MONTGOMERY COUNTY COMMUNITY COLLEGE

3 0473 00191312 1

W9-BWE-347

VAN NOSTRAND'S
ENCYCLOPEDIA OF CHEMISTRY

Fifth Edition

Edited by
Glenn D. Considine

WILEY-INTERSCIENCE

A John Wiley & Sons, Inc., Publication

Copyright © 2005 by John Wiley & Sons Inc. All rights reserved.

Published by John Wiley & Sons, Inc., Hoboken, New Jersey.
Published simultaneously in Canada.

No part of this publication may be reproduced, stored in a retrieval system, or transmitted in any form or by any means, electronic, mechanical, photocopying, recording, scanning, or otherwise, except as permitted under Section 107 or 108 of the 1976 United States Copyright Act, without either the prior written permission of the Publisher, or authorization through payment of the appropriate per-copy fee to the Copyright Clearance Center, Inc., 222 Rosewood Drive, Danvers, MA 01923, 978-750-8400, fax 978-646-8600, or on the web at www.copyright.com. Requests to the Publisher for permission should be addressed to the Permissions Department, John Wiley & Sons, Inc., 111 River Street, Hoboken, NJ 07030, (201) 748-6011, fax (201) 748-6008.

Limit of Liability/Disclaimer of Warranty: While the publisher and author have used their best efforts in preparing this book, they make no representations or warranties with respect to the accuracy or completeness of the contents of this book and specifically disclaim any implied warranties of merchantability or fitness for a particular purpose. No warranty may be created or extended by sales representatives or written sales materials. The advice and strategies contained herein may not be suitable for your situation. You should consult with a professional where appropriate. Neither the publisher nor author shall be liable for any loss of profit or any other commercial damages, including but not limited to special, incidental, consequential, or other damages.

For general information on our other products and services please contact our Customer Care Department within the U.S. at 877-762-2974, outside the U.S. at 317-572-3993 or fax 317-572-4002.

Wiley also publishes its books in a variety of electronic formats. Some content that appears in print, however, may not be available in electronic format.

Library of Congress Cataloging-in-Publication Data:

Van Nostrand's encyclopedia of chemistry.–5th ed. / edited by Glenn D. Considine.
　　p. cm.
　Rev. ed. of: Van Nostrand Reinhold encyclopedia of chemistry, 4th ed.c1984.
　Includes bibliographical references and index.
　ISBN 0-471-61525-0
　　1. Chemistry–Encyclopedias. I. Considine, Glenn D. II. Van Nostrand Reinhold encyclopedia of chemistry.

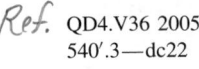 QD4.V36 2005
　　540′.3—dc22

2004057156

Printed in the United States of America.
10 9 8 7 6 5 4 3 2 1

PREFACE

The editors are pleased to introduce *Van Nostrand's Encyclopedia of Chemistry*, Fifth Edition, thus continuing a proud tradition of excellence. Previous editions (1956, 1966, 1973 and 1984) were published under the name *Van Nostrand Reinhold Encyclopedia of Chemistry*. This Fifth edition reflects progress and change during the past two decades with over 75% new text. The expansion of chemistry into new fields and the impact of related sciences on chemistry have required the reporting of many new topics and a considerable reorganization of the format to accommodate the rapidly growing interdisciplinary character of chemistry. The science of chemistry has continued in its trend toward greater compartmentalization and specialization. Concurrently, countless thousands of scientists, engineers, and technologists have been drawn to the knowledge of chemistry because of the importance of chemistry in diverse technologies—electronics and communications, energy sources and conservation, waste handling and pollution abatement, biotechnology, molecular biology, and the development of pharmaceuticals, biologicals, and chemotherapeutic methodologies. In addition to "classical" topics of chemistry, the new Encyclopedia covers nanotechnology, fuel cell technology, green chemistry, forensic chemistry, supramolecular chemistry, combinatorial chemistry, brief biographies on scores of scientists, and materials chemistry.

The essence of *Van Nostrand's Encyclopedia of Chemistry* is enduring and it remains a fine, concise, comprehensive, and accessible general chemistry work. Its intellectual scope ranges from the introductory to the highly technical. As has long been the case, the editors have designed the book to be approachable by students of many ages. An important feature continued in this work, is the progressive development of the discussion of each topic, beginning with a simple definition expressed in plain terms, developing into a more detailed treatment, and augmented by often-extensive Additional Reading suggestions and Internet references.

Scientists, engineers, technologists and contemporary readers can now turn to *Van Nostrand's Encyclopedia of Chemistry* for information about how their daily lives are increasingly affected by the sophistication of today's science and the complexity of modern chemistry. They will be reminded that knowledge and discovery exist in a continuum, and often, but not always, what is new depends entirely on what came before. As our esteemed, late editor of more than 20 years, Douglas M. Considine, was wont to say, "Science is history". With that mantra in mind, and noting that it has already been 20 years since the Fourth Edition in 1984, it is time to ask: What's new?

In designing this Fifth Edition, the editors have considered the foregoing observations. For example, to accommodate the much wider areas of interest in chemistry and to provide substantive coverage of the progress made during the past two decades, the number of specific entries in the volume has increased from just over 1250 to approximately 2750 topics. To avoid an unwieldy long volume, this has required tighter wording and condensation of the text and greater use of tabular and graphic presentations. The interior references in the book, where one article refers to another article that offers augmented or corollary coverage, and the visual aids, as well as the index, have been entirely overhauled; this will result in much greater ease in "navigating" the book. In addition to including topics, which the user of an encyclopedia of chemistry would normally expect to find, the editors have given particular emphasis to the following:

1. Advanced processes— catalytic conversion, cryogenics, dialysis, exomosis, freeze-concentrating, drying, and preserving, molecular distillation, photonuclear reactions, reverse osmosis, semipermeable membranes, molecular sieves, solvolysis, supercooling, superfluidity, thermoelectric cooling, and ultrafiltration.

2. Strategic raw materials— several hundred economic minerals and important raw materials used in chemical processing.

3. Chemistry of metals— greater stress on metallurgical phenomena and processes.

4. Energy sources and conversion— biomass, batteries, fuel cells and fuel cell technology, hydrogen as a fuel, liquid and gaseous fuels from coal, oil shale, tar sands, nuclear fission and fusion, lithium for thermonuclear reactors, insulating materials, and solar energy.

5. Wastes and pollution— carcinogens, cytotoxic chemicals, dioxin, biphenyls, air pollution, water treatment and pollution, radioactive waste handling.

6. Analytical instrumentation— new tools for determining ppm and ppb trace materials, as by chromatography.

7. Growing use of food chemicals— descriptions of all major food additives (anticaking agents, antimicrobial agents, bodying and bulking agents, coating materials, flavorings and flavor enhancers and potentiometers, humectants, intermediate-moisture food technology, polymeric food additives, among others).

8. Structure of matter— once in the province of chemistry and later annexed by physics, research in molecular biology and biotechnology has again narrowed the gap between chemistry and physics. Considerable emphasis is given in this volume to structure, including comprehensive coverage of subatomic particles at one end of the spectrum, to macromolecules at the other end.

9. New and improved materials— metal alloys, glass fibers for fiber optic communications, aerogels, plastics, graphite structures for aircraft, xanthan gum for the food industry, dyes, smart materials, YIGs and YAGs from the rare-earth metals for the electronics industry, cavitands, metallobiomolecules, metalloids, metalloproteins, electroconductive polymers, and superconductors, among others.

10. Plant chemistry— allopathic substances, anthocyanins, betalaines, gibberellic acid and gibberelin plant growth hormones, maleic hydrazide growth inhibitors, herbicides, insecticides, and other agricultural control chemicals, including new concepts in insect control.

11. Biochemistry and biotechnology— enzymes, coenzymes, antibiotics, fermentation, recombinant DNA, pharmaceuticals, hormones, contractile proteins, vitamins, enkephalins and endorphis, antimetabolites, immunochemistry, dietary minerals, vitamins, amino acids, and proteins are among topical coverage in this important topical area. Notable are detailed discussions of the effects of the chemical elements in biological systems, including calcium, chloride ion, cobalt, copper, fluorine, iodine, iron, magnesium, molybdenum, phosphorus, sodium, potassium, sulfur, and zinc.

Lastly, but certainly not least, the editors, faced with daunting amounts of information in highly specialized fields, have relied increasingly on the generous contributions of industry experts and scholars from all over all over North America and Europe. It has always been in the best tradition of the history of chemistry to share knowledge. It is therefore no mere coincidence that so many contributors are teachers at the university level, for they not only have deep knowledge in their respective fields, but they also can communicate that knowledge effectively. The great improvements to the substance of this book would not have been possible without them. Therefore, the editors have preserved the individual styles of the authors in keeping with the tradition of *Van Nostrand's Encyclopedia of Chemistry*, as an eminently personal, and one hopes, more accessible work of general chemistry and its related fields. We hope that this Fifth Edition, will be many things to many people and that it will serve you, the reader, as a useful chemical reference.

GLENN D. CONSIDINE, Editor
GREG GALLAGHER, Associate Editor
PETER H. KULIK, Associate Editor

CONTRIBUTORS

Several hundred scientists, engineers, and educators, located worldwide, have made this Fifth Edition of *Van Nostrand's Encyclopedia of Chemistry* a reality. Authorities from the various subdivisions of chemistry and related scientific disciplines assisted in the compilation of this work. Their inputs ranged from detailed information, graphics, and editorial counsel to the creation of comprehensive manuscripts on complex subjects. The editors and staff of this encyclopedia gratefully acknowledge their excellent cooperation and stress that the following abridged list of over 450 individuals and groups could be much longer. Although not mentioned there, grateful acknowledgement is also expressed for the earlier work of people who contributed to the prior editions of this book.

Special appreciation must be extended for the efforts of Drs. M. L. and W. L. Dilling, who skillfully summarized the complex world of organic chemistry, its nomenclature and equations.—Elmer Rowley, who made the coverage of mineralogy and crystallography in this encyclopedia truly outstanding.—Karl A. Gschneidner, Jr., who authored and arranged numerous entries on Rare-Earth elements.—Jeanne Maree Iacono, who authored and rendered invaluable assistance toward creating brief biographies on scores of scientists.—The staff of the *Kirk-Othmer Concise Encyclopedia of Chemical Technology* for their invaluable help in obtaining numerous entries. Without exaggeration, the list of such very special efforts could be extended by several additional paragraphs.

NOTE: In the cases of relatively short articles, the authors' initials may be used instead of their full names. In the following list, an asterisk indicates such authors.

Gurpreet S. Ahluwalia, *Gillette Research Institute, Gaithersburg, MD,* **Enzyme Therapeutic**

Satinder Ahuja, *Ahuja Consulting, Monsey, NY,* **Trace and Residue Analysis**

Jean-Francois Alary, *Universite de Sherbrooke, Sherbrooke, Quebec, Canada,* **Asbestos**

H. J. Albert, *Parr Instrument Co., Moline, IL,* http://www.parrinst.com/, **Calorimetry**

Lyle F. Albright, *Purdue University, West Lafayette, IN,* http://www.purdue.edu/, **Nitration**

P. S. Albright, *Wichita, KS,* **Electroplating**

Edward S. Amis, *University of Arkansas, Fayetteville, AR,* http://www.uark.edu/, **Activity Coefficient; Dissociation**

Frederick J. Antosz, *The Upjohn Company, Kalamazoo, MI,* http://www.pfizer.com/main.html, http://www.pearce-bennett.freeserve.co.uk/ancestors/upjohn/weupjo, **Antibiotics: Ansamacrolides**

Mohammad Aslam, *Hoechst Celanese Corporation, Corpus Christi, TX,* **Esterification**

Carmen Avendano, *Universidad Complutense, Madrid, Spain,* http://www.ucm.es/, **Hydantoin and its Derivatives**

Malcolm H. I. Baird, *McMaster University, Hamilton, Ontario, Canada,* http://www.mcmaster.ca/, **Extraction (Liquid-Liquid)**

Lawrence E. Ball, *BP Research, Cleveland, OH,* http://www.bp.com/, **SAN** (under **Acrylonitrile Polymers**)

Brian Bannister, *The Upjohn Company, Kalamazoo, MI,* http://www.pfizer.com/main.html, http://www.pearce-bennett.freeserve.co.uk/ancestors/upjohn/weupjo, **Antibiotics: Lincosaminides**

Cynthia S. Barcelon-Yang, *E. I. Du Pont de Nemours & Co., Inc., Wilmington, DE,* http://www.dupont.com/ag/, **Information Retrieval**

Jay A. Bardole, *Vincennes University, Vincennes, IN,* http://www.vinu.edu/, **Paint and Finish Removers**

Robert Q. Barr, *Director Technical Information, Climax Molybdenum Company, Greenwich, CT,* http://www.climaxmolybdenum.com/, **Molybdenum;** and **Technetium**

Roger G. Bates, *University of Florida, Gainesville, FL,* http://www.ufl.edu/, **Hydrogen-Ion Activity**

William Bauer, Jr., *Rohm and Haas Company, Spring House, PA,* http://www.rohmhaas.com/, **Acrylic Acid and Derivatives**

W. F. Beach, *Alpha Metals, Bridgewater, NJ,* **Xylene Polymers**

James Bellows, *Westinghouse Electric Corporation, Orlando, FL,* http://www.westinghouse.com/, **Steam**

J. R. Benke, *Westinghouse Electric Corporation, Pittsburgh, PA,* http://www.westinghouse.com/, **Fuel Cells**

Yvette Berry, *Reckitt & Colman Inc., Wayne, NJ,* http://www.reckitt.com/, **Odor Modification**

J. C. Bevington, *University of Lancaster, Lancaster, England,* http://www.lancs.ac.uk/, **Polymerization (Radical)**

Ernst Billig, *Union Carbide Chemicals and Plastics Company Inc., South Charleston, WV,* http://www.unioncarbide.com/, **Butyl Alcohols;** and **Oxo Process**

John Blackwell, *Case Western Reserve University, Cleveland, OH,* http://www.cwru.edu/, **Macromolecular Science**

Gary T. Blair, *Haarmann & Reimer Corporation, Springfield, NJ,* **Hydroxy Dicarboxylic Acids**

George Blomgren, *Eveready Battery Company, Inc., West Lake, OH,* http://www.eveready.com/, **Batteries: Primary Cells**

John E. Boliek, *E. I. Du Pont de Nemours & Co., Inc., Wilmington, DE,* http://www.dupont.com/ag/, **Fibers: Elastomeric**

A. A. Bondi, *Shell Development Co., Houston, TX,* http://www.shellus.com/sepco/, **Lubrication** (under **Lubricating Agents**)

Daniel P. Bonner, *Bristol-Myers Squibb Pharmaceutical Research Institute, Princeton, NJ,* http://www.bms.com/research/data/, **Antibiotics: Monobactams**

Donald Borders, *American Cyanamid Company, Pearl River, NY,* http://www.basf-corp.com/, **Antibiotics**

Barbara H. Bory, *Lever Company, Edgewater, NJ,* **Surfactants**

Judith M. Bradow, *U.S. Department of Agriculture, New Orleans, LA,* http://www.ars.usda.gov/, **Herbicides**

James F. Brazdil, *BP Research, Cleveland, OH,* http://www.bp.com/, **Acrylonitrile**

Thomas E. Breuer, *Humko Chemical Division of Witco Corporation, Memphis, TN,* **Dimer Acids**

James A. Brient, *Merichem Company, Houston, TX,* http://www.merichem.com/, **Naphthenic Acids**

Angelo Brisimitzakis, *GE Plastics, Pittsfield, MA,* http://www.geplastics.com/resins, **ABS Resins** (under **Acrylonitrile Polymers**)

Ralph Brodd, *Gould, Inc., East Lake, OH,* **Batteries**

Robert T. Brooker, *Olin Corporation, Charleston, TN,* http://www.olin.com/, **Perchloric Acid and Perchlorates**

Donald E. Brownlee, *University of Washington, Seattle, WA,* http://www.washington.edu/, **Extraterrestrial Materials**

Evelyn L. Brownlee, *E. I. Du Pont de Nemours & Co., Inc., Wilmington, DE,* http://www.dupont.com/ag/, **Information Retrieval**

Daniel J. Brunelle, *General Electric, Schenectady, NY,* http://www.geplastics.com/, **Polycarbonates**

David R. Bryant, *Union Carbide Chemicals and Plastics Company Inc., South Charleston, WV,* http://www.unioncarbide.com/, **Oxo Process**

Kathryn R. Bullock, *Johnson Controls, Inc.,* http://www.johnsoncontrols.com/bg/, **Batteries: Lead-Acid**

H. P. Burchfield, *Gulf South Research Institute, New Iberia, LA,* **Hydroquinones**

Joseph C. Burnett, *Huntsman Specialty Chemical Corporation, St. Louis, MO,* http://www.huntsman.com/, **Maleic Anhydride, Maleic Acid,** and **Furmaric Acid**

Timothy A. Calamari, Jr., *United States Department of Agriculture, New Orleans, LA,* **Flame Retardants for Textiles**

Narasimhan Calamur, *Amoco Chemical Company, Naperville, IL,* **Butylenes**

Emmett D. Calhoun, *E. I. Du Pont de Nemours & Co., Inc., Wilmington, DE,* http://www.dupont.com/ag/, **Information Retrieval**

William J. Cannella, *Chevron Research & Technology Company, Richmond, CA,* **Xylenes and Ethylbenze**

Robert R. Cantrell, *Union University, Jackson, TN,* http://www.uu.edu/, **Dicarboxylic Acids**

Bruno A. Caputo, *E. I. Du Pont de Nemours & Co., Inc., Wilmington, DE,* http://www.dupont.com/ag/, **Information Retrieval**

S. C. Carapella, Jr., *ASARCO Incorporated, South Plainfield, NJ,* http://www.asarco.com/, **Tellurium; Thallium**

F. Patrick Carr, *OMYA, Inc., Proctor, VT,* http://www.omya-na.com/, **Calcium Carbonate**

Martin E. Carrera, *Amoco Chemical Company, Naperville, IL,* **Butylenes**

Richard G. Carter, *Union Carbide Chemicals and Plastics Company Inc., South Charleston, WV,* http://www.unioncarbide.com, **Diamines and Higher Amines, Aliphatic**

B. Cavalleri, *Merrell-Dow Research Institute, Gerenzano, Italy,* **Antibiotics: Glycopeptides (Dalbaheptides)**

Peter Cervoni, *American Cyanamid Company, Pearl River, NY,* http://www.basf-corp.com/, **Diuretic Agents**

Peter S. Chan, *American Cyanamid Company, Pearl River, NY,* http://www.basf-corp.com/, **Diuretic Agents**

Shiou-Shan Chen, *Raytheon Engineers & Constructors, East Weymouth, MA,* http://www.mbendi.co.za/orgs/cnkl.htm, **Styrene**

Michael S. Cholod, *Rohm and Haas Company, Bristol, PA,* http://www.rohmhaas.com/, **Cyanohydrins**

A. J. Cofrancesco, *Consultant, Windward Meadows, Delanson, NY,* **Dyes: Natural**

Peter J. Collings, *Swarthmore College, Swarthmore, PA,* http://www.swarthmore.edu/, **Liquid Crystalline Materials**

Alan E. Comyns, *Solvay Interox, Chester, England,* www.solvay.com, **Peroxides and Peroxide Compounds (Inorganic)**

J. G. Converse, *Sterling Chemicals Inc., Texas City, TX,* http://www.sterlingchemicals.com/, **Chromatography**

B. Sharp Cook, *The University of Texas, El Paso, TX,* http://www.utep.edu/, **Fallout (Radioactive); Nuclear Fission; Nuclear Forces**

P. H. Cook, *Dow Chemical U.S.A., and operating unit of the Dow Chemical Company, Freeport, TX,* **Vinyl Ester Resins**

David A. Cooney, *National Cancer Institute, Bethesda, MD,* http://www.nci.nih.gov/, **Enzyme Therapeutic**

Cajetan F. Cordeiro, *Air Products and Chemicals, Inc., Allentown, PA,* http://www.airproducts.com/, **Vinyl Acetate Polymers**

Donald A. Corrigan, *Handy & Harman, Fairfield, CT,* http://www.handyharman.com/, **Gold;** and **Silver**

Joseph A. Cowfer, *The Geon Company, Avon Lake, OH,* http://www.geon.com/, **Vinyl Chloride**

Louise W. Crandall, *Lilly Research Laboratories, Indianapolis, IN,* http://www.lilly.com/research/, **Antibiotics: Polyethers**

Charles C. Cumbo, *E. I. Du Pont de Nemours & Co., Inc., Wilmington, DE,* http://www.dupont.com/ag/, **Information Retrieval**

Benedict S. Curatolo, *BP Research, Cleveland, OH,* http://www.bp.com/, **SAN** (under **Acrylonitrile Polymers**)

R. Marc Dahlgren, *Ivory Technical Center, The Procter & Gamble Company, Cincinnati, OH,* http://www.pg.com/, **Soaps**

Joseph P. Daniszewski, *E. I. Du Pont de Nemours & Co., Inc., Wilmington, DE,* http://www.dupont.com/ag/, **Information Retrieval**

Rathin Datta, *Consultant, Chicago, IL,* **Hydroxycarboxylic Acids**

Darwin D. Davis, *E. I. Du Pont de Nemours & Co., Inc., Victoria, TX,* http://www.dupont.com/ag/, **Adipic Acid**

Ann C. Debaldo, Ph.D., *College of Public Health, University of South Florida, Tampa, FL,* http://publichealth.usf.edu/, **Antitoxin; Bacteria; Genetics and Gene Science (Classical); Immune System and Immunochemistry;** and **Virus**

D. F. DeCrane, *Chemetals Corporation, Baltimore, MD,* http://www.chemetals.com, **Manganese**

Jeffrey J. DeFraties, *Haarmann & Reimer Corporation, Springfield, NJ,* **Hydroxy Dicarboxylic Acids**

Ernesto DeGuzman, *Sybron Chemicals Inc., Welford, SC,* http://www.sybronchemicals.com/, **Dye Carriers**

Phillip DeLassus, *The Dow Chemical Company, Midland, MI,* http://www.dow.com/Homepage/index.html, **Barrier Polymers**

P. T. DeLassus, *The Dow Chemical Company, Freeport TX,* **Vinylidene Chloride Monomer and Polymers**

Suman C. Desai, *Davy McKee Iron & Steel Division, Ashmore House, Stockton-on-Tees, England,* **Iron**

Stephen C. DeVito, *U.S. Environmental Protection Agency, Washington, D.C.,* http://www.epa.gov/, **Nitriles**

Earl D. Dietz, *Toledo, OH,* **Glass**

M. L., and **W. L. Dilling,** *The Dow Chemical Company, Midland, MI,* http://www.dow.com/, **Organic Chemistry**

Christopher P. Dionigi, *U.S. Department of Agriculture, New Orleans, LA,* http://www.ars.usda.gov/, **Herbicides**

Murice Dery, *Akzo Nobel Chemicals Inc., Dobbs Ferry, NY,* http://www.akzonobelusa.com/, **Quaternary Ammonium Compounds**

John C. Dobson, *Rohm and Haas Company, Spring House, PA,* http://www.rohmhaas.com/, **Methacrylic Acid and Derivatives**

T. Dombrowski, *Engelhard Corporation, Iselin, NJ,* http://www.engelhard.com/, **Clays**

Ronald L. Dotson, *Olin Corporation, Charleston, TN,* http://www.olin.com/, **Perchloric Acid and Perchlorates**

Arthur R. Doumaux, Jr., *Union Carbide Chemicals and Plastics Company Inc., South Charleston, WV,* http://www.unioncarbide.com, **Diamines and Higher Amines, Aliphatic**

Mary Noon Doyle, *Shepard Chemical Company, Cincinnati, OH,* http://www.shepchem.com/, **Naphthenic Acids**

Terrence W. Doyle, *Bristol-Meyers Squibb Company, Wallingford, CT,* http://www.bms.com/, **Chemotherapeutics, Anticancer**

Hans Dressler, *Koppers Company, Inc., Monroeville, PA,* http://www.koppers.com/, **Pyridine and Derivatives**

E. I. Dupont, *E. I. Du Pont de Nemours & Co., Inc., Wilmington, DE,* http://www.dupont.com/ag/, **Polyamide Resins**

Douglas J. Durian, *University of California, Los Angeles, Los Angeles, CA,* http://www.ucla.edu, **Foam**

Richard A. Durst, *Cornell University, Geneva, NY,* http://www.cornell.edu/, **Hydrogen-Ion Activity**

H. W. Earhart, *Consultant, Wichita, KS,* **Polymethylbenzenes**

Chris Eberspacher, *UNISUN, Newbury Park, CA,* http://members.aol.com/unisun/, **Photovoltaic Cells**

Douglas A. Eckel, *E. I. Du Pont de Nemours & Co., Inc., Wilmington, DE,* http://www.dupont.com/ag/, **Information Retrieval**

Varadaraj Elango, *Hoechst Celanese Corporation, Corpus Christi, TX,* **Esters, Organic**

George A Ellestad, *American Cyanamid Company, Pearl River, NY,* http://www.basf-corp.com/, **Antibiotics: Tetracyclines**

Stanley B. Elliott, *Bedford, OH,* **Saponification**

Kenneth R. Engh, *CELITE Corporation, Lompac, CA,* **Diatomite**

Mary G. Enig, *Enig Associates, Inc., Silver Springs, MD,* http://www.enig.com/, **Mineral Nutrients**

Lawrence J. Esposito, *Rhone-Poulene, Cranbury, NJ,* **Vanillin**

W. G. Etzkorn, *Union Carbide Chemicals & Plastics Company Inc, South Charleston, WV,* http://www.unioncarbide.com/, **Acrolein and Derivatives**

B. Evans, *Assistant Chemist, Rare-Earth Information Center, Energy and Mineral Resources Research Institute, Iowa State University, Ames, IA,* http://www.external.ameslab.gov/, **Cerium; Dysprosium; Erbium; Europium; Gadolinium; Holmium; Lanthanum; Lutetium; Neodymium; Rare-Earth Elements and Metals; Praseodymium; Samarium; Scandium; Terbium; Thulium; Ytterbium;** and **Yttrium**

Daniel F. Farkas, *Oregon State University, Corvallis, OR,* http://oregonstate.edu/, **Food Processing**

James P. Farr, *The Clorox Company, Pleasanton, CA,* http://www.clorox.com/, **Bleaching Agents**

William D. Faust, *Ferro Corporation, Cleveland, OH,* http://www.ferro.com/, **Enamels, Porcelain or Vitreous**

Rudolf Faust, *University of Massachusetts, Lowell, MA,* http://www.uml.edu/, **Initiators (Cationic)**

Timothy R. Felthouse, *Huntsman Specialty Chemical Corporation, St. Louis, MO,* http://www.huntsman.com/, **Maleic Anhydride, Maleic Acid, and Furmaric Acid**

T. E. Ferrington, *Clarksville, MD,* **Polymers (Inorganic); Polymers (Organic)**

L. F. and M. Fieser, *Harvard University, Cambridge, MA,* http://www.harvard.edu/, **Reformatsky Reaction; Willgerodt Reaction;** and **Wolff-Kishner Reaction**

R. Fikentscher, *BASF AG, Ludwigshafen, Germany,* **Imines, Cyclic**

K. Thomas Finley, *State University of New York, Brockport, NY,* http://www.suny.edu/, **Quinolines and Isoquinolines;** and **Quinones**

Barry A. J. Fisher, *Scientific Services Bureau, Los Angeles County Sheriff's Department, Los Angeles, CA,* **Forensic Chemistry**

William B. Fisher, *Allied Signal Inc., Petersberg, VA,* http://www.alliedsignal.com/, **Cyclohexanol-Cyclohexanone**

K. Formanek, *Rhone-Poulene, Cranbury, NJ*, **Vanillin**

David K. Frederick, *OMYA, Inc., Proctor, VT*, http://www.omya-na.com/, **Calcium Carbonate**

Stig E. Friberg, *Clarkson University, Potsdam, NY*, http://www.clarkson.edu/, **Emulsions**

Leslie J. Friedman, *Arthur D. Little, Inc., Cambridge, MA*, **Food Additives**

Leroy W. Fritch, Jr., *GE Plastics, Pittsfield, MA*, http://www.geplastics.com/resins, **ABS Resins** (under **Acrylonitrile Polymers**)

Ashit K. Ganguly, *Schering-Plough Corporation, Kenilworth, NJ*, http://www.sch-plough.com/, **Antibiotics: Oligosaccharides**

Richard G. Gann, *National Institute of Standards and Technology, Gaithersburg, MD*, http://www.nist.gov/, **Flame-Retarding Agents**

John Gannon, *Consultant, Danbury, CT*, **Epoxy Resins**

Roque A. Garcia, *Technicon Instruments Corporation, Tarrytown, NY*, **Automated Instrumentation: Hematology**

Charles F. Gay, *National Renewable Energy Laboratory, Golden, CO*, http://www.nrel.gov/, **Photovotaic Cells**

Chester H. Gelbert, *E. I. Du Pont de Nemours & Co., Inc., Louisville, KY*, http://www.dupont.com/ag/, **Latex Technology**

Stanley A. Gembicki, *UOP, Des Plaines, IL*, http://www.uop.com/, **Adsorption: Liquid Separation**

Paul Gherson, *Miles, Inc., Tarrytown, NY*, **Automated Instrumentation: Clinical Chemistry**

D. S. Gibbs, *The Dow Chemical Company, Midland, MI*, **Vinylidene Chloride Monomer and Polymers**

Paul R. Gifford, *Ovonic Battery Company, Troy, MI*, http://www.ovonic.com/, **Batteries: Other**

V. M. Girijavallabhan, *Schering-Plough Corporation, Kenilworth NJ*, http://www.sch-plough.com/, **Antibiotics: Oligosaccharides**

Kenneth H. Glaspey, *E. I. Du Pont de Nemours & Co., Inc., Wilmington, DE*, http://www.dupont.com/ag/, **Information Retrieval**

J. Edward Glass, *North Dakota State University, Fargo, ND*, http://www.chem.ndsu.nodak.edu/, **Water Soluble Polymers**

Brian Glover, *Zeneca Colours, Manchester, U. K.*, **Dyes: Application and Evaluation**

Louis H. Goodson, *Midwest Research Institute, Kansas City, MO*, http://www.mriresearch.org/, **Racemization**

Maximillian B. Gorensek, *The Geon Company, Avon Lake, OH*, http://www.geon.com/, **Vinyl Chloride**

Michael C. Grady, *E. I. Du Pont de Nemours & Co., Inc., Louisville, KY*, http://www.dupont.com/ag/, **Latex Technology**

Darlyn C. Green-Kocher, *E. I. Du Pont de Nemours & Co., Inc., Wilmington, DE*, http://www.dupont.com/ag/, **Information Retrieval**

C. Gail Greenwald, *Arthur D. Little, Inc., Cambridge, MA*, **Food Additives**

Peter Gregory, *Zeneca Specialties, Manchester, U.K.*, **Dye and Dye Intermediates**

Andrew W. Gross, *Rohm and Haas Company, Spring House, PA*, http://www.rohmhaas.com/, **Methacrylic Acid and Derivatives**

Marianne B. Gruber, *E. I. Du Pont de Nemours & Co., Inc., Wilmington, DE*, http://www.dupont.com/ag/, **Information Retrieval**

Karl. A. Gschneidner, Jr., *Senior Metallurgist, Rare-Earth Information Center, Energy and Mineral Resources Research Institute, Iowa State University, Ames, IA*, http://www.external.ameslab.gov/, **Cerium; Dysprosium; Erbium; Europium; Gadolinium; Holmium; Lanthanum; Lutetium; Neodymium; Rare-Earth Elements and Metals; Praseodymium; Samarium; Scandium; Terbium; Thulium; Ytterbium; and Yttrium**

Reigh C. Gunderson, *The Dow Chemical Co., Midland, MI*, http://www.dow.com/, **Synthetic Lubricants** (under **Lubricating Agents**)

C. E. Habermann, *Dow Chemical, USA, Midland, MI*, http://www.dow.com/, **Acrylamide**

Charles M. Hall, *Upjohn Company, Kalamazoo, MI*, **Analgesics, Antipyretics, and Antiinflammatory Agents**

Wilbur S. Hall, *Amchem Products, Inc., Ambler, PA*, **Autodeposition; Conversion Coatings**

Robert L. Hamill, *Lilly Research Laboratories, Indianapolis, ID*, http://www.lilly.com/research/, **Antibiotics: Polyethers**

William M. Hann, *Rohn and Haas Company, Spring House, PA*, http://www.rohmhaas.com/, **Dispersants**

Robert J. Harper, Jr., *United States Department of Agriculture, New Orleans, LA*, **Flame Retardants for Textiles**

Guy H. Harris, *University of California at Berkeley, Berkeley, CA*, **Xanthates**

James A. Harvey, *Hewlett-Packard Company, Oregon Graduate Institute of Science & Technology, Corvallis, OR*, **Smart Materials**

W. D. Hatfield, *Decatur, IL*, **Activated Sludge**

Makoto Hattori, *Sumitomo Chemical Company, Tokyo, Japan*, http://www.sumitomo-chem.co.jp/english/, **Dyes: Anthraquinone**

Alan S. Hay, *General Electric Company, Schenectady, NY*, **Polymerization (Oxidative-Coupling)**

B. W. Heinemeyer, *The Dow Chemical Company, Freeport, TX*, http://www.dow.com/, **Polyethylene**

Howard I. Heitner, *Cytec Industries, Stamford, CT*, http://www.cytec.com/, **Flocculating Agents**

L. L. Hench, *University of Florida, Gainesville, FL*, http://www.ufl.edu/, **Glass Blocks (under Glass); and Sol-Gel Technology**

Joel H. Hilderbrand, *University of California, Berkeley, CA*, http://www.cchem.berkeley.edu/, **Solutions**

Joseph J. Hlavka, *American Cyanamid Company, Pearl River, NY*, http://www.basf-corp.com/, **Antibiotics: Tetracyclines**

Stephen E. Hluchan, *Business Manager, Calcium Metal Products, Minerals, Pigments, Metals Division, Pfizer, Inc., Wallingford, CT*, **Barium; Calcium; and Strontium**

A. Hohn, *BASF AG, Lugwigshafen, Germany*, http://www.corporate.basf.com/en/, **Formamide**

Allen T. Hopper, *University of Wisconsin-Madison, Madison, WI*, http://www.wisc.edu/, **Pharmaceuticals, Chiral**

Fred H. Hoskins, *Washington State University, Pullman, WA*, http://www.wsu.edu/, **Food Toxicants, Naturally Occurring**

Robert A. Howell, *Central Michigan University, Mount Pleasant, MI*, http://www.cmich.edu/, **Vinylidene Chloride Monomer and Polymers**

Chia-Lung Hsieh, *Wyeth-Lederle Vaccine and Pediatrics, Pearl River, NY*, **Vaccine Technology**

James Hunter, *Eveready Battery Company, Inc., West Lake, OH*, http://www.eveready.com/, **Batteries: Primary Cells**

Charles D. Hurd, *Northwestern University, Evanston, IL*, http://www.northwestern.edu/, **Ketenes**

*Jeanne Maree Iacono, (J. M. I.), *Dammeron Valley, UT*, *Brief biographies on scores of scientists*

Margaret M. Isselmann, *E. I. Du Pont de Nemours & Co., Inc., Wilmington, DE*, http://www.dupont.com/ag/, **Information Retrieval**

Jipi Jaakkola, *Paper Machine Product Manager, Valmet Corporation, Charlotte, NC*, http://www.valmet.com/, **Papermaking and Finishing**

M. L. Jackson, *University of Wisconsin, Madison, WI*, http://www.chem.wisc.edu/, **Soil Chemistry**

E. E. Jaffe, *Ciba-Geigy Corporation, Newport, DE*, **Pigments (Organic)**

Edwin C. Jahn, *State College of Forestry, Syracuse, NY*, http://www.esf.edu/, **Wood**

Monique M. L. Janssens, *Janssen Research Foundation, Berse, Belgium*, **Histamine and Histamine Antagonists**

Hiremagular N. Jayaram, *Indiana University School of Medicine, Indianapolis, IN*, http://www.medicine.iu.edu/, **Enzyme Therapeutic**

A. Jayaraman, *AT&T Bell Laboratories, Murray Hill, NJ*, http://www.belllabs.com/, **Diamond Anvil High Pressure Cell**

Arnold W. Jensen, *E. I. Du Pont de Nemours & Co., Inc., Wilmington, DE*, http://www.dupont.com/ag/, **Fibers: Elastomeric**

Thomas C. Johns, *E. I. Du Pont de Nemours & Co., Inc., Wilmington, DE*, http://www.dupont.com/ag/, **Information Retrieval**

James A. Johnson, *UOP, Des Plaines, IL*, http://www.uop.com/, **Adsorption: Liquid Separation**

Richard M. Johnson, *U.S. Department of Agriculture, New Orleans, LA*, http://www.ars.usda.gov/, **Herbicides**

Carmel R. Jolicoeur, *Universite de Sherbrooke, Sherbrooke, Quebec, Canada*, **Asbestos**

Robert W. Johnson, *Union Camp Corporation, Savannah, GA*, **Dicarboxylic Acids**

Mark M. Jones, *Vanderbilt University, Nashville, TN*, http://www.vanderbilt.edu/, **Activation (Molecular)**

Steven Jones, *Clarkson University, Potsdam, NY*, http://www.clarkson.edu/, **Emulsions**

Alexy D. Kachikovski, *Institute of Organic Chemistry, National Academy of Sciences of the Ukraine*, **Polymethine Dyes**

Steven W. Kaiser, *Union Carbide Chemicals and Plastics Company Inc., South Charleston, WV,* http://www.unioncarbide.com, **Diamines and Higher Amines, Aliphatic**

John N. Kalberg, *Ivory Technical Center, The Procter & Gamble Company, Cincinnati, OH,* http://www.pg.com/, **Soaps**

Norbert W. Kaleta, *Natural Gypsum, Gold Bond Research Center, Buffalo, NY,* http://www.national-gypsum.com/, **Calcium Sulfate**

George J. Kaminsky, *The Procter & Gamble Company, Cincinnati, OH,* http://www.pg.com/, **Detergents**

Curtis R. Kates, *Advanced Aromatics, Inc., Baytown, TX,* www.advancedaromatics.com, **Naphthalene Derivatives**

Samuel Kaufman, *Naval Research Laboratory, Washington, D.C.,* http://www.nrl.navy.mil/, **Solubilization**

Thomas R. Keenan, *Kind & Knok Gelatine, Inc., Sioux, IA,* **Gelatin**

Donald R. Kemp, *E. I. Du Pont de Nemours & Co., Inc., Victoria, TX,* http://www.dupont.com/ag/, **Adipic Acid**

George L. Kenyon, *University of California, San Francisco,* **Enzyme Inhibitors**

G. Kientz, *Rhone-Poulene, Cranbury, NJ,* **Vanillin**

Alilcia P. King, *E. I. Du Pont de Nemours & Co., Inc., Wilmington, DE,* http://www.dupont.com/ag/, **Information Retrieval**

Larry W. Kingston, *Natural Gypsum, Gold Bond Research Center, Buffalo, NY,* http://www.national-gypsum.com/, **Calcium Sulfate**

H. A. Kirst, *Eli Lilly and Company, Indianapolis, IN,* http://www.lilly.com/, **Antibiotics: Macrolides**

Yury V. Kissin, *Mobil Chemical Company, Edison, NJ,* http://www.exxonmobilchemical.com/, **Polyethylene** (under **Olefin Polymers**); **High Density Polyethylene** (under **Olefin Polymers**); **Linear Low Density Polyethylene** (under **Olefin Polymers**); and **Polymers of Higher Olefins** (under **Olefin Polymers**)

Martin Klein, *Rutgers, University, Piscataway, NJ,* http://www.rutgers.edu/, **Batteries: Secondary Cells**

Jerome M. Klosowski, *Dow Corning Corporation, Midland, MI,* http://www.dowcorning.com/, **Sealants**

Edward A. Knaggs, *Consultant, Deerfield, IL,* **Sulfonation and Sulfation**

Jerome W. Knapczk, *Solutia, Indian Orchard, MA,* http://www.solutia.com/, **Vinyl Acetal Polymers**

Raymond S. Knorr, *Monsanto Company, Pensacola, FL,* http://www.monsanto.com/, **Fibers: Acrylic**

Charles J. Knuth, *Pfizer, Inc., New York, NY,* http://www.pfizer.com/main.html, **Plasticizers**

Edmund I. Ko, *Carnegie Mellon University, Pittsburgh, PA,* http://www.cmu.edu/, **Aerogels**

M. I. Kohan, *E. I. Du Pont de Nemours & Co., Inc., Wilmington, DE,* http://www.dupont.com/ag/, **Polyamide Resins**

Andrew P. Komin, *Koch Chemical Company, Wichita, KS,* **Polymethylbenzenes**

Kenneth D. Kopple, *Illinois Institute of Technology, Chicago, IL,* http://www.iit.edu/, **Polymers (Electroconductive)**

Gabe I. Kornis, *Pharmacia & Upjohn Inc., Kalamazoo, MI,* http://www.pharmacia.com/, **Pyrazoles, Pyrazolines, and Pyrazolones**

William H. Koster, *Bristol-Myers Squibb Pharmaceutical Research Institute, Princeton, NJ,* http://www.bms.com/research/data/, **Antibiotics: Monobactams**

Joseph Kozakiewicz, *American Cyanamid Company, Stamford, CT,* http://www.basf-corp.com/, **Acrylamide Polymers**

David M. Krentz, *E. I. Du Pont de Nemours & Co., Inc., Wilmington, DE,* http://www.dupont.com/ag/, **Information Retrieval**

Charles T. Kresge, *Mobile Research and Development Corporation, Paulsboro, NJ,* **Molecular Sieves**

Gerald A. Krulik, *Applied Electroless Concepts, Inc., El Toro, CA,* **Metallic Coatings**

Gunther H. Kuhl, *Consultant, Cherry Hills, NJ,* **Molecular Sieves**

Donald M. Kulich, *GE Plastics, Pittsfield, MA,* http://www.geplastics.com/resins, ABS Resins (under **Acrylonitrile Polymers**)

Ralf Kuriyel, *Millipore Coroporation, Bedford, MA,* **Ultrafiltration**

J. J. Kurland, *Union Carbide Chemicals & Plastics Company, Inc., South Charleston, WV,* http://www.unioncarbide.com/, **Acrolein and Derivatives**

Roger Kust, *Tetra Chemicals, Houston, TX; Tetra Technologies, Inc., The Woodlands, TX,* http://www.tetratec.com/nav.html, **Calcium Chloride**

Fluorence M. Kvalnes, *E. I. Du Pont de Nemours & Co., Inc., Wilmington, DE,* http://www.dupont.com/ag/, **Information Retrieval**

Michael R. Ladisch, *Laboratory of Renewable Resources Engineering and Department of Agricultural and Biological Engineering, Purdue University, West Lafayette, IN,* http://www.purdue.edu/, **Biotechnology (Bioprocess Engineering)**

Robert T. LaLonde, *State University of New York, Syracuse, NY,* **Terpenes and Terpenoids**

Joseph B. Lambert, *Northwestern University, Evanston, IL,* http://www.chem.northwestern.edu/, **Stereochemistry**

L. M. Landoll, *Hercules Inc., Wilmington, DE,* http://www.herc.com/, **Olefin Fibers**

Howard Lanza, *Miles, Inc., Tarrytown, NY,* http://www.bayer.com/, **Automated Instrumentation: Clinical Chemistry**

Walter C. Lapple, *Alliance, OH,* **Stoichiometry**

G. R. Lappin, *Albermarle Corp., Baton Rouge, LA,* **Olefins, Higher** (under **Olefin Polymers**)

Richard R. Lattime, *The Goodyear Tire & Rubber Company, Akron, OH,* http://www.goodyear.com, **Styrene-Butadiene Rubber**

G. G. Lauer, *Koppers Company, Inc.,Pittsburgh, PA,* http://www.koppers.com/, **Coal Tar and Derivatives**

M. E. Leaphart II, *University of South Carolina, Columbia, SC,* http://www.sc.edu/, **Nitrides**

H. David Leigh III, *Clemson University, Clemson, SC,* http://www.clemson.edu/, **Refractories**

George R. Lenz, *BOC Group Technical Center, Murray Hill, NJ,* http://www.boc.com/index.asp, **Anesthetics**

Patricia M. Lesko, *Rohm and Haas Company, Spring House, PA,* http://www.rohmhaas.com/, **Methacrylic Polymers**

Robert Leurs, *Leiden/Amsterdam Center for Drug Research,* **Histamine and Histamine Antagonists**

Cynthia L. Levinson, *University of California, San Francisco,* http://www.ucsf.edu/, **Enzyme Inhibitors**

B. A. Lewis, *Cornell University, Ithaca, NY,* http://www.cornell.edu/, **Fiber (Dietary)**

Linton Libby, *Chief Chemist, Simmons Refining Company, Chicago, IL,* **Rhodium; Ruthenium; Palladium;** and **Platinum and Platinum Group**

Richard B. Lieberman, *Montell Polyolefins, Elkton, MD,* **Polypropylene** (under **Olefin Polymers**)

Thomas A. Liederbach, *Electrode Corporation, Chardon, OH,* **Metal Anodes**

K. F. Lin, *Hercules Incorporated, Wilmington, DE,* http://www.herc.com/, **Alkyd Resins**

Klaus R. Lindner, *Bristol-Myers Squibb Pharmaceutical Research Institute, Princeton, NJ,* http://www.bms.com/landing/data/, **Antibiotics: Monobactams**

David Lipp, *American Cyanamid Company, Stamford, CT,* http://www.basf-corp.com/, **Acrylamide Polymers**

William F. Little, *University of North Carolina, Chapel Hill, NC,* http://www.unc.edu/, **Metallocenes**

The C. Lo, *T. C. Lo & Associates, Wayne, NJ,* **Extraction (Liquid-Liquid)**

Gerd Loebbert, *BASF Corporation, Holland, MI,* http://www.corporate.basf.com/en/, **Phthalocyanine Compounds; Pigment Dispersions**

Robert B. Login, *Sybron Chemicals, Inc., Wellford, SC,* http://www.sybronchemicals.com/, **Vinylamide Polymers;** and **Vinyl Ether Monomers and Polymers**

Gabriel Lopez, *University of New Mexico, Albuquerque, NM,* http://www.unm.edu/, **Nanotechnology (Molecular)**

Jose A. Lopez, *Baylor College of Medicine, Houston, TX,* http://www.bcm.tmc.edu/, **Nanotechnology (Molecular)**

Lucent Technologies, *Optical Fiber Solutions, Norcross, GA,* **Optical Fiber Systems**

John H. Lupinski, *General Electric Company, Schenectady, NY,* **Polymers (Electroconductive)**

Hugh M Lybarger, *The Goodyear Tire and Rubber Company, Akron, OH,* http://www.goodyear.com/, **Isoprene**

Jesse L. Lynn, Jr., *Lever Company, Edgewater, NJ,* **Surfactants**

Fred T. Mackenzie, *Northwestern University, Evanston, IL,* http://www.northwestern.edu/, **Ocean Water**

H. Maehr, *Hoffmann-LaRoche, Inc., Nutley, NJ,* http://www.rocheusa.com/, **Antibiotics: Elfamycins**

Douglas Magde, *University of California at San Diego, San Diego, CA,* http://www.ucsd.edu/, **Kinetic Measurements**

Henry E. Mahncke, *King of Prussia, PA*, **Lubricating Oils** (under **Lubricating Agents**)

Nenad V. Mandich, *HBM Engineering Company, Lansing, IL*, **Survey** (under **Metallic Coatings**)

Chien-Pei Mao, *Delavan Inc., West Des Moines, IA*, http://www.delavan.com/, **Sprays**

Eric J. Markel, *University of South Carolina, Columbia, SC*, http://www.sc.edu/, **Nitrides**

Kathleen Markey, *Synergen, Inc., Boulder, CO*, http://www.synergen.com/, **Electrophoresis**

F. Lennart Marten, *Air Products and Chemicals, Inc., Allentown, PA*, http://www.airproducts.com/, **Vinyl Acetate Polymers**

J. R. Mason, *(Oil & Chemicals) Ltd., London, England*, **Formaldehyde**

Robert T. Mason, *Koppers Industries, Inc., Pittsburgh, PA*, http://www.koppers.com/, **Naphthalene**

F. Mauger, *Rhone-Poulene, Cranbury, NJ*, **Vanillin**

V. Maureaux, *Rhone-Poulene, Cranbury, NJ*, **Vanillin**

Joseph A. McDonough, *Hoechst Celanese Corporation, Corpus Christi, TX*, **Esters, Organic**

Donald McGregor, *Bristol-Myers Squibb Company, Wallingford, CT*, http://www.bms.com/landing/data/, **Antibiotics: Aminoglycosides**

Davy McKee, *(Oil & Chemicals) Ltd., London, England*, **Formaldehyde**

Ronald J. McKinney, *E. I. Du Pont de Nemours & Co., Inc., Wilmington, DE*, http://www.dupont.com/ag/, **Nitriles**

George L. McNew, *Boyce Thompson Laboratories, Yonkers, NY*, **Fungicides**

Wayne A. McRae, *Consultant, Mannedork, Switzerland*, **Electrodialysis**

Sudhir K. Mendiratta, *Olin Corporation, Charleston, TN*, http://www.olin.com/, **Perchloric Acid and Perchlorates**

J. Carlos Menendez, *Universidad Complutense, Madrid, Spain*, http://www.ucm.es/, **Hydantoin and its Derivatives**

George H. Miller, *Schering-Plough Research, Bloomfield, NJ*, http://www.sch-plough.com/, **Antibiotics: Chloramphenicol and Analogues**

Luray M Minkiewicz, *E. I. Du Pont de Nemours & Co., Inc., Wilmington, DE*, http://www.dupont.com/ag/, **Information Retrieval**

Scott F. Mitchell, *Huntsman Specialty Chemical Corporation, St. Louis, MO*, http://www.huntsman.com/, **Maleic Anhydride, Maleic Acid, and Furmaric Acid**

Stephen Mitchell, *University of London, Birmingham, England*, http://www.lon.ac.uk/, **Aminophenols**

Irving Moch, Jr., *E. I. Du Pont de Nemours & Co., Inc., Wilmington, DE*, http://www.dupont.com/ag/, **Hollow-Fiber Membranes**

Victor M. Monroy, *General Tire, Inc., Akron, OH*, http://www.generaltire.com/, **Initiators (Anionic)**

Braja D. Mookherjee, *International Flavors & Fragrances, Union Beach, NJ*, http://www.iff.com/, **Benzyl Alcohol and β-Phenethyl Alcohol; and Oils, Essential**

L. Dow Moore, *PPG Industries, Pittsburgh, PA*, http://www.ppg.com/, **Fiber Glass**

Bradley P. Morgan, *Pfizer, Inc., Groton, CT*, http://www.pharmacia.com/, **Steroids**

Melinda S. Moynihan, *Pfizer, Inc., Groton, CT*, http://www.pharmacia.com/, **Steroids**

Michael J. Mummey, *Huntsman Specialty Chemical Corporation, St. Louis, MO*, http://www.huntsman.com/, **Maleic Anhydride, Maleic Acid, and Furmaric Acid**

Toru Murakami, *UBE Industries, Ltd., Tokyo, Japan*, http://www.ube.com/, **Oxalic Acid**

Angelika Muscate, *University of California, San Francisco*, **Enzyme Inhibitors**

Terry N. Myers, *Elf Atochem North America, Inc., Buffalo, NY*, http://www.elf-atochem.com/, **Initiators (Free-Radical); and Peroxides and Peroxide Compounds (Organic)**

Tattanahalli Nagabhushan, *Schering-Plough Research, Bloomfield, NJ*, http://www.sch-plough.com/, **Antibiotics: Chloramphenicol and Analogues**

Nobuyuki Nagato, *Showa Denko K. K., Tokyo, Japan*, http://www.sdk.co.jp/index_e.htm, **Allyl Alcohol and Monoallyl Derivatives**

John Nagy, *Beckman Industrial Corporation, Cedar Grove, NJ*, http://beckman.com/Default.asp?bhfv=6, **Electrolytic Conductivity and Resistivity Measurements**

Kurt Nassau, *Nassau Consultants, Lebanon, NJ*, **Gemstones**

Robert J. Naumann, *University of Alabama in Huntsville, Huntsville, AL*, http://www.atmos.uah.edu/, **Space Processing**

Behrooz Nazer, *E. I. Du Pont de Nemours & Co., Inc., Wilmington, DE*, http://www.dupont.com/ag/, **Information Retrieval**

William T. Nearn, *Weyehaeuser Company, Seattle, WA*, **Wood Preservative**

W. D. Neilsen, *Union Carbide Chemicals & Plastics Company Inc*, http://www.unioncarbide.com/, **Acrolein and Derivatives**

L. H. Nemec, *Albermarle Corp., Baton Rouge, LA*, **Olefins, Higher** (under **Olefin Polymers**)

Marshall J. Nepras, *Stepan Company, Northfield, IL*, http://www.stepan.com/, **Sulfonation and Sulfation**

Ronald, W. Novak, *Rohm and Haas Company, Spring House, PA*, http://www.rohmhaas.com/, **Acrylic Ester Polymers; Methacrylic Polymers**

Mirek Novotny, *Cerdec Corporation, Washington, PA*, http://www.cerdec.com/, **Pigments (Inorganic)**

Richard A Nugent, *Upjohn Company, Kalamazoo, MI*, **Analgesics, Antipyretics, and Antiinflammatory Agents**

B. E. Obi, *The Dow Chemical Company, Midland, MI*, **Vinylidene Chloride Monomer and Polymers**

J. T. O'Connor, *Loctite Corporation, Newington, CT*, http://www.loctite.com/, **Acrylic Ester Polymers**

Hauromi Oeda, *Ajinomoto Co., Inc., Kawasaki, Japan*, http://www.ajinomoto.com/, **Amino Acids**

E. A. Ogryzlo, *University of British Columbia, Vancouver, British Columbia, Canada*, http://www.ubc.ca/, **Orbitals**

Rodrigo Orefice, *University of Florida, Gainesville, FL*, http://www.ufl.edu/, **Sol-Gel Technology**

Anil R. Oroskar, *UOP, Des Plaines, IL*, http://www.uop.com/, **Adsorption: Liquid Separation**

Neal F. Osborne, *SmithKline Beecham Pharmaceuticals, Surrey, United Kingdom*, http://www.sb.com/, **Antibiotics: β-Lactams**

Lloyd Osipow, *New York, NY*, **Surface Chemistry**

Michael J. Owen, *Dow Corning Corporation, Midland, MI*, http://www.dowcorning.com/, **Release Agents**

S. Ted Oyama, *Center for Advanced Materials, Lawrence Berkeley Laboratory, The University of California Berkeley, CA*, http://www.lbl.gov/, **Catalysis**

E. Dickson Ozokwelu, *Amoco Chemical Company, Naperville, IL*, **Toluene**

John E. Page, *GE Plastics, Pittsfield, MA*, http://www.geplastics.com/resins, ABS Resins (under **Acrylonitrile Polymers**)

Richard A. Palmer, *Dow Corning Corporation, Midland, MI*, http://www.dowcorning.com/, **Sealants**

Anthony J. Papa, *Union Carbide Chemicals and Plastics Company, Inc., South Charleston, WV*, http://www.unioncarbide.com/, **Amyl Alcohols**

F. Parenti, *Merrell-Dow Research Institute, Gerenzano, Italy*, **Antibiotics: Glycopeptides (Dalbaheptides)**

Angela K. G. Parsons, *E. I. Du Pont de Nemours & Co., Inc., Wilmington, DE*, http://www.dupont.com/ag/, **Information Retrieval**

Lloyd W. Pebsworth, *Polyethylene Technology, Morris, IL*, **Low Density Polyethylene** (under **Olefin Polymers**)

Milton Pelavin, *Miles, Inc., Tarrytown, NY*, http://www.bayer.com/, **Automated Instrumentation: Clinical Chemistry**

Carol R. Perrotto, *E. I. Du Pont de Nemours & Co., Inc., Wilmington, DE*, http://www.dupont.com/ag/, **Information Retrieval**

Donald J. Peterson, *Natural Gypsum, Gold Bond Research Center, Buffalo, NY*, http://www.national-gypsum.com/, **Calcium Sulfate**

W. L. Peticolas, *University of Oregon, Eugene, OR*, http://www.uoregon.edu/, **Order-Disorder Theory and Applications**

A. W. Petrocelli, *Westerley, RI*, **Superoxides**

Alex Pettigrew, *Ethyl Technical Center, Baton Rouge, LA*, **Halogenated Flame Retardants**

John R. Pierson, *Johnson Controls, Inc., Milwaukee, WI*, http://www.johnsoncontrols.com/bg/, **Batteries: Lead-Acid**

William Pietro, *York University, North York, Ontario, Canada*, http://www.yorku.ca/, **Biosensors**

W. T. Plass, *Forest Service, Northeastern Forest Experimentation, Princeton, WV*, **Revegetation**

Alphonsus V. Pocius, *The 3 M Company, St. Paul, MN*, http://www.3m.com/US/, **Adhesives**

Peter Pollak, *Lonza Ltd., Basel, Switzerland*, http://www.lonza.com/group/en.html, **Malonic Acid and Derivatives**

Charles M. Pollock, *Union Camp Corporation, Savannah, GA,* **Dicarboxylic Acids**

R. J. Ponsford, *SmithKline Beecham Pharmaceuticals, Surrey, United Kingdom,* http://www.sb.com/, **Antibiotics: Penicillins and Others**

Howard W. Post, *Williamsville, NY,* **Silicone Resins**

Duane B. Priddy, *The Dow Chemical Company, Midland, MI,* http://www.dow.com/, **Carboxylic Acids**

Ralph D. Priester, Jr., *Dow Chemical USA, Freeport, TX,* http://www.dow.com, **Isocyantes, Organic**

Roger C. Prince, *Exxon Research and Engineering Company, Annandale, NJ,* http://www.exxonmobil.com/corporate/, **Bioremediation**

Salvatore Profeta, Jr., *Monsanto, St. Louis, MO,* http://www.monsanto.com/, **Molecular Modeling**

Roderic P. Quirk, *University of Akron, Akron, OH,* http://www.uakron.edu/, **Initiators (Anionic)**

Philip E. Rakita, *Elf Atochem Japan, Chuo-ku, Tokyo, Japan,* **Grignard Reactions**

D. Rastler, *Electric Power Research Institute, Palo Alto, CA,* http://www.epri.com/, **Fuel Cells**

Rita D. Ratliff, *E. I. Du Pont de Nemours & Co., Inc., Wilmington, DE,* http://www.dupont.com/ag/, **Information Retrieval**

Richard W. Rees, *E. I. Du Pont de Nemours & Co., Inc., Wilmington, DE,* http://www.dupont.com/ag/, **Ionomers**

Jill Rehmann, *Fordham University, Brooklyn, NY,* http://www.fordham.edu/, **Nucleic Acids**

Nancy L. Reichenbach, *Temple University School of Medicine, Philadelphia, PA,* http://www.medschool.temple.edu/, **Antibiotics: Nucleosides and Nucleotides**

Kenneth I. G. Reid, *Tetra Chemicals, Houston, TX;* http://www.tetratec.com/nav.html, **Calcium Chloride**

Abraham Reife, *CIBA-GEIGY Corporation, Toms River, NJ,* **Dyes: Environmental Chemistry**

Ganapathi R. Revankar, *Triplex Pharmaceutical Corporation, The Woodlands, TX,* **Antiviral Agents**

Reinhard H. Richter, *Dow Chemical USA, Freeport, TX,* http://www.dow.com, **Isocyantes, Organic**

J. A. Riddick, *Baton Rouge, LA,* **Associaton (Chemical)**

Mary B. Ritchey, *Wyeth-Lederle Vaccine and Pediatrics, Pearl River, NY,* **Vaccine Technology**

G. Robert, *Rhone-Poulene, Cranbury, NJ,* **Vanillin**

Garth Roberts, *FRS, Thorn EMI plc and University of Oxford,* **Molecular and Supermolecular Electronics**

John Roberts, *Hoffmann-LaRoche, Inc., Nutley, NJ,* http://www.rocheusa.com/, **Antibiotics: Cephalosporins**

Roland K. Robins, *ICN Nucleic Acid Research Institute, Costa Mesa, CA,* http://www.usu.edu/~iar/Brochure/brochure.html, **Antiviral Agents**

Gerard Romeder, *Lonza Ltd., Basel, Switzerland,* http://www.lonza.com/group/en.html, **Malonic Acid and Derivatives**

George R. Romovacek, *Koppers Company, Inc., Monroeville, PA,* http://www.koppers.com/, **Coal Tar and Derivatives**

Stephen L. Rosen, *University of Missouri-Rolla, Rolla, MO,* http://www.chem.umr.edu/web/, **Polymers**

Donald M. Ross, *Consulting Engineer, Lancaster, CA,* **Rocket Propellants**

Alex T. Rowland, *Gettysburg College, Gettysburg, PA,* http://www.gettysburg.edu/homepage/home.html, **Stereoisomerism**

Elmer B. Rowley, *F.M.S.A., formerly Mineral Curator, Department of Civil Engineering, Union College, Schenectady, NY,* http://www.union.edu/, **Beryl;** **Feldspar; Galena; Garnet; Hornblende; Magnetite; Mineralogy; Nepheline; Quartz; Serpentine; Spinel;** and **Tourmaline**

Aroop K. Roy, *Dow Coring Corporation, Midland, MI,* **Inorganic High-Polymers**

Scott Rudge, *Synergen, Inc., Boulder, CO,* http://www.synergen.com/, **Electrophoresis**

Charles V. Rue, *Norton Company, Worcester, MA,* www.nortonabrasives.com, **Abrasives**

Douglas M. Ruthven, *University of New Brunswick, Fredericton, New Brunswick, Canada,* http://www.unb.ca/, **Adsorption**

Alvin J. Salkind, *Rutgers, University, Piscataway, NJ,* http://www.rutgers.edu/, **Batteries: Secondary Cells**

Jose Sanchez, *Elf Atochem North America, Inc., Buffalo, NY,* http://www.elf-atochem.com/, **Initiators (Free-Radical);** and **Peroxides and Peroxide Compounds (Organic)**

S. J. Sansonetti, *Consultant, Reynolds Metals Company, Richmond, VA,* http://www.alcoa.com/, **Aluminum;** and **Cryolite**

Robert P. Santandrea, Ph.D., *Los Alamos National Laboratory, Los Alamos, NM,* http://www.lanl.gov/worldview/, **Microgravity and Materials Processing**

E. J. Sare, *PPG industries, Inc.,* http://www.ppg.com/, **Allyl Ester Resins**

Harold A. Sarvetnick, *Westfield, NJ,* **Polyvinyl Chloride (PVC)**

J. D. Sauer, *Albermarle Corp., Baton Rouge, LA,* **Olefins, Higher** (under **Olefin Polymers**)

Hiroyuki Sawada, *UBE Industries, Ltd., Tokyo, Japan,* http://www.ube.com/, **Oxalic Acid**

Joseph W. Schappel, *Avtex Fibers, Inc., Fort Royal, VA,* **Fibers: Acetate**

G. Scherr, *BASF AG, Ludwigshafen, Germany,* **Imines, Cyclic**

C. E. Schildknecht, *Gettysburg College, Gettysburg, PA,* http://www.gettysburg.edu/homepage/home.html, **Stereoregular Polymers**

Hollis G Schoepke, *Anaquest, Murray Hill, NJ,* **Anesthetics**

M. Schussler, *Fansteel, Inc., North Chicago, IL,* http://www.fansteel.com/, **Tantalum**

Glenn T. Seaborg, *University of California, Berkeley, Berkeley, CA,* http://www.berkeley.edu/, **Actinides and Transactinides**

Masaaki Sekino, *Toyobo Co., Ltd., Iwakuni, Yamaguchi-Pref, Japan,* **Desalination**

Curtis C. Selph, *Propellant Research Engineer, U.S. Air Force Rocket Propulsion Laboratory, Edwards, CA,* http://www.pr.afrl.af.mil/, **Rocket Propellants**

George A. Serad, *Hoechst Celanese Corporation, Charlotte, NC,* http://www.celanese.com/, **Fibers: Cellulose Esters**

Raymond B. Seymour, *University of Houston, Houston, TX,* http://www.uh.edu/, **Polymides; Polyvinyl Alkyl Ethers; Polyvinylidene Chloride**

Anand S. G. Sharangpani, *BASF Corporation, Holland, MI,* http://www.corporate.basf.com/en/, **Pigment Dispersions**

Winston G. Shequen, P.E., *Bausch & Lomb/ARL, Sunland, CA,* http://www.bausch.com/, **Ion Microprobe Mass Analyzer;** and **X-Ray Analysis**

John D. Sherman, *UOP, Tarrytown, NY,* http://www.uop.com/, **Adsorption: Gas Separation**

Gary J. Shiflet, *University of Virginia, Charlottesville, VA,* http://www.virginia.edu/, **Glassy Metals**

E. C. Shuman, *Consulting Engineer, State College, PA,* **Insulation (Thermal)**

Jeanette C. Sikes, *E. I. Du Pont de Nemours & Co., Inc., Wilmington, DE,* http://www.dupont.com/ag/, **Information Retrieval**

Gary S. Silverman, *Elf Atochem North America, King of Prussia, PA,* http://www.arcat.com/arcatcos/cos32/arc32238.cfm, **Grignard Reactions**

David E. Smith, *FMC Corporation, Princeton, NJ,* http://www.fmc.com/, **Carbon Disulfide**

Duane H. Smith, *Technical Solutions and West Virginia University, Morganstown, WV,* **Microemulsions**

Peter A. S. Smith, *University of Michigan, Ann Arbor, MI,* http://www.umich.edu/, **Nomenclature**

Roy E. Smith, *CIBA-GEIGY Corporation, Greensboro, NC,* **Dyes: Reactive**

William L. Smith, *The Clorox Company, Pleasanton, CA,* http://www.clorox.com/, **Bleaching Agents**

Asta Sokov, *Universite de Sherbrooke, Sherbrooke, Quebec, Canada,* **Asbestos**

Z. Solc, *University of Pardubice, Czech Republic,* http://www.upce.cz/indexan.htm, **Pigments (Inorganic)**

Gabor A. Somorjai, *Center for Advanced Materials, Lawrence Berkeley Laboratory, The University of California Berkeley, CA,* http://www.lbl.gov/, **Catalysis**

Robert Southgate, *SmithKline Beecham Pharmaceuticals, Surrey, United Kingdom,* http://www.sb.com/, **Antibiotics: β-Lactams**

Theodore C. Spaulding, *Anaquest, Murray Hill, NJ,* **Anesthetics**

Elmer Sperry, *Beckman Industrial Corporation, Cedar Grove, NJ,* http://beckman.com/Default.asp?bhfv=6, **Electrolytic Conductivity and Resistivity Measurements**

Apryll M. Stalcup, *University of Cincinnati, Cincinnati, OH,* http://www.uc.edu/, **Chiral Separations**

J. G. Stam, *Pfizer Central Research, Groton, CT,* http://www.pfizer.com/main.html, **Antibiotics: β-Lactamase Inhibitors**

Dale S. Steichen, *The Clorox Company, Pleasanton, CA,* http://www.clorox.com/, **Bleaching Agents**

U. Steuerle, *BASF AG, Ludwigshafen, Germany,* **Imines, Cyclic**

Norman S. Stoloff, *Rensselaer Polytechnic Institute,Troy, NY,* http://www.rpi.edu/, **High Temperature Alloys**

Ulrich P. Strauss, *Rutgers University, New Brunswick, NJ,* http://www.rutgers.edu/, **Polyelectrolytes**

David M Sturmer, *Eastman Kodak Company, Rochester, NY,* http://www.kodak.com/, **Dyes: Sensitizing**

Kyung W. Suh, *The Dow Chemical Company, Granville, OH,* http://www.dow.com/, **Foamed Plastics**

Robert J. Suhadolnik, *Temple University School of Medicine, Philadelphia, PA,* http://www.medschool.temple.edu/, **Antibiotics: Nucleosides and Nucleotides**

Edward A. Sullivan, *Morton International, Danvers, MA,* **Hydrides**

James W. Summers, *The Geon Company, Avon Lake, OH,* http://www.geon.com/, **Vinyl Chloride**

Richard J. Sundberg, *University of Virginia, Charlottesville, VA,* http://www.virginia.edu/, **Indoles**

Kenneth S. Suslick, *University of Illinois at Urbana-Champaign, Urbana, IL,* http://www.scs.uiuc.edu/chem/, **Sonochemistry**

Boyce Sutton, Jr., *Sybron Chemicals Inc., Welford, SC,* http://www.sybronchemicals.com/, **Dye Carriers**

Ladislav Svarovsky, *Consultant Enginners and Fine Particle Software, Heaton/Bradford/West, Yorkshire, U. K.,* **Filtration**

Henry F. Szepan, *(retired), Ingersoll-Rand Co, Impco Division, Nashua, NH,* **Pulp (Wood) Production and Processing**

Mannan Talukder, *Advanced Aromatics, Inc., Baytown, TX,* www.advancedaromatics.com, **Naphthalene Derivatives**

James T. Tanner, *North Carolina State University, Asheville, NC,*ncsu.edu www.engr.ncsu.edu/mrl, **Mica**

Roger Tate, *Delavan Inc., West Des Moines, IA,* http://www.delavan.com/, **Sprays**

Kwoliang D. Tau, *Hoechst Celanese Corporation, Corpus Christi, TX,* **Esters, Organic**

Dunbar G. Terry, *(retired), Ingersoll-Rand Co., Impco Division, Nashua, NH,* **Pulp (Wood) Production and Processing**

Dean Thetford, *Zeneca Specialties, Manchester, U.K.,* http://www.touchmanchester.co.uk/comdir/cditem.cfm/5036, **Triphenylmethane and Related Compounds**

Curt Thies, *Washington University, St. Louis, MO,* http://www.wustl.edu/, **Microencapsulation**

Hendrik Timmerman, *Leiden/Amsterdam Center for Drug Research,* **Histamine and Histamine Antagonists**

Norman S. Thompson, *Consultant, Appleton, WI,* **Hemicellulose**

Quentin E. Thompson, *Monsanto Company, Millstadt, IL,* http://www.monsanto.com/, **Biphenyl and Terphenyls**

Robert W. Timmerman, *FMC Corporation, Princeton, NJ,* http://www.fmc.com/, **Carbon Disulfide**

G. Paull Torrence, *Hoechst Celanese Corporation, Corpus Christi, TX,* **Esterification**

Irving Touval, *Touval Associates, Sparta, NJ,* **Antimony and Other Inorganic Flame Retardants**

David J. Triggle, *State University of New York at Buffalo, Buffalo, NY,* http://www.suny.edu/, **Pharmacodynamics**

M. Trojan, *University of Pardubice, Czech Republic,* http://www.upce.cz/indexan.htm, **Pigments (Inorganic)**

F. Truchet, *Rhone-Poulene, Cranbury, NJ,* **Vanillin**

Paul S. Tully, *Stepan Company, Northfield, IL,* http://www.stepan.com/, **Sulfonic Acids**

John Turrell, *Technicon Instruments Corporation, Tarrytown, NY,* **Automated Instrumentation: Hematology**

Henri Ulrich, *Consultant, Guilford, CT,* **Urethane Polymers**

Pamla R. Umberger, *Union Carbide Chemicals and Plastics Company Inc., South Charleston, WV,* http://www.unioncarbide.com, **Diamines and Higher Amines, Aliphatic**

Arthur M. Usmani, *Bridgestone/Firestone, Inc., Carmel, IN,* http://www.bridgestone-firestone.com/, **Medical Diagnostic Reagents**

Rajesh Vaidya, *University of New Mexico, Albuquerque, NM,* http://www.unm.edu/, **Nanotechnology (Molecular)**

Jan F. VanPeppen, *Allied Signal Inc., Petersberg, VA,* http://www.alliedsignal.com/, **Cyclohexanol-Cyclohexanone**

Kanwal J. Varma, *Schering-Plough Research, Bloomfield, NJ,* http://www.sch-plough.com/, **Antibiotics: Chloramphenicol and Analogues**

Manual G. Venegas, *The Procter & Gamble Company, Cincinnati, OH,* http://www.pg.com/, **Detergents**

***R. C. Vickery, M.D., Ph.D., D.Sc, (R. C. V),** Blanton/Dade City, FL,* **Carcinogens; Free Radical; Photochemistry and Photolysis; Photosynthesis; and Polypeptide**

R. Villalobos, *The Foxboro Company (A Siebe Company) Foxboro, MA,* **Chromatography**

Dan Vlastelica, *Miles, Inc., Tarrytown, NY,* http://www.bayer.com/, **Automated Instrumentation: Clinical Chemistry**

Karl S. Vorres, *Argonne National Laboratory, Argonne, IL,* http://www.anl.gov/, **Lignite and Brown Coal**

Dolatrai M. Vyas, *Bristol-Meyers Squibb Company, Wallingford, CT,* http://www.bms.com/, **Chemotherapeutics, Anticancer**

J. D. Wagner, *Albermarle Corp., Baton Rouge, LA,* **Olefins, Higher (under Olefin Polymers)**

Phillip A. Waitkus, *Plastics Engineering Company, Sheboygan, WI,* http://www.plenco.com/, **Phenolic Resins**

Richard J. Wakeman, *University of Exeter, Devon, U. K.,* http://www.ex.ac.uk/, **Extraction (Liquid-Solid)**

Kenneth A. Walsh, Ph.D., *Brush Wellman Inc., Elmore, OH,* http://www.brushwellman.com/, **Beryllium**

Timothy Ward, *Millsaps College, Jackson, MS,* http://www.millsaps.edu/, **Biopolymers**

Rosemary Waring, *University of Birmingham, Birmingham, England,* http://www.bham.ac.uk/, **Aminophenols**

Albin H. Warth, *Cape May, NJ,* **Waxes**

A. Dinsmoor Webb, *University of California, Davis,* http://www-chem.ucdavis.edu/, **Vinegar**

Byron H. Webb, *U.S. Department of Agriculture, Washington, D.C.,* **Milk and Milk Products**

Edwin Weber, *Technische Universitat Bergakademi Freiberg, Institut Fur organische Chemie, Germany,* **Inclusion Compounds; Molecular Recognition**

Amie H. Webster, *E. I. Du Pont de Nemours & Co., Inc., Wilmington, DE,* http://www.dupont.com/ag/, **Information Retrieval**

Edward D. Weil, *Polytechnic University, Brooklyn, NY,* http://www.poly.edu/, **Phosphorus Flame Retardants**

David A. Weitz, *Exxon Research & Engineering Company, Annandale, NJ,* http://www.exxon.com/, **Foam**

J. Y. Welsh, *Chemetals Corporation, Baltimore, MD,* http://www.chemetals.com, **Manganese**

Armin Wendel, *Rhone-Poulenc Rorer, Cologne, Germany,* http://www.rhone-poulenc.com/, **Lecithin**

Jack Wernick, *Murray Hill, NJ,* **Thin Films and Particles (under Magnetic Materials)**

R. A. Wessling, *The Dow Chemical Company, Midland, MI,* **Vinylidene Chloride Monomer and Polymers**

Peter J. Wessner, *Merichem Company, Houston, TX,* http://www.merichem.com/, **Naphthenic Acids**

David L. White, *Kwick Kleen Industrial Solvents, Inc., Vicennes, IN,* **Paint and Finish Removers**

Zeno W. Wicks, Jr., *Consultant, Las Cruces, NM,* **Drying Oils**

Pieter R. Wiederhold, *General Eastern Instruments Corporation, Watertown, MA,* **Hygrometry and Psychrometry**

Paul Wight, *Seneca Specialties, Manchester, U.K.* **Xanthene Dyes**

E. Williams, *Cobalt Information Centre, London, UK,* **Cobalt**

Richard A Wilsak, *Amoco Chemical Company, Naperville, IL,* **Butylenes**

Richard A. Wilson, *International Flavors & Fragrances, Union Beach, NJ,* http://www.iff.com/, **Benzyl Alcohol and β-Phenethyl Alcohol; Oils, Essential**

Edmund M. Wise, Jr., *The Wellcome Research Laboratories, Research Triangle Park, NC,* **Antibiotics: Peptides**

Donald T. Witlak, *University of Wisconsin-Madison, Madison, WI,* http://www.wisc.edu/, **Pharmaceuticals, Chiral**

Shuad Wojkowski, *U.S. Department of Agriculture, New Orleans, LA,* http://www.ars.usda.gov/, **Herbicides**

John A. Wojtowicz, *Consultant, Olin Corporation, Cheshire, CT,* **Ozone**

C.J. Wust, Jr., *Hercules Inc., Wilmington, DE,* http://www.herc.com/, **Olefin Fibers**

Carmen M Yon, *UOP, Tarrytown, NY,* http://www.uop.com/, **Adsorption: Gas Separation**

Raymond A. Young, *University of Wisconsin, Madison, WI,* http://www.wisc.edu/, **Fibers: Vegetable**

Aleksey Zaks, *Schering-Plough Research Institute, Union, NJ,* http://www.sch-plough.com/schering_plough/research/research_institute.jsp, **Enzymes in Organic Synthesis**

Paul Zanowiak, *Temple University, Philadelphia, PA,* http://www.temple.edu/, **Pharmaceuticals**

David Zelmanovic, *Technicon Instruments Corporation, Tarrytown, NY,* **Automated Instrumentation: Hematology**

Edward G/Zey, *Hoechst Celanese Corporation, Corpus Christi, TX,* **Esterification**

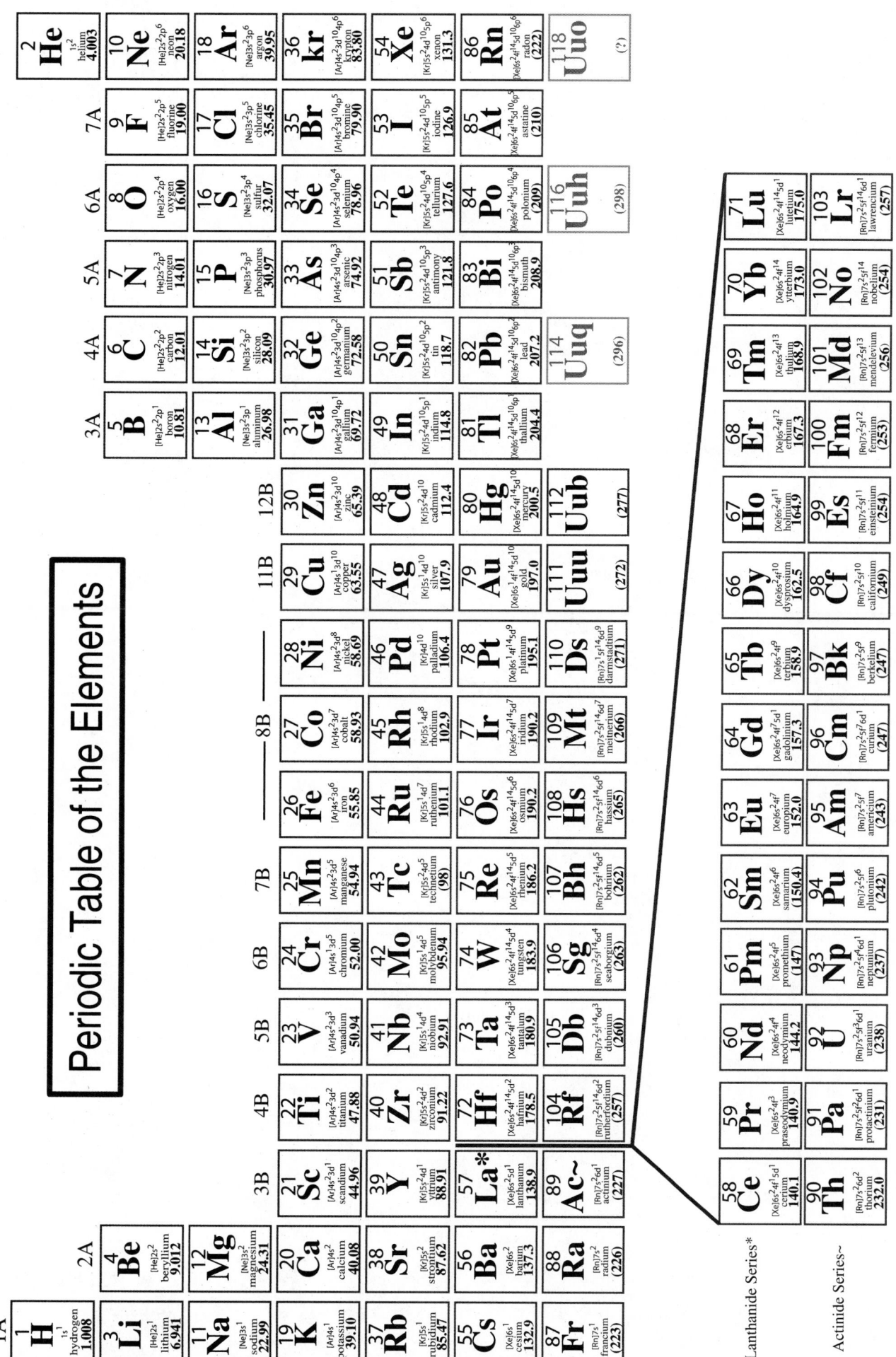

Periodic Table of the Elements

Los Alamos National Laboratory Chemistry Division

VAN NOSTRAND'S

ENCYCLOPEDIA OF CHEMISTRY

Fifth Edition

A

AAAS. The American Association for the Advancement of Science was founded in 1848 and incorporated in 1874. Its objectives are to further the work of scientists, to facilitate cooperation among them, to foster scientific freedom and responsibility, to improve the effectiveness of science in promoting human welfare, to advance education in science, and to increase public understanding and appreciation for the importance and promise of the methods of science in human progress. The AAAS head quarters is in Washington, DC. Additional information on the AAAS can be found at *http://www.aaas.org/* and *http://www.sciencemag.org/*.

ABACA. The sclerenchyma bundles from the sheathing leaf bases of *Musa textilis*, a plant closely resembling the edible banana plant. These bundles are stripped by hand, after which they are cleaned by drawing over a rough knife. The fiber bundles are now whitish and lustrous, and from six to twelve feet (1.8–3.6 meters) long. Being coarse, extremely strong and capable of resisting tension, they are much used in the manufacture of ropes and cables. Since the fibers swell only slightly when wet, they are particularly suited for rope that will be used in water. Waste manila fibers from rope manufacture and other sources are used in the making of a very tough grade of paper, known as manilla paper. The fibers may be obtained from both wild and cultivated plants, the latter yielding a product of better grade. The cultivated plants, propagated by seeds, by cuttings of the thick *rhizomes* or by suckers, are ready for harvest at the end of three years, after which a crop may be expected approximately every three years.

ABHERENT. Any substance that prevents adhesion of a material to itself or to another material. It may be in the form of a dry powder (a silicate such as talc, mica, or diatomaceous earth); a suspension (bentonite-water); a solution (soap-water); or a soft solid (stearic acid, tallow waxes). Abherents are used as dusting agents and mold washes in the adhesives, rubber, and plastics industries. Fats and oils are used as abherents in the baking industry. Fluorocarbon resin coatings on metals are widely used on cooking utensils.

ABLATING MATERIAL. A material, especially a coating material, designed to provide thermal protection to a body in a fluid stream through loss of mass. Ablating materials are used on the surfaces of some reentry vehicles to absorb heat by removal of mass, thus blocking the transfer of heat to the rest of the vehicle and maintaining temperatures within design limits. Ablating materials absorb heat by increasing in temperature and changing in chemical or physical state. The heat is carried away from the surface by a loss of mass (liquid or vapor). The departing mass also blocks part of the convective heat transfer to the remaining material in the same manner as transpiration cooling.

(1) Fibers made from white silica, fused in an oven, cut into blocks, and coated with borosilicate glass; these are extremely efficient at temperatures up to 2300F. (2) An all-carbon composite (called reinforced carbon-carbon) make by laminating and curing layers of graphite fiber previously coated with a resin, which is pyrolized to carbon. The resulting tile is then treated with a mixture of alumina, silicon, and silicon carbide. Such composites are used for maximum-temperature (nose cone) exposure up to 3000F. Both types are undamaged by the heat and are reusable. The tiles are adhered to the body of the spacecraft with a silicone adhesive. Ablative materials used on early spaceship trials were fluorocarbon polymers and glass-reinforced plastics, but these were wholly or partially decomposed during reentry.

ABLATION. The removal of surface material from a body by vaporization, melting, chipping, or other erosive process; specifically, the intentional removal of material from a nose cone or spacecraft during high-speed movement through a planetary atmosphere to provide thermal protection to the underlying structure. See also **Ablating Material**.

ABRASION. All metallic and nonmetallic surfaces, no matter how smooth, consist of minute serrations and ridges that induce a cutting or tearing action when two surfaces in contact move with respect to each other. This wearing of the surfaces is termed abrasion. Undesirable abrasion may occur in bearings and other machine elements, but abrasion is also adapted to surface finishing and machining, where the material is too hard to be cut by other means, or where precision is a primary requisite.

Temperature is a significant factor: friction may raise the temperature of the surface layers to the point where they become subject to chemical attack. Abrasion causes deterioration of many materials, especially of rubber (tire treads), where it can be offset by a high percentage of carbon black. Other materials subjected to abrasion in their service life are textiles (laundering), leather and plastics (shoe soles, belting), and house paints and automobile lacquers (airborne dust, grit, etc.).

See also **Abrasives**.

ABRASION pH. A term originated by Stevens and Carron in 1948 "to designate the pH values obtained by grinding minerals in water." Abrasion pH measurements are useful in the field identification of minerals. The pH values range from 1 for ferric sulfate minerals, such as coquimbite, konelite, and rhomboclase, to 12 for calcium-sodium carbonates, such as gaylussite, pirssonite, and shortite. The recommended technique for determining abrasion pH is to grind, in a nonreactive mortar, a small amount of the mineral in a few drops of water for about one minute. Usually, a pH test paper is used. Values obtained in this manner are given in the middle column of Table 1. Another method, proposed by Keller et al. in 1963, involves the grinding of 10 grams of crushed mineral in 100 milliliters of water and noting the pH of the resulting slurry electronically. Values obtained in this manner are given in the right-hand column.

TABLE 1. ABRASION pH VALUES OF REPRESENTATIVE MINERALS

Mineral	pH by Stevens-Carron Method	pH by Keller et al. Method
Coquimbite	1	
Melanterite	2	
Alum	3	
Glauconite	5	5.5[a]
Kaolinite	5, 6, 7	5.5[a]
Anhydrite	6	
Barite	6	
Gypsum	6	
Quartz	6, 7	6.5
Muscovite	7, 8	8.0
Calcite	8	8.4
Biotite	8, 9	8.5
Microcline	8, 9	8.0 9.0[a]
Labradorite		8.0 9.2[a]
Albite	9, 10	
Dolomite	9, 10	8.5
Hornblende	10	8.9
Leucite	10	
Diopside	10, 11	9.9
Olivine	10, 11	9.6[a]
Magnesite	10, 11	

[a] More recent values published in literature.

Additional Reading

Keller, W.D., W.D. Balgord, and A.L. Reesman: "Dissolved Products of Artificially Pulverized Silicate Minerals and Rocks," *Jrnl. Sediment. Petrol.*, **33**(1), 191–204 (1963).

ABRASIVES. An abrasive is a substance used to abrade, smooth, or polish an object. If the object is soft, such as wood, then relatively soft abrasive materials may be used. Usually, however, abrasive connotes very hard substances ranging from naturally occurring sands to the hardest material known, diamond.

There are three basic forms of abrasives: grit (loose, granular, or powdered particles); bonded materials (particles are bonded into wheels, segments, or stick shapes); and coated materials (particles are bonded to paper, plastic, cloth, or metal).

Properties of Abrasive Materials

Hardness. Table 1 lists the various scales of hardness used for abrasives.

Toughness. An abrasive's toughness is often measured and expressed as the degree of friability, the ability of an abrasive grit to withstand impact without cracking, spalling, or shattering.

Refractoriness (Melting Temperature). Instantaneous grinding temperatures may exceed 3500°C at the interface between an abrasive and the workpiece being ground. Hence melting temperature is an important property.

Chemical Reactivity. Any chemical interaction between abrasive grains and the material being abraded affects the abrasion process.

Thermal Conductivity. Abrasive materials may transfer heat from the cutting tip of the grain to the bond posts, retaining the heat in a bonded wheel or coated belt. The cooler the cutting point, the harder it is.

Fracture. Fracture characteristics of abrasive materials are important, as well as the resulting grain shapes. Equiaxed grains are generally preferred for bonded abrasive products and sharp, acicular grains are preferred for coated ones. How the grains fracture in the grinding process determines the wear resistance and self-sharpening characteristics of the wheel or belt.

Microstructure. Crystal size, porosity, and impurity phases play a major role in fixing the fracture characteristics and toughness of an abrasive grain.

Natural Abrasives

Naturally occurring abrasives are still an important item of commerce, although synthetic abrasives now fill many of their former uses. They include diamonds, corundum, emery, garnet, silica, sandstone, tripoli, pumice, and pumicite.

Manufactured Abrasives

Manufactured abrasives include silicon carbide, fused aluminum oxide, sintered aluminum oxide, sol–gel sintered aluminum oxide, fused zirconia–alumina, synthetic diamond, cubic boron nitride, boron carbide, slags, steel shot, and grit.

Sizing, Shaping, and Testing of Abrasive Grains

Sizing. Manufactured abrasives are produced in a variety of sizes that range from a pea-sized grit of 4 (5.2 mm) to submicrometer diameters.

Shaping. Desired shapes are obtained by controlling the method of crushing and by impacting or mulling. In general, cubical particles are preferred for grinding wheels, whereas high aspect-ratio acicular particles are preferred for coated abrasive belts and disks.

Testing. Chemical analyses are done on all manufactured abrasives, as well as physical tests such as sieve analyses, specific gravity, impact strength, and loose poured density (a rough measure of particle shape). Special abrasives such as sintered sol–gel aluminas require more sophisticated tests such as electron microscope measurement of α-alumina crystal size, and indentation microhardness.

Coated Abrasives

Coated abrasives consist of a flexible backing on which films of adhesive hold a coating of abrasive grains. The backing may be paper, cloth, open-mesh cloth, vulcanized fiber (a specially treated cotton rag base paper), or any combination of these materials. The abrasives most generally used are fused aluminum oxide, sol–gel alumina, alumina–zirconia, silicon carbide, garnet, emery, and flint.

A new form of coated abrasive has been developed that consists of tiny aggregates of abrasive material in the form of hollow spheres. As these spheres break down in use, fresh cutting grains are exposed; this maintains cut-rate and keeps power low.

Bonded Abrasives

Grinding wheels are by far the most important bonded abrasive product both in production volume and utility. They are produced in grit sizes ranging from 4, for steel mill snagging wheels, to 1200, for polishing the surface of rotogravure rolls.

Marking System. Grinding wheels and other bonded abrasive products are specified by a standard marking system which is used throughout most of the world. This system allows the user to recognize the type of abrasive, the size and shaping of the abrasive grit, and the relative amount and type of bonding material.

Bond Type. Most bonded abrasive products are produced with either a vitreous (glass or ceramic) or a resinoid (usually phenolic resin) bond.

Special Forms of Bonded Abrasives. Special forms of bonded abrasives include honing and superfinishing stones, pulpstone wheels, crush-form grinding wheels, and creep feed wheels.

Superabrasive Wheels

Superabrasive wheels include diamond wheels and cubic boron nitride (CBN) wheels.

Grinding Fluids

Grinding fluids or coolants are fluids employed in grinding to cool the work being ground, to act as a lubricant, and to act as a grinding aid. Soluble oil coolants in which petroleum oils are emulsified in water have been developed to impart some lubricity along with rust-preventive properties.

Loose Abrasives

In addition to their use in bonded and coated products, both natural and manufactured abrasive grains are used loose in such operations as polishing, buffing, lapping, pressure blasting, and barrel finishing.

Jet Cutting

High pressure jet cutting with abrasive grit can be used on metals to produce burn-free cuts with no thermal or mechanical distortion.

Health and Safety

Except for silica and natural abrasives containing free silica, the abrasive materials used today are classified by NIOSH as nuisance dust materials and have relatively high permissible dust levels.

CHARLES V. RUE
Norton Company

TABLE 1. SCALES OF HARDNESS

Material	Mohs' scale	Ridgeway's scale	Woodell's scale	Knoop hardness[a], kN/m^{2b}
talc	1			
calcite	3			
apatite	5			
vitreous silica		7		
topaz	8	9		13
corundum	9		9	20
fused ZrO_2/Al_2O_3[c]				16
SiC		13	14	24
cubic boron nitride				46
diamond	10	15	42.5	78

[a] At a 100-g load (K-100) average.
[b] To convert kN/m^2 to kgf/mm^2 divide by 0.00981.
[c] 39% ZrO_2 (NZ Alundum).

Additional Reading

Arpe, H.-J.: *Ullmann's Encyclopedia of Industrial Chemistry, Abrasives to Aluminum Oxide*, Vol. 1, 5th Edition, John Wiley & Sons, Inc., New York, NY, 1997.
Coes, L. Jr., *Abrasives*, Springer-Verlag, New York, NY, Vienna, 1971.

Ishikawa, T. *1986 Proceedings of the 24th Abrasive Engineering Society Conference,* Abrasive Engineering Society, Pittsburgh, PA, 1986, pp. 32–51.

Shaw, M.C.: *Principles of Abrasive Processing,* Oxford University Press, New York, NY, 1996.

Sluhan, C.A. *Lub. Eng.,* 352–374 (Oct. 1970).

ABSOLUTE 1. Pertaining to a measurement relative to a universal constant or natural datum, as absolute coordinate system, absolute altitude, absolute temperature. See also **Absolute Temperature**. 2. Complete, as in absolute vacuum.

ABSOLUTE TEMPERATURE. The fundamental temperature scale used in theoretical physics and chemistry, and in certain engineering calculations such as the change in volume of a gas with temperature. Absolute temperatures are expressed either in degrees Kelvin or in degrees Rankine, corresponding respectively to the centigrade and Fahrenheit scales. Temperatures in Kelvins are obtained by adding 273 to the centigrade temperature (if above $°C$) or subtracting the centigrade temperature from 273 (if below $°C$). Degrees Rankine are obtained by subtracting 460 from the Fahrenheit temperature.

ABSOLUTE ZERO. Conceptually that temperature where there is no molecular motion, no heat. On the Celsius scale, absolute zero is $-273.15°C$, on the Fahrenheit scale, $-459.67°F$; and zero Kelvin (0 K). The concept of absolute zero stems from thermodynamic postulations.

Heat and temperature were poorly understood prior to Carnot's analysis of heat engines in 1824. The Carnot cycle became the conceptual foundation for the definition of temperature. This led to the somewhat later work of Lord Kelvin, who proposed the Kelvin scale based upon a consideration of the second law of thermodynamics. This leads to a temperature at which all the thermal motion of the atoms stops. By using this as the zero point or absolute zero and another reference point to determine the size of the degrees, a scale can be defined. The Comit'e Consultative of the International Committee of Weights and Measures selected 273.16 K as the value for the triple point for water. This set the ice-point at 273.15 K.

From the standpoint of thermodynamics, the thermal efficiency E of an engine is equal to the work W derived from the engine divided by the heat supplied to the engine, $Q2$. If $Q1$ is the heat exhausted from the engine,

$$E = (W/Q2) = (Q2 - Q1)/Q2 = 1 - (Q1/Q2)$$

where W, $Q1$, and $Q2$ are all in the same units. A Carnot engine is a theoretical one in which all the heat is supplied at a single high temperature and the heat output is rejected at a single temperature. The cycle consists of two adiabatics and two isothermals. Here the ratio $Q1/Q2$ must depend only on the two temperatures and on nothing else. The Kelvin temperatures are then defined by the relation where $Q1/Q2$ is the ratio of the heats rejected and absorbed, and $T1/T2$ is the ratio of the Kelvin temperatures of the reservoir and the source. If one starts with a given size for the degree, then the equation completely defines a thermodynamic temperature scale.

$$\frac{Q1}{Q2} = \frac{T1}{T2}$$

A series of Carnot engines can be postulated so that the first engine absorbs heat Q from a source, does work W, and rejects a smaller amount of heat at a lower temperature. The second engine absorbs all the heat rejected by the first one, does work, and rejects a still smaller amount of heat which is absorbed by a third engine, and so on. The temperature at which each successive engine rejects its heat becomes smaller and smaller, and in the limit this becomes zero so that an engine is reached which rejects no heat at a temperature that is absolute zero. A reservoir at absolute zero cannot have heat rejected to it by a Carnot engine operating between a higher temperature reservoir and the one at absolute zero. This can be used as the definition of absolute zero. Absolute zero is then such a temperature that a reservoir at that temperature cannot have heat rejected to it by a Carnot engine which uses a heat source at some higher temperature.

ABSORPTIMETRY. A method of instrumental analysis, frequently chemical, in which the absorption (or absence thereof) of selected electromagnetic radiation is a qualitative (and often quantitative) indication of the chemical composition of other characteristics of the material under observation. The type of radiation utilized in various absorption-type instruments ranges from radio and microwaves through infrared, visible, and ultraviolet radiation to x-rays and gamma rays. See also **Analysis (Chemical)**; and **Spectro Instruments**.

ABSORPTION BAND. A range of wavelengths (or frequencies) in the electromagnetic spectrum within which radiant energy is absorbed by a substance. When the absorbing substance is a polyatomic gas, an absorption band actually is composed of a group of discrete absorption lines, which appear to overlap. Each line is associated with a particular mode of vibration or rotation induced in a gas molecule by the incident radiation. The absorption bands of oxygen and ozone are often referred to in the literature of atmospheric physics.

The important bands for oxygen are (1) the Hopfield bands, very strong, between about 670 and 1000 angstroms in the ultraviolet; (2) a diffuse system between 1019 and 1300 angstroms; (3) the Schumann-Runge continuum, very strong, between 1350 and 1760 angstroms; (4) the Schumann-Runge bands between 1760 and 1926 angstroms; (5) the Herzberg bands between 2400 and 2600 angstroms; (6) the atmospheric bands between 5380 and 7710 angstroms in the visible spectrum; and (7) a system in the infrared at about 1 micron.

The important bands for ozone are the Hartley bands between 2000 and 3000 angstroms in the ultraviolet, with a very intense maximum absorption at 2550 angstroms; the Huggins bands, weak absorption between 3200 and 3600 angstroms; the Chappius bands, a weak diffuse system between 4500 and 6500 angstroms in the visible spectrum; and the infrared bands centered at 4.7, 9.6 and 14.1 microns, the latter being the most intense.

See also **Absorption Spectrum**; and **Electromagnetic Spectrum**.

ABSORPTION COEFFICIENT 1. For the absorption of one substance or phase in another, as in the absorption of a gas in a liquid, the absorption coefficient is the volume of gas dissolved by a specified volume of solvent; thus a widely used coefficient is the quantity a in the expression $\alpha = V_0/V_p$, where V_0 is the volume of gas reduced to standard conditions, V is the volume of liquid, and p is the partial pressure of the gas.

2. In the case of sound, the absorption coefficient (which is also called the acoustical absorptivity) is defined as the fraction of the incident sound energy absorbed by a surface or medium, the surface being considered part of an infinite area.

3. In the most general use of the term, absorption coefficient, applied to electromagnetic radiation and atomic and subatomic particles, is a measure of the rate of decrease in intensity of a beam of photons or particles in its passage through a particular substance. One complication in the statement of the absorption coefficient arises from the cause of the decrease in intensity. When light, x-rays, or other electromagnetic radiation enters a body of matter, it experiences in general two types of attenuation. Part of it is subjected to scattering, being reflected in all directions, while another portion is absorbed by being converted into other forms of energy. The scattered radiation may still be effective in the same ways as the original, but the absorbed portion ceases to exist as radiation or is re-emitted as secondary radiation. Strictly, therefore, we have to distinguish the true absorption coefficient from the scattering coefficient; but for practical purposes it is sometimes convenient to add them together as the total attenuation or extinction coefficient.

If appropriate corrections are made for scattering and related effects, the ratio I/I_0 is given by the laws of Bouguer and Beer. Here, I_0 is the intensity or radiant power of the light incident on the sample and I is the intensity of the transmitted light. This ratio $I/I_0 = T$ is known as the transmittance. See also **Spectrochemical Analysis (Visible)**.

ABSORPTION (Process). Absorption is commonly used in the process industries for separating materials, notably a specific gas from a mixture of gases; and in the production of solutions such as hydrochloric and sulfuric acids. Absorption operations are very important to many air pollution abatement systems where it is desired to remove a noxious gas, such as sulfur dioxide or hydrogen sulfide, from an effluent gas prior to releasing the material to the atmosphere. The absorption medium is a liquid in which (1) the gas to be removed, i.e., absorbed is soluble in the liquid, or (2) a chemical reaction takes place between the gas and the absorbing liquid. In some instances a chemical reagent is added to the absorbing liquid to increase the ability of the solvent to absorb.

Wherever possible, it is desired to select an absorbing liquid that can be regenerated and thus recycled and used over and over. An example of absorption with chemical reaction is the absorption of carbon dioxide from a flue gas with aqueous sodium hydroxide. In this reaction, sodium carbonate is formed. This reaction is irreversible. However, continued absorption of the carbon dioxide with the sodium carbonate solution results in the formation of sodium acid carbonate. The latter can be decomposed upon heating to carbon dioxide, water, and sodium carbonate and thus the sodium carbonate can be recycled.

Types of equipment used for absorption include (1) a packed tower filled with packing material, absorbent liquid flowing down through the packing (designed to provide a maximum of contact surface), and gas flowing upward in a countercurrent fashion; (2) a spray tower in which the absorbing liquid is sprayed into essentially an empty tower with the gas flowing upward; (3) a tray tower containing bubble caps, sieve trays, or valve trays; (4) a falling-film absorber or wetted-wall column; and (5) stirred vessels. Packed towers are the most commonly used.

A representative packed-type absorption tower is shown in Fig. 1. In addition to absorption efficiency, a primary concern of the tower designer is that of minimizing the pressure drop through the tower. The principal elements of pressure drop are shown at the right of the diagram. Important to efficiency of absorption and pressure drop is the type of packing used. As shown by Fig. 2, over the years numerous types of packing (mostly ceramic) have been developed to meet a wide variety of operating parameters. A major objective is that of providing as much contact surface as is possible with a minimum of pressure drop. Where corrosion conditions permit, metal packing sometimes can be used. Of the packing designs illustrated, the berl saddles range in size from $\frac{1}{4}$ inch (6 millimeters) up to 2 inches (5 centimeters); raschig rings range from $\frac{1}{4}$ inch (6 millimeters) up to 4 inches (10 centimeters); lessing rings range from 1 inch (2.5 centimeters) up to 2 inches (5 centimeters); partition and spiral rings range from 3 inches (7.5 centimeters) up to 6 inches (15 centimeters).

In operation, the absorbing liquid is pumped into the top of the column where it is distributed by means of a weir to provide uniform distribution of the liquid over the underlying packing. Gas enters at the base of the tower and flows upward (countercurrent with the liquid) and out the top of the tower. The liquid may or may not be recycled without regeneration, depending upon the strength of the absorbent versus the quantity of material (concentration) in the gas to be removed. In a continuous operation, of course, a point is reached where fresh absorbing liquid must be added.

It is interesting to note that over 100,000 of the $\frac{1}{4}$-inch (6-millimeter) size packing shapes will be contained in each cubic foot (0.02832 cubic meter) of tower space if dense packing is desired.

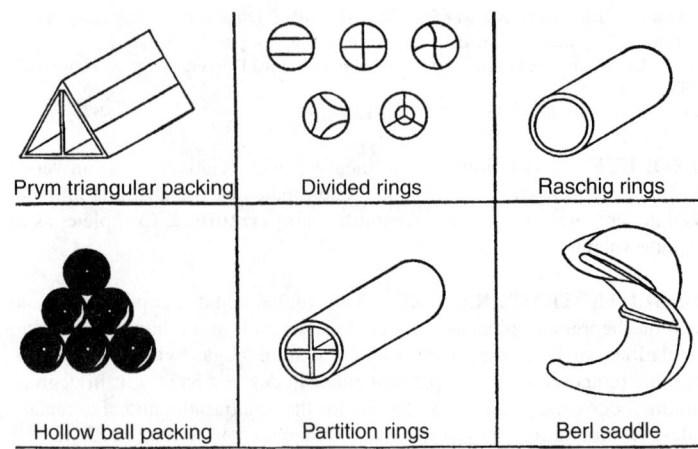

Fig. 2.   Types of packing used in absorption towers

In the purification of natural gas, the gas is fed into the bottom of an absorption tower where the gas is contacted countercurrently by a lean absorption oil. Hydrochloric acid is produced by absorbing gaseous hydrogen chloride in water, usually in a spray-type tower. Unreacted ammonia in the manufacture of hydrogen cyanide is absorbed in dilute sulfuric acid. In the production of nitric acid, ammonia is catalytically oxidized and the gaseous products are absorbed in water. The ethanolamines are widely used in scrubbing gases for removal of acid compounds. Hydrocarbon gases containing hydrogen sulfide can be scrubbed with monoethanolamine, which combines with it by salt formation and effectively removes it from the gas stream. In plants synthesizing ammonia, hydrogen and carbon dioxide are formed. The hydrogen can be obtained by countercurrently scrubbing the gas mixture in a packed or tray column with monoethanolamine which absorbs the carbon dioxide. The latter can be recovered by heating the monoethanolamine. In a nonliquid system, sulfur dioxide can be absorbed by dry cupric oxide on activated alumina, thus avoiding the disadvantages of a wet process. Sulfuric acid is produced by absorbing sulfur trioxide in weak acid or water.

See also **Coal**; **Ethanolamines**; **Chromatography**; and **Pollution (Air)**.

Additional Reading

Felder, T.D. and E.L. Garrett: *Process Technology Systems,* Pearson Education, Boston, MA, 2002.

Geankoplis, C.J. and P.R. Toliver: *Transport Processes and Separation Process Principles (Includes Unit Operations),* 4th Edition, Prentice Hall Professional Technical Reference, Upper Saddle River, NJ, 2003.

Thomas, W.J. and B. Crittenden: *Adsorption Technology and Design,* Elsevier Science & Technology Books, New York, NY, 2000.

Yang, R.T.: *Gas Separation by Adsorption Processes,* Vol. 1, World Scientific Publishing Company, Inc., Riveredge, NJ, 2000.

ABSORPTION SPECTROSCOPY.   An important technique of instrumental analysis involving measurement of the absorption of radiant energy by a substance as a function of the energy incident upon it. Adsorption processes occur throughout the electromagnetic spectrum, ranging from the γ region (nuclear resonance absorption of the Mossbauer effect) to the radio region (nuclear magnetic resonance). In practice, they are limited to those processes that are followed by the emission of radiant energy of greater intensity than that which was absorbed. All absorption process involve absorption of a photon by the substance being analyzed. If it loses the excess energy by emitting a photon of less energy than that absorbed, fluorescence or phosphorescence is said to occur, depending on the lifetime of the excited state.

The emitted energy is normally studied. If the source of radiant energy and the absorbing species are in identical energy states (in resonance), the excess energy is often given up by the nondirectional emission of a photon whose energy is identical with the absorbed.

Either adsorption or emission may be studied, depending upon the chemical and instrumental circumstances. If the emitted energy is studied, the term *resonance fluorescence* is often used. However, if the absorbing

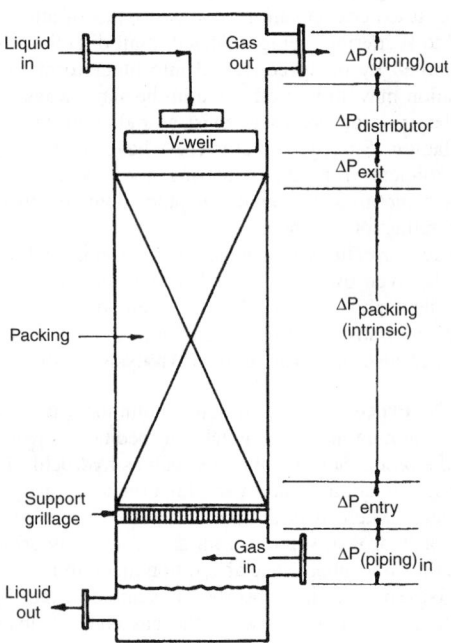

Fig. 1.   Section of representative packed absorption tower

species releases the excess energy in small steps by intermolecular collision or some other process, it is commonly understood that this phenomenon falls within the realm of absorption spectroscopy. The terms *absorption spectroscopy, spectrophotometry*, and *absorptimetry* are often used synonymously.

Most absorption spectroscopy is done in the ultraviolet, visible, and infrared regions of the electromagnetic spectrum.

See also **Emission Spectroscopy**; and **Infrared Radiation**.

ABSORPTION SPECTRUM. The spectrum of radiation that has been filtered through a material medium. When white light traverses a transparent medium, a certain portion of it is absorbed, the amount varying, in general, progressively with the frequency of which the absorption coefficient is a function. Analysis of the transmitted light may, however, reveal that certain frequency ranges are absorbed to a degree out of all proportion to the adjacent regions; that is, with a distinct selectivity. These abnormally absorbed frequencies constitute, collectively, the "absorption spectrum" of the medium, and appear as dark lines or bands in the otherwise continuous spectrum of the transmitted light. The phenomenon is not confined to the visible range, but may be found to extend throughout the spectrum from the far infrared to the extreme ultraviolet and into the x-ray region.

A study of such spectra shows that the lines or bands therein accurately coincide in frequency with certain lines or bands of the emission spectra of the same substances. This was formerly attributed to resonance of electronic vibrations, but is now more satisfactorily explained by quantum theory on the assumption that those quanta of the incident radiation which are absorbed are able to excite atoms or molecules of the medium to some (but not all) of the energy levels involved in the production of the complete emission spectrum.

A very familiar example is the spectrum of sunlight, which is crossed by innumerable dark lines, the Fraunhofer lines, much has been learned about the constitution of the sun, stars, and other astronomical objects from the Fraunhofer lines.

A noteworthy characteristic of selective absorption is found in the existence of certain anomalies in the refractive index in the neighborhood of absorption frequencies; discussed under Dispersion. See also **Absorption Band**; and **Electromagnetic Spectrum**.

Additional Reading

Baeyans, W.R.G., et al.: *Luminescence Techniques in Chemical and Biochemical Analysis,* in Practical Spectroscopy Series, Vol. 12, Marcel Dekker, New York, NY, 1991.

Burgess, C. and D.G. Jones: Spectrophotometry, Luminescence and Colour: Science and Compliance: Papers Presented at the Second Joint Meeting of the Uv Spectrometry Group of the u, Elsevier Science, Ltd, New York, NY, 1995.

Evans, N.J.: "Impedance Spectroscopy Reveals Materials Characteristics," *Adv. Mat. & Proc.,* **41** (November 1991).

Ewing, G.W., Editor: *Analytical Instrumentation Handbook,* 2nd Edition Marcel Dekker, New York, NY, 1997.

Grant, E.R. and R.G. Cooks: "Mass Spectrometry and Its Use in Tandem with Laser Spectroscopy," *Science,* **61** (October 5, 1990).

Robinson, J.W.: *Atomic Spectroscopy,* 2nd Edition Marcel Dekker, New York, NY, 1996.

Van Grieken, R. and A. Markowicz: *Handbook of X-Ray Spectrometry: Methods and Techniques,* Marcel Dekker, New York, NY, 1992.

Various: "Application Reviews (Chemical Instrumentation)" *Analytical Chemistry* (Special Issue), (June 15, 1991).

ABS RESINS (Acrylonitrile-Butadiene-Styrene). See **Resins (Acrylonitrile-Butadiene-Styrene)**.

ABUNDANCE. The relative amount (% by weight) of a substance in the earth's crust, including the atmosphere and the oceans.

(1) The abundance of the elements in the earth's crust is shown in Table 1.

(2) The percentages of inorganic compounds in the earth's crust, exclusive of water, are:

(1) SiO_2	55	(2) Al_2O_3	15	(3) $CaCO_3$	8.8
(4) MgO	1.6	(5) Na_2O	1.6	(6) K_2O	1.9

TABLE 1.

Rank	Element	% by wt.
1	Oxygen	49.2
2	Silicon	25.7
3	Aluminum	7.5
4	Iron	4.7
5	Calcium	3.4
6	Sodium	2.6
7	Potassium	2.4
8	Magnesium	1.9
9	Hydrogen	0.9
10	Titanium	0.6
11	Chlorine	0.2
12	Phosphorus	0.1
13	Manganese	0.1
14	Carbon	0.09
15	Sulfur	0.05
16	Barium	0.05
	all others	0.51

(3) The most abundant organic materials are cellulose and its derivatives, and proteins.

Note: In the universe as a whole, the most abundant element is hydrogen.

ACARICIDE. A substance, natural or synthetic, used to destroy or control infestations of the animals making up *Arachnida, Acarina*, mainly mites and ticks, some forms of which are very injurious to both plants and livestock, including poultry. There are numerous substances that are effective both as acaricides and insecticides; others of a narrower spectrum are strictly acaricides. See also **Insecticide**; and **Insecticide and Pesticide Technology**.

ACCELERATOR   1. A compound, usually organic, that greatly reduces the time required for vulcanization of natural and synthetic rubbers, at the same time improving the aging and other physical properties. See also **Rubber (Natural)**. Organic accelerators invariably contain nitrogen, and many also contain sulfur. The latter type are called *ultra-accelerators* because of their greater activity. The major types include amines, guanidines, thiazoles, thiuram sulfides, and dithiocarbamates. The amines and guanidines are basic, the others acidic. The normal effective concentration of organic accelerators in a rubber mixture is 1% or less depending on the rubber hydrocarbon present. Zinc oxide is required for activation, and in the case of acidic accelerators, stearic acid is required. The introduction of organic accelerators in the early twenties was largely responsible for the successful development of automobile tires and mechanical products for engineering uses. A few inorganic accelerators are still used in low-grade products, e.g., lime, magnesium oxide, and lead oxide.

2. A compound added to a photographic developer to increase its activity, such as certain quaternary ammonium compounds and alkaline substances.

3. A particle accelerator.

ACETALDEHYDE. [CAS: 75-07-0]. CH_3CHO, formula weight 44.05, colorless, odorous liquid, mp $-123.5°C$, bp $20.2°C$, sp gr 0.783. Also known as *ethanal*, acetaldehyde is miscible with H_2O, alcohol, or ether in all proportions. Because of its versatile chemical reactivity, acetaldehyde is widely used as a commencing material in organic syntheses, including the production of resins, dyestuffs, and explosives. The compound also is used as a reducing agent, preservative, and as a medium for silvering mirrors. In resin manufacture, paraldehyde $(CH_3CHO)_3$ sometimes is preferred because of its higher boiling and flash points.

In tonnage production, acetaldehyde may be manufactured by (1) the direct oxidation of ethylene, requiring a catalytic solution of copper chloride plus small quantities of palladium chloride, (2) the oxidation of ethyl alcohol with sodium dichromate, and (3) the dry distillation of calcium acetate with calcium formate.

Acetaldehyde reacts with many chemicals in a marked manner, (1) with ammonio-silver nitrate ("Tollen's solution"), to form metallic silver, either as a black precipitate or as an adherent mirror film on glass, (2) with alkaline cupric solution ("Fehling's solution") to form cuprous oxide, red to yellow precipitate, (3) with rosaniline (fuchsine, magenta), which has been decolorized by sulfurous acid ("Schiff's solution"), the pink color of rosaniline is restored, (4) with NaOH, upon warming, a yellow to brown resin of unpleasant odor separates (this reaction is given by aldehydes immediately following acetaldehyde in the series, but not by formaldehyde, furfuraldehyde or benzaldehyde), (5) with anhydrous ammonia, to form aldehyde-ammonia $CH_3 \cdot CHOH \cdot NH_2$, white solid, mp 97°C, bp 111°C, with decomposition, (6) with concentrated H_2SO_4, heat is evolved, and with rise of temperature, paraldehyde $(C_2H_4O)_3$ or

$$CH_3 \cdot CH \overset{OCH(CH_3)}{\underset{OCH(CH_3)}{\diagdown}} O$$

colorless liquid bp 124°C, slightly soluble in H_2O, is formed, (7) with acids, below 0°C, forms metaldehyde $(C_2H_4O)x$, white solid, sublimes at about 115°C without melting but with partial conversion to acetaldehyde, (8) with dilute HCl or dilute NaOH, aldol, $CH_3 \cdot CHOH \cdot CH_2CHO$ slowly forms, (9) with phosphorus pentachloride, forms ethylidene chloride, $CH_3 \cdot CHCl_2$, colorless liquid, bp 58°C, (10) with ethyl alcohol and dry hydrogen chloride, forms acetal, 1,1-diethyoxyethane $CH_3 \cdot CH(OC_2H_5)_2$, colorless liquid, bp 104°C, (11) with hydrocyanic acid, forms acetaldehyde cyanohydrin, $CH_3 \cdot CHOH \cdot CN$, readily converted into alphahydroxypropionic acid, $CH_3 \cdot CHOH \cdot COOH$, (12) with sodium hydrogen sulfite, forms acetaldehyde sodium bisulfite, $CH_3 \cdot CHOH \cdot SO_3Na$, white solid, from which acetaldehyde is readily recoverable by treatment with sodium carbonate solution, (13) with hydroxylamine hydrochloride forms acetaldoxime, $CH_3 \cdot CH:NOH$, white solid, mp 47°C, (14) with phenylhydrazine, forms acetaldehyde phenylhydrazone, $CH_3 \cdot CH:N \cdot NH \cdot C_6H_5$, white solid, mp 98°C, (15) with magnesium methyl iodide in anhydrous ether ("Grignard's solution"), yields, after reaction with water, isopropyl alcohol, $(CH_3)_2CHOH$, a secondary alcohol, (16) with semicarbazide, forms acetaldehyde semicarbazone, $CH_3 \cdot CH:N \cdot NH \cdot CO \cdot NH_2$, white solid, mp 162°C, (17) with chlorine, forms trichloroacetaldehyde ("chloral"), $CCl_3 \cdot CHO$, (18) with H_2S, forms thioacetaldehyde, $CH_3 \cdot CHS$ or $(CH_3 \cdot CHS)_3$. Acetaldehyde stands chemically between ethyl alcohol on one hand—to which it can be reduced—and acetic acid on the other hand—to which it can be oxidized. These reactions of acetaldehyde, coupled with its ready formation from acetylene by mercuric sulfate solution as a catalyzer, open up a vast field of organic chemistry with acetaldehyde as raw material: acetaldehyde hydrogenated to ethyl alcohol; oxygenated to acetic acid, thence to acetone, acetic anhydride, vinyl acetate, vinyl alcohol. Acetaldehyde is also formed by the regulated oxidation of ethyl alcohol by such a reagent as sodium dichromate in H_2SO_4 (chromic sulfate also produced). Reactions (1), (3), (14), and (16) above are most commonly used in the detection of acetaldehyde. See also **Aldehydes**.

ACETAL GROUP. An organic compound of the general formula RCH(OR')(OR'') is termed an *acetal* and is formed by the reaction of an aldehyde with an alcohol, usually in the presence of small amounts of acids or appropriate inorganic salts. Acetals are stable toward alkali, are volatile, insoluble in H_2O, and generally are similar structurally to ethers. Unlike ethers, acetals are hydrolyzed by acids into their respective aldehydes. $H(R)CO + (HO \cdot C_2H_5)_2 \longrightarrow H(R)C(OC_2H_5)_2 + H_2O$. Representative acetals include: $CH_2(OCH_3)_2$, methylene dimethyl ether, bp 42°C; $CH_3CH(OCH_3)_2$, ethylidene dimethyl ether, bp 64°C; and $CH_3CH(OC_2H_5)_2$, ethylidene diethyl ether, bp 104°C.

ACETAL RESINS. See **Resins (Acetal)**.

ACETATE DYE. One group comprises water insoluble azo or anthraquinone dyes that have be highly dispersed to make them capable of penetrating and dyeing acetate fibers. A second class consists of water-insoluble amino azo dyes that are made water soluble by treatment with formaldehyde and bisulfite. After absorption by the fiber, the resulting sulfonic acids hydrolyze and regenerate the insoluble dyes.

See also **Dye and Dye Intermediates**; and **Dyes: Anthraquinone**.

ACETATE FIBERS. See **Fibers: Acetate**.

ACETATES. See **Acetic Acid**; **Fibers: Acetate**.

ACETIC ACID. [CAS: 64-19-7]. CH_3COOH, formula weight 60.05, colorless, acrid liquid, mp 16.7°C, bp 118.1°C, sp gr 1.049. Also known as ethanoic acid or vinegar acid, this compound is miscible with H_2O, alcohol, and ether in all proportions. Acetic acid is available commercially in several concentrations. The CH_3COOH content of glacial acetic is approximately 99.7% with H_2O, the principal impurity. Reagent acetic acid generally contains 36% CH_3COOH by weight. Standard commercial aqueous solutions are 28, 56, 70, 80, 85, and 90% CH_3COOH. Acetic acid is the active ingredient in vinegar in which the content ranges from 4 to 5% CH_3COOH. Acetic acid is classified as a weak, monobasic acid. The three hydrogen atoms linked to one of the two carbon atoms are not replaceable by metals.

In addition to the large quantities of vinegar produced, acetic acid in its more concentrated forms is an important high-tonnage industrial chemical, both as a reactive raw and intermediate material for various organic syntheses and as an excellent solvent. Acetic acid is required in the production of several synthetic resins and fibers, pharmaceuticals, photographic chemicals, flavorants, and bleaching and etching compounds.

Early commercial sources of acetic acid included (1) the combined action of *Bacterium aceti* and air on ethyl alcohol in an oxidation-fermentation process: $C_2H_5OH + O_2 \longrightarrow CH_3COOH + H_2O$, the same reaction which occurs when weak alcoholic beverages, such as beer or wine, are exposed to air for a prolonged period and which turn sour because of the formation of acetic acid; and (2) the destructive distillation of wood. A number of natural vinegars still are made by fermentation and marketed as the natural product, but diluted commercially and synthetically produced acetic acid is a much more economic route to follow. The wood distillation route was phased out because of shortages of raw materials and the much more attractive economy of synthetic processes.

The most important synthetic processes are (1) the oxidation of acetaldehyde, and (2) the direct synthesis from methyl alcohol and carbon monoxide. The latter reaction must proceed under very high pressure (approximately 650 atmospheres) and at about 250°C. The reaction takes place in the liquid phase and dissolved cobaltous iodide is the catalyst. $CH_3OH + CO \longrightarrow CH_3COOH$ and $CH_3OCH_3 + H_2O + 2CO \longrightarrow 2CH_3COOH$. The crude acid produced first is separated from the catalyst and then dehydrated and purified in an azeotropic distillation column. The final product is approximately 99.8% pure CH_3COOH.

Acetic acid solution reacts with alkalis to form acetates, e.g., sodium acetate, calcium acetate; similarly, with some oxides, e.g., lead acetate; with carbonates, e.g., sodium acetate, calcium acetate, magnesium acetate; with some sulfides, e.g., zinc acetate, manganese acetate. Ferric acetate solution, upon boiling, yields a red precipitate of basic ferric acetate. Acetic acid solution attacks many metals, liberating hydrogen and forming acetate, e.g., magnesium, zinc, iron. Acetic acid is an important organic substance, with alcohols forming esters (acetates); with phosphorus trichloride forming acetyl chloride $CH_3CO \cdot Cl$, which is an important reagent for transfer of the acetyl (CH_3CO-) group; forming acetic anhydride, also an acetyl reagent; forming acetone and calcium carbonate when passed over lime and a catalyzer (barium carbonate) or when calcium acetate is heated; forming methane (and sodium carbonate) when sodium acetate is heated with NaOH; forming mono-, di-, trichloroacetic (or bromoacetic) acids by reaction with chlorine (or bromine) from which hydroxy- and amino-, aldehydic-, dibasic acids, respectively, may be made; forming acetamide when ammonium acetate is distilled. Acetic acid dissolves sulfur and phosphorus, is an important solvent for organic substances, and causes painful wounds when it comes in contact with the skin. Normal acetates are soluble, basic acetates insoluble. The latter are important in their compounds with lead, and copper ("verdigris").

A large number of acetic acid esters are important industrially, including methyl, ethyl, propyl, butyl, amyl, and cetyl acetates; glycol mono- and diacetate; glyceryl mono-, di-, and triacetate; glucose pentacetate; and cellulose tri-, tetra-, and pentacetate.

Acetates may be detected by formation of foul-smelling cacodyl (poisonous) on heating with dry arsenic trioxide. Other tests for acetate are the lanthanum nitrate test in which a blue or bluish-brown ring forms when a drop of 2.5% La(NO$_3$)$_3$ solution, a drop of 0.01-N iodine solution, and a drop of 0.1% NH$_4$OH solution are added to a drop of a neutral acetate solution; the ferric chloride test, in which a reddish color is produced by the addition of 1-N ferric chloride solution to a neutral solution of acetate; and the ethyl acetate test, in which ethyl alcohol and H$_2$SO$_4$ are added to the acetate solution and warmed to form a colorless solution.

Additional Reading

Agreda, V.H. and J. Zoeller: *Acetic Acid and Its Derivatives,* Marcel Dekker, Inc., New York, NY, 1992.

Behrens, D.: *DECHEMA Corrosion Handbook,* Vol. 6, John Wiley & Sons, Inc., New York, NY. 1997.

Dillon, C.P. and W.I. Pollock: *Materials Selector for Hazardous Chemicals: Formic, Acetic and Other Organic Acids,* Elsevier Science, New York, NY, 1998.

ACETOACTETIC ESTER CONDENSATION. A class of reactions occasioned by the dehydrating power of metallic sodium or sodium ethoxide on the ethyl esters of monobasic aliphatic acids and a few other esters. It is best known in the formation of acetoacetic ester:

$$2\,CH_3 \cdot COOC_2H_5 + 2\,CH_3 \cdot COOC_2H_5 + 2\,Na \longrightarrow$$

$$2\,CH_3 \cdot C(ONa) : CH \cdot COOC_2H_5 + 2\,C_2H_5OH + H_2$$

The actual course of the reaction is complex. By the action of acids the sodium may be eliminated from the first product of the reaction and the free ester obtained. This may exist in the tautomeric enol and keto forms (CH$_3$ · COH:CH · COOC$_2$H$_5$ and CH$_3$ · CO · CH$_2$ · COOC$_2$H$_5$).

On boiling ester with acids or alkalies it will split in two ways, the circumstances determining the nature of the main product. Thus, if moderately strong acid or weak alkali is employed, acetone is formed with very little acetic acid (ketone splitting). In the presence of strong alkalies, however, very little acetone and much acetic acid result (acid splitting). Derivatives of acetoacetic ester may be decomposed in the same fashion, and this fact is responsible for the great utility of this condensation in organic synthesis. This is also due to the reactivity of the · CH$_2$· group, which reacts readily with various groups, notably halogen compounds. Usually the sodium salt of the ester is used, and the condensation is followed by decarboxylation with dilute alkali, or deacylation with concentrated alkali.

$$CH_3 \cdot CO \cdot CHNa \cdot COOC_2H_5 + RI \longrightarrow CH_3 \cdot CO \cdot CHR \cdot COOC_2H_5$$

$$+ NaI$$

$$CH_3 \cdot CO \cdot CHR \cdot COOC_2H_5 \xrightarrow[\text{Dilute alkali}]{H_2O} CH_3 \cdot CO \cdot CH_2R$$

$$+ C_2H_5OH + CO_2$$

$$CH_3 \cdot CO \cdot CHR \cdot COOC_2H_5 \xrightarrow[\text{Concentrated alkali}]{2\,H_2O} HOOC \cdot CH_2 \cdot R$$

$$+ C_2H_5OH + CH_3COOH$$

ACETONE. [CAS: 67-64-1]. CH$_3$· CO· CH$_3$, formula weight 58.08, colorless, odorous liquid ketone, mp −94.6°C, bp 56.5°C, sp gr 0.792. Also known as dimethyl ketone or propanone, this compound is miscible in all proportions with H$_2$O, alcohol, or ether. Acetone is a very important solvent and is widely used in the manufacture of plastics and lacquers. For storage purposes, acetylene may be dissolved in acetone. A high-tonnage chemical, acetone is the starting ingredient or intermediate for numerous organic syntheses. Closely related, industrially important compounds are diacetone alcohol (DAA) CH$_3$· CO· CH$_2$· COH(CH$_3$)$_2$ which is used as a solvent for cellulose acetate and nitrocellulose, as well as for various resins and gums, and as a thinner for lacquers and inking materials. Sometimes DAA is mixed with castor oil for use as a hydraulic brake fluid for which its physical properties are well suited, mp −54°C, bp 166°C, sp gr 0.938. A product known as synthetic methyl acetone is prepared by mixing acetone (50%), methyl acetate (30%), and methyl alcohol (20%) and is used widely for coagulating latex and in paint removers and lacquers.

In older industrial processes, acetone is prepared (1) by passing the vapors of acetic acid over heated lime. In a first step, calcium acetate is produced, followed by a breakdown of the acetate into acetone and calcium carbonate:

$$CH_3 \cdot CO \cdot O \cdot Ca \cdot OOC \cdot CH_3 \longrightarrow CH_3 \cdot CO \cdot CH_3 + CaCO_3;$$

and (2) by fermentation of starches, such as maize, which produce acetone along with butyl alcohol. Modern industrial processes include (3) the use of cumene as a chargestock, in which cumene first is oxidized to cumene hydroperoxide (CHP), this followed by the decomposition of CHP into acetone and phenol; and (4) by the direct oxidation of propylene, using air and catalysts. The catalyst solution consists of copper chloride and small amounts of palladium chloride. The reaction: CH$_3$CH = CH$_2$ + 1/2 O$_2$ $\longrightarrow$ CH$_3$COCH$_3$. During the reaction, the palladium chloride is reduced to elemental palladium and HCl. Reoxidation is effected by cupric chloride. The cuprous chloride resulting is reoxidized during the catalyst regeneration cycle. The process is carried out under moderate pressure at about 100°C.

Acetone reacts with many chemicals in a marked manner: (1) with phosphorus pentachloride, yields acetone chloride (CH$_3$)$_2$CCl$_2$, (2) with hydrogen chloride dry, yields both mesityl oxide CH$_3$COCH:C(CH$_3$)$_2$, liquid, bp 132°C, and phorone (CH$_3$)$_2$C:CHCOCH: C(CH$_3$)$_2$, yellow solid, mp 28°C, (3) with concentrated H$_2$SO$_4$, yields mesitylene C$_6$H$_3$(CH$_3$)$_3$ (1,3,5), (4) with NH$_3$, yields acetone amines, e.g., diacetoneamine C$_6$H$_{12}$ONH, (5) with HCN, yields acetone cyanohydrin (CH$_3$)$_2$CHOH· CN, readily converted into alpha-hydroxy acid (CH$_3$)$_2$CHOH· COOH, (6) with sodium hydrogen sulfite, forms acetone-sodiumbisulfite (CH$_3$)$_2$ COH· SO$_3$Na white solid, from which acetone is readily recoverable by treatment with sodium carbonate solution, (7) with hydroxylamine hydrochloride, forms acetoxime (CH$_3$)$_2$C:NOH, solid, mp 60°C, (8) with phenylhydrazine, yields acetonephenyl-hydrazone (CH$_3$)$_2$C:NNHC$_6$ H$_5$· H$_2$O, solid, mp 16°C, anhydrous compound, mp 42°C, (9) with semicarbazide, forms acetonesemicarbazone (CH$_3$)C:NNHCONH$_2$, solid, mp 189°C, (10) with magnesium methyl iodide in anhydrous ether ("Grignard's solution"), yields, after reaction with H$_2$O, trimethylcarbinol (CH$_3$)$_3$COH, a tertiary alcohol, (11) with ethyl thioalcohol and hydrogen chloride dry, yields mercaptol (CH$_3$)$_2$C(SC$_2$H$_5$)$_2$, (12) with hypochlorite, hypobromite, or hypoiodite solution, yields chloroform CHCl$_3$, bromoform CHBr$_3$ or iodoform CHI$_3$, respectively, (13) with most reducing agents, forms isopropyl alcohol (CH$_3$)$_2$CHOH, a secondary alcohol, but with sodium amalgam forms pinacone (CH$_3$)$_2$COH· COH(CH$_3$)$_2$ (14) with sodium dichromate and H$_2$SO$_4$, forms acetic acid CH$_3$COOH plus CO$_2$. When acetone vapor is passed through a tube at a dull red heat, ketene CH$_2$:CO and methane CH$_4$ are formed.

ACETYL CHLORIDE. See **Chlorinated Organics**.

ACETYLENE. [CAS: 74-86-2]. CH:CH formula weight 26.04, mp −81.5°C, bp −84°C, sp gr 0.905 (air = 1.000). Sometimes referred to as ethyne, ethine, or gaseous carbon (92.3% of the compound is C), acetylene is moderately soluble in H$_2$O or alcohol, and exceptionally soluble in acetone (300 volumes of acetylene in 1 volume of acetone at 12 atmospheres pressure). The gas burns when ignited in air with a luminous sooty flame, requiring a specially devised burner for illumination purposes. An explosive mixture is formed with air over a wide range (about 3 to 80% acetylene), but safe handling is improved when the gas is dissolved in acetone. The heating value is 1455 Btu/ft^3 (8.9 Cal/m^3).

Although acetylene still is used in a number of organic syntheses on an industrial scale, its use on a high-tonnage basis has diminished because of the lower cost of other starting materials, such as ethylene and propylene. Acetylene has been widely used in the production of halogen derivatives, acrylonitrile, acetaldehyde, and vinyl chloride. Within recent years, producers of acrylonitrile switched to propylene as a starting material.

Commercially, acetylene is produced from the pyrolysis of naphtha in a two-stage cracking process. Both acetylene and ethylene are end products. The ratio of the two products can be changed by varying the naphtha feed rate. Acetylene also has been produced by a submerged-flame process from crude oil. In essence, gasification of the crude oil occurs by means of the flame, which is supported by oxygen beneath the surface of the oil. Combustion and cracking of the oil take place at the boundaries of the flame. The composition of the cracked gas includes about 6.3% acetylene and 6.7% ethylene. Thus, further separation and purification are required. Several years ago when procedures were developed for the safe handling of acetylene on a large scale, J. W. Reppe worked out a series of reactions that later became known as "Reppe chemistry." These reactions were particularly important to the manufacture of many high polymers and other synthetic products. Reppe and his associates were able to effect synthesis of chemicals that had been commercially unavailable. An example is the synthesis of cyclooctatetraene by heating a solution of acetylene under pressure in tetrahydrofuran in the presence of a nickel cyanide catalyst. In another reaction, acrylic acid was produced from CO and H_2O in the presence of a nickel catalyst: $C_2H_2 + CO + H_2O \longrightarrow CH_2 : CH \cdot COOH$. These two reactions are representative of a much larger number of reactions, both those that are straight-chain only, and those involving ring closure.

Acetylene reacts (1) with chlorine, to form acetylene tetrachloride $C_2H_2Cl_4$ or $CHCl_2 \cdot CHCl_2$ or acetylene dichloride $C_2H_2Cl_2$ or $CHCl:CHCl$, (2) with bromine, to form acetylene tetrabromide $C_2H_2Br_4$ or $CHBr_2 \cdot CHBr_2$ or acetylene dibromide $C_2H_2Br_2$ or $CHBr:CHBr$, (3) with hydrogen chloride (bromide, iodide), to form ethylene monochloride $CH_2:CHCl$ (monobromide, monoiodide), and 1,1-dichloroethane, ethylidene chloride $CH_3 \cdot CHCl_2$ (dibromide, diiodide), (4) with H_2O in the presence of a catalyzer, e.g., mercuric sulfate, to form acetaldehyde $CH_3 \cdot CHO$, (5) with hydrogen, in the presence of a catalyzer, e.g., finely divided nickel heated, to form ethylene C_2H_4 or ethane C_2H_6, (6) with metals, such as copper or nickel, when moist, also lead or zinc, when moist and unpurified. Tin is not attacked. Sodium yields, upon heating, the compounds C_2HNa and C_2Na_2. (7) With ammoniocuprous (or silver) salt solution, to form cuprous (or silver) acetylide C_2Cu_2, dark red precipitate, explosive when dry, and yielding acetylene upon treatment with acid, (8) with mercuric chloride solution, to form trichloromercuric acetaldehyde $C(HgCl)_3 \cdot CHO$, precipitate, which yields with HCl acetaldehyde plus mercuric chloride.

Additional Reading

Stang, P.J. and F. Diederich: *Modern Acetylene Chemistry*, John Wiley & Sons, Inc., New York, NY, 1995.

ACETYLENE SERIES.
A series of unsaturated hydrocarbons having the general formula C_nH_{2n-2}, and containing a triple bond between two carbon atoms. The series is named after the simplest compound of the series, acetylene HC:CH. In more modern terminology, this series of compounds is termed the *alkynes*. See also **Alkynes**.

ACETYLSALICYLIC ACID.
[CAS: 50-78-2]. $C_6H_4(COOH)CO_2CH_3$, formula wt, 180.06, mp 133.5°C, colorless, crystalline, slightly soluble in water, soluble in alcohol and ether, commonly known as aspirin, also called orthoacetoxybenzoic acid. The substance is commonly used as a relief for mild forms of pain, including headache and joint and muscle pain. The drug tends to reduce fever. Aspirin and other forms of salicylates have been used in large doses in acute rheumatic fever, but must be administered with extreme care in such cases by a physician. Commercially available aspirin is sometime mixed with other pain relievers as well as buffering agents.

See also **Aspirin**; and **Salicylic Acid and Related Compounds**.

ACHLORHYDRIA.
Lack of hydrochloric acid in the digestive juices in the stomach. Hydrochloric acid helps digest food. The low pH of the normal stomach contents is a barrier to infection by various organisms and, where achlorhydria develops—particularly in malnutrition—it renders the patient more susceptible to infection, such as by *Vibrio cholerae* and *Giardia lamblia*. The condition is relatively common among people of about 50 years of age and older, affecting 15 to 20% of the population in this age group. The acid deficiency also occurs in about 30% of patients with adult onset-type of primary hypogamma-globulinemia. A well-balanced diet of easily digestible foods minimizes the discomforting effects of complete absence of hydrochloric acid in the stomach. The condition does not preclude full digestion of fats and proteins, the latter being attacked by intestinal and pancreatic enzymes. In rare cases, where diarrhea may result from achlorhydria, dilute hydrochloric acid may be administered by mouth. Where this causes an increase in discomfort or even pain, the use of dexamethasone or mucosal coating agents is preferred.

Commonly, achlorhydria may not be accompanied by other diseases, but in some cases there is a connection. For example, achlorhydria is an abnormality that sometimes occurs with severe iron deficiency. Histalog-fast achlorhydria, resulting from intrinsic factor deficiency in gastric juice, may be an indication of pernicious anemia. Hyperplastic polyps are often found in association with achlorhydria.

Excessive alcohol intake can also lead to achlorhydria and it is said that the resistance for cyanide poisoning of the Russian mystic, Rasputin, was attributable to that effect. The great amount of vodka that he consumed led to achlorhydria and thus the ingested potassium cyanide did not liberate lethal hydrocyanic gas, nor was the potassium salt absorbed through the stomach walls.

Additional Reading

Holt, P.R. and R.M. Russell: *Chronic Gastritis-Achlorhydria in the Elderly*, CRC Press, LLC., Boca Raton, FL, 1993.

ACID-BASE REGULATION (Blood).
The hydrogen ion concentration of the blood is maintained at a constant level of pH 7.4 by a complex system of physico-chemical processes, involving, among others, neutralization, buffering, and excretion by the lungs and kidneys. This topic is sometimes referred to as *acid-base metabolism*. The clinical importance of acids and bases in life processes derives from several fundamental factors. (1) Most chemical reactions within the body take place in water solutions. The type and rate of such reactions is seriously affected by acid-base concentrations, of which pH is one indication. (2) Hydrogen ions are mobile charged particles and the distribution of such ions as sodium, potassium, and chloride in the cell environment are ultimately affected by hydrogen ion concentration (pH). (3) It also has been established that hydrogen ion concentration influences the three-dimensional configurations of proteins. Protein conformational changes affect the biochemical activity of proteins and thus can affect normal protein function. For example, enzymes, a particular class of proteins, exhibit optimal activity within a narrow range of pH. Most physiological activities, and especially muscular exercise, are accompanied by the production of acid, to neutralize which, a substantial alkali reserve, mainly in the form of bicarbonate, is maintained in the plasma, and so long as the ratio of carbon dioxide to bicarbonate remains constant, the hydrogen ion concentration of the blood does not alter. Any non-volatile acid, such as lactic or phosphoric, entering the blood reacts with the bicarbonate of the alkali reserve to form carbon dioxide, which is volatile, and which combines with hemoglobin by which it is transported to the lungs and eliminated by the processes of respiration. It will also be evident from this that no acid stronger than carbon dioxide can exist in the blood. The foregoing neutralizing and buffering effects of bicarbonate and hemoglobin are short-term effects; to insure final elimination of excess acid or alkali, certain vital reactions come into play. The rate and depth of respiration are governed by the level of carbon dioxide in the blood, through the action of the respiratory center in the brain; by this means the pulmonary ventilation rate is continually adjusted to secure adequate elimination of carbon dioxide. In the kidneys two mechanisms operate; ammonia is formed, whereby acidic substances in process of excretion are neutralized, setting free basic ions such as sodium to return to the blood to help maintain the alkali reserve. Where there is a tendency toward development of increased acidity in the blood, the kidneys are able selectively to re-absorb sodium bicarbonate from the urine being excreted, and to release into it acid sodium phosphate; where there is a tendency toward alkalemia, alkaline sodium phosphate is excreted, the hydrogen ions thus liberated are re-absorbed to restore the diminishing hydrogen ion concentration. See also **Achlorhydria**; **Acidosis**; **Alkalosis**; **Blood**; **pH (Hydrogen Ion Concentration)**; and **Potassium and Sodium (In Biological Systems)**.

ACIDIC SOLVENT.
A solvent which is strongly protogenic, i.e., which has a strong tendency to donate protons and little tendency to accept them. Liquid hydrogen chloride and hydrogen fluoride are acidic solvents, and in them even such normally strong acids as nitric acid do not exhibit acidic properties, since there are no molecules that can accept protons; but, on the contrary, behave to some extent as bases by accepting protons yielded by the dissociation of the HCl or the HF. See **Acids and Bases**.

ACIDIMETRY.
An analytical method for determining the quantity of acid in a given sample by titration against a standard solution of a base, or, more broadly, a method of analysis by titration where the end point is recognized by a change in pH (hydrogen ion concentration). See also **Analysis (Chemical)**; **pH (Hydrogen Ion Concentration)**; **Titration (Potentiometric)**; and **Titration (Thermometric)**.

ACIDITY.
The amount of acid present, expressed for a solution either as the molecular concentration of acid, in terms of normality, molality,

etc., or the ionic concentration (hydrogen ions or protons) in terms of pH (the logarithm of the reciprocal of the hydrogen ion concentration). The acidity of a base is the number of molecules of monoatomic acid which one molecule of the base can neutralize. See **Acids and Bases**.

ACID NUMBER. A term used in the analysis of fats or waxes to designate the number of milligrams of potassium hydroxide (KOH) required to neutralize the free fatty acids in 1 gram of substance. The determination is performed by titrating an alcoholic solution of the wax or fat with tenth or half-normal alkali, using phenolphthalein as indicator.

ACIDOSIS. A condition of excess acidity (or depletion of alkali) in the body, in which acids are absorbed or formed in excess of their elimination, thus increasing the hydrogen ion concentration of the blood, exceeding the normal limit of 7.4. The acidity-alkalinity ratio in body tissue normally is delicately controlled by several mechanisms, notably the regulation of carbon dioxide-oxygen transfer in the lungs, the presence of buffer compounds in the blood, and the numerous sensing areas that are a part of the central nervous system. Normally, acidic materials are produced in excess in the body, this excess being neutralized by the presence of free alkaline elements, such as sodium occurring in plasma. The combination of sodium with excess acids produces carbon dioxide which is exhaled. Acidosis may result from: (1) severe exercise, leading to increased carbon dioxide content of the blood, (2) sleep, especially under narcosis, where the elimination of carbon dioxide is depressed, (3) heart failure, where there is diminished ventilation of carbon dioxide through the lungs, (4) diabetes and starvation, in which organic acids, such as β-hydroxybutyric and acetoacetic acids, accumulate, (5) kidney failure, in which the damaged kidneys cannot excrete acid radicals, and (6) severe diarrhea, in which there is loss of alkaline substances. Nausea, vomiting, and weakness sometimes may accompany acidosis. See also **Acid-Base Regulation (Blood); Blood;** and **Potassium and Sodium (In Biological Systems)**.

ACID RAIN. Acid rain can be simply described as rain that is more acidic than normal. Acid rain is a complicated problem. Caused by air pollution, acid rain's spread and damage involve weather, chemistry, soil, and the life cycles of plants and animals on the land and in the water.

Scientists have discovered that air pollution from the burning of fossil fuels is the major cause of acid rain. Acidic deposition, or acid rain, as it is commonly known, occurs when emissions of sulfur dioxide (SO_2) and oxides of nitrogen (NO_X) react in the atmosphere with water, oxygen, and oxidants to form various acidic compounds. This mixture forms a mild solution of sulfuric acid and nitric acid. Sunlight increases the rate of most of these reactions.

These compounds then fall to the earth in either wet form (such as rain, snow, and fog or dry form (such as gas and particles). About half of the acidity in the atmosphere falls back to earth through dry deposition as gases and dry particles. The wind blows these acidic particles and gases onto buildings, cars, homes, and trees. In some instances, these gases and particles can eat away the things on which they settle. Dry deposited gases and particles are sometimes washed from trees and other surfaces by rain. When that happens, the runoff water adds those acids to the acid rain, making the combination more acidic than the falling rain alone. The combination of acid rain plus dry deposited acid is called *acid deposition*. See **Acid Deposition**, which is discussed in more detail later in this entry. Prevailing winds transport the compounds, sometimes hundreds of miles, across state and national borders.

Electric utility plants account for about 70 percent of annual SO_2 emissions and 30 percent of NO_X emissions in the United States. Mobile sources (transportation) also contribute significantly to NO_X emissions. Overall, over 20 million tons of SO_2 and NO_X are emitted into the atmosphere each year.

Acid rain causes acidification of lakes and streams and contributes to damage of trees at high elevations (for example, red spruce trees above 2,000 feet in elevation). In addition, acid rain accelerates the decay of building materials and paints, including irreplaceable buildings, statues, and sculptures that are part of our nation's cultural heritage. Prior to falling to the earth, SO_2 and NO_X gases and their particulate matter derivatives, sulfates and nitrates, contribute to visibility degradation and impact public health.

Implementation of the Acid Rain Program under the 1990 Clean Air Act Amendments will confer significant benefits on the nation. By reducing SO_2 and NO_X, many acidified lakes and streams will improve substantially so that they can once again support fish life. Visibility will improve, allowing for increased enjoyment of scenic vistas across our country, particularly in National Parks. Stress to the forests that populate the ridges of mountains from Maine to Georgia will be reduced. Deterioration of historic buildings and monuments will be slowed. Finally, reductions in SO_2 and NO_X will reduce sulfates, nitrates, and ground level ozone (smog), leading to improvements in public health.

Surface Waters

Acid rain primarily affects sensitive bodies of water, that is, those that rest atop soil with a limited ability to neutralize acidic compounds (called "buffering capacity"). Many lakes and streams examined in a National Surface Water Survey (NSWS) suffer from chronic acidity, a condition in which water has a constant low (acidic) pH level. The survey investigated the effects of acidic deposition in over 1,000 lakes larger than 10 acres and in thousands of miles of streams believed to be sensitive to acidification. Of the lakes and streams surveyed in the NSWS, acid rain has been determined to cause acidity in 75 percent of the acidic lakes and about 50 percent of the acidic streams. Several regions in the U.S. were identified as containing many of the surface waters sensitive to acidification. They include, but are not limited to, the Adirondacks, the mid-Appalachian highlands, the upper Midwest, and the high elevation West.

In some sensitive lakes and streams, acidification has completely eradicated fish species, such as the brook trout, leaving these bodies of water barren. In fact, hundreds of the lakes in the Adirondacks surveyed in the NSWS have acidity levels indicative of chemical conditions unsuitable for the survival of sensitive fish species.

Emissions from U.S. sources also contribute to acidic deposition in eastern Canada, where the soil is very similar to the soil of the Adirondack Mountains, and the lakes are consequently extremely vulnerable to chronic acidification problems. The Canadian government has estimated that 14,000 lakes in eastern Canada are acidic.

Streams flowing over soil with low buffering capacity are equally as susceptible to damage from acid rain as lakes are. Approximately 580 of the streams in the Mid-Atlantic Coastal Plain are acidic primarily due to acidic deposition. The New Jersey Pine Barrens area endures the highest rate of acidic streams in the nation with over 90 percent of the streams acidic. Over 1,350 of the streams in the Mid-Atlantic Highlands (mid-Appalachia) are acidic, primarily due to acidic deposition. Many streams in that area have already experienced trout losses due to the rising acidity.

Acidification is also a problem in surface water populations that were not surveyed in federal research projects. For example, although lakes smaller than 10 acres were not included in the NSWS, there are from one to four times as many of these small lakes as there are larger lakes. In the Adirondacks, the percentage of acidic lakes is significantly higher when it includes smaller lakes (26 percent) than when it includes only the target size lakes (14 percent).

The acidification problem in both the United States and Canada grows in magnitude if "episodic acidification" (brief periods of low pH levels from snowmelt or heavy downpours) is taken into account. Lakes and streams throughout the United States, including high-elevation western lakes, are sensitive to episodic acidification. In the Mid-Appalachians, the Mid-Atlantic Coastal Plain, and the Adirondack Mountains, many additional lakes and streams become temporarily acidic during storms and snowmelt. Episodic acidification can cause large-scale "fish kills."

For example, approximately 70 percent of sensitive lakes in the Adirondacks are at risk of episodic acidification. This amount is over three times the amount of chronically acidic lakes. In the mid-Appalachians, approximately 30 percent of sensitive streams are likely to become acidic during an episode. This level is seven times the number of chronically acidic streams in that area.

Acid rain control will produce significant benefits in terms of lowered surface water acidity. If acidic deposition levels were to remain constant over the next 50 years (the time frame used for projection models), the acidification rate of lakes in the Adirondacks that are larger than 10 acres would rise by 50 percent or more. Scientists predict, however, that the decrease in SO_2 emissions required by the Acid Rain Program will significantly reduce acidification due to atmospheric sulfur. Without the reductions in SO_2 emissions, the proportions of aquatic systems in sensitive ecosystems that are acidic would remain high or dramatically worsen.

The impact of nitrogen on surface waters is also critical. Nitrogen plays a significant role in episodic acidification and new research recognizes the importance of nitrogen in long-term chronic acidification as well. Furthermore, the adverse impact of atmospheric nitrogen deposition on estuaries and other large bodies of water may be significant. For example, 30 to 40 percent of the nitrogen in the Chesapeake Bay comes from atmospheric deposition. Nitrogen is an important factor in causing eutrophication (oxygen depletion) of water bodies.

Forests

Acid rain has been implicated in contributing to forest degradation, especially in high-elevation spruce trees that populate the ridges of the Appalachian Mountains from Maine to Georgia, including national park areas such as the Shenandoah and Great Smoky Mountain national parks. Acidic deposition seems to impair the trees' growth in several ways; for example, acidic cloud water at high elevations may increase the susceptibility of the red spruce to winter injury.

There also is a concern about the impact of acid rain on forest soils. There is good reason to believe that long-term changes in the chemistry of some sensitive soils may have already occurred as a result of acid rain. As acid rain moves through the soils, it can strip away vital plant nutrients through chemical reactions, thus posing a potential threat to future forest productivity.

Visibility

Sulfur dioxide emissions lead to the formation of sulfate particles in the atmosphere. Sulfate particles account for more than 50 percent of the visibility reduction in the eastern part of the United States, affecting our enjoyment of national parks, such as the Shenandoah and the Great Smoky Mountains. The Acid Rain Program is expected to improve the visual range in the eastern U.S. by 30 percent. Based on a study of the value national park visitors place on visibility, the visual range improvements expected at national parks of the eastern United States due to the Acid Rain Program's SO_2 reductions will be worth a billion dollars by the year 2010. In the western part of the United States, nitrogen and carbon also play roles, but sulfur has been implicated as an important source of visibility impairment in many of the Colorado River Plateau national parks, including the Grand Canyon, Canyonlands, and Bryce Canyon.

Materials

Acid rain and the dry deposition of acidic particles are known to contribute to the corrosion of metals and deterioration of stone and paint on buildings, cultural objects, and cars. The corrosion seriously depreciates the objects' value to society. Dry deposition of acidic compounds can also dirty buildings and other structures, leading to increased maintenance costs. To reduce damage to automotive paint caused by acid rain and acidic dry deposition, some manufacturers use acid-resistant paints, at an average cost of $5 for each new vehicle (or a total of $61 million per year for all new cars and trucks sold in the U.S.) The Acid Rain Program will reduce damage to materials by limiting SO_2 emissions. The benefits of the Acid Rain Program are measured, in part, by the costs now paid to repair or prevent damage—the costs of repairing buildings, using acid-resistant paints on new vehicles, plus the value that society places on the details of a statue lost forever to acid rain.

Health

Based on health concerns, SO_2 has historically been regulated under the Clean Air Act. Sulfur dioxide interacts in the atmosphere to form sulfate aerosols, which may be transported long distances through the air. Most sulfate aerosols are particles that can be inhaled. In the eastern United States, sulfate aerosols make up about 25 percent of the inhalable particles. According to recent studies at Harvard and New York Universities, higher levels of sulfate aerosols are associated with increased morbidity (sickness) and mortality from lung disorders, such as asthma and bronchitis. By lowering sulfate aerosol levels, the Acid Rain Program will reduce the incidence and the severity of asthma and bronchitis. When fully implemented by the year 2010, the public health benefits of the Acid Rain Program will be significant, due to decreased mortality, hospital admissions, and emergency-room visits.

Decreases in nitrogen oxide emissions are also expected to have positive health effects by reducing the nitrate component of inhalable particulates and reducing the nitrogen oxides available to react with volatile organic compounds (VOCs) and form ozone. Ozone impacts on human health include a number of morbidity and mortality risks associated with lung disorders.

Automotive Coatings

Since about 1990, reports of damage to automotive coatings have increased. The reported damage typically occurs on horizontal surfaces and appears as irregularly shaped, permanently etched areas. The damage can best be detected under fluorescent lamps, can be most easily observed on dark colored vehicles, and appears to occur after evaporation of a moisture droplet. In addition, some evidence suggests damage occurs most frequently on freshly painted vehicles. Usually the damage is permanent; once it has occurred, the only solution is to repaint.

The general consensus within the auto industry is that the damage is caused by some form of environmental fallout. "Environmental fallout," a term widely used in the auto and coatings industries, refers to damage caused by air pollution (e.g., acid rain), decaying insects, bird droppings, pollen, and tree sap. The results of laboratory experiments and at least one field study have demonstrated that acid rain can scar automotive coatings. Furthermore, chemical analyses of the damaged areas of some exposed test panels showed elevated levels of sulfate, implicating acid rain.

The popular term "acid rain" refers to both wet and dry deposition of acidic pollutants that may damage material surfaces, including auto finishes. These pollutants, which are released when coal and other fossil fuels are burned react with water vapor and oxidants in the atmosphere and are chemically transformed into sulfuric and nitric acids. The acidic compounds then may fall to earth as rain, snow, fog, or may join dry particles and fall as dry deposition. Automotive coatings may be damaged by all forms of acid rain, including dry deposition, especially when dry acidic deposition is mixed with dew or rain. However, it has been difficult to quantify the specific contribution of acid rain to paint finish damage relative to damage caused by other forms of environmental fallout, by the improper application of paint or by deficient paint formulations. According to coating experts, trained specialists can differentiate between the various forms of damage, but the best way of determining the cause of chemically induced damage is to conduct a detailed, chemical analysis of the damaged area.

Because evaporation of acidic moisture appears to be a key element in the damage, any steps taken to eliminate its occurrence on freshly painted vehicles may alleviate the problem. The steps include frequent washing followed by hand drying, covering the vehicle during precipitation events, and use of one of the protective coatings currently on the market that claim to protect the original finish. (However, data on the performance of these coatings are not yet sufficient.)

The auto and coatings industries are fully aware of the potential damage and are actively pursuing the development of coatings that are more resistant to environmental fallout, including acid rain. The problem is not a universal one—it does not affect all coatings or all vehicles even in geographic areas known to be subject to acid rain, which suggests that technology exists to protect against this damage. Until that technology is implemented to protect all vehicles or until acid deposition is adequately reduced, frequent washing and drying and covering the vehicle appear to be the best methods to minimize acid rain damage.

Acid Deposition

Sulfur and nitrogen oxides are emitted into the atmosphere primarily from the burning of fossil fuels. These emissions react in the atmosphere to form compounds that are transported long distances and are subsequently deposited in the form of pollutants such as particulate matter (sulfates and nitrates), SO_2, NO_2, nitric acid and when reacted with volatile organic compounds (VOCs) form ozone. The effects of atmospheric deposition include acidification of lakes and streams, nutrient enrichment of coastal waters and large river basins, soil nutrient depletion and decline of sensitive forests, agricultural crop damage, and impacts on ecosystem biodiversity. Toxic pollutants and metals also can be transported and deposited through atmospheric processes.

Both local and long-range emission sources contribute to atmospheric deposition. Total atmospheric deposition is determined using both wet and dry deposition measurements. Although the term "acid rain" is widely recognized, the dry deposition portion ranges from 20 to 60 percent of total deposition.

The United States Environmental Protection agency (EPA) is required by several Congressional and other mandates to assess the effectiveness of air pollution control efforts. These mandates include Title IX of the Clean Air Act Amendments (CAAA), the National Acid Precipitation Assessment Program (NAPAP), the Government Performance and Results

Act, and the U.S. Canada Air Quality Agreement. One measure of effectiveness of these efforts is whether sustained reductions in the amount of atmospheric deposition over broad geographic regions are occurring. However, changes in the atmosphere happen very slowly and trends are often obscured by the wide variability of measurements and climate. Many years of continuous and consistent data are required to overcome this variability, making long-term monitoring networks especially critical for characterizing deposition levels and identifying relationships among emissions, atmospheric loadings, and effects on human health and the environment.

For wet and dry deposition, these studies typically include measurement of concentration levels of key chemical components as well as precipitation amounts. For dry deposition, analyses also must include meteorological measurements that are used to estimate rate of the actual deposition, or "flux." Data representing total deposition loadings (e.g., total sulfate or nitrate) are what many environmental scientists use for integrated ecological assessments.

Primary Atmospheric Deposition Monitoring Networks

The National Atmospheric Deposition Program (NADP) and the Clean Air Status and Trends Network (CASTNET), described in detail below, were developed to monitor wet and dry acid deposition, respectively. Monitoring site locations are predominantly rural by design to assess the relationship between regional pollution and changes in regional patterns in deposition. CASTNET also includes measurements of rural ozone and the chemical constituents of $PM_{2.5}$. Rural monitoring sites of NADP and CASTNET provide data where sensitive ecosystems are located and provide insight into natural background levels of pollutants where urban influences are minimal. These data provide needed information to scientists and policy analysts to study and evaluate numerous environmental effects, particularly those caused by regional sources of emissions for which long-range transport plays an important role. Measurements from these networks are also important for understanding non-ecological impacts of air pollution such as visibility impairment and damage to materials, particularly those of cultural and historical importance.

National Atmospheric Deposition Network

The NADP was initiated in the late 1970s as a cooperative program between federal and state agencies, universities, electric utilities, and other industries to determine geographical patterns and trends in precipitation chemistry in the United States. Collection of weekly wet deposition samples began in 1978. The size of the NADP Network grew rapidly in the early 1980s when the major research effort by the NAPAP called for characterization of acid deposition levels. At that time, the network became known as the NADP/NTN (National Trends Network). By the mid-1980s, the NADP had grown to nearly 200 sites, where it stands today, as the longest running national deposition monitoring network.

The NADP analyzes the constituents important in precipitation chemistry, including those affecting rainfall acidity and those that may have ecological effects. The Network measures sulfate, nitrate, hydrogen ion (measure of acidity), ammonia, chloride, and base cations (calcium, magnesium, potassium). To ensure comparability of results, laboratory analyses for all samples are conducted by the NADP's Central Analytical Lab at the Illinois State Water Survey. A new subnetwork of the NADP, the Mercury Deposition Network (MDN) measures mercury in precipitation.

Clean Air Status and Trends Network

The CASTNET provides atmospheric data on the dry deposition component of total acid deposition, ground-level ozone, and other forms of atmospheric pollution. CASTNET is considered the nation's primary source for atmospheric data to estimate dry acidic deposition and to provide data on rural ozone levels. Used in conjunction with other national monitoring networks, CASTNET is used to determine the effectiveness of national emission control programs. Established in 1987, CASTNET now comprises over 70 monitoring stations across the United States. The longest data records are primarily at eastern sites. The majority of the monitoring stations are operated by EPA's Office of Air and Radiation; however, approximately 20 stations are operated by the National Park Service in cooperation with EPA.

Each CASTNET dry deposition station measures:

- weekly average atmospheric concentrations of sulfate, nitrate, ammonium, sulfur dioxide, and nitric acid.

- hourly concentrations of ambient ozone levels.
- meteorological conditions required for calculating dry deposition rates.

Dry deposition rates are calculated using atmospheric concentrations, meteorological data, and information on land use, vegetation, and surface conditions. CASTNET complements the database complied by NADP. Because of the interdependence of wet and dry deposition, NADP wet deposition data are collected at all CASTNET sites. Together, these two long-term databases provide the necessary data to estimate trends and spatial patterns in total atmospheric deposition. National Oceanic and Atmospheric Administration (NOAA) also operates a smaller dry deposition network called Atmospheric Integrated Assessment Monitoring Network (AIRMoN) focused on addressing research issues specifically related to dry deposition measurement.

Ozone Data Collection Network

Ozone data collected by CASTNET are complementary to the larger ozone data sets gathered by the State and Local Air Monitoring Stations (SLAMS) and National Air Monitoring Stations (NAMS) networks. Most air-quality samples at SLAMS/NAMS sites are located in urban areas, while CASTNET sites are in rural locations. Hourly ozone measurements are taken at each of the 50 sites operated by EPA. Data from these sites provide information to help characterize ozone transport issues and ozone exposure levels.

Integrated Monitoring, and AIRMoN

The Atmospheric Integrated Research Monitoring Network is an atmospheric component to the overall national integrated monitoring initiative that is currently evolving. AIRMoN is a relatively new program, constructed by combining and building upon pre-existing specialized wet deposition and dry deposition monitoring networks, and with two specific goals:

1. *To provide regular and timely reports on the atmospheric consequences of emission reductions, as imposed under the Clean Air Act Amendments.*
2. *To provide quantified information required to extend these observations of atmospheric effects to atmospheric deposition, both wet and dry.*

AIRMoN has two principal components: wet and dry deposition. All variables are measured in a manner that is designed to detect and properly attribute the benefits of emissions controls mandated under the Clean Air Act Amendments of 1990, and to reveal the actual deposition that occurred without fear of chemical (or other) contamination. It should be emphasized that conventional monitoring programs rely on statistical methods to extract small signals from imperfect and noisy data records. AIRMoN is designed to take a new step, that will remove much of the noise by integrating modern forecast technology into the monitoring process.

ARL presently focuses its research attention on:

- the measurement of precipitation chemistry with fine time resolution (AIRMoN-wet),
- the development of systems for measuring deposition, both wet and dry,
- the measurement of dry deposition using micrometeorological methods (AIRMoN-dry),
- the development of techniques for assessing air-surface exchange in areas (such as specific watersheds) where intensive studies are not feasible, and
- the extension of local measurements and knowledge to describe a real average exchange in numerical models.

Clean Air Act

The overall goal is to achieve significant environmental and public health benefits through reductions in emissions of sulfur dioxide (SO_2) and nitrogen oxides (NO_X), the primary causes of acid rain. To achieve this goal at the lowest cost to society, the program employs both traditional and innovative, market-based approaches for controlling air pollution. In addition, the program encourages energy efficiency and pollution prevention.

Title IV of the Clean Air Act Amendments of 1990 calls for a 10 million ton reduction in annual emissions of sulfur dioxide (SO_2) in the United States by the year 2010, which represents an approximately 40 percent reduction in anthropogenic emissions from 1980 levels. Implementation of Title IV is referred to as the Acid Rain Program; the primary motivation for this section of the Clean Air Act Amendments is to reduce acid precipitation and dry deposition. To achieve these reductions, the law requires a two-phase tightening of the restrictions placed on fossil-fuel-fired power plants.

The Act also calls for a 2 million ton reduction in NO_X emissions by the year 2000. A significant portion of this reduction will be achieved by coal-fired utility boilers that will be required to install low NO_X burner technologies and to meet new emissions standards.

Phase I began in 1995 and affects 263 units at 110 mostly coal-burning electric utility plants located in 21 eastern and midwestern states. An additional 182 units joined Phase I of the program as substitution or compensating units, bringing the total of Phase I affected units to 445. Emissions data indicate that 1995 SO_2 emissions at these units nationwide were reduced by almost 40% below their required level.

Phase II, which begins in the year 2000, tightens the annual emissions limits imposed on these large, higher emitting plants and also sets restrictions on smaller, cleaner plants fired by coal, oil, and gas, encompassing over 2,000 units in all. The program affects existing utility units serving generators with an output capacity of greater than 25 megawatts and all new utility units.

See also **Pollution (Air)**.

Additional Reading

Ellerman, A.D., R. Schmalensee, E.M. Bailey, et al.: *Markets for Clean Air: The U.S. Acid Rain Program,* Cambridge University Press, New York, NY, 2000.

Hocking, C., J. Barber, J. Coonrod, et al.: *Acid Rain,* University of California Press, Berkeley, CA, 2000.

Howells, G.P.: *Acid Rain and Acid Waters,* 2nd Edition, Prentice-Hall, Inc., Upper Saddle River, NJ, 1995.

Hunt, K.: *Changes in Global Environment-Acid Rain,* Kendall/Hunt Publishing Company, Dubuque, IA, 1997.

Hutterman, A. and D. Godbold: *Effects of Acid Rain on Forest Processes,* John Wiley & Sons, Inc., New York, NY, 1994.

Kosobud, R.F., D.L. Schreder and H.M. Biggs: *Emissions Trading: Environmental Policy's New Approach,* John Wiley & Sons, Inc., New York, NY, 2000.

Morgan, S.: *Acid Rain,* Franklin Watts, Danbury, CT, 1999.

Somerville, R.C.J.: *The Forgiving Air: Understanding Environmental Change,* University of California Press, Berkeley, CA, 1998.

Web References

http://www.epa.gov/ United States Environmental Protection Agency.

http://www.epa.gov/acidrain/ardhome.html United States Environmental Protection Agency Acid Rain Program.

http://www.epa.gov/acidrain/links.htm United States Environmental Protection Agency Links.

http://www.ec.gc.ca/acidrain/acidfact.html Environment Canada.

http://www.epa.gov/airsdata/ State and Local Air Monitoring Stations (SLAMS) and National Air Monitoring Stations (NAMS) networks.

http://www.arl.noaa.gov/ National Oceanic and Atmospheric Administration (NOAA).

http://www.arl.noaa.gov/research/themes/aq.html#3 Atmospheric Integra ted Assessment Monitoring Network (AIRMoN).

ACIDS AND BASES. The conventional definition of an acid is that it is an electrolyte that furnishes protons, i.e., hydrogen ions, H^+. An acid is sour to the taste and usually quite corrosive. A base is an electrolyte that furnishes hydroxyl ions, OH^-. A base is bitter to the taste and also usually quite corrosive. These definitions were formulated in terms of water solutions and, consequently, do not embrace situations where some ionizing medium other than water may be involved. In the definition of Lowry and Brnsted, an acid is a proton donor and a base is a proton acceptor. Acid-base theory is described later.

Acidification is the operation of creating an excess of hydrogen ions, normally involving the addition of an acid to a neutral or alkaline solution until a pH below 7 is achieved, thus indicating an excess of hydrogen ions. In *neutralization,* a balance between hydrogen and hydroxyl ions is effected. An acid solution may be neutralized by the addition of a base; and vice versa. The products of neutralization are a salt and water.

Some of the inorganic acids, such as hydrochloric acid, HCl, nitric acid, HNO_3, and sulfuric acid, H_2SO_4, are very-high-tonnage products and are considered very important chemical raw materials. The most common inorganic bases (or alkalis) include sodium hydroxide, NaOH, and potassium hydroxide, KOH, and also are high-tonnage materials, particularly NaOH.

Several classes of organic substances are classified as acids, notably the carboxylic acids, the amino acids, and the nucleic acids. These and the previously mentioned materials are described elsewhere in this volume.

Principal theories of acids and bases have included: (1) Arrhenius-Ostwald theory, which was proposed soon after the concept of the *ionization* of chemical substances in aqueous solutions was generally accepted. (2) Much later (1923), J.N. Brønsted defined an acid as a source of protons and a base is an acceptor of protons. (3) T.M. Lowry, working in the same time frame as Brønsted, developed a similar concept and, over the years, the concept has been referred to in the literature as the *Lowry-Brønsted theory.* It will be noted that this theory altered the definition of an acid very little, continuing to emphasize the role of the hydrogen ion. However, the definition of a base was extended beyond the role of the hydroxyl ion to include a wide variety of uncharged species, such as ammonia and the amines. (4) In 1938, G.N. Lewis further broadened the definition of Lowry-Brønsted. Lewis defined an acid as anything that can attach itself to something with an unshared pair of electrons. The broad definition of Lewis creates some difficulties when one attempts to categorize Lewis acids and bases. R.G. Pearson (1963) suggested two main categories—hard and soft acids as well as hard and soft bases. These are described in more detail by Long and Boyd (1983). (5) In 1939, M. Usanovich proposed still another theory called the *positive-negative* theory, also developed in detail by Long and Boyd.

In terms of the definition that an acid is a proton donor and a base is a proton acceptor, hydrochloric acid, water, and ammonia (NH_3) are acids in the reactions

$$HCl \rightleftharpoons H^+ + Cl^-$$

$$H_2O \rightleftharpoons H^+ + OH^-$$

$$NH_3 \rightleftharpoons H^+ + NH_2^-$$

Note that this definition is different in at least two major respects from the conventional definition of an acid as a substance dissociating to give H^+ in water. The Lowry-Brnsted definition states that for every acid there be a "conjugate" base, and vice versa. Thus, in the examples cited above, Cl^-, OH^-, and NH^--are the conjugate bases of HCl, H_2O, and NH_3.

Furthermore, since the equations given above should more properly be written

$$HCl + H_2O \rightleftharpoons H_3O^+ + Cl^-$$

$$H_2O + H_2O \rightleftharpoons H_3O^+ + OH^-$$

$$NH_3 + H_2O \rightleftharpoons H_3O^+ + NH_2^-$$

It can be seen that every acid-base reaction involving transfer of a proton will involve two conjugate acid-base pairs, e.g., in the last equation NH_3 and H_3O^+ are the acids and NH_2^- and H_2O the respective conjugate bases. On the other hand, in the reaction

$$NH_3 + H_2O \rightleftharpoons NH_4^+ + OH^-$$

H_2O and NH_4^- are the acids and NH_3 and OH^- the bases. In other reactions, e.g.,

Base$_1$	Acid$_2$		Acid$_1$	Base$_2$
$C_2H_3O_2^-$	$+ H_2O$	$\rightleftharpoons$	$HC_2H_3O_2$	$+ OH^-$
HCO_3^-	$+ HCO_3^-$	$\rightleftharpoons$	H_2CO_3	$+ CO_3^{-2}$
$N_2H_5^+$	$+ N_2H_5+$	$\rightleftharpoons$	$N_2H_6^{+2}$	$+ N_2H_4$
H_2O	$+ Cr(H_2O)_6^{+3}$	$\rightleftharpoons$	N_3O^+	$+ Cr(H_2O)_5OH^{2+}$

the conjugate acids and bases are as indicated. The theory is not limited to the aqueous solution; for example, the following reactions can be considered in exactly the same light:

Base$_1$	Acid$_2$		Acid$_1$	Base$_2$
NH_3	$+ HCl$	$\rightleftharpoons$	NH_4^+	$+ Cl^-$
CH_3CO_2H	$+ HF$	$\rightleftharpoons$	$CH_3CO_2H_2^+$	$+ F^-$
HF	$+ HClO_4$	$\rightleftharpoons$	H_2F^+	$+ ClO_4^-$
$(CH_3)_2O$	$+ HI$	$\rightleftharpoons$	$(CH_3)_2OH^+$	$+ I^-$
C_6H_6	$+ HSO_3F$	$\rightleftharpoons$	$C_6H_7^+$	$+ SO_3F^-$

Acids may be classified according to their charge or lack of it. Thus, in the reactions cited above, there are "molecular" acids and bases, such as HCl, H_2CO_3, $HClO_4$, etc., and N_2H_4, $(CH_3)_2O$, C_6H_6, etc., and also cationic acids and bases, such as H_3O^+, $N_2H_5^+$, $N_2H_6^{2+}$, NH_4^+, $(CH_3)_2OH^+$, etc., as well as anionic acids and bases, such as HCO_3^-, Cl^-, NH_2^-, NH_3^{-2} etc. In a more general definition, Lewis calls a base any substance with a free pair of electrons that it is capable of sharing with an electron pair acceptor, which is called an acid. For example, in the reaction:

$$(C_2H_5)_2O : + BF_3 \longrightarrow (C_2H_5)_2O{:}BF_3$$

the ethyl ether molecule is called a base, the boron trifluoride, an acid. The complex is called a *Lewis salt*, or *addition compound*.

Acids are classified as monobasic, dibasic, tribasic, polybasic, etc., according to the number (one, two, three, several, etc.) of hydrogen atoms, replaceable by bases, contained in a molecule. They are further classified as (1) organic, when the molecule contains carbon; (1a) carboxylic, when the proton is from a—COOH group; (2) normal, if they are derived from phosphorus or arsenic, and contain three hydroxyl groups: (3) ortho, meta, or para, according to the location of the carboxyl group in relation to another substituent in a cyclic compound; or (4) ortho, meta, or pyro, according to their composition.

Superacids. Although mentioned in the literature as early as 1927, superacids were not investigated aggressively until the 1970s. Prior to the concept of superacids, scientists generally regarded the familiar mineral acids (HF, HNO_3, H_2SO_4, etc.) as the strongest acids attainable. Relatively recently, acidities up to 10^{12} times that of H_2SO_4 have been produced.

In very highly concentrated acid solutions, the commonly used measurement of pH is not applicable. See also **pH (Hydrogen Ion Concentration)**. Rather, the acidity must be related to the degree of transformation of a base with its conjugate acid. In the *Hammett acidity function*, developed by Hammett and Deyrup in 1932,

$$H_0 = pK_{BH^+} - \log \frac{BH^+}{B}$$

where pkBH$^+$ is the dissociation constant of the conjugate acid (BH$^+$), and BH$^+$/B is the ionization ratio, measurable by spectroscopic means (UV or NMR). In the Hammett acidity function, acidity is a logarithmic scale wherein H_2SO_4 (100%) has an H_0 of -11.9; and HF, an H_0 of -11.0.

As pointed out by Olah et al. (1979), "The acidity of a sulfuric acid solution can be increased by the addition of solutes that behave as acids in the system: $HA + H_2SO_4 \rightleftharpoons H_3SO_4^+ + A^-$. These solutes increase the concentration of the highly acidic H_3SO_4 cation just as the addition of an acid to water increases the concentration of the oxonium ion. H_3O^+. Fuming sulfuric acid (oleum) contains a series of such acids, the polysulfuric acids, the simplest of which is disulfuric acid, $H_2S_2O_7$, which ionizes as a moderately strong acid in sulfuric acid: $H_2S_2O_7 + H_2SO_4 \rightleftharpoons H_3SO_4^- + HS_2O_7^-$. Higher polysulfuric acids, such as $H_2S_3O_{10}$ and $H_2S_4O_{13}$, also behave as acids and appear somewhat stronger than $H_2S_2O_7$."

Hull and Conant in 1927 showed that weak organic bases (ketones and aldehydes) will form salts with perchloric acid in nonaqueous solvents. This results from the ability of perchloric acid in nonaqueous systems to protonate these weak bases. These early investigators called such a system a superacid. Some authorities believe that any protic acid that is stronger than sulfuric acid (100%) should be typed as a superacid. Based upon this criterion, fluorosulfuric acid and trifluoro-methanesulfonic acid, among others, are so classified. Acidic oxides (silica and silica-alumina) have been used as solid acid catalysts for many years. Within the last few years, solid acid systems of considerably greater strength have been developed and can be classified as *solid superacids*.

Superacids have found a number of practical uses. Fluoroantimonic acid, sometimes called *Magic Acid*, is particularly effective in preparing stable, long-lived carbocations. Such substances are too reactive to exist as stable species in less acidic solvents. These acids permit the protonation of very weak bases. For example, superacids, such as Magic Acid, can protonate saturated hydrocarbons (alkanes) and thus can play an important role in the chemical transformation of hydrocarbons, including the processes of isomerization and alkylation. See also **Alkylation**; and **Isomerization**. Superacids also can play key roles in polymerization and in various organic syntheses involving dienone-phenol rearrangement, reduction, carbonylation, oxidation, among others. Superacids also play a role in

inorganic chemistry, notably in the case of halogen cations and the cations of nonmetallic elements, such as sulfur, selenium, and tellurium.

Free Hydroxyl Radical. It is important to distinguish the free radical · OH and the OH$^-$ ion previously mentioned. The free radical is created by complex reactions of so-called "excited" oxygen with hydrogen as the result of exposure to solar ultraviolet light. The radical has been found to be an important factor in atmospheric and oceanic chemistry. The life span of the radical is but a second or two, during which time it reacts with numerous atmospheric pollutants in a scavenging (oxidizing) manner. For example, it reacts with carbon monoxide, as commonly encountered in atmospheric smog. It also reacts with sulfurous gases and with hydrocarbons, as may result from incomplete combustion processes or that have escaped into the atmosphere (because of their volatility) from various sources.

Because of the heavy workload placed upon the hydroxyl radical through such "cleansing" reactions in the atmosphere, some scientists are concerned that the atmospheric content of · OH has diminished with increasing pollution, estimating the probable drop to be as much as 5–25% during the past three centuries since the start of the Industrial Revolution. Ironically, some of the very pollutants that are targets for reduction also are compounds from which the · OH radical is produced and, as they are reduced, so will the concentration of · OH be reduced. The fact that there is only one hydroxyl radical per trillion air molecules must not detract from its effectiveness as a scavenger.

Scientists at the Georgia Institute of Technology have devised a mass spectrometric means for testing the various theories pertaining to the chemistry of · OH.

The probable importance of · OH in the oceans also is being investigated. Researchers at Washington State University and the Brookhaven National Laboratory have confirmed the presence of · OH in seawater and now are attempting to measure its content quantitatively and to determine the sources of its formation. Dissolved organic matter is one highly suspected source. Tentatively, it has been concluded (using a method called flash photolysis) that · OH concentrations (as well as daughter radicals) range from 5 to 15 times higher in deep water than in open-ocean surface waters. This may indicate that · OH may have some impact on biota residing in deep water and may enhance the secondary production of bacterial growth, particularly in "carbon limited" oligotrophic waters, in upwelling waters, and in regions with high ultraviolet radiation.

See also specific acids and bases, such as sulfuric acid and sodium hydroxide, in alphabetically arranged entries throughout this *Encyclopedia*.

Additional Reading

Lide, D.R., Editor: *CRC Handbook of Chemistry and Physics,* 84th Edition, CRC Press LLC, Boca Raton, FL, 2003.

Long, F.A., and R.H. Boyd: "Acid and Bases," in *McGraw-Hill Encyclopedia of Chemistry*, McGraw-Hill Companies, Inc., New York, NY, 1983.

Olah, G.A., G.K. Surya Prakash, and J. Sommer: "Superacids," *Science*, **205**, 13–20 (1979).

Parker, P.: *McGraw Encyclopedia of Chemistry,* McGraw-Hill Companies, Inc., New York, NY, 1993.

Walling, C.: *Fifty Years of Free Radicals* (Profiles, Pathways, and Dreams); American Chemical Society, Washington, DC, 1994.

ACIDULANTS AND ALKALIZERS (Foods).   Well over 50 chemical additives are commonly used in food processing or as ingredients of final food products, essentially to control the pH (hydrogen ion concentration) of the process and/or product. An excess of hydrogen ions, as contributed by acid substances, produces a sour taste, whereas an excess of hydroxyl ions, as contributed by alkaline substances, creates a bitter taste. Soft drinks and instant fruit drinks, for example, owe their tart flavor to acidic substances, such as citric acid. Certain candies, chewing gums, jellies, jams, and salad dressings are among the many other products where a certain degree of tartness contributes to the overall taste and appeal.

Taste is only one of several qualities of a process or product that is affected by an excess of either of these ions. Some raw materials are naturally too acidic, others too alkaline—so that neutralizers must be added to adjust the pH within an acceptable range. In the dairy industry, for example, the acid in sour cream must be adjusted by the addition of alkaline compounds in order that satisfactory butter can be churned. Quite often, the pH may be difficult to adjust or to maintain after adjustment. Stability of pH can be accomplished by the addition of buffering agents that, within limits, effectively maintain the desired pH even when additional acid or alkali is added. For example, orange-flavored instant breakfast drink has just enough "bite" from the addition of potassium citrate (a buffering agent) to regulate the tart flavor imparted by another ingredient, citric acid. In some instances, the presence of acids or alkalies assists mechanical processing operations in food preparation. Acids, for example, make it easier to peel fruits and tubers. Alkaline solutions are widely used in removing hair from animal carcasses.

The pH values of various food substances cover a wide range. Plant tissues and fluids (about 5.2); animal tissues and fluids (about 7.0 to 7.5); lemon juice (2.0 to 2.2); acid fruits (3.0 to 4.5); fruit jellies (3.0 to 3.5).

Acidulants commonly used in food processing include: Acetic acid (glacial), citric acid, fumaric acid, glucono delta-lactone, hydrochloric acid, lactic acid, malic acid, phosphoric acid, potassium acid tartrate, sulfuric acid, and tartaric acid. Alkalies commonly used include: Ammonium bicarbonate, ammonium hydroxide, calcium carbonate, calcium oxide, magnesium carbonate, magnesium hydroxide, magnesium oxide, potassium bicarbonate, potassium carbonate, potassium hydroxide, sodium bicarbonate, sodium carbonate, sodium hydroxide, and sodium sesquicarbonate. Among the buffers and neutralizing agents favored are: Adipic acid, aluminum ammonium sulfate, ammonium phosphate (di- or monobasic), calcium citrate, calcium gluconate, sodium acid pyrophosphate, sodium phosphate (di-, mono-, and tri-basic), sodium pyrophosphate, and succinic acid.

See also **Buffer (Chemical)**; and **pH (Hydrogen Ion Concentration)**.

Functions of Acidulants

In the *baking industry*, acidulants and their salts control pH to inhibit spoilage by microbial actions to enhance the stability of foams (such as whipped egg albumin), to assist in leavening in order to achieve desired volume and flavoring, and to maximize the performance of artificial preservatives. A variety of the food acids previously mentioned is used. For example, citric acid traditionally has been favored by bakers for pie fillings. Baking powders (leavening agents) frequently will contain adipic acid, fumaric acid, and cream of tartar. Fumaric acid, in particular, has been the choice for leavening systems of cakes, pancakes, biscuits, waffles, crackers, cookies, and doughnuts. This acid also provides the desired characteristic flavor for sour rye bread—this eliminating fermentation of the dough to achieve desired flavor. Lactic acid and its salts sometimes are used as dough conditioners.

Acidulants are used in the *soft drink beverage industry* for producing a tart taste, improving flavor balance, modifying the "sweetness" provided by sugar and other sweeteners, extending shelf life by reducing pH value of final product, and improving the performance of antimicrobial agents. Specific acidulants preferred vary with the type of beverage—i.e. carbonated, non-carbonated, dry (reconstituted by addition of water), and low-calorie products.

In the production of *confections and gelatin desserts,* acidulants are used mainly for enhancing flavor, maintaining viscosity, and controlling gel formation. In confections, such as hard candies, acidulants are used to increase tartness and to enhance fruit flavors. Acidulants also contribute to the ease of manufacturing.

In *dairy products,* acidulants, in addition to achieving many of the foregoing functions, also help to process the products. As an example, adipic acid improves the texture and melting characteristics of processed cheese and cheese foods, where pH control is very important.

In *fruit and vegetable processing,* acidulants play somewhat different roles than previously described. These would include reducing process heating requirements through pH control, inactivating certain enzymes that reduce shelf life, and chelation of trace metals that may be present (through catalytic enzymatic oxidation). Citric acid is used widely in canned fruits, such as apricots, peaches, pears, cherries, applesauce, and grapes, to retain the firmness of the products during processing. The acid also provides a desirable tartness in the final products.

In the *processed meat field,* citric acid, along with oxidants, is used to prevent rancidity in frankfurters and sausages. Sodium citrate is used in processing livestock blood, which is used to manufacture some sausages and pet foods.

Acidulants and alkalizers, like other food additives, are controlled by regulatory bodies in most industrial nations. Some of the additives mentioned in this article are considered to be "Generally Regarded as Safe," having a GRAS classification. These include acetic, adipic, citric, glucone delta lactone, lactic, malic, phosphoric, and tartaric acids. Others are covered by the Code of Federal Registration (FDA) in the United States.

A very orderly and informative article (Dziezak 1990) is suggested as a source of detailed information on this topic.

Additional Reading

Dziezak, J.D.: "Acidulants: Ingredients That Do More than Meet the Acid Test," *Food Techy.*, 76 (January 1990).
Igoe, R.S.: *Dictionary of Food Ingredients,* Chapman & Hall, New York, NY, 1999.
Kirk, R.E. and D.F. Othmer: *Encyclopedia of Chemical Technology,* 4th Edition, Vol. 6, John Wiley & Sons, New York, NY, 1993.
Toledo, R.T.: *Fundamentals of Food Process Engineering,* 2nd Edition, Aspen Publishers, Inc., Gaithersburg, MD, 1999.

ACMITE-AEGERINE. Acmite is a comparatively rare rock-making mineral, usually found in nephelite syenites or other nephelite or leucite-bearing rocks, as phonolites. Chemically, it is a soda-iron silicate, and its name refers to its sharply pointed monoclinic crystals. Bluntly terminated crystals form the variety aegerine, named for Aegir, the Icelandic sea god.

Acmite has a hardness of 6 to 6.5, specific gravity 3.5, vitreous; color brown to greenish-black (aegerine), or red-brown to dark green and black (acmite). Acmite is synonymous with aegerine, but usually restricted to the long slender crystalline variety of brown color.

The original acmite locality is in Greenland. Norway, the former U.S.S.R., Kenya, India, and Mt. St. Hilaire, Quebec, Canada furnish fine specimens. United States localities are Magnet Cove, Arkansas, and Libby, Montana, where a variety carrying vanadium occurs.

ACREE'S REACTION. A test for protein in which a violet ring appears when concentrated sulfuric acid is introduced below a mixture of the unknown solution and a formaldehyde solution containing a trace of ferric oxide.

ACROLEIN AND DERIVATIVES. Acrolein (2-propenal), C_3H_4O, is the simplest unsaturated aldehyde (CH_2=CHCHO). The primary characteristic of acrolein is its high reactivity due to conjugation of the carbonyl group with a vinyl group. More than 80% of the refined acrolein that is produced today goes into the synthesis of methionine. Much larger quantities of crude acrolein are produced as an intermediate in the production of acrylic acid. More than 85% of the acrylic acid produced worldwide is by the captive oxidation of acrolein.

Acrolein is a highly toxic material with extreme lacrimatory properties. At room temperature acrolein is a liquid with volatility and flammability somewhat similar to acetone; but unlike acetone, its solubility in water is limited. Commercially, acrolein is always stored with hydroquinone and acetic acid as inhibitors. Special care in handling is required because of the flammability, reactivity, and toxicity of acrolein.

The physical and chemical properties of acrolein are given in Table 1.

Economic Aspects

Presently, worldwide refined acrolein nameplate capacity is about 113,000 t/yr. Degussa has announced a capacity expansion in the United States by building a 36,000 t/yr acrolein plant in Theodore, Alabama to support their methionine business. The key producers of refined acrolein are Union Carbide (United States), Degussa (Germany), Atochem (France), and Daicel (Japan).

Reactions and Derivatives

Acrolein is a highly reactive compound because both the double bond and aldehydic moieties participate in a variety of reactions, including oxidation, reduction, reactions with alcohols yielding alkoxy propionaldehydes,

TABLE 1. PROPERTIES OF ACROLEIN

Property	Value
Physical properties	
molecular formula	C_3H_4O
molecular weight	56.06
specific gravity at 20/20°C	0.8427
boiling point, °C at 101.3 kPa[a]	52.69
Chemical properties	
autoignition temperature in air, °C	234
heat of combustion at 25°C, kJ/kg[b]	5383

[a]To convert kPa to mm Hg, multiply by 7.5.
[b]To convert kJ to kcal, divide by 4.184.

acrolein acetals, and alkoxypropionaldehyde acetals, addition of mercaptans yielding 3-methylmercaptopropionaldehyde, reaction with ammonia yielding β-picoline and pyridine, Diels-Alder reactions, and polymerization.

Direct Uses of Acrolein

Because of its antimicrobial activity, acrolein has found use as an agent to control the growth of microbes in process feed lines, thereby controlling the rates of plugging and corrosion.

Acrolein at a concentration of <500 ppm is also used to protect liquid fuels against microorganisms.

W. G. ETZKORN
J. J. KURLAND
W. D. NEILSEN
Union Carbide Chemicals & Plastics Company Inc.

Additional Reading

Schulz, R.C.: in J.I. Kroschwitz, ed., *Encyclopedia of Polymer Science and Engineering,* 2nd Edition, Vol. 1, Wiley-Interscience, New York, NY 1985, pp. 160–169.

Ohara, T., T. Sato, N. Shimizu, G. Prescher, H. Schwind, and O. Weiberg: *Ullman's Encyclopedia of Industrial Chemistry,* 5th Edition, Vol. A1, 1985, pp. 149–160.

Smith, C.W. ed.: *Acrolein,* John Wiley & Sons, Inc., New York, NY 1962.

ACRYLAMIDE. Acrylamide (NIOSH No. A533250) has been commercially available since the mid-1950s and has shown steady growth since that time, but is still considered a small-volume commodity. Its formula, $H_2=CHCONH_2$ (2-propeneamide), indicates a simple chemical, but it is by far the most important member of the series of acrylic and methacrylic amides. Water-soluble polyacrylamides have the most important applications, including potentially large uses in enhanced oil recovery as mobility-control agents in water flooding, additives for oilwell drilling fluids, and aids in fracturing, acidifying, and other operations. Other uses include flocculants for waste-water treatment, the mining industry, and various other process industries, soil stabilization, papermaking aids, and thickeners. Smaller but nonetheless important uses include dye acceptors; polymers for promoting adhesion; additives for textiles, paints, and cement; increasing the softening point and solvent resistance of resins; components of photopolymerizable systems; and cross-linking agents in vinyl polymers.

Physical Properties

The physical properties of solid acrylamide monomer are summarized in Table 1. Typical physical properties of 50% solution in water appear in Table 2.

Chemical Properties

Acrylamide, C_3H_5NO, is an interesting difunctional monomer containing a reactive electron-deficient double bond and an amide group, and it undergoes reactions typical of those two functionalities. It exhibits both weak acidic and basic properties.

TABLE 1. PHYSICAL PROPERTIES OF SOLID ACRYLAMIDE MONOMER

Property	Value
molecular weight	71.08
melting point, °C	84.5 ± 0.3
boiling point, °C at 0.67 kPa[a]	103

[a]To convert kPa to mm Hg, multiply by 7.5.

TABLE 2. PHYSICAL PROPERTIES OF 50% AQUEOUS ACRYLAMIDE SOLUTION

Property	Value
pH	5.0–6.5
refractive index range, 25°C (48–52%)	1.4085–1.4148
viscosity, mPa (= cP) at 25°C	2.71
specific gravity, at 25°C	1.0412
boiling point at 101.3 kPa[a], °C	99–104

[a]To convert kPa to mm Hg, multiply by 7.5.

Manufacture

The current routes to acrylamide are based on the hydration of inexpensive and readily available acrylonitrile (C_3H_3N, 2-propenenitrile, vinyl cyanide, VCN, or cyanoethene) See also **Acrylonitrile.**

Health and Safety Considerations

Contact with acrylamide can be hazardous and should be avoided. The most serious toxicological effect of exposure to acrylamide monomer is as a neurotoxin. In contrast, polymers of acrylamide exhibit very low toxicity.

Economic Aspects

The largest production of acrylamide is in Japan; the United States and Europe also have large production facilities. The principal producers in North America are The Dow Chemical Company, American Cyanamid Company, and Nalco Chemical Company (internal use).

C. E. HABERMANN
Dow Chemical, USA

Additional Reading

Chemistry of Acrylamide, Bulletin PRC 109, Process Chemicals Department, American Cyanamid Co., Wayne, N.J., 1969.

Environmental and Health Aspects of Acrylamide, A Comprehensive Bibliography of Published Literature 1930 to April 1980, EPA Report No. 560/7-81-006, 1981.

MacWilliams, D.C.: in R.H. Yocum and E.B. Nyquist, eds., *Functional Monomers,* Vol. 1, Marcel Dekker, Inc., New York, 1973, pp. 1–197.

U.S. Pat. 3,597,481 (Aug. 3, 1971), B.A. Tefertiller and C.E. Habermann (to The Dow Chemical Co.); U.S. Pat. 3,631,104 (Dec. 28, 1971), C.E. Habermann and B.A. Tefertiller (to The Dow Chemical Co.); U.S. Pat. 3,642,894 (Feb. 15, 1972), C.E. Habermann, R.E. Friedrich, and B.A. Tefertiller (to The Dow Chemical Co.); U.S. Pat. 3,642,643 (Feb. 15, 1972), C.E. Habermann (to The Dow Chemical Co.); U.S. Pat. 3,642,913 (Mar. 7, 1972), C.E. Habermann (to The Dow Chemical Co.); U.S. Pat. 3,696,152 (Oct. 3, 1972), C.E. Habermann and M.R. Thomas (to The Dow Chemical Co.); U.S. Pat. 3,758,578 (Sept. 11, 1973), C.E. Habermann and B.A. Tefertiller (to The Dow Chemical Co.); U.S. Pat. 3,767,706 (Oct. 23, 1972), C.E. Habermann and B.A. Tefertiller (to The Dow Chemical Co.).

ACRYLAMIDE POLYMERS. Acrylamide,

$$(CH_2=CHC\overset{\overset{\displaystyle O}{\|}}{}-NH_2)$$

polymerizes in the presence of free-radical initiators to form polyacrylamide chains with the following structure:

$$(-CH_2-CH-)_n$$
$$|$$
$$CONH_2$$

In this article the term *acrylamide polymer* refers to all polymers which contain acrylamide as a major constituent. Consequently, acrylamide polymers include functionalized polymers prepared from polyacrylamide by postreaction and copolymers prepared by polymerizing acrylamide (2-propenamide, C_3N_5NO) with one or more comonomers.

Manufacturing processes have been improved by use of on-line computer control and statistical process control leading to more uniform final products. Production methods now include inverse (water-in-oil) suspension polymerization or polymerization in water on moving belts. Conventional azo, peroxy, redox, and gamma-ray initiators are used in batch and continuous processes.

Physical Properties

Solid Polymer. Completely dry polyacrylamide is a brittle white solid. The physical properties of nonionic polyacrylamide are listed in Table 1.

TABLE 1. PHYSICAL PROPERTIES OF SOLID POLYACRYLAMIDE

Property	Value
density, g/cm³	1.302
glass-transition temp, °C	188
chain structure	mainly heterotactic linear or branched, some head-to-head addition
crystallinity	amorphous (high mol wt)

Polymers in Solution. Polyacrylamide is soluble in water at all concentrations, temperatures, and pH values.

In general nonionic polyacrylamides do not interact strongly with neutral inorganic salts.

Flow Properties. In water, high molecular weight polyacrylamide forms viscous homogeneous solutions.

Chemical Properties

The preparation of polyacrylamides and postpolymerization reactions on polyacrylamides are usually conducted in water. Reactions on the amide groups of polyacrylamides are often more complicated than reactions of simple amides because of neighboring groups' effects.

Post-reactions of polyacrylamide to introduce anionic, cationic, or other functional groups are often attractive from a cost standpoint. This approach can suffer, however, from side reactions resulting in cross-linking or the introduction of unwanted functionality. Reactions include hydrolysis, sulfomethylation, methylol formation, reaction with other aldehydes, transamidation, Hoffman degradation, and reaction with chlorine.

Uses

Polyacrylamides are classified according to weight-average molecular weight ($\overline{m}_w$) as follows: high 15×10^6; low 2×10^5; and very low 2×10^3.

Most uses for high molecular weight polyacrylamides in water treating, mineral processing, and paper manufacture are based on the ability of these polymers to flocculate small suspended particles by charge neutralization and bridging. Low molecular weight polymers are employed as dispersants, crystal growth modifiers, or selective mineral depressants. In oil recovery, polyacrylamides adjust the rheology of injected water so that the polymer solution moves uniformly through the rock pores, sweeping the oil ahead of it. Other applications such as superabsorbents and soil modification rely on the very hydrophilic character of polyacrylamides.

Suppliers of polyacrylamide are listed in Table 2.

Safety and Health

Dry nonionic and cationic material caused no skin and minimal eye irritation during primary irritation studies with rabbits. Dry anionic polyacrylamide did not produce any eye or skin irritation in laboratory animals. Emulsion nonionic polyacrylamide produced severe eye irritation in rabbits, while anionic and cationic material produced minimal eye irritation in rabbits. Polyacrylamides are used safely for numerous indirect food packaging applications, potable water, and direct food applications.

Joseph Kozakiewicz
David Lipp
American Cyanamid Company

Additional Reading

Kulicke, W.M., R. Kniewske, and J. Klein: *Progr. Polym. Sci.* **8**, 373–468 (1982).

Kurenkov, V.F. and L.I. Abramova: *Polym.-Plast. Technol. Eng.* **31**(7,8), 659–704 (1992).

Kurenkov, V.F. and V.A. Myagchenkov: *Polym.-Plast. Technol. Eng.* **30**(4), 367–404 (1991).

Myagchenkov, V.A. and V.F. Kurenkov: *Polym.-Plast. Technol. Eng.* **30**(2,3), 109–135 (1991).

ACRYLATES AND METHACRYLATES. A wide range of plastic materials that date back to the pioneering work of Redtenbacher before 1850, who prepared acrylic acid by oxidizing acrolein

$$CH_2=CHCHO \xrightarrow{\text{O}} CH_2=CHCOOH.$$

At a considerably later date, Frankland prepared ethyl methacrylate and methacrylic acid from ethyl-α-hydroxyisobutyrate and phosphorus trichloride. Tollen prepared acrylate esters from 2,3-dibromopropionate esters and zinc. Otto Rohm, in 1901, described the structures of the liquid condensation products (including dimers and trimers) obtained from the action of sodium alkoxides on methyl and ethyl acrylate. Shortly after World War I, Rohm introduced a new acrylate synthesis, noting that an acrylate is formed in good yield from heating ethylene cyanohydrin and sulfuric acid and alcohol. A major incentive for the development of a clear, tough plastic acrylate was in connection with the manufacture of safety glass.

Ethyl methacrylate went into commercial production as early as 1933. The synthesis proceeded in the following steps:

TABLE 2. SUPPLIERS OF POLYACRYLAMIDE

Region	Companies
United States	Allied Colloids, Inc.
	American Cyanamid Co.
	Aqua Ben Corp.
	Betz Laboratories, Inc.
	Calgon Corp. (Merck & Co.)
	Chemtall, Inc. (SNF Floerger)
	Dearborn Chemical Co. (W. R. Grace & Co.)
	The Dow Chemical Company
	Drew Chemical Corporation (Ashland Chemical, Inc.)
	Exxon Chemical Co.
	Hercules, Inc.
	Nalco Chemical Co.
	Polypure, Inc.
	Secodyne, Inc.
	Stockhausen, Inc.
Europe	Allied Colloids, Ltd.
	American Cyanamid Co.
	BASF AG
	Chemische Fabrik Stockhausen & Cie
	The Dow Chemical Company
	SNF Floerger (France)
	Kemira Oy (Finland)
	Rohm GmbH
	Rhône-Poulenc Specialties Chimiques (France)
Japan	Dai-Ichi Kogyo Seiyaku Co., Ltd.
	Kurita Water Industries, Ltd.
	Kyoritsu Yuki Co., Ltd.
	Mitsubishi Chemical Industries, Ltd.
	Mitsui-Cyanamid, Ltd.
	Sankyo Kasei Co., Ltd.
	Sanyo Chemical Industries, Ltd.
	Takenaka Komuten Co., Ltd.
	Toa Gosei Chemical Industry Co., Ltd.

(1) Acetone and hydrogen cyanide, generated from sodium cyanide and acid, gave acetone cyanohydrin

$$HCN + CH_3COCH_3 \longrightarrow (CH_3)_2C(OH)CN$$

(2) The acetone cyanohydrin was converted to ethyl α-hydroxyisobutyrate by reaction with ethyl alcohol and dilute sulfuric acid

$$(CH_3)_2C(OH)CN + C_2H_5OH \xrightarrow{H_2SO_4} (CH_3)_2C(OH)COOC_2H_5$$

(3) The hydroxy ester was dehydrated with phosphorus pentoxide to produce ethyl methacrylate

$$(CH_3)_2C(OH)COOC_2H_5 \xrightarrow{P_2O_5} CH_2=C(CH_3)COOC_2H_5$$

In 1936, the methyl ester of methacrylic acid was introduced and used to produce an "organic glass" by cast polymerization. Methyl methacrylate was made initially through methyl α-hydroxyisobutyrate by the same process previously indicated for the ethyl ester. Over the years, numerous process changes have occurred and costs lowered, making these plastics available on a high production basis for many hundreds of uses. For example, the hydrogen cyanide required is now produced catalytically from natural gas, ammonia, and air.

As with most synthetic plastic materials, they commence with the monomers. Any of the common processes, including bulk, solution, emulsion, or suspension systems may be used in the free-radical polymerization or copolymerization of acrylic monomers. The molecular weight and physical properties of the products may be varied over a wide range by proper selection of acrylic monomer and monomer mixes, type of process, and process conditions.

In bulk polymerization, no solvents are employed and the monomer acts as the solvent and continuous phase in which the process is carried out. Commercial bulk processes for acrylic polymers are used mainly in the production of sheets, rods and tubes. Bulk processes are also used on a much smaller scale in the preparation of dentures and novelty items and in the preservation of biological specimens. Acrylic castings are produced by pouring monomers or partially polymerized sirups into suitably designed molds and completing the polymerization. Acrylic bulk polymers consist essentially of poly(methyl methacrylate) or copolymers with methyl methacrylate as the major component. Free radical initiators soluble in the monomer, such as benzoyl peroxide, are the catalysts for the polymerization. Aromatic tertiary amines, such as dimethylaniline, may be used as accelerators in conjunction with the peroxide to permit curing at room temperature. However, colorless products cannot be obtained with amine accelerators because of the formation of red or yellow colors. As the polymerization proceeds, a considerable reduction in volume occurs which must be taken into consideration in the design of molds. At 25°C, the shrinkage of methyl methacrylate in the formation of the homopolymer is 21%.

Solutions of acrylic polymers and copolymers find wide use as thermoplastic coatings and impregnating fluids, adhesives, laminating materials, and cements. Solutions of interpolymers convertible to thermosetting compositions can also be prepared by inclusion of monomers bearing reactive functional groups which are capable of further reaction with appropriate crosslinking agents to give three-dimensional polymer networks. These polymer systems may be used in automotive coatings and appliance enamels, and as binders for paper, textiles, and glass or nonwoven fabrics. Despite the relatively low molecular weight of the polymers obtained in solution, such products are often the most appropriate for the foregoing uses. Solution polymerization of acrylic esters is usually carried out in large stainless steel, nickel, or glass-lined cylindrical kettles, designed to withstand at least 50 psig. The usual reaction mixture is a 40–60% solution of the monomers in solvent. Acrylic polymers are soluble in aromatic hydrocarbons and chlorohydrocarbons.

Acrylic emulsion polymers and copolymers have found wide acceptance in many fields, including sizes, finishes and binders for textiles, coatings and impregnants for paper and leather, thermoplastic and thermosetting protective coatings, floor finishing materials, adhesives, high-impact plastics, elastomers for gaskets, and impregnants for asphalt and concrete.

Advantages of emulsion polymerization are rapidity and production of high-molecular-weight polymers in a system of relatively low viscosity. Difficulties in agitation, heat transfer, and transfer of materials are minimized. The handling of hazardous solvents is eliminated. The two principal variations in technique used for emulsion polymerization are the redox and the reflux methods.

Suspension polymerization also is used. When acrylic monomers or their mixtures with other monomers are polymerized while suspended (usually in aqueous system), the polymeric product is obtained in the form of small beads, sometimes called pearls or granules. Bead polymers are the basis of the production of molding powders and denture materials. Polymers derived from acrylic or methacrylic acid furnish exchange resins of the carboxylic acid type. Solutions in organic solvents furnish lacquers, coatings and cements, while water-soluble hydrolysates are used as thickeners, adhesives, and sizes.

The basic difference between suspension and emulsion processes lies in the site of the polymerization, since initiators insoluble in water are used in the suspension process. Suspensions are produced by vigorous and continuous agitation of the monomer and solvent phases. The size of the drop will be determined by the rate of agitation, the interfacial tension, and the presence of impurities and minor constituents of the recipe. If agitation is stopped, the droplets coalesce into a monomer layer. The water serves as a dispersion medium and heat-transfer agent to remove the heat of polymerization. The process and resulting product can be influenced by the addition of colloidal suspending agents, thickeners, and salts.

Product Groupings. The principal acrylic plastics are cast sheet, molding powder, and high-impact molding powder. The cast acrylic sheet is formable, transparent, stable, and strong. Representative uses include architectural panels, aircraft glazing, skylights, lighted outdoor signs, models, product prototypes, and novelties. Molding powders are used in the mass production of numerous intricate shapes, such as automotive lights, lighting fixture lenses, and instrument dials and control panels for autos, aircraft, and appliances. The high-impact acrylic molding powder yields a somewhat less transparent product, but possesses unusual toughness for such applications as toys, business machine components, blow-molded bottles, and outboard motor shrouds. The various acrylic resins find numerous uses, with varied and wide use in coatings. Acrylic latexes are composed mainly of monomers of the acrylic family, such as methyl methacrylate, butyl methacrylate, methyl

acrylate, and 2-ethylhexylacrylate. Additional monomers, such as styrene or acrylonitrile, can be polymerized with acrylic monomers. Acrylic latexes vary considerably in their properties, mainly affected by the monomers used, the particle size, and the surfactant system of the latex. Generally, acrylic latexes are cured by loss of water only, do not yellow, possess a good exterior durability, are tough, and usually have good abrasion resistance. The acrylic polymers are reasonably costly and some latexes do not have very good color compatibility. Acrylic latex paints can be used for concrete floors, interior flat and semigloss finishes, and exterior surfaces.

ACRYLIC ACID AND DERIVATIVES. [CAS: 79-10-7]. Acrylic acid (propenoic acid) was first prepared in 1847 by air oxidation of acrolein. Interestingly, after use of several other routes over the past half century, it is this route, using acrolein from the catalytic oxidation of propylene, that is currently the most favored industrial process.

Acrylates are primarily used to prepare emulsion and solution polymers. The emulsion polymerization process provides high yields of polymers in a form suitable for a variety of applications. Acrylate emulsions are used in the preparation of both interior and exterior paints, floor polishes, and adhesives. Solution polymers of acrylates, frequently with minor concentrations of other monomers, are employed in the preparation of industrial coatings. Polymers of acrylic acid can be used as superabsorbents in disposable diapers, as well as in formulation of superior, reduced-phosphate-level detergents.

The polymeric products can be made to vary widely in physical properties through controlled variation in the ratios of monomers employed in their preparation, cross-linking, and control of molecular weight. They share common qualities of high resistance to chemical and environmental attack, excellent clarity, and attractive strength properties.

Physical Properties

Physical properties of acrylic acid and representative derivatives appear in Table 1.

Reactions

Acrylic acid and its esters may be viewed as derivatives of ethylene, in which one of the hydrogen atoms has been replaced by a carboxyl or carboalkoxyl group. This functional group may display electron-withdrawing ability through inductive effects of the electron-deficient carbonyl carbon atom, and electron-releasing effects by resonance involving the electrons of the carbon−oxygen double bond. Therefore, these compounds react readily with electrophilic, free-radical, and nucleophilic agents.

Specialty Acrylic Esters

Higher alkyl acrylates and alkyl-functional esters are important in copolymer products, in conventional emulsion applications for coatings and adhesives, and as reactants in radiation-cured coatings and inks. In general, they are produced in direct or transesterification batch processes because of their relatively low volume.

Health and Safety Factors

The toxicity of common acrylic monomers has been characterized in animal studies using a variety of exposure routes. Toxicity varies with level, frequency, duration, and route of exposure. The simple higher esters of acrylic acid are usually less absorbed and less toxic than lower esters. In general, acrylates are more toxic than methacrylates.

TABLE 1. PHYSICAL PROPERTIES OF ACRYLIC ACID DERIVATIVES

Property	Acrylic acid	Acrolein	Acrylic anhydride	Acryloyl chloride	Acrylamide
molecular formula	$C_3H_4O_2$	C_3H_4O	$C_6H_6O_3$	C_3H_3OCl	C_3H_5ON
melting point, °C	13.5	−88			84.5
boiling point[a], °C	141	52.5	38[b]	75	125[c]
refractive index[d], n_D	1.4185[e]	1.4017	1.4487	1.4337	

[a] At 101.3 kPa = 1 atm unless otherwise noted.
[b] At 0.27 kPa.
[c] At 16.6 kPa.
[d] At 20°C, unless otherwise noted.
[e] At 25°C.

Current TLV/TWA values are provided in Material Safety Data Sheets provided by manufacturers on request.

WILLIAM BAUER, JR.
Rohm and Hass Company

Additional Reading

Acrylic and Methacrylic Monomers — Specifications and Typical Properties, Bulletin 84C2, Rohm and Haas Co., Philadelphia, PA, 1986.
Hydrocarbon Process. **60**(11), 124 (1981).
Snyder, T.P. and C.G. Hill, Jr., *Catal. Rev. Sci. Eng.* **31**, 43–95 (1989).

ACRYLIC ESTER POLYMERS.
Acrylic esters are represented by the generic formula

$$\begin{array}{ccc} H & & H \\ | & & | \\ C & = & C \\ | & & | \\ H & & COOR \end{array}$$

The nature of the R group determines the properties of each ester and the polymers it forms. Polymers of this class are amorphous and are distinguished by their water-clear color and their stability on aging. Acrylic monomers are extremely versatile building blocks. They are relatively moderate to high boiling liquids that readily polymerize or copolymerize with a variety of other monomers. Copolymers with methacrylates, vinyl acetate, styrene, and acrylonitrile are commercially significant. Polymers designed to fit specific application requirements ranging from soft, tacky adhesives to hard plastics can be tailored from these versatile monomers. Although the acrylics have been higher in cost than many other common monomers, they find use in high quality products where their unique characteristics and efficiency offset the higher cost.

Physical Properties

To a large extent, the properties of acrylic ester polymers (Table 1) depend on the nature of the alcohol radical and the molecular weight of the polymer. As is typical of polymeric systems, the mechanical properties of acrylic polymers improve as molecular weight is increased; however, beyond a critical molecular weight, which often is about 100,000 to 200,000 for amorphous polymers, the improvement is slight and levels off asymptotically.

Chemical Properties

Under conditions of extreme acidity or alkalinity, acrylic ester polymers can be made to hydrolyze to poly(acrylic acid) or an acid salt and the corresponding alcohol. However, acrylic polymers and copolymers have a greater resistance to both acidic and alkaline hydrolysis than competitive poly(vinyl acetate) and vinyl acetate copolymers.

Acrylic polymers are fairly insensitive to normal uv degradation since the primary uv absorption of acrylics occurs below the solar spectrum.

TABLE 1. PHYSICAL PROPERTIES OF ACRYLIC POLYMERS

Polymer[a]	Monomer molecular formula	T_g, °C
methyl acrylate	$C_4H_6O_2$	6
ethyl acrylate	$C_5H_8O_2$	−24
propyl acrylate	$C_6H_{10}O_2$	−45
isopropyl acrylate	$C_6H_{10}O_2$	−3
n-butyl acrylate	$C_7H_{12}O_2$	−50
sec-butyl acrylate	$C_7H_{12}O_2$	−20
isobutyl acrylate	$C_7H_{12}O_2$	−43
tert-butyl acrylate	$C_7H_{12}O_2$	43
hexyl acrylate	$C_9H_{16}O_2$	−57
heptyl acrylate	$C_{10}H_{18}O_2$	−60
2-heptyl acrylate	$C_{10}H_{18}O_2$	−38
2-ethylhexyl acrylate	$C_{11}H_{20}O_2$	−65
2-ethylbutyl acrylate	$C_9H_{16}O_2$	−50
dodecyl acrylate	$C_{15}H_{28}O_2$	−30
hexadecyl acrylate	$C_{19}H_{36}O_2$	35
2-ethoxyethyl acrylate	$C_7H_{12}O_3$	−50
isobornyl acrylate	$C_{13}H_{20}O_2$	94
cyclohexyl acrylate	$C_9H_{14}O_2$	16

[a]Density (g/cm³) and refractive index (n_D) for methyl acrylate, ethyl acrylate, and *n*-butyl acrylate: 1.22, 1.479; 1.12, 1.464; 1.08, 1.474. Density for isopropyl acrylate = 1.08 g/cm³.

TABLE 2. PHYSICAL PROPERTIES OF ACRYLIC MONOMERS

Acrylate	Molecular weight	bp, °C[a]	d^{25}, g/cm³
methyl	86	79–81	0.950
ethyl	100	99–100	0.917
n-butyl	128	144–149	0.894
isobutyl	128	61–63[b]	0.884
t-butyl	128	120	0.879
2-ethylhexyl	184	214–220	0.880

[a]At 101.3 kPa unless otherwise noted.
[b]At 6.7 kPa = 50 mm Hg.

Acrylic Ester Monomers

Some of the physical properties of the principal commercial acrylic esters are given in Table 2.

There are currently two principal processes used for the manufacture of monomeric acrylic esters: the semicatalytic Reppe process and the propylene oxidation process. The newer propylene oxidation process is preferred because of economy and safety.

The toxicities of acrylic monomers range from moderate to slight. In general, they can be handled safely and without difficulty by trained personnel following established safety practices.

Radical Polymerization

Usually, free-radical initiators such as azo compounds or peroxides are used to initiate the polymerization of acrylic monomers. Photochemical and radiation-initiated polymerizations are also well known. Methods of radical polymerization include bulk, solution, emulsion, suspension, graft copolymerization, radiation-induced, and ionic with emulsion being the most important.

The free-radical polymerization of acrylic monomers follows a classical chain mechanism in which the chain-propagation step entails the head-to-tail growth of the polymeric free radical by attack on the double bond of the monomer.

The vast majority of all commercially prepared acrylic polymers are copolymers of an acrylic ester monomer with one or more different monomers. Copolymerization greatly increases the range of available polymer properties and has led to the development of many different resins suitable for a broad variety of applications.

In general, acrylic ester monomers copolymerize readily with each other or with most other types of vinyl monomers by free-radical processes.

Health and Safety Factors

Acrylic polymers are considered to be nontoxic. In fact, the FDA allows certain acrylate polymers to be used in the packaging and handling of food.

Potential health and safety problems of acrylic polymers occur in their manufacture. During manufacture, considerable care is exercised to reduce the potential for violent polymerizations and to reduce exposure to flammable and potentially toxic monomers and solvents.

Uses

Acrylic ester polymers are used primarily in coatings, textiles, adhesives, and paper.

2-Cyanoacrylic Ester Polymers

The polymers of the 2-cyanoacrylic esters, more commonly known as the alkyl 2-cyanoacrylates, are hard glassy resins that exhibit excellent adhesion to a wide variety of materials. The polymers are spontaneously formed when their liquid precursors or monomers are placed between two closely fitting surfaces. The spontaneous polymerization of these very reactive liquids and the excellent adhesion properties of the cured resins combine to make these compounds a unique class of single-component, ambient-temperature-curing adhesive of great versatility (Table 3). The materials that can be bonded run the gamut from metals, plastics, most elastomers, fabrics, and woods to many ceramics.

The utility of these adhesives arises from the electron-withdrawing character of the groups adjacent to the polymerizable double bond, which accounts for both the extremely high reactivity or cure rate and their polar nature, which enables the polymers to adhere tenaciously to many diverse substrates.

At present, a number of manufacturers in the United States, Europe, Japan, and elsewhere marked extended lines of these adhesives all over the

TABLE 3. ADHESIVE BOND PROPERTIES OF 2-CYANOACRYLIC ESTERS WITH METALS AND VARIOUS POLYMERIC MATERIALS

Property	Methyl	Ethyl[a]	Butyl	Isobutyl	Methoxyethyl	Ethoxyethyl
			SET TIME, s			
steel	20	10	30	20	15	5
nitrile rubber	5	3	5	5	5	3
ABS	20	10	20	20	5	5
polycarbonate	20	10	20	20	60	20
PVC	5	3	2	5		10
phenolic resin	5	3	30	5	25	5
			BOND STRENGTH, kPa[b]			
steel	206	172	151	96	206	165
ABS	48	48	96	48	48	48
polycarbonate	69	69	90	69		41
PVC	96	96	62	83	55	69
phenolic resin	69	76	90	62	62	55

[a] Set times for allyl esters are similar to those for ethyl esters, as are bond strengths to steel, ABS, and PC.

[b] To convert kPa to psi multiply by 0.145.

world. Some of the major producers and their trademarks include Loctite (Prism and Superbonder), Toagosei (Aron Alpha, Krazy Glue), Henkel (Sicomet), National Starch (Permabond), Sumitomo (Cyanonond), Three Bond (Super Three), and Alpha Giken (Alpha Ace, Alpha Techno).

Manufacture and Processing. The cyanoacrylic esters are prepared via the Knoevenagel condensation reaction, in which the corresponding alkyl cyanoacetate reacts with formaldehyde in the presence of a basic catalyst to form a low molecular weight polymer. The polymer slurry is acidified and the water is removed. Subsequently, the polymer is cracked and redistilled at a high temperature onto a suitable stabilizer combination to prevent premature repolymerization. Strong protonic or Lewis acids are normally used in combination with small amounts of a free-radical stabilizer.

Adhesives formulated from the 2-cyanoacrylic esters typically contain stabilizers and thickeners, and may also contain tougheners, colorants, and other special property-enhancing additives.

Economic Aspects. Production of the 2-cyanoacrylic ester adhesives on a worldwide basis is estimated to be approximately 2400 metric tons. This amounts to only 0.02% of the total volume of adhesive produced but about 3% of the dollar volume.

Because of the high costs of raw materials and the relatively complex synthesis, the 2-cyanoacrylic esters are moderately expensive materials when considered in bulk quantities. In typical bonding applications, where single drops are adequate for bonding, the adhesives are very economical to use.

Health and Safety Factors. The 2-cyanoacrylic esters have sharp, pungent odors and are lacrimators, even at very low concentrations. The TLV for methyl 2-cyanoacrylate is 2 ppm and the short-term exposure limit is 4 ppm. Good ventilation when using the adhesives is essential.

Eye and skin contact should be avoided because of the adhesive's rapid tissue-bonding capabilities.

Both the liquid and cured 2-cyanoacrylic esters support combustion.

Uses. Some of the market segments served by these versatile materials include automotive, electronic, sporting goods, toys, hardware, morticians, law enforcement, cosmetics, jewelry, and medical devices. Although they are not approved for such use in the United States, their strong tissue bonding characteristics have led to their use as chemical sutures and hemostatic agents in other countries around the world.

RONALD W. NOVAK
Rohm and Haas Company

J. T. O'CONNOR
Loctite Corporation

Additional Reading

Billmeyer, F.W. Jr.: *Textbook of Polymer Chemistry,* Interscience Publishers, New York, 1957.

Coover, H. W., Dreifus, D. W., and O'Connor, J. T: in I. Skeist, ed., *Handbook of Adhesives,* 3rd ed., Van Nostrand Reinhold Co., Inc., New York, NY, 1990, Chapt. 27.

Lee, H. ed.: *Cyanoacrylate Resins—The Instant Adhesives Monograph,* Pasadena Technology Press, CA, 1986.

Riddle, E.H.: *Monomeric Acrylic Esters,* Reinhold Publishing Corp., New York, 1954.

ACRYLIC PAINT.   See **Paint and Finish Removers**.

ACRYLIC PLASTICS.   A wide range of plastic materials dates back to the pioneering work of Redtenbacher before 1850 who prepared acrylic acid by oxidizing acrolein

$$CH = CHCHO \xrightarrow{O} CH_2=CHCOOH$$

At a considerably later date, Frankland prepared ethyl methacrylate and methacrylic acid from ethyl α-hydroxyisobutyrate and phosphorus trichloride. Tollen prepared acrylate esters from 2,3-dibromopropionate esters and zinc. Otto Rohm, in 1901, described the structures of the liquid condensation products (including dimers and trimers) obtained from the action of sodium alkoxides on methyl and ethyl acrylate. Shortly after World War I, Rohm introduced a new acrylate synthesis, noting that an acrylate is formed in good yield from heating ethylene cyanohydrin and sulfuric acid and alcohol. A major incentive for the development of a clear, tough plastic acrylate was for use in the manufacture of safety glass.

Ethyl methacrylate went into commercial production in 1933. The synthesis proceeded in the following steps:

(1) Acetone and hydrogen cyanide, generated from sodium cyanide and acid, gave acetone cyanohydrin

$$HCN + CH_3COCH_3 \longrightarrow (CH_3)_2C(OH)CN$$

(2) The acetone cyanohydrin was converted to ethyl α-hydroxyiso butyrate by reaction with ethyl alcohol and dilute sulfuric acid

$$(CH_3)_2C(OH)CN + C_2H_5OH \xrightarrow{H_2SO_4} (CH_3)_2C(OH)COOC_2H_5$$

(3) The hydroxy ester was dehydrated with phosphorus pentoxide to produce ethyl methacrylate

$$(CH_3)_2C(OH)COOC_2H_5 \xrightarrow{P_2O_5} CH_2=C(CH_3)COOC_2H_5$$

In 1936, the methyl ester of methacrylic acid was introduced and used to produce an "organic glass" by cast polymerization. Methyl methacrylate was made initially through methyl α-hydroxyisobutyrate by the same process previously indicated for the ethyl ester. Over the years, numerous process changes have taken place and costs lowered, making these plastics available on a very high tonnage basis for thousands of uses. For example, the hydrogen cyanide required is now produced catalytically from natural gas, ammonia, and air.

As with most synthetic plastic materials, they commence with the monomers. Any of the common processes, including bulk, solution, emulsion, or suspension systems, may be used in the free-radical polymerization or copolymerization of acrylic monomers. The molecular weight and physical properties of the products may be varied over a wide range by proper selection of acrylic monomer and monomer mixes, type of process, and process conditions.

In bulk polymerization no solvents are employed and the monomer acts as the solvent and continuous phase in which the process is carried out. Commercial bulk processes for acrylic polymers are used mainly in the production of sheets, rods and tubes. Bulk processes are also used on a much smaller scale in the preparation of dentures and novelty items and in the preservation of biological specimens. Acrylic castings are produced by pouring monomers or partially polymerized syrups into suitably designed molds and completing the polymerization. Acrylic bulk polymers consist essentially of poly(methyl methacrylate) or copolymers with methyl methacrylate as the major component. Free radical initiators soluble in the monomer, such as benzoyl peroxide, are the catalysts for the polymerization. Aromatic tertiary amines, such as dimethylaniline, may be used as accelerators in conjunction with the peroxide to permit curing at room temperature. However, colorless products cannot be obtained with amine accelerators because of the formation of red or yellow colors. As the polymerization proceeds, a considerable reduction in volume occurs which must be taken into consideration in the design of molds. At 25°C, the shrinkage of methyl methacrylate in the formation of the homopolymer is 21%.

Solutions of acrylic polymers and copolymers find wide use as thermoplastic coatings and impregnating fluids, adhesives, laminating materials, and cements. Solutions of interpolymers convertible to thermosetting compositions can also be prepared by inclusion of monomers bearing reactive functional groups which are capable of further reaction with appropriate cross-linking agents to give three-dimensional polymer networks. These polymer systems may be used in automotive coatings and appliance enamels, and as binders for paper, textiles, and glass or non-woven fabrics. Despite the relatively low molecular weight of the polymers obtained in solution, such products are often the most appropriate for the foregoing uses. Solution polymerization of acrylic esters is usually carried out in large stainless steel, nickel, or glass-lined cylindrical kettles, designed to withstand at least 50 psig. The usual reaction mixture is a 40–60% solution of the monomers in solvent. Acrylic polymers are soluble in aromatic hydrocarbons and chlorohydrocarbons.

Acrylic **emulsion polymers and copolymers** have found wide acceptance in many fields, including sizes, finishes and binders for textiles, coatings and impregnants for paper and leather, thermoplastic and thermosetting protective coatings, floor finishing materials, adhesives, high-impact plastics, elastomers for gaskets, and impregnants for asphalt and concrete.

Advantages of emulsion polymerization are rapidity and production of high-molecular-weight polymers in a system of relatively low viscosity. Difficulties in agitation, heat transfer, and transfer of materials are minimized. The handling of hazardous solvents is eliminated. The two principal variations in technique used for emulsion polymerization are the redox and the reflux methods.

Suspension polymerization also is used. When acrylic monomers or their mixtures with other monomers are polymerized while suspended (usually in aqueous system), the polymeric product is obtained in the form of small beads, sometimes called pearls or granules. Bead polymers are the basis of the production of molding powders and denture materials. Polymers derived from acrylic or methacrylic acid furnish exchange resins of the carboxylic acid type. Solutions in organic solvents furnish lacquers, coatings and cements, while water-soluble hydrolysates are used as thickeners, adhesives, and sizes.

The basic difference between suspension and emulsion processes lies in the site of the polymerization, since initiators insoluble in water are used in the suspension process. Suspensions are produced by vigorous and continuous agitation of the monomer and solvent phases. The size of the drop will be determined by the rate of agitation, the interfacial tension, and the presence of impurities and minor constituents of the recipe. If agitation is stopped, the droplets coalesce into a monomer layer. The water serves as a dispersion medium and heat-transfer agent to remove the heat of polymerization. The process and resulting product can be influenced by the addition of colloidal suspending agents, thickeners, and salts.

Product Groupings

The principal acrylic plastics are cast sheet, molding powder, and high-impact molding powder. The cast acrylic sheet is formable, transparent, stable, and strong. Representative uses include architectural panels, aircraft glazing, skylights, lighted outdoor signs, models, product prototypes, and novelties. Molding powders are used in the mass production of numerous intricate shapes, such as automotive lights, lighting fixture lenses, and instrument dials and control panels for autos, aircraft, and appliances. The high-impact acrylic molding powder yields a somewhat less transparent product, but possesses unusual toughness for such applications as toys, business machine components, blow-molded bottles, and outboard motor shrouds. The various acrylic resins find numerous uses as previously mentioned, with varied and wide use in coatings. Acrylic latexes are composed mainly of monomers of the acrylic family, such as methyl methacrylate, butyl methacrylate, methyl acrylate, and 2-ethyl hexylacrylate. Additional monomers, such as styrene or acrylonitrile, can be polymerized with acrylic monomers. Acrylic latexes vary considerably in their properties, mainly affected by the monomers used, the particle size, and the surfactant system of the latex. Generally, acrylic latexes are cured by loss of water only, do not yellow, possess good exterior durability, are tough, and usually have good abrasion resistance. The acrylic polymers are reasonably costly and some latexes do not have very good color compatibility. Acrylic latex paints can be used for concrete floors, interior flat and semigloss finishes, and exterior surfaces. See also **Paints and Coatings**.

A developing and potentially large volume use for acrylics is in the video, audio, and data-storage disk markets. The properties, as well as ease of fabrication, have made acrylics a primary choice for these applications.

Additional Reading

Harper, C.: "Modern Plastics Handbook," *Modern Plastics Magazine*, McGraw-Hill Companies, Inc., New York, NY, 1999.

Modern Plastics Encyclopedia 97/Edition, Price Stern Sloan, Inc., Los Angeles, CA, 1997.

"Plastics Handbook," *Modern Plastics Magazine*, McGraw-Hill Companies, New York, NY, 1994.

Web References

http://www.socplas.org/ SPI The Society of the Plastics Industry/The Plastics Industry Trade Association.
http://www.4spe.org/ Society of Plastics Engineers.

ACRYLONITRILE. [CAS: 107-13-1]. Today over 90% of the approximately 4,000,000 metric tons of acrylonitrile (also called acrylic acid nitrile, propylene nitrile, vinyl cyanide, and propenoic acid nitrile) produced worldwide each year use the Sohio-developed ammoxidation process. Acrylonitrile is among the top 50 chemicals produced in the United States as a result of the tremendous growth in its use as a starting material for a wide range of chemical and polymer products. Acrylic fibers remain the largest use of acrylonitrile; other significant uses are in resins and nitrile elastomers and as an intermediate in the production of adiponitrile and acrylamide.

Physical Properties

Acrylonitrile (C_3H_3N, mol wt = 53.064) is an unsaturated molecule having a carbon–carbon double bond conjugated with a nitrile group. It is a polar molecule because of the presence of the nitrogen heteroatom. Tables 1 and 2 list some physical properties and thermodynamic information, respectively, for acrylonitrile.

Acrylonitrile is miscible in a wide range of organic solvents, including acetone, benzene, carbon tetrachloride, diethyl ether, ethyl acetate, ethylene cyanohydrin, petroleum ether, toluene, some kerosenes, and methanol.

Acrylonitrile has been characterized using infrared, Raman, and ultraviolet spectroscopies, electron diffraction, and mass spectroscopy.

Chemical Properties

Acrylonitrile undergoes a wide range of reactions at its two chemically active sites, the nitrile group and the carbon–carbon double bond. Detailed descriptions of specific reactions have been given.

Manufacturing and Processing

Acrylonitrile is produced in commercial quantities almost exclusively by the vapor-phase catalytic propylene ammoxidation process developed by Sohio.

$$C_3H_6 + NH_3 + \tfrac{3}{2}O_2 \xrightarrow{\text{catalyst}} C_3H_3N + 3H_2O$$

Economic Aspects

More than half of the worldwide acrylonitrile production is situated in Western Europe and the United States. In the United States, production is dominated by BP Chemicals, with more than a third of the domestic capacity. The export market has been an increasingly important outlet for U.S. production, exports growing from around 10% in the mid-1970s to 53% in 1987 and 43% in 1988.

TABLE 1. PHYSICAL PROPERTIES OF ACRYLONITRILE

Property	Value
appearance/odor	clear, colorless liquid with faintly pungent odor
boiling point, °C	77.3
freezing point, °C	−83.5
density, 20°C, g/cm³	0.806
vapor density (air = 1)	1.8
pH (5% aqueous solution)	6.0–7.5
viscosity, 25°C, mPa · s(= cP)	0.34

TABLE 2. THERMODYNAMIC DATA[a]

Property	Value
flash point, °C	0
autoignition temperature, °C	481
heat of combustion, liquid, 25°C, kJ/mol	1761.5
heat of vaporization, 25°C, kJ/mol	32.65

[a]To convert kJ to kcal, divide by 4.184.

Storage and Transport

Acrylonitrile is transported by rail car, barge, and pipeline. Department of Transportation (DOT) regulations require labeling acrylonitrile as a flammable liquid and poison.

Health and Safety Factors

Acrylonitrile is highly toxic if ingested, with an acute LDL_0 value for laboratory rats of 113 mg/kg. It is moderately toxic if inhaled (rat LCL_0 = 500 ppm/4 h), and it is extremely irritating and corrosive to skin and eyes. Acrylonitrile is categorized as a cancer hazard by OSHA.

Acrylonitrile will polymerize violently in the absence of oxygen if initiated by heat, light, pressure, peroxide, or strong acids and bases. It is combustible and ignites readily, producing toxic combustion products such as hydrogen cyanide, nitrogen oxides, and carbon monoxide.

Federal regulations (40 CFR 261) classify acrylonitrile as a hazardous waste, and it is listed as Hazardous Waste Number U009. Disposal must be in accordance with federal (40 CFR 262, 263, 264), state, and local regulations only at properly permitted facilities.

JAMES F. BRAZDIL
BP Research

Additional Reading

Chem. Mark. Rep. **235**, 50 (1989).

Dalin, M.A., I.K. Kolchin, and B.R. Serebryakov: *Acrylonitrile*, Technomic, Westport, CT, 1971.

The Chemistry of Acrylonitrile, 1st Edition, American Cyanamid Co., New York, 1951.

U.S. Pat. 3,193,480 (July 6, 1965), M.M. Baizer, C.R. Campbell, R.H. Fariss, and R. Johnson (to Monsanto Chemical Co.).

ACRYLONITRILE POLYMERS

SAN

Acrylonitrile has found its way into a great variety of polymeric compositions based on its polar nature and reactivity. Some of these areas include adhesives and binders, antioxidants, medicines, dyes, electrical insulations, emulsifying agents, graphic arts materials, insecticides, leather, paper, plasticizers, soil-modifying agents, solvents, surface coatings, textile treatments, viscosity modifiers, azeotropic distillations, artificial organs, lubricants, asphalt additives, water-soluble polymers, hollow spheres, cross-linking agents, and catalyst treatments.

SAN Physical Properties and Test Methods

Styrene–acrylonitrile (SAN) resins possess many physical properties desired for thermoplastic applications. They are characteristically hard, rigid, and dimensionally stable with load bearing capabilities. They are also transparent, have high heat distortion temperatures, possess excellent gloss and chemical resistance, and adapt easily to conventional thermoplastic fabrication techniques.

TABLE 1. PROPERTIES OF INJECTION-MOLDED COMMERCIAL SAN RESINS[a]

Property	Monsanto Lustran-35	Dow Tyril-880
tensile strength, MPa[b]	79.4	82.1
ultimate elongation, %	3.0	3.0
Izod impact strength, J/m[c]	24.0	26.7
hardness—Rockwell M	83	80
coefficient of linear thermal expansion, cm/(cm·°C)	6.8×10^{-5}	6.6×10^{-5}
flammability, cm/min		2.0
specific heat, J/(g·K)[d]		1.3
dielectric constant, kHz (MHz)		3.18 (3.02)
water absorption, % in 24 h	0.25	0.35
specific gravity	1.07	1.08

[a] Data taken from Monsanto and Dow product data sheets.
[b] To convert MPa to psi multiply by 145.
[c] To convert J/m to ftlb/in. divide by 53.39.
[d] To convert J to cal divide by 4.184.

SAN polymers are random linear amorphous copolymers. Physical properties are dependent on molecular weight and the percentage of acrylonitrile. An increase of either generally improves physical properties, but may cause a loss of processibility or an increase in yellowness. Various processing aids and modifiers can be used to achieve a specific set of properties. Modifiers may include mold release agents, uv stabilizers, antistatic aids, elastomers, flow and processing aids, and reinforcing agents such as fillers and fibers. Some typical physical properties are listed in Table 1.

SAN Chemical Properties and Analytical Methods

SAN resins show considerable resistance to solvents and are insoluble in carbon tetrachloride, ethyl alcohol, gasoline, and hydrocarbon solvents. They are swelled by solvents such as benzene, ether, and toluene. Polar solvents such as acetone, chloroform, dioxane, methyl ethyl ketone, and pyridine will dissolve SAN.

The properties of SAN are significantly altered by water absorption. The equilibrium water content increases with temperature while the time required decreases. A large decrease in T_g can result. Strong aqueous bases can degrade SAN by hydrolysis of the nitrile groups.

SAN Manufacture

Commercially, SAN is manufactured by three processes: emulsion, suspension, and continuous bulk.

Processing. SAN copolymers may be processed using the conventional fabrication methods of extrusion, blow molding, injection molding, thermoforming, and casting.

Other Copolymers

Acrylonitrile copolymerizes readily with many electron-donor monomers other than styrene. Hundreds of acrylonitrile copolymers have been reported, and a comprehensive listing of reactivity ratios for acrylonitrile copolymerizations is readily available.

Copolymers of acrylonitrile and methyl acrylate and terpolymers of acrylonitrile, styrene, and methyl methacrylate are used as barrier polymers. Acrylonitrile copolymers and multipolymers containing butyl acrylate, ethyl acrylate, 2-ethylhexyl acrylate, hydroxyethyl acrylate, methyl methacrylate, vinyl acetate, vinyl ethers, and vinylidene chloride are also used in barrier films, laminates, and coatings. Environmentally degradable polymers useful in packaging are prepared from polymerization of acrylonitrile with styrene and methyl vinyl ketone.

Economic Aspects

Since its introduction in the 1950s, SAN has shown steady growth. The combined properties of SAN copolymers such as optical clarity, rigidity, chemical and heat resistance, high tensile strength, and flexible molding characteristics, along with reasonable price have secured their market position. The largest portion of SAN (80%) is incorporated into ABS resins, and their markets are inexorably joined.

There are two major producers of SAN resin in the United States, Monsanto Chemical Company and The Dow Chemical Company, which market these materials under the names of Lustran and Tyril, respectively.

Health and Toxicology

SAN resins themselves appear to pose few health problems in that they have been approved by the FDA for food packaging use. The main concern is that of toxic residuals, e.g., acrylonitrile, styrene, or other polymerization components such as emulsifiers, stabilizers, or solvents. Each component must be treated individually for toxic effects and safe exposure level.

ABS Resins

Acrylonitrile–butadiene–styrene (ABS) polymers are composed of elastomer dispersed as a grafted particulate phase in a thermoplastic matrix of styrene and acrylonitrile copolymer (SAN). The presence of SAN grafted onto the elastomeric component, usually polybutadiene or a butadiene copolymer, compatabilizes the rubber with the SAN component. Property advantages provided by this graft terpolymer include excellent toughness, good dimensional stability, good processibility, and chemical resistance. Property balances are controlled and optimized by adjusting elastomer particle size, morphology, microstructure, graft structure, and SAN composition and molecular weight. Therefore, although the polymer is a relatively low cost engineering thermoplastic the system is structurally complex. This complexity is advantageous in that altering these structural and compositional parameters allows considerable versatility in the tailoring of properties to meet specific product requirements. This versatility may be even further enhanced by adding various monomers to raise the heat deflection temperature, impart transparency, confer flame retardancy, and, through alloying with other polymers, obtain special product features. Consequently, research and development in ABS systems is active and continues to offer promise for achieving new product opportunities.

Physical Properties. The range of properties typically available for general purpose ABS is illustrated in Table 2. Numerous grades of ABS are available including new alloys and specialty grades for high heat, plating, flaming-retardant, or static dissipative product requirements.

Chemical Properties. The behavior of ABS may be inferred from consideration of the functional groups present within the polymer.

Chemical Resistance. The polar character of the nitrile group reduces interration of the polymer with hydrocarbon solvents, mineral and vegetable oils, waxes, and related household and commercial materials. Good chemical resisatance provided by the presence of acrylonitrile as a comonomer combined with the relatively low water absorptivity (<1%) results in high resistance to the staining agents typically encountered in household applications.

Processing Stability. As with elastomers or other rubber modified polymers, the presence of double bonds in the elastomeric phase increases sensitivity to thermal oxidation either during processing or end use. Antioxidants are generally added at the compounding step to ensure retention of physical properties. Physical effects can also have marked effects on mechanical properties due to orientation, molded-in stress, and the agglomeration of dispersed rubber particles under very severe conditions. Proper drying conditions are essential to prevent moisture-induced splay. Discoloration can be minimized by reducing stock temperature during molding or extrusion.

Thermal Oxidative Stability. ABS undergoes autoxidation with the polybutadiene component more sensitive to thermal oxidation than the styrene–acrylonitrile component. Antioxidants substantially improve oxidative stability. Studies on the oven aging of molded parts have shown that oxidation is limited to the outer surface (<0.2 mm), i.e., the oxidation process is diffusion limited.

Photoxidative Stability. Unsaturation present as a structural feature in the polybutadiene component of ABS (also in high impact polystyrene, rubber-modified PVC, and butadiene-containing elastomers) also increases lability with regard to photooxidative degradation. Such degradation also only occurs in the outermost layer, and impact loss upon irradiation can be attributed to embrittlement of the rubber and possibly to scission of the grafted styrene–acrylonitrile copolymer. Applications involving extended outdoor exposure, especially in direct sunlight, require protective measures such as the use of stabilizing additives, pigments, and protective coatings and film.

Flammability. The general-purpose grades are usually recognized as 94 HB according to the requirements of Underwriters' Laboratories UL94 and also meet the requirements, dependent on thickness, of the Motor Vehicle Safety Standard 302.

Polymerization. In all manufacturing processes, grafting is achieved by the free-radical copolymerization of styrene and acrylonitrile monomers in the presence of an elastomer. Ungrafted styrene–acrylonitrile copolymer is formed during graft polymerization or added afterward.

Manufacturing. There are three commercial processes for manufacturing ABS: emulsion, mass, and mass-suspension. ABS is sold as an unpigmented product for on-line coloring using color concentrates during molding, or as precolored pellets matched to exacting requirements.

Analysis. Analytical investigations may be undertaken to identify the presence of an ABS polymer, characterize the polymer, or identify nonpolymeric ingredients. Fourier transfrom infrared (ftir) spectroscopy is the method of choice to identify the presence of an ABS polymer and determine the acrylonitrile–butadiene–styrene ratio of the composite polymer. Confirmation of the presence of rubber domains is achieved by electron microscopy. Comparison with available physical property data serves to increase confidence in the identification or indicate the presence of unexpected structural features. Phase-seperation techniques can be used to provide detailed compositional analyses.

Processing. Good thermal stability plus shear thinning allow wide flexibility in viscosity control for a variety of processing methods. ABS exhibits non-Newtonian viscosity behavior. ABS can be processed by all the techniques used for other thermoplastics: compression and injection molding, extrusion, calendering, and blow-molding. Clean, under-graded regrind can be reprocessed in most applications (plating excepted), usually at 20% with virgin ABS. Post-processing operations include cold forming; thermoforming; metal plating; painting; hot stamping; ultrasonic, spin, and vibrational welding; and adhesive bonding.

Applications

Its broad property balance and wide processing window have allowed ABS to become the largest selling engineering thermoplastic. ABS enjoys a unique position as a "bridge" polymer between commodity plastics and other higher performance engineering thermoplastics. Table 3 summarizes estimates for 1988 regional consumption of ABS resins by major use.

TABLE 2. MATERIAL PROPERTIES OF GENERAL PURPOSE AND HEAT DISTORTION RESISTANT ABS

Properties	High impact	Medium impact	Heat resistant
notched Izod impact at RT, J/m[a]	347–534	134–320	107–347
tensile strength, MPa[b]	33–43	30–52	41–52
tensile modulus, GPa[c]	1.7–2.3	2.1–2.8	2.1–2.6
flexural modulus, GPa[c]	1.7–2.4	2.2–3.0	2.1–2.8
Rockwell hardness	80–105	105–112	100–111
heat deflection[d], °C at 455 kPa[e]	99–107	102–107	110–118
coefficient of linear thermal expansion, $\times 10^5$ cm/cm·°C	9.5–11.0	7.0–8.8	6.5–9.2
dielectric strength, kV/mm	16–31	16–31	14–34
dielectric constant, $\times 10^6$ Hz	2.4–3.8	2.4–3.8	2.4–3.8

[a] To convert J/m to ft · lb/in. divide by 53.4.
[b] To convert MPa to psi multiply by 145.
[c] To convert GPa to psi multiply by 145,000.
[d] Annealed.
[e] To convert kPa to psi multiply by 0.145.

TABLE 3. MARKETS FOR ABS PLASTICS BY REGION IN 1988, 10³ T

	United states and canada	Western europe	Japan	Total	%
transportation	139	120	96	355	25
appliances	95	97	117	309	22
business machines	124	60	99	298	20
pipe and fittings	91	23		114	8
other	98	144	102	344	25
Total	*547*	*444*	*414*	*1405*	*100*

LAWRENCE E. BALL
BENEDICT S. CURATOLO
BP Research

DONALD M. KULICH
JOHN E. PACE
LEROY W. FRITCH, JR.
ANGELO BRISIMITZAKIS
GE Plastics

Additional Reading

Chemical Economics Handbook, SRI International, Menlo Park, CA, 1989, 580.0180D.

Cycolac Brand ABS Resin Design Guide, Technical Publication CYC-350, GE Plastics, Pittsfield, Mass., 1990.

Johnston, N.W. *J. Macromol. Sci. Rev. Macromol. Chem.* **C14**, 215 (1973).

Kulich, D. M., P. D. Kelley, and J. E. Pace: in J. I. Kroschwitz, ed., *Encyclopedia of Polymer Science and Engineering*," 2nd Edition, Vol. 1, Wiley-Interscience, New York, NY, 1985, p. 396.

Peng, F.M.: in J.I. Kroschwitz, ed. *Encyclopedia of Polymer Science and Engineering,* 2nd Edition, Vol. 1, Wiley-Interscience, New York, 1985, p. 463.

Pillichody, C.T. and P. D. Kelley: in I. I. Rubin, ed., *Handbook of Plastic Materials and Technology,* John Wiley & Sons, Inc., New York, NY, 1990. Chapt. 3.

Reithel, F.L.: in R. Juran, ed., *Modern Plastics Encyclopedia* 1989, **65**(11), McGraw-Hill Book Co., Inc., New York, p. 105.

The Chemistry of Acrylonitrile, 2nd Edition, American Cyanamid Co., Petrochemical Division, New York, 1959.

ACRYLONITRILE-BUTADIENE RUBBER (ABR).   See **Elastomers**.

ACTH.

ACTH. [CAS: 9002-60-2]. The adrenocorticotropic hormone of the anterior lobe of the pituitary gland, which specifically stimulates the adrenal cortex to secrete cortisone, and hence has effects identical with those of cortisone. ACTH differs in its chemistry, absorption, and metabolism from the other adrenal steroids. Chemically, it is a water-soluble polypeptide having a molecular weight of about 3500. Its complete amino acid sequence has been determined. It produces its peripheral physiological effects by causing discharge of the adrenocortical steroids into the circulation. ACTH has been extracted from pituitary glands. In purified form, ACTH is useful in treating some forms of arthritis, lupus erythematosus, and severe skin disorders. The action of ACTH injections parallels the result of large quantities of naturally formed cortisone if they were released naturally. See also **Steroids**.

ACTIN.

ACTIN. One of the two proteins that makes up the myofibrils of striated muscles. The other protein is myosin. The combination of these two proteins is sometimes spoken of as actinomyosin. The banded nature of the myofibrils is due to the fact that both proteins are present where the bands are dark and only one or the other is present in the light bands. Since these bands lie side by side in the different myofibrils that go to make up a muscle fiber, the entire muscle fiber shows a banded or striated appearance. See also **Contractility and Contractile Proteins**.

ACTINIDE CONTRACTION.

ACTINIDE CONTRACTION. An effect analogous to the Lanthanide contraction, which has been found in certain elements of the Actinide series. Those elements from thorium (atomic number 90) to curium (atomic number 96) exhibit a decreasing molecular volume in certain compounds, such as those which the actinide tetrafluorides form with alkali metal fluorides, plotted in Fig. 1. The effect here is due to the decreasing crystal radius of the tetrapositive actinide ions as the atomic number increases. Note that in the Actinides the tetravalent ions are compared instead of the trivalent ones as in the case of the Lanthanides, in which the trivalent state is by far the most common.

The behavior is attributed to the entrance of added electrons into an (inner) f shell ($4f$ for the Lanthanides, $5f$ for the Actinides) so that the increment they produce in atomic volume is less than the reduction due to the greater nuclear charge.

ACTINIDES AND TRANSACTINIDES

Actinides

The actinide elements are a group of chemically similar elements with atomic numbers 89 through 103, and their names, symbols, and atomic numbers are given in Table 1, see also **Radioactivity**; **Nuclear Power Technology**; **Plutonium**; **Thorium**; and **Uranium**. Each of the elements has a number of isotopes, all radioactive and some of which can be obtained in isotopically pure form.

Thorium, uranium, and plutonium are well known for their role as the basic fuels (or sources of fuel) for the release of nuclear energy. The importance of the remainder of the actinide group lies at present, for the most part, in the realm of pure research, but a number of practical

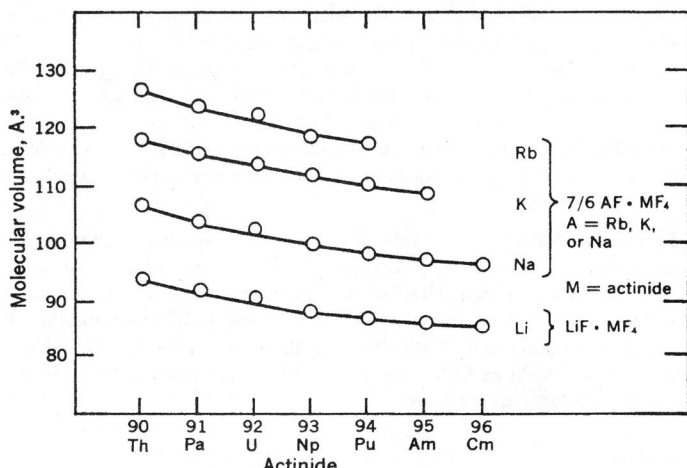

Fig. 1. Plot of molecular volume versus atomic number of the tetravalent Actinides

TABLE 1. THE ACTINIDE ELEMENTS

Atomic number	Element	Symbol	Atomic weight[a]
89	actinium	Ac	227
90	thorium	Th	232
91	protactinium	Pa	231
92	uranium	U	238
93	neptunium	Np	237
94	plutonium	Pu	242
95	americium	Am	243
96	curium	Cm	248
97	berkelium	Bk	249
98	californium	Cf	249
99	einsteinium	Es	254
100	fermium	Fm	257
101	mendelevium	Md	258
102	nobelium	No	259
103	lawrencium	Lr	260

[a]Mass number of longest-lived or most available isotope.

applications are also known. The actinides present a storage-life problem in nuclear waste disposal and consideration is being given to separation methods for their recovery prior to disposal, see also **Nuclear Power Technology**.

Source

Only the members of the actinide group through Pu have been found to occur in nature. Thorium and uranium occur widely in the earth's crust in combination with other elements, and, in the case of uranium, in significant concentrations in the oceans. With the exceptions of uranium and thorium, the actinide elements are synthetic in origin for practical purposes, i.e., they are products of nuclear reactions. High neutron fluxes are available in modern nuclear reactors, and the most feasible method for preparing actinium, protactinium, and most of the actinide elements is through the neutron irradiation of elements of high atomic number.

Experimental Methods of Investigation. All of the actinide elements are radioactive and, except for thorium and uranium, special equipment and shielded facilities are usually necessary for their manipulation.

The study of the chemical behavior of concentrated preparations of short-lived isotopes is complicated by the rapid production of hydrogen peroxide in aqueous solutions and the destruction of crystal lattices in solid compounds. These effects are brought about by heavy recoils of high energy alpha particles released in the decay process.

Special techniques for experimentation with the actinide elements other than Th and U have been devised because of the potential health hazard to the experimenter and the small amounts available. In addition, investigations are frequently carried out with the substance present in very low concentration as a radioactive tracer. Tracer studies offer a method for obtaining knowledge of oxidation states, formation of complex ions, and the solubility of various compounds. These techniques are not applicable to crystallography, metallurgy, and spectroscopic studies. Microchemical or ultramicrochemical techniques are used extensively in chemical studies of actinide elements.

Electronic Structure. Measurements of paramagnetic susceptibility, paramagnetic resonance, light absorption, fluorescence, and crystal structure, in addition to a consideration of chemical and other properties, have provided a great deal of information about the electronic configuration of the aqueous actinide ions in which the electrons are in the $5f$ shell. There are exceptions, such as U_2S_3, and subnormal compounds, such as Th_2S_3, where $6d$ electrons are present.

Properties

The close chemical resemblance among many of the actinide elements permits their chemistry to be described for the most part in a correlative way.

Oxidation States. The oxidation states of the actinide elements are summarized in Table 2. The most stable states are designated by boldface type and those which are very unstable are indicated by parentheses.

The actinide elements exhibit uniformity in ionic types. Corresponding ionic types are similar in chemical behavior, although the oxidation–reduction relationships and therefore the relative stabilities differ from element to element.

Hydrolysis and Complex Ion Formation. Of the actinide ions, the small, highly charged M^{4+} ions exhibit the greatest degree of hydrolysis and complex ion formation.

The degree of hydrolysis or complex ion formation decreases in the order $M^{4+} > MO_2^{2+} > M^{3+} > MO_2^+$. Presumably the relatively high tendency toward hydrolysis and complex ion formation of MO ions is related to the high concentration of charge on the metal atom. On the basis of increasing charge and decreasing ionic size, it could be expected that the degree of hydrolysis for each ionic type would increase with increasing atomic number.

Metallic State. The actinide metals, like the lanthanide metals, are highly electropositive. They can be prepared by the electrolysis of molten salts or by the reduction of a halide with an electropositive metal, such as calcium or barium. Their physical properties are summarized in Table 3.

Solid Compounds. Thousands of compounds of the actinide elements have been prepared, and the properties of some of the important binary compounds are summarized in Table 4.

Crystal Structure and Ionic Radii. Crystal structure data have provided the basis for the ionic radii (coordination number = CN = 6). For both M^{3+} and M^4 ions there is an actinide contraction, analogous to the lanthanide contraction, with increasing positive charge on the nucleus. As a consequence of the ionic character of most actinide compounds and of the similarity of the ionic radii for a given oxidation state, analogous compounds are generally isostructural.

Absorption and Fluorescence Spectra. The absorption spectra of actinide and lanthanide ions in aqueous solution and in crystalline form contain narrow bands in the visible, near-ultraviolet, and near-infrared regions of the spectrum.

TABLE 3. PROPERTIES OF ACTINIDE METALS

Element	Melting point, °C	Heat of vaporization, ΔH_v, kJ/mol (kcal/mol)	Boiling point, °C
actinium	1100 ± 50	293 (70)	
thorium	1750	564 (130)	3850
protactinium	1575		
uranium	1132	446.4 (106.7)	3818
neptunium	637 ± 2	418 (100)	3900
plutonium	646	333.5 (79.7)	3235
americium	1173	230 (55)	2011
curium	1345	386 (92.2)	3110
berkelium	1050		
californium	900 ± 30		
einsteinium	860 ± 30		

Transactinides

The elements beyond the actinides in the periodic table can be termed the transactinides. These begin with the element having the atomic number 104 and extend, in principle, indefinitely. Although only seven such elements, numbers 104–110 were definitely known in 2003, (Rutherfordium 104, Dubnium 105, Seaborgium 106, Bohrium 107, Hassium 108, Meitnerium 109, and Darmstadtium 110), there are good prospects for the discovery of a number of additional elements just beyond number 110 or in the region of larger atomic numbers. They are synthesized by the bomunderlinedment of heavy nuclides with heavy ions. See also **Chemical Elements**.

On the basis of the simplest projections it is expected that the half-lives of the elements beyond element 110 will become shorter as the atomic number is increased, and this is true even for the isotopes with the longest half-life for each element.

Turning to consideration of electronic structure, upon which chemical properties must be based, modern high speed computers have made possible the calculation of such structures. The calculations show that elements 104 through 112 are formed by filling the $6d$ electron subshell, which makes them, as expected, homologous in chemical properties with the elements hafnium ($Z = 72$) through mercury ($Z = 80$). Elements 113 through 118 result from the filling of the $7p$ subshell and are expected to be similar to the elements thallium ($Z = 81$) through radon ($Z = 86$).

It can be seen that elements in and near the island of stability based on element 114 can be predicted to have chemical properties as follows: element 114 should be a homologue of lead, that is, should be eka-lead; and element 112 should be eka-mercury, element 110 should be eka-platinum, etc. If there is an island of stability at element 126, this element and its neighbors should have chemical properties like those of the actinide and lanthanide elements.

GLENN T. SEABORG
University of California, Berkeley

TABLE 2. THE OXIDATION STATES OF THE ACTINIDE ELEMENTS

						Atomic number and element								
89	90	91	92	93	94	95	96	97	98	99	100	101	102	103
Ac	Th	Pa	U	Np	Pu	Am	Cm	Bk	Cf	Es	Fm	Md	No	Lr
						(2)			(2)	(2)	2	2	**2**	
3	(3)	(3)	3	3	3	**3**	**3**	**3**	**3**	**3**	3	3	3	**3**
	4	4	4	4	**4**	4	4	4	(4)					
		5	5	5	**5**	5								
			6	6	6	6								
				7	(7)									

Additional Reading

Hermann, G.: *Superheavy Elements, International Review of Science, Inorganic Chemistry,* Series 2, Vol. 8, Butterworths, London, and University Park Press, Baltimore, MD, 1975; G.T. Seaborg and W. Loveland, *Contemp. Physics* **28**, 233 (1987).

Herrmann, W.A.: *Synthetic Methods of Organometallic and Inorganic Chemistry,* Lanthanides and Actinides, Thieme Medical Publishers, New York, NY, 1997.

Katz, J.J., G.T. Seaborg, and L.R. Moss, *The Chemistry of the Actinide Elements,* 2nd Edition, Chapman & Hall, New York, NY, 1986.

Lide, D.R., *Handbook of Chemistry and Physics,* 84th Edition, CRC Press LLC, Boca Raton, FL, 2003.

Marks, T.J.: "Actinide Organometallic Chemistry," *Science,* **217**, 989–997 (1982).

Meyer, G. and L.R. Moss: *Synthesis of Lanthanide and Actinide Compounds,* Kluwer Academic Publishers, New York, NY, 1991.

Seaborg, G.T.: *The Transuranium Elements,* Yale University Press, New Haven, CT, 1958.

Seaborg, G.T. *Ann. Rev. Nucl. Sci.* **18**, 53 (1968); O.L. Keller, Jr., and G.T. Seaborg, *Ann. Rev. Nucl. Sci.* **27**, 139 (1977).

Web Reference

http://www.acs.org/ American Chemical Society.

TABLE 4. PROPERTIES AND CRYSTAL STRUCTURE DATA FOR IMPORTANT ACTINIDE BINARY COMPOUNDS

Compound	Color	Melting point, °C	Symmetry	Space group or structure type	Density, g/mL
AcH_2	black		cubic	fluorite ($Fm3m$)	8.35
ThH_2	black		tetragonal	$F4/mmm$	9.50
Th_4H_{15}	black		cubic	$I4\bar{3}d$	8.25
$\alpha\text{-}PaH_3$	gray		cubic	$Pm3n$	10.87
$\beta\text{-}PaH_3$	black		cubic	β-W	10.58
$\alpha\text{-}UH_3$	?		cubic	$Pm3n$	11.12
$\beta\text{-}UH_3$	black		cubic	β-W ($Pm3n$)	10.92
NpH_2	black		cubic	fluorite	10.41
NpH_3	black		trigonal	$P\bar{3}c1$	9.64
PuH_2	black		cubic	fluorite	10.40
PuH_3	black		trigonal	$P\bar{3}c1$	9.61
AmH_2	black		cubic	fluorite	10.64
AmH_3	black		trigonal	$P\bar{3}c1$	9.76
CmH_2	black		cubic	fluorite	10.84
CmH_3	black		trigonal	$P\bar{3}c1$	10.06
BkH_2	black		cubic	fluorite	11.57
BkH_3	black		trigonal	$P\bar{3}c1$	10.44
Ac_2O_3	white		hexagonal	La_2O_3 ($P\bar{3}m1$)	9.19
Pu_2O_3	?		cubic	$Ia3$ (Mn_2O_3)	10.20
Pu_2O_3	black	2085	hexagonal	La_2O_3	11.47
Am_2O_3	tan		hexagonal	La_2O_3	11.77
Am_2O_3	reddish brown		cubic	$Ia3$	10.57
Cm_2O_3	white to faint tan	2260	hexagonal	La_2O_3	12.17
Cm_2O_3			monoclinic	$C2/m$ (Sm_2O_3)	11.90
Cm_2O_3	white		cubic	$Ia3$	10.80
Bk_2O_3	light green		hexagonal	La_2O_3	12.47
Bk_2O_3	yellow-green		monoclinic	$C2/m$	12.20
Bk_2O_3	yellowish brown		cubic	$Ia3$	11.66
Cf_2O_3	pale green		hexagonal	La_2O_3	12.69
Cf_2O_3	lime green		monoclinic	$C2/m$	12.37
Cf_2O_3	pale green		cubic	$Ia3$	11.39
Es_2O_3	white		hexagonal	La_2O_3	12.7
Es_2O_3	white		monoclinic $C2/m$	$C2/m$	12.4
Es_2O_3	white		cubic	$Ia3$	11.79
ThO_2	white	ca 3050	cubic	fluorite	10.00
PaO_2	black		cubic	fluorite	10.45
UO_2	brown to black	2875	cubic	fluorite	10.95
NpO_2	apple green		cubic	fluorite	11.14
PuO_2	yellow-green to brown	2400	cubic	fluorite	11.46
AmO_2	black		cubic	fluorite	11.68
CmO_2	black		cubic	fluorite	11.92
BkO_2	yellowish-brown		cubic	fluorite	12.31
CfO_2	black		cubic	fluorite	12.46
Pa_2O_5	white		cubic	fluorite-related	11.14
Np_2O_5	dark brown		monoclinic	$P2_1/c$	8.18
$\alpha\text{-}U_3O_8$	black-green	1150 (dec)	orthorhombic	$C2mm$	8.39
$\beta\text{-}U_3O_8$	black-green		orthorhombic	$Cmcm$	8.32
$\gamma\text{-}UO_3$	orange	650 (dec)	orthorhombic	$Fddd$	7.80
$AmCl_2$	black		orthorhombic	$Pbnm$ ($PbCl_2$)	6.78
$CfCl_2$	red-amber		?		
$AmBr_2$	black		tetragonal	$SrBr_2$ ($P4/n$)	7.00
$CfBr_2$	amber		tetragonal	$SrBr_2$	7.22
ThI_2	gold		hexagonal	$P6_3/mmc$	7.45
AmI_2	black	ca 700	monoclinic	EuI_2 ($P2_1/c$)	6.60
CfI_2	violet		hexagonal	CdI_2 ($P\bar{3}m1$)	6.63
CfI_2	violet		rhombohedral	$CdCl_2$ ($R\bar{3}m$)	6.58
AcF_3	white		trigonal	LaF_3 ($P\bar{3}c1$)	7.88
UF_3	black	>1140(dec)	trigonal	LaF_3	8.95
NpF_3	purple		trigonal	LaF_3	9.12
PuF_3	purple	1425	trigonal	LaF_3	9.33
AmF_3	pink	1393	trigonal	LaF_3	9.53
CmF_3	white	1406	trigonal	LaF_3	9.85
BkF_3	yellow-green		orthorhombic	YF_3 ($Pnma$)	9.70
BkF_3	yellow-green		trigonal	LaF_3	10.15
CfF_3	light green		orthorhombic	YF_3	9.88
CfF_3	light green		trigonal	LaF_3	10.28
$AcCl_3$	white		hexagonal	UCl_3 ($P6_3/m$)	4.81
UCl_3	green	835	hexagonal	$P6_3/m$	5.50
$NpCl_3$	green	ca 800	hexagonal	UCl_3	5.60
$PuCl_3$	emerald green	760	hexagonal	UCl_3	5.71
$AmCl_3$	pink or yellow	715	hexagonal	UCl_3	5.87
$CmCl_3$	white	695	hexagonal	UCl_3	5.95
$BkCl_3$	green	603	hexagonal	UCl_3	6.02

(*continued overleaf*)

25

TABLE 4. (*continued*)

Compound	Color	Melting point, °C	Symmetry	Space group or structure type	Density, g/mL
α-CfCl$_3$	green	545	orthorhombic	TbCl$_3$ (*Cmcm*)	6.07
β-CfCl$_3$	green		hexagonal	UCl$_3$	6.12
EsCl$_3$	white to orange		hexagonal	UCl$_3$	6.20
AcBr$_3$	white		hexagonal	UBr$_3$(P6$_3$/m)	5.85
UBr$_3$	red	730	hexagonal	P6$_3$/m	6.55
NpBr$_3$	green		hexagonal	UBr$_3$	6.65
NpBr$_3$	green		orthorhombic	TbCl$_3$(*Cmcm*)	6.67
PuBr$_3$	green	681	orthorhombic	TbCl$_3$	6.72
AmBr$_3$	white to pale yellow		orthorhombic	TbCl$_3$	6.85
CmBr$_3$	pale yellow-green	625 ± 5	orthorhombic	TbCl$_3$	6.85
BkBr$_3$	light green		monoclinic	AlCl$_3$(C2/m)	5.604
BkBr$_3$	light green		orthorhombic	TbCl$_3$	6.95
BkBr$_3$	yellow green		rhombohedral	FeCl$_3$ (R$\bar{3}$)	5.54
CfBr$_3$	green		monoclinic	AlCl$_3$	5.673
CfBr$_3$	green		rhombohedral	FeCl$_3$	5.77
EsBr$_3$	straw		monoclinic	AlCl$_3$	5.62
PaI$_3$	black		orthorhombic	TbCl$_3$ (*Cmcm*)	6.69
UI$_3$	black		orthorhombic	TbCl$_3$	6.76
NpI$_3$	brown		orthorhombic	TbCl$_3$	6.82
PuI$_3$	green		orthorhombic	TbCl$_3$	6.92
AmI$_3$	pale yellow	ca 950	hexagonal	BiI$_3$ (R$\bar{3}$)	6.35
AmI$_3$	yellow		orthorhombic	PuBr$_3$	6.95
CmI$_3$	white		hexagonal	BiI$_3$	6.40
BkI$_3$	yellow		hexagonal	BiI$_3$	6.02
CfI$_3$	red-orange		hexagonal	BiI$_3$	6.05
EsI$_3$	amber to light yellow		hexagonal	BiI$_3$	6.18
ThF$_4$	white	1068	monoclinic	UF$_4$(C2/c)	6.20
PaF$_4$	reddish-brown		monoclinic	UF$_4$	6.38
UF$_4$	green	960	monoclinic	C2/c	6.73
NpF$_4$	green		monoclinic	UF$_4$	6.86
PuF$_4$	brown	1037	monoclinic	UF$_4$	7.05
AmF$_4$	tan		monoclinic	UF$_4$	7.23
CmF$_4$	light gray-green		monoclinic	UF$_4$	7.36
BkF$_4$	pale yellow-green		monoclinic	UF$_4$	7.55
CfF$_4$	light green		monoclinic	UF$_4$	7.57
α-ThCl$_4$	white		orthorhombic		4.12
β-ThCl$_4$	white	770	tetragonal	UCl$_4$(I4$_1$/amd)	4.60
PaCl$_4$	greenish-yellow		tetragonal	UCl$_4$	4.72
UCl$_4$	green	590	tetragonal	I4$_1$/amd	4.89
NpCl$_4$	red-brown	518	tetragonal	UCl$_4$	4.96
α-ThBr$_4$	white		tetragonal	I4$_1$/a	5.94
β-ThBr$_4$	white		tetragonal	UCl$_4$	5.77
PaBr$_4$	orange-red		tetragonal	UCl$_4$	5.90
UBr$_4$	brown	519	monoclinic	2/c-/-	
NpBr$_4$	dark red	464	monoclinic	2/c-/-	
ThI$_4$	yellow	556	monoclinic	P2$_1$/n	6.00
PaI$_4$	black				
UI$_4$	black				
PaF$_5$	white		tetragonal	I$\bar{4}$2d	
α-UF$_5$	grayish white		tetragonal	I4/m	5.81
β-UF$_5$	pale yellow		tetragonal	I$\bar{4}$2d	6.47
NpF$_5$			tetragonal	I4/m	
PaCl$_5$	yellow	306	monoclinic	C2/c	
α-UCl$_5$	brown		monoclinic	P2$_1$/n	3.81
β-UCl$_5$	red-brown		triclinic	P$\bar{1}$	
α-PaBr$_5$			monoclinic	P2$_1$/c	
β-PaBr$_5$	orange-brown		monoclinic	P2$_1$/n	
UBr$_5$	brown		monoclinic	P2$_1$/n	
PaI$_5$	black		orthorhombic		
UF$_6$	white	64.02[a]	orthorhombic	*Pnma*	5.060
NpF$_6$	orange	55	orthorhombic	*Pnma*	5.026
PuF$_6$	reddish-brown	52	orthorhombic	*Pnma*	4.86
UCl$_6$	dark green	178	hexagonal	P$\bar{3}m$1	3.62

[a] At 151.6 kPa; to convert kPa to atm, divide by 101.3.

ACTINIUM. [CAS: 7440-34-8]. Chemical element symbol Ac, at. no. 89, at. wt. 227 (mass number of the most stable isotope), periodic table group 3, classed in the periodic system as a higher homologue of lanthanum. The electronic configuration for actinium is

$$1s^2 2s^2 2p^6 3s^2 3p^6 3d^{10} 4s^2 4p^6 4d^{10} 4f^{14} 5s^2 5p^6 5d^{10} 6s^2 6p^6 6d^1 7s^2$$

The ionic radius (Ac^{3+}) is 1.11Å.

Presently, 24 isotopes of actinium, with mass numbers ranging from 207 to 230, have been identified. All are radioactive. One year after the discovery of polonium and radium by the Curies, A. Debierne found an unidentified radioactive substance in the residue after treatment of pitchblende. Debierne named the new material *actinium* after the Greek word for ray. F. Giesel, independently in 1902, also found a radioactive material in the rare-earth extracts of pitchblende. He named this material *emanium*. In 1904, Debierne and Giesel compared the results of their experimentation and established the identical behavior of the two substances. Until formulation of the law of radioactive displacement by Fajans and Soddy about ten years later, however, actinium definitely could not be classed in the periodic system as a higher homologue of lanthanum.

The isotope discovered by Debierne and also noted by Giesel was ^{227}Ac which has a half-life of 21.7 years. The isotope results from the decay of ^{235}U (AcU-*actinouranium*) and is present in natural uranium to the extent of approximately 0.715%. The proportion of Ac/U in uranium ores is estimated to be approximately 2.10^{-10} at radioactive equilibrium. O. Hahn established the existence of a second isotope of actinium in nature, ^{228}Ac, in 1908. This isotope is a product of thorium decay and logically also is referred to as *meso*-thorium, with a half-life of 6.13 hours. The proportion of mesothorium to thorium (MsTh$_2$/Th) in thorium ores is about 5.10^{-14}. The other isotopes of actinium were found experimentally as the result of bombarding thorium targets. The half-life of 10 days of ^{225}Ac is the longest of the artificially produced isotopes. Although occurring in nature as a member of the neptunium family, ^{225}Ac is present in extremely small quantities and thus is very difficult to detect.

^{227}Ac can be extracted from uranium ores where present to the extent of 0.2 mg/ton of uranium and it is the only isotope that is obtainable on a macroscopic scale and that is reasonably stable. Because of the difficulties of separating ^{227}Ac from uranium ores, in which it accompanies the rare earths and with which it is very similar chemically, fractional crystallization or precipitation of relevant compounds no longer is practiced. Easier separations of actinium from lanthanum may be effected through the use of ion-exchange methods. A cationic resin and elution, mainly with a solution of ammonium citrate or ammonium-a-hydroxyisobutyrate, are used. To avoid the problems attendant with the treatment of ores, ^{227}Ac now is generally obtained on a gram-scale by the transmutation of radium by neutron irradiation in the core of a nuclear reactor. Formation of actinium occurs by the following process:

$$^{226}Ra(n,\,\gamma)^{227}Ra \xrightarrow{\beta-} {}^{227}Ac$$

In connection with this method, the cross-section for the capture of thermal neutrons by radium is 23 barns (23×10^{-24} cm^2). Thus, prolonged radiation must be avoided because the accumulation of actinium is limited by the reaction ($\sigma = 500$ barns):

$$^{227}Ac(n,\,\gamma)^{228}Ac(MsTh_2) \longrightarrow {}^{228}Th(RdTh)$$

In 1947, F. Hageman produced 1 mg actinium by this process and, for the first time, isolated a pure compound of the element. It has been found that when 25 g of RaCO$_3$ (radium carbonate) are irradiated at a flux of 2.6×10^{14} ncm^{-2} s^{-1} for a period of 13 days, approximately 108 mg of ^{227}Ac (8 Ci) and 13 mg of ^{228}Th (11 Ci) will be yielded. In an intensive research program by the Centre d'Etude de l'Energie Nucléaire Belge, Union Minière, carried out in 1970–1971, more than 10 g of actinium were produced. The process is difficult for at least two reasons: (1) the irradiated products are highly radioactive, and (2) radon gas, resulting from the disintegration of radium, is evolved. The methods followed in Belgium for the separation of ^{226}Ra, ^{227}Ac, and ^{228}Th involved the precipitation of Ra(NO$_3$)$_2$ (radium nitrate) from concentrated HNO$_3$ after which followed the elimination of thorium by adsorption on a mineral ion exchanger (zirconium phosphate) which withstand high levels of radiation without decomposition.

Metallic actinium cannot be obtained by electrolytic means because it is too electropositive. It has been prepared on a milligram-scale through the reduction of actinium fluoride in a vacuum with lithium vapor at about 350°C. The metal is silvery white, faintly emits a blue-tinted light which is visible in darkness because of its radioactivity. The metal takes the form of a face-centered cubic lattice and has a melting point of $1050 \pm 50°$C. By extrapolation, it is estimated that the metal boils at about 3300°C. An amalgam of metallic actinium may be prepared by electrolysis on a mercury cathode, or by the action of a lithium amalgam on an actinium citrate solution (pH = 1.7 to 6.8).

In chemical behavior, actinium acts even more basic than lanthanum (the most basic element of the lanthanide series). The mineral salts of actinium are extracted with difficulty from their aqueous solutions by means of an organic solvent. Thus, they generally are extracted as chelates with

trifluoroacetone or diethylhexylphosphoric acid. The water-insoluble salts of actinium follow those of lanthanum, namely, the carbonate, fluoride, fluosilicate, oxalate, phosphate, double sulfate of potassium. With exception of the black sulfide, all actinium compounds are white and form colorless solutions. The crystalline compounds are isomorphic.

In addition to its close resemblance to lanthanum, actinium also is analogous to curium ($Z = 96$) and lawrencium ($Z = 103$), both of the group of trivalent transuranium elements. This analogy led G.T. Seaborg to postulate the actinide theory, wherein actinium begins a new series of rare earths which are characterized by the filling of the 5f inner electron shell, just as the filling of the 4f electron shell characterizes the Lanthanide series of elements. However, the first elements of the Actinide series differ markedly from those of actinium. Notably, there is a multiplicity of valences for which there is no equivalent among the lanthanides. See **Chemical Elements** for other properties of actinium.

Mainly, actinium has been of interest from a scientific standpoint. However, ^{227}Ac has been proposed as a source of heat in space vehicles. It is interesting to note that the heat produced from the absorption of the radiation emitted by 1 g of actinium, when in equilibrium with its daughters, is 12,500 cal/hour.

See also **Actinide Contraction**.

Additional Reading

Greenwood, N.N. and A. Earnshaw: *Chemistry of the Elements*, 2nd Edition, Butterworth-Heinemann, UK, 1997.

Hageman, F.: "The Chemistry of Actinium", in G.T. Seaborg and J.J. Katz (editors), *The Actinide Elements, National Nuclear Energy Series, IV-14A*, p. 14, McGraw-Hill, New York, NY, 1954.

Katz, J.J., G.T. Seaborg, and L.R. Moss, *The Chemistry of the Actinide Elements*, 2nd Edition, Chapman & Hall, New York, NY, 1986.

Lide, D.R., *Handbook of Chemistry and Physics*, 84th Edition, CRC Press LLC, Boca Raton, FL, 2003.

Web Reference

http://www.acs.org/ American Chemical Society.

ACTINOLITE. The term for a calcium-iron-magnesium amphibole, the formula being Ca$_2$(Mg,Fe)$_5$Si$_8$O$_{22}$(OH)$_2$ but the amount of iron varies considerably. It occurs as bladed crystals or in fibrous or granular masses. Its hardness is 5–6, sp gr 3–3.2, color green to grayish green, transparent to opaque, luster vitreous to silky or waxy. Iron in the ferrous state is believed to be the cause of its green color. Actinolite derives its name from the frequent radiated groups of crystals. Essentially it is an iron-rich tremolite; the division between the two minerals is quite arbitrary, with color the macroscopic definitive factor—white for tremolite, green for actinolite. Actinolite is found in schists, often with serpentine, and in igneous rocks, probably as the result of the alteration of pyroxene. The schists of the Swiss Alps carry actinolite. It is also found in Austria, Saxony, Norway, Japan, and Canada in the provinces of Quebec and Ontario. In the United States actinolite occurs in Massachusetts, Pennsylvania, Maryland, and as a zinc-manganese bearing variety in New Jersey. See also **Amphibole**; **Tremolite**; and **Uralite**.

ACTINON. The name of the isotope of radon (emanation), which occurs in the naturally occurring actinium, series being, produced by alpha-decay of actinium X, which is itself a radium isotope. Actinon has an atomic number of 86, a mass number of 219, and a half-life of 3.92 seconds, emitting an alpha particle to form polonium-215 (Actinium A). See also **Chemical Elements**; and **Radioactivity**.

ACTIVATED SLUDGE. This is the biologically active sediment produced by the repeated aeration and settling of sewage and/or organic wastes. The dissolved organic matter acts as food for the growth of an aerobic flora. This flora produces a biologically active sludge which is usually brown in color and which destroys the polluting organic matter in the sewage and waste. The process is known as the activated sludge process.

The activated sludge process, with minor variations, consists of aeration through submerged porous diffusers or by mechanical surface agitation, of either raw or settled sewage for a period of 2–6 hours, followed by settling of the solids for a period of 1–2 hours. These solids, which are made up of the solids in the sewage and the biological growths which develop, are returned to the sewage flowing into the aeration tanks. As this cycle is repeated, the aerobic organisms in the sludge develop until there is 1000–3000 ppm of suspended sludge in the aeration liquor. After a while more of the active sludge is developed than is needed to purify the incoming sewage, and this excess is withdrawn from the process and either dried for fertilizer or digested anaerobically with raw sewage sludge. This anaerobic digestion produces a gas consisting of approximately 65% methane and 35% CO_2, and changes the water-binding properties so that the sludge is easier to filter or dry.

The activated sludge is made up of a mixture of zoogleal bacteria, filamentous bacteria, protozoa, rotifera, and miscellaneous higher forms of life. The types and numbers of the various organisms will vary with the types of food present and with the length of the aeration period. The settled sludge withdrawn from the process contains from 0.6 to 1.5% dry solids, although by further settling it may be concentrated to 3–6% solids. Analysis of the dried sludge for the usual fertilizer constituents show that it contains 5–6% of slowly available N and 2–3% of P. The fertilizing value appears to be greater than the analysis would indicate, thus suggesting that it contains beneficial trace elements and growth-promoting compounds. Recent developments indicate that the sludge is a source of vitamin B_{12}, and has been added to mixed foods for cattle and poultry.

The quality of excess activated sludge produced will vary with the food and the extent of oxidation to which the process is carried. In general, about 1 part sludge is produced for each part organic matter destroyed. Prolonged or over-aeration will cause the sludge to partially disperse and digest itself. The amount of air or more precisely oxygen that is necessary to keep the sludge in an active and aerobic condition depends on the oxygen demand of the sludge organisms, the quantity of active sludge, and the amount of food to be utilized. Given sufficient food and sufficient organisms to eat the food, the process seems to be limited only by the rate at which oxygen or air can be dissolved into the mixed liquor. This rate depends on the oxygen deficit, turbulence, bubble size, and temperature and at present is restricted by the physical methods of forcing the air through the diffuser tubers and/or mechanical agitation.

In practice, the excess activated sludge is conditioned with 3–6% $FeCl_3$ and filtered on vacuum filters. This reduces the moisture to about 80% and produces a filter cake which is dried in rotary or spray driers to a moisture content of less than 5%. It is bagged and sold direct as a fertilizer, or to fertilizer manufacturers who use it in mixed fertilizer.

The mechanism of purification of sewage by the activated sludge is twofold i.e., (1) absorption of colloidal and soluble organic matter on the floc with subsequent oxidation by the organisms, and (2) chemical splitting and oxidation of the soluble carbohydrates and proteins to CO_2, H_2O, NH_3, NO_2, NO_3, SO_4, PO_4 and humus. The process of digestion proceeds by hydrolysis, decarboxylation, deaminization and splitting of S and P from the organic molecules before oxidation.

The process is applicable to the treatment of almost any type of organic waste waters which can serve as food for biological growth. It has been applied to cannery wastes, milk products wastes, corn products wastes, and even phenolic wastes. In the treatment of phenolic wastes a special flora is developed which thrives on phenol as food.

W. D. HATFIELD
Decatur, Illinois

ACTIVATION (Molecular). When a molecule which is a Lewis base forms a coordinate bond with a metal ion or with a molecular Lewis acid (such as $AlCl_3$ or BF_3), its electronic density pattern is altered, and with this, the ease with which it undergoes certain reactions. In some instances the polarization that ensues is sufficient to lead to the formation of ions by a process of the type

$$A{:}B{:} + M = A^+ + {:}B{:}M^-$$

More commonly the molecule is polarized by coordination in such a manner that A bears a partial positive charge and B a partial negative charge in a complex A:B:M. Where M is a reducible or oxidizable metal ion, an electron-transfer process may result in which a free radical is generated from A:B: or its fragments and the metal assumes a different oxidation state. In any case the resulting species is often in a state in which it undergoes one or more types of chemical reaction much more readily.

Theoretical Basis

The theoretical basis underlying these activation processes is in the description of the bonding which occurs between the ligand (Lewis base) and the coordination center. This consists of variable contributions from two types of bond. The first is from the sigma bond in which both the electrons in the bonding orbital come from the ligand. This kind of bonding occurs with ligands with available lone pairs (as NH_3, H_2O and their derivatives) and leads to a depletion of electronic charge from the substrate. The second is from the pi bond; here the electrons may come from the metal (if it is a transition metal with suitably occupied d orbitals) or from the ligand (where it has filled p orbitals or molecular orbitals of suitable symmetry). In this case the electronic shifts may partially compensate those arising from the sigma bond if the metal "back-donates" electrons to the ligand. Such shifts may also accentuate those of sigma bonding where ligand electrons are used for both bonds. It is generally found that the net drift of the electronic density is *away* from the ligand.

Activation of Electrophiles

The activation of electrophiles by coordination is a direct result of the weakening of the bond between the donor atom and the rest of the ligand molecule after the donor atom has become bonded to the coordination center. There is considerable evidence to support the claim that this bond need not be broken heterolytically prior to reaction as an electrophile. When this does not occur the literal electrophile is that portion of the ligand which bears a partial positive charge. In such a case the activation process can be more accurately represented by:

$$A{:}B{:} + M = \overset{\delta+\ \delta-}{A{:}B{:}M}$$

Examples of this type of process are:

A:B:	+	M	$\rightarrow$	$\overset{\delta+\ \delta-}{A{:}B{:}M}$
Cl:Cl:		$FeCl_3$		Cl:Cl:$FeCl_3$
NC:Cl:		$AlCl_3$		NC:Cl:$AlCl_3$
RC:Cl: ‖ O		$AlCl_3$		RC:Cl:$AlCl_3$ ‖ O
O_2N:Cl: O ‖		$AlCl_3$		O_2N:Cl:$AlCl_3$ O ‖
RS:Cl: ‖ O		$AlCl_3$		RS:Cl:$AlCl_3$ ‖ O
R:Cl:		BF_3		R:Cl:BF_3

The resultant electrophiles are effective attacking species and can be used to replace an aromatic hydrogen by the group A. When coordination is used to activate an electrophile which has additional donor groups not involved in the principal reaction, and if these are sufficiently effective as donors, they will react with the coordination centers initially added and stop the activation process. In these cases a larger amount of the Lewis acid must be added so that there is more than enough to complex with all the uninvolved donor groups. The extra reagent then provides the Lewis acid needed for the activation process. This is encountered in the Fries reaction and in Friedel-Crafts reactions where the substrate has additional coordination sites. This particular procedure is also used in the "swamping catalyst" procedure for the catalytic halogenation of aromatic compounds.

The usual activation of carbon monoxide by coordination appears to involve complexes in which the carbon atom bonded to the metal is rendered slightly positive, and thus more readily attacked by electron rich species such as ethylenic or acetylenic linkages. An example is seen in the reaction of nickel carbonyl and aqueous acetylene, which results in the production of acrylic acid.

Activation of Free Radicals

The formation of free radicals results from a very similar process when the species M can be oxidized or reduced in a one-electron step with the resultant heterolytic splitting of the A:B bond. The basic reaction in an oxidation reaction of this sort is

$$A:B: + M^{+x} = A \cdot + :B:M^{+x+1}$$

where A and B may be the same or different. The most thoroughly characterized of these reactions is the one found with Fenton's reagent:

$$Fe^{2+} + HOOH = HO \cdot + Fe(OH)^{2+}.$$

The resultant hydroxyl radicals are effective in initiating many chain reactions. The number of metal ions and complexes which are capable of activating hydrogen peroxide in this manner is quite large and is determined in part by the redox potentials of the activator. Related systems in which free radicals are generated by the intervention of suitable metallic catalysts include many in which oxygen is consumed in autoxidations. Cobalt(II) compounds which act as oxygen carriers can often activate radicals in such systems by reactions of the type:

$$Co(II)L + O_2 = Co(III)L + O_2^-, \text{ etc.}$$

Processes of this sort have been used for catalytic oxidations and in cases where a complex with O_2 is formed, the reversibility of the reaction has been studied as a potential process for separating oxygen from the atmosphere.

Radical generating systems of this sort may be used for the initiation of many addition polymerization reactions including those of acrylonitrile and unsaturated hydrocarbons. The information on systems other than those derived from hydrogen peroxide is very meager.

The activation of O_2 by low oxidation states of plantinum has also been demonstrated in reactions such as

$$2P(C_6H_5)_3 + O_2 \xrightarrow{Pt(P(C_6H_5)_3)_3} 2(C_6H_5)_3PO.$$

Ligands such as ethylene are Lewis basesby virtue of the availability of the electrons of their pi bonds to external reagents and the coordination of unsaturated organic compounds to species such as Cu(I), Ag(I), Pd(II), and Pt(II) is a well established phenomenon. The coordination process with such ligands usually involves a considerable element of back bonding from the filled d orbitals of the metal ion. The coordination process activates olefins towards cis-trans isomerizations and attach by reagents such as hydrogen halides. Coordination to palladium(II) facilitates attach of olefins by water via a redox process as seen in the Smid reaction:

$$PdCl_2 + H_2C{=}CH_2 + H_2O = Pd + CH_3CHO + 2HCl$$

$$Pd + 2CuCl_2 = 2CuCl + PdCl_2$$

$$2HCl + 2CuCl + \tfrac{1}{2}O_2 = 2CuCl_2 + H_2O$$

A similar reaction also occurs for carbon monoxide:

$$Pd^{2+} + CO + H_2O = Pd + CO_2 + 2H^+.$$

Activation of nucleophiles by coordination, best exemplified by various complexes used as catalysts for hydrogenation, and coordination assistance to photochemical activation, may be similarly demonstrated.

MARK M. JONES
Vanderbilt University
Nashville, Tennessee

ACTIVATOR. 1. A substance that renders a material or a system reactive; commonly, a catalyst. 2. A special use of this term occurs in the flotation process, where an activator assists the action of the collector. 3. An impurity atom, present in a solid, that makes possible the effects of luminescence, or markedly increases their efficiency. Examples are copper in zinc sulfide, and thallium in potassium chloride. See also **Enzyme**.

ACTIVE CENTER. Atoms which, by their position on a surface, such as at the apex of a peak, at a step on the surface or a kink in a step, or on the edge or corner of a crystal, share with neighboring atoms an abnormally small portion of their electrostatic field, and therefore have a large residual field available for catalytic activity or for adsorption.

ACTIVE DEPOSIT. The name given to the radioactive material that is deposited on the surface of any substance placed in the neighborhood of a preparation containing any of the naturally occurring radioactive chains (uranium, thorium, or actinium chains). This deposit results from deposition of the nongaseous products of the gaseous radon nuclides that have escaped from the parent substance. An active deposit can be concentrated on a negatively charged metal wire or surface placed in closed vessels containing the radon. See also **Radioactivity**.

ACTIVITY COEFFICIENT. A fractional number which when multiplied by the molar concentration of a substance in solution yields the chemical activity. This term provides an approximation of how much interaction exists between molecules at higher concentrations. Activity coefficients and activities are most commonly obtained from measurements of vapor-pressure lowering, freezing-point depression, boiling-point elevation, solubility, and electromotive force. In certain cases, activity coefficients can be estimated theoretically. As commonly used, activity is a relative quantity having unit value in some chosen standard state. Thus, the standard state of unit activity for water, a_w, in aqueous solutions of potassium chloride is pure liquid water at one atmosphere pressure and the given temperature. The standard state for the activity of a solute like potassium chloride is often so defined as to make the ratio of the activity to the concentration of solute approach unity as the concentration decreases to zero.

In general, the activity coefficient of a substance may be defined as the ratio of the effective contribution of the substance to a phenomenon to the actual contribution of the substance to the phenomenon. In the case of gases the effective pressure of a gas is represented by the fugacity f and the actual pressure of the gas by P. The activity coefficient, γ, of the gas is given by

$$\gamma = f/P. \tag{1}$$

One method of calculating fugacity and hence γ is based on the measured deviation of the volume of a real gas from that of an ideal gas. Consider the case of a pure gas. The free energy F and chemical potential μ changes with pressure according to the equation

$$dF = d\mu = V\,dP. \tag{2}$$

but by definition

$$d\mu = V\,dP = RT\,d\ln f \tag{3}$$

If the gas is ideal, the molal volume V_i is given by

$$V_i = \frac{RT}{P} \tag{4}$$

but for a nonideal gas this is not true. Let the molal volume of the nonideal gas be V_n and define the quantity α by the equation

$$\alpha = V_i - V_n = \frac{RT}{P} - V_n \tag{5}$$

Then V of Eq. (2) is V_n of Eq. (5) and hence from Eq. (5)

$$V = \frac{RT}{P} - \alpha \tag{6}$$

Therefore from Eqs. (2), (3), and (6)

$$RT\,d\ln f = dF = d\mu = RT\,D\ln P - \alpha\,dP \tag{7}$$

and

$$RT\ln f = RT\ln P - \int_0^P \alpha\,dP \tag{8}$$

Thus knowing PVT data for a gas it is possible to calculate f. The integral in Eq. (8) can be evaluated graphically by plotting α, the deviation of gas volume from ideality, versus P and finding the area under the curve out to the desired pressure. Also it may be found by mathematically relating α to P by an equation of state, or by using the method of least squares or other acceptable procedure the integral may be evaluated analytically for any value of P. The value of f at the desired value of P may thus be found and consequently the activity coefficient calculated. Other methods are available for the calculation of f and hence of γ, the simplest perhaps being the relationship

$$f = \frac{P^2}{P_i} \tag{9}$$

where P_i is the ideal and P the actual pressure of the gas.

In the case of nonideal solutions, we can relate the activity α_A of any component A of the solution to the chemical potential μ_A of that component by the equation

$$\mu_A = \mu_A^{\circ\prime} + RT \ln a_A \tag{10}$$

$$= \mu_A^{\circ\prime} + RT \ln \gamma_A X_A \tag{11}$$

where $\mu_A^{\circ\prime}$ is the chemical potential in the reference state where a_i is unity and is a function of temperature and pressure only, whereas γ_A is a function of temperature pressure and concentration. It is necessary to find the conditions under which γ_A is unity in order to complete its definition. This can be done using two approaches—one using Raoult's law which for solutions composed of two liquid components is approached as $X_A \to 1$; and two using Henry's law which applies to solutions, one component of which may be a gas or a solid and which is approached at $X_A \to 0$. Here X_A represents the mole fraction of component A.

For liquid components using Raoult's law

$$\gamma_A \to 1 \text{ as } X_A \to 1 \tag{12}$$

Since the logarithmic term is zero in Eq. (11) under this limiting condition, $\mu_A^{\circ\prime}$ is the chemical potential of pure component A at the temperature and pressure under consideration. For ideal solutions the activity coefficients of both components will be unity over the whole range of composition.

The convention using Henry's law is convenient to apply when it is impossible to vary the mole fraction of both components up to unity. Solvent and solute require different conventions for such solutions. As before, the activity of the solvent, usually taken as the component present in the higher concentration, is given by

$$\gamma_A \to 1 \text{ as } X_A \to 0 \tag{14}$$

Thus, $\mu_A^{\circ\prime}$ for the solute in Eq. (11) is the chemical potential of the solute in a hypothetical standard state in which the solute at unit concentration has the properties which it has at infinite dilution.

γ_A is the activity coefficient of component A in the solution and is given by the expression

$$\gamma_A = \frac{a_A}{X_A} \tag{15}$$

In Eq. (15) a_A is the activity or in a sense the effective mole fraction of component A in the solution.

The activity a_A of a component A in solution may be found by considering component A as the solvent. Then its activity at any mole fraction is the ratio of the partial pressure of the vapor of A in the solution to the vapor pressure of pure A. If B is the solute, its standard reference state is taken as a hypothetical B with properties which it possesses at infinite dilution.

The equilibrium constant for the process

$$B \text{ (gas)} \leftrightarrow B \text{ (solution)} \tag{16}$$

is

$$K = \frac{a_{\text{solution}}}{a_{\text{gas}}} \tag{17}$$

Since the gas is sufficiently ideal its activity a_{gas} is equivalent to its pressure P_2. Since the solution is far from ideal, the activity a_{solute} of the liquid B is not equal to its mole fraction N_2 in the solution. However,

$$K' = N_2 / P_2 \tag{18}$$

and extrapolating a plot of this value versus N_2 to $N_2 = 0$ one obtains the ratio where the solution is ideal. This extrapolated value of K' is the true equilibrium constant K when the activity is equal to the mole fraction

$$K = a_2 / P_2 \tag{19}$$

Thus a_2 can be found. The methods involved in Eqs. (16) through (19) arrive at the activities directly and thus obviate the determination of the activity coefficient. However, from the determined activities and known mole fractions γ can be found as indicated in Eq. (15).

In the case of ions the activities, a_+ and a_- of the positive and negative ions, respectively, are related to the activity, a, of the solute as a whole by the equation

$$a_+^p \times a_-^q \tag{20}$$

and the activity coefficients γ_+ and γ_- of the two charge types of ions are related to the molality, m, of the electrolyte and ion activities a_+ and a_- by the equations

$$\gamma_+ = \frac{a_+}{pm}; \quad \gamma_- = \frac{a_-}{qm} \tag{21}$$

Also the activity coefficient of the electrolyte is given by the equation

$$\gamma = (\gamma_+^p \times \gamma_-^q)^{(1/p+q)} \tag{22}$$

In Eqs. (20), (21) and (22) p and q are numbers of positive and negative ions, respectively, in the molecule of electrolyte. In dilute solutions it is considered that ionic activities are equal for uni-univalent electrolytes, i.e., $\gamma_+ = \gamma_-$.

Consider the case of $BaCl_2$.

$$\gamma = (\gamma_+ \times \gamma_-^2)^{(1/1+2)} = (\gamma_+ \times \gamma_-^2)^{1/3} \tag{23}$$

or

$$\gamma^3 = \gamma_+ \gamma_-^2 \tag{24}$$

also

$$a = a_+ \times a_-^2 = (m\gamma_+)(2m\gamma_-)^2 \tag{25}$$

$$= 4\, m^3 \gamma_+ \gamma_-^2 = 4\, m^3 \gamma^3 \tag{26}$$

Activity coefficients of ions are determined using electromotive force, freezing point, and solubility measurements or are calculated using the theoretical equation of Debye and Hückel.

The solubility, s, of AgCl can be determined at a given temperature and the activity coefficient γ determined at that temperature from the solubility and the solubility product constant K. Thus

$$K = a_+ a_- = \gamma_+ c_+ \gamma_- c_- \tag{27}$$

where c_+ and c_- are the molar concentrations of the positive silver and negative chloride ions, respectively. The solubility s of the silver chloride is simply $s = c_+ = c_-$. The expression for K is then

$$K = \gamma^2 s^2 \tag{28}$$

and

$$\gamma = \frac{K^{1/2}}{s} \tag{29}$$

By measuring the solubility, s, of the silver chloride in different concentration of added salt and extrapolating the solubilities to zero salt concentration, or better, to zero ionic strength, one obtains the solubility when $\gamma = 1$, and from Eq. (29) K can be found. Then γ can be calculated using this value of K and any measured solubility. Actually, this method is only applicable to sparingly soluble salts. Activity coefficients of ions and of electrolytes can be calculated from the Debye-Hückel equations. For a uni-univalent electrolyte, in water at 25°C, the equation for the activity coefficient of an electrolyte is

$$\log \gamma = -0.509 z_+ z_- \sqrt{\mu} \tag{30}$$

where z_+ and z_- are the valences of the ion and μ is the ionic strength of the solution, i.e.,

$$\mu = \tfrac{1}{2} \Sigma c_i z_i^2 \tag{31}$$

where c_i is the concentration and z_i the valence of the ith type of ion.

To illustrate a use of activity coefficients, consider the cell without liquid junction

$$\text{Pt, } H_2(g); \text{ HCl (m); AgCl, Ag} \tag{32}$$

for which the chemical reaction is

$$\tfrac{1}{2} H_2(g) + AgCl \text{ (solid)} = HCl \text{ (molality, m)} + Ag \text{ (solid.)} \tag{33}$$

The electromotive force, E, of this cell is given by the equation

$$E = E^\circ - \frac{2.303\, RT}{n\mathbf{F}} \log \frac{a_{\text{HCl}}}{P_{H_2}}$$

$$= E^\circ - 0.05915 \log m^2 \gamma^2 \tag{34}$$

where $E°$ is the standard potential of the cell, n is the number of electrons per ion involved in the electrode reaction (here $n = 1$), $\mathbf{F}$ is the coulombs per faraday, a (equal to $m^2\gamma^2$) is the activity of the electrolyte HCl, P_{H_2} is the pressure (1 atm) and is equal to the activity of the hydrogen gas, and AgCl (solid) and Ag (solid) have unit activities. Transferring the exponents in front of the logarithmic term in Eq. (34), the equation can be written,

$$E = E° - 0.1183 \log m - 0.1183 \log \gamma \qquad (35)$$

which by transposing the log m term to the left of the equation becomes

$$E + 0.1183 \log m = E° - 0.1183 \log \gamma \qquad (36)$$

For extrapolation purposes, the extended form of the Debye-Hückel equation involving the molality of a dilute univalent electrolyte in water at 25°C is used:

$$\log \gamma = -0.509\sqrt{m} + bm \qquad (37)$$

where b is an empirical constant.

Substitution of $\log \gamma$ from Eq. (37) into Eq. (36) gives

$$E + 0.1183 \log m - 0.0602\, m^{1/2}$$
$$= E° - 0.1183\, bm \qquad (38)$$

A plot of the left hand side of Eq. (38) versus m yields a practically straight line, the extrapolation of which to $m = 0$ gives $E°$ the standard potential of the cell. This value of $E°$ together with measured values of E at specified m values can be used to calculate γ for HCl in dilute aqueous solutions at 25° for different m − values. Similar treatment can be applied to other solvents and other solutes at selected temperatures.

Activity coefficients are used in calculation of equilibrium constants, rates of reactions, electrochemical phenomena, and almost all quantities involving solutes or solvents in solution.

EDWARD S. AMIS
University of Arkansas
Fayetteville, Arkansas

TABLE 1. STANDARD ELECTRODE POTENTIALS (25°C)

	Reaction		Volts
$Li^+ + e^-$	$\rightleftharpoons$	Li	−3.045
$K^+, +e^-$	$\rightleftharpoons$	K	−2.924
$Ba^{2+} + 2e^-$	$\rightleftharpoons$	Ba	−2.90
$Ca^{2+} + 2e^-$	$\rightleftharpoons$	Ca	−2.76
$Na^+ + e^-$	$\rightleftharpoons$	Na	−2.711
$Mg^{2+} + 2e^-$	$\rightleftharpoons$	Mg	−2.375
$Al^{3+} + 3e^-$	$\rightleftharpoons$	Al	−1.706
$2H_2O + 2e^-$	$\rightleftharpoons$	$H_2 + 2\,OH^-$	−0.828
$Zn^{2+} + 2e^-$	$\rightleftharpoons$	Zn	−0.763
$Cr^{3+} + 3e^-$	$\rightleftharpoons$	Cr	−0.744
$Fe^{2+} + 2e^-$	$\rightleftharpoons$	Fe	−0.41
$Cd^{2+} + 2e^-$	$\rightleftharpoons$	Cd	−0.403
$Ni^{2+} + 2e^-$	$\rightleftharpoons$	Ni	−0.23
$Sn^{2+} + 2e^-$	$\rightleftharpoons$	Sn	−0.136
$Pb^{2+} + 2e^-$	$\rightleftharpoons$	Pb	−0.127
$2H^+ + 2e^-$	$\rightleftharpoons$	H_2	0.000
$Cu^{2+} + 2e^-$	$\rightleftharpoons$	Cu	+0.34
$I_2 + 2e^-$	$\rightleftharpoons$	$2I^-$	+0.535
$Fe^{3+} + e^-$	$\rightleftharpoons$	Fe^{2+}	+0.77
$Ag^+ + e^-$	$\rightleftharpoons$	Ag	+0.799
$Hg^{2+} + 2e^-$	$\rightleftharpoons$	Hg	+0.851
$Br_2 + 2e^-$	$\rightleftharpoons$	$2Br^-$	+1.065
$O_2 + 4H^+ + 4e^-$	$\rightleftharpoons$	$2H_2O$	+1.229
$Cr_2O_7^{2-} + 14H^+ + 6e^-$	$\rightleftharpoons$	$2Cr^{3+} + 7H_2O$	+1.33
$Cl_2(gas) + 2e^-$	$\rightleftharpoons$	$2Cl^-$	+1.358
$Au^{3+} + 3e^-$	$\rightleftharpoons$	Au	+1.42
$MnO_4^- + 8H^+ + 5e^-$	$\rightleftharpoons$	$Mn^{2+} + 4H_2O$	+1.491
$F_2 + 2e^-$	$\rightleftharpoons$	$2F^-$	+2.85

ACTIVITY (Radioactivity). The activity of a quantity of radioactive nuclide is defined by the ICRU as $\Delta N/\Delta t$, where N is the number of nuclear transformations that occur in this quantity in time Δt. The symbol Δ preceding the letters N and t denotes that these letters represent quantities that can be deduced only from multiple measurements that involve averaging procedures. The special unit of activity is the curie, defined as exactly 3.7×10^{10} transformations per second. See **Radioactivity**.

ACTIVITY SERIES. Also referred to as the *electromotive series* or the *displacement series*, this is an arrangement of the metals (other elements can be included) in the order of their tendency to react with water and acids, so that each metal displaces from solution those below it in the series and is displaced by those above it. See Table 1. Since the electrode potential of a metal in equilibrium with a solution of its ions cannot be measured directly, the values in the activity series are, in each case, the difference between the electrode potential of the given metal (or element) in equilibrium with a solution of its ions, and that of hydrogen in equilibrium with a solution of its ions. Thus in the table, it will be noted that hydrogen has a value of 0.000. In experimental procedure, the hydrogen electrode is used as the standard with which the electrode potentials of other substances are compared. The theory of displacement plays a major role in electrochemistry and corrosion engineering. See also **Corrosion**; and **Electrochemistry**.

ACYL. An organic radical of the general formula, RCO−. These radicals are also called acid radicals, because they are often produced from organic acids by loss of a hydroxyl group. Typical acyl radicals are acetyl, CH_3CO-, benzyl, C_6H_5CO-, etc.

ACYLATION. A reaction or process whereby an acyl radical, such as acetyl, benzoyl, etc., is introduced into an organic compound. Reagents often used for acylation are the acid anhydride, acid chloride, or the acid of the particular acyl radical to be introduced into the compound.

ADAMANTINE COMPOUND. A compound having in its crystal structure an arrangement of atoms essentially that of diamond, in which every atom is linked to its four neighbors mainly by covalent bonds. An example is zinc sulfide, but it is to be noted that the eight electrons involved in forming the four bonds are not provided equally by the zinc and sulfur atoms, the sulfur yielding its six valence electrons, and the zinc, two. This is the structure of typical semiconductors, e.g., silicon and germanium.

ADAMS, ROGER (1889–1971). An American chemist, born in Boston; graduated from Harvard, where he taught chemistry for some years. After studying in Germany, he move to the University of Illinois in 1916, where he later became chairman of the department of chemistry (1926–1954). During his prolific career, he made this department one of the best in the country, and strongly influenced the development of industrial chemical research in the U.S. His executive and creative ability made him an outstanding figure as a teacher, innovator, and administrator. Among his research contributions were development of platinum-hydrogenation catalysts, and structural determinations of chaulmoogric acid, gossypol, alkaloids, and marijuana. He held many important offices, including president of the ACS and AAS, and was a recipient of the Priestley medal.

ADDITIVE COLOR PROCESS. An early system of color imagery in which the color synthesis is obtained by the addition of colors one to another in the form of light rather than as colorants. This color addition may take place (1) by the simultaneous projection of two or more (usually three) color images onto a screen, (2) by the projection of the color images in rapid succession onto a screen or (3) by viewing minutely divided juxtaposed color images.

In the case of a three-color process, three-color records are made from the subject recording, in terms of silver densities, the relative amounts of red, green, and blue present in various areas of the subject.

When the additive synthesis is to be made by simultaneous projection, positives are made from the color separation negatives and projected with a triple lantern onto a screen through red, green, and blue filters. The registered color images give all colors of the subject due to simple color addition, red plus green making yellow, red plus blue appearing magenta, etc.

When the additive synthesis is made by successive viewing, the same three-color images must be flashed onto the screen in such rapid succession that the individual red, green, and blue images are not apparent. Simple

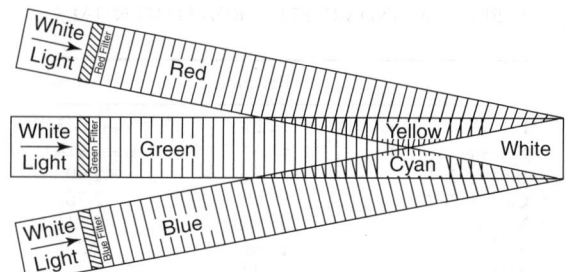

Fig. 1.   Mechanism of color addition

color addition is again obtained but this time use is made of the persistence of vision to "mix" the colors. See Fig. 1.

The third type of additive synthesis makes use of the fact that small dots of different colors, when viewed from such a distance that they are no longer individually visible, form a single color by simple color addition. The three-color images in this type of process are generally side by side in the space normally occupied by a single image. The red record image will be composed of a number of red dots or markings of differing density which, in total, will compose the red record image. Alongside the red markings will be green and blue markings, without any overlapping. When viewed at such a distance that the colored markings are at, or below, the limit of visual resolution, the color sensation from any given area will be the integrated color of the markings comprising the area—an additive color mixture.

ADENINE.   [CAS: 73-24-5]. A prominent member of the family of naturally occurring purines (see Structure 1). Adenine occurs not only in ribonucleic acids (RNA), and deoxyribonucleic acids (DNA), but in nucleosides, such as adenosine, and nucleotides, such as adenylic acid, which may be linked with enzymatic functions quite apart from nucleic acids. Adenine, in the form of its ribonucleotide, is produced in mammals and fowls endogenously from smaller molecules and no nutritional essentiality is ascribed to it. In the nucleosides, nucleotides, and nucleic acids, the attachment or the sugar moiety is at position 9.

$$NH_2$$

(1)

The purines and pyrimidines absorb ultraviolet light readily, with absorption peaks at characteristic frequencies. This has aided in their identification and quantitative determination.

ADENOSINE.   [CAS: 58-61-7]. An important nucleoside composed of adenine and ribose. White, crystalline, odorless powder, mild, saline, or bitter taste, Mp 229C, quite soluble in hot water, practically insoluble in alcohol. Formed by isolation following hydrolysis of yeast nucleic acid. The upper portion of Structure 1 represents the adenine moiety, and the lower portion of the pentose, D-ribose.

$$NH_2$$

(1)

ADENOSINE DI-AND TRIPHOSPHATE.   See **Carbohydrates**; **Phosphorylation (Oxidative)**; **Phosphorylation (Photosynthetic)**.

ADENOSINE PHOSPHATES.   The adenosine phosphates include *adenylic acid* (adenosine monophosphate, AMP) in which adenosine is esterified with phosphoric acid at the 5'-position; *adenosine diphosphate* (ADP) in which esterification at the same position is with pyrophosphoric acid,

$$HO_2-\overset{\overset{O}{\|}}{P}-O-\overset{\overset{O}{\|}}{P}-(OH)_2$$

and *adenosine triphosphate* (ATP) in which three phosphate residues

$$HO_2-\overset{\overset{O}{\|}}{P}-O-\overset{\overset{O}{\|}}{\underset{\underset{OH}{|}}{P}}-O-\overset{\overset{O}{\|}}{P}-(OH)_2$$

are attached at the 5'-position. Adenosine-3'-phosphate is an isomer of adenylic acid, and adenosine-2', 3'-phosphate is esterified in two positions with the same molecules of phosphoric acid and contains the radical.

$$-O-\overset{\overset{O}{\|}}{\underset{\underset{OH}{|}}{P}}-O-$$

ADHESIVES.   An *adhesive* is a material capable of holding together solid materials by means of surface attachment. *Adhesion* is the physical attraction of the surface of one material for the surface of another. An *adherend* is the solid material to which the adhesive adheres and the *adhesive bond* or *adhesive joint* is the assembly made by joining adherends together by means of an adhesive. *Practical adhesion* is the physical strength of an adhesive bond. It primarily depends on the forces of adhesion, but its magnitude is determined by the physical properties of the adhesive and the adherend, as well as the engineering of the adhesive bond.

The *interphase* is the volume of material in which the properties of one substance gradually change into the properties of another. The interphase is useful for describing the properties of an adhesive bond. The *interface*, contained within the interphase, is the plane of contact between the surface of one material and the surface of another. Except in certain special cases, the interface is imaginary. It is useful in describing surface energetics.

Theories of Adhesion

There is no unifying theory of adhesion describing the relationship between practical adhesion and the basic intermolecular and interatomic interactions which take place between the adhesive and the adherend either at the interface or within the interphase. The existing adhesion theories are, for the most part, rationalizations of observed phenomena, although in some cases, predictions regarding the relative ranking of practical adhesion can actually be made.

Diffusion Theory.   The diffusion theory of adhesion is mostly applied to polymers. It assumes mutual solubility of the adherend and adhesive to form an interphase.

Electrostatic Theory.   The basis of the electrostatic theory of adhesion is the differences in the electronegativities of adhering materials which leads to a transfer of charge between the materials in contact. The attraction of the charges is considered the source of adhesion.

Surface Energetics and Wettability Theory.   The surface energetics and wettability theory of adhesion is concerned with the effect of intermolecular and interatomic forces on the surface energies of the adhesive and the adherend and the interfacial energy between the two.

Mechanical Interlocking Theory.   A practical adhesion can be enhanced if the adhesive is applied to a surface which is microscopically rough.

Guidelines for Good Adhesion.   The various adhesion theories can be used to formulate guidelines for good adhesion:

1.   An adhesive should possess a liquid surface tension that is less than the critical wetting tension of the adherend's surface.

2. The adherend should be mechanically rough enough so that the asperities on the surface are on the order of, or less than, one micrometer in size.

3. The adhesive's viscosity and application conditions should be such that the asperities on the adherend's surface are completely wetted.

4. If an adverse environment is expected, covalent bonding capabilities at the interface should be provided.

For good adhesion, the adhesive and the adherend should, if possible, display mutual solubility to the extent that both diffuse into one another, providing an interphasal zone.

Advantages and Disadvantages of Using Adhesives

Adhesive Advantages. In comparison to other methods of joining, adhesives provide several advantages. First, a properly applied adhesive provides a joint having a more uniform stress distribution under load than a mechanical fastener which requires a hole in the adherend. Second, adhesives provide the ability to bond dissimilar materials such as metals without problems such as galvanic corrosion. Third, using an adhesive to make an assembly increases fatigue resistance. Fourth, adhesive joints can be made of heat- or shock-sensitive materials. Fifth, adhesive joining can bond and seal simultaneously. Sixth, use of an adhesive to form an assembly usually results in a weight reduction in comparison to mechanical fasteners since adhesives, for the most part, have densities which are substantially less than that of metals.

Adhesive Disadvantages. There are some limitations in using adhesives to form assemblies. The major limitation is that the adhesive joint is formed by means of surface attachment and is, therefore, sensitive to the substrate surface condition. Another limitation of adhesive bonding is the lack of a nondestructive quality control procedure. Finally, adhesive joining is still somewhat limited because most designers of assemblies are simply not familiar with the engineering characteristics of adhesives.

Mechanical Tests of Adhesive Bonds

The three principal forces to which adhesive bonds are subjected are a shear force in which one adherend is forced past the other, peeling in which at least one of the adherends is flexible enough to be bent away from the adhesive bond, and cleavage force. The cleavage force is very similar to the peeling force, but the former applies when the adherends are nondeformable and the latter when the adherends are deformable. Appropriate mechanical testing of these forces are used. Fracture mechanics tests are also typically used for structural adhesives.

Table 1 provides the approximate load-bearing capabilities of various adhesive types. Because the load-bearing capabilities of an adhesive are dependent upon the adherend material, the loading rate, temperature, and design of the adhesive joint, wide ranges of performance are listed.

Chemistry and Uses of Adhesives

Structural Adhesives. A structural adhesive is a resin system, usually a thermoset, that is used to bond high strength materials in such a way that the bonded joint is able to bear a load in excess of 6.9 MPa (1000 psi) at room temperature. Structural adhesives are the strongest form of adhesive and are meant to hold loads permanently. They exist in a number of forms. The most common form is the two-part adhesive, widely available as a consumer product. The next most familiar is that which is obtained as a room temperature curing liquid. Less common are primer—liquid adhesive combinations which cure at room temperature. Structural adhesive pastes which cure at 120°C are widely available in the industrial market.

Structural adhesives are formulated from epoxy resins, phenolic resins, acrylic monomers and resins, high temperature-resistant resins (e.g., polyimides), and urethanes. Structural adhesive resins are often modified by elastomers.

Natural-product-based structural adhesives include protein-based adhesives, starch-based adhesives, and cellulosics.

Pressure-Sensitive Adhesives. A pressure-sensitive adhesive, a material which adheres with no more than applied finger pressure, is aggressively and permanently tacky. It requires no activation other than the finger pressure, exerts a strong holding force, and should be removable from a smooth surface without leaving a residue.

Applications and Formulation. Pressure-sensitive adhesives are most widely used in the form of adhesive tapes. The general formula for a pressure-sensitive adhesive includes an elastomeric polymer, a tackifying resin, any necessary fillers, various antioxidants and stabilizers, if needed, and cross-linking agents.

Hot-Melt Adhesives. Hot-melt adhesives are 100% nonvolatile thermoplastic materials that can be heated to a melt and then applied as a liquid to an adherend. The bond is formed when the adhesive resolidifies. The oldest example of a hot-melt adhesive is sealing wax.

Solvent- and Emulsion-Based Adhesives. *Solvent-Based Adhesives.* Solvent-based adhesives, as the name implies, are materials that are formed by solution of a high molecular weight polymer in an appropriate solvent. Solvent-based adhesives are usually elastomer-based and formulated in a manner similar to pressure-sensitive adhesives.

Emulsion Adhesives. The most widely used emulsion-based adhesive is that based upon poly(vinyl acetate)–poly(vinyl alcohol) copolymers formed by free-radical polymerization in an emulsion system. Poly(vinyl alcohol) is typically formed by hydrolysis of the poly(vinyl acetate). This is also known as "white glue."

Economic Aspects

Although the manufacture and sale of adhesives is a worldwide enterprise, the adhesives business can be characterized as a fragmented industry. The 1987 Census of Manufacturers obtained reports from 712 companies in the United States, each of which considers itself to be in the adhesives or sealants business; only 275 of these companies had more than 20 employees. Phenolics, poly(vinyl acetate) adhesives, rubber cements, and hot-melt adhesives are the leading products in terms of monetary value. These products are used primarily in the wood, paper, and packaging industries. The annual growth rate of the adhesives market is 2.3%, and individual segments of the market are expected to grow faster than this rate.

An excellent review of "Adhesive Bonding" is contained in the *Modern Plastic Encyclopedia*, issued annually by Modern Plastics, Pittsfield, Massachusetts.

For further information, refer to Case Western Reserve University in Cleveland, Ohio, which maintains a fundamental research center for adhesives and coatings. http://www.cwru.edu/cse/eche/

ALPHONSUS V. POCIUS
The 3M Company

TABLE 1. LOAD-BEARING CAPABILITIES OF ADHESIVES[a]

Adhesive type	Shear load, MPa[b]	Peel load, N/m[c]
pressure sensitive	0.005–0.02[d]	300–600
rubber based	0.3–7	1000–7000
emulsion	10–14	
hot melt	1–15	1000–5000
natural product (structural)	10–14	
polyurethane	6–17	2000–10,000
acrylic	6–20	900–6000
epoxy	14–50	700–18,000
phenolic	14–35	700–9000
polyimide	13–17	350–1760

[a] Load bearing capabilities are dependent upon the adherend, joint design, rate of loading, and temperature. Values given represent the type of adherends normally used at room temperature. Lap shear values approximate those obtainable from an overlap of 3.2 cm^2.

[b] To convert from MPa to psi, multiply by 145.

[c] To convert from N/m to ppi, divide by 175.

[d] Pressure-sensitive adhesives normally are rated in terms of shear holding power, i.e., time to fail in minutes under a constant load.

Additional Reading

American Society for Testing Materials: ASTM, *Adhesives,* American Society for Testing and Materials, West Conshohocken, PA, 1999.

Budinski, K.G., and M.K. Budinski: *Engineering Materials: Properties and Selection,* Prentice-Hall Inc., Upper Saddle River, NJ, 1998.

Hartshorn, S.R. ed.: *Structural Adhesives: Chemistry and Technology,* Plenum, New York, NY, 1986.

Modern Plastics Encyclopedia 97/E: Price Stern Sloan, Inc., Los Angeles, CA, 1997.

Petrie, E.M.: *Handbook of Adhesives and Sealants,* The McGraw-Hill Companies, Inc., New York, NY, 1999.

Pocius, A.V.: *Adhesion and Adhesives Technology,* Hanser Gardner Publications, Cincinnati, OH, 1997.

Satas, D. ed.: *Handbook of Pressure Sensitive Adhesive Technology*, Van Nostrand Reinhold Co., Inc., New York, 1989.

Skeist, I.M. ed.: *Handbook of Adhesives*, 3rd Edition, Van Nostrand Reinhold Co., Inc. New York, 1990. A basic resource for practitioners of this technology.

Wu, S. *Polymer Interface and Adhesion*, Marcel Dekker, Inc., New York, NY, 1982. A basic textbook covering surface effects on polymer adhesion.

ADIABATIC PROCESS.   Any thermodynamic process, reversible or irreversible, which takes place in a system without the exchange of heat with the surroundings. When the process is also reversible, it is called *isentropic*, because then the entropy of the system remains constant at every step of the process. (In older usage, isentropic processes were called simply adiabatic, or quasistatic adiabatic; the distinction between adiabatic and isentropic processes was not always sharply drawn.)

When a closed system undergoes an adiabatic process without performing work (*unresisted expansion*), its internal energy remains constant whenever the system is allowed to reach thermal equilibrium. Such a process is necessarily irreversible. At each successive state of equilibrium, the entropy of the system S_i, has a higher value than the initial entropy, S_0. Example: When a gas at pressure p_0, temperature T_0, occupying a volume V_0 (see Fig. 1) is allowed to expand progressively into volumes $V_1 = V_0 + \Delta V$, etc., by withdrawing slides 1, 2, etc., one after another, it undergoes such a process if it is enclosed in an adiabatic container. After each withdrawal of a slide, the irreversibility of the process causes the system to depart from equilibrium; equilibrium sets in after a sufficiently long waiting period. At each successive state of equilibrium $U_1 = U_2 = \cdots = U_0$, but $S_0 < S_1 < S_2$, etc.

When an open system in steady flow undergoes an adiabatic process without performing external work, the enthalpy of the system regains its initial value at each equilibrium state, and the entropy increases as before. Example: Successive, *slow* expansions through porous plugs $P_1, P_2 \cdots$ (Fig. 2), when we have

$$H_1 = H_2 = \cdots = H_0$$

but

$$S_0 < S_1 < S_2, \text{ etc.}$$

This process is also necessarily irreversible.

A closed system cannot perform an isentropic process without performing work. Example (Fig. 3): A quantity of gas enclosed by an ideal, frictionless, adiabatic piston in an adiabatic cylinder is maintained at a pressure p by a suitable ideal mechanism, so that $Gl = pA$ (A being the area of piston). When the weight G is increased (or decreased) by an infinitesimal amount dG, the gas will undergo an isentropic compression (or expansion). In this case,

$$S = \text{constant}, \quad dS = 0$$

at any stage of the process, but

$$U \neq \text{constant}, \quad H \neq \text{constant}$$

During an isentropic process of a closed system between state 1 and 2, the change in internal energy equals *minus* the work done between the two states, or

$$U_2 - U_1 = -W_{12}$$

work is done "at the expense" of the internal energy.

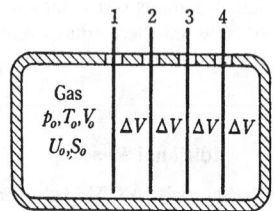

Fig. 1.   Successive adiabatic expansions of gas by withdrawing slides.

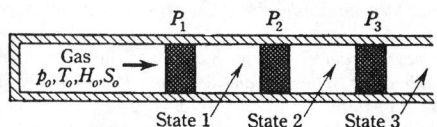

Fig. 2.   Successive, slow adiabatic expansions of gas through porous plugs.

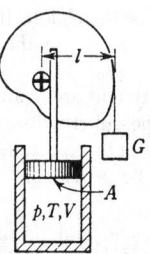

Fig. 3.   Isentropic compression (or expansion) in cylinder.

ADIPIC ACID.   [CAS: 124-04-9]. Adipic acid, hexanedioic acid, 1,4-butanedicarboxylic acid, mol wt 146.14, HOOCCH$_2$CH$_2$CH$_2$CH$_2$COOH, is a white crystalline solid with a melting point of about 152°C. Little of this dicarboxylic acid occurs naturally, but it is produced on a very large scale at several locations around the world. The majority of this material is used in the manufacture of nylon-6,6 polyamide, which is prepared by reaction with 1,6-hexanediamine.

Chemical and Physical Properties

Adipic acid is a colorless, odorless, sour-tasting crystalline solid. Its fundamental chemical and physical properties are listed in Table 1.

Chemical Reactions

Adipic acid undergoes the usual reactions of carboxylic acids, including esterification, amidation, reduction, halogenation, salt formation, and dehydration. Because of its bifunctional nature, it also undergoes several industrially significant polymerization reactions.

Manufacture and Processing

Adipic acid historically has been manufactured predominantly from cyclohexane and, to a lesser extent, phenol. During the 1970s and 1980s, however, much research has been directed to alternative feedstocks, especially butadiene and cyclohexene, as dictated by shifts in hydrocarbon markets. All current industrial processes use nitric acid in the final oxidation stage. Growing concern with air quality may exert further pressure for alternative routes as manufacturers seek to avoid NO$_x$ abatement costs, a necessary part of processes that use nitric acid.

Since adipic acid has been produced in commercial quantities for almost 50 years, it is not surprising that many variations and improvements have been made to the basic cyclohexane process. In general, however, the commercially important processes still employ two major reaction stages. The first reaction stage is the production of the intermediates cyclohexanone and cyclohexanol, usually abbreviated as KA, KA oil, ol-one, or anone-anol. The KA (ketone, alcohol), after separation from unreacted cyclohexane (which is recycled) and reaction by-products, is

TABLE 1. PHYSICAL AND CHEMICAL
PROPERTIES OF ADIPIC ACID

Property	Value
molecular formula	C$_6$H$_{10}$O$_4$
molecular weight	146.14
melting point, °C	152.1 ± 0.3
specific gravity	1.344 at 18°C (sol)
	1.07 at 170°C (liq)
vapor pressure, Pa[a]	
solid at °C	
18.5	9.7
47.0	38.0
liquid at °C	
205.5	1300
244.5	6700
specific heat, kJ/kg·K[b]	1.590 (solid state)
	2.253 (liquid state)
	1.680 (vapor, 300°C)
heat of fusion, kJ/kg[b]	115
melt viscosity, mPa·s (= cP)	4.54 at 160°C
heat of combustion, kJ/mol[b]	2800

[a] To convert Pa to mm Hg, divide by 133.3.
[b] To convert J to cal, divide by 4.184.

then converted to adipic acid by oxidation with nitric acid. An important alternative to this use of KA is its use as an intermediate in the manufacture of caprolactam, the monomer for production of nylon-6. The latter use of KA predominates by a substantial margin on a worldwide basis, but not in the United States.

Storage, Handling, and Shipping

When dispersed as a dust, adipic acid is subject to normal dust explosion hazards. The material is an irritant, especially upon contact with the mucous membranes. Thus protective goggles or face shields should be worn when handling the material.

The material should be stored in corrosion-resistant containers, away from alkaline or strong oxidizing materials.

Economic Aspects

Adipic acid is a very large-volume organic chemical. It is one of the top 50 chemicals produced in the United States in terms of volume. Demand is highly cyclic, reflecting the automotive and housing markets especially. Prices usually follow the variability in crude oil prices. Adipic acid for nylon takes about 60% of U.S. cyclohexane production; the remainder goes to caprolactam for nylon-6, export, and miscellaneous uses.

Toxicity, Safety, and Industrial Hygiene

Adipic acid is relatively nontoxic; no OSHA PEL or NIOSH REL have been established for the material.

DARWIN D. DAVIS
DONALD R. KEMP
E.I. du Pont de Nemours & Co., Inc.

Additional Reading

Castellan, A., J.C.J. Bart and S. Cavallaro: *Catalysis Today* **9**, 237–322 (1991).

Luedeke, V.D. "Adipic Acid", in *Encyclopedia of Chemical Processing and Design*, J. McKetta and W. Cunningham, eds., Vol. 2, Marcel Dekker, Inc., New York, 1977, pp. 128–146.

Suresh, A.K., T. Sridhar and O.E. Potter: *AIChE J.* **34**(1), 55–93 (1988).

Yen Y.C. and S.Y. Wu, *Nylon-6,6, Report No. 54B, Process Economics Program*, SRI International, Menlo Park, CA., 1987, pp. 1–148.

ADRENAL CORTICAL HORMONES. The hormones elaborated by the adrenal cortex are steroidal derivatives of cyclopentanoperhydrophenanthrene related to the sex hormones. The structural formulas of the important members of this group are shown in Fig. 1. With the exception of *aldosterone*, the compounds may be considered derivatives of *corticosterone*, the first of the series to be identified and named. The C_{21} steroids derived from the adrenal cortex and their metabolities are designated collectively as *corticosteroids*. They belong to two principal groups; (1) those processing an O or OH substituent at C_{11} (*corticosterone*) and an OH group at C_{17} (*cortisone* and *cortisol*) exert their chief action on organic metabolism and are designated as *glucocorticoids*; (2) those lacking the oxygenated group at C_{17} (*desoxycorticosterone* and *aldosterone*) act primarily on electrolyte and water metabolism and are designated as *mineralocorticoids*. In humans, the chief glucocorticoid is *cortisol*. The chief mineralocorticoid is *aldosterone*.

The *glucocorticoids* are involved in organic metabolism and in the organism's response to stress. They accelerate the rate of catabolism (destructive metabolism) and inhibit the rate of anabolism (constructive metabolism) of protein. They also reduce the utilization of carbohydrate and increase the rate of gluconeogenesis (formation of glucose) from protein. They also exert a lipogenic as well as lipolytic action, potentiating the release of fatty acids from adipose tissue. In addition to these effects on the organic metabolism of the basic foodstuffs, the glucocorticoids affect the body's allergic, immune, inflammatory, antibody, anamnestic, and general responses of the organism to environmental disturbances. It is these reactions which are the basis for the wide use of the corticosteroids therapeutically. See also **Immune System and Immunochemistry**.

Aldosterone exerts its main action in controlling the water and electrolyte metabolism. Its presence is essential for the reabsorption of sodium by the renal tube, and it is the loss of salt and water which is responsible for the acute manifestations of adrenocortical insufficiency. The action of aldosterone is not limited to the kidney, but is manifested on the cells generally, this hormone affecting the distribution of sodium, potassium, water, and hydrogen ions between the cellular and extracellular fluids independently of its action on the kidney.

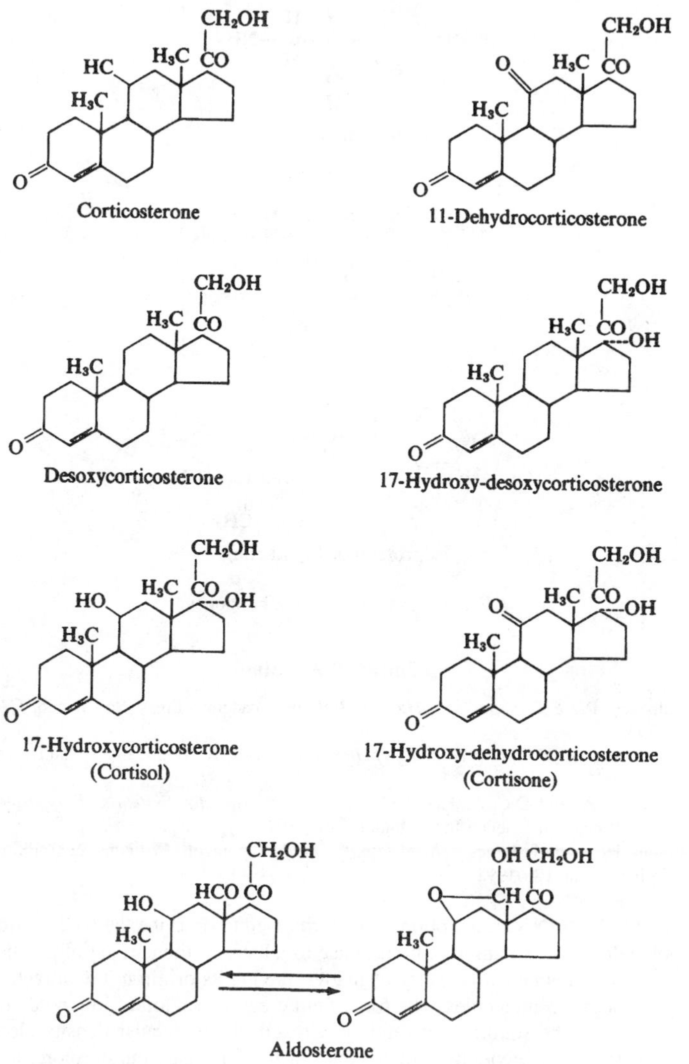

Fig. 1.   Adrenal cortical hormones

The differentiation in action of the glucocorticoids and the mineralocorticoids is not an absolute one. Aldosterone is about 500 times as effective as cortisol in its salt and water retaining activity, but is one-third as effective in its capacity to restore liver glycogen in the adrenalectomized animal. Cortisol in large doses, on the other hand, exerts a water and salt retaining action. Corticosterone is less active than cortisol as a glucocorticoid, but exerts a more pronounced mineralocorticoid action than does the latter. See also **Steroids**.

In addition to the aforementioned corticosteroidal hormones, the adrenal glands produce several oxysteroids and small amounts of testosterone and other androgens, estrogens, progesterone, and their metabolites.

ADRENAL MEDULLA HORMONES. *Adrenaline (epinephrine)* and its immediate biological precursor *noradrenaline* (*norepinephrine*, levarternol) are the principal hormones of the adult adrenal medulla. See Fig.1. Some of the physiological effects produced by adrenaline are: contraction of the dilator muscle of the pupil of the eye (*mydriasis*); relaxation of the smooth muscle of the bronchi; constriction of most small blood vessels; dilation of some blood vessels, notably those in skeletal muscle; increase in heart rate and force of ventricular contraction; relaxation of the smooth muscle of the intestinal tract; and either contraction or relaxation, or both, of uterine smooth muscle. Electrical stimulation of appropriate sympathetic (*adrenergic*) nerves can produce all the aforementioned effects with exception of vasodilation in skeletal muscle.

Noradrenaline, when administered, produces the same general effects as adrenaline, but is less potent. Isoproternol, a synthetic analogue of noradrenaline, is more potent than adrenaline in relaxing some smooth muscle, producing vasodilation and increasing the rate and force of cardiac contraction.

Noradrenaline

Adrenaline

Isoproterenol (Synthetic)

Fig. 1.   Adrenal medula hormones

Additional Reading

Dulbecco, R.: *Encyclopedia of Human Biology,* Academic Press, San Diego, CA, 1997.

Ramachandran, V.S.: *Encyclopedia of Human Behavior,* Academic Press, San Diego, CA, 1994.

Vinson, G.P., and D.C. Anderson: *Adrenal Glands Vascular System and Hypertension,* Blackwell Science Inc., Malden, MA, 1997.

Vivian, H., and T. James: *The Adrenal Gland,* Lippincott Williams & Wilkins, Philadelphia, PA, 1992.

ADSORPTION.   Adsorption is the term used to describe the tendency of molecules from an ambient fluid phase to adhere to the surface of a solid. This is a fundamental property of matter, having its origin in the attractive forces between molecules. The force field creates a region of low potential energy near the solid surface and, as a result, the molecular density close to the surface is generally greater than in the bulk gas. Furthermore, and perhaps more importantly, in a multicomponent system the composition of this surface layer generally differs from that of the bulk gas since the surface adsorbs the various components with different affinities. Adsorption may also occur from the liquid phase and is accompanied by a similar change in composition, although, in this case, there is generally little difference in molecular density between the adsorbed and fluid phases.

The enhanced concentration at the surface accounts, in part, for the catalytic activity shown by many solid surfaces, and it is also the basis of the application of adsorbents for low pressure storage of permanent gases such as methane. However, most of the important applications of adsorption depend on the selectivity, i.e., the difference in the affinity of the surface for different components. As a result of this selectivity, adsorption offers, at least in principle, a relatively straight-forward means of purification (removal of an undesirable trace component from a fluid mixture) and a potentially useful means of bulk separation.

Fundamental Principles

Forces of Adsorption. Adsorption may be classified as chemisorption or physical adsorption, depending on the nature of the surface forces. In physical adsorption the forces are relatively weak, involving mainly van der Waals (induced dipole–induced dipole) interactions, supplemented in many cases by electrostatic contributions from field–dipole or field–gradient–quadrupole interactions. By contrast, in chemisorption there is significant electron transfer, equivalent to the formation of a chemical bond between the sorbate and the solid surface. Such interactions are both stronger and more specific than the forces of physical adsorption and are obviously limited to monolayer coverage.

Selectivity. Selectivity in a physical adsorption system may depend on differences in either equilibrium or kinetics, but the great majority of adsorption separation processes depend on equilibrium-based selectivity. Significant kinetic selectivity is, in general, restricted to molecular sieve adsorbents—carbon molecular sieves, zeolites, or zeolite analogues.

Hydrophilic and Hydrophobic Surfaces. Polar adsorbents such as most zeolites, silica gel, or activated alumina adsorb water (a small polar molecule) more strongly than they adsorb organic species, and, as a result, such adsorbents are commonly called hydrophilic. In contrast, on a nonpolar surface where there is no electrostatic interaction, water is held only very weakly and is easily displaced by organics. Such adsorbents, which are the only practical choice for adsorption of organics from aqueous solutions, are termed hydrophobic.

Capillary Condensation. In a porous adsorbent the region of multilayer physical adsorption merges gradually with the capillary condensation regime, leading to upward curvature of the equilibrium isotherm at higher relative pressure. In the capillary condensation region the intrinsic selectivity of the adsorbent is lost.

Practical Adsorbents

To achieve a significant adsorptive capacity an adsorbent must have a high specific area, which implies a highly porous structure with very small micropores. Such microporous solids can be produced in several different ways. Adsorbents such as silica gel and activated alumina are made by precipitation of colloidal particles, followed by dehydration. Carbon adsorbents are prepared by controlled burn-out of carbonaceous materials such as coal, lignite, and coconut shells. The crystalline adsorbents (zeolite and zeolite analogues) are different in that the dimensions of the micropores are determined by the crystal structure and there is therefore virtually no distribution of micropore size. Although structurally very different from the crystalline adsorbents, carbon molecular sieves also have a very narrow distribution of pore size. The adsorptive properties depend on the pore size and the pore size distribution as well as on the nature of the solid surface.

Adsorption Equilibrium

Henry's Law. Like any other phase equilibrium, the distribution of a sorbate between fluid and adsorbed phases is governed by the principles of thermodynamics. Equilibrium data are commonly reported in the form of an isotherm, which is a diagram showing the variation of the equilibrium adsorbed-phase concentration or loading with the fluid-phase concentration or partial pressure at a fixed temperature. In general, for physical adsorption on a homogeneous surface at sufficiently low concentrations, the isotherm should approach a linear form, and the limiting slope in the low concentration region is commonly known as the Henry's law constant. The Henry constant is a thermodynamic equilibrium constant and the temperature dependence therefore follows the usual van't Hoff equation:

$$\lim p \to 0 \left(\frac{\delta q}{\delta p} \right) T \equiv K' = K_0' e^{-\Delta H_0/\mathrm{RT}} \tag{1}$$

in which $-\Delta H_0$ is the limiting heat of adsorption at zero coverage. Since adsorption, particularly from the vapor phase, is usually exothermic, $-\Delta H_0$ is a positive quantity and K' therefore decreases with increasing temperature.

Henry's law corresponds physically to the situation in which the adsorbed phase is so dilute that there is neither competition for surface sites nor any significant interaction between adsorbed molecules. At higher concentrations both of these effects become important and the form of the isotherm becomes more complex. The isotherms have been classified into five different types (Fig. 1). Isotherms for a microporous adsorbent are generally of type I; the more complex forms are associated with multilayer adsorption and capillary condensation.

Langmuir Isotherm. Type I isotherms are commonly represented by the ideal Langmuir model:

$$\frac{q}{q_s} = \frac{bp}{1 + bp} \tag{2}$$

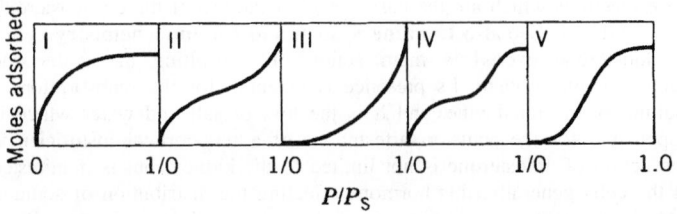

Fig. 1.   The Brunaner classification of isotherms (I–V)

where q_s is the saturation limit and b is an equilibrium constant which is directly related to the Henry constant ($K' = bq_s$).

Freundlich Isotherm. The isotherms for some systems, notably hydrocarbons on activated carbon, conform more closely to the Freundlich equation:

$$q = bp^{1/n}(n > 1.0) \qquad (3)$$

Adsorption of Mixtures. The Langmuir model can be easily extended to binary or multicomponent systems:

$$\frac{q_1}{q_{s1}} = \frac{b_1 p_1}{1 + b_1 p_1 + b_2 p_2 + \cdots}; \frac{q_2}{q_{s2}} = \frac{b_2 p_2}{1 + b_1 p_1 + b_2 p_2; + \cdots} \qquad (4)$$

Thermodynamic consistency requires $q_{s1} = q_{s2}$, but this requirement can cause difficulties when attempts are made to correlate data for sorbates of very different molecular size. For such systems it is common practice to ignore this requirement, thereby introducing an additional model parameter. This facilitates data fitting but it must be recognized that the equations are then being used purely as a convenient empirical form with no theoretical foundation.

Ideal Adsorbed Solution Theory. Perhaps the most successful general approach to the prediction of multicomponent equilibria from single-component isotherm data is ideal adsorbed solution theory. In essence, the theory is based on the assumption that the adsorbed phase is thermodynamically ideal in the sense that the equilibrium pressure for each component is simply the product of its mole fraction in the adsorbed phase and the equilibrium pressure for the pure component *at the same spreading pressure.* The theoretical basis for this assumption and the details of the calculations required to predict the mixture isotherm are given in standard texts on adsorption. Whereas the theory has been shown to work well for several systems, notably for mixtures of hydrocarbons on carbon adsorbents, there are a number of systems which do not obey this model. Azeotrope formation and selectivity reversal, which are observed quite commonly in real systems, are not consistent with an ideal adsorbed phase and there is no way of knowing *a priori* whether or not a given system will show ideal behavior.

Adsorption Kinetics

Intrinsic Kinetics. Chemisorption may be regarded as a chemical reaction between the sorbate and the solid surface, and, as such, it is an activated process for which the rate constant (k) follows the familiar Arrhenius rate law:

$$k = k_0 e^{-E/RT} \qquad (5)$$

Depending on the temperature and the activation energy (E), the rate constant may vary over many orders of magnitude.

In practice the kinetics are usually more complex than might be expected on this basis, since the activation energy generally varies with surface coverage as a result of energetic heterogeneity and/or sorbate—sorbate interaction. As a result, the adsorption rate is commonly given by the Elovich equation:

$$q = \frac{1}{k'} \ln(1 + k''t) \qquad (6)$$

where k' and k'' are temperature-dependent constants.

In contrast, physical adsorption is a very rapid process, so the rate is always controlled by mass transfer resistance rather than by the intrinsic adsorption kinetics. However, under certain conditions the combination of a diffusion-controlled process with an adsorption equilibrium constant that varies according to equation 1 can give the appearance of activated adsorption.

A porous adsorbent in contact with a fluid phase offers at least two and often three distinct resistances to mass transfer: external film resistance and intraparticle diffusional resistance. When the pore size distribution has a well-defined bimodal form, the latter may be divided into macropore and micropore diffusional resistances. Depending on the particular system and the conditions, any one of these resistances may be dominant, or the overall rate of mass transfer may be determined by the combined effects of more than one resistance. The magnitude of the intraparticle diffusional resistances, or any surface resistance to mass transfer, can be conveniently determined by measuring the adsorption or desorption rate, under controlled conditions, in a batch system.

Adsorption Column Dynamics

In most adsorption processes the adsorbent is contacted with fluid in a packed bed. An understanding of the dynamic behavior of such systems is therefore needed for rational process design and optimization. What is required is a mathematical model which allows the effluent concentration to be predicted for any defined change in the feed concentration or flow rate to the bed. The flow pattern can generally be represented adequately by the axial dispersed plug-flow model, according to which a mass balance for an element of the column yields, for the basic differential equation governing the dynamic behavior,

$$-D_L \frac{\delta^2 c_i}{\delta z^2} + \frac{\delta}{\delta z}(v c_i) + \frac{\delta c_i}{\delta t} + \left(\frac{1 - \varepsilon}{\varepsilon}\right) \frac{\delta \overline{q}_i}{\delta t} = 0 \qquad (7)$$

The term $\delta \overline{q}_i / \delta t$ represents the overall rate of mass transfer for component i (at time t and distance z) averaged over a particle. This is governed by a mass transfer rate expression which may be thought of as a general functional relationship of the form

$$\frac{\delta \overline{q}}{\delta t} = f(c_i, c_j, \cdots q_i, q_j, \cdots) \qquad (8)$$

This rate equation must satisfy the boundary conditions imposed by the equilibrium isotherm and it must be thermodynamically consistent so that the mass transfer rate falls to zero at equilibrium.

Equilibrium Theory. The general features of the dynamic behavior may be understood without recourse to detailed calculations since the overall pattern of the response is governed by the form of the equilibrium relationship rather than by kinetics. If the equilibrium isotherm is of "favorable" form (i.e., slope decreasing with increasing concentration as in Figure 1,I) the concentration front, for adsorption, will assume the form of a travelling shock wave, whereas for desorption the front will assume the form of a simple wave which spreads as it propagates through the column.

Constant Pattern Behavior. In a real system the finite resistance to mass transfer and axial mixing in the column lead to departures from the idealized response predicted by equilibrium theory. In the case of a favorable isotherm the shock wave solution is replaced by a constant pattern solution. The concentration profile spreads in the initial region until a stable situation is reached in which the mass transfer rate is the same at all points along the wave front and exactly matches the shock velocity. In this situation the fluid-phase and adsorbed-phase profiles become coincident. This represents a stable situation and the profile propagates without further change in shape—hence the term constant pattern.

Length of Unused Bed. The constant pattern approximation provides the basis for a very useful and widely used design method based on the concept of the length of unused bed (LUB). In the design of a typical adsorption process the basic problem is to estimate the size of the absorber bed needed to remove a certain quantity of the adsorbable species from the feed stream, subject to a specified limit (c') on the effluent concentration. The length of unused bed, which measures the capacity of the adsorber which is lost as a result of the spread of the concentration profile, is defined by

$$\text{LUB} = (1 - q'/q_o)L = (1 - t'/\overline{t})L \qquad (9)$$

where q' is the capacity at the break time t' and $\overline{t}$ is the stoichiometric time (see Fig. 2). The values of t', $\overline{t}$, and hence the LUB are easily determined from an experimental breakthrough curve since, by overall mass balance:

$$\overline{t} = \frac{L}{v}\left[1 + \left(\frac{1 - \varepsilon}{\varepsilon}\right)\left(\frac{q_0}{c_0}\right)\right] = \int_\infty^0 \left(1 - \frac{c}{c_0}\right) dt \qquad (10)$$

$$t' = \frac{L}{v}\left[1 + \left(\frac{1 - \varepsilon}{\varepsilon}\right)\left(\frac{q'}{c_0}\right)\right] = \int_{L'}^0 \left(1 - \frac{c}{c_0}\right) dt \qquad (11)$$

The length of column needed for a particular duty can then be found simply by adding the LUB to the length calculated from equilibrium considerations, assuming a shock concentration front.

Proportionate Pattern Behavior. If the isotherm is unfavorable (as in Fig. 1,III), the stable dynamic situation leading to constant pattern behavior can never be achieved. The equilibrium adsorbed-phase concentration then lies above rather than below the actual adsorbed-phase profile. As the mass transfer zone progresses through the column it broadens, but the limiting situation, which is approached in a long column, is simply local equilibrium at all points ($c = c^*$) and the profile therefore continues to

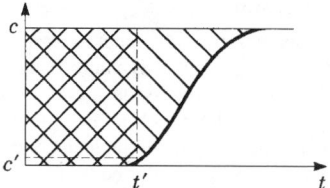

Fig. 2.   Sketch of breakthrough curve showing break time t' and the method of calculation of the stoichiometric time $\bar{t}$ and LUB. ⊠ = the integral of equation 10; ◺ = integral of equation 11.

spread in proportion to the distance traveled. This difference in behavior is important since the LUB approach to design is clearly inapplicable under these conditions.

Adsorption Chromatography. In a linear multicomponent system (several sorbates at low concentration in an inert carrier) the wave velocity for each component depends on its adsorption equilibrium constant. Thus, if a pulse of the mixed sorbate is injected at the column inlet, the different species separate into bands which travel through the column at their characteristic velocities, and at the outlet of the column a sequence of peaks corresponding to the different species is detected. Measurement of the retention time $(\bar{t})$ under known flow conditions thus provides a simple means of determining the equilibrium constant (Henry constant).

In an ideal system with no axial mixing or mass-transfer resistance, the peaks for the various components propagate without spreading. However, in any real system the peak broadens as it propagates and the extent of this broadening is directly related to the mass transfer and axial dispersion characteristics of the column. Measurement of the peak broadening therefore provides a convenient way of measuring mass-transfer coefficients and intraparticle diffusivities.

Applications

The applications of adsorbents are many and varied. They may be classified as "regenerative" and "nonregenerative". Most process applications, in which the adsorbent is used as a means of purifying or separating the components of a gas or liquid mixture are regenerative. The process operates in a cyclic manner so that the adsorbent is alternately saturated and regenerated. Nonregenerative applications include the use of adsorbents in cigarette filters, in some water purification systems, as deodorants in health care products and as desiccants in storage, packaging and dual-pane windows.

Adsorption Separation and Purification Processes. Adsorption processes can be classified according to the flow system (cyclic batch or continuous countercurrent) and the method by which the adsorbent is regenerated. The two basic flow schemes are illustrated in Figure 3. The cyclic batch scheme is simpler but less efficient. It is generally used where selectivity is relatively high. Countercurrent or simulated countercurrent schemes are more expensive in initial cost and are generally used only for difficult separations in which selectivity is limited or mass-transfer resistance is high.

The three common methods of regeneration are thermal swing, pressure swing, and displacement. The main factors governing this choice are summarized in Table 1.

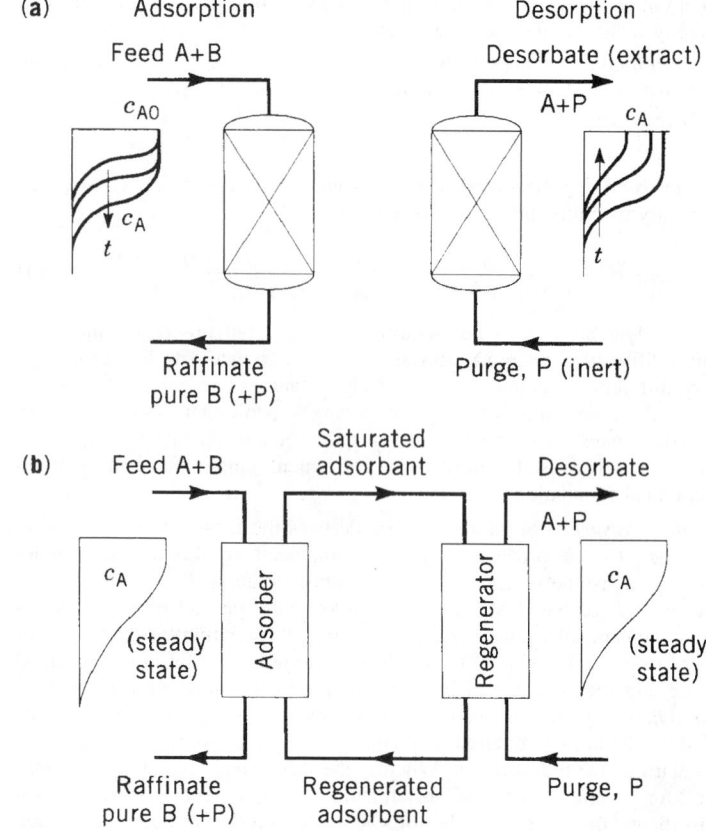

Fig. 3.   The two basic modes of operation for an adsorption process: (**a**) cyclic batch system; (**b**) continuous countercurrent system with adsorbent recirculation.

Notation

b = Langmuir equilibrium constant
c = sorbate concentration in fluid phase
c_0 = initial value of c
D_L = axial dispersion coefficient
E = activation energy
$-\Delta H_0$ = limiting heat of adsorption
K' = Henry's law constant
K_0' = preexponential factor
k = rate constant
k_0 = preexponential factor
k', k'' = constants in Elovich equation
L = bed length

Douglas M. Ruthven
University of New Brunswick, Canada

TABLE 1.   FACTORS GOVERNING CHOICE OF REGENERATION METHOD

Method	Advantages	Disadvantages
thermal swing	good for strongly adsorbed species; small change in T gives large change in q^*	thermal aging of adsorbent
	desorbate may be recovered at high concentration	heat loss means inefficiency in energy usage
		unsuitable for rapid cycling, so adsorbent cannot be used with maximum efficiency
	gases and liquids	in liquid systems the latent heat of the interstitial liquid must be added
pressure swing	good where weakly adsorbed species is required at high purity	very low P may be required
		mechanical energy more expensive than heat
	rapid cycling—efficient use of adsorbent	desorbate recovered at low purity
displacement desorption	good for strongly held species	product separation and recovery needed (choice of desorbent is crucial)
	avoids risk of cracking reactions during regeneration	
	avoids thermal aging of adsorbent	

Additional Reading

Annino, R., and R. Villalobos: *Process Gas Chromatography,* ISA, Research Triangle Park, NC, 1992.

Lang, K.R., and K. Lang: *Astrophysical Formulae: Radiation, Gas Processes and High Energy Astrophysics,* Vol. 1, Springer-Verlag Inc., New York, NY, 1998.

Levan, M.D.: *Fundamentals of Adsorption: Proceedings of the Fifth International Conference on Fundamentals of Adsorption,* Kluwer Academic Publishers, Norwell, MA, 1996.

Ruthven, D.M.: *Principles of Adsorption and Adsorption Processes,* Wiley-Interscience, New York, NY, 1984.

Staff: "Gas Process Handbook '92," in *Hydrocarbon Processing,* **85** (April 1992).

Suzuki, M.: *Adsorption Engineering,* Kodansha-Elsevier, Tokyo, 1990.

Suzuki, M.: *Fundamentals of Adsorption: Proceedings of the Fourth International Conference on Fundamentals of Adsorption,* Elsevier Science, New York, NY, 1993.

Szostak, R., *Handbook of Molecular Sieves,* Van Nostrand Reinhold, New York, NY, 1992.

Wankat, P. *Large Scale Adsorption and Chromatography,* CRC Press, Boca Raton, FL, 1986.

Yang, R.T.: *Gas Separation by Adsorption Processes,* Butterworths, New York, NY, 1997.

Yiacoumi, S.: *Kinetics of Metal Ion Adsorption from Aqueous Solutions: Models, Algorithms, and Applications,* Kluwer Academic Publishers, Norwell, MA, 1995.

ADSORPTION: GAS SEPARATION.

Gas-phase adsorption is widely employed for the large-scale purification or bulk separation of air, natural gas, chemicals, and petrochemicals (Table 1). In these uses it is often a preferred alternative to the older unit operations of distillation and absorption.

An adsorbent attracts molecules from the gas, and the molecules become concentrated on the surface of the adsorbent and are removed from the gas phase. Many process concepts have been developed to allow the efficient contact of feed gas mixtures with adsorbents to carry out desired separations and to allow efficient regeneration of the adsorbent for subsequent reuse. In nonregenerative applications, the adsorbent is used only once and is not regenerated.

Most commercial adsorbents for gas-phase applications are employed in the form of pellets, beads, or other granular shapes, typically about 1.5 to 3.2 mm in diameter. Most commonly, these adsorbents are packed into fixed beds through which the gaseous feed mixtures are passed. Normally, the process is conducted in a cyclic manner. When the capacity of the bed is exhausted, the feed flow is stopped to terminate the loading step of the process, the bed is treated to remove the adsorbed molecules in a separate regeneration step, and the cycle is then repeated.

The growth in both variety and scale of gas-phase adsorption separation processes, particularly since 1970, is due in part to continuing discoveries of new porous, high surface-area adsorbent materials (particularly molecular sieve zeolites) and, especially, to improvements in the design and modification of adsorbents. These advances have encouraged parallel inventions of new process concepts. Increasingly, the development of new

TABLE 1. COMMERCIAL ADSORPTION SEPARATIONS

Separation	Adsorbent
Gas bulk separations	
normal paraffins, isoparaffins, aromatics	zeolite
N_2/O_2	zeolite
O_2/N_2	carbon molecular sieve
CO, CH_4, CO_2, N_2, Ar, NH_3/H_2	zeolite, activated carbon
acetone/vent streams	activated carbon
C_2H_4/vent streams	activated carbon
H_2O/ethanol	zeolite
Gas purifications	
H_2O/olefin-containing cracked gas, natural gas, air, synthesis gas, etc	silica, alumina, zeolite
CO_2/C_2H_4, natural gas, etc	zeolite
organics/vent streams	activated carbon, others
sulfur compounds/natural gas, hydrogen, liquified petroleum gas (LPG), etc	zeolite
solvents/air	activated carbon
odors/air	activated carbon
NO_x/N_2	zeolite
SO_2/vent streams	zeolite
Hg/chlor—alkali cell gas effluent	zeolite

applications requires close cooperation in adsorbent design and process cycle development and optimization.

Adsorption Principles

The design and manufacture of adsorbents for specific applications involves manipulation of the structure and chemistry of the adsorbent to provide greater attractive forces for one molecule compared to another, or, by adjusting the size of the pores, to control access to the adsorbent surface on the basis of molecular size. Adsorbent manufacturers have developed many technologies for these manipulations, but they are considered proprietary and are not openly communicated. Nevertheless, the broad principles are well known.

Adsorption Forces. Coulomb's law allows calculations of the electrostatic potential resulting from a charge distribution, and of the potential energy of interaction between different charge distributions. Various elaborate computations are possible to calculate the potential energy of interaction between point charges, distributed charges, etc.

Adsorption Selectivities. For a given adsorbent, the relative strength of adsorption of different adsorbate molecules depends on the relative magnitudes of the polarizability α, dipole moment μ, and quadrupole moment Q of each. Often, just the consideration of the values of α, μ, and Q allows accurate qualitative predictions to be made of the relative strengths of adsorption of given molecules on an adsorbent or of the best adsorbent type (polar or nonpolar) for a particular separation.

Heats of Adsorption. The integral heat of adsorption is the total heat released when the adsorbate loading is increased from zero to some final value at isothermal conditions. The differential heat of adsorption δH_{iso} is the incremental change in heat of adsorption with a differential change in adsorbate loading. This heat of adsorption δH_{iso} may be determined from the slopes of adsorption isosteres (lines of constant adsorbate loading) on graphs of $\ln P$ vs $1/T$ (Fig. 1) through the Clausius-Clapeyron relationship:

$$\frac{d \ln P}{d(1/T)} = -\frac{\delta H_{iso}}{R}$$

where R is the gas constant, P the adsorbate absolute pressure, and T the absolute temperature.

Isotherm Models. *Thermodynamically Consistent Isotherm Models.* These models include both the statistical thermodynamic models and the models that can be derived from an assumed equation of state for the adsorbed phase plus the thermodynamics of the adsorbed phase.

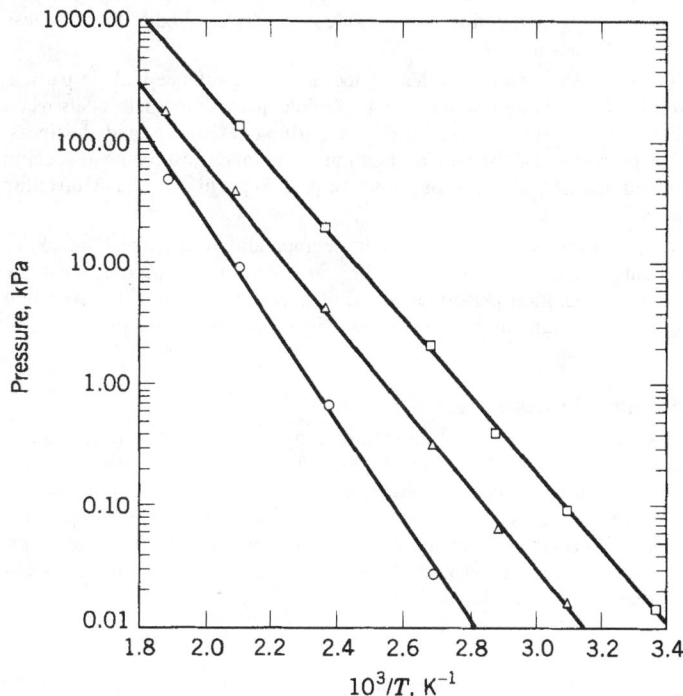

Fig. 1. Adsorption isosteres, water vapor on 4A (NaA) zeolite pellets. H_2O loading: □, 15 kg/100 kg zeolite; △, 10 kg/100 kg, ○, 5 kg/100 kg. To convert kPa to mm Hg, multiply by 7.5. Courtesy of Union Carbide

Statistical Thermodynamic Isotherm Models. These approaches were pioneered by Fowler and Guggenheim and Hill and this approach has been applied to modeling of adsorption in microporous adsorbents.

Semiempirical Isotherm Models. Some of these models have been shown to have some thermodynamic inconsistencies and should be used with due care. They include models based on the Polanyi adsorption potential (Dubinin-Radushkevich, Dubinin-Astakhov, Radke-Prausnitz, Toth, UNILAN, and BET).

Isotherm Models for Adsorption of Mixtures. Of the following models, all but the ideal adsorbed solution theory (IAST) and the related heterogeneous ideal adsorbed solution theory (HIAST) have been shown to contain some thermodynamic inconsistencies. They include Markham and Benton, the Leavitt loading ratio correlation (LRC) method, the ideal adsorbed solution (IAS) model, the heterogeneous ideal adsorbed solution theory (HIAST), and the vacancy solution model (VSM).

Adsorption Dynamics. An outline of approaches that have been taken to model mass-transfer rates in adsorbents has been given. Extensive literature exists on the interrelated topics of modeling of mass-transfer rate processes in fixed-bed adsorbers, bed concentration profiles, and breakthrough curves and the related simple design concepts of WES, WUB, and LUB for constant-pattern adsorption.

Reactions on Adsorbents. To permit the recovery of pure products and to extend the adsorbent's useful life, adsorbents should generally be inert and not react with or catalyze reactions of adsorbate molecules. These considerations often affect adsorbent selection or require limits be placed upon the severity of operating conditions to minimize reactions of the adsorbate molecules or damage to the adsorbents.

Adsorbent Principles

Principal Adsorbent Types. Commercially useful adsorbents can be classified by the nature of their structure (amorphous or crystalline), by the sizes of their pores (micropores, mesopores, and macropores), by the nature of their surfaces (polar, nonpolar, or intermediate), or by their chemical composition. All of these characteristics are important in the selection of the best adsorbent for any particular application. However, the size of the pores is the most important initial consideration because if a molecule is to be adsorbed, it must not be larger than the pores of the adsorbent.

Adsorption Properties. Not only do the more highly polar molecular sieve zeolites adsorb more water at lower pressures than do the moderately polar silica gels and alumina gels, but they also hold onto the water more strongly at higher temperatures. For the same reason, temperatures required for thermal regeneration of water-loaded zeolites are higher than for less highly polar adsorbents.

Physical Properties. Physical properties of importance include particle size, density, volume fraction of intraparticle and extraparticle voids when packed into adsorbent beds, strength, attrition resistance, and dustiness. These properties can be varied intentionally to tailor adsorbents to specific applications. See also **Adsorption: Liquid Separation**; and **Molecular Sieves**.

Deactivation. Gradual adsorbent degradation by chemical attack or physical damage commonly occurs in many uses, accompanied by declining separation performance. Allowance for this must be taken into account in design of the process and in scheduling the replacement of spent adsorbents.

Adsorption Processes

Adsorption processes are often identified by their method of regeneration. Temperature-swing adsorption (TSA) and pressure-swing (PSA) are the most frequently applied process cycles for gas separation. Purge-swing cycles and nonregenerative approaches are also applied to the separation of gases. Special applications exist in the nuclear industry. Others take advantage of reactive sorption. Most adsorption processes use fixed beds, but some use moving or fluidized beds.

Design Methods

Design techniques for gas-phase adsorption range from empirical to theoretical. Methods have been developed for equilibrium, for mass transfer, and for combined dynamic performance. Approaches are available for the regeneration methods of heating, purging, steaming, and pressure swing. Several broad reviews have been published on analytical equations

describing adsorption, on experimental adsorption processes, and on adsorption design considerations.

Future Directions

Advances in fundamental knowledge of adsorption equilibrium and mass transfer will enable further optimization of the performance of existing adsorbent types. Continuing discoveries of new molecular sieve materials will also provide adsorbents with new combinations of useful properties. New adsorbents and adsorption process will be developed to provide needed improvements in pollution control, energy conservation, and the separation of high value chemicals. New process cycles and new hybrid processes linking adsorption with other unit operations will continue to be developed.

JOHN D. SHERMAN
CARMEN M. YON
UOP

Additional Reading

Keller, G.E. II, R.A. Anderson, and C.M. Yon: in R.W. Rousseau, ed., *Handbook of Separation Process Technology,* John Wiley & Sons, Inc., New York, NY, 1987, pp. 644–696.
Barrer, R.M. *Zeolites and Clay Minerals as Adsorbents and Catalysts,* Academic Press, London, UK, 1978, pp. 164, 174, and 185.
Breck, D.W. *Zeolite Molecular Sieves—Structure, Chemistry, and Use,* John Wiley & Sons, Inc., New York, NY, 1974.
Macnair R.N. and G.N. Arons: in P.N. Cheremisinoff and F. Eleerbusch, eds., *Carbon Adsorption Handbook,* Ann Arbor Science, Ann Arbor, MI, 1978, pp. 819–859.

ADSORPTION: LIQUID SEPARATION. Liquid-phase adsorption has long been used for the removal of contaminants present at low concentrations in process streams. In most cases, the objective is to remove a specific feed component; alternatively, the contaminants are not well defined, and the objective is the improvement of feed quality defined by color, taste, odor, and storage stability.

In contrast to trace impurity removal, the use of adsorption for bulk separation in the liquid phase on a commercial scale is a relatively recent development. This article is devoted mainly to the theory and operation of these liquid-phase bulk adsorptive separation processes.

Adsorbate–Adsorbent Interactions

An adsorbent can be visualized as a porous solid having certain characteristics. When the solid is immersed in a liquid mixture, the pores fill with liquid, which at equilibrium differs in composition from that of the liquid surrounding the particles. These compositions can then be related to each other by enrichment factors that are analogous to relative volatility in distillation. The adsorbent is selective for the component that is more concentrated in the pores than in the surrounding liquid.

A significant advantage of adsorbents over other separative agents lies in the fact that favorable equilibrium-phase relations can be developed for particular separations; adsorbents can be produced that are much more selective in their affinity for various substances than are any known solvents. This selectivity is particularly true of the synthetic crystalline zeolites containing exchangeable cations.

An example of unique selectivity is provided by the use of 5A molecular sieves for the separation of linear hydrocarbons from branched and cyclic types. In this system only the linear molecules can enter the pores; others are completely excluded because of their larger cross section. Thus the selectivity for linear molecules with respect to other types is infinite. In the more usual case, all the feed components access the selective pores, but some components of the mixture are adsorbed more strongly than others. A selectivity between the different components that can be used to accomplish separation is thus established.

Practical Adsorbents

The search for a suitable adsorbent is generally the first step in the development of an adsorption process. A practical adsorbent has four primary requirements: selectivity, capacity, mass transfer rate, and long-term stability. The requirement for adequate adsorptive capacity restricts the choice of adsorbents to microporous solids with pore diameters ranging from a few tenths to a few tens of nanometers.

Traditional adsorbents such as silica, SiO_2; activated alumina, Al_2O_3; and activated carbon, C, exhibit large surface areas and micropore volumes.

TABLE 1. MOLECULAR SIEVE PORE STRUCTURES

Common name	Ring size, number of atoms	Free aperture, nm	Pore structure	Formula
faujasite	12	0.74	3-D	$(Ca, Mg, Na_2, K_2)_{29.5}[(AlO_2)_{59}(SiO_2)_{133}]\cdot235H_2O$
mordenite	8	0.29×0.57	1-D	$Na_8[(AlO_2)_8(SiO_2)_{40}]\cdot24H_2O$
	12	0.67×0.7	1-D	
L	12	0.71	1-D	$K_9[(AlO_2)_9(SiO_2)_{27}]\cdot22H_2O$
ZSM-5	10	0.54×0.56	1-D	$(Na, TPA^a)_3 [(AlO_2)_3(SiO_2)_{93}\cdot16H_2O]$
	10	0.51×0.56	1-D	
Erionite	8	0.36×0.52	2-D	$(Ca, Mg, Na_2, K_2)_{4.5}[(AlO_2)_9(SiO_2)_{27}]\cdot27H_2O$
A	8	0.42	3-D	$Na_{12} [(AlO_2)_{12}(SiO_2)_{12}]\cdot27H_2O$

[a] TPA = tetrapropylammonium.

The surface chemical properties of these adsorbents make them potentially useful for separations by molecular class. However, the micropore size distribution is fairly broad for these materials. This characteristic makes them unsuitable for use in separations in which steric hindrance can potentially be exploited.

In contrast to these adsorbents, zeolites offer increased possibilities for exploiting molecular-level differences among adsorbates. Zeolites are crystalline aluminosilicates containing an assemblage of SiO_4 and AlO_4 tetrahedra joined together by oxygen atoms to form a microporous solid, which has a precise pore structure. Nearly 40 distinct framework structures have been identified to date. Table 1 and Figure 1 summarizes some of those structures that have been widely used in the chemical industry. The versatility of zeolites lies in the fact that widely different adsorptive properties may be realized by the appropriate control of the framework structure, the silica-to-alumina ratio (Si/Al), and the cation form.

Commercial Processes

Industrial-scale adsorption processes can be classified as batch or continuous. In a batch process, the adsorbent bed is saturated and regenerated in a cyclic operation. In a continuous process, a countercurrent staged contact between the adsorbent and the feed and desorbent is established by either a true or a simulated recirculation of the adsorbent. The efficiency of an adsorption process is significantly higher in a continuous mode of operation than in a cyclic batch mode. For difficult separations, batch operation may require 25 times more adsorbent inventory and twice the desorbent circulation rate than does a continuous operation. In addition, in a batch mode, the four functions of adsorption, purification, desorption, and displacement of the desorbent from the adsorbent are inflexibly linked, whereas a continuous mode allows more degrees of freedom with respect to these functions, and thus a better overall operation.

Continuous Countercurrent Processes

The need for a continuous countercurrent process arises because the selectivity of available adsorbents in a number of commercially important separations is not high. In the p-xylene system, for instance, if the liquid around the adsorbent particles contains 1% p-xylene, the liquid in the pores contains about 2% p-xylene at equilibrium. Therefore, one stage of contacting cannot provide a good separation, and multistage contacting must be provided in the same way that multiple trays are required in fractionating materials with relatively low volatilities.

Since the 1960s the commercial development of continuous countercurrent processes has been almost entirely accomplished by using a flow scheme that simulates the continuous countercurrent flow of adsorbent and process liquid without the actual movement of the adsorbent. The idea of a simulated moving bed (SMB) can be traced back to the Shanks system for leaching soda ash.

Such a concept was originally used in a process developed and licensed by UOP under the name UOP Sorbex. The extent of commercial of Sorbex processes is shown in Table 2. Other versions of the SMB

system are also used commercially. Toray Industries built the Aromax process for the production of p-xylene. Illinois Water Treatment and Mitsubishi have commercialized SMB processes for the separation of fructose from dextrose.

Cyclic-Batch Processes

Continuous processes have wide application in different areas of the chemical industry. The separation efficiency of a continuous process is generally higher than that of a batch or cyclic-batch process. However, in some applications the cyclic-batch process may be preferred because of the complexity of design and the difficulty of controlling the continuous processes. Examples of commercial cyclic-batch adsorption processes operating in liquid phase include the UOP methanol recovery (UOP MRU) and oxygenate removal (UOP ORU) processes, which separate oxygenates from C_4 hydrocarbons; the UOP Cyclesorb process, which separates fructose from glucose; and ion-exclusion processes for recovering sucrose from molasses.

Liquid Chromatography

Conventional liquid chromatography has not attained great commercial significance in the area of large-scale bulk separations from the liquid phase. In analytical chromatography, the primary objective is to maximize the resolution between two components subject to some restrictions on the maximum time of elution. As a result, the feed pulse loading is minimized, and the number of theoretical plates is maximized. In preparative chromatography, the objective is to maximize production rate as well as reduce capital and operating costs at a given separation efficiency. The adsorption column is therefore commonly run under overload conditions with a finite feed pulse width. The choice of operating conditions for preparative chromatography has been discussed. In production chromatography, the optimal pulse sequence occurs when the successive pulses of feed are introduced at intervals such that the feed components are just resolved both within a given sample and between adjacent samples.

Outlook

Liquid adsorption processes hold a prominent position in several applications for the production of high purity chemicals on a commodity scale. Many of these processes were attractive when they were first introduced to the industry and continue to increase in value as improvements in adsorbents, desorbents, and process designs are made. The UOP Parex process alone has seen four generations of adsorbent and four generations of desorbent.

The value of many chemical products from pesticides to pharmaceuticals to high performance polymers, is based on unique properties of a particular isomer from which the product is ultimately derived. Often the purity requirement for the desired product includes an upper limit on the content of

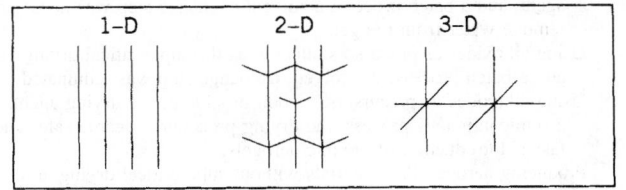

Fig. 1.   Schematic diagram of molecular sieve pore structure. See Table 1

TABLE 2. UOP SORBEX PROCESSES FOR COMMODITY CHEMICALS

UOP processes	Separation	Licensed units
Parex	p-xylene from C_8 aromatics	53
Molex	n-paraffins from branched and cyclic hydrocarbons	33
Olex	olefins from paraffins	6
Cymex	p- or m-cymene from cymene isomers	1
Cresex	p- or m-cresol from cresol isomers	1
Sarex	fructose from dextrose plus polysaccharides	5
Total		99

one or more of the other isomers. This separation problem is a complicated one, but one in which adsorptive separation processes offer the greatest chances for success.

STANLEY A. GEMBICKI
ANIL R. OROSKAR
JAMES A. JOHNSON
UOP

Additional Reading

Mantell, C.L. *Adsorption,* 2nd Edition, McGraw-Hill, Inc., New York, NY, 1951.

Broughton, D.B. *Chem. Eng. Prog.* **64**, 60 (1968).

Breck, D.W. *Zeolite Molecular Sieves,* John Wiley & Sons, Inc., New York, NY, 1974.

Ruthven, D.M. *Principles of Adsorption and Adsorption Processes,* John Wiley & Sons, Inc., New York, NY, 1984.

AERATION.   A process of contacting a liquid with air, often for the purpose of releasing other dissolved gases, or for increasing the quantity of oxygen dissolved in the liquid. Aeration is commonly used to remove obnoxious odors or disagreeable tastes from raw water. The principle of aeration is also used in the treatment of sewage by a method known as the activated sludge process. The sewage is allowed to flow into an aeration tank where it is mixed with a predetermined volume of sludge. Compressed air is introduced which agitates the mixture and furnishes oxygen that is necessary for certain biological changes. Sewage may also be aerated by mechanically actuated paddles that rotate the liquid and constantly bring a fresh surface in contact with the atmosphere.

Aeration is of importance in the fermentation industries. In the manufacture of baker's yeast, penicillin, and other antibiotics, an adequate air supply is required for optimum yields in certain submerged fermentation processes.

Aeration can be accomplished by allowing the liquid to fall in a thin film or to be sprayed in the form of droplets in air at atmospheric pressure; or the air, under pressure, may be bubbled into the liquid by means of a sparger, or other device that creates thousands of small bubbles, thus providing maximum contact area between the air and the liquid.

AEROGELS.   Aerogels are solid materials that are so porous that they contain mostly air. Almost all applications of aerogels are based on the unique properties associated with a highly porous network. Envision an aerogel as a sponge consisting of many interconnecting particles which are so small and so loosely connected that the void space in the sponge, the pores, can make up over 90% of its volume. The ability to prepare materials of such low density, and perhaps more importantly, to vary the density in a controlled manner, is indeed what make aerogels attractive in many applications.

Sol–Gel Chemistry

Inorganic Materials.   Sol–gel chemistry involves first the formation of a sol, which is a suspension of solid particles in a liquid, then of a gel, which is a diphasic material with a solid encapsulating a solvent. A detailed description of the fundamental chemistry is available in the literature. The chemistry involving the most commonly used precursors, the alkoxides $(M(OR)_m)$, can be described in terms of two classes of reactions:

Hydrolysis $-M-OR + H_2O \longrightarrow -M-OH + ROH$

Condensation $-M-OH + XO-M- \longrightarrow -M-O-M- + XOH$

where X can either be H or R, an alkyl group

The important feature is that a three-dimensional gel network comes from the condensation of partially hydrolyzed species. Thus, the microstructure of a gel is governed by the rate of particle (cluster) growth and their extent of crosslinking or, more specifically, by the *relative* rates of hydrolysis and condensation.

Acid- and base-catalyzed gels yield micro- (pore width less than 2 nm) and meso-porous (2–50 nm) materials, respectively, upon heating. An acid-catalyzed gel which is weakly branched and contains surface functionalities that promote further condensation collapses to give micropores. This example highlights a crucial point: *the initial microstructure and surface functionality of a gel dictates the properties of the heat-treated product.*

Besides pH, other preparative variables that can affect the microstructure of a gel, and consequently, the properties of the dried and heat-treated product include water content, solvent, precursor type and concentration, and temperature.

In the preparation of a two-component system, the minor component can either be a network modifier or a network former. In the latter case, the distribution of the two components, or mixing, at a molecular level is governed by the *relative* precursor reactivity. Qualitatively good mixing is achieved when two precursors have similar reactivities. When two precursors have dissimilar reactivities, the sol–gel technique offers several strategies to prepare well-mixed two-component gels. Two such strategies are prehydrolysis, which involves prereacting a less reactive precursor and chemical modification, which involves slowing down a more reactive precursor. The ability to control microstructure *and* component mixing is what sets sol–gel apart from other methods in preparing multicomponent solids.

Organic Materials.   The sol–gel chemistry of organic materials is similar to that of inorganic materials. The first organic aerogel was prepared by the aqueous polycondensation of resorcinol with formaldehyde using sodium carbonate as a base catalyst.

Resorcinol–formaldehyde gels are dark red in color and do not transmit light. The preparation of melamine–formaldehyde gels, which are colorless and transparent, is also aqueous-based. Since water is deleterious to a gel's structure at high temperatures and immiscible with carbon dioxide (a commonly used supercritical drying-agent), these gels cannot be supercritically dried without a tedious solvent-exchange step. In order to circumvent this problem, an alternative synthetic route of organic gels that is based upon a phenolic–furfural reaction using an acid catalyst has been developed. The solvent-exchange step is eliminated by using alcohol as a solvent. The phenolic–furfural gels are dark brown in color.

Carbon aerogels can be prepared from the organic gels mentioned above by supercritical drying with carbon dioxide and a subsequent heat-treating step in an inert atmosphere.

Despite these changes, the carbon aerogels are similar in morphology to their organic precursors, underscoring again the importance of structural control in the gelation step. Furthermore, changing the sol–gel conditions can lead to aerogels that have a wide range of physical properties.

Inorganic–Organic Hybrids.   One of the fastest growing areas in sol–gel processing is the preparation of materials containing both inorganic and organic components, because many applications demand special properties that only a combination of inorganic and organic materials can provide. In this regard, sol–gel chemistry offers a real advantage because its mild preparation conditions do not degrade organic polymers, as would the high temperatures that are associated with conventional ceramic processing techniques. The voluminous literature on the sol–gel preparation of inorganic–organic hybrids can be found in several recent reviews and the references therein.

Preparation and Manufacturing

Supercritical Drying.   The development of aerogel technology from the original work of Kistler to about late 1980s has been reviewed. Over this period, supercritical drying was the dominant method in preparing aerogels. Several advances, summarized in Table 1, have made possible the relatively safe supercritical drying of aerogels in a matter of hours. In recent years, the challenge has been to produce aerogel-like materials without using supercritical drying at all in an attempt to deliver economically competitive products.

Supercritical drying should be considered as part of the aging process, during which events such as condensation, dissolution, and reprecipitation

TABLE 1. IMPORTANT DEVELOPMENTS IN THE PREPARATION OF AEROGELS

Decade	Developments
1930	Using inorganic salts as precursors, alcohol as the supercritical drying agent, and a batch process; a solvent-exchange step was necessary to remove water from the gel.
1960	Using alkoxides as precursors, alcohol as the supercritical drying agent, and a batch process; the solvent exchange step was eliminated.
1980	Using alkoxides as precursors, carbon dioxide as the drying agent, and a semicontinuous process; the drying procedure became safer and faster. Introduction of organic aerogels.
1990	Producing aerogel-like materials without supercritical drying at all; preparation of inorganic–organic hybrid materials.

can occur. The extent to which a gel undergoes aging during supercritical drying depends on the structure of the initial gel network. A higher drying temperature changes the particle structure of base-catalyzed silica aerogels but not that of acid-catalyzed ones. Gels that have uniform-sized pores can withstand the capillary forces during drying better because of a more uniform stress distribution. Such gels can be prepared by a careful manipulation of sol–gel parameters such as pH and solvent or by the use of so-called drying control chemical additives (DCCA).

Carbon dioxide is the drying agent of choice if the goal is to stabilize kinetically constrained structure, and materials prepared by this low-temperature route are referred to by some people as *carbogels*. In general, carbogels are also different from aerogels in surface functionality, in particular hydrophilicity.

However, even with carbon dioxide as a drying agent, the supercritical drying conditions can affect the properties of a product. Other important drying variables include the path to the critical point, composition of the drying medium, and depressurization.

For some applications it is desirable to prepare aerogels as thin films that are either self-supporting or supported on another substrate. All common coating methods such as dip coating, spin coating, and spray coating can be used to prepare gel films.

In all the processes discussed above, the gelation and supercritical drying steps are done sequentially. Recently a process that involves the direct injection of the precursor into a strong mold body followed by rapid heating for gelation and supercritical drying to take place was reported. By eliminating the need of forming a gel first, this entire process can be done in less than three hours per cycle. Besides saving time, gel containment minimizes some stresses and makes it possible to produce near net-shape aerogels and precision surfaces. The optical and thermal properties of silica aerogels thus prepared are comparable to those prepared with conventional methods.

Ambient Preparations. Economic and safety considerations have provided a strong motivation for the development of techniques that can produce aerogel-like materials at ambient conditions, i.e., without supercritical drying. The strategy is to minimize the deleterious effect of capillary pressure which is given by:

$$P = 2\sigma \cos(\theta)/r$$

where P is capillary pressure, σ is surface tension, θ is the contact angle between liquid and solid, and r is pore radius.

The equation above suggests that one approach would be to use a pore liquid that has a low surface tension. In fact, with a pore liquid that has a sufficiently small surface tension, ambient pressure acid catalyzed aerogels with comparable pore volume and with bulk density to those prepared with supercritical drying (see Fig. 1) have been produced.

For base-catalyzed silica gels, it has been shown that modifying the surface functionality is an effective way to minimize drying shrinkage. In particular, surface hydroxyl groups, the condensation of which leads to pore collapse, can be "capped off" via reactions with organic groups such as tetraethoxysilane and trimethylchlorosilane. This surface modification approach (also referred to as surface derivatization), initially developed for bulk specimens, has recently been applied to the preparation of thin films.

In changing surface hydroxyls into organosilicon groups, surface modification has an additional advantage of producing hydrophobic gels. This feature, namely the immiscibility of surface-modified gel with water, has led to the development of a rapid extractive drying process shown in Figure 2. This ambient pressure process offers improved heat transfer rates and, in turn, greater energy efficiency without compromising desirable aerogel properties.

Another approach to produce aerogels without supercritical drying is freeze drying, in which the liquid–vapor interface is eliminated by freezing a wet gel into a solid and then subliming the solvent to form what is known as a *cryogel*. The limited data available on freeze drying suggest that it might not be as attractive as the above ambient approaches in producing aerogels on a commercial scale.

Properties

Table 2 summarizes the key physical properties of silica aerogels. A range of values is given for each property because the exact value is dependent on the preparative conditions and, in particular, on density.

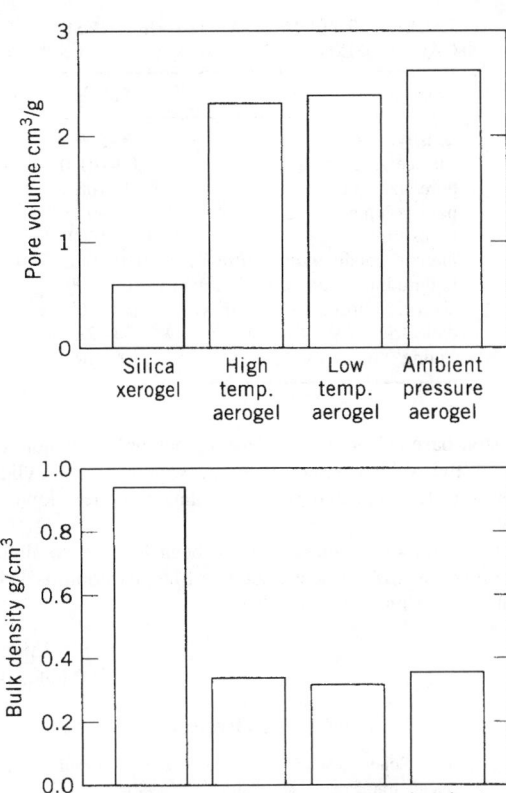

Fig. 1. Comparison of physical properties of silica xerogels and aerogels. Note the similar properties of the aerogels prepared with and without supercritical drying. Reproduced from C. J. Brinker and co-workers, *Mat. Res. Soc. Symp. Proc.* **271**, 567 (1992). Courtesy of the Materials Research Society

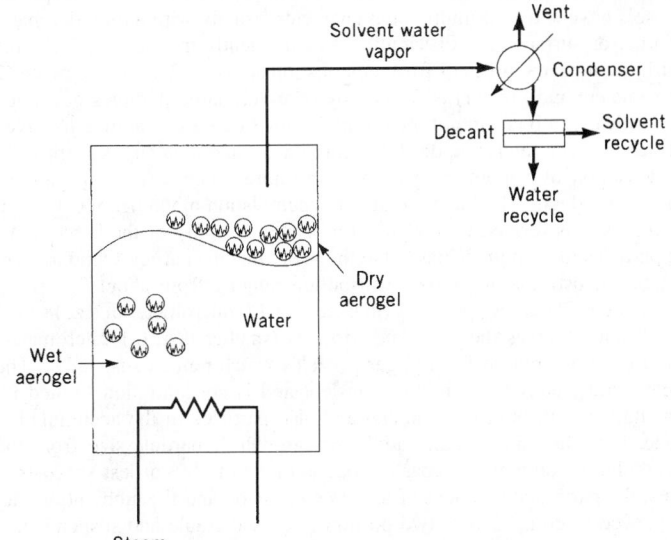

Fig. 2. Schematic diagram of an extractive drying process that produces aerogels at ambient pressure. Reproduced from D. M. Smith and co-workers, *Mat. Res. Soc. Symp. Proc.* **431**, 291 (1996). Courtesy of the Materials Research Society

Applications

Aerogels are used in thermal insulation, catalysis, detection of high energy particles, piezoceramic, ultrasound transducers, integrated circuits, and as dehydrating agents.

Summary

It has been hailed as the world's lightest solid, and a near-magic material. Yet more than 70 years after chemists first discovered the extraordinary form of matter called "aerogels," it's been used almost exclusively by space researchers and in niche markets.

TABLE 2. TYPICAL VALUES OF PHYSICAL PROPERTIES OF SILICA AEROGELS

Property	Values
density, kg/m^3	3–500
surface area, m^2/g	800–1000
pore sizes, nm	1–100
pore volume, cm^3/g	3–9
porosity, %	75–99.9
thermal conductivity, $W/(m \cdot K)$	0.01–0.02
longitudinal sound velocity, m/s	100–300
acoustic impedance, $kg/(m^2 \cdot s)$	10^3–10^6
dielectric constant	1–2
Young's modulus, N/m^2	10^6–10^7

Now, Boston-based Cabot Corp. is rolling out the first major commercial application of the silicon-based material: window and skylight panels that use aerogels for heat and sound insulation while allowing light to pass through.

While large commercial markets have been long in coming, aerogels have been used in NASA projects, such as Mars exploration vehicles and a space probe capturing comet-tail dust.

EDMUND I. KO
Carnegie Mellon University

Additional Reading

Brinker, C.J. and G.W. Scherer: *Sol-Gel Science: The Physics and Chemistry of Sol-Gel Processing,* Academic Press, New York, NY, 1990.
Fricke, J.: *Sci. Amer.* **256**(5), 92 (1988).
Livage, J.M. Henry, and C. Sanchez: *Prog. Solid State Chem.* **18**, 259 (1988).
Schneider, M. and A. Baiker: *Catal. Rev. - Sci. Eng.* **37**(4), 515 (1995).

AEROSOLS. A colloidal system in which a gas, frequently air, is the continuous medium, and particles of solids or liquid are dispersed in it. Aerosol thus is a common term used in connection with air pollution control. Studies of the particle size distribution of atmospheric aerosols have shown a multimodal character, usually with a bimodal mass, volume, or surface area distribution and frequently trimodal surface area distribution near sources of fresh combustion aerosols. The coarse mode (2 micrometers and greater) is formed by relatively large particles generated mechanically or by evaporation of liquid from droplets containing dissolved substances. The nuclei mode (0.03 micrometer and smaller) is formed by condensation of vapors from high-temperature processes, or by gaseous reaction products. The intermediate or accumulation mode (from 0.1 to 1.0 micrometer) is formed by coagulation of nuclei. Study of the behavior of the particles in each mode has led to the belief that the particles tend to form a stable aerosol having a size distribution ranging from about 0.1 to 1.0 micrometer. The larger particles (in excess of 1.0 micrometer in size) settle, or fall out, whereas the very fine particles (smaller than 0.1 micrometer) tend to agglomerate to form larger particles which remain suspended. The nuclei mode tends to be highly transient and is concentration limited by coagulation with both other nuclei and also particles in the accumulation mode. It further appears that additional growth of particle size from the accumulation mode to the coarse mode is limited to 5% or less (by mass). Thus, the particulate content of a source emission and the ambient air can be viewed as composed of two portions, i.e., settle-able and suspended.

Both settle-able and suspended atmospheric particles have deleterious effects upon the environment. The settle-able particles can affect health if assimilated and also can cause adverse effects on materials, crops, and vegetation. Further, such particles settle out in streams and upon land where soluble substances, sometimes including hazardous materials, are dissolved out of the particles and thus become pollutants of soils and surface and ground waters. Suspended atmospheric particulate matter has undesirable effects on visibility and, if continuous and of sufficient concentration, possible modifying effects on the climate. Importantly, it is particles within a size range from 2 to 5 micrometers and smaller that are considered most harmful to health because particles of this size tend to penetrate the body's defense mechanisms and reach most deeply into the lungs.

The term aerosol is also applied to a form of packaging in which a gas under pressure, or a liquefied gas that has a pressure greater than atmospheric pressure at ordinary temperatures, is used to spray a liquid. The result of the spraying process is to produce a mist of small liquid droplets in air, although not necessarily a stable colloidal system. Numerous products, such as paints, clear plastic solutions, fire-extinguishing compounds, insecticides, and waxes and cleaners, are packaged in this fashion for convenience. Food products, such as topping and whipped cream, also are packaged in aerosol cans.

For a number of years, chlorofluorocarbons were the most popular source of pressure for these cans. Because of concern in recent years over the reactions of chlorofluorocarbons in the upper atmosphere of the earth that appear to be leading to a deterioration of the ozone layer, some countries have banned their use in aerosol cans. Manufacturers have turned to other gases or to conveniently operated hand pumps. See also **Colloid Systems**; and **Pollution (Air)**.

Additional Reading

Hinds, W.W.: *Aerosol Technology: Properties, Behavior, and Measurement of Airborne Particles,* John Wiley & Sons Inc., New York, NY, 1999.
Spurny, K.R.: *Analytical Chemistry of Aerosols,* CRC Press LLC, Boca Raton, FL, 1999.
Willeke, K. and P.A. Baron: *Aerosol Measurement: Particles, Techniques, and Applications,* John Wiley & Sons Inc., New York, NY, 1997.

AFFINITY. The tendency of an atom or compound to react or combine with atoms or compounds of different chemical constitution. For example, paraffin hydrocarbons were so named because they are quite unreactive, the word *paraffin* meaning "very little affinity." The hemoglobin molecule has a much greater affinity for carbon monoxide than for oxygen. The free energy decrease is a quantitative measure of chemical affinity.

AGAR. Sometimes called *agar-agar,* this is a gelatine-like substance which is prepared from various species of red algae growing in Asiatic waters. The prepared product appears in the form of cakes, coarse granules, long shreds, or in thin sheets. It is used extensively alone or in combination with various nutritive substances, as a medium for culturing bacteria and various fungi. See also **Gums and Mucilages**.

AGATE. Agate is a variety of chalcedony, whose variegated colors are distributed in regular bands or zones, in clouds or in dendritic forms, as in moss agate.

The banding is often very delicate with parallel lines of different colors, sometimes straight, sometimes undulating or concentric. The parallel bands represent the edges of successive layers of deposition from solution in cavities in rocks that generally conform to the shape of the enclosing cavity.

As agate is an impure variety of quartz it has the same physical properties as that mineral. It is named from the river Achates in Sicily where it has been known from the time of Theophrastus.

Agate is found in many localities; India, Brazil, Uruguay, and Germany are notable for fine specimens.

Onyx is a variety of agate in which the parallel bands are perfectly straight and can be used for the cutting of cameos. Sardonyx has layers of dark reddish-brown carnelian alternating with light and dark colored layers of onyx.

See also **Chalcedony**; and **Quartz**.

AGENT ORANGE. Common name for a 50–50% mixture of the herbicides 2,4,5-T and 2,4-D, once widely used by the military as a defoliant. The mixture contains dioxin as a contaminant. See also **Dioxin**; and **Herbicides**.

AGGLOMERATION. This term connotes a gathering together of smaller pieces or particles into larger size units. This is a very important operation in the process industries and takes a number of forms. Specific advantages of agglomeration include: increasing the bulk density of a material; reducing storage-space needs; improving the handling qualities of bulk materials; improving heat-transfer properties; improving control over solubility; reducing material loss and lessening of pollution, particularly of dust; converting waste materials into a more useful form and reducing labor costs because of resulting improved handling efficiency.

The principal means used for agglomerating materials include (1) compaction, (2) extrusion, (3) agitation, and (4) fusion.

Tableting is an excellent example of compaction. In this operation, loose material, such as a powder, is compressed between two opposing surfaces, or compacted in a die or cavity. Some tableting machines use the action of two opposing plungers that operate within a cavity. Resulting tablets may range from $\frac{1}{8}$ to 4 inches (3 millimeters to 10 centimeters) in diameter. Uniformity and dimensional precision are outstanding. Numerous pharmaceutical products are formed in this manner, as well as some metallic powders and industrial catalysts.

Pellet mills exemplify the use of extrusion. In some designs the charge material is forced out of cylindrical or other shaped holes located on the periphery of a cylinder within which rollers and spreaders force the bulk materials through the openings. A knife cuts the extruded pellets to length as they are forced through the dies.

The *rolling drum* is the simplest form of aggregation using agitation. Aggregates are formed by the collision and adherence of the bulk particles in the presence of a liquid binder or wetting agent to produce what essentially is a "snowball" effect. As the operation continues, the spheroids become larger. The strength and hardness of the enlarged particles are determined by the binder and wetting agent used. The operation is followed by screening, with recycling of the fines.

The *sintering process* utilizes fusion as a means of size-enlargement. This process, used mainly for ores and minerals and some powdered metals, employs heated air that is passed through a loose bed of finely ground material. The particles partially fuse together without the assistance of a binder. Sintering frequently is accompanied by the volatilization of impurities and the removal of undesired moisture.

The *spray-type* agglomerator utilizes several principles. Loosely bound clusters or aggregates are formed by the collision and coherence of the fine particles and a liquid binder in a turbulent stream. The mixing vessel consists of a vertical tank, around whose lower periphery are mounted spray nozzles for introduction of the liquid. A suction fan draws air through the bottom of the tank and creates an updraft within the mixing vessel. Materials spiral downward through the mixing chamber, where they meet the updraft and are held in suspension near the portion of the vessel where the liquids are injected. The liquids are introduced in a fine mist. Individual droplets gather the solid particles until the resulting agglomerate overcomes the force of the updraft and falls to the bottom of the vessel as finished product.

Additional Reading

Elimelech, M., J. Gregory, R.A. Williams, and X. Jia: *Particle Deposition and Aggregation: Measurement, Modelling and Simulation,* Butterworth-Heinemann Inc., Woburn, MA, 1995.
Pietsch, W.: *Size Enlargement by Agglomeration,* John Wiley & Sons Inc., New York, NY, 1997.

AGGLUTINATION. The combination or aggregation of particles of matter under the influence of a specific protein. The term is usually restricted to antigen-antibody reactions characterized by the clumping together of visible cells such as bacteria orerythrocytes.

See also **Aggregation.**

AGGLUTININ. One of a class of substances found in blood to which certain foreign substances or organisms have been added or admixed. As the name indicates, agglutinins have the characteristic property of causing agglutination, especially of the foreign substances or organisms responsible for their formation.

See also **Agglutination.**

AGGREGATE. The solid conglomerate of inert particles which are cemented together to form concrete are called aggregate. A well-graded mixture of fine and coarse aggregates is used to obtain a workable, dense mix. The aggregate may be classed as fine or coarse depending upon the size of the individual particles. The specifications for the concrete on any project will give the limiting sizes that will distinguish between the two classifications. Fine aggregate generally consists of sand or stone screenings while crushed stone, gravel, slag or cinders are used for the coarse aggregate. The aggregates should be strong, clean, durable, chemically inert, free of organic matter, and reasonably free from flat and elongated particles since the strength of the concrete is dependent upon the quality of the aggregates as well as the matrix of cementing material. See also **Concrete.**

AGGREGATION. This overall operation may be considered to include the more specific designations of *agglutination, coagulation,* and *flocculation.* These terms imply some change in the state of dispersion of sols or of macromolecules in solution.

Agglutination generally refers to the aggregation of particulate matter mediated by an interaction with a specific protein. More specifically, the term refers to antigen-antibody reactions characterized by a clumping together of visible cells, such as bacteria or erythrocytes. The distinguishing feature of these reactions appears to be the presence of special areas where the orientation of active groups permits specific interaction of antigen with antibody. There is evidence that the forces involved may include hydrogen bonding, electrostatic attraction, and London-van der Waals forces. The clumping of the fat globules in milk has been described as an agglutination by the proteins of the euglobulin fraction. When the fat globules are dispersed in a dilute salt solution, the addition of this protein fraction induces normal creaming. Although this action is nonspecific, the same protein fraction causes agglutination of certain bacteria when they are added to milk.

Coagulation of a hydrophobic sol may be brought about by the addition of small amounts of electrolytes. Coagulation may be rapid, occurring in seconds, or slow, requiring months for completion. The resultant coagula contain relatively small proportions of the dispersion medium, in contrast with jellies formed from hydrophilic systems. Sometimes it is found that what at first appeared to be a homogeneous liquid becomes turbid and distinctly nonhomogeneous. Systems which are intermediate between true hydrophobic sols and hydrophilic sols are encountered frequently, so that in common usage the word *coagulation* is applied to such diverse phenomena as the clotting of blood by thrombin or the clotting of milk by rennin.

Flocculation is generally considered synonymous with coagulation, but is widely used in connection with certain kinds of applications. If one considers only hydrophilic systems, it is apparent that an important factor in flocculation is the solvation of the particles, despite the common presence of an electric charge. Since stability appears to depend upon solute-solvent interactions and solubility properties, flocculation can frequently be brought about by either of two pathways. The addition of salts may compress the double layer, leaving the macromolecules stabilized by a diffuse solvation shell. The addition of alcohol or acetone will dehydrate the particles, leading to instability and flocculation. Alternatively, the alcohol or acetone may be added first, which will convert the particle to one of hydrophobic character stabilized largely by the electric double layer. Such a sol can be coagulated by the addition of small amounts of electrolytes.

See also **Colloid Systems**

AIChE. The American Institute of Chemical Engineers was founded in Philadelphia, Pennsylvania, in 1908 to serve what, at that time, was an emerging new engineering discipline, *chemical engineering.* The general aim of the Institute is to promote excellence in the development and practice of chemical engineering through semiannual district meeting and an annual national meeting for the presentation and discussion of technical papers and the exhibition of equipment and materials used in chemical engineering projects. The Institute publishes several periodicals, including the *AIChE Journal, International Chemical Engineering,* and *Chemical Engineering Progress.* Technical divisions of the AIChE include Computer and Systems Technology, Engineering and Construction Contracting, Environmental Technology, Food, Pharmaceutical and Bioengineering, Forest Products, Fuels and Petrochemicals, Heat Transfer and Energy Conversion, Management, Materials Engineering and Sciences, Nuclear Engineering, Safety and Health, and Separations Technology. The Institute sponsors research projects in cooperation with corporate, governmental, and institutional sources, including the Center for Chemical Process Safety (CCPS), the Center for Waste Reduction Technologies (CWRT), the Design Institute for Emergency Relief Systems (DIERS), the Design Institute for Physical Property Data (DIPRR), the Process Data Exchange Institute (PDXI), and the Research Institute for Food Engineering. Headquarters of the AIChE is in New York City.

AIR. In addition to being the principal substance of the earth's atmosphere, air is a major industrial medium and chemical raw material. The average composition of dry air at sea level, disregarding unusual concentrations of certain pollutants, is given in Table 1. The amount of water vapor in the air varies seasonally and geographically and is a factor of large importance where air in stoichiometric quantities is required for reaction processes, or where water vapor must be removed in air-conditioning and compressed-air systems. The water content of air for varying conditions of temperature and pressure is shown in Table 2. The

TABLE 1. COMPOSITION OF AIR

Constituent	Percent by weight	Percent by volume
Oxygen (O_2)	23.15	20.95
Ozone (O_3)	1.7×10^{-6}	0.00005
Nitrogen (N_2)	75.54	78.08
Carbon dioxide (CO_2)	0.05	0.03
Argon (Ar)	1.26	0.93
Neon (Ne)	0.0012	0.0018
Krypton (Kr)	0.0003	0.0001
Helium (He)	0.00007	0.0005
Xenon (Xe)	5.6×10^{-5}	0.000008
Hydrogen (H_2)	0.000004	0.00005
Methane (CH_4)	trace	trace
Nitrous oxide (N_2O)	trace	trace

TABLE 2. WATER CONTENT OF SATURATED AIR

Temperature (F)	(C)	Water content (Pounds in 1 Pound of Air, or Kilograms in 1 Kilogram of Air)
40	4.44	0.00520
45	7.22	0.00632
50	10	0.00765
55	12.8	0.00920
60	15.6	0.01105
65	18.3	0.01322
70	21.1	0.01578
75	23.9	0.01877
80	26.7	0.02226
85	29.4	0.02634
90	32.2	0.03108
95	35.0	0.03662
100	37.8	0.04305
105	40.6	0.05052

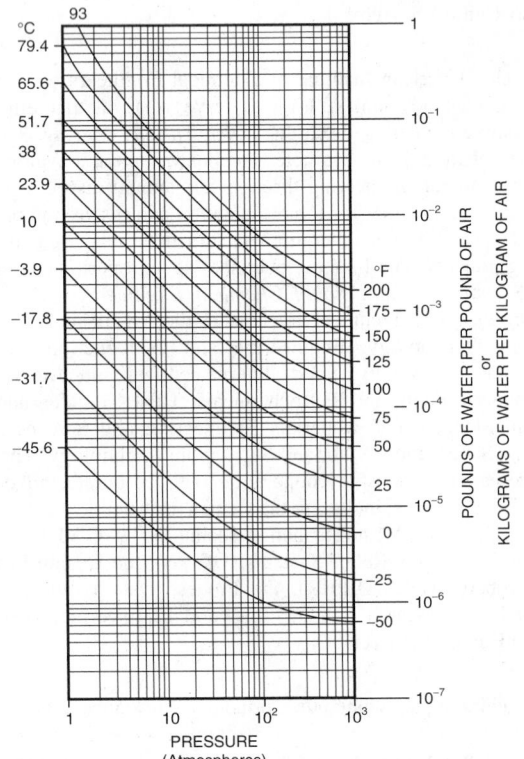

Fig. 1.   Water content of saturated air at various temperatures and pressures

water content of saturated air at various temperatures is shown in Fig. 1. See also **Oxygen**; **Nitrogen**; and **Pollution (Air)**.

ALABANDITE.   Manganese sulfide, MnS. Associated with pyrite, sphalerite, and galena in metallic sulfide vein deposits.

ALABASTER.   A fine-grained variety of the mineral gypsum, formerly much used for vases and statuary. It is usually white in color or may be of other light, pleasing tints.

The word alabaster is derived from the Greek name for this substance. See also **Gypsum**.

ALANINE.   See **Amino Acids**.

ALBERTITE.   An oxygenated hydrocarbon that differs from asphaltum slightly in that it is not completely soluble in turpentine, nor can it be perfectly fused. Specific gravity, 1.097, pitchy luster, dark brown to black color. Occurs in veins from 1 to 16 feet (0.3 to 5 meters) wide in the Albert Shale of Albert County, New Brunswick.

ALBUMIN.   An albumin is a member of a class of proteins which is widely distributed in animal and vegetable tissues. Albumins are soluble in water and in dilute salt solutions, and are coagulable by heat.

Albumin is of great importance in animal physiology; in man it constitutes about 50% of the plasma proteins (blood) and is responsible to a great extent for the maintenance of osmotic equilibrium in the blood. The high molecular weight (68,000) of the albumin molecule prevents its excretion in the urine; the appearance of albumin may indicate kidney damage.

ALCHEMY.   The predecessor of chemistry, practiced from as early as 500 BC through the 16th century. Its two principal goals were transmutation of the baser metals into gold and discovery of a universal remedy. Modern chemistry grew out of alchemy by gradual stages.

ALCOHOL.   A term commonly used to designate ethyl alcohol or ethanol. See **Ethyl Alcohol**. Also a class of organic compounds. See **Alcohols**.

ALCOHOLATE.   Replacement of the hydrogen in the hydroxyl group of an alcohol by a metal, particularly a metal that forms a strong base, results in formation of an alcoholate. An example is sodium ethylate, C_2H_5ONa.

ALCOHOLS.   The alcohols may be regarded as hydrocarbon derivatives in which the hydroxyl group (OH) replaces hydrogen on a saturated carbon. Alcohols are classified as *primary, secondary*, or *tertiary*, according to the number of hydrogen atoms that are bonded to the carbon atom with the hydroxyl substituent. Alcohols also may be regarded as alkyl derivatives of water. Thus, alcohols with a small hydrocarbon group tend to be more like water in properties than a hydrocarbon of the same number. Alcohols with a large hydrocarbon group are found to have physical properties similar to a hydrocarbon of the same structure. Some comparisons are given in Table 1. Structures are summarized by:

$$
\begin{array}{ccc}
\text{H} & \text{H} & \text{R} \\
| & | & | \\
\text{R'—C—OH} & \text{R—C—OH} & \text{R—C—OH} \\
| & | & | \\
\text{H} & \text{R} & \text{R} \\
\text{Primary} & \text{Secondary} & \text{Tertiary}
\end{array}
$$

where R' = H, alkyl, aryl; R = alkyl, aryl.

In addition to the basic classification as primary, secondary, or tertiary, alcohols may be further grouped according to other structural features. *Aromatic alcohols* contain an aryl group attached to the carbon having the hydroxyl function; *aliphatic alcohols* contain only aliphatic groups. The prefix *iso* usually indicates branching of the carbon chain.

Alcohols containing two hydroxyl groups are called *dihydric alcohols* or *glycols*. Ethylene glycol, $HOCH_2CH_2OH$, trimethylene glycol, $HOCH_2CH_2CH_2CH_2OH$, and 1,4-butanediol are examples of industrially important glycols. Glycerol, $HOCH_2CHOHCH_2OH$, has three hydroxyl groups per molecule and is a *trihydric* alcohol. Physical properties of alcohols containing more than one hydroxyl group can be estimated by considering the number of carbons for each hydroxyl group as in the case of simple alcohols.

TABLE 1. COMPARISON OF PHYSICAL PROPERTIES OF ALCOHOLS AND HYDROCARBONS

Alcohol	Hydrocarbon	Formula	Properties
Methanol		CH_3OH	Liquid, water soluble, bp. 65°C
	Methane	$CH_3{}^-H$	Gas, water insoluble
Ethanol		CH_3CH_3OH	Liquid, water soluble, bp. 78.5°C
	Ethane	$CH_3CH_2{}^-H$	Gas, water insoluble
Tetradecanol		$CH_3(CH_3)_{12}CH_2OH$	Liquid, water insoluble, bp. 263.2°C
	Tetradecane	$CH_3(CH_2)_{12}CH_2{}^-H$	Liquid, water insoluble, bp. 253.5°C

Reactions of Alcohols

Alcohols undergo a large number of reactions. However, these reactions may be grouped into a few general types. Reactions of alcohols may involve the O—H or C—O bonds. Ester formation and salt formation are examples of the former class, while conversion to halides is an example of the latter type.

$$-O-H \text{ Bond Cleavage}$$

$$ROH + CH_3COOH \underset{}{\overset{H_+}{\rightleftharpoons}} CH_3COOR + H_2O \tag{1}$$

$$ROH + K \longrightarrow RO^-K^+ + \tfrac{1}{2}H_2 \tag{2}$$

$$C-O \text{ Bond Cleavage}$$

$$RCH_2OH + HCl \xrightarrow[170°C]{ZnCL_2} RCH_2Cl + H_2O \tag{3}$$

Many industrially important substitution reactions of alcohols are conducted in the vapor phase over a catalyst. Only primary alcohols give satisfactory yields of product under these conditions.

$$CH_3OH + H_2S \xrightarrow{K_2WO_4} CH_3SH + H_2O \tag{4}$$

$$RCH_2OH + (CH_3)_2NH \xrightarrow{Al_2O_3} RCH_2N(CH_3)_2 + H_2O \tag{5}$$

Production of Alcohols

Lower alcohols (amyl and below) are prepared by (a) hydrogeneration of carbon monoxide (yields methanol), (b) olefin hydration (yields ethanol, isopropanol, secondary and tertiary butanol), (c) hydrolysis of alkyl chlorides, (d) direct oxidation, and (e) the OXO process.

$$C = O + 2 H_2 \longrightarrow CH_3OH \tag{6}$$

$$CH_3CH = CH_2 + H_2O \xrightarrow{H^+} CH_3CHOHCH_3 \tag{7}$$

$$C_5H_{11}Cl + H_2O \longrightarrow \underset{\text{(mixture of isomers)}}{C_5H_{11}OH} \tag{8}$$

Most higher alcohols (hexanol and higher) and primary alcohols of three carbons or more are synthesized by one of four general processes, or derived from a structurally related natural product. See also **Organic Chemistry**.

The OXO Process

An olefin may be hydroformylated to a mixture of aldehydes. The aldehydes are readily converted to alcohols by hydrogenation. Many olefins from ethylene to dodecenes are used in the OXO reaction. OXO alcohols are typically a mixture of linear and methyl branched primary alcohols. See also **Oxo Process**.

$$RCH = CH_2 + CO + H_2 \tag{9}$$

$$\longrightarrow [RCH_2CH_2CHO + RCH(CHO)CH_3]$$

Aldol Condensation

Aldehydes may also be dimerized by an aldol condensation reaction to give a branched unsaturated aldehyde. This may be converted to a branched alcohol by hydrogenation. See also **Aldol Condensation**.

$$2CH_3CH_2CH_2CHO \longrightarrow \tag{10}$$

$$CH_3CH_2CH_2CH = CCHO \xrightarrow{H_2} CH_3(CH_2)_3 CHCH_2 OH$$
$$\qquad\qquad\qquad |\qquad\qquad\qquad\qquad\qquad |$$
$$\qquad\qquad\qquad C_2H_5 \qquad\qquad\qquad\qquad\quad C_2H_5$$

Alcohols from an aldol reaction may be linear if acetaldehyde is a reactant, but usually aldol alcohols are branched primary alcohols. An aldol condensation sometimes is done with an OXO reaction. The combined process is called the ALDOX process.

Oxidation of Hydrocarbons

Using air, the oxidation of hydrocarbons generally results in a mixture of oxygenated compounds and is not a useful synthesis of alcohols except under special circumstances. Cyclohexanol may be prepared by air oxidation of cyclohexane inasmuch as only one isomer can result.

The yield of alcohol from normal paraffin oxidation may be improved to a commercially useful level by oxidizing in the presence of boric acid.

$$3 RH + \tfrac{3}{2}O_2 + H_3BO_3 \longrightarrow (RO)_3B + 3 H_2O \tag{12}$$

$$(RO)_3B + 3 H_2O \longrightarrow H_3BO_3 + 3 ROH \tag{13}$$

A borate ester is formed which is more stable to further oxidation than the free alcohol. This is easily hydrolyzed to recover the alcohol. These alcohols which are predominately secondary are used in surfactant manufacture.

Synthesis from Alkylaluminums

Fundamental work on organoaluminum chemistry by Prof. Karl Ziegler and co-workers at the Max Planck Institute provided the basis for a commercial synthesis of even-carbon-numbered straight chain primary alcohols. These alcohols are identical with products derived from naturally occurring fats. In this process, ethylene is reacted with aluminum triethyl to form a higher alkylaluminum which then is oxidized and hydrolyzed to give the corresponding alcohols.

$$(C_2H_5)_3Al \xrightarrow{CH^2 = CH^2} \tag{14}$$

$$[CH_3(CH_2CH_2)_nCH_2]_3Al \xrightarrow[(2)\ H_2O]{(1)\ O_2}$$

$$3 CH_3(CH_2CH_2)_nCH_2OH + Al(OH)_3$$

Commercialization of this route to higher alcohols is the most significant development in this area in recent years.

Synthesis from Natural Products

Many alcohols are prepared by reduction of the corresponding methyl esters which are derived from animal or vegetable fats. These alcohols are straight chain even-carbon-numbered compounds. Tallow and coconut oil are two major raw materials for higher alcohol manufacture.

$$(RCOO)_3C_3H_5 + 3 CH_3OH \longrightarrow 3 RCOOCH_3 \tag{15}$$

$$\textit{Triglyceri de}$$
$$+ CH_2OHCHOHCH_2OH$$

$$RCOOCH_3 + 2 H_2 \xrightarrow{Catalyst} RCH_2OH + CH_3OH \tag{16}$$

The production of ethyl alcohol for beverage, cosmetic, and pharmaceutical products is commonly accomplished by the natural process of fermentation. See **Ethyl Alcohol**; **Methyl Alcohol**; and **Fermentation**. Beer, wine, and whiskey production are extensively covered in the *Foods and*

Food Production Encyclopedia (D.M. and G.D. Considine, Eds.), Van Nostrand Reinhold, New York, 1982.

Additional Reading

Lide, D.R.: *Handbook of Chemistry and Physics,* 84th Edition, CRC Press LLC, Boca Raton, FL, 2003.

Lagowski, J.J.: *MacMillan Encyclopedia of Chemistry,* Vol. 1, MacMillian Library Reference, New York, NY, 1997.

Parker, S.P.: *McGraw-Hill Concise Encyclopedia of Science and Technology,* McGraw-Hill Companies, New York, NY, 1998.

Ziegler, K. In H. Zeiss: *Organometallic Chemistry,* ACS Monograph **147**, Van Nostrand Reinhold Company, New York, NY, 1960.

ALCOHOLYSIS. If a triglyceride oil is heated with a polyol, such as glycerol or pentaerythritol, mixed partial esters are produced in a reaction known as *alcoholysis.*

ALDEHYDES. The homologous series of aldehydes (like ketones) has the formula $C_nH_{2n}O$. The removal of two hydrogen atoms from an alcohol yields an aldehyde. Thus, two hydrogens taken away from ethyl alcohol $CH_3 \cdot C(H_2)OH$ yields acetaldehyde CH_3CH_2OH; and two hydrogens removed from propyl alcohol $C_2H_5 \cdot C(H_2)OH$ yields propaldehyde $C_2H_5 \cdot CHO$. The trivial names of aldehydes derive from the fatty acid which an aldehyde will yield upon oxidation. Thus, formaldehyde is named from formic acid, the latter being the oxidation product of formaldehyde. Similarly, acetaldehyde is oxidized to acetic acid. Or, the aldehyde may be named after the alcohol from which it may be derived. Thus, formaldehyde, which may be derived from methyl alcohol, may be named methaldehyde; or acetaldehyde may be named ethaldehyde since it may be derived from ethyl alcohol. In still another system, the aldehyde may take its name from the parent hydrocarbon from which it theoretically may be derived. Thus, prop*anal* (not to be confused with prop*anol*) may signify propaldehyde (as a derivative of propane).

Essentially aldehydes exhibit the following properties: (1) with exception of the gaseous formaldehyde, all aldehydes up to C_{11} are neutral, mobile, volatile liquids. Aldehydes above C_{11} are solids under usual ambient conditions; (2) formaldehyde and the liquid aldehydes have an unpleasant, pungent, irritating odor, (3) although the low-carbon aldehydes are soluble in H_2O, the solubility decreases with formula weight, and (4) the high-carbon aldehydes are essentially insoluble in H_2O, but are soluble in alcohol or ether.

The presence of the double bond (carbonyl group C:O) markedly determines the chemical behavior of the aldehydes. The hydrogen atom connected directly to the carbonyl group is not easily displaced. The chemical properties of the aldehydes may be summarized by: (1) they react with alcohols, with elimination of H_2O, to form *acetals*; (2) they combine readily with HCN to form *cyanohydrins*, (3) they react with hydroxylamine to yield *aldoximes*; (4) they react with hydrazine to form *hydrazones*; (5) they can be oxidized into *fatty acids*, which contain the same number of carbons as in the initial aldehyde; (5) they can be reduced readily to form primary alcohols. When benzaldehyde is reduced with sodium amalgam and H_2O, benzyl alcohol $C_6H_5 \cdot CH_2 \cdot OH$ is obtained. The latter compound also may be obtained by treating benzaldehyde with a solution of cold KOH in which benzyl alcohol and potassium benzoate are produced. The latter reaction is known as Cannizzaro's reaction.

In the industrial production of higher alcohols (above butyls), aldehydes play the role of an intermediate in a complete process that involves aldol condensation and hydrogenation. In the OXO process, olefins are catalytically converted into aldehydes that contain one more carbon than the olefin in the feedstock. Aldehydes also serve as starting materials in the synthesis of several amino acids. See also **Acetaldehyde**; **Aldol Condensation**; **Benzaldehyde**; and **Furfuraldehyde**.

ALDER, KURT (1902–1958). A German chemist who won the Nobel prize for chemistry along with Otto Diels in 1950 for a project involving a practical method for making ring compounds from chain compounds by forcing them to combine with maleic anhydride. This is known as the Diels-Alder reaction and provided a method for synthesis of complex organic compounds. He had degrees from the Universities of Berlin and Kiel.

ALDOL CONDENSATION. A reaction between aldehydes or aldehydes and ketones that occurs without the elimination of any secondary product and yields β-hydroxycarbonyl compounds. It is distinguished from polymerization by the fact that it occurs between aldehydes and ketones and is not generally reversible. In its simplest form it may be represented by the condensation of two molecules of acetaldehyde to aldol:

$$CH_3 \cdot CHO + CH_3 \cdot CHO \longrightarrow CH_3 \cdot CHOHCH_2 \cdot CHO$$

Weak alkalies and acids are employed to effect the condensation.

Researchers at the University of California, Berkeley, have accomplished acyclic stereocontrol through the aldol condensation. As observed by C.H. Heathcock (*Science*, **214**, 295–400, Oct. 23, 1981), one of the most difficult problems in the synthesis of complex organic compounds is that of controlling the relative stereochemistry, that is, establishing the correct configuration at the various chiral centers as the synthesis is carried out. In recent years, researchers have been attempting to find direct solutions to the problem, particularly in synthesizing acyclic and other conformation-flexible molecules. Heathcock and colleagues have found that aldol condensation, one of the oldest and most familiar organic reactions, can be a very effective tool for achieving stereocontrol.

See also **Aldehydes**; and **Ketones**.

ALDOSES. See **Carbohydrates**.

ALDOSTERONE. See **Steroids**.

ALDOXIMES. See **Hydroxylamine**.

ALEXANDRITE. A variety of chrysoberyl, originally found in the schists of the Ural Mountains. It absorbs yellow and blue light rays to such an extent that it appears emerald green by daylight but columbine-red by artificial light. It is used as a gem, and was named in honor of Czar Alexander II of Russia. See also **Chrysoberyl**.

ALGICIDE. A substance, natural or synthetic, used for destroying or controlling algae. The term is also sometimes used to describe chemicals used for controlling aquatic vegetation, although these materials are more properly classified as aquatic herbicides. See **Herbicides**.

ALGIN. A hydrophilic colloidal polysaccharide obtained from several species of brown algae. The term is used both in reference to the pure substance, alginic acid, extracted from the algae and also to the salts of this acid such as sodium or ammonium alginate, in which forms it is used commercially. The alginates currently find a large number of applications in the paint, rubber, pharmaceutical, food, and other industries. See also **Gums and Mucilages**.

ALIPHATIC COMPOUND. An organic compound that can be regarded as a derivative of methane, CH_4. Most aliphatic compounds are open carbon chains, straight or branched, saturated or unsaturated. Originally, the term was used to denote the higher (fatty) acids of the $C_nH_{2n}O_2$ series. The word is derived from the Greek term for oil. See also **Compound (Chemical)**; and **Organic Chemistry**.

ALKALI. A term that was originally applied to the hydroxides and carbonates of sodium and potassium but since has been extended to include the hydroxides and carbonates of the other alkali metals and ammonium. Alkali hydroxides are characterized by ability to form soluble soaps with fatty acids, to restore color to litmus which has been reddened by acids, and to unite with carbon dioxide to form soluble compounds. See also **Acids and Bases**.

ALKALI METALS. The elements of group 1 of the periodic classification. In order of increasing atomic number, they are hydrogen, lithium, sodium, potassium, rubidium, cesium, and francium. With the exception of hydrogen, which is a gas and which frequently imparts a quality of acidity to its compounds, the other members of the group display rather striking similarities of chemical behavior, all reactive with H_2O to form strongly alkaline solutions. The elements in the group, including hydrogen, are characterized by a valence of one, having one electron in an outer shell available for reaction. Because of their chemical similarities, these elements, along with *ammonium* and sometimes magnesium, are considered the sixth group in classical qualitative chemical analysis separations.

ALKALINE EARTHS. The elements of group 2 of the periodic classification. In order of increasing atomic number, they are beryllium, magnesium, calcium, strontium, barium, and radium. The members of the group display rather striking similarities of chemical behavior, including stable oxides and carbonates, with hydroxides that are less alkaline than those of group 1. The elements of the group are characterized by a valence of two, having two electrons in an outer shell available for reaction. Because of their chemical similarities, these elements are considered the fifth group in classical qualitative chemical analysis separations.

ALKALI ROCKS. Igneous rocks which contain a relatively high amount of alkalis in the form of soda amphiboles, *soda* pyroxenes, or felspathoids, are said to be alkaline, or alkalic. Igneous rocks in which the proportions of both lime and alkalis are high, as combined in the minerals, feldspar, hornblende, and augite, are said to be calcalkalic.

ALKALOIDS. The term, alkaloid, which was first proposed by the pharmacist, W. Meissner, in 1819, and means "alkali-like," is applied to basic, nitrogen-containing compounds of plant origin. Two further qualifications usually are added to this definition: (1) the compounds have complex molecular structures; and (2) they manifest significant pharmacological activity. Such compounds occur only in certain genera and families, rarely being universally distributed in larger groups of plants. Many widely distributed bases of plant origin, such as methyltrimethyl-and other open-chain simple alkylamines, the cholines, and the phenylalkylamines, are not classed as alkaloids. Alkaloids usually have a rather complex structure with the nitrogen atom involved in a heterocyclic ring. However, thiamine, a heterocyclic nitrogenous base, is not regarded as an alkaloid mainly because of its almost universal distribution in living matter. Colchicine, on the other hand, is classed as an alkaloid even though it is not basic and its nitrogen atom is not incorporated into a heterocyclic ring. It apparently qualifies as an alkaloid because of its particular pharmacological activity and limited distribution in the plant world.

Over 2000 alkaloids are known and it is estimated that they are present in only 10–15% of all vascular plants. They are rarely found in cryptogamia (exception, ergot alkaloids), gymnosperms, or monocotyledons. They occur abundantly in certain dicotyledons and particularly in the following families: *Apocynaceaae* (dogbane, quebracho, pereiro bark); *Papaveraceae* (poppies, chelidonium); *Papilionaceae* (lupins, butterfly-shaped flowers); *Ranunculaceae* (aconitum, delphinium); *Rubiaceae* (cinchona bark, ipecacuanha); *Rutaceae* (citrus, fagara); and *Solanaceae* (tobacco, deadly nightshade, tomato, potato, thorn apple). Well-characterized alkaloids have been isolated from the roots, seeds, leaves or bark of some 40 plant families. *Papaveraceae* is an unusual family, in that all of its species contain alkaloids.

Brief descriptions in alphabetical order of alkaloids of commercial or medical importance or of societal concern (alkaloid narcotics) are given later in this entry. See also **Amphetamine**; **Morphine**; and **Pyridine and Derivatives**.

The nomenclature of alkaloids has not been systemized, both because of the complexity of the compounds and for historical reasons. The two commonly used systems classify alkaloids either according to the plant genera in which they occur, or on the basis of similarity of molecular structure. Important classes of alkaloids containing generically related members are the aconitum, cinchona, ephedra, lupin, opium, rauwolfia, senecio, solanum, and strychnos alkaloids. Chemically derived alkaloid names are based upon the skeletal feature which members of a group possess in common. Thus, indole alkaloids (e.g., psilocybin, the active principle of Mexican hallucinogenic mushrooms) contain an indole or modified indole nucleus, and pyrrolidine alkaloids (e.g., hygrine) contain the pyrrolidine ring system. Other examples of this type of classification include the pyridine, quinoline, isoquinoline, imidazole, pyridine-pyrrolidine, and piperidine-pyrrolidine type alkaloids. Several alkaloids are summarized along these general terms in Table 1.

The beginning of alkaloid chemistry is usually considered to be 1805 when F.W. Sertürner first isolated morphine. He prepared several salts of morphine and demonstrated that it was the principle responsible for the physiological effect of opium. Alkaloid research has continued to date, but because most likely plant sources have been investigated and because a large number of synthetic drugs serve medical and other needs more effectively, the greatest emphasis has been placed upon the synthetics.

Sometimes, there is confusion between alkaloids and narcotics. It should be stressed that all alkaloids are not narcotics; and all narcotics are not

TABLE 1. GENERAL CLASSIFICATION OF ALKALOIDS

General Class	Examples
Derivatives of aryl-substituted amines	Adrenaline, amphetamine, ephedrine, phenylephrine tyramine
Derivatives of pyrrole	Carpaine, hygrine, nicotine
Derivatives of imidazole	Pilocarpine
Derivatives of pyridine and piperidine	Anabasine, coniine, ricinine
Containing fusion of two piperidine rings	Isopelletierine, pseudopelletierine
Pyrrole rings fused with other rings	Gelsemine, physostigmine, vasicine, yohimbine
Aporphone alkaloids	Apomorphine corydine, isothebaine
Berberine alkaloids	Berberine, emetine
Bis-benzylisoquinoline alkaloids	Bebeering, trilobine
Cinchona alkaloids	Cinchonine, quinidine, quinine
Cryptopine alkaloids	Cryptopine, protopine
Isoquinoline alkaloids	Anhalidine, pellotine, sarsoline
Lupine alkaloids	Lupanine, sparteine
Morphine and related alkaloids	Codeine, morphine, thebaine
Papaverine alkaloids	Codamine, homolaudanosine, papeverine
Phthalide isoquinoline alkaloids (also known as narcotine alkaloids)	Hydrastine, narceine, narcotine
Quinoline alkaloids	Dictamine, galipoline, lycorine
Tropine alkaloids	Atropine, cocaine, ecgonine, scopolamine, tropine
Other alkaloids	Brucine, sclanidine, strychnine

alkaloids. A narcotic has the general definition of a drug that produces sleep or stupor, and also relieves pain. Many alkaloids do not meet these specifications.

The molecular complexity of the alkaloids is demonstrated by Fig. 1. Alkaloids react as bases to form salts. The salts used especially for crystallization purposes are the hydrochlorides, sulfates, and oxalates, which are generally soluble in water or alcohol, insoluble in ether, chloroform, carbon tetrachloride, or amyl alcohol. Alkaloid salts unite with mercury, gold, and platinum chlorides. Free alkaloids lack characteristic color reactions but react with certain reagents, as follows, with (1) iodine in potassium iodide solution, forming chocolate brown precipitate; (2) mercuric iodide in potassium iodide solution (potassium mercuriiodide), forming precipitate; (3) potassium iodobismuthate, forming orange-red precipitate; (4) bromine-saturated concentrated hydrobromic acid forming yellow precipitate; (5) tannic acid, forming precipitate; (6) molybdophosphoric acid, forming precipitate; (7) tungstophosphoric acid, forming precipitate; (8) gold(III) chloride, forming crystalline precipitate of characteristic melting point; (9) platinum(IV) chloride, forming crystalline precipitate of characteristic melting point; (10) picric acid, forming precipitate; (11) perchloric acid, forming precipitate. Many alkaloids form more or less characteristic colors with acids, solutions of acidic salts, etc.

The function of alkaloids in the source plant has not been fully explained. Some authorities simply regard them as by-products of the plant metabolism. Others conceive of alkaloids as reservoirs for protein synthesis; as protective materials discouraging animal or insect attacks; as plant stimulants or regulators in such activities as growth, metabolism, and reproduction; as detoxifying agents, which render harmless (by processes such as methylation, condensation, and ring closure) substances whose accumulation might otherwise cause damage to the plant. While these theories are of interest, it is also of interest to observe that from 85–90% of all plants manage well without the presence of alkaloids in their structures.

Adrenaline®. See **Epinephrine** later in this entry.

Atropine, also known as daturine, $C_{17}H_{23}NO_3$ (see structural formula in accompanying diagram), white, crystalline substance, optically inactive, but usually contains levorotatory hyoscyamine. Compound is soluble in alcohol, ether, chloroform, and glycerol; slightly soluble in water; mp 114–116°C. Atropine is prepared by extraction from *Datura stramonium*, or synthesized. The compound is toxic and allergenic. Atropine is used in medicine and is an antidote for cholinesterase-inhibiting compounds, such as organophosphorus insecticides and certain nerve gases. Atropine is commonly offered as the sulfate. Atropine is used in connection with the treatment of disturbances of cardiac rhythm and conductance, notably in the therapy of sinus bradycardia and sick sinus syndrome. Atropine is also used in some cases of heart block. In particularly high doses, atropine may induce ventricular tachycardia in an ischemic myocardium. Atropine is frequently one of several components in brand name prescription drugs.

Caffeine, also known as theine, or methyltheobromine, 1,2,7-trimethyl xanthine (see structural formula in accompanying diagram), white, fleecy or long, flexible crystals. Caffeine effloresces in air and commences losing water at 80°C. Soluble in chloroform, slightly soluble in water and alcohol, very slightly soluble in ether, mp 236.8°C, odorless, bitter taste. Solutions are neutral to litmus paper.

Caffeine is derived by extraction of coffee beans, tea leaves, and kola nuts. It is also prepared synthetically. Much of the caffeine of commerce is a by-product of decaffeinized coffee manufacture. The compound is purified by a series of recrystallizations. Caffeine finds use in medicine and in soft drinks. Caffeine is also available as the hydrobromide and as sodium benzoate, which is a mixture of caffeine and sodium benzoate, containing 47–50% anhydrous caffeine and 50–53% sodium benzoate. This mixture is more soluble in water than pure caffeine. A number of nonprescription (pain relief) drugs contain caffeine as one of several ingredients. Caffeine is a known cardiac stimulant and in some persons who consume significant amounts, caffeine can produce ventricular premature beats.

Cocaine (also known as methylbenzoylepgonine), $C_{17}H_{21}NO_4$, is a colorless-to-white crystalline substance, usually reduced to powder. Cocaine is soluble in alcohol, chloroform, and ether, slightly soluble in water, giving a solution slightly alkaline to litmus. The hydrochloride is levorotatory, mp 98°C. Cocaine is derived by extraction of the leaves of coca (*Erythroxylon*) with sodium carbonate solution, followed by treatment with dilute acid and extraction with ether. The solvent is evaporated after which the substance is re-dissolved and subsequently crystallized. Cocaine also is prepared synthetically from the alkaloid ecgonine. Cocaine is *highly toxic* and *habit-forming*. While there are some medical uses of cocaine,

usage must always be under the direction of a physician. It is classified as a narcotic in most countries. Society's major concern with cocaine is its use (increasing in recent years) as a narcotic.

Cocaine has been known as a very dangerous material since the early 1900s. When use of it as a narcotic increased during the early 1970s, serious misconceptions concerning its "safety" as compared with many other narcotics led and continue to lead to many deaths from its use.

Addicts use cocaine intravenously or by snorting the powder. After intravenous injections, coma and respiratory depression can occur rapidly. It has been reported that fatalities associated with snorting usually occur shortly after the abrupt onset of major motor seizures, which may develop within minutes to an hour after several nasal ingestions. Similar results occur if the substance is taken by mouth. Treatment is directed toward ventilatory support and control of seizures—although in many instances a victim may not be discovered in time to prevent death. It is interesting to note that cocaine smugglers, who have placed cocaine-filled condoms in their rectum or alimentary tract, have died (Suarez et al., 1977). The structural formula of cocaine is given in Fig. 1.

Codeine, also known as methylmorphine, $C_{18}H_{21}NO_3 \cdot H_2O$, is a colorless white crystalline substance, mp 154.9°C, slightly soluble in water, soluble in alcohol and chloroform, effloresces slowly in dry air. Codeine is derived from opium by extraction or by the methylation of morphine. For medical use, codeine is usually offered as the dichloride, phosphate, or sulfate. Codeine is *habit forming*. Codeine is known to exacerbate *urticaria* (familiarly known as *hives*). Since codeine is incorporated in numerous prescription medicines for headache, heartburn, fatigue, coughing, and relief of aches and pains, persons with a history of urticaria should make this fact known to their physician. Codeine is sometimes used in cases of acute *pericarditis* to relieve severe chest pains in early phases of disease. Codeine is sometimes used in drug therapy of renal (kidney) diseases.

Colchicine, an alkaloid plant hormone, $C_{22}H_{25}NO_6$, is yellow crystalline or powdered, nearly odorless, mp 135–150°C, soluble in water, alcohol,

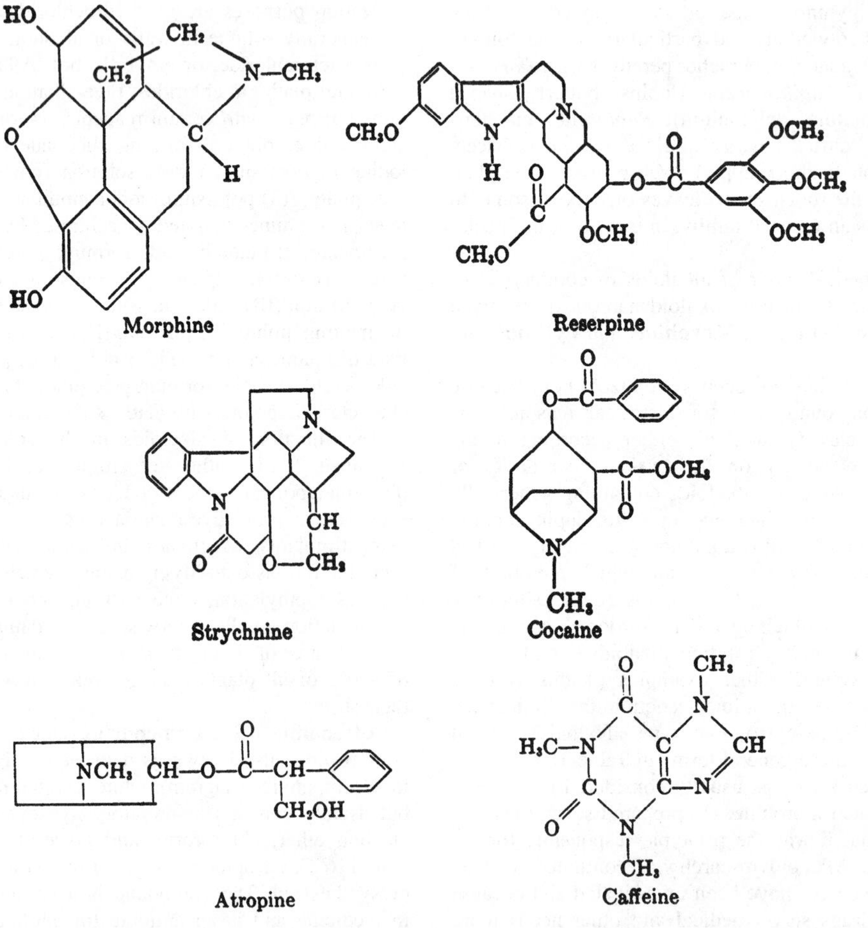

Fig. 1.   Structures of representative alkaloids. The carbon atoms in the rings and the hydrogen atoms attached to them are not designated by letter symbols. However, there is understood to be a carbon atom at each corner (except for the cross-over in the structure of morphine) and each carbon atom has four bonds, so that any bonds not shown or represented by attached groups are joined to hydrogen atoms

and chloroform, moderately soluble in ether. Solutions are levorotatory and deteriorate under light. The substance is highly toxic (0.02 gram may be fatal if ingested). Colchicine is extracted from the plant *Colchicum autumnale* after which it is crystallized. The compound also has been synthesized. Biologists have used colchicine to induce chromosome doubling in plants. Colchicine finds a number of uses in medicine.

Although colchicine has been known for many years, interest in the drug has been revitalized in recent years as the result of the discovery that it interferes with cell division by destroying the spindle mechanism. The two chromatids, which represent one chromosome at the metaphase stage, fail to separate and do not migrate to the poles (ends) of the cell. Each chromatid becomes a chromosome *in situ*. The entire group of new chromosomes now forms a resting nucleus and the next cell division reveals twice as many chromosomes as before. The cell has changed from the diploid to the tetraploid condition. Applied to germinating seeds or growing stem tips in concentrations of about 1 gram in 10,000 cubic centimeters of water for 4 or 5 days, colchicine may thus double the chromosome number of many or all of the cells, producing a tetraploid plant or shoot. Offspring from such plants may be wholly tetraploid and breed true. Tetraploid plants are larger than diploid plants and often more valuable. The alkaloid has also been used to double the chromosome number of sterile hybrids produced by crossing widely separated species of plants. Such plants, after colchicine treatment, contain in each cell two complete diploid sets of chromosomes, one from each of the parent species, and become fertile, pure-breeding hybrid species.

In medicine, colchicine is probably best known for its use in connection with the treatment of gout. Acute attacks of gout are characteristically and specifically aborted by colchicine. The response noted after administration of the drug also can be useful in diagnosing gout cases where synovial fluid cannot be aspirated and examined for the presence of typical urate crystals. However, colchicine does not affect the course of acute synovitis in rheumatoid arthritis.

Kaplan (1960) observed that colchicine may produce objective improvement in the periarthritis associated with *sarcoidosis* (presence of noncaseating granulomas in tissue). Colchicine is sometimes used in the treatment of *scleroderma* (deposition of fibrous connective tissues in skin or other organs); it may assist in preventing attacks of Mediterranean fever; and it is sometimes used as part of drug therapy for some renal (kidney) diseases.

Colchicine can cause diarrhea as the result of mucosal damage and it has been established that colchicine interferes with the absorption of vitamin B_{12}.

Emetine, an alkaloid from ipecac, $C_{29}H_{40}O_4N_2$, is a white powder, mp 74°C, with a very bitter taste. The substance is soluble in alcohol and ether, slightly soluble in water. Emetine darkens upon exposure to light. The compound is derived by extraction from the root of *Cephalis ipecacuanha* (ipecac). It is also made synthetically. Medically, ipecac is useful as an emetic (induces vomiting) for emergency use in the treatment of drug overdosage and in certain cases of poisoning. Ipecac should not be administered to persons in an unconscious state. It should be noted that emesis is not the proper treatment in all cases of potential poisoning. It should not be induced when such substances as petroleum distillates, strong alkali, acids, or strychnine are ingested.

Ephedrine, 1-phenyl-2-methylaminopropanol, $C_6H_5CH(OH)CH$ $(NHCH_3)$ CH_3, is a white-to-colorless granular substance, unctuous (greasy) to the touch, and hygroscopic. The compound gradually decomposes upon exposure to light. Soluble in water, alcohol, ether, chloroform, and oils, mp 33–40°C, by 255°C, and decomposes above this temperature. Ephedrine is isolated from stems or leaves of *Ephedra*, especially Ma huang (found in China and India). Medically, it is usually offered as the hydrochloride. In the treatment of bronchial asthma, ephedrine is known as a *beta agonist*. Compounds of this type reduce obstruction by activating the enzyme adenylate cyclase. This increases intracellular concentrations of cAMP (cyclic 3′5′-adenosine monophosphate) in bronchial smooth muscle and mast cells. Ephedrine is most useful for the treatment of mild asthma. In severe asthma, ephedrine rarely maintains completely normal airway dynamics over long periods. Ephedrine also has been used in the treatment of cerebral transient ischemic attacks, particularly with patients with vertabrobasilar artery insufficiency who have symptoms associated with relatively low blood pressure, or with postural changes in blood pressure. Ephedrine sulfate also has been used in drug therapy in connection with urticaria (hives).

Epinephrine, a hormone having a benzenoid structure, $C_9H_{13}O_3N$, also called adrenaline. It can be obtained by extraction from the adrenal glands of cattle and also prepared synthetically. Its effect on body metabolism is pronounced, causing an increase in blood pressure and rate of heartbeat. Under normal conditions, its rate of release into the system is constant, but emotional stresses, such as fear or anger rapidly increase the output and result in temporarily heightened metabolic activity. Epinephrine is used for the symptomatic treatment of bronchial asthma and reversible bronchospasm associated with chronic bronchitis and emphysema. The drug acts on both alpha and beta receptor sites. Beta stimulation provides bronchodilator action by relaxing bronchial muscle. Alpha stimulation increases vital capacity by reducing congestion of the bronchial mucosa and by constricting pulmonary vessels.

Epinephrine is also used in the management of anesthetic procedures in connection with noncardiac surgery of patients with active ischemic heart disease. The drug is useful in the treatment of severe urticarial (hives) attacks, especially those accompanied by angioedema.

Epinephrine has numerous effects on intermediary metabolism. Among these are promotion of hepatic glycogenolysis, inhibition of hepatic gluconeogenesis, and inhibition of insulin release. The drug also promotes the release of free fatty acids from triglyceride stores in adipose tissues. Epinephrine produces numerous cardiovascular effects. Epinephrine is particularly useful in treating conditions of immediate hypersensitivity—interactions between antigen and antibody. These mechanisms cause attacks of anaphylaxis, hay fever, hives and allergic asthma. Anaphylaxis can occur after bee and wasp stings, venoms, etc. Although the mechanism is not fully understood, epinephrine can play a lifesaving role in the treatment of acute systemic anaphylaxis.

In some instances, epinephrine can be a cause of a blood condition involving the leukocytes and known as neutrophilia. In very rare cases, an intramuscular injection of epinephrine can be a cause of clostridial myonecrosis (gas gangrene).

Heroin, diacetylmorphine $C_{17}H_{17}NO(C_2H_3O_2)_2$, is a white, essentially odorless, crystalline powder with bitter taste, soluble in alcohol, mp 173°C. Heroin is derived by the acetylization of morphine. The substance is highly toxic and is a habit-forming narcotic. One-sixth grain (0.0108 gram) can be fatal. Although emergency facility personnel in some areas during recent years have come to regard heroin overdosage as approaching epidemic statistics, it is nevertheless estimated that the majority of persons with heroin overdose die before reaching a hospital. The initial crisis of an overdose is a severe respiratory depression and sometimes *apnea* (cessation of breathing). In emergency situations, the victim may be ventilated with a self-inflating resuscitative bag with delivery of 100% oxygen. Then, an endotracheal tube attached to a mechanical ventilator may be inserted. Naloxone (*Narcan*®), a narcotic antagonist, then may be administered intraveneously, often with repeated dosages over short intervals, until an improvement is noted in the respiratory rate or sensorial level of the victim. If a victim does not respond, this is usually indication that the situation is not opiate-related, or that other drugs also have been taken. Inasmuch as the antagonizing action of naloxone persists for only a few hours, a heroin overdose patient should be observed in the hospital for an indeterminate period. In heroin overdose cases, pulmonary edema (as the result of altered capillary permeability) may occur. This is directly associated with the overdose and not with subsequent treatment. Aside from severe overdose, the drug causes or contributes to a number of ailments. These include chronic renal (kidney) failure and nephritic syndrome. Septic arthritis, caused by *Pseudomonas* and *Serratia* infections, is sometimes found as the result of intravenous heroin abuse. Drug-induced immune platelet destruction also may occur.

Morphine. See separate entry on **Morphine**.

Neo-Synephrine®. See **Phenylephrine hydrochloride** later in this entry.

Nicotine, beta-pyridyl-alpha-N-methylpyrrolidine, $C_5H_4NC_4H_7NCH_3$, is a thick, water-white levorotatory oil that turns brown upon exposure to air. The compound is hygroscopic, soluble in alcohol, chloroform, ether, kerosene, water, and oils, bp 247°C, at which point it decomposes. Specific gravity is 1.00924. Nicotine is combustible with an auto-ignition temperature of 243°C. Nicotine is derived by distilling tobacco with milk of lime and extracting with ether. Nicotine is used in medicine, as an insecticide, and as a tanning agent. Nicotine is commercially available as the dihydrochloride, salicylate, sulfate, and bitartrate. Nicotinic acid (pyridine-3-carboxylic acid) is a vitamin in the B complex. See also **Vitamin**.

PhenylephrineHydrochloride

l-1-(meta-hydroxyphenyl-2-methyl-) aminoethanol hydrochloride, $HOC_6H_4CH(OH)CH_2NHCNH_3$· HCl, white or nearly white crystalline substance, odorless, bitter taste. Solutions are acid to litmus paper, freely soluble in water and in alcohol, mp 140–145°C. Levorotatory in solution. Phenylephrine hydrochloride is used medically as a vasoconstrictor and pressor drug. It is chemically related to epinephrine and ephedrine. Actions are usually longer lasting than the latter two drugs. The action of phenylephrine hydrochloride contrasts sharply with epinephrine and ephedrine, in that its action on the heart is to slow the rate and to increase the stroke output, inducing no disturbance in the rhythm of the pulse. In therapeutic doses, it produces little if any stimulation of either the spinal cord or cerebrum. The drug is intended for the maintenance of an adequate level of blood pressure during spinal and inhalation anesthesia and for the treatment of vascular failure in shock, shock-like states, and drug-induced hypotension, or hypersensitivity. It is also used to overcome paroxysmal supraventricular tachycardia, to prolong spinal anesthesia, and as a vasoconstrictor in regional analgesia. Caution is required in the administration of phenylephrine hydrochloride to elderly persons, or to patients with hyperthyroidism, bradycardia, partial heart block, myocardial disease, or severe arteriosclerosis. The brand name *Neo-Synephrine*® is also used to designate another product (nose drops) which does not contain phenylephrine hydrochloride. The nose drops contain xylometazoline hydrochloride.

Quinine, $C_{20}H_{24}N_2O_2$· H_2O, a bulky, white, amorphous powder or crystalline substance, with very bitter taste. It is odorless and levorotatory. Soluble in alcohol, ether, chloroform, carbon disulfide, oils, glycerol, and acids; very slightly soluble in water. Quinine is derived from finely ground cinchona bark mixed with lime. This mixture is extracted with hot, high-boiling paraffin oil. The solution is filtered, shaken with dilute sulfuric acid and then neutralized while hot with sodium carbonate. Upon cooling, quinine sulfate crystallizes out. Pure quinine is obtained by treating the sulfate with ammonia. In addition to medical uses, quinine and its salts are used in soft drinks and other beverages.

Quinine derivatives are used in therapy for mytonic dystrophy (usually weakness and wasting of facial muscles); in the treatment of certain renal (kidney) diseases. Quinine and derivatives are best known for their use in connection with malaria. Acute attacks of malaria are usually treated with oral chloroquine phosphate. The drug is given intramuscularly to patients who cannot tolerate oral medication. Combined therapy is indicated for treating *P. falciparum* infections, using quinine sulfate and pyrimethamine. A weekly oral dose of chloroquinone phosphate is frequently prescribed for persons who travel in malarial regions. The drug is taken one week prior to travel into such areas and continued for six weeks after leaving the region. Chloroquine phosphate has not proved fully satisfactory in the treatment of babesiosis, a malaria-like illness caused by a parasite.

Strychnine, $C_{21}H_{24}ON_2$, hard, white crystals or powder of a bitter taste. Soluble in chloroform, slightly soluble in alcohol and benzene, slightly soluble in water and ether, mp 268–290°C, bp 270°C (5 millimeters pressure). Strychnine is obtained by extraction of the seeds of *Nux vomica* with acetic acid, followed by filtration, precipitation by an alkali, followed by final filtration. The compound is highly toxic by ingestion and inhalation. The phosphate finds limited medical use. Strychnine is also used in rodent poisons. Strychnine acts as a powerful stimulant to the central nervous system. At one time, strychnine was used in a very carefully controlled way in the treatment of some cardiac disorders. Acute strychnine poisoning resembles fully developed generalized tetanus.

Additional Reading

Bentley, K.W.: *The Isoquinoline Alkaloids,* Gordon and Breach Science Publishers, Newark, NJ, 1998.

Cordell, L.: *The Alkaloids,* Academic Press, Inc., San Diego, CA, 2000.

Gawin, F.H.: "Cocaine Addiction: Psychology and Neurophysiology," *Science,* 1580 (March, 29, 1991).

Gerstein, D.R. and L.S. Lewin: "Treating Drug Problems," *New Eng. J. Med.,* 844 (September, 20, 1990).

Gillin, J.C.: "The Long and the Short of Sleeping Pills," *New Eng. J. Med.,* 1735 (June, 13, 1991).

Gilpin, R.K. and L.A. Pachla: "Pharmaceuticals and Related Drugs," *Analytical Chemistry,* 130R (June, 15, 1991).

Gorrod, J.W. and J. Wahren: *Nicotine and Related Alkaloids, Adsorption, Distribution, Metabolism and Excretion,* Chapman & Hall, New York, NY, 1993.

Grobbee, D.E., et al.: "Coffee, Caffeine, and Cardiovascular Disease in Men," *New Eng. J. Med.,* 1026 (October, 11, 1990).

Holloway, M.: "Rx for Addiction," *Sci. Amer.,* 94 (March, 1991).

Jackson, J.F. and H.F. Linskens: *Alkaloids,* Springer-Verlag New York, Inc., New York, NY, 1994.

Kaplan, H.: "Sarcoid Arthritis wilth a Response to Colchicine: Report of Two Cases," *N. Eng. J. Med.,* **263**, 778 (1960).

Kroschwitz, J.I. and M.H. Grant: *Encyclopedia of Chemical Technology: A to Alkaloids,* 4th Edition, Vol. 1, John Wiley & Sons, New York, NY, 1991.

Masto, D.F.: "Opium and Marijuana in American History," *Sci. Amer.,* 40 (July, 1991).

Oates, J.A. and A.J.J. Wood: "Drug Therapy," *New Eng. J. Med.,* 1017 (October, 3, 1991).

Pelletier, S.W.: *Alkaloids, Chemical and Biological Perspectives,* Vol. 12, Elsevier Science, New York, NY, 1998.

Rahman, Atta-Ur and A. Basha: *Indole Alkaloids,* Gordan and Breach Science Publishers, Newark, NJ, 1998.

Roberts, M.F. and M. Wink: *Alkaloids, Biochemistry, Ecology, and Medicinal Applications,* Kluwer Academic/Plenum Publishers, New York, NY, 1998.

Suarez, C.A. et al.: "Cocaine-Condom Ingestion: Surgical Treatmen," *J. Amer. Med. Assn.,* **238**, 1391 (1977).

Tonnesen, P., et al.: "A Double-Blind Trial of a 16-Hour Transdermal Nicotine Patch in Smoking Cessation," *New Eng. J. Med.,* 311 (August, 1, 1991).

Wetli, C.V. and R.K. Wright: "Death Caused by Recreational Cocaine Use," *J. Amer. Med. Assn.,* **241**, 2510 (1979).

Winks: *Biochemistry of the Quinilizidine Alkaloid,* Chapman & Hall, New York, NY, 1999.

ALKALOSIS. A condition of excess alkalinity (or depletion of acid) in the body, in which the acid-base balance of the body is upset. The hydrogen ion concentration of the blood drops below the normal level, increasing the pH value of the blood above the normal 7.4. The condition can result from the ingestion or formation in the body of an excess of alkali, or of loss of acid. Common causes of alkalosis include: (1) overbreathing (hyperventilation), where a person may breathe too deeply for too long a period, consequently washing out carbon dioxide from the blood, (2) ingestion of excessive alkali, as for example an overdosage of sodium bicarbonate possibly taken for the relief of gastric distress, and (3) excessive vomiting, which leads to loss of chloride and retention of sodium ions. The usual, mild symptoms of alkalosis are restlessness, possible numbness or tingling of the extremities (hands and feet), and generally increased muscular irritability. Only in extreme cases, tetany (muscle spasm) and convulsions may be evidenced.

See also **Acid-Base Regulation (Blood)**; **Blood**; and **Potassium and Sodium (In Biological Systems)**.

ALKANE. One of the group of hydrocarbons of the paraffin series, e.g., methane, ethane, and propane. See also **Organic Chemistry**.

ALKENE. One of a group of hydrocarbons having one double bond and the type formula $C_n H_{2n}$, e.g., ethylene and propylene. See also **Organic Chemistry**.

ALKYD. See **Paint and Finish Removers**.

ALKYD RESINS. In spite of challenges from many new coating resins developed over the years, alkyd resins as a family have maintained a prominent position for two principal reasons, their high versatility and low cost.

Fundamental Reactions and Resin Structure

The main reactions involved in alkyd resin synthesis are polycondensation by esterification and ester interchange. Figure 1 uses the following symbols to represent the basic components of an alkyd resin.

As Figure 1 implies, there is usually some residual acidity as well as free hydroxyl groups left in the resin molecules.

Classification of Alkyd Resins

Alkyd resins are usually referred to by a brief description based on certain classification schemes. From the classification the general properties of the resin become immediately apparent. Classification is based on the nature of the fatty acid and oil length.

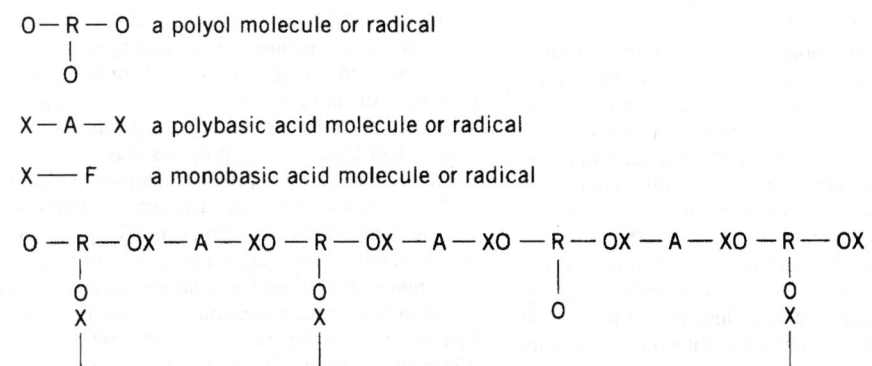

O — R — O   a polyol molecule or radical
     |
     O

X — A — X   a polybasic acid molecule or radical

X —— F   a monobasic acid molecule or radical

O — R — OX — A — XO — R — OX — A — XO — R — OX — A — XO — R — OX — A — X
    | | | |
    O   O   O   O
    X   X   X
    | | |
    F   F   F

Fig. 1.   Schematic representation of an alkyd resin molecule

TABLE 1. PROPERTY CHANGES WITH OIL LENGTH OF ALKYD RESINS[A]

Property	Long	Medium	Short
		Oil length	
requirement of aromatic/polar solvents	———————→		
compatibility with other film-formers	———————→		
viscosity	———————→		
ease of brushing	←———————		
air dry time, set-to-touch	←———————		
through-dry	←———— ———→		
film hardness	←———— ———→		
gloss	———————→		
gloss retention	———————→		
color retention	———————→		
exterior durability	←———————→		

[a]Primarily drying-type alkyds.

TABLE 2. POLYOLS FOR ALKYD SYNTHESIS

Type	Mol wt	Eq wt
pentaerythritol	136	34
glycerol	92	31[a]
trimethylolpropane	134	44.7
trimethylolethane	120	40
ethylene glycol	62	31
neopentyl glycol	104	52

[a]Because glycerol is usually supplied at 99% purity (1% moisture), its eq wt is commonly assumed to be 31 in recipe calculations.

Oil Length-Resin Property Relationship

The oil length of an alkyd resin has profound effects on the properties of the resin (Table 1).

Alkyd Ingredients

For each of the three principal components of alkyd resins, the polybasic acids, the polyols, and the monobasic acids, there is a large variety to be chosen from. The selection of each of these ingredients affects the properties of the resin and may affect the choice of manufacturing processes. Thus, to both the resin manufacturers and the users, the selection of the proper ingredients is a significant decision.

Polybasic Acids and Anhydrides. The principal polybasic acids used in alkyd preparation include phthalic anhydride (mol wt 148, eq wt 74), isophthalic acid (mol wt 166, eq wt 83), maleic anhydride (mol wt 98, eq wt 49), fumaric acid (mol wt 116, eq wt 58), adipic acid (mol wt 146, eq wt 73), azelaic acid (mol wt 160, eq wt 80), sebacic acid (mol wt 174, eq wt 87), chlorendic anhydride (mol wt 371, eq wt 185.5), and trimellitic anhydride (mol wt 192, eq wt 64).

Polyhydric Alcohols. The principal types of polyol used in alkyd synthesis are shown in Table 2.

Monobasic Acids. The overwhelming majority of monobasic acids used in alkyd resins are long-chain fatty acids of natural occurrence. They may be used in the form of oil or free fatty acid. Free fatty acids are usually available and classified by their origin, *viz*, soya fatty acids, linseed fatty acids, coconut fatty acids, etc. Fats and oils commonly used in alkyd resins include castor oil, coconut oil, cottonseed oil, linseed oil, oiticica oil, peanut oil, rapeseed oil, safflower oil, soyabean oil, sunflowerseed oil, and tung oil.

The drying property of fats and oils is related to their degree of unsaturation, and hence, to iodine values.

Linolenic acid is responsible for the high yellowing tendency of alkyds based on linseed oil fatty acids. Alkyds made with nondrying oils or their fatty acids have excellent color and gloss stability. They are frequently the choice for white industrial baking enamels and lacquers.

The Concept of Functionality and Gelation

The concept of functionality and its relationship to polymer formation was greatly expanded the theoretical consideration and mathematical treatment of polycondensation systems. Thus if a dibasic acid and a diol react to form a polyester, assuming there is no possibility of other side reactions to complicate the issue, only linear polymer molecules are formed. When the reactants are present in stoichiometric amounts, the average degree of polymerization, $\bar{x}_n$ follows the equation:

$$\bar{x}_n = 1/(1 - p) \tag{1}$$

where p is the fractional extent of reaction. Thus when the reaction is driven to completion, theoretically, the molecular weight approaches infinity and the whole mass forms one giant polymer molecule. Although the material should theoretically still be soluble and fusible the molecular weight would be so high that it would not be processible by any of the existing methods. For all practical purposes it is a gel; this is the sole example of difunctional monomers being polymerized to gelation.

The functionality of the system, f, is the sum of all of the functional groups, i.e., equivalents, divided by the total number of moles of the reactants present in the system. Thus, in the above equimolar reaction system,

$$f = (1 \times 2 + 1 \times 2)/(1 + 1) = 2 \tag{2}$$

Microgel Formation and Molecular Weight Distribution

The behavior of alkyd resin reactions often deviates from that predicted by the theory of Flory. To explain this, a mechanism of microgel formation by some of the alkyd molecules at relatively early stage of the reaction was proposed. The microgel particles are dispersed and stabilized by smaller molecules in the remaining reaction mixture. As polyesterification proceeds, more microgel particles are formed, until finally a point is reached where they can no longer be kept separated. The microgel particles then coalese or flocculate, phase inversion occurs, and the entire reaction mass gels. The drying capability of an alkyd resin comes primarily from the microgel fraction. For example, when the highest molecular weight fraction representing about 20% of the total was removed through fractionation, a residual linoleic alkyd lost all ability to air dry to a hard film.

Principles for the Designing of Alkyd Resins

The process of alkyd resin designing should begin with the question "What are the intended applications of the resin?" The application dictates property requirements, such as solubility, viscosity, drying characteristics, compatibility, film hardness, film flexibility, acid value water resistance, chemical resistance, and environmental endurance With the targets in mind, a selection of oil length and a preliminary list of alternative choices of ingredients can then be made. For commercial production, the raw material list is screened based considerations of material cost, availability, yield, impact on processing cost, and potential hazard to health, safety, and the environmen. The list may be further narrowed by limitations imposed by the production equipment or other considerations. Once the oil length and ingredients are chosen, the first draft of a detailed formulation for the resin can be made.

A simple molecular approach is favored by some alkyd chemists for deriving a starting formulation. The basic premise of this approach is that when the total number of moles of the polyols is equal to or slightly larger than that of the dibasic acids, and the hydroxyl groups are present in an empirically prescribed excess amount, the probability of gelation is very small. Table 3 lists the empirical requirements for excess hydroxyl groups at various oil (fatty acid) lengths of the alkyd.

Chemical Procedures for Alkyd Resin Synthesis

Different chemical procedures may be used for the synthesis of alkyd resins. The choice is usually dictated by the selection of the starting ingredients. Procedures include the alcoholysis process, the fatty acid process, the fatty acid–oil process, and the acidolysis process.

Alkyd Resin Production Processes

Depending on the requirements of the chemical procedures, the processing method may be varied with different mechanical arrangements to remove the by-product, water, in order to drive the esterification reaction toward completion. Methods include the fusion process and the solvent process.

Process Control. The progress of the alkyd reaction is usually monitored by periodic determinations of the acid number and the solution viscosity of samples taken from the reactor. The frequency of sampling is commonly every half-hour.

Safety and Environmental Precautions

The manufacturing of alkyd resins involves a wide variety of organic ingredients. Whereas most of them are relatively mild and of low toxicity, some, such as phthalic anhydride, maleic anhydride, solvents, and many of the vinyl (especially acrylic) monomers, are known irritants or skin sensitizers and are poisonous to humans. The hazard potential of the chemicals should be determined by consulting the Material Safety Data Sheets provided by the suppliers, and recommended safety precautions in handling the materials should be practiced.

With the ever-increasing awareness of the need of environment protection, the emission of solvent vapors and organic fumes into the atmosphere should be prevented by treating the exhaust through a proper scrubber. The solvent used for cleaning the reactor is usually consumed as part of the thinning solvent. Aqueous effluent should be properly treated before discharge.

Modification of Alkyd Resins by Blending With Other Polymers

One of the important attributes of alkyds is their good compatibility with a wide variety of other coating polymers. This good compatibility comes from the relatively low molecular weight of the alkyds, and the fact that the resin structure contains, on the one hand, a relatively polar and aromatic backbone, and, on the other hand, many aliphatic side chains with low polarity. An alkyd resin in a blend with another coating polymer may serve as a modifier for the other film-former, or it may be the principal film-former and the other polymer may serve as the modifier for the alkyd to enhance certain properties. Examples of compatible blends follow.

Nitrocellulose-based lacquers often contain short or medium oil alkyds to improve flexibility and adhesion. The principal applications are furniture coatings, top lacquer for printed paper, and automotive refinishing primers.

Amino resins are probably the most important modifiers for alkyd resins. Many industrial baking enamels, such as those for appliances, coil coatings, and automotive finishes (especially refinishing enamels), are based on alkyd-amino resin blends. Some of the so-called catalyzed lacquers for finishing wood substrate require very low bake or no bake at all.

Chlorinated rubber is often used in combination with medium oil drying-type alkyds. The principal applications are highway traffic paint, concrete floor, and swimming pool paints.

Vinyl resins, i.e., copolymers of vinyl chloride and vinyl acetate which contain hydroxyl groups from the partial hydrolysis of vinyl acetate or carboxyl groups, e.g., from copolymerized maleic anhydride, may be formulated with alkyd resins to improve their application properties and adhesion. The blends are primarily used in making marine top-coat paints.

Synthetic latex house paints sometimes contain emulsified long oil or very long oil drying alkyds to improve adhesion to chalky painted surfaces.

Silicone resins with high phenyl contents may be used with medium or short oil alkyds as blends in air-dried or baked coatings to improve heat or weather resistance; the alkyd component contributes to adhesion and flexibility. Applications include insulation varnishes, heat-resistant paints, and marine coatings.

Chemically Modified Alkyd Resins

Although blending with other coating resins provides a variety of ways to improve the performance of alkyds, or of the other resins, chemically combining the desired modifier into the alkyd structure eliminates compatibility problems and gives a more uniform product. Several such chemical modifications of the alkyd resins have gained commercial importance. They include vinylated alkyds, silicone alkyds, urethane alkyds, phenolic alkyds, and polyamide alkyds.

High Solids Alkyds

There has been a strong trend in recent years to increase the solids content of all coating materials, including alkyds, to reduce solvent vapor emission. In order to raise solids and still maintain a manageable viscosity, the molecular weight of the resin must be reduced. Consequently, film integrity must be developed through further chain extension or cross-linking of the resin molecules during the "drying" step. A high cross-linking density necessitated by the lower molecular weight of the resin builds high stress in the film and causes it to be prone to cracking. Therefore, adequate flexibility should be designed into the resin structure. Chain extension and cross-linking of high solids alkyd resins are typically achieved by the use of polyisocyanato oligomers or amino resins.

Water-Reducible Alkyds

Replacing solvent-borne coatings with water-borne coatings not only reduces solvent vapor emission, but also improves the safety against the fire and health hazards of organic solvents. Alkyd resins may be made water-reducible either by converting the resin into an emulsion form or by incorporating "water-soluble" groups in the molecules.

Economic Aspects

Alkyd resins, as a family, have remained the workhorse of the coatings industry for decades. The top alkyd resin manufacturers in the United States are Cargill, Reichhold, a subsidiary of Dainippon Ink & Chemicals, Inc., and Spencer Kellog, now a part of NL Industries, Inc.

Future Prospects. Because of the efforts of the coatings industry to reduce solvent emission, there has been a clear gradual decline in the market share of alkyds as a group relative to all synthetic coating resins. However, their versatility and low cost will undoubtedly maintain them as significant players in the coatings arena. Alkyds are much more amenable to development of higher solids compositions than most other coating resins. Great strides in the development of water-borne types have also been made

TABLE 3. EXCESS HYDROXYL CONTENT REQUIRED IN ALKYD FORMULATIONS

Oil length, fatty acid, %[a]	Excess OH based on diacid equivalents, %
62 or more	0
59–62	5
57–59	10
53–57	18
48–53	25
38–48	30
29–38	32

[a]Based on C-18 fatty acids with average eq wt of 280. If the average eq wt of the monobasic acids is significantly different, adjustment is necessary.

in recent years. Another good reason to remain optimistic about alkyds for the future is that a significant portion of their raw material, fatty acids, is renewable.

K. F. LIN
Hercules Incorporated

Additional Reading

Lanson, H.J.: in J.I. Kroschwitz, ed., *Encyclopedia of Polymer Science and Engineering*, 2nd Edition, Vol. 1, John Wiley & Sons, Inc., New York, NY, 1985, pp. 644–679.

Mraz, R.G. and R.P. Silver: in N.M. Bikales, ed., *Encyclopedia of Polymer Science and Technology*, Vol. 1, John Wiley & Sons, Inc., New York, NY, 1964, pp. 663–734.

Patton, T.C.: *Alkyd Resin Technology, Formulating Techniques and Allied Calculations*, Wiley-Interscience, New York-London, 1962.

ALKYL. A generic name for any organic group or radical formed from a hydrocarbon by elimination of one atom of hydrogen and so producing a univalent unit. The term is usually restricted to those radicals derived from the aliphatic hydrocarbons, those owing their origin to the aromatic compounds being termed "aryl."

ALKYLATION. Addition of an alkyl group. These reactions are important throughout synthetic organic chemistry; for example, in the production of gasoline with high antiknock ratings for automobiles or for use in aircraft.

The nature of the products of these reactions, as well as the yields, depend upon the catalysts and physical conditions. The reactions in Equation 1 have been written to show two combination reactions of two isobutene molecules, one yielding diisobutene, which reduces to isooctane, and the other yielding a trimethylpentane by a direct reduction reaction. Specifically, the term is applied to various methods, including both thermal and catalytic processes, for bringing about the union of paraffin hydrocarbons with olefins. The process is especially effective in yielding gasoline of high octane number and low boiling range (aviation fuels).

$$2CH_3-\underset{\underset{CH_3}{|}}{C}=CH_2 + CH_3-\underset{\underset{CH_3}{|}}{C}=CH_2 \longrightarrow CH_3-\underset{\underset{CH_3}{|}}{\overset{CH_3}{\underset{|}{C}}}-CH_2-\underset{CH_3}{C}=CH_2$$

Isobutene   Isobutene   Diisobutene

$$\xrightarrow{H_2} CH_3-\underset{\underset{CH_3}{|}}{\overset{CH_3}{\underset{|}{C}}}-CH_2-CH=CH_3$$

Isooctane

$$2CH_3-\underset{\|}{\overset{CH_3}{C}}CH_3 + CH=CH-CH_3 \xrightarrow{H_2} CH_3-\underset{\underset{CH_3}{|}}{\overset{CH_3}{\underset{|}{C}}}\ \ -\underset{CH_3}{CH}-CH_2-CH_3$$

Isobutene   Isobutene   2-2-3-Trimethylpentane

In the petroleum industry, catalytic cracking units provide the major source of olefinic fuels for alkylation. A feedstock from a catalytic cracking units is typified by a C_3/C_4 charge with an approximate composition of: propane, 12.7%; propylene, 23.6%; isobutane, 25.0%; *n*-butane, 6.9%; isobutylene, 8.8%; 1-butylene, 6.9%; and 2-butylene, 16.1%. The butylenes will produce alkylates with octane numbers approximately three units higher than those from propylene.

One possible arrangement for a hydrofluoric acid alkylation unit is shown schematically in Fig. 1. Feedstocks are pretreated, mainly to remove sulfur compounds. The hydrocarbons and acid are intimately contacted in the reactor to form an emulsion, within which the reaction occurs. The reaction is exothermic and temperature must be controlled by cooling water. After reaction, the emulsion is allowed to separate in a settler, the hydrocarbon phase rising to the top. The acid phase is recycled. Hydrocarbons from the settler pass to a fractionator which produces an overhead stream rich in isobutane. The isobutane is recycled to the reactor. The alkylate is the bottom product of the fractionater (isostripper). If the olefin feed contains propylene and propane, some of the isostripper overhead goes to a depropanizer where propane is separated as an overhead

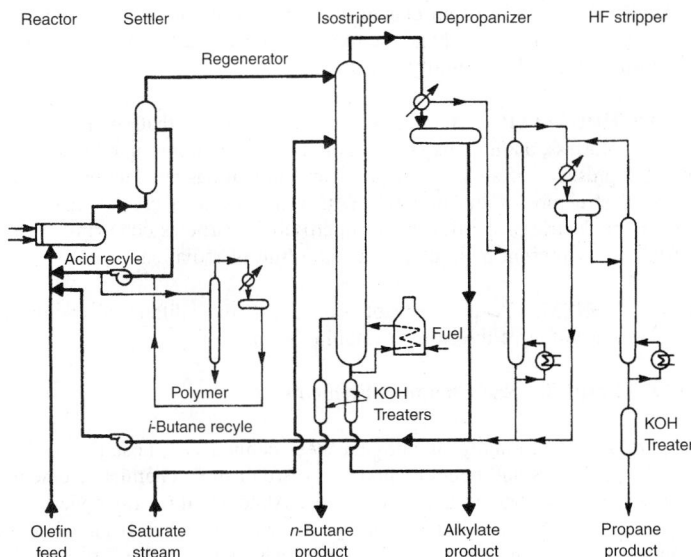

Fig. 1.   Hydrofluoric acid alkylation unit. (*UOP Process Division*)

product. A hydrofluoric acid (HF) stripper is required to recover the acid so that it may be recycled to the reactor. HF alkylation is conducted at temperature in the range of 24–38°C (75–100°F).

Sulfuric acid alkylation also is used. In addition to the type of acid catalyst used, the processes differ in the way of producing the emulsion, increasing the interfacial surface for the reaction. There also are important differences in the manner in which the heat of reaction is removed. Often, a refrigerated cascade reactor is used. In other designs, a portion of the reactor effluent is vaporized by pressure reduction to provide cooling for the reactor.

ALKYNES. A series of unsaturated hydrocarbons having the general formula C_nH_{2n-2}, and containing a triple bond between two carbon atoms. The simplest compound of this series is acetylene HC:CH. Formerly, the series was named after this compound, namely the *acetylene* series. The latter term remains in popular usage. Particularly, the older names of specific compounds, such as acetylene, allylene $CH_2C:CH$, and crotonylene $CH_3C:CCH_3$, persist. These compounds also are sometimes called *acetylenic hydrocarbons*. In the alkyne system of naming, the "yl" termination of the alcohol radical corresponding to the carbon content of the alkyne is changed to "yne." Thus, C_2H_2 (acetylene by the former system) becomes *ethyne* (the "eth" from ethyl (C_2); and C_4H_6 (crotonylene by the former system) becomes *butyne* (the "but" from butyl (C_4)). See also **Organic Chemistry**.

ALLANITE. Allanite is a rather rare monoclinic mineral of somewhat variable but quite complex chemical composition, perhaps represented satisfactorily by the formula $(Ce,Ca,Y)_2(Al,Fe)_3Si_3O_{12}(OH)$. The color of the fresh mineral is black but it is usually brown or yellow with a coating of some alteration product; often the altered crystals have the appearance of small rusty nails. It occurs characteristically in plutonic rocks like granite, syenite or diorite and is found in large masses in pegmatites. Localities in the United States are in New York, New Jersey, Virginia, and Texas. The slender prismatic crystals are sometimes called orthite. Allanite was named for its discoverer, T. Allan. Orthite was so named from the Greek word meaning straight, in reference to the straight prisms, a common habit of this mineral.

ALLELOPATHIC SUBSTANCE. A material contained within a plant that tends to suppress the growth of other plant species. The alkaloids present in several seed-bearing plants are believed to play an allelopathic role. Other suspected allelopathic substances contained in some plants include phenolic acids, flavonoids, terpenoid substances, steroids, and organic cyanides.

ALLOBAR. A form of an element differing in atomic weight from the naturally occurring form, hence a form of element differing in isotopic composition from the naturally occurring form.

ALLOCHROMATIC. With reference to a mineral that, in its purest state, is colorless, but that may have color due to submicroscopic inclusions, or to the presence of a closely related element that has become part of the chemical structure of the mineral. With reference to a crystal that may have photoelectric properties due to microscopic particles occurring in the crystal, either present naturally, or as the result of radiation.

ALLOMERISM. A property of substances that differ in chemical composition but have the same crystalline form.

ALLOTROPES. See **Chemical Elements**.

ALLOYS. Traditionally, an alloy has been defined as a substance having metallic properties and being composed of two or more chemical elements of which at least one is a metal (ASM). Although this still covers the general use of the word, in recent years alloy also has been used in connection with other, non-metallic, materials. Most metals are soluble in one another in their liquid state. Thus alloying procedures usually involve melting. However, alloying by treatment in the solid state without melting can be accomplished in some instances by such methods as powder metallurgy. When molten alloys solidify, they may remain soluble in one another, or may separate into intimate mechanical mixtures of the pure constituent metals. More often, there is partial solubility in the solid state and the structure consists of a mixture of the saturated solid solutions. Another important type of solid phase is the inter-metallic compound, which is characterized by hardness and brittleness and usually has only limited solid solubility with the other phases present. The interactions of two or more elements both in the liquid and solid state are effectively characterized by phase diagrams. Where only two principal materials are involved, *binary alloy* is the term used. Three principal ingredients are referred to as *ternary alloy*. Beyond three components, the material may be referred to as a multi-composition system or alloy.

The decade of the 1980s witnessed the development of hundreds of new alloys, involving not only the traditional metals, but much greater use of the less common chemical elements, such as indium, hafnium, etc. In this encyclopedia, alloys of a chemical element are discussed mainly under that particular element, or in an entry immediately following. Also check alphabetical index.

In addition to the appearance of numerous new alloys, sometimes called superalloys, recent developments in this field include many relatively new processes and methodologies, such as electron beam refining, rapid solidification, single-crystal superalloys, and metallic glasses, among others. Some of these are described in separate articles in this encyclopedia.

Motivation for the development of new alloys is found in nearly all consuming areas, but particular emphasis has been given to the expanding and increasingly demanding requirements of the aircraft and aerospace industries, including much attention directed to the lighter elements (titanium, aluminum, etc.); the needs of the military; the very difficult requirements of jet engine parts; and the electronics industry where much attention has been directed toward the less common metals. Within recent years, metallurgists also have come to appreciate that processing of alloys can be of as much importance as the elements they contain. New processes have been developed during the past decade or so, including rapid solidification, electron beam refining, and many others. Metal alloy research also has been impacted by the rapidly and continuously expanding science and art of making composites, often involving ceramics, graphite, organics, etc. in addition to metals. Knowledge of alloys not only must assist the applications of simply the alloys themselves, but also how the alloys perform in a composite part.

Predicting the Performance of Alloys

It is well known that many important alloy combinations have properties that are not easy to predict, simply on the basis of knowledge of the constituent metals. For example, copper and nickel, both having good electrical conductivity, form solid-solution type alloys having very low conductivity, or high resistivity, making them useful as electrical resistance

wires. In some cases very small amounts of an alloying element produce remarkable changes in properties, as in steel containing less than 1% carbon with the balance principally iron. Steels and the age-hardening alloys depend on heat treatment to develop special properties such as great strength and hardness. Other properties which can be developed to a much higher degree in alloys than in pure metals include corrosion-resistance, oxidation-resistance at elevated temperatures, abrasion- or wear-resistance, good bearing characteristics, creep strength at elevated temperatures, and impact toughness. However, solid state physics has been successful in explaining many of these properties of metal and alloys.

There are various types of alloys. Thus, the atoms of one metal may be able to replace the atoms of the other on its lattice sites, forming a substitutional alloy, or solid solution. If the sizes of the atoms, and their preferred structures, are similar, such a system may form a continuous series of solutions; otherwise, the miscibility may be limited. Solid solutions, at certain definite atomic proportions, are capable of undergoing an order-disorder transition into a state where the atoms of one metal are not distributed at random through the lattice sites of the other, but form a superlattice. Again, in certain alloy systems, inter-metallic compounds may occur, with certain highly complicated lattice structures, forming distinct crystal phases. It is also possible for light, small atoms to fit into the interstitial positions in a lattice of a heavy metal, forming an interstitial compound.

In this encyclopedia alloys of chemical elements of alloying importance are discussed under that particular element.

Alloy Phase Diagram Data Programme

Alloy phase diagrams have been known since 1829 when the Swedish scientist Rydberg, who observed the thermal effects that occur during the cooling of binary and ternary alloys from a molten condition. Gibbs many years later published a treatise on the theory of heterogeneous equilibria. The practical importance of phase diagrams awaited the development of the phase diagram for the iron-carbon system, which became central to the metallurgy of steel. See **Iron Metals, Alloys, and Steels**. With the development over the years of scores of binary alloys, considering the number of chemical elements involved, and then followed by ternary and much more complex alloys—with many hundreds of professionals in the metal sciences contributing knowledge—the problems of collecting and of easily locating such information took on formidable proportions. The start of an effective database was the publication of a compilation, by Hansen in 1936, of information gleaned from the literature on 828 binary systems, for which sufficient data were available to construct phase diagrams for 456 binary systems. An English version of the German works, updated to some extent, appeared in 1958. This compilation included 1324 binary systems and 717 binary phase diagrams. A supplementary volume by R.P. Elliott brought the number of binary phase diagrams to 2067 and a later work by F.A. Skunk (1969) included data on 2380 systems. These efforts became key reference works for metallurgists concerned with alloy development and alloy applications. For obvious reasons, information on ternary phase diagrams and other multicomponent systems was far less satisfactory.

To improve this important metallurgical database, the National Bureau of Standards (U.S.) and the American Society for Metals, after many prior deliberations, each signed a memorandum of agreement to proceed with a data programme for alloy phase diagrams, concentrating on binary systems. As early as 1975, T.B. Massalski (Carnegie-Mellon University), then chairman of the programme, observed that a knowledge of phase diagram data is basic for the technological application of metals and alloys; that any programme to provide critically evaluated data would have to be a worldwide enterprise because there is too much work for any institution or organization, or even any country, to accomplish the task alone; that the programme should deal with binary and multicomponent systems; that a computerized bibliographic database should be developed; that computer technology should be used to provide data for the generation of phase diagrams at remote terminals; and that funding for the programme should be sought. As of the late 1980s, many of these objectives had been achieved. Among organizations not previously mentioned, cooperation has been given by the Institute of Metals (U.K.), the U.K. Universities' Science Research Council, the Max-Planck Institut für Metallforschung (Stuttgart), among several other sources, including funding from various interested corporations.

The massive task required that data be sorted into categories by some thirty editors. The first comprehensive publication to be released thus far is from the American Society for Metals (ASM International), entitled "Binary Alloy Phase Diagrams," which contains up-to-date and comprehensive phase diagram information for more than 1850 alloy systems, representing the first major release of critically evaluated phase diagrams since 1969.

Importance of Phase Diagrams

As pointed out by Massalski and Prince (1986), phase diagrams are graphic displays of the thermodynamic relationships of one or more elements at different temperatures and pressures. It has been stated that phase diagrams are to the metallurgist what anatomy is to the medical profession or cartography to the explorer. To explain this analogy, reference is made to a specific phase diagram (Fig. 1). This gold-silicon (Au-Si) phase diagram is a two-dimensional mapping of the phases that form between Au and Si as a function of temperature and of alloy composition. In Fig. 1(a), the alloy composition is defined in terms of the percentage of atoms of Si in the alloy. It will be noted that a dramatic lowering of the freezing point of pure gold (1064.43°C) occurs upon the addition of silicon. Conversely, there is a more regular depression of the freezing point of Si (1414°C) upon the addition of Au.

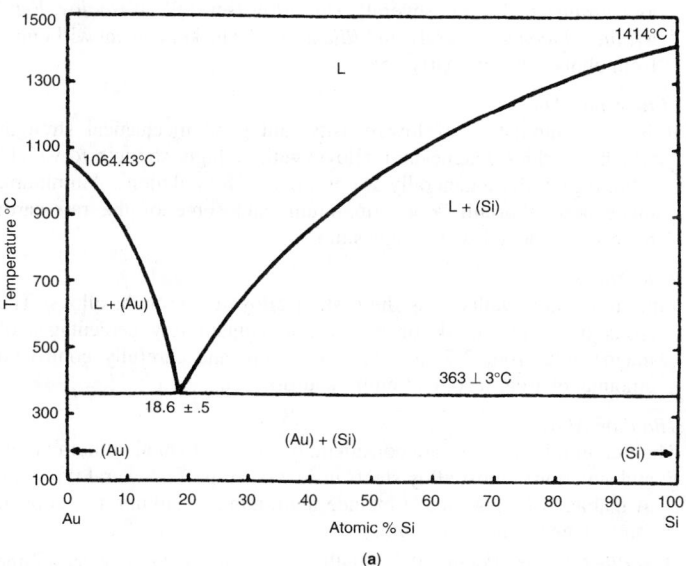

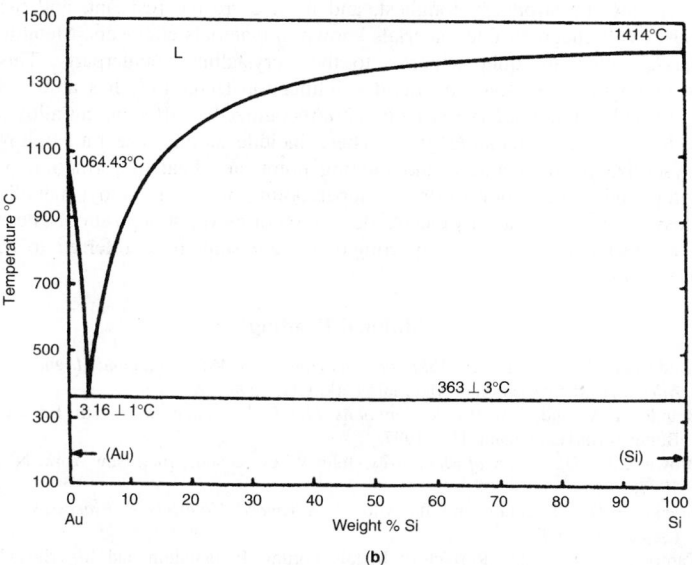

Fig. 1. Representative binary alloy phase diagrams: **(a)** the Au-Si phase diagram with compositions in atomic percent; **(b)** the Au-Si phase diagram with compositions in weight percent. (*ASM News*)

The two upper curves, termed the *liquidus curve*, define the temperatures at which Au-Si alloys begin to solidify. The curves meet at 363°C at an alloy composition containing 18.6 atomic percent Si. At this temperature, all Au-Si alloys, irrespective of composition, complete their solidification by the eutectic separation of a fine mixture of Au and Si from the liquid phase containing 18.5 atomic percent Si. The horizontal line at 363°C is called the *solidus* because below such a line all of the alloys are completely solid.

The effect of presenting the alloy composition in terms of the weight percentage of Si is shown in Fig. 1(b). The liquidus curves drop even more dramatically towards the Au-rich side of the phase diagram and the eutectic liquid at 363°C contains only 3.16 weight percent Si. The movement of the eutectic composition, the lowest-melting alloy composition, from 18.6 percent Si of Fig. 1(a) to 3.15 weight percent Si in Fig. 1(b) simply reflects the great difference in the atomic weights of Au and Si.

Massalski and Prince selected this particular phase diagram because of its simplicity for illustration and because this particular phase diagram is of considerable importance in the semiconductor device industry. Silicon chips are frequently bonded to a heat sink, using a gold alloy or more frequently a Au-Si alloy foil placed between them. Upon heating above 363°C, the Au reacts with Si to form a brazed joint between the silicon chip and the heat sink.

Phase diagrams are condensed presentations of a large amount of information. They provide quantitative information on the phases present under given conditions of alloy composition and temperature and, to the experienced metallurgist, a guide to the distribution of the phases in the microstructure of the alloy. They also dictate what alterations in phase constitution will occur with changing conditions, whether these be alteration of alloy composition, temperature, pressure or atmosphere in equilibrium with the material.

Further, the phases present in an alloy and their morphological distribution within the microstructure, define the mechanical, chemical, electrical, and magnetic properties that may be achievable. Thus, we have the essential link between the engineering properties of an alloy and its phase diagram. Indeed, a distinctive feature of metallurgy as a profession is that it is primarily concerned with the relationship between the constitution and the properties of alloys. Phase diagram data are key elements to understanding, and thereby controlling, the properties of alloys.

Broad Categories of Alloys

Although there are thousands of alloys, with many new alloys appearing each year, there are certain traditional alloys that serve the vast majority of materials needs. The bulk of new alloys, although extremely important, are frequently application-specific. The broad classes are described briefly as follows.

Cast Ferrous Metals.

Gray, Ductile, and High-Alloy Irons
In gray iron, most of the contained carbon is in the form of graphite flakes, dispersed throughout the iron. In ductile iron, the major form of contained carbon is graphite spheres, which are visible as dots on a ground surface. In white iron, practically all contained carbon is combined with iron as iron carbide (cementite), a very hard material. In malleable iron, the carbon is present as graphite nodules. High-alloy irons usually contain an alloy content in excess of 3%.

Malleable Iron
The two main varieties of malleable iron are ferritic and pearlitic, the former more machinable and more ductile; the latter stronger and harder. Carbon in malleable iron ranges between 2.30 and 2.65%. Ranges of other constituents are: manganese, 0.30 to 0.40%; silicon, 1.00 to 1.50%; sulfur, 0.07 to 0.15%; and phosphorus, 0.05 to 0.12%.

Carbon and Low-Alloy Steels
Low-carbon cast steels have a carbon content less than 0.20%; medium-carbon steels, 0.20 to 0.50%; and high-carbon steels have in excess of 0.50% carbon. Ranges of other constituents are: manganese, 0.50 to 1.00%; silicon, 0.25 to 0.80%; sulfur, 0.060% maximum; and phosphorus, 0.050% maximum.

Low-alloy steels have a carbon content generally less than 0.40% and contain small amounts of other elements, depending upon the desired end-properties. Elements added include aluminum, boron, chromium, cobalt, copper, manganese, molybdenum, nickel, silicon, titanium, tungsten, and vanadium.

High-Alloy Steels

When "high-alloy" is used to describe steel castings, it generally means that the castings contain a minimum of 8% nickel and/or chromium. Commonly thought of as stainless steels, nevertheless *cast grades* should be specified by ACI (Alloy Casting Institute) designations and not by the designations that apply to similar *wrought alloys*.

Wrought Ferrous Metals.

Carbon Steels

These steels account for over 90% of all steel production. There are numerous varieties, depending upon carbon content and method of production. In one classification, there are *killed* steels, *semikilled* steels, *rimmed* steels, and *capped* steels. These are described in considerable detail under **Iron Metals, Alloys, and Steels**.

High-Strength Low-Alloy Steels

There are several varieties, with high-yield strength depending mainly on the precipitation of martensitic structures from an austenitic field during quenching. Small additions of alloy elements, such as manganese and copper, are dissolved in a ferritic structure to obtain high strength and corrosion resistance.

Low and Medium-Alloy Steels

The two basic types are (1) *through* hardenable, and (2) *surface* hardenable. Subcategories of surface hardenable alloys include carburizing alloys, flame and induction-hardening alloys, and nitriding alloys.

Stainless Steels

A stainless steel is defined as iron-chromium alloy that contains at least 11.5% chromium. There are three major categories: (1) austenic, (2) ferritic, and (3) martensitic, depending upon the metallurgical structure. There are scores of varieties. Type 302 is the base alloy for austenitic stainless steels. Representative stainless steels in this category provide some insight into why so many varieties are made, and how rather small changes in composition and production can bring about significant differences in the final properties of the various stainless steels. A slightly lower carbon content improves weldability and inhibits carbide formation. An increase in nickel content lowers the work hardening. By increasing both chromium and nickel, better corrosion and scaling resistance is achieved. The addition of sulfur or selenium increases machinability. The addition of silicon increases scaling resistance at high temperature. Small amounts of molybdenum improve resistance to pitting corrosion and temperature strength.

High-Temperature, High-Strength, Iron-Base Alloys

There are two general objectives in making these alloys: (1) they can be strengthened by a martensitic type of transformation, and (2) they will remain austenitic regardless of heat treatment and derive their strength from cold working or precipitation hardening. Again, there are numerous types. Considering the main types, the carbon content may range from 0.05% to 1.10%; manganese, 0.20 to 1.75%; silicon, 0.20 to 0.90%; chromium, 1.00 to 20.75%; nickel, 0 to 44.30%; cobalt, 0 to 19.50%; molybdenum, 0 to 6.00%; vanadium, 0 to 1.9%; tungsten, 0 to 6.35%; copper, 0 to 3.30%; columbium (niobium), 0 to 1.15%; tantalum, 0 to <1%; aluminum, 0 to 1.17%; and titanium, 0 to 3%.

Ultrahigh-Strength Steels

Normally a steel is considered in this category if it has a yield strength of 160,000 psi or more. The first of these steels to be produced was a chromium-molybdenum alloy steel, shortly followed by a stronger chromium-nickel-molybdenum grade.

Free-Machining Steels

Normally, the carbon content is kept under 0.10%, but as much as 0.25% carbon has little deleterious effect on machinability. Aluminum and silicon are held to a minimum (aluminum not used as a de-oxidizer where machinability is extremely important). Lead, sulfur, bismuth, selenium, and tellurium (0.04%) improve machinability when in the proper combination. Sulfur improves machinability by combining with any manganese and oxygen present to form oxysulfides.

Nonferrous Metals.

Aluminum Alloys

These alloys are available as wrought or cast alloys. The principal metals alloyed with aluminum include copper, manganese, silicon, magnesium, and zinc. These alloys are discussed in considerable detail under **Aluminum Alloys**.

Copper Alloys

These alloys are available as wrought or cast alloys. The principal wrought copper alloys are the brasses, leaded brasses, phosphor bronzes, aluminum bronzes, silicon bronzes, beryllium coppers, cupronickels, and nickel silvers. The major cast copper alloys include the red and yellow brasses, manganese, tin, aluminum, and silicon bronzes, beryllium coppers, and nickel silvers. The chemical compositions range widely. For example, a leaded brass will contain 60% copper, 36 to 40% zinc, and lead up to 4%; a beryllium copper is nearly all copper, containing 2.1% beryllium, 0.5% cobalt, or nickel, or in another formulation, 0.65% beryllium, and 2.5% cobalt.

Nickel Alloys

Although nickel is present in varying amounts in stainless steel *commercially* a high-nickel stainless steel is not categorized as a nickel *alloy*, but rather as a stainless steel. Most nickel alloys are proprietary formulations and hence designated by trade names, such as *Duranickel, Monel* (several), *Hastelloy* (several), *Waspaloy, Rene 41, Inco, Inconel* (several), and *Illium* G. The nickel content will range from about 30% to nearly 95%.

Magnesium Alloys

It is the combination of low density and good mechanical strength which provides magnesium alloys with a high strength-to-weight ratio. Again, these generally are proprietary formulations. Aluminum, manganese, thorium, zinc, zirconium, and some of the rare-earth metals are alloyed with magnesium.

Zinc Alloys

Zinc alloys are available as die-casting alloys or wrought alloys. The sprincipal alloys used for die casting contain low percentages of magnesium, from 3.5 to 4.3% aluminum, and carefully controlled amounts of iron, lead, cadmium, and tin.

Titanium Alloys

The titanium-base alloys are considerably stronger than aluminum alloys and superior to most alloy steels in several respects. Several types are available. Alloying metals include aluminum, vanadium, tin, copper, molybdenum, and chromium.

Metallic Glasses. Potentially, metallic glasses and metastable crystalline alloys are the strongest, toughest, and most corrosive resistant, and the most easily magnetizable materials known to materials engineers. Metallic glasses often are quite superior to their crystalline counterparts. This is an important reason why rapid solidification technology has attracted worldwide attention. Several factors are recognized as affecting an alloy's ability to form a metallic glass. These include atomic size ratio, alloy crystallization temperature and melting point, and heat of formation of compounds. In addition to drastic supercooling of a metal alloy, metallic glasses have been made by electro-deposition or by vapor deposition. These non-crystalline metal or alloy compounds are sometimes referred to as *amorphous* alloys.

Additional Reading

Avedesian, M. and H. Baker: *ASM Specialty Handbook, Magnesium and Magnesium Alloys*, ASM International, Materials Park, OH, 1998.

Brandes, E.A. and G.B. Brook: *Smithells Metals Reference Book*, 7th Edition, Butterworth-Heinemann, UK, 1997.

Brown, C.D.: *Dictionary of Metallurgy*, John Wiley & Sons, Inc., New York, NY, 1997.

Bryskin, B.D.: "Rhenium and Its Alloys," *Advanced Materials & Processes*, **22** (September, 1992).

Cardonne, S.M. et al.: "Refractory Metals Forum: I: Tantalum and Its Alloys," *Advanced Materials & Processes*, 16 (September, 1992).

Clement, T.P., Parsonage, T.B., and M.B. Kuxhaus: "Ti$_2$AlNb = Based Alloys Outperform Conventional Titanium Aluminides," *Advanced Materials & Processes*, **37** (March 1992).

Davis, J.R.: *ASM Materials Engineering Dictionary*, ASM International, Materials Park, OH, 1992.

Davis, J.R.: *ASM Specialty Handbook, Aluminum and Aluminum Alloys*, ASM International, Materials Park, OH, 1993.

Davis, J.R.: *ASM Specialty Handbook, Stainless Steels*, ASM International, Materials Park, OH, 1994.

Davis, J.R.: *ASM Specialty Handbook, Carbon and Alloy Steels*, ASM International, Materials Park, OH, 1995.

Davis, J.R.: *ASM Specialty Handbook of Cast Iron*, ASM International, Materials Park, OH, 1996.

Davis, J.R.: *ASM Specialty Handbook, Heat-Resistant Materials*, ASM International, Materials Park, OH, 1997.

Eillenauer, J.P., T.G. Nieh, and J. Wadsorth: "Tungsten and Its Alloys," *Advanced Materials & Processes*, **28** (September, 1992).

Elliott, R.P.: *Constitution of Binary Alloys, First Supplement*, McGraw-Hill Companies, Inc., New York, NY, 1965.

Fremond, M. and S. Miyazaki: *Shape Memory Alloys*, Springer-Verlag New York, Inc., New York, NY, 1996.

Frick, J., Editor: *Woldman's Engineering Alloys*, 8th Edition, ASM International, Materials Park, OH, 1994.

Habashi, F.: *Alloys, Properties, Applications*, John Wiley & Sons, Inc., New York, NY, 1998.

Hansen, M.: *Der Aufbau Der Zweistoff-Legierungen*, Julis Springer-Verlag, Berlin, 1936.

Hansen, M. and K. Anderko: *Constitution of Binary Alloy*, McGraw-Hill Companies, Inc., New York, NY, 1958.

Jackman, L.A., G.E. Maurer, and S. Widge: "New Knowledge About 'White Spots' in Superalloys," *Advanced Materials & Processes*, 18 (May, 1993).

Kane, R.D.: "Super Stainless Steels Resist Hostile Environments," *Advanced Materials & Processes*, 16 (July, 1993).

Kane, R.D. and R.G. Taraborelli: "Selecting Alloys to Resist Heat and Corrosion," *Advanced Materials & Processes*, 22 (April, 1993).

Kutz, M.: *Mechanical Engineers Handbook*, 2nd Edition, John Wiley & Sons, Inc., New York, NY, 1998.

Lai, G.Y.: *High-Temperature Corrosion of Engineering Alloys*, ASM International, Materials Park, OH, 1990.

Lamb, S.: *Practical Handbook of Stainless Steels & Nickel Alloys*, Co-published by CASTI and ASM International, Materials Park, OH, 1999.

Massalski, T.B.: *Binary Alloy Phase Diagrams*, American Society for Metals, Metals Park, OH, 1986.

Massalski, T.B. and A. Prince: "An International Programme for the Critical Evaluation of Alloy Phase Diagram Data, *ASM News*, 4–5 (November, 1986).

McCaffrey, T.J.: "Combined Strength and Toughness Characterize New Aircraft Alloy," *Materials & Processes*, **47** (September, 1992).

Otsuka, K. and C.M. Wayman: *Shape Memory Materials*, Cambridge University Press, New York, NY, 1998.

Rahoi, D.: *Alloy Digest, Print Complete Set 1952 to 1998*, ASM International, Materials Park, OH, 1998.

Rioja, R.J. and R.H. Graham: "Al-Li Alloys Find Their Niche," *Advanced Materials & Processes*, 23 (June, 1992).

Schweitzer, P.A.: *Corrosion Resistance Tables*, 4th Edition, ASM International, Materials Park, OH, 1995.

Shields, J.A., Jr.: "Refractory Metals Forum II: Molybdenum and Its Alloys," *Advanced Materials & Processes*, **28** (October, 1992).

Skunk, F.A.: *Constitution of Binary Alloys, Second Supplement*, McGraw-Hill Companies, Inc., New York, NY, 1969.

Smith, W.F.: *Structure and Properties of Engineering Alloys*, 2nd Edition, McGraw-Hill Companies, New York, NY, 1993.

Staff: *ASM Engineered Materials Reference Book*, 2nd Edition, ASM International, Materials Park, OH, 1994.

Staff: *Properties and Selection: Irons, Steels, and High Performance Alloys*, CD-ROM Edition, ASM International, Materials Park, OH, 1999.

Staff: *Alloy Phase Diagrams*, 10th Edition, Vol. 003, ASM International, Materials Park, OH, 1992.

Staff: *Properties and Units for Engineering Alloys*, ASM International, Materials Park, OH, 1997.

Staff: "Advances in Aluminum and Alloys," *Advanced Materials & Processes*, ASM International, Materials Park, Ohio, **17** (January, 1992).

Staff: *Metals and Alloys in the Unified Numbering System*, Society of Automotive Engineers, Warrendale, PA, 1998.

Toropova, L.S., D.G. Eskin, M.L. Kharakterova, and T.V. Dobatkina: *Advanced Aluminum Alloys Containing Scandium*, Structure and Properties, Gordon and Breach Science Publishers, Newark, NJ, 1998.

Verrill, et al.: *Properties and Selection: Nonferrous Alloys and Special-Purpose Materials Metals Handbook*, 10th Edition, American Institute of Chemical Engineers (AIChE) New York, NY, 1999.

ALLYL ALCOHOL AND MONOALLYL DERIVATIVES. The technology of introducing a new functional group to the double bond of allyl alcohol was developed in the mid-1980s. Allyl alcohol is accordingly used as an intermediate compound for synthesizing raw materials such as epichlorohydrin and 1,4-butanediol, and this development is bringing about expansion of the range of uses of allyl alcohol.

Physical Properties

Allyl alcohol is a colorless liquid having a pungent odor; its vapor may cause severe irritation and injury to eyes, nose, throat, and lungs. It is also corrosive. Allyl alcohol is freely miscible with water and miscible with many polar organic solvents and aromatic hydrocarbons, but is not miscible with *n*-hexane. It forms an azeotropic mixture with water and a ternary azeotropic mixture with water and organic solvents. Allyl alcohol has both bacterial and fungicidal effects. Properties of allyl alcohol are shown in Table 1.

Chemical Properties

Addition Reactions. The C=C double bond of allyl alcohol undergoes addition reactions typical of olefinic double bonds.

Hydroformylation. Hydroformylation of allyl alcohol is a synthetic route for producing 1,4-butanediol, a raw material for poly(butylene terephthalate), an engineering plastic.

Substitution of Hydroxyl Group. The substitution activity of the hydroxyl group of allyl alcohol is lower than that of the chloride group of allyl chloride and the acetate group of allyl acetate. However, allyl alcohol undergoes substitution reactions under conditions in which saturated alcohols do not react. Reactions proceed in catalytic systems in which a π-allyl complex is considered as an intermediate.

Oxidation. The C=C double bond of allyl alcohol undergoes epoxidation by peroxide, yielding glycidol. This epoxidation reaction is applied in manufacturing glycidol as an intermediate for industrial production of glycerol.

Industrial Manufacturing Processes for Allyl Alcohol

There are four processes for industrial production of allyl alcohol. One is alkaline hydrolysis of allyl chloride. A second process has two steps. The first step is oxidation of propylene to acrolein and the second step is reduction of acrolein to allyl alcohol by a hydrogen transfer reaction, using isopropyl alcohol. At present, neither of these two processes is being used industrially. Another process is isomerization of propylene oxide. Until 1984, all allyl alcohol manufacturers were using this process. Since 1985 Showa Denko K.K. has produced allyl alcohol industrially by a new process which they developed. This process, which was developed partly for the purpose of producing epichlorohydrin via allyl alcohol as the intermediate, has the potential to be the main process for production of allyl alcohol. The reaction scheme is as follows:

$$CH_2{=}CHCH_3 + CH_3COOH + 1/2\ O_2 \xrightarrow{Pd} CH_2{=}CHCH_2O\overset{O}{\overset{\|}{C}}CH_3$$

$$CH_2{=}CHCH_2O\overset{O}{\overset{\|}{C}}CH_3 + H_2O \underset{\rightleftarrows}{\overset{H^+}{}} CH_2{=}CHCH_2OH + CH_3COOH$$

TABLE 1. PROPERTIES OF ALLYL ALCOHOL

Property	Value
molecular formula	C_3H_6O
molecular weight	58.08
boiling point, °C	96.90
freezing point, °C	−129.00
density, d^{20}_4	0.8520
refractive index, n^{20}_D	1.413
viscosity at 20°C, mPa·s(= cP)	1.37
flash point[a] °C	25
solubility in water at 20°C, wt %	infinity

[a] Closed cup.

The world's manufacturers of allyl alcohol are ARCO Chemical Company, Showa Denko K.K., Daicel Chemical Industries, and Rhône-Poulenc Chimie; total production is approximately 70,000 tons per year.

Monoallyl Derivatives

In this article, mainly monoallyl compounds are described. Diallyl and triallyl compounds used as monomers are covered elsewhere.

Reactivity of Allyl Compounds

Hydrosilylation. The addition reaction of silane

$$H—Si\diagup\diagdown$$

to the C=C double bond of allyl compounds is applied in the industrial synthesis of silane coupling agents.

π-Allyl Complex Formation. Allyl halide, allyl ester, and other allyl compounds undergo oxidative addition reactions with low atomic valent metal complexes to form π-allyl complexes.

Physical Properties of Derivatives

The physical properties of some important monoallyl compounds are summarized in Table 2.

Allyl Chloride

This derivative, abbreviated AC, is a transparent, mobile, and irritative liquid. It can be easily synthesized from allyl alcohol and hydrogen chloride. However, it is industrially produced by chlorination of propylene at high temperature.

Uses. Allyl chloride is industrially the most important allyl compound among all the allyl compounds. It is used mostly as an intermediate compound for producing epichlorohydrin, which is consumed as a raw material for epoxy resins.

Allyl Esters

Allyl Acetate. Allyl acetate is produced mostly for manufacturing allyl alcohol.

Allyl Methacrylate. At present, allyl methacrylate, AMA, is used mostly as a raw material for silane coupling agents.

Allyl Ethers

The C–H bond of the allyl position easily undergoes radical fission, especially in the case of allyl ethers, reacting with the oxygen in the air to form peroxide compounds.

Therefore, in order to keep allyl ether for a long time, it must be stored in an air-tight container under nitrogen.

Allyl Glycidyl Ether. This ether is used mainly as a raw material for silane coupling agents and epichlorohydrin rubber.

Allyl Amines

Allylamine. This amine can be synthesized by reaction of allyl chloride with ammonia at the comparatively high temperature of 50–100°C, or at lower temperatures using $CuCl_2$ or CuCl as the catalyst.

Dimethylallylamine. Dimethylallylamine is used in the production of insecticides and pesticides.

Safety and Handling

Most allyl compounds are toxic and many are irritants. Those with a low boiling point are lachrymators. Precautions should be taken at all times to ensure safe handling.

NOBUYUKI NAGATO
Showa Denko K.K.

Additional Reading

Allyl Alcohol, Technical Publication SC: 46-32, Shell Chemical Corp., San Francisco, CA, Nov. 1, 1946.
Schildknecht, C.E.: *Allylic Compounds and Their Polymers,* John Wiley & Sons, Inc., New York, NY, 1973.
Raech, H. Jr.: *Allylic Resins and Monomers,* Reinhold Publishing Corporation, New York, NY, 1965.

ALLYL CHLORIDE. See **Chlorinated Organics**; **Allyl Alcohol and Monoallyl Derivatives**.

ALLYL ESTER RESINS. The allyl radical ($CH_2CH=CH_2$) is the basis of the allyl family of resins. Allyl esters are based on monobasic and dibasic acids and are available as low-viscosity monomers and thermoplastic prepolymers. They are used as crosslinking agents for unsaturated polyester resins and in the preparation of reinforced thermoset molding compounds and high-performance transparent articles. All modern thermoset techniques may be used for processing allyl resins.

The most widely used allyls are the monomers and prepolymers of diallyl phthalate and diallyl isophthalate. These are readily converted into thermoset molding compounds and into preimpregnated glass cloths and papers.

Diethyleneglycol-*bis*-(allylcarbonate), marketed as *CR-39*™, is finding increasing use where optical transparency is required. It is the primary material used in the manufacture of *plastic lenses for eyewear* because of its light weight, dimensional stability, abrasion resistance, and dye-ability. Other applications for this product include instrument panel covers, camera filters, and myriad glazing uses. In these applications, the solvent and chemical resistance of the material are important.

Other allyl monomers of commercial significance are diallyl fumarate and diallyl maleate. These highly reactive trifunctional monomers contain two types of polymerizable double bonds.

Allyl methacrylate also exhibits dual functionality and finds use as both a crosslinking agent and as a monomer intermediate. Triallyl cyanurate has found use as a crosslinking agent in unsaturated polyester resins.

Most diallyl phthalate compounds are used in critical electrical/electronic applications requiring high reliability under long-term adverse environmental conditions. Compatibility with modern electronic finishing technology, such as vapor phase soldering, is inherent in these materials.

E. J. SARE
PPG Industries Inc.

ALPHA (α). 1. A prefix denoting the position of substituting atom or group in an organic compound. The Greek letters α, β, γ, etc., are usually not identical with the IUPAC numbering system 1, 2, 3, etc., since they

TABLE 2. PROPERTIES OF IMPORTANT ALLYL COMPOUNDS

Property	Allyl chloride	Allyl acetate	Allyl methacrylate	AGE[a]	Allyl amine	DMAA[b]
molecular formula	C_3H_5Cl	$C_5H_8O_2$	$C_7H_{10}O_2$	$C_6H_{10}O_2$	C_3H_7N	$C_5H_{11}N$
molecular weight	76.53	100.12	126.16	114.14	57.10	85.15
boiling point, °C	44.69	104	150	153.9	52.9	64.5
freezing point, °C	−134.5	−96	−60	−100	−88.2	
density, d^{20}_4	0.9382	0.9276	0.934	0.9698	0.7627	0.72
viscosity at 20°C, mPa · s(= cP)	3.36	0.52	13	1.20		0.44
flash point, °C	−31.7	6	33	57.2	−29	−23

[a] Allyl glycidyl ether.
[b] Dimethylallylamine.

do not start from the same carbon atom. However, α and β are used with naphthalene ring compounds to show the 1 and 2 positions, respectively. α, β, etc., are also used to designate attachment to the side chain of a ring compound.

2. Both a symbol and a term used for relative volatility in distillation.

3. Symbol for optical rotation.

4. A form of radiation consisting of helium nuclei.

5. The major allotropic form of a substance, especially of metals, e.g., α-iron.

ALPHA DECAY.   The emission of alpha particles by radioactive nuclei. The name *alpha particle* was applied in the earlier years of radioactivity investigations, before it was fully understood what alpha particles are. It is known now that alpha particles are the same as helium nuclei. When a radioactive nucleus emits an alpha particle, its atomic number decreases by $Z = 2$ and its mass number by $A = 4$. The process is a spontaneous nuclear reaction, and the radionuclide that undergoes the emission is known as an *alpha emitter*.

The entire energy released by the transition is carried away by the product nuclei. Therefore, a spectrum of alpha-particle numbers as a function of energy shows a series of distinct peaks, each corresponding to a single alpha-particle transition. To conserve both energy and momentum, the energy must be shared by the two product nuclei, with the daughter nucleus ($^{A-4}Z - 2$) recoiling away from the direction of emission of the alpha particle. If E_X and M_X are, respectively, the kinetic energy and mass of the alpha particle and E_R and M_R the kinetic energy and mass of the recoiling product nucleus, the transition energy is $Q = E_\alpha + E_R$; and the kinetic energy of the emitted alpha particle is $E_\alpha = [M_R(M_\alpha + M_R)]Q$.

Almost all radioactive nuclides that emit alpha particles are in the upper end of the periodic table, with atomic numbers greater than 82 (lead), but a few alpha-particle emitting nuclides are scattered through lower atomic numbers. The reason why alpha-particle emitters are limited to nuclides with larger mass numbers is that generally only in this region is alpha-particle emission energetically possible. Most radioactive nuclides with smaller mass numbers emit beta-particle radiation.

See **Particles (Subatomic)**; and **Radioactivity**.

ALPHA PARTICLE.   A helium nucleus emitted spontaneously from radioactive elements both natural and, man-made. Its energy is in the range of 4–8 MeV and is dissipated in a very short path, i.e., in a few centimeters of air or less than 0.005 mm of aluminum. It has the same mass (4) and positive charge (2) as the helium nucleus. Accelerated in a cyclotron alpha-particles can be used to bombard the nuclei of other elements.

See also **Beta Decay**; **Helium**; and **Radioactivity**.

ALTMAN, SIDNEY (1939–).   Awarded the Nobel prize in Chemistry in 1989 jointly with Thomas R. Cech for the discovery that RNA acts as a biological catalyst as well as a carrier of genetic information. He received his Doctorate in 1967 from the University of Colorado.

ALUM.   A series (*alums*) of usually isomorphous crystalline (most commonly octahedral) compounds in which sulfate is usually the negative ion. Alums have the general formula $M^I M^{III} (AX_4) \cdot 12H_2O$. The first is potassium or a higher alkali metal (sodium alums are rare) or thallium(I) or ammonium (or a substituted ammonium ion) and the second is a tripositive ion of relatively small ionic radius (0.5–0.7Å). The tripositive ions of larger ionic radius such as the Lanthanides form double sulfates, but not alums. The tripositive ions in alums, in roughly the order of number of known compounds, are aluminum, chromium, iron, manganese, vanadium, titanium, cobalt, gallium, rhodium, iridium, and indium.

The term *alum*, itself, refers to potassium alum, potassium aluminum sulfate, $KAl(SO_4)_2 \cdot 12H_2O$. Other common alums are ferric ammonium alum, $NH_4 Fe(SO_2)_2 \cdot 12H_2O$, and sodium chrome alum, $NaCr(SO_4)_2 12H_2O$.

ALUMINA.   See **Adsorption**; and **Bauxite**.

ALUMINUM.   [CAS: 7429-90-5]. Chemical element symbol Al, at. no. 13, at. wt. 26.98, periodic table group 13, mp $660 \pm 1°C$, bp $2452 \pm 15°C$, sp gr 2.699. In Canada and several other English-speaking nations, the spelling of the element is *aluminium*. The element is a silver-white metal, with bluish tinge, capable of taking a high polish. Commercial aluminum has a purity of 99% (minimum), is ductile and malleable, possesses good weldability, and has excellent corrosion resistance to many chemicals and most common substances, particularly foods. The electrical conductivity of aluminum (on a volume basis) is exceeded only by silver, copper, and gold. First ionization potential, 5.984 eV; second, 18.823 eV; third, 28.44 eV. Oxidation potentials, $Al \longrightarrow Al^{3+} + 3e^-$, V; $Al + 4OH^- \longrightarrow AlO_2^- + 2H_2O + 3e^-$, 2.35 V. Other important physical properties of aluminum are given under **Chemical Elements**.

Aluminum occurs abundantly in all ordinary rocks, except limestone and sandstone; is third in abundance of the elements in the earth's crust (8.1% of the solid crust), exceeded only by oxygen and silicon, with which two elements aluminum is generally found combined in nature; present in igneous rocks and clays as aluminosilicates; in the mineral cryolite in Greenland as sodium aluminum fluoride Na_3AlF_6; in the minerals corundum and emery, the gems ruby and sapphire, as aluminum oxide Al_2O_3; in the mineral bauxite in southern France, Hungary, Yugoslavia, Greece, the Guianas of South America, Arkansas, Georgia, and Alabama of the United States, Italy, Russia, New Zealand, Australia, Brazil, Venezuela, and Indonesia as hydrated oxide $Al_2O(OH)_4$; in the mineral alunite or alum stone in Utah as aluminum potassium sulfate $Al_2(SO_4)_3 \cdot K_2SO_4 \cdot 4Al(OH)_3$. See also **Bauxite**; and **Cryolite**.

Uses

Because of its corrosion resistance and relatively low cost, aluminum is used widely for food-processing equipment, food containers, food-packaging foils, and numerous vessels for the processing of chemicals. Because of the presence of aluminum in soils and rocks, there are natural traces of the element in nearly all foods. The processing of foods in copper vessels will cause destruction of vitamins, whereas aluminum does not accelerate the degradation of vitamins. Aluminum foil has been used to cover severe burns as a means to enhance healing. Because of its good electrical conductivity (exceeded only by gold, silver and copper), aluminum is used as an electrical conductor, particularly for high-voltage transmission lines. For the same conductance, the weight of aluminum required is about one-half that of annealed copper. The greater diameter of aluminum conductors also reduces corona loss. But, because aluminum has about $1.4\times$ the linear temperature coefficient of expansion as compared with annealed copper, the changes in sag of the cable are greater with temperature changes. For long spans requiring high strength, the center strand may be a steel cable, or supporting steel cables may be used. Aluminum is used for bus bars because of its large heat-dissipating surface available for a given conductance.

Aluminum alloys readily with copper, manganese, magnesium, silicon, and zinc. Many aluminum alloys are commercially available. See also **Aluminum Alloys and Engineered Materials**.

Because aluminum, particularly in alloys, combines strength with light weight, it has been a favorite construction material for transportation equipment, such as airplanes, the early dirigibles, and parts of rail cars. With the emphasis on fuel economy during the past decade or two, the automotive industry has become a major user of aluminum alloys. For example, important developments include cast-aluminum engine blocks with metallurgically bonded-in-place cylinder liners, cast-spun wheels, aluminum space frames, and auto radiators. The first mass production of aluminum radiators occurred in the 1980s (Ford Motor Co.) and it is predicted that nearly all Ford vehicles will feature aluminum radiators by the mid-1990s. Of course, other auto firms worldwide are also actively pursuing the metal because of its light weight and other attractive properties.

The applications of aluminum are growing larger and more varied with the introduction of advanced materials engineering techniques, including sandwich-type construction, laminates, aluminum powder metallurgy, and composite technology.

Discovery and Early Production

Although aluminum was predicted by Lavoisier (France) as early as 1782, when he was investigating the properties of aluminum oxide (alumina), the metal was not isolated until 1825 by H.C. Oersted (Denmark). Oersted obtained an impure aluminum metal by heating potassium amalgam with anhydrous aluminum chloride, followed by distilling off the mercury. Using similar methods, Woehler (Germany) produced an aluminum powder in 1827. In 1854, Deville (France) and Bunsen (Germany) separately but concurrently found that aluminum could be isolated by using sodium instead of potassium in the amalgam of Oersted and Woehler. Deville exhibited his product at the Paris Exposition of 1855, after which Napoleon III commissioned Deville to improve the process and lower the cost. Because of improved processing techniques, the price declined from $115 to $17 per pound ($254 to $38 per kilogram) by 1859, with numerous plants throughout France. The price was further reduced to $8 per pound ($17.60 per kilogram) by 1885. The first breakthrough for production of aluminum on a large scale and at a much lower cost occurred as the result of experimentation by Hall (Oberlin, Ohio) who found that metallic aluminum could be produced by dissolving Al_2O_3 (alumina) in molten $3NaF \cdot AlF_3$ (cryolite) at a temperature of above 960°C and then passing an electric current through the bath. Heroult (France) independently discovered the same process.

Contemporary Production Methods

For many years the electrolytic process has dominated. The major difference found in aluminum production plants relate to the design of the electrolytic cells. Basically, the cells are large carbon-lined steel boxes. The carbon lining serves as the cathode. Separate anodes are immersed in the bath. The electric circuit is completed by passage of current through the bath. Thus, the alumina is decomposed into aluminum and oxygen by action of the electric current. The molten aluminum collects at the bottom of the cell from which it is siphoned off periodically. The released oxygen combines with carbon at the electrodes to form CO_2. The bath is replenished with alumina intermittently. The carbon anodes must be replaced periodically. Additional cryolite also is required to make up for volatilization losses. The several reactions which occur during electrolysis still are not fully understood.

Energy requirements for aluminum production in modern plants approximate 6 to 8 kilowatt-hours per pound (13.2 to 17.6 kilowatt-hours per kilogram). Per weight unit of aluminum produced, 0.4 to 0.6 weight units of carbon, 1.9 weight units of alumina, and 0.1 weight unit of cryolite are required. One form of electrolytic cell is shown in Fig. 1.

An estimate (1987) showed that the aluminum industry accounts for nearly 1.5% of the annual world energy consumption. Because of the need to conserve energy and concurrently to alleviate (even to a small extent) the disposal of solid waste materials, a program was initiated in some industrial nations to collect and recycle the aluminum cans used mainly by the carbonated beverage industry. The program has proved to be more successful than originally predicted. As of 1990, over one-half of the aluminum beverage cans are recycled in the United States. See Figs. 2 and 3. Efforts also have been made to reduce the amount of aluminum needed for each beverage can. The thickness of the aluminum can body has been reduced from an average weight of 21 grams/can in 1972 to approximately 15 grams/can in 1992. Can manufacturers have developed new designs that retain strength while reducing the amount of metal needed. In one innovation, the top of the can was "necked in"—first one, then two, three, and four times. Each successive necking down of the end lowered the weight of the can. By 1990, the average weight/can had been reduced by at least 26%.

Because of production process improvements, the cost of aluminum has been reduced by many fold over the past century. The demographics of aluminum production have changed markedly in recent years. Since 1982, the virgin metal producers in the United States have closed down or abandoned seven of the older, less-efficient, and higher-cost reduction plants. American producers have gone to other countries to build new modern plants using the most recently developed technology for large electrolytic cells. New generation electrolytic cells operate at very high current densities (180,000–275,000 amperes). These cells will produce four times the output of many existing cells while using 25% less energy.

The principal aluminum-producing areas of the world, in addition to the United States and Europe, are Canada, Venezuela, Brazil, Norway, Australia, and the Republic of the Congo. In the United States the remaining plants are located in the states of Washington, Oregon, and New York, along the Mississippi, and where water power generation is still available at somewhat reduced cost.

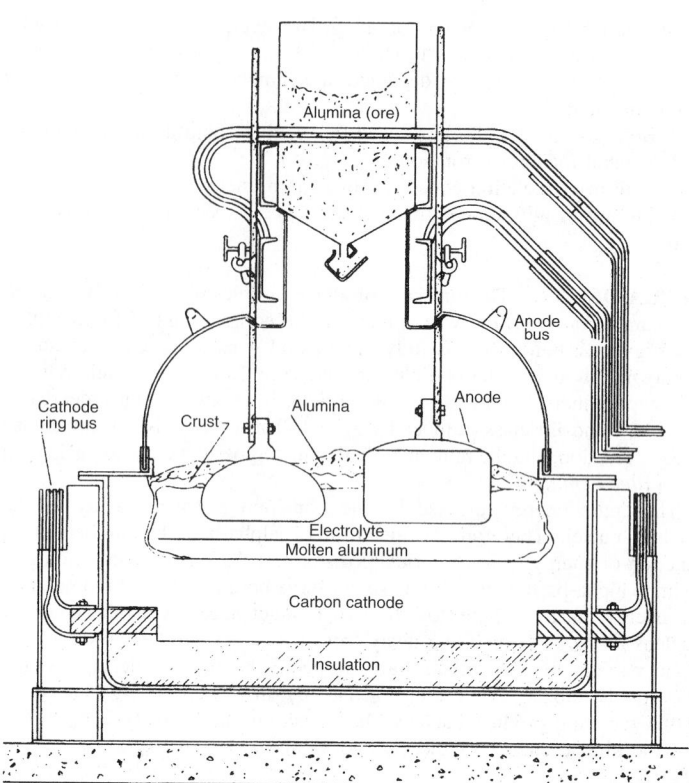

Fig. 1.　Aluminum electrolytic reduction cell of the pre-bake type. The anodes are constructed of separate blocks of carbon which have been pre-baked. There is a lead to the main bus bar from each block. (*Reynolds Metals Company*)

Fig. 2.　Recycling of aluminum beverage cans and other scrap saves nearly 95% of the energy required when producing the virgin metal from ore. Approximately 50% of the metal processed in 1990 was recycled from scrap. (*Reynolds Aluminum Recycling Company*)

In 1960, 90% of world capacity was held by private enterprises. That percentage had dropped to 60% in 1985. The trend is to greater governmental participation in Latin America. Africa, the Mideast, and Asia. Fifty percent of Western World capacity is government sponsored, primarily in Europe and Latin America. Six corporate groups account for 50% of aluminum production. These are Alcan, Alcoa, Reynolds, Kaiser, Pechiney, and Alusuisse. They possess 40% of the bauxite capacity for

Percent Recycled

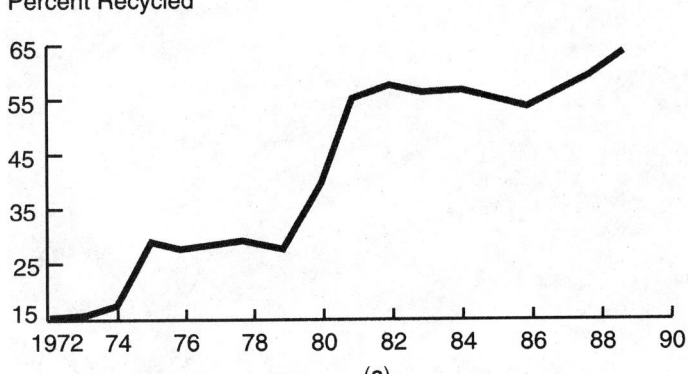

(a)

Average weight per can (grams)

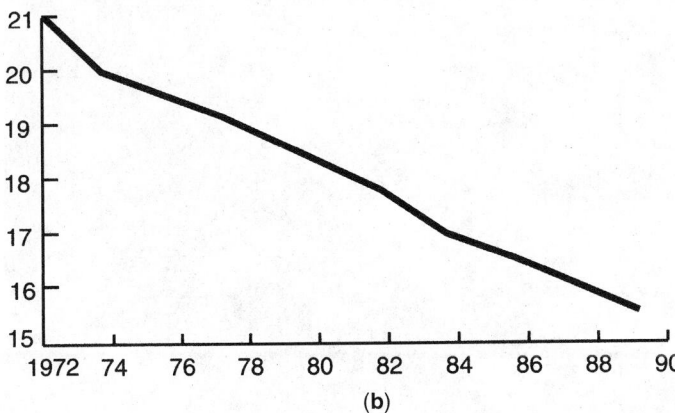

(b)

Fig. 3.   Steps to conserve aluminum and the electrical power to produce it: (a) Percent of aluminum cans recycled (1972–1990); (b) average weight of beverage can (grams) (1972–1990). (*Source: Aluminum Association*, Washington, DC)

alumina and aluminum. All are integrated from mine to ultimate consumer. Australia is the world's largest producer of bauxite.[1]

By 1995, Australia will probably be one of the world's three largest aluminum producers. At present, it supplies one-third of the world's bauxite. New smelter construction underway is expected to boost the world production by two million tons (1.8 million metric tons). New Zealand also has a good potential. Australia has coal reserves estimated at 1500 years.

It is also projected that Venezuela, because of its abundance of hydropower, will become a major aluminum producer. During the past decade, Brazil has attracted considerable interest because of its vast reserves of good quality bauxite and hydropower. The Tucurui hydroelectric power facility will double its output by 1995.

The People's Republic of China is modernizing with the help and technology of foreign companies to bring its production to over 500,000 metric tons.

Most of the increased capacity for virgin aluminum production for the past two and a half decades has occurred outside the continental United States. The annual growth rate for the aluminum industry during this period is estimated at about 1.6%. The projected U.S. growth rate is about 1% per year, as contrasted with $2\frac{1}{4}\%$ worldwide. U.S. consumption is about 50 pounds (22.5 kg) per person per year. World consumption has doubled about every 20 years. The developed nations (North America, Western Europe, Japan-Oceania) account for two-thirds of world consumption of aluminum (1991).

[1] Firms participating in the development of Australian production capacity include Alcan, Alcoa, Kaiser, Comalco, Alumax, Reynolds, Pechiney and several Japanese companies.

Thermal Reduction Process

Because of the large electric power requirements of the electrolytic process, an effort has been underway for several years to cut production costs. The thermal process, including the use of electric arc furnaces, make it possible to achieve a greater concentration of electric energy in the refining process. Commercial electric furnace processes are fed with bauxite and clays (kaolin or kyanite or partially refined bauxite mixed with carbon or charcoal in briquette form). An electric dc or ac arc is used. To date, this process has been practical only for the production of impure aluminum alloys, ranging in content from 65–70% aluminum, 25–30% silicon, and 1% or more of iron. Other impurities may include titanium and the carbides and nitrides of aluminum and iron. This impure product may be refined to yield pure aluminum by means of a subsequent distillation process in which the electric-furnace product is dissolved at high temperatures in zinc, magnesium, or mercury baths and then distilled. Aluminum with a purity of 99.99 + % may be produced in this manner.

A gas-reaction purification process also has attracted attention over the years. This is based on aluminum trichloride gas reacting with molten aluminum at about 1000°C to produce aluminum monochloride gas. Aluminum fluoride gas may be substituted for the aluminum monochloride gas in the process. Raw materials for the process may be scrap aluminum, aluminum from thermal-reduction processes, aluminum carbide, or aluminum nitride. As of the beginning of 1987, none of these other processes are fully commercial.

For the production of superpurity aluminum on a large scale, the Hoopes cell is used. This cell involves three layers of material. Impure (99.35 to 99.9% aluminum) metal from conventional electrolytic cells is alloyed with 33% copper (eutectic composition) which serves as the anode of the cell. A middle, fused-salt layer consists of 60% barium chloride and 40% $AlF_3 1.5NaF$ (chiolite), mp 720°C. This layer floats above the aluminum-copper alloy. The top layer consists of superpurity aluminum (99.995%). The final product usually is cast in graphite equipment because iron and other container metals readily dissolve in aluminum. For extreme-purity aluminum, zone refining is used. This process is similar to that used for the production of semiconductor chemicals and yields a product that is 99.9996% aluminum and is available in commercial quantities.

Continuing Research and Development

The industry continues to research improvements in the present production cells. Special attention is being focused on developing inert anodes and cathodes. Ferrites may find use as inert anodes, while titanium diboride may become the optimum material for cathodes. Before commercial use of inert electrodes can be achieved, cell sidewall materials must be developed which will withstand extremely reactive conditions and further improvements (i.e., less solubility of the anode and cathode materials are required). Over the past 15 years, American and Canadian aluminum producers have channeled nearly $1.5 billion into manufacturing technology research, the modernization and computerization of plant facilities, and new and better applications for the metal. Some of the results achieved thus far include:

- Energy efficiency—improved by almost 20%.
- New smelters with large, more efficient electrolytic cells.
- Recycling processes for all types of aluminum scrap, including aluminum beverage cans.
- Electromagnetic casting (EMC) for aluminum ingots in which an electromagnetic force field acts as the mold, thus producing smooth surfaces. In many cases, this eliminates wasteful scalping.
- Fluxing and degassing methods, combined with filtration, to give highest quality aircraft and other critical alloys.
- Continuous casting of filtered molten metal into giant coils (50,000 lb; 22,680 kg) of 0.250-inch (~6.4 mm) strip. This eliminates all hot rolling and intermediate annealing for rolling thin sheet products, such as foil, container stock, siding, etc.
- Aluminum particle rolling directly into continuous strip. Molten metal is sprayed into particles about the size of rice in a controlled environment, then brought to a desired rolling temperature and continuously rolled into "instant" coiled sheet This process can handle alloys which normally cannot be cast in ingot form.

- Computerized-automated cold rolling mills capable of 8000 ft (2438 m) per minute speeds. Multistand cold mills eliminate coil changes and wasted time. See Fig. 4.
- Foil rolling at speeds up to 8000 ft (2438 m) per minute with automatic gauge and shape control.
- Aluminum cans have captured over 90% of the beverage industry in the United States. The intent now is to capture a major share of the 30 billion cans food packaging market. (This will be challenged by the increasing use of plastics.)
- Increased use of aluminum in automotive vehicles. Approximately 200 pounds (90 kg) of aluminum per U.S.-built vehicle (i.e., wheels, engine blocks, bumpers, heat exchangers, and other parts).
- New alloys, such as aluminum-lithium alloys for aircraft applications that will permit 7 to 10% weight reduction per aircraft.
- Aluminum powder metallurgy continues to receive high research and development priority. A major objective is to produce final aluminum parts by direct pressing of rapidly solidified aluminum particles. The process handles high-strength aluminum alloy materials.
- Aluminum composites aim at combining high-strength aluminum alloys with alumina ceramic fibers and filaments to provide even higher-strength structural materials.
- Superplastic—high-strength aluminum alloys which can be plastically shaped into difficult designed parts used on cars and aircraft.
- Rheocasting technology for producing stronger castings by combining casting and forging of molten aluminum alloys in one step.
- Ultrasonic welding, used increasingly in the aerospace, automotive, electrical, and electronics industries. The process requires simpler surface preparation and less energy and allows welding of thinner sheets to thicker structural members.
- Inertia (friction) welding in the solid state, which is useful for joining dissimilar metals, i.e., aluminum to steel and stainless steel. At least one of the pieces to be welded must be circular and capable of being rotated at high velocities before being brought into contact with pressure to the second member.
- Vacuum brazing for producing heat exchangers.
- Aluminum as energy (fuel) source. The high energy content of aluminum has made it a logical choice as a sold fuel for boosting vehicles, such as the space shuttle, into orbit.
- Aluminum—air battery. A second potential application of this available energy is based on electrochemical oxidation of aluminum in air to produce electricity. In an aluminum—air battery, for example, thin coils of aluminum strip may be used as the fuel. No electric battery recharging would be required since the aluminum is consumed to generate the electricity directly. This fuel would not give off fumes or pollute and could be stored in solid form indefinitely. If this concept materializes into commercial viability, it will provide the energy needed for electric vehicles.

Some of the foregoing topics are described in more detail in the following article on **Aluminum Alloys and Engineered Materials**.

Aluminum and Space

The year 1986 marked the hundredth anniversary of the Hall-Heroult electrolytic process for producing aluminum economically. Jules Verne, the 19th century fiction writer, foretold the use of aluminum for space travel while it was still a mystery metal. Aluminum is the key metal in the aerospace industry. The Wright brothers recognized its value on their first flight in 1903 by using aluminum to build part of their engine. In military and commercial aircraft, aluminum became the metal of construction as early as 1930 for the U.S. Army P-12 and the Navy F-48, as well as an aluminum dirigible ZNC-2. World War II "light metals war" planes (i.e., B-17, B-25, B-29, etc.) became the forerunners of present aircraft. The U.S. Army Ballistic Missile Agency launched a program in 1950 for surface-to-surface missiles—the Redstone and Jupiter rocket programs. These programs prepared the way for NASA to launch unmanned satellites and to put men on the moon. Aluminum was the metal used for the rockets, the space modules, the moon rover (land vehicle), and most equipment associated with the programs. Today aluminum is used in the solid fuel booster rockets (some containing 0.5 mil lb-0.23 mil kg-of aluminum powder), in the space shuttles and planned space station. Some commercial airliners (jumbo

Fig. 4. High-speed, computer-controlled cold rolling mill can roll coils up to 40,000 pounds (18,150 kg), at speeds up to 5500 feet (1675 m) per minute. The system is equipped with automatic hydraulic gage control and shape control systems. (*Reynolds Metals Company*)

jets 747 and DC-10) contain more than 300,000 lb (~136,000 kg) of aluminum each.

Aluminum Chemistry and Chemicals

The CAS registry lists 5037 aluminum-containing compounds exclusive of alloys and intermetallics.

Since aluminum has only three electrons in its valence shell, it tends to be an electron acceptor. Its strong tendency to form an octet is shown by the tetrahedral aluminum compounds involving sp^3 hybridization. Aluminum halides include the trifluoride, an ionic crystalline solid, the trichloride and tribromide both of which are dimeric in the vapor and in nonpolar solvents, having the halogen atoms arranged in tetrahedra about each aluminum atom, giving a bridge structure; and the triiodide. They combine readily with many other molecules, especially organic molecules having donor groups, and aluminum chloride is an active catalyst. Certain of the hydrates have a saltlike nature; thus $AlCl_3 \cdot 6H_2O$ acts as a salt, even though Al_2Cl_6 is covalent.

Common water-soluble salts include the sulfate, selenate, nitrate, and perchlorate, all of which are hydrolyzed in aqueous solution. Double salts are formed readily by the sulfate and by the chloride.

There are many fluorocomplexes of aluminum. The general formula for the fluoroaluminates is $M_x^1 [Al_y F_{x+3y}]$, based upon AlF_6 octohedra, which may share corners to give other ratios of Al:F than 1:6. Chloroaluminates of the type $M^1[AlCl_4]$ are obtainable from fused melts. Aluminum ions form chloro-, bromo-, and iodo-complexes containing tetrahedral $[AlX_4]^-$ ions. However, in sodium aluminum fluoride $NaAlF_4$, the aluminum atoms are in the centers of octohedra of fluorine atoms in which the fluorine atoms are shared with neighboring aluminum atoms.

Aluminum hydroxide is about equally basic and acidic, the pK_b being about 12 and the pK_a being 12.6. Sodium aluminate seems to ionize as a uni-univalent electrolyte: $NaAl(OH)_4(H_2O)_2 \rightleftharpoons Na^+ + [Al(OH)_4(H_2O)_2]^-$. The high viscosity of sodium aluminate solutions is explained by hydrogen bonding between these hydrated ions, and between them and water molecules. By reaction of aluminum and its chloride or bromide at high temperature, there is evidence of the existence of monovalent aluminum. Here the aluminum atom is apparently in the sp state, with an electron pair on the side away from the chlorine atom, whereby the single pairs on the two chlorine atoms are shared to form two weak π bonds.

Aluminum forms a polymeric solid hydride, $(AlH_3)_x$, unstable above 100°C, extremely reactive, e.g., giving hydrogen explosively with H_2O and igniting explosively in air. From it are derived the tetrahydroaluminates, containing the ion AlH_4^-, which are important, powerful but selective reducing agents, e.g., reducing chlorophosphines R_2PCl to phosphines R_2PH; reactive alkyl halides and sulfonates to hydrocarbons, epoxides, and carboxylic acids; aldehydes and ketones to alcohols; nitrites, nitro compounds and N,N-dialkyl amides to amines. The salt most commonly used is lithium (tetra)hydroaluminate ("lithium aluminum hydride").

Aluminum compounds are generally made starting with bauxite, which is reactive with acids and with bases. With acids, e.g., H_2SO_4, any iron contained in the bauxite is dissolved along with the aluminum and silicon is left in the residue, whereas with bases, e.g., NaOH, any silicon is dissolved and iron left in the residue.

Acetate: Aluminum acetate $Al(C_2H_3O_2)_3$, white crystals, soluble, by reaction of aluminum hydroxide and acetic acid and then crystallizing. Used (1) as a mordant in dyeing and printing textiles, (2) in the manufacture of lakes, (3) for fireproofing fabrics, (4) for waterproofing cloth.

Aluminates: Sodium aluminate $NaAlO_2$, white solid, soluble, (1) by reaction of aluminum hydroxide and NaOH solution, (2) by fusion of aluminum oxide and sodium carbonate; the solution of sodium aluminate is reactive with CO_2 to form aluminum hydroxide. Used as a mordant in the textile industry, in the manufacture of artificial zeolites, and in the hardening of building stones. See silicates below and calcium aluminates.

Alundum: See oxide (below).

Carbide: Aluminum carbide Al_4C_3, yellowish-green solid, by reaction of aluminum oxide and carbon in the electric furnace, reacts with H_2O to yield methane gas and aluminum hydroxide.

Chlorides: Aluminum chloride $AlCl_3 \cdot 6H_2O$, white crystals, soluble, by reaction of aluminum hydroxide and HCl, and then crystallizing; anhydrous aluminum chloride $AlCl_3$, white powder, fumes in air, formed by reaction of dry aluminum oxide plus carbon heated with chlorine in a furnace, used as a reagent in petroleum refining and other organic reactions.

Fluoride: Aluminum fluoride AlF_3, white solid, soluble, by reaction of aluminum hydroxide plus hydrofluoric acid and then crystallizing $2AlF_3 \cdot 7H_2O$, used in glass and porcelain ware.

Hydroxide: Aluminum hydroxide $Al(OH)_3$, white gelatinous precipitate, by reaction of soluble aluminum salt solution and an alkali hydroxide, carbonate or sulfide (sodium aluminate is formed with excess NaOH but no reaction with excess NH_4OH), upon heating aluminum hydroxide the residue formed is aluminum oxide. Used as intermediate substance in transforming bauxite into pure aluminum oxide.

Nitrate: Aluminum nitrate $Al(NO_3)_3$, white crystals, soluble, by reaction of aluminum hydroxide and HNO_3, and then crystallizing.

Oleate: Aluminum oleate $Al(C_{18}H_{33}O_2)_3$, yellowish-white powder, by reaction of aluminum hydroxide, suspended in hot H_2O, shaken with oleic acid, and then drying, the product is used (1) as a thickener for lubricating oils, (2) as a drier for paints and varnishes, (3) in waterproofing textiles, paper, leather.

Oxide: Aluminum oxide, alumina Al_2O_3, white solid, insoluble, melting point 2020°C, formed by heating aluminum hydroxide to decomposition; when bauxite is fused in the electric furnace and then cooled, there results a very hard glass ("alundum"), used as an abrasive (hardness 9 Mohs scale) and heat refractory material. Aluminum oxide is the only oxide that reacts both in H_2O medium and at fusion temperature, to form salts with both acids and alkalis.

Palmitate: Aluminum palmitate $Al(C_{16}H_{31}O_3)_3$, yellowish-white powder, by reaction of aluminum hydroxide, suspended in hot H_2O, shaken with palmitic acid, and then drying, the product is used (1) as a thickener for lubricating oils, (2) as a drier for paints and varnishes, (3) in waterproofing textiles, paper, leather, (4) as a gloss for paper.

Silicates: Many complex aluminosilicates or silicoaluminates are found in nature. Of these, clay in more or less pure form, pure clay, kaolinite, kaolin, china clay $H_4Si_2AlO_2O_9$ or $Al_2O_3.2SiO_2.2H_2O$ is of great importance. Clay is formed by the weathering of igneous rocks, and is used in the manufacture of bricks, pottery, procelain, and Portland cement.

Stearate: Aluminum stearate $Al(C_{18}H_{35}O_2)_3$, yellowish-white powder, by reaction of aluminum hydroxide suspended in hot H_2O, shaken with stearic acid, and then drying the product, used (1) as a thickener for lubricating oils, (2) as a drier for paints and varnishes, (3) in waterproofing textiles, paper, leather, (4) as a gloss for paper.

Sulfate: Aluminum sulfate $Al_2(SO_4)_3$, white solid, soluble, by reaction of aluminum hydroxide and H_2SO_4, and then crystallizing, used (1) as a clarifying agent in water purification, (2) in baking powders, (3) as a mordant in dyeing, (4) in sizing paper, (5) as a precipitating agent in sewage disposal; aluminum potassium sulfate, see **Alum**; and **Alunite**.

Sulfide: Aluminum sulfide Al_2S_3, white to grayish-black solid, reactive with water to form aluminum hydroxide and H_2S, formed by heating aluminum powder and sulfur to a high temperature.

Aluminum in solution of its salts is detected by the reaction (1) with ammonium salt of aurin tricarboxylic acid ("aluminon"), which yields a red precipitate persisting in NH_4OH solution, (2) with alizarin red S, which yields a bright red precipitate persisting in acetic acid solution.

Organoaluminum Compounds: By the action of alpha olefins on AlH_3, the trialkyls of aluminum can be prepared:

$$AlH_3 + 3CH_2 = CHR \xrightarrow{120°} (CH_2CH_2R)Al$$

In the presence of ethers, magnesium-alloy alloys react with alkyl halides to yield trialkyls, R_3AlOEt_2. The trimethyl aluminum compound is a dimer, mp 15°C. Triethyl aluminum is a liquid. The alkyls react readily with H_2O:

$$Et_3Al + 3H_2O \longrightarrow Al(OH)_3 + 3EtH$$

When aluminum is reacted with diphenyl mercury $(C_6H_5)_2Hg$, triphenylaluminum $(C_6H_5)_3Al$ is yielded. The aluminum organometallic alkyls and aryls act as Lewis acids to form compounds with electron-donating substances. The resulting compounds are the organometallic basis for producing polymers of the Al-N type.

Zeolite Structures: These are crystalline, microporous solids that contain cavities and channels of molecular dimensions (3 Å to 10Å) and sometimes are called molecular sieves. Zeolites are used principally in catalysis, separation, purification, and ion exchange. The fundamental building block of a zeolite is a tetrahedron of four oxygen atoms surrounding a central silicon atom (i.e., $(SiO_4)^{4-}$). From the fundamental unit, numerous combinations of secondary building units (polygons) can be formed. The corners of these polyhedra may be Si or Al atoms.[2]

Different combinations of the same secondary building unit may yield numerous distinctive zeolites. As pointed out by D.E.W. Vaughan, few fields of chemistry offer such chemical and structural diversity. Although only about sixty structures are known, tens of thousands of theoretical structures are possible. The science of synthesizing zeolites, which achieves much more diversity than is the case with natural zeolites, has been under continuous development for several years. Synthetic zeolites often are cataloged by their Si/Al composition ratios, usually ranging from 1:1 to 10:500. Thus, aluminum plays a large role in this field. In the manufacture of synthetic zeolites, the common reactants are silica from sodium silicate (water-glass) or an aqueous colloidal silica, aluminum from sodium aluminate, alum, or a colloidal alumina (boehmite), and sodium or potassium hydroxide. For further detail, see the Vaughan (1988) reference.

S. J. SANSONETTI
Consultant, Reynolds Metals Company
Richmond, Virginia
(Updated by editorial staff).

Additional Reading

Note: See list of references at end of next article.

ALUMINUM ALLOYS AND ENGINEERED MATERIALS. Aluminum alloys have been used effectively for scores of years. Much

---

[2] Other atoms are used, but to a lesser extent, including Li, Be, B, Mg, Co, Mn, Zn, P, As, Ti, Ga, and Fe.

research during the past few decades has been directed not only to expand the list of alloys, but also for applying aluminum in other engineered forms, such as composites, sandwiched, and laminated materials. Aluminum fabrication and surface treatment have been improved. The techniques of powder metallurgy have been expanded to include the new engineering forms of aluminum. Aluminum quasi-crystals are a recent discovery. The "shape memory" of certain aluminum-base materials is being exploited. A process for using shock waves to shape aluminum also is among recent achievements.

Annually, sessions are held on *advanced aluminum alloys* as part of the conference on "Advanced Aerospace Materials/Processes and Exposition." Topics discussed have included rapid-solidification alloys, very low density alloys, laminates, new alloying elements, such as lithium, and such processes as diffusion bonding and means to improve fracture toughness and reduce fatigue and corrosion.

Aluminum is unique among the metals because it responds to nearly all of the known finishing processes. It can be finished in the softest, most delicate textures as exemplified by tableware and jewelry. Aluminum can be anodized and dyed to appear like gold. It can be made as specular as a silver mirror and jet black. The metal also can be anodized to an extremely hard, wear- and abrasion-resistant surface that approaches the hardness of a diamond. Aluminum is available in many convenient forms-shapes, sheet, plate, ingot, wire, rod and bar, foil, castings, forgings, powdered metals, and extrusions.

In the production of these products, new high-speed automated equipment has been developed over the past decade. Rolling speeds as high as 8000 feet (2438 meters) per minute are possible on some products. New continuous casting processes have been developed where molten metal can be converted directly into coiled sheet, using roll casting, belt casting, and articulated block casting. See Figs. 1 and 2. Many of the common alloys respond well to these new processes. Electromagnetic casting of the conventional giant ingots used for the conventional rolling mills has been developed in the fully automated mode. This casting process eliminates the need for scalping ingots prior to processing into final products, such as can stock, foil, auto body sheet, siding, and many other products. In this process, an electrical field rather than a solid mold wall is used to shape the ingot. The process is based upon the fact that liquids with good electrical conductivity, such as aluminum, can be constrained where an alternating magnetic field is applied. The liquid metal remains suspended by the electrical forces while it is being solidified.

The casting alloy products comprise sand castings, permanent mold castings, and die castings. Aluminum is the basic raw material for more than 20,000 businesses in the United States. Aluminum is an indispensable metal for aircraft, for example. See Fig. 3. Representative aluminum alloys for a broad classification of uses is given in Table 1.

Worldwide demand for aluminum over the past several decades has increased by approximately 5–6% per year. Aluminum is quite well established as an energy saver and a cost-effective material in numerous applications. Particularly strong growth is occurring in the automotive and transportation fields, housing, and food packaging. For example, a number of breakthroughs in foil pouch packaging have been achieved for replacing conventional cans for shelf-stable foods. Steady growth has

Fig. 2. Continuous rolling plant (Hot Spring, Arkansas) has cast nearly two billion pounds (0.9 billion kg) of aluminum foil feedstock since it began operations in 1979. That quantity of foil would wrap the earth almost 1600 times at the equator. The plant produces coils of feedstock for aluminum foil and flexible packaging products at a 40% energy saving compared to conventional rolling processes. (*Reynolds Metals Company.*)

Fig. 3. Final operations on machined aluminum plates for wings of large jet aircraft. (*Reynolds Metals Company.*)

continued in other areas, including aircraft and space applications, electrical wire and cable, marine, architectural, and consumer durable products. See Fig. 4.

Pure aluminum is soft and ductile. In its annealed form, the tensile strength is approximately 13,000 pounds per square inch (90 megapascals). This can be work hardened by rolling, drawing into wire, or by other cold-working techniques to increase its strength to approximately 24,000 pounds per square inch (165 megapascals). Pure aluminum is particularly useful in the food and chemical industries where its resistance to corrosion and high thermal conductivity are desirable characteristics, and in the electrical industry, where its electrical conductivity of about 62% that of copper and its lightweight make it desirable for wire, cable, and bus bars. Large quantities are rolled to thin foils for the packaging of food products, for collapsible tubes, and for paste and powders used for inks and paints.

Most commercial uses require greater strength than pure aluminum affords. Higher strength is achieved by alloying other metal elements with aluminum and through processing and thermal treatments. The alloys can be classified into two categories, non-heat-treatable and heat-treatable.

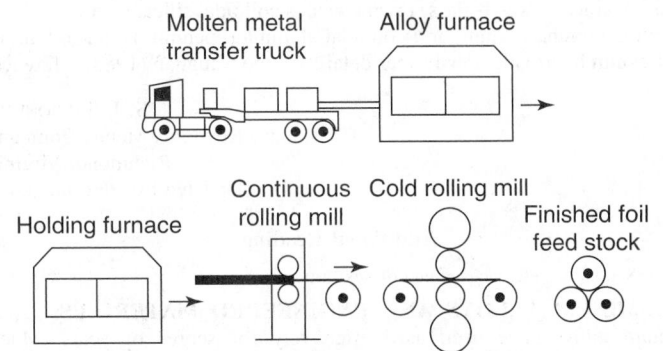

Fig. 1. Schematic portrayal of continuous aluminum roll casting process. (*Reynolds Metals Company.*)

TABLE 1. REPRESENTATIVE ALUMINUM ALLOYS

Product	Alloy	Form	Product	Alloy	Form
Architectural and Building Products					
Awnings	3003	Sheet	Cable	E.C., 5005, 6201	Wire
Fence Wire	6061	Wire	Conduit	6063	Tubing
Fittings	A514.0	Castings	Motors	319.0, 355.0, 360.0, 380.0	Castings
Gutters	Alc. 3004	Sheet			
Nails	5056, 6061	Wire	Transmission Towers and Substations	6061, 6063, 7005	Extrusions
Panels	3003, 6063	Sheet and Extrusions	*Consumer Durables*		
Roofing	Alc. 3004	Sheet	Appliances	3003, 4343, 5052, 5252, 5357, 5457	Sheet and Extrusions
Screens	Alc. 5056	Wire			6063, 6463
Siding	3003, Alc. 3004	Sheet	Sheet, Plate,	356.0, 390.0	Castings
Transportation			Cooking Utensils	1100, 3003, Alc. 3003,	Sheet 3004, Alc. 3004, 5052,
Aircraft	2014, 2024, 2048, 2090, 2124, 2219, 7075	Extrusions, and		5454	
	7079, 7175, 7178,	Forgings		B443.0, A514.0	Castings
	7179, 7475, 8090, 8091,		Furniture	3003, 3004, 5005, 6463	Sheet and Tubing
	8092. B295.0, 355.0,	Castings	Refrigerators	3003, Alc. 3003, 3004, Alc. 3004, 4343, 5005, 5050, 5052, 5252, 5457, 6061, 6463	Sheet and Extrusions
	356.0, 518.0,				
	520.0				
Automotive			Water Heaters	Alc. 6061, 3003	Sheet
Auto and Truck Bodies	2036, 2037, 5052, 5182, 5252, 5657, 6009, 6010, 6061, 6063, 6463, 7016, 7021, 7029, 7046, 7129	Sheet and Extrusions	*Machinery and Equipment*		
			Chemical Processing	1060, 1100, 3003, Alc. 3003, 3004, Alc. 3004, 5083, 5086, 5154, 5454, 6061, 6063	Sheet, Plate, Extrusions, and Tubes
Buses	2036, 5083, 5086, 5182, 5252, 5457, 5657, 6061, 6063, 6463, 7016, 7046	Sheet and Extrusions		356.0, 360.0, B514.0	Castings
	242.0, B295.0, 355.0, 356.0, 360.0, 380.0, 390.0, B443.0	Castings	Heat exchangers and Solar Panels	3003, Alc. 3003, 3004, Alc. 3004, 4343, 5005, 5052, 6061, 6951	Sheet and Tubing
Marine			Sheet for Vacuum Brazing Heat Exchangers	4003, 4004, 4005, 4044, 4104	
Barges, Small Craft, Ships, Tankers	5052, 5083, 5086, 5454, 5456, 6061, 6063	Sheet, Plate, Extrusions, and Tubing	Irrigation	3003, Alc. 3003, 3004, Alc. 3004, 5052, 6061, 6063	Tubing
	360.0, 413.0 B443.0	Castings	Sewage Plants	3003, Alc. 3003, 3004, Alc. 3004, 5052, 5083, 5086, 5454, 5456, 6061, 6063	Sheet, Extrusions, and Tubes
Railroad	Alc. 2024, 5052, 5083, 5086, 5454 5456, 6061, 6063, 7005	Sheet, Plate, and Extrusions		B514.0	Castings
	B295.0, 356.0, 520.0	Castings	Screw Machine Parts	2011, 2024, 6262	Rod and Bar
Containers and Packaging			Textile Machinery	2014, 2024, 6061, 6063	Extrusions, Sheet and Plate
Foils	1100, 1235, 3003, 5005, 8079, 8111	Foils	*Finished Pressed Parts*	7090, 7091	Powder Metallurgy
Cans	3003, 3004, 5182	Sheet			
Electrical					
Bus Bar	6063, 6201	Extrusions or Rolled Rod			

Non-Heat-Treatable Alloys

The initial strength of alloys in this group depend upon the hardening effect of elements such as manganese, silicon, iron, and magnesium, singly or in various combinations (see three alloys in Table 2). The non-heat-treatable alloys are usually designated therefore in the 1xxx, 3xxx, 4xxx, and 5xxx series (Table 3). Since these alloys are work hardenable, further strengthening is made possible by various degrees of cold work denoted by the H series tempers.

Heat-Treatable Alloys. This group of alloys includes the alloying elements, copper, magnesium, zinc, and silicon (−T tempered alloys in Table 2). Since these elements singly or in combination show increasing solid solubility in aluminum with increasing temperature, it is possible to subject them to thermal treatments that will impart pronounced strengthening. The first step called heat treatment or solution heat treatment, is an elevated temperature process designed to put the soluble element or elements in solid solution. This is followed by a rapid quenching which momentarily "freezes" the structure and for a short time renders the alloy very workable. At room or elevated temperatures, the alloys are not stable after quenching, and precipitation of the constituent from the supersaturated solid solution begins. After several days at room temperature, termed aging or room temperature precipitation, the alloy is considerably stronger. Many alloys approach a stable condition at room temperature but some alloys, especially those containing magnesium and silicon or magnesium and zinc continue to age harden for long periods of time at room temperature. By heating for a controlled time at slightly elevated temperatures, even further strengthening is possible and the properties are stabilized.

Clad Alloys. The heat-treatable alloys, in which copper or zinc are major alloying constituents, are less resistant to corrosive attack than a majority of the non-heat-treatable alloys. To increase the corrosion resistance of these alloys in sheet and plate form, they are often clad with a high-purity aluminum, a low-magnesium-silicon alloy, or an aluminum alloy containing 1% zinc.

The cladding, usually from $2\frac{1}{2}$ to 5% of the total thickness on each side, not only protects the composite due to its own inherent excellent corrosion resistance, but also exerts a galvanic effect that further protects the core alloy.

Special composites may be obtained such as clad non-heat-treatable alloys for extra corrosion protection, for brazing purposes, and for special surface finishes.

Composites. The most commonly used metal elements alloyed with aluminum are magnesium, manganese, silicon, copper, zinc, iron, nickel, chromium, titanium, and zirconium. The strength of aluminum can be tailored to the specific end applications ranging from the very soft ductile foils to the high-strength aircraft and space alloys equal to steel in the 90,000 pounds per square inch (621 megapascals) tensile strength range. In addition to the homogeneous aluminum alloys, dispersion hardened and advanced filament-aluminum composites provide even higher strengths. Boron filaments in aluminum can provide strengths in the range of 150,000–300,000 pounds per square inch (1034–2068 megapascals) tensile strength.

Superplasticity. Eutectoid and near eutectoid alloy chemistry research has resulted in alloys exhibiting superplasticity with unusual elongation (>1000%) and formability and with 60,000 pounds per square inch (414 megapascals) tensile strength. Three examples of these alloys are: 94.5% Al-5% Cu-0.5% Zr; 22% Al-78% Zn; and 90% Al, 5% Zn, 5% Ca.

Casting Alloys

During the last two decades, the quality of castings has been improved substantially by the development of new alloys and better liquid-metal treatment and also by improved casting techniques. Casting techniques include sand casting, permanent mold casting, pressure die casting, and others. Today sand castings can be produced in high-strength alloys and are weldable. Die casting permits large production outputs per hour on intricate pieces that can be cast to close dimensional tolerance and have excellent surface finishes; hence, these pieces require minimum machining. Since aluminum is so simple to melt and cast, a large number of foundry shops have been established to supply the many end products made by this method of fabrication. See Table 3.

Al_2O_3, Casting Semisolid Metal. A new casting technology is based on vigorously agitating the molten metal during solidification. A very different metal structure results when this metal is cast. The vigorously agitated liquid-solid mixture behaves as a slurry still sufficiently fluid (thixotropic) to be shaped by casting. The shaping of these metal slurries is termed "Rheocasting."

The slurry nature of "Rheocast" metal permits addition and retention of particulate nonmetal (e.g., Al_2O_3, SiC, T, C, glass beads) materials for cast composites. This new technology is beginning to be commercialized.

Alloy and Temper Designation Systems for Aluminum. The aluminum industry has standardized the designation systems for wrought aluminum alloys, casting alloys and the temper designations applicable. A system of four-digit numerical designations is used to identify wrought aluminum alloys. The first digit indicates the alloy group as shown in Table 4. The 1xxx series is for minimum aluminum purities of 99.00% and greater; the last two of the four digits indicate the minimum aluminum percentage; i.e., 1045 represents 99.45% minimum aluminum, 1100 represents 99.00%

Fig. 4.   Four aluminum sections of this type make up fuel tank for aerospace vehicle. Each section is produced from computer-machined plate and weighs 4750 pounds (2155 kg). (*Reynolds Metals Company*)

TABLE 2. NOMINAL CHEMICAL COMPOSITION[1] AND TYPICAL PROPERTIES OF SOME COMMON ALUMINUM WROUGHT ALLOYS PTV

	1100	3003	5052	2014T6[10]	2017T4[9]	2024T4[9]	6061T6[10]	7075T6[10]	6101T6[10]
Nominal chemical Composition[1]	99% Min. Alum.	1.2% Mn	2.5% Mg 0.25% Cr	4.4% Cu 0.8% Si 0.8% Mn 0.4% Mg	4.0% Cu 0.5% Mn 0.5% Mg	4.5% Cu 1.5% Mg 0.6% Mn	1.0% Mg 0.6% Si 0.25% Cu 0.25% Cr	5.5% Zn 2.5% Mg 1.5% Cu 0.3% Cr	0.5% Mg 0.5% Si
Tensile strength, psi	A 13,000[7] H 24,000[7]	16,000 29,000	28,000 42,000	— 70,000	— 62,000	— 68,000	45,000	83,000	32,000
Tensile strength, MPa	A 90 H 165	110 200	193 290	— 483	— 427	— 469	310	572	221
Yield strength, psi[2]	A 5000 H 22,000	6000 27,000	13,000 7,000	— 60,000	— 40,000	— 47,000	40,000	73,000	28,000
Yield strength, Mpa	A 34 H 152	41 186	90 255	— 414	— 276	— 324	276	503	193
Elongation percent in 2 in. (5.1 cm)11	A 45 H 15	A 40 H 10	A 30 H 8	13	22	19	17	11	15
Modulus of elasticity[3]	10	10	10.2	10.6	10.5	10.6	10	10.4	10
Brinnell hardness[8]	23–44	28–55	45–85	135	105	120	95	150	71
Melting range,°C	643–657	643–654	593–649	510–638	513–640	502–638	582–652	477–638	616–651
Melting range,°F	1190–1215	1190–1210	1100–1200	950–1180	955–1185	935–1180	1080–1250	890–1180	1140–1205
Specific gravity	2.71	2.73	2.68	2.80	2.79	2.77	2.70	2.80	2.70
Electrical resistivity[4]	2.9	3.4	4.93	4.31	5.75	5.75	4.31	5.74	3.1
Thermal conductivity[5]	0.53	A 0.46	A 0.33	0.37	0.29	0.29	0.37	0.29	0.52
SI units	221.9	A 192.6	A 138.2	154.9	121.4	121.4	154.9	121.4	217.7
Coefficient of expansion[6]	23.6	23.2	23.8	22.5	23.6	22.8	23.4	23.2	23

[1] Aluminum plus normal impurities is the remainder.
[2] 0.2% permanent set.
[3] Multiply by 10^6.
[4] Microhms per cm (room temperature).
[5] C.g.s. units (at 100°C).
[6] Per°C (20–100°C); multiply by 10^{-6}.
[7] A = annealed; H = hard.
[8] 500 kg load, 10 mm ball.
[9] Solution heat-treated and naturally aged.
[10] Solution heat-treated and artificially aged.
[11] Round specimens, $\frac{1}{2}$-m diameter.
Conversion factors used: 1 psi = 6.894757×10^{-3} megapascals (MPa).
C.g.s. = (cal) $(cm^2)/(sec)(cm)(°C)$.
SI unit = Watts/meter °K.
1 C.g.s. unit = 418.68 SI units.

minimum aluminum. The 2xxx through 8xxx series group aluminum alloys by major allowing elements. In these series the first digit represents the major alloying element, the second digit indicates alloy modification, while the third and fourth serve only to identify the different alloys in the group. Experimental alloys are prefixed with an X. The prefix is dropped when the alloy is no longer considered experimental.

Cast Aluminum Alloy Designation System. A four-digit number system is used for identifying aluminum alloys used for castings and foundry ingot (see Table 5). In the 1xx.x group for aluminum purity of 99.00% or greater, the second and third digit indicate the minimum aluminum percentage. The last digit to the right of the decimal point indicates the product form: 1xx.0 indicates castings and 1xx.1 indicates ingot. Special control of one or more individual elements other than aluminum is indicated by a serial letter before the numerical designation. The serial letters are assigned in alphabetical sequence starting with A but omitting I, O, Q, and X, the X being reserved for experimental alloys.

In the 2xx.x through 9xx.x alloy groups, the second two of the four digits in the designation have no special significance but serve only to identify the different aluminum alloys in the group. The last digit to the right of the decimal point indicates the product form: .0 indicates casting and .1 indicates ingot. Examples: Alloy 213.0 represents a casting of an aluminum alloy whose major alloying element is copper. Alloy C355.1 represents the third modification of the chemistry of an aluminum alloy ingot whose major alloying elements are silicon, copper, and magnesium.

Temper Designation System. A temper designation is used for all forms of wrought and cast aluminum alloys. The temper designation follows the alloy designation, the two letters being separated by a hyphen. Basic designations consist of letters followed by one or more digits. These designate specific sequences of basic treatments but only operations recognized as significantly influencing the characteristics of the product. Basic tempers are —F (as fabricated), —O annealed (wrought products only), —H strain-hardened (degree of hardness is normally quarter hard, half hard, three-quarters hard, and hard, designated by the symbols H12, H14, H16, and H18, respectively). —W solution heat-treated and —T thermally treated to produce stable tempers. Examples: 1100-H14 represents commercially pure aluminum cold rolled to half-hard properties. 2024-T6 represents an aluminum alloy whose principal major element is copper that has been solution heat treated and then artificially aged to develop stable full-strength properties of the alloy.

Contemporary Advancements and Future Potential

Highlighted in the following paragraphs are improvements in aluminum metallurgy that have occurred and have been available only relatively recently or that are promising but that still remain in a late phase of research or testing.

Aluminum-Lithium Alloys. Both private and government funding have been invested in Al—Li alloy research for several years. As of the early 1990s, exceptionally good results had been achieved by way of increasing the strength-to-weight ratio and the stiffness of Al—Li alloys. Low ductility in the short-transverse direction has been a difficult problem to solve. Wide usage awaits further problem solving and testing for critical applications. The Al—Li alloy 2091-T3 (Pechiney) is a medium-strength, lightweight alloy quite similar to the traditional alloy 2024-T3, which it is expected to

TABLE 3. NOMINAL CHEMICAL COMPOSITION[1] AND TYPICAL PROPERTIES OF SOME ALUMINUM CASTING ALLOYS

Properties for alloys 195, B195, 220, 355, and 356 are for the commonly used heat treatment

	413.0[2]	B443.0[2]	208.0[3]	308.0[4]	295.0[3]	B295.0[4]	514.0[3]	518.0[2]	520.0[3]	355.0[3]	356.0[3]	380.0[2]
Nominal chemical composition	12% Si	5% Si	4% Cu	5.5% Si 3% Si	4.5% Cu 4.5% Cu	4.5% Cu 0.8% Si	3.8% Mg 2.5% Si	8% Mg	10% Mg	5% Si	7% Si 0.5% Mg	8.5% Si
Tensile strength, psi[5]	37,000	19,000	21,000	28,000	36,000	45,000	25,000	42,000	46,000	35,000	33,000	45,000
Tensile strength, Mpa[5]	255	131	145	193	248	310	172	290	317	241	228	310
Yield strength, psi[5]	18,000	9000	14,000	16,000	24,000	33,000	12,000	23,000	25,000	25,000	24,000	25,000
Yield strength, Mpa[5]	124	62	97	110	165	228	83	159	174	174	165	174
Elongation, percent[5]	1.8	6	2.5	2	5	5	9	7	14	2.5	4	2
Brinnell hardness[6]	—	40	55	70	75	90	50	—	75	80	70	—
Melting range,°C	574–585	577–630	521–632	—	549–646	527–627	580–640	540–621	449–621	580–627	580–610	521–588
Melting range,°F	1065–1085	1070–1165	970–1170	—	1020–1195	980–1160	1075–1185	1005–1150	840–1150	1075–1160	1075–1130	970–1090
Specific gravity	2.66	2.69	2.79	2.79	2.81	2.78	2.65	2.53	2.58	2.70	2.68	2.76
Electrical resistivity	4.40	4.66	5.56	4.66	4.66	3.45	4.93	7.10	8.22	4.79	4.42	6.50
Thermal conductivity	70.37	0.35	0.29	0.34	0.35	0.45	0.33	0.24	0.21	0.34	0.36	0.26
SI units	154.9	146.5	121.4	142.4	146.5	188.4	138.2	100.5	87.9	142.4	150.7	108.9
Coefficient of expansion	820.0	22.8	22.8	22.7	23.9	22.8	24.8	24.0	25.4	22.8	22.8	20.0

[1] Remainder is aluminum plus minor impurities.
[2] Die cast.
[3] Sand cast.
[4] Permanent mold cast.
[5] For separately cast test bars.
[6] 500 kg/load, 10 mm, ball.
[7] C.g.s. units.
[8] Multiply by 10^{-6}. Per°C, for temperature range 20 to 200°C.
Conversion factors used: 1 psi = 6.894757×10^{-3} megapascals (MPa).
SI unit = Watts/meter °K.
1 C.g.s. unit = 418.68 SI units.

replace for aerospace applications. The new alloy has a 7% lower density and a 10% higher stiffness. The new alloy, like most Al alloys, is notch sensitive. An oxide film composed of MgO, LiO_2, $LiAlO_2$, Li_2CO_3, and LiOH tends to develop under normal production conditions. Cracks form in this film and tend to initiate cracks in the alloy's substrate and this reduces fatigue life. When the film is removed, in both longitudinal and long-transverse directions, the new alloy's fatigue properties are comparable with other aluminum alloys.

In late 1989, the availability of a proprietary family of weldable, high-strength (*Weldalite*) Al−Li products appeared. The material was claimed to be nearly twice as strong (100×10^3 psi) as other leading alloys then

TABLE 4. DESIGNATIONS FOR CAST ALUMINUM ALLOY GROUPS

Alloy No.	
Aluminium-99.00% minimum and greater	1xxx
MAJOR ALLOYING ELEMENT	
Aluminium Alloys Grouped by Major Alloying Elements { Copper	2xxx
Manganese	3xxx
Silicon	4xxx
Magnesium	5xxx
Magnesium and Silicon	6xxx
Zinc	7xxx
Other Element	8xxx
Unused Series	9xxx

(1) For codification purposes an alloying element is any element which is intentionally added for any purpose other than grain refinement and for which minimum and maximum limits are specified.
(2) Standard limits for alloying elements and impurities are expressed to the following places:

Less than 1/1000%	0.000X
1/1000 up to 1/100%	0.00X
1/100 up to 1/10%	
Unalloyed aluminum made by a refining process	0.0XX
Alloys and unalloyed aluminum not made by a refining process	0.0X
1/10 through 1/2%	0.XX
Over 1/2%	0.X, X.X, etc.

currently used for aerospace applications. The alloy was initially developed especially for space-launch systems. Specific advantages claimed include: (1) high strength over a broad temperature range, from cryogenic to highly elevated temperatures, (2) light weight, and (3) weldability—this property being of particular value for fabricating fuel and oxidizer tanks for space vehicles. *Weldalite* is produced in sheet, plate, extrusion, and ingot products.

Al−Li investment castings are gaining acceptance. Among aluminum alloying elements, lithium is one of the most soluble. About 4.2% Li can be dissolved in Al at the eutectic temperature, 1116°F (602°C). However, in commercial-size ingots, the maximum Li content that can be cast without cracking is about 2.7%. Lithium is a strengthening element because of the formation of small, coherent ordered Al_3Li precipitates during aging (secondary hardening when Li content exceeds 1.4%). The toughness of Al−Li alloys, unlike conventional Al alloys, does not increase with increasing aging temperature (beyond that point needed for peak strength).

Metal-Matrix Composites. Silicon carbide particles are contributing to easy-to-cast metal-matrix composites (MMCs). When compared with their non-reinforced counterparts, the SiCp/Al components are more wear resistant, stiffer, and stronger, accompanied by improved thermal stability. Additional advantages include lower density and lower cost. Nearly all prior aluminum MMCs required labor-intensive methods, such as powder metallurgy, diffusion bonding, squeeze casting, or thermal spraying.

The new SiC composites are available as foundry ingot or extrusion billets. A new process ensures complete wetting of the SiC particles by molten aluminum. A number of investment castings are now being made, including aircraft hydraulic components and other small parts. These composites have excellent prospects for use in a variety of small parts, including medical prostheses and golf club heads.

Sialons consist of three-dimensional arrays of (Si−Al) $(O,N)_4$ tetrahedra. These oxynitrides are traditionally fabricated with silicon nitride. An example is beta-sialon, where the O and Si are partially replaced by N and Al, respectively. Advanced sialons are now being researched to enhance fracture toughness and improved creep properties.

Aluminides. These are intermetallic compounds of aluminum. The potential of these products includes uses where low weight, high-temperature strength, and oxidation resistance are required. Traditionally, these products are made by way of powder metallurgy technology.

TABLE 5. DESIGNATIONS FOR CAST ALUMINUM ALLOY GROUPS

		Alloy No.
Aluminum	99.00% minimum and greater	1xx.x
	Major Alloy Element	
	Copper	2xx.x
Aluminum	Silicon, with added Copper	
Alloys	and/or Magnesium	3xx.x
Grouped	Silicon	4xx.x
By Major	Magnesium	5xx.x
Alloying	Zinc	7xx.x
Elements	Tin	8xx.x
	Other Element	9xx.x
Unused Series		6xx.x

(1) For codification purposes an alloying element is any element which is intentionally added for any purpose other than grain refinement and for which minimum and maximum limits are specified.

(2) Standard limits for alloying elements and impurities are expressed to the following places:

Less than 1/1000%	0.000X
1/1000 up to 1/100%	0.00X
1/100 up to 1/10%	
Unalloyed aluminum made by a refining process	0.0XX
Alloys and unalloyed aluminum not made by a refining process	0.0X
1/10 through 1/2%	0.XX
Over 1/2%	0.X, X.X, etc.

Powder consolidation has been affected by sintering and hot isostatic pressing, both methods requiring long processing at height temperature. They rely mainly on solid-state diffusion. In a more recent method, *dynamic consolidation* uses high-pressure shock waves traveling at several kilometers per second. Such shocks can be generated through the use of detonating explosives or a gun-fired projectile. Upon full development of the shock-wave technique, advantages predicted include: (1) the non-equilibrium microstructures produced in rapid-solidification processing of powders will be retained in the final compact, (2) composite materials may be fabricated with very thin reaction zones between matrix and reinforcement, thus minimizing brittle reaction products that distract from the composite properties, and (3) net shapes may be produced. Normally confined in the past to production of centimeter-size parts, an improved process may be scaled up to meter-size products. Further development is required to prevent the formation of cracks.

Shape-Memory Alloys. Stoeckel defines a shape-memory alloy as the ability of some plastically deformed metals (and plastics) to resume their original shape upon heating. This effect has been observed in numerous metal alloys, notably the Ni−Ti and copper-based alloys, where commercial utilization of this effect has been exploited. (An example is valve springs that respond automatically to change in transmission-fluid temperature.) Copper-based alloy systems also exhibit this effect. These have been Cu−Zn−Al and Cu−Al−Ni systems. In fact, the first thermal actuator to utilize this effect (a greenhouse window opener) uses a Cu−Zn−Al spring.

ARALL Laminates

Developed in the late 1970s, *AR*amid *AL*uminum *L*aminates were developed by Delft University and Fokker Aircraft Co. The laminate currently is used for the skin of the cargo door for the Douglas C-17 military transport aircraft, but additional aerospace applications are envisioned. In essence, the laminate comprises a prepreg (i.e., unidirectional aramid fibers embedded in a structural epoxy adhesive) sandwiched between layers of aircraft alluminum alloy sheet. The fibers are oriented parallel to the rolling direction of the aluminum sheet. Prior to lay-up and autoclave curing, the aluminum surfaces are anodized and primed to ensure good bond integrity and to inhibit corrosion of the metal in the event of moisture intrusion at the bond line.

Quasicrystals

In the early 1980s, D. Schechtman at NIST (U.S. Nat ional Institute for Standards and Technology) discovered quasicrystals in aluminum alloys. Since then, they also have been noted in other alloys, including those of copper, magnesium, and zinc. Quasicrystals contradict the traditional fundamentals of crystallography to the effect that the periodicity of a perfect crystal structure is not possible with pentagon shapes. Much pioneering research on quasicrystals also has been conducted at the Laboratoire de Science at Gènie des Matèriaux Mètalliques in France.

To date, little use has been found for quasicrystals in bulk, but they have proved very effective as coatings, notably in cookware. Recent cookware, with a different appearance and "feel," has appeared in the marketplace. These pots, pans, and so on, have a hardness equal to that of hardened alloy steel and thus are practically immune to scratching. They also are thermally stable and corrosion and oxidation resistant.

The coating is applied by using flame, supersonic, and plasma-arc spraying. The deposited material consists of a mixture of quasicrystals and crystalline phases. The quasicrystal content of the surface ranges from 30−70%.

In structure, the quasicrystal relates to the Penrose tile structures (polygon), originally proposed by Roger Penrose, a mathematician at Oxford University. See **Crystal**.

Advances in Powdered Metallurgy (PM) Aluminum Alloys

As noted by Frazier, materials for advanced airframe structures and propulsion systems must withstand increasingly high temperature exposure. For example, frictional heating can raise supersonic skin temperatures to a range of 555° to 625°F (290° to 330°C). Unfortunately, wrought age-hardening aluminum alloys lose strength above 265°F (130°C). Titanium alloys perform well under these conditions, but they are 67% denser than aluminum, constituting about 42% of the weight of contemporary turbofan engines. Replacement of half the titanium with aluminum would reduce engine weight by about 20%. The motivation for using PM products is cost reduction and improved performance. Advanced thermoplastic matrix composites under development are difficult to process and presently cost prohibitive. Thus, intensive research is underway to improve rapid solidification technology and other new PM processes to increase the alloy aluminum content, thus reducing weight and cost.

Aluminum Electroplating

Electroplated aluminum is growing in acceptance for use in automotive parts, electrical equipment, and appliances and for products in a marine environment. Markets may be extended as the result of a new galvano-aluminum electroplating process developed by Siemens Research Laboratory (Erlangen, Germany) and described in the Hans reference.

S. J. SANSONETTI
Consultant, Reynolds Metals Company
Richmond, Virginia
(Updated by Editorial Staff).

Additional Reading

Aluminum Association: "Aluminum Standards and Data" and "Aluminum Statistical Review" (issued periodically). http://www.aluminum.org/

Carter, G.F., and D.E. Paul: *Materials Science and Engineering,* ASM International, Materials Park, OH, 1991. http://www.asm-intl.org/

Cathonet, P.: "Quasicrystals at Home on the Range," *Adv. Mat. & Proc.,* 6 (June, 1991).

Davis, J.R.: *Corrosion of Aluminum and Aluminum Alloys,* ASM International, Materials Park, OH, 1999.

Frazier, W.E.: "PM Al Alloys: Hot Prospects for Aerospace Applications," *Adv. Mat. & Proc.,* 42, (November, 1988).

Frick, J., Editor: *Woldman's Engineering Alloys,* 8th Edition, ASM International, Materials Park, OH, 1994.

Gregory, M.A.: "ARALL Laminates Take Wing," *Adv. Mat. & Proc.,* 115 (April 1990).

Hans, R.: "High-Purity Aluminum Electroplating," *Adv. Mat. & Proc.,* 14 (June 1989).

Kaufman, J.G.: *Properties of Aluminum Alloys: Tensile, Creep, and Fatigue Data at High and Low Temperatures,* ASM International, Materials Park, OH, 1999.

Kaufman, J.G.: *Introduction to Aluminum Alloys and Tempers,* ASM International, Materials Park, OH, 2000.

Kennedy, D.O.: "SiC Particles Beef up Investment-Cast Aluminum," *Adv. Mat. & Proc.,* 42–46 (June 1991).

Kim, N.J., K.V. Jata, W.E. Frazier, and E.W. Lee: *Light Weight Alloys for Aerospace Applications,* The Minerals, Metals & Materials Society, Warrendale, PA, 1998. http://www.tms.org/

Lide, D.R.: *CRC Handbook of Chemistry and Physics,* 84th Edition, CRC Press, LLC., Boca Raton, FL, 2003.

Loffler, H.: *Structure and Structure Development of Al–Zn Alloys,* John Wiley & Sons, Inc., New York, NY, 1995.

Perry, R.H., and D. Green: *Perry's Chemical Engineers' Handbook,* 7th Edition, McGraw-Hill Companies, Inc., New York, NY, 1999.

Peterson, W.S.: *Hall-Heroult Centennial—First Century of Aluminum Process Technology—1886–1986,* The Metallurgical Society, London, 1986.

Rioja, R.J., and R.H. Graham: "Al–Li Alloys Find Their Niche," *Adv. Mat. & Pro.,* 23 (June 1992).

Samuels, L.E.: *Metals Engineering: A Technical Guide,* ASM International, Materials Park, OH, 1988.

Sousa, L.J.: "The Changing World of Metals," *Adv. Mat. & Proc.,* **27** (September 1988).

Staff: *Aluminum and Magnesium Alloys,* American Society for Testing & Materials, West Conshohocken, PA, 1999. http://www.astm.org/

Staff: "Aluminum, Steel Cans Make a Dent in the Market," *Adv. Mat. & Proc.,* **12** (June 1989).

Staff: "Sialons Produced by Combustion Synthesis," *Adv. Mat. & Proc.,* **11** (September 1989).

Staff: *Aluminum Data Sheets, #7450G,* ASM International, Materials Park, OH, 1990.

Staff: "Strength (Metals)," *Adv. Mat. & Proc.,* **19** (June 1990).

Staff: *Properties and Selection: Nonferrous Alloys and Special-Purpose Materials,* ASM International, Materials Park, OH, 1991.

Staff: "Audi To Get Aluminum Space Frame," *Adv. Mat. & Proc.,* **9** (January 1992).

Staff: "Forecast '92—Aluminum," *Adv. Mat. & Proc.,* **17** (January 1992).

Stoeckel, D.: "Shape–Memory Alloys Prompt New Actuator Designs," *Adv. Mat. & Proc.,* 33 (October 1990).

Strauss, S.: "Impossible Matter (Quasicrystals)," *Techy. Review (MIT),* 19 (January 1991).

Taketani, H.: "Properties of Al–Li Alloy 2091-T3 Sheet," *Adv. Mat. & Proc.,* 113 (April 1990).

Van Horn, K.R., Editor: *Aluminum,* Vol. 1–3, ASM International, Materials Park, OH, 1967. (A classic reference.)

Vassilou, M.S.: "Shock Waves Shape Aluminides," *Adv. Mat. & Proc.,* 70 (October 1990).

Vaughan, D.E.W.: "The Synthesis and Manufacture of Zeolites," *Chem. Eng. Prog.,* **25** (February 1988).

Webster, D., T.G. Haynes, III, and R.H. Fleming: "Al–Li Investment Castings Coming of Age," *Adv. Mat. & Proc.,* 25 (June 1988).

Webster, D., and C.G. Bennett: "Tough(er) Aluminum-Lithium Alloys," *Adv. Mat. & Proc.,* 49 (October 1989).

Winterbottom, W.L.: "The Aluminum Auto Radiator Comes of Age," *Adv. Mat. & Proc.,* 55 (May 1990).

ALUNITE. The mineral alunite, $KAl_3(SO_4)_2(OH)_6$, is a basic hydrous sulfate of aluminum and potassium; a variety called natroalunite is rich in soda. Alunite crystallizes in the hexagonal system and forms rhombohedrons with small angles, hence resembling cubes. It may be in fibrous or tabular forms, or massive. Hardness, 3.5–4; sp gr, 2.58–2.75; luster, vitreous to pearly; streak white; transparent to opaque; brittle; color, white to grayish or reddish.

Alunite is commonly associated with acid lava due to sulfuric vapors often present; it may occur around fumaroles or be associated with sulfide ore bodies. It has been used as a source of potash. Alunite is found in the Czech Republic and Slovakia, Italy, France, and Mexico; in the United States, in Colorado, Nevada, and Utah.

Alunite is also known as *alumstone.*

AMALGAM. 1. An alloy containing mercury. Amalgams are formed by dissolving other metals in mercury, when combination takes place often with considerable evolution of heat. Amalgams are regarded as compounds of mercury with other metals, or as solutions of such compounds in mercury. It has been demonstrated that products which contain mercury and another metal in atomic proportions may be separated from amalgams. The most commonly encountered amalgams are those of gold and silver. See also **Gold**; **Mercury**; and **Silver**.

2. A naturally occurring alloy of silver with mercury, also referred to as mercurian silver, silver amalgam, and argental mercury. The natural amalgam crystallizes in the isometric system; hardness, 3–3.5; sp gr, 13.75–14.1; luster, metallic; color, silver-white; streak, silver-white; opaque. Amalgam is found in Bavaria, British Columbia, Chile, the Czech Republic and Slovakia, France, Norway, and Spain. In some areas, it is found in the oxidation zone of silver deposits and as scattered grains in cinnabar ores.

AMBER. Amber is a fossil resin known since early times because of its property of acquiring an electric charge when rubbed. In modern times it has been used largely in the making of beads, cigarette holders, and trinkets. Its amorphous non-brittle nature permits it to be carved easily and to acquire a very smooth and attractive surface. Amber is soluble in various organic solvents, such as ethyl alcohol and ethyl ether.

It occurs in irregular masses showing a conchoidal fracture. Hardness, 2.25; sp gr, 1.09; luster, resinous; color, yellow to reddish or brownish; it may be cloudy. Some varieties will exhibit fluorescence. Amber is transparent to translucent, melts between 250 and 300°C.

Amber has been obtained for over 2,000 years from the lignite-bearing Tertiary sandstones on the coast of the Baltic Sea from Gdansk to Liepàja; also from Denmark, Sweden and the other Baltic countries. Sicily furnishes a brownish-red amber that is fluorescent.

The association of amber with lignite or other fossil woods, as well as the beautifully preserved insects that are occasionally in it, is ample proof of its organic origin.

AMBERGRIS. A fragrant waxy substance formed in the intestine of the sperm whale and sometimes found floating in the sea. It has been used in the manufacture of perfumes to increase the persistence of the scent.

AMBLYGONITE. A rather rare compound of fluorine, lithium, aluminum, and phosphorus, $(Li, Na)AlPO_4(F, OH)$. It crystallizes in the tri-clinic system; hardness, 5–5.6; sp gr 3.08; luster, vitreous to greasy or pearly; color, white to greenish, bluish, a yellowish or grayish; streak white; translucent to subtransparent.

Amblygonite occurs in pegmatite dikes and veins associated with other lithium minerals. It is used as a source of lithium salts. The name is derived from two Greek words meaning blunt and angle, in reference to its cleavage angle of 75°30′.

Amblygonite is found in Saxony; France; Australia; Brazil; Varutrask, Sweden; Karibibe, S.W. Africa; and the United States.

AMERICIUM. [CAS: 7440-35-9]. Chemical element, symbol Am, at no. 95, at. wt. 243 (mass number of the most stable isotope), radioactive metal of the actinide series, also one of the transuranium elements. All isotopes of americium are radioactive; all must be produced synthetically. The element was discovered by G.T. Seaborg and associates at the Metallurgical Laboratory of the University of Chicago in 1945. At that time, the element was obtained by bombarding uranium-238 with helium ions to produce ^{241}Am, which has a half-life of 475 years. Subsequently, ^{241}Am has been produced by bombardment of plutonium-241 with neutrons in a nuclear reactor. ^{243}Am is the most stable isotope, an alpha emitter with a half-life of 7950 years. Other known isotopes are ^{237}Am, ^{238}Am, ^{240}Am, ^{241}Am, ^{242}Am, ^{244}Am, ^{245}Am, and ^{246}Am. Electronic configuration is $1s^2 2s^2 2p^6 3s^2 3p^6 3d^{10} 4s^2 4p^6 4d^{10} 4f^{14} 5s^2 5p^6 5d^{10} 5f^7 6s^2 6p^6 7s^2$. Ionic radii are: Am^{4+}, 0.85 Å; Am^{3+}, 1.00Å.

This element exists in acidic aqueous solution in the (III), (IV), (V), and (VI) oxidation states with the ionic species probably corresponding to Am^{3+}, Am^{4+}, AmO_2^+ and AmO_2^{2+}.

The colors of the ions are: Am^{3+}, pink; Am^{4+}, rose; AmO_2^+, yellow; and AmO_2^{2+}, rum-colored.

It can be seen that the (III) state is highly stable with respect to disproportionation in aqueous solution and is extremely difficult to oxidize or reduce. There is evidence for the existence of the (II) state since tracer amounts of americium have been reduced by sodium amalgam and precipitated with barium chloride or europium sulfate as carrier. The (IV) state is very unstable in solution: the potential for americium(III)-americium(IV) was determined by thermal measurements involving solid AmO_2. Americium can be oxidized to the (V) or (VI) state with strong oxidizing agents, and the potential for the americium(V)-americium(VI) couple was determined potentiometrically.

In its precipitation reactions americium(III) is very similar to the other tripositive actinide elements and to the rare earth elements. Thus the fluoride and the oxalate are insoluble and the phosphate and iodate are only moderately soluble in acid solution, whereas the nitrates, halides, sulfates, sulfides, and perchlorates are all soluble. Americium(VI) can be precipitated with sodium acetate giving crystals isostructural with sodium uranyl acetate,

$$NaUO_2(C_2H_3O_2)_3 \cdot xH_2O$$

and the corresponding neptunium and plutonium compounds.

Of the hydrides of americium, both AmH_2 and Am_4H_{15} are black and cubic.

When americium is precipitated as the insoluble hydroxide from aqueous solution and heated in air, a black oxide is formed which corresponds almost exactly to the formula AmO_2. This may be reduced to Am_2O_3 through the action of hydrogen at elevated temperatures. The AmO_2 has the cubic fluorite type structure, isostructural with UO_2, NpO_2, and PuO_2. The sesquioxide, Am_2O_3 is allotropic, existing in a reddish brown and a tan form, both hexagonal. As in the case of the preceding actinide elements, oxides of variable composition between $AmO_{1.5}$ and AmO_2 are formed depending upon the conditions.

All four of the trihalides of americium have been prepared and identified. These are prepared by methods similar to those used in the preparation of the trihalides of other actinide elements. AmF_3 is pink and hexagonal, as is $AmCl_3$; $AmBr_3$ is white and orthorhombic; while a tetrafluoride, AmF_4 is tan and monoclinic.

In research at the Institute of Radiochemistry, Karlsruhe, West Germany during the early 1970s, investigators prepared alloys of americium with platinum, palladium, and iridium. These alloys were prepared by hydrogen reduction of the americium oxide in the presence of finely divided noble metals according to:

$$Am_2O_3 + 10Pt \xrightarrow[1100°C]{H_2} 2\,AmPt_5 + H_2O$$

The reaction is called a *coupled reaction* because the reduction of the metal oxide can be done only in the presence of noble metals. The hydrogen must be extremely pure, with an oxygen content of less than 10^{-25} torr.

See also **Chemical Elements**.

Industrial utilization of americium has been quite limited. Uses include a portable source for gamma radiography, a radioactive glass thickness gage for the flat glass industry, and an ionization source for smoke detectors.

Americium is present in significant quantities in spent nuclear reactor fuel and poses a threat to the environment. A group of scientists at the U.S. Geological Survey (Denver, Colorado) has studied the chemical speciation of actinium (and neptunium) in ground waters associated with rock types that have been proposed as possible hosts for nuclear waste repositories. Researchers Cleveland, Nash, and Rees (see reference list) concluded that americium (and neptunium) are relatively insoluble in ground waters containing high sulfate concentrations (90°C).

Additional Reading

Cleveland, J.M., K.L. Nash, and T.F. Rees: "Neptunium and Americium Speciation in Selected Basalt, Granite, Shale, and Tuff Ground Waters," *Science*, **221**, 271–273 (1983).

Fisk, Z. et al.: "Heavy-Electron Metals: New Highly Correlated States of Matter," *Science*, **33** (January 1, 1988).

Greenwood, N.N. and A. Earnshaw, Editors: *Chemistry of the Elements*, 2nd Edition, Butterworth-Heinemann, UK, 1997.

Lide, D.R.: *Handbook of Chemistry and Physics*, 84th Edition, CRC Press LLC, Boca Raton, FL, 2003.

Moss, L.R. and J. Fuger, Editors: *Transuranium Elements: A Half Century, American Chemical Society*, 1992.

Seaborg, G.T.: "The Chemical and Radioactive Properties of the Heavy Elements." *Chemical & Engineering News*, **23**, 2190–2193 (1945).

Seaborg, G.T. and W.D. Loveland: *The Elements Beyond Uranium*, John Wiley & Sons, New York, NY, 1990.

Seaborg, G.T., Editor: *Transuranium Elements*, Dowden, Hutchinson & Ross, Stroudsburg, PA, 1978.

Silva, R.J., G. Bidoqlio, M.H. Rand, and P. Robouch: *Chemical Thermodynamics of Americium (Chemical Thermodynamics, Vol. 2)* North-Holland, New York, NY, 1995.

AMERICAN ASSOCIATION OF SCIENTIFIC WORKERS (AASW). Founded in 1946. Scientists concerned with national and international relations of science and society and with organizational aspects of science. It is located at the School of Veterinary Medicine, University of Pennsylvania, Philadelphia, PA 19172.

AMERICAN ASSOCIATION OF TEXTILE CHEMISTS AND COLORISTS (AATCC). Founded in 1921. It has over 6500 members. A technical and scientific society of textile chemists and colorists in textile and related industries using colorants and chemical finishes. It is the authority for test methods. It is located at PO Box 12215, Research Triangle Park, NC 27709. http://www.aatcc.org/

AMERICAN CARBON SOCIETY (ACS). The present name of the American Carbon Committee, a group incorporated in 1964 to operate the Biennial American Carbon Conferences. The committee also has sponsored the international journal *Carbon*. Over 500 members (physicist, chemists, technicians, and other scientific personnel) worldwide focus on the physics and chemistry of organic crystals, polymers, chars, graphite, and carbon materials. It is located at the Stackpole Corporation, St. Mary's, PA 19174. http://www.americancarbonsociety.org/

AMERICAN CERAMIC SOCIETY (ACerS). Founded 1899. It has 12,000 members. A professional society of scientists, engineers, and plant operators interested in glass, ceramics-metal systems, cements, refractories, nuclear ceramics, white wares, electronics, and structural clay products. It is located at 65 Ceramic Dr., Columbus, OH 43214. http://www.acers.org/

AMERICAN CHEMICAL SOCIETY (ACS). Founded in 1876. It has over 150,000 members. The nationally chartered professional society for chemists in the U.S. One of the largest scientific organizations in the world. Its offices are at 1155 16th St., NW, Washington DC 20036. http://www.acs.org/

AMERICAN INSTITUTE OF CHEMISTS (AIC). Founded in 1923, it is primarily concerned with chemists and chemical engineers as professional people rather than with chemistry as a science. Special emphasis is placed on the scientific integrity of the individual and on a code of ethics adhered to by all its members. It publishes a monthly journal, *The Chemist*. It is located at 7315 Wisconsin Ave, NW, Bethesda, MD 20814. http://www.theaic.org/

AMERICAN NATIONAL STANDARDS INSTITUTE (ANSI). Founded in 1918. A federation of trade associations, technical societies, professional groups, and consumer organizations that constitutes the U.S. clearinghouse and coordinating body of voluntary standards activity on the national level. It eliminates duplication of standards activities and combines conflicting standards into single, nationally accepted standards. It is the U.S. member of the International Organization for Standardization and the International Electrotechnical Commission. Over 1000 companies are members of the ANSI. One of its primary concerns is safety in such fields as hazardous chemicals, protective clothing, welding, fire control, electricity and construction operations, blasting, etc. Its address is 1430 Broadway, New York, NY 10018. http://www.ansi.org/

AMERICAN OIL CHEMIST'S SOCIETY (AOCS). Founded in 1909. It has over 5000 members. These members are chemists, biochemists, chemical engineers, research directors, plant personnel, and persons concerned with animal, marine, and regular oils and fats and their extraction, refining, safety, packaging, quality control, and use. The address is 508 S. 6th St., Champaign, IL 61820. http://www.aocs.org/

AMERICAN PETROLEUM INSTITUTE (API). Founded in 1919. It has 5500 members. The members are the producers, refiners, marketers, and transporters of petroleum and allied products such as crude oil, lubricating oil, gasoline, and natural gas. The address is 1220 L Street, NW. Washington, DC 20005-4070. http://www.api.org

AMERICAN SOCIETY FOR METALS (ASM). Formally organized in 1935, this society actually had been active under other names since 1913, when the need for standards of metal quality and performance in the automobile became generally recognized. ASM has over 53,000 members and publishes *Metals Review* and the famous *Metals Handbook*, as well as research monographs on metals. It is active in all phases of metallurgical activity, metal research, education, and information retrieval. Its headquarters is at Metals Park, OH, 44073. http://www.asm.org/

AMERICAN SOCIETY FOR TESTING AND MATERIALS (ASTM). This society, organized in 1898 and chartered in 1902, is a scientific and technical organization formed for "the development of standards on characteristics and performance of materials, products, systems and services, and the promotion of related knowledge." There are over 31,000 members. It is the world's largest source of voluntary consensus standards. The society operates via more than 125 main technical committees that function in prescribed fields under regulations that ensure balanced representation among producers, users, and general-interest participants. Headquarters of the society is at 655 15th St., Washington DC 2005. http://www.astm.org/

AMETHYST. A purple- or violet-colored quartz having the same physical characteristics as quartz. The source of color is not definite but thought to be caused by ferric iron contamination. Oriental amethysts are purple corundum.

Amethysts are found in the Ural Mountains, India, Sri Lanka, Madagascar, Uruguay, Brazil, the Thunder Bay district of Lake Superior in Ontario, and Nova Scotia; in the United States, in Michigan, Virginia, North Carolina, Montana, and Maine.

The name amethyst is generally supposed to have been derived from the Greek word meaning not drunken. Pliny suggested that the term was applied because the amethyst approaches but is not quite the equivalent of a wine color.

See also **Quartz**.

AMIDES. An amide may be defined as a compound that contains the $CO \cdot NH_2$ radical, or an acid radical(s) substituted for one or more of the hydrogen atoms of an ammonia molecule. Amides may be classified as (1) *primary amides*, which contain one acyl radical, such as $-CO \cdot CH_3$ (acetyl) or $-CO \cdot C_6H_5$ (benzoyl), linked to the *amido* group ($-NH_2$). Thus, acetamide NH_2COCH_3 is a combination of the acetyl and amido groups; (2) *secondary amides*, which contain two acyl radicals and the *imido* group ($-NH_2$) Diacetamide $HN(COCH_3)_2$ is an example; and (3) *tertiary amides*, which contain three acyl radicals attached to the N atom. Triacetamide $N(COCH_3)_3$ is an example.

A further structural analysis will show that amides may be regarded as derivatives of corresponding acids in which the amido group substitutes for the hydroxyl radical OH of the carboxylic group COOH. Thus, in the instance of formic acid HCOOH, the amide is $HCOONH_2$ (formamide); or in the case of acetic acid CH_3COOH, the amide is CH_3CONH_2 (acetamide). Similarly, urea may be regarded as the amide of carbonic acid (theoretical) $O:C$, that is, NH_2CONH_2 (urea). The latter represents a dibasic acid in which two H atoms of the hydroxyl groups have been replaced by amido groups. A similar instance, malamide,

$$NH_2CO \cdot CH_2CH(OH) \cdot CONH_2$$

is derived from the dibasic acid, malic acid,

$$OHCO \cdot CH_2CH(OH) \cdot COOH.$$

Aromatic amides, sometimes referred to as *arylamides*, exhibit the same relationship. Note the relationship of benzoic acid C_6H_5COOH with benzamide $C_6H_5CONH_2$. *Thiamides* are derived from amides in which there is substitution of the O atom by a sulfur atom. Thus, acetamide $NH_2 \cdot CO \cdot CH_3$, becomes thiacetamide $NH_2 \cdot CS \cdot CH_3$; or acetanilide $C_6H_5 \cdot NH \cdot CO \cdot CH_3$ becomes thiacetanilide $C_6H_5 \cdot NH \cdot CS \cdot CH_3$. *Sulfonamides* are derived from the sulfonic acids. Thus, benzene-sulfonic acid $C_6H_5 \cdot SO_2 \cdot OH$ becomes benzene-sulfonamide $C_6H_5 \cdot SO_2 \cdot NH_2$. See also **Sulfonamide Drugs**.

Amides may be made in a number of ways. Prominent among them is the acylation of amines. The agents commonly used are, in order of reactivity, the acid halides, acid anhydrides, and esters. Such reactions are:

$$R'COCl + HNR_2 \longrightarrow R'C(=O)NR_2 + HCl$$

$$R'C(=O)OC(=O)R' + HNR_2 \longrightarrow R'C(=O)NR_2 + R'COOH$$

$$R'C(=O)OR'' + HNR_2 \longrightarrow R'C(=O)NR_2 + R''OH$$

The hydrolysis of nitriles also yields amides:

$$RCN + H_2O \xrightarrow{OH} RCONH_2$$

Amides are resonance compounds, having an ionic structure for one form:

$$R-C(=O)NR_2 \qquad R-C(-O^-):N^+R_2$$

Evidence for the ionic form is provided by the fact that the carbon-nitrogen bond (1.38 Å) is shorter than a normal $C-N$ bond (1.47 Å) and the carbon-oxygen bond (1.28 Å) is longer than a typical carbonyl bond (1.21 Å). That is, the carbon-nitrogen bond is neither a real $C-N$ single bond nor a $C-N$ double bond.

The amides are sharp-melting crystalline compounds and make good derivatives for any of the acyl classes of compounds, i.e., esters, acids, acid halides, anhydrides, and lactones.

Amides undergo hydrolysis upon refluxing in H_2O. The reaction is catalyzed by acid or alkali.

$$R-\underset{NR_2}{\overset{\overset{\textstyle O}{\|}}{C}} + HOH \xrightarrow{H_3O^+ \text{ or } OH^-} R-\overset{\overset{\textstyle O}{\|}}{\underset{OH}{C}} + R_2NH$$

Primary amides may be dehydrated to yield nitriles.

$$R-CONH_2 + C_6H_5SO_2Cl \xrightarrow[70°]{\text{pyridine}} R-CN + C_6H_5SO_3H + HCl$$

The reaction is run in pyridine solutions.

Primary and secondary amides of the type $RCONH_2$ and $RCONHR$ react with nitrous acid in the same way as do the corresponding primary and secondary amines.

$$RCONH_2 + HONO \longrightarrow RCOOH + N_2 + HOH$$

$$RCONHR + HONO \longrightarrow RCON(NO)R + HOH$$

When diamides having their amide groups not far apart are heated, they lose ammonia to yield imides. See also **Imides**.

AMINATION. The process of introducing the amino group ($-NH_2$) into an organic compound is termed *amination*. An example is the reduction of aniline, $C_6H_5 \cdot NH_2$, from nitrobenzene, $C_6H_5 \cdot NO_2$. The reduction may be accomplished with iron and HCl. Only about 2% of the calculated amount of acid (to produce H_2 by reaction with iron) is required because of the fact that H_2O plus iron in the presence of ferrous chloride solution (ferrous and chloride ions) functions as the primary reducing agent. Such groups as nitroso ($-NO$), hydroxylamine ($-NH \cdot NH-$), and azo ($-N:N-$) also yield amines by reduction. Amination also may be effected by the use of NH_3, in a process sometimes referred to as *ammonolysis*. An example is the production of aniline from chlorobenzene:

$$C_6H_5Cl + NH_3 \longrightarrow C_6H_5 \cdot NH_2 + HCl$$

The reaction proceeds only under high pressure. In the ammonolysis of benzenoid sulfonic acid derivatives, an oxidizing agent is added to prevent the formation of soluble reduction products, such as $NaNH_4SO_4$, which commonly form. Oxygen-function compounds also may be subjected to ammonolysis: (1) methanol plus aluminum phosphate catalyst yields mono-, di-, and trimethylamines; (2) β-naphthol plus sodium ammonium sulfite catalyst (Bucherer reaction) yields β-naphthylamine; (3) ethylene oxide yields mono-, di-, and triethanolamines; (4) glucose plus nickel catalyst yields glucamine; and (5) cyclohexanone plus nickel catalyst yields cyclohexylamine.

AMINES. An amine is a derivative of NH_3 in which there is a replacement for one or more of the H atoms of NH_3 by an alkyl group, such as $-CH_3$ (methyl) or $-C_2H_5$ (ethyl); or by an aryl group, such as $-C_6H_5$ (phenyl) or $-C_{10}H_7$ (naphthyl). Mixed amines contain at least one alkyl and one aryl group as exemplified by methylphenylamine $CH_3 \cdot N(H) \cdot C_6H_5$. When one, two, and three H atoms are thus replaced, the resulting amines are known as *primary, secondary*, and *tertiary*, respectively. Thus, methylamine, CH_3NH_2, is a primary amine; dimethylamine, $(CH_3)_2NH$, is a secondary amine; and trimethylamine, $(CH_3)_3N$, is a tertiary amine. Secondary amines sometimes are called *imines*; tertiary amines, *nitriles*.

Quaternary amines consist of four alkyl or aryl groups attached to an N atom and, therefore, may be considered substituted ammonium bases. Commonly, they are referred to in the trade as quaternary ammonium compounds. An example is tetramethyl ammonium iodide.

$$\underset{I}{\overset{H_3C}{\diagdown}} \underset{CH_3}{\overset{CH_3}{\diagup}} N \diagdown_{CH_3}^{CH_3}$$

The amines and quaternary ammonium compounds, exhibiting such great versatility for forming substitution products, are very important starting and intermediate materials for industrial organic syntheses, both on a small scale for preparing rare compounds for use in research and on a tonnage basis for the preparation of resins, plastics, and other synthetics. Very important industrially are the ethanolamines which are excellent absorbents for certain materials. See also **Ethanolamines**. Hexamethylene tetramine is a high-tonnage product used in plastics production. See also **Hexamine**. Phenylamine (aniline), although not as important industrially as it was some years ago, still is produced in quantity. Melamine is produced on a large scale and is the base for a series of important resins. See also **Melamine**. There are numerous amines and quaternary ammonium compounds that are not well known because of their importance as intermediates rather than as final products. Examples along these lines may include acetonitrile and acrylonitrile. See also **Acrylonitrile**.

Primary amines react (1) with nitrous acid, yielding (a) with alkyl-amine, nitrogen gas plus alcohol, (b) with warm arylamine, nitrogen gas plus phenol (the amino-group of primary amines is displaced by the hydroxyl group to form alcohol or phenol), (c) with cold arylamine, diazonium compounds, (2) with acetyl chloride or benzoyl chloride, yielding substituted amides, thus, ethylamine plus acetyl chloride forms N-ethylacetamide, $C_2H_5NHOCCH_3$, (3) with benzene-sulfonyl chloride, $C_6H_5SO_2Cl$, yielding substituted benzene sulfonamides, thus, ethylamine forms N-ethylbenzenesulfonamide, $C_6H_5SO_2-NHC_2H_5$, soluble in sodium hydroxide, (4) with chloroform, $CHCl_3$ with a base, yielding isocyanides (5) with HNO_3 (concentrated), yielding nitra-mines, thus, ethylamine reacts to form ethylnitramine, $C_2H_5-NHNO_2$.

Secondary amines react (1) with nitrous acid, yielding nitrosamines, yellow oily liquids, volatile in steam, soluble in ether. The secondary amine may be recovered by heating the nitrosamine with concentrated HCl, or hydrazines may be formed by reduction of the nitrosamines, e.g., methylaniline from methylphenylnitrosamine, $CH_3(C_6H_5)NNO$, reduction yielding unsymmetrical methylphenylhydrazine, $CH_3(C_6H_5)NHNH_2$, (2) with acetyl or benzoyl chloride, yielding substituted amides, thus, diethylamine plus acetyl chloride to form N, N-diethylacetamide $(C_2H_5)-NOCCH_3$, (3) with benzene sulfonyl chloride, yielding substituted benzene sulfonamides, thus, diethylamine reacts to form N, N-diethylbenzenesulfonamide, $C_6H_5SO_2N(C_2H_5)_2$, insoluble in NaOH.

Tertiary amines do not react with nitrous acid, acetyl chloride, benzoyl chloride, benzenesulfonyl chloride, but react with alkyl halides to form quaternary ammonium halides, which are converted by silver hydroxide to quaternary ammonium hydroxides. Quaternary ammonium hydroxides upon heating yield (1) tertiary amine plus alcohol (or, for higher members, olefin plus water). Tertiary amines may also be formed (2) by alkylation of secondary amines, e.g., by dimethyl sulfate, (3) from amino acids by living organisms, e.g., decomposition of fish in the case of trimethylamine.

AMINO ACIDS. The scores of proteins which make up about one-half of the dry weight of the human body and that are so vital to life functions are made up of a number of amino acids in various combinations and configurations. The manner in which the complex protein structures are assembled from amino acids is described in the entry on Protein. For some users of this book, it may be helpful to scan that portion of the protein entry that deals with the chemical nature of proteins prior to considering the details of this immediate entry on amino acids.

Although the proteins resulting from amino acid assembly are ultimately among the most important chemicals in the animal body (as well as plants), the so-called infrastructure of the proteins is dependent upon the amino acid building blocks. Although there are many hundreds of amino acids, only about 20 of these are considered very important to living processes, of which six to ten are classified as essential. Another three or four may be classified as quasi-essential, and ten to twelve may be categorized as nonessential. As more is learned about the fundamentals, protein chemistry, the scientific importance attached to specific amino acids varies. Usually, as the learning process continues, the findings tend to increase the importance of specific amino acids. Actually, the words *essential* and *nonessential* are not very good choices for naming categories of amino acids. Generally, those amino acids that the human body cannot synthesize at all or at a rate commensurate with its needs are called *essential amino acids* (EAA). In other words, for the growth and maintenance of a normal healthy body, it is essential that these amino acids be ingested as part of the diet and in the necessary quantities. To illustrate some of the indefinite character of amino acid nomenclature, some authorities classify histidine as an essential amino acid; others do not. The fact is that histidine is essential for the normal growth of the human infant, but to date it is not regarded as essential for adults. By extension of the preceding explanation, the term nonessential is

taken to mean those amino acids that are really synthesized in the body and hence need not be present in food intake. This classification of amino acids, although amenable to change as the results of new findings, has been quite convenient in planning the dietary needs of people as well as of farm animals, pets, and also in terms of those plants that are of economic importance. The classification has been particularly helpful in planning the specific nutritional content of food substances involved in various aid and related programs for the people in needy and underdeveloped areas of the world.

Food Fortification with Amino Acids. In a report of the World Health Organization, the following observation has been made: "To determine the quality of a protein, two factors have to be distinguished, namely, the proportion of essential to nonessential amino acids and, secondly, the relative amounts of the essential amino acids. . . The best pattern of essential amino acids for meeting human requirements was that found in whole egg protein or human milk, and comparisons of protein quality should be made by reference to the essential amino acid patterns of either of these two proteins." The ratio of each essential amino acid to the total sum is given for hen's egg and human and cow's milk in Table 1.

In the human body, tyrosine and cysteine can be formed from phenylalanine and methionine, respectively. The reverse transformations do not occur. Human infants have an ability to synthesize arginine and histidine in their bodies, but the speed of the process is slow compared with requirements.

Several essential amino acids have been shown to be the limiting factor of nutrition in plant proteins. In advanced countries, the ratio of vegetable proteins to animal proteins in foods is 1.4:1. In underdeveloped nations, the ratio is 3.5:1, which means that people in underdeveloped areas depend upon vegetable proteins. Among vegetable staple foods, wheat easily can be fortified. It is used as flour all over the world. L-Lysine hydrochloride (0.2%) is added to the flour. Wheat bread fortified with lysine is used in several areas of the world; in Japan it is supplied as a school ration.

The situation of fortification in rice is somewhat more complex. Before cooking, rice must be washed (polished) with water. In some countries, the cooking water is allowed to boil over or is discarded. This significant loss of fortified amino acids must be considered. L-Lysine hydrochloride (0.2%) and L-threonine (0.1%) are shaped like rice grain with other nutrients and enveloped in a film. The added materials must hold the initial shape and not dissolve out during boiling, but be easily freed of their coating in the digestive organs.

The amino acids are arranged in accordance with essentiality in Table 2. Each of the four amino acids at the start of the table are all limiting factors of various vegetable proteins. Chick feed usually is supplemented with fish meal, but where the latter is in limited supply, soybean meals are substituted. The demand for DL-methionine, limiting amino acid in soybean meals, is now increasing. When seed meals, such as corn and sorghum, are used as feeds for chickens or pigs, L-lysine hydrochloride must be added for fortification. Lysine production is increasing upward to the level of methionine.

TABLE 1. REPRESENTATIVE ESSENTIAL AMINO ACID PATTERNS *A/E RATIO (MILLIGRAMS PER GRAM OF TOTAL ESSENTIAL AMINO ACIDS)

	Hen's egg (Whole)	Human milk	Cow's milk
Total "aromatic" amino acids	195	226	197
Phenylalanine	(114)	(114)	(97)
Tyrosine	(81)	(112)	(100)
Leucine	172	184	196
Valine	141	147	137
Isoleucine	129	132	127
Lysine	125	128	155
Total "S"	107	87	65
Cystine	(46)	(43)	(17)
Methionine	(61)	(44)	(48)
Threonine	99	99	91
Tryptophan	31	34	28

Source: World Health Organization; FAO Nutrition Meeting Report Series, No. 37, Geneva, 1965.

* A/E Ratio equals ten times percentage of single essential amino acid to the total essential amino acids contained.

TABLE 2. IMPORTANT NATURAL AMINO ACIDS AND PRODUCTION

Amino acid	World annual production, tons	Present mode of manufacture	Characteristics
		ESSENTIAL AMINO ACIDS	
DL-Methionine	10^4	Synthesis from acrolein and mercaptan	First limiting amino acid for soybean
L-Lysine. HCl	10^3	Fermentation (AM)*	First limiting amino acid for cereals
L-Threonine	10	Fermentation (AM)	Second limiting amino acid for rice
L-Tryptophan	10	Synthesis from acrylonitrile and resolution	Second limiting amino acid for corn
L-Phenylalanine	10	Synthesis from phenyl-acetaldehyde and resolution	
L-Valine	10	Fermentation (AM)	Rich in plant protein
L-Leucine	10	Extraction from protein	
L-Isoleucine	10	Fermentation (WS)**	Deficient in some cases
		QUASI-ESSENTIAL AMINO ACIDS	
L-Arginine. HCl	10^2	Synthesis from L-ornithine Fermentation (AM)	
L-Histidine. HCl	10	Extraction from protein	Essential to human infants
L-Tyrosine	10	Enzymation of phenol and Serine	Limited substitute for phenylalanine
L-Cysteine L-Cystine	10	Extraction from human hair	Limited substitute for methionine
		NONESSENTIAL AMINO ACIDS	
L-Glutamic acid	10^5	Fermentation (WS) Synthesis from acrylonitrile and resolution	MSG, taste enhancer
Glycine	10^3	Synthesis from formaldehyde	Sweetener
DL-Alanine	10^2	Synthesis from acetaldehyde	
L-Aspartic acid	10^2	Enzymation of fumaric acid	Hygienic drug
L-Glutamine	10^2	Fermentation (WS)	Anti-gastroduodenal ulcer drug
L-Serine	<10	Synthesis from glycolonitrile and resolution	Rich in raw silk
L-Proline	<10	Fermentation (AM)	
L-Hydroxyproline	<10	Extraction from gelatin	Rich in gelatin
L-Asparagine	<10	Synthesis from L-aspartic acid	Neurotropic metabolic regulator
L-Alanine	<10	Enzymation of L-aspartic acid	Rich in degummed white silk
L-Dihydroxy-phenylalanine	10^2	Synthesis from piperonal, vanillin, or acrylonitrile and resolution	Specific drug for Parkinson's disease
L-Citrulline	<10	Fermentation (AM)	Ammonia detoxicant
L-Ornithine	<10	Fermentation (AM)	

*AM, artificial mutant;
**WS, wild strain. MSG, monosodium glutamate.

Early Research and Isolation of Amino Acids. Because of such rapid studies made within the past few decades in biochemistry and nutrition, these sciences still have a challenging aura about them. But, it is interesting to note that the first two natural amino acids were isolated by Braconnot in 1820. As shown by Table 3, these two compounds were glycine and leucine. Bopp isolated tyrosine from casein in 1849. Additional amino acids were isolated during the 1880s, but the real thrust into research in this field commenced in the very late 1800s and early 1900s with the work of Emden, Fischer, Mörner, and Hopkins and Cole. It is interesting to observe that Emil Fischer (1852–1919), German chemist and pioneer in the fields of purines and polypeptides, isolated three of these important compounds, namely, proline from gelatin in 1901, valine from casein in 1901, and hydroxyproline from gelatin in 1902. As an understanding of the role of amino acids in protein formation and of the function of proteins in nutrition progressed, the pathway was prepared for further isolation of amino acids. For example, in 1907, a combined committee representing the American Society of Biological Chemists and the American Physiological Society, proposed a formal classification of proteins into three major categories: (1) simple proteins, (2) conjugated proteins, and (3) derived proteins. The last classification embraces all denatured proteins and hydrolytic products or protein breakdown and no longer is considered as a general class.

Very approximate annual worldwide production of amino acids, their current method of preparation (not exclusive), and general characteristics are given in Table 2.

The *isoelectric point* is very important in the preparation and separation of amino acids and proteins. Protein solubility varies markedly with pH and is at a minimum at the isoelectric point. By raising the salt concentration and adjusting pH to the isoelectric point, it is often possible to obtain a precipitate considerably enriched in the desired protein and to crystallize it from a heterogenous mixture.

Chemical Nature of Amino Acids

In a very general way, an amino acid is any organic acid that incorporates one or more amino groups. This definition includes a multitude of substances of most diverse structure. There are seemingly limitless related compounds of differing molecular size and constitution which incorporate varying kinds and numbers of functional groups. Most extensive study has centered on the relatively small group of alpha-amino acids that are combined in amide linkage to form proteins. With few exceptions, these compounds possess the general structure $NH_2-CHRCO_2-H$, where the amino group occupies a position on the carbon atom alpha to that of the carboxyl group, and where the side chain R may be of diverse composition and structure.

Few products of natural origin are as versatile in their behavior and properties as are the amino acids, and few have such a variety of biological duties to perform. Among their general characteristics would be included:

(a) Water-soluble and amphoteric electrolytes, with the ability to form acid salts and basic salts and thus act as buffers over at least two ranges of pH (hydrogen ion concentration).

(b) Dipolar ions of high electric moment with a considerable capacity to increase the dielectric constant of the medium in which they are dissolved.

(c) Compounds with reactive groups capable of a wide range of chemical alterations leading readily to a large variety of degradation, synthetic, and transformation products, such as esters, amides, amines, anhydrides, polymers, polypeptides, diketopiperazines, hydroxy acids, halogenated acids, keto acids, acylated acids, mercaptans, shorter- or longer-chained acids, and pyrrolidine and piperidine ring forms.

(d) Indispensable components of the diet of all animals including humans.

(e) Participants in crucial metabolic reactions on which life depends, and substrates for a variety of specific enzymes in vitro.

(f) Binders of metals of many kinds.

Optical Properties. With the exception of glycine ($NH_2-CH_2-CO_2-H$) and amino-malonic acid [$NH_2-CH(CO_2H)_2$], all α-amino acids which are classifiable according to the general formula previously given exist in at least two different optically isometric forms. The optical isomers of a given amino acid possess identical empirical and structural formulas, and are indistinguishable from each other on the basis of their chemical and physical properties, with the singular exception of their effect on plane

TABLE 3. FIRST ISOLATION OF AMINO ACIDS

Abbreviation	Name and formula	First isolation and (Source)	Isoelectric point
	NEUTRAL AMINO ACIDS-ALIPHATIC TYPE		
Ala	Alanine $CH_3-CH-COOH$ $\quad\quad\;\; | \quad$ $\quad\quad\;\; NH_2$	1879 by Schutzenberger 1888 by Weyl (silk fibroin)	6.0
Gly	Glycine NH_2-CH_2-COOH	1820 by Braconnot (gelatin)	6.0
Ile	Isoleucine $\quad\quad H$ $\quad\quad |$ $C_2H_5-C-CH-COOH$ $\quad\quad | \quad |$ $\quad\; H_3C \quad NH_2$	1904 by Ehrlich (fibrin)	6.0
Leu	Leucine $(CH_3)_2CH-CH_2-CH-COOH$ $\quad\quad\quad\quad\quad\quad	$ $\quad\quad\quad\quad\quad\quad NH_2$	1820 by Braconnot (muscle fiber; wool)	6.0
Val	Valine $(CH_3)_2CH-CH-COOH$ $\quad\quad\quad\quad\;\; |$ $\quad\quad\quad\quad\;\; NH_2$	1901 by Fischer (casein)	6.0
	NEUTRAL AMINO ACIDS—HYDROXY TYPE		
Ser	Serine $HO-CH_2-CH-COOH$ $\quad\quad\quad\quad\; |$ $\quad\quad\quad\quad\; NH_2$	1865 by Cramer (sericine)	5.7
Thr	Threonine $CH_3-CH-CH-COOH$ $\quad\quad\; | \quad |$ $\quad\quad\; OH \quad NH_2$	1925 by Gortner and Hoffman 1925 by Schryver and uston oat protein)	6.2
	NEUTRAL AMINO ACIDS—SULFUR-CONTAINING TYPE		
Cys	Cysteine $HS-CH_2-CH-COOH$ $\quad\quad\quad\quad |$ $\quad\quad\quad\quad NH_2$	— — —	5.1
Cys Cys	Cystine $(-SCH_2-CH-COOH)_2$ $\quad\quad\quad\quad	$ $\quad\quad\quad\quad NH_2$	1899 by Mörner (horn) 1899 by Emden	4.6
Met	Methionine $CH_3-S-CH_2-CH_2-CH-COOH$ $\quad\quad\quad\quad\quad\quad\quad\quad |$ $\quad\quad\quad\quad\quad\quad\quad\quad NH_2$	1922 by Mueller (casein)	5.7
	NEUTRAL AMINO ACIDS—AMIDE TYPE		
Asn	Asparagine $H_2NOC-CH_2-CH-COOH$ $\quad\quad\quad\quad\quad\; |$ $\quad\quad\quad\quad\quad\; NH_2$	1932 by Damodaran (edestin)	5.4
Gln	Glutamine $H_2NOC-CH_2-CH_2-CH-COOH$ $\quad\quad\quad\quad\quad\quad\quad	$ $\quad\quad\quad\quad\quad\quad\quad NH_2$	1932 by Damodaran, Jaaback, and Chibnall (gliadin)	5.7
Phe	Phenylalanine $C_6H_5-CH_2-CH-COOH$ $\quad\quad\quad\quad\quad	$ $\quad\quad\quad\quad\quad NH_2$	1881 by Schulze and Barbieri (lupine seedings)	5.5

TABLE 3. (*continued*)

Abbreviation	Name and Formula	First Isolation and (Source)	Isoelectric Point
Trp	Tryptophan 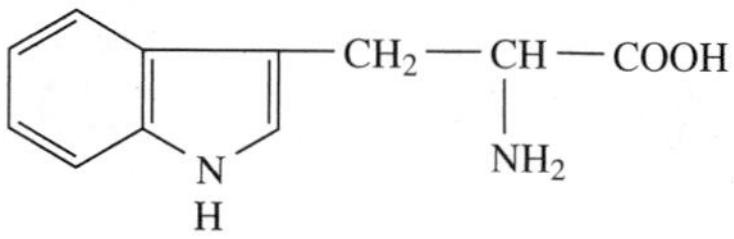 $CH_2-CH-COOH$ / NH_2 / (indole ring with N-H)	1902 by Hopkins and Cole (casein)	5.9
Tyr	Tyrosine 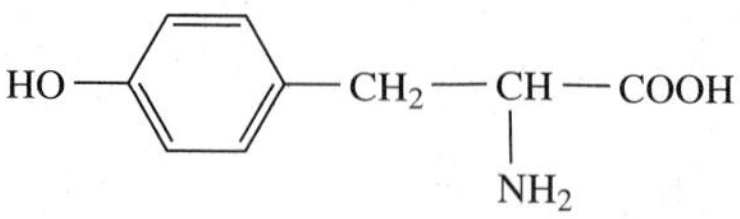 $HO-$(benzene ring)$-CH_2-CH-COOH$ / NH_2	1849 by Bopp (casein)	5.7

ACIDIC AMINO ACIDS

Asp	Aspartic Acid $HOOC-CH_2-CH-COOH$ / NH_2	1868 by Ritthausen (conglutin; legumin)	2.8
Glu	Glumatic Acid $HOOC-CH_2-CH_2-CH-COOH$ / NH_2	1866 by Ritthausen (gluten-fibrin)	3.2

BASIC AMINO ACIDS

Arg	Arginine $H_2N-C-NH(CH_2)_3-CH-COOH$ / $\parallel$NH / NH_2	1895 by Hedin (horn)	11.2
His	Histidine 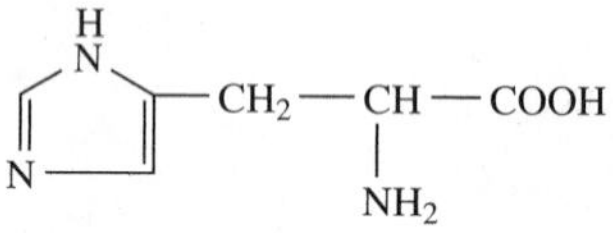 $CH_2-CH-COOH$ / NH_2 / (imidazole ring)	1896 by Kossel (sturine) 1896 by Hedin (various protein ydrolysates)	7.6
Lys	Lysine $H_2N-(CH_2)_4CH-COOH$ / NH_2	1889 by Dreschel (casein)	9.7

IMINO ACIDS

Hyp	Hydroxyproline 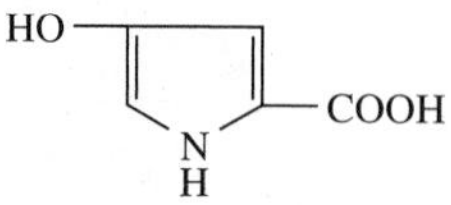 $HO-$(pyrrolidine ring)$-COOH$ / N-H	1902 by Fischer (gelatin)	5.8
Pro	Proline 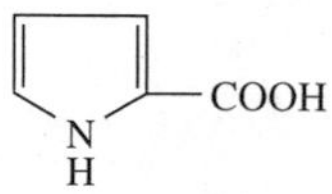 (pyrrolidine ring)$-COOH$ / N-H	1901 by Fischer (casein)	6.3

polarized light. This may be illustrated with the two optically active forms as shown by

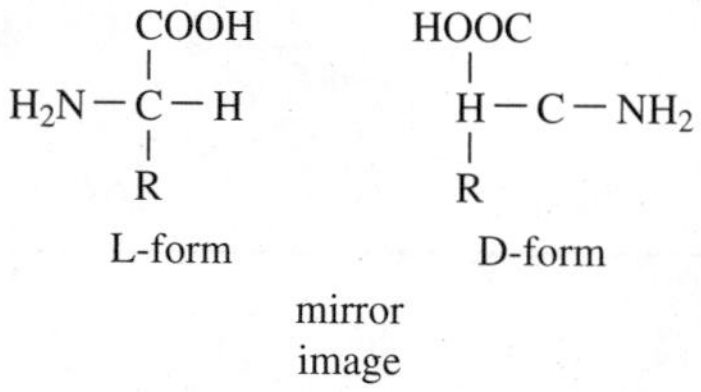

$$H_2N-\underset{R}{\underset{|}{\overset{COOH}{\overset{|}{C}}}}-H \qquad H-\underset{R}{\underset{|}{\overset{HOOC}{\overset{|}{C}}}}-NH_2$$

L-form   D-form

mirror
image

One form (L-form) exhibits the ability to rotate the plane of polarization of plane polarized light to the left (levorotatory), whereas the other form (D-form) rotates the plane to the right (dextrorotatory). Although the direction of optical rotation exhibited by these optically active forms is different, the magnitude of their respective rotations is the same. If equal amounts of *dextro* and *levo* forms are admixed, the optical effect of each isomer is neutralized by the other, and an optically inactive product known as a *racemic modification* or *racemate* is secured.

The ability of the alanine molecule, for example, to exist in two stereo-isomeric forms can be attributed to the fact that the α-carbon atom of this compound is attached to four different groups which may vary in their three-dimensional spatial arrangement. Compounds of this type do not possess complete symmetry when viewed from a purely geometrical standpoint and hence are generally referred to as *asymmetric*. As a consequence of this molecular asymmetry, the four covalent bonds of an asymmetric carbon atom can be aligned in a manner such that a regular tetrahedron is formed by the straight lines connecting their ends. Hence, two different tetrahedral arrangements of the groups about the asymmetric carbon atom can be devised so that these structures relate to one another as an object relates to its mirror image, or as the right hand relates to the left hand. Molecules of this type are endowed with the property of optical activity and, together with their nonsuperimposable mirror images, are generally referred to as *enantiomorphs, enantiomers, antimers,* or *optical antipodes*.

Classification. In accordance with the structure of the R-group, the amino acids of primary importance can be classified into eight groups. Additional amino acids composing protein are not included in this classification, because they occur infrequently. See Table 4.

Normally, amino acids exist as dipolar ions. $RCH(NH_3^+)COO^-$, in a neutral state, where both amino and carboxyl groups are ionized. The dipolar form, $RCH(NH_2)COOH$ may be considered, but the dipolar form predominates for the usual monoamino monocarboxylic acid and it is estimated that these forms occur 10^5 to 10^6 times more frequently than the non-polar forms. Amino acids decompose thermally at what might be considered a relatively high temperature ($200-300°C$). The compounds are practically insoluble in organic solvents, have low vapor pressure, and do not exhibit a precisely defined melting point.

The ionic states of a simple α-amino acid are given by

$$RCH(NH_3^+)COOH \quad \underset{+H^+}{\overset{-H+(K_1)}{\rightleftharpoons}}$$
(Cationic form; acidic)

$$RCH(NH_3^+)COO^- \quad \underset{+H^+}{\overset{-H+(K_2)}{\rightleftharpoons}} \quad RCH(NH_2)COO^-$$
(Dipolar form; neutral) (Anionic form; basic)

In accordance with the change of the ionic state, dissociation constants are

$$K_1(COOH) = \frac{[H^+][RCH(NH_3^+)COO^-]}{[RCH(NH_3^+)COOH]}$$

$$K_2(NH_3^+) = \frac{[H^+][RCH(NH_2)COO^-]}{[RCH(NH_3^+)COO^-]}$$

In as much as $pK = -\log K$, the values for glycine are $pK_1 = 2.34$ and $pK_2 = 9.60$ (in aqueous solution at $25°C$). The homologous amino acids indicate similar values. The pH at which acidic ionization balances basic ionization is termed the *isoelectric point* (pH_1), (corresponding to

$$[RCH(NH_3^+)COOH] = [RCH(NH_2)COO^-]$$

TABLE 4. STRUCTURAL CLASSIFICATION OF AMINO ACIDS

NEUTRAL AMINO ACIDS

Aliphatic-type	*Hydroxy-type*	*Sulfur-containing*
Glycine	Serine	Cysteine
Alanine	Threonine	Cystine
Valine		Methionine
Leucine		
Isoleucine		
Amide-type	*Aromatic-type*	
Asparagine	Phenylalanine	
Glutamine	Tryptophan	
	Tyrosine	

ACIDIC AMINO ACIDS

Aspartic acid
Glutamic acid

BASIC AMINO ACIDS

Histidine
Lysine
Arginine

IMINO ACIDS

Proline
Hydroxyproline

Thus, from these formulas, the pH_1 is

$$pH_1 = \tfrac{1}{2}(pK_1 + pK_2)$$

Formation of Salts. Amino acids have certain characteristics of both organic bases and organic acids because they are amphoteric. As amines, the amino acids form stable salts, such as hydrochlorides or aromatic sulfonic acid salts. These are used as selective precipitants of certain amino acids. As organic acids, the amino acids form complex salts with heavy metals, the less soluble salt being used for amino acid separation.

Esters. When heated with the equivalent amount of a strong acid, usually hydrochloric acid in absolute alcohol, amino acids form esters. These are obtained as hydrochlorides.

Acylation. In alkaline solution, amino acids react with acid chlorides or acid anhydrides to form acyl compounds of the type

$$\begin{array}{c} RCH-COONa \\ | \\ NHCOR' \end{array}$$

Van Slyke Reaction (Deamination). With excess nitrous acid, -amino acids react to form-hydroxyl acids on a quantitative basis. Nitrogen gas is generated.

$$\begin{array}{cc} RCH_2-COOH + HNO_2 \longrightarrow & RCH_2-COOH + N_2 + H_2O \\ | & | \\ NH_2 & OH \end{array}$$

The reaction is completed within five minutes at room temperature. Thus, measurement of the volume of nitrogen generated can be used in amino acid determinations.

Decarboxylation. When heated with inert solvents, such as kerosene, amino acids form amines

$$\begin{array}{cc} RCH-COOH \longrightarrow & RCH_2 + CO_2 \\ | & | \\ NH_2 & NH_2 \end{array}$$

Decarboxylative enzymes may react specifically with amino acids having free polar groups at the ω position. Cadaverine can be produced from lysine, histamine from histidine, and tyramine from tyrosine.

Formation of Amides. When condensed with ammonia or amines, amino acid esters form acid amides:

$$\begin{array}{c} RCH-CONH_2 \\ | \\ NH_2 \end{array}$$

Oxidation. Oxidizing agents easily decompose α-amino acids, forming the corresponding fatty acid with one less carbon number:

$$\begin{array}{c} RCH-\Psi OOH \overset{o}{\longrightarrow} RCHO \overset{o}{\longrightarrow} RCOOH + NH_3 + CO_2 \\ | \\ NH_2 \end{array}$$

Ninhydrin Reaction. A neutral solution of an amino acid will react with ninhydrin (triketohydrindene hydrate) by heating to cause oxidative decarboxylation. The central carbonyl of the triketone is reduced to an alcohol. This alcohol further reacts with ammonia formed from the amino acid and causes a red-purplish color. Since the reaction is quantitative, measurement of the optical density of the color produced is an indication of amino acid concentration. Imino acids, such as hydroxyproline and proline, develop a yellow color in the same type of reaction.

Maillard Reaction. In amino acids, the amino group tends to form condensation products with aldehydes. This reaction is regarded as the cause of the browning reaction when an amino acid and a sugar coexist. A characteristic flavor, useful in food preparations, is evolved along with the color in this reaction.

Ion-exchange Separations. Because amino acids are amphoteric, they behave as acids or bases, depending upon the pH of the solution. This makes it possible to adsorb amino acids dissolved in water on either a strong-acid cation exchange resin; or a strong-base anion exchange resin. The affinity varies with the amino acid and the solution pH. Ion-exchange resins are widely used in amino acid separations.

Production of Amino Acids. There are three means available for making (or separating) amino acids in large quantity lots: (1) extraction from natural protein; (2) fermentation; and (3) chemical synthesis. During the early investigations of amino acids, the first method was widely used and still applies to four amino acids. See Table 3.

L-Leucine is easily extracted in quantity from almost any type of vegetable protein hydrolyzates. Cystine is extracted from the human-hair hydrolyzate. L-Histidine is obtainable from the blood of animals, but future yields may stem from fermentation inasmuch as some artificial mutants of bacteria have been discovered. Gelatin is the prime source of L-hydroxyproline.

Natural amino acids, normally not contained in proteins, but which are effective in medicine, include citrulline, ornithine, and dihydroxyphenylalanine. These are not listed in Tables 2 and 3. Citrulline (Cit) with an isoelectric point of 5.9 was isolated by Koga in 1914; by Odake in 1914; and by Wada in 1930. It has the formula

$$H_2NC-NH-(CH_2)_3CH-COOH$$
$$\overset{\|}{O} \qquad\qquad\qquad \overset{|}{NH_2}$$

Dihydroxyphenylalanine (Dopa) with an isoelectric point of 5.5 was isolated by Torquati in 1913; and by Guggenheim in 1913. It has the formula

Ornithine (Orn) with an isoelectric point of 9.7 was isolated by Riesser in 1906 from arginine. It has the formula

$$CH_2-CH_2-CH_2-CH-COOH$$
$$\overset{|}{NH_2} \qquad\qquad\qquad \overset{|}{NH_2}$$

Fermentation Methods. Numerous microorganisms can synthesize the amino acids required to support their life from a simple carbon source and an inorganic nitrogen source, such as ammonium or nitrate salts, or nitrogen gas.

Japanese microbiologists, in 1956, first succeeded in developing industrial production of L-glutamic acid by a microbiological process. As of the present, nearly all common amino acids can be produced on a low cost industrial scale by fermentation. From microbiological studies, it has been ascertained that some microbial stains isolated from natural sources serve to excrete and accumulate a large amount of a particular amino acid in the cultural broth under carefully controlled conditions. The production of glutamic acid is produced by adding a selected bacterial strain and culturing aerobically for one to two days in a chemically defined medium which contains carbon sources, such as sugar or acetate, and nitrogen sources, such as ammonium salts. About 50% (wt) of the carbon sources can be converted to glutamate.

Genetic techniques have been used to improve the ability of microorganisms to accumulate amino acids. Several amino acids are manufactured from their direct precursors by the use of microbially produced enzymes. For example, bacterial L-aspartate β-carboxylase is used for the production of L-alanine from L-aspartic acid.

In isolating the amino acids from the fermentation broth, chromatographic separations using ion-exchange resins are the most important commercial method. Precipitation with compounds which yield insoluble salts with amino acids are also used. Purification is possible by crystallization through careful adjustment of the isoelectric point, at which point the amino acid is least soluble.

There are several laboratory-size methods for synthesizing amino acids, but few of these have been scaled up for industrial production. Glycine and DL-alanine are made by the Strecker synthesis, commencing with formaldehyde and acetaldehyde, respectively. In the Strecker synthesis, aldehydes react with hydrogen cyanide and excess ammonia to give amino nitriles which, in turn, are converted into α-amino acids upon hydrolysis.

$$RCH \xrightarrow{HCN} RCH-CN \xrightarrow{NH_3} RCH-CN \xrightarrow{NaOH} RCH-COONa$$
$$\overset{\|}{O} \qquad\quad \overset{|}{OH} \qquad\qquad \overset{|}{NH_2} \qquad\qquad \overset{|}{NH_2}$$

The Hydantoin Process. Hydantoins are produced by reacting aldehydes with sodium cyanide and ammonium carbonate. Upon hydrolysis, α-amino acids will be yielded

$$RCHO \xrightarrow{NaCN,\ (NH_4)_2\,CO_3} \underset{HN\quad NH}{\overset{RCH-CO}{|\qquad|}} \xrightarrow{NaOH}$$
$$\underset{CO}{|}$$

$$RCH-COONa + (NH_4)HCO_3$$
$$\overset{|}{NH_2}$$

The production of α-amino acids by chemical synthesis yields a mixture of DL forms. The D-form of glutamic acid has no flavor-enhancing properties and thus requires transformation into the optically active form insofar as monosodium glutamate is concerned. The three methods for separating the optical isomers are: (1) preferential inoculation method; (2) the diastereoisomer method; and (3) the acylase method.

HAUROMI OEDA
Ajinomoto Co., Inc.
Kawasaki, Japan

Additional Reading

Alger, B.E. and Hanns Mohler: *Pharmacology of Inhibitory Amino Acid Transmitters,* Springer-Verlag Inc., New York, NY, 2000.

Barrett, G.C.: *Amino Acid Derivatives: A Practical Approach,* Oxford University Press, Inc., New York, NY, 2000.

Barrett, G.C. and D.T. Elmore: *Amino Acids and Peptides,* Cambridge University Press, New York, NY, 1999.

Baxevanis, A.D. and B.F. Ouellette: *Bioinformatics: A Practical Guide to the Analysis of Gene and Proteins,* Vol. 39, John Wiley & Sons, Inc., New York, NY, 1998.

Bishop, M.J.: *Genetic Databases,* Academic Press, Inc., San Diego, CA, 1999.

Bork, Peer and D.E. Eisenberg: *Analysis of Amino Acid Sequences,* Academic Press, Inc., San Diego, CA, 2000.

Durbin, R., R. Eddy, A. Krogh, and G. Mitchison: *Biological Sequence Analysis: Probabilistic Models of Proteins and Nucleic Acids,* Cambridge University Press, New York, NY, 1998.

El-Khoury, A.E.: *Methods for Investigation of Amino Acid and Protein Metabo lism,* CRC Press, LLC., Boca Raton, FL, 1999.

Goodfriend, G.A., M. Collins, J.F. Wehmiller, M. Fogel, and S. Macko: *Perspectives in Amino Acid and Protein Geochemistry,* Oxford University Press, Inc., New York, NY, 2000.

Harris, R.A. and J.R. Sokatch: *Branched Chain Amino Acids: Part B,* Vol. 324, Academic Press, Inc., San Diego, CA, 2000.

Kellner, R., F. Lottspeich, and H.E. Meyer: *Microcharacterization of Proteins,* John Wiley & Sons, Inc., New York, NY 1999.

Seeburg, P.H., L. Turski, and I. Bresink: *Excitatory Amino Acids: From Genes to Therapy,* Springer-Verlag Inc., New York, NY, 1999.

Singh, B.K.: *Plant Amino Acids: Biochemistry and Biotechnology,* Marcel Dekker, Inc., New York, NY, 1999.

Wheal, H. and A. Thomson: *Excitatory Amino Acids and Synaptic Transmission,* Academic Press, Inc., San Diego, CA, 1997.

AMINO RESINS. A family of resins resulting from an addition reaction between formaldehyde and compounds, such as aniline, ethylene urea, dicyandiamide, melamine, sulfonamide, and urea. The resins are thermosetting and have been used for many years in such products as textile-treating agents, laminating coatings, wet-strength paper coatings, and wood adhesives. The urea and melamine compounds are the most widely used. Both of these basic resins are water white (transparent). However, the resins readily accept pigments and opacifying agents. The addition of cellulose filler can be used to reduce light transmission. Where color is unimportant, various materials are added to the melamine resin compounds, including macerated fabric, glass fiber, and wood flour. Wood flour frequently is added to the urea resins to yield a low cost industrial material.

Advantages claimed for amino resins include: (1) good electrical insulation characteristics, (2) no transfer of tastes and odors to foods, (3) self-extinguishing burning characteristics, (4) resistance to attack by oils, greases, weak alkalis and acids, and organic solvents, (5) abrasion resistance, (6) good rigidity, (7) easy fabrication by economical molding procedures, (8) excellent resistance to deformation under load, (9) good subzero characteristics with no tendency to become brittle, and (10) marked hardness.

Amino resins are fabricated principally by transfer and compression molding. Injection molding and extrusion are used on a limited scale. Urea resins are not recommended for outdoor exposure. The resins show rather high mold shrinkage and some shrinkage with age. The melamines are superior to the ureas insofar as resistance to heat and boiling water, acids, and alkalis is concerned.

Some of the hundreds of applications for amino resins include: closures for glass, metal, and plastic containers; electrical wiring devices; appliance knobs, dials, handles, and push buttons; lamp shades and lighting diffusers; organ and piano keys; dinnerware; food service trays; food-mixer housings; switch parts; decorative buttons; meter blocks; aircraft ignition parts; heavy duty switch gear; connectors; and terminal strips. Not all of the urea or melamine amino resins are suited to all of the foregoing uses. Because of the large number of fillers and additives available, the overall range of use of this family of resins is large.

AMINOPHENOLS. Aminophenols and their derivatives are of commercial importance, both in their own right and as intermediates in the photographic, pharmaceutical, and chemical dye industries. They are amphoteric and can behave either as weak acids or weak bases, but the basic character usually predominates. 3-Aminophenol (2) is fairly stable in air unlike 2-aminophenol (1) and 4-aminophenol (3) which easily undergo oxidation to colored products. The former are generally converted to their acid salts, whereas 4-aminophenol is usually formulated with low concentrations of antioxidants which act as inhibitors against undesired oxidation.

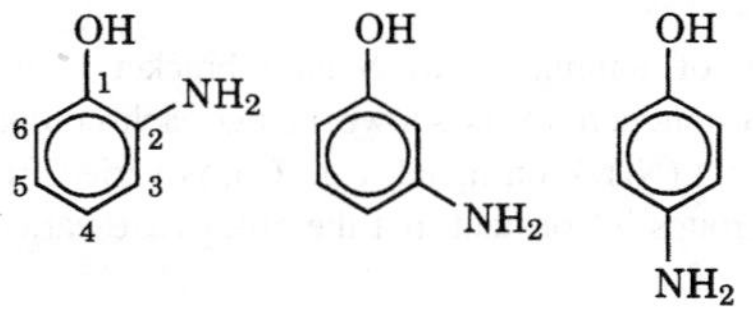

(1)   (2)   (3)

Physical Properties

The simple aminophenols exist in three isomeric forms depending on the relative positions of the amino and hydroxyl groups around the benzene ring. At room temperature they are solid crystalline compounds. In the past the commercial-grade materials were usually impure and colored because of contamination with oxidation products, but now virtually colorless, high purity commercial grades are available. General properties are listed in Table 1.

2-Aminophenol. This compound forms white orthorhombic, bipyramidal needles when crystallized from water or benzene, which readily become yellow-brown on exposure to air and light.

3-Aminophenol. This is the most stable of the isomers under atmospheric conditions. It forms white prisms when crystallized from water or toluene.

4-Aminophenol. This compound forms white plates when crystallized from water. The base is difficult to maintain in the free state and deteriorates rapidly under the influence of air to pink-purple oxidation products.

Chemical Properties

The chemical properties and reactions of the aminophenols and their derivatives are to be found in detail in many standard chemical texts. The acidity of the hydroxyl function is depressed by the presence of an amino group on the benzene ring; this phenomenon is most pronounced with 4-aminophenol. The amino group behaves as a weak base, giving salts with both mineral and organic acids. The aminophenols are true ampholytes, with no zwitterion structure; hence they exist either as neutral molecules (4), or as ammonium cations (5), or phenolate ions (6), depending on the pH value of the solution.

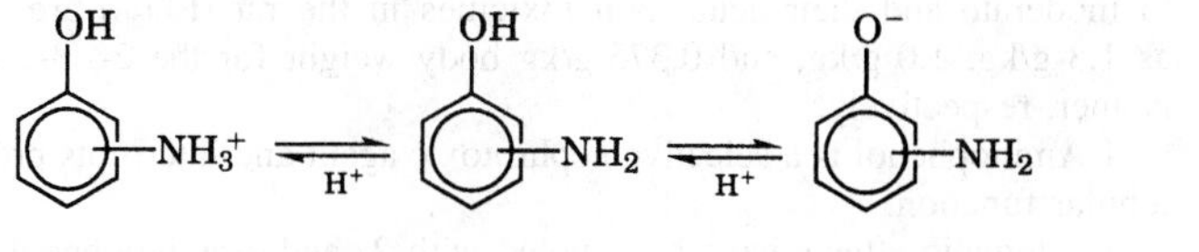

(5)   (4)   (6)

The aminophenols are chemically reactive, undergoing reactions involving both the aromatic amino group and the phenolic hydroxyl moiety, as well as substitution on the benzene ring. Oxidation leads to the formation of highly colored polymeric quinoid structures. 2-Aminophenol undergoes a variety of cyclization reactions. Important reactions include alkylation, acylation, diazonium salt formation, cyclization reactions, condensation reactions, and reactions of the benzene ring.

Manufacture and Processing

Aminophenols are either made by reduction of nitrophenols or by substitution. Reduction is accomplished with iron or hydrogen in the presence of a catalyst. Catalytic reduction is the method of choice for the production of 2- and 3-aminophenol. Electrolytic reduction is also under industrial consideration and substitution reactions provide the major source of 3-aminophenol.

Purification

Contaminants and by-products which are usually present in 2- and 4-aminophenol made by catalytic reduction can be reduced or even removed completely by a variety of procedures. These include treatment with 2-propanol, with aliphatic, cycloaliphatic, or aromatic ketones, with aromatic amines, with toluene or low mass alkyl acetates, or with phosphoric acid, hydroxyacetic acid, hydroxypropionic acid, or citric acid. In addition, purity may be enhanced by extraction with methylene chloride, chloroform, or nitrobenzene. Another method employed is the treatment of aqueous solutions of aminophenols with activated carbon.

Economic Aspects

Production figures for the aminophenols are scarce, the compounds usually being classified along with many other aniline derivatives. Most production of the technical grade materials (95% purity) occurs on-site, as they are chiefly used as intermediate reactants in continuous chemical syntheses. World production of the fine chemicals (99% purity) is probably no more than a few hundred metric tons yearly.

Storage

Under atmospheric conditions, 3-aminophenol is the most stable of the three isomers. Both 2- and 4-aminophenol are unstable; they darken on exposure to air and light and should be stored in brown glass containers, preferably in an atmosphere of nitrogen. The use of activated iron oxide

TABLE 1. GENERAL PROPERTIES OF AMINOPHENOLS

Property	2-Aminophenol	3-Aminophenol	4-Aminophenol
alternative names	2-hydroxyaniline 2-amino-1-hydroxybenzene	3-hydroxyaniline 3-amino-1-hydroxybenzene	4-hydroxyaniline 4-hydroxy-1-aminobenzene
molecular formula	C_6H_7NO	C_6H_7NO	C_6H_7NO
molecular weight	109.13	109.13	109.13
melting point, °C	174	122–123	189–190[a]
boiling point, °C			
1.47 kPa	153[b]	164	174
101.3 kPa			284
ΔH_f, kJ/mol[c]	-191.0 ± 0.9	-194.1 ± 1.0	-190.6 ± 0.9[d]

[a] Decomposes.
[b] Sublimes. To convert kPa to mm Hg, multiply by 7.5.
[c] In the crystalline state. To convert kJ to kcal, divide by 4.184.
[d] -179.1 is also noted.

in a separate cellophane bag inside the storage container, or the addition of stannous chloride or sodium bisulfite inhibits the discoloration of aminophenols. The salts, especially the hydrochlorides, are more resistant to oxidation and should be used where possible.

Health and Safety Factors

In general, aminophenols are irritants. Their toxic hazard rating is slight to moderate and their acute oral toxicities in the rat (LD_{50}) are quoted as 1.3 g/kg, 1.0 g/kg, and 0.375 g/kg body weight for the 2-, 3-, and 4-isomer, respectively.

4-Aminophenol is a selective nephrotoxic agent and interrupts proximal tubular function.

Teratogenic effects have been noted with 2- and 4-aminophenol in the hamster, but 3-aminophenol was without effect in the hamster and rat.

Obviously, care should be taken in handling these compounds, with the wearing of chemical-resistant gloves and safety goggles; prolonged exposure should be avoided.

The addition of slaked lime and the initiation of polymerization reactions with H_2O_2 and ferric or stannous salts are techniques employed to remove aminophenols from wastewaters.

Uses

The aminophenols are versatile intermediates and their principal use is as synthesis precursors; their products are represented among virtually every class of stain and dye.

Derivatives

The derivatives of the aminophenols have important uses in both the photographic and pharmaceutical industries. They are also extensively employed as precursors and intermediates in the synthesis of more complicated molecules, especially those used in the staining and dye industry. All of the major classes of dyes have representatives that incorporate substituted aminophenols; those compounds produced commercially as dye intermediates have been reviewed. Details of the more commonly encountered derivatives of the aminophenols can be found in standard organic chemistry texts.

STEPHEN MITCHELL
University of London

ROSEMARY WARING
University of Birmingham, England

Additional Reading

Beilstein's Handbuch der Organischen Chemie, Julius Springer, Berlin, 1918, Section 13, pp. 354–549 and Section 13(2), pp. 164–308.
Coffey, S. ed.: *Rodd's Chemistry of Carbon Compounds,* 2nd Edition, Vol. 3A, Elsevier Publishing Co., Amsterdam, The Netherlands, 1971, pp. 352–363.
Colour Index, 3rd Edition, The Society of Dyers and Colorists, Bradford, England, Vol. 4, 1971, pp. 4001–4863 and Vol. 6, 1975, pp. 6391–6410.
Forester, A.R. and J.L. Wardell in S. Coffey, ed., *Rodd's Chemistry of Carbon Compounds,* 2nd Edition, Vol. 3A, Elsevier Publishing Co., Amsterdam, The Netherlands, 1971, Chapt. 4, pp. 352–363.

AMMINES. Dry ammonia gas reacts with dehydrated salts of some of the metals to form solid ammines. Ammines, upon warming, evolve ammonia, sometimes with final decomposition of the salt itself, in a manner analogous to the decomposition of certain hydrates. The ammines of chromium(III)(Cr^{3+}), cobalt(III), platinum(IV), and other metals have been studied in detail. Two series of ammines are shown in Fig. 1 the first is one in which the neutral ammonia group is replaced step by step by the negative nitro group (NO_2^-), the second is one in which the neutral ammonia group is replaced, step by step, by the neutral H_2O group.

The neutral group of the complex may be replaced step by step by the following negative groups: Cl^-, Br^-, I^-, F^-, OH^-, NO_2^-, NO_3^-, CN^-, CNS^-, SO_4^{2-}, CO_3^{2-}, $C_2O_4^{2-}$; or by the following neutral groups: H_2O, NO, NO_2, SO_2, S, N_2H_4, H_2NOH, CO, C_2, H_5OH, C_6H_6. All neutral groups are of substances capable of independent existence.

In the ammines, trivalent metals, such as cobalt(III) and chromium(III) and iron(III), possess a coordination number of 6, this number being the sum of the unit replacements on the metal in the complex ion. Since a regular octahedron has six corners equidistant from the center, it is assumed that the metal occupies the center and each of the six replacing groups

$[Co(NH_3)_6]Cl_3$
410

$[Co(NH_3)_5(NO_2)]Cl_2$
240

$[Co(NH_3)_4(NO_2)_2]Cl$
95

$[Co(NH_3)_3(NO_2)_3]$
1.5

$K[Co(NH_3)_2(NO_2)_4]$ $[Cr(NH_3)_6]X_3$
95                          $[Cr(NH_3)_5H_2O)]X_3$
                            $[Cr(NH_3)_4H_2O_2]X_3$

$K_2[Co(NH_3)(NO_2)_5]$ $[Cr(NH_3)_3(H_2O)_3]X_3$
240                        $[Cr(NH_3)_2(H_2O)_4]X_3$

$K_2[Co(NO_3)_6]$
420   X = unit anion

(a)   **(b)**

Fig. 1. Two series of ammines: (**a**) Square bracket contains the ion. The equivalent electrical conductivity is shown below each compound. The number of neutral groups, e.g., (NH_3), on metal, e.g., Co, is varied from 6 to 0. (**b**) The number of neutral groups is constant, but the groups are varied

occupies a corner of a regular octahedron. Support for this assumption is offered by the x-ray examination of these ammines. When there is only one of the six groups replaced by a second group, as in $[Co(NH_3)_5(NO_2)]Cl_2$, and in $[Cr(NH_3)_5(H_2O)]X_3$ the octahedral placement of groups supplies only one form, but when two of the six groups are replaced by a second group, as in $[Co(NH_3)_4(NO_2)_2]Cl$, and in $[Cr(NH_3)_4(H_2O)_2]X_3$, two different octahedral corner arrangements are possible depending upon whether the two replacing groups are adjacent (cis-form) or opposite (trans-form). Two substances differing in physical properties and corresponding to these two forms are known. Further, when three divalent groups, e.g., $3C_2O_4^2A$ are present in the complex, two arrangements—not identical but mirror-images of each other—are possible. Two optically active substances are known in such cases corresponding to these two stereo-isomeric forms.

Six is the ordinary coordination number for metallic ammines and similar complexes. Additional examples are $K_2[Pt(NH_3)_2(CN)_4]$, $[Ni(NH_3)_6]Cl_2$, $K_4[Fe(CN)_6]$, $K_3[Fe(CN_6)]$, $K_2[Fe(CN)_5(NO)]$, $K_2[SiF_6]$, $[Ca(NH_3)_6]Cl_2$. But, for the elements boron, carbon, and nitrogen four is the coordination number, e.g., $[BH_4]Cl$, CH_4, NH_4Cl, and in these substances the groups are assumed to occupy the corners of a regular tetrahedron; in $[K_4Mo(CN)_8]$ and $[Ba(NH_3)_8]Cl_2$ the coordination number is eight, and the groups are assumed to occupy the corners of a cube.

AMMONIA. [CAS: 7664-41-7]. Known since ancient times, ammonia, NH_3 has been commercially important for well over 100 years and has become the second largest chemical in terms of tonnage and the first chemical in value of production. The first practical plant of any magnitude was built in 1913. Worldwide production of NH_3 as of the early 1980s is estimated at 100 million metric tons per year or more, with the United States accounting for about 14% of the total production. A little over three-fourths of ammonia production in the United States is used for fertilizer, of which nearly one-third is for direct application. An estimated 5.5% of ammonia production is based in the manufacture of fibers and plastics intermediates.

Properties. At standard temperature and pressure, NH_3 is a colorless gas with a penetrating, pungent-sharp odor in small concentrations which, in heavy concentrations, produces a smothering sensation when inhaled. Formula weight is 17.03, mp $-77.7°C$, bp $-33.35°C$, and sp gr 0.817 (at $-79°C$) and 0.617 (at $15°C$). Ammonia is very soluble in water, a saturated solution containing approximately 45% NH_3 (weight) at the freezing temperature of the solution and about 30% (weight) at standard conditions. Ammonia dissolved in water forms a strongly alkaline solution of ammonium hydroxide, NH_4OH. The univalent radical NH_4^+ behaves in many respects like K^+ and Na^+ in vigorously reacting with acids to form salts. Ammonia is an excellent nonaqueous electrolytic solvent, its ionizing power approaching that of water. Ammonia burns with a greenish-yellow flame.

Ammonia derives its name from sal ammoniac, NH_4Cl, the latter material having been produced at the Temple of Jupiter Ammon (Libya) by distilling camel dung. During the Middle Ages, NH_3 was referred to as the spirits of hartshorn because it was produced by heating the hoofs and horns of oxen. The composition of ammonia was first established by Claude Louis Berthollet (France, ca. 1777). The first significant commercial source of NH_3 (during the 1880s) was its production as a byproduct in the making of manufactured gas through the destructive distillation of coal. See also **Coal Tar and Derivatives**.

Nitrogen fixation is a term assigned to the process of converting nitrogen in the air to nitrogen compounds. Although some bacteria in soil are capable of this process, N_2 as an ingredient of fertilizer is required for soils that are depleted by crop production. The production of synthetic NH_3 is the most important industrial nitrogen-fixation process. See also **Fertilizer**.

Synthesis of Ammonia

The first breakthrough in the large-scale synthesis of ammonia resulted from the work of Fritz Haber (Germany, 1913), who found that ammonia could be produced by the direct combination of two elements, nitrogen and hydrogen, ($N_2 + 3H_2 \rightleftharpoons 2NH_3$) in the presence of a catalyst (iron oxide with small quantities of cerium and chromium) at a relatively high temperature ($550°C$) and under a pressure of about 200 atmospheres, representing difficult processing conditions for that era. Largely because of the urgent requirements for ammonia in the manufacture of explosives during World War I, the process was adapted for industrial-quality production by Karl Bosch, who received one-half of the 1931 Nobel Prize for chemistry in recognition of these achievements. Thereafter, many improved ammonia-synthesis systems, based on the Haber-Bosch process, were commercialized, using various operating conditions and synthesis-loop designs.

The principal features of an NH_3 synthesis process system are the converter designs, operating conditions, method of product recovery, and type of re-circulation equipment. Most current systems operate at or above the pressure used in the original Haber-Bosch process. Converter designs have either a single continuous catalyst bed, which may or may not have heat-exchange cooling for controlling reaction heat, or several catalyst beds with provision for temperature control between the beds.

Claude Process. The original Claude process was one of the first systems to use a high operating pressure (1000 atmospheres), achieving 40% conversion without recycling. This system used multiple converters in a series-parallel arrangement. The present Claude process[1] operates at 340–650 atmospheres, using a single converter with continuous catalyst-charged tubes externally cooled to remove the heat of reaction. Approximate hydrogen conversion is 30–34 mole percent per pass. The pressure is increased gradually to compensate for catalyst aging and loss in activity. Product recovery is by simple condensation in a water-cooled condenser. Nonreacted gas is recycled by compressor.

Casale Process. This is another high-pressure conversion system, using synthesis pressures of 450–600 atmospheres, which also permits hydrogen conversions in the 30 mole percent range. As in the Claude process, the high pressure allows NH_3 to be recovered from the converter effluent by water cooling. The Casale converter uses a single catalyst bed with internal heat-exchange surfaces. Reaction rate and temperature rise across the catalyst are controlled by the internal exchanger and retaining 2–3 mole percent NH_3 in the converter feed. An ejector is used to remove nonreacted gas. This eliminates the need for a mechanical recycle compressor, but requires high feed-gas pressures to supply the energy required for the ejector.

Low-Pressure Processes. Several systems use low synthesis pressures with hydrogen conversion below 30 mole percent and product recovery by water and refrigeration.

Synthesis-Gas-Production Processes. These processes were improved and developed as a result of changes in feedstock availability and economics. Before World War II, most NH_3 plants obtained H_2 by reacting coal or coke with steam in the water-gas process. A small number of plants used water electrolysis or coke-oven byproduct hydrogen. The subsequent low-cost availability of natural gas brought about steam-hydrocarbon reforming as the major source of H_2 for the NH_3 synthesis gas.

Partial oxidation processes to produce H_2 from natural gas and liquid hydrocarbons were also developed after World War II and accounted for 15% of the synthetic NH_3 capacity by 1962. The steam-hydrocarbon reforming process[2] was developed in 1930. In this process, methane was mixed with an excess of steam at atmospheric pressure, and the mixture reformed inside nickel-catalyst-filled alloy furnace tubes. The heat of reaction was supplied by externally heating the catalyst-filled tubes to about $871°C$. Since the late 1950s, improvements in the tubular-reforming technology and metallurgy have brought about the utilization of high-pressure (>24 atmospheres) reforming, which cut synthesis-gas-compression costs and increased heat recovery. The first pressure reformer[3] was built in 1953. In addition, the higher pressures allowed improvements in the efficiency of synthesis-gas-purification systems. High-pressure steam-reforming technology also has been extended to cover heavier hydrocarbon gases, including propane, butane, reformer gases, and streams containing a high amount of olefins. In 1962, a process[4] for reforming straight-run liquid distillates (naphthas) was commercialized. This process is based on the use of an alkali oxide-promoted nickel catalyst[5] which permits reforming of desulfurized naphthas at low (~3.5:1) steam-to-carbon ratios, without significant carbon deposition problems.

Noncatalytic partial oxidation processes designed to produce H_2 from a wide range of hydrocarbon liquids, including heavy fuel oils, crudes, naphthas, coal tar, and pulverized bituminous coal, were commercialized in 1954[6] and 1956.[7] In both these processes, the hydrocarbon feed is oxidized and reformed in a refractory-lined pressure vessel. The required oxygen usually is supplied by an air separation plant from which nitrogen also is used as feed for the synthesis gas. The main differences between the two processes are in the reactor design, feeding method, burner design, and carbon and heat recovery. The partial oxidation processes and the steam-naphtha reforming process are favored in areas with short supplies of natural gas.

The source of nitrogen for the synthesis gas has always been air, either supplied directly from a liquid-air separation plant or by burning a small amount of the hydrogen with air in the H_2 gas. The need for air separation plants has been eliminated in modern ammonia plants by use of secondary reforming, where residual methane from the primary reformer is adiabatically reformed with sufficient air to produce a 3:1 mole ratio hydrogen-nitrogen synthesis gas.

Most ammonia plants built since the early 1960s are in the 600–1500 short tons/day (540–1350 metric tons/day) range and are based on new integrated designs that have cut the cost of ammonia manufacture in half. The plants of the early 1960s, in fact, have reached the best combination in terms of plant overall efficiency and cost by combining all the separate units (e.g., synthesis-gas preparation, purification, and ammonia synthesis) in one single train. High-pressure reforming has reduced the synthesis-gas compression load and front end plant equipment size. This compactness in design has also led to increased plant size at reduced investment and operating costs.

Use of Multistage Centrifugal Compressors. One of the major factors contributing to the improved economics of ammonia plants is the application of multistage centrifugal compressors, which have replaced the reciprocating compressors traditionally used in the synthesis feed and recycle service. A single centrifugal compressor can do the job of several banks of reciprocating compressors, thus reducing equipment cost, floor space, supporting foundations, and maintenance.

The use of multistage centrifugal compressors was made possible by redesigning the synthesis loop to operate at low pressures (150–240 atmospheres) and by increasing plant capacity to above the compressor's minimum-flow restriction in order to obtain a reasonable compressor efficiency. (Most synthesis loops using reciprocating compressors had been operating at intermediate pressures of 300–350 atmospheres.) Centrifugal compressors capable of developing pressures up to 340 atmospheres already

[1] Developed by Grande Pariosse and L'Air Liquide.

[2] Originally developed by Standard Oil Company of New Jersey.
[3] Built by M.W. Kellogg for Shell Chemical Corp. (Ventra, California).
[4] W. Kellogg and Imperial Chemical Industries.
[5] Developed by M.W. Kellogg.
[6] Texaco partial oxidation process.
[7] Shell gasification process.

are being offered and used in some large-capacity (1000 short tons/day; 900 metric tons/day) plants, where the increasing compressor horsepower is partially offset by reduction of the refrigeration horsepower requirement.

An operating ammonia plant using the aforementioned improvements is shown schematically in Fig. 1. This plant[8] has a capacity of 1000 short tons/day (900 metric tons/day) and uses natural gas as feedstock. The plant can be divided into the following integrated-process sections: (a) synthesis-gas preparation; (b) synthesis-gas purification; and (c) compression and ammonia synthesis. A typical (Kellogg designed) ammonia plant is shown in Fig. 2.

Synthesis Gas Preparation. The desulfurized natural gas mixed with steam is fed to the primary reformer, where it is reacted with steam in nickel-catalyst-filled tubes to produce a major percentage of the hydrogen required. The principal reactions taking place are[9]

$$CH_4 + H_2O \rightleftharpoons CO + 3H_2 \qquad (1)$$

$$\Delta H_{298} = 49.3 \text{kcal/mole}$$

$$CO + H_2O \rightleftharpoons CO_2 + H_2 \qquad (2)$$

$$\Delta H_{298} = -9.8 \text{kcal/mole}$$

Reaction (1) is the principal reforming reaction, and reaction (2) is the water-gas shift reaction. The net reactions are highly endothermic. The partially reformed gas leaves the primary reformer containing approximately 10% methane, on a mole dry-gas basis, at 27–34 atmospheres and up to 816°C. The required heat of reaction is supplied by natural-gas-fired arch burners, which are designed to also burn purge and flash gases from the synthesis section. Waste heat from the primary reformer flue gas is recovered by generating high-pressure superheated

steam, which along with waste-heat process boilers and an appended auxiliary boiler assure a steam system that is always in balance, while providing high-pressure steam to compressor turbine drivers and low-pressure steam to pump drivers. Further waste heat is recovered by preheating the natural-gas-steam feed mixture, steam-air for secondary reforming, and fuel.

The primary reforming step is followed by conversion of the residual methane to hydrogen and carbon oxides over a bed of high-temperature chrome and nickel catalysts in the secondary reformer. The secondary reforming step not only achieves a great degree of overall reforming economically possible, but also reduces fuel-gas input and overall reforming costs by shifting part of the required hydrocarbon conversion from the high-cost primary reformer to the lower-cost secondary reformer. It also permits an increase in the residual methane level at the primary effluent, which results in lower operating temperatures, reduced steam requirements, and milder tube-metal conditions.

Process waste-heat boilers then cool the reformed gas to about 371°C while generating high-pressure steam. The cooled gas-stream mixture enters a two-stage shift converter. The purpose of shift conversion is to convert CO to CO_2 and produce an equivalent amount of H_2 by the reaction: $CO + H_2O \rightleftharpoons CO_2 + H_2$. Since the reaction rate in the shift converter is favored by high temperatures, but equilibrium is favored by low temperatures, two conversion stages, each with a different catalyst provide the optimum conditions for maximum CO shift. Gas from the shift converter is the raw synthesis gas, which, after purification, becomes the feed to the NH_3 synthesis section.

Purification of Synthesis Gas. This involves the removal of carbon oxides to prevent poisoning of the NH_3 catalyst. An absorption process is used to remove the bulk of the CO_2, followed by methanation of the residual carbon oxides in the methanator. Modern ammonia plants use a variety of CO_2-removal processes with effective absorbent solutions. The principal absorbent solutions currently in use are hot carbonates and ethanolamines. Other solutions used include methanol, acetone, liquid nitrogen, glycols, and other organic solvents.

[8] Designed by The M.W. Kellogg Company, Houston, Texas, for which Kellogg received the 1967 Kirkpatrick Chemical Engineering Achievement Award.

[9] Heats of reaction at 198°K (25°C), 1 atmosphere pressure, gaseous substances in ideal state.

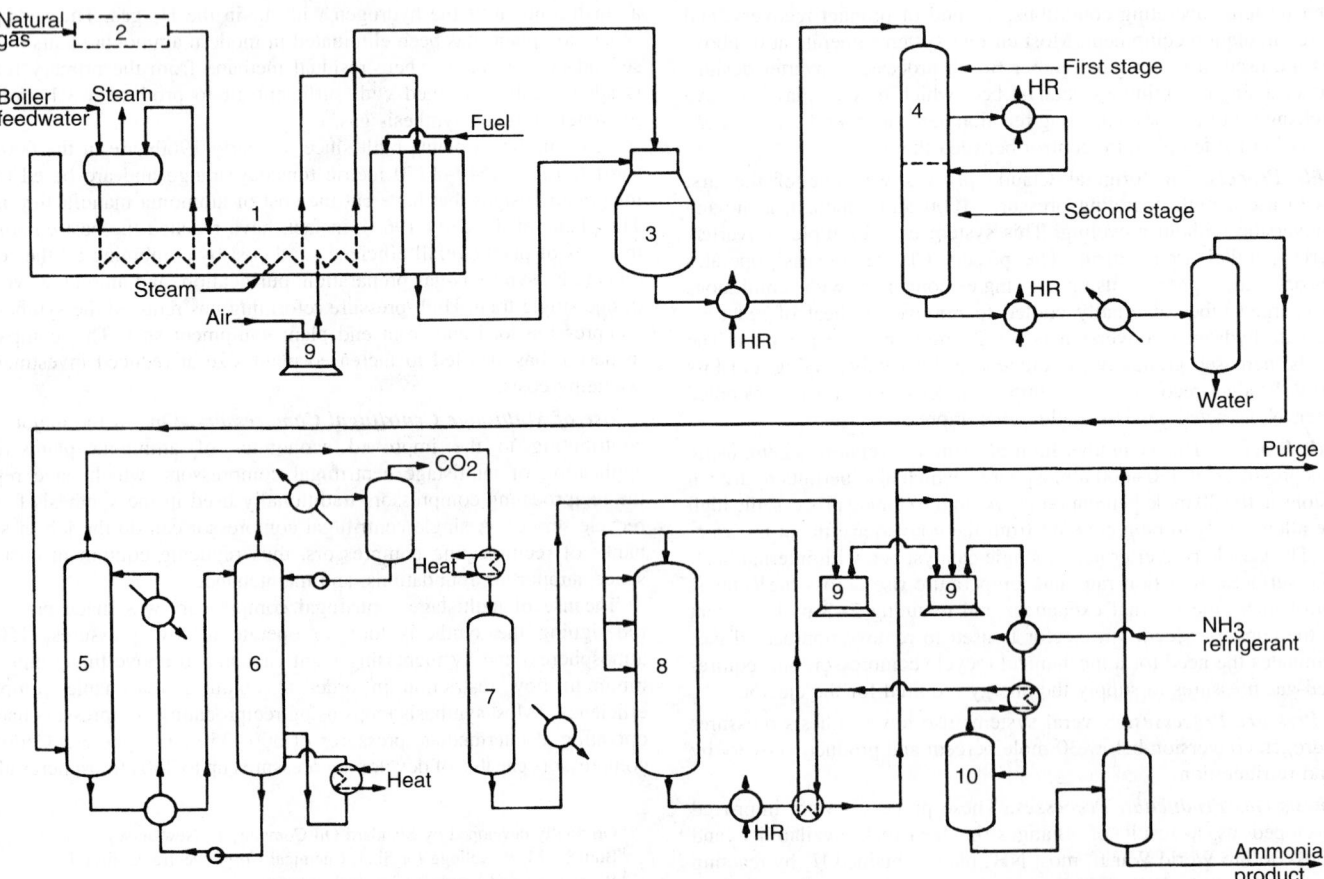

Fig. 1.   Ammonia production process: (1) Primary reformer, (2) desulfurization, (3) secondary reformer, (4) CO shift converter (in two stages), (5) CO_2 absorber, (6) CO_2 stripper, (7) methanator, (8) NH_3 converter, (9) compressor, (10) separator. HR=heat recovery. (*M.W. Kellogg*)

Fig. 2.   Two 1000 short tons/day (900 metric tons/day) Kellogg-designed modern ammonia plants. (*M.W. Kellogg*)

The partially purified synthesis gas leaves the CO_2 absorber containing approximately 0.1% CO_2 and 0.5% CO. This gas is preheated at the methanator inlet by heat exchange with the synthesis-gas compressor interstage cooler and the primary-shift converter effluent and reacted over a nickel oxide catalyst bed in the methanator. The methanation reactions are highly exothermic and are equilibrium favored by low temperatures and high pressures.

$$CO + 3H_2 \rightleftharpoons CH_4 + H_2O$$

$$CO_2 + 4H_2 \rightleftharpoons CH_4 + 2H_2O$$

The methanator effluent is cooled by heat exchange with boiler feedwater and cooling water. The synthesis gas leaves the methanator containing less than 10 parts per million (ppm) of carbon oxides.

Compression and Synthesis. The purified synthesis gas, containing H_2 and N_2 in a 3:1 mole ratio and with an inert gas (methane and argon) content of about 1.3 mole percent, is delivered to the suction of the synthesis-gas compressor. Anhydrous ammonia is catalytically synthesized in the converter. The effluent from the converter, after taking off a small purge stream, is recycled for eventual conversion to ammonia. Reaction takes place at approximately 370–482°C. Ammonia liquid, separated from the loop in the separator and from the purge, contains dissolved synthesis gas, which is released when the combined stream is flashed into the letdown drum. The flashed gas is then separated in the letdown drum and combined with the vapors from the purge separator to form a stream of purge fuel gases. Liquid ammonia in the letdown drum still contains some dissolved gases that must be disengaged. This liquid ammonia is let down to the refrigeration system where the dissolved gases are flashed and released to fuel. A centrifugal compressor is used to provide the refrigeration for the ammonia condensing. Ammonia product is withdrawn from the ammonia refrigeration system.

Modern Ammonia Plant. Since the development of the single train ammonia plant in the 1960s, many improvements have been made to reduce energy consumption. The plant of the 1980s has achieved striking results of reducing energy consumption by 20 to 30%—to less than 25 MMBTU(LHV)/ST (million Btus of low heating value fuel per short ton) of ammonia. This achievement represents a constant effort of development in seeking out more energy efficient design. Those developments have been centered around the 1960s basic process scheme with modifications to improve efficiency. Therefore, the basic process steps have not changed in any major way. The modern ammonia plants (after 1980) have incorporated energy-saving features, including: (1) more efficient furnace design to reduce fuel consumption, (2) more efficient drivers, compressors, and reduced power consumption, (3) low energy consumption in carbon dioxide removal system, (4) more efficient waste heat recovery and utilization, and (5) more efficient synthesis loop design, such as make-up gas drying, purge gas hydrogen recovery, and intercooled ammonia converter. The trend toward greater energy conservation is expected to continue.

Future Considerations in Ammonia Production. In addition to continued emphasis on energy efficiency, alternate feedstocks will continue to be a primary area of ammonia technology.

In the past, coal or heavy hydrocarbon feedstock ammonia plants were not economically competitive with plants where the feedstocks were light hydrocarbons (natural gas to naphtha). Because of changing economics, however, plants that can handle heavy hydrocarbon feedstock are now attracting increasing attention. In addition, the continuous development and improvement of partial oxidation processes at higher pressure have allowed reductions in equipment size and cost. Therefore, the alternate feedback ammonia plants based on a partial oxidation process may become economically competitive in the near future.

The ultimate goal of any ammonia process will be the direct fixation of nitrogen by reaction of water with air, $1.5H_2O + 0.5N_2 \rightleftharpoons NH_3 + 0.75O_2$. The theoretical energy requirement of the reaction is about 18 MMBTU/ST of ammonia. This feed energy requirement is the same as a natural gas feed ammonia plant. The major difference is that there is no short supply of water, air, and solar energy. However, the technology required for this route is not expected to be available any time soon. It is believed that, in the near future, ammonia plant designs will be based essentially on the present-day process with modifications to reduce energy consumption.

Additional Reading

Aika, K. and A. Neilsen: *Ammonia, Catalysis and Manufacture*, Springer-Verlag New York, Inc., New York, NY, 1995.

Appl, M.: *Ammonia, Principles and Industrial Practice*, John Wiley & Sons, Inc., New York, NY, 1999.

Staff: *Ammonia Plant Safety & Related Facilities*, Vol. 39, American Institute of Chemical Engineers, New York, NY, 1999.

AMMONIUM CHLORIDE. [CAS: 12125-02-9]. NH_4Cl, formula weight 53.50, white crystalline solid, decomposes at 350°, sublimes at 520°C under controlled conditions, sp gr 1.52. Also known as *sal ammoniac*, the compound is soluble in H_2O and in aqueous solutions of NH_3; slightly soluble in methyl alcohol. Ammonium chloride is a high-tonnage chemical, finding uses as an ingredient of dry cell batteries, as a soldering flux, as a processing ingredient in textile printing and hide tanning, and as a starting material for the manufacture of other ammonium chemicals. The compound can be produced by neutralizing HCl with NH_3 gas or with liquid NH_4OH, evaporating the excess H_2O, followed by drying, crystallizing, and screening operations. The product also can be formed in the gaseous phase by the reacting of hydrogen chloride gas with NH_3. Ammonium chloride generally is not attractive as a source of nitrogen for fertilizers because of the build-up and damaging effects of chloride residuals in the soil. See also **Nitrogen**.

AMMONIUM COMPOUNDS. Several of the principal ammonium compounds are described in separate entries in this volume. See also **Ammonium Chloride**; **Ammonium Nitrate**; **Ammonium Phosphates**; and **Ammonium Sulfate**. The important aspects of several other ammonium compounds are summarized below.

Acetate: Ammonium acetate [CAS: 631-61-8] $NH_4C_2H_3O_2$, white solid, soluble, formed by reaction of ammonia or NH_4OH and acetic acid, reacts upon heating to yield acetamide.

Alum: Ammonium alums are those alums, such as aluminum ammonium sulfate $Al_2(NH_4)_2(SO_4)_4 \cdot 24H_2O$, ferric ammonium sulfate $Fe_2(NH_4)_2(SO_4)_4 \cdot 24H_2O$, chromium ammonium sulfate $Cr_2(NH_4)_2(SO_4)_4 \cdot 24H_2O$ where ammonium sulfate is crystallized with the heavier metal sulfate.

Benzoate: Ammonium benzoate [CAS: 1863-63-4] $NH_4C_7H_5O_2$, white solid, soluble, formed by reaction of NH_4OH and benzoic acid. Used (1) as a food preservative, (2) in medicine.

Borate: Ammonium borate, [CAS: 11128-98-6] ammonium tetraborate

$$(NH_4)_2B_4O_7 \cdot 4H_2O,$$

white solid, soluble, formed by reaction of NH_4OH and boric acid. Used (1) in fireproofing fabrics, (2) in medicine.

Bromide: Ammonium bromide [CAS: 12124-97-9] NH_4Br white solid, soluble, sublimes at 542°C, formed by reaction of NH_4OH and hydrobromic acid. Used in photography.

Carbonates: Ammonium carbonate, [CAS: 506-87-6] $(NH_4)_2CO_3$, volatile, white solid, soluble, formed by reaction of NH_4OH and CO_2 by crystallization from dilute alcohol, loses NH_3, CO_2, and H_2O at ordinary temperatures, rapidly at 58°C; ammonium hydrogen carbonate, ammonium bicarbonate, ammonium acid carbonate NH_4HCO_3, white solid, soluble, formed by reaction of NH_4OH and excess CO_2. This salt is the important reactant in the ammonia soda process for converting sodium chloride in solution into sodium hydrogen carbonate solid.

Chloroplatinate: Ammonium chloroplatinate [CAS: 16919-58-7] $(NH_4)_2PtCl_6$, yellow solid, insoluble, formed by reaction of soluble ammonium salt solutions and chloroplatinic acid. Used in the quantitative determination of ammonium.

Cobaltinitrite: Diammonium sodium cobaltinitrite,

$$(NH_4)_2NaCo(NO_2)_6 \cdot H_2O$$

golden-yellow precipitate, formed by reaction of sodium cobaltinitrite solution in acetic acid with soluble ammonium salt solution. Used in the detection of ammonium.

Cyanate: Ammonium cyanate NH_4CNO, white solid, soluble, formed by fractional crystallization of potassium cyanate and ammonium sulfate (ammonium cyanate is soluble in alcohol), when heated changes into urea.

Dichromate: Ammonium dichromate [CAS: 7789-09-5] $(NH_4)_2Cr_2O_7$, red solid, soluble, upon heating evolves nitrogen gas and leaves a green insoluble residue of chromic oxide.

Fluoride: Ammonium fluoride [CAS: 12125-01-8] NH_4F, white solid, soluble, formed by reaction of NH_4OH and hydrofluoric acid, and then evaporating. Used (1) as an antiseptic in brewing, (2) in etching glass; ammonium hydrogen fluoride, ammonium bifluoride, ammonium acid fluoride NH_4F_2, white solid, soluble.

Iodide: Ammonium iodide [CAS: 12027-06-4] NH_4I, white solid, soluble, formed by reaction of NH_4OH and hydriodic acid, and then evaporating. Used (1) in photography, (2) in medicine.

Linoleate: Ammonium linoleate $NH_4C_{18}H_{31}O_2$. Used (1) as an emulsifying agent, (2) as a detergent.

Nitrite: Ammonium nitrite [CAS: 13446-48-5] NH_4NO_2 when ammonium sulfate or chloride and sodium or potassium nitrite are heated, the mixture behaves like ammonium nitrite in yielding nitrogen gas.

Oxalate: Ammonium oxalate [CAS: 1113-38-8] $(NH_4)_2C_2O_4$, white solid, soluble, formed by reaction of NH_4OH and oxalic acid, and then evaporating. Used as a source of oxalate; ammonium binoxalate $NH_4HC_2O_4 \cdot H_2O$, white solid, soluble.

Perchlorate: Ammonium perchlorate [CAS: 7790-98-9] $(NH(4)CLO(4))$ is a white crystalline substance. It is a powerful oxidizing material. AP is the oxidizer used in the solid rocket boosters on the space shuttle. It is stable in pure form at ordinary temperature, but decomposes at a temperature of 150 degrees C or above. It becomes an explosive when mixed with finely divided organic materials. AP exhibits the same explosive sensitivity to shock as picric acid (Class A explosive). Sensitivity to shock and friction may be great when contaminated with small amounts of some impurities such as sulfur, powdered metals and carbonaceous materials. AP may explode when involved in fire. *Periodate*: Ammonium periodate NH_4IO_4, white solid, moderately soluble.

Persulfate: Ammonium persulfate [CAS: 7727-54-0] $(NH_4)_2S_2O_8$, white solid, soluble, formed by electrolysis of ammonium sulfate under proper conditions. Used (1) as a bleaching and oxidizing agent, (2) in electroplating, (3) in photography.

Phosphomolybdate: Ammonium phosphomolybdate [CAS: 12026-66-3]

$$(NH_4)_3PO_4 \cdot 12MoO_3$$

(or similar composition), yellow precipitate, soluble in alkalis, formed by excess ammonium molybdate and HNO_3 with soluble phosphate solution. Used as an important test for phosphate (similar product and reaction when arsenate replaces phosphate).

Salicylate: Ammonium salicylate $NH_4C_7H_5O_3$, white solid, soluble, formed by reaction of NH_4OH and salicylic acid, and then evaporating. Used in medicine.

Sulfide: Ammonium sulfide [CAS: 12124-99-1] $(NH_4)_2S$, colorless to yellowish solution, formed by saturation with hydrogen sulfide of one-half of a solution of NH_4OH, and then mixing with the other half of the NH_4OH. Dissolves sulfur to form ammonium polysulfide, yellow solution. Used as a reagent in analytical chemistry; ammonium hydrogen sulfide, ammonium bisulfide, ammonium acid sulfide NH_4HS, colorless to yellowish solution, formed by saturation with H_2S of a solution of NH_4OH.

Tartrate: Ammonium tartrate [CAS: 3164-29-2] $(NH_4)_2C_4H_4O_6$, white solid, moderately soluble, formed by reaction of NH_4OH and tartaric acid, and then evaporating. Used in the textile industry; ammonium hydrogen tartrate, ammonium bitartrate, ammonium acid tartrate $NH_4HC_4H_4O_6$, white solid, slightly soluble, formation sometimes used in detection of ammonium or tartrate.

Thiocyanate: Ammonium thiocyanate, [CAS: 1762-95-4] ammonium sulfocyanide, ammonium rhodanate NH_4CNS, white solid, soluble, absorbs much heat on dissolving with consequent marked lowering of temperature, mp 150°C, formed by boiling ammonium cyanate solution with sulfur, and then evaporating. Used (1) as a reagent for ferric, (2) in making cooling solutions, (3) to make thiourea.

Ammonium compounds liberate NH_3 gas when warmed with NaOH solution.

AMMONIUM HYDROXIDE. [CAS: 1336-21-6]. NH$_4$OH, formula weight 35.05, exists only in the form of an aqueous solution. The compound is prepared by dissolving NH$_3$ in H$_2$O and usually is referred to in industrial trade as aqua ammonia. For industrial procurements, the concentration of NH$_3$ in solution is normally specified in terms of the specific gravity (degrees Baumé, °Be). Common concentrations are 20 °Be and 26 °Be. The former is equivalent to a sp gr of 0.933, or a concentration of about 17.8% NH$_3$ in solution; the latter is equivalent to a sp gr of 0.897, or a concentration of about 29.4% NH$_3$. These figures apply at a temperature of 60°F (15.6°C). Reagent grade NH$_4$OH usually contains approximately 58% NH$_4$OH (from 28 to 30% NH$_3$ in solution).

Ammonium hydroxide is one of the most useful forms in which to react NH$_3$ (becoming the NH$_4^+$ radical in solution) with other materials for the creation of ammonium salts and other ammonium and nitrogen-bearing chemicals. Ammonium hydroxide is a direct ingredient of many products, including saponifiers for oils and fats, deodorants, etching compounds, and cleaning and bleaching compounds. Because aqua ammonia is reasonably inexpensive and a strongly alkaline substance, it finds wide application as a neutralizing agent. See also **Nitrogen**.

AMMONIUM NITRATE. [CAS: 6484-52-2]. NH$_4$NO$_3$, formula weight 80.05, colorless crystalline solid, occurs in two forms:

α-NH$_4$NO$_3$, tetragonal crystals, stable between $-16°$C and $32°$C, sp gr 1.66.
β-NH$_4$NO$_3$, rhombic or monoclinic crystals, stable between $32°$C and $84°$C, sp gr 1.725.

The melting point generally ascribed to the alpha form is 169.6°C, with decomposition occurring above 210°C. Upon heating, ammonium nitrate yields nitrous oxide (N$_2$O) gas and can be used as an industrial source of that gas. Ammonium nitrate is soluble in H$_2$O, slightly soluble in ethyl alcohol, moderately soluble in methyl alcohol, and soluble in acetic acid solutions containing NH$_3$.

As shown in Fig. 1 in making ammonium nitrate on a large scale, NH$_3$, vaporized by waste steam from neutralizer, is sparged along with HNO$_3$ into the neutralizer. A ratio controller automatically maintains the proper proportions of NH$_3$ and acid. The heat of neutralization evaporates a part of the H$_2$O and gives a solution of 83% NH$_4$NO$_3$. Final evaporation to above 99% for agricultural prills or to approximately 96% for industrial prills is accomplished in a falling-film evaporator located at the top of the prilling tower. The resultant melt flows through spray nozzles and downward through the tower. Air is drawn upward by fans at the top of the tower. The melt is cooled sufficiently to solidify, forming round pellets or prills of the desired range of sizes. The prills are removed from the bottom of the tower and fed to a rotary cooler. Where industrial-type prills are produced, a pre-drier and drier precede the cooler. Fines from the rotary drums are collected in wet cyclones. This solution eventually is returned to the neutralizer. After cooling, the prills are screened to size and the over and undersize particles are sent to a sump and returned to the neutralizer. Intermediate or product-size prills are dusted with a coating material, usually diatomaceous earth,

in a rotary coating drum and sent to the bagging operation. The process can be adapted to other types of materials and mixtures of ammonium nitrate and other fertilizer materials. Mixtures include the incorporation of limestone and ammonium phosphates.

Ammonium nitrate is a very high tonnage industrial chemical, finding major applications in explosives and fertilizers, and additional uses in pyrotechnics, freezing mixtures (for obtaining low temperatures), as a slow-burning propellant for missiles (when formulated with other materials, including burning-rate catalysts), as an ingredient in rust inhibitors (especially for vapor-phase corrosion), and as a component of insecticides.

Amatol, an explosive developed by the British, is a mixture of ammonium nitrate and TNT. A special explosive for tree-trunk blasting consists of ammonium nitrate coated with TNT. In strip mining, an explosive consisting of ammonium nitrate and carbon black is used. The explosive ANFO is a mixture of ammonium nitrate and fuel oil. ANFO accounts for about 50% of the commercial explosives used in the United States. Slurry explosives consist of oxidizers (NH$_4$NO$_3$ and NaNO$_3$), fuels (coals, oils, aluminum, other carbonaceous materials), sensitizers (TNT, nitrostarch, and smokeless powder), and water mixed with a gelling agent to form a thick, viscous explosive with excellent water-resistant properties. Slurry explosives may be manufactured as cartridged units, or mixed on-site. Although Nobel introduced NH$_4$NO$_3$ into his dynamite formulations as early as 1875, the tremendous explosive power of the compound was not realized until the tragic Texas City, Texas disaster of 1947 when a shipload of NH$_4$NO$_3$ blew up while in harbor. See also **Explosives**.

As a fertilizer, NH$_4$NO$_3$ contains 35% nitrogen. Because of the explosive nature of the compound, precautions in handling are required. This danger can be minimized by introducing calcium carbonate into the mixture, reducing the effective nitrogen content of the product to 26%. In as much as NH$_4$NO$_3$ is highly hygroscopic, clay coatings and moisture-proof bags are means used to preclude spoilage in storage and transportation. See also **Fertilizer**; and **Nitrogen**.

AMMONIUM PHOSPHATES. There are two ammonium phosphates, both produced on a very high-tonnage scale; monoammonium phosphate, NH$_4$H$_2$PO$_4$, white crystals, sp gr 1.803 formula weight 115.04, N = 12.17%; P$_2$O$_5$ = 61.70% diammonium phosphate, (NH$_4$)$_2$HPO$_4$, white crystals, sp gr 1.619 formula weight 132.07, N = 21.22%, P$_2$O$_5$ = 53.74%

Both compounds are soluble in H$_2$O; insoluble in alcohol or ether. A third compound, triammonium phosphate (NH$_4$)$_3$PO$_4$, does not exist under normal conditions because, upon formation, it immediately decomposes, losing NH$_3$ and reverting to one of the less alkaline forms.

Large quantities of the ammonium phosphates are used as fertilizers and in fertilizer formulations. The compounds furnish both nitrogen and phosphorus essential to plant growth. The compounds also are used as fire retardants in wood building materials, paper and fabric products, and in matches to prevent afterglow. Solutions of the ammonium phosphates sometimes are air dropped to retard forest fires, serving the double purpose of fire fighting and fertilizing the soil to accelerate new plant growth. The compounds are used in baking powder formulations, as nutrients in

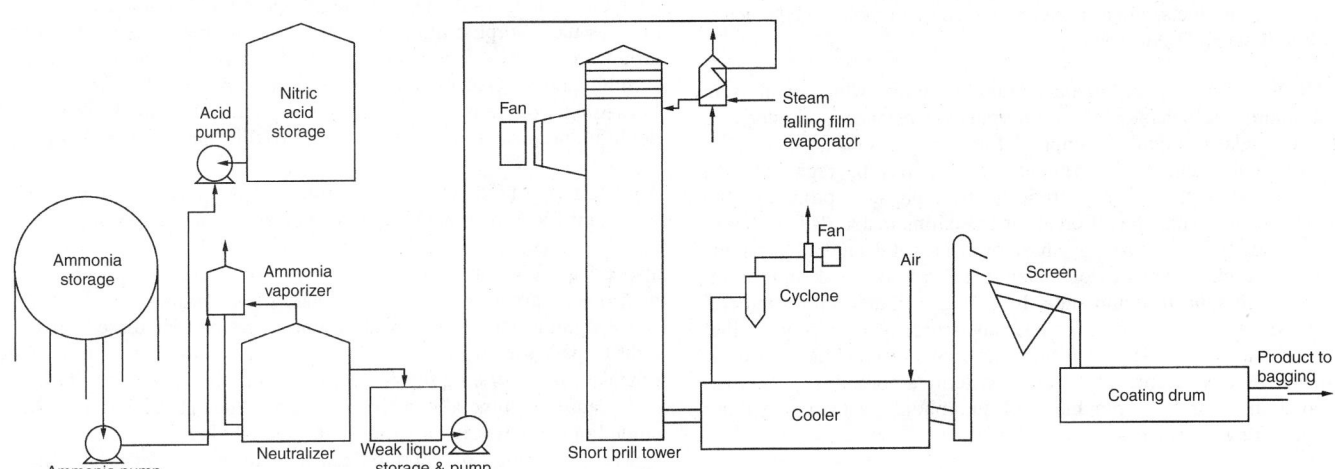

Fig. 1.   Process for making ammonium nitrate on a large scale

the production of yeast, as nutritional supplements in animal feeds, for controlling the acidity of dye baths, and as a source of phosphorus in certain kinds of ceramics.

Ammonium phosphates usually are manufactured by neutralizing phosphoric acid with NH_3. Control of the pH (acidity/alkalinity) determines which of the ammonium phosphates will be produced. Pure grades can be easily made by crystallization of solutions obtained from furnace grade phosphoric acid. Fertilizer grades, made from wet-process phosphoric acid, do not crystallize well and usually are prepared by a granulation technique. First, a highly concentrated solution or slurry is obtained by neutralization. Then the slurry is mixed with from $6\times$ to $10\times$ its weight of previously dried material, after which the mixture is dried in a rotary drier. The dry material is then screened to separate the desired product size. Oversize particles are crushed and mixed with fines from the screen operation and then returned to the granulation step where they act as nuclei for the production of further particles. Other ingredients often are added during the granulation of fertilizer grades. The ratio of nitrogen to phosphorus can be altered by the inclusion of ammonium nitrate, ammonium sulfate, or urea. Potassium salts sometimes are added to provide a 3-component fertilizer (N, P, K). A typical fertilizer grade diammonium phosphate will contain 18% N and 46% P_2O_5 (weight). See also **Fertilizer**; and **Nitrogen**.

There has been a trend toward the production of ammonium phosphates in powder form. Concentrated phosphoric acid is neutralized under pressure, and the heat of neutralization is used to remove the water in a spray tower. The powdered product then is collected at the bottom of the tower. Ammonium nitrate/ammonium phosphate combination products can be obtained either by neutralizing mixed nitric acid and phosphoric acid, or by the addition of ammonium phosphate to an ammonium nitrate melt.

AMMONIUM SULFATE. [CAS: 7783-20-2]. $(NH_4)_2SO_4$, formula weight 132.14, colorless crystalline solid, decomposes above 513°C, sp gr 1.769. The compound is soluble in H_2O and insoluble in alcohol. Ammonium sulfate is a high-tonnage industrial chemical, but frequently may be considered a byproduct as well as intended end-product of manufacture. Large quantities of ammonium sulfate result from a variety of industrial neutralization operations required for alleviation of stream pollution by free H_2SO_4. The ammonium sulfate so produced is not always recovered and marketed. A significant commercial source of $(NH_4)_2SO_4$ is its creation as a byproduct in the manufacture of caprolactam, which yields several tons of the compound per ton of caprolactam made. See also **Caprolactam**. Ammonium sulfate also is a byproduct of coke oven operations where the excess NH_3 formed is neutralized with H_2SO_4 to form $(NH_4)_2SO_4$. However, as a major fertilizer and ingredient of fertilizer formulations, additional production is required, largely depending upon the proximity of consumers to byproduct $(NH_4)_2SO_4$ sources. In the Meresburg reaction, natural or byproduct gypsum is reacted with ammonium carbonate:

$$CaSO_4 \cdot 2H_2O + (NH_4)_2CO_3 \longrightarrow CaCO_3 + (NH_4)_2SO_4 + 2\,H_2O$$

The product is stable, free-flowing crystals. As a fertilizer, $(NH_4)_2SO_4$ has the advantage of adding sulfur to the soil as well as nitrogen. By weight, the compound contains 21% N and 24% S. Ammonium sulfate also is used in electric dry cell batteries, as a soldering liquid, as a fire retardant for fabrics and other products, and as a source of certain ammonium chemicals. See also **Fertilizer**; and **Nitrogen**.

AMORPHOUS. As opposed to a crystalline substance which exhibits an orderly structure, the behavior of an amorphous substance is similar to a very viscous, inelastic liquid. Examples of amorphous substances include amber, glass, and pitch. An amorphous material may be regarded as a liquid of great viscosity and high rigidity, with physical properties the same an all directions (may be different for crystalline materials in different directions). Usually, upon heating, an amorphous solid gradually softens and acquires the characteristics of a liquid, but without a definite point of transition from solid to liquid state. In geology, an amorphous mineral lacks a crystalline structure, or has an internal arrangement so irregular that there is no characteristic external form. This does not preclude, however, the existence of any degree of order. The term amorphous is used in connection with amorphous graphite and amorphous peat, among other naturally occurring substances.

AMOSITE. Amosite is a long-fiber gray or greenish asbestiform mineral related to the cummingtonite-grunerite series, and is of economic importance. It occurs within both regional and contact metamorphic rocks in the Republic of South Africa. The name *amosite* is a product of the initial letters of its occurrence at the Asbestos Mines of South Africa. See also **Asbestos**

AMPEROMETER. An instrument for the chemical analysis of electro-reducible or oxidizable ions, molecules, or dissolved gases in solution. Included are the majority of metal ions and many organic substances that contain oxidizable or reducible groups. The range of concentration measurements of which the instrument is capable is from 0.01 to 1,000 ppm. This instrumental method is used for determining free available or total available chlorine, particularly in connection with water-chlorination control. In this application, the range of chlorine concentration is from 0 to 50 ppm. Free iodine also may be determined with an amperometer. The identification of unknown substances is by inference from noting the fixed potential between polarized microelectrode and reference electrode that causes oxidation or reduction of a given composition sought. Where identity is firmly established, the concentration also may be determined because the concentration is proportional to the diffusion-limited current that flows in the electrode circuit. In this latter respect, amperometry is similar to polarography. See also **Analysis (Chemical)**.

AMPHETAMINE. Also called methylphenethylamine; 1-phenyl-2-aminopropane; Benzedrine; formula $C_6H_5CH_2CH(NH_2)CH_3$; *amphetamine* is a colorless, volatile liquid with a characteristic strong odor and slightly burning taste. Boils and commences decomposition at 200–203°C. Low flash point, 26.7°C. Soluble in alcohol and ether; slightly soluble in water. Amphetamine is the basis of a group of hallucinogenic, habit-forming drugs that affect the central nervous system. The drug also finds medical application, notably in appetite suppressants. It should be emphasized that administration of amphetamines for prolonged periods in connection with weight-reduction programs may lead to drug dependence. Professionals must pay particular attention to the possibility of persons obtaining amphetamines for nontherapeutic use or distribution to others.

AMPHIBOLE. This is the name given to a closely related group of minerals all showing in common a prismatic cleavage of 54–56° as well as similar optical characteristics and chemical composition.

The amphiboles may be said to represent chemically a series of metasilicates corresponding to the general formula $RSiO_3$ where R may be calcium, magnesium, iron, aluminum, titanium, sodium, or potassium. The crystals of the amphibole family group fall within both the monoclinic and orthorhombic systems.

There is a clear parallelism between the amphiboles and the pyroxenes. There are two basic differences between the minerals of these two family groups; amphiboles with cleavage angles of 56° and 124°, with essential OH groups in their structure; pyroxenes with cleavage angles of 87° and 93°, and being anhydrous, with no OH content. Amphibole crystals are usually long and slender and tend to be simple while pyroxene crystals tend to be complex, short, and stout prisms.

Amphibole is common in both lavas and deep-seated rocks, though less so in the basic lavas than pyroxene. Many of the amphiboles may be developed as metamorphic minerals. The following members of the amphibole group are described under their own headings: actinolite, anthophyllite, cummingtonite, glaucophane, grunerite, hornblende, riebeckite and tremolite. Amphibole was so named by Haüy from the Greek word, meaning doubtful, because of the many varieties of this mineral. See also **Pyroxene**.

AMPHIBOLITE. The amphibolites form a large group of rather important rocks of metamorphic character. As the name implies they are made up very largely of minerals of the amphibole group. There may be also a variety of other minerals present, such as quartz, feldspar, biotite, muscovite, garnet, or chlorite in greater or lesser amounts.

Depending upon the particular amphibole present these rocks may be light to dark green or black, the amphibole usually being in long slender prisms or laths, often quite coarse, sometimes in acicular or fibrous forms.

Because the mineral constituents are arranged parallel to the schistosity, amphibolites may have a strongly developed cleavage.

The occurrence of amphibolites accompanying gneisses, schists, and other metamorphic rocks of probable sedimentary origin strongly suggests a similar derivation. Yet some amphibolites cut other metamorphic rocks in the manner of dikes or sills. It is very likely that they have been derived from both original igneous and sedimentary rocks. Large masses of amphibolite suggest gabbroic stocks. Well-known areas in which amphibolites are found are New England, New York State, Canada, Scotland, and the Alps.

AMPHIPATHY. The simultaneous attraction and repulsion in a single molecule or ion consisting of one or more groups having an affinity for the phase in which they are dissolved together with groups that tend to be expelled by the medium.

AMPHIPHILIC. Molecule having a water-soluble polar head (hydrophilic) and a water-insoluble organic tail (hydrophobic), e.g., octyl alcohol, sodium stearate. Such molecules are necessary for emulsion formation and for controlling the structure of liquid crystals.

See also **Emulsions**; and **Liquid Crystals**.

AMPHIPROTIC. Capable of acting either as an acid or as a base, i.e., as a proton donor or acceptor, according to the nature of the environment. Thus, aluminum hydroxide dissolves in acids to form salts of aluminum, and it also dissolves in strong bases to form aluminates. Solvents like water, which can act to give protons or accept them, are amphiprotic solvents. See **Acids and Bases**; and **Salt**.

AMPHOLYTE. A substance that can ionize to form either anions or cations and thus may act as either an acid or a base. An ampholytic detergent is cationic in acid media and anionic in base media. Water is an ampholyte.

AMPOULE. Sometimes spelled ampule, a small sealed glass container for drugs that are to be given by injection. As they are completely sealed, the contents are kept in their original sterile condition.

AMYGDALOID. A vesicular rock, commonly lava, whose cavities have become filled with a secondary deposit of mineral material such as quartz, calcite, and zeolites. The term is derived from the Greek word meaning almond in reference to the frequent almondlike appearance of the filled vesicles, which are called amygdales or amygdules.

AMYL. The 5-carbon aliphatic group C_5H_{11}, also known as pentyl. Eight isometric arrangements (exclusive of optical isomers) are possible. The amyl compounds occur (as in fusel oil) or are formed (as from the petroleum pentanes) as mixtures of several isomers, and, since their boiling points are close and their other properties similar, it is neither easy nor usually necessary to purify them.

See also **Amyl Alcohols**.

AMYL ALCOHOLS. Amyl alcohol describes any saturated aliphatic alcohol containing five carbon atoms. This class consists of three pentanols, four substituted butanols, and a disubstituted propanol, i.e., eight structural isomers $C_5H_{12}O$: four primary, three secondary, and one tertiary alcohol. In addition, 2-pentanol, 2-methyl-1-butanol, and 3-methyl-2-butanol have chiral centers and hence two enantiomeric forms.

The odd-carbon structure and the extent of branching provide amyl alcohols with unique physical and solubility properties and often offer ideal properties for solvent, surfactant, extraction, gasoline additive, and fragrance applications. Amyl alcohols have been produced by various commercial processes in past years. Today the most important industrial process is low pressure rhodium-catalyzed hydroformylation (oxo process) of butenes.

Mixtures of isomeric amyl alcohols (1-pentanol and 2-methyl-1-butanol) are often preferred because they are less expensive to produce commercially; also, the different degree of branching of the mixture imparts a more desirable combination of properties. One such mixture is a

TABLE 1. THE AMYL ALCOHOLS AND SOME OF THEIR PHYSICAL PROPERTIES

Properties	1-Pentanol	2-Pentanol	3-Pentanol	2-Methyl-1-butanol	3-Methyl-1-butanol	2-Methyl-2-butanol	3-Methyl-2-butanol	2,2-Dimethyl-1-propanol
common name	*n*-amyl alcohol	*sec*-amyl alcohol			isoamyl alcohol	*tert*-amyl alcohol		neopentyl alcohol
critical temperature, °C	315.35	287.25	286.45	291.85	306.3	272.0	300.85	276.85
critical pressure, kPa[a]	3868	3710	3880	3880	3880	3880	3960	3880
critical specific volume, mL/mol	326.5	328.9	325.3	327	327	327	327	327
critical compressibility	0.25810	0.26188	0.27128	0.27009	0.26335	0.27992	0.27133	0.27745
boiling point at pressure, °C								
101.3 kPa[a]	137.8	119.3	115.3	128.7	130.5	102.0	111.5	113.1
40 kPa	111.5	93.8	90.9	103.5	105.6	78.3	87.2	89.0
1.33 kPa	44.6	32.0	27.7	40.2	43.0	21.0	26.0	25.0
vapor pressure[b], kPa[a]	0.218	0.547	0.761	0.274	0.200	1.215	0.810	0.929
melting point, °C	−77.6	−73.2	−69.0	< −70	−117.2	−8.8	forms glass	54.0
heat of vaporization at normal boiling point, kJ/mol[c]	44.83	43.41	42.33	44.75	43.84	40.11	41.10	41.35
ideal gas heat of formation[d], kJ/mol[c]	−298.74	−313.80	−316.73	−302.08	−302.08	−329.70	−314.22	−319.07
liquid density[b], kg/m³	815.1	809.4	820.3	819.1	810.4	809.6	818.4	851.5[e]
liquid viscosity[b], mPa·s(= cP)	4.06	4.29	6.67	5.11	4.37	4.38	3.51[d]	2.5[e]
surface tension[b], mN/m(= dyn/cm)	25.5	24.2	24.6	25.1[d]	24.12	22.7	23.0[d]	14.87[e]
refractive index[d]	1.4080	1.4044	1.4079	1.4086	1.4052	1.4024	1.4075	1.3915
solubility parameter[d], $(MJ/m^3)^{0.5f}$	22.576	21.670	21.150	22.274	22.322	20.758	21.607	19.265[e]
solubility in water[b], wt%	1.88	4.84	5.61	3.18	2.69	12.15	6.07	3.74
solubility of water in[b], wt%	9.33	11.68	8.19	8.95	9.45	24.26	11.88	8.23

[a] To convert kPa to mm Hg, multiply by 7.5.
[b] At 20°C unless otherwise noted.
[c] To convert kJ/mol to cal/mol, multiply by 239.
[d] At 25°C.
[e] At the melting point.
[f] To convert (MJ/m^3) to $(cal)^{0.5}$, divide by 2.045.

commercial product sold under the name Primary Amyl Alcohol by Union Carbide Chemicals and Plastics Company Inc.

Physical Properties

With the exception of neopentyl alcohol (mp 53°C), the amyl alcohols are clear, colorless liquids under atmospheric conditions, with characteristic, slightly pungent and penetrating odors. They have relatively higher boiling points than ketonic or hydrocarbon counterparts and are considered intermediate boiling solvents for coating systems (Table 1).

Commercial primary amyl alcohol is a mixture of 1-pentanol and 2-methyl-1-butanol, in a ratio of ca 65 to 35 (available from Union Carbide Chemicals and Plastics Company Inc. in other ratios upon request). Typical physical properties of this amyl alcohol mixture are listed in Table 2.

Like the lower alcohols, amyl alcohols are completely miscible with numerous organic solvents and are excellent solvents for nitrocellulose, resin lacquers, higher esters, and various natural and synthetic gums and resins. However, in contrast to the lower alcohols, they are only slightly soluble in water. Only 2-methyl-2-butanol exhibits significant water solubility.

Chemical Properties

The amyl alcohols undergo the typical reactions of alcohols which are characterized by cleavage at either the oxygen—hydrogen or carbon—oxygen bonds. Important reactions include dehydration, esterification, oxidation, amination, etherification, and condensation.

Manufacture

Three significant commercial processes for the production of amyl alcohols include separation from fusel oils, chlorination of C-5 alkanes with subsequent hydrolysis to produce a mixture of seven of the eight isomers (Pennsalt), and a low pressure oxo process, or hydroformylation, of C-4 olefins followed by hydrogenation of the resultant C-5 aldehydes.

The oxo process is the principal one in practice today; only minor quantities, mainly in Europe, are obtained from separation from fusel oil. *tert*-Amyl alcohol is produced on a commercial scale in lower volume by hydration of amylenes (Dow, BASF).

Health and Safety Factors

The main effects of prolonged exposure to amyl alcohols are irritation to mucous membranes and upper respiratory tract, significant depression of the central nervous system, and narcotic effects from vapor inhalation or oral absorption. All the alcohols are harmful if inhaled or swallowed, appreciably irritating to the eyes and somewhat irritating to uncovered skin on repeated exposure. Prolonged exposure causes nausea, coughing, diarrhea, vertigo, drowsiness, headache, and vomiting. 3-Methyl-1-butanol has demonstrated carcinogenic activity in animal studies.

All of the amyl alcohols are TSCA and EINECS (European Inventory of Existing Commercial Chemical Substances) registered.

The amyl alcohols are readily flammable substances; *tert*-amyl alcohol is the most flammable (closed cup flash point, 19°C). Their vapors can form explosive mixtures with air.

TABLE 2. PHYSICAL PROPERTIES OF PRIMARY AMYL ALCOHOL, MIXED ISOMERS

Property	Value[a]
molecular weight	88.15
boiling point at 101.13 kPa[b], °C	133.2
freezing point, °C	−90[c]
specific gravity 20/20°C	0.8155
absolute viscosity at 20°C, mPa·s(= cP)	4.3
vapor pressure at 20°C, kPa[b]	0.27
flash point (closed cup), °C	45
solubility at 20°C, by wt%	
in water	1.7
water in	9.2

[a] 65/35 blend, i.e., a mixture of 1-pentanol and 2-methyl-1-butanol, 65/35 wt%, respectively.

[b] To convert kPa to mm Hg, multiply by 7.5.

[c] Sets to glass below this temperature.

Shipping and Storage

Amyl alcohols are best stored or shipped in either aluminum, lined steel, or stainless steel tanks.

Economic Aspects

All eight amyl alcohol isomers are available from fine chemical supply firms in the United States. Five of them, 1-pentanol, 2-pentanol, 2-methyl-1-butanol, 3-methyl-1-butanol, and 2-methyl-2-butanol (*tert*-amyl alcohols) are available in bulk in the United States; in Europe all but neopentyl alcohol are produced.

ANTHONY J. PAPA
Union Carbide Chemicals and Plastics Company Inc.

Additional Reading

Bieber, H.: *Encycl. Chem. Process Des.* **3**, 278 (1977).
Kirkpatrick Chemical Engineering Achievement Award, *Chem. Eng.* **84**, 110 (Dec. 5, 1977).

AMYLOSE.   See **Starch**.

ANABOLIC AGENTS.   See **Steroids**.

ANAEROBE.   An organism that can grow in the absence of free oxygen and referred to as an anaerobic organism. Subdivided into *facultative anaerobes*, which can grow and utilize oxygen when it is present; and obligatory anaerobes, which cannot tolerate even a trace of oxygen in their surroundings. Yeast is an example of a facultative anaerobe. Yeast can grow and utilize oxygen in its metabolism, in which case it utilizes all the energy in a carbohydrate and yields water and carbon dioxide. In the absence of oxygen, the yeast cells turn to fermentation and anaerobic metabolism wherein the carbohydrate is converted into alcohol and carbon dioxide. The bacterium which produces botulism in "preserved" foods is an obligatory anaerobe (*Clostridium botulinum*).

In addition to botulism, other species of Clostridium are the etiological agents of tetanus, gas gangrene, and a variety of animal diseases including struck, lamb dysentery, pulpy kidney, and enterotoxemia. The pathogenesis of the clostridia depends largely on the production of potent toxins. Generally, these are heat labile exotoxins and often several pharmacologically and immunologically different toxins may be produced by the same species. In recent years, certain of the toxins have been purified to the point of crystalline purity and revealed to be high-molecular-weight proteins. The availability of the pure toxins has permitted more precise studies both of the chemical nature of the toxin and also of their properties as enzymes. From such toxins, it is possible to produce efficient toxoids for use in prophylactic immunization measures for tetanus and botulism. Some clostridia, in contrast, are beneficial. In addition to the nitrogen-fixing ability of certain species, *C. sporogenes* is active in the decomposition of proteins and contributes to biodegradability of substances, particularly of plant remains. *C. acetobutyleium* is active in production of acetone and butyl alcohol by fermentation.

Glucose can be broken down through glycolysis into pyruvic acid, but in the absence of oxygen as a final acceptor for the hydrogen, the pyruvic acid cannot enter the tricarboxylic acid cycle for further breakdown. Instead the pyruvic acid itself serves as an acceptor for the hydrogen split off in glycolysis. Hence, much less energy is obtained from food in anaerobic metabolism. Lactic acid, alcohol, and some extremely poisonous substances are products of anaerobic metabolism.

In addition to the one-celled anaerobes, certain parasitic worms, such as *Ascaris*, are thought to use anaerobic metabolism, to a certain extent at least. Living in the intestine of higher animals these worms have little access to free oxygen.

The muscles of higher animals are also known to use anaerobic metabolism when the demands on the muscles for energy are greater than can be supplied through the available oxygen. Lactic acid is generated.

The diverse mechanisms that permit animal life in the absence of oxygen are described in considerable detail by P.W. Hochachka ("Living without Oxygen," Harvard Univ. Press, Cambridge, Massachusetts, 1980).

Additional Reading

Blankenship, R.E., M.T. Madigan, and C.E. Bauer: *Anoxygenic Photosynthetic Bacteria*, Kluwer Academic Publishing, Norwell, MA, 1995.

Carlile, M.J. and S. Watkinson: *The Fungi*, Academic Press, San Diego, CA, 1996.

Dusenbery, D.B.: *Life at Small Scale, The Behavior of Microbes*, Scientific American Library, W.H. Freeman and Company, New York, NY, 1996.

Evans, I.H.: "Yeast Protocols," *Methods in Cell and Molecular Biology*, Vol. 53, Humana Press, Totowa, NJ, 1996.

Hauschild, A.H. and K.L. Dodds, *Clostridium Botulinum*, Marcel Dekker, Inc. New York, NY, 1992.

Mountfort, D.O. and C.G. Orpin: *Anaerobic Fungi, Biology, Ecology and Function*, Marcel Dekker, Inc., New York, NY, 1994.

Spencer, J.F.T. and D.M. Spencer: *Yeasts, Their Lives in Natural and Artificial Habitats*, Springer-Verlag, Inc., New York, NY, 1997.

Williams, J.: *Bioinorganic Chemistry, Trace Element Evolution from Anaerobes to Aerobes*, Springer-Verlag, Inc., New York, NY, 1997.

ANALCIME. A common zeolite mineral, $NaAlSi_2O_6 \cdot H_2O$, a hydrous soda-aluminum silicate. It crystallizes in the isometric system, hardness, 5–5.5; specific gravity, 2.2; vitreous luster; colorless to white; but may be grayish, greenish, yellowish, or reddish. Its trapezohedral crystal resembles garnet but is softer; it is distinguished from leucite only by chemical tests.

There are many excellent European localities. Magnificent crystals occur at Mt. St. Hilaire, Quebec, Canada; in the United States at Bergen Hill and West Paterson, New Jersey, Keweenaw County, Michigan, and Jefferson County, Colorado. Nova Scotia furnishes beautiful specimens.

Analcime is a relatively common mineral and occurs with other zeolites in cavities and fissures in basic igneous rocks, occasionally in granites or gneisses. It seems to occur as a replacement and perhaps in some cases as a primary mineral crystallizing from a magma rich in soda and water vapor under pressure. The name analcime is derived from the Greek word meaning weak, in reference to the weak electric charge developed when heated or subjected to friction.

ANALGESICS, ANTIPYRETICS, AND ANTIINFLAMMATORY AGENTS. Pain, pyresis, and inflammation are distinct physiological responses which can occur independently; they are often associated as the body mounts a response to an injury or insult. Each is an important signal from injured tissue and the signal's continued presence can guide a physician in diagnosing and treating the condition which led to the occurrence.

The twentieth century has seen considerable progress in the understanding of pain and inflammation and the relationship of one to the other. It is increasingly apparent that the central and peripheral nervous systems are capable of causing the production of mediators which can attract and activate inflammatory cells, thereby initiating or amplifying an inflammatory response. At the same time, inflammatory cells are also capable of releasing substances which not only respond to the original insult, but also stimulate the nervous systems.

A number of interdependent physiological mechanisms, each providing new targets for therapeutic intervention, and allowing for the development of new treatments for pain, pyresis, and inflammation, have been discovered. At least eight distinct types of opioid receptors have been identified and the corresponding individual functions are beginning to be understood. Moreover, a great deal has been learned about the role of lipid mediators in the inflammatory response and how antiinflammatory agents control their production.

Opioid Agonists

Opium is the dried, powdered sap of the unripe seed pod of *Papaver somniferum*, a poppy plant indigenous to Asia Minor. More than 20 different alkaloids (qv) of two different classes comprise 25% of the weight of dry opium. The benzylisoquinolines, characterized by papaverine (1.0%), a smooth muscle relaxant, and noscapine (6.0%), an antitussive agent, do not have any analgesic effects. The phenanthrenes, the second group, are the more common and include 10% morphine (**1**, R′ = R = H), 0.5% codeine, $C_{18}H_{21}NO_3$, (**1**, R′ = H, R = CH$_3$), and 0.2% thebaine, $C_{19}H_{21}NO_3$, (**2**).

(1)

(2)

Morphine, mol wt 285.3, effectively relieves pain and increases an individual's ability to tolerate a painful experience. It also produces a remarkably broad range of other effects, including drowsiness, mood changes, respiratory depression, nausea, decreased gastrointestinal motility, and vomiting. Morphine behaves as a receptor agonist, acting preferentially at the μ receptor, but also exhibiting appreciable affinity for other opioid receptors. A standard therapeutic dose is 10–15 mg, usually administered subcutaneously. Analgesia peaks in about one hour and lasts for four to five hours. An important feature of morphine and related drugs is the development of physical dependence on, and tolerance to, some of the effects.

Codein, mol wt 299.3, is a significantly less potent analgesic than morphine, requiring 60 mg (0.20 mmol) to equal the effectiveness of 10 mg (0.04 mmol) of morphine. However, codeine is orally effective, and it is less addictive and associated with less nausea than morphine.

Introduced in 1898, heroin (**1**, R = R′ = COCH$_3$) was heralded as a nonaddictive alternative to morphine. Subsequent clinical experience showed it to be highly addictive and preferred by addicts over morphine.

Synthetic and Semisynthetic Agonists

In attempts to discover drugs demonstrating fewer undesirable side effects than morphine, many synthetic analogues have been prepared. Some of these are shown in Table 1.

Several common structural features necessary for opioid, analgesic activity have been identified from the action of the analogues. Systematic simplification demonstrated that much of the morphine ring structure could be modified or even eliminated. The piperidine ring, in a chair conformation, and in particular the nitrogen atom of that ring, appears essential to pharmacological activity. The nitrogen is believed to attach to an anionic center in the receptor. Also crucial is the presence of the phenyl ring which, through van der Waals forces, binds to the hydrophobic portion of the receptor.

Opioid Antagonists and Partial Agonists

The replacement of the *N*-methyl group on the nitrogen atom of the piperidine ring of morphine and analogues, by allyl, isopropyl, or methyl cyclopropyl, an isopropyl isostere, results in compounds which antagonize opioid responses, especially respiratory depression. Naloxone (**7**), $C_{19}H_{21}NO_4$, and naltrexone (**8**), $C_{20}H_{23}NO_4$, are both derived from oxymorphone (Table 1) and exhibit agonist activity only at doses that are of little clinical significance. In the absence of opioid drugs, naloxone does not cause analgesia, respiratory depression, or sedation. However, when administered with an opioid analgesic, the effects produced by the opioid agonist are promptly reversed. The ability to antagonize opioids at all of the different opioid receptors makes naloxone useful for the treatment of opioid overdose. Naltrexone has a similar profile, but it is orally active and has a significantly longer half-life.

(7)

$R = CH_2CHCH_2$

TABLE 1. SYNTHETIC MORPHINE ANALOGUES

Name	Molecular formula	Structure
levorphanol	$C_{17}H_{23}NO$	

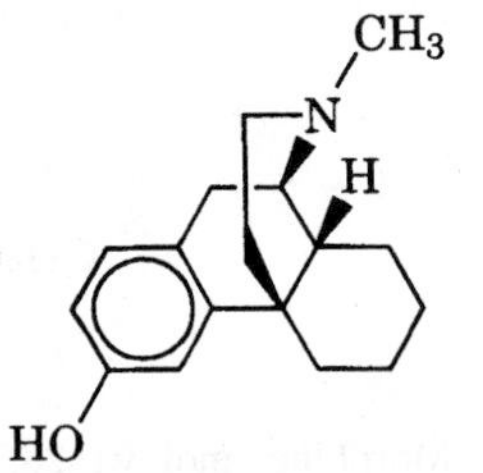

(3)

| hydromorphone | $C_{17}H_{19}NO_3$ | |

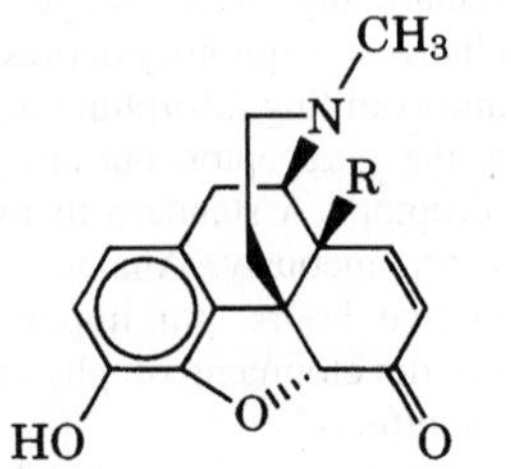

(4, R = H) (4, R = OH)

| oxymorphone | $C_{17}H_{19}NO_4$ | |
| phenazocine | $C_{22}H_{27}NO$ | |

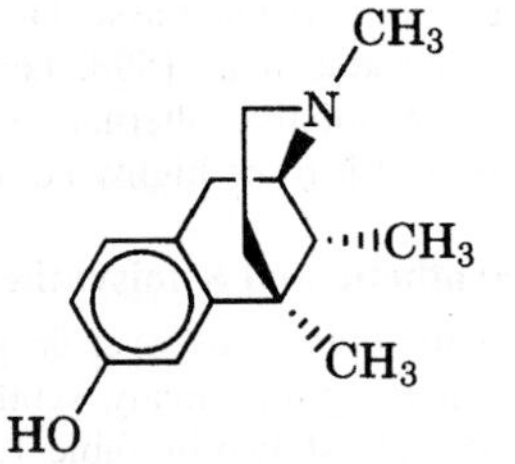

(5)

| fentanyl | $C_{22}H_{29}NO_2$ | |

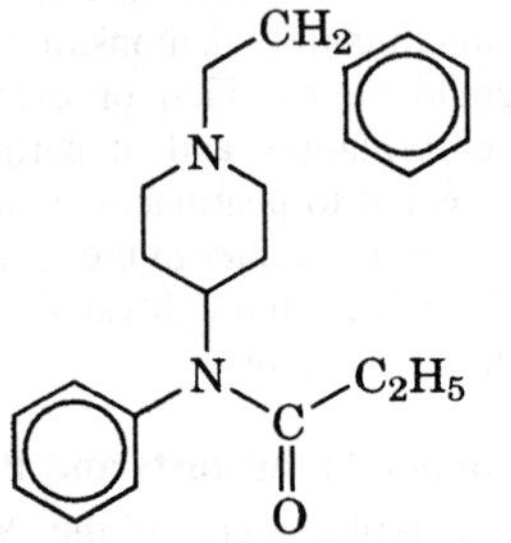

(6)

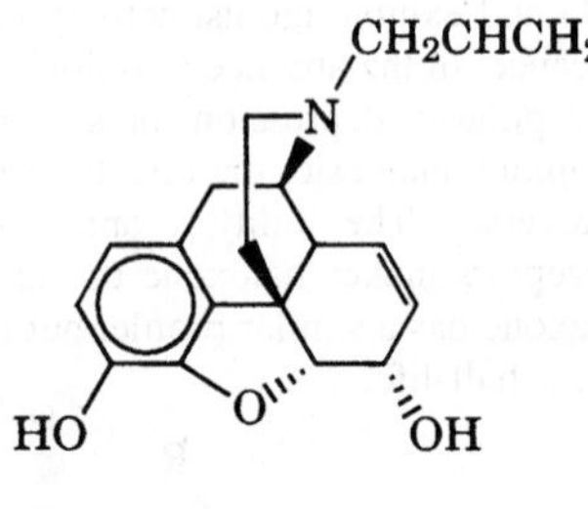

(8)

$$R = CH_2 - \triangleleft$$

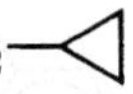

The quest for compounds that combined the analgesic properties of morphine, were nonaddictive and lacked side effects, led to the development of the drugs that have both agonist and antagonist activities. Nalbuphine and buprenorphine are semisynthetic materials derived from oxymorphone and thebaine respectively, whereas pentazocine and butorphanol are benzomorphan and morphan derivatives. Although structurally similar, they display different receptor affinities: pentazocine is a weak μ-antagonist, but a strong agonist of the κ-receptor; nalbuphine is a competitive antagonist for the μ-receptor, blocking the effects of the morphinelike drugs, but is a partial agonist for the κ and ς receptors; and buprenorphine is a partial agonist for the μ-receptor.

Pharmacologically, the effects of these drugs resemble those of opioid agonists. All four have analgesic potency equal to or greater than morphine and, like morphine, they cause respiratory depression. A ceiling effect is reached, however, above which increased doses do not increase respiratory depression or do not produce proportionally greater depression.

Other Analgesic Agents

Most analgesic agents rely on agonism of the μ receptor for their activity, however the ability of the κ and δ receptors to induce analgesia is also well documented. Some nonmorphine analgesics which may preferentially bind to κ and δ receptors are found in Table 2.

Antiinflammatories and Antipyretics

Most of the time, the powerful analgesia supplied by morphine and the other opioid analgesics is not needed. Rather, a mild analgesic, such as aspirin, the most commonly employed analgesic agent, can be used for the treatment of simple pain associated with headaches, minor muscle pain, mild trauma, arthritis, cold and flu symptoms, and fever.

Aspirin, the oldest of the nonsteroidal antiinflammatory drugs (NSAIDs), is a member of the salicylate group.

Aspirin's pain relief results, not through direct action on the central nervous system, but rather through peripheral action. It has been proposed that aspirin and aspirinlike drugs inhibit the enzymatic production of prostaglandins, a group of endogenous agents which are well known to cause erythema, edema, pain, and fever. Aspirin does not act as a prostaglandin receptor antagonist; rather it blocks the cyclooxygenase enzymecatalyzed conversion of arachidonic acid, $C_{20}H_{32}O_2$, to cyclic endoperoxides, a prostaglandin precursor.

The action of endogenous pyrogens on the hypothalamus produces fever, because of a readjustment in the central set point controlling the body's internal temperature. Salicylates and other NSAIDs achieve their antipyretic effect by controlling the prostaglandin-induced release of pyrogens.

A second class of NSAIDs, the so-called coal-tar analgesics, are derived from acetanilide. Although it is no longer used therapeutically, its analogues, phenacetin and the active metabolite, acetaminophen, are effective alternatives to aspirin. They have analgesic and antipyretic effects that do not differ significantly from aspirin, but they do not cause the gastric irritation, erosion, and bleeding that may occur after salicylate treatment.

TABLE 2. NONMORPHINE ANALGESICS

Name	Molecular formula	Structure
acetorphan	$C_{21}H_{23}NO_4S$	

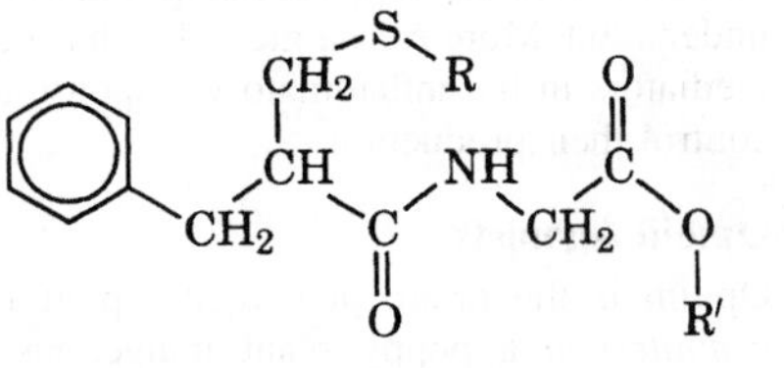

(9, R = COCH$_3$, R' = CH$_2$C$_6$H$_5$) (9, R = H, R' = H)

| thiorphan | $C_{12}H_{15}NO_3S$ | |
| spiradoline | $C_{22}H_{30}Cl_2N_2O_2$ | |

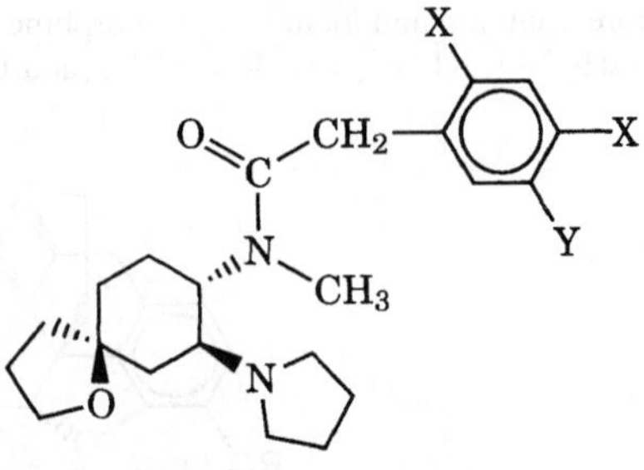

(10, X = H, Y = Cl)

In contrast to aspirin, however, they are not cyclooxygenase inhibitors and have no antiinflammatory properties. Clinically, acetaminophen is preferred over phenacetin, because it has less overall toxicity.

A more recently introduced, nonprescription analgesic is the aryl propionic acid, ibuprofen, which offers significant advantages over aspirin. Ibuprofen, a cyclooxygenase inhibitor, displays good antiinflammatory activity. It is more potent than aspirin and has a lower incidence of gastrointestinal irritation, although at high doses or chronic exposure, gastric irritation, as well as some renal toxicity, has been observed. Ibuprofen is more effective than propoxyphene in relieving episiotomy pain, pain following dental extractions, and menstrual pain.

The adrenal cortex produces steroidal hormones that are associated with carbohydrate, fat, and protein metabolism, electrolyte balance, and gonadal functions. One of these, cortisone, $C_{21}H_{28}O_5$ (11), demonstrated a remarkable ability to relieve the symptoms of inflammatory conditions. Other glucocorticoid steroids, such as dexamethasone, $C_{22}H_{29}FO_5$ (12, R = F, R' = CH_3), and prednisolone, $C_{21}H_{28}O_5$ (12, R = R' = H), also have antiinflammatory properties.

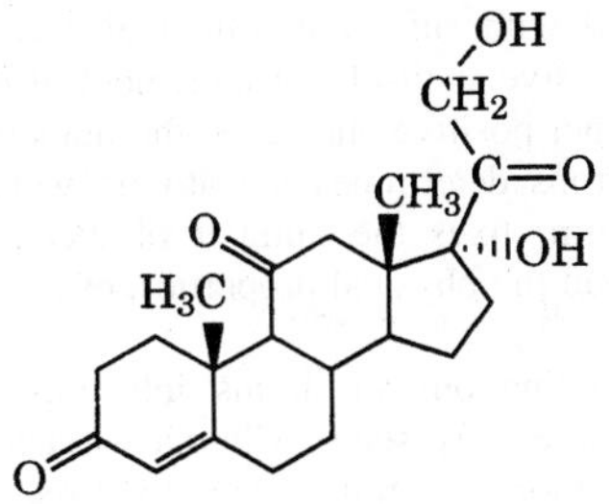

(11)

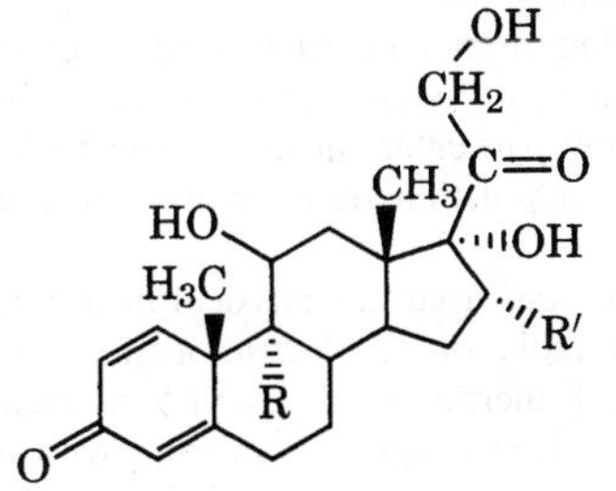

(12)

These steroids are capable of preventing or suppressing the development of the swelling, redness, local heat, and tenderness which characterize inflammation.

Unfortunately steroids merely suppress the inflammation while the underlying cause of the disease remains. Another serious concern about steroids is that of toxicity. The abrupt withdrawal of glucocorticoid steroids results in acute adrenal insufficiency. Long-term use may induce osteoporosis, peptidic ulcers, the retention of fluid, or an increased susceptibility to infections. Because of these problems, steroids are rarely the first line of treatment for any inflammatory condition, and their use in rheumatoid arthritis begins after more conservative therapies have failed.

Economic Aspects

Analgesics and antiarthritics represent significant worldwide pharmaceutical markets. Table 3 lists trade names and producers of some of the more commercially important agents.

See also **Alkaloids; Aspirin** and **Morphine**.

RICHARD A. NUGENT
CHARLES M. HALL
Upjohn Company

Additional Reading

Buschmann, H., and T. Christoph: *Analgesics: From Chemistry and Pharmacology to Clinical Application*, John Wiley & Sons, Inc., New York, NY, 2002.

Gallin, J.I.I. M. Goldstein, and R. Snyderman, eds., *Inflammation: Basic Principles and Clinical Correlates*, Raven Press, New York, NY, 1992.

Herz, A. ed.: Opioids I, in *Handbook of Experimental Pharmacology*, Vol. 104, Springer-Verlag, New York, NY, 1993.

TABLE 3. TRADE NAMES AND PRODUCERS OF ANALGESICS, ANTIPYRETICS, AND ANTIINFLAMMATORY DRUGS

Compound name	Trade name	Producer
levorphanol	Dromoran	Hoffmann-LaRoche
phenazocine	Prinadol	Smith Kline & Beecham
meperidine	Demerol	Winthrop
propoxyphene	Darvon	Eli Lilly
naloxone	Narcan	Du Pont
naltrexone	Trexan	Du Pont
nalbuphine	Nubain	Du Pont
buprenorphine	Buprenex	Norwich Eaton
pentazocine	Talwin	Winthrop
butorphanol	Stadol	Bristol
acetaminophen	Tylenol	McNeil
ibuprofen	Motrin	Upjohn
naproxen	Naprosyn	Syntex
ketoprofen	Orudis	Wyeth
flurbiprofen	Ansaid	Upjohn
indomethacin	Indocin	Merck Sharpe & Dohme
sulindac	Clinoril	Merck Sharpe & Dohme
diclofenac	Voltaren	CIBA-GEIGY
meclofenamate	Meclomen	Parke-Davis
piroxicam	Feldene	Pfizer
diflunisal	Dolobid	Merck Sharpe & Dohme

Herz, A. ed.: "Opioids II," in *Handbook of Experimental Pharmacology*, Vol. 104, Springer-Verlag, New York, NY, 1993.

Jacobs, B.L.: *Aspirin Alternatives: The Top Natural Pain-Relieving Analgesics*, B L Publications, Aliso Viejo, CA, 1999.

Lombardino, J.G. ed.: *Nonsteroidal Antiinflammatory Drugs*, John Wiley & Sons, Inc., New York, NY, 1985.

Omoigui, S.: *Sota Omoigui's Pain Drugs Handbook*, 2nd Edition, Blackwell Science, Inc. Malden, MA, 1999.

Staff: *Drug Facts and Comparisons*, 54th Edition, Facts And Comparisons, St. Louis, MO, 1999.

ANALYSIS (Chemical). Analytical chemistry is that branch of chemistry concerned with the detection and identification of the atoms, ions, or radicals (groups of atoms which react as a unit) of which a substance is composed, the compounds which they form, and the proportions of these compounds present in a given substance. The work of the analyst begins with sampling, since analyses are performed upon small quantities of material. The validity of the result depends upon the procurement of a sample that is representative of the bulk of material in question (which may be as large as a carload or tankload).

The Revolution in Analytical Chemistry—A Perspective

During the last few decades, not many branches of any science have undergone so much change in the equipment and procedures used as has the field of *chemical analysis*. This revolution also has impacted on how the principles of chemistry are taught today. The revolution in analysis also has had wide influence on technology in general because of the far greater accuracy with which chemical determinations can be made. Just a few years ago, analyses that would yield reliable data in the range of a few parts per million (*ppm*) were considered excellent. The accurate reporting of parts per billion (*ppb*) was achieved in the 1960s. With modern analytical instrumentation available beginning in the 1990s, a part per trillion ($\frac{1}{10^{12}}$) sensitivity was achieved for some routine analyses, as, for example, in determinations of the dangerous pollutant, *dioxin*. A special tandem-accelerator mass spectrometer now can detect three atoms of ^{14}C in the presence of 10^{16} atoms of ^{12}C in a radiocarbon age dating procedure. (It is interesting to note that a pinhead would occupy a part per trillion of the area of a road from New York to California; 10^{12} molecules of molecular weight 600 weigh only 10^{-9} gram.)

While the impact of vastly improved chemical analysis has been felt by essentially all phases of science, dramatically more precise data have been of notable significance in the area of pollutants and pharmaceuticals. The effects of minute impurities, beyond detection just a few years ago, now can be determined and become the basis for pollution and drug legislation, litigation, etc. In some cases, unfortunately, the long-term effects of impurities in substances and in the environment remains a pseudoscience of statistics. Consequently instrumentally yielded analytical chemical information requires caution and prudence in its application to decision making at a policy level.

Traditional Analytical Chemistry

In the interest of putting modern analytical chemistry in perspective, it is in order to review that long time period (essentially prior to the 1940s) when the subject was divided into two readily understood areas:
1. *Qualitative chemical analysis*, in which one is concerned simply with the identification of the constituents of a compound or components of a mixture, sometimes accompanied by observations (rough estimates) of whether certain ingredients may be present in major or trace proportions.

2. *Quantitative chemical analysis*, which is concerned with the amounts (to varying degrees of precision) of all, or frequently of only some, specific ingredients of a mixture or compound. Classically, quantitative chemical analysis is divided into (a) *gravimetric analysis* wherein weight of sample, precipitates, etc., is the underlying basis of calculation, and (b) *volumetric analysis* (titrimetric analysis) wherein solutions of known concentration are reacted in some fashion with the sample to determine the concentration of the unknown. Obviously, the Figure from either gravimetric or volumetric determinations are convertible and the two methodologies frequently are combined in a multistep analytical procedure.

Classical laboratory, manual methods conducted on a macro-scale where sample quantities are in the range of grams and several milliliters. These are the techniques that developed from the earliest investigations of chemistry and which remain effective for teaching the fundamentals of analysis. However, these methods continue to be widely used in industry and research, particularly where there is a large variety of analytical work to be performed. The equipment, essentially comprised of analytical balances and laboratory glassware, tends to be of a universal nature and particularly where budgets for apparatus are limited, the relative modest cost of such equipment is attractive.

A Gradual Break from Tradition

One of the first breaks from traditional analytical chemistry was the addition of *microchemical methods*. These methods essentially extended macro-scale techniques so that they could be applied for determinations involving very small (milligram) quantities of samples. These methods required fully new approaches or extensive modifications of macro-scale equipment. Consequently, the apparatus usually was sophisticated, relatively costly, and required, greater manipulative skills. Nevertheless, microchemical methods opened up entirely new areas of research, making possible the determination of composition where the availability of samples, as in many areas of biochemistry, was confined to very small quantities.

A second major break was the introduction of *semi-automated analytical apparatus* that introduced an interim step between (a) macro-scale and microchemical analysis techniques on the one hand and (b) fully instrumented and automated analytical methods on the other hand. Significant design changes in chemical balances that greatly increased the speed of weighing samples and reagents and automatic and self-refilling burettes are examples of ways in which an analytical procedure could be "tooled" to conserve technician power, reduce drudgery, and often contribute to more reliable and precise results.

Another major break was the introduction of *process analyzers*, which moved the chemical control laboratory from a central location in a materials-manufacturing plant to the use of chemical analyzers *on-line*. With this concept, quality control no longer depended upon grab-samples, analyzed periodically and thus always behind (time lag) conditions actually occurring in the process itself at any given instant. Many analytical instruments today, at least in principle, are applicable to on-line installation. While thousands of on-line analyzers are in place, usage throughout the processing and manufacturing industries is far from universal. Difficulties in designing and protecting sensitive instrumentation from the very rugged environments encountered on-line continue. Thus, chemical composition is commonly inferred from other related measurements, such as temperature, pressure, and careful chemical analysis of raw materials at the input side and similar analyses of products on the output side. Numerous techniques used for on-line instrumentation are described in this encyclopedia.

Energy—Matter Interactions in Analytical Instrumentation

Modern chemical analyzers, ranging from research and laboratory applications to process control, developed in a rather chaotic manner over several decades. There indeed was no master plan and, in fact, it was not until the late 1950s that a concerted attempt was made to classify analytical instruments in a scientific way. Table 1 is an updated, but abridged version of a summary prepared and first published in 1957[1]. The thrust of the summary is directed toward industrial instrumentation, although it embraces the principal laboratory instruments as well.

Chemical-composition variables are measured by observing the interactions between matter and energy. That such measurements are possible stems from the fundamental that all known matter is comprised of complex, but systematic arrangements of particles which have mass and electric charge. Thus, there are neutrons that have mass but no charge; protons that have essentially the same mass as neutrons with a unit positive charge; and electrons that have a negligible mass with a unit negative charge. The neutrons and protons comprise the nuclei of atoms. Each nucleus ordinarily is provided with sufficient orbital electrons, in what is often visualized as a progressive shell-like arrangement of different energy levels, to neutralize the net positive charge on the nucleus. The total number of protons plus neutrons determines the atomic weight. The number of protons, which, in turn, fixes the number of electrons, determines the chemical properties and the physical properties, except mass, of the resulting atom.

The chemical combinations of atoms into molecules involve only the electrons and their energy states. Chemical reactions involving both structure and composition generally occur by loss, gain, or sharing of electrons among the atoms. Thus, every configuration of atoms in a molecule, crystal, solid, liquid, or gas may be represented by a specific system of electron energy states. Also, the particular physical state of the molecules, as resulting from their mutual arrangement, also is reflected upon these energy states. Fortunately, these energy states, characteristic of the composition of any particular substance, can be inferred by observing the consequences of interaction between the substance and an external source of energy.

External energy sources used in analytical instrumentation include:
1. electromagnetic radiation; 2. electric or magnetic fields; 3. chemical affinity or reactivity; 4. thermal energy; and 5. mechanical energy.

The interaction of electromagnetic radiation with matter yields fundamental information as the result of the fact that photons of electromagnetic radiation are emitted or absorbed whenever changes take place in the quantized energy states occupied by the electrons associated with atoms and molecules. X-rays (photons or electromagnetic wave packets with relatively high energy) penetrate deeply into electron orbits of an atom and provide, upon absorption, the large quantity of energy required to excite one of the innermost electrons. Thus, the pattern of x-ray excitation or absorption is relative to the identity of those atoms whose orbital electrons are excited, ideally suiting x-ray techniques for determining atoms and elements in dense samples. But, because of the penetrating power of x-rays, they are not suited to the excitation of low-energy states that correspond to outer-shell or valence electrons; or of the inter-atomic bonds which involve vibration or rotation.

In contrast, the relatively longer wavelengths of infrared radiation (photons having relatively low energy) correspond to the energy transformations involved in the vibration of atoms in a molecule as resulting from stretching or twisting of the inter-atomic bonds. Thus, because the penetrating power of electromagnetic radiation varies over the total spectrum, an instrumental irradiation technique can be developed for almost any analytical instrumentation requirement.

The interaction of matter with electric or magnetic fields is widely applied for determining chemical composition. The mass spectrometer, for example, which uses a combination of electric and magnetic fields to sort out constituent ions in a sample, takes full advantage of this interaction. A simple electric-conductivity apparatus determines ions in solution as the result of applying an electric potential difference across an electrolyte.

---

[1] Albright, C.M., Jr., "Chemical Composition," in *Process Instruments and Controls Handbook*, D.M. Considine, Editor, McGraw-Hill, New York, 1957. The book is now in its 5th edition (1999).

TABLE 1. INTERACTIONS BETWEEN ENERGY AND MATTER UTILIZED IN ANALYTICAL INSTRUMENTATION

GROUP I—INTERACTIONS WITH ELECTROMAGNETIC RADIATION

Measurement of the quantity and quality of electromagnetic radiation emitted, reflected, transmitted, or diffracted by the sample.

Electromagnetic radiation varies in energy with radiation frequency, that of the highest frequency or shortest wavelength having the highest energy and penetration into matter. Radiation of the shortest wavelengths (gamma rays) interacts with atomic nuclei; x-rays with the inner shell electrons; visible and ultraviolet light with valence electrons and strong interatomic bonds; and infrared radiation and microwaves with the weaker interatomic bonds and with molecular vibrations and rotation. Most of these interactions are structurally related and unique. They may be used to detect and measure the elemental or molecular composition of gas, liquid, and solid substances within the limitations of available equipment.

Emitted Radiation
Thermally Excited: Optical emission spectro-chemical analysis
  Flame photometry
Electro-magnetically Excited: Fluorescence
  Raman spectro-photometry
  Induced radioactivity
  X-ray fluorescence
Transmitted and Reflected Radiation
  X-ray analysis
  Ultraviolet spectro-photometry
  Ultraviolet absorption analysis
  Conventional photometry—transmission colorimetry
  Colorimetry
  Light scattering techniques
  Optical rotation-polarimetry
  Refractive index
  Infrared spectro-photometry
  Infrared process analyzers
  Microwave spectroscopy
  Gamma ray spectroscopy
  Nuclear quadrupole moment

GROUP II—INTERACTION WITH CHEMICALS

Measurement of the results of reaction with other chemicals in terms of amount of sample or reactant consumed, product formed, or thermal energy liberated, or determination of equilibrium attained.

The selectivity inherent in the chemical affinity of one element or compound for another, together with their known stoichiometric and thermodynamic behavior, permits positive identification and analysis under many circumstances. In a somewhat opposite sense, the apparent dissociation of substances at equilibrium in chemical solution gives rise to electrically measurable valence potentials, called oxidation-reduction potentials, whose magnitude is indicative of the concentration and composition of the substance. While individually all the above effects are unique for each element or compound, many are readily masked by the presence of more reactive substances so they can be applied only to systems of known composition limits.

Consumption of Sample or Reactant
  Orsat analyzers
  Automatic titrators
Measurement of Reaction Products
  Impregnated paper-tape devices
  Photometric reaction product analyzers
Thermal Energy Liberation
  Combustion-type analyzers
  Total combustibles analyzers (hydrocarbons and carbon monoxide analyzers)

Equilibrium Solution Potentials
  Redox potentiometry
  pH (hydrogen ion concentration)
  Metal ion equilibria

GROUP III—REACTION TO ELECTRIC AND MAGNETIC FIELDS

Measurement of the current, voltage, or flux changes produced in energized electric and magnetic circuits containing the sample.

The production of net electric charge on atoms or molecules by bombardment with ionizing particles or radiation or by electrolysis or dissociation in solution or the induction of dipoles by strong fields establishes measurable relationships between these ionized or polarized substances and electric and magnetic energy. Ionized gases and vapors can be accelerated by applying electric fields, focused or deflected in magnetic fields, and collected and measured as anelectric current in mass spectroscopy. Ions in solution can be transported, and deposited if desired, under the influence of various applied potentials for coulometric or polarographic analysis and for electrical conductivity measurements. Inherent and induced magnetic properties give rise to specialized techniques, such as oxygen analysis based on its paramagnetic properties and nuclear magnetic resonance, which is exceedingly precise and selective for determination of the compounds of many elements.

Mass Spectroscopy
  Quadrupole mass spectrometry
Electrochemical
  Reaction product analyzers
Electrical Properties
  Electrical conductivity/electrical resistivity
  Dielectric constant and loss factor
  Oscillometry
  Gaseous conduction
Magnetic Properties
  Paramagnetism
  Nuclear magnetic resonance
  Electron paramagnetic resonance

GROUP IV—INTERACTION WITH THERMAL OR MECHANICAL ENERGY

Measurement of the results of applying thermal or mechanical energy to a sample in terms of energy transmission, work-done, or changes in physical state.

The thermodynamic relationship involving the physical state and thermal energy content of any substance permits analysis and identification of mixtures of solids, liquids, and gases to be based on the determination of freezing or boiling points and on the quantitative measurement of physically separated fractions. Useful information can often be derived from thermal conductivity and viscosity measurements, involving the transmission of thermal and mechanical energy, respectively.

Effects of Thermal Energy
  Thermal conductivity
  Melting and boiling point determinations
  Ice point-humidity instrumentation, among others
  Dew point-humidity instrumentation, among others
  Vapor pressure
  Fractionation
  Chromatography
  Thermal expansion
Effects of Mechanical Energy or Forces
  Viscosity
  Sound velocity
  Density and specific gravity

The use of chemical reactions in analytical instrumentation essentially extends the fundamental techniques of older laboratory analytical methods.

Numerous analytical instrumentation techniques involve interactions between mechanical and thermal energy with matter. All of these interactions are summarized in the accompanying table.

Targets of Analytical Instrumentation

Some authorities feel that less emphasis and even abandonment by some educational institutions of the traditional qualitative-analysis (wet basis) course represents the loss of a great learning experience in the fundamentals of chemistry. Generally, for teaching purposes, the course is limited to inorganic substances. Practically all of the fundamentals of inorganic chemistry are called upon in the execution of qualitative analysis. Thus, in addition to serving as an effective analytical procedure, the method is an effective teacher.

The first important step is that of putting the sample (unknown) into solution. For metals and alloys, strong acids, such as HCl, HNO$_3$, or aqua regia may be used. If the material is not fully dissolved by these acids, it should be fused, either with sodium carbonate (alkaline fusion) or potassium acid sulfate (acid fusion). Care should be exercised to make certain that no portion of the unknown is volatilized and thus lost during these procedures.

The next step is the detection of the cations of the metals. For this purpose, the solution should be treated with HNO$_3$, by evaporation and re-dissolving if necessary, to remove other acid radicals, so that nitrate is the only anion present in the solution. Then, a systematic procedure is followed for separation of groups of the cations. Such schemes of separation have been devised for all the metals found in nature. A shortened plan, which applies to 24 of the commonly occurring metals and ammonium, has been known and practiced for many years. This plan consists of the separation of the 24 metals into five groups. Further details of the procedure are given in the 6th Edition of this encyclopedia.

Criteria for Selecting Appropriate Analytical Method

These include (1) sensitivity, (2) specificity, (3) speed, (4) sampling methodology re quired, (5) simplicity (translated into terms of expertise needed), and, of course, cost as traded off against the other criteria.

Sensitivity, as previously mentioned, has improved almost astoundingly over the past few years. Of course, great sensitivity is not always needed, with the ppm and ppb levels well serving many industrial and laboratory requirements. Most frequently, sensitivity is closely related to cost. Sensitivity is also closely related to accuracy, i.e., sensitivity of a reliable, repeatable nature. Sensitivity depends on how well the target of measurement can be transduced into some reliable signal from which the instrument can create a display and/or record. The absorption or emission of photons is the basis of many spectroscopic analytical methods, such as x-ray, ultraviolet, infrared, nuclear magnetic resonance, Raman, Mossbauer, etc., as well as of charged particles, which serve as the basis for electron and mass spectrometry and of the electrochemical, flame ionization, etc. used in chromatography. Through the use of intense energy sources, such as lasers, synchrotron radiation, and plasmas, the efficiency of converting (transducing) the objective parameter of analysis (*analyte*) is greatly improved. McLafferty reports that efficiencies approaching 100% have been experienced for resolution-enhanced multi-photon ionization of atomic and molecular species. Multiplier detectors can respond to the arrival of a single photon or ion. Such methods can detect, for example, a single cesium atom or naphthalene molecule.

Specificity is an ever-present criterion because there are indeed few analytical techniques that detect single species without careful tuning. Frequently, filtering techniques must be used as a means of narrowing the range of detection. See **Infrared Radiation**; and **Ultraviolet Spectrometers**.

Speed. The rapidity with which an analysis can be performed and utilized (including interpretation, whether manual or automatic) is particularly important in industrial chemical analysis. In a laboratory setting, this may not be quite so urgent, but even then time is a major criterion where, in most cases, special personnel are held up in other activities, awaiting the results of an analysis. Frequently higher cost can be justified on the basis of less time and lower personnel costs per analysis made.

One of the principal contributions of electronic data processing over the past several years in terms of chemical analysis is the saving of manual effort in interpreting analytical data. Special techniques, such as Fourier transform, have increased speed (as well as sensitivity) by orders of magnitude in connection with infrared, nuclear magnetic resonance, and mass spectroscopy. Of course, for on-line process analyses, essentially instantaneous interpretation is required to provide the proper error signal that is used to position the final control element (valve, feeder, damper, etc.).

Sampling for analysis is sometimes considered a secondary criterion in the selection of an analysis system. In the laboratory or on the process, sampling often is the practical key to success. The entire result can depend upon obtaining a *truly representative* sample and the sampling methodology used varies widely with the materials to be analyzed. Sampling of solids, such as coal, metals, etc. differs markedly from sampling for fluids. Particularly in process analyses, where the environment varies and contrasts markedly with the usual laboratory conditions, a gas or liquid will require filtering and temperature and pressure conditioning—so that the composition detector will consistently be exposed to the material in question under the same physical conditions.

Additional Reading

Adamovics, J.A., Editor: *Chromatographic Analysis of Pharmaceuticals,* 2nd Edition, Marcel Dekker, Inc., New York, NY, 1999.

Adlard, E.R.: *Chromatography in the Petroleum Industry,* Elsevier Science, New York, NY, 1995.

Alfassi, Z.B.: *Instrumental Multi-Element Chemical Analysis,* Kluwer Academic Publishers, New York, NY, 1998.

Altgelt, K.H. and M.M. Boduszynski: *Composition and Analysis of Heavy Petroleum Fractions, Chemical Industries,* Marcel, Inc., New York, NY, 1999.

Anderson, D.J.: "Analysis in Clinical Chemistry," *Analytical Chemistry,* **165R** (June 15, 1991).

Bequette, W.B.: *Process Dynamics,* Modeling, Analysis and Simulation, Prentice-Hall, Inc., Upper Saddle River, NJ, 1997.

Bialkowski, S.E., Editor: *Photothermal Spectroscopy Methods for Chemical Analysis,* John Wiley & Sons, Inc., New York, NY, 1995.

Brettell, T.A. and R. Saferstein: "Analysis in Forensic Science," *Analytical Chemistry,* 148R (June 15, 1991).

Bruno, T.J., and P.D. Svoronos: *Handbook of Basic Tables for Chemical Analysis,* 2nd Edition, CRC Press, LLC., Boca Raton, FL, 2003.

Chalmers, J.M.: *Industrial Analysis with Vibrational Spectroscopy,* American Chemical Society, Washington, DC, 1997.

Cherry, R.H.: "Thermal Conductivity Gas Analyzers," in *Process/Industrial Instruments and Controls Handbook,* D.M. Considine, Editor, McGraw-Hill, New York, NY, 1993.

Converse, J.G.: "Sampling for On-Line Analyzers," in *Process/Industrial Instruments & Controls Handbook,* D.M. Considine, Editor, 4th Edition, McGraw-Hill, New York, NY, 1993.

Converse, J.G.: "Process Chromatography," in *Process/Industrial Instruments & Controls Handbook,* D.M. Considine, Editor, 4th Edition, McGraw-Hill, 1993.

Dean, J.A., Editor: *Analytical Chemistry Handbook,* McGraw-Hill, New York, NY, 1995.

Duncan, T.M. and J.A. Reimer: *Chemical Engineering Design and Analysis, An Introduction,* Cambridge University Press, New York, NY, 1998.

Eberhart, J.P.: *Structural and Chemical Analysis of Materials,* John Wiley & Sons, Inc., New York, NY, 1999.

Ewing, G.W.: *Analytical Instrumentation Handbook,* 2nd Edition, Marcel Dekker, Inc., New York, NY, 1997.

Fawcett, A.H.: *Polymer Spectroscopy,* John Wiley & Sons, Inc., New York, NY, 1996.

Foucault, A.P., Editor: *Centrifugal Partition Chromatography,* Maecel Dekker, Inc., New York, NY, 1995.

Glajch, J.L. and L.R. Snyer, Editors: *Computer-Assisted Method Development for High-Performance Liquid Chromatography,* Elsevier Science, New York, NY, 1990.

Hardcastle, W.A.: *Qualitative Analysis,* A Guide to Best Practice, Springer-Verlag Inc., New York, NY, 1998.

Harman, J.N. and D.M. Gray: "pH and Redox Potential Measurements," in *Process/Industrial Instruments & Controls Handbook,* D.M. Considine, Editor, 4th Edition, McGraw-Hill, New York, NY, 1993.

Harris, D.C.,: *Exploring Chemical Analysis,* 3rd Edition, W.H. Freeman & Co., New York, NY, 2004.

Harris, D.C.,: *Quantitative Chemical Analysis,* 5th Edition, W.H. Freeman & Co., New York, NY, 1998.

Hatakeyama, T. and F.X. Quinn: "Thermal Analysis," *Fundamentals and Applications to Polymer Science,* 2nd Edition, John Wiley & Sons, Inc., New York, NY, 1999.

Hollas, J.M.: *Modern Spectroscopy,* 4th Edition, John Wiley & Sons, Inc., Hoboken, NJ, 2004.

Holtzelaw, H.F., W.R. Robinson, and J.D. Odom: *General Chemistry with Qualitative Analysis,* 10th Edition, Houghton Mifflin Co. (College Division), New York, NY, 1996.

Ito, Y., W.D. Conway, and Y. Ato, Editors: *High-Speed Countercurrent Chromatography* (Chemical Analysis, Vol. 132) John Wiley & Sons, New York, NY, 1995.

Kenneth, W.W., R.E. Davis, and P.M. Larry: *General Chemistry with Qualitative Analysis,* 6th Edition, Saunders College Publishing, Philadelphia, PA, 1999.

Kohlmann, F.: "Electrical Conductivity Measurements," in *Process/Industrial Instruments & Controls Handbook,* D.M. Considine, Editor, 4th Edition, McGraw-Hill, New York, NY, 1993.

Levie, R.De: *Principles of Quantitative Chemical Analysis,* McGraw-Hill Companies, New York, NY, 1996.

Lex, D.: Turbidity Measurement, in *Process/Industrial Instruments & Controls Handbook,* D.M. Considine, Editor, 4th Edition, McGraw-Hill, New York, NY, 1993.

McLafferty, F.W.: "Trends in Analytical Instrumentation," *Science,* **226,** 251–253 (1984).

Montaudo, G. and P.P. Lattimer: *Mass Spectrometry,* CRC Press, LLC, Boca Raton, FL, 1999.

Palumbo, S., P. Fratamico, and M. Tunick: *New Techniques in the Analysis of Foods,* Kluwer Academic/Plenum Publishers, New York, NY, 1999.

Pare, J.R. and J.M. Belanger: *Instrumental Methods in Food Analysis,* Techniques and Instrumentation in Analytical Chemistry, Vol. 18, Elsevier Science, New York, NY, 1997.

Patnaik, P.: *Handbook of Environmental Analysis, Chemical Pollutants in Air, Water, Soil, and Soil Wastes,* Lewis Publishers, New York, NY, 1997.

Patnaik, P.: *Dean's Analytical Chemistry Handbook,* 2nd Edition, The McGraw-Hill Companies, Inc., New York, NY, 2004.

Poppiti, J.A.: *Practical Techniques for Laboratory Analysis,* Lewis Publishers, New York, NY, 1999.

Rubinson, J.F. and K.A. Rubinson: *Contemporary Chemical Analysis,* Prentice-Hall, Inc., Upper Saddle River, NJ, 1998.

Saltzman, R.S.: "Gas and Process Analyzers," in *Process/Industrial Instruments & Controls Handbook*, D.M. Considine, Editor, 4th Edition, McGraw-Hill, New York, NY, 1993.

Settle, F.A.: *Handbook of Instrumental Techniques for Analytical Chemistry*, Prentice-Hall, Inc., Upper Saddle River, NJ, 1997.

Sibilla, J.P.: *Guide to Materials Characterization and Chemical Analysis*, 2nd Edition, John Wiley & Sons, Inc., New York, NY, 1996.

Staff: *Polymer Analysis/Polymer Physics*, Springer-Verlag Inc., New York, NY, 1996.

Tipping, F.T.: "Oxygen Determination," in *Process/Industrial Instruments and Controls Handbook*, D.M. Considine, Editor, 4th Edition, McGraw-Hill, New York, NY, 1993.

Tranter, R.: *Design and Analysis in Chemical Research*, CRC Press, LLC, Boca Raton, FL, 2000.

Yazbak, G.: "Refractomers," in *Process/Industrial Instruments & Controls Handbook*, D.M. Considine, Editor, 4th Edition, McGraw-Hill, New York, NY, 1993.

Zerbi, G. and H.W. Siesler: *Modern Polymer Spectroscopy*, John Wiley & Sons, Inc., New York, NY, 1999.

ANALYSIS (Organic Chemical).

Various techniques are used in the chemical analysis of organic substances both in microanalysis and macro laboratory procedures. As contrasted with the determination of total carbon content or the amounts of other specific chemical elements, the representative analytical techniques described here are directed toward the determination of presence and amount of various functional groups (radicals). These groups also are described elsewhere in this volume and, in several instances, additional analytical procedures are related.

(1) The *carboxyl group* is determined by titration with standard sodium hydroxide solution, using phenolphthalein as the indicator, by the reaction

$$RCOOH + NaOH \longrightarrow RCOONa + H_2O$$

(2) The *hydroxyl group* is determined by reaction with acetic anhydride on heating in a sealed tube by the reaction

$$ROH + (CH_3CO)_2O \longrightarrow ROOCCH_3 + CH_3COOH$$

The amount of hydroxyl group present is found by titrating the resulting acetic acid (CH_3COOH) with standard sodium hydroxide, as in (1).

(3) The *acyl group* ($-COOR$) in esters and amides is determined by hydrolysis in alcoholic sodium hydroxide solution, followed by ion exchange with an acidic resin. The carboxylic acid formed is then titrated with standard sodium hydroxide, as in (1). The reactions are

$$\underset{\text{Ester}}{RCOOR'} + NaOH \longrightarrow RCOONa + R'OH$$

or

$$\underset{\text{Amide}}{RCONHR'} + NaOH \longrightarrow RCOONa + R'NH_2 + H_2O$$

$$RCOONa + Resin\text{-}SO_3H \longrightarrow RCOOH + Resin\text{-}SO_3Na$$

(4) The *carbonyl group* is determined by a reaction with 2,4-dinitrophenyl hydrazine which precipitates the 2,4-dinitrophenyl hydrazone of the aldehyde or ketone, which is then filtered off, dried, and weighed. The reaction is

$$RR'CO + H_2NNHC_6H_3(NO_2)_2 \longrightarrow$$
$$RR'C{:}NNHC_6H_3(NO_2)_2 + H_2O$$

(5) The *peroxy group* is determined by treatment with sodium iodide. The liberated iodine is then titrated with standard sodium thiosulfate solution. The reaction is

$$RCOO_2OCR + 2\ NaI \longrightarrow I_2 + 2\ R'COONa$$

(6) The *primary amino group* is determined by treatment with nitrous acid and measurement of the nitrogen (gas) produced by the reaction

$$RNH_2 + HNO_2 \longrightarrow N_2 + ROH + H_2O$$

(7) The aromatic *nitro group* is determined by its reduction with excess titanium(III) chloride. After the reaction, the unused titanium(III) ions (Ti^{3+}) are determined by titration with iron(III) sulfate or iron alum solution:

$$RNO_2 + 6\ TiCl_3 + 6\ HCl \longrightarrow RNH_2 + 6\ TiCl_4 + 2\ H_2O$$
$$Ti^{3+} + Fe^{3+} \longrightarrow Ti^{4+} + Fe^{2+}$$

(8) The *hydrazino group* is determined by oxidation with copper(II) sulfate solution, and measurement of the nitrogen (gas) formed. The reaction is

$$RNHNH_2 + 4\ CuSO_4 + H_2O \longrightarrow$$
$$N_2 + ROH + 2\ Cu_2SO_4 + H_2SO_4$$

(9) The *sulfhydryl group* is determined by reaction with iodine, which is produced in the vessel from potassium iodide, added in excess to the solution, and potassium iodate, added from a buret until the completion of the reaction is shown by the permanent appearance of the blue color of starch–iodine.

$$2\ RSH + I_2 \longrightarrow RSSR + 2\ HI$$

(10) *Unsaturated groups* are determined by addition of bromine, by the reaction

$$R_2C{=}CR'_2 + Br_2 \longrightarrow R_2CBr{-}CBrR'_2$$

The term *functional group analysis* sometimes is used to describe the foregoing kinds of analyses.

Ultimate Analysis. This term, generally limited to organic chemical analysis, denotes the determination of the proportion of each element in a given substance. The primary determination is that of carbon and hydrogen, which is conducted by mixing the sample with copper(II) oxide and heating it in a stream of oxygen to a temperature of 700 to 800°C. The carbon is converted to carbon dioxide and the hydrogen to water. These products are then absorbed by suitable reagents. For example, magnesium perchlorate dehydrate may be used to absorb water and sodium hydroxide to absorb carbon dioxide. Although the fundamental procedure is simple, a rather elaborate train of apparatus, involving both temperature and flow control, is required. The traditional procedure for determining nitrogen is the Kjeldahl method. The Unterzaucher method is used for determining oxygen in organic substances, the sample is heated to a high temperature (approximately 1120°C) in an atmosphere of nitrogen. Under these conditions, the oxygen present combines with part of the carbon content to form carbon dioxide and with part of the hydrogen content to form water. The gases then are passed over hot carbon (1150°C), whereupon both the carbon dioxide and water are converted to carbon monoxide. The latter gas upon leaving the furnace is passed over iodine pentoxide I_2O_5 at about 110°C to form iodine by the reaction: $5CO + I_2O_5 \longrightarrow I_2 + 5CO_2$. The freed iodine is titrated with a standard sodium thiosulfate solution.

ANALYZER (Reaction-Product).

Chemical composition may be determined by the measurement of a reaction product—in an automatic fashion utilizing the basic principles of conventional qualitative and quantitative chemical analysis. Two steps usually are involved in this type of instrumental analysis: (1) the formation of a target chemical reaction, and (2) the determination of one or more of the reaction products.

Determination of a constituent in a process stream or sample by measurement of a reaction product can be represented by: $C + R \rightarrow P$, where C = constituent to be determined; R = reactant; and P = reaction product to be measured. If reactant R already is present in the sample, it is only required to expose the sample to suitable reaction conditions to form P. Under normal instrument operating conditions, the reaction of C and R may be spontaneous. In other instances, suitable conditions may have to be established either (1) to promote the desired reaction (for example, setting the proper temperature and pressure, or using a catalyst) or (2) to assure a suitable reaction rate.

Frequently, it is possible to measure the reaction product as it forms in the reaction zone. In some instances, the products and sample residue must be removed from the reaction zone before a measurement can be made. Also, the reaction product may be measured directly; or its presence may have to be inferred from a secondary reaction. In one example, carbon monoxide in air or oxygen may be determined by combustion to carbon dioxide. The latter may be measured directly as by thermal-conductivity methods; or inferentially by absorbing the carbon dioxide in a solution and then measuring the change of that solution by electrolytic conductance.

ANALYZER (Reagent-Tape).   The key to chemical analysis by this method is a tape (paper or fabric) that has been impregnated with a chemical substance that reacts with the unknown to form a reaction product on the tape which has some special characteristic, e.g., color, increased or decreased opacity, change in electrical conductance, or increased or lessened fluorescence. Small pieces of paper treated with lead acetate, for example, have been used manually by chemists for many years to determine the presence of hydrogen sulfide in a solution or in the atmosphere. This basic concept forms the foundation for a number of sophisticated instruments that may pretreat a sample gas, pass it over a cyclically advanced tape, and, for example, photo-metrically sense the color of the exposed tape, to establish a relationship between color and gas concentration. Depending upon the type of reaction involved, the tape may be wet or dry and it may be advanced continuously or periodically. Obviously, there are many possible variations within the framework of this general concept.

ANAMORPHISM.   A term proposed by Van Hise in 1904 to designate the deep-seated constructive processes of metamorphism by which new complex (metamorphic) minerals are formed from the preexisting simpler minerals, as contrasted with the surface alteration of rocks due to weathering and cementation, termed katamorphism.

ANAPHYLAXIS.   State of supersensitivity that may develop after a first injection of a foreign protein, such as a therapeutic or prophylactic serum. See also **Alkaloids**.

ANA-POSITION.   The position of two substituent groups on atoms diagonally opposite, in α-positions on symmetrical fused rings, as the 1,5 or the 4,8 positions (which are identical) of the naphthalene ring.

ANATASE.   The mineral anatase, TiO_2 crystallizing in the tetragonal system is a relatively uncommon mineral. It occurs as a tri-morphous form of TiO_2 with rutile and brookite. Rutile and anatase have tetragonal crystallization; brookite, orthorhombic. Hardness, 5.5–6; sp. gr. 3.82–3.97; brittle with sub-conchoidal fracture; color, shades of brown, into deep blue to black; also colorless, grayish, and greenish; also transparent to opaque with adamantine luster.

Anatase occurs as an accessory mineral in igneous and metamorphic rocks, gneisses, and schists. Fine crystals have been found in Arkansas in the United States, and in Switzerland.

ANDALUSITE.   An aluminum silicate corresponding to the formula Al_2SiO_5, and is one of a three-member polymorphous group consisting of andalusite, sillimanite, and kyanite. Andalusite occurs in contact-metamorphic shales, and in rocks of regional metamorphic origin in association with sillimanite and kyanite. Andalusite crystallizes in the orthorhombic system, developing coarse prisms of approximately square cross section, but may be massive or granular. It shows a distinct cleavage parallel to the prism; hardness 6.5–7.5; sp. gr., 3.13–3.16; vitreous luster; colorless to white, gray, brown, greenish, or reddish; streak, white; transparent to opaque.

This mineral is named for its original locality, Andalusia, Spain. A variety of andalusite, chiastolite, has carbonaceous impurities so oriented that they produce a cross or a tesselated figure at right angles to the prism. Chiastolite comes from the Greek word meaning a cross. Localities are the Urals, the Alps, the Tyrol, the Pyrenees, Australia, and Brazil; in the United States, at Standish, Maine; Sterling and Lancaster, Massachusetts; Delaware County, Pennsylvania; and Madera County, California.

When clear it is used as a gemstone, and it has also been used to manufacture porcelain for spark plugs.

ANDESITE.   A term originally applied to a porphyritic lava from the Andes Mountains by Leopold Van Buch. In modern terminology andesite is an extrusive igneous rock, the surface equivalent of diorite. In other words, it is composed chiefly of plagioclase, corresponding in chemical composition to oligoclase or andesine together, with biotite, hornblende, or pyroxene in varying quantities.

Andesites are of rather widespread occurrence, being found in the Rocky Mountains, California. Alaska, South America, and in many other localities.

ANDROGENS.   The relation between the testis and the male secondary sex characteristics has long been known. The evidence that a chemical substance present in the testis could elicit androgenic effects was not achieved until 1908 by Walker, who prepared an aqueous glycerol extract of bull testis tissue that caused growth of the capon's comb. More active extracts from bull testes were prepared in 1927 by McGee and Koch by using organic solvents. The extracts were assayed quantitatively by measuring the increase in area of the capon's comb. The discovery of androgenic activity in urine made possible the isolation of the first biologically active crystalline androgens by Butenandt et al. in 1931–1934. Androsterone and dehydroisoandrosterone were isolated from male urine. In 1935, David et al. isolated a crystalline hormone from bull testis extract. This possessed a higher biological activity than either androsterone or dehydroisoandrosterone. It was named testosterone. Testosterone was also prepared from cholesterol within months of its isolation from testular extract. See Fig. 1.

As pointed out by Liddle and Melmon (1974), adrenal androgen production also carried out in the *zona fasciculata* and in the *zona reticularis*, varies greatly at different stages of life. The fetus makes significant amounts of adrenal androgen, whereas the child makes very little. Beginning with puberty, adrenal androgen production increases, reaches a peak in early adulthood, and then declines to rather low levels beyond age 50. On the other hand, the secretion of ACTH, the only known control of adrenal androgen biosynthesis, shows no age-related fluctuations. The full regulation of adrenal androgen production is not understood. Adrenal androgens are relatively weak, but some serve as precursors for hepatic conversion to testosterone. Hyperfunction of this pathway in the female may lead to significant masculinization. See also **Adrenal Cortical Hormones**; and **Adrenal Medulla Hormones**.

The androgens stimulate the development of the male secondary structures, such as the penis, scrotum, seminal vesicles, prostate gland, vas deferens and epididymis. The deepening of the voice, the growth of pubic, axillary, body, and facial hair, as well as the development of the characteristic musculature of the human male, are also under the influence of testosterone. If the testes fail to develop or are removed prior to puberty, these changes do not occur. Thus, testosterone is essential for reproductive function of the male.

The adrenal cortex produces hydroisoandrosterone, which is largely found in blood and urine conjugated as the sulfate ester. The amounts of androgen secreted by the normal adrenal cortex are insufficient to maintain reproductive function in the male. The normal human ovary and placenta also produce small amounts of androgenic steroids that serve as precursors for the estrogens in these tissues. In the human, little testosterone is excreted into the urine and virtually none into the feces. The principal metabolic transformation products are androsterone and 5β-androsterone, with small amounts of other reduced compounds. These substances are excreted in the urine in the form of esters with sulfuric acid or glycosides with glucuronic acid.

Like all other classes of steroid hormones, the androgens are synthesized from acetyl coenzyme A *via* mevalonic acid, isopentenyl pyrophosphate, farnesyl pyrophosphate, squalene, lanosterol, and cholesterol. Enzyme

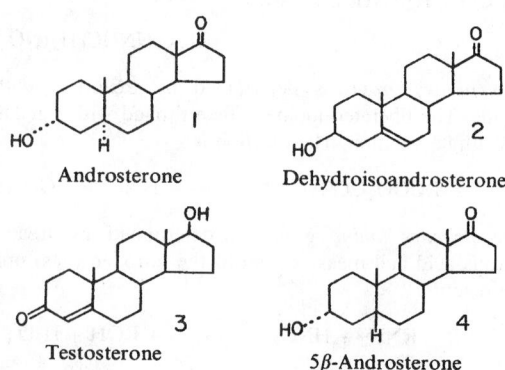

Androsterone   Dehydroisoandrosterone

Testosterone 5β-Androsterone

Fig. 1.   Androsterone and related hormones

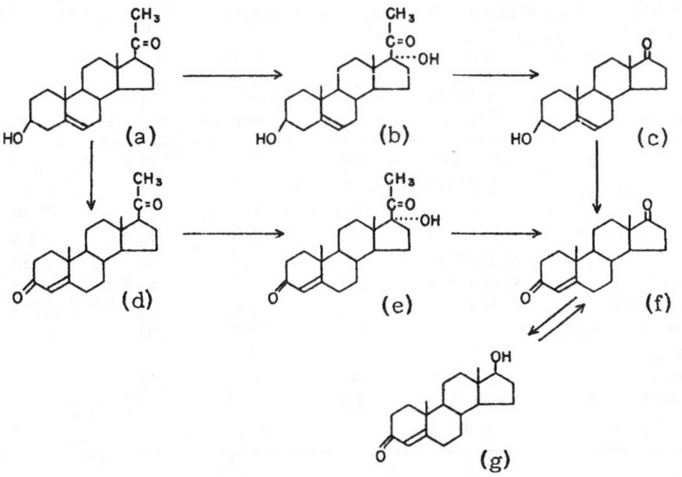

Fig. 2. Biosynthesis of testosterone: (**a**) Pregnenolone; (**b**) 17-hydroxy pregnenoline; (**c**) dehydroisoandrosterone; (**d**) progesterone; (**e**) 17-hydroxypro ges terone; (**f**) androstenedione; (**g**) testosterone

systems in the testis then catalyze the cleavage of the side chain of cholesterol to pregnenolone which can give rise to testosterone by the two pathways shown in Fig. 2.

Testosterone is formed by the interstitial or Leydig cells of the testes which develop under the influence of gonadotrophic hormones discharged into the bloodstream by the anterior pituitary gland. In pituitary insufficiency, this hormonal stimulus is lacking and, as a consequence, the Leydig cells do not secrete testosterone. In such instances, the male secondary sex characteristics fail to develop. However, interstitial cell tumors may occur, leading to excessive androgen production and precocious puberty. In women, tumors or excessive function of the adrenal cortex and, rarely, of the ovary, result in the production of large amounts of androgens with associated virilization.

Acne vulgaris, a chronic skin disorder, is related to androgen production. The post-pubescent development of the sebaceous glands and the onset of acne are dependent upon the presence of androgens, but are not related to testosterone blood levels. Sansone et al. (1971) observed that the hypothesis of increased end-organ sensitivity, which is supported by the heightened ability of skin with acne to metabolize testosterone, may explain the lack of correlation between levels of circulating androgens and occurrence and severity of the disease.

There is androgen involvement in polycystic ovary syndrome (PCO) and hyperthecosis. In 1935, Stein and Leventhal defined a condition with hirsutism, secondary amenorrhea, and enlarged ovaries—a syndrome now referred to as PCO. In this condition, the female usually shows signs of androgen excess, including increased body hair, but true virilism, with balding and deepening of the voice, is less common. Usually, one or both ovaries are enlarged. In many patients the ovaries are cystic, with thickened capsules, yet not palpably enlarged. It has been postulated that the development of PCO commences when luteinizing hormone (LH) triggers an increase in ovarian androgen, which is converted to estrogen, causing estrogen levels (particularly estrone) to increase. This is followed by an anterior pituitary response to luteinizing hormone-releasing hormone (LRH). This completes the cycle by creating exaggerated pulsatile but surgeless LH levels. The initiating lesion remains obscure. Wedge resection of one or both ovaries has largely been replaced by the administration of the antiestrogen clomiphene. This was the first drug known to trigger ovulation in women. Patients with anovulation arising from PCO are treated with the drug primarily when fertility is desired. The hirsutism found with PCO has been difficult to treat. Wood and Boronow (1976) reported that long intervals of anovulation, such as occur in PCO, may be a prelude to endometrial carcinoma. As postulated, the link may either be continuous exposure of the endometrium to estrogen unopposed by progesterone. There is also the postulation that estrone, the estrogen that appears to be high in this disorder, may be a causative factor. A carcinogenic role has been alleged for estrone.

In a somewhat related disorder, hyperthecosis, androgen excess tends to be greater. In this condition, there is prominent luteinization of the theca, whereas the cystic development and capsular thickening of the PCO are absent. Ovarian tumors making androgen can produce the features of PCO, but they tend to have a course with sharper onset and clearer progression.

Androgens also have been used in the management and treatment of agnogenic myeloid metaplasia, aplastic anemia, breast cancer, hereditary angiodema, osteoporosis, paroxysmal nocturnal hemoglobinuria, and sideroblastic anemia.

Oral androgens as may be used in hormone therapy for the management of metastatic breast cancer are effective in about 30% of women regardless of age, but virilizing doses of the hormone are usually required. The oral androgens carry the additional risk of toxic hepatitis.

Androgens may explain some of the differences between heart diseases of males and females.

Additional Reading

Austin, C.R. and R.V. Short, Eds.: *Hormonal Control of Reproduction,* 2nd Edition, Cambridge University Press, New York, NY, 1984.

Azziz, R. and J.E. Nestler, MD, Editors: *Androgen Excess Disorders in Women,* Lippincott Williams & Wilkens, Philadelphia, PA, 1997.

Barbieri, R.L. and K. Schiff, Eds.: *Reproductive Endocrine Therapeutics;* Alan R. Liss, New York, NY, 1989.

Eldar-Geva, T. et al.: "Secondary Biosynthetic Defects in Women with Late-Onset Congenital Adrenal Hyperplasia," *New Eng. J. Med.,* 855 (September 27, 1990).

Griffin, J.E.: "Androgen Resistance—The Clinical and Molecular Spectrum," *New Eng. J. Med.,* 611 (February 27, 1992).

Liddle, G.W. and K.L. Melmon: *The Adrenals: Text book of Endocrinology,* 5th Edition (R.H. Williams, editor), Saunders, Philadelphia, PA, 1974.

Makin, H.: "Mass Spectra and GC Data of Steroids," *Androgens and Estrogens,* John Wiley & Sons, Inc., New York, NY, 1999.

Oddens, B.J. and A. Vermeulen, Editors: Androgens and the Aging Male: Proceedings of a Meeting Held in Geneva, 1996, Parthenon Publishing Group, New York, NY, 1997.

Sansone, G. and R.M. Reisner: *J. Invest. Dermatol.,* **56**, 366 (1971).

Shackman, J., G. Ptacek, and K.C. Ullis, Editors: Super't: *The Complete Guide to Creating an Effective Safe, and Natural Testosterone Supplement Program for Men and Women,* Fireside, Columbus, OH, 1999.

Spieler, J.M., C. Wang, C. Kelly, R.S. Swerdloff, H.L. Gabelnick, and S. Bhasin: "Pharmacology, Biology, and Clinical Applications of Androgens," *Current Status and Future Prospects,* John Wiley & Sons, Inc., New York, NY, 1996.

Stein, I.F. and M.L. Leventhal: "Amenorrhea Associated with Bilateral Polycystic Ovaries," *Am. J. Obstet. Gynecol.,* **29**, 181 (1935).

Wilson, J.D., D.W. Foster, Editors: *Williams Textbook of Endocrinology,* 9th Edition, W.B. Saunders Co. Philadelphia, PA, 1998.

Yen, S.S.C. and R.B. Jaffe: *Reproductive Endocrinology: Physiology, Pathophysiology and Clinical Management,* 4th Edition, W.B. Saunders Co, Philadelphia, PA, 1999.

ANDROSTERONE. See **Steroids**.

ANESTHETICS. The term anesthesia comes from the Greek *anaisthai* or insensibility and constitutes a state in which perception of noxious events such as surgical procedures is imperceptible. This state may or may not be accompanied by loss of consciousness. A complete or general anesthetic given by the inhalation or intravenous routes produces hypnosis (profound sleep), analgesia, muscle relaxation, and protection against the increase in blood pressure and heart rate resulting from surgical stress (maintains homeostasis). An anesthetic which blocks the neural transmission of painful stimuli through afferent nerves and does not affect the level of consciousness can be classified as a local anesthetic.

The induction of general anesthesia produces a progressive deepening of the anesthetic state and represents, in the anatomical sense, a descending desensitization of the central nervous system (CNS). A progression of clinical signs are useful for estimating the depth of anesthesia. These signs vary somewhat for each anesthetic, but in general four stages can be defined: *(1)* state of altered consciousness and analgesia indicative of action on cerebral canticle areas begin; *(2)* loss of consciousness, often accompanied by delirium and excitement, occurs. Irregular respiration and motor movement result from depression of higher motor inhibitory centers with the release of lower motor mechanisms; *(3)* stage of surgical anesthesia is reached in which spinal cord and spinal reflexes are abolished, providing relaxation of skeletal musculature. Within the four planes in this state are loss of corneal, conjunctival, pharyngeal, and laryngeal reflexes as the depth of anesthesia progresses; *(4)* onset of respiratory paralysis occurs resulting from significant depression of the medullary respiratory center. Subsequent cardiovascular collapse ensues.

Most signs which require a skeletal muscle reflex would not be apparent after treatment with a neuromuscular blocking drug or would be altered significantly by preoperative drugs. In these cases central nervous system monitoring via an electroencephalogram (EEG), and hemodynamic and blood gas monitoring, help assess the depth of anesthesia. The potency of inhaled agents is expressed as the minimum alveolar concentration (MAC) that is required to prevent spontaneous movement in response to a surgical or equivalent stimulus in 50% of patients.

The onset of action is fast (within 60 s) for the intravenous anesthetic agents and somewhat slower for inhalation and local anesthetics. The induction time for inhalation agents is a function of the equilibrium established between the alveolar concentration relative to the inspired concentration of the gas. Onset of anesthesia can be enhanced by increasing the inspired concentration to approximately twice the desired alveolar concentration, then reducing the concentration once induction is achieved. The onset of local anesthetic action is influenced by the site, route, dosage (volume and concentration), and pH at the injection site.

Theories of General Anesthesia

Although the modern practice of anesthesia is exceedingly sophisticated, identification of the basic molecular mechanism underlying it is still lacking. Even whether the target sites of the inhalational anesthetics are lipid or protein remains unknown, although it is likely that proteins are intimately involved. The inhalational anesthetics are very diverse chemically and extrapolation from the effect of a particular agent to the physiological state of anesthesia is problematic. However, anesthesia theories may be roughly divided into two categories: lipid theories and protein theories.

Lipid Theories. Although there are many varieties of the lipid theory all postulate that inhalational anesthetics exert their primary effects by dissolving in the lipid portions of nerves, thereby altering the conductivity. The primary site of action is postulated to be the lipid matrix of cell membranes.

Protein Theories. The direct interaction of inhalation anesthetics and proteins has been proposed as the cause of anesthesia. An inhalation agent, whether a noble gas or a fluorinated ether, could dissolve asymmetrically in a protein. Resultant conformational changes in the protein, if these changes occur, could then cause changes in biological activity.

Anesthetic Agents

Inhalation Agents. An ideal inhalation anesthetic would exhibit physical, chemical, and pharmacological properties allowing safe usage in a variety of surgical interventions. The agent should be odorless, nonflammable at concentrations which are likely to be used in the operating room, and stable both on storage and to soda lime, which is used as the CO_2 absorber in the anesthetic circuit. Induction of, and recovery from, anesthesia should be rapid, and minimal side effects should be observed on the cardiovascular (depression and epinephrine compatibility) or central nervous systems (EEG activation). See also **Neuroregulators**. The drug should not be metabolized.

A number of inhalation anesthetics have been introduced to clinical practice, some of which are listed in Table 1. All agents introduced after 1950, except ethyl vinyl ether, contain fluorine. Agents such as ether, chloroform, trichloroethylene (Trilene), cyclopropane, and fluoroxene (Fluoromar), which were once used, have been displaced by the newer fluorinated anesthetics.

Intravenous Anesthetic. The intravenous (iv) anesthetic agents are of two types: those which are used to induce, but not maintain, anesthesia, and those which are useful not only for induction, but also for maintenance. The period of induction is perhaps the most crucial part of the anesthesia. The need is for an anesthetic agent having an extremely fast rate of onset and limited side effects. Fast onset minimizes the stress and agitation which could arise during a more lengthy induction. Various classes of compounds have been used. Among these are: barbiturates, opioids, steroids, benzodiazepines, and hindered phenols. But the ideal agent for both induction and maintenance has not yet been found. For this reason, balanced anesthesia is often used: a potent opioid is combined with an inhalation agent.

A major difference between the inhalational agents and the iv anesthetics is that the former probably exert their biological activity through physical effects while the latter, in most cases, function through a biological receptor-mediated pathway.

TABLE 1. PROPERTIES AND PARTITION COEFFICIENTS OF INHALATION ANESTHETICS

Agent	Molecular formula	Partition coefficient Oil/gas	Partition coefficient Blood/gas	Boiling point, °C	Year introduced
ethyl ether	$C_4H_{10}O$	65	12.1	35	1842
chloroform	$CHCl_3$	394	8.4	61	1847
trichloroethylene	C_2HCl_3	200–250	9.0	87	1934
fluoroxene	$C_4H_5F_3O$	47	1.37	43.1	1960
halothane	$C_2HBrClF_3$	224	2.3	50.2	1956
isoflurane	$C_3H_2ClF_5O$	90.8	1.4	48.5	1980
enflurane	$C_3H_2ClF_5O$	96.5	1.9	56.5	1972
methoxyflurane	$C_3H_4Cl_2F_2O$	970	12.0	105	1962
sevoflurane	$C_4H_3F_7O$	42	0.6	58	1989
desflurane	$C_3H_2F_6O$	18.7	0.42	23.5	
nitrous oxide	N_2O	1.4	0.46	−89.5	1850s
cyclopropane	C_3H_6			−33	1934

Opioid Reversal Agents. At the end of a surgery, when the decision is made to reverse opioid-induced activity, the primary concern is to eliminate the respiratory depressant effects inherent in the potent opiates used as anesthetics. This reversal, brought about by opioid reversal agents, can also diminish the analgesic effects of the opiates. There are two types of reversal agents: pure antagonists and mixed agonist–antagonists. The antagonists naloxone and naltrexone are the most commonly used, although the use of the agonist–antagonist nalbuphine is increasing because it maintains a moderate level of analgesia while reversing the side effects. Nalmefene, a long-acting reversal agent, is in clinical trials.

Local Anesthetics. Nerve impulses are initiated by membrane depolarization, effected by the opening of a sodium-ion channel and an influx of sodium ions. Local anesthetics act by inhibiting the channel's opening; they bind to a receptor located in the channel's interior. The degree of blockage on an isolated nerve depends not only on the amount of drug, but also on the rate of nerve stimulation.

Local anesthetic activity is usually demonstrated by compounds which possess both an aromatic and an amine moiety separated by a lipophilic hydrocarbon chain and a polar group. In the clinically useful agents (Table 2) the polar group is an ester or an amide. Activity may be maintained, however, when the polar function is an ether, thioether, ketone, or thioester.

Pharmacological Profile. The profile of the ideal local anesthetic agent depends largely on the type and length of the surgical procedure for which it is applied. Procedures could include neuraxial (spinal and epidural) anesthesia, nerve and plexus blocks, or field blocks (local infiltration). In general, the ideal agent should have a short onset of anesthesia and be useful for multiple indications such as infiltration, nerve blocks,

TABLE 2. CLINICALLY USED LOCAL ANESTHETIC AGENTS

Agent	Method of application	Comment
	Amino esters	
procaine	infiltration, spinal	slow onset, short duration
chloroprocaine	peripheral nerve and obstetric extradural	fast onset, short duration, low systemic toxicity
tetracaine	spinal	slow onset, long duration, high systemic toxicity
	Amino amides	
lidocaine	infiltration, iv regional, peripheral nerve block, extradural block, spinal and topical	fast onset, moderate duration, low systemic toxicity, most versatile agent
mepivacaine	infiltration, peripheral nerve block, extradural	similar to lidocaine
prilocaine	similar to lidocaine	safest amino amide, methemoglobinemia at high doses
bupivacaine	infiltration, peripheral nerve block, extradural	moderate onset, long duration, potential for cardiovascular side effects, sensory/motor separation
etidocaine	infiltration, peripheral nerve block, extradural	fast onset, long duration, profound motor block

intravenous, extradural, spinal, and topical administration. The therapeutic indexes for systemic CNS and cardiovascular toxicity should be high and the agent should be compatible with the vasoconstrictor epinephrine.

GEORGE R. LENZ
BOC Group Technical Center

HOLLIS G. SCHOEPKE
THEODORE C. SPAULDING
Anaquest

Additional Reading

Barash, P. G., R. K. Stoelting, and B. F. Cullen: *Clinical Anesthesia,* 4th Edition, Lippincott Williams & Wilkins, Philadelphia, PA, 2000.

Chung, D. C., A. M. Lam, and L. Day: *Essentials of Anesthesiology,* 3rd Edition, Elsevier Science, New York, NY, 1996.

Covino, B. G. and H. G. Vasallo: *Local Anesthetics: Mechanism of Action and Clinical Use,* Grune and Stratton, New York, NY, 1976.

Evers, A. S., and M. Maze: *Anesthetic Pharmacology: Physiologic Principles and Clinical Practice,* Elsevier Science, New York, NY, 2003.

Gilman, A. G., L. S. Goodman, T. W. Rall, and F. Murad, eds.: *The Pharmacological Basis of Therapeutics,* Macmillan Publishing Company, New York, NY 1985.

Lenz, G. R., S. M. Evans, D. E. Walters, and A. J. Hopfinger: *Opiates,* Academic Press, Orlando, FL, 1986.

Miller, K. W. and M. E. Wolff, eds.: *Burger's Medicinal Chemistry,* 4th Edition, Vol. 3, John Wiley & Sons, Inc., New York, NY, 1981, p. 623.

Roizen, M. F., and L. A. Fleischer: *Essence of Anesthesia Practice,* 2nd Edition, Elsevier Science, New York, NY, 2001.

Stoelting, R. K.: *Pharmacology and Physiology in Anesthetic Practice,* 3rd Edition, Lippincott Williams & Wilkins, Philadelphia, PA, 2003.

Tetzlaff, J. E.: *Clinical Pharmacology of Local Anesthetics,* Elsevier Science, New York, NY, 1999.

ANFINSEN, CHRISTIAN B. (1916–1995). An American biochemist who won the Nobel prize for chemistry in 1972 for his work on ribonuclease, especially concerning the connection between the amino acid sequence and the biologically active conformation. He shared the Nobel prize with Stanford Moore and William H. Stein. His doctorate was granted from Harvard.

ANGLESITE. Naturally occurring lead sulfate ($PbSO_4$), which crystallizes in the orthorhombic system and may be found mixed with galena, from which it is usually formed by oxidation. Hardness, 3; specific gravity, 6.12–6.39; luster, adamantine to vitreous or resinous; transparent to opaque; streak, white; colorless to white or green, but rarely may be yellow or blue. This mineral is used as a source of lead.

Anglesite, whose name derives from Anglesey, England, is found in many European localities; in the United States it has been found in large crystals in the Wheatley Mine, Phoenixville, Pennsylvania, and also in Missouri, Utah, Arizona, and Idaho.

ÅNGSTROM (Å). A unit of length almost one non-hundred millionth (10^{-8}) centimeter. The Ångstrom is defined in terms of the wavelength of the red line of cadmium (6438.4696 Å). Used in stating distances between atoms, dimensions of molecules, wavelengths of short-wave radiation, etc.

ANHYDRIDE. A chemical compound derived from and acid by elimination of a molecule of water. Thus sulfur trioxide, SO_3, is the anhydride of sulfuric acid; CO_2 is the anhydride of carbonic acid; phthalic acid, $C_6H_4(CO_2H)_2$ minus water gives phthalic anhydride. The term should not be confused with *anhydrous*.

ANHYDRITE. The mineral anhydrous calcium sulfate, $CaSO_4$, occurs in granular, scaly, or fibrous masses, is rarely crystallized in orthorhombic tabular or prismatic forms. Hardness, 3–3.5; sp gr, 2.9–2.98; translucent to opaque; streak white; color, white, gray, bluish, or reddish. Anhydrite has three cleavages at right angles to one another. It is similar to gypsum and occurs under the same conditions, often with the latter mineral. It is usually found in sedimentary rocks associated with limestones, salt, and gypsum, into which it changes slowly by the absorption of water. See also **Gypsum**.

Anhydrite is found in Poland, Saxony, Bavaria, Württemberg, Switzerland, and France; in the United States, in South Dakota, New Mexico, Texas, New Jersey, and Massachusetts; in Canada, in Nova Scotia, New Brunswick, and exceptional specimens from the Faraday Uranium Mine near Bancroft, Ontario.

ANHYDROUS. Descriptive of an inorganic compound that does not contain water either adsorbed on its surface or combined as water of crystallization. Do not confuse with *anhydride*.

ANILINE. Aniline, phenylamine, aminobenzene, $C_6H_5NH_2$, is a colorless, odorous liquid, an amine, with melting point−6°C, boiling point 184°C, is slightly soluble in water, miscible in all proportions with alcohol or ether, poisonous, turns yellow to brown in the air, is a weak base forming salts with acids, e.g., anilinehydrochloride ("aniline salt," $C_6H_5NH_2 \cdot HCl$) from which aniline is reformed by addition of sodium hydroxide solution. Aniline reacts (1) with hypochlorite solution, to form a transient violet coloration, (2) with nitrous acid (a) warm, to form nitrogen gas plus phenol, (b) cold, to form diazonium salt (benzene diazonium chloride, C_6H_5N-Cl), (3) with acetyl chloride, acetic anhydride, or acetic acid glacial, to form N-phenylacetamide

acetanillide, "antifebrin," C_5H_6N

benzanilide, C_5H_6N

(4) with benzoyl chloride, to form N-phenylbenzamide

(5) with benzenesulfonyl chloride, to form N-phenylbenzene sulfonamide, $C_6H_5SO_2NHC_6H_5$, soluble in sodium hydroxide, (6) with chloroform, $CHCl_3$, plus alcohol plus sodium hydroxide, to form phenyl isocyanide, C_6H_5NC, very poisonous, (7) with H_2SO_4 at 180° to 200°C, to form para -aminobenzene sulfonic acid (sulfanilic acid, $H_2N \cdot C_6H_4 \cdot SO_2H$ (1,4)), (8) with HNO_3, when the amine group is protected, e.g., using acetanilide, to form mainly paranitroacetanilide, $CH_3CONH \cdot C_6H_4 \cdot NO_2$ (1.4), from which paranitroaniline, $H_2N \cdot C_6H_4 \cdot NO_2(1,4)$ is obtained by boiling with concentrated hydrochloric acid, (9) with chlorine in an anhydrous solvent, such as chloroform or acetic acid glacial, to form 2,4,6-trichloroaniline (1) $H_2N \cdot C_6H_2Cl_3(2,4,6)$, (10) with bromine water, to form white solid 2,4,6-tribromoaniline, (1)$H_2N \cdot C_6H_2Br_3(2,4,6)$, (11) with potassium dichromate in sulfuric acid, to form aniline black dye, and, by further oxidation, benzoquinone, $O:C_6H_4:O$ (1,4), (12) with potassium permanganate in sodium hydroxide, to form azobenzene, $C_6H_5N:NC_6H_5$, along with some azoxybenzene $C_6H_5NO:NC_6H_5$, (13) with reducing agents, to form aminohexahydrobenzene (cyclohexylamine, $H_2N \cdot C_6H_{11}$), (14) with alkyl halides or alcohols heated, to form alkyl anilines, e.g., methylaniline, $C_6H_5NHCH_3$, dimethylaniline, $C_6H_5N(CH_3)_2$.

Aniline may be made (1) by the reduction, with iron or tin in HCl, of nitrobenzene, and (2) by the amination of chlorobenzene by heating with ammonia to a high temperature corresponding to a pressure of over 200 atmospheres in the presence of a catalyst (a mixture of cuprous chloride and oxide). Aniline is the end-point of reduction of most mono-nitrogen substituted benzene nuclei, as nitrosobenzene, beta-phenylhydroxylamine, azoxybenzene, azobenzene, hydrazobenzene. Aniline is detected by the violet coloration produced by a small amount of sodium hypochlorite.

Aniline is used (1) as a solvent, (2) in the preparation of compounds as illustrated above, (3) in the manufacture of dyes and their intermediates, (4) in the manufacture of medicinal chemicals. See also **Amines**.

ANION. A negatively-charged atom or radical. In electrolysis, an anion is the ion which deposits on the anode; that portion of an electrolyte which carries the negative charge and travels against the conventional direction of the electric current in a cell. Within the category of anions are included the nonmetallic ions and the acid radicals, as well as the hydroxyl ion, OH^-. In electrochemical reactions, they are designated by the minus sign placed above and after the symbol, such as Cl^- and SO_4^{2-}, the number of the minus sign indicating the magnitude, in electrons, of the electrical charge carried by the anion. In a battery, it is the deposition of negative anions that makes the anode negative. See also **Ion**.

ANISE. Of the family *Umbelliferae* (carrot family), the anise plant (*Pimpinella anisum*) is native to the Mediterranean region and is cultivated in Egypt, Malta, Spain, and Syria, but also in other areas of the world, such as Germany and the United States.

The aromatic, warm, and sweetish odor and taste of the seed, leaves, and stem arises from the presence of a volatile oil that contains anethole (*p*-propenyl phenylmethyl ether, $C_3H_5C_6H_4OCH_3$), the derivatives of which (anisole and anisaldehyde) are used in food flavoring, particularly bakery, liqueur, and candy products, as well as ingredients for perfumes. For commercial production of anise oil, the seeds and the dried, ripe fruit of the plant are used. Anise oil, a colorless to pale-yellow, strongly refractive liquid of characteristic odor and taste, is prepared by steam distillation of the seed and fruit. The oil contains choline, which finds use in medicine as a carminative and expectorant.

ANISODESMIC STRUCTURE. A type of ionic crystal in which some of the ions tend to form tightly bound groups, e.g., nitrate and chlorate.

ANISOTROPIC MEDIUM. An anisotropic medium has different optical or other physical properties in different directions. Wood and calcite crystals are anisotropic, while fully annealed glass and, in general, fluids at rest are isotropic.

ANNABERGITE. The mineral annabergite is a rather rare nickel arsenate with the formula $Ni_3(AsO_4)_2 \cdot 8H_2O$, crystallizing in the monoclinic system. It is of secondary origin, resulting from the alteration of preexisting nickel minerals, commonly found as surface alteration crust on nickeline. Annabergite has been found in Saxony, France, including Annaberg, from which its name is derived, and as exceptional crystals at Laurium, Greece, and in Cobalt, Ontario, Canada.

ANNATTO FOOD COLORS. These colors are natural carotenoid colorants derived from the seed of the tropical annatto tree (*Bixa orellana*). The surface of the seeds contains a highly colored resin, consisting primarily of the carotenoid *bixin*. The bixin is extracted from the seed by a special process to produce a pure, soluble colorant. Bixin, one of the relatively few naturally occurring *cis* compounds, has a chemical structure similar to the nucleus of carotene with a free and esterified carboxyl group as end groups. Its formula is $C_{25}H_{30}O_4$. (See Fig. 1).

$$H_3COOC-CH \overset{CH_3}{\underset{CH}{\diagdown}} \overset{}{\underset{CH}{C}} \overset{CH_3}{\underset{}{\diagdown}} CH \overset{}{\underset{CH}{\diagdown}} CH \overset{}{\underset{CH}{\diagdown}} CH \overset{H}{\underset{}{C}}=\overset{H}{\underset{C}{C}} \overset{CH_3}{\underset{}{\diagdown}} CH-COOH$$

Fig. 1.   Structure of bixin

Bixin is an oil-soluble, highly stable coloring ingredient. The saponification of the methyl ester group to form the dicarboxylic acid yields the water-soluble form of bixin, sometimes called *norbixin*. Annatto colorants date back into antiquity. The colorant has been used for centuries in connection with various textiles, medicinals, cosmetics, and foods. Annatto colors have also been used to color cheese, butter, and other dairy products for over a century. See also **Carotenoids**.

Processors make annatto colors available as a refined powder, soluble in water at pH values above 4.0 (solubility about 10 grams in 100 milliliters of distilled water at 25°C), in an acid-soluble form, in an oil-soluble form, in a water- and oil-soluble form, and in a variety of hues ranging from delicate yellows to hearty orange. Annatto extract is frequently mixed with turmeric extract to obtain various hues.

ANNEALING. The process of holding a solid material at an elevated temperature for a specified length of time in order that any metastable condition, such as frozen-in stains, dislocations, and vacancies may go into thermodynamic equilibrium. This may result in re-crystallization and polygonization of cold-worked materials.

Annealing generally falls into the technology of heat treatment and varies with materials and the intended end uses of the materials, as well as the prior processing of them. In the case of nonferrous alloys, annealing is primarily a heat treatment for the purpose of removing the hardening due to cold work. Annealing also may be used with nonferrous precipitation

hardening alloys to cause softening through agglomeration of the hardening constituent into fewer and larger particles.

Ferrous Metallurgy. In the case of ferrous materials, the term annealing usually implies full annealing. This heat treatment involves a change of phase inasmuch as the metal is heated into the austenitic region. Cooling slowly back to room temperature then develops a softened structure of pearlite and ferrite. The annealing of cold-worked metal is termed *process annealing*, wherein a change of phase is not involved. Annealing takes several forms in terms of the time-temperature relationships imposed upon the materials. *Box annealing, isothermal annealing, normalizing, patenting, spheroidize annealing,* and *stress relieving* are described under **Iron Metals, Alloys, and Steels**.

Annealing of Cold-Worked Metals. Ductile metals hardened by cold-working may be softened by annealing. Annealing is often an important intermediate step in producing metals by cold deformation. Thus, in the formation of fine wires through wire drawing, several intermediate anneals may be required. Annealing may also be the last production step when metal objects are desired in a final softened condition.

In general, cold working increases many-fold the dislocation density of a metal. A severely cold-worked metal may easily have a dislocation density 10^6 times greater than in the same unworked metal. Since each dislocation is surrounded by a strain field extending over long distances on an atomic scale, each dislocation contributes to the strain energy of the metal and, accordingly, to its free energy. When the metal is annealed, the free energy associated with the dislocations resulting from cold work furnishes a driving force that can effectively reduce the dislocation density back to the value that existed before deformation.

Three basic stages are generally recognized as occurring during the annealing of cold worked metals. These are recovery, re-crystallization, and grain growth.

In recovery, the strain energy is lowered by the recombination of dislocations of opposite sign, or by rearrangements of dislocations into configurations of lower strain energy. A simple well-known example of this latter is the polygonization of the dislocations in a bent crystal. When a crystal is bent, the curved shape is the result of the accumulation, upon the slip planes of the crystal, of a large number of edge dislocations of the same sign. During recovery, these dislocations move from their more or less random positions along the slip planes into a set of vertical walls normal to the slip planes. This movement is accomplished by both slip and dislocation climb. The walls of dislocations that are formed in this manner constitute a form of grain boundary across which the crystal lattice is slightly rotated by the order of minutes of arc. Such boundaries are better known as sub-grain boundaries. The crystalline material between these sub-boundaries is effectively free of dislocations. It is thus apparent that polygonization transforms a highly strained bent crystal into a set of small sub-grains that are nearly strain-free. (See Fig. 1.)

The rate of recovery is normally highest at the start of an isothermal annealing cycle because the driving force is largest at that time. As recovery continues, the driving force diminishes as the available strain energy is used up and the rate of recovery falls continuously toward zero. A plot of the rate of recovery as a function of time yields a curve that is somewhat similar in appearance to an exponential decay curve. The rate of recovery is also temperature dependent and may be expressed, in a number of cases, by a simple empirical equation of the form

$$1/t = Ae^{-Q/RT}$$

where t is the time to attain a certain fixed amount of recovery, A is a constant, R, the universal gas constant, T, the absolute temperature, and Q, an empirical activation energy. Because the reactions that occur during

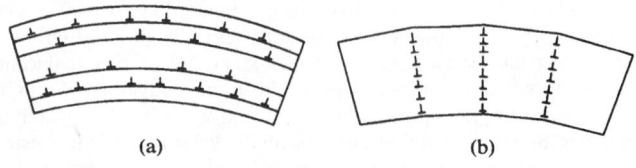

(a)   (b)

Fig. 1.   Realignment of edge dislocations during polygonization: **(a)** excess edge dislocations that remain on active slip planes after a crystal is bent: **(b)** arrangement of dislocations after polygonization

recovery are complex, it is usually not possible to attach a simple meaning to the activation energy for recovery.

Re-crystallization is the process whereby the distorted grains or crystals of a cold-worked metal are reconverted into new (essentially) strain-free grains. It occurs by the nucleation of minute submicroscopic crystals that grow out into and consume the strained material surrounding them. Re-crystallization is, therefore, a nucleation and growth phenomenon and, characteristically, the rate of re-crystallization starts slowly, builds up to a maximum, and then diminishes back to zero. The increase in the rate at the early stages of re-crystallization is due primarily to continued nucleation of new grains while the older ones continue to grow. The final falling off in the rate is the result of the progressive consumption of the material available for re-crystallization. A metal is said to be completely re-crystallized when all of the original deformed structure has been eliminated.

Re-crystallization, like recovery, is thermally activated and occurs at a rate that grows very rapidly with increasing temperature, as may be seen in the accompanying diagram where the amount of re-crystallization in copper is plotted as a function of the time for six different temperatures. (See Fig. 2.)

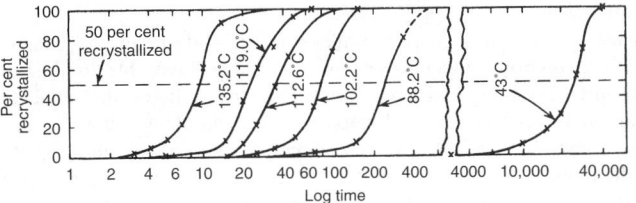

Fig. 2.   Complex polygonized structure in a single silicon-iron crystal deformed 8% by cold rolling before being annealed 1 hour at 1100°C

The driving force for re-crystallization also comes from the strain energy of the excess dislocations created by cold work. It is therefore apparent that recovery and re-crystallization are competitive processes. Usually a metal may undergo a considerable degree of recovery before visible evidence of re-crystallization is obtained. However, since the re-crystallized grains grow from very small beginnings, the re-crystallization process undoubtedly is occurring long before it can be detected visually. Also, there is reason to believe that the nuclei of the re-crystallized grains may be formed as a result, at least in some cases, of processes related to recovery. It should also be noted that recovery phenomena may continue to occur during re-crystallization in those grains not yet consumed by the re-crystallization process. The degree to which the manifestations of re-crystallization and recovery appear to overlap is a function of the metal concerned and of the nature of the deformation that it has received. Under certain conditions, it is possible to have recovery occur without re-crystallization. This is particularly true when the amount of deformation is insufficient to cause re-crystallization, or when the type of deformation, although extensive, is very simple, as in the case of a zinc or magnesium crystal deformed only by slip on the basal plane. Examples have been observed where single crystals of these metals have been deformed in this manner by as much as 700% and still failed to re-crystallize on annealing.

After a metal has undergone re-crystallization it can still undergo grain growth. The driving force in this case comes from the surface energy of the grain boundaries. A close analogy exists between the growth of grains in a metal during annealing and the growth of soap bubbles in soap froth. In the soap froth, a small bubble that finds itself surrounded by larger neighbors will normally have but a few sides convex toward the bubble center. This curvature produces a small but finite excess gas pressure inside the small bubble, which causes gas to diffuse through the bubble wall into the neighboring larger bubbles. The smaller bubble, consequently, grows smaller and disappears, while the larger ones surrounding it grow in size. At the same time, the average bubble in the froth must also grow in size. The same basic phenomenon occurs during grain growth in metals. In this case, the atoms from the smaller grains move across the grain boundaries and become part of the crystals of the larger grains. At all times a geometrically similar distribution of grain sizes exists in the metal, ranging from small to large, which promotes continued grain growth. However, as the average size increases, there is a corresponding decrease in the growth rate. This is easily understood in terms of the soap froth analogy because with an increase in bubble size there is a corresponding decrease in the average bubble wall curvature and in the pressure difference across bubble walls.

It may also be shown that in soap froth the average bubble size should increase as the square root of the time. This one-half power law, however, is seldom observed during grain growth in a metal. Usually the grain growth exponent (power to which the time is raised) is much smaller than one half, signifying that the empirical growth rates are much lower than would be expected from the soap froth analogy. Several reasons may be proposed in explanation of this fact. The motion of grain boundaries is easily influenced by impurity atoms in solid solution, or present as small inter-metallic inclusions. In either case, the grain boundary mobility is lowered with a corresponding decrease in the value of the grain growth exponent.

Under the proper conditions, a limiting grain size may be attained in a metal at which point grain growth ceases. This is often true in very thin specimens when the average grain diameter approaches the thickness of the specimen. At this time, the grain boundary geometry becomes two- instead of three-dimensional which reduces the average curvature and hinders further growth. Alternatively, it is possible for grain boundaries to become so held up by the nonmetallic inclusions that further growth is prevented. This condition is only achieved after a critical grain size has been achieved.

Associated with the limiting grain size effect mentioned above is a phenomenon known as secondary re-crystallization. Sometimes, after a limiting grain size has been attained, a few grains may begin to grow again and may obtain very large sizes. This is actually not a true re-crystallization but rather an unusual manifestation of grain growth. This formation of a new set of very large grains in material where growth had apparently ceased is known as secondary re-crystallization.

Annealing of Glass. As with metals, glass is fabricated at high temperatures and is annealed to relieve stresses that would develop if the glass were permitted to cool in an uncontrolled fashion. If not annealed, products made from high-expansion glasses can break spontaneously as they cool freely in air. In annealing glasses, they are raised to an annealing point temperature and then cooled gradually to a temperature that is somewhat below the strain point. Usually, the rate of cooling within this range determines the magnitude of residual stresses after the glass arrives at room temperature. Once below the strain point, the cooling rate is limited only by any transient stresses that may develop. A typical time-temperature glass-annealing curve is shown in Fig. 3. Normally, glass for optical purposes is annealed much more slowly than commercial glassware to improve the optical homogeneity of the material. In the annealing process, glass normally is heated and held at a temperature slightly higher than the annealing point temperature and controlled cooling is effected to a temperature slightly below the strain point to accommodate for differences in materials and as an extra safeguard. Controlled cooling may occur over two periods as indicated by the diagram.

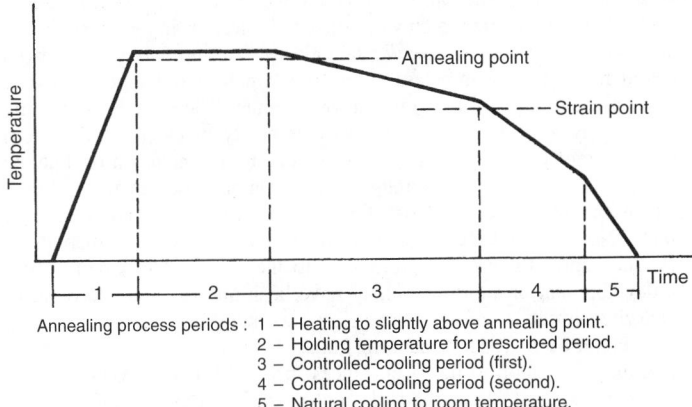

Annealing process periods : 1 – Heating to slightly above annealing point.
2 – Holding temperature for prescribed period.
3 – Controlled-cooling period (first).
4 – Controlled-cooling period (second).
5 – Natural cooling to room temperature.

Fig. 3.   Typical glass annealing curve

Additional Reading

Azencott, R.: *Simulated Annealing, Parallelization Techniques,* John Wiley & Sons, Inc., New York, NY, 1991.

Bauccio, M., Editor, American Society for Metals, *ASM Metals Reference Book,* 3rd Edition, ASM International, Materials Park, OH, 1993.

Brooks, C.R.: *Principles of the Heat Treatment of Plain Carbon and Low Alloy Steel,* ASM International, Materials Park, OH, 1996.

Bryson, B., *Heat Treatment, Selection, and Application of Tool Steels,* Gardner Publications, Inc., Cincinnati, OH, 1997.

Chandler, H.E.: *Heat Treater's Guide, Practices and Procedures for Irons and Steels*, ASM International, Materials Park, OH, 1995.

Davis, J.R., Editor: *ASM International Handbook Committee, Metals Handbook: Desk Edition*, ASM International, Materials Park, OH, 1999.

Dossett, J. and R. Juetje: Heat Treating, Proceedings of the 16th Conference, ASM International, Materials Park, OH, 1996.

Humphreys, F.J. and M. Hatherly, Editors: *Recrystallization and Related Annealing Phenomena*, Elsevier Science, New York, NY, 1996.

Kalivas, J.H.: *Adaption of Simulated Annealing to Chemical Optimization Problems*, Elsevier Science, New York, NY, 1995.

Staff: Heat Treating, *ASM Handbook* 10th Edition, Vol. 004, ASM International, Materials Park, OH, 1991.

Totten, G.E., and M. A. Howes: *Steel Heat Treatment Handbook*, Marcel Dekker, Inc., New York, NY, 1997.

ANODE. In the most general sense, an anode is the electrode via which current enters a device. The anode is the positively charged electrode of an electrolytic cell. See **Electrochemistry**. The anode (also frequently called the plate) is the principal electrode for collecting electrons in an electron tube, and is, therefore, operated at a positive potential with respect to the cathode.

ANODIC OXIDATION. Oxidation is defined not only as reaction with oxygen, but as any chemical reaction attended by removal of electrons. Therefore, when current is applied to a pair of electrodes so as to make them anode and cathode, the former can act as a continuous remover of electrons and hence bring about oxidation (while the latter will favor reduction since it supplies electrons). This anodic oxidation is utilized in industry for various purposes. One of the earliest to be discovered (H. Kolbe, 1849) was the production of hydrocarbons from aliphatic acids, or more commonly, from their alkali salts. Many other substances may be produced, on a laboratory scale or even, in some cases, on an economically sound production scale, by anodic oxidation. The process is also widely used to impart corrosion-resistant or decorative (colored) films to metal surfaces. For example, in the anodization or Eloxal process, the protection afforded by the oxide film ordinarily present on the surface of aluminum articles is considerably increased by building up this film by anodic oxidation.

ANODIZE. This term means to place a protective film on a metal surface by electrolytic or chemical action in which the metal surface is made the anode in an electrochemical process. Aluminum and magnesium parts of electronics equipment are frequently anodized.

ANODIZED COATINGS. See **Conversion Coatings**.

ANORTHOSITE. The name anorthosite was given by T. Sterry Hunt to rocks of gabbroid nature which were essentially free from pyroxene, hence almost wholly plagioclase *usually* labradorite. The term is derived from the French word for plagioclase, anorthose. Small quantities of pyroxene may be present as well as magnetite or ilmenite. The rock is commonly white to gray, bluish, greenish, or perhaps nearly black. A variety from the Province of Quebec is purplish-brown due to the inclusion of ilmenite dust within the feldspars. Although not a common rock in the ordinary sense of the word, occurrences of great territorial extent are known in Canada, Norway, and Russia and in the United States in northern New York State and Minnesota. Opinions as to the origin of this rock differ. The development of anorthosite may have been due to the settling out of labradorite crystals from a gabbro magma as many believe, or there may have been an original anorthosite magma.

A study of anorthosite occurrences brings out two very curious circumstances, first, that there is no extrusive (lava) equivalent of anorthosite, and second, that most anorthosite masses seem to be of pre-Cambrian age.

ANTACIDS. These are formulations widely used in the treatment of excessive gastric secretions and peptic ulcer. Several factors determine the efficacy of antacids, including (1) the ability and capacity of the stomach to secrete acid; (2) the duration of time the antacid is retained in the stomach; and (3) the nature of the gastric response upon eating.

Five principal active ingredients are used in antacid preparations: (1) *Sodium bicarbonate* is a rapid and effective neutralizer. The compound does yield large amounts of absorbable sodium, undesirable in some persons (heart disease; hypertension). The compound also may induce milk-alkali syndrome. (2) *Calcium carbonate* is a strong, effective neutralizer, but can cause constipation, hypercalcemia, acid rebound, and milk-alkali syndrome. (3) *Aluminum hydroxide* provides slow and not potent action. The compound causes constipation, absorbs phosphates, as well as certain drugs, such as tetracyclines. (4) *Magnesium hydroxide* which provides a slow and prolonged action with no major side reactions. (5) *Magnesium trisilicate* which acts like magnesium hydroxide, but which is poorly absorbed and acts as an osmotic laxative. In cases of renal insufficiency, the serum magnesium should be monitored.

The foregoing compounds are frequently used in combination and, in some, simethicone is added to relieve flatulence. There are striking differences of commercial antacids in terms of their neutralizing capacity.

The physician is concerned with at least three factors when prescribing antacids: (1) Acid rebound (associated with calcium carbonate); (2) milk-alkali syndrome (caused by ingestion of large quantities of alkali); and (3) phosphorus depletion (by aluminum salts). The mechanism of acid rebound, especially in the long-term use of calcium carbonate, is poorly understood. It has been established that there is an excessive re-acidification of the antrum (pyloric gland area) a number of hours after ingestion of calcium carbonate.

The ingestion of a quart of milk or more while taking large amounts of alkali, as from antacids, sets up conditions favorable to milk-alkali syndrome. Generally, with withdrawal of the milk or the antacid, the condition is self-correcting. Symptoms of milk-alkali syndrome include nausea, vomiting, anorexia, weakness, polydipsia, and polyuria. Abnormal calcifications also may occur in the chronic stage and other symptoms include mental changes, asthenia, aching muscles, band keratopathy, and nephocalcinosis. Symptoms of milk-alkali syndrome sometimes tend to mimic hyperparathyroidism and vitamin D intoxication.

ANTHOCYANINS. A group of water-soluble pigments that account for many of the red, pink, purple, and blue colors found in higher plants. Most plants contain more than one of these pigments and they occur most prevalently as glycosides. Several hundred different anthocyanins are known. Anthocyanins have been isolated and some have been found to be acylated with substituted cinnamic acids. The site of attachment of these acids to the anthocyanins has not been fully defined. The natural role of the anthocyanins in plants to date has not been related to any factor of plant metabolism; many authorities believe that the pigments play more of an ecological role with regard to pollination and seed dispersal through their ability to act as an insect and bird attractant.

The anthocyanins are part of the larger group of aromatic oxygen-containing, heterocyclic compounds, known as flavonoids, most of which have a 2-phenylbenzopyran skeleton as their basic ring system. Although widely distributed among higher plants, including ferns and mosses, they are not found in algae, fungi, bacteria, or lichens.

There has been considerable interest and research activity in connection with anthocyanins during the past decade or so, stemming principally from the tighter restrictions, including banning, of several synthetic colorants. See also **Colorants (Foods)**. Representative of the food processing industry's desire to find colorants that are beyond suspicion as health deterrents, scientists have been investigating various sources of anthocyanins. They have found that pigments from roselle plants (*Hibiscus sabdariffa*) native to the West Indies can be used for coloring apple and pectin jellies. A cranberry pomace extract has been found useful in coloring cherry pie filling. The potential of blueberry as a source of anthocyanin pigments also has been investigated. The berry is rich in non-acylated anthocyanins, but presently appears to be too costly as a coloring substitute.

Grape anthocyanins have been intensely investigated and have been found reasonably satisfactory, for example, in carbonated beverages. Although to date the grape anthocyanins are not as stable as Red No. 2, research continues, encouraged by the large amounts of grape wastes produced in the production of wine and grape juice. Red cabbage also has been seriously considered as a source of anthocyanin pigments. Anthocyanins are most stable at a pH range of 1.0 to 4.0, and this acidity dictates the products in which they can be used.

Much more detail on this topic can be found in the references listed below. See also **Colorants (Foods)**; **Glycosides (Steroid)**; and **Pigmentation (Plants)**.

Additional Reading

Bullock: *Colorant Formulation and Applications,* Routledge, New York, NY, 1999.

Considine, D.M. and G.D. Considine: *Foods and Food Production Encyclopedia,* Van Nostrand Reinhold Company, Inc., New York, NY, 1982.

Hendry, G.F. and J.D. Houghton: *Natural Food Colorants,* Routledge, New York, NY, 1999.

Marmion, D.M.: *Handbook of U.S. Colorants, Foods, Drugs, Cosmetics, and Medical Devices,* 3rd Edition, John Wiley & Sons, Inc., New York, NY, 1993.

ANTHOPHYLLITE. The mineral anthophyllite is an orthorhombic amphibole essentially $(Mg, Fe)_7Si_8O_{22}(OH)_2$ with aluminum sometimes present. This mineral corresponds to enstatite and hypersthene in the pyroxene group. It has a prismatic cleavage; hardness, 5.5–6; sp gr, 2.8–3.57; luster, vitreous; color, gray, yellow, brown, green or brownish-green; transparent to translucent; probably always a metamorphic mineral in magnesium-rich rocks, often associated with talc; very common in schist. Found in Norway, Austria, Greenland, Pennsylvania, Georgia, and elsewhere. The name is derived from the Latin *anthophyllum,* clove, because of its usual brownish shades. See also **Amphibole.**

ANTHRACENE. [CAS: 120-12-7]. A colorless solid; melting point 218°C, blue fluorescence when pure; insoluble in water, slightly soluble in alcohol or ether, soluble in hot benzene, slightly soluble in cold benzene; transformed by sunlight into para-anthracene $(C_{14}H_{10})_2$.

Anthracene reacts: (1) With oxidizing agents, e.g., sodium dichromate plus sulfuric acid, to form anthraquinone, $C_6H_4(CO)_2C_6H$. (2) With chlorine in water or in dilute acetic acid below 250°C to form anthraquinol and anthraquinone, at higher temperatures 9,10-dichloroanthracene. The reaction varies with the temperature and with the solvent used. The reaction has been studied using, as solvent, benzene, chloroform, alcohol, carbon disulfide, ether, glacial acetic acid, and also without solvent by heating. Bromine reacts similarly to chlorine. (3) With concentrated sulfuric acid to form various anthracene sulfonic acids. (4) With nitric acid, to form nitroanthracenes and anthraquinone. (5) With picric acid $(1)HO \cdot C_6H_2(NO_2)_3(2,4,6)$ to form red crystalline anthracene picrate, melting point 138°C.

$$C_{14}H_{10} \text{ or}$$

Anthracene is obtained from coal tar in the fraction distilling between 300° and 400°C. This fraction contains 5–10% anthracene, from which, by fractional crystallization followed by crystallization from solvents, such as oleic acid, and washing with such solvents as pyridine, relatively pure anthracene is obtained. It may be detected by the formation of a blue-violet coloration on fusion with mellitic acid. Anthracene derivatives, especially anthraquinone, are important in dye chemistry.

ANTHRAQUINONE. [CAS: 84-65-1]. Anthraquinone (9,10) is a yellow solid, melting point 286°C; can be sublimed;

forms monoxime, melting point 224°C, by heating under pressure at 180°C with hydroxylamine chloride; forms no phenylhydrazone with phenylhydrazine; with strong oxidizing agents reacts with difficulty to yield phthalic acid $C_6H_4(COOH)_2(1,2)$; with reducing agents, such as sodium hyposulfite, zinc in sodium hydroxide solution, tin or stannous chloride in hydrochloric acid (but not sulfurous acid), is reduced to anthraquinol, anthrone, dianthrol and dianthrone, depending on the conditions.

Anthraquinone is obtained by oxidation of anthracene using sodium dichromate plus sulfuric acid, and is purified by dissolving in concentrated sulfuric acid at 130°C and pouring into boiling water, whereupon anthraquinone separates as pure solid, and is recovered by filtration. Further purification

may be accomplished by sublimation or crystallization from nitrobenzene, aniline or tetrachloroethane. Anthraquinone is used as the material from which many dyes are made, notably alizarin $C_6H_4(CO)_2C_6H_2(OH)_2$ and related substances. These are vat dyes, that is, insoluble colored substances which are readily reduced to a substance having marked affinity for the fiber to be dyed and which upon exposure to the air are readily re-oxidized to the original dye. Anthraquinone may be detected by the appearance of a red color on treatment with alkali, zinc powder, and water. See also **Coal Tar and Derivatives.**

ANTHRAXOLITE. A coal-like metamorphosed bitumen, often closely associated with igneous rocks. Commonly associated with "Herkimer Diamond" type quartz crystals in dolomitic limestones in Herkimer and Montgomery counties in New York State.

ANTIBIOTICS. Antibiotics are chemical substances produced by microorganisms and other living systems that are capable in low concentrations of inhibiting the growth of bacteria or other microorganisms. This inhibitory effect can be *in vitro* or *in vivo.* Antibiotics having both *in vivo* activity and low mammalian toxicity have been extremely valuable in treating infectious diseases. There are over 10,000 antibiotics produced by microorganisms that have been reported in the scientific literature, approximately 6100 of which have been characterized and have had molecular structures assigned. These substances range from the very simple to extremely complex, but most antibiotics are in the 300–800 mol wt range. Many thousands of semisynthetic variations of the naturally occurring antibiotics have been prepared. Only relatively few (~200) have become commercial products for human and veterinary uses.

The mechanism of action of a number of antibiotics with regard to the inhibition of bacteria, fungi, or other organisms has been established. The more common mechanisms include inhibition of bacteria cell wall biosynthesis, inhibition of protein, RNA, or DNA synthesis, and damaging of membranes. Cell wall biosynthesis is a target present in bacteria but not in mammalian cells. Thus the β-lactams, which are very effective against bacteria, are relatively nontoxic to humans. In contrast, antibiotics that damage DNA, like adriamycin, are relatively toxic to both types of cells. However, adriamycin, which was found to be more toxic to rapidly proliferating tumor cells than to most normal cells with slower turnover rates, shows significant selectivity against tumor cells to find clinical application as an antitumor agent.

The number of naturally occurring antibiotics increased from about 30 known in 1945 to 150 in 1949, 450 in 1953, 1,200 in 1960, and to 10,000 by 1990. Table 1 lists the years of historical importance to the development of antibiotics used for treatment in humans. Most of the antibiotics introduced since the 1970s have been derived from synthetic modifications of the β-lactam antibiotics.

Classification of Antibiotics

A chemical classification of some of the commercially more important antibiotic families is given here.

Aminoglycosides. Antibiotics in the aminoglycoside group characteristically contain amino sugars and deoxystreptamine or streptamine. This family of antibiotics has frequently been referred to as aminocyclitol aminoglycosides. Representative members are streptomycin, neomycin, kanamycin, gentamicin, tobramycin, and amikacin. These antibiotics all inhibit protein biosynthesis.

Ansamacrolides. Antibiotics in the ansamacrolide family are also referred to as ansamycins. They are benzenoid or naphthalenoid aromatic compounds in which nonadjacent positions are bridged by an aliphatic chain to form a cyclic structure. One of the aliphatic–aromatic junctions is always an amide bond. Rifampin is a semisynthetically derived member of this family and has clinical importance. It has selective antibacterial activity and inhibits RNA polymerase.

β-Lactams. All β-lactams are chemically characterized by having a β-lactam ring. Substructure groups are the penicillins, cephalosporins, carbapenems, monobactams, nocardicins, and clavulanic acid. Commercially this family is the most important group of antibiotics used to control bacterial infections. The β-lactams act by inhibition of bacterial cell wall biosynthesis.

TABLE 1. YEAR OF DISCOVERY OR MARKET INTRO-
DUCTION OF SOME OF THE MORE IMPORTANT
ANTIBIOTICS

Antibiotic	Year	
	Discovery	Introduction
penicillin	1929	
tyrothricin	1939	
griseofulvin	1939	
streptomycin	1944	
bacitracin	1945	
chloramphenicol		1947
polymyxin	1948	
chlortetracycline		1948
cephalosporin C,N,P	1948	
neomycin	1949	
oxytetracycline		1950
nystatin	1950	
erythromycin		1952
novobiocin	1955	
kanamycin		1957
ampicillin[a]		1962
fusidic acid		1961
cephalothin[a]	1962	
lincomycin		1963
gentamicin		1963
carbenicillin[a]	1964	
cephalexin[a]	1966	
clindamycin[a]		1967
cephaloxidine and cephalothin[a,b]		1969
minocycline[a]		1971
amoxycillin[a]		1972
cefoxitin[a,c]		1978
tricarcillin[a]		1979
mezlocillin[a]		1980
piperacillin[a]		1980
cefotaxime[a]		1980
moxalactam[a]		1981
augmentin[d]		1984
aztreonam[e]		1984
imipenem[a,f]		1985

[a] Semisynthetic products.
[b] First oral cephalosporins.
[c] First commercial cephamycin.
[d] First β-lactamase inhibitor combination.
[e] First monobactam and a synthetic product.
[f] First carbapenem.

Chloramphenicol. Only chloramphenicol and a few closely related analogues fall into this group. Chloramphenicol, a nitro benzene derivative of dichloroacetic acid, inhibits protein biosynthesis.

Glycopeptides. Vancomycin, avoparcin, and teicoplanin are examples of glycopeptide antibiotics. This family has cyclic peptide structures and biphenyl containing amino acids. Sugars are attached to the peptide unit resulting in compounds frequently in the molecular weight range of 1400–2000. These antibiotics inhibit bacterial cell wall biosynthesis by binding to D-alanyl-D-alanine units found in the cell walls.

Lincomycin. The lincomycins and celesticetins are a small family of antibiotics that have carbohydrate-type structures. Clindamycin, a chemical modification of lincomycin, is clinically superior. Antibiotics in this family inhibit gram-positive aerobic and anaerobic bacteria by interfering with protein biosynthesis.

Macrolides. Antibiotics in the macrolide group are macrocyclic lactones that can be further classified into two main subgroups: (1) polyene macrolides that are antifungal agents and include compounds like nystatin and amphotericin B; and (2) antibacterial antibiotics represented by erythromycin and tylosin. A number of other subfamilies of antibacterial and antifungal antibiotics fall into the broad category of macrolides.

Polyethers. Antibiotics within this family contain a number of cyclic ether and ketal units and have a carboxylic acid group. They form complexes with mono- and divalent cations that are soluble in nonpolar organic solvents. They interact with bacterial cell membranes and allow cations to pass through the membranes causing cell death. Because of this property they have been classified as ionophores. Monensin, lasalocid, and maduramicin are examples of polyethers that are used commercially as anticoccidial agents in poultry and as growth promotants in ruminants.

Tetracyclines. The tetracyclines are a small group of antibiotics characterized as containing a polyhydronaphthacene nucleus. Commercially the tetracyclines are important. They have been used clinically against gram-positive and gram-negative bacteria, spirochete, mycoplasmas, and rickettsiae and have veterinary applications in promoting growth and feed efficiency. The mode of action is inhibition of protein synthesis. Some of the more important members of this family are tetracycline, minocycline, and doxycycline.

Production

Most of the microorganisms used to produce antibiotics were isolated from soil samples. These microorganisms occur as heterogeneous populations and generally inhabit the top few centimeters of soil. Families demonstrated to produce antibiotics include actinomycetes, bacteria, and fungi. Actinomycetes are in numbers the most productive for antibiotics.

To obtain reproducible antibiotic production by fermentation, it is necessary to obtain a pure culture of the producing organism. Pure cultures are isolated from mixed soil sample populations by various streaking and isolation techniques on nutrient media. Once a pure culture has been found that produces a new antibiotic typically on a mg/L scale, improvement in antibiotic yield is accomplished by modification of the fermentation medium or strain selection and mutation of the producing organism. Production of g/L quantities may take years to accomplish.

The vast majority of new antibiotics result from screening soil microorganisms or by semisynthetic modification of naturally occurring antibiotics. Genetic engineering technology has begun to evolve that allows modifications of a microorganism's DNA so that it will produce new antibiotics.

Commercial fermentations are conducted in large bioreactors which are usually referred to as fermentors and are designed for operation in batch, fed-batch, or continuous fermentation modes. The batch and fed-batch procedures are used for most commercial antibiotic fermentations.

Uses

Antibiotics are used as antibacterial agents, anticancer agents, antituberculin agents, antifungal agents, antiviral agents, and in veterinary products and animal feed supplements for growth promotion.

DONALD BORDERS
American Cyanamid Company

Additional Reading

Bérdy, J. A. Aszalos, and K.L. McNitt, *CRC Handbook of Antibiotic Compounds,* Vol. 14, CRC Press, Inc., Boca Raton, FL, 1987.

Conte, J.E.: *Manual of Antibiotics and Infectious Diseases,* 8th Edition, Lippincott Williams & Wilkins, Philadelphia, PA, 1995.

County NatWest WoodMac, *International Pharmaceutical Service,* County NatWest Securities Limited, London, 1989; Part 4, 1990.

Cuevas, C.F.: *Antibiotic Resistance, From Molecular Basics to Therapeutic Options,* Chapman & Hall, New York, NY, 1997.

Gilespie, S.H.: *Antibiotic Resistance Methods and Protocols,* Humana Press, Totowa, NJ, 2000.

Goode J. and D.J. Chadwick: *Antibiotic Resistance, Origins, Evolution, Selection and Spread,* Vol. 207, John Wiley & Sons, Inc., New York, NY, 1997.

Gottfried, T.: *Alexander Fleming, Discover of Penicillin,* Grolier Publishing, Danbury, CT, 1997.

Higton, A.A. and A.D. Roberts, in Bycroft, B.W. ed., *Dictionary of Antibiotics and Related Substances,* Chapman and Hall, New York, NY 1988.

Jacoby, G.A. and G.L. Archer: "New Mechanisms of Bacterial Resistance to Antimicrobial Agents," *N. Eng. J. Med.,* **601** (February 28, 1991).

Kucers, A. N.Mck. Bennett, and R.J. Kemp, *The Use of Antibiotics,* 4th Edition, William Heinemann Medical Books, London, UK 1987, pp. 914, 1418–1528.

Kucers, A., S.M. Crowe, M.L. Grayson, and J.F. Hoy: *The Kucer's Use of Antibiotics, A Clinical Review of Antibacterial Antifungal and Antiviral Drugs,* 5th Edition, Butterworth-Heinemann, Inc., Woburn, MA, 1997.

Lorian, V.: *Antibiotics in Laboratory Medicine,* 4th Edition, Lippencott Williams & Wilkins, Philadelphia, PA, 1996.

Mark, A.L.: "Cyclosporine, Sympathetic Activity, and Hypertension," *N. Eng. J. Med.,* 746 (September 13, 1990).

Moberg, C.L.: "Penicillin's Forgotten Man: Norman Heatley," *Science,* 734 (August 16, 1991).

Moberg, C.L. and Z.A. Cohn: "Rene Jules Dubos," *Sci. Amer.,* 66 (May 1991).

Neu, H.C.: "The Crisis in Antibiotic Resistance," *Science,* 1064 (August 21, 1992).

Parenti, F. and G.G. Gallo: *Antibiotics, A Multidisciplinary Approach,* Plenum Publishing, New York, NY, 1995.

Roberts, A.D., A.A. Higton, and B.W. Bycroft: *Dictionary of Antibiotics and Related Subs,* CRC Press LLC, Boca Raton, FL, 1999.

Zinner, S.H., L.S. Young, H.C. Neu, and J.F. Acar: *Expanding Indications for the New Macrolides, Azalides and Streptogramins,* Vol. 21, Marcel Dekker, Inc., New York, NY, 1997.

ANTIBIOTICS: AMINOGLYCOSIDES.

The term *aminoglycoside* is commonly used to refer to members of the class of antibacterial antibiotics, the structures of which are derived from D-streptamine (**1**, R=OH), D-2-deoxystreptamine (**1**, R=H), or closely related compounds. The term *aminocyclitol* is also sometimes used to identify this group of compounds. A typical member of this class, tobramycin, has the structure (**2**).

(1)

(2)

Aminoglycosides in Medical Usage

In 1991, the most widely used aminoglycosides in medical practice were gentamicin, tobramycin, amikacin, and netilmicin. Other aminoglycosides used to a lesser extent include dibekacin, isepamicin, neomycin, astromicin, spectinomycin, kanamicin A, sisomicin, and streptomycin.

Medical and Biological Properties

General Antibacterial Properties. In general, the aminoglycosides are useful for the treatment of serious infections involving aerobic or facultative gram-negative bacilli, especially in the compromised host. Particular advantages of the aminoglycosides include the findings that, in general, the bactericidal concentration is not significantly higher than the growth inhibitory concentration, and that the bactericidal effect is rapid and concentration-dependent. Clinical usage has been extensively reviewed in the medical literature.

Bacterial Resistance Mechanisms. The most common resistance mechanism involves the inactivation of the aminoglycoside by reactions catalyzed by plasmid-borne enzymes. In general, amikacin and isepamicin tend to be least susceptible to inactivation by this mechanism, while netilmicin and dibekacin are intermediate and gentamicin and tobramycin are most susceptible. Less common resistance mechanisms include decreased affinity for the antibiotic by the bacterial ribosome, and decreased rate of transport into the bacterial cytoplasm.

Pharmacokinetics. The aminoglycosides are not reliably absorbed following oral dosing, so they are administered primarily by intravenous infusion or intramuscular injection. Distribution throughout the vascular and interstitial space is fairly rapid, but intracellular, cerebrospinal fluid, and bronchial secretion levels are generally low. Most of the administered dose is eliminated unmetabolized in the urine.

Toxicology. Potential toxicity is a primary limiting factor in the clinical use of aminoglycosides. The most important toxicities are nephrotoxicity, ototoxicity, and to a lesser extent, neuromuscular blockade. Although there is some variation, all the aminoglycosides in medical practice are capable of causing these adverse events. The effect of this potential is to prevent the use of significantly increased doses to cover difficult infections. In addition, serum aminoglycoside concentrations can reach toxic levels after normal dosing due to variations in glomerular filtration efficiency.

Mechanism of Antibacterial Action. The bactericidal mechanisms employed by aminoglycosides are incompletely understood. Initially, the cationic aminoglycoside binds nonspecifically to anionic groups on the bacterial cell surface. Passage through the cell wall is via porins or self-promoted defects. Uptake into the cell cytoplasm appears to be a two-step, energy-requiring process. In the cytoplasm, ribosomal binding leads to misreading of the genetic code and production of abnormal proteins and, at higher concentrations, protein synthesis inhibition. The cause of cell death is uncertain.

Aminoglycoside Biosynthesis. The biosynthesis of aminoglycosides has been extensively studied. Probably the most interesting aspect is the biosynthesis of 2-deoxystreptamine, in which the C-1 and C-6 of a D-glucose molecule become the C-1 and C-2 of 2-deoxysptreptamine by way of the intermediate 2-deoxy-*scyllo*-inosose.

Structure-Activity Relationships Among Aminoglycoside Derivatives

The aminoglycosides possess properties which make them valuable for the control of bacterial infectious disease, but they also have distinct limitations, especially in the areas of toxicity and susceptibility to bacterial resistance mechanisms. Thus, a large amount of research has been conducted aimed at reducing the limitations while maintaining the advantages by modification of the aminoglycoside molecular structure. The principal approaches to novel structural variations have been (*1*) a search for new microorganisms which produce novel aminoglycosides directly (e.g., tobramycin); (*2*) chemical modification of available aminoglycosides (semisynthesis) (e.g., amikacin); and (*3*) generation of microorganism mutants which require a modifiable exogenous substrate for the biosynthesis of the aminoglycoside. Overall, structural modification has been successful in reducing susceptibility to bacterial inactivation mechanisms. It has not been possible, however, to substantially dissociate the toxicity potential from the antibacterial activity.

DONALD MCGREGOR
Bristol-Myers Squibb Company

Additional Reading

Keuhl, F.A., R.L. Peck, C.E. Hoffhine, Jr., and K. Folkers: *J. Am. Chem. Soc.* **70,** 2325 (1948).

Neidle, S., D. Rogers, and M.B. Hursthouse: *Tetrahedron Lett.* 4725 (1968).

Schatz, A., E. Bugie, and S.A. Waksman: *Proc. Soc. Exp. Biol. Med.* **55,** 66 (1944).

Umezawa, S., Y. Takahasi, T. Usui, and T. Tsuchiya: *J. Antibiot.* **27,** 997 (1974).

ANTIBIOTICS: ANSAMACROLIDES.

The ansamacrolides or ansamycins are a family of antibiotics characterized by an aliphatic ansabridge that connects two nonadjacent positions of an the aromatic nucleus. Ansamacrolides can be divided into two groups based on the nature of the aromatic nucleus. One group contains a naphthoquinoid nucleus and includes the streptovaricins, the rifamycins, tolypomycin, the halomycins, the naphthomycins, actamycin, the diastovaricins, kanglemycin, awamycin, and ansathiazin. The other group contains a benzoquinoid nucleus and includes geldanamycin, the maytansinoids, the herbimycins, the macbecins, the mycotrienins, the trienomycins, the ansatrienins, and the ansamitocins. Table 1 summarizes the biological activity of these antibiotics.

Naphthoquinoids

Streptovaricins. The streptovaricins are produced by *Streptomyces Spectablis* n. sp. and are isolated as a crude complex.

Chemical Properties and Derivatives. All of the streptovaricins except streptovaricin D react with one mole of sodium periodate to yield the corresponding streptovals. The streptovaricins undergo thermal isomerization to the corresponding atropisostreptovaricins. In the natural streptovaricins the ansa-bridge lies above the aromatic nucleus but in the atropisostreptovaricins this bridge lies below the aromatic nucleus. Most spectral properties of the isomers are nearly identical, but the optical rotations, although of approximately equal magnitude, are of opposite sign.

Assay Methods. The primary assay for the streptovaricins is the microbiological assay using the agar diffusion method or a turbidimetric procedure. The streptovaricins can also be identified by paper or thin-layer chromatography.

Rifamycins. The rifamycins were first isolated from a broth of *Nocardia mediterranei* (the producing organism was originally identified as *Streptomyces mediterranei*). The rifamycins were originally designated as rifomycins. Only rifamycin B, which accounts for 10–15% of the crude complex, can be isolated easily as a stable crystalline compound.

The structures of the rifamycins were arrived at by chemical degradation studies and confirmed by x-ray crystallography. The absolute configuration of the ansa-bridge is 6(S), 7 (S), 8 (R), 9 (R), 10 (R), 11 (S), 12 (R), and 13 (S). Studies of ^{13}C nmr and ir have been reported and mass spectra of the rifamycins have been obtained.

Chemical Properties and Derivatives. There have been thousands of rifamycin derivatives prepared in an attempt to obtain a broader-spectrum antibiotic having good oral absorption. Rifamycins B, O, and S have served as starting materials for the preparation of numerous classes of derivatives. Several of the semisynthetic derivatives are more active, have a broader spectrum of biological activity, and are therapeutically more effective than the parent antibiotics.

Manufacture and Processing. Although fermentation procedures have not been reported, assumptions concerning fermentation media and optimal conditions have been made. The transformation of the biologically inactive rifamycin B to the biologically active rifamycin S is usually accomplished chemically. Several rifamycin B oxidases have been isolated that can enzymatically transform rifamycin B to rifamycin O, which is hydrolyzed in the fermentation medium to rifamycin S. The enzymes from *Monocillium spp.* ATC 20621 and *Humicold spp.* ATCC 20620 are intracellular, whereas the enzyme from *Curvularia lunata* var. *aeri* is extracellular. The use of a fluidized bed reactor containing immobilized whole cells of *Humicola* for the transformation of rifamycin B to rifamycin S has been described. Rifamycin SV-producing strains have been isolated, but it is not known if these strains are used commercially.

Assay Methods. A large number of assays exist for the determination of the various rifamycins. Rifamycin SV and rifampicin can be determined by a microbiological assay using *Sarcina lutea* ATTC 9341 as the test organism, and rifampicin can be determined using *S. aureus* 560. Rifamycins B, S, and SV can be separated by electrophoresis on agar gel and determined microbiologically using *B. subtilis* or *S. lutea*. Spectrophotometric assays exist for the rifamycins and for rifampicin. Rifamycins B, O, S, and SV can be determined via polarography or by amperometric titration. Rifamycins B, O, S, and SV can be separated by thin-layer chromatography on silica gel or by paper chromatography. Fluorimetric assays exist for rifamycin B and rifampicin. High performance liquid chromatographic (hplc) procedures exist for rifamycins B, O, S, and SV, for rifampicin in formulations, in body fluids, in mixtures of antibiotics, and for rifapentine in plasma.

Tolypomycins. The addition of small amounts of iron salts to the fermentation medium increases the production of tolypomycin Y, the structure of which was arrived at by chemical degradation and confirmed by x-ray crystallographic analysis.

Tolypomycin Y shows strong antibacterial activity against gram-positive bacteria and *Neisseria gonorrheae*.

A differential bioassay was developed to distinguish tolypomycin Y from rifamycin B.

Halomycins. The halomycins are a group of four antibiotics produced by *Micromonospora halophytica* and separated by partition chromatography on Chromosorb W coated with formamide. Further purification was accomplished using preparative tlc.

The halomycins are active against gram-positive bacteria. The halomycin complex exhibited high activity against bacterial strains resistant to penicillin G.

Naphthomycins, Naphthoquinomycins, Actamycin, and Diastovaricins. The naphthomycins are a group of closely related antibiotics differing in the substituent at C-2 and C-30, and in the geometry about the C-4 and C-6 double bonds. The naphthoquinomycins, diastovaricins, and actamycin are all closely related to the naphthomycins. Naphthomycin A is isolated from a fermentation beer of *Streptomyces collinus* (Tü 105).

Naphthomycin B is produced by *Streptomyces galbus* (Tü 353) whereas naphthomycin C is produced by *Streptomyces diastatochromogenes* (Tü 1892).

Naphthoquinomycins A and B are isolated from *Streptomyces S-1998* and their structures are assigned on the basis of spectral data. Actamycin is obtained from *Streptomyces* sp. E/784, and its structure arrived at on the basis of spectral data and degradation studies.

Diastovaricins I and II are produced by *Streptomyces diastochromogenes*. Diastovaricins I and II are active against Friend mouse leukemia cells. Spectral data are used to determine the structures.

Kanglemycin. Kanglemycin is isolated from the fermentation broth filtrate of *Nocardia mediterranei* var *kanglensis* and its structure determined by x-ray crystallographic studies. The antibiotic is active against gram-positive bacteria.

Awamycin and Ansathiazin. Awamycin and ansathiazin are produced by *Streptomyces* sp. No. 80-217. The structures for awamycin and ansathiazin were assigned on the basis of spectral data. Both antibiotics are active against gram-positive bacteria, and awamycin is reported to have antitumor activity.

Benzoquinoids

Geldanamycin. Geldanamycin is isolated from the filtered beer of *Streptomyces hygroscopicus* var. *geldanus* var. *nova*. This organism also produces nigericin nocardamine, and a libanamycinlike activity depending on the composition of the fermentation medium. The structure of geldanamycin was assigned in great part on the basis of nmr studies. Unlike the naphthoquinoid ansamacrolides, geldanamycin has little antibacterial activity, being primarily active against protozoa and fungi, especially *Tetrahymena pyriformis* and *Crithidia fasciculata*. Geldanamycin also has herbicidal activity.

TABLE 1. BIOLOGICAL ACTIVITY OF THE ANSAMACROLIDES

Ansamacrolide	Biological activity
streptovaricins	antibacterial (gram-positive and mycobacteria), antiviral, inhibitors of reverse transcriptase
rifamycins	antibacterial (gram-positive, gram-negative, and mycobacteria), antiviral, inhibitors of reverse transcriptase
tolypomycins	antibacterial (gram-positive)
halomycins	antibacterial (gram-positive)
naphthomycins	antibacterial (gram-positive), vitamin K antagonist
actamycins	inhibitors of fatty acid synthesis
diastovaricins	antileukemic
kanglemycins	antibacterial (gram-positive)
awamycins	antibacterial (gram-positive), antitumor
ansathiazins	antibacterial (gram-positive)
geldanamycins	antiprotozoal, herbicidal, inhibitors of reverse transcriptase
herbimycins	herbicidal, antitumor, antiviral, inhibitors of tyrosine kinase
macbecins	antibacterial (gram-positive), antiprotozoal, antifungal, antitumor
mycotrienins	antifungal
trienomycins	antitumor
ansatrienins	antifungal
maytansinoids	antileukemic, antitumor
ansamitocins	antiprotozoal, antifungal, antitumor

Herbimycins. Herbimycins A, B, and C along with some derivatives, are isolated from the fermentation broth of *Streptomyces hygroscopicus* AM-3672. The structure of herbimycin A was assigned on the basis of spectral data and confirmed by x-ray crystallographic studies. The structures for herbimycins B and C were derived by comparing spectral data to those for herbimycin A. The herbimycins possess strong herbicidal activity and exhibit some antitumor and antiviral activity.

Several derivatives of herbimycin A have been prepared that possess greater antitumor activity than the parent.

Macbecins. Macbecin I and II are isolated from the fermentation broth of *Nocardia sp.* C-14919. The structures were assigned on the basis of ^{1}H NMR studies on the intact antibiotics as well as on several degradation products. The assigned structures were confirmed by x-ray crystallographic studies. The macbecins are active against gram-positive bacteria, fungi, and protozoa and exhibit *in vitro* antitumor activity against murine leukemia P 388 and melanoma B 16.

Mycotrienins, Mycotrienols, Trienomycins, and Ansatrienins. The mycotrienins are produced by *Streptomyces rishiriensis* T-23. The structures for mycotrienins I and II were assigned primarily on the basis of NMR spectral analysis.

Streptomyces rishiriensis T-23 also produces mycotrienols I and II, the structures of which were based on spectral analysis. The mycotrienins possess no antibacterial activity but are active against fungi and yeasts, and exhibit weak antitumor activity. The mycotrienols are of an order of magnitude less active than the mycotrienins, suggesting that the cyclohexanecarbonylalanine group is important for biological activity.

The trienomycins are isolated from *Streptomyces* sp. 83-16. The assigned structures were based on spectral data. The trienomycins have no antimicrobial activity but have good antitumor activity. Trienomycin A is the most active, exhibiting good *in vivo* antitumor activity against sarcoma 180 and P 388 leukemia in mice.

The ansatrienins are produced by *Streptomyces collinus* Tü 1982. The structures were assigned on the basis of spectral data of the intact antibiotics as well as several derivatives. The ansatrienins are active against fungi.

Maytansinoids and Maytansides (Ansamitocins)

Isolation and Structure Proof. The maytansinoids were the first ansamacrolides to be found in plants. The term maytansinoids refers to those ansamacrolides related to maytansine, whereas the term maytansides refers to maytansinoids lacking the ester side chain at C-3 as well as the corresponding elimination products. Maytansine was first isolated from the alcoholic extract of *Maytenus ovatus* Loes. Several other maytansinoids and maytansides have been isolated from this species. The structure of maytansine was established by x-ray crystallographic analysis, and the structures of the other maytansinoids and maytansides were arrived at by comparative nmr studies using maytansine. The absolute configuration of maytansine is 3(*S*), 4(*S*), 5(*S*), 6(*R*), 7(*S*), 10(*R*), and 2′(*S*).

Colubrinol and colubrinol acetate are isolated from *Colubrina texensis* Gray (Rhamnaceae) along with maytanbutine. Colubrinol is also isolated from *Trewia nudiflora* (Euphorbiaceae). The structures for colubrinol and colubrinol acetate were established by high resolution ms and the comparison of their nmr spectra with that of the known maytanbutine.

Normaytansine is isolated from *Maytenus buchananii*, and the maytansinoids trewiasine, dehydrotrewiasine, and demethyltrewiasine are isolated from the ethanolic extract of the seed from *Trewia nudiflora* L. (Euphorbiaceae). Also isolated from *Trewia nudiflora* are the maytansinoids treflorine, trenudine, and *N*-methyltrenudone, all of which contain an additional macrocyclic ring linking C-3 to the aromatic amide nitrogen.

Another large group of maytansinoids are produced by the microorganism *Nocardia* sp. C-25003 (N-1) and are designated ansamitocins. The structures of the ansamitocins were determined by spectral analysis. By comparison of reported physical data, it was concluded that ansamitocins P-0, P-1, and P-2 were identical to maytansinol, maytanacine, and maytansinol propionate, respectively.

Chemical Properties and Derivatives. Procedures for the total synthesis of several of the maytansinoids have been thoroughly reviewed. A variety of bacteria, actinomycetes, yeasts, and fungi were screened for their ability to modify the ansamitocins.

Several semisynthetic maytansinoids have been prepared by acylating the C-3 hydroxyl group of maytansinol. Some of these derivatives have antiprotozoal and antitumor activity similar to maytansine and

ansamitocin P-3. 3-Epimaytansinoids have been synthesized and were not biologically active.

Biological Activity. The maytansinoids possess antitumor activity, particularly against P 388 lymphocytic leukemia, B 16 melanocarcinoma, and Lewis lung carcinoma. A number of semisynthetic esters of maytansinol have been prepared and exhibit good antileukemic activity. The maytansides lack antitumor activity, indicating that the ester at C-3 is a requirement for activity. The carbinolamide also appears to be necessary for antitumor activity. The maytansinoids do not inhibit bacterial RNA polymerase as do the other ansamacrolides. Besides having antitumor activity, the ansamitocins have antiprotozoal and antifungal activity. Maytansine has undergone Phase I and II clinical studies and does not appear to be effective.

Mode of Action and Biosynthesis

The mode of action of the naphthoquinoid ansamacrolides was established through studies using the rifamycins and streptovaricins. The ansamacrolides inhibit bacterial growth by inhibiting RNA synthesis. This is accomplished by forming a tight complex with DNA-dependent RNA polymerase.

The ansamacrolides form no such complex with mammalian RNA polymerase and thus have low mammalian toxicity.

The antiviral activity of the ansamacrolides does not result from inhibition of RNA polymerase but rather from the inhibition of the assembly of the virus particles.

The antitumor activity of geldanamycin and its derivatives appears to result from inhibition of DNA synthesis, whereas RNA synthesis is not affected. The antitumor activity of the maytansinoids also appears to result from the inhibition of DNA synthesis.

The ansa-chain of the ansamycins streptovaricins, rifamycins, geldanamycin, and herbimycin has been shown to be polyketide in origin, being made up of propionate and acetate units with the *O*-methyl groups coming from methionine. The remaining aromatic C_7N portion of the ansamacrolides is derived from 3-amino-5-hydroxybenzoic acid, which is formed via shikimate precursors. Based on the precursors of the rifamycins and streptovaricins isolated from mutant bacteria strains, a detailed scheme for the biosynthesis of most of the ansamacrolides has been proposed.

Commercially Available Ansamacrolides

Rifampicin, the only commercially available ansamacrolide, is manufactured by Merrell Dow under the trade name Rifadin, and by CIBA under the trade name Rimactane. Rifampicin is also supplied in combination with isoniazid or pyrazinamide. The rifampicin–isoniazid combination is known as Rifamate (Merrell Dow), Rifinah (Merrell Dow), and Rimactazid (CIBA); the rifampicin–pyrazinamide as Rifater (Merrell Dow). Several other rifamycin derivatives including rifabutin and rifapentine are undergoing clinical studies.

FREDERICK J. ANTOSZ
The Upjohn Company

Additional Reading

Ghisalba, O., J.A.L. Auden, T. Schupp, and J. Nüsch: in E.J. VanDamme, ed., *Biotechnology of Industrial Antibiotics,* Marcel Dekker, New York, NY, 1984, Chapt. 9.

Lancini, G.: in H. Pape and H.-J. Rehm, eds., *Biotechnology: Microbial Products II,* Vol. 4, VCH, New York, NY, 1986, Chapt. 14.

Rinehart, K.L. Jr. and L.S. Shield: *Fortschr. Chem. Org. Naturst.* **33**, 231 (1976).

Smith, C.R. Jr. and R.G. Powell: in S.W. Pelletier, ed., *Alkaloids: Chemical and Biological Perspectives,* John Wiley & Sons, Inc., New York, NY, 1984, Chapt. 4.

ANTIBIOTICS: β-LACTAMASE INHIBITORS. The antibacterial effectiveness of penicillins, cephalosporins, and other β-lactam antibiotics depends on selective acylation and consequently, inactivation, of transpeptidases involved in bacterial cell wall synthesis. This acylating ability is a result of the reactivity of the β-lactam ring (**1**). Bacteria that are resistant to β-lactam antibiotics often produce enzymes called β-lactamases that inactivate the antibiotics by catalyzing the hydrolytic opening of the β-lactam ring to give products (2) devoid of antibacterial activity.

(1) **(2)**

active inactive

Based on sequence data, it has been suggested that β-lactamases evolved from the enzymes involved in bacterial cell wall synthesis.

One approach to combating antibiotic resistance caused by β-lactamase is to inhibit the enzyme. See also **Enzyme Inhibitors**. Effective combinations of enzyme inhibitors with β-lactam antibiotics such as penicillins or cephalosporins result in a synergistic response, lowering the minimal inhibitory concentration (MIC) by a factor of four or more for each component. However, inhibition of β-lactamases alone is not sufficient. Pharmacokinetics, stability, ability to penetrate bacteria, cost, and other factors are also important in determining whether an inhibitor is suitable for therapeutic use. Almost any class of β-lactam is capable of producing β-lactamase inhibitors. Several reviews have been published on β-lactamase inhibitors, detection, and properties.

Table 1 shows the clinically most important bacteria that produce β-lactamases, separated into gram-positive and gram-negative organisms. The prevalence of β-lactamase is indicated as is the origin and Richmond-Sykes classification. Based on these data the most important β-lactamases to inhibit clinically are the gram-positive penases, the gram-negative TEM, which are Richmond-Sykes type III, and the gram-negative chromosomal cephalosporinases–cephases which are Richmond-Sykes type I. These enzymes are subsequently referred to as penase, TEM(III), and cephase(I).

Mechanistic Aspects of β-Lactamase Inhibition

The clinically important β-lactamases, e.g., the penases, TEM(III), and cephases(I), are serine proteases that form an acyl enzyme intermediate with β-lactam substrates and β-lactam-derived β-lactamase inhibitors. Mechanistic studies using several β-lactamase inhibitors have been extensively reviewed and a general inhibition scheme is illustrated in Figure 1.

Active site-directed β-lactam-derived inhibitors have a competitive component of inhibition, but once in the active site they form an acyl enzyme species which follows one or more of the pathways outlined in Figure 1.

β-Lactam-Based Inhibitors

Penicillins, Cephalosporins, and Monobactams. Early attempts at inhibiting β-lactamases using inorganics or penicillin fragments were not successful. The discovery that cephalosporin C and methicillin inhibited β-lactamases resulted in the screening of numerous antibiotics, and a number of β-lactamase-resistant penicillins and cephalosporins were found to be β-lactamase inhibitors. These compounds act by forming a transiently inhibited acyl enzyme species as a result of conformational change and are inhibitory substrates (Fig. 1, route C). No clinically useful inhibitors have been identified from this class. These efforts have been extensively reviewed.

Modern β-lactamase-resistant cephalosporins have been reported to be inhibitors of type I cephases. β-Lactamase inhibition occurs through the transiently inhibited enzyme species, which requires a good C-3 leaving group.

Several monobactams have been reported to be inhibitory substrates for type I cephases.

Clavulanic Acid Class of β-Lactamase Inhibitors. Clavulanic acid has only weak antibacterial activity, but is a potent irreversible inhibitor for many clinically important β-lactamases, including penases and Richmond-Sykes types II, III, IV, V, VI (Bacteroides).

Carbapenem-β-Lactamase Inhibitors. Carbapenems are another class of natural product β-lactamase inhibitors discovered about the same time as clavulanic acid.

Penem β-Lactamase Inhibitors. The synthesis and antibacterial properties of penems have been reviewed. Like the closely related carbapenems, many of the penems are potent antibacterials. Additionally, penems are also susceptible to degradation by renal dipeptidase, but to a lesser extent.

Penam Sulfone β-Lactamase Inhibitors. Natural product discoveries stimulated the rational design of β-lactamase inhibitors based on the readily accessible penicillin nucleus. An early success was penicillanic acid sulfone, $(2(S)$-cis)-3,3-dimethyl-7-oxo-4,4-dioxide-4-thia-1-azabicyclo [3.2.0]heptane-2-carboxylic acid (sulbactam) $(R = R^1 = H, R^2 = R^3 = CH_3)$, $C_8H_{11}NO_5S$. The synthesis, microbiology, and clinical use of sulbactam have been reviewed. Sulbactam, with minor exceptions, is a weak antibacterial, but is a potent irreversible inactivator of many β-lactamases, including penases and Richmond-Sykes type II, III, IV, V, and VI (Bacteroides) β-lactamases. Sulbactam is better than clavulanic acid against type I cephases, and synergy is observed for combinations of many penicillins and cephalosporins. Because sulbactam is not well absorbed orally, prodrug forms have been developed. Numerous other penicillin sulfones have been reported to be β-lactamase inhibitors.

Penam β-Lactamase Inhibitors. Penam is the trivial name of 4-thia-1-azabicyclo[3.2.0]heptane. The report that 6-β-bromopenicillanic acid, $[2(S)$-$(2\alpha, 5\alpha, 6\beta)]$-6-bromo-3,3-dimethyl-7-oxo-4-thia-1-azabicyclo [3.2.0]heptane-2-carboxylic acid, $(R = Br, R^1 = H, R^2 = R^3 = CH_3)$ is a potent inhibitor led to intense study both of this compound and analogues. The microbiology profile of 6-β-bromopenicillanic acid has been reported and the compound has progressed to clinical trials. Mechanistic studies have demonstrated that the dihydrothiazine derivative is responsible for inactivation of β-lactamases.

Other Unusual β-Lactam-Based Inhibitors. There are a number of other unusual β-lactams reported to have β-lactamase inhibition activity.

TABLE 1. β-LACTAMASE-PRODUCING BACTERIA

Organism	β-Lactamase-producers, %[a]	Enzyme origin, %		Richmond-Sykes classification
		Chromosomal	Plasmid	
Gram-positive				
Staphylococcus aureus	80		penase	
Staphylococcus epidermidis	80		penase	
Gram-negative				
Escherichia coli	25(16–76)	15	TEM, 80; OXA-1, 7.5	I,III,V
Haemophilus influenzae	25–60		TEM, 92; ROB-1, 8	III
Neisseria gonorrhea	1–10		TEM, 100	III
Salmonella	6		TEM, 100	III
Shigella			TEM	III
Klebsiella pneumoniae	60–90		TEM, 24; SHV-1, 76	III,IV
Enterobacter	25–30	73	TEM, 27	I,III
Citrobacter	20–50	77	TEM, 23	I,III
Pseudomones aeruginosa	23	44	TEM, 9; PSE, 10; others	I,III
Bacillus catarrhalis	87	90+	BRO-1	
Bacillus fragilis	87	~100		VI

[a] Clinically resistant bacteria, virtually 100% of the *Enterobacteriaceae*, produce a low level of chromosomal enzyme that can clinically be selected for higher levels.

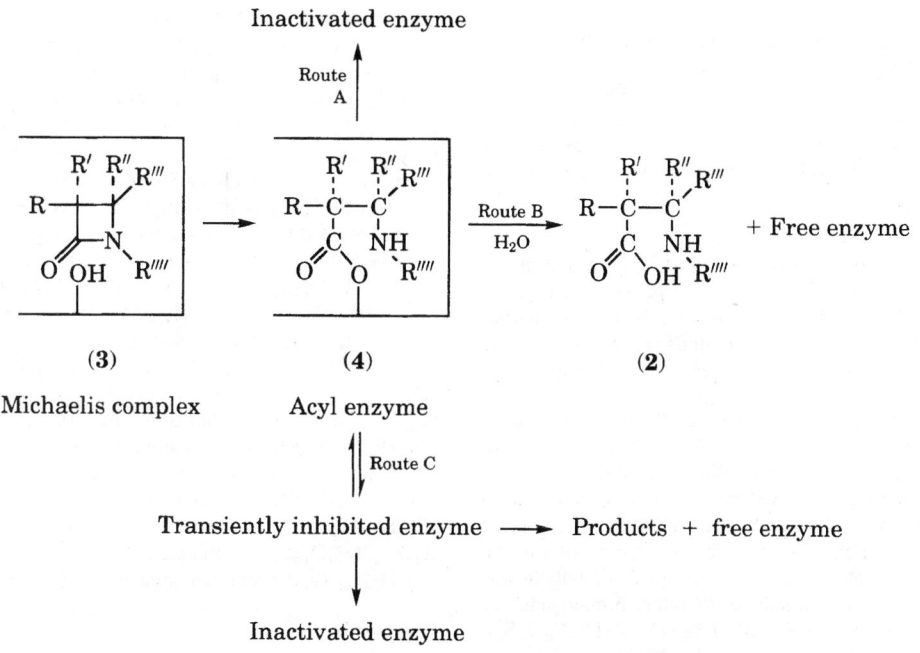

Fig. 1.   Scheme for the interaction of β-lactamase inhibitors and β-lactamases where the enzyme is represented by

In general these compounds are not very potent and are not irreversible inhibitors. Data are also very limited.

Economic Aspects

Although a broad range of β-lactamase inhibitors has been discovered, only clavulanic acid and sulbactam have been commercialized. Clavulanic acid manufactured by SmithKline Beecham is sold as an oral and parenteral product in combination with amoxicillin under the trade name Augmentin. A parenteral product in combination with ticarcillin, $C_{15}H_{16}N_2O_6S$, has the trade name, Timentin.

Sulbactam is produced by Pfizer. The oral version of sulbactam in combination with ampicillin is called Unasyn Oral, which is the mutual prodrug sultamicillin. Two sulbactam parenteral products are sold, a combination product with ampicillin called Unasyn and a combination with cefoperazone called Sulperazon. In addition, sulbactam is sold alone for parenteral use with any β-lactam antibiotic as Betamaze.

J. G. STAM
Pfizer Central Research

Additional Reading

Bush, K. and R.B. Sykes: *J. Antimicrob. Chemother.* **11**, 97 (1983)
Hamilton Miller, J.M.T. and J.T. Smith, eds.: *Beta-Lactamases*, Academic Press, New York, 1979.
Knowles, J.R.: *Acc. Chem. Res.* **18**, 97 (1985).
Reading, C. and M. Cole: *J. Enzyme Inhib.* **1**, 83 (1986).

ANTIBIOTICS: β-LACTAMS.

In the period up to 1970 most β-lactam research was concerned with the penicillin and cephalosporin group of antibiotics. Since that time, however, a wide variety of new mono- and bicyclic β-lactam structures have been described. The carbapenems, characterized by the presence of the bicyclic ring system (**1**, X =CH$_2$) originated from natural sources; the penem ring (**1**, X =S) and its derivatives are the products of the chemical synthetic approach to new antibiotics. The chemical names are: 7-oxo-(R)-1-azabicyclo[3.2.0]hept-2-ene-2-carboxylic acid, $C_7H_7NO_3$, and 7-oxo-(R)-4-thia-1-azabicyclo[3.2.0]hept-2-ene-2-carboxylic acid, $C_6H_5NO_3S$, respectively.

(**1**)

Carbapenems

Carbapenems include thienamycin ($C_{11}H_{16}N_2O_4S$), MM 4550 ($C_{13}H_{16}N_2O_9S_2$), MM 13902 ($C_{13}H_{16}N_2O_8S_2$), MM 17880 ($C_{13}H_{18}N_2O_8S_2$), PS-5 ($C_{13}H_{18}N_2O_4S$), carpetimycin A ($C_{14}H_{18}N_2O_6S$), asparenomycin A ($C_{14}H_{16}N_2O_6S$), and pluracidomycin A ($C_9H_{11}NO_{10}S_2$).

Occurrence, Fermentation, and Biosynthesis. Although a large number of *Streptomyces* species have been shown to produce carbapenems, only *S. cattleya* and *S. penemfaciens* have been reported to give thienamycin. Generally the antibiotics occur as a mixture of analogues or isomers and are often coproduced with penicillin N and cephamycin C.

Properties. Thienamycin is isolated as a colorless, hygroscopic, zwitterionic solid, although the majority of carbapenems have been obtained as sodium salts and, in the case of the sulfated olivanic acids, as disodium salts. Concentrated aqueous solutions of the carbapenems are generally unstable, particularly at low pH. All the substituted natural products have characteristic uv absorption properties that are often used in assay procedures. The ir frequency of the β-lactam carbonyl is in the range 1760–1790 cm^{-1}.

Structure Determinations. The structural elucidation of the early carbapenems, thienamycin, and the olivanic acids, followed a fairly similar sequence making use of both spectroscopic and degradation studies. Infrared absorption spectra suggested the presence of a β-lactam ring (ν_{max} 1765 cm^{-1}) and in the case of thienamycin a trans-arrangement of β-lactam protons was indicated by the small coupling constant ($J_{5,6} < 3$ Hz) for the β-lactam hydrogens in the nmr. For the sulfated olivanic acids, the coupling constant ($J_{5,6} \approx 6$ Hz) indicated the more familiar *cis-β*-lactam stereochemistry found in the penicillins and cephalosporins.

Reactions. Although carbapenems are extremely sensitive to many reaction conditions, a wide variety of chemical modifications have been carried out. Many derivatives of the amino, hydroxy, and carboxy group of thienamycin have been prepared primarily to study structure–activity relationships.

Synthesis. One consequence of the discovery of the carbapenem natural products has been the development of new synthetic methods, the impetus for which was provided by the exceptional antibacterial potential of the compounds coupled with the extremely poor fermentation yields. Only chemical synthesis could provide the quantities of *N*-formimidoyl thienamycin (MK 0787) necessary for clinical trials and commercial production.

A synthetic approach that involves the [3,4] bond formation using a carbene insertion reaction has been highly successful and is illustrated by the enantioselective synthesis of (+-thienamycin) starting from L-aspartic acid, $C_4H_7NO_4$.

A second method makes use of the lactone from acetone dicarboxylate and for which a synthesis form (−)-carvone has been reported.

Biological Properties. Thienamycin, the olivanic acids, and the majority of carbapenems are highly active broad-spectrum antibiotics having good stability to β-lactamases. Of the natural products, thienamycin is the most potent, having a spectrum of activity encompassing both aerobic and anaerobic gram-positive and gram-negative bacteria, including *Pseudomonas* species.

Penems

Historically, the development of penems is contemporary with that of the naturally occurring carbapenems, and the direction of penem research has clearly been influenced by the structures of the closely related natural products. The origins of the two groups of compounds are, however, quite different. Unlike carbapenems, no penems have been found in nature.

Synthesis

Woodward's Phosphorane Route. The first penem synthesis utilized an intramolecular Wittig reaction to form the [2, 3] double bond of the thiazoline ring. Reductive acylation of the penicillin-derived disulfide gave the thioester. Ozonolysis of the latter provided the oxalimide which on mild methanolysis gave the azetidinone. Well-established methods were applied to convert to the phosphorane which underwent thermal cyclization to the penem ester. Catalytic hydrogenation gave the penem acid, which was shown to possess antibacterial activity in spite of its rather limited stability Penems include SCH 29482 ($C_{10}H_{13}NO_4S_2$), SCH 34343 ($C_{11}H_{14}N_2O_6S_2$), FCE 221201 ($C_{10}H_{12}N_2O_6S$), FCE 22891 ($C_{13}H_{16}N_2O_8S$), HRE 664 ($C_{15}H_{14}N_2O_6S$), SUN 5555 ($C_{12}H_{15}NO_5S$), CGP 31608 ($C_9H_{12}N_2O_4S$), CP 65207 ($C_{12}H_{15}NO_5S_3$), and FCE 25199 ($C_{15}H_{17}NO_7S$).

Biological Properties. In marked contrast to the antibacterially inactive penicillanic and cephalosporanic acids, 6-unsubstituted penems exhibit good activity against both gram-positive and gram-negative bacteria.

Economic Aspects

Extensive carbapenem and penem antibiotic research has been ongoing since thienamycin was discovered in 1978. However, only the imipenem–cilastatin combination has become a commercial product. Meropenem was expected to be the second carbapenem on the market by 1998.

ROBERT SOUTHGATE
NEAL F. OSBORNE
SmithKline Beecham Pharmaceuticals (U.K.)

Additional Reading

Brown, A.G., M.J. Pearson, and R. Southgate: *Comprehens. Med. Chem.* **2**, 655 (1990).

Ratcliffe, R.W. and G. Albers-Schönberg: "The Chemistry of Thienamycin and Other Carbapenem Antibiotics" and I. Ernest, "The Penems" in R.B. Morin and M. Gorman, eds., *Chemistry and Biology of β-Lactam Antibiotics*, Academic Press, New York, 1982, Chapts. 4 and 5.

Southgate, R. and S. Elson: *Progr. Chem. Org. Nat. Prod.* **47**, 1 (1985).

ANTIBIOTICS: CEPHALOSPORINS.

The cephalosporins, a subgroup of β-lactam antibiotics, consist of a 4-membered lactam ring fused through the nitrogen and the adjacent tetrahedral carbon atom to a second heterocycle forming a 6-membered dihydrothiazine ring. Other structural features common to all the cephalosporins are a carboxyl group on the dihydrothiazine ring on the carbon next to the ring nitrogen and a functionalized amino group on C-7, the carbon of the β-lactam ring opposite the nitrogen. These features are evidenced in 7-aminocephalosporanic acid (7-ACA), $C_{10}H_{12}N_2O_5S$ (**1**). Cephalosporins, like all β-lactam antibiotics, exert their antibacterial effect by interfering with the synthesis of the bacterial cell wall. These antibiotics tend to be "irreversible" inhibitors of cell wall biosynthesis and they are usually bactericidal at concentrations close to their bacteriostatic levels. Cephalosporins are widely used for treating bacterial infections. They are highly effective antibiotics and have low toxicity.

(**1**)

Nomenclature and Stereochemistry

Naturally occurring cephalosporins, cephamycins, and the 7-formamido cephalosporins are deacetoxycephalosporin C ($C_{14}H_{19}N_3O_8S$), deacetyl-cephalosporin C ($C_{14}H_{19}N_3O_7S$), cephalosporin C ($C_{16}H_{12}N_2O_5S$), O-carbamoyldeacetylcephalosporin C ($C_{15}H_{19}N_4O_8S$), 3'-methylthiodeacetoxycephalosporin C (F_1) ($C_{15}H_{21}N_3O_6S_2$), 3'-sulfothiodeacetoxycephalosporin C (F_2), ($C_{14}H_{19}N_3O_9S_3$), C43-219 ($C_{19}H_{28}N_4O_8S_2$), 7α-methoxycephalosporin C ($C_{17}H_{23}N_3O_9S$), cephamycin C ($C_{16}H_{22}N_4O_9S$), cephamycin A ($C_{25}H_{29}N_3O_{14}S_2$), cephamycin B ($C_{25}H_{29}N_3O_{11}S$), Takeda C2801X ($C_{25}H_{29}N_3O_{12}S$), SF-1623 ($C_{15}H_{21}N_3O_{10}S_3$), SQ 28, 516 ($C_{36}H_{55}N_{11}O_{15}S_1$), SQ 28, 517 ($C_{36}H_{56}N_{12}O_{14}S_1$), cephabacin F_1 or chitinovorin A ($C_{26}H_{41}N_9O_{11}S$), cephabacin F_2 or chitinovorin B ($C_{29}H_{46}N_{10}O_{12}S$), cephabacin F_3 ($C_{32}H_{51}N_{11}O_{13}S$), cephabacin F_4 ($C_{26}H_{41}N_9O_{12}S$), cephabacin F_5 ($C_{29}H_{46}N_{10}O_{14}S$), cephabacin F_6 ($C_{32}H_{51}N_{11}O_{15}S$), cephabacin F_7 ($C_{26}H_{41}N_7O_{12}S$), cephabacin F_8 ($C_{29}H_{46}N_8O_{14}S$), cephabacin F_9 ($C_{32}H_{51}N_9O_{14}S$), cephabacin H_1 ($C_{25}H_{40}N_8O_{10}S$), cephabacin H_2 ($C_{28}H_{45}N_9O_{11}S$), cephabacin H_3 ($C_{31}H_{50}N_{10}O_{12}S$), cephabacin H_4 ($C_{25}H_{40}N_8O_{11}S$), cephabacin H_5 ($C_{28}H_{45}N_9O_{13}S$), cephabacin H_6 ($C_{31}H_{50}N_{10}O_{14}S$), cephabacin M_1 ($C_{31}H_{50}N_8O_{13}S$), cephabacin M_2 ($C_{34}H_{55}N_9O_{15}S$), cephabacin M_3 ($C_{37}H_{60}N_{10}O_{16}S$), cephabacin M_4 ($C_{41}H_{69}N_{11}O_{15}S$), cephabacin M_5 ($C_{44}H_{74}N_{12}O_{17}S$), and cephabacin M_6 ($C_{47}H_{79}N_{13}O_{18}S$).

Biogenesis

The biosynthesis of cephalosporins and penicillins both start from the amino acids and proceed via δ-(L-α-aminoadipyl)-L-cysteinyl-D-valine (LLD-ACV), $C_{14}H_{25}N_3O_6S$, often referred to as the Arnstein tripeptide. Because LLD-ACV is not transported into intact cells, a cell-free system was required to determine that this intermediate is the precursor of the penicillins. The cell-free system, obtained from *C. acremonium*, converts LLD-ACV into isopenicillin N, $C_{14}H_{21}N_3O_6S$. Isopenicillin N (IPN) synthetase, the enzyme which catalyzes this conversion, requires oxygen, Fe^{2+}, a reducing agent such as ascorbate, and a thiol group such as that of dithiothreitol for high activity. Isopenicillin N synthetase is present in *P. chrysogenum* and in species of *Streptomyces* as well as in *C. acremonium*. In the *Cephalosporium* species and the *Streptomyces* species, an epimerase is present that converts isopenicillin N into penicillin N, $C_{14}H_{21}N_3O_6S$, which then undergoes a ring expansion to deacetoxycephalosporin C. The ring expansion enzyme (REX) from *C. acremonium* appears to be bifunctional; it also catalyzes the subsequent hydroxylation of deacetoxycephalosporin C to deacetylcephalosporin C, itself a precursor of cephalosporin C. The introduction of the methoxyl group is also a two-step process involving molecular oxygen. Using cephalosporin C or the corresponding carbamoyl derivative as a substrate, another dioxygenase catalyzes the incorporation of a 7α-hydroxy function. The resulting 7α-hydroxycephalosporin is then methylated using S-adenosylmethionine to form the corresponding 7α-methoxycephalosporin. The details of these steps are discussed in depth in the literature. Rapid advances in this area have been made possible by the successful cloning and expression of isopenicillin N synthetase (IPNS) and ring expansion-hydroxylase (REX).

Physical Properties

Most cephalosporin antibiotics are white, off-white, tan, or pale yellow solids that are usually amorphous, but can sometimes be obtained crystalline. The cephalosporins do not usually have sharp melting points, but rather decompose upon heating at elevated temperatures. The acid strength, pK_a, of the carboxyl group on the dihydrothiazine ring depends on environment. Representative pK_a values are given in Table 1, as are other physical properties.

One of the distinguishing physical characteristics of the cephalosporins is the infrared stretching frequency of the β-lactam carbonyl. This absorption occurs at higher frequencies (1770–1815 cm^{-1}) than those of either normal secondary amides (1504–1695 cm^{-1}) or ester carbonyl groups (1720–1780 cm^{-1}).

Chemical Properties

Much of the chemical reactivity of the β-lactam antibiotics is associated with the β-lactam moiety. The geometry and the accompanying increased ring strain results in very little, if any, amide-resonance stabilization leading to a marked increase in chemical reactivity when compared to a normal amide. In fact, in many instances the reactivity of the lactam carbonyl is

TABLE 1. PROPERTIES OF CEPHALOSPORINS

Name	pK$_a$[b]	β-Lactam IR stretching frequency, cm^{-1}[c]	UV absorption, λ max, nm (ε, cm^{-1} M^{-1})[d]	NMR absorption, ppm[a]		
				H-7[e]	H-6[f]	$J_{6,7}$, Hz[g]
7-ACA	1.75, 4.63	1806	261(8500)	5.53d 4.83d	5.13d	4.5
cephalosporin C	<2.6, 3.1, 9.8	1780	260(8900)	5.66d	5.15d	4.7
cephalothin	5.0	1760	265(9000)	5.70d	5.14d	4.5
cephalexin	5.2, 7.3	1775	260(7750)	6.10d	5.45d	4.2
cephamycin C	4.2, 5.6, 10.4[h]	1770	264(6900) 242(5700)		5.19s	

[a] In D$_2$O relative to external standard tetramethyl silane (TMS); s = singlet, d = doubet.
[b] Values for aqueous solutions unless otherwise noted, either by direct determination or by extrapolation from mixed solvents.
[c] Nujol mull.
[d] In aqueous solution.
[e] Range 5.23–6.21.
[f] Range 4.24–5.46.
[g] $J_{6,7}(cis)$ = 4–5 Hz, $J_{6,7}(trans)$ = 1.5–2 Hz, J_{7-NH} = 8–11 Hz.
[h] In 66% DMF.

analogous to that of a carboxylic acid anhydride. Fused β-lactam antibiotics are readily attacked by nucleophiles with resultant ring opening and loss of biological activity. The cephalosporins are more resistant to ring opening than the penicillins.

Biological Properties

The clinical effectiveness of the cephalosporins depends on a number of properties. The antibiotic must inhibit, or preferably kill, bacteria at acceptable concentrations of the drug (*in vitro* activity); it must be capable of achieving host serum and tissue levels greater than those required to inhibit the pathogenic organism; and the selective toxicity profile must allow for safe administration to the host.

Classification. As of 1991, there were approximately fifty different cephalosporins in clinical use or at an advanced stage of evaluation and development. These include oral cephalosporins [cephalexin (C$_{16}$H$_{17}$N$_3$O$_4$S), cefaclor (C$_{15}$H$_{14}$N$_3$O$_4$S$_1$Cl), cephradine (C$_{16}$H$_{19}$N$_3$O$_4$S), cefadroxil (C$_{16}$H$_{17}$N$_3$O$_5$S), cefixime (C$_{16}$H$_{15}$N$_5$O$_7$S$_2$), ceftibuten (C$_{15}$H$_{14}$N$_4$O$_6$S$_2$), ceprozil (C$_{18}$H$_{19}$N$_3$O$_5$S), and C$_{15}$H$_{15}$N$_5$O$_6$S$_2$]; oral cephalosporins-prodrugs [cefuroximeaxetil (C$_{20}$H$_{22}$N$_4$O$_{10}$S$_1$), cefpodoximeproxetil (C$_{21}$H$_{27}$N$_5$O$_9$S$_2$), and (C$_{18}$H$_{19}$N$_5$O$_7$S$_2$)]; cephamycins [cefoxitin (C$_{15}$H$_{15}$N$_3$O$_6$), cefmetazole (C$_{14}$H$_{15}$N$_7$O$_4$S$_3$), cefminox (C$_{15}$H$_{19}$N$_7$O$_6$S$_3$), and cefotetan (C$_{16}$H$_{15}$N$_7$O$_7$S$_4$)]; parenteral cephalosporins [cephalothin (C$_{16}$H$_{16}$N$_2$O$_6$S$_2$), cephacetrile (C$_{13}$H$_{13}$N$_3$O$_6$S), cephapirin (C$_{17}$H$_{17}$N$_3$O$_6$S$_2$), cefamandole (C$_{11}$H$_{18}$N$_6$O$_5$S$_2$), cefonicid (C$_{18}$H$_{17}$N$_6$O$_8$S$_3$Na), cefazolin (C$_{14}$H$_{14}$N$_8$O$_4$S$_3$), ceforanide (C$_{20}$H$_{19}$N$_7$O$_6$S$_2$), cefoperazone (C$_{25}$H$_{27}$N$_9$O$_8$S$_2$), cefuroxime (C$_{16}$H$_{16}$N$_4$O$_8$S), cefotaxime (C$_{16}$H$_{17}$N$_5$O$_7$S$_2$), ceftizoxime (C$_{13}$H$_{13}$N$_5$O$_5$S$_2$), cefmenoxime (C$_{16}$H$_{17}$N$_9$O$_5$S$_3$), ceftriaxone (C$_{18}$H$_{17}$N$_8$O$_7$S$_3$Na), ceftazidime (C$_{22}$H$_{22}$N$_6$O$_7$S$_2$), cefsulodin (C$_{22}$H$_{20}$N$_4$O$_8$S$_2$), cefpiramide (C$_{25}$H$_{24}$N$_8$O$_7$S$_2$), cefpirome (C$_{22}$H$_{22}$N$_6$O$_5$S$_2$), cefpimizole (C$_{28}$H$_{26}$N$_6$O$_{10}$S$_2$), cefepime (C$_{19}$H$_{24}$N$_6$O$_5$S$_2$), C$_{28}$H$_{21}$N$_9$O$_{11}$S$_3$Na$_2$, and C$_{31}$H$_{31}$N$_8$O$_8$S$_2$F$_3$]; and oxadethiacephalosporins [moxalactam latamoxef (C$_{20}$H$_{20}$N$_6$O$_8$S) and flomoxef (C$_{15}$H$_{18}$N$_6$O$_6$S$_2$F$_2$)]. Cephalosporins may be classified for convenience by their clinical pharmacology, β-lactamase resistance, chemical structure, or their antibacterial spectrum. The most common classification, which is somewhat arbitrary, divides the cephalosporins into three groups or generations, based primarily on their antibacterial spectrum. First-generation cephalosporins are characterized by good gram-positive activity and modest to weak gram-negative activity. Third-generation cephalosporins have an expanded gram-negative spectrum and are the most active against enteric gram-negative bacilli, including penicillinase-producing strains, as well as *Serratia* and *Citrobacter*.

Some other compounds tentatively labeled as examples of the as yet undefined "fourth generation" have appeared in the literature, but these are probably best thought of as third-generation cephalosporins having slight advantages over earlier examples of this group. This group includes cefpirome, cefepime, and others undergoing clinical trials.

In Vitro Antibacterial Activity and Structure-Activity Relationships. The *in vitro* antibacterial activity of any particular cephalosporin is a combination of the degree and type of activity at the target site,

the ease with which it can penetrate to the target, and the ability to resist the attack of destructive enzymes. The nature and complexity of the biochemical target(s) for β-lactam antibiotics is fairly well-established and the mechanisms of penetration are also understood to some extent. However, many factors are involved, including pharmacokinetic and pharmacodynamic properties, and the antibiotic may not perform as predicted.

Structure–activity relationships can be inferred by comparison of the antibacterial properties of the clinical agents and related compounds. Different acyl side chains can result in significant changes in the antibacterial activity, both with respect to potency and to breadth of spectrum. The highest activities are observed when the acylamino side chain at C-7 is a substituted acetic acid. Homologation of the acetic acid moiety lowers activity dramatically as exemplified by the naturally occurring cephalosporins, which all have weak activity.

One of the principal deficiencies of the older cephalosporins was the lack of resistance to β-lactamases. Compounds with improved β-lactamase resistance have one or more of the following characteristics: a second, monovalent substituent on the α-carbon of the C-7 acyl group such as is found in cefamandole and cefoperazone; a *syn*-oxime substituent, e.g., cefuroxime and ceftriaxone; a methoxy, or formamido, substituent on the β-lactam ring at the 7 α-position such as in cefoxitin. However, these various substituents have different effects, and increased resistance to one enzyme does not indicate resistance to all β-lactamases.

Intrinsic Activity. β-Lactam antibiotics affect sensitive bacteria by inhibiting late stages in the biosynthesis of their cell wall peptidoglycan.

Resistance. Resistance to the cephalosporins may result from the alteration of target penicillin-binding sites (PBPs), decreased permeability of the bacterial cell wall and outer membrane, or by inactivation via enzyme-mediated hydrolysis of the lactam ring. This resistance can be either natural or acquired.

Transport and Cell Penetration. One of the causes of bacterial resistance to the cephalosporins is poor transport of the antibiotic through the outer membrane of gram-negative bacteria. This lipid-bilayer membrane carries receptor proteins for the recognition and transport of essential nutrients, but provides an effective barrier to large molecules. In the case of the cephalosporins there can be a considerable difference between the concentration required to inhibit intact cells and the concentrations required to saturate the target enzymes in broken cell preparations.

Pharmacokinetics. The pharmacokinetic properties of the cephalosporins depend to a large extent on the substituent at C-3. The 3′-acetoxy group is metabolized in the body to the less active 3′-carbamate, most other substituents, including the 3′-carbamate, are metabolically stable. Most cephalosporins are eliminated rapidly, having serum half-lives of 1–2 h.

Manufacture and Chemical Synthesis

At present all of the cephalosporins are manufactured from one of four β-lactams, cephalosporin C, penicillin V, penicillin G, and cephamycin C, which are all produced in commercial quantities by fermentation. The manufacturing process consists of three steps: fermentation, isolation, and chemical modification.

Nuclear Analogues of Cephalosporins

In the search for improved antibacterials not only has the effect produced by the variation of the C-7 amido side chain and the 3′ substituent been studied, but so also has the more synthetically challenging question of the effect of changes in the cephem nucleus. Nuclear analogues have been studied since the early 1970s but only the oxacephem class has reached the marketplace.

Cephalosporins With Special Properties

Chromogenic Cephalosporins. A 3-substituted pyridinium cephalosporin known as PADAC, $C_{27}H_{26}N_6O_4S_2$, is purple in color but on hydrolysis or treatment with β-lactamase releases the 3′-pyridinium group with concomitant loss of the purple color. PADAC is an example of the chromogenic cephalosporins which are useful in studying interactions with β-lactamases.

Uses

The cephalosporins are used for treating infectious diseases of bacterial origin in both humans and animals. First-generation cephalosporins such as cephalothin and cephalexin are the most active against staphylococci and nonenterococcal streptococci and are effective alternatives to the penicillins in patients with endocarditis, osteomyelitis, septic arthritis, and cellulitis. They are especially useful for treating patients who are allergic to the penicillins or who have mixed infections from gram-positive and gram-negative bacteria. Although these drugs have proved useful in treating infections such as bacteremias, urinary tract infections, and pneumonias, caused by gram-negative bacilli, their use as single agents in this regard is not recommended because activity against gram-negative organisms is somewhat weak and unpredictable. The first-generation cephalosporins have been widely used for prophylaxis in cardiovascular, orthopedic, biliary, pelvic, and intraabdominal surgery.

Whereas third-generation cephalosporins do have some coverage against gram-positive infections, they are not the agents of choice. Similarly, most community-acquired infections are better treated with drugs other than the third-generation cephalosporins. The treatment of meningitis is an important exception.

The third-generation cephalosporins are effective in the treatment of bacteremias, pneumonias, urinary tract infections, intraabdominal infections, and skin and soft tissue infections.

Toxicity

The cephalosporins generally cause few side effects. Thrombophlebitis occurs as a result of intravenous administration of all cephalosporins. Hypersensitivity reactions related to the cephalosporins are the most common side effects observed, but these are less common than found with the penicillins.

Although immediate reactions of anaphylaxis, bronchospasm, and urticaria have been reported, most commonly patients exhibiting an adverse reaction develop a maculopapular rash, usually after several days of therapy. They may also develop fever and eosinophilia.

JOHN ROBERTS
Hoffmann-LaRoche

Additional Readings

Baldwin, J.E.: in P.H. Bentley and R. Southgate, John Roberts, Hoffmann- La Roche eds., *Recent Advances in the Chemistry of β-Lactam Antibiotics: Proceedings of the Fourth International Symposium,* The Royal Society of Chemistry, London, 1989, p. 1.

Donowitz, G.R. and G.L. Mandell: *N. Engl. J. Med.* **318,** 490 (1988).

Mandell, G.L.: in G.L. Mandell, R.G. Douglas, Jr., and J.E. Bennett, *Anti-infective Therapy.* John Wiley & Sons, Inc., New York, NY, 1985, p. 76.

Squires, E. and R. Cleedland: *Microbiology and Pharmacokinetics of Parenteral Cephalosporins,* Roche Laboratories, Nutley, NJ, 1985.

ANTIBIOTICS: CHLORAMPHENICOL AND ANALOGUES.

Chloramphenicol (**1**, R = NO$_2$), $C_{11}H_{12}Cl_2N_2O_5$, is a commercially significant antibacterial agent and its status in clinical practice has been reviewed. Although widespread use of this antibiotic declined in the United States in the 1960s because of reports of serious toxic effects, this situation changed a decade later when ampicillin-resistant *Hemophilus influenzae* emerged on the clinical scene. The appearance of *Bacteroides* species and of *Streptococcus pneumoniae* resistant to β-lactam antibiotics contributed further to

the resurgence. In the 1970s, chloramphenicol also became important in the treatment of serious *Salmonella* invasive gastroenteritis in infants less than three months of age. Because chloramphenicol crosses the blood brain barrier, it is indicated in infections of the central nervous system caused by susceptible organisms.

(1)

(2)

The emergence of quinolones and other antibiotics is expected to curtail the use of chloramphenicol in the future, but this drug is relatively inexpensive, orally active, and the toxicity, except for the rare idiosyncratic aplastic anemia, can be managed through monitoring of blood levels by sensitive modern analytical procedures. However, clinical use is being further curtailed by the emergence of chloramphenicol-resistant organisms. In Table 1, the median *in vitro* susceptibilities of chloramphenicol, thiamphenicol, and a fluoroanalogue, florfenicol (**2**), $C_{12}H_{14}Cl_2FNO_4S$, against a host of chloramphenicol-sensitive and -resistant organisms are given. Bacteria resistant to chloramphenicol are also resistant to thiamphenicol.

Both chloramphenicol and thiamphenicol cause reversible bone marrow suppression. The irreversible, often fatal, aplastic anemia, however, is only seen for chloramphenicol. This rare (1 in 10,000–45,000) chloramphenicol toxicity has been linked to the nitroaromatic function. Thiamphenicol, which is less toxic than chloramphenicol in regard to aplastic anemia, lacks potency and has never found much usage in the United States. An analogue of thiamphenicol having antimicrobial potencies equivalent to chloramphenicol was sought. Florfenicol (**2**) was selected for further development from a number of closely related structures.

Bacterial Resistance of Amphenicol

Of the many mechanisms of bacterial resistance to chloramphenicol and thiamphenicol, the plasmid-mediated transmissible resistance conferred by the presence in resistant bacteria of chloramphenicol-acetyltransferases (CAT) is the most important.

Structure–Activity Relationship of Chloramphenicol

Structure–activity and mechanism of action studies indicate that the requirements for chloramphenicol activity are the D-threo-configuration, the 1,3-propanediol moiety, and a strong electron-withdrawing group on the aromatic ring. The L-threo, the mirror image of (**1**), and the D-erythro and L-erythro isomers are not biologically active. Thus the speculation arose that certain specific intramolecular dipolar attractive interactions must exist in chloramphenicol leading to greater stabilization of one particular conformer over the others where biological activity results from the most stable conformer. The three basic conformational isomers of chloramphenicol are shown in Figure 1.

Fluoroanalogues

Because the lack of biological activity of 3-substituted chloramphenicols reported previously might result from the inability to exist in the "active" (**a**) type conformation, it was speculated that the size and nature of the C-3 substituent, maintenance of a low barrier to rotation about the C-2—C-3 bond, and the length of the carbon-substituent atom bond at C-3 were highly critical for achieving a conformational preference of the (**a**) type. Thus, on the basis of the van der Waals radii of fluorine and oxygen being the same (0.14 nm) and the average C—O and C—F bond lengths being close (0.131 nm and 0.138 nm, respectively), the C-3-hydroxyl group

TABLE 1. *IN VITRO* SUSCEPTIBILITIES OF AMPHENICOLS

Organism strain	No. of strains tested	Susceptibility[b]	MIC, μg/mL[a]		
			Chloramphenicol	Thiamphenicol	Florfenicol
Enterobacter	4	S	4	64	4
	14	R	512	1024	8
Citrobacter	3	S	4	32	8
	3	R	512	1024	128
E. coli	9	S	4	64	8
	20	R	256	1024	8
Klebsiella	9	S	4	64	8
	20	R	512	1024	4
Providencia	4	S	16	128	8
	12	R	128	1024	8
Pseudomonas	13	R	128	128	256
Serratia	6	S	16	512	64
	18	R	512	1024	64
Salmonella	15	S	4	32	8
	7	R	256	1024	8
Shigella	9	S	1	2	2
Proteus	23	R	256	512	8
Acinetobacter	4	R	64	512	128
Staphylococcus aureus	9	S	4	8	8
	7	R	64	512	8
Streptococcus pneumonae	3	R	8	64	4

[a] Agar dilution, 24 h, Mueller-Hinton agar.

[b] S = susceptible; R = resistant.

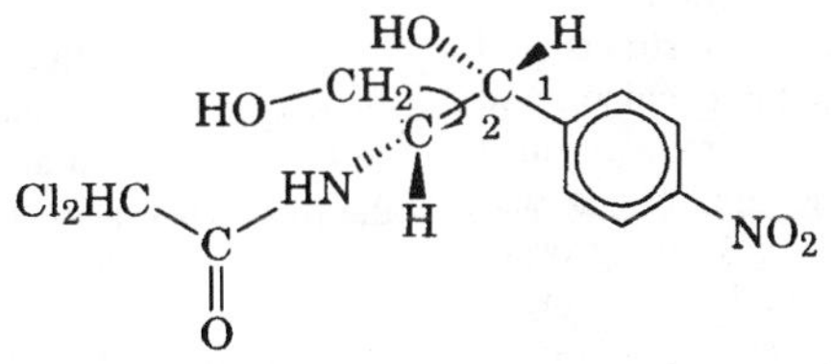

(1)

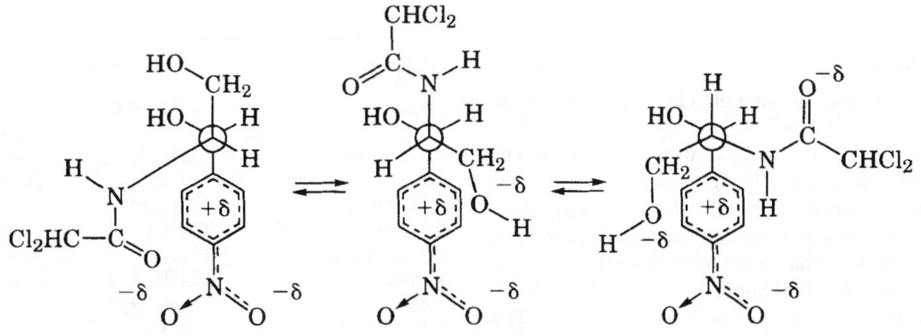

Fig. 1.   Conformers (**a**), (**b**), and (**c**) of chloramphenicol (**1**, R = NO$_2$)

of chloramphenicol was replaced by a fluorine atom. Optical rotation measurements of 3-fluoro-chloramphenicol, C$_{11}$H$_{11}$Cl$_2$FN$_2$O$_4$, in ethanol, gave $[\alpha]_D = +24.4°$ and in dimethylformamide gave $-23.4°$. Thus, as in the case of chloramphenicol, the optical rotation changed from a positive to a negative value on going from a protic to a dipolar aprotic solvent. The solid-state conformation was determined by single crystal x-ray structure analysis and the crystals contained two rotameric structures in the asymmetric unit.

Biological Activity. The biological activity of 3-fluoro-chloramphenicol against chloramphenicol-sensitive and -resistant organisms was determined. Potencies against sensitive strains are similar to those of chloramphenicol. Additionally, this fluoroderivative is highly active against chloramphenicol-resistant organisms having MICs ranging from 1 to 16 μg/L. This result prompted the synthesis and biological evaluation of a number of amphenicols containing a fluorine atom at the 3-position. The most

promising florfenicol (**2**), 3-fluorothiamphenicol, is not only active against the chloramphenicol–thiamphenicol-resistant strains, but the potency of florfenicol against sensitive organisms is also superior to any of the other amphenicols.

Veterinary Potential of Florfenicol. The absolute ban on the use of chloramphenicol in food-producing animals in the United States and Canada has accentuated the need for an effective broad-spectrum antibiotic in animal food medicine. Florfenicol and other antibiotics commonly used in veterinary medicine have been evaluated *in vitro* against a variety of important veterinary and aquaculture pathogens. Florfenicol was broadly active, having MICs lower than those of chloramphenicol in each of the genera tested. Florfenicol was also superior to chloramphenicol, thiamphenicol, oxytetracycline, ampicillin, and oxolinic acid against the most commonly isolated bacterial pathogen of fish in Japan.

Structure–Activity Relationships of 3-Fluoro-Amphenicols. A number of analogues of 3-fluorochloramphenicol and florfenicol (2) have been synthesized and the biological activities examined. Replacement of the dichloroacetyl group by a difluoroacetyl function in both series led basically to retention of potency and the spectrum of activity of the parent structures. However, changing the difluoroacetyl to a trifluoro- or a chlorodifluoroacetyl group abolished the antimicrobial activity almost completely. Reduced level of potency was also seen when the dichloroacetyl group was changed to a chlorofluoroacetyl group. Other amide functions such as methoxyacetyl or methylsulfonylacetyl did not give any appreciable activity. In the florfenicol structure, changing the methylsulfonyl group to methylsulfoxide greatly reduced the potency and the methylthio analogue was practically inactive.

Mechanism of Action of Florfenicol. The inhibitory activities of chloramphenicol $(1, R = NO_2)$, thiamphenicol $(1, R = SO_2CH_3)$, and florfenicol (2) against a sensitive *E. coli* strain have been studied. In two different liquid media, both chloramphenicol and florfenicol allowed only 20–30% residual growth at a drug concentration of 2 mg/L, whereas a thiamphenicol concentration of 25 mg/L was required to produce a similar effect. Florfenicol was also found to be a selective inhibitor of prokaryotic cells. At concentrations of 1 mg/L chloramphenicol and florfenicol, and at a concentration of 25 mg/L, thiamphenicol, inhibited protein synthesis.

In Vivo Effects of Florfenicol. Florfenicol is similar to thiamphenicol in acute toxicity by oral and subcutaneous (sc) administration, but is comparable to chloramphenicol by intraperitoneal (ip) and intravenous (iv) routes.

The efficacy of florfenicol *in vivo* was determined by measuring the dose required to obtain values for protection from infection in 50% of the animals (PD_{50}) against 10 chloramphenicol-resistant strains and two chloramphenicol-sensitive isolates. Florfenicol, chloramphenicol, and thiamphenicol were evaluated concurrently against each strain. Against sensitive *Enterobacter*, PD_{50} by the subcutaneous and oral routes was similar for florfenicol and chloramphenicol (25 mg/kg sc, 5 mg/kg oral), but higher for thiamphenicol (30 mg/kg sc, 20 mg/kg oral). A dramatic effect was seen for florfenicol against *Shigella* (3 mg/kg sc, 2 mg/kg oral) as compared to chloramphenicol and thiamphenicol (100 mg/kg by both routes). Against resistant strains of *Enterobacter, Klebsiella, Providencia, Serratia, Salmonella,* and *Staphylococcus,* the PD_{50} values for florfenicol ranged from 5 to 60 mg/kg whereas chloramphenicol and thiamphenicol were practically ineffective.

Pharmacokinetics in Nonrodents

The pharmacokinetic disposition of florfenicol (2) was studied in preruminant veal calves after administration of a single 22-mg/kg dose intravenously, orally after a 12-h fast, and orally 5 min postfeeding. The disposition of florfenicol in veal calves following a single iv dose was adequately described by a two-compartment open model where there was no significant effect of the animal's age on the pharmacokinetic parameters. Calves given the oral doses had a complex absorption pattern and delayed absorption. Administering florfenicol with milk delayed the onset of absorption and therefore the time to peak concentration. The disposition of the serum concentration of florfenicol in veal calves given by either oral method could be adequately described by a one-compartment pharmacokinetic model with first-order drug absorption and first-order drug elimination. The bioavailability of florfenicol was significantly less when given with milk replacer than when given on an empty stomach: after a 12-h fast median bioavailability was 88% of the dose; and given 5 min post-feeding, median bioavailability of the drug was 65%.

The elimination half-life of florfenicol after a single iv dose of 22 mg/kg (138–204 min) compares well with the elimination half-life of chloramphenicol $(1, R = NO_2)$ reported in cattle, except in very young calves.

Florfenicol concentrations in tissues and body fluids of male veal calves were studied after 11 mg/kg intramuscular doses administered at 12-h intervals. Concentrations of florfenicol in the lungs, heart, skeletal muscle, synovia, spleen, pancreas, large intestine, and small intestine were similar to the corresponding serum concentrations, indicating excellent penetration of florfenicol into these tissues. Because the florfenicol concentration in these tissues decreased over time as did the corresponding serum concentrations, it was deemed that florfenicol equilibrated rapidly between these tissues and the blood. Thus serum concentrations of florfenicol can be used as an indicator of drug concentrations in these tissues.

Florfenicol has a wide tissue distribution, similar to that reported for chloramphenicol in calves and thiamphenicol in humans.

TATTANAHALLI NAGABHUSHAN
GEORGE H. MILLER
KANWAL J. VARMA
Schering-Plough Research

Additional Reading

Nagabhushan, T.L., D. Kandasamy, H. Tsai, W.N. Turner and G.H. Miller: *Proc. 11th ICC 19th ICAAC Am. Soc. Microbiol.* **1,** 442–443 (1980); U.S. Pat. 4,235,892 (1980), T.L. Nagabhushan.
Nagabhushan, T.L. and A.T. McPhail: unpublished data, 1979.
Neu, H.C., K.P. Fu and Kong, K.: *Curr. Chemother. Infect. Dis.* **1,** 446–447 (1980).
Schafer, T.W., E.L. Moss, T.L. Nagabhushan and G.H. Miller: *Curr. Chemother. Infect. Dis.* **1,** 444–446 (1980).

ANTIBIOTICS: ELFAMYCINS. The elfamycins are so named because they exhibit antimicrobial activity through the inhibition of protein biosynthesis via binding to the *el*ongation *fac*tor Tu. All of the known elfamycins are listed in Table 1. These antibiotics are distinguished by low mammalian toxicity, narrow-range antimicrobial activity, and positive effects on feed utilization and growth promotion in farm animals. Elfamycins also improve milk production in lactating ruminants.

Elfamycins are natural products. Aurodox and efrotomycin have been synthesized chemically.

Properties

Elfamycins are slightly acidic because of the 4-hydroxy-2-pyridone or the carboxylic acid moiety. They are soluble in most polar organic solvents and the alkali and ammonium salts are water-soluble. The extractability of the free acids from aqueous solution into solvents such as dichloromethane and ethyl acetate is utilized in their isolation from fermentation broths.

Production, Biosynthesis, and Chemistry

The production of elfamycins has been described in the literature. Fermentation yield improvements with aurodox has proved difficult

TABLE 1. ELFAMYCINS

Antibiotic	Molecular formula	Producing organism
aurodox	$C_{44}H_{62}N_2O_{12}$	*Streptomyces goldiniensis* var. *goldiniensis*
kirromycin[a]	$C_{43}H_{60}N_2O_{12}$	*S. collinus* Tü 365
azdimycin	unknown	*S. diastatochromogenes* ATCC 31013
efrotomycin	$C_{59}H_{88}N_2O_{20}$	*S. lactamdurans* NRRL 3802
dihydromocimycin	$C_{43}H_{62}N_2O_{12}$	*S. ramocissimus* CBS190.69
heneicomycin	$C_{44}H_{62}N_2O_{11}$	*S. filipinensis* NRRL 11044
kirrothricin	$C_{44}H_{64}N_2O_{10}$	*S. cinnamomeus* Tü 89
factumycin	$C_{44}H_{62}N_2O_{10}$	*S. lavendulae* ATCC 31312
MSD A63A	unknown	*Streptoverticillum hiroshimense*
L681,217	$C_{36}H_{53}N_1O_{10}$	*S. cattleya* ATCC 39203
SB22484, factor 3	$C_{41}H_{56}N_2O_{11}$	*S.* strain NRRL 15496
SB22484, factor 4	$C_{42}H_{58}N_2O_{11}$	*S.* strain NRRL 15496
phenelfamycin A	$C_{51}H_{71}N_1O_{15}$	*S. violaceoniger* str. AB999F-80, AB1047T-33
phenelfamycin B	$C_{61}H_{71}N_1O_{15}$	*S. violaceoniger* str. AB999F-80, AB1047T-33
phenelfamycin C	$C_{58}H_{83}N_1O_{18}$	*S. violaceoniger* str. AB999F-80, AB1047T-33
phenelfamycin D	$C_{58}H_{83}N_1O_{18}$	*S. violaceoniger* str. AB999F-80, AB1047T-33
phenelfamycin E	$C_{65}H_{95}N_1O_{21}$	*S. violaceoniger* str. AB999F-80, AB1047T-33
phenelfamycin F	$C_{65}H_{95}N_1O_{21}$	*S. violaceoniger* str. AB999F-80, AB1047T-33
unphenelfamycin	$C_{43}H_{65}N_1O_{14}$	*S. violaceoniger* str. AB999F-80, AB1047T-33
LL-E19020α	$C_{65}H_{95}N_1O_{21}$	*S. lydicus* sp. *tanzanius* NRRL 18036
LL-E19020β	$C_{65}H_{95}N_1O_{21}$	*S. lydicus* sp. *tanzanius* NRRL 18036
UK-69,753	$C_{58}H_{86}N_2O_{18}$	*Amycolatopsis orientalis* ATCC 53550
N-demethylefrotomycin	$C_{58}H_{86}N_2O_{20}$	*Nocardia* ATCC 53758

[a] Identical to mocimycin which was isolated from *S. ramocissimus* CB5190.69.

because of feedback inhibition. Aurodox-resistant strains, however, respond positively to conventional mutagenic methods, leading to yield increases from 0.4 to 2.5 g/L. Scale-up of efrotomycin fermentations was found to be particularly sensitive to small changes in sterilization conditions of the oil-containing medium used.

Biological Properties

Mode of Action. Elfamycins block bacterial protein biosynthesis at the level of elongation factor Tu (EF-Tu).

Antimicrobial Activity. The elfamycins' antimicrobial specificity and lack of toxicity in animals can be explained in view of species-dependent specificity of elfamycin binding to EF-Tu. Inefficient cellular uptake or the presence of a nonresponding EF-Tu were cited as responsible factors for the natural resistance in *Halobacterium cutirubrum, Lactobacillus brevis,* and in actinomycetes. The low activity of elfamycins against *S. aureus* was also attributed to an elfamycin-resistant EF-Tu system. However, cross-resistance with other antibacterial agents has not been observed.

Elfamycins have similar *in vitro* antimicrobial spectra and the activity against *Moraxella, Pasteurella, Yersinia, Haemophilus, Streptococcus, Corynebacterium,* and *Neisseria* appears to be common.

Growth Promotion. Elfamycins, in general, enhance the growth of farm animals. Growth improvement and feed conversion was studied using aurodox in chicks and turkeys. Efrotomycin is being developed as a growth-promoting agent for swine.

Stable growth promoting feed additives were prepared by granulation of the drug with alginic acid and magnesium hydroxide and adding it to oiled rice hulls or by adsorption of elfamycins onto corn cob grits and coating with 10% tristearin. A molecular efrotomycin dispersion with (2-vinyl-pyridine)-styrene copolymer served as a postrumen effective dosage form for oral administration.

Improvement of Lactation. Qualitative and quantitative improvements in milk production. See also **Milk and Milk Products**. Also are dependent on changes of ruminal volatile fatty acid (VFA) production and VFA composition. Increased propionate production at the expense of acetate and butyrate is responsible for growth enhancement and is achieved by additives such as polyether antibiotics. Relatively high levels of acetate and butyrate, however, are required to maintain adequate fat content in milk. Oral administration of elfamycins at concentrations of 1–10 mg/kg of animal body weight per day has been shown to increase milk volume from 2–15% relative to untreated animals without compromising fat content. The methods of administration to ruminants such as dairy cows and goats were similar to those employed for improving feed utilization and growth promotion.

Economic Aspects

The potential usefulness of elfamycins as growth promoters and feed-conversion enhancers is now generally recognized. Low original fermentation yields and difficulties in yield improvements discouraged early attempts to develop aurodox and mocimycin (kirromycin) commercially. A development program for efrotomycin, however, is ongoing as of this writing. Some of the newer elfamycins, such as the LL-E19020 pair, are considerably more active growth promoters than either aurodox or mocimycin, pointing toward the emergence of a second generation of elfamycins.

H. MAEHR
Hoffmann-LaRoche Inc.

Additional Reading

Miller, D.L. and H. Weissbach: *Arch. Biochem. Biophys.* **141**, 26 (1970).
Parmeggiani, A. and G.W.M. Swart: *Ann. Rev. Microbiol.* **39**, 557 (1985).
Parmeggiani, A. and G. Sander: *Top. Antibiot. Chem.* **5**, 159 (1980).
U.S. Pat. 4,808,412 (Feb. 28, 1989), E.P. Smith, and S.H. Wu.
U.S. Pat. 4,336,250 (June 22, 1982), C.C. Scheifinger.

ANTIBIOTICS: GLYCOPEPTIDES (Dalbaheptides).

The vancomycin–ristocetin family of glycopeptides is a subclass of linear sugar-containing peptides composed of seven amino acids cross-linked to generate a specific stereochemical configuration. This configuration forms the basis of a particular mechanism of action, e.g., complexation with the D-alanyl-D-alanine terminus of bacterial cell wall components. Because the mechanism of action is the distinguishing feature of these peptides, the term dalbaheptide, from D-al (anyl-D-alanine)b(inding)a(ntibiotics) having

hept(apept)ide structure, has been proposed to distinguish them within the larger and diverse groups of glycopeptide antibiotics.

About 40 different naturally produced dalbaheptides have been reported (Table 1). They correspond to a larger number of chemical entities, because many are groups of strictly related factors called complexes. Among them, vancomycin and teicoplanin are used clinically as a result of high activity against gram-positive pathogens such as many coagulase-negative *Staphylococci* (CNS), corynebacteria, *Clostridium difficile*, multiresistant *Staphylococcus aureus*, and highly gentamicin-resistant *Enterococci* which are refractory to established drugs. Eremomycin is under clinical evaluation. Many patents claim feed-utilization efficiency increase and growth promotion in domestic animals for several dalbaheptides but only avoparcin (Avotan) is commercially used. The platelet aggregation ability of ristocetin A renders it a suitable diagnostic agent for von Willebrand's disease, a hematologic disorder of genetic origin.

Knowledge of the mechanism of action and investigations on the physicochemical characteristics of the therapeutically used dalbaheptides has stimulated the transformation of natural antibiotics into new derivatives using both chemical and biosynthetic modification.

Screening. All the dalbaheptides are produced by strains belonging to the order of *Actinomycetales*. They were at first isolated by conventional screening procedures. More recently specific discovery assays have been used. For example, differential inhibition of a vancomycin-resistant *S. aureus* strain and its susceptible parent, and an assay based on antagonism of the antibacterial activity by N,N-diacetyl-L-Lys-D-Ala-D-Ala, a tripeptide analogue of the dalbaheptides receptor.

Fermentation. Dalbaheptides are produced by submerged fermentation and this topic has been reviewed. In a few cases addition of a biosynthetic precursor of the core aglycones such as tyrosine and *p*-hydroxyphenylglycine increased the yields.

Recovery and Purification. The dalbaheptides are present in both the fermentation broth and the mycelial mass, from which they can be extracted with acetone or methanol, or by raising the pH of the harvested material. A detailed review on the isolation of dalbaheptides has been written. Recovery from aqueous solution is made by ion pair (avoparcin) or butanol (teicoplanin) extraction.

Physical Properties. The molecular weight of dalbaheptides ranges from about 1150 to 2200. Pure dalbaheptides are obtained as colorless or whitish amorphous powders that usually retain water and solvents. Dalbaheptides are generally water-soluble. Teicoplanin can be obtained as an internal salt or as a partial monoalkaline (sodium) salt depending on the pH of the aqueous solution in the final purification step. Other dalbaheptides are obtained as acidic salts, such as hydrochlorides (vancomycin, actaplanin) or sulfates (ristocetin A, avoparcin, eremomycin). The presence of amino, carboxyl, benzylic, and phenolic hydroxyl functions, sugars, and aliphatic chains influences both water solubility and total charge.

Dalbaheptides are levorotatory. The absolute configuration of vancomycin was determined by x-ray analysis of degradation product CDP-I.

Biosynthesis. Biochemical studies on dalbaheptides have been reviewed. Experiments with ^{13}C and ^{2}H have shown that in vancomycin, D-tyrosine is the precursor of D-*p*-hydroxyphenylglycine and β-hydroxy-*m*-chlorotyrosine, and acetate the precursor of the two m,m'-dihydroxyphenylglycines. Similar results using either ^{13}C or radioactively labeled material have been reported for avoparcin, ristocetin, ardacin, and A47934.

Kibdelins are converted into the corresponding ardacins by cultures of *Kibdelosporangium aridum*, indicating that the oxidation of carbon-6 of glucosamine to a carboxyl group is the last biosynthetic step.

Biological Properties

Mechanism of Action. The basis of the antibacterial action of dalbaheptides is the ability to form a complex with the terminal D-Ala-D-Ala residues of growing peptidoglycan chains, thus preventing transglycosylation, e.g., chain elongation, and transpeptidation, e.g., cross-linking. The consequent defective cell wall stops bacterial growth and eventually leads to cell death. The mechanism of action has been reviewed both at the molecular and at the biochemical level.

Antibacterial Activity

In Vitro Properties. The antibacterial spectrum of most dalbaheptides is known. Vancomycin or teicoplanin is generally introduced as a reference

TABLE 1. NATURALLY OCCURRING DALBAHEPTIDES

Antibiotic	Producing organism	Year[a]	Company or institute
ristocetin A, B	*Nocardia lurida* NRRL 2430	1953	Abbott
vancomycin	*Streptomyces orientalis* NRRL 2450	1955	Lilly
actinoidin A, B	*Proactinomyces actinoides*	1956	Research Institute for the Discovery of New Antibiotics, Moscow
K-288	*Streptomyces haranomachiensis* n. sp.	1961	Tohoku University, Sendai
ristomycin A, B	*Proactinomyces fructiferi* var. *ristomycini*	1962	Research Institute for the Discovery of New Antibiotics, Moscow
avoparcin (LL-AV290 complex)	*Streptomyces candidus* NRRL 3218	1966	American Cyanamid
AM374	*Streptomyces eburosporeus* NRRL 3582	1969	American Cyanamid
A477	*Actinoplanes* sp. NRRL 3884	1971	Lilly
actaplanin (A4696 complex)	*Actinoplanes missouriensis* ATCC 23342	1971	Lilly
AB-65	*Saccharomonospora viride* T-80	1973	Dainippon Pharmaceuticals
teicoplanin (teichomycin complex)	*Actinoplanes teichomyceticus* ATCC 31121	1975	Lepetit
A35512 complex	*Streptomyces candidus* NRRL 8156	1976	Lilly
OA-7653A, B	*Streptomyces hygroscopicus* subsp. *hiwasaensis* ATCC 31613	1978	Otsuka Pharmaceutical
A51568A (*N*-demethyl-vancomycin), B	*Nocardia orientalis* NRRL 15232	1982	Lilly
A41030 complex	*Streptomyces toyocaensis* NRRL 15156	1982	Lilly
A47934	*Streptomyces toyocaensis* NRRL 15009	1982	Lilly
ardacin (AAD-216 complex, aridicin)	*Kibdelosporangium aridum* ATCC 39323	1983	Smith, Kline and French
chloropolysporin A, B, C	*Faenia interjecta* sp. nov. FERM BP-538	1983	Sankyo
A40926 complex	*Actinomadura* sp. ATCC 39727	1984	Lepetit
M43 complex	*Nocardia orientalis* NRRL 2450	1984	Lilly
kibdelin (AAD-609 complex)	*Kibdelosporangium aridum* subsp. *largum* ATCC 39922	1985	Smith, Kline and French
izupeptin A, B	*Nocardia* sp. FERM P-8656	1986	Kitasato Institute, Tokyo
parvodicin (AAJ-271)	*Actinomadura parvosata* ATCC 53463	1986	Smith, Kline and French
synmonicin A, B, C (CWI-785)	*Synnemomyces mamnoorii* ATCC 53296	1986	Eskayef Ltd.; Smith, Kline and French
actinoidin A$_2$	*Nocardia* sp. SKF-AAJ-193	1987	Smith, Kline and French
orienticin (PA 42867-A, B, C, D)	*Nocardia orientalis* PA-42867	1987	Shionogi
eremomycin	*Actinomyces* sp. INA-238	1987	Institute of New Antibiotics, Academy of Medical Sciences, Moscow
A42867	*Nocardia* sp. nov. ATCC 53492	1987	Lepetit
A82846 A, B, C	*Amycolatopsis orientalis* NRRL 18098	1987	Lilly
UK-68,597	*Actinoplanes* sp. ATCC 53533	1987	Pfizer
helvecardin A, B	*Pseudonocardia compacta* subsp. *helvetica* SANK 65185	1988	Sankyo
chloroorienticins A, B, C, D, E (PA-45052)	*Amycolatopsis orientalis* PA-45052	1988	Shionogi
A80407 A, B	*Kibdelosporangium philippinensis* NRRL 18198	1989	Lilly
MM 45289, MM 47756	*Amycolatopsis orientalis* NCIB 12531	1989	Beecham
MM 47761, MM 49721	*Amycolatopsis orientalis* NCIB 12608	1989	Beecham
MM 47766, MM 47767, MM 55256, MM 55260	*Amycolatopsis orientalis* NCIB 40011	1989	Beecham
MM 49728, MM 55266, MM 55267, MM 55268	*Amycolatopsis* sp. NCIB 40089	1989	Beecham
decaplanin	*Actinomyces* sp. DSM 4763	1990	Hoechst
UK-72,051	*Amycolatopsis orientalis* n.sp.	1990	Pfizer

[a] Year of patent or first publication.

compound. Most aerobic and anaerobic gram-positive bacteria are sensitive to the action of dalbaheptides. Gram-negative bacteria are generally insensitive with the partial exception of *Neisseria gonorrhoeae*.

In Vivo Properties. The efficacy of dalbaheptides has been assessed in various models of experimental infections. In general there was good correlation between the ED$_{50}$s in the septicemia model in mice and the MICs on test strains.

Pharmacokinetics. Pharmacokinetic studies in mice via iv administration have been done mainly on vancomycin, eremomycin, and lipo-dalbaheptides. A systematic investigation of the relationship between pI, lipophilicity, and pharmacokinetic parameters of ardacin and its hydrolysis products in comparison to those of ristocetin, teicoplanin, and vancomycin has been carried out. Ardacin and its pseudoaglycone having pIs ≈ 3.8 yielded high and prolonged serum concentrations and the half-life ranged from 226 to 492 min. In contrast, vancomycin and ristocetin, which have pIs ≈ 8, had $t_{1/2} = 20$ and 62 min, respectively; teicoplanin and ardacin aglycone, pIs ≈ 5, had intermediate elimination rates, $t_{1/2}$ ranged from 118 to 155 min. For dalbaheptides having similar pIs, clearance decreased and half-life increased as lipophilicity increased.

Mechanism of Resistance. Resistance to high levels of dalbaheptides has been described in enterococci. In some isolates, resistance is inducible. Transfer of resistance, in some cases plasmid-mediated, has been described. More recently, strains highly resistant to vancomycin but sensitive to teicoplanin have been isolated in the United States. Among the resistant enterococci, the VanA strains have been the most intensively studied and a reasonable picture of the resistance mechanism is emerging.

Chemical Modifications

Although it has been known for some time that hydrolysis of ristocetin leads to derivatives having different or increased antimicrobial activity, little chemical modification work was reported in the literature until the mid-1980s. Selective removal of sugar moieties and glycosylation, deacylation, deamination, dechlorination, introduction of bromine or chlorine atoms, esterification or amidation of the terminal carboxyl, acylation or alkylation of the terminal (or sugar) amino groups have all since been described, mainly in patent applications. Some of the aglycones and pseudoaglycones showed a weak activity against selected gram-negative bacteria. Among the most interesting compounds produced were SKF 104662 and MDL 62,873. SKF 104662 was obtained from the B$'$ epimer of synmonicin. The *in vitro* activity and therapeutic efficacy in experimental infection in mice for SKF 104662 were similar to those for vancomycins or teicoplanins. SKF 104662 was less toxic than vancomycin in mice, and the pharmacokinetic profile in mice, rats, and dogs indicated that the half-life was also longer than that of vancomycin. MDL 62,873 is the $-NH(CH_2)_3N(CH_3)_2$ amide of teicoplanin factor A2-2. The combined effect of a moderate basicity and a slightly increased lipophilicity at neutral pH probably led to a better penetration through the cell wall. MDL 62,873 was consistently more active than teicoplanin against CNS clinical isolates.

Economic Aspects

Only three dalbaheptides are commercialized: vancomycin and teicoplanin for human health, and avoparcin for animal usage.

Clinical Use. Vancomycin and teicoplanin as formulated drugs are lyophilized powders to be reconstituted with sterile water for injection.

B. CAVALLERI
F. PARENTI
Merrell-Dow Research Institute

Additional Reading

Campoli-Richards, D.M., R.N. Brogden, and D. Faulds: *Drugs* **40**, 449 (1990).

Cassani, G.: in M.E. Bushell and U. Gräfe, eds., *Bioactive Metabolites from Microorganisms, Progress in Industrial Microbiology*, Vol. 27, Elsevier, Amsterdam, the Netherlands, 1989, p. 221.

Cassetta, A., E. Bingen, and N. Lambert-Zechovsky: *Pathol. Boil.* **39**, 700 (1991).

Lancini, G.C. and B. Cavalleri: in H. Kleinkauf and H. von Doehren, eds., *Biochemistry of Peptide Antibiotics, Recent Advances in the Biotechnology of β-Lactams and Microbial Bioactive Peptides*, W. de Gruyter and Co., Berlin, 1990, pp. 159–178.

ANTIBIOTICS: LINCOSAMINIDES. Lincomycin ($\mathbf{1}$, R = OH, R' = H), $C_{18}H_{34}N_2O_6S$, the first lincosaminide antibiotic to which a structure was assigned, is defined chemically as methyl 6,8-dideoxy-6-(1-methyl-*trans*-4-propyl-L-pyrrolidin-2-ylcarbonylamino)-1-thio-D-erythro-D-galacto-octopyranoside. Both lincomycin and the semisynthetic clindamycin ($\mathbf{1}$, R = H, R' = Cl), $C_{18}H_{33}ClN_2O_5S$, are widely used in clinical practice. The trivial name of the sugar fragment of this antibiotic, methyl α-thiolincosaminide, has lent itself to the other members of this family, whether produced as secondary metabolites of soil microorganisms or derived semisynthetically by chemical modification.

Thus celesticetin ($\mathbf{2}$, R = OC—[benzene ring]—HO), $C_{24}H_{36}N_2O_9S$, and desalicetin ($\mathbf{2}$, R = H), $C_{17}H_{32}N_2O_7S$, are also lincosaminides.

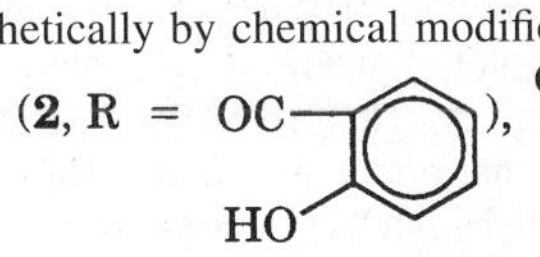

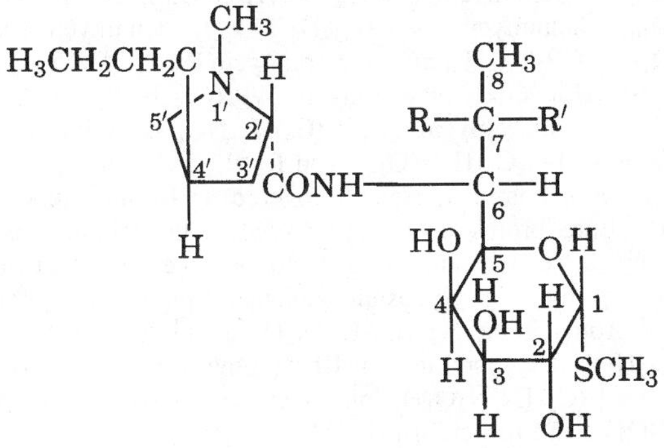

(1)

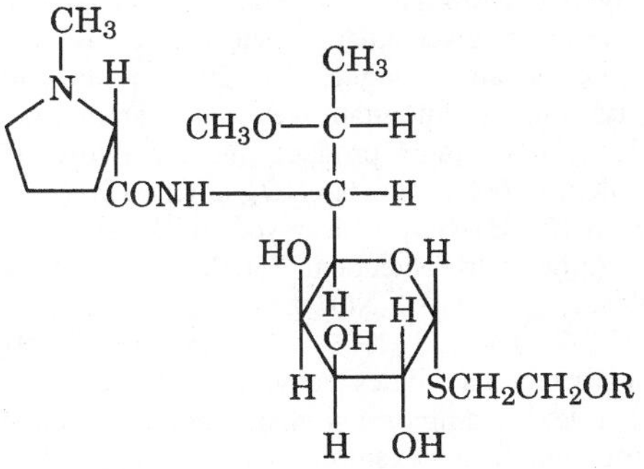

(2)

Lincomycin

The discovery and biological properties of lincomycin ($\mathbf{1}$, R = OH, R' = H) were described in 1962. This antibiotic is active *in vitro* and *in vivo* against most of the common gram-positive pathogens. Resistance by Staphylococci is developed slowly in a stepwise manner, based on *in vitro*

serial subculture experiments, and its activity is not influenced by body fluids up to concentrations of 50% in the assay medium.

Lincomycin has been produced by a variety of *Streptomyces* strains and by strain 1146 of *Actinomyces roseolus*. Extraction by standard procedures using *Sarcina lutea* as the assay organism on agar trays leads to a crystalline hydrochloride having molecular formula $C_{18}H_{34}N_2O_6S \cdot HCl \cdot \frac{1}{2}H_2O$.

Biosynthesis. The terminal *C*-methyl of the propyl side chain, the *S*-methyl, and the *N*-methyl groups are derived from methionine. *trans*-4-Propyl-L-proline was shown to accumulate when *Streptomyces lincolnensis* is grown in media deficient in sulfur, and the addition of L-tyrosine or L-dihydroxyphenylalanine (DOPA) was shown to stimulate this production.

Mechanism of Action. The earliest studies on the mechanism of action of lincomycin showed that lincomycin had the immediate effect on *Staphylococcus aureus* of complete inhibition of protein synthesis. This inhibition results from the blocking of the peptidyltransferase site of the 50S subunit of the bacterial ribosome. Little effect on DNA and RNA synthesis was observed.

Resistance to Lincomycin. Resistance to lincomycin is developed slowly, and is usually caused by modification of 23S ribosomal RNA, which leads to coresistance to macrolide, lincosaminide, and streptogramin B antibiotics.

Pharmacology and Uses. Lincomycin hydrochloride (Lincocin) is available in oral dosage forms and as a sterile solution for injection.

Lincomycin has found use in the treatment of diseases of the ear, throat, nose, respiratory tissue, skin and soft tissue, bone, joint, dental, and septicemic infections caused by staphylococci, pneumonococci, and streptococci (other than enterococci). It has also been used in the treatment of diphtheria and a variety of anaerobic infections, including actinomycosis.

Celesticetin and Other Lincosaminide Metabolites

The production and isolation of the antibiotic celesticetin (**2**), a salicylate ester of the β-hydroxyethylthio-substituent, was reported as early as the 1950s, although its structure was not determined until 1968.

Antibiotic Bu-2545, 7-*O*-methyl-4'-depropyllincomycin (**1**, R = OCH, R = H'3 but lacking the 4'-propyl group), $C_{16}H_{30}N_2O_6S$, produced by *Streptomyces* strain No. H 230-5, possesses structural features in common with both celesticetin and lincomycin.

Accompanying lincomycin in *S. lincolnensis* fermentations is a small amount of the analogue 4'-depropyl-4'-ethyllincomycin, $C_{17}H_{32}N_2O_6S$, which has considerably lower antibacterial activity than the parent compound. Extension of the normal six-day fermentation to twelve days resulted in the formation of lincomycin sulfoxide, $C_{18}H_{34}N_2O_7S$, and 1-demethylthio-1-hydroxylincomycin (lincomycose), $C_{17}H_{32}N_2O_7$, both of which have greatly reduced antibacterial activity.

Clindamycin

Clindamycin, 7(*S*)-7-chloro-7-deoxylincomycin, (**1**, R = H, R = Cl'), also known as Cleocin, first resulted from the reaction of lincomycin and thionyl chloride; improved synthetic methods involve the reaction of lincomycin and triphenylphosphine dichloride or triphenylphosphine in carbon tetrachloride. Clindamycin is significantly more active than lincomycin against gram-positive bacteria *in vitro*, and is absorbed rapidly following oral administration. Clindamycin 2-palmitate, the 2-palmitate ester of clindamycin, is tasteless and antibacterially inactive. However, following oral administration, esterase cleavage occurs to give good blood levels of clindamycin, and this ester has been developed as a pediatric formulation of the antibiotic Cleocin Pediatric.

Uses. Clindamycin has found use in the treatment of common infections caused by gram-positive cocci. It is also efficacious in the treatment of anaerobic infections, including antinomycosis. Clindamycin has been shown to be active against strains of *Plasmodium* in animals.

Resistance to Clindamycin. Cross-resistance between lincomycin and clindamycin is complete, and co-resistances of lincomycin also apply to clindamycin. However, the inactivation of clindamycin by clinical isolates of *Staphylococcus haemolyticus* and *Staphylococcus aureus* is caused by adenylylation at the 4-position to form clindamycin 4-(5'-adenylate) in contrast to the lincomycin 3-(5'-adenylate) that forms.

Biomodification of Clindamycin. When added to fermentations of *Streptomyces punipalus*, clindamycin is converted into de-*N*-methylclindamycin. However, when clindamycin is incubated with *Streptomyces armentosus*, clindamycin sulfoxide, $C_{18}H_{33}ClN_2O_6S$, which has low antibacterial activity, is formed. Clindamycin 3-phosphate, antibacterially inactive *in vitro*, and the ribonucleotides clindamycin 3-(5'-cytidylate), clindamycin 3-(5'-adenylate), clindamycin 3-(5'-uridylate), and clindamycin 3-(5'-guanylate), all inactive *in vitro*, can be generated. All of these derivatives protect mice infected with *Staphylococcus aureus*, however, presumably because of biotransformation into clindamycin.

Other Changes in the Amino Acid Fragment

The influence of the size of the cyclic amino acid in analogues of clindamycin has been examined. The (2-(*S*)-*cis*)-4-ethylpipecolic acid analogue (6-ring), $C_{17}H_{31}ClN_2O_5S$, was highly active both *in vitro* and *in vivo*, but the *cis*-(*R*)-isomer had minimal activity. Significant activity was shown for the azepine (7-ring) analogue, $C_{16}H_{29}ClN_2O_5S$, but the azetidine (4-ring) analogue, $C_{13}H_{23}ClN_2O_5S$, showed little activity.

Economic Aspects

The composition of matter patents in the United States issued to The Upjohn Company on clindamycin phosphate and hydrochloride expired at the end of 1986 and in early 1987, respectively. Since then, these compounds have been available generically from more than two dozen companies in the United States alone.

BRIAN BANNISTER
The Upjohn Company

Additional Reading

Fass, R.J.: in E.M. Kagan, ed., *Antimicrobial Therapy*, 2nd Edition, W. B. Saunders Co., Philadelphia, PA, 1974, p. 83.

ANTIBIOTICS: MACROLIDES.

Macrolide antibiotics are well-established antimicrobial agents in both clinical and veterinary medicine. These agents can be administered orally and are generally used to treat infections in the respiratory tract, skin and soft tissues, and genital tract caused by gram-positive organisms, *Mycoplasma* species, and certain susceptible gram-negative and anaerobic bacteria.

The macrolide class is large and structurally diverse. Macrolides are produced by fermentation of soil microorganisms. Additionally, structural modifications using both chemical and microbiological means have yielded biologically active semisynthetic derivatives.

The term macrolide was introduced to denote the class of substances produced by *Streptomyces* species containing a macrocyclic lactone ring. The generalized structure is a highly substituted monocyclic lactone (aglycone) to which is attached one or more saccharides glycosidically linked to hydroxyl groups on either the aglycone or another saccharide. The aglycones are derived via similar polyketide biosynthetic pathways and thus share many structural features in terms of pattern and stereochemistry of substituents. Traditional macrolide antibiotics are divided into three families according to the size of the aglycone, which can be 12-, 14-, or 16-membered.

Because one or more aminosugars are usually present, these compounds are basic and can form acid addition salts. In addition, one or more neutral sugars are often present. A few macrolides possess no aminosugar. The saccharides share some common features: they tend to be highly deoxygenated and *N*- or *O*-methylated; and the amino groups are located at either position 3 or 4.

12-Membered Ring Macrolides

12-Membered ring macrolides include methymycin ($C_{25}H_{43}NO_7$), neomethymycin ($C_{25}H_{43}NO_7$), and YC-17 ($C_{25}H_{43}NO_6$).

14-Membered Ring Macrolides

Natural Products. Naturally occurring 14-membered macrolides include erythromycin A ($C_{37}H_{67}NO_{13}$), erythromycin B ($C_{37}H_{67}NO_{12}$), erythromycin C ($C_{36}H_{65}NO_{13}$), erythromycin D ($C_{36}H_{65}NO_{12}$), erythromycin F ($C_{37}H_{67}NO_{14}$), erythromycin E ($C_{37}H_{65}NO_{14}$), oleandomycin ($C_{35}H_{61}NO_{12}$), oleandomycin Y ($C_{34}H_{59}NO_{12}$), pikromycin ($C_{28}H_{47}NO_8$), narbomycin ($C_{28}H_{47}NO_7$), 5-*O*-mycaminosylnarbonolide ($C_{28}H_{47}NO_8$), 10,11-dihydropikromycin ($C_{28}H_{49}NO_8$), kayamycin ($C_{28}H_{49}NO_8$), 12-deoxykromycin ($C_{20}H_{30}O_4$), kromycin ($C_{20}H_{30}O_5$), megalomicin A

($C_{44}H_{80}N_2O_{15}$), megalomicin B ($C_{46}H_{82}N_2O_{16}$), megalomicin C_1 ($C_{48}H_{84}N_2O_{17}$), megalomicin C_2 ($C_{49}H_{86}N_2O_{17}$), XK-41-B_2 ($C_{47}H_{84}N_2O_{16}$), lankamycin (kujimycin B) ($C_{42}H_{72}O_{16}$), kujimycin A ($C_{40}H_{70}O_{15}$), 15-dehydrolankamycin ($C_{42}H_{70}O_{16}$), 15-*O*-(4-*O*-acetylarcanosyl)-lankamycin ($C_{52}H_{88}O_{20}$), 3''-*O*-demethyl-2'',3''-anhydrolankamycin ($C_{41}H_{68}O_{15}$), and 23672 RP ($C_{48}H_{82}O_{20}$).

Semisynthetic Derivatives. Erythromycin has been the principal subject of modification of 14-membered macrolides; some of the derivatives are being commercially launched. Derivatives of erythromycin and oleandomycin include 2'-*O*-acetylerythromycin ($C_{39}H_{69}NO_{14}$), 2'-*O*-propionylerythromycin ($C_{40}H_{71}NO_{14}$), erythromycin ethyl carbonate ($C_{40}H_{71}NO_{15}$), erythromycin ethyl succinate ($C_{43}H_{75}NO_{16}$), tri-*O*-acetyloleandomycin ($C_{41}H_{67}NO_{15}$), erythromycin-11,12-carbonate ($C_{38}H_{65}NO_{14}$), roxithromycin ($C_{41}H_{76}N_2O_{15}$), 9(*S*)-erythromycylamine ($C_{37}H_{70}N_2O_{12}$), dirithromycin ($C_{42}H_{78}N_2O_{14}$), azithromycin ($C_{38}H_{72}N_2O_{12}$), clarithromycin ($C_{38}H_{69}NO_{13}$), and flurithromycin ($C_{37}H_{66}FNO_{13}$).

16-Membered Ring Macrolides

Natural Products. 16-Membered macrolides are divided into leucomycin- and tylosin-related groups, which differ in the substitution pattern of their aglycones. Multifactor complexes are usually produced and some compounds have been isolated from culture broths of different organisms and then been given different names. Natural products include leucomycin A_1 ($C_{40}H_{67}NO_{14}$), leucomycin A_5 ($C_{39}H_{65}NO_{14}$), leucomycin A_7 ($C_{38}H_{63}NO_{14}$), leucomycin A_9 ($C_{37}H_{61}NO_{14}$), leucomycin V ($C_{35}H_{59}NO_{13}$), leucomycin A_3 (josamycin) ($C_{42}H_{69}NO_{15}$), leucomycin A_4 ($C_{41}H_{67}NO_{15}$), leucomycin A_6 ($C_{40}H_{65}NO_{15}$), leucomycin A_8 ($C_{39}H_{63}NO_{15}$), leucomycin U ($C_{37}H_{61}NO_{14}$), platenomycin A_1 ($C_{43}H_{71}NO_{15}$), midecamycin A_2 ($C_{42}H_{69}NO_{15}$), midecamycin A_1 (espinomycin A_1) (platenomycin B_1), ($C_{41}H_{67}NO_{15}$), platenomycin C_2 (espinomycin A_3) ($C_{40}H_{65}NO_{15}$), DHP ($C_{38}H_{63}NO_{14}$), espinomycin A_2 ($C_{42}H_{69}NO_{15}$), platenomycin A_0 ($C_{44}H_{73}NO_{15}$), maridomycin I (platenomycin C_3) ($C_{43}H_{71}NO_{16}$), maridomycin VII ($C_{42}H_{69}NO_{16}$), maridomycin III (platenomycin C_1) ($C_{41}H_{67}NO_{16}$), maridomycin IV ($C_{40}H_{65}NO_{16}$), maridomycin VI ($C_{39}H_{63}NO_{16}$), carbomycin A (deltamycin A_4) ($C_{42}H_{67}NO_{16}$), deltamycin A_3 ($C_{41}H_{65}NO_{16}$), deltamycin A_2 ($C_{40}H_{63}NO_{16}$), deltamycin A_1 ($C_{39}H_{61}NO_{16}$), deltamycin X (EOA) ($C_{37}H_{59}NO_{15}$), EOP ($C_{38}H_{61}NO_{15}$), carbomycin B ($C_{42}H_{67}NO_{15}$), platenomycin W_1 ($C_{43}H_{69}NO_{15}$), platenomycin W_2 ($C_{44}H_{71}NO_{15}$), niddamycin ($C_{39}H_{63}NO_{14}$), midecamycin A_4 ($C_{42}H_{67}NO_{15}$), midecamycin A_3 ($C_{41}H_{65}NO_{15}$), DOA ($C_{37}H_{59}NO_{14}$), and DOP ($C_{38}H_{61}NO_{14}$).

The spiramycin complex, also discovered as foromacidine, was isolated from culture broths of *S. ambofaciens*. The spiramycins are distinguished by a second aminosugar, forosamine, attached to the 9-hydroxyl group by a β-glycosidic linkage. Spiramycins include spiramycin I (foromacidine A) ($C_{43}H_{74}N_2O_{14}$), spiramycin II (foromacidine B) ($C_{45}H_{76}N_2O_{15}$), spiramycin III (foromacidine C) ($C_{46}H_{78}N_2O_{15}$), neospiramycin I ($C_{36}H_{62}N_2O_{11}$), spiramycin U ($C_{42}H_{71}NO_{16}$), spiramycin S ($C_{42}H_{73}NO_{16}$), and forocidin I ($C_{28}H_{47}NO_{10}$).

A second large group of 16-membered macrolides differs from the leucomycins in the substitution pattern of the aglycone. One difference is a methyl or hydroxymethyl group at C-14. If hydroxymethyl is present, it may be glycosidically substituted by a neutral sugar such as mycinose or an analogue. The most prominent member of this group is tylosin, an important veterinary antibiotic produced by *S. fradiae*. Tylosin and related products include tylosin ($C_{46}H_{77}NO_{17}$), relomycin (20-dihydrotylosin) ($C_{46}H_{79}NO_{17}$), macrocin ($C_{45}H_{75}NO_{17}$), *O*-demethylmacrocin ($C_{44}H_{73}NO_{17}$), desmycosin ($C_{39}H_{65}NO_{14}$), lactenocin ($C_{38}H_{63}NO_{14}$), *O*-demethyllactenocin (DOML) ($C_{37}H_{61}NO_{14}$), 23-*O*-demycinosyltylosin ($C_{38}H_{63}NO_{13}$), 23-(demycinosyloxy)tylosin ($C_{38}H_{63}NO_{12}$), GS-77-1 ($C_{38}H_{65}NO_{11}$), and GS-77-3 ($C_{38}H_{65}NO_{12}$).

Many 16-membered macrolides possess a tylosin-type aglycone with only one saccharide, an aminosugar that is either mycaminose or desosamine. Rosaramicin is an example. These compounds also differ in their degree of oxidation at C-20, C-23, and C-9 to C-13 (dienone or epoxyenone). Hydrolysis of both neutral sugars from tylosin yields 5-*O*-mycaminosyltylonolide (OMT). However, this compound is more conveniently obtained by hydrolysis of mycarose from the fermentation-derived demycinosyltylosin. Analogous hydrolyses of mycarose from DMOT, GS-77-3, and GS-77-1 yield, respectively, 23-deoxy-5-*O*-mycaminosyltylonolide (DOMT), 20-deoxo-5-*O*-mycaminosyltylonolide (GS-77-4), and 5-*O*-mycaminosyltylactone (GS-77-2).

The mycinamicin complex, also found as AR-5 complex, was isolated from culture broths of *M. griseorubida*. These compounds contain a 2,3-double bond and a methyl group rather than a two-carbon substituent at C-6 of the aglycone. A hydroxyl group is present at C-14 in mycinamicins II and V. X-ray crystallography established the stereochemistry and absolute configuration of the aglycone as identical to tylosin.

A few neutral 16-membered macrolides have been isolated. The structure of chalcomycin, produced by *S. bikiniensis*, was established as a 16-membered lactone having two neutral sugars, D-chalcose and D-mycinose. Its aglycone resembles that of the mycinamicins, but differs by a hydroxyl group, the stereochemistry of which is uncertain, at C-8 and a methyl group at C-15.

The aldgamycins, produced by *S. lavendulae*, contain a novel bicyclic sugar, aldgarose.

As is the case for the leucomycin group, several aglycones exist in the tylosin group which differ in the pattern of oxidation and substitution of the lactone. If the aglycone contains an aldehyde, cleavage of the aminosugar yields the hemiketal, such as tylonolide and rosaranolide.

Semisynthetic Derivatives. 3''-O-Acyl derivatives have not been found via fermentation, but chemical acylation of the 3''-hydroxyl group yields products having good antibiotic activity and better pharmacokinetics than the parent macrolides. Two such compounds have been developed: 3''-O-propionyl-leucomycin A$_5$ (rokitamycin) $C_{42}H_{69}NO_{15}$, formerly TMS-19-Q,; and 9,3''-di-O-acetylmidecamycin (miokamycin) $C_{45}H_{71}NO_{17}$. At least part of the *in vivo* improvement was attributed to slower elimination of active metabolites from serum.

The enhanced activity from acylation of the 3- and 4''-hydroxyl groups of leucomycin prompted analogous studies of tylosin. Bioconversion of tylosin by *S. thermotolerans* yielded 3- or 4''-O-acyl derivatives possessing increased activity against certain resistant microorganisms and higher concentrations of antibiotic in serum after oral administration. From this study, 3-O-acetyl-4''-O-isovaleryltylosin (AIV-tylosin), $C_{53}H_{87}NO_{19}$, was developed as a new veterinary antibiotic.

Tilmicosin, $C_{46}H_{80}N_2O_{13}$, was selected as a therapeutic agent to treat pneumonia in cattle and pigs because of its activity against *Pasturelle* species, oral bioavailability, and prolonged concentrations *in vivo*.

Hybrid Macrolides

Other macrolides have been prepared which represent hybrids of structures within the 14-membered family, within the 16-membered family, or between the two families. These hybrids have been made by chemical, bioconversion, or genetic manipulations.

Aglycones obtained from one macrolide have been bioconverted by microorganisms producing either the same or a different macrolide; such studies have been important for production of new compounds and elucidation of biosynthetic pathways.

The advent of molecular biology has opened new possibilities for producing hybrid macrolides. Insertion of DNA from the oleandomycin-producer into a mutant strain of the erythromycin-producer produced derivatives of erythromycin lacking the 2-methyl group. Genetic manipulations of biosynthetic pathways in macrolide-producing microorganisms complement traditional chemical and microbiological approaches. For example, inactivation of the *Ery F* gene coding for a 6-hydroxylase in *Sa. Erythraea* produced 6-deoxyerythromycin.

Biological Properties

Antimicrobial Properties. Macrolides inhibit growth of gram-positive bacteria, *Mycoplasma* species, and certain gram-negative and anaerobic bacteria. Susceptible gram-positive bacteria include many species of *Staphylococcus* and *Streptococcus*; susceptible gram-negative bacteria include *Bordetella pertussis, Legionella pneumophila, Moraxella catarrhalis* (formerly *Branhamella*), and *Haemophilus ducreyi*. Although erythromycin has some *in vitro* activity against *Haemophilus influenzae*, it does not exhibit a high level of *in vivo* efficacy. An important susceptible anaerobe is *Propionibacterium acnes*. Comparative evaluations have been published, and monographs devoted to particular macrolides and reviews are available.

Azithromycin expanded the traditional *in vitro* spectrum because of increased activity against gram-negative bacteria, including *H. influenzae*.

Macrolides inhibit growth of bacteria by inhibiting protein synthesis on ribosomes. Bacterial resistance to macrolides is often accompanied by cross-resistance to lincosamide and streptogramin B antibiotics (MLS-resistance), which can be either inducible or constitutive. 14-Membered macrolides generally induce resistance to themselves, whereas 16-membered macrolides do not; consequently, one advantage of the latter is their activity against bacteria which are inducibly resistant to erythromycin. Both 14- and 16-membered macrolides lack activity against constitutively resistant strains.

Bacterial resistance to antibiotics usually results from modification of a target site, enzymatic inactivation, or reduced uptake into or increased efflux from bacterial cells.

Pharmacokinetics and Pharmacology. Older macrolides such as erythromycin exhibit relatively low serum concentrations, short *in vivo* half-lives, highly variable oral absorption, and low oral bioavailability. Improvements in these pharmacokinetic parameters have been accomplished for newer derivatives. The principal side effects of macrolides are gastrointestinal problems, such as pain, indigestion, diarrhea, nausea, and vomiting.

Biosynthetic Patterns and Conformational Analysis

Macrolides are obtained by controlled submerged aerobic fermentations of soil microorganisms. Although species of *Streptomyces* have dominated, species of *Saccharopolyspora, Micromonospora,* and *Streptoverticillium* are also well represented. New techniques such as enzyme-linked immunosorbent assay (ELISA) may prove beneficial for discovering new structures.

Economic Aspects

Macrolide antibiotics are used clinically to treat infections resulting from susceptible organisms in the upper and lower respiratory tract, skin and soft tissues, and genital tract. They are generally used orally, although

TABLE 1. COMMERCIAL MACROLIDE PRODUCTS

Macrolide	Trade name[a]	Clinical route	Manufacturer
erythromycin[b]	Ery-tab	oral	Abbott
erythromycin[c]	Erythromycin base filmtab	oral	Abbott
erythromycin[d]	Erythromycin delayed-release capsules	oral	Abbott
erythromycin[e]	E-Mycin	oral	Boots
erythromycin[b]	Ilotycin	oral	Dista
erythromycin[d]	Eryc	oral	Parke-Davis
erythromycin topical solution	Eryderm 2%	topical	Abbott
erythromycin topical solution	ETS-2%	topical	Paddock
erythromycin ophthalmic ointment	Ilotycin ointment	topical	Dista
erythromycin stearate	Erythrocin stearate	oral	Abbott
erythromycin lactobionate	Erythrocin lactobionate-iv	iv	Abbott
	Erythrocin piggyback	iv	Abbott
erythromycin gluceptate	Ilotycin gluceptate	iv	Dista
erythromycin ethyl succinate	E.E.S.	oral	Abbott
	Eryped	oral	Abbott
erythromycin estolate	Ilosone	oral	Dista
erythromycin acistrate	Erasis (Finland)	oral	Orion
erythromycin stinoprate	Erythrocist (Italy)	oral	Refarmed
propionylerythromycin mercaptosuccinate	Zalig (Italy)	oral	Pierrel
erythromycin-11,12-carbonate	Davercin (Poland)	oral	Tarchomin
roxithromycin	Rulid (France)	oral	Roussel-Uclaf
clarithromycin	Biaxin	oral	Abbott
azithromycin	Zithromax	oral	Pliva
triacetyloleandomycin	TAO	oral	Roerig
josamycin	Josamycin (Japan)	oral	Yamanouchi
miokamycin	Miocamycin (Japan)	oral	Meiji Seika
rokitamycin	Ricamycin (Japan)	oral	Toyo Jozo

[a] Country in parenthesis represents launch outside the United States.
[b] Enteric-coated tablets.
[c] Film-coated tablets.
[d] Enteric-coated pellets.
[e] Coated tablets.

they can be given intravenously. For the latter purpose, a water-soluble salt is used, because macrolides are poorly soluble in water as free bases. Commonly employed acid-addition salts include the lactobionate, gluceptate, tartrate, and phosphate. For oral administration, acid-stable esters, salts, formulations, and coatings are necessary for erythromycin, which is unstable as its unprotected free base under acidic conditions such as those in the stomach. They are not administered by intramuscular injection because of severe pain.

Reported worldwide sales of macrolides are estimated as $800,000,000 annually; approximately 25% of this figure are U.S. sales. Table 1 provides a partial list of products commercially available for clinical use.

Macrolides are regarded as among the safest of antibiotics.

Relatively few macrolides are used in veterinary medicine. The most important is tylosin (Tylan, Elanco Products), which is used to control chronic respiratory disease caused by *Mycoplasma gallisepticum* in poultry. It is also used to treat and control infections in pigs resulting from gram-positive bacteria and *Mycoplasma*, and as a growth promotant for pigs and poultry. Other macrolides in veterinary use include erythromycin, olean-domycin, and spiramycin. Newer macrolides being developed for veterinary applications are 3-*O*-acetyl-4″-*O*-isovaleryltylosin and tilmicosin. The pharmacokinetics of macrolides in animals have been reviewed, and a book devoted to pharmacokinetics of veterinary antibiotics has been published.

H. A. KIRST
Eli Lilly and Company

Additional Reading

Bryskier, A., J.P. Butzler, H.C. Neu, and P.M. Tulkens, eds.: *The Macrolides*, Arnette-Blackwell Publishers, Oxford, UK, 1993.
Kirst, H.A.: *Prog. Med. Chem.* **30** (1993).
Kirst, H.A.: *Prog. Med. Chem.* **31** (1994)
Neu, H.C., L.S. Young, and S.H. Zimmer, eds.: *The New Macrolides, Azalides, and Streptogramins: Pharmacology and Clinical Applications*, Marcel Dekker, Inc. New York, NY, 1993.

ANTIBIOTICS: MONOBACTAMS. β-Lactam antibiotics are one of the best established classes of antimicrobial agents for the treatment of infectious diseases.

By screening strains of bacteria that were specifically responsive to β-lactam antibiotics, monocyclic β-lactams, i.e., monobactams, varying in substitution at the C-3 position, were identified. The naturally occurring monobactams include SQ 26,180 ($C_6H_{10}N_2O_6S \cdot K$), SQ 26,445 ($C_{12}H_{20}N_4O_9S$), sulfazecin ($C_{12}H_{20}N_4O_9S$), and isosulfazecin ($C_{12}H_{20}N_4O_9S$). More recently, the 4β-methyl analogue of SQ 26,445/sulfazecin was isolated using a differential antibacterial assay.

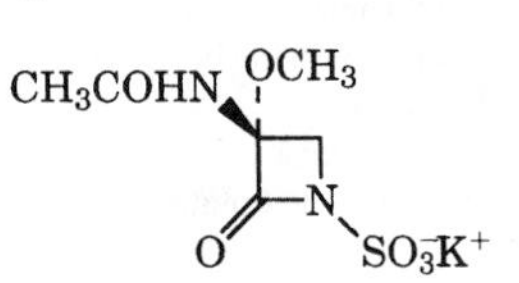

SQ 26, 180

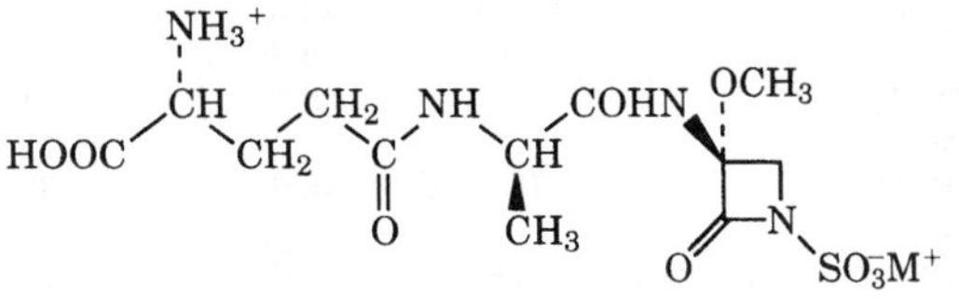

SQ 26, 445/ sulfazecin

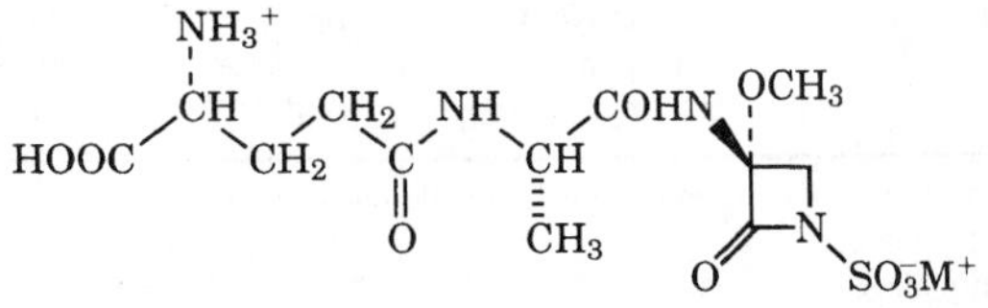

isosulfazecin

Synthesis

Initial syntheses employed the sulfonation of an N-1 unsubstituted azetidinone as the key step. The natural product SQ, 26,180 as well as other methoxylated monobactams were synthesized, starting from either 7-aminocephalosporanic acid (7-ACA) or 6-aminopenicillanic acid (6-APA), via sulfonation of the N-1 unsubstituted degradation products. Subsequently, many more C-3 side-chain analogues were prepared by this method.

A second, conceptually distinct chiral synthesis of monobactams was developed from β-hydroxy amino acids. Cyclization of the acylsulfamate of an amino-protected *O*-mesylserine derivative leads directly to the monobactam nucleus. This methodology was also applied to the synthesis of 4α- and 4β-methyl monobactams from L-threonine and allothreonine, respectively.

The monobactams, like penicillins and cephalosporins, interfere with the synthesis of bacterial cell walls. β-Lactam antibiotics bind to a series of penicillin-binding proteins (PBPs) on the cytoplasmic membrane and their antibacterial effect is believed to result from inhibition of a subset of these PBPs known as peptidoglycan transpeptidases.

The biosynthetic origin of monobactams has been elucidated by fermentation experiments using radioactively labeled amino acids. The monocyclic ring of the naturally occurring C-4 unsubstituted monobactams is derived from serine. Similar techniques have shown that the methyl moiety of the methoxyl group in 3α-methoxylated monobactams is derived from methionine.

All of the naturally occurring monobactams discovered as of this writing have exhibited poor antibacterial activity. However, as in the case of the penicillins and cephalosporins, alteration of the C-3 amide side chain led to many potent new compounds.

Aztreonam. Aztreonam is a totally synthetic compound having an antibacterial spectrum that is unique among β-lactam antibiotics. It exhibits potent and specific activity against a wide range of both β-lactamase-producing and nonproducing aerobic gram-negative bacteria, including *Pseudomonas aeruginosa*, but displays minimal inhibition against anaerobic and gram-positive aerobic bacteria, e.g., staphylococci and streptococci.

Overall, aztreonam appears to be a safe agent having toxicity side effects similar to those of other β-lactams. The safety profile suggests that aztreonam may be useful as a replacement for aminoglycoside therapy. The biological properties of aztreonam have been extensively reviewed.

Alternative N-1 Activating Groups

β-Lactam antibiotics exert their antibacterial effects via acylation of a serine residue at the active site of the bacterial transpeptidases. Critical to this mechanism of action is a reactive β-lactam ring having a proximate

TABLE 1. BIOLOGICALLY ACTIVE MONOBACTAM SUBCLASSES

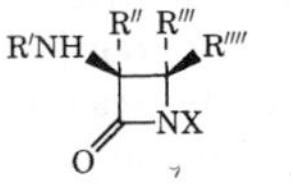

Subclass	Substituent, X
monobactams	$-SO_3H$
monophosphams	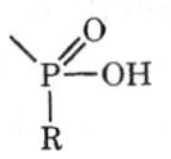 $-\overset{O}{\underset{R}{P}}-OH$
monosulfactams	$-OSO_3H$
monophosphatams	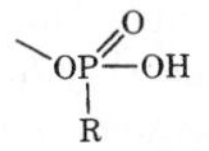 $-O\overset{O}{\underset{R}{P}}-OH$
monoxacetams (oxamazins)	$-OCH_2COOH$
monocarbams	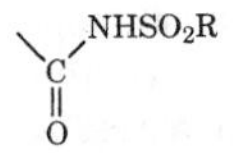 $-\overset{NHSO_2R}{\underset{O}{C}}$
tetrazole-activated monobactams	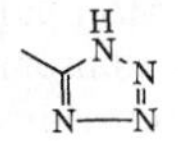 (tetrazole group)

anionic charge that is necessary for positioning the ring within the substrate binding cleft.

All of the naturally occurring monobactams and aztreonam are characterized by the presence of the N-1 sulfonate group, which serves a dual function. The electron-withdrawing sulfonate moiety renders the β-lactam ring more reactive toward nucleophilic attack and at the same time provides the anionic charge necessary for binding. A variety of monobactam subclasses bearing N-1 activating groups having the necessary physical properties for antibacterial activity are listed in Table 1.

Economic Aspects

Two monobactams were in clinical use as of 1990. Aztreonam, manufactured by Bristol-Myers Squibb, has the worldwide trademark of Azactam. The global experience with aztreonam and its clinical acceptance signifies that the monobactams are recognized as an important new class of antibacterial agents. Carumonam, manufactured by Takeda, received approval for human usage in 1988 in Japan. Carumonam has the trademark Amasulin.

KLAUS R. LINDNER
DANIEL P. BONNER
WILLIAM H. KOSTER
Bristol-Myers Squibb Pharmaceutical Research Institute

Additional Reading

J.F. Acar and H.C. Neu, eds., *Rev. Infect. Disease*, **1** (Suppl. 4) (1985).
Koster, W.H., C.M. Cimarusti, and R.B. Sykes: in R.B. Morin and M. Gorman, eds., *Chemistry and Biology of β-Lactam Antibiotics,* Vol. 3, Academic Press, New York, NY, 1982, pp. 339–375.
Koster, W.H. and D.P. Bonner: in H. Umezawa ed., *Frontiers of Antibiotic Research,* Academic Press, New York, NY 1987, pp. 211–226.
Parker, W.L., J. O'Sullivan, and R.B. Sykes: *Advances in Applied Microbiology,* Vol. 31, Academic Press, New York, 1986, pp. 181–205.

ANTIBIOTICS: NUCLEOSIDES AND NUCLEOTIDES.

The naturally occurring nucleoside and nucleotide antibiotics exist as either the C- or N-glycosides (see **Nucleic Acids**). They include ezomycin A_1 ($C_{26}H_{38}N_8O_{15}S$), ezomycin B_1 ($C_{26}H_{39}N_7O_{17}S$), ezomycin C_1 ($C_{26}H_{37}N_7O_{16}S$), ezomycin A_2 ($C_{19}H_{26}N_6O_{12}$), ezomycin B_2 ($C_{19}H_{25}N_5O_{13}$), ezomycin C_2 ($C_{19}H_{25}N_5O_{13}$), showdomycin ($C_9H_{11}NO_6$), isoshowdomycin ($C_9H_{11}NO_6$), maleimycin ($C_7H_7NO_3$), oxazinomycin ($C_9H_{11}NO_7$), pyrazomycin (pyrazofurin) ($C_9H_{13}N_3O_6$), formycin ($C_{10}H_{13}N_5O_4$), formycin B ($C_{10}H_{12}N_4O_5$), oxoformycin B ($C_{10}H_{12}N_4O_6$). These antibiotics contain a variety of purine and pyrimidine rings, including the diazepin, maleimide, indole, imidazole, pyrrolopyrimidine, pyrazolopyrimidine, pyrazole, oxazine, triazene, hydantoin, and the purine ring having a 6-N-phosphoramidate substituent. The structures of the carbohydrate moieties also vary. Some nucleosides have additional carbon atoms attached to the ribosyl moiety and in some cases ribose has been replaced by other sugars such as allulose, olefinic sugars, 2-, 3-, 4-, or 5-amino sugars, 4-fluororibose, 4-aminohexoses, aminouronic acids, disaccharides, or a tricyclic dodecose.

The naturally occurring nucleoside/nucleotide antibiotics, which have been isolated from bacteria, fungi, blue-green algae, and marine sponges, have proven to be useful biochemical probes in eucaryotic, procaryotic, viral, fungal, and plant systems. Some excellent reviews on the nucleoside/nucleotide antibiotics are available.

C-Nucleosides

The naturally occurring C-nucleosides containing C-glycosyl linkages include ezomycins, showdomycin, oxazinomycin, pyrazomycin, and pyrazolopyrimidine nucleosides.

N-Nucleosides

The naturally occurring nucleoside analogues contain the N-glycosyl linkage and either purine, pyrimidine, imidazole, diazepin, or indole rings. The purine nucleosides inhibit protein synthesis, RNA and DNA synthesis, and methyltransferases; they have antimycoplasmal, antiviral, hypotensive, antifungal, antimycobacterial, and antitumor activities and induce sporulation. The pyrimidine nucleosides inhibit protein synthesis, virus replication, RNA and DNA synthesis, and cAMP phosphodiesterase. The imidazole nucleosides inhibit nucleic acid synthesis. The diazepin nucleosides inhibit adenosine deaminase (ADA). The indole nucleosides inhibit bacteria, yeast, fungi, and viruses.

Purine N-Nucleosides. The purine N-nucleoside antibiotics include 2′-amino-2′-deoxyadenosine ($C_{10}H_{14}N_6O_3$), 2′-amino-2′-deoxyguanosine ($C_{10}H_{14}N_6O_4$), 3′-amino-3′-deoxyadenosine ($C_{10}H_{14}N_6O_3$), cordycepin (3′-deoxyadenosine) ($C_{10}H_{13}N_5O_3$), puromycin ($C_{22}H_{29}N_7O_5$), homocitrullylaminoadenosine ($C_{17}H_{27}N_9O_5$), lysylaminoadenosine ($C_{16}H_{26}N_8O_4$), chryscandin ($C_{20}H_{23}N_7O_6$), A201A ($C_{37}H_{50}N_6O_{14}$), psicofuranine ($C_{11}H_{15}N_5O_5$), decoyinine (angustmycin A) ($C_{11}H_{13}N_5O_4$), arabinofuranosyladenine (ara-A, vidarabine) ($C_{10}H_{13}N_5O_4$), sinefungin ($C_{15}H_{23}N_7O_5$), A9145A ($C_{15}H_{23}N_7O_5$), A9145C ($C_{15}H_{21}N_7O_5$), herbicidins A, B, E, F, and G ($C_{23}H_{30}N_5O_{10}$), doridosine ($C_{11}H_{15}N_5O_5$), aristeromycin ($C_{11}H_{15}N_5O_3$), neplanocin A ($C_{11}H_{13}N_5O_3$), neplanocin B ($C_{11}H_{13}N_5O_4$), neplanocin C ($C_{11}H_{13}N_5O_4$), neplanocin D ($C_{11}H_{12}N_4O_4$), neplanocin F ($C_{11}H_{13}N_5O_3$), nucleocidin ($C_{10}H_{13}FN_6O_6S$), ascamycin ($C_{13}H_{18}ClN_7O_7S$), nebularine ($C_{10}H_{12}N_4O_4$), crotonoside ($C_5H_5N_5O$), griseolic acid ($C_{14}H_{13}N_5O_8$), griseolic acid B ($C_{14}H_{13}N_5O_9$), griseolic acid C ($C_{14}H_{15}N_5O_7$), oxetanocin A ($C_{10}H_{13}N_5O_3$), and oxanosine ($C_{10}H_{12}N_4O_6$).

Pyrrolopyrimidine Nucleosides. The pyrrolopyrimidine N-nucleoside antibiotics include tubercidin ($C_{11}H_{14}N_4O_4$), toyocamycin ($C_{12}H_{13}N_5O_4$), sangivamycin ($C_{12}H_{15}N_5O_5$), cadeguomycin ($C_{12}H_{14}N_4O_7$), kanagawamicin ($C_{13}H_{17}N_5O_6$), mycalisine A ($C_{13}H_{13}N_5O_3$), and mycalisine B ($C_{13}H_{12}N_4O_4$).

Diazepin Nucleosides. Four naturally occurring diazepin nucleosides, coformycin ($C_{11}H_{16}N_4O_5$), 2′-deoxycoformycin ($C_{11}H_{16}N_4O_4$), adechlorin or 2′-chloro-2′-deoxycoformycin ($C_{11}H_{15}ClN_4O_4$), and adecypenol ($C_{12}H_{16}N_4O_4$), have been isolated.

Bredinin, Neosidomycin, and SF-2140. Bredinin ($C_{19}H_{13}N_3O_6$) isolated from the culture filtrates of *Eupenicillium brefeldianum*, inhibits the multiplication of L5178Y, HeLa S3, RK-13, mouse L-cells, and Chinese hamster cells.

Neosidomycin ($C_{17}H_{20}N_2O_6$) and SF-2140 ($C_{18}H_{20}N_2O_6$) are indole N-glycosides produced by *S. hygroscopicus* and *Actinomadura*, respectively. Both show activity against gram-positive bacteria, yeast, fungi, and viruses.

Pyrimidine-Nucleoside Antibiotics

Fatty acyl Nucleosides. The nucleoside antibiotics with fatty acyl groups containing adenine, uracil, and dihydrouracil aglycons, the tunicamycins, streptovirudins, corynetoxins include tunicamycin [I (A_0) ($C_{36}H_{58}N_4O_{16}$), II (C) ($C_{37}H_{60}N_4O_{16}$), III (A_2) ($C_{37}H_{60}N_4O_{16}$), IV (B_2) ($C_{38}H_{62}N_4O_{16}$), V (A) ($C_{38}H_{62}N_4O_{16}$), VI ($C_{38}H_{62}N_4O_{16}$), VII (B) ($C_{39}H_{64}N_4O_{16}$), VIII (C_2), ($C_{39}H_{64}N_4O_{16}$), IX (D_1) ($C_{40}H_{66}N_4O_{16}$), and X (D) ($C_{40}H_{66}N_4O_{16}$)], streptovirudins ($A_1 - D_1$) [A_1 ($C_{35}H_{58}N_4O_{16}$), B_1 ($C_{36}H_{60}N_4O_{16}$), B_{1a} ($C_{36}H_{60}N_4O_{16}$), C_1 ($C_{37}H_{62}N_4O_{16}$), D_1 ($C_{38}H_{64}N_4O_{16}$)], streptovirudins ($A_2 - D_2$) [A_2 ($C_{35}H_{56}N_4O_{16}$), B_2 ($C_{36}H_{58}N_4O_{16}$), B_{2a} ($C_{36}H_{58}N_4O_{16}$), C_2 ($C_{37}H_{60}N_4O_{16}$), and D_2 ($C_{38}H_{62}N_4O_{16}$)], and corynetoxins [H16i ($C_{39}H_{66}N_4O_{17}$), H18i ($C_{41}H_{70}N_4O_{17}$), H17a ($C_{40}H_{68}N_4O_{17}$), H19a ($C_{42}H_{72}N_4O_{17}$), U16i ($C_{39}H_{64}N_4O_{16}$), U18i ($C_{41}H_{68}N_4O_{16}$), U17a ($C_{40}H_{66}N_4O_{16}$), U19a ($C_{42}H_{70}N_4O_{16}$), S16i ($C_{39}H_{66}N_4O_{16}$), S18i ($C_{41}H_{70}N_4O_{16}$), S15a ($C_{38}H_{64}N_4O_{16}$), S17a ($C_{40}H_{68}N_4O_{16}$), and S19a ($C_{42}H_{72}N_4O_{16}$)].

Peptidyl N-Nucleoside Antibiotics

Polyoxins and Neopolyoxins. The polyoxins and neopolyoxins are peptidylpyrimidine nucleoside antibiotics that have achieved use as agricultural fungicides. See **Fungicides**.

Mureidomycins and Pacidamycins. The mureidomycins and pacidamycins are listed in Table 1. The four peptidylnucleosides, mureidomycins A-D, are produced by *S. flavidovirens*.

4-Aminohexose Nucleosides. The 4-aminohexose nucleosides include gougerotin ($C_{16}H_{25}N_7O_8$), blasticidin S ($C_{17}H_{26}N_8O_5$), amicetin ($C_{29}H_{42}N_6O_9$), bamicetin ($C_{28}H_{40}N_6O_9$), oxamicetin ($C_{29}H_{42}N_6O_{10}$), plicacetin ($C_{25}H_{35}N_5O_7$), norplicacetin ($C_{24}H_{33}N_5O_7$), hikizimycin (anthelmycin) ($C_{21}H_{37}N_5O_{14}$), bagougeramine A ($C_{17}H_{28}N_{10}O_7$), bagougeramine B ($C_{24}H_{44}N_{12}O_7$), arginomycin ($C_{18}H_{28}N_8O_5$), mildiomycin ($C_{19}H_{30}N_8O_9$), and SCH 36605 (rodaplutin) ($C_{24}H_{39}N_9O_7$). A biosynthetic relationship between the 4-aminohexose peptidyl nucleoside antibiotics and the pentopyranines has been proposed. The 4-aminohexose pyrimidine nucleoside antibiotics block peptidyl transferase activity and inhibit transfer of amino acids from aminoacyl-tRNA to polypeptides.

Nikkomycins. The nikkomycins, isolated from *S. tendae*, are nucleoside–peptide antibiotics. They include Z ($C_{20}H_{25}N_5O_{10}$), X ($C_{20}H_{25}N_5O_{10}$), J ($C_{25}H_{32}N_6O_{13}$), I ($C_{25}H_{32}N_6O_{13}$), B_z, B_x ($C_{21}H_{26}N_4O_{10}$), K_z ($C_{19}H_{23}N_5O_9$), K_x ($C_{19}H_{23}N_5O_9$), O_z ($C_{19}H_{23}N_5O_{10}$), O_x ($C_{19}H_{23}N_5O_{10}$),

TABLE 1. THE MUREIDOMYCINS AND THE PACIDAMYCINS

Name	Molecular formula
Mureidomycins	
mureidomycin A	$C_{38}H_{48}N_8O_{12}S$
mureidomycin B[a]	$C_{38}H_{50}N_8O_{12}S$
mureidomycin C	$C_{40}H_{51}N_9O_{13}S$
mureidomycin D[a]	$C_{40}H_{53}N_9O_{13}S$
Pacidamycins	
pacidamycin 1	
pacidamycin 2	$C_{39}H_{49}N_9O_{12}$
pacidamycin 3	$C_{39}H_{49}N_9O_{13}$
pacidamycin 4	
pacidamycin 5	$C_{36}H_{44}N_8O_{11}$
pacidamycin 6	
pacidamycin 7	$C_{38}H_{47}N_9O_{12}$

[a] Positions 5, 6 of the uracil moiety are saturated to give dihydrouracil.

Q_z $(C_{24}H_{32}N_6O_{12})$, Q_x $(C_{24}H_{32}N_6O_{12})$, P_x $(C_{20}H_{25}N_5O_9)$, R_z $(C_{25}H_{32}N_6O_{12})$, R_x $(C_{25}H_{32}N_6O_{12})$, W_z $(C_{19}H_{21}N_4O_9)$, W_x $(C_{19}H_{21}N_4O_9)$, pseudo-Z $(C_{20}H_{25}N_5O_{10})$, and pseudo-J $(C_{25}H_{32}N_6O_{13})$.

Glycosyl Nucleosides. There are five glycosyl antibiotics with either the adenine, uracil, or 7-deazaguanine aglycon. They are dapiramicin A $(C_{21}H_{29}N_5O_{10})$, epidapiramicin A $(C_{21}H_{29}N_5O_{10})$, dapiramicin B $(C_{21}H_{29}N_5O_{11})$, capuramycin $(C_{23}H_{31}N_5O_{12})$, and adenomycin $(C_{25}H_{39}N_7O_{18}S)$.

Octosyl Acids. Three octosyl uronic acid nucleosides are produced by *S. cacaoi* sub sp. *asoensis.*

Arabinosylpyrimidine Nucleosides. 1-β-D-Arabinofuranosylthymine (ara-T), $C_{10}H_{14}N_2O_6$, and 1-β-D-arabinofuranosyluracil (ara-U), $C_9H_{12}N_2O_6$, also known as spongouridine, were first isolated from the sponge *C. crypta.*

5-Azacytidine. 5-Azacytidine, $C_8H_{12}N_4O_5$, 4-amino-1-β-D-ribofuranosyl-*S*-triazine, 2-(1*H*-one), was chemically synthesized in 1964 and subsequently isolated from culture filtrates of *S. ladakanus.*

5,6-Dihydro-5-azathymidine. 5,6-Dihydro-5-azathymidine, $C_9H_{15}N_3O_5$, contains the *s*-triazine ring and is isolated from the culture filtrates of *S. platensis* var. *clarensis.*

Clitocine. Clitocine, $C_9H_{13}N_5O_6$, isolated from the mushroom, *Clitocybe inversa*, is 5-nitro-4-(β-D-ribofuranosylamino) pyrmidine-6-amine. The crystal structure indicates that the nitro group is hydrogen bonded to the 4-amino hydrogen atom. The base is in the anti conformation and the sugar moiety is disordered. Clitocine shows strong insecticidal activity against *Pectinophoro gossypiella* and is a substrate and inhibitor of adenosine kinase.

Uridine Analogues. 5-Formyloxymethyluridine, $C_{11}H_{14}N_2O_8$, produced by *Serratia plymuthica*, inhibits bacterial growth. 3-Methylpseudouridine, $C_{10}H_{14}N_2O_6$, has been isolated and identified from the culture filtrates of *Nocardia lactamdurans*. 1-Methylpseudouridine was previously isolated from *S. platensis*. 4-Thiouridine, $C_9H_{12}N_2O_5S$, has been isolated from an actinomycete resembling *S. hygroscopicus.*

Hydantocidin. Hydantocidin, $C_7H_{10}N_2O_6$, is elaborated by *S. hygroscopicus*. It is unique in that the anomeric carbon of the ribosyl moiety forms the spiro bond of hydantoin. The ribofuranose moiety which has been reported to be in a C_2-*endo* conformation has been synthesized. Hydantocidin is a herbicidal nucleoside with activity against monocotyledenous and dicotyledenous plants.

N-Nucleotides

The *N*-nucleotide antibiotics include agrocin 84 $(C_{21}H_{34}N_6O_{16}P_2)$, thuringiensin $(C_{22}H_{30}N_5O_{19}P)$, phosmidosine $(C_{16}H_{24}N_7O_8P)$, fosfadecin $(C_{13}H_{19}N_5O_{10}P_2)$, and fosfocytocin $(C_{12}H_{20}N_4O_{13}P_2)$. Agrocin 84 is an adenine 6-*N*-phosphoramidate nucleotide analogue that contains adenine, phosphate, and D-glucose in a 1:2:1 ratio. It is produced by *Agrobacterium radiobacter* strain K-84.

Thuringiensin, produced by *B. thuringiensis*, is a β-exotoxin that exerts its toxic action on insects and mammals through the inhibition of RNA polymerases.

The structure of the nucleotide antibiotic, phosmidosine, isolated from the culture filtrates of *Streptomyces* sp. RK-16 has been elucidated. Phosmidosine inhibits pore formation of *Botrytis cinerea* and *Aspergillus niger.*

Fosfadecin and fosfocytocin are adenine and cytosine nucleotide antibiotics isolated from culture filtrates of *Pseudomonas* species. Fosfadecin and fosfocytocin inhibit gram-positive and gram-negative bacteria.

ROBERT J. SUHADOLNIK
NANCY L. REICHENBACH
Temple University School of Medicine

Additional Reading

Isono, K.: *J. Antibiot.* **41**, 1711 (1988).
Isono, K.: *Pharmacol. Therapeut.*, **52**, 269 (1991).
Suhadolnik, R.J.: *Nucleoside Antibiotics*, John Wiley & Sons, Inc., New York, NY, 1970.
Suhadolnik, R.J.: *Nucleosides as Biological Probes*, John Wiley & Sons, Inc., New York, NY, 1979.

ANTIBIOTICS: OLIGOSACCHARIDES. Oligosaccharide antibiotics represented by the everninomicins [everninomicin B $(C_{66}H_{99}Cl_2NO_{36})$, everninomicin C $(C_{63}H_{93}Cl_2NO_{34})$, everninomicin D $(C_{66}H_{99}Cl_2NO_{35})$, everninomicin 2 $(C_{58}H_{86}Cl_2O_{31})$, everninomicin 3 $(C_{66}H_{98}Cl_2O_{33})$, everninomicin 7 $(C_{66}H_{100}Cl_2O_{34})$, everninomicin-13-384-Component 1 $(C_{70}H_{97}Cl_2NO_{38})$, and everninomicin-13-384-Component 5 $(C_{70}H_{99}Cl_2O_{36})$], flambamycins [flambamycin $(C_{61}H_{88}Cl_2O_{33} \cdot H_2O)$], avilamycins [avilamycin A $(C_{61}H_{88}Cl_2O_{32})$, avilamycin A_1, avilamycin B $(C_{59}H_{84}Cl_2O_{32})$, avilamycin C $(C_{61}H_{90}Cl_2O_{32})$, avilamycin D_1 $(C_{57}H_{82}Cl_2O_{31})$, avilamycin D_2, avilamycin E, avilamycin F $(C_{60}H_{87}ClO_{32})$, avilamycin G $(C_{62}H_{90}Cl_2O_{32})$, avilamycin H $(C_{61}H_{89}ClO_{32})$, avilamycin I $(C_{60}H_{86}Cl_2O_{32})$, avilamycin J $(C_{60}H_{86}Cl_2O_{32})$, avilamycin K $(C_{61}H_{88}Cl_2O_{33})$, avilamycin L $(C_{60}H_{86}Cl_2O_{32})$, avilamycin M $(C_{60}H_{86}Cl_2O_{32})$, and avilamycin N $(C_{60}H_{86}Cl_2O_{32})$], and curamycins [curamycin A $(C_{59}H_{84}Cl_2O_{32})$] have complex and unique structural features. They are sometimes referred to as orthosomycins because characteristically these oligosaccharides possess two acid-sensitive ortho ester linkages, cleavage of which results in the complete loss of antibiotic activity. Another structural feature common to all the oligosaccharide antibiotics is the presence of a substituted phenol ester derived from dichloroisoeverninic acid attached to the sugar ring B at C-13. This acidic phenolic group, which could form salts with organic or inorganic bases, e.g., the sodium or *N*-methylglucamine salt, is also essential for the antibiotic activity.

Properties

Physical Properties. Oligosaccharide antibiotics are colorless solids, which are often crystalline and have defined melting points and optical rotations.

Chemical Properties and Structure. Oligosaccharide antibiotics are sensitive to acid pH because of the ortho ester linkages. Yet all of them have an acidic phenolic group which makes the molecule relatively unstable to handling conditions. The ortho ester connecting the C and D rings is comparatively more sensitive to acid pH than the one linking the G and H residues. Other chemically reactive groups in these compounds are the glycosidic linkages, the hydroxyl and carbonyl groups, and any nitro groups present.

Everninomicins

Everninomicin D is the principal component from cultures of *Micromonospora carbonacae*. Chemical modification of the nitro group in everninomicin D has resulted in the formation of amino, mono- and dialkylamino, *N*-acylamino, and *N*-hydroxylamino (and its nitrone derivatives) everninomicin D, all of which possess great antibiotic activity against gram-positive bacteria.

Everninomicin B, though a secondary component, has been extensively investigated because of its improved pharmacokinetic properties over those of everninomicin D.

Flambamycin, Curamycin, and Avilamycins

Flambamycin. Flambamycin, produced by *Streptomyces hygroscopicus* DS 23230, also has undergone substantial chemical degradative experiments during structure elucidation. The structure, chemical reactions, and biological properties have been thoroughly reviewed; it melts at 202–203°C.

Curamycin. Curamycin A is the primary component of the culture *Streptomyces curacoi*. Curamycin A melts at 192–199°C.

Avilamycins. At least 16 avilamycins have been reported. Avilamycins A (mp 181–182°C) and C (mp 188–189°C) are the primary components produced by the strain *Streptomyces viridochromogenes*.

Other Oligosaccharides and Relevant Work

Sporacuracin A, $C_{63}H_{94}Cl_2O_{35}$, mp 145–148°C, $[\alpha]_D - 26.3$, and sporacuracin B, $C_{63}H_{94}Cl_2O_{35}$, mp 162–165°C, $[\alpha]_D - 18.8$, are other probable members of the oligosaccharide antibiotic family.

Biological Properties. The *in vitro* activity is such that oligosaccharides, in general, are highly potent, but are narrow-spectrum antibiotics. Everninomicins are active against a wide variety of gram-positive aerobes and anaerobes; *Neisseria*, *Mycoplasma*, and some *Mycobacteria*. Comparatively, flambamycin is less potent than everninomicin D.

In Vivo Activity. Potential for oligosaccharide antibiotics in both human and animal health care has been claimed in various patents.

V. M. GIRIJAVALLABHAN
ASHIT K. GANGULY
Schering-Plough Corporation

Additional Reading

Ganguly, A.K.: in P. Sammes, ed.: *Oligosaccharide Antibiotics, Topics in Antibiotic Chemistry*, Vol. 2, Ellis Horwood Publishers, Chichester, UK 1978, p. 49.
Ganguly, A.K., O.Z. Sarre, D. Greeves, and J. Morton: *J. Amer. Chem. Soc.* **97**, 1982 (1975).
Ganguly, A.K. et al.: *Heterocycles* **28**, 83 (1989).

ANTIBIOTICS: PENICILLINS AND OTHERS.

The basic structural features of the penicillin nucleus (**1**) include a β-lactam ring fused through nitrogen and the adjacent tetrahedral carbon to a second heterocycle which, in natural penicillin is a 5-membered thiazolidine ring. Biologically active penicillins are generally characterized by a functionalized amino group in the 6β-position of the β-lactam ring and a carboxyl group in the 3-position in the thiazolidine ring. The unsubstituted bicyclic ring system of the penicillins is designated as penam, C_5H_7NOS (**2**) and the penicillins (**1**) are generally 6-acylamino-2,2-dimethylpenam-3-carboxylic acids. A further simplification is the use of the term penicillanic acid, $C_8H_{11}NO_3S$, to designate the penicillin ring system having the substituents indicated in (**3**).

(**1**)

(**2**)

(**3**)

In general, penicillins exert their biological effect, as do the other β-lactams, by inhibiting the synthesis of essential structural components of the bacterial cell wall. These components are absent in mammalian cells so that inhibition of the synthesis of the bacterial cell wall structure occurs with little or no effect on mammalian cell metabolism. Additionally, penicillins tend to be irreversible inhibitors of bacterial cell-wall synthesis and are generally bactericidal at concentrations close to their bacteriostatic levels. Consequently penicillins have become widely used for the treatment of bacterial infections and are regarded as one of the safest and most efficacious classes of antibiotics. Penicillins in clinical use include limited spectrum penicillins [benzyl penicillin (penicillin G) ($C_{16}H_{18}N_2O_4S$), phenoxymethyl penicillin (penicillin V) ($C_{16}H_{18}N_2O_5S$), and phenethicillin ($C_{17}H_{20}N_2O_5S \cdot K$)], β-lactamase stable penicillins [methicillin ($C_{17}H_{20}N_2O_6S \cdot Na$), oxacillin ($C_{19}H_{19}N_3O_5S$), cloxacillin ($C_{19}H_{18}ClN_3O_5S$), dicloxacillin ($C_{19}H_{17}Cl_2N_3O_5S$), flucloxacillin ($C_{19}H_{17}ClFN_3O_5S$), and nafcillin ($C_{21}H_{22}N_2O_5SNa$)], broad spectrum penicillins [ampicillin ($C_{16}H_{19}N_3O_4S$), hetacillin ($C_{19}H_{23}N_3O_4S$), pivampicillin ($C_{22}H_{29}N_3O_6S$), talampicillin ($C_{24}H_{23}N_3O_6S$), bacampicillin ($C_{21}H_{27}N_3O_7S$), ciclacillin ($C_{15}H_{23}N_3O_4S$), amoxicillin ($C_{16}H_{19}N_3O_5S$), carbenicillin ($C_{17}H_{18}N_2O_6S$), ticarcillin ($C_{15}H_{16}N_2O_6S_2$), sulbenicillin ($C_{16}H_{18}N_2O_7S_2$), azlocillin ($C_{20}H_{23}N_5O_6S$), mezlocillin ($C_{21}H_{25}N_5O_8S_2$), and piperacillin ($C_{23}H_{27}N_5O_7S$)], and directed spectrum β-lactamase stable penicillins [temocillin ($C_{16}H_{18}N_2O_7S$)].

Physical Properties

Penicillins have several properties that are characteristic of β-lactam antibiotics. They are obtained in relatively pure form as off-white, tan, or yellow freeze-dried or spray-dried solids that are usually amorphous. Alternatively they are sometimes obtained as crystalline solids, often as hydrates. Penicillins do not usually have sharp melting points, but decompose upon heating to elevated temperatures. Most natural members have a free carboxyl group and commercial preparations are generally either supplied as salts, most frequently as sodium salts, or in zwitterionic form as hydrates, e.g., amoxicillin trihydrate. The acid strength of the carboxyl group in aqueous solution varies from $pK_{a1} = 2.73$ for oxacillin to $pK_{a1} = 3.06$ for carbenicillin.

Spectral Characteristics. The infrared stretching frequency of the penicillin β-lactam carbonyl group normally occurs at relatively high frequencies (1770–1815 cm^{-1}), as compared to the absorptions for the secondary amide (1504–1695 cm^{-1}) and ester (1720–1780 cm^{-1}) carbonyl groups. There is little difference between solution and KBr spectra. The nuclear magnetic resonance spectrum of penicillins invariably provides information about the integrity of the ring, attachments to the ring, and the stereochemistry of those attachments. The disposition of the C-5 and C-6 resonances is also characteristic. More recently NMR spectroscopy has also proved to be a valuable tool for detecting and characterizing penicillin metabolites in biofluids.

Stereochemistry. The absolute stereochemistry of the penicillins is 3(*S*):5(*R*):6(*R*) and the stereochemistry of the substituents attached to the ring is designated by the α and β notations. Thus the β-lactam hydrogens are α, the acylamino group is β, and the penicillin carboxyl is α as in (**1**).

Chemical Properties

Penicillin to Other β-Lactam Conversions. Penicillin G (**1**, R $=C_6H_5CH_2$), penicillin V (**1**, R$=C_6H_5OCH_2$), and 6-APA (**1**, RCO $=$H) have proven to be cheap and versatile starting materials for a number of conversions to novel β-lactam systems including 7-aminodeacetoxycephalosporanic acid (7-ADCA), the deacetoxycephalosporins, cephalexin and cephradin, moxalactam and related oxacephems, novel penems, and the carbapenem antibiotic thienamycin.

Synthesis

The only penicillins used in their natural form are benzylpenicillin (penicillin G) and phenoxymethylpenicillin (penicillin V). The remainder of penicillins in clinical use are derived from 6-APA and most penicillins having useful biological properties have resulted from acylation of 6-APA using standard procedures.

A variety of coupling methods have been employed including: acid chorides, mixed anhydrides, mixed sulfonic acid anhydrides, *N,N*-dicyclohexylcarbodiimide and similar condensing agents, activated esters with *N*-hydroxysuccinimide, and *N*-hydroxybenzotriazole together with other acylating agents commonly used in peptide synthesis.

The use of protecting groups is common in penicillin chemistry: the amino function is normally protected by a trityl, benzyloxycarbonyl, *p*-nitrobenzyloxycarbonyl, trichloroethyloxycarbonyl, or trimethylsilyl group; and the carboxylic acid is usually protected as a benzyl, *p*-nitrobenzyl, *p*-methoxybenzyl, or trichloroethyl ester. Acylations may thus be carried out in aqueous or nonaqueous media with subsequent removal of the protecting group as required.

Chemical Modification. Chemical modification of most positions in the penicillin nucleus have been carried out. Apart from acylation of 6-APA, few of these modifications have proven profitable in terms of improving the biological properties of the derived penicillins. However, one of the modifications that has led to beneficial properties is substitution at the 6α-position.

Degradation. Penicillins are rapidly hydrolyzed by aqueous alkali to the corresponding penicilloic acids which are stable as salts, but which decarboxylate on acidification to yield penilloic acids. Penicillins are also degraded by aqueous acids via initial reaction of the side-chain carbonyl group with the β-lactam.

Biological Properties

Structure–Activity Relationships. Biological evaluation of penicillins yields information such as *in vitro* and *in vivo* antibacterial activities, minimum inhibitory concentration (MIC), minimum bactericidal concentration (MBC), protective effectiveness in laboratory animals (PD₅₀), and pharmacokinetic characteristics including efficiency of absorption, serum levels, tissue distribution, urinary excretion, recycling, etc. Penicillins are also tested for ability to resist inactivation by β-lactamase produced by both gram-positive and gram-negative bacteria.

Penicillin G remains probably the most active penicillin against gram-positive organisms. However, the majority of *Staphylococcus aureus* strains are resistant to penicillin by virtue of β-lactamase production. The β-lactamase-resistant penicillins such as oxacillin, cloxacillin, flucloxacillin, and nafcillin are active against most penicillin-resistant *Staphylococci* but lack activity against methicillin-resistant *Staphylococci* (MRSA) and the majority of gram-negative organisms.

Ampicillin and its congeners amoxicillin, bacampicillin, and ciclacillin, have largely similar antibacterial spectra, exhibiting activity against both penicillin-sensitive gram-positive and gram-negative microorganisms. The susceptibility of ampicillin or amoxicillin to β-lactamase may be overcome by combination with a β-lactamase inhibitor. Clavulanic acid and sulbactam are two such β-lactamase inhibitors in clinical use. The products Augmentin, a combination of clavulanic acid and amoxycillin for oral use primarily, and Unasyn, a combination of sulbactam and ampicillin for use by injection, are highly active against both gram-positive aerobic and anaerobic organisms in addition to many important gram-negative pathogens. Carbenicillin and ticarcillin possess moderate broad-spectrum activity and were the first penicillins to show activity against *Pseudomonas* species. Azlocillin, mezlocillin, and piperacillin possess largely the same characteristics but have slightly greater potency against many gram-negative species. Temocillin is the first penicillin to possess a high level of activity against gram-negative organisms, notably the *Enterobacteriaceae*, as well as excellent β-lactamase stability. The introduction of the 6α-methoxy group and the concomitant β-lactamase stability compromises activity against both gram-positive organisms and *Pseudomonas* species. Similarly BRL 36650, having the 6α-formamido substituent, exhibits high bacterial β-lactamase stability. Some activity is retained against most species of *Streptococci* but the compound is inactive against *Staphylococci*. BRL 36650 is highly potent against most gram-negative organisms including refractory gram-negative species such as *Acinetobacter* and *Pseudomonas*. The level of potency is much greater than either that of ticarcillin or piperacillin.

The pharmacology of penicillins differs markedly from compound to compound. The majority of derivatives, including penicillin G and the antipseudomonal penicillins, are unstable in gastric acid and are not available orally. The isoxazolyl penicillins are relatively acid-stable but

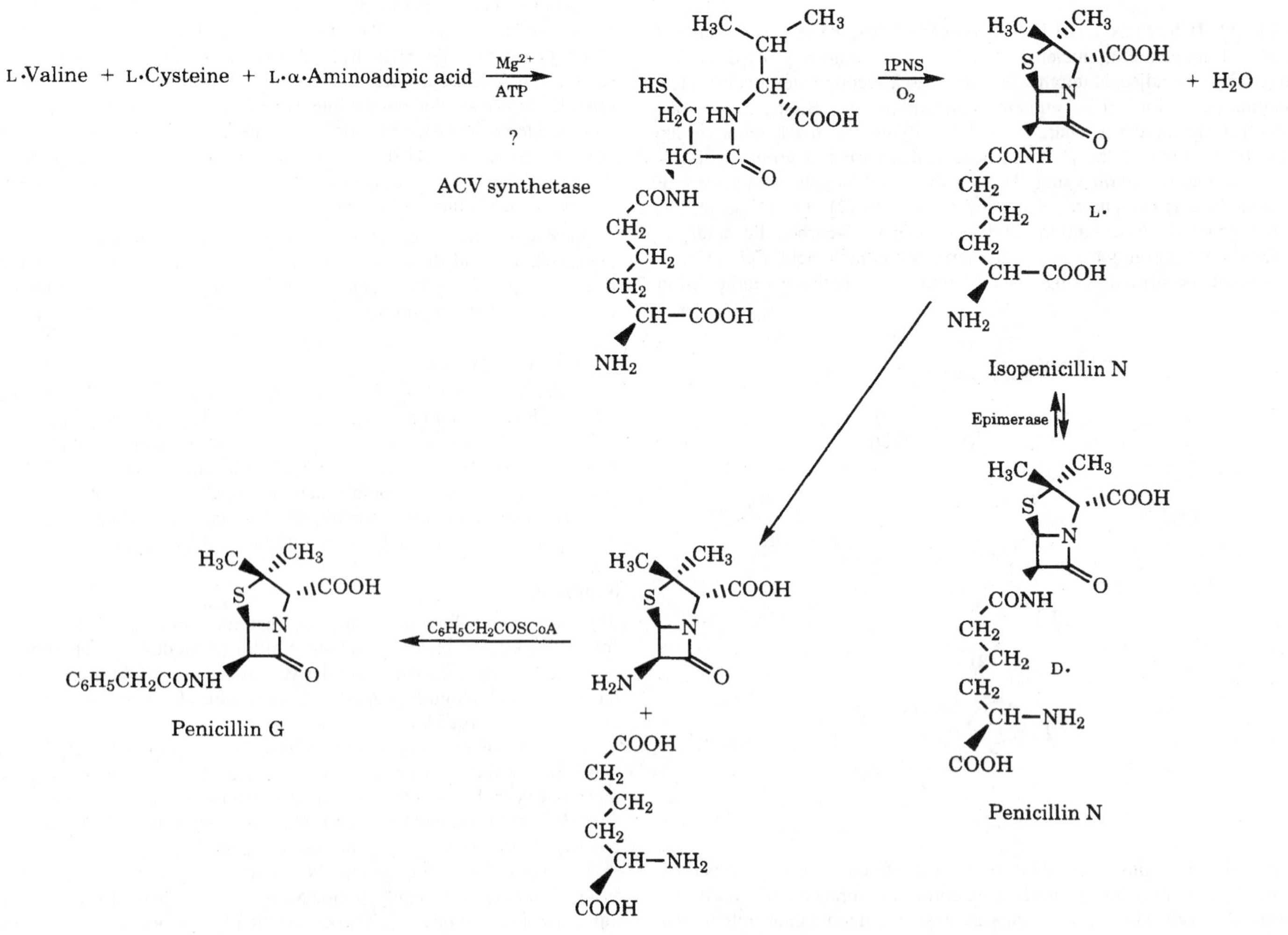

Fig. 1. Biosynthesis of penicillins when ACV is aminoadipoly cysteinyl valine and IPNS is isopenicillin N synthase and C₆H₅CH₂COSCoA represents benzyl coenzyme A. ACV synthetase is thought to catalyze the first step of this reaction sequence. Courtesy of J. E. Baldwin, Royal Society of Chemistry

not consistently well absorbed by the oral route. Nafcillin and oxacillin are poorly absorbed orally; cloxacillin, dicloxacillin, and flucloxacillin are more reliable. Penicillin V, ampicillin, and particularly amoxicillin are relatively well absorbed orally. Esters of ampicillin such as bacampicillin, pivampicillin, and talampicillin improve the level of oral absorption of ampicillin to that achieved by amoxicillin. Absorption can be diminished by food after oral administration, however, and peak blood levels, usually achieved after 1 to 2 h, are somewhat delayed after ingestion of food.

The penicillins in general are renowned for their lack of toxicity. The most common adverse effect of the use of penicillins is an allergic reaction which can change from a mild rash to fatal anaphylactic shock in rare cases. All penicillins cross the placenta and are excreted in maternal milk. However, the relative freedom from toxicity renders these compounds valuable agents during pregnancy and lactation.

Biosynthesis. The microbial synthesis of penicillins from eukaryotes such as *C. acremonium* and *P. chrysogenum* has been comprehensively reviewed. In essence the biosynthesis of penicillins is described in Figure 1 although certain stages have yet to be fully characterized. Products are isopenicillin N, penicillin G, and penicillin N, $C_{14}H_{21}N_3O_6S$.

Mode of Action. Penicillins exert their antibacterial effect by inhibiting the high molecular weight penicillin-binding proteins (PBPs) that are implicated in the final stages of peptidoglycan synthesis.

Production, Manufacture, and Processing

Most methods used for the production of the commercially important α-amino penicillins, such as ampicillin and amoxicillin, are based on modifications of an enamine process employing the appropriate phenylglycine and methylacetoacetate followed by coupling with 6-APA.

R. J. PONSFORD
SmithKline Beecham Pharmaceuticals

Additional Reading

Long, A.A.W., J.H.C. Nayler, H. Smith, T. Taylor, and N. Ward: *J. Chem. Soc.* (C), 1920 (1971).
Neway, J.O. ed.: *Fermentation Process Development of Industrial Organisms,* Marcel Dekker Inc., New York, NY, 1989.
Parry, M.F.: *Med. Clin. North Am., Update on Antibiotics 1, The Penicillins* **71**, 1093 (1987).
Pratt, W.B. and R. Fekety, eds.: *The Antimicrobial Drugs,* Oxford University Press, London, UK, 1986.

ANTIBIOTICS: PEPTIDES.

Peptide antibiotics are classified according to their overall shape, which can be linear or cyclic, and by the nature of the bonds joining the constituent amino acids and other carboxylic acids, which can be all amide bonds or amide plus ester bonds. Most peptide antibiotics are cyclic peptides that do not contain disulfide linkages.

Peptide antibiotics differ in many respects from proteins and from peptides having hormonal or other functions in higher animals and plants. Among the significant differences are low molecular weight, where the vast majority of peptide antibiotics have molecular weights in the 500–1500 range. The average protein has a mol wt of 40,000. Many peptide antibiotics have unusual fatty acids and amino acids, such as D-amino acids, N-methyl amino acids, or imino acids, and they usually lack methionine and histidine. Additionally, peptide antibiotics often have other nonamino acid moieties, such as the chromophore of dactinomycin. Ring closure in cyclic peptide antibiotics is by amide or ester bonds, not by disulfide bonds as in normal proteins; or, if the peptide antibiotic has a tail, a diamino acid or a hydroxy amino acid provides the amide or ester bonds for ring closure. Peptide antibiotics are normally resistant to the usual proteases and peptidases. Also they are usually synthesized on multienzyme complexes much as are fatty acids, and not on ribosomes as are proteins. Nisin and related antibiotics are exceptions. Most peptide antibiotics are synthesized, and are often marketed, as groups of closely related structures, usually reflecting a lack of specificity of the biosynthetic enzymes. Even the small fraction of peptide antibiotics that have therapeutic usefulness are quite toxic. For these useful peptide antibiotics the therapeutic index, the ratio of the minimum toxic dose to the maximum effective dose, is smaller than for most nonpeptide antibiotics.

Although historically most useful antibiotics have come from spore forming microorganisms, marine organisms have yielded the candidate antitumor peptide didemnin B and cytostatic peptides such as the patellamides. Many of the marine peptides have little or no antimicrobial activity. Antibacterial peptides called magainins are found in frog skin and antibacterial polypeptides called defensins are found in mammalian white blood cells and other tissues.

Peptide antibiotics are not often the drugs of first choice for systemic therapy of important human disease. However, the World Health Organization, which chooses drugs especially for Third World use based on efficacy, safety, quality, price, and availability, includes as essential such peptide antibiotics as bleomycin, dactinomycin, and bacitracin (as an ointment containing neomycin), plus several β-lactams. See also **Antibiotics, Antibiotics: β-Lactams**. Systemic use of peptide antibiotics is many times limited by nephrotoxicity and other toxicities. Semisynthesis or complete chemical synthesis of analogues of peptide antibiotics has most often not resulted in improved drugs.

The complex structure of peptide antibiotics adds considerably to the problems of synthesis, but more recent efforts toward improved peptide antibiotics are encouraging. Methods of bioassay and other laboratory use of economic antibiotics are available. Table 1 is a list of important peptide antibiotics. More extensive listings of minor peptide antibiotics have been published.

EDMUND M. WISE, JR.
The Wellcome Research Laboratories

Additional Reading

Kleinkauf, H. and H. von Döhren, eds.: *Biochemistry of Peptide Antibiotics,* W. de Gruyter, Berlin, 1990.
Lorian, V. ed.: *Antibiotics in Laboratory Medicine,* 3rd Edition, Williams & Wilkins, Baltimore, Md., 1991.
Bycroft, B.W. ed.: *Dictionary of Antibiotics and Related Substances,* Chapman and Hall, London, 1988.
Laskin, A.I. and H.A. Lechevalier, eds.: *CRC Handbook of Microbiology,* 2nd Edition, Vol. 9, Parts A and B, CRC Press, Boca Raton, Fla., 1988.

ANTIBIOTICS: POLYETHERS.

The polyether antibiotics were first recognized as a separate class with the publication of the structure of monensin in 1967. Several members of the group have since found commercial application as anticoccidials in poultry farming and in improvement of feed efficiency for ruminants.

These antibiotics are characterized by multiple tetrahydrofuran and tetrahydropyran rings connected by aliphatic bridges, direct C–C linkage, or spiro linkage. Other features include a free carboxyl function, many lower alkyl groups, and a variety of functional oxygen groups. These structural features enable the molecule to form a cyclic conformation with the oxygen functions at the center and the alkyl groups on the outer surface. This conformation results in lipid solubility, even for the salt forms, enabling transport of cations across lipid membranes.

Individual polyethers exhibit varying specificities for cations. Some polyethers have found application as components in ion-selective electrodes for use in clinical medicine or in laboratory studies involving transport studies or measurement of transmembrane electrical potential.

Properties

The polyether class can be subdivided based on the number of carbon atoms in the backbone. The carbon backbone or skeleton refers to the longest chain of contiguous carbons between the carboxyl group and the terminal carbon. The 30 C skeleton group accounts for about 60% of the polyethers for which structures have been determined. Most of these contain one or two sugar moieties, usually 2,3,6-trideoxy-4-O-methyl-D-*erythro* pyranose. Monensin (**1**), widely used as an anticoccidial and feed efficiency enhancer, is an example of the 26 C backbone class. Maduramicin (**2**), another commercially important polyether, is shown as an example of the 30 C group with a sugar moiety.

(**1**) Monensin A

TABLE 1. PEPTIDE ANTIBIOTICS OF SPECIAL INTEREST

Name	Related to	Producing microbe	Antibiotic activity[a]	Mol wt (number of amino acids)	Number in family
From bacteria of the genus Bacillus					
bacillomycin F		*B. subtilis*	F	1080 (8)	2
bacilysin		*B. subtilis, B. pumilis*	P, N	270 (2)	
bacitracin (ayfivin)		*B. licheniformis*	P	1410 (11)	5
circulins	polymyxins	*B. circulans*	N	1150 (10)	2
colistin (polymyxin E)	polymyxins, circulins	*B. colistinus*	N	1150 (10)	10
edeine A$_1$		*B. brevis*	P, N, F	755 (5)	6
gramicidins, linear, Dubos		*B. brevis*	P	1900 (15)	6
gramicidin S		*B. brevis*	P	1141 (10)	3
iturins A	bacillomycin B	*B. subtilis*	F	1050 (8)	9
micrococcin P$_1$	thiocillins	*B. pumilis*	P	1144 (5)	4
mycobacillin		*B. subtilis*	F	1528 (13)	1
polymyxins	circulins	*B. polymyxa*	N	1200 (10)	10
subtilin	nisin	*B. subtilis*	P	3321 (26)	1
tyrocidines		*B. brevis*	P, N	1300 (10)	5
tyrothricin		*B. brevis*	P, N	mixture	
From higher bacteria of the genus Streptomyces					
albomycin		*S. subtropicus, S. griseus*	P	950 (5)	5
amphomycin (glumamycin)		*S. canis, S. violaceus, S. albogriseolus*	P	1290 (10)	5
bialaphos (SF 1293)	phosphinothricin	*S. hydroscopicus*, other *S.* species	P, N,[b]	323 (3)	2
bicozamycin (bicyclomycin)		*S. sapporonensis, S. aizunensis*	N	302 (2)	
bleomycins	phleomycin, peplomycin, liblomycin	*S. verticillis*	P, N, M, F, T	1416 (12)	
bottromycin		*S. bottropensis*	P, N	810 (6)	5
capreomycins	viomycin, enviomycin, tuberoactinomycin	*S. capreolus*	P, N, M	670 (5)	4
dactinomycin (actinomycin D)	cactinomycin	*S. parvullus*	P, T	1255 (10)	30
daptomycin (deptomycin, LY 146032, semisynthetic)		from A21978C of *S. roseosporus*	P	1619 (14)	6
distamycin A (stallimycin)	netropsin	*S. distallicus*	F, T, V	482 (4)	
enduracidins (enramycin)		*S. fungicidicus*	P, M	2355 (17)	4
enviomycin (tuberactinomycin N)	capreomycin	*S. griseoverticilatus* var. *tuberacticus*	P, N, M	686 (6)	3
ferrimycin A	sideromycins	*S. griseoflavus*, other *S.* species	P, N	978 (6)	4
globomycin		*S. halstedii, S. cinnamomeus*	N	657 (5)	
ilamycin A (rufomycin A)		*S. islandicus, S. atratus*	P, M	1026 (7)	5
neoviridogrisein II (etamycin B)		*S. griseoviridis*	P	863 (7)	3
netropsin	distamycin	*S. netropsis, S. chromogens*, others	P, N, M, T, V	430 (2)	8
quinomycin C	triostin	*S. aureus*	P, N, T, V	1141 (8)	3
stendomycin		*S. antimycoticus*	P, N, F	1850 (14)	3
streptogramin A (pristinamycin IIA, virginiamycin M1)	virginiamycin M2	*S. virginiae*, others	P	527 (3)	2
streptogramin B (pristinamycin IA)	virginiamycin S$_1$	*S. virginiae*, others	P	867 (6)	12
streptothricin A (polymycin A)	noureothricins, racemomycins	*S. lavendulae*, others	P, N, M, V	1143 (6)	7
telomycin		*S. canus*, others	P	1272 (11)	1
thiopeptin		*S. tateyamensis*	P	1550 (6)	8
thiostrepton (bryamycin)		*S. aureus*	P	1665 (9)	1
valinomycin		*S. fulvissimus, S. tsusimaensis*	P, N, F, T, V	1111 (6)	
viridogrisein (etamycin A)		*S. lavendulae*	P	879 (7)	2
zinostatin (neocarzinostatin)		*S. carcinostaticus*	P, T	11400 (113)	1
From other bacteria					
nisin	subtilin	*Streptococcus lactis*	P	3354 (34)	
From fungi					
alamethicin		*Trichoderma viridis*	F	1970 (19)	2
cilofungin (semisynthetic)	aculeacin B, sporiofungin	from echinocandin B of *Aspergillus* spp.	F	1029 (4)	
cyclosporine (cyclosporin A)[c]	other cyclosporins	*Cylindrocarpon lucidum, Tolypocladium inflatum*	F[c]	1203 (11)	25
destruxin B		*Oospora destructor*, others	[d]	584 (5)	
echinocandin B		*Aspergillus rugulosus, A. nidulans* var. *roseus*	F	1060 (4)	8
enniatin A		*Fusarium orthoceras, F. sciroi*	P, M, F	682 (6)	3
tentoxin		*Alternaria tenuis, A. alternata*	[b]	415 (4)	2

[a] Activity against gram-positive bacteria = P; gram-negative bacteria = N; mycobacteria = M; fungi = F; tumors = T; and viruses = V.
[b] Active as a herbicide.
[c] Active against the immune system.
[d] Active as an insecticide.

(2) Maduramicin alpha

Purification and Production

Polyethers are usually found in both the filtrate and the mycelial fraction, but in high yielding fermentations they are mostly in the mycelium because of their low water-solubility. The high lipophilicity of both the free acid and the salt forms of the polyether antibiotics lends these compounds to efficient organic solvent extraction and chromatography on adsorbents such as silica gel and alumina. Many of the production procedures utilize the separation of the mycelium followed by extraction using solvents such as methanol or acetone. A number of the polyethers can be readily crystallized, either as the free acid or as the sodium or potassium salt, after only minimal purification.

Polyethers such as monensin, lasalocid, salinomycin, and narasin are sold in many countries in crystalline or highly purified forms for incorporation into feeds or sustained-release bolus devices (see **Controlled-Release Technology**). There are also mycelial or biomass products, especially in the United States. The mycelial products are generally prepared by separation of the mycelium and then drying by azeotropic evaporation, fluid-bed driers, continuous tray driers, flash driers, and other types of commercial driers. In countries allowing biomass products, crystalline polyethers may be added to increase the potency of the product.

Biological Activities

The polyether antibiotics exhibit a broad range of biological, antibacterial, antifungal, antiviral, anticoccidial, antiparasitic, and insecticidal activities. They improve feed efficiency and growth performance in ruminant and monogastric animals. Only the anticoccidial activity in poultry and cattle, and the effect on feed efficiency in ruminants such as cattle and sheep are of commercial interest.

The discovery of the activity of the polyethers against *Eimeria* sp. has greatly altered the prevention and control of coccidiosis. It is estimated that the polyether ionophores constitute more than 80% of the total worldwide usage of anticoccidials. The enhancement of feed efficiency in ruminants correlates with increased production of propionic acid in the rumen. A decrease in the production of acetic acid, butyric acid, and methane often accompanies the increased propionic acid production. These volatile fatty acids are produced in the rumen by the degradation of carbohydrates by microorganisms. The polyethers apparently selectively inhibit certain of the microorganisms to achieve greater propionate production. Propionate is thought to be more effectively utilized in energy metabolism in the host animal than acetate or butyrate.

Biosynthesis

Bacteria belonging to the order Actinomycetales are the organisms reported to produce all of the polyethers. Most are secondary metabolites of *Streptomyces* sp. with the species *hygroscopicus* and *albus* accounting for about one-third of the antibiotics. Other genera represented are *Streptoverticillium*, *Dactylosporangium*, *Actinomadura*, *Nocardia*, and *Nocardiopsis*. The taxonomy of these producing organisms has been reviewed.

LOUISE W. CRANDALL
ROBERT L. HAMILL
Lilly Research Laboratories

Additional Reading

Robinson, J.A.: "Chemical and Biochemical Aspects of Polyether–Ionophore Antibiotic Biosynthesis" in W. Herz, G.W. Kirby, W. Steglich, Ch. Tamm, eds., *Progress in the Chemistry of Organic Natural Products*, Springer-Verlag, Wien, New York, NY, 1991, pp. 1–82.

Westley, J.W. ed.: *Polyether Antibiotics,* Vols. 1 and 2, Marcel Dekker, Inc., New York, NY, 1982.

Yonemitsu, O. and K. Horita: "Total Synthesis of Polyether Antibiotics" in G. Lukacs and M. Ohno, eds. *Recent Progress in the Chemical Synthesis of Antibiotics,* Springer Verlag, Berlin, Heidelberg, 1990, pp. 448–466.

ANTIBIOTICS: TETRACYCLINES.

The tetracyclines are a group of antibiotics having an identical 4-ring carbocyclic structure as a basic skeleton and differing from each other chemically only by substituent variation. Figure 1 shows the principal tetracycline derivatives now used commercially.

The first tetracycline discovered was produced by a soil organism, *Streptomyces aureofaciens*, and is now known as chlortetracycline (2), $C_{22}H_{23}ClN_2O_8$. This compound ushered in a new era in antibacterial chemotherapy because it was effective orally and against a broad range of gram-positive and gram-negative bacteria.

The three tetracyclines most recently marketed were made by a semisynthetic pathway. The first of these were methacycline (6-methylene oxytetracycline) (5), $C_{22}H_{22}N_2O_8$, and its reduction product doxycycline (6), $C_{22}H_{24}ClN_2O_8$. The latter compound is a potent antibiotic which is well-absorbed and slowly excreted, thus allowing small and infrequent (once or twice a day) dosage schedules. Finally, the most recent addition to the commercial tetracyclines is minocycline (7), $C_{23}H_{27}N_3O_7$, which is also well-absorbed and slowly excreted.

Physical Properties

In general, the tetracyclines are yellow crystalline compounds that have amphoteric properties (Fig. 2). They are soluble in both aqueous acid and aqueous base.

The tetracyclines are strong chelating agents. This ability to chelate to metals, such as calcium, results in tooth discoloration when tetracycline is administered to children.

Semisynthetic Modifications

The tetracycline molecule (1) presents a special challenge with regard to the study of structure–activity relationships. The difficulty has been to devise chemical pathways that preserve the BCD ring chromophore and its antibacterial properties. The lability of the 6-hydroxy group to acid and base degradation, plus the ease of epimerization at position 4, contribute to chemical instability under many reaction conditions.

Although many of the tetracycline derivatives showed useful *in vivo* activity against a wide spectrum of pathogenic organisms, few showed any significant improvement in overall activity when compared to tetracycline. The exception is minocycline (7) which exhibits superior activity against tetracycline-sensitive organisms and many tetracycline-resistant strains of gram-positive bacteria.

Structure–Activity Correlations

There are a number of tetracycline structural features that are prerequisites for biological activity. The linear arrangement of the rings, coupled with the phenolic β-diketone system, is essential. Any structure variation at the 11a position results in loss of activity. The C-11 to C-12 β-diketone system has exceptional chelating qualities, and probably is involved in the binding of the tetracyclines to ribosomes, in the interactions with bacterial repressor proteins, and in transport of tetracyclines into the bacterial cell. The amide hydrogen can be replaced by a methyl group, but larger residues, if not rapidly cleaved in water, bring about a reduction in activity.

The configuration at the chiral centers C-4a, C-5a, and C-12a determine the conformation of the molecule. In order to retain optimum *in vitro* and *in vivo* activity, these centers must retain the natural configuration. The hydrophobic part of the molecule from C-5 to C-9 is open to modification in many ways without losing antibacterial activity. However, modification at C-9 may be critical because steric interactions or hydrogen bonding with the oxygen atom at C-10 may be detrimental to the activity.

Manufacture

Most of the fermentation and isolation processes for manufacture of the tetracyclines are described in patents. Manufacture begins with the cultivated growth of selected strains of *Streptomyces* in a medium chosen to produce optimum growth and maximum antibiotic production. Some clinically useful tetracyclines (2–4) are produced directly in these fermentations; others (5–7) are produced by subjecting the fermentation

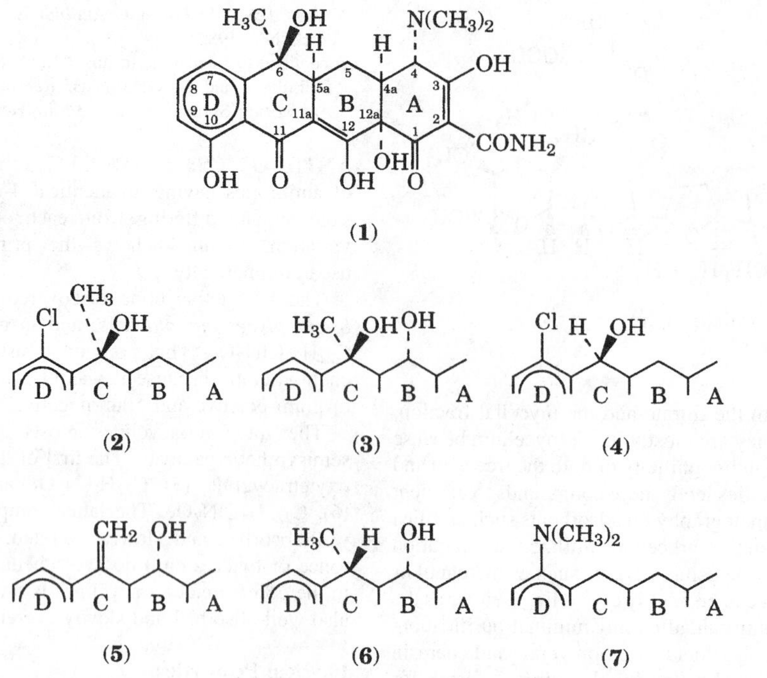

Fig. 1. Tetracycline (1) and its derivatives: chlortetracycline (7-chlorotetracycline) (2); oxytetracycline (5-hydroxytetracycline) (3); demeclocycline (6-demethyl-7-chlorotetracycline) (4); methacycline (6-demethyl-6-deoxy-5-hydroxy-6-methylenetetracycline) (5); doxycycline (6α-deoxy-5-hydroxytetracycline) (6); and minocycline (6-demethyl-6-deoxy-7-dimethylamino tetracycline) (7). Substituents at positions not specifically shown or mentioned remain as in (1). Courtesy of Blackwell Scientific Publications, Ltd

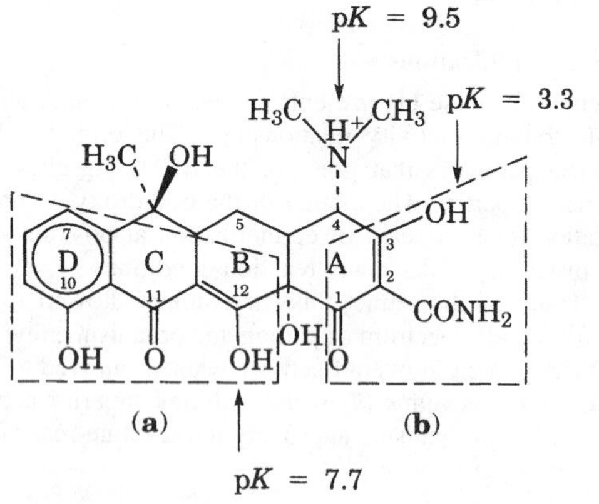

Fig. 2. Tetracycline (1) indicating the titratable hydrogens and showing (a) the BCD-chromophore and (b) the A-chromophore

TABLE 1. TETRACYCLINES USED FOR THE THERAPY OF INFECTIOUS DISEASES

Generic name	Trade name	Year of discovery
chlortetracyline	Aureomycin	1948
oxytetracyline	Terramycin	1948
tetracycline	Achromycin	1953
demeclocycline	Declomycin	1957
methacycline	Rondomycin	1965
doxycycline	Vibramycin	1967
minocycline	Minocin	1972

of protein synthesis results primarily from disruption of codon–anticodon interaction between tRNA and mRNA so that binding of aminoacyl–tRNA to the ribosomal acceptor (A) site is prevented. The precise mechanism is not understood. However, inhibition is likely to result from interaction of the tetracyclines with the 30S ribosomal subunit because these antibiotics are known to bind strongly to a single site on the 30S subunit.

Uses

Clinical Uses. The emergence of bacterial resistance to tetracyclines has limited the use of these agents as the drugs of first choice in the treatment of many infections for which they were previously effective. Nevertheless, they are still the treatment of first choice in the following cases: *(1)* for bacterial infections causing brucellosis, cholera, chancroid, granuloma inguinale, and Lyme disease; *(2)* for rickettsial infections; *(3)* for chlamydial infections; *(4)* in the treatment of nonspecific urethritis because of *Chlamydia* or *Ureaplasmas;* and *(5)* in the treatment of acne vulgaris and rosacea.

Tetracyclines are used as alternative drugs in a variety of circumstances when the patient is unable to take the drug of choice, e.g., in patients allergic to penicillin.

Veterinary Uses. Tetracyclines are widely used for veterinary therapy. The types of pathogens encountered are frequently different from those for which tetracyclines are used in humans. Tetracyclines are also used in animal husbandry as growth promoters.

Resistance to Tetracyclines

Mechanisms of Resistance. Three distinct biochemical mechanisms of resistance to tetracyclines have been identified. The energy-dependent efflux of antibiotic mediated by resistance proteins located in the bacterial

products to one or more chemical alterations. The purified antibiotic produced by fermentation is used as the starting material for a series of chemical transformations.

Economic Aspects

The development of the semisynthetic β-lactam antibiotics and emergence of resistance to the tetracyclines has steadily diminished the clinical usefulness of tetracyclines.

In the United States, the manufacturers of fermentation-derived tetracyclines (1), (2), and (3) are the Lederle Laboratories, a division of American Cyanamid Company, Charles Pfizer Inc., Bristol Laboratories, and Rachelle Laboratories. There are also several manufacturers abroad. Tetracycline is now sold generically by many companies. Pfizer's doxycycline (6) and Lederle's minocycline (7), both semisynthetic tetracyclines, are the only members of the group that have increasing sales. Table 1 lists the commercial tetracyclines and the corresponding trade names.

Biological Aspects

It has been known for some time that tetracyclines are accumulated by bacteria and prevent bacterial protein synthesis. Furthermore, inhibition of protein synthesis is responsible for the bacteriostatic effect. Inhibition

cytoplasmic membrane is one mechanism. The intracellular tetracycline concentration remains too low for effective binding to ribosomes. Ribosomal protection, whereby tetracyclines no longer bind productively to the bacterial ribosome, is a second mechanism. In a tetracycline-resistant cell, tetracycline accumulation within the cell is similar to that in the sensitive cell, but the ribosome is modified so that tetracycline no longer binds productively to the ribosome. Finally, chemical alteration of the tetracycline molecule by a reaction in the cytoplasm that requires oxygen renders the drug inactive as an inhibitor of protein synthesis. The altered tetracycline then diffuses out of the cell.

JOSEPH J. HLAVKA
GEORGE A. ELLESTAD
IAN CHOPRA
American Cyanamid Company

Additional Reading

Blackwood, R.K. et al: *J. Amer. Chem. Soc.* **85**, 3943 (1963).
Broschard, R.W. et al: *Science* **109**, 199 (1949).
Duggar, B.M. *Ann. N.Y. Acad. Sci.*, **51**, 177 (1948).
Hlavka, J.J. and J.H. Boothe, eds.: *Handbook of Experimental Pharmacology*, Vol. 78, Springer-Verlag, New York, 1985, p. 86.
Martell, M.J. Jr. and J.H. Boothe: *J. Med. Chem.* **10**, 44 (1967).
Redin, G.S. in G.L. Hobby, ed., *Antimicrobial Agents and Chemotherapy*, Williams & Wilkins, Baltimore, Md., 1966, p. 371.
Stephens, C.R. et al: *J. Amer. Chem. Soc.* **85**, 2643 (1963).

ANTIBODY. This article gives a generalized description of antibodies and their role in the body's immune system. More details in terms of the most recent findings in this field are given in the entry on **Immune System and Immunochemistry**. In medicine and physiology, immunity is the ability of the body to resist invasion by pathogenic organisms and substances. Immunity may be initially in place, that is, genetically ordered for a given species. Humans are naturally immune to canine distemper; dogs are immune to measles; rats are immune to diphtheria; and domestic fowls are immune to anthrax. Many other examples could be cited. Immunity may be acquired as the result of exposure to an invasive pathogen, triggering the immune system to construct cells that will be in reserve to resist subsequent invasions by the same pathogen. Immunity also may be acquired artificially through the use of preventive immunization techniques. Immunity is effected through antibodies.

Any substance that can provoke a response by the body's immune system is called an *antigen*. This property of an antigen is referred to as *immunogenicity*. Although the first antigens to be investigated were microorganisms and proteins foreign to the body, research during recent decades has been directed toward understanding the immune response at the molecular level—for it is at this level that the actions and reactions of the immune system occur. At the molecular level, numerous previously unsuspected complexities of the immune process have been revealed and still others are only partially understood, if at all. It has been discovered that several cell types, in addition to the lymphocytes, act cooperatively in effecting what might be called the total immune response. Although the lymphocytes appear to play the dominant role in the immune system, several other cells are now known to cooperate with the lymphocytes, and the functions of these other cells are no longer considered of secondary importance. Study of the antigens at the molecular level also has contributed to a much better understanding of the immune response.

Numerous molecules can evoke an immune response, sometimes when the responses from the standpoint of protecting body functions are not immediately obvious, other than that such molecules do not meet the criteria of "self" and "nonself" described later. In a general way, it may be observed that the immune system tends to have a bias toward suspicion and may, on occasion, overreact, as in cases of autoimmunity. Currently, it is generally hypothesized that recognition of antigens at the receptor sites of the lymphocytes is based upon the shapes of molecules, reminiscent of some of the current hypotheses concerning taste and odor receptors in the tongue and nasal membranes. Until the mechanism occurring at the receptor sites is more fully explained, numerous questions as regards what molecules do and do not evoke immune response will remain unanswered.

Considerable research with synthetic polymers comprised of various amino acids has been undertaken in an effort to determine the requirements of immunogenicity. It has been established, for example, that tyrosine as well as some other aromatic amino acids will confer immunogenicity to certain polypeptides which in they are not or are only slightly antigenic. Further, it has been found in such cases that the antibody is directed against the polypeptide and not the amino acid. Research with synthetic polymers led investigators to the finding that immune response is under genetic control, this based upon the observations that different animal strains and species respond differently to a given polymer. However, in a descriptive fashion, this principle had been demonstrated by the different reactions to antigens by various species many decades ago.

Although not catalytic in the usual sense, certain substances, known as *adjuvants*, are capable of enhancing the immunogenicity of certain antigens. Among the adjuvants are aluminum salts, bacterial endotoxins, bacillus Calmette-Guèrin (BCG), *Bordetella pertussis*, and mycobacteria. These materials and this phenomenon have been important in immunity research. Sometimes adjuvants are used clinically in connection with certain immunizations, such as against tetanus.

Where antigens are introduced into the body intravenously, they usually travel rapidly to the spleen, followed by the fast production of an antibody. Subcutaneous or intradermal injection of antigens most frequently localizes in the lymph nodes and antigens that are inhaled favor local sensitization. In some cases, such as tetanus immunization, toxin produced by the bacteria may be slow and insufficient to provoke a significant immunologic reaction. Thus, the requirement for properly timed booster injections.

Clinical Use of Antigens

Without the benefit of understanding the complexities of the immune system, particularly at the molecular level, much progress was made over the years in taking advantage of certain antigens and a qualitative or descriptive understanding of the immune response. This was the early development of vaccines and antitoxins.

The history of the development of antitoxins in combating bacterial infection dates back to the early beginnings of organized bacteriology. Behring was the first to show that animals that were immune to diphtheria contained, in their serum, factors which were capable of neutralizing the poisonous effects of the toxins derived from the diphtheria bacillus. While this work was carried out in 1890 (prior to many of the great discoveries of mass immunization, and, prior to the much later discovery of antibiotics), it is interesting to note that a relatively limited place remains for antitoxins in the modern medical treatment, or prophylaxis, of a few diseases, such as tetanus, botulism, and diphtheria. In the case of diphtheria, equine antitoxin is the only specific treatment available. However, it is only reasonably effective if used during the first 48 hours of the onset of the disease. Trivalent (ABE) antitoxin is used in the treatment of botulism. In the treatment of tetanus, human tetanus immune globulin is preferred, but when it is not available, equine antitoxin is substituted. See also **Antitoxin**.

For many years the preferred approach to immunity to infectious disease has been by development of active immunity through the injection of a vaccine. The vaccine may be either an attenuated live infectious agent, or an inactivated or killed product. In either case, protective substances called antibodies are generated in the bloodstream; these are described in the next section. Vaccines for a number of diseases have been available for many years and have assisted in the eradication of some diseases, such as smallpox. As new strains of bacteria and viruses are discovered, additional vaccines becomes available from time to time. See also **Vaccine Technology**.

Antibodies

Antigens are excluded from the body by skin and mucous membranes. If these barriers are penetrated, the foreign organism may be ingested by phagocytic cells (monocytes, polymorphs, macrophages) and subsequently destroyed by cytoplasmic enzymes. Some time after a foreign macromolecule has entered the body, induced mechanisms come into play. There are two basic biological manifestations of the immune reaction: (1) Immunity to infectious agents; and (2) specific hypersensitivity. Hypersensitivity, or the heightened response to an agent, can be divided into anaphylactic, allergic, and bacterial. Anaphylaxis, which can be produced by either active or passive sensitization, is a laboratory tool for studying the fundamental nature of hypersensitivity. The amounts of antigen and antibody involved, as well as the nature and source of the antibody, govern the extent of the reaction.

In immunity to infectious agents, some time after a foreign macromolecule has entered the body, induced mechanisms come into play, which result in the synthesis of specially adapted molecules (*antibodies*) capable of combining with the foreign substances which have elicited them. Most macromolecules (proteins, carbohydrates, nucleic acids, etc.) can function as antigens, provided that they are different in structure from autologous macromolecules, i.e., from the macromolecules of the responding organism.

Antibodies are proteins with a molecular weight of 150,000–1,000,000 and with electrophoretic mobility predominantly of gamma globulins. The combination between antigen and antibody results in inhibition of the biological activity of the antigen and leads to increased rate of ingestion (*opsonization*) of the antigen by phagocytic cells. In addition, combination of antigen and antibody results in the activation of a complex chain of interacting constitutive molecules—the *complement system*—leading to lysis of the cell membranes to which antibody, directed against cellular antigens, is attached.

Biochemical Individuality

This is a unique quality, genetically determined, for each individual and is exhibited with respect to: (1) the composition of blood, tissues, urine, digestive juices, cerebrospinal fluid, etc.; (2) the enzyme levels in tissues and in body fluids, particularly the blood; (3) the pharmacological responses to specific drugs and poisons; (4) the biochemical responses to bacteria, fungi, and other microorganisms; (5) the quantitative needs for specific nutrients—minerals, amino acids, vitamins, etc.—and in a number of other ways, including reactions of taste and smell and the effects of heat, cold, and electricity. Although individual *similarities* are readily apparent at the macro level, each individual must possess a highly distinctive pattern, since the differences between individuals with respect to measurable items in a potentially long list are by no means trifling. Out of these relatively small, but distinct differences has risen the concept of the so-called normal individual, against which high and low levels of response are compared when considering the "chemistry" of a given individual.

Autoimmunity

A very important characteristic of the body's immune system is a capacity to distinguish between *self* and *non-self*. In terms of the immune system, the ability to make such distinctions proceeds at the biochemical level without conscious awareness of the individual. Less than a century ago, the ability of the body to distinguish self from non-self (foreign), at the biochemical level, was considered impossible. Over the years, however, much descriptive information accumulated which, without ample explanation, proved that the body does remember at the biochemical level. Seldom, for example, have medical records shown a second infection with mumps, measles, or smallpox, once an individual survived the first attack. The question, how does the body remember? remained a mystery for many decades until the concept of the immune system was first outlined in a very general way.

When the immune system functions property, which is the normal situation, antibodies to parts of the same body are not produced. But a condition known as *auto-immunity* can sometimes occur. In such circumstances, antibodies and sensitized (antigen-reactive) cells may be produced and directed against "self" antigens. To treat auto-immunity is one of the challenges of modern medicine. This auto-immunity mechanism may trigger some asthmatic paroxysms and is presently being considered as a suspect mechanism by cancer researchers—with considerable study being directed to tumor immunology. Some investigators hypothesize that the body system has the capacity to recognize neoplastic cells and to destroy them, but that in some individuals the process becomes inoperative, allowing the cells to multiply.

The ability of the body to tolerate self-antigens and thus to preclude auto-immunity is known as *immunological tolerance*. As early as 1959, some scientists suggested that the antigen-specific lymphocytes which interact with self-antigens are eliminated during the prenatal state. In recent years, the concept of *suppressor mechanisms* has been well received. In this concept, the immune system, responding to an antigen, in addition to producing antibodies and sensitized cells to combat foreign substances, also initiates various suppressor mechanisms, causing some mediation of the process. Such substances are termed *mediators*. This may explain why, in some clinical situations, an early and strong immune response may be due to deficiencies in the suppressor mechanisms; or, in contrast, a deficient immune response may not be caused by the lack of particular lymphocytes, but rather by over-reactive suppressor mechanisms.

A study of the body's immune system over the years has made possible the preparation of antitoxins and vaccines for preventing and treating many diseases. These studies have assisted in dealing with the problems of allergy and hypersensitivity and in finding partial or effective solutions to problems that arise from malperformance of the immune systems. Further progress is expected in the understanding and treatment of immune-system-related disorders, such as amyloidosis, macroglobulinemia, multiple myeloma, systemic lupus erythematosus, asthma, and various allergies, among others.

See also **Immune System and Immunochemistry**.

Additional Reading

Delves, P.J.: *Antibody Applications, Essential Techniques*, John Wiley & Sons, Inc., New York, NY, 1995.
Frank, M.M. and J.E. Volanakis: *The Human Complement System in Health Disease*, Marcel Dekker, Inc., New York, NY, 1998.
Harris: *Antibody Identification Protocols*, American Association of Blood Banks, Bethesda, MD, 1999.
Hue, K.M. and J.L. Bidwell: *Handbook of HLA Typing Techniques*, CRC Press, LLC, Boca Raton, FL, 1993.
Leong: *Antibody Manual*, Greenwich Medical Media, New York, NY, 1999.
Lillehoj, E.P. and V.S. Malik: *Antibody Techniques*, Academic Press, San Diego, CA, 1995.
Shoenfeld, Y.: *The Decade of Autoimmunity*, Elsevier Science, New York, NY, 1999.

ANTICAKING AGENTS.   Some products, particularly food products that contain one or more hygroscopic substances, require the addition of an *anticaking agent* to inhibit formation of aggregates and lumps and thus retain the free-flowing characteristic of the products. Calcium phosphate, for example, is commonly used in instant breakfast drinks and lemonade and other soft-drink mixes.

The general function of an anticaking agent can be described by using silica gel as an example. Generally anticaking agents are available as very small particles (ranging from 2 to 9 micrometers in diameter). A typical application for silica gel is admixture with orange-juice crystals to assure a free-flowing product, avoiding formation of crystal cakes and hard lumps. The very high adsorption properties of the anticaking substance removes moisture that can cause fusion. The billions of extremely fine, inert particles coat and separate each grain of powder (product) to keep it free flowing. Many anticaking agents, including silica gel, also act as dispersants for powdered products. Many food products, when stirred into water, tend to form lumps that are difficult to disperse or dissolve. The agent not only improves flow properties, but also increases speed of dispersion by keeping the food particles separated and permitting water to wet them individually instead of forming lumps. As is true with so many food additive chemicals, anticaking agents serve multiple functions. In addition to acting as an anticaking and dispersing agent, silica gel also can be used as a moisture scavenger and carrier. Some additives, when they are capable of serving several functions, may be called *conditioning agents*.

Anticaking agents commonly used include: calcium carbonate, phosphate, silicate, and stearate; cellulose (microcrystalline); kaolin; magnesium carbonate, hydroxide, oxide, silicate, and stearate; myristates; palmitates; phosphates; silica (silicon dioxide); sodium ferrocyanide; sodium silicoaluminate; and starches.

ANTICOAGULANTS.   These are substances that prevent coagulation of the blood. For blood investigations made outside the body, sodium or potassium citrates, oxalates, and fluorides are sometimes used. For blood that is to be used for transfusions, sodium citrate is used.

Organic anticoagulants are used in vivo in the treatment of numerous conditions where blood coagulation can be dangerous, as in cerebral thrombosis and coronary heart disease, among others which will be described later. The main anticoagulants used are heparin and coumarin compounds, such as warfarin.

Heparin

A complex organic acid (mucopolysaccharide) present in mammalian tissues and a strong inhibitor of blood coagulation. Although the precise formula and structure of heparin are uncertain, it has been suggested that the formula for sodium heparinate, generally the form of the drug used in anticoagulant therapy, is $(C_{12}H_{16}N_2Na_3)_{20}$ with a molecular weight of about 12,000. The commercial drug is derived from animal livers or lungs.

Heparin is considered a hazardous drug. Heparin may be the leading cause of drug-related deaths in hospitalized patients who are relatively well (Porter and Jick, 1978). It has been reported (Bell, et al., 1976) that some patients who receive continuously infused intravenous heparin develop *thrombocytopenia* (condition where the platelet count is less than 100,000 per cubic millimeter). Some authorities believe that the risk of thrombocytopenia associated with porcine heparin may be less than the risk associated with heparin of bovine origin.

Heparin, in addition to inhibiting reactions that lead to blood clotting, also inhibits the formation of fibrin clots, both in vitro and in vivo. Heparin acts at multiple sites in the normal coagulation system. Small amounts of heparin in combination with antithrombin III (heparin co-factor) can prevent the development of a hyper-coagulable state by inactivating activated factor X, preventing the conversion of prothrombin to thrombin. Once a hyper-coagulable state exists, larger doses of heparin, in combination with antithrombin II, can inhibit the coagulation process by inactivating thrombin and earlier clotting intermediates, thus preventing the conversion of fibrinogen to fibrin. Heparin also prevents the formation of a stable fibrin clot by inhibiting the activation of the fibrin, stabilizing factor. The half-life of intravenously administered heparin is about 90 minutes.

Coumarin

Oral anticoagulants can be prepared from compounds with coumarin as a base. Coumarin has been known for well over a century and, in addition to its use pharmaceutically, it is also an excellent odor-enhancing agent. However, because of its toxicity, it is not permitted in food products in the United States (Food and Drug Administration). One commercial drug is 3-(alpha-acetonyl-4-nitrobenzyl)-4-hydroxycoumarin. This drug reduces the concentration of prothrombin in the blood and increases the prothrombin time by inhibiting the formation of prothrombin in the liver. The drug also interferes with the production of factors VII, IX, and X, so that their concentration in the blood is lowered during therapy. The inhibition of prothrombin involves interference with the action of vitamin K, and it has been postulated that the drug competes with vitamin K for an enzyme essential for prothrombin synthesis.

Another commercial drug is bis-hydroxy-coumarin, $C_{19}H_{12}O_6$. The actions of this drug are similar to those just described.

Warfarin

This compound is also of the coumarin family. The formula is 3-(alpha-acetonylbenzyl)-4-hydroxycoumarin. In addition to use in anticoagulant therapy in medicine, the compound also has been used as a major ingredient in rodenticides, where the objective is to induce bleeding and, when used in heavy doses, is thus lethal. The compound can be prepared by the condensation of benzylidene-acetone and 4-hydroxycoumarin.

The anticoagulant action of warfarin is through interference of the gamma-carboxylation of glutamic acid residues in the polypeptide chains of several of the vitamin K-dependent factors. The carboxylation reaction is required for the calcium-binding activity of the K-dependent factors. Because of the reserve of pro-coagulant proteins in the liver, usually several days are required to effect anticoagulation with warfarin.

Warfarin antagonists include vitamin K, barbiturates, glutethimide, rifampin, and cholestyramine. Warfarin potentiators include phenylbutazone, oxyphenbutazone, anabolic steroids, clofibrate, aspirin, hepatotoxins, disulfiram, and metronidazole. In patients undergoing anticoagulation therapy with warfarin, it has been found that cimetidine (used in therapy of duodenal ulcer) may increase anticoagulant blood levels and consequently prolong the prothrombin time.

Anticoagulation Therapy

Prior to administration of anticoagulant drugs, patients must be carefully evaluated. Anticoagulant drugs are to be avoided if any of the following conditions prevail: a history of abnormal bleeding, recent corticosteroid therapy, recent intra-ocular or intracranial bleeding, recent pericarditis, and recent peptic ulcer or esophageal bleeding. A history of the individual's use of antiplatelet agents, such as aspirin, dipyridamole, phenylbutazone, and indomethacin, should be obtained and evaluated. Anticoagulant drugs should be administered with particular care during pregnancy. Heparin does not anticoagulate the fetus because it does not cross the placenta. Warfarin, on the other hand, anticoagulates both the mother and the fetus. Problems may arise from the administration of warfarin during pregnancy, particularly during the first trimester.

The bile sequestrant cholestyramine is frequently used in the treatment of familial hypercholesterolemia. This drug not only binds cholesterol, but also a number of other drugs, including anticoagulants.

Deep Vein Thrombosis and Pulmonary Embolism

Prompt administration of intravenous heparin is indicated in the treatment of this condition. Heparin is fast-acting, prevents further thrombus formation, and when used in therapeutic doses, also prevents the release of serotonin and thromboxane A_2 from platelets that adhere to thrombi that embolize to the lungs. The size of the dose required varies with a number of patient conditions. Several authorities are convinced that the continuous (pump-driven) infusion method is superior to intermittent injections.

Heparin is usually administered for a period ranging from 7 to 10 days. Frequently, during the last half of this period of heparin therapy, oral anticoagulation will be commenced with warfarin. The time during which oral anticoagulation administration should be continued may be three months or longer after clinical evidence that the venous thrombosis has subsided; and for one year after pulmonary embolism.

Cerebral Thrombosis

Among the specific modes of treatment that have been used in anticoagulation therapy. Many authorities suggest the use of anticoagulants for an evolving stroke in an effort to arrest the propagation of thrombus. In this procedure, a lumbar puncture is usually prepared first. If there is a presence of red blood cells in the spinal fluid, this infers a hemorrhagic infarction, in which case anticoagulants are withheld for a minimum of 48 hours. If the fluid is clear, heparin can be administered by continuous intravenous drip.

Cerebral Transient Ischemic Attack (TIA)

Aspirin, as a platelet-inhibiting agent, has been found effective in the medical management of TIA. As studied by the Joint Committee for Stroke Facilities in the late 1970s, the results of anticoagulant therapy for TIA were reported as vague, but possibly this is due to poorly planned tests. However, anticoagulant therapy is considered a proper mode of treatment for persons who cannot tolerate aspirin, with warfarin the drug of choice. Oral anticoagulants are not given to persons with gastrointestinal ulcerations, severe hypertension, bleeding tendencies, or renal or hepatic failure.

Coronary Heart Disease

Long-term preventive anticoagulation with warfarin and similar drugs in patients with coronary artery disease has decreased in popularity during the last few years because evidence collected over a long period of time has not shown, in a convincing way, that the therapy is of value. As of the late 1980s, the therapy of choice included the use of platelet-inhibiting drugs, notably acetylsalicylic acid (aspirin), sulfinpyrazone, and dipyridamole.

Prevention of Thromboembolism

Anticoagulation for prevention of thromboembolism is not used to the extent that it was once employed in the 1950s. Some professionals have reexamined the data of that period, however, and have concluded that anticoagulant therapy does have value, but their observations have not been widely accepted.

In present times, because of early mobilization and shorter stays in hospital, venous thrombosis in the legs and resulting pulmonary embolism has declined to a large degree. In persons with acute myocardial infarction, prophylactic low-dose heparin has reduced the incidence of venous thrombosis in the legs. It is considered as a reasonable alternative to warfarin in selected patients. Preventive anticoagulation may be indicated in some cases to prevent strokes due to left ventricular mitral thrombi embolizing in the brain.

Prosthetic Valve Endocarditis

Anticoagulants are sometimes used in the overall treatment of PVE even though there are risks of intracerebral hemorrhage or hemorrhagic infarction. Countering this risk, however, is the risk of major thromboembolic complications involving the central nervous system that may occur in the absence of continued anticoagulant therapy.

Anticoagulant therapy is also sometimes used in cases of congestive heart failure and in the treatment of polycythemia vera (elevation of the packed cell volume or the hemoglobin level) where not contraindicated.

Massive Venous Occlusion

This may be described as a surgical emergency. Immediately after diagnosis, intravenous heparinization is started and continued during the thrombectomy.

Mini-Dose Heparin

Small subcutaneous doses of heparin have been found to be effective in high-risk post-surgical patients and in patients with acute myocardial infarction. The preventive treatment is commenced a few hours before an operative procedure and continued postoperatively for 4 to 5 days. As the result of a study in 1975, low-dose heparin prophylaxis in high-risk patients who undergo abdomino-thoracic surgery has become a widely accepted practice. However, preventive anticoagulant therapy, to date, has been unsatisfactory and controversial in the instances of hip surgery or prostatectomy.

Additional Reading

Editor's Note : The following references have been selected to provide the interested reader with more detail on the pharmacologic complexities of anticoagulants. The article by Edwin W. Salzman, M.D. is a short, but excellent summary of the status of antithrombotic drugs as of early 1992. The article on heparin by Jack Hirsh, M.D. and the article on warfarin by the same author provide important fundamental background information on the pharmacokinetics and pharmacodynamics of the most widely used anticoagulant drugs. Brandjes, D.P.M., et al.: "Acenocoumaral and Heparin Compared with Acenocoumarol Alone in the Initial Treatment of Proximal-Vein Thrombosis," *N. Eng. J. Med.*, 1485 (November 19, 1992).

Ansell, J.E.: "Managing Oral Anitcoagulation Therapy," *Clinical and Operational Guidelines*, Aspen Publishers, Inc., Gaithersburg, MD, 1997.

Bell, W.R., et al.: "Thrombocytopenia Occurring During the Administration of Heparin," *Ann. Inter. Med.*, **85**, 155 (1976).

Chesebro, J.H., V. Fuster, and J.L. Halperin: "Atrial Fibrillation—Risk Marker for Stroke," *N. Eng. J. Med.*, 1556 (November 29, 1990).

Cheesebro, J.H. and V. Foster: "Thrombosis in Unstable Angina," *N. Eng. J. Med.*, 192 (July 16, 1992).

Coccheri: *Guide to Oral Anticoagulant Therapy*, Karger Libri, Farmington, CT, 1999.

Conrad, H.E.: *Heparin-Binding Proteins*, Academic Press, San Diego, CA, 1997.

Doutremepuich, C.: *Anticoagulation*, Springer-Verlag, Inc., New York, NY, 1994.

Ezekowitz, M.D., et al.: "Warfarin in the Prevention of Stroke Associated with Nonrheumatic Atrial Fibrillation," *N. Eng. J. Med.*, 1406 (November 12, 1992).

Gold, H.K.: "Conjunctive Antithrombotic and Thrombolytic Therapy for Coronary-Artery Occlusion," *N. Eng. J. Med.*, 1483 (November 22, 1990).

Green, D.: *Anticoagulants, Physiologic, Pathologic, Pharmacologic*, CRC Press, LLC., Boca Raton, FL, 1994.

Hirsh, J.: "Heparin," *N. Eng. J. Med.*, 1565 (May 30, 1991).

Hirsh, J.: "Drug Therapy: Oral Anticoagulant Drugs," *N. Eng. J. Med.*, 1865 (June 27, 1991).

Hirsh, J. and L. Poller: *Oral Anticoagulants*, Oxford University Press, New York, NY, 1996.

Hull, R.D., et al.: "Heparin for 5 Days as Compared with 10 Days in the Initial Treatment of Proximal Venous Thrombosis," *N. Eng. J. Med.*, 1260 (May 3, 1990).

Hull, R.D., et al.: "Subcutaneous Low-Molecular-Weight Heparin Compared with Continuous Intravenous Heparin in the Treatment of Proximal-Vein Thrombosis," *N. Eng. J. Med.*, 975 (April 9, 1992).

Poller, L. and F.R.C. Path: "The Effect of Low-Dose Warfarin on the Risk of Stroke in Patients with Nonrheumatic Atrial Fibrillation," *New Eng. J. Med.*, 129 (July 11, 1992).

Porter, J. and H. Jick: "Drug-related Deaths among Medical Inpatients," *J. Amer. Med. Assn.*, **237**, 879 (1977).

Salzman, E.W.: "Low-Molecular-Weight Heparin and Other New Antithrombotic Drugs," *N. Eng. J. Med.*, 1017 (April 6, 1992).

Saour, J.N., et al.: "Trial of Different Intensities of Anticoagularion in Patients with Prosthetic Heart Valves," *N. Eng. J. Med.*, 428 (February 15, 1990).

Stroke Prevention in Atrial Fibrillation Study Group of Investigators: "Special Report," *N. Eng. J. Med.*, 863 (March 22, 1990).

Theroux, P. et al.: "Reactivation of Unstable Angina after the Discontinuation of Heparin," *N. Eng. J. Med.*, 141 (July 16, 1992).

Thomas, D.P.: "Low-Molecular-Weight Heparin," *N. Eng. J. Med.*, 817 (September 1, 1992).

Thornes, R.D. and R. O'Kennedy: *Coumarins, Biology Applications and Mode of Action*, John Wiley & Sons, Inc., New York, NY, 1997.

ANTIDOTE. An agent that inhibits or counteracts the action of a poison. There is a wide variety of poisons, such as the *corrosives* (strong acids and alkalis) that cause local destruction of tissues; irritants that produce congestion of the organ they contact; the neurotoxins that affect the nerves or some of the basic processes within the cell; hemotoxins; hepatotoxins, and nephrotoxins. Consequently, the list of effective antidotes is long and reasonably complex, and usually much less lifesaving than making immediate efforts to have the poison victim vomit and thus expel as much of the poison as may be possible. Unless the appropriate antidote is selected, administration can be harmful rather than helpful. To illustrate this, if a sleep-producing drug, such as opium or morphine, has been taken in overdose, it is best to keep the patient awake by giving strong coffee. In contrast, in the instance of strychnine poisoning, no stimulants should be given and the patient should be kept as quiet as possible. In every type of poisoning, immediate medical aid is essential. Most local health departments have lists of antidotes for common poisons.

ANTIFOULING AGENTS. Various chemical substances added to paints and coatings to combat mildew and crustacean formations, such as barnacles on the hull of a ship. In the past, large quantities of mercury compounds have been used in this manner. With growing environmental concern over possible mercury pollution, manufacturers have been turning to other, sometimes less efficacious compounds. Research continues to find compounds of a less toxic, but equally effective power of the mercury compounds. Bis(tributyltin) fluoride has been used on ship bottoms. See also **Mercury**.

ANTIFREEZE AGENTS. A substance that lowers the freezing point (essentially of water). At one time, sodium chloride and magnesium chloride were widely used, but their extremely corrosive properties made them a liability in automotive and industrial cooling systems. Methyl alcohol, which requires about 27% by volume to protect against freezing to $0°F$ $(-17.8°C)$, has a tendency to evaporate rapidly at operating temperatures and, coupled with its flammability and low boiling point of $147°F$ $(64.9°C)$ limits its practical use in cooling systems. Many years ago, alcohols were replaced by glycol derivatives, which are relatively noncorrosive, nonflammable, have very low evaporation rates, and are effective heat-exchange media. A concentration of about 35% ethylene or propylene glycol antifreeze provides protection against freezing to $0°F$ $(17.8°C)$. Because of their overall properties, it is now common practice to retain the antifreeze concentrations in cooling systems the year around.

Antifreeze agents are also used in fuels where severe environmental conditions are encountered. For example, a mixture of methyl alcohol, isopropyl alcohol, TM "Drygas," and sometimes proprietary substances are used to inhibit the formation of ice from water vapor in hydrocarbon fuels. It is added directly to the gasoline. These additives are toxic and flammable.

See also **Coolant**.

ANTIGEN. A substance, usually a protein, a polysaccharide, or a lipoid, which when introduced into the body stimulates the production of antibodies. Bacteria, their toxins, red blood corpuscles, tissue extracts, pollens, dust, and many other substances may act as antigens. See **Antibody**.

ANTIKNOCK AGENTS. Any of a number of organic compounds that increase the octane number of a gasoline when added in low percentages by reducing knock, especially in high-compression engines. Knock is caused by spontaneous oxidation reactions in the cylinder head, resulting in loss of power and characteristic ignition noise. Branched-chain hydrocarbon gasolines ameliorate this problem, and antiknock additives virtually eliminate it. Tetraethyllead, the most effective of these, has been used for many years, but its contribution to air pollution has almost eliminated its use in automotive fuels. Lead-free gasolines are now used in conjunction with catalytic converters. The antiknock agents used in them are nonmetallic compounds such as methyl-*tert*-butyl ether (MTBE) or a mixture of methanol and *tert*-butyl alcohol.

See also **Octane Number**; and **Petroleum**.

ANTIHISTAMINE. A synthetic substance essentially structurally analogous to histamine, the presence of which in minute amounts prevents or counteracts the action of excess histamine formed in body tissues. See also **Histamine and Histamine Antagonists**. Antihistamines are usually complex amines of various types. They find a number of medical uses.

In immediate hypersensitivity situations (reaction between antigen and antibody as encountered in hay fever, hives (urticaria), allergic (extrinsic) asthma, bites, drug injections, among others), antihistamines can be part of the effective therapy. Although widely used, antihistamines and steroids are not always the drugs of choice. In atopic dermatitis (chronic skin disorder), antihistamines may assist in breaking the itch-scratch cycle, particularly in persons whose sleep may be interrupted by pruritus. Antihistamine compounds for urticaria are also effective for atopic dermatitis therapy. Frequently, shifting from one antihistamine to another is effective and helps to reduce side effects of the drugs. Antihistamines are sometimes effective in the treatment of autoerythrocyte purpura, a rare disease. Antihistamines are also used in connection with mild penicillin reactions, and in cases of penicillin desensitization procedures. Certain antihistamines find application to control mild Parkinsonism.

Some antihistamines are particularly effective in alleviating the onset of motion sickness. Some antihistamines have been found helpful in relieving persistent, unproductive coughs that frequently accompany bronchitis or coughs associated with allergy. They are used in connection with perennial and seasonal allergic rhinitis.

Most antihistamines have anticholinergic (drying) and sedative side effects, sometimes producing marked drowsiness and reduction of mental alertness and thus should not be used by persons who operate machinery, drive vehicles, or otherwise must react quickly. Because of their similar structure, antihistamines appear to compete with histamine for cell receptor sites. Although conventional antihistaminic drugs, such as mepyramine, block the allergic and smooth muscle effects caused by histamine, the structure of these drugs is not sufficiently similar to histamine to inhibit histamine-stimulated gastric acid secretion. However, during the last few years, so-called histamine-blocking drugs have been developed which appear to be effective. It has been found that such compounds must contain the imidazole ring of histamine, with their potency enhanced by extension of the side chain. Among these new drugs are metiamide and cimetidine.

Some drugs in the antihistamine series play markedly different roles. Hydroxyzine hydrochloride and hydroxyzine pamoate have been used in the total management of anxiety, tension, and psychomotor agitation in conditions of emotional stress, usually requiring a combined approach of psychotherapy and chemotherapy. Hydroxyzine has been found to be particularly useful for making the disturbed patient more amenable to psychotherapy in long-term treatment of the psychoneurotic and the psychotic. The drug is not used as the only treatment of psychosis or of clearly demonstrated cases of depression. Hydroxyzine has also been found useful in alleviating the manifestations of anxiety and tension in acute emotional problems and in such situations as preparation for dental procedures. Hydroxyzine therapy has been used in treatment of chronic alcoholism where anxiety withdrawal symptoms or delirium tremens may be present. Hydroxyzine may potentiate narcotics and barbiturates.

Most conventional antihistamines are available for both oral and intravenous or intramuscular administration. In serious cases of urticaria (hives), for example, the injection rather than oral route is most effective. The major excretion route for most antihistamines is hepatic (liver), occurring within 4 to 15 hours.

In addition to the side effects previously mentioned, some antihistamines may cause neutropenia (neutrophil count in the blood is less than 1800 per cubic millimeter). Some antihistamines also may cause a modification of normal platelets in the blood.

Some of the more commonly used antihistamine compounds are listed below:

Ethanolamines:
    Diphenhydramine hydrochloride (Benadryl®)
    Dimenhydrinate (Dramamine®)
Ethylenediamines:
    Tripelennamine hydrochloride (Pyribenzamine®)
Alkylamines:
    Chlorpheniramine maleate (Chlor-Trimeton®)
Piperazines:
    Cyclizine hydrochloride (Marezine®)
Phenothiazines:
    Promethazine hydrochloride (Phenergan®)
Others:
    Cyproheptadine hydrochloride (Periactin®)
    Hydroxyzine hydrochloride (Atarax®)
    Hydroxyzine pamoate (Vistaril®)

Additional Reading

Anderson, K.N., L.E. Anderson, and W.D. Glanze: *Mosby's Medical Dictionary,* 5th Edition, Mosby-Year Book, Inc., St. Louis, MO, 1997.
Budavari, S., A. Smith, P. Heckelman, J. Kinneary, and M.J. O'Neill: "The Merck Index," *An Encyclopedia of Drugs, Chemicals, and Biologicals,* 12th Edition, Merck Research Laboratories, Whitehouse Station, NJ, 1996.
Hodgson, B.B. and R.J. Kizior: *Saunders Nursing Drug Handbook 2000,* W.B. Saunders Company, Philadelphia, PA, 1999.
Leurs, R. and H. Timmerman: *Histamine H3 Receptor, A Target for New Drugs,* Elsevier Science, New York, NY, 1999.
Simons, F.E.: *Histamine and H1-Receptor Antagonists in Allergic Disease,* Vol. 7, Marcel Dekker, Inc., New York, NY, 1996.
Staff: *Nursing 2000 Drug Handbook,* Springhouse Publishing Company, Springhouse, PA, 1999.
Staff: *Drug Facts and Comparisons 2000,* 54th Edition, Facts and Comparisons, St. Louis, MO, 1999.
Stedman, T.: *Stedman's Medical Dictionary,* 27th Edition, Lippincott Williams & Wilkins, Philadelphia, PA, 2000.

ANTIMETABOLITES.

These substances fall into the general class of cytotoxic chemicals, i.e., agents that damage cells to which they are applied. Antimetabolites are so similar to normal enzymatic substrate molecules or metabolites as to gain entry into the cellular machinery of intermediary metabolism, but once there they differ enough to cause enzymatic inhibition. If incorporated into protein, nucleic acids, or coenzymes, for example, they will diminish the biological worth of those substances. Spectacular agents of this sort include the antifolic acids, such as aminopterin and amethopterin, various other vitamin analogs, and analogs of the naturally occurring purines, pyrimidines, nucleosides, and amino acids. Effective action against the integrity of the cell appears to be exerted at a number of points of intermediary metabolism in these multifarious antimetabolites. Of particular interest with many of them is an interference in normal nucleic acid metabolism. 5-Fluoro-2′-deoxyuridine, for example, acts to inhibit the synthesis of thymidylate, a necessary precursor of DNA, and the related 5-bromo-2′-deoxyuridine is actually incorporated into new DNA in the place of thymidine. Both of these agents increase the frequency of chromosomal disturbances. Various other base analogs, if incorporated into DNA, can lead to gene mutation by alteration of the normal sequence of nucleotides during replication through incorrect base pairing. 2-Aminopurine is an example of such a mutagen. 8-Azaguanine can be incorporated into ribonucleic acids, which are thus rendered defective. Among the actions of 6-mercaptopurine is an interference in the biochemical activity of coenzyme A, with resultant mitochondrial damage. Such amino acid analogs as b-fluorophenylalanine can effectively halt cellular activities by being incorporated into new proteins, which thereupon fail to attain their proper enzymatic or other functions.

Advantage is taken of the properties of antimetabolites in chemotherapy. In cancer chemotherapy, several antimetabolites are used. These include methotrexate, 6-mercaptopurine, 6-thioguanine, 5-fluorouracil, and cystine arabinoside. In the chemotherapy of metastatic breast cancer, 5-fluorouracil and methotrexate, in combination with cyclophosphamide, have been used. Antimetabolites, sometimes along with corticosteroids, are used in the therapy of various autoimmune diseases, such as thrombocytenic purpura, thyroiditis, Goodpasture's syndrome, among others.

Metabolites are implicated as agents that produce marrow aplasia as found in leukemia.

Additional Reading

Note: Check references listed in articles on **Biotechnology (Bioprocess Engineering)**; **Genetics and Gene Science (Classical)**; and **Industrial Biotechnology**.

ANTIMICROBIAL AGENTS (Foods).

Frequently substances are added to food products or applied to the surface of some foods while in transit or storage for the purpose of inhibiting the growth of certain destructive microorganisms. Such organisms affect a large percentage of fresh fruits and vegetables, as well as meat that has been cut and packaged. In addition to spoilage of the foods themselves, inadvertent consumption of certain molds can cause human disease. See also **Food Toxicants, Naturally Occurring**.

Antimicrobials for foods act against microorganisms by (1) adversely affecting the cellular membranes of the destructive molds, yeasts, bacteria, etc.; (2) by interfering with the genetic mechanisms of the offending microorganisms; and (3) by interfering with cellular membranes of the microorganisms. The chemicals used for such agents function against the microorganisms not unlike the actions of antibiotics and other antimicrobial drugs in preventing or arresting microbially caused infections in humans and other animals.

For several years, the use of antimicrobial agents in connection with food products has been strictly regulated in the United States and most of the other developed countries. Consequently, this is a field that is subject to constant change. This is evidenced by continuing controversy concerning, for example, the use of nitrates and nitrites and, more recently, the use of sulfur dioxide and sulfites. In the United States, the best source for current regulations is the Food and Drug Administration, Washington, DC.

The principal antimicrobial agents still permitted, in most countries, for use in foods are benzoic acid and sodium benzoate; the parabens; sorbic acid and sorbates; propionic acid and propionates; acetic acid and acetates; nitrates and nitrites; sulfur dioxide and sulfites; diethyl pyrocarbonate; epoxides; hydrogen peroxide; and phosphates. These agents are not used uniformly—some are applied to the food products near point of sale or consumption; others are introduced directly into the products during processing; still others are used only in connection with controlling microorganisms that may collect in the processing equipment. For the latter purpose, inasmuch as the equipment is thoroughly washed after control chemicals are used, many stronger and toxic (to humans in high concentrations) can be used, including hypochlorites and strong detergents. Because the need to maintain food processing equipment in ultra-clean condition at all times, microbial decontamination is practiced on a shift or daily basis. To alleviate many of the problems and loss of time required by such cleaning operations, much food processing equipment has been designed for cleaning in place (CIP), where the cleaning is essentially automatic and does not require disassembly and re-assembly of critical parts. Antimicrobial agents are also used in the packaging and wrapping of a number of food products, notably in connection with certain fruits and vegetables. Numerous studies have been made and, consequently, regulations established concerning the possible migration of such control chemicals from the packaging materials to the consumable product. Also, very powerful antimicrobial agents are permitted and used for special purposes, such as fumigating grain elevators and bulk food storage warehouses, where the objective is not simply that of killing microorganisms, but also to kill insects, rodents, and the like.

Although food irradiation has been proved effective and could ultimately significantly alter the need for antimicrobial chemical agents, there remain a number of regulatory as well as consumer acceptance questions that require further resolution.

Benzoic Acid and Sodium Benzoate

These compounds are most active against yeasts and are less effective against molds. They are best suited for foods with a natural or adjusted pH below 4.5. The average dosage in foods ranges between 0.05–0.1% (weight), depending upon product. Benzoic acid occurs naturally in cinnamon, ripe cloves, cranberries, greengage plums, and prunes. See also **Benzoic Acid**. For a number of years the sodium salt has been preferred over the acid by food processors.

Common applications for these compounds include carbonated and non-carbonated beverages, but excluding beers and wines because of their action against yeasts. They are also used in salted margarine, jams, jellies, and preserves, pie fillings, salads and salad dressings, pickles, relishes, and other condiments, as well as olives and sauerkraut. In terms of human metabolism of these substances, some authorities have suggested that benzoate is conjugated with glycine to produce hippuric acid, which is excreted, possibly accounting for 65–95% of benzoate ingested. It has been postulated that the remainder is detoxified by conjugation with glycuronic acid.

Parabens

These compounds include the methyl, ethyl, propyl, and butyl esters of para-hydroxybenzoic acid. In the United States and a number of other countries, the methyl and propyl esters are preferred, while European food processors favor the ethyl and butyl esters. The parabens were first described in 1924 as having antimicrobial activity and initially were used in cosmetic and pharmaceutical products.

The parabens are most effective against molds and yeasts, but less active against bacteria, particularly gram-negative bacteria. The antimicrobial activity of the parabens is directly related to the molecular chain length (methyl is weakest; butyl is strongest). However, the solubility of these compounds is in inverse relationship with chain length. These characteristics give rise to the use of two or more esters in combination and sometimes in combination with entirely different antimicrobial agents, such as sodium benzoate. Below a pH of 7, the parabens are only weakly effective.

The parabens are used in carbonated beverages and other soft drinks, including cider. In lieu of pasteurizing or using Millipore filtration, some brewers use the parabens for controlling secondary yeast formation. Because of their activity against yeasts, they are not used in bread and rolls, but they find use in other bakery products, such as pie crusts, certain pastries, icings, toppings, fillings, and cakes. The parabens are particularly effective in preserving fruit cakes. Usage also includes creams and pastes, fruit products, flavor extracts, pickles and olives, and artificially sweetened jams, jellies, and preserves. Average dosage ranges from 0.03–0.6% (weight).

Sorbic Acid and Sorbates

Sorbic acid and its potassium and sodium salts are effective against molds, but less effective against bacteria. These compounds may be incorporated directly into the food product, but they are frequently applied by spraying, dipping, or coating. The compounds are effective up to a pH of about 6.5. This is higher than propionates and sodium benzoate, but not so high as the parabens. Metabolism in humans parallels that of other fatty acids.

Because sorbates affect yeasts, the compounds are not directly useful in yeast-raised goods. Sorbates are particularly favored for use in chocolate syrups. They can be used in wine production in conjunction with sulfur dioxide against bacteria and are effective in inhibiting development of unwanted yeasts. They are also used in artificially sweetened jellies, jams, and preserves; in pickles and related products; in nonsalted margarines; in dried and smoked fish products; in semimoist pet foods; in dry sausage castings; in fruit-filled toaster pastries; and in cheese and cheese products. In the latter products, the agents usually are applied by dipping or spraying. Wrappers also may be impregnated with sorbates.

Propionic Acid and Propionates

The antimicrobial properties of propionic acid and its calcium and sodium salts were first noted in 1913. Today, the calcium and sodium salts are most commonly used. These compounds are more active against molds than sodium benzoate, but have little if any activity against yeasts. The propionates are well known for their effectiveness against *Bacillus messentericus*, a "rope"—forming microorganism. These compounds are effective up to a pH of 5 or slightly higher. Metabolism in the human body parallels that of other fatty acids.

An early application for the propionates was that of dipping cheddar cheese in an 8% propionic acid solution. This increased mold-free life by 4 to 5 times more than when no preservative was added. For pasteurized process cheese and cheese products, propionates can be added before or with emulsifying salts. Research has indicated that propionate-treated parchment wrappers provide protection for butter.

Use of propionates in breads can extend mold-free life by 8 days or more. Propionates are favored by bakers because of their effectiveness against ropy mold in breads up to pH levels of 6. For cakes and unleavened bakery goods, the sodium salt is usually preferred; for bread, the calcium salt is favored. This additive also contributes to the mineral enrichment of the product.

Acetic Acid and Acetates

Acetic acid (pure and as vinegar) and calcium, potassium, and sodium acetates, as well as sodium diacetate, serve as antimicrobial agents. In the United States, vinegar can contain no less than 4 grams of acetic acid per 100 milliliters of product. Acetic acid and calcium acetate are most effective against yeasts and bacteria, and to a lesser extent, molds. The diacetate is effective against both rope and mold in bread. It is interesting to note that the antimicrobial effectiveness of acetic acid and its salts is increased as the pH is lowered.

Optimal pH range varies with products and target microorganisms, but generally falls between 3.5 and 5.5. These agents are particularly effective against *Salmonella aertrycke. Staphylococcus aureus, Phytomonas phaseoli, Bacillus cereus, B. mesetericut. Saccharomyces cerevisiae,* and *Aspergillus niger.*

Unfortunately, to be effective against microorganisms in bakery products, acetic acid concentrations must be so high that an overly sour taste is imparted to the products. Sodium diacetate, however, can be used in small concentrations in bread and rolls to control rope and molds. Traditional concentrations of the acetate are 0.4 part to 100 parts of flour. During recent years, the propionates have largely displaced sodium diacetate for this use.

Vinegar or acetic acid is used in a number of products as much for its sour taste as for its antimicrobial properties. Such products include catsup, mayonnaise, pickles, salad dressing, and various condiment sauces. These agents also have been used to a lesser extent in malt syrups and concentrates, cheeses, and in the treatment of parchment wrappers for products, such as butter, to inhibit mold.

Nitrates and Nitrites

For many decades, sodium nitrate and nitrite, and potassium nitrate and nitrite have been used to cure, preserve, and provide a characteristic flavor to such meats as bacon, corned beef, frankfurters, ham, and various sausages. This tradition continued into the late 1980s, but was seriously threatened in the mid- and late 1970s. Some researchers reported that N-nitrosopyrolidine (NPyr) formed in bacon upon application of heat during preparation for consumption. It was observed that there was a greater concentration of the NPyr in adipose tissue than in the lean portion. A connection was proposed that involved serious implications of the ultimate carcinogenic risk involved in meat treated with the nitrates and nitrites. Numerous tests proceeded. One of the main factors learned during this period was how little knowledge food scientists had concerning the fate of the nitrates and nitrites. Although the precursors of the nitrosamines formed under certain conditions of cooking were known, the mechanism of formation was unknown.

Sulfur Dioxide and Sulfites

The use of sulfur dioxide gas and with it the production of sulfites differ somewhat from the other antimicrobial agents thus far described. Historical records show that burning sulfur to produce sulfur dioxide (SO_2 gas) dates back to the ancient Egyptians and Romans who used it in connection with wine making. See **Sulfur**. Action by sulfur dioxide is accomplished in the gaseous phase. The effect of SO_2 is markedly determined by concentration and pH conditions of the target product. Research has demonstrated that most bacteria are inhibited by HSO_3^- at concentrations of 200 parts per million (ppm) or less. With few exceptions, yeasts are also similarly inhibited. There are, however, some strains of molds that are considerably more resistant. The sulfite salts tend to be unstable and oxidize during long periods of storage, thus decreasing the availability of SO_2. This process is aggravated by the presence of moisture.

The most effective range for optimal microbial inhibition with sulfites is a pH of 2.5 to 3. It has been found that from 2 to 4 times greater concentrations of SO_2 are needed to inhibit the growth of microorganisms at a pH of 3.5 than at a pH of 2.5. It also has been demonstrated that, at a pH of 7, SO_2 has little if any effect on yeasts and molds, even at concentrations up to 1000 ppm. The inhibitory effects against bacteria are also considerably less when pH rises above 3.5. Some researchers believe that at higher pH levels, penetration of cell walls is much more difficult.

Because residual levels of sulfites in excess of 500 ppm impart a noticeable taste to food substances, this fact, regardless of any regulations toward limiting concentration, requires SO_2 levels to be controlled.

The use of SO_2 for preserving fruit juices, syrups, concentrates, and purees is particularly attractive in regions with warm climates and where products must be stored in bulk prior to processing. In these situations, the SO_2 concentration will range between 350 and 600 ppm. High sugar concentrations require higher levels of SO_2. For optimal effectiveness, the pH of some products has to be reduced.

It is a common practice to expose many fruits to SO_2 prior to dehydration. The SO_2 also extends storage life of raw fruit prior to dehydration. The optimal temperature for exposure to the gas is from 43 to 49°C. Unlike fruit, vegetables are usually dipped in solutions of neutral sulfites and bisulfites.

Many countries do not allow use of SO_2 or sulfite salts on meats, fish, processed meat and fish products, or fresh fruits and vegetables. Where permitted, sulfite is helpful in eliminating "black spot" formation in shrimp. End-use application of sulfite related compounds as, for example, to prevent discoloration of leafy vegetables on salad bars, is now subject to regulation in the United States.

A major use of sulfites is in wine making. It is used for sanitizing equipment and, prior to fermenting, the grape *musts* have to be treated with sulfites to inhibit the growth of any natural microbial flora present. This is done prior to the addition of pure cultures of the appropriate wine-making yeasts. While fermenting SO_2 also can function as an antioxidant, clarifier, and dissolving agent. Sulfur dioxide is often used after fermentation to prevent undesirable post-fermentation alterations by various microorganisms. Levels of SO_2 during fermentation range from 50 to 100 ppm, depending upon condition of the grapes, temperature, pH, and sugar concentration. The wine industry uses sulfur dioxide dissolved in water, vaporized SO2, and sulfite salts. An SO_2 level of 50–75 ppm assists the prevention of bacterial spoilage during the bulk storage of wine after fermentation.

Antibiotics

Much attention has been given to the use of antibiotics in food-associated applications since the introduction of penicillin in the 1940s. Their use has been limited. The use of antibiotics in animal foodstuffs continues to remain a topic of controversy in many countries; in other countries they have been banned. Antibiotics carry into the meat produced and further into human diets, thus possibly reducing their effectiveness in the treatment of human diseases.

Diethyl Pyrocarbonate

The preservative qualities of this compound were not recognized until the late 1930s and research on the compound continues to date. Also called pyrocarbonic acid diethyl ester, the compound is extremely effective against yeasts. It is also active against bacteria, such as *Lactobacillus pastorianus*, and various molds. The substance is generally used in still wines, fermented malt beverages, and non-carbonated soft drinks, as well as fruit-based beverages. Regulations on its use vary from one country to the next. Effective inhibition by diethyl pyrocarbonate is largely confined to acid products of low microorganism count. Some researchers point out that the pH should be less than 4 and that the microorganism count should not exceed 500 per milliliter. Some authorities observe that because of its rapid hydrolysis, no toxicity or residue problems should occur in products where the compound is permitted.

Epoxides

Two compounds are included in this category of antimicrobials. One is the gas, *ethylene oxide*; the other, *propylene oxide*, a colorless liquid with a boiling point of 35°C. Ethylene is highly reactive and must be used carefully and only with proper equipment. Somewhat less hazardous from an explosion standpoint, propylene oxide also has an explosive range of 2–22%. Consequently, these materials are usually mixed with inert substances, such as carbon dioxide or organic diluents.

Ethylene oxide is a universal antimicrobial in that it is lethal to all microorganisms. However, it is not universal from the standpoint of application. Propylene oxide is considered a broad-range microbiocide. In practically all aspects, propylene oxide is a considerably less effective agent, requiring longer exposures and greater concentrations because of its low penetrating power. However, propylene oxide is less toxic to humans.

The use of these gases (propylene oxide is volatilized) has been called "cold sterilization" and is frequently useful in sterilizing a number of low-moisture ingredients that end up on high-moisture foods. This prior sterilization lessens the total load on later thermal processing. The ability of these gases to kill microorganisms in low-moisture foods is an outstanding advantage. At the same time, macro-organisms also are killed. The gases find application in connection with spices, starches, nut meats, dried prunes, and glacè fruit. They are not used on peanuts (groundnuts).

Hydrogen Peroxide

Although usually regarded as a bleaching and oxidizing agent, this compound, H_2O_2, can be an effective antimicrobial and can be particularly useful in sterilizing processing equipment and packaging materials, notably prior to the aseptic packaging process. Regulations regarding the use of hydrogen peroxide vary from one country to the next.

Phosphates

The various phosphates are an effective multipurpose food additive chemical and functions other than their antimicrobial properties are usually given the greatest stress. The antimicrobial properties of the phosphates have been investigated over the years and are reasonably well documented.

This topic is discussed in detail in the "Foods and Food Production Encyclopedia," (D.M. and G.D. CONSIDINE, Eds.), Van Nostrand Reinhold, NY, 1982.

Additional Reading

Davidson, P.M. and A.L. Branen: *Antimicrobials in Foods,* 2nd Edition, Marcel Dekker, Inc., New York, NY, 1993.

Dillon, V.M. and R.G. Board: *Natural Antimicrobial Systems and Food Preservation,* CAB International, New York, NY, 1994.

Jacoby, G.A. and G.L. Archer: "New Mechanisms of Bacterial Resistance to Antimicrobial Agents," *N. Eng. J. Med.,* 601 (February 28, 1991).

Jay, J.M.: *Modern Food Microbiology,* 6th Edition, Aspen Publishers, Inc., Gaithersburg, MD, 2000.

Luck, E.M., Jager, and S.F. Laichena (Translator): *Antimicrobial Food Additives, Characteristics, Uses, Effects,* 2nd Edition, Springer-Verlag, Inc., New York, NY, 1997.

Ray, B.: *Fundamental Food Microbiology,* CRC Press, LLC., Boca Raton, FL, 1998.

Robinson, R.K., C.A. Batt, and P. Patel: *Encyclopedia of Food Microbiology,* Academic Press, San Diego, CA, 2000.

Sofos, J.N.: *Naturally Occurring Antimicrobials in Food,* Vol. 132, Council for Agricultural Science & Technology, Ames, IA, 1998.

ANTIMONY. [CAS: 7440-36-0]. Chemical element, symbol Sb, at. no. 51, at. wt. 121.75, periodic table group 15, mp 630.5°C, bp 1950°C, sp gr 6.62 (vacuum-distilled solid at 20°C) and 6.73 (single crystal). Naturally occurring isotopes are ^{121}Sb and ^{123}Sb. Antimony metal is a lustrous, silvery, blue-white solid, extremely brittle, and exhibiting a scale-like or flaky crystalline texture. The metal is easy to pulverize. The pure metal has a hardness of 3.0–3.3 on the Mohs scale and 55 on the Brinell scale. Of the more common metals, antimony is the poorest conductor (4.5 on a scale of 100 for copper). From careful studies, it has been observed that Sb contracts upon solidification rather than expanding. The element was first described by Th "olden (Valentine) in 1450.

There are two natural isotopes, ^{121}Sb and ^{123}Sb; and ten radioactive isotopes, ^{116}Sb through ^{120}Sb, ^{122}Sb, and ^{124}Sb through ^{127}Sb. ^{124}Sb is used as a radiation source in industrial instruments for the measurement of flow of slurries and interface measurements in pipelines. See also **Radioactivity**.

First ionization potential 8.64 eV; second 16.5 eV; third 25.3 eV; fourth 44.1 eV; fifth 56 eV. Oxidation potentials $Sb + H_2O \longrightarrow SbO^+ + 2H^+ + 3e^-$, -0.212 V,

$$2 SbO^+ + 3 H_2O \longrightarrow Sb_2O_5 + 6 H^+ + 4 e^-,$$

-0.581 V, $Sb + 4OH^- \longrightarrow SbO_2^- + 2H_2O + 3e^-$, 0.66 V. Other important physical properties of antimony are given under **Chemical Elements**.

Antimony exists in a number of allotropic forms. Gray or metallic antimony, density 6.79 g/per cm^3, is the stable form, forming rhombohedral crystals. Its vapor is that of Sb$_4$ up to 800°C, where dissociation to Sb$_2$ commences. Yellow antimony, Sb$_4$, density 5.3 g/cm^3 is less stable than yellow arsenic. It is produced by oxidation of stibine (see below) at very low temperatures, above which it is unstable. It changes even in the dark to black antimony at -90°C (in the light at -180°C). Black antimony, produced most readily by cooling antimony vapor or oxidizing stibine at 40°C, density 5.3, is metastable with respect to the gray form. It is also more reactive, igniting in air at room temperatures or above. Explosive antimony is produced by rapid electrodeposition of antimony from its halides. When heated or scratched, it undergoes an exothermic transformation to gray antimony. Its structure is amorphous, and differs somewhat from that of gray antimony.

Antimony is used in alloys, with lead for storage battery plates, with lead and tin in type metals and body solders, with tin and copper in bearing or antifriction metals. Antimony occurs chiefly as the sulfide (stibnite, Sb$_2$S$_3$) which is produced mainly in China, only small amounts in Mexico and Bolivia. Stibnite is (1) melted and reduced to antimony by iron metal and separated from fused ferrous sulfide (See also **Stibnite**); (2) roasted in air, and sublimed antimonous oxide collected and reduced by heating to fusion with carbon and sodium carbonate.

Antimony is also leached from tetrahedryte ore and recovered by electrowinning.

Antimony is scarcely tarnished in dry air but oxidizes slowly in moist air; burns at a red heat in air or oxygen with incandescence forming antimonous oxide; it is insoluble in HCl and is converted by HNO$_3$ into antimonous oxide or antimonic oxide, depending upon the concentration of acid; by chlorine into trichloride or pentachloride, by NaOH solution into antimonite.

Stibine. SbH$_3$, is formed by hydrolysis of some metal antimonides or reduction (with hydrogen produced by addition of zinc and HCl) of antimony compounds, as in the Gutzeit test. It is decomposed by aqueous bases, in contrast with arsine. It reacts with metals at higher temperatures to give the antimonides. The antimonides of elements of group 1a, 2a, and 3a usually are stoichiometric, with antimony trivalent. With other metals, the binary compounds are essentially intermetallic, with such exceptions as the nickel series, Ni$_2$Sb$_3$, NiSb, Ni$_5$Sb$_2$ and Ni$_4$Sb.

Trihalides

SbF$_3$, SbCl$_3$, SbBr$_3$, and SbI$_3$, are solids, and have pyramidal structures. Except for the fluoride, which is not hydrolyzed, they undergo partial hydrolysis only (in contrast with the phosphorus trihalides) on contact with water to yield insoluble oxyhalides, either of composition SbOX or varying somewhat from this composition to give such compounds as Sb$_4$O$_5$Cl$_2$. The antimony pentahalides, SbF$_5$ and SbCl$_5$ can be prepared, but the pentabromide exists only in double compounds, known as bromoantimonates, those for monovalent metals being of the type MSbBr$_6$, plus water of hydration, and yielding SbBr$_6^-$ ions. SbCl$_6^-$ and SbF$_6^-$ ions are also known. Mixture of antimony(III) chloride, SbCl$_3$, in HCl solution with antimony(V) chloride, SbCl$_5$, in equimolar proportions yields a dark-colored solution. While antimony(IV) chloride cannot be isolated from it, compounds such as cesium antimony(IV) chloride, Cs$_2$SbCl$_6$ are formed by addition of cesium chloride, CsCl, and they are isomorphous with similar compounds of lead, tin, and other metals. However, tetravalent antimony should be paramagnetic because of the unpaired electron, whereas compounds of SbCl$_6^{2-}$ are diamagnetic. Therefore it may be that these compounds contain equimolar mixtures of SbCl$_6^-$ and SbCl$_6^{-3}$. The existence of these higher halide complexes with tin (and bismuth), but not with phosphorus or arsenic, may be due to steric considerations.

Antimony(III) oxide

[CAS: 1309-64-4]. Sb$_2$O$_3$ or Sb$_4$O$_6$, is formed by melting antimony in air, or from the hydroxide Sb(OH)$_3$. The Sb$_2$O$_3$ of commerce is produced from the oxidation of stibnite ore. Antimony is below arsenic in the periodic table, and Sb(OH)$_3$ is more definitely amphiprotic than As(OH)$_3$, forming not only antimony(III) salts and antimonites (containing the ion SbO$_2^-$ or Sb(OH)$_4^-$), but also basic salts, especially the antimonyl salts, containing the ion SbO$^+$. Antimony(V) oxide, formed by oxidation of the metal with HNO$_3$, is less soluble in H$_2$O than As$_2$O$_5$. Antimonic acid cannot be obtained by hydration, and the product resulting upon hydrolysis of pentahalides has a variable H$_2$O content. The salts of the acid, the antimonates, are of the type M^I Sb(OH)$_6$, as Pauling showed to be necessary to conform to accepted ionic radius ratios. Although the strength of antimonic acid has not been accurately determined, it appears to be comparable to acetic acid.

Antimony(IV) oxide

Obtained by heating in air the trioxide or the hydrated pentoxide.

There is a marked structural difference between the phosphates and the antimonates. Thus sodium pyroantimonate, Na$_2$H$_2$Sb$_2$O$_7 \cdot$ 5H$_2$O contains the ion Sb(OH)$_6^-$ rather than Sb$_2$O$_7^{4-}$, and the magnesium compound (hydrated) which has a 12:1 ratio of oxygen to antimony, and would thus be a hexahydroxyantimonate, has the (X-ray determined) structure [Mg(H$_2$O)$_6$][Sb(OH)$_6$]$_2$.

Sulfides

Sb$_2$S$_3$ and Sb$_2$S$_5$, which may be obtained from the elements or by precipitation, respectively, of Sb(III) and Sb(V) solutions with H$_2$S. The Sb$_2$S$_3$ dissolves in alkaline solutions to form thioantimonites, containing the ion SbS$_3^{3-}$, or Sb(SH)$_6^-$, while Sb$_2$S$_5$ forms the thioantimonates, containing SbS$_4^{3-}$. The latter is probably present as [SbS$_2$(SH)$_{21}$(OH)$_2$]$^{3-}$ or [SbS$_4$(H$_2$O)$_2$]$^{3-}$.

In alloys, antimony is easily detected by its formation of a white solid upon treatment with concentrated HNO$_3$ and subsequent separation from tin, which is the only other metal thus forming a white solid.

Both trivalent and pentavalent antimony form several organic antimony compounds. Some of these include methylstibine CH$_3$SbH$_2$

TABLE 1. ANTIMONY CONTENT OF REPRESENTATIVE ANTIMONY-CONTAINING ALLOYS

Hard lead	Up to 12% Sb
Antimony reduces mp of Pb and hardens resulting alloy. Alloy has better abrasion resistance than chemical Pb at temperatures below 140°C. Alloy is age-hardenable.	
Tin-lead solders	Up to 1% Sb
Type metals	3–19% Sb
These Pb-base alloys also contain from 3–9% Sn.	
Lead-base diecasting alloys:	
[a]ASTM No. 4	14–6% Sb
ASTM No. 5	9.25–10.75% Sb
Bearing alloy	15% Sb
CT metal	12.5% Sb
Tin-free alloy	10% Sb
Babbitt (bearing) metals:	
[b]SAE 10	4–5% Sb
SAE 11	6–7.5% Sb
SAE 12	7–8.5% Sb
SAE 13	9.25–10.25% Sb
SAE 14	14–16% Sb
SAE 15	14.5–16% Sb
Britannia metal	5% Sb
This alloy also contains 93% Sn and 2% Cu. Very useful for spinning utensils.	
Pewter	Up to 7% Sb
Pewter also contains up to 20% Pb and 4% Cu with the remainder made up by Sn.	

[a] American Society for Testing and Materials
[b] Society of Automotive Engineers

and the substitution product, methyldichlorostibine CH_3SbCl_2; phenylstibine $C_6H_5SbH_2$ and the substitution product, phenyldichlorostibine $C_6H_5SbCl_2$; methylantimony tetrachloride CH_3SbCl_4; phenylantimony tetrachloride $C_6H_5SbCl_4$; sodium methylantimonate $Na[CH_3Sb(OH)_5]$; sodium trifluoromethyl antimonate $Na[(CF_3)_3Sb(OH)_3]$; triethylstibine sulfide $(C_2H_5)_3SbS$; tetraphenylstibonium tetraphenylborate $[(C_6H_5)_4Sb]$ $[B(C_6H_5)_4]$; stibiobenzene $C_6H_5Sb = SbC_6H_5$ and lithium hexaphenylantimonate $LiSb(C_6H_5)_6$.

Uses

Representative alloys containing antimony are described in the Table 1.

Metallic antimony is an effective pearlitizing agent for producing pearlitic cast iron. The principal use of antimony, however, is in the form of the oxide. Its major application is as a flame retardant for plastics and textiles. Other applications of importance are in glass, pigments, and catalysts.

Toxicity

The threshold limit value of antimony and its compounds is 0.5 milligram/cubic meter (as Sb). Antimony and its compounds used under conditions giving rise to dust, fume, and vapor should be carried out under proper ventilation. In handling antimony and its compounds, appropriate hygienic practices and good housekeeping should be observed. Stibine, SbH_3, requires extreme caution in handling because it is very toxic. When using antimony and its compounds, reducing conditions, which may give rise to the undesired formation of stibine, must be avoided.

Additional Reading

Buch, A.: *Pure Metals Properties,* A Scientific and Technical Handbook, ASM International, Materials Park, OH, 1999.

King: *Encyclopedia of Inorganic Chemistry,* 2nd Edition, John Wiley & Sons, Inc., New York, NY, 1997.

Lagowski, J.J.: *MacMillan Encyclopedia of Chemistry,* Vol. 1, Macmillan Library Reference, New York, NY, 1997.

Lide, D.R.: *CRC Handbook of Chemistry and Physics,* 84th Edition, CRC Press, LLC., Boca Raton, FL, 2003.

Norman, N.C.: *Chemistry of Arsenic, Antimony and Bismuth,* Kluwer Academic Publishers, Norwell, MA, 1998.

Perry, R.H., D. Green, and J.O. Maloney: *Perry's Chemical Engineers' Handbook,* 7th Edition, McGraw-Hill, Companies, New York, NY, 1999.

ANTIOXIDANTS. Usually an organic compound added to various types of materials, such as rubber, natural fats and oils, food products, gasoline, and lubricating oils, for the purposes of retarding oxidation and associated deterioration, rancidity, gum formation, reduction in shelf life, etc.

Rubber antioxidants are commonly of an aromatic amine type, such as dibeta-naphthyl-para-phenylenediamine and phenyl-beta-naphthylamine. Usually, only a small fraction of a percent affords adequate protection. Some antioxidants are substitute phenolic compounds (butylated hydroxyanisole, di-tert-butyl-para-cresol, and propyl gallate).

When used in foods, antioxidants are highly regulated to extremely small percentages in most countries—down to the low fractions of one percent. Composition of the substrate, processing conditions, impurities, and desired shelf life are among the most important factors in selecting the best antioxidant system for a given food product. The desirable features of antioxidants may be summarized as (1) effectiveness at low concentrations; (2) compatibility with the substrate; (3) nontoxic to consumers; (4) stability in terms of conditions encountered in processing and storage, including temperature, radiation, pH, etc.; (5) non-volatility and non-extractability under the conditions of use; (6) ease and safety in handling; (7) freedom from off-flavors, off-odors, and off-colors that might be imparted to the food products; and (8) cost effectiveness.

Mechanism of Oxidative Degradation

It could appear that inasmuch as oxidative degradation occurs in a variety of organic materials that are dissimilar in appearance and have entirely different applications and different properties, with degradation producing different effects, the oxidation mechanism itself might be different. Current knowledge indicates, however, that the mechanism of oxidative degradation is the same for all organic substances. They appear to degrade by the same free-radical mechanism.

Common examples of food oxidative degradation include products that contain oils and fats. For example, some antioxidants have made it possible to store groundnuts (peanuts) and other nuts, maize (corn) products, and bakery and cereal products on the shelf for periods well in excess of the four months that was considered the traditional limiting period prior to the appearance of such additives. Other examples of food products that tend to become rancid by way of oxidation include various meat-flavored stuffing mixes, cake mixes, unbaked cheesecake mix, and essentially all foods that incorporate lipids. The stability of natural fats and oils present in raw materials varies over a wide range and hence the amount of antioxidant required must be tailored to each product situation. Enzymatic "browning" is another example of oxidative degradation. The enzymes in fruits and vegetables cause apples, apricots, and potatoes, among others, to darken when they are exposed to air after being cut, bruised, or allowed to overmature. Some antioxidants can prevent or delay enzymatic browning much in the same manner as dipping freshly cut fruits in lemon, orange, or pineapple juice. Limonene and ascorbic acid naturally present in these juices serve as antioxidants. Oxidative changes may affect carbohydrate, protein, and fat substances, the primary building blocks of foodstuffs, but generally the oxidative rancidity problem results mainly from the *autoxidative* degradation of fatty (glyceridic) components.

Some authorities describe oxidation as a free radical, chain-type reaction. At usual processing temperatures and more slowly at room temperature, organic free radicals (R·) are formed. These react with oxygen to form peroxy radicals (ROO·), which can abstract a hydrogen atom from the affected substance to form a hydroperoxide (ROOH) and another organic free radical. The cycle repeats itself with the addition of oxygen to the new free radical. The unstable hydroperoxides left along with the substance are the major source of degradation. Under the influence of heat, light, and any metals if present, the hydroperoxides decompose to form carbonyl groups. When this happens, the organic molecule breaks and splits off another organic free radical. Ultimately, this type of degradation can lead to rancidity and color deterioration in oils and fats.

An antioxidant ties up the peroxy radicals so that they are incapable of propagating the reaction chain or to decompose the hydroperoxides in such a manner that carbonyl groups and additional free radicals are not formed. The former, which are called *chain-breaking antioxidants, free-radical scavengers,* or *inhibitors,* are usually hindered phenols or amines. The latter, called *peroxide decomposers,* are generally sulfur compounds or organophosphites. A number of antioxidants useful in rubber and plastics, for example, are not suited to food products because of their toxicity.

A mixture of two antioxidants often will display synergism. Probably the most generally effective mixtures of antioxidants are those in which one compound functions as a decomposer of peroxides (sulfides, thiodiproprionate) and the other as an inhibitor of free radicals (hindered phenols, amines). Although the latter retards the formation of reaction chains, some hydroperoxide is nevertheless formed. If this hydroperoxide then reacts with a decomposer of peroxides, instead of decomposing into free radicals, the two antioxidants act together to complement each other. Moreover, the peroxide decomposer may itself be subject to oxidation by peroxy radicals, and its efficiency will therefore be increased in the presence of an inhibitor of free radicals. In the case of phenolsulfide mixtures, the sulfide (peroxide decomposer) also continuously regenerates the phenol (radical scavenger) to accentuate the synergistic nature of the mixture. Metal chelators or deactivators, such as citric and phosphoric acids, of pro-oxidant metals (iron, copper, nickel, tin), ultraviolet-light absorbers (carbon black, substitute benzophenones, benzotriazoles, and salicylates), and antiozonants (substituted phenylenediamines) also develop synergistic effects with antioxidants.

Applications of Antioxidants

The use of antioxidants in foods, pharmaceuticals, and animal feeds (direct feed additives), as well as their use in food-contact surfaces (indirect additives) is closely regulated by the governments of several countries. Antioxidants are approved only after extensive extraction, toxicological, and feeding studies. The list is relatively limited. Although antioxidants have been used for several decades and some occur naturally in food substances, intensive research in continuing, partly accelerated by the growing use of unsaturated oils in numerous food products.

Butylated hydroxyanisole (BHA) was first used in food products in 1940. This continues as one of the commonly used antioxidants, sometimes in combination with butylated hydroxytoluene (BHT), propyl gallate, citric, or phosphoric acids, to obtain a synergistic effect. In foodcontact surfaces, BHT has been used by itself or in combination with thiodipropionates and/or phosphoric acids, to obtain a synergistic effect. Well over $50 million of antioxidants are produced per year commercially in the United States alone.

The value of antioxidant protection by way of natural food sources has been pointed out in the literature with considerable frequency. Among the components of soy flour known to have some antioxidant properties are isoflavones and phospholipids. Amino acids and peptides in soybean flour also possess some antioxidant activity. There also may be some antioxidant impact from aromatic amines and sulfhydryl compounds.

Rosemary and sage have been shown to have effective antioxidant properties. The extracts in the past have been of strong odor and bitter taste and thus unsuited for use in most food products. However, solvent extraction procedures have been developed to produce purified antioxidants from rosemary and sage.

For many years, in connection with certain food products, a barrier to freeze-drying has been the problems associated with the storage stability of foods that are susceptible to lipid oxidation. In order for such foods to have a reasonable shelf life and acceptable flavor characteristics, protective additives, which retard oxidation, are often added before dehydration. Such antioxidants must carry through the process and not be lost because of volatilization. For these applications, BHA, BHT, and tert-butylhydroquinone (TBHQ) have been found quite effective.

Additional Reading

Bigelow, S.W.: "Food Chemicals Codex: A Progress Report," *Food Technology*, 88 (May 1991).

Burdock, G.A. et al.: "GRAS Substances," *Food Technology*, 78 (February 1990).

Cadenas, E. and L. Packer: *Handbook of Antioxidants*, Vol. 8, Marcel Dekker, Inc., New York, NY, 2001.

Colgan, M.: *Antioxidants, The Real Story*, Apple Publishing Company, New York, NY, 1999.

Denisov, E.T. and T.G. Denisov: *Handbook of Antioxidants Bond Dissociation Energies, Rate Constants, Activation Energies, and Enthalpies of Reactions*, 2nd Edition, CRC Press, LLC., Boca Raton, FL, 2000.

Hiramatsu, M. and M. Inoue: *Food and Free Radicals*, Kluwer Academic/Plenum Publishers, New York, NY, 1997.

Madhavi, D.L., D.K. Salunkhe, and S.S. Deshpande: *Food Antioxidants, Toxicological and Health Perspectives*, Vol. 71, Marcel Dekker, Inc., New York, NY, 1995.

Packer, L. and E. Cadenas: *Handbook of Synthetic Antioxidants*, Vol. 4, Marcel Dekker, Inc., New York, NY, 1997.

Packer, L. and C. Colman: *The Antioxidant Miracle, Put Lipoic Acid, Coq 10, Pycnogenol, and Vitamins E and C to Work for You*, John Wiley & Sons, Inc., New York, NY, 1999.

Shriner, R.L., T.C. Morrill, and C. Hermann: *The Systematic Identification of Organic Compounds*, 7th Edition, John Wiley & Sons, Inc., New York, NY, 1997.

Yoshikawa, T., M. Hiramatsu, and L. Packer: *Antioxidant Food Supplements in Human Health*, Academic Press, Inc., San Diego, CA, 1999.

ANTIPARTICLES. One of the great discoveries of modern physics is that for every type of elementary entity of matter and radiation (particle), there exists a corresponding conjugate type of entity (antiparticle). In the antiparticle, certain of the particle-defining properties are identical (*conjugation-invariant*) and others are reversed in sign (*conjugation-reversing*). The reversed sign in a conjugation-reversing property allows one to maintain a conservation law for that property in the dramatic processes of *pair creation* and *pair annihilation* in which an antiparticle is observed to appear and disappear together with the particle to which it is conjugate. In those cases where all the conjugation-reversing properties occur with zero values, the antiparticle is identical with the particle. The progressive recognition of the existence of antiparticles was initiated by Dirac's relativistic antielectron theory in 1931, and by Anderson's independent experimental discovery of the antielectron (the positron) in 1932. See also **Particles (Subatomic)**.

ANTIPROTON. An elementary particle having a mass equal to that of the proton, differing from the proton only in the sign of its charge, which is negative, and a magnetic moment oppositely directed with respect to its spin. Positive identification of the antiproton was first made at the University of California. In 1959 Segre and Chamberlain received the Nobel Prize in physics for this discovery. Protons that had been accelerated to an energy of 6200 MeV in the bevatron, the proton synchrotron of the University of California Radiation Laboratory at Berkeley, were allowed to collide with a copper target. Negatively charged particles coming out in a forward direction from this collision were selected and separated in momentum by a focusing and analyzing magnet system to provide a beam of negative particles of known momentum. After a time of flight of about one-tenth of a microsecond, this beam may be expected to consist mainly of negative pions and muons, with some negative kaons (mass about 965 electron masses) and possibly negative protons. These particles were then distinguished both by measurement of their time of flight from the target (since particles of different mass have different velocities for given momentum) and by means of a device measuring the velocity of each particle passing through by the angle of its Cerenkov radiation. In this way the presence of negative particles with protonic mass (within about 10%) and distinct from the known kaons and hyperons was established. Their rate of production for the momentum and direction of this experiment was about one negative proton for every 50,000 negative pions with the same momentum and direction.

Antiprotons were first captured (1987) by researchers working with the Low-Energy Antiproton Ring (LEAR), which is a part of the large European Laboratory for Particle Physics (CERN) located in Geneva, Switzerland. The team of scientists made up from physicists of the University of Washington and the University of Mainz (West Germany) captured antiprotons from a high-energy accelerator and stored them for several minutes in an electromagnetic ion trap. The scientists forecast that with improvements in the trapping process it will facilitate the precise measurement of the inertial mass of the antiproton. They plan to ascertain the mass from the frequency of the circular motion of the antiprotons around the magnetic field lines in the trap (cyclotron resonance frequency). This will require slowing down the axial motion in the trap from the kiloelectron volt energies of the present equipment to a maximum of approximately 5×10^{-4} eV, i.e., the thermal energy associated with the ambient temperature of 4.2 K. To make the most meaningful measurement, there should be just one antiproton cooled to 4.2 K and the trapping time should be at least one day, compared with the minutes of the first experiment. Much greater vacuum and cryogenic cooling with helium is planned for future experiments. The scientists envision a number of interesting uses for trapped antiprotons:—determination of the gravitational constant of an antiproton; and possible use of large numbers of antiprotons as an energy source for space and military applications, although the scientists in the late 20th Century admitted that the idea of collecting antiprotons in macroscopic quantities stretched the imagination. See also **Particles (Subatomic)**; and **Proton**.

ANTIPYRETIC.   Any physical agent or drug that lowers the temperature of the body. Among antipyretics used are aspirin, antipyrine, acetanilid, and phenacetin. See also **Analgesics, Antipyretics, and Antiinflammatory Agents**.

ANTITOXIN.   (1) A substance made and elaborated in the body to neutralize a specific bacterial, plant, or animal toxin; (2) one of the class of specific antibodies. See also **Antibody**.

The history of the development of antitoxins in combating bacterial infection dates back to the early beginnings of organized bacteriology. Behring was the first to show that animals that were immune to diphtheria contained, in their serum, factors which were capable of neutralizing the poisonous effect of the toxins derived from the diphtheria bacillus. While this work was carried out in 1890, prior to many of the great discoveries of mass immunization, and much later the antibiotics, there yet remains a place for antitoxins in medical treatment or prophylaxis for some diseases, such as tetanus and botulism.

The more important approach to immunity to infectious disease now is the development of *active immunity* by injection of a vaccine. The vaccine may be either an attenuated live infectious agent, or an inactivated or killed product. In either case, protective substances generated in the bloodstream, called antibodies, help to neutralize the infectious agent when it is introduced. The principle of *passive immunization*, on the other hand, involves the development of the antibodies in another host and most frequently a different species as well. The antiserum or antitoxin (from the other host) is employed in preventing the onset of the disease, or in actual treatment of the active infection in subjects who have not had the advantage of becoming actively immunized due either to neglect or to the fact that an effective vaccine was not available. The use of antitoxins prepared in another species (for example, horse) is not without some element of risk.

Antitoxins are prepared by injecting the donor animals with frequent and increasing doses of toxin while maintaining a level at each injection that the animal can tolerate. The initial doses are critical since these toxins may be among the most poisonous agents known. In one technique, the toxin is diluted so that the first injection contains less than the minimum lethal dose. Other programs used toxins that are inactivated with formaldehyde so that they are no longer poisonous, but may still elicit an immune response and result in antitoxin that will neutralize the unaltered toxin. This method of inactivation also was developed in the nineteenth century for preparing many of the important vaccines against diseases, such as influenza, tetanus, and diphtheria.

Small laboratory animals are helpful in carrying out research in this field, but in commercial production, larger animals are required. The horse proved to be very satisfactory for large-scale production of antitoxins, and is still used for the major portion of antitoxin production, although some material is also derived from cattle. The management of the animals and the schedules and dosages are perhaps as much an art as an exact science. The selection of the horses that have the best potential as antitoxin producers is also a critical factor. When a horse is receiving the maximum level of toxin in the hyper-immunization program, it may be injected in a single dose with enough toxin to be fatal for 100 million mice if the material was suitably distributed.

A means for enhancing the potency of the antitoxin in horses is to use an agent called an *adjuvant*. Several adjuvants have been used, including tapioca, mineral oil, and aluminum hydroxide. The mechanism by which these agents increase the intensity of the immune response is not fully understood, but local inflammatory reaction and the resulting slower release of the injected material from the original site appear to play a role. Adjuvants also are sometimes used with vaccines for human use. It has been recorded that one horse, during an eleven-year period, gave 657 gallons (~ 25 hectoliters) of blood from which tetanus antitoxin and, at different times, pneumococcus antiserum, was prepared. The volume of serum removed from the horse can be increased by the return of the red cells after removal of the plasma or liquid component. It has been shown that if the red cell level can be maintained in human donors of special sera, they can safely give as much as one liter of serum per week.

The injection of serum components of another species into human patients has not been without problems. A condition known as serum sickness develops in an alarmingly high proportion of those treated in this way, which is apparently due to a generalized sensitivity that develops to the foreign serum protein. The onset of illness is usually delayed for several days after the injection of antitoxin and symptoms may be quite severe.

See also **Immune System and Immunochemistry**.

ANN C. DEBALDO, Ph.D.
College of Public Health, University of South Florida
Tampa, Florida

Additional Reading

Landon, J. and T. Chard: *Therapeutic Antibodies,* Springer-Verlag Inc., New York, NY, 1994.
Staff: Institute of Medicine Committee: *Dietary Reference Intakes: Review of Dietary Antitoxins,* National Academy Press, Washington, DC, 1999.

ANTIVIRAL AGENTS.   Viral infections are among the greatest causes of human morbidity, and it is estimated that in developed countries more than 60% of all the episodes of human illness result from viral infections. High virus infection rates also occur among pets, livestock, and plants. The high morbidity and the resulting economic loss caused by these infections have generated tremendous efforts in recent years to develop means to combat viral infections using antiviral agents.

Since viruses propagate only within living cells, the development of antiviral drugs which would disrupt the viral replication without affecting the metabolism of the host cell was initially believed to be difficult, if not impossible. However, dramatic progress in viral molecular biology has now made it possible to identify enzymatic processes which are unique to virus-infected cells. As a consequence, it is becoming feasible to design chemical compounds which identify infected cells, block a specific step in viral replication, and leave uninfected cells unharmed.

The majority of antiviral drugs which are under clinical development today generally interrupt viral nucleic acid synthesis. These compounds often do not affect host cell metabolism and possess considerable selectivity against virus-induced enzymes. This article discusses agents exhibiting significant antiviral activity against viral infections in animal model systems.

Agents Active Against DNA Viruses

One of the simplest molecules found to inhibit the replication of DNA viruses in animals is phosphonoformic acid (PFA), CH_3O_5P (**1**). Both PFA (as the trisodium salt CNa_3O_5P, foscarnet) and its homologue phosphonoacetic acid (PAA), $C_2H_5O_5P$ (**2**), were developed by Astra Pharmaceuticals and show selective inhibition of DNA polymerase in various herpes viruses.

(1)

(2)

PFA has recently undergone clinical evaluation in humans for the treatment of recurrent genital herpes, hepatitis B viral infection, and acquired immunodeficiency syndrome (AIDS), as well as cytomegalovirus (CMV) infection of bone marrow and renal transplant patients.

Levamisole   (6-phenyl-2,3,5,6-tetrahydroimidazol[2,1-*b*]thiazole), $C_{11}H_{12}N_2S$ (**3**), was found to be effective against herpes virus infections in humans.

(3)

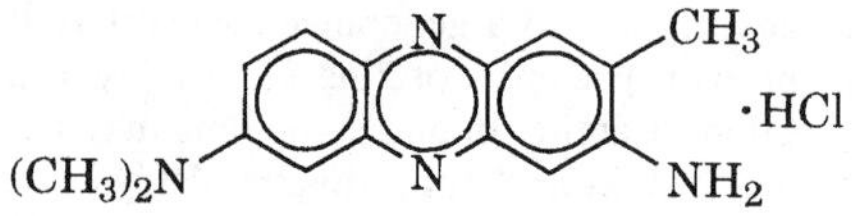

(4)

Certain heterocyclic dyes such as neutral red, $C_{15}H_{17}ClN_4$ (4), acridine orange, and proflavine, which act by binding to viral nucleic acid, absorbing visible light, and inactivating the virus by oxidation, have shown antiviral activity of clinical potential against cutaneous herpes virus infections. However, concern has been raised that photoinactivation of herpes viruses may be clinically dangerous because the inactivated virus may be capable of transforming normal cells to a malignant state. Moreover, these dyes have a strong affinity for the nuclei of normal cells and may damage their genetic makeup.

Several pyrimidine bases have been found to inhibit herpes virus-induced keratitis in rabbits, e.g., 2-amino-4,6-dichloropyrimidine, $C_4H_3Cl_2N_3$ (5), and 1-allyl-6-chloro-3,5-diethyluracil, $C_{11}H_{15}ClN_2O_2$ (6).

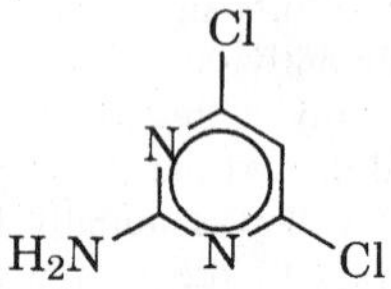

(5)

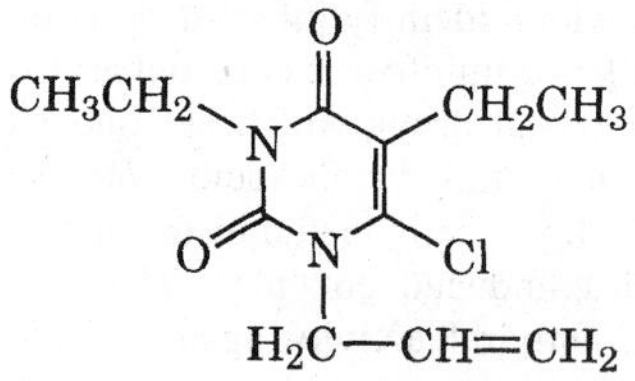

(6)

The most successful clinical antiviral agents belong to the nucleoside category. Nucleoside analogues with potent antiviral activity have been known since idoxuridine and trifluridine were shown to be efficacious against herpes keratitis more than 30 years ago. However, despite their antiviral efficacy the therapeutic usefulness of these early nucleosides was limited because of mutagenic, teratogenic, carcinogenic, cytostatic, or cytotoxic side effects. Recently the specificity of antiviral action of this class of compounds has been significantly improved; potent and highly selective nucleoside antiviral agents have now been developed. These are either antiviral nucleosides which are only activated by virus-infected cells, or antiviral agents which specifically inhibit virus-induced enzymes required for viral replication.

Cytosine   Derivatives. 1-β-D-Arabinofuranosylcytosine   (ara-C), $C_9H_{13}N_3O_5$ reportedly has had significant therapeutic effects in patients with localized herpes zoster, herpes eye infections, and herpes encephalitis, although several negative results have also been reported. Ara-C, also known as cytarabine, is quite toxic and is only recommended for very severe viral infections. A number of derivatives of ara-C have been prepared in an effort to improve on antiviral activity and to reduce the toxicity.

Purine Nucleoside Derivatives. A number of purine nucleoside analogues are also found to be active against several DNA viruses (Fig. 1). They include ara-A (9-β-D-arabinofuranosyladenine, vidarabine), $C_{10}H_{13}N_5O_4$ (7), ara-HxMP, $C_{10}H_{13}N_4O_8P$ (8), cyclaradine, $C_{11}H_{15}N_5O_3$ (9) and the xylofuranosyl analogue of tubercidin, $C_{11}H_{14}N_4O_4$ of tubercidin, (10).

Adenine Derivatives. These include (S)-9-(3-hydroxy-2-phosphonyl-methoxypropyl) adenine [(S)-HPMPA], $C_9H_{14}N_5O_5P$ (11), the cyclic phosphonate of (S)-HPMPA, (S)-cHPMPA, $C_9H_{12}N_5O_4P$ (12), (S)-DHPA, $C_8H_{11}N_5O_2$ (13), and adeninylhydroxypropanoic acid alkyl esters [(R,S)-AHPA esters] (14) (Fig. 2).

Acyclic Purine Nucleosides. As a consequence of earlier studies involving adenosine deaminase as a model enzyme for examining structural features required for antiviral activity, a number of acyclic purine nucleosides have emerged as antiviral agents (Fig. 3). They include acyclovir [9-(2-hydroxyethoxymethyl) guanine], $C_8H_{11}N_5O_3$ (15), 6-deoxyacyclovir, $C_8H_{11}N_5O_2$ (16), DHPG [9-(1,3-dihydroxy-2-propoxymethyl)guanine, ganciclovir], $C_9H_{13}N_5O_4$ (17) DHPG cyclic phosphate $C_9H_{11}N_5O_6P \cdot$ Na (18), 3,4-dihydroxybutylguanine (DHBG), $C_9H_{13}N_5O_3$ (19), (±)-(1α, 2β, 3α)-9-[2,3-bis(hydroxymethyl) cyclobutyl] guanine [(± − BHCG)], $C_{11}H_{15}N_5O_3$ (20), the ara-carbocyclic analogue of 7-deazaguanosine (±)-2-amino-3,4-dihydro-7-[(1α, 2α, 3β, 4α)-2,3-dihydroxy-4-(hydroxymethyl)-1-cyclopentyl] pyrrolo[2,3-d]-pyrimidin-4-one, $C_{12}H_{16}N_4O_4$ (21), and 9-β-D-+xylofuranosylguanine (xylo-G), $C_{10}H_{13}N_5O_5$ (22).

Agents Active Against RNA Viruses. A large number of α-hydroxybenzylbenzimidazole (HBB), $C_{14}H_{12}N_2O$ (23), derivatives have been prepared and extensively studied as selective inhibitors of the RNA-containing enteroviruses. Although none of these derivatives have shown any antiviral activity in animals, 1,2-bis(5-methoxy-2-benzimidazol-2-yl)-1,2-ethanediol, $C_{18}H_{18}N_4O_4$ (24), was found to be active against an experimentally induced rhino virus infection in chimpanzees. However, the *in vivo* antiviral efficacy was accompanied by significant toxicity.

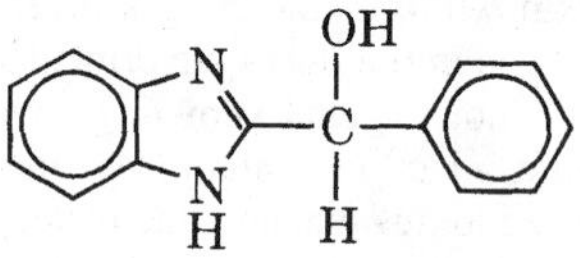

(23)

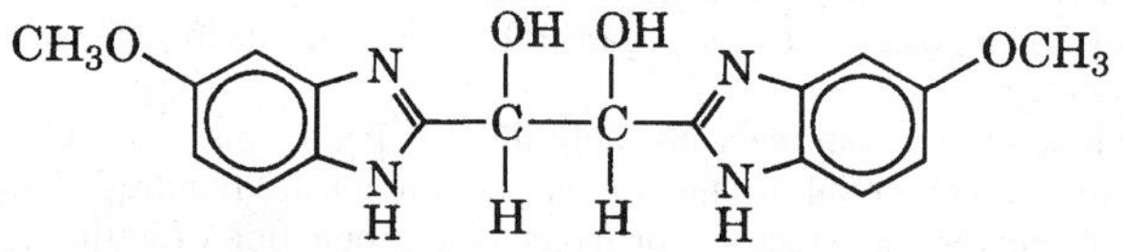

(24)

Amantadine   hydrochloride   (1-adamantanamine   hydrochloride), $C_{10}H_{17}N \cdot$ HCl (25), is a good example of a narrow-spectrum agent active only against influenza A virus. It became the first antiviral drug available for systemic use in the United States when it was approved by the FDA in 1966 for use against Asian influenza. A structurally related drug, rimantadine hydrochloride, (α-methyl-1-adamantane-methylamine hydrochloride),

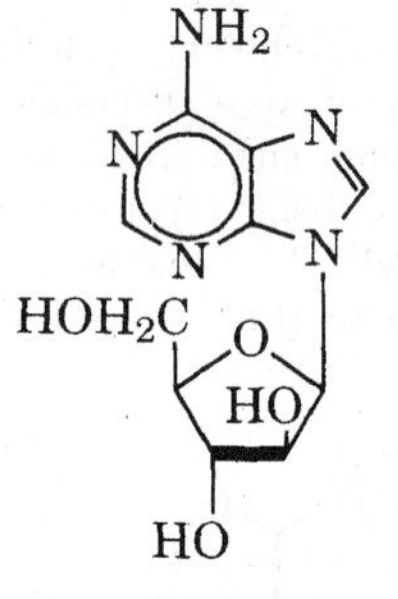

(7) 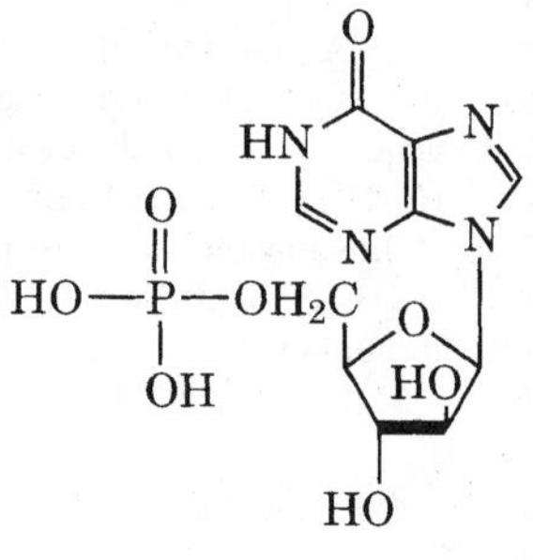   **(8)** 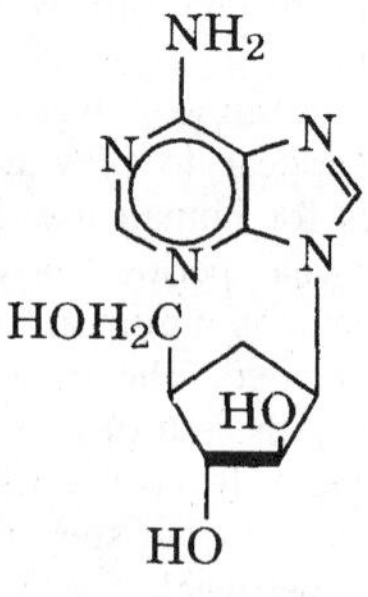   **(9)** 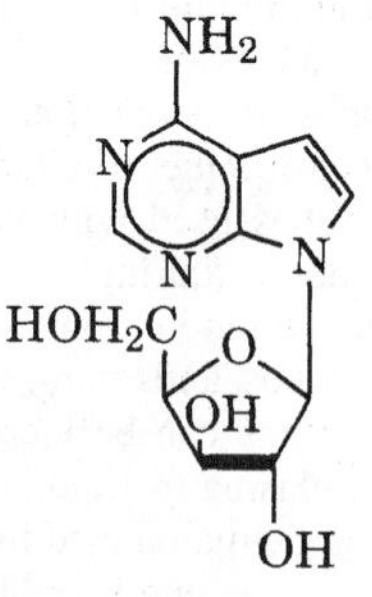   **(10)**

Fig. 1.   Purine nucleoside analogues found to be active against DNA viruses

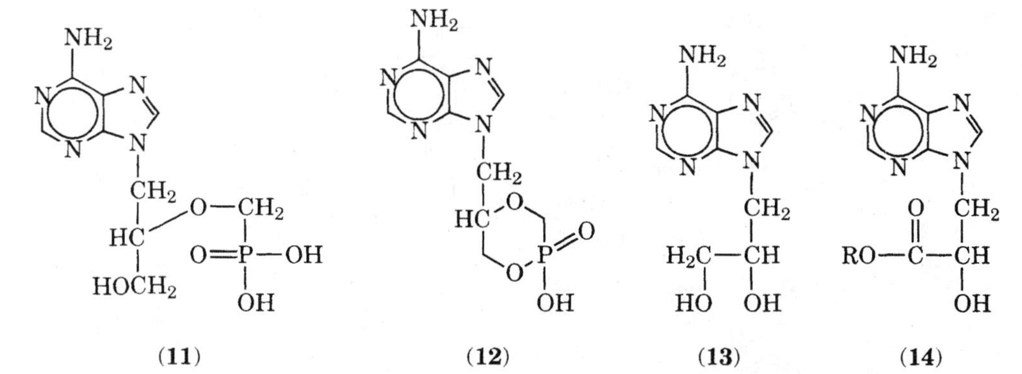

(11) (12) (13) (14)

Fig. 2. Aliphatic adenosine analogues with broad-spectrum antiviral activity

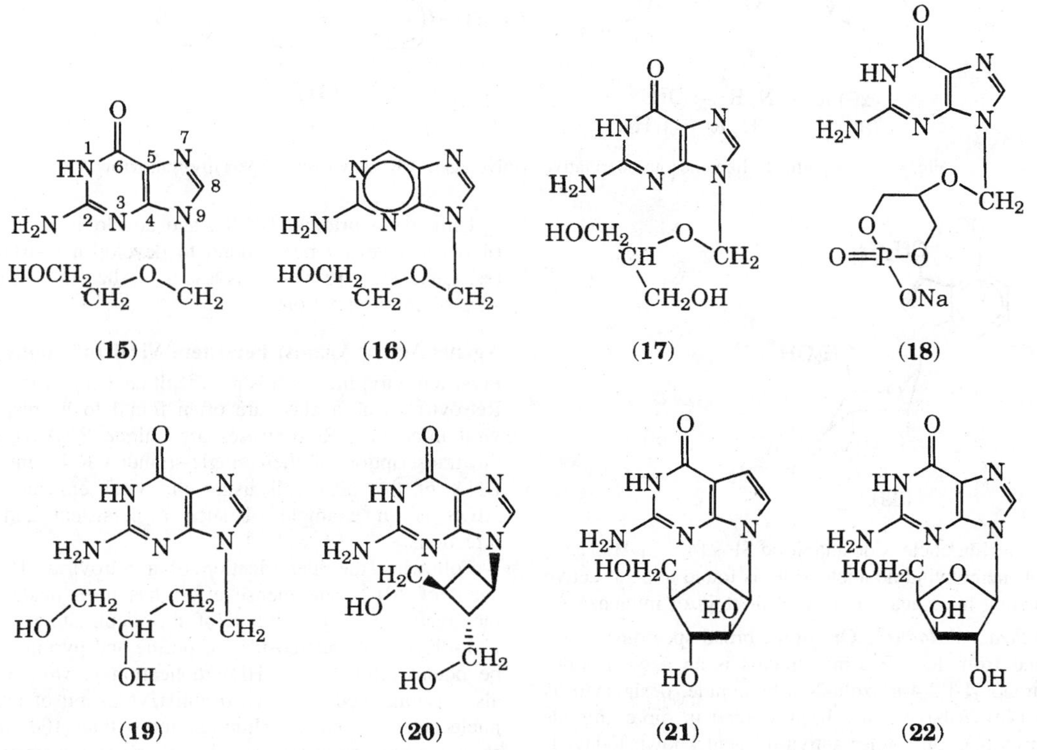

(15) (16) (17) (18)

(19) (20) (21) (22)

Fig. 3. Acyclovir and its analogues

$C_{12}H_{21}N \cdot HCl$ (**26**), is widely used in Russia to treat influenza A virus.

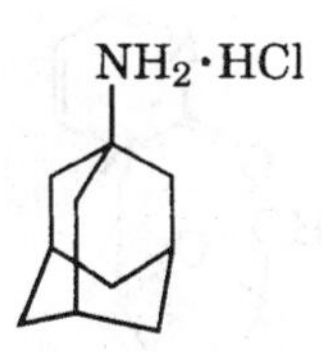

(**25**)

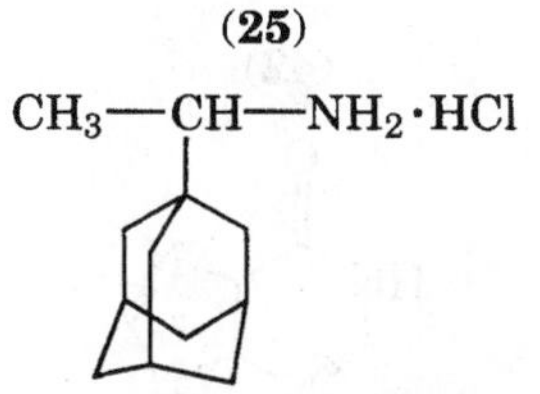

(**26**)

Lipophilic β-Diketones. Arildone, $C_{20}H_{29}ClO_4$ (**27**), and several other lipophilic β-diketones (Fig. 4) have exhibited significant *in vivo* activity against a number of RNA viruses. These include WIN 51711, $C_{20}H_{26}N_2O_3$ (**28**), enviroxime, $C_{17}H_{18}N_4O_3S$ (**29**), the enviroxime surrogate enviradene, $C_{19}H_{21}N_3O_2S$ (**30**), and 3-methoxy-6-[4-(3-methylphenyl)-1-piperazinyl]pyridazine (R61837), $C_{16}H_{20}N_4O$ (**31**).

Adenosine and Guanosine Analogues. Several adenosine analogues have been found to be active against both RNA and DNA viruses (Fig. 5). They include 3-deazaadenosine $C_{11}H_{14}N_4O_4$ (**32**), the carbocyclic analogue of 3-deazaadenosine, [(±)-3-deazaaristeromycin], $C_{12}H_{16}N_4O_4$ (**33**), neplanocin A, $C_{11}H_{13}N_5O_3$ (**34**), sinefungin $C_{15}H_{23}N_7O_5$ (**35**), 3-deazaguanine, $C_6H_6N_4O$ (**36**), and 3-deazaguanosine, $C_{11}H_{14}N_4O_5$ (**37**).

The naturally occurring nucleoside antibiotic pyrazofurine, $C_9H_{13}N_3O_6$ (**38**), shows broad-spectrum antiviral activity and a broad safety margin against both RNA and DNA viruses *in vitro*.

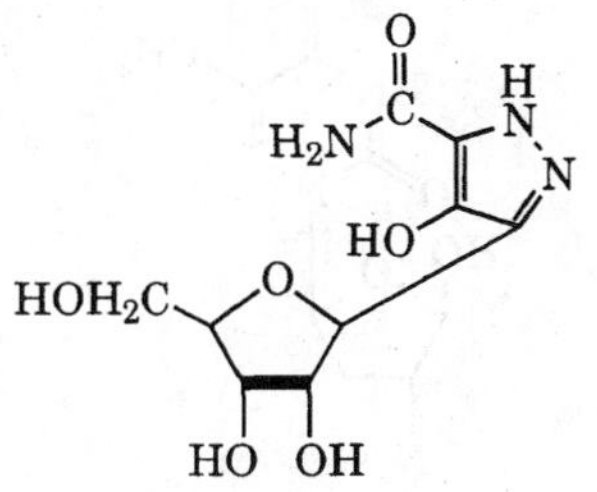

(**37**)

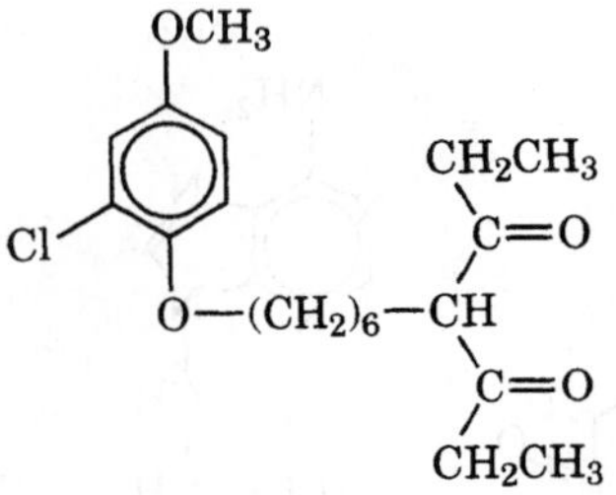

(27)

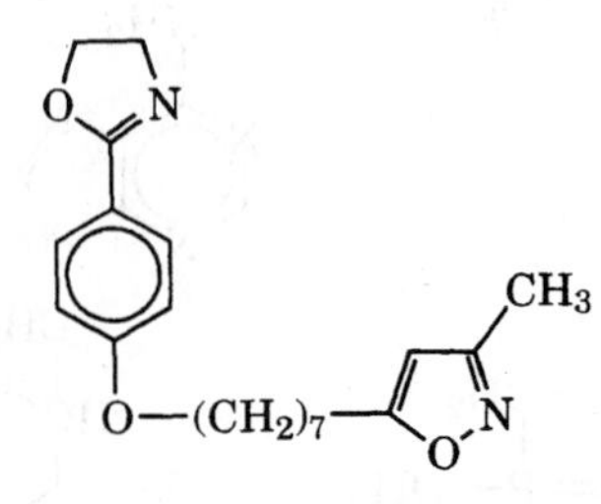

(28)

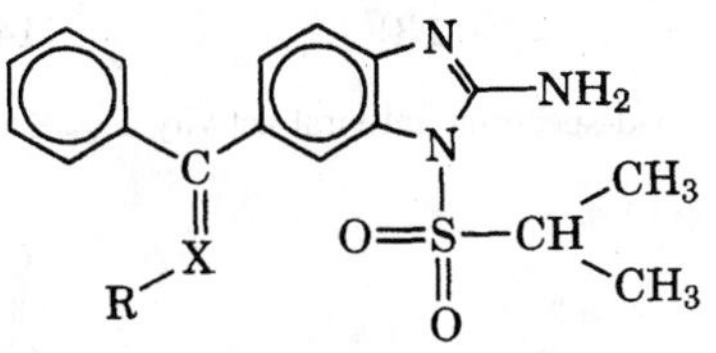

(29) X = N, R = OH
(30) X = CH, R = CH₃

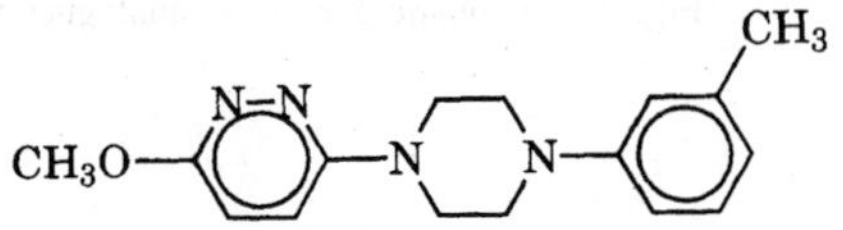

(31)

Fig. 4.   Lipophilic β-diketones and derivatives active against RNA viruses, especially rhinovirus

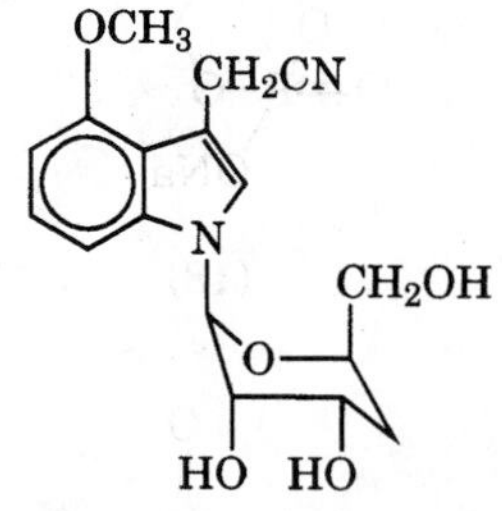

(39)

Another naturally occurring nucleoside antibiotic SF-2140, $C_{17}H_{20}N_2O_5$ (39) a 3-cyanomethyl-4-methoxyindole nucleoside, is found to be as active as amantadine in protecting mice against an APR-8 strain of influenza A.

Ribavirin and Structural Analogues. One of the broad-spectrum antiviral agents that emerged from ICN Pharmaceuticals is an azole ribonucleoside, 1-β-D-ribofuranosyl-1,2,4-triazole-3-carboxamide, designated as ribavirin, $C_8H_{12}N_4O_5$ (40). Ribavirin has been studied in more animals and against more viruses than any other antiviral agent known today. It is active in cell culture against approximately 85% of all viruses studied, although it is inactive against polio virus, certain coxsackie viruses, most corona viruses, pseudorabies virus, and HBV.

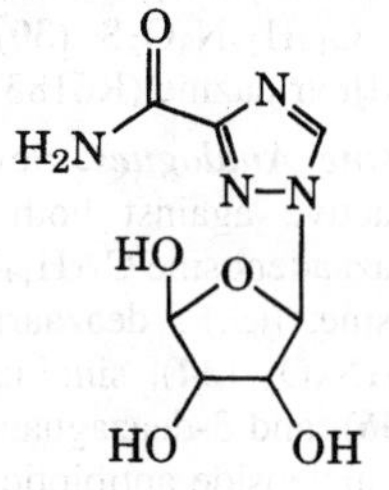

Unlike idoxuridine, BVdU, and acyclovir, viral strains susceptible to ribavirin have not been found to develop a resistance to the drug. The resistance against ribavirin is less likely because the drug exhibits multiple sites of antiviral action.

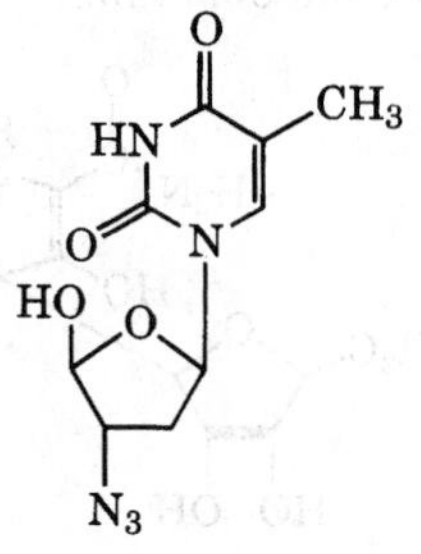

Agents Active Against Persistent Viral Infections

Persistent viral infection is a difficult challenge for antiviral chemotherapy. Retroviruses as a class are often found to be responsible for persistent viral infections. Retroviruses are unique RNA viruses characterized by the transcription of their single-stranded RNA into the double-stranded DNA of the host cell using the viral enzyme reverse transcriptase. AIDS is an example of such a persistent and latent human viral infection.

Following the identification of a retrovirus, HIV, as the etiological agent of AIDS, an intense effort has been made to identify drugs for the treatment or prevention of this debilitating, lethal disease. Several 2′,3′-dideoxyribonucleosides of purine and pyrimidine were discovered to be potent inhibitors of HIV replication *in vitro*. Considerable data has also accumulated on *in vitro* antiHIV testing of acyclic and carbocyclic nucleoside analogues. Although more than 100 nucleosides have been shown to have an *in vitro* antiHIV activity commensurate with development to clinical trials, only 3′-azido-3′-deoxythymidine (retrovir, zidovudine, AZT), $C_{10}H_{13}N_5O_4$ (41) 2,′3′-dideoxycytidine (DDC), $C_9H_{13}N_3O_3$ (42), and 2′,3′-dideoxyinosine (DDI), $C_{10}H_{12}N_4O_3$ (43), have become widely

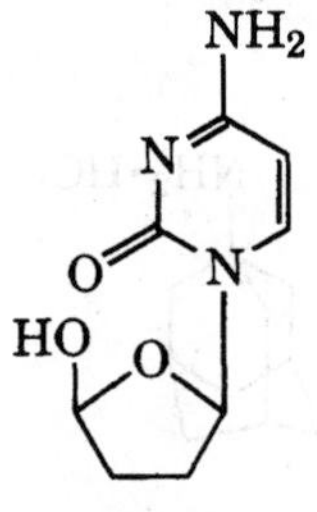

(40)

(41)

(42)

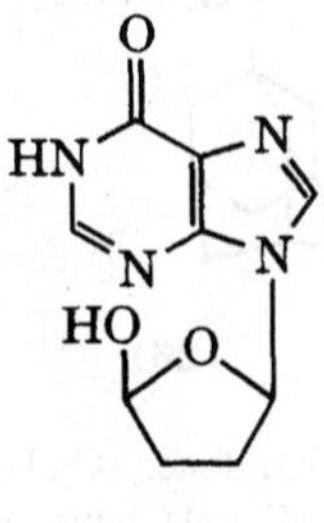

(43)

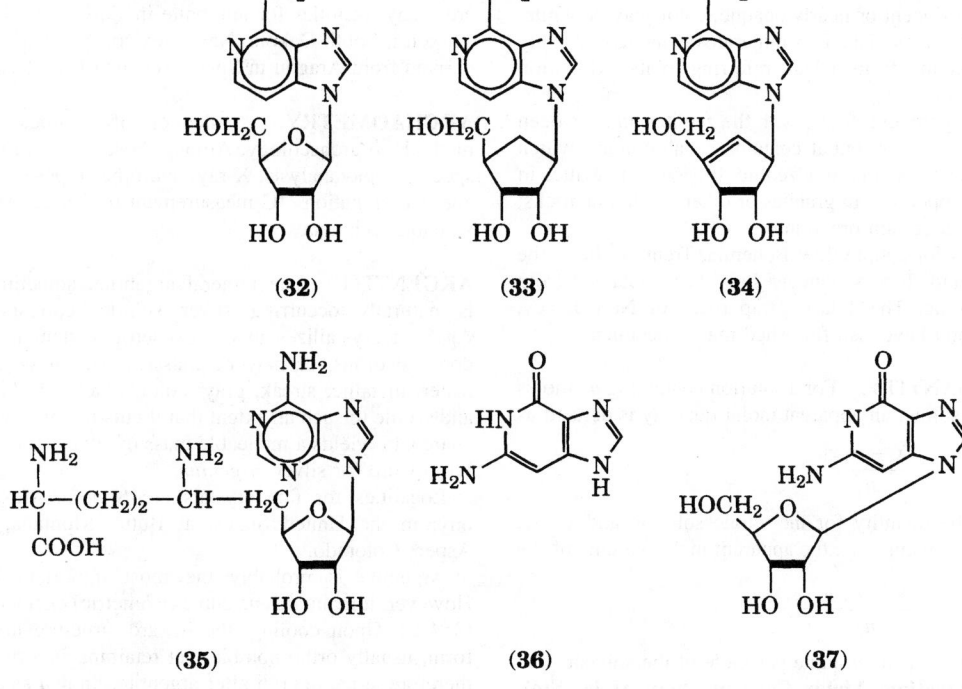

(32)　　　(33)　　　(34)

(35)　　　(36)　　　(37)

Fig. 5.   Adenosine and guanosine analogues with both RNA and DNA antiviral effectiveness

available for the treatment of AIDS and approved by the FDA for treatment of advanced AIDS cases.

GANAPATHI R. REVANKAR
Triplex Pharmaceutical Corporation

ROLAND K. ROBINS
ICN Nucleic Acid Research Institute

Additional Reading

Adams, J. and V.J. Merluzzi: *Search for Antiviral Drugs, Case Histories from Concept to Clinic*, Birkhauser Verlag, Cambridge, MA, 1994.

Billups, N.F. and S.M. Billups: *American Drug Index 2000*, 44th Edition, Facts And Comparisons, St. Louis, MO, 1999.

Budavari, S., A. Smilth, P. Heckelman, J. Kinneary, and M.J. O'Neill: *The Merck Index, An Encyclopedia of Drugs, Chemical, and Biologicals*, 12th Edition, Merck Research Laboratories, Whitehouse Station, NJ, 1996.

De Clercq, E. ed.: in *Design of Anti-Aids Drugs*, Elsevier, New York, NY, 1990, pp. 1–24.

Declercq, E. and D.J. Jeffries: *Antiviral Chemotherapy*, John Wiley & Sons, Inc., New York, NY, 1995.

Elion, G.B.: *Am. J. Med.* **73**, 7 (1982).

Richman D.D.: *Antiviral Drug Resistance*, John Wiley & Sons, Inc., New York, NY, 1996.

Robins, R.K.: *Chem. Eng. News* **64**, 28 (1986).

Roth, L.S.: *Mosby's Nursing Drug Reference 2000*, Harcourt Health Sciences Group, San Diego, CA, 1999.

Schinazi, R.F. and D. Kinchington: *Antiviral Methods and Protocols*, Vol. 24, Humana Press, Totowa, NJ, 1999.

Sidwell, R.W. G.R. Revankar, and R.K. Robins: in D. Shugar, ed., *Intl. Enc. Pharmacol. Ther.*, Sec. 116, Vol. 2, Pergamon Press, New York, NY, 1985, pp. 49–108.

Staff: *Drug Facts and Comparisons w 2000*, 54th Edition, Facts And Comparisons, St. Louis, MO, 1999.

ANTLERITE.   Antlerite is a relatively uncommon mineral found within the oxidized zones of copper deposits in arid regions. It is a basic sulfate of copper, $Cu_3(SO_4)(OH)_4$, crystallizing in the orthorhombic system. Hardness of 3.5, specific gravity 3.88, with vitreous luster, and emerald-green to black-green color.

Originally found in Arizona. It is the principal copper ore mineral at Chuquicamata, Chile.

ANTONOFF RULE.   The tension at the interface between two saturated liquid layers that are in equilibrium is equal to the difference between the individual surface tensions against air or vapor of the two saturated solutions. This rule is approximate only and a number of exceptions are known.

APATITE.   The mineral apatite is a phosphate of calcium with either fluorine or chlorine or sometimes both, hence the distinction between fluorapatite and chlorapatite. Sometimes both fluorine and chlorine are present. Most apatite is, however, fluorapatite, $Ca_5(PO_4)_3F$.

Apatite crystallizes in the hexagonal system in prismatic and tabular forms. Hardness, 4.5–5; specific gravity, 3.17–3.23; luster, vitreous to resinous; transparent to opaque; streak, white; cleavage, imperfect basal and prismatic; color, white, green, yellow, red, brown and purple; sub-conchoidal fracture. The variety called asparagus stone is yellow-green and manganapatite, which is a dark bluish-green, may contain as much as 10% manganese dioxide replacing the calcium. Werner devised the name *apatite* from the Greek word meaning "to deceive", because it was frequently mistaken for beryl and other species. Apatite has been found widely distributed both geographically and petrologically as it occurs in many sorts of rocks, metamorphic limestones, gneisses, schists, granites and syenites, pegmatite veins and even with iron ores. It has been prepared artificially. It has been mined for the manufacture of fertilizers and to a slight extent for jewelry.

Apatite occurs extensively in Europe and America, especially in New England, New Jersey, New York, North Carolina, California, and in the provinces of Ontario and Quebec in Canada.

API GRAVITY.   See **Petroleum**; **Specific Gravity**.

APLITE.   This term is applied to fine-grained, sometimes sugary-textured igneous rocks, composed almost wholly of quartz and feldspar. Except for size of grain, aplites resemble permatites both in mineral composition and in mode of occurrence in dikes and veins, save that the rare minerals often present in pegmatites are wanting here. The word aplite is derived from the Greek word meaning simple, referring to its ordinarily simple mineral composition.

APOPHYLLITE.   The mineral is a hydrous silicate of potassium, calcium, and fluorine, corresponding to the formula, $KCa_4Si_8O_{20}(F,OH)\cdot 8H_2O$. The true crystallographic symmetry is evident on crystals by the luster difference between the basal pinacoid facial planes and other crystal faces. Also, prism faces show vertical striations; basal planes do not. It crystallizes in the tetragonal system in square prisms resembling cubes terminated by based pinacoid or pyramids, often with both. Prism faces show vertical striations; basal pinacoid either dull or rough. Cleavage is perfect, parallel to the base; hardness, 4.5–5; sp gr, 2.3–2.4; luster, vitreous to pearly; transparent to translucent or nearly opaque; color may be white, grayish, greenish, yellowish, or reddish. This mineral was named by Haüy from the Greek words meaning from a leaf, referring to its exfoliation when heated with the blow pipe.

Apophyllite is a secondary mineral found with the zeolites and has been classed with them by some writers, but it contains no aluminum, which element is understood to be an essential in a zeolite. It occurs in cavities in basalts and less often filling openings in granites or other crystalline rocks; it also is a gangue mineral in certain ore veins.

There are many localities for apophyllite: Bohemia, Trentino, Italy, the Hartz Mountains, and Iceland. Fine specimens have been obtained from the Ghats Mountains in India. The Triassic trap rocks of New Jersey, Connecticut, and Nova Scotia have also furnished many specimens.

APPARENT MOLAR QUANTITY.

For a solution containing n_1 moles of solvent and n_2 moles of solute, an apparent molar quantity is defined as

$$\frac{X - n_1 x_1}{n_2}$$

where X is the value of the quantity for the whole solution and x_1 the molar quantity for the pure solvent; e.g., the apparent molar volume of the solute is

$$\frac{V - n_1 v_1}{n_2}$$

V being the total volume and v_1 the volume per mole of the solvent.

See also **Molal Concentration**; **Molar Concentration**; **Mole (Stoichiometry)**; **Mole Fraction**; and **Mole Volume**.

APPLIED RESEARCH.

The experimental investigation of a specific practical problem for the immediate purpose of creating a new product, improving an older one, or evaluating a proposed ingredient. The experimental program is set up to answer the question "What happens?" rather than "Why does it happen?" Examples are the determination of the value of a new rubber antioxidant, the substitution of one drying oil for another in a paint formulation, and the development of a new synthetic product. While applied research has produced the multitude of new materials that have revolutionized industry, agriculture, and medicine in the last 50 years or so, its achievements have usually been practical outgrowths of prior fundamental research.

AQUAMARINE.

A form of the gem beryl. See **Beryl**.

AQUA REGIA.

Also known as nitrohydrochloric acid, aqua regia is made up of three parts hydrochloric acid and one part nitric acid, each of the usual concentrated laboratory form. Aqua regia will dissolve all metals except silver. The latter is converted to silver chloride. The reaction of metals with nitrohydrochloric acid typically involves oxidation of the metal to a metallic ion and the reduction of the nitric acid to nitric oxide. Aqua regia also dissolves the common oxides and hydroxides of metals with the exception of silver, the ignited oxides of tin, aluminum, chromium, and iron, and the higher oxides of lead, cobalt, nickel, and manganese, the latter dissolving effectively in hydrochloric acid alone.

ARACHIDIC ACID.

[CAS: 506-30-9]. Also known as eicosanoic acid, formula $CH_3 (CH_2)_{18}COOH$. A widely distributed, but minor, component of the fats of certain edible vegetable oils. Shining, white crystalline leaflets; soluble in ether, slightly soluble in water. Sp gr, 0.8240 (100/4°C); mp 75.4°C; bp 205°C (1 millimeter pressure). Decomposes at 328°C. Commercial product derived from groundnut (peanut) oil. Used in organic synthesis, lubricating greases, waxes, and plastics. Source of arachidyl alcohol. See also **Vegetable Oils (Edible)**.

ARAGONITE.

The mineral aragonite is calcium carbonate, $CaCO_3$, chemically identical with calcite but crystallizing in the orthorhombic system, with acicular crystals. By repeated twinning, pseudo-hexagonal forms result. Aragonite may be columnar or fibrous, occasionally in branching stalactitic forms called flosferri (flowers of iron) from their association with the ores at the Carinthian iron mines. Its hardness is 3.5–4; specific gravity, 2.93–2.95; luster, vitreous to resinous; colors, white, gray, green-yellow or purple; transparent to translucent. Aragonite forms at temperatures of 80–100°C and is relatively unstable at ordinary temperatures and pressures. It alters to calcite, although very slowly. There are many localities for aragonite in Europe, Bolivia, Pennsylvania, Iowa, Missouri, South Dakota, New Mexico, Arizona and Colorado. Its name is derived from Aragon in Spain. See also **Calcite**.

ARCHAOMETRY.

Application of chemical and physical analytical methods to archaeology. Among those used are microanalytical methods, spectroscopic analysis, X-ray, and other types of nondestructive tests. For age determination ^{14}C measurement (chemical dating) is one of the most valuable techniques.

ARGENTITE.

The mineral argentite, sometimes called silver glance, is naturally occurring silver sulfide, corresponding to the formula Ag_2S. It crystallizes in the isometric system in cubes, octahedrons and dodecahedrons, or may be massive. Hardness, 2–2.5; sp gr, 7.2–7.34; luster, metallic; streak, gray; color, black, blackish-gray or gray; opaque and sectile to such an extent that it cuts like wax with a knife. Heated upon charcoal it yields a malleable mass of silver. The name is derived from the Latin word for silver, *argentum*.

Localities for fine crystals are Sonora, Mexico, and Freiberg, Saxony; in the United States, at Butte, Montana; Tonopah, Nevada; and Aspen, Colorado.

Argentite is probably the most important primary silver mineral. However, it maintains its cubic (isometric) characteristic only above 179°C (354°F). Upon cooling, the inward structure inverts to a non-isometric form, usually orthorhombic, yet retaining its original outward form. It is, therefore, a paramorph after argentite, known as acanthite.

ARGILLITE.

A dense, fine-grained, hard, sedimentary rock of various colors (usually white, gray or red). Composed of minute grains of both clay and quartz. Certain types of argillites are easily confused with certain types of fine-grained acid lava flows, such as felsites, unless studied microscopically.

ARGININE.

See **Amino Acids**.

ARGON.

[CAS: 7440-37-1]. Chemical element, symbol Ar, at. no. 18, at. wt. 39.948, periodic table group 18 (inert or noble gases), mp −189.2°C, bp −185.7°C, density 1.78 g/cm^3 (solid at −233°C). Solid argon has a face-centered cubic crystal structure. At standard conditions, argon is a colorless, odorless gas and does not form stable compounds with any other element. Because of its low valence forces, argon is unable to form diatomic molecules, except in discharge tubes. It does form compounds under highly favorable conditions, as excitation in discharge tubes, or pressure in the presence of a powerful dipole. As an example of the first, argon forms amorphous compounds of the type FeA in a discharge tube having iron electrodes. An example of the second is furnished by the hydrates, which argon forms with H_2O at 150 atmospheres and 0°C. Argon forms compounds, possibly clathrates, with a number of organic substances, such as a compound with hydroquinone containing 9% argon, in which the amount of argon may vary from this proportion. The compounds are made by crystallization of the aqueous solution of the hydroquinone under argon gas pressure on the order of 40 atmospheres.

Argon occurs in the atmosphere to the extent of approximately 0.935%. In terms of abundance, argon does not appear on lists of elements in the earth's crust because it does not exist in stable compounds. However, argon is 2.5 × more soluble in H_2O than nitrogen and thus is found in seawater to the extent of approximately 2800 tons per cubic mile (605 metric tons per cubic kilometer). Commercial argon is derived from air by liquefaction and fractional distillation. There are three natural isotopes, ^{36}Ar, ^{38}Ar, and ^{40}Ar, and four radioactive isotopes, ^{35}Ar, ^{37}Ar, ^{39}Ar, and ^{41}Ar. The lengths of half-lives of the isotopes vary widely, the shortest ^{35}Ar with a half-life of about 2 seconds; the longest ^{39}Ar with a half-life of about 260 years. The first ionization potential of Ar is 15.755 eV; second 27.76 eV; third 40.75 eV. Other important physical characteristics of argon are given under **Chemical Elements**.

The presence of argon in air was suspected by Cavendish as early as 1785, but was not positively identified until 1894 by Lord Rayleigh and Sir William Ramsay. Argon exhibits a characteristic series of lines in the red end of the spectrum. Commercially, argon gas is used in incandescent lamps and fluorescent lamps as an inert gas to minimize vaporization of the filaments and, for this, is preferable to nitrogen. The gas also is used for shielding electrodes in arc welding. A gas of about 99.995% purity is required for lamps. Argon also has found effective use in certain lasers.

Regarding argon in meteorites, see **Krypton**.

Ar Isotopes in Paleochronology. Isotopes of Ar and their ratios have proven useful in geological period dating and thus are of assistance in establishing databases for paleoclimatology and estimating periods of mass extinction on Earth. For example, major changes in Earth's climate are known to have occurred during transition between the Eocene, q.v., and Oligocene, q.v., periods. This was a major transition from the essentially tropical environment of the Mesozoic, q.v., to the start of the galacial world. This is considered the most significant climatic alteration since the prior demise of the dinosaurs. However, it has been difficult to establish the timing of prehistoric events, for lack of evidence.

Since the mid-1960s, $^{40}K–^{40}Ar$ dates have been used to estimate time periods in terms of Ma (millions of years), and some adjustments were made to prior estimates of the Chadronian boundary (36–32 Ma). Other isotopic data (^{40}Ar, $^{40}Ar/^{39}Ar$) and Rb-Sr have been used for time estimating.

As pointed out by Swisher and Prothero (1990) relatively recent advances (1989–1990) in mass spectrometric techniques and the development of laser-fusion $^{40}Ar/^{39}Ar$ dating techniques have resulted in the ability to date individual volcanic crystals. Multiple analysis and the ability to date single crystals allow the identification of multiple-age components, due to detrital contamination, and thus permit improved precision and accuracy. To date, studies have been directed to North American chronology, notably minerals (biotite, anor, plag) found in Nebraska and Wyoming.

Additional Reading

Hawley, G.G. and Lewis, R.J.: *Hawley's Condensed Chemical Dictionary,* 13th Edition, John Wiley & Sons, Inc., New York, NY, 1996.

Kent, J.A.: *Riegel's Handbook of Industrial Chemistry,* Chapman & Hall, New York, NY, 1992.

Lagowski, J.J.: *MacMillan Encyclopedia of Chemistry,* MacMillan Library Reference, New York, NY, 1997.

Lide, D.R.: *CRC Handbook of Chemistry and Physics,* 84th Edition, CRC Press, LLC., Boca Raton, FL, 2003.

Parker, S.P.: *McGraw-Hill Concise Encyclopedia of Science and Technology,* 4th Edition, McGraw-Hill Companies, New York, NY, 1998.

Perry, R.H., D.W. Green, and J.O. Maloney: *Perry's Chemical Engineers' Handbook,* 7th Edition, McGraw-Hill Companies, New York, NY, 1999.

Swisher, C.C., III, and D.R. Prothero: "Single-Crystal 40Ar/39Ar Dating of the Eocene-Oligocene Transition in North America," *Science,* 760 (August, 17, (1990)).

AROMATIC COMPOUND. An organic compound that incorporates a closed-chain or (ring) nucleus in its structure. This is in contrast with the aliphatic compound, which is comprised of an open-chain structure. The classical example of an aromatic compound is benzene. Aromatic compounds also are sometimes referred to as benzenoids. Some ring-type compounds are not classified as aromatic. These include the cycloparaffins and cycloolefins which are considered to be derivatives of methane. See also **Compound (Chemical)**; and **Organic Chemistry**.

ARSENIC. [CAS: 7440-38-2]. Chemical element, symbol As, at. no. 33, at. wt. 74.9216, periodic table group 15, mp 817°C (24 atmospheres), sublimes at 613°C, density 5.72 g/cm^3. One naturally-occurring stable isotope: ^{75}As. Various studies indicate that arsenic exists in several allotropic forms. The metallic form has a steel-gray color in the crystalline form and is brittle. Although the red form of arsenic sulfide As_2S_2 was observed by Aristotle as early as 400 B.C., the first attempt to isolate the metal was not made until 1250 by Albert Magnus. Later documentation on the preparation of the element was given by J. Schroder and N. Lèmery in the 1600s. First ionization potential 9.8 eV; second 18.63 eV; third 28.34 eV; fourth 50.1 eV; fifth 62.5 eV. Oxidation potentials $As \longrightarrow As + 3H^+ + 3e^-$, 0.54 V; $As + 2H_2O \longrightarrow HAsO_2 + 3H^+ + 3e^-$, −0.2475 V; $HAsO_2 + 2H_2O \longrightarrow H_3AsO_4 + 2H^+ + 2e^-$, −0.559 V; $AsO_2^- + 4OH^- \longrightarrow AsO_4^{3-} + 2H_2O + 2e^-$, 0.71 V; $As + 4OH^- \longrightarrow AsO_2^- + 2H_2O + 3e^-$, 0.68 V. Other important physical properties of arsenic are given under **Chemical Elements**.

Gray or metallic arsenic, density 5.73 g/cm^3, which sublimes on heating, and has the vapor composition As_4, becoming As_2 at higher temperatures, is the ordinary variety. On rapid cooling, the vapor condenses to yellow arsenic, density 1.97 g/cm^3, which reverts to the gray variety on warming.

An intermediate form in the transition is black amorphous β arsenic, density 4.6–5.2, also obtained by the thermal decomposition of arsine. Brown arsenic, density 3.7–4.2, obtained by reduction of acid solutions of trivalent arsenic, is probably a finely divided form of black arsenic.

Arsenic sublimes on heating and is unchanged in dry air, but a film of oxide is formed in moist air; heated in air at 180°C, arsenic forms arsenic trioxide of the odor of garlic, poisonous; insoluble in HCl but soluble in concentrated HNO_3 or concentrated H_2SO_4 to form arsenic acid; soluble in hot NaOH solution; heated with chlorine forms arsenic trichloride; heated with metals forms metallic arsenides. When arsenic is heated in a tube and the vapor cooled (1) slowly (that is, in the hot part of the tube) black arsenic is formed, and this form is converted into the gray at 360°C, (2) rapidly (that is, in the cold part of the tube) yellow arsenic is formed, and this form is quickly converted into the gray by the action of light. Yellow arsenic is soluble in CS_2.

Arsenic occurs in nature as the arsenide of iron, cobalt, nickel, and as the mineral sulfides, *realgar* (arsenic monosulfide, AsS), red colored; *orpiment* (arsenic trisulfide, As_2S_3), yellow colored—these two minerals when powdered once were used as paint pigments—*arsenopyrite, mispickel* (iron arsenosulfide, FeAsS); *enargite,* Cu_3AsS_4; and *tennantite,* $Cu_8As_2S_7$.

The primary arsenic-containing material is arsenious oxide obtained by separation from roaster or smelter flue gases. Metallic arsenic is obtained as sublimate by heating the oxide with carbon.

Arsine

[CAS: 7748-42-1]. AsH_3, is formed by hydrolysis of arsenides, or reduction (by zinc and HCl or aluminum and NaOH) of arsenic compounds, as in the Gutzeit test. It reacts with metals at higher temperatures or in solution to give the arsenides. Diarsine, As_2H_4, is produced by reduction of arsenic trichloride, $AsCl_3$, by lithium aluminum hydride, $LiAlH_4$ in ether at −190°C. It melts below −50°C, but begins to decompose into AsH_3 and brown polymeric $(AsH)_x$ about −100°C. It is more stable in the gas phase than in the solid or liquid phases.

Arsenides

These are prepared by fusion from the elements. Their properties vary across the periodic table, those of the alkalies and alkaline earths being readily hydrolyzed by H_2O or acids and are stoichiometric, while the arsenides of the other metals show an increasingly intermetallic character and resist hydrolysis.

Gallium Arsenide

Devices using gallium arsenide are extremely important in contemporary electronic equipment. For example, some of the fastest digital integrated circuits built to date are made of gallium arsenide. They are biphase clock flip-flops configured to perform frequency division. They operate at frequencies up to 5.77 GHz, which is about the highest division speed yet reported for integrated circuits operating at room temperature. These circuits have been fabricated by electron-beam lithography to produce gate lengths of 0.5 micrometer in the MESFET (metal-semiconductor field-effect transistor) switching transistors. Gallium arsenide devices can be used in very-high-frequency signal processing or as interfaces to more complex chips, including VHSIC (very high speed integrated circuits). (*Note*: This and the following two paragraphs were prepared by staff.)

In special tests carried out by several research groups (AT&T Bell Laboratories, IBM, Arizona State University, and Japanese investigators) on gallium arsenide devices, experimental evidence indicates that possibly electrons can travel through a semiconductor without being slowed by collisions, that is, ballistically. As pointed out by Robinson (January 1986), at least in theory, the faster the electrons travel through a transistor, the faster the device can switch on and off. Although, to date, there has been no demonstration of how to take advantage of ballistic transport in a practical transistor, the concept does excite visions of ballistic electrons traveling at nearly the speed of light and providing the basis for transistors that can switch trillions of times per second.

Gallium arsenide and silicon transistors each have their own specific advantages. GaAs transistors switch faster than Si transistors and they also emit near-infrared and visible light, a property of value when both optical and electrical functions are combined in one chip. In many other respects, the GaAs devices are inferior to their silicon counterparts. Researchers have recently found how to effect epitaxial growth of crystalline GaAs layers on silicon wafers and thus combine the properties of both semiconductors, particularly in the manufacture of high-speed microelectronic chips.

The potential of GaAs and other III-V semiconductors are well portrayed by Yablonovitch (see list).

The trifluoride and trichloride, AsF_3 and $AsCl_3$, are liquids at room temperature and the tribromide and triiodide, $AsBr_3$ and AsI_3 are solids, although the former melts at $31°C$. Like the analogous phosphorus compounds, they have pyramidal structures. Their hydrolysis in aqueous solution is not quite complete, consistent with their greater ionic character (than the phosphorus halides), as is the fact that As^{3+} is precipitated from their solutions as the sulfide. The only stable binary pentahalogen compound of arsenic is the pentafluoride, AsF_5, a colorless gas, which like the trihalides, is less readily hydrolyzed than the corresponding phosphorus compound. A very unstable pentachloride, $AsCl_5$, has been reported. The mixed halide AsF_3Cl_2 can be made by passing chlorine into ice-cold arsenic trifluoride.

Arsenic (III) Oxide

As_4O_6, exists as tetraarsenic hexoxide in the solid state and in the vapor to above $800°C$, where dissociation to As_2O_3 commences. It is somewhat soluble in H_2O (about 20 g/l at $25°C$), and its solutions have some acidic properties, although the acid has not been isolated and its formula is probably not $As(OH)_3$, the form used for convenience in writing reactions. It is an amphiprotic substance, since, as stated above, As^{3+} is precipitated by H_2S from acid solutions as the sulfide, while the salts, the arsenites (containing the ion AsO_3^{3-}), are readily formed. Their solubility in H_2O varies across the periodic table, those of the alkali metals being very soluble, those of the alkaline earth metals less so, and those of the heavy metals essentially insoluble. Arsenite ion probably exists as $As(OH)_4^-$ in solution.

Arsenic (V) Oxide

As_4O_{10} is a white solid, decomposes at $315°C$, isomorphous with phosphorus pentoxide, P_2O_5, but not produced by a simple oxidation of As_2O_3. It is made by dehydration of arsenic acid or $As_4O_{10} \cdot 4H_2O$. It hydrates to give arsenic acid, $H_3AsO_4 \cdot \frac{1}{2}H_2O$. This acid is only slightly weaker than phosphoric acid, which it resembles in forming a wide variety of polyacids. It also forms primary, secondary, and tertiary (ortho) arsenates. Raman spectral studies of concentrated arsenic acid solutions in H_2O have a strong band assigned to the $-OH$ group, whence it is inferred that the acid is present in different forms in concentrated and dilute solutions. Many arsenates are converted by ignition into pyro-arsenates, e.g., calcium pyroarsenate, $Ca_2As_2O_7$, and meta-arsenates are also known.

Direct fusion of the elements yields a number of arsenic sulfides, including As_4S_3, As_4S_4, As_2S_3, and As_2S_5, the last two being obtained also by precipitation from arsenic (III) and arsenic (V) solutions, respectively. The trisulfide dissolves in alkali sulfide solutions to form thio-arsenites:

$$As_2S_3 + 3S_2^{2-} \longrightarrow 2AsS_3^{3-} + S$$

while with polysulfides it forms thioarsenates:

$$As_2S_3 + 2S_2^{2-} + S \longrightarrow 2AsS_4^{2-}$$

Organoarsenic Compounds

The largest group of organic arsenic-containing compounds is the arsenic acids $RAsO(OH)_2$, where the R may be alkyl, aryl, or heterocyclic groups and their salts. In addition to specific compounds mentioned under the uses of arsenic, some organo-arsenic compounds include methylarsine CH_3AsH_2; methylarsine tetrachloride CH_3AsCl_4; diphenylarsenic peroxide $(C_6H_5)_2 AsOOAs (C_6H_5)_2$; triphenylarsenic dihydroxide

$$(C_6H_5)_3As(OH)_2;$$

dimethylarsine borane $(CH_3)_2AsHBH_3$; and ethoxydichloroarsine $C_2H_5OAsCl_2$.

Uses

Future production of As_2O_3 will be influenced by the ability to handle ores in compliance with environmental restrictions. As_2O_3 is available in two grades: (1) crude, 95% As_2O_3; and (2) refined arsenic, 99% As_2O_3. Domestic supplies of the United States are supplemented by imports from Sweden, France, and Mexico. Commercial arsenic metal is produced chiefly by the United States and Sweden.

Arsenic trioxide finds major use in the preparation of other compounds, notably those used in agricultural applications. The compounds monosodium methylarsonate, disodium methylarsonate, methane arsenic acid (cacodylic acid) are used for weed control, while arsenic acid, H_3AsO_4, is used as a desiccant for the defoliation of cotton crops. Other compounds once widely used in agriculture are calcium arsenate for control of boll weevils, lead arsenate as a pesticide for fruit crops, and sodium arsenite as a herbicide and for cattle and sheep dip. In some areas, arsenilic acid has been used as a feed additive for swine and poultry. Restrictions on these compounds vary from one country and region to the next.

Refined arsenic trioxide is used both as a fining and decolorizing agent in glass. As_2O_5 and arsenic acid are used in the manufacture of chromated copper arsenate, which is used extensively as a wood preservative.

Indium arsenide, gallium arsenide, and gallium arsenide phosphide find use as semiconductors. For these materials, the starting arsenic source must be extremely pure. Arsenic trichloride and arsenic hydride (very high purity) find application in the production of epitaxial gallium arsenide. Also, in various combinations with iodine, germanium, selenium, sulfur, tellurium, and thallium, arsenic will form a group of glasses with very low melting points.

The applications of arsenic as a metal are quite limited. Metallurgically, it is used mainly as an additive. The addition of from $\frac{1}{2}$ to 2% of arsenic improves the sphericity of lead shot. Arsenic in small quantities improves the properties of lead-base bearing alloys for high-temperature operation. Improvements in hardness of lead-base battery grid metal and cable-sheathing alloys can be obtained by slight additions of arsenic. Very small additions (0.02–0.05%) of arsenic to brass reduce dezincification.

Toxicity

Although metallic arsenic and arsenic trisulfide may be handled, as in the case of most arsenical compounds, skin contact should be avoided. Arsine requires extreme caution in handling because of its very high toxicity. In handling arsenic and its compounds, reducing conditions should be avoided because these may give rise to the undesired formation of arsine. Wherever arsenic and its compounds are present as dusts or vapors, proper ventilation and respirators are mandatory. Good housekeeping and appropriate hygienic practices should also be observed.

Arsenic is commonly found in small amounts in the tissues of plants and animals. A human body may contain as much as 20 mg (As_2O_3). No role in natural biological phenomena has been found for As. Although the element may be present in seawater to the extent of 0.006–0.03 ppm, it may be ten times as high in estuaries. Shellfish tend to accumulate the arsenic from the large amount of seawater with which they come in contact. Oysters may contain 3–10 ppm. However, shellfish of the same species grown in different localities show wide variations in arsenic content, suggesting that it is an accidental constituent, which the organisms learn to tolerate. Its lack of function in the human body is suggested by the fact that it tends to accumulate in the hair and nails, which are essentially nonliving.

As and Gene Amplification

Research by Te-Chang Lee and associated scientists have reported that arsenic acts specifically in the progression phase of carcinogenicity. Findings indicate that arsenic may be related to its ability to cause gene amplification, but not gene mutations. Data collected regarding humans occupationally exposed to arsenic indicates that exposure to As appears to act at a late stage in the carcinogenic process. Thus, the scientists postulate that amplification of an altered or activated oncogene may be a late stage in neoplastic progression. The hypothesis would explain why arsenic is not an effective complete carcinogen, initiator, or tumor promotor. There may be other substances, in addition to As, that act in this manner.

Additional Reading

Blakemore, J.S.: *Gallium Arsenide,* Springer-Verlag, Inc., New York, NY, 1998.

Burch, A.: *Pure Metals Properties, A Scientific and Technical Handbook,* ASM International, Materials Park, OH, 1999.

Calderon, A.: *Arsenic, Exposure and Health Effects,* Vol. 2, Chapman & Hall, New York, NY, 1997.

Carapella, S.C., Jr.: "Properties of Pure Arsenic," in *Metals Handbook,* ASM International, Materials Park, OH. (Published periodically.)

Chang, L.W., T. Suzuki, and L. Magos: *Toxicology of Metals,* Vol. 1, Lewis Publishers, New York, NY, 1998.

Ecobichon, D.J.: *The Basic Toxicity Testing,* 2nd Edition, CRC Press, LLC., Boca Raton, FL, 1997.

Herrmann, W.A. and H.H. Karsch: *Synthetic Methods of Organometallic and Inorganic Chemistry, Phosphorus, Arsenic, Antimony, and Bismuth*, Vol. W, Thieme Medical Publishers, Inc., New York, NY, 1996.

Lee, Te-Chang, et al.: "Induction of Gene Amplification by Arsenic," *Science*, 79 (October 20, 1989).

Lide, D.R.: *CRC Handbook of Chemistry and Physics*, 84th Edition, CRC Press, LLC., Boca Raton, FL, 2003.

Norman, N.C.: *Chemistry of Arsenic, Antimony and Bismuth*, Kluwer Acanemic/Plenum Publishers, New York, NY, 1998.

Perry, R.H., D.W. Green, and J.O. Maloney: *Perry's Chemical Engineers' Handbook*, 7th Edition, McGraw-Hill Companies, New York, NY, 1997.

Robinson, A.L.: "Silicon Solution of Gallium Arsenide IC's," *Science*, **232**, 826–828 (1986).

Yablonovitch, E.: "The Chemistry of Solid-State Electronics," *Science*, 347 (October 20, 1989).

ARSENOPYRITE. The mineral arsenopyrite is a sulfarsenide of iron corresponding to the formula FeAsS. A variety in which some of the iron is replaced by cobalt is known as danaite. It crystallizes in the monoclinic system but twinning produces pseudo-orthorhombic crystals. Its hardness is 5.5–6; sp gr, 6.07; luster, metallic color, silvery-white to steel gray, but usually with a yellow to gray tarnish; streak, black. Arsenopyrite is a common mineral with tin and lead ores and in pegmatites, probably having been deposited by action of both vapors and hydrothermal solutions. It is a widespread mineral, well-known deposits occurring in Austria, Saxony, Switzerland, Sweden, Norway; Cornwall and Devonshire, England; Bolivia; in the United States at Roxbury, Connecticut; Franklin, New Jersey; Paris, Maine; Emery, Montana; and Leadville, Colorado. Danaite was first found in Franconia, New Hampshire, by J.D. Dana, for whom it was later named. Arsenopyrite also is known as *mispickel*, an old German term whose exact derivation is unknown.

ASBESTOS. The term asbestos is a generic designation referring usually to six types of naturally occurring mineral fibers which are or have been commercially exploited. These fibers are extracted from certain varieties of hydrated alkaline silicate minerals comprising two families: serpentines and amphiboles. The serpentine group contains a single fibrous variety: chrysotile; five fibrous forms of amphiboles are known: anthophyllite, amosite, crocidolite, tremolite, and actinolite.

These fibrous minerals share several properties which qualify them as asbestiform fibers: (*1*) they are found in large clusters which can be easily separated from the host matrix or cleaved into thinner fibers; (*2*) the fibers exhibit high tensile strengths; (*3*) they show high length:diameter ratios, from a minimum of 20 up to >1000; (*4*) they are sufficiently flexible to be spun; and (*5*) macroscopically, they resemble organic fibers such as cellulose.

Since asbestos fibers are all silicates, they exhibit several other common properties, such as incombustibility, thermal stability, resistance to biodegradation, chemical inertia toward most chemicals, and low electrical conductivity.

The usual definition of asbestos fiber excludes numerous other fibrous minerals which could be qualified as asbestiform following the criteria listed above. However, it appears the term asbestos has traditionally been attributed only to those varieties which are commercially exploited.

The fractional breakdown of the recent world production of the various fiber types shows that the industrial applications of asbestos fibers have now shifted almost exclusively to chrysotile. Two types of amphiboles, commonly designated as amosite and crocidolite are still being used, but their combined production is currently less than 2% of the total world production. The other three amphibole varieties, anthophyllite, actinolite, and tremolite, have no significant industrial applications presently. This statement excludes asbestiform amphiboles which may occur in other industrial minerals.

Geology and Fiber Morphology

The genesis of asbestos fibers as mineral deposits required certain conditions with regard to chemical composition, nucleation, and fiber growth; such conditions must have prevailed over a period sufficiently long and perturbation-free to allow a continuous growth of the silicate chains into fibrous structures. Some of the important geological or mineralogical features of the industrially significant asbestos fibers are summarized in Table 1.

Crystal Structure of Asbestos Fibers

The microscopic and macroscopic properties of asbestos fibers stem from their intrinsic, and sometimes unique, crystalline features. As with all silicate minerals, the basic building blocks of asbestos fibers are the silicate tetrahedra which may occur as double chains $(SiO)^{6-}114$, as in the amphiboles, or in sheets $(SiO)^{4-}104$, as in chrysotile.

Chrysotile. In the case of chrysotile, an octahedral brucite layer having the formula $(MgO_4(OH)_8)^{4-}6$ is intercalated between each silicate tetrahedra sheet.

Amphiboles. The crystalline structure common to amphibole minerals consists of two ribbons of silicate tetrahedra placed back to back.

Properties of Asbestos Fibers

Asbestos fibers used in most industrial applications consist of aggregates of smaller units (fibrils). This is most evident with chrysotile which exhibits an inherent, well-defined unit fiber. Typical diameters of fibers in bulk industrial samples may reach several tens of micrometers; fiber lengths are on the order of one to ten millimeters.

The mechanical processes employed to extract the fibers from the host matrix, or to further separate (defiberize, open) the aggregates, can impart significant morphological alterations to the resulting fibers. Typically, microscopic observations on mechanically opened fibers reveal fiber bends and kinks, partial separation of aggregates, fiber end-splitting, etc. The resulting product thus exhibits a wide variety of morphological features. The consequences of the peculiar morphology of fiber shapes are difficult to assess, but it is quite obvious that a proper dimensional characterization of these fibers requires a shape factor in addition to diameter and length.

The morphological variance appears more important with chrysotile than with amphiboles. The intrinsic structure of chrysotile, its higher flexibility, and interfibril adhesion allow a variety of intermediate shapes when fiber aggregates are subjected to mechanical shear. Amphibole fibers are generally more brittle and accommodate less morphological deformation during mechanical treatment.

Fiber Length Distribution. For industrial applications, the fiber length and length distribution are of primary importance because they are closely related to the performance of the fibers in matrix reinforcement. Representative distributions of fiber lengths and diameters can be obtained through measurement and statistical analysis of microphotographs; fiber length distributions have also been obtained recently from automated optical analyzers.

Physico-Chemical Properties. The industrial applications of chrysotile fibers were developed taking advantage of their particular combination of properties: fibrous morphology, high tensile strength, resistance to heat and corrosion, low electrical conductivity, and high friction coefficient. In many applications, the surface properties of the fibers also play an important role; in such cases, a distinction between chrysotile and amphiboles can be

TABLE 1. GEOLOGICAL OCCURRENCE OF ASBESTOS FIBERS

	Chrysotile	Amosite	Crocidolite	Tremolite
mineral species	chrysotile	cummingtonite-grunerite	riebeckite	tremolite
structure	as veins in serpentine	lamellar, coarse to fine, fibrous and asbestiform	fibrous in ironstones	long, prismatic, and fibrous aggregates
origin	alteration and metamorphism of basic igneous rocks rich in magnesium silicates	metamorphic	regional metamorphism	metamorphic
essential composition	hydrous silicates of magnesia	hydroxy silicate of Fe and Mg	hydroxy silicate of Na, Mg, and Fe	hydroxy silicate of Ca and Mg

TABLE 2. PHYSICAL AND CHEMICAL PROPERTIES OF ASBESTOS FIBERS

Property	Chrysotile	Amosite	Crocidolite	Tremolite
color	usually white to grayish green; may have tan coloration	yellowish gray to dark brown	cobalt blue to lavender blue	gray-white, green, yellow, blue
luster	silky	vitreous to pearly	silky to dull	silky
hardness, Mohs	2.5–4.0	5.5–6.0	4.0	5.5
specific gravity	2.4–2.6	3.1–3.25	3.2–3.3	2.9–3.2
optical properties	biaxial positive parallel extinction	biaxial positive parallel extinction	biaxial oblique extinction	biaxial negative oblique extinction
refractive index	1.53–1.56	1.63–1.73	1.65–1.72	1.60–1.64
flexibility	high	fair	fair to good	poor, generally brittle
texture	silky, soft to harsh	coarse but somewhat pliable	soft to harsh	generally harsh
spinnability	very good	fair	fair	poor
tensile strength, MPa[a]	1100–4400	1500–2600	1400–4600	<500
resistance to:				
acids	weak, undergoes fairly rapid attack	fair, slowly attacked	good	good
alkalies	very good	good	good	good
surface charge, mV (zeta potential)	+13.6 to +54[b]	−20 to −40	−32	
decomposition temperature, °C	600–850	600–900	400–900	950–1040
residual products	forsterite, silica, eventually enstatite	Fe and Mg pyroxenes, magnetite, haematite, silica	Na and Fe pyroxenes, haematite, silica	Ca, Mg, and Fe pyroxenes, silica

[a] To convert MPa to psi, multiply by 145.
[b] Chrysotile fibers tend to become negative after weathering and/or leaching.

observed because of their differences in chemical composition and surface microstructure. Technologically relevant physical and chemical properties of asbestos fibers are given in Table 2.

Analytical Methods and Identification of Asbestos Fibers

In a general way, the identification of asbestos fibers can be performed through morphological examination, together with specific analytical methods to obtain the mineral composition and/or structure. Morphological characterization in itself usually does not constitute a reliable identification criterion. Hence, microscopic examination methods and other analytical approaches are usually combined.

Fiber Classification and Standard Testing Methods

In the production, or industrial applications, of asbestos fibers, several parameters are considered critically important and are used as standard evaluation criteria: length (or length distribution), degree of opening and surface area, performance in cement reinforcement, and dust and granule content. The measurement of fiber length is important since the length determines the product category in which the fibers will be used and, to a large extent, their commercial value.

Dry Classification Method. The most widely accepted method for chrysotile fiber length characterization in the industry is the Quebec Standard test (QS).

Wet Classification Method. A second industrially important fiber-length evaluation technique is the Bauer-McNett (BMN) classification.

Other classification techniques have been developed which provide some insight on fiber lengths, typically the Ro-Tap test, the Suter-Webb Comb, and the Wash test.

Industrial Applications

Asbestos fibers have been used in a broad variety of industrial applications. In the peak period of asbestos consumption in industrialized countries, some 3000 applications, or types of products, have been listed. Because of recent restrictions, many of these applications have now been abandoned and others are pursued under strictly regulated conditions.

The main characteristic properties of asbestos fibers that can be exploited in industrial applications are their thermal, electrical, and sound insulation; nonflammability; matrix reinforcement (cement, plastic, and resins); adsorption capacity (filtration, liquid sterilization); wear and friction properties (friction materials); and chemical inertia (except in acids). These properties have led to several main classes of industrial products or applications: fire protection and heat or sound insulation,

fabrication of papers and felts for flooring and roofing products, pipeline wrapping, electrical insulation, thermal and electrical insulation, friction products in brake or clutch pads, asbestos–cement products, reinforcement of plastics, fabrication of packings and gaskets, friction materials for brake linings and pads, reinforcing agents, vinyl or asphalt tiles, and asphalt road surfacing.

Alternative Industrial Fibers and Materials

Table 3 lists some of the materials and fibers that have been suggested or used in the development of asbestos-free products.

Health and Safety

The relationship between workplace exposure to airborne asbestos fibers and respiratory diseases is one of the most widely studied subjects of modern epidemiology.

TABLE 3. ASBESTOS SUBSTITUTES AND RELATIVE COSTS[a]

Minerals	Synthetic mineral fibers	Synthetic organic fibers
	<2 $/kg[b]	
attapulgite	mineral wool	
diatomite	glass wool	
mica		
perlite		
sepiolite		
talc		
vermiculite		
wollastonite		
asbestos, grades 3–7	2–10 $/kg	
	steel fibers	polypropylene (PP)
	continuous filament glass	poly(vinyl alcohol) (PVA)
	alkali-resistant glass	polyacrylonitrile (PAN)
	aluminosilicates	
	10–20 $/kg	
	continuous filament glass	polytetrafluoroethylene (PTFE)
	<20 $/kg	
	alumina fibers	polybenzimidazole (PBI)
	silica fibers	aramid fibers
		pitch and PAN carbon fibers

[a] In U.S. $, 1989.
[b] The natural organic fiber, cellulose (pulp), also falls in the <$2/kg range.

The research efforts resulted in significant consensus in some areas, although strong controversies remain in other areas. Typically, it is widely recognized that the inhalation of long (considered usually as $> 5\ \mu m$), thin, and durable fibers can induce or promote lung cancer. It is also widely accepted that asbestos fibers can be associated with three types of diseases: asbestosis: a lung fibrosis resulting from long-term, high level exposures to airborne fibers; lung cancer: usually resulting from high level exposures and often correlated with asbestosis; mesothelioma: a rare form of cancer of the lining of the thoracic and abdominal cavities (mesothelium).

A further consensus developed within the scientific community regarding the relative carcinogenicity of the different types of asbestos fibers. There is strong evidence that the genotoxic and carcinogenic potentials of asbestos fibers are not identical; in particular, mesothelial cancer is mostly, if not exclusively, associated with amphibole fibers.

Regulation. The identification of health risks associated with asbestos fibers, together with the fact that huge quantities of these minerals were used ($\approx 5 \times 10^6$ t/yr) in a variety of applications, has prompted strict regulations to limit the maximum exposure of air-borne fibers in workplace environments.

See also http://www.epa.gov/opptintr/asbestos/index.htm

CARMEL R. JOLICOEUR
JEAN-FRANCOIS ALARY
ASTA SOKOV
Université de Sherbrooke, Canada

Additional Reading

Beard, M.E. and H.L. Rook: *Advancements in Environmental Measurement Methods for Asbestos,* American Society for Testing & Materials, West Conshohocken, PA, 2000. http://www.astm.org/

Hodgson, A.A. in L. Michaels and S.S. Chissick, eds.: *Asbestos: Properties, Applications and Hazards,* Vol. 1, John Wiley & Sons, Inc., New York, NY, 1979; A.A. Hodgson, *Scientific Advances in Asbestos, 1967 to 1985,* Anjalena Publication, Crowthorne, UK, 1986.

Hodgson, A.A. ed.: *Alternatives to Asbestos, The Pros and Cons,* John Wiley & Sons, Inc., New York, NY, 1989, p. xi.

Holden, C.: "Asbestos Regulations to be Re-Examined," *Science,* 1639 (March 27, 1992).

International Symposium on Man-Made Mineral Fibers in the Working Environment, Copenhagen, Denmark, Oct. 28–29, 1986; *Ann. Occup. Hyg.* **31,** 4B (1987).

Mossman, B.T., et al.: "Asbestos: Scientific Developments and Implications for Public Policy," *Science,* 294 (January 19, 1990).

Oberta, A.F.: *Manual on Asbestos Control: Removal, Management and the Visual Inspection Process,* American Society for Testing & Materials, West Conshohocken, PA, 1995.

Rom, W.N. and A. Upton: "Asbestos-Related Diseases," *N. Eng. J. Med.,* 129 (January 11, 1991).

Ross, M. R.A. Kuntze, and R.A. Clifton: in B. Levadie, ed., *Definition for Asbestos and Other Health Related Silicates,* ASTM STP 834, American Society for Testing and Materials, Philadelphia, PA, 1984, pp. 139–147; W. J. Campbell and co-workers, *Selected Silicates Minerals and Their Asbestiform Varieties,* IC 8751, U.S. Bureau of Mines, Washington, DC, 1977, pp. 5–17, 33.

Skinner, H.C.W. M. Ross, and C. Frondel: *Asbestos and Other Fibrous Materials,* Oxford Press, New York, 1988, pp. 21–23, 25, 31, 34, 35.

Stone, R.: "No Meeting of the Minds on Asbestos," *Science,* 928 (November 15, 1991).

Stone, R.: "Fiber Flap: Refractory Ceramic Fibers," *Science,* 1356 (March 13, 1992).

Tweedale, G., P. Hansen, and T. Newall: *Magic Mineral to Killer Dust,* Oxford University Press, Inc., New York, NY, 2000.

ASCORBIC ACID (Vitamin C).

[CAS: 50-81-7]. Infrequently referred to as the anti-scorbutic vitamin and earlier called cevitamic acid or hexuronic acid, the present terms, *ascorbic acid* and *vitamin C,* are synonymous. Ascorbic acid was one of the first (if not the first) nutrients to be associated with a major disease. Lind first described *scurvy* in 1757. However, this vitamin C deficiency disease had been recognized by Hippocrates in the 13th century and was a curse during the time of the Crusaders. In time of war, the disease killed untold numbers in armies, navies, and besieged towns. During the early days of the sailing ships, often requiring months between port calls accompanied by a lack of fresh food for long periods, the disease affected the crew as a plague. Scurvy was of some importance as recently as World War II. Currently, the disease is of prime concern in pediatrics. It is rarely seen in breast-fed children, but pasteurization of cow's milk degrades the vitamin and an addition to the diet of ascorbic acid must be provided for infants under 1 year of age.

Scurvy was first produced experimentally by Holst and Frolich in 1907. About 6 months were required to produce scurvy experimentally, as individual susceptibility and the quantity of vitamin C previously stored in the body affects the onset of scurvy. The earliest sign of scurvy is usually a sallow or muddy complexion, a feeling of listlessness, general weakness, and mental depression. Soon the bones are affected and increasing pain and tenderness develop. Teeth easily decay and become loose and often fall out, while the gums bleed easily and are sore. Changes in the blood vessels occur, producing hemorrhages in different parts of the body. In infants, irritability, loss of appetite, fever, and anemia also occur. An infant between 6 and 12 months of age, who has not had sufficient intake of vitamin C (as from fruit juices, supplements, etc.) may show abnormal irritability and tenderness and pain in the legs, often accompanied by pain and swelling of joints (elbows and knees). Immediate administration of vitamin C is indicated in such cases.

In 1928, Zilva first described antiscorbutic agents in lemon juice, although the importance of fresh fruit or vegetables for preventing scurvy had been established a century or more earlier. Also in 1928, Szent-Györgyi isolated hexuronic acid (vitamin C) from lemon juice. In 1932, Waugh and King identified hexuronic acid as an antiscorbutic agent. Haworth, in 1933, established the configuration of hexuronic acid and, in that same year, Reichstein first synthesized hexuronic acid. Later in that year, Haworth and Szent-Györgyi changed the name of hexuronic acid to ascorbic acid.

In 1950, King et al., by the use of glucose labeled with radiocarbon in known positions, traced glucose through intermediate steps in the formation of ascorbic acid in plant and animal tissues; then, by using ascorbic acid with radiocarbon-labeled positions, it was possible to determine with considerable accuracy the metabolic distribution, storage, and chemical changes characteristic of the vitamin molecule. That experimentation made it clear that the carbon atoms in glucose or galactose all retain their original positions, along the carbon chain in the vitamin when it is formed biologically. No rupture or replacement in the chain during conversion was noted. It was also found that the synthesis can be considerably enhanced by feeding livestock small amounts of *chloretone* or any of a score or more of organic compounds. Reactions just described are indicated below.

Biological Role of Ascorbic Acid

Apparently all forms of life, both plant and animal, with the possible exception of simple forms, such as bacteria that have not been studied thoroughly, either synthesize the vitamin from other nutrients or require it as a nutrient. Dormant seeds contain no measurable quantity of the vitamin, but after a few hours of soaking in water, the vitamin is formed.

Ascorbic acid is easily oxidized to dehydroascorbic acid. The latter is less stable than ascorbic acid and tends to yield products, such as oxalate, threonic acid, and carbon dioxide. When administered to animals or consumed in foods, dehydroascorbic acid has nearly the same antiscorbutic activity as ascorbic acid, and it can be quantitatively reduced to ascorbic acid.

In its biochemical functions, ascorbic acid acts as a regulator in tissue respiration and tends to serve as an antioxidant in vitro by reducing oxidizing chemicals. The effectiveness of ascorbic acid as an antioxidant when added to various processed food products, such as meats, is described in entry on **Antioxidants**. In plant tissues, the related glutathione system of oxidation and reduction is fairly widely distributed and there is evidence that electron transfer reactions involving ascorbic acid are characteristic of animal systems. Peroxidase systems also may involve reactions with ascorbic acid. In plants, either of two copper-protein enzymes are commonly involved in the oxidation of ascorbic acid.

In animal tissues, it is easily demonstrated that, as the vitamin content of tissues is depleted, many enzyme systems in the body are decreased in activity. Full explanation of these decreased activities still requires further research. In the total animal and in isolated tissues from animals with scurvy, there is an accelerated rate of oxygen consumption even though the animal becomes very weak in mechanical strength and many physiologic functions are disorganized. With the onset of scurvy, the most conspicuous tissue change is the failure to maintain normal collagen. Sugar tolerance is decreased and lipid metabolism is altered. There is also marked structural disorganization in the odontoblast cells in the teeth and in bone-forming cells in skeletal structures. In parallel with the foregoing changes, there is a decrease in many hydroxylation reactions. The hydroxylation of organic compounds is one of the most characteristic features disturbed by a vitamin C deficiency. These reactions relate to the vitamin's regulation of respiration, hormone formations, and control of collagen structure.

A partial list of physiological functions that have been determined to be affected by vitamin C deficiencies includes: (1) absorption of iron; (2) cold tolerance, maintenance of adrenal cortex; (3) antioxidant; (4) metabolism of tryptophan, phenylalanine, and tyrosine; (5) body growth; (6) wound healing; (7) synthesis of polysaccharides and collagen; (8) formation of cartilage, dentine, bone, and teeth; and (9) maintenance of capillaries.

Requirements

Species known to require exogenous sources of ascorbic acid include the primates, guinea pig, Indian fruit bat, red vented bulbul, trypanosomes, and yeast. Species capable of endogenous sources include the remainder of vertebrates, invertebrates, plants, and some molds and bacteria. The estimation of requirements of vitamin C by humans has been approached in several ways: (1) direct observation in human studies; (2) analogy to experimentation with guinea pigs; (3) analogy to experimental studies in monkeys and other primates; and (4) analogy to animals, such as the albino rat, that normally synthesize the vitamin in accordance with physiological need. It is relatively easy to maintain intakes at recommended levels by use of mixed practical dietaries that include nominal quantities of fresh, canned, or frozen vegetables or fruits. Generally, ascorbic acid is considered as nontoxic to humans. Possible exceptions include kidney stones (in gouty individuals); inhibitory in excess doses on cellular level (mitosis inhibition) possible damage to beta-cells of pancreas and decreased insulin production by dehydroxyascorbic acid.

Distribution and Sources

Natural sources of vitamin C include the following:

> *High ascorbic acid content (100–300 milligrams/100 grams)*
> Broccoli, Brussels sprouts, collards, currant (black), guava, horse-radish, kale, parsley, pepper (sweet), rose hips, turnip greens, walnut (green English)
> *Medium ascorbic acid content (50–100 milligrams/100 grams)*
> Beet greens, cabbage, cauliflower, chives, kohlrabi, lemon, mustard, orange, papaya, spinach, strawberry, watercress
> *Low ascorbic acid content (25–50 milligrams/100 grams)*
> Asparagus, bean (lima), cantaloupe, chard, cowpea, currant (red and white), dandelion greens, fennel, grapefruit, kumquat, lime, loganberry, mango, melon (honeydew), mint, okra, onion (spring), passion fruit, potato, radish, raspberry, rutabaga, soybean, spring greens, squash (summer), tangerine, tomato, turnip

For the species where the ascorbic acid is synthesized endogenously, the precursors include *d*-mannose, *d*-fructose, glycerol, sucrose, *d*-glucose, and *d*-galactose. Intermediates include uridine diphosphate glucose, *d*-glucuronic acid, gulon acid, *l*-gulonolactone, (Mn^{2+} cofactor). Production sites in animals are the kidney and liver in most instances. In rats, it is the intestinal bacterial supply. In plants, the production sites are found in green leaves and fruit skins. Cell sites include microsomes, mitochondria, and golgi.

Supplements

Commercially available ascorbic acid still includes isolation from natural sources, such as rose hips, but large-scale production will involve the microbiological approach, i.e., *Acetobacter suboxidans* oxidative fermentation of calcium *d*-gluconate; or the chemical approach, i.e., the oxidation of *l*-sorbose.

Bioavailability of Ascorbic Acid

The general causes of reduced availability of vitamin C include damage to adrenal cortex, presence of antagonists, and food preparation practices (oxidation, storage, leaching, cooking). Excepting the use of supplements, the almost universal requirement for fresh foods as a source of vitamin C is readily explained by the sensitivity of the vitamin to destruction by reaction with oxygen. This is accelerated by the presence of minute quantities of enzymes that occur in most living tissues, in which copper or iron is combined with a protein to form a catalyst for the oxidation reaction. Other chemicals, such as quinones or high-valence salts of manganese, chromium, and iodine can also oxidize the vitamin readily in aqueous solutions. Most of these reactions increase rapidly in proportion to exposure to air and rising temperature. In the dry crystalline state, however, and in many dried plant tissues, particularly if acidic in reaction, the vitamin is quite stable at room temperature over a period of several months.

Freshly cut oranges or their juices may be exposed in an open glass for several hours without appreciable loss of the vitamin because of the protective effect of the acids present and the practical absence of enzymes that catalyze its destruction. In potatoes, when baked or boiled, there is a slight loss of the vitamin, but if they are whipped up with air while hot, as in the production of mashed potatoes, a large fraction of the initial vitamin content usually will be lost. In freezing foods, it is common practice to dip them in boiling water or to treat them briefly with steam to inactivate enzymes, after which they are frozen and stored at very low temperatures. In this state, the vitamin is reasonably stable. Vitamin C degradation in dehydrated food systems is described shortly.

Factors which increase the bioavailability of ascorbic acid include the presence of antioxidants and synergists in the diet.

Numerous studies have been conducted concerning vitamin C degradation during food processing, including dehydrated food systems. In the latter, the degradation is dependent upon water activity, moisture content, and storage temperature (for example, in containers with no headspace). Ascorbic acid destruction is dramatically increased in the presence of oxygen. See Structure (**1**).

Additional Reading

Considine, D.M. and G.D. Considine: *Foods and Food Production Encyclopedia*, Van Nostrand Reinhold, Company, Inc., New York, NY, 1982.

Harris, J.R.: *Ascorbic Acid, Biochemistry and Biomedical Cell Biology*, Kluwer Academic/Plenum Publishers, New York, NY, 1996.

King, C.G.: "Ascorbic Acid (Vitamin C) and Scurvy," in (R.J. Williams and E.M. Lansford, Jr., Editors), *The Encyclopedia of Biochemistry*, Van Nostrand Reinhold Company, Inc., New York, NY, 1967.

Sies, H., J. Bug, E. Grossi, A. Poli, and R. Paoletti: "Vitamin C," *The State of the Art Disease Prevention Sixty Years after the Nobel Prize*, Springer-Verlag, Inc., New York, NY, 1999.

D-Glucose   D-Glucuronic acid lactone   L-Gulonic acid lactone   L-Ascorbic acid   L-Dehydro ascorbic acid

(1)

ASH. (1) In analytical chemistry the residue remaining after complete combustion of a material. It consists of mineral matter (silica, alumina, iron oxide, etc.), the amount often being a specification requirement. (2) The end product of large-scale coal combustion as in power plants; now said to be the sixth most plentiful mineral in the U.S. It consists principally of fly ash, bottom ash, and boiler ash. Some of its values are recoverable, and there are a number of industrial uses of fly ash, e.g., in cement products and road fill.

ASPARAGINE. See **Amino Acids**.

ASPARTIC ACID. See **Amino Acids**.

ASPHALT (or Asphaltuim). [CAS: 8052-42-4]. A semisolid mixture of several hydrocarbons, probably formed because of the evaporation of the lighter and more volatile constituents. It is amorphous, of low specific gravity, 1–2, with a black or brownish-black color and pitchy luster. Notable localities for asphaltum are the Island of Trinidad and the Dead Sea region, where Lake Asphaltites were long known to the ancients. See also **Coal Tar and Derivatives**; and **Petroleum**.

ASPIRIN. [CAS: 50-78-2]. A drug used for nearly a century to relieve headaches and general aches and pains and to reduce the swelling and pain associated with joints (gout, ague, rheumatoid arthritis). In recent years, attention to aspirin for its apparent role in reducing heart attacks (coronary thrombosis) and strokes has increased. Trial studies also are underway for its use in reducing the risk of fatal colon cancer.

As early as 1763, the Rev. Edward Stone of Chipping-Norton, Oxfordshire (England), reported to the Royal Society that the bark of the willow tree (*Salix alba*) was found to be effective by his local constituents for treating ague. He reported his findings to the Royal Society, creating much interest. Medical historians also report that Hippocrates as well as some North American Indian tribes were aware of the analgesic effects of the bark of certain trees. Ways were sought to prepare what chemists at the time referred to as *salicin* by extracting the active ingredient from the willow bark. To produce very small amounts of salicin required several pounds of bark, causing the price to be quite high. However, with further efforts, the extraction process was improved, lowering the price. Ague and gout occurred widely, and the extract market became quite large—sufficiently large to interest early European pharmaceutical firms to find a way to synthesize the product. The Germans, who during the mid-1800s excelled in organic chemistry due to their synthesizing important dye chemicals, finally found a way to synthesize salicylic acid. Ironically, however, chemists in France were first to name the product, *l'acide salicylique* (salicylic acid).

It was found that salicylic acid in its pure form had a number of deficiencies, and for a number of years chemists sought a salicylic acid–based compound that would be effective yet less harsh and that could counteract pain with smaller dosages. This process ended in 1898 with the introduction by Bayer of acetylsalicylic acid, which has the formula $C_6H_4(COOH)CO_2CH_3$ and since then has been commonly referred to as *aspirin*. See also **Acetylsalicylic Acid**; and **Salicylic Acid and Related Compounds**.

The market for aspirin grew at a rapid rate, with sales in the United States reaching $2 billion/year in 1990. This represents 1600 tons of the drug, or 80 million tablets. Within recent years, some aspirin has been formulated with other materials. These include buffers for reducing stomach irritation experienced by some people who consume aspirin. Also within the last decade or so, other nonsteroidal anti-inflammatory drugs (NSAIDs) have been introduced into this highly competitive marketplace.

In recent years aspirin has been subjected to some negative publicity. In a minority of the population, aspirin can induce hypersensitivity syndrome. Obviously, persons who exhibit an allergic reaction to the drug are not candidates for it and should turn to other NSAIDs. Needless to say, aspirin must be used with moderation. A most unusual situation was reported by Thibault (1992) in one of the publications listed. A middle-aged man, who had a psychiatric history, complained at an emergency center of nausea, vomiting, shortness of breath, and hallucinatory hearing. It was learned that the patient had consumed four aspirin tablets every 2 to 4 hours for a period of 2 weeks. Obviously, this defines *extreme immoderation* in terms of the drug's use. However, with treatment, the patient's symptoms disappeared within 2 days. Fortunately, such findings of the abuse of aspirin are rare, but the case does emphasize, in perspective, the relative safety of the drug when properly administered.

Probably the most negative situation involving aspirin arose about a decade ago, in connection with the appearance of Reye's syndrome. The biochemistry of this connection has not been fully elucidated, but based upon clinical findings, the general medical community stipulates that aspirin not be used (particularly with young children) where there any symptoms or suspicions that influenza may be present. Also, as a result of this incident, aspirin was removed from the World Health Organization's list of essential drugs, representing a decision that was not universally accepted by medical professionals.

Biochemistry of Aspirin

The biochemical paths and actions by which aspirin and other salicylates achieve their therapeutic effects were poorly understood until at least a partial mechanism was proposed by Sir John Vane in 1971. Vane, who later received a Nobel Prize for his efforts (1982), found that NSAIDs, including aspirin, block the production of prostaglandins by cells and tissues. During the same time frame, Vane and other researchers also confirmed the inhibitory effects of aspirin on platelet aggregation, this caused by interference with the ability of platelets to synthesize prostaglandins, notably thromboxane A_2. The complexities of the topic go well beyond the scope of this volume, but are well ventilated in the Vane (1971), the Smith-Willis (1971), and the Weissmann (1991) articles listed. See also **Prostaglandins**.

Much current research relating to aspirin and heart attacks and strokes is going forward, principally in the form of trial study groups, with emphasis on the effects of dosage. The findings of aspirin's advantage in connection with fatal colon cancer are in their early and debatable study phases.

See also **Analgesics, Antipyretics, and Antiinflammatory Agents**.

Additional Reading

Abramson, S., et al.: "Modes of Action of Aspirin-Like Drugs," *Proceedings, National Academy of Sciences* **82**(21) 7227 (November 1985).

Bashein, G., et al.: "Preoperative Aspirin Therapy and Reoperation for Bleeding After Coronary Artery Bypass Surgery," *Arch. Intern. Med.,* **114**, 835–9 (1991).

Dutch TIA Trial Study Group: "A Comparison of Two Doses of Aspirin (30 mg vs. 283 mg a day) in Patients After a Transient Ischemic Attack or Minor Ischemic Stroke," *N. Eng. J. Med.,* 1261 (May 7, 1992).

Ferreira, S.H. and J.R. Vane: *Annual Review of Pharmacology,* Vol. 14, 57 (1974).

Mills, J.D.: "Aspirin, The Ageless Remedy?" *N. Eng. J. Med.,* 1303 (October 31, 1991).

Pederson, A.K. and G.A. FitzGerald: "Dose-Related Kinetics of Aspirin: Presystemic Acetylation of Platelet Cyclooxygenase," *N. Eng. J. Med.,* 1206 (November 6, 1984).

Smith, J.B. and A.L. Willis: "Aspirin Selectivity Inhibits Prostaglandin Production in Human Platelets," *Nature-New Biology,* **231**, 235 (1971).

Thibault, G.E.: "The Landlady Confirms the Diagnosis (Aspirin Overdose)," *N. Eng. J. Med.,* 1272 (May 7, 1992).

Vane, J.R.: "Inhibition of Prostaglandin Synthesis As a Mechanism of Action for Aspirin-Like Drugs," *Nature-New Biology,* **231**(25), 232 (June 23, 1971).

Weissmann, G.: "Aspirin," *Sci. Amer.,* 84 (January 1991).

ASSOCIATION (Chemical). The combination of molecules of the same substance to form larger aggregates consisting of two or more molecules. See also **Elastomers**; and **Molecule**.

Association was first thought of as a reversible reaction between like molecules that distinguished it from polymerization, which is not reversible. Association is characterized by reversibility or ease of disassociation, low energy of formation (usually about 5 and not more than 10 kcal per mole), and the coordinate covalent bond which Lewis called the acid-base bond. Association takes place between like and unlike species. The most common type of this phenomenon is hydrogen bonding. Association of like species is demonstrable by one or more of the several molecular weight methods. Association between unlike species is demonstrable by deviation of the system from Raoult's law.

The strength of the coordinate covalent bond is a function of polarity of the associating molecules. Hence, associated molecules vary in stability from very unstable to very stable. The argon-boron trifluoride complex is quite unstable, whereas calcium sulfate dihydrate (gypsum) is very stable. The bond strength associated with stability has been measured for a number of combinations. The strengths of some hydrogen bond types decrease in the order FHF, OHO, OHN, NHN, CHO, but they are dependent upon the geometry of the combination and upon the acid-base characteristics of the group. Steric effects can have a marked effect on the strength of the coordinate covalent bond. This was demonstrated by a stude strength of the series NH_3, $C_2H_5 NH_2$, $(C_2H_5)_2NH$, $(C_2H_5)_3N$ as bases toward an acid in solution and in the gaseous state and the comparison of the base strength of triethylamine and quinuclidine. The latter is, in effect, triethylamine in which the two-carbon atoms of each ethyl group are tied together by another carbon. The geometry of the ethyls around the nitrogen is drastically changed, and the cyclic is a stronger base than the triethyl compound. The factors affecting the strength of the hydrogen bond also influence the degree of association.

Association within the same species accounts for the high boiling points of water, ammonia, hydrogen fluoride, alcohols, amines, and amides. Ethyl ether and butanol contain the same number of atoms of each element, but butanol has a boiling point of 83°C above that of ethyl ether as a result of more extensive hydrogen bonding. Some substances associate completely to two or more formula weights per molecule. Carboxylic acid, by a hydrogen-to-oxygen association, form dimers with a six-membered ring. N-unsaturated amides dimerize in the same manner, whereas N-substituted amides dimerize in a chain form in a *trans* configuration.

Hydrogen bonding is so common that coordinate bonds between other elements are sometimes overlooked. Antimony(III) halides form very few complexes with other halides, whereas aluminum halides readily form complexes. The octet of electrons is complete in all atoms of the antimony halides, but is incomplete in the aluminum atom of aluminum halides:

(a)   (b)   (c)

Aluminum can accept two electrons to complete its octet. The pair of electrons is available from the halogen. An alkali halide can supply the electrons and form a complex (c), or the electron pair may come from the halogen of another aluminum chloride. Association with other aluminum halides accounts for the higher melting point of aluminum halides over antimony(III) halides which have a formula weight of 95 or more. The association of aluminum sulfate, alkali metal sulfate, and water to form the stable alums is one of the more complex examples.

The formation of solvates is association between unlike species. Solvation is more frequent between substances of high polarity than those of low polarity. This is illustrated by the decrease in the tendency to form solvates with decrease in dipole moment and dielectric constant (shown in parentheses) for N-methylacetamide (3.59; 172), to water (1.84; 78.4), to ethanol (1.70; 24.6); to ammonia (1.48; 78.4); to ethanol (1.70; 24.6); to ammonia (1.48; 17.8); to methylcyclohexane (0; 2.02) for which few associations are known.

J. A. Riddick
Baton Rouge, Louisiana

ASSOCIATION OF OFFICIAL ANALYTICAL CHEMISTS (AOAC). Formerly called the Association of Official Agricultural Chemists. It was founded in 1884. There are 3000 members. These scientists develop, test and study methods for analyzing fertilizer, foods, feeds, drugs, pesticides, cosmetics, and other products related to agricultural and public health. It is located at 1111 N. 19th St., Ste. 210, Arlington, VA, 22209. http://www.aoac.org/

ASTATINE. [CAS: 7440-68-8]. Chemical element, symbol At, at. no. 85, at. wt. 210 (mass number of the most stable isotope), periodic table group 17, classed in the periodic system as a halogen, mp 302°C, bp 337°C. All isotopes are radioactive. This element occurs in nature only in minute amounts, as a result of minor branching in the naturally occurring alpha decay series: ^{218}At($t_{1/2}$ = ca. 2 sec); it is produced to the extent of 0.03% by the beta decay of ^{218}Po(radium A), 99.97% going by alpha decay to ^{214}Pb(RaB); ^{216}At($t_{1/2} = 3 \times 10^{-4}$ sec.) and 0.013% by beta decay from ^{216}Po(thorium A);

$$^{215}At(t_{1/2} = 0.018 \text{ sec.})$$

0.0005% by beta decay from ^{215}Po (actinium A). Astatine-217 ($t_{1/2}$ = 0.020 sec.) is a principal member of the neptunium ($4n + 1$) series, all members of which occur only to that extent to which the parent ^{237}Np is produced by naturally occurring slow neutrons from uranium.

The first isotope to be discovered was ^{211}At made by Carson, Mackenzie, and Segrè by bombardment of a bismuth target with α-particles from the 60-inch cyclotron at Berkeley in 1940. The reaction is ^{209}Bi($\alpha, 2n$)^{211}At. The half-life of ^{211}At is 7.2 hr. It decays in two modes, 60% by K-electron capture and 40% by α-particle emission. The longest-lived isotope is ^{210}At($t_{1/2}$ = 8.3 hr.); other isotopes having half-lives longer than 1 hr are 206, 207, 208, and 209. Some of the collateral radioactive series involving bombardment reactions contain other astatine isotopes, such as ^{214}At and ^{216}At. All these isotopes have half-lives that are only fractions of a second. The total number of isotopes is at least nineteen, including spallation reaction products as well as bombardment ones. They also include two short-lived isotopes, ^{215}At and ^{218}At, occurring in very small amounts in the branched β-disintegration of ^{215}Po(actinium A) and ^{218}Po(radium A), respectively, as noted above.

The chemistry of astatine determined by tracer techniques, is in keeping with the regular transition of properties of the halogens. The acid properties of astatine are less marked than are those of iodine, while its electropositive character is more marked than that of iodine. After reduction by SO_2 or metallic zinc, the astatine activity is carried by silver iodide or thallium iodide, so it evidently forms insoluble silver and thallium salts. This represents astatine in the univalent negative state characteristic of the halogens. However, astatine is very readily oxidized by bromine and ferric ions, giving indications of two higher oxidation states. Although there is no evidence from migration experiments of the presence of positive ions in the solution, astatine deposits on the cathode, as well as on the anode, in the electrolysis of oxidized solutions. Elemental astatine can be volatilized, although not so readily as iodine, and it has a specific affinity for metallic silver. The similarity to iodine is also shown by the observation that astatine concentrates in the thyroid glands of animals.

Additional Reading

Kent, J.A.: *Riegel's Handbook of Industrial Chemistry,* 9th Edition, Chapman & Hall, New York, NY, 1992.
Krebs, R.E.: *The History and Use of Our Earth's Chemical Elements,* A Reference Guide, Greenwood Publishers Group, Inc., Westport, CT, 1998.
Lide, D.R.: *CRC Handbook of Chemistry and Physics,* 84th Edition, CRC Press, LLC., Boca Raton, FL, 2003.
Lagowski, J.J.: *MacMillan Encyclopedia of Chemistry,* Vol. 1, MacMillan Library Reference, New York, NY, 1997.
Parker, S.P.: *McGraw-Hill Concise Encyclopedia of Science and Technology,* 4th Edition, McGraw-Hill Companies, New York, NY, 1998.
Stwertka, A. and E. Stwertka: *A Guide to the Elements,* Oxford University Press, New York, NY, 1998.

ASTON, FRANCIS WILLIAM (1877–1945). Aston was an English scientist who studied chemistry at Mason College and worked there with P.F. Frankland on optical rotations. In 1910, Aston entered the Cavendish Laboratory, Cambridge, to work under J.J. Thomson, who at the time was examining the positive rays produced in discharge tubes. Aston left his research during the war but returned to Cambridge in 1919. He began using positive-ray analysis to study isotopes. One of Aston's greatest contributions is his invention of the mass spectrograph which he used to examine the isotopic conditions of over 50 (fifty) elements. Today, an improved mass spectrograph is used in the study of nuclear physics, chemistry, and organic chemistry. Aston had the honor of being Nobel Laureate in chemistry in 1922. See also **Aston Whole Number Rule**; and **Spectroscope**.

J.M.I.

ASTON WHOLE NUMBER RULE. The atomic weights of isotopes are (very nearly) whole numbers when expressed in atomic weight units, and the deviations from the whole numbers of the atomic weights of the elements are due to the presence of several isotopes with different weights.

ASTROCHEMISTRY. Application of radioastronomy (microwave spectroscopy) to determination of the existence of chemical entities in the gas clouds of interstellar space and of elements and compounds in celestial bodies, including their atmospheres. Such data are obtained from spectrographic study of the light from the sun and stars, from analysis of meteorites, and from actual samples from the moon. Hydrogen is by far the most abundant element in interstellar space, with helium a distant second.

Over 25% of the elements, including carbon, have been identified, as well as molecules of water, carbon monoxide, carbon dioxide, ammonia, ethane, methane, acetylene, formaldehyde, formic acid, methyl alcohol, hydrogen cyanide, and acetonitrile. When applied to the planets only, the science is called chemical planetology.

ASYMMETRIC TOP. A model of a molecule which has no three-fold or higher-fold axis of symmetry, so that during rotation all three principal moments of inertia are in general different. Example, the water molecule.

ASYMMETRY (Chemical). Asymmetry involves the presence of four different atoms or substituent groups bonded to an atom. Its existence was discovered in 1815 by the French physicist, J.B. Biot (1774–1867). Biot found that oil of turpentine and solutions of sugar, camphor, and tartaric acid all rotate the plane of plane-polarized light when placed between two Nicol prisms. This phenomenon is called *optical rotation* and is indicated in symbols, such as: $[\alpha]_D^{20°} = +53.4$ aq., signifying that the substance gives a rotation of 53.4° to the right (clockwise, or plus) in water solution at 20°C using sodium D line as the light source. Substances in solution that rotate light to the right are designated *d* and are called *dextrorotatory*; substances rotating light to the left are designated *l* and are called *levorotatory*. See also **Isomerism**.

ATACAMITE. This mineral is a basic chloride of copper corresponding to formula $Cu_2Cl(OH)_3$. Crystallizes in thin, orthorhombic prisms, may occur massive. Hardness, 3–3.5; sp gr, 3.76–3.78; luster, adamantine to vitreous; color, green, streak, green; transparent to translucent.

It is a secondary mineral found associated with malachite and cuprite; originally found at Atacama, Chile, whence its name. Other localities are Bohemia, South Australia, and in the United States in Arizona, Utah, and Wyoming. See also **Cuprite**; and **Malachite**.

ATHERMAL TRANSFORMATION. A reaction that occurs without thermal activation. Such a reaction also takes place without diffusion and can occur with great rapidity under the influence of a sufficiently high driving force. The martensite transformation that occurs in steel is primarily athermal, so that the amount of austenite transformed to martensite depends primarily on the temperature to which the steel is cooled and not upon the rate of cooling or the length of time the metal is held at the quenching temperature. It is necessary to note the difference between an isothermal transformation and an athermal transformation. In the former, the reaction occurs at constant temperature and depends, in general, on both diffusion and thermal activation. The transformation of austenite to pearlite can occur isothermally, with carbon atoms diffusing out of the austenite and into the cementite lamellae. See also **Iron Metals, Alloys, and Steels**.

ATMOLYSIS. The separation of a mixture of gases by means of their relative diffusibility through a porous partition, as burned clay. The rates of diffusion are inversely proportional to the square roots of the densities of the gases. Hydrogen, thus, is the most diffusible gas.

ATMOSPHERE (Earth). An envelope (actually a series of envelopes) in the form of imperfect spherical shells of various materials that are bound to the earth by gravitational force. Consisting of gases, vapors, and suspended matter, the total mass of the earth's atmosphere is estimated at approximately 5.1×10^{15} tons, or somewhat less than one-millionth part of the total mass of the earth. One-half of this total mass lies below about 5500 meters (18,000 feet). More than three-fourths of the atmosphere exists below about 10,700 meters (~35,000 feet). The composition of the lower layers of the atmosphere is assumed for purposes of most engineering calculations as 76.8% nitrogen and 23.2% oxygen by weight; 79.1% nitrogen and 20.9% oxygen by volume. A more precise composition of this mixture of gases, including minor constituents, is given in entry on **Air**.

The earth's atmosphere extends some 600 to 1500 kilometers into space. Two factors are involved in this great extension of the atmosphere. First, above about 100 kilometers, the atmospheric temperature increases rapidly with altitude, causing an outward expansion of the atmosphere far beyond that which would occur were the temperature within the bounds observed at the earth's surface. Second above this distance, the atmosphere is sufficiently rarefied so that the different atmospheric constituents attain diffusive equilibrium distributions in the gravitational field; the lighter

constituents then predominate at the higher altitudes and extend farther into space than would an atmosphere of more massive particles. This effect is enhanced by the dissociation of some molecular species into atoms.

Atmosphere-Altitude (Pressure-Temperature) Relationships

The composition of the atmosphere does not change much up to 100 kilometers; there is a region of maximum concentration of ozone (still a very minor constituent) near 20 to 30 kilometers; the relative concentration of water vapor falls markedly from its average sea-level value up to 10 or 15 kilometers, and the relative abundance of atomic oxygen begins to become appreciable on approaching 100 kilometers, due to photodissociation of oxygen by ultraviolet sunlight. Above 200 kilometers, atomic oxygen is the principal atmospheric constituent for several hundred kilometers. However, helium is even lighter than atomic oxygen, so its concentration falls less rapidly with altitude, and it finally replaces atomic oxygen as the principal atmospheric constituent above some altitude which varies with the sunspot cycle between 600 and 1500 kilometers. At still higher altitudes, atomic hydrogen finally displaces helium as the principal constituent. The hydrogen extends many earth radii out into space and constitutes the telluric hydrogen corona, or *geocorona*.

The temperature of the upper atmosphere, and hence its density, varies with the intensity of solar ultraviolet radiation and this, in turn, varies with the sunspot cycle and with solar activity in general. The solar radio-noise flux is a convenient index of solar activity, since it can be monitored at the earth's surface. The minimum nighttime temperature of the upper atmosphere above 300 kilometers has been expressed in terms of the 27-day average of the solar radio-noise flux at 8-centimeter wavelength. This varies from about 600 K near the minimum of the sunspot cycle to about 1400 K near the maximum of the cycle. The maximum daytime temperature is about one-third larger than the nighttime minimum.

Various properties of the earth's atmosphere are described in Tables 1 through 5 and by Figs. 1 and 2. The several layers of the atmosphere are indicated in Table 1, along with the relationship between atmospheric

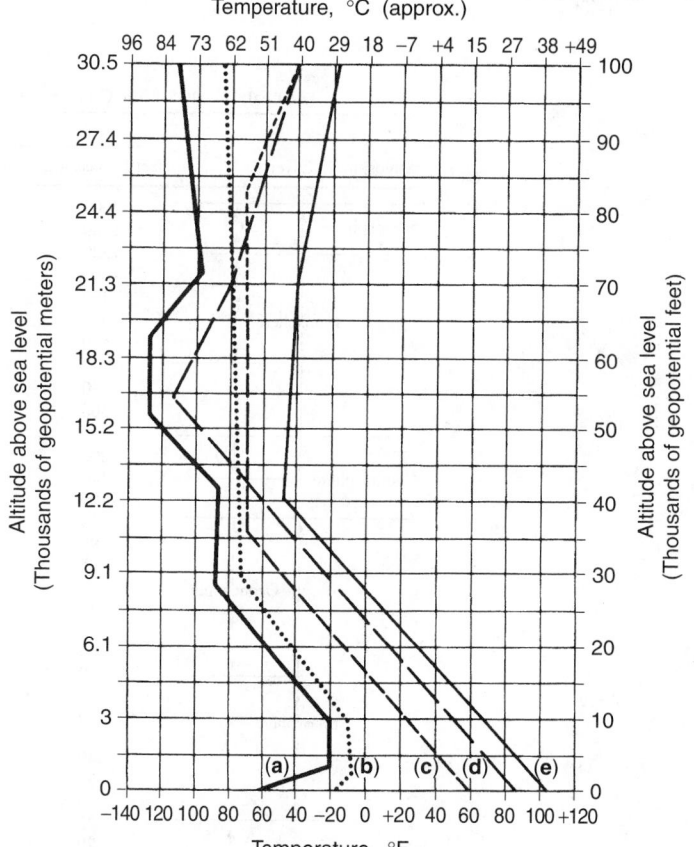

Fig. 1. Relationship between temperature and altitude: (**a**) cold and (**e**) hot are the composites of extremes of cold and hot atmospheres; (**b**) arctic and (**d**) tropical are the composites of the arctic and tropical regions: (**e**) is the standard atmosphere upon which altimetry is based

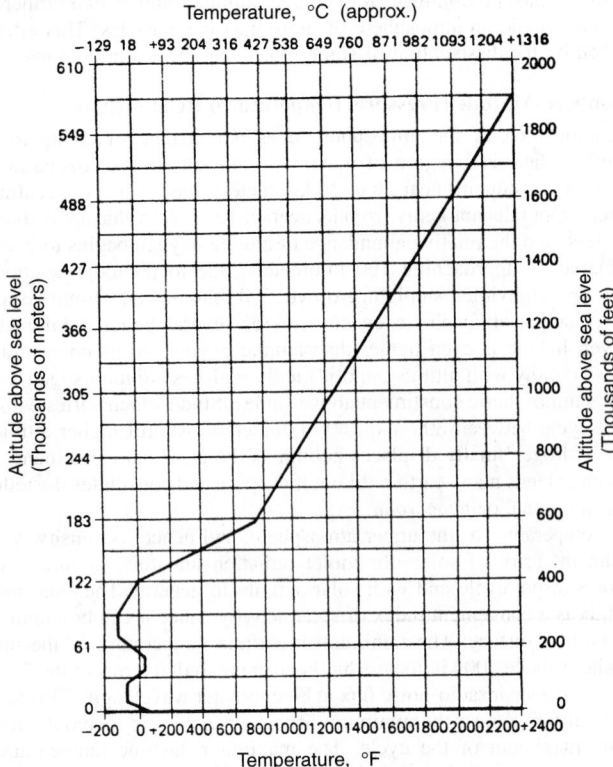

Temperature, °C (approx.)

Fig. 2. Real kinetic temperature of the atmosphere, a measure of the kinetic energy of the molecules and atoms constituting the atmosphere is plotted against altitude above sea level here. Numerical values are determined by the assumed molecular weight of the air (see Table 4), as well as assumed values of the temperature lapse rate

TABLE 2. ATMOSPHERIC DENSITY VERSUS ALTITUDE ABOVE SEA LEVEL

Altitude above sea level		Specific weight	
Feet (thousands)	Meters (thousands)	pounds per cubic foot	kilograms per cubic meter
2000	609.60	1.614×10^{-15}	25.856×10^{-13}
1000	304.80	2.374×10^{-13}	38.031×10^{-13}
100	30.48	0.00101	0.016
95	28.96	0.00129	0.021
90	27.43	0.00166	0.027
85	25.91	0.00214	0.034
80	24.38	0.00275	0.044
75	22.86	0.00350	0.056
70	21.34	0.00445	0.071
65	19.81	0.00566	0.091
60	18.29	0.00720	0.115
55	16.76	0.00915	0.146
50	15.24	0.01164	0.186
45	13.72	0.01480	0.237
40	12.19	0.01883	0.302
35	10.67	0.02370	0.380
30	9.14	0.02861	0.458
25	7.62	0.03427	0.549
20	6.10	0.04075	0.653
15	4.57	0.04812	0.771
10	3.05	0.05648	0.905
5	1.52	0.06590	1.056
0	0	0.07648	1.225

Conversion factors used: 1 foot = 0.3048 meter
1 pound/cubic foot = 16.02 kilograms/cubic meter.

pressure and altitude. Atmospheric density versus altitude are given in Table 2. Geopotential altitude as related with actual altitude and the acceleration due to gravity is given in Table 3. It is interesting to note that the energy required to lift an object 2 million geometric feet is only

TABLE 1. ATMOSPHERIC PRESSURE VERSUS ALTITUDE ABOVE SEA LEVEL

Atmospheric Layer		Altitude Above Sea Level		Pressure	
		feet (thousands)	meters (thousands)	inches of mercury	millibars
Mesosphere	G	2000	609.60	$7:959 \quad 10^{12}$	$269:524 \quad 10^{12}$
		1920	585.22	10^{11}	$3.4 \quad 10^{10}$
		1320	402.34	10^{10}	$3.4 \quad 10^{9}$
	F	1000	304.80	$5:256 \quad 10^{10}$	$177:989 \quad 10^{10}$
Ionosphere		900	274.32	10^{9}	$3:4 \quad 10^{8}$
		640	195.07	10^{8}	$3:4 \quad 10^{7}$
		480	146.30	10^{7}	$3:4 \quad 10^{6}$
		400	121.92	10^{6}	$3:4 \quad 10^{5}$
	E	340	103.63	10^{5}	$3:4 \quad 10^{4}$
		300	91.44	10^{4}	$3:4 \quad 10^{3}$
		260	79.25	10^{3}	$3:4 \quad 10^{2}$
	D	200	60.96	10^{2}	$3:4 \quad 10^{1}$
Chemosphere		140	42.67	10^{1}	3.4
		100	30.48	0.32	10.8
		95	28.96	0.4	13.5
		90	27.43	0.5	17.1
		85	25.91	0.64	21.6
Ozonosphere		80	24.38	0.81	27.4
		75	22.86	1.03	34.9
		70	21.34	1.31	44.4
		65	19.81	1.67	56.6
		60	18.29	2.12	71.8
		55	16.76	2.69	91.1
Stratosphere		50	15.24	3.42	115.8
		45	13.72	4.35	147.3
		40	12.19	5.54	187.6
		35	10.67	7.04	238.4
		30	9.14	8.89	301.1
		25	7.62	11.10	375.9
		20	6.10	13.75	465.6
Troposphere		15	4.57	16.89	572.0
		10	3.05	10.58	696.9
		5	1.52	24.90	843.2
		0	0	29.92	1013.3

Conversion factors used: 1 foot   0:3048 meter.

TABLE 3. ACCELERATION DUE TO GRAVITY AND GEOPOTENTIAL ALTITUDE VERSUS ACTUAL ALTITUDE ABOVE SEA LEVEL

Actual altitude above sea level		Geopotential altitude		Acceleration due to gravity	
Feet (thousands)	Meters (thousands)	Feet (thousands)	Meters (thousands)	Feet/second /second	meters/second /second
2000	609.6	1825	556.26	26.79	8.17
1800	548.64	1657	505.05	27.26	8.31
1600	487.68	1485	452.63	27.75	8.46
1400	426.72	1310	399.29	28.25	8.61
1200	365.76	1132	345.03	28.77	8.77
1000	304.8	950	289.56	29.3	8.93
800	243.8	766	233.48	29.84	9.1
600	182.88	579	176.48	30.4	9.27
400	121.92	389	118.57	30.97	9.44
200	60.96	196	59.74	31.57	9.62
0	0	0	0	32.17	9.81

Conversion factors used: 1 foot = 0.3048 meter
1 foot/sec/sec = 0.3048 meter/sec/sec.

1.824 million times that required to lift it 1 foot above sea level—this because of the decrease in the acceleration due to gravity with altitude.

Reduction of molecular weight, indicating the change in composition of the atmosphere with increasing altitude, is shown in Table 4. The molecular weight of air is assumed essentially constant from sea level up to about 300,000 feet (91,440 meters). At altitudes higher than this, lower molecular weight is largely attributed to the dissociation of oxygen. Above an altitude of about 590,000 feet (179,832 meters), the lower molecular weight is also affected by the diffusive separation and dissociation of nitrogen.

The percent water vapor content of air at saturation versus representative temperatures and pressure altitudes is given in Table 5.

The layers of the earth's atmosphere of interest to meteorologists are the *troposphere* and the *stratosphere*. The troposphere is a thermal atmospheric region, extending from the earth's surface to the stratosphere and characterized by decreasing temperature with height, appreciable vertical wind motion, appreciable water vapor content, and containing nearly all clouds, storms, and pollutants. The thickness of the troposphere varies from as little as about 7–8 kilometers in the cold polar regions to more than 13 kilometers in the warmer, equatorial regions. Temperatures decrease to the interface between the troposphere and stratosphere. This interface is termed the *tropopause*. At the tropopause, polar temperatures average around −55°C, in equatorial regions, −80°C. Above the stratosphere are the *mesosphere* and *ionosphere*, and the outermost layer, the exosphere, gradually fades into the plasma continuum between earth and sun.

In these higher layers of the atmosphere, complex interactions between the fluxes of electromagnetic radiation of various wavelengths and

TABLE 4. MOLECULAR WEIGHT OF ATMOSPHERE VERSUS ALTITUDE

Altitude above sea level		Molecular Weight
feet (thousands)	meters (thousands)	
2000	609.6	15.67
1900	579.12	15.80
1800	548.64	15.96
1700	518.16	16.13
1600	487.68	16.33
1500	457.20	16.56
1400	426.72	16.82
1300	396.24	17.14
1200	365.76	17.51
1100	335.28	17.97
1000	304.8	18.54
900	274.32	19.27
800	243.8	20.24
700	213.36	21.59
600	182.88	23.60
500	152.4	24.09
400	121.92	24.76
300	91.44	28.89
0	0	28.97

Conversion factor used: 1 foot = 0.3048 meter.

TABLE 5. PERCENT WATER VAPOR CONTENT OF AIR AT SATURATION VERSUS REPRESENTATIVE TEMPERATURES AND PRESSURE ALTITUDES

Temperature °C	1,000 Millibars 370 Feet (113 Meters)	850 Millibars 4,780 Feet (1,457 Meters)	700 Millibars 9,880 Feet (3,011 Meters)	500 Millibars 18,280 Feet (5,572 Meters)
40	4.97%	5.93%	7.35%	—
30	2.76	3.28	4.03	5.79
20	1.49	1.76	2.16	3.06
10	0.77	0.91	1.12	1.57
0	0.38	0.45	0.55	0.77
−10	0.18	0.21	0.26	0.36
−20	0.08	0.09	0.11	0.16
−30	0.03	0.04	0.05	0.06
−40	0.01	0.01	0.02	0.02

corpuscular radiation from the sun on one side and the low-density concentrations of atmosphere gases on the other side take place. The particulate radiations are also governed by the earth's magnetic field. Radiations of short wavelength cause a variety of photochemical reactions, the most notable of which is the creation of a layer of ozone acting as an effective absorber of solar ultraviolet and thus causing a warm layer at 30 kilometers in the atmosphere. See also **Aerosols**; and **Oxygen**. The upper atmosphere, as an absorber of primary cosmic rays, shows many interesting nuclear reactions and is an important natural source of radioactive substances, including tritium and carbon 14 which are used as tracers of atmospheric motions and as criteria of age.

Thermodynamics of the Atmosphere

In meteorological calculations, the ideal gas law is a satisfactory approximation for the derivation of formulas for the mixture of gases that constitute the atmosphere. The derivation is:

$$PV = RT$$

where V = volume
P = pressure
R = universal gas constant
T = absolute temperature

For one gram, this becomes

$$PV = \frac{RT}{m}$$

where m is the molecular weight of the gas. For G grams, this becomes,

$$PV = \frac{GRT}{m}$$

This equation is valid for each of the constituent gases of the atmosphere. For nitrogen,

$$P_n V = \frac{G_n RT}{m_n}$$

For oxygen,

$$P_0 V = \frac{G_0 RT}{m_0}$$

For argon,

$$P_a V = \frac{G_a RT}{m_a}$$

For water vapor,

$$P_w V = \frac{G_w RT}{m_w}$$

When there is no water vapor present in the atmosphere, these equations can be combined as follows:

$$P_t V = (P_n + P_0 + P_a)V = RT \left[\frac{G_n}{m_n} + \frac{G_0}{m_0} + \frac{G_a}{m_a} \right]$$

$$= RT \left(\frac{G_t}{m_t} \right)$$

In these equations, P_t is the total pressure of the nitrogen, oxygen, and argon. Also, G_t is the total mass of the gases; and m_t is the molecular weight of the mixture, with a numerical value of 28.97.

Because water vapor is always present in varying quantities in the atmosphere, corrections in the equation of state must be made in accordance with the amount of water vapor present. Procedure is as follows:

$$PV = (P_t + P_w)V = RT \left[\frac{G_t}{m_t} + \frac{G_w}{m_w} \right]$$

$$= RT \left[\frac{G}{m_t} + \frac{G_w}{m_t} \frac{G_w}{m_w} \right]$$

In these equations, P is the total pressure of the air gases plus the water vapor, and G is the total mass of the air gases plus the water vapor.

This equation can be rearranged and simplified:

$$PV = RT \frac{G}{m_t} \left[1 - \frac{G_w}{G} \left(1 - \frac{m_t}{m_w} \right) \right]$$

where m_t has a value of 28.97 and m_w has a value of 18.00. This equation is easily reduced to

$$PV = RT \frac{G}{m_t} \left[1 + 0.6 \frac{G_w}{G} \right]$$

Virtual temperature of the air is defined as

$$T' = T \left[1 + 0.6 \frac{G_w}{G} \right]$$

Virtual temperature is, in effect, the temperature of a mass of dry air having the same density of another mass of air containing water vapor. Virtual temperature is always greater than real temperature, except when G_w is nil.

The equation of state for real air becomes

$$PV = \frac{RT'G}{m_t}$$

If R, the universal gas constant, is made into a specific gas constant for air by letting $R/m_t = R_a$, then for one gram of air, the equation of state becomes $PV = R_a T'$.

The hydrostatic state of equilibrium of the atmosphere varies with the type of atmosphere that is under consideration.

Standard atmosphere is a term used in the following references:

1. A hypothetical vertical distribution of atmosphere temperature, pressure, and density, which, by international agreement, is taken to be representative of atmosphere for purposes of pressure altimeter calibrations, aircraft performance calculations, aircraft and missile design, ballistic tables, etc. The air is assumed to obey the ideal gas law and the hydrostatic equation, which, taken together, relate temperature, pressure, and density variations in the vertical. It is further assumed that the air contains no water vapor, and that the acceleration of gravity does not change with height. The current standard atmosphere was adopted in 1952 by the International Civil Aeronautical Organization (ICAO) and supplants the U.S. Standard

Atmosphere prepared in 1925. The parametric assumptions and physical constants used in preparing the ICAO Standard Atmosphere are as follows:

(a)   Zero pressure altitude corresponds to that pressure which will support a column of mercury 760 mm high. This pressure is taken to be 1.013250×106 dynes/cm^2, or 1013.250 mb, or 101.325 kPa (and is known as one standard atmosphere or one atmosphere).
(b)   The gas constant for dry air is 2.8704×10^6 erg/gm K.
(c)   The ice point at one standard atmosphere pressure is 273.16 K.
(d)   The acceleration of gravity is 980.665 cm/sec^2.
(e)   The temperature at zero pressure altitude is 15°C or 288.16 K.
(f)   The density at zero pressure altitude is 0.0012250 gm/cm^3.
(g)   The lapse rate of temperature in the troposphere is 6.5°C/km.
(h)   The pressure altitude of the tropopause is 11 km.
(i)   The temperature at the tropopause is −56.5°C.

2. A standard unit of atmospheric pressure; the 45° atmosphere, defined as the pressure exerted by a 760 mm column of mercury at 45° latitude at sea level at temperature 0°C (acceleration of gravity = 980.616 cm/sec^2). One 45° atmosphere equals 760 mm Hg(45°); 29.9213 in. Hg(45°); 1013.200 mb; 101.325 kPa.

3. Any environmental gas or mixture of gases, e.g., and atmosphere of nitrogen of an inert atmosphers.

ATOM. An atom is a basic structural unit of matter, being the smallest particle of an element that can enter into chemical combination. Each atomic species has characteristic physical and chemical properties that are determined by the number of constituent particles (protons, neutrons, and electrons) of which it is composed; especially important are the number Z of protons in the nucleus of each atom. To be electrically neutral the number of electrons in an atom must also be Z. The arrangement of these electrons in the internal structure of an atom determines its chemical properties. All atoms having the same atomic number Z have the same chemical properties, but differ in greater or lesser degree from atoms having any other value of Z. Thus, for example, all atoms of sodium ($Z = 11$) exhibit the same characteristic properties and undergo those reactions which chemists have found for the element sodium. Although these reactions are similar in some degree to those reactions characteristics of certain other elements, such as potassium and lithium, they are not exactly the same and hence can be distinguished chemically (see **Chemical Elements**), so sodium has properties distinctly different from those of all other elements. Individual atoms can usually combine with other atoms of either the same or another species to form molecules.

As explained in the entry on **Chemical Elements; Atomic Structure of the Elements**, atoms having the same atomic number may differ in their neutron numbers or in their nuclear excitation energies.

The term *atom* has a long history, which goes back as far as the Greek philosopher Democritus. The concept of the atomic nature of matter was revived near the beginning of the nineteenth century. It was used to explain and correlate advancing knowledge of chemistry and to establish many of the basic principles of chemistry, even though conclusive experimental verification for the existence of atoms was not forthcoming until late in the nineteenth century. It was on the basis of this concept that Mendeleev first prepared a periodic table. See also **Periodic Table of the Elements**.

Several qualifying terms are used commonly to refer to specific types of atoms. Examples of some of the terms are given in the following paragraphs.

An *excited atom* is an atom that possesses more energy than a normal atom of that species. The additional energy commonly affects the electrons surrounding the atomic nucleus, raising them to higher energy levels.

An *ionized atom* is an ion, which is an atom that has acquired an electric charge by gain or loss of electrons surrounding its nucleus.

A *labeled atom* is a tracer which can be detected easily, and which is introduced into a system to study a process or structure. The use of those labeled atoms is discussed at length in the entry **Isotope**.

A *neutral atom* is an atom that has no overall, or resultant, electric charge.

A *normal atom* is an atom which has no overall electric charge, and in which all the electrons surrounding the nucleus are at their lowest energy levels.

A *radiating atom* is an atom that is emitting radiation during the transition of one or more of its electrons from higher to lower energy states.

A *recoil atom* is an atom that undergoes a sudden change or reversal of its direction of motion as the result of the emission by it of a particle or radiation in a nuclear reaction.

A *stripped atom* is an atomic nucleus without surrounding electrons; also called a nuclear atom. It has, of course, a positive electric charge equal to the charge on its nucleus.

Subatomic particles and organization of the atom are discussed in the entry **Particles (Subatomic)**.

Additional Reading

Delone, N.B. and V.P. Krainov: *Multiphoton Processes in Atoms,* 2nd Edition, Springer-Verlag, Inc., New York, NY, 2000.
Rossotti, H.: *Diverse Atoms,* Oxford University Press, Inc., New York, NY, 1998.
Silverman, M.P.: *Probing the Atom,* Princeton University Press, Princeton, NJ, 2000.

ATOMIC DISINTEGRATION.   The name sometimes given to radioactive decay of an atomic nucleus and occasionally to the breakup of a compound nucleus formed during a nuclear reaction. See also **Radioactivity**.

ATOMIC ENERGY.   1. The constitutive internal energy of the atom, which would be released when the atom is formed from its constituent particles, and absorbed when it is broken up into them. This is identical in magnitude with the total binding energy and is proportional to the mass defect. 2. Sometimes this term is used to denote the energy released as the result of the disintegration of atomic nuclei, particularly in large-scale processes, but such energy is more commonly called nuclear energy. See also **Nuclear Power Technology**.

ATOMIC ENERGY LEVELS.   1. The values of the energy corresponding to the stationary states of an isolated atom. 2. The set of stationary states in which an atom of a particular species may be found, including the ground state, or normal state, and the excited states.

ATOMIC FREQUENCY.   The vibrational frequency of an atom, used particularly with respect to the solid state.

ATOMIC HEAT.   The product of the gram-atomic weight of an element and its specific heat. The result is the atomic heat capacity per gram-atom. For many solid elements, the atomic heat capacity is very nearly the same, especially at higher temperatures and is approximately equal to $3R$, where R is the gas constant (Law of Dulong and Petit).

ATOMIC HEAT OF FORMATION.   Of a substance, the difference between the enthalpy of one mole of that substance and the sum of the enthalpies of its constituent atoms at the same temperature; the reference state for the atoms is chosen as the gaseous state. The atomic heat of formation at 0 K is equal to the sum of all the bond energies of the molecule, or to the sum of all the dissociation energies involved in any scheme of step-by-step complete dissociation of the molecule.

ATOMIC MASS (Atomic Weight).   As of the late 1980s, the current and internationally accepted unit for atomic mass is 1/12*th of the mass of an atom of the* ^{12}C *nuclide and the official symbol is u*. The SI symbol *u* was selected so that it would indicate measurements made on the unified scale.[1]

It is interesting to note that prior to 1961, *two* atomic mass scales were used. Chemists preferred a scale based on the assignment of *exactly* 16, which experience had shown as the *average* mass of oxygen atoms as they are found in nature. On the other hand, physicists preferred to base the scale on a single isotope of oxygen, namely, ^{16}O (oxygen-16). The two

scales differed because oxygen has three stable isotopes, ^{16}O, ^{17}O, and ^{18}O (as well as three identifiable radioactive isotopes, ^{14}O, ^{15}O, and ^{19}O).

Long before an understanding of the structure of the atom had been established and before the existence of isotopes was evidenced, several pioneers proposed what have become known as the concepts (laws) of:

Combining Volumes—under comparable conditions of pressure and temperature, the volume ratios of gases involved in chemical reactions are simple whole numbers

Combining Weights—if the weights of elements that combine with each other be called their 'combining weights,' then elements always combine either in the ratio of their combining weights or of simple multiples of these weights.

This then led to the establishment of the basic principle that the *combining weight of an element or radical is its atomic weight divided by its volume*.

Although the tables of atomic weights published today embrace all of the known chemical elements, it should be pointed out that the concept of combining weights stemmed exclusively from very early experiments strictly with *gases*. The kinetic theory of gases, which was developed from a line of logic that did not require the innermost understanding of the atom as we know it today, served as the early basis of how atoms react in quantitative proportions with each other to form compounds.

Boyle (1662) observed that at constant temperature the volume of a sample of gas varies inversely with pressure, but Boyle did not explain why this was so. Somewhat later, Charles (1787) refined the observation to the effect that the volume of any sample of a gas varies directly with the *absolute temperature* provided that the pressure is held constant. A few years later, Gay-Lussac (1808), in reporting the results of his experiments with reacting gases, observed that volumes of gases that are used or produced in a chemical reaction can be expressed in ratios of small whole numbers—a concept to become known as Gay-Lussac's law of combining volumes. It should be noted that the foregoing concepts proposed by Boyle, Charles, and Gay-Lussac were based upon experimental observations, not on theory.

An explanation for the law of combining volumes was given by Avogadro (1811) in which he proposed that equal volumes of all gases at the same pressure and temperature contain the same number of molecules. This, obviously, was an extension of Bernoulli's earlier thinking.[2] Avogadro's observations were essentially ignored and it remained for Cannizzaro (1858–1864) to develop, in a practical way, a method for computing the combining weights for gaseous compounds. This work led to the universal acceptance[2] of Avogadro's principle. Cannizzaro used gas densities to assign atomic and molecular weights, basing his atomic weight scale on hydrogen. The hydrogen atom was assigned a value of 1 (approximately its assigned value today). The molecular weight of hydrogen was 2.

Much further research and careful experimentation was required to convey the principle to solid compounds. See entries on **Chemical Composition**; and **Chemical Formula**.

In returning to the attractive simplicity of combining weights in terms of ratios of small whole numbers, why then is it necessary, considering the standard for comparison ($^{12}C = 12$), to extend the atomic weight values to four and more decimal places? The principal answer is the presence of isotopes. Isotopes were unknown in the days of Boyle and other early pioneers. In essence, the atomic weight of an element is a *weighted average* of the atomic masses of the natural isotopes. The weighted average is determined by multiplying the atomic mass of each isotope by its fractional abundance and adding the values thus obtained. A fractional abundance is the decimal equivalent of the percent abundance. However, for the standard of comparison, obviously a specific isotope was selected rather than a weighted average for that element.

One might also query—since we know so much today about the masses of the protons, neutrons, and electrons comprising an atom,[3] why not

---

[1] It should be stressed that *u*, as the standard for comparing the masses (weights) of all chemical elements in all kinds of chemical compounds, refers not simply to the carbon atom, but rather to one very specific isotope, carbon-12. There are two stable isotopes of carbon, ^{12}C and ^{13}C, and four known radioactive isotopes, ^{10}C, ^{11}C, ^{14}C, and ^{15}C.

[2] An attempt was made by Daniel Bernoulli (1738) to explain Boyle's law on the basis of what later became known as the *kinetic theory of gases*. Bernoulli introduced the concept that the pressure of a gas results from the collisions of gas molecules within the walls of the gas container. This established a connection between the *numbers* of gas molecules present and their kinetic energy present at any given temperature.

[3] Mass of proton is 1.007277 *u*; of neutron is 1.008665 *u*; mass of electron is 0.0006486 *u*.

simply add up these specific values for a given atom? This, of course, still would not relieve the isotope problem, but it is not accurate to do so mainly because of Einstein's equation $E = mc^2$. As pointed out by Mortimer, with the exception of $_1{}^1H$, the sum of the masses of the particles that make up a nucleus will always differ from the actual mass of the nucleus. If the required nucleons were brought together to form a nucleus, some of their mass would be converted into energy. Called the *binding energy*, this is also the amount of energy required to pull the nucleus apart.

For practical purposes, a majority of elements have a constant mixture of natural isotopes. For example, mass spectrometric studies of chlorine show that the element consists of 75.53% $_{17}{}^{35}Cl$ atoms (mass = 34.97 u) and 24.47% $_{17}{}^{37}Cl$ atoms (mass = 36.95 u). Experience has shown that any sample of chlorine from a natural source will consist of these two isotopes in this proportion.

Additional Reading

Lide, D.R.: *CRC Handbook of Chemistry and Physics,* 84th Edition, CRC Press, LLC., Boca Raton, FL, 2003.
Perlmutter, A., S.L. Mintz, and B.N. Kursunoglu: *Physics of Mass,* Kluwer Academic/Plenum Publishers, New York, NY, 1999.
Sherrill, B.M., D.J. Morrissey, and C.N. Davis: *Enam98, Nuclei and Atomic Masses,* American Institute of Physics, College Park, MD, 1999.

ATOMIC NUMBER.   The number of protons (positively charged mass units) in the nucleus of an atom, upon which its structure and properties depend. This number represents the location of an element in the periodic table. It is normally the same as the number of negatively charged electrons in the shells. Thus, an atom is electrically neutral except in an ionized state, when one or more electrons have been gained or lost. Atomic numbers range from 1, for hydrogen, to 110 for darmstadtium.

See also **Atomic Mass (Atomic Weight)**; and **Periodic Table of the Elements**.

ATOMIC ORBITALS.   See **Orbitals**.

ATOMIC PERCENT.   The percent by atom fraction of a given element in a mixture of two or more elements.

ATOMIC PLANE.   A plane passed through the atoms of a crystal space lattice, in accordance with certain rules relating its position to the crystallographic axes. See also **Mineralogy**.

ATOMIC RADIUS.   See **Chemical Elements**.

ATOMIC SPECIES.   A distinctive type of atom. The basis of differentiation between atoms is (1) mass, (2) atomic number, or number of positive nuclear charges, (3) nuclear excitation energy. The reason for recognizing this third class is because certain atoms are known, chiefly among those obtained by artificial transmutation, which have the same atomic (isotopic) mass and atomic number, but differ in energetics.

ATOMIC SPECTRA.   An atomic spectrum is the spectrum of radiation emitted by an excited atom, due to changes within the atom; in contrast to radiation arising from changes in the condition of a molecule. Such spectra are characterized by more or less sharply defined "lines," corresponding to pronounced maxima at certain frequencies or wavelengths, and representing radiation quanta of definite energy.

The lines are not spaced at random. In the spectrum of hydrogen, for example, there is a prominent red line (H_α) and, far from it, another (H_β) in the greenish-blue, then after a shorter wavelength interval a blue-violet line (H_γ), and after a still shorter interval another violet line (H_δ), etc. One has only to plot the frequencies of these lines as a function of their ordinal number in the sequence to get a smooth curve, which shows that they are spaced in accordance with some law. In 1885, Balmer studied these lines, now called the Balmer series, and arrived at an empirical formula which in modern notation reads

$$v = Rc \left(\frac{1}{n_1^2} - \frac{1}{n_2^2} \right)$$

It gives the frequency of successive lines in the Balmer series if R is the Rydberg constant, c the velocity of light, $n_1 = 2, n_2 = 3, 4, 5, \ldots$. As n_2 becomes large, the lines become closer together and eventually reach the series limit of $v = Rc/4$. Ritz, as well as Rydberg, suggested that other

series might occur where n_1 has other integral values. These, with their discoverers and the spectral region in which they occur are as follows:

Lyman series, far ultraviolet,	$n_2 = 2, 3, 4, \ldots, n_1 = 1$
Paschen series, far infrared,	$n_2 = 4, 5, 6, \ldots, n_1 = 3$
Brackett series, far infrared,	$n_2 = 5, 6, 7, \ldots, n_1 = 4$
Pfund series, far infrared,	$n_2 = 6, 7, 8, \ldots, n_1 = 5$

See also **Energy Level**.

ATOMIC SPECTROSCOPY.   Chemical analysis by atomic absorption spectrometry involves converting the sample, at least partially, into an atomic vapor and measuring the absorbance of this atomic vapor at a selected wavelength which is characteristic for each element. The measured absorbance is proportional to concentration, and analyses are made by comparing this absorbance with that given under the same experimental conditions by reference samples of known composition. Several methods of vaporizing solids directly can be used in analytical applications. One of the first methods used was spraying a solution of the sample into a flame, giving rise to the term "absorption flame photometry." When a flame is used, the atomic absorption lines are usually so narrow (less than 0.05 Å) that a simple monochromator is not sufficient to obtain the desired resolution. Commercial atomic absorption spectrophotometers overcome this difficulty by using light sources which emit atomic spectral lines of the element to be determined under conditions which ensure that the lines in the spectrum are narrow, compared with the absorption line to be measured. With this arrangement, peak absorption can be measured, and the monochromator functions only to isolate the line to be measured from all other lines in the spectrum of the light source. See Fig. 1.

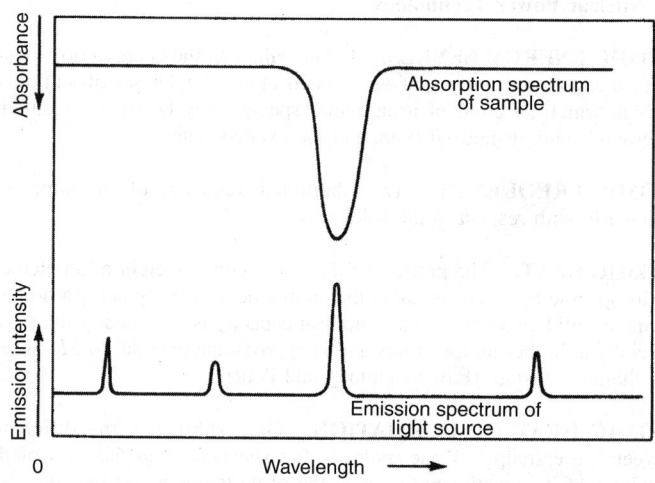

Fig. 1.   Lines emitted by the light source are much narrower than the absorption line to be measured

Atomic spectra, which historically contributed extensively to the development of the theory of the structure of the atom and led to the discovery of the electron and nuclear spin, provide a method of measuring ionization potentials, a method for rapid and sensitive qualitative and quantitative analysis, and data for the determination of the dissociation energy of a diatomic molecule. Information about the type of coupling of electron spin and orbital momenta in the atom can be obtained with an applied magnetic field. Atomic spectra may be used to obtain information about certain regions of interstellar space from the microwave frequency emission by hydrogen and to examine discharges in thermonuclear reactions.

ATTRITION MILLS.   Equipment of this type is used in the process industries to reduce the size of various feeds. Attrition connotes a rubbing action, although this action usually is combined with other forces, including shear and impact. Attrition mills also are referred to as disk mills and normally comprise two vertical disks mounted on horizontal shafts, with adjustable clearance between the vertical disks. In some designs, one vertical disk may be stationary. In other designs, the two vertical disks rotate at differential speeds or in opposite directions. Material is fed to the mill so that it is subjected to a tearing or shredding action. Because of the frictional nature of the operation, temperatures build up and heat-sensitive materials cannot be size-reduced in this type of equipment. Attrition mills sometimes are used principally as mixers to provide an intimate blending of powders. Special plates are used to permit intensive blending with a minimum of grinding action. Throughput rates per horsepower required are high.

AUGITE. This mineral is a common monoclinic variety of pyroxene whose name is derived from the Greek word meaning "luster," in reference to its shining cleavage faces. Chemically it is a complex metasilicate of calcium, magnesium, iron and aluminum. Color, dark green to black, may be brown or even white; hardness, 5–6; specific gravity, 3.23–3.52. Augite is important as a primary mineral in the igneous rocks and also as secondary mineral. The white augite is called leucaugite from the Greek word meaning "white." Chemical analysis reveals this variety as containing little or no iron. Augite is of widespread occurrence.

See also **Pyroxene**.

AURIC AND AUROUS. Prefixes often used in the naming of gold salts of valence +3 (ic) and +1 (ous). Thus, auric chloride, aurous nitrate, and so on.

AUSTENITE. The solid solution based upon the face-centered cubic form of iron. The most important solute is usually carbon, but other elements may also be dissolved in the austenite. See also **Iron Metals, Alloys, and Steels**.

AUTOCATALYSIS. A word used to describe the experimentally observable phenomenon of a homogeneous chemical reaction that shows a marked increase in rate with time, reaching a peak at about 50% conversion and then dropping off. The temperature has to remain constant and all ingredients mixed at the start for proper observation.

This definition excludes those exothermic reactions which shown and increase in rate with time (like explosions) caused by the rapidly rising temperature.

AUTODEPOSITION. A generic term coined to describe a fairly recent (late 1970s) development in which conversion coatings and organic coatings are applied to metal substrates in a single stage. The process is analogous to the electrodeposition of organic coatings, but in autodeposition, the coatings are deposited by means of chemical action rather than under the influence of an electric current and there is no need for a separate conversion coating stage. See also entry on **Conversion Coatings**.

Autodeposition baths are comprised of colloidal dispersions of film-forming coating materials, such as latexes or polymer emulsions, acids and oxidizers. Cleaned metal surfaces are immersed in the coating composition where the substrate metal is lightly attacked by the activating system of acid and oxidizer. This results in the formation of metal ions which overcome the stabilizing charges on the polymer particles and cause them to deposit on the metal surface. The thickness and type of conversion coating which forms simultaneously is determined by the kind and degree of activation.

Autodeposition is characterized by the growth of the coating with time of immersion and the ability to withstand water rinsing immediately upon removal from the coating bath and before fusion or curing of the coating by heat without loss of coating.

The elimination of the conversion coating sequence used in conventional industrial coating processes means that autodeposited coatings can be applied in fewer steps and in smaller finishing areas. There are no solvents in the commercial baths. This reduces air pollution to virtually zero and completely eliminates fire hazard. The coatings exhibit excellent adhesion, impact resistance, flexibility, and chemical resistance. There is no "throwing power" limit, such as that in electrodeposition. The coatings can be applied to any partially enclosed surface that can be contacted by the bath and are not attacked by "solvent wash." As of the early 1980s, the largest application of autodeposition systems is in the automotive industry.

WILBUR S. HALL
Amchem Products, Inc., Ambler, Pennsylvania

AUTOIONIZATION (or Preionization). Some bound states of atoms have energies greater than the ionization energy. An atom that is in a discrete energy state above the ionization point can ionize itself automatically with no change in its angular momentum vectors if there is a continuum with exactly the same characteristics. This process is called autoionization.

AUTOLYSIS. The energy derived from biological oxidations in living cells serves to promote anabolic processes, i.e., to produce relatively complex, highly ordered molecules and structures, and thus normally keeps living cells in a steady state remote from equilibrium. In organisms that lack cellular nutrients or oxygen (or in dead organisms or cells that have been disrupted so as to destroy much subcellular organization), the opposing catabolic tendency toward equilibrium, including the tendency toward degradation of macromolecules to simpler monomeric subunits, is not counterbalanced. These degradative processes, many of them enzymatically catalyzed, are collectively termed autolysis. Autolytic processes may include, for example, hydrolysis of proteins catalyzed by proteolytic enzymes or hydrolysis of nucleic acids catalyzed by nucleases. Autolysis of tissues (e.g., liver homogenate) has sometimes been used as a method for releasing bound molecules (e.g., vitamins or coenzymes) into free soluble form.

AUTOMATED INSTRUMENTATION: CLINICAL CHEMISTRY. Clinical chemistry, initiated in the 1940s, involves the biochemical testing of body fluids to provide objective information on which to base clinical diagnosis. The ever-increasing demand for high quality, routine clinical testing stimulated the development of automated techniques, and as early as the 1960s automation in the clinical laboratory was the rule rather than the exception.

Although growth was initially driven by automation, in the 1990s the growth in U.S. laboratory testing may be attributed to several factors. Among them are the discovery of new diseases and the introduction of new therapies, as well as better understanding of body chemistry and the aging of the U.S. population, which increases the risk of contracting age-related illnesses requiring clinical chemistry analyses.

Assay Automation

Clinical chemistry analyzers are automated instruments used for measuring concentrations of the various chemical constituents of blood or other body fluids. For a discussion of the related category of instruments used for the measurement of blood cell parameters. See also **Automated Instrumentation: Hematology**.

Before the advent of automation, the steps a clinical technologist had to follow in performing a single clinical chemistry assay included sample preparation, identification, centrifugation, and filtering; reagent preparation; manual metering and addition of sample and reagents into a reaction vessel; mixing and incubation in the reaction vessel; optical measurement of the mixture in a separate cuvette; and calculation and recording of the results. With the exception of the sample and reagent preparation procedure, which is sometimes done separately, these steps are all performed by an automated analyzer such as that shown in Figure 1.

Technology

An automated system for clinical analysis consists of the instrument (hardware), the reagents, and the experimental conditions (time, temperature, etc) required for each determination. The reagents plus the experimental conditions are sometimes referred to as the chemistry of the system. The chemistry employed is generally similar to that used in manual assays because most automated assay methods have been adapted from the manual ones. However, automated analyzers rarely afford the flexibility of experimental procedure that is possible in manual analyses. Chemical determinations available on automated analyzers include those for albumin, calcium, chloride, creatinine, iron, total bilirubin, total protein, alanine aminotransferase (ALT), alkaline phosphatase, aspartate transaminase (AST), carbon dioxide, cholesterol, creatine kinase (CK), gamma glutamyl transferase (GGT), glucose, lactate dehydrogenase (LD), triglycerides, and urea nitrogen.

Essential features of an automated method are the specificity, i.e., the assay should be free from interference by other serum or urine constituents, and the sensitivity, i.e., the detector response for typical sample concentration of the species measured should be large enough compared to the noise level to ensure assay precision. Also important are

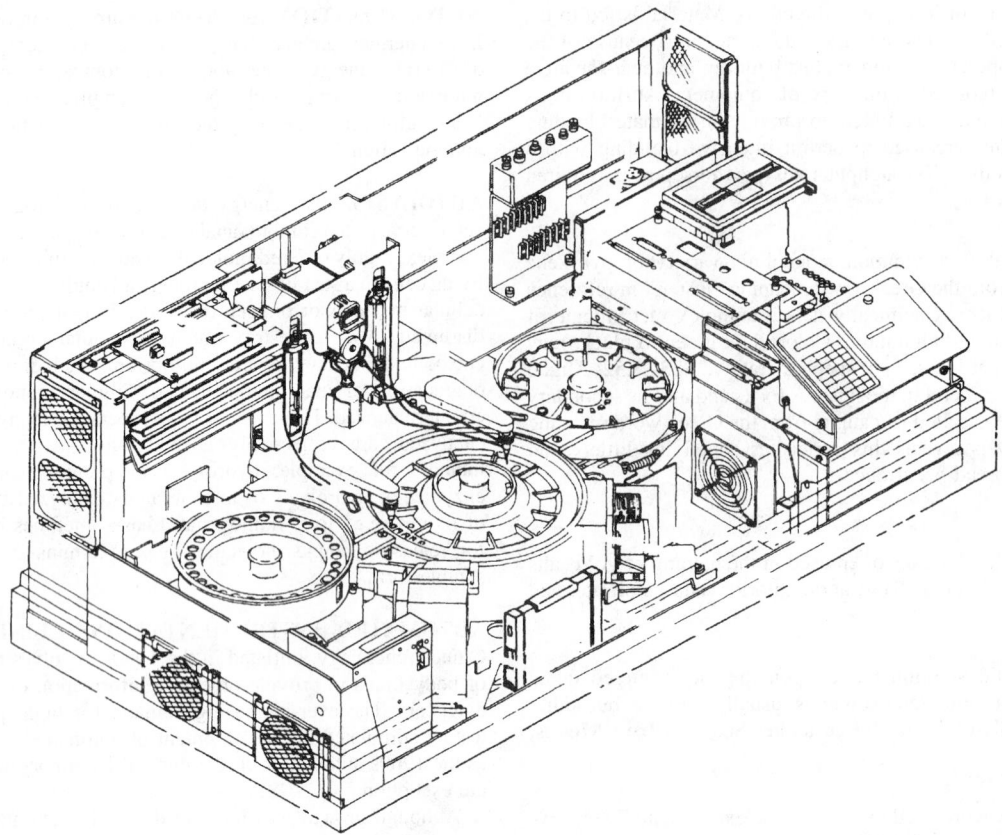

Fig. 1.   Internal view of a clinical chemistry analyzer (Technicon RA-XT). Courtesy of Miles Inc

the speed, i.e., the reaction should occur within a convenient time interval (for fast analysis rates), and adequate range; the result for most samples should fall within the allowable range of the assay.

The assay methods for the various biochemical species can be classified according to reaction rate behavior, e.g., end point vs kinetic methods, blanking schemes, or reaction principle and type of reagents employed.

Measurement Methods

The majority of the various analyte measurements made in automated clinical chemistry analyzers involve optical techniques such as absorbance, reflectance, luminescence, and turbidimetric and nephelometric detection means.

Classification of Clinical Chemistry Analyzers

Automated clinical analyzers can be divided into discrete and continuous-flow systems. In discrete systems, the sample–reagent mixtures from different specimens are kept in individual reaction cuvettes during incubation and optical reading. In continuous-flow analyzers, the sample–reagent mixtures form liquid segments flowing through a tube. Adjacent segments are separated by air bubbles. Clinical analyzers can also be classified according to their degree of flexibility. Most of the modern systems are random access analyzers, for which the tests on various specimens are performed in any order programmed by the operator.

The throughput range, number of assays performed per hour, of clinical analyzers reflects the diversity of the market. Systems that perform less than 400 tests per hour are usually referred to as small analyzers; medium analyzers cover the range from about 500 tests per hour to 1,500 tests per hour, while high throughput analyzers can process up to 10,000 tests per hour. Table 1 lists a number of representative automated systems and their characteristics.

Functional Elements of Clinical Chemistry Analyzers

Sample Handling System. Venous or capillary blood, urine, and cerebrospinal fluid are specimens routinely used in medical diagnostic testing. Of these biological fluids, the use of venous blood is by far the most prevalent. Collection devices such as syringes and partial vacuum test tubes, e.g., Vacutainer, are used to draw 10 mL or less of venous blood. At collection time, the test tubes are carefully labeled for later identification.

The specimen, as drawn, contains cells, platelets, fibrin, and particulates. For most chemistry tests, e.g., determination of glucose, cholesterol, etc, it is necessary to first separate the cellular blood fraction which, if present during the assay, would interfere with the determination and adversely affect the accuracy of the measurement. The cellular components are separated by centrifugation.

After centrifugation, the specimens are placed in a holder specifically designed to work with a particular instrument. Once loaded, the sample holder is placed onto the carrier transport mechanism of the system, which is referred to as the sampler. The function of the sampler is to bring each specimen into a position where a sample of it can be taken by the aspiration probe, then to transport it further to the unloading area.

Sample Aspiration and Dispensing. Precisely metered amounts of the sample have to be aspirated rapidly and without allowing intersample contamination, known as sample carry-over. Generally, pipetting is accomplished by motor-driven syringes connected through a fluid line to a thin aspiration probe (see Fig. 1)

In most analyzers, sample carry-over is minimized by one or more of the following techniques: sensing the liquid level in the sample container, and limiting the probe immersion in the liquid to a few millimeters; or rinsing the aspiration probe inside and out after each immersion.

Reagent Handling System. Many analyzers can be programmed to preform a wide variety of assays. However, reagents for only a limited number of tests, usually referred to as resident tests, are available on the instrument at any one time. The main reason for this limitation is that, for infrequently requested tests, the time period until reagent depletion may exceed the chemical stability time limit.

Precise metering of the amount of liquid reagent needed for a test is generally done using a motor-driven syringe.

Two notable technological developments related to the reagent handling system occurred in the late 1980s. One of these was the introduction of the dry slide reagent technology where the reagents needed for a particular assay are deposited in thin layers on a slide.

Another unique development, which significantly reduced the amount of reagent necessary for an assay, is an extension of the continuous-flow analyzer technology called capsule chemistry. The core of the system is a capillary Teflon tube coated on the inside with a thin, flowing film of

TABLE 1. FEATURES OF AUTOMATED ANALYZERS

Manufacturer and system	Throughput per hour		Number of resident methods	Amount of sample needed per test, μL	Incubator temperature, °C	Optical system[a]	Distinctive features
	Samples	Tests					
Roche COBAS MIRA	variable	125	30	2–95	37	AMW,P	disposable cuvettes
Technicon RA-1000	variable	240	12	2–30	30, 37	AMW,P	no probe wash
Ames CLINISTAT	up to 80	up to 80		10	37	R	dry chemistry
Abbott SPECTRUM	variable	400–600	23	1.25–25	25, 30, 37	AMW,P,T	polychromatic sample blanking
Beckman SYNCHRON CX3	up to 75	600	8	122/8 tests	37	AMW,P	STAT analyzer[b] Beckman liquid calibrators and controls
Kodak EKTACHEM 700	300 max	600	26	10, 11	37	R	dry chemistry slides, routine/STAT[b]
Baxter PARAMAX	240	720	32	2–50	37	AMW,P	closed container sampling option; unit dose dry tablet reagent
BM Hitachi 737	300	1,200	23	3–20	30, 37	AMW,P	
Technicon CHEM 1	variable	1,800	35	1	30, 37	AMW,P	continuous-flow capsule chemistry
BM Hitachi 736–50	300	8,100	27	10	30, 37	AMW,P	
Technicon DAX = 96	300	10,200	34	3–67	37	AMW,P	

[a] AMW = absorption, multiple wavelength per test; N = nephelometry; T = turbidimetry; P = potential; and R = reflectance.

[b] A specimen that has to be processed with priority vs other specimens is known as STAT sample.

fluorocarbon oil. Sample and liquid reagents, alternating with air bubbles, are aspirated into the tube, forming a segmented flow.

Optical Detection System. The optical detection system includes dedicated or time-multiplexed arrangements of the various elements such as the light source, cuvette or flowcell, wavelength isolation elements, and detector. A common layout may have a single light source with light distribution via fiber optics to many cuvettes or flowcell read stations. On the output side of the reaction cuvette or flowcell, fiber optics pipe the light signals to an optical chopper, through dedicated interference filters, and to a common multiplexed photomultiplier. A variation of this arrangement uses a rotating mirror device to multiplex the light output of many cuvettes into the input of a single grating–array photodiode.

Computer System. The brain of the modern clinical chemistry analyzer is its computer system. The part of the computer system that controls the functional aspects of the analyzer is known as the process control computer or analytical processor (AP); the test results are handled by the data management computer, also known as the results processor (RP).

Economic Aspects

It is estimated that the worldwide clinical chemistry diagnostics market is about $3 billion. This amount includes an estimated $700 million in instrument sales, and $2.3 billion in sales of reagents and consumables. Some of the principal instrument manufacturers are Hitachi (Japan), Miles Laboratories (United States), E. I. du Pont de Nemours (United States), Beckman Instruments (United States), Eastman Kodak (United States), Abbott Laboratories (United States), Olympus (Japan), Toshiba (Japan), Hoffmann-La Roche (Switzerland), and Ciba Corning Diagnostics (United States).

PAUL GHERSON
HOWARD LANZA
MILTON PELAVIN
DAN VLASTELICA
Miles Inc.

Additional Reading

Tietz, N.W. ed.: *Textbook of Clinical Chemistry,* W. B. Saunders Co., Philadelphia, Pa., 1986.

Kaplan, L.J. and A.J. Pesce: *Clinical Chemistry, Theory, Analysis, and Correlation,* The C. V. Mosby Co., 1989.

Kessler, G. and L.R. Snyder: *Technicon International Congress,* Vol. 1, Mediad Press, Tarrytown, NY, 1977, pp. 28–35.

Kolthoff, I.M. P.J. Elving, and E.J. Meehan: *Optical Methods of Analysis, Vol. 8, Treatise on Analytical Chemistry,* Part I, John Wiley & Sons, Inc., New York, NY, 1986.

AUTOMATED INSTRUMENTATION: HEMATOLOGY. Hematology analyzers provide information about blood cells and their constituents. The three basic blood cell types are erythrocytes or red blood cells, leukocytes or white blood cells, and thrombocytes or platelets. Hemoglobin is the principal nonaqueous component of red blood cells. Its physiological importance gives it the status of a primary hematological constituent.

Aperture Impedance Instruments

The aperture impedance principle of blood cell counting and sizing, also called the Coulter principle, exploits the high electrical resistivity of blood cell membranes. Red blood cells, white blood cells, and blood platelets can all be counted. In the aperture impedance method, blood cells are first diluted and suspended in an electrolytic medium, then drawn through a narrow orifice (aperture) separating two electrodes (Fig. 1). In the simplest form of the method, a d-c current flows between the electrodes, which are held at different electrical potentials. The resistive cells reduce the current as the cells pass through the aperture, and the current drop is sensed as a change in the aperture resistance.

Aperture impedance counters are flow cytometers, as are all current automated cell counters. Flow cytometers measure cells as the cells flow hydraulically, one at a time, through sensing zones.

An alternative approach is to model the probability of coincidence as a function of either sample concentration or the fraction of time the sensing zone is occupied by passing cells (dead time). The model produces a coincidence-correction factor which is applied to the count. Still another approach is to reduce the size of the sensing zone, which reduces the probability of coincidence. In practice, commercial counters combine the latter two approaches.

Flow cytometer cell counts are much more precise and more accurate than hemocytometer counts. Hemocytometer cell counts are subject both to distributional and sampling errors.

Aperture impedance counters provide cell volume information as well as cell counts. They also provide mean red cell volume (MCV) and mean platelet volume (MPV) as well as cell volume distribution information because they measure volume on a cell-by-cell basis.

Aperture impedance counters provide cell volume information as well as cell counts.

In summation, aperture impedance counters provide information on WBC, RBC, HCT, MCV, PLT, MPV, RDW, PDW, and three-part Diff, for d-c current only, or four-part Diff using a d-c/r-f combination.

Light-Scattering Instruments

The light-scattering principle of cell counting is based on the observation that microscopic particles, such as blood cells, scatter into small (0–15°) angles, most of the visible light incident upon them. This principle is used to count red blood cells, white blood cells, and platelets. In the basic form of the light-scattering method, a dilute suspension of cells and a sheathing

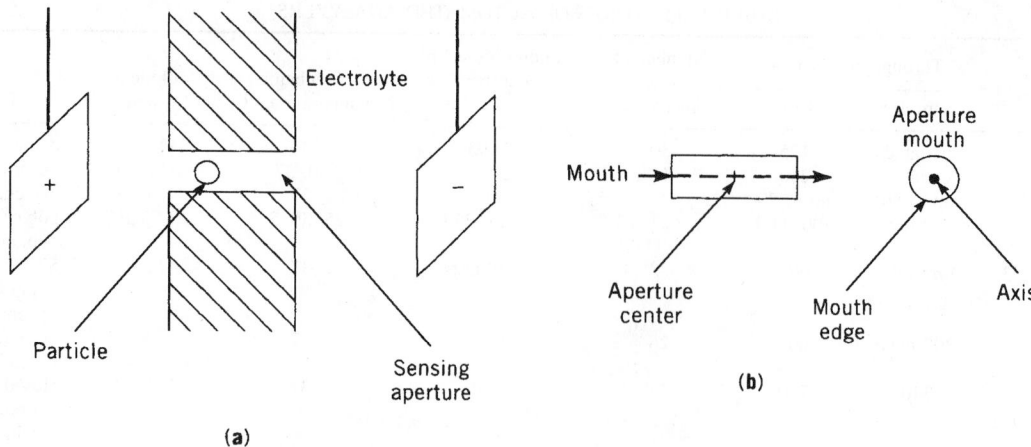

Fig. 1.   (**a**) Schematic diagram of the electrical impedance method for counting and sizing blood cells. The electrodes are represented as charged rectangles. As particles flow through an aperture, the impedance and, hence, the flow of current are modified; (**b**) schematic diagram of the sensing aperture of a flow cytometer

fluid having a matching refractive index flow concentrically through an optical flow cell onto which light is focused. The cells scatter the incident light as they pass through the flow cell, and the light scattered into a small-angle interval, typically $1-3°$, is sensed by an optical detector. The purpose of the sheath is to narrow the cell stream and reduce the cell passage to single file while centering the stream in the flow cell. The unscattered component of focused light is much larger than the scattered component and must be blocked from the detector so that it does not swamp the desired small-angle scatter signal.

The basic single-angle interval light-scattering method cannot accurately measure individual red blood cell or platelet volumes, but it can provide MCV and MPV. Red cells are bi-concave disks, and platelets are rod to disk shaped. Scattering intensities depend on the orientation in the flow cell. Because the cells can interrupt the optical path in random orientations, individual scattering intensities are not proportional to cell volume. However, because thousands of cells of each type pass through the flow cell, the effects of orientation can be averaged.

Light-scattering measurements of sphered cells are not subject to orientation effects, and a method for the rapid sphering and fixing of red cells for the purpose of measuring them in a light-scattering flow cytometer has been developed.

Light-scattering measurements of red cell volume are accurate only if the measurements account for the effects of cell refractive index on scattering intensity. Basic single-angle interval light-scattering measurements of red cell volume, even for sphered and fixed red cells, are only accurate to within approximately 10%. Mie Scattering Theory predicts that the angular scattering intensity pattern for sphered red cells depends on cell refractive index as well as cell size. Red cell refractive index is linearly related to cellular hemoglobin concentration. Therefore, two normal red cells of the same volume but of different cellular hemoglobin concentration scatter light differently. A sphered cell's volume and the hemoglobin concentration can both be accurately determined from Mie Scattering Theory calculations by making simultaneous measurements of light-scatter intensity over two suitably chosen angle interval. Two suitable angle intervals for red cells are $2-3°$, known as low angle scatter (SL), and $5-15°$, high angle scatter (SH). In contrast to the aperture impedance method, this method for determining red cell volume does not suffer from inaccuracies as a result of cellular hemoglobin concentration variability.

The two-angle interval scattering method can provide CBC, RDW, PDW, PCT, and HDW.

The single-angle interval light-scattering intensity counts white cells but cannot provide five-part Diffs. White cell light-scattering intensity depends on cell size, cell refractive index, and internal cell structure. White cell size ranges overlap and the cells differ in refractive index. Scattering measurements can distinguish among cell types based on differences in internal structure alone, but only if the cell types have the same overall size and refractive index.

Light-scattering measurements made over two suitable chosen angle intervals can combine with depolarized light-scattering measurements to provide five-part Diffs.

Cytochemistry

Cytochemical techniques can be combined with light-scattering and absorption measurements to provide five-part Diffs. Cytochemistry concerns the chemical reactions of cell components. The reactions for automated white blood cell differential analysis include those that bind chromophores to the granules of white cell types, based on the presence of various substrates and enzymes in the granules. These reactions yield products suitable for light-scattering and optical absorption measurements. Other reactions exploit the differential resistance of white cell types to cytoplasmic stripping by the lysing action of surfactants. The reaction products are suitable for light-scattering and aperture impedance measurements.

Various method combinations can be used to produce five-part Diffs including d-c/r-f aperture impedance plus light scattering, two-angle interval light-scattering plus depolarized light scattering; light scattering plus cytochemistry; and d-c/r-f aperture impedance plus cytochemistry. Flow cytometric methods produce five-part Diffs that are generally more precise and accurate than manual Diffs. However, manual methods can provide more information.

Photometric Hemoglobinometry

Most automated hemoglobinometry methods are derivatives of the manual ICSH reference method. The manual method follows three steps: (*1*) The blood sample is diluted in a reagent containing Triton X-100, a nonionic surfactant. The surfactant lyses the RBC membrane, releasing hemoglobin into solution. (*2*) Ferricyanide present in the diluent diffuses into the hemoglobin molecule and oxidizes heme from the Fe_{2+} to the Fe_{3+} form, yielding methemoglobin. (*3*) Cyanide diffuses into methemoglobin's interior and reacts with heme to yield cyanmethemoglobin (HiCN), which has a characteristic absorption spectrum that is measured spectrophotometrically.

The automated method differs from the ICSH method chiefly in that oxidation and ligation of heme iron occur after the hemes have been released from globin. Therefore, ferricyanide and cyanide need not diffuse into the hemoglobin and methemoglobin, respectively. Because diffusion is rate-limiting in this reaction sequence, the overall reaction time is reduced from approximately three minutes for the manual method to $3-15$ s for the automated method.

DAVID ZELMANOVIC
ROQUE A. GARCIA
JOHN TURRELL
Technicon Instruments Corporation

Additional Reading

Grant J.L., et al.: *Amer. J. Clin. Path.* **33**, 138–143 (1960).
Malin M.J., et al.: NY, *J. Clin. Path.* **92**(3), 286–294 (1989).
van de Hulst H.C. *Light Scattering by Small Particles,* John Wiley & Sons, Inc., New York, 1957, Chapt. 8.
Wales M., et al.: *Rev. Sci. Inst.* **32**, 1132–1136 (1961).

AUTOXIDATION. A word used to describe those spontaneous oxidations, which take place with molecular oxygen or air at moderate temperatures (usually below 150°C) without visible combustion. Autoxidation may proceed through an ionic mechanism, although in most cases the reaction follows a free radical-induced chain mechanism. The reaction is usually autocatalytic and may be initiated thermally, photochemically, or by addition of either free radical generators or metallic catalysts. Being a chain reaction, the rate of autoxidation may be greatly increased of decreased by traces of foreign material.

Many organic and a variety of inorganic compounds are susceptible to autoxidation.

AUTUNITE. This mineral is a hydrous phosphate of calcium and uranium, crystallizing in the tetragonal system, usually in thin tabular crystals. Good basal cleavage; hardness, 2–2.5; specific gravity, 3.1; luster, subadamantine to pearly on the base; color, lemon yellow; streak, yellow; transparent to translucent; strongly fluorescent.

Originally from near Autun in France, whence the name, it is a secondary mineral associated commonly with uraninite. In the United States, it occurs sparsely in the pegmatites of Connecticut, New Hampshire and North Carolina. Autunite also is known as *calco-uranite*. See also **Uraninite.**

AVOGADRO CONSTANT. The number of molecules contained in one mole or gram-molecular weight of a substance. The most recent value is $6.0220943 \times 10^{23} \pm 6.3 \times 10^{17}$. In measurements made by scientists at the National Bureau of Standards (Gaithersburg, Maryland) and announced in late 1974, the uncertainty (as compared with previous determinations) of the number was reduced by a factor of 30.

AVOGADRO LAW. The well-recognized principle known by this name was originally a hypothesis suggested by the Italian physicist Avogadro, in 1811, to explain the puzzling rule of proportional volumes observed in chemical reactions of gases and vapors. It states simply that equal volumes of all gases and vapors at the same temperature and pressure contain the same number of molecules. Though this assumption accords with the facts and aids the kinetic theory of gases, just why it should be true is by no means self-evident, unless one starts with the much more recent Maxwell-Boltzmann law of equipartition of energy, which also requires proof. That Avogadro's law is true cannot be said to have been positively established until the experiments of J.J. Thomson, Millikan, Rutherford, and others determined the value of the electron as an electric charge and thereby made it possible to count the number of atoms of different elements in a gram. The actual number of molecules contained in one mole (gram-molecular weight) of a substance is the Avogadro constant.

At any fixed temperature and pressure, the density of carbon dioxide gas, for example, is approximately 22 times greater than the density of hydrogen gas. Thus, the mass of 1 liter of carbon dioxide is 22 times the mass of 1 liter of hydrogen gas. According to Avogadro's principle, the number of molecules in 1 liter of carbon dioxide is the same as the number of molecules in 1 liter of hydrogen. Thus, it follows that a carbon dioxide molecule must have a mass that is 22 times larger than the mass of a hydrogen molecule. Since the molecular weight of hydrogen (H_2) was set equal to 2, carbon dioxide was assigned a molecular weight of 22×2, or 44. Cannizzaro was the first to use gas densities to assign atomic and molecular weights. Avogadro's principle also may be used to assign molecular weights in a slightly different way. At standard temperature and pressure, the volume of a mole of any gas is 22.4 liters. The molecular weight of a gas, therefore, is the mass (in grams) of 22.4 liters of the gas under standard conditions. For most gases, the deviation from this ideal value is less than 1%. See also **Avogadro Constant;** and **Combustion (Fuels).**

AXINITE. This mineral is an aluminum-boron-calcium silicate with iron and manganese, $(Ca, Mn, Fe)_3Al_2BSi_4O_{15}(OH)$. It crystallizes in the triclinic system, yielding broad sharp-edged forms, which has led to its name, derived from the Greek word meaning axe. It breaks with a conchoidal fracture; hardness, 6.5–7; specific gravity, 3.22–3.31; luster, vitreous; colors, brown, blue, yellow and gray; transparent to translucent.

Axinite occurs in granites or more basic rocks along contacts and in cavities in Saxony, Switzerland, France, England, Tasmania, and Japan; in the United States, in New Jersey, Pennsylvania, and California.

AZEOTROPIC SYSTEM. A system of two or more components that has a constant boiling point at a particular composition. If the constant boiling point is a minimum, the system is said to exhibit *negative azeotropy*, if it is a maximum, *positive azeotropy*.

Consider a mixture of water and alcohol in the presence of the vapor. This system of two phases and two components is divariant (see Phase Rule). Now choose some fixed pressure and study the composition of the system at equilibrium as a function of temperature. The experimental results are shown schematically in Fig. 1.

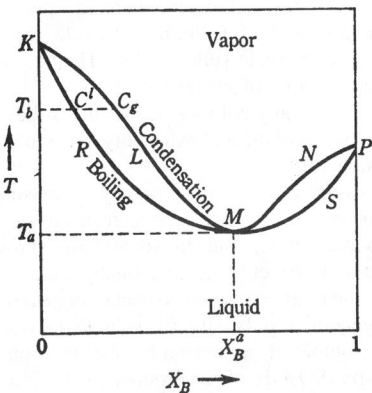

Fig. 1. Azeotropic system

The vapor curve KLMNP gives the composition of the vapor as a function of the temperature T, and the liquid curve KRMSP gives the composition of the liquid as a function of the temperature. These two curves have a common point M. The state represented by M is that in which the two states, vapor and liquid, have the same composition x_B^a on the mole fraction scale. Because of the special properties associated with systems in this state, the Point M is called an azeotropic point and the system is said to form an azeotrope. In an azeotropic system, one phase may be transformed to the other at constant temperature, pressure and composition without affecting the equilibrium state. This property justifies the name azeotropy, which means a system that boils unchanged.

AZIDES. The salts of hydrazoic acid are termed *azides*. Metallic azides can be prepared from barium azide and the metal sulfate, or from potassium azide and the metal perchlorate.

Soluble azides react with iron (III) salt solutions to produce a red color, similar to that of iron(III) thiocyanate. Sodium azide is not explosive, even on percussion, and nitrogen may be evolved upon heating. With iodine dissolved in cold ether, silver azide forms iodine azide (IN_3), a yellow explosive solid.

Sodium azide is a slow oxidizing agent. It has a selective action in inhibiting the growth of gram-negative organisms. It has been used as a component in selective media such as azide glucose broth or azide blood agar base for the isolation of mastitis and fecal *streptococci*.

A number of alkyl and aryl azides are known, such as CH_3N_3, $C_2H_5N_3$ and $C_6H_5N_3$. The nonmetallic inorganic azides include ClN_3, an explosive gas, BrN_3, an orange liquid, mp $-45°C$, and IN_3, a yellow solid, decomposing above $-10°C$. The gas FN_3 is more stable than ClN_3, decomposing only slowly at room temperature.

Lead and silver azides are widely used as initiating, or primary explosives because they can be readily detonated by heat, impact, or friction. As such, these materials, particularly lead azide, are used in blasting caps, percussion caps, and delay initiating devices. The function of the azides is similar to that of mercury fulminate or silver fulminate.

AZINES. The products of the reaction between an aldehyde or a ketone with hydrazine are termed *azines*. A number of dyestuffs and complex members of the pyridine family of compounds also are termed *azines*. See also **Pyridine and Derivatives.**

AZO AND DIAZO COMPOUNDS. Characteristically, these are compounds containing the group–N:N–(azo) or >N : N (diazo). They are closely related to the substituted hydrazines. The N_2 group may be covalently attached to other groups at both ends, as in the azo compounds, or at only one end, as in the diazo compounds or diazonium salts. Although organic chemistry furnishes the most numerous examples, many inorganic azo compounds also exist.

Compounds related to aniline, either directly or by oxidation, and to nitrobenzene by reduction, are numerous and important. When nitrobenzene is reduced in the presence of hydrochloric acid by tin or iron, the product is aniline (colorless liquid); in the presence of water by zinc, the product is phenylhydroxylamine (white solid); in the presence of methyl alcohol by sodium alcoholate or by magnesium plus ammonium chloride solution, the product is azoxybenzene (pale yellow solid); by sodium stannite, or by water plus sodium amalgam, the product is azobenzene (red solid); in the presence of sodium hydroxide solution by zinc, the product is hydrazobenzene (pale yellow solid). The behavior of other nitro-compounds is similar to that of nitrobenzene.

Diazonium salts are usually colorless crystalline solids, soluble in water, moderately soluble in alcohol, and when dry are violently explosive by percussion or upon heating.

The simplest azo-dyes are yellow, but by increasing the number of auxochrome groups or by increasing the percentage of carbon, the color darkens to red, violet, blue, and in some cases brown. Naphthalene residues darken to red, violet, blue and finally black. These aminoazo-dyes, together with the hydroxyazo-dyes (containing auxochrome hydroxyl-group–OH), are generally only slightly soluble in water. In order that the dye may be soluble it is desirable that it contain one or more sulfonic acid groups–SO_2OH. This group may be introduced either by treating the dye with concentrated sulfuric acid, or by using sulfonic acid derivatives in preparing the dye, e.g., methyl orange, sodium dimethyl-*para*-aminoazobenzene-*para*-sulfonate

$$(4)(CH_3)_2NC_6H_4N{:}NC_6H_4SO_2ONa(4)$$

from dimethylaniline and diazotized sulfanilic acid (*para*-amino-benzene sulfonic acid), $(1)H_2N{\cdot}C_6H_4{\cdot}SO_2OH(4)$, and then the sodium salt is made from the product. Other azo-dyes are

$$\text{chrysoidine}\left(C_6H_5N{:}NC_6H_3\begin{array}{c}NH_2(2)\\[4pt]NH_2(4)\end{array}\right)$$

$$\text{Bismarck brown}\left((3)H_2N{\cdot}C_6H_4N{:}NC_6H_3\begin{array}{c}NH_2(2)\ \cdot HCl\\[4pt]NH_2(4)\end{array}\right)$$

AZURITE. This mineral is a basic carbonate of copper, crystallizing in the monoclinic system, with the formula $Cu_3(CO_3)_2(OH)_2$, so called from its beautiful azure-blue color. It is a brittle mineral with; a conchoidal fracture; hardness, 3.5–4; sp gr, 3.773; luster, vitreous, color and streak, blue; transparent to translucent. Azurite, like malachite, is a secondary mineral, but far less common than malachite. It is formed by the action of carbonated waters on compounds of copper or solutions of copper compounds.

B

BABINGTONITE. This mineral is a relatively rare calcium-iron-manganese silicate, occurring in small black triclinic crystals, found in Italy, Norway and in the United States at Somerville and Athol, Massachusetts, and in Passaic County, New Jersey. It was named for Dr. William Babington.

BACK-GOUDSMIT EFFECT. An effect closely related to the Zeeman effect. It occurs in the spectrum of elements having a nuclear magnetic and mechanical moment. See also **Hyperfine Structure**; and **Paschen-Back Effect**.

BACKSCATTERING. The deflection of particles or of radiation by scattering processes through angles greater than 90° with respect to the original direction of motion.

One particular application of backscattering is the use of beta rays (electrons) to determine properties of substances. This phenomenon has been known for many years, but more recently its variations in value with differences in the atomic and molecular composition of the scattering substance have been found to give results that can be correlated with the atomic numbers of the atoms. In some cases where identical backscattering is obtained from two or more compounds (because of an accidental agreement between the total scattering of their atoms) differences in beta-ray absorption between them are usually available to provide another clue to their composition. Backscattering has also found application in measuring the thickness of coatings on materials such as paper, plastic films, and strip steel.

BACTERIA. Microscopic, unicellular cells bounded by a membrane-wall complex and containing a variety of inclusions. Depending upon the species and cultural conditions, bacteria occur as individual cells or in clumps or chains of sister cells. Bacteria lie at the lower limits of resolution of the optical microscope. The average length lies within the range of 2 to 5 micrometers, although some are as small as 0.2 micrometer, or as large as 100 micrometers in length.

Bacteria are classified, somewhat arbitrarily, by a descriptive array of features, one of the most common being shape.

Bacilli

In terms of shape, the first of these are rod-shaped and are called *bacilli* (singular, *bacillus*). The bacilli often have small, whip-like structures known as flagella, with which they are able to move about. Some bacilli have oval, egg-shaped, or spherical bodies in their cells, known as spores. Under adverse conditions, such as dehydration, and in the presence of disinfectants, the bacteria may die, but the spores may be able to live on. The spores germinate when the conditions become favorable, and form new bacterial cells. Some are so resistant that they can withstand boiling and freezing temperatures and prolonged desiccation. See Fig. 1.

Cocci

A second type of bacteria is the *cocci* (singular, *coccus*) which are spherical or ovoid in shape. The individual bacterial cells of this group may occur singly (*Micrococcus*), in chains (*Streptococcus*), in pairs (*Diplococcus*), in irregular bunches (*Staphylococcus*), and in the form of cubical packets (*Sarcina*). The coccus does not form spores and usually is nonmotile. See Fig. 2.

Curved or Bent Rods

A third group of bacteria are the curved or bent rods. Of these, the genus *Vibrio* is composed of bacteria that are comma-shaped; and the genus *Spirillum* consists of those that are twisted and spiral in form. All members

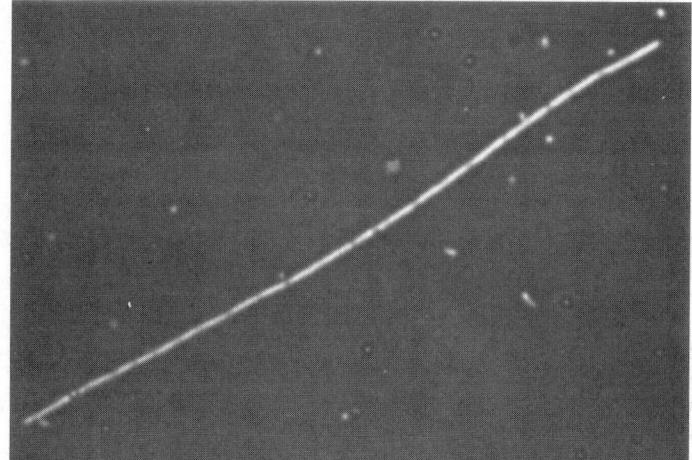

Fig. 1.   Bacillus. (*A.M. Winchester*)

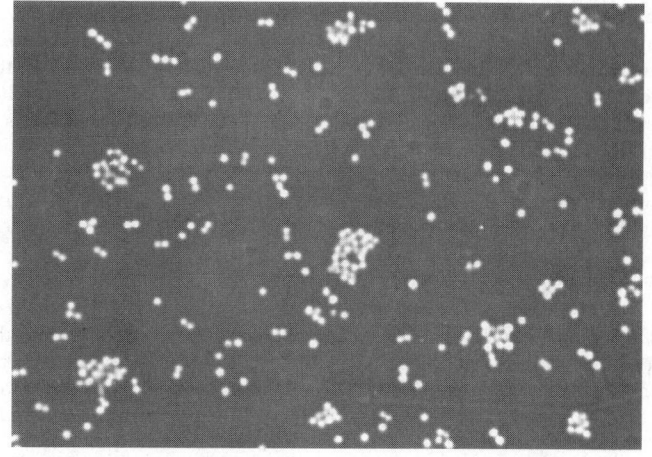

Fig. 2.   Coccus. (*A.M. Winchester*)

of this group are motile, but none form spores. However, some of these bacteria form a gelatinous capsule or covering by which they are probably protected from adverse environmental conditions. Still another group of spiral-shaped bacteria are known as the *spirochetes*, one of which is the cause of syphilis. See Fig. 3.

Other Bases for Classifying Bacteria

Bacteria also may be classified on the basis of their requirements of free atmospheric oxygen. Those requiring atmospheric oxygen are called *aerobic* (air-living); those which cannot live in the presence of atmospheric oxygen are called *anaerobic*; those that do well with oxygen, but can get along without it are termed *facultative anaerobes*.

Bacteria are dependent upon the proper temperature for life and reproduction; and the various species of bacteria may differ widely in their temperature requirements. Most of the disease-producing (*pathogenic*) bacteria thrive best at body temperatures; others may live and multiply in much cooler temperatures; while still others live in hot springs. Freezing, as a rule, does not destroy bacteria, but prevents their reproduction.

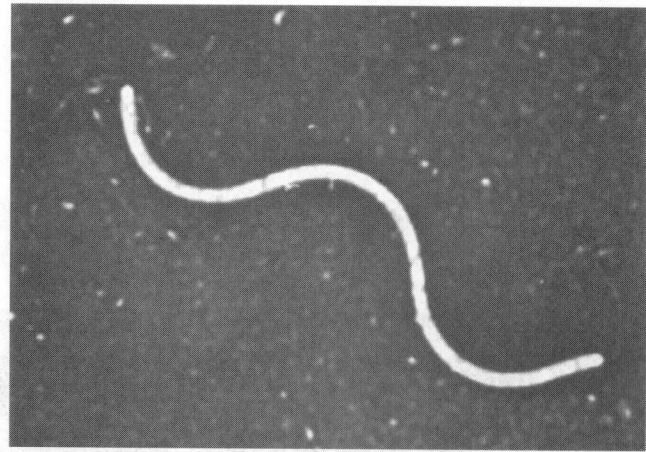

Fig. 3.   Spirillum. (*A.M. Winchester*)

High temperature, conversely, quickly kills many bacteria. Most disease-producing organisms in milk, for example, may be killed by raising the temperature to 143°F (61.6°C) in the pasteurizing process. Most non-spore-forming, disease-producing microorganisms are destroyed by boiling water. In the spore stage, some bacteria must be heated to 240°F (116°C) for a considerable period of time, in order that the spores be destroyed. These high temperatures are best obtained by steam under pressure.

A further taxonomic characteristic of bacteria, which is of some importance in disease diagnosis and treatment, is based upon their staining reactions—specifically upon their response to the Gram staining technique. In this reaction, a basic dye (crystal violet) is first applied, then a solution of iodine. All bacteria will be stained blue at this point. The cells are then treated with alcohol. Gram-positive bacteria retain the crystal violet-iodine complex and remain blue. Gram-negative cells are completely decolorized by the alcohol. A red counterstain is then applied so that the decolorized Gram-negative cells will take on a contrasting color. The basis of the differential Gram reaction lies in the structure of the cell wall. Another staining approach distinguishes acid-fast bacteria (which retain carbolfuchsin dye even when decolorized by hydrochloric acid) from non-acid-fast bacteria. Capsule staining also can be used, but the Gram reaction is of primary importance. Table 1 shows which organisms are Gram negative and which are Gram positive.

Fundamental Structure of Bacteria

For many years, before the concept of *recombinant DNA*, scientists postulated that bacteria were unicellular microbes. In more recent years, the unicellular structure and, notably, the unicellular behavior of bacteria has been subjected to serious questions by a number of scientists. As observed by Shapiro, "Investigators are finding that in many ways an individual bacterium is more analogous to a component cell of a multicellular organism than it is to a free-living autonomous organism." There is

TABLE 1. GRAM REACTION CATEGORIES OF BACTERIA

Gram-positive	Gram-negative
Actinomyces bovis	Aerobacter aerogenes
Bacillus anthracis	Brucella abortus
Clostridium butyricum	Brucella suis
Clostridium septicum	Eberthella typhi
Clostridium sordelli	Escherichia coli
Clostridium tetani	Hemophilus pertussis
Clostridium welchii	Klebsiella pneumoniae
Corynebacterium diphtheriae	Neisseria gonorrheae
Diplococcus pneumoniae	Neisseria intracellularis
Erysipelothrix muriseptica	Pasteurella pestis
Mycobacterium tuberculosis	Proteus vulgaris
Staphylococcus aureus	Pseudomonas aeruginosa
Streptococcus fecalis	Salmonella enteriditis
Streptococcus hemolyticus	Shigella paradysenteriae
Streptococcus lactis	Vibrio comma
Streptococcus salivarius	
Streptococcus viridans	

evidence that complex communities of bacteria hunt prey, sometimes leaving chemical trails for the guidance of thousands of individuals. Scientists have observed such bacterial communal activity in *Rhizobium* microorganisms that fix nitrogen in the roots of leguminous plants. See also **Nitrogen**. Some bacteria have been observed in distinct colonies in petri dishes, and it has also been observed that *photosynthetic* bacteria (*Cyanobacteria*) grow as connected chains or intertwined mats with definitive configurations. Such reconfigurations, obvious in numerous bacteria, project to some scientists a manifestation of DNA rearrangement. With reference to the most morphologically complex of all bacteria, Shapiro observes, "Their elaborate fruiting bodies rival those of fungi and slime molds and have long been an object of scientific curiosity." Much research along these lines has been conducted by Hans Reichenbach (Society for Biotechnological Research, Braunschweig) and other researchers at the Institute for Scientific Film (Göttingen, Germany). Shapiro poses an interesting question, "What practical value, if any, do these findings have?" They may serve as insights to researchers who are seeking ways to produce various biochemicals via genetically engineered bacteria, or to those scientists in the medical field who are seeking improved drugs for handling bacterial infections.

In a scholarly paper, Magasanik reviews the role of biological research into the genetics, biochemistry, and physiology of bacteria during the past four decades and stresses the importance of these efforts in learning more about cells of all types at the molecular level. Magasanik stresses the advantages of using bacteria in genetic studies, including in most instances their simple non-compartmentalized structures and the accessibility of their genetic material. The importance of studying *Escherichia coli* for over a half-century is reviewed. Of notable interest is Magasanik's observation, "Yet, less than 50 years ago, in 1954, Kluyver and Van Niel, two eminent microbiologists, found it necessary to devote five lectures at Harvard University to convince their audience that the study of microbes could make a major contribution to biology."

Bacterial Genetics. The deoxyribonucleic acid (DNA) of bacteria is predominantly located in masses of variable shape, nuclearbodies (nucleoplasm or genophore), unbounded by a nuclear membrane. Bacteria thus are classified as procaryoids, in contrast to higher organisms containing nuclear membranes, the eucaryoids. In general, nongrowing, stationary-phase bacteria contain one nuclear body per cell, whereas exponentially growing, log-phase bacteria contain two or more nuclear bodies per cell. These nuclei are the sister products of a preceding nuclear division.

When bacteria are inoculated into growth medium, there is a delay (lag phase) before division and exponential growth ensue. The rate of exponential growth is a characteristic of the bacterial strain, the temperature, and the nutritional environment. The amount of DNA per nuclear body remains constant at various growth rates, although cell mass and average number of nuclei per cell are functions of the growth rate.

Most of the genetic information of bacteria is contained in a single structure of fixed DNA content, a giant circular DNA molecule that replicates semi-conservatively. The enzymatic reactions involved in the biologically fundamental processes of DNA biosynthesis and genetic recombination are being elucidated in studies with bacterial systems.

Chemical and Therapeutic Measures

In addition to the preventive measures described, for well over a century people have turned to chemical substances to assure clean environments and numerous drugs for treating bacterial infections once acquired.

A majority of bacteria may be killed by the action of chemical disinfectants. Sometimes a substance that prevents infection or inhibits the growth of microorganisms may be referred to as an antiseptic. Over the years, phenol and related compounds have proven effective when used with care (to avoid chemical injury). Free chlorine gas is an excellent disinfectant, as are the hypochlorites, for use in sterilizing structures. Tincture of iodine used on some cuts and other wounds has good disinfecting power, as does hydrogen peroxide in some cases, but these materials should be used with professional guidance. At one time, a number of mercury-containing compounds were used effectively and widely prior to environmental concerns associated with the element mercury.

There has been much emphasis during the last few decades on *bacteriostatic* agents, which prevent or slow down the rate of bacterial growth and reproduction, so that the natural protective mechanisms of the body can overcome the infection. These chemicals include the sulfonamide group, such as sulfathiazole, sulfadiazine, sulfanilamide, sulfasuxidine, and sulfaguanidine. Although valuable in the treatment of certain diseases, the drugs should not be taken indiscriminately, nor in conjunction with bactericidal agents. See also **Sulfonamide Drugs**.

Antibiotics, which are produced by other living organisms, inhibit the growth of bacteria or destroy them (bactericidal). There are few known bacterial diseases, the effects of which cannot be mitigated if the proper antibiotic is used early in the course of the disease. Tetanus and botulism are exceptions. These diseases are the manifestation of extremely potent toxins produced by the bacteria, rather than symptoms caused by infections of the microorganisms themselves. See also **Antibiotics**.

Remarkable Survivability of Some Bacteria

Plastic pipes, even when flushed out with the most powerful disinfectants and germicides, have proven to be safe havens for some bacterial strains. Bacteria-resistant piping is of major importance in pharmaceutical manufacture. Research is underway to find plastic piping that will reject the adhesion of bacterial slimes. Currently, alloy steels are widely used. The adherence of slimes to plastic pipes permits colonies of bacteria to multiply. A similar problem exists when patients are furnished with plastic implants or prostheses. Hospital water supplies must be continuously monitored.

Beneficial Bacteria

As with insects—of which there are numerous species that are damaging to life processes (crop infestation, human discomfort, etc.), but also many that are beneficial (honeybee, lady beetle, etc.)—so are there both helpful and harmful bacteria. Bacteria play a major and constructive role in numerous processes that support life, such as food digestion and synthesis of vitamins. They serve as a basis for manufacturing antibiotics and as tools of genetic research. This list easily could be expanded a thousandfold. Some of the lesser-known examples are cited here.

Oil-Eating Bacteria. For several years, studies have been conducted to determine the ability of specialized bacteria. Some of these microorganisms are capable of converting oil into fatty acids, the result of which makes the oil products more water-soluble. Special strains have been grown by oceanologists at the University of Texas. These bacteria were tested in connection with the oil slick in the Gulf of Mexico after a supertanker (Mega Borg) spilled nearly 4 million tons of crude oil. Tests with other bacteria were tested earlier in connection with the Exxon Valdez spill in Alaska. See **Water Pollution**.

Bifidobacterium in Food Products. Commonly referred to as bifidobacteria, these microorganisms were discovered by Tissier (Pasteur Institute) in 1900 in the feces of infants. These bacteria are not true lactic acid bacteria, such as *Lactococcus* or *Pediococcus*, because they produce both acetic and lactic acids. Early research was difficult because of the lack of effective laboratory procedures. Considerable research since the mid-1950s, however, has been conducted. Hughes and Hoover (University of Delaware) reported in 1991 on the beneficial qualities of bifidobacteria and the possibility of their use in "Bifid"-amended food products, notably dairy products. These therapeutic effects include:

1. Maintenance of normal intestinal microflora balance;
2. Improvement of lactose tolerance of milk products;
3. Anti-tumorigenic activity;
4. Reduction of serum cholesterol levels; and
5. Synthesis of B-complex vitamins.

Products fermented with the bifid culture have a mild acidic flavor, similar to that of yogurt. A bifidus milk was developed for therapeutic use as early as the 1940s. By the 1960s it was found that it was possible to positively modify intestinal flora with bifidum cultures. By the late 1980s, in Japan, it was found that yogurt sales nearly doubled with bifid-containing products. Similar increases in popularity occurred in France. As of the early 1990s, bifidus products are marketed in Brazil, Canada, England, Italy, Poland, and some of the Balkan countries. Because of health benefits and a good track record in other countries, no major barriers for expanding its use in the United States are foreseen by the experts.

Archaebacteria

Sometimes referred to as the "Third Kingdom of Life," the archaebacteria differ markedly from other bacteria. In fact, most scientists do not consider this form of life as a bacterium in any sense. The topic is included here because the association with bacteria is often made, since this microorganism has been misnamed, and most readers seeking information on it would most likely turn to this topic initially.

The archaebacteria were discovered by Woese (Univ. of Illinois, Urbana) as recently as 1977. These microorganisms differ markedly from eukaryotes, which have visible nuclei and are found in plants and animals, and differ as well from prokaryotes, which principally are found in bacteria and blue-green algae. See also **Genetics and Gene Science (Classical)**.

In the area of a volcano, one may observe hot muds and polluted areas of water and air and quickly reach the conclusion that no life could possibly be present in such an environment. But microbial forms of life may be present, as typified by archaebacteria, which resemble ordinary bacteria, but which some scientists suggest may be another form of living material. There are numerous species of *Archaebacterium*, including: (1) *thermophiles*, which can survive up to temperatures of boiling water and greater; (2) *halophiles*, which tolerate extremely salty substances (greater, for example, than would be encountered in the Dead Sea); (3) *acidophiles*, one species that accommodates great acidity (ph = 1) and high temperatures (96°C) and is aptly named *Acidanus infernus*, and (4) barophiles, which can withstand tremendous deep-sea pressures and simulated laboratory pressures (Scripps Institution of Oceanography) up to 1300 to 1400 atmospheres. It is interesting to note that acidophiles maintain their interiors at a neutral pH of 7.0, the mechanism of which remains to be discovered. In fact, the manner in which these organisms alter their molecular structure to withstand such trying conditions thus far has defied logical explanation.

A scientist at the Woods Hole Oceanographic Institution observes that it is fortunate that deep-sea organisms can adapt to such extreme pressures—otherwise, dead plant and animal debris that falls to the ocean bottom probably would not decay. It is surmised that barophilic bacteria participate in recycling organic materials in the ocean.

To date, archaebacteria have been positively identified in volcanic areas, such as Iceland, Italy, and Yellowstone Park (U.S.), and in the vicinity of hydrothermal vents in deep oceanic depths. Limited research to date has been conducted by Woese, previously mentioned, by researchers in oceanology, and by Stretter (Univ. of Regensburg, Germany). Generally, it has been found that the growth of most species stops at about 110°C, but that optimal growth occurs at about 100°C. It has been surmised to date that these microorganisms convert various organic materials by combining C with H_2 to form methane (CH_4). Other species appear to combine S and H to form H_2S.

Some scientists currently forecast that research on the archaebacteria may lead to a better understanding of catalytic enzymes and, because of this property, lead to catalysts that can participate at higher temperatures and thus accelerate chemical reaction time.

Additional Reading

Ball, A.S.: *Bacterial Cell Culture,* John Wiley & Sons, Inc., New York, NY, 1997.

Barton, L.L.: *Sulfate-Reducing Bacteria,* Plenum Publishing Corporation, New York, NY, 1995.

Blankenship, R.E., M.T. Madigan, and C.E. Bauer: *Anoxygenic Photosynthetic Bacteria,* Kluwer Academic Publishers, Morwell, MA, 1995.

Cossart, P.F., R. Rappuoli, and P. Boquet: *Cellular Microbiology,* ASM Press, Washington, DC, 1999.

Dale, J.W.: *Molecular Genetics of Bacteria,* 3rd Edition, John Wiley & Sons, Inc., New York, NY, 1998.

Denny, F.W.: "The Streptococcus Saga Continues," *N. Eng. J. Med.,* **127** (July 11, 1991).

Donowitz, L.G.: "Hospital-Acquired Infections in Children," *N. Eng. J. Med.,* **1836** (December 27, 1990).

Dunny, G.M. and S.C. Winans: *Cell-Cell Signaling in Bacteria,* ASM Press, Washington, DC, 1999.

Edwards, C. and J.M. Walker: *Environmental Monitoring of Bacteria,* Vol. 12, Humana Press, Totowa, NJ, 1999.

Fischetti, V.A., J.J. Ferretti, R.P. Novick, and R.K. Tweten: *The Gram-Positive Pathogens,* ASM Press, Washington, DC, 2000.

Fletcher, M.M.: *"Bacterial Adhesion," Molecular and Ecological Diversity,* John Wiley & Sons, Inc., New York, NY, 1996.

Gray, G.C. et al.: "Hyperendemic Streptococcus pyogenes Infections Despite Prophylaxis with Penicillin G Benzathine," *N. Eng. J. Med.,* **92** (July 14, 1991).

Henderson, B., M. Wilson, and A.J. Lax: *"Cellular Microbiology," Bacteria-Host Interactions in Health and Disease,* John Wiley & Sons, Inc., New York, NY, 1999.

Hess, D.J.: *"Can Bacteria Cause Cancer?" Alternative Medicine Confronts Big Science*, New York University Press, New York, NY, 1997.

Hoekstra, W.P., B.A. Van der Zeijst, and A.J. Alphen: *Ecology of Pathogenic Bacteria*, Elsevier Science, New York, NY, 1997.

Hughes, D.B. and D.G. Hoover: "Bifidobacteria: Their Potential for Use in American Dairy Products," *Food Tech.*, **74** (April 1991).

Isberg, R.R.: "Discrimination Between Intracellular Uptake and Surface Adhesion of Bacterial Pathogens," *Science*, **934** (May 17, 1991).

Kaiser, A.B.: "Surgical Wound Infections," *N. Eng. J. Med.*, **123** (January 10, 1991).

Klein, J.O.: "From Harmless Commensal to Invasive Pathogen," *N. Eng. J. Med.*, **339** (August 2, 1990).

Kluyver, A.J. and C.B. Van Niel: *The Microbe's Contribution to Biology*, Harvard Univ. Press, Cambridge, MA, 1956.

Lowry, P.W. et al.: "A Cluster of Legionella Sternal-Wound Infections Due to Postoperative Topical Exposure to Contaminated Tap Water," *N. Eng. J. Med.*, **109** (January 10, 1991).

Lunt, G.G., M.J. Danson, and D.W. Hough: *"Archaebacteria,"* Biochemistry and Biotechnology, Ashgate Publishing Company, Brookfield, VT, 1992.

MacFaddin, J.F.: *Biochemical Tests for Identification of Medical Bacteria*, 3rd Edition, Lippincott Williams Wilkins, Philadelphia, PA, 2000.

Mann, J. and J.C. Crabbe: *Bacteria and Antibacterial Agents*, Oxford University Press, Inc., New York, NY, 1998.

Magasanik, B.: "Research on Bacteria in the Mainstream of Biology," *Science*, **1435** (June 10, 1988).

Marrack, P. and J. Kappler: "The Staphylococcal Enterotoxins and Their Relatives," *Science*, **705** (May 11, 1990).

Moffat, A.S.: "Nitrogen-Fixing Bacteria Find New Partners," *Science*, **910** (November 16, 1990).

Neidhardt, F.C., J.L. Ingraham, and M. Schaechter: "Physiology of the Bacterial Cell," *A Molecular Approach*, Sinauer Associates, Inc., Sunderland, MA, 1997.

Patten, C.L., G. Holguin, D.M. Penrose, and B.R. Glick: *Biochemical and Genetic Mechanisms Used by Plant Growth Promoting Bacteria*, World Scientific Publishing Company, Inc., Riveredge, NJ, 1999.

Pool, R.: "Pushing the Envelope of Life," *Science*, **158** (January 12, 1990).

Prusiner, S.B.: "Molecular Biology of Prion Diseases," *Science*, **1515** (June 14, 1991).

Richet, H.M. et al.: "A Cluster of Rhodococcus (Gordona) bronchialis Sternal-Wound Infections after Coronary-Artery Bypass Surgery," *N. Eng. J. Med.*, **104** (January 10, 1991).

Rietschel, E.T. and H. Brade: "Bacterial Endotoxins," *Sci. Amer.*, **54** (August 1992).

Robb, F.T. and A.R. Place: *"Archaea," A Laboratory Manual: Thermophiles/with 1999 Biosupplynet Source Book*, Harbor Laboratory Press, Cold Spring Harbor, New York, NY, 1995.

Satin, M.: *"The Food Alert,"* The Ultimate Source Book for Food Safety, Facts on File, Inc., New York, NY, 1999.

Schaechter, M., G. Medoff, B.I. Eisenstein, and N.C. Engleberg: *Mechanisms of Microbial Disease,* 3rd Edition, Lippincott Williams Wilkins, Philadelphia, PA, 1998.

Shapiro, J.A.: *Bacteria as Multicellular Organism*, Oxford University Press, Inc., New York, NY, 1996.

Shapiro, J.A.: "Organization of Developing Escherichia coli Colonies Viewed by Scanning Electron Microscopy," *J. of Bacteriology*, **169**, 142 (January 1987).

Snyder, L. and W. Champness: *Molecular Genetics of Bacteria*, ASM Press, Washington, DC, 1997.

Tetz, V.V., and A.A. Totolian: *Molecular Biology of Bacteria*, NOVA Science Publishers, Inc., Huntington, NY, 1996.

Vreeland, R.H. and L.I. Hochstein: *The Biology of Halophilic Bacteria*, CRC Press, LLC., Boca Raton, FL, 1992.

Wood, B.J.: *The Lactic Acid Bacteria in Health and Disease,* Aspen Publishers, Inc. Gaithersburg, MD, 1999.

Wright, K.: "Bad News Bacteria," *Science*, **22** (July 6, 1990).

Major portions of this entry were prepared by
ANN C. DEBALDO, Ph.D.
Assoc. Prof., College of Public Health, University of South Florida,
Tampa, Florida. Other portions and updating by Staff.

BAEKELAND, L. H. (1863–1944).

Born in Ghent, Belgium. He did early research in photographic chemistry and invented Velox paper (1893). After working for several years in electrolytic research, he under took fundamental study of the reaction products of phenol and formaldehyde, which culminated in his discovery in 1907 of phenol-formaldehyde polymers originally called "Bakelite." The reaction itself had been investigated by Bayer in 1872, but Baekeland was the first to learn how to control it to yield dependable results on a commercial scale. The Bakelite Co. (now a division of Union Carbide) was founded in 1910.

BAFFLE.

An object, usually a partition, placed for some specific purpose in the flow path of a fluid, causing the fluid to take some prearranged and circuitous path. Thus, baffles are found in steam boilers to direct the hot gas properly back and forth over the tubes so that the gas will give up its heat to the required degree and will not short-circuit directly from the furnace to the stack. For this service the baffle is composed of refractory material similar to firebrick and will be found in longitudinal or transverse arrangement. Transverse baffling is made by building the baffle perpendicular to the tubes. Longitudinal baffles are usually precast and laid upon the tubes of the boiler, forming a baffle whose surface is parallel to the tubes.

Baffles are built in coagulation basins to impede the flow of liquids, and are also found in exhaust mufflers, where their purpose is to mix the flow of gases in adjacent exhaust puffs so that they may emerge from the muffler in a silent steady stream.

BAGASSE.

In the manufacture of sugar from sugar cane, the crushed fibers from which the sap has been expressed are called bagasse. Its principal use is as a fuel to run the mills that crush the cane. For this purpose bagasse is mixed with petroleum oil. It is also used as a fertilizer and to some extent in manufacturing heavy insulation board and coarse paper.

BAINITE.

A product of the decomposition of austenite that usually occurs at temperatures between those that produce pearlite and those that produce martensite. Its structure consists of finely divided carbide particles in a matrix of ferrite. See also **Austenite**.

BALANCE.

(1) Exact equality of the number of atoms of various elements entering into a chemical reaction and the number of atoms of those elements in the reaction products. For example, in the reaction $NaOH + HCl \rightarrow NaCl + H_2O$, the atoms in the input side are $H[2]$, $Na[1]$, $O[1]$, and $Cl[1]$. Each of these is also present in the products, though in different combination. The atoms of catalysts (when present) do not enter into reactions and therefore are not involved. The balance of chemical reactions follows the law of conservation of mass.

The term *material balance* is used by chemical engineers in designing processing equipment. It denotes a precise list of all the substances to be introduced into a reaction and all those that will leave it in a given time, the two sums being equal.

(2) A precision instrument designed for weighing extremely small amounts of material with high accuracy. An analytical balance or microbalance for weights from about 1 g to 0.1 mg is standard equipment in chemical laboratories. Its essential feature is a one-piece metal beam (lever) pivoted on a knife-edge or flexure at its exact center (fulcrum) so that it is free to oscillate. From it are suspended two scale pans approximately 2 inches in diameter, each of which is also positioned on a knife-edge on the lower arms of the beam. Exact balance is indicated by a pointer attached to the beam. Either an aluminum rider or a chain and vernier is provided for maximum accuracy. Highly sophisticated balances operating electronically with built-in microprocessors have become available in recent years.

BALL, PEBBLE, AND ROD MILLS.

Basically, all of these mills used for the size reduction of materials are comprised of a rotating drum which operates on a horizontal axis and is filled partially with a free-moving grinding medium which is harder and tougher than the material to be ground. The tumbling action of the grinding medium crushes and grinds the material by combination of attrition and impact. The grinding medium used may be a large number of round metal balls, operating in a drum with a metal lining. A conical ball mill is shown in Fig. 1. In the case of a pebble mill, a nonmetallic medium, such as flint, pebbles, or even large pieces of the material being ground, is used. Instead of a metallic lining, the lining may comprise flint or porcelain blocks. In a rod mill, the grinding medium consists of a series of metallic rods essentially as long as the mill cylinder. These rods rotate freely like balls or pebbles as the mill turns. Like ball mills, rod mills have metallic linings. Feed for rotating drums varies—from a maximum of $1\frac{1}{2}$-inch (3.8 centimeter) ring size downward. These units can be operated either in a batch or continuous mode. Grinding generally requires several hours to assure the necessary fineness within particle-size limits. Usually, oversize material will be returned continuously or handled in a subsequent batch. A rod mill is shown in Fig. 2.

Fig. 1.   Conical ball mill

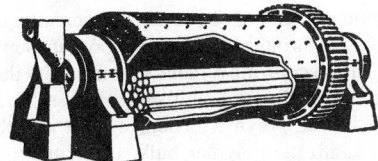

Fig. 2.   Rod mill

BANTING, SIR FREDERICK (1891–1941). A native of Ontario, Canada, Banting did his most important work in endocrinology. His brilliant research culminated in the preparation of the antidiabetic hormone that he called insulin, derived from the isles of Langerhans in the pancreas. He received the Nobel prize in medicine for this work together with John MacLeod of the University of Toronto. In 1930, the Banting Institute was founded in Toronto. He was killed in an airplane crash.

BARBITURATES. Sedative drugs derived from barbituric acid. These drugs depress the central nervous system and act especially on the sleep center in the brain, thus their sedative and sometimes hypnotic effects. Barbital and phenobarbital are relatively long lasting in effects. Other drugs are more powerfully hypnotic and have a shorter action. In the treatment of epilepsy, phenobarbital is sometimes used. It has been shown to be an effective anticonvulsant, but produces drowsiness when given in large amounts. Barbiturates have been used in sleeping pills, but in recent years several other compounds also have been introduced for this purpose. Barbiturates induce a feeling of relaxation, usually followed by sleep. The drugs have been used to provide temporary respite in times of unusual emotional stress, but they must not be taken regularly as a substitute for a cure in chronic nervous tension. The drugs will only prolong the stress and encourage the patient to continue reliance on drugs instead of seeking a solution to emotional problems.

A few years ago, the Expert Committee on Drug Addiction of the World Health Organization advised the United Nations that barbiturates "must be considered drugs liable to produce addiction." Some persons develop a physical dependence on barbiturates; others may be able to stop using the drugs voluntarily. As in the use of other psychological supports, the need for continued barbiturates lies in the underlying personality disorder.

Because phenobarbital is frequently considered the drug of choice for treating young children with febrile seizures, a study was made, from November 1982 through December 1985, involving over 200 children between 8 and 36 months of age who had had at least one febrile seizure and represented a heightened risk of further seizures. The purpose of the study was to determine possible behavioral and cognitive side effects of the drug. The report concluded: "This study found a depression of cognitive performance associated with phenobarbital, with indications of a disadvantage that outlasted the administration of the drug by several months and did not demonstrate a countervailing benefit." The report is discussed in some detail in the *N. Eng. J. Med*, 364 (February 8, 1990).

Additional Reading

Ator, N. and J.E. Henningfield: *Barbiturates: Sleeping Potion of Intoxicant?,* Chelsea House Publishers, Broomall, PA, 1988.

Clayton, L.: *Barbiturates and Other Depressants,* Rosen Publishing Group, Inc., New York, NY, 1998.

Houle, M.M.: *Tranquilizer, Barbiturate, and Downer Drug Dangers,* Enslow Publishers, Inc., Berkeley Heights, NJ, 2000.

Web Reference

WebMD Health. *http://my.webmd.com/*

BARDHAN-SENGUPTA SYNTHESIS. Phosphorus pentoxide and other powerful dehydrating agents act upon 2-beta-phenethyl-1-cyclohexanol to form octahydrophenanthrene compounds.

BARFF PROCESS. A process for oxidizing the surface of metals, by the action of superheated steam, to increase their resistance to corrosion.

BARITE. The mineral barite is barium sulfate, $BaSO_4$ crystallizing in the orthorhombic system. It may occur as tabular crystals, in groups, or lamellar, fibrous and massive. Barite has two perfect cleavages, basal and prismatic; hardness, 3–3.5; specific gravity, 4.5, which has led to the term heavy spar, occasionally used for this mineral. Its luster is vitreous; streak, white; color, white to gray, yellowish, blue, red and brown; transparent to opaque. It sometimes yields a fetid odor when broken or when pieces are rubbed together, due probably to the inclusion of carbonaceous matter. It is used as a source of barium compounds.

Barite is a frequently occurring gangue mineral and is found also in large masses in sedimentary rocks. It occurs in many places in Europe, including the Czech Republic and Slovakia, Germany, France, Spain and England; in the United States, New York, Connecticut, Pennsylvania, Virginia, Michigan, Missouri, New Mexico, Oklahoma, Utah, Colorado, South Dakota, Georgia and Tennessee. In Canada it occurs in Ontario and in Nova Scotia.

The name of this mineral derives from the Greek word meaning heavy.

BARIUM. [CAS: 7440-39-3]. Chemical element symbol Ba, at. no. 56, at. wt. 137.33, periodic table group 2 (alkaline earths), mp 725°C, bp 1640°C, density 3.5 g/cm^3 (20°C). Body-centered cubic crystal form. Naturally occurring isotopes are ^{130}Ba, ^{130}Ba, ^{134}Ba, ^{135}Ba, ^{136}Ba, ^{137}Ba, and ^{138}Ba. Barium metal is comparatively soft and ductile and capable of mechanical working. Barium metal and all barium compounds are highly toxic to humans, although barium sulfate (because of its insolubility in H_2O and body fluids) can be ingested without harm and is widely used as an opaque medium in x-ray diagnostic studies of the body. First ionization potential 5.21 eV; second 9.95 eV. Oxidation potentials $Ba \rightarrow Ba^{2+} + 2e^-$, 2.90 V, $Ba + 2OH^- + 8H_2O \rightarrow Ba(OH)_2 \cdot 8H_2O + 2e^-$, 2.97 V. Other important physical properties of barium are given under **Chemical Elements**.

Barium occurs chiefly as sulfate (barite, barytes, heavy spar, $BaSO_4$), and, of less importance, carbonate (witherite, $BaCO_3$). Georgia and Tennessee are the principal producing states. The sulfate is transformed into chloride, and the electrolysis of the fused chloride yields barium metal. See also **Barite**; and **Witherite**. Barium ores are mined chiefly as a source of barium compounds because very little metallic barium is consumed commercially. The metal is obtained by thermal reduction of the oxide, using aluminum metal at a high temperature and under vacuum in a closed retort: $4BaO + 2Al \rightarrow BaOAl_1O_3 + 3Ba$. The gaseous barium produced is recovered by condensation.

As is to be expected from its high electrode potential (2.90 V) barium, like strontium and calcium, reacts readily with the halogens, oxygen and sulfur to form halides, oxide, and sulfide, as well as with nitrogen and hydrogen at higher temperatures to form the nitride and hydride. In all its stable compounds it is divalent. It reacts vigorously with water, displacing hydrogen to form the hydroxide. Barium peroxide is formed on treatment of the hydroxide with hydrogen peroxide in the cold and also by direct combination of oxygen and barium oxide or metal. The peroxide prepared in the latter way is frequently paramagnetic because of the presence of some superoxide, $Ba(O_2)_2$. Barium exhibits little tendency to form complexes, the amines formed with NH_3 being unstable and the β-diketones and alcoholates are not well characterized. Barium metal solutions in liquid NH_3 solution yield $Ba(NH_3)_6$ upon evaporation. Common compounds of barium are:

Barium acetate. [CAS: 543-80-6]. $Ba(C_2H_3O_2)_2$, white crystals, solubility 76.4 g/100 ml H_2O at 26°C, formed by reaction of barium carbonate or hydroxide and acetic acid.

Barium carbide (acetylide), BaC_2 black solid, by reaction of barium oxide and carbon at electric furnace temperatures, reacts with H_2O; yielding acetylene gas and barium hydroxide.

Barium carbonate. [CAS: 513-77-9]. $BaCO_3$, white solid, insoluble ($K_{sp} = 5.13 \times 10^{-9}$), formed (1) by reaction of barium salt solution and sodium carbonate or bicarbonate solution; (2) by reaction of barium hydroxide solution and CO_2. With excess CO_2 barium hydrogen carbonate, $Ba(HCO_3)_2$, solution is formed. Barium carbonate decomposes at 1450°C.

Barium chloride. [CAS: 10361-37-2]. $BaCl_2 \cdot 2H_2O$, white crystals, solubility 31 g/100 ml H_2O at 0°C, formed by reaction of barium carbonate or hydroxide and HCl.

Barium chromate. [CAS: 10294-40-3]. $BaCrO_4$, yellow precipitate, $K_{sp} = 1.17 \times 10^{-10}$, formed by reaction of barium salt solution and potassium chromate solution.

Barium cyanamide, $BaCN_2$, formed in a mixture with barium cyanide, [CAS: 542-62-1], $Ba(CN)_2$, by heating barium carbide at 800°C with nitrogen gas. Fusion of the cyanamide-cyanide mixture with sodium carbonate converts it entirely to cyanide.

Barium nitrate. [CAS: 10022-31-8]. $Ba(NO_3)_2$, white crystals, solubility 8.7 g/100 ml H_2O at 20°C, formed by reaction of barium carbonate or hydroxide and HNO_3.

Barium oxide. [CAS: 1304-28-5]. BaO, white solid, mp about 1900°C, reactive with H_2O to form barium hydroxide. Barium peroxide, $BaO_2 \cdot 8H_2O$, white precipitate, formed by reaction of barium salt solution and hydrogen or sodium peroxide, yields anhydrous barium peroxide upon heating at 100°C in a current of dry air. Anhydrous barium peroxide is also formed by heating barium oxide in air or oxygen under pressure (at somewhat over one atmosphere pressure) and temperature of 400°C.

Barium oxalate. [CAS: 516-02-9]. BaC_2O_4, white precipitate, $K_{sp} = 1.1 \times 10^{-7}$, formed by reaction of barium salt solution and ammonium oxalate solution.

Barium sulfate. [CAS: 7727-43-7]. $BaSO_4$, white precipitate, $K_{sp} = 8.7 \times 10^{-11}$, formed by reaction of barium salt solution and H_2SO_4 or sodium sulfate solution, insoluble in acids; by heating with carbon yields barium sulfide.

Barium sulfide. [CAS: 21109-95-5]. BaS, grayish-white solid, formed by heating barium sulfate and carbon, reactive with H_2O to form barium hydrosulfide, $Ba(SH)_2$, solution. The latter is also made by saturation of barium hydroxide solution with H_2S. Barium polysulfides are formed by boiling barium hydrosulfide with sulfur.

Uses of Barium. The major use of barium metal for a number of years has been as a getter for oxygen in electronic vacuum tubes. A layer of the metal is deposited inside the glass envelope of the tube. Minute quantities of gases that leak into the tube react with the barium layer to form compounds. If the gases remained free, they would alter the conductance of the tube and cause deterioration of its performance.

STEPHEN E. HLUCHAN
Business Manager, Calcium Metal Products, Minerals, Pigments Metals Division, Pfizer Inc. Wallingford, Connecticut

Additional Reading

Carter, G.F. and D.E. Paul: *Materials Science and Engineering,* ASM International, Materials Park, OH, 1991.

Lewis, R.S. et al.: "Barium Isotopes in Allende Meteorite: Evidence Against an Extinct Superheavy Element," *Science,* **222,** 1013–1015 (1983).

Lide, D.R.: *CRC Handbook of Chemistry and Physics,* 84th Edition, CRC Press, LLC., Boca Raton, FL, 2003.

Parker, P.: "Barium" in *McGraw-Hill Encyclopedia of Chemistry,* 2nd Edition, McGraw-Hill Companies, Inc., New York, NY, 1993.

Perry, R.H. and D. Green: *Perry's Chemical Engineers' Handbook,* 7th Edition, McGraw-Hill Companies, Inc., New York, NY, 1999.

Staff: *ASM Handbook—Properties and Selection: Nonferrous Alloys and Special-Purpose Materials,* ASM International, Materials Park, OH, 1990.

BARLOW RULE. The volumes of space occupied by the various atoms in a given molecule are approximately proportional to the valences of the atoms; whenever an element exhibits more than one kind of valence the lowest value is generally selected.

BAROMETRIC PRESSURE. The pressure of the air at a particular point on or above the surface of the earth. At sea level, this pressure is sufficient to support a column of mercury approximately 29.9 inches in height (760 mm), equivalent to 14.7 lbs/inch^2 absolute (psia) or 1 atm.

BARRIER LAYER. An electrical double layer formed at the junction, or surface of contact, between a metal and a semiconductor, or between two metals, for various purposes.

BARRIER (Moisture). Any substance that is impervious to water or water vapor. Most effective are high-polymer materials such as vulcanized rubber, phenolformaldehyde resins, polyvinyl chloride, and polyethylene, which are widely used as packaging films. The chief factors involved are polarity, crystallinity, and degree of cross-linking. Water-soluble surfactants and protective colloids increase the susceptibility of a film to water penetration. Any pigments and fillers must be completely wetted by the polymer. Properly formulated paints are effective moisture barriers.

BARRIER POLYMERS. Barrier polymers are used for many packaging and protective applications. As barriers they separate a system, such as an article of food or an electronic component, from an environment.

The Permeation Process

Barrier polymers limit movement of substances, hereafter called permeants. The movement can be through the polymer or, in some cases, merely into the polymer. The overall movement of permeants through a polymer is called permeation, which is a multistep process. First, the permeant molecule collides with the polymer. Then, it must adsorb to the polymer surface and dissolve into the polymer bulk. In the polymer, the permeant "hops" or diffuses randomly as its own thermal kinetic energy keeps it moving from vacancy to vacancy while the polymer chains move. The random diffusion yields a net movement from the side of the barrier polymer that is in contact with a high concentration or partial pressure of the permeant to the side that is in contact with a low concentration of permeant. After crossing the barrier polymer, the permeant moves to the polymer surface, desorbs, and moves away.

Permeant movement is a physical process that has both a thermodynamic and a kinetic component. For polymers without special surface treatments, the thermodynamic contribution is in the solution step. The permeant partitions between the environment and the polymer according to thermodynamic rules of solution. The kinetic contribution is in the diffusion. The net rate of movement is dependent on the speed of permeant movement and the availability of new vacancies in the polymer.

Small Molecule Permeation

Permanent Gases. Table 1 lists the permeabilities of oxygen, nitrogen, and carbon dioxide for selected barrier and nonbarrier polymers at 20°C and 75% rh. The effect of temperature and humidity are discussed later. For many polymers the permeabilities of nitrogen, oxygen, and carbon dioxide are in the ratio 1:4:14.

The traditional definition of a barrier polymer required an oxygen permeability less than 2 nmol/(m·s·GPa) (originally, less than (1 cc·mil)/(100 in.^{2}·d·atm)) at room temperature. This definition was based partly on function and partly on conforming to the old commercial unit of permeability. The old commercial unit of permeability was created so that the oxygen permeability of Saran Wrap brand plastic film, a trademark of The Dow Chemical Company, would have a numerical value of 1.

TABLE 1. PERMEABILITIES OF SELECTED POLYMERS

Polymer	Gas permeability, nmol/m·s·GPa[d]		
	Oxygen	Nitrogen	Carbon dioxide
vinylidene chloride copolymers	0.02–0.30	0.005–0.07	0.1–1.5
ethylene-vinyl alcohol copolymers, dry	0.014–0.095		
at 100% rh	2.2–1.1		
nylon-MXD6[a]	0.30		
nitrile barrier polymers	1.8–2.0		6–8
nylon-6	4–6		20–24
amorphous nylon (Selar[b] PA 3426)	5–6		
poly(ethylene terephthalate)	6–8	1.4–1.9	30–50
poly(vinyl chloride)	10–40		40–100
high density polyethylene	200–400	80–120	1200–1400
polypropylene	300–500	60–100	1000–1600
low density polyethylene	500–700	200–400	2000–4000
polystyrene	500–800	80–120	1400–3000

[a] Trademark of Mitsubishi Gas Chemical Co.
[b] Trademark of E. I. du Pont de Nemours & Co., Inc.

TABLE 2. DIFFUSION AND SOLUBILITY COEFFICIENTS FOR OXYGEN AND CARBON DIOXIDE IN SELECTED POLYMERS AT 23°C, DRY

	Oxygen		Carbon dioxide	
S, nmol/(m³·GPa)[a]	D, m²/s	S, nmol/(m³·GPa)[a]	Polymer	D, m²/s
vinylidene chloride copolymer	1.2×10^{-14}	1.01×10^{13}	1.3×10^{-14}	3.2×10^{13}
ethylene-vinyl alcohol copolymer[b]	7.2×10^{-14}	2.4×10^{12}		
acrylonitrile barrier polymer	1.0×10^{-13}	1.0×10^{13}	9.0×10^{-14}	4.4×10^{13}
poly(ethylene terephthalate)	2.7×10^{-13}	2.8×10^{13}	6.2×10^{-14}	8.1×10^{14}
poly(vinyl chloride)	1.2×10^{-12}	1.2×10^{13}	8.0×10^{-13}	9.7×10^{13}
polypropylene	2.9×10^{-12}	1.1×10^{14}	3.2×10^{-12}	3.4×10^{14}
high density polyethylene	1.6×10^{-11}	7.2×10^{12}	1.1×10^{-11}	4.3×10^{13}
low density polyethylene	4.5×10^{-11}	2.0×10^{13}	3.2×10^{-11}	1.2×10^{14}

[a] Solubility coefficient in cc(STP)/(cm³·atm) × (4.04×10^{14}) = solubility coefficient in nmol/(m³·GPa).
[b] 42 mol% ethylene.

Poly(ethylene terephthalate) (PET), with an oxygen permeability of 8 nmol/(m·s·GPa), is not considered a barrier polymer by the old definition; however, it is an adequate barrier polymer for holding carbon dioxide in a 2-L bottle for carbonated soft drinks. The solubility coefficients for carbon dioxide are much larger than for oxygen. For the case of the PET soft drink bottle, the principal mechanism for loss of carbon dioxide is by sorption in the bottle walls as 500 kPa (5 atm) of carbon dioxide equilibrates with the polymer. For an average wall thickness of 370 μm (14.5 mil) and a permeability of 40 nmol/(m·s·GPa), many months are required to lose enough carbon dioxide (15% of initial) to be objectionable.

The diffusion and solubility coefficients for oxygen and carbon dioxide in selected polymers have been collected in Table 2.

Polymers With Good Barrier to Permanent Gases. The polymers that are good barriers to permanent gases, especially oxygen, have important commercial significance.

Vinylidene chloride copolymers are available as resins for extrusion, latices for coating, and resins for solvent coating. Comonomer levels range from 5 to 20 wt %. Common comonomers are vinyl chloride, acrylonitrile, and alkyl acrylates. The permeability of the polymer is a function of type and amount of comonomer. As the comonomer fraction of these semicrystalline copolymers is increased, the melting temperature decreases and the permeability increases. The permeability of vinylidene chloride homopolymer has not been measured.

Vinylidene chloride copolymers are marketed under a variety of trade names. Saran is a trademark of The Dow Chemical Company for vinylidene chloride copolymers. Other trade names include Daran (W.R. Grace), Amsco Res (Union Oil), and Serfene (Morton Chemical) in the United States; and Haloflex (Imperial Chemical Industries, Ltd.), Diofan (BASF), Ixan (Solvay and Cie SA), and Polyidene (Scott-Bader) in Europe.

Hydrolyzed ethylene-vinyl acetate copolymers, commonly known as ethylene-vinyl alcohol (EVOH) copolymers, are usually used as extrusion resins, although some may be used in solvent-coating applications.

Copolymers of acrylonitrile are used in extrusion and molding applications. Commercially important comonomers for barrier applications include styrene and methyl acrylate.

Polyamide polymers can provide a good-to-moderate barrier to permeation by permanent gases.

Two often-used polymers have adequate properties for some applications. Poly(ethylene terephthalate) (PET) is used to make films and bottles. Poly(vinyl chloride) (PVC) is a moderate barrier to permanent gases. Plasticized poly(vinyl chloride) is used as a household wrapping film. The plasticizers greatly increase the permeabilities.

Water Vapor Transmission. Table 3 lists water vapor transmission (WVTR) values for selected polymers. Comparison of Tables 1 and 3 shows that often there is a reversal of roles. Those polymers that are good oxygen barriers are often poor water-vapor barriers and vice versa. This can be rationalized as follows. Barrier polymers often rely on dipole-dipole interactions to reduce chain mobility and, hence, diffusional movement of permeants. These dipoles can be good sites for hydrogen bonding. Water molecules are attracted to these sites. Polymer molecules without dipole—dipole interactions, such as polyolefins, dissolve very little water and have low. WVTR and permeability values. The low values of S more than compensate for the naturally higher values of D.

TABLE 3. WATER-VAPOR TRANSMISSION RATES OF SELECTED POLYMERS[a]

Polymer	WVTR, nmol/(m·s)
vinylidene chloride copolymers	0.005–0.05
high density polyethylene (HDPE)	0.095
polypropylene	0.16
low density polyethylene (LDPE)	0.35
ethylene-vinyl alcohol, 44 mol % ethylene[b]	0.35
poly(ethylene terephthalate) (PET)	0.45
poly(vinyl chloride) (PVC)	0.55
ethylene-vinyl alcohol, 32 mol % ethylene[b]	0.95
nylon-6,6, nylon-11	0.95
nitrile barrier resins	1.5
polystyrene	1.8
nylon-6	2.7
polycarbonate	2.8
nylon-12	15.9

[a] At 38°C and 90% rh unless otherwise noted.
[b] Measured at 40°C.

Large Molecule Permeation

The permeation of flavor, aroma, and solvent molecules in polymers follows the same physics as the permeation of small molecules. However, there are two significant differences. For these larger molecules, the diffusion coefficients are much lower and the solubility coefficients are much higher. This means that steady-state permeation may not be reached during the storage time of some packaging situations. Hence, large molecules from the environment might not enter the contents, or loss of flavor molecules would be limited to sorption into the polymer. However, since the solubility coefficient is large, the loss of flavor could be important solely from sorption in the polymer. Furthermore, the large solubility coefficient can lead to enough sorption of the large molecule that plasticization occurs in the polymer, which can increase the diffusion coefficient.

Table 4 contains some selected permeability data, including diffusion and solubility coefficients for flavors in polymers used in food packaging. Generally, vinylidene chloride copolymers and glassy polymers such as polyamides and EVOH are good barriers to flavor and aroma permeation, whereas the polyolefins are poor barriers. Comparison to Table 2 shows that the large-molecule diffusion coefficients are 1000 or more times lower than the small-molecule coefficients.

Physical Factors Affecting Permeability

Several physical factors can affect the barrier properties of a polymer. These include temperature, humidity, orientation, and cross-linking.

Temperature. The temperature dependence of the permeability arises from the temperature dependencies of the diffusion coefficient and the solubility coefficient. Typically, the permeability increases 5 to 10% for every increase of 1°C.

TABLE 4. EXAMPLES OF PERMEATION OF FLAVOR AND AROMA COMPOUNDS IN SELECTED POLYMERS AT 25°C[a], DRY[b]

Flavor/aroma compound	Permeant formula	P, MZU[c]	D, m^2/s	S, kg/(m·Pa^3)
Vinylidene chloride copolymer				
ethyl hexanoate	$C_8H_{16}O_2$	570	8.0×10^{-18}	0.71
ethyl 2-methylbutyrate	$C_7H_{14}O_2$	3.2	1.9×10^{-17}	1.7×10^{-3}
hexanol	$C_6H_{14}O$	40	5.2×10^{-17}	7.7×10^{-3}
trans-2-hexenal	$C_6H_{10}O$	240	1.8×10^{-17}	0.14
d-limonene	$C_{16}H_{16}$	32	3.3×10^{-17}	9.7×10^{-3}
3-octanone	$C_8H_{16}O$	52	1.3×10^{-18}	0.40
propyl butyrate	$C_7H_{14}O_2$	42	4.4×10^{-18}	9.4×10^{-2}
dipropyl disulfide	$C_6H_{14}S_2$	270	2.6×10^{-18}	1.0
Ethylene–vinyl alcohol copolymer				
ethyl hexanoate		0.41	3.2×10^{-18}	1.3×10^{-3}
ethyl 2-methylbutyrate		0.30	6.7×10^{-18}	4.7×10^{-4}
hexanol		1.2	2.6×10^{-17}	4.6×10^{-4}
trans-2-hexenal		110	6.4×10^{-17}	1.8×10^{-2}
d-limonene		0.5	1.1×10^{-17}	4.5×10^{-4}
3-octanone		0.2	1.0×10^{-18}	2.0×10^{-3}
propyl butyrate		1.2	2.7×10^{-17}	4.5×10^{-4}
Low density polyethylene				
ethyl hexanoate		4.1×10^6	5.2×10^{-13}	7.8×10^{-2}
ethyl 2-methylbutyrate		4.9×10^5	2.4×10^{-13}	2.3×10^{-2}
hexanol		9.7×10^5	4.6×10^{-13}	2.3×10^{-2}
trans-2-hexenal		8.1×10^5		
d-limonene		4.3×10^6		
3-octanone		6.8×10^6	5.6×10^{-13}	1.2×10^{-1}
propyl butyrate		1.5×10^6	5.0×10^{-13}	3.0×10^{-2}
dipropyl disulfide		6.8×10^6	7.3×10^{-14}	9.3×10^{-1}
High density polyethylene				
d-limonene		3.5×10^6	1.7×10^{-13}	2.5×10^{-1}
menthone	$C_{10}H_{18}O$	5.2×10^6	9.1×10^{-13}	4.7×10^{-1}
methyl salicylate	$C_8H_8O_3$	1.1×10^7	8.7×10^{-14}	1.6
Polypropylene				
2-butanone	C_4H_8O	8.5×10^3	2.1×10^{-15}	4.0×10^{-2}
ethyl butyrate	$C_6H_{12}O_2$	9.5×10^3	1.8×10^{-15}	5.3×10^{-2}
ethyl hexanoate		8.7×10^4	3.1×10^{-15}	2.8×10^{-1}
d-limonene		1.6×10^4	7.4×10^{-16}	2.1×10^{-1}

[a] Values for vinylidene chloride copolymer and ethylene–vinyl alcohol are extrapolated from higher temperatures.
[b] Permeation in the vinylidene chloride copolymer and the polyolefins is not affected by humidity; the permeability and diffusion coefficient in the ethylene–vinyl alcohol copolymer can be as much as 1000 times greater with high humidity.
[c] MZU = $(10^{-20}$ kg·m)/(m·s·Pa)^2.

Humidity. When a polymer equilibrates with a humid environment, it absorbs water. This can plasticize the polymer and increase the permeability.

Orientation. The effect of orientation on the permeability of polymers is difficult to assess because the words orientation and elongation or strain have been used interchangeably in the literature. Diffusion in some polymers is unaffected by orientation; in others, increases or decreases are observed.

Cross-Linking. Cross-linking has been shown in a few cases to decrease the diffusion coefficient.

Barrier Structures

Barrier polymers are often used in combination with other polymers or substances. The combinations may result in a layered structure either by coextrusion, lamination, or coating. The combinations may be blends that are either miscible or immiscible. In each case, the blend seeks to combine the best properties of two or more materials to enhance the value of a final structure.

Predicting Permeabilities

Reasonable prediction can be made of the permeabilities of low molecular weight gases such as oxygen, nitrogen, and carbon dioxide in many polymers. The diffusion coefficients are not complicated by the shape of the permeant, and the solubility coefficients of each of these molecules do not vary much from polymer to polymer. Hence, all that is required is some correlation of the permeant size and the size of holes in the polymer matrix. Reasonable predictions of the permeabilities of larger molecules such as flavors, aromas, and solvents are not easily made. The diffusion coefficients are complicated by the shape of the permeant, and the solubility coefficients for a specific permeant can vary widely from polymer to polymer.

The permachor method is an empirical method for predicting the permeabilities of oxygen, nitrogen, and carbon dioxide in polymers. In this method a numerical value is assigned to each constituent part of the polymer. An average number is derived for the polymer, and a simple equation converts the value into a permeability. This method has been shown to be related to the cohesive energy density and the free volume of the polymer. The model has been modified to liquid permeation with some success.

For larger molecules, independent predictions of the diffusion coefficients and the solubility coefficients are required. Figure 1 shows how the diffusion coefficient varies as a function of permeant size in poly(vinyl chloride) (PVC). The two sets of data represent glassy PVC, which is below its glass-transition temperature and plasticized PVC, which is above its glass-transition temperature. Other glassy polymers show the steep slope, and other rubbery polymers show the shallow slope. The points near the lines represent spherical permeants whereas the points above the lines represent linear permeants. Predicting the diffusion coefficient for a permeant in a polymer requires knowing one other diffusion coefficient in the polymer.

The solubility coefficients are more difficult to predict. Although advances are being made, the best method is probably to use a few known solubility coefficients in the polymer to predict others with a simple plot of S vs $(\delta_{poly} - \delta_{perm})^2$ where δ_{poly} and δ_{perm} are the solubility parameters of the polymer and permeant, respectively. When insufficient data are available, S at 25°C can be estimated with equation 1 where $\kappa = 1$ and the resulting units of cal/cm^3 are converted to kJ/mol by dividing by the

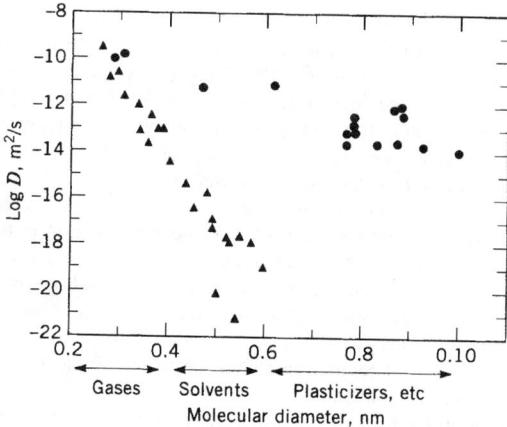

Fig. 1. Diffusivities of penetrants in rigid (▲) and plasticized (●) poly(vinyl chloride) versus molecular diameter at 30°C

polymer density and multiplying by the molecular mass of the permeant and by 4.184.

$$S = \kappa(\delta_{poly} - \delta_{perm})^2 \qquad (1)$$

The boiling temperature of a permeant can be used to predict the solubility coefficient only when the solubility coefficients of other permeants of the same chemical family are known.

Measuring Barrier Properties

Measuring the barrier properties of polymers is important for several reasons. The effects of formulation or process changes need to be known, new polymers need to be evaluated, data are needed for a new application before a large investment has been made, and fabricated products need to have performance verified.

Oxygen Transport. The most widely used methods for measuring oxygen transport are based upon the Ox-Tran instrument (Modern Controls, Inc.). Several models exist, but they all work on the same principle. The most common application is to measure the permeability of a film sample.

Water Transport. Two methods of measuring water-vapor transmission rates (WVTR) are commonly used. The newer method uses a Permatran-W (Modern Controls, Inc.). The other method is the ASTM cup method.

Carbon Dioxide Transport. Measuring the permeation of carbon dioxide occurs far less often than measuring the permeation of oxygen or water. A variety of methods are used; however, the simplest method uses the Permatran-C instrument (Modern Controls, Inc.).

Flavor and Aroma Transport. Many methods are used to characterize the transport of flavor, aroma, and solvent molecules in polymers. Each has some value, and no one method is suitable for all situations. Any experiment should obtain the permeability, the diffusion coefficient, and the solubility coefficient. Furthermore, experimental variables might include the temperature, the humidity, the flavor concentration, and the effect of competing flavors.

Applications

The primary application for barrier polymers is food and beverage packaging. Barrier polymers are also used for packaging medical products, agricultural products, cosmetics, and electronic components and in moldings, pipe, and tubing.

Safety and Health Factors

The use of safe materials is vital for barrier applications, particularly for food, medical, and cosmetics packaging. Suppliers of specific barrier polymers can provide the necessary details, such as material safety data sheets, to ensure safe processing and use of barrier polymers.

PHILLIP DE LASSUS
The Dow Chemical Company

Additional Reading

Comyn, J. ed.: *Polymer Permeability*, Elsevier Applied Science Publishers, Ltd., Barking, U.K., 1985.
Karos, W. J. ed.: *Barrier Polymers and Structures*, ACS Symposium Series No. 423, American Chemical Society, Washington, DC, 1990.
Risch, S. A. and J. H. Hotchkiss, eds.: *Food and Packaging Interactions II*, ACS Symposium Series, No. 473, American Chemical Society, Washington, DC, 1991.
Vieth, W. R.: *Diffusion In and Through Polymers*, Hanser Publishers, Munich, Germany 1991.

BARTON, DEREK H. R. (1918–1998). An English organic chemist who won the Nobel prize for chemistry in 1969 with Odd Hassel. The field of conformational analysis in organic chemistry was initiated through his research in the terpene and steroid fields. He did extensive research in the area of carbanion autoxidations. He was instrumental in research concerning the relationship of molecular rotation to structure in complex organic molecules. His education took place in London, France, and Ireland.

BARYONS. A class of subatomic particles including the proton, the neutron, and several heavier particles, such as the lambda, the sigma (plus, minus, and neutral), and the omega (minus) particles. Baryons are particles that interact with the strong nuclear force. Each baryon is given a baryon number 1, each corresponding antibaryon is given a baryon number −1, while the light particles (photons, electrons, neutrinos, muons, and mesons) are given baryon number 0. The total baryon number in a given reaction is found by algebraically adding up the baryon numbers of the particles entering into the reaction. During any reaction among particles, the baryon number cannot change. This rule ensures that a proton cannot change into an electron, even though a neutron can change into a proton. Similarly, to create an antiproton in a reaction, one must simultaneously create a proton or other baryon. Baryon conservation ensures the stability of the proton against decaying into a particle of smaller mass. See also **Neutron**; **Particles (Subatomic)**; and **Proton**.

BARYTOCALCITE. This mineral is a carbonate of barium and calcium; it crystallizes in the monoclinic system but occurs massive as well. It has a perfect cleavage parallel to the prism and one, less perfect, parallel to the base; fracture, sub-conchoidal; brittle; hardness, 4; specific gravity, 3.66–3.71; luster, vitreous; color, white or gray or may be greenish or yellowish; transparent to translucent. Barytocalcite is found in Cumberland, England, associated with barite and fluorite.

BASAL METABOLISM. The metabolism of a living cell or organism refers to the total turnover of chemical material and energy. It consists of *anabolism*, or assimilation, mostly of substances of high potential energy (primarily protein, fat, and carbohydrate), and *catabolism* or dissimilation. In common speech, metabolism refers to the oxidation of major foodstuffs and the concomitant release of energy. Metabolic rate refers to the metabolism in a given period of time. The "basal" metabolic rate refers to the fundamental energy requirement for maintenance and continued functioning of the organism (aside from external muscular work and work of digestion), such as respiration, contraction of the heart, function of the kidney, the liver, and of all cells in general. Basal metabolic rate (BMR) in humans refers to the determination of metabolic rate under certain standardized conditions, including complete physical rest (but not sleep), a fasting state, and an ambient temperature that does not require energy expenditure for physiological temperature regulation. Actually the BMR refers not to a "basal" rate but to a determination under these standard conditions. The BMR is below normal in sleep, starvation, anesthesia, and certain endocrine disturbances (hypothyroidism), and is elevated in fever, athletic training, under the influence of drugs (e.g., caffeine) and endocrines (adrenaline, thyroid hormones).

In studies of animals it becomes technically difficult to make observations under standard conditions which include rest. Restraining an animal increases the metabolic rate and inactivation through anesthesia lowers it; ruminants and other plant eaters cannot be brought into a fasting state unless they are deprived of food for prolonged periods of time; small animals (e.g., shrews) have such high metabolic rates that they must eat almost continuously to sustain a normal metabolic rate, etc.

Methods of Determination

In principle, the metabolic rate is determined in three different ways. (a) Determination of the energy value of all food less the energy value of excreta (mainly feces and urine) should give the energy turnover of the organism. However, the result must be corrected for any change in the composition of the body, mainly deposition or utilization of body fats. The method is cumbersome and is accurate only if the period of observation is sufficiently long. (b) Measurement of total heat production of the organism. This is fundamentally the most accurate method. The value obtained must be corrected for any external work performed, including such items as heating of the foodstuffs taken in, vaporization of water, etc. The determinations are made with the organism in a calorimeter, technically a rather difficult procedure, but it yields very accurate results. (c) The amount of oxygen used in oxidation processes can be used to determine the metabolic rate. (In theory, the carbon dioxide production could also be used, but it is less accurate, mainly because there is a large pool of carbon dioxide in the organism that undergoes changes relatively easily.) The reason that oxygen can be used is that similar amounts of heat are produced for each liter of oxygen, irrespective of whether fat, carbohydrate or protein is oxidized. The figures are: fat 4.7 kcal; carbohydrate, 5.0 kcal; and protein, 4.5 kcal, per liter oxygen. It is customary to use an average value, 4.8 kcal/liter oxygen consumed. The use of oxygen consumption for the determination of metabolic rate is so common that the two concepts have become practically synonymous. Obviously, the oxygen consumption cannot be used for determinations of metabolic rate in, for example, anaerobic organisms.

Temperature Effects

Animals whose body temperature changes with that of the environment (poikilothermic or cold-blooded animals) have a metabolic rate which depends on their temperature. In general, the metabolic rate increases, within the range tolerated by the organism, some two- or threefold for a temperature increase of 10°C. This change, designated as Q_{10}, is a term preferred by most physiologists and biologists over the use of the Arrhenius constant, which is a thermodynamically more correct way of expressing temperature dependence. Because of the temperature effect, information about metabolic rate in cold-blooded animals is meaningful only if the temperature is known.

Mammals and birds maintain a relatively constant body temperature within a wide range of ambient temperatures, and are called warm-blooded or homothermic animals. When the ambient temperature falls below a certain critical level, their metabolic rate increases so that the increased heat loss is balanced by increased heat production. Most of the increased heat production is due to involuntary muscle contractions (shivering).

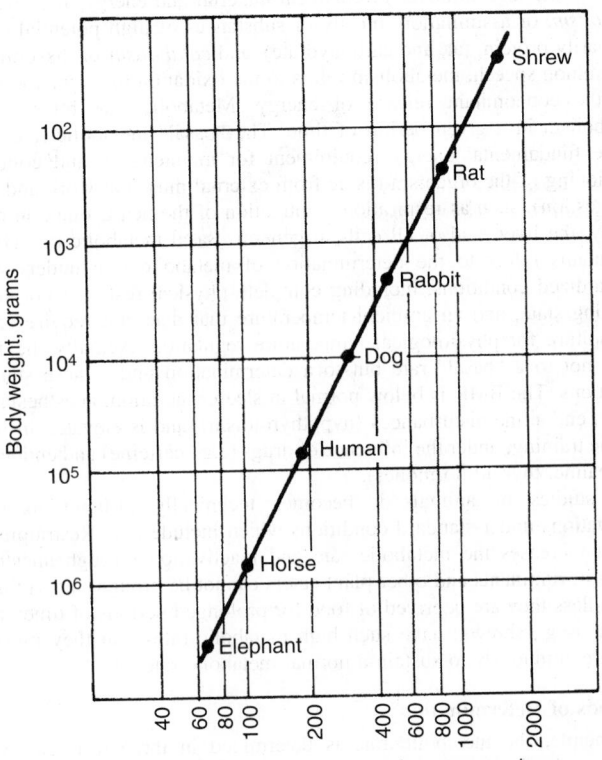

Fig. 1.   Body weight versus metabolic rate plotted on logarithmic coordinates

Metabolic Rate in Relation to Body Size

If a uniform group of animals, such as mammals, is used for a comparison of metabolic rates, an interesting relationship is revealed. The smaller the animal, the higher is the metabolic rate per gram of body weight. If, on logarithmic coordinates, the metabolic rate is plotted against body size, we obtain a straight line (see Fig. 1) that corresponds to the equation: log metabolic rate = $k + 0.74$ log body weight (k being a constant whose numerical value depends on the units used). It has been suggested that this relationship expresses the need for a higher heat production in the smaller animal, which, because of its larger relative surface, must produce heat at a higher rate than a large animal in order to maintain its body temperature. However, similar relationships between metabolic rate and body size have been found in numerous groups of cold-blooded animals as well as plants, where the need for heat regulation cannot be the fundamental explanation of this interesting relationship.

See also **Metabolism**.

Additional Reading

Coffee, C.J.: *Metabolism: Quick Look Series,* Fence Creek Publishing, LLC., Madison, CT, 1999.
Felig, P., J.D. Baxter, and L.A. Frohman: *Endocrinology and Metabolism,* 3rd Edition, McGraw-Hill Companies, Inc., New York, NY, 1995.
Groff, J.L. and S.S. Gropper: *Advanced Nutrition and Human Metabolism,* 3rd Edition, Wadsworth Publishing Company, Belmont, CA, 1999.
Salway, J.G.: *Metabolism at a Glance,* 2nd Edition, Blackwell Science, Inc., Malden, MA, 1999.
Stephanopoulos, G.N., J. Nielsen, and A.A. Aristidou: *Metabolic Engineering: Principles and Methodologies,* Academic Press, Inc., San Diego, CA, 1999.

BASE (Chemistry).   See **Acids and Bases.**

BASIC OXIDE.   An oxide that is a base or that forms a hydroxide when combined with water and/or that will neutralize acidic substances. Basic oxides are all metallic oxides, but there is a great variation in the degree of basicity. Some basic oxides, such as those of sodium, calcium, and magnesium, combine with water vigorously or with relative ease and also neutralize all acidic substances rapidly and completely. The oxides of the heavy metals are only weakly basic, do not dissolve or react with water to any extent, and neutralize only the more strongly acidic substances. There is a gradual transition from basic to acidic oxides, and certain oxides, such as aluminum oxide, show both acidic and basic properties.

See also **Acids and Bases.**

BASIC SALT.   A compound belonging to the categories of both salts and bases, because it contains OH (hydroxyl) or O (oxide) as well as the usual positive and negative radicals of normal salts. Among the best examples are bismuth subnitrate, often written $BiONO_3$; and basic copper carbonate, $Cu_2(OH)_2CO_3$. Most basic salts are insoluble in water and many are of variable composition.

BASTNASITE.   A wax-yellow reddish-brown, greasy mineral of the composition (Ce, La) (CO$_3$) F, usually found in contact zones or associated with zinc lodes. Sometimes spelled *bastnaesite*.

See also **Rare-Earth Elements and Metals.**

BATTERIES.   Batteries, storehouses for electrical energy "on demand," range in size from large house-sized batteries for utility storage, cubic meter-sized batteries for automotive starting, lighting, and ignition, down to tablet-sized batteries for hearing aids and paper-thin batteries for memory protection in electronic devices.

In bulk chemical reactions, an oxidizer (electron acceptor) and fuel (electron donor) react to form products resulting in direct electron transfer and the release or absorption of energy as heat. By special arrangements of reactants in devices called batteries, it is possible to control the rate of reaction and to accomplish the direct release of chemical energy in the form of electricity on demand, without intermediate processes.

Figure 1 schematically depicts an electrochemical reactor in which the chemical energy stored in the electrodes is manifested directly as a voltage and current flow. The electrons involved in the chemical reactions are transferred from the active materials undergoing oxidation to the oxidizing agent by means of an external circuit. The passage of electrons through this external circuit generates an electric current, providing a direct means for energy utilization without going through heat as an intermediate step. As

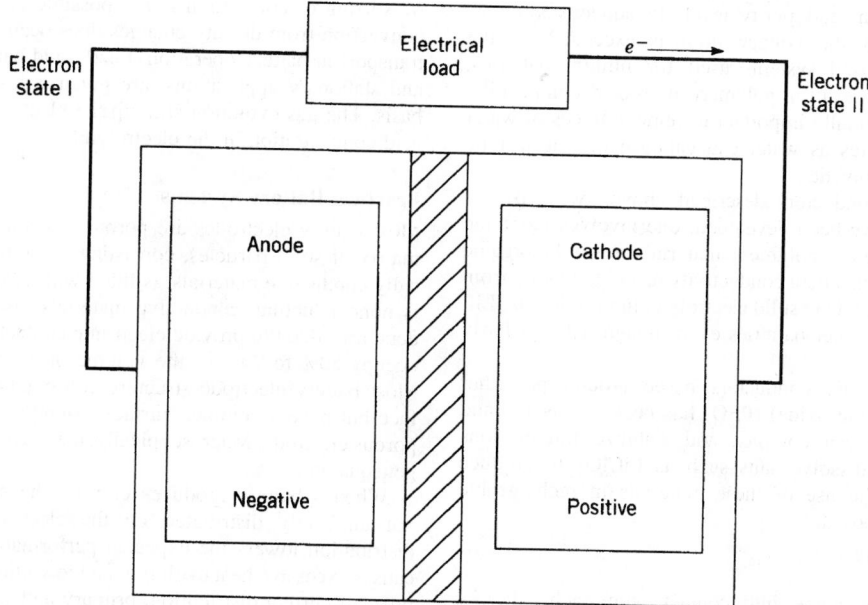

Fig. 1. Schematic representation of a battery system also known as an electrochemical transducer where the anode, also known as electron state I, may be comprised of lithium, magnesium, zinc, cadmium, lead, or hydrogen, and the cathode, or electron state II, depending on the composition of the anode, may be lead dioxide, manganese dioxide, nickel oxide, iron disulfide, oxygen, silver oxide, or iodine

a result, electrochemical reactors can be significantly more efficient than Carnot Cycle heat engines.

The three main types of batteries are primary, secondary, and reserve. A primary battery is used or discharged once and discarded. Secondary or rechargeable batteries can be discharged, recharged, and used again. Reserve batteries are normally special constructions of primary battery systems that store the electrolyte apart from the electrodes, until put into use. They are designed for long-term storage before use. Fuel cells are not discussed herein.

Economic Aspects

The U.S. primary battery market is usually divided according to the chemical system used in the batteries, whereas the secondary battery market is usually divided according to usage. The lead–acid battery accounts for over 85% of the secondary battery market.

Thermodynamics

Batteries are miniature chemical reactors that convert chemical energy into electrical energy on demand. The thermodynamics of battery systems follow directly from that for bulk chemical reactions. For the general reaction

$$aA + bB \rightleftharpoons cC + dD \tag{1}$$

the basic thermodynamic equations for a reversible electrochemical transformation are given as

$$\Delta G = \Delta H - T\Delta S \tag{2}$$

$$\Delta G^\circ = \Delta H^\circ - T\Delta S^\circ \tag{3}$$

where ΔG is the Gibbs free energy, or the energy of a reaction available for useful work, ΔH is the enthalpy, or the energy released by the reaction, ΔS is the entropy, or the heat associated with the organization of material, and T is the absolute temperature. The superscript $^\circ$ is used to indicate that the value of the function is for the material in the standard state at 25°C and unit activity. Although the Helmholtz free energy ΔA is used to describe constant volume situations found in battery systems, the use of the Gibbs free energy ΔG is adequate to describe practical battery systems.

The terms ΔG, ΔH, and ΔS are state functions and depend only on the identity of the materials and the initial and final state of the reaction.

Because ΔG is the net useful energy available from a given reaction, in electrical terms, the net available electrical energy from a reaction is given by

$$-\Delta = nFE \tag{4}$$

and

$$-\Delta^\circ = nFE^\circ \tag{5}$$

where n is the number of electrons transferred in the reaction, F is Faraday's constant, E is the voltage or electromotive force (emf) of the cell, and E° is that voltage at 25°C and unit activity. The voltage is unique for each group of reactants comprising the battery system. The amount of electricity produced is determined by the total amount of materials involved in the reaction. The voltage may be thought of as an intensity factor, and the term nF may be considered a capacity factor.

The more negative the value of ΔG, the more energy or useful work can be obtained from the reaction. Reversible processes yield the maximum output. In irreversible processes, a portion of the useful work or energy is used to help carry out the reaction. The cell voltage or emf also has a sign and direction. Spontaneous processes have a positive emf; the reaction, written in a reversible fashion, goes in the forward direction.

Electrolytes

Electrolytes are a key component of electrochemical cells and batteries. Electrolytes are formed by dissolving an ionogen into a solvent. When salts are dissolved in a solvent such as water the salt dissociates into ions through the action of the dielectric, water. Strong electrolytes, i.e., salts of strong acids and bases, are completely dissociated in solution into positive and negative ions. The ions are solvated but positive ions tend to interact more strongly with the solvent than do the anions. The ions of the electrolyte provide the path for the conduction of electricity by movement of charged particles through the solution. The electrolyte also provides the physical separation of the positive and negative electrodes needed for electrochemical cell operation.

Transport properties of the electrolyte, as well as electrode reactions, have a significant impact on battery operation. The electrode reactions and ionic transference that occur during discharge result in considerable modifications to the solution composition at each electrode compartment. The negative and positive electrode compartments can lose or gain electrolyte and solvent depending on the transference numbers of the ions and the electrode reactions. The composition of the electrolyte in the separator between the two compartments generally remains unchanged.

Battery electrolytes are concentrated solutions of strong electrolytes and the Debye-Hückel theory of dilute solutions is only an approximation. Typical values for the resistivity of battery electrolytes range from about 1 ohm · cm for sulfuric acid, H_2SO_4, in lead–acid batteries and for potassium hydroxide, KOH, in alkaline cells to about 100 ohm · cm for organic electrolytes in lithium Li, batteries.

Each electrolyte is stable only within certain voltage ranges. Exceeding these limits results in decomposition. The stable range depends on the solvent, electrolyte composition, and purity level. In aqueous systems, hydrogen and oxygen form if the voltage limit is exceeded. In the nonaqueous organic solvent-based systems used for lithium batteries, exceeding the voltage limit can result in polymerization or decomposition of the solvent system. It is especially important to remove traces of water from the nonaqueous electrolytes as water can catalyze the electrolytic decomposition of the organic solvent.

In addition to the liquid conductors described above, two types of solid-state ionic conductors have been developed; one involves inorganic compounds and the other is based on polymeric materials. Several inorganic solids have been found to have excellent conductivity resulting wholly from ionic motion in the crystal lattice. One solid electrolyte, lithium iodide, LiI, has found application in heart-pacer batteries even though it has a fairly low conductivity.

A second type of solid ionic conductors based around polyether compounds such as poly(ethylene oxide) (PEO) has been discovered and characterized. The polyethers can complex and stabilize lithium ions in organic media. They also dissolve salts such as $LiClO_4$ to produce conducting solid solutions. The use of these materials in rechargeable lithium-batteries has been proposed.

Electrical Double Layer

When two conducting phases come into contact with each other, a redistribution of charge occurs as a result of any electron energy level difference between the phases. If the two phases are metals, electrons flow from one metal to the other until the electron levels equilibrate. When an electrode, i.e., electronic conductor, is immersed in an electrolyte, i.e., ionic conductor, an electrical double layer forms at the electrode–solution interface resulting from the unequal tendency for distribution of electrical charges in the two phases. Because overall electrical neutrality must be maintained, this separation of charge between the electrode and solution gives rise to a potential difference between the two phases, equal to that needed to ensure equilibrium.

On the electrode side of the double layer the excess charges are concentrated in the plane of the surface of the electronic conductor. On the electrolyte side of the double layer the charge distribution is quite complex. The potential drop occurs over several atomic dimensions and depends on the specific reactivity and atomic structure of the electrode surface and the electrolyte composition. The electrical double layer strongly influences the rate and pathway of electrode reactions.

Electrically, the electrical double layer may be viewed as a capacitor with the charges separated by a distance of the order of molecular dimensions. The measured capacitance ranges from about two to several hundred microfarads per square centimeter, depending on the structure of the double layer, the potential, and the composition of the electrode materials.

Kinetics and Transport

Activation Processes. To be useful in battery applications reactions must occur at a reasonable rate. The rate or ability of battery electrodes to produce current is determined by the kinetic processes of electrode operations, not by thermodynamics, which describes the characteristics of reactions at equilibrium when the forward and reverse reaction rates are equal. Electrochemical reaction kinetics follow the same general considerations as those of bulk chemical reactions. Two differences are a potential drop that exists between the electrode and the solution because of the electrical double layer at the electrode interface, and the reaction that occurs at a two-dimensional interfaces rather than in three-dimensional space.

Transport Processes. The velocity of electrode reactions is controlled by the charge-transfer rate of the electrode process, or by the velocity of the approach of the reactants, to the reaction site. The movement or transport of reactants to and from the reaction site at the electrode interface is a common feature of all electrode reactions. Transport of reactants and products occurs by diffusion, by migration under a potential field, and by convection. The complete description of transport requires a solution to the transport equations. A full account is given in texts and discussions on hydrodynamic flow. Molecular diffusion in electrolytes is relatively slow. Although the process can be accelerated by stirring, enhanced mass transfer by stirring or convection is not possible in most battery designs. Natural convection from density changes does occur but does not greatly enhance transport in battery operation. Lead–acid batteries, used for motive power and stationary applications, are given a gassing overcharge on a regular basis. The gas evolution stirs up the electrolyte and equalizes the sulfuric acid concentration in the electrolyte.

Practical Battery Systems

Most battery electrodes are porous structures in which an interconnected matrix of solid particles, consisting of both nonconductive and electronically conductive materials, is filled with electrolyte. When the active mass is nonconducting, conductive materials, usually carbon or metallic powders, are added to provide electronic contact to the active mass. The solids occupy 50% to 70% of the volume of a typical porous battery electrode. Most battery electrode structures do not have a well-defined planar surface but have a complex surface extending throughout the volume of the porous electrode. Macroscopically, the porous electrode behaves as a homogeneous unit.

When a battery produces current, the sites of current production are not uniformly distributed on the electrodes. The nonuniform current distribution lowers the expected performance from a battery system, and causes excessive heat evolution and low utilization of active materials. Two types of current distribution, primary and secondary, can be distinguished. The primary distribution is related to the current production based on the geometric surface area of the battery construction. Secondary current distribution is related to current production sites inside the porous electrode itself. Most practical battery constructions have nonuniform current distribution across the surface of the electrodes. This primary current distribution is governed by geometric factors such as height (or length) of the electrodes, the distance between the electrodes, and the resistance of the anode and cathode structures; by the resistance of the electrolyte; and by the polarization resistance or hindrance of the electrode reaction processes.

Cell geometry, such as tab/terminal positioning and battery configuration, strongly influence primary current distribution. The monopolar construction is most common. Several electrodes of the same polarity may be connected in parallel to increase capacity. The current production concentrates near the tab connections unless special care is exercised in designing the current collector. Bipolar construction, wherein the terminal or collector of one cell serves as the anode and cathode of the next cell in pile formation, leads to greatly improved uniformity of current distribution.

Whereas current-producing reactions occur at the electrode surface, they also occur at considerable depth below the surface in porous electrodes. Porous electrodes offer enhanced performance through increased surface area for the electrode reaction and through increased mass-transfer rates from shorter diffusion path lengths. The key parameters in determining the reaction distribution include the ratio of the volume conductivity of the electrolyte to the volume conductivity of the electrode matrix, the exchange current, the diffusion characteristics of reactants and products, and the total current flow. The porosity, pore size, and tortuosity of the electrode all play a role.

Mathematical formulations of the models of primary and secondary current distribution permit rapid optimization in the design of new battery configurations. Models to describe and predict porous electrode performance in the lead–acid battery system have been developed. The high rate performance of the present starting, lighting, and ignition (SLI) automotive batteries have evolved directly from coupling collector designs with the porous electrode compositions identified from modeling studies.

The positive electrode in a battery system is most often a metal oxide, but it may also be a metal sulfide or halide. Generally, these materials are relatively poor electrical conductors and exhibit extremely high ohmic polarizations (impedances) if not combined with supporting electronic conductors such as graphite, lead, silver, copper, or nickel in the form of powder, rod, mesh, wire, grid, or other configurations. In almost all cases the negative electrode is a metallic element of sufficient conductivity to require only minimal supporting conductive structures. Exceptions are the oxygen (air) positive and hydrogen gas negative electrodes, which require

TABLE 1. COMMERCIAL PRIMARY BATTERY SYSTEMS

Common name	Cell reaction	Nominal voltage	Energy content[a]		Comments	Manufacturers
			$W \cdot h/mL$	$W \cdot h/kg$		
Leclanché	$Zn + 2MnO_2 + 2 NH_4Cl \rightarrow$ $Zn(NH_3)_2Cl_2 + H_2O + Mn_2O_3$	1.5	0.20	80	low cost; general purpose, wide range of sizes	Eveready
zinc chloride	$4Zn + 8MnO_2 + ZnCl_2 + 9H_2O \rightarrow$ $8MnOOH + ZnCl_2 \cdot 4ZnO \cdot 5H_2O$	1.5	0.20	125	intermediate cost and performance	Rayovac Bright Star Eveready
alkaline	$2Zn + 3MnO_2 \rightarrow 2ZnO + Mn_3O_4$	1.5	0.25	130	sets standard for cylindrical cells	Rayovac Duracell Eveready
silver	$Zn + Ag_2O \rightarrow 2Ag + ZnO$	1.6	0.5	200	good pulse, higher voltage than mercury or Zn–air cells	Rayovac Duracell
mercury	$Zn + HgO \rightarrow Hg + ZnO$	1.35	0.5	165	sets standard for button cells; environmental problems	Eveready Rayovac Duracell
air	$2Zn + O_2(air) \rightarrow 2ZnO$	1.4	1.0	300	twice capacity of mercury and silver cells, limited active stand	Rayovac Alexander Duracell
Li–CuO	$2Li + CuO \rightarrow Li_2O + Cu$	1.5	0.6	300	potential replacement for Leclanché and zinc chloride	Rayovac Panasonic
Li–FeS	$2Li + FeS \rightarrow Li_2S + Fe$	1.6	0.4	160	replacement for mercury and silver cells, no mercury	Eveready
Li–CF$_x$	$xLi + CF_x \rightarrow xLiF + C$	2.7	0.5	300	high voltage, long shelf life, wide operating temperature	Rayovac
Li–MnO$_2$	$Li + MnO_2 \rightarrow LiMnO_2$	2.8	0.5	330	high voltage, long shelf life, wide operating temperature	Eagle Picher Duracell
Li–I$_2$	$2Li + I_2 \rightarrow 2LiI$	2.8	0.9	290	solid electrolyte, 10+ year life, welded construction, used in most of heart pacers, iodine charge transfer complex	Eveready Power Conversion Medtronic
Li–SO$_2$	$2Li + 2SO_2 \rightarrow Li_2S_2O_4$	2.8	0.4	330	military battery, low temperature, excellent storage	Catalyst Wilson Greatbatch SAFT
Li–SOCl$_2$	$4Li + 2SOCl_2 \rightarrow 4LiCl + SO_2 + S$	3.6	1.0	530	high voltage, high energy density	Power Conversion Electrochem Industries Eagle Picher Power Conversion

[a] Approximate values.

a substantial conductive, catalytically active surface support, which also serves as current collector.

Although there are a multitude of chemical reactions that release energy, only a few reactions have the characteristics requisite for use in commercial batteries. A set of criteria can be established to characterize reactions suitable for battery development. The principal features necessary for battery reactions are as follow. *(1) Mechanical and chemical stability.* The reactants or active masses and cell components must be stable over time (5 years or more) in the operating environment and must reform in their original condition on recharge. *(2) Energy content.* The reactants must have sufficient energy content to provide a useful voltage and current level, measured in $W \cdot h/L$ or $W \cdot h/kg$. *(3) Power density.* The reactants must be capable of reacting at rates sufficient to deliver useful rates of electricity, measured in terms of W/L or W/kg. *(4) Temperature range.*

The reactants must be able to maintain energy, power, and stability over a normal operating environment. The military often specifies −50 to 75°C. The average consumer has a less severe range of operating requirements, usually −10 to 50°C. *(5) Safety.* The battery must be safe in the normal operating environment as well as under mild abusive conditions. *(6) Cost.* The reactants and the materials of construction should be inexpensive and in good supply.

Tables 1 and 2 contain characteristics of various primary and secondary battery systems, respectively. Table 3 contains performance parameters for promising rechargeable battery systems in various stages of research and commercial development.

RALPH BRODD
Gould Inc.

TABLE 2. COMMERCIAL RECHARGEABLE BATTERY SYSTEMS

Common name	Cell reaction	Nominal voltage	Energy content[a] W·h/L	W·h/kg	Comments	Manufacturers
lead–acid	$Pb + PbO_2 + 2H_2SO_4 \rightleftarrows$ $2PbSO_4 + 2H_2O$	2.10	80	35	lowest cost, largest sales, available sealed	Johnson Controls Delco Exide GNB
nickel–cadmium	$Cd + 2NiOOH + 2H_2O \rightleftarrows$ $Cd(OH)_2 + 2Ni(OH)_2$	1.35	80	38	high rate, available sealed	Gates SAFT
nickel–metal hydride	$H_2(M) + 2NiOOH \rightleftarrows$ $2Ni(OH)_2 + M$	1.35	160	55	hydrogen absorbing alloy, good cycle life, high self-discharge	Ovonics
nickel–iron	$Fe + 2NiOOH + 2H_2O \rightleftarrows$ $Fe(OH)_2 + 2Ni(OH)_2$	1.25	90	30	limited production, very long cycle life, almost indestructible, old technology	Gates SAB
nickel–hydrogen	$H_2(g) + 2NiOOH \rightleftarrows 2Ni(OH)_2$	1.35	90	45	special space battery, very long cycle life, high self-discharge	Eagle Picher Hughes
silver–zinc	$Zn + 2AgO \rightleftarrows ZnO + Ag_2O$	1.86	200	100	two-step discharge, limited cycle life, high energy density	Eagle Picher Yardney
lithium–MoS₂	$Zn + Ag_2O \rightleftarrows ZnO + 2Ag$ $Li + MoS_2 \rightleftarrows LiMoS_2$	1.60 2.3	150	80	under development, small sealed cell	Moli Energy
lithium–MnO₂	$Li + MnO_2 \rightleftarrows LiMnO_2$	3.2	140	55	sealed coin cell	Sanyo
lithium–V₂O₅	$Li(C) + V_2O_5 \rightleftarrows LiV_2O_5 + C$	3.0	75	30	sealed coin cell	Toshiba
lithium–carbon	$Li + C \rightleftarrows Li(C)$ (intercalate)	3.0	6	2.1	sealed coin cell, "supercapacitor"	Panasonic

[a] Approximate values.

TABLE 3. RECHARGEABLE BATTERY SYSTEMS IN VARIOUS STAGES OF RESEARCH AND DEVELOPMENT

Common name	Cell reaction	Nominal voltage	Energy content[a] W·h/L	W·h/kg	Comments
nickel–zinc	$Zn + 2NiOOH + H_2O \rightleftarrows$ $ZnO + 2Ni(OH)_2$	1.65	95	60	limited cycle life, high rate capability
zinc–bromine	$Zn + Br_2 \rightleftarrows ZnBr_2$	1.85	75	65	bromine complex with quaternary ammonium salt, circulating electrolyte
lithium–FeS	$2LiAl + FeS \rightleftarrows$ $Li_2S + 2Al + Fe$	1.33	90	95	low cost, high temperature fused salt, second step possible
sodium–sulfur	$2Na + 3S \rightleftarrows Na_2S_3$	2.1	120	160	solid β-Al_2O_3 separator, high temperature operation
aluminum–air	$4Al + 3O_2 + 2H_2O \rightleftarrows$ $2Al_2O_3 \cdot H_2O$	1.6	360	250	circulating electrolyte, low cost, low energy efficiency, replaceable negative electrode
lithium–V₆O₁₃	$8Li + V_6O_{13} \rightleftarrows Li_8V_6O_{13}$ (intercalate)	2.5	200	200	solid polymer electrolyte

[a] Approximate values.

Additional Reading

Bard, A.J. and L.R. Faulkner: *Electrochemical Methods, Fundamentals and Applications,* John Wiley & Sons, Inc., New York, NY, 1980.

Bennett, P.D. and S. Gross: *Aqueous Batteries,* The Electrochemical Society, Inc., Pennington, NJ, 1997.

Berndt, D.: *Maintenance-Free Batteries: Lead-Acid, Nickel/Cadmium, Nickel/Metal Hydride,* John Wiley & Sons, Inc., New York, NY, 1997.

Besenhard, J.O. and D.O. Besenhard: *Handbook of Battery Materials,* John Wiley & Sons, Inc., New York, NY, 1999.

Broadhead, J. and B. Scrosati: *Lithium Polymer Batteries,* The Electrochemical Society, Inc., Pennington, NJ, 1997.

Carcone, J.: "Rechargeable Lithium Batteries Ideally Suited for Memory Backup," *Electronic Products,* **41** (January 1990).

Chin, S.: "Batteries," *Electronic Products,* **47** (May 1991).

Crompton, T.R.: *Battery Reference Book,* Butterworth-Heinemann, Inc., Woburn, MA, 2000.

Ingram, W.J.: *New Technology Batteries Guide,* DIANE Publishing Company, Collingdale, PA, 2000.

Julien, C. and Z. Stoynov: *Materials for Lithium-Ion Batteries,* Kluwer Academic Publishers, Norwell, MA, 2000.

Julien, C. and Z. Stoynov: *Materials for Lithium-Ion Batteries: Proceedings of the NATO Advanced Study Inst. On Materials for Lithium-Ion Batteries,* Kluwer Academic Publishers, Norwell, MA, 2000.

Koryta, J. and J. Dvorak: *Principles of Electrochemistry,* John Wiley & Sons, Inc., New York, NY, 1987.

Levy, S.C. and P. Bro: *Battery Hazards and Accident Prevention,* Kluwer Academic Publishers, Norwell, MA, 1994.

Linden, D. ed.: *Handbook of Batteries and Fuel Cells,* McGraw-Hill Book Co., Inc., New York, NY, 1984.

McKeefry, H.L.: "The Heat's On for Battery Makers," *Electronic Buyers News,* **30** (June 1, 1992).

Munshi, M.Z.A.: *Handbook of Solid State Batteries and Capacitors,* World Scientific Publishing Company, Inc., Riveredge, NJ, 1995.

Overshinsky, S.R., M.A. Fetcenko, and J. Ross: "A Nickel Metal Hydride Battery for Electric Vehicles," *Science*, **176** (April 9, 1993).

Reasbeck, P. and J.G. Smith: *Batteries for Automotive Use,* John Wiley & Sons, Inc., New York, NY, 1997.

Richter, A.: "Battery Developments Slow to Come as Cost Battles Function in Market," *Electronic Buyers News*, **32** (June 18, 1990).

Shulman, S.: "Plotting Revolutions in Electricity Storage," *Technology Review* (*MIT*). **19** (November/December 1992).

Staff: "New Technology Offers Longer Battery Life," *Today's Chemist at Work*, **18** (April 1992).

Staff: *Rechargeable Batteries Applications Handbook, Technical Marketing Staff of Gates Energy Products,* Butterworth-Heinemann, Inc., Woburn, MA, 1997.

Stix, G.: "Electric Car Pool," *Sci. Amer.*, **126** (May 1992).

Surampudi, S. and R. Marsh: *Lithium Batteries,* The Electrochemical Society, Inc., Pennington, NJ, 1999.

Vincent, C.A., B. Scrosati, M. Lazzari, and F. Bonino: *Modern Batteries,* Edward Arnold, Ltd., London, 1984.

Vincent, C.A. and B. Scrosati: *Modern Batteries: An Introduction to Electrochemical Power Sources,* Butterworth-Heinemann, Inc., Woburn, MA, 1997.

Wakihara, M. and O. Yamamoto: *Lithium-Ion Batteries: Fundamentals and Performance,* John Wiley & Sons, Inc., New York, NY, 1998.

Young, J. and A. Richter: "Nickel/Metal Hydroxide Batteries Taking Shape," *Electronic Buyers News*, **14** (May 25, 1992).

Web References

Eveready Battery Company, Inc. http://www.energizer.com/flashed.html
Battery Directory.com. http://www.batterydirectory.com/cons.php3
Battery-Index.com. http://www.battery-index.com/
Exide Industries. http://www.exideindustries.com/
Interstate Batteries. http://www.interstatebatteryofdet.com/
Rayovac. http://Rayovac.com/index.shtml

BATTERIES: LEAD-ACID.

The lead–acid battery is one of the most successful electrochemical systems and the most successful storage battery developed. About 80% of the lead, Pb, consumption in the United States was for batteries in that year.

The lead–acid battery consists of a number of cells in a container. These cells contain positive (PbO_2) and negative (Pb) electrodes or plates, separators to keep the plates apart, and sulfuric acid, H_2SO_4, electrolyte. The battery reactions are highly reversible, so that the battery can be discharged and charged repeatedly. The number of charge–discharge cycles that can be obtained depends strongly on the use mode and can vary from several hundred to thousands of cycles.

Each cell has a nominal voltage of 2 V and capacities typically vary from 1 to 2000 ampere-hours. Lead–acid cells can be operated with coulombic efficiencies as high as 95% and with energy efficiencies greater than 80%. The many cell designs available for a wide variety of uses can be divided into three main categories: automotive, industrial, and consumer. Automotive batteries, starting, lighting, and ignition (SLI) for cranking of internal combustion engines accounted for nearly 73% of the lead–acid battery sales in 1988. Industrial batteries are used for heavy-duty application such as motive and standby power. More recently, the use of batteries for utility peak shaving has been increasing. Consumer batteries are used for emergency lighting, security alarm systems, cordless convenience devices and power tools, and small engine starting. This is one of the fastest growing markets for the lead–acid battery.

In Figure 1, the cutaway view of the automotive battery shows the components used in its construction. Automotive and industrial motive power batteries have the standard free electrolyte systems and operate only in the vertical position.

Two types of batteries having immobilized electrolyte systems are also made. They are most common in consumer applications, but their use in industrial and SLI applications is increasing. Both types have low maintenance requirements and usually can be operated in any position. They are sometimes called valve regulated or recombinant batteries because they are equipped with a one-way pressure relief vent and normally operate in a sealed condition with an oxygen recombination cycle to reduce water loss.

In the gelled electrolyte battery, the sulfuric acid electrolyte has been immobilized by a thixotropic gel. This is made by mixing an inorganic powder such as silicon dioxide, SiO_2, with the acid. Other cells use a highly absorbent separator to immobilize the electrolyte.

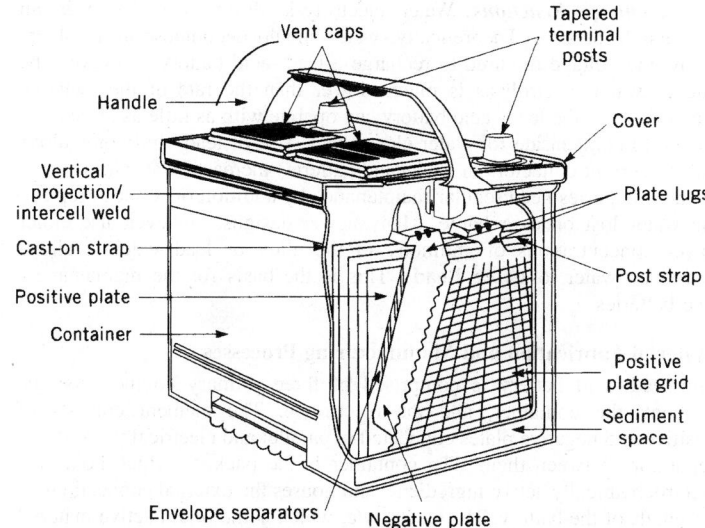

Fig. 1. Cutaway view of an automotive SLI lead–acid battery container and cell element. Courtesy of Johnson Controls, Inc

Cell Thermodynamics

The chemical reaction of the lead–acid battery was explained as early as 1882. The double sulfate theory has been confirmed by a number of methods as the only reaction consistent with the thermodynamics of the system. The thermodynamics of the lead–acid battery has been reviewed in great detail.

Lead sulfate is formed as the battery discharges, and sulfuric acid is regenerated as the battery is charged. The open circuit voltage of the lead–acid battery is the function of the acid concentration and temperature. A review of this subject is available. The Nernst equation may be used to calculate the open circuit cell voltage. The battery voltage is then obtained by multiplying the cell voltage by the number of cells.

Lead Grid Corrosion

The corrosion of the lead grid at the lead dioxide electrode is one of the primary causes of lead–acid battery failure. The mechanisms of lead corrosion in sulfuric acid have been studied and good reviews of the literature are available.

Charge–Discharge Processes

An excellent review covers the charge and discharge processes in detail and ongoing research on lead–acid batteries may be found in two symposia proceedings. Detailed studies of the kinetics and mechanisms of lead–acid battery reactions are published continually.

At high discharge rates, such as those required for starting an engine, the voltage drops sharply primarily because of the resistance of the lead current collectors. This voltage drop increases with the cell height and becomes significant even at moderate discharge rates in large industrial cells. Researchers have measured this effect in industrial cells and have developed a model which has been used to improve grid designs for automotive batteries.

Self-Discharge Processes. The shelf life of the lead–acid battery is limited by self-discharge reactions, which proceed slowly at room temperature. High temperatures reduce shelf life significantly. The reactions which can occur are well defined, and self-discharge rates in lead–acid batteries having immobilized electrolyte and limited acid volumes have been measured.

The lead current collector in the positive lead–dioxide plate corrodes and the compounds which form are a function of the acid concentration and positive electrode voltage. Other reactions which take place at the positive electrode are oxygen evolution, oxidation of organics, sulfation of PbO (in new cells), and oxidation of additives such as antimony, in the grid alloy.

Similar reactions can be written for other metallic additives. At the negative electrode two more reactions can occur, hydrogen evolution and oxygen recombination.

Overcharge Reactions. Water electrolysis during overcharge is an irreversible process. Theoretically, water should decompose at a voltage below the voltage required to recharge a lead–acid battery. However, the rate of water electrolysis is much slower than the rate of the recharge reaction. Thus the lead–acid battery can operate with as little as 5% excess charge to compensate for water electrolysis. Use of lead–antimony alloys for the current collectors in lead–acid batteries increases water loss. Some of these batteries need regular maintenance by addition of water to replace the water lost on overcharge. Many newer designs, however, use either lower concentrations of antimony in the alloy or lead–calcium alloys to reduce water loss (see **Lead**). This is the basis for the maintenance-free batteries.

Material Fabrication and Manufacturing Processes

The lead–acid battery is comprised of three primary components: the element, the container, and the electrolyte. The element consists of positive and negative plates connected in parallel and electrically insulating separators between them. The container is the package which holds the electrochemically active ingredients and houses the external connections or terminals of the battery. The electrolyte, which is the liquid active material and ionic conductor, is an aqueous solution of sulfuric acid (see Fig. 1).

Economic Aspects

Whereas automotive batteries have the majority of the market, other types of lead–acid batteries, such as sealed and small maintenance-free varieties, are making inroads into various applications. The automotive battery's operating environment has changed substantially in the last 10 years. Underhood temperature has risen and electrical loads have increased. This trend is expected to continue as car manufacturers reevaluate their design strategies and objectives. Battery design is changing to meet these needs.

KATHRYN R. BULLOCK
JOHN R. PIERSON
Johnson Controls, Inc.

Additional Reading

Bode, H.: *Lead-Acid Batteries* (trans. R. J. Brodd and K. V. Kordesch), John Wiley & Sons, Inc., New York, NY, 1977.
"SAE Standard Test Procedure for Storage J537—June '86," *SAE Recommended Practices,* SAE, New York, NY, June 1986.
Storage Battery Technical Service Manual, 10th Edition, Battery Council International, Chicago, Ill., 1987.
Vinal, G. W.: *Storage Batteries,* 4th Edition, John Wiley & Sons, Inc., New York, NY, 1955.

BATTERIES: OTHER.

The proliferation of portable electronic devices has fueled rapid market growth for the rechargeable battery industry. Miniaturization of electronics coupled with consumer demand for lightweight batteries providing ever longer run times continues to spur interest in advanced battery systems. Interest also continues to run strong in electric vehicles (EVs) and the large auto manufacturers continue to develop prototype EVs. It is clear that advances in battery technology are required for a widely acceptable EV. Advanced batteries continue to play a strong role in other applications such as load leveling for the electric utility industry and satellite power systems for aerospace.

Ambient Temperature Lithium Systems

Traditionally, secondary battery systems have been based on aqueous electrolytes. Whereas these systems have excellent performance, the use of water imposes a fundamental limitation on battery voltage because of the electrolysis of water, either to hydrogen at cathodic potentials or to oxygen at anodic potentials. The application of nonaqueous electrolytes affords a significant advantage in terms of achievable battery voltages. By far the most actively researched field in nonaqueous battery systems has been the development of practical rechargeable lithium batteries. These are systems that are based on the use of lithium metal, Li, or a lithium alloy, as the negative electrode. See also **Lithium**.

The use of lithium as a negative electrode for secondary batteries offers a number of advantages. Lithium has the lowest equivalent weight of any metal and affords very negative electrode potentials when in equilibrium with solvated lithium ions, resulting in very high theoretical energy densities for battery couples. These high theoretical energy densities have prompted a wealth of research activity in a wide variety of experimental battery systems. However, realization of the technology to commercialize these systems has been slow.

A key technical problem in developing practical lithium batteries has been poor cycle life attributable to the lithium electrode. The highly reactive nature of freshly plated lithium leads to reactions with electrolyte and impurities to form passivating films that electrically isolate the lithium metal.

The choice of battery electrolyte is of paramount importance in achieving acceptable cycle life because of the high reducing power of the metallic lithium. The formation of surface films on the lithium electrode imparts the apparent stability of the electrolyte to the electrode. It is critical to determining lithium cycling efficiency. In addition to providing a stable film in the presence of lithium, the electrolyte must satisfy additional requirements, including good conductivity, being in the liquid range over the battery operating temperature, and electrochemical stability over a wide voltage range. Solubility of the electrolyte salt in the solvent system is important in achieving good conductivity. In order to satisfy the various electrolyte system requirements, the use of mixed solvent electrolytes has become common in practical cells. Examples are tetrahydrofuran, C_4H_8O,-based electrolytes or ethylene carbonate $C_2H_4O_3$,−propylene carbonate, $C_4H_6O_3$, mixed solvent systems.

A second class of important electrolytes for rechargeable lithium batteries are solid electrolytes. Of particular importance is the class known as solid polymer electrolytes (SPEs). SPEs are polymers capable of forming complexes with lithium salts to yield ionic conductivity. The best known of the SPEs are the lithium salt complexes of poly(ethylene oxide) (PEO), $-(CH_2CH_2OH)_n-$, and poly(propylene oxide) (PPO).

The lithium or lithium alloy negative electrode systems employing a liquid electrolyte can be categorized as having either a solid positive electrode or a liquid positive electrode. Systems employing a solid electrolyte employ solid positive electrodes to provide a solid-state cell. Another class of lithium batteries are those based on conducting polymer electrodes. Several of these systems have reached advanced stages of development or initial commercialization such as the Seiko Bridge-stone lithium polymer coin cell.

The most important rechargeable lithium batteries are those using a solid positive electrode within which the lithium ion is capable of intercalating. These intercalation, or insertion, electrodes function by allowing the interstitial introduction of the Li^+ ion into a host lattice. A large number of inorganic compounds have been investigated for their ability to function as a reversible positive electrode in a lithium battery. Intercalation electrodes have found wide application in systems employing both solid or liquid electrolytes.

High Temperature Systems

Lithium–Aluminum/Metal Sulfide Batteries. The use of high temperature lithium cells for electric vehicle applications has been under development since the 1970s. Advances in the development of lithium alloy–metal sulfide batteries have led to the Li–Al/FeS system, where the following cell reaction occurs.

$$2\ LiAl_x + FeS \Longleftrightarrow Fe + Li_2S + 2x\ Al$$

The cell voltage is 1.33 V to give a theoretical energy density of 458 W·h/kg. The cell employs a molten salt electrolyte, most commonly a lithium chloride/potassium chloride, LiCl–KCl eutectic mixture. The cell is generally operated at 400–500°C. The negative electrode is composed of lithium–aluminum alloy, which operates at about 300 mV positive of pure lithium. The positive electrode is composed of iron sulfide mixed with a conductive agent such as carbon or graphite. Electrodes are constructed by cold pressing powder onto current collectors.

Development of practical and low cost separators has been an active area of cell development. Cell separators must be compatible with molten lithium, restricting the choice to ceramic materials. Early work employed boron nitride, BN, but a more desirable separator has been developed using magnesium oxide, MgO, or a composite of MgO powder–BN fibers.

Li–Al/FeS cells have demonstrated good performance under EV driving profiles and have delivered a specific energy of 115 W·h/kg for advanced cell designs. Cycle life expectancy for these cells is projected to be about 400 deep discharge cycles. This system shows considerable promise for use as a practical EV battery.

A similar system under development employs iron disulfide, FeS_2, as the positive electrode. Whereas this system offers a higher theoretical energy density than does Li–Al/FeS, the FeS_2 cell is at a lower stage of development.

Sodium–Sulfur. The best known of the high temperature batteries is the sodium–sulfur, Na–S, battery. The cell reaction is best represented by the equation:

$$2\,Na + 3\,S \rightleftharpoons Na_2S_3$$

occurring at a cell voltage of 1.74 V, to give a specific energy of 760 W·h/kg. The cell is constructed using a solid electrolyte typically consisting of β-alumina, β-Al_2O_3, ceramic, although borate glass fibers have also been used. These materials have high conductivities for the sodium ion. The negative electrode consists of molten sodium metal and the positive electrode of molten sulfur. Because sulfur is not conductive, a current collection network of graphite is required. The cell is operated at about 350°C.

The Na–S battery couple is a strong candidate for applications in both EVs and aerospace. Projected performance for a sodium–sulfur-powered EV van is shown in Table 1 for batteries having three different energies.

The Na–S system is expected to provide significant increases in energy density for satellite battery systems. In-house testing of Na–S cells designed to simulate midaltitude (MAO) and geosynchronous orbits (GEO) demonstrated over 6450 and over 1400 cycles, respectively.

Difficulties with the Na–S system arise in part from the ceramic nature of the alumina separator: the specific β-alumina is expensive to prepare, and the material is brittle and quite fragile. Separator failure is the leading cause of early cell failure. Cell failure may also be related to performance problems caused by polarization at the sodium/solid electrolyte interface. Lastly, seal leakage can be a determinant of cycle life. In spite of these problems, however, the safety and reliability of the Na–S system has progressed to the point where pilot plant production of these batteries is anticipated for EV and aerospace applications.

A battery system closely related to Na–S is the Na–metal chloride cell. The cell design is similar to Na–S; however, in addition to the β-alumina electrolyte, the cell also employs a sodium chloroaluminate, $NaAlCl_4$, molten salt electrolyte. The positive electrode active material consists of a transition metal chloride such as iron(II) chloride, $FeCl_2$, or nickel chloride, $NiCl_2$, in lieu of molten sulfur. This technology is in a younger state of development than the Na–S.

Miscellaneous Systems

Rechargeable cells employing aluminum, Al, as a negative electrode in room temperature molten salts have been investigated.

Redox flow batteries, under development since the early 1970s, are still of interest primarily for utility load leveling applications. Unlike other batteries, the active materials are not contained within the battery itself but are stored in separate tanks. The reactants each flow into a half-cell separated from the other by a selective membrane. An oxidation and reduction electrochemical reaction occurs in each half-cell to generate current. Examples of this technology include the iron–chromium, Fe–Cr, battery and the vanadium redox cell.

Other flow batteries investigated for both electric vehicle applications and utility load leveling include zinc–chlorine, Zn–Cl$_2$, and zinc–bromine, Zn–Br$_2$, batteries.

Economic Aspects

As of this writing, there is little commercialization of advanced battery systems.

Efforts to develop commercially viable EV versions of advanced battery systems continue. The ultimate goal is to develop battery technology suitable for practical, consumer-acceptable electric vehicles. The United States Advanced Battery Consortium (USABC) has been formed with the express purpose of accelerating development of practical EV batteries.

Paul R. Gifford
Ovonic Battery Company

Additional Reading

Bockris, J. and co-eds.: *Comprehensive Treatise of Electrochemistry,* Vol. 3, Plenum Press, New York, NY, 1981.

Matsuda, Y. and C. Schlaikjer, eds.: *Practical Lithium Batteries,* JEC Press, Inc., 1988.

Pletcher, D. *Industrial Electrochemistry,* Chapman and Hall Ltd., UK, 1984, pp. 272–274.

Sudworth, J. L. and A. R. Tilley: *The Sodium Sulfur Battery,* Chapman and Hall Ltd., UK, 1985, and references therein.

BATTERIES: PRIMARY CELLS.

Primary cells are galvanic cells designed to be discharged only once, and attempts to recharge them can present possible safety hazards. The cells are designed to have the maximum possible energy in each cell size because of the single discharge. Thus, comparison between battery types is usually made on the basis of the energy density in W·h/cm^3. The specific energy, W·h/kg, is often used as a secondary criterion for primary cells, especially when the application is weight-sensitive, as in space applications. The main categories of primary cells are carbon–zinc, known as heavy-duty and general purpose; alkaline, cylindrical, and miniature; lithium; and reserve or specialty cells.

Carbon–Zinc Cells

Carbon–zinc batteries are the most commonly found primary cells worldwide and are produced in almost every country. Traditionally there are a carbon rod, for cylindrical cells, or a carbon-coated plate, for flat cells, to collect the current at the cathode and a zinc anode. There are two basic versions of carbon–zinc cells: the Leclanché cell and the zinc chloride, $ZnCl_2$, or heavy-duty cell. Both have zinc anodes, manganese dioxide, MnO_2, cathodes, and include zinc chloride in the electrolyte. The Leclanché cell also has an electrolyte saturated with ammonium chloride, NH_4Cl. Additional undissolved ammonium chloride is usually added to the cathode, whereas the zinc chloride cell has at most a small amount of ammonium chloride added to the electrolyte. Both types are dry cells, in the sense that there is no excess liquid electrolyte in the system. The zinc chloride cell is often made using synthetic manganese dioxide and gives higher capacity than the Leclanché cell, which uses inexpensive natural manganese dioxide for the active cathode material. The MnO_2 is only a modest conductor. Thus the cathodes in both types of cell contain 10–30% carbon black in order to distribute the current. Because of the ease of manufacture and the long history of the cell, this battery system can be found in many sizes and shapes.

Performance. Carbon–zinc cells perform best under conditions of intermittent use, and many standardized tests have been devised that are appropriate to such applications as light and heavy flashlight usage, radios, cassettes, and motors (toys). The most frequently used tests are American National Standards Institute (ANSI) tests. The tests are carried out at constant resistance and the results reported in minutes or hours of service. Figure 1 shows typical results under a light load for different size cells.

To compare one battery with another, it is useful to compute the energy density from these data. Because the voltage declines with capacity, the average voltage during the discharge is used to compute an average current, which is then multiplied by the service in hours to give the ampere-hours of capacity. Watt-hours of energy can be obtained by multiplying again by the average voltage.

Cylindrical Alkaline Cells

Primary alkaline cells use sodium hydroxide or potassium hydroxide as the electrolyte. They can be made using a variety of chemistries and physical constructions. The alkaline cells of the 1990s are mostly of the limited electrolyte, dry cell type. Most primary alkaline cells are made using zinc as the anode material; a variety of cathode materials can be used. Primary alkaline cells are commonly divided into two classes, based on type of construction: the larger, cylindrically shaped batteries, and the miniature, button-type cells. Cylindrical alkaline batteries are mainly produced using zinc–manganese dioxide chemistry, although some cylindrical zinc–mercury oxide cells are made.

TABLE 1. ELECTRIC VEHICLE BATTERY PERFORMANCE

Parameter	Lead–acid		Battery Sodium–sulfur	
battery energy, kW·h	40.0	40.0	60.0	85.0
range, km	84.0	113.0	169.0	242.0
max payload, t	0.9	1.7	1.6	1.6
battery weight, kg	1250.0	330.0	424.0	580.0

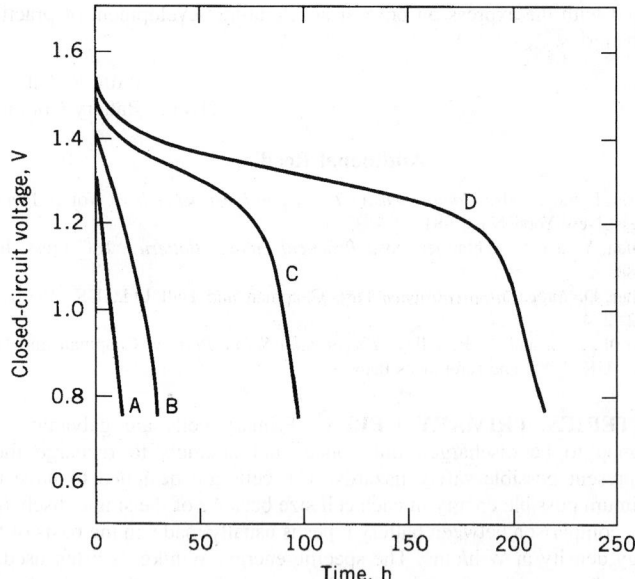

Fig. 1.   Hours of service on 40-Ω discharge for 4 h/d radio test at 21°C for A, RO3 "AAA"; B, R6 "AA"; C, R14 "C"; and D, R20 "D" paperlined, heavy-duty zinc chloride cells

Cylindrical alkaline cells are zinc–manganese dioxide cells having an alkaline electrolyte, which are constructed in the standard cylindrical sizes, R20 "D", R14 "C", R6 "AA", RO3 "AAA", as well as a few other less common sizes. They can be used in the same types of devices as ordinary Leclanché and zinc chloride cells. Moreover, the high level of performance makes them ideally suited for applications such as toys, audio devices, and cameras.

Performance. Alkaline manganese dioxide batteries have relatively high energy density, as can be seen from Table 1. This results in part from the use of highly pure materials, formed into electrodes of near optimum density. Moreover, the cells are able to function well with a rather small amount of electrolyte. The result is a cell having relatively high capacity at a fairly reasonable cost.

Miniature Alkaline Cells

Miniature alkaline cells are small, button-shaped cells which use alkaline NaOH or KOH electrolyte and generally have zinc anodes, but may have a variety of cathode materials. They are used in watches, calculators, cameras, hearing aids, and other miniature devices.

Cylindrical alkaline cells are made in only a few standard sizes and have only one important chemistry. In contrast, miniature alkaline cells are made in a large number of different sizes, using many different chemical systems. Whereas the cylindrical alkaline batteries are multipurpose batteries, used for a wide variety of devices under a variety of discharge conditions, miniature alkaline batteries are highly specialized, with the cathode material, separator type, and electrolyte all chosen to match the particular application.

Zinc–Mercuric Oxide Batteries. Miniature zinc–mercuric oxide batteries have a zinc anode and a cathode containing mercuric oxide, HgO.

Miniature zinc–mercuric oxide batteries function efficiently over a wide range of temperatures and have good storage life.

Although the zinc–mercuric oxide battery has many excellent qualities, increasing environmental concerns have led to a deemphasis in the use of this system. The main environmental difficulty is in the disposal of the cell. Both the mercuric oxide in the fresh cell and the mercury reduction product in the used cell have long-term toxic effects.

Zinc–Silver Oxide Batteries. Miniature zinc–silver oxide batteries have a zinc anode, and a cathode containing silver oxide, Ag_2O. Miniature zinc–silver oxide batteries are commonly used in electronic watches and in other applications where high energy density, a flat discharge profile, and a higher operating voltage than that of a mercury cell are needed. These batteries function efficiently over a wide range of temperatures and are comparable to mercury batteries in this respect. Miniature zinc–silver oxide batteries have good storage life.

Divalent Silver Oxide Batteries. It is possible to produce a silver oxide in which the silver has a higher oxidation state, approaching a composition of AgO. This material can provide both higher capacity and higher energy density than Ag_2O. Alternatively, a battery can be made with the same capacity as a monovalent silver cell, but with cost savings. However, some difficulties with regard to material stability and voltage regulation must be addressed.

Zinc–Manganese Dioxide Batteries. The combination of a zinc anode and manganese dioxide cathode, which is the dominant chemistry in large cylindrical alkaline cells, is used in some miniature alkaline cells as well. Overall, this type of cell does not account for a large share of the miniature cell market. It is used in cases where an economical power source is wanted and where the devices can tolerate the sloping discharge curve shown in Figure 2.

Zinc–Air Batteries. Zinc–air batteries offer the possibility of obtaining extremely high energy densities. Instead of having a cathode material placed in the battery when manufactured, oxygen from the atmosphere is used as cathode material, allowing for a much more efficient design. The construction of a miniature air cell is shown in Figure 3. From the outside, the cell looks like any other miniature cell, except for the air access holes in the can. On the inside, however, the anode occupies much more of the internal volume of the cell. Rather than the thick cathode pellet, there is a thin layer containing the cathode catalyst and air distribution passages. Air enters the cell through the holes in the can and the oxygen reacts at the surface of the cathode catalyst. The air access holes are often covered with a protective tape, which is removed when the cell is placed in service.

The performance level of air cells is exceptional, but these are not general-purpose cells. They must be used in applications where the usage

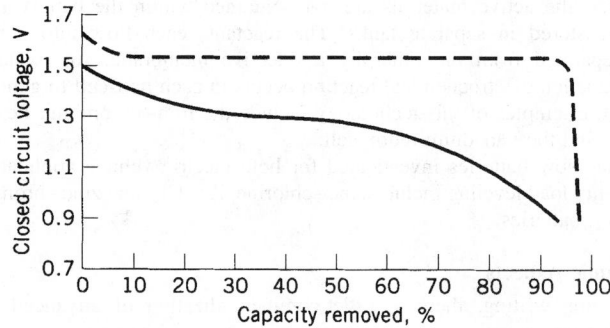

Fig. 2.   Discharge curves for miniature zinc–silver oxide batteries (– – –), and zinc–manganese dioxide batteries (—). Courtesy of Eveready Battery Co

Parameter	Carbon–zinc (Zn/MnO_2)	Alkaline manganese dioxide ($Zn–MnO_2$)	Mercuric oxide ($Zn–HgO$)	Silver oxide ($Zn–Ag_2O$)	Zinc–air ($Zn–O_2$)
nominal voltage, V	1.5	1.5	1.35	1.5	1.25
working voltage, V	1.2	1.2	1.3	1.55	1.25
specific energy, W·h/kg	40–100	80–95	100	130	230–400
energy density, W·h/mL	0.07–0.17	0.15–0.25	0.40–0.60	0.49–0.52	0.70–0.80
temperature range, °C					
storage	−40–50	−40–50	−40–60	−40–60	−40–50
operating	−5–55	−20–55	−10–55	−10–55	−10–55

TABLE 1. CHARACTERISTICS OF AQUEOUS PRIMARY BATTERIES

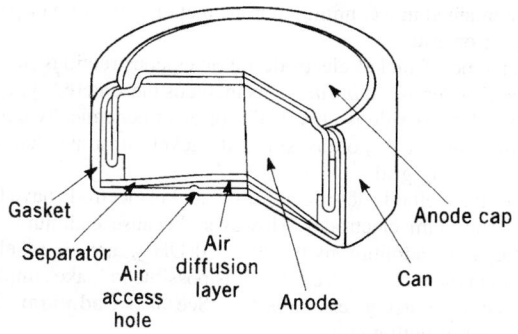

Fig. 3. Cutaway view of a miniature air cell battery. Courtesy of Eveready Battery Co.

is largely continuous, and where the discharge level is relatively constant and well-defined. The reasons for these limitations lies in the fact that the cell must be open to the atmosphere and the holes that allow oxygen into the cell also allow other gases to enter or leave the cell.

Miniature air cells are mainly used in hearing aids, where they are required to produce a relatively high current for a relatively short time period such as a few weeks. In this application they provide exceptional performance compared to other batteries.

Lithium Cells

Cells having lithium anodes are generally called lithium cells regardless of the cathode. They can be conveniently separated into two types: cells having solid cathodes and cells having liquid cathodes. Cells having liquid cathodes also have liquid electrolytes and in fact, at least one component of the electrolyte solvent and the active cathode material are one and the same. Cells having solid cathodes may have liquid or solid electrolytes but, except for the lithium–iodine system, those having solid electrolytes are not yet commercial.

All of the cells take advantage of the inherently high energy of lithium metal and its unusual film-forming property.

Much analytical study has been required to establish the materials for use as solvents and solutes in lithium batteries. Among the best organic solvents are cyclic esters, such as propylene carbonate (PC), $C_4H_6O_3$, ethylene carbonate (EC), $C_3H_4O_3$, and butyrolactone, $C_4H_6O_2$, and ethers, such as dimethoxyethane (DME), $C_4H_{10}O_2$, the glymes, tetrahydrofuran (THF), and 1,3-dioxolane, $C_3H_6O_2$. Among the most useful electrolyte salts are lithium perchlorate, $LiClO_4$, lithium trifluoromethanesulfonate, $LiCF_3SO_3$, lithium tetrafluoroborate, $LiBF_4$, and lithium hexafluoroarsenate, $LiAsF_6$. A limitation of these organic electrolytes is the relatively low conductivity, compared to aqueous electrolytes. This limitation, combined with the generally slow kinetics of the cathode reactions, has forced the use of certain designs, such as thin electrodes and very thin separators, in all lithium batteries. This usage led to the development of coin cells rather than button cells for miniature batteries and jelly or Swiss roll designs rather than bobbin designs for cylindrical cells.

Many of the cylindrical cells have glass-to-metal hermetic seals, although this is becoming less common because of the high cost associated with this type of seal. Alternatively, cylindrical cells have compression seals carefully designed to minimize the ingress of water and oxygen and the egress of volatile solvent. These construction designs are costly and the high price of the lithium cell has limited its use. However, the energy densities are superior.

Solid Cathode Cells. Solid cathode cells include lithium–manganese dioxide cells, lithium–carbon monofluoride cells, lithium–iron disulfide cells, and lithium–iodine cells.

Liquid Cathode Cells. Liquid cathode cells include lithium–sulfur dioxide cells and lithium–thionyl chloride cells.

Reserve Batteries

Reserve batteries have been developed for applications that require a long inactive shelf period followed by intense discharge during which high energy and power, and sometimes operation at low ambient temperature, are required. These batteries are usually classified by the mechanism of activation which is employed. There are water-activated batteries that utilize fresh or seawater; electrolyte-activated batteries, some using the complete electrolyte, some only the solvent; gas-activated batteries where the gas is used as either an active cathode material or part of the electrolyte; and heat-activated or thermal batteries which use a solid salt electrolyte activated by melting on application of heat.

Activation of these batteries involves adding the missing component which can be done in a simple way, such as pouring water into an opening in the cell, for water-activated cells, or in a more complicated way by using pistons, valves, or heat pellets activated by gravitational or electric signals for the case of the electrolyte- or thermal-activation types. Such batteries may be stored for 10–20 yr while awaiting use. Reserve batteries are usually manufactured under contract for various government agencies such as the U.S. Department of Defense, although occasional industrial or safety uses have been found. Many of the electrochemical systems involved in these batteries are beyond the scope of this article.

The lithium–thionyl chloride, or the lithium–sulfur dioxide, system is often used in a reserve battery configuration in which the electrolyte is stored in a sealed compartment which upon activation may be forced by a piston or inertial forces into the interelectrode space. Most applications for such batteries are in mines and fuse applications in military ordnance.

One variant of the liquid cathode reserve battery is the lithium–water cell in which water serves as both the liquid cathode and the electrolyte. A certain amount of corrosion occurs, but sufficient lithium is provided to compensate. These cells are mostly used in the marine environment where water is available or compatible with the cell reaction product. Common applications are for torpedo propulsion and to power sonobuoys and submersibles.

The last type of reserve cell is the thermally activated cell. The older designs use calcium or magnesium anodes; newer types use lithium alloys as anodes.

The heat pellet used for activation in these batteries is usually a mixture of a reactive metal such as iron or zirconium, and an oxidant such as potassium perchlorate. An electrical or mechanical signal ignites a primer, which then ignites the heat pellet which in turn melts the electrolyte. Sufficient heat is given off by the high current to sustain the necessary temperature during the lifetime of the application. Many millions of these batteries have been manufactured for military ordnance and employed in rockets, bombs, missiles, etc.

GEORGE BLOMGREN
JAMES HUNTER
Eveready Battery Company, Inc.

Additional Reading

Brodd, R. J. *Batteries for Cordless Appliances,* John Wiley & Sons, Inc., New York, NY, 1987.

Crompton, T. R. *Battery Reference Book,* Butterworths, London, UK, 1990.

Heise, G. W. and N. C. Cahoon, eds.: *The Primary Battery,* Vol. 1 and 2, John Wiley & Sons, Inc., New York, NY, 1971, 1976.

Linden, D. ed.: *Handbook of Batteries and Fuel Cells,* McGraw-Hill, New York, NY, 1984.

BATTERIES: SECONDARY CELLS.

Alkaline electrolyte storage battery systems are more suitable than others in applications where high currents are required, because of the high conductivity of the electrolyte. Additionally, in almost all of these battery systems, the electrolyte which is usually an aqueous solution containing 25–40% potassium hydroxide, KOH, does not enter into the chemical reaction. Thus concentration and cell resistance are invariant with state of discharge and these battery systems give high performance and have long cycle life. The annual production value of alkaline storage batteries is growing: it was over 2×10^9 worldwide in 1990, representing approximately 20% of the value for all secondary batteries.

Positive electrode active materials have been made from the oxides or hydroxides of nickel, silver, manganese, copper, mercury, and from oxygen. Negative electrode active materials have been fabricated from various geometric forms of cadmium, Cd, iron, Fe, and zinc, Zn, and from hydrogen. Two different types of hydrogen electrode designs are common: those used in space, which employ hydrogen as a gas, and those used in consumer batteries, where the hydrogen is used as a metallic hydride. As indicated in Table 1, nine electrode combinations exist in some scale of commercial production. Five system combinations are in the research and development stage as of this writing, and two have been abandoned before

TABLE 1. RECHARGEABLE ALKALINE STORAGE BATTERY SYSTEMS

System[a]	Historical name	Voltage, V	Production[b]
nickel–cadmium	Jungner	1.30	vl
nickel–iron	Edison	1.37	s
nickel–zinc	Drumm	1.70	vs
nickel–hydrogen		1.30	l
silver–cadmium		1.38 and 1.16[c]	vs
silver–iron	Jirsa	1.45 and 1.23[c]	s
silver–zinc	Andre	1.86 and 1.60[c]	s
silver–hydrogen		1.38 and 1.16[c]	vs
manganese–zinc[d]		1.52	vs
mercury–cadmium		0.92	r
air (oxygen)–zinc		1.60	r
air (oxygen)–iron		1.40	r
air (oxygen)–aluminum			r
copper–lead		1.20	r
copper–cadmium	Darrieus	0.45	n
copper–zinc	Waddell-Entz, Edison-LeLande, Lelande-Chaperon	0.85	n

[a] The substance named first represents the positive electrode; the substance named second is the negative electrode. In all cases except for air (oxygen) systems, the active electrode material is the oxide or the hydroxide of the named species.
[b] vl => 100×10^6 3A·h/yr product; l => 25×10^6 A·h/yr; s => 5×10^6 A·h/yr; vs =< 5×10^6 A·h/yr; r = research and development phase; and n = no longer in production.
[c] silver electrodes have two voltages plateaus.
[d] secondary system.

or after commercial production for reasons such as short life, high cost, low voltage, low energy density, and excessive maintenance.

The annual production value of small, sealed nickel–cadmium cells is over $\$1.2 \times 10^9$. However, environmental considerations relating to cadmium are necessitating changes in the fabrication techniques, as well as recovery of failed cells. Battery system designers are switching to nickel–metal hydride (MH) cells for some applications, typically in "AA"-size cells, to increase capacity in the same volume and avoid the use of cadmium.

Many of the most recent applications for alkaline storage batteries require higher energy density and lower cost designs than were previously available. Materials such as foam or fiber nickel, Ni, mats as substrates, and new processing techniques including plastic bounded, pasted, or electroplated electrodes, have enabled the alkaline storage battery to meet these new requirements, while reducing environmental problems in the manufacturing plants. In addition, substantial technical efforts have been devoted to the recovery of used batteries. The most recent innovations in materials relate to the development of metal–hydride alloys for the storage and electrochemical utilization of hydrogen. Modifications to the chemical structure or the cell design of manganese dioxide MnO_2, electrodes have resulted in sufficient improvement to allow the reintroduction of the rechargeable MnO_2–zinc cell to the market as a lower cost, albeit lower performance, alternative to nickel–cadmium consumer size cells. Improvements in materials science and electrical circuits have led to better separators, seals, welding techniques, feedthroughs, and charging equipment.

Nickel–Cadmium Cells

Electrodes. A number of different types of nickel oxide electrodes have been used. The term nickel oxide is common usage for the active materials that are actually hydrated hydroxides at nickel oxidation state 2+, in the discharged condition, and nickel oxide hydroxide, NiO·OH, nickel oxidation state 3+, in the charged condition. Nickelous hydroxide, $Ni(OH)_2$, can be precipitated from acidic solutions of bivalent nickel either by the addition of sodium hydroxide or by cathodic processes to cause an increase in the interfacial pH at the solution–electrode surface. See also **Nickel**.

The many varieties of practical nickel electrodes can be divided into two main categories. In the first, the active nickelous hydroxide is prepared in a separate chemical reactor and is subsequently blended, admixed, or layered with an electronically conductive material. This active material mixture is afterwards contained in a confining porous metallic structure or pasted onto a metallic mat or grid.

The other type of nickel electrode involves constructions in which the active material is deposited *in situ*. This includes the sintered-type electrode in which nickel hydroxide is chemically or electrochemically deposited in the pores of a 80–90% porous sintered nickel substrate that may also contain a reinforcing grid.

Almost all the methods described for the nickel electrode have been used to fabricate cadmium electrodes. However, because cadmium, cadmium oxide, CdO, and cadmium hydroxide, $Cd(OH)_2$, are more electrically conductive than the nickel hydroxides, it is possible to make simple pressed cadmium electrodes using less substrate. See also **Cadmium**. These are commonly used in button cells.

Electrochemistry and Crystal Structure. The solid-state chemistry of the nickel electrode is complex. Nickel hydroxide in the discharged state has a hexagonal layered lattice, where planes of Ni^{2+} ions are sandwiched between planes of OH^-. This structure, similar to that of cadmium iodide, CdI_2, is common to seven metal hydroxides, including those of cadmium and cobalt. There are various hydrated and nonhydrated nickel hydroxides that have slightly different crystal habitats and electrochemical potentials. The most common form of charged material observed in batteries is NiOOH, density = 4.6 g/mL. In comparison, $Ni(OH)_2$ has a density of 4.15 g/mL. Thus the theoretical change in density on charge–discharge is only 9%, and the kinetics involve only a proton transfer.

The chemistry, electrochemistry, and crystal structure of the cadmium electrode is much simpler than that of the nickel electrode. The overall reaction is generally recognized as:

$$Cd + 2OH^- \underset{charge}{\overset{discharge}{\rightleftharpoons}} Cd(OH)_2 + 2e^-$$

However, there is a strong likelihood of a soluble intermediate in the formation of $Cd(OH)_2$. Cadmium has an appreciable solubility in alkaline solutions: $\sim 2 \times 10^{-4}$ mol/L in 8 M potassium hydroxide at room temperature. In general, it is believed that the solution process consists of anodic dissolution of cadmium ions in the form of complex hydroxides.

Sealed Cells. Most sealed cells are based on the principles appearing in patents of the early 1950s where the virtues of limited electrolyte, a separator that would absorb and retain electrolyte, and providing free passage for the oxygen from the positive to the negative plate were described. Although both pocket and sintered electrodes of the nickel–cadmium type have been used in sealed-cell construction, the preponderant majority of cells in commercial production use sintered positive (nickel) electrodes, and either sintered or pasted negative (cadmium) electrodes.

Cell Fabrication Methods

Pocket Cells. The essential steps of positive (nickel) electrode construction are (1) cold-rolled steel ribbon is cut to proper width and is perforated using either needles or rolls; (2) the perforated steel ribbon is nickel-plated and usually annealed in hydrogen. The ribbon is formed into a trough shape, is filled with active material by either a briquetting or a powder-filling technique; (3) a second strip is formed into a lid that covers and locks with the filled trough; (4) the filled strips are cut to length and are arranged to form an electrode sheet by interleaving. This operation, carried out by means of rollers in a forming roll, is often combined with the pressing of a pattern into the electrode sheet in order to ensure good contact between ribbon and active material and to add mechanical strength to the construction; and (5) the electrode sheet is then cut to pieces of appropriate size, and side bedding and lugs are attached to form a metallic frame. The frame material is usually also cold-rolled steel ribbon.

The pockets are usually arranged horizontally in the electrodes, but in a few cases vertical pockets are used. No significant difference has been observed between the two arrangements.

Pocket-type cadmium electrodes are made by a procedure similar to that described for the positive electrode. Because cadmium active material is more dense than nickel active material, and because cadmium has a 2+ valence, cadmium electrodes, when fabricated to equal thicknesses, have almost twice the working capacity of the nickel electrode.

After the individual pocket electrodes are fabricated, they are assembled into electrode groups. Electrodes of the same polarity are electrically and mechanically connected to each other and to a pole bolt. Plates of opposite polarity are interleaved with separators.

To complete the assembly of a cell, the interleaved electrode groups are bolted to a cover and the cover is sealed to a container. Originally, nickel-plated steel was the predominant material for cell containers, but more recently plastic containers have been used for a considerable proportion of pocket nickel–cadmium cells. Polyethylene, high impact polystyrene, and a copolymer of propylene and ethylene have been the most widely used plastics.

Tubular Cells. Although the tubular nickel electrode invented by Edison is almost always combined with an iron negative electrode, a small quantity of cells is produced in which nickel in the tubular form is used with a pocket cadmium electrode. This type of cell construction is used for low operating temperature environments, where iron electrodes do not perform well or where charging current must be limited.

Sintered Cells. The fabrication of sintered electrode batteries can be divided into five principal operations: preparation of sintering-grade nickel powder; preparation of the sintered nickel plaque; impregnation of the plaque with active material; assembly of the impregnated plaques (often called plates) into electrode groups and into cells; and assembly of cells into batteries.

Other Cells. Other methods to fabricate nickel–cadmium cell electrodes include those for the button cell, used for calculators and other electronic devices. This cell, the construction of which is illustrated in Figure 1, is commonly made using a pressed powder nickel electrode mixed with graphite that is similar to a pocket electrode. The cadmium electrode is made in a similar manner. The active material, graphite blends for the nickel electrode, are almost the same as that used for pocket electrodes, i.e., 18% graphite.

Lower cost and lower weight cylindrical cells have been made using plastic bound or plated active material pressed into a metal screen. These cells suffer slightly in utilization at high rates compared to a sintered-plate cylindrical cell, but they may be adequate for most applications.

Applications. Nickel–cadmium cells represent almost 20% of the market for all storage batteries, including lead–acid, manufactured in the world. Uses are divided into three categories: pocket cells are used in emergency lighting, diesel starting, and stationary and traction applications where the reliability, long life, medium-high rate capability,

and low temperature performance characteristics warrant the extra cost over lead–acid storage batteries; sintered, vented cells are used in extremely high rate applications, such as jet engine and large diesel engine starting as well as some electric vehicle designs; and sealed cells, both the sintered and button types, are used in computers, phones, cameras, portable tools, electronic devices, calculators, cordless razors, toothbrushes, carving knives, flashguns, and in space applications, where nickel–cadmium is optimum because it can be recharged a great number of cycles and not suffer from prolonged trickle overcharge. Cells of this category are generally made in sizes comparable to conventional dry cells, such as "D," "C," "AA," etc.

Charger Technology. Alkaline storage batteries are commonly charged from rectified d-c equipment, solar panels, or other d-c sources and have fairly good tolerance to ripple and transient pulses. Advances in electronic control circuitry in charges and redesign of small sealed nickel–cadmium cells permit the rapid (15-min) recharge of special designs.

Nickel–Iron Cells. There has been renewed interest in the system for electric vehicles (EV). The EV design is based on a high rate, usually sintered, iron electrode as well as high rate nickel electrodes.

Electrochemistry and Kinetics. The electrochemistry of the nickel–iron battery and the crystal structures of the active materials depends on the method of preparation of the material, degree of discharge, the age (life cycle), concentration of electrolyte, and type and degree of additives, particularly the presence of lithium and cobalt. A simplified equation representing the charge–discharge cycle can be given as:

$$2NiOOH^* + \alpha - Fe + 2H_2O \rightleftharpoons 2Ni(OH)_2^* + Fe(OH)_2$$

where the asterisks indicate adsorbed water and KOH.

Electrode Structures. The classical iron active material for pocket and pasted iron electrodes was formed by roasting recrystallized ferrous sulfate, $FeSO_4$, in an oxidizing atmosphere to ferric oxide, Fe_2O_3, and then reducing the latter in hydrogen. The α-iron formed was then heated to a mixture of Fe_3O_4 and Fe. As such it was pure enough to be used for pharmaceutical purposes. For battery use, a small amount of sulfur, as FeS, was added, as were other additives which were believed to increase the cycle life by acting as depassivating agents, i.e., helping to reduce the tendency of iron to evolve hydrogen upon standing in alkaline electrolyte.

A study of sintered iron electrodes claimed advantages of high rate capability, long life, and low hydrogen evolution.

Sintered nickel electrodes used in nickel iron cells are usually thicker than those used in Ni–Cd cells. These result in high energy density cells, because very high discharge rates are usually not required.

Performance Characteristics. The sintered nickel-sintered iron design battery has outstanding power characteristics at all states of discharge, making them attractive to the design of electric vehicles (EV) which must accelerate with traffic even when almost completely discharged.

Silver–Zinc Cells

The silver–zinc battery has the highest attainable energy density of any rechargeable system in use as of this writing. In addition, it has an extremely high rate capability coupled with a very flat voltage discharge characteristic. Its use, in the early 1990s, is limited almost exclusively to the military for various aerospace applications such as satellites and missiles, submarine and torpedo propulsion applications, and some limited portable communications applications. The main drawback of these cells is the rather limited lifetime of the silver–zinc system. Life is normally limited to less than 200 cycles with a total wet-life of no more than about two years. The silver–zinc system also carries a very high cost and applications are justified only where cost is a minor factor. The high cost of silver battery systems is attributable to the cost of the active silver material used in the positive electrodes.

Cellophane or its derivatives have been used as the basic separator for the silver–zinc cell since the 1940s. The cellophane is the principal limitation to cell life. Oxidation of the cellophane in the cell environment degrades the separator and within a relatively short time short circuits may occur in the cell.

A second lifetime limitation is the zinc anode. In spite of the separator and cell designs, some zinc material is solubilized during the

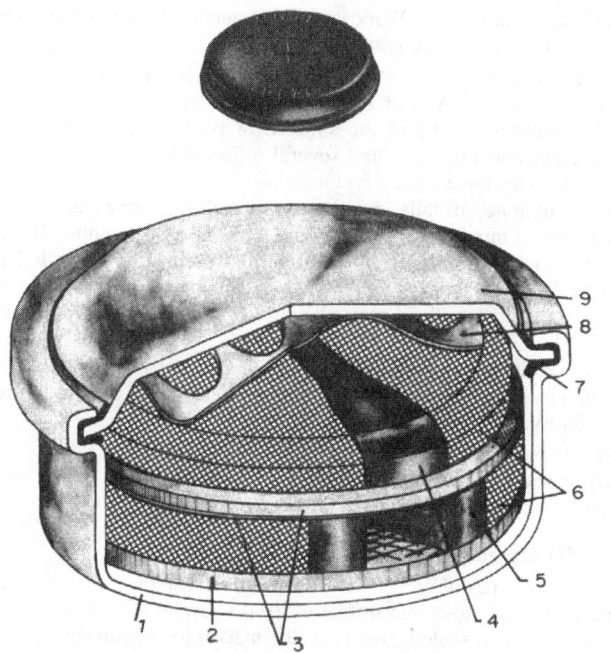

Fig. 1. Section of disk-type cell where: 1 is the cell cup; 2, the bottom insert; 3, the separator; 4, the negative electrode; 5, the positive electrode; 6, the nickel wire gauze; 7, the sealing washer; 8, the contact spring; and 9, the cell cover

charge–discharge reaction. Over a period of cycling there is a shift of active material, originally distributed evenly over the face of the electrode, to the center and bottom areas of the electrode. This shape change limits the life of the cell as exemplified by a fading of the capacity and a build-up of internal pressure that may eventually lead to a short circuit.

Electrochemistry. Silver–zinc cells have some unusual thermodynamic properties. The equations indicate that the higher valence silver oxide is AgO, silver(II) oxide. However, in the crystallographic unit cell, which is monolithic, there are four silver atoms and four oxygen atoms, and none of the Ag–O bonds conforms to a silver(II) bond length. Instead there are two Ag–O bonds of 0.218 nm corresponding to silver(I) and two Ag–O bonds of 0.203 nm corresponding to silver(III). This structure has also been proposed on the basis of magnetic and semiconductor properties and confirmed using neutron diffraction.

Electrodes. All of the finished silver electrodes have certain common characteristics: the grids or substrates used in the electrodes are made exclusively of silver, although in some particular cases silver-plated copper is used.

There are three methods of silver electrode fabrication: (*1*) the slurry pasting of monovalent or divalent silver oxide to the grid, drying, reducing by exposure to heat, and then sintering to agglomerate the fine particles into an integral, strong structure; (*2*) the dry processing of fine silver powders by pressing in a mold or by a continuous rolling operation onto a silver grid followed by sintering; and, (*3*) the use of plastic-bonded active material formed by embedding the active material (fine silver powder) in a plastic vehicle such as polyethylene, which can then be milled into flexible sheets. These sheets are cut to size, pressed in a mold on both sides of a conductive grid, and the pressed electrode subjected to sintering where the plastic material is fired off, leaving the metallic silver.

Silver electrodes prepared by any of the three methods are almost always subjected to a sintering operation prior to cell or battery assembly.

Zinc electrodes for secondary silver–zinc batteries are made by one of three general methods: the dry-powder process, the slurry-pasted process, or the electroformed process. The active material used in any of the processes for the manufacture of electrodes is a finely divided zinc oxide powder, USP grade 12.

Electrolyte. The electrolyte in silver–zinc cells is 30–45% KOH. The lower concentrations in this range have higher conductivities and are preferred for high rate cells. Higher concentrations have a less deleterious effect on cellulosic separators and are preferable for extended life characteristics.

Cell Hardware. Cell jars are constructed almost exclusively of injection-molded plastics, which are resistant to the strong alkali electrolyte.

Cell terminal connections are usually brought out by two-threaded terminals that protrude through the cell jar cover. They are usually steel, brass, or copper with a hollow construction.

Performance

Charging. Charging of silver–zinc cells can be done by one of several methods. The constant-current method which is most common consists of a single rate of current, usually equivalent to a full input within the 12–16-h period.

Discharge. Silver–zinc cells have one of the flattest voltage curves of any practical battery system known, although there are two voltage steps caused by the two different valence states of silver oxide.

Performance of silver–zinc cells is normally considered to be adequate in the temperature range of 10–38°C. If a wider temperature range is desired silver–zinc cells and batteries may be used in the range 0–71°C without any appreciable derating.

Cell Life. Silver–zinc cells are usually manufactured as either low or high rate cells. Approximately 10–30 cycles can be expected for high rate cells, depending on the temperature of use, the rate of discharge, and methods of charging. Low rate cells have been satisfactorily used for 100–300 cycles under the proper conditions. In general, the overall life of the silver–zinc cells with the separator systems normally in use is approximately 1–2 yr.

Other Silver Positive Electrode Systems

Silver–Cadmium Cells. In satellite applications the nonmagnetic property of the silver–cadmium battery is of utmost importance because magnetometers were used on satellites to measure radiation and the effects of

magnetic fields of energetic particles. Satellites have to be constructed of nonmagnetic components in sealed batteries.

Silver–cadmium satellite batteries have been used in cyclic periods of five hours or more with discharge times of 30–60 min. Operational and test programs have shown cycle life periods of 3 yr at low temperatures. At temperatures of 40°C and 50°C, the cycle life is 1 yr and 0.2 yr, respectively. The cycle life at intermediate temperatures is 1.4–2.0 yr.

Another application for silver–cadmium batteries is propulsion power for submarine simulator-target drones.

Silver–Iron Cells. The silver–iron battery system combines the advantages of the high rate capability of the silver electrode and the cycling characteristics of the iron electrode. Commercial development has been undertaken to solve problems associated with deep cycling of high power batteries for ocean systems operations.

Cells consist of porous sintered silver electrodes and high rate iron electrodes. The latter are enclosed with a seven-layered, controlled porosity polypropylene bag which serves as the separator. The electrolyte contains 30% KOH and 1.5% LiOH.

Applications have been found for these batteries in emergency power applications for telecommunications systems in tethered balloons. Unfortunately, the system is expensive because of the high cost of the silver electrode. Applications are, therefore, generally sought where recovery and reclamation of the raw materials can be made.

Applications have been found for these batteries in emergency power applications for telecommunications systems in tethered balloons. Unfortunately, the system is expensive because of the high cost of the silver electrode. Applications are, therefore, generally sought where recovery and reclamation of the raw materials can be made.

Nickel–Zinc Cells

Nickel–zinc cells offer potential advantages over other rechargeable alkaline systems. The single-level discharge voltage, 1.60–1.65 V/cell is approximately 0.35–0.45 V/cell higher than nickel–cadmium or nickel–iron and approximately equal to that of silver–zinc. In addition, the use of zinc as the negative electrode should result in a higher energy density battery than either nickel–cadmium or nickel–iron and a lower cost than silver–zinc. In fact, nickel–zinc cells having energy densities in the range of 40–60 W·h/kg have been successfully demonstrated.

A commercial nickel–zinc battery is considered to be the most likely candidate for electric vehicle development. If the problems of limited life and high installation cost ($100–150/kW·h) are solved, a nickel–zinc EV battery could provide twice the driving range for an equivalent weight lead–acid battery. Work is developmental; there is no commercial production of nickel–zinc batteries.

Cell Construction. Nickel–zinc batteries are housed in molded plastic cell jars of styrene, SAN, or ABS material for maximum weight savings. Nickel electrodes can be of the sintered or pocket type; however, these types are not cost-effective, and several different types of plastic-bonded nickel electrodes have been developed.

Nickel hydrate, usually 5–10% cobalt added, serves as the active material and is mixed with a conductive carbon, e.g., graphite. The active mass is mixed with an inert organic binder such as polyethylene or poly(tetrafluoroethylene) (TFE). The resultant mass is rolled into sheets on a compounding mill or pressed into electrodes as a dry powder on a nickel grid.

Negative electrodes are fabricated of zinc oxide by any of the methods (pasting, pressing, etc) described. Binders, usually TFE, are used to reduce the solubility of the electrode in KOH.

Separators are both of the organic and inorganic type.

Performance. The limited life of nickel–zinc batteries is the principal drawback to widespread use.

Nickel–Hydrogen Cells

There are two types of nickel–hydrogen cells: those that employ a gaseous H_2 electrode and those that utilize a metal hydride, MH. The use of MH electrodes in small-scaled. However, the market for consumer-type cells is rapidly growing. The system has also been proposed for electric vehicles.

Other Cell Systems

Silver–Hydrogen Cells. With the development of the nickel–hydrogen system limited attention was directed to the development of a silver–hydrogen cell. The main characteristics of interest were the potential

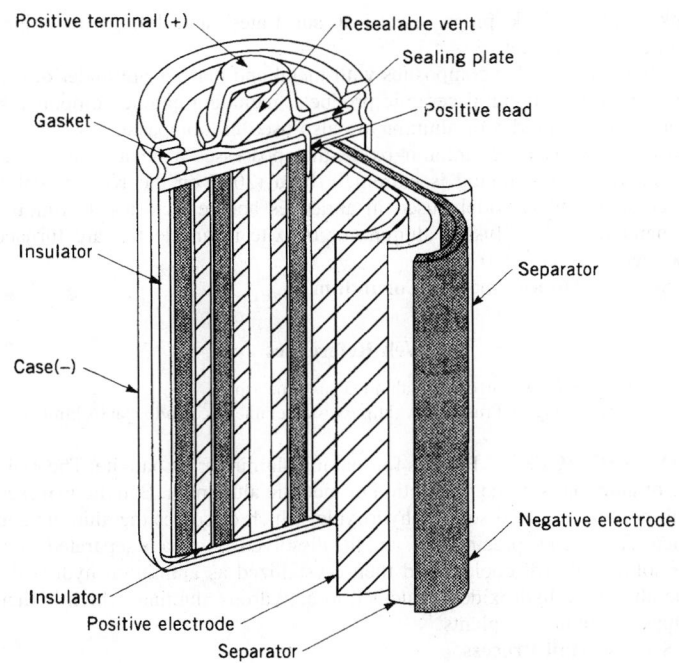

Fig. 2.   Schematic diagram of a Ni–H (MH) cell

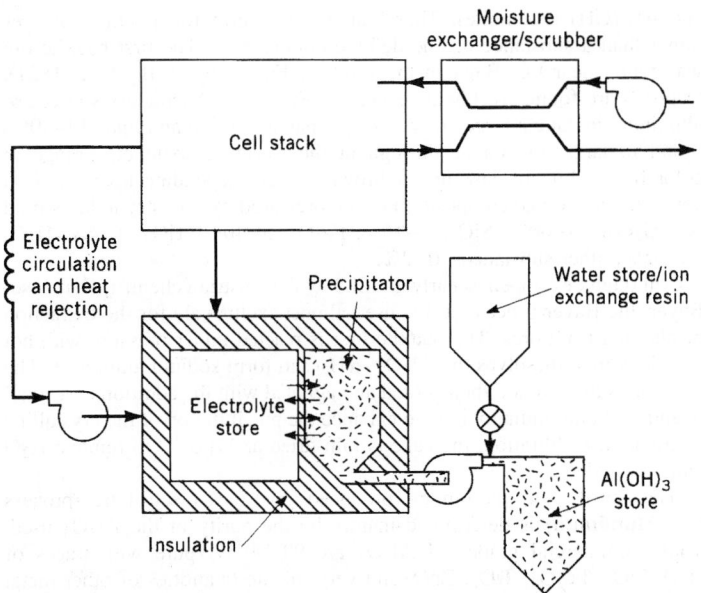

Fig. 3.   Aluminum–air power cell system. The design provides for forced convection of air and electrolyte, heat rejection, electrolyte concentration control via Al(OH)$_3$ precipitation, and storage for reactants and products

for a higher gravametric energy density based on the lighter weight of the silver electrode vs that of the nickel. The packaging approach utilized for this battery is similar to that of nickel–hydrogen single cylindrical cells as shown in Figure 2. The silver electrode is typically the sintered type used in rechargeable silver–zinc cells. The hydrogen electrode is a Teflon-bonded platinum black gas diffusion electrode.

Because the silver oxide electrode is slightly soluble in the potassium hydroxide electrolyte the separator is of a barrier type to minimize silver diffusion to the opposite electrode.

Zinc–Oxygen Cells. On the basis of reactants the zinc–oxygen or air system is the highest energy density system of all the alkaline rechargeable systems with the exception of the H$_2$·O$_2$ one. The reactants are cheap and abundant and therefore a number of attempts have been made to develop a practical rechargeable system.

Iron–Air Cells. The iron–air system is a potentially low cost, high energy system being considered mainly for mobile applications. The iron electrode, similar to that employed in the nickel–iron cell, exhibits long life and therefore this system could be more cost effective than the zinc–air cell.

Hydrogen–Oxygen Cells. The hydrogen–oxygen cell can be adapted to function as a rechargeable battery, although this system is best known as a primary one. See also **Fuel Cells**.

Mechanically Rechargeable Batteries. To avoid the time required for electric recharge, the problems of *in situ* electric recharge, or to utilize anodes that are not electrically rechargeable in aqueous electrolytes, mechanically rechargeable batteries have been studied. These systems are metal–air couples. The anodes that have received attention are zinc, lithium, and aluminum.

The most significant results with these battery types focus on aluminum as the anode. Figure 3 shows an aluminum–air cell being developed for electric vehicle applications. The aluminum hydroxide reaction product would be returned to the factory to be reprocessed into fresh aluminum anodes. One set of anodes could yield up to 800 km range before replacement. However, the corrosion reaction of the aluminum with the electrolyte is still a problem. Additionally, the system is complex and it is anticipated that replacing the anodes repeatedly will be problematic. The system has poor energy efficiency when consideration is given to the full cycle of electric generation, aluminum production, battery efficiency, and reprocessing of the battery reaction product back to aluminum.

Electrolyte

Potassium hydroxide is the principal electrolyte of choice for the above batteries because of its compatibility with the various electrodes, good conductivity, and low freezing point temperature.

Safety and Disposal

The potassium hydroxide electrolyte used in alkaline batteries is a corrosive, hazardous chemical. It is a poison and if ingested attacks the throat and stomach linings. Immediate medical attention is required. It slowly attacks skin if not rapidly washed away. Extreme care should be taken to avoid eye contact that can result in severe burns and blindness. Protective clothing and face shields or goggles should be worn when filling cells with water or electrolyte and performing other maintenance on vented batteries.

Alkaline batteries generate hydrogen and oxygen gases under various operating conditions. This can occur during charge, overcharge, open circuit stand, and reversal. In vented batteries free ventilation should be provided to avoid hydrogen accumulations surrounding the battery.

Alkaline batteries are capable of high current discharges and accidental short circuits should be avoided. Spontaneous low resistance internal short circuits can develop in silver–zinc and nickel–cadmium batteries.

Because of increasing environmental concerns, the disposal of all batteries is being reviewed. Traditionally silver batteries were reclaimed for the silver metal and all other alkaline batteries were disposed of in landfills or incinerators. Some aircraft and industrial nickel–cadmium batteries are rebuilt to utilize the valuable components.

To reduce or eliminate the scattering of cadmium in the environment, the disposal of nickel–cadmium batteries is under study. Already a large share of industrial batteries are being reclaimed for the value of their materials.

ALVIN J. SALKIND
MARTIN KLEIN
Rutgers University

Additional Reading

Andersson, B. and L. Ojefors: *11th International Power Sources Symposium,* Brighton, U.K., 1978; *J. Electrochem. Soc.* **123,** 824 (1976).

Klein, M.: in D. H. Collins, ed., *Power Sources,* Vol. 5, Academic Press, New York, NY, 1974.

Ojefors, L.: *Proceedings of the 9th International Society for Electrochemistry,* Marcoussis, France, 1975.

Sulkes, M.J.: Nickel-Zinc Secondary Batteries, Proceedings of the 23rd Annual Power Sources Conference, 1969.

Swaroop, R.: "Reports on Electric Vehicle Batteries", *Electric Power Research Institute (EPRI)*, Palo Alto, CA, 1989–1991.

BAUME SCALE. See **Specific Gravity**.

BAUXITE. There are two main types of bauxite ores used as the primary sources for aluminum metal and aluminum chemicals: Al(OH)$_3$ (gibbsite) and AlO(OH) (boehmite). Thus, bauxite is a term for a family of ores rather than a substance of one definite composition. The first bauxite ore was found near Les Baux in the south of France by P. Berthier (1821). Deposits are found worldwide except in Antarctica. Secondary sources of aluminum include a large array of clays that are rich in alumina (30–40%) found in large abundance throughout the world. Bauxite ores range in color from white to dark red or brown, largely depending upon the iron content. An average composition of the ores used by industry today would be: Al_2O_3, 35–60%; SiO_2, 1–15%; Fe_2O_3, 5–40%; TiO_2, 1–4%, H_2O; 10–35%; other substances, 0–2%.

Although developed as early as 1888 by the Austrian chemist, Karl Josef Bayer, the Bayer process still is used almost exclusively for the extraction of alumina from ores. The bauxite first is reacted under pressure with hot caustic, which dissolves the $Al_2O_3 \cdot xH_2O$ to form sodium aluminate. The solution is filtered hot, then cooled and agitated with the addition of a small quantity of aluminum hydrate to enhance the precipitation of the crystalline hydrate. After filtration, the cake is kiln-dried at 1100 °C to remove H_2O and yield Al_2O_3.

The purity of aluminum produced by the electrolytic process (see **Aluminum**) is determined mainly by the purity of the Al_2O_3 used. Thus, commercial grades of Al_2O_3 are 99–99.5% pure with traces of H_2O, SiO_2, Fe_2O_3, TiO_2, ZnO, and very minute quantities of other metal oxides.

As of 1991, there were nine alumina plants in the United States and territories. Six plants were located on the Gulf Coast, two plants were in Arkansas, and one plant was in the Virgin Islands. Total capacity was estimated at 7.7 million short tons (approximately 6.9 million metric tons). The aluminum industry in the United States has depended on imports from the Caribbean, South America, and Australia. Caribbean exporting countries have high levies on bauxite exports and, politically, have moved to nationalize and expropriate the bauxite mines developed by United States industry. It is doubtful that new Bayer alumina plants will be built in the United States. The technology in the United States is focusing on extracting alumina from abundant alumina-containing clays, but these programs are being de-emphasized because of great new bauxite finds. New Bayer plants in Australia and Venezuela will be key future suppliers.

The industry has maintained a continuous exploration program, which has been very successful. There has been a fivefold increase in world reserves of bauxite since 1965; at that time the world reserves were 6 billion tons, whereas in 1981 there were 29 billion tons in reserve. The source of that information is the *J. of Metals*, February 1986, page 19, article by A.S. Russell, "Aluminum Technology Responds to Change."

Uses and Grades of Alumina

Although aluminum production is a major consumer of alumina, the compound is used widely elsewhere. The properties of alumina and hydrated aluminas may be varied, ranging from a talc-like softness to the hardness of a ruby or sapphire. Some of the uses for alumina include water purification, glassmaking, production of steel alloys, waterproofing of textiles, coatings for ceramics, abrasives and refractory materials, cosmetics, and electronics. *Hydrated aluminas* may be represented: α-$Al_2O_3 \cdot H_2O$ or α-$Al(OH)_3$. The compounds are dry, snow-white, free-flowing crystalline powders and may be obtained in a wide range of particle sizes. The compounds are widely used in the production of aluminum salts because of their reactions with strong acids and alkalies. Some of the salts prepared in this manner include aluminum chloride, aluminum phosphate, and aluminum sulfate. *Activated aluminas* are very porous aluminum oxide. λ-Al_2O_3 is made by heating the hydrate to drive off nearly all of the combined water. The final products are granules or fine powders with a large surface area to provide absorptive capacity per unit volume. Applications of the aluminas are enhanced by virtue of their chemical inertness and nontoxic qualities. They are extensively used for drying gases and for dehydrating liquids, such as alcohol, benzol, carbon tetrachloride, ethyl acetate, gasoline, toluol, and vegetable and animal oils. They also are used as filter aids in the manufacture of lubricating and other oil products. Their large surface area qualifies the aluminas as catalysts for numerous reactions. The compounds also find extensive application in ceramics, particularly in abrasive and cutting wheels, polishing compounds, additives to glass,

tank linings, spark plugs, electrical substrates, and linings for high-temperature furnaces.

Alumina fibers for composites with metal and plastics are under development as are structural ceramic products, including engine components. Fine, specialty grades of alumina are also used in toothpaste.

Corundum is an aluminum oxide that possesses a hexagonal crystal structure. The compound is extremely hard (2000 on the Knoop scale), sp gr 3.95, and is widely used in abrasives and refractories. Corundum is manufactured by fusing alumina or bauxite in an electric arc furnace operated at about 2200°C.

See also **Aluminum**; and **Corundum**.

Web References

Alcoa, Inc. http://www.alcoa.com/alumina/en/home.asp
C-E Minerals, King of Prussia, PA. http://www.ceminerals.com/loc_aGA.html

BAYER PROCESS. Process for making alumina from bauxite. The main use of alumina is in the production of metallic aluminum. Bauxite is mixed with hot concentrated sodium hydroxide, which dissolves the alumina and silica. The silica is precipitated, and the dissolved alumina is separated from the solids, diluted, cooled, and then crystallized as aluminum hydroxide. The aluminum hydroxide is calcined to anhydrous alumina, which is then shipped to reduction plants.

See also **Hall Process**.

BECKE TEST. A microscope of moderate or high magnification is used to compare the indices of refraction of two contiguous minerals (or of a mineral and a mounting medium or immersion liquid), in a thin section or other mount. When the two substances differ substantially in refractive index, they are separated by a bright line, called the *Becke line*. The line moves toward the less refractive of two materials when the tube of the microscope is lowered.

BECKMANN, ERNST (1853–1923). Beckmann was a German chemist who discovered in 1886 the arrangement of oximes of ketones into acid amides or anilides, named the Beckmann molecular transformation. He was the inventor of two pieces of apparatus used in determining freezing and boiling points of solutions. The Beckmann thermometer is used for determining molecular weights in solutions.

See also **Beckmann Method**.

J.M.I.

BECKMANN METHOD. A method of measuring elevation of the boiling point or depression of the freezing point of a solution. It may be used to measure concentration if the nature of the solute is known, or the molecular weight of the solute if the volume concentration is known.

See also **Analysis (Chemical)**; and **Freezing-Point Depression**.

BECQUEREL, ANTOINE HENRI (1852–1908). Becquerel, a Frenchman, represented the third generation in his family to become a physicist. He was trained as an engineer beginning his research career at the Ecole Polytechnique in 1875. His early research was work was concerned with the plane polarization of light, with the phenomenon of phosphorescence and with the absorption of light by crystals (his doctorate thesis). He also worked on the subject of terrestrial magnetism.

Becquerel invented a phosphoroscope to study phosphorescence. He is most famous for his discovery of natural radioactivity, a property of uranium, in 1896. His discovery lead eventually to the atomic and nuclear age and the research work of Marie and Pierre Curie. Becquerel shared the Nobel Prize in Physics in 1903 with Marie and Pierre Curie for "his discovery of spontaneous radioactivity".

See also **Becquerel Effect**.

J.M.I.

BECQUEREL EFFECT. A photographic effect discovered by E. Becquerel (1895). Experimenting with the daguerreotype process, Becquerel found that a plate will produce a direct (positive) image if exposed first to diffuse daylight. See also **Photochemistry and Photolysis**; and **Photography and Imagery**.

BEER'S LAW. If the Bouguer law is applied to a solution of fixed thickness, b cm. and concentration c, the result is $\log I_0/I = abc$ which is the fundamental equation of quantitative absorptimetry. Here I/I_0 is the transmittance and the constant a, the absorptivity depends on the nature of the absorbing material, the wavelength of the incident radiation, the nature of the solvent, the temperature, and perhaps other controllable experimental conditions. When c is given in moles per liter, a is called the molar absorptivity. If other concentration units are more convenient, the numerical value of a must be changed accordingly. The product $A = abc$ is the absorbance of the sample and it equals $1/T$, the reciprocal of the transmittance. Experimental verification of Beer's law will succeed only if appropriate corrections are made or can be neglected as with the Bouguer law.

BEGGIATOA. A genus of filamentous sulfur bacteria which is capable of converting hydrogen sulfide to sulfuric acid. Sulfur granules are stored in the cells, and may be oxidized to supply the cell with the necessary energy for life. The bacteria are autotrophic in form, using chemosynthesis to obtain energy.

BEILSTEIN, F. P. (1838–1906). A German chemist noted for his compilation "Handbuch der Organischen Chemie," the first edition of which appeared in 1880. A multivolume compendium of the properties and reactions of organic compounds, it has been revised several times and remains a unique and fundamental contribution to chemical literature.

BEILSTEIN'S TEST. A test to detect halogens in organic compounds. Copper gauze is heated in a flame until the flame shows no green color; if the addition of an organic compound produces a green flame, a halogen is present.

BÉNARD CONVECTION CELLS. When a layer of liquid is heated from below, the onset of convection is marked by the appearance of a regular array of hexagonal cells, the liquid rising in the center and falling near the wall of each cell. The criterion for the appearance of the cells is that the Rayleigh number should exceed 1700 (for rigid boundaries).

BENEDICT SOLUTION. In its original, classical form, this was an alkaline solution of copper hydroxide and sodium citrate in sodium carbonate used either as a mild oxidizing agent or as a test for easily oxidizable groups such as aldehyde groups. The formation of cuprous oxide is a positive test, its color red, but often yellow at first. Many other forms of this solution have been developed. Glucose reacts with Benedict solution to form cuprous oxide.

See also **Fehling's Solution**.

BENTONITE. The term applied to altered fine-grained volcanic ashes which have been blown considerable distance from their origin and deposited in marine waters. The resulting material is usually a white, but sometimes a colored, clay-like sediment which may contain bits of volcanic glass but is composed mainly of colloidal silica which will absorb large quantities of water. Since bentonites are wind-blown deposits they are useful as definite datum planes in stratigraphy, especially in helping to determine the contemporaneity of the different facies of marine sediments.

BENZALDEHYDE. [CAS: 100-52-7]. C_6H_5CHO, formula weight 106.12, colorless liquid, mp bp $-26°C$, bp $179°C$, sp gr 1.046. Sometimes referred to as artificial almond oil or "oil of bitter almonds," benzaldehyde has a characteristic nutlike odor. The compound is slightly soluble in H_2O, but is miscible in all proportions with alcohol or ether. On standing in air, benzaldehyde oxidizes readily to benzoic acid. Commercially, benzaldehyde may be produced by (1) heating benzal chloride $C_6H_5CHCl_2$ with calcium hydroxide, (2) heating calcium benzoate and calcium formate:

$$(C_6H_5COO)_2Ca + (HCOO)_2Ca \longrightarrow 2C_6H_5CHO + 2CaCO_3,$$

or (3) boiling glucoside amygdalin of bitter almonds with a dilute acid.

Benzaldehyde reacts with many chemicals in a marked manner: (1) with ammonio-silver nitrate ("Tollen's solution") to form metallic silver, either as a black precipitate or as an adherent mirror film on glass (but does not reduce alkaline cupric solution, "Fehling's solution"); (2) with rosaniline (fuchsine, magenta) that has been decolorized by sulfurous acid ("Schiff's

solution"), restoring the pink color of rosaniline; (3) with NaOH solution, yielding benzyl alcohol and sodium benzoate; (4) with NH_4OH, yielding tribenzaldeamine (hydrobenzamide, $(C_6H_5CH)_3N_2$), white solid, mp $101°C$, (5) with aniline, yielding benzylideneaniline ("Schiff's base" $C_6H_5CH:NC_6H_5$); (6) with sodium cyanide in alcohol, yielding benzoin C_6H_5. $CHOHCOC_6H_5$, white solid, mp $133°C$; (7) with hydroxylamine hydrochloride, yielding benzaldoximes $C_6H_5CH:NOH$, white solids, antioxime, mp $35°C$, syn-oxime, mp $130°C$; (8) with phenylhydrazine, yields benzaldehyde phenylhydrazone $C_6H_5CH:NNHC_6H_5$, pink solid, mp $156°C$; (9) with concentrated HNO_3, yields metanitrobenzaldehyde $NO_2·$ C_6H_4CHO, white solid, mp $58°C$; (10) with concentrated H_2SO_4 yields metabenzaldehyde sulfonic acid $C_6H_4CHO(SO_3H)_2$, (11) with anhydrous sodium acetate and acetic anhydride at $180°C$, yielding sodium benzoate $C_6H_5CHOONa$ (12) with sodium hydrogen sulfite, forming benzaldehyde sodium bisulfite $C_6H_5CHOHSO_3Na$, a white solid, from which benzaldehyde is readily recoverable by treatment with sodium carbonate solution; (13) with acetaldehyde made slightly alkaline with NaOH, yielding cinnamic aldehyde $C_6H_5CH:CHCHO$, (14) with phosphorus pentachloride, yielding benzylidine chloride $C_6H_5CHCl_2$.

Benzaldehyde may be detected by the appearance of a blue color on treating with acenaphthene and H_2SO_4, followed by heating. Benzaldehyde is used (1) as a flavoring material, (2) in the production of cinnamic acid, (3) in the manufacture of malachite green dye.

BENZENE. [CAS: 71-43-2]. C_6H_6, formula weight 78.11, colorless, highly flammable liquid that burns with a smoky flame, mp $5.5°C$, bp $80.1°C$, sp gr 0.879 ($20°C$ referred to H_2O at $4°C$). Sometimes called *benzol, phenyl hydride*, or *cyclohexatriene*, benzene is practically insoluble in water (0.07 part in 100 parts at $22°C$); and fully miscible with alcohol, ether, and numerous organic liquids. Benzene is of large importance industrially, mainly as a starting ingredient of many reactions and as a solvent. The compound also is of much theoretical interest, being the simplest hydrocarbon of the aromatic group. Forming an explosive mixture with air, benzene can be used as a fuel or fuel component for internal combustion engines. Benzene is the first member of a homologous series of compounds, C_nH_{2n-6}. Methylbenzene or toluene, C_6H_5. CH_3, is the only homolog with the formula C_7H_8. Next in order of the homologous series, C_8H_{10}, exists in four isomeric forms, namely, ethylbenzene, C_6H_5. C_2H_5, and ortho-, meta-, and paradimethylbenzene, $C_6H_4(CH_3)_2$. Of the formula C_9H_{12}, eight isomerides are possible. Possible isomerides increase rapidly as the carbon count increases.

Many of the C_nH_{2n-6} series of hydrocarbons occur in coal gas and coal tar, from which they may be extracted. These were the early sources for benzene and related compounds. Much of the benzene currently is produced from petroleum sources. The separation of benzene from coal tar is difficult because of the presence of scores of isomerides with close boiling points. In one process for making high-purity benzene (99.94% or higher) from coke-oven light oil (the cut boiling between $60-150°C$), the light oil and a stream of hydrogen are heated to reaction temperature and passed through fixed-bed reactors that contain a proprietary catalyst. In this reaction, the non-aromatics present are converted to light hydrocarbon gases. Sulfur compounds present are converted to H_2S. Some de-alkylation of the higher aromatics present also produces benzene in addition to that contained in the feedstock. Vapors from the reactor are cooled and passed to a stabilizer tower where dissolved H_2S and light hydrocarbons (with boiling points lower than benzene) are removed. The bottoms from the stabilizer containing benzene, toluene, and xylene, are clay-treated. Then follows a series of fractionations to produce benzene, toluene, and xylene, in addition to higher-boiling hydrocarbons. If a portion of the hydrocarbons in the product fuel gas is reformed, no external hydrogen is required.

The synthetic production of benzene generally involves the de-alkylation of toluene. In one non-catalytic process, a hydrogen-rich gas is mixed with liquid toluene feed and preheated prior to charging to the reactor. Toluene reacts with the hydrogen to form benzene and methane. The reaction is exothermic. Operating conditions approximate $500-1000$ psi and $595-760°C$. The process provides about 98% yield of benzene. The toluene is recycled.

In a catalytic de-alkylation process, toluene or C_8 aromatics (alkylbenzenes) are fed to a reactor, together with a hydrogen-containing gas. See Fig. 1. The hydrogen source is not critical and may be manufactured hydrogen or off-gas from a reforming or other refining unit. Effluent from the reactor, after cooling, is charged to a separator, from which hydrogen is removed and recycled to the reactor. Liquid phase from the separator is stripped of hydrocarbons (boiling lower than benzene) in a stabilizing column. One further fractionating step yields product benzene overhead. The bottoms from the tower are recycled to the reactor for de-alkylation. Yields of 98% of theoretical are claimed.

In another process, mixtures of aromatics and nonaromatic hydrocarbons comprise the charge. In a first step, aromatics are continuously extracted from the feed by using an aqueous solution of N-methylpyrollidone. A multistage countercurrent extraction tower is used. The operation is carried out at modest temperatures and pressures. The rich aromatic extract phase then proceeds to a stripper, where pentane and a part of the benzene are removed overhead and recycled to the extractor. The bottoms from the stripper are free of nonaromatics and enter a second stripper for further separation. The distillate from the second stripper contains aromatic-free solvent, which is returned to the extractor. One or more further fractionations yield benzene, toluene, and xylenes of desired specification. A typical feedstock may contain an mixture of aromatics in following ranges: benzene, 26–60%; toluene, 14–22%, xylenes plus ethylbenzene, 15–50%. A similar process uses dimethyl sulfoxide as a solvent.

Styrene is a major consumer of benzene, followed by the production of cyclohexane. At one time, phenol production was the second largest consumer of benzene. Benzene and cyclohexene are closely related economically because cyclohexane can be produced by reacting benzene with hydrogen. Although the foregoing represent the major tonnage uses of benzene, the compound is critically important to the production of hundreds of other compounds. The halogen derivatives of benzene are particularly important. See also **Chlorinated Organics**. These include chlorobenzene, C_6H_5Cl; bromobenzene, C_6H_5Br; benzal chloride, $C_6H_5 \cdot CHCl_2$; benzyl chloride, $C_6H_5 \cdot CH_2Cl$; and benzotrichloride, $C_6H_5 \cdot CCl_3$. Important nitro derivatives of benzene include nitrobenzene, $C_6H_5 \cdot NO_2$; and metadinitrobenzene, $C_6H_4(NO_2)_2$. Amino compounds of large importance derived from benzene include aminobenzene, $C_6H_5 \cdot NH_2$ (aniline); diaminobenzene, $C_6H_4(NH_2)_2$; and triaminobenzene, $C_6H_3(NH_2)_3$.

Phenol, C_6H_5OH, is hydroxybenzene. Resorcinol, catechol, and quinol, $C_6H_4(OH)_2$, may be considered to be dihydroxybenzenes. Pyrogallol and phloroglucinol, $C_6H_3(OH)_3$, may be considered as trihydroxybenzenes. The benzene-related alcohols, aldehydes, and ketones include benzyl alcohol, $C_6H_5 \cdot CH_2 \cdot OH$; benzaldehyde, $C_6H_5 \cdot CHO$; benzoin, $C_6H_5 \cdot CO \cdot CH(OH) \cdot C_6H_5$; salicyaldehyde, $C_6H_4(OH) \cdot CHO$; anisaldehyde, $C_6H_4(OHC_3) \cdot CHO$; acetophenone, $C_6H_5 \cdot CO \cdot CH_3$; benzophenone, $C_6H_5 \cdot CO \cdot C_6H_5$; and quinone, $C_6H_4O_2$. The benzene-related acids and salts include benzoic acid, $C_6H_5 \cdot COOH$; ethyl benzoate, $C_6H_5 \cdot COOC_2H_5$, benzoyl chloride, $C_6H_5 \cdot COCl$; benzoic anhydride, $(C_6H_4 \cdot CO)_2O$; benzamide, $C_6H_5 \cdot CO \cdot NH_2$; benzonitrile, $C_6H_5 \cdot CN$; anthranilic acid, $C_6H_4(NH_2) \cdot COOH$; phthalic acid, $C_6H_4(COOH)_2$; phthalic anhydride, $C_6H_4(CO)_2O$; phthalimide, $C_6H_4(CO)_2NH$; isophthalic acid, $C_6H_4(COOH)_2$; terephthalic acid, $C_6H_4(COOH)_2$; benzenehexacarboxylic acid, $C_6H_4(COOH)_6$; and phenylacetic acid, $C_6H_5 \cdot CH_2 \cdot COOH$.

The structure of benzene with its six C−H groups has been known for decades. However, arriving at the ring structure of the compound required much research and imagination in the early days of organic chemistry. Both Kekulè and Dewar and, a bit later, Claus proposed ingenious structures.

Benzene reacts (1) with chlorine, to form (a) substitution products (one-half of the chlorine forms hydrogen chloride) such as chlorobenzene, C_6H_5Cl; dichlorobenzene, $C_6H_4Cl_2(1,4)$ and $(1,2)$; trichlorobenzene, $C_6H_3Cl_3(1,2,4)$; tetrachlorobenzene $(1,2,3,5)$; and (b) addition products, such as benzene dichloride $C_6H_6Cl_2$; benzene tetrachloride, $C_6H_6Cl_4$; and benzene hexachloride, $C_6H_6Cl_6$. The formation of substitution products of the benzene nucleus, whether in benzene or its homologues, is favored by the presence of a catalyzer, e.g., iodine, phosphorus, iron; (2) with concentrated HNO_3, to form nitrobenzene, $C_6H_5NO_2$; 1,3-dinitrobenzene, $C_6H_4(NO_2)_2$ $(1,3)$, 1,3,5-trinitrobenzene, $C_6H_5(NO_2)_3$ $(1,3,5)$; (3) with concentrated H_2SO_4, to form benzene sulfonic acid, $C_6H_5SO_3H$, benzene disulfonic acid, $C_6H_4(SO_3H)_2(1,3)$, benzene trisulfonic acid, $C_6H_3(SO_3H)_3$ $(1,3–5)$; (4) with methyl chloride plus anhydrous aluminum chloride (Friedel-Crafts reaction) to form toluene, monomethyl benzene, $C_6H_5CH_3$; dimethyl benzene $C_6H_4(CH_3)_2$; trimethyl benzene, $C_6H_3(CH_3)_3$; (5) with acetyl chloride plus anhydrous aluminum chloride (Friedel-Crafts reaction) to form acetophenone (methylphenyl ketone), $C_6H_5COCH_3$. See also **Organic Chemistry**.

Additional Reading

Bailey, P.S. and C.A. Bailey: *Organic Chemistry: A Brief Survey,* 6th Edition, Prentice-Hall, Inc., New York, NY, 1999.
Brown, W.H. and C.S. Foote: *Organic Chemistry,* Harcourt College Publishers, Philadelphia, PA, 1999.
Bruice, P.Y.: *Organic Chemistry,* Prentice-Hall, Inc., New York, NY, 1998.
Jones, L. and P.W. Atkins: *Chemistry: Molecules, Matter and Change,* W.H. Freeman and Company, New York, NY, 1999.
Lide, D.R.: *CRC Handbook of Chemistry and Physics,* 84th Edition, CRC Press, LLC., Boca Raton, FL,. 2003.
Morrison, R.T. and R.N. Boyd: *Organic Chemistry,* 7th Edition, Prentice-Hall, Inc., New York, NY, 2000.
Solomons, G. M. M. Shenkman and C. Fryhle: *Organic Chemistry,* 8th Edition, John Wiley & Sons, Inc., New York, NY, 2003.
Sorrell, T.N.: *Organic Chemistry,* University Science Books, Sausalito, CA, 1999.
Vollhardt, K., C. Peter, and Neil Eric, Schore: *"Organic Chemistry,"* 3rd Edition, W.H. Freeman and Company, New York, NY, 1999.
Wade, L.G. Jr.: *Organic Chemistry,* Prentice-Hall, Inc., Upper Saddle River, NJ, 1999.

BENZIDINE REARRANGEMENT. See **Rearrangement (Organic Chemistry)**.

BENZIL REARRANGEMENT. See **Rearrangement (Organic Chemistry)**.

BENZINE. A product of petroleum boiling between 120°F and 150°F (49–66°C) and composed of aliphatic hydrocarbons. Not to be confused with benzene, which is a single chemical compound and an aromatic hydrocarbon.

BENZOIC ACID. [CAS: 65-85-0]. $C_6H_5 \cdot COOH$, formula weight 122.12, white crystalline solid, mp 121.7°C, bp 249.2°C, sublimes readily at 100°C and is volatile in steam, sp gr 1.266. Sometimes referred to as phenylformic acid, the compound is insoluble in cold H_2O, but readily soluble in hot H_2O, or in alcohol or ether. Commercially, benzoic acid finds major use as a starting or intermediate material in various industrial organic syntheses, notably in the preparation of the high-tonnage chemical, terephthalic acid. See also **Terephthalic Acid**. Benzoic acid forms benzoates: e.g., sodium benzoate, calcium benzoate which, when heated with calcium oxide, yields benzene and calcium. With phosphorus trichloride, benzoic acid forms benzoyl chloride C_6H_5COCl, an important agent for the transfer of the benzoyl group (C_6H_5CO-). Benzoic acid reacts with chlorine to form m-chlorobenzoic acid and reacts with HNO_3 to form m-nitrobenzoic acid. Benzoic acid forms a number of industrially useful esters, including methyl benzoate, ethyl benzoate, glycol dibenzoate, and glyceryl tribenzoate.

Although benzoic acid occurs naturally in some substances, such as gum benzoin, dragon's blood resin, Peru and Tolu balsams, cranberries, and the urine of the ox and horse, the product is made on a large scale by synthesis from other materials. Benzoic acid can be prepared from toluene and air in a process that takes place in the liquid-phase in a continuous oxidation reactor operated at moderate pressure and temperature: $C6H5 \cdot CH3 + 1\frac{1}{2} O2 \rightarrow C6H5 \cdot COOH + 2H2O$. The acid also can be obtained as a byproduct of the manufacture of benzaldehyde from benzal chloride or benzyl chloride. See also **Antimicrobial Agents (Foods)**.

BENZOIN. The term benzoin is used in botany to denote a tree (*Styrax benzoin*) and a resin obtained from it. The former is a tall, quick-growing tree native to Sumatra and Java. The tree has alternate entire leaves, the

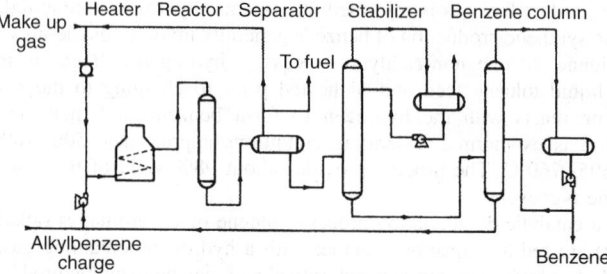

Fig. 1. Process for converting toluene or C_8 aromatic hydrocarbons (catalytic hydrodealkylation) to high-purity benzene. (*UOP, Inc*)

lower surface of which is soft and hairy, the upper smooth. The flowers are borne in compound axillary racemes; the fruit is a drupe. The resin is obtained by making incisions in the bark. From these a thick white juice exudes and hardens. This is scraped off. It is a soft fragrant substance either white or of yellowish color.

Frequently confused with this is the North American shrub *Benzoin aestivale*, often called Spice bush, which blossoms very early in the spring. All parts of the shrub contain an aromatic substance, which is very noticeable when the plant is bruised.

In chemistry, benzoin is a compound obtained from the resin described above, or obtained synthetically. When benzaldehyde C_6H_5CHO is warmed with sodium cyanide dissolved in alcohol, benzoin $C_6H_5CO-CHOHC_6H_5$ white solid, 137°C, is formed, and has the characteristics of a ketone and a secondary alcohol.

BENZOYL CHLORIDE.   See **Chlorinated Organics**.

BENZYL ALCOHOL AND β-PHENETHYL ALCOHOL.   Benzyl
alcohol [CAS: 100-51-6]. (**1**) β-phenethyl alcohol (**2**) (2-phenylethanol) are the simplest of the aromatic alcohols, and, as such, are chemically similar. Their physical properties are given in Table 1.

(1)

(2)

Benzyl Alcohol

Benzyl alcohol [CAS: 100-51-6]. (**1**) occurs widely in essential oils both as the free alcohol, and, more importantly from a fragrance standpoint, in the form of various esters.

Although benzyl alcohol itself is rather bland in odor, combined with its much more fragrant esters it is an important part of the door of jasmine, ylang-ylang, gardenia, some rose varieties, narcissus and peony, as well as castoreum, balsams of peru and tolu, and propolis. Benzyl alcohol occurs primarily in flower oils and tree exudates.

Benzyl alcohol readily undergoes the reactions characteristic of a primary alcohol, such as esterification and etherification, as well as halide formation. In addition, it undergoes ring substitution. Catalytic oxidation over copper oxide yields benzaldehyde; benzoic acid is obtained by oxidation with chromic acid or potassium permanganate.

TABLE 1. PHYSICAL PROPERTIES OF BENZYL ALCOHOL AND β-PHENETHYL ALCOHOL

Property	Benzyl alcohol	β-Phenethyl alcohol
molecular formula	C_7H_8O	$C_8H_{10}O$
mp, °C	−15	−25.8
bp at 101.3 kPa[a]	205.4–205.7	219.5–220
bp at 1.33 kPa[a], °C	89.0–89.5	99–100
d_{25}^{25}	1.0441	1.017
flash point, closed cup, °C	100.4	
open cup, °C	104.4	
autoignition temp, °C	436	
vapor density (air = 1)	3.7	
vapor pressure at 58°C, kPa[a]	0.133	0.133
100°C, kPa[a]	2.02	1.33
surface tension at 20°C, mN/m(= dyn/cm)	39	
80°C, mN/m(= dyn/cm)	33	
viscosity at 25°C, mPa · s(= cP)	5.05	7.58
solubility at 25°C, water	1 g/25 mL	1 g/51 mL
50% ethanol	1 g/1.5 mL	1 g/1.7 mL

[a] To convert kPa to mm Hg, multiply by 7.5., °C

Manufacture. Today benzyl alcohol is almost universally manufactured from toluene which is first chlorinated to give benzyl chloride. This is then hydrolyzed to benzyl alcohol by treatment with aqueous sodium carbonate.

World Consumption and Uses. In the soap, perfume, and flavor industries benzyl alcohol is primarily used in the form of its aliphatic esters. Benzyl alcohol is commercially available in five grades.

The largest proportion of benzyl alcohol is for use in the photographic and textile industries although the latter use has been declining. The textile grade is used as a dyeing assistant for wool and nylon.

The pharmaceutical industry makes use of benzyl alcohol's local anesthetic, antiseptic, and solvent properties. It is used in nail lacquers and as a color developer in hair dyes by the cosmetics industry, and in acne treatment preparations.

Because of its strong polarity and limited water solubility, the technical grade of benzyl alcohol is used in rug cleaners as a degreasing agent, in leather dyeing, in ballpoint inks, as a cleaner for soldering, and as an extractive distillation solvent for xylenes and cresols. It is used as a stabilizer in insecticidal formulations by the agriculture industry and in treating fruits and vegetables. In addition, benzyl alcohol is used extensively in the polymer industry and in the manufacture of automobile tires.

Health and Safety. This material has a Generally Recognized As Safe (GRAS) status indicated by the Flavor and Extract Manufacturers' Association for use in flavors and by the Council of Europe for use as a flavor. Benzyl alcohol satisfies the most current guidelines published by the International Fragrance Association (IFRA) which governs the use of fragrance materials.

β-Phenethyl Alcohol

Of all the aromatic organic molecules β-phenethyl alcohol (PEA) (**2**) is probably the most prestigious aroma chemical in the world of perfumery. This is because of its exquisite odor of natural rose petals.

Physical properties of PEA are shown in Table 1. The compound undergoes the usual chemical reactions of alcohols or aromatic compounds.

In insect control, PEA has been considered as a mosquito repellant, and its acetate has been used as an ingredient in Japanese beetle bait. The alcohol also has bacteriostatic action and antifungicidal properties, and it has been claimed as a surface-active agent.

Because of factors of low cost, stability, and odor quality, PEA is ideally suited for use in bar soap fragrances, where its use can be up to 30–50% of the fragrance.

Manufacture. Current commercial methods for making PEA include Grignard synthesis, Friedel-Crafts process, and catalytic hydrogenation of styrene oxide.

Future methods of production include catalytic air oxidation and microbiological oxidation.

Purification. Purification problems are primarily solved by two methods: continuous vacuum fractionation and chemical combination to yield a high boiling ester, separation of the noncombining impurities by distillation, and hydrolysis of the ester.

World Consumption. Approximately 85% of the PEA is employed for fragrance use.

Health and Safety. The use of β-phenethyl alcohol generally presents no health problems. This material has Generally Recognized As Safe (GRAS) status as indicated by the Flavor and Extract Manufacturers Association and is approved by the U.S. Food and Drug Administration and the Council of Europe for use in flavors. PEA satisfies the most current guidelines published by the International Fragrance Association (IFRA), which governs the use of fragrance materials.

BRAJA D. MOOKHERJEE
RICHARD A. WILSON
International Flavors & Fragrances

Additional Reading

Bedoukian, P.Z. *Perfumery and Flavoring Synthetics,* 2nd Revised Edition, Elsevier Publishing Co., New York, NY, 1967, p. 55.
Bujold, A. *Chem. Econ. Handbook Program,* Stanford Research Institute, Menlo Park, CA, 1990.

BENZYL BENZOATE.   [CAS: 120-51-4]. A water-white liquid with
formula $C_6H_5CH_2OOCC_6H_5$. Sharp, burning taste with faint aromatic

odor. Supercools easily. Insoluble in water and glycerin; soluble in alcohol, chloroform, and ether. B.p. 325°C, mp 18.8°C, flash point, ~150°C. This compound is produced by the Cannizzaro reaction from benzaldehyde, by esterifying benzyl alcohol with benzoic acid, or by treating sodium benzoate with benzyl chloride. It is purified by distillation and crystallization. Benzyl benzoate is used as a fixative and solvent for musk in perfumes and flavors, as a plasticizer, miticide, and in some external medications. The compound has been found effective in the treatment of *scabies* and *pediculosis capitis* (head lice, *Pediculus humanus* var. *capitis*).

BENZYL CHLORIDE.   See Chlorinated Organics.

BENZYNE. The concept of benzyne intermediates has largely stemmed from work by Georg Wittig of the University of Heidelberg, Germany, and J.D. Roberts of the California Institute of Technology. Roberts postulated that benzynes form when a substituted benzene (such as bromobenzene) reacts with a nucleophilic reagent, such as potassium amide in liquid ammonia. He and other workers have shown that strong nucleophiles add readily to arynes. If the nucleophile is attached to a side chain on the aryne, a new ring fused to the original aromatic nucleus forms by intramolecular addition. This was shown by Bunnett and B.F. Hrutfiord and also by R. Huisgen and co-workers of the University of Munich, Germany. In this work, heterocyclic compounds were usually obtained. However, in later work J.F. Bunnett and J.A. Skorez, then at Brown University, used the synthesis to obtain homocyclic ring closures, producing derivatives of such compounds as benzocyclobutene, indane, tetralin and benzocycloheptane. Their type reaction may be written as where $n = 1, 2, 3,$ or 4 $X =$ cyano, acyl, carbethoxy or sulfonyl.

Intermediate benzyne
compound

Lester Friedman and Francis M. Logullo prepared substituted benzynes by diazotizing substituted anthranilic acid. This is a mild, room-temperature reaction which permits simultaneous reactions of the benzynes with suitable acceptors to prepare halogen. $-NO_2$, $-CH_3$, and $-OCH_3$ derivatives. They have also prepared new heterocyclic arynes, such as 3-pyridine from 3-amino-isonicotinic acid.

BERGAMOT OIL. An essential oil, Brownish-yellow to green liquid, agreeable odor, bitter taste, produced from the rind of the fruit of *Citrus aurantium* or *C. bergamia*, relatives of the orange and lemon. The small trees are cultivated in southern Europe. The oil is expressed from the skin of the small yellow fruits and sometimes is used as a scent for cosmetics. The oil also is used sometimes as a clearing agent in the preparation of material for microscopic examination.

BERGIUS, FREDERICK (1884–1949). A German chemist who won the Nobel prize in 1931 with Carl Bosch for chemical high-pressure methods. He invented a method of converting coal dust into oil via pressurized hydrogen. He also invented a method for production of cattle feed and sugar from wood by hydrolysis. He was educated in Poland and Germany.

BERGIUS PROCESS. Formation of petroleum-like hydrocarbons by hydrogenation of coal at high temperatures and pressures (e.g., 450 C and 300 atm) with or without catalysts; production of toluene by subjecting aromatic naphthas to cracking temperatures at 100 atm with a low partial pressure of hydrogen in the presence of a catalyst.

BERGIUS-WILLSTATTER SACCHARIFICATION PROCESS. Process for industrial production of fermentable sugar from wood by hydrolysis of tannin and xylan-free cellulose with 40–45% hydrochloric acid. The use of concentrated acid requires acid-resistant equipment and recovery of acid. The sugar produced must be rehydrolyzed prior to fermentation.

BERGMANN AZLACTONE PEPTIDE SYNTHESIS. Conversion of an acetylated amino acid and an aldehyde into an azlactone with an alkylene side chain; reaction with a second amino acid with a second amino acid with ring opening and formation of an acylated unsaturated dipeptide, followed by catalytic hydrogenation and hydrolysis to the dipeptide.

BERGMANN DEGRADATION. Stepwise degradation of polypeptides involving benzoylation, conversion to azides, and treatment of the azides with benzyl alcohol; this treatment yields, via rearrangement to isocyanates, carbobenzoxy compounds which undergo catalytic hydrogenation and hydrolysis to the amide of the degraded peptide.

BERG, PAUL (1926–). An American molecular biologist who won the Nobel prize for chemistry in 1980 with Frederick Sanger and Walter Gilbert. Berg's research concerned the biochemistry of nucleic acid, particularly regarding recombinant DNA, that is combining a molecule DNA's from different species. His Doctorate was attained at Western Reserve, and later he performed research at Stanford University.

BERKELIUM. [CAS: 7440-40-6]. Chemical element, symbol Bk, at. no. 97, at. wt. 247 (mass number of the most stable isotope), radioactive metal of the *Actinide* series, also one of the *Transuranium* elements. All isotopes of berkelium are radioactive; all must be produced synthetically. The element was discovered by G.T. Seaborg and associates at the Metallurgical Laboratory of the University of Chicago in 1949. At that time, the element was produced by bombarding ^{241}Am with helium ions. ^{247}Bk is an alpha-emitter and may be obtained by alpha-bombardment of ^{244}Cm, ^{245}Cm, or ^{246}Cm. Other nuclides include those of mass numbers 243–246 and 248–250. Probable electronic configuration:

$$1s^2 2s^2 2p^6 3s^2 3p^6 3d^{10} 4s^2 4p^6 4d^{10} 4f^{14} 5s^2 5p^6 6d^{10} 5f^9 6s^2 6p^6 7s^2$$

Ionic radius Bk^{+3}, 0.99 Å. Longest-lived isotope, ^{247}Bk($t_{1/2} = 700$ years).

Berkelium is known to exist in aqueous solution in two oxidation states, the (III) and the (IV) states, and the ionic species presumably correspond to Bk^{+3} and Bk^{+4}. The oxidation potential for the berkelium(III)-berkelium(IV) couple is about -1.6 V on the hydrogen scale (hydrogen-hydrogen ion couple taken as zero).

The solubility properties of berkelium in its two oxidation states are entirely analogous to those of the actinide and lanthanide elements in the corresponding oxidation states. Thus in the tripositive state such compounds as the fluoride and the oxalate are insoluble in acid solution, and the tetrapositive state has such insoluble compounds as the iodate and phosphate in acid solution. The nitrate, sulfate, halides, perchlorate, and sulfide of both oxidation states are soluble.

The first compound of berkelium of proven molecular structure was isolated in 1962 by Cunningham and Wallman. A small quantity (0.004 microgram) of berkelium (as berkelium-249) dioxide was used to determine structure by x-ray diffraction.

Berkelium metal exists in two crystal modifications. Melting point of crystals has been estimated at 986°C. The first production of berkelium required the solution to several problems in high energy physics. These included the necessity to synthesize the highly radioactive target elements used; safe handling procedures for the high levels of radioactivity encountered had to be developed; new systematics for predicting the modes of decay and half-lives of the still undiscovered isotopes were needed. See also **Chemical Elements**.

Additional Reading

Emsley, J.: *The Elements,* 3rd Edition, Oxford University Press, Inc., New York, NY, 1998.

Fuger, J. and L.R. Moss: "Transuranium Elements," *A Half Century,* American Chemical Society, Washington, DC, 1992.

Krebs, R.E.: *"The History and Use of Our Earth's Chemical Elements," A Reference Guide,* Greenwood Publishers Group, Inc., Westport, CT, 1998.

Lagowski, J.J.: *MacMillian Encyclopedia of Chemistry,* Vol. 1, MacMillian Library Reference, New York, NY, 1997.

Lide, D.R.: *CRC Handbook of Chemistry and Physics,* 84th Edition, CRC Press, LLC., Boca Raton, FL, 2003.

Perry, R.H., D. Green, and J.O. Maloney: *Perry's Chemical Engineers' Handbook,* 7th Edition, McGraw-Hill, New York, NY, 1997.

Seaborg, G.T. (editor): *Transuranium Elements,* Dowden, Hutchinson Ross, Stroudsburg, Pennsylvania, 1978 (Classical reference.)

Seaborg, G.T. and W.D. Loveland: *The Elements beyond Uranium,* John Wiley & Sons, Inc., New York, NY, 1990.

Stwertka, A. and E. Stwertka: *A Guide to the Elements,* Oxford University Press, Inc., New York, NY, 1998.

BERTHELOT EQUATION.

A form of the equation of state, relating the pressure, volume, and temperature of a gas, and the gas constant R. The Berthelot equation is derived from the Clausius equation and is of the form

$$PV = RT \left(1 + \frac{9PT_c}{128 P_c T} \left[1 - 6\frac{T_c^2}{T^2} \right] \right)$$

in which P is the pressure, V is the volume, T is the absolute temperature, R is the gas constant, T_c is the critical temperature, and P_c the critical pressure.

BERTHELOT, PIERRE EUGENE MARCELLIN (1827–1907).

A French Chemist, Berthelot, was a professor at the Ecole Superieure de Pharmacie (1859) and at the College de France from 1865. He did important work in organic synthesis. Berthelot proved that hydrocarbons and other organic compounds could be synthesized from inorganic materials. In later years, he did valuable research in thermochemistry. Much of Berthelot's research involved dyes and explosives.

See also **Berthelot Equation**.

J.M.I.

BERTHOLLET, CLAUDE LOUIS (1748–1822).

A French chemist, born in Talloires, near Annecy, France. Followed Antoine Lavoisier, but did not accept the latter's contention that oxygen is the characteristic constituent of acids. He was the first to propose chlorine as a bleaching agent. His essay on chemical physics (1803) was the first attempt to explain this subject. His speculations on stoichiometry, especially as regards relative masses of reacting atoms, profoundly affected later theories of chemical affinity.

BERTHOLLIDE COMPOUNDS.

See **Chemical Composition**; **Compound (Chemical)**.

BERYL.

The mineral beryl is a silicate of beryllium (glucinium) and aluminum corresponding to the formula $Be_3Al_2Si_6O_{18}$. Crystallizing in the hexagonal system the 6-sided prisms of beryl may be very small or range up to several feet (meters) in length and 3 feet (1 meter) or so in diameter. Terminated crystals are relatively rare. Its fracture is conchoidal; hardness, 7.5–8; specific gravity, 2.6–2.9; colors, emerald green, green, blue-green, blue, yellow, red, white and colorless; luster, vitreous; transparent to translucent.

Beryl has long been used as a gem, the emeralds being a rich green variety, colored probably by minute amounts of some chromium compound. A beautiful bluish sort is called aquamarine; morganite is pink, and the golden beryl is a clear bright yellow. Other shades like honey yellow and yellowish-green are common. Metallic beryllium is obtained from beryl. Its lightness and strength make it very valuable for industrial purposes.

Beryl is found in granite rocks and especially in pegmatites, but it occurs also in mica schists in the Urals. In addition to the many European localities, including Austria, Germany, and Ireland, beryls of gem quality are found in Africa, Madagascar (especially for morganite), and Brazil. The most famous place in the world for emeralds is at Muso, Colombia, South America, where they form a unique occurrence in limestones. Emeralds are also obtained in the Transvaal and near Mursinsk, in Siberia. In the United States, New England has furnished much beryl from its pegmatites, and for a long time the huge crystals from Acworth and Grafton, New Hampshire, were the largest known. Later, however, giant crystals even larger than those from New Hampshire were discovered in Albany, Maine, the largest of which was 18 by 4 feet (5.4 by 1.2 meters), and weighed about 36,000 pounds (16,330 kilograms). Other localities are Paris and, elsewhere, in Oxford County, Maine; Royalston, Massachusetts; North Carolina; Colorado; South Dakota; and California.

ELMER B. ROWLEY
F.M.S.A., formerly Mineral Curator
Department of Civil Engineering
Union College
Schenectady, New York

BERYLLIUM.

[CAS: 7440-41-7]. Chemical element symbol Be. at. no. 4, at. wt. 9.0122, periodic table group 2, mp $1287 - 1292 \pm 3°C$, bp $2970 \pm 5°C$. The vapor pressure at the melting point calculates to be 55 N/m² from the equations for the vapor pressure.

1. Solid $\log P_{(bar)} = 6.266 + 1.473 \times 10^{-4} \, T - 16,950 \, T^{-1}$
2. Liquid $\log P_{(bar)} = 6.578 - 11,860 \, T^{-1}$

Considerable variation exists in the reported specific heat data. The following appear to be representative:

TEMPERATURE °C	SPECIFIC HEAT kJ/(kg·K)
−13	1630
+25	1970
100	2130
200	2340

Similar scatter exists in the reported thermal conductivity data. An average value lies around 125 kW/(m·K).

The density of beryllium is 1.847 g/cm³ based upon average values of lattice parameters at 25°C (a = 22.856 nm and c = 35.832 nm). Beryllium products generally have a density around 1.850 g/cm³ or higher because of impurities, such as aluminum and other metals, and beryllium oxide. The crystal structure is close-packed hexagonal. The alpha-form of beryllium transforms to a body-centered cubic structure at a temperature very close to the melting point.

First ionization potential 9.32 eV; second 18.4 eV. Oxidation potentials $Be \rightarrow Be^{2+} + 2e^-$, 1.70 V; $2Be + 6OH^- \rightarrow Be_2O_3^{2-} + 3H_2O + 4e^-$, 2.28 V.

All naturally occurring beryllium compounds are made up of the ^9Be isotope. Artificially produced isotopes occur during some nuclear reactor operations and include ^6Be, ^7Be, ^8Be, and ^{10}Be.

The thermal neutron absorption cross section is 0.0090 barn/atom.

The electrical conductivity of beryllium is dependent upon both temperature and metal purity. It varies at room temperature between 38–42% (International Annealed Copper Standard). Electrical resistivity of $4.266' \, 10^{-8}$ ohm· m at 25°C has been reported.

Background

In 1797, Vauquelin discovered beryllium to be a constituent of the minerals beryl and emerald. Soluble compounds of the new element tasted sweet, so it was first known as glucinium from the corresponding Greek term. Quarrels over the name of the element were perpetuated by the simultaneous and independent isolations of metallic beryllium in 1828 by Wohler and Bussy. Both reduced beryllium chloride with metallic potassium in a platinum crucible. The name beryllium and symbol Be were officially recognized by the IUPAC in 1957.

Hope for the emergence of beryllium beyond the laboratory curiosity status resulted from publication of the work of the French scientist Lebeau in 1899. His paper described the electrolysis of fused sodium fluoberyllate to produce small hexagonal crystals of beryllium. Lebeau also reported the direct reduction of a beryllium oxide-copper oxide mixture with carbon to yield a beryllium-copper alloy. In 1926, Lebeau's alloy was rediscovered and found to have remarkable age-hardenable mechanical properties. A copper-beryllium alloy was first marketed in 1931, and this market remains important today.

Commercial development of beryllium in the United States was begun in 1916 by Hugh S. Cooper with the production of the first significant metallic beryllium ingot. This was followed by formation of the Brush Laboratories Company, which started its development work under the direction of Dr. C.B. Sawyer in 1921. In Germany, the Siemens-Halske Konzern began commercial development work in 1923.

Occurrence

A few years ago, when present theories concerning the formation of the universe were proposed, cosmologists suggested that only hydrogen and helium and also lithium, in very small concentrations, were present in the primordial matter. It was postulated that all other elements were produced as the result of subsequent star formation through nuclear reactions or cosmic ray radiation, thus creating all of the elements in the Periodic Table.

Studies of old stars, such as HD140283, have been made quite recently. This star is considered to be so old that it has only about 1% of the oxygen and other heavier elements that the sun has. About 1000 times more beryllium has been found than possibly could be attributed to cosmic radiation. These observations were essentially confirmed by one of the early experiments using the Hubble Space Telescope. These observations will contribute to further unraveling the remaining problems pertaining to the origin of the universe.

Occurrences of beryllium in the earth's crust are widely distributed and estimates of the amount fall in the 4–6 ppm range. Forty-five beryllium-containing minerals have been identified. Only two are commercially important—beryl, $3BeO \cdot Al_2O_3 \cdot 6SiO_2$, for its high beryllium content, and bertrandite, $Be_4Si_2O_7 (OH)_2$, for its large quantities located in the United States.

In 1959, beryllium was found in the rhyolitic tuffs of Spor Mountain, Utah, containing from 0.1 to 1.0% beryllium oxide. The practical processing limit requires an average beryllium oxide content of the ore of 0.6% to compete with beryl ore processing of material with more than 10% beryllium oxide. Deposits of this processable grade are adequate for the industrial requirements of the United States for several decades at present levels of consumption. Although the ore grade is much lower than that of beryl ore, the beryllium values in the rhyolitic tuffs are acid-soluble and recoverable by established processing technology.

In pure form, beryl mineral contains nearly 14% beryllium oxide, as found in its precious forms, emerald and aquamarine. Industrial grades of the mineral contain 10–12% beryllium oxide. Beryl occurs as a minor constituent of pegmatic dikes and is mined primarily as a by-product of feldspar, spodumene, and mica operations. Only the relatively large crystals are recovered by handpicking or cobbing to supply the industrial requirements of about 4000 tons (3600 metric tons) per year. Principal suppliers have been Argentina, Brazil, China, and Russia. A mill that processes beryllium from tertrandite-bearing ores also operates in Utah, thus somewhat reducing the demand for raw beryl.

Extractive and Process Metallurgy

The production of metallic beryllium, its alloys, or its ceramic products centers around the recovery of an intermediate partially purified concentrate from ore processing. The usual intermediate is beryllium basic carbonate or hydroxide. The mill in Utah is the only one in the Western world that extracts beryllium from its ores. The processes used to extract the beryllium are based on sulfuric acid. The sulfate solutions from beryl or bertrandite sources are partially purified by solvent extraction before yielding beryllium hydroxide as the end product. The hydroxide is converted to beryllium fluoride by reaction with ammonium bifluoride. Thermal reduction with magnesium metal forms beryllium pebbles. Final purification is accomplished by vacuum melting the beryllium pebbles to remove fluorides and magnesium impurities and casting into graphite molds. Standard powder metallurgy processes are generally used to convert the cast billets to solid shapes. The prevalent final consolidation step is hot pressing.

Important Commercial Properties

Beryllium has several unique properties which have given it a position of commercial significance. Its low atomic mass, low absorption cross section, and high scattering cross section are neutronic properties of importance. These properties spurred the expansion of beryllium production beyond the pilot scale immediately after the formation of the United States Atomic Energy Commission and the initiation of nuclear reactor development programs. About 1960, structural applications using beryllium began to utilize its modulus of elasticity of 2.93×10^5 mPa, its low density, and its relatively high melting point. Beryllium has good thermal conductivity and excellent thermal capacity properties, which gave rise to its use as a thermal barrier and heat sink for re-entry vehicles and other aerospace applications. The latter properties have been coupled with favorable ductility properties at elevated temperatures for the development of aircraft brakes.

Commercial and Aerospace Applications

Although the early applications of beryllium took advantage of the element's nuclear characteristics, structural uses of beryllium in aircraft and aerospace have developed because no other known material exceeds beryllium's modulus-to-weight ratio, while still retaining significant ductility. Often, where stiffness-to-weight is a problem, engineers will turn to beryllium. Beryllium has the stiffness to contain both inertial and vibratory loads, as well as the thermal conductivity to prevent undesirable heat gradients. The density of beryllium does not penalize the payload.

Applications over the past several years have included guidance system parts, such as gimbals, gyroscopes, stable platforms, housings, mirrors, aircraft brakes, and accelerometers. In advanced (U.S.) land-based nuclear missiles, there is a precisely machined beryllium sphere, floating, warmed and protected, in a fluid bath.

This makes it possible to achieve exacting guidance without reference to external benchmarks. The beryllium sensor minutely determines all changes in acceleration and orientation and is claimed to have improved striking accuracy by some twenty times.

Applications of beryllium within the recent past have included structures that are loaded in compression. Wrought products with yield strengths approaching 690 MPa (1,000,000 psi) and 20% elongation at room temperature have been achieved.

Recent beryllium processing, such as the production of near-net shapes by way of cold and hot isostatic pressing, is reducing the cost of beryllium parts by as much as 35%. Hot isostatic processing also is being used to produce entirely new families of beryllium products, such as beryllide intermetallic compounds for rocket nozzles and other high-temperature needs, as well as metal-matrix composites for hypersonic aircraft components, and strong, lightweight aluminum alloys containing as much as 40% Be.

Researchers at the Brush-Wellman Laboratory (Elmore, Ohio) have developed a bench-scale, inert-gas atomization method for producing ultraclean, spherical powders, which then can be hot isostatic processed.

For several years, additions of Be to commercial copper- and nickel-based alloys have enabled these materials to be precipitation-hardened to strengths approaching those of heat-treated steels. Yet Cu-Be alloys retain the corrosion resistance, electrical and thermal conductivities, and spark resistance of copper-based alloys.

Very low electrical conductivity and high transparency to microwaves in microelectronic substrate applications have proven very advantageous.

Chemical Properties

Many chemical properties of beryllium resemble aluminum, and to a lesser extent, magnesium. Notable exceptions include solubility of alkali metal fluoride-beryllium fluoride complexes and the thermal stability of solutions of alkali metal beryllates.

All of the common mineral acids attack beryllium metal readily with the exception of nitric acid. It is also attacked by sodium hydroxide and potassium hydroxide, but not by ammonium hydroxide.

Beryllium interacts with most gases. Polished beryllium surfaces retain their brilliance for years on exposure to air at ambient temperatures. The oxidation rate in air increases parabolically at temperatures above 850°C with the formation of a loosely adherent, white oxide.

Compounds of Beryllium

Ammonium beryllium carbonate solutions are prepared by dissolving the hydroxide or the basic carbonate in warm (50°C) aqueous mixtures of NH_4HCO_3 and $(NH_4)_2CO_3$. After filtering to remove insoluble impurity hydroxides and adding a chelating agent, heating above 88°C evolves

NH_3 and CO_2 and precipitates a high-purity, basic beryllium carbonate. If the aqueous system has the stoichiometry of $(NH_4)_4Be(CO_3)_3$, analogous to the ammonium uranyl carbonate system, the basic beryllium carbonate product of hydrolysis is $2BeCO_3 \cdot Be(OH)_2$. This compound is readily dissolved in all mineral acids, making it a valuable starting material for laboratory synthesis of beryllium salts of high purity.

Beryllium hydroxide, [CAS: 13327-32-7], $Be(OH)_2$ is precipitated as an amorphous, gelatinous material by addition of ammonia or alkali to a solution of a beryllium salt at slightly basic pH values. A pure hydroxide can be prepared by pressure hydrolysis of a slurry of beryllium basic carbonate in water at 165°C. All forms of beryllium hydroxide begin to decompose in air or water to beryllium oxide at 190°C.

Beryllium sulfate, [CAS: 13510-49-1], $BeSO_4 \cdot 4H_2O$, is an important salt of beryllium used as an intermediate of high purity for calcination to beryllium oxide powder for ceramic applications. A saturated aqueous solution of beryllium sulfate contains 30.5% $BeSO_4$ by weight at 30°C and 65.2% at 111°C.

Beryllium fluoride, [CAS: 7787-49-7], BeF_2, is readily soluble in water, dissolving in its own water of hydration as $BeF_2 \cdot 2H_2O$. The compound cannot be crystallized from solution and is prepared by thermal decomposition of ammonium fluoberyllate, $(NH_4)_2BeF_4$.

Beryllium chloride, [CAS: 7787-47-5], $BeCl_2$, with a melting point of 440°C, is used as a component of molten salt baths for electrowinning or electrorefining of the metal. The compound hydrolyzes readily with atmospheric moisture, evolving HCl, so protective atmospheres are required during processing.

Basic beryllium acetate, [CAS: 543-81-7], $Be_4O(C_2H_3O_2)_6$, is the best known of the beryllium salts of organic acids which can be divided into normal beryllium carboxylates, $Be(RCOO)_2$, and beryllium oxide carboxylates, $Be_4O(RCOO)_6$. The basic acetate is soluble in glacial acetic acid and can readily be crystallized therefrom in very pure form. It is also soluble in chloroform and other organic solvents. It has been used as a source of pure beryllium salts.

Biology and Toxicology

Beryllium can be handled safely with reasonable controls, but it can cause serious illness if these controls are not observed. Skin and respiratory reactions can be experienced. There is, however, no ingestion problem.

The hazards are generally classified as (1) acute respiratory disease, (2) chronic pulmonary disease, and (3) dermatitis.

Dermatitis is produced by skin contact with soluble salts of beryllium, especially the fluoride. It is controlled by a program of good personal hygiene, frequent washing of the exposed parts of the body, as well as a program where clothing is laundered on the plant site.

Acute pulmonary disease is due exclusively to inhalation of soluble beryllium salts and is not caused by exposure to the oxide, the metal, or its alloys. The exact forms of beryllium causing the chronic pulmonary disease and the degree of exposure necessary to induce it are not precisely known. It is known that under the completely uncontrolled conditions existing in beryllium extraction plants before the establishment of air-count standards in 1949, when beryllium air-counts were in milligrams per cubic meter of air rather than micrograms, only about 1% of the exposed workers became ill. This would indicate a sensitivity of a limited number of individuals to beryllium.

Investigations by medical, toxicological, and engineering personnel led to the promulgation of safe limits of exposure by the Atomic Energy Commission in 1949. The disease is believed to be avoidable when air-counts are held within average limits of 2 micrograms per cubic meter of air for an 8-hour exposure, with a maximum at any time of 25 micrograms per cubic meter of air.

The Occupational Safety and Health Administration several years later issued a proposed new occupational standard for beryllium air-counts. This proposal was highly controversial.

Local exhaust ventilation is the major engineering control used to limit concentrations of airborne beryllium. Modern air cleaners allow control within recommended outplant levels of 0.01 microgram beryllium per cubic meter of air, averaged over 1-month periods.

Because of the increasing use of beryllium in a growing diversity of end products, a new study of beryllium disease was undertaken by the U.S. Department of Defense and the U.S. Department of Energy, as of December 1991. This study was precipitated by finding that nuclear-related weapons workers at Oak Ridge National Laboratory (Tennessee) were diagnosed has having chronic beryllium disease. The study will probe beryllium worker health statistics back to the mid-1980s.

KENNETH A. WALSH, Ph.D.
Brush Wellman Inc.
Elmore, Ohio

Additional Reading

Brush, Wellman Inc.: "Properties and Applications of Beryllium and Beryllium Alloys," *Metal Progress*, **128**(6), 56 (November 1985).

Bunn, M.: "Birth of the Beryllium Baby," *Techy. Rev. (MIT)*, 75 (August/September 1991).

Carter, G.F. and D.E. Paul: *Materials Science and Engineering,* ASM International, Materials Park, Ohio, 1991.

Copley, S.M.: "Applied General and Nonferrous Physical Metallurgy," *Encyclopedia of Materials Science and Engineering,* MIT Press, Cambridge, MA, 1986.

Gibbons, A.: "In the Beginning, Let There Be Beryllium," *Science*, 162 (January 10, 1992).

Holden, C.: "Beryllium Disease," *Science*, **1724** (December 20, 1991).

Lide, D.R.: *CRC Handbook of Chemistry and Physics,* 84th Edition, CRC Press, LLC., Boca Raton, FL, 2003.

Perry, R.H. and D. Green: *Perry's Chemical Engineers' Handbook,* 7th Edition, McGraw-Hill Companies, Inc., New York, NY, 1999.

Staff: "Beryllium—HIP Helps Spark Surge in Beryllium Applications," *Adv. Mat. Proc.*, **24** (January 1991).

Staff: *ASM Handbook—Properties and Selection: Nonferrous Alloys and Special-Purpose Materials,* ASM International, Materials Park, Ohio, 1990.

BERZELIUS, J. J. (1779–1848). A native of Sweden, Berzelius was one of the foremost chemists of the 19th century. He made many contributions to both fundamental and applied chemistry; coined the words *isomer* and *catalyst*; classified minerals by chemical compound. He recognized organic radicals which maintain their identity in a series of reactions; discovered selenium and thorium, and isolated silicon, titanium, and zirconium; did pioneer work with solutions of proteinaceous materials which he recognized as being different from "true" solutions.

BESSEL FUNCTION. The differential equation

$$x^2 y'' + xy' + (x^2 - n^2)y = 0, \; n = \text{const.}$$

is called Bessel's equation of order n. Certain of its solutions (see below) are called Bessel functions. The general solution is

$$y = AJ_n(x) + BY_n(x)$$

where

$$J_n(x) = \sum_{k=0}^{\infty} \frac{(-1)^k}{\Gamma(k+1)\Gamma(k+n+1)} \left(\frac{x}{2}\right)^{n+2k}$$

and $Y_v(x) = J_{-n}(x)$ if n is not an integer. These functions are called Bessel functions of the first kind. If n is an integer, then $J_{-n}(x)(-1)^n J_n(x)$, so that $Y_n(x)$ is defined as

$$Y_n(x) = \lim_{k \to \infty} \frac{J_k(x)\cos kx - J_{-k}(x)}{\sin k\pi}$$

which is called a Bessel function of the second kind. The functions, much used in physics,

$$H_n^{(1)}(x) = J_n(x) + iY_n(x), \quad i = \sqrt{-1}$$
$$H_n^{(2)}(x) = J_n(x) - iY_n(x)$$

are Bessel functions of the third kind; they are also called Hankel functions of the first and second kind, respectively. Other combinations of Bessel functions are also given names. These functions have certain standard properties of recurrence, orthogonality, etc.

BEST, CHARLES H. (1899–1978). Born in Maine, Best was educated at the University of Toronto, where he distinguished himself as a student of biochemistry. He collaborated with the late Dr. Frederick Banting in the isolation of the hormone insulin. He later became head of the insulin division of the Connaught Laboratories of the University as well as of the Banting and Best Research Institute. He also developed histaminase (an antiallergic enzyme) and the anticoagulant heparin.

See also **Banting, Sir Frederick (1891–1941)**.

BETA DECAY.   The process that occurs when beta particles are emitted by radioactive nuclei. The name *beta particle* or beta radiation was applied in the early years of radioactivity investigations, before it was fully understood what beta particles are. It is known now, of course, that beta particles are electrons. When a radioactive nuclide undergoes beta decay its atomic number Z changes by $+1$ or -1, but its mass number A is unchanged. When the atomic number is increased by 1, negative beta particle (negatron) emission occurs; and when the atomic number is decreased by 1, there is positive beta particle (positron) emission or orbital electron capture.

Because atomic nuclei contain only protons and neutrons, beta particles must be created at the moment of emission, just as photons are created at the time of emission of electromagnetic radiation. Because of this creation process, the amount of energy equal to the rest energy, $m_e c^2$ of an electron, must be consumed when beta decay occurs. Any remaining energy can be given to the beta particle as kinetic energy. The nuclear transitions producing beta decay are between discrete energy states differing by a definite amount of energy W_0, so we expect the total energy of a beta-decay transition to be W_0. However, emitted beta particles are experimentally found to have a continuous range of total (rest plus kinetic) energies W of such magnitude that $m_e c^2 < W < W_0$, rather than all having a single energy W_0. This distribution as a function of energy (or momentum) forms what is known as a beta-ray spectrum. The shape of the spectrum depends on the sign of the charge on the beta particle (positive or negative), the energy W_0, and the degree to which the transition is forbidden (explained below). Unless energy and momentum are not conserved in the process, the energy not carried away by the beta particle must be given to some other particle. Furthermore, since the beta particle has a spin quantum number $\frac{1}{2}$, angular momentum cannot be conserved unless another $\frac{1}{2}$ unit of angular momentum can be disposed of. Both of these possible discrepancies in the conservation laws have been taken care of in the Fermi theory of beta decay through postulation of a massless particle, a neutrino or an antineutrino, which has a spin quantum number $\frac{1}{2}$ and also carries away the remaining energy and momentum. Neutrinos were difficult to find experimentally but, even before they were experimentally detected, so much evidence had been developed to show their existence that the Fermi theory of beta decay was generally accepted.

Beta-decay processes are classified as allowed or forbidden but, as in many other physical processes, the term forbidden does not mean nonoccurrence, just a significant retardation relative to the rate for allowed transitions. The degree of forbiddenness is determined by the magnitude of the difference in angular momentum between the initial and final nuclear states as well as the parity of these states. If more than one unit of angular momentum must be carried away by the decay products ($\frac{1}{2}$ unit by the beta particle and $\frac{1}{2}$ unit by the neutrino or antineutrino), the transition must be forbidden. Allowed transitions give straight-line Fermi plots, as do some forbidden transitions, but some forbidden transitions have distinct shapes other than straight lines for their Fermi plots.

A negatron emitted during beta decay has its spin aligned away from the direction of its emission (its angular momentum vector is antiparallel to its momentum vector) and hence has a negative helix, but an emitted positron has positive helix. It is because of the absence of beta particles with both positive and negative helix in both types of beta-emission processes that parity is not conserved in beta decay.

See also **Particles (Subatomic)**; and **Radioactivity**.

BETA-LACTAM RING.   See **Antibiotics**.

BETALAINES.   See **Dyes: Natural**

BETA-RAY CHEMICAL ANALYZERS.   Instrumental beta-ray absorption techniques can be used for determining H_2 in hydrocarbons and, consequently, the hydrogen-carbon ratio. The range of concentration measurements is from 0 to 100%, although the presence of sulfur and oxygen may interfere, resulting in high H_2 readings. The measurement principle is based upon the fact that H_2 has twice as many electrons per unit weight as other atoms and, accordingly, has twice the beta-ray absorbency as carbon per unit weight. Beta-ray absorbency of the sample is measured by a null-balance ion chamber to indicate readings that are proportional to (Weight of Carbon/ml) $+ 2 \times$ (Weight of H_2/ml). A simultaneous density reading is proportional to (Weight of Carbon/ml) $+$ (Weight of H_2/ml). These expressions thus permit calculation, based upon empirical calibration

of the absorbency scale, of (Weight of H_2/ml) and the consequent determination of weight percent or the hydrogen-carbon ratio. Determinations require about 5 minutes, although readings can be made continuously. Less than 50 milliliters of sample are required. A convenient beta-ray source is strontium-90, with a half-life of 25 years. See also **Analysis (Chemical)**.

Beta-ray backscattering techniques also are applied to determine (1) the average atomic number of a sample having a fixed thickness, and (2) the thickness of coatings having fixed composition. Elements of high atomic number scatter beta rays more intensely than those of low atomic number. The beta rays from a radioactive source strike the sample. Scattered beta rays re-emitted from the same side of the sample are detected in an ion chamber, shielded from the source, to produce a small current proportional to the backscattering effect. The sample thickness must be controlled. Sample windows must be thin and of low-atomic-number material to avoid contribution to the measured effect.

BETTENDORF'S REAGENT.   A reagent used for the detection of arsenic in presence of bismuth and antimony compounds. It consists of a concentrated solution of stannous chloride in fuming hydrochloric acid.

BETTERTON-KROLL PROCESS.   A process for obtaining bismuth and purifying desilverized lead that contains bismuth. Metallic calcium or magnesium is added to the molten lead to cause formation of high-melting intermetallic compounds with bismuth. These separate as a surface scum and are skimmed off. The excess calcium and magnesium are removed from the lead by use of chlorine gas as mixed molten chlorides of lead or zinc. Bismuth of 99.995% purity is produced in this way.

BILE.   A bitter alkaline fluid secreted by the liver into the duodenum, which aids in the digestion of food. The chief components of bile are bile salts and bile pigments. Because of its strong alkalinity, bile neutralizes the acid coming into the duodenum from the stomach. The bile not only performs important functions in the process of digestion, but also serves as a vehicle for the excretion of waste products from the body.

Bile salts help in the breakdown of fat in the intestines and in fat absorption through the intestinal wall. The bile salts are injected into the digestive canal at the duodenum. They are not excreted, but are almost totally absorbed through the walls of the intestine, to be used over and over again. Bile pigments are derived from the hemoglobin of broken-down red blood cells and are excreted with the feces. When the pigments appear in excessive amounts in the blood, the mucous membranes and conjunctiva of the eye become stained a pale yellow, and the patient is said to be jaundiced.

Bile is continually secreted by the liver and stored in the gallbladder. Here the bile is concentrated by the absorption of water through the walls of the gallbladder. Bile is released from the gallbladder into the intestine when food passes through the pyloric valve from the stomach into the small intestine. Gallstones are formed of constituents of the bile which have settled out of solution. The stones vary in size, color, and structure, according to the materials composing them.

An inadequate supply of bile contributes to vitamin A deficiency because of disturbances of the intestinal track which prevent the effective absorption of the vitamin. In an average adult, from one-half to one liter of bile is secreted every 24 hours, the quantity depending upon the amount and kind of food eaten. The absence or lack of secretion of bile is known as *acholia*.

Part of the cholesterol newly synthesized in the liver is excreted into bile in a free non-esterified state (in constant amount). Cholesterol in bile is normally complexed with bile salts to form soluble cholic acids, Free cholesterol is not readily soluble and with bile stasis or decreased bile salt concentration may precipitate as gallstones. Most common gallstones are built of alternating layers of cholesterol and calcium bilirubin and consist mainly (80–90%) of cholesterol. Normally, 80% of hepatic cholesterol arising from blood or lymph is metabolized to cholic acids and is eventually excreted into the bile in the form of bile salts.

The C_{24} bile acids arise from cholesterol in the liver after saturation of the steroid nucleus and reduction in length of the side chain to a 5-carbon acid; they may differ in the number of hydroxyl groups on the sterol nucleus. The four acids isolated from human bile include *cholic acid* (3,7,12-trihydroxy), as shown in Fig. 1; *deoxycholic acid* (2,12-dihydroxy); *chenodeoxycholic acid* (3,7-dihydroxy); and *lithocholic acid* (3-hydroxy). The bile acids are not excreted into the bile as such, but are conjugated through the C_{24} carboxylic acid with glycine or

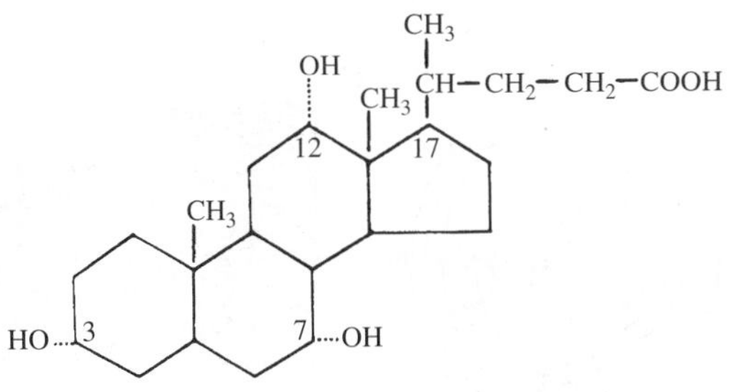

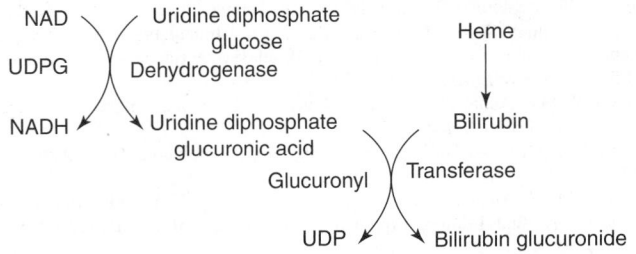

Fig. 1. Cholic Acid

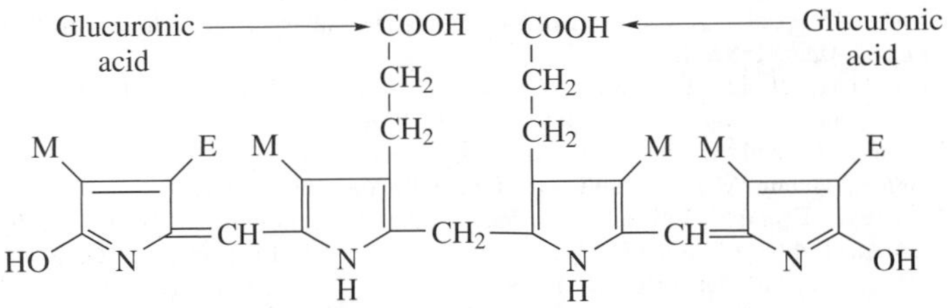

Fig. 2. Structure of *direct* bilirubin

Fig. 3. Conjugation of bilirubin

taurine, $NH_2-CH_2-CH_2-SO_3H$. This esterification of the bile acids to soluble conjugates occurs in the microsomes and requires coenzyme A, magnesium ion, and ATP (adenosine triphosphate). Although taurocholic acid predominates at birth, the most abundant of the bile acids in the adult is glycocholic acid. In alkaline bile, the conjugated bile acids exist in their ionized form as the bile salts, glycocholate or taurocholate. Bile salts can function as effective product feedback inhibitors of hepatic cholesterol synthesis. Because of their detergent action, bile salts play an important role in the absorption of cholesterol, fats, and fat-soluble vitamins. The bile salts are believed to facilitate absorption of these compounds by the formation of micelles or aggregates of low osmotic pressure. The bile salts themselves are not absorbed during this process. Their absorption from the intestine occurs at a different site and at an entirely different rate from that of the lipids. Approximately 95% of the bile salts are reabsorbed, enter the enterohepatic circulation, and are ultimately re-excreted into the bile for further utilization in lipid absorption.

Most of the hormones that are normally conjugated in the liver to form glucuronides or sulfates, such as the steroids, thyroxine, epinephrine, and norepinephrine, are secreted into the bile, but to a varying degree, may be re-absorbed in the intestine and eventually excreted in the urine. The 17-hydroxysteroids, including cortisol, are secreted into the bile primarily as reduced glycuronide conjugates. More than 70% of these conjugates enter the enterohepatic circulation and are eventually excreted into the urine; less than 30% are found in the feces. Progesterone, after its conversion to pregnanediol, is also excreted into the bile primarily as the glucuronide, some 75% of which is eventually excreted in the feces. Most androgens are excreted as sulfates in the urine, part of which are of nonhepatic or nonbiliary origin. Significant amounts of estrogens are excreted into bile as estriol glucuronide or estrone sulfates. Many derivatives of epinephrine and norepinephrine are eventually conjugated with either glucuronide or sulfate at the 4-hydroxy position and excreted into the bile. Thyroxine is predominantly conjugated with glucuronic acid and is excreted as such into the bile. The bile, however, is not a significant route for the net disposal of thyroxine, since this hormone is rapidly reabsorbed, enters the enterohepatic circulation and is eventually excreted as urinary metabolites.

The major components of bile, the bile pigments, can account for 15–20% of the total solids. *Bilirubin* comes primarily from the degradation of heme in the reticuloendothelial system in the spleen, bone marrow, and to a lesser extent, the liver. The initial step in the metabolism of heme is the cleavage of the porphyrin ring and elimination of the alpha methylene carbon to produce an open tetrapyrrole. This may exist as a complex with iron and globin called choleglobin. After removal of the iron and globin, the resulting tetrapyrrole, biliverdin, is rapidly reduced to bilirubin, the major pigment in human bile. Not all bilirubin results from the breakdown of hemoglobin from mature red cells. The early appearance of labeled bilirubin after injection of precursor glycine-^{14}C indicates that some bilirubin (approximately 10%) may arise from: (1) the rapid breakdown of immature red cells in the bone marrow; (2) from heme that had not entered hemoglobin; or (3) from the destruction of newly formed red cells in the peripheral circulation. This "shunt" pathway for bilirubin formation may predominate in pernicious anemia and some porphyrias. A small amount of bilirubin may also arise from other heme pigments, such as myoglobin or the cytochromes. The bilirubin that enters the blood is rapidly and solely bound to albumin. Normal circulating levels of bilirubin are less than 1 mg/100 ml. Free bilirubin, which readily crosses the blood-brain barrier in the newborn, and to a lesser extent in the adult, is an effective uncoupler of oxidative phosphorylation in the brain and is highly toxic.

The hepatic transport of bilirubin from plasma to bile involves three independent, but related mechanisms, i.e., uptake, conjugation, and secretion. Plasma bilirubin is dissociated from plasma albumin in the liver and is rapidly concentrated in the cytoplasm of the hepatic cells by an unknown mechanism which precedes and is relatively independent of any subsequent hepatic conjugation. After concentration in the liver, bilirubin is conjugated with 2 moles of glucuronic acid to form bilirubin diglucuronide, the glucuronic acid moieties being attached in ester linkage to the carboxyl groups on the propionic acid side chains. See Figs. 2 and 3.

Glucuronyl transferase, the enzyme catalyzing the final step, is located in the smooth endoplasmic reticulum of liver and to a lesser extent in kidney and gastric mucosa, where a small amount of extrahepatic conjugation may occur. This enzyme has not been purified and it is unclear whether it nonspecifically catalyzes glucuronide conjugation of many non-bilirubinoid substrates, or is bilirubin specific and a member of a large group of closely related glucuronyl transferases. Its activity can be induced by a variety of drugs and can be inhibited with steroids or steroid glucuronides found in plasma of pregnant women.

Crigler-Najjar's disease in humans is characterized by increased levels of unconjugated bilirubin in the serum. A genetic impairment of glucuronyl transferase, the enzyme responsible for the transfer of glucuronic acid from uridine diphosphate glucuronic acid, exists not only in the liver, but in the kidney as well. Gilbert's disease is characterized by a mild increase of unconjugated bilirubin in the plasma, which may result from a partial impairment of glucuronyl transferase, from a defect in bilirubin transport in the blood, or a defect in hepatic uptake. In subjects with Dubin-Sprinz or Dubin-Johnson disease, the serum contains high levels of both unconjugated and conjugated bilirubin, and an unidentified brown pigment is present in the liver. A defect in the secretion of the bilirubin conjugates from the hepatic cell is a probable causative factor. Rotor's disease is also characterized by increased serum levels of both unconjugated and conjugated bilirubin, but it differs from Dubin-Sprinz disease in that the hepatic brown pigment is not found. The foregoing syndromes are sometimes collectively referred to as *idiopathic hyperbilirubinemia*.

The mild nonhemolytic jaundice often present in the newborn (physiological jaundice) or the more severe jaundice and kernicterus in premature infants may result in part from an inability of the immature liver to conjugate bilirubin; low hepatic levels of both glucuronyl transferase and uridine diphosphate glucuronic acid dehydrogenase (the enzyme that catalyzes the synthesis of uridine diphosphate glucose glucuronic acid from uridine diphosphate glucose) are found in fetus and newborn. Hepatic secretion of conjugated bilirubin may also be impaired.

Additional Reading

Alpers, D.H., C. Owyang, F.E. Silverstein, W.L. Hasler, D.W. Powell, and T. Yamada: *Handbook of Gastroenterology,* Lippincott Williams & Wilkins, Philadelphia, PA, 1998.

Alpers, D.H., C. Owyang, L. Laine, D.W. Powell, and T. Yamada: *Textbook of Gastroenterology,* 3rd Edition, Lippincott Williams & Wilkins, Philadelphia, PA, 1999.

Edward, C.S. et al. "Endoscopic Biliary Drainage for Severe Acute Cholangitis," *N. Eng. J. Med.*, **1582** (June 11, 1992).

Logan, G.M. et al.: "Bile Porphyrin Analysis in the Evaluation of Variegate Porphyria," *N. Eng. J. Med.*, **1408** (May 16, 1991).

McNulty, J.G. and A. Chua: *Minimally Invasive Therapy of the Liver and Biliary System,* Thieme Medical Publishers, Inc., New York, NY, 1994.

Morrissey, J.F. and M. Reichelderfer: "Gastrointestinal Endoscopy—Part I," *N. Eng. J. Med.*, 1142–1149 (October 17, 1991); "Part II," 1214–1291 (October 24, 1991).

Rizzetto, M., Jean-Pierre Benhamou, N. Mcintyre, J. Rodes, and J. Dircher: *Oxford Textbook of Clinical Hepatology,* 2 Vol. Set, 2nd Edition, Oxford University Press, Inc., New York, NY, 1999.

Sherlock, S. and J. Dooley: *Diseases of the Liver and Biliary System,* 10th Edition, Blackwell Scientific, Boston, MA, 1991.

Stiehl, A., W. Gerok, and G. Paumgartner: Falk Symposium Staff, "Bile Acids and the Hepatobiliary System-from Basic Science to Clinical Practice," Proceedings of the 68th Galk Symposium Held in Basel, Switzerland, Kluwer Academic Publishers, Norwell, MA, 1993.

Steinberg, W.M.: "Acute Pancreatitis—Never Leave a Stone Unturned," *N. Eng. J. Med.*, **635** (February 27, 1992).

Yamada, T., Editor: *Textbook of Gastroenterology,* Lippincott, Hagerstown, MD, 1991.

Yamada, T., D.H. Alpers, C. Owyang, L. Laine, and D.W. Powell, *Atlas of Gastroenterology,* 2nd Edition, Lippincott Williams & Wilkins, Philadelphia, PA, 1999.

Zakim, D., T.D. Boyer, and R. Zorab: *Hepatology: A Textbook of Liver Disease,* 3rd Edition, W.B. Saunders, Co., Philadelphia, PA, 1996.

BIMETAL THERMOMETER.

Thermostatic bimetal can be defined as a composite material, made up of strips of two or more metals fastened together, which, because of the different expansion rates of the components, tends to change its curvature when subjected to a change in temperature.

With one end of a straight strip fixed, the other end deflects in direct proportion to the temperature change and the square of the length, and inversely as the thickness, throughout the linear portion of the deflection characteristic curve. If a strip of bimetal is wound into a helix or spiral and one end is fixed, the other end will rotate when heat is applied. The angular deflection varies directly with the temperature change and the length of the strip, and inversely with the thickness of the material, over the linear parts of the deflection characteristic curve. Bimetals show uniform deflection only over part of the deflection characteristic curve, as shown in Fig. 1. The three types of elements most commonly used in thermometers are shown in Fig. 2.

Bimetal thermometers are made in ranges from +1000°F (538°C) down to −300°F (−184°C) to and lower. However, at low temperatures, the rate of deflection drops off quite rapidly. Because of its long-term instability at high temperatures, the maximum temperature for continuous use is about 800°F (427°C). However, special bimetal thermometers can be obtained for continuous use up to 1200°F (649°C). Good bimetal thermometers retain their accuracy indefinitely. Usually industrial bimetal thermometers read with an accuracy of ±1% at any point on the scale. The speed of response of bimetal thermometers is generally about the same as that for liquid-in-glass thermometers in similar ranges.

The thermostatic bimetal approach is used widely in a variety of thermostatic-type temperature-control situations, as found in heating and air-conditioning systems and in automotive cooling systems, among others. Bimetals are also used in thermal type time-delay relays and switches.

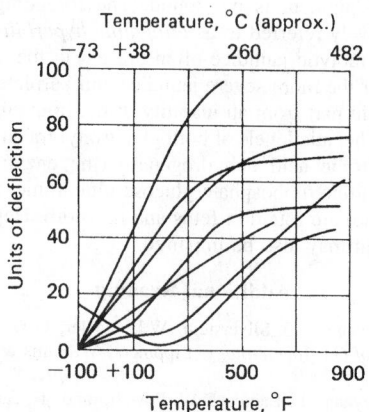

Fig. 1. Deflection characteristics of various bimetals

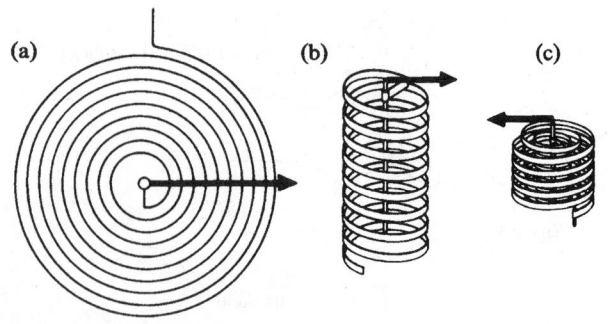

Fig. 2. Principal configurations of elements used in bimetal thermometers: **(a)** flat spiral, **(b)** single helix, **(c)** multiple helix

BINARY.

Descriptive of a system containing two and only two components. Such a system may be a chemical compound composed of two elements, an element and a group (hydroxyl, methyl, etc.), or two groups, (e.g., oxalic acid). It may also be a two-component solution or alloy.

BINDING ENERGY.

This term is used in atomic physics with two closely related meanings: the binding energy of a particle (or other entity) is the energy required to remove the particle from the system to which it belongs; the binding energy of a system is the energy required to disperse the system into its constituent entities. Explicit definitions are obviously necessary.

Some explicit definitions for the binding energies of particles are the following:

(1) The *electron binding energy* is the energy necessary to remove an electron from an atom. It is identical with the ionization potential.

(1a) The *total electron binding energy* is the energy necessary to remove all the electrons from an atom to infinite distances, so that only the nucleus remains. It is equal to the sum of the successive ionization potentials of that atom.

(2) The *proton binding energy* is the energy necessary to remove a single proton from a nucleus. Most known proton binding energies are in the range 5–12 MeV, although that for ^{2}H is 2.23 MeV, that for ^{4}He is 19.81 MeV, and those for ^{5}Li and ^{9}Be are negligible.

(3) The *neutron binding energy* is the energy required to remove a single neutron from a nucleus. Most known neutron binding energies are in the ranges 5–8 MeV, though that for ^{2}H is 2.23 MeV, that for ^{9}Be is 1.67 MeV, and that for ^{12}C is 18.7 MeV.

(4) The *alpha-particle binding energy* is the energy required to remove an alpha-particle from a nucleus. For most light nuclides the alpha-particle binding energy is positive and is equal to several MeV. For nuclides of mass number about 125, it is approximately zero. For nuclides of mass number about 150 to 200, it is negative by about 1 to 3 MeV, but the magnitude of the Coulomb potential barrier at the nucleus is sufficiently large that penetration by an alpha particle is so improbable that lifetimes for alpha disintegration are generally too long for detection of alpha activity. For most nuclides of mass number exceeding 200, the alpha-particle binding energy is negative by about 4 to 8 MeV, which is a negative binding energy of sufficiently large magnitude to give a measurable probability of penetration of the potential barrier by an alpha particle, hence an observable alpha activity.

Some explicit definitions for the binding energies of systems are:

(1) The *nuclear binding energy* is the energy that would be necessary to separate an atom of atomic number Z and mass number A into Z hydrogen atoms and $A-Z$ neutrons. This energy is the energy equivalent of the difference between the sum of the masses of the product hydrogen atoms and neutrons, and the mass of the atom; it includes the effect of electronic binding. (See *total electron binding energy* above.)

(2) The binding energy of a solid is the energy required to disperse a solid into its constituent atoms, against the forces of cohesion. In the case of ionic crystals, it is given by the Born-Mayer equation. See **Crystal**.

Although the concept of the atom has not been in serious question for nearly a half-century, full understanding of the forces that hold the neutrons and protons together has not yet been achieved. In 1927, Aston found that experimentally measured isotopic weights differed slightly from whole numbers. See also **Aston Whole Number Rule.** From this he was led to the concept of the *packing fraction*, which is defined as the algebraic difference between the isotopic weight and the mass number, divided by the mass number. Although the theoretical significance of the packing fraction is difficult to assess, it does lead to some interesting conclusions with respect to nuclear stability. A negative packing fraction derives from a situation where the isotopic weight is less than the mass number, inferring that in the formation of the nucleus from its constituent particles, some mass is converted into energy. Since an equivalent amount of energy would be necessary to break up the nucleus into its constituent particles again, a negative packing fraction suggests a high order of nuclear stability. By the same reasoning, a positive packing fraction indicates nuclear instability. Stable elements with mass numbers above about 175 and below about 25 have positive packing fractions. It is interesting to note that the packing fractions of both hydrogen and uranium are positive.

Actually, a comparison of the isotopic weight with the mass number (as is done in determining the packing fraction) is somewhat artificial. A rigorous determination of the mass-energy interconversion in the formation of an atom would seem to require a calculation of the difference between the sum of the masses of the constituent particles of the atom and the experimentally measured isotopic weight. The value of the mass difference thus obtained is the *mass defect*. The energy equivalent of this mass difference as derived from the Einstein equation yields a measure of the binding energy of the nucleus. Division of the binding energy of a nucleus by the number of nucleons (the total number of protons and neutrons) therein yields the binding energy per nucleon. In stable isotopes, the binding energy per nucleon decreases with increasing mass number, a fact that is important in nuclear fission. Secondly, the binding energy per nucleon derived in the manner just described is an average value, whereas each additional nucleon added to the nucleus has a binding energy less than those that preceded it. Thus, the most recently added nucleons are bound less tightly than those already present.

Additional considerations regarding nuclear stability may be gleaned from a consideration of the odd or even nature of the numbers of protons and neutrons in the nucleus. According to the Pauli exclusion principle, no two extranuclear electrons having an identical set of quantum numbers can occupy the same electron energy state. See also **Pauli Exclusion Principle**. The application of this principle to the nucleus leads to conclusions which at least are not at variance with observations of nuclear stability. Thus, it is inferred that no two nucleons possessing an identical set of quantum numbers may occupy the same nuclear energy state. It would appear, then, that both protons and neutrons that differ only in their angular momenta or spins may exist in a nuclear state. The exclusion principle requires, therefore, that only protons having opposite spins can exist in the same state. The same consideration applies to neutrons. Accordingly, two protons and two neutrons might occupy the same nuclear energy state provided the nucleons in each pair have opposite spins. Such two-proton-two-neutron groupings are termed "closed shells," and by virtue of their proton-neutron interaction, they confer exceptional stability to nuclei they inhabit. The nuclear forces in closed shells are said to be "saturated", which means that the nucleons therein interact strongly with each other, but weakly with those in other states. Since like particles tend to complete an energy state by pairing of opposite spins, two neutrons of opposite spin, or a single neutron or proton also might exist in a particular energy state.

Any of the foregoing conditions may be achieved when the nucleus contains an even number of both protons and neutrons, or an even number of one and an odd number of the other. Since there is an excess of neutrons over protons for all but the lowest atomic number elements, in the odd-odd situation there is a deficiency of protons necessary to complete the two-proton-two-neutron quartets. It might be expected that these could be provided by the production of protons via beta decay. However, there exist only four stable nuclei of odd-odd composition, whereas there are 108 such nuclei in the even-odd form and 162 in the even-even series. It will be seen that the order of stability, and presumably the binding energy per nucleon, from greatest to smallest, seems to be even-even, even-odd, odd-odd.

Although the existence of binding energies holding the nucleus together has been demonstrated, the problem of defining the nature of these forces presents itself. Clearly, repulsive electrostatic forces must exist between protons. These are "long range" in effect. To achieve nuclear stability then, compensating attractive forces also must exist. It has been concluded that "short range" attractive forces exist between protons, between neutrons, and between protons and neutrons. The $(p - n)$ attractive forces are considered to be of the greatest magnitude, while the $(n - n)$ and $(p - p)$ forces are of less intensity, with the latter decreased by virtue of electrostatic repulsion. When the number of protons in a nucleus is greater than 20, it is found that the ratio of neutrons to protons exceeds unity. The additional short-range attractive forces provided by the excess neutrons, therefore, may be considered as compensating for the long-range electrostatic repulsive forces between the protons. Nevertheless, when the number of protons exceeds about 50, the short-range forces are insufficient to counteract the electrostatic forces completely, with the result that the binding energy per each additional nucleon decreases.

The nature of the short-range attractive forces between nucleons requires further investigation. An interpretation of them has been presented by Heisenberg in terms of wave-mechanical exchange forces. Thus, if the basic difference between the proton and neutron in a system composed of these two particles is considered to be that the former is electrically charged while the latter is not, then the transfer of the electric charge from the proton to the neutron results in an exchange of individual identity, but not a change in the system. That is to say, the system still is composed of a proton and neutron, despite the fact that the particles have exchanged their identities. Since the system itself has the same composition, it must possess the same energy after the exchange as it did before. One of the principles of wave mechanics is that, if a system may be represented by two states, each of which has the same energy, then the actual state of the system is a result of the combination, i.e., resonance, of the two separate states and is more stable than either. In the proton-neutron system, the energy difference between the "combined" state and the individual states may be considered as the "exchange energy" or "attractive force" between the particles. In an extension of Heisenberg's proposal, Yukawa postulated that the exchange energy is carried by a particle termed the *meson*. Particles having the properties attributed by Yukawa to mesons have been identified in cosmic rays.

With such concepts of nuclear structure and stability, however imperfect, the process of nuclear fission of uranium can be considered. Although fast neutrons (greater than 0.1 MeV) can cause fission in both uranium-235 and uranium-238, thermal neutrons (about 0.03 MeV) are effective only with uranium-235. Uranium-238 is unsatisfactory as a fissionable material for most purposes, however, since it has a high probability for "resonance capture" of fast neutrons, which is a non-fission process. It is instinctive to ponder why uranium-235 undergoes fission with thermal neutrons and uranium-238 does not. It will be recalled that the binding energy for an even-even nucleus exceeds that for an even-odd. Consequently, the addition of a neutron to uranium-235, which yields an even-even compound nucleus, will contribute a greater binding energy than in the case of uranium-238 where an even-odd compound nucleus would be produced. Calculations yield a value of 6.81 MeV for the additional neutron in the former case, and 5.31 MeV in the latter. Using Bohr and Wheeler's calculations, it is found that the activation energy for fission is 5.2 MeV for uranium-235 and 5.9 MeV for uranium-238. Thus, the binding energy for an additional neutron in uranium-235 exceeds its fission activation energy, whereas it is less in the case of uranium-238. It can be seen, then, that uranium-235 fission is energetically feasible with thermal neutrons while the fission of uranium-238 is not.

In considering the physical forces acting in fission, use may be made of the Bohr liquid drop model of the nucleus. Here it is assumed that in its normal energy state, a nucleus is spherical and has a homogeneously distributed electrical charge. Under the influence of the activation energy furnished by the incident neutron, however, oscillations are set up which tend to deform the nucleus. In the ellipsoid form, the distribution of the protons is such that they are concentrated in the areas of the two foci. The electrostatic forces of repulsion between the protons at the opposite ends of the ellipse may then further deform the nucleus into a dumbbell shape. From this condition, there can be no recovery, and fission results.

It will be recalled that the binding energy per nucleon decreases with increasing mass number, that is, a greater amount of energy is released in the formation of nuclei of intermediate mass number from their constituent nucleons than is the case of nuclei of high mass number. Thus, energy is released in fission because the binding energy of the high mass number uranium-236 compound nucleus is less than that of the intermediate mass number fission products which are produced. The total energy thus liberated in fission is about 200 MeV. Of this, the kinetic energy of the fission products accounts for 160 MeV. These fragments, being of significantly lower atomic number, require fewer neutrons for stability than they actually contain immediately after fission. These excess neutrons, therefore, are "boiled off" the fission fragments, the process occurring in two distinct phases. In the first phase, "prompt" neutrons of about 2-MeV energy are released within 10^{-2} seconds after fission occurs and take up about 7% of the fission energy. Subsequently, after several seconds, additional "delayed" neutrons with about 0.5-MeV energy are boiled off the fission products. See also **Energy**; and **Particles (Subatomic)**.

Additional Reading

Bernal, J.D.: "A History of Classical Physics," *From Antiquity to the Quantum*, Barnes & Noble Books, New York, NY, 1997.

Davies, P.C.W.: *The New Physics,* Press Syndicate of the University of Cambridge New York, NY, 1992.

Ellis, P.J. and Y.C. Tang, Editors: *Trends in Theoretical Physics,* Addison-Wesley, Redwood City, CA, 1990.

Falomir, H., R.E. Gomboa, and F.A. Schaposnki: *Trends in Theoretical Physics,* American Institute of Physics, College Park, MD, 1998.

Gutbrod, H. and H. Stocker: "The Nuclear Equation of State," *Sci. Amer.,* **58** (November 1991).

Haber, H.E. and G.L. Kane: "Is Nature Supersymmetric?" *Sci. Amer.,* **253**(6), 52–60 (June 1986).

Hamilton, D.P.: "A Tentative Vote for Supersymmetry," *Science,* **272** (July 19, 1991).

Heller, M.: *The New Physics and a New Theology,* University of Notre Dame Press, Notre Dame, IN, 1997.

Imry, Y. and R.A. Webb: "Quantum Interference and the Aharonov-Bohm Effect," *Sci. Amer.,* **56** (April 1989).

Korepin, V.E., A.G. Izergin, and N.M. Bogoliubov: *Quantum Inverse Scattering Method and Correlation Functions,* Cambridge University Press, New York, NY, 1997.

Landau, R.H.: "Quantum Mechanics ll," *A Second Course in Quantum Theory,* 2nd Edition, John Wiley & Sons, Inc., New York, NY, 1995.

Messiah, A.: *Quantum Mechanics,* Dover Publications, Inc., Mineola, NY, 2000.

Omnes, R.: "Quantum Philosophy," *Understanding and Interpreting Contemporary Science,* Princeton University Press, Princeton, NJ, 1999.

Ruthen, R.: "Quantum Pinball Machine," *Sci. Amer.,* **38** (November 1991).

BIOCHEMICAL INDIVIDUALITY.

The possession of biochemical distinctiveness by individual members of a species, whether plant, animal, or human. The primary interest in such distinctiveness has centered in the human family, and in the distinctiveness within animal species as it might illuminate some of the questions on human biochemistry.

While it has been known for centuries that bloodhounds, for example, can tell individuals apart even by the attenuated odors from their bodies left on a trail, the first scientific work which hinted at the existence of substantial biochemical distinctiveness in human specimens was the discovery of blood groups by Landsteiner about 1900.

A few years later Garrod noted what he called "inborn errors of metabolism"—rare instances where individuals gave evidence of being abnormal biochemically in that they were albinos (lack of ability to produce pigment in skin, hair and eyes), or excreted some unusual substance in the urine or feces. To Garrod these observations suggested the possibility that the biochemistry of all individuals might be distinctive.

About 50 years later serious attention to the phenomenon of biochemical individuality resulted in the publication of several articles and a book on this subject. (Williams, R. J.: *Biochemical Individuality*, Wiley, New York, 1956.) These reported evidence indicating that every human being, including all those designated as "normal," possesses a distinctive metabolic pattern which encompasses everything chemical that takes place in his or her body. That these patterns, like the abnormalities discussed by Garrod, have genetic roots is indicated by the pioneer explorations of Beadle and Tatum in the field of biochemical genetics in which they established the fact that the potentiality for producing enzymes resides in the genes.

Biochemical individuality, which is genetically determined, is accompanied by, and in a sense based upon, anatomical individuality, which must also have a genetic origin. Substantial differences, often of large magnitude, exist between the digestive tracts, the muscular systems, the circulatory systems, the skeletal systems, the nervous systems, and the endocrine systems of so-called normal people. Similar distinctiveness is observed at the microscopic level, for example in the size, shape and distribution of neurons in the brain and in the morphological "blood pictures," i.e., the numbers of the different types of cells in the blood.

Individuality in the biochemical realm is exhibited with respect to (1) the composition of blood, tissues, urine, digestive juices, cerebrospinal fluid, etc.; (2) the enzyme levels in tissues and in body fluids, particularly the blood; (3) the pharmacological responses to numerous specific drugs; (4) the quantitative needs for specific nutrients—minerals, amino acids, vitamins—and in miscellaneous other ways including reactions of taste and smell and the effects of heat, cold, electricity, etc. Each individual must possess a highly distinctive pattern, since the differences between individuals with respect to the measurable items in a potentially long list are by no means trifling. Often a specific value derived from one "normal" individual of a group will be several times as large as that derived from another.

BIOCHEMICAL OXYGEN DEMAND (BOD).

A standardized means of estimating the degree of contamination of water supplies, especially those which receive contamination from sewage and industrial wastes. It is expressed as the quantity of dissolved oxygen (in mg/L) required during stabilization of the decomposable organic matter by aerobic biochemical action. Determination of this quantity is accomplished by diluting suitable portions of the sample with water saturated with oxygen and measuring the dissolved oxygen in the mixture both immediately and after a period of incubation, usually five days.

See also **Biodegradability**.

BIOCHEMISTRY.

The dawn of biochemistry may have been the discovery of the first enzyme, diastase, in 1833 by Anselme Payen. In 1828, Friedrich Wöhler published a paper about the synthesis of urea, proving that organic compounds can be created artificially, in contrast to the common belief of the time that organic compounds can only be made by living organisms. Since then, biochemistry has advanced, especially since the mid-20th century, with the development of new techniques such as chromatography, X-ray diffraction, NMR, radioisotope labelling, electron microscopy and molecular dynamics simulations. These techniques allowed for the discovery and detailed analysis of many molecules and metabolic pathways of the cell, such as glycolysis and the Krebs cycle.

Originally a subdivision of chemistry but now an independent science, biochemistry includes all aspects of chemistry that apply to living organisms. Thus, photochemistry is directly involved with photosynthesis, and physical chemistry with osmosis—two phenomena that underlie all plant and animal life. Other important chemical mechanisms that apply directly to living organisms are catalysis, which takes place in biochemical systems by the agency of enzymes; nucleic acid and protein constitution and behavior, which are known to control the mechanism of genetics; colloid chemistry, which deals in part with the nature of cell walls, muscles, collagen, etc,; acid-base relations, involved in the pH of body fluids; and such nutritional components as amino acids, fats, carbohydrates, minerals, lipids, and vitamins, all of which are essential to life. The chemical organization and reproductive behavior of microorganisms (bacteria and viruses) and a large part of agricultural chemistry are also included in biochemistry. Particularly active areas of are nucleic acids, cell surfaces (membranes), enzymology, peptide hormones, molecular biology, and recombinant DNA.

See also **Biotechnology (Bioprocess Engineering)**; **Carbohydrates**; **Colloid Systems**; **Genetics and Gene Science (Classical)**; and **Industrial Biotechnology**.

BIODEGRADABILITY.

The susceptibility of a substance to decomposition by microorganisms, specifically the rate at which detergents and pesticides and other compounds may be chemically broken down by bacteria and/or natural environmental factors. Branched-chain alkylbenzene sulfonates (ABS) are much more resistant to such decomposition than are linear alkylbenzene sulfonates (LAS), in which the long, straight alkyl chain is readily attacked by bacteria. If the branching is at the end of a long alkyl chain (isoalkyls), the molecules are about as biodegradable as the normal alkyls. The alcohol sulfate anionic detergents and most of the nonionic detergents are biodegradable. Among pesticides, the organophosphorus types, while highly toxic, are more biodegradable than DDT and its derivatives. Tests on a number of compounds gave results as follows. Easily biodegraded: *n*-propanol, ethanol, benzoic acid, benzaldehyde, ethyl acetate. Less easily biodegraded: ethylene glycol, isopropanol, *o*-cresol, diethylene glycol, pyridine, triethanolamine. Resistant to biodegration: aniline, methanol, monothanolamine, methyl ethyl ketone, acetone. Additives that accelerate biodegradation of polyethylene, polystyrene, and other plastics are available.

See also **Detergents**.

BIOELECTROCHEMISTRY.   Application of the principles and techniques of electrochemistry to biological and medical problems. It includes such surface and interfacial phenomena as the electrical properties of membrane systems and processes, ion adsorption, enzymatic clotting, transmembrane pH and electrical gradients, protein phosphorylation, cells, and tissues.

BIOGEOCHEMISTRY.   A branch of geochemistry dealing with the interactions between living organisms and their mineral environment. It includes, among other studies, the effect of plants on weathering of rocks, of the chemical transformations that produced petroleum and coal, of the concentration of specific elements in vegetation at some time in the geochemical cycle (iodine in sea plants, uranium in some form of decaying organic matter), and of the organic constituents of fossils.

BIOINORGANIC CHEMISTRY.   Study of the mechanisms involved in the behavior of metal-containing molecules in living organisms, e.g., biological transport of iron, the effect of copper on nucleic acid and nucleoproteins, molybdenum, and manganese complexes, etc.

BIOLOGICAL ENERGY TRANSFER.   When ionization occurs in a substance such as a protein, the net charge produced in the protein probably migrates throughout a large region of the molecule with various probabilities favoring its occurrence in one part of the molecule or another. Eventually, after approximately 10^{-14} seconds, the excess (or deficiency) of charge probably settles in an s–s bond or in the hydrogen atom attached to the carbon of the peptide bond which is opposite to one or other of the amino acid residues. Thus, regardless of the site of the original ionization in the molecule, there is considerable transfer of energy throughout a large portion of the molecule. However, the phrase *energy transfer* is generally meant to include those cases where it might occur in addition to this; for example, intermolecularly either between adjacent protein molecules or between protein and solvent molecules. It can also apply to excitation.

BIOLOGY.   The science of life. As with several of the fundamental sciences, over the last several decades, biology has been segmented into a number of fields of specialization. These include biochemistry, bioengineering, biomedicine, biophysics, cell biology, developmental biology, ecogenetics, evolutionary biology, marine biology, microbiology, and molecular biology, among others. Convenient umbrella terms sometimes used include the *biological sciences* and the *life sciences*.

There are hundreds of entries of varying length included throughout this encyclopedia that relate to the biological sciences. Many of these entries include lists of references for further reading.

BIOLUMINESCENCE.   Many living organisms exhibit the unique property of producing visible light, a phenomenon referred to as bioluminescence. Known light-emitting organisms have either oxidative or peroxidative enzymes that couple the chemical energy released from the enzyme reaction to give electronic excitation of a luminescent compound. The compound that is oxidized with subsequent light emission is usually referred to as *luciferin* and the enzyme that catalyzes the reaction as *luciferase*. Most luciferins and luciferases that have been isolated from unrelated species are different in molecular structure. With one known exception, combinations of luciferin and luciferase from different species do not exhibit bioluminescence.

The light-producing reaction in a number of organisms can be represented simply by: Luciferin + O_2 $\xrightarrow{\textit{Lusiferase}}$ Light. Some luminous organisms catalyzing this reaction are: (1) *Cypridina* (a crustacean); (2) *Apogon* (a fish), and (3) *Gonyaulax* (a protozoan). The latter organism is mainly responsible for the phosphorescence (so-called) of the sea.

In other instances, some luciferins must first undergo a luciferase-catalyzed activation reaction prior to their being catalytically oxidized by the enzyme to produce light. There are two well-known cases:

(1)   The firefly:

Luciferin + Adenosine Triphosphate (ATP)

$$\xrightarrow{\textit{Luciferase;}Mg^{2+}} \text{Activate Luciferin}$$

Activated Luciferin + O_2 $\xrightarrow{\textit{Luciferase}}$ Light

(2)   The sea pansy (*Renilla*):

Luciferin + 3′, 5′-Diphosphoadenosine (DPA)

$$\xrightarrow{\textit{Luciferase;}Ca^{2+}} \text{Activate Luciferin}$$

Activated Luciferin + O_2 $\xrightarrow{\textit{Luciferase}}$ Light

Both of these activation reactions are linked to adenine-containing nucleotides of great biological importance. Since the measurement of light can be made an extremely sensitive and rapid technique, the most sensitive and rapid assays known have been developed for ATP and DPA, using the foregoing luminescent systems. Nucleotide concentrations of less than 1×10^{-9} M are easily detectable using electronic instrumentation. Firefly luciferase-luciferin preparations for ATP assays are commercially available.

The structure of firefly luciferin has been confirmed by total synthesis. The firefly emits a yellow-green luminescence, and luciferin in this case is a benzthiazole derivative. Activation of the firefly luciferin involves the elimination of pyrophosphate from ATP with the formation of an acid anhydride linkage between the carboxyl group of luciferin and the phosphate group of adenylic acid forming luciferyl-adenylate.

All other systems that have been extensively studied emit light in the blue-green region of the spectrum. In these cases, the luciferins appear to be indole derivatives.

Some animals, such as the marine acorn worms (Balanoglossus), produce light via a peroxidation reaction and appear not to require molecular oxygen for luminescence. The luciferase in this case is a peroxidase of the classical type and catalyzes the reaction: Luciferin + H_2O_2 $\xrightarrow{\textit{Lusiferase}}$ Light.

Commercially available horseradish peroxidase (crystalline) will substitute for luciferase in the foregoing reaction. In addition, a compound of known structure, 5-amino-2, 3-dihydro-1, 4-phthalazinedione (also known as *luminol*), will substitute for luciferin. The mechanisms appear to be the same regardless of the way in which the crosses are made. Thus, a model bioluminescent system is available and can be used as a sensitivity assay for H_2O_2 at neutral pH. The identification of luciferase as a peroxidase is of interest since this represents the only demonstration of a bioluminescent system in which the catalytic nature of a luciferase molecule has been defined.

Most of the luminescent systems mentioned appear to be under some nerve control. Normally, a luminous flash is observed after mechanical or electrical stimulation of most of the aforementioned species. A number of these also exhibit a diurnal rhythm of luminescence.

Among the lower forms of life, there are two well-known examples of luminescence which are not under nerve control, giving a continuous glow of visible light. These are the luminous bacteria, frequently found growing on dead fish, and luminous fungi, which grow abundantly on rotting wood. These cells apparently depend upon the oxidation of an organic molecule and hydrogen that is transferred through diphosphopyridine nucleotide (DPN; also termed NAD, nicotinamide adenine dinucleotide) and the enzyme system to drive the luminescent reaction. Known details of these luminescent reactions are represented as follows. For bacteria:

$$\text{DPNH} + \text{H}^+ + \text{Flavin Mononucleotide (FMN)} \xrightleftharpoons{\textit{Oxidase}} \text{FMNH}_2 + \text{DPN}$$

$$\text{FMNH}_2 + \text{Long-chain Aliphatic Aldehyde} + O_2 \xrightarrow{\textit{Lusiferase}} \text{Light and}$$

for fungi

$$\text{DPNH} + \text{H}^+ + \text{Unknown compound (X)} \xrightleftharpoons{\textit{Oxidase}} \text{XH}_2 + \text{DPN}$$

$$\text{XH}_2 + O_2 \xrightarrow{\textit{Lusiferase}} \text{Light}$$

Both of these systems are apparently closely linked to respiratory processes and in this sense are analogous to one another. Luciferase from a luminous bacterium, *Photobacterium fischeri*, has been made into-crystal in high yield.

See also **Luminescence**.

Additional Reading

Baretta-Bekker, J.G., E.K. Duursma, and B.R. Kuipers: *Encyclopedia of Marine Sciences,* Springer-Verlag Inc., New York, NY, 1992.

Hanneke, J.G., B.R. Kuipers, and H.J. Baretta-Bekker: *Encyclopedia of Marine Sciences,* 2nd Edition, Springer-Verlag, Inc., New York, NY, 1998.

Hastings, J.W., L.J. Kricka, and P.E. Stanley: Bioluminescence and Chemiluminescence, Proceedings of the 9th International Symposium, John Wiley & Sons, Inc., New York, NY, 1998.

Muller, W. and GyForgy E. *Muller: "Signaling Mechanisms in Protozoa and Invertebrates,"* Vol. 17, Springer-Verlag, Inc., New York, NY, 1996.

Roda, A., L.J. Kricka, P.E. Stanley, and M. Pazzagli: *Bioluminescence and Chemiluminescence—Perspectives for the 21st Century: Proceedings of 10th International Symposium, 1998,* John Wiley & Sons, Inc., New York, NY, 1999.

Stanely, P.E. and L.J. Kricka: Bioluminescence and Chemiluminescence, *Fundamentals of Applied Aspects,* John Wiley & Sons, Inc., New York, NY, 1996.

Ziegler, M.M. and T.O. Baldwin: *Bioluminescence and Chemiluminescence, Part C,* Vol. 305, Academic Press, Inc., San Diego, CA, 2000.

BIOMASS. See **Wastes and Pollution**; **Wastes as Energy Sources**; **Water**; **Water Pollution**.

BIOMIMETIC CHEMISTRY. An interdisciplinary approach to biochemistry including both organic and inorganic aspects of this field. The term means imitation or mimicry of natural organic processes in living systems, and encompasses such subjects as enzyme systems, vitamin B_{12} and flavins, oxygen binding and activation, bioorganic mechanisms, and nitrogen and small-molecule fixation. The technique was utilized in the synthesis of the bleomycin molecule. A notable example of biomimetic chemistry is the development of model synthetic catalysts that imitate the action of natural enzymes. The behavior of chymotrypsin has been duplicated by a manufactured catalyst that can accelerate certain reaction rates by the incredible factor of 100 billion.

BIOPOLYMERS. Biopolymers are the naturally occurring macromolecular materials that are the components of all living systems. There are three principal categories of biopolymers, proteins; nucleic acids; and polysaccharides. See also **Carbohydrates**. Biopolymers are formed through condensation of monomeric units; i.e., the corresponding monomers are amino acids, nucleotides, and monosaccharides for proteins, nucleic acids, and polysaccharides, respectively. The term biopolymers is also used to describe synthetic polymers prepared from the same or similar monomer units as are the natural molecules.

In addition to being necessary for all forms of life, biopolymers, especially enzymes (proteins), have found commercial applications in various analytical techniques. See also **Automated Instrumentation: Clinical Chemistry**; **Automated Instrumentation: Hematology**; and **Biosensors**. In synthetic processes (see also **Enzymes in Organic Synthesis**); and in prescribed therapies (See also **Enzyme Therapeutic**; and **Vitamin**). Other naturally occurring biopolymers having significant commercial importance are the cellulose derivatives, e.g., cotton and wood, which are complex polysaccharides.

Analytical Techniques

Analytical techniques that utilize biopolymers, i.e., natural macromolecules such as proteins, nucleic acids, and polysaccharides that compose living substances, represent a rapidly expanding field. The number of applications is large and thus uses herein are limited to chiral chromatography, immunology, and biosensors.

Biopolymers in Chiral Chromatography. Biopolymers have had a tremendous impact on the separation of nonsuperimposable, mirror-image isomers known as enantiomers. Enantiomers have identical physical and chemical properties in an achiral environment except that they rotate the plane of polarized light in opposite directions. Thus separation of enantiomers by chromatographic techniques presents special problems. Direct chiral resolution by liquid chromatography (lc) involves diastereomeric interactions between the chiral solute and the chiral stationary phase. Because biopolymers are chiral molecules and can form diastereomeric

interactions with chiral solutes, they are ideal for use as chiral stationary phases. This property has led to a rapid growth of chromatographic stationary phases utilizing biopolymers to separate chiral molecules. They include cyclodextrin chromatographic phases, α_1-acid glycoprotein chromatographic phases, bovine serum albumin chromatographic phases, and cellulose triacetate and cellulose derivatives.

Biopolymers in Immunology. Biopolymers are employed in many immunological techniques, including the analysis of food, clinical samples, pesticides, and in other areas of analytical chemistry. Immunoassays are specific, sensitive, relatively easy to perform, and usually inexpensive. For repetitive analyses, immunoassays compare very favorably with many conventional methods in terms of both sensitivity and limits of detection.

Antigens. One condition that must be met for the application of an immunochemical method is that the analyte must be capable of stimulating an immune response leading to the formation of antibodies in the immunized animal. These antibodies can then be isolated and used as highly specific analytical reagents (immunoassays). Analytes that can combine with the corresponding antibodies are called antigens. There are physical and chemical restrictions on the types of analytes that may be used as immunoassay antigens. In general, large, rigid, chemically complex molecules make good antigens.

Antibodies. Antibodies are proteins, found in many body fluids such as tears, saliva, and urine, that are present in highest concentrations in blood serum. Because antibodies are proteins (qv), they may be characterized by such physical properties as solubility, electrostatic charge, isoelectric point, and molecular weight. The particular proteins which exhibit antibody activity are the immunoglobulins (Ig). The principal immunoglobulin in blood serum is immunoglobulin G (IgG), the structure of which is similar to the other immunoglobins.

There are estimated to be approximately 10 million potential combinations of antigen-binding specificities resulting from light-and heavy-chain combinations in the immunoglobulins. The possibility of utilizing all of these combinations as reagents in immunochemical methods is highly interesting though improbable.

Biopolymers as Biosensors. Selectivity is an important consideration in analytical chemistry. Biologically derived polymers can be used as highly selective immobilized reagents in analytical applications.

Immobilized Enzymes. The immobilized enzyme electrode is the most common immobilized biopolymer sensor, consisting of a thin layer of enzyme immobilized on the surface of an electrochemical sensor. The enzyme catalyzes a reaction that converts the target substrate into a product that is detected electrochemically. The advantages of immobilized enzyme electrodes include minimal pretreatment of the sample matrix, small sample volume, and the recovery of the enzyme for repeated use. Several reviews and books have been published on immobilized enzyme electrodes.

Enzyme Immunosensors. Enzyme immunosensors are enzyme immunoassays coupled with electrochemical sensors. These sensors require multiple steps for analyte determination, and either sandwich assays or competitive binding assays may be used. Both of these assays use antibodies for the analyte of interest attached to a membrane on the surface of an electrochemical sensor.

Economic Aspects

Enantiomeric separations are expected to continue to have a considerable economic impact on the development of new drugs and therapy in the biomedical field.

The importance of immunoassays for food monitoring and in the detection of diseases is expected to continue to grow as techniques and detection limits improve.

The development of biosensors is expected to benefit monitoring therapeutic drug levels, office testing, and implantable devices because of the advantages of cost-saving automation and improved data handling.

TIMOTHY WARD
Millsaps College

Additional Reading

Armstrong, D.W. and S.M. Han: *CRC Crit. Rev. Analyt. Chem.* **19**(3), 175 (1988).

Diamandis, E.P. and T.K. Christopoulos: *Anal. Chem.* **62**, 1149A (1990).

Steinbuchel, A.: *Biopolymers: Cumulative Index,* John Wiley & Sons, Inc., New York, NY, 2003.

Stinshoff, E., W. Stein, W.G. Wood, and P. Laska: *Anal. Chem.* **59**, 339R (1987).

Thompson, M. and U.J. Krull: *Anal. Chem.* **63**, 393A (1991).

BIOREMEDIATION. Bioremediation is the process of judiciously exploiting biological processes to minimize an unwanted environmental impact; usually it is the removal of a contaminant from the biosphere.

Bioremediation is already a commercially viable technology, with estimates of aggregate bioremediation revenues of $2–3 billion for the period 1994–2000. There are significant opportunities to enlarge upon this success. Bioremediation has applications in the gas phase, in water, and in soils and sediments. For water and soils, the process can be carried out *in situ*, or after the contaminated medium has been moved to some sort of contained reactor (*ex situ*). The former is generally rather cheaper, but the latter may result in such a significant increase in rate that the additional cost of manipulating the contaminated material is overshadowed by the time saved. Bioremediation may explicitly exploit bacteria, fungi, algae, or higher plants. Each, in turn, may be part of a complex food-web, and optimizing the local ecosystem may be as important as focusing solely on the primary degraders or accumulators.

Bioremediation usually competes with alternative approaches to achieving an environmental goal. It is typically among the least expensive options, but an additional important consideration is that in many cases bioremediation is a permanent solution to the contamination problem, since the contaminant is completely destroyed or collected. Some of the alternatives technologies, such as thermal desorption and destruction of organics, are also permanent, solutions, but the simplest, removing the contaminant to a dump site, merely moves the problem, and may well not eliminate the potential liability. Furthermore, by its very nature bioremediation addresses the bioavailable part of any contamination. The same cannot necessarily be said for nonbiological technologies, which may leave bioavailable contaminants at low levels.

Bioremediation also has the advantage that it can be relatively nonintrusive, and can sometimes be used in situations where other approaches would be severely disruptive. On the other hand, bioremediation is usually slower than most physical techniques, and may not always be able to meet some very strict cleanup standards.

General Technological Aspects

Table 1 lists some of the technologies in use today in bioremediation.

TABLE 1. SOME TECHNOLOGICAL DEFINITIONS RELEVANT TO BIOREMEDIATION

Technology	Description
air sparging; aquifer sparging; biosparging	injection of air to stimulate aerobic degradation; may also stimulate volatilization
air stripping	injection of air to stimulate volatilization
aquifer bioremediation	*in situ* bioremediation in an aquifer, usually by adding nutrients or co-substrates
aquifer sparging	injection of air into a contaminated aquifer to stimulate aerobic degradation, may also stimulate volatilization
batch reactor	a bioreactor loaded with contaminated material, and run until the contaminant has been consumed, then emptied, and the process is repeated
bioactive barrier; bioactive zone; biowall	a zone, usually subsurface, where biodegradation of a contaminant occurs so that no contaminant passes the barrier
bioaugmentation	addition of exogenous bacteria with defined degradation potential (or rarely indigenous bacteria cultivated in a reactor and reapplied)
biofilm reactor	a reactor where bacterial communities are encouraged on a high surface area support, biofilms often have a redox gradient so that the deepest layer is anaerobic while the outside is aerobic
biofiltration	usually an air filter with degrading organisms supported on a high surface area support such as granulated activated carbon
biofluffing	augering soil to increase porosity
bioleaching	extracting metallic contaminants at acid pH
biological fluidized bed; fluidized-bed bioreactor	bioreactor where the fluid phase is moving fast enough to suspend the solid phase as a fluid-like phase
biopile; soil heaping	an engineered pile of excavated contaminated soil, with engineering to optimize air, water, and nutrient control
bioslurping	vacuum extraction of the floating contaminant, water, and vapor from the vadose zone; the air flow stimulates biodegradation
biostimulation	optimizing conditions for the indigenous biota to degrade the contaminant
biotransformation	the biological conversion of a contaminant to some other form, but not to carbon dioxide and water
biotrickling filter	a reactor where a contaminated gas stream passes up a reactor with immobilized microorganisms on a solid support, while nutrient liquor trickles down the reactor
bioventing	vacuum extraction of contaminant vapors from the vadose zone, thereby drawing in air that stimulates the biodegradation of the remainder
borehole bioreactor; in-well bioreactor	the addition of nutrients and electron acceptor to stimulate the biodegradation *in situ* in a contaminated aquifer
closed-loop bioremediation	groundwater recovery, a bioreactor, and low-pressure reinjection to maximize nutrient use, and maintain temperature in cold climates
composting	addition of biodegradable bulking agent to stimulate microbial activity; optimal composting generally involves self-heating to 50–60°C
constructed wetland	artificial marsh for bioremediation of contaminated water
continuous stirred tank reactor (CSTR)	a completely mixed bioreactor
digester	usually an anaerobic bioreactor for digestion of solids and sludges that generates methane
ex-situ bioremediation	usually the bioremediation of excavated contaminated soil in a biopile, compost system or bioreactor
fixed-bed bioreactor	bioreactor with immobilized cells on a packed column matrix
land-farming; land treatment	application of a biodegradable sludge as a thin layer to a soil to encourage biodegradation; the soil is typically tilled regularly
natural attenuation; intrinsic bioremediation	unassisted biodegradation of a contaminant
phytoextraction	the use of plants to remove and accumulate contaminants from soil or water to harvestable biomass
phytofiltration	the use of completely immersed plant seedlings, to remove contaminants from water
phytoremediation	the use of plants to effect bioremediation
phytostabilization	the use of plants to stabilize soil against wind and water erosion
pump and treat	pumping groundwater to the surface, treating, and reinjection or disposal
rhizofiltration	the use of roots to immobilize contaminants from a water stream
rotating biological contractor	bioreactor with rotating device that moves a biofilm through the bulk water phase and the air phase to stimulate aerobic degradation
sequencing batch reactor	periodically aerated solid phase or slurry bioreactor operated in batch mode
soil-vapor extraction	vacuum-assisted vapor extraction

Organic Contaminants

Hydrocarbons

Constituents. Hydrocarbons get into the environment from biogenic and fossil sources. Methane is produced by anaerobic bacteria in enormous quantities in soils, sediments, ruminants and termites, and it is consumed by methanotrophic bacteria on a similar scale. Submarine methane seeps support substantial oases of marine life, with a variety of invertebrates possessing symbiotic methanotrophic bacteria. Thus, methanotrophic bacteria are ubiquitous in aerobic environments. Plants generate large amounts of volatile hydrocarbons, including isoprene and a range of terpenes. These compounds provide an abundant substrate for hydrocarbon-degrading organisms.

Crude oil has been part of the biosphere for millennia, leaking from oil seeps on land and in the sea. Crude oils are very complex mixtures, primarily of hydrocarbons although some components do have heteroatoms such as nitrogen (e.g., carbazole) or sulfur (e.g., dibenzothiophene). Chemically, the principal components of crude oils and refined products can be classified as aliphatics, aromatics, naphthenics, and asphaltic molecules. When crude oils reach the surface environment the lighter molecules evaporate, and are either destroyed by atmospheric photooxidation or are washed out of the atmosphere in rain, and are biodegraded. Some molecules, such as the smaller aromatics (benzene, toluene, etc.) have significant solubilities, and can be washed out of floating slicks, whether these are at sea, or on terrestrial water tables. Fortunately the majority of molecules in crude oils, and refined products made from them, are biodegradable, at least under aerobic conditions.

Biodegradation. Methane and the volatile plant terpenes are fully biodegradable by aerobic organisms, and most refined petroleum products are essentially completely biodegradable under aerobic conditions. The least biodegradable material, principally polar molecules and asphaltenes, lacks the "oily" feel and properties that are associated with oil. These are essentially impossible to distinguish from more recent organic material in soils and sediments, such as the humic and fulvic acids, and appear to be biologically inert.

Numerous bacterial and fungal genera have species able to degrade hydrocarbons aerobically and the pathways of degradation of representative aliphatic, naphthenic and aromatic molecules have been well characterized in at least some species. It is a truism that the hallmark of an oil-degrading organism is its ability to insert oxygen atoms into the hydrocarbon, and

Linear alkanes

Cycloalkanes

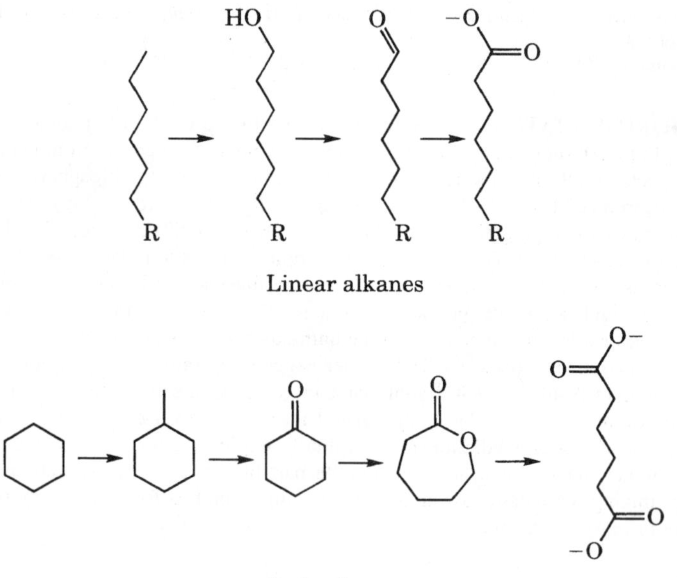

Fig. 1.   Initial steps in the biodegradation of linear and cyclic alkanes

there are many ways in which this is achieved. Figures 1 and 2 show the most well-studied.

In recent years it has become clear that at least some hydrocarbons are oxidized by bacteria under completely anaerobic conditions, where the oxygen is probably coming from water. Limited hydrocarbon biodegradation has now been shown under sulfate-, nitrate-, carbon dioxide- and ferric iron-reducing conditions (Table 2). Figure 3 shows the intermediates identified in anaerobic toluene degradation in different organisms. It is noteworthy that while organisms capable of aerobic oil biodegradation seem to be ubiquitous, organisms capable of the anaerobic degradation of hydrocarbon have to date only been found in a few places.

Although the majority of molecules in crude oils and refined products are hydrocarbons, the U.S. Clean Air Act amendment of 1990 mandated the addition of oxygenated compounds to gasoline in many parts of the United States. The requirement is usually that 2% (w/w) of the fuel be oxygen,

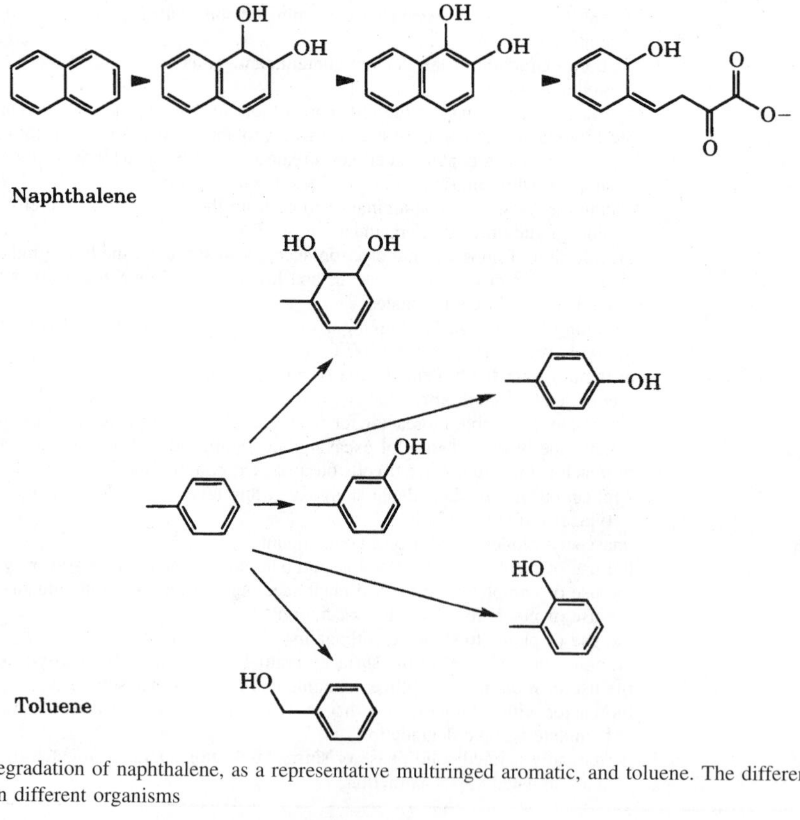

Naphthalene

Toluene

Fig. 2.   Initial steps in the aerobic degradation of naphthalene, as a representative multiringed aromatic, and toluene. The different initial steps of toluene degradation are examples of the diversity found in different organisms

TABLE 2. HYDROCARBONS THAT HAVE BEEN SHOWN TO
BE BIODEGRADED UNDER ANAEROBIC CONDITIONS

Electron acceptor	Substrate
nitrate (to nitrogen)	heptadecene
	toluene, ethylbenzene, xylene
	naphthalene
	terpenes
iron(III) (to iron(II))	toluene
manganese(IV) (to Mn(II))	toluene
sulfate (to sulfide)	hexadecane, alkylbenzenes
	benzene
	naphthalene, phenanthrene
CO_2 (to methane)	toluene, xylene

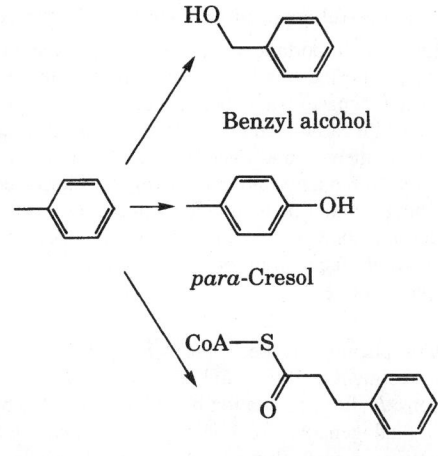

Benzyl alcohol

para-Cresol

β-PhenylpropionylCoA

Fig. 3.   Proposed initial steps in the anaerobic biodegradation of toluene in different organisms

which requires that 5–15% (v/v) of the gasoline be an oxygenated additive (e.g., methanol, ethanol, methyl *tert*-butyl ether (MTBE), etc.). Although methanol and ethanol are readily degraded under aerobic conditions, the degradability of MTBE remains something of an open question. The compound was previously very rare in the environment, but now it is one of the major chemicals in commerce. At first it seemed that the compound was completely resistant to biodegradation, but complete mineralization has now been reported. Whether biodegradation can be optimized for effective bioremediation remains to be seen.

Bioremediation. Crude oil and refined products are readily biodegradable under aerobic conditions, but they are only incomplete foods since they lack any significant nitrogen, phosphorus, and essential trace elements. Bioremediation strategies for removing large quantities of hydrocarbon must therefore include the addition of fertilizers to provide these elements in a bioavailable form.

Air. Hydrocarbon vapors in air are readily treated with biofilters. These are typically rather large devices with a very large surface area provided by bulky material such as a bark or straw compost. The contaminated air, perhaps from a soil vapor-extraction treatment, or from a factory using hydrocarbon solvents, is blown through the filter, and organisms, usually indigenous to the filter material or provided by a soil or commercial inoculum, grow and consume the hydrocarbons.

Sea. Significantly more oil reaches the world's oceans from municipal sewers than widely covered crude oil spills. Physical collection of spilled oil is the preferred remediation option, but if skimming is unable to collect the oil, biodegradation and perhaps combustion or photooxidation are the only routes for eliminating of the spill. One approach to stimulating biodegradation is to disperse the oil with chemical dispersants. Modern dispersants and application protocols stimulate biodegradation by increasing the surface area of the oil available for microbial attachment, and perhaps providing nutrients to stimulate microbial growth.

Shorelines. The successful bioremediation of shorelines affected by the spill from the *Exxon Valdez* in Prince William Sound, Alaska, was perhaps the largest project to date. Bioremediation focused on the addition

of nitrogen and phosphorus fertilizers to partially remove the nutrient-limitation on oil degradation. Of course the addition of fertilizers was complicated by the fact that oiled shorelines were washed by tides twice a day. Two fertilizers were used in the full-scale applications; one, an oleophilic product known as Inipol EAP22 (trademark of CECA, Paris, France), was a microemulsion of a concentrated solution of urea in an oil phase of oleic acid and trilaurethphosphate, with butoxyethanol as a cosolvent. This product was designed to adhere to oil, and to release its nutrients to bacteria growing at the oil-water interface. The other fertilizer was a slow-release formulation of inorganic nutrients, primarily ammonium nitrate and ammonium phosphate, in a polymerized vegetable oil skin. This product, known as Customblen (trademark of Grace-Sierra, Milpitas, California), released nutrients with every tide, and these were distributed throughout the oiled zone as the tide fell. Fertilizer application rates were carefully monitored so that the nutrients would cause no harm, and the rate of oil biodegradation was stimulated between two- and five-fold.

Areas where there are currently few remediation options but where bioremediation may provide an option include oiled marshes, mangroves, and coral reefs. Bioremediation also offers options for dealing with oiled material, such as seaweed, that gets stranded on shorelines; composting has been shown to be effective.

Groundwater. Spills of refined petroleum product on land, and leaking underground storage tanks, sometimes contaminate groundwater. Stand-alone bioremediation is an option for these situations, but "pump and treat" is the more usual treatment. Contaminated water is brought to the surface, free product is removed by flotation, and the cleaned water re-injected into the aquifer or discarded. Adding a bioremediation component to the treatment, typically by adding oxygen and low levels of nutrients, is an appealing and cost-effective way of stimulating the degradation of the residual hydrocarbon not extracted by the pumping. This approach is becoming widely used.

Typically only small aromatic molecules, the infamous BTEX (benzene, toluene, ethylbenzene, and xylenes), are soluble enough to contaminate groundwater. With the advent of oxygenated gasolines, it is expected that these oxygenates [ethanol, methanol, MTBE (methyl-*tert*-butyl ether) etc] will also be found in groundwater. These contaminants are biodegradable, and some biodegradation is probably already occurring when the contamination is discovered. The cheapest approach to remediation is, thus, to allow this intrinsic process to continue.

Intrinsic bioremediation is becoming an acceptable option in locations where the contaminated groundwater poses little threat to environmental health. Nevertheless, it may not be the lowest cost option if there are extensive monitoring and documentation costs involved for several years. In such cases it may well be more cost effective to optimize conditions for biodegradation.

One approach is to optimize the levels of electron acceptors. Slow release formulations of inorganic peroxides, such as magnesium peroxide, have recently been used with success. Nitrate may be added, although there are sometimes regulatory limitations on the amount of this material that may be added to groundwater. Ferric iron availability may be manipulated by adding ligands.

If there are significant amounts of both volatile and nonvolatile contaminants, remediation may be achieved by a combination of liquid and vapor extraction of the former, and bioremediation of the latter. This combination has been termed "bioslurping".

The majority of remediation operations include stopping the source of the contamination, but in some cases this is impossible, either because of the location of the spill, or because it is over a large area, and not a point source. In these situations it may be possible to intercept the flow of contaminated groundwater on-site, and ensure that no contamination passes. Approaches include biowall, trench biosparge, funnel and gate, bubble curtain, sparge curtain and engineered trenches and gates. Both aerobic and anaerobic designs have been successfully installed.

Where there are large volumes of contaminated water under a small site, it is sometimes most convenient to treat the contaminant in a biological reactor at the surface.

Of course the presence of a liquid phase of hydrocarbon in a soil gives rise to vapor contamination in the vadose zone above the water table. This can be treated by vacuum extraction, and the passage of the exhaust gases through a biofilter (see above) can be a cheap and effective way of destroying the contaminant permanently.

Soil. Spills from production facilities and pipelines often involve both oil and brine. Successful bioremediation strategies must therefore include remediating the brine. In wet regions the salt is eventually diluted by rainfall, but in arid regions, and to speed the process in wetter regions, gypsum is often added to restore soil porosity.

Many hydrocarbons bind quite tightly to soil components, and are thereby less available to microbial degradation. Intrinsic biodegradation occurs, but it usually only removes the lightest refined products, such as gasoline, diesel and jet fuel. Active intervention is typically required. Usually the least expensive approach is *in situ* remediation, typically with the addition of nutrients, and the attempted optimization of moisture and oxygen by tilling.

Deeper contamination may be remedied with bioventing, where air is injected through some wells, and extracted through others to both strip volatiles and provide oxygen to indigenous organisms. Fertilizer nutrients may also be added. This is usually only a viable option with lighter refined products.

A recent suggestion has been to use plants to stimulate the microbial degradation of the hydrocarbon (hydrocarbon phytoremediation). The plants are proposed to help deliver air to the soil microbes, and to stimulate microbial growth in the rhizosphere by the release of nutrients from the roots. The esthetic appeal of an active phytoremediation project can be very great.

When soil contamination extends to some depth it may be preferable to excavate the contaminated soil and put it into "biopiles" where oxygen, nutrient and moisture levels are more easily controlled. Composting by the addition of readily degradable bulking agents is also a useful option for relatively small volumes of excavated contaminated soil.

Slurry bioreactors offer the most aggressive approach to maximizing contact between the contaminated soil and the degrading organisms. Slurry bioreactors are usually the most expensive bioremediation option because of the large power requirements, but under some conditions this cost is offset by the rapid biodegradation that can occur.

In all these cases it is important to bear in mind that although the majority of hydrocarbons are readily biodegraded, some are very resistant to microbial attack. It is thus important to run laboratory studies to ensure that the contaminant is sufficiently biodegradable that clean-up targets can be met.

Halogenated Organic Solvents

Constituents. Halogenated organic solvents are widely used in metal processing, electronics, dry cleaning and paint, paper and textile manufacturing and are fairly widespread contaminants. Unlike the hydrocarbons, the halogenated solvents typically have specific gravities greater than 1, and they generally sink to the bottom of any groundwater, and float on the bedrock. For this reason they are sometimes known as DNAPLs for dense nonaqueous phase liquids.

Biodegradation. Halogenated solvents are degraded under aerobic and anaerobic conditions. The anaerobic process is typically a reductive dechlorination that progressively removes one halide at a time.

The simplest chlorinated alkanes, alkenes, and alcohols (e.g., chloromethane, dichloromethane, chloroethane, 1,2-dichloroethane, vinyl chloride, and 2-chloroethanol) serve as substrates for aerobic growth for some bacteria, but the majority of halogenated solvents cannot support growth. Nevertheless these compounds are mineralized under aerobic conditions.

Bioremediation

Air. Biofilters are an effective way of dealing with air from industrial processes that use halogenated solvents that support aerobic growth. Both compost-based dry systems and trickling filter wet systems are in use. Similar filters could be incorporated into pump-and-treat operations.

Groundwater and Soil. Pumping out the liquid phase is an obvious first step if the contaminant is likely to be mobile, but *in situ* bioremediation is a promising option. Thus, the U.S. Department of Energy is investigating the use of anaerobic *in situ* degradation of carbon tetrachloride with nitrate as electron acceptor, and acetate as electron donor.

Trichloroethylene, the most frequent target of remediation, is only metabolized co-metabolically. Remediation operations thus incorporate the addition of co-metabolized substrate.

Plants may have a role to play in enhancing microbial biodegradation of halogenated solvents, for it has recently been shown that mineralization of radiolabelled trichlorothylene is substantially greater in vegetated rather than unvegetated soils, indicating that the rhizosphere provides a favorable environment for microbial degradation of organic compounds.

Methane has been used in aerobic bioreactors that are part of a pump-and-treat operation, and toluene and phenol have been used as co-substrates at the pilot scale. Anaerobic reactors have also been developed for treating trichloroethylene.

Groundwater contaminated with other halogenated solvents can also be treated in aboveground reactors. Aerobic reactors are useful for those compounds that can support growth. Sequential anaerobic and aerobic reactors are capable of mineralizing tetrachloroethylene.

Halogenated Organic Compounds

Constituents. Complex halogenated organic compounds have been widely used in commerce in the last fifty years. Representative examples are pentachlorophenol, (2,4-dichlorophenoxy)acetate, DDT and polychlorinated biphenyls (PCBs). They may not seem a good target for bioremediation but some successful applications have been developed.

Biodegradation. An important characteristic of degradation is the cleavage of carbon–chlorine bonds, and the enzymes that catalyze these reactions, the dehalogenases, are being characterized. The reductive dechlorination seen with carbon tetrachloride and tetrachloroethylene seems to be a general phenomenon, and even compounds as persistent as DDT and the polychlorinated biphenyls are reductively dechlorinated under some conditions, particularly under methanogenic conditions. Some compounds, such as pentachlorophenol, can be completely mineralized under anaerobic conditions, but the more recalcitrant ones require aerobic degradation after reductive dehalogenation.

Bioremediation

Soil. Pentachlorophenol has been the target of bioremediation at a number of wood-treatment facilities, and good success has been achieved in several applications. *In situ* degradation has been stimulated by bioventing. Just as with the halogenated solvents, it seems that plants stimulate microbial degradation of pentachlorophenol in the rhizosphere.

The kinetics of such *in situ* degradation are rather slow, however, and more active bioremediation is usually attempted. For example, contaminated soil at the Champion Superfund site in Libby, Montana, was placed into 1-acre land treatment units in 6-in. layers, and irrigated, tilled, and fertilized. Under these conditions, the half-lives of pentachlorophenol, pyrene, and several other polynuclear aromatic hydrocarbons, initially present at around 100–200 ppm, were on the order of 40 days. Composting, and bioremediation focusing on the use of white-rot fungi, has also met with success at the pilot scale. Others have used fed-batch or fluidized-bed bioreactors to stimulate the biodegradation of pentachlorophenol. This allows significant optimization of the process and increases in rates of degradation by tenfold.

A major concern when remediating wood-treatment sites is that pentachlorophenol was often used in combination with metal salts, and these compounds, such as chromated-copper-arsenate, are potent inhibitors of at least some pentachlorophenol degrading organisms. Sites with significant levels of such inorganics may not be suitable candidates for bioremediation.

The phenoxy-herbicide, 2,4-D, has been successfully bioremediated in a soil contaminated with such a high level of the compound (710 ppm) that it was toxic to microorganisms. Success relied on washing a significant fraction of the contaminant off the soil and adding bacteria enriched from a less contaminated site. Success was achieved in remediating both soil washwater and soil in a bioslurry reactor. 2,4-D is also effectively degraded in composting, with about half being completely mineralized, and the other half becoming incorporated in a nonextractable form in the residual soil organic matter.

Cultures are being found that can degrade both polychlorinated biphenyls and petroleum hydrocarbons. There is also interest in the role of rhizosphere organisms in polychlorinated biphenyl degradation, particularly since some plants exude phenolic compounds into the rhizosphere that can stimulate the aerobic degradation of the less chlorinated biphenyls.

Groundwater. A successful groundwater bioremediation of pentachlorophenol is being carried out at the Libby Superfund site described above. A shallow aquifer is present at 5.5 to 21 m below the surface, and a contaminant plume is nearly 1.6 km in length. Nutrients and hydrogen peroxide were added at the source area and approximately half way along the plume, and pentachlorophenol concentrations decreased from 420 ppm to 3 ppm where oxygen concentrations were successfully raised.

Pentachlorophenol is readily degraded in biofilm reactors, so bioremediation is a promising option for the treatment of contaminated groundwater brought to the surface as part of a pump-and-treat operation.

River and Pond Sediments. Much of the work on polychlorinated biphenyls has focused on the remediation of aquatic sediments, particularly from rivers, estuaries, and ponds. Harkness and co-workers have successfully stimulated aerobic biodegradation in large caissons in the Hudson River by adding inorganic nutrients, biphenyl, and hydrogen peroxide, but found that repeated addition of a polychlorinated-biphenyl degrading bacterium (*Alcaligenes eutrophus* H850) had no beneficial effect. Whether this approach can be scaled-up for large-scale use, with a net environmental benefit, remains to be seen.

Nonchlorinated Pesticides and Herbicides

Constituents. It is unusual for these compounds to become contaminants where they are applied correctly, but manufacturing facilities, storage depots and rural airfields where crop-dusters are based have had spills that can lead to long lasting contamination.

Biodegradation. The vast majority of pesticides, herbicides, fungicides, and insecticides in use today are biodegradable, although the intrinsic biodegradability of individual compounds is one of the variables used in deciding which compound to use for which task.

Compounds with organophosphate moieties, such as Diazinon, Methyl Parathion, Coumaphos and Glyphosate are usually hydrolyzed at the phosphorus atom. Indeed several *Flavobacterium* isolates are able to grow using parathion and diazinon as sole sources of carbon.

Very few pure cultures of microorganisms are able to degrade triazines such as Atrazine, although some *Pseudomonads* are able to use the compound as sole source of nitrogen in the presence of citrate or other simple carbon substrates. The initial reactions seem to be the removal of the ethyl or isopropyl substituents on the ring, followed by complete mineralization of the triazine ring.

Nitroaromatic compounds, such as Dinoseb, are degraded under aerobic and anaerobic conditions. The nitro group may be cleaved from the molecule as nitrite, or reduced to an amino group under either aerobic or anaerobic conditions. Alternatively, the ring may be the subject of reductive attack. Recent work has isolated a *Clostridium bifermentans* able to anaerobically degrade dinoseb cometabolically in the presence of a fermentable substrate. The dinoseb was degraded to below detectable levels, although only a small fraction was actually mineralized to CO_2.

Carbamates such as Aldicarb undergo degradation under both aerobic and anaerobic conditions.

Bioremediation

Groundwater. Atrazine dominated the world herbicide market in the 1980s, and contamination of groundwater has been reported in several locations in the U.S., Europe, and South Africa. Successful biodegradation has been achieved with indigenous organisms in laboratory mesocosms after a lag phase, and once activity was found, it remained. It is clear that intrinsic remediation is likely to lead to the disappearance of atrazine from groundwaters.

If more active treatment is required, such as pump and treat, it is possible that biological reactors will be a cost-effective replacement for activated carbon filters.

Marsh and Pond Sediments. Herbicides and pesticides are detectable in marsh and pond sediments, but intrinsic biodegradation is usually found to be occurring.

Soil. Herbicides and pesticides are of course metabolized in the soil to which they are applied, and there are many reports of isolating degrading organisms from such sites. Little work has yet been presented where the biodegradation of these compounds has been successfully stimulated by a bioremediation approach.

It is a general observation that herbicide degradation occurs more readily in cultivated than fallow soil, suggesting that rhizosphere organisms are effective herbicide degraders. Whether this can be effectively exploited in a phytoremediation strategy remains to be seen.

Military Chemicals

Constituents. The military use a range of chemicals such as explosives and propellants that are sometimes termed "energetic molecules." Generally speaking, modern explosives are cyclic, often heterocyclic, composed of carbon, nitrogen and oxygen, eg; 2,4,6-trinitrotoluene; RDX (Royal Demolition eXplosive; hexahydro-1,3,5-trinitro-1,2,3-triazine); HMX (High Melting eXplosive; octahydro-1,3,5,7-tetranitro-1,3,5,7-tetrazocine); and N,N-dimethylhydrazine. These compounds are sometimes present at quite high levels in soils and groundwater on military bases and production sites. Bioremediation is a promising new technology for treating sites contaminated with such compounds.

Bioremediation may also be an appropriate tool for dealing with chemical agents such as the mustards and organophosphate neurotoxins, but little work on actual bioremediation has been published.

Biodegradation. Nitrosubstituted compounds are subject to a variety of degradative processes. Under anaerobic conditions TNT is readily reduced to the corresponding aromatic amines and subsequently deaminated to toluene. As shown in the section on hydrocarbons, the latter can be mineralized under anaerobic conditions, leading to the potentially complete mineralization of TNT in the absence of oxygen.

Under aerobic conditions TNT can be mineralized by a range of bacteria and fungi, often co-metabolically with the degradation of a more degradable substrate. There is even evidence that some plants are able to deaminate TNT reductively.

RDX and HMX are rather more recalcitrant, especially under aerobic conditions, but there are promising indications that biodegradation can occur under some conditions, especially composting.

Little work has been reported on the biodegradation of dimethylhydrazine.

Bioremediation

Groundwater. Nitrotoluenes have been detected in groundwater in some areas, and intrinsic remediation may be occurring at some sites by anaerobic degradation.

A commercial technology, the SABRE process, treats contaminated water and soil in a two-stage process by adding a readily degradable carbon and an inoculum of anaerobic bacteria able to degrade the contaminant. An initial aerobic fermentation removes oxygen so that the subsequent reduction of the contaminant is not accompanied by oxidative polymerization.

Soil. Composting of soils contaminated by high explosives is being carried out at the Umatilla Army Depot near Hermiston, Oregon. If this is successful, there are 30 similar sites on the National Priority List that could be treated in a similar way.

Other Organic Compounds. The majority of organic compounds in commerce are biodegradable, so bioremediation is a potential option for cleaning up after industrial and transportation accidents.

Inorganic Contaminants

Nitrogen Compounds

Constituents. Nitrate levels are regulated in groundwater because of concerns for human and animal health. Ammonia is regulated in streams and effluents as a potential fish toxicant, and any nitrogenous contaminant is a potential problem in water because of its stimulatory effect on the growth of algae. Other nitrogenous contaminants include cyanides in mine waters. Fortunately, all are amenable to biological treatment.

Biodegradation. The biological mineralization of fixed nitrogen is well studied; ammonia is oxidized to nitrite, and nitrite to nitrate, by autotrophic bacteria, and nitrate is reduced to nitrogen by anaerobic bacteria. On the other hand, ammonia and nitrate are essential nutrients for plant and bacterial growth, so one option is to use these organisms to take up and use the contaminants.

Bioremediation

Surface Water. One example of exploiting biology to handle excess nitrogen in a surface water is the case of the Venice Lagoon in Italy. About a million tons of sea lettuce (*Ulva*) grows in the lagoon annually because of the high levels of nitrogenous nutrients in this relatively landlocked bay. This material is harvested, composted and sold as a low-cost remediation of this problem.

A more constrained opportunity for nitrate bioremediation arose at the U.S.-DoE Weldon Spring Site near St. Louis, Missouri which had been a uranium and thorium processing facility. Two pits had nitrate levels that required treatment before discharge. Bioremediation by the addition of calcium acetate as a carbon source successfully treated more than 19 million liters of water at a reasonable cost.

Groundwater. One approach to minimizing the environmental impact of excess nitrogen in groundwater migrating into rivers and aquifers is to intercept the water with rapidly growing trees, such as poplars, that will use the contaminant as a fertilizer.

An alternative approach is to add a readily degradable substrate to the contaminated aquifer, in the absence of oxygen, to stimulate bacterial denitrification.

Metals and Metalloids. A wide range of metallic and nonmetallic elements are present as contaminants at industrial and agricultural sites throughout the world, both in ground and surface water, and in soils. They pose a quite different problem from that of organic contaminants, since they cannot be degraded so that they disappear. Some metal and metalloid elements have radically different bioavailabilities and toxicities depending on their redox state, so one option is stabilization by converting them to their least toxic form. This can be a very effective way of minimizing the environmental impact of a contaminant, but if the contaminant is not removed from the environment, there is always the possibility that natural processes, biological or abiological, may reverse the process. Removing the contaminants from water phases is relatively straightforward, and the wastewater treatment industry practices this on an enormous scale. Pump-and-treat systems that mimic wastewater treatments are already being used for several contaminants, and less complex systems involving biological mats are a promising solution for less demanding situations.

In the past, removing metal and metalloid contaminants from soil has been impossible, and site clean-up has meant excavation and disposal in a secure landfill. An exciting new approach to this problem is phytoextraction, where plants are used to extract contaminants from the soil and harvested.

Bioremediation

Water. Groundwater can be treated in anaerobic bioreactors that encourage the growth of sulfate reducing bacteria, where the metals are reduced to insoluble sulfides, and concentrated in the sludge.

Phytoremediation is not yet being used commercially, but results at several field trial suggest that commercialization is not far away. Perhaps the biggest success to date is the successful rhizofiltration of radionuclides from a Department of Energy site at Ashtabula, Ohio, where uranium concentrations of 350 ppb were reduced to less than 5 ppb, well below groundwater standards, by Sunflower roots.

Mine Drainage. In recent years it has become clear that the environmental impact of acid mine drainage can be minimized by the construction of artificial wetlands that combine geochemistry and biological treatments. These systems are being designed for a range of wastewaters, most of which fall outside the scope of this article.

Soil. The results of the first reported field trial of the use of hyperaccumulating plants to remove metals from a soil contaminated by sludge applications were positive. However, the rates of metal uptake suggest a time scale of decades for complete cleanup. Trials with higher biomass plants, such as *B. juncea*, are underway.

The bacterial reduction of Cr(VI) to Cr(III) is also being used to reduce the hazards of chromium in soils and water.

Conclusions

Bioremediation has many advantages over other technologies, both in cost and in effectively destroying or extracting the pollutant. An important issue is thus when to consider it, and a series of questions may lead to the appropriate answer (Table 3).

If the answers to the questions in Table 3 lead to the selection of bioremediation, it then becomes important to assess the success of the bioremediation strategy in achieving the cleanup criteria. A major disadvantage of bioremediation is that it is typically rather slower than competing technologies such as thermal treatments. How can regulators and responsible parties gain confidence during this time that success will indeed be achieved? The National Research Council has recently addressed this issue, and suggested a three-fold strategy for "proving" bioremediation: (*1*) a documented loss of contaminants from the site; (*2*) laboratory tests showing the potential of endogenous microbes to catalyze the reactions of interest; and (*3*) some evidence that this potential is achieved in the field.

Finally a caveat. Despite its documented success in many situations, bioremediation may not always be able to meet current clean up criteria for a particular site. Some standards are so tight that they are essentially "detection limit" standards, and it is not clear that biological processes will be able to remove contaminants to such low levels. Bioremediation will be

TABLE 3. WILL BIOREMEDIATION BE A SUITABLE TREATMENT FOR A SITE CONTAMINATED WITH ORGANIC, NITROGENOUS, OR ORGANIC CONTAMINANTS?

ORGANIC

Is the contaminant biodegradable? If the contaminant is a complex mixture of components, are the individual chemical species biodegradable? If the contaminant has been at the site for some time, biodegradation of the most readily degradable components may have already occurred. Is the residual contamination biodegradable?

Are degrading organisms present at the site?

What is limiting their growth and activity? Can this be added effectively?

Are the levels of contaminant amenable to bioremediation? Are they toxic to microorganisms? Are they so abundant that even substantial microbial activity will take too long to clean the site?

Are the clean-up standards reasonable? Are biological processes known to degrade substrates down to the levels required?

NITROGENOUS

Are appropriate microorganisms present at the site?

What is limiting their growth and activity? Can this be added effectively?

Are the levels of contaminant amenable to bioremediation? Are they toxic to microorganisms? Are they so abundant that even substantial microbial activity will take too long to clean the site?

Can the nitrogenous compound be used by plants?

Are the clean-up standards reasonable? Are biological processes known to degrade substrates down to the levels required?

INORGANIC

Can the contaminant be made less hazardous by changing its redox state?

Can the contaminant be brought to a reactor or constructed wetland where biological systems, microbial or plant, can extract and immobilize the contaminant?

Can plants extract the contaminant from the soil matrix?

Are the clean-up standards reasonable? Are biological processes known to accumulate contaminants down to the levels required?

more likely to fulfill its promise as an important tool in contaminated site remediation if there is progress towards standards based on bioavailability and net environmental benefit from the clean up, rather than on arbitrary absolute standards.

ROGER C. PRINCE
Exxon Research and Engineering Company

Additional Reading

Alexander, M.A.: *Biodegradation and Bioremediation*, Academic Press, New York, 1994.

Atlas, R.M. and T. Hazen: *Bioremediation*, ASM Press, Washington, DC, 2004.

Bioremediation; when does it work?, National Research Council, National Academy Press, 1993.

Crawford, R.L. and D.L.: *Bioremediation: Principles and Practice*, Cambridge University Press, 1996.

Evans, G.M., and J.C. Furlong: *Environmental Biotechnology: Theory and Application*, John Wiley & Sons, Inc., New York, NY, 2002.

Singh, A., and O.P. Ward: *Applied Bioremediation and Phytoremediation*, Springer-Verlag New York, LLC., New York, NY, 2004.

Young, L.Y. and C.E. Cerniglia, eds: *Microbial Transformation and Degradation of Toxic Organic Compounds*, Wiley-Liss, New York, NY, 1995.

BIOSENSORS. The detection of trace and low levels of biologically active substances is among the most significant and challenging analytic technologies. See also **Trace and Residue Analysis.** Any discrete sensing device that relies on a biologically derived component as an integral part of its detection mechanism is known as a biosensor. See also **Biopolymers.**

Basic Components

All biosensors are composed of two basic parts, the molecular recognition component and the transducer component. The molecular recognition component is typically a complex chemical system usually extracted or derived directly from a biological organism. The function of the molecular recognition component is to interact specifically with a target compound, ie, the chemical to be detected. The molecular recognition component must be capable of discerning the presence of a target in a solution containing possibly hundreds of other compounds, some of which may have molecular structures closely resembling that of the target. The term selectivity is used to quantify the degree to which a molecular recognition system responds specifically to a target while rejecting compounds having related molecular structures. There are three principal classes of molecular recognition components used in biosensors: enzymes or catalytic proteins; immunoglobins, which are biological macromolecules that selectively bind to foreign substances invading an organism; and chemoreceptors, biomolecular units responsible for sensory reception in living organisms. Each class has advantages and disadvantages, and each class comprises a vast number of highly selective possible molecular recognition components.

Once the molecular recognition component interacts with its target, a user-readable signal must be generated. This is the purpose of the transducer component. Ideally, the generated signal should not only report the presence of target but should also relay information concerning the amount or concentration of target compound in the test solution. As of this writing, there were four main types of analytical transducer components employed in biosensors: optical, electrochemical, field effect transistor, and piezoelectric.

The term sensitivity is used to quantify the ability of a biosensor to reliably report target level. There are two separate uses of the term. First, the detection limit of a biosensor is the lowest concentration of target for which the biosensor provides a reliable, reproducible, and unambiguous response. Increasing the sensitivity of a biosensor frequently refers to decreasing the biosensor's detection limit. Second, sensitivity can also refer to the smallest change in concentration from some nominal level of target compound that the biosensor can unambiguously report. Increasing sensitivity in these latter terms is usually a more difficult task than decreasing the detection limit.

Biosensor design involves consideration of many factors. Molecular recognition components are frequently large complex biomolecules or biological macroassemblies and usually perform efficiently within narrow ranges of pH, temperature, and ionic strength. Degradation and the limited lifetimes of biologically derived components can be important. Additionally, immobilization of enzymes sometimes results in decreased target activity. Factors influencing the selection of an appropriate transducer component include compatibility with the molecular recognition component and the environment in which the biosensor is to be used. Development and optimization of biosensor techniques is a rapidly growing research area; as of the early 1990s, publications on fundamental and applied aspects of biosensing were appearing at a rate approaching 400 papers per year.

Factors Affecting Biosensor Detection Limits

Biosensor designers are striving for lower detection limits, higher sensitivity, greater selectivity, and lower occurrences of false-positive signals. The detection limit for a biosensor is related to the nature of both the molecular recognition component and the transducer component. In general, antibodies and chemoreceptors are well-suited for low detection limit biosensors because of the high binding constants for target molecules.

In general, low level detection is masked by the noise level inherent in any measuring device. Electrochemical methods are susceptible to electrical interference from external sources, variations in reference electrode parameters resulting from aging or contamination, and interference from redox contaminants in the test solution.

An ingenious method of decreasing detection limits in biosensors that is frequently employed is enzyme amplification. An example is enzyme multiplied immunoassay technology (emit).

WILLIAM PIETRO
York University

Additional Reading

Cooper, J. and T. Cass: *Biosensors,* 2nd Edition, Oxford University Press, New York, NY, 2004.

Diamond, D.: *Principles of Chemical and Biological Sensors,* John Wiley & Sons, Inc., New York, NY, 1998.

Eggins, B. R.: *Chemical Sensors and Biosensors,* John Wiley & Sons, Inc., New York, NY, 2002.

Heller, A. and Y. Degani: in G. Dryhurst and K. Niki, eds., *Redox Chemistry and Interfacial Behavior of Biological Molecules,* Plenum Publishing Corp., New York, NY, 1988, pp. 151–170.

Recent Advances in the Development and Analytical Application of Biosensing Probes, Vol. 20, Chemical Rubber Co. Press, Boca Raton, FL. 1988.

Turner, A. P. F., I. Karube, and G. S. Wilson, eds.: *Biosensors: Fundamentals and Applications,* Oxford University Press, New York, NY, 1987.

BIOTECHNOLOGY (Bioprocess Engineering).   Biotechnology is the application of biological systems to the manufacture of products ranging from food-grade sweeteners and fuel alcohol to recombinant proteins for treatment of heart disease and cancer. The application of biotechnology also encompasses the use of chemicals to modify the behavior of biological systems, the genetic modification of organisms to confer new traits, and the science by which foreign DNA may be inserted into people to compensate for genes whose absence cause life-threatening conditions. Bioprocess engineering, the subject of this article, translates biotechnology into unit operations, biochemical processes, equipment and facilities for manufacturing bioproducts.

An array of engineering fundamentals is required to translate the discoveries of biotechnology into tangible commercial products, thereby putting biotechnology to work. This article presents the principles of the life sciences and engineering, the practice of key biotechnology manufacturing techniques, and the economic characteristics of the industries and manufacturing processes that encompass:

(1) Production of human and animal health-care products, food products, biologically active proteins, and chemicals.
(2) Design and scale-up of bioreactors to generate large quantities of transformed microbes or cells and the products which they generate.
(3) Separation and purification of biological molecules.

The principles and practice of bioprocess engineering are based in the biological sciences. The key technologies based on the biosciences are

(1) Identification of genes, and the products which result from them, for purposes of disease prevention, remediation, and development of new medicines.
(2) Application of molecular biology to obtain transformed microorganisms, cells, or animals having new and/or enhanced capabilities to generate bioproducts.
(3) Development of biological sensors coupled to microprocessors or computers for process control and monitoring of biological systems (including humans).

The pioneers of the biotechnology industry had a grasp not only of the science, but also an understanding of the financial aspects of taking a new technology from test tube to market in less than seven years. In the late 1970s and early 1980s, the technical success of the insulin project, and the apparent availability of venture capital for risky enterprises, converged to promote the industry. Venture capital was available in part due to government policy that promoted limited partnerships and relieved taxes on capital gains, although the laws have since changed. Some of the pioneers also had a sense that the large scale production, clinical testing, obtaining government regulatory approval, and marketing, required the infrastructure of a large pharmaceutical company. The first recombinant product approved by the FDA, human insulin, resulted from the cooperation of the biotechnology company, Genentech, and the pharmaceutical company, Eli Lilly. Genentech ultimately merged with another major international pharmaceutical company, Hoffmann-LaRoche, in 1990 (see timeline in Table 1).

Although the industry grew rapidly, profits were slow in coming. This is not surprising, given the seven to ten year time required to bring a new product to market. Financial success for some of the companies has occurred as the twenty-fifth birthday of the new biotechnology field is approached. Some companies have consolidated when faced with large costs of clinical trials, regulatory approval, and marketing of biotechnology derived products. In the meantime, the realization has grown that there are other products amenable to bioprocessing approaches. These products include polypeptides (animal growth hormones), amino acids, enzymes, and food products. As the price categories and profit margins of the products decrease, there is an increased need for bioprocess engineering to design efficient and cost-effective production systems.

TABLE 1. TIME-LINE OF MAJOR DEVELOPMENTS IN THE BIOTECHNOLOGY INDUSTRIES[a]

Date	Event
6000 B.C.	Bread making (involves yeast fermentation).
3000 B.C.	Moldy soybean curd used to treat skin infections in China.
2500 B.C.	Malting of barley, fermentation of beer in Egypt.
1857 A.D.	Pasteur proved yeasts are living cells that cause alcohol fermentation.
	Birth of microbiology.
1877	Pasteur and Joubert discovered some bacteria can kill anthrax bacilli.
1896	Gosio discovered mycophenolic acid; an antibacterial substance produced by microbes (too toxic for use as antibiotic).
1902	*Bacillus thuringiensis* first isolated from silk worm culture by Ishiwata.
1908	Ikeda identified MSG as flavor enhancer in Konbu.
	Invertase absorbed onto charcoal, i.e., first example of immobilized enzyme.
1909	Ajinomoto (Japan) initiates commercial production of sodium glutamate from wheat gluten and soybean hydrolysates.
1900–1920	Ethanol, glycerol, acetone, and butanol produced by fermentation.
1922	Banting and Best treated human with insulin obtained from animal pancreatic extractions.
1923	Citric acid fermentation plant using *Aspergillus niger* by Charles Pfizer.
1928	Alexander Fleming discovered penicillin from *Penicillum notatum*.
1943	Submerged culture of *P. chrysogenum* opened way for large-scale production of penicillin.
1945	Production through fermentation process scaled up to make enough penicillin to treat 100,000 patients per year.
	Beginning of rapid development of antibiotic industry. During World War II, research driven by 85% tax on "excess" profits, encouraged investment in research and development of antibiotics this led to their postwar growth.
1953	DNA structure and function elucidated.
	Xylose isomerase discovered.
1957	Commercial production of natural (L) amino acids via fermentation facilitated the discovery of *Micrococcus glutamicus* (later renamed *Corynebacterium glutamicum*).
	Glucose isomerizing capability of xylose isomerase reported.
1960	Lysine produced on a technical scale.
1961	First commercial production of MSG via fermentation.
1965	Corn bran and hull replaced xylose as inducer of glucose (xylose) isomerase in *Streptomyces phaeochromogenius*.
	Phenyl methyl ester of aspartic acid and phenylalanine (Aspartame) synthesized at G.D. Searle Co.
1967	Clinton Corn Processing ships first enzymatically produced fructose syrup.
1970	Smith et al., reported restriction endonuclease from *Haemophilis influenzae* that recognizes specific DNA target sequences.
1971	Cetus founded.
1973	Cohen and Boyer reported genetic engineering technique (EcoRI enzyme).
	Aspartase in immobilized *E. coli* cells catalyzes L-aspartic acid production from fumarate and ammonia.
1973–1974	Oil price increase (Yom Kippur war).
	HFCS market about 500–600 $\times 10^6$ lb/yr in U.S. sugar price peaked at 30/lb.
1975	Kohler and Milstein reported monoclonal antibodies.
	Basic patent coverage for xylose (glucose) isomerase lost in lawsuit; opened up development of new HFCS processes.
1976	Genentech, Inc. was founded in by venture capitalist Robert A. Swanson and biochemist Dr. Herbert W. Boyer. In the early 1970s, Boyer and geneticist Stanley Cohen pioneered a new scientific field called recombinant DNA technology.
1977	Genentech, Inc., reported the production of the first human protein manufactured in a bacteria: somatostatin, a human growth hormone-releasing inhibitory factor. For the first time, a synthetic, recombinant gene was used to clone a protein. Many consider this to be the advent of the Age of Biotechnology.

Date	Event
	Sixteen bills introduced in Congress to regulate recombinant DNA research. The bills called for the development of bacteria and plasmids that could be prevented from escaping the laboratory environment. None of these bills passed.
	Bill Rutter and Howard Goodman isolated the gene for rat insulin.
	Walter Gilbert and Allan Maxam at Harvard University devised a method for sequencing DNA using chemicals rather than enzymes.
1978	Genentech, Inc. and The City of Hope National Medical Center announced the successful laboratory production of human insulin using recombinant DNA technology.
	Harvard researchers used genetic engineering techniques to produce rat insulin.
	Studies by David Botstein and others found that when a restrictive enzyme is applied to DNA from different individuals, the resulting sets of fragments sometimes differ markedly from one person to the next. Such variations in DNA are called restriction fragment length polymorphisms, or RFLPs, and they are extremely useful in genetic studies.
	William J. Rutter's laboratory at UCSF cloned a coat protein of the virus that causes hepatitis B.
	Biogen formed, developed interferons.
	Eli Lilly licensed recombinant insulin technology from Genentech.
1979	3.5 $\times 10^9$ lb HFCS produced in U.S.
	John Baxter reported cloning the gene for human growth hormone.
	Another major oil price increase (OPEC). Sugar price at 12/lb.
	Energy saving method for drying ethanol using corn (starch) and cellulose based adsorbents reported.
1980	The U.S. Supreme Court ruled in that genetically altered life forms can be patented a Supreme Court decision in 1980 allowed the Exxon oil company to patent an oil-eating microorganism. This ruling opened up enormous possibilities for commercially exploiting genetic engineering.
	Researchers successfully introduced a human gene-one that codes for the protein interferon-into a bacterium.
	Kary Mullis and others at Cetus Corporation in Berkeley, California, invented a technique for multiplying DNA sequences *in vitro* by, the polymerase chain reaction (PCR). PCR has been called the most revolutionary new technique in molecular biology in the 1980s. Cetus patented the process, and in the summer of 1991 sold the patent to Hoffmann-La Roche, Inc. for $300 million.
	Amgen founded.
1981	Genentech, Inc. cloned interferon gamma.
	Bill Rutter and Pablo Valenzuela published a report in Nature on a yeast expression system to produce the hepatitis B surface antigen.
	Scientists at Ohio University produced the first transgenic animals by transferring genes from other animals into mice.
	Mary Harper and two colleagues mapped the gene for insulin. Mapping by *in situ* hybridization became a standard method.
	Hoechst AG, a West German chemical company, gave Massachusetts General Hospital, a teaching facility of Harvard Medical School, $70 $\times 10^6$ to build a new Department of Molecular Biology in return for exclusive rights to any patent licenses that might emerge from the facility.
	Congressman Al Gore held a series of hearings on the relationship between academia and commercialization in the arena of biomedical research. He focused on the effect that the potential for huge profits from intellectual property and patent rights could have on the research environment at universities. Jonathan King, a professor at MIT speaking at the Gore hearings, reminded the biotech industry that "the most important long-term goal of biomedical research is to discover the causes of disease in order to prevent disease."

TABLE 1. (*continued*)

Date	Event
	AIDS cases identified and reported in San Francisco.
	Aspartame approved for food use by FDA.
	Chiron founded.
	Gene synthesizing machines developed.
	Sugar price at 30/lb.
1982	Genentech, Inc. received approval from the Food and Drug Administration to market genetically engineered human insulin. 1982 The U.S. Food and Drug Administration approves the first genetically engineered drug, a form of human insulin produced by bacteria.
	Applied Biosystems, Inc. introduced the first commercial gas phase protein sequencer, dramatically reducing the amount of protein sample needed for sequencing.
	Lindow requested government permission to test genetically engineered bacteria to control frost damage to potatoes and strawberries.
	Michael Smith at the University of British Columbia, Vancouver, developed a procedure for making precise amino acid changes anywhere in a protein.
	First transgenic mouse, rat gene transferred to mouse.
1983	Eli Lilly received a license to make insulin.
	Syntex Corporation received FDA approval for a monoclonal antibody-based diagnostic test for Chlamydia trachomatis.
	Stanford Research Institute International filed for a patent for an *E. coli* expression vector.
	Jay Levy's laboratory at UCSF isolated the AIDS virus at almost the same moment it was isolated at the Pasteur Institute in Paris and at the NIH.
	U.S. patents were granted to companies genetically engineering plants.
	A study of an extended family in Venezuela with Huntington's chorea demonstrated that family members with the disease show a distinct and characteristic pattern of restriction fragment lengths, leading to a new screening test. The same methods of investigation revealed patterns for cystic fibrosis, adult polycystic kidney disease, Duchenne muscular dystrophy, and others.
	Marvin Carruthers at the University of Colorado devised a method to construct fragments of DNA of predetermined sequence from five to about 75 base pairs long. He and Leroy Hood at the California Institute of Technology invented instruments that could make such fragments automatically.
	Aspartame came on market sold by G.D. Searle as Nutrasweet®.
	Worldwide antibiotic sales at about $8 billion.
	First product sales of recombinant insulin.
	HIV virus identified as cause of AIDS
1984	Cal Bio scientists described in *Nature* the isolation of a gene for anaritide acetate, which helps to regulate blood pressure and control salt and water excretion.
	The Plant Gene Expression Center was established.
	Stanford University received a product patent for prokaryote DNA.
	Chiron Corp. announced the first cloning and sequencing of the entire human immunodeficiency virus (HIV) genome.
	Charles Cantor and David Schwartz developed pulsed-field gel electrophoresis.
	British geneticist Alec Jeffreys introduced technique for DNA fingerprinting to identify individuals.
	Oil at $40/barrel.
	Process for industrial drying of fuel alcohol using a corn based adsorbent in place of azeotropic distillation demonstrated on an industrial scale.
	Transgenic pig, rabbit, and sheep by micro injection of foreign DNA into egg nuclei.
	PRC developed at Cetus.
1985	Genetic fingerprinting entered the courtroom.
	Axel Ullrich reported the sequencing of the human insulin receptor in Nature. Bill Rutter's UCSF team described the sequencing in Cell two months later.
	Cal Bio cloned the gene that encodes human lung surfactant protein, a major step toward reducing a premature birth complication.

Date	Event
	Cetus Corporation's developed GeneAmp polymerase chain reaction (PCR) technology, which could generate billions of copies of a targeted gene sequence in only hours.
	Genetically engineered plants resistant to insects, viruses, and bacteria were field tested for the first time.
	The NIH approved guidelines for performing experiments in gene therapy on humans.
	Genetic Sciences surreptitiously performed the first deliberate release experiment, injecting genetically engineered microbes into trees growing on the company's roof, while waiting for approval from the EPA to conduct a different deliberate release experiment involving strawberry plants.
1986	UC Berkeley chemist Peter Schultz described how to combine antibodies and enzymes (creating "abzymes") to create pharmaceuticals.
	Orthoclone OKT3® (Muromonab-CD3) approved for reversal of acute kidney transplant rejection.
	A regiment of scientists and technicians at Caltech and Applied Biosystems, Inc., invented the automated DNA fluorescence sequencer.
	The FDA granted a license for the first recombinant vaccine (for hepatitis) to Chiron Corp.
	The EPA approved the release of the first genetically engineered crop, gene-altered tobacco plants.
	Phase down of lead as octane booster in gasoline in U.S. Created demand for nonleaded octane booster for liquid fuels.
	Ethanol production at 500×10^6 gallon/year for use as octane booster.
	Merck licensed Chiron's recombinant hepatitis B vaccine.
1987	Genentech received FDA approval to market rt-PA (genetically engineered tissue plasminogen activator) to treat heart attacks.
	Calgene, Inc. received a patent for the tomato polygalacturonase DNA sequence, used to produce an antisense RNA sequence that can extend the shelf-life of fruit.
	Advanced Genetic Sciences, Inc. conducted a field trial of a recombinant organism, a frost inhibitor, on a Contra Costa County strawberry patch.
	Maynard Olson and colleagues at Washington University invented "yeast artificial chromosomes," or YACs, expression vectors for large proteins.
	Recombivax-HB® (recombinant hepatitis B vaccine) approved.
	Interleukin (IL-2) by Cetus undergoes clinical trials.
	Aspartame sales at $500 million.
	Oil at $15 to $20/barrel.
1988	Harvard molecular geneticists Philip Leder and Timothy Stewart awarded the first patent for a genetically altered animal, a mouse that is highly susceptible to breast cancer.
	SyStemix Inc. received a patent for the SCIDHU Mouse, an immune-deficient mouse with a reconstituted human immune system. The mouse was engineered for AIDS research.
	Genencor International, Inc. received a patent for a process to make bleach-resistant protease enzymes to use in detergents.
	Hoffmann-La Roche, Inc. and Cetus Corp. negotiated a licensing agreement for two anti-cancer drugs, Interleukin-2 and Polyethylene Glycol Modified IL-2. This agreement became the prototype for cross-licensing between companies with parallel patents.
	Tissue plasminogen activator (TPA) introduced by Genetech.
	Human genome project started
	Ethanol production at 800×10^6 gallons/year.
	Cetus requested approval for IL-2 to treat advanced kidney cancer
1989	UC Davis scientists developed a recombinant vaccine against the deadly rinderpest virus, which had wiped out millions of cattle in developing countries.
	Creation of the National Center for Human Genome Research, headed by James Watson, which will oversee the $3 billion U.S. effort to map and sequence all human DNA by 2005.

(continued overleaf)

TABLE 1. (*continued*)

Date	Event
	Epogen® (Epoetin alfa) a genetically engineered protein introduced, providing a means to help patients with kidney failure.
	American Home Products purchased A.H. Robbins.
	Amgen introduced EPO (produced in 2-L roller bottles).
	Merrill Dow combined with Marion Laboratories for $7.7 billion.
	Bristol Myers and Squibb merged ($12 billion).
1990	UCSF and Stanford University were issued their 100th recombinant DNA patent license. By the end of fiscal 1991, both campuses had earned $40 $\times 10^6$ from the patent.
	Actimmune® (Interferon gamma-1b) approved for treatment of chronic granulomatous disease.
	Adagen® (adenosine deaminase) approved for treatment of severe combined immunodeficiency disease.
	The first successful field trial of genetically engineered cotton plants, was conducted by Calgene Inc. The plants had been engineered to withstand use of the herbicide Bromoxynil.
	The FDA licensed Chiron's hepatitis C antibody test to help ensure the purity of blood bank products.
	Michael Fromm, molecular biologist at the Plant Gene Expression Center, reported the stable transformation of corn using a high-speed gene gun.
	Mary Claire King, epidemiologist at UC-Berkeley, reported the discovery of the gene linked to breast cancer in families with a high degree of incidence before age 45.
	GenPharm International, Inc. created the first transgenic dairy cow. The cow was used to produce human milk proteins for infant formula.
	The first gene therapy took place, on a four-year-old girl with an immune-system disorder called ADA deficiency. The therapy appeared to work, but set off a fury of discussion of ethics both in academia and in the media.
	The Human Genome Project, the international effort to map all of the genes in the human body, was launched. Estimated cost: $13 $\times 19^9$. 1990 Formal launch of the international Human Genome Project.
	FDA rejected IL-2 application of Cetus.
	Hoffmann-LaRoche acquired 60% of Genetech for $2.1 $\times 10^6$.
	Genetech's TPA earned $210 $\times 10^6$.
	Protropin® (human growth hormone) earned $157 $\times 10^6$.
	First attempt at human gene therapy.
	Amgen's EPO sales at $300 $\times 10^6$/yr, in U.S. Licensed by Kilag (Johnson and Johnson in Europe) and Kirin (Japan).
	HFCS world sales estimated at about $3 $\times 10^9$ (17 $\times 10^9$ lb).
1991	The celebrated reference work "Mendelian Inheritance in Man," was made available through an on-line computer network. The catalogue lists some 5,600 genes known or thought on good evidence to be inherited in Mendelian patterns. http://www3.ncbi.nlm.nih.gov/Omim/
	Analyzing chromosomes from women in cancer-prone families, Mary-Claire King, of the University of California, Berkeley, found evidence that a gene on chromosome 17 causes the inherited form of breast cancer, and also increases the risk of ovarian cancer.
	Amgen EPO sales exceeded $293 $\times 10^6$ by August 1991.
	Genetics Institute lawsuit against Amgen for American rights to EPO.
	American Home Products bought $666 $\times 10^6$ (60%) share in Genetics Institute.
	Chiron purchased Cetus for $650 $\times 10^6$.
1992	The U.S. Army began collecting blood and tissue samples from all new recruits as part of a "genetic dog tag" program aimed at better identification of soldiers killed in combat.
	American and British scientists unveiled a technique for testing embryos in vitro for genetic abnormalities such as cystic fibrosis and hemophilia.
	IL-2 (now owned by Chiron) approved for further testing by FDA.
	TPA (from Genentech) under competitive pressure from less expensive product by Swedish Kabi.
	Policy guidelines for the agricultural biotechnology established.

Date	Event
1993	Kary Mullis won the Nobel Prize in Chemistry for inventing the technology of polymerase chain reaction (PCR).
	Chiron's Betaseron (a β-interferon) was approved as the first treatment for multiple sclerosis in 20 years. Drug marketed by Berlex (owned by Schering AG).
	The FDA declared that genetically engineered foods are "not inherently dangerous" and do not require special regulation.
	The Biotechnology Industry Organization was created by merging two smaller trade associations.
	George Washington University researchers cloned human embryos and nurtured them in a Petri dish for several days. The project provoked protests from ethicists, politicians, and critics of genetic engineering.
	An international research team, led by Daniel Cohen, of the Center for the Study of Human Polymorphisms in Paris, produced a rough map of all 23 pairs of human chromosomes.
	Genentech launched Access Excellence, a $10 $\times 10^6$ nationwide communications network program designed to enable high school biology teachers across the country to access their peers as well as experts.
	Health care reform proposals created uncertainty in biotechnology industry.
	Merck acquired Medco containment services for $6.6 $\times 10^9$.
1994	The first genetically engineered food product, the Flavr Savr tomato, gained FDA approval.
	The BRCA1 gene, previously implicated in the development of rare familial forms of breast cancer, also found to appear to play a role in much more common types of noninherited breast cancers.
	A multitude of genes, human and otherwise, were identified and their functions described. These included:
	• Ob, a gene predisposing to obesity.
	• BCR, a breast cancer susceptibility gene.
	• BCL-2, a gene associated with apoptosis (programmed cell death) hedgehog genes (so named because of their shape, these produce proteins which guide cell differentiation in advanced organisms).
	• Vpr, a gene governing reproduction of the HIV virus.
	Linkage studies identified genes for a variety of ailments including: bipolar disorder, cerulean cataracts, melanoma, hearing loss, dyslexia, thyroid cancer, sudden infant death syndrome, prostate cancer, and dwarfism.
	Genetic researchers successfully transferred the CFTR (cystic fibrosis transmembrane conductance regulator) gene into the intestines of mice. This appeared to be a major step towards gene therapy for patients with cystic fibrosis. Researchers reported early success with a liposomal method for delivering the CFTR gene in humans.
	The approval of genetically engineered version of human DNAase, which breaks down protein accumulation in the lungs of CF patients. It was the first new therapeutic drug for the management of cystic fibrosis in over 30 years.
	Another group of researchers reported the first successful systemic selective inhibition of gene expression using antisense oligonucleotides.
	Centocor's ReoPro cleared for marketing in United States by the FDA and by the European Union's regulatory body, CPMP for patients undergoing high-risk balloon angioplasty.
	Genzyme's Ceredase®/Cerezyme® (alglucerase/recombinant alglucerase) approved for type 1 Gaucher's disease.
	The first crude but thorough linkage map of the human genome appeared (see *Science*, **265**, (Sept.30, '1994), for the full color pull-out).
	The year also saw an increase in squabbling over who owns what parts of the genome. The scientists and research corporations have worked out a way to share access to a computerized database detailing 35,000 human genes.

TABLE 1. (*continued*)

Date	Event
	Researchers at the University of Texas reported that the enzyme telomerase appears to be responsible for the unchecked growth of cells seen in human cancers. The discovery could lead to many new diagnostic and therapeutic applications.
	Recombinant GM-CSF approved for chemotherapy-induced neutropenia.
	American Home Products bought American Cyanimid.
	Roche Holding, parent of Swiss Drug company Hoffmann-LaRoche bought Syntes for $5.3 billion.
	SmithKline Beecham merged with Diversified Pharmaceutical Services ($2.3 billion) and Sterling Winthrop ($2.9 $\times 10^9$).
	Eli Lilly acquired PCS Health Systems for $4.1 $\times 10^9$.
	Bayer purchased Sterling Winthrop NA for $1.0 $\times 10^9$.
	Based on corn adsorbent dries half of fermentation ethanol in U.S. (750 $\times 10^6$ gallons/yr).
1995	A European research team identified a genetic defect which appears to underlie the most common cause of deafness Kok et al. *Science*, **267**, (Feb. 3, 1995).
	Researchers at Duke University Medical Center transplanted hearts from genetically altered pigs into baboons, proving that cross-species operations are possible. Later, the first baboon-to-human bone marrow transplant was performed on an AIDS patient.
	The first full gene sequence of a living organism other than a virus was completed for the bacterium Hemophilus influenzae. See: Smith et al., *Science* **269**, 495–511 (July 28, 1995): and associated commentary, pp.468–470.
	Former football player O.J. Simpson was found not guilty in a high-profile double-murder trial in which PCR and DNA fingerprinting played a prominent, but apparently unpersuasive role.
	Investigators at the Centers for Disease Control and Prevention confirmed that the Ebola virus is behind outbreak of hemorrhagic fever in Zaire.
	Hitherto unrecognized properties of RNA added further support to the idea that RNA was the central molecule in the origin of life.
	Leptin, a protein product of the recently identified obesity gene (ob) appeared to cause weight loss in experimental animals. *Journal Nature Medicine* (1995).
	A new gene mapping technique, STS gene mapping, thought to greatly speed the work of geneticists involved in the international Human Genome Project. *Nature Genetics.* (1995).
	A single gene identified that appears to control the growth and development of eyes throughout the animal kingdom. Halder et al., *Science*, **267**, 1788–1792 (1995).
	A new transgenic mouse carrying a gene for human Alzheimer's disease developed. Games et al. *Nature*, **373**, 476–477, 523–537 (1995).
	Gene therapy, immune system modulation and genetically engineered antibodies entered the clinic in the war against cancer. The *Journal of Clinical Oncology*. (1995).
	UpJohn and Pharmacia merged to form Pharmacia-UpJohn in a $6 $\times 10^9$ stock swap.
	Hoescht/Marion Merrell Dow merged for $7.1 $\times 10^9$.
	Glaxo/Wellcome merged in a $15 $\times 10^9$ deal.
1996	Biogen's recombinant interferon drug. Avonex® approved for the treatment of multiple sclerosis.
	A collaboration of scientists reported sequencing of the complete genome of a complex organism, *Saccharomyces cerevisiae*, otherwise, known as baker's yeast. The achievement marked the complete sequencing of the largest genome to date: more than 12 million base pairs of DNA. I
	Analysis of a small meteorite that landed on Antarctica some 15 $\times 10^6$ years ago sparked what may be the greatest scientific discovery ever, possible evidence of life on Mars.
	The sequencing of the genome of ancient organisms, archaea, found in inhospitable climates deep in thermal vents under the sea thought to greatly advance understanding of the evolution of life on Earth. The microorganisms are neither eukaryotes nor prokaryotes. Science; (1996).

Date	Event
	A new inexpensive diagnostic biosensor test for the first time allowed instantaneous detection of the toxic strain of E. coli (E. coli strain 0157: H7), the bacteria responsible for several recent food-poisoning outbreaks.
	The discovery of a gene associated with Parkinson's disease provided an important new avenue of research into the cause and potential treatment of the debilitating neurological ailment. *Science*; (1996).
	Monsanto purchased Ecogen for $25 $\times 10^6$, Dekalb Genetics for $160 $\times 10^6$. Agracetus for $150 $\times 10^6$; and 49.9% of Calgene.
	American Home Products purchased the remaining 40% stake in Genetics Institute, Inc. for $1.25 $\times 10^9$.
	Biogen introduced Avonex, to compete with Berlex's Betaseron for MS sufferers.
	$27 $\times 10^9$ merger of Ciba-Geigy AG and Sandoz AG to form Norvatis approved. Estimated annual sales of $27.3 $\times 10^9$. U.S. Federal Trade Commission (FTC) prevented monopoly that does not yet exist by requiring Norvatis to provide access to key genetic-research discoveries. The goal of the FTC: to prevent company from dominating gene-therapy research.
	Hoechst Schering AgrEvo purchased PGS International for $730 $\times 10^6$, thereby obtaining access to technology for constructing transgenic plants resistant to an AgrEvo broad spectrum herbicide, and for introducing Bt toxin gene.
1997	Monsanto completed purchase of Calgene for $320 $\times 10^6$. Agreed to buy Asgrow Agronomics Soybean business for $240 $\times 10^6$, and Holden's Foundation Seeds (corn, sales of $50 $\times 10^6$/yr) for $1.02 billion.
	Proctor and Gamble paid Regeneron $135 $\times 10^6$. to carry out research on small-molecule drugs.
	SmithKline Beecham formed joint venture with Incyte to enter genetic-diagnostics business.
	Schering-Plough Corporation acquired Mallinckrodt Animal-Health Unit for $405 $\times 10^6$.
	Merck, Rhône-Poulenc formed animal health care 50–50 joint venture (Merial Animal Health). Estimated annual sales were $1.7 billion).
	Novartis purchased Merck and Co.'s insecticide and fungicide business (sales of $200 $\times 10^6$/yr) for $910 $\times 10^6$.
	Roche Holding, parent of Swiss Drug Company Hoffmann-LaRoche bought Boehringer Mannheim for about $11 $\times 10^9$.
	Proctor and Gamble signed $25 $\times 10^6$ agreement with Gene Logic to identify genes associated with onset and progression of heart failure.
	Dupont purchased Protein Technologies (division of Ralston Purina) for $1.5 $\times 10^9$ as part of business plan to develop soy protein foods.
	American Home Products discussed $60 $\times 10^9$ merger with SmithKline Beecham PLC.
	Monsanto spun off chemicals unit and becomes Monsanto Life Sciences.
	Researchers at Scotland's Roslin Institute reported that they had cloned a sheep-named Dolly-from the cell of an adult ewe. Polly, the first sheep cloned by nuclear transfer technology bearing a human gene, appeared later.
	Artificial human chromosomes created for the first time. *Nature Genetics*. (1997).
	Follistim, a recombinant follicle-stimulating hormone, approved for treatment of infertility.
	Scientists at the Oregon Regional Primate Research Center, cloned two Rhesus monkeys from the early stage embryos using the nuclear transfer method.
	Leading geneticists expressed shock and dismay as word spread of the U.S. Patent and Trademark Office announcement that it would allow patents on expressed sequence tags (ESTs), short sequences of human DNA that have proven useful in genome mapping.
	Orasure, a bloodless HIV-antibody test using cells from the patient's gums approved, *JAMA*, (1997).

(*continued overleaf*)

TABLE 1. (continued)

Date	Event
	Researchers at Northwestern University used a kind of reverse-knockout approach to identify the Clock gene. The first gene providing the circadian rhythm of mammalian life has now been identified.
	Using a bit of DNA and some commonplace biological laboratory techniques, researchers engineered the first DNA computer "hardware" ever: logic made of DNA. The research was announced at the First International Conference on Computational Molecular Biology in Santa Fe, NM.
	A new *E. coli* vaccine for prevention of urinary tract infections developed.
	The complete genome of the Lyme disease pathogen, *Borrelia burgdorferi*, sequenced, along with the genomes for *E. coli* and *H pylori*. *Nature*; (1997).
	A new DNA technique combined PCR, DNA chips, and computer programming providing a new tool in the search for disease-causing genes without the necessity of costly DNA sequencing.
	The FDA approved Rituxan, the first antibody-based therapy for cancer (for patients with non-Hodgkin's lymphoma).
1998	University of Hawaii scientists cloned three generations of mice from nuclei of adult ovarian cumulus cells. *Nature*; (July 23, 1998).
	The FDA granted marketing clearance to RemicadeTM (infliximab), a novel monoclonal antibody for treatment of Crohn's disease.
	Research teams from The University of Wisconsin and Johns Hopkins University succeeded in growing embryonic stem cells, the long sought grail of molecular biology. *Science*; (Nov 5, 1998) and *Proceedings of the National Academy of Sciences*, (Nov 1998.)
	Scientists at Japan's Kinki University cloned eight identical calves using cells taken from a single adult cow.
	Favorable results with a new antibody therapy against breast cancer, HER2neu (Herceptin), heralded a new era of treatment based on molecular targeting of tumor cells
	Fomivirsen became the first approved therapeutic agent developed with antisense medical technology.
	The first complete animal genome the *C. elegans* worm was sequenced. *Science*; (Dec 11, 1998).
	A rough draft of the human genome map was produced, showing the locations of more than 30,000 genes, *Science* **282**, 744–746; (1998).
	Smith Kline Beecham broke off talks with American Home Products.
	Glaxo entered merger discussions with SmithKline Beecham in a deal valued at $65–70 10^9. Merger discussions driven by the successful hunt for human genes and opportunities for exploiting these findings for development of new pharmaceuticals.
1999	A new medical diagnostic test for the first time allowed quick identification of BSE/CJD a rare but devastating form of neurologic disease transmitted from cattle to humans,
	A new technique based on unique individual antibody profiles offered an alternative to current DNA fingerprinting methods. The method is simple to use and has attracted considerable attention from law enforcement.
	Genentech to remain independent company, Roche intended to exercise its call option and publicly reissue Genentech shares.
	Genentech issued new t-PA patent and files patent infringement suit against Centocor's thrombolytic agent.
	FDA approved Nutropin DepotTM, first long-acting dosage form of recombinant growth hormone for growth hormone deficiency in children.
	FDA approves efficacy results for spine bone mineral density in childhood-onset growth hormone-deficient adults.
	COR Therapeutics, Schering-Plough and Genentech announce plans to study INTEGRILIN eptifibatide, injection in combination with TNKase–tenecteplase, in heart attack patients.
	Genentech and Immunex announce patent license agreement for immuno-adhesin technology for ENBREL.
	Immunex Corporation and Genentech, Inc. join forces to develop TRAIL/Apo2L in cancer.

Date	Event
2000	Actelion and Genentech sign licensing agreement for Tezosentan in the U.S.
	Genentech announces clinical trials testing TNKase™ (Tenecteplase) with leading antithrombotics for treatment of heart attack.
	Genentech and Novartis submit application for FDA approval of anti-IgE antibody.
	DNA sequencing of Human Genome roughly completed.
2001	OSI Pharmaceuticals, Genentech and Roche to develop and commercialize OSI-774, OSI's lead cancer drug.
	COR, Schering-Plough, and Genentech announce U.S. collaboration INTEGRILIN® (eptifibatide) injection, TNKase™ (Tenecteplase) and Activase® (Alteplase, recombinant).

^a *Note*: "the field of biotechnology" continues to evolve and grow at a breath-taking rate, therefore some of the entries in this table will have resulted in different outcomes than initially thought.

New Biotechnology

The directed manipulation of genes distinguishes the new biotechnology from prior biotechnology. Biotechnology was broadly defined in 1991 by the U.S. Congress as "any technique that uses living organisms (or parts of organisms) to make or modify products, to improve plants or animals, or to develop microorganisms for specific uses" (OTA). This technology was probably first applied about 8000 years ago in the making of bread. Over the last 60 years it has helped to catalyze the growth of the pharmaceutical, food, agricultural-processing and specialty-product sectors of the U.S. economy to the point where sales exceed $100 billion/year.

A new sector of the biotechnology industry has grown significantly in economic importance since about 1990. It is sometimes referred to as the "new" biotechnology which represents a technology for manipulating genetic information and manufacturing products that are of biological origin or which affect biological activity. It is based on methods, introduced since 1970, and applied on an industrial scale since 1979, that enable directed manipulation of the cell's genetic machinery through recombinant DNA techniques. Recombinant DNA is defined by a "DNA molecule formed *in vitro* by ligating DNA molecules that are not normally joined "(see Walker and Cox)." A recombinant technique is a method "that helps to generate new combinations of genes that were originally present" in different organisms.

The new biotechnology, which occurred only recently on the time-line of bioprocessing, enables production of mammalian or plant proteins and other biomolecules having therapeutic value in quantities required for practical use (see Table 1). Recombinant methodologies also have potential for dramatically improving production efficiency of products already derived by fermentation through the directed modification of cellular metabolism, i.e., metabolic engineering. As summarized in the preface of a 1992 National Research Council Report, "scientists and engineers can now change the genetic make-up of microbial, plant, and animal cells to confer new characteristics. Biological molecules, for which there is no other means of industrial production, can now be generated. Existing industrial organisms can be systematically altered (i.e., engineered) to enhance their function and to produce useful products in new ways." Since the alteration of the cells is carried out by inserting new genes into the DNA of the organisms in a rational, rather than random, manner, the process has popularly been referred to as "genetic engineering." This report, and another one that had preceded it, had recognized that the new biotechnology was setting the stage for unprecedented growth in products, markets, and expectations, and that there was a need for substantial manufacturing capability to bring about the application of biotechnology for the benefit of society.

The U.S. industry based on the new biotechnology is projected to grow to $50 billion by the 21st century. This industry has the potential to surpass the computer industry in size and importance because of the pervasive role of biologically produced substances in everyday life. The transition from basic discovery to production through scale-up of bioprocesses will be a key element in the growth of the industry that will impact products and production methods in the chemical, fermentation, and food processing industries. By 1998, the impact of the new biotechnology on the pharmaceutical industry was becoming profound. Its potential effect on food processing and production agriculture was resulting in multibillion dollar investments by some of the world's largest chemical companies (Fritsch and Kilman, Kilman, 1996, 1998). In 1997, Monsanto sold off its chemical business with sales of about 4×10^9 ($10^9 = 1$ billion) to form an entity known as Solutia, in order to focus on its more profitable agricultural, food ingredients, and pharmaceutical businesses with sales of over 5×10^9 (Fritsch). Dupont, Dow Chemical, Norvatis, and Monsanto were investing heavily or acquiring food technology, plant biotechnology, and seed companies at costs of billions of dollars by early 1998. Bioengineered foods is one target of these acquisitions. Fat-free pork, vegetarian meat, bread enriched with cancer fighting compounds and corn products that fight osteoporosis are part of these companies' vision and future profits of biotechnology (Kilman, 1998).

New biotechnology methods enable rapid identification of genes and their protein products, and facilitate new drug candidates from combinatorial chemistry. The discovery of biochemical nature of diseases, and identification of ways by which drugs can interfere with the disease process has been greatly accelerated by the new biotechnology and by the identification of genes that represent underlying mechanisms of disease. The new genes are patentable, hence giving a twenty year monopoly to the companies first to identify and patent them (Kinoshita). This capability also speeds up drug discovery. The resulting candidates hold the promise of new and more potent medicines.

Although genomics (i.e., gene-identifying technology) provides information on the targets or sites upon which drugs can act, combinatorial chemistry and the automation of chemical synthesis make it possible to generate thousands of candidates rapidly. There conducted could be tested against targets identified through genomic methods. This in turn is driving the merger of drug companies to make resources and multibillion dollar research budgets available to screen, develop, and test the new drugs resulting from the combination of computerized gene-discovery and automated synthesis of molecules with potential as drugs. The largest merger proposed to date (in early 1998) was that of Smith Kline Beecham PLC and Glaxo Wellcome PLC, in a deal valued at between $65 and 70 \times 10^9$. Annual sales of the combined companies would be on the order of 25×10^9 (Tanouye and Langreth).

Products from the new plant biotechnology are changing the structure of large companies that sell agricultural chemicals. The chemical, food and agricultural industries also began to experience major changes by the mid 1990s, catalyzed by advances in plant biotechnology and by the ability to identify and manipulate traits of economically important crops at the cellular level. The first commercial success of plant biotechnology came in 1996, after about 20 years of research and development in the genetic engineering of crops. In this case, Monsanto had shown that soybean seeds, genetically engineered to resist the herbicide, Round-up, gave higher yields. The herbicide could be applied to kill all other plants in a field while having no effect on the soybean plants. The introduction of genetically engineered seeds fostered concerns about transgenic supercrops. Could these breed super weeds if genes from widely planted and cultivated, weed resistant plants were to migrate to wild relatives (Kling). The risk is perceived to be small. However, France, a major corn producer, prohibited planting (but not consumption) of transgenic corn (Balter).

Another genetically altered product by Monsanto, introduced in 1996, was cotton seed that contained a gene from bacterium, *Bacillus thuringiensis*. The gene enables the bacterium, and now the cotton plants, to make a protein that is toxic to bollworms and tobacco budworms, thereby reducing the need for pesticides that are otherwise needed to control insect infestations (Fritsch and Kilman). The *Bacillus thuringiensis* gene has also been incorporated into potato so that it produces a protein toxic to the Colorado potato beetle. The potato has been approved for use in Canada (Yanchinski). Dow-Elanco through its investment in Mycogen (a San Diego based biotechnology company) has marketed transgenic corn with *Bacillus thuringiensis*-based resistance to the European Corn Borer (Thayer, 1996).

There were 30 crop species that, by 1996, had been engineered to express *Bacillus thuringiensis* endotoxins. These proteins are highly toxic to specific insect pests. There are concerns of the evolution of insects that are resistant to these toxins since millions of acres are now being planted with these crops. (Ives). If insect species develop that are resistant to *Bacillus thuringiensis* endotoxin, strategies for overcoming this resistance will need to be developed, resulting in a cycle of further manipulation of the plants genes to alter the toxins (Ives). Nonetheless, the use of this class of bioproduct is rapidly growing. Success of the transgenic soybean

seeds, insect resistant cotton seeds, and other genetically engineered seeds have resulted in predictions, from market analysis, that sales would grow to 6.5×10^9, annually, within 10 years (Fritsch and Kilman).

The current growth of the agricultural biotechnology sector may, in part, have benefited from a 1992 ruling in which policy guidelines were defined on how the Food and Drug Administration would oversee new plant varieties, included those developed through biotechnology. Their basis was "that products would be judged by their risks ... not by the process which produces them." Hence if the composition of a food product is not significantly altered, the FDA's premarket approval would not be needed, and special labeling would not be required. However, if "genetic manipulation were to lead to a product that has increased allergenicity, or that contains new proteins, fats, or carbohydrates, it will have to pass rigorous premarket review. And if approved its new traits will have to be labeled," according to Ember. At the time of this announcement in 1992, overall sales of biotechnology were estimated to be 4×10^9 annually, and expected to grow to 50×10^9 by the end of the 20th century. Of these revenues, the agricultural market was projected to be about 28×10^9 (Ember). The widespread tests and apparent initial success of several genetically engineered crops, about four to six years later, may have benefitted from this policy, proposed under the Bush administration. The timeline between 1992 and 1997 shows that a combination of, technical, business, and regulatory factors coincided at about the same time, and culminated in the first significant commercialization of plant biotechnology products by 1997 (see timeline in Table 1).

Hileman summarized the status of regulation, concisely in 1995, as follows. The statutes divide agricultural products into six categories: plants engineered for (1) herbicide resistance, (2) pesticide resistance, (3) delayed (tomato) ripening; (4) plants modified to make products otherwise obtained from other crops; (5) plants engineered so their crop can be processed more easily; and (6) bacteria enhanced to fix nitrogen or control insects. Genetically modified plants and animals for purposes of producing pharmaceuticals are not included in these categories. Although few products from transgenic plants require agency approval, most companies have voluntarily consulted the FDA to obtain its stamp of approval. Transgenic plants to be grown on a large scale are regulated by the U.S. Department of Agriculture (USDA). The U.S. Environmental Protection Agency (U.S. EPA), regulates gene-modified organisms that express a pesticide or function as a pesticide. Genetically engineered microorganisms, such as a bacterium engineered to produce ethanol, would also be regulated by the U.S. EPA. *Bacillus thuringiensis* corn is regulated by all three agencies: insecticidal properties by the EPA; large-scale growing by the USDA; and corn as a food product by the FDA.

Growth of the Fermentation Industries

The history of bioprocessing probably started with bread making, which utilized yeast, about 6000 years ago. Fermentation of grains and fruits to alcoholic beverages was carried out in Egypt, and other parts of the ancient world about 2500 B.C. Other types of food fermentation practiced for thousands of years included the transformation of milk into cheeses, and fermentation of soybeans to give sauces and cheeses. However, it was not until 1857, that Pasteur proved that alcoholic fermentation was caused by living cells, i.e., yeasts. The intentional manipulation of microbial fermentations to obtain food products, solvents, and beverages, and later, substances having therapeutic value as antimicrobial agents for fighting infections (Aiba et al., Evans), continued to give rise to a large fermentation industry over the next 100 years.

Submerged Fermentations. Submerged fermentations represent a technology in which microorganisms are grown in large agitated tanks filled with liquid fermentation media consisting of sugar, vitamins, and minerals. Many of the fermentations are aerated, with vigorous bubbling of air providing the oxygen needed for microbial growth. Since the microorganisms are in a liquid slurry, they are considered to be submerged, compared to microbial growth that occurs on a surface (such as on a moldy fruit or piece of bread).

The fermentation industry produced food and beverage products, and some types of oxygenated chemicals by submerged fermentations prior to the first submerged antibiotic fermentations in 1943. Products in the early 20th century included ethanol, glycerol, acetone, and butanol. However the growth of an efficient petrochemical industry, in the years following the Second World War, made some of these fermentation products economically unattractive. Petroleum sources supplanted many of the large volume fermentation products, with the exception of ethanol produced with government subsidies, yeasts for baking, vinegar, carboxylic acids, and amino acids for use as food additives and in the formulation of animal feeds. These large volume fermentations depended on the availability of molasses or glucose from corn (starch) as the fermentation substrates as well as the ability to grow organisms in large tanks (Hacking).

Oil Prices and the Fermentation Industry. The low prices of oil prior to 1973 made the manufacture of many oxygenated chemicals by fermentation uneconomical. The low oil prices stimulated research and development on single cell protein and attempts to commercialize it. Large, aerated fermentations utilized methane or methanol (from petrochemical processing) as the main substrate in order to grow the cells. The goal was to propagate microorganisms, whose protein content would make them attractive as a food source, using a relatively inexpensive substrate derived from petroleum. Since these fermentations were based on the growth of unicellular microorganisms, the source of the protein was appropriately named single-cell protein. However, concerns about carry-over of harmful substances from the petroleum source into the fermentation, where it would be incorporated into the edible cellular biomass, coupled with a sharp rise in oil prices in 1973, and subsequent decreases in soybean prices (a vegetable source of protein), made single-cell protein processes unattractive.

At the same time that the prices of crude oil quadrupled in late 1973 and 1974, in part triggered by a war in the Middle East, the combined use of restriction enzymes (that cut DNA in a directed manner) and ligases (i.e., enzymes that join foreign genes with the DNA of the host cell) was demonstrated. Stanley Cohen and Herbert Boyer showed that DNA could be cut and rejoined in new arrangements in a directed manner. Their work gave birth to the field and the industry based on new forms of organisms obtained through the sequencing, removal, insertion, and amplification of genes across different species of organisms. While the old, high volume fermentation industry based on inexpensive petroleum was lost, a new high-value, lower volume fermentation industry was evolving (Hacking, Olsen).

Value-added Products. The economic impact of value-added chemicals derived through fermentations is currently modest, but the potential is clear. For example, in 1997, Dupont and Genencor developed an inexpensive and environmentally friendly fermentation for making polyester building blocks from glucose. This process is based on a proprietary, recombinant microorganism which combines bacterial and yeast pathways so that the fermentable sugar is converted to glycerol, and the glycerol to trimethylene glycol, which in turn is polymerized to polyester. The resulting polymer is reported to have properties, which are superior to existing polyesters derived from petrochemical building blocks (balter).

Another example is a genetically engineered bacteria that can convert naphthalene (historically, priced much lower than indigo, as low as $1.00/kg) to indigo (priced at $8–17/kg). Indigo is used to dye blue jeans, has an annual market of $250 × 10^6, and accounts for 3% of all of the dye colors currently used. This dye, given the name, "bio-indigo," was proposed to replace dye currently made using cyanide and formaldehyde (O'Hamilton). The market for this biotechnology product is not yet clear although the potential benefits of this material could cause it to evolve into an economic success (O'Hamilton). The enzyme that enables this conversion and the genetically engineered bacteria that gave the result were discovered at Amgen. The technology was sold to Genencor International Inc., in 1989, when one of Amgen's biopharmaceuticals began to take-off and caused Amgen to drop its research on industrial enzymes (see timeline Table 1).

Growth of the Antibiotic/Pharmaceutical Industry

The genesis of the antibiotic industry, which preceded the current era of engineered organisms, was catalyzed by government incentives and support during World War II. The motivation was both the discovery of the beneficial effects of penicillin and the difficulty of producing penicillin through chemical synthesis. Once the fermentation route was chosen over chemical synthesis, events progressed rapidly. At first, production was tried by growing the *Penicillum* on the surface of moist bran. This did not work well because it was difficult to control temperature and sterility. The second approach was to grow the microorganism on the surface of quiescent liquid growth medium in milk bottles and other types of containers. Yields were high, but potential for mass production was limited. This led to fermentations in large, agitated vessels that held a liquid volume of 7000 gallons or more (Shuler and Kargi).

The rapid scale-up of production required development of a cost-effective medium on which the penicillin producing microorganism, *Penicillin chrysogenum*, was grown, design of large-scale-aerated tanks that could be operated under sterile conditions, and recovery and purification of the final product. The first submerged fermentations were carried out on corn steep liquor–lactose based medium. Recovery of the labile penicillin molecule was achieved through a combination of pH shifts and rapid liquid–liquid extraction (Shuler and Kargi). According to Hacking, "the discovery of penicillin, which has now virtually passed into folklore, has been one of the major milestones of the twentieth century, and arguably one of the most beneficial discoveries of all time. Its impact on biotechnology, and the public perception of biotechnology should not be underestimated." Hacking also points out that the development of cost-effective processes for manufacture of penicillin provides an example of the impact of government support in starting up a new technology with the potential of widespread benefits for society. "During the Second World War, the U.S. government imposed an excess profits tax of 85% as one measure to pay for the war effort. As many pharmaceutical companies were liable for this tax, additional research cost them effectively only 15% of its total cost, "the 15 dollar." Several of the early large antibiotic screens and process development of penicillin were funded on this basis (Hacking)." This led to the growth of a fermentation based antibiotic industry (see 1943 in timeline Table 1).

Penicillin. In 1877, Pasteur and Joubert discovered that anthrax bacilli were killed by other bacteria. In 1896, Gosio isolated mycophenolic acid from *Penicillum brevi-compactum*. Mycophenolic acid inhibited *Bacillus anthracis*, but was too toxic to be used as a therapeutic agent. By 1917 Grieg-Smith showed actinomycetes give substances with antibacterial activity. In 1928, Alexander Fleming showed staphylococcus cultures were inhibited by growing colonies of *Penicillum notatum*. (Unlike many of the other antibiotics at that time, penicillin was later found to be effective as well as suitable for systemic use). By 1937, actinomycetin, an antibacterial agent, had been isolated from a streptomycete culture, but was again too toxic for use as a therapeutic agent. In 1939, Dubos obtained gramacidin and tyrocidin from the spore-bearing soil bacillus, *Bacillus brevis*. These agents proved to be effective for treatment of skin infections (topical use) but again were too toxic for internal systemic use.

Penicillin was the first antibiotic suitable for human systemic use. Experience with penicillin during World War II showed it to be one of the few antibiotics, available at that time, which was suitable for systemic use to fight infections and save thousands of lives (Evans). Florey and Chain catalyzed the rediscovery of Fleming's penicillin, starting in 1939, during studies of compounds that would lyse bacteria. Their work was carried out in search of new therapeutic agents for use in the World War. Fleming's strain of *P. notatum* did not produce large amounts of penicillin, and penicillin itself was relatively unstable. Thus, significant efforts were carried out to develop isolation and purification procedures that minimized product loss. In a pattern, which continues to be repeated to this day, the researchers developed methods for isolating small quantities of penicillin for trials with mice. The penicillin was shown to protect the mice against *Streptomyces haemolyticus* while being relatively nontoxic. Clinical trials in 1941 were most successful, and scale-up to obtain larger amounts of the antibiotic were then initiated by joint Anglo-American efforts that included the involvement of Merck, Pfizer, and Squibb with the help of government laboratories (Aiba et al.).

Discovery, selection, and development of microorganisms suitable for submerged fermentations made penicillin's mass production possible. Penicillin was at first produced by surface culture, with the mold being grown on the surface of a nutrient medium in 1–2 liter batches in vessels resembling milk bottles. Antibiotic yield was improved by discovery of higher yielding strains of *P. notatum* and *P. chrysogenum* and mutants obtained by ultraviolet and x-ray irradiation of several *Penicillum* strains.

Scale-up from 1 to 2 L to 100,000 L capacity was made possible by the discovery of a strain of *P. chrysogenum* (on a moldy cantaloupe in Peoria, Illinois) which could be grown in submerged culture, and by the development of submerged fermentation technology for this microbe in large agitated vessels by USDA's Northern Regional Laboratories in Peoria. Mass production was now possible, since the microorganism could be grown in large liquid volumes of nutrient media to which sterile air was provided to satisfy oxygen requirements for growth (Aiba et al. and Evans). Prices of penicillin quickly decreased because of productivity gains made possible by submerged fermentation.

Acler Actiluatics. Other antibiotics were quickly discovered after the introduction of penicillin. A different type of antibiotic, streptomycin, which is active against a wider range of pathogens than penicillin and is a potent inhibitor of *Mycobacterium tuberculosis*, was isolated from a strain of actinomycete from the throat of a chicken by Waksman at Rutgers in 1944 (Aiba and Evans). Actinomycetes are commonly found in soil and are intermediate between fungi and bacteria. Since 1944, numerous other metabolic products of actinomycetes have been isolated and have made a transition from the bench scale to wide therapeutic use.

Between 1945 and 1965, about 30 new antibiotics gained the status of established therapeutic agents. By 1981, about 5500 antibiotics were identified with only 100 on the market. The number grew to about 160 variations of 16 basic compounds by 1996, together with an alarming rise in antibiotic and antibacterial resistant bacterial infections (Stinson). This array of products was based on the discovery of thousands of new antibiotics, of which only a few hundred were characterized, during the same time period (Aiba et al., Hacking). This illustrates how successful development of therapeutic agents derived from microbial sources requires tremendous screening efforts. There is a continuing need for antibiotic discovery and development as drug-resistant infections have begun to appear, with drug-resistant bacteria now being viewed as a threat that is just as serious as the AIDS (HIV) epidemic (Stinson and Tanouye).

Stinson gives a clear description of the difference between antibiotics and antibacterial drugs. Antibiotics are drugs that kill bacteria and are derived from natural sources such as molds. Chemically synthesized antibacterial drugs, whose structures are the same as antibacterials from natural sources are also antibiotics. Hence, penicillin G from *Penicillum chrysogenum*, or organically synthesized ceftriaxone with the same 7-aminocephasosporanic acid nucleus as compounds derived from *Cephalosporium* molds are antibiotics. "Antibiotics that are toxic to human cells, and to malignant human cells more so than normal ones," are anticancer antibiotics. Examples of antibacterial drugs are quinolones, oxazalidinones, and sulfa drugs. Some antibacterial drugs are also active against fungal, viral, or parasitic microorganisms. For example, *Helicobacter pylori* infections are now treated with antibacterial drugs. *H. pylori* infections of the stomach wall are found in 95% of patients with duodenal ulcers, and 80% of patients with stomach ulcers (excludes patients who take nonsteroidal, antiinflammatory drugs).

Discovery and Scale-up. This synopsis of penicillin and antibiotic development illustrates how discovery and scale-up are synergistic. Following discovery, enough of the new substance must be obtained for testing with animals, and later, if warranted, for treating humans. The ability to scale-up production of a promising therapeutic agent rapidly complements the discovery process by providing sufficient material so that it can be characterized in detail on what a new compound can do. This help to provide input on where future discovery efforts might be directed. The second role of scale-up is to be able to provide larger quantities of pure drugs for clinical trials. Last, but not least, if the trials are successful, scale-up is needed to provide manufacturing capability to make the new drug available on a wide basis.

Pharmaceutical Industry Growth. The pharmaceutical industry has invested heavily in research to invent, screen, develop, and test new products. The industry has historically invested between 16 and 20% of its earnings in research, and has been continually concerned about developing new products, and keeping its pipeline full of new pharmaceuticals. Although the production of pharmaceuticals is sometimes perceived to be principally a fermentation- and biotechnology-based industry, many drugs are obtained through chemical synthesis or chemical modification of fermentation-derived molecules. It is estimated that pharmaceuticals accounted for half of the 45×10^9 fine chemical sales, worldwide, in 1991. Custom drug discovery, synthesis and marketing is carried out on a time-scale similar to biopharmaceuticals, with 10 years often being required to develop a product from discovery to routine production, where routine production is classified as 500 kg/year or more (Stinson). The combination of fermentation derived and chemically synthesized drugs, show that there has been a significant increase in the overall sales of the top ten companies, as well as a major change in their relative ranking based on sales (see Table 2). While the U.S. sales of drugs by US pharmaceutical companies exceeded 32×10^9 in 1991, the U.S. sales of protein biopharmaceuticals, was only 1.2×10^9. By 1996 the sales value of the biopharmaceuticals had grown to 7.5×10^9 (Anonymous, 1992; Thayer, 1991).

TABLE 2. SALES OF TOP 15 PHARMACEUTICAL COMPANIES IN 1991[a]

Companies	Approximate Pharmaceutical Sales (1991)[b]	Estimated 1997[c]
U.S. Pharmaceuticals		
Baxter International	NA	6.0
Merck and Co.	6.37	22.9
Bristol-Myers Squibb	5.26	16.2
Eli Lilly	3.70	7.9
American Home Products	3.46	14.2
Johnson and Johnson	3.30	22.9
Pfizer	3.23	11.8
Abbott Laboratories	3.16	11.8
Warner-Lambert	3.08	7.5
Schering-Plough	NA	6.6
Sub-Total	31.6	127.8
International		
Glaxo (UK)	6.05	
Hoeschst-Roussel (Germany)	4.99	
Bayer (Germany)	4.96	
Ciby-Geigy (Switzerland)	4.58	
SmithKline Beechem (UK/US)	4.24	
Sandoz (Switzerland)	4.09	
Hoffmann-LaRoche (Switzerland)	3.46	
Sub-Total	32.4	
TOTAL of 15.	64.0	

[a] From Anonymous, 1992.

[b] NA = not available.

[c] Estimated from first half sales and earnings reported by Thayer (1997), and multiplying by two.

Despite the large increase in sales, the cost of prescription drugs as a fraction of health care costs remained constant at 4 to 5% of health care expenditures during this period (Reisch, 1993). Health care costs, which as of 1998 were contained, had risen very rapidly in the period from 1982 to 1992 from about 300×10^9/year to 820×10^9/year, respectively. The rapid increase during this ten year period led to close examination of medical costs, including costs of prescription drugs. One report suggested that the industry priced their drugs 4.3% higher than needed to recover research and development costs. At about the same time, the Pharmaceutical Manufacturer's Association (PMA), which represents companies that account for 90% of US pharmaceutical sales, proposed to limit aggregate price increases to no greater than the general rate of inflation as measured by the government's consumer price index. The PMA did not wish to appear to act in restraint of trade and hence first sent the proposal to the antitrust division of the Department of Justice (Reisch, 1992).

The estimated 1997 sales in Table 2, in part, reflect the growth of revenues from areas other than nonprescription drugs. Despite constraints on prices, the need and markets for new pharmaceuticals is large, and will continue to grow as the population grows and ages. The fit of new biotechnology products into this growing market, and the pending loss of patent protection of existing drugs, help to explain the increasing pace of consolidation in the pharmaceutical industry since about 1995 (Thayer, 1995). The pending loss of patent protection will affect, or has affected, major products from Merck & Co., Schering-Plough, Eli Lilly, Glaxo, Bristol-Myers Squibb, Syntex (now a part of Roche), UpJohn (now Pharmacia-UpJohn) and Astra AB. An example of the effect of loss of patent protection is given by heart drug Capoten (Bristol-Myers Squibb), whose US sales fell from 146×10^6 to 25×10^6 within a year after coming off patent. The price of the medicine dropped from 57/pill to 3/pill when generic versions of the drug entered the market. Other examples of drugs that will come off patent between 2000 and 2002, given by Tanouye and Langreth, have total current U.S. sales of 6.8×10^9 among four U.S. companies. Hence, both economics and technology are likely to play a role of the expansion of the products and processes, based on the new biotechnology, into the pharmaceutical industry.

Growth of the Amino Acid/Acidulant Fermentation Industry

The production of α-amino acids has a long history, starting with the isolation of asparagine from asparagus juice in 1806. The other α-amino acids were isolated from a variety of natural substances with threonine from fibrin being the last of these acids to be isolated and identified in 1935 (DeGussa, 1985). Glutamic acid, in the form of monosodium glutamate (MSG), was identified by Ikeda in 1908 as the agent responsible for the flavor enhancing properties of konbu, a traditional seasoning (food) additive in Japan. This discovery helped to initiate the development of an amino acid production industry starting in 1909, when amino acid hydrolysates of wheat gluten (protein) or soybean protein were used by Ajinomoto (Japan) to (Hirose et al.) commercially produce MSG. The discovery of the glutamic acid producing bacteria *Micrococcus glutamicus* (later named *Corynebacterium glutamicum*) in Japan in 1957 (Kinoshita) marked the beginning of a 40 year period of growth in high volume, fermentation derived bioproducts.

Monosodium Glutamate (MSG). Production of monosodium glutamate (MSG) via fermentation was not practiced until 1961. The discovery of the L-glutamic acid producer, *M. glutamicus* in 1957 (Hirose et al., Kinoshita) and the realization that a critical concentration of biotin (ca. 2 ppb), or biotin together with an antibiotic or non-ionic detergent, was needed to maximize L-glutamic acid production (Aiba et al., Hirose et al.), marks the transition of an amino acid industry based on processing of natural materials, to one dominated by fermentation based technology for production of food and feed additives, and products used in therapeutic applications. At about the same time, the first industrial production of another α-amino acid, L-lysine, was also initiated (DeGussa, 1985). By 1972, the world production of glutamate was 180,000 tons, of which 90% was made by fermentation (Aiba et al.), and by 1976, Japan produced two-thirds of the world's amino acids with an estimated annual sales value of 3×10^9 and a volume of 200,000 tons/year. Half of this volume was MSG, which was used almost exclusively as a food seasoning. A small amount was employed in the production of a leather substitute (Hirose et al.). Just as is the case for antibiotics, some amino acids are produced by chemical synthesis. In particular DL-methionine, DL-alanine, and glycine are obtained via chemical synthesis.

The impact of glutamic acid bacteria on monosodium glutamate cost was dramatic. In 1950, MSG was obtained from natural products (i.e., soybean) and cost $4/kg. When the first fermentation process started up in 1961, the price dropped to $2/kg. By 1970 the price settled to about $1/kg, and in 1983 approximately 220,000 tons of MSG/year were being produced by fermentation, worldwide, with a sales value of about 550×10^6.

The discovery of glutamic acid bacteria resulted in a fundamental change in the industry, not only for MSG production, but also for other amino acids, which could now be obtained via fermentation. Mutation of glutamic acid bacteria with ultraviolet light, X-rays, γ-rays and chemicals resulted in amino acid producing auxotropic and regulatory mutants for the L-forms of lysine, threonine, tyrosine, phenylalanine, L-ornithine, proline, leucine, citrulline, and homoserine, as well as glutamic acid. The multiplicity of names given to glutamic acid bacteria discovered after 1957 was a source of confusion. Based on conventional taxonomy, and the 73 to 100% DNA homology reported for several strains of *Corynebacterium, Brevibacterium*, and *Microbacterium* in 1980, the glutamic acid bacteria were reasonably regarded as a single species in the genus *Corynebacterium* (Kinoshita).

Although MSG remains the single largest fermentation derived amino acid, there are a number of other valuable amino acids obtained from mutants or related microorganisms of the genus *Corynebacterium*. Lysine (an animal feed additive), is the second largest amino acid obtained by fermentation. It had a worldwide production of 40,000 tons/year in 1983 compared to MSG at 220,000 tons/year. More recent valuations place the worldwide lysine market at 600×10^6/year in 1996, with half of the sales attributed to Archer Daniels Midland. Some lysine producers also manufacture the carboxylic acid, citric acid, with annual sales of 1×10^9 in 1996. Citric acid is a widely used flavoring agent and acidulant in beverages (Kilman, 1996a, b). Most other amino acids are small volume (10 to 300 tons/year), high value products obtained by fermentation, microbial conversion of intermediates, or enzyme synthesis (Hirose et al.). Dominant suppliers of amino acids are Ajinomoto, Takeda, and Kyowa (in Japan) (Hacking) and Archer-Daniels Midland (abbreviated ADM) (Kilman 1996a, b). The corn milling division of Cargill, a privately owned company entered lysine production. A letter of intent with the German Company, Degussa, resulted in an agreement "to build a 165×10^6 lb/year synthetic lysine production facility at Cargill's Blair, Nebraska complex," with production to begin in the second half of 1999. (Anonymous, 1997). The announcement stated that "Degussa says it is the world's leading supplier of methionine, and a leading supplier of threonine. Lysine, synthetically produced by corn-fed microorganisms,

is used as an additive for feeds for hogs and poultry." The syntax of this announcement is almost as interesting as its content. The term "Synthesis," which has previously been associated with *in vitro* chemical methods, is now used to describe an *in vivo* microbial method. This gives an indication of how the distinction between chemistry and biotechnology as a means of production is becoming blurred.

Some amino acids are derived by chemical routes, or through enzyme synthesis and microbial production starting with an intermediate. The largest volume, chemically synthesized amino acid is DL-methionine (also an animal feed additive) which is obtained from propylene, hydrogen sulfide, methane, and ammonia (DeGussa, Tanner and Schmidtborn). Another major chemically synthesized amino acid is a glycine chemical building block which is an intermediate in the manufacture of Monsanto's herbicide, Round-up®. Round-up® was estimated to generate about 2.2×10^9 in sales in 1995, and sales were expected to continue to grow with the successful introduction of "Round-up ready" soybean seeds in 1996. Commercial quantities of these herbicide resistant seeds were planted in 1996 and generated sales of about 45×10^6. A survey of 1058 farmers (about 10% of the farmers who planted the seed) showed that 90% of those surveyed felt the product met or exceeded expectations (Fritsch). These seeds were genetically engineered by inserting a gene from petunia that confers herbicide resistance to soybean plants (Fritsch and Kilman). This allows the fields to be sprayed with the herbicide, thereby killing all other vegetation, followed by planting of soybean seeds. The soybeans are able to grow since they are not effected by the herbicides.

These examples illustrate some of the cross-over and cross-disciplinary characteristics of biotechnology when applied to a specialty chemical business.

By late 1996, Monsanto had planned to divide itself into two entities: Monsanto Life Sciences and a chemical entity (later given the name Solutia) (Reisch). This occurred within a year, during which Monsanto purchased a plant biotechnology company (Calgene) that held patents for improved fresh produce, cotton seeds, specialty industrial and edible oils (derived from seeds) and plant varieties (Anonymous, 1997). Based on analysis of 1995 sales, the life sciences products accounted for annual sales of $5.3 billion based on the major products of Round-up herbicide, Round-up-resistant soybeans, Bollguard® insect-protected cotton, Nutrasweet® sweetner, and prescription drugs for arthritis and insomnia.

The incorporation of the *Bacillus thuringiensis* gene into cotton (Bollguard®) enabled the cotton to produce proteins that are toxic to cotton bollworm and budworm. The first trials of the transgenic cotton were a qualified success (Thayer, 1997). Approximately 40% of the Bollguard planted fields had to be sprayed although only one to two pesticide applications were required compared with a more typical four to five. Cotton growers reported an average 7% improvement in yield, as well as reducing or avoiding the use of pesticides. The farmers reported a $33/acre cost advantage of Bollguard cotton compared with insecticide treated non-Bollguard cotton, after the technology licensing fee of $32/acre and supplemental pesticide applications were accounted for.

Approximately 3% of Texas' three million acres of cotton, and 60–70% of Alabama's five hundred thousand acres of cotton were planted with *Bacillus thuringiensis* cotton in 1996. The potential economic and environmental benefits are evident when the size of the annual US cotton harvest (9×10^9 pounds, worth about 7.2×10^9), and volume of insecticides ($400–500 \times 10^6$/year) is considered. Bollguard plantings for 1997 were estimated at 2.5 million acres out of 14 million acres of cotton (Thayer, 1997).

Prices and Volumes. The price of biotechnology products (including, in the example at hand: Vitamin B_{12}, cephalosporin, penicillin, lysine, xanthan, glutamic acid, citric acid, gluconic acid, lactic acid, ethanol, and high fructose syrup) decreases as the total sales volume increases, with the relationship between price and volume falling on a characteristic line when the data are plotted on a double log scale (Hacking). This type of plot, previously used for correlating the cost of thermoplastics to their volume, is relevant for biotechnology products according to Hacking. Although the price and volume data are from 1986, and there is scatter in the data, the plot illustrates the necessity of reducing production costs to achieve bulk sales. This is both a function of the technology which must be available if production costs are to be reduced, and market demand that would drive the production of larger volumes of the product. In the case of monosodium glutamate, technology enabled a reduction in cost, which in turn increased the volume used. The essential amino acid, Tryptophan, would find uses in animal feed, and therefore offer larger volume markets, if its price were to decrease from \$100/kg to \$10/kg (based on 1986 prices, per Hacking). The dramatic decrease in the cost of penicillin within 12 years after its introduction illustrates that the volume–price relationship will also affect a high value drug. Penicillin was a wonder-drug when first introduced in 1941–1945, but then decreased in price from \$500/g in 1943 to \$0.10/g by 1956 (Hacking) as advances in manufacturing technology, improvements in productivity and product concentration, ramp-up of its volume, and economics of scale drove the price down (Hacking). The production cost of penicillin, while not a key issue in 1945, is certainly an important factor now.

The economics that currently drive the biopharmaceutical industry may be different, although it is too early to tell. Large volume uses of biotherapeutic products have not yet developed since the industry is still starting-up, and has yet to reach a steady-state growth phase. While the costs of production for a high value therapeutic contribute as little as 10% to the overall price of the product, production economics are secondary only to the timeline of drug development, competitive pressures, and marketing issues normally associated with introduction of a new product.

Biochemical Engineers and Microbial Fermentation. Biochemical engineering was defined by Aiba, Humphrey and Millis, in 1973, as "conducting biological processes on an industrial scale, providing the link between biology and chemical engineering," with its heart being "the scale-up and management of cellular processes" (Aiba et al.). It was viewed as the interaction of two disciplines. The definition was extended by Bailey and Ollis in 1977 to encompass "the domain of microbial and enzyme processes, natural or artificial," including "wastewater treatment as well as industrial fermentation and enzyme utilization." Shuler and Kangi described biochemical engineering as "the extension of chemical engineering principles to systems using a biological catalyst to bring about desired chemical transformations," and as being "subdivided into bioreaction engineering and bioseparations."

Challenges that a biochemical engineer faces are the prevention of contamination (sterility); control and promotion of microbial growth and productivity through media composition; control of growth conditions (pH, temperature, product precursors); and oxygen transfer (for aerobic fermentation). The engineer must devise efficient and robust methods for recovering, and purifying an often, unstable product. These same disciplines and skills apply to manufacture of products from organisms that form the basis of the biopharmaceuticals industry.

Growth of the Biopharmaceutical Industry

Traditional biotechnology is based on indirect methods of selecting, improving, and propagating organisms. Its transition to a new biotechnology, based on genetic engineering, could be interpreted as an abrupt change in technology when viewed from an historical (8000 year) perspective. Actually, the technological change in the industry followed an evolutionary path in both methodology and perceptions that occurred over a period of about 40 years. Microorganisms, cells, plants, and animals (collectively referred to as organisms) with enhanced characteristics for production of bioproducts and antibiotics have been selected, manipulated and improved since the beginning of the industry. Genetic structures have been purposely altered using mutagens or radiation, selective pressures, breeding, and strain improvement. The discovery and development of the structure and function of DNA over a period of a century, allowed genes to be identified, and traits to be correlated to an organism's characteristics. These findings advanced both the fundamental understanding and practical applications of organisms for generating bioproducts (food, beverages, medicines, and chemicals). The developments of recombinant technology, that now benefits the pharmaceutical industry was catalyzed by, or was "a direct result of generous governmental funding for basic biomedical research since World War II (Olsen)."

The discovery of enzymes that cut or ligate DNA at specific locations, and development of practical methods for achieving alterations in a cell's genetic make-up changed the field. These enzymes allowed genetic manipulations, and the determination of the effects of these alterations, to be achieved in a matter of months, rather than years or decades. In addition to rapidity, these methods could be carried out in a directed and rational manner, thereby allowing introduction of completely new traits by design, rather than by chance, into the host organisms. This gave rise to the expression, "genetic engineering," i.e., genes could now be engineered and altered in a predetermined manner to confer a specified end result.

U.S. government policy in the 1980s regarding limited partnerships in risky ventures may have encouraged start-ups. An estimated, 1000 biotechnology companies were formed during 1980 to 1990. The major product groups addressed by these companies were: (1) diagnostics (human and/or animal health); (2) therapeutics (human and/or animal health); (3) Ag-bio (plant genetics and/or microbial crop protectants; and (4) suppliers to the industry (according to a survey by the Arthur Young Company, Burrill, 1988). By 1996 this number increased to 1300, of which about 200 are public. The growth in the number of companies is appearing to level-off in the U.S. (Thayer, 1997a) while growth in the number of companies, internationally, was continuing at a rapid pace. For example, the number of Canadian biotechnology companies doubled over a three year period, reaching 224 by 1997, with combined revenues of 837×10^6 (Thayer, 1997a).

Sales of biotechnology products grew from zero in 1980 to 4×10^9 by 1991, and 25×10^9 by 1996 (NAS, Olsen), with about 15×10^9 due to biopharmaceutical products. Of this amount, about 10×10^9 was attributed to clearly identifiable biotechnology companies.

The projections and financial performance of the biotechnology industry does not uniformly coincide with companies that manufacture and sell biotechnology products. For example, Eli Lilly and Company and Novo Nordisk are major suppliers of biosynthetic human insulin derived from recombinant microorganisms, yet these companies are sometimes identified as drug or industrial enzyme companies, not biotechnology companies. Consequently, a complete accounting of their sales derived from recombinant technology may not be possible. The overall accounting of the industry has included plant biotechnology products from a major chemical company such as Monsanto, even though it did not become a biotechnology company until September 1, 1997 (Anonymous, 1997). The internally consistent comparison of revenues and earnings of specific sectors of the bioindustry for estimating growth over different time periods can be challenging as well since large consolidations of companies have occurred (see timeline after 1992 in Table 1). Hence, revenues will vary significantly and depend on a particular survey's definition of the biotechnology industry.

This is well within the projections made for growth rate in 1991, where market estimates (in 1991) placed sales of biopharmaceuticals, alone, at 8×10^9 within 10 years (Thayer, 1998). The analysis of these sales in the context of the growth potential of biopharmaceuticals is somewhat complicated since the effect of major pharmaceutical companies buying smaller biotechnology companies, and the merging of sales figures is difficult to place into categories. The adaptation of biotechnology as a means of production by some of the large and long-established chemical, agricultural, and food companies, further obscures the definition of a biotechnology company. Furthermore, a significant volume of pharmaceuticals is manufactured by chemical synthesis, and these are not considered to be biotechnology products.

This sector is expected to continue to show strong growth with 284 new biotechnology drugs under development in 1996 (Thayer, 1997a Anonymous, 1996). However, this is still an industry that is very much in its early growth phase. A complete set of earnings figures would show that only 10 out of the 1300 biotechnology companies reported profits in 1997. All the others had negative earnings as of the first half of 1997 (Thayer, 1997a). Nonetheless, the professional services firm of Ernst and Young, which had analyzed and reported on the biotechnology industry, annually, for more than 11 years, characterized the period between July 1, 1995 to June 30, 1996 as showing "stunning results" for the industry. This is based on sales of biotechnology-related products, which rose 16% to 10.8×10^9/yr.

Biotechnology products, as defined here, are products sold by the biotechnology companies and exclude many bioproducts sold by established pharmaceutical, food or chemical industries, even though these products are biotechnology related. While pharmaceutical companies report sales, the biotechnology companies report revenues, which includes sales as well as income received for services and contract research. This method of accounting helps the financial analysis of the industry to track the progress of companies clearly started up within the last 20 years with 80% of these started up since 1986. The distinction is clearly described by Burrill and Lee in 1991. "Biotech companies will transform traditional industries. Pharmaceutical companies will become biopharmaceutical companies, being themselves transformed as they partner with biotech companies, acquire biotech companies and their technologies, and marry traditional pharmaceutical R&D methodologies with biotechnology's understanding of cellular

TABLE 3. ANNUAL SALES OF MAJOR THERAPEUTIC BIOPHARMACEUTICALS TOP 6.8×10^9 IN 1995.[a]

Biopharmaceutical[a]	Disease	Annual sales[b] ($ millions)
Anti-CD3 Mab (Monoclonal Antibody)	Transplant rejection	$80
Dnase	Cystic fibrosis	111
Erythropoietin	Anemia	1,650
Factor VIII	Hemophilia	250
GCSF (Granulocyte Colony Stimulating Factor)	Neutropenia	936
Glucocerebrosidase	Gaucher's disease	215[c]
GMCSF (Granulocyte Macrophage Colony Stimulating Factor)	Bone marrow transplant	41
GPIIb/IIIa Mab	Blood clots in angioplasty	130
Hepatitis B Vaccine	Hepatitis B	1,000
Human Growth Hormones	Growth deficiency, renal insufficiency	450
Insulin	Diabetes	700
Interleukin-2	Kidney cancer	40
α-Interferon	Cancers, hepatitis	700
β-Interferon	Multiple sclerosis	255
γ-Interferon	Granulomatous disease	4
Tissue Plasminogen Activator	Heart attack/embolism, stroke	300

[a] *Sources*: Biotechnology Industry Organization, Pharmaceutical Research & Manufacturers of America, company results, analyst reports.
[b] May include different companies' products approved for several uses.
[c] Includes sales of isolated natural protein. Thayer (1995).

behavior." "... we will see emergence of bioagricultural, biochemical, and bioenergy companies. Biotech companies will continue to exist with traditional companies transformed through their partnership."

The top 10 recombinant drugs accounted for nearly 5×10^9 in sales and the top 16 had sales of about 6.8 billion (Thayer, 1996), as shown in Table 3. Product sales, revenues, research and development expenses, and market value increased significantly for the industry between 1991 through 1996 as shown in Table 4 (Thayer, 1997c), although the industry as a whole would show a negative earnings and a deferred profit margin. This is consistent with the characteristics of the "present phase" of the biotechnology industry's life cycle. See Fig. 1. As of the first half of 1997, six leading biotechnology companies dominated the earnings figures. The single major contributing company was Amgen (201×10^6 earnings for the first half of 1997). The combination of Biochem Pharma, Biogen, Chiron, Genentech, and Genzyme contributed another 98×10^6. The remaining 24 firms of the top 30 combined for a net loss of 120×10^6. The accounting of costs helps to communicate the stage at which the industry finds itself. However, an understanding of the technology, and insights into potential applications and markets is needed if the industry's tremendous potential for growth is to be understood and windows of opportunity identified. This article attempts to provide a brief

TABLE 4. BIOTECHNOLOGY INDUSTRY SUMMARY: JULY 1 TO JUNE 30[a]

Values in $\$ \times 10^9$	1990–1991	1994–1995	1995–1996	% Change
Product sales	4.0	9.3	10.5	16
Revenues	5.8	12.7	14.6	15
Research and development expense	2.9	7.7	7.9	3
Net loss	10.9	4.6	4.5	−2
Market value		52.0	83.0	60
Number of companies (Industry total)	1,100	1,308	1,287	−2
Number of employees	70,000	108,000	118,000	9

[a] Adapted from Thayer based on data of Ernst & Young; 1990–1991 data from Burrill and Lee.

introduction to the possible importance of selected molecules. These will be given as examples during presentation of bioseparations and bioprocess engineering principles.

Published analyses of the revenues and earnings segregate the biotechnology industry from the large pharmaceutical companies, which also market biotechnology products but are not counted as part of the startup biotechnology industry. Comparison of the pharmaceutical industry's earnings compared to the combined earnings of 30 major biotechnology industries during the first half of 1997 illustrate the large differences in sales (revenues), earnings, and profit margins (see Table 5). The current pharmaceutical industry is much larger than the combined earnings of the 30 most significant biotechnology companies.

The biopharmaceutical industry is in the early part of its life cycle. The assessment of an industry doing well when it is losing money may seem unusual to a student in engineering or biological sciences. However, this is consistent with the start-up of an entire new sector of the economy, where negative cash flow during the beginning years of the individual companies cannot be offset until approved products reach the market and sales are generated. It is only in the later phase of the industry's life cycle that the overall industry becomes profitable.

The key phases of the biotechnology industry's life cycle are schematically represented by Burrill and Lee. The data in Table 4 (for 1994–1995 and 1995–1996) are based on companies that were started up during the prior 10 to 15 years. All of them were at a similar stage of the product development cycle, and few had positive earnings at this point. This differs from an established industry where most companies have positive earnings, with only a few start-ups that might be experiencing a negative cash flow. The rapid changes between negative earnings and profitability can be illustrated by the earnings history of the six largest biotechnology companies with acquisitions and restructuring contributing to the volatility. The biotechnology industry, while risky, also has an enormous earnings potential. Hence, there are investors willing to risk capital to get the industry started.

The importance of new developments is often difficult to gauge at times relatively close to their discovery. Nonetheless, it would appear that an important transition occurred approximately 45 years after the discovery of penicillin, and 30 years after its first large scale manufacture through submerged fermentation. The first therapeutics, were microbial substances made by microbes for activity against other microorganisms. Biologically active proteins, referred to as biopharmaceuticals, were made possible by advances in molecular biology since microorganisms could now express mammalian proteins having therapeutic use. Perhaps more importantly, molecular biology and genetic techniques have made it practical to discover new knowledge about how biological systems function, and how various molecules, not just proteins, moderate the biological chemistry of living cells. New types of therapeutic approaches are made practical by this expanded knowledge of biological systems. The means by which molecules for therapeutic use are manufactured may prove to be a key technological shift that differentiates the current fermentation and pharmaceutical industry from the nascent biotechnology industry.

Discovery of Type II Restriction Endonucleases. In 1970, Kelly and Smith, and Smith and Wilcox, discovered enzymes, known as Type II restriction endonucleases, which recognize a particular target sequence in duplex DNA. This enabled use of the enzyme to obtain discrete DNA fragments of defined length and sequence, and therefore facilitate a key step in carrying out genetic engineering (Old and Primrose). Unlike mutation and screening procedures, genetic engineering entails insertion of a foreign gene into the genetic material of a microbial cell, so that the cell can

TABLE 5. COMPARISON OF THE BIOTECHNOLOGY AND PHARMACEUTICAL COMPANIES FIRST HALF SALES AND EARNINGS, 1997[a]

	Industry	
	First half of 1997, all pharmaceutical companies	First half of 1997, 30 major biotechnology companies
Sales (revenues), $\$ \times 10^9$	63.0	3.73
Earnings, $\$ \times 10^9$	11.0	0.269
Profit margin[b], %	17.3	7.2

[a] (From Thayer, (1997d)).
[b] After tax earnings as a percent of sales.

Biotechnology Industry Life Cycle and Value Drivers

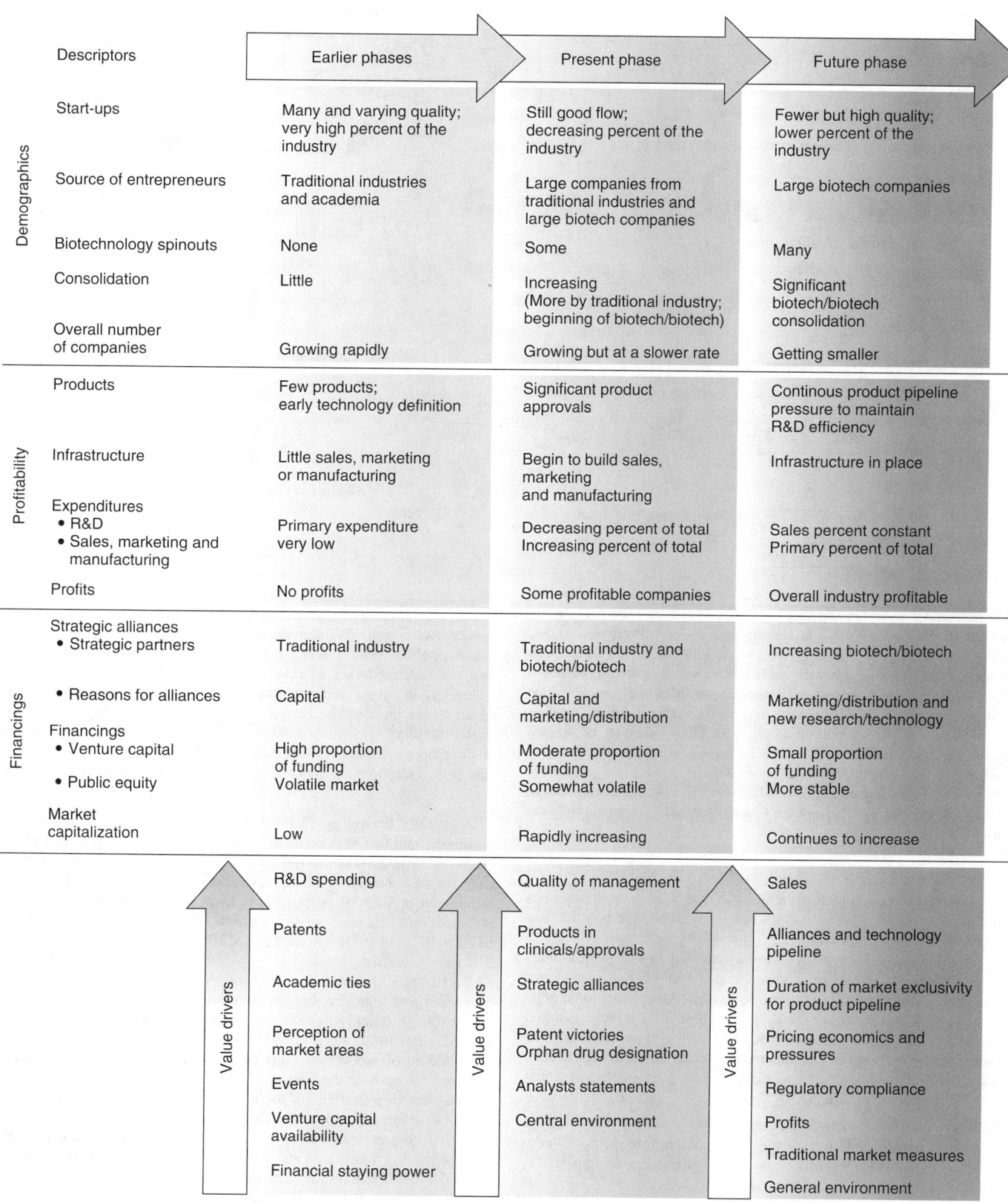

Descriptors	Earlier phases	Present phase	Future phase
Demographics			
Start-ups	Many and varying quality; very high percent of the industry	Still good flow; decreasing percent of the industry	Fewer but high quality; lower percent of the industry
Source of entrepreneurs	Traditional industries and academia	Large companies from traditional industries and large biotech companies	Large biotech companies
Biotechnology spinouts	None	Some	Many
Consolidation	Little	Increasing (More by traditional industry; beginning of biotech/biotech)	Significant biotech/biotech consolidation
Overall number of companies	Growing rapidly	Growing but at a slower rate	Getting smaller
Profitability			
Products	Few products; early technology definition	Significant product approvals	Continous product pipeline pressure to maintain R&D efficiency
Infrastructure	Little sales, marketing or manufacturing	Begin to build sales, marketing and manufacturing	Infrastructure in place
Expenditures • R&D • Sales, marketing and manufacturing	Primary expenditure very low	Decreasing percent of total Increasing percent of total	Sales percent constant Primary percent of total
Profits	No profits	Some profitable companies	Overall industry profitable
Financings			
Strategic alliances • Strategic partners	Traditional industry	Traditional industry and biotech/biotech	Increasing biotech/biotech
• Reasons for alliances	Capital	Capital and marketing/distribution	Marketing/distribution and new research/technology
Financings • Venture capital	High proportion of funding	Moderate proportion of funding	Small proportion of funding
• Public equity	Volatile market	Somewhat volatile	More stable
Market capitalization	Low	Rapidly increasing	Continues to increase
Value drivers	R&D spending	Quality of management	Sales
	Patents	Products in clinicals/approvals	Alliances and technology pipeline
	Academic ties	Strategic alliances	Duration of market exclusivity for product pipeline
	Perception of market areas	Patent victories Orphan drug designation	Pricing economics and pressures
	Events	Analysts statements	Regulatory compliance
	Venture capital availability	Central environment	Profits
	Financial staying power		Traditional market measures
			General environment

Fig. 1.   Conceptual representation of biotechnology industry life cycle. The early phase coincides with period of approximately 1975 to 1990. The present phase started in about 1990 and is likely to continue through the beginning of the 21st century (figure from Burrill and Lee) [1991].

produce proteins that would otherwise not be found in the cell. The impact and commercial possibilities of this technique are credited to Cohen, Chang, and Boyer, who in 1974, established practical utility of constructing biologically functional plasmids *in vitro* and transforming *E. coli* to express foreign protein (Ryu and Lee, MacQuitty). The basis of the technique is illustrated in Figure 2.

Kohler and Milstein are responsible for the third significant development during this time. In 1975 they demonstrated that cells derived from mouse *B lymphocytes*, which secrete antibodies, were fused with mouse myeloma tumor cells from a mouse, which can grow indefinitely in culture (see Fig. 3). The resulting fused cell, is a *hybrid-myeloma* or hybridoma cell that can grow in cell culture and produce large quantities of chemically

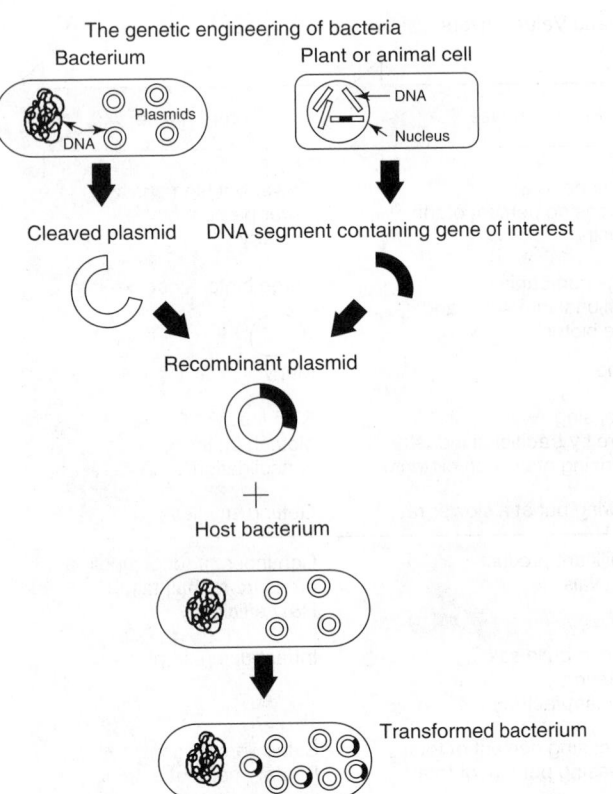

Fig. 2. One common way to genetically engineer bacteria involves the use of small, independently replicating loops of DNA known as plasmids. Certain enzymes can cleave these plasmids at specific sequences in their genetic codes. DNA from other organisms that has been treated with the same enzymes can then be spliced into the plasmids with enzymes that join the cut ends of DNA. These recombinant plasmids are reinserted into bacteria, where they can reproduce themselves many times over. At the same time, the bacteria can divide, creating millions of copies of the introduced DNA. This DNA can then be studied through analytical techniques, or, if a gene within the introduced DNA can be made to produce the same protein it did in its original location, the genetically engineered bacteria can be used as microbial factories to make large quantities of the protein (reprinted with permission from Olsen, 1986, p. 17, Copyright National Academy Press)

identical antibodies, i.e., monoclonal antibodies, which are excellent probes for diagnostic purposes (Olsen) and are being developed for possible use as a drug delivery system (see Fig. 4).

The restriction enzymes and hybridomas enabled new types of pharmaceutical products The developments associated with restriction enzymes (facilitate genetic engineering) and hybridomas (produce monoclonal antibodies) opened up vistas for a new range of potential products. Isolated from humans directly, these can now be produced in microorganisms or cell culture, and be used in large concentrations to enhance the body's ability to fight heart disease, cancer, and viral infections against which most antibiotics have little or no effect.

Regulatory Issues. A new drug must move through three stages of testing and review, adding to the time required to bring a new drug to market. These stages follow an investigational new drug (IND) application (Thayer, 1991):

Phase I Clinical Trials: Safety and pharmacological activity of a drug is established using 100 or less, healthy human volunteers. These require about one year;

Phase II Chemical Trials: Controlled studies are carried out on 200 to 300 volunteer patients to test drug efficacy together with animal and human studies for safety. This stage requires about two years.

Phase III Clinical Trials: Efficacy results are confirmed using about 10 times more patients than used in Phase II trials. Low-incidence adverse effects are also identified. This phase requires about three years.

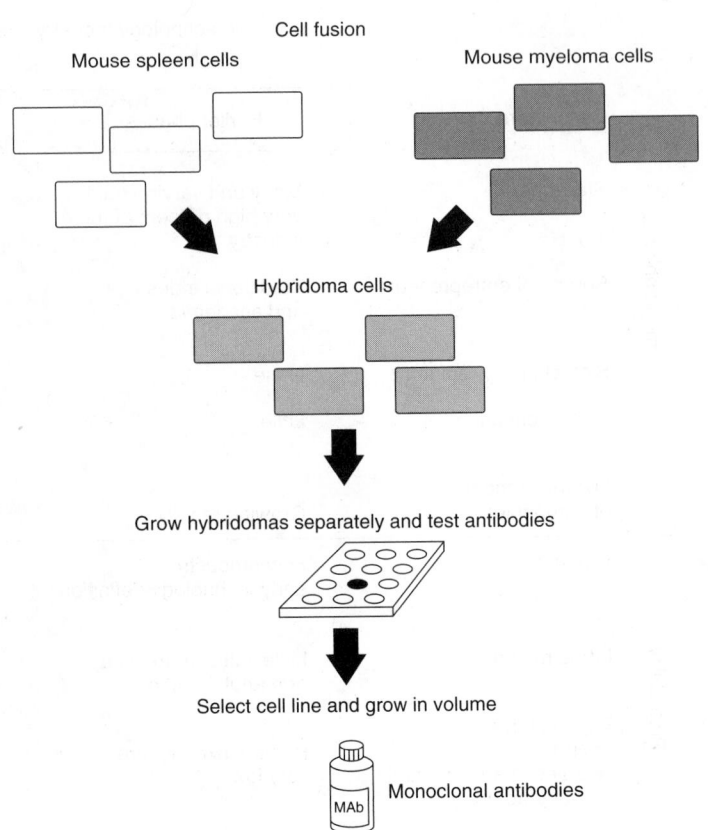

Fig. 3. To produce monoclonal antibodies, antibody-producing spleen cells from a mouse that has been immunized against an antigen are mixed with mouse myeloma cells. Under the proper conditions, pairs of the cells fuse to form antibody-producing hybrid-myeloma ("hybridoma") cells, which can live indefinitely in culture. Individual hybridomas are grown in separate wells, and the antibodies they produce are tested against the antigen. When an effective cell line is identified, it is grown either in culture or in the body cavities of mice to produce large quantities of chemically identical, monoclonal antibodies (reprinted with permission from Olsen, 1986, p. 26, Copyright, National Academy Press)

A product license application is then filed with the FDA, which had required about two to two and one half years to review up until 1993, when efforts were undertaken to reduce the time required. Now, some approvals can now be achieved in 15 months. Thus, if approved, a new product may require up to 7 to 10 years to pass from preclinical trials to commercial introduction. This process, for one new drug, is estimated to cost at least 120×10^6 and as much as 240×10^6. For serious or life-threatening diseases, promising Phase I results can lead to combined Phase II and Phase III trials, thus reducing the time by two or three years. The approval process is thorough, lengthy, and expensive. Bioprocess engineering plays a key role in quickly developing means of production for supplying the needed amounts of pure material required for these trials. In this case, the scale-up of production must follow a time-scale that is less than that required for each of the trials. Otherwise, completion of the trials can be delayed due to a shortage of the drug being tested.

The challenge for bioprocess engineers transcends scale-up issues and the drive for a company to be the first to market. As articulated by Bailey and Ollis, "... modifications in the organism or in the process require some degree of reiteration of earlier tests, or re-examination of the system by regulatory authorities. Therefore, a premium exists for engineering strategies that provide effective *a priori* guidance ... in organism, product, and process development." Because of the associated regulatory costs, incremental process improvements may not be commercially significant until relatively late in the product life cycle, when there are several suppliers and competition for a given drug develops. A similar analysis was also presented by Wheelwright (1991) in the context of purifying proteins in a commercial setting.

Reduction in the time and expense for approving new biopharmaceuticals may accelerate development of new products and processes. The increase in the number of new biotechnology drugs and other products that

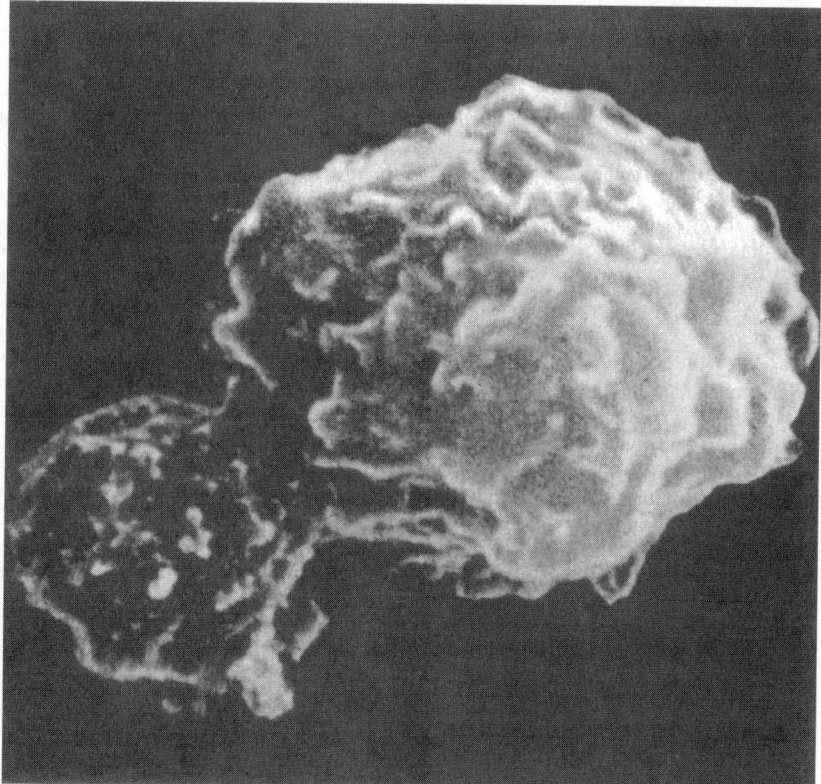

Hybritech

Fig. 4. A mouse spleen cell and tumor cell fuse to form a hybridoma. As the hybridoma divides, it gives rise to a "clone" of identical cells, giving the name "monoclonal" to the antibodies those cells produce (reprinted with permission from Olsen, 1986, p. 27, Copyright, National Academy Press)

are subject to review and approval by the FDA coincided with changes at the FDA starting in about 1990. Practices were initiated to decrease the amount of time between development of a new drug or biologic and its approval. A concurrent process was instituted to move safe, efficacious drugs, that were not fully tested, to terminally ill patients (Ember). However, for many new pharmaceuticals, a key issue was the length of time required for the review process which, in 1993, required about 30 months. By 1996, the time was reduced to 15 months (Stone). Only 15 biotechnology drugs had been approved from 1983 to 1992 year period. However, there were approximately 300 products in clinical trials in 1992. Many of these were likely to reach the FDA for review (Ember). Consequently, the review process needed to be handled more quickly.

Fifty additional reviewers (for a total of 250) and the appointment of Kathryn C. Zoon, as a new director for the Center for Biologics Evaluation and Research, occurred in 1992, under the leadership of FDA commissioner David Kessler, despite the lack of additional funding for the FDA to carry this out (Ember). By 1996, 500 new scientific staff (from microbiologists to statisticians) had been hired to review applications. The funds for the additional staffing resulted from authorization by Congress for the FDA to collect user fees from the industry.

The effect of user fees, and the additional staff that it enabled, was a reduction in the time required to process a new drug application. It was 30 months in 1992 and 17 months in 1995, with 15 months projected for 1996. Furthermore, the time required for anticancer and AIDS drugs was as little as six months. There were still industry complaints, however, that FDA reviewers would often ask for additional, time-consuming tests during the later part of the review process, thereby causing extra research and development expenses. Overall costs for developing, producing, and testing a new drug was estimated to be 500×10^6 in 1996 (Stone).

The renewal of the user fee arrangement for an additional five years by the U.S. Congress in November 1997, was viewed as "landmark legislation that codifies many new FDA regulations, revises its policies, and updates a number of outmoded practices. The bill also reauthorizes the Prescription Drug User Fee Act (abbreviated PDUFA) for another five years, an objective that pushed the entire package through the legislative labyrinth (Wechsler)". Some key provisions, other than user-fee contributions from the pharmaceutical industry, were fast-track approval of critical drugs;

expanded access to investigational therapies for patients with AIDS, cancer, and other serious diseases; and six months of additional exclusivity for manufacturers that conduct pediatric studies for new or approved drugs or manufacturers that develop new antibiotics.

Another key provision could encourage, as well as speed-up, development of new biopharmaceuticals. The FDA may now approve a new drug or biotech therapy based on pilot or small-scale manufacturing data. Preapproval requirements were reduced for certain post-approval manufacturing changes for drugs and biologics (Wechsler). Although the regulations on these policies would require another two years (until 1999) to be published (Wechsler), one implication was that some types of biologics (such as recombinant vaccines) derived through recombinant organisms or recombinant technology might be regulated more like drugs. Another practical effect may be that test quantities of new recombinant protein products can be produced in a facility other than the one used for full-scale production. Previously a manufacturing facility would be built (before the product was approved) to produce test lots of a new product. It may soon be possible to contract out manufacture of the product (to an existing facility), rather than build a new plant in-house. A new plant, based on an untested product, had previously been required due to the constraints of the FDA approval process. This is very expensive and inhibited product development, since such a plant would face a risk of being "moth-balled" if clinical trials were unsuccessful. While the equipment in such a facility could be used elsewhere, the investment in engineering, systems design, process control, and investor's capital would be lost. Changes in the guidelines would ease this restriction on many types (but not all) biotechnology products. A facility designed specifically for manufacturing the product would not need to be built before the product is approved. This could reduce the time and expense of introducing a new product, as well as moderate the risk of developing a new product.

The goal of the FDA is to reduce the time needed to review and act on new drug applications, product and biologic license applications, efficacy and manufacturing supplements, and resubmissions. According to the synopsis given by Wechsler in November, 1997 (Wechsler): "the agency's goal will be to review and act on 90% of standard submissions and efficacy supplements within 10 months of receipt, priority submissions and efficacy supplements within 6 months, manufacturing supplements within 4–6 months, and resubmissions within 2–4 months." The practical consequence of these modernized policies and shorter turn around time for supplements may be to promote development and installation of process improvements. This should have a positive effect on the development on new types of bioseparations for both new and existing products, where overall purification costs account for 50 to 90% of manufacturing costs. This change in regulation should create opportunities for bioprocess engineers, conversant in bioseparations, to invent, develop, design, and operate improved purification processes that cost less.

Growth of Industrial Enzymes

Catalysis in biotechnology, biocatalysis, has traditionally been associated with enzymes. At first glance the enzyme market would appear to be large at 650×10^6 (worldwide) in 1996. About 40% of the enzymes are used in detergents, 25% in starch conversion, and 19% in the dairy (cheese) industry. Consequently only a small fraction of the enzymes are employed as catalysts in reactions other than hydrolysis of starches (amylases) protein (proteases), and milk (chymosin and related proteases) (Layman). An example of how the new biotechnology has affected the enzyme business is given by the production of the milk clotting enzyme, chymosin, through a recombinant yeast, and improving cellulase enzyme titers in fungal fermentations. The recombinant chymosin is in commercial use in the dairy industry, where it supplants an enzyme previously derived from an animal source. Cellulase enzyme systems have been used in the clarification of citrus juices, and more recently to impart a "stone-washed" appearance to cotton fabrics by limited hydrolysis of the cellulose containing cotton fabric (Tyndall). These are derived from fermentation of recombinant microorganisms. Technology for cost-effectively converting cellulosic wastes to fermentable sugars could some day be enabled by application of cellulase enzymes, and thereby create a very large market for these enzymes. In the interim, a number of smaller, specialty niches will make up the demand for these enzymes, although the impact of such products in processing bulk agricultural commodities into value-added products should not be overlooked.

Glucose Isomerase and Sugar Production from Corn. Glucose isomerase is an example of an enzyme that catalyzed the development of a new industry, as well as the isomerization reaction for converting glucose (from starch) into fructose. Fructose containing corn syrups had grown to 17×10^9 pounds/year in 1991, worldwide, since being introduced in 1967 and 21×10^9 pounds in the U.S. in 1997 (Corn Refiners Assoc., Lastick and Spencer). The industry is based on an enzyme known as glucose isomerase that catalyzes the conversion of glucose (dextrose) to its chemical isomer (fructose). The equilibrium mixture of glucose with 42% fructose has a sweetness which approaches that of sugar derived from sucrose (cane sugar). Development of large-scale separation technology for partially separating glucose from fructose to give a product with 55% fructose, enabled this sugar to be used in consumer products ranging from soft drinks to donuts. Since the glucose was derived from corn, which is a less expensive agricultural commodity than sucrose from sugar cane, high fructose corn syrups displaced sucrose in the majority of high volume applications. The applications had previously been based on sucrose that had been hydrolyzed to glucose and fructose to obtain the same level of sweetness as high fructose corn syrup.

Glucose isomerase was initially isolated and identified in 1953 as a xylose isomerase (Hochester and Watson). Four years later the glucose isomerizing capability of xylose isomerase from *Pseudomonas hydrophilia* was discovered although arsenate was required to enhance the reaction. This enzyme was thus impractical for food production (Marshall and Kooi).

A key development for commercial prospects occurred in 1961, when it was found that not all glucose isomerases required arsenate, and that at least one of the arsenate requiring enzymes was in fact a glucose phosphate isomerase which converted glucose-6-phosphate to fructose-6-phosphate. While *Escherichia intermedia* had glucose phosphate isomerase (Natake and Yoshimura, Natake 1966, 1968) several other organisms (*Lactobacillus brevis, Aerobacter cloacae,* and *Streptomyces phaeochromogenus*) yielded isomerases, which did not require arsenate (Yamanaka, Tsumura and Sato, 1965a, b, 1970). Glucose and xylose isomerase activities were proposed to be due to the same enzyme (Yamanaka 1963b, 1968). Evaluation of the activities of a highly purified crystallized glucose isomerase from *Bacillus coagulans* was consistent with this hypothesis (Danno). It has been suggested that two enzymes, xylose isomerase (requires xylose as an inducer, but no arsenate), and glucose phosphate isomerase (does not

require xylose as an inducer, but needs arsenate for activity) were co-purified in the early work, hence giving the observed properties (Tsumura et al. 1967).

The identification of a thermally stable glucose isomerase and an inexpensive inducer was needed for an industrial process. The discovery of enzyme from *S. phaeochromogenus*, which was thermally stable at 90 °C in the presence of substrate and Co^{++} (Tsumura and Sato 1965a, b, Tsumura et al. 1967), was important to setting the stage for commercialization. A thermally stable enzyme reduces cost by increasing its useful lifetime, and therefore, its overall productivity. The last major barrier to large-scale production of the enzyme was replacement of the expensive xylose media on which *S. phaeochromogenus* had to be grown to obtain glucose (xylose) isomerase enzyme, with a less expensive carbohydrate source. This turned out to be xylan containing wheat bran, corn bran, and corn hull (Takasaki, Takasaki et al.). Later it was found that D-xylose could induce glucose isomerase activity in *Bacillus coagulans* cells previously grown on glucose (Danno). The stage was now set for a new biocatalyst to transform the corn wet-milling industry.

The demand for high fructose corn syrup (HFCS) resulted in large scale use of immobilized enzymes and liquid chromatography. In 1965–1966, Clinton Corn Processing Company of Iowa (at that time a division of Standard Brands, Inc.) entered into an agreement with the Japanese government to develop the glucose isomerase technology for commercial use in the U.S. (Tsumura et al., Lloyd and Horvath). The glucose isomerase was really a xylose isomerase derived from *Streptomyces rubiginosus* and had the properties of thermal stability and high activity which made it amenable for use at industrially relevant conditions for use in batch reactions. The first enzymatically produced fructose corn syrup, was shipped by Clinton Corn in 1967. A.E. Staley licensed the technology from Clinton Corn and entered the market.

The basic patent coverage for use of xylose isomerase to convert glucose to fructose was lost in 1975 as the result of a civil action suit between CPC International and Standard Brands. This enabled development of alternate processes. By 1978, the estimated US production volume was 3.5×10^9 lb and consisted mostly of syrups containing 42% D-fructose sold at a 71% solids level (Antrim et al.). The introduction of large scale liquid chromatographic purification of the fructose enabled production of a 55% fructose which could be used in soft drinks in place of invert from sucrose.

The growth of the industry both promoted and followed development of large-scale technologies for carrying out the conversion in reactors packed with immobilized glucose isomerase, and large-scale liquid-chromatographic separations of the resulting sugars. Immobilized enzyme technology took advantage of the heat stability of the glucose isomerase by fixing the enzyme within, or on the surface of, solid particles. These particles, when packed in large vessels, enabled the enzyme, which was otherwise water soluble, to be held immobile while the glucose solution (at concentrations of up to 40% solids) was passed through the reactor. This made it possible to continuously process an aqueous glucose solution by passing it through a fixed bed, biocatalytic reactor to convert it to fructose. However, equilibrium limited the conversion to 42% fructose. Large-scale liquid chromatography was therefore needed to upgrade the 42% fructose directly obtainable from enzyme conversion, to the 55% required for sweetness equivalent to invert (a hydrolyzed form of sucrose). Liquid-chromatographic separations technology (Sorbex) for continuous sorptive separation of xylenes in a simulated moving bed was adapted to glucose/fructose separation by UOP to give the SAREX process for enriching the fructose content. Part of the effluent from the immobilized enzyme reactor is diverted to the chromatography column. The glucose-rich stream can be recycled to the immobilized bed reactor to be converted to 42% fructose, as shown in Figure 5. The stream high in fructose is mixed with the remaining 42% fructose stream to give the 55% product. The combination of an enzyme reaction with separation made the 55% HFCS product economically and technically feasible in about 1975. Subsequently, HFCS −42% and −55% syrups gradually displaced sucrose for commercially produced beverages and baked products.

Rapid growth of HFCS market share was enabled by large-scale liquid chromatography and propelled by record high sugar prices. The "Pepsi" taste challenge showed consumers preferred soft drinks that were made from 55% HFCS compared to a competing product made from sucrose. Large-scale liquid chromatography made it possible to upgrade the high fructose corn syrups from 42% fructose to 55% fructose. By 1984, the

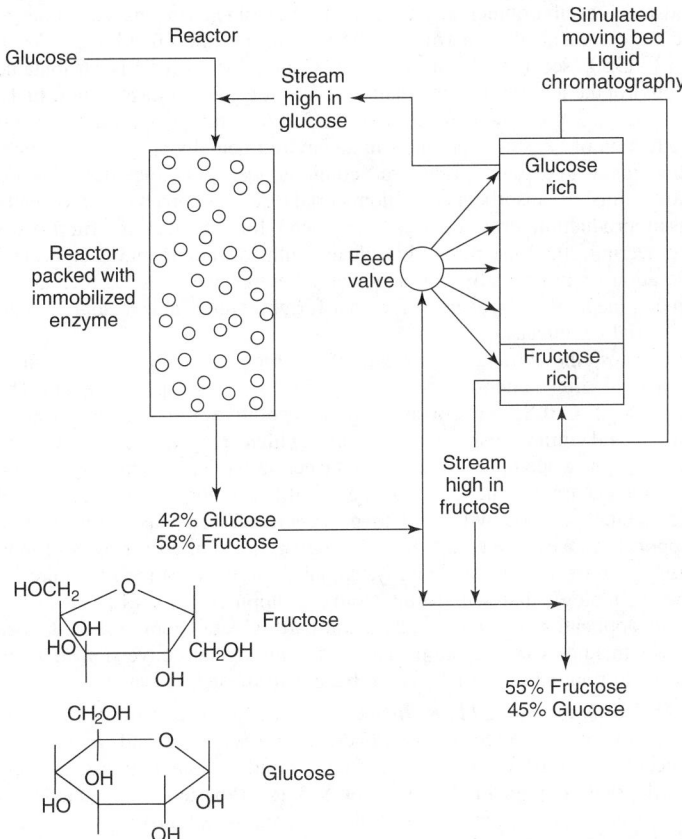

Fig. 5. Schematic diagram of combined immobilized enzyme reactor and simulated moving bed chromatography for producing 55% HFCS. The immobilized reactor is filled with pellets or particles in which the glucose isomerase is either entrapped or attached. Glucose solution, which also contains ions required for enzyme activity and about 3% residual soluble starch fragments (maltodextrins). The glucose feed is passed through the reactor, which is maintained at 50 to 60 °C. The immobilized enzyme converts the glucose to fructose. The product leaving the reactor contains about 42% fructose as well as the other ions and maltodextrins that do not react. The composition shown in the figure is normalized for the total glucose and fructose content, and does not reflect other dissolved solids. The effluent from the reactor enters a very large liquid chromatography column, packed with an adsorbent that selectively, but weakly, retains the fructose over glucose. The 42% fructose feed is introduced to different parts of the column in a time varying manner, through a special feed valve, in a manner that simulates a moving bed, so that the use of adsorbent is maximized. Sections of the bed that are rich in glucose and fructose are formed and maintained a steady state operation. This system allows a feed of 42% fructose to be continuously separated. The glucose that is separated out is returned to the reactor so that it will be converted to fructose

42% and 55% syrups exceeded 8×10^9 pounds annually, and accounted for 30% of all nutritive sweeteners in the U.S. (Lloyd and Horvath). The 55% fructose was quickly displacing industrial sugars previously derived from sucrose and accounted for over 50% of the HFCS shipped in the US in 1984. By 1990 the world market for HFCS had grown to about 17×10^9 lb/yr (Lastick and Spencer). The U.S. production of high fructose corn syrup −42% (denotes 42% fructose) was 8.57×10^9 lb pounds while high fructose corn syrup −55% + was 12.5×10^9 lb for a total of 21×10^9 in 1997 (Corn Refiners Assoc.). The value of the xylose (glucose) isomerase enzyme used in these processes is estimated at between 15×10^6 (Lastick and Spencer) to 70×10^6 dollars/yr. (Petsko), with most glucose isomerase being produced for captive or contract use.

The rapid adaption of the combined enzyme reactor/product separation technology in the wet milling industry coincided with dramatic, but short-lived, increases in world sugar (sucrose) prices. The demand for industrial sugars exceeded the supply. This short fall not only raised prices, but also provided a window of opportunity for a competing product, HFCS, to enter the market. The properties of HFCS were the same as dissolved invert (from sucrose) in applications where sucrose would otherwise be used. The

cost was less. By positioning the price of the HFCS slightly below that of sucrose (invert), HFCS producers were able to retain market share. When the prices increased by extraordinary values, the price of HFCS was about double of its cost of production. Consequently, investments in building new plants were quickly paid off, making further rounds of expansion possible, and further entrenching the competing product in the market. This is not typical of other biotechnology products, where higher prices are based on a unique product, protected by a patent.

Stabilization Sugar Prices. Immobilized enzymes and large scale liquid chromatography may have helped to stabilize sugar prices. Sugar prices are among the most unstable prices in international trade. World sugar price spikes were characterized by a pattern of high prices for one or two years followed by five to ten years of low prices. The price instability reflects the relatively small share of the world sugar production that is freely traded. Thus world crop changes and shifts in government sugar policies exert a disproportionate effect. The inability for sugar producers to adjust output rapidly in response to changing economic conditions helped to propagate this cycle. The delay between planting and harvesting is eight months for sugar beets and up to two years for sugar cane. Consequently, increases in production capacity during the high-price phase of the sugar cycle would take several seasons to be absorbed by a relatively steady, but slow, increase in consumption. High prices also promoted expansion or construction of new plants to process the sugars. As explained in the Agriculture Information Bulletin No. 478 (Petrides et al.), "Processing facilities are expensive to construct. They require large size to capture scale economies. Consequently once a plant is in place, there is a strong incentive for the plans to be fully utilized to spread out fixed costs. After five to ten years of low prices and slow growth in consumption, world sugar production typically catches up with processing capacity. At this point, a disruption to production triggers an explosive price rise, and the sugar cycle begins a new"(Petrides et al.).

This cycle has occurred five times: 1950–1951, 1957, 1963–1964, 1974–1975, and 1980–1981 (Barry et al.). Domestic (U.S.) sugar prices were also higher than the world price during several periods of low prices due to government policies. The major price surges of 1979/80 and 1980/81 resulted from bad weather in the USSR, India and Thailand; crop disease in Cuba; and reduced sugarcane acreage in Brazil. These periods coincided with major installation of plants for converting corn to high fructose corn syrup, with immobilized enzyme technology available in 1975, and combined enzyme conversion/liquid chromatography units being introduced in about 1977. Rapid payback of the capital investment was possible since the high world sugar prices helped to support the price of HFCS (Barry et al.). The technical ability to substitute HFCS for sugar in a wide range of products also promoted demand. The cost of production of HFCS, including normal returns on capital was estimated to be about 14/lb dry basis (Barry et al.) (This estimate, is based on costs of 4/lb net starch (equivalent to $2.60/bushel). The relatively low net starch costs are due to the value of corn wet milling by-products of oil, gluten feed and meal, with by-product values paying for approximately half of the corn costs. During the 1980s, enzyme costs also decreased, the scale of production increased, and the by-products enabled the plants to be fully utilized.) The lower production cost enabled the HFCS to be priced strategically below refined sugar prices. The prices of HFCS followed changes in sugar prices, but at a 10 to 30% discount. By 1989, HFCS had captured about half of the sugar market (Barry et al.).

Although per capita (U.S.) consumption increased to 130 lb of sugars/year, the price spike behavior was moderated. The acceptance of starch-derived and low-caloric sweeteners being widely accepted as sugar substitutes was a factor. The introduction of Aspartame, in 1981, and several other low-calorie, high intensity sweeteners coincided with a significant increase in the U.S. per capita consumption of low-calorie sweeteners, with low-calorie sweetener use being equivalent to 20 lb per capita in sugar-sweetness-equivalent (SSE). In 1980, this was about 6% of the total sugar consumption compared to 13% in 1988. Consequently, starch-based sweeteners, and to a less extent-low caloric sweeteners, are poised to take advantage of sugar shortfalls and high-prices, should they occur (Barry et al.). The recent history of sugar prices illustrates the complexity interactions between technology, commodity prices, and market demand for a product derived from renewable resources.

Cost-effective Enzymes. Submerged fermentation was the key to obtaining cost effective enzymes. As was the case for penicillin, success in submerged fermentation of microbes that produced the intracellular enzyme glucose isomerase was a key step in obtaining a cost-effective enzyme. When combined with development of processes for utilizing the enzyme at a commercial scale and an economic separation technology, a new product resulted. HFCS grew from zero to 17×10^9 lb annually (worldwide) in less than 25 years. The enzyme glucose isomerase also set the stage for the U.S. fuel alcohol industry, since the major HFCS producing companies had the infrastructure and large-scale processing know-how to establish it. The fermentation ethanol industry grew from about 100×10^6 gal in 1980 to over 1.5×10^9 gal/year in 1997.

Biocatalysts Used in Fine Chemical Manufacture. Although the total U.S. enzyme market in the mid-1990s was about $\$650 \times 10^6$ dollars/year, only about 5% was for biocatalysts used in fine chemical manufacture. The main biocatalysts were hydantoinase, β-lactamases, acylases/esterases, and steroid transforming enzymes (Polastro et al.). The discovery and development of catalytic antibodies may expand the role of biocatalysts, since this type of biocatalyst combines specificity of binding of an antibody together with the ability to transform a target molecule chemically. Hybridoma technology allows generation of antibodies to haptens or stabilized transition state analogues that have catalytic side chains in an antibody combining site. Hybridoma technology facilitates synthesis of gram quantities of the catalysts with very high specificity, compared to years required by other design and organic synthesis approaches. Catalysis using these types of antibodies are likely to occur since the antibodies are able to catalyze a large number of different reactions, including at least one type (Diels-Alder) for which an enzyme has yet to be isolated (Schultz et al.).

Enzymes in low-water environments can carry out selective transesterification reactions. Examples in the literature include polymerization of phenols in an organic solvent, and the formation of the biodegradable polymer sucrose polyester (Ryu et al., Patil et al.). This is currently a developing field, but holds promise of applications in organic synthesis, since some enzymes have been shown to be active in organic solvents. Fundamental studies have shown the enzyme subtilisin Carlsberg, which ordinarily hydrolyzes proteins, will carry out transesterification of N-acetyl-L-phenylalanine 2-chloro ethyl ester (NAPCE) with 1-propanol, when the reactants and enzyme are mixed in tetrahydrofuran containing a small amount of water (Affleck et al.).This research and work that has preceded it, gives a first indication of the potential of proteins, i.e. enzymes, as catalysts in organic synthesis.

The Polymerase Chain Reaction (PCR). Another significant development in enzyme technology was the polymerase chain reaction (abbreviated PCR). Since about 1985, this method has significantly increased the ease and speed of DNA sequence isolation *in vitro*. PCR was developed by Cetus Corporation scientists in 1984 and 1985. It is an enzyme catalyzed reaction that facilitates gene isolation and avoids the complex process of cloning which requires the *in vivo* replication of a target DNA sequence integrated into a cloning vector in a host organism. PCR is initiated by DNA denaturation, followed by primer annealing, and then addition of a DNA polymerase and deoxynucleoside triphosphates to form a new DNA strand across the target sequence. This cycle, when repeated n times, produces 2^n times as much target sequence as was initially present. Thus 20 cycles of the PCR yields about a millionfold increase or amplification of the DNA. Applications include comparison of altered, uncloned genes to cloned genes, diagnosis of genetic diseases, and retrospective analysis of human tissue (Arnheim et al.). Methodology for using PCR to clone DNA has also been recently described, although unwanted mutagenesis poses limitations (Higuchi).

The PCR is an example of a biotechnology that touches other scientific disciplines (and even the movie industry). Consider the study of ancient DNA for tracking evolution. Several researchers have referred to this as molecular archeology, since preselected DNA, which has survived in ancient tissue for 45,000 years or more, can be sufficiently amplified to provide quantities that permit direct sequencing of the DNA. This approach, in effect, provides a time machine that enables students of molecular evolution "to retrieve and study ancient DNA molecules and thus to catch evolution red-handed" (Pääbo et al.).

Growth of Renewable Resources. There are already large industries, associated with corn processing and food manufacture that utilize enzymes and microbial fermentation on an extremely large scale. In these cases, production and substrate costs can be 70% of the total product costs, and cost efficient engineering becomes paramount. The development of the industry that produces fuel grade ethanol, used as a nonleaded octane booster and environmentally acceptable fuel oxygenate, had its roots in the development of high fructose corn syrup and the oil price shocks of 1973–1980 (see timeline in Table 1). Continuing growth is attributable to environmental regulations relating to clean air and transportation fuels, as well as a continued government subsidy. Future potential exists for production of consumer products in an environmentally acceptable manner from renewable and agricultural commodities. This includes plastics, paints, finishes, adsorbents and biomaterials. As the impact of biologically based production grows; so will the need for engineering. Bioprocess engineering, the discipline that deals with development, design, and operation of biologically based processes, will have a major role in the development of bioprocesses for transforming renewable resources into industrial chemicals.

Perhaps the ultimate application of biotechnology is to control ecological life support systems (or CELSS) for space travel. The concept of CELSS encompasses bioregenerative life support systems that would mimic earth's ecosystem, which recycles C, H, O and N, through a complex and interconnecting series of biological and biochemical transformations (Averner et al.). For long term space travel and habitation, food will need to be generated, and wastes recycled to support life (Westgate et al.). Such biological life support is needed if long term space travel, and particularly, colonization of Mars is to be attained. The technology that will result from development of CELSS also has direct application to understanding the earth's ecosystem, and addresses environmental issues. Bioremediation, the biological conversion of toxic wastes to nontoxic materials, could benefit from such research.

Technologies Needed to Reduce Costs. The future growth of renewable resources as a source of value-added biochemicals and oxygenated molecules will be driven by technologies rather than markets, as long as oil prices remain in the range of $15 to 20/barrel. There may be a few cases where a fermentation process or enzyme transformation could provide a specific product at an equivalent or lower cost compared to a chemical process, while achieving a reduced environmental impact. In such a case, a biotechnology process might be used. An example is the production of fumaric acid for use as a food additive by fermentation, compared to fumaric acid by chemical synthesis for other purposes (Cao et al.). Some fermentation processes may offer attractive economics if there were a sustained increase in oil prices. Temporary spikes in oil prices could also offer windows of opportunity, although long term cost advantages would be needed to sustain growth of a renewable resource-based industry.

The attractiveness of renewable resources, aside from being renewable, lies in their ability to sequester and/or recycle CO_2, and their potential to be processed into useful products in an environmentally friendly manner. This has already been illustrated by corn. A renewable resource, it is used for food, animal feed, specialty starches, and as a source of sugars (mainly glucose) that can be fermented or converted to numerous value added products. See Fig. 6. Fats and oils derived from plants and animals/are examples of another type of renewable resource. These are the basis of industries that already produce a number of products that range from soaps to lubricants. See Fig. 7. The focus here is materials generated from sugars and polysaccharides. A thorough analysis of industrial uses of agricultural materials shows their potential (USDA). Current products range from fuel ethanol to starch adhesives, although the production of fuel alcohol from corn requires a government subsidy. Emerging technologies in ethanol production have decreased cost. The total energy required to process one gallon of ethanol with an energy value of 76,000 Btu/gallon was as high as 120,000 Btu when the industry started in 1977. Even when an energy credit of 32,000 Btu was assigned to the co-products of ethanol production (corn oil, gluten feed, gluten meal, and CO_2), there was still a net energy loss of 12,000. Research, and the rapid implementation of the resulting process improvements on an enormous scale by some of the ethanol producers, decreased energy costs. Process integration and implementation of new process technologies improved energy efficiency (USDA). (An example was a new anhydrous process at ADM for removing water from alcohol, which was installed in 1983). According to a stockholder's report, the company said that the process "will reduce total alcohol distillation energy by 35% or 4/gallon, and after shakedown is complete will probably be installed in all of our alcohol plants" (ADM). By 1993, the total energy requirement for an average-efficiency corn farm and an average-efficiency ethanol plant was 75,811 Btu/gal. With a co-product energy credit of 24,950 Btu, a net energy *gain* of Btu was achieved for modern, large-scale corn to ethanol routes. (Ryu and Lee, USDA).

Processing starches and sugars into industrial and consumer products

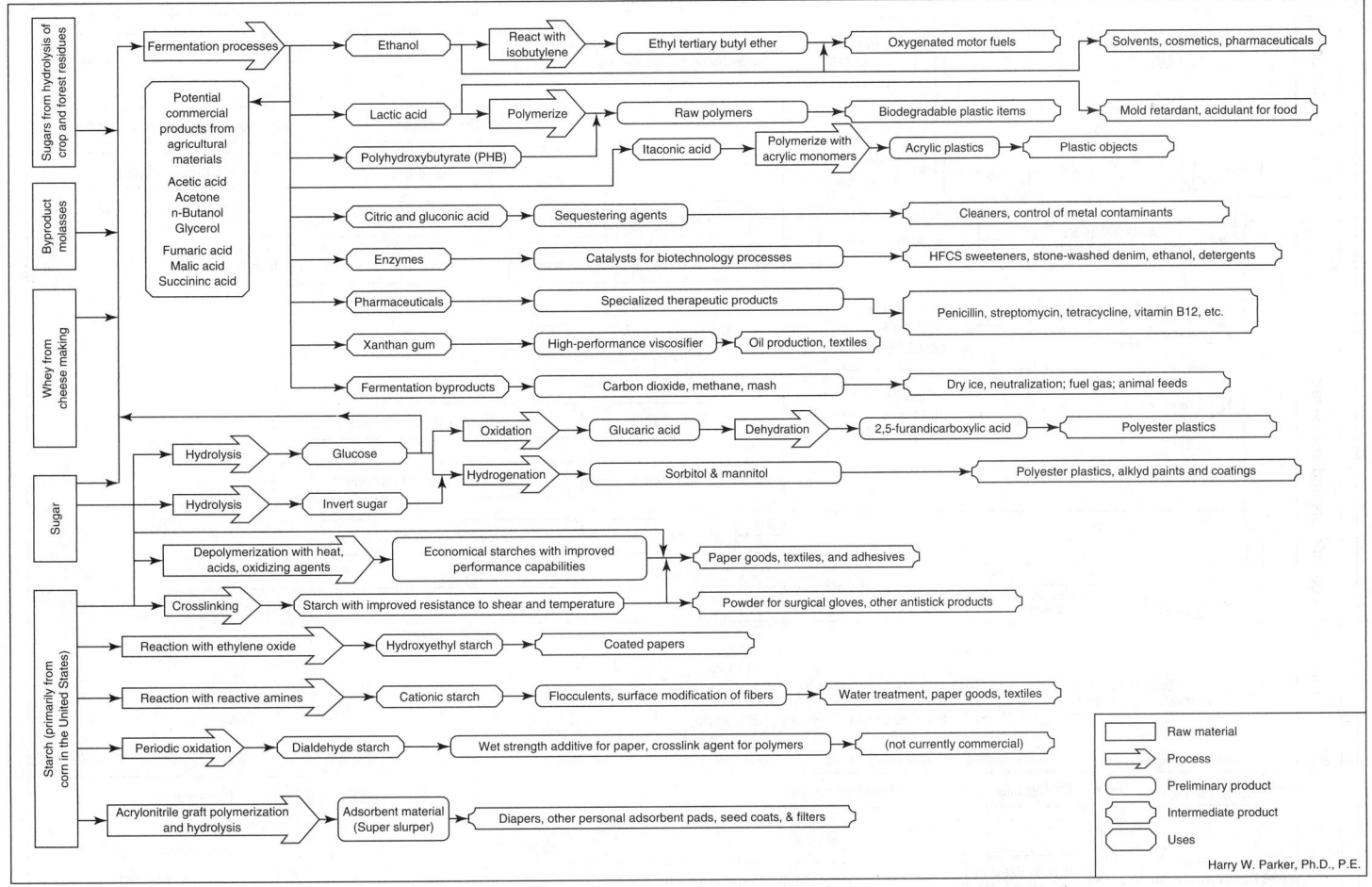

Fig. 6. Chart showing industrial chemicals derived from starches and sugars. (*USDA*)

Further improvements and decreased costs were anticipated due to better enzymes, bioreactor designs, and membranes for enhanced recovery of high-value co-products (Hohmann and Rendleman). Further significant reductions in cost, however, require lower feedstock prices. Thus, for example, the conversion of corn fiber, a cellulose containing co-product of ethanol production, to fermentable sugars could reduce costs by 3.0 to 7.5/gallon. (These cost comparisons were made relative to the cost of producing fermentation ethanol from corn in 1992, which was at $1.25/gallon, down from $1.35 to $1.45/gallon in 1980.) This is significant relative to the net feedstock (corn) costs of at least 44 per gallon. Cellulosic residues have the potential as being a less expensive substrate, but their composition and physical structure is different from starch. The cellulose structure is crystalline and resistant to being hydrolyzed. Consequently, a pretreatment is necessary to soften the cellulose prior to subjecting it to hydrolysis. Less expensive enzymes (that hydrolyze cellulose to glucose) are also needed to achieve a cost-effective hydrolysis step. Pentoses and hexoses occur in approximately 4:5 ratio, in cellulosic materials, compared to a 1:10 to 1:15 ratio in corn. Conversion of sugars from the pentosan and hexosan fractions of cellulosic materials will require microorganisms that ferment both pentoses and hexoses, to a single product (Hohmann and Rendleman).

Progress has been made in pretreatment (Weil et al., 1994a, Kohlmann et al., 1995a, Ladisch et al.) and enzyme hydrolysis (Weil et al., 1994b, Kohlmann et al. 1995b) as well as obtaining genetically engineered microorganisms that convert both pentoses and hexoses to ethanol (Lynd et al., Toon et al.). However, an economically viable process has not yet been achieved (Gulati et al.). Major improvements are still needed for all three technologies. Once these occur, their integration into existing ethanol plants is possible, as suggested by Figure 8. (Hohmann and Rendleman). The emergence of a cellulose-based industry would open the way for production of many other types of oxygenated chemicals and lead to the growth of a fermentation-based, renewable resources industry. However,

this development is likely to be incremental, at first, and subject to government policies, much as was the case for other biotechnologies and in particular, HFCS. The plants which produce HFCS are also the major source of fermentation ethanol, large volumes of amino and carboxylic acids, and high-volume specialty products. These plants are well positioned to implement improvements in cellulose conversion, as another source of fermentable sugars, and then utilizing these sugars in an optimal manner for production of a diverse product mix. The future of the industry is likely to evolve from being based only on hexoses to an industry that utilizes and transforms both hexoses and pentoses to value-added products. Bioprocess engineering will play a major role in implementing technology for this transition, while the current and massive restructuring and consolidation of the high volume bioproducts sector will shape the business alliances needed for the evolution of a biobased industry.

Phytochemicals. Renewable resources are a source of natural plant chemicals. Chemicals derived from plant products are referred to as phytochemicals. These chemicals are formed as either primary or secondary metabolites. Balandrin et al. proposed that chemicals from primary metabolism could be classified as high volume-low value bulk chemicals with values less than $2/lb. Examples are vegetable oils, fatty acids (for making soaps or detergents), and carbohydrates (sucrose, starch, pectin or cellulose). Secondary metabolites are classified as high value with values ranging from $100/lb to $5000/gram. Examples are pyrethins and rotenone (used as pesticides); steroids and alkaloids such as *Digitalis* glycosides, anticancer alkaloids, and scopolamine. Secondary metabolites in plants often have an ecological role as pollination attractants or for mounting chemical defenses against microorganisms and insects. Secondary natural products often consist of complex structures with many chiral centers that are important to the biological activity of the materials. Plant derived proteins can also be classified as high value products. For example, papain is a proteolytic enzyme used as a meat tenderizer.

Processing fats and oils into industrial and consumer products

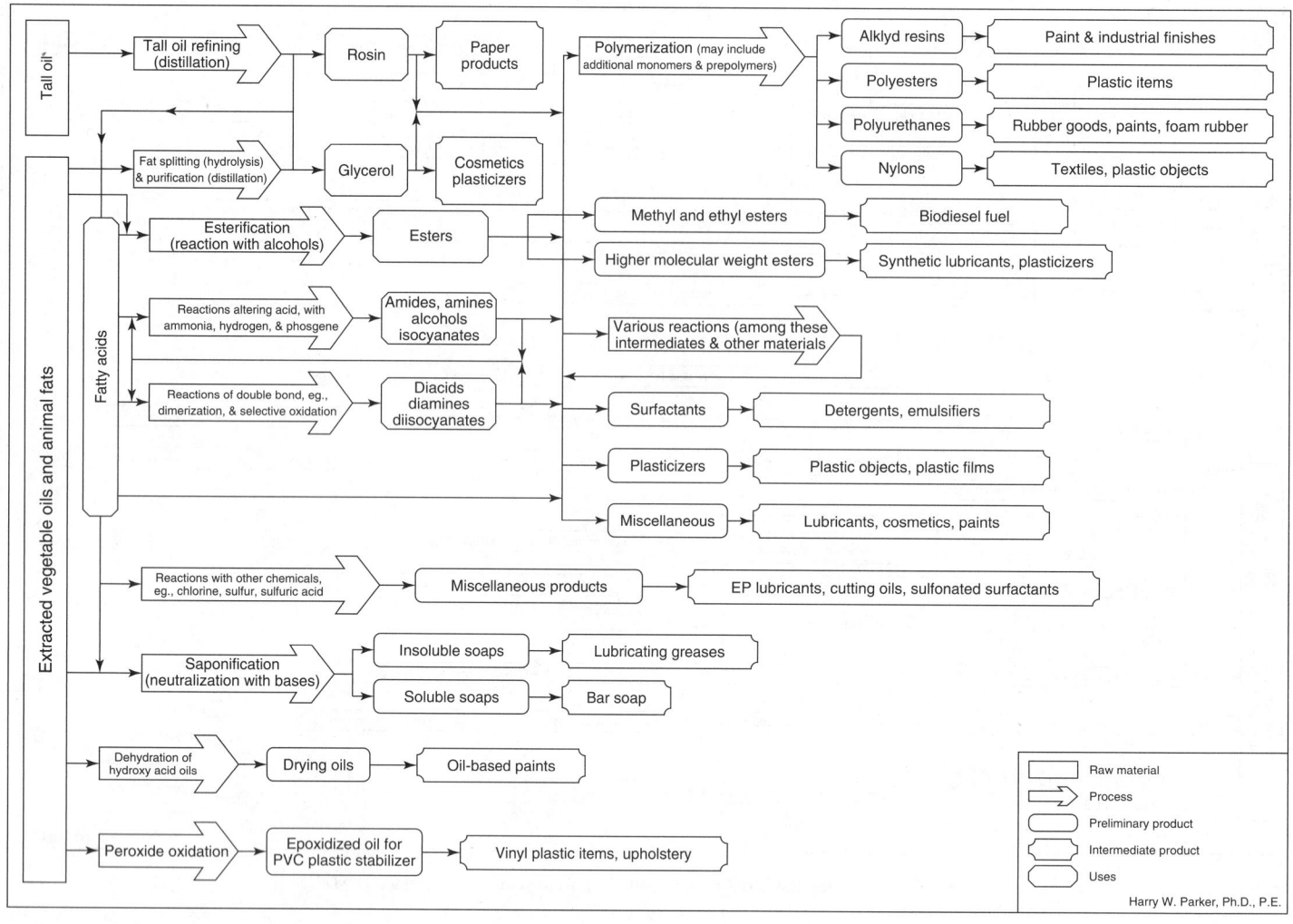

Fig. 7. Chart showing products derived from renewable sources of fats and oils. (*USDA*)

Bromelain (from pineapple) has applications in milk clotting. Malt extract from barley has amylolytic enzymes used in beer brewing.

The summary of valuable products derived from plant materials by Balandrin et al., indicate that the recovery of bioactive compounds, classified as secondary metabolites, will continue to be an important source of medicines, pesticides, and specialty chemicals. The naturally occurring compounds could be numerous. These include: pesticides (pyrethrum from flowers, rotenone from roots, and nicotine); growth inhibitors of soil microorganisms (allelochemicals) plant growth regulators (brassinolide promotes growth at 1 gram per 5 acres); insect growth regulators; and medicinals (anti-cancer and analgesic drugs).

A recent example is given by the insecticidal properties of the neem tree (Stone). Neem seeds have chemicals that ward off more than 200 species of insects, while exhibiting less toxicity to humans than synthetic pesticides, and having little effect on predators of the insects. Neem seeds were valued at $300/ton in 1992. However, a lower price is needed if the extracts are to find widespread use as insecticides. The economic potential for naturally derived insecticides is large considering that about 2×10^9 of synthetic insecticides are sold annually in the U.S. (Stone).

Another class of plant-derived compounds are known as saponins. Saponins consists of a aglycone consisting of a titerpene, a steroid, or a steroidal alkaloid attached to a one or more sugar chains consisting of up to 11 monosaccharide units. One saponin derivative (*Abrusoside E*) is 150 times sweeter than sugar. Another group of saponins from *Yucca schidigera*, a medicinal plant used by Native North Americans, have antifungal activity and could be used as preservatives for processed foods. The plant itself is reported to be recognized by the FDA as safe for human use. A saponin from a member of the lily family, *Ornithogalum*

saundersiae, has been found to exhibit little toxicity with respect to human cells, but is "remarkably toxic to malignant tumor cells" (Rouhi). Saponins from the South American tree, *Q. saponaria*, make animal vaccines more effective. Extracts from two types of African trees have been used to control Schistosomiasis, by eliminating host snails that harbor the disease (Rouchi).

Genetic engineering of bacteria or yeast to produce complex secondary metabolites remains a challenge, since these are formed via the joint action of many enzymes (i.e., many gene products). The culturing of plant cells, or plant tissues, to produce secondary metabolites could be used to produce some of the more valuable and bioactive compounds, and has been proposed for obtaining the chemotherapy agent, Taxol. It was estimated (in 1985) that a yield of 1 ggproduct/liter culture for compounds exceeding $1/gram in value is needed to obtain an economically viable process. Such an estimate may be optimistic. Even if successful, such cultures would still owe their start to the plants from which the compounds were identified, and the cells derived.

The advances in methods for genetic engineering of plants and propagating the transformed species could someday make use of the knowledge on the structure and function of plant metabolites. Attempts to determine plant genomes may soon be initiated. The combination of these factors could lead to development of other types of transgenic, pest resistant species, and perhaps the development of plant species grown to generate pharmaceuticals that are difficult or too costly to obtain by other means. These possibilities give a strong rationale for preserving the biodiversity of Earth's plant life, estimated between 250,000 and 750,000 species, since only 5 to 15% have been surveyed for biologically active compounds (Balandrin).

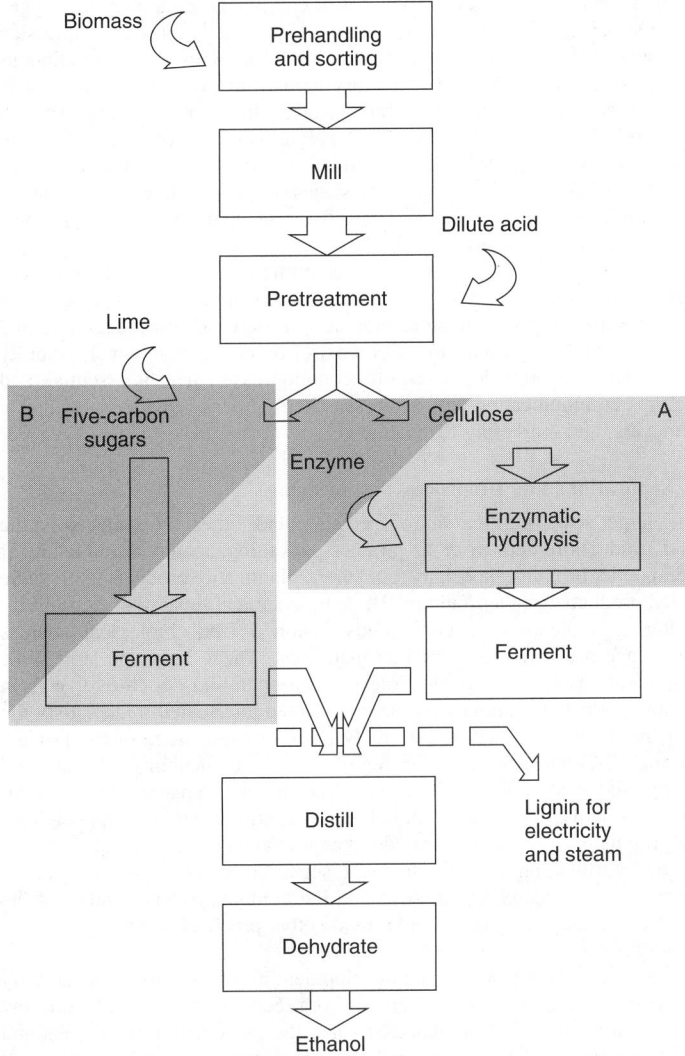

Flowchart of yard waste biomass processing plant
Biomass conversion requires a separate process to break down
complex five-carbon sugars.

Fig. 8.   Flow chart showing process for converting a cellulosic material (yard wastes i.e., grass clippings) to ethanol (from Hohmann and Rendleman)

Bioseparations. Bioseparations are important to the extraction, recovery, and purification of plant derived products. The recovery and processing of plant derived molecules and the processing of renewable resources can be classified by the type of product. Primary metabolites are generally biopolymers. Natural rubber, condensed tannins and lignin, and high molecular weight polysaccharides, pectin, starch, cellulose, hemicellulose, and gums, are included in this category (Balandrin). Their processing into high-volume, relatively low-value products, are subject to economics of scale (bigger is better). The separation of the plant material into value-added components may require a combination of physical treatments combined with a change in the material's structure. For example, the processing of cellulose (in the cell walls of wood or agricultural residues) requires the cellulose be hydrolyzed to its monomer, glucose. This can only be readily achieved if the crystallinity of cellulose structure and its close association with lignin is disrupted and an enzyme or acid catalyst is used to depolymerize the cellulose polymer into its constituent monosaccharide. This processing, referred to as pretreatment/hydrolysis must be carried out in a very inexpensive manner, if it is to be practical. A commercially feasible process is yet to be demonstrated (Weil et al., 1994a, Kohlmann et al., 1995a, Ladisch et al.).

The recovery of secondary metabolites can be achieved by steam distillation, aqueous extraction, or extraction with an organic solvent. However, these bioproducts, whose molecular weights are below 2000, may be chemically unstable or only sparingly soluble in the extractant, thereby making their processing difficult. An example is given by annonaeous acetogenins from the Indiana banana (paw paw) tree (McLaughlin). These compounds have potential as chemotherapy agents against breast cancer, and are isolated from 100 samples of natural materials. These are chromatographed over a 25 cm × 2 m long silica gel column, in a gradient method that requires over a week to complete. The challenge is to increase the rate of preparative-scale chromatography so that samples, can be isolated and tested more quickly.

Impacts of Biotechnology on Drug Discovery and Biology

There are a number of disciplines that affect biotechnology. The fields of biology, molecular biology, biochemistry, chemistry, and biochemical engineering have made individual contributions to the rapid advance of biotechnology in a commercial sense. The demonstration that a gene from human pancreas cells could be introduced into a β-galactosidase operon, which in turn was introduced into *Escherichia coli*, so that the *E. coli* would produce human insulin, was a key event. The first steps in initiating this chain of events include the development of techniques for transgenosis, the artificial transfer of genetic information from bacterial cells to eucaryotic cells by transducing phages; isolation of a restriction enzyme that specifically cleaves certain parts of a small circular DNA known as a plasmid; the first cloning experiments in 1972 to 1973; electrophoresis and blotting techniques; and *in vitro* translation systems for converting mRNA into proteins (Old and Primrose, Alberts et al.). Other important enzymes were those that could ligate pieces of DNA (i.e., ligases), and would otherwise allow manipulation of DNA strands to obtain predetermined DNA sequences, i.e., genes, to be introduced into a bacterial plasmid.

The plasmid pBR322 was one of the most versatile artificial plasmid vectors since it contains ampicillin (Ap^R) and tetracycline (Tc^R) resistant genes, as well as desirable replication elements. These genes, when introduced into the cell, enable the bacteria to make enzymes that hydrolyze the two antibiotics. Hence, when a foreign gene is purposely engineered into the plasmid, and the plasmid is placed back into the bacterial host cell to obtain a transformed cell, the transformed cell will also be able to resist the two antibiotics as well as express a foreign protein corresponding to the gene. Transformed cells will grow in the presence of the antibiotic, while all other cells will be killed. Since transformations are not 100% efficient, the presence of ampicillin and tetracycline resistant genes in a recombined plasmid enable the researcher to select for the transformed cells by growing them in a medium that contains the antibiotics.

The pBR322 plasmid is 4362 base pairs long, and is completely sequenced (Old and Primrose). Plasmid pBR322 was the starting point for constructing the plasmid pSomII-3 containing a chemically synthesized gene for the small hormone, somatostatin (Itakura et al.). These developments resulted in a new language, as well as a new technology. Consequently, bioprocess engineers involved in reducing recombinant technology to commercial practice now must learn a new vocabulary to communicate with bioscientists.

At a meeting convened by Eli Lilly and Co. in Indianapolis, in 1976, these tools, together with a method for producing copious quantities of mRNA for insulin by a special tumor cell known as an insulinoma apparently helped to encourage a race for the cloning of the human insulin gene into *E. coli*. The ensuing three years of intense activity (about 1976 to 1979) make up the fascinating story told by Hall.

The realization that there is an interface where biology, fermentation, and recovery technologies all affect each other (Swartz), gave rise to new challenges. Questions that may arise as a process is designed, scale-up, and operated are if recovery is difficult, can the protein be modified to make it easier to purify? If protein refolding gives low yields, can the molecular biologist clone the gene to obtain a proactive form? If test tube scale production gives low yields, can the biochemical engineer optimize conditions to give a better yield? If the product is intracellular, how can the cell best be programmed to respond to a step input to the fermentation, to turn on protein production machinery? How can a separation system be modified to handle complex mixtures containing low titers of the final product? Another important question in a practical context: How much will this cost and how long will it take?

The response to these types of questions has fostered evolution of a team or cross-disciplinary approach to problem solving in the biotechnology industry.

Biotechnology and Biological Research. Genetic engineering techniques have facilitated studies of proteins in plant and animal cells, which otherwise would not have been possible since these proteins (such as DNA polymerases, reverse transcriptase, and ribozymes) are normally present in very small amounts. Their study led, in turn, to their use as tools in further understanding cellular function and genetics, as well as practical applications such as PCR (requires polymerase) and development of antiviral drugs (a possible application of ribozymes) (Pääbo et al., Gibbons). Genetic engineering techniques have been used to amplify mammalian proteins in bacterial systems, so that sufficient protein becomes available for classical study using established enzymology techniques. Some examples are neural peptides, which impact hypertension, and co-reductases implicated in heart disease.

Drug Discovery. Drug discovery was thought to be the area that would benefit enormously from techniques developed through genetic engineering. Proteins having possible therapeutic value are often difficult to study because their *in vivo* concentration in mammalian systems is very low, or because they are difficult to isolate due to limited specimen availability. Once a gene can be identified, molecular biology techniques can be used to produce larger quantities of the protein, thus facilitating its study. The interleukins are an example of this approach, where it is estimated that more interleukin was produced during several years of its discovery phase than was generated since the beginning of humanity. These cytokines are now being examined or used in treatment of wound healing, cancer, and immune deficiencies (Thayer, 1991).

Site specific mutations can also be achieved using recombinant techniques. This facilitates study of a protein's activity if given amino acids are replaced in its structure. Thus, the protein's binding properties with respect to inhibitors, where the inhibitors represent an experimental drug, can be studied.

Biotechnology for Value Added Products from Renewable and Nonrenewable Resources. Biotechnology is usually identified with the microbial production of pharmaceuticals (including vitamins, hormones, alkaloids, antitumor drugs, and interferon). The proliferation of applied molecular biology to microbial, animal, and plant cell modifications, together with a heightened awareness by the chemical industry of the potential of biotechnology is adding a new dimension. Genetic modification techniques are resulting in new types of plant systems that hold the promise of increased productivity and enhanced disease and pest tolerance in the context of integrated agricultural and land management techniques. This is changing the types of agricultural chemicals needed, and hence, is affecting an important sector of the chemical industry. Engineering expertise is needed to translate research advances of biochemists, microbiologists, and agricultural scientists to commercial scale practice.

Agriculture will be an essential component of the emerging biotechnology industry, both as an end-user of technological advances and supplier of the carbohydrates and other biological materials that will be needed as feed stocks by a rapidly growing industry. Development of crops, with enhanced levels of natural pesticides, are beginning to reduce the need for chemical pesticides. Such crops are already undergoing trials.

Bioprocess Engineering and Economics

Evaluation of the engineering economics of bioprocesses is becoming increasingly important as new products move from the research and development phase into commercial production and test marketing. Products of current interest include biological molecules for human or animal health, monoclonal antibodies for diagnostic purposes, enzymes, biochemical intermediates, antibiotics and amino acids. Engineers and scientists sometimes need to obtain a first estimate of relative costs to choose process alternatives and establish magnitude or order product costs. Since there is such a broad range of products with different economic criteria, a single type of cost calculation approach cannot be used for all these cases. Consequently, there appears to be a need for procedures for estimating the relative economics for each type of product on an internally consistent basis. This article addresses the basic concepts of a thought process which Taylor calls "economic decision making."

Bioprocess economics require the engineer or scientist to be able to quickly recognize limiting technical parameters that are likely to be encountered upon scale-up. Processing schemes that minimize operational difficulties and costs must be selected accordingly. Hence, process conceptualization is an important part of the evaluation process and must reflect regulatory constraints as well as transport and kinetic phenomena required to size key pieces of equipment. Examples are oxygen transfer, mixing, and heat removal in a fermentation (i.e., compression and heat exchange costs); maintaining sterile conditions in a continuous bioreactor (productivity and regulatory issues); or predicting resolution in a chromatography column (separation costs). In some cases it is difficult to carry out calculations since physical properties (for example, broth viscosities and heat capacities, or diffusion coefficients of biomolecules) are not well defined during the early stages of a research and development program. Hence, approaches and heuristic rules that might be used in this situation are also needed.

Disparate elements, which affect economic decisions for bioprocesses, can be combined into a logical approach. An attempt must be made to combine fundamentals of bioreactor design, downstream processing, and properties of biomolecules in the context of process evaluation. The ability to set technical priorities based on economic assessment is essential, and requires communication between researchers, process engineers, technical managers, and marketing managers.

Bioseparations and Bioprocess Engineering

The large scale purification of proteins and other bioproducts is the final production step, prior to product packaging, in the manufacture of therapeutic proteins, specialty enzymes, diagnostic products and value added products from agriculture. These biochemical separation steps purify biological molecules or compounds obtained from biological sources and hence are referred to as bioseparations. The essence of large-scale bioseparations is both art and science. Bioseparations often evolve from laboratory scale techniques. These are adapted and scaled-up to satisfy the need for larger amounts of extremely pure, test quantities of the product for analysis, characterization, testing of efficacy, clinical or field trials, and finally, full scale commercialization. The uncompromising standards for product quality, driven by commercial competition, end-use applications, and regulatory oversight, provide many challenges to the scale-up of protein purification and biochemical separations. The rigorous quality control of manufacturing practices and the complexity and liability of the macromolecules being processed provide other practical issues, which must be addressed.

The emphasis of research and development in bioseparations is currently on the purification of proteins derived from recombinant fermentations and cell culture. This is to be expected since the purification of proteins and other bioproducts is a critical and expensive part of most biotechnology based manufacturing processes, and may account for 50% or more of production costs (NAS). While overall production costs have been considered to be secondary to being the first to market, this perspective is changing as the price, and value, of new bioproducts is decreasing. When the volume of the products is small and the price is high (Table 6), being the first to market, together with attaining high product quality (in terms of purity, activity, dependability, or flexibility) are the major competitive advantages (NAS, Wheelwright, Bisbee). Although the data in Table 6 will change with time as prices change, the extraordinarily high values illustrate

TABLE 6. UNIT VALUES AND RELATIVE PRODUCTION QUANTITIES FOR SELECTED APPROVED BIOPHARMACEUTICALS

Product	Year Approved	Approximate selling price $/g	Amount of product for $200 million in sales, kg
Human Insulin	1982	375[b]	530.
Tissue Plasminogen Activator	1987	23,000	8.7
Growth Hormone	1985	35,000	5.7
Erythropoetein (Epogen)	1989	840,000	0.24
GM CSF	1991	384,000	0.52
G-CSF	1991	450,000	0.44

[a] Adapted from Table 2–1, National Academy of Sciences, 1992 (NAS) and data of Bisbee.

[b] A 1996 estimate places the selling price at $130/g, thus indicating a significant decrease in price (Petrides).

the difference between these protein biopharmaceuticals and other drugs, or larger volume bioproducts.

Bioseparations are important in assuring product quality, but manufacturing cost is likely to be secondary for these types of products. When the scale of production of new bioproducts increases from kilograms to tons, the need for cost-effective purification schemes is also increased in importance. High volume products could range from serum (blood) proteins produced by recombinant organisms, to organic acids, enzymes, and food additives obtained from large scale fermentations or from enzyme transformations.

A major technical challenge in the production of biopharmaceuticals is the "development of high-resolution protein purification technologies that are relatively inexpensive, are easily scaled-up and have minimal waste-disposal requirements (NAS)." Separation processes for bioprocessing of renewable resources and agricultural products will benefit from development of "more efficient separations for recovering fermentation products, sugars, and dissolved materials from water," and in particular, lowering the cost of separating water from the product in the fermentation broth (NAS). The bioprocess engineer must therefore be familiar with principles of a wide range of bioseparations techniques, and the practice of the dominant methods likely to be used in the industry.

The most popular method for purifying protein biopharmaceuticals is some form of liquid chromatography due to its resolving powers. Large-scale liquid chromatography can also be used to purify high volume bioproducts including sugars, antibiotics, vitamins, nucleosides, organic acids, and alcohols. However, chromatography is only one of the many steps needed to purify bioproducts. Recovery and purification steps preceding chromatography may include centrifugation or filtration in order to remove solids; membrane filtration or fractionation to separate large molecules from small ones; and precipitation, crystallization, extraction, and/or adsorption to concentrate the product.

It is difficult to predict which combinations of separation methods might be used to purify a new product on the process scale. It is hoped that the combination of principles, practice, and a sense of economics that drive the industry, will assist process engineers and scientists in selecting time-efficient and cost-effective purification methods.

The engineering of separation processes for biotechnology products, also requires a fundamental understanding of the biological materials being processed. The range of products derived from genetically modified organisms will have a diverse range of properties, but a common requirement that they be recovered, concentrated, and purified to give a safe and, in many cases, biologically active product. Bioseparations engineering combines the disciplines of engineering, life sciences, chemistry, and medicine in order to match the unique properties of biomolecules with the most appropriate techniques for their large scale purification. The rapid development of new methods, organisms, and molecules means that these products will be continually changing, as will the methods by which they are purified. Consequently, the design of bioseparation processes requires that the engineering principles of individual separation steps be understood, so that they can be combined into sequences of steps that result in the needed product purity at a cost which is clearly defined and calculated.

Parts of this section taken directly, or paraphrased from Ladisch, which was presented as a plenary lecture in the Separations Technology Program of the 1997 Annual meeting of the American Institute of Chemical Engineers.

MICHAEL R. LADISCH
Laboratory of Renewable Resources Engineering and
Department of Agricultural and Biological Engineering
Purdue University
West Lafayette, Indiana

Additional Reading

ADM, Archer Daniels Midland Company Stockholder's Report, November 28, 1983.

Affleck, R., et al.: "Enzymatic Catalysis and Dynamics in Low-Water Environments", *Proc. Natl. Acad. Sci. USA*, **89**, 1100–1104 (1992).

Aiba, S., A.E. Humphrey, and N.F. Millis, *Biochemical Engineering,* Academic Press, (1973), 3–6.

Alberts, B., et al.: *Molecular Biology of the Cell,* 2nd Edition, Garland Publishing, New York, NY, 1989, pp. 258–284.

Anonymous: Advertisement, *Wall Street Journal*, **A15** (Dec. 18, 1996).

Anonymous: "Biotechnology Therapeutic Medicines and Vaccines Under Development", *Genetic Eng. News*, **16**(16), 29–34 (1996).

Anonymous: "Cargill, Degussa to Form Lysine Venture," *Chem. Eng. News*, **75**(28), 17 (1997).

Anonymous: "Dupont Takes Stake in Pioneer Hi-Bred," *Chem. Eng. News*, **75**(32), 14 (1997).

Anonymous: "Facts and Figures for the Chemical Industry", *Chem. Eng. News*, **73**(26), 60, 61 (1995).

Anonymous: "Facts and Figures for the Chemical Industry", *Chem. Eng. News*, **59**, 60 (1997).

Anonymous: "Monsanto Offers to Buy Remainder of Calgene," *Chem. Eng. News*, **75**(6), 17 (1997).

Anonymous: "Monsanto Sets Spinoff Date," *Chem. Eng. News*, **75**(33), 21 (1997).

Anonymous: "Recession Proof," *Science*, **255**(5043), 407 (1992).

Antrim, R.L., W. Colilla, and B.J. Schnyder: "Glucose Isomerase Production of High Fructose Syrups," *Appl. Biochem. Bioeng.*, **2**, New York, NY, 98–154 (1979).

Arnheim, N. and C.H. Levenson: "Polymerase Chain Reaction," *Chem. Eng. News*, **68**(40), 38–47 (1990).

Averner, M., M. Karel, and R. Radner: Problems Associated with Utilization of Algae in Bioregenerative Life Support Systems, NASA Contractor Report 166615, 1984.

Bailey, J.E. and Ollis, D.F.: *Biochemical Engineering Fundamentals,* McGraw-Hill Book Company, New York, NY, (1977), pp. xiii, 3–23, 45–57, 225.

Balandrin, M.F., et al.: "Natural Plant Chemicals: Sources of Industrial and Medicinal Materials," *Science*, **228**, 1154–1160 (1985).

Balter, M.: "Transgenic Corn Sparks a Furor," *Science*, **275**(5303), 1063 (1997).

Barry, R.D., et al.: Sugar, Background for 1990 Farm Legislation, AGES 9006, USDA Economic Research Service, Feb. 1990, pp. 19–35.

Bisbee, C.A.: "Current Perspectives on Manufacturing and Scale-up of Biopharmaceuticals," *Genetic Eng. News*, **13**(14), 8–9 (1993).

Burrill G. *Steven with the Arthur Young High Technology Group: Biotech 89: Commercialization,* Mary Ann Liebert Publishers, New York, NY, 1988.

Burrill, G.S. and K.B. Lee, Jr.,: Biotech '92: Promise to Reality—An Industry Annual Report, Ernst & Young, San Francisco, 1991.

Cao, N., et al.: "Production of Fumaric Acid by Immobilized Rhizopus Using Rotary Biofilm Contractor," *Appl. Biochem. Biotechnol.*, **63–65**, 387–394 (1997).

Corn Refiners Association, Inc., 1997 Corn Annual, Washington, DC, 1997.

Council on Competitiveness, Office of the Vice President, reportReport on National Biotechnology Policy, Washington, DC, 1991.

Danno, G.: *Agr. Biol. Chem.*, **34**(11), 1658–1667 (1970).

Danno, G.: *Agr. Biol. Chem.*, **34**(12), 1805–1814 (1970b).

DeGussa, *Amino Acids for Animal Nutrition,* A.G. DeGussa, 1985.

Economic Research Service, U.S. Department of Agriculture, Sugar, Background for 1985 Farm Legislation, Agriculture Information Bulletin No. 478, 21–24 (Sept, 1984).

Ember, L.: "FDA Chief Aims to Revitalize Agency with New, Tough Agenda," *Chem. Eng. News*, **69**(10), 21–22 (1991).

Ember, L.: "FDA to Speed Approval of Biotechnology Drugs," *Chem. Eng. News*, **70**(11), 6 (1992).

Ember, L.: "No Special Rules Needed for Biotech Food Products," *Chem. Eng. News*, **70**(22), 5–6 (1992).

Evans, R.M.: *The Chemistry of Antibiotics Used in Medicine,* Pergamon Press, Oxford, (1965).

Fritsch, P. and S. Kilman: "Seed Money: Huge Biotech Harvest Is A Boon For Farmers—and For Monsanto", *Wall Street Journal*, **A1, A8** (Oct. 24, 1996).

Fritsch, P.: "Biotech Boosts Monsanto Stock to Record Height," *Wall Street Journal*, **A3** (Oct. 4, 1996).

Fritsch, P.: "Monsanto May Shed Its Chemical Unit", *Wall Street Journal*, **A3** (Oct. 11, 1996).

Gibbons, A.: "Molecular Scissors: RNA Enzymes Go Commercial," *Science*, **251**(4993), 521 (1991).

Goetz, R.: *Change in the Air,* Forefront, Office of Agriculture Research Programs, Purdue University, **6**(27), 3 (1995).

Gulati, M., et al.: "Assessment of Ethanol Production Options for Corn Products," *Bioresource Technology*, **58**, 253–264 (1996).

Hacking, A.J., *Economic Aspects of Biotechnology,* Cambridge University Press, Cambridge, (1986), pp. 6–8, 18–20, 39–72.

Hall, S.S.: *Invisible Frontiers, The Race to Synthesize a Human Gene,* Atlantic Monthly Press, New York, NY, 1987.

Hamilton, J.O'C.: "A Gene to Make Greener Blue Jeans", *Business Week*, **3520**, 82–83 (1997).

Hamilton, J.O.'C.: "Heartbreak and Triumph in Biotech Land," *Business Week*, **3250**, 33 (1992).

Higuchi, R.: *Using PCR to Engineer DNA,* in PCR Technology, H.A. Erlich, ed., Stockton Press, New York, NY, 1989, pp. 61–70.

Hileman, B.: "Views Differ Sharply Over Benefits, Risks of Agricultural Biotechnology," *Chem. Eng. News*, **73**(34), 8–17 (1995).

Hirose, Y., K. Sano, and H. Shibai: *Amino Acids,* in Annual Reports on Fermentation Processes, Academic Press, New York, NY, Vol. 2, (1978), pp. 155–189.

Hochester, R.M. and R.W. Watson: *J. Am. Chem. Soc.*, **75**, 3284–3285 (1953).

Hohmann, N. and C.M. Rendleman: Emerging Technologies in Ethanol Production, Agriculture Information Bulletin Number 663.

Itakura, K., et al.: "Expression in Escherichia coli of a Chemically Synthesized Gene for the Hormone Somatostatin," *Science*, **198**, 1056–1063 (1977).

Ives, A.R.: Technical Comments—"Evolution of Insect Resistance to Bacillus thuringiensis-Transformed Plants," *Science*, **273**(5280), 1412 (1996).

Johannes, L.: "New Drug Aims to Win Over MS Sufferers," *Wall Street Journal*, B7 (May 20, 1996).

Kilman, S.: ADM Settles Two Suits For $65 Million, Raising Hopes for Plea in Federal Probe, *Wall Street Journal*, A1, A6 (Sept. 30, 1996a).

Kilman, S.: Ajinomoto Pleads Guilty to Conspiring With ADM, Others to Fix Lysine Prices, *Wall Street Journal*, B5 (Nov. 15, 1996b).

Kilman, S.: "Green Genes, If Fat-Free Pork Is Your Idea of Savory, Its a Bright Future- Dupont, Others Are Plowing Billions Into A Bioengineered Menu," *Wall Street Journal* A1, A10 (Jan 29, 1998).

Kilman, S.: "Green Genes: If Fat-free Pork Is Your Idea of Savory, It's a Bright Future", *Wall Street Journal*, A1, A10 (Jan. 29, 1998).

Kinoshita, S.: *Glutamic Acid Bacteria,* in Biology of Industrial Microorganisms, A.L. Demain, and N.A. Solomon, eds., Benjamin/Cummings, Menlo Park, CA, 1985, pp. 115–142.

Kirschner, E.: "Norvatis Merger Approved—FTC Requires Licensing of Gene-Therapy Technology, Sale of two Sandoz Units," *Chem. Eng. News*, **74**(52), 4 (1996).

Kling, J.: "Agricultural Ecology: Could Transgenic Supercrops One Day Breed Superweeds?" *Science* **274**, 180–181 (1996).

Kohlmann, K.L., et al., *Enhanced Enzyme Activities on Hydrated Lignocellulosic Substrates,* in Enzymatic Degradation of Insoluble Carbohydrates, ACS Symp. Ser. No. 618, J.N. Saddler and M.H. Penner, eds., 1995, pp. 237–255.

Kohlmann, K.L., et al.: *Enhanced Enzyme Activities on Hydrated Lignocellulosic Substrates in Enzymatic Degradation of Insoluble Carbohydrates,* ACS Symp. Ser. No. 618, J.N. Saddler and M.H. Penner, eds., 1995b, pp. 238–255.

Ladisch, M.R. and K. Dyck: "Dehydration of Ethanol: New Approach Gives Positive Energy Balance," *Science*, **205**(4409), 898–900 (1979).

Ladisch, M.R.: Current Status and Future Directions In Biochemical Separations," lecture at plenary session on Separation Science and Technologies: New Developments and Opportunities II, paper 2 cc1997 AIChE Annual Meeting, Los Angeles CA, Nov. 17, 1997.

Ladisch, M.R. and K.L. Kohlmann: Recombinant Human Insulin, *Biotechnol. Progr.*, **8**, 469–478 (1992).

Ladisch, M.R., et al.: "Process Considerations in Enzymatic Hydrolysis of Biomass," *Enz. Microb. Technol.*, **6**, 82–100 (1983).

Landgreth, R. and S.D. Moore: "Behind Merger Talks, Two CEO's at a Crossroads," *Wall Street Journal*, B1, B8 (Jan. 21, 1998).

Langreth, R. and P. Thomas: "P&G to Pay Regeneron $135 Million In Research Effort for Several Ailments", *Wall Street Journal*, B10 (May 14, 1997).

Langreth, R.D.: "Schering-Plough Corp. to Acquire Mallinckrodt Animal-Health Unit," *Wall Street Journal*, B6 (May 30, 1997).

Lastick, S.M. and G.T. Spencer: *Xylose-Glucose Isomerases, Structure, Homology, Function,* in Enzymes in Biomass Conversion, G.F. Leatham and M.E. Himmel, eds., ACS Symposium Ser., **460**, (1991), pp. 487–500.

Layman, P.: "Promising New Markets Emerging for Enzymes," *Chem. Eng. News*, **68**(39), 17–18 (1990).

Lipin, J.S., S.D. Moore, and T.M. Burton: "UpJohn and Pharmacia Sign $6 Billion Merger—Accord Would Link 2 Second-Tier Players In the Drug Industry," *Wall Street Journal*, A3 (Aug. 21, 1995).

Lloyd, N.E. and R.O. Horvath: *Biotechnology and Development of Enzymes for the HFCS Industry,* in Bio-Expo 85, O. Zaborsky, ed., Cahners Exposition Group, Stanford, CT, 1985, pp. 116–134.

Lynd, L.R., et al., "Fuel Ethanol from Cellulosic Biomass," *Science*, **251**(4999), 1318–1323 (1991).

MacQuitty, J.J.: *The Impact of Chemistry on Biotechnology,* in ACS Symp. Ser. 362, M. Phillips, et al., eds., (1988), pp. 11–29.

Marshall, R.O. and E.R. Kooi: *Science*, **25**, 648–649 (1957).

McLaughlin, J.L.: Annonaceous Acetogenins: Antitumor and Pesticidal Principles of the Paw Paw Tree, Abstract—Purdue University Chromatography Workshop '95, W. Lafayette, IN, 1995, 47.

Moore, S.D.: "Norvatis Jumps Last Regulatory Hurdle and Major Drug-Industry Force Is Born," *Wall Street Journal*, B4 (Dec. 18, 1996).

Moore, S.D.: "SmithKline Beecham Enters Genetics-Diagnostics Race—Joint Venture With Incyte Bolsters A Promising but Troubled Industry", *Wall Street Journal*, B4 (Sept. 3, 1997).

NAS, Committee on Bioprocess Engineering, National Research Council, *Putting Biotechnology to Work: Bioprocess Engineering,* National Academy of Sciences, Washington, DC, (1992).

Natake, M.: *Agr. Biol. Chem.*, **30**(9), 887–895 (1966).

Natake, M.: *Agr. Biol. Chem.*, **32**(3), 303–313 (1968).

Natake, M. and S. Yoshimura: *Agr. Biol. Chem.*, **28**(8), 505–509 (1964).

Narisetti, R.: "P&G Signs Accord With Gene Logic to Develop Drugs," *Wall Street Journal*, B5 (July 13, 1997).

Old, R.W. and S.B. Primrose: *Principles of Gene Manipulation–An Introduction to Genetic Engineering,* Studies in Microbiology, Vol. 2, University of California Press, Berkeley, (1981), pp. 1–47.

Olsen, S.: *Biotechnology: An Industry Comes of Age,* National Academy Press, Washington, DC, 1986.

OTA, reportOffice of Technology Assessment, U.S. Congress, *Biotechnology in A Global Economy*, B. Brown, ed., OTA-BA-494, U.S. Government Printing Office, Washington, DC, 1991.

Pääbo, S., R.G. Higuchi, and A.C. Wilson: "Ancient DNA and the Polymerase Chain Reaction," *J. Biol. Chem.* **264**(17), 9709–9712 (1989).

Patil, D.R., D.G. Rethwisch, and J.S. Dordick: "Enzymatic Synthesis of a Sucrose—Containing Linear Polyester in Nearly Anhydrous Organic Media", *Biotechnol. Bioeng.*, **37**, 639–646 (1991).

Petrides, D.P., J. Calandranis, and C.L. Cooney: "Bioprocess Optimization via CAPD and Simulation for Product Commercialization," *Genetic Eng. News*, **16**(16), 24, 28 (1996).

Petsko, G.A.: *Protein Engineering,* in Biotechnology and Materials Science-Chemistry for the Future," M.L. Good, ed., American Chemical Society, Washington, DC, 1988, pp. 53–60.

Polastro, E.T., A. Walker, and H.W.A. Teeuwen: "Enzymes in the Fine Chemicals Industry: Dreams and Realities," *Bio/Technology*, **7**(12), 1238–1241 (1989).

Potera, C.: "Genencor and Dupont Create Green Polyester," *Genetic Eng. News*, **17**(11), 17 (1997).

Reisch, M.: Drug Price Controls: Manufacturers Prescribe Self-Regulation," *Chem. Eng. News*, **71**(12), 6 (1992).

Reisch, M.: "Drug Price Controls, Manufacturers Prescribe Self-Regulation," *Chem. Eng. News*, **71**(12), 6 (1993).

Reisch, M.: "Merck, Rhône-Poulenc form Animal Health Care Joint Venture," *Chem. Eng. News*, **75**(1), 8 (1997).

Reisch, M.: "Monsanto Plans to Split in Two: Chemical and Life Sciences Firms will be Formed, Up to 2,500 Jobs May Be Cut", *Chem. Eng. News,* **74**(51), 9 (1996).

Rouhi, A.M.: "Researchers Unlocking Potential of Diverse, Widely Distributed Saponins," *Chem. Eng. News*, **73**(37), 28–35 (1995).

Ryu, D.D.Y. and S.-B. Lee: in *Horizons of Biochemical Engineering,* S. Aiba, ed., Oxford University Press, Oxford, 1988, pp. 97–124.

Ryu, K., D.R. Stafford, and J.S. Dordick: "Peroxidase-Catalyzed Polymerization of Phenols: Kinetics of *p*-Cresol Oxidation in Organic Media, in *Biocatalysis* in *Agricultural Biotechnology,* J.R. Whitake, ed. ACS Symp. Ser. No. 389.

Schultz, P.G., R.A. Lerner, and S.J. Benkovic: "Catalytic Antibodies," *Chem. Eng. News*, **68**(22), 1–56 (1990).

Shuler, M.L. and F. Kargi: *Bioprocess Engineering. Basic Concepts,* Vol. **2**, Prentice Hall PTR, Englewood Cliffs, NJ, 1992, pp. 4–8.

Standard & Poor's Research Report, Amgen Inc. (Dec. 19, 1995).

Stinson, S.C.: "Custom Chemical Producers Set to Meet Industry Standards," *Chem. Eng. News*, **70**(5), 25–50 (1992).

Stinson, S.C.: "Drug Firms Restock Antibacterial Arsenal, Growing Bacterial Resistance, New Disease Threats Spur Improvements to Existing Drugs and Creation of New Classes," *Chem. Eng. News*, **74**(39), 75–100 (1996).

Stinson, S.: "Drug Industry Consolidations, Roche Buys Boehringer Mannheim, Becomes Worlds Leading Diagnostics Player," *Chem. Eng. News*, **75**(22), 6 (1997).

Stone, R.: "A Biopesticidal Tree Begins to Blossom, Neem Seed Oil has Insect Toxicologists Buzzing About Its Potential As a Source of Natural Insecticides," *Science*, **255**(5048), 1070–1071.

Stone, R.: "Kessler's Legacy: Unfinished Reform," *Science*, **274**(5293), 1603 (1996).

Swartz, J.: at the 1988 Engineering Foundation Conference coined the phrase "Fermentation/Recovery Interface', ' April, 1988.

Takasaki, Y.: *Agr. Biol. Chem.*, **30**(12), 1247–1253 (1966).

Takasaki, Y., K. Yoshiju, and A. Kanbagashi: in *Fermentation Advances,* D. Perlman, ed., Academic Press, New York, NY, 1969, pp. 561–589.

Tanner, H. and H. Schmidtborn: *DL-Methionine—The Amino Acid for Animal Nutrition,* A.G. DeGussa, 1981.

Tanouye, E. and R. Langreth: "Time's Up—With Patents Expiring on Big Prescriptions, Drug Industry Quakes: T Gird For Onslaught from Generics, Scramble to Develop New Products—Bet on Fewer Blockbusters," *Wall Street Journal*, A1, A6 (Aug. 12, 1997).

Tanouye, E.: "Drug Makers Go All Out to Squash Superbugs," *Wall Street Journal*, B1, B6 (June 25, 1996).

Tanouye, E. and R. Langreth: "Genetic Giant: Cost of Drug Research Is Driving Merger Talks of Glaxo, SmithKline," *Wall Street Journal*, A1, A8 (Feb. 2, 1998).

Tanouye, E. and R. Langroth, "Genetic Giant: Cost of Drug Research Is Driving Merger Costs of Glaxo SmithKline—New Colassus Could Afford Molecular Engineering Needed to Stay on Top," *Wall Street Journal*, A1, A8 (Feb. 2, 1998).

Tanouye, E. and S.D. Moore: "Novartis to Pay $910 Million for Merck Business," *Wall Street Journal*, A3 (May 14, 1997).

Taylor, G.A.: *Managerial and Engineering Economy, Economic Decision Making*, D. Nostrand Co., New York, NY, 1966.

Thayer, A.M.: "Betting the Transgenic Farm, First Plantings of Engineered Crops from Monsanto and Others Are Serving as Agbiotech's Proving Ground", *Chem. Eng. News*, **75**(17), 15–19 (1997).

Thayer, A.M.: "Biopharmaceuticals Go Global," *Chem. Eng. News*, **75**(33), 14–19 (1997).

Thayer, A.M.: "Biopharmaceuticals Overcoming Market Hurdles," *Chem. Eng. News*, **69**(8), 27–48 (1991).

Thayer, A.: "Biotech Industry Breeding Growth," *Chem. Eng. News*, **75**(1), 15–16 (1997a).

Thayer, A.: "Chemical Firms Extend Positions in Ag-Bitoech", *Chem Eng. News*, **74**(29), 22–25 (1996).

Thayer, A.: Consolidation Reshapes Drug Industry, Shifts Employment and R&D Outlook: In Response to Global Pressures on the Industry, Companies Are Merging to Create Entities With the Critical Mass to Compete, *Chem. Eng. News*, **73**(42), 10 (1995).

Thayer, A.: "Drug, Biotech Firms Strong in the First Half," *Chem. Eng. News*, **75**(34), 18–19 (1997d).

Thayer, A.M.: "Drug, Biotech Firms Strong In First Half," *Chem. Eng. News*, **75**(34), 18–19 (1997).

Thayer, A.M.: "Many Biotech Companies Enter 90's With Sharply Higher Revenues", *Chem. Eng. News*, **68**(14), 9–11 (1990).

Thayer, A.M.: "Market Investor Attitudes Challenge Developers of Biopharmaceuticals," *Chem. Eng. News*, **74**(33), 13–22 (1996).

Thayer, A.M.: "Monsanto Gets all of Calgene," *Chem. Eng. News,* **75**(14), 11–12 (1997).

Toon, S.T., et al.: "Enhanced Cofermentation of Glucose and Xylose by Recombinant Saccharomyces Yeast Strains in Batch and Continuous Operating Modes", *Appl. Biochem. Biotechnol.*, **63–65**, 243–255 (1997).

Tsumura, N., et al.: *Agr. Biol. Chem.*, **31** (8), 902–907 (1967).

Tsumura, N. and T. Sato: *Agr. Biol. Chem.*, **29** (12), 1123–1128 (1965a).

Tsumura, N. and T. Sato: *Agr. Biol. Chem.*, **29** (12), 1129–1134 (1965b).

Tsumura, N. and T. Sato: *Bull. Agr. Chem. Soc., Japan*, **24** (3), 326–327 (1970).

Tyndall, R.M.: Application of Cellulase Enzymes to Cotton Fabrics and Garments, *AATCC Book of Papers*, 259–273 (1991).

USDA, Economic Research Service Report, Industrial Uses of Agricultural Materials, reportnoReport IUS-1, 10–35, June, 1993.

Walker, J.W. and M. Cox: with A. Whitaker as a contributor, The Language of Biotechnology, A Dictionary of Terms, ACS Professional Reference Book, American Chemical Society, Washington, DC, (1988), pp. 204.

Wheelwright, Protein Purification, Hansen Publishers, Munich, 1991.

Wechsler, J.: "Congress Modernizes FDA," *Pharmaceutical Technol.*, **21**(12), 16–26 (1997).

Weil, J., et al.: "Cellulose Pretreatments of *Lignocellulosic Substrates*, *Enzyme Microb. Technol.*, **16**, 1002–1004 (1994a).

Weil, J., et al.: "Cellulosic Pretreatments of Lignocellulosic Substrates," *Enz. Microb. Technol.*, **16**, 1002–1004 (1994b).

Werth, B.: "Biomania," *Worth*, **1**(1), 104–107 (1992).

Westgate, P.J., et al.: "Bioprocessing in Space," *Enz. Microb. Technol.*, **14**(1), 76–79 (1992).

Wilke, J.R.: U.S. Forces New Drug Giant to Share Genetic Research, In Sandoz, Ciba Merger, FTC Prevents Monopoly That Doesn't Yet Exist, *Wall Street Journal*, **B4**(Dec. 18, 1996).

Yanchinski, S.: "Canadian Health Agency OK's Monsanto's Genetically Engineered Potatoes", *Genetic Eng. News*, **17**(4), 24 (1996).

Yamanaka, K.: *Agr. Biol. Chem.*, **27**(4), 265–270 (1963a).

Yamanaka, K.: *Agr. Biol. Chem.*, **27**(4), 271–278 (1963b).

Yamanaka, K.: *Biochimica Biophysica Acta*, **151**, 670–680 (1968).

BIOTIN. Infrequently referred to as Bios IIB, protective factor X, vitamin H, egg white injury factor, and CoR. Biotin, required by most vertebrates, invertebrates, higher plants, and most fungi and bacteria, falls into the general classification of vitamins. In certain species, a deficiency of this substance is a cause of: desquamation of the skin; lassitude, somnolence, and muscle pain; hyperesthesia; seborrheic dermatitis; alopecia, spastic gait and kangaroo-like posture (rats and mice); dermatitis and perosis (chicks and turkeys); progressive paralysis, and K^+ deficiency (dogs); alopecia, spasticity of hind legs (pigs); and thinning and depigmentation of hair (monkeys).

Biotin reacts with an oxidized carbon fragment (denoted as CO_2) and an energy-rich compound, adenosine triphosphate (ATP), to form carboxy biotin, which is "activated carbon dioxide." Biotin is firmly bound to its enzyme protein by a peptide linkage. Biotin and carboxy biotin are:

Biotin

"Activated" carboxy-biotin

Biotin enzymes are believed to function primarily in reversible carboxylation-decarboxylation reactions. For example, a biotin enzyme mediates the carboxylation of propionic acid to methylmalonic acid, which is subsequently converted to succinic acid, a citric acid cycle intermediate. A vitamin B_{12} coenzyme and coenzyme A are also essential to this overall reaction, again pointing out the interdependence of the B vitamin coenzymes. Another biotin enzyme-mediated reaction is the formation of malonyl-CoA by carboxylation of acetyl-CoA ("active acetate"). Malonyl-CoA is believed to be a key intermediate in fatty acid synthesis.

Bios, from the Greek, meaning life, was a word coined to describe a growth-promoting substance for yeast and discovered by Wildiers in 1901. When added in small amounts to sugar and salts medium, it permitted rapid growth of yeast even from a small seeding. Subsequent investigations proved that there was not merely a single substance involved, but that depending upon the strain of yeast and the circumstances of basting, a number of different substances could act, often synergistically, to promote the rapid growth of yeast. Pantothenic acid, biotin, inositol, thiamine, and pyridoxine all have "bios" properties when appropriately tested. Even an amino acid may be a limiting factor for yeast growth when other needs are supplied. The term "bios" has fallen from use in the literature.

Biotin required for growth and normal function by animals, yeast, and many bacteria is seldom found in deficiency in humans because the intestinal bacteria synthesize it in sufficient quantity to meet requirements. Biotin deficiency does occur, however, in animals fed raw whites of eggs. The egg white contains a protein, *avidin*, which combines with biotin, and this complex is not broken down by enzymes of the gastrointestinal tract. Hence, a deficiency develops.

Biotin was first isolated in pure form in 1936 by two Dutch chemists, Koegel and Tonnis, who obtained 1.1 milligrams from 250 kilograms of dried egg yolk. They showed that the compound was necessary for the growth of yeast and gave it the name, biotin. Five years later, in America, György and co-workers found that the same compound prevented the toxicity of raw egg white in animals and, in 1942, du Vigneaud and collaborators determined the structure of the compound.

Distribution and Sources

Natural sources of biotin include the following:

High biotin content (100–400 micrograms/100 grams)

Lamb liver, pork liver, soyal jelly, yeast

Medium biotin content (10–100 micrograms/100 grams)

Grains: Barley, corn (maize), oats, rice, wheat

Meat and Fish: Beef liver, chicken, eggs, mackerel, salmon, sardines

Nuts: Almonds, filberts, hazelnuts, groundnuts (peanuts), pecans, walnuts

Vegetables: Cauliflower, chick-peas, cowpeas, lentils, soybeans

Other: Chocolate, mushrooms

Low biotin content (0–10 micrograms/100 grams)

Dairy: Cheese, milk

Fruits: Apple, avocado, banana, cantaloupe, grape, grapefruit, orange, peach, strawberry, watermelon

Meat and Fish: Beef, halibut, lamb, oyster, pork, tuna, veal

Vegetables: Bean (lima), beet, beet greens, cabbage, carrot, lettuce, onion, pea, spinach, sweet corn, sweet potato, tomato

Biotin can be produced commercially by using, for starting the synthesis, meso-diamino succinic acid derivative of fumaric acid.

Determination of Biotin

Bioassay methods include the (1) rat and chick method (growth response after biotin deficiency); (2) microbiological with *L. arabinosus*. Physicochemical methods make use of polarography.

Bioavailability of Biotin

Factors which cause a decrease in bioavailability include: (1) presence of avidin in food; (2) cooking losses; (3) presence of antibiotics; (4) presence of sulfa drugs; and (5) binding in foods (such as yeast). Availability can be increased by stimulating synthesis by intestinal bacteria.

Antagonists of biotin include desthiobiotin in some forms, ureylene phenyl, homobiotin, urelenecyclohexyl butyric and valeric acid, norbiotin, avidin, lysolecithin, and biotin sulfone. Synergists include vitamins B_2, B_6, B_{12}, folic acid, pantothenic acid, somatotrophin (growth hormone), and testosterone.

Precursors for the biosynthesis of biotin include pimelic acid, cysteine, and carbamyl phosphate. Desthiobiotin acts as an intermediate. In plants, the production sites are seedlings and leaves. In most animals, production is in the intestine. Storage site is the liver.

Some of the unusual features of biotin noted by investigators include: (1) the binding and inactivation by avidin protein found in egg white; (2) fetal tissues and cancer tissues higher in biotin than normal adult tissues; (3) biotin deficiency increasing the severity and duration of some diseases, notably some of the protozoan infections; and (4) oleic acid and related compounds replacing biotin as unspecific stimulatory compounds in bacteria.

BIOTITE. A common silicate mineral containing potassium, magnesium, iron, and aluminum. Biotite is found in granitic rocks, gneisses, and schists. Although actually monoclinic, it often assumes a pseudohexagonal form. Like others of the mica group, it shows a highly perfect basal cleavage. Hardness, 2.5–3; sp gr, 2.7–3.4; luster, pearly to vitreous or sometimes sub-metallic when very black in color; cleavage sheets are elastic; color, greenish to brown or black; transparent to opaque. A general formula is $K(Mg,Fe)_3, (Al,Fe)Si_3O_{10} (OH,F)_2$.

Biotite is occasionally found in large sheets, especially in pegmatite veins. It also occurs as a contact metamorphic mineral or the product of the alteration of hornblende, augite, wernerite, and similar minerals.

Biotite is found in the lavas of Vesuvius, at Monzoni, and in many other European localities; in the United States, especially in the pegmatites of New England, Virginia, and North Carolina, and the granite of Pikes Peak, Colorado. The mineral was named in honor of the French physicist, J.B. Biot. Biotite is also known as *iron mica*.

BIPHENYL AND TERPHENYLS. Biphenyl (diphenyl, phenylbenzene) and terphenyl are the lowest members of a family of polyphenyls in which benzene rings are attached one to another in a chainlike manner, $C_6H_5(C_6H_4)_mC_6H_5$. Many higher polyphenyls are known, but only biphenyl and the terphenyls are of commercial significance.

Physical Properties

Pure biphenyl is a white crystalline solid that separates from solvents as plates or monoclinic prismatic crystals. Commercial samples are often slightly yellow or tan in color. Similarly, pure terphenyls are white crystalline solids whereas commercial grades are somewhat yellow or

TABLE 1. PHYSICAL PROPERTIES OF BIPHENYL

Property	Value
melting point, °C	69.2
freezing point commercial grades, °C	68.5–69.4
boiling point at 101.3 kPa[a] °C	255.2 ± 0.2
critical properties	
temperature, °C	515.7
pressure, MPa[a]	4.05
density, g/mL	0.314
flash point, °C	113.0
fire point, °C	123.0
autogenous ignition temperature, °C	560.0
heat of combustion, kJ/mol[b]	6243.2
heat of fusion, kJ/mol[b]	18.60[c,d]

[a] To convert kPa to mm Hg, multiply by 7.5.
[b] To convert MPa to psi, multiply by 145.
[c] To convert J to cal, divide by 4.184.
[d] To convert W/(cm · K) to (cal·cm)/(s·cm^2· °C), divide by 4.184.

TABLE 2. PHYSICAL PROPERTIES OF PURE TERPHENYL ISOMERS

Property	Ortho-	Meta-	Para-
melting point, °C	56.2	87.5	212.7
boiling point at 101.3 kPa[a], °C	332.0	365.0	376.0
heat of vaporization at 101.3 kPa[a]	253.0	279.0	272.0
flash point, °C	171.0	206.0	210.0
fire point, °C	193.0	229.0	238.0
auto ignition temperature, °C	530.0	555.0	555.0
vapor pressure, kPa[a]			
93°C	0.01172	0.00165	
315.6°C	64.40	27.3	
density of liquid, g/L			
93°C	1022.0	1039.0	solid
315.6°C	842.0	871.0	879.0
heat capacity of liquid, kJ/kg[b]			
93°C	1.007	0.970	
398.9°C	1.400	1.397	1.116
viscosity of liquid, mPa · s(= cP)			
100°C	4.34	3.87	solid
300°C	0.30	0.40	0.43
thermal conductivity of liquid, W/(m·K)[c]			
100°C	0.1316	0.1347	
210°C	0.1206	0.1356	0.1359
heat of vaporization, J/g at 252°C	280.0	298.0	305.0
heat of fusion, kJ/kg[b]	55.2	73.7	146.5
critical temperature, K	891.0	927.0	926.0
critical pressure, MPa[d]	3.903	3.503	3.330

[a] To convert kPa to mm Hg, multiply by 7.5.
[b] To convert J to cal, divide by 4.184.
[c] To convert W/(m · K) to (cal·cm)/(s·cm^2·°C), divide by 418.4.
[d] To convert MPa to psi, multiply by 145.

tan. Physical and chemical constants for biphenyl and the three isomeric terphenyls, respectively, are given in Tables 1 and 2.

Chemical Properties

Biphenyl and terphenyls may be regarded as substituted benzenes that undergo acylation, alkylation, halogenation, nitration, sulfonation, and other reactions common to benzene. The points of initial attack on chlorination, nitration, and sulfonation of biphenyl occur at the 2- and 4-positions; the latter group predominates.

Terphenyls, like biphenyl, undergo the usual reactions of aromatic hydrocarbons. The *ortho-* and *para-* isomers nitrate initially at the 4-position.

Manufacture

Dow, Monsanto, and Koch Chemical Company are the principal biphenyl producers, with lesser amounts coming from Sybron Corporation and Chemol, Inc. With the exception of Monsanto, the above suppliers recover biphenyl from high boiler fractions that accompany the hydrodealkylation of toluene to benzene.

High purity biphenyl is currently produced by Monsanto in the United States and United Kingdom by direct dehydrocondensation of benzene. Terphenyls are also obtained from the higher boiling polyphenyl by-products that accompany the biphenyl.

Shipping

By-product biphenyl is usually sold as a dye carrier in the molten state in tank truck or tank car lots. Grades of higher purity are also sold in the molten state or as flakes in 22.7-kg bags.

Biphenyl is defined as a toxic chemical under, and subject to, reporting requirements of Section 313 of Title III of the Superfund Amendments and Reauthorization Act (SARA) of 1986 and 40 CFR, Part 372, under the name biphenyl. It is identified as a hazardous chemical under criteria of the OSHA Hazard Communication Standard (29 CFR 1910.1200).

The small amount of mixed terphenyls that are sold as such are shipped in the form of flaked solids in 22.7-kg multiwall bags. The U.S. freight classification is Plastics, synthetic other than liquid, NOIBN. Like biphenyl, mixed terphenyls fall under the hazardous chemical criteria of the OSHA Hazard Communication Standard (29 CFR 1910.1200).

Economic Aspects

Reliable estimates of annual production of biphenyl in the United States are difficult to obtain. About 10% of the biphenyl derived from HDA sources is consumed as 93–95% grade in textile dye carrier applications. The remainder is used for alkylation or upgraded to ≥99.9% grades for heat-transfer purposes. Essentially all of the high purity biphenyl produced by dehydrocondensation of benzene is used as alkylation feedstock or is utilized directly in heat-transfer applications.

As in the case of biphenyl, current worldwide production figures for terphenyls are not readily obtainable. Currently, most of the terphenyl produced is converted to a partially hydrogenated form.

Health and Safety Factors

Although biphenyl and the terphenyls fall under the hazardous chemical criteria of the OSHA Hazard Communications Standard, the products themselves are fairly low in toxicity and do not constitute a serious industrial hazard.

Because biphenyl is often transported in the molten state, a moderate fire hazard does exist under these circumstances.

Environmental Considerations

The widespread use of biphenyl and methyl-substituted biphenyls as dye carriers in the textile industry has given rise to significant environmental concern because of the amount released to the environment in wastewater effluent. Although biphenyl and simple alkyl-biphenyls are themselves biodegradable, the prospect of their conversion by chlorination to PCBs in the course of wastewater treatment has been a subject of environmental focus. Despite the fact that the lower chlorinated biphenyls are also fairly biodegradable continued environmental concern has resulted in decreased use of biphenyl as a dye carrier.

Terphenyls in heat-transfer applications are used in relatively smaller quantities with negligible release to the environment. They are sufficiently biodegradable so as not to constitute an environmental threat.

Derivatives

Short-chain alkylated biphenyls are the principal biphenyl derivatives in commercial use. They are generally produced by liquid-phase Friedel-Crafts alkylation of biphenyl with ethylene, propylene, or mixed butenes.

Ortho- and *para*-phenylphenols are commercially significant biphenyl derivatives that do not involve biphenyl as a starting material. Both are produced as by-products from the hydrolysis of chlorobenzene with aqueous sodium hydroxide. *o*-Phenylphenol, i.e., 1,1-biphenyl-2-ol, particularly as its sodium salt, is widely used as a germicide or fungicide. *para*-Phenylphenol with formaldehyde forms a resin used in surface coatings.

Several functionalized biphenyls either are, or show promise of becoming, commercially significant polymer building blocks.

QUENTIN E. THOMPSON
Monsanto Company

Additional Reading

Dasgupta, R. and B. Maiti: *Ind. Eng. Chem. Process Des. Dev.* **25**(2), 381 (1986).
Mandel, H.: *Heavy Water Organic Cooled Reactor, Physical Properties of Some Polyphenyl Coolants*, AEC Report A 1-CE-15, Apr. 15, 1966.
Meylan, W. M. and P. H. Howard: Chemical Market Input/Output Analysis of Selected Chemical Substances to Assess Sources of Environmental Contamination: Task II. Biphenyl and Diphenyl Oxide, EPA Contract No. 68-1-3224, Syracuse, N.Y., 1976, p. 2.
Taylor, W. E.: in R. F. Makens, ed., *Organic Coolant Summary Report, AEC Accession No. 15554, Report No. IDO-11401*, Idaho Operations Office, U.S. Atomic Energy Commission, CFSTI, Washington, D.C., 1964, pp. 9–38.
U.S. Pat. 2,143,509 (1939), C. Conover and A. E. Huff (to Monsanto Co.).

BISMALEIMIDE POLYMERS. These relatively new polymeric materials were developed to serve the increasing requirements for materials of high strength in high-temperature applications. Currently, a high percentage of the bismaleimides produced are used for printed circuit boards (PCBs). The materials usually are cured with aromatic amines and then compression molded into the PCBs. Future uses include aircraft structural components where bismaleimides may prove superior for high-temperature skin surface applications as compared with present epoxy composites.

Bismaleimides are produced by the condensation reaction of a diamine, such as methylenedianiline, with maleic anhydride. The reaction product tends to be crystalline with a high melting point. Eutectic blends of different bismaleimides reduce the melting point. However, a co-reactant generally is required to improve the processing properties of the material. Bismaleimides owe their reactivity to the double bonds on each end of the molecule, which can react with themselves or with other compounds containing functional groups (vinyls, allyls, or amines). A typical bismaleimide structure is shown by:

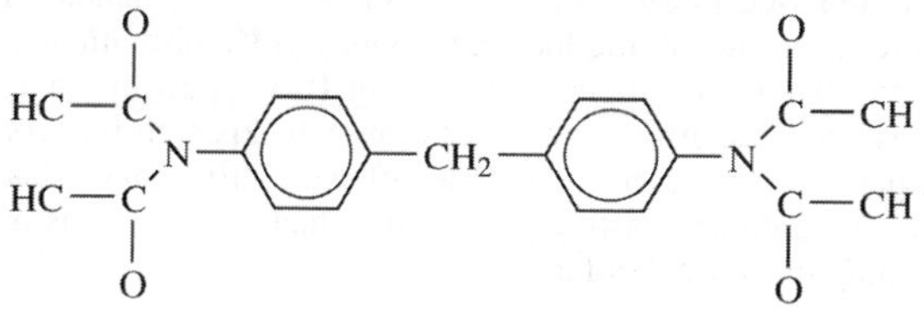

Bismaleimides require an initial cure of from 350 to 450 °F (177 to 232 °C) for one to four hours, followed by a postcure at 450 °F (232 °C) for four hours, if the full properties are to be developed. The glass transition temperature of bismaleimides generally exceeds 500 °F (260 °C). The materials generally have a continuous-use temperature of from 400 to 450 °F (204 to 232 °C).

Compounds based on allyl phenols, such as diallyl bisphenol A, are a recent development. These compounds have superior mechanical properties, processing, and toughness. Some of these compounds are liquids that can dissolve the bismaleimide and thus result in a resin system that is suitable for filament winding and casting in addition to fiber impregnation. When allyl phenols are used as co-curing agents with bismaleimides, the gains in strength and toughness at room and elevated temperatures are marked. Hot acid resistance is also outstanding. Coating applications are developing where resistance to acids and high temperatures are required.

BISMUTH. [CAS: 7440-69-9]. Chemical element, symbol Bi, at. no. 83, at. wt. 208.981, periodic table group 15, mp 271.3 °C, bp 1555–1565 °C, density 9.75 g/cm^3 (20°C). Elemental bismuth has a rhombohedral crystal structure. The metal is of a silvery-white color with limited ductility. Like gallium, bismuth is one of the few metals that increases its volume (3.32%) upon solidifying from the molten state. It is the most diamagnetic of all the metals. All isotopes of the element (^{205}Bi through ^{215}Bi) are radioactive. See also **Radioactivity**. However, the naturally occurring isotope ^{209}Bi generally is not regarded in this category because of its extremely long half-life (2×10^{17} years). Although described by Basil Valentine in the fifteenth century, the element was not defined as a new element until its characteristics were published in 1753 by C. Geoffroy and T. Bergman.

First ionization potential 7.287 eV; second 16.6 eV; third 25.56 eV; fourth 45.1 eV; fifth 55.7 eV. Oxidation potentials Bi + H$_2$O → BiO$^+$ + 2H$^+$ + 3e$^-$, −0.32 V; Bi + 3OH$^-$ → BiOOH + H$_2$O + 3e$^-$, 0.46 V. Other important physical characteristics of bismuth are given under **Chemical Elements**.

Bismuth occurs as native bismuth in Bolivia and Saxony and frequently is associated with lead, copper, and tin ores—the sulfide (bismuthinite, bismuth glance, Bi_2S_3) is also found in nature. Separation of bismuth from lead takes place during the electrolytic refining of the latter with bismuth remaining in the anode mud, or by prometallurgical methods by which it is removed from the lead as a calcium-magnesium compound. See also **Bismuthinite**.

Alloys

Metallurgically, bismuth is used in the production of low melting point fusible alloys and as an additive to steel, cast iron, and aluminum. The fusible alloys contain about 50% bismuth in combination with lead, tin, cadmium, and indium and are used in a variety of ways, including fire-protection devices, joining and sealing hardware, and short-life dies. Because of the special volume-increase property with solidification, bismuth is used to manufacture alloys with a zero liquid-to-solid volume change. Alloy compositions are given in Table 1. The addition of about 0.2% bismuth, along with a similar quantity of lead, improves the machineability of aluminum. Very small quantities (0.02%) of bismuth are used in the production of malleable cast iron for stabilization of carbides upon solidification, particularly desirable for castings with heavy cross sections. Combinations of bismuth and tin and bismuth and cadmium have found use as counterelectrode alloys in the manufacture of selenium rectifiers. Bismuth telluride Bi_2Te_3 and bismuth selenide Bi_2Se_3 display thermoelectric properties. With modification, these compounds are used for certain commercial and military solid-state devices, including small units for portable power generation and refrigeration.

In 1912, a number of bronze artifacts from Late Horizon times (A.D. 1476–1534) were recovered at the Inca city of Machu Picchu in Peru. These were among the first artifacts ever to be subjected to metallographic studies. Researchers Gordon and Rutledge (Kline Geology Laboratory, Yale University) reported in 1984 that the decorative bronze handle of a tumi (small knife) excavated at the Inca city contains 18% bismuth and appears to be the first known example of the use of Bi with Sn to make bronze. The alloy is not made brittle by the Bi because the bismuth-rich constituent does not penetrate the grain boundaries of the matrix phase. The use of Bi facilitated the duplex casting process by which the tumi was made and forms an alloy of unusual color.

Chemistry and Compounds

Generally, the chemical behavior of bismuth parallels that of arsenic and antimony, but bismuth is the most metallic of the group. Bismuth is not soluble in cold H_2SO_4 or cold HCl, but is attacked by these acids when hot and also by cold aqua regia. Elemental bismuth is not attacked by cold alkalies. The metal is soluble in HNO_3 and forms nitrates. When heated with chlorine, bismuth yields a chloride.

Some of the salts of bismuth are used in medicines for the relief of digestive disorders because of the smooth, protective coating the compounds impart to irritated mucous membranes. Like barium, bismuth also is used as an aid in x-ray diagnostic procedures because of its opacity to x-rays. At one time, certain bismuth compounds were used in the treatment of syphilis. Bismuth oxychloride, which is pearlescent, has found use in cosmetics, imparting a frosty appearance to nail polish, eye shadow, and lipstick, but may be subject to increasing controls. Bismuth phosphomolybdate has been used as a catalyst in the production of acrylonitrile for use in synthetic fibers and paints. Bismuth oxide and subcarbonate are used as fire retardants for plastics.

Bismuth trihalides exhibit an increased tendency toward hydrolysis, usually forming bismuthyl compounds, also called bismuth oxyhalides, which are often assumed to contain the ion BiO^+. This is not a discrete ion, however, and the crystal lattices of the "bismuthyl" compounds actually are comprised of Bi(III), O(−II) and X(−I) units. For example, BiOCl has the same crystal structure as PbFCl. The trihalides also form halobismuthates, with halogen ions, such as the chlorobismuthates, which contain the ions $BiCl_4^-$ and $BiCl_5^{2-}$. The BiI_4^{2-} ion is precipitated analytically as the cinchonine salt.

Bismuth(III) oxide, Bi_2O_3, is the compound produced by heating the metal, or its carbonate, in air. It is definitely a basic oxide, dissolving readily in acid solutions, and unlike the arsenic or antimony compounds, not amphiprotic in solution, although it forms stoichiometric addition compounds on heating with oxides of a number of other metals. It exists in three modifications, white rhombohedral, yellow rhombohedral, and gray-black cubical. Bismuth(II) oxide, BiO, has been produced by heating the basic oxalate.

Bismuth(III) hydroxide also is not significantly amphiprotic in solution, dissolving only in acids. Its formula is given as $Bi(OH)_3$ but it is difficult to isolate, due to adsorption of acid anions and to its dehydration to BiO(OH). The action of strong oxidants in concentrated alkalies on the hydroxide yields alkali bismuthates, such as $NaBiO_3$, sodium metabismuthate, from which $NaBi(OH)_6$ is initially produced. Other metal bismuthates may be made from them or directly from the oxides and Bi_2O_3, and bismuth(V) oxide is obtained by the action of HNO_3 on the alkali bismuthates; however, some oxygen is lost and the product is a mixture of Bi_2O_5 and BiO_2.

Bismuth(III) sulfide, Bi_2S_3, is precipitated by H_2S from bismuth solutions. Complex sulfide ions form only slowly, so bismuth sulfide may be separated from the arsenic and antimony sulfides by this difference in properties. Like the oxide, bismuth sulfide forms double compounds with the sulfides of the other metals.

Bismuth vanadate, $BiVO_4$, exhibits a ferroelastic-paraelastic phase transition and had been the subject of considerable investigation. This is reported in some detail in the entry on **Vanadium**.

Bismuth forms a number of complex compounds, including the sulfatobismuthates, e.g., $NaBi(SO_4)_2$ and $Na_3Bi(SO_4)_3$; and the thiocyanatobismuthates, e.g., $Na_3Bi(SCN)_6$, by the interaction of sodium thiocyanate and $Bi(SCN)_3$. The salts of bismuth tend to lose part of their acid readily, especially on heating, to form basic salts.

True pentavalent compounds of bismuth are rare, but include bismuth pentafluoride, BiF_5 (subl. 550 °C), and $KBiF_6$; pentaphenylbismuth, $(C_6H_5)_5$ Bi; various compounds $(C_6H_5)_3$ BiX_2, where X = F, Cl, Br, N_3, NCO, CH_3CO_2, $\frac{1}{2}CO_3$; and tetraphenylbismuthonium salts, $[(C_6H_5)_4$ Bi]X, where X = Cl, $-[B(C_6H_5)_4]$, etc.

Organobismuth Compounds

Numerous bismuth organic compounds have been prepared. Some of these include methylbismuthine CH_3BiH_2; phenyldibromobismuthine $C_6H_5BiBr_2$; potassium diphenylbismuthide $K[Bi(C_6H_5)_2]$; triphenylbismuthdihydroxide $(C_6H_5)_2Bi(OH)_2$; tetraphe nylbismuthonium tetraphenylborate $[(C_6H_5)_4Bi] [B(C_6H_5)_4]$; and pentaphenylbismuth $(C_6H_5)_5Bi$.

Additional Reading

Carter, G.F. and D.E. Paul: *Materials Science and Engineering,* ASM International, Materials Park, OH, 1991.

Gordon, R.B. and J.W. Rutledge: "Bismuth Bronze from Machu Picchu, Peru," *Science,* **223,** 585–588 (1984).

Greenwood, N.N. and A. Earnshaw: *Chemistry of the Elements,* 2nd Edition, Butterworth-Heinemann, Inc., Woburn, MA, 1997.

Hermann, W.A. and H.H. Karsch: "Synthetic Methods of Organometallic and Inorganic Chemistry, Phosphorus, Arsenic, Antimony, and Bismuth*, Vol. 3, Thieme Medical Publishers, Inc., New York, NY, 1996.

Krebs, R.E.: "The History and Use of Our Earth's Chemical Elements," A Reference Guide, Greenwood Publishers Group, Inc., Westport, CT, 1998.

Lide, D.R.: *CRC Handbook of Chemistry and Physics,* 84th Edition, CRC Press, LLC., Boca Raton, FL, 2003.

Norman, N.C.: *Chemistry of Arsenic and Bismuth,* Kluwer Academic Publishers, Norwell, MA, 1998.

Patai, S.E.: *Chemistry of Organic Arsenic, Antimony, and Bismuth Compounds,* John Wiley & Sons, Inc., New York, NY, 1994.

Perry, R.H., D. Green, and J.O. Maloney: *Perry's Chemical Engineers' Handbook,* 7th Edition, McGraw-Hill, New York, NY, 1997.

Staff: *ASM Handbook—Properties and Selection: Nonferrous Alloys and Special-Purpose Materials,* ASM International, Materials Park, OH, 1990.

TABLE 1. SOME REPRESENTATIVE LOW-MELTING-POINT ALLOYS CONTAINING BISMUTH

Fusible alloy, melting at 96°C	53% Bi	32% Pb	15% Sn	
Fusible alloy, melting at 91.5°C	52% Bi	40% Pb	8% Cd	
Fusible alloy, melting at 100°C	50% Bi	30% Sn	20% Pb	
Fusible alloy, melting at 70°C. (Wood's metal)	50% Bi	25% Pb	12.5% Sn	
			12.5% Cd	
Fusible alloy, melting at 70°C. (Lipowitz' alloy)	50% Bi	27% Pb	13% Sn	
			10% Cd	
Rose metal	50% Bi	27% Pb	23% Sn	
Bismuth solder, melting at 111°C.	40% Bi	40% Pb	20% Sn	

BISMUTHINITE. A mineral containing a sulfide of bismuth, Bi_2S_3, and sometimes copper and iron; a variety from Mexico contains about 8% antimony. Bismuthinite is orthorhombic although its thin needlelike crystals are rare as it usually occurs in foliated or fibrous masses. It has one good cleavage parallel to the prism; hardness, 2; specific gravity, 6.78; metallic luster; streak, lead gray; color, similar but often with iridescent tarnish; opaque.

Bismuthinite is a rather rare mineral although somewhat widely distributed. European localities are in Norway, Sweden, Saxony, Rumania, and England. It is found also in Bolivia, Australia, and in the United States in Utah. It is used as an ore of bismuth. Bismuthinite also is known as *bismuth glance*.

BISULFITE PROCESS. See **Pulp (Wood) Production and Processing.**

BITTER PATTERNS. A method for detecting domain boundaries at the surface of ferromagnetic crystals. If a drop of a colloidal suspension of ferromagnetic particles is placed on the surface of the crystals, the particles will collect along the domain boundaries where the field is strongest.

BITUMEN. Natural flammable substances of a wide range of color, hardness, and volatility, constituted mainly of a mixture of hydrocarbons and essentially free from oxygenated bodies. Petroleums, asphalts, natural mineral waxes, and asphaltites are considered bitumens. See also **Tar Sands.**

BIXIN. See **Annatto Food Colors.**

BLACK BODY. This term denotes an ideal body which would, if it existed, absorb all and reflect none of the radiation falling upon it; its reflectivity would be zero and its absorptivity would be 100%. Such a body would, when illuminated, appear perfectly black, and would be invisible except its outline might be revealed by the obscuring of objects beyond. The chief interest attached to such a body lies in the character of the radiation emitted by it when heated and the laws that govern the relations of the flux density and the spectral energy distribution of that radiation with varying temperature.

The total emission of radiant energy from a black body takes place at a rate expressed by the Stefan-Boltzmann (fourth-power) law; while its spectral energy distribution is described by Wien's laws, or more accurately by Planck's equation, as well as by a number of other empirical laws and formulas. See also **Thermal Radiation.**

The nearest approach to the ideal black body, experimentally, is not a sooty surface, as might be supposed, but an almost completely closed cavity in an opaque body, such as a jug. The laboratory type is usually a somewhat elongated, hollow metal cylinder, blackened inside, and completely closed except for a narrow slit in one end. When such an enclosure is heated, the radiation escaping through the opening closely resembles the ideal black-body radiation; light or other radiation entering by the opening is almost completely trapped by multiple reflection from the walls, so that the opening usually appears intensely black.

BLAGDEN LAW. The depression of the freezing point of a solution is, for small concentrations, proportional to the concentration of the dissolved substance.

BLEACHING AGENTS. A bleaching agent is a material that lightens or whitens a substrate through chemical reaction. The bleaching reactions usually involve oxidative or reductive processes that degrade color systems. These processes may involve the destruction or modification of chromophoric groups in the substrate as well as the degradation of color bodies into smaller, more soluble units that are more easily removed in the bleaching process. The most common bleaching agents generally fall into two categories: chlorine and its related compounds (such as sodium hypochlorite) and the peroxygen bleaching agents, such as hydrogen peroxide and sodium perborate. Reducing bleaches represent another category. Bleaching agents are used for textile, paper, and pulp bleaching as well as for home laundering.

Chlorine-Containing Bleaching Agents

Chlorine-containing bleaching agents are the most cost-effective bleaching agents known. They are also effective disinfectants, and water disinfection is often the largest use of many chlorine-containing bleaching agents. They may be divided into four classes: chlorine, hypochlorites, N-chloro compounds, and chlorine dioxide.

The first three classes are called available chlorine compounds and are related to chlorine by the equilibria in equations 1–4. These equilibria are rapidly established in aqueous solution, but the dissolution of some hypochlorite salts and N-chloro compounds can be quite slow.

$$Cl_2(gas) \rightleftharpoons Cl_2(aq) \tag{1}$$

$$Cl_2(aq) + H_2O \rightleftharpoons HOCl + H^+ + Cl^- \tag{2}$$

$$HOCl \rightleftharpoons H^+ + OCl^- \tag{3}$$

$$RR'NCl + H_2O \rightleftharpoons HOCl + RR'NH \tag{4}$$

The total concentration or amount of chlorine-based oxidants is often expressed as available chorine, or less frequently as active chlorine. Available chlorine is the equivalent concentration or amount of Cl_2 needed to make the oxidant according to equations 1–4. Active chlorine is the equivalent concentration or amount of Cl atoms that can accept two electrons. This is a convention, not a description of the reaction mechanism of the oxidant. Because Cl_2 only accepts two electrons as does HOCl and monochloramines, it only has one active Cl atom according to the definition. Thus the active chlorine is always one-half of the available chlorine. The available chlorine is usually measured by iodometric titration. The weight of available chlorine can also be calculated by equation 5, where 70.9 represents the mol wt of Cl_2 and moles of oxidant can be represented wt oxidant/mol wt of oxidant.

$$\text{weight available} = 70.9 \times \text{moles of oxidant}$$

$$\times \frac{\text{number active Cl atoms}}{\text{molecule}} \tag{5}$$

In solutions, the concentration of available chlorine in the form of hypochlorite or hypochlorous acid is called free-available chlorine.

Commercially important solid available chlorine bleaches are usually more stable than concentrated hypochlorite solutions. They decompose very slowly in sealed containers. But most of them decompose quickly as they absorb moisture from air or from other ingredients in a formulation. This may release hypochlorite that destroys other ingredients as well.

Chlorine. Except to bleach wood pulp and flour, chlorine itself is rarely used as a bleaching agent.

Hypochlorites. The principal form of hypochlorite produced is sodium hypochlorite, NaOCl.

Other hypochlorites include calcium hypochlorite, bleach liquor, bleaching powder and tropical bleach, dibasic magnesium hypochlorite, lithium hypochlorite, chlorinated trisodium phosphate, and hypochlorous acid.

N-Chloro Compounds

Chlorinated Isocyanurates. The principal solid chlorine bleaching agents are the chlorinated isocyanurates. The one used most often for bleaching applications is sodium dichloro-isocyanurate dihydrate, with 56% available chlorine.

Other N-chloro compounds include halogenated hydantoins, sodium N-chlorobenzenesulfonamide (chloramine B), sodium N-chloro-p-toluene-sulfonamide (chloramine T), N-chlorosuccinimide, and trichloromelamine.

Chlorine Dioxide. Chlorine dioxide, ClO_2, is a gas that is more hazardous than chlorine. Large amounts for pulp bleaching are made by several processes in which sodium chlorate is reduced with chloride, methanol, or sulfur dioxide in highly acidic solutions by complex reactions. For most other purposes chlorine dioxide is made from sodium chlorite.

Peroxygen Compounds

Peroxygen compounds contain the peroxide linkage ($-O-O-$) in which one of the oxygen atoms is active.

Hydrogen Peroxide. Hydrogen peroxide is one of the most common bleaching agents. See also **Hydrogen Peroxide**. It is the primary bleaching agent in the textile industry, and is also used in pulp, paper, and home laundry applications. In textile bleaching, hydrogen peroxide is the most common bleaching agent for protein fibers, and is also used extensively for cellulosic fibers.

Solid Peroxygen Compounds. Hydrogen peroxide reacts with many compounds, such as borates, carbonates, pyrophosphates, sulfates, silicates, and a variety of organic carboxylic acids, esters, and anhydrides to give peroxy compounds or peroxyhydrates. A number of these compounds are stable solids that hydrolyze readily to give hydrogen peroxide in solution.

Compounds include perborates, sodium carbonate peroxyhydrate, and peroxymonosulfate.

Peracids. Peracids are compounds containing the functional group $-OOH$ derived from an organic or inorganic acid functionality. Typical structures include $CH_3C(O)OOH$ derived from acetic acid and $HOS(O)_2OOH$ (peroxymonosulfuric acid) derived from sulfuric acid. Peracids have superior cold water bleaching capability versus hydrogen peroxide because of the greater electrophilicity of the peracid peroxygen moiety. Lower wash temperatures and phosphate reductions or bans in detergent systems account for the recent utilization and vast literature of peracids in textile bleaching.

Peracids can be introduced into the bleaching system by two methods. They can be manufactured separately and delivered to the bleaching bath with the other performance components or as a separate product. Peracids can also be formed *in situ* utilizing the perhydrolysis reaction shown in equation 6.

$$\underset{\displaystyle R-\overset{\displaystyle O}{\overset{\|}{C}}-Z}{} + {}^-OOH \longrightarrow RCOOH + Z^- \qquad (6)$$

Peracid Precursor Systems. Compounds that can form peracids by perhydrolysis are almost exclusively amide, imides, esters, or anhydrides. Two compounds were commercially used for laundry bleaching as of 1990. Tetraacetylethylenediamine (TAED) is utilized in over 50% of Western European detergents. Nonanoyloxybenzene sulfonate (NOBS) is used in detergent products in the United States and Japan.

Preformed Peracids. Peracids can be generated at a manufacturing site and directly incorporated into formulations without the need for *in situ* generation. Two primary methods are utilized for peracid manufacture. The first method uses the equilibrium shown in equation 7 to generate the peracid from the parent acid.

$$\underset{\displaystyle RCOH}{\overset{\displaystyle O}{\overset{\|}{}}} + H_2O_2 \rightleftharpoons \underset{\displaystyle RCOOH}{\overset{\displaystyle O}{\overset{\|}{}}} + H_2O \qquad (7)$$

The equilibrium is shifted by removal of the water or removal of the peracid by precipitation. Peracids can also be generated by treatment of an anhydride with hydrogen peroxide to generate the peracid and a carboxylic acid.

$$\underset{\displaystyle RCOCR}{\overset{\displaystyle O\quad O}{\overset{\|\quad\|}{}}} + H_2O_2 \longrightarrow \underset{\displaystyle RCOOH}{\overset{\displaystyle O}{\overset{\|}{}}} + \underset{\displaystyle RCOH}{\overset{\displaystyle O}{\overset{\|}{}}} \qquad (8)$$

The latter method (eq. 8) typically requires less severe conditions than the former because of the labile nature of the organic anhydride. Both of these reactions can result in explosions, and significant precautions should be taken prior to any attempted synthesis of a peracid.

Peracid Decomposition. Peracids, whether preformed or formed *in situ* via the perhydrolysis reaction, are susceptible to decomposition in an aqueous bleaching bath. The decomposition is caused by the occurrence of one of four reactions. The peracid can decompose as a result of oxidation of the bleachable material. Transition metals present even at extremely low concentration in the bath from the incoming water can decompose the peracid catalytically. To minimize this effect, metal-sequestering agents have been proposed to prevent the degradation of the peracid in solution. Peracids can also hydrolyze to the parent acid and hydrogen peroxide because of the large excess of water present in the aqueous bleaching bath. This is generally a kinetically slow process. A final decomposition mechanism involves the reaction of two moles of peracid generating two moles of parent acid and a mole of oxygen.

Reducing Bleaches

The reducing agents generally used in bleaching include sulfur dioxide, sulfurous acid, bisulfites, sulfites, hydrosulfite (dithionites), sodium sulfoxylate formaldehyde, and sodium borohydride. These materials are used mainly in pulp and textile bleaching.

The Mechanism of Bleaching

Bleaching is a decolorization or whitening process that can occur in solution or on a surface. The color-producing materials in solution or on fibers are typically organic compounds that possess extended conjugated chains of alternating single and double bonds and often include heteroatoms, carbonyl, and phenyl rings in the conjugated system. The portion of molecule that absorbs a photon of light is referred to as the chromophore (Greek: *color bearer*). For a molecule to produce color the conjugated system must result in sufficiently delocalized electrons such that the energy gap between the ground and excited states is small enough so that photons in the visible portion of the light spectrum are absorbed.

Bleaching and decolorization can occur by destroying one or more of the double bonds in the conjugated chain, by cleaving the conjugated chain, or by oxidation of one of the other moieties in the conjugated chain. The result of any one of the three reactions is an increase in the energy gap between the ground and excited states, so that the molecule then absorbs light in the ultraviolet region, and no color is produced. Bleaching may also increase the water solubility of organic compounds after reaction. Conversion of an olefin to a vicinal diol, for example, dramatically increases the polarity and consequently the water solubility of the compound. A variety of bleaching agents can affect this transformation. The increased solubility allows actual removal of the bleached substance from a surface.

Chlorine bleaches react with more chromophores than oxygen bleaches. They react irreversibly with aldehydes, alcohols, ketones, carbon–carbon double bonds, acidic carbon–hydrogen bonds, nitrogen compounds, sulfur compounds, and aromatic compounds.

The mechanism of bleaching of hydrogen peroxide is not well understood. It is generally believed that the perhydroxyl anion (HOO^-) is the active bleaching species since both the concentration of this anion and the rate of the bleaching process increase with increasing pH. Hydrogen peroxide and other peroxygen compounds can destroy double bonds by epoxidation. This involves addition of an oxygen atom across the double bond usually followed by hydrolysis of the epoxide formed to 1,2-diols under bleaching conditions.

Peracids undergo a variety of reactions which result in bleaching. Peracids can add an oxygen across a double bond to give an epoxide, which can undergo further reactions including hydrolysis to give a vicinal diol. Peracids can oxidize aldehydes to acids, sulfur compounds to sulfoxides and sulfones, and nitrogen compounds to amine oxides, hydroxylamines, and nitro compounds. Peracids can also oxidize α-diketone compounds to anhydrides and ketones to esters.

Reducing agents are thought to work by reduction of the chromophoric carbonyl groups in textiles or pulp.

Applications of Bleaching Compounds

Laundering and Cleaning

Home and Institutional Laundering. The most widely used bleach in the United States is liquid chlorine bleach, an alkaline aqueous solution of sodium hypochlorite. This bleach is highly effective at whitening fabrics and also provides germicidal activity at usage concentrations. Liquid chlorine bleach is sold as a 5.25% solution, and 1 cup provides 200 ppm of available chlorine in the wash. Liquid chlorine bleaches are not suitable for use on all fabrics. Dry and liquid bleaches that deliver hydrogen peroxide to the wash are used to enhance cleaning on fabrics. They are less efficacious than chlorine bleaches but are safe to use on more fabrics. The dry bleaches typically contain sodium perborate in an alkaline base whereas the liquid peroxide bleaches contain hydrogen peroxide in an acidic solution. Detergents containing sodium perborate tetrahydrate are also available. See also **Detergents**; and **Oxidation and Oxidizing Agents**.

The worldwide decreasing wash temperatures, which decrease the effectiveness of hydrogen peroxide-based bleaches, have stimulated research to identify activators to improve bleaching effectiveness. Tetraacetylethylenediamine is widely used in European detergents to compensate for the trend to use lower wash temperatures. TAED generates peracetic acid in the wash in combination with hydrogen peroxide. TAED has not been utilized in the United States, where one activator nonanoyloxybenzene sulfonate (NOBS) has been commercialized and incorporated into several detergent products. NOBS produces pernonanoic acid when combined with hydrogen peroxide in the washwater and is claimed to provide superior cleaning in contrast to perborate bleaches.

In industrial and institutional bleaching, either liquid or dry chlorine bleaches are used because of their effectiveness, low cost, and germicidal properties. Dry chlorine bleaches, particularly formulated chloroisocyanurates, are used in institutional laundries.

Hard Surface Cleaners and Cleansers. Bleaching agents are used in hard surface cleaners to remove stains caused by mildew, foods, etc, and to disinfect surfaces. Disinfection is especially important for many industrial uses. Alkaline solutions of 1–5% sodium hypochlorite that may contain surfactants and other auxiliaries are most often used for these purposes. These are sometimes thickened to increase contact times with vertical surfaces. A thick, alkaline cleaner with 5% hydrogen peroxide is also sold in Europe. Liquid abrasive cleansers with suspended solid abrasives are also available and contain about 1% sodium hypochlorite. Powdered cleansers often contain 0.1–1% available chlorine and they may contain abrasives. Sodium dichloroisocyanurate is the most common bleach used in powdered cleansers, having largely replaced chlorinated trisodium phosphate. Calcium hypochlorite is also used. Dichloroisocyanurates are also used in effervescent tablets that dissolve quickly to make cleaning solutions. In-tank toilet cleaners use calcium hypochlorite, dichloroisocyanurates, or *N*-chloro compounds to release hypochlorite with each flush.

Automatic Dishwashing and Warewashing. The primary role of bleach in automatic dishwashing and warewashing is to reduce spotting and filming by breaking down and removing the last traces of adsorbed soils. They also remove various food stains such as tea. All automatic dishwashing and warewashing detergents contain alkaline metal salts or hydroxides.

Textile Bleaching. Many textiles are bleached to remove any remaining soil and colored compounds before dyeing and finishing (see TEXTILES). Bleaching is usually preceded by washing in hot alkali to remove most of the impurities in a process called scouring. Bleaching is usually done as part of a continuous process, but batch processes are still used. Bleaching conditions vary widely, depending on the equipment, the bleaching agent, the type of fiber, and the amount of whiteness required for the end use.

Cotton and Cotton–Polyester. Cotton is the principal fiber bleached today, and almost all cotton is bleached. About 80–90% of all cotton and cotton–polyester fabric is bleached with hydrogen peroxide.

Other Cellulosics. Rayon is bleached similarly to cotton but under milder conditions since the fibers are more easily damaged and since there is less colored material to bleach.

Synthetic Fibers. Most synthetic fibers are sufficiently white and do not require bleaching. When needed, synthetic fibers and many of their blends are bleached with sodium chlorite solutions. Solutions of 0.1% peracetic acid are also used.

Wool and Silk. Wool must be bleached carefully, in order to avoid fiber damage. It is usually bleached with 1–5% hydrogen peroxide solutions. Silk is bleached similarly, but at slightly higher temperatures.

Bleaching of Other Materials

Hair. Hydrogen peroxide is the most satisfactory bleaching agent for human hair.

Fur. The coloring matter in fur is usually bleached using hydrogen peroxide stabilized with sodium silicate.

Foodstuffs, Oils. Sulfur dioxide is used to preserve grapes, wine, and apples; the process also results in a lighter color. During the refining of sugar, sulfur dioxide is added to remove the last traces of color. Flour can be bleached with a variety of chemicals including chlorine, chlorine dioxide, oxides of nitrogen, and benzoyl peroxide. Bleaching agents such as chlorine dioxide or sodium dichromate are used in the processing of nonedible fats and fatty oils for the oxidation of pigments to colorless forms. See also **Food Processing**.

JAMES P. FARR
WILLIAM L. SMITH
DALE S. STEICHEN
The Clorox Company

Additional Reading

Coons, D. M.: *J. Amer. Oil Chemist's Soc.* **55**, 104 (1978).
Household and Industrial Bleach Systems, North America Forecast to 2000, Colin A. Houston and Associates, Mamaroneck, NY, 1988, pp. 2,3.
Sheltmire, W. H.: in J. S. Sconce, ed., *Chlorine: Its Manufacture, Properties And Uses,* American Chemical Society Monograph Series 154, Reinhold Publishing Co., New York, NY, 1962, pp. 512–542.

BLENDING. See **Mixing and Blending**.

BLOOD. Classified as a major tissue of the human body, blood is a characteristically red, mobile fluid with an average specific gravity of about 1.058. Slightly sticky and somewhat viscous, blood has a viscosity between 4.5 and 5.5 times greater than that of water at the same temperature. Thus, blood flows somewhat more sluggishly than water. The odor of blood is characteristic; the taste is slightly saline. The pH of blood ranges between 7.35 and 7.45. The complex acid-base regulatory system of the blood is described in entry on **Acid-Base Regulation (Blood)**. Under normal conditions, the blood circulates through the body at a temperature of 100.4°F (38.0°C). This is slightly higher than the body temperature as determined by mouth, 98.6°F (37.0°C). An adult human of average age and size has just over 6 quarts (5.7 liters) of blood.

In very general terms, blood serves as a chemical transport and communications system for the body (i.e., it carries chemical messengers as well as nutrients, wastes, etc.). Circulated by the heart through arteries, veins, and capillaries, blood carries oxygen and a variety of chemicals to all cells, acting as a delivery agent to serve the needs of the cells. Blood also takes away waste products, including carbon dioxide, from the various tissues to organs such as the kidneys and lungs which ultimately dispose these wastes to the environment. Thus, the blood serves as a collecting agent. Unlike a simple liquid, such as water, or a simple solution, such as salt water, blood is a complex fluid made up of several components, each of which is, in turn, extremely complex and even today not fully understood. Many of these substances are solids in suspension. Unlike most simple liquids that are not easily changed when exposed to air or to slight alterations in their environment, the physical and biochemical properties of blood undergo marked changes (Hemostatic responses) when blood is taken from the body's circulatory system and, for example, placed in a test tube. Separation of blood from its usual environment immediately initiates biochemical processes that alter its properties and cause it to release its components. When so removed, blood shortly becomes viscid and forms a soft, jelly-like substance, then soon separates into a firm solid mass (clot) and liquid (serum). This extremely important property of clotting is unique to blood among known inorganic and organic fluids and solutions. Were it not for this property, a person would bleed (hemorrhage) to death if a blood vessel were opened by accident or as the consequence of disease. Thus, blood may be described as a living fluid and most accurately as a living tissue, like the other tissues of the body.

Illustrative of the complex constitution of blood is the list of Table 1.

Principal Components of Human Blood

Not considering the numerous substances, other than oxygen, carried by the blood where the main function of the blood is one of transport, the main functional components of the blood are indicated in Fig. 1.

Erythrocytes. It is estimated that an adult man will have about 5 million erythrocytes per cubic millimeter of blood. This is equivalent to about 82 billion erythrocytes per cubic inch of blood. In an adult woman, there are about 4.5 million erythrocytes per cubic millimeter. Erythrocytes are homogeneous circular disks with no nucleus. These red cells are about 0.0077 millimeter in diameter. When viewed singly by transmitted light, the erythrocyte has a yellowish red tinge, but when viewed in great numbers, the erythrocytes have the distinctly blood red coloration. Erythrocytes possess a certain degree of elasticity, so that they can pass through tiny apertures and passages on their way to reach tissue supplied by the capillaries.

The prime function of the erythrocytes is to deliver oxygen to peripheral tissues. This oxygen is furnished to these cells by an exchange-diffusion system brought about in the lungs. The color of the erythrocytes is derived from a red iron-containing pigment called hemoglobin. This is a conjugated protein that consists of a globin (a protein) and hematin (a non-protein pigment), the latter containing iron. Hemoglobin contains 0.33% iron. When hemoglobin combines with oxygen, oxyhemoglobin is formed. When oxygen is given up to the tissues, it is then reduced back to hemoglobin. The erythrocytes also carry some carbon dioxide from the tissues and function to maintain a normal acid-base balance (pH) of the blood. When the hemoglobin has its full complement of oxygen, it is a bright red. This scarlet blood is found in the arteries that carry the blood to organ tissues throughout the body. As the oxyhemoglobin gives up oxygen, it takes on a darker crimson hue, and this is found in the veins which return the blood to the lungs for re-oxygenation. See **Hemoglobin**.

Megaloblasts. Cells that are the precursors of erythrocytes, are noted in the blood islands of the yolk sac of the human embryo. By the end

TABLE 1. REPRESENTATIVE CONSTITUENTS OF HUMAN BLOOD (VALUES ARE PER 100 MILLILITERS)

Constituent	Plasma of serum	Whole blood
Adenosine	1.09 mg	
Adenosine triphosphate (total)		31–57 mg
Amino acids (total)		38–53 mg
Ammonia N	0.1–1.1 mg	0.1–0.2 mg
Ascorbic acid	0.7–1.5 mg	0.1–1.3 mg
Base (total)	145–160 meq/liter	
Bicarbonate	25–30 meq/liter	19–23 meq/liter
Bile acids		0.2–3.0 mg
Biotin		0.7–1.7 μg
Blood volume		2990–6980 ml
—adult men	33.7–43.7 ml/kg	66.2–97.7 ml/kg
—adult women	32.0–42.0 ml/kg	46.3–85.5 ml/kg
—infants	36.3–46.3 ml/kg	79.7–89.7 ml/kg
Carbon dioxide		
—Arterial blood (total)		45–55 vol%
—Venous blood (total)		50–60 vol%
Cholesterol (total)	120–250 mg	115–225 mg
Cholesterol esters	75–150 mg	48–115 mg
Cholesterol (free)	30–60 mg	82–113 mg
Choline (total)	26–35 mg	11–31 mg
Fat (neutral)	25–260 mg	85–235 mg
Fatty acids	190–450 mg	250–390 mg
Fibrinogen	200–400 mg	120–160 mg
Fructose	7–8 mg	0–5 mg
Glucose (adult)	65–105 mg	80–120 mg
Hemoglobin	trace	14.8–15.8 g
Histamine		6.7–8.6 μg
Ketone bodies (total)	0.15–1.36 mg	0.23–1.00 mg
Lactic acid	30–40 mg	5–40 mg
Lecithin	100–225 mg	110–120 mg
Lipids (total)	400–700 mg	445–610 mg
Mucopolysaccharides	175–225 mg	
Mucoproteins	86.5–96 mg	
Nicotinic acid	0.02–015 mg	0.5–0.8 mg
Nitrogen (total)		3.0–3.7 g
Non-protein nitrogen	18–30 mg	25–50 mg
Nucleotide (total)		31–52 mg
Nucleotide phosphorus		2–3 mg
Oxygen (arterial)		17–22 vol%
Oxygen (venous)		11–16 vol%
PH	7.38–7.42	7.36–7.40
Pantothenic acid	6–35 μg	15–45 μg
Polysaccharides (total)	73–131 mg	
Protein (total)	6.0–8.0 g	19–21 g
Protein (albumin)	4.0–4.8 g	
Protein (globulin)	1.5–3.0 g	
Purines (total)		9.5–11.5 mg
Pyruvic acid	0.7–1.2 mg	0.5–1.0 mg
Riboflavin	2.6–3.7 μg	15–60 μg
Ribonucleic acid	4–6 mg	50–80 mg
Sphingomyelin	10–47 mg	150–185 mg
Thiamine	1–9 μg	3–10.7 μg
Urea	28–40 mg	20–40 mg
Urea N	8–28 mg	5–28 mg
Uric acid (male)	2.5–7.2 mg	0.6–4.9 mg
Vitamin A (caretenol)	15–60 μg	9–17 μg
Vitamin A (carotene)	40–540 μg	20–300 μg
Vitamin B$_{12}$ (cyanocobalamin)	0.01–0.07 μg	0.06–0.14 μg
Vitamin D$_2$ (as calciferol)	1.7–4.1 μg	
Vitamin E	0.9–1.9 mg	
Water	93–95 g	81–86 g

Note: Plasma is the liquid portion of whole blood. *Serum* is the liquid portion of blood after clotting, the fibrinogen having been removed.

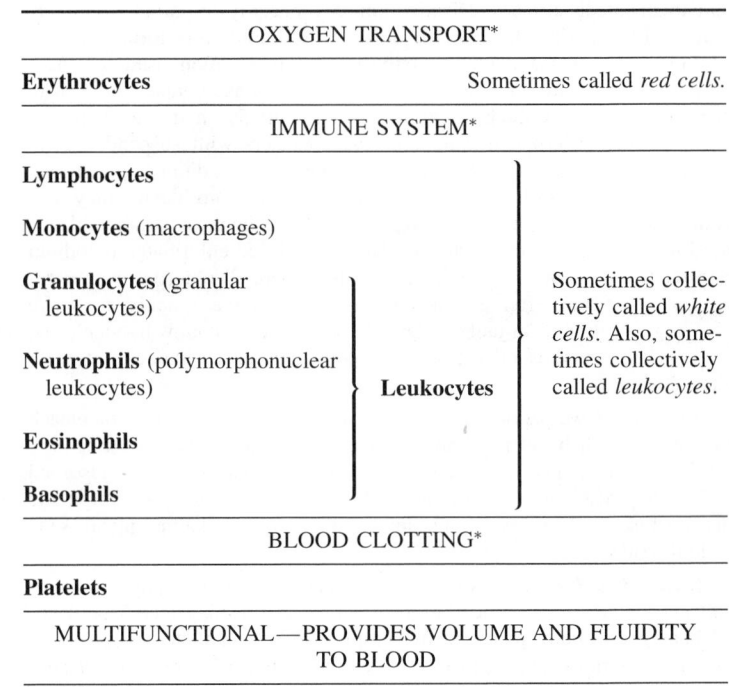

Fig. 1.   Basic components of blood

of the embryo's second month of life, manifesting the second step in the erythrocytic series, erythroblasts are found in the liver and spleen. These cells are somewhat smaller and possess a smaller nucleus than the megaloblasts. At about the fifth month, centers of blood formation appear in the middle regions of the bones, with an accompanying progressive expansion of the marrow cavities. At this stage the marrow assumes nearly exclusively the function of producing the erythrocytes (red cells) required by the body—a process which continues throughout the life of the individual. At the time of birth, essentially all bone marrow is engaged

in blood formation (not exclusively red cells). As the individual progresses toward maturity, much of the marrow of the long bones is converted into a fatty tissue in which blood cell formation (hematopoiesis) is no longer apparent. In adults, bone marrow active in the formation of blood cells is found in the ribs, vertebrae, skull, and the proximal ends of the humerus (upper arm) and femur (upper thigh).

Once erythrocytes enter the blood, it is estimated that they have an average lifetime of about 120 days. In an average person, this indicates that about 1/120th or 0.83% of the red cells are destroyed each day. At least three important mechanisms are involved in the death of erythrocytes: (1) Phagocytosis, defined as the ingestion of solid particles by living cells—in this case, by cells of the reticuloendothelial system. (2) Hemolysis by specific agents in the blood plasma. The erythrocytes are protected by a membrane. If this membrane is broken, the hemoglobin goes into solution in the plasma. Numerous substances (hemolytic agents) may cause this action and these include hypotonic solutions, foreign blood serums, snake venom, various bacterial metabolites, chloroform, bile salts, ammonia and other alkalis, among others. In this condition, the erythrocytes no longer can serve as oxygen carriers. (3) Mechanical damage and destruction, brought about by simple wear and tear as the reasonably fragile red cells circulate and recirculate through the body.

The stimulus for production of new red cells is provided by erythropoietin, a hormone that is apparently produced by the kidneys. The actual production is accomplished almost entirely by the red portions of the bone marrow, but certain substances necessary for their manufacture must be furnished by the liver. Surplus red cells, needed to meet an emergency, are stored in the body, mainly in the spleen. The spleen also breaks down old and worn red cells, conserving the iron during the process. When a sudden loss of a large amount of blood occurs, the spleen releases large numbers of red cells to make up for the loss, and the bone marrow is stimulated to increase its rate of manufacture of blood cells. When a donor gives a pint of blood, it usually requires about seven weeks for the body reserve of red corpuscles to be replaced, although the circulating red cells may be back almost to normal within a few hours. Repeated losses of blood within a short time, however, may easily deplete the red cell reserves.

In addition to hemoglobin, it has been found that there are least two other alternative oxygen carriers—hemerythrin and hemocyanin. In overall terms, as presently understood, these carriers are minor. Unlike hemoglobin, these two blood proteins do not incorporate an iron-porphyrin ring. The three blood proteins are strikingly colored in their oxygenated states—the familiar red of hemoglobin; the unusual reddish-tinted violet of hemerythrin; and the cupric-bluish color of hemocyanin. Klotz and colleagues (Northwestern University and other locations) have made a detailed study of the alternative oxygen carriers and suggest that an understanding of the three-dimensional structure of hemerythrin and of the electronic state of the active site is approaching, in refinement, that which is currently known about hemoglobin.

White Cells. There are several types of white cells, which are sometimes collectively called leukocytes, although some authorities reserve that term to identify only the granulocytes. White cells are irregular in shape and size, but generally are larger than the red cells. They differ from the red cells in that each white cell contains a nucleus. Adult humans have from 5000 and 9000 leucocytes per cubic millimeter of blood. In infants, the number is essentially doubled. There is roughly a ratio of 1 white to every 700 red cells. When white cells increase in number, the condition is called leukocytosis, a situation that is presented in pneumonia, appendicitis, and abscesses, among other conditions. A decrease in the number of leukocytes below normal is called leukopenia. In leukemia, there is an uncontrolled increase in the number of leucocytes.

In general terms, the white cells, each type with a specific function, accomplish the following actions: (1) Protection of the body from pathogenic organisms; and (2) participation in tissue repair and regeneration. Over the years, an increasingly detailed understanding of the white cells has occurred. These matters are described in some detail in the entry on **Immune System and Immunochemistry**. Generally, whenever bacteria or other foreign substances enter the tissues, large numbers of white cells immediately travel through the walls of the blood vessels and to the site of disturbances. They take the bacteria and any other foreign materials into their own bodies, where they are digested. White cells are able to break up and carry away even as large an object as a splinter or thorn in the skin. They also help in carrying away dead tissue and blood clots that remain after a wound. Pus is largely composed of white cells that have been drawn to the infected area, as well as the dead and disintegrating tissue and bacteria. During severe infections, the white cells may be increased in the blood five- or tenfold. Because of this, a white cell count is made on the blood in order to confirm diagnosis in many infections.

Lymphocytes generally comprise between 25–30% of the white cells in human blood. These immunologically active cells are comprised of several classes, each of which has specific properties and functions. Lymphocytes are derived from stem cells located in the yolk sac and fetal liver. Later, some stem cells originate from the bone marrow. These cells then differentiate into lymphocytes in the primary lymphoid organs, principally the thymus and lymph nodes.

Monocytes (macrophages) are part of the mononuclear phagocytic system. They are large, mononuclear cells and comprise 3–8% of the leukocytes found in the peripheral blood. Monocytes originate in the bone marrow. When the mature cells enter the peripheral blood, they are called monocytes; when they leave the blood and infiltrate tissues, they are called macrophages. These cells play an important role in induction of the immune response. They present antigen to the lymphocytes that bear specific receptors for the antigen and also act as effector cells, attacking certain microorganisms and neoplastic cells.

Granulocytes contain specifically identifiable granules, including the neutrophils, eosinophils, and basophils. The neutrophils comprise 60–70% of all leukocytes in the blood. Neutrophils arise from precursors in the bone marrow and have a half-life of 4 to 8 hours in the blood, with about a day of life in the tissues. Neutrophils hasten to inflammatory sites by a number of different and poorly understood chemotactic (response to chemical stimulation) factors. Neutrophils have a marked capacity to phagocytize and destroy microorganisms. These cells also contain a number of degradative enzymes and small proteins. The cells are endowed with receptors for IgC and for a complement component (C3b). See also **Immune System and Immunochemistry**. The eosinophils are named by virtue of the fact that the granules of cytoplasm are stainable with acid dyes, such as eosin. These cells are present in small numbers (2–4% of the blood), but under certain pathological conditions they show a marked increase. The exact function of eosinophils has been a mystery for many years. Some studies commenced in the mid-1970s have indicated a number of different functions. Many eosinophils have been found in tissues at sites of immune reactions that have been triggered by IgE antibodies (as found in nasal polyps or in the bronchial wall of some patients with asthma). Eosinophils have been found to contain several enzymes that can degrade mediators of immediate hypersensitivity, such as histamine, suggesting that they may control or diminish some hypersensitivity reactions. These cells have been found associated with infections caused by helminths (worms). The basophils are formed in the bone marrow and have a polymorphic nucleus. They occur only to the extent of about 1% of the leukocytes. The function of these cells is poorly understood. They are known to play a role in immediate hypersensitivity reactions and in some cell-mediated delayed reactions, such as contact hypersensitivity in humans and skin graft or tumor rejections and hypersensitivity to certain microorganisms in animals.

Platelets are the smallest of the formed elements of the blood. Every cubic millimeter of blood contains about 250 million platelets, as compared with only a few thousand white cells. There are about a trillion platelets in the blood of an average human adult. Platelets are not cells, but are fragments of the giant bone-marrow cells called megakaryocytes. When a megakaryocyte matures, its cytoplasm (substance outside cell nucleus) breaks up, forming several thousand platelets. Platelets are roughly disk-shaped objects between one-half and one-third the diameter of a red cell, but containing only about one-thirteenth the volume of the red cell. Platelets lack DNA and have little ability to synthesize proteins. When released into the blood, they circulate and die in about ten days. However, they do possess an active metabolism to supply their energy needs.

Because platelets contain a generous amount of contractile protein (actomyosin), they are prone to contract much as muscles do. This phenomenon explains the shrinkage of a fresh blood clot after it stands for only a few minutes. The shrinkage plays a role in forming a hemostatic plug when a blood vessel is cut. The primary function of platelets is that of forming blood clots. When a wound occurs, numbers of platelets are attracted to the site where they activate a substance (thrombin) which starts the clotting process. Prothrombin is the precursor of thrombin. Thrombin, in addition to converting fibrinogin into fibrin, also makes the platelets sticky. Thus, when exposed to collagen and thrombin, the platelets aggregate to form a plug in the hole of an injured blood vessel. Persons with a low platelet count (thrombocytopenia) have a long bleeding time. Platelet counts may be low because of insufficient production in the bone marrow (from leukemia or congenital causes, or from chemotherapy used in connection with cancer), among other causes. Also, individuals may manufacture antibodies to their own platelets to the point where they are destroyed at about the same rate they are produced. A major symptom of this disorder is purpura. Aspirin may aggravate this condition. The bleeding of hemophilia results from a different cause. Transfusion of blood is a major therapy used in treating platelet disorders.

Platelets not only tend to stick to one another, but to the walls of blood vessels as well. Obviously because they promote clotting, they have a key role in forming thrombi. Many attempts have been made to explain the process of atherogenesis, that is, the creation of plaque, which narrows arteries and, of particular concern, the coronary arteries. Recently, there has been increasing interest in the possible role of platelets in atherosclerosis. Evidence from experimentation with laboratory animals has provided some evidence of a role for platelets in this process. This is covered in some detail by Zucker (1980).

As reported by Turitto and Weiss (1980), red blood cells may have a physical and chemical effect on the interaction between platelets and blood vessel surfaces. Under flow conditions in which primarily physical effects prevail, it has been found that platelet adhesion increases fivefold as *hematocrit*[1] values increase from 10 to 40%, but undergoes no further increase from 40 to 70%, implying a saturation of the transport-enhancing capabilities of red cells. For flow conditions in which platelet surface reactivity is more dominant, platelet adhesion and thrombus formation increase monotonically as hematocrit values increase from 10 to 70%. Thus, the investigators suggest that red cells may have a significant influence on hemostasis and thrombosis; the nature of the effect is apparently related to the flow conditions.

[1] A hematocrit is a tube calibrated to facilitate determination of the volume of erythrocytes (red cells) in centrifuged, oxalated blood, expressed as corpuscular volume percent.

Human von Willebrand Factor (vWF). In an excellent technical discussion, Ginsburg et al. review human factor VIII-von Willebrand factor. vWF is a large, multimeric glycoprotein that plays a central role in the blood coagulation system, serving both as a carrier for factor VIIC (antihemophilic factor) and as a major mediator of platelet-vessel wall interaction. Diminished or abnormal vWF activity results in von Willebrand's disease (vWD), a common and complex hereditary bleeding disorder. In the article, Ginsburg and colleagues describe how they have isolated a nearly full-length cDNA for human vWF and initial characterization of the vWF genetic locus. Such studies shed new knowledge on how the hemostatic system has evolved to minimize blood loss following vascular injury. In higher vertebrates, including humans, the system is complex and requires the interaction of circulating platelets, a series of plasma coagulation proteins, endothelial cells, and components of the vascular subendothelium. The initial and critical event in hemostasis is the adhesion of platelets to the subendothelium, a process that occurs within seconds of injury and provides a location for platelet plug assembly and fibrin clot formation.

Doolittle (1981), in an excellent paper on fibrinogen and fibrin, presents a detailed pictorial model of fibrinogen and develops the amino acid sequence of the fibrinogen molecule. In the paper, the author demonstrates how knowledge of the amino acid sequence of fibrinogen bolsters some long-standing notions about the protein's three-dimensional structure and general behavior. The sequence data complete a model in which two large terminal domains are connected to a central region by sets of three-strand ropes, giving rise to a trinodular structure similar to the one that electron microscopists proposed nearly 30 years ago. This extended polydomainal structure, as stressed by Doolittle, is exquisitely suited to a series of consecutive operations—polymerization, stabilization, and fibrinolysis—the first processes that stop bleeding and then clear away the clot to prevent blood vessel blockage. It is expected that this knowledge will be useful in helping patients whose blood tends to clot under the wrong circumstances.

Molecular Defects in Interactions of Platelets with Vessel Wall. As reviewed by J.N. George and colleagues (see reference), it was shown nearly a century ago that blood platelets are required for hemostasis. In that era, it was also learned that a congenital hemorrhagic disease can result from abnormal platelet functions. Over the years, many additional disorders of this type have been noted. Relatively recent analytical techniques have been used to identify the molecular abnormalities causing the defects and to define the mechanisms of platelet-vessel-wall interactions. It has been shown that platelet function requires specific receptors on the platelet surface that interact with macromolecules on the blood vessel wall, or with proteins in plasma. Some of these proteins are secreted by the platelet. George refers to these as "contact interactions." These reactions may include the adhesion of platelets to subendothelial tissue exposed at the cut end of a divided vessel, the recruitment of adjacent platelets to form a cohesive aggregate, and the generation of thrombin on the platelet surface to form the fibrin network that provides stability for the initial hemostatic plug.

Abnormalities in platelet function are placed by George and colleagues into four classes:

(1) Defects of platelet adhesion to subendothelium—these causing (a) the Bernard-Soulier syndrome, a rare, autosomal recessive trait which results in severe or even fatal hemorrhagic disease; (b) von Willebrand's disease, which presents mucocutaneous problems of bruising, epistaxis, and gingival bleeding; and (c) pseudo-von Willebrand's disease, in which platelets bind an increased amount of plasma vWF.

(2) Defects of platelet aggregation—these causing (a) Glanzmann's thrombasthenia and (b) congenital afibrogenemia.

(3) Defects of platelet secretory granules—the cause of gray-platelet syndrome.

(4) Defects in platelet coagulant activity.

Plasma. Normal blood plasma is a clear, slightly yellowish fluid, which is approximately 55% of the total volume of the blood. The plasma is a water solution in which are transported the digested food materials from the walls of the small intestine to the body tissues, as well as the waste materials from the tissues to the kidneys. Consequently, this solution contains several hundred different substances. In addition, the plasma carries antibodies, which are responsible for immunity to disease, and hormones. The plasma transports most of the waste carbon dioxide from the tissues back to the lungs. Plasma consists of about 91% water, 7% protein material, and 0.9% various mineral salts. The remainder consists of substances already mentioned. The salts and proteins are important in keeping the proper balance between the water in the tissues and in the blood. Disturbances in this ratio may result in excessive water in the tissues (swelling or edema). The mineral salts in the plasma all serve

TABLE 2. INORGANIC CONSTITUENTS OF HUMAN PLASMA OR SERUM

Constituent	Value/100 ml
Aluminum	45 μg
Bicarbonate	24–31 meq/liter
Bromine	0.7–1.0 μg
Calcium	9.8 (8.4–11.2) mg
Chloride	369 (337–400) mg
Cobalt	10 (3.7–16.6) μg
Copper	8–16 μg
Fluorine	109 (75–145) μg
Iodine, total	7.1 (4.8–8.6) μg
Protein bound I	6.0 (3.5–8.4) μg
Thyroxine I	4–8 μg
Iron	105 (39–170) μg
Lead	2.9 μg
Magnesium	2.1 (1.6–2.6) mg
Phosphorus, total	11.4 (10.7–12.1) mg
Inorganic P	3.5 (2.7–4.3) mg
Organic P	8.2 (7–9) mg
ATP P	0.16 (0–6.4) mg
Lipid P	9.2 (6–12) mg
Nucleic acid P	0.54 (0.44–0.65) mg
Potassium	16.0 (13–19) mg
Rubidium	0.11 mg
Silicon	0.79 mg
Sodium	325 (312–338) mg
Sulfur	
Ethereal S	0.1 (0–0.19) mg
Inorganic S	0.9 (0.8–1.1) mg
Non-protein S	2.8 (2.4–3.6) mg
Organic S	1.7 (1.4–2.6) mg
Sulfate S	1.1 (0.9–1.3) mg
Tin	4 μg
Zinc	300 (0–613) μg

other vital functions in the body and must be supplied through diet. See Table 2.

Some of the blood plasma, as well as some of the white cells, filters through the walls of the blood vessels and out into the tissues. This filtered plasma (lymph) is a clear and colorless fluid that returns to the blood through a series of canals referred to as the lymphatic system. This system contains filters (lymph nodes) which remove bacteria and other debris from the lymph. These nodes, especially those located in the neck, armpit, and groin, may become swollen when an infection occurs in a nearby site. Blood clots do not occur normally while the blood is in the vessels. But in an injury, one of the plasma proteins (fibrin) forms a mesh in which the blood cells are trapped, and this mesh is the clot. Blood serum is the yellowish fluid left after the cells and fibrin have been removed from the blood.

Blood Osmotic Pressure. The presence of solute molecules and ions in relatively high concentrations in blood establishes an osmotic pressure, which tends to transport water from the exterior, through the semipermeable membranes of the blood vessel walls, into the bloodstream. This osmotic transport of water inward is opposed by the effect of hydrostatic pressure within the blood vessels, tending to force water (and soluble substances) out through the capillary walls. The loss through leakage of some of these solutes is indirectly restored through the action of the lymphatic system. Among the blood constituents important in maintaining blood osmotic pressure (and thus helping to regulate the volume of fluid in the blood) are the blood proteins. Among these, the protein fraction termed albumins, being relatively low in molecular weight, makes the greatest contribution to the total osmotic effect.

Blood Processing and Transfusion Therapy

Blood transfusion practice has changed markedly in recent years. At one time, units of whole blood were administered to patients with a variety of requirements stemming from different conditions. These conditions ranged from acute blood loss (hemorrhage, bleeding from injuries, etc.) to aplastic anemia, among other blood-related problems.

The outer portion of the erythrocyte (red cell) is a very complex material composed of proteins, polysaccharides, and lipids, many of which are antigens, sometimes referred to as blood group substances. The presence of most, if not all, of these antigens is genetically determined, and their number is such that there may be few, if any, individuals in the world with an identical set of antigens on the red cells—monozygotic twins excepted. These differences in whole blood were learned early in the development of transfusion technology. Fundamental to the refinement of the technology was the discovery by Karl Landsteiner (Nobel Prize winner in 1930) for his observations of the four hereditary blood groups. Landsteiner developed the ABO blood-typing system, which serves as a principal guideline in determining the suitability of donors and recipients. This system consists of three allelic genes, dividing all humans into four groups, A, AB, B, and O. In a few rare individuals, the presence of a suppressor gene may prevent the expression of the A, B, O group character. The products of these genes are the A, B, and O antigens or substances. These antigens not only are located on red cells, but are widely distributed in the body, occurring in the endothelium of capillaries, veins, and arteries, and in numerous cells throughout the body. In addition to a cell-associated form, these antigens occur in soluble form in many body fluids, such as the saliva, gastric juice, urine, amniotic fluid, and in very high concentrations is pseudomucinous ovarian cyst fluid. All individuals possess cell-associated A, B, and O antigens. The presence of the soluble form, however, is governed by a recessive gene called the secretor gene, which exists as two alleles, *Se* and *se*. Individuals who possess at least one *Se* gene secrete the antigens, while those with two se genes do not. The A, B, and O antigens are not uniquely human, but are quite widely distributed in nature. They are found on primate erythrocytes and in the stomach lining of pigs and horses. Intensive investigation has produced considerable information concerning the chemical composition of these antigens. They are extremely stable substances, which is attested by the fact that they can be extracted from Egyptian mummies, thus making it possible to obtain the blood groups of this ancient people. Specific antigenic activity is associated with the carbohydrate moiety, and since the A, B, and O substances possess the same four sugars, the difference between them lies in their arrangement. Analysis of purified A, B, and O substances reveals that about three-fourths of the weight is accounted for by four sugars: L-fructose, D-galactose, *N*-acetyl-D-glucosamine, and *N*-acetyl-D-galactosamine. The remainder consists of amino acids. See also **Immune System and Immunochemistry**.

Cross-Matching. Upon receipt of a tube of clotted blood at a blood bank for typing and cross-matching, procedures are undertaken to determine which antigens are on patient's red cells and which antibodies against red cell antigens are present in the patient's serum. Typing is routine for red blood cell antigens in the ABO system and for a single specificity in the Rh system, namely, the D phenotype.

The Rh group, so denoted because the antigen was first found in the red cells of Rhesus monkeys, is very complex, consisting of perhaps 20 antigens. The Rh_0D antigen is the most important of these antigens because of its possible involvement in the induction of *hemolytic disease of the newborn*. Today, Rh_0D immune globulin (RhoGam® and Gamulin-Rh®, among others) is available to alleviate this danger. This danger is brought about when an Rh-negative woman and an Rh-positive man have an Rh-positive child. There is the grave risk that the woman will become sensitized to the Rh factor in her infant's blood and begin to produce anti-Rh antibodies. The first child is not usually affected, but with subsequent pregnancies, the mother may send sufficient damaging antibodies into the child's blood to threaten its life. When this occurs, in the absence of using the Rh_0D immune globulin, an exchange blood transfusion with almost complete replacement of the infant's blood by Rh-negative blood of the proper ABO group is necessary.

Component Therapy. Frequently, patients do not require all of the blood components and, in fact, their presence can cause many problems. From experience with whole blood therapy over a number of years, *component therapy* emerged. In component therapy, which has many advantages, the patient is given specifically what is needed by way of blood components. Further, separate blood fractions can be stored under those special conditions best suited to assure their biological activity at the time of transfusion. Component therapy also avoids the introduction of foreign antigens and antibodies. It is seldom that fresh whole blood is the treatment of choice providing that specific components are readily available within the time needed.

Processing of Donor Blood. When donor blood is received at a processing center, it is first tested for syphilis and hepatitis B antigen. One unit of whole blood is 500 milliliters. It is then separated into: (1) a unit of *packed erythrocytes* (volume of 300 milliliters and hematocrit value of 70 to 90); this substance is storable in citrate-phosphate-dextrose at 4°C (39.2°F) for up to 3 weeks; (2) a unit of *platelets* (packet) with a volume of 50 milliliters containing about 80 billion platelets). This substance is storable (while being gently mixed) at room temperature for 2 or 3 days. (3) A unit of *cryoprecipitate* (volume of about 10 milliliters containing from 80 to 120 units of Factor VIII and from 300 to 400 milligrams of fibrinogen). This substance is stable in a frozen condition for about one year. (4) One unit of *plasma* (a volume of about 200 milliliters, from which about half of the fibrinogen has been removed). This substance contains platelets, Factor VIII, as well as all remaining procoagulants, albumin, salt, and antibodies that were a part of the original plasma. This substance may be (a) stored in the frozen state, (b) refrigerated, or (c) further processed for individual globulin classes and albumin.

Thus, somewhat analogous to obtaining specific drugs for different conditions, the physician can order up specifically those blood components required for a given need. Platelets, which survive poorly in whole blood stored in acid-citrate-dextrose solution under refrigeration, can be obtained in platelet packets as previously mentioned, or as washed platelets. Factor VIII, for the management of hemophilia, can be obtained in a lyophilized or other purified form. In whole blood, by contrast, procoagulants stored in whole blood decay so that, after a few days, the availability of Factor VIII and other allied components is extremely low. As the result of additional processing, blood centers can furnish prothrombin complex concentrate (*Proplex*), each batch bearing a specific analysis. Preparations of peripheral white blood cells are available from centers with specific blood processing equipment. Additionally, substances for expanding plasma volume, such as fresh frozen plasma, albumin solutions, and Dextran, among others, are available.

Frozen Red Cells. In the early 1950s, the concept of using previously frozen red cells was considered to be one of great potential for the medical profession. One early assumption was that infectious agents would not be passed on because of their destruction by the freezing process. This was disproved in 1978, however, when it was demonstrated that hepatitis B virus was not rendered noninfectious by such processing. The unique contribution of the freezing technology was claimed to be the ability to guarantee the availability of rare blood types for transfusion to recipients who were sensitized to single or multiple high-frequency red-cell antigens. Storage at −80°C made it possible to stockpile such blood for approximately 3 years, but no more than 10 years. Unfortunately the shelf life of a thawed deglycerolized unit was only about 24 hours. As reported by Chaplin, it was predicted that the future large-scale use of previously frozen red cells would depend mainly on the superiority of the product in broad areas of transfusion practice, and secondarily on simplifying the technology, reducing the cost, and extending the post-thawing shelf life. The declining use of previously frozen red cells over the intervening years reflected a lack of progress on all fronts.

Statistics indicate that in late 1978, the American Red Cross reported a demand for thawed, deglycerolized human red cells of nearly 100.000 units per year. By May 1983, the demand was only approximately 42.000 units. Thus, it now appears that the need for frozen red cells was mainly found in filling rare donor blood types. Chaplin summarizes—future developments may yet prove previously frozen red cells to be the sleeping giant of red-cell replacement therapy; for the present, we must be content with the minor but crucial role in which they have proved their worth beyond doubt.

Impact of AIDS on Blood Supply. Acquired immunodeficiency syndrome was first diagnosed as a specific disease entity in 1981 and, by 1983, health officials were aware that the disease could be transmitted via infected blood. In the spring of 1985, kits became available for testing blood for antibodies to the AIDS virus. Since then, virtually all official and approved blood bank organizations have used the test for screening donated blood. Prior to invoking this test, the Centers for Disease Control (Atlanta, Georgia) estimated that well over 400 people had developed AIDS because they received infected blood or blood products. It was later estimated by the CDC that the persons so infected represented less than 2% of persons who had developed AIDS. A major drawback of the ELISA (enzyme-linked immunosorbent assay) test is that it will sometimes indicate a positive result even though an individual may not be infected with the AIDS virus. The test has since been refined and improved. It is the normal practice of blood bank operators to make three ELISA tests on a blood sample before rejection. A more accurate test, known as the Western blot test, may then be used. This latter test is much more likely to be antibody-positive due to infection by the AIDS virus. It should be noted that the ELISA test can, but rarely, yield a "false negative" result. Timing is also a factor. Persons who are in very early stages of AIDS infection may not contain detectable levels of antibodies simply because their immune system has not had sufficient time to produce the antibodies. A celebrated case along these lines occurred in Colorado in 1986—where a newly infected person was not positively detected as having AIDS and the blood was used for transfusion in two patients. One of the persons contracted AIDS, and the infection was attributed to that particular transfusion.

The proposed use of *autologous* blood where an individual places blood in a bank for possible future personal needs has received mixed reactions among the professionals. There is general agreement that this is a good procedure where pre-operative knowledge of surgery is established, but for long-term storage, many authorities feel that the procedure is logistically impractical. The public media have thoroughly explored the psychosocial implications of the AIDS virus relationship to the national blood supply.

Blood Transfusion and Athletics. At the time of the 1984 Los Angeles Olympic games, it was reported that 7 members of a 24-member cycling team, including 4 medalists, had received blood transfusions in an effort to enhance their performance. Team officials report that the athletes were given transfusions of whole blood, collected from both relatives and from unrelated donors, in a motel room. The initial public reaction was negative and cries for disqualification were heard. The medical profession spoke out forcefully against the practice. Be that as it may, Klein (see reference) asks some interesting questions: Do blood transfusions afford world-class athletes a substantial competitive advantage? Is the practice safe? Is it ethical to use blood as a recreational drug? It has been well established, including a test conducted over 50 years ago, that the capacity to perform sustained muscular activity depends on the ability to transport oxygen to the contracting muscle cell. Relating exercise capacity to the maximal oxygen uptake is a widely accepted measure of physical fitness. Transfusion increases oxygen delivery to exercising muscle by increasing the amount of the carrier protein hemoglobin. Red-cell mass and maximal oxygen uptake are generally well correlated. Thus, as reasoned by Horstman et al. (1976), if the metabolic limit of muscle is not exceeded, an increase in the hemoglobin concentration should result in increased oxygen consumption and muscle performance. The elevated hemoglobin concentration induced by hypoxia is one rationale for the widely accepted technique of high-altitude endurance training. This training increases the oxidative capacity of muscles as well. Thus, transfusion would seem least likely to benefit the sprinter, whose muscles generate energy primarily by anaerobic metabolism, and most likely to benefit endurance athletes, whose work capacity depends on a ready supply of transported oxygen. In a 1980 test by Buick et al., of the re-infusion of autologous red cells (previously frozen and stored) in subjects, with the elevation of the circulating red-cell mass as 1 g per deciliter above control values; this resulted in improved treadmill endurance, a lower heart rate during exercise, and less accumulation of blood lactate, all measures which contribute to performance. There was a mean overall increase in maximal oxygen uptake of only 5%. One conclusion that may be drawn—red-cell infusions can improve performance of world-class athletes, but the advantage may be slight.

The general conclusions are aptly expressed by Klein—blood is a drug. Collection, storage, and compatibility testing of blood for transfusion are carefully prescribed by the Food and Drug Administration in the United States and by similar organization in a number of countries. Facilities for blood collection and transfusion are registered, licensed, and inspected for compliance. Like other drugs, blood should be given only for medical indications. As early as 1976, the Medical Commission of the International Olympic Committee formally condemned the practice of blood transfusion for athletes in good health. It has been suggested that even stronger regulations should be formulated.

Blood Substitutes. Researchers in Japan (Fukushima Medical Center) and in other institutions in Europe and North America have been investigating substances that, in major characteristics, may serve as a substitute for blood, particularly in emergency situations where rare blood types are not immediately available to severely ill patients who require transfusions. For example, in early 1979, a Japanese patient with a rare O-negative blood was given an infusion of one liter of a new, oxygenated perfluorocarbon emulsion. This compound carried oxygen through the patient's circulatory system until the rare blood could be obtained.

Later that year, eight additional patients survived infusions with artificial blood. As early as 1966, investigators at the University of Cincinnati demonstrated that life could be sustained when rodents were immersed in perfluorochemicals for long periods. This class of chemicals can dissolve as much as 60% oxygen by volume, as contrasted with whole blood (20%), or salt water or blood plasma (3%). Initially, a major problem existed because pure perfluorochemicals are not miscible with blood. In the late-1960s, researchers (University of Pennsylvania; Harvard School of Public Health) demonstrated that perfluorochemicals could be emulsified. Research is continuing along these lines. Also, new chemicals of this class are being sought. Initially, the research was done with perfluorobutyltetrahydrofuran and perfluorotripopylamine, both superior carriers of oxygen, but prone to concentrate in some organs of the body, notably the liver and spleen. In 1973, perfluorodecalin was found to be completely eliminated from the body. The approach in Japan has differed somewhat, in that research has been directed to add other chemicals that will increase the half-life of the chemicals in the body.

Reports show the synthesis of artificial red cell prototypes that meet the six essential specifications for such cells are: (1) the micro-capsule membrane must be biodegradable and physiologically compatible; (2) the encapsulation process must avoid significant hemoglobin (Hb) degradation; (3) when encapsulated, the oxygen affinity of Hb must be reduced relative to that of free human Hb; (4) the encapsulated Hb must be sufficiently concentrated, that is, more than 33% of that in erythrocytes; (5) there should be no evidence of overt intravascular coagulopathy; and (6) the artificial cells must be small enough to pass unrestricted through normal capillaries. These prototypal artificial red cells are called *neohemocytes* (NHC). The researchers point out that a nontoxic resuscitation fluid that combines the functions of a plasma expander with the ability to carry and deliver oxygen to tissues could prove useful in treatment of trauma, as a temporary substitute for red cells, and for the treatment of tissue ischemia.

Blood Recycling. In a process known as *autotransfusion*, introduced in the early 1980s, blood lost during operative procedures, particularly in heart surgery, is recycled back to the patient. Some reports indicate the need for donor blood can be reduced by as much as 60%. Instead of discarding blood lost during surgery, as has been the traditional practice, the blood is collected in a plastic bag with a special filter to cleanse impurities before the blood is returned to the patient. The procedure has many advantages, including costs of transfusion and elimination of risks from hepatitis, errors in mismatching blood types, and other complications that may arise with donor blood. Although results appear to be positive thus far, a few additional years may be required before the procedure is fully accepted as standard practice.

Blood as an Indicator of Disorders and Diseases

Since the blood performs many services for all parts of the body, it will reflect disturbances that occur as the result of many widely divergent diseases. This had led to the development of a variety of blood tests, either to confirm a diagnosis or to follow the effectiveness of treatment in the patient. *Immunological* or *serological* tests are performed to confirm the diagnosis of selected types of infectious diseases, and are based upon the principle that in certain diseases there appear in the blood specific substances (antibodies) which are produced by the body in resisting invasion by specific disease-producing media. One of the more widely used tests is the Kolmer test for syphilis. Blood typing tests are also serological in nature. A second group of blood tests are known as *hematological*. These tests determine the number of each type of circulating blood cell (*blood count*), the total volume of red cells in a blood sample (*hematocrit*), and the hemoglobin content of the blood. A *differential* blood count is one in which selected dyes are used to distinguish better the different kinds of white blood cells. These tests are important in diagnosing and treating illnesses, such as infections, the anemias, and the leukemias.

Another group of blood tests involves *bacteriological* techniques. Blood and bone marrow samples are obtained under aseptic precautions and introduced into a variety of artificial culture media, with subsequent isolation and identification of the specific microorganism responsible for the illness. Relative susceptibility of the specific strain of bacteria to the available chemotherapeutic and antibiotic agents may then be determined and the effectiveness of such agents in sterilizing the blood-stream can be determined by further blood cultures.

Many *chemical* tests are performed on blood samples to determine the quantitative relationships between circulating globulins, albumin, sugar, non-protein nitrogen, minerals, and other normal and abnormal constituents. Such chemical tests are important in diabetes, kidney diseases, the failing heart, and in pancreatic and liver diseases. In all of these disorders, pronounced changes in the relative amounts of the various chemical constituents of the blood occur. Chemical tests also may be performed on urine, spinal fluid, and saliva for some special purpose, and since most of these fluids are derived from the blood plasma, their chemical analysis frequently reflects changes in the blood itself. During prolonged therapy with certain drugs, it may be desirable to measure chemically the concentration of the drug in the blood plasma.

Sophisticated instrumentation and procedures are used in research involving blood and its functions. Phase contrast microscopy has the advantage that living cells can be studied for long periods of time; chromatin, mitochondria, centrosomes and specific granules can be seen and photographed at magnifications of 2, 500×. The method is excellent for the study of granules of the matrix of cells which is unseen in traditionally fixed and stained cells. It is an excellent aid for those who wish to use the electron microscope, because areas demonstrated by light can be compared with those visualized by the electron beam. The study of blood by motion pictures (*microcinematography*) has been used for many years. With the invention of the phase microscope, this approach to the study of blood cells has been an important tool. Studies of the movements of the lymphocytes in rats showed a softening of the membrane at the forward moving end, and pseudopod formation; contractions of the cell force the inner plasma forward, while the external plasma gel remains fixed except at the posterior end, then it becomes softer and passes through the stiffer ring of plasma gel to become more gelated at the anterior end. This is an example of the type of detailed investigation that can be made with microcinematography. Using speed photography at 3,200 frames per second, the red blood cell has been observed to have an interior velocity of 30× that of water at 38°C. In the dog's mesentery, red blood cells passing into capillaries from larger arterioles take the form of an inverted cap or parachute; when blood flow is stopped, they become biconcave disks. The cup shape is suggested as bringing more surface close to the capillary endothelium.

Other blood research techniques include the use of physical and chemicals agents, ultracentrifugation, cytochemical methods, microincineration, and autoradiography.

Occult Blood as an Indicator of Cancer. A good number of physicians encourage asymptomatic patients who are over 40 to undergo annual testing for occult blood in stool as part of a screening program, with the hope that colon cancers may be detected at an early, curable stage or that adenomatous polyps may be found and removed in an attempt to prevent cancer. This concept remains controversial and is unproven, but because it is noninvasive and simple and relatively low in cost, many physicians regard the test as quite useful. Sampling errors do occur. The test is not always positive with patients who have colon cancer because some cancers do not bleed, or bleed intermittently. Vitamin C may inhibit the oxidation of guaiac (a colorless phenolic compound that is converted into a colored quinone when contacted by hemoglobin). Some stools may yield a positive result when fresh, but a negative result after drying on the *Hemoccult* card. Also, the test can be positive when colon cancer is not present—the result caused by bleeding from some other lesion, such as salicylate gastritis or hemorrhoids. Dietary substances (red meat, uncooked peroxidase-rich vegetables, elemental iron, etc.) may cause a "false positive" result. Persons undergoing such screening are directed to avoid red meat in their diet for several days prior to the test.

More recently, a new test (*HemoQuant*) has been developed. This test involves the chemical conversion of stool heme to porphyrins that can be assayed fluorometrically. The test also detects porphyrins present in stool as a result of bacterial and enzymatic degradation of hemoglobin as it travels through the intestines. Thus, this new test provides a quantitative measure of all blood that enters the gastrointestinal tract. It appears to be biochemically sound and is considered a methodological breakthrough.

Additional Reading

Babior, B.M. and T.P. Stossel: *Hematology: A Pathophysiological Approach,* Churchill Livingstone, New York, NY, 1990.

Bain, B.J.: "Blood Cells," *A Practical Guide,* Blackwell Science, Inc., Malden, MA, 1996.

Barnard, D.L., McVerry and D.R. Norfolk: *Clinical Haematology,* Oxford University Press, New York, NY, 1989.

Beck, W.S., Ed.: *Hematology,* 5th Edition, MIT Press, Cambridge, MA, 1991.

Buetler, E., M.A. Lichtman, and U. Seligsohn: *Williams Hematology,* McGraw-Hill Companies, Inc., New York, NY, 2000.

Buick, F.J. et al.: "Effect of Induced Erthrocythemia on Aerobic Work Capacity," *J. Appl. Physiol.,* **48**, 636–642 (1980).

Brookes, M.: *Blood Supply of Bone," Scientific Aspects,* Springer-Verlag, Inc., New York, NY, 1998.

Carr, J.H. and B.F. Rodak: *Clinical Hematology Atlas,* W.B. Saunders Co., Philadelphia, PA, 1998.

Chaplin, H., Jr.: "Frozen Red Cells Revisited," *N. Eng. J. Med.,* **311**(26) 1696–1698 (December **27**, 1984).

Davie, E.W.: *Blood Coagulation, Fibrinolysis, and Platelets,* Springer-Verlag, Inc., New York, NY, 1996.

Deeg, H.J.: *A Guide to Blood and Marrow Transplantation,* Springer-Verlag, Inc., 1999.

Delamore, I.W. and J.A. Liu Yin, Eds.: *Haematological Aspects of Systemic Disease,* W.B. Saunders, Philadelphia, PA, 1990.

Dixon, B.: "Of Different Bloods," *Science* **84**, **5**(9), 65–70 (November 1984).

Doolittle, R.F.: "Fibrinogen and Fibrin," *Sci. Amer.,* **245**(6) 126–135 (1981).

Furie, B. and B.C. Furie: "Molecular and Cellular Biology of Blood Coagulation," *N. Eng. J. Med.,* **800** (March **19**, 1992).

Garratty, G.: *Applications of Molecular Biology to Blood Transfusion American Medicine,* American Association of Blood Banks, Bethesda, MD, 1997. *http://www.aabb.org/.*

George, J.N. and S.J. Shattil: "The Clinical Importance of Acquired Abnormalities of Platelet Function," *N. Eng. J. Med.,* **27** (January 3, 1991).

Ginsberg, D. et al.: "Human von Willebrand Factor (vWF); Isolation of Complementary DNA (cDNA) Clones and Chromosomal Localization," *Science,* **228**, 1401–1406 (1985).

Handin, R.I.: *"Blood," Principles and Practice of Hematology,* Lippincott Williams Wilkins, Philadelphia, PA, 1995.

Hillis, L.D. and R.A. Lange: "Serotonin and Acute Ischemic Heart Disease," *N. Eng. J. Med.,* **688** (March 7, 1991).

Hillman, R.S. and C.A. Finch: *Red Cell Manual,* 7th Edition, F.A. Davis Co., Philadelphia, PA, 1996.

Hoffer, E.P.: *"Arterial Blood Gases,"* Version 5, Lippincott Williams Wilkins, Philadelphia, PA, 1997.

Hoffman, R., E.J. Benz, Jr., S.J. Shattil et al.: *"Hematology," Basic Principles and Practice,* 3rd Edition, Churchill Livingstone, New York, NY, 1999.

Horstman, D.H., M. Gleser, and J. Delehunt: "Effects of Altering O_2 Delivery on VO_2 of Isolated, Working Muscle," *Amer. J. Physio.,* **230**, 327–334 (1976).

Issitt, P.D. and D.J. Anstee: *Applied Blood Group Serology,* Montgomery Scientific Publications, Durham, NC, 1998.

Issitt, L.: *"Blood Groups," Refresher and Update,* American Association of Blood Banks, Bethesda, MD, 1996.

Jackson, J.B. et al.: "Absence of HIV Infection in Blood Donors with Indeterminate Western Blot Tests for Antibody to HIV-1," *N. Eng. J. Med.,* **217** (January **25**, 1990).

Jandl, J.H.: *Blood Textbook of Hematology,* Lippincott Williams Wilkins, Philadelphia, PA, 1996.

Klein, H.G.: "Blood Transfusion and Athletics," *N. Eng. J. Med.,* **312**(13), 854–856 (March **28**, 1985).

Klotz, I.M. et al.: "Hemerythrin: Alternative Oxygen Carrier," *Science,* **192**, 335–344 (1976).

Krieger, H.: *Blood Conservation in Cardiac Surgery,* Springer-Verlag Inc., New York, NY, 1998.

Kulig, K.: "Cyanide Antidotes and Fire Toxicology," *N. Eng. J. Med.,* 1801 (December 19, 1991).

Lake, C.L.: *"Blood," Hemostasis, Transfusion, and Alternatives in the Perioperative Period,* Lippincott Williams Wilkins, Philadelphia, PA, 1995.

Larsen, M.L., Horder, M., and E.F. Mogensen: "Effect of Long-term Monitoring of Glycosylated Hemoglobin Levels in Insulin-Dependent Diabetes Mellitus," *N. Eng. J. Med.,* 1021 (October 11, 1990).

Lichtman, M., E. Henderson, and S.J. Shattil: *Hematology,* Academic Press Inc., San Diego, CA, 1999.

Loffler, H.: *Atlas of Clinical Hematology,* Springer-Verlag Inc., New York, NY, 2000.

Majerus, P.W., R.M. Perlmutter, and H. Varmus: *The Molecular Basis of Blood Disease,* W.B. Saunders Company, Philadelphia, PA, 2000.

Martin, L.: *All You Really Need to Know to Interpret Arterial Blood Gases,* 2nd Edition, Lippincott Williams Wilkins, Philadelphia, PA, 1999.

McCurdy, K.: *Blood Bank Regulations: A to Z,* American Association of Blood Banks, Bethesda, MD, 1999.

McCullough, M.D.: *Blood Transfusion: A Practical Guide,* McGraw Hill Companies, Inc., New York, NY, 1997.

Nathan, D.M.: "Hemoglobin Alc—Infatuation or the Real Thing?" *N. Eng. J. Med.,* 1062 (October 11, 1990).

Peterson, W.L. and J.S. Fordtran: "Quantitating the Occult," *N. Eng. J. Med.,* **312**(22), 1448–1450 (May **30**, 1985).

Petz, L.D. and L. Calhoun: "Changing Blood Types and Other Immunohematologic Surprises," *N. Eng. J. Med.,* **888** (March 26, 1992).

Redman, C.W.G.: "Platelets and the Beginnings of Preeclampsia," *N. Eng. J. Med.,* **478** (August 16, 1990).

Reid, M.E. and S.J. Nance: *Red Cell Transfusion: A Practical Guide,* Humana Press, Totowa, NJ, 1997.

Reiffers, J.: *Blood Stem Cell Transplantation,* Harcourt Inc., San Diego, CA, 1998.

Ruiz, A.L.: *Blood Circulation,* Sterling Publishing Co., Inc., New York, NY, 1997.

Shinton, N.K.: *Desk Reference for Hematology,* CRC Press, LLC., Boca Raton, FL, 1998.

Silberstein, L.: *Autoimmune Disorders of the Blood,* American Association of Blood Banks, Bethesda, MD, 1996.

Turgeon, M.L.: *Clinical Hematology: Theory and Procedures,* Lippincott Williams Wilkins, Philadelphia, PA, 1998.

Turitto, V.T. and J.J. Weiss: "Reed Blood Cells: Their Dual Role in Thrombus Formation," *Science,* **207,** 541–543 (1980).

Winkelstein, A. and R.A. Sacher: *White Cell Manual,* F.A. Davis Company, Philadelphia, PA, 1998.

Yawata, Y.: *Atlas of Blood Diseases: Cytology and Histology,* Blackwell Science, Inc., Malden, MA, 1996.

Zon, L.I.: *Hematopoiesis,* Oxford University Press, Inc., New York, NY, 2001.

BLOODSTONE. A massive variety of quartz of greenish color with small spots of red jasper somewhat resembling blood drops. It is used as a semiprecious stone. When placed in water in full sunlight bloodstone will frequently give a general reddish reflection, hence the term heliotrope, derived from the Greek words meaning sun and to turn. See also **Chalcedony**; and **Quartz.** Bloodstone also is known as *heliotrope.*

BLOOM. In surface-coating technology, bloom is a whitish, filmy layer that appears on films of paints, varnishes, or lacquers due to contamination from the atmosphere. The term is also applied to a filmy layer deposited on a photographic plate by tap water, which can be removed by rubbing the plate with wet cotton. The term bloom is used in metallurgy to denote a mass of malleable iron from which the slag has been removed. See also **Iron Metals, Alloys, and Steels.**

BLUE GLOW. A type of luminescence emitted by certain metallic oxides, when heated. A blue glow is normally seen in electron tubes containing mercury vapor, arising from the ionization of the molecules in the mercury vapor.

BODYING AND BULKING AGENTS (Foods). These terms tend to be self-defining. Additives in these classifications are frequently described together because many substances will serve one or both purposes.

Bodying Agents

The *body* of a food substance is generally associated with the textural qualities of the substance, notably with mouth-feel or chewiness. Some food products, particularly those of a fabricated nature, may possess a full complement of desirable consumer appeals (taste, odor, color, nutritive value, etc.) and yet lack the desirable textural quality of body. Thus, soups, gravies, sauces, cheese foods and spreads, dressings, snack dips, and margarines, among others, can be improved through the addition of bodying agents. For example, formulations for frozen desserts can be improved in this respect by the addition of low levels of a material such as microcrystalline cellulose (about 0.25% weight), in combination with soluble hydrocolloids, such as guar, locust bean gum, alginates, or carrageenans. See also **Gums and Mucilages.**

Bulking Agents

These substances are added to semiliquid and solid food products to add bulk to the end product over and beyond the bulk resulting from the strict use of conventional ingredients.

For example, the natural sugars (sucrose, fructose, etc.) are best known for their contribution of sweetness to food products. The sugars, however, also perform other useful functions, including their natural preservative qualities and, in many foods such as baked goods, contribute considerable bulk to the finished product. As described under **Sweeteners**, a number of artificial, non-carbohydrate compounds serve as excellent sweeteners, but they lack the ability to achieve the desired bulk.

Microcrystalline Cellulose. This additive achieves about the same degree of body and substance in frozen desserts that is normally achieved only in well-emulsified products with a 2–4% higher fat content. This is the result of the ability of microcrystalline cellulose to stabilize the serum solid. Microcrystalline cellulose imparts body and smoothness to

ice cream and ice milk, and tends to make them less "cold tasting." There are no off-flavors associated with the substance and frozen desserts melt to smooth, creamy consistencies. Bodying agents play an effective, if not exclusive role, in improving freeze-thaw properties of numerous products. In another example, when whey or sugar solids are used to reduce or replace portions of the milk solids nonfat (MSNF), there is a definite loss of functionality of the mix, resulting in reduced body and texture. Problems, such as stickiness, gumminess, and weak body can be corrected by the addition of a bodying agent, such as microcrystalline cellulose, in a very small amount (0.25–0.4% weight).

When microcrystalline cellulose is used to add bulk to the end product, the bulk not only may be compensated for the lack of a natural bulking agent, but may add extra bulk. In addition to achieving the natural characteristics expected of a product, there are two additional advantages: (1) The cellulose increases the fiber content of a food product, and (2) weight is added, thus reducing the effective caloric content (caloric density) of a given weight serving of a product. However, it should be stressed that microcrystalline cellulose is only a partial substitute for fat, which is needed for air entrapment, or for flour, which provides the elastic gluten structure.

In the currently very important field of manufacturing low-calorie foods, a bulking agent essentially can be considered as a diluent even though it may play other important roles. Thus, the diet-conscious consumer can eat cookies, doughnuts, or portions of cake of traditional size and yet consume considerably fewer calories. The important factor in selecting a bulking agent for low-calorie foods is that of finding a substance that combines non-caloric qualities with other functional capabilities so that lower amounts of relatively high-calorie ingredients can be reduced or replaced without detracting drastically from the consumer appeals of the finished product.

Isomalt. An odorless, white, sweet-tasting, crystalline, practically non-hygroscopic substance is used in a wide variety of confectionery products, such as chocolates, caramels, hard candy, tablets, pan-coated products, and chewing gum. The low hygroscopicity simplifies packaging requirements. For baking, isomalt can be substituted for sucrose on a 1:1 basis. Although used in fruit-flavored products, it has not been used in traditional jelly and preserves. Isomalt can be used as a substitute for sucrose in ice cream, ice milk, yogurt, and a variety of desserts and fillings. For tabletop sweeteners, isomalt can be used in combination with saccharin, cyclamates, aspartame, acesulfame-K, and some other artificial sweeteners.

Isomalt (*Palatnit^R*) was developed in Germany and has been well accepted in Europe. A sucrose-glucosylfructose-mutase from *Protamino bacter rubrum,* a nonpathogenic organism found in beet sugar factories, is used to transform sucrose into the reducing sugar isomaltose. The properties of isomalt coincide well with existing food processing equipment and procedures.

Oat Bran. Fat replacement is a very timely topic in the food processing industry as of the early 1990s, and is also described in other articles of this *Encyclopedia.* Check Alphabetical Index.

As pointed out by Pszczola, certain fat replacers, when used with 90% fat-free ground beef or pork sausage, can provide the texture, flavor, and juiciness of full-fat meat products. One fat replacer is specially processed oat bran, with added flavorings and seasonings. It is estimated that up to 20% of the ground beef sold in the near future will be of the low-fat variety. (Total consumption of ground beef in the United States is estimated at about 7 billion pounds (3.2 billion kg) annually!) It is reported that, after 3 years of research by the Webb Technical Group (consultants sponsored by the Beef Industry Council of the National Live Stock and Meat Board), oat brain was selected above other substances studied. These included wheat bran, psyllium husk, rice bran, barley bran, vegetable protein, soy fiber, cane fiber, and carrageenan. As reported, the advantages of oat bran include: 1. Keeps meats from drying out when cooked; 2. Good mouth-feel that imitates fat; 3. Lack of a cereal flavor; 4. Retention of natural meat flavoring; and 5. A holding time superior to other fat replacers.

Thus far, all ingredients in the replacer are considered GRAS (Generally Regarded as Safe) and have FDA (U.S.) approval.

Classification of Fat Substitutes. The development of suitable fat substitutes or replacers will remain a very active field for the foreseeable future. Experts have classified fat substitutes into several categories:

1. *Protein-based substitutes*, the present major limitation being that such substitutes do not lend themselves to use in cooking oils or with products that require frying or baking, because excessive heat causes the ingredients to coagulate, with loss of fat-like mouthfeel.

2. *Synthetic, fat-like substances* that are resistant to hydrolysis by digestive enzymes. One type is a mixture of the hexa-to-octaesters of sucrose; others are esterified propoxylated glycerols and dialkyl dihexadecylmalonate (DDM), which has been used in potato and tortilla chips; trialkoxytricarballate-tricarballic acid esterified with fatty alcohols, currently under trial for use in margarine- and mayonnaise-type products.

3. *Carbohydrate-based substitutes*, which include *Gums*, sometimes referred to as hydrophilic colloids or hydrocolloids. These are long-chain, high–molecular weight polymers that dissolve or disperse in water, providing a thickening, sometimes gelling, effect. The period of usage of these substances dates back to the early 1980s. They include:

   a. *Corn starch maltodextrin*, which is a non-sweet saccharide polymer produced by a limited hydrolysis of corn starch;

   b. *Potato starch maltodextrin*, rather widely used for bakery products, dips, salad dressings, frosting, frozen desserts, mayonnaise-like products, meat products, and confections;

   c. *Tapioca dextrins*, used in a number of products and, more recently, in microwavable cheese sauces;

   d. *Konjac Flour*, a product of the konjac root traditionally has been used by the Japanese and other Far East nations for over a thousand years to make gels and noodles that are stable in boiling water. Products range from chewy desserts to colored soup dumplings. Konjac was first used in the United States in the early 1900s. Konjac flour is the dried, pulverized, and winnowed tubers of the perennial herb *Amorphophallus konjac*. The dried tuber contains up to 60–80% konjac flour. Because of its interaction with carrageenan and starches to form heat-stable gels, the potential of konjac flour as a fat substitute is promising.

It is important to mention that combinations of some of the aforementioned products are available to food processors. Over the years, for certain special products, including pharmaceuticals, glycerin, methylcellulose, polyvinylpyrolidon (PVP), sodium carboxymethycellulose, and whey solids have been used.

Additional Reading

Carroll, L.E.: "Functional Properties and Applications of Stabilized Rice Bran in Bakery Products," *Food Technology*, **74** (April 1990).

Carroll, L.E.: "Stabilizer Systems Reduce Texture Problems in Multicomponent Foods and Bakery Products," *Food Technology*, **94** (April 1990).

Considine, D.M. and G.D. Considine: *Foods and Food Production Encyclopedia*, Van Nostrand Reinhold, New York, NY, 1982.

Imeson, A.: *Thickening and Gelling Agents for Food*, Aspen Publishers, Gaithersburg, MD, 1999.

Irwin, W.E.: "Isomalt—A Sweet, Reduced-Calorie Bulking Agent," *Food Technology*, **128** (June 1990).

Pszczola, D.E.: "Oat-Bran-Based Ingredient Blend Replaces Fat in Ground Beef and Pork Sausage," *Food Technology*, **60** (November 1991).

Staff: "Fat Substitute Update," *Food Technology*, **92** (March 1990).

Taki, G.H.: "Functional Ingredient Blend Produces Low-Fat Meat Products to Meet Consumer Expectations," *Food Technology*, **70** (November 1991).

Tye, R.J.: "Konjac Flour: Properties and Applications," *Food Technology*, **82** (March 1991).

BOHRIUM. See **Chemical Elements**.

BOILING CURVE AND CONDENSATION CURVE. Consider the phase diagram (Fig. 1) of a binary system forming a liquid and a vapor phase at constant pressure. Curve I is the boiling curve, which gives the coexistence temperature as a function of liquid composition; and curve II is the condensation curve, which gives the coexistent temperature as a function of the composition of the vapor phase. If the temperature is increased, vaporization begins when the boiling curve is crossed.

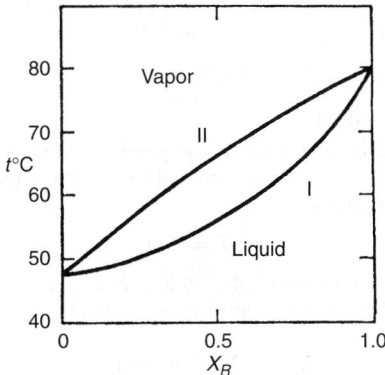

Fig. 1. Temperature composition of a liquid-vapor system at constant pressure

Inversely, condensation begins when the temperature is decreased below the condensation curve.

BOILING POINT. The normal boiling point of a liquid is the temperature at which its maximum or "saturated" vapor pressure is equal to the normal atmospheric pressure, 760 mm of mercury. If the pressure on the liquid varies, the actual boiling point varies in accordance with the relation between the vapor pressure and the temperature for the liquid in question. See also **Vapor**. Water, for example, with a normal boiling point of 100°C or 212°F, boils at ordinary room temperature when the pressure is reduced to about 17 mm; and inhabitants of elevated regions often find difficulty in cooking food by boiling, because of the low boiling point. On the other hand, the boiling water and steam in a "pressure cooker" are so hot that such foods as meat and rice are cooked tender in a very short time. If a solid is dissolved in the liquid, or if another, less volatile liquid is mixed with it, the boiling point is raised to a degree expressed by the boiling point laws of Van't Hoff, Raoult, and others.

A liquid does not necessarily begin boiling when the temperature reaches the boiling point. If kept perfectly quiet, and especially if covered with a film of oil, water may be raised several degrees above its normal boiling point, before it suddenly boils with explosive violence; it then returns to its true boiling point.

To prevent this superheating, it is customary to add to laboratory distillation flasks small pieces of inert material having sharp corners; the latter favor the formation of "bubbles" of vapor when the liquid reaches the boiling point.

The *maximum boiling point* is that temperature corresponding to a definite composition of a two-component or multicomponent system at which the boiling point of the system is a maximum. At this temperature the liquid and vapor have the same composition and the solution distills completely without change in temperature. Binary liquid systems that show negative deviations from Raoult's law have maximum boiling points. See **Raoult's Law**; and **Van't Hoff Law**.

The *minimum boiling point* is that temperature corresponding to a definite composition of a two-component or multicomponent liquid system at which the boiling temperature is the lowest for that particular system. At the minimum boiling point the liquid and vapor have the same composition.

BOILING POINT CONSTANT. Consider a dilute solution in which all solute species may be regarded as nonvolatile. The vapor in equilibrium with the solution is then formed from the solvent only. Call T^0 the boiling point of the pure solvent at the pressure concerned, and T the boiling point of the solution. For a dilute solution, the difference

$$\theta = T - T^0$$

will be small compared with T^0.

If the solution is also ideal, one has

$$\theta = \frac{R(T^0)^2}{\Delta_e h^0}\frac{M_1}{1000}\sum_s m_s = \theta_e \sum_s m_s$$

M_1 is the molar mass of the solvent, $\Delta_e h^0$ its latent heat of vaporization in kcal per mole at temperature T^0, m_s the molality of solute s; θ_e is called the *boiling point constant*, or *ebullioscopic constant*. It depends only on the properties of the solvent. For water,

$$\theta_e = 0.51°C$$

Boiling Point Elevation

The boiling point of a solution is, in general, higher than that of pure solvent, and the elevation is proportional to the active mass of the solute for dilute (ideal) solutions.

$$\Delta T = Km$$

where ΔT is the elevation of the boiling point, K is the *boiling point constant* or the *ebullioscopic constant* and m, the molality of the solution.

BOILING POINT ELEVATION. The boiling point of a solution is, in general, higher than that of pure solvent, and the elevation is proportional to the active mass of the solute for dilute (ideal) solutions.

$$\Delta T = Km$$

where ΔT is the elevation of the boiling point, K is the *boiling point constant* or the *ebullioscopic constant* and m, the molality of the solution.

There are several methods for measuring elevation: In the Beckmann method a Beckmann-type thermometer is immersed in a weighed amount of solvent and the boiling point determined by gentle heating until a steady temperature is reached. A weighed amount of solute is then added and the boiling point redetermined. The difference gives the elevation of the boiling point. The glass vessel containing the liquid is provided with a platinum wire sealed through the bottom to promote steady boiling and to prevent overheating, and reflux condensers are used to minimize loss of liquid.

In the Landsberger method, vapor from boiling solvent is passed through the solvent contained in another vessel and by giving up its latent heat will eventually raise the liquid to the boiling point. At this stage a weighed amount of solute is added to the second vessel and the boiling point is again determined.

In the Cottrell method, the thermometer is placed in the vapor phase above the surface of the liquid and the apparatus so designed that boiling liquid is pumped continuously over the bulb of the thermometer.

BOILING WATER REACTOR.   See **Nuclear Reactor**.

BOLTZMANN, LUDWIG (1844–1906). Boltzmann was born in Vienna. He was a great theoretical physicist and is known for his contributions to mathematical analyses of physical phenomena, and to the kinetic theory of gases, especially in regard to viscosity and diffusion.

Unique to Boltzmann is the H-theorem that resolves the problem of microscopic reversibility and macroscopic irreversibility. He is recognized for the Maxwell-Boltzmann integro differential equation governing the change in distribution of atoms due to collisions. Throughout his life, Boltzmann used his intuition and this intuition led him to consider the probability of energy distributions in gases. He believed that thermodynamic phenomena were the macroscopic reflection of atomic phenomena regulated by mechanical laws and by chance. He was ridiculed for his idea, and this ridicule may have led to his depression and suicide. We now acknowledge Boltzmann as the creator of statistical mechanics.

See also **Boltzmann's Distribution Law**; and **Boltzmann Transport Equation**.

J.M.I.

BOLTZMANN'S DISTRIBUTION LAW. Consider a system composed of molecules, of one or more kinds, able to exchange energy at collisions but otherwise independent of one another. Evidently we cannot say anything useful or interesting about the state of a particular molecule at a particular time. We can however make useful statements about the average fraction of molecules of a given kind in a given state, or, what is the same thing, the fraction of time spent by each molecule of a given kind in a given state. If the system is maintained at a definite temperature, then the fraction of molecules of a given kind in a given state is determined by the energy of this state and by the temperature.

In particular if we denote by i and k two completely defined states of a molecule of a given kind and by E_i and E_k the energies of these two states then the average numbers N_i and N_k of molecules in these two states are related by

$$N_i/N_k = \exp\{-\beta(E_i - E_k)\} \tag{1}$$

where β is a parameter having a positive value determined entirely by the thermostat; i.e., β has the same value for all states of a given kind of molecule and for all kinds of molecules. In other words β has all the characteristics of temperature except that it decreases as temperature increases. If we write

$$\beta = 1/kT \tag{2}$$

then it can be shown that T is identical with thermodynamic (or absolute) temperature and k is a universal constant whose value determines the unit of T called the degree. When k is given the value 1.38041×10^{-23} joule/degree, the temperature scale becomes the Kelvin scale defined by $T = 273.16$ K at the triple point of water. Substitution of Eq. (2) into Eq. (1) leads to

$$N_i/N_k = \exp\{-(E_i - E_k)/kT\} \tag{3}$$

This fundamental relation is called Boltzmann's distribution law after the creator of statistical mechanics, Ludwig Boltzmann (1844–1906), Professor of Physics in Leipzig, and k is called Boltzmann's constant.

We must now discuss the meaning of the words used above, "completely defined state." These words have one meaning in classical mechanics and a different, but related, meaning in quantum mechanics. Since the quantum definition is the simpler we shall discuss it first. We begin by considering a system of highly abstract "molecules" having only a single degree of freedom, for example linear oscillators. The quantum states form a simple series specified by consecutive integers called the quantum numbers. In this simple example there is no ambiguity in the meaning of "completely defined state"; each state i is completely defined by the integral value of a single quantum number. Let us now consider a "molecule" with three degrees of freedom such as a structureless particle moving in three-dimensional space. The complete specification of this particle's state requires not one but three integral quantum numbers. If the particle moves freely in a cubical box, the three quantum numbers may be associated with motion along the three directions normal to the faces of the box. The subscript labels i and k in the previous formulas are abbreviations for sets of three quantum numbers. For example i might mean (2, 5, 1) and k might mean (3, 4, 2). There can now be several states having the same energy. For a particle moving freely in a cubical box, it follows from symmetry that the states (2, 5, 1), (1, 2, 5), (5, 1, 2), (1, 5, 2), (2, 1, 5), and (5, 2, 1) all have the same energy; such an energy level is called sixfold degenerate. (One should *not* speak of a p-fold degenerate *state*, but of a p-fold degenerate energy level.) It is sometimes desirable to consider the fraction of molecules having a given energy rather than the fraction in a given state. If N_r and N_s denote the average number of molecules of a given kind having energy E_r and E_s then evidently

$$N_r/N_s = (p_r/p_s) \exp\{-(E_r - E_s)/kT\} \tag{4}$$

Alternatively, if f_r denotes the average fraction of molecules of a given kind having energy E_r, then

$$f_r = p_r \exp(-E_r/kT) \Big/ \sum_s p_s \exp(-E_s/kT) \tag{5}$$

The sum $\sum_s$ occurring in the denominator is called the partition function.

It may happen that certain degrees of freedom are completely independent of other degrees of freedom. We call such degrees of freedom "separable." The partition function can then be separated into factors relating to the several sets of separable degrees of freedom, and Boltzmann's distribution law is applicable separately to each set of separable degrees of freedom. For example, for an electron moving freely in a rectangular box, the translational motions normal to the three pairs of faces and the fourth degree of freedom due to spin are all separable.

We shall now consider briefly the meaning of completely specified state according to classical mechanics. We know that classical mechanics is merely an approximation, sometimes good but sometimes bad, to quantum mechanics. Motion in each separable degree of freedom can be described classically by a coordinate x and its conjugate momentum p_x. If x and p_x are plotted as Cartesian coordinates, the diagram is called the phase plane. There is a simple correlation between the quantum definition and the classical descriptions: the density of quantum states is one per area h (Planck's constant) in the phase plane. This may be extended to several degrees of freedom. If there are f degrees of freedom, the motion is described by f coordinates $q_1, \ldots q_f$ are the conjugate momenta $p_1, \ldots p_f$. We can imagine these plotted in a $2f$ dimensional Cartesian space called phase space. There is then one quantum state per $2f$ dimensional volumes

h^f of phase space. In the classical as in the quantum description there can be degenerate energy values and there can be separable degrees of freedom. The classical description is a good approximation to the quantum description when the spacing between energy levels is small compared with kT. An example of an effectively classical separable degree of freedom is the motion in a given direction of a free particle. If the linear coordinate is denoted by x and the linear momentum by p_x, then the fraction of molecules at a position between x and $x + dx$ and having a momentum between p_x and $p_x + dp_x$ is

$$\frac{\exp(-p_x{}^2/2mkT)dxdp_x}{\int dx \int_{-\infty}^{\infty} dp_x \exp(-p_x{}^2/2mkT)} \tag{6}$$

where m is the mass of the particle so that its (kinetic) energy is $p_x{}^2 p/2m$. In the classical treatment, the kinetic and potential factors are separable. Consequently the fraction of molecules, anywhere or everywhere, having momentum between p_x and $p_x + dp_x$ is

$$\exp(-p_x{}^2/2mkT)dp_x \Big/ \int_{-\infty}^{+\infty} \exp(-p_x{}^2/2mkT)dp_x \tag{7}$$

Equation (7) is called Maxwell's distribution law after Clerk Maxwell (1831–79), Professor of Physics at Cambridge (England), who obtained it in 1860, before Boltzmann in 1871 obtained his wider distribution law. Maxwell derived his distribution law from the conservation of energy together with the assumption that the motion is separable in three mutually orthogonal directions. The latter assumption was violently attacked by mathematicians, but we now recognize that the assumption is both reasonable and true.

In conclusion we must mention that a necessary condition for the validity of Eq. (3), and consequently of other formulas derived from Eq. (3) is that $N_i \ll 1$ for the state (or states) of lowest energy and *a fortiori* for all other states. When this inequality does not hold, Boltzmann's distribution law must be replaced by a more general and more precise distribution law, either that of Fermi and Dirac or that of Bose and Einstein according to the nature of the molecules. See also **Statistical Mechanics**.

BOLTZMANN TRANSPORT EQUATION. The fundamental equation describing the conservation of particles which are diffusing in a scattering, absorbing, and multiplying medium. It states that the time rate of change of particle density is equal to the rate of production, minus the rate of leakage and the rate of absorption, in the form of a partial differential equation such as

$$\frac{\partial n}{\partial t} = \text{production} - \text{leakage} - \text{absorption}$$

$$= S + D\nabla^2\phi + \phi \sum_a$$

in which S in the cgs system is in units of neutrons $\text{cm}^{-3} \text{ sec}^{-1}$, D is the diffusion coefficient in units of cm, $\phi = nv$ is the neutron fluence in units of neutrons $\text{cm}^{-2} \text{ sec}^{-1}$, Σ_s is the absorption cross section per unit volume in units of cm^{-1}, and ∇ is del, the vector differential operator.

BOND CALORIMETER. See **Calorimetry**.

BOND (Chemical). An attractive force between atoms strong enough to permit the combined aggregate to function as a unit. A more exact definition is not possible because attractive forces ranging upward from 0 to those involving more than 250 kcal/mole of bonds are known. A practical lower limit may be taken as 2–3 kcal/mole of bonds, the work necessary to break approximately 1.5×10^{24} bonds by separating their component atoms to infinite distance.

All bonds appear to originate with the electrostatic charges on electrons and atomic nuclei. Bonds result when the net coulombic interactions are sufficiently attractive. Different principal types of bonds recognized include metallic, covalent, ionic, and bridge.

Metallic bonding is the attraction of all the atomic nuclei in a crystal for the outer shell electrons that are shared in a delocalized manner among all available orbitals. Metal atoms characteristically provide more orbital vacancies than electrons for sharing with other atoms.

Covalent bonding results most commonly when electrons are shared by two atomic nuclei. Here the bonding electrons are relatively localized in the region of the two nuclei, although frequently a degree of delocalization occurs when the shared electrons have a choice of orbitals. The conventional *single* covalent bond involves the sharing of two electrons. There may also be *double* bonds with four shared electrons, *triple* bonds with six shared electrons, and bonds of intermediate multiplicity.

Covalent bonds may range from *nonpolar*, involving electrons evenly shared by the two atoms, to extremely *polar*, where the bonding electrons are very unevenly shared. The limit of uneven sharing occurs when the bonding electron spends full time with one of the atoms, making the atom into a negative ion and leaving the other atom in the form of a positive ion. Ionic bonding is the electrostatic attraction between oppositely charged ions.

Bridge bonds involve compounds of hydrogen in which the hydrogen bears either a + or − charge. When hydrogen is attached by a polar covalent bond to one molecule, it may attract another molecule, bridging the two molecules together. If the hydrogen is +, it may attract an electron pair of the other molecule. This is called a *protonic bridge*. If the hydrogen is −, it may attract through a vacant orbital the nucleus of an atom of a second molecule. This is called a *hydridic bridge*. Such bridges are at the lower range of bond strength but may have a significant effect on the physical properties of condensed states of those substances in which they are possible.

See also **Chemical Elements**.

BORACITE. This mineral is a magnesium borate containing some chlorine, $Mg_3 B_7 O_{13} Cl$. It appears to be isometric but probably becomes so only at 265°C, below which temperature it is believed to be orthorhombic. Its hardness is 7; specific gravity, 2.9; luster, vitreous; color, white to gray, sometimes yellow or green; translucent to subtransparent. It occurs in beds with gypsum and salt in Germany, particularly at Stassfurt in Saxony.

BORANE. One of a series of boron hydrides (compounds of boron and hydrogen). The simplest of these, BH_3, is unstable at atmospheric pressure and becomes diborane (B_2H_6) as gas at normal pressures. This is converted to higher boranes, i.e., penta-, deca-, etc., by condensation. This series progresses through a number of well-characterized crystalline compounds. Hydrides up to $B_{20}H_{26}$ exist. Most are not very stable and readily react with water to yield hydrogen. Many react violently with air. As a rule, they are highly toxic. Their properties have suggested investigation for rocket propulsion, but they have not proved satisfactory for this purpose. There are also a number of organoboranes used as reducing agents in electroless nickel-plating of metals and plastics. Some of the compounds used are di- and triethylamine borane and pyridine borane. See also **Carbonate**; and **Organoborane**.

BORAX. This hydrated sodium borate mineral, $Na_2B_4O_7 \cdot 10H_2 O$, is a product of evaporation from shallow lakes and plays. Borax crystallizes in the monoclinic system, usually in short prismatic crystals. Its color grades from colorless through gray, blue to greenish. Vitreous to resinous luster of translucent to opaque character. Hardness of 2–2.5, and specific gravity of 1.715.

Borax from the salt lakes of Kashmir and Tibet has been known since early history. India, the former U.S.S.R., and Persia possess small deposits. Extensive deposits are known in the United States, notably in Lake, San Bernardino, Inyo, and Kern Counties in California, and Esmeralda and Dona Ana Counties in New Mexico.

It is used in antiseptics and medicines, as a flux in smelting, soldering and welding operations, as a deoxidizer in nonferrous metals, as a neutron absorber for atomic energy shields, in rocket fuels, and as extremely hard abrasive boron carbide (harder than corundum). See also **Boron**.

BORIC ACID. See **Boron**.

BORNITE. Named for the German mineralogist of the eighteenth century, Ignatius von Born, this mineral is a sulfide of copper and iron corresponding to the formula $Cu_5 FeS_4$. It is isometric with a cubic habit, although crystals are rare, usually occurring as granular or compact masses. Its fracture is conchoidal to uneven; brittle; hardness, 3; specific gravity, 5.079; color, copper-red to reddish-brown (hence the name horseflesh ore) when freshly fractured; it soon assumes an iridescent tarnish (hence the name peacock ore); luster, metallic; streak, grayish-black; opaque.

Bornite as a primary mineral has been observed in pegmatite veins and in igneous rocks and is also a common secondary mineral.

Bornite crystals have been obtained in Austria and England. As an ore it is important in Tasmania, Chile, Peru and in Montana. In the United States, bornite also has been found in Connecticut, and in Canada, in the Province of Quebec.

Bornite also is known as *peacock ore* and *horseflesh ore*.

BORN, MAX (1882–1970).

A German-born British physicist, Max Born studied mathematics and physics and in 1904 became David Hilbert is private assistant for. While at the University of Breslau, he won a competition on the stability of elastic wires and it became the dissertation for his Ph.D. After graduate school, he studied special relativity for a while, then became interested in the physics of crystals. In 1912, he published the Born-Karman theory of specific heats and his work on crystals is a cornerstone of solid-state theory.

Born coined the term "Quantum mechanics" and in 1925 devised a system called matrix mechanics, which accounted mathematically for the position and momentum of the electron in the atom. He devised a technique called the Born approximation in scattering theory for computing the behavior of subatomic particles which is used in high-energy physics. Also, interpretation of the wave function for Schrodinger's wave mechanics was solved by Born who suggested that the square of the wave function could be understood as the probability of finding a particle at some point in space. For this work in quantum mechanics, Max Born received the Nobel Prize in Physics in 1954.

J.M.I.

BORN-OPPENHEIMER APPROXIMATION.

An argument for calculating the force constants between atoms in a molecule or solid, based on the observation that the motion of the electrons is so rapid compared with that of the heavier nuclei that it can be assumed that the electrons follow the motion of the nuclei adiabatically. That is, one calculates the eigenvalues of energy for the electrons with the nuclei in fixed positions; the variation of this electronic energy with the configuration of the nuclei may then be treated as a contribution to the potential energy of the interatomic forces.

BORON.

[CAS: 7440-42-8]. Chemical element, symbol B, at. no. 5, at. wt. 10.81, periodic table group 13, mp 2079°C, sublimes at approximately 2550°C, density 2.35 g/cm^3 (amorphous form). There are four principal crystal modifications of boron: (1) α-rhombohedral, (2) β-rhombohedral, (3) I-tetragonal, and (4) II-tetragonal. There are two natural isotopes, ^{10}B and ^{11}B. In 1807, Davy first produced elemental boron in amorphous form by electrolyzing boric acid. A year later, Gay-Lussac and Thènard produced elemental boron by reducing boric acid with potassium. However, it was not until 1892 that boron with a purity of over 90% was produced by Moissan, who reduced the element from B_2O_3. Moissan observed that the produced substances earlier claimed to be elemental boron were in effect compounds of boron. First ionization potential 8.296 eV; second 23.98 eV; third 37.75 eV. Oxidation potential $B + 3H_2O \rightarrow H_3BO_3 + 3H^+ + 3e^-$, 0.73 V; $B + 4OH^- \rightarrow H_2BO_3^- + H_2O + 3e^-$, 2.5 V. Other important physical properties of boron are given under **Chemical Elements**.

Boron is (1) a yellowish-brown crystalline solid and (2) an amorphous greenish-brown powder. Both forms are unaffected by air at ordinary temperatures but when heated to high temperatures in air form oxide and nitride. Crystalline boron is unattacked by HCl or HNO$_3$, or by NaOH solution, but with fused NaOH forms sodium borate and hydrogen; reacts with magnesium but not with sodium.

Boron occurs as rasorite or kernite (sodium tetraborate tetrahydrate, $Na_2B_4O_7 \cdot 4H_2O$) and colemanite (calcium borate, $Ca_2B_6O_{11} \cdot 5H_2O$) in California, as sassolite (boric acid, H_3BO_3) in Tuscany, Italy, and also locally in Chile, Turkey, and Tibet. See also **Colemanite**; **Kernite**; and **Ulexite**.

Production: Commercial boron is produced in several ways. (1) Reduction with metals from the abundant B_2O_3, using lithium, sodium, potassium, magnesium, beryllium, calcium, or aluminum. The reaction is exothermic. Magnesium is the most effective reductant. With magnesium, a brown powder of approximately 90–95% purity is produced. (2) By reduction with compounds, such as calcium carbide or tungsten carbide, or with hydrogen in an electric arc furnace. The starting boron source may be B_2O_3 or BCl$_3$. (3) Reduction of gaseous compounds with hydrogen. In an atmosphere of a boron halide, metallic filaments or bars at a surface temperature of about 1200°C will receive depositions of boron upon admission of hydrogen to the process atmosphere. Although the deposition rate is low, boron of high purity can be obtained because careful control over the purity of the starting ingredients is possible. (4) Thermal decomposition of boron compounds, such as the boranes (very poisonous). Boranes in combination with oxygen or H$_2$O are very reactive. In this process, boron halides, boron sulfide, some borides, boron phosphide, sodium borate and potassium borate also can be decomposed thermally. (5) Electrochemical reduction of boron compounds where the smeltings of metallic fluoroborates or metallic borates are electrolytically decomposed. Boron oxide alkali metal oxide–alkali chloride compounds also can be decomposed in this manner.

Both chemical methods and float zoning are used to purify the boron product from the foregoing processes. In the latter method, a boron of 99.99% purity can be obtained.

Although the chemistry of boron is extremely interesting, there is no substantial market for elemental boron. Some boron compounds are high-tonnage products. Elemental boron has found limited use to date in semiconductor applications, although it does possess current-voltage characteristics that make it suitable for use as an electrical switching device. In a limited way, boron also is used as a dopant (*p*-type) for *p-n* junctions in silicon. The principal problem deterring the larger use of boron as a semiconductor is the high-lattice defect concentration in the crystals currently available.

Uses of Boron

As early as 1959, boron filaments were introduced as the first of a family of high-strength, high-modulus, low-density reinforcements developed for advanced aerospace applications. A process was engineered by Avco Specialty Materials (Lowell, Massachusetts) and the U.S. Air Force to manufacture boron filaments that had high strength and high stiffness, but low density and, hence, low weight. During the interim, advanced boron fibers have been used as a reinforcement in resin-matrix composites. Boron aluminum has been used for tube-shaped truss members, for reinforcing space vehicle structures, and has also been considered as a fan blade material for turbofan jet engines. However, boron's rapid reaction with molten metals, such as aluminum, and the degradation of its mechanical properties when diffusion-bonded at temperatures above 480°C have been difficult to surmount. These shortcomings led to the development of silicon-carbide (SiC) fibers for some applications.

The principal use of boron filaments is in the form of continuous boron-epoxy pre-impregnated tape, commonly known as *prepreg*. The boron filaments are unidirectionally arranged and occupy about 50% of the composite volume. Typically, there are about 200 filaments/inch (8/cm). Usually, the resin content is about 30–35% (weight). Boron composites have been used in military aircraft, including helicopters. In addition to aircraft, boron-epoxy composites have been used in tennis, racquetball, squash, and badminton rackets, fishing rods, skis, and golf club shafts, for improving strength and stiffness.

Boron, which is extremely hard (3300–3500 on the Knoop scale; 9.5 on the Mohs scale), has been used in cutting and grinding tools. Boron is 30–40% harder than silicon carbide and almost twice as hard as tungsten carbide. Boron also has interesting microwave polarization properties. Research (Southern Illinois University) has shown that a single ply of boron epoxy will transmit 98.5% and reflect 0.6% of the incident microwave power when the angle between the grain and the E-field is 90°. This property has been useful in the design of spacecraft antennas and radomes.

As described by Buck, a chemical vapor deposition (CVD) process is used to form boron fibers. A small-diameter substrate wire is run through a glass reactor tube and suitable gases are introduced. The substrate is heated by electrical resistance, causing the gases to react and allowing boron to deposit on the heated wire, thus forming the filament. In the process, elemental boron is obtained through the reaction of boron trichloride and hydrogen gases.

Various boron compounds have been used as rocket fuels, diamond substitutes, and additives to aluminum alloys to improve electrical and thermal conductivity, as well as for grain refining. Boron hydrides are sensitive to shock and can detonate easily. Boron halides are corrosive and toxic.

Biological Functions: Although boron is required by plants, there is little solid evidence to date that it is required for the nutrition of livestock or humans. Boron deficiency may alter the levels of vitamins or sugars in plants owing to the effect of boron upon the synthesis and translocation of these compounds within the plant. The addition of boron to some boron-deficient soils has increased the carotene or provitaminA concentration in carrots and alfalfa.

Like several of the other trace elements, while concentrations of very low levels are desirable, high levels of boron are toxic to plants. Different plant species vary widely in their requirement for this element and in their tolerance for high levels. Application of boron-containing fertilizer can be carefully adjusted for different crops. An application of boron-containing fertilizer to improve the yields of alfalfa or beets may be toxic to such boron-sensitive crops as tomatoes and grapes. In the southwestern United States, serious boron toxicity to plants has resulted from using irrigation waters that are high in boron.

Boric Acid: Boric oxide, [CAS: 1303-86-2], B_2O_3, is acidic. It exists in two forms, a glassy form obtained by high temperature dehydration of boric acid, and crystalline form obtained by slow heating of metaboric acid.

The oxyacids of boron are of two types: (A) the boric acids, based upon boric oxide, and (B) the lower oxyacids based upon boron-to-boron structural linkages.

The really acidic boric acids consist essentially of metaboric acid (HBO_2), a polymer, and boric or orthoboric acid, H_3BO_3 ($pK_a = 9.24$). There is no compound corresponding to the formula for tetraboric acid, $H_2B_4O_7$, although there are a number of salts that may be based upon this composition. Sometimes called *boracic acid*, H_3BO_3, is a high-tonnage material, the main uses being in the medical and pharmaceutical fields. A saturated solution of H_3BO_3 contains about 2% of the compound at 0°C, increasing to about 39% at 100°C. The compound also is soluble in alcohol. In preparations, solutions of boric acid are nonirritating and slightly astringent with antiseptic properties. Although no longer used as a preservative for meats, boric acid finds extensive use in mouthwashes, nasal sprays, and eye-hygiene formulations. Boric acid (sometimes with borax) is used as a fire-retardant. A commercial preparation of this type (*Minalith*) consists of diammonium phosphate, ammonium sulfate, sodium tetraborate, and boric acid. The tanning industry uses boric acid in the deliming of skins where calcium borates, soluble in H_2O, are formed. As sold commercially, boric acid is $B_3O_3 \cdot 3H_2O$, prepared by adding HCl or H_2SO_4 to a solution of borax.

Borates: Sodium tetraborate, $Na_2B_4O_7 \cdot 10H_2O$, is a very-high-tonnage material. Natural borax has a hardness of 2–2.5, mp 75°C, sp gr 1.75. An aqueous solution of borax is mildly alkaline and antiseptic. The compound finds many uses, including: (1) cleaning compounds of numerous types; (2) important ingredient of glass and ceramics, notably for heat-resistant glass where as much as 40 pounds of borax may be required per 100 pounds of finished glass; (3) source of elemental boron and other boron compounds; (4) flux for soldering and welding; (5) constituent of fertilizers; (6) filler in paper and paints; and (7) corrosion inhibitor in antifreeze formulations. Borax also is used in fire retardants.

Chemistry of Boron and Other Boron Compounds: In 1901, the German chemist Alfred Stock stated, "It was evident that boron, the close neighbor of carbon in the periodic system, might be expected to form a much greater variety of interesting compounds than merely boric acid and the borates, which were almost the only ones known." In 30 years of research that followed that statement, Stock synthesized almost all of the important *boranes* (hydrogen and boron). Some of these compounds now find use in glass, ceramics, synthetic lubricants, and as ingredients of high-energy rocket fuels and jet-engine and automotive fuels. Further pioneering of borohydride chemistry was carried on by Schlesinger and Burg of the University of Chicago in the late 1940s. Boron carbide, B_4C, is used as neutron-absorbing material in nuclear reactors. Sodium borohydride, $NaBH_4$, is applied as a reducing agent in the manufacture of certain synthetics. Although not ultimately selected, because of the greater volatility of uranium hexafluoride, UF_6, both uranium borohydride, $U(BH_4)_4$, and its methyl derivative, $U(CH_3BH_3)_4$, were considered for use in separating the isotopes of uranium during the Manhattan Project. ^{10}B is used in brain tumor research. When injected intravenously, borax concentrates in the areas of tumors and its presence can be detected by radiation techniques. With further research, the tendency of boron to link with itself may comprise the foundation of future inorganic polymeric materials. Although they have poor mechanical strength, boron-phosphorus polymers, prepared by reacting diborane with phosphone derivatives, do exhibit excellent heat-resistance.

X-ray diffraction studies show five general types of structures in solid borates:

1. Discrete anions containing individual BO_3^{3-} groups, or a limited number of other groups combined by sharing oxygen atoms. (The simplest is $B_2O_5^{4-}$, which is called pyroborate.)

2. Extended anions in which individual BO_3 groups are linked into rings or chains, such as $B_3O_6^{3-}$ or $B_2O_4^{2-}$ (metaborate).

3. Sheet structures in which all the oxygen atoms are shared between borate groups, as in $B_5O_{10}^{5-}$ (pentaborate).

4. Structures containing the tetrahedral $B(OH)_4^-$ ion, which is the principal ion found in alkaline aqueous solutions.

5. Extended anions containing tetrahedral BO_4 units, usually linked with triangular BO_3 groups.

The lower oxyacids of boron may be derived from the various boron hydrides, whence their boron-boron linkages result. These compounds include the hypoborates, which may be produced by reactions of tetraborane with strong alkali, and which may be formulated from the structure $H_2[H_6B_2O_2]$; the subborates, derived from $H_4[B_2O_4]$, which is called subboric acid; and the borohydrates, which are derived from acids of various compositions, such as $H_2[B_4O_2]$, $H_2[B_2O_2]$ and $H_2[H_4B_2O_2]$. The last of these compounds contains a double-bonded boron-boron linkage, and exhibits *cis-trans* isomerism.

The borides are binary compounds of boron with metals or electropositive elements in general. Except in isolated cases their compositions depart from the stoichiometry of trivalent boron compounds and are determined more by the requirements of metal and boron lattices than by valencies. On the basis of composition, they may be classified into types based respectively upon zigzag chains (MB) represented by CoB; isolated boron atoms (M_2B) represented by Co_2B; double chains (M_3B_4) represented by Mo_3B_4; hexagonal layers (MB_2) represented by CoB_2; three-dimensional frameworks (MB_6 or MB_{12}) represented by SiB_6 or UB_{12}. It is apparent that these borides are interstitial compounds existing primarily with the metals of main groups, II, III, IV, V, and VI.

There are at least six definitely characterized boron hydrides, as follows: diborane(6), B_2H_6; tetraborane(10), B_4H_{10}; pentaborane(9) (stable), B_5H_9; pentaborane(11) (unstable), B_5H_{11}; hexaborane(10), B_6H_{10}; and decaborane(14), $B_{10}H_{14}$. In these names, note that the prefix denotes the number of boron atoms, while the figure in parentheses denotes the number of hydrogen atoms. In addition to these compounds, which are all gases or volatile liquids except decaborane(14), decomposition of the lower boron hydrides yields colorless or yellow solid boron hydrides, ranging in composition from $(BH_{1.5})_x$ to $(BH)_x$. This readiness to polymerize is evidence of the reactivity of these borane compounds, which readily form additional products with ammonia, with the amalgams of the active metals, and with many organic compounds, as well as with CO.

In addition to BH_4^- there exist a number of hydroborate anions, which may be derived from real or hypothetical boron hydrides by addition of hydride ion. These include $B_2H_7^-$, formed by the reaction of B_2H_6 and BH_4^- in organic solvents, and the extremely stable ions $B_{10}H_{10}^{2-}$ and $B_{12}H_{12}^{2-}$ unaffected by either acidic or alkaline aqueous solutions or by atmospheric oxygen. Free halogens merely cause substitution of halogen for hydrogen. The structure of $B_{10}H_{10}^{2-}$ is based on the square antiprism, while that of $B_{12}H_{12}^{2-}$ is a regular icosahedron.

In 1976, the Nobel Prize for Chemistry was awarded to William Nunn Lipscomb, Jr., of Harvard University, for original research on the structure and bonding of boron hydrides and their derivatives. As pointed out by Grimes (1976), the insight into electron-deficient borane structures originally provided by Lipscomb carries over not only to the carboranes, but also to their organic cousins, the so-called "nonclassical" carbonium ions. The three-center bond descriptions given by Lipscomb to B_5H_9 and B_6H_{10} can as easily be applied to their hydrocarbon analogs, the pyramidal ions $C_5H_5^+$ and $C_6H_6^{2+}$, both presently known as alkyl derivatives. Also, molecules usually not so considered, such as metallocenes, organometallics, such as $(C_4H_4)Fe(CO)_3$ or $[(CO)_3Fe]_5C$, metal clusters and others, can be considered from the perspective of borane analogs. The boranes, once considered peculiar, over the years have provided insight to many cluster-type molecules, for which classical Lewis bond descriptions do not fit. Lipscomb's lecture given in Stockholm on December 11, 1976 provides an excellent overview of the boranes and their relatives.

In 1979, the Nobel Prize for Chemistry was received by Herbert C. Brown (shared with Georg Witting for research in another field) of Purdue University for the discovery of the *hydroboration reaction*. This reaction, depicted below, has made the organoboranes readily available as chemical intermediates. The boron atom adds to the less substituted carbon atom. As pointed out by Brewster and Negishi (1980), depending upon steric factors, mono-, di-, or trialkylboranes may be formed. These products comprise synthetically useful reactions whereby the boron atom is replaced, but the mono- and dialkylboranes are also useful as reducing or hydroborating agents.

$$RCH{=}CH_2 + B_2H_6 \longrightarrow (RCH_2CH_2{-})_3B \qquad (1)$$

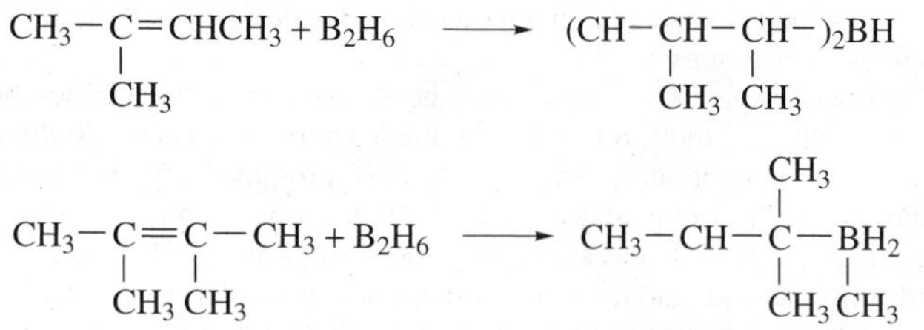

$$CH_3{-}\underset{\underset{CH_3}{|}}{C}{=}CHCH_3 + B_2H_6 \longrightarrow (\underset{\underset{CH_3}{|}}{CH}{-}\underset{\underset{CH_3}{|}}{CH}{-}CH{-})_2BH \qquad (2)$$

$$CH_3{-}\underset{\underset{CH_3}{|}}{C}{=}\underset{\underset{CH_3}{|}}{C}{-}CH_3 + B_2H_6 \longrightarrow CH_3{-}CH{-}\underset{\underset{CH_3}{|}}{\overset{\overset{CH_3}{|}}{C}}{-}BH_2 \qquad (3)$$

Among the other inorganic compounds of boron are the following:

Borides: Carbon boride, CB_6, and silicon borides SiB_3 and SiB_6 are hard, crystalline solids, produced in the electric furnace; magnesium boride, Mg_3B_2, brown solid, by reaction of boron oxide and magnesium powder ignited, forms boron hydrides with HCl; calcium boride, Ca_3B_2, forms boron hydrides and hydrogen gas with HCl.

Nitride: Boron nitride, BN, white solid, insoluble, reacts with steam to form NH_3 and boric acid, formed by heating anhydrous sodium borate with ammonium chloride, or by burning boron in air.

Sulfide: Boron sulfide, B_2S_3, white solid, unpleasant odor, irritating to the eyes, reactive with water to form boric acid and hydrogen sulfide, formed by reaction of boron oxide plus carbon heated in a current of CS_2 at red heat.

The great number of compounds of boron is due to the readiness with which boron atoms form, to some extent, chain structures with other boron atoms, and, to a far greater extent, cyclic compounds, both with other boron atoms, and with atoms of carbon, oxygen, nitrogen, phosphorus, arsenic, the halogens, and many other elements. Examples of them are shown below, beginning with the two pentaboranes B_5H_9 and B_5H_{11}:

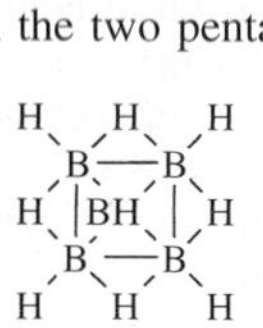

B_5H_9 Pentaborane (9)

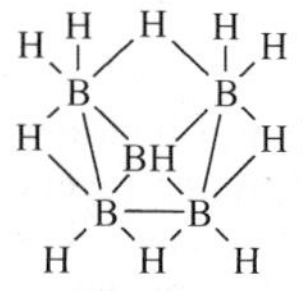

B_5H_{11} Pentaboranze (11)

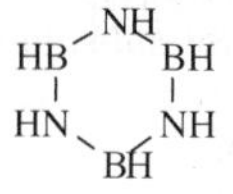

hexahydro-*s*-triazatriborine (also called *borazine*)

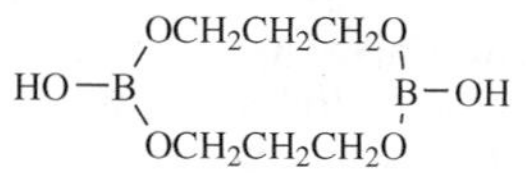

2,8-dihydroxy-1,3,7,9-tetroxa-2,8-diboracyclododecane

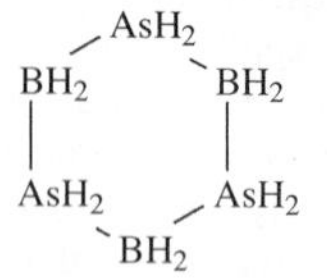

dodecahydro-*s*-triarsatriborine (also called *s*-triphosphatriborane or *borarsane*)

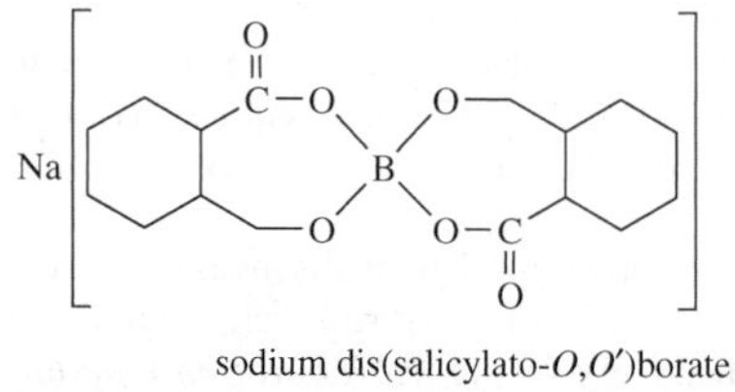

sodium dis(salicylato-*O*,*O'*)borate (1–)

Halides: Since simple boron compounds have only three electron pairs in the valence shell of boron, they tend to be electron acceptors. Its simple molecules are formed by sp^2 hybrid sigma bonds lying in a plane. Its strong tendency to form an octet is shown by the tetrahedral boron compounds involving sp^3 hybridization. Boron halides include the trifluoride, BF_3, the trichloride, BCl_3, the tribromide, BBr_3 and the triiodide, BI_3, which range in mp from -127 to $+43°C$. Typical methods of forming the boron halides are: treatment of boron oxide with hot concentrated H_2SO_4 in a reaction mixture with calcium fluoride to produce BF_3, and by heating boron, or boron oxide plus carbon with chlorine to produce the chloride.

In addition to the simple halides, boron forms fluorine complexes containing the fluoroborate ion $(BF_4{}^-)$ Subhalides of boron are known (B_2X_4) of the structure:

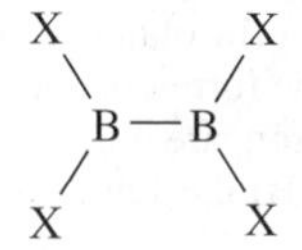

and B_4X_4 of the structure

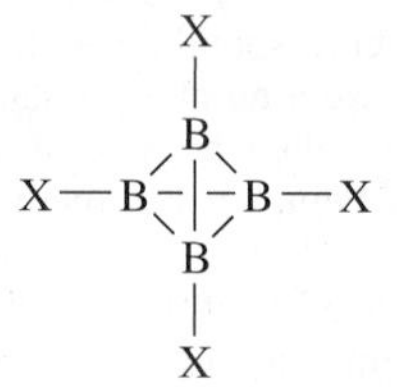

Additional Reading

Barton, L.: *Introduction to The Inorganic Chemistry of Boron,* John Wiley & Sons, Inc., New York, NY, 2000.

Brewster, J.H. and E. Negishi: "The 1979 Nobel Prize for Chemistry," *Science,* **207,** 44 (1980). (A classic reference.)

Carter, G.F. and D.E. Paul: *Materials Science and Engineering,* ASM International, Materials Park, OH, 1991.

Grew, E.S. and L.M. Anovitz: *Boron: Mineralogy Petrology and Geochemistry,* Mineralogical Society of America, Washington, DC, 1996.

Grimes, R.N.: "The 1979 Nobel Prize in Chemistry," *Science,* **194,** 709 (1979). (A classic reference.)

Hawley, G.G. and R.J. Lewis: *Hawley's Condensed Chemical Dictionary,* 13th Edition, John Wiley & Sons, Inc., New York, NY, 1999.

King, R.B.: *Boron Chemistry At The Millennium,* Elsevier Science, New York, NY, 1999.

Lide, D.R.: *Handbook of Chemistry and Physics,* 84th Edition, CRC Press, LLC., Boca Raton, FL, 2003.

Lipscomb, W.N.: "The Boranes and Their Relatives," *Science,* **196,** 1047–1055 (1977).

Perry, R.H. and D. Green: *Perry's Chemical Engineers' Handbook,* 7th Edition, McGraw-Hill Companies, Inc., New York, NY, 1997.

Rogl, P. and G. Effenberg: *Phase Diagrams of Ternary Metal-Boron-Carbon Systems,* ASM International, Materials Park, OH, 1998.

Siebert, W.: *Advances in Boron Chemistry,* Springer-Verlag, Inc., New York, NY, 1997.

Staff: *Boron-Environmental Aspects,* World Health Organization, Washington, DC, 2000.

Web Reference

Mineralogical Society of America, *http://www.minsocam.org/*

BOSONS. Those elementary particles for which there is symmetry under intra-pair production. They obey Bose-Einstein statistics. Included are photons, pi mesons, and nuclei with an even number of particles. (Those particles for which there is antisymmetry are *fermions.*) See **Mesons; Particles (Subatomic);** and **Photon and Photonics**.

BOSTONITE. A rather rare rock type, dense, with an occasional feldspar phenocryst and grayish in color. It is composed almost wholly of alkaline feldspar, being analogous to aplites. The type locality is Salem Neck, Massachusetts, close to Boston, for which it was named.

BOUGUER AND LAMBERT LAW. In homogeneous materials, such as glass or clear liquids, the fractional part of intensity or radiant energy absorbed is proportional to the thickness of the absorbing substance.

Summing over a series of thin layers or integration over a finite thickness gives the relation

$$\log I_0/I = k_1 b$$

where I_0 is the intensity or radiant power incident on a sample b centimeters thick and I is the intensity of the transmitted beam. The constant k_1 depends on the wavelength of the incident radiation, the nature of the absorbing material and other experimental conditions. Verification of the law fails unless appropriate corrections are made for reflection, convergence of the light beam and spectral slit width, as well as possible scattering, fluorescence, chemical reaction, nonhomogeneity, and anisotropy of the sample. Formerly, the constant k_1 was called the absorption coefficient. It is now preferable to avoid this term and to call the ratio I/I_0 the transmittance. The law was first expressed by Bouguer in 1729 but it is often attributed to Lambert, who restated it in 1768.

BOULANGERITE. A mineral compound of lead-antimony sulfide, $Pb_5Sb_{4s}S_{11}$. Crystallizes in the monoclinic system; hardness, 2.5–3; specific gravity, 6.23; color, lead gray.

BOURNONITE. An antimony-copper-lead sulfide corresponding to the formula $PbCuSbS_3$. It is orthorhombic, and repeated twinning often produces crosses or wheel-shaped crystals. It is brittle; fracture, subconchoidal; hardness, 2.5–3; specific gravity, 5.83; luster, metallic; color and streak, dark gray to black; opaque.

Bournonite is found with galena, chalcopyrite, and sphalerite. There are many European localities; it was first found in Cornwall, England, by Count Bournon, for whom it was later named. Bournonite occurs in Bolivia and Peru and in the United States in Arizona, Montana, Nevada and Utah. Bournonite is also known as *wheel ore*.

BOYLE-CHARLES LAW. This law states that the product of the pressure and volume of a gas is a constant which depends only upon the temperature. This law may be stated mathematically as

$$p_2 v_2 = p_1 v_1 [1 + a(t_2 - t_1)]$$

where p_1 and v_1 are the pressure and volume of a body of gas at temperature t_1, p_2 and v_2 are the pressure and volume of the same body of gas at another temperature t_2, and a is the volume coefficient of expansion of the gas. If the temperature is expressed in degrees absolute, this expression becomes

$$\frac{p_2 v_2}{T_2} = \frac{p_1 v_1}{T_1}$$

which is the ideal gas law, so-called because all real gases depart from it to a greater or lesser extent. See also **Characteristic Equation**.

BOYLE, ROBERT (1627–1691). Boyle was an Anglo-Irish scientist who is known for his research in chemistry, physics, and medicine. He is also remembered for his inventions and also for his study of philosophy and theology. He was educated at Eton College from 1635 to 1639. Beginning in 1644, he began doing experimental scientific work and he is considered one of the founders of modern science.

Boyle was one of the founders and key members of the Royal Society, a group of scientists who conducted experiments under the patronage of King Charles the Second.

In 1667, Boyle repeated the air pump experiments of von Guericke with an improved version of the apparatus so that it created a nearly perfect vacuum in a container. When he pumped air out of a pipe that was placed in a container of water, he was able to raise the water a little over 10 m, but no more. He continued further experiments of his own and published his results in *"New Experiments Physico-Mechanical,"* a later edition which includes his famous gas law. Boyle's Law states that the pressure of a given mass of gas is inversely proportional to the volume. Boyle is credited for realizing the true nature of elements. He believed all matter was made up from extremely small particles, which were combined together in varying ways to form the different chemical substances. In "Skeptical Chymist," he expressed his views. He attacked the old theories of the alchemists. He was the first to suggest that matter can be classified in terms of similar chemical properties. Boyle is remembered as a scientist who gave important weight to the atomic theory of matter. He is recognized for his development of the experimental method, his careful observation, recording, and reporting of results.

Boyle was a very religious man and is also known as a devout Christian. Interestingly, he wrote many articles on theology. His Boyle Lectures defended Christianity against atheists.

See also **Boyle's Law.**

J.M.I.

BOYLE'S LAW. This law, attributed to Robert Boyle (1662) but also known as Mariotte's law, expresses the isothermal pressure-volume relation for a body of ideal gas. That is, if the gas is kept at constant temperature, the pressure and volume are in inverse proportion, or have a constant product. The law is only approximately true, even for such gases as hydrogen and helium; nevertheless it is very useful. Graphically, it is represented by an equilateral hyperbola (see Fig. 1). If the temperature is not constant, the behavior of the ideal gas must be expressed by the Boyle-Charles law.

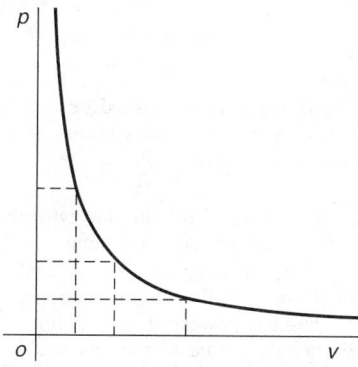

Fig. 1. Equilateral hyperbola representing Boyle's law. The rectangular areas (PV) are all equal

The Boyle temperature is that temperature, for a given gas, at which Boyle's law is most closely obeyed in the lower pressure range. At this temperature, the minimum point (of inflection) in the pV-T curve falls on the pV axis. See **Compression (Gas)**; and **Ideal Gas Law**.

B. P. An abbreviation of "Before Present." This term is an indication of time calculation, used especially when referring to radiometric dating.

BRAGG'S CURVE. There are two types of curves to which Bragg's name is occasionally given: 1. A graph for the average number of ions per unit distance along a beam of initially monoenergetic alpha particles, or other ionizing particles, passing through a gas. 2. A graphical relationship between the average specific ionization of an ionizing particle of a particular kind, and some other variable, such as the kinetic energy, the residual range, or the velocity of the particle.

BRAGG'S LAW. The law expressing the condition under which a crystal will reflect a beam of x-rays with maximum distinctness, at the same time giving the angle at which the reflection takes place. For x-ray reflection it is customary to use the complement of the angle of incidence and reflection, that is, the angle which the incident or the reflected beam makes with the crystal planes, rather than with the normal. Let this "Bragg angle" be θ. If the planes or layers of atoms are spaced at a distance d apart, and if λ is the wavelength of the x-rays, Bragg's law is expressed by the equation

$$\sin \theta = \frac{n\lambda}{2d}$$

The condition for an intensity maximum is that n must be a whole number. For example if the planes of rock salt parallel to the natural cubical faces are spaced at $d = 2.814 \times 10^{-8}$ centimeters or 2814 x-units, and if the incident rays have a component of wavelength $\lambda = 714$ x-units, the above equation gives sin; $\theta = 0.1269n$. Then if the crystal is rotated slowly, there will be a distinct reflection where θ reaches $7°17'$ ($n = 1$), again at $14°42'$ ($n = 2$), also at $22°23'$ ($n = 3$), etc. See also **Crystal**.

BRAGG SPECTROMETER. An instrument for the x-ray analysis of crystal structure, in which a homogeneous beam of x-rays is directed on the known face of a crystal, C, and the reflected beam detected in a suitably placed ionization chamber, E. As the crystal is rotated, the angles at which

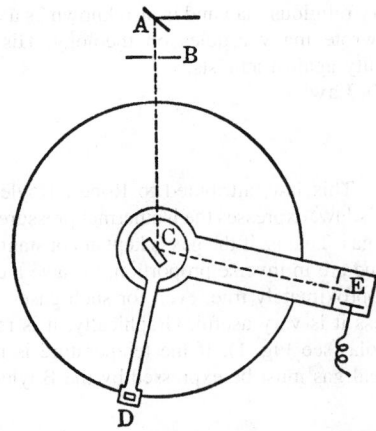

Fig. 1.   Bragg spectrometer

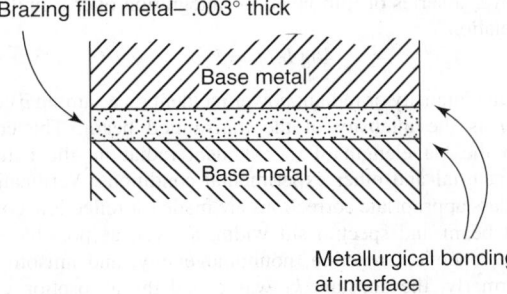

Fig. 1.   The principle by which the filler metal is drawn through the joint to create a bond between the base metals is capillary action. In brazing, heat is applied broadly to the base metals. The filler metal is then brought into contact with the heated parts, whereupon the filler metal melts instantly and is drawn completely through the joint. (*Lucas-Milhaupt, Inc., A Handy & Harman Company*)

the equation expressing Bragg's law is satisfied are identified as sharp peaks in the ionization current. See Fig. 1. This is one of the early, classical instruments in the laboratory field.

BRAGG'S RULE.   An empirical relationship whereby an elements for mass stopping power of an element of alpha particles is inversely proportional to the one-half power of the atomic weight. This relationship is also stated in the form that the atomic stopping power is directly proportional to the one-half power of the atomic weight. The wide usefulness of the Bragg rule is due to the fact that it leads to relations between the stopping powers of different elements for alpha particles. It also applies to other charged particles as well as alpha particles, and to the same degree of approximation, which is about ±15%.

See **Particles (Subatomic)**.

BRAGG, WILLIAM LAWRENCE SIR (1890–1971).   Bragg was an Australian physicist. When he was a small boy, he shattered his elbow and an X-ray was taken. William Bragg's X-ray was the first medical X-ray in Australia. He received his early education at St. Peter' College in his birthplace, proceeding to Adelaide University to take his degree in mathematics with first-class honors in 1908. He came to England with his father in 1909 and entered Trinity College, Cambridge, as an Allen Scholar, taking first-class honors in the Natural Science Tripos 1912. Bragg's father, William Henry Bragg, was also a professor of physics and the two men worked together. They determined the crystal structures of basic substances such as diamonds, zinc blende, and calcite to mention a few. In 1915, he received with his father, the Nobel Prize in Physics for work in X-ray crystallography.

See also **Bragg's Law**; **Bragg Spectrometer**; and **Bragg's Rule**.

J.M.I.

BRAIN CHEMICALS.   See **Enkephalins and Endorphins**.

BRASS.   See **Copper**.

BRAVAIS-MILLER INDICES.   See **Crystal**.

BRAZING.   Brazing may be defined as the joining of metals through the use of heat and a filler metal whose melting temperature is above 840°F (450°C), but below the melting point of the metals being joined. A more exact name for many brazing processes would be "silver brazing," since the filler metal used most often is a silver alloy. Brazing may be the most versatile method of metal joining today. Brazed joints are strong-on nonferrous metals and steels, the tensile strength of a properly made joint will often exceed that of the metals joined. Brazed joints are ductile, able to withstand considerable shock and vibration. Brazing is essentially a one-operation process. There is seldom any need for grinding, filing, or mechanical finishing after the joint is complete. In comparing brazing with welding, it should be noted that welding, by its nature, presents problems in automation. A resistance weld joint made at single point is relatively easy to automate, but once the point becomes a line (a *linear joint*), the line has to be *traced*.

In contrast, a brazed joint is made in a completely different way from a welded joint. The first big difference is in temperature. Brazing does not melt the base metals; therefore, brazing temperatures are invariably lower than the melting points of the base metals, and they are always lower than welding temperatures for the same base metals. Brazing joins metals by creating a *metallurgical bond* between the filler metal and the surfaces of the two metals being joined. See Fig. 1. The principle by which the filler metal is drawn through the joint to create this bond is *capillary action*. In brazing, heat is applied broadly to the base metals. The filler metal is then brought into contact with the heated parts. It is melted instantly by the heat in the base metals. Because of this action, brazing joins almost any configuration with equal ease. See Fig. 2.

The six basic steps in brazing are: (1) *Good fit and proper clearance*—Because brazing depends upon capillary action to distribute the molten filler metal between the surfaces of the base metals, care must be taken to make certain that the clearance between the base metals is right, which in most cases can be described as a *close clearance*. (2) *Cleaning the metals*—because oil, grease, rust, scale, dirt, etc. form barriers between the base metal surfaces and the brazing materials. (3) *Fluxing the parts*—With few exceptions, flux is applied to the joint surfaces before brazing. A coating of flux on the joint area shields the surfaces from air, preventing oxide formation. The flux also dissolves and absorbs any oxides that form during heating, or that were not completely removed in the cleaning procedure. (4) *Assembly for brazing*—After cleaning and fluxing, the parts must be held in firm position for brazing. The simplest way to hold parts together is by gravity, providing the shape and weight of the parts permit. Where there are several assemblies to braze and their configurations are too complex for self-support or clamping, a brazing support fixture is indicated. (5) *Brazing the assembly*—which involves the application of heat. Commonly a torch is used to furnish the heat. Well suited to automation (providing other variables permit) is furnace brazing, a method that has been used quite successfully in the manufacture of heavy-duty electrical contacts. If the furnace has an inert atmosphere, fluxing can be eliminated. (6) *Cleaning the brazed joint*.

The physical properties of the filler metal are based on its metallurgical properties. The composition will determine whether the filler metal is compatible with the metals being joined—capable of wetting them and flowing completely through the joint area. There are also special

Brazing joins all these configurations with equal ease

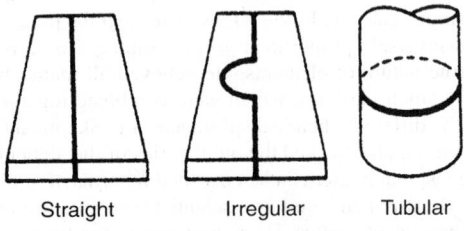

Straight   Irregular   Tubular

Fig. 2.   Representative brazing joint configurations. (*Lucas-Milhaupt, Inc., A Handy & Harman Company*)

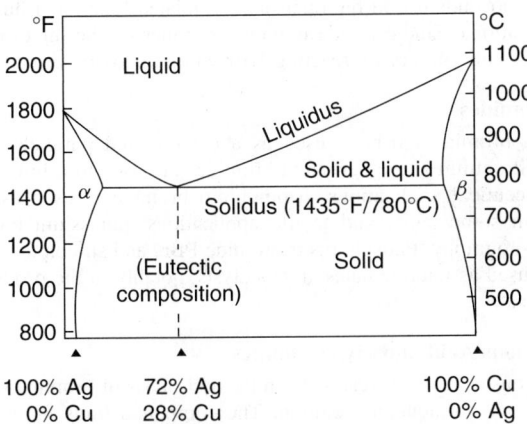

Fig. 3.   Silver-copper equilibrium diagram. For a 72% silver-28% copper alloy, liquidus and solidus temperatures are the same. Alloys to the left or to the right of this eutectic composition do not go directly from a solid to a liquid state, but pass through a "mushy" range where the alloy consists of both solid and liquid states. Some brazing alloys are formulated to melt in a narrow temperature range. They are very fluid when melted and thus flow easily into close-clearance joints. Other brazing alloys are formulated for a wide melting range. Their relatively sluggish flow is desirable for filling wide gaps, or for building up stress-distribution fillets at the joint edges. (*Lucas-Milhaupt, Inc., A Handy & Harman Company*)

requirements, such as brazing in a vacuum where a filler metal free of any volatile elements, such as cadmium or zinc, must be selected. Some electronic components require filler metals of exceptionally high purity. Corrosion-resistant joints require filler metals that are both corrosion-resistant and compatible with the base metals jointed. The melting behavior of the filler metal is based on its metallurgical composition. Since filler metals are alloys, they usually do not melt in the same manner as pure metals, which go from a solid to a liquid state at one temperature. An important exception is the eutectic alloys which do melt in the same way as pure metals. One such eutectic composition is a simple silver-copper alloy (72% Ag and 28% Cu), also known as *Harman's Braze 720*, which melts completely at a single temperature, 780°C (1435°F). The melting behavior is shown by Fig. 3.

In all brazing applications, a critical factor is the "flow point" of the brazing filler material. This is the temperature above which the filler metal is liquid and flows readily, as distinguished from the melting point, when melting begins. Since in brazing the base metals must not be melted, a filler metal whose flow point is lower than the melting point of either of the base metals being joined must be used. A practical problem sometimes arises wherein there are two brazed joints in relatively close proximity (Fig. 4). So that the second brazing operation will not adversely affect the

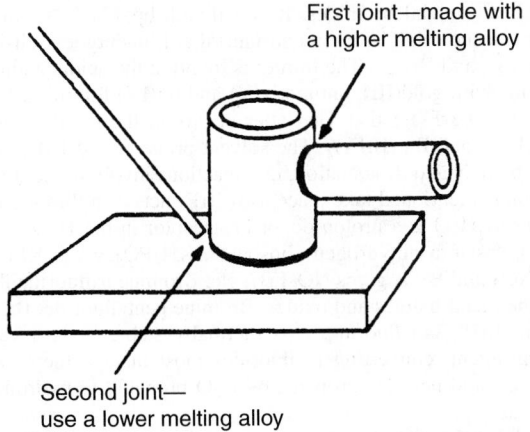

Fig. 4.   Demonstration of the use of a higher melting alloy for first joint and a lower melting alloy for an adjacent second joint so that first joint will not be disturbed when the second joint is made. (*Lucas-Milhaupt, Inc., A Handy & Harman Company*)

first operation, the filler metal selected for the second joint will have a flow point that is lower than that used for the first joint.

Brazing is widely used in assembling heat exchangers, piping systems, electrical products, cutting tools, bicycles, and control instruments, among many other applications.

BREEDER REACTOR.   See Nuclear Reactor.

BREWSTER ANGLE.   The Brewster angle, or polarizing angle, of a dielectric is that angle of incidence for which a wave polarized parallel to the plane of incidence is wholly transmitted (no reflection). An unpolarized wave incident at this angle is therefore resolved into a transmitted partly-polarized component and a reflected perpendicularly-polarized component. See Fig. 1.

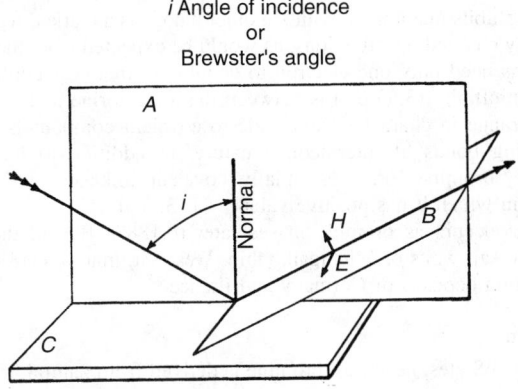

Fig. 1.   Brewster angle. *A*, incident plane (plane of polarization or plane of magnetic vector, after reflection); *B*, plane of vibration (plane of electric vector, after reflection); *C*, reflecting surface (dielectric)

BRIGHTENERS.   See Detergents.

BRILLOUIN ZONE.   See Fermi Surface.

BRINELL HARDNESS TEST.   The standard method of measuring the hardness of metals. The smooth surface of the metal is dented by a steel ball under force. The standard load and time are 500 kg for 60 seconds for soft metals and 3000 kg for 30 seconds for steel and other hard metals. The size (diameter) of the resulting dent is measured, and the hardness determined from a chart or formula.

BRIX SCALE.   See Specific Gravity.

BROCHANTITE.   A mineral composed of basic copper sulfate corresponding to the formula $Cu_4(SO_4)(OH)_6$, crystallizing in the monoclinic system in needle-like prisms, or forming druses or masses. Hardness, 3.5–4; specific gravity, 3.9; vitreous luster; color, green; streak, green; transparent to translucent.

Brochantite is a secondary mineral occurring in the oxidized zones with other copper minerals, and is found in the Ural Mountains, in Rumania, in Sardinia; Cornwall, England; Chile. In the United States this mineral has been found at Bisbee, Arizona, Utah, in the Tintic District and in Inyo County, California. Brochantite was named for Brochant de Villiers.

BROMINE.   [CAS: 7726-95-6]. Chemical element, symbol Br, at. no. 35, at. wt. 79.904, periodic table group 17 (halogens), mp −7.2°C, bp 58.8°C, density 3.12 g/cm³ (20°C). Bromine is one of the few elements that is liquid at standard conditions. The element volatilizes readily at room temperature to form a red vapor that is very irritating to the eyes and throat. Liquid bromine causes painful lesions upon contact with the flesh. Bromine has two stable isotopes ^{79}Br and ^{81}Br. Elemental bromine finds limited application as a chemical intermediate and as a sanitizing, disinfecting, and bleaching agent. Both the inorganic and organic compounds of the element find extensive commercial usage. Bromine was discovered in 1826 by Antoine-Jérôme Balard, who identified the element as a component of seawater bitterns. Electronic configuration $1s^2 2s^2 2p^6 3s^2 3p^6 3d^{10} 4s^2 4p^5$. Ionic radius Br^- 1.97 A, Br^{7+} 0.39 Å. Covalent radius 1.193$_5$. First ionization potential 11.84 eV; second 19.1 eV; third 25.7 eV. Oxidation potentials $2Br^- \rightarrow Br_2(1) + 2e^-$, -1.065 V; $2Br^- \rightarrow Br_2(aq) + 2e^-$, -1.087 V; $Br^- + H_2O \rightarrow HBrO + H^+ + 2e^-$, -1.33 V; $Br^- + 3H_2O \rightarrow BrO_3^- + 6H^+ + 6e^-$, -1.44 V; $\frac{1}{2}Br_2 + 3H_2O \rightarrow BrO_3^- + 6H^+ + 5e^-$, -1.52 V; $\frac{1}{2}Br_2 + H_2O \rightarrow HBrO + H^+ + e^-$, -1.59 V; $Br^- + 6OH^- \rightarrow BrO_3^- + 3H_2O + 6e^-$, -0.61 V; $Br^- + 2OH^- \rightarrow BrO^- + H_2O + 2e^-$, -0.70 V.

Other important physical properties of bromine are described under **Chemical Elements**.

Bromine is only moderately soluble in H_2O (3.20 g/100 ml) but markedly so in nonpolar solvents, e.g., carbon tetrachloride, as is consistent with the covalent character of the Br—Br bond. It dissolves more readily in alkali bromide solutions due to the formation of the tribromide ion (Br_3^-), and in certain associated solvents, such as concentrated H_2SO_4 and ethyl alcohol. Its aqueous solution is more stable than that of chlorine, since the tendency of Br_2 to hydrolyze to unstable hypobromous acid and hydrogen bromide is less than the corresponding reaction for chlorine. Bromine exhibits in common with the other halogens a marked readiness to form singly charged negative ions, as would be expected from the fact that these atoms need only one electron to acquire an inert gas configuration. Its electron affinity (3.53 eV) is between that of chlorine and iodine. The bromides range in character from ionic to covalent compounds, many of them having bonds of intermediate nature. In addition to its negative univalence, bromine forms essentially covalent linkages with negative elements, in which it has positive valences 1, 3, and 5.

Bromine occurs as bromide in seawater (0.188% Br), in the mother liquor from salt wells of Michigan, Ohio, West Virginia, Arkansas, and in the potassium deposits of Germany and France.

Production

In the United States, nearly all bromine is derived from natural brines. The Arkansas brines which contain a minimum of 4000 ppm bromide account for over half of this production. Recovery is effected by a *steaming-out* process. After heating fresh brine, the solution is fed to the top of a tower. Chlorine and steam are injected at the bottom of the tower. The chlorine oxidizes the bromide and displaces one resultant bromine from solution. For brines of lower concentration, air instead of steam is used to sweep out the bromine vapors after chlorination.

Hydrogen Bromide and Hydrobromic Acid

HBr, is formed directly from the elements, effectively when catalyzed by sunlight, by heated charcoal or platinum, or more conveniently by hydrolysis of phosphorus tribromide. Treatment of bromides with H_2SO_4 yields mixtures of HBr and bromine. The H—Br bond is considered to be partly covalent. Hydrobromic acid is a strong acid in aqueous solution. Its salts are the bromides, all of which are water-soluble except those of copper(I), silver, gold(I), mercury(I), thallium(I) and lead(II), the divalent ions of the elements of the second and third transition series, and the salts of the heavy alkali ions with many bromo-complex anions, e.g., Cs_2PtBr_6, $RbAuBr_4$, etc. The main uses of HBr and hydrobromic acid are in the production of alkyl bromides (by replacement of alcoholic hydroxyl groups or by addition to olefins) and inorganic bromides.

Sodium Bromide

This is a high-tonnage chemical and one of the most important of the bromide salts commercially. High-purity grades are required in the formulation of silver bromide emulsions for photography. The compound, usually in combination with hypochlorites, is used as a bleach, notably for cellulosics. The production of sodium bromide simply involves the neutralization of HBr with NaOH or with sodium carbonate or bicarbonate.

Calcium Bromide

Because of its ready solubility, calcium bromide forms solutions of high density which when properly formulated are finding increasing use as functional fluids in oil well completion and packing applications.

Lithium Bromide

LiBr finds use as a desiccant in industrial air conditioning systems.

Zinc Bromide

$ZnBr_2$ is used as a rayon-finishing agent, as a catalyst, as a gamma-radiation shield in nuclear reactor viewing windows, and as an absorbent in humidity control. It too finds use in high density formulated functional fluids in oil well applications. Zinc bromide is prepared either by the direct reduction of bromine with zinc, or by reacting HBr with zinc oxide or carbonate.

Other Bromides

Aluminum bromide $AlBr_3$ is used as a catalyst and parallels $AlCl_3$ in this role. Strontium and magnesium bromides are used to a limited extent in pharmaceutical applications. Ammonium bromide is used as a flame retardant in some paper and textile applications; potassium bromide is used in photography. Phosphorus tribromide PBr_3 and silicon tetrabromide $SiBr_4$ are used as intermediates and catalysts, notably in the production of phosphite esters.

Hypobromous Acid and Hypobromites

Hypobromous acid HOBr results from the hydrolysis of bromine with H_2O and exists only in aqueous solution. The compound finds limited use as a germicide and in water treatment; also it can be used as an oxidizing or brominating agent in the production of certain organic compounds. Although hypobromous acid is low in bromine content, concentrated hypobromite solutions can be formed by adding bromine to cooled solutions of alkalis.

Bromic Acid and Bromates

Bromic acid, $HBrO_3$, can exist only in aqueous solution. Bromic acid and bromates are powerful oxidizing agents. Bromic acid decomposes into bromine, oxygen, and water. Many oxidizing agents, e.g., hydrogen peroxide, hypochlorous acid, and chlorine convert Br_2 or Br^- solutions to bromates. The decomposition reactions of bromates vary considerably. Lead(II) bromate and copper(II) bromate give the metal oxides and Br^-; silver, mercury(II) and potassium bromates give the metal ion, Br^- and oxygen, while zinc, magnesium, and aluminum bromates give the metal oxide, Br_2 and oxygen.

Halogen Compounds

Bromine forms a number of compounds with the other halogens. Its binary iodine compounds are discussed under iodine; other interhalogen compounds of bromine include bromine monochloride, bromine monofluoride, bromine trifluoride, and bromine pentafluoride. The nonexistence of higher chlorides of bromine, differing from iodine, can readily be explained in terms of the oxidation potential of Br(III) and Br(V). The monochloride, bromine chloride, BrCl, exists in pure state only at very low temperatures in the solid form. Dissociation in the gas phase is approximately 40% at 25 °C and increases slowly with increasing temperature; less than 20% occurs in the liquid phase. With many substrates, bromine chloride reacts much more rapidly than does bromine itself to introduce bromine substituents. Bromine monofluoride, BrF, is also somewhat unstable, decomposing spontaneously at 50 °C to Br_2, BrF_3, and BrF_5. It has never been prepared pure since it is always in equilibrium with Br_2 and BrF_5. It is a gas at room temperature, reacting readily with water, phosphorus, and the heavy metals. Bromine trifluoride, BrF_3, is much more stable than the monofluoride. It is obtained directly from the elements at 10 °C, or by fluorination of univalent heavy metal bromides. It is a liquid, bp 127.6 °C, mp 8.8 °C. There is evidence (high Trouton constant) that it undergoes self-ionization to form BrF_2^+ and BrF_4^-. The former is found in the acidic addition products it forms with gold(III), antimony(V) and tin(IV) fluorides, BrF_2AuF_4, BrF_2SbF_6 and $(BrF_2)_2SnF_6$. The latter occurs in the tetrafluorobromates, such as $KBrF_4$ and Ba $(BrF_4)_2$. The solvent properties of BrF_3 are consistent with its indicated dissociation, i.e., reactions involving the two classes of compounds mentioned take place as if BrF_3 acts as a fluoride ion donor or acceptor as H_2O is a proton donor or acceptor in the H_2O system. For example, potassium dihydrogen phosphate, KH_2PO_4 gives KPF_6, a mixture of HNO_3 and B_2O_3 gives NO_2FB_4, etc. Bromine trifluoride fluorinates many of the metal halides and oxides. Bromine pentafluoride, BrF_5, is prepared from BrF_3 and fluorine. It is thermally stable. It is a very active fluorinating agent, converting to fluorides most metals, their oxides and other halides, and being hydrolyzed by H_2O probably to hydrofluoric and bromic acids.

The polyhalide complexes of bromine include $(PBr_4)(IBr_2)$ formed by reaction of phosphorus pentabromide and iodine monobromide, and dissociating in certain organic polar solvents to the ions PBr_4^+ and IB_2^-. Other polyhalides include NH_4IBr_2, $[(CH_3)_4N][IBr_2]$, $Cs[IFBr]$, $Rb[IClBr]$, and $Cs[IClBr]$. Most of these compounds hydrolyze readily, ionize in polar nonreacting solvents to the corresponding polyhalide ions, and decompose on heating to give the metal halide of greatest lattice energy.

Oxides

In binary combination with oxygen, bromine forms at least three compounds. Bromine(I) oxide, Br_2O, is a dark brown solid that is stable only in the dark below $-40°C$. It is prepared by passing dry gaseous bromine through dry mercury(II) oxide and sand. Bromine(I) oxide, in carbon tetrachloride at low temperatures, reacts with alkali hydroxide to give hypobromites. Bromine(IV) oxide, BrO_2, is obtained by reaction of the elements in a cooled electric discharge tube. It is yellow, and is stable only at low temperatures. The compound appearing in the older literature as tribromine octoxide, Br_3O_8, is actually bromine trioxide, BrO_3, or dibromine hexoxide, Br_2O_6 (cf. chlorine). It is obtained from the low temperature, low pressure reaction of ozone and bromine; it is stable only at low temperatures and is soluble in H_2O with decomposition. Bromine(VII) oxide may be present among the decomposition products of BrO_2, or BrO_3, but no other evidence for its existence has been found.

Organic Bromine Compounds

Commercially important organic bromine compounds include: (1) methyl bromide CH_3Br formed by reacting methanol with HBr or hydrobromic acid. The compound is a highly toxic gas at standard conditions. Because of its toxicity, it is used as a soil and space fumigant. In many organic syntheses, the compound is used as a methylating agent; (2) ethylene dibromide (1,2-dibromoethane) is used in combination with lead alkyls as an antiknock agent for gasoline. The compound also is used as a fumigant; (3) methylene chlorobromide (bromochloromethane) is a low-boiling liquid of low toxicity and is useful as a fire-extinguishing agent in portable equipment and aircraft; (4) bromotrifluoromethane is increasingly employed as a fire extinguishant in permanently installed systems protecting high-cost installations, such as computer rooms, where its low toxicity and especially its freedom from corrosivity are important considerations; (5) acetylene tetrabromide (1,1,2,2-tetrabromoethane) is made by adding bromine to acetylene. The compound is comparatively dense and finds use as a gage fluid and in specific gravity separations of solids. It is also used as part of the catalyst system for the oxidation of p-xylene to terephthalic acid; (6) tris(2,3-dibromopropyl) phosphate may be prepared by the reaction of phosphorus oxychloride with 2,3-dibromopropanol, or by the addition of bromine to triallyl phosphate. This viscous fluid was used as a flame retardant in a number of polymer systems, but has been displaced from most, or all, of these uses, because it is a mutagen and suspect carcinogen; (7) tetrabromobisphenol A is produced by the direct bromination of bisphenol A. The compound is used extensively as a flame retardant and usually is incorporated into the polymer backbone structure of epoxy resins, unsaturated polyesters, and polycarbonates; (8) tetrabromophthalic anhydride, made by the catalytic bromination of phthalic anhydride in fuming H_2SO_4, also finds use as a reactive flame retardant in the formulation of polyol systems (for polyurethane foams) and in unsaturated polyesters; (9) decabromodiphenyl ether, and others of the lower brominated diphenyl ethers, are finding increasing use as flame retardants in a variety of thermoplastic polymer systems; (10) vinyl bromide has also found major use in flame-retarding modacrylic textile fibers when introduced as a co-monomer in the synthesis of the polymer itself.

Alkanes and arenes, e.g., ethane and benzene, respectively, react with bromine by *substitution* of bromine for hydrogen (hydrogen bromide also formed)—ethane to yield ethyl bromide C_2H_5Br plus further substitution products; benzene, in the presence of a catalyst, e.g., iodine, phosphorus, iron, to yield bromobenzene C_6H_5Br plus further substitution products; toluene, under like conditions to benzene, to yield orthobromotoluene and parabromotoluene $CH_3C_6H_4Br$ plus further substitution products, but at the boiling temperature, in sunlight, dry, and in the absence of a catalyst, to yield alkyl side-chain substitution products, benzyl bromide $C_6H_5CH_2Br$, benzal bromide $C_6H_5CHBr_2$, and benzotribromide, $C_6H_5CBr_3$.

Alkenes, alkynes, and arenes, e.g., ethylene, acetylene and benzene, respectively, react (1) with bromine by *addition*, e.g., ethylene dibromide $C_2H_4Br_2$ (1,2), acetylene tetrabromide $C_2H_2Br_4$ 1,1,2,2, hexabromocyclohexane $C_6H_6Br_6$; also carbon monoxide yields carbonyl bromide $COBr_2$; (2) with hypobromous acid by *addition*, e.g., olefins form, for example,

ethylene bromohydrin $CH_2Br·CH_2OH$; (3) with hydrogen bromide by *addition*, to form, for example, ethyl bromide $CH_3·CH_2Br$ from ethylene. When the two olefin carbons have unequal numbers of hydrogens, the carbon to which one bromide or one hydroxyl attaches can be controlled by the reaction conditions.

Oxygen-function compounds, e.g., ethyl alcohol, acetaldehyde, acetone, acetic acid, react (1) with bromine, to form *bromo-substituted* corresponding or related *compounds*, e.g., ethyl alcohol or acetaldehyde to yield bromal $CBr_3·CHO$, acetone to yield bromoacetone $CH_2Br·CO·CH_3$; acetic acid to yield, at the boiling temperature, dry, and in the absence of a catalyst, monobromoacetic acid $CH_2Br·COOH$, dibromoacetic acid $CHBr_2·COOH$, tribromoacetic acid $CBr_3·COOH$, the substitution taking place on the alpha-carbon (the carbon next to the carboxyl-group-COOH), (2) with phosphorus bromides, to form corresponding *bromides*, e.g., ethyl bromide C_2H_5Br, ethylidene dibromide CH_3CHBr_2, acetone bromide $(CH_3)_2CBr_2$, acetyl bromide CH_3COBr, (3) with hydrobromic acid, concentrated, alcohol forms the corresponding bromide.

Bromoform is made by reaction of acetone or ethyl alcohol with sodium hypobromite; carbon tetrabromide by reaction of CS_2 plus bromine Br_2 in the presence of iron, heated; or by one reaction of bromoform with aqueous hypobromite solutions. Use is made of the diazo-reaction to introduce bromine into aryl compounds.

Many of the bromo-compounds are used as reagents or as intermediate compounds in organic chemistry. When alkyl bromocompounds are treated (1) with NaOH dissolved in alcohol, hydrogen bromide is removed, e.g., ethyl bromide $CH_3·CH_2Br$ yields ethylene $CH_2:CH_2$, ethyl dibromide $CH_2Br·CH_2Br$ yields acetylene $CH:CH$; (2) with magnesium or zinc and alcohol, bromine is removed, e.g., ethylene dibromide $CH_2Br·CH_2Br$ yields ethylene $CH_2:CH_2$, acetylene tetrabromide $CHBr_2:CHBr_2$, yields acetylene $CH:CH$.

Additional Reading

Bell, C.H., N. Price, and B. Chakrabarti: *The Methyl Bromide Issue,* John Wiley & Sons, Inc., New York, NY, 1997.

Cheremisinoff, N.P.: *Handbook of Industrial Toxicology and Hazardous Materials,* Marcel Dekker, Inc., New York, NY, 1999.

Hathaway, G.J., N.H. Proctor, and J.P. Hughes: *Proctor and Hughes' Chemical Hazards of the WorkPlace,* 4th Edition, John Wiley & Sons, Inc., New York, NY, 1997.

Hawley, G.G. and R.L. Lewis: *Hawley's Condensed Chemical Dictionary,* 13th Edition, John Wiley & Sons, Inc., New York, NY, 1999.

Hermann, W.A.: *Synthetic Methods of Organometallic Inorganic Chemistry: Halogen Compounds Rare Metals,* Thieme, New York, NY, 1999.

Krebs, R.E.: *The History and Use of Our Earth's Chemical Elements: A Reference Guide,* Greewood Publishers Group, Inc., Westport, CT, 1998.

Lefevre, M.J. and S. Conibear: *First Aid Manual for Chemical Accidents,* 2nd Edition, John Wiley & Sons, Inc., New York, NY, 1997.

Lewis, R.J.: *SAX's Dangerous Properties of Industrial Materials,* 10th Edition, John Wiley & Sons, Inc., New York, NY, 1999.

Lide, D.R.: *CRC Handbook of Chemistry and Physics,* 84th Edition, CRC Press, LLC., Boca Raton, FL, 2003.

Newton, D.E.: *Chemical Elements: From Carbon to Krypton,* Gale, Group, New York, NY, 1998.

Perry, R.H. and D.W. Green, et al.: *Perry's Chemical Engineers' Handbook,* 7th Edition, McGraw-Hill Companies, Inc., New York, NY, 1997.

Roche, L.P.: *The Chemical Elements: Chemistry, Physical Properties, and Uses in Science and Industry,* Prentice-Hall, Inc., Upper Saddle River, NJ, 1997.

Williams, P.L., R.C. James, et al.: *Principles of Toxicology: Environmental and Industrial Applications,* 2nd Edition, John Wiley & Sons, Inc., New York, NY, 2000.

BROOKITE. Brookite, composed of titanium dioxide, TiO_2, is an orthorhombic mineral of the same chemical composition as rutile and octahedrite. It was named for the English mineralogist H.J. Brooke. See also **Rutile**; and **Titanium Dioxide**.

BRONZE. See **Copper**.

BROWN, HERBET C. (1912–). An English-born chemist who was the recipient of the Nobel prize for chemistry with Wittig, Georg in 1979. Via his work in organic synthesis, he discovered new routes to add substituents to olefins selectively. His early education was irregular and disjointed as a result of family circumstances and the economic depression of the 1930s. He eventually received his Ph.D. from the University of Chicago. The reduction of carbonyl compounds with diborane was the topic of his thesis. The bulk of his career has been spent at Purdue University. http://www.chem.purdue.edu/faculty.htm

See also **Boron**.

BROWNIAN MOTION. The random movement observed among microscopic particles suspended in a fluid medium. The phenomenon was observed in 1827 with suspensions in liquids, colloids, by Robert Brown, English botanist, who is said to have attributed it to living organisms. Not until the kinetic theory was developed was it generally understood to be due to the thermal agitation of the suspending medium. A smoke particle floating in the air, for example, is battered on all sides by the high-speed air molecules. The resultant displacement is for the most part nearly zero, but there are statistical inequalities which now and then reach such magnitude as to produce motions visible in a high-powered microscope, and which result in an irregular migration of the particle. In fact, such particles may be regarded essentially as huge molecules, with mean square speeds of thermal motion proportionately smaller as their masses are larger than that of the true molecules of the surrounding medium.

In a series of papers published from 1905 to 1908, Einstein successfully incorporated the suspended particles into the molecular-kinetic theory of heat. He treated the suspended particles as being in every way identical to the suspending molecules except for the vast difference of their size. He set forth several relationships that were capable of experimental verification and he invited experimentalists to "solve" the problem.

Several workers undertook this task. The most notable of these was Perrin. Perrin's special success was due to his technique for preparing particles to suspend that were of uniform and known size. The uniformity was achieved by fractional centrifuging, and the size was established by noting that they could be coagulated into "chains" whose length could be measured and whose "links" could be counted. The microscopic observation of these uniform particles enabled Perrin and his students to verify the Einstein results and to make four independent measurements of Avogadro's number. See Fig. 1. These results not only established an understanding of Brownian movement, but also they silenced the last critics of the atomic view of matter.

Probably the simplest example of Perrin's experiments was his test of the Law of Atmospheres. If it is assumed that air is at rest and has the same temperature from ground level upward, it can be shown that the pressure (and concentration) of the air falls off exponentially with increasing altitude. For particles of mass m and density ρ suspended in a medium of density ρ' at absolute temperature T, the ratio of the particle concentrations n_1 to n_2 at heights h_1 and h_2 is given by

$$\frac{n_1}{n_2} = \exp\left[-\frac{mg(\rho - \rho')N_0(h_1 - h_2)}{\rho RT}\right]$$

where N_0 is Avogadro's number, g is the acceleration of gravity, and R is the universal gas constant. Although the concentration of air varies slowly with height, the concentration of the relatively heavy particles varied significantly over a height change of a few millimeters. By observing the concentration variation as a function of height, all quantities in the given equation were known except Avogadro's number, which could, therefore, be determined.

An exceptionally interesting and complete paper on Brownian motion was prepared in 1985 by B.H. Lavenda (University of Camerino, Italy). In a very workmanlike manner the author reviews the impact of the Brownian motion concept on other important topics of science, as previously mentioned, and continuing to the present time. In recent years, study of Brownian motion has led to the invention of mathematical techniques for the general investigation of probabilistic processes. For example, such techniques have been applied in the control of electromagnetic "noise," and they have contributed to the comprehension of the dynamics of star clustering, and the development and adaptation of ecological systems, not to mention studies of stock and commodity prices.

In another paper, R. Kuho (Keio University, Japan) illustrates in a rather technical and mathematical fashion the relationship between Brownian motion and non-equilibrium statistical mechanics. In this paper, the author describes the linear response theory, Einstein's theory of Brownian motion, coarse-graining and stochastization, and the Langevin equations and their generalizations.

Additional Reading

Borodin, A.N. and P. Salminen: *Handbook of Brownian Motion: Facts and Formulae,* Birkhauser Verlag, Cambridge, MA, 1996.

Chung, Kai Lai, and Z. Zhao: *From Brownian Motion to Schrodinger's Equation,* Springer-Verlag, Inc., New York, NY, 1995.

Karatzas, I.: *Brownian Motion and Stochastic Calculus,* Springer-Verlag, Inc., New York, NY, 1998.

Kazuaki, T.: *Brownian Motion and Index Formulas for the de Rham Complex,* John Wiley & Sons, Inc., New York, NY, 1999.

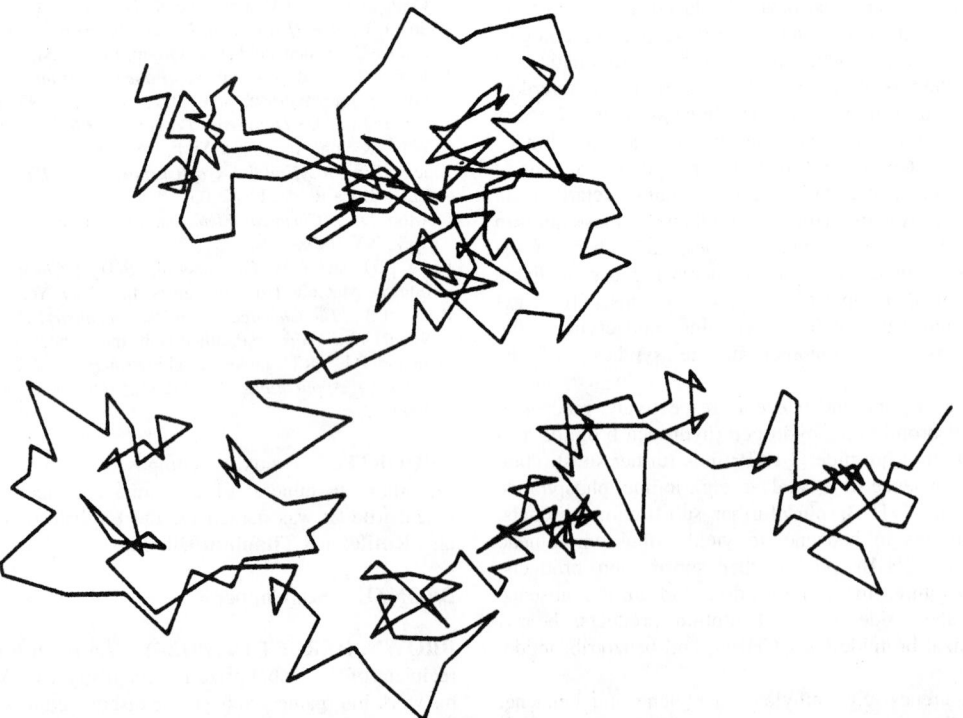

Fig. 1. Replica of plot made by Jean Baptiste Perrin (France) in 1912 of a microscopic particle suspended in water. The position of the particle was recorded at half-minute intervals. At the time, Perrin observed that "only a very meager idea of the extraordinary discontinuity of the actual trajectory" can be obtained in this experimental fashion

Kubo, R.: "Brownian Motion and Non-equilibrium Statistical Mechanics," *Science*, **223**, 330–334 (1986).

Lavenda, B.H.: "Brownian Motion," *Sci. Amer.*, **252**(2), 70–85 (February 1985).

Neuenschwander, D.: *Probabilities on the Heisenberg Group: Limit Theorems and Brownian Motion*, Vol. 163," Springer-Verlag, Inc., New York, NY, 1996.

Perrin, J.: *Atoms*, Ox Bow Press, Woodbridge, MA, 1990.

Sznitman, Alain-Sol: *Brownian Motion, Obstacles, and Random Media*, Springer-Verlag, Inc., New York, NY, 1998.

BRUCITE. The mineral brucite is magnesium hydroxide corresponding to the formula $Mg(OH)_2$; iron and manganese may occasionally be present. The crystals are usually tabular rhombohedrons of the hexagonal system; it may also occur fibrous or foliated. Brucite has one perfect cleavage parallel to the prism base; hardness, 2.5; specific gravity, 2.39; luster, pearly to vitreous; commonly white but may be gray, bluish or greenish; transparent to translucent. Brucite is a secondary mineral found with serpentine and metamorphic dolomites. It has been found in Italy, Sweden, and the Shetland Islands; and in the United States in New York, Pennsylvania, Nevada and California. Brucite was named in honor of Archibald Bruce, an American physician.

BUBBLE CAP. See **Distillation**.

BUCHNER, EDUARD (1860–1917). A German chemist who was awarded the Nobel prize for chemistry in 1907. His works included the synthesis of diiodoacetamid through alcoholic fermentation caused by enzymes, as well as the discovery of zymase, the first enzyme to be isolated. He received his Ph.D. at the University of Munich, where he became a lecturer. Later, he taught and performed research at Tübingen, Berlin, and Würzburg.

BUCKMINSTERFULLERENE (Buckyballs). C_{60}. Spherical aromatic molecule with a hollow truncated-icosahedron structure, similar to a soccer ball. First reported in the mid 1980s. Capable of enclosing ions or atoms in a host-guest relationship. See also **Carbon Compounds**.

BUFFER (Chemical). When acid is added to an aqueous solution, the pH (hydrogen ion concentration) falls. When alkali is added, it rises. If the original solution contains only typical salts without acidic or basic properties, this rise or fall may be very large. There are, however, many other solutions that can receive such additions without a significant change in pH. The solutes responsible for this resistance to change in pH, or the solutions themselves, are known as *buffers*. A weak acid becomes a buffer when alkali is added, and a weak base becomes a buffer upon the addition of acid. A simple buffer may be defined, in Brønsted's terminology, as a solution containing both a weak acid and its conjugate weak base.

Buffer action is explained by the mobile equilibrium of a reversible reaction:

$$A + H_2O \rightleftarrows B + H_3O^+$$

in which the base B is formed by the loss of a proton from the corresponding acid A. The acid may be a *cation*, such as NH_4^+, a *neutral molecule*, such as CH_3COOH, or an *anion*, such as $H_2PO_4^-$. When alkali is added, hydrogen ions are removed to form water, but so long as the added alkali is not in excess of the buffer acid, many of the hydrogen ions are replaced by further ionization of A to maintain the equilibrium. When acid is added, this reaction is reversed as hydrogen ions combine with B to form A.

The pH of a buffer solution may be calculated by the mass law equation

$$pH = pK' + \log \frac{C_B}{C_A}$$

in which pK' is the negative logarithm of the apparent ionization constant of the buffer acid and the concentrations are those of the buffer base and its conjugate acid.

A striking illustration of effective buffer action may be found in a comparison of an unbuffered solution, such as 0.1 M NaCl with a neutral phosphate buffer. In the former case, 0.01 mole of HCl will change the pH of 1 liter from 7.0 to 2.0, while 0.01 mole of NaOH will change it from 7.0 to 12.0. In the latter case, if 1 liter contains 0.06 mole of Na_2HPO_4 and 0.04 mole of NaH_2PO_4, the initial pH is given by the equation:

$$pH = 6.80 + \log \frac{0.06}{0.04} = 6.80 + 0.18 = 6.98$$

After the addition of 0.01 mole of HCl, the equation becomes:

$$pH = 6.80 + \log \frac{0.05}{0.05} = 6.80$$

while, after the addition of 0.01 mole of NaOH, it is

$$pH = 6.80 + \log \frac{0.07}{0.03} = 6.80 + 0.37 = 7.17.$$

The buffer has reduced the change in pH from ±5.0 to less than ±0.2.

Figure 1 shows how the pH of a buffer varies with the fraction of the buffer in its more basic form. The buffer value is greatest where the slope of the curve is least. This is true at the midpoint, where $C_A = C_B$ and $pH = pK'$. The slope is practically the same within a range of 0.5 pH unit above and below this point, but the buffer value is slight at pH values more than 1 unit greater or less than pK'. The curve has nearly the same shape as the titration curve of a buffer acid with NaOH or the titration curve of a buffer base with HCl. Sometimes buffers are prepared by such partial titrations instead of by mixing a weak acid or base with one of its salts. Certain "universal" buffers, consisting of mixed acids partly neutralized by NaOH, have titration curves that are straight over a much wider pH interval. This is also true of the titration curves of some polybasic acids, such as citric acid, with several pK' values not more than 1 or 2 units apart. Other polybasic acids, such as phosphoric acid, with pK' values farther apart, yield curves having several sections, each somewhat similar to the accompanying curve. At any pH, the buffer value is proportional to the concentration of the effective buffer substances or groups. See also **pH (Hydrogen Ion Concentration)**.

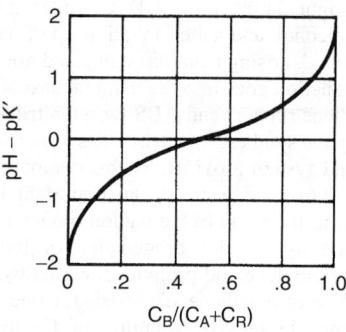

Fig. 1. The pH of a simple buffer solution. Abscissas represent the fraction of the buffer in its more basic form. Ordinates are the difference between pH and pK'

Table 1 gives approximate pK' values, obtained from the literature, for several buffer systems.

TABLE 1. REPRESENTATIVE BUFFER SOLUTIONS

Constituents	pK'
H_3PO_4; KH_2PO_4	2.1
HCOOH; HCOONa	3.6
CH_3COOH; CH_3COONa	4.6
KH_2PO_4; Na_2HPO_4	6.8
HCl; $(CH_2OH)_3CNH_2$	8.1
$Na_2B_4O_7$; HCl or NaOH	9.2
NH_4Cl; NH_3	9.2
$NaHCO_3$; Na_2CO_3	10.0
Na_2HPO_4; NaOH	11.6

Buffer substances which occur in nature include phosphates, carbonates, and ammonium salts in the earth, proteins of plant and animal tissues, and the carbonic-acid-bicarbonate system in blood. See also **Acid-Base Regulation (Blood)**.

Buffer action is especially important in biochemistry and analytical chemistry, as well as in many large-scale processes of applied chemistry. Examples of the latter include the manufacture of photographic materials, electroplating, sewage disposal, agricultural chemicals, and leather products.

BULKING AGENTS (Foods). See **Bodying and Bulking Agents (Foods)**.

BUNSEN, ROBERT WILHELM (1811–1899). Born in Germany, Bunsen is remembered chiefly for his invention of the laboratory burner named after him. He engaged in a wide range of industrial and chemical research, including blast-furnace firing, electrolytic cells, separation of metals by electric current, spectroscopic techniques (with Kirchhoff), and production of light metals by electrical decomposition of their molten chlorides. He also discovered two elements, rubidium and cesium.

BURETTE. A long, slender, graduated glass tube used in volumetric analysis, particularly titrations. An average-size burette will contain 50 milliliters of reagent and will be graduated down to tenths of a milliliter. The amount of reagent required to effect a given reaction, such as neutralization of an acid with a base, is determined by taking the difference between the starting and ending readings of the graduations. So that the end-point can be determined with maximum precision, the bottom portion of the burette is in the form of a tapered, narrow tip, such that the exiting droplets are quite small. Flow of reagent from the burette is controlled by a ground glass stop cock which is an intimate part of the total glass assembly. For laboratories where numerous titrations requiring the same reagent are made routinely, a so-called automatic burette is available. Essentially, this simplifies the reagent refilling operation.

Other types of microanalytical apparatus, such as chromatography, have replaced manual titration methods, particularly where a procedure can be tooled for making tens of thousands of similar determinations automatically.

BUTADIENE. [CAS: 106-99-0]. $CH_3CH:C:CH_2$, 1,3-butadiene (methylallene), formula weight 54.09, bp −4.41°C, sp gr 0.6272, insoluble in H_2O, soluble in alcohol and ether in all proportions. Butadiene is a very reactive compound, arising from its conjugated double-bond structure. Most butadiene production goes into the manufacture of polymers, notably SBR (styrene-butadiene rubber) and ABS (acrylonitrile-butadiene-styrene) plastics. Several organic syntheses, such as Diels-Alder reaction, commence with the double-bond system provided by this compound.

Butadiene came into prominence as an important industrial chemical during World War II as the result of the natural rubber shortage. Originally, butadiene was made by the dehydrogenation of butylenes. Later, the naphtha cracking for ethylene and propylene, with a byproduct C_4 stream, created another source of butadiene. The basis for one butadiene recovery process is the change in relative volatility of C_4 hydrocarbons in the presence of acetonitrile solvent. The latter makes the separation easier. The C_4 mixed charge goes to an extractive distillation column where it is separated in a solvent environment into a solvent-butadiene stream and a byproduct butane-butylenes stream (overhead). The acetonitrile is recovered from the butane-butylenes. Butadiene is stripped from the fat solvent, after which it goes to a postfractionator for recovery as a 99.5% pure product.

Other solvents used in the extractive distillation include n-methyl pyrrolidone, dimethyl formamide, furfural, and dimethyl acetamide.

BUTANE. See **Organic Chemistry**.

BUTYL ALCOHOLS. Butyl alcohols encompass the four structurally isomeric 4-carbon alcohols of empirical formula $C_4H_{10}O$. One of these, 2-butanol, can exist in either the optically active R(−) or S(+) configuration or as a racemic (±) mixture.

Physical and Chemical Properties

The butanols are all colorless, clear liquids at room temperature and atmospheric pressure with the exception of t-butyl alcohol which is a low melting solid (mp 25.82°C); it also has a substantially higher water miscibility than the other three alcohols. Physical constants of the four butyl alcohols are given in Table 1.

Physical constants for the optically pure stereoisomers of 2-butanol have been reported as follows:

	$d^t{}_4$	n_D^{20}	$[\alpha]^{27}D$
(S)-(+)-2-butanol	0.8025^{27}	1.3954	+13.52
(R)-(−)-2-butanol	0.8042^{25}	1.3970	−13.52

Butyl alcohol liquid vapor pressure–temperature responses, which are important parameters in direct solvent applications, are presented in Figure 1.

The butanols undergo the typical reactions of the simple lower chain aliphatic alcohols.

Manufacture

The principal commercial source of 1-butanol is n-butyraldehyde, obtained from the Oxo reaction of propylene.

Uses

The largest-volume commercial derivatives of 1-butanol are n-butyl acrylate and methacrylate. These are used principally in emulsion polymers

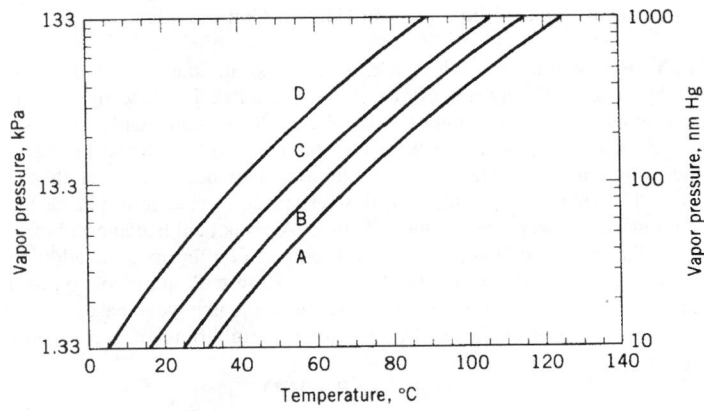

Fig. 1.   Vapor pressure of butyl alcohols: A, n-butyl; B, isobutyl; C, sec-butyl; D, t-butyl. To convert kPa to mm Hg, multiply by 7.5

TABLE 1. PHYSICAL PROPERTIES OF THE BUTYL ALCOHOLS (BUTANOLS)

Common Name	n-Butyl alcohol	Isobutyl alcohol	sec-Butyl alcohol	t-Butyl alcohol
systematic name	1-butanol	2-methyl-1-propanol	2-butanol	2-methyl-2-propanol
formula	$CH_3(CH_2)_3OH$	$(CH_3)_2CHCH_2OH$	$CH_3CH(OH)C_2H_5$	$(CH_3)_3COH$
critical temperature, °C	289.90	274.63	262.90	233.06
critical pressure, kPa[a]	4423	4300	4179	3973
normal boiling point, °C	117.66	107.66	99.55	82.42
melting point, °C	−89.3	−108.0	−114.7	25.82
heat of fusion, kJ/mol[b]	9.372	6.322	5.971	6.703
heat of vaporization at normal boiling point, kJ/g[b]	43.29	41.83	40.75	39.07
liquid density, kg/m³ at 25°C	809.7	801.6	806.9	786.6[c]
refractive index at 25°C	1.3971	1.3938	1.3949	1.3852
flash point, closed cup, °C	28.85	27.85	23.85	11.11
dielectric constant, ε	17.5^{25}	17.93^{25}	16.56^{20}	12.47^{30}
solubility in water at 30°C, % by weight	7.85	8.58	19.41	miscible
solubility of water in alcohol at 30°C, % by weight	20.06	16.36	36.19	miscible

[a] To convert kPa to mm Hg, multiply by 7.50.
[b] To convert kJ to kcal, divide by 4.184.
[c] For the subcooled liquid below melting point.

for latex paints, in textile applications and in impact modifiers for rigid poly(vinyl chloride).

Isobutyl alcohol has replaced *n*-butyl alcohol in some applications where the branched alcohol appears to have preferred properties and structure.

Health, Safety, and Environmental Considerations

All four butanols are thought to have a generally low order of human toxicity. However, large dosages of the butanols generally serve as central nervous system depressants and mucous membrane irritants.

All four butanols are registered in the United States on the Environmental Protection Agency Toxic Substances Control Act (TSCA) Inventory, a prerequisite for the manufacture or importation for commercial sale of any chemical substance or mixture in quantities greater than 454 kg (1000 lbs.). Additionally, the manufacture and distribution of the butanols in the United States are regulated under the Superfund Amendments and Reauthorization Act (SARA), Section 313, which requires that anyone handling at least 4545 kg (10,000 lbs.) a year of a chemical substance report to both the U.S. EPA and the state any release of that substance to the environment.

Storage and Handling

The C-4 alcohols are preferably stored in baked phenolic-lined steel tanks. However, plain steel tanks can also be employed provided a fine-porosity filler is installed to remove any contaminating rust.

Storage under dry nitrogen is also recommended since it limits flammability hazards as well as minimizing water pickup.

ERNST BILLIG
Union Carbide Chemicals and Plastics Company Inc.

Additional Reading

Cropley, J. B., L. M. Burgess, and R. A. Loke: *Chemtech.* **14**, 374–380 (1984).

Chemical Economics Handbook, SRI International, Menlo Park, Calif.

George, F. E. C. and D. Clayton, eds.: *Patty's Industrial Hygiene and Toxicology,* Vol. 2C, John Wiley & Sons, Inc., New York, NY, 1982, pp. 4571–4586.

BUTYLATED HYDROXYANISOLE (BHA). See **Antioxidants**.

BUTYLATED HYDROXYTOLUENE (BHT). See **Antioxidants**.

BUTYLENES. Butylenes are C_4H_8 mono-olefin isomers: 1-butene, *cis*-2-butene, *trans*-2-butene, and isobutylene (2-methylpropene). These isomers are usually coproduced as a mixture and are commonly referred to as the C-4 fraction. These C-4 fractions are usually obtained as by-products from petroleum refinery and petrochemical complexes that crack petroleum fractions and natural gas liquids. Since the C-4 fractions almost always contain butanes, it is also known as the B−B stream. The linear isomers are referred to as butenes.

Physical Properties

For any industrial process involving vapors and liquids, the most important physical property is the vapor pressure. Table 1 presents values for the constants for a vapor-pressure equation and the temperature range over which the equation is valid for each butylene.

TABLE 2. PHYSICAL PROPERTIES OF THE BUTYLENES

Property	Values			
	1-Butene	*cis*-2-Butene	*trans*-2-Butene	Isobutylene
molecular weight	56.11	56.11	56.11	56.11
melting point, K	87.80	134.23	167.62	132.79
boiling point, K	266.89	276.87	274.03	266.25
critical temperature, K	419.60	435.58	428.63	417.91
critical pressure, MPa[a]	4.023	4.205	4.104	4.000
critical volume, L/mol	0.240	0.234	0.238	0.239
critical compressibility factor	0.277	0.272	0.274	0.275
Flammability limits, vol % in air				
lower limit	1.6	1.6	1.8	1.8
upper limit	9.3	9.7	9.7	8.8
autoignition temperature, K	657	598	597	738

[a] To convert MPa to atm, multiply by 9.869.

Table 2 presents other important physical properties for the butylenes. Thermodynamic and transport properties can also be obtained from other sources.

Chemical Properties

The carbon–carbon double bond is the distinguishing feature of the butylenes and, as such, controls their chemistry. The carbon–carbon bond, acting as a substitute, affects the reactivity of the carbon atoms at the alpha positions through the formation of the allylic resonance structure. This structure can stabilize both positive and negative charges. Thus allylic carbons are more reactive to substitution and addition reactions than alkane carbons. Therefore, reactions of butylenes can be divided into two broad categories: (*1*) those that take place at the double bond itself, destroying the double bond; and (*2*) those that take place at alpha carbons.

The electron-rich carbon–carbon double bond reacts with reagents that are deficient in electrons, e.g., with electrophilic reagents in electrophilic addition, free radicals in free-radical addition, and under acidic conditions with another butylene (cation) in dimerization.

Manufacture

The C-4 isomers are almost always produced commercially as by-products in a petroleum refiner–petrochemical process.

There are other commercial processes available for the production of butylenes. However, these are site- or manufacturer-specific, e.g., the Oxirane process for the production of propylene oxide; the disproportionation of higher olefins; and the oligomerization of ethylene. Any of these processes can become an important source in the future. More recently, the Coastal Isobutane process began commercialization to produce isobutylene from butanes for meeting the expected demand for methyl-*tert*-butyl ether (MTBE).

New Technology. Several technologies are emerging for the production of isobutylene to meet the expected demand for isobutylene: (*1*) deep catalytic cracking; (*2*) superflex catalytic cracking; (*3*) dehydrogenation of butanes; and (*4*) the Coastal process of thermal dehydrogenation of butanes.

Separation and Purification of C-4 Isomers. 1-Butene and isobutylene cannot be economically separated into pure components by conventional

TABLE 1. VAPOR-PRESSURE EQUATION CONSTANTS FOR THE BUTANES, BUTYLENES, AND BUTADIENES[a]

	A	B	C	D	N	Temperature range, K
n-butane	61.5623	−4259.90	−6.20315	3.07575×10^{-7}	2.5	135–423
isobutane	66.7163	−4237.62	−7.08156	4.00506×10^{-7}	2.5	129–408
1-butene	78.8760	−4713.65	−9.05743	1.28654×10^{-5}	2.0	126–416
cis-2-butene	71.9534	−4681.34	−7.87527	1.00237×10^{-5}	2.0	203–358
trans-2-butene	74.3950	−4648.45	−8.33977	1.20897×10^{-5}	2.0	195–358
isobutylene	83.8683	−4822.95	−9.90214	1.51060×10^{-5}	2.0	194–359
1,2-butadiene	49.49.5031	−4021.95	−4.28893	5.13547×10^{-6}	2.0	200–284
1,3-butadiene	73.0016	−4547.77	−8.11105	1.14037×10^{-5}	2.0	164–425

[a] $\ln P = A + B/T + C^*\ln T + D^*T^{**}N$ where P is in Pa and T is in K.

TABLE 3. BUTYLENE CONSUMPTION, % OF SUPPLY

Use	United States	Western Europe	Japan
Fuel			
alkylate	85.6	22.1	2.0
methyl *tert*-butyl ether	4.5	8.2	0.0
polygas, LPG, blending	3.4	50.3	69.2
Chemicals			
sec-butyl alcohol/MEK	1.5	5.2	7.5
polyethylene copolymer	0.9	0.6	1.6
heptene, octene	0.7	3.4	1.3
butadiene	0.1	0.0	0.0
maleic anhydride	0.0	0.0	1.6
polybutene-polyisobutylenes	1.9	3.1	2.5
butyl rubbers	1.0	3.7	4.8
di- and triisobutylenes	~0.0	2.1	2.7
methyl methacrylate	0.0	0.0	4.4
other	0.4	1.3	2.4

distillation because they are close boiling isomers. 2-Butene can be separated from the other two isomers by simple distillation. There are four types of separation methods available: (*1*) selective removal of isobutylene by polymerization and separation of 1-butene; (*2*) use of addition reactions with alcohol, acids, or water to selectively produce pure isobutylene and 1-butene; (*3*) selective extraction of isobutylene with a liquid solvent, usually an acid; and (*4*) physical separation of isobutylene from 1-butene by absorbents.

There are three important processes for the production of isobutylene; (*1*) the extraction process using an acid to separate isobutylene; (*2*) the dehydration of *tert*-butyl alcohol, formed in the Arco's Oxirane process; and (*3*) the cracking of MTBE.

Handling and Analysis

Storage and Transportation. Handling requirements are similar to liquefied petroleum gas (LPG). Storage conditions are much milder.

Butylenes are stored as liquids at temperatures ranging from 0 to 40°C and at pressures from 100 to 400 kPa (1–4 atm). Their transportation is also similar to LPG; they are shipped in tank cars, transported in pipelines, or barged.

Health and Safety

Butylenes are not toxic. The effect of long-term exposure is not known; hence, they should be handled with care. They are volatile and asphyxiants. Care should be taken to avoid spills because they are extremely flammable. Physical handling requires adequate ventilation to prevent high concentrations of butylenes in the air.

Commercial Utilization

Pricing of butylenes determines the end use of butylenes in different geographic areas. The use pattern of butylenes in the United States, Europe, and Japan is shown in Table 3 as a percentage of supply in 1984.

NARASIMHAN CALAMUR
MARTIN E. CARRERA
RICHARD A. WILSAK
Amoco Corporation

Additional Reading

DIPPR, Project 801, Data Compilation (July 1990).

Dorman, D. E., M. Jantelot, and J. D. Roberts: *J. Org. Chem.* **36**, 2157 (1971);

Hirana, I., O. Kikuchi, and K. Suzuki: *Bull. Chem. Soc. Jpn.* **49**, 3321 (1976).

Hobson, G. D. and W. Pohl: *Modern Petroleum Technology*, 5th Edition, John Wiley & Sons, Inc., New York, NY, 1984, p. 517.

Morrison, R. T. and R. N. Boyd: *Organic Chemistry*, 4th Edition, Allyn and Bacon, Boston, Mass., 1983;

Verscheren, K. ed.: *Handbook of Environmental Data on Organic Chemicals*, Van Nostrand Reinhold Co., New York, 1983, pp. 304, 317.

BUTYL RUBBER. See **Elastomers**.

BUTYRATE PLASTICS. See **Cellulose Ester Plastics (Organic)**.

C

CADMIUM. [CAS: 7440-43-9]. Chemical element, symbol Cd, at. no. 48. at. wt. 112.41, periodic table group 12, mp 321°C, bp 765°C, density 8.65 g/cm³ (20°C). Elemental cadmium has a hexagonal crystal structure. Cadmium is a silver-white metal, malleable and ductile, but at 80°C becomes brittle. It remains lustrous in dry air and is only slightly tarnished by air or H_2O at standard conditions. The element may be sublimed in a vacuum at a temperature of about 300°C, and when heated in air burns to form the oxide. Cadmium dissolves slowly in hot dilute HCl or H_2SO_4 and more readily in HNO_3. The element first was identified by M. Stromeyer in 1817. Naturally occurring isotopes 106, 108, 110–114, 116. ^{113}Cd is unstable with respect to beta decay (0.3 MeV) into ^{113}In ($t_{1/2} \geq 10^{13}$ years). Electronic configuration $1s^2 2s^2 2p^6 3s^2 3p^6 3d^{10} 4s^2 4p^6 4d^{10} 5s^2$. Ionic radius Cd^{2+} 0.99 Å. Metallic radius 1.489 Å. First ionization potential, 8.99 eV; second, 16.84 eV; third, 38.0 eV. Oxidation potentials $Cd \longrightarrow Cd^{2+} + 2e^-$, 0.402 V; $Cd + 2OH^- \longrightarrow Cd(OH)_2 + 2e^-$, 0.915 V; $Cd + 4CN^- \longrightarrow Cd(CN)_4 + 4e^-$, 0.90 V. Other important physical properties of cadmium are given under **Chemical Elements**.

Although ranking 57th in abundance in the earth's crust (0.15 ppm), cadmium is not encountered alone, but is always associated with zinc. The only known cadmium minerals are greenockite (sulfide) and otavite (carbonate), both minor constituents of sphalerite (zinc oxide) and smithsonite (zinc carbonate), respectively. See also **Greenockite**; **Smithsonite**; and **Sphalerite Blende**.

Production

Two major processes are used for producing cadmium: (1) pyro-hydro-metallurgical and (2) electrolytic. Zinc blende is roasted to eliminate sulfur and to produce a zinc oxide calcine. The latter is the starting material for both processes. In the pyro-hydro-metallurgical process, the zinc oxide calcine is mixed with coal, pelletized, and sintered. This procedure removes volatile elements such as lead, arsenic, and the desired cadmium. From 92–94% of the cadmium is removed in this manner, the vapors being condensed and collected in an electrostatic precipitator. The fumes are leached in H_2SO_4 to which iron sulfate is added to control the arsenic content. The slurry then is oxidized, normally with sodium chlorate, after which it is neutralized with zinc oxide and filtered. The cake goes to a lead smelter, while the filtrate is charged with high-purity zinc dust to form zinc sulfate or zinc carbonate and cadmium sponge. The latter is briquetted to remove excess H_2O and melted under caustic to remove any zinc. The molten metal then is treated with zinc ammonium chloride to remove thallium, after which it is cast into various cadmium metal shapes. The process just described is known as the *melting under caustic process*. In a *distillation process*, regular rather than high-purity zinc is used to make the sponge. Then, after washing and centrifuging to remove excess H_2O, the sponge is charged to a retort. The heating and distillation process is under a reducing atmosphere. Lead and zinc present in the vapors contaminate about the last 15% of the distillate. Thus, a redistillation is required. The cadmium vapors produced are collected and handled as previously described.

Reactions that occur in the foregoing processes are: (Leaching): $CdO + H_2SO_4 \longrightarrow CdSO_4 + H_2O$; (Oxidation): $3As_2O_3 + 2NaClO_3 \longrightarrow 3As_2O_5 + 2NaCl$; and $6FeSO_4 + NaClO_3 + 3H_2SO_4 \longrightarrow 3Fe_2(SO_4)_3 + NaCl + 3H_2O$; (Neutralization): $Fe_2(SO_4)_3 + As_2O_5 + 3ZnO + 8H_2O \longrightarrow 2FeAs(OH)_8 + 3ZnSO_4$; (Cadmium Precipitation): $CdSO_4 + Zn \longrightarrow Cd + ZnSO_4$; (Melting Under Caustic): $Zn + 2NaOH + \frac{1}{2}O_2 \longrightarrow Na_2ZnO_2 + H_2O$.

In the electrolytic process, the calcine first is leached with H_2SO_4. Charging the resultant solution with zinc dust removes the cadmium and other metals that are more electronegative than zinc. The sponge that results is digested in H_2SO_4 and purified of all contaminants except zinc.

Nearly pure cadmium sponge is precipitated by the addition of high-purity, lead-free zinc dust. The cadmium sponge then is redigested in spent cadmium electrolyte, after which the cadmium is deposited by electrolysis onto aluminum cathodes. The metal is then stripped from the electrodes, melted, and cast into various shapes. Reactions which occur during the electrolytic process are: (Roasting): $ZnS + 1\frac{1}{2}O \longrightarrow ZnO + SO_2$; (Leaching): $ZnO + H_2SO_4 \longrightarrow ZnSO_4 + H_2O$; (Neutralization): $Fe_2(SO_4)_3 + 3ZnO + 3H_2O \longrightarrow 2Fe(OH)_3 + 3ZnSO_4$; (Cadmium Precipitation): $CdSO_4 + Zn \longrightarrow Cd + ZnSO_4$; (Electrolysis): $CdSO_4 + H_2O \longrightarrow Cd + H_2SO_4 + O_2$.

Industrial specifications normally require that impurities in cadmium metal not exceed the following: zinc, 0.035%; copper, 0.015%; lead, 0.025%; tin, 0.01%; silver, 0.01%; antimony, 0.001%; arsenic, 0.003%; and tellurium, 0.003%. The metal is available in numerous forms. Electroplaters generally prefer balls 2 inches (5 centimeters) in diameter.

Uses

A major use of cadmium is for electroplating steel to improve its corrosion resistance. It is also used in low-melting-point alloys, brazing alloys, bearing alloys, nickel-cadmium batteries, and nuclear control rods, and as an alloying ingredient to copper to improve hardness. Cadmium, unfortunately, is limited in its usefulness because fumes and dusts containing cadmium are quite toxic. Melting and handling conditions that create dust or fumes must be equipped with exhaust ventilation systems. See also specific Cd compounds in this article.

Biological Properties

Over the last several years, concern over the poisonous nature of cadmium, particularly of Cd powder and chips from Cd plating, has increased. For example, cadmium-plated hardware has not been used for food processing equipment for a decade or more. A part of this concern pertains to the incineration of waste materials that may contain cadmium, for fear of introducing Cd particles into the atmosphere. Also, although the quantity of cadmium used in pigments by artists is indeed very small, many artists are adamant concerning possible legislation. One painter has said, "Losing cadmiums would be like a composer losing the use of several keys." As another artist has pointed out, "Van Gogh could not have painted his 'Sunflowers' without cadmium." The jewelry industry also uses cadmium as an ingredient of low-melting silver solders.

In February 1990 the Occupational Safety and Health Administration (U.S.) published a report that summarizes the history of cadmium regulation, studies of health problems, and risk calculations for cancer, kidney damage, and other disorders. This report represents a formal step toward implementation of stricter limits on cadmium exposure in the workplace. Some authorities admit that considerably more research is required.

The battery industry also has been plagued with metal pollution problems. These problems began with lead storage batteries several years ago and was at least partially solved by a lead recycling program based upon manufacturers' recalling "spent" car batteries. This was followed in the 1980s by the grossly reduced quantities of mercury, which is used to coat the electrodes in alkaline batteries. Initially, mercury accounted for about 1% of a battery's weight. By 1993 this has been reduced to 0.025% of battery weight.

In the early 1990s it is estimated that nearly 300 million nickel-cadmium batteries were sold in the United States. A large percentage of these were embedded in a variety of cordless appliances, such as power tools, small vacuum cleaners, and even toothbrushes. It was recently estimated that nearly 2,000 tons of cadmium appeared in the industrial waste stream as the result of equipment "junked" during the mid-1980s. Legislation directed toward keeping cadmium out of landfills and incinerators already has been passed in Connecticut and Minnesota. Other states considering similar legislation include New Jersey, Vermont, Michigan, California, and Oregon. Battery makers are investigating suitable substitutes for Cd, including nickel–nickel hydride batteries. Metal hydrides, which are porous compounds capable of storing hydrogen, ultimately may suffice for low-power devices (toys, photoflash devices), but presently do not look promising for high-power devices, such as motorized hand tools.

Chemistry and Compounds

In virtually all of its compounds, cadmium exhibits the +2 oxidation state, although compounds of cadmium(I) containing the ion Cd_2^{2+}, have occasionally been reported. Cadmium hydroxide is more basic than zinc hydroxide, and only slightly amphiprotic, requiring very strong alkali to dissolve it, and forming $Cd(OH)_3^-$ or $Cd(OH)_4^{2-}$ depending upon the pH.

Cadmium is found in metallothioneins, which are low-molecular-weight, cysteine-rich proteins that bind metal ions. Metallothioneins and their genes have several potential kinds of physiological activity. See Metallothioneins. Furey et al.(1986) report on a thorough investigation of the crystal structure of Cd, Zn metallothionein.

Cadmium Oxide

[CAS: 1306-19-0]. CdO, formed by burning the metal in air or heating the hydroxide or carbonate, is soluble in acids, ammonia, or ammonium sulfate solution, and is more readily reduced on heating with carbon, carbon monoxide or hydrogen than zinc oxide. Cadmium suboxide, Cd_2O, formed by thermal reduction of cadmium oxalate with carbon monoxide, is believed to be a mixture of CdO and finely divided cadmium. CdO_2 and Cd_4O have been reported. Sodium hydroxide solution precipitates cadmium hydroxide, $Cd(OH)_2$, from solutions of the sulfate or nitrate, but with the chloride the $Cd(OH)_2$ precipitate is mixed with CdOHCl and other hydroxychlorides. $Cd(OH)_2$ exists in two forms, an "active" and an "inactive" one, which have different solubility products. Cadmium(I) hydroxide $Cd_2(OH)_2$, prepared by hydrolysis of Cd_2Cl_2, is, like Cd_2O, believed to be a mixture of the metal and the divalent compound. The Cd_2^{2+} ion is definitely established, however, in such compounds as $Cd_2(AlCl_4)_2$.

Cadmium Halides

These compounds can be prepared by the action of the corresponding hydrohalic acids upon the carbonate; or by direct union of the elements. If bromine water is used, some hydrobromic acid must be added to prevent hydrolysis of the bromide to the oxybromide, CdOHBr.

In general, the cadmium halides show in their crystal structure the relation between polarizing effect and size of anion. The fluoride has the smallest and least polarizable anion of the four and forms a cubic structure, while the more polarizable heavy halides have hexagonal layer structures, increasingly covalent and at increasing distances apart in order down the periodic table. In solution the halides exhibit anomalous thermal and transport properties, due primarily to the presence of complex ions, such as CdI_4^{2-} and $CdBr_4^{2-}$, especially in concentrated solutions or those containing excess halide ions.

Cadmium Sulfide

[CAS: 1306-23-6]. CdS is the most extensively used of cadmium compounds and generally is prepared by precipitation from cadmium salts. The wide range of colors, varying from lemon yellow through the oranges and deep red, coupled with the stability and intensity of these colors, qualify CdS as a most desirable pigment for paints, plastics, and other products. The range of colors of CdS precipitates results from differing conditions in their formulation, including the temperature and acidity of the salt solutions from which they are precipitated. The particular salt, such as nitrate, chloride, sulfate, etc., also affects the resulting color. The rate of addition of hydrogen sulfide to the liquor affects particle size and color of the precipitates. Cadmium sulfide is insoluble in H_2O, is dimorphous, and sublimes at 1,350°C. Several crystalline forms exist. When precipitated from normal H_2SO_4 and HNO_3 solutions, the crystals are cubic. From other media, stable alpha hexagonal and unstable beta cubic forms may be formed, these ranging in specific gravity from 3.9 to 4.5, respectively. Pigment colors are not due to crystal form, but rather

derive from the particle size and dispersion of the precipitates. Of total cadmium production, pigments account for 20–25% of the total.

Other cadmium compounds used as pigments in ceramics, glass, and paints include cadmium nitrate, selenide, sulfoselenide, and tungstate. Cadmiopone ($BaSO_4$ plus CdS) ranges from yellow to crimson and is used for coloring plastics and rubber goods. Cadmium stearate, when combined with barium stearate, is widely used as a stabilizer in thermosetting plastics and accounts for well over 20% of the total cadmium produced.

Cadmium Carbonate

[CAS: 513-78-0]. $CdCO_3$, $pK_{sp} = 11.3$, is formed by the hydroxide upon absorption of CO_2, or upon precipitation of a cadmium salt with ammonium carbonate. With alkali carbonates, the oxycarbonates are produced.

Cadmium nitrate tetrahydrate, solubility 215 g/100 ml H_2O at 0°C, is obtained by action of HNO_3 upon the carbonate. It is ionized completely only in solutions weaker than about tenth molar. However, it does not form hydroxy compounds as readily as the zinc salt, requiring the action of NaOH, which in moderate concentration gives $Cd(NO_3)_2 \cdot 3Cd(OH)_2$ and $Cd(NO_3)_2 \cdot Cd(OH)_2$; excess sodium hydroxide precipitates the hydroxide.

Cadmium forms a wide variety of other salts, many by reaction of the metal, oxide, or carbonate with the acids, although some can be obtained only by fusion of the oxides or hydroxides. They include the antimonates (pyro- and meta-), the arsenates (ortho-, meta-, and pyro-, including acid salts as well as normal), the arsenites, the borates ($Cd(BO_2)_2$, $Cd_2B_6O_{11}$, $Cd_3(BO_3)_3$ and $Cd_3B_2O_6$ have been identified), the bromate, the bicarbonate, the chlorate, the chlorite, chromates and dichromates, the cyanide, the ferrate, the iodate, the molybdate, $CdMoO_4$, the nitrate, the perchlorate, various periodates, the permanganate, various phosphates (ortho, meta, and para, including acid salts as well as normal), the selenates and selenites, various silicates, the stannate, the sulfate (which reacts with limited amounts of NaOH or NH_3 solution to give various hydroxy sulfates), the thiosulfate, the titanate, the tungstate, and the uranate.

Cadmium arsenide, nitride, selenide, and telluride are known, the first and third obtainable from the elements, while the nitride is obtained by heating the amide (obtained by reaction of cadmium thiocyanate and potassium amide in liquid NH_3), and the telluride is obtainable by reduction of the tellurate with hydrogen. Cadmium arsenide is used as a semiconductor.

One of the features of the chemistry of cadmium is that it forms a relatively large number of complexes. A number of solid double halides of compositions $MCdX_3$, M_2CdX_4, M_3CdX_5 and M_4CdX_6 where M is an alkali metal and X a halogen are known, the last two probably existing only in the solid state. Conductance studies of solutions indicate the presence of such ions as CdX^+, CdX_3^- and CdX_4^{2-}. The donor ability of oxygen is less toward cadmium than toward zinc, fewer oxygen complexes and organic oxygen-linked complexes being known. Sulfur is a better donor than oxygen; additives of the type $(R_2S)_2 \cdot CdX_2$ are formed from di-alkyl sulfides and cadmium halides. The ready reactions with NH_3, as with amines, give large numbers of complexes; those with ammonia include tetrammines and hexamines, containing $[Cd(NH_3)_4]^{2+}$ and $[Cd(NH_3)_6]^{2+}$, respectively. Ethylenediamine forms 6-coordinate compounds containing $[Cd(en)_3]^{2+}$. Prominent among the carbon donor complexes are the cyanides, principally compounds of $Cd(CN)_4^{2-}$, although $Cd(CN)_3^-$ is also known. Other carbon donor compounds are the organometallic compounds CdR_2, where R may be methyl, ethyl, propyl, butyl, isobutyl, isoamyl, amylthio, phenyl, octylthio, decylthio, and higher organic radicals.

Additional Reading

Amato, I.: "Singing the Cadmium Blues," *Science News,* **168** (September 15, 1990).

Berndt, D.: *Maintenance-Free Batteries: Lead-Acid, Nickel/Cadmium, Nickel/Metal Hydride,* 2nd Edition, John Wiley & Sons, Inc., New York, NY, 1997.

Carter, G.F. and D.E. Paul: *Materials Science and Engineering,* ASM International, Materials Park, OH, 1991.

Erickson, D.: "Cadmium Charges," *Sci. Amer.,* **122** (May 1991).

Furey, W.F. et al.: "Crystal Structure of Cd, Zn Metallothionein," *Science,* **231,** 704–710 (1986).

Greenwood, N.N. and A. Earnshaw: *Chemistry of the Elements,* 2nd Edition, Butterworth-Heinemann, Inc., Woburn, MA, 1997.

Hawley, G.G. and R.J. Lewis: *Hawley's Condensed Chemical Dictionary,* 13th Edition, John Wiley & Sons, Inc., New York, NY, 1999.

Jackson, T. and A. MacGillivary: *Accounting for Cadmium,* Gordon & Breach Science Publishers, Newark, NJ, 1996.

Klepper, G., P. Michaelis, and G. Mahlau: *Industrial Metabolism: A Case Study on the Economics of Cadmium Control,* University of Michigan Press, Ann Arbor, MI, 1995.

Lagowski, J.J.: *MacMillan Encyclopedia of Chemistry,* Vol. 1, MacMillan Library Reference, New York, NY, 1997.

Lide, D.R.: *CRC Handbook of Chemistry and Physics,* 84th Edition, CRC Press, LLC., Boca Raton, FL, 2003.

Perry, R.H., D.W. Green, and J.O. Maloney: *Perry's Chemical Engineers' Handbook,* 7th Edition, McGraw-Hill Companies, Inc., New York, NY, 1997.

Sax, N.I. and R.J. Lewis, Sr.: *Dangerous Properties of Industrial Materials,* 10th Edition, John Wiley & Sons, Inc., New York, NY, 1999.

CAFFEINE.   See **Alkaloids.**

CAIRNGORM STONE.   The name given to the smoky brown variety of quartz, particularly when transparent, from Cairngorm, Scotland, a well-known locality. See also **Quartz.**

CALAVERITE.   A gold telluride, $AuTe_2$, associated with quartz in low-temperature veins. A valuable gold ore from Kalgorrlie, Western Australia and the Cripple Creek region of Colorado. The ore occurs in bladed to lath-like monoclinic crystals with striations parallel to the long axis of the crystals. The ore has a metallic luster of brass-yellow to silver-white color, a hardness of 2.5 to 3, a specific gravity of 9.24 to 9.31, and a yellowish to greenish-gray streak.

CALCINATION.   The subjection of a substance to a high temperature below its fusion point, often to make the substance friable. Calcination frequently is carried out in long, rotating, cylindrical vessels, known as kilns. Material so treated may (1) lose moisture, e.g., the heating of silicic acid or ferric hydroxide resulting in the formation of silicon oxide or ferric oxide, respectively, (2) lose a volatile constituent, e.g., the heating of limestone (calcium carbonate) resulting in the formation of carbon dioxide gas and calcium oxide residue—destructive distillation of many organic substances is of this type—(3) be oxidized or reduced, e.g., the heating of pyrite (iron disulfide) in air resulting in the formation of sulfur dioxide gas and ferric oxide residue. When the calcination involves oxidation, as in the preceding case, the operation is termed roasting. When heating involves reduction of metals from their ores with separation from the gangue of the liquid metal and slags, the process is termed smelting.

CALCITE.   The mineral calcite, carbonate of calcium corresponding to the formula $CaCO_3$, is one of the most widely distributed minerals. Its crystals are hexagonal-rhombohedral although actual calcite rhombohedrons are rare as natural crystals. However, they show a remarkable variety of habit including acute to obtuse rhombohedrons, tabular forms, prisms, or various scalenohedrons. It may be fibrous, granular, lamellar or compact. The cleavage in three directions parallel to rhombohedron is highly perfect; fracture, conchoidal but difficult to obtain; hardness, 3; specific gravity, 2.7; luster, vitreous in crystallized varieties; color, white or colorless through shades of gray, red, yellow, green, blue, violet, brown, or even black when charged with impurities; streak, white; transparent to opaque; it may occasionally show phosphorescence or fluorescence.

Calcite is perhaps best known because of its power to produce strong double refraction of light such that objects viewed through a clear piece of calcite appear doubled in all of their parts. A beautifully transparent variety used for optical purposes comes from Iceland, for that reason is called Iceland spar.

Acute scalenohedral crystals are sometimes referred to as dogtooth spar. Calcite represents the stable form of calcium carbonate; aragonite will go over to calcite at 470°C (878°F). Calcite is a common constituent of sedimentary rocks, as a vein mineral, and as deposits from hot springs and in caves as stalactites and stalagmites.

Localities that produce fine specimens in the United States include the Tri-State area of Missouri, Oklahoma, and Kansas, as well as Wisconsin, Tennessee, and Michigan with inclusions of native copper; several areas in Mexico, notably Charcas and San Luis Potosi; Iceland; Cumberland and Durham regions in England; and at various regions in S.W. Africa,

notably Tsumeb. The exceptionally fine sand-calcite crystals from South Dakota and Fontainebleau in France are well known.

CALCIUM.   [CAS: 7440-70-2]. Chemical element, symbol Ca, at. no. 20, at. wt. 40.08, periodic table group 2 (alkaline earths), mp 837–841°C, bp 1,484°C, density 1.54 g/cm³ (single crystal). Elemental calcium has a face-centered cubic crystal structure when at room temperature, transforming to a body-centered cubic structure at 448°C.

Calcium is a silver-white metal, somewhat malleable and ductile; stable in dry air, but in moist air or with water reacts to form calcium hydroxide and hydrogen gas; when heated burns in air to form calcium oxide emitting a brilliant light. Discovered by Davy in 1808.

There are six stable isotopes, ^{40}Ca, ^{42}Ca, ^{43}Ca, ^{44}Ca, ^{46}Ca, and ^{48}Ca, with a predomination of ^{40}Ca. In terms of abundance, calcium ranks fifth among the elements occurring in the earth's crust, with an average of 3.64% calcium in igneous rocks. In terms of content in seawater, the element ranks seventh, with an estimated 1,900,000 tons of calcium per cubic mile (400,000 metric tons per cubic kilometer) of seawater. Electronic configuration $1s^2 2s^2 2p^6 3s^2 3p^6 4s^2$. Ionic radius Ca^{2+} 1.06 Å. Metallic radius 1.874 Å. First ionization potential 6.11 eV; second, 11.82 eV; third, 50.96 eV. Oxidation potentials $Ca \longrightarrow Ca^{2+} + 2e^-$, 2.87 V; $Ca + Ca(OH)_2 + 2e^-$, 3.02 V.

Other important physical properties of calcium are given under **Chemical Elements.**

Calcium occurs generally in rocks, especially limestone (average 42.5% CaO) and igneous rocks; as the important minerals limestone (calcium carbonate, $CaCO_3$), gypsum (calcium sulfate dihydrate, $CaSO_4 \cdot 2H_2O$), phosphorite, phosphate rock (calcium phosphate, $Ca_3(PO_4)_2$), apatite (calcium phosphate-fluoride, $Ca_3(PO_4)_2$ plus CaF_2), fluorite, fluorspar (calcium fluoride, CaF_2); in bones and bone ash as calcium phosphate, and in egg shells and oyster shells as calcium carbonate. See also **Apatite; Calcite; Fluorite;** and **Gypsum.**

In the United States and Canada, calcium metal is produced by the thermal reduction of lime with aluminum. Before World War II, most elemental calcium was made by electrolysis of fused calcium chloride. In the thermal reduction process, lime and aluminum powder are briquetted and charged into high-temperature alloy retorts, which are maintained at a vacuum of 100 μm or less. Upon heating the charge to 1,200°C, the reaction takes place slowly, releasing Ca vapor. The latter is removed continuously by condensation, thus permitting the reaction to proceed to completion. High-purity lime is required as a starting ingredient if resulting calcium metal of high purity is desired. Aluminum contamination of the resulting calcium is removed by an additional vacuum-distillation step. Other impurities also are reduced by this distillation step.

Uses of Elemental Calcium

The very active chemical nature of calcium accounts for its major uses. Calcium is used in tonnage quantities to improve the physical properties of steel and iron. Tonnage quantities are also used in the production of automotive and industrial batteries. Other major uses include refining of lead, aluminum, thorium, uranium, samarium, and other reactive metals.

Calcium treatment of steel results in improved yields, cleanliness, and mechanical properties. Because it is a very strong deoxidizer and sulfide former, calcium will improve the deoxidation and desulfurization of steel. In addition, it alters the morphology and size of inclusions, reduces internal and surface defects, and reduces macrosegregation. Hydrogen-induced cracking of line pipe steels by high-sulfur fuels is reduced with calcium treatment. Several grades of calcium-treated steel are used in automotive, industrial, and aircraft applications. Oil line pipe, heavy plate, and deep drawing sheet were first treated in Japan. Additional uses have been developed in the United States and Europe.

The high vapor pressure and reactivity of calcium limited its use in steel and iron making prior to the development of injection systems and mold nodularization processes. There are two types of injection systems. One consists of the use of a holding furnace, a sealed vessel, a carrier gas, and a lance through which calcium or calcium compounds are blown into the molten metal. This system is effective for massive desulfurization of large quantities of steel. It is a ladle process. The second type of injection process is wire feeding. A steel-jacketed calcium-core wire is fed through a delivery system that drives the composite wire below the surface of the liquid metal bath. The steel jacket protects the solid metallic calcium from reacting at the surface and allows it to penetrate deep into the bath. Because the reaction occurs below the surface, high and reproducible calcium recoveries are possible. This process is used in both ladle additions and in tundish additions for continuous casting. It provides shape control, deoxidation, final desulfurization and reduction of macrosegregation.

Ladle and mold processes using calcium ferroalloys are important in the production of nodular iron castings. The principal calcium alloy used is magnesium ferrosilicon. Calcium reduces the reactivity of the alloy; with the molten iron it enhances nucleation and improves morphology. The calcium content of the alloy is proportional to the magnesium content, typically in the range of 15–50% of magnesium content. In ladle or sandwich treatment techniques, pieces of the ferroalloy are placed in a pocket cut in the refractory lining of the ladle and the molten iron is then poured into the ladle. The treated, nodularized iron is then cast from the ladle into molds.

In the mold addition process, a granular form of the alloy is placed in a small reaction chamber in the mold. The nodularization treatment occurs in the mold when the iron is cast, rather than in the ladle. The reaction is contained in the mold, and high recoveries result. The production of nodular iron castings is over three million tons per year.

A calcium lead alloy is used in maintenance-free automotive and industrial batteries. The use of calcium reduces gassing and improves the life of the battery. From 0.1 to 0.5% calcium is alloyed with the lead prior to the fabrication of the battery plates either by casting or through the production of coiled sheet. With calcium present, these lead-acid batteries can be sealed and do not require the service of conventional batteries. The batteries have a higher energy-to-weight ratio. Of the battery market in the United States, over 50 million batteries per year, 40% are maintenance-free types.

Calcium is used in refining battery-grade lead for removing bismuth. Calcium is also used as an electrode material in high-energy thermal batteries.

The production of samarium cobalt magnets requires the use of calcium. The reaction is

$$3\ Sm_2O_3 + 10\ Co_3O_4 + 49\ Ca\ (vapor) \xrightarrow[\Delta]{850-1150°C} 6\ SmCo_5 + 49\ CaO$$

$$0.75\ \text{weight units of Ca} \longrightarrow 1\ \text{weight unit of SmCo}_5$$

Samarium cobalt magnets have three to six times greater magnetic energy than alnico magnets.

Calcium serves as a reductant for such reactive metals as zirconium, thorium, vanadium, and uranium. In zirconium reduction, zirconium fluoride is reacted with calcium metal. The high heat of the reaction melts the zirconium. The zirconium ingot resulting is remelted under vacuum for purification. Thorium and uranium oxides are reduced with an excess of calcium in reactors or trays under an atmosphere of argon. The resulting metals are leached with acetic acid to remove the lime.

Calcium is also used in aluminum alloys and as an addition in a magnesium alloy used for etching. An alloy of 80% Ca-20% Mg is used to deoxidize magnesium castings. The metal also is used in the production of calcium pantothenate, a B-complex vitamin.

Chemistry and Compounds

Calcium exhibits a valence state of +2 and is slightly less active than barium and strontium in the same series. Calcium reacts readily with all halogens, oxygen, sulfur, nitrogen, phosphorus, arsenic, antimony, and hydrogen to form the halides, oxide, sulfide, nitride, phosphide, arsenide, antimonide, and hydride. It reacts vigorously with water to form the hydroxide, displacing hydrogen. Calcium oxide (quicklime) adds water readily and with the evolution of much heat (slaked lime) to form the hydroxide. Calcium hydroxide forms a peroxide on treatment with hydrogen peroxide in the cold. Calcium exhibits little tendency to form complexes; the ammines formed with ammonia are unstable, although a solid of composition $Ca(NH_3)^6$ can be isolated from solutions of the metal in liquid ammonia.

Calcium Acetate. [CAS: 62-54-4], $Ca(C_2H_3O_2)_2 \cdot H_2O$, white solid, solubility: at 0°C, 27.2 g; at 40°C, 24.9 g, at 80°C, 25.1 g of anhydrous salt per 100 g saturated solution, formed by reaction of calcium carbonate or hydroxide and acetic acid.

Calcium Aluminates. [CAS: 065997-16-2]. Four in number, have been prepared by high-temperature methods and identified, $3CaO \cdot Al_2O_3$, at 1,535°C, decomposes with partial fusion; $5CaO \cdot Al_2O_3$, mp 1,455°C, $CaO \cdot Al_2O_3$, mp 1,590°C, $3CaO \cdot Al_2O_3$, mp 1,720°C.

Calcium Aluminosilicates. Two in number, have been prepared by high-temperature methods and identified: $2CaO \cdot Al_2O_3 \cdot SiO_2$, gehlinite; $CaO \cdot Al_2O_3 \cdot 2SiO_2$, anorthite.

Calcium Arsenate. [CAS: 7778-44-1]. $Ca_3(AsO_4)_2$, white precipitate, formed by reaction of soluble calcium salt solution and sodium arsenate solution. $pK_{sp} = 18.17$.

Calcium Arsenite. [CAS: 52740-16-6]. $Ca_3(AsO_3)_2$, white precipitate, formed by reaction of soluble calcium salt solution and sodium arsenite solution.

Calcium Borates. Found in nature as the minerals colemanite, $Ca_2B_6O_{11} \cdot 5H_2O$, borocalcite, $CaB_4O_7 \cdot 4H_2O$, and pandermite $Ca_2B_6O_{11} \cdot 3H_2O$. See also **Colemanite**.

Calcium Bromide. [CAS: 7789-41-5]. $CaBr_2 \cdot 6H_2O$, white solid, solubility 1,360 g/100 ml H_2O at 25°C, formed by reaction of calcium carbonate or hydroxide and hydrobromic acid.

Calcium Carbide. [CAS: 75-20-7]. CaC_2, grayish-black solid, reacts with water yielding acetylene gas and calcium hydroxide, formed at electric furnace temperature from calcium oxide and carbon.

Calcium Carbonate. [CAS: 1317-65-3]. $CaCO_3$, found in nature as calcite, Iceland spar, marble, limestone, coral, chalk, shells of mollusks, aragonite. $pK_{sp} = 8.32$. It is (1) readily dissolved by acids forming the corresponding calcium salts, (2) converted to calcium oxide upon heating. Aragonite is an unstable form at room temperature, although no change is observable until heated, when, at 470°C, it is quickly converted into calcite; calcium hydrogen carbonate, calcium bicarbonate, $Ca(HCO_3)$, known only in solution, formed by reaction of calcium carbonate and carbonic acid. See also **Aragonite**; **Calcite**; and **Calcium Carbonate**.

Calcium Chloride. [CAS: 10043-52-4]. $CaCl_2 \cdot 6H_2O$, white solid, solubility 536 g/100 g H_2O at 20°C, absorbs water from moist air, formed by reaction (1) of calcium carbonate or hydroxide and HCl, (2) of calcium hydroxide and ammonium chloride. See also **Calcium Chloride**.

Calcium Chromate. [CAS: 13765-19-0]. $CaCrO_4$, yellow solid, formed by the reaction of chrome ores and calcium oxide heated to a high temperature in a current of air. $pK_{sp} = 3.15$.

Calcium Citrate. $Ca_3(C_6H_5O_7)_2 \cdot 4H_2O$, white solid, solubility: at 18°C 0.085 g/100 g H_2O, formed by reaction of calcium carbonate or hydroxide and citric acid solution.

Calcium Cyanamide. [CAS: 156-62-7]. $CaCN_2$, white solid, formed (1) by heating cyanamide or urea with calcium oxide, sublimes at 1,050°C, (2) by heating calcium carbide at 1,100–1,200°C in a current of nitrogen. Decomposes in water with evolution of NH_3.

Calcium Fluoride. [CAS: 7789-75-5]. CaF_2, white precipitate, formed by reaction of soluble calcium salt solution and sodium fluoride solution. $pK_{sp} = 10.40$. See also **Fluorite**.

Calcium Formate. [CAS: 544-17-2]. $Ca(CHO_2)_2$, white solid, solubility at 0°C 13.90 g, at 40°C 14.56 g, at 80°C 15.22 g of anhydrous salt per 100 g saturated solution, formed by reaction of calcium carbonate or hydroxide and formic acid. Calcium formate, when heated with a calcium salt of a carboxylic acid higher in the series, yields an aldehyde.

Calcium Furoate. $Ca(C_4H_3O \cdot COO)_2$, formed by reaction of calcium carbonate or hydroxide and furoic acid.

Calcium Hydride. [CAS: 7789-78-8]. CaH_2, white solid, reacts with water yielding hydrogen gas and calcium hydroxide; when electrolyzed in fused potassium lithium chloride, hydrogen is liberated at the anode.

Calcium Hypochlorite. [CAS: 7778-54-3]. $CaOCl_2$ or $Ca(ClO)_2 \cdot 4H_2O$, white solid, contains 60%–65% "available chlorine" and sufficient calcium hydroxide to stabilize, formed by reaction of calcium hydroxide and chlorine. Very soluble in water.

Calcium Hypophosphite. [CAS: 7789-79-9]. $Ca(H_2PO_2)_2$, white solid, solubility 15.4 g/100 g H_2O at 25°C, formed (1) by boiling calcium hydroxide suspension in water and yellow phosphorus, (2) by reaction of calcium carbonate or hydroxide and hypophosphorous acid.

Calcium Iodide. [CAS: 10102-68-8]. CaI_2, yellowish-white solid, solubility 66 g/100 g H_2O at 10°C, formed by reaction of calcium carbonate or hydroxide and hydriodic acid. The hexahydrate, $CaI_2 \cdot 6H_2O$, is soluble to the extent of 1.680 g/100 g H_2O at 30°C.

Calcium Lactate. [CAS: 814-80-2]. $Ca(C_3H_5O_3)_2 \cdot 5H_2O$, white solid, solubility at 0°C 3.1 g, at 30°C 7.9 g of anhydrous salt per 100 g H_2O, formed by reaction of calcium carbonate or hydroxide and lactic acid.

Calcium Malate. [CAS: 17482-42-7]. $CaC_4H_4O_5 \cdot 2H_2O$, white solid, solubility at 0°C 0.670 g, at 37.5°C 1.011 g of anhydrous salt per 100 g saturated solution. Formed (1) by reaction of calcium carbonate or hydroxide and malic acid, (2) by precipitation of soluble calcium salt solution and sodium malate solution.

Calcium Nitrate. [CAS: 10124-37-5]. $Ca(NO_3)_2 \cdot 4H_2O$, white solid, solubility 660 g/100 g H_2O at 30°C, formed by reaction of calcium carbonate or hydroxide and HNO_3.

Calcium Oxalate. CaC_2O_4, white precipitate, insoluble in weak acids, but soluble in strong acids, formed by reaction of soluble calcium salt solution and ammonium oxalate solution. Solubility at 18°C 0.0056 g anhydrous salt per liter of saturated solution.

Calcium Oxide. [CAS: 1305-78-8]. CaO (quicklime), white solid, mp 2,570°C, reacts with H_2O to form calcium hydroxide with the evolution of much heat; reacts with H_2O vapor and CO_2 of the atmosphere to form calcium hydroxide and carbonate mixture (slaked lime); formed by heating limestone at high temperature (800°C) and removal of CO_2. This process is conducted industrially in a lime kiln.

Tricalcium Phosphate. $Ca_3(PO_4)_3$, white solid, insoluble in water; reactive with silicon oxide and carbon at electric furnace temperature yielding phosphorus vapor; reactive with H_2SO_4 to form, according to the proportions used, phosphoric acid, or dicalcium hydrogen phosphate, $CaHPO_4$, white solid, insoluble; or calcium dihydrogen phosphate, $Ca(H_2PO_4)_2 \cdot H_2O$, white solid, soluble. $pK_{sp} = 28.70$. See also **Apatite**.

Calcium Silicates. Four have been prepared by high-temperature methods and identified: $3CaO \cdot SiO_2$, prepared by heating the constituents to a temperature below the mp (mp is 1,700°C but substance unstable); $2CaO \cdot SiO_2$, mp 2,080°C, but upon slow cooling changes to forms of different volume; $3CaO \cdot 2SiO_2$, mp 1,475°C; and $CaO \cdot SiO_2$, wollastinite, mp approximately 1,400°C. See also **Clinozoisite**; **Datolite**; **Diopside**; **Feldspar**; **Lawsonite**; **Tremolite**; and **Wollastonite**.

Calcium Sulfate. [CAS: 10101-41-4]. Gypsum, $CaSO_4 \cdot 2H_2O$, plaster of Paris, $CaSO_4 \cdot \frac{1}{2}H_2O$ anhydrite $CaSO_4$, white solid, slightly soluble (about 0.2 g per 100 ml of H_2O), formed by reaction of soluble calcium salt solution with a sulfate solution. pK_{sp} of $CaSO_4 = 4.6_{25}$. See also **Anhydrite**; **Gypsum**; and **Calcium Sulfate**.

Calcium Sulfide. [CAS: 20548-54-3]. CaS, grayish-white solid, reactive with H_2O, formed by reaction of calcium sulfate and carbon at high temperatures. Calcium hydrogen sulfide, $Ca(HS)_2$, formed in solution by saturating calcium hydroxide suspension with H_2S. pK_{sp} of $CaS = 7.24$.

Calcium Sulfite. [CAS: 10257-55-3]. $CaSO_3 \cdot 2H_2O$, white precipitate, $pK_{sp} = 7.9$, formed by reaction of soluble calcium salt solution and sodium sulfite solution, or by boiling calcium hydrogen sulfite solution; calcium hydrogen sulfite, $Ca(HSO_3)_2$, formed in solution by saturating calcium hydroxide or carbonate suspension with sulfurous acid.

Calcium Tartrate. $CaC_4H_4O_6 \cdot H_2O$, white solid, solubility: at 0°C 0.0875, at 80°C 0.180 g anhydrous salt in 100 ml saturated solution, formed by reaction of calcium carbonate or hydroxide and tartaric acid, or by precipitation of Ca^{2+} with a tartrate solution.

For the role of calcium in biological systems, see **Calcium (In Biological Systems)**.

STEPHEN E. HLUCHAN
Pfizer Inc.
Wallingford, Connecticut

Additional Reading

Carter, G.F. and D.E. Paul: *Materials Science and Engineering,* ASM International, Materials Park, OH, 1991.

Considine, D.M. and G.D. Considine: *Van Nostrand Reinhold Encyclopedia of Chemistry,* 4th Edition, Van Nostrand Reinhold Company, New York, NY, 1984. (A Classic Reference).

Kent, J.A.: *Riegel's Handbook of Industrial Chemistry,* 9th Edition, Chapman & Hall, New York, NY, 1992.

Lewis, R.J. and N.I. Sax: *Sax's Dangerous Properties of Industrial Materials,* John Wiley & Sons, Inc., New York, NY, 1999.

Lide, D.R.: *CRC Handbook of Chemistry and Physics,* 84th Edition, CRC Press, LLC, Boca Raton, FL, 2003.

Meyers, R.A.: *Handbook of Chemicals Production Processes,* McGraw-Hill, New York, NY, 1986.

Parker, P.: *McGraw-Hill Encyclopedia of Chemistry,* 2nd Edition, McGraw-Hill Companies, Inc., New York, NY, 1993.

Perry, R.H. and D.W. Green: *Perry's Chemical Engineers' Handbook,* 7th Edition, McGraw-Hill Companies, Inc., New York, NY, 1999.

Staff: *ASM Handbook—Properties and Selection: Nonferrous Alloys and Special-Purpose Materials,* ASM International, Materials Park, OH, 1990.

CALCIUM CARBONATE. [CAS: 1317-65-3]. Calcium carbonate, [CAS: 471-34-1]. $CaCO_3$, mol wt 100.09, occurs naturally as the principal constituent of limestone, marble, and chalk. Powdered calcium carbonate is produced by two methods on the industrial scale. It is quarried and ground from naturally occurring deposits and in some cases beneficiated. It is also made by precipitation from dissolved calcium hydroxide and carbon dioxide. The natural ground calcium carbonate and the precipitated material compete industrially based primarily on particle size and the characteristics imparted to a product.

Calcium carbonate is one of the most versatile mineral fillers and is consumed in a wide range of products, including paper, paint, plastics, rubber, textiles, caulks, sealants, and printing inks. High purity grades of both natural and precipitated calcium carbonate meet the requirements of the *Food Chemicals Codex* and the *United States Pharmacopeia* and are used in dentifrices, cosmetics, foods, and pharmaceuticals.

Properties

Calcium carbonate occurs naturally in three crystal structures: calcite, aragonite, and, although rarely, vaterite. Calcite is thermodynamically stable; aragonite is metastable and irreversibly changes to calcite when heated in dry air to about 400°C.

The commercial grades of calcium carbonate from natural sources are either calcite, aragonite, or sedimentary chalk. In most precipitated grades aragonite is the predominant crystal structure. The essential properties of the two common crystal structures are shown in Table 1.

TABLE 1. PROPERTIES OF CALCIUM CARBONATE

Property	Calcite	Aragonite
specific gravity	2.60–2.75	2.92–2.94
hardness, Mohs'	3.0	3.5–4.0
solubility at 18°C, g/100 g H_2O	0.0013	0.0019
melting point, °C	1339[a] dec 900	[b]
index of refraction		
α		1.530
β		1.680
γ		1.685
ω	1.658	
ε	1.486	

[a] At 10.38 MPa (102.5 atm).
[b] Decomposes to calcite at temperatures >400°C

Economic Aspects

The principal U.S. producers of ground calcium carbonate are Columbia River Carbonates, ECC International, Franklin Limestone Company, Genstar Stone Products, Georgia Marble Company, J.M. Huber Corporation, Calcium Carbonates Division, James River Limestone Company, Inc., OMYA Inc. (Pluess-Staufer), and MTI Inc. The principal U.S. producers of precipitated calcium carbonate are Mississippi Lime Company and MTI Inc.

Health and Safety Factors

Calcium carbonate is listed as a food additive and not considered a toxic material. The exposure to dust is regulated and a Threshold Limit Value–Time-Weighted Average (TLV–TWA) of 10 mg/m^3 is set. Both natural ground and precipitated calcium carbonates can contain low levels of impurities that are regulated.

F. PATRICK CARR
DAVID K. FREDERICK
OMYA, Inc.

Additional Reading

Katz, H. S. and J. V. Milewski: *Handbook of Fillers for Plastics,* Van Nostrand Reinhold Co., New York, NY, 1987, p. 123.
Klein, C. and C. S. Hurlbut, Jr.: *Manual of Mineralogy,* John Wiley & Sons, Inc., New York, NY, 1985, pp. 328, 335.
O'Driscoll, M.: *Industrial Fillers* **276**, 21 (Sept. 1990).
Reeder, R. J. ed.: *Carbonates, Mineralogy and Chemistry,* Mineralogical Society of America, Washington, DC, 1990, p. 191.

CALCIUM CHLORIDE.   Calcium chloride, [CAS: 10043-52-4]. CaCl$_2$, is a white, crystalline salt that is very soluble in water. Solutions containing 30–45 wt % CaCl$_2$ are used commercially. Of the alkaline-earth chlorides it is the most soluble in water. It is extremely hygroscopic and liberates large amounts of heat during water absorption and on dissolution. It forms a series of hydrates containing one, two, four, and six moles of water per mole of calcium chloride (Table 1). Another hydrate, CaCl · 0.33H$_2$O, has been identified, mol wt 116.98; 94.8 wt % CaCl$_2$; heat of solution in water to infinite dilution, −71.37 kJ/mol (−17.06 kcal/mol).

Commercial applications of calcium chloride and its hydrates exploit one or more of its properties with regard to aqueous solubility, hygroscopic nature, the heat gained or lost when one hydrated phase changes to another, and the depressed freezing point of the eutectic solution at a composition of about 30% by weight calcium chloride.

Properties

The properties of calcium chloride and its hydrates are summarized in Table 1.

Calcium Chloride Solutions. Because of high solubility in water, calcium chloride is used to obtain solutions having relatively high densities.

Viscosity is an important property of calcium chloride solutions in terms of engineering design and in application of such solutions to flow-through porous media. Data and equations for estimated viscosities of calcium chloride solutions over the temperature range of 20–50°C are available.

Production and Consumption

Significant quantities of calcium chloride are produced in the United States, Canada, Mexico, Germany, Belgium, Sweden, Finland, Norway, and Japan. In the United States the principal route for making calcium chloride is by the evaporation of underground brines. Additional commercial material is available by the action of hydrochloric acid on limestone.

Uses

Calcium chloride, manufactured for over 100 years, has been used for a variety of purposes. The primary CaCl$_2$ markets have not changed since the 1950s. Significant markets in the United States are for deicing during the winter and roadbed stabilization, and as a dust palliative during the summer. Use as an accelerator in the ready-mix concrete industry is sizable but there is concern about chloride usage because of the possible corrosion of steel in highways and buildings. Calcium chloride is also used in oil and gas well drilling. The size of that market is dependent on the state of the worldwide oil and gas industry.

Food. Food-grade calcium chloride is used in cheese making to aid in rennet coagulation and to replace calcium lost in pasteurization. In the canning industry it is used to firm the skin of fruit such as tomatoes, cucumbers, and jalapenos. It acts as a control in many flocculation, coagulation systems. Food-grade calcium chloride is used in the brewing industry both to control the mineral salt characteristics of the water and as a basic component of certain beers.

Toxicity and Environmental

Above certain levels chloride is toxic to plants and animals. Thus, when considering calcium chloride, potentially large concentrations of calcium ion can be tolerated, but at these concentrations the chloride ion becomes toxic.

Calcium chloride solutions, typically employed at 2–5% concentration, are used as antispasmodics, diuretics, and in the treatment of tetany. Concentrated solutions of calcium chloride cause erythema, exfoliation, ulceration, and scarring of the skin. Injections into the tissue may cause necrosis. If given orally calcium chloride can cause irritation to the gastrointestinal tract unless accompanied by a demulcent. There is no published information on mutagenicity or carcinogenicity caused by calcium ions or calcium chloride. Calcium chloride has been give a toxicity or hazard level of 3. Materials in this classification typically have LD$_{50}$ below 400 mg/kg or an LC$_{50}$ below 100 ppm.

KENNETH I. G. REID
ROGER KUST
Tetra Chemicals

Additional Reading

Chemical Economics Handbook—Chlorine and Alkali Chemicals, SRI International, Menlo Park, CA, 1990.
Meissingset, K. K. and F. Gronvold: *J. Chem. Thermodynam.* (18), 159–173 (1986).
Sax, N. I. and R. J. Lewis: *Dangerous Properties of Industrial Materials,* 7th Edition, Vol. I, Von Nostrand Reinhold, New York, NY, 1989, p. 678.
Sinke, G. C., E. H. Mossner, and J. L. Curnutt: *J. Chem. Thermodynam.* (17), 893–899 (1985).

CALCIUM HYPOCHLORITE.   See **Bleaching Agents.**

CALCIUM (In Biological Systems).   The biological role and, consequently, the importance of calcium in foods for humans and feedstuffs for

TABLE 1. PROPERTIES OF CALCIUM CHLORIDE HYDRATES

Property	CaCl · 6H$_2$O	CaCl · 4H$_2$O	CaCl · 2H$_2$O	CaCl · H$_2$O	CaCl$_2$
mol wt	219.09	183.05	147.02	129.00	110.99
composition, wt % CaCl$_2$	50.66	60.63	75.49	86.03	100.00
mp, °C	30.08	45.13	176	187	772
sp gravity, d^{25}_4	1.71	1.83	1.85	2.24	2.16
heat of fusion or transition, kJ/mol[a]	43.4	30.6	12.9	17.3	28.5
heat of solution in water[b], kJ/mol[a]	15.8	−10.8	−44.05	−52.16	−81.85
heat of formation, at 25°C, kJ/mol[a]	−2608	−2010	−1403	−1109	−795.4
heat capacity, at 25°C, J/(g · ° C)[a]	1.66	1.35	1.17	0.84	0.67

[a] To convert J to cal, divide by 4.184.
[b] To infinite dilution.

livestock is well established. Although about 99% of the calcium in the bodies of animals is found in bones and teeth, the element is an essential constituent of all living cells.

Various calcium salts and organic compounds fall into this category of dietary supplements and are frequently used in feeds and foods. Some of the more important additives include calcium carbonate, calcium glycerophosphate, calcium phosphate (di- and monobasic), calcium pyrophosphate, calcium sulfate, and calcium pantothenate.

Limestone is frequently used to augment animal feedstuffs. When used, it must be low in fluorine. Calcite limestone is preferred. Calcium is also supplied in the form of crushed oyster shells, marl, gypsum (calcium sulfate), bone meal, and basic slag. In compounding feedstuffs, the specific selection of calcium source is dependent upon the species to be fed. The requirements differ, for example, between cattle, swine, and poultry. The quantity required also varies with the life stage of the animal. For example, laying hens require a much higher percentage of calcium in their diet than starting poultry.

In the mammalian body, calcium is required to insure the integrity and permeability of cell membranes, to regulate nerve and muscle excitability, to help maintain normal muscular contraction, and to assure cardiac rhythmicity. Calcium plays an essential role in several of the enzymatic steps involved in blood coagulation and also activates certain other enzyme-catalyzed reactions not involved in any of the foregoing processes. Calcium is the most important element of bone salt. Together with phosphate and carbonate, calcium confers on bone most of its mechanical and structural properties.

Calcium Metabolism

The aggregate of the various processes by which calcium enters and leaves the body and its various subsystems can be summarized by the term *calcium metabolism*. The principal pathways of calcium metabolism are intake, digestion and absorption, transport within the body to various sites, deposition in and removal from bone, teeth, and other calcified structures, and excretion in urine and stool.

Pathways

The principal pathways involve three subsystems of the body: (1) the oral cavity where ingestion occurs and the gastrointestinal tract where digestion and absorption take place and from which the feces is excreted; (2) the body fluids, including blood, which transport calcium, and the soft tissues and body organs to which calcium is transported and where many of its physiological functions are carried out (some of the organs, like the kidney, the liver, and sweat glands, are also responsible for calcium excretion); and (3) the skeleton, including the teeth, where calcium is deposited in the form of bone salt and from where it is removed (resorbed) after destruction of the bone salt.

Calcium Intake

This varies in different populations and is related to the food supply and to the cultural and dietary patterns of a given population. The intake of a substantial fraction of the world population falls between 400 and 1,100 mg/day, but a range encompassing 95% of all people would undoubtedly be even wider. Most populations derive half or more of their calcium intake from milk and dairy products. Calcium intakes of domestic and laboratory animals are higher than are those of humans. For example, rats typically ingest 250 mg Ca/kg body weight, and cattle 100 mg/kg, whereas humans ingest only 10 mg/kg. Ingestion falls with age in all species. The average percentage concentration of minerals in the lean body mass of vertebrates ranges from 1.1 to 2.2%.

Calcium Absorption

In most animals, including the human body, this occurs mainly in the upper portion of the small intestine. The amount and, therefore, the fraction of calcium absorbed from the gut are a function of intake, age, nutritional status, and health. Generally, the fraction absorbed decreases with age and intake and as the nutritional status improves. The absolute amount absorbed increases with intake and may or may not decrease with age. The mechanisms by which calcium is absorbed are not well understood. Active transport of the ion against an electrochemical gradient seems to be involved, but not all of the calcium appears to be absorbed by ways of this process, because

calcium absorption continues under conditions when active transport is severely depressed, as in vitamin D deficiency. Calcium absorption can be enhanced by the administration of large doses of vitamin D and is depressed in vitamin D deficiency. There is uncertainty regarding the effect on calcium absorption of the parathyroid hormone, the major endocrine control of the blood calcium level. Patients with hyperparathyroidism have been shown to have higher than normal absorption and patients with hypoparathyroidism to have lower than normal absorption. Similar effects have been observed in acute animal experiments, but in most of these instances a possible indirect effect has not been excluded.

Effects of Microgravity

Experience to date with humans who have lived under microgravity conditions in spacecraft has indicated possible "demineralization" of bone structure. Research has been difficult because the time spans of exposure have been so short. More must be known, however, as plans for programs requiring living under microgravity conditions for months and years are getting underway. Some analytic marker, which can return a record of changes that have occurred during space travel, is needed. A marker isotope, calcium-48, is now being seriously considered. The isotope is not abundant in nature and must be produced in the laboratory. Researchers contemplate that, by using a laser to excite calcium-48 at its resonance frequency, it will be possible to extract the isotope from samples.

Interrelationship with Phosphorus and Vitamin D

The interdependence of calcium, phosphorus, and vitamin D is exemplary of how synergistic effects can occur from combinations of feed and food components, either with a positive or negative result in the animal body. The relative concentrations (proportions) of each component in such a combination can be quite critical. Much research has gone into these particular interrelationships; much further research is required. The relationship between phosphorus and calcium nutrition has been known since the early 1840s, when Chossat in France first discovered that pigeons develop a poor bone structure when fed diets low in calcium. A few years later, the fundamental relationship of calcium and phosphorus in animal diets was developed by French and German researchers. It was not until 1922, however, with the discovery of vitamin D, that a triangular relationship was observed. See also **Phosphorus**; and **Vitamin D**.

Calcium in Blood Plasma

The concentration of calcium in the blood plasma of most mammals and many vertebrates is quite constant at about 2.5 mM (10 milligrams per 100 milliliters plasma). In the plasma, calcium exists in three forms: (1) as the free ion. (2) bound to proteins, and (3) complexed with organic (e.g., citrate) or inorganic (e.g., phosphate) acids. The free ion accounts for about 47.5% of the plasma calcium; 46% is bound to proteins; and 6.5% is in complexed form. Of the latter, phosphate and citrate account for half.

The mechanism involved in the regulation of the plasma calcium level is not fully understood. The parathyroid glands regulate both level and constancy; when these glands are removed, the plasma level drops and tends to stabilize at about 1.5 mM, but variations in calcium intake may induce fairly wide fluctuations in the plasma level. In the intact organism, wide variations in intake produce essentially no variations in the plasma calcium value, which is stabilized at about 2.5 mM. The equilibrium between bone and plasma is believed to determine the level of the plasma calcium in parathyroidectomized animals, but this reasonable hypothesis requires further experimental support. See also **Blood**. The problem of whether parathyroid regulation is due to a single hormone with hypercalcemic properties or to two hormones, one hypocalcemic, termed calcitonin, the other hypercalcemic, termed parathyroid hormone, continues under investigation.

When the calcium ion concentration is lowered in the fluids bathing nerve axons (fluids which are in very rapid equilibrium with the blood plasma) the electrical resistance of the axon membrane is lowered, there is increased movement of sodium ions to the inside, and the ability of the nerve to return to its normal state following a discharge is slowed. Thus, on the one hand, there is hyperexcitability. But, the ability for synaptic transmission is inhibited because the rate of acetylcholine liberation is a function of the calcium ion concentration. The neuromuscular junction is affected in a similar fashion; hence, the end plate potential is lowered before the muscle membrane potential and the muscle membrane is in a hyperexcitable state. These events are reversed when the calcium ion concentration is raised above the normal in the blood plasma and in the fluids bathing muscle and nerve. It is for these reasons that hypocalcemia is associated with hyperexcitability and ultimately tetany and hypercalcemia with sluggishness and bradycardia.

Muscular Contraction and Relaxation

The role of calcium in this function is not fully understood. Some researchers have proposed that calcium is the link between the electrical and mechanical events in contraction. It has been shown *in vitro* that when calcium ions are applied locally, muscle fibers can be triggered to contract. It has further been postulated that relaxation of muscle fibers is brought about by an intracellular mechanism for reducing the concentration of calcium ions available to the muscle filaments. Others postulate that contraction occurs because calcium inactivates a relaxing substance, which is released from the sarcoplasmic reticulum in the presence of ATP (adenosine triphosphate).

Bone

This is the most important reservoir of calcium in the animal body. Accounting for the largest portion of the body's calcium, bone calcium also constitutes about 25% (weight) of fat-free, dried bones. Calcium occurs in bone mostly in the form of a complex, apatitic salt, so named for its structural resemblance to a family of calcium phosphates of which hydroxyapatite $[Ca_{10}(PO_4)_6(OH)_2]$ is the best-known mineralogical example. Since calcium occurs also as the carbonate, there is discussion as to whether bone salt contains the carbonate as a separate phase, whether some of the surface phosphate in apatite has been substituted for by carbonate, or whether bone mineral is a carbonato-apatite, such as dahlite. It is important to recognize that the crystal lattice of the bone mineral, when first laid down, does not and probably cannot have all possible calcium positions occupied. Whether stability is derived from hydrogen and/or organic bonds to which the mineral may be attached is not fully determined. It has been proposed that bone salt is a lamellar mixture of octocalcium phosphate and hydroxyapatite. This hypothesis has to account for the amount of pyrophosphate formed when bone salt is heated and also for its evolution with age, i.e., the increase with age in the calcification of bone and the corresponding drop in its induced pyrophosphate content, observations for which the apatitic structure can account. The proponents of the octocalcium phosphate hypothesis explain this by showing that octocalcium phosphate breaks down to apatite and anhydrous dicalcium phosphate, which upon further heating give rise to pyrophosphate. Finally, it is postulated that octocalcium phosphate may be present in young and presumably newly formed bone, whereas in older bone an apatitic phosphate admittedly dominates the equilibrium.

Calcium enters and remains in bone as a result of calcification processes that involve two steps: (1) deposition of bone salt of a minimum calcium content and specific gravity, which occurs by way of nucleation, probably an epitactic process on the collagen fibers, with the ground substance (mostly mucopolysaccharides) between the fibers exerting either a positive or an inhibitory effect on the nucleation process; and (2) subsequent further mineralization of the bone mineral, leading to an increase in its calcium content and its specific gravity.

Calcium removal, in contrast, involves destruction of the calcified structure *in toto*. There is no evidence that only particular structures are resorbed, e.g., those with a given degree of mineralization.

The amount of calcium deposited in bone at any moment may be determined from experiments with radioactive calcium. In growing individuals, it exceeds the amount removed by bone destruction. In adults, it is about the same as the amount removed. Such individuals are considered to be in "zero" calcium balance. In older persons, the amount deposited is less than the amount removed.

Calcium's Role in Postmenopausal Women

The effectiveness of calcium supplementation in retarding the rate of bone loss in older, postmenopausal women, continued to be debated in the 1990s. Some studies demonstrated that calcium could reduce the rate of bone loss; other studies were not fully convincing, particularly with regard to slowing bone loss from the spine and hip. Dowson-Hughes and a group of researchers (Tufts University) conducted a double-blind, placebo-controlled, random trial to determine the effect of calcium on bone loss from the spine, femoral neck, and radius in over 300 healthy postmenopausal women. Conclusions: Healthy postmenopausal women whose usual dietary calcium intake is low should increase their calcium intake to 800 mg per day (essentially consistent with most RDAs). In the study calcium citrate maleate was found to be a better source of calcium than calcium carbonate for dietary augmentation.

A 1990 study by Sheikh and Fordtran (Baylor University Medical Center) indicated that there are important differences in the bioavailability of calcium from different calcium-containing compounds. The ability to dissolve a preparation in dilute acid is a major factor that contributes to bioavailability. Currently, the FDA (Food and Drug Administration, U.S.) does not require commercially available products to meet specific dissolution standards.

R.L. Prince and a group of investigators (Sir Charles Gairdner Hospital, Nedlands, Western Australia and King Edward Memorial Hospital, Subiaco, Western Australia) researched the effects of exercise, calcium supplementation, and hormone replacement therapy over a two-year period involving 120 postmenopausal women. General conclusions of the study: "In postmenopausal women with low bone density, bone loss can be slowed or prevented by exercise plus calcium supplementation, or prevented by exercise plus calcium supplementation or estrogen-progesterone replacement. Although the exercise-estrogen regimen was more effective than exercise and calcium supplementation in increasing bone mass, it also caused more side effects.

Preclampsia

During but mainly at the end of pregnancy, a syndrome referred to as *preclampsia* may develop during labor or in the immediate puerperium. The condition is relatively common and poses a danger to mother and baby. With current knowledge, the condition is unpredictable in its onset and progression. Presently, the only known treatment is to terminate the pregnancy.

Although there is no specific diagnostic test, certain abnormalities, including hypertension and proteinuria, may be detected. Zemel et al., as the result of conducting a study of over 50 women during each trimester of pregnancy, have found that an increase in the sensitivity of platelet calcium to arginine vasopressin may be an early predictor of subsequent preclampsia.

Excretion of Calcium

The principal routes of excretion are stool and urine. Calcium in the stool may be considered as made up of unabsorbed food calcium and non-reabsorbed digestive juice calcium. The latter is termed the fecal endogenous calcium. The proportion of fecal endogenous calcium to urinary calcium varies in different species. It is approximately 1:1 in humans and 10:1 in the rat and in cattle. The calcium in the urine may have a dual origin—calcium that was filtered at the glomerulus and failed to get reabsorbed along the length of the nephron, and calcium that may have originated from trans-tubular movement in certain regions of the nephron. The amount of calcium that may be lost in sweat can be large, but there is no convincing evidence that sweat is a habitual route of significant loss.

Natural Availability of Calcium

The soils of humid regions are commonly low in calcium; thus, ground limestone usually is applied to add the element, reduce the toxicity of aluminum and manganese, and correct soil acidity. The soils of dry areas are frequently rich in calcium. There is little evidence to indicate a strong relationship between human nutrition and calcium excesses or deficiencies in the soil. Even with farm livestock, most calcium deficiencies are not related to levels of available calcium in the soil. The reason for this anomaly is evident when one examines some of the controls over the movement of calcium in the food chain.

At the step in the food chain when calcium moves from the soil to the plant, controls based upon the genetic nature of the plant are very important.

Because of these controls, certain plant species always accumulate fairly high concentrations of calcium; while other plants accumulate rather low concentrations. Among the forage crops, red clover grown, for example, on the low-calcium soils of the northeastern United States, contains more calcium than grasses grown on the high-calcium soils of the western United States. Among the food crops, snap beans and peas normally contain about three to five times as much calcium as corn (maize) and tomatoes. Thus, the level of calcium in the diets of people or of animals depends more on what kinds of plants are included in the diet than it does on the supply of available calcium in the soil where these plants are grown.

Adding limestone to soils to correct soil acidity and to supplement available calcium will, of course, indirectly affect human and calcium nutrition, but this is a difficult quantity to measure.

Additional Reading

Amjad, Z.: *Calcium Phosphates in Biological and Industrial Systems,* Kluwer Academic Publishers, Norwell, MA, 1997.

Carafoli, E. and C.B. Klee: *Calcium as a Cellular Regulator,* Oxford University Press, Inc., New York, NY, 1999.

Carafoli, E. and J.R. Krebs: *Calcium Homeostasis,* Springer-Verlag Inc., New York, NY, 2000.

Dawson-Hughes, B., et al.: "A Controlled Trial of the Effect of Calcium Supplementation on Bone Density in Postmenopausal Women," *N. Eng. J. Med.,* **878** (September 27, 1990).

Fleisch, H.: *Biophosphonates in Bone Disease: From the Laboratory to the Patient,* Academic Press, Inc., San Diego, CA, 2000.

Kostyuk P.G. and A. Verkhratsky: *Calcium Signalling in the Nervous System,* John Wiley & Sons, Inc., New York, NY, 1996.

Nuccitelli, R., American Society for Cell Biology: *A Practical Guide to the Study of Calcium in Living Cells,* Vol. 40, Academic Press, Inc., San Diego, CA, 2000.

Peterson, Ole, H.: *Measuring Calcium and Calmodulin Inside and Outside Cells,* Springer-Verlag Inc., New York, NY, 2000.

Prince, R.L., et al.: "Prevention of Postmenopausal Osteoporosis," *N. Eng. J. Med.,* **1189** (October 24, 1991).

Putney, J.W.: *Calcium Signaling,* CRC Press, LLC., Boca Raton, FL, 1999.

Redman, C.W.G.: "Platelets and the Beginnings of Preclampsia," *N. Eng. J. Med.,* **478** (August 16, 1990).

Romanini, C. and A.L. Tranquilli: *Calcium Antagonists in the Treatment of Hypertension in Pregnancy,* Parthenon Publishing Group, New York, NY, 1999.

Sheikh M.S. and J.S. Fordtran: "Calcium Bioavailability from Two Calcium Carbonate Preparations" (correspondence), *N. Eng. J. Med.,* **921** (September 27, 1990).

Sotelo J.R. and J.C. Benech: *Calcium and Cellular Metabolism Transport and Regulation,* Kluwer Academic Publishers, Norwell, MA, 1997.

Vedral J.L., Institute of Medicine: *Dietary Reference Intakes: For Calcium, Phosphorus, Magnesium, Vitamin D, and Fluoride,* National Academy Press, Washington, DC, 1999.

Verkhratsky, A. and E.C. Toescu: *Integrative Aspects of Calcium Signalling,* Plenum Publishing Corporation, New York, NY, 1998.

Watterson, D.M. and L.J. Van Eldik: *Calmodulin and Signal Transduction,* Academic Press, Inc., San Diego, CA, 1998.

Zemel, M.B., et al.: "Altered Platelet Calcium Metabolism as an Early Predictor of Increased Peripheral Vascular Resistance and Preclampsia in Urban Black Women," *N. Eng. J. Med.,* **434** (August 16, 1990).

CALCIUM SULFATE. [CAS: 10101-11-4]. Calcium sulfate, $CaSO_4$, in mineral form is commonly called and occurs abundantly in many areas of the world. In natural deposits, the main form is the dihydrate. Some anhydrite is also present in most areas, although to a lesser extent. Mineral composition can be found in Table 1. The hemihydrate is normally produced by heat conversion of the dihydrate from which $\frac{3}{2}H_2O$ is removed as vapor. The resulting powder is also known as plaster of Paris. Stucco has the greatest commercial significance of these materials. It is the primary constituent used to fabricate products and in formulated plasters used in job or shop-site applications.

About 23×10^6 t of gypsum are consumed annually. About 80% is processed into the commercially usable hemihydrate. Uses of gypsum are in fabricated or formulated building materials, Portland cement set regulation, and agricultural soil conditioning.

Properties

Table 2 lists the physical properties of calcium sulfate.

TABLE 1. GYPSUM FORMS AND COMPOSITION

Common name	Molecular formula	Composition, wt %		
		CaO	SO₃	Combined H₂O
anhydrite	$CaSO_4$	41.2	58.8	
gypsum	$CaSO_4 \cdot 2H_2O$	32.6	46.5	20.9
stucco	$CaSO_4 \cdot \frac{1}{2}H_2O$	38.6	55.2	6.2

TABLE 2. PHYSICAL PROPERTIES OF CALCIUM SULFATE

Property	Dihydrate	Hemihydrate	Anhydrite
mol wt	172.17	145.15	136.14
transition point, °C	128[a]	163[b]	
	163[b]		
mp[c] °C	1450	1450	1450
specific gravity	2.32		2.96
solubility at 25°C, g/100 g H₂O	0.24	0.30	0.20
hardness, Mohs'	1.5–2.0		3.0–3.5

[a] Hemihydrate is formed.
[b] Anhydrous material is formed.
[c] Compound decomposes.

Sources

The natural, or mineral, form of gypsum is most widely extracted by mining or quarrying and used commercially.

Gypsum is also obtained as a by-product of various chemical processes. The main sources are from processes involving scrubbing gases evolved in burning fuels that contain sulfur, and the chemical synthesis of chemicals, such as sulfuric acid, phosphoric acid, titanium dioxide, citric acid, and organic polymers.

Decomposition Thermodynamics

The thermodynamic properties of gypsum decomposition, which involve two distinct steps, have been the subject of much theoretical and practical study. Two forms of the hemihydrate, α and β, have been identified.

$$CaSO_4 \cdot 2H_2O \xrightarrow{\Delta} CaSO_4 \cdot \frac{1}{2}H_2O + 1\frac{1}{2}H_2O$$

$$CaSO_4 \cdot \frac{1}{2}H_2O \xrightarrow{\Delta} CaSO_4 + \frac{1}{2}H_2O$$

Manufacture

Natural Gypsum. Gypsum rock from the mine or quarry is crushed and sized to meet the requirements of future processing or removed for direct marketing of the dihydrate as a cement retarder. Once subjected to a secondary crusher, calcining, and drying, the product is fine-ground. Fine-ground dihydrate is commonly called land plaster, regardless of its intended use. The degree of fine grinding is dictated by the ultimate use. The majority of fine-ground dihydrate is used as feed to calcination processes for conversion to hemihydrate.

β-Hemihydrate. The dehydration of gypsum, commonly referred to as calcination in the gypsum industry, is used to prepare hemihydrate, or anhydrite. Kettle calcination continues to be the most commonly used method of producing β-hemihydrate.

α-Hemihydrate. Three processing methods are used for the production of α-hemihydrate. One, developed in the 1930s, involves charging lump gypsum rock 1.3–5 cm in size into a vertical retort, sealing it, and applying steam at a pressure of 117 kPa (17 psi) and a temperature of about 123°C. After calcination under these conditions for 5–7 h the hot moist rock is quickly dried and pulverized.

Another method, first reported in the 1950s, has lower water demand. The dihydrate is heated in a water solution containing a metallic salt such as $CaCl_2$ at pressures not exceeding atmospheric. A third method, developed in 1967, prepares very low water-demand α-hemihydrate by autoclaving powdered gypsum in a slurry. A crystal-modifying substance such as succinic acid or malic acid is added to the slurry in the autoclave to produce large squat crystals.

Anhydrite. In addition to kettle calcination, soluble anhydrite is commercially manufactured in a variety of forms, from fine powders to granules 4.76 mm (4 mesh) in size, by low temperature dehydration of gypsum.

Production and Trade

Crude gypsum is the principal form of calcium sulfate shipped in international trade, although the 1980s saw an increase in the volume of fabricated products moved across international borders. See also **Gypsum**.

DONALD J. PETERSEN
NORBERT W. KALETA
LARRY W. KINGSTON
Natural Gypsum

Additional Reading

Kelly, K. K., J. C. Southard, and C. T. Anderson: *U.S. Bureau of Mines Technical Papers*, Technical Paper 625, 1941.
Mineral Industry Surveys, U.S. Dept. of Interior, Bureau of Mines, Washington, DC, Jan. 1990.

CALICHE (Nitrate).   The gravel, rock, soil, or alluvium cemented with soluble salts of sodium in the nitrate deposits of the Atacama Desert of northern Chile and Peru. The material contains from 14 to 25% sodium nitrate, 2 to 3% potassium nitrate, and up to 1% sodium iodate, plus some sodium chloride, sulfate, and borate. At one time, this was an important natural fertilizer.

CALIFORNIUM.   [CAS: 7440-71-3]. Chemical element, symbol Cf, at. no. 98. at. wt. 251 (mass number of the most stable isotope), radioactive metal of the *Actinide* series, also one of the *Transuranium* elements. All isotopes of californium are radioactive; all must be produced synthetically. See also **Radioactivity**. The isotope ^{245}Cf was first produced by S.G. Thompson, K. Street, Jr., A. Ghiorso, and G.T. Seaborg at the University of California at Berkeley in 1950 by bombarding microgram quantities of ^{242}Cm with helium ions. The reaction: ^{242}Cm $(\alpha, n) \rightarrow {}^{245}$Cf. The isotope has a half-life of 44 min. A number of other isotopes of Cf have been made, one of which, ^{254}Cf, half-life 55 days, is of interest because it decays predominantly by spontaneous fission. The longest-lived isotope is ^{251}Cf ($t_{1/2}$ = about 700 yrs), the next is ^{249}Cf ($t_{1/2}$ = 470 yrs). Except for ^{250}Cf ($t_{1/2}$ = 10 yrs), and ^{252}Cf ($t_{1/2}$ = 2.2 yrs), all other isotopes have half-lives less than one year. Several other isotopes (246, 248, 249, 250, 252) also decay by spontaneous fission, but with fission half-lives much longer than the half-lives for alpha-decay. Californium is considered to occur in its compounds only in the tripositive state.

Studied through the use of tracer quantities, the chemical properties of californium indicate that its chemical properties are analogous to those of the tripositive actinides and lanthanides, showing the fluoride and the oxalate to be insoluble in acid solution, and the halides, perchlorate, nitrate, sulfate and sulfide to be soluble.

Probable electronic configuration:

$$1s^2 2s^2 2p^6 3s^2 3p^6 3d^{10} 4s^2 4p^6 4d^{10} 4f^{14} 5s^2 5p^6 5d^{10} 5f^{10} 6s^2 6p^6 7s^2$$

Ionic radius: Cf^{3+} 0.98 Å.

In 1960, Cunningham and Wallmann isolated 0.3 microgram of californium (as californium-249) oxychloride. The best isotope for the study of californium is ^{249}Cf, which can be isolated in pure form through its beta particle-emitting parent, ^{249}Bk.

Californium-252 is an intense neutron source. One gram emits 2.4 × 10^{12} neutrons per second. This isotope shows promise for applications in neutron activation analysis, neutron radiography, and as a portable source for field use in mineral prospecting and oil well logging.

See also **Chemical Elements**.

Additional Reading

Choppin G.R., G.S. Thompson, A. Ghiorso, and B.G. Harvey: "Nuclear Properties of Some Isotopes of Californium, Elements 99 and 100," *Phys. Rev.*, **94**, 4, 1080–1081 (1954). (A classic reference.)
Fuger, J. and L.R. Moss: *Transuranium Elements: A Half Century,* American Chemical Society, Washington, DC, 1992.
Greenwood N.N. and A. Earnshaw: *Chemistry of the Elements,* 2nd Edition, Butterworth-Heinemann, Inc., Woburn, MA, 1997.
Hawley G.G. and R.J. Lewis: *Hawley's Condensed Chemical Dictionary,* 13th Edition, John Wiley & Sons, Inc., New York, NY, 1999.
Hulet E.K., Thompson, S.G., Ghiorso, A., and K. Street, Jr.: "New Isotopes of Berkelium and Californium," *Phys. Rev.*, **84**, 2, 366–367 (1951). (A classic reference.)
Krebs, R.E.: *The History and Use of Our Earth's Chemical Elements: A Reference Guide,* Greenwood Publishers Group, Inc., Westport, CT, 1998.
Lagowski, J.J.: *MacMillan Encyclopedia of Chemistry,* Vol. 1, MacMillan Library Reference, New York, NY, 1997.
Lide, D.R.: *CRC Handbook of Chemistry and Physics,* 84th Edition, CRC Press, LLC., Boca Raton, FL, 2003.
Loretta, J. and P.W. Atkins: *Chemistry: Molecules, Matter and Change,* W.H. Freeman and Company, New York, NY, 1999.
Parker, S.P.: *McGraw-Hill Concise Encyclopedia of Science and Technology,* 4th Edition, The McGraw-Hill Companies, Inc., New York, NY, 1998.
Seaborg, G.T. and W.D. Loveland: *The Elements beyond Uranium,* John Wiley & Sons, Inc., New York, NY, 1990.
Stwertka, A. and E. Stwertka: *A Guide to the Elements,* Oxford University Press, Inc., New York, NY, 1998.
Wierzbicki, J.G. and Staff: North Atlantic Treaty Organization: *Californium-252: Isotope for 21st Century Radiotherapy,* Kluwer Academic Publishers, Norwell, MA, 1997.

CALORESCENCE.   A term designating the production of visible light by means of energy derived from invisible radiation of frequencies below the visible range. Tyndall found it possible to raise a piece of blackened platinum foil to a red heat by focusing upon it infrared radiation from an arc or from the sun, the visible wavelengths having been filtered out. It is to be noted that the transformation is indirect, the light being produced by heat and not by any direct stepping up of the infrared frequency. A somewhat analogous phenomenon is the production of visible sparks or the glowing of a fine platinum wire in a resonant circuit energized by long-wave Hertzian radiation.

CALORIMETRY.   Calorimetry is one of the oldest reported scientific measurement techniques. Calorimetry is derived from the Latin *calor* meaning heat, and the Greek *metry* meaning to measure. All calorimetric techniques are based on the measurement of heat that may be generated (exothermic process) or consumed (endothermic process) by a sample or system. The approaches to measuring such heat transfers are numerous. Since calorimetry's advent in the late 18th century, a large assortment of techniques has been developed. Initially calorimetric techniques were based on simple temperature measurement methods. More recently, advances in electronics and control have added a new dimension to calorimetry, enabling users to collect data and maintain samples under conditions that were previously unattainable.

Any process that results in heat being generated and/or exchanged with the environment is a candidate for a calorimetric study. As a result, it is not surprising to discover that calorimetry has a very broad range of applicability, with examples ranging from characterizing the heating value of fuel materials to drug design in the pharmaceutical industry, to quality control of process streams in the chemical industry, and the study of metabolic rates in biological systems.

Traditionally, calorimeters have been classified according to the degree of heat transfer occurring between the reacting system and its surroundings. At one extreme, the calorimeter is isolated as fully as possible from its surroundings so that heat transfer is minimized. Work added or energy converted by a chemical process causes a change in temperature of the calorimeter and its contents. For an exothermic reaction, the ideal situation is approximated by heating the jacket that surrounds the calorimeter in order to nullify the temperature differential between the calorimeter cell and its surroundings during a test. This type of calorimeter is known as an adiabatic calorimeter.

The second extreme case uses a good, systematic path for heat to flow between the calorimeter system and its surroundings. In this type of calorimeter, there is no net change in temperature between the calorimeter cell and its surroundings. In this conduction type or isothermal calorimeter, the transfer of heat from the calorimeter to the surroundings is typically measured by integrating the voltage output of a thermoelectric transducer situated in the heat flow path between the calorimeter and a heat sink. Another variation of this type of calorimeter operates on the heat compensation principle. Either Joule heating or Peltier cooling compensation of the reaction enthalpy achieves isothermal conditions.

An important variation of the adiabatic principle is isoperibol calorimetry. Well-defined heat leaks, minimized by efficient calorimeter construction and experiment design, are compensated for by calculation and/or extrapolation. The isoperibol design holds the temperature of the immediate environment surrounding the calorimeter constant. The word isoperibol literally means "constant temperature environment."

Calorimetry is used to determine the thermodynamic properties of materials as well as to measure the thermal effects associated with physical and chemical processes. These properties are valuable, for example, in choosing practical manufacturing techniques in the chemical process industries, optimizing yields of reaction products, making energy balances, and as diagnostic or analysis tools. Calorimetry is also used to determine the important properties of complex or poorly defined materials. One example of this is in determining the heating value of fuels such as coal and coke, petroleum products and gaseous fuels. In recent years, the heating value of incinerated waste materials and other refuse-derived fuels have become important. The calorimetry of hazardous and explosive materials has direct relevance to the transport and safe handling of chemicals.

Interest in the use of calorimetry as a routine diagnostic or analysis tool has gained significant momentum only in the last 50 years. This interest has lead to the development of popular procedures such as differential thermal analysis (DTA) and differential scanning calorimetry (DSC). A wide variety of solution calorimetric techniques exist today. These techniques include thermometric titration, injection and flow enthalpimetry. The major growth of commercial instrumentation for calorimetry has occurred to address applications in routine analysis and the rapid characterization of materials.

The following discussion will use an oxygen combustion calorimeter to help illustrate some of the basic concepts of calorimetry. One of the fundamental characteristics of any fuel is the amount of energy released as it is burned. This value is referred to as the heat of combustion or the calorific value of the fuel. This value is usually expressed in British thermal units per pound (Btu/lb), calories per gram (cal/g), or megajoules per kilogram (MJ/kg). The heat of combustion of fuels is routinely determined in order to establish the price of the fuel as well as to serve as a basis for calculating the overall efficiency of a power-generating facility or engine.

To determine the heat of combustion of a fuel, a representative sample (typically one gram) is burned in a high-pressure oxygen atmosphere within a metal "bomb" or pressure vessel. The energy released by this combustion is adsorbed by the calorimeter and results in a temperature rise. The heat of combustion of the sample is the product of the temperature rise and the predetermined energy equivalent or calibration factor for the instrument. Burning a sample with a known heat of combustion and recording the temperature rise of the calorimeter determines the energy equivalent of the calorimeter. This factor is usually expressed as the amount of heat necessary to raise the calorimeter one degree Celsius. Benzoic acid is used almost exclusively as a reference material for fuel calorimetry because it is nonhydroscopic and is readily available in a very pure form.

Any oxygen bomb calorimeter consists of four essential parts: (1) a bomb or vessel in which the sample is burned; (2) a bucket or container which holds the bomb as well as a precisely measured quantity of water to absorb the heat released from the bomb and a stirring device to aid in achieving rapid thermal equilibrium; (3) a jacket for protecting the bucket from transient thermal stresses; and (4) a thermometer for measuring temperature changes within the bucket. The cross section of a simple calorimeter is shown in Fig. 1.

The bomb consists of a strong, thick-walled, metal vessel that can be opened for inserting the sample, for cleaning, and for recovering the products of combustion. Valves are provided for filling the bomb with oxygen and releasing the residual gases after the test is complete. Electrodes to carry the ignition current to a fuse wire that ignites the sample are also provided. Pressures up to 100 bar are developed during the combustion. As a result, most bombs are constructed to safely withstand pressures of at least double this value.

The calorimeter bucket contains the bomb plus a sufficient quantity of water to completely immerse the bomb in order to absorb the heat released from the combustion. A stirrer is used in the bucket to facilitate bringing the bucket and its contents to thermal equilibrium.

The jacket that contains the bucket with its bomb provides a thermal shield to control the heat transfer between the calorimeter bucket and its surroundings. In an isoperibol calorimeter, it is not necessary to prevent this transfer, as long as a means of precisely determining the amount of heat transferred during the determination can be established.

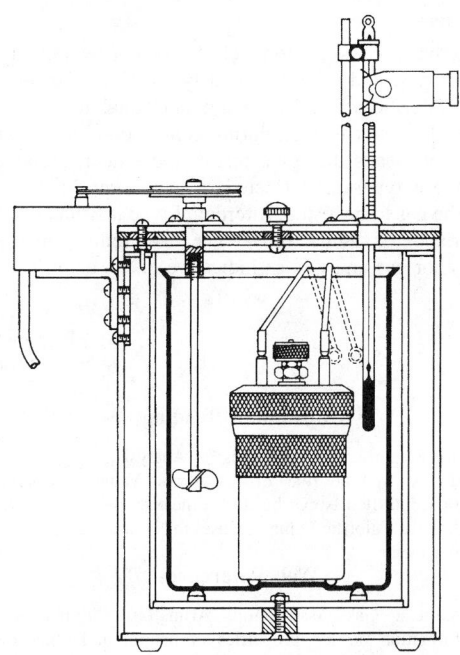

Fig. 1.   Cross-section of plain jacket oxygen bomb calorimeter. (*Parr Instrument Co.*)

The calorimetric thermometer measures temperature changes within the calorimeter bucket. It must be able to provide excellent resolution and repeatability. High single-point accuracy is not required since it is the change in temperature that is important in fuel calorimetry. Mercurial thermometers, platinum resistance thermometers, quartz oscillators, and thermistor systems have all been successfully used as calorimetric thermometers.

The American Society for Testing and Materials (ASTM) has developed a series of standard test methods for both solid and liquid fuels in oxygen bomb calorimeters. Advanced combustion calorimeters are capable of performing 8 to 10 tests per hour with a precision of 0.1%.

Calorimeters of Historical and Special Interest

Around 1760 Black realized that heat applied to melting ice facilitates the transition from the solid to the liquid state at a constant temperature. For the first time, the distinction between the concepts of temperature and heat was made. The mass of ice that melted, multiplied by the heat of fusion, gives the quantity of heat. Others, including Bunsen, Lavoisier, and Laplace, devised calorimeters based upon this principle involving a phase transition. The heat capacity of solids and liquids, as well as combustion heats and the production of heat by animals were measured with these calorimeters.

A steam calorimeter was perfected by J. Joly (1886) and used for the accurate determination of specific heats of solids, liquids, and gases. In principle this apparatus consists of a balance, with the specimen hung from one pan and surrounded by an enclosure that can be flooded with steam. The mass of moisture condensing on the specimen, multiplied by the heat of vaporization of water, gives the quantity of heat imparted to the specimen.

The Nernst calorimeter is a calorimeter for the measurement of specific heat capacities at low temperatures. The sample to be measured is suspended in a glass or metal envelope that can be evacuated. The sample is heated by means of a platinum wire located in a bore inside the sample. The wire also serves as a resistance thermometer. The specific heat capacity is determined by recording the temperature rise in the sample for a given delivery of energy.

Differential scanning calorimetry (DSC) is a technique determining the variation in the heat flow given out or taken in by a sample when it undergoes temperature scanning in a controlled atmosphere. Any transformation taking place in a material during the temperature scan is accompanied by an exchange of heat. DSC enables the temperature of this transformation to be determined and the heat from it to be quantified. The DSC is used to measure specific heat capacity and heats of transition as well as to detect the temperature of phase changes and melting points.

Gas Calorimeters

There are three basic classifications: (1) total calorific value types, (2) net calorific value types, and (3) inferential types. Net calorific value is less than the total calorific value by an amount equal to the latent heat of vaporization of the water formed during combustion. A net calorific value instrument uses a means that give results more nearly related to the net value. Thus, these types are affected by gas composition and must be calibrated for the gas to be tested. Inferential-type instruments depend upon such characteristics as flame appearance, maximum flame temperature, specific gravity, or gas analysis as indicative of the calorific value.

H. J. ALBERT
Parr Instrument Co.
Moline, Illinois

Additional Reading

D2013 Method for Preparing Coal Samples for Analysis.
D1989 Standard Test Method for Gross Calorific Value of Coal and Coke by Microprocessor Controlled Isoperibol Calorimeters.
D1826 Test Method for Calorific Value of Gases in Natural Gas Range by Continuous Recording Calorimeter.

Web References

AGA: The American Gas Association, Arlington, Virginia is an excellent source of information on the properties of natural and other fuel gases and their measurement, including calorimetry. Publications are frequently updated. *http://www.aga.org/*
ASTM: The American Society for Testing and Materials, Philadelphia, Pennsylvania has established standards and methodologies for testing fuels of all types. The following ASTM methods and standards, periodically revised, are of particular relevance. *http://www.astm.org/*

CALORIZING. Production of a protective coating of iron-aluminum alloy on iron or steel. The articles are ordinarily coated by heating to a high temperature in a closed container packed with powdered aluminum. Other processes include impregnation at high temperature with an aluminum chloride vapor and spraying with molten aluminum from a spray gun and then heating to a high temperature. When the aluminum coating is held at high temperatures, an iron-aluminum alloy forms which is resistant to oxidation and corrosion by hot combustion gases, especially those containing sulfur compounds which are particularly corrosive to bare iron or steel.

Steel sheets are aluminized by a hot-dip process similar to galvanizing. The principal applications for such a product are furnaces and ovens, automobile mufflers, and other equipment requiring heat and corrosion resistance. When a sheet which has been coated with aluminum by a hot-dip process is exposed to a temperature over 1,000°F (538°C), the aluminum forms an iron-aluminum alloy which is heat- and corrosion-resistant.

CALVIN, MELVIN (1911–1997). An American chemist who won the Nobel prize for chemistry in 1961. Muck of his work involved the study of photosynthesis, biophysics, and application of physics and chemistry of molecules to some of the basic problems of biology. His doctorate was from the University of Minnesota. He did postgraduate work in England and at Northwestern University and the University of Notre Dame.

CAMPHOR *(Cinnamonum camphora; Lauraceae).* A crystalline compound occurring in various parts of the wood and leaves of the camphor tree, a large evergreen tree with light green leaves growing in many warm regions of southeastern Asia, notably Taiwan. Camphor, $C_{10}H_{16}O_7$, is a white solid, mp 179°C, bp 209°C, of a characteristic pleasant odor, insoluble in H_2O, soluble in alcohol or ether. Camphor may be produced synthetically by converting pinene into bornyl chloride with HCl, thence to isobornyl acetate, thence to isoborneol, and finally oxidizing borneol to camphor. Camphor has found use in medicines, insecticides and moth preventives. Earlier uses included the manufacture of plastics and lacquers.

As reported by American Forests, a champion camphor tree growing in Darby Florida was selected in 1992. Dimensions of the tree: circumference (at $4\frac{1}{2}$ feet; 1.4 meters above ground level) = 422 inches (1070 centimeters); height = 67 feet (20.4 meters); spread = 103 feet (31.4 meters).

CANCER (Drugs). See **Chemotherapeutics, Anticancer.**

CANCRINITE. The mineral cancrinite is a complex hydrous silicate (see also **Silicon**) corresponding approximately to the formula $(Na, K, Ca)_{6-8}(Al, Si)_{12}O_{24} (SO_4, CO_3, Cl)_{1-2} \cdot nH_2O$. It is hexagonal, with prismatic cleavage; hardness, 5–6; specific gravity, 2.42–2.50; color, white to gray or may be greenish, bluish, yellow, or flesh red; colorless streak; luster, subvitreous to greasy; transparent to translucent. Cancrinite is found only in the nephelite-syenites and related rock types and is commonly associated with sodalite. It is believed to be in part primary, having crystallized direct from the magma, and in part secondary as a result of alteration of nephelite by solutions of calcium carbonate. It is found in the Ilmen Mountains of the former U.S.S.R., in Rumania, in Norway, in Canada in Hastings County, Ontario, and in the United States in Kennebec County, Maine. This mineral was named for Count Georg Cancrin, a Russian statesman who died in 1845.

CANNIZZARO REACTION. Base catalyzed dismutation of aromatic aldehydes or aliphatic aldehydes with no α-hydrogen into the corresponding acids and alcohols. When the aldehydes are not identical, the reaction is called the "crossed Cannizzaro reaction."

CANNIZZARO, STANISLAO (1826–1910). Born in Italy, he extended the research of Avogadro on the molecular concentration of gases and thus was able to prove the distinction between atoms and molecules. His investigations of atomic weights helped to helped to make possible the discovery of the periodic law by Mendeleyev. His research in organic chemistry led to the establishment of the Cannizzaro reaction involving the oxidation reduction of an aldehyde in the presence of concentrated alkali.

CAPILLARITY. The name given to a class of phenomena, of which the elevating or depression of liquids in fine tubes is representative. When the interface between a liquid and a gas, or between two liquids, is intercepted by a solid surface, an equilibrium is established at the junction among the forces acting along the three surfaces of contact. For example, let a plate of solid S be dipped into a liquid L having gas G above it. See Fig. 1. A molecule at the junction O is acted upon by the adhesive attraction P, by the forces which give rise to the three surface tensions along the interfaces OH, OE, and OD, and by the reaction R of the plate S against which it is drawn by the adhesion. (Its weight may be considered negligible.) The flexible interface OH adjusts itself so that these forces come into equilibrium; unless, indeed, one of them, E, exceeds the sum of D and C, in which case the liquid "creeps" indefinitely along the surface as oil does over a glass or tin container. The equilibrium polygon at the right is labeled in each case to correspond with the figure representing the surfaces. The "angle of contact" α, between the liquid surface at O and the solid surface OD, is determined by the aforesaid forces acting at O. For most liquids against glass it is acute; for mercury against glass it is obtuse. See Fig. 2. In special cases it may be 90°, and in others it reduces to zero.

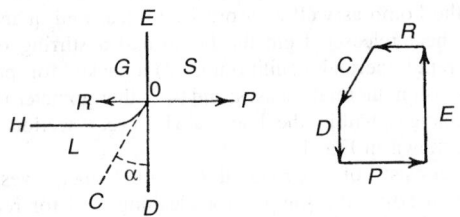

Fig. 1. Capillary force, large adhesion

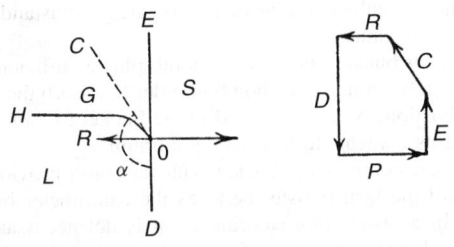

Fig. 2. Capillary force, small adhesion

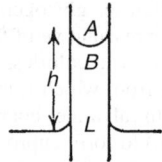

Fig. 3. Rise of liquid in capillary tube

If the interface between two media A and B (Fig. 3) is curved, A being on the concave side, the pressure in A is greater than in B on account of the surface tension; much as the pressure inside a rubber balloon is greater than outside. We can now understand why water rises in a capillary tube. In order to secure equilibrium, the liquid must rise until the pressure inside the surface at B, plus the pressure due to gravity at depth h, makes the pressure at L equal to that at the surface level outside; that is, to the atmospheric pressure. Similar reasoning applies to the depression of mercury in a glass tube. For a circular tube of internal radius r, the distance h to which capillarity will elevate (or depress) a liquid of density ρ and surface tension T (against air) is readily shown to be

$$h = \frac{2T \cos \alpha}{r \rho g}$$

where g is gravity. See also **Electrocapillarity**.

CAPILLARY. 1. Hair-like, especially in application to fine tubes. 2. A minute thin-walled blood vessel intervening between the arteries and veins. 3. A cylindrical space of small radius, or a tube containing such a space. The numerous uses of such tubes has given rise to a number of derived terms. Thus, the capillary correction is a correction applied to mercury barometers, widebore thermometers, etc., for the effect of capillarity on the height of the column. Capillary pressure is a pressure due to capillary force. See also **Capillarity**.

Capillary rise is the elevation of liquid in a capillary tube above the general level. Capillary separation is the separation of gases by flow through a porous medium. In a theory of this process based on the concept of momentum transfer, the actual porous medium is treated as equivalent to a bundle of parallel capillary tubes.

CAPRIC ACID. [CAS: 334-48-5]. Also called decanoic, decoic, and decyclic acid, formula $CH_3(CH_2)_8COOH$. The acid occurs as a glyceride in natural oils. Usual form is white crystals having an unpleasant odor. Soluble in most organic solvents and dilute nitric acid; insoluble in water. Specific gravity 0.8858 (40°C); mp 32.5°C; bp 270°C. Combustible. A component of some edible vegetable oils. See also **Vegetable Oils (Edible)**. Capric acid is derived from the fractional distillation of coconut oil fatty acids. The acid is used in esters for perfumes; fruit flavors; a base for wetting agents; as an intermediate in organic synthesis; plasticizer; resins; and used in food-grade additives.

CAPROIC ACID. [CAS: 142-62-1]. Also called hexanoic, hexylic, or hexoic acid, formula $CH_3(CH_2)_4COOH$. Present in milk fats to extent of about 2%. Also a constituent of some edible vegetable oils. See also **Vegetable Oils (Edible)**. The acid is oily, colorless or slightly yellow, and liquid at room temperature. Odor is that of Limburger cheese. Soluble in alcohol and ether; slightly soluble in water. Specific gravity 0.9276 (20.4°C); mp −4.0°C; bp 205°C. Combustible. Caproic acid is derived from the crude fermentation of butyric acid; or by fractional distillation of natural fatty acids. Used in various flavorings; manufacture of rubber chemicals; varnish dryers; resins; pharmaceuticals.

CAPROLACTAM. [CAS: 105-60-2]. $NH(CH_2)_5CO$, formula weight 112.15, liquid ingredient used in the manufacture of type 6 nylon. See also **Fibers**. Several hundred million pounds of the compound are produced annually. There are a number of proprietary processes for caprolactam production.

In one process, the chargestock is nitration-grade toluene, air, hydrogen, anhydrous NH_3, and H_2SO_4. The toluene is oxidized to yield a 30% solution of benzoic acid, plus intermediates and byproducts. Pure benzoic acid, after fractionation, is hydrogenated with a palladium catalyst in stirred reactors operated at about 170°C under a pressure of 10 atmospheres. The resultant product, cyclohexanecarboxylic acid, is mixed with H_2SO_4

and then reacted with nitrosylsulfuric acid to yield caprolactam. The nitrosylsulfuric acid is produced by absorbing mixed nitrogen oxides N_2O_3 in H_2SO_4: $N_2O_3 + H_2SO_4 \longrightarrow SO_3 + 2NOHSO_4$. The resulting acid solution is neutralized with NH_3 to yield $(NH_4)_2SO_4$ and a layer of crude caprolactam which is further purified.

A later process utilizes a photochemical reaction in which cyclohexane is converted into cyclohexanone oxime hydrochloride.

The yield of cyclohexanone is estimated at about 86% by weight. Then, in a Beckmann rearrangement, the cyclohexanone oxime hydrochloride is converted to ∈-caprolactam:

$$C_6H_{10}NOH \cdot 2HCl \xrightarrow{H_2SO_4} CH_2{-}(CH_2)_4{-}C{=}O + 2HCl$$
$$\underset{\hspace{3em}NH\hspace{3em}}{\rule{6em}{0.4pt}}$$

CARAT. See **Diamond**.

CARBAMATES. Derivatives of the hypothetical carbamic acid, H_2NCOOH, which does not exist. The ethyl derivative urethane is prepared by heating urea in alcohol under pressure, by the reaction $H_2NC({=}O)NH_2 + C_2H_5OH \longrightarrow H_2NCOOC_2H_5 + NH_3$. The structures of representative carbamates are shown below:

$$OC \Big\langle \begin{matrix} OCH_3 \\ NH_2 \end{matrix} \Big\} \qquad OC \Big\langle \begin{matrix} OC_2H_5 \\ NH_2 \end{matrix} \Big\} \qquad OC \Big\langle \begin{matrix} SC_2H_5 \\ NH_2 \end{matrix} \Big\}$$

Methyl carbamate   Methyl carbamate (urethane)   Thiourethane

CARBAMIC ACID. See **Herbicides; Insecticide**.

CARBANION. An ion of the general formula

$$\begin{matrix} & A & \\ & | & \\ B & {-}C{:}^- , & \\ & | & \\ & D & \end{matrix}$$

where A, B and D are substituent groups. Their importance in elucidating the mechanism of organic reactions is because a considerable proportion of all organic reactions involve carbanions, as others do carbonium ions and carbon free radicals (including carbene radicals). Many carbanion reactions involve removal of a proton from a carboxylic acid to form a carbanion. Many electrophilic substitution reactions involve carbanions. Carbanions are strong bases or nucleophiles. Many electrophilic substitution reactions that have carbanion intermediates are base-catalyzed since the basic reagent produces the basic carbanion. Because of the negative charge on carbanions, their structures are affected by cations, by attached substituents and particularly by the solvent.

CARBENE. The name quite generally used for the methylene radical,: CH_2. It is formed during a number of reactions. Thus the flash photochemical decomposition of ketene ($CH_2{=}C{=}O$) has been shown to proceed in two stages. The first yields carbon monoxide and :CH_2, the latter then reacting with more ketene to form ethylene and carbon monoxide. Carbene reacts by insertion into a C−H bond to form a C−CH$_3$ bond. Thus carbene generated from ketene reacts with propane to form n-butane and isobutane. Carbene generated by pyrolysis of diazomethane reacts with diethyl ether to form ethylpropyl ether and ethylisopropyl ether.

Substituted carbenes are also known; chloroform reacts with potassium t-butoxide to form dichlorocarbene:CCl_2, which adds to double or triple carbon-carbon bonds to form cyclopropane derivatives.

CARBIDES. A binary solid compound of carbon and another element. The most familiar carbides are those of calcium, tungsten, silicon, boron, and iron (cementite). Two factors have an important bearing on the properties of carbides: (1) the difference in electronegativity between carbon and the second element, and (2) whether the second element is a transition metal. Saltlike carbides of alkali metals are obtained by reaction with acetylene. Those obtained from silver, copper, and mercury salts are explosive. See also **Carbon**; and **Iron Metals, Alloys, and Steels**.

CARBOCYCLIC COMPOUNDS. See **Organic Chemistry**.

CARBOHYDRATES. These are compounds of carbon, hydrogen, and oxygen that contain the saccharose grouping (below), or its first reaction product, and in which the ratio of hydrogen to oxygen is the same as in water.

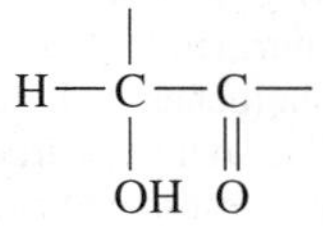

$$H - \underset{\underset{OH}{|}}{C} - \underset{\underset{O}{\|}}{C} -$$

Carbohydrates are the most abundant class of organic compounds, representing about three-fourths of the dry weight of all vegetation. Carbohydrates are also widely distributed in animals and lower life forms. These compounds comprise one of the three major components (others are protein and fat) of the human diet, and indeed that of most other animals. In a nutrition-conscious era, advocates for both more and fewer carbohydrate calories in the human diet can be found.

Classification of Carbohydrates

Because carbohydrates as components of foods and feedstuffs are not limited to just a few specific classes or types, but essentially run the gamut of the carbohydrate spectrum, it is in order here to review briefly the organization of carbohydrate chemistry, with some examples from the various classes. See also entry on **Organic Chemistry**.

Elementary Terminology

A term synonymous with carbohydrate is *saccharide* (sometimes *saccharose*). When referring to saccharides, the basic molecular formula is considered to be $C_6H_{12}O_6$. Compounds with this general formula, such as glucose, mannose, and galactose, are known as *monosaccharides* because they contain one $C_6H_{12}O_6$. A *disaccharide*, as typified by sucrose, lactose, and maltose, has the general molecular formula, $C_{12}H_{22}O_{11}$ and may be considered as containing two $C_6H_{12}O_6$ groupings that have been joined by one atom of oxygen, with the elimination of one molecule of water. Similarly, the *trisaccharides*, such as raffinose, have the molecular formula, $C_{18}H_{32}O_{16}$. Any larger molecules of the $C_x(H_2O)_y$ configuration are termed *polysaccharides*, and include the starches, celluloses, dextrin, and glycogen. See also **Starch**. An *oligosachharide* is a carbohydrate containing from two up to ten simple sugars linked together (e.g., sucrose, composed of dextrose and fructose). Beyond ten, the term *polysaccharide* is used. Gums and mucilages are complex carbohydrates. See also **Gums and Mucilages**.

Both the terms carbohydrate and saccharide are significant only by way of classifying these compounds, because neither term appears in whole or in part in any of the widely used names of these compounds. About the only point of nomenclature enjoyed in common by several of the saccharides is the termination *-ose*, as found, for example, in cellulose, dextrose, sucrose, and glucose. Any saccharides having the structure of an aldehyde is termed as *aldose*; any saccharide with the structure of a ketone is termed a *ketose*. For those saccharides that contain 4–6 carbons, the number of carbons forms a nomenclature base, as a *tetrose*, $C_4H_8O_4$, a *pentose*, $C_5H_{10}O_5$, and a *hexose*, $C_6H_{12}O_6$.

To be consistent with the relationship between a mono- and a disaccharide, some authorities do not term a tetrose or a pentose a monosaccharide. By combining the *ald-* and *ket-* prefixes, certain compounds then may be called aldohexoses, such as glucose and galactose, or ketohexoses, such as fructose and sorbose.

The mono-, di-, and trisaccharides are also commonly termed *sugars*. A sugar generally is considered to possess the properties of a crystalline solid with a relatively low melting point (below 150°C), of being soluble in water, and of possessing a sweet taste. Thus, the common names of several saccharides incorporate the term sugar, preceded by the common raw source of the substance, as glucose (grape sugar), sucrose (cane or beet sugar), maltose (malt sugar), and lactose (milk sugar). The crosscurrents of the nomenclature employed for the carbohydrates will be evident from Table 1.

Important Carbohydrates in Foods and Biological Systems

The properties of several carbohydrates that are of particular importance in foods and biological systems are described in the following paragraphs.

Glucose. This may be considered the key carbohydrate. It is the leading member of the aldohexose group, and is formed as one of the products or the only product when the following carbohydrates are hydrolyzed,

sucrose, lactose, maltose, cellulose, glycogen. In many of its properties and its structural forms, it is representative of the sugars, and it is therefore discussed in detail here. Glucose is a colorless solid ($C_6H_{12}O_6$), less sweet than sucrose, soluble in water from which it may be crystallized $C_6H_{12}O_6$ H_2O. Glucose reacts (1) with alkaline cupric salt solution (Fehling's solution or Benedict's solution) to form cuprous oxide, (2) with ammonio-silver salt solution (Tollens' solution) to form finely divided or mirror film of silver, (3) with phenylhydrazine in acetic acid, to form glucose phenylhydrazone $CH_2OH(CHOH)_4CH:NNHC_6H_5$, white solid, melting point alpha 159–160°C, beta 140–141°C, with excess phenylhydrazine to form glucosazone

$$CH_2OH(CHOH)_3C : (NNHC_6H_5) \cdot CH : NNHC_6H_5$$

yellow solid, melting point 205°C decom., (4) with acetic anhydride, to form glucose pentacetate $C_5H_6(OOCCH_3)_5CHO$, melting point alpha 112 to 113°C, beta 131 to 134°C, (5) with sodium amalgam, to form sorbitol $CH_2OH(CHOH)_4CH_2OH$, (6) with hydriodic acid, to form 2-iodo-normal-hexane $CH_3(CH_2)_3CHICH_3$, (7) with sodium hydroxide solution, to form yellowish-brown solutions upon warming, (8) with calcium hydroxide solution, to form calcium glucosate $CH_2OH (CHOH)_4COCa(OH)$, slightly soluble solid from which glucose is recoverable by action of carbon dioxide (calcium carbonate formed simultaneously). Strontium hydroxide and barium hydroxide react similarly. Any of these three reactions may be utilized to recover glucose, with the limitation that barium soluble compounds are poisonous, (9) with hydroxylamine hydrochloride, to form glucoseoxime $CH_2OH(CHOH)_4CH:NOH$, melting point 138°C, (10) with hydrocyanic acid, to form glucosecyanhydrin

$$CH_2OH(CHOH)_4CHOHCN,$$

(11) by oxidation, to yield with bromine gluconic acid $CH_2OH(CHOH)_4$ $COOH$, and with nitric acid saccharic acid $COOH(CHOH)_4COOH$, (12) with alpha-naphthol dissolved in chloroform and then forming a layer of concentrated sulfuric acid beneath the mixture, to form a red coloration at the junction of the two liquid layers (Molisch's test for carbohydrates). Upon standing, the color changes to purple. (13) With methyl alcohol in the presence of hydrogen chloride, to form methyl glucoside (methyl ether of glucose). See also **Glycosides**.

If a sample of glucose is recrystallized from water, it is found that a freshly prepared aqueous solution of this sample has a specific rotation of +113°, and upon standing, the value steadily changes to +52° and remains there. On the other hand, if a sample of the same glucose is recrystallized from pyridine, a freshly prepared aqueous solution has a specific rotation of +19°, which steadily increases upon standing and levels off at a constant value of +52°. This changing of optical rotation with time is referred to as mutarotation. The fact that the two portions of glucose when recrystallized from different solvents mutarotate and stop at the same position suggests the formation of some equilibrium mixture.

To explain this situation, it must be recognized that glucose contains an aldehyde (−CHO) group and four alcohol groups (−OH). These two kinds of groups can react to form a hemiacetal just as if they were present in different molecules. See Fig. 1.

Glucose and fructose are present in sweet fruits, such as grapes and figs, and in honey. These two are the only hexoses found in nature in the free state. Glucose is normally present in human urine to the extent of about 0.1%, but in the case of those suffering from diabetes glucose is excreted in large amounts. Glucose is formed, as previously mentioned, by the reaction of polysaccharides and water, the reaction with starch in the

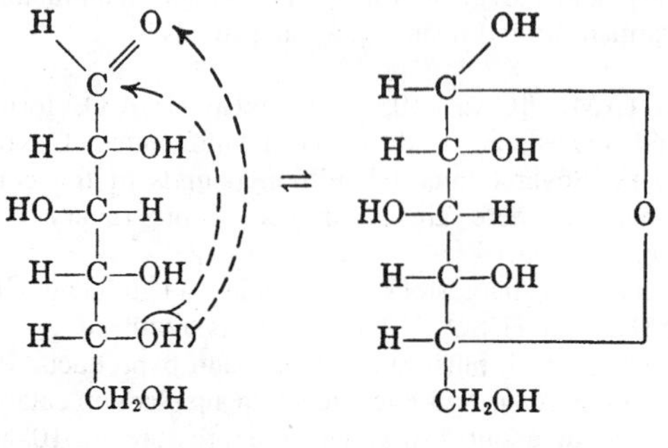

Fig. 1. Mutarotational aspects of glucose

TABLE 1. CLASSES OF CARBOHYDRATES (with examples)

Monosaccharides (sugars):

    crystalline solids, soluble in water, sweet taste; those that occur in nature are hydrolyzed by certain enzymes.

    Tetrose, $C_4H_8O_4$

        1. Erythrose

    Pentoses, $C_5H_{10}O_5$

        2. Arabinose
            By boiling gum arabic, cherry gum, corn pith, elder pith with dilute sulfuric acid.

        3. Xylose
            By boiling substances mentioned under arabinose above.

        4. Ribose

        5. Lyxose

Hexoses, $C_6H_{12}O_6$
    Aldohexoses

        6. Glucose, dextrose ("grape sugar"), melting point 146°C (anhyjeniferdrous).
            With the enzyme zymase (of yeast) yields ethyl alcohol plus carbon dioxide. Specific rotatory power—see glucose below.

        7. Galactose
            Specific rotatory power +83.9°.

        8. Mannose
            Specific rotatory power +14.1°.

        9. Glucose

      10. Idose

      11. Talose

      12. Altrose

      13. Allose

    Ketohexoses

      14. Fructose, levulose ("fruit sugar"), melting point 95°C. Specific rotatory power −88.5°.

      15. Sorbose

      16. Tagatose

Disaccharides (sugars), $C_{12}H_{22}O_{11}$: crystalline solids, soluble in water, sweet taste.

      17. Sucrose ("cane sugar," "beet sugar"), melting point 170–186°C (decomposes). With the enzyme invertase, yields glucose plus fructose. Specific rotatory power +66.4°.

      18. Lactose ("milk sugar"), melting point 202°C (anhydrous). With the enzyme lactase yields glucose plus galactose. Specific rotatory power +52.4°.

      19. Maltose ("malt sugar"), melting point of $C_{12}H_{22}O_{11} \bullet H_2O$: 100°C. With the enzyme maltase yields glucose plus glucose. Specific rotatory power +138.5°.

      20. Melibiose
           With enzymes or dilute acid yields glucose plus galactose.

      21. Cellobiose
           With the enzymes maltase, or cellase, yields glucose plus glucose.

      22. Trehalose

Trisaccharide, $C_{18}H_{32}O_{16}$: crystalline solid, soluble in water, tasteless.

      23. Raffinose, melitose, melting point 118°C (anhydrous). With the enzyme invertase, yields fructose plus melibiose. With the enzyme emulsin, yields sucrose plus galactose.

Polysaccharides (non-sugars), $(C_6H_{10}O_5)_n$: noncrystalline solids, insoluble in water, tasteless.

      24. Starches
           With the enzyme diastase yield maltose.

      25. Celluloses
           With hydrochloric acid, heated, yield glucose.
           With acetic anhydride plus concentrated sulfuric acid, yield cellobiose.

      26. Dextrin
           With the enzyme diastase yields maltose.
           With the enzyme maltase or with acids yields glucose.

      27. Inulin, melting point 178°C (decom.) $(C_6H_{10}O_5)_n$.
           With the enzyme inulase (but not with diastase) yields fructose.

      28. Glycogen, melting point 240°C.
           With the enzyme diastase (or ptyalin), yields glucose plus maltose.

      29. Pentosans

presence of very dilute hydrochloric acid serving as the industrial source (the hydrochloric acid acts as a catalyzer, and the small percentage present is later neutralized to form sodium chloride). The solution is evaporated to a syrup or to crystallization, and is used in the manufacture of sweets, and (usually) alcohol, and in foods. The reaction of glucosides with water, by enzymes or acids, produces glucose as one of the products. With sodium hydroxide, under carefully defined conditions, glucose forms lactic acid. Glucose is used as food and for the production of alcohol (wines) from fruit juices. Glucose may be detected by formation of glucosazone, and determination of its melting point.

Industrial process for converting starch into dextrose (glucose) are described under **Starch**.

Fructose. This sugar is present with glucose in sweet fruits and honey, and may be obtained free by reaction of insulin of dahlia tubers or artichokes with water, and with glucose by reaction of sucrose with water, the product being known as invert sugar. Fructose differs from glucose in structure in being a pentahydroxy-2-ketone,

$$CH_2OH(CHOH)_3COCH_2OH$$

instead of aldehyde. The specific rotary power of fructose is $-88.5°$. Fructose forms the same identical osazone as glucose, and sorbitol plus mannitol by reduction. Fructose may be used as sugar by diabetic patients to advantage instead of glucose or sucrose. Fructose is detected by the violet color its alkaline solution gives with meta-dinitrobenzene.

Sucrose. This is a colorless solid which when heated melts at 170–186°C, and upon cooling forms barley sugar, which gradually crystallizes. Upon heating above the melting point, it forms caramel, a brown liquid, with decomposition. Caramel is used in confectionery, and in coloring beverages and foods. At higher temperatures decomposition into gaseous and tarry substances occurs, finally leaving a residue of carbon ("sugar charcoal"). Other sugars behave similarly. Sugars are also carbonized by concentration sulfuric acid. Sucrose is very soluble in water, and is obtained from solution by crystallization, usually by vacuum evaporation. The solution has a specific rotatory power of $+66.4°$, does not exhibit mutarotation, but is converted by acids or invertase into invert sugar (glucose plus fructose), specific rotatory power $-19.7°$. Sucrose forms with calcium hydroxide calcium sucrosate, a 1% solution of sugar dissolves about 18 times as much calcium hydroxide as does pure water. This behavior is utilized to recover sugar from solutions, as in the case of glucose, and also to determine free calcium oxide in burnt lime, due to the reactivity of calcium hydroxide and non-reactivity of calcium carbonate. Sucrose is nonreactive with dilute sodium hydroxide, with phenylhydrazine, with ammonio-silver salt solution, but, when inverted to glucose plus fructose, these reactions may be obtained. Sucrose forms with acetic anhydride sucrose octaacetate. The suggested structural formula is as shown in Fig. 2. Sucrose is an important food preservative, food flavor, and a raw material for confectionery and for industrial alcohol.

Sucrose is extensively distributed in the seeds and leaves of plants, and is the most abundant of the sugars. The commercial sources of sucrose are the stems of sugar-cane (11 to 16% sucrose, average 13%), the root of the sugar-beet (average 16% sucrose, selection having raised the sucrose content from 5% to a maximum of 20%), the sap of the sugar maple, and the stems of sorghum-cane. Sucrose is pressed from the stems of sugar cane or sorghum cane, and extracted with the water from the sliced roots

of sugar beets. The solutions are purified, evaporated and crystallized to such a degree that commercial sucrose is practically chemically pure (about 99.8% sucrose). The purity of sugar and the concentration or strength of sugar solutions is determined by the rotatory power of the solution, and the special polariscope usually used is called a saccharimeter. Sucrose is reduced with Fehling's solution only after inversion. See also **Sugar**.

The sugar content of some common fruits have been reported by Kulisch:

	SUCROSE	HEXOSES
Apple	1.0–5.4	7.0–13.0
Apricot	6.0	2.7
Banana, ripe	5.0	10.0
Pineapple	11.3	2.0
Strawberry	6.3	5.0

Lactose. This sugar is obtained from the residual water solution (whey) of milk after removal of fat and casein for making butter and cheese. Milk contains about 4.5% of lactose. Lactose forms hard gritty crystals ("sand sugar") $C_{12}H_{22}O_{11} \cdot H_2O$, loses water at 140°C, melting point 202°C (anhydrous) with decomposition; is less sweet than sucrose, reduces ammoniocupric salt solution, ammoniosilver salt solution, forms osazone, melting point 200°C, turns yellow when warmed with sodium hydroxide solution. Lactose is the source of galactose, and undergoes, with the proper enzymes, fermentation into lactic acid and butyric acid.

Maltose. This sugar is found in soybean, and is produced by the action of the enzyme diastase of germinated barley (malt) on starch at 50°C, and is thus an intermediate product in the transformation of starch into alcohol. Maltose $C_{12}H_{22}O_{11} \cdot H_2O$, melting point 100°C, when rapidly heated, may be crystallized from the concentrated malt syrup after removal of proteins and insoluble material. Maltose reduces ammonio-cupric salt solution, and forms osazone.

Starch. This is a white powder, odorless and tasteless, insoluble in cold water, forming an emulsion ("starch paste") or gel with hot water, the consistency of which depends upon the ratio of starch to water used. When boiled starch emulsion is cooled and treated with a solution of iodine in alcohol or potassium iodide, a blue coloration is produced, which is a sensitive and characteristic test. The blue color is associated with the adsorption of iodine on the surface of the starch, and disappears in the presence of alkalis. When boiled with dilute acid, starch is first changed into a soluble gummy mixture known as dextrin, and finally into glucose. When starch, either alone or in the presence of a slight amount of nitric acid, is heated to 120° to 200°C, dextrin is formed; at higher temperatures starch behaves similarly to sucrose. With concentrated nitric acid, starch forms esters, similar to cellulose nitrates. By the action of the enzyme diastase, starch is converted into maltose, which, with the enzyme maltase, yields glucose. Starch is nonreactive with ammonio-cupric salt solution, and with phenylhydrazine. See also **Starch**.

Dextrin. This is a white-to-yellow solid, forming an adhesive with water, nonreactive with ammonio-cupric salt solution, reactive with iodine in alcohol or potassium iodide, usually forming red, brown, or blue color. Formed when starch is (1) heated to 120° to 200°C either alone or in the

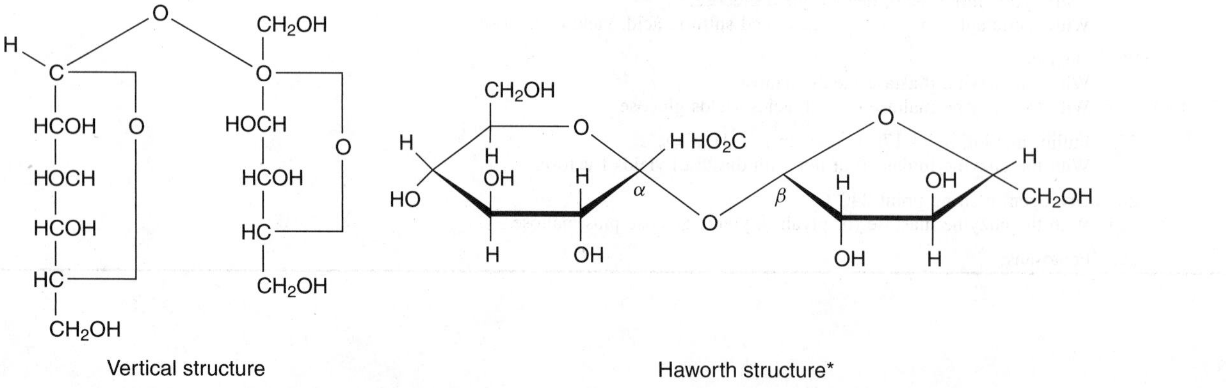

Vertical structure   Haworth structure*

Fig. 2.   Direction of angular momentum vector is perpendicular to plane formed by radial vector **r** and momentum vector **p**.

presence of a slight amount of nitric acid. Dextrin is formed when bread is toasted and is present in well-baked bread crust, and on the surface of starched goods that have been ironed hot. Dextrin is used in adhesives.

Inulin. This is a white solid, soluble in warm water, specific rotatory power −40°, with iodine in alcohol or potassium iodide gives yellow color. Inulin is present in tubers of dahlia to the extent of about 10%. Inulin reacts with water in the presence of the enzyme inulase or of acids to form fructose. The enzyme diastase does not produce this change.

Glycogen. Also known as *animal starch*, this is a white solid, soluble in water, specific rotatory power +197°, with iodine in alcohol or potassium iodide solution, forming brown color. Glycogen is found as reserve carbohydrates in the animal body, more particularly in the liver. Horse flesh, oysters and beef are sources of glycogen.

Pentosans. These compounds are polysaccharides which may be considered as anhydrides of pentose sugars, after the manner of the hexosans, sucrose, starch, from glucose, fructose. When pentosans or pentoses are heated with hydrochloric or sulfuric acid, furfural $C_4H_3O \cdot CHO$ is formed, and addition of aniline produces a red color. Pentosans are present in gummy carbohydrates, in bran of wheat seed, and in woods.

By means of the cyanohydrin reaction, higher sugars of the heptose, octose, and nonose types have been prepared. A monosaccharide such as an aldohexose may be converted into the next lower monosaccharide, such as an aldopentose, by oxidation to the acid, which corresponds to the aldohexose, then treating the calcium salt solution of this acid with a solution of ferrous acetate plus hydrogen peroxide. Carbon dioxide is evolved and aldopentose formed.

For a description of cellulose, see **Cellulose**.

Carbohydrate Metabolism

Carbohydrates are utilized by the cells as a source of energy and as precursors for the manufacture of many of their structural and metabolic components. In the mammal, for example, D-glucose is the carbohydrate primarily used for this purpose. certain microorganisms, in contrast, can grow on a medium containing some other hexose or a pentose as the principal source of carbon. Green plants obtain their carbohydrates by photosynthesis, while animals receive most of their carbohydrates by ingestion and digestion. See also **Photosynthesis**.

The complete oxidation of glucose to carbon dioxide and water yields 689 kcal of heat per mole of glucose. When this oxidation occurs in a cell, the energy is not all dissipated as heat. Some of the evolved energy is conserved in biochemically utilizable form of "high-energy" phosphates, such as adenosine triphosphate (ATP) and guanosine triphosphate (GTP). In addition to enzymes concerned with energy metabolism, there are enzymes in biological systems which catalyze the transformation of glucose into various carbohydrates, fatty acids, steroids, amino acids, nucleic acid components, and other necessary biochemical substances. The entire network of reactions involving compounds that interconvert carbohydrates constitutes *carbohydrate metabolism*. By convention, some reactions involving compounds which are not carbohydrates, but which are derived from them, may also be included in this area of metabolism.

Anaerobic Oxidation of Glucose. Historically, the first system of carbohydrate metabolism to be studied was the conversion by yeast of glucose to alcohol (fermentation) according to the equation: $C_6H_{12}O_6 \rightarrow 2CH_3CH_2OH + 2CO_2$. The biochemical process is complex, involving the successive catalytic actions of 12 enzymes and known as the *Embden-Meyerhof pathway*. This series of reactions is summarized in the entry on **Glycolysis**.

In order for the cell to carry out a "controlled" oxidation of D-glucose and conserve some of the energy derived from the process, it is first necessary to add phosphate to the hexose with the expenditure of energy. The necessary energy and the phosphate per se is supplied by ATP in two separate reactions of the system. Since each molecule of glucose can yield two molecules of triose phosphate for oxidation, the conversion of glucose to pyruvic acid nets two molecules of ATP per molecule of hexose utilized.

Approximately 30% of the evolved energy is conserved as ATP, but only about 8% of the total energy in glucose is made available in this anaerobic oxidation of glucose to pyruvic acid. Since nicotinamide adenine dinucleotide (NAD$^+$), also called diphosphoryidine nucleotide (DPN$^+$), which is involved in the oxidation of glyceraldehyde-3-phosphate, is present in the cell in small quantities only, this coenzyme must constantly be regenerated for the oxidative process to continue. This regeneration is accomplished by the reduction of *acetaldehyde* to *ethanol*.

Since oxygen plays no role in this process, the system can obviously proceed anaerobically. In fact, the presence of oxygen decreases the net disappearance of glucose (*Pasteur effect*).

Fermentation occurs in many microorganisms, but not all organisms reoxidize the reduced nicotinamide adenine dinucleotide (NADH) through the formation of ethanol. In certain organisms, for example, *pyruvic acid* is converted to *acetoin*, which is then reduced with NADH to 2,3-butylene glycol. In other organisms and in animal tissues, NADH is oxidized in the reduction of *pyruvic acid* to *lactic acid*. In insects, and possibly in some animal tissues, the reduction of *dihydroxyacetone phosphate* to *alpha-glycerol phosphate* may serve to regenerate NAD$^+$. The conversion of glucose to lactic acid in animal tissues is termed *glycolysis*. This term arose from the initial understanding that this process was markedly different from the microbial fermentation process. Fermentation and glycolysis are now known to differ primarily in the further anaerobic utilization of pyruvic acid.

Aerobic Oxidation of Pyruvic Acid. Pyruvic acid can be oxidized completely to carbon dioxide and water in a cyclic enzymatic system known as the *Krebs citric acid cycle*, or the *tricarboxylic acid cycle (TCA cycle)*. In this system, a two-carbon unit in the form of acetyl coenzyme A (acetyl = CoA), derived from the NAD$^+$ mediated oxidative decarboxylation of pyruvic acid in the presence of coenzyme A, is condensed with oxalacetic acid to form citric acid. This tricarboxylic acid is then converted back to oxalacetic acid in a stepwise manner with the formation of $2CO_2$ and $2H_2O$. In addition to this formation of CO_2, one reduced nicotinamide adenine dinucleotide phosphate (NADPH), two NADH, one reduced flavin, and one GTP arise per two-carbon unit oxidized in the cycle. Since in the aerobic oxidation of the reduced flavin and the reduced nicotinamide adenine nucleotides, ATP is formed, the oxidation of a molecule of "acetate" results in the conservation of energy in the form of 12 molecules of triphosphate. In the complete oxidation of glucose through glycolysis and the citric acid cycle, about 40% of the energy originally present in the glucose can be retained as triphosphate. The ubiquitous distribution of this cycle in nature suggests that the citric acid cycle is a major energy-yielding pathway in biological systems.

Certain microorganisms have a modification of this cycle in which isocitric acid is cleaved to succinic acid and glyoxylic acid. The latter acid is condensed with acetyl-CoA to form malic acid. In this modification (the *glyoxylic acid cycle*), oxalsuccinic acid and alpha-ketoglutaric acid are not involved. This is sometimes referred to as the "glyoxylate shunt" pathway.

Since in the citric acid cycle there is no net production of its intermediates, mechanisms must be available for their continual production. In the absence of a supply of oxalacetic acid, "acetate" cannot enter the cycle. Intermediates for the cycle can arise from the carboxylation of pyruvic acid with CO_2 (e.g., to form malic acid), the addition of CO_2 to phosphenolpyruvic acid to yield oxalacetic acid, the formation of succinic acid from propionic acid plus CO_2, and the conversion of glutamic acid and aspartic acid to alpha-ketoglutaric acid and oxalacetic acid, respectively. See Fig. 3.

The utilization of carbohydrate intermediates for the biosynthesis of amino acids, fatty acids, steroids, etc. occurs at various stages of the cycle and its related reactions. See Fig. 4. See also **Coenzymes**.

Other Carbohydrate Interconversions. Two systems, as shown in Fig. 5, are available for the synthesis of ribose-5-phosphate, a precursor of the pentose moiety of ribonucleic acid, ATP, and other substances. The formation of ribose-5-phosphate from glucose-6-phosphate by formation and decarboxylation of 6-phosphogluconic acid and isomerization of the resulting ribulose-5-phosphate is termed the *hexose monophosphate oxidative pathway*. The scheme, together with the system involving the enzymes *transketolase* and *transaldolase* (which also can synthesize pentose) that act to form hexose phosphate from pentose phosphate, is called the *pentose phosphate cycle*. This cycle represents an alternative pathway to glycolysis for the formation of triose phosphate from glucose-6-phosphate. The relative importance of the two pathways seems to be different among the various organisms and tissues.

In a certain group of bacteria, still another pathway (*Entner-Doudoroff pathway*) for the utilization of glucose has been studied. Here glucose-6-phosphate is oxidized to 6-phosphogluconic acid which is dehydrated to 2-keto-3-deoxy-6-phosphogluconic acid. This substance is then split to pyruvic acid and glyceraldehyde-3-phosphate (which also can be converted to pyruvic acid).

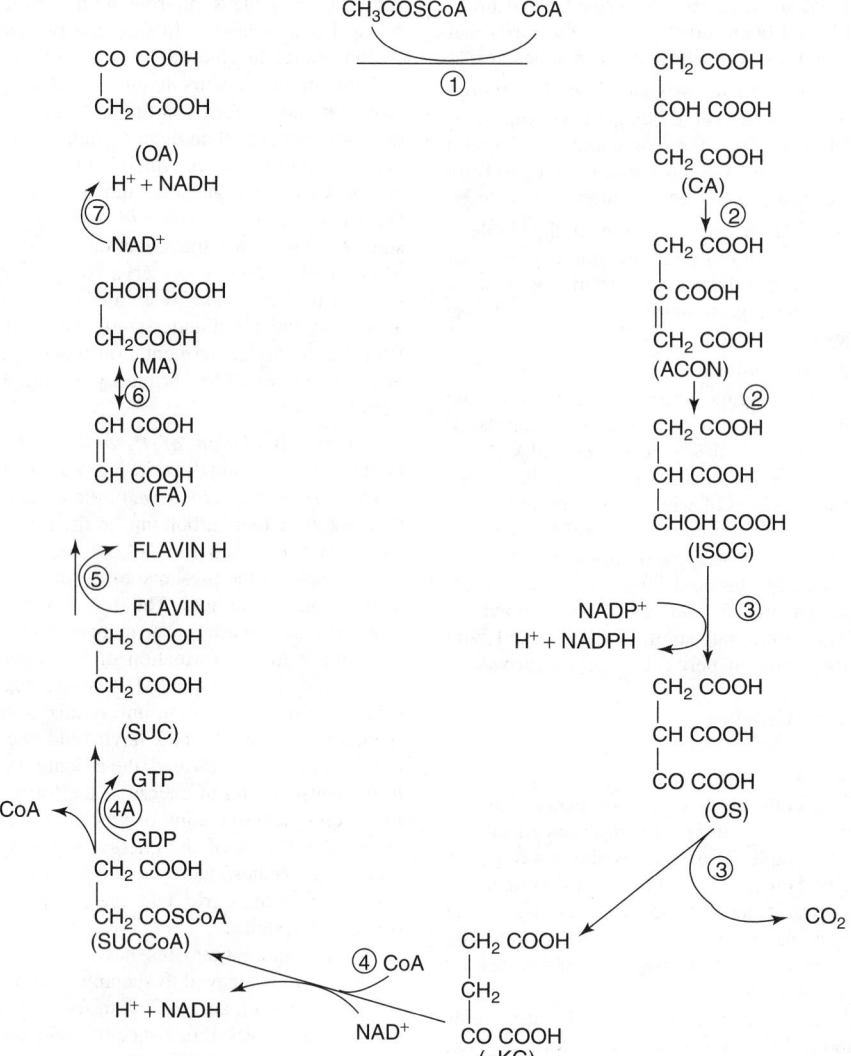

Fig. 3.   Krebs citric acid cycle. *Enzymes involved*: (1) Condensing enzyme; (2) aconitase; (3) isocitric acid; (4) α-ketoglutaric acid dehydrogenase; (4) α-succinic acid thiokinase; (5) succinic acid dehydrogenase; (6) fumarase; (7) malaic acid dehydrogenase. *Abbreviations*: CA = citric acid; ACOM = *cis*-aconitic acid; KG = α-ketoglutaric acid; SIC = succinic acid; FA = fumaric acid; MA = malic acid; OA = oxalacetic acid

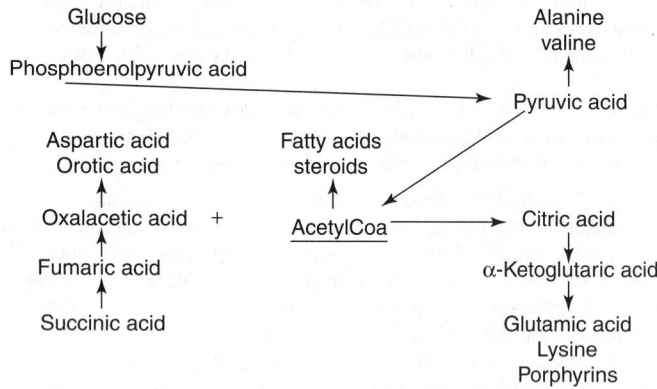

Fig. 4.   Representative conversions of carbohydrates to other substances

The formation of deoxyribose, the pentose moiety of deoxyribonucleic acid, can occur directly from ribose while the latter is in the form of a nucleotide diphosphate. Deoxyribose-5-phosphate can also be formed by condensation of acetaldehyde and glyceraldehyde-3-phosphate.

Transglycosylation. An enzymatic process, transglycosylation, plays an important role in carbohydrate metabolism. Figure 6 represents the formation of the disaccharide, sucrose, as an example of this mechanism. In the upper reaction of Fig. 6, glucose-1-phosphate is the glycosyl donor and

fructose is the acceptor. In the lower reaction, the sugar nucleotide, uridine diphosphoglucose (UDP-glucose) is the glycosyl donor. With UDP-glucose as donor and glucose-6-phosphate as acceptor, trehalose-6-phosphate may be formed. Polysaccharides may also be formed by this process. The donor residues provided by sugar nucleotides are added to preexisting polysaccharide chains (known as "primers") acting as glycosyl acceptors. In the formation of glycogen, for example, UDP-glucose donates the glucose moiety which is added to the end of a previously synthesized chain by a 1,4-linkage, thereby lengthening the chain by one glucose unit.

Digestion, Absorption and Storage of Carbohydrates

In the mammal, complex polysaccharides which are susceptible to such treatment, are hydrolyzed by successive exposure to the amylase of the saliva, the acid of the stomach, and the disaccharidases (e.g., maltase, invertase, amylase, etc.) by exposure to juices of the small intestine. The last mechanism is very important. Absorption of the resulting monosaccharides occurs primarily in the upper part of the small intestine, from which the sugars are carried to the liver by the portal system. The absorption across the intestinal mucosa occurs by a combination of active transport and diffusion. For glucose, the active transport mechanism appears to involve phosphorylation. The details are not yet fully understood. Agents which inhibit respiration (e.g., azide, fluoracetic acid, etc.) and phosphorylation (e.g., phlorizin), and those which uncouple oxidation from phosphorylation (e.g., dinitrophenol) interfere with the absorption of glucose. See also **Phosphorylation (Oxidative)**. Once the various monosaccharides pass through the mucosa, interconversion of the other

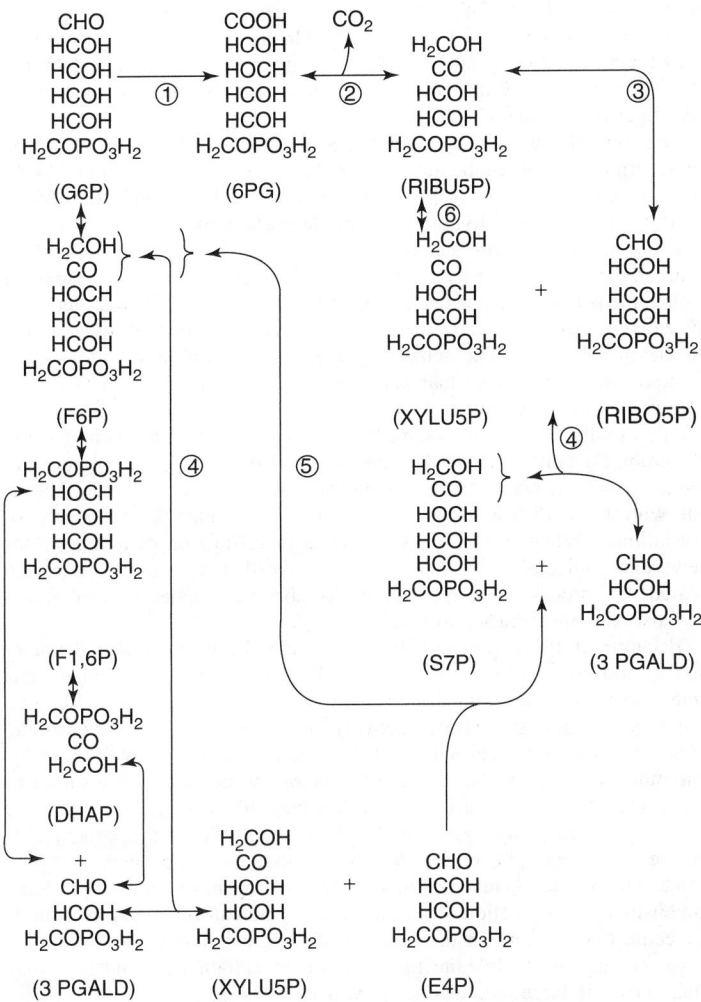

Fig. 5. Pentose phosphate cycle. *Enzymes involved*: (1) Glucose-6-phosphate dehydrogenase; (2) 6-phosphogluconic acid dehydrogenase; (3) pentose phosphate isomerase; (4) transketolase; (5) transaldolase; (6) pentose phosphate epimerase. *Abbreviations*: G6P = glucose-6-phosphate; 6PG = 6-phospho gluconic acid; RIBIUP = ribulose-5-phosphate; 3PGALD = glyceraldehyde-3-phosphate; E4P = erythrose-4-phosphate; F1,6P = fructose-1-6-diphosphate; DHAP = dihydroxyacetone phosphate; F6P = fructose-6-phosphate. Enzymes not named are those of glycolysis. NADP$^+$ is reduced in reactions (1) and (2)

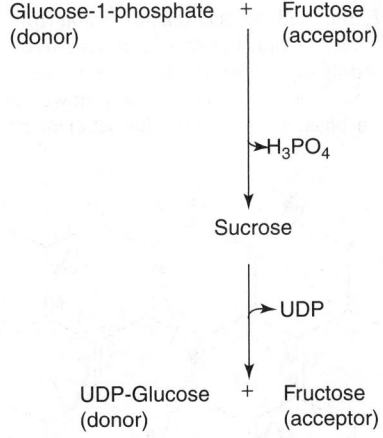

Fig. 6.   Examples of transglycosylation

sugars to glucose can begin, although the liver is probably the chief site for such conversions. Even though many organs and tissues store carbohydrates as glycogen for their own use, the liver provides the main source of glucose for all tissues, through conversion of its glycogen (and

other substances) to glucose-6-phosphate, hydrolysis of this ester by the specific liver glucose-6-phosphatase, and transport of the free glucose in the bloodstream throughout the body.

A common cause of *osmotic diarrhea* is the ingestion of carbohydrates that a person cannot digest. Retention of a disaccharide, such as lactose or sucrose, within the intestinal lumen occurs because of the absence of the appropriate disaccharidase at the intestinal surface membrane. Unless they are converted to monosaccharides, these sugars cannot be transported. Their retention in the lumen can cause a significant diarrheal water loss per day. As an added complication, bacteria in the lower small intestine and colon may catabolize the 12-carbon sugars to 3-carbon fragments, further aggravating the osmotic effect. Infants and very young children usually have sufficient lactase and sucrase, but there is a tendency among some people to become lactase deficient between the ages of 3 and 14 years. Once such a condition is fully recognized, the ingestion of milk and other dairy products should be eliminated. It should be pointed out, however, that the so-called *irritable bowel syndrome* is attributable to lactase deficiency in a relatively small percentage of cases. Carbohydrates that can cause diarrhea in some persons include: lactose and sucrose, already mentioned; stachyose and raffinose, contained in many legumes; mannitol and sorbitol, contained in artificial sweeteners (which contain sugar alcohols); glucose and galactose, present in all dietary sugars; and lactulose, contained in nondietary disaccharides as parts of certain medications.

Endocrine Influences. A number of hormones are known to influence carbohydrate metabolism in the mammal. Insulin seems to increase oxidation of glucose, lipogenesis, and glycogenesis. Its primary mode of action may be to facilitate the entry of glucose into the cell.

Vitamin Influences. The involvement of NAD$^+$ and NADP$^+$ in many carbohydrate reactions explains the importance of nicotinamide in carbohydrate metabolism. Thiamine, in the form of thiamine pyrophosphate (cocarboxylase), is the cofactor necessary in the decarboxylation of pyruvic acid, in the *trans*-ketolase-catalyzed reactions of the pentose phosphate cycle, and in the decarboxylation of alpha-ketoglutaric acid in the citric acid cycle, among other reactions. Biotin is a bound cofactor in the fixation of carbon dioxide to form oxalacetic acid from pyruvic acid. Pantothenic acid is a part of the CoA molecule. There are separate alphabetical entries in this volume on the various specific vitamins as well as a review entry on **Vitamin**.

Photosynthesis. The formation of carbohydrates in green plants by the process of photosynthesis is described in the entry on **Photosynthesis**. The synthetic mechanism involves the addition of carbon dioxide to ribulose-1,5-diphosphate and the subsequent formation of two molecules of 3-phosphoglyceric acid which are reduced to glyceraldehyde-3-phosphate. The triose phosphates are utilized to again from ribulose-5-phosphates by enzymes of the pentose phosphate cycle. Phosphorylation of ribulose-5-phosphate with ATP regenerates ribulose-1,5-diphosphate to accept another molecule of carbon dioxide. See also **Phosphorylation (Photosynthetic)**.

Carbohydrates in Foods

Sugar is discussed in several entries in this volume, including **Fiber (Dietary)**; **Gums and Mucilages**; and **Sugar**.

Statistics on the carbohydrate content of diets of various peoples throughout the world have not been very reliable because of the scores of variables involved, the great difficulties in establishing reliable sampling procedures, lack of past records, among other factors. One summary, for example, that breaks down food energy from protein, fat, and carbohydrates shows a downward trend for carbohydrates in the American diet—from 56% in 1911 to 46% in the mid-1970s. These figures were based upon U.S. Department of Agriculture statistics of food disappearance at the retail level, but they do not take into consideration food spoilage, cooking waste, plate waste, and other factors that affect actual consumption. Since protein remained quite constant at 11–12% throughout this time span, the drop in carbohydrates was made up by an increase in fats—from 32% in 1911 to 42% in the mid-1970s. In another study, of the 46% carbohydrate energy intake as of 1977, 24% is attributed to sugar and 22% to complex carbohydrates. In a controversial U.S. government study that attempted to set new dietary goals for the nation, it was suggested that the traditional 12% protein be retained, but that fat be reduced from 42% and carbohydrates upped to 58%, but with a major difference, namely, cutting the sugar portion of carbohydrates from 24% to 15%. Thus, the dietary goal would require 40–45% complex carbohydrates in the diet. It has been suggested that to achieve the projected carbohydrate goals, there would have to be a 66% increase in the consumption of grain products; a 25% increase of vegetables and fruit; and a 50% reduction in sugar and sweets.

Even though much visibility had been given by the various news media to the dietary role of sugar, it was obvious that, as of the late 20th Century, a great deal of fundamental research remained to be done to prove or disprove many conclusions, often conflicting and confusing, in order to establish reliable dietary guidance in this area.

Additional Reading

Alvarez, J. and L.C. Polopolus: *Marketing Sugar and Other Sweeteners,* Elsevier, New York, NY, 1991.

Appl, R.C.: "Confectionary Ingredients from Starch," *Food Technology,* **148** (March 1991).

Bednarski, M.D. and E.S. Simon, Editors: *Enzymes in Carbohydrate Synthesis,* ACS Symposium Series, American Chemical Society, Washington, DC, 1991.

Binkley, E.R. and R.W. Binkley: *Carbohydrate Photochemistry,* American Chemical Society, Columbus, OH, 1998.

Bols, M.: *Carbohydrate Building Blocks,* John Wiley & Sons, Inc., New York, NY, 1995.

Cho, S.S.: *Complex Carbohydrates in Foods,* Marcel Dekker, Inc., New York, NY, 1999.

El, R.: *Carbohydrate Analysis,* Elsevier Science, New York, NY, 1995.

Farrar, J.F.: *Fructose Polymers in Plants and Micro-Organisms: New Phytologist,* Cambridge University Press, New York, NY, 1997.

Finch, P.: *Carbohydrates: Structures, Syntheses and Dynamics,* Kluwer Academic Publishers, Norwell, MA, 1999.

Fox, P.F.: *Advanced Dairy Chemistry: Lactose, Water, Salts and Vitamins,* Aspen Publishers, Gaithersburg, MD, 1997.

Freeman, T.P. and D.R. Shelton: "Microstructure of Wheat Starch: From Kernel to Bread," *Food Technology,* **162** (March 1991).

Gould, G.W.: *Facilitative Glucose Transporters,* landes Bioscience Publishers, Austin, TX, 1997.

Hecht, S.M.: *Bioorganic Chemistry: Carbohydrates,* Oxford University Press, New York, NY, 1998.

Higley, N.A. and J.S. White: "Trends in Fructose Availability and Consumption in the United States," *Food Technology,* **118** (October 1991).

Horton, D.: *Advances in Carbohydrate Chemistry and Biochemistry,* Vol. 55, Academic Press, Inc., San Diego, CA, 2000.

Houts, S.S.: "Lactose Intolerance," *Food Technology,* **110** (March 1988).

Huber, G.R.: "Carbohydrates in Extrusion Processing," *Food Technology,* **160** (March 1991).

Kulp, K., K. Lorenz, and M. Stone: "Functionality of Carbohydrate Ingredients in Bakery Products," *Food Technology,* **138** (March 1991).

Lehman, J.: *Carbohydrates Structure and Biology,* Tieme Medical Publishers, Inc., New York, NY, 1999.

MacDonald, G.A. and T.C. Lanier: "Carbohydrates as Cryopectants for Meats and Surimi," *Food Technology,* **150** (March 1991).

Osborn, H., and L.M. Harwood: *Carbohydrates,* Elsevier Science & Technology, New York, NY, 2003.

Pennington, N.L. and C.W. Baker: *A User's Guide to Sucrose,* Van Nostrand Reinhold, New York, NY, 1990.

Scherz, H., and G. Bonn: *Analytical Chemistry of Carbohydrates,* John Wiley & Sons, Inc., New York, NY, 2002.

Scholz, C. and R. Gross, et al.: *Polymers from Renewable Resources: Carbohydrates and Agroproteins,* Oxford University Press, New York, NY, 2000.

Sivak, M.N. and J. Preiss: *Starch: Basic Science to Biotechnology,* Academic Press, Inc., San Diego, CA, 1997.

Walter, R.H.: *The Chemistry and Technology of Pectin,* Academic Press, Inc., San Diego, CA, 1991.

Walters, D.E., F.T. Orthoefer, and G.E. DuBois, Editors: *Sweeteners: Discovery, Molecular Design, and Chemoreception,* ACS Symposium Series, American Chemical Society, Washington, DC, 1991.

CARBON. [CAS: 7440-44-0]. Chemical element symbol C, at. no. 6, at. wt. 12.011, periodic table group 14, mp 3,550°C (approximate), bp 4,289°C (approximate), density 3.52 g/cm³ (diamond at 20°C), 2.25 g/cm³ (graphite at 20°C). The specific gravity of amorphous carbon at 20°C ranges from 1.8 to 2.1. There are two stable isotopes of the element, ^{12}C and ^{13}C, and four known radioactive isotopes, ^{10}C, ^{11}C, ^{14}C, and ^{15}C. Because the half-life (about 5,760 years) of ^{14}C has been established, this isotope is useful for dating ancient documents and materials.

The first ionization potential of carbon is 11.264 eV; Second, 24.28 eV; third, 47.7 eV. Other important physical characters of carbon are given in article on **Chemical Elements**.

Traditionally, the principal forms of carbon have (1) *diamond*, with its tetrahedral arrangement of atoms; (2) *graphite*, whose structure resembles layers of chicken wire; and sometimes (3) *amorphous*, a poorly defined grouping of carbons. This latter classification was chosen more out of convenience than grounded scientifically. However, by recent consensus, a third form of carbon is now officially recognized: *fullerene*, of which the C_{60} so-called *buckminsterfullerene* or "buckyball" is the most thoroughly investigated example of its class.

Diamond, the hardest of natural materials, consists of a lattice of carbon atoms arranged in a tetrahedral structure at equal distances apart (1.544 Å) and bonded by electron pairs in localized molecular orbitals formed by overlapping of the sp^3 hybrids. See article on **Diamond**.

Graphite, a very soft material, consists of carbon atoms arranged in laminar sheets, 3.40 Å apart and composed of carbon atoms in hexagonal arrangement 1.42 Å apart. Each atom is bonded to three others in its sheet by electron pairs in localized molecular orbitals formed by overlapping of the sp^2 hybrids. The remaining p-electrons form a mobile system of nonlocalized pi bonds that permit electrical conductivity between the lamina. See also **Graphite**.

The familiar carbon blacks are formed by such methods as combustion of carbon-containing materials with sufficient oxygen. These carbons are found to have x-ray diffraction patterns that are suggestive of graphite, but with more diffuse rings, thus indicating a much lower degree of crystallinity. When carbon black is heated, its diffraction pattern develops new rings indicative of a structure more like that of graphite. When so heated, the properties of the carbons as absorbent materials deteriorate. See also **Carbon Black and Coal**.

Because of the comparatively recent demonstration of the geometry of C_{60} and cousins of both lower and higher carbon atom content, this giant molecule is ushering in a new concept to the chemistry of carbon and organic chemistry. This discovery, which later may have profound implications for practical industrial and scientific usage, provides insights into other scientific fields, such as astrochemistry, and serves as a source of yet unknown chemical derivatives; it has been likened by some scientists to the first practical suggestion by Kekulè (1825) on the structure of the benzene ring. The carbon atoms in Kekulè's design have dangling bonds which usually are accommodated by hydrogen. In contrast the three-dimensional configuration of C_{60} has exactly 60 carbon atoms in a single molecule that is inert, in the absence of dangling bonds. The molecule, however, appears to tolerate the insertion of certain ions and thus may make possible large numbers of derivatives.

Traditional Carbon Chemistry

The probable importance to high-temperature behavior of the $-C \equiv C-$ triple bond, most familiarly encountered in acetylene, was proposed in the late 1960s. The essence of the proposal is that high-temperature carbon forms are made up of chains of triple-bonded carbon atoms, termed *carbynes*. See Fig. 1. It will be noted that at high temperatures, a single bond in the structure may break. This shifts an electron into each of the adjacent double bonds, forming a triple bond. Completion of the process transforms the sheet of atoms into a chain of carbynes. The chains can be variously stacked. In 1973, at least five such forms had been reported. Researchers in 1977 reported that the transformation from the carbyne form to graphite involves a reaction between acetylene-like molecules (acting rapidly and exothermically), whereas the reverse reaction (breaking of single bonds) can be expected to be a much slower process. Thus, the conventional carbon phase diagram may be deficient because it does not

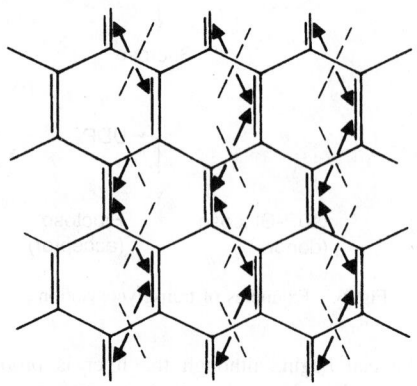

Fig. 1.   Mechanisms suggested for transformation of a graphite basal plane sheet of atoms into carbyne chains

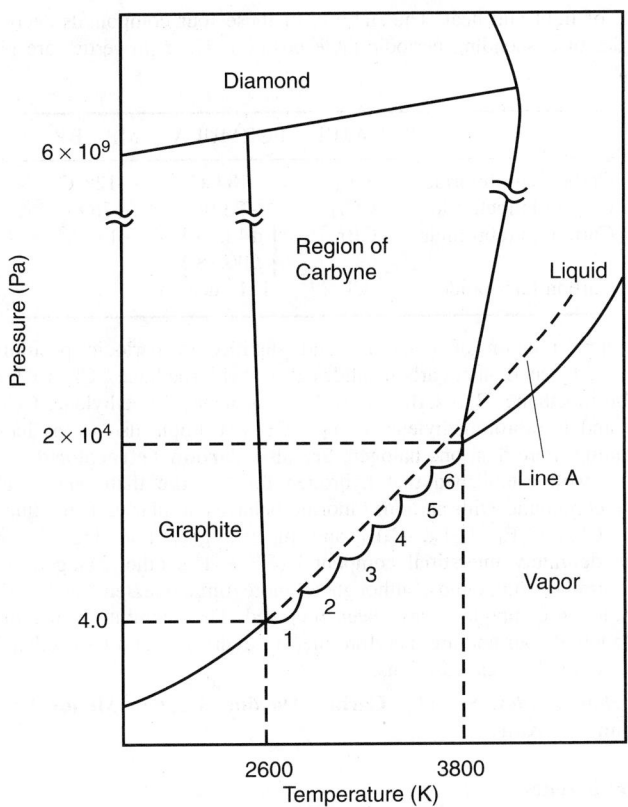

Fig. 2. Carbon phase diagram (suggested by Whittaker) to accommodate region of carbyne. Dashed line A is vapor pressure graphite would have if it were the stable form above 2600 K. Whittaker notes: (1) graphite is not stable above 2600 K at any pressure; (2) the solid-liquid-vapor triple point occurs at 3800 K and 2×

consider the carbyne forms. One scientist pointed this out because it had been difficult to reconcile high-pressure results with the low-pressure data available on the vapor pressure of carbon. See Fig. 2.

Research pertaining to carbynes, in way, was a prelude to further molecular carbon research and the now well-known buckyball.

Carbon Compounds

With the exception of hydrogen, carbon forms more compounds than any of the other chemical elements. Traditionally, carbon compounds fall into two fundamental classes: (1) *inorganic* compounds and (2) *organic* compounds.

The main subclasses of inorganic carbon compounds include:

1. The *carbon oxides*, notably CO (carbon monoxide) and CO_2 (carbon dioxide).
2. The *carbonates* $-CO_3$, which occur widely in nature—minerals, rocks, ores, and mineral waters—and include such compounds as Na_2CO_3, $CaCO_3$ (limestone), $MgCO_3$ (magnesite), $MgCa(CO_3)_2$ (dolomite), etc.
3. Some carbon-sulfur compounds, such as CS_2 (carbon disulfide), the thiocyanides, and thiocyanates $-CNS$, such as HCNS (thiocyanic acid), $Pb(CNS)_2$ (lead thiocyanate), etc.
4. The *carbides*, such as Na_2C_2, Cu_2C_2, WC, ZrC, etc.
5. The *carbonyls* $-CO$, such as $Cr(CO)_6$, $Fe(CO)_5$, $Ni(CO)_4$, etc.
6. The *halides*, such as CCl_4 (carbon tetrachloride), CBr_4, etc.

The subclasses of organic compounds comprise the realm of organic chemistry and are described under **Organic Chemistry**. There are several subclasses of organic compounds that include oxygen along with hydrogen and carbon in their structure—e.g., acid anhydrides, alcohols, aldehydes, carbohydrates, carboxylic acids, esters, ethers, fatty acids, furans, ketones, lactides, lactones, phenols, quinones, and terpenes. Some of the main subclasses of nitrogen-bearing organic compounds include the amides, amines, amino acids, anilides, azo and diazo compounds, carbamates, cyanamides,

hydrazines, polypeptides, proteins, purines, pyridines, pyrroles, quaternary ammonium compounds, semicarbazones, ureas, and ureides. The addition of the halogens to the structure yields chlorine organics, brominated compounds, fluorocarbons, etc. Most of the metals combine with carbon compounds to form organometallics. Sulfur-bearing organics include the sulfonic acids, sulfonyls, sulfones, thioalcohols, thioaldehydes, sulfoxides, etc. Silicones are silicon-bearing carbon compounds. See also **Chlorinated Organics**.

Carbides. As might be expected from its position in the periodic table, carbon forms binary compounds with the metals in which it exhibits a negative valence, and binary compounds with the non-metals in which it exhibits a positive valence. A convenient classification of the binary compounds of carbon is into ionic or salt-like carbides, intermediate carbides, interstitial carbides, and covalent binary carbon compounds.

The ionic or salt-like carbides are formed directly from the elements, or from metallic oxides and carbon, carbon monoxide, or hydrocarbons. This last reaction is reversible, and this group of carbides may be further subdivided into acetylides, e.g., Li_2C_2, Na_2C_2, K_2C_2, Rb_2C_2, Cs_2C_2, Cu_2C_2, Ag_2C_2, Au_2C_2, BeC_2, Mg_2C_2, CaC_2, SrC_2, BaC_2, ZrC_2, CdC_2, Al_2C_6, Ce_2C_6, and ThC_4: methanides, e.g., Be_2C and Al_4C_3; and the allylides, primarily magnesium allylide, Mg_2C_3, according to the hydrocarbon or the principal hydrocarbon formed upon hydrolysis. By the term intermediate carbides is meant compounds intermediate in character between the ionic carbides and the interstitial carbides. The intermediate carbides, such as Cr_3C_2, Mn_3C, Fe_3C, Co_3C, and Ni_3C are similar to the ionic carbides in that they react with water or dilute acids to give hydrocarbons, and they resemble the interstitial carbides in their electrical conductivity, opacity, and metallic luster. The interstitial carbides have these properties, and are uniformly chemically inert. They include those having cubic close-packed structures, such as TiC, ZrC, HfC, VC, NbC, TaC, MoC, and WC, and those having hexagonal close-packed structures such as V_2C, Mo_2C, and W_2C. In both, the carbon atoms occupy interstitial positions in the crystal lattices of the metals, giving hardness, high melting points, and chemical inertness, as well as electrical conductivity with a positive temperature coefficient and metallic luster. The covalent binary compounds of carbon range in character from hard, chemically inert solids, such as silicon carbide, SiC, to volatile liquids, such as carbon disulfide and carbon tetrachloride, CS_2 and CCl_4, and even to gases such as carbon tetrafluoride, carbon dioxide and methane, CF_4, CO_2 and CH_4, varying in thermal stability. With several of these elements carbon forms a series of compounds, or as with hydrogen, a number of series of hydrocarbons, consisting of both compounds based upon chains and branched chains of carbon atoms, variously saturated (i.e., joined by single, double, or triple bonds), and also of ring connected carbon atoms, with or without side chains, with varying degrees of saturation, and capable of replacement of the hydrogen atoms with other atoms or radicals.

Carbonates. Carbonic acid H_2CO_3 is present to the extent of 0.27% of the total CO_2 present in the solution that is formed by dissolving CO_2 in H_2O at room temperature. The CO_2 may be expelled fully upon boiling. The solution reacts with alkalis to form carbonates, e.g., sodium carbonate, sodium hydrogen carbonate, calcium carbonate, calcium hydrogen carbonate. The acid ionization constant usually cited for carbonic acid (4.2×10^{-7}) is actually for the equilibrium $CO_2(aq) + H_2O \longleftrightarrow H^+ + HCO_3^-$. The true ionization constant, i.e., for the equilibrium $H_2CO_3 \longleftrightarrow H^+ + HCO_3^-$ is about 1.5×10^{-4}. The carbonate ion is a resonance hybrid of the three structures shown a, b, and c as well as structures of the type d which give a partial ionic character to bonds. This resonance is somewhat inhibited in the acid and its esters, but is complete, or much more nearly complete, in many other derivatives and in the carbonate ion. Esters of both metacarbonic, $(RO)_2CO$, and orthocarbonic acid, $(RO)_4C$, are known. The esters also exhibit resonance.

Metallic carbonates are (1) soluble in H_2O, e.g., sodium carbonate, potassium carbonate, ammonium carbonate (2) insoluble in H_2O and excess alkali carbonate, e.g., calcium carbonate, strontium carbonate, barium carbonate, magnesium carbonate, ferrous carbonate (3) insoluble in H_2O but soluble in excess alkali carbonate forming carbonate complexes, e.g., compounds of uranium and ytterbium $U(CO_3)_2$, UO_2CO_3, $Yb_2(CO_3)_3$. Metallic bicarbonates are known in solution and on warming are converted into ordinary or normal carbonates, e.g., bicarbonates of sodium, potassium, calcium, barium. These are preferably named as "hydrogen carbonates," e.g., $NaHCO_3$ = sodium hydrogen carbonate. Basic carbonates are important in such cases as lead ("white lead"), zinc, magnesium, and copper. Carbonates of very weak bases, such as aluminum, iron(III), and chromium(III), are now known.

The carbonates are found in nature as the carbonates, calcite, iceland spar, limestone and various forms of impure calcium carbonate $CaCO_3$, as magnesite (magnesium carbonate, $MgCO_3$), as dolomite (various compositions of calcium and magnesium carbonates), as witherite $SrCO_3$, as strontianite $SrCO_3$, as azurite and malachite (various compositions of cupric hydroxycarbonates), in various natural waters as carbonic acid, calcium and magnesium hydrogen carbonates, in blood, as sodium hydrogen carbonate.

Many esters of carbonic acid are known, e.g., diethyl carbonate, ethyl ester of metacarbonic acid, $(C_2H_5O)_2CO$, made by reaction of ethyl alcohol and carbonyl chloride; dimethyl carbonate, $(CH_3O)_2CO$; methyl ethyl carbonate, $(CH_3O)CO(OC_2H_5)$; dipropyl carbonate, $(C_3H_7O)_2CO$; tetraethyl carbonate, ethyl ester of orthocarbonic acid, $(C_2H_5O)_4C$, bp 158°C.

Peroxycarbonic acid exists only in its compounds. Alkali peroxycarbonates are obtained by electrolysis of concentrated solutions of the carbonates, the anodic reaction being written as

$$2CO_3^{2-} \longrightarrow C_2O_6^{2-} + 2_e{}^-$$

The peroxycarbonates are relatively stable only in concentrated alkaline solutions. On dilution they decompose to give the bicarbonate and hydrogen peroxide

$$Na_2C_2O_6 + 2H_2O \longrightarrow 2NaHCO_3 + H_2O_2$$

when acidified, the peroxycarbonate ion gives, correspondingly, CO_2 and hydrogen peroxide

$$C_2O_6^{2-} + 2H^+ \longrightarrow 2CO_2 + H_2O_2$$

Carbonyls. The metal carbonyls are strongly covalent in character, as shown by their volatility, their solubility in many nonpolar solvents, and their insolubility in polar solvents. They also behave in many reactions like mixtures of carbon monoxide, CO, and the metal. Those of group 6b elements, $Cr(CO)_6$, $Mo(CO)_6$, and $W(CO)_6$ are more stable and less reactive than the others, especially those of group 8 elements. Group 7b carbonyls are $Mn_2(CO)_{10}$, $Tc_2(CO)_{10}$, and $Re_2(CO)_{10}$, while group 8 elements form $Fe(CO)_5$, $Fe_2(CO)_9$, $Fe_3(CO)_{12}$, $Co_2(CO)_8$, $Co_4(CO)_{12}$, $Ni(CO)_4$, $Ru(CO)_5$, $Ru_2(CO)_9$, $Ru_2(CO)_{12}$, $Rh_2(CO)_8$, $Rh_3(CO)_9$, (and multiples), $Rh_4(CO)_{14}$, (and multiples), $Os(CO)_5$, $Os_2(CO)_9$, $Ir_2(CO)_8$, and $Ir_3(CO)_9$ (and multiples). The carbonyls form a wide variety of addition compounds; they are dissolved in alcoholic potassium hydroxide or other strong alkalies to form hydrides which are acids, and can be used to form a wide variety of more complex compounds. Although $H_2Fe(CO)_4$ is a moderately weak acid, $pK_1 = 4.44$, $pK_2 = 14.0$, $HCo(CO)_4$ appears to be comparable with HCl in acidity. The carbonyl compounds have zero charge number on the metal. The mononuclear carbonyls are spin-paired complexes, and are formed only by metals having even atomic numbers. However, metals having odd atomic numbers can form carbonyl compounds with other atoms or radicals, as exemplified by the nitrosyl compound of cobalt carbonyl, $Co(CO)_2NO$, where the $-NO$ radical contributes the electron necessary to complete the $3d$ level of the cobalt atom. More than one NO group may occur in a metal carbonyl, as, for example, in $Fe(CO)_2(NO)_2$. This is isostructural with $Co(CO)_3NO$ and $Ni(CO)_4$.

Halides. The four tetrahalides of carbon are symmetrical, planar compounds, with the general property of marked stability to chemical reactions, although the tetraiodide undergoes slow hydrolysis in contact with water to form iodoform and iodine. It also decomposes under the

action of light and heat. The stability of these four compounds decreases in order of descending periodic table position. Their properties are given below:

NAME	FORMULA	MP	BP
Carbon tetrafluoride	CF_4	−184°C	−128°C
Carbon tetrachloride	CCl_4	−23.0°	−76.8°
Carbon tetrabromide	CBr_4	$\begin{cases} \alpha 48.48 \\ \beta 90.18 \end{cases}$	−189.5°
Carbon tetraiodide	CI_4	171° dec	

The same relation of reactivity and stability to periodic position is exhibited by such other carbon halides as hexachloroethane $CCl_3 \cdot CCl_3$ and hexabromoethane, $Br_3 \cdot CBr_3$, as well as by hexachloroethylene, $CCl_2 = CCl_2$ and hexabromoethylene, $CBr_2 = CBr_2$. Carbon also forms halides containing more than one halogen. See also **Carbon Tetrachloride**.

It is well established that hydrogen forms more than one covalent binary compound with carbon. Fluorine behaves similarly. Thus, fluorine forms CF_4, C_2F_4, C_2F_6, C_3F_8 and many higher homologs, as well as the definitely interstitial compound $(CF)_n$. The other halogens form some similar compounds, although to more limited extent, and various polyhalogen compounds have been prepared. They exhibit the maximum covalency of four and are therefore inert to hydrolysis and most other low temperature chemical reactions.

Carbon Oxides. See also **Carbon Dioxide**; **Carbon Monoxide**; and **Carbon Suboxide**.

The Fullerenes

The less-than-scientific ring ascribed to the comparatively recent discovery of a third form of carbon, the *fullerenes*, is reminiscent of *flavors* used a few years ago to describe the various kinds of quarks in the field of high-energy physics. The technical literature on fullerenes, as of early 1994, features such terms as bucky-ball, buckminsterfullerene, buckytube, carbon cage, dopey ball, hairy ball, Russian doll, et al., some of which terms are synonymous; others have specific connotations. Considered as an entity, fullerene chemistry constitutes a major breakthrough in the science of physics and chemistry of materials at the molecular level.

The absence of a formal nomenclature at this juncture is accompanied by a somewhat fuzzy chronology pertaining to the discovery and early research on the fullerenes. However, the isolation and confirmation of the C_{60} all-carbon molecule sans any dangling bonds, as first conjectured in 1985, was pivotal to subsequent research.

Setting the Stage for Carbon 60 Research. The pathways that ultimately led to the geometric visualization of the C_{60} molecule were several and varied.

(1) A growing interest in cluster configurations extends back to the 1950s. The sophistication of instrumentation for investigating once exotic substances has improved many times over during recent years, and numerous schemes of molecular geometry have been proposed. Thus, in retrospect, the efforts made to visualize the structure of the C_{60} molecule were not exclusively of a pioneering bent.

(2) For many years, astrophysicists have been interested in the role of carbon molecules, both as building blocks and as photofragments of carbonaceous materials. As early as 1972, *polyyne* chain molecules $(\cdots C \equiv C - C \equiv C - C \equiv C \cdots)$ were proposed as being present in interstellar space and in the atmospheres of carbon stars. This material, in the form of HC_5N was prepared in the laboratory and later was found, by way of radio astronomy, to exist in space. The structural geometry of the compound, however, could not be explained satisfactorily.

(3) Over a number of years, researchers engaged in the study of carbon (fuel) combustion reactions sought a better understanding of such reactions at the molecular level and, in particular, those processes that produced soot (carbon particles). Thus, carbon particles became a major target of their research. Because of concerns with air pollution, efforts were made to make carbon particle determination an exact science.

Carbon Sixty Research Chronology

In 1984, scientists (Rohlfing, Cox, and Caldor at Exxon Research and Engineering) created clusters of carbon (soot) by the laser vaporization of a carbon target rod in connection with a supersonic nozzle. By means of mass spectroscopy, the researchers determined the relative abundance of the carbon clusters produced. Small, 20- to 40-atom clusters of carbon were expected inasmuch as these had been produced a number of times by earlier investigators working on the soot problem. In such experiments, an interesting but unexplained question always arose—Why were only even-numbered carbon clusters produced in the complete absence of odd-numbered clusters? See Fig. 3.

In 1985, similar experiments were conducted at Rice University. In a 1988 paper, Curl and Smalley (Rice University) outlined their experiments with carbon cluster beams, essentially using the cluster-generating apparatus previously described by the Exxon researchers. Initially, this experimentation was motivated by an interest that had been shown by the astrophysicist, Kroto (University of Sussex), who had been modeling the formation of carbon molecules in circumstellar shells. As a consequence, the Rice University team concentrated its studies on the smaller (2- to 30-atom) carbon clusters. As pointed out in the Curl-Smalley paper, the objective was to "determine if some or all of the species had the same form as the long linear carbon chains known to be abundant in interstellar space."

Over time, the research interests of the Rice University team and of Kroto were directed increasingly to developing a suitable structural explanation of the even-numbered carbon clusters and, notably, of the C_{60} molecule.

It has been reported that, over a period of at least several months, Kroto and the Rice University team had formed a sort of research camaraderie, which developed out of their common interests in learning more about the structure of C_{60}. There are, however, some differences in opinion as to how the buckyball was visualized initially.

In a 1988 article, Kroto observed, "Initially, cluster reactions were probed which showed that C_n ($n > 30$) clusters did indeed react with H and N to form polyynes, which had been detected in space, a result satisfyingly consistent with the idea of a stellar source of interstellar chains. The larger clusters were totally inert, and as the experiments progressed, it became impossible to ignore the antics of the C_{60} peak which varied from relative insignificance to total dominance, depending on the clustering conditions.

"After much discussion we conjectured that the bizarre behavior, particularly of the dominance of C_{60}, could be the result of stabilization by closure of a graphite net into a hollow chicken-wire cage similar to the geodesic domes of Buckminster Fuller.[1] Such closure would eliminate all 20 or so reactive edge bonds of a 60-atom sheet. This led to the realization that there was a most elegant and, at the time, overwhelming solution—the *truncated icosahedron cage.*" See Fig. 4.

"The structure necessitated the throwing of all caution to the wind (the Greek icosahedron) and it was proposed immediately by Kroto, Heath, O'Brien, Curl, and Smalley: *Nature,* **318**, 162 (1987). After all, it was surely too perfect a solution to be wrong. We named C_{60} after Buckminster Fuller, which has turned out to be a highly appropriate name."

The diagram (Fig. 5) shows a full accounting of the 60 carbon atoms that fully close the cage without any dangling bonds. Because the molecule resembles a soccer ball, it has been called the "buckyball."

Smalley's description of how the geometry of C_{60} was revealed varies somewhat from Kroto's accounting. This is explained partially by Philip Yam (reference listed), who writes in a short biography of Smalley, "Neither individual probably would have discovered buckyballs had they not collaborated, and both agree that it was a serendipitous finding." Ironically, it is interesting to note that David Jones (writing under the pseudonym, Daedalus) previously had proposed such cages as early as 1982.

Ensuing Fullerene Research

In the 1990s, fullerene research continued apace. Hundreds of new papers appeared each quarter pertaining to the properties of C_{60} and its cousins, and the prospects for developing new materials based upon this new dimension of carbon chemistry. (A sampling list of additional reading is given at the end of this article.)

Researchers J.M. Hawkins, et al. (University of California, Berkeley), for example, studied infrared, Raman, ^{13}C nuclear magnetic resonance, and photoelectron spectra of C_{60}, and found the data to be consistent with

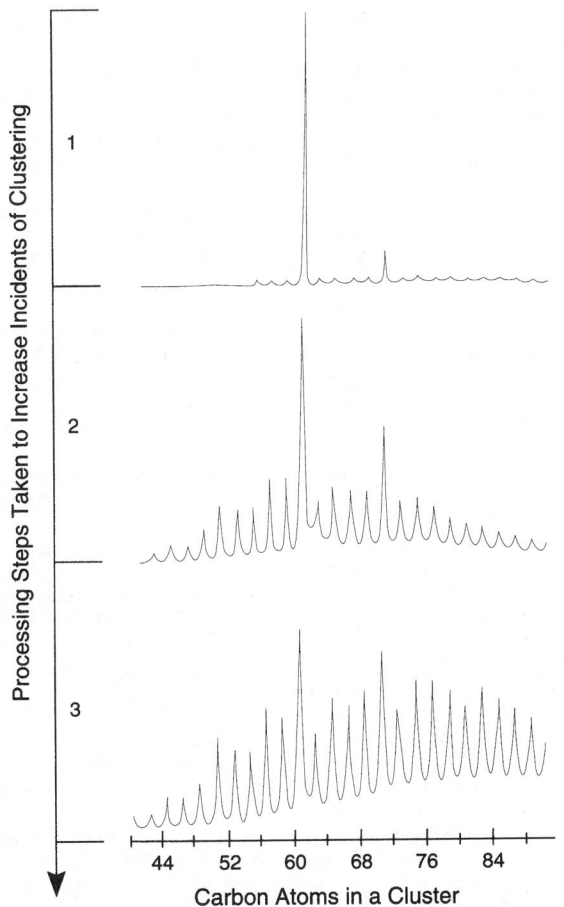

Fig. 3.   Reasonable facsimiles of mass spectra produced by laser vaporization of carbon in a supersonic beam, indicating three stages in the process for increasing the extent of clustering. Experiment was carried out by Rohlfing, Cox, and Kaldor (Exxon Research and Engineering). Original diagrams were featured in *Nature* (1985)

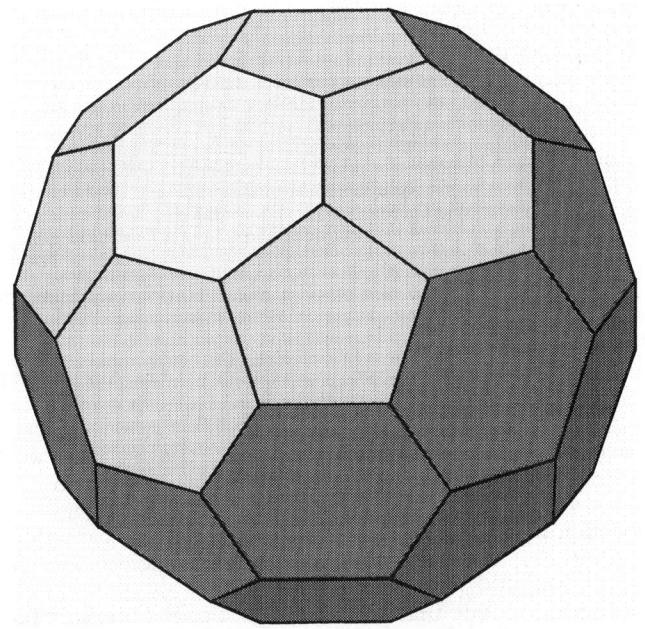

Fig. 4.   Frontal view of truncated icosahedral structure of C_{60} cluster

[1] Architect, Buckminster Fuller, probably is most famous for his design of the United States exhibit building for the 1967 Exposition in Montreal.

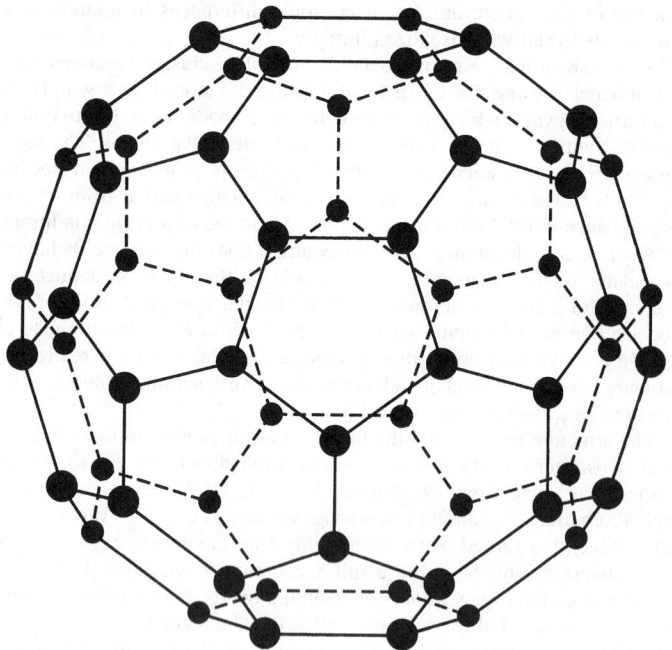

Fig. 5. C_{60} shown as transparent to indicate all sixty carbon atoms

icosahedral symmetry and thus highly supportive of the original proposed structure. However, the researchers did not strictly prove the soccer ball framework or provide atomic positions through the studies of spectra. Therefore, the investigators added an osmyl unit to C_{60} in order to break its pseudospherical symmetry and give an ordered crystal. The crystal structure of this derivative, C_{60} (OsO_4) (4-*tert*-butylpyridine)$_2$, revealed atomic positions within the carbon cluster, thus essentially confirming the soccer ball framework.

Scientists Y.Z. Li, R.E. Smalley, et al. (Rice University) have used scanning tunneling microscopy to study monolayer and multilayer structures of C_{60}. Detailed studies of potassium incorporation in crystalline C_{60} show highly ordered structures in the K_3C_{60} metallic state, but disordered non-metallic structures for high potassium concentrations.

Researcher R.C. Haddon (AT&T Bell Laboratories) reports that there seems to be no doubt that the C_{60} molecule is highly electronegative, but that recent research has characterized C_{60} as that of an electron-deficient polyalkene without significant delocalization. Fullerenes are without boundary conditions; just as in an ideal graphite sheet, there are no peripheral atoms to serve as sites of preferred activity. "Without curvature, the fullerenes would be no more reactive than an ideal graphite sheet. The chemistry of the fullerenes is best described as that of a class of strained and continuous aromatic molecules. C_{60} is of ambiguous magnetic properties but with the reactivity of a continuous aromatic molecule moderated only by the tremendous strain inherent in the spheroidal structure."

J.E. Fischer and a team of researchers (University of Pennsylvania) report, "The recent discovery of an efficient synthesis of C_{60} and C_{70} ... has facilitated the study of a new class of molecular crystals ('fullerites') based on these molecules ('fullerenes')." In the Fischer reference listed, a study of the compressibility of solid C_{60} is described.

In the S. Chakravarty (University of California, Los Angeles) et al. reference, "A theory of the electronic properties of doped fullerenes is proposed in which electronic correlation effects, within single fullerene molecules play a central role and qualitative predictions are made, which, if verified, would support this hypothesis. Depending on the effective intra-fullerene electron-electron repulsion and the interfullerene hopping amplitudes (which would depend on the dopant species, among other things), the calculations indicate the possibilities of singlet superconductivity and ferromagnetism."

As described by V.P. Dravid (Northwestern University) et al., "Transmission electron microscopy (TEM) observations of graphite tubules (*buckytubes*) and their derivatives have revealed not only the previously reported buckytube geometries but also additional shapes of the buckytube derivatives. Detailed cross-sectional TEM images reveal the cylindrical cross section of buckytubes and the growth pattern of buckytubes as well as

their derivatives.... Based on the TEM observations, it is proposed that buckytubes act as precursors to closed-shell fullerene (buckyball) formation."

The NEC Corporation, when examining deposits of soot on a carbon electrode used for generating fullerenes, found miniscule (up to a micron long) fibers which were tiled in hexagonal arrays. The arrays appear to tightly bind the carbon atoms and terminate in faceted, conical caps. The fibers immediately were referred to as "buckytubes." Iijama observes, "It (buckytube) could be the strongest fiber that can exist. Its strength flows from the nature of carbon-carbon bonds, on the one hand, and the nearly flawless structure of the tubular crystals, on the other." A scientist at the Massachusetts Institute of Technology has observed, "Buckyfibers have very few defects and so in that sense are better than graphite."

Materials engineers are becoming very interested in buckytubes because they may perform better than graphite in carbon-carbon composite materials, as currently used in aircraft.

Fullerenes in nature have been reported by a team of scientists, including P.R. Buseck (Arizona State University). High-resolution transmission electron microscopy images of poorly graphitized material in carbon-rich (coaly) rock, taken from an area near Karelia, Russia, are similar to those produced by synthetic fullerenes. The presence of C_{60} and C_{70} was confirmed by mass spectrometry. Needless to say, the finding was a surprise because the natural conditions for producing the fullerenes differ so much from the high-temperature processes of the laboratory. This finding may contribute in some way to future fullerene research.

Additional Reading

Aldersey-Williams, H.: *The Most Beautiful Molecule: The Discovery of the Buckyball,* John Wiley & Sons, Inc., New York, NY, 1997.

Alers, G.B., et al.: "Existence of an Orientational Electric Dipolar Response in C_{60} Single Crystals," *Science,* **511** (July 24, 1992).

Amato, I.: "Buckyballs Get Their First Major Physical," *Sci. News,* **357** (December 8, 1990).

Amato, I.: "Buckeyball, Hairyballs, Dopeyballs," *Sci. Amer.,* **646** (May 3, 1991).

Amato, I.: "A Transforming Look at C_{60}," *Science,* **1785** (June 28, 1991).

Amato, I.: "Doing Chemistry in the Round," *Science,* **30** (October 4, 1991).

Amato, I.: "First Sighting of Buckyballs in the Wild," *Science,* **167** (July 10, 1992).

Andreoni, W.: *The Chemical Physics of Fullerenes 10: NATO Advanced Research Workshop,* Kluwer Academic Publishing, Norwell, MA, 1996.

Bandyopadhyay, J.K. and K.L. Gauri: *Carbonate Stone: Chemical Behavior, Durability, and Conservation,* John Wiley & Sons, Inc., New York, NY, 1999.

Benning, P.J., et al.: "Electronic States of K_xC_{60}: Insulating, Metallic, and Superconducting Character," *Science,* **1417** (June 7, 1991).

Blashfield, J.F.: *Carbon,* Raintree Steck-Vaughn Publishers, Orlando, FL, 1998.

Buseck, P.R., S.J. Tsipursky, and R. Hettich: "Fullerenes from the Geological Environment," *Science,* **215** (July 10, 1992).

Chakravarty, S., Gelfand, M.P., and S. Kivelson: "Electronic Correlation Effects and Superconductivity in Doped Fullerenes," *Science,* **970** (November 15, 1991).

Culotta, E. and D.E. Koshland, Jr.: "Buckyballs: Wide Open Playing Field for Chemists," *Science,* **1706** (December 20, 1991).

Curl, R.F. and R.E. Smalley: "Probing C_{60}," *Science,* **7** (November 18, 1988).

Curl, R.F. and R.E. Smalley: "Fullerenes," *Sci., Amer.,* **54** (October 1991).

Daly, T.K., et al.: "Fullerenes from a Fulgurite," *Science,* **1599** (March 12, 1993).

Ebert, L.B.: "Is Soot Composed Predominantly of Carbon Clusters?" *Science,* **1468** (March 23, 1990).

Delhaes, P. and H. Kuzmany: *Fullerens and Carbon Based Materials,* Elsevier Science Ltd., New York, NY, 1998.

DeMeijere, A.: *Carbon Rich Compounds II: MacRocyclic Oligoacetylenes and Other Linearly Conjugated Systems,* Springer-Verlag, Inc., New York, NY, 1999.

Diederich, F., et al.: "Fullerene Isomerism," *Science,* **1768** (December 20, 1991).

Dravid, V.P., et al.: "Buckytubes and Derivatives: Their Growth and Implications for Buckyball Formation," *Science,* **1601** (March 12, 1993).

Dresselhaus, M.S., P.C. Eklund, and G. Dresselhaus: *Science of Fullerenes and Carbon Nanotubes,* Academic Press, Inc., San Diego, CA, 1996.

Fischer, J.E., et al.: "Compressibility of Solid C_{60}," *Science,* **1288** (May 31, 1991).

Flam, F.: "Buckyballs: A Little Like Basketballs—Only Smaller," *Science,* **29** (April 5, 1991).

Gogotsi, Y.G. and R.A. Andrievski: *Materials Science of Carbides, Nitrides and Borides,* Kluwer Academic Publishers, Norwell, MA, 1999.

Greenwood, N.N. and A. Earnshaw: *Chemistry of the Elements,* 2nd Edition, Butterworth-Heinemann, Inc., Woburn, MA, 1997.

Guo, T., et al.: "Uranium Stabilization of C28: A Tetravalent Fullerene," *Science,* **1661** (September 18, 1992).

Haddon, R.C.: "Chemistry of the Fullerenes: The Manifestation of Strain in a Class of Continuous Aromatic Molecules," *Science,* **1545** (September 17, 1991).

Hawkins, J.M., et al.: "Crystal Structure of Osmylated C_{60}: Confirmation of the Soccer Ball Framework," *Science,* **312** (April 12, 1991).

Hawley, G.G. and R.J. Lewis: *Hawley's Condensed Chemical Dictionary,* 13th Edition, John Wiley & Sons, Inc., New York, NY, 1999.

Hedberg, K., et al.: "Bond Lengths in Free Molecules of Buckminsterfullerene, C_{60} from Gas-Phase Electron Diffraction," *Science,* **410** (October 18, 1991).

Holden, C.: "Buckyballs for Sale," *Science,* **516** (February 1, 1991).

Hunter, J., J. Fye, and M.F. Jarrold: "Annealing C_{60+}; Synthesis of Fullerenes and Large Carbon Rings," *Science,* **784** (May 7, 1993).

Jones, J.: *Core Carbonyl Chemistry,* Oxford University Press, New York, NY, 1997.

Jones, D.: *The Inventions of Daedalus,* Freeman, Oxford, 1982.

Jones, L. and P.W. Atkins: *Chemistry: Molecules, Matter and Change,* W.H. Freeman and Company, New York, NY, 1999.

Kadish, K.: *Fullerenes: Chemistry, Physics, & Technology,* John Wiley & Sons, Inc., New York, NY, 1999.

Krebs, R.E.: *The History and Use of Our Earth's Chemical Elements: A Reference Guide,* Greenwood Publishers Group, Inc., Westport, CT, 1998.

Kroto, H.: "Space, Stars, C_{60} and Soot," *Science,* **1139** (November 25, 1988).

Kroto, H.: *Fullerenes: The First International Interdisciplinary Colloquium on the Science and Technology of the Fullerenes,* Pergamon Press, New York, NY, 1993.

Kroto, H.W. and D.R.M. Walton: *The Fullerenes: New Horizons for the Chemistry, Physics and Astrophysics of Carbon,* Cambridge University Press, New York, NY, 1994.

Li, Y.Z., et al.: "Order and Disorder in C_{60} and $K_x C_{60}$ Multilayers: Direct Imaging with Scanning Tunneling Microscopy," *Science,* **429** (July 26, 1991).

Lide, D.R.: *CRC Handbook of Chemistry and Physics,* 84th Edition, CRC Press, LLC., Boca Raton, FL, 2003.

Meijere, A. De: *Carbon Rich Compounds I,* Springer-Verlag, Inc., New York, NY, 1998.

Moffat, A.S.: "Chemists Cluster in Chicago to Confer on Cagey Compounds," *Science,* **400** (October 16, 1992).

Newton, D.E. and L.W. Baker: *Chemical Elements: From Carbon to Krypton,* UxI, Inc., Campbell, CA, 1998.

Olson, J.R., Topp, K.A., and R.O. Pohl: "Specific Heat and Thermal Conductivity of Solid Fullerenes," *Science,* **1145** (February 19, 1993).

Otera, J.: *Modern Carbonyl Chemistry,* VCH Publishers, Inc., New York, NY, 2000,

Pasquarello, A., M. Schulter, and R.C. Haddon: "Ring Currents in Icosahedral C_{60}," *Science,* **1660** (September 18, 1992).

Pennisi, E.: "Hot Times for Buckyball Superconductors," *Sci. News,* **84** (August 10, 1991).

Pennisi, E.: "Buckyballs' Supercool Spring Surprise," *Sci. News,* **244** (April 20, 1991).

Pennisi, E.: "Buckyballs Still Charm," *Sci. News,* **120** (August 24, 1991).

Pennisi, E.: "Buckyballs Shine as Optical Materials," *Sci. News,* **127** (August 24, 1991).

Poirier, D.M., et al.: "Formation of Fullerides and Fullerene-Based Heterostructures," *Science,* **646** (August 9, 1991).

Ross, P.E.: "Buckyballs: Fullerenes Open New Vistas in Chemistry," *Sci. Amer.,* **114** (January 1991).

Ross, P.E.: "Buckytubes: Fullerenes May Form the Finest, Toughest Fibers Yet," *Sci. Amer.,* **24** (December 1991).

Ross, P.E.: "Billions of Buckytubes," *Sci. Amer.,* **115** (October 1992).

Ross, P.E.: "Faux Fullerenes," *Sci. Amer.,* **24** (February 1993).

Sainsbury, M.: *Second Supplements to the Second Edition of Rodd's Chemistry of Carbon Compounds,* Elsevier Science Ltd., New York, NY, 1999.

Saunders, M.: "Buckminsterfullerene: The Inside Story," *Science,* **330** (July 19, 1991).

Shengzhong, L., et al.: "The Structure of the C_{60} Molecule: X-Ray Crystal Structure Determination of a Twin at 110 K," *Science,* **408** (October 18, 1991).

Staff: *Refractories, Carbon and Graphite Products, Activated Carbon, Advanced Ceramics,* American Society for Testing and Materials, West Conshohocken, PA, 1999.

Stevenson, F.J.: *Cycles of Soil; Carbon, Nitrogen, Phosphorus, Sulfur, Micronutrients,* John Wiley & Sons, Inc., New York, NY, 1999.

Stwertka, A. and E. Stwertka: *A Guide to the Elements,* Oxford University Press, Inc., New York, NY, 1998.

Thrower, P.A.: *Chemistry and Physics of Carbon: A Series of Advances,* Marcel Dekker. Inc., New York, NY, 1999.

Weber, E.: *Fullerenes and Related Structures,* Springer-Verlag, Inc., New York, NY, 1998.

Wigley, T.M.L. and D.S. Schimel: *The Carbon Cycle,* Cambridge University Press, New York, NY, 1999.

Yam, P.: "The All-Star of Buckyball," *Sci. Amer.,* **46** (September 1993).

Zhou, O., et al.: "Compressibility of $M_3 C_{60}$ Fullerene Superconductors," *Science,* **833** (February 14, 1992).

CARBONADO. The mineral carbonado is an opaque massive black variety of diamond, often crystalline to granular or compact and without cleavage. In thin splinters it appears greenish-black by transmitted light. It is found chiefly in Bahia, Brazil. Carbonado is used for rock-drilling apparatus.

Carbonado also is known as *black diamond.*

CARBONATE. A compound resulting from the reaction of either a metal or an organic compound with carbonic acid. The reaction with a metal yields a salt (calcium carbonate) and that with an aliphatic or aromatic compound forms an ester, e.g., diethylcarbonate, diphenyl carbonate. The latter are liquids used as solvents and in synthesizing polycarbonate resins. See also **Carbon.**

CARBON BLACK. [CAS: 1333-86-4]. Finely divided carbonaceous pigments of a wide variety are termed carbon blacks. Over 90% of the carbon black manufactured is consumed as reinforcing and compounding agents for rubber, mainly for motor vehicle and aircraft tires. Most users of tires do not realize that the effective use of these agents extends the life of a tire in normal usage by eight to ten times. The addition of as little as 1 to 2% carbon black to plastics greatly minimizes the effects of sunlight in degrading the materials. Most carbon blacks are derived from the pyrolysis of hydrocarbon gases and oils. The permanent and penetratingly deep black coloration obtainable with carbon blacks also makes the materials attractive for paints, inks, protective coatings, and as colorants for paper and plastics.

Two properties of carbon blacks are most significant for commercial applications: (1) particle size and (2) surface area. The particle sizes range from 100 to 5,000 micrometers. Surface areas will range from 6 to 1,100 m^2/g of material. Under electron microscopic examination, the carbon particles appear as rough spheres, usually as clusters of spheres rather than as individual spheres. The clustering characteristics stem from both chemical and physical bonding forces. Classically, the arrangement of the carbon particles may be likened to hexagonal nets of carbon atoms, which are *paracrystalline* in nature. The particle size and surface area characteristics essentially are at the microscopic level-hence control over carbon black production is exacting. In terms of coloration, for example, the human eye can resolve 260 shades of blackness. The blackest of commercially produced carbon particles will have a diameter of about 100 micrometers. The grayest particle will have a diameter of about 5,000 micrometers. The blackness characteristic sometimes is referred to as *masstone* (mass-tone). The particles with the smaller diameters and hence greater surface area exhibit the highest masstone.

Lampblacks have been made for many centuries. Early methods involved the burning of petroleum-like substances or coal-tar residues with a minimum of air, thus producing large amounts of unoxidized carbon particles. The earlier settling chambers in which the particles collected have been replaced by cyclones, bag filters, or electrical precipitators. Modern installations use oil furnaces to create the particles.

Channel or *impingement carbons* are produced from burning natural gas (sometimes containing oil vapors) in many hundreds of small burners. The flames from the burners impinge upon flat surfaces called channels. The carbon deposits are periodically removed by scraping into a collector. The burning equipment is contained within a large burner house, which has means for carefully regulating bottom and top drafting of air.

Thermal blacks also are derived from natural gas, but by thermal decomposition completely in the absence of air. Large furnaces first are preheated to a temperature ranging from 1,100–1,650°C. When the checkerwork is at the proper temperature, natural gas is bled into the furnace, whereupon the gas decomposes into carbon and hydrogen. This is a batch process, requiring pairs of furnaces, one furnace preheating, while the other furnace is decomposing the gas feed. Frequently, the hydrogen byproduct is recycled as fuel to heat the furnaces. Where very fine thermal blacks are produced, the byproduct hydrogen is used as a diluent for the gas feed.

Furnace carbons also are derived from natural gas, but in a process in which a slight excess of air is introduced to support combustion. The hydrocarbon feedstock or liquid oil is injected into the furnace at a location where the so-called blast-flame gases are circulating at their greatest velocity. Injection of the feed at this point causes an instant high rise in temperature, which results in practically instantaneous decomposition of the feed into carbon black. For coarse particles, the oil/air ratio is greater, furnace gas velocities are lower, and residence time in the furnace is longer. There is a wide range of furnace carbon particle sizes. The very fine particles go into tire treads, whereas the coarser particles are used in tire carcasses.

Acetylene black is derived from feeding acetylene into high-temperature retorts whereupon the acetylene dissociates into carbon and hydrogen. This reaction is exothermic (other carbon black processes are endothermic). Temperature control of the furnace is effected by throttling the acetylene feed.

Additional Reading

Donnet, Jean-Baptiste, R.C. Bansal, et al.: *Carbon Black: Science and Technology,* Marcel Dekker Inc., New York, NY, 1993.

Staff: *Rubber, Natural and Synthetic-General Test Methods, Carbon Black,* American Society for Testing and Materials, West Conshohocken, PA, 1999.

CARBON COMPOUNDS. See **Organic Chemistry**.

CARBON DATING. See **Radioactivity**.

CARBON DIOXIDE. [CAS: 124-38-9]. CO_2, formula weight 44.01, colorless, odorless, nontoxic gas at standard conditions. High concentrations of the gas do cause stupefaction and suffocation because of the displacement of ample oxygen for breathing. Density 1.9769 g/l (0°C, 760 torr), sp gr 1.53 (air = 1.00), mp −56.6°C (5.2 atmospheres), solid CO_2 sublimes at −79°C (760 torr), critical pressure 73 atmospheres, critical temperature 31°C. Carbon dioxide is soluble in H_2O (approximately 1 volume CO_2 in 1 volume H_2O at 15°C, 760 torr), soluble in alcohol, and is rapidly absorbed by most alkaline solutions. The solubility of CO_2 in H_2O for various pressures and temperatures is given in Table 1.

Carbon dioxide plays several roles: (1) as a *raw material* for several processes, as in the Solvay process for the manufacture of sodium bicarbonate and sodium carbonate. (2) as a *byproduct* from many processes, notably as a product of combustion of fossil fuels, (3) as an *ingredient* of products, for example, carbonated beverages, (4) as a *product* for direct consumption, for example, CO_2 fire extinguishers and dry ice refrigerants, and (5) as a *pollutant* of the atmosphere. Carbon dioxide is useful in all three of its physical phases-gas, liquid, and solid. Although not toxic, the presence of CO_2 in the atmosphere disturbs the environmental energy balance. The latter aspects of CO_2 are discussed under **Pollution (Air)**. Normally, CO_2 is present in the air at sea level to the extent of about 0.05% by weight.

Transportation Uses

Solid carbon dioxide (dry ice) is an effective refrigerant for transportation uses. Refrigeration of moving vehicles may be derived from (1) mechanical systems which, of course, require a continuous input of energy, (2) water ice and ice-salt mixes which require water (often briny) removal, and are corrosive and subject to algae formations, and (3) dry ice, the end-product of which is simply gaseous CO_2, which is easily removed. To maintain a cool temperature in a railroad refrigerator car for a trip between California and New York, about 1,000 pounds (~454 kg) of dry ice would be required. To maintain the same conditions with water ice and salt would require 10,000 pounds of ice.

Specially designed rail cars have replaced on-board diesel-powered refrigeration units, with a CO_2 injection system and ceiling-mounted bunker. These bunkers carry sufficient quantities of dry ice snow to provide sufficient refrigeration for long trips. There are similar applications where perishables are moved by truck. Particularly in truck shipments, CO_2 systems not only refrigerate the cargo, but the inert atmosphere (CO_2 in gaseous phase) retards bacterial growth and thus prevents spoilage. The system is widely used for local route deliveries where frequent and lengthy door openings are needed. Automatic temperature controllers are used. Airlines, hotels, and restaurants keep prepared foods fresh during transport by dispensing CO_2 snow into the bunker portion of customized food service carts.

TABLE 1. SOLUBILITY OF CARBON DIOXIDE IN WATER

Pressure (atmospheres)	Parts (Weight) CO_2 Soluble in 100 Parts Water				
	18°C	35°C	50°C	75°C	100°C
25	3.7	2.6	1.9	1.4	1.1
50	6.3	4.4	4.0	2.5	2.0
75	6.7	5.5	4.5	3.4	2.8
100	6.8	5.8	5.1	4.1	3.5
200	—	6.3	5.8	5.3	5.1
300	7.4	—	6.2	5.8	5.7
400	7.8	7.1	6.6	6.3	6.4
700	—	—	7.6	7.4	7.6

Fire-Fighting Uses

The fact that CO_2 is heavier than air makes it particularly effective for fighting fires in low places, such as pipe trenches and hard-to-reach low corners and basements, where the CO_2 tends to roll under the air required to maintain combustion. Both manually and automatically controlled CO_2 fire-fighting systems are available. These can be actuated by heat-sensitive systems—just as a conventional water-sprinkling system. CO_2 is effective for fires involving electrical and electronic gear because, if a fire is not fully out-of-hand, the CO_2 often can quickly quench the fire source without leaving any residual damage, as often is the disastrous consequences of using water or sand.

Food Industry Uses

Large quantities of CO_2 are used in food processing, ranging over a wide variety of cooling and freezing operations. A number of freezer designs have been developed, including tunnel, cabinet, spiral, flighted, and drum designs. For example, wide usage of CO_2 in the baking industry includes chilling pneumatically conveyed dry ingredients, such as flour and powdered sugar, to controlling the temperature of dough during the mixing process.

Carbon dioxide is used for carbonating soft drinks. The wine industry also uses CO_2 to add effervescence to sparkling burgundies, rose wines, and some champagne.

The use of CO_2 atmosphere systems in greenhouses has been found to increase plant growth. During winter months, heating costs are markedly reduced and crop yields are increased.

Oil Production Enhancement

For a number of years, depending upon the geopolitics of crude oil production, considerable interest has been shown in the use of carbon dioxide for increasing the recovery of oil from old wells. In the United States alone, it is estimated that there are more than 300 billion barrels of oil left in known formations, which are incapable of recovery through the use of traditional recovery enhancement techniques, such as steam flooding and the use of surfactants. Supercritical fluid carbon dioxide is an impressive solvent for fats and hydrocarbons. The problems of geological formations underground and their varying characteristics (permeability, etc.) present difficulties as with past methods, but it has been established that the dense fluid CO_2 will contribute to recovery wherever it contacts oil. Consequently, some major oil firms already have expended large sums to ready pipelines and other facilities for bringing CO_2 to oil fields as, for example, those in the Permian basin of western Texas and New Mexico. Although carbon dioxide has been a useful material for other purposes, oil recovery usage may require the gas in huge quantities not heretofore contemplated. The target, of course, is to capture the needed CO_2 mainly from wastes to the atmosphere, as from power plants. Although authorities still consider oil recovery as a long-range goal, the short-term pace is affected by the fluctuating price of crude oil on world markets. More detail concerning the use of supercritical CO_2 for this purpose is given in entry on **Petroleum**.

Sources of Commercial Carbon Dioxide

Although carbon dioxide must be generated on site for some processes, there is a trend toward CO_2 recovery where it is a major reaction byproduct and, in the past, vented to the atmosphere. For example, very large quantities of CO_2 are generated by various fermentation processes and in cement production. If the CO_2 must be removed from stack gases because of pollution control regulations, it is only one more step to purify the gas and sell it, usually in compressed liquid form. There are, of course, several economic tradeoffs that must be considered. Where the gas is recovered, it usually is first absorbed in sodium or potassium carbonate solutions, followed by steam-heating the solutions to free a reasonably pure CO_2. The last step is compression of the gas into steel cylinders. The ethanolamines also are excellent absorbents of CO_2.

Carbon Dioxide in Biological Systems

Carbon dioxide, which is a byproduct of the metabolic activity of all cells, is one of the most important chemical regulators in the human body. It can be said that human life without carbon dioxide would be impossible. In less specialized forms of life, carbon dioxide is essentially a waste product. In the more highly developed animals, such as humans, the gas is used to regulate the activity of the heart, the blood vessels, and the respiratory system.

As mentioned, CO_2 is normally present in air at sea level at about 0.05% (weight). A poorly ventilated room may contain as much as 1% (volume). Concentrations of the gas from about 0.1–1% (volume) induce languor and headaches; concentrations of 8–10% (volume) bring about death by asphyxiation. High concentrations of the gas are toxic. See also **Basal Metabolism**.

As a general rule, the respiration of individual cells decreases as the concentration of carbon dioxide in the medium increases. Fish show a lessened capacity to extract oxygen from their environment with increasing amounts of carbon dioxide present. On the other hand, many invertebrates show marked increases in respiratory rate (or ventilation) with increased amounts of the gas in their surroundings.

Photosynthetic and autotrophic bacteria reduce carbon dioxide, which is assimilated into complex molecules for use in synthesizing various cellular constituents. The gas is apparently assimilated, at least to a small extent, by the heterotrophic bacteria. Certainly it is required for any growth in these forms. Many pathogenic bacteria required increased carbon dioxide tension for growth immediately after they are isolated from the body. The production of hemolysins and like substances is greatly enhanced by adding 10–20% of CO_2 in the air that comes in contact with the cultures.

The oxygen dissociation curve for blood is shifted to the right when the partial pressure of carbon dioxide is increased. This is referred to as the "Bohr Effect." It means that for a given partial pressure of oxygen, hemoglobin holds less oxygen at high concentration of carbon dioxide than at a lower concentration. It is evident, then, that the production of carbon dioxide by actively metabolizing tissues favors the release of oxygen from the blood to the cells where it is urgently needed. Moreover, at the alveolar surfaces in the lungs, the blood is losing carbon dioxide rapidly, which loss favors the combination of oxygen with hemoglobin. In males, the average amount of CO_2 in the alveolar air is about 5.5% (volume); during the breathing cycle, this concentration varies only slightly. In females and children, somewhat lower mean values obtain.

In every 100 milliliters of arterial blood, there is a total of 48 milliliters of free and combined CO_2. In venous blood of resting humans, there is about 5 milliliters more than this. Only about 1/20 of the carbon dioxide is uncombined, a fact which indicates that there is a specialized mechanism, aside from simple solution, for the transport of CO_2 in the blood.

About 20% of the CO_2 in the blood is carried in combination with hemoglobin as *carbaminohemoglobin*. The balance of the combined carbon dioxide is carried as bicarbonate. A CO_2 dissociation curve for blood can be prepared just as for oxygen, but the shape is not the same as for the latter. As the partial pressure of CO_2 in the air increases, the amount in the blood increases; the increase is practically linear in the higher ranges. Oxygen exerts a negative effect on the amount of CO_2 which can be taken up by the blood.

In working muscles large amounts of CO_2 are produced. This causes local vasodilation. The diffusion of some of the CO_2 into the bloodstream slightly raises the concentration there. It circulates through the body and the capillaries of the vasoconstrictor center, where it excites the cells of the center, resulting in an increase of constrictor discharges. Regarding the stimulating effect of CO_2 on cardiac output, it is evident that a most effective mechanism exists for increasing circulation through active muscles: more blood is pumped by the heart per minute, and the arterial pressure is increased by the general vasoconstriction; blood is forced from the inactive regions, under increased pressure, through the widely dilated vessels of the active muscles.

The partial pressure of CO_2 is important in connection with a number of physiological problems. For example, respiratory acidosis is the result of an abnormally high $p \ldots CO_2$. The value of arterial pCO_2 varies directly with changes in the metabolic production of CO_2 and indirectly with the amount of alveolar ventilation. The problem is more commonly the result of decreased alveolar ventilation caused by abnormally low CO_2 excretion by the lungs (alveolar *hypo*ventilation).

On the other hand, primary respiratory alkalosis occurs as a result of alveolar *hyper*ventilation. This condition is associated with a number of pulmonary diseases, but also may appear during pregnancy, liver disease, and salicylate intoxication, among others. The sequence of events proceeds along these lines: (1) Ventilation removes CO_2 faster than the gas is produced by metabolism, causing a decrease in pCO_2 in the blood and body fluids, including a reduction of venous pCO_2. This reduces the gradient

for excretion of CO_2 by the lungs. (2) Pulmonary excretion and metabolic production ultimately balance out at a lower pCO_2 level for all body fluids. (3) The lower pCO_2 level causes a lower carbonic acid concentration and consequently an increase in pH. The latter is relative to the reduced level of pCO_2, but the pH change also alters bicarbonate concentration. The steplike process is quite complex. See also **Blood**.

Narcosis due to CO_2 is characterized by mental disturbances which may range from confusion, mania, or drowsiness to deep coma, headache, sweating, muscle twitching, increased intracranial pressure, pounding pulse, low blood pressure, hypothermia, and sometimes papilloedema. The basic mechanisms by which carbon dioxide induces narcosis is probably through interference with the intracellular enzyme systems, which are all sensitive to pH changes.

See also **Photosynthesis**.

Carbon Dioxide and Enzymes

Dr. Harland Wood (Case Western Reserve University) has made major contributions to the understanding of carbon dioxide cycles and enzyme reactions within living organisms. While investigating the process of bacterial fermentation, Wood discovered that some heterotrophic organisms (non-plant forms that require organic compounds for growth) can use carbon dioxide along with organic compounds to build essential compounds. This was in 1935, when it was considered that only plants could use carbon dioxide and that, in heterotrophs, carbon dioxide was a waste product. Wood also researched the role of carbon dioxide in the metabolism of carbohydrates, fats, and amino acids by forming the required intermediate compounds. In 1985 Wood found that certain bacteria produce organic compounds entirely from carbon dioxide by a pathway that differs from that of photosynthesis. Certain parts of the cycle involving use of carbon dioxide and hydrogen were found to exist within many organisms.

Wood also has worked with transcarboxylase, a complex, biotin-containing enzyme important in the use of carbon dioxide within heterotrophs. As of the early 1990s this pathway had not been fully delineated.

Additional Reading

Berliner, L.J. and Pierre-Marie. Robitaille: *Biological Magnetic Resonance, Vol. 15: In Vivo Carbon-13 NMR,* Plenum Publishing Corporation, New York, NY, 1998.

Branden, Carl-Ivar and G. Schneider: *Carbon Dioxide Fixation and Reduction in Biological and Model Systems: Proceedings of the Royal Swedish Academy of Sciences Nobel Symposium 1991,* Oxford University Press, New York, NY, 1994.

Halmann, M.M. and M. Steinberg: *Greenhouse Gas Carbon Dioxide Mitigation: Science and Technology,* Lewis Publishing, Cherry Hill, NJ, 1998.

Halmann, M.M.: *Chemical Fixation of Carbon Dioxide: Methods for Recycling CO_2 into Useful Products,* CRC Press, LLC., Boca Raton, FL, 1993.

Hawley, G.G. and R.J. Lewis: *Hawley's Condensed Chemical Dictionary,* 13th Edition, John Wiley & Sons, Inc., New York, NY, 1999.

Lide, D.R.: *CRC Handbook of Chemistry and Physics,* 84th Edition, CRC Press, LLC., Boca Raton, FL, 2003.

Luo, Y. and H.A. Mooney: *Carbon Dioxide and Environmental Stress,* Academic Press, Inc., San Diego, CA, 1999.

Perry, R.H., D.W. Green, and J.O. Maloney: *Perry's Chemical Engineers' Handbook,* 7th Edition, The McGraw-Hill Companies, Inc., New York, NY, 1997.

Pradier, Jan Paul, and Claire-Marie, Pradier: *Carbon Dioxide Chemistry: Environmental Issues,* Lewis Publishing, Cherry Hill, NJ, 1994.

Staff: *Carbon-Dioxide Fire Extinguishers, UI 154,* 8th Edition, Laboratories Incorporated Underwriters, Northbrook, IL, 1995.

Staff: *Carbon Dioxide,* Liquid Carbonic Corporation, Chicago, Illinois (1990).

Williams, A.: *Concerted Organic and Bio-Organic Mechanisms,* CRC Press, LLC., Boca Raton, FL, 1999.

Wittwer, S.H.: *Food, Climate, and Carbon Dioxide: The Global Environment and World Food Production,* CRC Press, LLC., Boca Raton, FL, 1995.

CARBON DISULFIDE. [CAS: 75-15-0]. Carbon disulfide (carbon bisulfide, dithiocarbonic anhydride), CS_2, is a toxic, dense liquid of high volatility and flammability. It is an important industrial chemical and its properties are well established. Low concentrations of carbon disulfide naturally discharge into the atmosphere from certain soils, and carbon disulfide has been detected in mustard oil, volcanic gases, and crude petroleum. Carbon disulfide is an unintentional by-product of many combustion and high temperature industrial processes where sulfur compounds are present.

Commercial uses grew rapidly from about 1929 to 1970, when the principal applications included manufacturing viscose rayon fibers, cellophane, carbon tetrachloride, flotation aids, rubber vulcanization accelerators, fungicides, and pesticides. Production of carbon disulfide in the United States has declined in recent years. Other chemical fibers and films, as well as environmental and toxicity considerations related to carbon tetrachloride, have had significant impact on the demand for carbon disulfide. Worldwide annual production capacity in 1991 was approximately 1.3 million tons, with actual production estimated at about one million metric tons.

Physical Properties

Pure carbon disulfide is a clear, colorless liquid with a delicate, ether-like odor. Carbon disulfide is slightly miscible with water, but it is a good solvent for many organic compounds. Thermodynamic constants, vapor pressure, spectral transmission, and other properties of carbon disulfide have been determined. Principal properties are listed in Table 1.

Carbon disulfide is completely miscible with many hydrocarbons, alcohols, and chlorinated hydrocarbons. Phosphorus and sulfur are very soluble in carbon disulfide.

Chemical Properties

The low flash point temperature of $-30°C$ at atmospheric pressure and wide flammability range of carbon disulfide deserve special attention. The flash point is lowered if the pressure is decreased or the oxygen content enriched. The flammability limits or explosive ranges depend on conditions of temperature, pressure, and geometry of the enclosure. Flammability limits of 1.06–50.0 vol % carbon disulfide in air are reported for upward propagation and 1.91–35.0 vol % for downward propagation in a 75-mm diameter glass tube.

Carbon disulfide chemistry is thoroughly described in several publications which include many references.

Manufacture

The earliest method for manufacturing carbon disulfide involved synthesis from the elements by reaction of sulfur and carbon as hardwood charcoal in externally heated retorts. Safety concerns, short lives of the retorts, and low production capacities led to the development of an electric furnace process, also based on reaction of sulfur and charcoal. The commercial use of hydrocarbons as the source of carbon was developed in the 1950s, and it was still the predominate process worldwide in 1991. That route, using methane and sulfur as the feedstock, provides high capacity in an economical, continuous unit.

Potential Processes. Sulfur vapor reacts with other hydrocarbon gases, such as acetylene or ethylene, to form carbon disulfide. Light gas oil was reported to be successful on a semiworks scale. In the reaction with hydrocarbons or carbon, pyrites can be the sulfur source. With methane and iron pyrite the reaction products are carbon disulfide, hydrogen sulfide, and iron or iron sulfide. Pyrite can be reduced with carbon monoxide to produce carbon disulfide.

Toxicology, Health, and Safety Factors

Care must be exercised in handling carbon disulfide because of both health concerns and the danger of fire or explosions. Occupational exposure potentially may involve as many as 20,000 workers in the United States. Ingestion is rare, but a 10 mL dose can prove fatal. Contact usually occurs by inhalation of vapor. However, vapor and liquid can be absorbed through intact skin and poisoning may occur by the dermal route. Repeated contact of liquid carbon disulfide with the skin can cause inflammation and cracking because carbon disulfide removes protective waxes and oils. Extended skin contact results in blistering and possibly second- and third-degree burns. Precautions should be taken to avoid breathing of vapors or mists that may contain carbon disulfide. Contact with skin or eyes should also be avoided, and adequate safety gear should be worn, including goggles, impervious gloves, and appropriate clothing.

The odor threshold of carbon disulfide is about 1 ppm in air but varies widely depending on individual sensitivity and purity of the carbon disulfide. However, using the sense of smell to detect excessive concentrations of carbon disulfide is unreliable because of the frequent co-presence of hydrogen sulfide that dulls the olfactory sense.

Immediate effects of overexposure to carbon disulfide vapors range from headache, dizziness, nausea, and vomiting to life-threatening convulsions, unconsciousness, and respiratory paralysis. For an exposure time of 30 min, 1150 ppm carbon disulfide in air results in serious symptoms, 3210 ppm is dangerous to life, and 4815 ppm is fatal. Prolonged and repeated exposure to carbon disulfide vapor can affect both the central and peripheral nervous systems. In recent years, previously unrecognized and more subtle toxic effects of repeated lower level exposures became evident. This led OSHA in 1989 to reduce permissible concentration limits to 4 ppm (12 mg/m^3) maximum time-weighted average for 8-h exposure and 12 ppm (36 mg/m^3) maximum for 15-min short-term exposure. Analysis of urine specimens for carbon disulfide metabolites by an iodine-azide test and other methods can indicate overexposure.

Health hazards linked to carbon disulfide are extensively covered in the literature. Also available are epidemiological studies, general reviews containing many references, and a Material Safety Data Sheet.

Uses

United States consumption of carbon disulfide totaled about 108,000 t in 1990 according to SRI International, with the following distribution by end use application: 46,000 t for rayon; 33,000 t for carbon tetrachloride; 12,000 t for rubber; 5,000 t for cellophane; and 12,000 t for agricultural and miscellaneous uses. During 1991 the carbon tetrachloride application disappeared entirely, thereby reducing the annualized carbon disulfide usage to an estimated 75,000 t. Net exports are around 6,000 t, and are expected to increase in the future.

DAVID E. SMITH
ROBERT W. TIMMERMAN
FMC Corporation

Additional Reading

Dunn, A. D. and W. D. Rudorf: *Carbon Disulphide in Organic Chemistry,* Ellis Horwood Ltd., Chichester, UK; Halsted Press, div. of John Wiley & Sons, Inc., New York, NY, 1989.
O'Brien, L. J. and W. J. Alford: *Ind. Eng. Chem.* **43**, 506 (1951).
Patty, F. A.: *Industrial Hygiene and Toxicology,* Vol. II, 2nd. rev. ed., John Wiley & Sons, Inc., New York, NY, 1962, pp. 901–904.
U.S. Pat. 2,568,121 (Sept. 18, 1951), H. O. Folkins, C. A. Porter, E. Miller, and H. Hennig (to The Pure Oil Co.).

CARBON GROUP (The). The elements of group 14 of the periodic classification sometimes are referred to as the Carbon Group. In order of increasing atomic number, they are carbon, silicon, germanium, tin, and lead. The elements of this group are characterized by the presence of four electrons in an outer shell. The similarities of chemical behavior among the elements of this group are less striking than that for some of the other groups. e.g., the close parallels of the alkali metals or alkaline earths. However, as more knowledge is gained of silicon, including the element's ability to form "carbon-like" chains with alternating silicon and

TABLE 1. PROPERTIES OF CARBON DISULFIDE

Property	Values
General	
melting point, K	161.11
latent heat of fusion, kJ/kg[a]	57.7
boiling point at 101.3 kPa[b], °C	46.25
flash point at 101.3 kPa[b], °C	−30
ignition temperature in air, °C	
10-s lag time	120
0.5-s lag time	156
critical temperature, °C	273
critical pressure, kPa[b]	7700
critical density, kg/m^3	378
solubility H$_2$O in CS$_2$	
at 10°C, ppm	86
at 25°C, ppm	142
dielectric constant	2.641
Thermochemical data at 298 K[a]	
heat capacity, C$°_p$, J/(mol · K)[a]	45.48
entropy, S°, J/(mol · K)[a]	237.8
heat of formation, H$°_f$, kJ/mol[a]	117.1
free energy of formation, G$°_f$, kJ/mol[a]	66.9

[a] To convert J to cal, divide by 4.184.
[b] To convert kPa to atm, divide by 101.3.

oxygen atoms, to polymerize, and to form silicones, silanes, etc., the similarity of silicon and carbon emerges more sharply. The semiconductor properties of silicon and germanium in this group are striking, but such properties are not limited to elements in this group. Although some of the elements of the group have valences in addition to $+4$, all do have the $+4$ valence in common. Unlike the alkali metals or alkaline earths, for example, the elements of the carbon group are not so similar chemically that they comprise a separate group in classical qualitative chemical analysis separations.

CARBONITRIDING. A surface hardening process for steels involving the introduction of carbon and nitrogen into steels by heating in a suitable atmosphere containing various combinations of hydrocarbons, ammonia, and carbon monoxide followed by a quenching to harden the case.

CARBONIZATION (Coal). See **Coal**.

CARBONIUM ION. An ion of the general formula

$$
\begin{array}{c}
\text{A} \\
| \\
\text{B}\!-\!\overset{+}{\text{C}} \\
| \\
\text{D}
\end{array}
$$

where A, B and D are substituent groups. It is important in elucidating the mechanism of organic reactions because a considerable proportion of all organic reactions involve carbonium ions, as others do carbanions and carbon free radicals (including carbene radicals). Nucleophilic substitution at saturated carbon atoms includes most of carbonium ion chemistry. Carbonium ions are usually powerful acids or electrophiles, and thus many nucleophilic substitution reactions that involve carbonium ions are acid-catalyzed. For example, the tertiary-butyl carbonium ion offers a clear understanding of the probable course of the conversion of isobutylene to its dimers and trimers.

$$(CH_3)_2C\!=\!CH_2 + H^+ \longleftrightarrow (CH_3)_3C^+$$

$$(CH_3)_3C^+ + (CH_3)_2C\!=\!CH_2 \rightleftharpoons (CH_3)_2C\!-\!CH_2C(CH_3)_3$$

The larger carbonium ion thus formed cannot continue to exist, but may depolymerize, unite with the catalyst, or stabilize itself by the attraction of an electron pair from a carbon atom adjacent to the electronically deficient carbon (C^+) with its proton. This establishes a double bond involving the formerly deficient atom. Thus a proton is expelled to the catalyst or attracted to the catalyst. If this takes place with one of the methyl groups, the product

$$
\begin{array}{c}
CH_2\!=\!C\!-\!CH_2C(CH_3)_3 \\
| \\
CH_3
\end{array}
$$

is If the methylene group is involved, the product is $(CH_3)_2C\!=\!CHC(CH_3)_3$.

CARBON MONOXIDE. [CAS: 630-08-0]. CO, formula weight 28.01, colorless, odorless, very toxic gas at standard conditions, density 1.2504 g/l ($0°C$, 760 torr), sp gr 0.968 (air = 1.000), mp $-207°C$, bp $-192°C$, critical temperature $-139°C$, critical pressure 35 atmospheres. Carbon monoxide is virtually insoluble in H_2O (0.0044 part CO in 100 parts H_2O at $50°C$). The gas is soluble in alcohol or solutions of cupric chloride. Because carbon monoxide has an affinity for blood hemoglobin that is 300 times that of oxygen, exposure to the gas greatly reduces or fully hinders the ability of hemoglobin to carry oxygen throughout the body, causing death in excessive concentrations. Engines and stoves in poorly ventilated areas are especially hazardous.

Carbon monoxide plays several roles: (1) as a raw material for chemical processes (a) particularly as an effective reducing agent in various metal smelting operations, (b) in the manufacture of formates: $CO + NaOH \longrightarrow HCOONa$, (c) in the production of carbonyls, such as $Ni(CO)_4$ and $Fe(CO)_5$, which are useful intermediate compounds in the separation of certain metals, (d) in combination with chlorine to form $COCl_2$ (phosgene), (e) as an ingredient of several synthesis gases, as for the production of methanol and ammonia; (2) as a *fuel* where CO is a major ingredient of such artificial fuels as coal gas, producer gas, blast-furnace gas, and water gas; (3) as a byproduct of numerous chemical

reactions, notably combustion processes where there is insufficient oxygen for complete combustion—the fumes from internal-combustion engines may contain in excess of 7% CO, and (4) as a dangerous air pollutant, particularly in industrial areas and where there are high concentrations of automotive vehicles and aircraft. The latter aspects of CO are discussed under **Pollution (Air)**

Summary of Chemical Reactivity

Chemically, carbon monoxide is (1) reactive with oxygen to form CO_2 accompanied by a transparent blue flame and the evolution of heat, but the fuel value is low (320 Btu per ft^3), (2) reactive with chlorine, forming carbonyl chloride $COCl_2$ in the presence of light and a catalyzer, (3) reactive with sulfur vapor at a red heat, forming carbonyl sulfide COS, (4) reactive with hydrogen, forming methyl alcohol, CH_3OH or methane CH_4 in the presence of a catalyzer, (5) reactive with nickel (also iron, cobalt, molybdenum, ruthenium, rhodium, osmium, and iridium) to form nickel carbonyl, $Ni(CO)_4$ (and carbonyls of the other metals named), (6) reactive with fused NaOH, forming sodium formate, HCOONa, (7) reactive with cuprous salt dissolved in either ammonia solution or concentrated HCl, which solutions are utilized in the estimation of carbon monoxide in mixtures of gases, e.g., flue gases of combustion, coal gas, exhaust gases of internal combustion engines, (8) reactive with iodine pentoxide at $150°C$. For the reaction of carbon monoxide with oxygen to form CO_2 finely divided iron or palladium wire is used as a catalyzer; for the reaction of carbon monoxide with H_2O vapor to form CO_2 plus hydrogen ("water gas reaction") important studies have been made of the conditions; and for the reaction of CO_2 plus carbon (hot) similar important studies have been made (at $675°C$, 50% CO_2 plus 50% CO; at $900°C$, 5% CO_2 plus 95% CO). The reaction of carbon plus oxygen at such a temperature as produces carbon monoxide (say $900°C$, 95% CO plus 5% CO_2) and *evolves heat*; while the reaction of carbon plus CO_2, producing carbon monoxide at the same temperature *absorbs heat*. Accordingly it is possible to arrange the oxygen (free or as air) and CO_2 supply ratio in such a way that the desired temperature may be continuously maintained. The reduction of CO_2 by iron forms carbon monoxide plus ferrous oxide.

In valence bond terms, carbon monoxide is considered as a resonance compound with the structures

$$\overset{+}{:}C:\overset{..}{\underset{..}{O}}:^{-} \quad :C::\overset{..}{O}: \quad :C::\underset{..}{O}: \quad {}^{-}:C:::O:^{+}$$

In molecular orbital terms the CO molecule is described as $CO(KK(z\sigma)^2 (y\sigma)^2(x\sigma)^2(w\pi)^4)$, one $(z\sigma)$ pair being formed from the oxygen $2s$ electrons, and one $(y\sigma)$ pair held by the carbon sp hybrid. This $(y\sigma)^2$ pair offsets the dipole moment of the π electrons, and also accounts for the readiness with which the CO molecule coordinates with metals to form the carbonyls.

Additional Reading

Cargill, R.W.: *Carbon Monoxide*, Pergamon Press, Mineola, NY, 1990.

Hawley, G.G. and R.J. Lewis: *Hawley's Condensed Chemical Dictionary*, 13th Edition, John Wiley & Sons, Inc., New York, NY, 1999.

Hirschler, M.M.: *Carbon Monoxide and Human Lethality: Fire and Non-Fire Studies*, Elsevier Science, New York, NY, 1993.

Lewis, R.J. and N.I. Sax: *Dangerous Properties of Industrial Materials*, 10th Edition, John Wiley & Sons, Inc., New York, NY, 1999.

Lide, D.R.: *CRC Handbook of Chemistry & Physics*, 84th Edition, CRC Press, LLC., Boca Raton, FL, 2003.

Parker, S.P.: *McGraw-Hill Concise Encyclopedia of Science and Technology*, 4th Edition, The McGraw-Hill Companies, Inc., New York, NY, 1998.

Penney, D.G.: *Carbon Monoxide*, CRC Press, LLC., Boca Raton, FL, 1996.

CARBON SUBOXIDE. C_3O_2, formula weight 68.03, colorless, toxic, gas at room temperature, very unpleasant odor, sp gr 2.10 (air = 1.00), 1.24 (liquid at $-87°C$), mp $-107°C$, bp $7°C$ (760 torr), burns with a blue smoky flame, producing CO_2. When condensed to liquid, the oxide slowly changes at ordinary temperature to a dark red solid, soluble in water to a red solution. Reacts with water to form malonic acid, with hydrogen chloride to form malonyl chloride, with ammonia to form malonamide. Made by heating malonic acid or its ester at $300°C$ under diminished pressure, and separation from simultaneously formed carbon dioxide and ethylene by condensation and fractional distillation.

Carbon suboxide has a linear structure, probably a resonance of four structures of which the last two below probably make a smaller contribution to the normal state of the molecule than the first two.

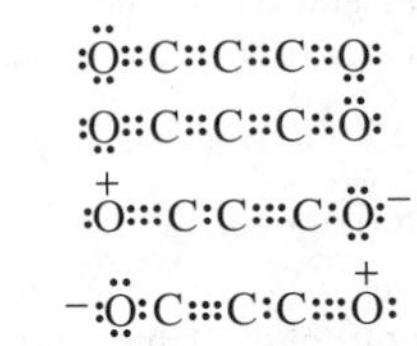

$$:\ddot{O}::C::C::C::\ddot{O}:$$
$$:\ddot{O}::C::C::C::\ddot{O}:$$
$$\overset{+}{:}\ddot{O}:::C:C:::C:\ddot{O}:^{-}$$
$$^{-}:\ddot{O}:::C:::C:C:::\overset{+}{O}:$$

CARBON TETRACHLORIDE. [CAS: 56-23-5]. CCl_4, formula weight 82.82, heavy, colorless, nonflammable, noncombustible liquid, mp $-23°C$, bp $76.75°C$, sp gr 1.588 ($25°C/25°C$), vapor density 5.32 (air = 1.00), critical temperature $283.2°C$, critical pressure 661 atmospheres, solubility 0.08 g in 100 g H_2O, odor threshold 80 ppm. Dry carbon tetrachloride is noncorrosive to common metals except aluminum. When wet, CCl_4 hydrolyzes and is corrosive to iron, copper, nickel, and alloys containing those elements. About 90% of all CCl_4 manufactured goes into the production of chlorofluorocarbons:

$$2CCl_4 + 3HF \xrightarrow{\text{catalyst}} CCl_2F_2 + CCl_3F + 3HCl.$$

Carbon tetrachloride was first made by chlorinating chloroform (1839). Later, CCl_4 was made by chlorinating carbon disulfide, CS_2, in the first commercial process, developed by Müller and Dubois (1893). Large-scale production commenced in the early 1900s at which time carbon tetrachloride became a popular metal-degreasing solvent, dry-cleaning fluid, fabric-spotting fluid, grain fumigant, and fire extinguishing fluid. In many of these uses, it has now been displaced by other less toxic chlorinated hydrocarbons. The carbon disulfide process consists of: (1) $3C + 6S \longrightarrow 3CS_2$; (2) $2CS_2 + 6Cl_2 \longrightarrow 2CCl_4 + 2S_2 Cl_2$; (3) $CS_2 + 2S_2 Cl_2 \longrightarrow CCl_4 + 6S$. The reaction must be carried out in a lead-lined reactor in a solution of CCl_4 at $30°C$ in the presence of iron filings as catalyst. The chlorination of methane is now the principal production route to CCl_4: $CH_4 + Cl_2 \longrightarrow CH_3 Cl + CH_2 Cl_2 + CHCl_3 + CCl_4 + HCl + $ excess CH_4. The reaction is carried out in the liquid phase at about $35°C$. Ultraviolet light is used as a catalyst. The same reaction can be carried out at $475°C$ without catalyst. The unreacted methane and partially chlorinated products are recycled to control the yield of CCl_4.

Toxicity.   The experimental exposure of laboratory animals to the vapors of CCl_4 has shown it to be very toxic by inhalation at concentrations easily obtainable at ambient temperatures. An overexposure to carbon tetrachloride has been known to cause acute but temporary loss of renal function.

CARBONYLS. See **Carbon**.

CARBONYLS (Chlorinated). See **Chlorinated Organics**.

CARBORANE.   A crystalline compound composed of boron, carbon, and hydrogen. It can be synthesized in various ways, chiefly by the reaction of a borane (penta-or deca-) with acetylene, either at high temperature in the gas phase or in the presence of a Lewis base. Alkylated derivatives have been prepared. Carboranes have different structural and chemical characteristics and should not be confused with hydrocarbon derivatives or boron hydrides. The predominant structures are the cage type, the nest type, and the web type, these terms being descriptive of the arrangement of atoms in the crystals. Active research on cargorane chemistry has been conducted under sponsorship of the U.S. Office of Naval Research. http://www.onr.navy.mil/

CARBORUNDUM. See **Silicon**.

CARBOXYLIC ACIDS.   The general formula for a carboxylic acid is

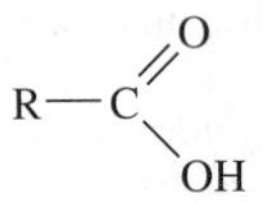

In terms of structure, a carboxylic acid may be aliphatic, carbocyclic, or heterocyclic:

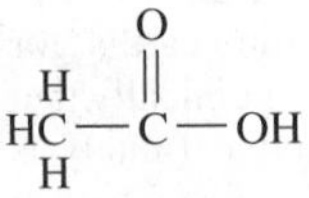

| Aliphatic acetic acid | Carbocyclic or aromatic benzoic acid | Heterocyclic pyromucic or furoic acid |

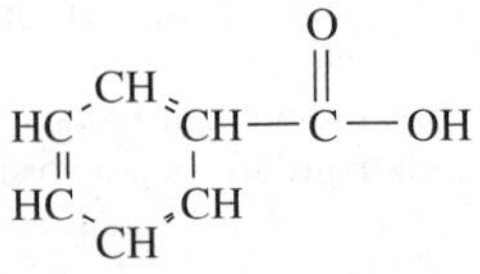

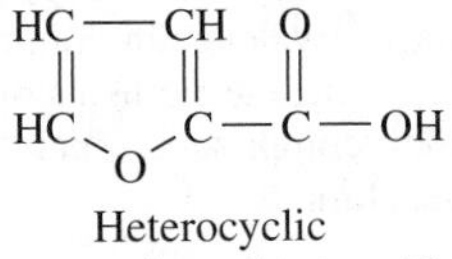

Or, a carboxylic acid may be classified in terms of the number of carboxyl ($-COOH$) groups which it contains. If one carboxyl group, it is designated as *mono*carboxylic; if two groups, as *di*carboxylic; if three groups, as *tri*carboxylic; and if four groups, as *tetra*carboxylic:

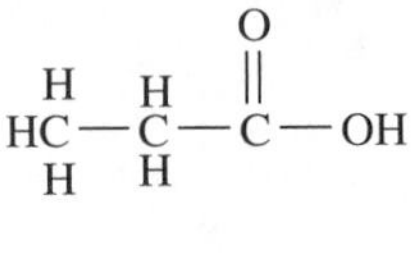

Propionic acid

(mono)

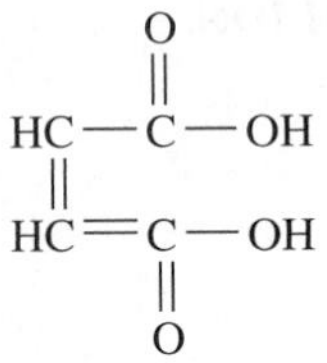

Maleic acid or *cis*-ethylene dicarboxylic acid

(di)

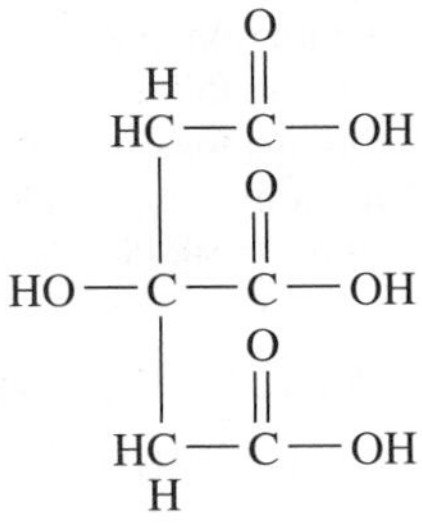

Citric acid

(tri)

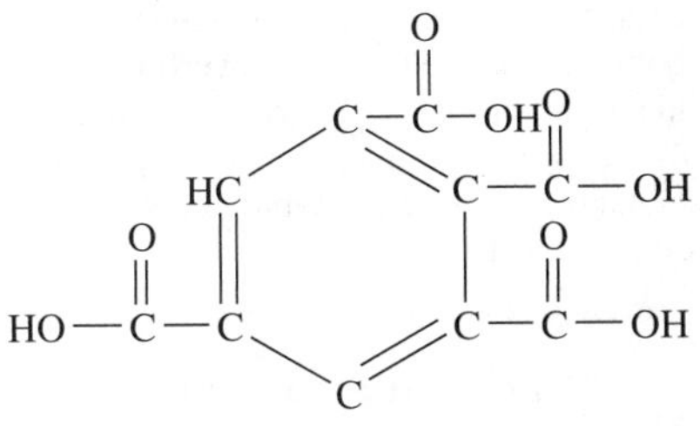

1,2,3,5-Benzenetetracarboxylic acid or mellophanic acid

(tetra)

When a carboxylic acid contains a hydroxyl group in addition to that of the principal $-COOH$ grouping, the term *hydroxy* is sometimes used. If there is only one additional hydroxyl group, the acid may be designated simply as a *hydroxycarboxylic* acid; if two groups, a *di*hydroxycarboxylic acid; if three groups, a *tri*hydroxycarboxylic acid.

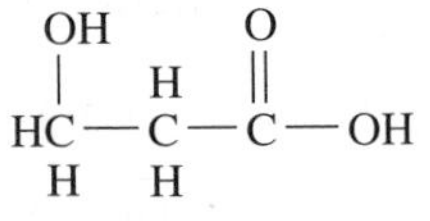

Hydracrylic acid or *β*-hydroxypropionic acid A hydroxymono-carboxylic acid

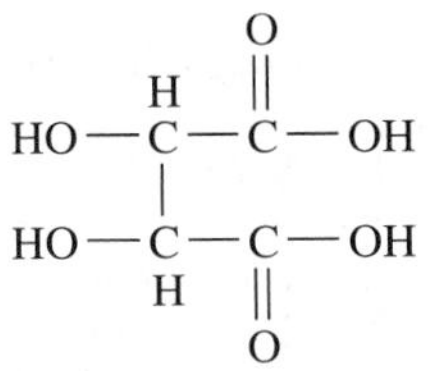

Tartaric acid A dihydroxycarboxylic acid

A carboxylic acid may be classified in accordance with the number of available hydrogens for salt formation. If only one hydrogen is available, the acid is *monobasic*; if two hydrogens are available, the acid is *dibasic*; if three or more hydrogens are available, the acid is *polybasic*.

A carboxylic acid also may be classified from the standpoint of other groups it contains. An *aldehydic* carboxylic acid contains the CHO group. An example is glyoxalic acid, $CHO \cdot COOH$. An *amino* carboxylic acid contains the NH_2 group. An example is carbamic or amino-formic acid, NH_2COOH. A *ketonic* carboxylic acid contains the CO group. An example is benzoylacetic acid, $C_6H_5 \cdot CO \cdot CH_2 \cdot COOH$. In the case of a *phenolic* carboxylic acid, the acid is structurally derived from benzoic acid, with uniting of the OH group with a carbon of the nucleus.

$$O=C-OH$$
$$\mid$$
$$CH$$
$$HC \quad\; C$$
$$\parallel$$
$$HO-C \quad\; C-OH$$
$$C=OH$$

Gallic or pyrogallol carboxylic acid
or 3,4,5-trihydroxybenzoic acid
A trihydroxymonocarboxylic acid

There are several homologous series of carboxylic acids, including:

$C_nH_{2n}O_2$	Saturated monobasic fatty acids
$C_nH_{2n-2}O_2$	Unsaturated monobasic fatty acids
$C_nH_{2n-4}O_2$	Propioloic acid series
$C_nH_{2n}(COOH)_2$	Dicarboxylic acids, where $n = 0$ for oxalic acid
$C_nH_{2n}(OH)(COOH)$	Hydroxymonocarboxylic acids, where $n = 0$ for carbonic acid

Fatty Acids

The simplest or lowest member of the fatty acid series is formic acid, HCOOH, followed by acetic acid, CH_3COOH, propionic acid with three carbons, butyric acid with four carbons, valeric acid with five carbons, and upward to palmitic acid with sixteen carbons, stearic acid with eighteen carbons; and melissic acid with thirty carbons. Fatty acids are considered to be the oxidation product of saturated primary alcohols. These acids are stable, being very difficult (with the exception of formic acid) to convert to simpler compounds; they easily undergo double decomposition because of the carboxyl group; they combine with alcohols to form esters and water; they yield halogen-substitution products; they convert to acid chlorides when reacted with phosphorus pentachloride; and their acidic qualities decrease as their formula weight increases.

Monohydroxy Fatty Acids

Structurally, these acids may be considered as the monohydroxy derivatives of the fatty acids. Included among these acids are hydroxyacetic acid (glycollic acid) and β-hydroxypropionic acid (β-lactic acid). These acids generally are syrupy liquids that tend to give up water readily and form crystalline anhydrides; they decompose when volatilized, and they are soluble in water and usually in alcohol and ether.

Polyhydric Monobasic Acids

Structurally, these acids are considered to be the oxidation products of polyhydric alcohols. However, a number of them can be formed from the oxidation of sugars. The careful oxidation of glycerol will yield a syrupy liquid, glyceric acid, an example of a dihydroxymonobasic carboxylic acid.

Aromatic Carboxylic Acids

In many ways, these acids are similar to the fatty acids. Generally, they are crystalline solids that are only slightly soluble in water, but most often they dissolve easily in alcohol or ether. The simpler aromatic acids may be distilled (or sublimed) without decomposition. The more complex acids, such as the phenolic and polycarboxylic aromatic acids, break down when heated, yielding carbon dioxide and a simpler compound. As an example, salicylic acid degrades to carbon dioxide and phenol. In nature, the aromatic acids are found in balsams, animal organisms, and resins.

The monobasic saturated aromatic acids include benzoic, hippuric, toluic acids (three structures), phenylacetic, phenylchloracetic, and dimethylbenzoic acid. Among the monobasic unsaturated acids are cinnamic, atropic, and phenylpropionic acids. The saturated phenolic acids include gallic and salicylic acids. The alcohol acids include amygdalic, tropic, and mandelic acids. One example of an unsaturated monobasic phenolic acid is coumaric acid.

Formation of Carboxylic Acids

Commercially, these acids are produced in several ways: (1) oxidation of relevant alcohol—e.g., acetic acid from ethyl alcohol; (2) oxidation of relevant aldehyde—e.g., acetic acid from acetaldehyde; (3) bacterial fermentation of dilute alcohols; (4) reacting a methyl ketone with sodium hypochlorite (haloform reaction); (5) carbonation of Grignard reagents; (6) hydrolysis of nitriles; (7) malonic ester synthesis route; (8) oxidation of relevant alkylaromatic—e.g., benzoic acid from toluene; (9) reaction of an alkali metal phenolic with carbon dioxide; and (10) hydrocarboxylation of olefins—e.g., butyric acid from propylene.

See also **Organic Chemistry**.

DUANE B. PRIDDY
The Dow Chemical Company
Midland, Michigan

Additional Reading

Bingham, E., C. Powell, and B. Cohrssen: *Organic Halogenated Hydrocarbons and Aliphatic Carboxylic Acid Compounds,* Vol. 5, John Wiley & Sons, Inc., New York, NY, 2000.

Kreysa, G., and R. Eckermann: *DECHEMA Corrosion Handbook: Corrosive Agents and Their Interaction with Materials, Carboxylic Acid Esters, Drinking Water, Nitric Acid,* John Wiley & Sons, Inc., New York, NY, 1992.

Lewis, R.J., and N.I. Sax: *Sax's Dangerous Properties of Industrial Materials,* 10th Edition, John Wiley & Sons, Inc., New York, NY, 1999.

Lide, D.R.: *CRC Handbook of Chemistry and Physics,* 84th Edition, CRC Press, LLC., Boca Raton, FL, 2003.

Wolfe, J.F., and M.A. Ogliaruso: *Synthesis of Carboxylic Acids, Esters and Their Derivatives–Updates,* John Wiley & Sons, Inc., New York, NY, 1991.

CARBURETION. The fuel for an internal combustion engine must be well mixed with the air required for combustion. This is particularly true of the Otto cycle engine, inasmuch as thorough distribution of particles of fuel in the air is essential to rapid and complete explosive combustion of the fuel in that cycle. One of the most effective means of mixing the particles of a liquid fuel with air is by vaporization. The vaporizing and mixing of a liquid fuel with air in the correct proportions is called *carburetion*; the device used is called a *carburetor*.

By using multiple jets, adjustable orifices, and other intricacies, commercial carburetors attain mixture control approximating the desirable performance for a gasoline engine as shown in Fig. 1. Increasing air pollution standards have exacted greater demands on carburetor performance and most of these demands have been met by utilizing solid-state controls over carburetor optimization.

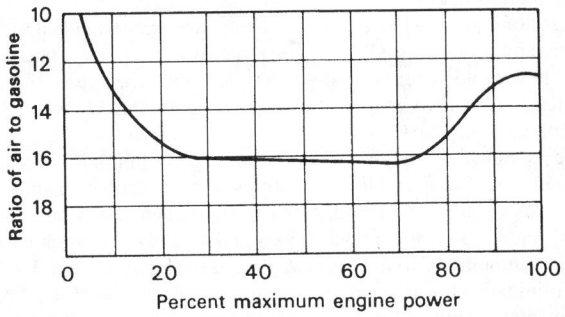

Fig. 1.   Optimal carburetor performance for gasoline engine

CARBURIZING. Machine parts requiring high strength, hardness, and toughness can often be made by either of two methods, one based on the use of a medium-carbon steel (0.30–0.50% carbon) heat treated to the required properties, and the other based on the use of a low-carbon steel (0.08–0.25% carbon) carburized to give a high-carbon surface layer and then heat treated. The carburized part will have a harder, more wear-resistant surface and a tougher core than the heat-treated medium-carbon steel. Transmission gears, camshafts, and piston pins are typical parts which can be made advantageously of carburizing grade steels.

The process consists of heating the fully machined part in an atmosphere rich in carbon monoxide or hydrocarbon gases at a temperature in the range 1650–1800°F (899–982°C). Reactions at the surface of the metal liberate atomic carbon which is readily dissolved by the steel and diffuses inward from the surface. In a typical carburized case a depth of penetration of 0.05 inch (0.13 centimeter) was obtained in 4 hours at 1700°F (927°C). The maximum carbon content at the surface was 1.10%. Shallow cases under 0.02 inch (0.05 centimeter) are useful for many purposes and very deep cases over 0.10 inch (0.25 centimeter) thick are required for gears for heavy machinery and for armor plate.

The process is most often carried out in sealed containers in which the parts are packed in carburizing compound consisting of a mixture of charcoal, coke, and other carbonaceous solids, together with barium carbonate and other compounds which act as energizers. At high temperatures these solids burn slowly, maintaining a supply of carbon monoxide. Carburizing is also carried out in batch-type and continuous-type furnaces in an atmosphere of natural gas, propane, butane, or specially mixed gases. Liquid baths consisting mainly of molten cyanide and chloride salts are also used for surface hardening. These baths supply both nitrogen and carbon to the surface of the steel, and where nitrogen is the principal hardener the process is known as cyaniding. Nitrogen hardens steel by forming hard compounds with iron and with certain alloying elements that may be present such as aluminum, chromium, and vanadium. See also **Nitriding**. In general, the salt-bath methods give shallower but harder cases than regular solid-pack carburizing. The pieces are quenched for hardening directly from the bath.

Carburized steels may also be quenched in oil or water directly from the box or furnace, or they may be cooled and reheated for hardening. A low temperature tempering treatment is given for relief of quenching stresses. A surface hardness of 60 Rockwell "C" is readily obtained, and when medium alloy steels of fine grain size are used, the strength and ductility of the core is exceptionally high, for example, 165,000 psi (11,224 atmospheres) tensile strength and 18% elongation.

CARCINOGENS.

A carcinogen may be defined as a substance, normally not present in the body that, when absorbed by the body in some manner (breathing, eating, drinking, injecting, skin contact, etc.), will induce the formation of malignant neoplasms (cancers); that is, a carcinogen initiates and nurtures tumor growth.

Progress of the ensuing carcinogenesis is dependent upon many factors, such as the frequency of exposure to the carcinogen (single, multiple, continuous), the concentration of the carcinogen when absorbed (ranging from parts per billion to parts per million and greater), as well as the poorly understood "natural resistance" of individual organisms to expel a given carcinogen. Very important is the total length of time over which exposure has occurred (ranging from seconds to years). Because of these extreme variations, group studies of environmental carcinogenicity are difficult and frequently unreliable. Consequently, the dangers of carcinogens can be over- or underestimated.

In terms of exposure to carcinogens by the average individual, particular attention should be given (1) to the *habitat* (water and air contamination; use of household chemical products), (2) the *workplace* (industrial chemicals), and, of course, (3) the *general environment*, particularly in industrialized urban areas.

There is much the average person can do to minimize exposure to carcinogens as, for example, carefully selecting garden chemicals and using gloves to prevent exposure of the skin when hazardous materials, including paint solvents, are handled—and in assuring good ventilation and air conditioning of living quarters to remove airborne particles. Within practical limitations, it is good practice to consider the local neighborhood environmental quality to avoid locating near known sources of air pollution. A check on possible radon pollution at a given site may be in order. See **Radon**.

Millions of people are avoiding exposure to the carcinogenic substances in tobacco. Millions of others are checking their dietary intake to avoid any substances that are suspected of promoting forms of carcinogenesis within the body, which often are of a long-term nature.

Until the early 1950s, the concept prevailed that the activity of carcinogenic chemicals was somehow related to the fact that they were synthetic "unnatural" substances that, since they are not present in the natural environment, were not factors of selection during developing life processes; hence, contemporary living organisms were not equipped for effective metabolic "detoxification" of these compounds. In the intervening years, however, a number of carcinogenic compounds of plant and fungal origin have been identified, including: safrole in sassafras; capsicine in chili peppers; various tannins; cycasin in the cycad groundnut; parasorbic acid in mountain ash berry; pyrollizidine alkaloids in *Senecio* shrubbery; and patulin, griseofulvin, penicillin G, and actinomycin produced by various molds. It also appears that liver cancer in Africa and the Peoples Republic of China is caused by interaction of aflatoxin and the co-carcinogen, hepatitis B virus. The number and variety of identified naturally occurring carcinogens continues to increase at a rapid rate.

Grouping of Carcinogens

Chemically identified carcinogens may be grouped in many ways, including a division into inorganic ions and organic compounds. The inorganic carcinogens contain the elements beryllium, cadmium, iron, cobalt, nickel, silver, lead, zinc, and possibly arsenic; these can form coordination compounds and/or react with sulfhydryl groups. Also asbestos powder is a powerful carcinogen toward the lung upon inhalation (asbestos cancer of miners). In recent years, the characteristics of asbestos and related substances have caused much controversy in connection with the pollution of certain waters, notably Lake Superior, by taconite processing waste products which contain fibers that have been compared with asbestos fibers. Most likely this situation will require some years for full scientific and legal resolution. Distinction must be made, however, between the supposed carcinogenic properties of asbestos materials and the silicosis engendered by the inhalation of other silicate dusts. There appears no reason to believe that in this latter instance the lesions are related to lung cancer per se.

The organic carcinogens may be subdivided in several ways, including: (a) condensed polycyclic aromatic hydrocarbons and heteroaromatic polycyclic compounds; (b) aromatic amines and N-aryl hydroxylamines; (c) amino-azo dyes and di-arylazo compounds; (d) aminostilbenes and stilbene analogues of sex hormones. Further breaking down the aliphatic carcinogens, these include: (1) alkylating agents (such as sulfur and nitrogen mustards, derivates of ethyleneimine, lactones, epoxides, alkane-α-ω-*bis*-methanesulfonates, certain dialkylnitrosamines, and ethionine; (b) lipophilic agents and hydrogen-bond reactors; this class comprises a wide variety of agents, such as chlorinated hydrocarbons (chloroform, carbon tetrachloride, and compounds used as pesticides under the names aldrin and dieldrin), bile acids, certain water-soluble high polymers, certain phenols, urethane and some of its derivatives, thiocarbonyls, and cycloalkynitrosamines; (c) naturally occurring carcinogens.

Until the mid-1970s, most studies in chemical carcinogenesis were experimental, i.e., suspect materials were placed continuously on the skin or in the diets of laboratory animals who were then observed to see if any neoplasms developed. While such work was invaluable in identifying materials which should be removed from the environment, or otherwise avoided, it did not provide any major understanding of the basis of chemical carcinogenesis. Indeed, many of the dermal tests merely indicated allergic reactions and many of the dietary tests showed that some animals thrived on trace additives.

Over the past decade, however, emphasis has been placed upon the molecular biology of carcinogenesis and it has been demonstrated that, in the first steps of carcinogen interaction, most carcinogens must be activated by the host cell's metabolism. Cell culture techniques have demonstrated that normal healthy cells can be transmuted into malignancies by certain chemicals. This in vitro work, however, has yet to transform human cells in the same way.

There is some evidence that the form of the chemical carcinogen that ultimately reacts with cellular macromolecules must contain a reactive electrophilic center, that is, an electron-deficient atom that can attack the numerous electron-rich centers in polynucleotides and proteins. As examples, significant electrophilic centers include free radicals, carbonium ions, epoxides, the nitrogen in esters of hydroxylamines and hydroxamic acids, and some metal cations. It is believed that carcinogens, which in themselves are not electrophiles, are metabolized to electrophilic derivatives that then become the "ultimate" carcinogens.

In this context, oxygen free radicals have been linked to many diseases other than cancer. There is much evidence that such free radicals may be developed in any kind of inflamed tissues and in chronically irritated organs the free radicals produced may convert exogenous chemicals to active carcinogens.

Workplace Carcinogens

During the early 1990s, an extensive study of occupational medicine was undertaken by the Yale-New Haven Occupational Medicine Program, Yale University School of Medicine, and the Occupational Medicine Program, University of Washington School of Medicine. M.R. Cullen, M.G. Cherniack, and L. Rosenstock, members of the Yale-Washington team, reported, "The success of epidemiologists in the 1960s and 1970s in establishing the excess cancer risk for workers exposed to several widespread workplace agents, most notably asbestos, benzene, and benzidine dyes, raised the possibility that cancer overall might largely be attributable to exposure in the workplace. Ecologic (sic) data showed some congruity between regions with high rates of cancer and high levels of industrial activity, and an unpublished government document purporting to show that 20 to 38 percent of all cancers were attributable to workplace exposure received circulation and attention. The past decade (1980s) has witnessed a considerable sobering and refinement of the prevailing views. Although over 300 compounds have been shown to have carcinogenic potential on the basis of their effects in laboratory animals, no new class of compounds has been added to the list of previously established human carcinogens." See Tables 1 and 2.

Carcinogen Mechanisms

The biochemical pathways in the cell are closely interconnected and are in a state of dynamic equilibrium (homeostasis). This equilibrium is maintained by feedback relationships existing between a great number of pathways. Chemical communication between subcellular organelles, such as the nucleus (within which the chromosomes contain the genetic blueprints for cell reproduction and the synthetic processes of cell life), the mitochondria (the powerhouse of the cell, which assures the synthesis of the universal cellular fuel, ATP, through the metabolism of carbohydrates and fatty acids), and the endoplasmic reticulum (synthesizing the proteins of the cell and assuring the metabolic breakdown—detoxification—of a multitude of endogenous and foreign compounds), depends on the constant interchange of a large variety of metabolic products and inorganic ions between them. There are probably a very great number of loci (receptor sites) upon which these regulatory chemical "stimuli" act. The receptor sites are of an enzymatic and nucleic acid nature. Other control points of protein character regulate the morphology of the intracellular lipoprotein membranes, which serve as "floor space" to the organized arrangements of multi-enzyme systems. The specificity of compounds of chemical control toward given receptor sites is due to a three-dimensional geometric "fit" following the lock and key analogy. Such is the general scheme of functional interrelationships in monocellular organisms which, hence, in a favorable medium multiply unchecked to the limit of the availability of nutrients.

In multicellular organisms, the subordination of the individual cells to the whole is assured by the existence of additional receptor sites which enable the cells to be response to chemical "stimuli" emitted by neighboring cells in the tissue and to hormonal regulation by the endocrine system in higher organisms. Hence, depending on the requirements of the moment, cells may remain stationary or may undergo cell division because of the need for repair of tissue injury, they may secrete different products, or they may perform some other specialized function depending on the nature of the particular tissue.

Carcinogenic substances are nonspecific cell poisons that cause the alterations and hence functional deletion of a large number of metabolic control sites. Present evidence suggests that these alterations are produced by the accumulation of the carcinogen in subcellular organelles, by covalent binding of the carcinogen to cellular macromolecules (proteins and nucleic acids) through metabolism, and by denaturation (i.e., destruction of the three-dimensional geometry) of the control sites through secondary valence interactions (hydrogen bonds, hydrophobic bonding, etc.) with the carcinogen. Early stages of tumor induction generally coincide with extensive cell death (necrosis) in the target tissues because a number of the biochemical lesions cause the irreversible blocking of metabolic pathways essential for cell life. However, because of the random distribution of the biochemical lesions in the cell population, in a small number of cells vital pathways are only slightly damaged and the lesions involve those sites and pathways which are not essential for cell life proper, but are necessary for organismic control. Thus, due to the action of the carcinogen, these cells escape physiological control and revert to a simpler, less specialized cell type (i.e., dedifferentiate). Such cells respond to continuous nutrition with continuous growth, which is an essential characteristic of malignant tumor cells.

The high incidence of skin cancer in coal tar workers was recognized as early as 1880. The carcinogenic activity of coal tar was demonstrated in 1915, when Yamagiwa and Ichikawa obtained epitheliomas (malignant tumor originating from epithelial cells) by its prolonged application to the ears of rabbits. Identification of the active material (in 1933) as

TABLE 1. DEFINITELY ESTABLISHED WORKPLACE CARCINOGENS

(Adapted from Yale-Washington Occupational Medicine Program)

Operations/processes carcinogen	Primary body organ where encountered	Affected
Para-aminodiphenyl	Chemical processing	Urinary bladder
Asbestos	Construction, asbestos mining and milling, production of friction products (brake Linings, etc.), and cement	Pleura and bronchus (lungs) and peritoneum
Arsenic	Copper mining and smelting	Skin, bronchus, liver
Alkylating agents (mechlorethamine hydrochloride) and bis [chloromethyl] ether	Chemical processing	Bronchus
Benzene	Chemical and rubber processing, petroleum refining	Bone marrow
Benzidine, beta naphthylamine and derived dyes	Dye and textile production	Urinary bladder
Chromium and chromates	Tanning, pigment making	Nasal sinus, bronchus
Ionizing radiation	Nuclear industry, health care settings bone marrow	Skin, thyroid, bronchus,
Radon	Uranium and hematite mining	Bronchus
Radium	Watch painting	Bone
Nickel	Nickel plating	Nasal sinus, bronchus
Polynuclear aromatic hydrocarbons (from coke, coal tar, shale, mineral oils and creosote)	Steelmaking, roofing, chimney cleaning	Skin, scrotum, bronchus
Vinyl chloride monomer	Chemical processing	Liver
Wood dust	Cabinetmaking, carpentry	Nasal sinus

TABLE 2. SUSPECTED (WIDELY USED) CARCINOGENS

(Adapted from Yale-Washington Occupational Medicine Program)

Operations/Processes Agent	Primary Body Organ Where Encountered	Affected
Beryllium	Beryllium processing, aircraft manufacturing, electronics, secondary smelting	Bronchus
Cadmium	Smelting, battery making, welding	Bronchus
Ethylene oxide	Hospitals, hospital supply manufacturing	Bone marrow
Formaldehyde	Plastics, textile and chemical processing; health care	Nasal sinus, bronchus
Synthetic mineral fibers (fiberglass)	Fiber manufacturing and installation	Bronchus
Polychlorinated biphenyls (PCBs)	Electrical equipment manufacturing/maintenance	Liver
Organochlorine pesticides (e.g., chlordane, dieldrin)	Pesticide manufacture and agricultural applications	Bone marrow
Silica	Casting, mining, refracting	Bronchus

the polycyclic aromatic hydrocarbon 3,4-benzopyrene (III, Fig. 1) is due to Cook, Kennaway, Hieger and their co-workers. This discovery was followed up by the synthesis and testing of a considerable variety of polycyclic aromatic hydrocarbons. All compounds of this class may be regarded as composed of condensed benzene rings. The arrangement of the hexagonal rings in various patterns results in a variety of compounds having different physical, chemical, and biological properties. However, not all polycyclic aromatic hydrocarbons possess carcinogenic activity; certain requirements of molecular geometry must be met.

For maximum activity, the molecule must have (Fig. 1): (a) an optimum size; (b) a coplanar molecular configuration, meaning that all hexagonal rings must lie flatly in one plane; in fact, hydrogenation of many of the active hydrocarbons results in buckled molecular conformation and this is concomitant with partial or total loss of activity; and (c) at least one meso-phenanthrenic double bond, also called the K-region (indicated by arrows in Fig. 1) of high π-electron density (i.e., of high chemical reactivity). In addition to III, 1,2,5,6-dibenzanthracene (IV), and 20-methylcholanthrene (V) are commonly used to study the experimental induction of tumors. The activity of most hydrocarbon carcinogens was tested on the skin of mice and the subcutaneous connective tissue of mice and rats. There is a vast body of evidence indicating that 3,4-benzopyrene and other carcinogenic hydrocarbons are formed during pyrogenation or incomplete burning of almost any kind of organic material. For example, carcinogenic hydrocarbons have been identified in overheated fats, broiled and smoked meats, coffee, burnt sugar, rubber, commercial paraffin oils and solids, soot, the tar contained in the exhaust fumes of internal combustion engines, cigarette smoke, etc.

It must be pointed out, however, that direct evidence of human involvement is lacking. For example, despite much publicity, N-nitrosamines have *not* been proven to be causative agents on induction of human cancer. They are, however, patent carcinogens in *experimental animals*.

Attention to the carcinogenic aromatic amines was drawn by the high incidence of urinary bladder tumors in dye works exposed to 2-naphthylamine (VII, Fig. 2), and benzidine (IX). The carcinogenic activity of VII, IX, and 4-aminobiphenyl (X) toward the bladder of the dog and the mouse has been demonstrated. In the rat, however, there is a change in target specificity, and tumors are induced by IX and X in the liver, mammary gland, ear duct, and small intestine. Carcinogenic activity is considerably heightened in 2-acetylaminofluorene (XI) without change of target specificity. Increased activity is due to the fact that XI is more coplanar than X, because of the internuclear methylene ($-CH_2-$) bridge in the former. 2-Acetylaminofluorene was proposed as an insecticide before its carcinogenic activity was accidentally discovered; it is a ubiquitous, potent carcinogen in a variety of species. Changing the internuclear bridge of XI

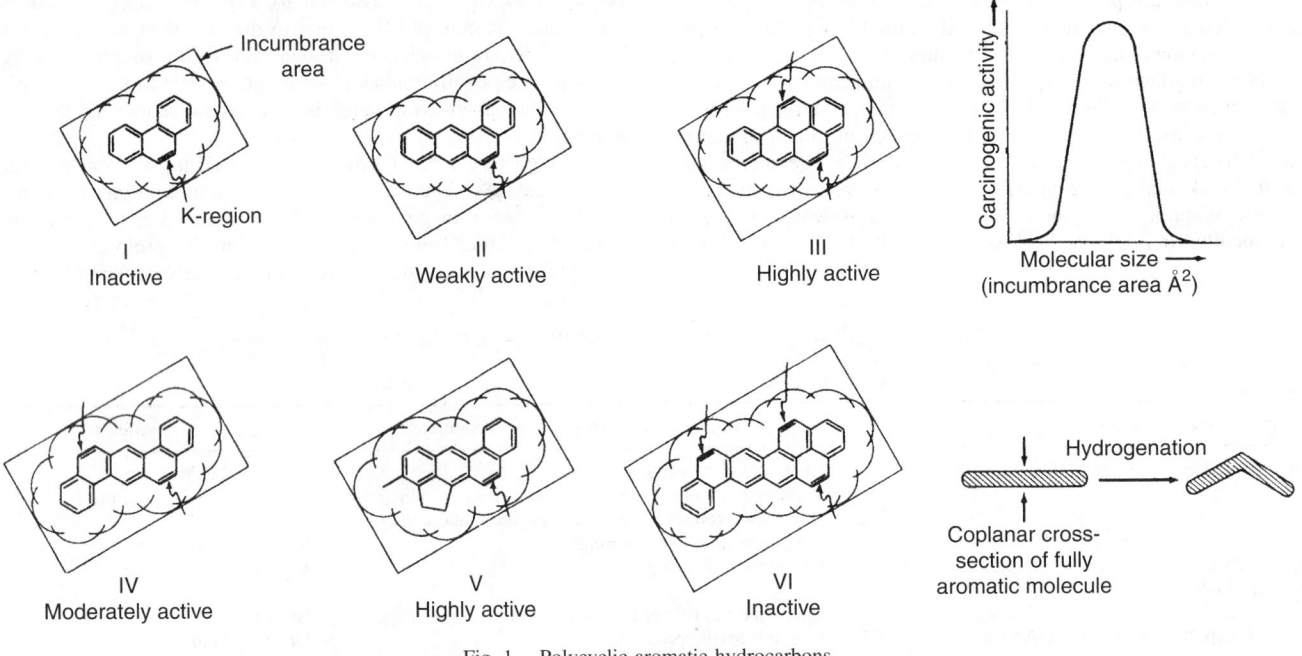

Fig. 1. Polycyclic aromatic hydrocarbons

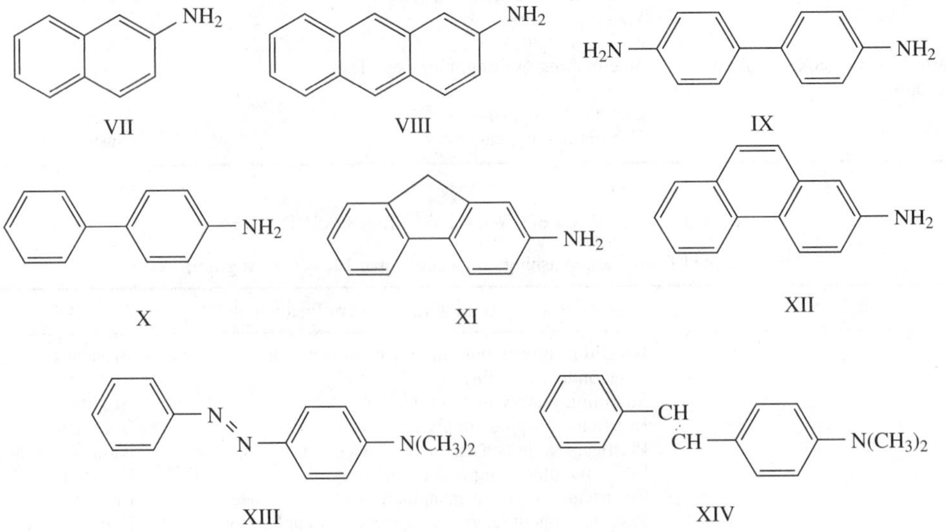

Fig. 2. Carcinogenic aromatic amines

to −CH=CH−, as in 2-aminophenanthrene (XII), causes a shift in target specificity; thus, in the rat, XII is inactive toward the liver, but in addition to inducing tumors in the mammary gland, ear duct, and small intestine, it produces leukemia. Compound XII represents a structural link between the aromatic amine and polycyclic hydrocarbon carcinogens (compare XII with I); it is also interesting in this respect that 2-amino-anthracene (VIII), which is a higher homologue of VII, is inactive toward the bladder, but is able to induce skin tumors in rats.

4-Dimethylamino-azobenzene (XIII) is the parent compound of the amino-azo dye carcinogens; it is also known in the earlier literature as Butter Yellow, because it was used to color butter and vegetable oils before its carcinogenic activity was discovered. Many derivatives of XIII have been prepared and tested for carcinogenic activity. In the rat, the amino-azo dye carcinogens, administered in the diet, specifically induce hepatomas. Tumor induction by most of the amino-azo dyes is delayed or inhibited by high dietary levels of riboflavin (vitamin B$_2$) or protein. Replacement of the −N=N−azo linkage by −CH=CH−, as in 4-dimethylaminostilbene (XIV), results in widening the target tissue spectrum; XIV induces tumors in the liver, mammary gland, and ear duct. Mice are much more resistant than rats to the carcinogenic activity of both amino-azo dyes and aminostilbenes.

Figure 3 illustrates some aliphatic carcinogens. N-Methyl-bis-β-chloro ethylamine (XV), a nitrogen mustard, produces local sarcomas, lung, mammary, and hepatic tumors upon injection in mice; because of its tumor-inhibitory properties, XV has also been used in the therapeutic treatment of certain types of human cancers. Bisepoxybutane (XVI), β-propiolactone (XVII), and N-lauroylethyleneimine (XVIII) produce local sarcomas in rats upon injection. Ethylcarbamate (XIX), the parent compounds of several hypnotic drugs used in humans, produces malignant lung adenomas in rats, mice, and chickens. Dimethylnitrosamine (XX) is a potent carcinogen toward the liver, lung, and kidney, and ethionine (XXI) toward the liver of the rat; the former is an intermediate in the manufacture of the rocket-fuel component, dimethylhydrazine (CH$_3$)$_2$N−NH$_2$, while the latter is the S-ethyl analogue of the natural amino acid methionine.

Testing

Because of their short life span (average 3 years) small rodents (mice, rats, and hamsters) are frequently used for the testing of chemicals for carcinogenic activity; occasionally testing is done with rabbits, dogs, fowls, monkeys, etc. While a great variety of ways of administration has been used, a common method is to introduce substances to be tested in the following ways: (a) skin painting; small volumes of solution of the substance in an inactive solvent (e.g., benzene) are applied to the shaved surface of the skin (generally of mice, in the interscapular region) daily or at longer intervals; (b) subcutaneous injection of the pure substance or its solution (once or at repeated intervals); (c) feeding; the substance is mixed in the diet at given levels, or dissolved in the drinking water. Testing of new substances for possible carcinogenic activity is conducted for a minimum of 1 year to be meaningful. At the end of the testing period, all animals are necropsied, and all tumors and dubious tissues examined histopathologically.

Within the past few years, the costs of carrying out long-term bioassays have increased markedly to the point at which it is too costly for small or medium-size manufacturers to fund a study of the long-term toxicity or carcinogenicity of any chemicals proposed for commercial development. Therefore, many short-term tests are being developed to predict which compounds would more likely be carcinogenic. A prominent one is

a test for mutagenicity in various strains of the bacterium *Salmonella typhimurium*. The tests generally involve addition of the compound under test into a culture dish, which is seeded with the bacterium. The medium is deficient or lacking in histidine; therefore, no bacterial growth occurs, unless mutation by the test chemical yields a form of bacteria that does not require histidine. A count of the bacterial colonies can therefore be used as a measure of the mutagenicity of the test compound. The test, although quite rapid, does suffer from certain deficiencies. For one, a fair number of compounds are not mutagenic even though they are strong carcinogens in animals. These are usually compounds that require metabolic activation to demonstrate their carcinogenicity. Several variations on the test conditions have been proposed and, therefore, with these variations and parallel tests in other systems it is possible to establish whether the chemical under test may be a potential hazard.

A carcinogen that is highly active in one species may be totally inactive in another species, and vice versa. The susceptibility of a species to a given carcinogen also depends on the genetic strain, sex, and dietary conditions. Moreover, carcinogenic substances generally show a rather selective specificity toward certain target tissues; e.g., certain compounds produce exclusively hepatomas in the susceptible species. For these reasons, no chemical compound may be stated safely to be devoid of carcinogenic activity toward humans unless it has been found inactive when tested in a variety of mammalian species and by a variety of routes of administration for a length of time corresponding to half the life span of each species.

The foregoing observation emphasizes one of the main problems in cancer research, namely, *extrapolation* of research findings. For a given population to be fully safe, literally tens of thousands of commonly encountered natural materials and synthetic materials, covering the complete spectrum of products with which people are in contact over their lives, would have to be tested and regarded suspect until thoroughly tested on the species of most importance to humans, namely, the testing of reactions among people themselves. But this alone would not suffice because, as stressed throughout biochemical studies of human systems, there is individuality. The problem of developing improved (vastly improved) testing systems and attention to the problem of extrapolation of findings rivals in difficulty the basic problem of identifying the nature of cancer itself.

Many lists have appeared indicating the toxicity and/or carcinogenicity of specific chemicals. It would be invidious to attempt to insert these lists in this article because the number is extensive. Reference can be made to the sources cited at the end of this article.

The interesting observation has been made that, because of the vast amounts of money going into cancer research, the field has become a large business for numerous suppliers. Animal cells now can be procured by the kilogram from suppliers. Because of this availability (mostly two kinds, 3T3 and W138), much information has been accumulated concerning the biology of these cells. But is much of what has been learned in this regard meaningful? Researchers have found that viruses and chemicals can transform 3T3 cells to a neoplastic state and that these cells can produce tumors when inoculated in suitable hosts. It should be noted, however, that the tumors so produced are sarcomas (derived from fibroblasts), which are very rare in human beings. Ninety percent of human tumors are carcinomas. With the exception that epithelial cells and fibroblasts are both animal cells, they have little in common. They stem from two embryonic sources with different functions, and the tumors they produce are different as well.

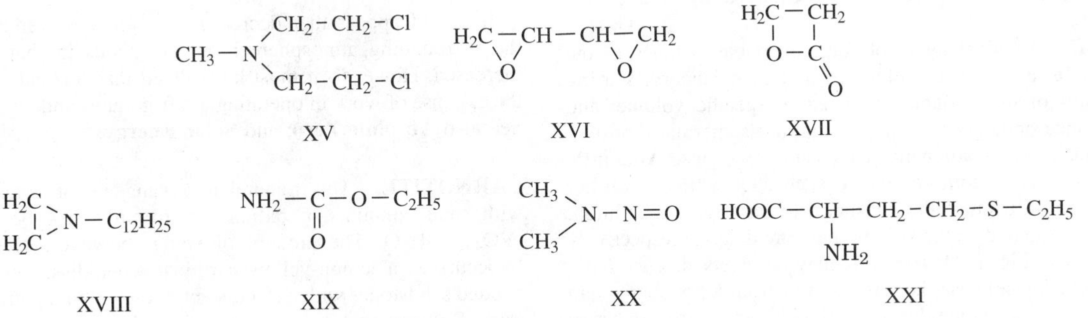

Fig. 3.   Aliphatic carcinogens

Greater Knowledge of Carcinogenesis Anticipated

To comprehend the process of carcinogenesis is to understand, at the molecular level, the nature and workings of the cells that constitute life itself. Biochemists and geneticists are making excellent progress toward understanding life at the molecular level and out of this research, possibly by the early 2000s, the identification and role of carcinogens, as described here, will be subject to revision in the light of new knowledge.

See also **Agent Orange**; **Asbestos**; **Biphenyl and Terphenyls**; and **Dioxin**.

Major portions of this article were contributed by
R. C. VICKERY, M.D., Ph.D., D.Sc., Blanton/Dade City, Florida

Additional Reading

Bartone, J.C.: *Human Carcinogens: Index of New Information with Authors, Subjects and References,* Abbe Publishing Association of Washington DC, Washington, DC, 2000.

Cullen, M.R., M.G. Cherniack, and L. Rosenstock: "Occupational Medicine," *N. Eng. J. Med.,* **675** (March 8, 1990).

Greim, H.: *Occupational Toxicants: Critical Data Evaluation of MAK Values and Classification of Carcinogens,* Vol. 14, John Wiley & Sons, Inc., New York, NY, 2000.

Lehnert, G. and D. Henschler: *Biological Exposure Values for Occupational Toxicants and Carcinogens Substances: Critical Data Evaluation for Bat and Eka Values,* Vol. 3, John Wiley & Sons, Inc., New York, NY, 1998.

Lewis, R.J. and N. Irving Sax: *'Sax's Dangerous Properties of Industrial Materials,* 10th Edition, John Wiley & Sons, Inc., New York, NY, 1999.

McLachlan, J.A., R.M. Pratt, and C.L. Markert: *Developmental Toxicology; Mechanisms and Risk,* Cold Spring Harbor Laboratory, Cold Spring Harbor, New York, NY, 1987.

Muller-Hermelink, H.K., H.G. Neumann, and W. Dekant: *Risk and Progression Factors in Carcinogenesis,* Springer-Verlag Inc., New York, NY, 1997.

Nago, M. and T. Sugimura: *Food Borne Carcinogens: Heterocyclic Amines,* John Wiley & Sons, Inc., New York, NY, 2000.

NAS: *Food Chemicals Codex,* Institute of Medicine, National Academy of Sciences, Washington, DC, 1993.

Sperber, W.H.: "The Modern Hazard Analysis and Critical Control Point System," *Food Technology,* **115** (June 1991).

Staff: National Research Council, *Carcinogens and Anticarcinogens in the Human Diet: A Comparison of Naturally Occurring and Synthetic Substances,* National Academy Press, Washington, DC, 1996.

Tisler, J.M.: "The Food and Drug Administration's Perspective on the Modern Hazard Analysis and Critical Point System," *Food Technology,* **125** (June 1991).

Zeckhauser, R.J. and W.K. Viscusi: "Risk within Reason," *Science,* **559** (May 4, 1990).

CARNALLITE. This mineral is a product of evaporation of saline deposits rich in potash content, as a hydrated chloride of potassium and magnesium, $KMgCl_3 \cdot 6H_2O$. Hardness, 2.5; specific gravity, 1.602. It crystallizes in the orthorhombic system usually as massive, granular aggregates. Luster greasy, with indistinct cleavage and conchoidal fracture. Color grades from colorless to white, into reddish from included hematite scales. Transparent to translucent with bitter taste, deliquesces readily in moist environment.

Found associated with sylvite, halite and polyhalite at Stassfurt, Germany; Abyssinia; the former U.S.S.R.; and in southeastern New Mexico and adjacent areas in Texas. It is an important source of potash for use in fertilizers. See also **Potassium**.

CARNELIAN. The mineral carnelian is a red or reddish-brown chalcedony; the word is derived from the Latin word meaning flesh, in reference to the flesh color sometimes exhibited. See also **Chalcedony**.

CARNOT CYCLE. An ideal cycle of four reversible changes in the physical condition of a substance, useful in thermodynamic theory. Starting with specified values of the variable temperature, specific volume, and pressure, the substance undergoes, in succession, an isothermal (constant temperature) expansion, an adiabatic expansion (see also **Adiabatic Process**), and an isothermal compression to such a point that a further adiabatic compression will return the substance to its original condition. These changes are represented on the volume-pressure diagram respectively by *ab, bc, cd,* and *da* in Fig. 1. Or the cycle may be reversed: *a d c b a.*

In the forward (clockwise) case, heat is taken in from a hot source and work is done by the hot substance during the high-temperature expansion *ab*; also additional work is done at the expense of the thermal energy of

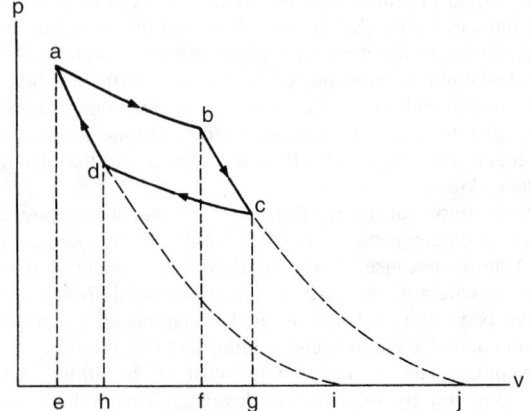

Fig. 1. Carnot cycle on v p diagram: *ab* and *cd,* isothermals: *bc* and *da,* adiabatics which, for some theoretical purposes, are produced to infinity

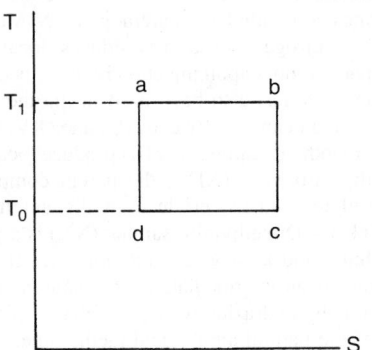

Fig. 2. Carnot cycle temperature-entropy diagram

the substance during the further expansion *bc.* Then a less amount of work is done on the cooled substance, and a less amount of heat discharged to the cool surroundings, during the low-temperature compression *cd*; and finally, by the further application of work during the compression *da,* the substance is raised to its original high temperature. The net result of all this is that a quantity of heat has been taken from a hot source and a portion of it imparted to something colder (a "sink"), while the balance is transformed into mechanical work represented by the area *abcd.* Thus, the forward Carnot cycle can be used for the production of power. If the cycle takes place in the counterclockwise direction, heat is transferred from the colder to the warmer surroundings at the expense of the net amount of energy which must be supplied during the process (also represented by area *abcd*). It can thus serve as a refrigerating cycle.

The temperature-entropy diagram for the Carnot cycle, corresponding to the pressure-volume diagram is shown in Fig. 2.

It should be noted that the efficiency of the forward cycle is highest when T_1 is as high as possible. Since, in practice, T_0 will always be fixed by the temperature of the surrounding atmosphere, a high efficiency corresponds to a large difference $T_1 - T_0$. In contrast, a high coefficient of performance, or a high effectiveness of a heat pump corresponds to a small difference $T_1 - T_0$.

It would appear that decreasing T_0 for a power cycle below that of the surrounding atmosphere is advantageous in that the efficiency η is increased. However, it must be realized that this can only be achieved at the expense of work in operating a refrigerator, and no advantage is gained. See also **Absolute Zero**; and **Solar Energy**.

CARNOTITE. This mineral is a vanadate of potassium and uranium with small amounts of radium. Its formula may be written $K_2 (UO_2)_2 (VO_4)_2 \cdot 3H_2O$. The amount of water, however, seems to be variable. It occurs as a lemon-yellow earthy powder disseminated through cross-bedded sandstones with rich concentrations around petrified and carbonized trees. Soft; sp gr 4.7. It was mined in Colorado and Utah as a source of radium. Other localities are in Arizona, Pennsylvania, and Zaire.

CARNOT THEOREMS. (1) No heat engine operating between two given temperatures can be more efficient than a perfectly reversible engine operating between the same temperatures. (2) The efficiency of any reversible heat engine working between two temperatures is independent of the nature of the engine and the working substances and depends only on the temperatures.

See also **Carnot Cycle**.

CAROTENOID PIGMENTS.   See **Pigmentation (Plants)**.

CAROTENOIDS.   See **Dyes: Natural**.

CAROTHERS, WALLACE H. (1896–1937).   Born in Iowa, Carothers obtained his doctorate in chemistry at the University of Illinois. He joined the research staff of Du Pont in 1928, where he undertook the development of polychloroprene (later called neoprene) that had been initiated by Nieuland" research on acetylene polymers. Carother's crowning achievement was the synthesis of nylon, the reaction product of hexamethylenetetramine and adipic acid. Carother's work in the polymerization mechanisms of fiber like synthetics of cyclic organic structures was brilliant and productive, and he is regarded as one of the most original and creative American chemists of the early 20th century.

CARRIER.   (1) A neutral material such as diatomaceous earth used to support a catalyst in a large-scale reaction system. (2) A gas used in chromatography to convey the volatilized mixture to be analyzed over the bed of packing that separates the components. (3) An atomic tracer carrier; a stable isotope or a natural element to which radioactive atoms of the same element has been added for purposes of chemical or biological research.

CARRIER (Food Additive).   A substance well named because its primary function is that of conveying and distributing other substances throughout a food substance. The role parallels that of a carrier in paint, wherein vehicle (carrier) holds and distributes pigment throughout the entire paint product. Silica gel and magnesium carbonate serve as carriers in food substances. For example, the high porosity of silica gel enables it to adsorb internally up to three times its own weight of many liquids. This property is used to convert various liquid ingredients, such as flavors, vinegar, oils, vitamins, and other nutritional additives, into easy-to-handle powders. These powders, in turn, can be measured easily and blended effectively with other constituents to provide a uniform food substance. Advantage is taken of the properties of carriers in the convenience food field, where flavors remain entrapped inside silica particles until the food product is mixed with water, at which time the flavors are released just prior to consumption, giving the product an aura of richness and freshness.

CAS.   Chemical Abstracts Registry or Chicago Academy of Science. See also Chemical Abstract Service Registry Number (CAS).

CASCADE.   (1) Any connected arrangement of separative elements whose result is to multiply the effect, such as isotope separation, created by the individual elements. A bubble plate-tower is a cascade whose elements are the individual plates; a plant consisting of many towers in series and parallel is similarly a cascade whose elements may be considered to be either the towers or the individual plates. Similarly, an amplifier in which each stage except the first has as its input the output of the preceding stage is spoken of as a cascade amplifier. A stage of a grounded-cathode vacuum-tube amplifier is defined as the section from a point just before the grid of one tube to that just before the grid of the next. Similarly, for grounded-emitter transistor amplifiers, it is defined as the section from a point just before the base of one transistor to that just before the base of the next. (2) Coined term used to describe a large number of compounds derived from a common source, e.g., the arachidonic acid cascade.

CASCADE COOLING.   See **Cryogenics**; **Natural Gas**.

CASE HARDENING.   Hardening of the surface layer or case of a ferrous alloy while leaving the core or center in a softer, tougher condition. There are two basic methods of case hardening. In the first, gaseous elements such as carbon or nitrogen are introduced into the surface layer, thereby forming a hardening or hardenable alloy at the surface. Examples are carburizing, nitriding, and carbonitriding. Alternatively, the surface may be given a hardening heat treatment that does not affect the core. This may be accomplished by flame hardening or induction heating, whereby the surface is rapidly heated into the austenite range and the specimen quenched before the center has obtained a temperature high enough to allow it to be hardened.

CASEIN.   [CAS: 9005-46-3]. Casein is the phosphoprotein of fresh milk; the rennin-coagulated product is sometimes called paracasein. British nomenclature terms the casein of fresh milk caseinogen and the coagulated product casein. As it exists in milk it is probably a salt of calcium.

Casein is not coagulated by heat. It is precipitated by acids and by rennin, a proteolytic enzyme obtained from the stomach of calves. Casein is a conjugated protein belonging to the group of phosphoproteins. The enzyme trypsin can hydrolyze off a phosphorus-containing peptone.

The commercial product also known as casein is used in adhesives, binders, protective coatings, and other products.

The purified material is a water-insoluble white powder. While it is also insoluble in neutral salt solutions, it is readily dispersible in dilute alkalis and in salt solutions such as those of sodium oxalate and sodium acetate.

CASSITERITE.   The mineral cassiterite, chemically tin dioxide, SnO_2, is almost the sole ore of tin. It is a noticeably heavy mineral crystallizing in the tetragonal system, as low pyramids, prisms, often very slender, and as twinned forms. It is a brittle mineral, hardness, 6.0–7.0; specific gravity, 6.99; luster, adamantine; color; generally brown to black, but may be red, gray to white, or yellow; streak whitish, grayish, or brownish; may be almost transparent to opaque. A fibrous variety somewhat resembling wood is called wood tin. Cassiterite occurs in widely scattered areas, but deposits of a size to be commercially important are few. It is associated with granites and rhyolites.

Cassiterite is heavily concentrated in bands and layers of varying thickness, forming economically valuable deposits, such as those found in the Malay States of southeastern Asia; Bolivia, Nigeria, and the Congo are also major producers of tin ore. Cassiterite is also known as tin stone.

CASTING.   A process for producing specific shapes of materials by pouring the material, while in fluid form, into a shaped cavity (mold) where the material solidifies in the desired shape. The resulting shape is also called a *casting*. In terms of metals, the art of casting is one of the oldest methods for making metal parts and is still used extensively even though numerous other methods for producing shaped metal products, such as forging, rolling, and extruding, have been developed. In terms of plastic materials, casting is also widely practiced.

Metal Casting

Production of a metal casting involves the use of a pattern, usually of wood or metal, which is similar in shape to the desired finished piece and slightly larger in all dimensions to allow for shrinkage of the metal upon solidification. The pattern is bedded down in a special damp sand by an operation called molding. When the pattern is removed it leaves an impression of the shape of the desired casting. This impression is completely surrounded by sand and provided with openings called gates through which the molten metal enters. After pouring and cooling the mold is broken open and the casting removed. All adhering sand particles together with any extraneous projections such as those left by the gate system are removed after which the casting is machined to the required finish.

The term is also applied to the casting of pig iron in blast furnace practice and the casting of ingots in steel-mill practice.

Centrifugal casting is applicable to the production of pipe and tubing, wheels, gear blanks, and other castings having rotational symmetry. While the mold is rotated on a horizontal axis for pipe and tubing, and on a vertical axis for wheels and gear blanks, a measured amount of molten metal is added. The mold may be sand or water-cooled metal for more rapid solidification. Centrifugal castings have good structure and density.

Metal molds are also used for making die castings and permanent mold castings. In the latter process a permanent metal mold is filled by gravity in the usual manner, while in die casting considerable pressure is exerted on the molten metal, insuring rapid and complete filling of the mold. Die-casting machines are highly mechanized for rapid and nearly automatic operation. The product is characterized by high dimensional accuracy and clear reproduction of mold details including screw threads, holes, and intricate sections, all of which greatly reduces the machining required. The process is limited in its application by the high cost of making alloy steel dies or molds. The lower melting zinc alloys and aluminum alloys are most successfully die cast; however, certain brasses and bronzes can also be die cast. Tin- and lead-base alloys are easily die cast but have limited application.

The zinc-base die-casting alloys are the most widely used. A typical composition is 1.0% copper, 3.9% aluminum, 0.06% magnesium, balance zinc. This alloy has a strength of about 45,000 psi (3,061 atmospheres) with 3% elongation in 2 inches (5 centimeters). Typical applications are carburetors, fuel pumps, tools, typewriter frames, instrument cases, and hardware often finished by chromium plating.

The investment or "lost wax" process has lately been revived as a method of making precision castings of metals such as steel and zinc having too high a melting point for die casting. A wax pattern is made in a die-casting machine, sprayed with a highly refractory slurry, dried, and embedded in sand. The mold passes through a furnace where the wax is melted or burned out, and the mold baked. The casting is then poured into the cavity left by the melting out of the wax, resulting in castings that rival die castings for dimensional accuracy.

Vacuum Casting

Although considered theoretically possible for many years, the commercialization of vacuum casting of metals was not demonstrated until the late 1980s. Vacuum casting offers an alternative position between investment and conventional shell-mold or green sand casting. Advantages of vacuum casting include thin-wall, near-net shape; multiple-core, complex shapes; and metallurgical integrity. Costs for vacuum castings appear to be competitive with green sand casting methods.

In the well-established *gravity-pour* process, molten metal is poured into the mold at atmospheric pressure. In the vacuum process, the molten alloy is drawn into the mold through gates in the bottom by a pressure differential between the atmospheric pressure of the melting furnace and a partial vacuum produced in the mold. This increase in molding pressure makes it possible to produce components having wall sections as thin as 1.75 mm (0.07 in.) in near-net shapes with increased metallurgical integrity and consistency. To make the vacuum process cost effective, however, very careful control of all conditions must be maintained, as by controls that have become available through the use of computers and microprocessors.

Heat- and corrosion-resistance materials have not always been compatible when traditional casting methods are used. Vacuum casting, on the other hand, is readily adaptable to a wide range of materials, such as low-nickel heat- and corrosion-resistant alloys. It should be noted that these materials frequently rely on combinations of silicon, chromium, and manganese as alternatives to the higher-cost nickel. These different materials, in the past, have added to casting difficulties. The vacuum process has been well received for making automotive and machinery parts.

Solidification Processing

The microstructure (arrangements of electrons, ions, space lattices, defects, and phases and their morphology) affects an alloy's ultimate properties and performance. As pointed out by Ahmed (Youngstown State University) and a team of research metallurgists, "Phase morphology of a particular microstructure is established during solidification, which essentially is a thermally activated nucleation and growth process requiring simultaneous control of several dependent and independent parameters to achieve a specific end result."

The solidification process is comprised of two principal phases: (1) the *nucleation stage*, the most important parameters of which include: changes in chemical-free energy between the solid and liquid phases, the surface free energy of the solid/liquid interface, the elastic strain energy, the amount of superheating and undercooling, the latent heat of solidification, the thermal conductivities of the phases, and the interdependence among these parameters; and (2) the *growth stage*, which tends to be of even greater complexity. It is during the growth stage that physical defects, such as chemical nonhomogeneities, dislocations, voids, and unwanted phases appear. The aforementioned research team has developed (patent pending) for solidification processing in an applied electric field. It is claimed that this method produces homogeneous nucleation and eliminates porosity.

Improved Melting Practices

Many new alloys were developed to meet the requirements of aircraft-engine manufacture. The principal flurry of activities occurred in the 1940s

through the 1970s. As engines' specifications grew tighter to cope with the need for greater performance, the needs for improving the quality of the earlier alloys became evident and, thus, during the 1980s and 1990s, was termed, "cleaner" alloys, was a major goal for metallurgists. Most of these improvements could be achieved during the melting process.

As observed by C.H. White and a team of metallurgists (Inco Alloys Ltd., Hereford, England), "It is well established that the presence of small-scale inclusions limits the maximum stresses at which the material can operate because these inclusions can cause premature failure." To meet these objectives, the principal objectives of melting have been formulated and include: (1) providing adequate deoxidation (via magnesium, calcium, cerium, or zirconium additions) to ensure alloy cleanness and good workability; (2) refining the metal to remove metalloid (sulfur, lead, and bismuth) and gaseous (oxygen, nitrogen, and hydrogen) impurities; (3) minimizing nonmetallic contamination; (4) obtaining a homogeneous mixing of the constituent-alloy ingredients (nickel, iron, cobalt, chromium, tungsten, molybdenum, titanium, aluminum, and niobium) within specified limits; and (5) casting into an ingot suitable for further processing.

A number of melting processes are in use or under consideration. Vacuum induction melting (VIM) substantially reduces gases present in the melt during the melt cycle. Methods under development to improve VIM include gas purging, melt filtering, continuous monitoring of temperatures and pressure, continuous monitoring of furnace atmospheres (residual-gas analysis) by mass spectrometry, and automatic operation and data storage using process-control computers. Remelting also is practiced. Two consumable electrode remelting processes in use are vacuum arc remelting (VAR) and electroslag remelting (ESR). These processes are detailed in the White reference.

Defects that must be appraised continuously include "white spot" and "freckle." White spot is an area of alloy depleted in the lower melting-point alloying additions. Freckle is a mid-radial channel segregation resulting from a deep melt pool and a steeply sloping liquidus profile. This segregation develops in the liquid before solidification.

The C.H. White research group has observed, "Melting without the use of refractories is the only way future cleanness-level requirements are likely to be achieved. Electron-beam cold-hearth refining (EBCHR) possibly is the best candidate for such a melting system. Plasma melting and refining also is being evaluated."

Plastic Casting

Several families of thermoplastic materials are capable of taking form by casting, although the process differs considerably from that used for metals just described. Some plastic casting processes depend upon melting and solidifying, as with metals; others depend upon solubility, as in the case of *solvent casting*.

Acrylic castings usually consist of poly(methyl methacrylate) or copolymers of this ester as the major components, with small amounts of other monomers to modify the material properties. Incorporating acrylates or higher methacrylates, for example, lowers the heat deflection temperature and hardness and improves thermoformability and solvent cementing capability, but with some loss to weathering resistance. Dimethacrylates or other crosslinking monomers increase the resistance to solvents and moisture. Castings are made by pouring the monomers or partially polymerized syrups into suitably designed molds and heating to complete the polymerization. A large reduction in volume, sometimes exceeding 20%, takes place during the cure. The reaction is also accompanied by liberation of substantial heat. At conversion, the polymerization may become auto-accelerated, and the rate of conversion may increase rapidly until about 85% conversion is achieved. Thereafter, the reaction slows down and post-curing may be required to complete the polymerization. On the other hand, with certain materials combinations, a violent runaway polymerization can occur.

The syrups made prior to casting (and final polymerization) can be stored safely at a controlled temperature until required. The preparation of syrups in advance shortens the time in the mold, decreases the tendency for leakage from the molds, and greatly minimizes the chance of dangerous runaways.

The majority of acrylic casting is in the manufacture of sheet. Cast sheet generally is made in a batch process within a mold or cell, but the process can be continuous through the use of stainless steel belts. Molds consist of two pieces of polished (or tempered) plate glass slightly larger in area than the desired finished sheet. The mold (or cell) is held together by spring clips that respond to the contraction of the acrylic material during the cure. The plates are separated by a flexible gasket of plasticized polyvinyl chloride tubing that controls the thickness of the product. Once filled, the mold is moved to an oven for cure. Thin sheet is cured in a forced-draft oven using a programmed temperature cycle, starting at about 45°C and ending at 90°C. The curing cycle requires several hours, the period increasing with the size of the sheet.

In continuous casting, a viscous syrup is cured between two highly polished moving stainless steel belts. Distance between the belts determines the thickness of the sheets. Although less versatile, the continuous process eliminates a number of problems in handling and breakage of large sheets of plate glass used for the batch process. Continuous processing produces sheets of more uniform thickness and essentially eliminates warping.

Nylon casting is a four-step process: (1) melting the monomer, (2) adding catalyst and activator, (3) mixing the melts, and (4) casting. Molds must be capable of containing a low-viscosity liquid at temperatures of 200°C and must allow for normal shrinkage. Two-piece molds are commonly used for simple shapes. More complex shapes require molds that can be disassembled to remove the cast shape. Stresses that develop during the casting can be controlled by very slowly cooling the casting over a period of 24 hours or longer.

Solvent casting is sometimes used, as in the case of polyvinyl chloride (PVC) film. In this process, resins, plasticizers and other ingredients are added to a solvent (tetrahydrofuran) in an inert, gas-blanketed mixing tank. Thorough mixing and degassing are critical for producing high-quality film. The mixture, below the boiling point, is pumped to a casting tank. The solution is filtered to a particle size not exceeding 5 micrometers. The solution is cast onto a stainless steel belt, which then enters an oven where solvent is evaporated from the film. After cooling, the film is stripped from the belt and wound into rolls. The gage of the film is controlled by the die opening, the pumping pressure, and the speed of the belt, all variables that can be carefully monitored and controlled. Films made by this process have good clarity, low strains, and freedom from pinholes.

Additional Reading

Ahmed, S., R. Bond, and E.C. McKannan: "Solidification Processing Superalloys in an Electric Field," *Advanced Materials & Processes*, **30–37** (October 1991).

Blackburn, R.D.: "Advanced Vacuum Casting," *Advanced Materials & Processes*, **17** (February 1990).

Blair, M., T.L. Stevens, and B. Linskey: *Steel Castings Handbook*, 6th Edition, ASM International, Materials Park, OH, 1995.

Cervellero, P.: "Levitation-Melting Method Intrigues Investment Casters," *Advanced Materials & Processes*, **41** (March 1991).

Daniels, J.A. and J.A. Douthett: "New Alloys Cut Auto-Casting Costs," *Advanced Materials & Processes*, **20** (February 1990).

Emmons, J.B.: "Component Design from Systems Design," *Advanced Materials & Processes*, **21** (February 1990).

Heine, R.: *Principles of Metal Casting,* The McGraw-Hill Companies, Inc., New York, NY, 1998.

Hicks, C.T.: *Casting of Acrylic,* Modern Plastics Encyclopedia, Price Stern Sloan Publishing, Los Angeles, CA, 1997.

Katgerman, L.: *Continuous Casting of Aluminum,* Ashgate Publishing Company, Brookfield, VT, 1998.

Lane, M.J.: "Investment-Cast Superalloys Challenge Wrought Material," *Advanced Materials & Processes*, **107** (April 1990).

Molloy, W.J.: "Investment-Cast Superalloys a Good Investment," *Advanced Materials & Processes*, **23** (October 1990).

Poirier, D.R. and G.H. Geiger: *Heat Transfer Fundamentals for Metal Casting,* Minerals Metals and Materials Society, Warrendale, PA, 1998.

Staff: "Rapid-Solidification Processing Improves Metal-Matrix Composites," *Advanced Materials & Processes*, **71** (November 1990).

Staff: *Advances in Aluminum Casting Technology: Proceedings from Material Solutions Conference 98 on Aluminum Casting Technology,* ASM International, Materials Park, OH, 1998.

Staff: *Ferrous Castings, Ferroalloys,* American Society for Testing and Materials, West Conshohocken, PA, 1999.

Thorp, J.: *Casting of Nylon,* Modern Plastics Encyclopedia, Price Stern Sloan Publishing, Los Angeles, CA, 1997.

Wallace, J.F.: "Casting," *Advanced Materials & Processes*, **53** (January 1990).

Weeks, R.A.: *Casting of Film,* Modern Plastics Encyclopedia, Price Stern Sloan Publishing, Los Angeles, CA, 1997.

White, C.H., P.M. Williams, and M. Morley: "Cleaner Superalloys Via Improved Melting Practices," *Advanced Materials & Processes*, **53** (April 1990).

CAST IRON.   Generic term for a group of metals that basically are alloys of carbon and silicon with iron. Relative to steel, cast irons are high in carbon and silicon, carbon ranging from 0.5 to 4.2% and silicon from 0.2 to 3.5%. All these metals may contain other alloys added to modify their properties.

See also **Iron Metals, Alloys, and Steels**.

CASTOR OIL (*Ricinus communis; Euphorbiaceae*).   [CAS: 8001-79-4]. Castor oil is obtained from a short-lived perennial tree that occurs wild in tropical Africa and perhaps in India. Cultivation of the tree is widespread not only in the tropics but also in temperate regions, where it is often grown as an ornamental plant. In the tropics it becomes a tree 36 feet tall, with large coarse leaves often of reddish color, and green flowers. An annual herbaceous variety is grown widely and produces a superior oil. The seeds, borne three in each of the smooth or prickly capsules, have a hard mottled shell. These seeds are ejected violently from the mature fruit.

The principal use of the plant is for the oil contained in the seeds. This oil is pressed out without heating the seeds. The particular properties make this oil valuable for specialized uses, such as low temperature lubrication. It is an important constituent of hydraulic brake fluid and other fluids where the degree of compressibility is important. Castor oil also finds medical uses, as an ingredient of special soaps, and in the preparation of some textile dyes. Ricin, an alkaloid present in castor oil, also has been used in insecticides. Prior to the preparation of refined castor oil for medical purposes, ricin must be removed.

CATALYSIS.   Catalysts have been employed since antiquity in such activities as wine, bread and cheese making. In many cases it was found that the addition of a small portion from a previous batch, a "starter," was necessary to begin the next production. In 1835, Berzelius published an account which tied together earlier observations by chemists, such as Thènard, Davy, and Döbereiner, by suggesting that minute amounts of a foreign substance were able to greatly affect the course of chemical reactions, both inorganic and biological. Berzelius attributed a mysterious force to the substance which he called *catalytic*. In 1894, Ostwald proposed that catalysts are substances that accelerate the rate of chemical reactions without themselves being consumed during the reactions. This definition is still applicable today.

The scope of catalysis is enormous. Catalysts are widely used in the commercial production of fuels, chemicals, foods and medicines. They also play an essential role in processes in nature, like nitrogen fixation, metabolism and photosynthesis.

Classification of Catalysts

Catalysts can be protons, ions, atoms, molecules, or larger assemblages. Traditionally, catalysts have been classified as homogeneous, heterogeneous, and enzymatic, reflecting an increasing hierarchy of complexity.

Homogeneous Catalysts: The first of the aforementioned species may be considered examples of *homogeneous* catalysts. In addition, metal complexes and organometallic compounds are important members of this class of catalysts. As the name implies, these catalysts are uniformly dispersed or dissolved in a gas or liquid phase together with the reactant of the reaction.

Heterogeneous Catalysts: In contrast to homogeneous catalysts, *heterogeneous* catalysts are usually solid surfaces, attached to solid surfaces, or part of insoluble matrices, such as polymers, and are thus phase-separated from the fluid medium surrounding them. Regardless of their form, the active catalytic component is located at the interface between the solid and the fluid and may consist of a wide diversity of species. Examples are: One or two atoms of the total surface; a larger ensemble of such surface atoms; an organometallic compound attached to the surface atoms; an organometallic compound attached to the surface by covalent bonds; or a molecular cluster lying on the surface.

Enzymatic Catalysts: These are like homogeneous catalysts in being dissolved in liquid media, but enzymatic catalysts are of biological origin and possess the highest level of complexity among the three types. Ironically, as mentioned in the opening sentence of this article, they were probably the first catalysts to be utilized commercially and industrially. Enzymatic catalysts are proteins composed of repeating units of amino acids, often twisted into helices, and in turn folded into 3-dimensional structures. The protein structures often surround a central organometallic structure. See also **Enzyme**.

Fundamentals of Catalysis

The action of catalysis can be illustrated by an example-the water gas shift reaction catalyzed by iron and chromium oxides.

$$H_2O + CO \longrightarrow H_2 + CO_2$$

This reaction is used in the production of hydrogen in several commercial processes. It is an example of a heterogeneous catalytic reaction, but the principles derived from it are also applicable to homogeneous and enzymatic catalytic reactions. A simplified scheme for the reaction is given as follows:

$$H_2O +{}^* \longrightarrow H_2 + O^*$$

$$CO + O^* \longrightarrow CO_2 +{}^*$$

In the first step, one of the reactants, H_2O, reacts with an empty catalytic site, denoted by * to produce a product, H_2, and a reactive intermediate consisting of an oxygen atom associated with the site, denoted by O^*. In the second step, the other reactant, CO, reacts with the intermediate to produce the product, CO_2, and regenerating the catalytic site, *. The energetics associated with this process are given in Fig. 1. A key aspect of this scheme is that it represents a cycle that occurs many times as the reaction proceeds. Each repetition of the cycle is called a turnover. A good catalyst will have millions of turnovers. In contrast, a stoichiometric reactant will have only one. Several important points are to be made concerning the energetics and scheme just presented.

1. The energy level diagram shows that the catalyzed reaction has a *lower activation barrier* than the uncatalyzed thermal reaction. This is the origin of the enhancement in the rate and it applies both in the forward and reverse directions of the reaction.
2. Regardless of the details of the mechanism and the energetics of the transformation of R into P, their relative energies, as shown by $\Delta H^{\circ}_{reaction}$, do not change. [*Strictly speaking, it is a free energy of reaction*, ΔG°. The equilibrium constant is given by $K = \exp(-\Delta G / RT)$.] This means that the thermodynamic equilibrium between them does not change. Catalysts increase the *rate of approach to equilibrium*, but do not alter the thermodynamic equilibrium.
3. As shown by the overall reaction stoichiometry, there is no net consumption or production of the catalytic site, *. The reaction proceeds by repetition of the catalytic cycle or chain, with the catalytic species remaining *unchanged* at the end. This explains the observation noted earlier that miniscule amounts of catalyst can give rise to very large amounts of product.
4. The intermediate, O^*, must be neither too stable nor too unstable. If it is too stable, it will not decompose to form the product; if it is too unstable, it will not form in the first place.

Nomenclature of Catalysis

The performance of catalysts is generally described by their activity, conversion, selectivity and yield. *Activity* is a measure of the rate at which the catalyst is able to transform reactants into products and is given in terms of an extensive property of the catalyst, such as mass, volume or number of moles. *Active sites* are the atomic or molecular species responsible for catalytic activity (represented by the symbol* as mentioned previously). Their identity and number are in general very difficult to measure. Various examples of the type of entities they might be are given in the definitions of the three kinds of catalysts. *Turnover frequency*, also known as turnover number or turnover rate, is the most fundamental measure of the activity, and represents the rate at which the catalytic cycle proceeds. It is equivalent to the number of molecules undergoing transformation per active site per unit time. The term *conversion* refers to the percentage of a reactant that is reacted to form all products. The term *selectivity* is applied to a specific product and refers to the percentage of that product among the total products formed. Equivalently, it is equal to the percentage of the product formed of the total reactant consumed. A high selectivity implies little waste of reactant. *Yield* is the product of conversion and selectivity, and is a measure of the efficiency of carrying out a particular transformation. *Specificity* is used mainly with enzymatic catalysts and describes their propensity to carry out only one type of reaction or to act upon only one isomer of a particular compound.

Other terms chiefly pertain to industrial applications of catalysts. *Stability* and *lifetime* refer to the ability and length of time that a catalyst is able to maintain the conversion and selectivity necessary to run a process. *Deactivation* refers to the loss of catalytic function by any of a number of causes, such as decline in surface area, decomposition of active species, or poisoning. *Denaturization* describes the deactivation of an enzyme by the loss of its 3-dimensional folded structure. This is generally caused by extremes in temperature or pH. *Poisoning* is a type of deactivation caused by the strong binding of a foreign substance to the active site of a catalyst in competition with the reactant. *Regenerability* refers to the ability to chemically or physically treat a catalyst that has lost its activity.

Industrial Usage of Catalysts

The most important catalysts employed commercially are listed in Tables 1 and 2. The remainder of this article is devoted to specific industrial uses of catalysts. The segment dealing with *fuels* covers the major operations used in the refining of petroleum. This is followed by descriptions of a few of the major processes used to produce *industrial chemicals*. The segment covering *foods and medicines* deals exclusively with enzymes.

Fuels. Catalytic Cracking: Catalysts are used to refined a moderately heavy crude oil fraction known as gas oil to gasoline. The net result of the process is a lighter product with a high content of branched-chain and aromatic hydrocarbons, the species responsible for raising gasoline octane levels. The transformations are complex, but can be considered to involve the following major acid-catalyzed reactions:

1. C−C *bond breaking*: $\underset{\text{paraffin}}{C_{18}H_{38}} \longrightarrow \underset{\text{paraffin}}{C_{10}H_{22}} + \underset{\text{olefin}}{C_8H_{16}}$

TABLE 1. PRINCIPAL USES FOR CATALYSTS

Petroleum refining	% of Total
Catalytic cracking	7.9
Reforming	≪1
Hydrocracking	≪1
Hydrotreating	≪1
Alkylation	91.2

Chemical production	
Polymerization	44.3
Alkylation	12.1
Hydrogenation	8.0
Dehydrogenation	2.4
Oxidation, ammoxidation, and oxychlorination	21.1
Ammonia, hydrogen, and methanol production	12.1

Note: It is estimated that in 1991, approximately 2400 million kilograms of catalysts were consumed by the petroleum industry; approximately 110 million kilograms were consumed for chemical production. Approximately 650 million kilograms were used strictly for emission and pollution control by various industries.

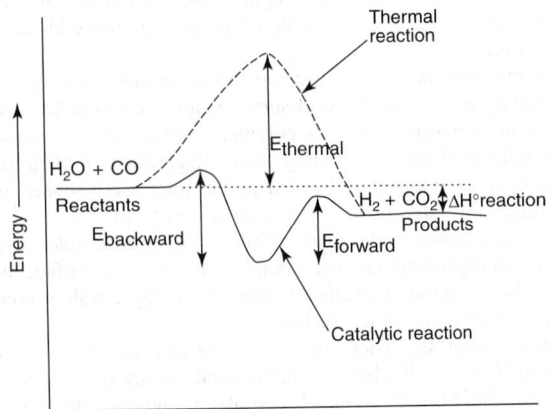

Fig. 1.   Energy level diagram for the hypothetical catalytic and thermal water gas shift reaction. The overall heat of reaction is given by $\Delta H^{\circ}_{reaction}$, the activation barriers in the forward and backward direction by $E_{forward}$ and $E_{backward}$, respectively; and the activation energy for the thermal reaction by $E_{thermal}$

TABLE 2. PRINCIPAL USES FOR ENZYME (CATALYSTS)

	% of Total
Alkaline protease	53.6
Glucose isomerase	9.7
Rennets	18.3
Glucoamylase	12.2
Other amylases	6.2

Note: It is estimated that in 1991, approximately 2 million kilograms of enzymes were consumed by the chemical and food production industries. Note included in the foregoing figures are enzymes for leather-bating, papain, pectinase, bromelain, and several others.

2. *Dealkylation*: $ArC_4H_9 \longrightarrow C_4H_8 + ArH$
   alkylaromatic   olefin   aromatic

3. *Hydrogen transfer*: $C_6H_{12} + 3\,C_8H_{16} \longrightarrow C_6H_6 + 3\,C_8H_{18}$
   cycloparaffin   olefin   aromatic   paraffin

4. *Isomerization*: $n\text{-}C_{10}H_{20} \longrightarrow i\text{-}C_{10}H_{20}$
   olefin   isoolefin

The heterogeneous catalysts employed in cracking are acidic materials composed of 3 to 25% (wt) of zeolites embedded in a silica-alumina matrix. Zeolites are crystalline aluminosilicates possessing a network of uniform pores whose walls hold the catalytically active acid sites. The reactant molecules pass through the pores and react within the zeolites.

Reforming: The catalysts are used to treat naphtha, a fraction of crude oil somewhat lighter than gas oil and containing large amounts of straight-chain paraffins. Several examples of typical reactions carried out by these catalysts are given below. The result of these reactions is to reconstruct or "reform" the hydrocarbons in the feed so as to increase the octane level. The catalysts used here differ from cracking catalysts because they tend not to alter the carbon number of the reactants and also because they produce a substantial amount of byproduct hydrogen gas.

1. *Isomerization:*

$n\text{-}C_7H_{16}$ n-heptane $\longrightarrow$ 2,2-dimethylpentane

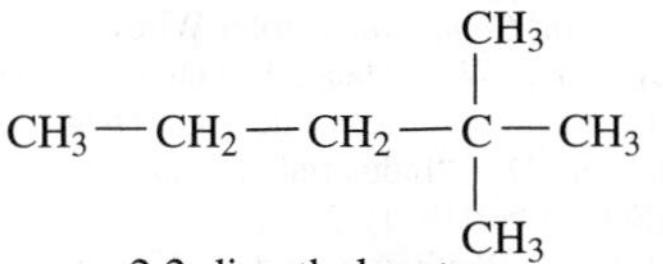

2. *Dehydrocyclization:*

$n\text{-}C_7H_{16}$ n-heptane $\longrightarrow$ toluene $+ H_2$ hydrogen

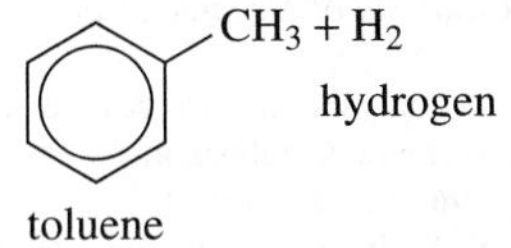

3. *Aromatization:*

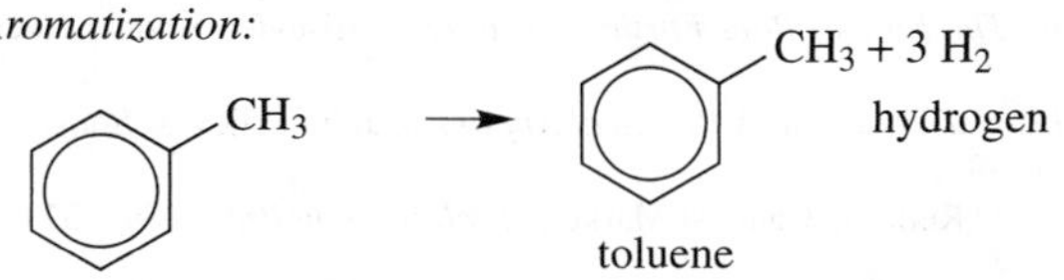

methylcyclohexane $\longrightarrow$ toluene $+ 3\,H_2$ hydrogen

These heterogeneous catalysts consist of multimetallic clusters, containing metals, such as platinum, iridium, or rhenium, supported on porous acidic oxide supports, such as alumina. The catalysts are said to be bifunctional because both the metal and the oxide play a part in the reactions. The metal is believed to carry out reversible dehydrogenation of paraffins to olefins, while the oxide is believed to carry out isomerization.

Hydrocracking. In hydrocracking, catalysts are used to reduce the molecular weight of a feedstock. A typical use is the conversion of light gas oil to naphtha for gasoline production through reforming. An example of a characteristic reaction is given as follows.

$$n\text{-}C_{16}H_{34} + H_2 \longrightarrow n\text{-}C_7H_{16} + i\text{-}C_9H_{20}$$
n-hexadecane   n-heptane   iso-nonane

These heterogeneous catalysts contain nickel, cobalt, molybdenum, tungsten, platinum, or palladium on acidic aluminum silicate or zeolite supports. As with reforming catalysts, the catalysts here are also believed to be *bifunctional*—with the metal component carrying out the reversible dehydrogenation of paraffins to olefins. Hydrocracking is carried out in the presence of hydrogen and produces saturated products.

Hydrotreating. This process comprises a mild hydrogenolysis of nitrogen, oxygen and sulfur compounds prior to catalytic cracking. The reactions carried out in this step are as follows.

1. *Desulfurization*: $R{-}SH + H_2 \longrightarrow RH + H_2S$

2. *Denitrogenation*: $R{-}NH_2 + H_2 \longrightarrow RH + NH_3$

Hydrotreating catalysts are composed of cobalt or nickel molybdate or nickel tungstate on an alumina or zeolite support. The materials are sulfided with hydrogen sulfide (H_2S) before use, but the final catalysts may retain some oxide and be of complex composition.

Alkylation: This process converts isobutane and butylenes produced in the catalytic cracking step into a mixture of dimers known as alkylate. This product is a gasoline blending stock of high octane value. Alkylation catalysts are homogeneous liquid catalysts, either sulfuric or hydrofluoric acids.

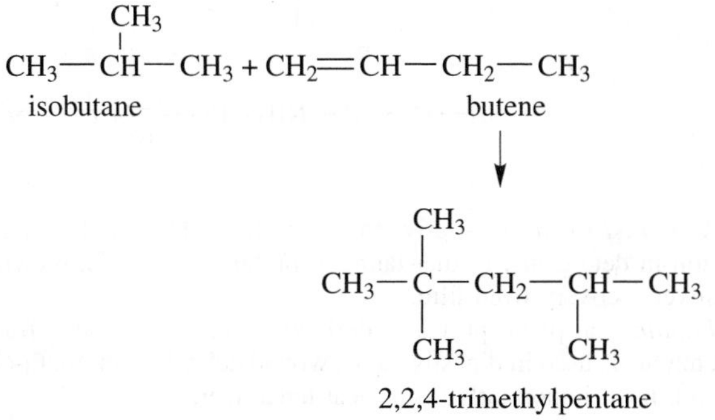

isobutane + butene $\longrightarrow$ 2,2-trimethylpentane

Chemicals. *Polymerization*: Catalysts are used in the production of polymers, such as linear and low-density polyethylene (LLDPE). An example of these catalysts are Ziegler-Natta catalysts, which are combinations of titanium halides with aluminium and magnesium alkyls.

$$n\text{-}CH_2{=}CH_2 \longrightarrow (-CH_2{-}CH_2)_n-$$
ethylene   LLDPE

Alkylation: Catalysts are used to make carbon-carbon bonds, as in the liquid phase alkylation of benzene to ethylbenzene, a styrene precursor. The catalyst used in this case is aluminum chloride.

$$C_6H_6 + CH_2{=}CH_2 \longrightarrow C_6H_5{-}CH_2{-}CH_3$$
benzene   ethylene   ethylbenzene

Hydrogenation: These catalysts are used to add hydrogen to unsaturates, as in the hydrogenation of vegetable oils to form hardened oils. Most catalytic systems consist of nickel or a noble metal on a support.

$$CH_3(CH_2)_4\,CH{=}CH{-}CH_2{-}CH{=}CH(CH_2)_7COOH \longrightarrow$$
linoleic acid

$$CH_3(CH_2)_7CH{=}CH(CH_2)_7\,COOH$$
oleic acid

Dehydrogenation: Catalysts are used to remove hydrogen from hydrocarbons. Many catalysts have been developed, including metals and oxides. An example of the latter is chromia-alumina used in the dehydrogenation of butane.

$$CH_3{-}CH_2{-}CH_2{-}CH_3 \longrightarrow CH_2{-}CH{-}CH{=}CH_2$$
butane   butadiene

Oxidation, Ammoxidation, and Oxychlorination: Numerous catalysts have been developed for a number of processes in this category. Examples are supported vanadium oxide, complex multimetallic oxides, and supported cupric chloride, used respectively for the following reactions:

1. *Butane oxidation*: $C_4H_{10} + O_2 \rightarrow C_4H_2O_3$
   butane   maleic anhydride

2. *Ammoxidation*: $C_3H_6 + O_2 + NH_3 \rightarrow CH_2{=}CH{-}CH$
   propylene   acrylonitrile

3. *Oxychlorination*: $C_2H_4 + Cl_2 + O_2 \rightarrow ClCH_2{-}CH_2Cl$
   ethylene   1,2-dichloroethane

Ammonia, Hydrogen, and Methanol Production: The ammonia synthesis catalyst is metallic iron promoted with Al_2O_3, K_2O, MgO, and CaO. The hydrogen-producing (methane reforming) catalyst is supported nickel. The methanol synthesis catalyst is ZnO promoted with Cr_2O_3 or Cu(I)–ZnO promoted with Cr_2O_3 or Al_2O_3. The respective reactions are cited as follows.

1. *Ammonia synthesis*: $N_2 + 3\ H_2 \longrightarrow 2\ NH_3$
2. *Methane reforming*: $CH_4 + H_2O \longrightarrow CO + 3\ H_2$
3. *Methanol synthesis*: $CO + 2\ H_2 \longrightarrow CH_3OH$

Foods, Medicines, and Other Products. *Proteases*: The function of these enzymes is to hydrolyze the peptide bond in proteins. Considerable variety exists in source, specificity, and reaction conditions for these enzymes. An example follows.

1. *Alkaline Proteases*—derived from bacteria. They find wide application in detergents, leather tanning, protein hydrolysis, brewing, and silver recovery from film.
2. *Papain*—a plant protease derived from the papaya fruit. The enzyme is used in digestive aids, wound debridement, tooth-cleaning and, most importantly, as a meat tenderizer.
3. *Bromelain*—a plant protease with uses similar to those of papain. Bromelain is obtained from stumps left over from pineapple harvest.
4. *Rennet or rennin*—an animal protease derived from the stomachs of calves as well as from microorganisms. Rennet is used in the manufacture of cheese to clot milk.

Glucose Isomerase: This enzyme is found in many organisms and, in practice, is used in the form of entrapped cells or bound to ion-exchange resins. Glucose isomerase converts glucose to fructose, one of the principal components of table sugar.

Leather Bating Enzymes: Enzymes used in leather manufacture to remove flesh from hides. The enzymes generally are derived from hog and beef pancreas and consist of mixtures of enzymes that attack both proteins and lipids.

Amylases. These enzymes hydrolyze the D-glycosidic linkage in starch.

1. *Glucoamylase*—found in blood, molds, and bacteria. This enzyme produces glucose by removing the end glucose unit in long-chain carbohydrates, such as starch, glycogen, dextrins, and maltoses. The main commercial use of glucoamylase is in the production of glucose syrup, glucose paste, and crystalline glucose.
2. *Other amylases*, constituting a large family of enzymes that act on different substrates, are found in saliva, animal tissues, plants, yeast, and other microorganisms. They find wide use in the manufacture of glue, starchy syrups, and in various steps in the production of brewery and bakery products.

Pectinases: These enzymes carry out the hydrolytic degradation of the D-glycosidic linkage in pectins. The latter substances, also known as pectic substances, are polymeric components of plant cell walls and, like starch, are composed of sugar residues linked by glycosidic bonds. The chemistry is the same as that shown for the amylases previously described. The main application of pectinases is in the production of fruit juices, wines, and certain other food products.

Major portions of this article were prepared by S. Ted Oyama and Prof. Gabor A. Somorjai, Center for Advanced Materials, Lawrence Berkeley Laboratory, The University of California. Berkeley, California.

Additional Reading

Adams, R.D. and F.A. Cotton: *Catalysis by Di- and Polynuclear Metal Cluster Complexes,* John Wiley & Sons, Inc., New York, NY, 1998.

Anderson, J.R. and M. Boudart: *Catalysis: Science and Technology,* Vol. 10, Springer-Verlag Inc., New York, NY, 1996.

Anpo, M., M. Onaka, and H. Yamashita: *Science and Technology in Catalysis,* Elsevier Science & Technology, New York, NY, 2003.

Baerns, M.: *Basic Principles in Applied Catalysis,* Springer-Verlag New York, LLC., New York, NY, 2004.

Buchmeiser, M.R.: *Polymeric Materials in Organic Synthesis and Catalysis,* John Wiley & Sons, Inc., New York, NY, 2003.

Chorkendorff, I., and J.W. Niemantsverdriet: *Concepts of Modern Catalysis and Kinetics,* John Wiley & Sons, Inc., New York, NY, 2003.

Cornils, B. and W.A. Hermann: *Applied Homogeneous Catalysis with Organo metallic Compounds,* John Wiley & Sons, Inc., New York, NY, 1999.

Eley, D.D., W.O. Haag, B. Gates, and H. Knozinger: *Advances in Catalysis,* Vol. 42, Academic Press, Inc., San Diego, CA, 1998.

Erickson, D.: "Industrial Immunology: Catalytic Antibodies," *Sci. Amer.,* **174** (September 1991).

Ertl, G., J. Weitkamp, and H. Knoezinger: *Handbook of Heterogeneous Catalysis,* John Wiley & Sons, Inc., New York, NY, 1997.

Ford, M.E.: *Catalysis of Organic Reactions,* Marcel Dekker, Inc., New York, NY, 2000.

Friend, C.M.: "Catalysis on Surfaces," *Sci. Amer.,* **74** (April 1993).

Gross, A.: "Enzymatic Catalysis in the Production of Novel Food Ingredients," *Food Technology,* **96** (January 1991).

Gschneidner, K.A. Jr. and L. Eyring: *Handbook on the Physics and Chemistry of Rare Earths: The Role of Rare Earths in Catalysis,* Elsevier Science, New York, NY, 2000.

Hagen J.: *Industrial Catalysis: A Practical Approach,* John Wiley & Sons, Inc., New York, NY, 1999.

Hoffman, H.J.L.: "Refining Catalyst Market," *Hydrocarbon Processing,* **37** (February 1991).

Johnson, A.D. et al.: "The Chemistry of Bulk Hydrogen: Reaction of Hydrogen Embedded in Nickel with Adsorbed CH_3," *Science,* **223** (July 10, 1992).

Kurosawa, H., and A. Yamamoto: *Fundamentals of Molecular Catalysis,* Elsevier Science & Technology Books, New York, NY, 2003.

Lerner, R.A., S.J. Benkovic, and P.G. Schultz: "At the Crossroads of Chemistry and Immunology: Catalytic Antibodies," *Science,* **659** (May 3, 1991).

Masel, R.I.: *Chemical Kinetics and Catalysis,* John Wiley & Sons, Inc., New York, NY, 2001.

McLean, J.B. and E.L. Moorehead: "Steaming Affects FCC Catalyst," *Hydrocarbon Processing,* **41** (February 1991).

Moulijn, J.A. and R.A. Van Santen: *Catalysis: An Integrated Approach,* Elsevier Science, New York, NY, 2000.

Niemantsverdriet, J.W.: *Spectroscopy in Catalysis: An Introduction,* John Wiley & Sons, Inc., New York, NY, 2000.

Ojima, I.: *Catalytic Asymmetric Synthesis,* John Wiley & Sons, Inc., New York, NY, 2000.

Rase, H.F.: *Handbook of Commercial Catalysts,* CRC Press, LLC., Boca Raton, FL, 2000.

Rosso, J.P.: "Maximize Precious-Metal Recovery from Spent Catalysts," *Chem. Eng. Progress,* **66** (December 1992).

Scott, D.L. et al.: "Interfacial Catalysis: The Mechanism of Phospholipase A2," *Science*, **1541** (December 14, 1990).

Sheldon, R.A. and H.V. Bekkum: *Fine Chemicals through Heterogenous Catalysis*, John Wiley & Sons, Inc., New York, NY, 2001.

Staff: "Microcalorimeter Studies Uncover Multistate Catalysts," *Chem. Eng. Progress*, **12** (October 1991).

Staff: "Single Site Catalysts Get Commercial Tryout," *Chem. Eng. Progress*, **21** (October 1991).

Staff: "New Catalyst Boosts Aromatics Yields," *Chem. Eng. Progress*, **30** (November 1991).

Thomas, J.M., Sir: "Solid Acid Catalysts," *Sci. Amer.*, **112** (April 1992).

Waldrop, M.M.: "Catalytic RNA Wins Chemistry Nobel," *Science*, **325** (October 20, 1989).

Waldrop, M.M.: "The Reign of Trial and Error Draws to a Close: Designing Catalysts at the Molecular Level," *Science*, **28** (January 5, 1990).

Weitkamp, J. and L. Puppe: *Catalysis and Zeolites: Fundamentals and Applications*, Springer-Verlag Inc., New York, NY, 1999.

Worstell, J.J.: "Succeed at Catalyst Upgrading," *Chem. Eng. Progress*, **33** (June 1992).

CATALYTIC CONVERTER (Internal Combustion Engine). A combination of the Clean Air Act Amendments of 1970 and the Energy Policy and Conservation Act of 1975 (United States Congress) has promoted the widespread use of catalytic aftertreatment to control automotive exhaust emissions with a concomitant increase in fuel economy. The catalytic converter, comprised of a ceramic catalyst and the necessary stainless steel hardware to ensure that the exhaust gases pass through the catalyst, permits the conventional spark-ignition automobile engine to run at near optimum efficiency to afford good fuel economy. The catalyst itself has the capability of promoting (or accelerating) the rate at which reactions occur. In the case of an oxidation catalyst, the function is to cause the carbon monoxide (CO) and hydrocarbons (HC) which result from incomplete combustion to be converted to CO_2 and water. In the case of a three-way catalyst, the oxidation reactions (HC and CO) are promoted as well as the reduction reaction of oxides of nitrogen (NO_x).

Converters now in use contain noble metals on a ceramic substrate (e.g., platinum dispersed on alumina). The converter is typically located in the exhaust system in one of two general locations: an underfloor location, or a close-coupled location near the manifold. The operating temperature range for noble metal catalyst is from 600 to 1200°F (316 to 649°C), which is similar to the exhaust pipe skin temperature range normally encountered on standard automobile engines.

Catalytic materials can be physically supported on either pelleted or monolithic substrates. In the case of the pelleted catalyst, the support is an activated alumina. A typical monolithic catalyst is composed of a channeled ceramic (cordierite) support having, for example, 300 to 400 square channels per square inch on which an activated alumina layer is applied. The active agents (platinum, palladium, rhodium, etc.) are then highly dispersed on the alumina.

In the case of pelleted catalyst, the pellets are confined by screens (Fig. 1); the monolithic-type catalyst (Fig. 2); being a single rigid material,

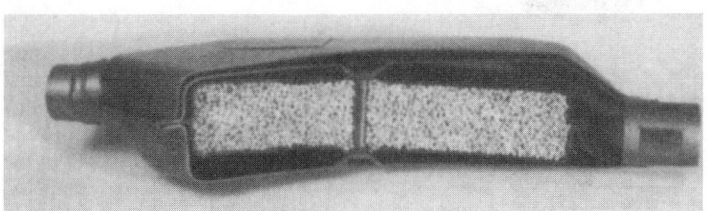

Fig. 1. Converter to use pelleted catalyst

Fig. 2. Converter to use monolithic catalyst

needs no such confinement. The arrangement within the container, regardless of which type of catalyst is used, is intended to ensure that the exhaust gases pass through the catalyst bed without bypassing it or "channeling" along outside walls of the catalyst.

Exhaust emission standards since the 1981 model year vehicles have required the use of three-way catalysts, either alone or in combination with an oxidation catalyst. Three-way catalysts are designed to operate in a very narrow range about the stoichiometric air/fuel ratio. In this range the HC and CO are subject to oxidation and the NO_x compounds undergo reduction. The downstream oxidation catalyst in a dual bed system is generally used as a "clean-up" catalyst to further control HC and CO emissions. The most common catalytic combination in three-way uses is platinum/rhodium. Current production applications use these elements in a relatively rich proportion of 5:1 to 10:1, whereas the respective mine ratio is about 19:1.

Since the introduction of catalytic converters on passenger cars in the 1973 model year in the United States and in the 1974 model year in Japan, the demand for improved air quality has grown worldwide. With that demand has grown converter usage to control mobile source emissions. Several European countries are in the process of drafting and enacting legislation that requires catalytic converters. Australia uses catalytic converters, Korea will soon require them, Brazil is also developing emission control strategies, and several other countries are studying the need for emission controls.

Technical Staff, Allied-Signal Catalyst Company
Catoosa, Oklahoma

CATENATION COMPOUND. See **Compound (Chemical)**.

CATHODE. 1. In general, the electrode at which positive current leaves a device which employs electrical conduction other than that through solids. 2. In an electron tube, the electrode through which a primary stream of electrons enters the inter-electrode space. 3. The negative terminal of an electroplating cell (i.e., the electrode from which electrons enter the cell, and thus at which positively charged ions (cations) are discharged). 4. The positive terminal of a battery. See also **Batteries**.

CATHODIC PROTECTION. See **Corrosion**.

CATION. A positively charged ion. Cations are those ions that are deposited, or which tend to be deposited, on the cathode. They travel in the nominal direction of the current. In electrochemical reactions they are designated by a dot or a plus sign placed above and behind the atomic or radical symbol as H• or H^+, the number of dots or plus signs indicating the valence of the ion.

In electrolysis, the cathode is negative, and attracts cations. In a battery, the transfer of charges of cations to the cathode makes it the positive terminal.

CAT'S-EYE. This name is applied to varieties of several mineral and gemstone species that enclose fine fibers or cellular structures in parallel arrangement, causing, particularly when cut and polished *en cabochon*, a band of reflected light to play on the surface of it. Because of fancied resemblance to the eyes of cats, such stones are called cat's-eyes, and the effect is referred to as chatoyancy. The stone is said to be chatoyant. True cat's-eye is a variety of chrysoberyl, but tourmaline and quartz are also found which show this same effect. Ordinary quartz cat's-eyes are a pale yellowish or greenish, but a beautiful golden-yellow sort, known from South Africa and called tiger's-eye, probably represents a replacement of crocidolite by quartz.

When the term *tiger's eye* is used, this applies only to chrysoberyl. Other gemstones that exhibit this phenomenon include sillimanite, scapolite, cordierite, orthoclase, albite, and beryl. See also **Chrysoberyl**; and **Crocidolite (Blue Asbestos)**.

CAUSTIC (Chemical). A corrosive substance, almost always of an alkaline nature, such as sodium hydroxide, NaOH; potassium hydroxide, KOH; or calcium oxide, CaO. Such substances attack many metals, plastics, and other materials, including human tissue, and generally fall in the category of *corrosives*.

CAUSTIC POTASH. See **Potassium**.

CAUSTIC SODA. See **Sodium**.

CAVENDISH, HENRY (1731–1810). Cavendish was an English chemist and physicist who performed numerous scientific investigations with experiments on electricity. He measured the strength of a current by shocking himself and then estimating the amount of pain.

Cavendish investigated "fixed air" and isolated "inflammable air", in 1776 he discovered hydrogen. He produced dew (water) by exploding a mixture of hydrogen and air with an electric spark. He also found that a small volume of air did not combine with nitrogen while using electric sparks. Later, Ramsay repeated Cavendish's work and discovered argon.

Cavendish is also known for a sensitive torsion balance, known as the Cavendish balance, to measure the value of the gravitational constant, which defines the strength of the force of gravity.

J.M.I.

CAVITAND. As defined by Cram (1983), a cavitand is a synthetic organic compound that contains an enforced cavity of dimensions at least equal to those of the smaller ions, atoms, or molecules. As Cram points out, if organic compounds that contain such rigid cavities are to be designed and prepared, they must be composed of units that are concave on parts of their surfaces. Very few organic compounds have concave surfaces of any size. Among the most studied of the naturally occurring compounds that contain rigid cavities are the cyclodextrins. In these cyclic oligomers of the 1,4-glucopyranoside unit, from 6 to 8 monosaccharide units are contained in a torus-shaped cavity. Organic hosts are now being designed and synthesized which contain enforced cavities sufficiently large to complex and even surround simple inorganic or organic guest compounds. This new field of investigation is of interest in enzyme and catalytic systems.

Additional Reading

Cram, D.J.: "Cavitands: Organic Hosts with Enforced Cavities," *Science,* **219,** 1177–1183 (1983).
Rouvray, D.H.: "Predicting Chemistry from Topology," *Sci. Amer.,* **255**(3), 40–47 (September 1983).

CAVITATION. Cavities may form, grow, and collapse in a liquid when variational tensile stresses are superimposed on the prevailing ambient pressure. Pure liquids have theoretical tensile strengths, which are estimated on various grounds to be of the order of 300 to 1500 atmospheres (bars), but the observed tensile strengths of real liquids are much lower. It is presumed, therefore, that the observed tensile strength is a measure of the stress required to enlarge the minute cavities, or cavitation nuclei, which already exist in the liquid rather than the stress required to form new interior surfaces.

The transient cavities formed by tensile stress are unstable and would grow indefinitely if the stress were maintained. After the cavitation nuclei have been expanded to many times their original size, however, they may collapse violently if the stress is reduced or removed. The kinetic energy of the liquid that follows each inwardly collapsing interface becomes highly concentrated as the cavity collapses. If such transient cavities contain very little permanent gas, the peak pressures at collapse may reach thousands of bars, the temperature may reach thousands of degrees, and strong shock waves may be radiated to a distance of several cavity radii. Similar cavities formed in saturated liquids will usually contain more gas and their collapse will be less violent, but the peak pressures attained are sufficient to produce unique mechanical effects, such as the corrosion and pitting of metallic surfaces (as in marine propellers and sonar projectors) and the beneficial removal of embedded dirt (as in ultrasonic cleaners). In the latter case, the soil to be removed provides a prolific source of cavitation nuclei at exactly the sites where cavitation is desired.

In hydrodynamic cavitation, the tensile stress is of relatively long duration and plenty of cavitation nuclei are usually available. As a result, cavitation occurs when the total net pressure, or the stagnation pressure, becomes approximately equal to the vapor pressure of the liquid. In acoustic cavitation, the cyclic pressure required to produce cavitation is a function of the frequency, the partial pressure of any dissolved gas, and the population of cavitation nuclei. For frequencies above about 200,000 Hz, the threshold pressure for cavitation increases with the square of the frequency and is almost independent of the degree of gas saturation. For frequencies below 200,000 Hz, the threshold pressure is a function of the partial pressure of the dissolved gas. In saturated liquids at sound pressures less than a few bars, stable bubbles can grow from cavitation nuclei by the process of rectified diffusion. At higher levels of acoustic excitation, transient cavities can be formed. The threshold sound pressure at which they appear and the violence of their collapse increase as the partial pressure of the dissolved gas is lowered.

Cavitation in Process Control Valves

Serious cavitation problems are sometimes encountered in valves designed to automatically control the flow of fluids in chemical and other processing plants. See Fig. 1. Much research has been directed by valve manufacturers and users to the alleviation of these problems. While the damage rate and the total damage from cavitation in valves are not highly predictable, a number of strategies may be used to reduce hardware damage. These measures involve system design, material selection, and using anticavitation products.

Consideration of potentially damaging cavitation conditions at the time a system is designed is the primary and preferred strategy. For example, the placement of control valves in high back-pressure locations, when possible, reduces the tendency of a valve to cavitate. When placement is not flexible, downstream "breakdown" orifices may be inserted to "artificially" increase the back-pressure. This is not a preferred method because the effective flow velocities through the breakdown orifice may be so low as to eliminate its effectiveness and, at high flow rates, the orifice plate may become the primary restriction and, in turn, limit or completely choke off the flow. If properly sized to a particular valve, the downstream orifice may prevent the valve from cavitating, but the orifice itself may cavitate. In another strategy, the use of a sacrificial member may be considered. In some situations, economics may warrant the installation of a lower-cost valve which is allowed to cavitate. Immediately downstream of such a valve where cavitation damage is likely to occur, a comparatively inexpensive pipe or fitting may be installed and periodically replaced. This method does not afford relief from the other side effects of cavitation, such as noise,

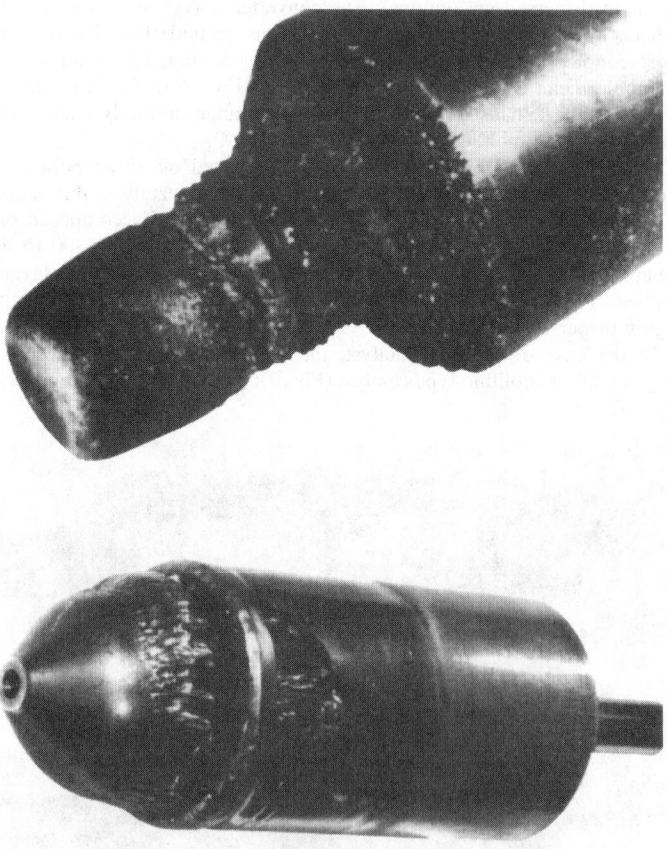

Fig. 1. (*Top*) Typical appearance of Cavitation damage caused by repeated attack by millions of micro-jets and shock waves. (*Bottom*) The term *cavitation* also is used to describe a variety of related phenomena that involve bubble or cavity formation. Examples include effervescence or outgassing, the boiling of a liquid, and flashing, the results of which are shown in bottom view. (*Fisher Controls*)

vibration, and choking, but in certain cases does permit more economical control of damage.

A second category is material selection. However, this does not offer a full solution because no material is completely immune to cavitation damage. Mechanical attack interacts synergistically with corrosion to create a different situation for nearly every application. In general, metals with greater hardness, ultimate resilience, or strain energy to failure offer better resistance to cavitation. Elastomers and compliant surfaces, in general, exhibit an ability to withstand levels of cavitation attack greater than their standard structural indicators may suggest. This apparently results from an interaction between the surface and the bubble, which orients the microjet away from the surface, thus reducing mechanical attack. The temperature and pressure limitations of these materials, however, place restrictions on their use.

The most effective means for controlling cavitation is the use of cavitation control equipment. Valve manufacturers in recent years have employed two basic strategies along these lines: (1) control energy transformations, and (2) isolate cavities. The former is usually attained by *staging*. This approach to damage control routes the flow through several restrictions in series, as opposed to a single restriction. Each restriction dissipates a certain amount of available energy and progressively lowers the inlet pressure to each succeeding stage. This enables the valve to take a large pressure differential, yet maintain the minimum fluid pressure above the vapor pressure of the liquid, and thus may eliminate cavitation. Traditional valve design, on the other hand, was targeted for maximum efficiency and not for cavitation prevention.

In other modern valve designs, the objective may not be that of preventing cavitation fully, but to control the cavitation that does occur. A mechanical component of attack is always present in cavitation damage, taking two forms—microjet impact and shock wave impact—both of which must occur in close proximity to the surface in order to be damaging. Thus, it follows that, if they occur far enough away from the surface, no mechanical attack will occur, thus controlling damage.

Additional Reading

Brekke, H., C.G. Duan, R.K. Fisher, and R. Schilling: *Hydraulic Machinery and Cavitation,* World Scientific Publishing Company, Inc., Riveredge, NJ, 1998.

Furuya, G.: *Cavitation and Multiphase Flow Forum 1992,* American Society of Mechanical Engineers," New York, NY, 1992.

Grist, E.: *Cavitation and the Centrifugal Pump: A Guide for Pump Users,* Taylor and Francis, Dallas, TX, 1999.

Margulis, M.A.: *Sonochemistry and Cavitation,* Gordon and Breach Science Publishers, Newark, NJ, 1995.

Riveland, M.L.: "Control Valve Cavitation—An Overview," in *Process/Industrial Instruments and Controls Handbook,* D.M. Considine and G.K. McMillan, Editors, The McGraw-Hill Companies, Inc., New York, NY, 1999.

Robertson, J.M.: "Cavitation Today—An Introduction," in *Cavitation—State of Knowledge,* American Institute of Mechanical Engineers, New York, NY, June 1969.

Shah, Y.T., A.B. Brandit, et al.: *Cavitation Reaction Engineering,* Kluwer Academic Publishers, Norwell, MA, 1999.

Tullis, J.P.: *Hydraulics of Pipelines: Pumps, Valves, Cavitation, Transients,* John Wiley & Sons, Inc., New York, NY, 1990.

Yatish, T.S., A.B. Pandit, V.S. Moholkar, et al.: *Cavitation Reaction Engineering,* Plenum Publishing Corporation, New York, NY, 1999.

Young, F.R.: *Cavitation,* Imperial College Press, London, UK, 2000.

CECH, R. THOMAS (1947–). Awarded the Nobel prize for chemistry in 1989 jointly with Sidney Altman for the discovery that RNA acts as a biological catalyst, as well as a carrier of genetic information. His Doctorate was awarded in 1975 by the University of California.

CELESTITE. The mineral celestite (also known as celestine) is composed of strontium sulfate, $SrSO_4$, occasionally with calcium and barium. It crystallizes in the orthorhombic system in tabular or prismatic crystals. More rarely it may be pyramidal or simply fibrous or granular. Two essentially perfect cleavages may be observed, one parallel to the base, the other parallel to the prism. Its fracture is uneven; hardness, 3–3.5; specific gravity, 3.97; luster, vitreous; color, white, but may be slightly reddish or bluish; transparent to translucent.

Celestite may occur with gypsum and salt associated with beds of limestone, or by itself in large, commercially important veins. It sometimes occurs with sulfur in volcanic localities and is often a gangue mineral in veins of galena, sphalerite and similar metallic minerals. In Europe there are many localities for fine crystals, especially in England. In the United States celestite is found in New York, Pennsylvania, West Virginia, Tennessee, Kansas, Colorado, and California. The first celestite described was the delicate blue material from Blair County, Pennsylvania. Its "celestial" tints suggested the name. Celestite resembles barite.

CELLULOSE. [CAS: 9004-34-6]. The formula for cellulose is sometimes given as $(C_6H_{10}O_5)_n$. This is an oversimplification inasmuch as the cellulose present in natural substances, such as wood and cotton fibers, usually is combined with other constituents, such as fats and gums. Cellulose is found almost exclusively in plants and accounts for about 30% of all vegetable matter. Cellulose is the principal substance of which the walls of vegetable cells are constructed. The term *cellulose* is derived from the Latin *cellula,* meaning little cell. Relatively pure cellulose can be obtained from cotton fibers (90% cellulose) and flax fibers. Very small amounts of cellulose are found in insects and not in other animal tissues. Digestive juices and enzymes present in animal systems do not appear to attack cellulose and thus ingestion by humans is relatively limited. By means of other biological processes, such as those involving amoeboid protozoa present in the digestive tract, herbivora and insects digest and absorb some cellulose.

Cellulose is a polysaccharide of glucose. See Fig. 1. Cellulose is a white solid, odorless and tasteless, insoluble in cold or hot water, and chemically nonreactive except when treated with strongly corrosive materials. If heated with water at 260°C and under rather high pressure, however, cellulose dissolves, but with decomposition. Concentrated sulfuric acid dissolves cellulose, the solution upon dilution and boiling yielding glucose. When treated with sodium hydroxide (15 to 25% NaOH) cellulose fibers swell up and upon washing and drying possess a lustrous appearance. This is the mechanism of *mercerization.* With iodine in potassium iodide solution plus zinc chloride (Schulze's solution), cellulose produces a dark blue color. When treated with an 80% sulfuric acid solution and rapidly washed and dried, cellulose yields a parchment-like surface.

In biochemical terms, cellulose is the name given both to a specific polysaccharide, consisting of βD-glucose residues joined end-to-end by linkage through $-O-$ of C_1 of one residue to C_4 of the next (see Fig. 2); and to a resistant family of polysaccharides (containing cellulose in the strict sense) isolated from plants by specific chemical treatment. Cellulose appears to be always associated in plants with other polysaccharides and polysaccharide derivatives, such as mannan, xylan, araban, galactan, polygalacturonic acid, and, in woody plants, with lignin.

The mechanism of cellulose synthesis and microfibril orientation is not fully understood. Synthesis probably occurs from a glucose/phosphate precursor, which may be guanosine diphosphate glucose. The enzyme system involved can be extracted from the plant (e.g., from *Acetobacter xylinum*) and synthesis by cell-free extracts has been achieved. The synthetic mechanism in organisms higher than bacteria is thought to be located on the cell surface, and both granular aggregates and microtubules seen in the electron microscope are considered as possible sites.

The "brown rots" (e.g., *Coniphora casebella, Poria monticola*) and the "white rots" (e.g., *Polystictus versicola*) are both basidiomycetes. Both attack cellulose, but only the latter takes lignin to any large extent. "Soft rot fungi," recognized relatively recently as important in this regard are members of the *Ascomycetes* and *Fungi imperfecti* (e.g., *Chaetomium globosum*). They all attack the cellulose of wood, producing characteristic angular cavities. The evidence is that all of these fungi attack the paracrystalline component of cellulose more rapidly than the crystalline component.

Compound celluloses are widely distributed in plants, the two principal types being:

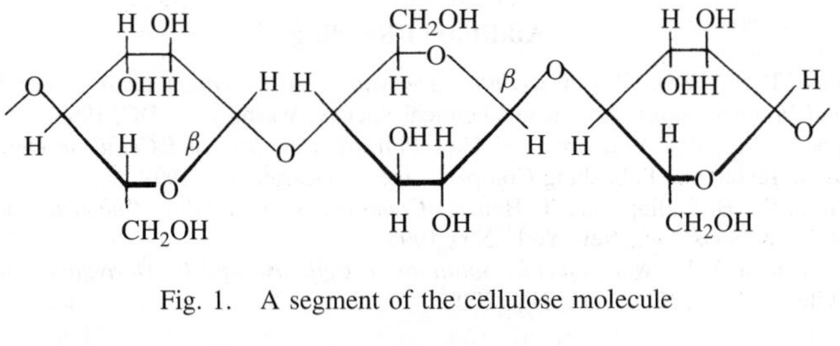

Fig. 1.   A segment of the cellulose molecule

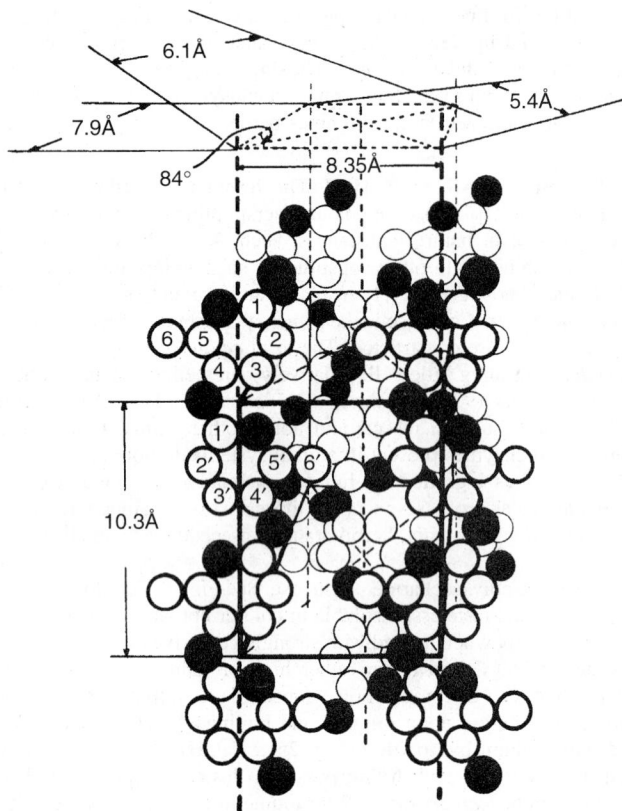

Fig. 2.   Diagrammatic representation of the unit cell of cellulose. The monoclinic cell, dimensions 10.3 Å × 8.35 Å = 84°, is delineated by solid lines with one cellulose chain at each vertical edge and one (antiparallel) in the center. Open circles = carbon atoms; solid circles = oxygen atoms. For clearness of the diagram, the hydroxyl groups on carbons 2, 3, and 6 are omitted; and hydrogen atoms are omitted. Two spacing at 6.1 Å and 5.4 Å are included, inasmuch as these are strongly represented in the x-ray diffraction diagram

(a) Lignocelluloses, of woods, cereal straws, jute. These cellulose materials yield lignin by treatment (1) with 43% hydrochloric acid, cold, for 12 hours, (2) with 8 to 12% sodium hydroxide at 140 to 160°C for 6 to 10 hours, (3) with 72% sulfuric acid at ordinary temperature for 18 hours (the common method). Wood yields about 25% of lignin by the last treatment.

(b) Pectocelluloses, of flax, hemp, ramine. These cellulose materials yield pectic substances by treatment with oxalic acid or ammonium oxalate at 85°C for 24 hours, followed by carefully defined treatment with alcohol, acetic acid and calcium chloride. Pectic substances are most abundant in leaves, e.g., ivy, sycamore, and in apples or oranges, especially the white peel of the latter.

Cellulose dissolves in Schweitzer's reagent, an ammoniacal solution of cupric oxide. After treatment with an alkali, the addition of carbon disulfide causes formation of sodium xanthate, a process used in the production of rayon. See also **Fibers**. The action of acetic anhydride in the presence of sulfuric acid produces cellulose acetates, the basis for a line of synthetic materials. See also **Cellulose Ester Plastics (Organic)**. Nitrocelluloses are produced by the action of nitric acid and sulfuric acid on cellulose, yielding compounds that are highly flammable and explosive. See also **Explosives**.

On a heavy tonnage basis, cellulose is important as a raw material in the production of wood pulp and paper. See also **Papermaking and Finishing**.

Additional Reading

Heinze, T. and W.G. Glasser: *Cellulose Derivatives: Modification, Characterization, and Nanostructures,* American Chemical Society, Washington, DC, 1998.

Kennedy, J.F., P.A. Williams, and G.O. Phillips: *Cellulose and Cellulose Derivatives,* Technomic Publishing Company, Inc., Lancaster, PA, 1999.

Klemm, D., B. Philipp, and T. Heinze: *Comprehensive Cellulose Chemistry,* John Wiley & Sons, Inc., New York, NY, 1998.

Myasoedova, V.V.: *Non-Aqueous Solutions of Cellulose and Its Derivatives,* John Wiley & Sons, New York, NY, 2000.

CELLULOSE ACETATE.   See **Acetic Acid**; **Fibers: Acetate**.

CELLULOSE ESTER PLASTICS (Organic).   The cellulosics are unique among the plastics in that the basic materials used in their manufacture are not synthetic polymers. Rather, they are derivatives of a natural polymer, *cellulose*. See also **Cellulose**. The preparation of an organic cellulose ester plastic involves the formation of a suitable cellulose derivative, followed by processing steps that convert the cellulose derivative into a plastic.

Cellulose, with its many hydroxyl groups, can react with organic reagents such as acids, anhydrides, and acid chlorides to form organic esters. The first reported organic ester of cellulose was cellulose acetate, prepared by Schützenberger in 1865 by heating cotton and acetic anhydride to about 180°C in a sealed tube until the cotton dissolved. Franchimont, in 1879, accomplished this reaction at a lower temperature with the aid of sulfuric acid as a catalyst. The product in both cases was very nearly the triester. Miles, in 1903, first described partially hydrolyzed (generally called "secondary") cellulose acetate and distinguished it from the triacetate by its acetone solubility. The solubility of secondary cellulose acetate in such inexpensive and relatively nontoxic solvents as acetone contributed greatly to the development and commercialization of this material.

Cellulose esters of the 2-, 3-, and 4-carbon acids are readily prepared by the cellulose-anhydride reaction; the acetate ester and the mixed acetate butyrate and acetate propionate esters are manufactured and used in large amounts. Esters of higher acids require different synthesis techniques and tend to be prohibitively expensive except as specialty products. Some are in commercial production, however. Cellulose acetate phthalate, for example, is manufactured for use as an enteric coating on pills.

Most commercial preparations of cellulose esters still follow, basically, the methods described by Franchimont and Miles—esterification with sulfuric acid catalyst followed by hydrolysis. The principal steps in this process are shown in Fig. 1.

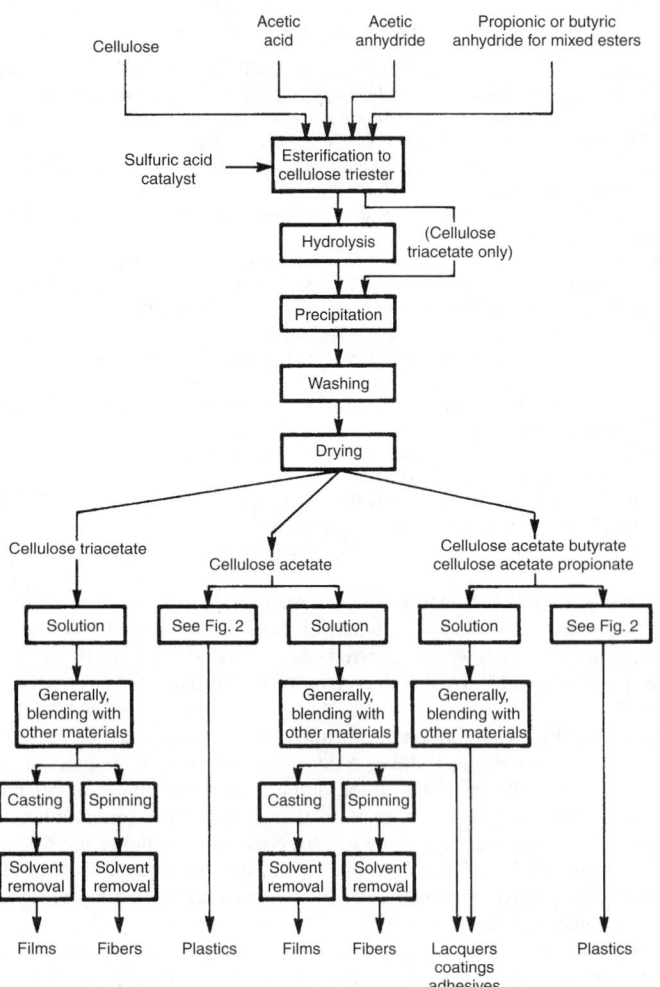

Fig. 1.   Production and end-uses of organic cellulose ester

Esterification

The nature of cellulose is such that its esterification does not occur randomly. Even when the DS (degree of substitution—the average number of hydroxyl groups replaced per anhydroglucose unit in the cellulose chain) is approaching 3 (complete reaction), many anhydroglucose units have a DS of zero. If the ester is recovered before the reaction is complete, the product will not be homogeneous and it will be hazy. Regardless of the desired DS of the final product, therefore, the reaction must be allowed to proceed to virtual completion if a homogeneous material is to be produced. In most processes, the ester dissolves in the reaction mixture as the reaction approaches completion.

The cellulose used to manufacture cellulose esters is highly purified cotton linters or wood pulp. It is generally treated to reduce its crystallinity and make it more reactive, then agitated at somewhat elevated temperatures with the appropriate acids, anhydrides, and catalyst until it dissolves. Some of the polymeric chains of the cellulose are broken during the reaction, and consequently the molecular weight decreases; thus, the catalyst concentration, reaction temperature, and reaction time must be controlled very carefully to give a product of the desired molecular weight.

Some acetate ester is recovered at the completion of reaction and marketed as commercial cellulose triacetate; it has a DS very close to 3. Other triesters have found no commercial applications.

Hydrolysis

The compatibility of a cellulose ester with other materials is influenced by the acid or acids used in its preparation, but it is controlled primarily by the hydroxyl content of the ester, which varies reciprocally with the degree of substitution. Cellulose triacetate, for example, does not dissolve in acetone; its solution requires very strong solvents such as methylene chloride. Plasticizers can be added to the triacetate in solution and will remain in the ester if the solvent is removed. Plasticizer added in this manner will affect some of the properties of the triester, but they will not appreciably reduce its softening temperature, which is higher than its decomposition temperature. The hydroxyl groups in secondary cellulose acetate, however, have for other polar materials an affinity that is lacking in the triacetate. The ester will dissolve in acetone, and plasticizers will both affect its mechanical properties and reduce its softening temperature sufficiently for the mixture to be processed as a thermoplastic. It is necessary, then, that cellulose esters to be used for plastics contain a significant number of hydroxyl groups, and these are produced by partial hydrolysis of the triester formed in the reaction step. Since the triester is in solution, hydrolysis is random and produces a homogeneous product.

Hydrolysis is initiated by the addition of water and stopped at the desired point by neutralization of the catalyst.

Precipitation, Washing, and Drying

The cellulose ester in solution in the reaction mixture is precipitated by the addition of water. Some precipitation processes produce a flake precipitate; some produce a powder. The precipitate is removed from the slurry, washed with water until it is free of acids, and dried.

Cellulose Esters

The cellulose esters that result from the process described are chemical raw materials and are used by many branches of the chemical industry. They are generally characterized by acyl content in weight percent and viscosity in seconds, the viscosity being obtained by timing the fall of a steel ball through a solution of the cellulose ester in accordance with ASTM Method D 1343. Low-viscosity esters are generally used in solution processes; high-viscosity esters are used in the production of plastics.

Cellulose Triacetate

It has already been implied that cellulose triacetate will not produce a thermoplastic, as its softening point cannot be reduced appreciably by plasticizers. It is used in solution processes, however, to produce films and fibers. Triacetate films absorb less water than films of secondary cellulose acetate, and they are therefore more dimensionally stable in environments where the humidity is not controlled. Triacetate fibers, with a similar resistance to water, impart to fabrics wrinkle resistance, dimensional stability, and the ability to dry rapidly. Under United States federal regulations, a fiber must be made from a cellulose acetate having

at least 92% of its hydroxyl groups acetylated if it is to be called "triacetate."

Cellulose Acetate

Like the triacetate, secondary cellulose acetate (CA) is used in solution processes to produce fibers and films. CA fibers were originally called "rayon," the name that was already in use for regenerated cellulose fibers. In 1951, however, the regulatory authorities formally acknowledged the chemical distinction between CA and cellulose, and the term "rayon" was reserved for fibers of regenerated cellulose. CA fibers are officially called "acetate," and they are used in a wide variety of fabrics. They also are used for cigarette filters. However, the majority of CA produced is used for manufacture of plastics.

Cellulose Acetate Butyrate and Cellulose Acetate Propionate

These two cellulose esters are somewhat similar in properties and applications. Cellulose acetate butyrate is commonly referred to in the chemical industry as CAB, while cellulose acetate propionate is simply termed "cellulose propionate" and referred to as CAP or as CP.

The major usage of CAB and CP is in plastics. Additionally, CAB and CP are mixed with a variety of synthetic polymers to produce lacquers for wood, metal, and plastics. CAB finds relatively small usage in hot-metal coatings and in optical-grade cast film used in the manufacture of sunglass lenses.

Organic Cellulose Ester Plastics

Although the first cellulose plastic (cellulose nitrate plastic-based on an *inorganic* ester of cellulose) was developed in 1865, the first *organic* cellulose ester plastic was not offered commercially until 1927. In that year, cellulose acetate plastic became available as sheets, rods, and tubes. Two years later, in 1929, it was offered in the form of granules for molding. It was the first thermoplastic sufficiently stable to be melted without excessive decomposition, and it was the first thermoplastic to be injection molded. Cellulose acetate butyrate plastic became a commercial product in 1938 and cellulose propionate plastic followed in 1945. The latter material was withdrawn after a short time because of manufacturing difficulties, but it reappeared and became firmly established in 1955.

Since the cellulose esters CA, CAB, and CP are chemical raw materials, the word "plastic" was used in the preceding paragraph to differentiate the product from the raw material. The cellulose ester plastics are commonly called simply "acetate," "butyrate," and "propionate," and these names will be used in the text that follows.

Commercial Production

In the manufacture of cellulose ester plastic, the appropriate ester is blended with plasticizer and other additives, such as stabilizers, ultraviolet inhibitors, dyes, and pigments, commonly in a large sigma-blade mixer. The mixture thus obtained is heated to its softening temperature and kneaded until it is homogeneous. This is done on hot milling rolls, in a compounding extruder, or in a Banbury mixer. The molten mass of plastic that results is formed into small rods or strips that are then cut into cylindrical or cubical pellets, which ordinarily have dimensions of about $\frac{1}{8}$ inch (3 millimeters). See Fig. 2.

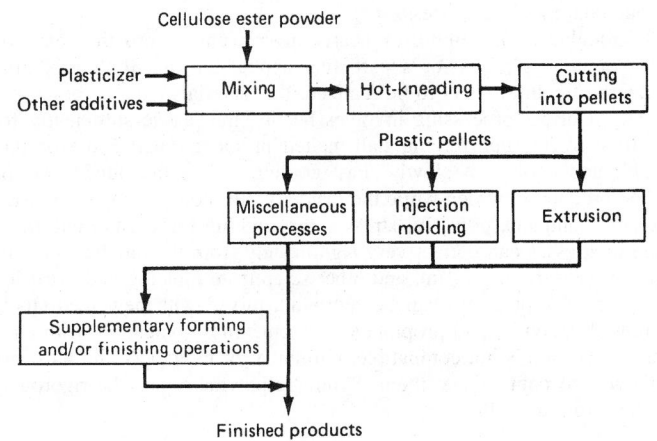

Fig. 2.   Manufacture of cellulose ester plastics

Properties

The concentration of plasticizer in a cellulose ester plastic determines its "flow temperature," which in turn determines its "flow designation," as defined by ASTM Method D 569. Flow designations range from various degrees of hardness (H4, H3, H2, H) through medium-hard (MH), medium (M), and medium-soft (MS) to various degrees of softness (S, S2, S3, S4, etc.). At any given flow designation, the characteristics of the plastic will vary somewhat with the identity of the plasticizer used. Some plasticizers, for example, give very hard materials, some give very low water absorption, and some permit unusual ease of processing. The flow temperature is used for quality control and the corresponding flow designation is a part of the purchase specification when material is ordered.

Depending upon plasticizer content, cellulose ester plastics range from soft, extremely tough materials to hard, strong, stiff compositions that still retain a considerable degree of toughness over a wide range of temperatures. They are basically transparent and virtually colorless, which makes it possible for them to be manufactured in almost any desired transparent, translucent, or opaque color. They are resistant to water and aqueous salt solutions, but they are attacked by aqueous solutions that are strongly acidic or basic. They resist several types of organic solvents, such as ethers and aliphatic hydrocarbons, but they are dissolved or swollen by strongly polar liquid organic compounds such as aromatic hydrocarbons, chlorinated hydrocarbons, ketones, and esters. The susceptibility of the plastics to attack decreases as the molecular weight of the attacking compound increases.

All three cellulose ester plastics are available in formulations that meet the regulatory requirements for use in contact with food.

Butyrate and propionate are available in special formulations for continuous use outdoors, where they generally remain useful for several years. Formulations other than the special outdoor-type materials should not be used in this manner. Acetate formulations are not suggested for outdoor use, as CA does not respond well to the addition of protective compounds.

Although acetate, butyrate, and propionate resemble each other in many ways, there are a number of significant differences among them. Butyrate and propionate are generally easier to process than acetate, and this factor, also, subtracts from the price advantage of acetate. Acetate is available in very hard flows, so it can be obtained with higher stiffness, hardness, and tensile strength than can the other two cellulose ester plastics. Butyrate is available in the softest flows, so it can be obtained in the toughest and easiest-processing formulations. Butyrate and propionate are generally considered to be tougher than acetate, even though in some instances the measured impact strengths may be similar. Butyrate retains its toughness better than does propionate at low temperatures. Butyrate and propionate use higher-boiling, less-water-soluble plasticizers than does acetate, which leads to better retention of plasticizer by butyrate and propionate when exposed to elevated temperatures or to the leaching action of water. Better plasticizer retention, in turn, leads to better permanence characteristics in the plastic—i.e., smaller changes in dimensions and properties with time.

Processing

The organic cellulose ester plastics are versatile materials and can be processed by almost any hot-processing technique used for thermoplastics. The principal techniques for all three plastics are injection molding and extrusion. Blow molding is also possible. Butyrate and propionate powder are used in fluidized-bed and electrostatic coating processes, as well as in the rotational molding process.

The toughness of cellulosics nearly always enters into the selection of one of these plastics for a particular application, but if the potential toughness of cellulosics is to be realized, the materials must be processed correctly. Correct processing involves heating the plastic sufficiently for it to flow freely (not forcing half-melted plastic through a die or into a mold) and cooling it slowly. Fast cooling causes the outside of the finished product to harden while the inside is still molten. When the inside cools and contracts, powerful stresses form within the plastic, and these frozen-in stresses can detract very significantly from the toughness of the plastic. Sprues, runners, trim, and other scrap from molding and extrusion operations, if kept clean, can be reground, mixed with new feed stock, and reused. Butyrate and propionate are sometimes compatible with each other, but acetate is not compatible with either of the others and must not be allowed to contaminate them. Synthetic polymers must be rigorously excluded from all cellulosics.

R. P. RICH
Eastman Chemical Products, Kingsport, Tennessee

[*Neither Eastman Kodak Company nor its marketing affiliates shall be responsible for the use of this information, or of any product, method or apparatus mentioned, and you must make your own determination of its suitability and completeness for your own use, for the protection of the environment, and for the health and safety of your employees and purchasers of your products. No warranty is made of the merchantability or fitness of any product; and nothing herein waives and of the Seller's conditions of sale.*]

CELLULOSICS (Applications). The cellulose ester plastics described in the prior article are suitable for a wide range of uses because of their exceptional balance of properties. Cellulose acetate and cellulose acetate butyrate are widely used for making tool handles because of their toughness, torque strength, and machinability for obtaining a good grip. Tool parts can be solvent vapor-polished to a smooth finish.

Acetate and propionate are widely used in eyeglass frames because their properties are well adapted to the post-finishing operations needed. Cellulosics are frequently used for a variety of home furnishing products, such as table edging and venetian blind wands, and fixtures. The clarity of the finished plastic is desirable for such applications.

Increasing numbers of retail goods, such as small hardware pieces, electronic parts, and some apparel, including T-shirts, underwear, and pantyhose, are packaged in cellulosic tubes. Because of their clarity and toughness, butyrate and propionate are commonly used. Tooth and hair brushes are commonly made of butyrate, again where clarity and toughness are required. Other major uses for cellulosics include extruded tape, audio tape, pressure-sensitive tape, gold-stamped foils, and face shields. Printed signs for interior use frequently are made from butyrate sheet. When coated with a weather-resistant formulation, butyrate is often used for heavy-gage outdoor signs.

In the automotive industry, propionate and butyrate find many over-foil uses, where the plastics are coated over foil strips for vehicle trim.

CELSIUS. Temperature scale proposed by Swedish astronomer Anders Celsius in 1742. A mixture of ice and water is zero on the scale; boiling water is designated as 100 degrees. A degree is defined as one hundredth of the difference between the two reference points, resulting in the original term, "centigrade" (100th part). To convert Celsius to Fahrenheit: multiply the Celsius temperature by 1.8 and add 32 degrees. $F = 9/5\ C + 32$. To convert Fahrenheit to Celsius: subtract 32 degrees from the Fahrenheit temperature and divide the quantity by 1.8. $C = (F-32)/1.8$. See also **Units and Standards**.

CELSIUS, ANDER (1701–1744). Celsius was a Swedish astronomer, mathematician, and physicist. He is best known for his development of a temperature scale based on the decimal system with fixed points separated by one hundred degrees. He proposed all scientific measurements of temperature should be based on the boiling point and the freezing point of water.

Celsius was a professor of astronomy at Uppsala University. He is noted for trying to determine the magnitude of the stars in the constellation Aries. He published 316 of his observations on the aurora borealis.

See also **Celsius**; and **Heat**.

J.M.I.

CEMENT. Cement is a finely powdered substance that possesses strong adhesive powers when combined with water. Gypsum plaster (see also **Calcium**), common lime, hydraulic limes, Puzzolan, natural and Portland cements are a few of the materials which are used for cementing purposes.

Portland cement, [CAS: 68475-76-3], which is the most important of these materials since it is a basic ingredient of concrete, was first manufactured in England in the early part of the nineteenth century. It derived its name from the fact that this newly discovered cement resembled a building stone that was quarried near Portland, England.

There are three fundamental stages in the process of manufacture of Portland cement, namely, (1) preparation of the raw mixture, (2) production of the clinker, (3) preparation of the cement. Whether the process used is wet or dry, the raw materials are selected, analyzed, and mixed so that, after treatment, the product, or clinker, has a desired, narrowly specified composition. A factory analysis of slurry, where the wet process is in use, is as follows: calcium oxide 44%, aluminum oxide 3.5%, silicon oxide

14.5%, ferric oxide 3%, magnesium oxide 1.6%, loss on ignition about 33% (largely carbon dioxide), showing that the composition of the resulting burned clinker is essentially a calcium aluminosilicate. The system calcium oxide-aluminum oxide-silicon oxide has been determined by Rankin and co-workers. In some places the composition of the rock is practically of the desired composition, and in other places clay and limestone are mixed in the desired proportions.

The raw mixture is heated in a continuously operated, long, almost horizontal, slowly rotated furnace or kiln at a high temperature. The temperature is regulated so that the product consists of sintered but not fused lumps. This is clinker. Too low a temperature causes insufficient sintering, and too high a temperature results in a molten mass or glass, the product in either of these cases being valueless for cement purposes. Clinker is unaffected by water, and may be stored indefinitely without detriment. In 1824, Joseph Aspin, an English bricklayer, took out a patent for the manufacture of an improved cement, which he called Portland cement. The cement thus produced was not what is known now as Portland cement as the temperature of burning was not sufficiently high. The value of burning at a temperature sufficiently high to cause incipient fusion was soon afterwards discovered.

In order to obtain the desired setting qualities in the finished cement, there is added to the clinker about 2% of gypsum (calcium sulfate, $CaSO_4 \cdot 2H_2O$), and the mixture is pulverized very finely. For every ton of Portland cement shipped, over two and one-half tons of raw materials *and* cement clinker must be ground very finely. See Table 1.

The wet process for making cement is shown in the flowsheet in Fig. 1. Selective quarry of raw materials for specific cement requirements is based upon chemical data. A television receiver, mounted on the quarry control panel, oversees these operations. Rocks sent to the crushing plant are reduced to pieces smaller than 2 inches. These are dropped onto a reversible shuttle conveyor that stockpiles the materials according to chemical composition in the raw materials storage located above a reclaiming tunnel. From the stockpiles, vibrating feeders then withdraw specified amounts of raw materials and discharge them over a 1230-foot (375-meter) belt conveyor that passes through the reclaiming tunnel. A television camera at the tunnel entrance observes the materials on their way to the next crushing stage. A second television camera supervises the screening operation, and a third camera observes the unloading of crushed material into the storage silos.

Raw materials, withdrawn from the silos, are conveyed to the raw-grinding mill, and water is added. A sonic system which "listens" to the sound of steel balls impacting the inside of the mill indicates the amount and fineness of the material being ground. Feedback from this system controls the feed rate to the mill. Raw-mill discharge (a slurry of about two-thirds solids content) is screened over a 50-mesh screen cloth and pumped to two slurry basins where it is homogenized. Each basin (44 feet high × 85 feet diameter) (13.2 meters high × 25.5 meters diameter) holds about a 3-day supply for the kiln.

The rotary kiln is 510 feet (153 meters) long and is fired by oil. Maximum daily output of the kiln is 1,760 tons (1,584 metric tons) of clinker, equivalent to 9,700 barrels of cement. Operations are automatically controlled, and two television cameras, one at each end of the kiln, observe all material flow.

TABLE 1. ANALYSIS OF MATERIALS USED FOR MANUFACTURE OF LIME AND CEMENT

Material	SiO_2	Al_2O_3	Fe_2O_3 CaO	MgO	CO_2	SO_3	Used for
Limestone	0.36	0.45	54.45	0.54	43.24		Portland cement
Limestone	3.30	1.30	52.15	1.58	40.98		"
Limestone	0.74	0.13	52.94	1.87	43.68		"
Marl	1.78	1.21	49.55	1.30	40.35		"
Cement rock	13.44	6.60	41.84	1.94	32.94		"
Cement rock	11.11	6.31	42.51	2.89	36.57		"
Clay	61.09	26.97	2.51	0.65	—	1.42	"
Clay	58.44	26.50	1.70	1.88	—		"
Cement rock	15.37	11.38	25.50	12.35	34.20		Natural cement
Limestone	0.89	0.47	54.68	0.32	43.44		Lime
Limestone	0.78	0.48	31.15	20.78	45.76		"
Oyster shells	3.30	0.25	52.14	0.25	41.61		"

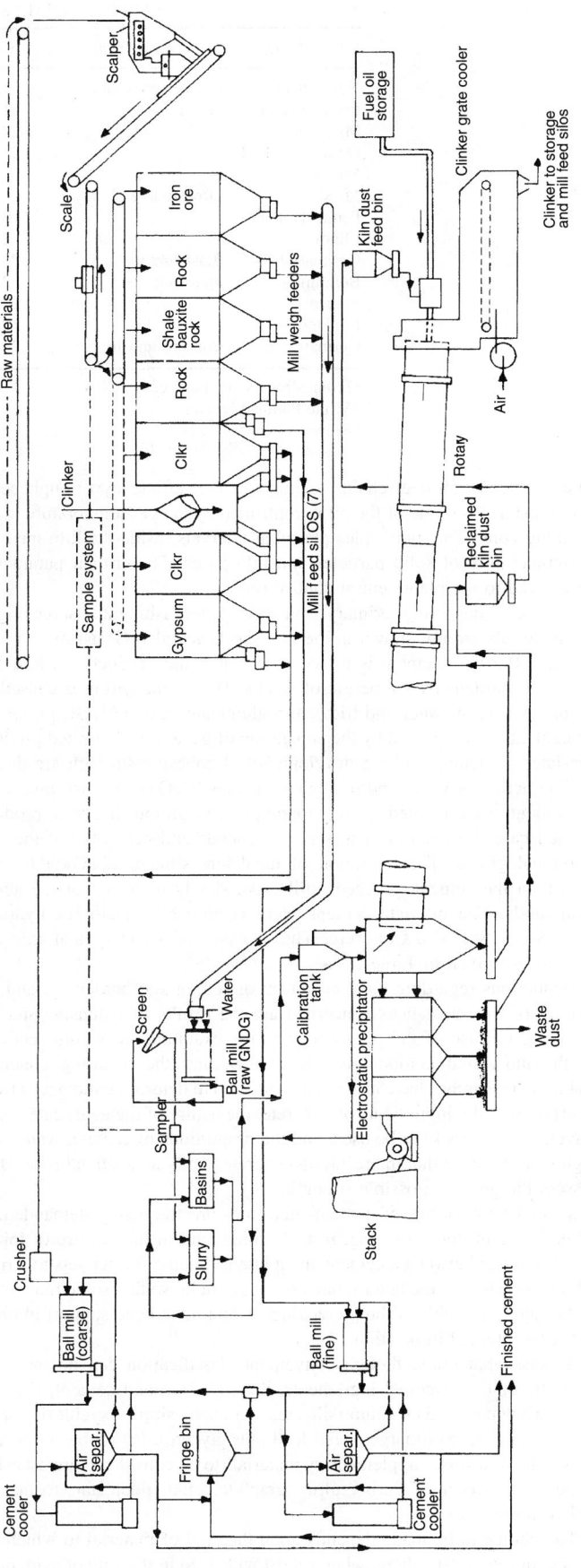

Fig. 1.   Wet process for making cement

In its downward path through the kiln, the slurry passes first into a 91-foot long drying zone (maintained at 2,000°F) (1,093°C) and then into a hotter (above 2,800°F) (1,538°C) calcining and burning section. The drying

TABLE 2. ANALYSIS OF PORTLAND CEMENTS[a]

Where Made	Made from	SiO$_2$	Fe$_2$O$_3$	Al$_2$O$_3$	CaO	MgO	SO$_3$	Loss
New Jersey	Cement rock and	21.82	2.51	8.0	362.19	2.71	1.02	1.05
Pennsylvania	limestone	21.94	2.37	6.87	60.25	2.78	1.38	3.55
Michigan	Marl and clay	22.71	3.54	6.71	62.18	1.12	1.21	1.58
Ohio		21.86	2.45	5.91	63.09	1.16	1.59	2.98
Virginia		21.31	2.81	6.54	63.01	2.71	1.42	2.01
Missouri	Limestone and clay	23.12	2.49	6.18	63.47	0.88	1.34	1.81
Pennsylvania[b]		23.56	0.30	5.68	64.12	1.54	1.50	2.92
Illinois		22.41	2.51	8.12	62.01	1.68	1.40	1.02
Germany	Blast furnace slag	20.48	3.88	7.28	64.03	1.76	2.46	
Belgium	limestone	23.87	2.27	6.91	64.49	1.04	0.88	
France		22.30	3.50	8.50	62.80	0.45	0.70	
England		19.75	5.01	7.48	61.39	1.28	0.96	
Germany[c]	Iron ore and limestone	20.5	11.0	1.5	63.5	1.5	1.0	

[a] From Meade's "Portland cement."

[b] White Portland cement.

[c] Seawater cement.

zone is fitted with steel chains to improve fuel efficiency through better heat transfer. Feed rate of the oil is controlled by the gas temperature in the calcining zone. Thermocouples placed at intervals inside the kiln measure the temperature of solid particles and kiln gases. The critical parameters are relayed to the plant central control room.

The initial quarrying, primary, and secondary crushing and screening of raw materials are not shown in the flowsheet. See also **Gypsum**.

When Portland cement is mixed with water, the product sets in a few hours and hardens over a period of weeks. The initial setting is caused by the interaction of water and tricalcium aluminate 3CaO•Al$_2$O$_3$, present in the cement, accompanied by the separation of gelatinous hydrated product. The later hardening and the development of cohesive strength are due to the interaction of water and tricalcium silicate 3CaO•SiO$_2$, also present in the cement, accompanied by the separation of gelatinous hydrated product. In each case the gelatinous material surrounds and cements together the individual grains. The hydration of dicalcium silicate 2CaO•SiO$_2$, also present in the cement, proceeds still more slowly than that of the above compounds. The ultimate cement agent is probably gelatinous hydrated silica SiO$_2$. See also **Concrete**. The analyses of some typical Portland cements are given in Table 2.

Deductions regarding the mechanism of setting and hardening, and the identity of the substances concerned are the results of extensive studies, involving the use of the microscope in the examination of thin sections, on the individual compounds, the clinker, and the resulting concrete. Elaborate researches have also been conducted to determine the best way of incorporating the ingredients of concrete, the nature of the aggregate (sand, gravel, crushed rock) to be used, and the proportions of cement, water, and aggregate, in order that the resulting concrete, really an artificial rock, shall possess the greatest possible strength.

Special applications of cements and lutes are frequently demanded, as for example in floor covering, in tank lining, and in the closure of joints. The difference between a cement and a lute is that the former sets to a rigid solid mass whereas the latter retains some plasticity so that some movement of the lute is possible without cracking. A lute must have support in order that it be retained in position.

A somewhat crude though convenient classification can be made on the basis of the principal ingredients, thus, (1) Portland cement, (2) high alumina cement, (3) sodium silicate, (4) magnesium oxychloride plus copper powder, (5) litharge or red lead plus glycerol, (6) rubber latex, and (7) synthetic resins. Supplementary materials to be considered are asbestos, white lead, plaster of Paris, sulfur, graphite, sand, pitch, tar, rosin, and boiled linseed oil.

The choice to be made depends upon the kind of material to which the cement or lute must adhere; what it must withstand in the way of acid, base, sulfate, or organic liquid; also what temperature is involved; and finally the matter of resistance to vibration and shock.

Portland and high alumina cements do not withstand acids but are resistant to bases. High alumina cement attains its maximum strength more quickly than Portland, and has the extra advantage that it withstands solutions of sulfates.

Sodium silicate cement does not withstand bases, but is resistant to acids except hydrofluoric. This cement sets to a very rigid solid, so that when subjected to mechanical shock or to temperature change it is liable to crack.

A cement containing 90% magnesium oxychloride and 10% copper powder is strong, resistant to abrasion, and can be bonded to Portland cement.

Rubber latex cement withstands dilute acids and dilute bases, and adheres well to ceramic materials such as stoneware. This cement remains somewhat pliable, thus resisting mechanical shock and temperature change. Organic liquids in general attack this cement.

Synthetic resin cements withstand hydrochloric acid, dilute nitric acid, dilute sulfuric acid, and dilute bases, and are frequently more resistant to organic liquids than is rubber latex cement. The adherence to ceramic materials is good, and the liability to cracking less than for sodium silicate cement.

As for the miscellaneous ingredients mentioned, some are used as fillers or extenders as in the case of sand, and some are used in their own right as when pitch, tar, rosin, molten sulfur, or packed asbestos can be used. See also **Adhesives**.

Additional Reading

Brandt, A.M.: *Cement-Based Composites; Material, Properties and Performance,* Chapman & Hall, New York, NY, 1994.

Derucher, K.N.: *Durability of Concrete,* John Wiley & Sons, Inc., New York, NY, 2000.

Gani, J.: *Cement and Concrete,* Chapman & Hall, New York, NY, 1997.

Hewlett, P.C.: *Lea's Chemistry of Cement and Concrete,* John Wiley & Sons, Inc., New York, NY, 1997.

Odler, I.: *Special Inorganic Cements,* Routledge, New York, NY, 2000.

Peray, K.E.: *Cement Manufacturer's Handbook,* California Historical Society, San Francisco, CA, 1998.

Staff: *Portland Cement Plaster Stucco Manual,* Portland Cement Association, Skokie, IL, 1996. http://www.portcement.org/index.asp

Staff: *Annual Book of ASTM Standards, 1997: Cement; Lime; Gypsum,* American Society for Testing and Materials, West Conshohocken, PA, 1997. http://www.astm.org/

Struble, L.J.: *Cement Research Progress 1993,* American Ceramic Society, Westerville, OH, 1997. http://www.acers.org/

CEMENTITE. Iron carbide, Fe$_3$C, is a compound present at room temperature in nearly all iron-carbon alloys such as steel and cast iron. It is very hard and brittle and weakly magnetic and has an orthorhombic *crystal* structure. In commercial steels, the chemical composition of cementite is usually changed by the presence of manganese and similar carbide forming elements in the steels, which replace iron atoms in the compound. Cementite is a metastable phase and, under the proper conditions, decomposes to form carbon (graphite) and iron. In ordinary steels, this decomposition rarely occurs. Most cast iron, however, contains graphite. This is because the higher silicon content of the cast iron makes cementite less stable. See also **Iron Metals, Alloys, and Steels**.

CENTER FOR THE HISTORY OF CHEMISTRY (CHOC). In 1982 the Center for the History of Chemistry (CHOC) was launched as a pilot project of the University of Pennsylvania and the American Chemical

Society (ACS). In 1984 the American Institute of Chemical Engineers (AIChE) became the third sponsor.

In 1987 the Center was incorporated as a not-for-profit organization, now named the National Foundation for the History of Chemistry, by joint action of the ACS and AIChE. The Foundation rented space in a new building at the University of Pennsylvania and restructured its activities into those of the Beckman Center for the History of Chemistry (established 1987) and the Othmer Library of Chemical History (established 1988).

In 1992 the Chemical Heritage Foundation (CHF) assumed its present name to better reflect the interdisciplinary nature of the sciences and industries it serves and the widening public scope of its activities. In 1995 CHF purchased its permanent home, the First National Bank building at the Independence National Historical Park in Philadelphia, and moved into its new location a year later. Today CHF enjoys the endorsement and support of 29 affiliated organizations.

It is located at 315 Chestnut Street, Philadelphia, PA 19106 www.chemheritage.org.

CENTRIFUGATION. A separation technique based upon the application of centrifugal force to a mixture or suspension of materials of closely similar densities. The smaller the difference in density, the greater is the force required. The equipment used (centrifuge) is a chamber revolving at high speed to impart a force up to 17,000 times that of gravity (much higher in the *ultracentrifuge*). The materials of higher density are thrown toward the outer portion of the chamber, while those of lower density are concentrated at or near the inner portion. A common cream separator is a type of centrifuge in which the flow is continuous. The centrifugal force throws the heavier milk into a different chamber from the lighter cream. Many hundreds of products, particularly in the chemical and food processing industries, require centrifugation at some stage in their manufacture. For example, important applications are found in the beet sugar and sugar cane industries, in the processing of corn (maize) products, in soybean processing, and in milk and dairy products manufacture.

Basic Principle of Centrifuging

The force acting on a particle within a centrifugal field is defined by Newton's fundamental force equation: $FM = ma$. Acceleration acting upon the particle, directed toward the center of rotation is $a = rw^2$. Therefore, the centrifugal force acting on the particle is $F = mrw^2$, or expressed as multiples of gravity,

$$F = 14.2 \times 10^{-6} DN^2 \quad (D \text{ in inches})$$
$$= 5.59 \times 10^{-6} DN^2 \quad (D \text{ in centimeters})$$

where m = mass of particle, g
$\quad a$ = acceleration, cm/s^2
$\quad r$ = radical distance of a particle in a centrifugal field
$\qquad$ from axis of rotation, cm
$\quad \omega$ = angular velocity, rad/s
$\quad D$ = inner diameter of centrifugal bowl
$\quad N$ = bowl speed, rpm

A particle and a mixture introduced, confined, and rotated within a circular enclosure accelerates as it moves from a neutral center toward the maximum diameter (inner periphery) of the enclosure. Thus, if a mixture is introduced to the center of a 24-inch (61-centimeter) diameter solid-bowl centrifuge rotating at 1500 rpm, the particle will be caused to move at a speed of 32.2 feet (9.8 meters) per second at the center. At the maximum diameter, the particle will have a terminal velocity of 766.8 × 32.2 feet per second, or 24,690 feet (7525 meters) per second. Essentially, the separation occurs at 766.8 × g.

Industrial Designs

Industrial centrifuges are available in a variety of designs. Bowl sizes range from less than 0.5 to over 4 feet (15.2 to 122 centimeters) in diameter. Forces applied generally range from about 500 to 14,200 times the force of gravity. Operating temperatures range up to 260°C or higher; pressures up to about 8.5 atmospheres. Liquid flow rates range from 5 to 500 gallons (19 to 1893 liters) per minute; solid rates from 0.2 to 68 metric tons per hour. Depending upon design, the range of size of particles that can be handled is from 1 micrometer to as large as 0.6 centimeter. Centrifuges often are equipped with filters to provide a highly clarified effluent. The

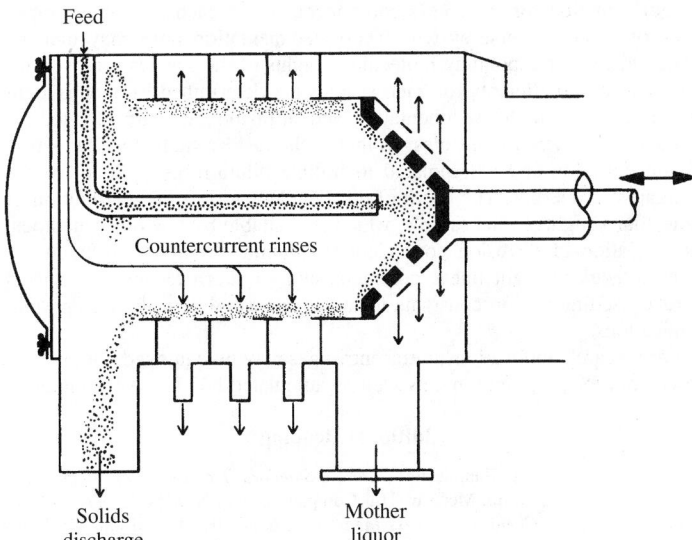

Fig. 1. Continuous horizontal, pusher (reciprocating) filter centrifuge. Double arrow at right indicates motion of ram

majority of industrial centrifuges are designed for continuous operation, although batch designs are obtainable.

One of several types of industrial designs is shown in Fig. 1. This is a continuous horizontal, pusher (reciprocating) filter centrifuge. The feed is continuously introduced through a stationary pipe. The solids settle to the screen surface and form a filter bed through which the liquid passes. The liquid continues through the basket and is discharged from the machine, while the solids, having formed a uniform filter-bed thickness in the cylinder basket, are assisted to discharge by a pusher ram. Washing of cake solids, following the initial dewatering, is a feature of this design.

Among other types are the gas centrifuges, which have been used for the separation of isotopes. In the concurrent type, one or more streams of gas enters at one end of the centrifuge and the partially separated isotopes are removed in two or more streams at the other end. In the countercurrent type, countercurrent circulation is established in a centrifuge either thermally or mechanically. By the circulation of the gas the radial concentration gradient is converted into an axial gradient. The evaporation type operates on volatile liquids, which evaporate within the apparatus. Two streams of vapor are removed from a point near the axis of the centrifuge, having been separated by diffusion through the centrifugal field.

Large industrial centrifuges can cause serious vibration problems. Some centrifuges as found, for example, in separating water from caustic potash slurries may accommodate loads of up to 1 ton. To avoid unduly vibrating an entire building that may house up to twelve of the large centrifuges, some means for damping the vibrations is required. Heavy-duty isolators must be installed. One solution is the use of multistranded steel wire rope. The ropes are coiled into helical assemblies. Under load, the wire rope helices assume an oval shape and deflect in any direction, providing isolation in compression, tension, shear, or roll. In most cases the helices provide up to 90% isolation at high frequencies and amplitudes.

Ultracentrifuge

Centrifuges which operate at very high speeds and which find applications in colloid chemistry and biochemical research are sometimes called *ultracentrifuges*. In research laboratories where proteins, polymers, and other substances with high molecular weights are studied, the ultracentrifuge is effectively used. The sedimentation of large molecules in a strong centrifugal field enables the determination of both average molecular weights and the distribution of molecular weights in various systems. When a solution containing polymer or other large molecules is centrifuged at forces up to 250,000 times gravity, the molecules begin to settle, leaving pure solvent above a boundary, which progressively moves toward the bottom of the cell. This boundary is a rather sharp gradient of concentrations for molecules of uniform size, such as globular proteins. For polydisperse systems, the boundary is diffuse, the lowest molecular weights lagging behind the larger molecules. An optical system can be provided for viewing this boundary, and a study as a function of time of centrifuging yields the rate of sedimentation for the single component, or for each of many components of a polydisperse system. These sedimentation rates may then be related to the corresponding molecular weights of the species present after the diffusion coefficients for each species are determined by independent experiments. Both the sedimentation and diffusion rates are affected by interactions between molecules, so that each must be studied as a function of concentration and extrapolated to infinite dilution, as is done for the colligative properties. The result of this detailed work is the distribution of molecular weights in the sample, which is available by few other methods. Extrapolation of diffusion coefficients to infinite dilution is difficult for high-molecular-weight linear polymers, and so alternate means are used to relate sedimentation constants to molecular weights in these important applications.

For research applications, ultracentrifuges may be equipped with microprocessors for programming, as well as automated handling of samples.

Additional Reading

Avallone, E.A., and T. Baumiester: *Marks' Standard Handbook of Mechanical Engineers,* 10th Edition, McGraw-Hill Companies, Inc., New York, NY, 1996.

Basta, N.: *Shreve's Chemical Process Industries Handbook,* 6th Edition, McGraw-Hill Companies, Inc., New York, NY, 1993.

Craig, W.H., R.G. James, and A.N. Schofield: *Centrifuges in Soil Mechanics,* A.A. Balkema Publishers, Rotterdam, Netherlands, 1998.

Erich, F.F.: *Handbook of Rotordynamics,* 2nd Edition, Krieger Publishing Company, Melbourne, FL. 1998.

Leung, W.W.: *Industrial Centrifugation Technology,* McGraw-Hill Companies, Inc., New York, NY, 1998.

Lide, D.R.: *CRC Handbook of Chemistry and Physics,* 84th Edition, CRC Press, LLC., Boca Raton, FL, 2003.

Perry, R.H. and D.W. Green: *Perry's Chemical Engineers' Handbook,* 7th Edition, McGraw-Hill Companies, Inc., New York, NY, 1997.

Rousseau, R.W.: *Handbook of Separation Process Technology,* John Wiley & Sons, Inc., New York, NY, 1987.

Schweitzer, P.A.: *Handbook of Separation Techniques for Chemical Engineers,* 4th Edition, McGraw-Hill Companies, Inc., New York, NY, 1997.

CERAMICS.

Derived from the Greek word *keramos* ("burnt stuff"), ceramics comprise a wide variety of materials that constitute a major industry. The principal facets of the ceramic industry, in order of increasing value of annual production, are: (1) abrasives; (2) porcelain enamel coatings; (3) refractories; (4) whitewares; (5) structural clay products; (6) electronic and technical ceramic products; and (7) glass. Glass accounts for about 45% of all ceramics produced. See also **Glass.**

Porcelain enamels are used to protect and decorate steel and aluminum metals. Actually, they are glasses especially designed to have high thermal expansions to match the base metal and to mature (to become glassy) at temperatures low enough to prevent distortion of the underlying metal sheet. Glazes perform a similar function on ceramic substrates, again, as special glasses matching the thermal expansion of the base and maturing at the desired temperature. Although not relatively large in terms of production value, the preparation of ceramic composites is important and growing. This area covers a variety of combinations, such as sapphire, Al_2O_3, whiskers in metals, metal-bonded carbides used in the machine tool industry, and directionally solidified two-phase ceramic systems. In a system of this type, an oriented fibrous second phase is grown in a primary matrix phase to maximize in a selected direction various important characteristics, such as minimum long term, high-temperature creep. An example of use is in gas turbine rotor blades.

Conventional Ceramics (Structurals)

The essential raw material is clay. Clay is essentially a hydrated compound of aluminum and silicon $H_2Al_2Si_2O_9$, containing more or less foreign matter such as (1) ferric oxide Fe_2O_3, which contributes the reddish color frequently associated with clay, (2) silica SiO_2 as sand, (3) calcium carbonate $CaCO_3$ as limestone. Since clay is formed by the decomposition of igneous rocks, followed by transportation of the fine particles by running water and later deposition of these particles by sedimentation when the flow of water diminishes in speed, the quality of clays shows a wide range. When clay is wet it is plastic and can be shaped according to the desire and skill of the operator. The shape is retained on drying, and subsequent heating produces a coherent, hard mass, which suffers in the process more or less shrinkage and deformation depending upon the composition of the raw materials, and the method and temperature of treatment. Common bricks are made of crude materials without careful regulation of the conditions of treatment.

Bricks and plain clay products possess an earthy surface and fracture, are porous, and the strength depends upon the materials and treatment. Porcelain, on the other hand, possesses a glasslike or vitreous surface and fracture, and is not porous. Porcelain is made by mixing with the clay some powdered feldspar mineral, potassium aluminosilicate ($KAlSi_3O_8$ approximately). At the temperature of firing, feldspar undergoes a gradual change from the crystalline to the glassy state, the rate depending upon the time of heating and the temperature to which it is subjected. The fusion point of feldspar is of the order of 1300°C, whereas that of kaolin (pure clay) is of the order of 1700°C. Subjection of the porcelain raw material to the latter temperature would result in the formation of a glass. But when the temperature used is below the melting point of the clay portion and about the melting point of the feldspar, the latter produces a glass cement which binds together the particles of the former. When ground quartz SiO_2 is added to the original clay mixture the shrinkage of the material in the processes of drying and firing is reduced, the resistance to deformation during firing is increased, and the temperature coefficient of expansion of the product is affected.

The range of clay, feldspar, and quartz, as to the ratios in the mixture and as to individual composition of each (see Tables 1 and 2), as well as the available range of temperature of firing makes possible the production of products of a wide variety of physical structure. There has been proposed an arbitrary line of demarcation, namely that the unglazed product, such as has been described, which absorbs not more than 1% of its weight upon and after immersion in water, shall be termed porcelain, otherwise it shall be called earthenware. Such a nonporous material as porcelain, which includes chinaware, is also distinctly translucent in thicknesses of a few millimeters, whereas earthenware is nontranslucent and somewhat porous.

Materials that are to be glazed are dipped in a slip (the mixture of raw materials and water), dried, and refired. The glaze mixture is made up so that its fusion temperature is lower than that of the body of the ware, and the firing temperature is such that a surface of glass is formed over the body of the ware.

Designs and colors may be placed, as is commonly done, on the glaze and refired, or, as less commonly and more recently done with fine effect, placed directly on the body under the glaze, in which case the glaze when produced covers and protects both the body of the ware and the decoration.

The properties of ceramics depend mainly on how the atoms are arranged and the interatomic bonds they form. The most important bonding force in the crystalline phases in most ceramics is ionic bonding, the metallic atoms losing an outer electron to become positive ions; and the nonmetallic atoms gaining an outer electron to form a negative ion. Ionic crystals are brittle and hard, melt at high temperatures, and have low electrical conductivity at room temperature. Compounds of metals with oxygen ions that are largely ionic are MgO, Al_2O_3, and ZrO_2. Covalent bonding is also found in ceramic crystalline materials. In this case, a pair of electrons is shared by two atoms. Covalent crystals, such as diamond and silicon carbide, have high hardness, high melting points, and low electrical conductivities at low temperatures.

TABLE 1

Raw materials	Chinaware	Earthenware
Total clay, usually blended	46.5%	50.5%
Feldspar	15	13.5
Quartz	36	36
Dolomite	2.5	—
Porosity average	0.5	8

TABLE 2. ANALYSES OF CERTAIN CERAMIC RAW MATERIALS

	Silica	Alumina	Lime	Magnesia	Iron oxides	Soda	Potash
	%	%	%	%	%	%	%
Kaolin	58	29	0.2	0.3	1	1	1
Fire clay	61	26	0.3	0.4	1	1	1
Common brick clay	58	14	7	1.5	4	3	3
Feldspar	71	16	0.3	0.0	0.5	4	7
Quartz	100	—	—	—	—	—	—

The basic building block for the silicate crystal structures is the silicon-oxygen tetrahedron with a silicon atom at the center and four oxygen atoms at the corners. The silicates are classed by the types of bonding existing between the tetrahedra in their crystal structures. In orthosilicates, the tetrahedra are independent of each other. These structures make good refractories because of their high melting points. This group includes the olivine minerals, garnets, zircon, kyanite, and mullite. When the tetrahedra are joined at only one corner (oxygen atom), they form pyrosilicates, which are rare. In metasilicates, the tetrahedra share two corners to form a variety of ring or chain structures. Minerals of this type include the pyroxines, such as spodumene, and the amphiboles, such as asbestos. Sharing three corners, the tetrahedra form disilicates, which exist as sheets or planes, forming such minerals as mica. In the various forms of silica, such as quartz and crystobalite, all four tetrahedron corners are shared.

Most ceramic shapes do not consist of one single crystal, but are composed of numerous crystals joined together to form polycrystalline structures. The characteristics of the grain boundaries between crystals can influence the strength, chemical stability, and electrical properties as much as do the crystalline structures within the individual grains.

Glass is a very important part of ceramics. The glass industry is the largest single element of the entire ceramic industry, and the glassy portions of many ceramic bodies are the bonds that hold many ceramics together. Probably the majority of the ceramics produced are a mixture of crystalline grains and a glassy phase. The glass frequently acts as the bond. This is the basis of the vitrified-grinding wheel industry and much of the structural and whiteware branches of the ceramics industry.

The rare-earth elements have found application in the ceramics field. In one example, a mixture of about 90% yttrium oxide powder and 10% thorium oxide powder is pressed into the desired shape and then sintered at about 2200°C. This heat treatment removes the microscopically small pores from between the powder particles. The result is a single-phase, polycrystalline material with a grain size normally between 10 and 50 micrometers in diameter. Yttrium oxide has a cubic crystal structure, thus light is not scattered at grain boundaries. This property, combined with the absence of a second phase and pores, imparts exceptional transparency (with polishing) to visible and infrared light. The transmission cutoff in the ultraviolet range occurs at 0.24 micrometers and in the infrared range at about 9 micrometers. Although the index of refraction of the ceramic is high, about 1.91 at the sodium D line wavelength, the optical dispersion of the material is very low. See also **Rare-Earth Elements and Metals**.

High-Technology Ceramics

Development of these materials stems from the need for improved performance in terms of thermal, wear, and corrosion resistance, superior electrical insulation properties, high magnetic permeability, and, sometimes, unusual optical properties. Through precise control of composition, particle shape, and particle size distribution, some materials can be engineered and processed to provide specific properties that may be required of a an application for which readily available standard materials are inadequate. The demanding conditions of engines and turbines are frequently mentioned to illustrate applications where complete satisfaction with alloyed metals has not been achieved. In some instances, engineered ceramics can replace higher-cost materials, such as the use of cobalt, in jet aircraft engine applications. In addition to high-performance engines and machines, advanced ceramics are finding uses in the electronics industry for their electrical and optical behavior.

The ceramic materials most successfully applied to date are high-researching monoli purity oxides, nitrides, carbides, and borides in combination with silicon, tungsten, titanium, tantalum, zirconium, zinc, aluminum, and magnesium. Currently, the oxides dominate the market, commanding about 85% of the critical uses. The application of high-technology ceramics in aircraft and some automotive engines includes turbocharger rotors, rocker arms and cam followers, valves, valve guides, and valve seats, pistons, piston pins, piston rings, cylinder liners, and exhaust port liners.

In the United States, as early as 1987, a program known as "Ceramic Technology for Advanced Heat Engines" was established at Oak Ridge National Laboratory. Some of the research targets of this program included: (1) to develop more reliable monolithic ceramics, toughened ceramic composites, and ceramic coatings via improved and unique synthesis and fabrication technology; (2) to achieve reliable ceramic attachments by new and improved joining technology; (3) to understand better the physical and chemical mechanisms that control the friction and wear of ceramic materials and coatings under heat-engine operating conditions; and (4) to identify and explain the mechanisms that control the long-term mechanical reliability of structural ceramics in advanced-engine environments.

The program also addressed the need to develop tough ceramic-matrix composites (CMCs) with much greater resistance to brittle fracture. Early in the program, researchers found that the chemical structure that imparts superior thermal and mechanical properties to ceramics also results in negative attributes, particularly of brittleness, which easily can lead to catastrophic failure.

In addition to research monolithic parts, ceramic coatings also are being investigated. Although there are numerous problems to be solved, including very extensive testing of materials under standard operating conditions, most professionals in the field forecast that monolithic ceramics and ceramic coatings will be found in conventional automotive engines by the early 2000s. The full acceptance of ceramic parts in jet aircraft engines may not occur until well into the 2000s. For example, C.T. Sims (Rensselaer Polytechnic Institute) observed in mid-1991, "The chronicling of the imminent demise of superalloys in gas-turbine engines appears to have been premature. The lack of toughness in monolithic ceramics and CMCs remains a serious obstacle to their use in gas-turbine engines."

Manufacturers of ceramic components, often composites, must commence fabrication with exquisitely pure raw materials, most frequently in the form of powders. Some of the new ceramic powders include the lesser-known elements, such as niobium and hafnium. Ceramic powders fall into three fundamental classes—the carbides, the nitrides, and the oxides. The powders are synthesized by a number of means, including (1) carbothermal reduction; (2) solid-solid, solid-gas combustion; (3) vapor-phase synthesis; (4) laser synthesis; and (5) plasma synthesis. At the late-1980s state of the art, plasma synthesis, an efficient simple and continuous process, appears in the lead. However, because of some of its disadvantages, including high power demand, large capital and operating costs, health hazards, and sometimes low powder yields, among other negative factors, all of the other methods are also being intensively researched. As pointed out by Sheppard (April 1987), in plasma synthesis, the powders are formed in a vapor-phase reaction under a high-temperature gas ($\sim$10,000 K) generated by the plasma. Ultrafine (<100 micrometers), ultrapure powders, often in a metastable high-temperature phase, are produced by rapid quenching of the hot gases. No milling or grinding is usually required. By comparison, the RF-inductively-coupled plasma requires no electrodes and has the ability to inject the reactants axially. The dc plasma, formed by a high current flow between two electrodes, has high thermal efficiency, and higher plasma temperatures, among other factors.

Some of the ceramic powders now being researched or made in quantity by plasma synthesis are indicated in Table 3.

Ceramic materials also are being used in a limited fashion by the chemical process industries for such applications as high-temperature (1,100°C) heat exchangers, as absorption and distillation column packing, catalyst supports, and high-temperature refractory linings.

Ceramic-Fiber Reinforced Metal-Matrix Composites (MMCs)

These engineered materials provide good strength at high temperatures, good structural rigidity and dimensional stability, light weight, and good fabricability. The incorporation of ceramic fibers in these materials greatly increases their endurance at high temperatures.

- Graphite fibers have been incorporated with aluminum, magnesium, lead, and copper for use in satellite, missile, and helicopter structures; storage battery plates, and electrical contacts and bearings.
- Boron fibers have been used in connection with aluminum, magnesium, and titanium for such applications as compressor blades and structural supports, antenna structures, and jet-engine fan blades.
- Borsic fibers have been used with aluminum and titanium for jet-engine fan blades and high-temperature structures.
- Alumina fibers have been incorporated with aluminum, lead, and magnesium for superconductor restraints in fusion power reactors, storage battery plates, and helicopter transmission structures.
- Silicon carbide fibers have been used with aluminum, titanium, and cobalt-based superalloys for high-temperature structures and engine components.

TABLE 3. SOME CERAMIC POWDERS PREPARED BY PLASMA SYNTHESIS

Ceramic compound	Starting materials	Type of plasma used
CARBIDES		
SiC	CH_3SiCl_3	RF/Arc
SiC	$SiO_x + CH_4$	Arc
SiC	$SiH_4 + CH_4$	RF
WC	$W + C/W + CH_4$	Arc
WC	$W_3O + CH_4$	Arc
TiC	$TiCl_4 + CH_4 + H_2$	Arc
TaC	$Ta + CH_4$	RF
TaC	$TaCl_5 + CH_4 + H_2$	Arc
B_4C	$BCl_3 + CH_4 + H_2$	RF
NITRIDES		
Si_3N_4	$SiCl_4 + NH_3 + H_2$	RF/Arc
Si_3N_4	$SiH_4 + NH_3$	Rf
Si_3N_4	$Si + N2/NH3$	RF/Arc
AlN	$AlNH_3$	RF
TiN	$TiCl_4 + N_2 + H_2$	RF
TiN	$Ti + N_2$	RF
ZrN	$Zr + N_2$	RF/Arc
TaN	$Ta + N_2$	Arc
MgN	$Mg + N_2$	Arc
NbN	$Nb + N_2$	Arc
VN	$V + N_2$	Arc
HfN	$HfCl_4 + N_2 + H_2$	RF
BN	$BCl_3 + N_2 + H_2$	RF
OXIDES		
Al_2O_3	$Al/AlCl_3 + O_2$	RF/Arc
Al_2O_3/Cr_2O_3	Al Halide $+ O_2 + CrO_2Cl_2$	RF
SiO_2	$SiCl_4 + O_2$	RF
SiO_2/Al_2O_3	$Si + Al + O_2$	RF
$TiO_2, TiO_2/Cr_2O_3$	$TiCl_4 + O_2 + CrO_2Cl_2$	RF
ZnO, Sb_2O_3, BaO SiO_2, MgO	Oxides	Arc
MgO	$Mg(NO_3)_2(aq)$	Rf
$ZrO_2, ZrO_2/Al_2O_3$	$Zr(NO_3)_2(aq) + Al(NO_3)_3(aq)$	RF
ZrO_2/SiO_2	$Zr(NO_3)_2(aq) + $ Silicone Oil	RF

Information Source: Los Alamos National Laboratory, New Mexico.

Molybdenum and tungsten fibers also have been used with superalloys for a variety of applications.

A process known as *squeeze casting* (solidification of liquid metal under pressure) contributes to producing defect-free castings with improved metallurgical properties. See also **Casting**.

Preforms for squeeze-cast composites are comprised of ceramic fibers, such as Al_2O_3 and SiO_2, which are bound into near-net (finally desired dimensions) shapes.

Some researchers have found that the reinforcement of ceramics with microscopic "whiskers" as opposed to continuous fibers, increases fracture toughness, even though the basic material nominally remains brittle. It has been reported that the mechanisms by which whiskers toughen the composite include both whisker bridging and whisker pullout within the region immediately beyond the tip of a crack (fracture). Toughness appears to be affected by the volume fraction of whiskers, elastic properties of the matrix, relative fraction energy of the fiber/matrix interface, and whisker diameter and strength.

As observed by D. Johnson and J. Stiegler, "Polymer-precursor routes for fabricating ceramics offer one potential means of producing reliable, cost-effective ceramics. Pyrolysis of polymeric metalloorganic compounds can be used to produce a wide variety of ceramic materials." Silicon carbide and silicon oxycarbide fibers have been produced and sol gel methods have been used to prepare fine oxide ceramic powders, such as spherical alumina, as well as porous and fully dense monolithic forms.

D. Johnson and J. Stiegler also observe, "Polysilazines, which contain the (−Si−N) unit, typify the type of product that can be expected to result from this merging of ceramic science with organic- and inorganic-preparation chemistry."

Liquid-Ceramic Process

A process for producing ceramic or ceramic-matrix-composite (CMC) parts via a liquid route, instead of the conventional power method, was announced in mid-1992. Silicon nitride, for example, is produced by reacting a silicon-based liquid with ammonia, removing byproducts through supercritical fluid extraction, and then heat-treating. Carbide ceramics, such as silicon carbide, also can be made by using a hydrocarbon as one of the reactants. Developers of the process claim that the procedure is less damaging to reinforcing fibers for CMC applications. Other advantages include better adaptability of the process to mass production, considerably lower costs than traditional powder processing, and the manufacture of more complex shapes.

As described by A.J. Klein and T.M. Sullivan, "The Sullivan process uses supercritical fluid extraction to remove carbon and halide impurities from liquid ceramic that has been reacted with ammonia or another reactant under supercritical temperatures and pressures. Supercritical temperatures and pressures are those above the critical point, or the point where two phases (gas and liquid) of a material are continually approximating each other, and become identical to form one phase. In the process, inexpensive raw materials are converted under pressure to amorphous ceramics by reaction, distillation, fluid extraction, and densification. First, by-products of the reacted liquid ceramic are extracted in the supercritical fluid. The fluid then is decompressed and cooled in a separator to precipitate solids. Gases are trapped or condensed and then are recycled. Slight changes in temperature and pressure in the critical region cause large changes in solvent density and dissolving power. The supercritical fluid extraction process has wide flexibility for extractive separation by varying temperature, pressure, choice of solvent used, and entrainers (additives). Otherwise insoluble polymers may dissolve in supercritical fluids to an extent that there may be two to seven orders of magnitude in excess of that predicted by the ideal gas law.

In one application (production of aluminum beverage can lids), a monolithic silicon nitride punch has demonstrated a useful life greater than ten times that of a cemented carbide tool.

Additional Reading

Bach, H. and D. Krause: *Analysis of the Composition and Structure of Glass and Glass Ceramics,* Springer-Verlag Inc., New York, NY, 1999.

DiSalvo, F.J.: "Solid-State Chemistry: A Rediscovered Chemical Frontier," *Science,* **649** (1990).

Green, D.J.: *An Introduction to the Mechanical Properties of Ceramics,* Cambridge University Press, New York, NY, 1998.

Grisaffe, S.J.: "Ceramic-Matrix Composites," *Advanced Materials & Processes,* **43** (January 1990).

Huckins, H.A.: "Apply Advanced Ceramics More Widely in Chemical Processes," *Chem. Eng. Progress,* 57 (February 1991).

Johnson, D.R. and J.O. Stiegler: "Structural Ceramics R & D," *Advanced Materials & Processes,* 55 (September 1990).

King, A.G.: *Ceramic Technology and Processing,* Noyes Publications, Park Ridge, NJ, 1999.

Klein, A.J. and T.M. Sullivan: "Liquid-Ceramic Process Makes Better Components," *Advanced Materials & Processes,* 35 (August 1992).

Lehman, R.L.: "Primer on Engineering Ceramics," *Advanced Materials & Processes,* **31** (June 1992).

Mostaghaci, H.: *Advanced Ceramic Materials: Applications of Advanced Material in a High-Tech Society I,* Trans Technical Publishing, Zuerich, Switzerland, 1997.

Munz, D. and T. Feet: *Ceramics: Mechanical Properties, Failure Behaviour, Materials Selection,* Springer-Verlag Inc., New York, NY, 1999.

Richerson, D.W.: *Magic of Ceramics,* The American Ceramic Society," Westerville, OH, 2000.

Riedel, R.: *Handbook of Ceramic Hard Materials,* VCH Publisher Inc., New York, NY, 2000.

Shackelford, J.F.: *Bioceramics: Applications of Ceramic and Glass Materials in Medicine,* Technomic Publishing Company, Lancaster, PA, 2000.

Sheppard, L.M.: "Vapor-Phase Synthesis of Ceramics," *Advanced Materials & Processes,* 46–51 (April 1987).

Sims, C.T.: "Nonmetallic Materials for Gas Turbine Engines," *Advanced Materials & Processes,* 32 (June 1991).

Staff: "Ceramics," *Advanced Materials & Processes,* 43 (January 1991).

Staff: *Ceramics and Ceramic Composites: Materialographic Preparation,* Elsevier Science, New York, NY, 2000.

Sundaram, S.K., D.F. Bickford, and E.J. Hornyak, Jr.: *Electrochemistry of Glass and Ceramics,* The American Ceramic Society, Westerville, OH, 1998.

Switzer, J.A., Shane, M.J., and R.J. Phillips: "Electrodeposited Ceramic Superlattices," *Science*, **444** (January 26, 1990).

Thevenot, E.: *Ceramics Processing,* VCH Publisher Inc., New York, NY, 2000.

Upadhyaya, G.S.: *Sintered Metallic & Ceramic Materials: Preparation, Properties & Applications,* John Wiley & Sons, Inc., New York, NY, 2000.

Urguhart, A.W.: "Molten Metals Sire MMCs and CMCs," *Advanced Material and Processes,* **25** (July 1991).

Wachtman J.B. Jr.: *Mechanical Properties of Ceramics,* John Wiley & Sons, Inc., New York, NY, 1996.

Zweben, C.: "Metal Matrix Composites," *Advanced Materials & Processes,* **28** (January 1994).

Web Reference

The American Ceramic Society. http://www.acers.org/

CERIUM. [CAS: 7440-45-1]. Chemical element, symbol Ce, at. no. 58, at. wt. 140.12, first in the Lanthanide Series in the periodic table, mp 798°C, bp 3433°C, density 6.770 g/cm³ (20°C). Elemental cerium has a face-centered cubic crystal structure at 25°C. Cerium is the most abundant element of the rare-earth group and is 28th in ranking of the naturally occurring elements in the earth's crust. The element is a silver-gray metal which oxidizes readily at room temperature, particularly in moist air, to form the oxide CeO_2, which is of a pale yellowish-green color. Above 300°C, the element may ignite and burn with a bright red glow. Of the nineteen isotopes of cerium, only four occur in nature, ^{136}Ce, ^{138}Ce, ^{140}Ce, and ^{142}Ce. The thermal neutron-absorption cross section of the element is low. The element has a low toxicity rating. Electronic configuration is $1s^2 2s^2 2p^6 3s^2 3p^6 3d^{10} 4s^2 4p^6 4d^{10} 4f^1 5s^2 5p^6 5d^1 6s^2$. Ionic radius Ce^{3+} 1.034 Å, Ce^{4+} 0.92 Å Metallic radius 1.825 Å. First ionization potential 5.47 eV; second 10.85 eV. Other important physical properties of cerium are given under **Rare-Earth Elements and Metals**. See also **Chemical Elements**.

Cerium was first identified by M.H. Klaproth in 1803 and, independently, in the same year by J.J. Berzelius and W. Hisinger. The element occurs in four source minerals, allanite, bastnasite, cerite, and monazite. Bastnasite, which is a rare-earth fluorocarbonate, is found in southern California. Monazite, a phosphate that contains thorium and the light lanthanides, is distributed widely throughout the world. See also **Bastnasite**; and **Monazite**. Cerium is recovered from the minerals through an extractive process using H_2SO_4, followed by precipitation with oxalic acid, which separates the light lanthanides from thorium, yttrium, and the heavy lanthanides. Cerium metal is produced from its salts, such as CeF_3 or $CeCl_3$, by thermal reduction in a tantalum or molybdenum crucible. Alternative processes include the electrolysis of $CeCl_3$ or CeO_2. The latter compounds are soluble in a complex molten halide flux. The Ce^{3+} is reduced to metal at a molybdenum electrode. The process is carried out at from 800 to 1,000°C.

A major use for CeO_2 is in decolorizing soda-lime container glass. The compound also is used for polishing gemstones and glass, notably precision optical glasses. Cerium is particularly useful in glass that is subject to α-, γ-, and x-radiation, and the impingement of light and electrons because the cerium prevents discoloration that may arise from the presence of Fe(II) by oxidizing the Fe(II) as it is formed to Fe(III). This is an important factor in color television tubes. Cerium dioxide also is used in cathodes, capacitors, phosphors, ceramic coatings, refractory oxides, semiconductors, and photochromic glasses. The compound also is used as a catalyst and as an opacifying agent in porcelain enamels. Because of its low nuclear cross section, CeO_2 may be applied as a diluent in oxide nuclear fuels.

Cerium metal finds wide application in mischmetal, which is a rare-earth metal comprised of 50% Ce, 25% La, 18% Nd, 5% Pr, and 2% other rare earths. This alloy is used in shell linings for military projectiles, as an alloying agent for improving the malleability of ductile iron, and in lighter "flints" where the alloy is compounded with a 30% iron alloy. The pyrophoric and incendiary nature of cerium are evident when cerium-base alloys are machined. Mischmetal also improves the creep resistance of magnesium alloys, the resistance to oxidation of nickel alloys, the hardness of copper alloys, and the strength of aluminum alloys. Both cerium metal and mischmetal are used as getters to remove traces of oxygen in vacuum tubes and equipment. When alloyed with cobalt, cerium is gaining importance as a magnet material. $CeCo_5$, as a permanent magnet material, has properties that exceed those of the alnicos and ferrites. Mixed rare-earth oxides and fluorides containing up to 50% cerium are used as cores for carbon arcs which, for illuminating purposes, have much greater intensity

and color balance. The mixed oxides with cerium also are used as catalysts (petroleum cracking and chemical oxidation reactions) and in a variety of waterproofing agents, fungicides, and polishing materials.

Note: This entry is based upon a prior article furnished by K.A. Gschneidner, Jr., Director, and B. Evans, Assistant Chemist, Rare-Earth Information Center, Energy and Mineral Resources Research Institute, Iowa State University, Ames, Iowa.

Additional Reading

Lide, D.R.: *CRC Handbook of Chemistry and Physics,* 84th Edition, CRC Press, LLC., Boca Raton, FL, 2003.

Staff: *ASM Handbook—Properties and Selection of Nonferrous Alloys and Pure Metals,* ASM International, Materials Park, Ohio, 1990.

CERMET. See **Chromium**; **Nuclear Reactor**.

CERUSSITE. The mineral cerussite, lead carbonate, $PbCO_3$, is orthorhombic with tabular, prismatic and pyramidal crystals, with twinned forms very common. If not in crystal aggregates it may occur in granular or compact masses. Cerussite is very brittle with a conchoidal fracture; hardness, 3–3.5; specific gravity, 6.55 (a heavy mineral); luster, adamantine but may be vitreous to resinous, pearly or even submetallic. Its color is variable, white to gray, grayish-black or blue or green, transparent to translucent.

Cerussite is of secondary origin, being found associated with other lead minerals, and is widely distributed. There are many European and American localities. Fine crystals have been obtained from Phoenixville, Pennsylvania; Joplin, Missouri; Leadville, Colorado; Pima County, Arizona, and Dona Ana County, New Mexico. It is an ore of lead, and frequently carries values of silver. Derived from the Latin cerussa, white lead.

CESIUM. [CAS: 7440-46-2]. Chemical element, symbol Cs, at. no. 55, at. wt. 132.905, periodic table group 1, mp 28.40°C, bp 669°C, density 1.88 g/cm³ (20°C). Elemental cesium has a body-centered cubic crystal structure. Cesium is a silver-white, very soft metal, one of the softest of all metals. The element tarnishes instantly on exposure to air, soon igniting spontaneously with flame to form the oxide. Generally, the element is preserved under kerosene. Cesium reacts vigorously with H_2O, forming cesium hydroxide and hydrogen gas. The element first was identified by Bunsen and Kirchhoff in 1860 through spectroscopic observations. Cesium occurs in nature as the ^{133}Cs isotope. There are 15 radioactive isotopes ^{125}Cs through ^{132}Cs and ^{134}Cs through ^{139}Cs. The half-life of ^{137}Cs is 33 years. This isotope is used as a source of gamma radiation, particularly in radiography and therapy. See also **Radioactivity**. First ionization potential, 3.89 eV; second, 23.4 eV. Oxidation potential $Cs \longrightarrow Cs^+ + e^-$, 3.02 V. Other important physical characteristics of cesium are given under **Chemical Elements**.

The main source of cesium is carnallite $KCl \cdot MgCl_2 \cdot 6H_2O$ which contains a small percentage of cesium compounds. See also **Carnallite**. Cesium also occurs in pollucite (cesium aluminosilicate, 35% Cs_2O) and lepidolite (lithium aluminosilicate). See also **Lepidolite**; and **Pollucite**. In early processes, cesium metal was obtained by the reduction of cesium salts, such as the hydroxide or chloride. In current practice, the metal is produced by electrolyzing the cyanide. The latter compound usually is fused cesium barium cyanide mixture.

The uses for cesium and its compounds are limited. Cesium is used in photoelectric devices because of its high sensitivity to light, finding applications in television, motion picture, radar, and instrumentation equipment. Cesium also has been used in luminescent tubes and screens. Certain processes for the manufacture of synthetic resins, such as chloroprene, use cesium as a catalyst. Some interest has been indicated in cesium as a fuel for ion-propulsion engines of low thrust for spacecraft. Like sodium, cesium also has been considered as a heat-transfer medium for special applications. The function of cesium in time measurement is important. As officially defined in 1967 by the International Bureau of Weights and Measures, the atomic second is equivalent to 9,192, 631,770 oscillations of the atom of ^{133}Cs. This value expresses the ephemeris time (ET) second as closely as practical in terms of an atomic standard. To derive this value, scientists at Great Britain's National Physical Laboratory and the United States Naval Observatory used a dual-rate moon-position camera and a cesium-beam clock.

Cesium forms several solid solutions with rubidium. These alloys are used as *getters* for eliminating residual gases from vacuum tubes and systems. Because of their extreme reactivity in air, the alloys are difficult to apply. For easier handling, cesium can be alloyed with calcium, barium, or strontium. The ternary alloys of cesium, aluminum, and barium or strontium are employed in photoelectric cells. Cesium alloyed with antimony, silver, bismuth, and gold also displays photoelectric properties.

Chemistry and Compounds

Cesium is more electropositive than rubidium (or the lower alkali metals) as is consistent with its position in group 1.

Because of the ease of removal of its single 6s electron (3.89 eV) and the difficulty of removing a second electron (23.4 eV) cesium is monovalent in its compounds, which are ionic.

In its solutions in liquid NH_3, cesium is like the other alkali metals, a powerful reducing agent, so that in such solutions, titrations of cesium polysulfide with cesium are made by electrometric methods. The solubility of cesium salts in liquid NH_3 increases markedly with the radius of an anion (the chloride, CsCl, 0.0227 moles per kg, the bromide, CsBr, 0.215 moles per kg, and the iodide, CsI, 5.84 moles per kg), though the values are less than for the corresponding rubidium compounds.

As reported by Knittle and Jeanloz (1984), cesium iodide, a simple ionic salt at low pressures, undergoes a second-order transformation at 40 gigapascals (400 kilobars) from the cubic B2 (cesium chloride-type) structure to the body-centered tetragonal structure. Also, the energy gap between valence and conduction bands decreases from 6.4 eV at zero pressure to about 1.7 eV at 60 gigapascals, transforming cesium iodide from a highly ionic compound to a semiconductor. The structural transition increases the rate at which the band gap closes, and an extrapolation suggests that cesium iodide becomes metallic near (or somewhat above) 100 gigapascals. It is noted that similar changes in bonding character are apt to occur in other alkali halides at pressures exceeding 100 gigapascals.

As in the case of the other alkali metals, cesium forms compounds generally with the inorganic and organic anions. For a general discussion of these compounds (see also **Sodium**) because the sodium compounds differ principally in their greater extent of hydration and greater number of hydrates. However, cesium coordinates with large organic molecules, such as salicylaldehyde, even though it does not with H_2O.

One respect in which cesium (and rubidium) are outstanding among the alkali metals is the readiness with which it forms alums. Cesium alums are known for all of the trivalent cations that form alums, Al^{3+}, Cr^{3+}, Fe^{3+}, Mn^{3+}, V^{3+}, Ti^{3+}, Co^{3+}, Ga^{3+}, Rh^{3+}, Ir^{3+}, and In^{3+}.

As in the case of potassium and rubidium, cesium forms a superoxide on reaction of the metal with oxygen. The compound is orange in color and paramagnetic because it contains the O_2^- ion with an odd electron in an antibonding orbital, and has the formula CsO_2. On heating, this compound loses oxygen to form black Cs_2O_3, which contains both CsO_2 and Cs_2O_2 (peroxide), which is the product of further heating. A series of suboxides of cesium is known, Cs_7O, Cs_4O (uncertain), Cs_7O_2, Cs_3O and Cs_2O. Moreover the normal oxide, Cs_2O, can be prepared by heating cesium nitrite with metallic cesium. It reacts explosively with oxygen to form CsO_2.

Cesium hydroxide, [CAS: 21351-79-1], CsOH, is the strongest of the five alkali metal hydroxides, as would be expected from its position in the periodic table (francium hydroxide, when prepared, would be expected to be stronger). For the same reason, it has the lowest lattice energy of the five (135.6 kcal per mole).

The most numerous organic compounds of cesium are the oxygen-connected ones, such as the salts of organic acids, and alkoxy and aryloxy compounds (alcoholates, phenates, etc.). Among the carbon-connected compounds, an ethyl cesium, CsC_2H_5, and a phenyl cesium, CsC_6H_5, have been reported.

Later studies by investigators (Alberts et al., 1979) have shown that ^{137}Cs introduced into a watershed is attached to soil particles, which are removed by erosion and runoff. Some of the eroded soil particles comprise the sediments of the catchment basins in the watersheds and act as "sinks" for [137]Cs. Other investigators have reported an almost irreversible fixation of this element in clay interlattice sites in freshwater environments, and, that it is unlikely that this nuclide will be removed from these sediments under normal environmental conditions other than by exposure to solutions of high ionic strength, such as may occur in estuarine environments. Studies of ^{137}Cs have been important because the element can be introduced into a water system from a leak in a nuclear fuel element. These findings are reported in some detail by Alberts et al. in *Science*, **203**, 649–651 (1979).

It has been estimated that the most serious long-term threat to health and the environment as the result of the Chernobyl (former U.S.S.R) nuclear power plant disaster (1986) may come from radioactive cesium, which has a half-life of 33 years. Exposure to cesium-137 could increase the death rate from cancer in western Russia by a maximum of 0.4% over the next 70 years. That would equate with almost 40,000 excess deaths.

Additional Reading

Knittle, E. and R. Jeanloz.: "Structural and Bonding Changes in Cesium Iodide at High Pressure," *Science*, **223**, 53055 (1984).
Lide, D.R.: *CRC Handbook of Chemistry and Physics,* 84th Edition, CRC Press, LLC., Boca Raton, FL, 2003.
Staff: *ASM Handbook—Properties and Selection of Nonferrous Alloys and Pure Metals,* ASM International, Materials Park, Ohio, 1990.
Walker, B.A. and K. Ravikumar: "The Gramicidin Pore: Crystal Structure of a Cesium Complex," *Science*, **183** (July 8, 1988).

CETANE NUMBER.   See **Petroleum**.

CHABAZITE.   The mineral chabazite is a member of that group of hydrous silicates, the zeolites, and corresponds to the formula $CaAl_2Si_4O_{12}\cdot 6H_2O$ with sodium sometimes replacing a part of the calcium. Potassium, barium and strontium may be present in very small amounts. Chabazite is hexagonal, usually in rhombohedrons that tend to resemble cubes. It has a rhombohedral cleavage; is brittle; hardness 4–5; specific gravity 2.05–2.10; luster vitreous; color white to flesh-red; streak white; translucent to transparent. Chabazite is found in the amygdaloidal cavities of basalts often associated with other zeolites. It is occasionally found in such crystalline rocks as syenites, gneisses and schists. Chabazite is a rather common zeolite, being found in many localities in Europe. In the United States it occurs in the Triassic traps of New Jersey and Maryland. The Triassic lavas of Nova Scotia have yielded fine specimens. The name chabazite is derived from the Greek word meaning a precious stone.

CHADWICK, JAMES (1891–1974).   Chadwick was a distinguished English physicist. After receiving a Master's, he went to Germany to work with Hans Geiger. New research here showed that beta particles possess a range of energies up to some maximum value. During World War I he was interned in the Zivilgefangenenlager, Ruhleben. After the war, he was released and returned to England. He worked with Rutherford at the Cavendish Laboratory, Cambridge, on the bombardment of elements with alpha particles. He discovered the neutron, a particle in the nucleus of an atom that carries no electric charge. He received the Nobel Prize in 1935 for Physics. His work was a tool for investigating the atom and also led to the production of a variety of new radioisotopes.

See also **Neutron**.

J.M.I.

CHAIN REACTION (Free Radical).   See **Free Radical**.

CHALCANTHITE.   This mineral of triclinic crystallization is found only as a rare secondary mineral in the oxidized zones of sulfide copper ores within arid regions. It is a hydrous copper sulfate, $CuSO_4 \cdot 5H_2O$, and is a most unstable mineral in moist atmosphere environments, altering readily to a powder-blue dust. The mineral possesses a vitreous luster of deep azure-blue color, ranging from transparent to translucent. Chalcanthite is found in abundance only in Chuquicamata and other arid regions of Chile where it is an important copper ore.

CHALCEDONY.   One of the cryptocrystalline varieties of the mineral quartz, having a waxy luster. It may be semitransparent or translucent and is usually white to gray or grayish-blue or some shade of brown, sometimes nearly black. Light colored clear red chalcedony is known as carnelian; deep reddish brown as sardonyx; a green variety colored by nickel oxide is called chrysoprase. Prase is a dull green. Plasma is a bright to emerald-green chalcedony sometimes found with small spots of jasper resembling blood drops; it is then referred to as blood stone or heliotrope.

Chalcedony and agate are essentially porous, which permits their being dyed various colors by artificial means. Red color is produced by iron nitrate solution; nickel nitrate produces vivid green color; ammonium bichromate produces blue-green; and ferrocyanide salts a vivid blue. The black onyx used extensively in rings is a product of soaking chalcedony in sugar solutions and later in sulfuric acid.

The term chalcedony is derived from the Greek name Chalkedon, a town in the Middle East.

CHALCOCITE.

This mineral is cuprous sulfide, Cu_2S, crystallizing in the orthorhombic system, often in pseudohexagonal forms. Above a temperature of 91°C, chalcocite changes into an isometric form. It has conchoidal fracture; hardness, 2.5–3; specific gravity, 5.5–5.8; metallic luster; color, dark gray to blackish-gray, frequently with bluish-green tarnish. Chalcocite is of widespread occurrence and a valuable copper ore. It seems in some cases to be definitely secondary in origin, in other cases primary. It may have been formed from bornite by the action of alkaline solutions. It sometimes carries valuable amounts of silver.

Among the many European localities might be mentioned Cornwall, England, the Ural Mountains, and Rumania. It occurs also in the Congo, South West Africa, Peru, Mexico, and Alaska. In the United States it is found at Bristol, Connecticut, in fine crystals, Montana, Tennessee, Arizona, Nevada, and California.

The word *chalcocite* is derived from the Greek word meaning "copper." Chalcocite also is known as *copper glance*.

CHALCOPYRITE.

The mineral chalcopyrite (also known as copper pyrites) is a sulfide of copper and iron corresponding to the formula $CuFeS_2$. Its tetragonal crystals are often complex with repeated twinning; massive chalcopyrite is common. It has an uneven fracture; is brittle; hardness, 3.5–4; specific gravity, 4.1–4.3; luster, metallic; color, brass-yellow, may be iridescent from tarnish; streak, greenish-black; opaque. Chalcopyrite is the most common copper-bearing mineral known and it is the most important ore of copper. It is a primary mineral in many igneous rocks and from it a host of secondary copper minerals have been derived.

Among the many localities where fine specimens of this mineral have been obtained might be mentioned: Freiburg, Saxony; Alsace; Rio Tinto, Spain; Cornwall, England; Australia; Chile, Peru, and Bolivia, South America; and in the United States, Ellenville, New York; Chester County, Pennsylvania; Joplin, Missouri; Gilpin County, Colorado; Arizona, Montana, Utah, Nevada, California, New Mexico and Tennessee. In Canada there are notable deposits of chalcopyrite in the Provinces of British Columbia, Ontario, and Quebec. The name *chalcopyrite* is derived from the Greek word meaning "copper," and the word *pyrites*.

CHALK.

Chalk is a soft, porous limestone of white, grayish-white or buff color made up of the minute shells of foraminifera and fragments of cocospheres. It occurs extensively in England and France and less so in the United States.

Chalk consists almost entirely of calcite, formed principally by shallow-water accumulation of (1) calcareous tests of floating microorganisms and (2) comminuted remains of calcareous algae. The most widely distributed chalks are of Cretaceous age, as exemplified by the cliffs on both sides of the English Channel. Although an unaltered deposit, chalk masses may contain nodules of chert and pyrite.

CHARACTERISTIC EQUATION.

1. A class of equations connecting those variables, such as temperature, pressure, and volume, which define the physical condition of a given substance and are called variables of state.

The ideal gas law and the Boyle-Charles law represent approximately the behavior of all gases, but if one wishes to be accurate, some modification of these must be sought which will take into account the differences between individual gases. The best known characteristic equation for gases is that of van der Waals. Using the same notation as for the ideal gas law, this may be written

$$\left(p + \frac{n^2 a}{v_2}\right)(v - nb) = nRT$$

where n is the number of moles of gas, and a and b are constants characteristic of the gas in question. They are very small; if they were zero we should have the ideal gas law. Following are their approximate values for certain gases, where a is expressed in atmosphere (liter/gram-mole)^2 and b in liter/gram-mole:

Gas	a	B
Ammonia	4.170	0.03707
Helium	0.034	0.03412
Hydrogen	0.244	0.02661
Nitrogen	1.390	0.03913
Oxygen	1.360	0.03183

Characteristic equations of this sort are also known as equations of state. See also **Berthelot Equation**.

2. Equations that have solutions, subject to particular boundary conditions, only for certain specific parameters occurring in them. In differential equations, the complete solution includes the characteristic solution and the particular solution. The characteristic solution is obtained from the roots of the characteristic equation, and defines the transient or time response of the system. The particular solution is obtained from the forcing function or input signal and defines the steady-state response.

3. An equation in the linearized theory of hydromagnetics whose solutions show the frequencies and modes of the initial perturbations that will decay or grow exponentially in time for any given system. The solutions to this equation indicate the regions of stability for various hydromagnetic systems. See also **Hydromagnetic Equations**.

CHARDONNET, H. (1839–1924).

A native of France, he has been called the father of rayon because of his successful research in producing what was then called artificial silk from nitrocellulose. He was able to extrude fine threads of this semi-synthetic material through a spinnerette-like nozzle, and the textile product was made on a commercial scale in several European countries. He was awarded the Perkin medal for his work

CHARGE-MASS RATIO.

This term refers to the relationship between the electric charge of a particle and its mass, so important in the physics of electrons, ions, and other electrified bodies of molecular orders.

The earliest information on the subject followed from the researches of Faraday on electrochemical equivalents. From his results it appears that, in the electrolysis of chlorine, for example, 1 coulomb of negative electricity is carried by 0.00037 gram of this element, and hence that the carriers or ions have a charge-mass ratio of about 2,700 coulombs, or 8.1×10^{12} electrostatic units of electricity to the gram. Similarly, 1 coulomb of positive electricity is carried by 0.0000104 gram of hydrogen, which gives about 95,700 coulombs or 2.87×10^{14} electrostatic units to the gram for hydrogen ions. This is 35 times the ratio for chlorine ions. But the atomic masses of hydrogen and chlorine are in the ratio 1:35, which means that if the carriers are atoms, the charge per carrier is the same for both elements. Bivalent elements, on the other hand, carry twice this charge per ion.

When J.J. Thomson applied a magnetic field to a stream of hydrogen canal rays, and then neutralized the resulting deflection by means of an electric field, he was able to calculate the charge-mass ratio of these particles from the curvature of the magnetically deflected stream and the values of the two field intensities. This he found to be either 95,700 coulombs per gram as in the electrolysis of hydrogen, or $\frac{1}{2}$ that value, which indicated that some of the ions were atoms and some were molecules carrying the some charge as the atoms. But when a similar test was applied to the cathode rays in a Crookes tube, the ratio was found to be about 5.303×10^{17} electrostatic units per gram, or about 1,850 times that for hydrogen atoms, whatever the nature of the cathode. We know now that this enormous difference is one of mass, not of charge, and, that these experiments were the first direct revelation of the identity of the electron, the mass of which is now known to be approximately 1/1,836 of the mass of the proton (nucleus of hydrogen atom).

CHARLES LAW.

Although the coefficients of expansion of different solids or of different liquids are notably different, the coefficients of expansion of all gases are nearly the same, namely, about $\frac{1}{273}$ of the volume at 0°C per centigrade degree. The law, stated by Charles in 1787 and independently by Gay-Lussac in 1802 (hence sometimes called Gay-Lussac's law) is not strictly true. Regnault obtained the following values of the volume coefficient for various gases:

Air	0.0036706
Hydrogen	0.0036613
Carbon dioxide	0.0037099
Sulfur dioxide	0.0039028
Carbon monoxide	0.0036688
Nitrous oxide	0.0037195
Cyanogen	0.0038767

None of these is far from $\frac{1}{273} = 0.003663$, which is therefore commonly taken as the expansion coefficient for gases; especially as the value for hydrogen, commonly used in the standard gas thermometer, is very near it. If the pressure as well as the volume is allowed to vary, the behavior of the ideal gas must be expressed by the Boyle-Charles law or the ideal gas law; and the behavior of a real gas by one of the other equations of state. See also **Ideal Gas Law**.

CHARNOCKITE. Charnockite is a granular variety of hypersthene granite, first described from the gravestone of Job Charnock, who founded the city of Calcutta, India, whence the derivation of the name charnockite.

CHELATES AND CHELATION. Chelation compounds are coordination compounds in which a single ligand occupies more than one coordination position. Such ligands are called chelating agents (the word being derived from the Greek meaning *crab's claw*). Thus ethylenediamine, $H_2N-CH_2-CH_2-NH_2$, abbreviated as *en*, forms a $Cr(en)_3^{3+}$ ion having three molecules of ethylenediamine, each occupying two coordination positions.

Ethylenediamine is therefore called a bidentate group, as are many other ligands, such as the β-diketones, which form chelation compounds of the type where M is a metal ion.

Although bidentate ligands are more common, there are polydentate ligands, which occupy more than two coordination positions; ethylenediaminetetraacetic acid is such a polydentate ligand.

While ethylenediamine, and many other chelating agents, form only covalent bonds, there are others which attach by both covalent and ionic bonds. Thus glycine forms with cupric ions (Cu^{2+}) the compound copper bisaminoacetate.

A number of synthetic chelating agents have been developed. They are substances like ethylenediaminetetraacetic acid (EDTA) (Structure **1**) and N-hydroxyethylethylenediaminetriacetic acid (HEDTA) and their salts, usually sodium salts. Many of these compounds and mixtures of these compounds are sold under trademarks.

(1)

Chelating agents are being used in increasing amounts for a number of important purposes. These uses may be put into two important categories: first, artificial trace metal carriers and, second, sequestering agents.

As artificial carriers for trace metals, chelating agents can be used as aids in agriculture by supplying the metals for soils which are deficient in their trace metal content. Both EDTA and HEDTA are adequate iron carriers and EDTA can be used as the carrier for bivalent copper, zinc, manganese, and cobalt. By use of such carriers certain plant deficiency diseases can be controlled. Another example of artificial carrier use is the employment of chelating agents such as the EDTA derivatives or mixtures of such derivatives with pyrophosphates or a mixture of EDTA and the sodium salt of N,N-di(2-hydroxyethyl) glycine in controlling polymerization reactions in synthetic rubber manufacture by the controlled release of trace metal catalysts.

As sequestering agents, chelating compounds have a wide variety of uses, for instance: for water softening in both soaps and synthetic detergents; in textile processing, as in kier boiling operations, where iron, copper, zinc, etc., ions are inactivated, so that discoloration of cloth is prevented; in the stabilization of hydrogen peroxide; and in boiler and heat exchanger cleaning.

Chelation in Biological Systems

Chemical reactions in biological systems are usually mediated by selective catalysts called enzymes. The high efficiencies and stereospecificities achieved require that enzymes have definite and characteristic geometries, whereby specific functional groups coordinated to the metal ion are held in definite spatial positions relative to each other and relative to the substances on which they exert their catalytic effects. The incorporation of metal ions into enzyme structures can assist in the maintaining of a definite geometrical relationship between ionic and polar groups, through the geometric requirements of the coordinate bonds of the metal ion. Certain metal ions may also participate in the catalytic properties of enzymes through ionic and coordinate bonding between the metal ion and electron donating groups of the enzyme and substrate, and through the ability of the metal ion to initiate oxidation-reduction reactions. Because of these chemical and steric effects, coordinated metal ions in the complex compounds that catalyze biological reactions frequently are found. See also **Metalloproteins**.

Most of the metal ions that have biological functions have a coordination number of six, with the donor groups arranged in an octahedral fashion. There are a few metals, such as Mg^{2+} and Zn^{2+}, that frequently coordinate only four donor groups tetrahedrally, and Cu^{2+}, which has four coordinations directed to the corners of a square plane with the metal ion at the center of the plane.

Many simple acid-base reactions are catalyzed by both metal ions and hydrogen ions. Because of small size, the electronic interaction of the hydrogen ion with a substrate is much greater than that of a metal ion. The latter, however, has properties not possessed by hydrogen ions, which are useful in catalysis, i.e., the ability to coordinate a large number of electron donor groups simultaneously, the specific geometric orientation of the coordinate bonds of certain metal ions, and the ability of metal ions to undergo oxidation-reduction reactions. Many of these reactions are models of the more complex catalytic effects that occur in biological systems. Since these reactions of simple coordination compounds aid in the understanding of biological reactions, a few of the more common examples are given in Table 1.

The function of the metal ions in the reactions listed is to attract electrons from the substrate. When this effect takes the form of simple polarization of the functional groups of the substrate, charge variations and electron shifts in these groups facilitate the chemical reactions listed under solvolysis and acid catalysts. When the metal ion removes completely one or more electrons from the substrate, the first step in an oxidation reaction occurs.

TABLE 1. METAL ION AND METAL CHELATE CATALYSIS OF CHEMICAL REACTIONS

Solvolysis and other reactions involving acid catalysis by the metal ion reaction type

Reaction type	Substrate	Catalyst
Solvolysis	Amino acid esters, peptides, and amides	Cu^{2+}, Co^{2+}, Mn^{2+}
	Phosphate esters	La^{3+}, Cu^{2+}, VO^{2+}
	Fluorophosphates	Cu^{2+}, UO_2^{2+} diamine-Cu(II) complexes
	Polyphosphates	Ca^{2+}, Mg^{2+}
	Schiff bases	Cu^{2+}, Ni^{2+}
Transamination	Schiff bases of pyridoxal and α-amino acids	Fe^{3+}, Cu^{2+}, Al^{3+}, Zn^{2+}, Ni^{2+}, Co^{2+}
Decarboxylation	α-Keto polycarboxylic acids (e.g., oxalacetic and oxalsuccinic acids)	Cu^{2+}, Zn^{2+}, Ni^{2+}, Co^{2+}, Mn^{2+}, Fe^{2+}
Acylation	Acetylacetone	Co(III), Rh(III) or Cr(III) chelates of acetylacetone

Catalysis of Oxidation Reactions by Electron Exchange with Metal Ions or Metal Complexes

Reaction	Substrate	Metal Ion or Complex
Oxidation by molecular O_2	Ascorbic acid, catechols, quinoline, salicylic acid	Fe(III), Fe(III)-EDTA, Cu(II), Cu(II)-EDTA, V(IV)
Oxidation by H_2O_2	Phenol, anisole	Fe(II) (Fenton's reagent), Fe(II)-hydroquinone Fe(II)-EDTA-ascorbic acid
Formation of oxygen	Hydrogen peroxide	Fe^{3+}, Fe(III)-phthalocyanine chelate
Formation of disulfides From mercaptides	Thioglycolic acid	Fe^{3+}, Cu^{2+}

This type of catalysis can be accomplished only by metals capable of existing in more than one valence state.

There is a saturation effect in the coordination of a metal ion by donor groups of both the enzyme and the substrate. Therefore, one would expect that the interaction of a free metal ion with the substrate would be greater than that of the metalloenzyme (in which the metal is already partially coordinated). If this were true, the metal ion would have a greater catalytic effect than the metalloenzyme. The reverse is always the case; thus far, no metal ions, or metal complex enzyme models, have been found to approach the catalytic activities of the corresponding enzyme. This high activity of the enzyme is ascribed to the special environment of the substrate around the active site of the enzyme, through which additional binding of the substrate by adjacent organic groups of the enzyme takes place.

The enzyme aconitase, which contains the Fe^{2+} ion at the reactive center, catalyzes the interconversion of citric, isocitric, and aconitic acids. The reaction has been shown to occur through the formation of a single intermediate carbonium ion structure in which the Fe^{2+} ion is always bound to the same donor atoms, while the interconversion of the substrate occurs through the migration of only protons and electrons.

Some of the more important biological reactions that are catalyzed by metal ions are summarized in Table 2.

Chelates in Food Processing

The most commonly used sequestrants in the food field include:

Calcium acetate	Potassium phosphate (di- and monobasic)
Calcium chloride	Sodium acid pyrophosphate
Calcium citrate	Sodium citrate
Calcium gluconate	Sodium diacetate
Calcium phosphate, monobasic	Sodium gluconate
Citric acid	Sodium metaphosphate
Disodium EDTA	Sodium tartrate
Glucono delta-lactone	Sodium thiosulfate
Oxystearin	Sorbitol
Phosphoric acid	Tartaric acid
Potassium citrate	Triethyl citrate

Phosphates have the ability to combine with metal ions, such as calcium, magnesium, iron, and copper, and so render the metals nonactive. Calcium and magnesium are primarily responsible for the hardness of water. The addition of tripolyphosphate or hexametaphosphate will bind these elements and produce soft water. In a similar manner, sequestration is used to soften the skins of fruits and vegetables for faster cooking, and to increase the extraction and recovery of pectin in fruit. Calcium pectinates, which are

TABLE 2. BIOLOGICALLY ACTIVE METAL CHELATES

Metal	Metalloenzyme	Other Biological Functions
Mg	Polynucleotide phosphorylase, ATPase, choline acylase, deoxyribonuclease, acetate kinase, adenosine phosphokinase, fructokinase, glyceric kinase, hexokinase	Chlorophyll
Ca	α-Amylase, aldehyde dehydrogenase, lipase	
V		Green algae, blood of marine worm (ascidian)
Cr		Glucose tolerance factor
Mn	Arginase, carnosinase, prolinase, enolase, isocitricdehydrogenase, 3-phospho-glycerate kinase, glucose-1-P kinase	
Fe	Aconitase, formic hydrogenylase, phenylalanine hydroxylase, peroxidase, catalase, cytochromes	Hemoglobin, ferritin, hemosiderin, siderophilin
Co	Aspartase, acetylornithinase	Vitamin B12
Cu	Lactase, phenolase, tyrosinase, uricase	Ceruloplasmin, cytochrome
Zn	Carbonic anhydrase, carboxypeptidase, alcohol dehydrogenase, glutamic dehydrogenase, acylase	
Mo	Nitrate reductase, xanthine oxidase	

insoluble, are converted into sodium pectinates, which are soluble and readily extracted.

Pyrophosphates are especially effective sequestrants for iron, which catalyzes oxidative darkening of fruits and vegetables. Potatoes, in particular, turn dark after cooking unless the iron in the potato is sequestered. Iron or copper is also responsible for catalyzing oxidative rancidity in meat, poultry, and fish. The effectiveness of combinations of an antioxidant, chelating agents, and polyphosphates in retarding the chemical and organoleptic deterioration of mechanically deboned flounder meat during frozen storage has been investigated. Treatment with polyphosphates, usually applied for moisture binding, will also inhibit rancidity and prolong storage life of various meats. Canned fish frequently develops crystals of a compound known as *struvite*, which appears to the user to be pieces of glass. Although not harmful, these substances are the cause for rejection. Struvite formation is effectively prevented by the addition of a small amount of pyrophosphate to the canned fish.

Additional Reading

Crompton, R.: *Determination of Organic Compounds in Soils, Sediments and Sludges,* Routledge, New York, NY, 2000.

Garnovskii, A.D.: *Direct Synthesis Coordination and Organometallic Compounds,* Elsevier Science, New York, NY, 1999.

Lide, D.R.: *Properties of Organic Compounds, Crcnetbase 2000,* CRC Press, LLC., Boca Raton, FL, 2000.

Sigel, A. and H. Sigel: *Metal Ions in Biological Systems: Interrelations Between Free Radicals and Metal Ions in Life Process,* Marcel Dekker, Inc., New York, NY, 1999.

Walker, M.D. and S. Hitendrah: *Everything You Should Know about Chelation Therapy,* 2nd Edition, Keats Publications, Chicago, IL, 1997.

CHEMICAL ABSTRACT SERVICE REGISTRY NUMBER (CAS).

This universally used number permits use and comparison of data on a given material no matter the synonym with which it might be published. It will, in fact, permit absolute identification of a compound with all of its synonyms. CAS numbers also facilitate extraction of information from computerized data bases. See also **Chemicals (Number of)**.

CHEMICAL AFFINITY.

The entropy production due to a chemical reaction has the form

$$\frac{d_i S}{dt} = \frac{1}{T} \mathbf{A} v \geq 0 \qquad (1)$$

where $\mathbf{A}$ is the chemical affinity and v, the reaction rate. $\mathbf{A}$ is related to the characteristic functions U, H, A, G, and to the chemical potentials μ by the relations:

$$\mathbf{A} = -\left(\frac{\partial U}{\partial \xi}\right)_{S,V} = -\left(\frac{\partial H}{\partial \xi}\right)_{S,p}$$

$$= -\left(\frac{\partial A}{\partial \xi}\right)_{T,V} = -\left(\frac{\partial G}{\partial \xi}\right)_{T,p} \qquad (2)$$

$$= -\sum_i v_i \mu_i$$

when ξ is the extent of reaction and v_i the stoichiometric coefficient.

The basic properties of the affinity $\mathbf{A}$ are that it is always of the same sign as the reaction rate, and that if the affinity is zero the reaction rate is also zero, i.e., the system is in equilibrium.

This definition of affinity is essentially due to De Donder and is called De Donder's fundamental inequality. In the notation used by G.N. Lewis and his school, it is supposed that ξ increases by unity; therefore the relations of (2) are written in the form:

$$\mathbf{A} = -(\Delta U)_{S,V}, = -(\Delta H)_{S,p} = -(\Delta A)_{T,V} = -(\Delta G)_{T,p}, \qquad (3)$$

Note that in this entry, $\mathbf{A}$ is the affinity and A, the Helmholtz function (work function).

See also **Chemical Reaction Rate**.

CHEMICAL COMPOSITION.

Matter is composed of the chemical elements, which may be in the free or elementary state, or in combination. In the former case, as exemplified by iron, tin, lead, sulfur, iodine, and the rare gases, matter commonly exhibits the properties of the atoms of the particular element, including the chemical properties whereby they combine to form molecules. Molecules may (1) be monoatomic; (2) they may consist of atoms of one element only, such as nitrogen or hydrogen molecules (N_2 or H_2), (3) they may be composed of atoms of more than one element, called compounds, which usually have distinctive properties.

The molecular formulas of gaseous compounds are obtained from a study of the composition by elements and the density, by a method introduced by the Italian chemist, Cannizzaro, in 1858. Later, in 1872, in the course of his Faraday Lecture before the Chemical Society (London) on the subject, "Some Points in the Teaching of Chemistry", Cannizzaro stated that "Symbols and formulas, in my opinion, constitute the introduction, preparation, and base of the study of the transformations of matter, which is the true object of our science." The simplest way to understand the *Cannizzaro method* is to arrange in tabular form: (1) the individual gases; (2) the weight in grams of 1 liter (at 0°C, 760 millimeters of mercury pressure) of each gas; and (3) the weight in grams of *each element* present in the above volume (1 standard liter) found by exact analysis (percentage composition by chemical elements using the methods of analytical chemistry). See Table 1.

TABLE 1. CANNIZZARO METHOD OF COMPOUND COMPUTATION

Gas	Grams per Standard liter	Percentage composition by Chemical elements		Grams per standard liter by chemical elements					
				Hydrogen	Oxygen	Carbon	Nitrogen	Sulfur	Chlorine
1. Hydrogen chloride	1.639	{Hydrogen	2.76%}	0.045					1.594
		{Chlorine	97.24}						
2. Ammonia	0.771	{Hydrogen	17.75}	0.137			0.634		
		{Nitrogen	82.25}						
3. Carbon dioxide	1.977	{Oxygen	72.73}		1.438	0.539			
		{Carbon	27.27}						
4. Carbon monoxide	1.250	{Oxygen	57.14}		0.714	9.536			
		{Carbon	42.86}						
5. Methane	0.717	{Hydrogen	25.14}	0.180		0.537			
		{Carbon	74.86}						
6. Ethylene	1.260	{Hydrogen	14.38}	0.181		1.079			
		{Carbon	85.62}						
7. Acetylene	1.173	{Hydrogen	7.75}	0.091		1.082			
		{Carbon	92.25}						
8. Oxygen	1.429	Oxygen	100.00		1.429				
9. Hydrogen	0.090	Hydrogen	100.00	0.090					
10. Nitrogen	1.251	Nitrogen	100.00				1.251		
11. Chlorine	3.214	Chlorine	100.00						3.214
12. Sulfur dioxide	2.927	{Oxygen	49.95}		1.462			1.465	
		{Sulfur	50.05}						
13. Hydrogen sulfide	1.539	{Hydrogen	5.91}	0.091				1.448	
		{Sulfur	94.09}						
14. Nitrous oxide	1.978	{Oxygen	36.35}		0.719		1.259		
		{Nitrogen	63.65}						
15. Nitric oxide	1.340	{Oxygen	53.32}		0.715		0.625		
		{Nitrogen	46.68}						
Minimum weight (approximate)				0.045	0.715	0.538	0.626	1.45	1.60

Note: Data are displayed in this table to illustrate the Cannizzaro method of arriving at the symbol and symbol weight of chemical elements; and the formula and formula weight of chemical compounds.

Careful examination of the figures in the last six columns reveals the experimental fact that (1) in each separate vertical column the figures represent a minimum weight or a small multiple (approximately) of this weight, and (2) the smallest of the six minimum weights is that for hydrogen, namely, 0.045 gram in 1 standard liter of hydrogen chloride gas.

The next step involves changing 0.045 gram of hydrogen to exactly 1.000 gram and finding arithmetically the volume of hydrogen chloride containing this weight (1.000 gram hydrogen). The volume is found to be 22.2 standard liters.

Therefore, 1.000 gram minimum weight of hydrogen is contained in 22.2 standard liters of hydrogen chloride.

Using this standard volume of 22.2 liters, the next step is to ascertain the minimum weight of the other elements in this volume.

Chemical element	Approximate minimum weight in grams of each of the six chemical elements in the standard volume, 22.2 liters
Hydrogen	1
Oxygen	16
Carbon	12
Nitrogen	14
Sulfur	32
Chlorine	35.5

Then, the abbreviation is introduced by the representation:

COMMONLY USED SYMBOL WEIGHTS OF EACH ELEMENT BY THE SYMBOLS

1 gram of hydrogen by the symbol H
16 grams of oxygen by the symbol O
12 grams of carbon by the symbol C
14 grams of nitrogen by the symbol N
32 grams of sulfur by the symbol S
35.5 grams of chlorine by the symbol Cl

By setting up again the second half of the table for the 15 gases, this time for 22.2 standard liters instead of 1 standard liter, the results obtained may be observed in Table 2.

Thus, it is seen, the chemical formulas and formula weights (last column) of 15 gaseous chemical compounds have been arrived at, using the Cannizzaro method, by purely experimental and rational means, involving no theoretical considerations. Extension of the method serves to ascertain the chemical formula of all gases and vaporizable substances. For compounds which are neither gases nor vaporizable, other methods are available. Of these the most used are those of Raoult depending upon the depression of the freezing point or the elevation of the boiling point of a compound dissolved in a given solvent.

It remains to be noted that, when there is no method available for ascertaining the formula weight of a compound, the *simplest* formula, based on chemical analysis and the use of symbol weights of the contained elements, is used, e.g., ferric oxide, Fe_2O_3, ferroferric oxide, Fe_3O_4, ferrous oxide, FeO, cupric oxide (black copper oxide), CuO, cuprous oxide (red copper oxide), Cu_2O. The customary formula of water is H_2O, which is correct at temperatures above $100°C$—actually, liquid water is mainly dihydrol $(H_2O)_2$.

It should be understood from the above discussion that a chemical formula is no chance throwing together of chemical symbols, but represents the results of careful analysis, and the scrutiny and deduction of the most skillful workers in the field. On this score alone, chemical formulas demand the greatest respect in understanding and use.

Symbol weights and atomic weights are used synonymously, as are formula weights and molecular weights. Unless otherwise stated, symbol weights and formula weights are expressed in grams, and the numbers used are those taken from the accepted list of atomic weights. See **Chemical Elements**.

One formula volume of a gas is 22.242 liters. It is necessary to state that actual gases under ordinary conditions show some variation from this value, so that for accurate work the records should be consulted in each case.

Summarizing, the formula "HCl" states that "36.5 grams of hydrogen chloride gas occupies a standard volume of 22.2 liters and is composed of 1 gram of hydrogen element chemically united with 35.5 grams of chlorine element." The reason for the formulas of the simple gases, oxygen, O_2, hydrogen, H_2, nitrogen, N_2, chlorine, Cl_2, is apparent from the general method of deduction. The formula O_2 represents 22.2 liters or 32 grams of oxygen *gas*, whereas O represents 16 grams of oxygen *element* in any substance, or more precisely, 15.9994 grams.

Non-Stoichiometric Compounds

It has become customary in chemical literature to use the formula of a substance as an accepted abbreviation for the name of the substance, especially in cases of frequent repetition.

Up to this point, the discussion in this entry has related to substances that are either elements, or single compounds of elements combined in proportions that can be represented by the ratio of small whole numbers. Such compounds are called *stoichiometric compounds* or *Daltonide compounds* (after the British chemist Dalton). There exist, however, some compounds in which the ratios of the amounts of elements present are not integral. Such compounds are called *non-stoichiometric compounds* or *Berthollide compounds* (after the French chemist Berthollet), and are exemplified by some oxides of the transition elements, by many intermetallic compounds, by the copper sulfide $Cu_{1.7}S$, the copper selenide $Cu_{1.6}Se$ and the cerium hydride $CeH_{2.7}$. Some such compounds vary over a range of composition, depending upon their method of preparation.

In spite of these departures of some compounds from whole number formulas, the fact remains that the great majority of compounds with which the chemist is concerned do contain their constituent elements in integral multiples of their atomic weights. In fact, there is even a further uniformity in the behavior of many of the elements. Thus the great majority of the compounds of the alkali elements (Group 1 in the periodic table) contain equal atomic weight proportions of hydrogen or its equivalent in other elements. Thus the hydrides of this group have compositions corresponding to the formulas LiH, NaH, KH, etc.; the halogen compounds of the group have the compositions, LiF, NaF, KF, LiCl, NaCl, KCl, etc.; while their simple sulfur compounds (since in many of its compounds sulfur combines with two hydrogen equivalents as represented by the formula H_2S) have the compositions Li_2S, Na_2S, K_2S, etc. However, there also exist more complex binary sulfur compounds of these elements, which contain higher proportions of sulfur, so that they combine with sulfur in more than one atomic proportion. Thus this relative combining power, which is called valence, has more than one value for many elements, but is still useful in organizing the data of chemistry. It is discussed at length in the entry on valence, and is explained in structural terms in the entry on molecule.

TABLE 2. DERIVATION OF FORMULAS AND FORMULA WEIGHTS OF GASES

Gas (symbol weight) (symbol)	1 g. H	16 g. O	12 g. C	14 g. N	32 g. S	35.5 g. Cl	Formula of gas	Grams of same gas in 22.2 liters
1. Hydrogen chloride	1					1	HCl	36.5
2. Ammonia	3			1			NH_3	17
3. Carbon dioxide		2	1				CO_2	44
4. Carbon monoxide		1	1				CO	28
5. Methane	4		1				CH_4	16
6. Ethylene	4		2				C_2H_4	28
7. Acetylene	2		2				C_2H_2	26
8. Oxygen		2					O_2	32
9. Hydrogen	2						H_2	2
10. Nitrogen				2			N_2	28
11. Chlorine						2	Cl_2	71
12. Sulfur dioxide		2			1		SO_2	64
13. Hydrogen sulfide	2				1		H_2S	34
14. Nitrous oxide		1		2			N_2O	44
15. Nitric oxide		1		1			NO	30

Note: Derivation assumes data available on the percentage composition by chemical elements of each gas and the symbols and symbol weights of the elements contained.

Radicals

In many chemical compounds there are groups of two or more elements that frequently have the properties of or enter into chemical reaction as a unit. Of those which are of outstanding importance the following are cited:

1. Ammonium NH_4—behaves as a unit in ammonium compounds and in some of these compounds is very similar to potassium K—in potassium compounds.
2. Hydroxyl—OH which behaves as a unit in bases (e.g., sodium hydroxide, NaOH), alcohols (e.g., methyl alcohol, CH_3OH), and phenols (e.g., phenol, C_6H_5OH).
3. Anion-groups of acids, their salts and their esters: Sulfate $>SO_4$, sulfite $>SO_3$, nitrate $-NO_3$, nitrite $-NO_2$, phosphate $\rightarrow PO_4$, perchlorate $-ClO_4$, chlorate $-ClO_3$, chlorite $-ClO_2$, hypochlorite $-OCl$, carbonate $> CO_3$, formate $-CHO_2$, acetate $-C_2H_3O_2$, palmitate $-C_{16}H_{31}O_2$, stearate $-C_{18}H_{35}O_2$, oleate $-C_{18}H_{33}O_2$, oxalate $> C_2O_4$, lactate $-C_3H_5O_3$, malate $> C_4H_4O_5$, tartrate $> C_4H_4O_6$, citrate $\rightarrow C_6H_5O_7$, benzoate $-C_7H_5O_2$, cinnamate $-C_9H_7O_2$, phthalate $> C_8H_4O_4$, salicylate $-C_7H_5O_3$.
4. Alkyl- and aryl-groups of alcohols, phenols, their esters and their alcoholates and phenolates: (a) Alkyl (non-benzenoid)-methyl CH_3-, ethyl C_2H_5-, propyl C_3H_7-, butyl $C_4H_8-$and similar radicals of alcohols; (b) Aryl (benzenoid)-phenyl C_6H_5-, tolyl C_7H_7-, xylyl C_8H_9-, naphthyl $C_{10}H_7-$and similar radicals of phenols.
5. Acyl-groups of organic acids: acetyl CH_3CO-, benzoyl C_6H_5CO-.
6. Miscellaneous radicals, for example, cacodyl $(CH_3)_2As-$, celebrated on account of the investigations by Bunsen (1838).

All of the above radicals are associated with a corresponding radical or element in a compound. While a radical frequently and rather generally enters into chemical reaction as a unit, it is not implied that this is always so, the stability in each case is characteristic of each radical and each reaction in which it is involved. Thus, ammonium hydroxide NH_4OH yields ammonia gas NH_3 and water H_2O at room temperature; ammonium nitrate NH_4NO_3 is decomposed, upon heating, with the accompanying disruption of both the ammonium and nitrate radicals to yield nitrous oxide N_2O gas and water H_2O.

Radicals enter widely into reactions involving electrolytic dissociation of salts, acids, bases in water solution.

Radicals exist most commonly in combination with atoms or other radicals. However, they can be produced "free," and can so exist for a finite period. Even when it is very short, the radical itself is often of great interest in elucidating reaction mechanisms. The first free radical discovered was triphenylmethyl.

Gomberg, by treating triphenylmethyl chloride in carbon dioxide, with zinc, silver, or mercury, obtained the free radical, triphenylmethyl. On dissolving the colorless solid in organic solvents a yellow solution is obtained, and the reactivity (due to unsaturation) of the yellow solution is marked towards oxygen, dissolved iodine, ether. Triphenylmethyl is present in solution in two forms, (1) monomolecular $(C_6H_5)_3C$ yellow, in equilibrium with (2) dimolecular $((C_6H_5)_3C)_2$ colorless. But tribiphenylmethyl $(C_6H_5-C_6H_4)_3C$ occurs only in the monomolecular form, purple. The action of alkali metals on ketones in some cases produces metallic ketyl (Schlenk, 1913) thus:

$$\begin{array}{c} R' \\ \diagdown \\ C-ONa \\ \diagup \\ R'' \end{array}$$

which is a free radical, or contains trivalent carbon as does monomolecular triphenylmethyl. Many other free radicals are known. See also **Free Radical**.

This entry has dealt with two types of chemical composition—elements and compounds. Many materials, including the great majority of those found in nature, are mixtures of compounds and often elements. Practically all biochemical materials and rocks are complex mixtures. Obviously the first step in the determination of the composition of such substances is their separation into the individual compounds, and elements if any, which they contain.

Additional Reading

Alfassi, Z.B.: *The Chemistry of N-Centered Radicals,* John Wiley & Sons, Inc., New York, NY, 1998.
Alfassi, Z.B.: *S-Centered Radicals,* John Wiley & Sons, Inc., New York, NY, 1999.
Baskin, S.I.: *Oxidants, Antioxidants, and Free Radicals,* Taylor & Francis, Dallas, TX, 1997.
Rosen, G.M.: *Free Radicals: Biology and Detection by Spin Trapping,* Oxford University Press, New York, NY, 1999.

CHEMICAL ELEMENTS. A chemical element may be defined as a collection of atoms of one type which cannot be decomposed into any simpler units by any chemical transformation, but which may spontaneously change into other units by radioactive processes. A chemical element is a substance that is made up of but one kind of atom. Of the over 100 chemical elements known, only 90 are found in nature. The remaining elements have been produced in nuclear reactors and particle accelerators. Theoretical physicists do not all agree, but some believe that fission-stable nuclei should exist at atomic numbers 109, 114, and 126. Claims thus far have been made for the discovery, isolation, or creation of elements up to 110. The element with the highest atomic number officially named and entered into the formal table of atomic weight is darmstadtium (Ds) with an atomic number of 110.

The chemical element group numbering system was officially changed in the mid-1980s. The new notations are used throughout this encyclopedia. For clarification of differences between former IUPAC numbers and prior CAS versions, consult table in the entry on the periodic table.

Some of the principal characteristics of the elements are given in Table 1. All of the 110 elements described on this table also are explained in further detail under individual alphabetical entries throughout this volume. The lanthanide series elements also are described further under **Rare-Earth Elements and Metals**. The platinum group metals are further detailed under **Platinum and Platinum Group**; the refractory metals are detailed under **Niobium**. The great versatility of carbon is described under **Organic Chemistry**. In 1817, Johann Dobereiner noticed that strontium had similar chemical properties to calcium and barium, and that its molecular weight fell midway between the two. Additional work led him to discovering a similar trend for another triad of elements: chlorine, bromine and iodine along with lithium, sodium and potassium. In 1829 he proposed the "Law of Triads" which states that, " nature contains triads of elements where the middle element has properties that are an average of the other two members of the triad when ordered by the atomic weight. It was later noticed that these patterns extended beyond that of a triad and eventually led to the "Law of Octaves" discovered by John Newlands in 1863. This discovery originated from Newlands to classifying the current 56 elements into 11 groups based on similar physical properties. He noticed that many pairs of similar elements existed with atomic weights that differed by a multiple of eight. The Law of Octaves states that any given element will exhibit analogous behaviour to the eighth element following it in the 11 groups. Not much later in 1869 did Dimitri Mendeleev publishing a table from elements with symbols and atomic weights on them, arranged in a periodicity of eight. The resulting matrix arrangement is called the Periodic Table. See also **Periodic Table of the Elements**.

The first listing of the elements is generally attributed to Lavoisier in 1789. Of the twenty elements listed, the discovery of five was the result of research conducted by Scheele of Gothenberg. With the development of nuclear physics and the application of these principles to astronomy and cosmology, in recent years the chemical elements have been viewed from new vantage points with much concentration on physical and nuclear characteristics as well as chemical properties.

Origin of the Elements

An excellent summary of progress made in the study of the origin of the elements was given by Penzias in a lecture delivered in Stockholm when he received the Nobel Prize in Physics in 1978. See reference listed.

Another outstanding account was given in the lecture by William A. Fowler (Kellogg Radiation Laboratory, California Institute of Technology) in his acceptance of the 1983 Nobel Prize in Physics (shared with Chandrasekhar). Although life on Earth depends upon the energy of sunlight (originating from nuclear fusion of hydrogen and helium in the solar interior), the sun did not produce the chemical elements found in the earth. Rather, it is theorized by some scientists that these first two elements and their stable isotopes emerged from the first few minutes of the early

TABLE 1. PRINCIPAL CHARACTERISTICS OF CHEMICAL ELEMENTS

Name	Symbol	Atomic number	Atomic weight	Periodic group[b]	Valency (oxidation state)	Specific gravity (20°C)	Melting point, °C	Boiling point, °C	Discovery (year)
Actinium	Ac	89	227[a]	3	3	10.1	1050	2800–3500	1899
Aluminum	Al	13	26.98	13	3	2.699	660	2467	1827
Americium	Am	95	243[a]	Actinides	3, 4, 5, 6	13.67	990–998	2607	1944
Antimony	Sb	51	121.75	15	−3, +3, 5	6.68	630.7	1950	Early
Argon	Ar	18	39.948	18	0	1.78	−189.2[fp]	−185.7	1894
Arsenic	As	33	74.9216	15	−3, +3, 5	(c)	817(24[atm])	613[(sub)]	Early
Astatine	At	85	210[a]	17	1, 3, 5, 7[est]	—	302	337	1940
Barium	Ba	56	137.33	2	2	3.5	725	1640	1808
Berkelium	Bk	97	247[a]	Actinides	3, 4	14[est]	—	—	1949
Beryllium	Be	4	9.012	2	2	1.848	1273–1283	2970	1798
Bismuth	Bi	83	208.891	15	3.5	9.75	271.3	1555–1565	1753
Boron	B	5	10.81	13	3	(d)	2079	2550[(sub)]	1808
Bromine	Br	35	79.904	17	−1, 1, 5	(e)	−7.2	58.8	1826
Cadmium	Cd	48	112.41	12	2	8.65	321	765	1817
Calcium	Ca	20	40.08	2	2	1.54	837–841	1484	1808
Californium	Cf	98	251[a]	Actinides	3	—	—	—	1950
Carbon	C	6	12.011	14	−4, +2, +4	(f)	(f)	(f)	Early
Cerium	Ce	58	140.12	Lanthanides	3, 4	6.66	799	3426	1801
Cesium	Cs	55	132.905	1	1	1.88	28.4	669	1860
Chlorine	Cl	17	35.453	17	−1, +1, 5, 7	(g)	−101	−34.6	1774
Chromium	Cr	24	51.996	6	2, 3, 6	7.2	1837–1877	2672	1797
Cobalt	Co	27	58.9332	9	2, 3	8.832	1495	2870	1735
Copper	Cu	29	63.546	11	1, 2	8.92	1083	2567	Early
Curium	Cm	96	247[a]	Actinides	3, 4	13.5	1310–1370	—	1944
Dysprosium	Dy	66	162.50	Lanthanides	3	8.551	1412	2567	1886
Einsteinium	Es	99	252[a]	Actinides	3[est]	—	—	—	1952
Erbium	Er	68	167.26	Lanthanides	3	9.066	1529	2868	1843
Europium	Eu	63	151.96	Lanthanides	2, 3	5.244	822	1529	1896
Fermium	Fm	100	257[a]	Actinides	—	—	—	—	1952
Fluorine	F	9	18.9984	17	−1	(h)	−219.62	188.1	1886
Francium	Fr	87	223[a]	1	1	2.4	26.28	676–678	1939
Gadolinium	Gd	64	157.25	Lanthanides	3	7.901	1313	3273	1880
Gallium	Ga	31	69.72	13	2, 3	6.0	29.78	2403	1875
Germanium	Ge	32	72.59	14	2, 4	5.32	937	2830	1886
Gold	Au	79	196.967	11	1, 3	19.32	1064.43	3080	Early
Hafnium	Hf	72	178.49	4	4	13.3	2207–2247	4601–4603	1923
Helium	He	2	4.0026	18	0	(i)	−272.2	−268.93	1895
Holmium	Ho	67	164.93	Lanthanides	3	8.795	1474	2695	1878
Hydrogen	H	1	1.008	1	1	(j)	−259.14	−252.87	1766
Indium	In	49	114.82	13	1, 2, 3	7.31	156.6	2078–2082	1863
Iodine	I	53	126.9045	17	1, 3, 5, 7	4.94	113.5	184.35	1811
Iridium	Ir	77	192.22	9	3, 4	22.42	2410	4130	1803
Iron	Fe	26	55.847	8	2, 3, 4, 6	7.874	1535	2750	Early
Krypton	Kr	36	83.80	18	0	(k)	−156.6	−152.3	1898
Lanthanum	La	57	138.91	Lanthanides	3	6.146	918	3464	1839
Lawrencium	Lr	103	260[a]	Actinides	3	—	—	—	1961
Lead	Pb	82	207.2	14	2,4	11.35	327.5	1740	Early
Lithium	Li	3	6.941	1	1	0.534	180.54	1342	1817
Lutetium	Lu	71	174.98	Lanthanides	3	9.841	1663	3402	1907
Magnesium	Mg	12	24.305	2	2	1.74	649	1090	1755
Manganese	Mn	25	54.9380	7	1, 2, 3, 4, 7	(l)	1241–1247	1962	1774
Mendelevium	Md	101	257[a]	Actinides	2, 3	—	—	—	1955
Mercury	Hg	80	200.59	12	1, 2	13.546	−38.84	356.58	Early
Molybdenum	Mo	42	95.94	6	2, 3, 4?, 5?, 6	10.22	2617	4612	1782
Neodymium	Nd	60	144.24	Lanthanides	3	7.004	1021	3074	1885
Neon	Ne	10	20.179	18	0	(m)	−248.68	−246.01	1898
Neptunium	Np	93	237.0482	Actinides	3, 4, 5, 6	20.25	640	3902	1940
Nickel	Ni	28	58.69	10	2, 3	8.9	1453	2732	1751
Niobium	Nb	41	92.906	5	3, 5	8.6	2458–2468	4742	1801
Nitrogen	N	7	14.0067	15	−1, 2, 3, +1, 2, 3, 4, 5	(o)	−209.86	−195.8	1772
Nobelium	No	102	259[a]	Actinides	2, 3	—	—	—	1957
Osmium	Os	76	190.2	8	3, 4	22.6	3015–3075	4927–5127	1803
Oxygen	O	8	15.9994	16	−2	(p)	−218.4	−182.96	1774
Palladium	Pd	46	106.42	10	2, 4	12.02	1554	2970	1803
Phosphorus	P	15	30.9738	15	−3, +3, 5	(q)	44.1[white]	280[white]	1669
Platinum	Pt	78	195.09	10	2, 4	21.4	1772	3727–3927	1735
Plutonium	Pu	94	244[a]	Actinides	3, 4, 5, 6	19.8[Alpha]	640	3232	1940
Polonium	Po	84	210[a]	16	2, 4	9.3[Alpha]	254	962	1898
Potassium	K	19	39.098	1	1	0.86	63.3	760	1807

(continued overleaf)

TABLE 1. (*continued*)

Name	Symbol	Atomic number	Atomic weight	Periodic group[b]	Valency (oxidation state)	Specific gravity (20°C)	Melting point, °C	Boiling point, °C	Discovery (year)
Praseodymium	Pr	59	140.91	Lanthanides	3	6.773	931	3520	1879
Promethium	Pm	61	145[a]	Lanthanides	3	7.264	1042	3000	1945
Protactinium	Pa	91	231.036	Actinides	4, 5	15.4	−1600	—	1918
Radium	Ra	88	226.025	2	2	5est	700	1140	1898
Radon	Rn	86	222[(a)]	18	0	[(r)]	−71	−61.8	1900
Rhenium	Re	75	186.2	7	4, 6, 7	21.0	3180	5627	1925
Rhodium	Rh	45	102.906	9	3	12.4	1963–1969	3627–3827	1803
Rubidium	Rb	37	85.468	1	1	[(s)]	38.9	686	1861
Ruthenium	Ru	44	101.07	8	3	12.41	2310	3900	1844
Samarium	Sm	62	150.35	Lanthanides	3	7.520	1074	1794	1879
Scandium	Sc	21	44.956	3	3	2.985	1541	2831	1876
Selenium	Se	34	78.96	16	−2, 4, 6	[(t)]	217[Gray]	685[Gray]	1817
Silicon	Si	14	28.086	14	−4, +2, 4	2.3	1408–1412	2355	1824
Silver	Ag	47	107.868	11	1	10.50	961.93	2212	Early
Sodium	Na	11	22.9898	1	1	0.971	97.82	882.9	1807
Strontium	Sr	38	87.62	2	2	2.54	769	1384	1808
Sulfur	S	16	32.064u	16	−2, 4, 6	[(v)]	[(w)]	444.7	Early
Tantalum	Ta	73	180.948	5	5	16.65	2996	5325–5525	1802
Technetium	Tc	43	98.906	7	4, 6, 7	11.5	2172	4877	1937
Tellurium	Te	52	127.60	16	−2, 4, 6	6.25	450	690	1782
Terbium	Tb	65	158.92	Lanthanides	3	8.230	1365	3230	1843
Thallium	Tl	81	204.38	13	1, 3	11.85	303.5	1447–1467	1861
Thorium	Th	90	232.038	Actinides	4	11.7	1750	4790	1828
Thulium	Tm	69	168.93	Lanthanides	3	9.321	1545	1950	1879
Tin	Sn	50	118.69	14	2, 4	[(x)]	231.97	2270	Early
Titanium	Ti	22	47.9	4	2, 3, 4	4.5	1650–1670	3287	1791
Tungsten	W	74	183.85	6	6	19.3	3390–3420	5660	1781
Uranium	U	92	238.03	Actinides	3, 4, 5, 6	18.9	1131–1133	3818	1789
Vanadium	V	23	50.942	5	2, 3, 4, 5	6.1	1880–2000	3380	1830
Xenon	Xe	54	131.30	18	0	[(y)]	−112	−107.1 ± 2.5	1898
Ytterbium	Yb	70	173.04	Lanthanides	2, 3	6.966	819	1196	1878
Yttrium	Y	39	88.9058	3	3	4.469	1522	3338	1794
Zinc	Zn	30	65.38	12	2	7.1	419.58	907	1746
Zirconium	Zr	40	91.22	4	4	6.5	1853 ± 1.0	4377	1789

TRANSACTINIDE ELEMENTS*

Name	Symbol	Atomic number	Atomic weight	Periodic group[b]	Valency (oxidation state)	Specific gravity (20°C)	Melting point, °C	Boiling point, °C	Discovery (year)
Rutherfordium	Rf	104	261[(a)]	4	4	—	—	—	1964
Dubnium	Db	105	262[(a)]	5	—	—	—	—	1967
Seaborgium	Sg	106	266	6	—	—	—	—	1974
Bohrium	Bh	107	264	7	—	—	—	—	1981
Hassium	Hs	108	269	8	—	—	—	—	1984
Meitnerium	Mt	109	268	9	—	—	—	—	1982
Darmstadtium	Ds	110	271	10	—	—	—	—	1994

[a] Mass number of isotope of longest known half-life (or a better known one for Bk, Cf, Po, Pm and Tc).
[b] New Periodic Group notation.
[c] Specific gravity, arsenic: Yellow orpiment form, 1.97; gray metallic form, 5.73.
[d] Specific gravity, boron: Crystalline, 2.34; amorphous, 2.37.
[e] Density, bromine: Gas, 7.59 g/l; liquid, 3.12 g/ml.
[f] Specific gravity, carbon: Amorphous, 1.8 to 2.1; graphite, 1.9 to 2.3; diamond, 3.15 to 3.53 (gem). Melting point, carbon: Natural, 3550°C: graphite sublimes, 3342–3392°C (at 12–13 GPa). Triple point, graphite-liquid-gas, 3577–3677°C (at 10.1 MPa); graphite-diamond-liquid, 3830–3930°C (at 12–13 GPa)
[g] Density, chlorine: 3.2 g/l. Specific gravity (−33.6°C), 1.56.
[h] Density, fluorine: 1.696 g/l (0°C, 1 atm). Specific gravity, 1.108 (at boiling point).
[i] Density helium: 0.1785 g/l (0°C, 1 atm).
[j] Density, hydrogen: Gas, 0.08988 g/l; liquid (−253°C), 70.8 g/l; solid (−262°C), 70.6 g/l.
[k] Density, krypton: 3.7 g/l (0°C).
[l] Specific gravity, manganese: 7.2–7.4, depending on allotropic form.
[m] Density, neon: Gas (0°C. 1 atm), 0.8999 g/l; liquid (at boiling point), 1.207 g/cm^3.
[n] Niobium is sometimes called columbium in the U.S. and a few other countries.
[o] Density, nitrogen: 1.25 g/l. Specific gravity, liquid (−195.8°C), 0.808; solid (−252°C), 1.026.
[p] Density, oxygen (0°C), 1.429 g/l. Specific gravity, liquid (−182.96°C), 1.4.

high-temperature, high-density stage of the expanding Universe (so-called big bang concept). Although it has been theorized that a small quantity of lithium (element 3 in the periodic table) was also produced during the big bang, it has been proposed that the remainder of the Li and all of the beryllium and boron (elements 4 and 5, respectively) were produced by the spallation of still heavier elements in the interstellar medium by the process of cosmic radiation. The fact that these latter elements are relatively rare is consistent with this concept.

Early 1990s studies of very old stars, such as HD140283, brought some surprises. This star, considered so old that it only has about 1% of the oxygen and other heavy elements of the sun, was found by one of the first observations of the Hubbel Space Telescope to contain about 1000 times more beryllium than could be attributed to cosmic radiation. See **Beryllium**.

As pointed out by Fowler, the question remains pertaining to the origin of the heavier elements, namely, those ranging upward from carbon (6) to radioactive uranium (92). The origin of these elements has been attributed to nuclear processes in the interior of stars, particularly in the stars of our own Galaxy and prior to the formation of the solar system (some $4\frac{1}{2}$ billion years ago). The general process is believed to have occurred during a period of some 5 to 15 billion years prior to formation of the solar system and is related to star formation and degradation in our Galaxy during that long period. It is believed that old stars, during their giant stage of stellar evolution, were sources of the heavier elements. Possibly, greater activity occurred during outbursts (*novae*) or stellar explosions (*supernovae*). The current commonly held concept is that the planets of the solar system, including Earth, condensed under the forces of gravitation and rotation from a gaseous solar nebula in the interstellar medium, this medium containing hydrogen and helium (from the big bang) and admixed with the heavier elements synthesized in the earlier generations of galactic stars. It is interesting to note that the abundance of heavy elements in the older galactic stars is less than 1 percent of that found in the solar system. Element building in the manner described is sometimes referred to as the *nucleosynthesis* in stars. It is assumed, of course, that a similar process has occurred and continues to occur in galaxies other than our own, such as the Andromeda Nebula, and that the process is thus universal.

The principal targets of nuclear astrophysics are: (1) an understanding of the energy generation in the sun and other stars at all stages of stellar evolution, and (2) a comprehension of the nuclear processes which caused the relative abundances of the elements and their isotopes in nature. Suess and Urey, in 1956, first systematized the terrestrial, meteoritic, solar, and stellar data that are related to the origins and abundances of the elements. During the interim, these data have been periodically updated by Cameron; major work in the experimental measurement of atomic transition rates required to determine solar and stellar abundances has been contributed by Whaling. Check references at the end of this article. See also Fig. 1(**a**) and (**b**).

The remainder of the Fowler article is of great interest, but in detail that is beyond the scope of this encyclopedia. Topics covered by Fowler include early research on element synthesis; stellar reaction rates from laboratory cross sections; hydrogen burning in main-sequence stars and the solar neutrino problem; synthesis of ^{13}C and ^{16}O and

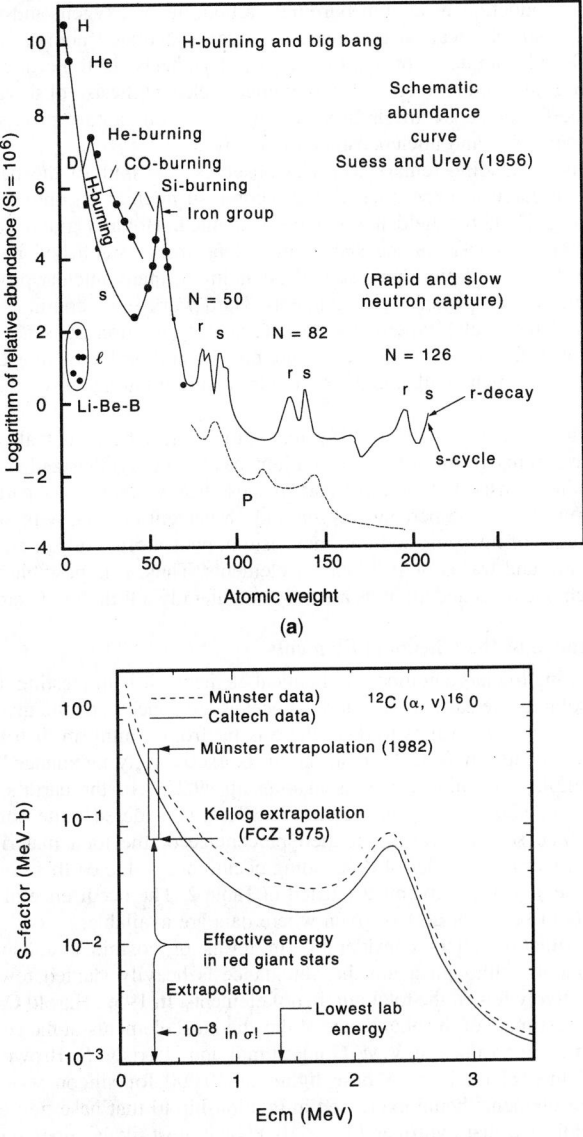

Fig. 1. (**a**) Atomic abundances relative to Si $= 10^6$ versus atomic weight for the sun and similar main-sequence stars. (**b**) Cross-section factor S (in MeV-varns) versus center-of-momentum energy (in MeV) for ^{12}C(α, γ)O Dashed and solid curves are theoretical extrapolations of Münster and Kellog Caltech data by Langanke and Koonin (see references). [*After paper on "The Quest for the Origin of the Elements" by William A. Fowler, presented in December 1983, when author received the Nobel Prize for Physics. Complete article in Science,* **226**, *922–935 (November 23, 1984) and in "Les Prix Nobel en 1983," Elsevier, New York, 1984*]

TABLE 1. (*continued*)

q Specific gravity, phosphorus: White, 1.82; red, 2.20; black, 2.25 to 2.9. 1.14.

r Density, radon: Gas, 9.73 g/l; liquid ($-62°$C), 4.4.

s Specific gravity, rubidium: Solid (20°C), 1.532; liquid (39°C), 1.475.

t Specific gravity, selenium: Gray. 4.79; vitreous, 4.28.

u Atomic weight of sulfur varies slightly because of naturally occurring isotopes 32, 33, 34, and 36, the total possible variation amounting to ±0.003.

v Specific gravity, sulfur: Rhombic, 2.07; monoclinic, 1.957.

w Melting point, sulfur: Rhombic, 112.8°C; monoclinic, 119.0°C.

x Specific gravity, tin: Gray, 5.75; white, 7.31.

y Density, xenon: Gas, 5.89 g/l. Specific gravity (liquid at $-109°$C), 3.52.

Abbreviations used in table:

est	Estimated
sub	Sublimation temperature
atm	Unit of pressure
fp	Freezing point

Note: Terms such as *alpha, gray,* and *white* refer to specific forms of element.

Values for all rare-earth elements furnished by Rare-Earth Information Center, Energy and Mineral Resources Research Institute, Iowa State University, Ames, Iowa.

neutron production in helium burning; carbon, neon, oxygen, and silicon burning; astrophysical weak interaction rates; calculated abundances for $A \lesssim 60$ and comments on explosive nucleosynthesis; isotopic anomalies in meteorites and evidence for ongoing nucleosynthesis; observational evidence for nucleosynthesis in supernovae; neutron-capture processes in nucleosynthesis; and nucleocosmochronology.

In his concluding remarks, Fowler observes; "In spite of the past and current research in experimental and theoretical nuclear astrophysics, the ultimate goal of the field has not been attained. Hoyle's grand concept of element synthesis in the stars will not be truly established until we attain a deeper and more precise understanding of many nuclear processes operating in astrophysical environments. Hard work must continue on all aspects of the cycle: experiment, theory, and observation. It is not just a matter of filling in the details. There are puzzles and problems in each part of the cycle which challenge the basic ideas underlying nucleosynthesis in stars."

Fowler further concludes: "My major theme has been that all of the heavy elements from carbon to uranium have been synthesized in stars. Our bodies consist for the most part of these heavy elements. Apart from hydrogen, we are 65 percent oxygen and 18 percent carbon, with smaller percentages of nitrogen, sodium, magnesium, phosphorus, sulfur, chlorine, potassium, and traces of still heavier elements. Thus, it is possible to say that each one of us and all of us are truly and literally a little bit of stardust."

Abundance of the Chemical Elements

Considering the large number of chemical elements, it is interesting to note that insofar as the earth's crust and the oceans are concerned, and given our knowledge of the cosmos to date, there is far from a uniform distribution of the elements. In fact, the distribution is exceedingly unbalanced with, for example, only nine elements making up 99.25% of the earth's crust. In terms of abundance, many of the materials considered quite common are, in fact, scarce in terms of their percentage of the total materials in the earth's crust. In order of descending occurrence in the earth's crust, all but the very scarce elements are listed in Table 2. The occurrence of these elements in seawater also is given where data are available.

The situation alters considerably in terms of cosmic abundance of the elements, although again the abundance is heavily slanted toward a comparatively few of the total number of elements. In 1952, Harold C. Urey made an estimate of the abundance of the chemical elements in the cosmos, wherein earlier values of V.M. Goldschmidt and Harrison S. Brown were used in the calculations. A base figure of 10,000 for silicon was used, the other elements being expressed in relationship to that base figure. The study indicated that hydrogen (3.5×10^8) is the most likely superabundant element in the cosmos, closely followed by helium (3.5×10^7). Other elements high on the list include oxygen (220,000), nitrogen (160,000), carbon (80,000), and neon (with a range estimated between 9,000 and 240,000). Magnesium, silicon, and iron also are relatively high.

Researchers at California Institute of Technology (Pasadena, California) have been studying the occurrence of various isotopes in the earth's crust as a lead to understanding the mechanics of continental crust formation. A means for determining the time of formation of new crustal segments is paramount to an understanding of how the continental crust was evolved. In this work, they have studied samarium–neodymium and rubidium–strontium isotopic systematics. In a study by Anderson (Seismological Laboratory, California Institute of Technology), more recent estimates of solar composition pose new questions. Compared with earlier findings, the study shows an enrichment of Fe and Ca relative to Mg, Al, and Si. The Fe/Si and Ca/Al atomic ratios are some 30–40% higher than chondritic values as found in chondrules (meteoric stones assumed to be of cosmic origin). Anderson suggests that the new data may require a revision in the estimates of cosmic abundance and the composition of nebula from which the planets accreted-inasmuch as the latter data previously have been based upon chondritic values. If so applied, the earth's mantle would contain about 15% (weight) more of FeO and more $CaMgSi_2O$ than once calculated.

Critical Importance of Certain Elements

The uneven distribution of the chemical elements in the earth's crust causes severe imbalances in their availability for industrial manufacturing uses. Frequently, the largest consuming nations are long distances from elemental sources, as in the case of the United States This clearly indicates the need for the various political entities of the world to stockpile and to project their

TABLE 2. ABUNDANCE OF THE CHEMICAL ELEMENTS (In Grams/Metric Ton)[a]

Element	Terrestrial Abundance	Occurrence in Seawater	Element	Terrestrial Abundance	Occurrence in Seawater
Oxygen	466,000	850,000	Beryllium	6	—
Silicon	277,200	2.98	Praseodymium	5.53	—
Aluminum	81,300	0.01	Arsenic	5	0.003
Iron	50,000	0.01	Scandium	5	4×10^{-5}
Calcium	36,300	404	Hafnium	4.5	—
Sodium	28,300	10,550	Dysprosium	4.47	—
Potassium	25,900	380	Uranium	4	—
Magnesium	20,900	1,290	Boron	3	4.9
Titanium	4,400	0.001	Thallium	3	—
Hydrogen	1,300	108,200	Ytterbium	2.66	—
Phosphorus	1,180	0.07	Erbium	2.47	—
Manganese	1,000	0.002	Tantalum	2.1	—
Fluorine	900	1.27	Bromine	1.62	66
Sulfur	520	894	Holmium	1.15	—
Carbon	320	27.6	Europium	1.06	—
Chlorine	314	19,050	Antimony	1	0.0005
Rubidium	310	0.121	Terbium	0.91	—
Strontium	300	8.1	Lutetium	0.75	—
Barium	250	0.006	Mercury	0.50	3×10^{-5}
Zirconium	220	—	Iodine	0.30	0.05
Chromium	200	5×10^{-5}	Thulium	0.20	—
Vanadium	150	0.002	Bismuth	0.20	0.0002
Zinc	132	0.01	Cadmium	0.15	6×10^{-5}
Nickel	80	0.0005	Silver	0.10	0.0003
Copper	70	0.003	Indium	0.10	0.02
Tungsten	69	0.0001	Selenium	0.09	4×10^{-5}
Lithium	65	0.2	Argon	0.04	0.595
Nitrogen	46.3	0.51	Palladium	0.01	—
Cerium	46.1	0.00038	Gold	0.005	4×10^{-6}
Tin	40	0.003	Osmium	0.005	—
Yttrium	28.1	0.0003	Platinum	0.005	—
Niobium (Columbium)	24	5×10^{-6}	Ruthenium	0.004	—
Neodymium	23.9	—	Tellurium	0.002	0.00001
Cobalt	23	0.0005	Rhodium	0.001	—
Lanthanum	18.3	0.0003	Iridium	0.001	—
Lead	16	0.003	Neon	7×10^{-5}	0.0003
Gallium	15	3×10^{-5}	Radium	13×10^{-6}	996×10^{-13}
Molybdenum	15	0.01	Krypton	9.8×10^{-6}	0.0003
Thorium	11.5	0.0007	Xenon	1.2×10^{-6}	0.0001
Cesium	7	0.0005	Protactinium	8×10^{-7}	0.003
Germanium	7	6×10^{-5}	Actinium	3×10^{-10}	—
Samarium	6.47	—	Polonium	3×10^{-10}	—
Gadolinium	6.36	—			

[a] Presented in order of diminishing terrestrial abundance.

needs many years in advance. Hence, for obvious reasons, these needs are commonly reflected in foreign policies.

Nuclides, Isotopes, and Isobars

A nuclide may be defined as a species of atoms, with specified atomic number and mass number. The term *nuclide* should be used, *not* isotope. Different nuclides having the same atomic number *are* isotopes. Different nuclides having the same mass number are *isobars*.

A comprehensive listing of the nonradioactive nuclides of the chemical elements is given in Table 3. The isotopic abundance and mass number are given. In all but a few instances, the isotopes listed are stable. The elements also have a number of radioisotopes each, the number varying considerably from one element to the next. Extensive tables also listing the radioactive isotopes, giving lifetime, modes of decay, decay energy, particle energies, particle intensities, and thermal neutron capture cross section, among other factors, can be found in the literature.

An important fact pertaining to naturally occurring elements is that many of them consist of several isotopes and that these isotopes are present in nearly all cases in the same proportion by weight. These constant isotopic compositions have enabled scientists over the years to analyze materials by weight, making reference to a table of atomic weights. In fact, for many years the atomic weight of an element was regarded as its most distinguishing characteristic, even though some discrepancies were noticed in the periodic table. It was not until the work of Mosley on characteristic

TABLE 3. THE NUCLIDES (ISOTOPES AND ISOBARS)[a]

Element	Mass No. A	Isotopic abundance %	Element	Mass No. A	Isotopic abundance %	Element	Mass No. A	Isotopic abundance %	Element	Mass No. A	Isotopic abundance %
H	1	99.985		66	27.8	Sn	112	0.96		162	25.5
	2	0.015		67	4.1		114	0.66		163	<25.0
He	3	1.3×10^{-4}		68	18.6		115	0.35		164	28.2
	4	~100		70	0.63		116	14.30	Ho	165	100
Li	6	7.5	Ga	69	60.1		117	7.61	Er	162	0.136
	7	92.5		71	39.9		118	24.03		164	1.56
Be	9	100	Ge	70	20.52		119	8.58		166	>33.4
B	10	18.7		72	27.43		120	32.85		167	22.9
	11	81.3		73	7.76		122	4.72		168	27.1
C	12	98.9		74	36.54		124	5.94		170	14.9
	13	1.1		76	<7.76	Sb	121	57.25	Tm	169	100
N	14	99.62	As	75	100		123	42.75	Yb	168	0.14
	15	0.38	Se	74	0.87	Te	120	0.088		170	3.03
O	16	99.76		76	9.02		122	2.83		171	14.3
	17	0.04		77	7.58		123	0.85		172	>21.8
	18	0.20		78	23.52		124	4.59		173	16.2
F	19	100		80	49.82		125	6.93		174	>31.8
Ne	20	90.8		82	9.19		126	18.71		176	12.7
	21	<0.3	Br	79	50.54		128	<31.86	Lu	175	97.5
	22	8.9		81	49.46		130	<34.52		176	2.5
Na	23	100	Kr	78	0.342	I	127	100	Hf	174	0.18
Mg	24	77.4		80	2.23	Xe	124	0.094		176	5.2
	25	11.5		82	11.50		126	0.088		177	>18.4
	26	11.1		83	11.48		128	1.92		178	27.1
Al	27	100		84	57.02		129	>26.23		179	13.8
Si	28	92.21		86	<17.43		130	4.05		180	35.3
	29	4.70	Rb	85	72.2		131	>21.14	Ta	180	0.012
	30	3.09		87	27.8		132	26.93		181	99.988
P	31	100	Sr	84	0.56		134	10.52	W	180	0.122
S	32	95.0		86	9.86		136	8.93		182	26.20
	33	<0.8		87	7.02	Cs	133	100		183	14.26
	34	4.2		88	82.56	Ba	130	0.101		184	<30.74
	36	<0.02	Y	89	100		132	0.097		186	<28.82
Cl	35	75.4	Zr	90	51.5		134	2.42	Re	185	37.1
	37	24.6		91	11.2		135	6.59		187	62.9
Ar	36	0.337		92	17.1		136	7.81	Os	184	0.018
	38	0.061		94	17.4		137	>11.32		186	1.59
	40	99.602		96	2.8		138	>71.66		187	1.64
K	39	93.1	Nb	93	100	La	138	0.089		188	13.3
	40	<0.012	Mo	92	15.84		139	99.911		189	16.1
	41	<6.9		94	9.04	Ce	136	0.19		190	26.4
Ca	40	96.96		95	15.72		138	0.26		192	<41.0
	42	0.64		96	16.53		140	88.47	Ir	191	38.5
	43	0.15		97	9.46		142	11.08		193	61.5
	44	2.06		98	23.78	Pr	141	100	Pt	190	0.012
	46	0.0033		100	9.63	Nd	142	27.11		192	0.78
	48	<0.019	Tc	(all	radioactive)		143	12.17		194	32.8
Sc	45	100	Ru	96	5.51		144	23.85		195	>33.7
Ti	46	7.93		98	1.87		145	8.30		196	25.4
	47	7.28		99	12.72		146	17.22		198	7.2
	48	73.94		100	12.62		148	5.73	Au	197	100
	49	5.51		101	17.07		150	5.62	Hg	196	0.15
	50	5.34		102	31.61	Pm	(all	radioactive)		198	10.1
V	50	0.25		104	18.58	Sm	147	15.0		199	17.0
	51	99.75	Rh	103	100		148	11.2		200	23.3
Cr	50	4.31	Pd	102	1.0		149	13.8		201	13.2
	52	83.76		104	11.0		150	7.4		202	<29.6
	53	9.55		105	22.2		152	26.8		204	6.7
	54	2.38		106	27.2		154	22.7	Tl	203	29.5
Mn	55	100		108	26.8	Eu	151	47.8		205	70.5
Fe	54	5.82		110	11.8		153	52.2	Pb	204	1.37
	56	91.66	Ag	107	51.4	Gd	152	0.2		206	26.26
	57	2.19		109	48.6		154	2.15		207	20.8
	58	0.33	Cd	106	1.23		155	<14.7		208	>51.55
Co	59	100		108	0.88		156	20.5	Bi	209	100
Ni	58	67.88		110	12.32		157	15.7			
	60	26.22		111	12.67		158	<24.9			
	61	1.18		112	24.15		160	21.9			
	62	3.66		113	12.21	Tb	159	100			
	64	<1.08		114	28.93	Dy	156	0.052			
Cu	63	69.09		116	7.61		158	0.090			
	65	30.91	In	113	4.2		160	2.29			
Zn	64	<48.9		115	95.8		161	>18.9			

[a] Following elements in increasing mass number are all radioactive: Po, At, Rn, Fr, Ra, Ac, Th, Pa, U, Np, Pu, Am, Cm, Bk, Cf, Es, Fm, Md, No, and Element 104 and heavier.

x-ray spectra and the development of positive-ray analysis that the nuclear charge was recognized as the fundamental chemical characteristic of an element. The use of mass spectrography, with other methods of determining the masses of the atoms in a given element and the proportions in which they are present, has permitted the determination of the isotopic composition of the elements.

Sulfur is one of the few exceptions to the constancy of isotopic proportions, in that there is sufficient variation, dependent upon the source of the sulfur, to cause a variation in its atomic mass by approximately ±0.01%. For normal stoichiometric calculations, however, this small variation is unimportant.

Naturally occurring elements which do not display any isotopic behavior include aluminum, arsenic, beryllium, bismuth, cobalt, fluorine, gold, helium, holmium, iodine, manganese, niobium (columbium), phosphorus, praseodymium, rhodium, scandium, sodium, terbium, thulium, and yttrium. Elements that have one predominating isotope (in excess of 98%) include argon, carbon, lanthanum, lutetium, nitrogen, oxygen, tantalum, and vanadium. Elements that have several isotopes and in which no one isotope is in excess of 80% of the total include antimony, barium, bromine, chlorine, copper, dysprosium, erbium, gadolinium, gallium, germanium, hafnium, iridium, krypton, lead, magnesium, mercury, molybdenum, nickel, osmium, palladium, platinum, rhenium, rubidium, ruthenium, selenium, silver, tellurium, tin, titanium, tungsten, xenon, ytterbium, zinc, and zirconium. Tin leads with a total of ten isotopes; xenon has nine isotopes; cadmium and tellurium each have eight isotopes.

Radioactive Elements

There are (1) naturally occurring radioactive and (2) artificially produced radioactive elements. There are three series of naturally occurring radioactive elements:

The Actinium Series. This series commences with ^{235}U and ends with the stable isotope ^{207}Pb. The decay scheme is represented by:

$$^{235}U \xrightarrow{\alpha}$$

$$^{231}Th \xrightarrow{\beta} \,^{231}Pa \xrightarrow{\alpha} \,^{227}Ac \xrightarrow{\beta \text{ and } \alpha} \,^{227}Th \xrightarrow{\alpha} \,^{223}Fr \xrightarrow{\beta} \,^{223}Ra$$

$$\xrightarrow{\alpha} \,^{219}Rn \xrightarrow{\alpha} \,^{215}Po \xrightarrow{\alpha \text{ and } \beta} \,^{211}Pb \xrightarrow{\beta} \,^{215}At \xrightarrow{\alpha} \,^{211}Bi \xrightarrow{\beta \text{ and } \alpha}$$

$$^{211}Po \xrightarrow{\alpha} \,^{207}Tl \xrightarrow{\beta} \,^{207}Pb \text{ (stable)}.$$

The Thorium Series. This series commences with ^{232}Th and ends with the stable isotope ^{208}Pb. The decay scheme is represented by:

$$^{232}Th \xrightarrow{\alpha}$$

$$^{228}Ra \xrightarrow{\beta} \,^{228}Ac \xrightarrow{\beta} \,^{228}Th \xrightarrow{\alpha} \,^{224}Ra \xrightarrow{\alpha} \,^{220}Rn \xrightarrow{\alpha} \,^{216}Po$$

$$\xrightarrow{\alpha} \,^{212}Pb \xrightarrow{\beta \text{ and } \alpha} \,^{216}At \xrightarrow{\alpha} \,^{212}Bi \xrightarrow{\beta \text{ and } \alpha} \,^{212}Po \xrightarrow{\alpha} \,^{208}Tl \xrightarrow{\beta}$$

$$^{208}Pb \text{ (stable)}.$$

The Uranium-Radium Series. This series commences with ^{238}U and ends with the stable isotope ^{206}Pb. The decay scheme is represented by:

$$^{238}U \xrightarrow{\alpha} \,^{234}Th \xrightarrow{\beta} \,^{234}Pa \xrightarrow{\beta} \,^{234}U \xrightarrow{\alpha} \,^{230}Th \xrightarrow{\alpha} \,^{226}Ra$$

$$\xrightarrow{\alpha} \,^{222}Ra \xrightarrow{\alpha} \,^{218}Po \xrightarrow{\alpha \text{ and } \beta} \,^{214}Pb \xrightarrow{\beta} \,^{218}At \xrightarrow{\alpha} \,^{214}Bi \xrightarrow{\beta \text{ and } \alpha}$$

$$^{214}Po \xrightarrow{\alpha} \,^{210}Tl \xrightarrow{\beta} \,^{210}Pb \xrightarrow{\beta} \,^{210}Bi \xrightarrow{\beta \text{ and } \alpha} \,^{210}Po \xrightarrow{\alpha} \,^{206}Tl$$

$$\xrightarrow{\beta} \,^{206}Pb \text{ (stable)}.$$

In the foregoing series, the type of radiation given off during the decay process is indicated above the arrows.

The production of artificially produced radioactive elements dates back to the early work of Rutherford in 1919 when it was found that alpha particles reacted with nitrogen atoms to yield protons and oxygen atoms. Curie and Joliot found (1933) that when boron, magnesium, or aluminum were bombarded with alpha particles from polonium, the elements would emit neutrons, protons, and positrons. They also found that upon cessation of bombardment the emission of protons and neutrons stopped, but that the emission of positrons continued. The targets remained radioactive. They also found that the radiation emitted dropped off exponentially as would be expected from a naturally occurring radioactive element. Further investigation indicated that nuclear reactions lead to the formation of radioactive isotopes. This and subsequent work by several investigators led to the formulation of another series of radioactive elements, namely, the Neptunium Series, which commences with ^{245}Cm (curium) and ends with the stable isotope 209Bi. The decay scheme is represented by: ^{245}Cm

$$\xrightarrow{\alpha} \,^{241}Pu \xrightarrow{\beta} \,^{241}Am \xrightarrow{\alpha} \,^{237}Np \xrightarrow{\alpha} \,^{233}Pa \xrightarrow{\beta} \,^{233}U$$

$$\xrightarrow{\alpha} \,^{229}Th \xrightarrow{\alpha} \,^{225}Ra \xrightarrow{\beta} \,^{225}Ac \xrightarrow{\alpha} \,^{221}Fr \xrightarrow{\alpha} \,^{217}At \xrightarrow{\alpha}$$

$$^{213}Bi \xrightarrow{\beta \text{ and } \alpha} \,^{213}Po \xrightarrow{\alpha} \,^{209}Tl \xrightarrow{\beta} \,^{209}Pb \xrightarrow{\beta} \,^{209}Bi \text{ (stable)}.$$

A *radioactive element* is an element that disintegrates spontaneously with the emission of various rays and particles. Most commonly, the term denotes radioactive elements such as radium, radon (emanation), thorium, promethium, uranium, which occupy a definite place in the periodic table because of their atomic number. The term *radioactive element* is also applied to the various other nuclear species, (which are produced by the disintegration of radium, uranium, etc.) including the members of the uranium, actinium, thorium, and neptunium families of radioactive elements, which differ markedly in their stability, and are isotopes of elements from thallium (atomic number 81) to uranium (atomic number 92), as well as the partly artificial actinide group, which extends from actinium (atomic number 89) to lawrencium (atomic number 103), and includes the following transuranic elements: neptunium (atomic number 93), plutonium (atomic number 94), americium (atomic number 95), curium (atomic number 96), berkelium (atomic number 97), californium (atomic number 98), einsteinium (atomic number 99), fermium (atomic number 100), mendelevium (atomic number 101), nobelium (atomic number 102). The radioactive nuclides produced from nonradioactive ones are discussed under **Radioactivity**.

A radioactive element may be designated as being in a *collateral series*. In addition to the three main natural and one artificial disintegration series of radioactive elements, each has been found to have at least one parallel or collateral series. The main series and the collateral series have different parents, but become identical in the course of disintegration, when they have a member in common.

Superheavy and Transactinide Elements

Those elements with an atomic number above 103 are sometimes referred to as the *transactinide elements*; and those with atomic (proton) numbers $Z \geq 110$ are sometimes called *superheavy elements* (SHEs). Considerable research has been concentrated on synthesizing elements heavier than 103. The last of the elements to be synthesized with positive proof of identity, timing of research, and place and persons associated with discovery—and thus relatively little controversy pertaining to the naming of the element—was lawrencium (103), discovered by Ghiorso, Sikkeland, Laesh, and Latimer in March 1961 at Berkeley. Research in this direction had been underway in the early 1960s at the Joint Nuclear Research Institute at Dubna (Russia). The half-lives of the Lr isotopes range from 8 to 35 seconds, enabling researchers to use solvent extraction techniques for determining the chemical characteristics and atomic number of the element. However, with the possible exception of a predicted *island of stability*, as one goes up the scale of heavier elements, the half-lives appear to become progressively shorter, making chemical separation, needed for identifying the atomic number of any laboratory-produced superheavy nucleus, increasingly difficult. Much research has been directed toward improving these techniques and, as of the late 1980s, methodology is available to cope with nuclei having a half-life of 1 second or greater. Separation techniques include the ion exchange behavior of the bromide complexes of the elements; and the ease with which the elements coprecipitate with cupric sulfide (Seaborg-Loveland-Morrissey, 1979).

Applying modern theories of nuclear structure, scientists working in the late 1960s and early 1970s made calculations that showed for superheavy elements in the vicinity of $Z = 114$ (proton number) and $N = 184$ (neutron number), ground states of nuclei were stabilized against fission. As pointed out in the aforementioned reference, "This stabilization was due to the complete filling of major proton and neutron shells in this region and is analogous to the stabilization of chemical elements, such as the noble gases by the filling of their electronic shells." Some of the calculations indicated that the half-lives of some of these superheavy nuclei might be on the order of the age of the universe. This observation, of course, was stimulating to further research. Calculations indicated that there should be an island of relative stability, which would extend above the Z and N figures previously given. The calculations also showed that between the presently known elements and these stable superheavy elements, there would be an intervening region of instability.

Although beyond the scope of this volume, in the early 1930s a new mechanism for the interaction of heavy ions was discovered. The method, known as *deep inelastic scattering*, involves a massive transfer of energy and nucleons between the projectile and the target.

A team of researchers from the Oak Ridge National Laboratory, the University of California (Davis, California), and Florida State University reported an x-ray spectrum in June 1976 that appeared to confirm the existence of superheavy elements with atomic numbers near 126. For a short period, this finding created much interest in the scientific community—mainly because the atomic numbers reported were much greater than might be expected at that stage of research, and, possibly of even greater interest, because the findings were based upon the elements being part of monazite crystals believed to be about 1 billion years old, thus giving rise to the previous mention of their life on the order of the age of the universe. Further confirmatory proof was lacking, capped by the finding of a researcher at Florida State University who showed that a gamma ray with the same energy as the x-ray peak for "element 126" is emitted when an excited praseodymium nucleus relaxes after being created from cerium during bombardment by protons. It is noteworthy that cerium is a major constituent of monazite.

Element 104

[CAS: 53850-36-5]. Rutherfordium. Researchers at Dubna (Russia), in 1964, bombarded plutonium with accelerated 113–115 MeV neon ions. During this process, an isotope that decayed by spontaneous fission was observed. It was reported that the isotope had a half-title of 0.3 ± 0.1 second and it was reasoned that the isotope was $^{260}104$, resulting from $_{94}^{242}Pu + _{10}^{22}Ne \longrightarrow {}^{260}104 + 4n$. Although subsequent work toward chemically separating the new element from all others has not been conclusive, considerable evidence for evaluation has been obtained. Estimates of the half-life of the element have been reduced from the prior 0.3 second to 0.15 second. Ghiorso, Nurmia, Harris, K. Eskola, and P. Eskola (University of California, Berkeley) reported in 1969 the positive identification of two and possibly three isotopes of the element. The Berkeley discovery resulted from bombarding a target of ^{249}Cf with ^{12}C nuclei of 71 MeV, and ^{13}C nuclei of 69 MeV. The first combination resulted in the instant emission of four neutrons to produce $^{257}104$. The isotope was reported to have a half-life of 4–5 seconds. Decay was by emission of an alpha particle into ^{253}No with a half-life of 105 seconds. In further research, several thousand atoms of $^{257}104$ and $^{259}104$ were produced. Thus far, the Dubna workers have proposed the name *kurchatovium* (Ku): and the Berkeley group has suggested *rutherfordium* (Rf). The name *rutherfordium* (Rf) has been adopted for element 104 in honor of Ernest R. Rutherford.

Element 105

[CAS: 53850-35-4]. Dubnium. In 1967, workers at Dubna (Russia) reported producing a few atoms of element $^{260}105$ and $^{261}105$, as the result of bombarding ^{243}Am with ^{22}Ne. Appropriate confirmations of identification, however, were lacking, the evidence being based upon time-coincidence measurements of alpha energies. In 1970, it was reported that the Dubna researchers had investigated all the types of decay of the new element and had determined its chemical properties. As of that time, the Dubna group had not proposed a name for element 105. Ghiorso, Nurmia, Harris, K. Eskola, and P. Eskola (University of California, Berkeley) reported a positive identification of element 105. This resulted from bombarding a target of ^{249}Cf with 84-MeV nitrogen nuclei. Upon absorption of ^{15}N nuclei by a ^{249}Cf nucleus, four neutrons are emitted, forming element $^{260}105$, with a half-life of 1.6 seconds.

In October 1971, Ghiorso and co-workers at Berkeley, using the heavy ion linear accelerator, announced production of element $^{261}105$ by bombarding ^{250}Cf with ^{15}N and by bombarding ^{249}Bk with ^{16}O. The isotope emits 8.93-MeV alpha particles, decaying to ^{257}Lr with a half-life of about 1.8 second. Element $^{262}105$ was produced by bombarding ^{249}Bk with ^{18}O. This isotope emits 8.45-MeV alpha particles and decays to ^{258}Lr, with a half-life of about 40 seconds. Element 105 has been name Dubnium (Db) after the location of the Joint Institute for Nuclear Research where the element was created.

Element 106

[CAS: 54038-81-2]. Seaborgium. In late 1974, two groups announced the synthesis of the 14th transurium element, namely, element number 106 (ekatungsten). The Russian group at Dubna bombarded a target of lead atoms with ions of various weights. Argon ions were used to form a short-lived isotope of fermium which decayed by spontaneous fission. After this trial, the Soviet group used ions of titanium to produce a similarly short-lived isotope of element 104. Then, the Soviet scientists finally used ions of chromium to produce what they believe to be element 106. This presumed element is described as having 151 or 152 neutrons, and decays by spontaneous fission with a half-life of about 4 to 10 milliseconds. The Russian scientists claim that the chromium and lead will combine to form element 106.

An American group at the University of California's Lawrence Berkeley Laboratory, using the modified super-HILAC accelerator (heavy ion linear accelerator) to bombard a target of californium-249 with ions of oxygen-18, caused the oxygen ions to combine with molecules in the target, releasing four neutrons per collision and producing eka-tungsten-263. It is reported that this isotope of element 106 has a half-life of 0.9 second. Contrary to earlier predictions, it does not decay by spontaneous fission, but emits an alpha particle with an energy of 9.06 MeV to become an isotope of element 104. The previously observed daughter isotope emits an alpha particle with an energy of 8.8 MeV, becoming nobelium-255. The latter, in turn, emits an alpha particle with an energy of 8.11 MeV. The American scientists believe that observation of this complete sequence of transmutations is conclusive proof of the formation of eka-tungsten. The American team first observed element 106 in 1970, but lead impurities in the target presented difficulties in providing conclusive evidence of the existence of the new element. The HILAC accelerator was shut down shortly after the experiment and two years were required for the improvements that resulted in the super-HILAC. Then other experiments were given a higher priority. Thus there was an approximately four-year delay in the American experiments. The isotope of element 106 synthesized at Berkeley contains 157 neutrons.

Element 106 has been named Seaborgium (Sb) in honor of the Glenn T. Seaborg, the father of nuclear chemistry and discoverer of plutonium.

Element 107

[CAS: 54037-14-8]. Bohrium is a chemical element in the periodic table that has the symbol Bh and atomic number 107. It is a synthetic element whose most stable isotope, Bh-262, has a half-life of 102 ms. It was synthesized in 1976 by a Soviet team led by Y. Oganessian at the Joint Institute for Nuclear Research at Dubna, who produced isotope 261 Bh with a half-life of 1–2 ms (later data give a half life of around 10 ms). They did this by bombarding bismuth-204 with heavy nuclei of chromium-54. In 1981 a German research team led by P. Armbruster and G. Münzenberg at the Gesellschaft für Schwerionenforschung (Institute for Heavy Ion Research) at Darmstadt were also able to confirm the Soviet team's results and produce bohrium, this time the longer-lived Bh-262. The Germans suggested the name nielsbohrium to honor the Danish physicist Niels Bohr. The Soviets had suggested this name be given to element 105 (dubnium). There was an element naming controversy as to what the elements from 101 to 109 were to be called; thus IUPAC adopted unnilseptium (symbol Uns) as a temporary name for this element. In 1994 a committee of IUPAC recommended that element 107 be named bohrium. While this conforms to the names of other elements honoring individuals, where only the surname is taken, it was opposed by many who were concerned that it could be confused with boron. Despite this, the name bohrium for element 107 was recognized internationally in 1997.

Element 108

[CAS: 54037-57-9]. Hassium is a chemical element in the periodic table that has the symbol Hs and atomic number 108. It is a synthetic element whose most stable isotope is Hs-265, with a half-life of 2 ms. It was first synthesized in 1984 by a German research team led by Peter Armbruster and Gottfried Münzenberg at the Institute for Heavy Ion Research at Darmstadt. The name hassium was proposed by them, derived from the

Latin name for the German state of Hessen where the institute is located. There was an element naming controversy as to what the elements from 101 to 109 were to be called; thus IUPAC adopted unniloctium (symbol Uno) as a temporary name for this element. In 1994 a committee of IUPAC recommended that element 108 be named hahnium. The name hassium was adopted internationally, however, in 1997.

Element 109

[CAS: 54038-01-6]. Meitnerium is a chemical element in the periodic table that has the symbol Mt and atomic number 109. It is a synthetic element whose most stable isotope is Mt-266 with a half-life of 3.4 ms. Meitnerium was first synthesized on August 29,1982 by a German research team led by Peter Armbruster and Gottfried Münzenberg at the Institute for Heavy Ion Research at Darmstadt. The team did this by bombing a target of bismuth-209 with accelerated nuclei of iron-58. The creation of this element demonstrated that nuclear fusion techniques could be used to make new, heavy nuclei. The name meitnerium was suggested in honor of the Austrian-Swedish physicist and mathematician Lise Meitner, but there was an element naming controversy as to what the elements from 101 to 109 were to be called; thus IUPAC adopted unnilennium (symbol Une) as a temporary, systematic element name. However in 1997 they resolved the dispute and adopted the current name.

Element 110

[CAS: 54083-77-1]. Darmstadtium (formerly Ununnilium) is a chemical element in the periodic table that has the symbol Ds and atomic number 110. It has an atomic weight of 271 making it one of the super-heavy atoms. It is a synthetic element and decays in thousandths of a second. Due to its presence in Group 10 it is believed to be likely to be metallic and solid. It was first created on November 9,1994 at the Gesellschaft für Schwerionenforschung (GSI) in Darmstadt, Germany. It has never been seen and only a few atoms of it have been created by the nuclear fusion of isotopes of lead and nickel in a heavy ion accelerator (nickel atoms are the ones accelerated and bombarded into the lead). Scientists are not always serious, so some suggested the name *policium* for the new element, because 110 is the telephone number of the German police. The element was named after the places of its discovery, Darmstadt (actually, the GSI is located in Wixhausen, a small suburb north of Darmstadt). The new name was given to it by the IUPAC in August 2003.

Allotropes. Some of the elements exist in two or more modifications distinct in physical properties, and usually in some chemical properties. Allotropy in solid elements is attributed to differences in the bonding of the atoms in the solid. Various types of allotropy are known. In *enantiomorphic allotropy*, the transition from one form to another is reversible and takes place at a definite temperature, above or below which only one form is stable, e.g., the alpha and beta forms of sulfur. In *dynamic allotropy*, the transition from one form to another is reversible, but with no definite transition temperature. The proportions of the allotropes depend upon the temperature. In *monotropic allotropy*, the transition is irreversible. One allotrope is metastable at all temperatures, e.g., explosive antimony.

Examples of allotropes include:

Arsenic with four forms, metallic, yellow, gray, and brown.
Boron with two forms, crystalline and amorphous.
Carbon with fiveforms, amorphous, diamond, graphite, buckmeister-fullerenes, and nanotubes.
Phosphorus with four forms, two white forms, a violet, and a black form. Red phosphorus is a mixture of the white and violet forms.
Selenium with four forms, amorphous, two crystalline monoclinic forms (red), and the stable, crystalline gray metallic form.
Sulfur with two forms, alpha-rhombic sulfur with a density of 2.07 and a mp of 112.8°C, and beta-monoclinic sulfur with a density of 1.96 and a mp of 119°C. The beta form changes to the alpha form below 96°C.

In the case of some of the less common elements, impure forms have been mistaken in the past as allotropic forms. A number of elements that once were considered to exist in both crystalline and amorphous forms have been found to exist in only one form when perfectly pure.

Gaseous Elements. Several gaseous elements (at standard conditions of temperature and pressure) form molecules of two atoms each. These are known as *diatomic gases*. Included in this category are hydrogen, H_2, nitrogen, N_2, oxygen, O_2, and chlorine, Cl_2. The inert gases, helium, neon, argon, krypton, and xenon, are *monatomic gases* and their symbols do not carry a subscript.

Atomic Structure of the Elements

The internal structure of an atom consists of electrons moving within a region having a diameter slightly greater than 10^{-8} cm and of protons and neutrons that are confined to a nucleus at the center of the electron distribution in a region having a diameter of about 10^{-12} cm. The electrons of a neutral atom are sufficient in number so that their total negative charge is equal to the positive charge on the nucleus. For example, atoms of the element hydrogen have in their neutral state a single electron moving about a nucleus, which has a positive charge equal to the negative charge of an electron. Helium, which has two electrons moving about its nucleus, has a positive charge on its nucleus equal to twice one electronic charge and lithium, which has three electrons moving about its nucleus, has a positive charge on its nucleus of three electronic charges. One basis for the classification of atoms is by those numbers which correspond to the number by which the charge on the hydrogen nucleus must be multiplied to equal the nuclear charge of the atom in question. These numbers, ranging from one, for hydrogen, to 110, for darmstadtium, are called atomic numbers.

The atomic number may be defined as the number of protons in an atomic nucleus, or the positive charge of the nucleus, expressed in terms of the electronic charge. Atomic number usually is denoted by the symbol Z. In the symbolic designation of individual nuclides, the atomic number sometimes is written as a subscript to the left of the chemical symbol of the atomic species, such as $_8{}^{16}O$ for the oxygen isotope of mass number 16. This usage is redundant, in that the chemical symbol per se specifies the atomic number of the nuclide.

Besides nuclear charge, atoms also differ in their masses. The mass is determined by the number of protons Z and the number of neutrons N in the atomic nucleus. The total number of nuclear particles, $Z + N$, in an atomic species is known as its mass number A.

The nuclear model for an atom is of relatively recent origin for, at the beginning of the twentieth century, J.J. Thomson's "plum-pudding" model of an atom was the more generally accepted version. In this model Thomson supposed that the positive charge forms a plasma that is distributed throughout the atomic volume and that the electrons are mixed into this plasma with a relatively uniform distribution. E. Rutherford proposed the nuclear model on the basis of experimental work by H. Geiger and E. Marsden in which they observed that a small number of alpha particles from a naturally occurring radioactive source are scattered through angles greater than 90° by thin foils of gold and silver. Although some small-angle scattering is predicted by the Thomson model, such large-angle scattering is not at all expected. Large-angle scattering is, however, completely consistent with the idea that the alpha particles interact with point positively charged objects of large mass at the center of each gold and silver atom.

Subsequently (1913), N. Bohr found that he could use the Rutherford nuclear model to explain in almost complete detail the observed spectrum of hydrogen (see **Atomic Spectra**). Bohr proposed that, in a neutral hydrogen atom, a single electron revolves in a stationary orbit around a point nucleus that has a charge $+e$. This electron is held in its orbit by the electrostatic force between the positive charge on the nucleus and its own negative charge. The stationary-orbit assumption was classically unsatisfactory but necessary to account for the behavior of the electron. If the laws of classical electrodynamics, according to which accelerated charges must be sources of electromagnetic radiation, were strictly obeyed, the electron would gradually lose energy; hence it would revolve in orbits of smaller and smaller radii and eventually fall into the nucleus. Furthermore, Bohr postulated that the magnitude of the orbital angular momentum of each stationary orbit is $L = nh/2\pi = n\hbar$, where n is an integer, h is Planck's constant, and $n\hbar$ is simply an abbreviated form for/2π.

The angular momentum of an electron moving in an orbit of the type described by Bohr is an axial vector $\mathbf{L} = \mathbf{r} \times \mathbf{p}$, formed from the radial distance $\mathbf{r}$ between electron and nucleus and the linear momentum $\mathbf{p}$ of the electron relative to a fixed nucleus. Figure 2 shows the customary method used to illustrate the axial vector $\mathbf{L}$ in terms of the orbital motion of any object, of which the electron of the Bohr atom is only one example. Although Bohr's planetary model needed only circular orbits to explain the spectral lines observed in the spectrum of a hydrogen atom, subsequent

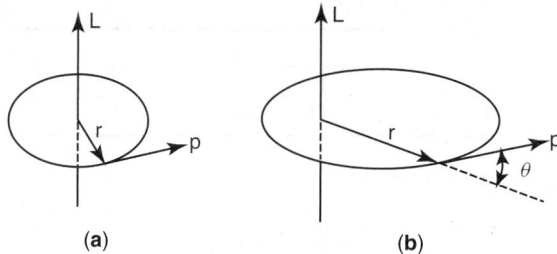

Fig. 2.   Direction of angular momentum vector is perpendicular to plane formed by radial vector **r** and momentum vector **p**

TABLE 4. IDENTIFICATION OF ATOMIC SHELLS

X-ray notation	Corresponding quantum numbers			X-ray notation	Corresponding quantum numbers		
	n	l	j		n	l	j
K	1	0	$\frac{1}{2}$	M_{V}	3	2	$\frac{5}{2}$
L_{I}	2	0	$\frac{1}{2}$	N_{I}	4	0	$\frac{1}{2}$
L_{II}	2	1	$\frac{1}{2}$	N_{II}	4	1	$\frac{1}{2}$
L_{III}	2	1	$\frac{1}{2}$	N_{III}	4	1	$\frac{3}{2}$
M_{I}	3	0	$\frac{1}{2}$	N_{IV}	4	2	$\frac{3}{2}$
M_{II}	3	1	$\frac{1}{2}$	N_{V}	4	2	$\frac{5}{2}$
M_{III}	3	1	$\frac{3}{2}$	N_{VI}	4	3	$\frac{5}{2}$
M_{IV}	3	2	$\frac{3}{2}$	N_{VII}	4	3	$\frac{5}{2}$

development of similar models for other atoms containing more than a single electron needed elliptically shaped orbits to explain the observed spectra. For such orbits **r** and **p** are usually not perpendicular to each other, so that the magnitude $\mathbf{L} = \mathbf{r} \times \mathbf{p}$ is $rp \sin \theta$.

A significant change in the theoretical treatment of atomic structure occurred in 1924 when Louis de Broglie proposed that an electron and other atomic particles simultaneously possess both wave and particle characteristics and that an atomic particle, such as an electron, has a wavelength $\lambda = h/p = h/mv$. Shortly thereafter, C.J. Davisson and L.H. Germer showed experimentally the validity of this postulate. De Broglie's assumption that wave characteristics are inherent in every atomic particle was quickly followed by the development of quantum mechanics. In its most simple form, quantum mechanics introduces the physical laws associated with the wave properties of electromagnetic radiation into the physical description of a system of atomic particles. By means of quantum mechanics a much more satisfactory explanation of atomic structure can be developed.

The quantity n introduced by Bohr in his description of the hydrogen atom is what is called a quantum number. These numbers enter quite naturally from quantum-mechanical descriptions of atomic energy states, including nuclear states. In quantum-mechanical descriptions of atoms, the number n characterizes a limited number of electron states that have very nearly the same energy. A group of electrons in an atom with a common value of n are usually said to be in a single shell of the atom. In the simple two-body problem of the hydrogen atom, all states having the same value of n have the same energy, but in multielectron atoms there are interactions between individual pairs of electrons as well as between electrons and the atomic nucleus. The result is a limited spread in energy between the most-tightly bound and the least-tightly bound electron with the same value of n. Except for $n = 1$, more than one orbital angular momentum state is possible for each shell. Each such state is described by a quantum number l, which can have only whole-number values between zero and $n - 1$. The number n is usually called the principal quantum number and l the orbital angular momentum, or sometimes azimuthal, quantum number. In addition to its characteristic angular momentum, each electron spins on an axis, such that it also has a spin angular momentum, described by a quantum number s. Because all electrons are identical, s has only one value, $\frac{1}{2}$.

In the development of the concepts of atomic structure much of the experimental evidence came from optical and x-ray spectroscopy. From this work certain notations have arisen that are now an accepted part of the language. For example, the $n = 1$ shell is sometimes known as the K-shell, the $n = 2$ shell as the L-shell, the $n = 3$ shell as the M-shell, etc., with consecutively following letters of the alphabet being used to designate those shells with successively higher principal quantum numbers. A Roman numeral subscript further subdivides the shells in accordance with the n, l, and j quantum numbers of the electrons, as shown in Table 4.

A letter notation has been given to the different orbital angular momentum states by optical spectroscopists. In this notation an $l = 0$ state is called an s state (not to be confused with the s used as a quantum number for spin, and which can usually be distinguished from the way it is used), an $l = 1$ state is a p state an $l = 2$ a d state, an $l = 3$ state an f state, with consecutively higher letters above f in the alphabet designating each succeeding value of l. A number often accompanies the state designation to indicate the appropriate value of n. For example, a $3p$ level is an $n = 3$, $l = 1$ state.

Electrically neutral atoms with nuclear charge $Z > 1$ are not hydrogen-like, but have more than a single orbital electron. With more than one electron in an atom, it is necessary to determine the relationship of one

electron to its neighbors. A clue to this relationship is found from the observation that the energy required to remove a second electron from an atom is greater than that required to remove the first electron and the energy so required increases with each succeeding electron that is removed. In other words, the potential energy that holds some electrons in an atom is much greater than the energy that holds other electrons in the same atom. From the results of the detailed observations, Pauli formulated a principle, the Pauli Exclusion Principle, which states that no two electrons within the same atom may have exactly the same wave function and, on this basis, that no two electrons in the same atom may have the same set of quantum numbers. Thus, if a particular configuration state in an atom is occupied by an electron, that state is forbidden to all other electrons. If each electron falls into the lowest possible energy state, the next electron to enter must remain in some higher energy state. Acceptance of this principle allows us to classify observed spectral characteristics of the radiation emitted by atoms in such a way that we can determine within reasonable limits the number of electrons in each shell or subshell of the atom producing the radiation. The electron configurations in the known atomic species are shown in Table 5, in which the atoms are ordered in accordance with their atomic numbers Z. Generally the innermost (lower quantum number) shells are filled with all the electrons allowed. For example, neon has two $1s$ electrons, two $2s$ electrons, and six $2p$ electrons, such that its K ($n = 1$) and L ($n = 2$) shells are both filled. An atom with Z having a magnitude one greater than neon, which is sodium, also has both the K and L shells filled; thus there is no space for another electron to go into either of these shells, and its additional electron must go into the M ($n = 3$) shell. Shells with higher values of the orbital angular momentum quantum number l are sometimes partially shielded electrically from the nucleus by other electrons in such a way that their binding energies are not as great as the lower orbital angular momentum subshells associated with the next higher principal quantum number. Thus, as can be seen in Table 5, the $5s$ subshell is filled in atomic species of lower Z than any of those that have any $4f$ electrons.

The periodic table of elements can be described in terms of the similarities and differences of the angular momentum characteristics of the various atomic species. For example, similar but not identical chemical properties are observed for a number of elements that have all but one of their electrons in filled shells, with the extra electron being a single s electron. In the periodic table, lithium, sodium, potassium, rubidium, and cesium all have this characteristic and thus are placed in a single column. Other groups of atomic species that have similar electron configurations in their outermost shell are arranged in the periodic table in the same chemical grouping. See also **Periodic Table of the Elements**.

Motions of charged particles are expected to establish magnetic fields that interact with each other. As a result, coupling between spin and orbital angular momenta of a single electron or coupling between angular momenta of different electrons in a single atom is expected. The spin-orbit coupling of a single electron results in a total angular momentum state, described by a total quantum number j, such that either $j = l + \frac{1}{2}$ or $j = l - \frac{1}{2}$. Quantum-mechanical solutions show that the magnitude of the orbital angular momentum is $l^*h = [l(l + 1)]1/2h$, not l, as would be expected if we had followed the simpler assumption of the Bohr model. Similarly, the magnitude of the spin and total angular momenta are $s^* = [s(s + 1)]1/2$

TABLE 5. ELECTRON STRUCTURE OF ATOMS (NORMAL STATE)

Element	Atomic number	Chemical symbol	K	L		M			N				O				
			1s	2s	2p	3s	3p	3d	4s	4p	4d	4f	5s	5p	5d	5f	5g
Hydrogen	1	H	1														
Helium	2	He	2														
Lithium	3	Li	2	1													
Beryllium	4	Be	2	2													
Boron	5	B	2	2	1												
Carbon	6	C	2	2	2												
Nitrogen	7	N	2	2	3												
Oxygen	8	O	2	2	4												
Fluorine	9	F	2	2	5												
Neon	10	Ne	2	2	6												
Sodium	11	Na	2	2	6	1											
Magnesium	12	Mg	2	2	6	2											
Aluminum	13	Al	2	2	6	2	1										
Silicon	14	Si	2	2	6	2	2										
Phosphorus	15	P	2	2	6	2	3										
Sulfur	16	S	2	2	6	2	4										
Chlorine	17	Cl	2	2	6	2	5										
Argon	18	Ar	2	2	6	2	6										
Potassium	19	K	2	2	6	2	6		1								
Calcium	20	Ca	2	2	6	2	6		2								
Scandium	21	Sc	2	2	6	2	6	1	2								
Titanium	22	Ti	2	2	6	2	6	2	2								
Vanadium	23	V	2	2	6	2	6	3	2								
Chromium	24	Cr	2	2	6	2	6	5	1								
Manganese	25	Mn	2	2	6	2	6	5	2								
Iron	26	Fe	2	2	6	2	6	6	2								
Cobalt	27	Co	2	2	6	2	6	7	2								
Nickel	28	Ni	2	2	6	2	6	8	2								
Copper	29	Cu	2	2	6	2	6	10	1								
Zinc	30	Zn	2	2	6	2	6	10	2								
Gallium	31	Ga	2	2	6	2	6	10	2	1							
Germanium	32	Ge	2	2	6	2	6	10	2	2							
Arsenic	33	As	2	2	6	2	6	10	2	3							
Selenium	34	Se	2	2	6	2	6	10	2	4							
Bromine	35	Br	2	2	6	2	6	10	2	5							
Krypton	36	Kr	2	2	6	2	6	10	2	6							
Rubidium	37	Rb	2	2	6	2	6	10	2	6			1				
Strontium	38	Sr	2	2	6	2	6	10	2	6			2				
Yttrium	39	Y	2	2	6	2	6	10	2	6	1		2				
Zirconium	40	Zr	2	2	6	2	6	10	2	6	2		2				
Niobium	41	Nb	2	2	6	2	6	10	2	6	4		1				
Molybdenum	42	Mo	2	2	6	2	6	10	2	6	5		1				
Technetium	43	Tc	2	2	6	2	6	10	2	6	(5)		(2)				
Ruthenium	44	Ru	2	2	6	2	6	10	2	6	7		1				
Rhodium	45	Rh	2	2	6	2	6	10	2	6	8		1				
Palladium	46	Pd	2	2	6	2	6	10	2	6	10						
Silver	47	Ag	2	2	6	2	6	10	2	6	10		1				
Cadmium	48	Cd	2	2	6	2	6	10	2	6	10		2				
Indium	49	In	2	2	6	2	6	10	2	6	10		2	1			
Tin	50	Sn	2	2	6	2	6	10	2	6	10		2	2			
Antimony	51	Sb	2	2	6	2	6	10	2	6	10		2	3			
Tellurium	52	Te	2	2	6	2	6	10	2	6	10		2	4			
Iodine	53	I	2	2	6	2	6	10	2	6	10		2	5			
Xenon	54	Xe	2	2	6	2	6	10	2	6	10		2	6			

Element	Atomic number	Chemical symbol	K	L	M	N				O					P						Q
						4s	4p	4d	4f	5s	5p	5d	5f	5g	6s	6p	6d	6f	6g	6h	7s
Cesium	55	Cs	2	8	18	2	6	10		2	6				1						
Barium	56	Ba	2	8	18	2	6	10		2	6				2						
Lanthanium	57	La	2	8	18	2	6	10		2	6	1			2						
Cerium	58	Ce	2	8	18	2	6	10	2	2	6				2						
Praseodymium	59	Pr	2	8	18	2	6	10	3	2	6				2						
Neodymium	60	Nd	2	8	18	2	6	10	4	2	6				2						
Promethium	61	Pm	2	8	18	2	6	10	5	2	6				2						
Samarium	62	Sm	2	8	18	2	6	10	6	2	6				2						
Europium	63	Eu	2	8	18	2	6	10	7	2	6				2						
Gadolinium	64	Gd	2	8	18	2	6	10	7	2	6	1			2						
Terbium	65	Tb	2	8	18	2	6	10	8 or 9	2	6	1 or 0			2						
Dysprosium	66	Dy	2	8	18	2	6	10	10	2	6				2						

TABLE 5. (*continued*)

Element	Atomic number	Chemical symbol	K	L	M	N				O							P				Q
						4s	4p	4d	4f	5s	5p	5d	5f	5g	6s	6p	6d	6f	6g	6h	7s
Holmium	67	Ho	2	8	18	2	6	10	11	2	6				2						
Erbium	68	Er	2	8	18	2	6	10	12	2	6				2						
Thulium	69	Tm	2	8	18	2	6	10	13	2	6				2						
Ytterbium	70	Yb	2	8	18	2	6	10	14	2	6				2						
Lutetium	71	Lu	2	8	18	2	6	10	14	2	6	1			2						
Hafnium	72	Hf	2	8	18	2	6	10	14	2	6	2			2						
Tantalum	73	Ta	2	8	18	2	6	10	14	2	6	3			2						
Tungsten	74	W	2	8	18	2	6	10	14	2	6	4			2						
Rhenium	75	Re	2	8	18	2	6	10	14	2	6	5			2						
Osmium	76	Os	2	8	18	2	6	10	14	2	6	6			2						
Iridium	77	Ir	2	8	18	2	6	10	14	2	6	7			2						
Platinum	78	Pt	2	8	18	2	6	10	14	2	6	9			1						
Gold	79	Au	2	8	18	2	6	10	14	2	6	10			1						
Mercury	80	Hg	2	8	18	2	6	10	14	2	6	10			2						
Thallium	81	Tl	2	8	18	2	6	10	14	2	6	10			2	1					
Lead	82	Pb	2	8	18	2	6	10	14	2	6	10			2	2					
Bismuth	83	Bi	2	8	18	2	6	10	14	2	6	10			2	3					
Polonium	84	Po	2	8	18	2	6	10	14	2	6	10			2	4					
Astatine	85	At	2	8	18	2	6	10	14	2	6	10			2	5					
Radon	86	Rn	2	8	18	2	6	10	14	2	6	10			2	6					
Francium	87	Fr	2	8	18	2	6	10	14	2	6	10			2	6					(1)
Radium	88	Ra	2	8	18	2	6	10	14	2	6	10			2	6					(2)
Actinium	89	Ac	2	8	18	2	6	10	14	2	6	10			2	6	(1)				(2)
Thorium	90	Th	2	8	18	2	6	10	14	2	6	10			2	6	(2)				(2)
Protactinium	91	Pa	2	8	18	2	6	10	14	2	6	10	(2)		2	6	(1)				(2)
Uranium	92	U	2	8	18	2	6	10	14	2	6	10	(3)		2	6	(1)				(2)
Neptunium	93	Np	2	8	18	2	6	10	14	2	6	10	(5)		2	6					(2)
Plutonium	94	Pu	2	8	18	2	6	10	14	2	6	10	(6)		2	6					(2)
Americium	95	Am	2	8	18	2	6	10	14	2	6	10	(7)		2	6					(2)
Curium	96	Cm	2	8	18	2	6	10	14	2	6	10	(7)		2	6	(1)				(2)
Berkelium	97	Bk	2	8	18	2	6	10	14	2	6	10	(9)		2	6					(2)
Californium	98	Cf	2	8	18	2	6	10	14	2	6	10	(10)		2	6					(2)
Einsteinium	99	Es	2	8	18	2	6	10	14	2	6	10	(11)		2	6					(2)
Fermium	100	Fm	2	8	18	2	6	10	14	2	6	10	(12)		2	6					(2)
Mendelevium	101	Mv	2	8	18	2	6	10	14	2	6	10	(13)		2	6					(2)
Nobelium	102	No	2	8	18	2	6	10	14	2	6	10	(14)		2	6					(2)
Lawrencium	103	Lw	2	8	18	2	6	10	14	2	6	10	(14)		2	6	(1)				(2)
Dubnium	105	Db	2	6	18	2	6	10	14	2	6	10	(14)		2	6	(3)				(2)
Seaborgium	106	Sg	2	6	18	2	6	10	14	2	6	10	(14)		2	6	(4)				(2)
Bohrium	107	Bh	2	6	18	2	6	10	14	2	6	10	(14)		2	6	(5)				(2)
Hassium	108	Hs	2	6	18	2	6	10	14	2	6	10	(14)		2	6	(6)				(2)
Meitnerium	109	Mt	2	6	18	2	6	10	14	2	6	10	(14)		2	6	(7)				(2)
Darmstadtium	110	Ds	2	6	18	2	6	10	14	2	6	10	(14)		2	6	(9)				(1)

and $j^* = [j(j+1)]1/2$, respectively. Coupled spin and orbital angular momenta thus do not align themselves in a linear pattern but in such a way that the spin and orbital angular momenta precess around the direction of the total angular momentum, as shown in Fig. 3. A similar precession around the direction of the resultant angular momentum state is found, as we shall find later, in the coupling of any two or more angular momentum states into a single common system.

Without some external influence, angular momentum states have no preferred orientation in space because the only interacting magnetic fields are entirely within the individual atom. An external magnetic field, however, may couple to the magnetic field of an atomic angular momentum state of the type described in the preceding paragraphs. In coupling to the total angular momentum, for example, only a limited number of orientations of j^*h are possible, these being such that their projections in the direction of the magnetic field have magnitudes m_j in which m_j, a magnetic quantum number, can have only those whole-number values for which $m_j \leq j$. Thus $2j + 1$ orientations, as illustrated in Fig. 4 for $j = \frac{3}{2}$, are possible. Note that the vector representing the total angular momentum is not parallel to the direction of the applied magnetic field. As a result, the total angular momentum vector processes around the direction of the magnetic field such that the component of the total angular momentum that is perpendicular to the direction of the magnetic field has a time-averaged value of zero and the only component that can be observed with an external detecting device is that component parallel to the direction of the external field. A magnetic field of this type splits a set of levels characterized by a quantum number j, all of which initially have the same energy, into $2j + 1$ components. The nature of this level splitting is deduced from observations of the splitting of characteristic line spectra, in which case an originally monoenergetic radiation is split by the magnetic field into several components that are usually relatively closely spaced in wave length. The observed effect is known as the Zeeman effect. See also **Zeeman Effect**.

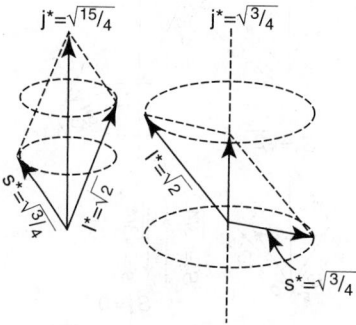

Fig. 3. Magnitudes and directions of the angular momentum vectors for an $l = 1$, $s = \frac{1}{2}$ electron in an atomic energy state

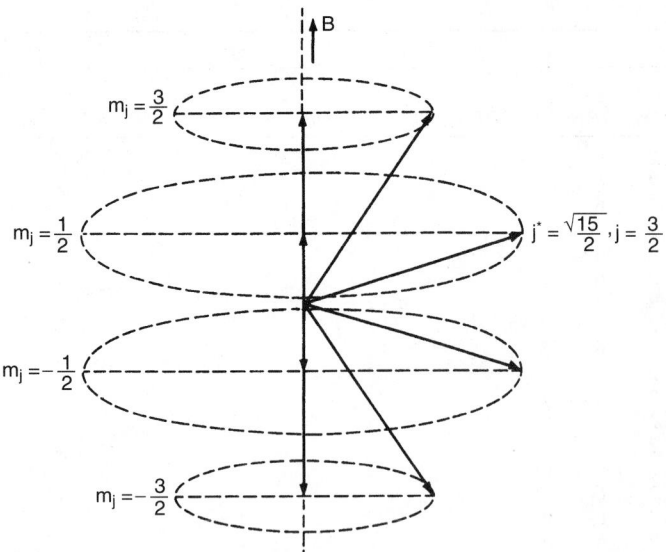

Fig. 4.   Possible orientations of the total angular momentum vector **i** relative to the direction of an externally applied magnetic field B and the magnitudes of the associated magnetic quantum state vectors **m**$_j$

The use of axial vectors to describe states of an atom in terms of a coupling between the angular momentum inherent in the electrons of the atom forms what is commonly called the *vector model* of the atom. The type of coupling just described forms the basis for the *j-j* coupling model. Such coupling is found between electrons that produce the optical radiation in the higher Z atomic systems. In these systems the total angular momentum j^* formed by the coupling of the spin and orbital angular momentum states of individual electrons are further coupled within any single shell of an atom to form a total angular momentum j^*, characterized by a quantum number J, which then is the angular momentum for the system of several electrons. According to customary usage, a lower case letter as a designator for a quantum number indicates that the quantum number is that of a single electron, while a capital letter indicates that the quantum number represents a state formed by several electrons.

For the lower Z part of the periodic table of elements, the appropriate coupling system for angular momentum states in an atom is the L-S, or Russell-Saunders, coupling. In this description, the orbital angular momenta of individual electrons in any single shell of an atom are coupled to form a resultant orbital angular momentum, described by the quantum number L, and the spin angular momenta of the same individual electrons are coupled to form a resultant spin angular momentum, described by the quantum number S. Coupling of the individual orbital angular momentum states

is not shown but possible coupling schemes for the spins of 2, 3, and 4 electrons are shown in Fig. 5. Note that as many possible couplings exist as there are electrons. Only one of these possible schemes will be the lowest energy state, the others being at higher energies. The resultant spin and orbital angular momentum vectors for a group of electrons in a single shell of an atom can then be used to describe the coupling that gives the total angular momentum J^* for these electrons. In Fig. 6 is shown the coupling according to the vector model between the orbital and spin angular momenta for which $L = 2$ and $S = 1$. There are $2J + 1$ possible states. All $2J + 1$ states have the same energy unless under the influence of an external magnetic field, in which case they are broken into components, each of which is designated by a magnetic quantum number M_j. In principle this splitting is the same as for the splitting of the total angular momentum states for a single electron, as shown in Fig. 3, except for the use of capital letters M_j and J.

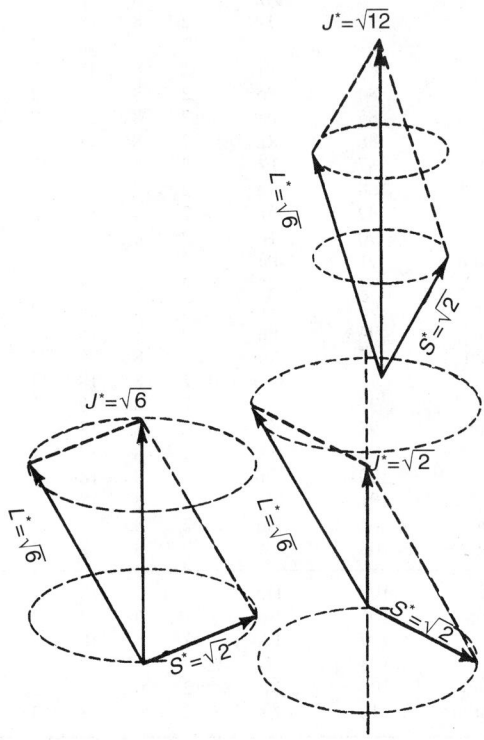

Fig. 6.   Vector model coupling of spin and orbital angular momenta for which $L = 2$, $S = 1$

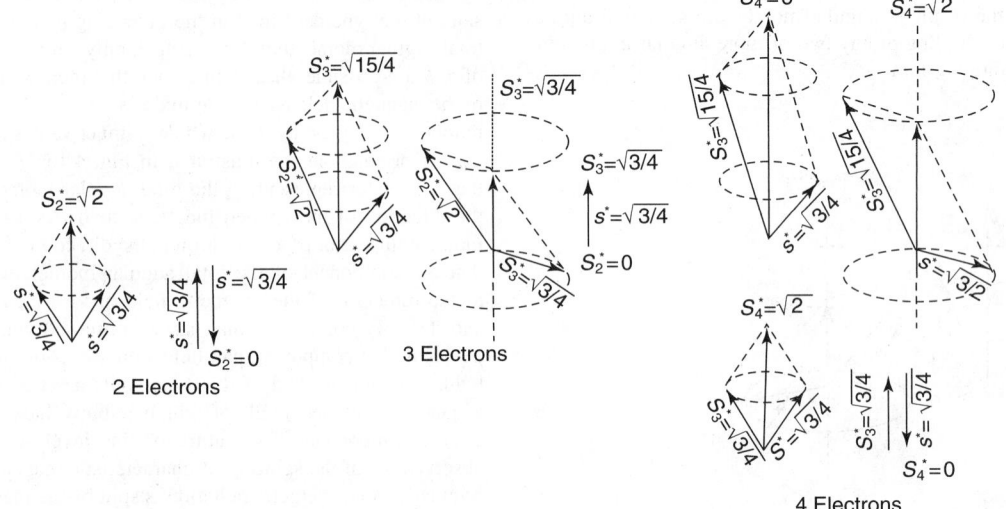

Fig. 5.   Vector model coupling of the spin angular momenta of two, three, and four electrons

The Pauli Exclusion Principle states that no two electrons of any single atom may simultaneously occupy a state described by only a single set of quantum numbers. Five such numbers are needed to describe fully the quantum-mechanical conditions of an electron. For j-j coupling this set is generally n, l, s, j, m_j, and for L-S it is n, l, s, m_l, m_s. From the coupling of the angular momentum associated with the latter sets a full description of the multielectron state, described by n, L, S, J, M_j, is determined.

Part of the outgrowth of the determination that atomic particles have wave properties, and of the subsequent development of quantum mechanics, is the Heisenberg Uncertainty Principle, which states that an electron cannot be located exactly in terms of both its space and momentum coordinates, or in terms of both its time and energy coordinates. See also **Uncertainty Principle**. If the energy of a particular atomic state is precisely defined, the time at which an electron is found in that energy state cannot be so defined. Likewise, if the momentum of an electron is precisely defined, its position in space cannot be so defined. On the other hand, the probability for finding an electron in a particular location relative to the atomic nucleus can be determined. After substitution of the appropriate potential energy terms needed to describe the atomic system, these probability density distributions may be found from solutions of the Schroedinger wave equation, which is a quantum-mechanical description of the system. This distribution, which is directly related to the wave function that describes the atomic state in a quantum-mechanical manner, must be distributed through a region of space and is hence a function of all three spatial variables, which are r, θ and ϕ in the spherical coordinate system. Because only two variables are available on the plane of a page to describe these functions, they are usually described by a series of graphs, which must then be assembled in the imagination of the reader to picture the complete three-dimensional distribution. The radial part of the distribution is dependent only on the quantum numbers n and l and is usually represented by the terminology R_{nl}^2. The radial distribution for selected states of a hydrogen atom are shown in Fig. 7. The dashed lines are proportional to the probability of finding the electron in the appropriate nl state in an incremental volume dv of constant magnitude at the indicated radial distance from the atomic nucleus. The solid lines are proportional to the probability of finding the electron in an incremental shell between the radial distances r and $r + dr$, with a volume $4\pi r^2 dr$ at the indicated radius. Radial distances in Fig. 7 are given in units of Bohr radii, the distance from the nucleus of the $n = 1$ orbit in the Bohr planetary model of the hydrogen atom.

In the quantum-mechanical description of a hydrogen atom, the radial portion of the probability density distribution is the same in all directions from the nucleus, but only for the case $l = 0$ is the magnitude of the distribution the same in all radial directions. For all other values of l, the magnitude of the distribution is a function of the angular direction, defined by the coordinates θ and ϕ. However, as in the case of the discussion of the vector model of an atom, we cannot define an angular direction unless an axis exists to provide a reference direction. To provide this axis some force or torque external to the electron configuration must be found. The only external force strong enough to interact with the electron configuration is that provided by certain magnetic fields. The interaction with the magnetic moment of the electron configuration can then be described in terms of a magnetic quantum number m_l, which is the magnetic quantum number associated with orbital angular momentum quantum number. When $l = 1$, m_l may have any of three values, $+1$, 0, or -1. Three possible angular distributions are then possible for the electrons described by a quantum number $l = 1$, but the distributions for $m_l = +1$ and $m_l = -1$ are identical, thereby providing some simplification. If we define $\theta = 0$ in the direction of the applied magnetic field, the probability density distribution in the $r\theta$-plane for an electron of a hydrogen atom described by $l = 1$, $m_l = 0$ is given in the $2p$, $m = 0$ part of Fig. 8. This distribution is symmetric in ϕ; it may thus be rotated around an axis perpendicularly directed through the center of the figure (the $\theta = 0$ axis) to give the full three-dimensional distribution. For either $m_l = +1$ or $m_l = -1$ the distribution is given by that part of Fig. 8 labeled $2p$, $m = 1$. When rotated about the $\theta = 0$ axis the $m_l = 0$ distribution gives a dumbbell-like distribution and the $m_l = 1$ distributions give ring-like distributions, all of which fade away to negligible magnitude at large distances from the center of the distribution. For the $2p$, $m_l = 0$ configuration, the angular distribution in the $r\theta$-plane is given by the equation $\frac{3}{2}\cos;^2\theta$, and for the $2p$, $m_l = \pm 1$ configuration by the equation $\frac{3}{4}\sin;^2\theta$. Since two $m_l = 1$ configurations exist for each $l = 1$, the distributions for all three m_l configurations, when summed, give a resultant distribution that is independent of the angle θ, as well as of the angle ϕ. This condition is reached when all possible electron states in the $2p$ subshell are filled. A similar situation, a complete symmetry of the electron distribution, exists for all filled subshells. Hence an external magnetic field interacts only with partially filled shells. In multielectron atoms only one shell is usually partially filled; thus interactions with external magnetic fields, such as those introduced artificially, by the atomic nucleus, or by neighboring atoms, occur only in one shell of the atom, sometimes called in chemical terminology the valence shell. For specific elements, the unfilled shell can be determined from Table 5. Probability density distributions for several other electron configurations besides $2p$ is given in Fig. 8.

The radial probability density distributions for individual electron configurations in atoms other than hydrogen are generally similar to those of Fig. 7, but, since the nuclear charge Z of these atoms is larger than for hydrogen, the electrostatic attractive force exerted by the nucleus on the innermost electron is stronger than for hydrogen and hence pulls its distribution closer to the nucleus. Outer electrons, however, are partially shielded electrostatically from the nucleus by the inner electrons and hence are influenced by a weaker force than that which would be provided by a bare nucleus. The least tightly-bound electron is held by a force that has an effective Z of about 1 but for the more tightly bound electrons the effective Z is much higher. The probability density distribution for the least tightly bound electron then extends to a distance comparable to the distribution for hydrogen, but the main part of the distribution is much closer to the nucleus. A typical distribution for all the electrons of a multielectron atom, in this case rubidium, is shown in Fig. 9.

As stated earlier in this entry, there are three p orbitals, with magnetic quantum numbers of 0, 1 and -1. The orbital for which $m = 0$ corresponds to the direction of the applied magnetic field, so that its effect is zero. This direction is conventionally taken as that of the z-axis, so that p_z, which is symmetrical about that axis, coincides with p_0. The other p orbitals, p_x and p_y, are perpendicular to p_z and to each other and correspond to standing waves built up by mixing of p_{+1} and p_{-1} which are oppositely directed waves.

An analogous situation for the d orbitals, for which $l = 2$, and hence $m_l = 2, 1, 0, -1$ and -2, gives rise to five d orbitals, designated as d_{xy}, d_{yz}, d_{x2-y2} and d_{z2}.

For further discussion of atomic orbitals see also **Orbitals**.

It follows from the earlier discussion that each orbital may be occupied, in accordance with the Pauli Exclusion Principle, by two electrons of opposed spin.

See also **Pauli Exclusion Principle**.

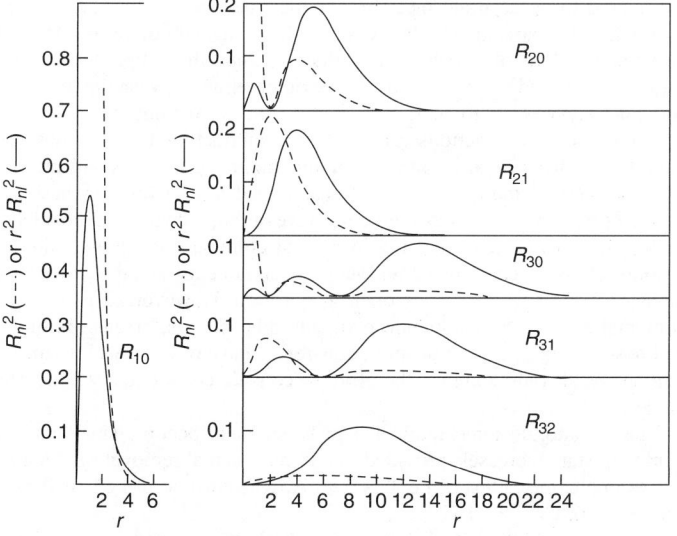

Fig. 7. Probability density distributions as a function of radial distance from the nucleus for several states of a hydrogen atom. The dashed lines are proportional to the probability of finding the electron in an incremental volume dv at the indicated radial distance. The solid lines are proportional to the probability for finding the electron in an incremental shell of volume $4\pi r^2 dr$ at the indicated radius

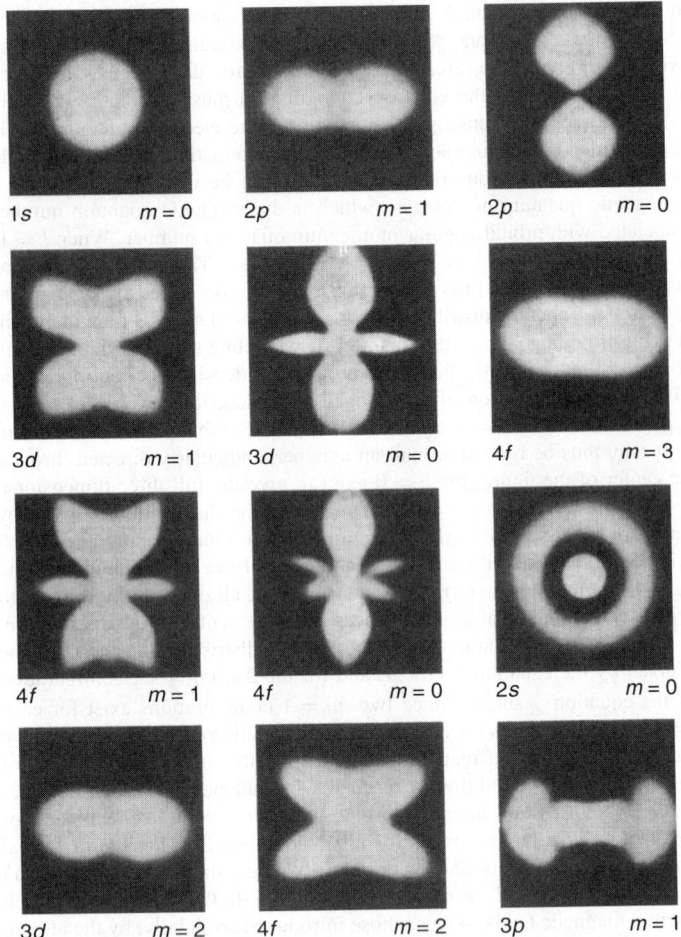

1s   m = 0

2p   m = 1

2p   m = 0

3d   m = 1

3d   m = 0

4f   m = 3

4f   m = 1

4f   m = 0

2s   m = 0

3d   m = 2

4f   m = 2

3p   m = 1

Fig. 8.   Electron wave density figures representing the single electron states of the hydrogen atom

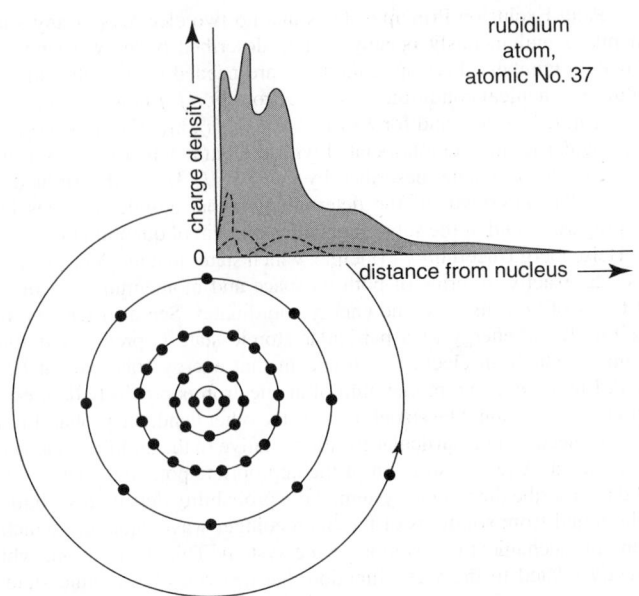

rubidium atom, atomic No. 37

charge density

0

distance from nucleus ⟶

Fig. 9.   Radial probability density distribution, derived from quantum-mechanical predictions, for rubidium, along with Bohr planetary model for the same atom

Atomic Radius

The radius of an atom may be defined as the distance of closest approach to another atom; it is the distance at which the mutual repulsion of the electron clouds and the mutual attraction of the nuclear charge of each for the electrons of the other are in equilibrium under specified circumstances. If the two atoms are the same, the radius of each is one half the internuclear distance; if they are unlike, the internuclear distance is the sum of the individual radii. Atomic radii fall roughly into four categories, which, however, merge into one another. These are Van der Waals and ionic radii, which are radii of equilibrium approach for mutually nonbonded atoms (neutral and charged, respectively), and covalent and metallic radii, which are for mutually bonded atoms (where the bonding electrons are, respectively, largely localized between the bonded atoms, or highly delocalized). Radii of all types vary in essentially the same manner from one part of the periodic table to another. In general, the radii decrease from the beginning to the end of any period, the rate of decrease being less the higher the number of the period. There are discontinuities in the decrease at points at which the quantum levels become half or completely filled. For example, Mn^{2+} ($r = 0.83$ Å) is larger than either Cr^{2+} (0.80 Å) or Fe^{2+} (0.80 Å), and Zn^{2+} (0.75 Å) is larger than Cu^{2+} (0.72 Å). (In this description, angstroms, Å, are used for convenience of notation. 1 angstrom = 10^{-10} meter.) Again, in general, the radii increase going down any column in the periodic table except in the region of hafnium. This last exception is caused by the lanthanide contraction occurring for the series of 14 elements between lanthanum and hafnium, which results in a far greater decrease between these two elements (average metallic radii: La 1.87 Å, Hf 1.58 Å) than between yttrium and zirconium (Y 1.80 Å, Zr 1.60 Å), where no such series intervenes. Because of the slower rate of decrease of the radius in the sixth period as compared to the fifth, however, the radii in the sixth period soon become larger again than those in the fifth (cf. Nb 1.429 Å, Ta 1.43 Å; Mo 1.36 Å, W 1.37 Å).

Van der Waals Radius

This is the radius of closest approach of one atom to another with which it does not form a chemical bond. It represents the distance at which the mutual repulsion of the nonbonding electrons of each exactly balances the attraction of the nucleus of each for the electrons of the other.

Ionic Radius

This is the radius of closest approach of a charged atom or group (i.e., an ion) to another atom or group with which it does not form a covalent bond. It represents the distance at which the mutual repulsion of the nonbonding electrons of each exactly balances the mutual attraction of oppositely charged ions, or the attraction of a positive ion for the electrons of a neutral atom or group or the attraction of a negative ion for the nucleus or nuclei of a neutral atom or group. Ionic radii for the cations of the least electronegative metals and of the most electronegative nonmetals are clearly defined, but for the elements of intermediate electronegativity (including the transition elements) they are much more doubtful because of the varying degrees of covalency in the compounds of these elements. Published ionic radii for these latter elements can therefore not be considered to be accurate measures of the true sizes of the ions, but are useful for comparison of relative sizes. In particular, there are certainly no simple cations of charge greater than 4+, and the simple tetrapositive cations are limited to Th^{4+}, probably the other tetrapositive actinide cations, and possibly Ce^{4+}. "Ionic radii" cited for such "cations" as Si^{4+}, P^{5+}, S^{6+}, Cl^{7+}, etc., are fictions arrived at by subtracting from an observed interatomic distance an arbitrary anionic radius for the second atom or else are extrapolated or theoretically calculated radii. However, inasmuch as all compounds of such elements have a high degree of covalency, such radii have no real meaning in normal compounds. Simple anions of absolute charge greater than 3 undoubtedly do not exist and the existence of monatomic trinegative ions is open to question. However, in their binary compounds with the lanthanide elements, phosphorus, arsenic, antimony and possibly nitrogen and bismuth apparently have radii very close to their Van der Waals radii and may therefore be considered essentially ionic. (See Table 6.)

The monatomic ionic radii of a given element become smaller as the oxidation state increases, provided this implies actual removal of electrons. For example, the radius of Fe^{2+} is 0.80 Å, whereas that of Fe^{3+} is 0.67 Å; $Cu^+ = 0.96$, $Cu^{2+} = 0.72$.

The ionic radii discussed above are properly "crystal radii," i.e., the radii exhibited by the ions in ionic crystals. Although these radii are probably reasonable representations of the radii of contact of the ions in solution with the nearest atoms of solvate molecules, especially for ions of low charge density, nevertheless, most ions in solution have far larger effective radii because they carry with them a sheath of solvent molecules, the tenacity

TABLE 6. IONIC CRYSTAL RADII (IN ANGSTROM UNITS)*

Element	Ion	Radius	Element	Ion	Radius	Element	Ion	Radius	Element	Ion	Radius
Actinium	Ac^{3+}	1.11	Copper	Cu^{+}	0.96	Mercury	Hg^{2+}	1.12	Scandium	Sc^{3+}	0.83
Aluminum	Al^{3+}	0.57		Cu^{2+}	0.72	Molybdenum	Mo^{4+}	0.68	Selenium	Se^{2-}	1.960
Americium	Am^{3+}	1.00	Curium	Cm^{3+}	1.00		(Mo^{6+})	0.65		(Se^{6+})	0.42
	Am^{4+}	0.85	Dysprosium	Dy^{3+}	0.908	Neodymium	Nd^{3+}	0.995	Silicon	(Si^{4+})	0.40
Antimony	Sb^{3-}	2.170	Einsteinium	Es^{3+}	0.97	Neptunium	Np^{3+}	1.02	Silver	Ag^{+}	0.97
	Sb^{3+}	0.90	Erbium	Er^{3+}	0.881		Np^{4+}	0.88	Sodium	Na^{+}	1.00
	(Sb^{5+})	0.62	Europium	Eu^{2+}	1.137	Nickel	Ni^{2+}	0.74	Strontium	Sr^{2+}	1.18
Arsenic	As^{3-}	1.991		Eu^{3+}	0.950	Niobium	Nb^{4+}	0.67	Sulfur	S^{2-}	1.855
	(As^{3+})	0.69	Fermium	Fm^{3+}	0.97		(Nb^{5+})	0.70?		(S^{6+})	0.29
	(As^{5+})	0.47	Fluorine	F^{-}	1.36	Nitrogen	N^{3-}	1.56	Tantalum	(Ta^{5+})	0.73
Astatine	At^{-}	2.2		(F^{7+})	0.07		(N^{5+})	0.11	Technetium	Tc^{4+}	0.5
Barium	Ba^{2+}	1.38	Francium	Fr^{+}	1.9	Nobelium			Tellurium	Te^{2-}	2.21
Berkelium	Bk^{3+}	0.99	Gadolinium	Gd^{3+}	0.938	Osmium	Os^{4+}	0.65		(Te^{4+})	0.84
Beryllium	Be^{2+}	0.31	Gallium	Ga^{3+}	0.65	Oxygen	O^{2-}	1.40		(Te^{6+})	0.56
Bismuth	Bi^{3-}	2.217	Germanium	Ge^{2+}	0.65		(O^{6+})	0.09	Terbium	Tb^{3+}	0.923
	Bi^{3+}	1.20		(Ge^{4+})	0.55	Palladium	Pd^{2+}	0.50	Thallium	Tl^{+}	1.50
	(Bi^{5+})	0.74	Gold	Au^{+}	1.37	Phosphorus	P^{3-}	1.920		Tl^{3+}	0.95
Boron	(B^{3+})	0.20	Hafnium	Hf^{4+}	0.86		(P^{5+})	0.34	Thorium	Th^{4+}	0.95
Bromine	Br^{-}	1.97	Holmium	Ho^{3+}	0.894	Platinum	Pt^{2+}	0.52	Thulium	Tm^{3+}	0.869
	(Br^{7+})	0.39	Hydrogen	H^{+}	2.08		(Pt^{4+})	0.55	Tin	Sn^{2+}	1.02
Cadmium	$Cd2+$	0.99	Indium	$In3+$	0.95	Plutonium	$Pu3+$	1.01		$(Sn4+)$	0.65
Calcium	$Ca2+$	1.06	Iodine	$I-$	2.16		$Pu4+$	0.86	Titanium	$Ti2+$	0.76
Californium	$Cf3-$	0.98		$(I7+)$	0.50	Polonium	$(Po4+)$	0.9		$(Ti4+)$	0.60
Carbon	$(C4-)$	2.60	Iridium	$Ir4+$	0.66	Potassium	$K+$	1.33	Tungsten	$W4+$	0.68
	$(C4+)$	0.15	Iron	$Fe2+$	0.80	Praseodymium	$Pr3+$	1.013		$(W6+)$	0.65
Cerium	$Ce3+$	1.034		$Fe3+$	0.67		$Pr4+$	0.87	Uranium	$U3+$	1.04
	$Ce4+$	0.941	Lanthanum	$La3+$	1.071	Promethium	$Pm3+$	0.98		$U4+$	0.89
Cesium	$Cs+$	1.70	Lead	$Pb2+$	1.18	Protactinium	$Pa3+$	1.06	Vanadium	$V2+$	0.82
Chlorine	$Cl-$	1.81		$(Pb4+)$	0.70	Radium	$Ra2+$	1.42		$V3+$	0.75
	$(Cl7+)$	0.26	Lithium	$Li+$	0.70	Rhenium	$(Re6+)$	0.52		$(V5+)$	0.59
Chromium	$Cr2+$	0.80	Lutetium	$Lu3+$	0.83	Rhodium	$Rh3+$	0.75	Ytterbium	$Yb2+$	1.02
	$Cr3+$	0.70	Magnesium	$Mg2+$	0.75		$Rh4+$	0.65		$Yb3+$	0.858
	$(Cr6+)$	0.52	Manganese	$Mn2+$	0.83	Rubidium	$Rb+$	1.52	Yttrium	$Y3+$	0.910
Cobalt	$Co2+$	0.78		$Mn3+$	0.52	Ruthenium	$Ru4-$	0.60	Zinc	$Zn2+$	0.75
	$Co3+$	0.65		$(Mn7+)$	0.46	Samarium	$Sm2+$	1.143	Zirconium	$Zr4+$	0.80
			Mendelevium	$Mv3+$	0.96		$Sm3-$	0.964			

* 1 angstrom = 10^{-10} meter.

and thickness of which is a function of the charge density and electronic structure of the ion. In consequence, ions of small crystal radius (e.g., Li^{+}) may act larger (e.g., have lower mobility) in solution than ions of large crystal radius (e.g., Cs^{+}).

The effective radius of an ion changes with coordination number. An ion of coordination number 4 (tetrahedral) has a radius 0.93–0.95 times as large as the same ion with coordination number 6, while an ion of coordination number 8 is about 1.03 times as large as the same ion with coordination number 6.

Effective spherical crystal radii for some polyatomic ions are given in Table 7.

Metallic Radius

Metals may be considered to be composed of cations bonded together by a cement of mobile electrons which are located in the conduction bands. Since the number of available energy levels in the conduction bands is a function of the number of available orbitals of the atoms making up the metal, and since the electron population of the nonbonding orbitals of the atoms and of the conduction band is a function of the number of valence electrons of the metal, the number and distance of nearest neighbors and the strength of the metal-metal bond varies in a fairly regular way across the periodic table. In particular, bond lengths are shortest and bond strengths are greatest in the vicinity of cobalt, rhodium and iridium, where the number of valence electrons and the number of available orbitals (nine) exactly match. Below this number there are too few electrons and above, too many, for maximum sharing and use of the bonding orbitals. The interatomic distances and coordination numbers of the elements in their metallic states are given in Table 8.

Covalent Radius. This is the radius of closest approach for atoms bonded together by electrons that are localized in the region between the atoms. It represents the distance at which the attraction of each nucleus for the bonding electrons is in equilibrium with the mutual repulsion of the two nuclei and the repulsion of the inner electrons of each atom for the inner electrons of the other.

The length of the covalent radius is a function of several factors, among which are (1) bond order (multiplicity), (2) electronegativity, (3) hybridization, (4) orbital overlap, (5) steric factors, and (6) special electronic effects.

(1) The effect of bond order is exemplified by the familiar shortening of the carbon-carbon bond in ethane (1.543 Å), graphite (1.4210 Å), benzene (1.397 Å), ethylene (1.353 Å), and acetylene (1.207 Å) as the bond order goes from 1 to $1\frac{1}{3}$ to $1\frac{1}{2}$ to 2 to 3. Bond orders affect the covalent radii of other elements similarly.

The determination of the standard single bond radius is not always a simple matter. In those cases where two like atoms are joined by an unquestionable single bond (as the carbon atoms in ethane) the standard single bond radius is one-half the internuclear distance. Similarly the single bond radii for nitrogen, oxygen, and fluorine are half the interatomic distances in N_2H_4, H_2O_2 and F_2. On the other hand, acceptance of half the interatomic distances in P_4, H_2S_2 and Cl_2 for the corresponding single bond radii is open to question, since in these and similar cases the possibility exists of multiple bond formation by overlap of the electron-filled p-orbitals

TABLE 7. CRYSTAL RADII OF POLYATOMIC IONS (Å)*

NH_4^{+}	1.48	OH^{-}	1.53
OH_3^{+}	1.35	SH^{-}	2.00
PH_4^{+}	1.61	SeH^{-}	2.15
BH_4^{-}	2.08	SiH_3^{-}	2.25
CN^{-}	1.92	TeH^{-}	(2.35)
NO_3^{-}	2.3		

* 1 angstrom = 10^{-10} meter.

TABLE 8. INTERATOMIC DISTANCES IN METALS (IN ANGSTROM UNITS)*

Element	Interatomic Distances	Coordination Number
Actinium (room temp.)	3.756	12
Aluminum (25°C)	2.8635	12 (f.c.c.)
Americium		
Antimony (25°C)	2.90, 3.36	3,3 (rhombohedral)
Arsenic	2.49₅, 3.33	3,3 (rhombohedral)
Barium (room temp.)	4.347	8 (b.c.c.)
Berkelium		
Beryllium (α-form, 20°C)	2.2260, 2.2856	6.6 (c.p. hex.)
Bismuth (25°C)	3.095, 3.47	3,3 (rhombohedral)
Boron	1.75–1.80	
Cadmium (21°C)	2.9788, 3.2933	6.6 (c.p. hex.)
Calcium (α-form, 18°C)	3.947	12 (f.c.c.)
(γ-form, 500°C)	3.877	8 (b.c.c.)
Californium		
Cerium (room temp.)	3.650	12 (f.c.c.)
	3.620, 3.652	6.6 (c.p. hex?)
(15,000 atm.)	3.42	(f.c.c.)
Cesium (−100°C)	5.264	8 (b.c.c.)
(−10°C)	5.309	8
Chromium (α-form, 20°C)	2.4980	8 (b.c.c.)
(β-form, >1850°C)	2.61	12 (f.c.c.)
Cobalt (18°C)	2.5061	12 (f.c.c.)
(room temp.)	2.505–2.498, 2.505–2.507	6.6 (c.p. hex.)
Copper (20°C)	2.5560	12 (f.c.c.)
Curium		
Dysprosium (room temp.)	3.503, 3.590	6.6 (c.p. hex.)
Einsteinium		
Erbium (room temp.)	3.468, 3.559	6.6 (c.p. hex.)
Europium (room temp.)	3.989	8 (b.c.c.)
Fermium		
Francium		
Gadolinium (20°C)	3.573, 3.636	6.6 (c.p. hex.)
Gallium (20°C)	2,44₂, 2.71₂, 2.74₂, 2.80	1,2,2,2 (orthorhombic)
Germanium (20°C)	2.4498	4 (diamond)
Gold (25°C)	2.8841	12 (f.c.c.)
Hafnium (α-form, 24°C)	3.1273, 3.1947	6.6 (c.p. hex.)
Holmium (room temp.)	3.486, 3.577	6,6 (c.p. hex.)
Indium (20°C)	3.2511, 3.3730	4,8 (f.c.t.)
Iridium (room temp.)	2.714	12 (f.c.c.)
Iron (α-form, 20°C)	2.4823	8 (b.c.c.)
(γ-form, 916°C)	2.578	12 (f.c.c.)
(δ-form, 1394°C)	2.539	8 (b.c.c.)
Lanthanum (α-form, room temp.)	3.739, 3.770	6,6 (c.p. hex.)
(β-form, room temp.)	3.745	12 (f.c.c.)
Lead (25°C)	3.5003	12 (f.c.c.)
Lithium (20°C)	3.0390	8 (b.c.c.)
(78°K)	3.111, 3.116	6,6 (c.p. hex.)
Lutetium (room temp.)	3.435, 3.503	6,6 (c.p. hex.)
Magnesium (25°C)	3.1971, 3.2094	6,6 (c.p. hex.)
Manganese (γ-form, 1095°C)	2.7311	12 (f.c.c.)
(δ-form, 1134°C)	2.6679	8 (b.c.c.)
Mendelevium		
Mercury (−46°C)	3.005	6 (rhombohedral)
Molybdenum (20°C)	2.7251	8 (b.c.c.)
Neodymium (room temp.)	3.628, 3.658	6,6(?) (modified c.p. hex.)
Neptunium (α-form, 20°C)	2.60–2.64	4 (orthorhombic)
(β-form, 313°C)	2.76	4 (tetragonal)
(δ-form, 600°C)	3.05	8 (b.c.c.)

Element	Interatomic Distances	Coordination Number
Nickel (18°C)	2.4916	12 (f.c.c.)
Niobium (20°C)	2.8584	8 (b.c.c.)
Nobelium		
Osmium (20°C)	2.6754, 2.7354	6,6 (c.p. hex.)
Palladium (25°C)	2.7511	12 (f.c.c.)
Phosphorus (black)	2.18,—	3,3 (orthorhombic)
Platinum (20°C)	2.7746	12 (f.c.c.)
Plutonium (γ-form, 235°C)	3.026, 3.159, 3.287	4,2,4 (f.c.c.)
Polonium (α-form, 10°C)	3.345	6 (cubic)
(β-form, 75°C)	3.359	6 (rhombohedral)
Potassium (78°K)	4.544	8 (b.c.c.)
Praseodymium (α-form, room temp.)	3.640, 3.673	6 (hexagonal)
(β-form, room temp.)	3.649	12 (f.c.c.)
Promethium		
Protactinium (room temp.)	3.212, 3.238	8,2 (b.c.t.)
Radium		
Rhenium (room temp.)	2.741, 2.760	6.6 (c.p. hex.)
Element	Interatomic Distances	Coordination Number
Rhodium (20°C)	2.6901	12 (f.c.c.)
Rubidium (20°C)	4.95	8 (b.c.c.)
(−196°C)	4.860	
Ruthenium (25°C)	2.6502, 2.7058	6.6 (c.p. hex.)
Samarium		
Scandium (room temp.)	3.256, 3.309	6,6 (c.p. hex.)
(room temp.)	3.212	12 (f.c.c.?)
Selenium (20°C)	2.321, 3.464	2,4 (hexagonal)
Silicon (20°C)	2.3517	4 (diamond)
Silver (25°C)	2.8894	12 (f.c.c.)
Sodium (20°C)	3.7157	8 (b.c.c.)
Strontium (α-form, 25°C)	4.3026	12 (f.c.c.)
(β-form, 248°C)	4.32, 4.324	6.6 (c.p. hex.)
(γ-form, 614°C)	4.20	8 (b.c.c.)
Tantalum (20°C)	2.86	8 (b.c.c.)
Technetium (room temp.)	2.703, 2.735	6.6 (c.p. hex.)
Tellurium (25°C)	2.864, 3.468	2,4 (hexagonal)
Terbium (room temp.)	3.525, 3.601	6,6 (c.p. hex.)
Thallium (α-form, 18°C)	3.4076, 3.4566	6,6 (c.p. hex.)
(β-form, 262°C)	3.362	8 (b.c.c.)
Thorium (α-form, 25°C)	3.595	12 (f.c.c.)
(β-form, 1450°C)	3.56	8 (b.c.c.)
Thulium (room temp.)	3.447, 3.538	6,6 (c.p. hex.)
Tin (α-form, 20°C)	2.8099	4 (diamond)
(β-form, 25°C)	3.022, 3.181	4,2 (tetragonal)
Titanium (α-form, 25°C)	2.8956, 2.9505	6,6 (c.p. hex.)
(β-form, 900°C)	2.8636	8 (b.c.c.)
Tungsten (25°C)	2.7409	8 (b.c.c.)
Uranium (α-form, room temp.)	2.77, 2.86, 3.28, 3.37	2, 2, 4, 4
(γ-form, 805°C)	3.058	8 (b.c.c.)
Vanadium (30°C)	2.6224	8 (b.c.c.)
Ytterbium (room temp.)	3.880	12 (f.c.c.)
Yttrium (room temp.)	3.551, 3.647	6,6 (c.p. hex.)
Zinc (25°C)	2.6649, 2.9129	6.6 (c.p. hex.)
Zirconium (α-form, 25°C)	3.1790, 3.2313	6,6 (c.p. hex.)
(β-form, 862°C)	3.1254	8 (b.c.c.)

* 1 angstrom = 10^{-10} meter.

of each atom with the empty d-orbitals of the other. This will result in an increase in strength and a decrease in length for these bonds compared with what they would have if they were exactly single bonds. In support of this idea, it can be seen from Table 9 that, although the homoatomic bond energy for silicon (where no electrons are available for multiple bonding) is less than that for carbon, the bond energies for phosphorus, sulfur, and chlorine are significantly greater than for nitrogen, oxygen, and fluorine, respectively. Representative bond strength s are given in Table 10.

In consequence of this effect, the single bond radii for such elements are better derived from the alkyl derivatives (with an electronegativity correction) in which only single bonding is possible. The single bond radii in Table 11, Covalent Radii of the Elements, were derived insofar as possible from such compounds. The multiple bond radii were derived similarly. For example, the oxygen double bond radius may be obtained from acetone by using the double bond radius for carbon taken from ethylene and applying the electronegativity correction discussed in the following paragraph.

(2) When two atoms of different electronegativity are connected by a covalent bond, the bond length is always shorter than the sum of the individual homoatomic covalent radii for the atoms in question. The shortening is proportional to the difference in electronegativity of the two elements and is expressed by the relationship due to Stevenson and Schomaker.

$$R = (r_A - r_B) - 0.09|x_A - x_B|$$

where

R = observed bond length

r_A, r_B = standard covalent radii of atoms A and B

x_A, x_B = electronegativities of A and B.

(3) It has been pointed out by a number of workers that for bonds of the same multiplicity, the lower the average value of the l quantum number in a hybrid bonding orbital, the shorter the bond should be. For example, the carbon-carbon single bond in $HC\equiv C-C\equiv CH$ which involves orbitals of sp hybridization is 1.37 Å long compared with the carbon-carbon bond in ethane (sp^3 hybridization) which is 1.543 Å long. Similarly the B—C bond

in $B(C_6H_5)_3$ ((sp^2) is shorter than in $B(C_6H_5)_4^-$ (sp^3). Thus it is found that a bond of given multiplicity has a length which is characteristic not only of the bond order, but also of the individual states of hybridization of the atoms. For example, the C—C bond in CH_3CN ($sp^3 + sp$) (1.46 Å) is almost exactly the average of the bonds in CH_3CH_3 (sp^3) and $N\equiv C-C\equiv N$ (sp) (1.37 Å). Though the data for other elements are less extensive than for carbon, and the interpretation frequently is much more complicated, the same general principles seem to apply to other elements as well.

(4) The effectiveness of overlap of bonding orbitals of the same symmetry appears to decrease as the principal quantum number increases and as the difference between the principal quantum numbers increases. This is reflected in the bond strengths shown in Table 10, The covalent radius of hydrogen is especially subject to effects of this kind, and has the values 0.3707, 0.362, 0.306, 0.284 and 0.293 Å respectively in H2, HF, HCl, HBr and HI. The apparent anomaly of the P—P, S—S, and Cl—Cl bonds being stronger than the N—N, O—O, and F—F bonds has been considered in paragraph (1).

(5) In the case of very large atoms or groups bonded to small atoms, the spatial requirements of the large groups may result in a bond lengthening. For example, it is probable that the C—I bond in Cl_4 is longer and weaker than in CH_3I because of the steric repulsions of the large iodine atoms. Again, the N—N bond in $[(CH_3)_3NN(CH_3)_3]^{2+}$ may be longer than in $H_3 NNH_3^{2+}$.

(6) Special effects of electronic or orbital structure may result in either lengthening or shortening a bond. The first situation is exemplified by such compounds as O_2N-NO_2, O_2N-X, $(C_6H_5)_3 C-C (C_6H_5)_3$ and the like, where the long bond is indicated in the formula. Cases of this sort involve molecules having a pi-electron system capable of accepting additional electrons (frequently in low-lying antibonding orbitals) so that the electrons required for the bond in question are partially drained away from it, leaving the bond weak and long. Thus the N—N bond in N_2O_4 has a length of 1.75 Å and a dissociation energy of 12.9 kcal compared with 1.47 Å and 60 kcal for N_2H_4.

Bond shortening, on the other hand, may occur in compounds of the most electronegative elements, notably fluorine. Typical examples are the fluoromethanes, in which the C—F bond lengths are: CH_3F 1.391 Å, CH_2F_2 1.358 Å, CHF_3 1.332 Å, and CF_4 1.323 Å. This has been explained in terms of electronegativity; i.e., that the polar C—F bond requires a high degree of p-character, thus releasing the s-orbital for the bonds to the less electronegative atoms, making them shorter and stronger. As more fluorine atoms are added, the s-orbital is more equally divided among the bonds, resulting in a regular shortening of the C—F bonds. This theory has been used to explain the supposed shortening of the C—C bond from 1.543 Å in C_2H_6 to 1.52 Å in C_2F_6. However, the experimental error attached to the latter value does not allow it to be considered really different from the former (a more recent value is 1.56 Å), and furthermore, the effect is not observed in other halo-substituted ethanes for which more accurate data are available: e.g., the C—C bond length in CF_3CN (1.464 Å) is if anything slightly longer than that in CH_3CN (1.458 Å), and microwave determinations on, $C_2H_5 Br$, C_2H_5Cl and C_2H_5 give 1.5508, 1.5508 and 1.540, respectively, for the C—C bonds. In addition, the C—H bond lengths in the fluoromethanes appear to be essentially constant: CH_4 1.092, CH_3F 1.109, CH_2F_2 1.092 and CHF_3 1.093. It should also be noted that the C—C stretching force constants in the two cyanides (4.50×10^{-5} and 4.55×10^{-5} dyne cm^{-1}, respectively) do not differ appreciably.

A theory that more satisfactorily explains all the known facts has been suggested by J.F.A. Williams (*Trans. Faraday Soc.*, **57**, 2089 (1961)). This proposes that the highly electronegative fluorine atom drains electron density away from the carbon atom in a C—F group sufficiently to make the lobe of the σ-antibonding (σ^*) orbital which is concentrated beyond the carbon atom available for π-bonding. In CH_3F where the hydrogen atoms have no nonbonding electrons to interact with the σ^* orbital, there is little or no effect. However, in CH_2F_2, where each fluorine atom has nonbonding electrons in p-orbitals of favorable disposition, the p-electrons of each interact with the σ^* orbital associated with the other to give a $p_\pi - \sigma_\pi^*$ bond, which results in a strengthening and shortening of both bonds.

This effect is observed in the shortening of such bonds as C—N in CCl_3NO_2 and $(CF_3)_3N$, C—P in $(CF_3)_3P$, C—S in $(CF_3)_2S$ and so forth.

TABLE 9. HOMOATOMIC SINGLE BOND ENERGIES (KCAL/MOLE)

C—C	78.9	Si—Si	53
N—N	39	P—P (in P$_4$)	48
O—O	35	S—S	58.1
F—F	38	Cl—Cl	57.87

TABLE 10. REPRESENTATIVE SINGLE-BOND ENERGIES (In kcal/mole)

H—H	104.18	O—Sb	71	C—Si	72	Cl—As	70
H—B	ca 93	O—I	ca 48	C—P	63	Cl—Se	58
H—C	98.7	F—F	38	C—S	65.6	Cl—Br	52.7
H—N	93.4	F—Si	135	C—Cl	78.2	Cl—Sn	76
H—O	110.6	F—P	117	C—Zn	40	Cl—Sb	74
H—F	135	F—S	68	C—Ge	ca 44	Cl—I	51
H—Si	76	F—Cl	ca 61	C—As	48	Cl—Hg	54
H—P	ca 77	F—As	111	C—Se	58	Cl—Bi	67
H—S	83	F—Se	68	C—Br	68	K—K	12.6
H—Cl	103.1	F—Br	61	C—Cd	32	Ge—Ge	45
H—As	ca 59	F—Te	80	C—Sn	54	Ge—Br	66
H—Se	ca 66	F—I	63	C—Sb	47	Ge—I	51
H—Br	87.4	Na—Na	18.4	C—I	51	As—As	35
H—Te	ca 57	Si—Si	53	C—Hg	23	As—Br	58
H—I	71.4	Si—S	60.9	C—Pb	31	As—I	43
Li—Li	27.2	Si—Cl	91	C—Bi	31	Se—Se	41
B—C	89	Si—Br	74	N—N	39	Br—Br	46.08
B—N	106.5	Si—I	56	N—O	48	Br—Sn	65
B—O	128	P—P	48	N—F	65	Br—I	43
B—F	154	P—Cl	78	N—Cl	46	Br—Hg	44
B—Cl	109	P—Br	63	O—O	35	Rb—Rb	11.5
B—Br	90	P—I	44	O—F	45.3	Sn—Sn	39
C—C	78.9	S—S	58.1	O—Si	108	Sn—I	65
C—N	72.8	S—Cl	61	O—P	ca 80	Sb—Sb	ca 29
C—O	85.5	S—Br	ca 52	O—Cl	ca 49	I—I	36.06
C—F	116	Cl—Cl	57.87	O—As	72	I—Hg	35
C—Al	61	Cl—Ge	81	O—Br	ca 48	Cs—Cs	10.4

TABLE 11. COVALENT RADII OF THE ELEMENTS (IN ANGSTROM UNITS)*

Element	Valence	Bond order	Hybridization	Radius	Element	Valence	Bond order	Hybridization	Radius
H	1	1	s	0.3754			1	d^2sp^3	1.43
Li	1	1	s	1.336			1	sp	1.34
Be	2	1	s	0.86		3	1	dsp^2	1.39
		1	sp^3	1.07	Zn	2	1	sp	1.15(?)
B	1	1	p	0.79			1	sp^3	1.34
	3	1	sp^2	0.84			1	sp^3d^2	1.46
		1	sp^3	0.92	Ga	1	1	p	1.29
C	4	1	sp	0.691		3	1	sp^2	1.23
		1	sp^2	0.74			1	sp^3	1.27
		1	sp^3	0.772	Ge	4	1	sp^3	1.225
		2	sp	0.643			1	sp^3d^2	1.3
		2	sp^2	0.666	As	3	1	p	1.218
		3	p	0.60		5	1	sp^3	1.19
		3	sp	0.602			1	sp^3d^2	1.33(?)
N	3	1	p	0.73_7	Se	2	1	p	1.21
	5	1	sp	0.700		4	1	pd	1.38
		1	sp^2	0.727		6	1	sp^3d^2	1.22(?)
		1	sp^3	0.74		2	2	p	1.075
	3	2	p	0.61	Br	1	1	p	1.193
	5	2	sp	0.617		5	1	pd	1.28
	3	3	p	0.60	Rb	1	1	s	2.06
	5	3	sp	0.638	Sr	2	1	s	1.49
O	2	1	p	0.745	Y	3			
		2	p	0.654	Zr	4	1	d^3s	1.42
	(6)	2	sp	0.58			1	d^5s	1.53
		3	p	0.599	Nb	5	1	d^4s	1.37(av.)
F	1	1	p	0.709			1	d^5s	1.5
Na	1	1	s	1.539	Mo	5	1	d^4s	1.30(av.)
Mg	2	1	s	1.20		6	1	d^3s	1.31
Al	1	1	p	1.22			1	d^5s	1.27
	3	1	sp^2	1.24			2	d^3s	1.23
		1	sp^3	1.26	Tc				
		1	sp^3d^2	1.44	Ru	4	1	d^2sp^3	1.38
Si	4	1	sp^3	1.176		7	2	d^3s	1.23
		1	sp^3d^2	1.31		8	2	d^3s	1.14
		2	sp^3	1.00	Rh	3	1	d^2sp^3	1.48
P	3	1	p	1.113	Pd	2	1	dsp^2	1.30
	5	1	sp^2	1.11	Ag	1	1	s	1.32
		1	sp^3	1.12			1	sp	1.42
		1	pd	1.23			1	sp^3	1.48
	5	1	sp^3d^2	1.20	Cd	2	1	sp	1.47
		2	sp^3	0.872			1	sp^3d^2	1.62
S	2,4	1	p	1.06	In	1	1	p	1.47
	6	1	sp^3	1.03		3	1	sp^2	1.47
		1	sp^3d^2	1.10			1	sp^3	1.43
	2	2	$p(p\pi)$	0.914			1	sp^3d^2	1.66
	4	2	$p(pd\pi)$	0.868	Sn	2	1	p	1.45
	6	2	sp^3	0.757		4	1	sp^3	1.405
Cl	1	1	p	1.050			1	sp^3d^2	1.47
	3	1	pd	1.135	Sb	3	1	p	1.376
	7	2	sp^3	0.681		5	1	sp^2	1.43
K	1	1	s	1.962			1	sp^3d^2	1.42(?)
Ca	2	1	s	1.39			1	pd	1.48
Sc					Te	2	1	p	1.39
Ti	4	1	d^3s	1.25		4	1	p^3d	1.34(?)
		1	d^5s	1.43			1	p^3d	1.55
V	4	1	d^3s	1.10		6	1	sp^3d^2	1.36(?)
	5	1	d^3s	1.17		2	2	p	1.279
		2	d^3s	1.05	I	1	1	p	1.360
Cr	3	1	d^2sp^3	1.45		5	1	pd	1.42
	6	1	d^3s	1.16	Cs	1	1	s	2.18
		2	d^3s	1.03	Ba	2	1	s	1.52
Mn	2	1	d^2sp^3	1.6	La				
	4	1	d^2sp^3	1.20	Hf				
	7	1	d^3s	1.13	Ta	5	1	d^4s	1.37(av.)
		2	d^3s	1.02			1	d^5sp	1.47(av.)
Fe	3	1	d^2sp^3	1.39			1	d^5sp^2	1.48(av.)
Co	2	1	sp^3	1.55	W	6	1	d^5s	1.32
	3	1	d^2sp^3	1.35	Re	4	1	d^2sp^3	1.44
Ni	2	1	dsp^2	1.28		6	1	d^2sp^3	1.37
		1	d^2sp^3	1.54		7	1	d^3s	1.25

(continued overleaf)

TABLE 11. (*continued*)

Element	Valence	Bond order	Hybridization	Radius	Element	Valence	Bond order	Hybridization	Radius
Cu	1	1	s	1.22			2	d^3s	1.22
		1	sp^3	1.38	Os	4	1	d^2sp^3	1.40
	2	1	dsp^2	1.29		8	2	d^3s	1.171
Ir	4	1	d^2sp^3	1.40	Pb	2	1	p	1.50
Pt	2	1	dsp^2	1.35		4	1	sp^3	1.44
	4	1	d^2sp^3	1.34	Bi	3	1	p	1.53
Au	1	1	s	1.24			1	p^3d^3	1.58
	4	1	sp^3d^2	1.51	Po	4	1	p^3d^3	1.58
Hg	2	1	$s(Hg_2{}^{2+})$	1.27	Th	4	1	d^3s	1.69
		1	sp	1.33	U	6	1	d^2sp^3	1.50
		1	sp^3	1.54			2	dp	1.41
Tl	1	1	p	1.54	Pu	4	1	d^2sp^3	1.72
	3	1	sp^3d^2	1.58	Am	5	2	dp	1.42

* 1 angstrom — 10^{-10} meter.

Such a mechanism, on the other hand, could not result in a shortening of the C—C bond in C_2F_6.

Table 11 gives standard covalent radii for most of the elements of the periodic table. Most of these have been calculated from the best data available in the literature using the considerations of paragraphs 1–3 above. Thus when radii of the appropriate multiplicity and hybridization are added and corrected for difference in electronegativity by the Stevenson and Schomaker relationship, the observed bond length will be obtained. For example, the O—F bond in OF_2 would have the value $0.745 + 0.709 - 0.9(0.5) = 1.409$ Å. The experimental value is 1.41 Å.

The sp^3 double bond radii for Si, P, S and Cl are taken from the paper of D.W.J. Cruickshank, *J. Chem. Soc.*, **5486** (1961). A calculation of the Cl<C:/eqbond>O bond length gives

$$0.681 + 0.654 - 0.9(0.5) = 1.290A$$

while for the Cl—O bond we calculate

$$1.050 + 0.745 - 0.9(0.5) = 1.750A$$

The observed Cl—O bond length in $ClO_4{}^-$ (1.48 Å) indicates that it has a bond order somewhat greater than 1.5.

The hybridization designations given in the table are not intended to be exact. In particular it must be recognized that the nitrogen bonding orbitals in the ammonia molecule, for example, have considerable s character. Nevertheless, such orbitals, for simplicity's sake, have been designated merely as p. This practice has been used uniformly when a nonbonding pair of electrons is found in the valence level. In the case of "sp^3d" hybridization, the radii are given as "sp^3d" if only average lengths are available, but are separated into sp^2 and pd if the data differentiated the two types of bonds. The values given for radii of the transition elements and for the less common hybridization states of the other elements, especially where the bond order is not accurately known, must be considered to be only approximate.

Valence

Valence is the capacity of an atom to combine with other atoms to form a molecule. Valence is specified as the number of hydrogen atoms or twice the number of oxygen atoms with which one atom of the element under question will combine. Thus, nitrogen has the valence 3,2,4,5 in the compounds NH_3, NO, NO_2, N_2O_5. A further distinction is made by considering positive and negative valences. If the hydrogen is assigned the valence of plus one, and oxygen that of minus two, and if the valences in a compound are made to total up to zero, we have a formal scheme of positive and negative valences. In ammonia, NH_3, the three hydrogen atoms each with a valence of plus one exactly balance the one nitrogen atom with the valence of negative three. Many atoms possess more than one valence but the principal valence is correlated with the periodic table and the atomic structure of the atom. The principal positive valence is the number of the group in which the element falls in the periodic table. Thus hydrogen is one, lithium also one, boron three, etc. The negative valence is eight minus the number of the group in the periodic table. Negative valences greater than four do not occur. For example, oxygen has the valence of eight minus six, that is two negative in H_2O (water): and nitrogen has the valence of eight minus five, that is three negative in NH_3 (ammonia).

On the basis of contemporary electronic theory of atomic structure we can classify the different types of valence. The guiding principle is that the atoms tend to assume an inert gas electronic structure of eight electrons in the outer shell (in the case of hydrogen it is two). To do this, the atom either loses to, gains from, or shares with other atoms, electrons. This process leads to molecule formation. The following are the principal types of valences and their electronic interpretation.

Electrovalence or polar valence is associated with a transfer of an electron from one element to the other in order to complete by such a transfer the octet of each element. Thus in sodium chloride the sodium atom has one valence electron outside a closed octet of eight. By loss of this electron the sodium atom becomes positively charged sodium ion because the nuclear positive charge exceeds that of the electrons by one. On the other hand, the chlorine atom has a grouping of seven electrons in the outer shell. It picks up another electron to complete its outer shell to an octet, but in so doing obtains a total charge of one minus, becoming a chloride ion. The result is that in sodium chloride we are not dealing with sodium atoms and chlorine atoms but with sodium and chloride ions. This is experimentally substantiated. The forces holding the ions together are the electrostatic forces, which are equal to the product of the electronic charges on the ions divided by the product of the separation squared times the dielectric constant of the medium. Thus when the sodium chloride crystal is placed in solvent of high dielectric constant such as water, the forces between the ions are weakened and the ions float away from each other. In other words, electrolytic dissociation takes place. It must be noted that polar valences have no specific directional effects in space. The electrostatic attraction is best satisfied by a close packing of the ions. Inasmuch as there are large stray electric fields present in polar compounds, they possess a high melting point and considerable hardness.

Homopolar or covalent bonds are formed by a different mechanism. Here again we have as the basis the tendency of each atom to complete its outer shell of electrons to eight, or in the case of hydrogen to a doublet. In contrast to polar valence, in covalence we have no direct transfer of electrons, but merely a sharing. In the case of molecular hydrogen each hydrogen atom with its one electron shares this electron with the other hydrogen. The result is that each atom in the molecule has at least part of the time a complete shell of two electrons. The electrons can be visualized as traveling in orbits encompassing the two hydrogen nuclei. It is a property of the covalent bond that it is not weakened by electrolytic solvents and that it has a definite direction in space. These directional effects of covalent bonds are expressed in stereochemistry. Thus, for example, the four valence bonds of the carbon atoms are arranged to extend from the center of a tetrahedron to the four corners. Furthermore, since there is a one-to-one saturation of the electron forces, the stray electric fields are negligible, the melting points are low, and the crystals are soft.

Intermediate in properties between the electrovalent and covalent bonds discussed above is the *semi-covalent bond* (also called *dative* or *polarized ionic bond*). It is formed when both electrons that constitute the bonding pair are supplied by one of the atoms. An example is the formation of amine oxides between tertiary amines and oxygen, in which both electrons are donated by the nitrogen atom. Such bonds naturally exhibit electrical polarity. They are members of the large class of heteropolar bonds characterized by an unequal distribution of charge due to a displacement of the electron-pair so that the effect of the bond is to make the atoms differ in polarity. In fact, atomic bonds are best described, not qualitatively, but in terms of bond angles and distances. In water, for example, the bond angle is 109.5°, indicating that the lines joining the two hydrogen atoms to the oxygen atom meet at this angle. However, there are two special types of bonds that deserve individual mention.

The *hydrogen bond* is actually two bonds, whereby two electronegative atoms are joined through a hydrogen atom. Since a stable hydrogen atom cannot be associated with more than two electrons, the hydrogen bond may be regarded as a resonance phenomenon, whereby the hydrogen atom is periodically attached to each of the two other atoms in turn, so that its behavior is a composite of the two structures.

Another type of bond that occurs in solids is the metallic bond. It can be considered as an extreme case of sharing of electrons in that an electron gas (present in the crystal lattice) is shared not by two ions but by all the ions in the lattice. This electron gas is responsible for the metallic properties of certain solids, especially for thermal and electrical conductivity.

As a consequence of the fact that many valence bonds leave residual electrical fields, many molecules in which the "primary" valences are satisfied can combine further with other molecules or with atoms. These higher combinations enter into many important areas of chemical science. They are the basis of the formation of coordination compounds, discussed under that heading. They cause molecular association. They are responsible for the formation of hydrates. They are in many cases the binding forces in nonionic solids, and are of great importance in explaining the structure of larger material aggregates.

The foregoing discussion of valence is, of course, a simplified one. From the development of the quantum theory and its application to the structure of the atom, there has ensued a quantum theory of valence and of the structure of the molecule, discussed in this book under **Molecule**. Topics that are basically important to modern views of molecular structure include, in addition to those already indicated: the Schroedinger wave equation; the molecular orbital method (introduced in the article on **Molecule**) as well as directed valence bonds; bond energies, hybrid orbitals, the effect of Van der Waals forces; and electron-deficient molecules. Some of these subjects are clearly beyond the space available in this book and its scope of treatment. Even more so is their use in interpretation of molecular structure. (However, see **Crystal Field Theory**; and **Ligand**.)

There are a number of terms used in describing the individual valence bonds. The *bond angle* is the angle between two bonds in a molecule, e.g., in water the bond angle is 109.5°, indicating that the lines joining the two hydrogen atoms to the oxygen atom meet at this angle.

The term *bond direction* arises from the fact that certain covalent bonds prefer to lie in particular directions with respect to the bonded atoms. For example, the bonds from carbon point from the center to the vertices of a regular tetrahedron.

Atoms sharing the two pairs of electrons are described as connected by a *double bond*. The bond energy of the C−C bond is 80 kcal and that of the C=C bond is 145 kcal. The second bond is formed by *p*-electrons and, while its energy effect is considerable, it does not produce a double bond having twice the energy of the single bond. Moreover, its electrons, being less firmly held between the carbon atoms, are available for addition reactions. These bonds between *p*-electrons, or π-bonds, tend to delocalize in many cases, i.e., the electronic charges "spread" over other atoms than those furnishing them.

The diagram below, which shows the bonding of the carbon atoms in the ethylene molecule, shows the carbon atoms connected by one of the sp^2 hybrid bonds (solid line), the other two being used for the hydrogen atoms. The dotted lines show the π-bond formed between the two *p*-electrons. Since they occupy *p*-orbitals that are perpendicular to the plane of the sp^2 bonds, they cannot form a bond without considerable overlapping. From the figures of 80 and 145 kcal for single and double bonds, the π-bond accounts for 44% of the energy of the double bond, indicating extensive overlapping.

Conjugated double bonds are two double bonds in positions connecting alternate pairs of carbon atoms. For example, the compound

$$CH_2{=}CH{-}CH{=}CH_2$$

has conjugated double bonds. In addition reactions, the conjugated double bond system commonly changes to a single double bond between the second and third carbon atoms, accompanied by the addition of atoms or groups to the first and fourth carbon atoms.

Triple Bond

A single C−C bond involves *sp-sp* overlapping of orbitals, while a C−H bond is the result of *sp-s* overlapping. The other two valences on carbon atoms are represented by two remaining π-electrons, occupying mutually perpendicular *p*-orbitals. In the case of *triple bonding*, −C≡C−, overlapping of these four *p*-orbitals gives two π molecular orbitals. Thus the carbon-carbon *triple bond* is conveniently represented as−C:C−, in which the π-electrons are shown occupying positions on the periphery of the carbon-carbon single bond. Their mutual repulsion reduces their bonding effectiveness, so that the bond energies for single, double and triple carbon-carbon bonds are 80, 145, and 198 calories, respectively.

Bond Energies

It has been suggested that ΔH^0_{298}, the heat of formation of a molecule from its constituent atoms (see also **Atomic Heat of Formation**), could be computed from a table of average bond energies, and the assumption of additivity:

$$\Delta H^0_{298} = \sum_{\substack{\text{all types} \\ \text{of bonds}}} n_{Xi-Xj} \cdot E_{Xi-Xj}$$

where n_{Xi-Xj} is the number of bonds between the two atomic species, X_i and X_j, in the molecule. E_{Xi-Xj} is the average bond energy associated with each of these bonds. Fairly accurate predictions of the heats of formations of organic molecules can be made in this way, particularly for the *larger* hydrocarbons, alcohols and other aliphatic derivatives.

The differences between the observed heats of formation of *cis-trans* isomers and of branched and unbranched hydrocarbon chains show that the additivity rule is not strictly rigorous. Various improvements have been suggested.

Single bond energies are given directly by the heat of dissociation of the corresponding molecules into neutral atoms. In cases where the molecule has no independent existence, other data may often be used. For example, the complete dissociation energy of a binary compound containing more than two atoms may be divided by the number of bonds broken in the dissociation, that is, the energy of the A−B bond may be taken as $\frac{1}{2}$ the dissociation energy of A−B−A into 2A and B, or as $\frac{1}{3}$ the dissociation energy of

into 3A and B. This multiple bond calculation yields, of course, an average value for the energy of the bonds involved. In the simple case of the H_2O molecule, the dissociation energies of the two successive steps $H_2O \longrightarrow H + OH \longrightarrow H + H + O$ differ by about 10%.

Representative single bond values were given in Table 10. Summation of such energies to obtain *average* values for molecules applies only when the constituent atoms exhibit their normal covalences and is subject to the exceptions already stated.

Pseudopotential Theory

It has been assumed for many years that the properties of matter (the chemical elements and the compounds that are built up from the atoms of the specific chemical elements) will follow the laws of quantum mechanics. Although the fundamental principles have been acknowledged, until recently it has been difficult to scale up these principles to aggregates of particles and to effectively apply fundamental theory to the prediction of materials properties, such as electrical conductivity, optical reflectivity, hardness, malleability, elasticity, and a number of chemical qualities. With a simple atom like hydrogen where its single electron is obviously a valence electron, calculations can be based on the potential energy of the electron. However, for atoms that contain two or more electrons, quantum scientists have found a mathematical approach impractical because the potential energy of a valence electron is affected by interactions with the more tightly bound core electrons of the atom. Further, these interactions are governed by the potential energy of each core electron. Consider, for example, the calculations that would be needed in connection with an electronic structure that may contain upward of 10^{20} particles or more. Some researchers have found that the interactions of the core and valence electrons have little effect outside the core. Thus, there is no requirement to know the true potential energy of all the electrons.

What is needed is a good approximation to the configuration of the valence electrons. Thus, as pointed out in a scholarly and detailed paper, Cohen, Heine, and Phillips observe, "The new method regards the core electrons and the atomic nucleus as if they constitute a single particle without internal structure. The method is called the *pseudopotential theory*." In their article, the authors review early computational strategy and problems, electron waves and standing wave patterns, the Fermi momentum, quantum momentum change, scattering in crystals, and applications to elemental solids and binary compounds. The authors conclude, "Perhaps the most remarkable aspect of pseudopotential theory is its capacity for steady growth through increasing physical and mathematical sophistication. The cumulative development has enabled pseudopotential theorists to keep basic physical issues as well as technical details in mind in describing the quantum structure of materials."

Among recent practical accomplishments from applying the theory are: (1) explanation of the properties of interfaces between metals and semiconductors and between two semiconductors; (2) calculation of the total energy of semiconducting materials with an accuracy of about 0.2 eV per atom for a variety of densities or volumes; and (3) forecasting the properties of light or heavy elements and covalent, ionic or metallic compounds throughout the periodic table more accurately, more rapidly, and more reliably than appraisal of any one light element a quarter of a century ago.

In the late twentieth century, a number of new instrumental techniques were developed for determining atomic properties with increased precision and reliability. Of marked importance is the increased facility for measuring minute dimensions and units of time at the respective nanometer and nanosecond levels. Laboratory techniques include laser atom probes, cold neutron research, scanning-tunneling microscopy, and atom trapping, among others.

Additional Reading

Amato, I.: "Mapping the Periodic Landscape of Elements," *Science News*, **390** (December 16, 1989).

Anderson, D.L.: "Composition of the Earth," *Science*, **367** (January 20, 1989).

Arnett, E.M., et al.: "Chemical Bond-Making, Bond-Breaking, and Electron Transfer in Solution," *Science*, **423** (January 26, 1990).

Atkins, P.W.: *General Chemistry: Solution Manual,* 6th Edition, W.H. Freeman Company, Salt Lake City, UT, 1997.

Bader, R.F.W.: *Atoms in Molecules: A Quantum Theory,* Oxford University Press, New York, NY, 1990.

Banjanin, M.: *Progressive Periodic Table of Elements Booklet & Chart,* Blackwell Science, Inc., Malden, MA, 1999.

Bard, A.J.: *Encyclopedia of Electrochemistry of the Elements,* Marcel Dekker, Inc., New York, NY, 1999.

Bauschlicher, C.W. Jr. and S.R. Langhoff: "Quantum Mechanical Calculations to Chemical Accuracy," *Science*, **394** (October 18, 1991).

Bonin, K.D. and V.V. Kresin: *Electric-Polarizabilities of Atoms, Molecules and Clusters,* World Scientific Publishing Company Inc., River Edge, NJ, 1997.

Buch, A.: *Pure Metals Properties: A Scientific and Technical Handbook,* ASM International, Materials Park, OH, 1999.

Burke, P.G. and C.J. Joachain: *Photon and Electron Collisions with Atoms and Molecules,* Plenum Publishing Corporation, New York, NY, 1997.

Cherfas, J.: "Proton Microbeam Probes the Elements," *Science*, **1500** (September 28, 1990).

Chu, W.: "Laser Manipulation of Atoms and Particles," *Science*, **861** (August 23, 1991).

Cohen-Tannoudji, C., Dupont-Roc, J., and G. Grynberg: *Atom-Photon Interactions: Basic Processes and Applications,* John Wiley & Sons, Inc., New York, NY, 1998.

Corcoran, E.: "Dimensioning Dimensions," *Sci. Amer.,* **122** (November 1990).

Crim, F.F.: "State- and Bond-Selected Unimolecular Reactions," *Science*, **1387** (September 21, 1990).

Delone, N.B., V.P. Krainov, et al.: *Multiphoton Processes in Atoms,* 2nd Edition, Springer-Verlag Inc., New York, NY, 2000.

DiSalvo, F.J.: "Solid-State Chemistry: A Rediscovered Chemical Frontier," *Science*, **649** (February 9, 1990).

Ebbing, D.D. and S.D. Gammon: *General Chemistry,* Houghton Mifflin Company, New York, NY, 2000.

Emsley, J.: *The Elements,* 3rd Edition, Oxford University Press, Inc., New York, NY, 1998.

Fowler, W.A.: "The Quest for the Origin of the Elements," *Science*, **226**, 922–935 (1984).

Gillard, R.D.: "Nature's Loaded Dice," Review (University of Wales), **21**(7), (1991).

Greenwood, N.N. and A. Earnshaw: *Chemistry of Elements,* 2nd Edition, Butterworth-Heinemann, Inc., Woburn, MA, 1997.

Greiner, W. and A. Sandulescu: "New Radioactivities," *Sci. Amer.,* **58** (March 1990).

Hill, J.W. and R.H. Petrucci: *General Chemistry: An Integrated Approach,* Prentice-Hall, Inc., Upper Saddle River, NJ, 1998.

Hinchliffe, A.: *Chemical Modeling: From Atoms to Liquids,* John Wiley & Sons, Inc., New York, NY, 1999.

Krebs, R.E.: *The History and Use of Our Earth's Chemical Elements: A Reference Guide,* Greenwood Publishers Group, Inc., Westport, CT, 1998.

Lide, D.R.: *CRC Handbook of Chemistry and Physics,* 84th Edition, CRC Press, LLC., Boca Raton, FL, 2003.

Lof, P.: *Elsevier's Periodic Table of the Elements,* Elsevier Science, New York, NY, 1990.

Ohrn, Y.N.: *Elements of Molecular Symmetry,* John Wiley & Sons, Inc., New York, NY, 2000.

Parker, P.: *McGraw-Hill Encyclopedia of Chemistry,* McGraw-Hill Companies, Inc., New York, NY, 1993.

Penzias, A.A.: *The Origin of the Elements,* Les Prix Nobel en 1978, Nobel Foundation, Stockholm; also "Nobel Lectures," (in English), Elsevier, Amsterdam and New York (1979); also reprinted in *Science,* **205**, 549–554 (1979).

Pool, R.: "Basic Measurements Lead to Physics Nobel," *Science,* **327** (October 20, 1989).

Roche, L.P.: *The Chemical Elements: Chemistry, Physical Properties, and Uses in Science and Industry,* Prentice-Hall, Inc., New Jersey, 1997.

Rossotti, H.: *Diverse Atoms: Profiles of the Chemical Elements,* Oxford University Press, Inc., New York, NY, 1998.

Saltpeter, E.E.: "The 1983 Nobel Prize in Physics," *Science,* **222**, 881–885 (1983).

Scott, P. and N. Kaltsoyannis: *The F Elements,* Oxford University Press, Inc., New York, NY, 1999.

Seaborg, G.T. and W.D. Loveland: *The Elements beyond Uranium,* John Wiley & Sons, Inc., New York, NY, 1990.

Seaborg, G.T.: *Modern Alchemy,* World Scientific Publishing Company, Inc., River Edge, NJ, 1994.

Servos, J.W.: *Physical Chemistry from Ostwald to Pauling,* Princeton University Press, Princeton, NJ, 1996.

Staff: *ASM Handbook—Properties and Selection: Nonferrous Alloys and Pure Metals,* ASM International, Materials Park, OH, 1990.

Wieberg, K.B. et al.: "The Response of Electrons to Structural Changes," *Science,* **1266** (May 31, 1991).

Yaws, C.L.: *Handbook of Thermal Conductivity: Inorganic Compounds and Elements,* Gulf Publishing Company, Houston, TX, 1997.

CHEMICAL EQUATION. By means of chemical formulas, the changes occurring during a chemical reaction can be expressed as an equation. Thus the reaction of 1 mole of sulfur with 1 mole of oxygen to produce 1 mole of sulfur dioxide is written as

$$S + O_2 \longrightarrow SO_2$$

The arrow is preferred to the equality sign, which does not emphasize the direction of the reaction. In addition to the identity of the reactants and products, the equation shows the number of atoms entering into the reaction, either in the atomic state or as constituents of molecules. It also shows the number of moles of each reactant and product, so that by use of the table of atomic weights, the relative masses can be computed.

Since the principle of conservation of masses applies to chemical reactions, coefficients must often be used in writing chemical reactions so that the number of atoms of products is equal to the number of atoms of reactants. An example is the reaction of *two* moles of hydrogen with *one* mole of oxygen to form *two* moles of water

$$2\,H_2 + O_2 \longrightarrow 2\,H_2O$$

In writing such equations, a convenient procedure is to write first an expression containing only the formulas, and then to add the smallest coefficients that will give the same number of atoms of products as of reactants. This operation is called balancing the equation.

Equilibrium reactions, such as that of acetic acid and ethyl alcohol to form ethyl acetate and water, which is cited in the entry on **Chemical Reaction Rate**, are indicated by use of the double arrow:

$$CH_3COOH + C_2H_5OH \rightleftharpoons HOH + CH_3COOC_2H_5$$

Reactions that result in the precipitation of a solid or the evolution of a gas are sometimes denoted by vertical arrows:

$$AgNO_3 + NaCl \longrightarrow AgCl \downarrow + NaNO_3$$

$$Na_2CO_3 + 2\,HCl \longrightarrow CO_2 \uparrow + H_2O + 2\,NaCl$$

This information about the state of the products may also be denoted by writing after their formulas the expressions (s), (l), or (g).

In some cases, as in reactions in electrochemical cells or other reactions involving oxidation-reduction, the half reactions of the ions are useful. Consider the Daniell cell, which consists of a zinc electrode in a zinc sulfate solution, and a copper electrode in a copper solution, the two solutions being separated by a porous partition. The half reactions are

$$Zn \longrightarrow Zn^{2+} + 2e^-$$

$$Cu^{2+} + 2e^- \longrightarrow Cu$$

so that the overall reaction is

$$CuSO_4 + Zn(s) \longrightarrow ZnSO_4 + Cu(s)$$

The more difficult oxidation-reduction equations can often be written more easily by use of the Stock system of oxidation numbers, which are positive or negative valences or charges. Consider the reaction of potassium dichromate, $K_2Cr_2O_7$, with potassium sulfite, K_2SO_3, in acid solution to form chromium(III) sulfate, $Cr_2[SO_4]_3$, and potassium sulfate, K_2SO_4. The unbalanced expression for the ionic reaction is

$$Cr_2O_7{}^{2-} + SO_3{}^{2-} \longrightarrow 2\,Cr^{3+} + SO_4{}^{2-}$$

Since the oxidation number of the combined oxygen atom is $2-$ throughout, that of the chromium atom in $Cr_2O_7{}^{2-}$ is $6+$, that of the Cr^{3+} ion is obviously $3+$, that of the sulfur atom in $SO_3{}^{2-}$ is $4+$, and that of the sulfur atom in $SO_4{}^{2-}$ is $6+$. The total loss in oxidation number by the two chromium atoms is therefore $(2 \times 6) - (2 \times 3) = 6+$. Since this loss must be offset by a gain made by the sulfur atoms, and since one sulfur atom gains $2+$, the reaction must require 3 sulfur atoms. Therefore, the next partially balanced equation is written as

$$Cr_2O_7{}^{2-} + 3\,SO_3{}^{2-} \longrightarrow 2\,Cr^{3+} + 3\,SO_4{}^{2-}$$

Counting the charges in this expression shows that there are 8 negative charges on the left-hand side and a net total of 0 charges on the right-hand side. Therefore, since the reaction occurs in acid solution, requiring that hydrogen ions be present, $8H^+$ are added to the right-hand side to balance the expression electronically, giving

$$Cr_2O_7{}^{2-} + 3\,SO_3{}^{2-} + 8\,H^+ \longrightarrow 2\,Cr^{3+} + 3\,SO_4{}^{2-}$$

Now it is balanced in number of atoms by counting the hydrogen ions (8 on the left-hand side), and the oxygen atoms (an excess of 4 on the left-hand side). Therefore $4\,H_2O$ is added to the right-hand side:

$$Cr_2O_7{}^{2-} + 3\,SO_3{}^{2-} + 8\,H^+ \longrightarrow 2\,Cr^{3+} + 3\,SO_4{}^{2-} + 4\,H_2O$$

If the molecular equation is wanted, it can be written by grouping the ions, and adding those that did not enter into the ionic equations, i.e., the potassium ions of the salts and the anions of the acid:

$$K_2Cr_2O_7 + 3\,K_2SO_3 + 4\,H_2SO_4 \longrightarrow Cr_2(SO_4)_3 + 4\,H_2O + 4\,K_2SO_4$$

CHEMICAL EQUILIBRIUM. The fundamental law of chemical equilibrium was enunciated by Le Châtelier (1884), and may be stated as follows: If any stress or force is brought to bear upon a system in equilibrium, the equilibrium is displaced in a direction which tends to diminish the intensity of the stress or force. This is equivalent to the principle of least action. Its great value to the chemist is that it enables him to predict the effect upon systems in equilibrium of changes in temperature, pressure, and concentration.

The chemical system hydrogen-nitrogen-ammonia furnishes a notable example of the application of the principle:

$$\underbrace{\underset{1\,vol.}{nitrogen} + \underset{3\,vol.}{hydrogen}}_{4\,vol.} \rightleftharpoons \underset{2\,vol.}{ammonia} + \underset{per\,mole\,ammonia}{12,000\,carlories}\,heat$$

At the temperature $700°C$ and pressure 1 atmosphere, the equilibrium percentage of ammonia is 0.03 in the above system, and at 100 atmospheres

2.5. Increase of pressure shifts the equilibrium toward the side of the smaller total volume, at a constant temperature. Decrease of pressure shifts the equilibrium toward the side of the larger total volume, at a constant temperature. Systems of the same initial and final volumes are unaffected, as to equilibrium amounts of materials, by change of pressure.

At the pressure 100 atmospheres, and temperature $700°C$, the equilibrium percentage of ammonia is 2.5 in the above system, at $600°C$ it is 5, at $500°C$ it is 10. Increase of temperature shifts the equilibrium in the direction which absorbs heat, at a constant pressure. Decrease of temperature shifts the equilibrium in the direction which evolves heat (van't Hoff's principle, 1884).

At constant pressure and temperature, the equilibrium is shifted away from the side subjected to an increase in concentration of any constituent, or toward the side subjected to a decrease in concentration of any constituent. (See also **Chemical Reaction Rate.**) For the qualitative effect of temperature change, one may visualize the heat of an equilibrium reaction as material, and an increase of temperature (heat intensity) as operating to increase the concentration of "heat material," thus shifting the equilibrium away from the side of its increased concentration, and conversely. It is possible, knowing the heat of reaction, Q, on the assumption that the heat of reaction is constant between two given (absolute) temperatures, T_1 and T_2, to calculate the equilibrium constant K_2 (at T_2) when the equilibrium constant K_1 (at T_1) and the gas constant, R (equals 2 calories per mole) are known, by the application of van't Hoff's equation:

$$\log_{10} K_2 - \log_{10} K_1 = \frac{Q}{2.3 \times R}\left(\frac{1}{T_2} - \frac{1}{T_1}\right)$$

In this way the quantitative effect of temperature change on the state of equilibrium may be calculated.

In reactions of the ammonia synthesis type, to which sulfur trioxide from sulfur dioxide plus oxygen also belongs, the rate of reaction decreases with lowering of the temperature as the conversion is increased. There is, in such types of reactions, a limit to the practicable lowering of the temperature. The finding of a positive catalyzer for a given reaction of this sort permits the operation to gain the advantage of equilibrium conversion at the lower temperature as well as the increased rate of reaction at that temperature due to the presence of the catalyzer. (See also **Chemical Reaction Rate.**) The time yield of product is, therefore, very important, and, with a catalyzer, the space-time yield.

Systems in equilibrium are divided into two great divisions, according to whether they are (A) homogeneous, that is, chemically and physically uniform throughout, or (B) heterogeneous, that is, not uniform throughout but consisting of two or more phases. Each phase is a homogeneous, physically distinct, and mechanically separable portion of a system. For example, ice, water, water vapor are three different phases (solid, liquid, gas) of the substance water. There can be only one gas phase of a system, and only one liquid phase where a *single* homogeneous solution is present. But the number of liquid and of solid phases in general is limited by the number of components (not constituents) of a system. The number of components is the least number of constituents *independently* variable and requisite to compose each and every phase. For example, the system consisting of saturated solution in water, H_2O, of sodium sulfate, Na_2SO_4, plus solid sodium sulfate decahydrate, $Na_2SO_4•10H_2O$, plus water vapor consists of three phases, (a) gas, (b) solution, (c) solid sodium decahydrate. The *least* number of constituents independently variable in amount *and* requisite to compose each and every phase is two, namely, Na_2SO_4 and H_2O. These, therefore, are the two components of this system. Since zero and negative as well as positive amounts of compounds are permitted in expressing the composition of each phase of any system, the three phases of this system are composed of the following components:

gas phase, zero	Na_2SO_4 plus H_2O
liquid phase,	Na_2SO_4 plus H_2O
solid phase,	Na_2SO_4 plus H_2O

The number of components in the ice-water–water-vapor system is one, namely, H_2O.

To systems in which equilibrium depends solely upon the variables, (1) composition of each and every phase, (2) temperature, and (3) pressure, the phase rule (Willard Gibbs, 1874) applies: The number of variables, that is (1) the number of components, C, plus (2) temperature plus (3) pressure, above, equals the number of phases, P, plus the number of degrees of freedom, F. The number of degrees of freedom of a system is the least number of the above variables which must be arbitrarily fixed in order to define the condition of the system:

$$C + 2 = P + F$$

The phase rule applies to true equilibrium systems, where the equilibrium can be reached from either side, and, furthermore, takes no account of the time involved to attain equilibrium. The phase rule is a qualitative statement, whereas the law of mass action (concentration effect) is quantitatively applicable to those equilibrium systems where the reaction which occurs may be considered to take place in a homogeneous system, e.g., gas phase, or solution phase. (See also **Chemical Reaction Rate**.).

In a one-component system, $P + F = 3$, and physical changes only occur. When only one phase is present, for example, liquid water (no vapor, no solid), the system is bivariant. That is, two variables—temperature and pressure—may be independently changed over a range. When a second phase, either vapor or solid, appears through a sufficient change of temperature or pressure or both, or when two phases are originally present, the system is univariant. That is, one variable—either temperature or pressure—may be independently changed over a range. When the third phase appears or when three phases are originally present, the system is invariant. That is, a change of either temperature or pressure destroys the equilibrium, and the disappearance of one of the phases occurs. A system of one component in three phases is invariant and the conditions are represented by a point known as the triple point. The triple point for water is 0.007°C, 4.6 millimeters mercury pressure. When the total pressure is one atmosphere (760 millimeters) the equilibrium temperature of water-ice is 0.000°C, and when the water vapor pressure is one atmosphere the equilibrium temperature of water–water vapor is 100.000°C.

If, in dealing with any system, the gas phase or pressure may be neglected, on account of constancy or slightness of effect, the phase rule is simplified for practical purposes to $C + 1 = P + F$, and, if both may be neglected, to $C = P + F$.

Many two- and three-component systems have been studied and recorded in detail. The iron-carbon system is one that has attracted much attention and been of great value in iron metallurgy.

In 1977, Professor Ilya Prigogine of the Free University of Brussels, Belgium, was awarded the Nobel Prize in chemistry for his central role in the advances made in irreversible thermodynamics over the last three decades. Prigogine and his associates investigated the properties of systems far from equilibrium where a variety of phenomena exist that are not possible near or at equilibrium. These include chemical systems with multiple stationary states, chemical hysteresis, nucleation processes which give rise to transitions between multiple stationary states, oscillatory systems, the formation of stable and oscillatory macroscopic spatial structures, chemical waves, and the critical behavior of fluctuations. As pointed out by I. Procaccia and J. Ross (*Science*, **198**, 716–717, 1977), the central question concerns the conditions of instability of the thermodynamic branch. The theory of stability of ordinary differential equations is well established. The problem that confronted Prigogine and his collaborators was to develop a thermodynamic theory of stability that spans the whole range of equilibrium and nonequilibrium phenomena.

CHEMICAL FORMULA.

The formulas of chemistry constitute a shorthand notation used to represent the composition by weight, the molecular properties, the characteristic chemical reactions or at times even the ordering of the atoms in space of the elements which go to make up the chemical compound. Chemical formulas are classified into empirical, molecular, structural, or configurational, the order given being that of increasing content of information. The meaning of empirical and the molecular formulas is explained in the entry on **Chemical Composition**, which also describes methods for determining the formulas for some simple compounds. See also **Atomic Mass (Atomic Weight)**. Their determination for compounds in general, especially if they are present in mixtures, requires considerable experimental work. The first step consists of the isolation of a pure chemical compound. Chemical purification can be obtained by methods such as crystallization, distillation, adsorption, and sublimation. Some of the criteria of purity a substance must satisfy are constancy and sharpness of melting point and boiling point on repeated purification. As an example, let us assume that we have succeeded in purifying a solid compound which we shall call tartaric acid and whose formula we wish to determine.

The second step consists in a qualitative and quantitative analysis of the compound. In the case of tartaric acid, qualitative analysis tells us that

the compound contains carbon, oxygen and hydrogen, while quantitative analysis shows that the proportions are 48 parts by weight of carbon, 96 of oxygen, and 6 of hydrogen. To obtain the empirical formula, one divides each proportion by the atomic weight of the particular element, obtaining in this way a set of numbers that can be represented by a ratio of small integers. The simplest ratio of integers is commonly used to indicate as subscripts on the right of the chemical symbol of the element to represent the empirical formula. In the case of tartaric acid, the atomic weights are approximately 12 for carbon, 16 for oxygen, and 1 for hydrogen. Dividing the percentages as determined by analysis by the atomic weights, we get:

$$\text{carbon } \frac{48}{12} = 4.00$$

$$\text{oxygen } \frac{96}{16} = 6.00$$

$$\text{hydrogen } \frac{6}{1} = 6.00$$

The set of numbers is 4,6,6 and can be presented, in this case, by the ratio of integers 2:3:3. The empirical formula is therefore $C_2O_3H_3$. Empirical formula is thus only a convenient method for representing the percentage composition by weight of the different elements in the compound. The third step is the determination of the molecular weight of the compound in question. This allows us to assign to the compound a molecular formula. The molecular weight can be determined in a variety of methods, such as by the determination of the weight of 22.242 liters of the vapor of the substance at 1 atmosphere pressure and 0°C, temperature. Other methods are based on the differences in the boiling point or freezing point of solutions of known concentration and those of the pure solvent. To determine the molecular formula from the knowledge of the empirical formula and the molecular weight, the following procedure must be followed: Multiply the atomic weight of each element by its subscript, as indicated in the empirical formula, and add the result. On comparison of such a sum with the molecular weight it will be found that the molecular weight is equal to the sum times an integer. To obtain the molecular formula, multiply each subscript in the empirical formula by this integer and obtain a new set of subscripts. We found the empirical formula of tartaric acid was $C_2O_3H_3$. The sum mentioned above is

$$12 \times 2 + 16 \times 3 + 1 \times 3 = 75$$

The molecular weight determined experimentally is 150. The integer multiple is 2, and the molecular formula becomes $C_4O_6H_6$.

The molecular weight of the compound can be obtained from the molecular formula by summing the products obtained by multiplication of the atomic weights of the elements times their subscripts in the molecular formula. The latter contains all the information that the empirical formula contains but in addition specifies the number of atoms in the molecule and also the molecular weight of the substance.

Important as the molecular formula is, it does not describe fully the properties, or even in some cases the identity, of chemical compounds. For example, there are two compounds that have the molecular formula C_2H_6O. They are different in all their properties, both chemical and physical. This difference is due to a difference in the manner in which the atoms are connected in the molecules of the two substances. These differences can be shown only by the use of structural formulas, such as those shown in Fig. 1, in which the valence bonds between the atom are shown. These structural formulas are determined circumstantially, that is, by the chemical reactions into which the compounds enter. (However, their arrangements have been confirmed in many cases by a direct instrumental means such as spectrometric methods, x-ray studies, etc.) These reactions differ markedly for ethyl alcohol and methyl ether. Such compounds which have the same molecular formula but differ due to the arrangements or positions of their atoms are called isomers, and the type just cited, in which the difference is in the grouping of the atoms, are called *functional isomers*. These, and many other types of isomers, are treated in the entry on **Isomerism**.

The structural formula is also a shorthand notation for the important chemical reactions of the compound. It can be considered as being built up of a group of organic radicals, i.e., groups of atoms that retain their individuality in the course of certain reactions. Each radical has reactions which are characteristic of its presence in the molecule.

For instance, the carboxyl radical will react with alkali such as sodium hydroxide to form salts with phosphorus pentachloride to form acid

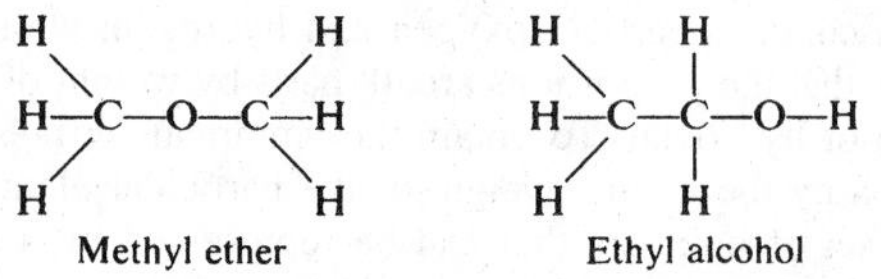

Fig. 1.   Examples of structural formulas

chlorides with alcohols to form esters; with reducing agents under certain conditions to form successively the aldehyde radical and the alcohol radical. Any pound that undergoes such reactions is said to contain a carboxyl group. The number of such carboxyl groups in a molecule can be determined by studying the above reactions quantitatively. On the other hand, if the compound will react: with sodium to give off hydrogen; with phosphorus trichloride to give a halogen substitution product which can be reduced to hydrocarbon; with an oxidizing agent to give an aldehyde or ketone, with organic acids to form esters; and with alcohols to form ethers; then the molecule is said to contain a hydroxyl group—OH. Analogously, there are similar characteristic reactions for a variety of radicals. It often happens that the presence of one type of a radical near another type mutually influences their reactivity, but one can consider to the first approximation that the radicals act independently of each other. The structural formula is considered completely established if one can synthesize the compound by simple clear-cut reactions involving no rearrangements on the basis of the proposed formula.

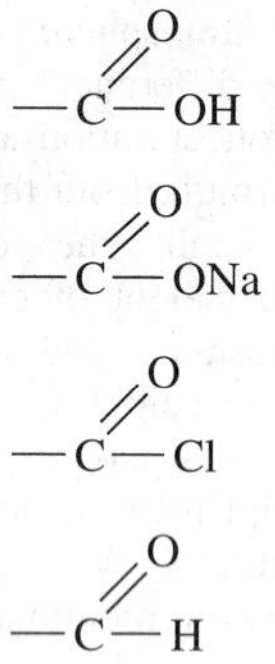

Just as it was stated above that two compounds may have the same molecular formula and yet have quite different structural formulas and properties, so there are also many instances in chemistry of compounds which have the same planar structural formula and yet differ in properties. In such cases their differences in structure can be shown only by three-dimensional formulas or their projections which portray differences in the arrangement in space of the atoms or radicals that make up the molecules of the two compounds.

Thus, in cases where four different atoms or groups are attached to the same atom, it is possible to have two arrangements in space that cannot be made to coincide geometrically. This situation can be demonstrated by use of a special type of formula, shown in Fig. 2 for the two forms of the compound fluorochlorobromomethane. This existence of two forms due to a difference in orientation in space is called *stereoisomerism*, and is discussed in the entry on **Isomerism**. It also follows that for compounds containing more than one atom bonded to four unlike groups, the number of different forms increases rapidly, as is shown by the three possible forms of tartaric acid, HOOC−CH−OH−CHOH−COOH, as portrayed by the three formulas shown in Fig. 3.

Still other types of formulas showing the spatial positions of atoms and groups are perspective and projection formulas, as shown in Fig. 4 for the

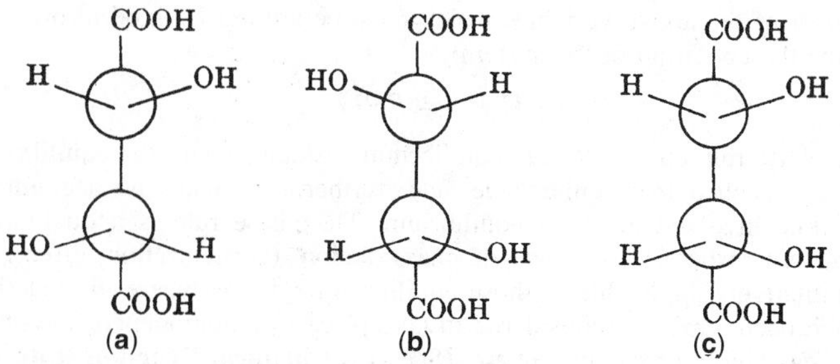

Fig. 3.   Formulas to demonstrate stereoisomerism

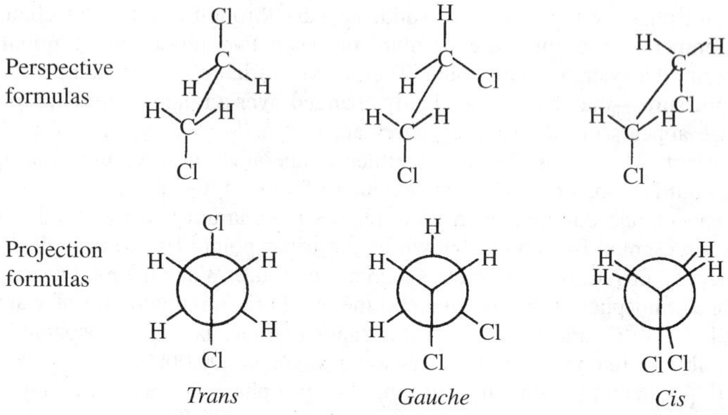

Fig. 4.   Perspective and projection type formulas

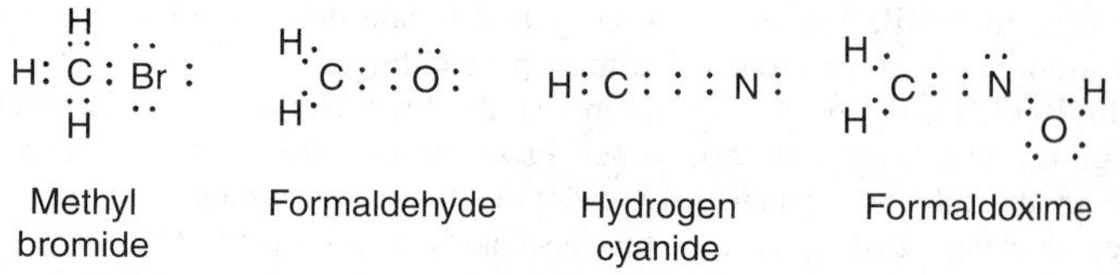

Fig. 5.   Formulas to demonstrate bonding

compound 1,2-dichloroethane. These differences are due to differences in conformation, that is, to the various configurations of a molecule, which differ in space by the rotation of two atoms about a single bond.

Another type of formula often written for compounds is the electronic formula, showing the distribution of the valence electrons among the atoms of the molecule, as shown in Fig. 5 and explained under valence and molecule.

CHEMICAL MANUFACTURERS ASSOCIATION (CMA).   Founded in 1863, this nonprofit trade association of chemical manufacture represents more than 90% of the chemical industry of the U.S. and Canada. It is located at 2501 M St., N. W., Washington, DC 20037. It is particularly active in the fields of packaging, transportation, and safe handling of hazardous chemicals and wording of precautionary labels, as well as in general chemical education and all aspects of pollution control. It has instituted an emergency telephone information service called *ChemTrec* to provide instant information for safety precautions in accidents involving chemicals.

CHEMICAL POTENTIALS.   Chemical potentials are defined in terms of the entropy by the relationship

$$\mu_i = -T \left(\frac{\partial S}{\partial n_i} \right)_{U,V} \qquad (1)$$

Apart from the factor T (the absolute temperature) the chemical potential is equal to the change of the entropy due to the introduction of the mole number **i** into the system, at constant total energy U and volume V. The parentheses in the above equation contain the partial derivative representing this rate of change.

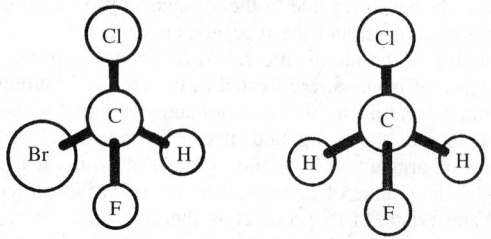

Fig. 2.   Formulas to indicate spatial geometry of compounds

Other independent variables are often much more convenient. One has also

$$\mu_i = \left(\frac{\partial U}{\partial n_i}\right)_{S,V} = \left(\frac{\partial G}{\partial n_i}\right)_{S,p}$$
$$= \left(\frac{\partial A}{\partial n_i}\right)_{T,V} = \left(\frac{\partial G}{\partial n_i}\right)_{T,p} \tag{2}$$

The last member of Equation (2) shows that μ_i is the partial molar quantity associated with the Gibbs free energy, G. Euler's theorem gives then

$$G = \sum_i n_i \mu_i \tag{3}$$

The relation to chemical affinity is also very direct

$$A = -\sum_i v_i \mu_i \tag{4}$$

Note that the roman capital A represents chemical affinity, while the italic capital A symbolizes the Helmholtz free energy (also called work function). It follows that the condition for chemical equilibrium is

$$\sum_i v_i \mu_i = 0 \tag{5}$$

where the v_i are stoichiometric coefficients. This formula expresses the law of mass action. Similarly the condition for two phases α and β to be in equilibrium with respect to species i is

$$\mu_i^{\alpha} - \mu_i^{\beta} \tag{6}$$

The chemical potential has then the same value in the two phases. See also **Thermodynamics**.

CHEMICAL REACTION RATE.

The chemical composition of a substance is subject to various changes under various conditions, depending upon (1) the nature of the specific substance, (2) the nature of other substances present, and (3) the environment in which it exists (the physical and chemical ambient conditions). Similarly, chemical reaction rates are affected.

The majority of reactions take place between two substances—occasionally one substance only, and sometimes three or more different substances. There are many cases when simple contact of the substances is sufficient to bring about the chemical change, e.g., the rusting of iron in oxygen. In many other cases, the change is not spontaneous, but must be induced, frequently by raising the temperature, as in the burning of fuels. The conditions that are considered important and fundamental are (1) temperature, (2) pressure, (3) medium, if any, (4) catalyzer, if any (5) electric direct current, (6) light. In a given reaction, the change in composition of the substance or substances involved is inherently connected with a change in energy. Thermal, electrical or light energy of a certain potential or intensity and in definite amounts, is requisite to initiate and carry on the reaction, and thermal, electrical or light energy of definite amount is liberated or consumed in the reaction. Every reaction, properly speaking, has both a matter and an energy aspect. While the energy aspect is frequently neglected directly, the conditions must always be in accord with the energy demand, even if apparently not considered. Chemical changes require consideration of three topics, namely, (1) natural rate of chemical reactions, (2) acceleration of the natural rate in the presence of a catalyzer, and (3) the endpoint of chemical reactions. See also **Electrochemistry**; **Photochemistry and Photolysis**; and **Thermochemistry**.

Contemporary Investigations of Reactions and Reaction Rates

The near term and continuing into the early 2000s will witness outstanding progress in the development of a better understanding of chemical reactions and the time frames within which they occur. In this excellent paper (1985), the late Frederick Kaufman (University of Pittsburgh) stated, "The experimental science of measuring the rates of elementary chemical reactions has advanced to a point where hundreds of such rates either have been determined or are within reach. In a sense, these data give us a global view of chemical reactivity for gas-phase reactions. What this information lacks in detailed, state-to-state insight it more than makes up for in chemical diversity. It is essential that this research be pursued for its own sake rather than just for the sake of providing input data for atmospheric and combustion chemistry (traditional areas of interest), however important these applications are. The existing data bases of elementary-reaction rate constants have not yet been systematically examined for what they tell us about chemical reactivity. In the light of present advances in the techniques for the generation and detection of reactive species, the number of accessible reactions and the range of temperatures and pressures being studied are rapidly increasing and the accuracy of rate measurements is rapidly improving. Strong interaction between theorists and experimentalists has developed but needs to be expanded. What is needed more than anything else is the full realization by the academic community that elementary reaction kinetics is an intellectually exciting field that gives insight into why, how, and how fast chemical reactions take place."

Modern experimental techniques for measuring rate parameters of elementary reactions have transformed the field of gas-phase reaction kinetics from one of indirect inference to one of direct determination. Two general kinds of experimental methods may be cited: (1) In the flash- or laser-photolysis (FP) method, reactive species are produced in a short (10^{-6} to 10^{-8} second) light flash, and their decay resulting from a reaction is monitored in real time by spectroscopic means. (2) In the flow reactor (flowing afterglow) discharge-flow (DF) method, the reactive species are produced continuously in steadily flowing carrier gas (helium or argon) and mixed with other reactants. The rate is determined either by changing the distance over which the reactants are in contact (reaction time) or by changing the concentration of one reactant at constant reaction time. Detailed diagrams of the apparatus used for these experimental methods are beyond the scope of this encyclopedia. They are carefully depicted in the Kaufman (1985) paper listed under additional reading.

As reported by Leone (1985), available laser sources for selective excitation and detection makes it possible to infer many things concerning the dynamics of reactions, such as the particular motions that molecules are likely to undergo during a reaction. The technique makes it possible to interrogate or probe the specific forms of excitation that best lead to chemical reaction, such as vibrational, rotational, or translational motion, or electronic and photo-excitation.

Under ideal circumstances, reactions under investigation are carried out under what is called "single-collision" conditions. These typically require low pressures, so that a product molecule born in a particular state is not modified by subsequent collisions before the time of interrogation has expired. Very short laser pulses usually are used, thus making the time between excitation and interrogation less than 10^{-8} second. However, at this speed, higher pressures can be used so that collisions occurring between formation of a product and its detection and identification can be minimized or essentially eliminated.

One of the fastest reactions reported to date resulted from research at Pennsylvania State University in 1986. It was reported that the time required for the rhodium helide ion ($RhHe^{2+}$) to dissociate into Rh and He was approximately 10^{-13} second, i.e., eight ten-trillionths of a second. Because this kind of interval cannot be resolved by present electronic timer technology, the investigators modified their apparatus to stretch out the reaction time by orders of magnitude. One part of the solution was to use a 4.2 meter evacuated flight tube that lead into an ion detector. Time was determined on the basis of the slight difference in kinetic energy (velocity) of the rhodium ions detected at the end of the flight tube.

Natural Rate of Chemical Reactions

Various factors operate to affect the rate of chemical reactions. By natural rate is understood the rate of a reaction in the absence of a catalyzer. Excluding electrochemical and photochemical reactions, and giving attention to thermochemical reactions only, there are four factors or conditions to be considered, namely, (1) concentration of constituents, (2) temperature, and (3) pressure—important where a gas is involved, (4) nature of the medium, if any.

The general mathematical definition of the rate of a chemical reaction v is

$$v = \frac{d\xi}{dt} \tag{1}$$

where ξ (Greek letter xi) is the extent of reaction, t is time, and the derivative thus represents the rate of change of the extent of reaction.

1. Relation between concentration of reactants and rate of reaction. The rate of a given reaction, at constant temperature and pressure under stated conditions of concentration of the reacting substances, is quantitatively

expressible by a velocity constant, which is the fraction of the substances transformed in a unit of time. Many reactions occur instantaneously—true for most reactions in solution in inorganic chemistry—and many others are complicated in subsidiary reactions, so that the velocity constant is measurable in comparatively few cases. The principle, however, holds as stated, whether or not the desired value can be ascertained experimentally.

A simple reaction that was studied by Wilhelmy, and since then by various investigators, is the transformation (hydrolysis) of sucrose $C_{12}H_{22}O_{11}$ in water solution into glucose ($C_6H_{12}O_6$, a polyhydroxy aldehyde) plus fructose ($C_6H_{12}O_6$, a polyhydroxy ketone), which proceeds at a measurable, steady rate in the presence of acid (hydrogen ion). The rate of reaction at any instant is found to be proportional to the amount of sucrose present at that instant.

When a dilute water solution of an ester, such as methyl acetate, is similarly hydrolyzed in the presence of hydrogen ion, the reaction is of the same type. And this statement also applies to the decay of radioactive elements. One of the important radioactive constants is the half-life, that is, the time required for the decay of one-half of the element present at a given instant.

The preceding cases are instances of *first order reactions*, that is, reactions in which the rate depends only upon the concentration of a single molecular or atomic species. They are also *monomolecular reactions*, that is, reactions in which the initial reactant is only of one species, that is, sucrose, or an ester, or a radioactive element. Note that first order reactions are not necessarily monomolecular; thus $H_2 + D_2 \rightleftharpoons 2HD$, is a first order reaction, even though it is called bimolecular because a hydrogen-1 molecule reacts with a hydrogen-2 (deuterium) molecule to form hydrogen deuteride molecules.

We can now introduce a general treatment of the concept of the order of a chemical reaction. A chemical reaction is said to be of the nth order if its rate is directly proportional to the product of n concentrations. Therefore the decomposition of A, if described by the equation

$$\frac{dC_A}{dt} = -kC_A \tag{2}$$

is a *first order reaction*. Similarly if it is described by the equation

$$\frac{dC_A}{dt} = -kC_A C_B \quad \text{or} \quad \frac{dC_A}{dt} = -kC_A^2 \tag{3}$$

it is a *second order reaction*.

The coefficient k, which appears in (2) or (3), is called the *rate constant*. Generally the temperature variation of a rate constant may be expressed by

$$k = Pe^{-Q/kT} \tag{4}$$

where k is the rate constant and k is the Boltzmann constant. This equation is called the *Arrhenius equation*.

We now introduce a general treatment of rate of reaction, which is often called the *absolute reaction rate theory*, because its purpose is to calculate the rate in terms of molecular quantities only.

Consider the reaction

$$A + BC \longrightarrow ; \; AB + C \tag{5}$$

To simplify the discussion, assume that A, B, and C always remain in a straight line. The course of the reaction may then be followed by noting the values of the two interatomic distances r_{AB} and r_{BC}. At the beginning of the reaction r_{AB} is large and r_{BC} is small while at the end of the reaction r_{AB} is small and r_{BC} is large.

Let us introduce the potential energy surface. The representative point of the system moves on this surface along the so-called *reaction coordinate*. The potential energy along the reaction coordinate is represented schematically in Fig. 1. The maximum of the curve corresponds to a situation where three atoms are very close to one another. Moreover this point is a *maximum* along the reaction coordinate but a *minimum* for the direction normal to the reaction coordinate. Indeed the most probable path is the path involving the minimum potential energy in going from the initial to the final state.

Therefore the point considered corresponds to a *saddle point* of the energy surface. It is called the *activated complex*.

One may now assume that the reaction rate is the product of the following three factors: (1) the average number of activated complexes; (2) the characteristic frequency of the activated complex (that is, the inverse of its lifetime); and (3) the *transmission coefficient*, K, which is

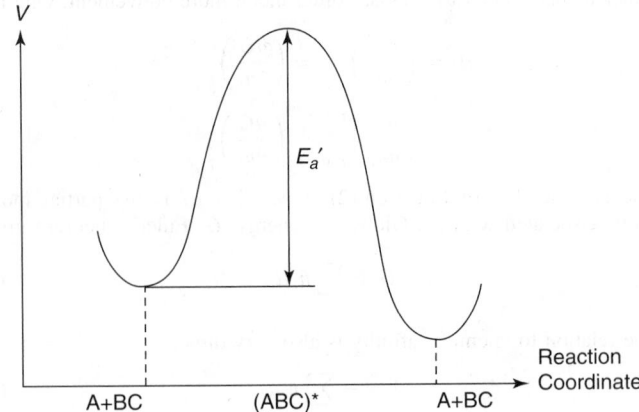

Fig. 1.   Potential energy along the reaction coordinate

the probability that a chemical reaction takes place after the system has reached the activated state.

Moreover, the number of activated complexes is calculated by the equilibrium assumption.

Using this description of the reaction process one derives the following expression for the reaction constant

$$k = K \frac{\phi(T)}{\phi_A(T)\phi_{BC}(T)} \frac{kT}{h} \exp\left(-\frac{E^x}{kT}\right) \tag{6}$$

Here the ϕ terms are the partition function $f(T, V)$, the volume factor being removed

$$f = V\phi \tag{7}$$

ϕ_t corresponds to the activated complex, the degree of freedom associated with the reaction coordinate being removed; k is Boltzmann's constant, h is Planck's constant, E^x is the energy associated with activated complex, or *activation energy* of the reaction.

This expression may also be written in the thermodynamic form

$$\begin{aligned} k &= K \frac{kT}{h} \exp\left(-\frac{\Delta G^{\ddagger}}{kT}\right) \\ &= K \frac{kT}{h} \exp[-(\Delta H^{\ddagger} - T\Delta S^{\ddagger})/kT] \end{aligned} \tag{8}$$

where $\Delta G^{\ddagger}$ is a suitable free energy of activation and $\Delta H^{\ddagger}$, $\Delta S^{\ddagger}$ the corresponding enthalpy and entropy of activation.

When a reaction involves two different phases, that is, when the system is not homogeneous but heterogeneous, as in reactions between a solid phase, such as zinc or calcium carbonate, and a liquid phase, such as hydrochloric acid solution, the rate of reaction involves consideration of (1) the area of the surface of contact of the solid with the solution, and (2) the rate of diffusion from the surface of the solid, as well as (3) the concentration of hydrogen ion of the acid solution.

When the rate of a chemical process is dependent upon (1) two or more *consecutive* reactions, the observed rate is limited by the rate of the slowest reaction in the series, or (2) two or more *concurrent* reactions, the products are in the same ratio at any instant only when the reactions themselves are of the same rate.

2. Relation between temperature of reactants and rate of reaction. The rate of chemical reaction is increased by an increase in temperature, as is evident from the fact that temperature occurs in the numerators in the foregoing equations.

3. Relation between pressure of reactants, if gaseous, and rate of reaction. Since pressure changes amount to concentration changes in such systems, the behavior is as described above under concentration.

4. Relation between nature of the medium and rate of reaction. Very slight changes in the nature of the medium greatly affect the rate of a chemical reaction, but attempts to relate any physical property of a solvent with the effect observed on the rate of a given reaction appear to have proved unsuccessful.

One should mention here that in reactions involving ions, the effects of electrolytes can be put into two principal categories: (a) primary salt effect and (b) secondary salt effects.

Primary salt effects refer to the effects of electrolyte concentration on the activity coefficients. Secondary salt effects are those concerned with the actual changes in concentration of the reacting species resulting from the addition of electrolytes.

BrØnsted has shown that the variation in specific rate with ionic strength depends on the magnitude and sign of the ionic charges. Thus three main types of primary salt effect in reactions of two species can be distinguished. If the products of the signs of the charges are positive, the velocity of the reaction increases with increasing ionic strength:

$$Co[(NH_3)_5Br]^{2+} + Hg^{2+} (+2 \times +2 = +4)$$

or

$$S_2O_8^{2-} + I^- \quad (-2 \times -1 = +2)$$

or

$$BrCH_2COO^- + S_2O_3^{2-} \quad (-1 \times -2 = +2)$$

If the products of the charges are negative the velocity of the reaction decreases with increasing ionic strength:

$$Co[(NH_3)_5Br]^{2+} + OH^- (+2 \times -1 = -2)$$

or

$$H_2O_2 + H^+ + Br^- (+1 \times -1 = -1)$$

In the third category where the product of the ionic charges is zero, the ionic concentration has no or very little effect on the velocity of the reaction, particularly in dilute solution, as in the case of

$$[Cr(NH_2CONH_2)_6]^{3+} + 6H_2O \longrightarrow [Cr(H_2O)_6]^{3+}$$
$$+ 6 NH_2CONH_2 (+3 \times 0 = 0)$$

BrØnsted showed the inadequacy of both the classical and the activity theories of rate of certain reactions. He proposed a new theory postulating that when ions or molecules react, they first form an unstable critical complex, which then decomposes to give the reaction products. The reaction that determines the velocity of a chemical change consists in the formation of that unstable critical complex.

Summarizing, the rates of chemical reactions are subject to highly specific influences in each case, as has been abundantly demonstrated by experimental investigations, and recognized in numerous legal battles in chemical patent suits.

Acceleration of the Natural Rate of Chemical Reactions in the presence of a positive or negative catalyzer. When, in the presence of a given substance, the natural rate of a chemical reaction is changed, either increased or decreased, the given substance is called a catalyzer. Examples are numerous. (1) When a gas-lighter of the type known as platinum black, the active part of which consists of very finely divided platinum, is held in a stream of hydrogen or city gas, the gas is ignited in air. Platinum is a catalyzer for this reaction, and causes ignition to take place at a temperature much lower than by subjecting to fire. (2) The changing of sulfur dioxide into sulfur trioxide is accomplished by passing a mixture of sulfur dioxide and air (one-fifth oxygen) over asbestos coated with finely divided platinum. The temperature required is much lower by the use of platinum catalyzer than without its use. (3) Solutions of sulfites are subject to oxidation to sulfates by oxygen upon allowing to stand in air. The addition of sugar or glycerol retards the speed of this reaction. These substances act in this case as negative catalyzers. (4) The combination of nitrogen and hydrogen gases under high pressure to form ammonia gas is accomplished at a lower temperature in the presence of a catalyzer than in its absence, thus increasing the yield of ammonia (see also **Equilibrium**). One of the catalyzers is composed of iron, intimately mixed with 1% aluminum oxide and 1% potassium oxide. Iron is a catalyzer for this reaction, but is more active as such in the presence of aluminum oxide and potassium oxide, which are spoken of as promoters, a sort of catalyzer of a catalyzer. (5) The hydrogenation of liquid fatty oils and of oleic acid is conducted in the presence of finely divided nickel as a catalyzer. (6) Enzymes are very specific catalyzers, "the most selective and delicate of all known catalysts (Hilditch)," at ordinary temperatures, say 25 to 30°C. Dextroglucose is converted into ethyl alcohol in the presence of the enzyme (zymase) of yeast, and ethyl alcohol into acetic acid (vinegar) in the presence of the enzyme of *Mycoderma aceti*. (7) Nitric acid reacts slowly with copper metal, but the rate of reaction is accelerated more and more as nitrogen tetroxide (catalyzer) is formed in the solution. This is an example of autocatalysis, wherein the reaction brings about the formation of its own catalyzer. (8) Arsenic-containing substances are extreme negative catalyzers, called inhibitors or poisons, of platinum catalyzer.

When the catalyzer is a solid substance, the greatest difficulty in use is to maintain a clean surface. The presence of a positive catalyzer enables a reaction to proceed more rapidly at a lower temperature than corresponds to the natural rate of the reaction. This increases the amount of substances converted in a given time, decreases the demands as to temperature resistance of materials of construction of the apparatus, and frequently makes possible a state of equilibrium more favorable to the yield of desired material.

The End-point of Chemical Reactions

If a chemically reactive system is isolated from the rest of the universe at a constant temperature and pressure, a definite end-point is often attained short of the complete transmutation of reactants into resultants. In order to be certain that this end-point (short of complete transmutation) is what is known as the equilibrium point, the equilibrium must be approached from both directions, e.g., $A + B \longrightarrow C + D$ and $C + D \longrightarrow A + B$. If the equilibrium constant (see treatment below) is the same when approached from both directions, then the reaction is one of true chemical equilibrium. Such equilibrium reactions are also referred to as balanced or reversible reactions. In such reactions the extent of the chemical change is proportional to the concentrations of all the reactants—reactants and resultants being interchangeable, depending upon the direction of the reaction. (Generalization of Guldberg and Waage, 1864, called Law of Mass Action, or more correctly Law of Concentration Effect. Reaction studied by Guldberg and Waage (1867): Barium sulfate plus potassium carbonate plus barium carbonate plus potassium sulfate.)

A classical case, frequently cited, is that investigated by Berthelot in 1963. When 1 mole (60 grams) of acetic acid CH_3COOH and 1 mole (46 grams) of ethyl alcohol C_2H_5OH, both of which substances are soluble in water, are mixed, a reaction takes place which results in the formation of water and ethyl acetate ester, which is likewise in the ratio of 1 mole (18 grams) of water, and 1 mole (88 grams) of ethyl acetate $CH_3COOC_2H_5$. On the other hand, when 1 mole of water and 1 mole of ethyl acetate ester are mixed, a reaction takes place which results in the formation of acetic acid and ethyl alcohol in the ratio of 1 mole of acetic acid and 1 mole of ethyl alcohol. Three important observations have resulted from the detailed study of this reaction, namely: (1) the reaction between acetic acid and ethyl alcohol as reactants proceeds at such a rate that the fraction $\frac{0.005}{5}$ of the amount present at any instant reacts, at 6 to 9°C, in 1 day to form equivalent amounts of water and ethyl acetate ester; (2) the reaction between water and ethyl ester as reactants proceeds at such a rate that the fraction 0.00144 of the amount present at any instant reacts, at 6 to 9°C, in 1 day to form equivalent amounts of acetic acid and ethyl alcohol; and (3) the end-point of each reaction is the same, that is, the reaction is one of true chemical equilibrium, and the resulting equilibrium mixture contains, in each case, 0.33 mole acetic acid plus 0.33 mole ethyl alcohol plus 0.67 mole water plus 0.67 mole ethyl acetate ester. This system attains practical equilibrium, at 6 to 9°C in about 1 year, at 100°C in about 8 days, and at 200°C in about 24 hours.

The equilibrium constant is calculated numerically as follows:

Equation:	CH_3COOH	$+ C_2H_5OH$	$\rightleftharpoons$	$HOH +$	$CH_3COOC_2H_5$
Reaction weights:	60	46		18	88
Molar ratio at equilibrium:	0.33	0.33		0.67	0.67
Weights at equilibrium:	0.33×60	0.33×46		0.67×18	0.67×88

$$\left.\begin{array}{r}\text{Equilibrium}\\\text{constant at } 9°C\end{array}\right\} = \frac{\text{conc. HOH} \times \text{conc. } CH_3COOC_2H_5}{\text{conc. } CH_3COOH \times \text{conc. } C_2H_5OH}$$

$$= \frac{0.67 \times 0.67}{0.33 \times 0.33}$$

Knowing the equilibrium constant at any stated temperature enables one to calculate the equilibrium end-point at that temperature for any ratio of reactants. Thus, when 1 mole (60) grams of acetic acid and 10 moles (460) grams of ethyl alcohol at 9°C are taken:

$$\text{Equilibrium constant at } 9°C = 4 = \frac{X \times X}{(1 - X) \times (10 - X)}$$

where X is the number of moles of water and also the number of moles of ethyl acetate ester formed (1 mole of each is formed by reaction of 1 mole acetic acid plus 1 mole ethyl alcohol). Solution of this equation shows $X = 0.97$. Therefore, by taking the above ratio of acetic acid (1 mole) to ethyl alcohol (10 moles) 0.97 (or 97%) of the acetic acid, the excess reactant being ethyl alcohol (9 moles), is converted at equilibrium into water plus ethyl acetate ester. In practice, the reaction is conducted by the use of a catalyzer, e.g., sulfuric acid concentrated, zinc chloride.

In cases where one of two resultants can be separated from the reactants and the other resultant, by precipitation as a solid, by condensation as a liquid, or by volatilization as a gas or vapor, the yield of the desired substance from a given amount of reactants can sometimes be materially increased. In the case of heterogeneous systems, whenever a solid participant is present, the *concentration* of said solid is considered constant. The precipitation and solution of solids are in this category, as well as the reactions between a gas and a solid, e.g., the system ferroferric oxide plus hydrogen gas plus iron plus water vapor.

The effect of change of temperature on a system in chemical equilibrium is that the equilibrium point is shifted (1) toward the side *away* from that which evolves heat when the temperature is *raised*, and (2) toward the side which evolves heat when the temperature is lowered. It is *as if* the amount of heat were a *material* reactant and its concentration (temperature or intensity of heat) increased, in respect to the *direction* of the shift of the equilibrium point. The amount of the shift at constant pressure can be calculated in cases where one possesses the proper data.

The effect of change of pressure on a system in chemical equilibrium is that the equilibrium point is shifted (1) towards the side possessing the smaller aggregate volume when the pressure is increased, and (2) towards the side possessing the larger aggregate volume when the pressure is decreased. The amount of the shift at constant temperature can be calculated by means of the equilibrium constant (above), recalling that increase of pressure is equivalent to increase of concentration of gases (temperature constant). When the volume of resultants equals the volume of reactants, no effect is produced on the equilibrium point by change of pressure. See also **Chemical Equilibrium**.

In some chemical reactions, the use of concentrations does not give calculated results that agree with those observed, because of the departure from ideality of real gases and solutions. In such reactions, concentrations are replaced by apparent effective concentrations, or activities, as explained in that entry.

Additional Reading

Amato, I.: "Unreal Reactions Elucidate Energy Flow," *Sci. News*, **53** (January 27, 1990).

Armentrout, P.B.: "Chemistry of Excited Electronic States," *Science*, **175** (January 11, 1991).

Arnett, E.M. et al.: "Chemical Bond-Making, Bond-Breaking, and Electron Transfer in Solution," *Science*, **423** (January 26, 1990).

Crim, F.F.: "State- and Bond-Selected Unimolecular Reactions," *Science*, **1387** (September 21, 1990).

Fodor, S.P.A. et al.: "Light-Directed, Spatially Addressable Parallel Chemical Synthesis," *Science*, **767** (February 15, 1991).

Jarrold, M.F.: "Nanosurface Chemistry on Size-Selected Silicon Clusters," *Science*, **1085** (May 14, 1991).

Kaufman, F.: "Rates of Elementary Reactions: Measurement and Applications," *Science*, **230**, 393–399 (1985).

Leone, S.R.: "Laser-Probing of Chemical Reaction Dynamics," *Science*, **227**, 889–895 (1985).

Moffat, A.S.: "Controlling Chemical Reactions with Laser Light," *Science*, **1643** (March 27, 1992).

Pool, R.: "Understanding the Simplest Reaction," *Science*, **411** (January 26, 1990).

Stucky, G.D. and J.E. MacDougall: "Quantum Confinement and Host/Guest Chemistry: Probing a New Dimension," *Science*, **669** (February 9, 1990).

Truhlar, D.G. and M.S. Gordon: "From Force Fields to Dynamics: Classical and Quantal Paths," *Science*, **491** (August 3, 1990).

Warren, W.S., Rabiz, H., and M. Dahleh: "Coherent Control of Quantum Dynamics: The Dream is Alive," *Science*, **1581** (March 12, 1993).

Williams, E.D. and N.C. Bartelt: "Thermodynamics of Surface Morphology," *Science*, **393** (January 25, 1991).

CHEMICALS (Number of). With over 100 chemical elements from which chemical compounds can be built, the large numbers of atoms which may be present in various compounds, and the many ways in which the atoms can be linked (straight chains, branching chains, rings, half-rings, etc.), the number of "possible" chemical compounds is a very high number indeed, and, of course, is much larger than the millions of "known" compounds, which can be described fully in terms of constituents and structure. Keeping track of so many compounds—in the interest of fundamental research and, in more recent years, the identification of chemical compounds in terms of both beneficial and adverse effects on biological systems, health, and the environment—commenced many years ago. Thousands of compounds are described in numerous handbooks, various societies have compiled lists and tabulations, as for example the Chemical Abstracts Service (CAS) of the American Chemical Society, and special encyclopedias which describe inorganic and organic compounds. Possibly the most outstanding example is the "Handbuch der Organische Chemie," first undertaken in Germany in 1918. The first 27 volumes of this book were published between 1928 and 1938. Supplements have been released periodically since that time. Part II, consisting of 29 volumes, was published between 1941 and 1957; Part III, 14 volumes, was published between 1958 and 1973. Part IV was commenced in 1972 and continues. Publisher is Springer-Verlag, Berlin. The "Dictionary of Organic Compounds," 5th Edition (J. Buckingham, Executive Editor), published by Chapman and Hall Ltd, London, describes some 150,000 compounds. The book is published in seven volumes, which comprise a total of approximately 7900 printed pages. The "Main Work" was published in October 1982, with supplements scheduled for publication periodically. The First Supplement was released in 1938.

The CAS registry listings run into the millions of compounds, a very high percentage of which contain carbon. The majority are synthesized for specific research use. It is estimated that about three-fourths of the compounds listed are mentioned only once or very few times in the literature. Large numbers of listed materials are coordination compounds, the structure of which has not been fully defined. The registry also lists alloys, polymers, and mixtures with definite names. The registry embraces only papers published since 1965.

The ACS also offers the "Beilstein Online Database." This is claimed to be the most complete and systematic collection of evaluated data on organic compounds. The database is reviewed in a ten-chapter, "The Beilstein Online Database: Implementation, Content, and Retrieval," available from American Chemical Society, Washington, DC.

CHEMISORPTION. The binding of a liquid or gas on the surface or in the interior of a solid by chemical bonds or forces.

CHEMOTHERAPEUTICS, ANTICANCER. Cancer is second only to cardiovascular disease as the principal cause of human mortality. As the median age of populations has risen, total deaths from cancer have increased. Treatment of cancer includes surgery, radiation, and chemotherapy, the last encompassing the use of both cytotoxic agents and relatively nontoxic hormonal agents for the control of tumor growth.

Drugs used in cancer chemotherapy or clinical trials are classified according to primary underlying mechanisms of action. However, many drugs operate through multiple mechanisms. Mechanisms include those of antimetabolites, DNA alkylating and/or cross-linking agents, DNA binding/cleaving agents, DNA topoisomerase interactive compounds, agents that act on tubulin structure, and hormones (Fig. 1). In addition to those drugs already approved by the FDA, a number of investigational drugs are undergoing clinical evaluation and many others are in the pipeline.

Antimetabolites. Antimetabolites, which represent one of the earliest groups of anticancer agents, are listed in Table 1.

DNA Alkylating/Cross-Linking Agents. This category includes compounds of diverse chemical classes (Table 2).

DNA Binding/Cleaving Agents. DNA binding and/or cleaving agents that have anticancer activity are listed in Table 3. All of the natural products and analogues of natural products in this category (Table 3) are able to bind DNA either as intercalators or as minor groove binders, hence inhibiting DNA-dependant RNA synthesis. Both bleomycin and esperamicin A_1 cleave DNA by forming free radicals in the immediate vicinity of the sugar—phosphate backbone. Activity as antitumor agents is related to the ability to induce irreparable lesions in DNA. Bleomycin generates oxygen free-radical species whereas esperamicin A_1 and a number of related natural products that include neocarzinostatin, dynemicin, and the calicheamicins generate aryl diradical species, which abstract hydrogen

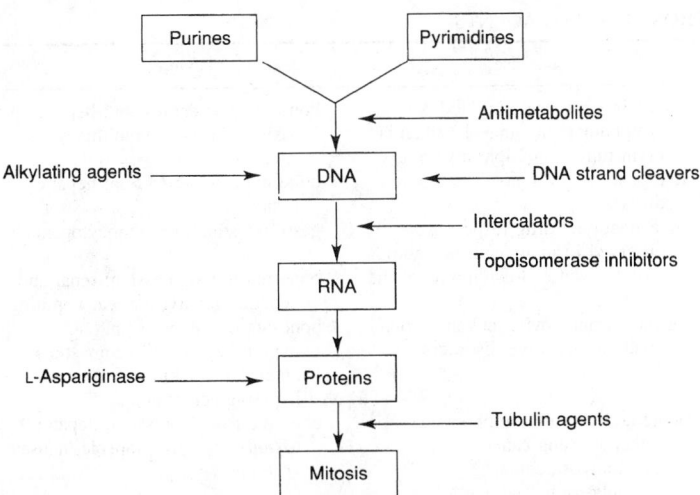

Fig. 1. Schematic of nucleic acid and protein synthesis and the steps leading to mitosis showing the common mechanisms of action and various classes of chemotherapeutic agents

atoms directly from the deoxyribose backbone. An analogue of the natural product CC1065 has the unique property of being a DNA alkylating agent which recognizes poly-AT regions of the minor groove of DNA. It remains

to be seen if the high potency and unique modes of action ascribed to these novel classes of agents can translate into clinically useful drugs.

Topoisomerase Interactive Drugs. Topoisomerases I and II have emerged as interesting targets for the design of new anticancer agents (Table 4).

Tubulin Active Drugs. Tubulin active drugs are listed in Table 5.

Hormones. Although not strictly cytotoxic, hormones have been used to control the environment of hormone-dependent tumors such as those of the prostate, breast, and endometrium, i.e., androgens are used to control the growth of estrogen-dependent breast tumors, whereas estrogens control androgen-dependent tumors of the prostate. Hormones that have anticancer activities are listed in Table 6.

Miscellaneous Agents. Those chemotherapeutic agents that do not fit into any of the classifications discussed are listed in Table 7.

Toxicity

As a result of the life-threatening nature of cancer and the general lack of therapeutically effective drugs for most cancers, doses of chemotherapeutic drugs in Phase I clinical trials are escalated until the emergence of a dose-limiting toxicity. The efficacy of these compounds in one or more tumor types is then established in Phase II/III clinical trials. The commonly observed dose-limiting toxicities include myelosuppression, gastrointestinal upset, and renal, hepatic, and cardiotoxicities.

TABLE 1. ANTIMETABOLITES

Drug (trade name)	Molecular formula	Molecular weight	Disease	Toxic effects
5-azacitidine[a] (Mylosar)	$C_8H_{12}N_4O_5$	244.21	acute myelogenous leukemia	nausea, vomiting; hepatic dysfunction; myelosuppression
cytarabine USP[a] (Cytosar)	$C_9H_{13}N_3O_5$	243.22	acute granulocytic leukemia (adults); acute lymphocytic leukemia (children); Hodgkin's disease	bone marrow depression; hepatic toxicity; megaloblastosis; nausea; vomiting; diarrhea
gemcitabine[b]	$C_9H_{11}F_2N_3O_4$	263.20	investigational drug; responses seen in Phase I trials in colon and nonsmall cell lung cancer	myelosuppression observed as dose-limiting toxicity
floxuridine USP[c] (FUDR)	$C_9H_{11}FN_2O_5$	246.21	palliative treatment of gastrointestinal adenocarcinoma with liver metastases	severe hematological toxicity; gastrointestinal hemorrhage; nausea; vomiting; diarrhea; enteritis; stomatitis; erythema
fluorouracil USP[e] (Fluorouracil)	$C_4H_3FN_2O_2$	130.08	palliative treatment of carcinoma of colon, rectum, breast, stomach, and pancreas	bone marrow depression; dermatitis; alopecia; nausea; vomiting; diarrhea; stomatitis; anorexia; GI ulcers; skin pigmentation
mercaptopurine USP[d] (Purinethol)	$C_5H_4N_4S$	152.19	acute leukemia (more effective in children than in adults); chronic granulocytic leukemia	bone marrow depression; hepatic toxicity; anemia; gastrointestinal (GI) ulceration; nausea; vomiting
thioguanine USP[d] (Tabloid)	$C_5H_5N_5S \cdot XH_2O$	167.19	acute leukemia; chronic granulocytic leukemia	bone marrow depression; stomatitis; anorexia; nausea; vomiting
methotrexate USP[e] (Methotrexate)	$C_{20}H_{22}N_8O_5$	454.46	acute lymphocytic leukemia; meningeal leukemia; choriocarcinoma; chorioadenoma destruens; lymphosarcoma; osteogenic sarcoma; cancer of lung, neck, head, cervix; mycosis fungoides; hydatidiform mole high dose MTX followed by leucovorin rescue in nonmetastatic osteosarcoma	bone marrow depression; renal and hepatic toxicity; enteritis; stomatitis; alopecia; abdominal distress; erythematous rash; oral and GI ulceration; diarrhea; nausea; vomiting
leucovorin[e] (calcium USP)	$C_{20}H_{21}CaN_7O_7$	511.51	high dose methotrexate rescue therapy in osteosarcoma	allergic sensitization
DDATHF[b] (Ly237147) (Lometrexol sodium)	$C_{21}H_{23}N_5Na_2O_6$	487.42	investigational drug	
trimetrexate[f]	$C_{19}H_{23}N_5O_3$	369.42	investigational drug; partial remissions in soft tissue sarcomas observed	myelosuppression; mucositis; nausea; vomiting; skin rash
hydroxyurea USP[g] (Hydrea)	$CH_4N_2O_2$	76.05	chronic granulocytic leukemia; melanoma; cancer of ovary, head, neck	vomiting; anorexia; fever; bone marrow depression; nausea; diarrhea

[a] Upjohn.
[b] Lilly.
[c] Hoffmann-La Roche.
[d] Burroughs Wellcome.
[e] Lederle.
[f] Parke-Davis.
[g] Bristol-Myers Squibb.

TABLE 2. DNA ALKYLATING/CROSS-LINKING AGENTS

Drug (trade name)	Molecular formula	Molecular weight	Disease	Toxic effects
carmustine USP[a] (BiCNU)	$C_5H_9Cl_2N_3O_2$	214.05	Hodgkin's disease; non-Hodgkin's lymphomas; meningeal leukemia; brain tumor; multiple myeloma	bone marrow depression; hepatic toxicity; nausea; vomiting
lomustine USP[a] (CeeNU)	$C_9H_{16}ClN_3O_2$	233.70	malignant brain tumors; Hodgkin's disease	bone marrow depression; hepatic toxicity
tauromustine[b]	$C_7H_{15}ClN_4O_4S$	286.73	investigational drug responses observed in malignant melanoma	gastrointestinal; thrombocytopenia
streptozocin USP[c] (Zanosar)	$C_8H_{15}N_3O_7$	265.22	metastatic islet cell carcinoma of the pancreas	bone marrow depression; renal and hepatic toxicity; nausea; vomiting
busulfan USP[d] (Myleran)	$C_6H_{14}O_6S_2$	246.29	chronic granulocytic leukemia; other myeloproliferative disorders	bone marrow depression; hyperuricemia; gynecomastia; amenorrhea; skin hyperpigmentation
cyclophosphamide USP[a] (Cytoxan)	$C_7H_{15}Cl_2N_2O_2P$	279.10	acute and chronic lymphocytic leukemia; lung cancer; rhabdomyosarcoma; neuroblastoma; ovarian and mammary carcinoma; multiple myeloma; lymphosarcoma; Burkitt's lymphoma; Hodgkin's disease; retinoblastoma; mycosis fungoides	bone marrow depression; hepatic toxicity; cystitis; alopecia; nausea; vomiting
ifosfamide USP[a] (Ifex)	$C_7H_{15}Cl_2N_2O_2P$	261.09	germ cell testicular cancer; used in combination with mesna	myelosuppression; urotoxicity; alopecia; nausea; vomiting; CNS toxicities
mesna USP[a] (Mesnex)	$C_2H_5NaO_3S_2$	164.17	prophylactic prevention of hemorrhagic cystitis	
mechlorethamine hydrochloride USP[e] (Mustargen)	$C_5H_{11}Cl_2N \cdot HCl$	192.52	Hodgkin's disease; non-Hodgkin's lymphomas; lymphosarcoma; cancer of breast, ovary, lung; neoplastic effusion	bone marrow depression; nausea; vomiting; anorexia; diarrhea; local irritation
chlorambucil USP[d]	$C_{14}H_{19}Cl_2NO_2$	304.23	chronic lymphocytic leukemia; cancer of ovary, breast, testis; Hodgkin's disease; non-Hodgkin's lymphomas	bone marrow depression; nausea; vomiting
melphalan USP[d] (Alkeran)	$C_{13}H_{18}Cl_2N_2O_2$	305.20	multiple myeloma; plasmacytic myeloma; cancer of breast and ovary	bone marrow depression; nausea; vomiting; anorexia
thiotepa USP[f] (Thiotepa)	$C_6H_{12}N_3PS$	189.21	cancer of breast, ovary, lung, bladder; Hodgkin's disease; non-Hodgkin's lymphomas; neoplastic effusion	bone marrow depression; amenorrhea; anorexia; nausea; vomiting
mitomycin C USP[a] (Mutamycin)	$C_{15}H_{18}N_4O_5$	334.33	chronic myelogenous leukemia; reticulum cell sarcoma; Hodgkin's disease; non-Hodgkin's lymphomas; cancer of stomach, pancreas, lung; epithelial tumors	bone marrow depression; renal toxicity; alopecia; stomatitis; anorexia; nausea; vomiting
BMY-25067[a]	$C_{23}H_{25}N_5O_7S_2$	547.60	investigational drug	
KW2149[g]	$C_{24}H_{34}N_6O_8S_2$	598.7	investigational drug	
cisplatin USP[a] (Platinol)	$Cl_2H_6N_2Pt$	300.06	metastatic testicular tumors; metastatic ovarian tumors; advanced bladder cancer	nephrotoxicity; ototoxicity; myelosuppression; nausea; vomiting; allergic reaction
carboplatin USP[a] (Paraplatin)	$C_6H_{12}N_2O_4Pt$	371.25	recurrent ovarian carcinoma	bone marrow suppression; emesis; allergic reactions
dacarbazine USP[h] (DTIC)	$C_6H_{10}N_6O$	182.18	malignant melanoma; Hodgkin's disease; soft tissue sarcomas	bone marrow depression; flulike syndrome; alopecia; nausea; vomiting; anorexia

[a] Bristol-Myers Squibb.
[b] Pharmacia.
[c] Upjohn.
[d] Burroughs Wellcome.
[e] Merck Sharp & Dohme.
[f] Lederle.
[g] Kyowa-Hakko.
[h] Dome.

Toxicity Amelioration

Research efforts to address the problem of toxicity amelioration has progressed in several directions. The three most prominent areas are analogue synthesis, chemoprotection, and drug targeting.

Drug Resistance

The most recognized and studied mechanisms of drug resistance are attributed to multidrug resistance (MDR), gene amplification, DNA repair, topoisomerase II activity, and glutathione and metallothionein levels. Even with advances in understanding the biology and mechanism of drug resistance among different classes of antitumor agents, no real breakthrough appears imminent.

Economics of Cancer Chemotherapy

The increased use of chemotherapy as a modality in the treatment of cancer has caused a corresponding increase in the market for anticancer

TABLE 3. DNA INTERACTIVE AGENTS

Drug (trade name)	Molecular formula	Molecular weight	Disease	Toxic effects
daunorubicin hydrochloride USP[a] (Cerubidine)	$C_{27}H_{29}NO_{10} \cdot HCl$	563.99	acute lymphocytic and granulocytic leukemia; lymphomas	bone marrow depression; cardiac toxicity; alopecia; stomatitis; GI disturbance
doxorubicin USP[b] (Adriamycin)	$C_{27}H_{29}NO_{11}$	543.53	soft-tissue and osteogenic sarcomas; Hodgkin's disease; non-Hodgkin's lymphomas; acute leukemia; cancer of thyroid, breast, lung, genitourinary (GU) tract; Wilm's tumor; neuroblastoma	bone marrow depression; cardiac toxicity; alopecia; stomatitis; GI disturbance
idarubicin hydrochloride[b] (Idamycin)	$C_{26}H_{27}NO_9 \cdot HCl$	533.96	acute myeloid leukemia in adults	bone marrow suppression; cardiotoxicity; nausea; vomiting; alopecia
mitoxanthrone hydrochloride USP[c] (Novantrone)	$C_{22}H_{28}N_4O_6 \cdot 2HCl$	517.41	acute nonlymphocytic leukemia, including myelogenous promyelocytic, monocytic, and erythroid acute leukemias	nausea; vomiting; alopecia; mucositis; stomatitus; myelosuppression; cardiotoxicity; allergic reaction; phlebitis
bleomycin sulfate USP[d] (Blenoxane)	mixture of bleomycin A_2, B_2 as primary components		squamous cell carcinoma of head, neck, esophagus, skin, GU tract; testicular tumor; Hodgkin's lymphomas	pulmonary fibrosis; skin reactions; alopecia; nausea; vomiting; anorexia; fever; stomatitis
esperamicin A_1[d]	$C_{59}H_{80}N_4O_{22}S_4$	1324.41	investigational drug	
Adozelesin[e] (U73,975)	$C_{30}H_{22}N_4O_4$	502.30	investigational drug	
dactinomycin USP[f] (Cosmegen)	$C_{62}H_{86}N_{12}O_{16}$	1255.43	Wilm's tumor; Ewing's tumor; choriocarcinoma; testicular carcinoma; rhabdomyosarcoma; neuroblastoma; melanoma; soft-tissue and osteogenic sarcomas	bone-marrow depression; renal and hepatic toxicity; alopecia; mental depression; stomatitis; nausea; vomiting; diarrhea; anorexia; local irritation
plicamycin USP[a] (Mithracin)	$C_{52}H_{76}O_{24}$	1085.16	testicular tumors; hypercalcemia and hypercalciuria associated with advanced malignancies	bone marrow depression; hepatic and renal toxicity; hypocalcemia; hemorrhage; stomatitis; nausea; vomiting; anorexia; diarrhea
procarbazine hydrochloride USP[g] (Matulane)	$C_{12}H_{19}N_3O \cdot HCl$	257.76	Hodgkin's disease; non-Hodgkin's lymphomas; lung cancer	bone marrow depression; neurological and dermatological toxicity; nausea; vomiting

[a] Wyeth-Ayerst.
[b] Adria.
[c] Lederle.
[d] Bristol-Myers Squibb.
[e] Merck Sharp & Dohme.
[f] UpJohn.
[g] Hoffmann-La Roche.

TABLE 4. TOPOISOMERASE INTERACTIVE DRUGS

Drug (trade name)	Molecular formula	Molecular weight	Disease	Toxic effects
etoposide USP[a] (Vepesid)	$C_{29}H_{32}O_{13}$	588.56	refractory testicular tumors; small cell lung cancer	myelosuppression; mild to moderate nausea and vomiting; transient hypotension; allergic reactions; alopecia
etoposide phosphate[a]	$C_{29}H_{33}O_{16}P$	712.51	investigational drug; prodrug of etoposide	prodrug of etoposide
teniposide[a]	$C_{32}H_{32}O_{13}S$	656.67	refractory acute lymphocytic leukemia in children	myelosuppression; mild to moderate nausea and vomiting; transient hypotension; allergic reactions; alopecia
CPT-11[b]	$C_{33}H_{38}N_4O_6 \cdot HCl$	622.78	investigational drug; topoisomerase I inhibitor	
topotecan hydrochloride[c]	$C_{23}H_{23}N_3O_5 \cdot HCl$	457.91	investigational drug; topoisomerase I inhibitor	
elsamitrucin tartrate[a]	$C_{33}H_{35}NO_{13} \cdot C_4H_6O_4$	771.37	investigational drug	
amsacrine[d] (Amsidyl)	$C_{21}H_{19}N_3O_3S$	393.46	investigational drug	

[a] Bristol-Myers Squibb.
[b] Yakult Honsha.
[c] Smith Kline Beecham.
[d] Parke-Davis.

agents. Only a few anticancer drugs have achieved sales in excess of $100 million per year and many of the drugs discussed herein sell less than $10 million per year. Those agents that have achieved greater economic importance are newer, frequently used in combination chemotherapy, and of use in the treatment of solid tumors which comprise the bulk of reported cancer incidence. A number of the agents in clinical trials, such as taxol and camptothecin analogues, are expected to have considerable economic impact based on activity in the treatment of the more common human

TABLE 5. TUBULIN ACTIVE DRUGS

Drug (trade name)	Molecular formula	Molecular weight	Disease	Toxic effects
vinblastin sulfate USP[a] (Velban)	$C_{46}H_{58}N_4O_9 \cdot H_2SO_4$	909.06	Hodgkin's disease; lymphosarcoma; reticulum cell sarcoma; neuroblastoma; choriocarcinoma; carcinoma of breast, lung, oral cavity, testis, bladder; acute and chronic leukemia; histiocytosis; mycosis fungoides	leukopenia; neurological toxicity (paresthesias, mental depression, loss of deep tendon reflexes, etc); dysfunction of autonomic nervous system (ileus, constipation, urinary retention, etc); alopecia; stomatitis; nausea; vomiting; local irritation
vincristin sulfate USP[a] (Oncovin)	$C_{46}H_{56}N_4O_{10} \cdot H_2SO_4$	923.04	acute leukemia in children; lymphocytic leukemia; Hodgkin's disease; non-Hodgkin's lymphomas; Wilm's tumor; neuroblastoma; rhabdomyosarcoma.	neurological toxicity (paresthesias, foot drop, double vision, etc); constipation; ileus; alopecia; leukopenia (occasional);
vindesine sulfate[a] (Eldisine)	$C_{43}H_{55}N_5O_7 \cdot H_2SO_4$	852.01	investigational drug	
navelbine[b] (Vinorelbine)	$C^{45}H^{54}N^4O_8$	778.45	investigational drug; nonsmall cell lung cancer	
taxol[c] (Paclitaxol)	$C_{47}H_{51}NO_{14}$	853.92	refractory ovarian cancer; refractory breast cancer; melanoma; lung cancer; head and neck cancer	alopecia; neutropenia; hypersensitivity; mucositis; neuropathy
taxotere[d] (Docetaxol)	$C_{43}H_{53}NO_{14}$	807.43	investigational drug	

[a] Lilly.
[b] Pierre Fabre.
[c] Bristol-Myers Squibb.
[d] Rhône-Poulenc.

TABLE 6. HORMONAL THERAPY

Drug (trade name)	Molecular formula	Molecular weight	Disease	Toxic effects
tamoxifen citrate USP[a] (Nolvadex)	$C_{26}H_{29}NO \cdot C_6H_8O_7$	563.65	breast cancer	visual disturbances
diethylstilbestrol diphosphate USP[b] (Stilphostrol)	$C_{18}H_{22}O_8P_2$	428.31	prostatic carcinoma	fluid retention; hypercalcemia; common side effects of steroids
chlorotrianisene USP[b] (TACE)	$C_{23}H_{21}ClO_3$	380.87	androgen dependent carcinoma of the prostate	fluid retention; hypercalcemia; common side effects of steroids
estradiol USP[c] (Estrace)	$C_{18}H_{24}O_2$	272.39	breast cancer; prostatic carcinoma	fluid retention; hypercalcemia; common side effects of steroids
estramustine phosphate sodium USP[d] (Emcyt)	$C_{23}H_{30}Cl_2N \cdot Na_2O_6P$	564.35	prostatic carcinoma	side effects because of estradiol; increased dyspnea; nausea; vomiting
medroxyprogesterone acetate[e] (USP Depo-provera)	$C_{24}H_{34}O_4$	386.53	metastatic endometrial carcinoma; renal carcinoma	fluid retention; hypercalcemia; common side effects of steroids
megestrol acetate USP[c] (Megace)	$C_{24}H_{32}O_4$	384.51	carcinoma of the breast or endorometrium	fluid retention; hypercalcemia; common side effects of steroids
testolactone USP[c] (Teslac)	$C_{19}H_{24}O_3$	300.40	breast cancer	fluid retention; hypercalcemia; common side effects of steroids
formestane[f] (Lentaron)	$C_{19}H_{26}O_3$	302.41	investigational drug; post-menopausal breast cancer	
goserelin USP[a] (Zoladex)	$C_{59}H_{84}N_{18}O_{14}$	1269.43	prostatic carcinoma	bone pain; common hormonal side effects
leuprolide acetate USP[g] (Leupron)	$C_{59}H_{84}N_{16}O_{12} \cdot C_2H_4O_2$	1269.47	prostatic carcinoma	common hormonal side effects
octreotide acetate USP[h] (Sandostatin)	$C_{49}H_{66}N_{10}O_{10}S_2 \cdot xC_2H_4O_2$	1019.24	mestastatic carcinoid tumors; vasoactive intestinal peptide-secretory tumors	nausea; diarrhea; loose stools; vomiting; abdominal pain; pain on injection
flutamide[i] (Eulexin)	$C_{11}H_{11}F_3N_2O_3$	276.21	metastatic prostatic carcinoma in combination with LHRH agonist	

[a] ICI.
[b] Marion-Merrill Dow.
[c] Bristol-Myers Squibb.
[d] Pharmacia.
[e] Upjohn.
[f] CIBA-GEIGY.
[g] TAP.
[h] Sandoz.
[i] Schering.

TABLE 7. CHEMOTHERAPEUTIC AGENTS

Drug (trade name)	Molecular formula	Molecular weight	Disease	Toxic effects
mitotane USP[a] (Lysodren)	$C_{14}H_{10}Cl_4$	320.05	palliative treatment of inoperable adrenal cortical carcinoma	skin toxicity; vertigo; lethargy; somnolence; anorexia; nausea; vomiting; diarrhea
isotretinoin USP[b] (Accutane)	$C_{20}H_{28}O_2$	300.44	investigational drug	
sulofenur[c]	$C_{16}H_{15}ClN_2O_3S$	350.82	investigational drug; refractory ovarian carcinoma response see in Phase I	anemia; methemoglobinemia
asparaginase[d] (Elspar)			acute lymphocytic leukemia	hepatic, renal, and pancreatic toxicity; neurological effects; hypersensitivity reactions; clotting abnormalities; nausea

[a] Bristol-Myers Squibb.
[b] Hoffmann-La Roche.
[c] Lilly.
[d] Merck Sharp & Dohme.

malignancies in, e.g., lung, breast, colon, and ovary. In addition to the market for cancer chemotherapeutic drugs, there is also a growing market for biologicals such as interferons.

TERRENCE W. DOYLE
DOLATRAI M. VYAS
Bristol-Meyers Squibb Company

Additional Reading

Carter, S. K., M. T. Bakowski, and K. Hellman: *Chemotherapy of Cancer,* John Wiley & Sons, Inc., New York, NY, 1987.
Fisher, J. D. and P. A. Aristoff: in *Progress in Drug Research,* Vol. 32, Birkhauser Verlag, Basel, Switzerland, 1988, pp. 411–484.
Boyd, M. R.: in *Current Therapy in Oncology,* B. C. Decleer, Inc., Philadelphia, PA., 1993, pp. 11–22.
Woolley, P. V. III and K. D. Tew, eds.: *Mechanism of Drug Resistance" in Neoplastic Cells,* Academic Press, Inc., New York, NY, 1988.

CHERT. An impure, flinty hard rock composed chiefly of cryptocrystalline silica. Chert varies in color from gray through brown to black according to the kind and amount of coloring matter. It occurs principally as concretions, nodules or bands in limestones and dolomites, and unlike flint its fracture tends to be splintery instead of conchoidal. A great deal has been written on the occurrence and origin of chert and there is no doubt but that it may be formed in several different ways. Many of the nodular and concretionary cherts have grown around siliceous sponge spicules or radiolaria. Chert may be either sygenetic or epigenetic. The former type is supposed by some authors to be chemically precipitated from river waters on the bottom of the sea as a colloid contemporaneously with the limestones or dolomites. On the other hand certain cherts are obviously secondary although they may have been formed previous to the final lithification of the formations in which they occur. Cherts containing relatively large amounts of iron are called Jasper.

CHINA CLAY. A commercial term, more or less identical with kaolin, as applied to the relatively pure clay concentrated by washing from a thoroughly kaolinized granite. England is the chief exporter of china clay. France has unique clays from which are made the famous Sévres and Limoges potteries.

China-clay rock is a kaolinized granite made up chiefly of quartz and kaolin, with sometimes the presence of muscovite and tourmaline. The rock crumbles easily in the fingers. *China stone* is (1) a partially kaolinized granite, which contains quartz, kaolin, and sometimes mica and fluorite, is harder than china-clay rock and is used as a glaze in the production of china; or (2) a fine-grained, compact mudstone or limestone found in England and Wales.

CHIRAL SEPARATIONS. Chiral separations are concerned with separating molecules that can exist as nonsuperimposable mirror images. Examples of these types of molecules, called *enantiomers* or *optical isomers,* are illustrated in Figure 1. Although chirality is often associated with compounds containing a tetrahedral carbon with four different substituents, other atoms, such as phosphorus or sulfur, may also be chiral. In addition,

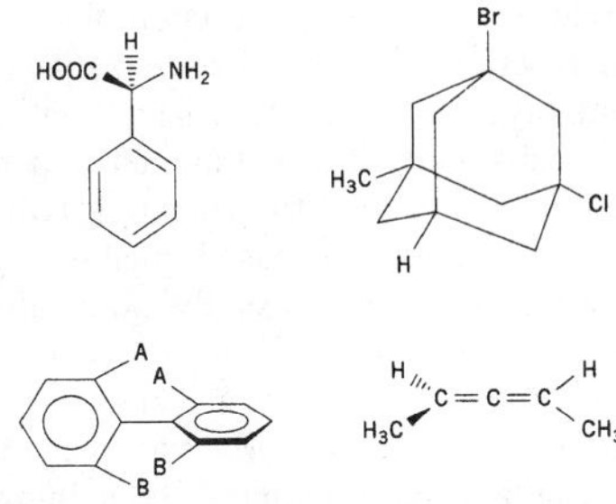

Fig. 1. Examples of chiral molecules

molecules containing a center of asymmetry, such as hexahelicene, tetrasubstituted adamantanes, and substituted allenes or molecules with hindered rotation, such as some 2, 2' disubstituted binaphthyls, may also be chiral. Compounds exhibiting a center of asymmetry are called *atropisomers.*

Although scientists have known since the time of Louis Pasteur that optical isomers can behave differently in a chiral environment (e.g., in the presence of polarized light), it has only been since about 1980 that there has been a growing awareness of the implications arising from the fact that many drugs are chiral and that living systems constitute chiral environments. Hence, the optical isomers of chiral drugs may exhibit different bioactivities and/or biotoxicities.

In the case of enantiomerically pure chiral drugs, the possibility of racemization or inversion either *in vivo* or during storage cannot be ruled out. Ibuprofen is an example of a chiral drug which undergoes rapid inversion *in vivo.* In addition, there are several examples of achiral (or *prochiral*) drugs being biotransformed into chiral entities. In some cases, the enantiomeric ratios produced by laboratory animals may differ from that produced in humans. This raises the question of the suitability of laboratory animals as appropriate test models for a certain drug.

For those drugs that are administered as the racemate, each enantiomer needs to be monitored separately yet simultaneously, since metabolism, excretion or clearance may be radically different for the two enantiomers. Further complicating drug profiles for chiral drugs is that often the pharmacodynamics and pharmacokinetics of the racemic drug is not just the sum of the profiles of the individual enantiomers.

Although a great deal of the work currently being done in chiral separations is related to pharmaceuticals, the agricultural and the food and beverage industries are affected as well. For instance, several chiral pesticides are used commercially. It is possible that the enantiomers may differ in their persistence in the environment and their effectiveness against specific pests. In the food and beverage industry, many of the constituents that confer flavor or aroma in foods and beverages are chiral. For instance, the configuration of the 4-alkyl-substituted γ-lactones responsible for much of the flavor in fruits is almost exclusively *R.* Often, the two enantiomers have very different aromas or flavors. The presence of any of the "unnatural" enantiomer may confer an "off-flavor" to the substance and may be indicative of racemization under adverse storage conditions, adulteration, or formulation from nonnatural sources.

The growing awareness of the implications of chirality to the pharmaceutical industry has spurred tremendous effort toward stereoselective synthetic strategies and the development of new chiral catalysts. However, the enantiomeric purity of these substances or their chiral precursors needs to be determined. Also, there are many chiral compounds for which no stereospecific synthetic pathways have been devised. Thus, there is a tremendous need not only for analytical scale (<5–10 mg), but bulk-scale chiral separations as well.

Whether analyzing drugs or synthetic precursors for enantiomeric purity, monitoring biological or environmental samples for chiral discrimination or trying to enantioresolve kilogram quantities of a racemic drug, there are a variety of reasons for performing chiral separations. The purpose of the separation dictates, to some extent, the method employed.

Traditionally, chiral separations have been considered among the most difficult of all separations.

A variety of strategies have been devised to obtain them. Although the focus of this article is on chromatographically based chiral separations, other methods include crystallization and stereospecific enzymatic-catalyzed synthesis or degradation. In crystallization methods, racemic chiral ions are typically resolved by the addition of an optically pure counterion, thus, forming diastereomeric complexes.

Enzymatically based methods depend upon the stereospecificity of an enzyme-catalyzed reaction, such as lipase-catalyzed esterification, to degrade enantioselectively the unwanted enantiomer or to produce the desired enantiomer. Because only one enantiomer undergoes the reaction, the subsequent separation is reduced to separating two different species. One disadvantage of enzymatically based methods is that only one enantiomer is obtained and there is usually no analogous method for producing the opposite enantiomer.

An alternative method of creating a chiral environmental is to derivatize a chiral analyte with an optically pure reagent, thus, producing diastereomers. The resultant diastereomers, containing more than one chiral center, have slightly different melting and boiling points and can often be separated using conventional methods. A number of chiral derivatizing agents, as well as the types of compounds for which they are useful, have been developed and are listed in Table 1. Limitations of this approach include lack of suitable functionality in the analyte that can be derivatized with an appropriate enantiomerically pure derivatizing agent, unavailability of a suitable derivatizing agent of sufficiently high or at least known optical purity, difficulty of removing the derivatizing group after the desired separation has been accomplished, enantiodiscrimination during derivatization, potential racemization either during derivatization or removal or the chiral derivatizing group (which is not always possible), and the additional validation required to confirm that the enantiomeric ratio of the final product corresponds to the original enantiomeric ratio.

Use of Chiral Additives

Another method for creating a chiral environment is to add an optically pure chiral selector to a bulk liquid phase. Chiral additives have several advantages over chiral stationary phases and continue to be the predominant mode for chiral separations by tlc and capillary electrophoresis (ce). First of all, the chiral selector added to a bulk liquid phase can be readily changed. The use of chiral additives allows chiral separations to be done using less expensive, conventional stationary phases. A wider variety of chiral selectors are available to be used as chiral additives than are available as chiral stationary phases, thus, providing the analyst with considerable flexibility. Finally, the use of chiral additives may provide valuable insight into the chromatographic conditions and/or likelihood of success with a potential chiral stationary-phase chiral selector. This is particularly important for the development of new chiral stationary phases because of the difficulty and cost involved.

Chiral additives, however, do pose some unique problems. Many chiral agents are expensive or are not commercially available, and therefore, must be synthesized. The presence of the chiral additive in the bulk liquid phase may also interfere with detection or recovery of the analytes. Finally, the presence of enantiomeric impurity in the chiral additive may add analytical complications.

Thin-Layer Chromatography. Thin-layer chromatography (tlc) offers several advantages for chiral separations and in the development of new chiral stationary phases. Besides being inexpensive, tlc can be used to screen mobile-phase conditions rapidly (i.e., organic modifier content, pH, etc.), chiral selectors, and analytes. Several different analytes may be run simultaneously on the same plate. Usually, no preequilibration of the mobile phase and stationary phase is required. In addition, only small

TABLE 1. ANALYTE FUNCTIONAL GROUPS AND CHIRAL DERIVATIZING REAGENTS

Analyte functional group	Derivatizing agent	Product	Examples of derivatizing agents
carboxylic acid (acid or base catalyzed)	alcohol	ester	(−)-menthol
	amine	amide	1-phenylethylamine
amine (1°)	aldehyde	isoindole	1-(1-naphthyl)ethylamine *o*-phthaldialdehyde–2-mercaptoethanol
amine (1° and 2°)	anhydrides	amide	γ-butyloxycarbonyl-L-leucine anhydride *O,O*-dibenzoyltartaric anhydride
	acyl halides	amide	(*R*)-(−)-methylmandelic acid chloride α-methoxy-α-trifluoromethylphenylacetyl chloride
	isocyanates	urea	α-methylbenzyl isocyanate 1-(1-naphthyl)ethyl isocyanate
	isothiocyanate	thiourea	2,3,4,6-tetra-*O*-acetyl-β-D-glucopyranosyl isothiocyanate α-methylbenzyl isothiocyanate
(1°, 2°; can N-dealkylate 3°)	chloroformates	carbamate	(−)-menthyl chloroformate (+)-1-(9-fluorenyl)ethylchloroformate
alcohols	acyl halides	ester	(−)-menthoxy acid chloride (*S*)-*O*-propionylmandelyl chloride (*S,S*)-tartaric anhydride
	anhydrides	ester	
	chloroformate	carbonate	(−)-menthyl chloroformate
	isocyanate	carbamate	α-methylbenzyl isocyanate

amounts of mobile phase, and therefore, chiral mobile-phase additive, are required. Another significant advantage is that the analyte can always be unambiguously found on the tlc plate.

Two mechanisms for chiral separations using chiral mobile-phase additives, analogous to models developed for ion-pair chromatography, have been proposed to explain the chiral selectivity obtained using chiral mobile-phase additives. In one model, the chiral mobile-phase additive and the analyte enantiomers form "diastereomeric complexes" in solution. As noted previously, diastereomers may have slightly different physical properties such as mobile phase solubilities or slightly different affinities for the stationary phase. Thus, the chiral separation can be achieved with conventional columns.

An alternative model has been proposed in which the chiral mobile-phase additive is thought to modify the conventional, achiral stationary phase *in situ*, thus, dynamically generating a chiral stationary phase. In this case, the enantioseparation is governed by the differences in the association between the enantiomers and the chiral selector in the stationary phase.

Several different types of chiral additives have been used including (1*R*)-(−)-ammonium-10-camphorsulfonic acid, cyclodextrins, proteins, and various amino acid derivatives such as *N*-benzoxycarbonyl-glycyl-L-proline as well as macrocyclic antibiotics.

Chiral separation validation in tlc may be accomplished by recovering the individual analyte spots from the plate and subjecting them to some type of chiroptical spectroscopy such as circular dichroism or optical rotary dispersion. Alternatively, the plates may be analyzed using a scanning densitometer.

Capillary Electrophoresis. Capillary electrophoresis (ce) or capillary zone electrophoresis (cze), a relatively recent addition to the arsenal of analytical techniques, has also been demonstrated as a powerful chiral separation method. Its high resolution capability and lower sample loading relative to hplc makes it ideal for the separation of minute amounts of components in complex biological mixtures.

In a ce experiment, a thin capillary is filled with a run buffer and a voltage is applied across the capillary. The underlying impetus for separations in ce is, in general, derived from the fact that charged species migrate in response to an applied electric field proportionately to their charge and inversely proportionately to their size.

Chiral separations by ce have been performed almost exclusively using chiral additives to the run buffer. The advantages of this approach are identical to the advantages mentioned previously with regard to using chiral mobile-phase additives in tlc. Many of the chiral selectors used successfully as mobile-phase additives in tlc and as immobilized ligands in hplc have been used successfully in ce including proteins, native and functionalized cyclodextrins, various carbohydrates, assorted functionalized amino acids, chiralion pairing agents, and macrocyclic antibiotics.

Although chiral ce is most commonly performed using aqueous buffers, there has been some work using organic solvents such as methanol, formamide, N-methylformamide or N,N-dimethylformamide with chiral additives such as quinine or cyclodextrins. Nonaqueous ce requires that the background electrolyte be prepared using organic acids (e.g., citric acid or acetic acid) and organic bases (e.g., tetraalkylammonium halides or tris(hydroxymethyl)aminomethane).

Chiral Stationary Phases

Most chiral chromatographic separations are accomplished using chromatographic stationary phases that incorporate a chiral selector. The chiral separation mechanisms are generally thought to involve the formation of transient diastereomeric complexes between the enantiomers and the stationary phase chiral ligand. Differences in the stabilities of these complexes account for the differences in the retention observed for the two enantiomers. Often, the use of a chiral stationary phase allows for the direct separation of the enantiomers without the need for derivatization. One advantage offered by the use of chiral stationary phases is that the chiral selector need not be enantiomerically pure, only enriched. In addition, for chiral stationary phases having a well understood chiral recognition mechanism, assignment of configuration (e.g., R or S) may be possible even in the absence of optically pure standards. However, chiral stationary phases have some limitations. The specificity required for chiral discrimination limits the broad applicability of most chiral stationary phases; thus there is no "universal" chiral stationary phase. The cost of most chiral columns are typically much higher ($\sim 3\times$) than for conventional columns. In contrast to conventional chromatographic columns, chiral stationary phases are generally not as robust, require more careful handling than conventional columns and usually, once column performance has begun to deteriorate, cannot be returned to their original performance levels. In many cases, chromatographic column choice or mobile phase optimization for chiral stationary phases is not as straightforward as with conventional stationary phases. For many of the chiral stationary phases, adequate chiral recognition models, used to guide selection of the appropriate column for a given separation, have yet to be developed. Column selection, therefore, is often reduced to identifying structurally similar analytes for which chiral resolution methods have been reported in the scientific literature or chromatographic supply catalogues and adapting a reported method for the chiral pair to be resolved.

An additional complication, sometimes arising with the use of chiral stationary phases, may occur when the analytes either exist as *conformers* or can undergo inversion during the chromatographic analysis.

Thin-Layer Chromatography. Chiral stationary phases in tlc have been primarily limited to phases based on normal or microcrystalline cellulose, triacetylcellulose sorbents or silica-based sorbents that have been chemically modified or physically coated to incorporate chiral selectors such as amino acids or macrocyclic antibiotics into the stationary phase. The cost of many chiral selectors, as well as the accessibility and success of chiral additives, may have inhibited widespread commercialization.

Of the silica-based materials, only the ligand-exchange phases are commercially available (Chiralplate, tlc plates are available through Alltech Associates, Inc.) Supelco, Inc., the Aldrich Chemical Company, and Bodman Industries are all based on ligand exchange. Typically in the case of the ligand-exchange type tlc plates, the ligand-exchange selector is comprised of an amino acid residue to which a long

hydrocarbon chain has been attached (e.g., ($2S$, $4R$, $2'RS$)4-hydroxy-1-(2-hydroxydodecyl)proline). The hydrocarbon chain of the functionalized amino acid is either chemically bonded to the substrate or intercalates in between the chains of a reversed phase-stationary phase thus immobilizing the chiral selector. The bidentate amino acid chiral selector is thought to reside close to the surface of the stationary phase and participates as a ligand in the formation of a bi-ligand complex with a divalent metal ion (e.g., Cu^{2+}) and the chiral bidentate analyte. Analytes enantioresolvable using ligand exchange are usually restricted to 1,2-diols, α-amino acids, α-amino alcohols, and α-hydroxyacids. Again, differences in the stabilities of the diastereomeric complexes thus formed give rise to the chiral separation.

High Performance Liquid Chromatography. The last decade has seen the commercialization of a large number of different types of chiral stationary phases including the cyclodextrin phases, the chirobiotic phases, the π-π interaction phases, the protein phases as well as the cellulosic and amylosic phases and chiral crown ether phases. Currently, there are over 50 different chiral columns that are commercially available for hplc. Table 2 briefly summarizes the types of columns available as well as typical applications and mobile-phase conditions. Each of these chiral stationary phases are very successful at separating large numbers of enantiomers, which in many cases, are unresolvable using any of the other chiral stationary phases. Unfortunately, despite the large number and variety of chiral stationary phases currently available, there remains a large number of enantiomeric compounds that are unresolvable by any of the existing chiral, stationary phases. In addition, incomplete understanding of the chiral recognition mechanisms of many of these chiral stationary phases limits the realization of the full potential of the existing chiral stationary phases and hampers development of new chiral stationary phases.

Ligand-Exchange Phases

Among the earliest reports of chiral separations by liquid chromatography were based on work done by Davankov using ligand exchange. These types of columns are available from Phenomenex, J. T. Baker, and Regis Technologies, Inc. Although almost any amino acid can form the basis for the chiral selector, proline and hydroxyproline exhibit the most widespread utility. Also, although other metals can be used, copper(II) is usually the metal of choice and is added to the aqueous buffer mobile phase.

The dependence of chiral recognition on the formation of the diastereomeric complex imposes constraints on the proximity of the metal binding sites, usually either an hydroxy or an amine α to a carboxylic acid, in the analyte. Principal advantages of this technique include the ability to: assign configuration in the absence of standards; enantioresolve

TABLE 2. CLASSES OF HPLC CHIRAL STATIONARY PHASES

Column chiral selector	Typical mobile phase conditions	Typical analyte features required
pirkle	nonpolar organic; 2-propanol-hexane	π-acid or π-basic moieties for charge transfer complex; hydrogen-bonding or dipole stacking capability near chiral center
protein	phosphate buffers	aromatic near chiral center; organic acids or bases; cationic drugs
cyclodextrin	aqueous buffers; polar organic	good "fit" between chiral cavity or chiral mouth of cyclodextrin and hydrophobic moiety; hydrogen-bonding capability near chiral center
ligand exchange	aqueous buffers	α-hydroxy or α-amino acids near chiral center; can do nonaromatic
chiral crown ether	0.01 N perchloric acid	primary amines near chiral center; can do nonaromatic
macrocyclic antibiotics	aqueous buffers, nonpolar and polar organic	amines, amides, acids, esters; aromatic; hydrophobic moiety
cellulosic and amylosic	nonpolar organic	aromatic

nonaromatic analytes; use aqueous mobile phases; acquire a stationary phase with the opposite enantioselectivity; and predict the likelihood of successful chiral resolution for a given analyte based on a well-understood chiral recognition mechanism.

Pirkle Phases

Of all of the commercially available chiral stationary phases for liquid chromatography, the chiral recognition mechanism for the "Pirkle" phases are among the best understood. Chiral recognition on Pirkle phases is thought to depend upon complimentary interactions between the analyte and the selector. These interactions may be π-π, steric, hydrogen-bonding, or dipole—dipole interactions and contribute to the overall stability of the diastereomeric association complexes that form between the individual enantiomers and the chiral selector in the stationary phase.

Nonpolar organic mobile phases, such as hexane with ethanol or 2-propanol as typical polar modifiers, are most commonly used with these types of phases. Under these conditions, retention seems to follow normal phase-type behavior (e.g., increased mobile phase polarity produces decreased retention). The normal mobile phase components only weakly interact with the stationary phase and are easily displaced by the chiral analytes thereby promoting enantiospecific interactions. Some of the Pirkle-types of phases have also been used, to a lesser extent, in the reversed phase mode.

Reciprocity, an important concept introduced by Pirkle, exploited the notion that analytes that were well resolved using a particular chiral selector would likely be good candidates for chiral selectors to enantioresolve analytes similar to the original chiral selector. This insight spawned a second generation of Pirkle phases based on N-(2-naphthyl)-α-amino acids. These phases were very successful at enantio-resolving analytes containing a 3,5-dinitrobenzoyl group, such as 3,5-dinitrophenyl carbamates, and ureas of chiral alcohols and amines. These columns are available through a variety of sources including Phenomenex, Regis Technologies, Inc., J. T. Baker, Inc., and Supelco, Inc.

The structure of the Whelk-O-1 phase, the most recent addition to this type of chiral stationary phase, is illustrated in Figure 2. The presence of both π-acid and π-base features, as well as the inherent rigidity of the chiral selector, confers greater versatility than any of the previous Pirkle-type phases, imposing fewer constraints on both analyte structural features required for successful enantioresolution and mobile phase conditions. Indeed, this chiral stationary phase has demonstrated considerable chiral selectivity for naproxen, warfarin, and its *p*-chloro analogue under nonaqueous reversed-phase conditions and reversed-phase conditions. An additional advantageous feature of this phase is its availability with either the (R,R) or (S,S) configuration, thus, permitting the enantiomeric elution order to be readily changed. The small size of the chiral selector also promotes fairly high bonded ligand densities in the stationary phase, which coupled with the high enantioselectivities often achieved with these phases, facilitates their use for preparative-scale separations.

Cyclodextrin Phases

Cyclodextrins are macrocyclic compounds comprised of D-glucose bonded through 1,4-α-linkages and produced enzymatically from starch. The greek letter which proceeds the name indicates the number of glucose units incorporated in the CD (e.g., $\alpha = 6$, $\beta = 7$, $\gamma = 8$, etc). Cyclodextrins are toroidal shaped molecules with a relatively hydrophobic internal cavity (Fig. 3).

Among the most successful of the liquid chromatographic reversed-phase chiral stationary phases have been the cyclodextrin-based phases, introduced by Armstrong and commercially available through Advanced

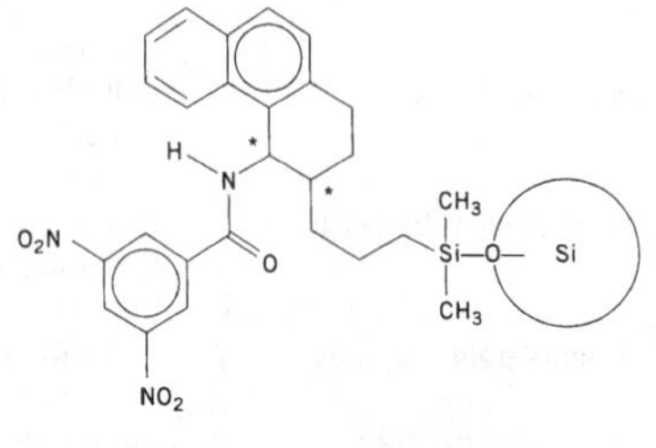

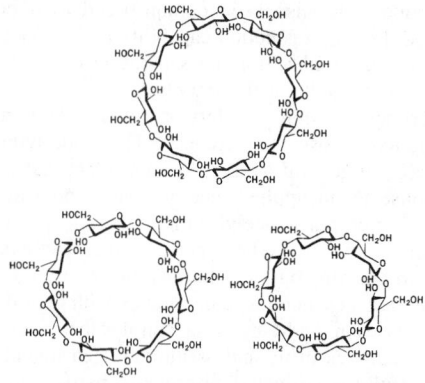

Fig. 3.   The structure of the three most common cyclodextrins

Separation Technologies, Inc. or Alltech Associates. The most commonly used cyclodextrin in hplc is the β-cyclodextrin. In the bonded phases, the cyclodextrins are thought to be tethered to the silica substrate through one or two spacer ligands. The mechanism thought to be responsible for the chiral selectivity observed with these phases is based upon the formation of an inclusion complex between the hydrophobic moiety of the chiral analyte and the hydrophobic interior of the cyclodextrin cavity. Preferential complexation between one optical isomer and the cyclodextrin through stereospecific interactions with the secondary hydroxyls which line the mouth of the cyclodextrin cavity results in the enantiomeric separation. Unlike the Pirkle-type phases, enantiospecific interactions between the analyte and the cyclodextrin are not the result of a single, well-defined association, but more of a statistical averaging of all the potential interactions with each interaction weighted by its energy or strength of interaction.

Vast amounts of empirical data suggest that chiral recognition on cyclodextrin phases in the reversed phase mode require the presence of an aromatic moiety that can fit into the cyclodextrin cavity, that there be hydrogen bonding groups in the molecule, and that the hydrophobic and hydrogen-bonding moieties should be in close proximity to the stereogenic center. Chiral recognition seems to be enhanced if the stereogenic center is positioned between two π-systems or incorporated in a ring.

Most of the chiral separations reported to date using the native cyclodextrin-based phases have been accomplished in the reversed-phase mode using aqueous buffers containing small amounts of organic modifiers. However, polar organic mobile phases have gained in popularity recently because of their ease of removal from the sample and reduced tendency to accelerate column degradation relative to the hydroorganic mobile phases. In these cases, because the more nonpolar component of the mobile phase is thought to occupy the cyclodextrin cavity, the analyte is thought to sit atop the mouth of the cyclodextrin much like a "lid".

Limitations with the chiral selectivity of the native cyclodextrins fostered the development of various functionalized cyclodextrin-based chiral stationary phases, including acetylated, sulfated, 2-hydroxypropyl, 3,5-dimethylphenylcarbamoylated and 1-naphthylethylcaarbamoylated cyclodextrin. Each of the glucose residues contribute three hydroxyl groups to which a substituent may be appended; thus, each cyclodextrin contributes multiple sites for derivatization. The substituents of these functionalized cyclodextrins seem to play a variety of roles in enhancing chiral recognition.

Cellulosic and Amylosic Phases

Cellulose and amylose are comprised of the same glucose subunits as the cyclodextrins. In the case of cellulose, the glucose units are attached through 1,4-β-linkages resulting in a linear polymer. In the case of amylose, the 1,4-α-linkages, as are found in the cyclodextrins, are thought to confer helicity to the polymeric chain.

Cellulosic phases as well as amylosic phases have been used extensively for enantiomeric separations recently. Most of the work in this area has been with various derivatives of the native carbohydrate. The enantioresolving abilities of the derivatized cellulosic and amylosic phases are reported to be very dependent upon the types of substituents on the aromatic moieties that are appended onto the native carbohydrate. Table 3 lists some of the cellulosic and amylosic derivatives that have been used.

Fig. 2.   The structure of the chiral selector in the Whelk-O-1 chiral stationary phase

These columns are available through Chiral Technologies, Inc. and J. T. Baker, Inc.

Protein-Based Phases

Proteins, amino acids bonded through peptide linkages to form macromolecular biopolymers, used as chiral stationary phases for hplc include bovine and human serum albumin, α_1-acid glycoprotein, ovomucoid, avidin, and cellobiohydrolase. The bovine serum albumin column is marketed under the name Resolvosil and can be obtained from Phenomenex. The human serum albumin column can be obtained from Alltech Associates, Advanced Separation Technologies, Inc., and J. T. Baker. The α_1-acid glycoprotein and cellobiohydrolase can be obtained from Advanced Separation Technologies, Inc. or J. T. Baker, Inc.

In most cases, the protein is immobilized onto γ-aminopropyl silica and covalently attached using a cross-linking reagent such as N,N'-carbonyldiimidazole. The tertiary structure or three dimensional organization of proteins are thought to be important for their activity and chiral recognition. Therefore, mobile phase conditions that cause protein "denaturation" or loss of tertiary structure must be avoided.

Typically, the mobile phases used with the protein-based chiral stationary phases consist of aqueous phosphate buffers. Often small amounts of organic modifiers, such as methanol, ethanol, propanol, or acetonitrile, are added to reduce hydrophobic interactions with the analyte and to improve enantioselectivity. In some cases, dramatic changes in chiral recognition occur when small amounts of organic modifiers, such as N,N-dimethyloctylamine or octanoic acid are added to the mobile phase. As in the case of the cyclodextrin and amylosic and cellulosic phases, the chiral recognition mechanism for these protein-based phases is not well understood. Optimization of chromatographic conditions and selection of analytes that can be successfully resolved on these phases is usually done empirically. In addition, the large molecular weight of these biopolymers dictates that the amount of chiral selector that can be immobilized on the column packing material is very small.

An interesting application of the protein-based phases is various protein binding and displacement experiments which can be done fairly routinely. For instance, differences in the enantioselectivity, toward a particular drug, of a column derived from human serum albumin and a column derived from some other animal serum albumin might be indicative that a particular species might not be a good animal model during drug development, thus, obviating the need for animal testing. Chiral separations on protein-based phases may also provide useful information on drug interactions.

Chirobiotic Phases

The chirobiotic chiral stationary phases are based on macrocyclic antibiotics such as vancomycin and teicoplanin. These chiral selectors, originally used as chiral additives in capillary zone electrophoresis, incorporate aromatic and carbohydrate, as well as peptide and ionizable moieties. The presence of aromatic groups, allowing for π-π interactions, and the macrocyclic rings, offering potential inclusion complexation, give these phases some of the advantages of the protein-based phases (e.g., peptide and hydrogen bonding sites) and the carbohydrate-based phases but with greater sample capacity and greater mobile phase flexibility. Indeed, these phases seem to be truly "multimodal" in that they have demonstrated chiral selectivity in the normal, polar organic, and reversed-phase modes. In addition, the use of such well-defined chiral selectors facilitate method development and optimization. These columns are commercially available through Advanced Separation Technologies, Inc. and Alltech Associates.

TABLE 3. CARBOHYDRATE DERIVATIVES USED AS HPLC CHIRAL STATIONARY PHASES

Cellulosic	Amylosic
triacetate	
tribenzoate	
tribenzylether	
tricinnamate	
triphenylcarbamate	triphenylcarbamate
tris-3,5-dichlorophenylcarbamate	
tris-3,5-dimethylphenylcarbamate	tris-3,5-dimethylphenylcarbamate
tris-1-phenylethylcarbamate	tris-1-phenylethylcarbamate

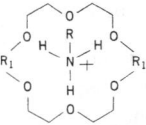

Fig. 4. An inclusion complex formed between a protonated primary amine and a chiral crown ether

Chiral Crown Ether Phases

Chiral crown ethers based on 18-crown-6 (Fig. 4) can form inclusion complexes with ammonium ions and protonated primary amines. Immobilization of these chiral crown ethers on a chromatographic support provides a chiral stationary phase which can resolve most primary amino acids, amines and amino alcohols. However, the stereogenic center must be in fairly close proximity to the primary amine for successful chiral separation. Significantly, the chiral crown ether phase is unique in that it is one of the few liquid chromatographic chiral stationary phases that does not require the presence of an aromatic ring to achieve chiral separations.

Mobile phases used with this stationary phase are typically 0.01 N perchloric acid with small amounts of methanol or acetonitrile. One significant advantage of these phases is that both configurations of the chiral stationary phase are commercially available and can be obtained from J. T. Baker Inc. and Chiral Technologies, Inc. (Crownpak CR).

Chiral Synthetic Polymer Phases

Chiral synthetic polymer phases can be classified into three types. In one type, a polymer matrix is formed in the presence of an optically pure compound to molecularly *imprint* the polymer matrix. Subsequent to the polymerization, the chiral template is removed, leaving the polymer matrix with chiral cavities. The selectivities achieved with these phases are generally excellent, thus, facilitating semipreparative separations. However, the applicability of these chiral stationary phases are generally limited to the analyte upon which the phase is based and a limited number of analogues. In addition, these types of phases generally exhibit poor efficiency in large part because the polymeric matrix contributes to nonsterespecific binding. Advantages of this approach include the ability to prepare reciprocal phases and the predictability of the enantiomeric elution order.

Another type of synthetic polymer-based chiral stationary phase is formed when chiral catalyst are used to initiate the polymerization. Columns of this type (e.g., Chiralpak OT) are available from Chiral Technologies, Inc., or J. T. Baker Inc.

A third type of synthetic polymer-based chiral stationary phase, developed by Blaschke, is produced when a chiral selector is either incorporated within the polymer network or attached as pendant groups onto the polymer matrix. Both are analogous to methods used to produce polymeric chiral stationary phases for gc.

In general, the synthetic polymeric phases seem to have polarities analogous to diol-type phases and a wide range of mobile phase conditions have been used including hexane, various alcohols, acetonitrile, tetrahydrofuran, dichloromethane and their mixtures, as well as aqueous buffers.

Chiral Separation Validation for Hplc

Chiral separations present special problems for validation. Typically, in the absence of spectroscopic confirmation (e.g., mass spectral or infrared data), conventional separations are validated by analyzing "pure" samples under identical chromatographic conditions. Often, two or more chromatographic stationary phases, which are known to interact with the analyte through different retention mechanisms, are used. If the pure sample and the unknown have identical retention times under each set of conditions, the identity of the unknown is assumed to be the same as the pure sample. However, often the chiral separation that is obtained with one type of column may not be achievable with any other type of chiral stationary phase. In addition, "pure" enantiomers are generally not available.

Most commonly, uv or uv–vis spectroscopy is used as the basis for detection in hplc. When using a chiral stationary phase, confirmation of a chiral separation may be obtained by either monitoring the column effluent at more than one wavelength or by running the sample more than once. Although not absolute proof of a chiral separation, this approach does provide strong supporting evidence.

As in tlc, another method to validate a chiral separation is to collect the individual peaks and subject them to some type of optical spectroscopy, such as, circular dichroism or optical rotary dispersion. Alternatively, a chiroptical spectroscopy can be used as the basis for detection on-line using commercially available optical rotary dispersion or circular dichroism-based detectors. Another method for validating chiral separations by lc is to couple the chromatographic system to a mass spectrometer.

Chiral Stationary Phases for Gas Chromatography

Gc chiral stationary phases can be broadly classified into three categories: diamide, cyclodextrin, and metal complex.

Diamide Chiral Separations. The first commercially available chiral column was the Chiralsil-val, which was introduced in 1976 for the separation of amino acid type compounds by gas chromatography. It is based on a polysiloxane polymer containing chiral side chains incorporating L-valine-*t*-butylamide. The polysiloxane backbone improved the thermal stability of these chiral stationary phases relative to the original coated columns and extended the operating temperatures up to 220°C. The column is effective for the separation of perfluoroacylated and esterified amino acids, amino alcohols, and some chiral sulfoxides. Another polysiloxane-based chiral stationary phase incorporating L-valine-(R)-α-phenylethylamide appended onto hydrolyzed XE-60 was found to be particularly successful at resolving perfluoroacetylated amino alcohol derivatives. Through judicious choice of derivatizing agent, chiral separations were obtained for a wider range of compounds, including amino alcohols, α-hydroxy acids, diols and ketones, than had previously been obtainable using these types of stationary phases.

Metal Complex. Complexation gas chromatography was first introduced by V. Schurig in 1980 and employs transition metals (e.g., nickel, cobalt, manganese or rhodium) complexed with chiral terpenoid ketoenolate ligands such as 3-trifluoroacetyl-1R-camphorate (**1**), 1R-3-pentafluoro-benzoylcamphorate or 3-heptafluorobutanoyl-(1R, 2S)-pinanone-4-ate. This class of chiral columns is particularly adept at enantioresolving some olefins and oxygen-containing compounds such as ketones, ethers, alcohols, spiroacetals, oxiranes, and esters.

(**1**)

Cyclodextrins. As indicated previously, the native cyclodextrins, which are thermally stable, have been used extensively in liquid chromatographic chiral separations, but their utility in gc applications was hampered because their highly crystallinity and insolubility in most organic solvents made them difficult to formulate into a gc stationary phase. However, some functionalized cyclodextrins form viscous oils suitable for gc stationary-phase coatings and have been used either neat or diluted in a polysiloxane polymer as chiral stationary phases for gc.

Chiral Separation Validation for Gas Chromatography

The special problems for validation presented by chiral separations can be even more burdensome for gc because most methods of detection (e.g., flame ionization detection or electron capture detection) in gc destroy the sample. Even when nondestructive detection (e.g., thermal conductivity) is used, individual peak collection is generally more difficult than in lc or tlc. Thus, off-line chiroptical analysis is not usually an option. Fortunately, gc can be readily coupled to a mass spectrometer and is routinely used to validate a chiral separation.

APRYLL M. STALCUP
University of Cincinnati

Additional Reading

Kuzel, R. A. S. K. Bhasin, H. G. Oldham, L. A. Damani, J. Murphy, P. Camilleri, and A. J. Hutt: *Chirality* **6**, 607 (1994).
Pasteur, L.: *Comptes Rendus de l'Academie des Sciences* **26**, 535 (1848).
Tobert, S. A.: *Clin. Pharmacol. Ther.* **29**, 344 (1981).
Wechter, W. J. D. G. Loughhead, R. J. Reischer, G. J. Van Giessen, and D. G. Kaiser: *Biochem. Biophys. Res. Comm.* **61**, 833 (1974).

CHLORAL. See **Chlorinated Organics**.

CHLORARGYRITE. Also known as horn silver, chlorargyrite is silver chloride, AgCl. The mineral crystallizes in the isometric system but is usually massive, appearing like wax or horn, hence the name. It has no cleavage, is highly sectile, yielding bright surfaces; hardness, 2.5; specific gravity, 5.55; luster, resinous to adamantine; color, gray, white to colorless. May be blue, violet-brown after exposure to light; transparent to translucent. Chlorargyrite is largely a secondary mineral, usually associated with other silver minerals as well as with compounds of lead, zinc, and copper. Saxony and the Harz Mountains are European localities. The Broken Hill district of New South Wales is a well-known occurrence, but probably the most important deposits are found in Atacama, Chile. The mineral also is found in Bolivia and Mexico. In the United States, chlorargyrite comes from Colorado, Idaho, Utah, Nevada, Arizona, and New Mexico.

CHLORIDE (Biological Aspects). Sodium chloride, potassium chloride, and other chloride salts, when ingested by animals from feedstuffs and humans from various food substances, reduce to a consideration of the cation involved (Na^+, K^+, etc.) and the Cl^- (chloride) ion. Generally, in terms of animal and human nutrition, more research has been conducted and more is known about the role of cations in metabolism than that of the chloride ion. Some physiologists and nutritionists in the past have described chloride as playing a "passive role" in maintaining the body's ionic and fluid balance. With exception of the "chloride shift" in venous blood, the movements of chloride have usually been considered secondary to those of the cations.

Much is known, of course, concerning the effects of excessive sodium chloride and of deficient sodium chloride in human and animal diets, but the physiological and nutritional roles of chloride have not been thoroughly studied and fully explained. C.E. Coppock, Department of Animal Science, Texas A & M University, College Station Texas; and M.J. Fettman, Cornell University College of Veterinary Medicine, undertook a study of chloride (targeted to chloride as a required nutrient for lactating dairy cows) and also carefully reviewed the prior work in this area of other researchers. (Interested readers are referred to the bibliography at end of article, "Chloride as a Required Nutrient for Lactating Dairy Cows," *Feedstuffs* (February 10, 1978).)

Despite important physiological functions and its presence in milk at about 0.11%, chloride is a neglected element in large animal nutrition. The practice of adding sodium chloride to concentrate mixtures and free-choice feeding seems to have precluded the possibility of a practical deficiency problem. When salt was omitted from the diet, researchers found that under the conditions used in their study, sodium was the first limiting element. This was true because sodium is present in most natural ingredients at much lower levels, relative to the cow's requirements, than is chloride.

Those who formulate diets usually ignore sodium levels in the natural ingredients (which are usually low in forages, but may be appreciable in certain concentrate ingredients) because of the traditional value of salt as a condiment. In addition, other sodium salts are often included in concrete mixtures: sodium sulfate as a sulfur replacement, the sodium phosphates as phosphorus supplements, and sodium bicarbonate as a buffer. Even when these supplements are used, salt is still often included because of tradition. Under many conditions, gross overfeed of both sodium and chloride occurs. High dietary levels of sodium and chloride do not increase the levels of these elements in milk. Excess will be excreted in urine and manure. Excess salt intake may result in greater water consumption, waste transport, bedding requirements, and transfer of sodium and chloride to the soil.

Of the seven macro mineral elements required by dairy cattle, five can be considered fertilizer elements (potassium, calcium, phosphorus, magnesium, and sulfur), but sodium and chloride are both toxic to plants at high concentrations and present practical problems in areas with saline soils. High salt intakes have also been shown to increase udder edema in heifers. Because of the importance of chloride in nutrition and metabolism, research is needed to define the chloride requirements of lactating cows and clarify mineral relationships, especially between chloride and potassium plus sodium.

Chloride and Plants

Chloride is one of the most recent elements to be shown essential for plant growth. Evidence shows that tomato plants grown in a low-chloride solution developed wilting of leaflet blade tips, which progressed to chlorosis, bronzing, and necrosis. Growth was proportional to chloride concentration (up to a point) in the culture medium. Chloride additions to the medium prevented the deficiency symptoms and caused their disappearance in deficient plants. Later, the same team showed that often, other species, including barley and alfalfa, displayed severe deficiency symptoms when grown in a low-chloride medium. Although buckwheat, corn (maize), and beans did not exhibit these obvious symptoms, yield effects were apparent.

Despite the presence of 250 parts per million (ppm) in the leaves of chloride-deficient tomato plants compared to a 0.1 ppm for molybdenum, chloride was classified as a micronutrient by plant physiologists. The essentiality of chloride escaped detection for so long because of its wide distribution in nature, the high solubility of most chloride salts, and difficulties of purification. Because chloride is a principal ion of sea water, it is picked up by winds from sea spray and carried far inland. For example, Geneva, New York has been estimated to receive 18 kilograms/hectare of chloride annually; Mount Vernon, Iowa, 73 kilograms/hectare annually. Deposits of more than 45 kilograms/hectare have been reported near coastlines. It was also suggested that leaf structures were capable of capturing airborne chloride. The popularity of potassium chloride as a potassium fertilizer and manure from cows fed excessive levels of chloride relative to their requirement are additional sources of soil chloride.

Chloride differs from other nutrient elements present in native rocks because it is not fixed by colloids; it is repelled by negatively charged clay surfaces, and all chloride compounds formed in soils are highly soluble. In addition to leaching, chloride is lost from soils through crop removal.

Several factors affect chloride uptake by plants: (1) age or advancing maturity shown in avocado, apricot, and grape leaves; (2) chloride concentration in the soil; (3) soil oxygen levels; (4) plant species; and (5) competition from other anions. Research has shown a specific antagonism between nitrate-nitrogen and chloride; increasing chloride levels in the growth medium reduces nitrogen uptake, but this reduction is primarily in the nitrate-nitrogen fraction, with little if any effect on the protein-nitrogen fraction. The reciprocal effect was also observed between chloride and sulfate accumulation. Chloride is essential in the plant for photosynthetic reactions in chloroplasts, which produce oxygen.

Chloride is also treated as a toxic element as well as a part of saline toxicity. Over the toxic range, the reduction in growth is approximately linear. For example, at about 3500 ppm chloride in the cultural medium, alfalfa growth will be depressed to about 60% that of normal. Obviously, in areas with saline soils, there is concern about excessive levels of salt returned to the soil via manure.

Gastrointestinal Absorption of Chloride

Since 1952, when in goats, sheep, and dairy cattle, the observation was made that chloride could be absorbed from ruminal fluid into the blood against a tenfold concentration difference (normal rumen fluid chloride concentrations may range from 10 to 30 mEq/l, while those of the plasma may range from 100 to 110 mEq/l), numerous researchers have attempted to describe accurately and explain the processes by which chloride might move across the reticulorumen epithelium against its apparent chemical gradients. For a number of years, it was assumed by some workers that the observed electrical potential difference across the forestomach epithelium, making the plasma approximately 30 mV positive to the contents of the reticulorumen, could adequately account for the otherwise anomalous movements. If chloride's movement into the blood was truly attributable solely to the combined electrochemical gradient acting upon it, then its distribution across the gastric epithelium should have been describable by the Nernst equation.

In rumen-fistulated experimental animals, it has been observed that for certain distribution ratios of chloride in the ruminal fluids and blood plasma, the calculated equilibrium potential for chloride is relatively the same as that measured directly with KCl–agar bridges and calomel electrodes. However, in many circumstances, the calculated and measured values have been found to be significantly different, an observation that could only be accounted for by the presence of an active transport mechanism responsible for the movement of chloride out of its equilibrium distribution.

The active transport of chloride has also been demonstrated across the wall of the frog stomach, rat ileum, dog ileum, and the human ileum. Experiments have produced a double exchange model in which bicarbonate secretion and chloride absorption are linked by an isoelectric mechanism to hydrogen ion secretion and sodium absorption across the human ileum. In 1972, a group of researchers proposed a similar model of coupled transport across the reticulorumen epithelium, and further concluded that active anion and cation transport in the rumen cannot function efficiently unless both components are intact, i.e., the net transport of the body's major cations from the rumen to the blood, appeared to be dependent in part on the activity of chloride in the system.

Chloride in Cerebrospinal Fluid

A similar story has unfolded concerning the distribution and movement of chloride between the blood and cerebrospinal fluid (CSF). The first indication for the possible existence of an active mechanism responsible for the maintenance of chloride levels in the CSF in the mid-20th Century when, Hiatt demonstrated in dogs the persistence of CSF chloride concentrations at 44% of normal, despite the reduction of chloride levels to 30% of normal in all body fluids during a nitrate-induced diuresis. Over the next 20 years, researchers recorded chloride ion concentration and electrical potential differences across CSF-ECF (extracellular fluid) and CSF-blood barriers ranging from 15 to 20% and from 5 to 30 mV, respectively, in such varied subjects as dogfish, rats, cats, dogs, monkeys, and humans. In all cases, the CSF was both higher in chloride ion concentration and negative in potential with respect to the reference body fluid.

In 1970, researchers studied the distribution and kinetics of chloride in cats following the iso-osmotic replacement of body chloride with isethionate via extracorporeal hemodialysis. They found that when the plasma chloride concentration was reduced by approximately 93%, the cerebral cortex, corpus callosum, and CSF chloride concentrations were reduced by approximately only 26.5, 35, and 21%, respectively. Other body tissues and fluids showed reductions in chloride concentration closer to those of the plasma (skeletal muscle and liver, 73%). The influx of chloride into the CSF at various plasma chloride concentrations was then plotted as a Lineweaver-Burk plot, and was shown to behave as a carrier-mediated process, as described by Michaelis-Menten kinetics. This information, combined with the observations made by Abbot et al. that reduction of the plasma chloride concentration by isethionate replacement did not produce a change in electrical potential "commensurate with or even in the same direction" as that expected by Nernst equation predictions, led workers in the field to ascribe the bulk of chloride movement from the blood into the CSF to an active transport process. It is possible that control of the rate of chloride transport is a factor in the regulation of the secretion of CSF, the medium that bathes, protects, and nourishes the central nervous system.

Chloride in the Humoral Regulation of Sodium and Potassium

Conventional presentations of the regulatory mechanisms involved in body fluid and electrolyte homeostasis usually have considered maintenance of the sodium/potassium ratio in the ECF both the prime means and end toward a functional electrolyte balance. Certainly, the sodium/potassium ratio provides the axis about which the body's humoral mechanism of electrolyte homeostasis revolves, represented mainly by the renin-angiotensin-aldosterone system. Aldosterone increases the activity of sodium retaining processes in the body. These include the active uptake of sodium from the gastrointestinal tract and the reabsorption of sodium from the renal tubes in exchange for potassium or hydrogen ion.

Given chloride's role in ruminal absorption and CSF secretion, perhaps its participation in the aldosterone mechanism of regulating the sodium/potassium "axis" should come as no surprise.

Upon detection of decreased blood pressure, volume, and/or sodium, the juxtaglomerular apparatus in the kidney secretes renin, an enzyme which then cleaves a decapeptide, angiotensin I, from a plasma a 2-globulin, angiotensinogen. A converting enzyme present in the plasma, and most abundant in the pulmonic circulation, cleaves a dipeptide from angiotensin I, thus forming angiotensin II. The effect of angiotensin II, potentiated by ACTH and high plasma potassium, is to induce the secretion of aldosterone by cells in the zona glomerulosa (arcuata) of the adrenal gland cortex. Aldosterone than exerts its effects on the sweat glands, salivary glands, intestinal mucosa, and distal convoluted tubules of the kidneys, promoting sodium absorption and retention. Because of its potent vasoconstrictive properties (40 times that of norepinephrine) angiotensin II can effectively reduce both renal blood flow and glomerular filtration rate, leading to an immediate decrease in excretion of water and electrolytes, before aldosterone can affect tubular sodium reabsorption.

Research has demonstrated that chloride is not only responsible in part for angiotensin II formation, but also for its deactivation or catabolism by the major angiotensinase of the body. Furthermore, chloride ion's relation to angiotensin II may not be its only route to affecting aldosterene secretion and sodium/potassium balance. Chloride's role in the metabolism of ACTH has led to support for the hypothesis that fluid and electrolyte homeostatic mechanisms may revolve not just around the sodium/potassium ratio, but also around the levels of chloride in the body.

See also **Blood**; **Sodium**; and **Sodium Chloride**.

Additional Reading

Adroque, H.J. and Donald E. Wesson: *Salt and Water,* Blackwell Science, Inc., Malden, MA, 1994.

Behrens, D.: *DECHEMA Corrosion Handbook,* John Wiley & Sons, Inc., New York, NY, 1990.

Kozlowski, R.Z.: *Chloride Channels,* Isis Medical Media, Oxford, UK, 2000.

Pollock, W.I.: *Materials Selector for Hazardous Chemicals: Hydrochloric Acid, Hydrogen Chloride and Chlorine,* Elsevier Science, New York, NY, 1999.

CHLORINATED ORGANICS.

Organic compounds containing chlorine are valued as reagents and intermediates in chemical synthesis and for their commercial and industrial importance. Several are produced in high tonnages. The large-volume market for these compounds is in plastics, including vinyl chloride for polyvinyl chloride (PVC), or as a copolymer with vinyl acetate; vinylidene chloride for *Saran*; and chloroprene for neoprene. Other important uses include agricultural chemicals, solvents, plasticizers, and medicines. Uses as intermediates to produce other chemicals also are important and varied. The largest-volume chlorine organic is ethylene dichloride (EDC). About 14 billion pounds (6.2 billion kilograms) per year of EDC are produced, but over half of this is consumed by the producers to make vinyl chloride monomer. Methyl chloride is the intermediate for many chemicals, as are benzyl chloride, phosgene, and chloroform.

The chlorine on certain compounds is used as a facile leaving group for the introduction of another functional group. The displacement of a halogen atom by a cyano group to form a nitrile is one of the most useful reactions of halogen compounds. This opens a route to carboxylic acids having one carbon atom more than the original halide, aside from the importance of the nitriles themselves. Adiponitrile, used in the manufacture of nylon, can be made by treatment of 1,4-dichlorobutane with cyanide:

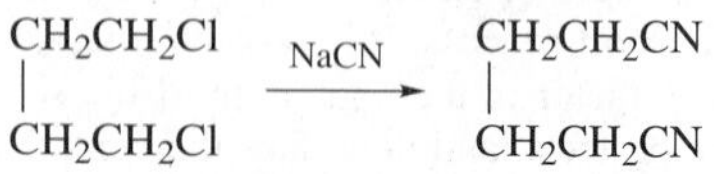

$$\begin{array}{ccc}
CH_2CH_2Cl & & CH_2CH_2CN \\
| & \xrightarrow{\text{NaCN}} & | \\
CH_2CH_2Cl & & CH_2CH_2CN
\end{array}$$

Long-chain alkyl chlorides can be used for the synthesis of various amines, while benzyl chloride is used for production of quaternary ammonium compounds. Alkyl chlorides are used for the formation of organometallics, including the Grignard reagents as well as for alkylation of aromatics. One of the important reactions of phosgene is with diamines for production of diisocyanates (polyurethanes).

Synthesis of Chlorinated Organics

Chlorine derivatives of organic compounds are obtained by substitution, addition, or displacement. Substitution reactions of Cl on hydrocarbons involve radical attack to remove a hydrogen, forming the hydrocarbon radical as an intermediate: $R-H + Cl \longrightarrow R + HCl$. Since a tertiary carbon radical $-\overset{|}{\underset{|}{C}}\cdot$ is most stable, it chlorinates more readily than a secondary carbon $-\overset{|}{CH_2}$ and that more readily than a methyl group. Due to inductive effects, the presence of chlorine in a molecule reduces the activity of the hydrogens on adjacent carbons more than on the chlorinated carbon. Thus, a second radical (Cl·) will preferentially attack a hydrogen on the same carbon. For example, chlorination of ethyl chloride will produce nearly twice as much 1,1-dichloroethane as 1,2-dichloroethane. Specificity of this sort is decreased at higher temperatures, leading to more random substitution. Hydrogens further away on a longer-chained molecule are essentially unaffected by the first chlorine, therefore little selectivity occurs for subsequent substitutions.

The wide range of organic compounds that can be produced through chlorination is illustrated by the very abridged listing in Table 1.

Chlorine can be substituted onto an aromatic ring in the presence of a catalyst, such as ferric chloride, $FeCl_3$, or aluminum chloride, $AlCl_3$. The simplest case would be chlorination of benzene. Substitution of a second Cl onto the ring preferentially goes to the para position, but the ortho and meta isomers can be formed with the latter least favored. If the chlorination

TABLE 1. PROPERTIES OF SOME CHLORINATED ORGANICS

Compound	Specific gravity	Melting point, °C	Boiling point, °C
Acetyl chloride	1.105	−112	51
Acetylene dichloride	1.291	−80	60
Allyl chloride	0.938	−136	45
Benzoyl chloride	1.212	−5	97
Benzyl chloride	1.100	−39	179
Carbon tetrachloride	1.595	−23	77
Chloroacetic acid	1.580	61	190
Chloral	1.505	−57	98
Chloral hydrate	1.619	52	98 (decomposes)
Chlorobenzene (mono)	1.107	−45	213
Chloroform	1.489	−64	61
Chlorophenol (ortho)	1.241	7	175
Cholorophenol (meta)	1.268	32	214
Chlorophenol (para)	1.306	42	217
Dichlorobenzene (ortho)	1.305	−18	179
Dichlorobenzene (meta)	1.288	−25	172
Dichlorobenzene (para)	1.458	53	174
Epichlorohydrin	1.204	—	−94
Ethyl chloride	0.917	−139	13
Ethylene dichloride	1.255	—	−84
Methyl chloride	0.952	−98	−24
Methylene chloride	1.336	−97	−40
Nitrochlorotoluene	—	−38	238
Pentachloroethane	1.671	−22	162
Vinyl chloride	0.908	−160	−12

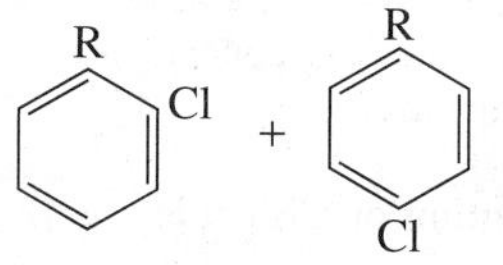

is carried out in the presence of a radical source, such as ultraviolet light, addition occurs instead. See *Chlorinated Aromatics* described later in this entry. When a functional group is present on the aromatic ring, Cl attack will depend upon the type of group present. Phenol and benzoic acid will chlorinate on the ring to give chlorophenol and p-chlorobenzoic acid. Alkyl benzenes will chlorinate on the alkyl group if a radical source is present. In the presence of an iron catalyst, the product is a mixture of the ortho- and para-chloroalkyl-benzenes:

With higher aromatics, such as naphthalene, chlorine successively substitutes all of the hydrogens. The first product is α-chloronaphthalene and the final compound is perchloronaphthalene.

Chlorine addition occurs on unsaturated hydrocarbons having double or triple bonds. Addition can occur by use of Cl_2, HCl, or HOCl:

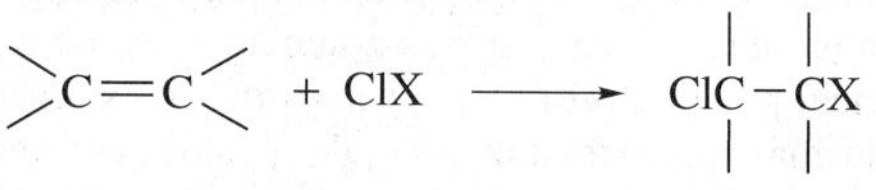

where X=Cl, H, or OH.

With ethylene, the products would be 1,2-dichloroethane, ethyl chloride, or ethylene chlorohydrin. When the unsaturated molecule has three or more carbons, HCl will add, preferentially, with the Cl on the carbon having the fewest hydrogens. For addition of HOCl, the opposite is favored.

Displacement occurs when a functional group is replaced. Chlorine can displace groups, such as hydroxyl, OH, in an acid-catalyzed reaction. For example, methyl chloride can be prepared from the reaction of HCl on methanol. Other alkyl chlorides can be made from their corresponding alcohols. Another type of displacement would be the exchange of one halogen for another, e.g., Cl can be substituted for bromine or iodine in a molecule. An acyl chloride can be formed by the reaction of a strong dehydrating Cl carrier, such as PCl_3, PCl_5, $POCl$, or $SOCl_2$ with an organic acid or its salt.

Industrially, chlorinations are carried out in five ways: (1) radical substitution of hydrogens (protons); (2) addition across unsaturated (double or triple) bonds, using molecular Cl, HCl, or HOCl; (3) HCl reaction with an alcohol; (4) chlorinolysis, in which the oxidative power of Cl is used at high temperature to cleave hydrocarbons, resulting in formation of smaller chlorinated fragments; and (5) oxychlorination. The latter reaction is similar to producing molecular chlorine, *in situ*, from HCl and air in the presence of a catalyst. An example of this is the production of ethylene dichloride, 1,2-dichloroethane, from ethylene, HCl, and oxygen.

Characteristics of Chlorinated Organics

The presence of Cl in an organic molecule increases the density, viscosity, and chemical reactivity, while decreasing the specific heat, solubility in water, and flammability. Chlorine is normally an excellent leaving group, particularly in base-catalyzed reactions, which makes it important for syntheses. Toxicity is the principal hazard. Threshold Limit Values, established by the American Conference of Governmental Industrial Hygienists, for tetra- and pentachloroethane are 5 ppm (vol.) in the atmosphere. Corresponding values for CCl_4, $CHCl_3$, and perchloroethylene are 10, 50, and 100 ppm, respectively. Chloroacetylenes are highly explosive, especially in contact with caustic, e.g., NaOH.

Safety and Handling. Chlorinated organics are absorbed through the skin and lungs and can seriously damage vital organs, especially the liver. Therefore, they should be handled with rubber gloves and in well-ventilated areas. When these materials are subject to burning, they have the potential of forming hydrochloric acid and phosgene, $COCl_2$, besides carbon monoxide. In highly chlorinated compounds, such as carbon tetrachloride and perchloroethylene, there is some danger of forming phosgene in a fire, or from high heat. Compounds that have sufficient hydrogen to combine with any Cl released, such as methyl chloride, vinyl chloride, and ethyl chloride, will form large amounts of hydrochloric acid. Although some small amounts of phosgene may be produced, the hydrogen chloride will naturally drive personnel away from such a fire.

Dioxin (2,3,7,8 tetrachlorodibenzo-*p*-dioxin [1746-01-61]), shown below,

is the extremely toxic byproduct of manufacture of certain chemicals, notably 2,4,5-T. Unfortunately, the name *dioxin* is used synonymously for other, non-chlorinated compounds, such as 2,6 dimethyl-1,3-dioxin-4-ol acetate [828-00-2], which do not deserve the notoriety.

Types or Families of Chlorinated Organics

Chlorinated Paraffins. Cl will displace one, two, three, or more hydrogens from the paraffins. These substitution products are referred to as *mono* ($C_nH_{2n+1}Cl$), *di* ($C_nH_{2n}Cl_2$), *tri* ($C_nH_{2n-1}Cl_3$), and so on. *Monochloro* derivatives include methyl chloride, CH_3Cl, ethyl chloride, C_2H_5Cl, and propyl chloride, C_3H_7Cl. These are also called *alkyl chlorides*. Examples of *dichloro* compounds include methylene dichloride, CH_2Cl_2, and ethylene dichloride, $C_2H_4Cl_2$. Chloroform, $CHCl_3$, and 1,1,1-trichloroethane, $C_2H_3Cl_3$, are *trichloro* derivatives, while carbon tetrachloride, CCl_4, is a *tetrachloro* molecule. When all the hydrogens are substituted by Cl, the term *perchloro* is sometimes used.

Chlorinated Carbonyls. Chlorination of an aldehyde or ketone occurs most readily on a carbon next to the carbonyl function. This is due to proton interaction with the carbonyl and is acid catalyzed. Reaction of Cl with acetone yields chloroacetone. Substitution of a second Cl on chloroacetone occurs with no preference for sites. Thus, equal amounts of 1,1-dichloro- and 1,3-dichloroacetone are produced. (The opposite is true when brominating, since it is possible to form nine parts of 1,3-dibromo-to one of 1,1-dibromoacetone.) Chloral is produced from acetaldehyde. It is also produced by hydrolysis of trichlorodiethyl ether. Acrolein reacts with dry HCl at low temperatures to give β-chloropropionaldehyde.

The chlorinated acetones are strong lachrymators. Tear gas contains chloroacetophenone, which is also a component of the nonlethal disabling spray chemical *Mace*.

Chlorination of diketene yields α-chloroacetoacetyl chloride:

This product is both a vesicant (blistering agent) and a lachrymator.

Chlorinated Fatty Acids. Chlorination of carboxylic acids is much more difficult because the contribution of the carbonyl group toward proton removal is offset by the electron donation effect from the hydroxyl group. This hindrance is obviated by reaction with the acid chloride or anhydride. Chlorination is normally accomplished by use of a catalyst, such as phosphorus trichloride. Monochloroacetic acid is an important industrial chemical. Dichloro- and trichloroacetic acids can be produced by further chlorination, although the latter can be produced conveniently by nitric acid oxidation of chloral. Higher chlorinated fatty acids can be produced by treatment of the hydroxy carboxylic acid or ester with HCl or PCl_5:

$$CH_3CHOHCH_2COOR + PCl_5 \longrightarrow CH_3CHClCH_2COOR$$

Amino fatty acids can be treated with a mixture of nitric oxide and chlorine to produce the corresponding chloroacid. Mono- and dichlorosuccinic acids are examples of chlorinated dicarboxylic acids.

Chlorinated Ethers. Ethylene chlorohydrin reacts with sulfuric acid to form β,β'-dichloroethyl ether. It is a byproduct of ethylene glycol production. The chlorines on this ether are inert, making it a good solvent. Further chlorination at 20–30°C gives α,β,β'-trichloro diethyl ether which hydrolyzes to chloroacetaldehyde and ethylene chlorohydrin. Ethylene and sulfur monochloride react to give β,β'-dichlorodiethyl sulfide (mustard gas), which is a thioether.

Chlorinated Aromatics. Chlorination of benzene in the presence of a catalyst ($FeCl_3$ or $AlCl_3$) yields chlorobenzene as the first product. Substitution with a second Cl yields ortho, para, or meta dichlorobenzene. Eventually all the hydrogens can be substituted to give hexachlorobenzene, C_6Cl_6. In the presence of ultraviolet light, the chlorination of benzene yields benzene hexachloride, $C_6H_6Cl_6$, a derivative of cyclohexane. Under the same conditions toluene chlorinates on the methyl group to give one, two or three substitutions (benzyl chloride, benzal chloride or benzotrichloride), while in the presence of an iron catalyst, one obtains ortho- and parachlorotoluene.

Chlorinated Heterocyclics. Substitution in pyridine is more difficult than in benzene but Cl will enter the β position slowly. Chlorine will not add to furan to give stable addition products but substitution occurs to give 2-chloro- or 3-chlorofuran, 2,5-dichlorofuran, and 2,3,5-trichlorofuran.

Important Specific Chlorinated Organic Compounds

Several thousand chlorine-containing compounds are known and have been synthesized. A select group is included for description here to provide a cross section of the most important of these compounds. The number in brackets following each heading, where appropriate, is the *Chemical Abstracts Service Registration* number.

Acetyl Chloride . [CAS: 75-36-5]. Acetyl chloride can be prepared by treatment of acetic acid with various reagents, such as PCl_3, $SOCl_2$ or $COCl_2$. It can be prepared by chlorination of acetic anhydride in several different ways, by reaction of methyl chloride with carbon monoxide in the presence of catalysts, by reaction of ketene ($H_2C=C=O$) with HCl, or by partial hydrolysis of 1,1,1-trichloroethane. Acetyl chloride hydrolyzes in the presence of water to give acetic acid. It reacts with ammonia and amines to give acetamides:

Reaction with alcohols gives the corresponding acetate esters. Acetyl chloride will add across unsaturated bonds in the presence of suitable catalysts to give halogenated ketones:

Allyl Chloride (3-chloropropene-1). [CAS: 107-05-1]. Allyl chloride can be synthesized by reaction of allyl alcohol with HCl or by treatment of allyl formate with HCl in the presence of a catalyst ($ZnCl_2$). Commercial production is by chlorination of propylene at high temperatures, about 500°C, using a large excess of propylene. It is used in the synthesis of glycerol, allyl alcohol and epichlorohydrin. Since the chlorine is situated alpha to a double bond, it is particularly reactive. Thus, hydrolysis to allyl alcohol occurs rapidly in dilute caustic at about 150°C. Addition of HOCl followed by treatment with an alkali yields epichlorohydrin:

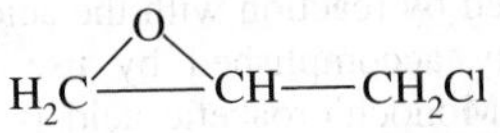

$$H_2C \overset{O}{\underset{\diagdown}{\diagup}} CH\!-\!\!-CH_2Cl$$

Addition of HBr in the presence of an oxidizing agent yields 1-chloro-3-bromopropane, which is used to prepare cyclopropane. Allyl chloride is one of the most toxic of the chlorinated organics.

Benzoyl Chloride. [CAS: 98-88-4].

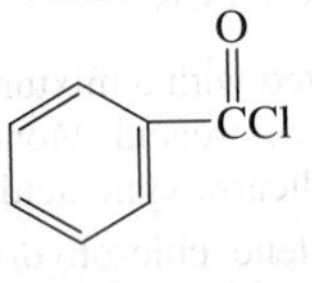

Benzoyl chloride can be prepared from benzoic acid by reaction with PCl_5 or $SOCl_2$, from benzaldehyde by treatment with $POCl_3$ or SO_2Cl_2, from benzotrichloride by partial hydrolysis in the presence of H_2SO_4 or $FeCl_3$, from benzal chloride by treatment with oxygen in a radical source, and from several other miscellaneous reactions. Benzoyl chloride can be reduced to benzaldehyde, oxidized to benzoyl peroxide, chlorinated to chlorobenzoyl chloride and sulfonated to m-sulfobenzoic acid. It will undergo various reactions with organic reagents. For example, it will add across an unsaturated (alkene or alkyne) bond in the presence of a catalyst to give the phenylchloroketone:

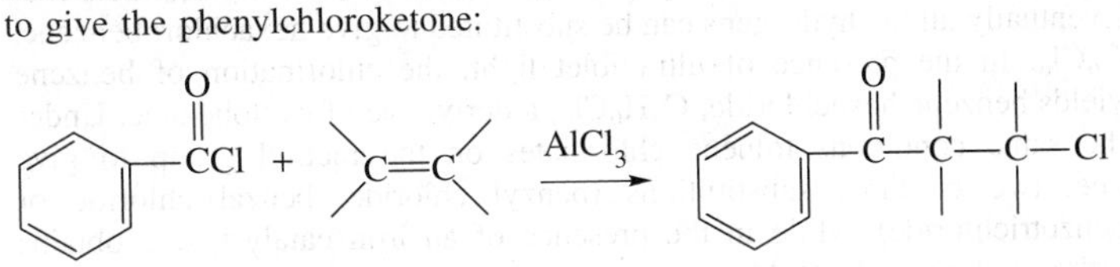

Reaction with benzene yields benzophenone while toulene gives phenyl-p-tolyl ketone. Reaction of benzoyl chloride with monohydric alcohols gives the corresponding alkyl ester, but with phenols the product can either be the phenylbenzoate or a phenolic ketone:

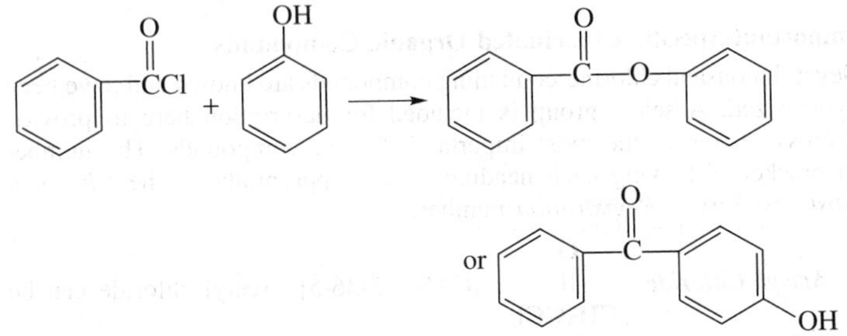

With ammonia and various primary and secondary amines the corresponding amide is formed.

Benzyl Chloride (α-chlorotoluene). [CAS: 100-44-7]. Benzyl chloride can be synthesized by chloromethylation of benzene in the presence of a catalyst ($ZnCl_2$) or by treatment of benzyl alcohol with SO_2Cl_2. Commercially it is produced by chlorination of boiling toluene in the presence of light. Benzyl chloride can be oxidized to benzoic acid or benzaldehyde, or substituted to give the halogenated, sulfonated or nitrated product:

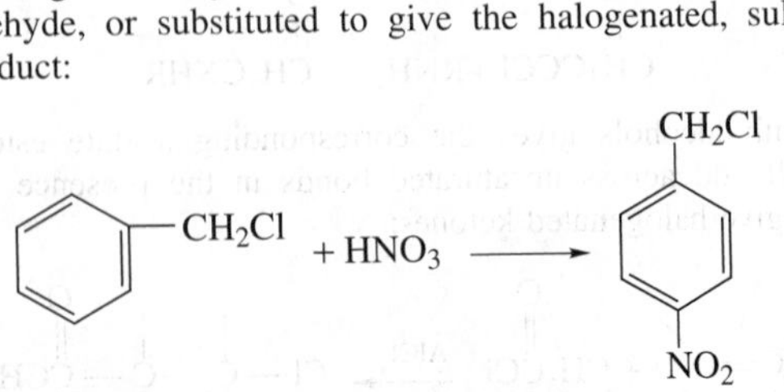

With NH_3 it yields mono-, di- or tribenzyl amine. With alcohols in base the benzylalkyl ether is formed

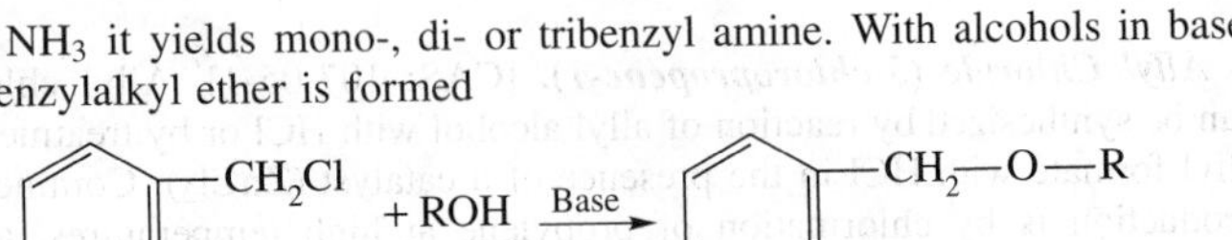

With phenols either the phenolic or nuclear hydrogens can react to give benzylaryl ether or benzylated phenols. Reaction with NaCN gives benzyl cyanide (phenylacetonitrile); with aliphatic primary amines the product is the N-alkylbenzylamine, and with aromatic primary amines N-benzylaniline is formed. Benzyl chloride is converted to butyl benzyl phthalate plasticizer and other chemicals.

Carbon Tetrachloride (tetrachloromethane). [CAS:56-23-5]. Carbon tetrachloride can be synthesized by the chlorination of CS_2, acetylene and other higher hydrocarbons but the primary source is the exhaustive chlorination of methane. It can be pyrolyzed to yield hexachloroethane, oxidized to phosgene and carbonylated with CO in the presence of $AlCl_3$ to give trichloroacetylchloride:

$$CCl_4 + CO \xrightarrow{AlCl_3} Cl_3C\!-\!\overset{O}{\overset{\|}{C}}Cl$$

Some of the more important commercial uses involve fluorine displacement to yield chlorofluoromethane refrigerants, such as trichlorofluoromethane (R-11) and dichlorodifluoromethane (R-12).

Chloroacetic Acid ($ClCH_2COOH$). [CAS: 79-11-8]. Chloroacetic acid can be synthesized by the radical chlorination of acetic acid, treatment of trichloroethylene with concentrated H_2SO_4, oxidation of 1,2-dichloro ethane or chloroacetaldehyde, amine displacement from glycine, or chlorination of ketene. It behaves as a very strong monobasic acid and is used as a strong acid catalyst for diverse reactions. The Cl function can be displaced in base-catalyzed reactions. For example, it condenses with alkoxides to yield alkoxyacetic acids: $ClCH_2COOH$ + KOR $\longrightarrow$ $ROCH_2COOH$. Oxidation of chloroacetic acid leads to formation of methylene chloride. Treatment with ammonia yields glycine ($ClCH_2COOH \xrightarrow{NH_3} H_2NCH_2COOH$) while use of amines leads to formation of substituted glycines. Commercially, chloroacetic acid is an intermediate in the production of herbicides (2,4-D, 2,4,5-T and others) and cellulose ethers.

Chloroacetylene ($HC \equiv CCl$). [CAS: 593-63-51]. This compound is a gas with a very unpleasant odor. It ignites spontaneously in air and may detonate during handling. It can be synthesized by dehydrochlorination of dichloroethylenes with a strong base. It will react with silver or mercury to give explosive salts. Addition occurs across the unsaturated bond-for example, bromination yields 1-chloro-1,1,2,2-tetrabromoethane.

Chloral (trichloroacetaldehyde). [CAS: 75-87-6]. Chloral can be prepared by action of Cl_2 on ethanol, chlorination of acetaldehyde, oxidation of 1,1,2-trichloroethylene in the presence of a catalyst ($FeCl_3$, $AlCl_3$, $TiCl_4$ or $SbCl_3$, and by reaction of CCl_4 with formaldehyde. Chloral can be reduced either at the $-CCl_3$ group or the $-CHO$ group. In the first case the product is acetaldehyde while the second gives Cl_3C-CH_2OH. Oxidation of chloral gives trichloroacetic acid. Polymerization in the presence of H_2SO_4 leads to metachloral or parachloral, depending on the temperature. It undergoes various condensation reactions with alcohols to yield hemiacetals. With organic acids and other functional groups a multitude of reactions are possible. It is used in medicine as a hypnotic.

Chlorobiphenyls. These compounds can be synthesized by direct chlorination of biphenyl in the presence of iron or other catalysts. Other means of preparation include reaction of diazotized aminobiphenyl with copper chloride. Treatment of chlorobiphenyls at elevated temperatures (300–400°C) with strong caustic yields hydroxybiphenyls. Various reactions, normal to aromatic systems, will occur—usually on the unsubstituted ring.

Chloroform (trichloromethane). [CAS: 67-66-3]. Although chloroform can be prepared by various means it is almost exclusively produced by the chlorination of methane. It can be oxidized to phosgene, substituted with various halogens, nitrated to chloropicrin (Cl_3CNO_2), hydrolyzed to formic acid ($\overset{O}{\overset{\|}{HCOH}}$) and carbonylated to dichloroacetic acid. It will react with unsaturated halohydrocarbons in the presence of $AlCl_3$:

$$ClHC = CHCl + CHCl_3 \xrightarrow{AlCl_3} Cl\!-\!\underset{H}{\overset{Cl}{C}}\!-\!\underset{H}{\overset{Cl}{C}}\!-\!\underset{H}{\overset{Cl}{C}}Cl$$

With aromatic aldehyde or ketones base catalyzed additions occur, while it condenses with primary amines to yield isocyanides: $C_2H_5NH_2$ + $CHCl_3 \xrightarrow{NaOH_3} C_2H_5N \equiv C$. A special type of addition can occur when chloroform is reacted with an unsaturated molecule in the presence of potassium alkoxide or sodium hydroxide in a polymer medium. The strong base removes HCl and produces dichlorocarbene (:CCl$_2$) which adds across the double bond:

Commercially, about 60% of chloroform is used in production of fluorocarbon refrigerants and propellants. Cl is replaced by treatment with fluorinated antimony pentachloride. The product, CHClF$_2$ (R-22), is used for home air-conditioning units. It is also used as a feed for production of tetrafluoroethylene, which polymerizes to Teflon. Total demand for chloroform approximates 600 million pounds (136 million kilograms) per year.

Chloronaphthalenes. These compounds can be prepared by direct chlorination of naphthalene in the liquid or vapor phase. They also can be synthesized from naphthylamines via diazotization reactions or from naphthols by treatment with PCl$_5$. Chloronaphthalenes can be further substituted in normal aromatic reactions e.g., halogenation, nitration and alkylation. The chloro group can be displaced to yield a naphthol, an amine, or a nitrile:

Chlorparaffins. These are produced by the random chlorination of various mixed long-chain paraffins. They are used as secondary plasticizers for polyvinyl chloride, lubricating oil additives, resinous materials for coatings, and in flame-retardants.

A particularly valuable use for chlorparaffins is in preparation of linear, primarily internal olefins as feedstock for long-chain synthetic oxoalcohols. Typically, *n*-paraffins (C$_{11}$–C$_{14}$) are chlorinated in a fluidized bed at about 300°C. Conversion is maintained low to limit multiple chlorination. After separation of the monochlorinated alkanes by distillation, dehydrochlorination over nickel acetate at 300°C yields the desired internal olefins. Unreacted paraffins are recycled.

Chloroprene (2-chlorobutadiene-1,3). [CAS: 126-99-8]. Chloroprene can be synthesized by addition of HCl to vinyl acetylene H$_2$C=CH–C≡CH + HCl ⟶ H$_2$C=CH–CCl=CH$_2$, and by dehydrochlorination of dichloro butenes or 2,2,3-trichlorobutane. It undergoes the normal addition reactions across the double bond and readily polymerizes or copolymerizes with other unsaturated compounds. These polymers resemble natural rubber but are superior in some respects, such as oil resistance (neoprene). Almost all the chloroprene produced is used for the manufacture of these polychloroprene rubbers. Chloroprene is a volatile, toxic, flammable liquid and is especially susceptible to oxidation and poly merization.

Chlorostyrene (chlorovinylbenzene). [CAS: 1331-28-8]. The alpha isomer can be prepared by PCl$_5$ reaction on acetophenone

by heating of acetophenone dichloride, or by hydrolysis of styrene dichloride in aqueous NaOH.

Dichlorobenzenes. [CAS: 95-50-1] [106-46-7] [541-73-1]. Dichlorobenzenes are primarily produced by the chlorination of benzene in the presence of a catalyst (FeCl$_3$ or AlCl$_3$) although there are other possible synthetic routes. The two commercially important isomers are the ortho- and para-dichlorobenzenes. Further chlorination yields 1,2,4-trichlorobenzene. Dichlorobenzenes participate in normal aromatic substitution and alkylation reactions. In the presence of CuCl$_2$, ammonia will react with the dichlorobenzenes to yield chloroanilines. The ortho-dichlorobenzene is used for pesticides, moth control, as a solvent and for dyestuff manufacture. About half of the para is used as a space odorant.

Epichlorohydrin (γ-chloropropylene Oxide). [CAS: 106-89-8]. This compound can be prepared from 1,3-dichloropropanol-2, 2,3-dichloropropanol-1, or allyl chloride. Commercially it is prepared as an intermediate in glycerol synthesis via alkaline hydrolysis of glycerol dichlorohydrin. Both come from allyl chloride. Epichlorohydrin reacts with monohydric alcohols to give ethers by opening the oxide ring. It will react with ethers, aldehydes, ketones, organic acids and amines to give a wide variety of useful syntheses.

Commercially the most important use is production of glycerine. Large volumes are consumed in nonglycerine areas, which largely consist of the various epoxy resins. It has use as a solvent and in the production of epichlorohydrin rubber.

Ethyl Chloride (chloroethane). [CAS: 75-00-3]. This compound can be synthesized by treatment of ethyl alcohol with HCl, cleavage of diethylether with HCl in the presence of a catalyst (ZnCl$_2$), chlorination of ethane or hydrochlorination of ethylene. The latter is the choice of industry. The reaction is carried out at 125°F and 125 psi in the presence of AlCl$_3$, which is dissolved in ethyl chloride. It will undergo all the reactions of a typical alkyl chloride—halogenation, hydrolysis, amination, alkylation—and will form the magnesium Grignard reagent. The compound is used in production of tetraethyllead (TEL) by reaction with sodium-lead alloy:

$$4PbNa + 4C_2H_5Cl \longrightarrow Pb(C_2H_5)_4 + 3Pb + 4NaCl$$

Ethyl cellulose is produced by treating alkali cellulose (cotton linter digested in dilute caustic) with ethyl chloride. Up to three ethyl ether stages can be made, giving various grades. These are used as synthetic gums and thickeners in the lacquer and plastics industries. Ethyl chloride is also used in the Friedel-Crafts alkylation of benzene and other aromatics. Additional uses include solvent, refrigerant, heat-transfer medium, aerosol propellant and anesthetic. Much is used captively by the producers.

Ethylene Dichloride (1,2-dichloroethane). [CAS:107-06-2]. Ethylene dichloride (EDC) is produced by reacting ethylene and chlorine in the presence of ferric chloride, using the liquid product as solvent. It is also produced by oxychlorination—ethylene, hydrogen chloride, and air are reacted at about 250°C with a copper chloride catalyst. This latter is the reaction of choice only when cheap by-product HCl is available. EDC reacts with Cl$_2$ to give derivatives of ethylene or ethane, depending on conditions and catalysts. It will dehydrochlorinate to give vinyl chloride, which is its principal commercial use. EDC hydrolyzes to ethylene glycol and reacts with aormatic hydrocarbons in the presence of AlCl$_3$ to give polyarylethylene plastics. The largest use is for vinyl chloride; next is its use as a solvent intermediate. Other uses include the manufacture of ethylenediamine and succinic acid, by way of the nitrile. Reaction of EDC with sodium tetrasulfide is used to produce thiokol rubbers.

Methyl Chloride. [CAS: 74-87-3]. This compound is produced by direct chlorination of methane. Since methyl chloride adds chlorine faster than methane, the yield of methyl chloride is increased by using a large excess of methane in the feed, i.e., about ten volumes of methane to one volume of chlorine. The reaction is carried out at about 450°C with very short contact times. Methyl chloride is also commercially produced by reaction of HCl on methanol in the presence of zinc chloride. Methyl chloride is mainly used in the production of silicone resins and rubbers. Silicon is reacted with an excess of methyl chloride at 300°C in the presence of a copper catalyst. The product includes mono-, di-, and trichloromethyl silanes. Hydrolysis of the chloro groups converts them into the corresponding hydroxymethylsilanes. There are then polymerized to silicones. Nearly equal amounts of methyl chloride are used in making these rubbers and the other principal user, production of tetramethyllead. Production of methyl chloride approximates 440 million pounds per year.

Methylene Chloride (dichloromethane). [CAS: 75-09-2]. As with the other members of the methyl series of chlorinated hydrocarbons, methylene chloride can be produced by direct chlorination of methane. The usual procedure involves a modification of the simple methane process. The product from the first chlorination passes through aqueous zinc chloride, contacting methanol at about 100°C. Thus, HCl from chlorination is used to displace the alcohol group, producing additional methyl chloride. This is further chlorinated to methylene chloride. Methylene chloride reacts violently in the presence of alkali or alkaline earth metals and will hydrolyze to formaldehyde in the presence of an aqueous base. Alkylation reactions occur at both functions, thus di-substitutions result. For example, reaction with benzene plus $AlCl_3$ yields diphenyl methane:

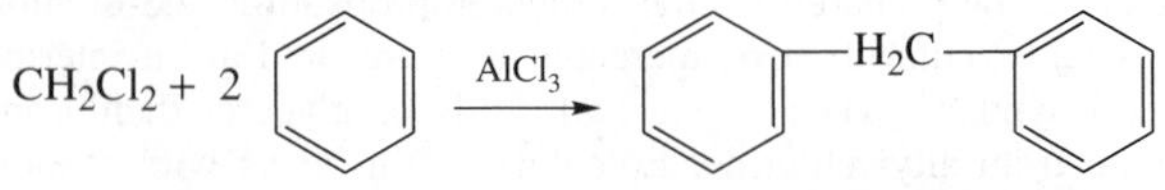

$$CH_2Cl_2 + 2 \text{ (benzene)} \xrightarrow{AlCl_3} \text{ diphenyl methane}$$

Catalyzed carbonylation (CO) reactions lead to formation of either chloroacetyl chloride or malonyl dichloride.

Methylene chloride is used in refrigeration, aerosol propellants, paint stripping, urethane foam-blowing agents, adhesive, and food extractants. It has low toxicity compared with other chlorinated hydrocarbons and has been shown to be neither mutagenic nor carcinogenic toward humans.

Monochlorobenzene (phenyl chloride). [CAS:108-90-7]. Benzene is chlorinated at 80°C in the presence of $FeCl_3$ catalyst. By using low conversions very little dichlorobenzene is produced. The chlorine on this compound is quite inactive, but hydrolysis can be effected by use of a strong caustic at high temperature and pressure, especially in the presence of a catalyst. This has been an important commercial route to phenol. With concentrated aqueous ammonia heated at high temperatures in the presence of a copper catalyst, aniline or diphenylaniline can be synthesized. Reaction with nitric acid yields chloronitrobenzenes-the para isomer predominates. Treatment with hot sulfuric acid leads to formation of *p*-chlorobenzensulfonic acid:

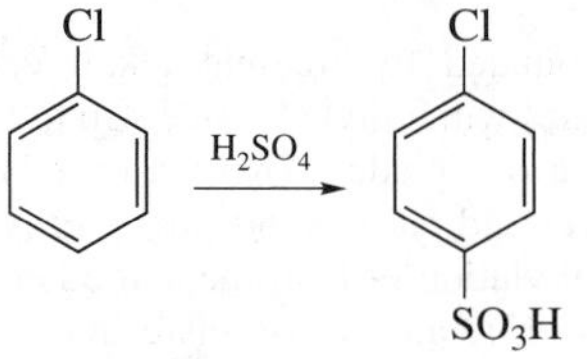

Monochlorobenzene is used commercially as a solvent and to produce phenol and nitrochlorobenzenes.

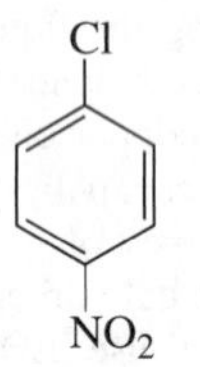

p-Nitrochlorobenzene. [CAS:100-00-5]. This compound is made by the nitration of chlorobenzene and is largely used to produce *p*-nitrophenol with smaller production of *p*-nitroaniline:

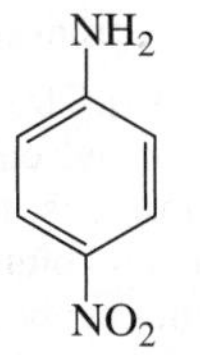

Various agricultural pesticides, rubber chemicals, phenacetin, and *p*-aminophenol consume about 30% of the total. Most of the production is used captively as an intermediate in the production of other chemicals.

Pentachlorophenol. [CAS: 87-86-5]. This compound can be produced by the chlorination of phenol in the presence of $AlCl_3$, or by hydrolysis of hexachlorobenzene with NaOH in methanol. Pentachlorophenol is used as a wood preservative for poles, crossarms, and pilings and thus competes with creosote.

Vinyl Chloride. [CAS:75-01-4]. This compound is produced by alkaline dehydrochlorination of ethylene dichloride, or by thermal cracking of EDC, or 1,1-dichloroethane. Vinyl chloride is polymerized in various ways to polyvinyl chloride (PVC). It is also copolymerized with various other monomers to make a variety of useful resins. The copolymers with about 3 to 20% vinyl acetate are the most important. Demand for vinyl chloride is high, approximating 8 billion pounds (3.6 billion kilograms) per year.

Additional Reading

Alleman, B.C. and A. Leeson: *Natural Attenuation of Chlorinated Solvents, Petroleum Hydrocarbons and Other Organic Compounds,* Battelle Press, Columbus, OH, 1999.

Carey, J., W. Owens, P. Cook, et al.: *Ecotoxicological Risk Assessment of the Chlorinated Organic Chemicals,* Society of Environmental Toxicology & Chemistry, Pensacola, FL, 1998. http://www.setac.org/
Heller, S.R., Editor: *The Beilstein Outline Database: Implentation, Content, and Retrieval,* American Chemical Society, Washington, DC, 1990.
Lewis, R.J. and N.I. Sax: *Sax's Dangerous Properties of Industrial Materials,* 10th Edition, John Wiley & Sons, Inc., New York, NY, 1999.
Lide, D.R.: *CRC Handbook of Chemistry and Physics,* 84th Edition, CRC Press, LLC., Boca Raton, FL, 2003.
Loehr, R.C., B. Smith, and W.C. Anderson: *Environmental Availability of Chlorinated Organics, Explosives, and Heavy Metals in Soils,* American Academy of Environmental Engineers, Inc., Annapolis, MD, 1999. http://www.enviro-engrs.org/
Ramamoorthy, S. and S. Ramamoorthy: *Chlorinated Organic Compounds in the Environment, Regulatory and Monitoring Assessment,* Lewis Publishers, Boca Raton, FL, 1997.
Staff: Chemical Abstracts Service, American Chemical Society, Washington, DC (current). http://www.acs.org/

CHLORINATION (Process). See **Chlorinated Organics**.

CHLORINATION (Water). A principal means for disinfecting municipal water supplies as well as public swimming pools, and some municipal and industrial wastes is by liquid- or gas-phase chlorination. Liquid chlorine is packaged in several types of containers to accommodate a wide range of uses, which may vary from a few hundred pounds during a season (in the case of a swimming pool) to many thousands of tons per year for water supplies. Liquid chlorine is obtainable in pressurized 100- and 150-pound (~45- and 68-kilogram) cylinders, 1-ton (0.9-metric-ton) containers, and for large users, is shipped by railroad tank cars, tank barges, and tank trailers. Large users, of course, must provide local storage means.

Chlorine cylinders are equipped with a single valve. Gas is delivered when the tank is in an upright position; liquid when the cylinder is in an inverted position. However, liquid withdrawal from cylinders is not usually practiced. In the case of ton containers, two valves are provided, permitting easy withdrawal of either gaseous or liquid chlorine. Bulk shipments almost always are unloaded in the liquid phase.

In the case of gaseous withdrawal, the vaporization of the liquid chlorine lowers the temperature surrounding the valve and hence withdrawal rates are limited, ranging up to a maximum of about 1.75 pounds (0.8 kilogram) per hour for a 150-pound (~68-kilogram) cylinder; 15 pounds (6.8 kilograms) per hour for a ton container. Sometimes, cylinders are manifolded to increase the capacity of the system. In the case of liquid withdrawal, the rate ranges up to 400 pounds (181 kilograms) per hour for ton containers; up to 7,000 pounds (3175 kilograms) per hour for a tank car where discharge is from one valve. Usually the liquid is forced out of the container or tank by its own vapor pressure. However, air pressure up to 200 psi (13.6 atmospheres) may be superimposed to increase withdrawal rates.

In municipal water and wastewater treatment installations, the chlorine usually is introduced into the main water system by way of a concentrated water solution of chlorine. Chlorine is metered under a vacuum created by the ejector. The chlorine is dissolved in water in the ejector, and then discharged into the water system as a high-strength solution. Frequently, the chlorine feeding is done automatically in proportion to the flow of water being chlorinated.

In the control of chlorine disinfectant systems, the effective use of the chlorine for its intended purpose is assumed if the treated water considerably downstream from the chlorinator contains a residual of chlorine. Depending upon use, full-contact time may be assumed after ten minutes, or the interval may be extended to several hours. The systems also are usually carefully monitored by bacteriological testing. Normally a dose of 1 to 2 milligrams of chlorine per liter is adequate to destroy all bacteria and leave an effective residual. Residuals of 0.1 to 0.2 milligrams per liter are usually maintained in the effluent streams from water-treatment plants as a factor of safety for consumers.

Surface waters require in most instances more extensive treatment, including chlorination, than do groundwaters. By the time some river water reaches some consuming communities it will have received large inputs of organics. Because of its great oxidizing power, chlorine is highly reactive and can combine in a variety of ways with both inorganic and organic pollutants. Thus, there is concern over the possible formation of carcinogens or otherwise harmful compounds in waters that are heavily chlorinated, particularly waters that have been recycled a number of times along a waterway. Major halogenated compounds found in water supplies suspected of posing health hazards to humans include: (1) chloro-esters, such as *bis*-(2-chloroethyl)ether and *bis*-(2-chloroisopropyl)ether; (2) halobenzenes, such as chlorobenzenes, bromobenzenes, and chloro-bromo benzenes; and (3) haloforms, such as chloroform, bromodichloromethane, dibromochloromethane, and bromoform. More information on this topic can be found in "Chlorine in the Marine Environment," by J.C. Goldman, *Oceanus*, **22**, 2, 36–43 (1979).

CHLORINE. [CAS: 7782-50-5]. Chemical element, symbol Cl, at. no. 17, at. wt. 35.435, periodic table group 17 (halogens), mp −101°C, bp −34.6°C, density (chlorine gas) 3.209 grams/liter (0°C and 1 atmosphere pressure). Chlorine gas is approximately 2.5 times heavier than air at standard conditions. Chlorine in the gaseous phase is diatomic (mol. wt. 70.906), pale greenish yellow of marked odor, irritating to the eyes and throat, poisonous. At 10°C and one atmosphere pressure 9.8 grams of Cl_2 will dissolve in one liter of water; at 30°C and one atmosphere pressure 5.6 grams will dissolve. Critical pressure is 1118.4 psia (7.7 mPa), critical temperature 144°C. CAS Registry No. 7782-50-5. Other important physical characteristics of chlorine are given under **Chemical Elements**.

Chlorine was discovered by Scheele in 1774 and confirmed as an element by Davy in 1810. It is a high-tonnage industrial chemical with many uses.

Naturally occurring isotopes 35, 37. Electronic configuration $1s^2 2s^2 2p^6 3s^2 3p^5$. Ionic radius Cl^{7+} 0.26 Å, Cl^- 1.81 Å. Covalent radius 1.050 Å. First ionization potential 13.01 eV; second, 23.70 eV; third, 39.69 eV; fourth, 53.16 eV; fifth, 67.4 eV. Oxidation potential $ClO_3^- + H_2O \longrightarrow ClO_4^- + 2H^+ + 2e^-$, −1.00 V; $HClO_2 + H_2O \longrightarrow ClO_3^- + 3H^+ + 2e^-$, −1.23 V; $\frac{1}{2}Cl_2 + 4H_2O \longrightarrow ClO_4^- + 8H^+ + 7e^-$, −1.34 V; $Cl^- \longrightarrow \frac{1}{2}Cl_2 + e^-$, −1.3583 V; $Cl^- + 3H_2O \longrightarrow ClO_3^- + 6H^+ + 6e^-$, −1.45 V; $\frac{1}{2}Cl_2 + 3H_2O \longrightarrow ClO_3^- + 6H^+ + 5e^-$, −1.47 V; $Cl^- + H_2O \longrightarrow HClO + H^+ + 2e^-$, −1.49 V; $Cl^- + 2H_2O \longrightarrow HClO_2 + 3H^+ + 4e^-$, −1.56 V; $\frac{1}{2}Cl_2 + H_2O \longrightarrow HClO + H^+ + e^-$, −1.63 V; $\frac{1}{2}Cl_2 + 2H_2O \longrightarrow HClO_2 + 3H^+ + 3e^-$, −1.67 V; $ClO_3^- + 2OH^- \longrightarrow ClO_4^- + H_2O + 2e^-$, −0.17 V; $ClO_2^- + 2OH^- \longrightarrow ClO_3^- + H_2O + 2e^-$, −0.35 V; $ClO^- + 2OH^- \longrightarrow ClO_2^- + H_2O + 2e^-$, −0.59 V; $Cl^- + 6OH^- \longrightarrow ClO_3^- + 3H_2O + 6e^-$, −0.62 V; $Cl^- + 4OH \longrightarrow ClO_2^- + 2H_2O + 4e^-$, −0.76 V; $Cl^- + 2OH^- \longrightarrow ClO^- + H_2O + 2e^-$, −0.94 V; $ClO_2^- \longrightarrow ClO_2 + e^-$, −1.15 V.

Production

Most of the chlorine produced in the world is manufactured by electrolysis of sodium chloride brine. Two processes are in common use: the mercury cell process and the diaphragm cell process. Since 1969 when international concern about the effect of mercury in the environment became widespread, some mercury cell plants have been shut down. Most expansion of chlorine production has been in diaphragm cell plants. Indeed, all mercury cell plants in Japan must now be converted to diaphragm cell plants. However, it should be noted that existing mercury cell plants operate well within strict standards as to mercury discharge both into the air and into waterways. They produce chlorine (and caustic soda) of the most exacting quality suitable for all uses including food preparation. There is no reason to believe that mercury cell technology is obsolescent in this country or, generally speaking, worldwide.

Technology. In production of chlorine by the diaphragm cell process, salt is dissolved in water and stored as a saturated solution. Chemicals are added to adjust the pH and to precipitate impurities from both the water and the salt. Recycled salt solution is added. The precipitated impurities are removed by settling and by filtration. The purified, saturated brine is then fed to the cell, which typically is a rectangular box. It uses vertical anodes (ruthenium dioxide with perhaps other rare metal oxides deposited on an expanded titanium support). The cathode is perforated metal, which supports the asbestos diaphragm. This is vacuum deposited in a separate operation. The diaphragm serves to separate the anolyte (the feed brine) from the catholyte (brine containing caustic soda). Chlorine is evolved at the anode. It is collected under vacuum, washed with water to cool it, dried with concentrated sulfuric acid, and further scrubbed, if necessary. It is then compressed and sent to process as a gas or liquefied and sent to storage for transfer to shipping containers and, ultimately, shipment to consumers.

A cell of this type is called a monopolar cell. In a cell bank, several cells have their negative electrodes and their positive electrodes connected by means of external bus bars. Some companies use a bipolar cell in which the electrodes are internally connected. This results in a configuration like a plate and frame filter press.

The latest development in cell technology is the so-called membrane cell. This uses a cation exchange membrane in place of an asbestos diaphragm. It permits the passage of sodium ions into the catholyte but effectively excludes chloride ions. Thus the concept permits the production of high-purity, high-concentration sodium hydroxide directly. The chlorine side of the cell is identical to existing technology. Research in membrane cells is proceeding at a rapid rate. Some companies are known to be operating this process commercially, but as of 1981 not all problems have been solved.

In the mercury cell process chlorine is liberated from a brine solution at the anodes which are, today, typically metal anodes (Dimensionally Stable Anodes or DSA). Collection and processing of the chlorine is similar to the techniques employed when diaphragm cells are used. However, the cathode is a flowing bed of mercury. When sodium is released by electrolysis it is immediately amalgamated with the mercury. The mercury amalgam is then decomposed in a separate cell to form sodium hydroxide and the mercury is returned for reuse.

Uses

The principal use of chlorine is in the production of organic compounds. Of these the production of PVC (polyvinylchloride) is probably the single largest consumer although chlorinated solvents as a class account for larger tonnage. See also **Chlorinated Organics**; and **Polyvinyl Chloride (PVC)**.

In many cases chlorine is used as a route to a final product which contains no chlorine. For instance propylene oxide has traditionally been manufactured by the chlorohydrin process. Modern technology permits abandoning this route in favor of direct oxidation, thus eliminating a need for chlorine.

Large quantities of chlorine are used in bleaching. Pulp bleaching for paper manufacture consumes about 13% of all chlorine produced in the United States. Since none of the chlorine used for this purpose winds up in the finished product, it must all be discharged as chlorides or chlorinated organics or be reprocessed. At the present time there is no proven, wholly satisfactory technique for removing chlorine compounds from pulp mill bleach plant wastes. It is doubtful that existing mills will be converted to a bleaching technique that does not require chlorine, but future mills may be designed to minimize the use of chlorine.

Substantial quantities of chlorine go into household bleaches. It is used also in laundry and other commercial bleaches. It is the active element in most swimming pool sanitizers.

Large quantities of chlorine are used for treating municipal and industrial water supplies and this use will probably continue. However, some concern has been felt that traces of organic compounds in all water supplies react with the chlorine to form chlorinated organics, which are suspected of being carcinogenic. Further the usefulness of chlorination of municipal wastes has been questioned in some quarters in the light of the fact that such treatment adds chlorinated organics to the waterways. See also **Chlorination (Water)**.

Safety and Handling

Although chlorine is a hazardous substance, it can be handled safely. All persons who handle chlorine should be thoroughly trained in its properties, in correct use of safety equipment, and in the operation of all other equipment including containers. The Chlorine Institute, 342 Madison Ave., New York, NY 10017, publishes the *Chlorine Manual* (available from the Institute at nominal cost) which provides useful information on these matters. In addition the Chlorine Institute has designed emergency kits capable of capping off certain types of leaks which can occur in chlorine containers.

There have been several recent studies of the physiologic effects of chlorine. These have considered chlorine both as an occupational exposure and as an environmental pollutant (see references). The National Institute of Occupational Safety and Health study recommended an 0.5 ppm concentration of chlorine in air for any 15-minute sampling period as the maximum permissible ceiling value. This contrasts with the generally accepted value of 1 ppm TLV (time weighted average for an eight hour exposure).

Chlorine is primarily a respiratory irritant. When the concentration in the air is sufficient, chlorine irritates the mucous membranes, the respiratory system, and the skin. It causes irritation of the eyes, coughing, and labored breathing. It may cause vomiting. In extreme cases, the difficulty of breathing may increase to the point where death can occur from suffocation. Liquid chlorine in contact with the eyes or skin will cause local irritation or severe burns.

Persons who have been overcome by chlorine should be removed to an uncontaminated area, their contaminated clothing should be removed, and they should be kept warm. Medical help should be provided. If breathing appears to have ceased artificial respiration should begin immediately. If breathing is labored, the administration of oxygen may be helpful.

Chlorine Chemistry

Chlorine exhibits in common with the other halogen elements a marked readiness to form singly charged negative ions, as would be expected from the fact that these atoms need only one electron to acquire an inert gas configuration. Thus, chlorine behaves in its normal chemical reactions as an electron acceptor. While there are many compounds in which chlorine has a positive valence, there are no simple compounds of positively charged chlorine (contrast the I^+ of iodine). The positively charged chlorine forms part of a radical, as in combination with oxygen. The electron affinity of chlorine (4.02 eV) is the greatest of all the halogens, and is greater than that of oxygen.

Chlorine reacts readily with hydrogen to form hydrogen chloride, with metals and many non-metals to give chlorides, with metal oxides to give chlorides or oxychlorides, and with many salts of metals to give chlorides. These include the iodides and bromides, whose halogen is displaced by chlorine.

Four isolatable oxides of chlorine are known, Cl_2O, ClO_2, Cl_2O_6 ($\rightleftharpoons$ $2ClO_3$), and Cl_2O_7. Chlorine(I) oxide, Cl_2O, obtained by passing Cl_2 over mercury(II) oxide and sand, is a gas, bp 2°C, somewhat soluble in H_2O to form hypochlorous acid. The $Cl-O-Cl$ bond angle is 111° and the $Cl-O$ distance 1.71 A. Cl_2O is an active oxidizing agent. Chlorine(II) oxide, ClO, is produced by reaction of Cl_2O with atomic chlorine, or as an intermediate product in the decomposition of the Cl_2O. The ClO then decomposes into chlorine and oxygen. In view of this instability the properties of the compound are not established. Chlorine(IV) oxide, ClO_2, is produced by treatment of sodium chlorate, $NaClO_3$, with mixed HCl, oxalic acid or other mild reducing agent, and H_2SO_4 (and H_2O). Cl(IV) oxide is a greenish yellow gas, having an odd electron in its molecule and is consequently paramagnetic. Electron diffraction studies indicate its structure to be

$$:Cl: \ddot{O}$$
$$:\ddot{O}:$$

with the odd electron in an antibonding orbital. The $O-Cl-O$ bond angle is 116.5° and $C-O$ distance 1.49 Å. It is readily hydrolyzed, but is stable when dry. The mechanism of its hydrolysis is complex, yielding all four of the oxychloric acids, and it is widely used as a heavy-duty oxidizing agent. It reacts with metal hydroxides to give the mixture of chlorate and chlorite, with metal peroxides to give chlorite and oxygen and with metals to give chlorite alone. ClO_2 is photosensitive, decomposing when illuminated at about 8°C, to give some Cl_2O_6, bp 3.5°C. Chlorine hexoxide, Cl_2O_6, has a molecular weight corresponding to the formula ClO_3-ClO_3. Its vapor pressure in the liquid state is 0.31 mm at 0°C against values of 23.7 for Cl_2O_7, 490 for ClO_2 and 699 mm for Cl_2O. This is consistent with a bi-trigonal-pyramidal structure in which the two pyramids have three oxygen atoms at their base corners, and are joined by the two chlorine atoms at the apices. In contrast, the additional oxygen atom of Cl_2O_7 would separate the two chlorine atoms, preventing close packed structure. Chlorine heptoxide, Cl_2O_7, the anhydride of perchloric acid, is obtained by heating the latter with phosphorus pentoxide, and consists of two chlorine atoms, each bonded to three oxygen atoms, and jointly bonded to the seventh. All of the oxides of chlorine are thermodynamically unstable with respect to decomposition into the elements.

Hypochlorous acid, [CAS: 7790-92-3], $HClO$, is formed by hydrolysis of chlorine(I) oxide. It is present in aqueous solutions of chlorine because of the equilibrium

$$Cl_2 + 2\,H_2O \rightleftharpoons HClO + H_3O^+ + Cl^-$$

and can be freed by the addition of any substance that combines with the Cl^-, such as mercury(II) oxide, or with the H^+, such as calcium carbonate or other weak bases which do not react with $HClO$. $HClO$ is a weak acid ($K = 3 \times 10^{-8}$ at 25°C). It reacts with hydrochloric acid to give chlorine

and H_2O. On warming or irradiation it undergoes this reaction as well as two other decompositions, i.e., to oxygen, H^+ and Cl^-, and to ClO_3^- and H^+. The presence of oxygen favors the last reaction. $HClO$ and its salts are strong oxidizing agents, oxidizing iodine and bromine to iodates and bromates. Covalent hypochlorites are known, such as the alkyl esters, $ROCl$. In common with other esters of oxidizing acids, these are unstable if R is a primary or secondary alkyl group. However, t-butyl hypochlorite is quite stable. Reduction of chlorine dioxide, ClO_2, with hydrogen peroxide, yields oxygen and chlorous acid, $HClO_2$, which exists only in solution. It is stronger than hypochlorous acid ($K = 1.01 \times 10^{-2}$ at 23°C). It is also a strong oxidizing agent and its sodium salt is widely used for this purpose, generally as a source of ClO_2.

Chloric acid, [CAS: 7790-93-4], $HClO_3 \bullet 7H_2O$, is readily prepared by passing chlorine into hot caustic solutions, since these conditions favor the formation of chlorate. Chloric acid is a more active oxidizing agent than $HClO$, reacting explosively with organic matter. Its alkali salts undergo on heating two modes of decomposition, one (catalyzed, e.g., by manganese dioxide) into the chloride and oxygen, and the other (uncatalyzed) into the chloride and perchlorate.

Perchloric acid, [CAS: 7601-90-3], $HClO_4$, is obtained in anhydrous form from perchlorates by H_2SO_4 distillation, or from ammonium perchlorate by aqua regia distillation. Perchlorates are also obtained by electrolysis of chlorides or chlorates. Perchloric acid is explosive unless properly handled; it is of course a powerful oxidizing agent, but has a higher activation energy than the lower acids.

The chlorides range in character from ionic to covalent compounds, many of them having bonds of intermediate character. There are also a number of interhalogen compounds containing chlorine. Those of iodine and bromine are discussed under those entries. With fluorine, chlorine forms ClF, chlorine monofluoride, which is also obtained (along with ClF_3), when mixtures of the elements are subjected to spark discharge. It may also be obtained by heating a mixture of chlorine and chlorine trifluoride. It is a reactive gas, bp -100.8°C, and with its bond having 20–30% ionic character. Chlorine trifluoride, ClF_3, is also obtained from the elements or from chlorine monofluoride and fluorine, by varying the conditions. It is a gas, bp of liquid 11.3°C, and is a more powerful fluorinating agent than the monofluoride. Present views on its structure suggest a trigonal bipyramid having chlorine in the center, a fluorine atom at each apex and the third fluorine and two nonbonding pairs of electrons in the three equatorial positions, giving a T-shaped molecule. ClF_3 reacts with all elements except the noble gases, nitrogen, chromium, and certain noble metals, although some metals (e.g., copper) require elevated temperature. It does not react with oxides or salts as readily as fluorine, but nevertheless ignites such materials as asbestos.

Chlorine Products Used in Food Industry

Chlorine compounds are used effectively to sanitize food-processing equipment and food containers, for washing and conveying raw food products, and for cooling heat-sterilized cans of foods. The use of chlorinated-water sprays at selected locations in a plant reduces or prevents accumulation of microorganisms and off-odors. For sanitizing equipment, solutions of 100–200 micrograms of hypochlorite per milliliter usually are held in or circulated through equipment for at least two minutes. In addition to hypochlorite, chlorinated trisodium phosphates and chloramine compounds have antimicrobial properties. The effectiveness of chlorine-type sanitizers is dependent upon pH, temperature, exposure time, types of organisms, presence of organic matter, and concentration of compound used. Biocidal activity of free-available chlorine compounds is greater if the presence of organic matter is minimized. Further information is given in the Foregoing (1983) reference.

See also **Halides**; **Hypochlorites**; and **Sodium Chloride**.

Additional Reading

Lewis, R.J. and N.I. Sax: *Sax's Dangerous Properties of Industrial Materials,* 10th Edition, John Wiley & Sons, Inc., New York, NY, 1999.

Lide, D.R.: *CRC Handbook of Chemistry and Physics,* 84th Edition, CRC Press, LLC., Boca Raton, FL, 2003.

Monastersky, R.: "U.S. Skies Harbor Ozone Destroyer (Chlorine Monoxide)," *Science News,* **84** (February 9, 1991).

Perry, R.H. and D.W. Green: *Perry's Chemical Engineers' Handbook,* 7th Edition, McGraw-Hill Companies, Inc., New York, NY, 1999.

Pollock, W.I.: *Materials Selector for Hazardous Chemicals: Hydrochloric Acid, Hydrogen Chloride and Chlorine,* Elsevier Science, New York, NY, 1999.

Schmittinger, P.: *Chlorine: Principles and Industrial Practice*, John Wiley & Sons, Inc., New York, NY, 2000.

Somerville, R.L.: "Reduce Risks of Handling Liquified Toxic Gas (Chlorine)," *Chem. Eng. Progress*, **64** (December 1990).

Staff: *Chlorine Manual*, The Chlorine Institute, New York, NY, (updated periodically). *http://www.cl2.com/*

Turoski, V.: *Chlorine and Chlorine Compounds in the Paper Industry*, CRC Press, LLC., Boca Raton, FL, 1998.

CHLORINITY. A measure of the chloride content of seawater. Originally, chlorinity was defined as the weight of chlorine in grams per kilogram of seawater after the bromides and iodides had been replaced by chlorides. To make the definition independent of atomic weights, chlorinity is now defined as 0.3285233 times the weight of silver equivalent to all the halides.

CHLORITE. Chlorite is a ubiquitous mineral usually a product of secondary origin from the alteration of silicates containing aluminum, ferrous iron, and magnesium. Pyroxenes, amphiboles, biotite garnet, and idocrase within rocks that have undergone metamorphism are common source minerals for chlorite. Distinct crystals are extremely rare; more often found as foliated masses or fine scaly aggregates. Color includes various shades of green. Hardness of 2–2.5, and specific gravity of 2.6–2.9, with vitreous to pearly luster. Individual folia characterized by flexible, not elastic property. A general formula is $(Mg, Fe^{2+}, Fe^{3+}, Mn)_6$ $AlSi_3O_{10}(OH)_8$.

CHLORITOID. A mineral that occurs as tabular crystals, probably triclinic, foliated masses or scattered scales and plates of a greenish-gray to greenish-black color. It is characteristic of the less intensely altered metamorphic rocks such as phyllites and quartzites. Chemically it is a hydrous iron-aluminum silicate, $Fe_2Al_4Si_2O_{10}(OH)_4$. Ottrelite contains some manganese as well. Chloritoid was originally noted as from the Ural Mountains and named for its greenish color from the Greek word meaning green. Ottrelite was named from Ottrez in Luxemburg.

CHLOROACETIC ACID. See **Chlorinated Organics**.

CHLOROACETYLENE. See **Chlorinated Organics**.

CHLOROBIPHENYLS. See **Chlorinated Organics**.

CHLOROFORM. See **Anesthetics**; **Chlorinated Organics**.

CHLOROFLUOROCARBONS (CFC). A family of compounds of chlorine, fluorine, and carbon, entirely of industrial origin. CFCs include refrigerants, propellants for spray cans (this usage is banned in the U.S., although some other countries permit it) and for blowing plastic-foam insulation, styrofoam packaging, and solvents for cleaning electronic circuit boards. The compounds' lifetimes vary over a wide range, exceeding 100 years in some cases.

CFCs' ability to destroy stratospheric ozone through catalytic cycles is contributing to the depletion of ozone worldwide. Because CFCs are such stable molecules, they do not react easily with other chemicals in the lower atmosphere. One of the few forces that can break up CFC molecules is ultraviolet radiation, however the ozone layer protects the CFCs from ultraviolet radiation in the lower atmosphere. CFC molecules are then able to migrate intact into the stratosphere, where the molecules are bombarded by ultraviolet rays, causing the CFCs to break up and release their chlorine atoms. The released chlorine atoms participate in ozone destruction. A single atom of chlorine is able to destroy ozone molecules over and over again.

International attention to CFCs resulted in a meeting of diplomats from around the world in Montreal in 1987. They forged a treaty that called for drastic reductions in the production of CFCs. In 1990, diplomats met in London and voted to strengthen the Montreal Protocol significantly by calling for a complete elimination of CFCs by the year 2000.

See also **Carbon**; **Chlorine**; **Fluorine**; and **Refrigeration**.

CHLOROPHYLLS. A group of closely related green pigments occurring in leaves, bacteria, and organisms capable of photosynthesis. The major chlorophylls in land plants are designated *a* and *b*. Chlorophyll *c* occurs in certain marine organisms. Because of the overwhelming percentage of the total photosynthesis which is performed by marine organisms, it is possible that chlorophyll *c* is equivalent in importance to chlorophyll *b*. Chlorophyll *a* is several times as abundant as chlorophyll *b*. See also **Photosynthesis**.

The canonical form for chlorophyll *a* is R = CH_3 when substituted in the following formula. For chlorophyll *b*, R = CHO. These structures have been established by a long series of degradation studies mainly by R. Willstätter, Hans Fischer and their collaborators and by synthetic studies by Fischer.

The biological significance of the chlorophylls stems from their role in photosynthesis, the process by which plants fix the sun's energy in the form of organic matter. This process corresponds to the reversal of the combustion of hydrogen. The oxygen liberated is set free in the air. Under special conditions, some organisms are also capable of liberating the hydrogen, but usually this is used for chemical reductions in the plant. Atmospheric carbon dioxide is fixed enzymatically and is thus used as the source of the carbon in the synthetic process, but is not reduced directly. The path of the carbon from carbon dioxide in photosynthesis has been elucidated largely by the studies of Calvin and his collaborators. See references.

While it is known that most of the energy fixed in photosynthesis is absorbed originally by the chlorophylls, the exact reactions which they undergo to initiate the process of reduction are not fully understood. The photosynthetic sequence requires a high degree of organization within the plant cells where it occurs, and destruction of the organization of the chloroplasts by processes like grinding is sufficient to bring photosynthesis to a stop, even when the chlorophyll and the soluble enzymes participating in the process are still presumably intact.

Chlorophyll derivatives with the phytyl group intact are oil soluble and form a series of green dyes used in the coloring of oils and waxes. The chlorophyll soaps, resulting from combined saponification and cleavage of the isocyclic ring, form "water soluble" dyes useful in the coloring of soaps and similar products. Both the medical and cosmetic literature are replete with claims of therapeutic or physiological activity of "chlorophyll." The substances utilized in this work range from partially purified chloroplasts to mixtures of materials that have undergone deep-seated chemical alteration. Some of the types of activity claimed can be shown to be due to incidental impurities. The field for investigation of the action of pure chemical individuals produced by the action of various reagents on chlorophyll or its derivatives is largely unexplored. It is known, however, that neither chlorophyll nor hemoglobin in the diet is utilized by the body in the formation of the physiologically active pyrrole pigments. These are derived, instead, from such simple building blocks as glycine and acetate ion. Only the iron in dietary blood pigment can be utilized by the body.

The work of Granick has shown that, in the physiological processes of plants, chlorophyll is formed from protoporphyrin, which can be obtained in the laboratory by the removal of iron from hemin. The pathways to heme and to chlorophyll diverge at protoporphyrin. To form heme, an organism introduces iron into protoporphyrin. To form chlorophyll from protoporphyrin, an oxidation, a reduction, a ring closure and esterifications are performed and the magnesium is introduced. The end product of the enzymatic synthetic chain is presumably protochlorophyll, the magnesium derivative of the porphyrin corresponding in structure to chlorophyll. The addition of the two hydrogens necessary to convert protochlorophyll to chlorophyll is accomplished under the influence of light.

Phytochrome is a blue-green protein that occurs in minute quantities in plant tissues. Phytochrome pigment is a photoreceptor for photomorphogenetic and photoperiodic responses in plants, and is found in many higher plants. It has been detected both in monocotyledonous plants, such as oats, barley, maize (corn), and at least one aquatic plant. It also occurs in dicolyledonous plants, such as sunflower, beans, peas, parsnip, soybean, lettuce, and radish. It is present in the aerial parts of plants (leaves, stems, buds, and inflorescences), such as in cauliflower. Phytochrome also occurs in some roots and in bryophytes, such as red algae and green algae. The structure of phytochrome is shown below.

Historically, lettuce seeds and cocklebur were the first two plants in which phytochrome was detected. Before isolation, the existence of phytochrome as a specific pigment was implied by physiological work with red and far-red light. The effect of red light (640–670 nm), which promotes germination in lettuce seeds, is reversible by far-red (710–740 nm). Thus, far-red radiation nullifies the promoting effect of the red light. It had been noted as early as 1935 that red light stimulated germination of lettuce seeds and that light of somewhat longer wavelengths inhibited germination. This knowledge was not elaborated further until 1952, when more exacting laboratory experiments were conducted.

Additional Reading

Anderson, B., J. Barber, and H. Salter: *Molecular Genetics of Photosynthesis,* Oxford University Press, Inc., New York, NY, 1998.

Dailey, H.A.: *Biosynthesis of Heme and Chlorophylls,* McGraw-Hill Companies, Inc., New York, NY, 1990.

Hall, D.O. and K. Rao: *Photosynthesis,* Cambridge University Press, New York, NY, 1999.

Ksenzhek, O.S. and A.G. Volkov: *Plant Energetics,* Kaufmann Publishers, San Diego, CA, 1997.

Sage, R.F.: *C4 Plant Biology,* Academic Press, Inc., San Diego, CA, 1998.

Scheer, H.: *Chlorophylls,* CRC Press, Inc., Boca Raton, FL, 1991.

Staff: CIBA Foundation, *Biosynthesis of the Tetrapyrrole Pigments,* Vol. 180, John Wiley & Sons, Inc., New York, NY, 1994.

CHOLEIC ACIDS. See **Bile.**

CHOLESTEROL. [CAS: 57-88-5]. Also known as cholesterin or 5-cholesten-3-betaol, cholesterol is the most common animal sterol, a monohydric secondary alcohol of the cyclopentenophenanthrene (4-ring fused) system containing one double bond. It occurs in part as the free sterol and in part esterified with higher fatty acids as a lipid in the human blood system. The primary precursor in biosynthesis appears to be acetic acid or sodium acetate. Cholesterol itself in the animal system is the precursor of bile acids, steroid hormones, and provitamin D$_3$. Cholesterol is a white, or faintly yellow, almost odorless substance and may take the form of pearly granules or crystals. The substance is affected by light; mp 148.5°C; bp 360°C, but tends to decompose at lower temperatures. Specific gravity 1.067 (20/4°C); insoluble in water; slightly soluble in alcohol; soluble in fat solvents, vegetable oils, and in aqueous solutions of bile salts. Cholesterol occurs in egg yolk, liver, kidneys, saturated fats and oils. In addition to its importance in medicine and biology in general, cholesterol is used as an emulsifying agent in cosmetic and pharmaceutical products. It is the source of estradiol.

Research over the years has shown that cholesterol is carried in the bloodstream in complexes with other lipids and proteins. Based upon their density, there are four classes of lipoproteins; the chylomicrons; the VLDLs (very low density); the LDLs (low density); and the HDLs (high density). It is estimated that about 80% of the total blood cholesterol is carried by the LDLs. Most of the remainder is carried by the HDLs. Large quantities of triglycerides, but very little cholesterol, are carried by the chylomicrons and VLDLs.

As early as 1951, researchers at Cornell University Medical College observed that in males with coronary heart disease, the HDL concentrations were low. There were several confirmations of this during the 1950s and 1960s. However, for many years, the principal criteria in connection with heart attack and stroke with relation to cholesterol was considered to be total cholesterol and LDL levels, with little attention given to the HDLs. It was not until 1975 researchers at the Royal Infirmary, Edinburgh, Scotland, reported an inverse correlation between blood concentration of HDLs and total body cholesterol. Miller hypothesized that HDLs may lessen body cholesterol by facilitating its excretion. Epidemiological studies were made shortly thereafter on Japanese-Hawaiian males, Israeli males, and black sharecroppers in Georgia. These studies showed that risk of heart attack increases as blood HDL level decreases. The average value for HDL levels in human males is 45 milligrams per deciliter (55 milligrams for females). The generally higher level of HDLs in females may partially account for their lower heart attack rates. For both sexes, it has been estimated that a 5 milligram drop in the aforementioned HDL levels may increase the risk of heart attack by about 25%.

See also **Steroids.**

Additional Reading

Challem, J. and V. Dolby: *Homocysteine: The New Cholesterol,* Keats Publishing, Inc., Chicago, IL, 1999.

Chang, T.Y. and D.A. Freeman: *Intracellular Cholesterol Trafficking,* Kluwer Academic Publishers, Norwell, MA, 1999.

Connor, W.E.: "Dietary Fiber—Nostrum or Critical Nutrient?" *N. Eng. J. Med.,* **193** (January 18, 1990).

Ginsberg, H.N. et al.: "Reduction of Plasma Cholesterol Levels in Normal Men on an American Heart Association Step 1 Diet or a Step 1 Diet with Added Monounsaturated Fat," *N. Eng. J. Med.,* **574** (March 1, 1990).

Gotto, A.M.: "Cholesterol Intake and Serum Cholesterol Level," *N. Eng. J. Med.,* **912** (March 28, 1991).

Manson, J.E. et al.: "The Primary Prevention of Myocardial Infarction," *N. Eng. J. Med.,* **1406** (May 21, 1992).

Mensink, R.P. and M.B. Katan: "Effect of Dietary Trans Fatty Acids on High-Density and Low-Density Lipoprotein Cholesterol Levels in Healthy Subjects," *N. Eng. J. Med.,* **439** (August 16, 1990).

Nesto, R.W. and L. Christenson: *Cholesterol-Lowering Drugs: Everything You and Your Family Need to Know,* Morrow, William & Company, New York, NY, 2000.

Pekkanen, J. et al.: "Ten-Year Mortality from Cardiovascular Disease in Relation to Cholesterol Level among Men with and without Preexisting Cardiovascular Disease," *N. Eng. J. Med.,* **1700** (June 14, 1990).

Raloff, J.: "Cholesterol: Up in Smoke," *Sci. News,* **60** (July 27, 1991).

Rossouw, J.E., Lewis, B., and B.M. Rifkind: "The Value of Lowering Cholesterol after Myocardial Infarction," *N. Eng. J. Med.,* **1112** (October 18, 1990).

Sacks, F.M. and W.W. Willett: "More on Chewing the Fat—The Good Fat and the Good Cholesterol," *N. Eng. J. Med.,* **1740** (December 12, 1991).

Small, D.M., Oliva, C., and A. Tercyak: "Chemistry in the Kitchen—Making Ground Meat More Healthful," *N. Eng. J. Med.,* **73** (January 10, 1991).

Swain, J.F., I.L. Rouse, Curley, C.B., and F.M. Sacks: "Comparison of the Effects of Oat Bran and Low-Fiber Wheat on Serum Lipoprotein Levels and Blood Pressure," *N. Eng. J. Med.,* **147** (January 18, 1990).

Web References

HealthlinkUSA. http://www.healthlinkusa.com/C.html

MDchoice.com. http://www.mdchoice.com/Pt/consumer/1.asp

National Heart, Lung, and Blood Institute. http://www.nhlbi.nih.gov/index.htm

National Institutes of Health. http://www.nih.gov/health/syh-hbc/chap1/

CHOLESTERIC LIQUID CRYSTALS. See **Liquid Crystals.**

CHOLIC ACID. See **Bile.**

CHOLINE AND CHOLINESTERASE. An enzyme (acetylcholinesterase) is specific for the hydrolysis of acetylcholine to acetic acid and choline in the animal body. It is found in the brain, nerve cells and red blood cells and is important in the mechanism of nerve action. Acetylcholine was first synthesized in 1867. It consists of a combination of choline and acetic acid in an ester linkage. The component parts of the acetylcholine molecule are both normal constituents of the animal body. Acetylcholine has the structure:

Acetylcholine assumed no importance to biologists until 1899 when Hunt identified the presence of choline in extracts of the adrenal glands and suggested that some derivative of choline was capable of causing a fall in blood pressure. This stimulated interest in studying the physiological effects of various choline derivatives and, in 1906, Hunt and Taveau found that acetylcholine was 100,000 times more effective than choline in causing a fall in blood pressure. Shortly thereafter, acetylcholine was identified in extracts of ergot, a fungus that grows on rye and other cereal grains. The first real proof of the role of acetylcholine in transmitting the effects of nerve stimulation did not come until 1921. The acetylcholine-cholinesterase system has served as the basis for the development of a number of drugs needed to alter the activity of the autonomic nervous system in certain disease states. Inhibitors of cholinesterase are used in insecticides. Parathion and malathion are examples of organic phosphate cholinesterase inhibitors that are effective for this purpose. Cholinesterase inhibitors are capable of producing poisoning and death in humans and domestic animals by the same mechanism.

Choline is an essential metabolic substance for building and maintaining cell structure. Choline is usually described along with B complex vitamins, although it is essentially a structural component of tissue rather than a metabolic catalyst. Choline is a part of the structure of phospholipids and acetylcholine.

Choline participates in normal fat metabolism and interrelates with methionine in a biochemical manipulation referred to as transmethylation. Choline, when in adequate quantity, can replace the essential amino acid methionine when the latter is in limited quantity; or the reverse may occur, that is, methionine can be dismantled to replace choline. Choline deficiencies result in numerous degradative physiologic changes in livestock. The usual dietary supplements are choline bitartrate and choline chloride.

Human Disease Implications

In Alzheimer's disease, neurochemical studies of cerebral cortex have shown specific reduction in choline acetyltransferase (CAT), the enzyme required for synthesis of the neurotransmitter acetylcholine (ACh). The degree of reduction of ACh correlates with the severity of dementia. This finding may reflect loss of the neurons in the substantia innominata of the basal forebrain that form the cholinergic projection to the hippocampus and cerebral cortex. Degeneration of the terminals of these neurons may initiate the formation of amyloid plaques.

In Huntington's disease, there is a progressive degenerative disease of the basal ganglia that is inherited as an autosomal dominant trait. In terms of the pathologic findings of this disease, there is a marked atrophy of the corpus striatum and depletion of interneurons in the striatum. The major neurochemical abnormalities in the basal ganglia are reduction of GABA (gamma-aminobutyric acid) and its synthesizing enzyme, glutamic acid decarboxylase; reduction of CAT; and reduction of postsynaptic receptors for these neurotransmitters.

Additional Reading

Bland, J. and R.A. Passwater: *Choline, Lecithin, Inositol and Other "Accessory" Nutrients,* Keats Publishing, Inc., Chicago, IL, 1999.

Bradford, H.F.: *Chemical Neurobiology,* W.H. Freeman and Company, New York, NY, 1986.

Staff: Institute of Medicine, *Dietary Reference Intakes: Folate, Other B Vitamins and Choline,* National Academy Press, Washington, DC, 2000.

Wurtman, R.J. and J.J. Wurtman: *Nutrition and the Brain,* Lippincott Williams & Wilkins, Philadelphia, PA, 1990.

Zeisel, S.H. and B.F. Szuhaj: *Choline, Phospholipids, Health, and Disease,* AOCS Press, Champaign, IL.

CHONDRODITE. Chondrodite, a magnesium fluosilicate mineral $Mg_5(SiO_4)_2(F,OH)_2$, crystallizing in the monoclinic system is a product of metasomatic origin in metamorphosed dolomitic limestones. Crystals are uncommon, usually occurring as discrete grains within the limestone, of light yellow to red color. Vitreous luster, translucent, with hardness of 6–6.5, and specific gravity of 3.1–3.2. This mineral is the most prominent member of minerals falling within the chondrodite group. These are norbergite, chondrodite, humite and clinohumite. Individual members of this group require optical evaluation for positive identification.

Noteworthy world occurrences include Mt. Somma, Italy; Pargas, Finland; Kafveltorp, Sweden; and in the United States at Brewster, and Warwick and Orange Counties, in New York.

CHROMATIN. This term was given by Flemming to denote that substance in cell nuclei which, in the usual treatment with nuclear dyes, takes up the color. In nondividing nuclei, chromatin is distributed throughout the entire nucleus (euchromatin), but in nuclei undergoing cell division, chromatin is confined to the chromosomes (heterochromatin). In Flemming's time, the chemistry of the nucleus was entirely unknown. Even in view of recent knowledge, it is not fully understood as to what particular substance(s) have the special affinity for the dyes. In isolation of chromatin from pea embryos by differential centrifugation and purified by sucrose gradient centrifugation, the composition of chromatin was found to be deoxyribonucleic acid (DNA), 31%; ribonucleic acid (RNA), 17.5%; histone protein, 33%; and the remainder, nonhistone protein.

CHROMATOGRAPHY. An instrumental procedure based on physical absorption principles for separating various components from a mixture of chemical substances. In its broad interpretation, chromatography is a combination of separation, identification, and quantitative measurements. As of the late 1980s, chromatography is a giant among the numerous techniques for chemical analysis; high-performance liquid chromatography (HPLC) and gas chromatography are two of the leaders in the field of separation science. Modern chromatography plays three major roles in science and industry, separately or in combination. (1) As a means for identifying and quantifying the ingredients of a mixture of chemical substances—*as used in the laboratory* for quality control, for example, detecting impurities in chemicals, pharmaceuticals, foodstuffs, etc.; assisting researchers: in the study of biochemistry; measuring and identifying unsafe materials in the environment; and as an aid in forensic science, among scores of other specific applications. (2) As a means for *separating materials* where separation (not analysis per se) is the principal target. In this manner, specific chemicals, not easily obtainable in other ways, can be isolated and then, in turn, used to participate in the synthesis of other chemicals. Simply stated, chromatographic separations yield raw materials, notably for research in organic syntheses. (3) As an important, widely acclaimed process—stream chemical composition controller. For such industrial applications, a chromatograph is a complex transducer which not only puts out a signal that identifies the types and amounts of a given substance, but also first separates the target substances from a stream that may contain numerous other substances, some of which may be closely related physically (close boiling points, specific gravities, etc.) or chemically. The transducer output can be used for off-line quality control, or on-line so that the chromatograph becomes part of the total control loop.

In laboratory chemical analysis, where only minute samples may be available, only milligrams of material or less are required for some forms of chromatography. Chromatography is widely used in pollution control technology.

Beginnings of Chromatography

In 1906, a Russian botanist, Mikhail Tswett, first used the chromatographic principle to separate plant pigments. Tswett filled a vertical glass tube with an adsorbent. As a sample of the pigments was washed through the tube with a solvent, a series of colored *adsorption bands* was produced, in essence forming a colorgraphic display. Thus, the apt term, *chromatography,* but no longer appropriate because frequently materials separated are colorless and the adsorption spectra are presented by way of electronic instrumentation.

Thin-Layer Chromatography (TLC). This format represented the first widespread employment of chromatography as an analytical tool. Surprisingly, even though decades old, the method (refined and improved) still persists and can be found in some pharmaceutical laboratories for drug testing, by medical laboratories, and in the chemical industry. TLC is a simple, low-cost technique. TLC, of course, has been replaced for a host of applications by the more sophisticated chromatographic apparatus available today that offers numerous advantages, disallowing cost, such as the conservation of laboratory workers' time, better performance, and speed. See Fig. 1. Highly engineered, sophisticated chromatographic apparatus did not become easily available until the 1950s and since that time has continually expanded apace. All of the advantages of computer and display technology have been applied to chromatography today.

Fundamentals of Chromatography

In chromatography, the *carrier* or *moving phase* may be a gas, liquid, or supercritical fluid. The separation is effected by distributing the mixture

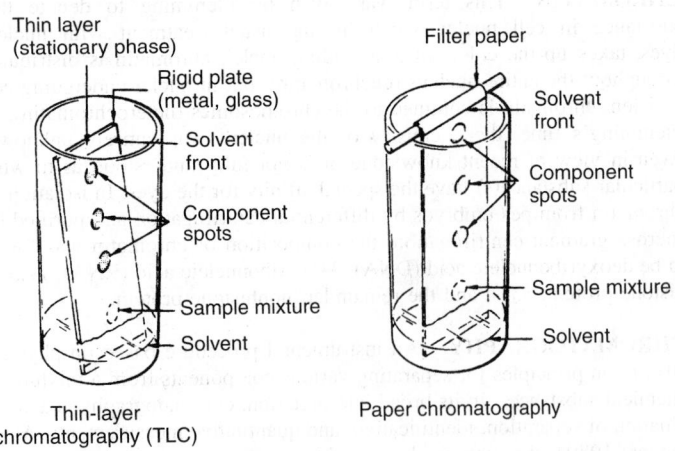

Fig. 1. Thin-layer chromatography (TLC), shown at left, and paper chromatography, shown at right, are similar in apparatus and technique. The sorbent bed is in the form of a thin layer or thin sheet of paper of a finely divided sorbent material. In TLC, the sorbent material is deposited on a supporting metal, glass, or plastic plate. The sample is spotted near one end of the bed, which is then brought in contact with a source of solvent. As the solvent moves through the bed by capillary action, the sample components are washed through the bed at different rates and are separated into spots of pure compound, which can be recovered after the solvent is allowed to evaporate. Spots can be detected visually if the substances are colored, or with ultraviolet light if they are fluorescent. Components can be reacted to give colored or fluorescent derivatives by spraying with reagents. Conventional quantitative determinations can be made after recovery of the substance of interest with solvent, or roughly estimated by comparison with spots produced with known quantities and concentrations. Recording densitometers may be used for quantitative measurements *in situ*. The paper may be drawn automatically past a slit in front of a photocell, and the transmitted or reflected light recorded. Radioactive compounds are detected by contact exposure with x-ray film, which after development can be measured with a densitometer. Quantitative measurements *in situ* can be made with apparatus similar to a densitometer in which a radiation detector is used in place of the photocell

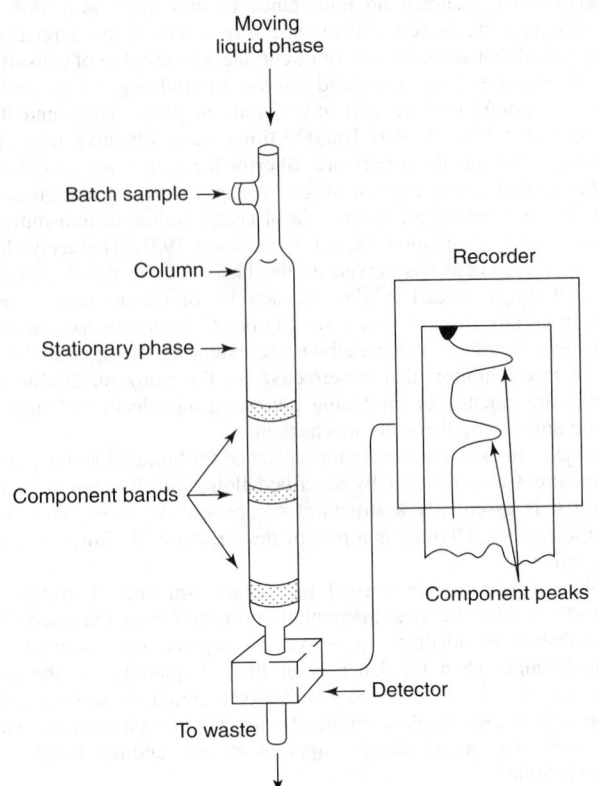

Fig. 2. Schematic of liquid-column chromatograph

between the fixed, or stationary, phase in a column and the carrier. The stationary phase may be a solid or a liquid-coated solid packed into the column, or it may be attached to the walls of a capillary. Liquid samples that can be vaporized can be separated with a gas carrier. High-boiling liquids and unstable compounds can be separated with a liquid carrier. Some materials can be handled better in a supercritical fluid and may separate faster. This latter approach, however, requires much more skill, is expensive, and thus, at present, is limited. A liquid-column is shown in Fig. 2.

Components of the sample are retained in the column for different lengths of time due to adsorption-desorption, solution-dissolution, chemical affinity, size exclusion, and other mechanisms of varying nature. Various components are continually washed from one part of the stationary phase and recaptured by another by the moving phase. Different components elute in groups from the column with respect to time from injection. Dispersion in the system causes the bands of components to emerge with

a Gaussian distribution or a distorted peak-shaped curve. A simplified diagram of the process is shown in Fig. 3. It becomes immediately obvious that chromatography not only is an analytical method, but also a separation method for substances that are difficult to separate by more conventional means.

Flow injection techniques are utilized to prepare samples, not only for chromatography, but also for numerous other analytical sensors that use thermal conductivity, flame ionization, and spectrometric principles. As important as these detectors may be, they must not overshadow the critical need to obtain a reliable sample. Flow injection analysis (FIA) is a technique for introducing a discrete sample into a flowing carrier, passing it through a conditioning "operator" system, measuring the concentration of one or more components in the modified sample, and displaying the results. Many different types of operators may be used, but component separation generally is associated with column chromatography. Flow injection is a procedure that was originally developed to automate wet lab test methods, but now has become a universal sample-preparation technique.

Advancements in Chromatography

When it first appeared as an industrial tool, the chromatograph was a single box with all of the components packaged together. Then it was considered

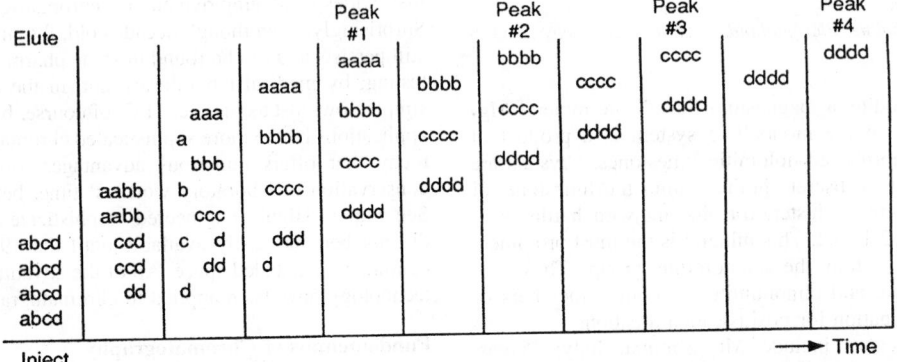

Fig. 3. Chromatographic separation process showing movement through column with time

an analyzer unto itself and required a separate sample conditioning system. Process engineers concerned with analysis began to develop techniques to modify the laboratory practice of using a single column. Multiple columns connected by switching valves were applied to backflush, foreflush, and cut components from the normal sequential elution scheme.

Some years later, laboratory chemists started to apply such techniques to specific industrial and process applications. Remote discrete sampling was developed in 1978. This allowed one to inject the small quantity of sample into a capillary hundreds of feet (tens of meters) from the "analyzer" and transport it to the column as a packet. The transfer line was viewed as a long oven and no longer constrained to a single box. Components of the original box now could be expanded into three-dimensional space by using several boxes with connecting capillaries. The programmer, which controls the system operation, became a main feature of the system.

In more recent years the capillary column, which now plays a major role in process applications, appeared. These columns produce significantly higher resolution of components and, in many cases, are faster and easier to use. Outer-surface-clad fused silica columns have been developed which are less fragile than those made of laboratory glass. Inert inner-surface metal capillary columns with cross-linked and chemically bound stationary phases also are available. The practice of technicians making their own columns has vanished into antiquity. Because there are numerous column types, details cannot be provided here. However, there are several buyers' guides available for the asking. The literature also is rich in terms of such factors as specific column selection, sample introduction techniques, analysis detectors, and overall performance qualities of a chromatographic system.

Control of chromatographic system operation has progressed from electromechanical and electro-optical devices to fully digital electronic components, which also incorporated computational capability. Microprocessors, personal computers, mini- and mainframe computers, as well as programmable logic controllers can be effectively integrated into chromatographic systems. However, for simple operations one still may use cam and digital timers.

A major improvement came with digital electronics, which allows high-resolution timing (0.1 second) and exact repeatability. Precise timing of sequential events is essential to multidimensional sample preparation utilizing column separation operators and switching valves.

Electronics for controlling instrument-operating parameters, such as temperature and pressure, have contributed to more stable operation. This produces better repeatability of the system for multiple cycles leading to higher precision of component isolation. Advantage of improved measurement devices can be realized once the sample preparation function is resolved. Computer hardware and software provide expanded capability for signal processing and data manipulation. Stable and precise sample preparation and measurement systems allow accurate analytical information provided that adequate maintenance and calibration practices are applied.

The wide range of measurement devices currently available makes almost any process chemical analysis possible. Column chromatographic techniques make interference-free determinations a reality.

Gas Chromatography

Almost any organic or inorganic compound that can be vaporized can be separated and analyzed with a gas chromatograph. As shown in Fig. 4, the minimum requirements for a system include: (1) a column that contains the substrate or stationary phase; (2) a supply of inert carrier gas (moving phase), which is continually passed through the columns; (3) a means for maintaining pressure and flow constant; (4) a means of admitting or injecting the sample into the carrier gas stream; (5) a detector which senses the sample components as they elute; and (6) a recorder. The carrier gas may be any gas that does not react with the sample or adversely affect the detector. Helium, hydrogen, nitrogen, and argon are most commonly used.

Microprocessor-Based Automation. Automatic systems now are used to carry out functions previously performed manually by the operator. These systems may take many forms: (1) integral to and dedicated to a single chromatograph and (2) as a separate unit which controls one or several chromatographs. The latter are also available for upgrading older manually operated chromatographs to provide automated data reduction and output.

Systems are equipped with read-only memory (ROM) for the operating system and random-access memory (RAM) for storage of user-specific

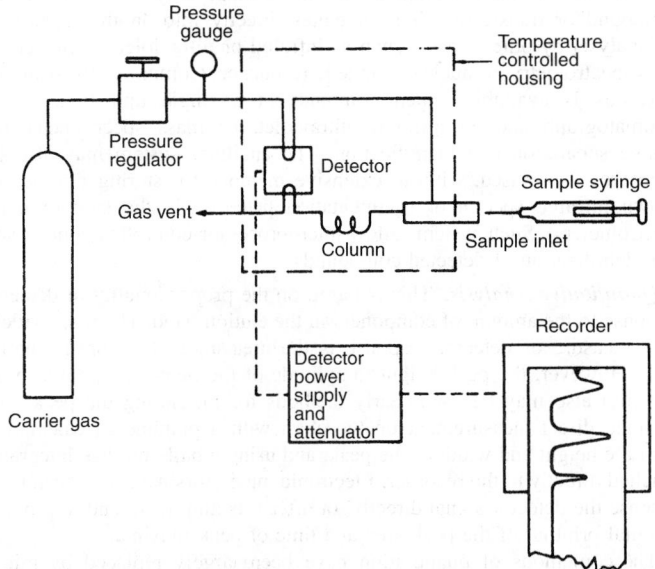

Fig. 4.   Minimum requirements for a gas chromatographic system include: (1) a column which contains the substrate or stationary phase, (2) a supply of inert carrier gas (moving phase) which is continually passed through the columns, (3) a means for maintaining pressure and flow constant, (4) a means of admitting or injecting the sample into the carrier gas stream, (5) a detector which senses the sample components as they elute, and (6) a display (recorder). The carrier gas may be any gas that does not react with the sample or adversely affect the detector. Helium, hydrogen, nitrogen, and argon are often used

methods, operating instructions, and analytical data acquisition and manipulation.

Depending on user requirements, one or more of the following functions can be provided:

1. Individual temperature control and indication of column, ovens, detectors, and sample inlets
2. Dynamic control of programmed temperature profiles, starting and ending temperatures, rate of temperature increase with one or more time-controlled segments of isothermal temperature, and cooling-down cycle.
3. Automatic integration of peak areas or peak height measurements, with a choice of several calibration methods for calibration of results.
4. Raw data storage with post-run calculation and manipulation of data for optimizing variables and methods development.
5. Integral printer-plotter for hard copy of chromatograms and analytical results.
6. Keyboard for input of analytical parameters and operator-selected variables.
7. Video terminal for display of chromatograms and analytical results.
8. Cassette or floppy disk extended memory for storage of files, methods, and user-specific applications programs.
9. High-level language for user programming of special calculations such as physical properties and comparison with alarm levels.
10. Serial communications via RS-232 ports with a central computer or other peripheral devices.
11. Timed output control signals to peripheral and ancillary devices such as sample valves, column switching valves, and other external functions which are part of the automated procedure.
12. On-board diagnostics to aid in troubleshooting and isolating and identifying failures.

Chromatograph Data Reduction—Qualitative Analysis. For a given substrate, under given conditions, each compound has a characteristic retention time, which can be used for tentative identification. However, two or more compounds may have the same elution time on a particular column. In such cases the compound may be rerun on a different column with other characteristics to reduce ambiguity. Extensive compilations of individual compound retention times on different substrates are available for reference. Positive identification can be made only by collecting the compound or transferring it as it elutes directly into another apparatus for analysis by other means, such as infrared or ultraviolet spectroscopy, mass spectrometry, or nuclear magnetic resonance. Commercially available apparatus is available which combines in a single unit both a gas chromatograph and an infrared, ultraviolet, or mass spectrometer for routine separation and identification. The ancillary system may also be microprocessor-based, with an extensive memory for storing libraries of known infrared spectra or fragmentation patterns (in the case of mass spectrometers). Such systems allow microprocessor-controlled comparison and identification of detected compounds.

Quantitative Analysis. This is based on the proportionality of detector response to the amount of component in the elution band. The most widely used measure of detector response is the area under the chromatogram peak. However, the peak height (amplitude of the detector signal at peak maxima) also may be used. Early methods for measuring the peak area included direct measurement on the chart with a planimeter, calculating from the height and width of the peak, and using a ball-and-disk integrator attached directly to the recorder. Electronic integrators have also been used to sense the detector signal directly, or after it is amplified, and to provide a digital printout of the peak area and time of peak maxima.

These methods of quantitation have been largely replaced by micro processor-based data systems. The detector signal is directly coupled to a high-speed analog-to-digital converter which samples it at predetermined rates as high as 40 Hz. The digitized values are stored in memory for subsequent manipulation. Slope sensing algorithms find the beginning and end of peaks, peak maxima, and valley points between incompletely resolved peaks and can differentiate between peaks and the baseline (absence of peaks) to correct the digitized signal for baseline drift. Tangent skimming algorithms can locate and digitize small "rider" peaks on tailing edges of larger peaks.

The microprocessor sums the digitized values for a given peak to obtain the peak area. Results are then calculated by various methods:

1. *Area Percentage, Normalized.* In this method, the area of each peak is determined as a percentage of the total of the areas of all peaks. Results must add up to 100 percent to indicate that all components have been detected.
2. *Concentration by Relative Response Factors, Normalized.* This is similar to area percentage, but each area is corrected by a relative response factor characteristic of a given compound.
3. *Internal Standard.* The results are calculated relative to a standard added to the sample in a known amount. Results are independent of sample size. The internal standard must be a compound not normally present in the sample and well separated from other components in the sample.
4. *Standard Addition.* The sample is analyzed with and without the addition of a known amount of a compound that is also in the sample (spiking). The concentration is calculated from the observed increase in area.
5. *External Standard.* The area of each component is compared to the area of a separately run standard with a known concentration of each component.

Process Gas Chromatography

This is a system for continuous, repetitive, and fully automatic on-line analysis of process streams; it is similar to the laboratory chromatograph in all essential elements of basic technique but different in design and appearance. Factors affecting design include the need (1) to comply with the National Electrical Code for operation in hazardous atmospheres, (2) to automate the procedure, and (3) for ready adaptability to closed-loop process control and communication with process control systems and computers. The demand for maximum reliability and minimum maintenance has emphasized simplicity of hardware and methodology. Emphasis is placed on analyzing for a few rather than a number of components and on minimizing analysis time. These design targets have resulted in the extensive use of multicolumn techniques for rapid separation of selected components, with large portions of the sample being discarded. As shown in Fig. 5 the major components of a microprocessor-based process gas chromatograph system are (1) the analyzer, (2) the data processor, (3) the sample conditioner system, (4) one or more recorders,

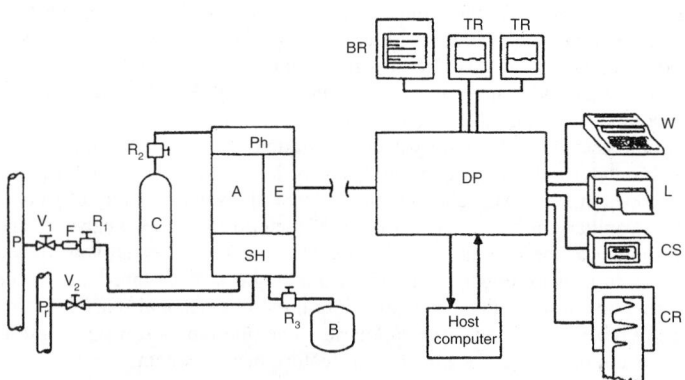

Fig. 5.   Basic elements of process gas chromatograph. The vapor sample is continuously withdrawn at a high rate from process line P, circulated through sample conditioner SH, and returned to lower pressure point P, through shutoff valves V_1 and V_2. Particulate matter is removed by filter F_1 and the pressure reduced to a constant low level be regulator R_1. The sample conditioner contains flow control and other conditioning components and a valve for switching to synthetic calibration blend B through pressure regulator R_3. A sample slipstream is circulated to the sample valve in analyzer A, which also contains columns, detectors, and a temperature-control system. Carrier gas C is controlled by regulator R_2 and pneumatics control section P. A microprocessor in electronics module E stores an analytical program in RAM and controls analyzer functions and data acquisition and reduction. Analytical results are transmitted over a serial link to data processor DP, which converts results to an analog signal for presentation to bar graph recorder BR and as many as 30 to 40 trend recorders TR. Real-time constructed chromatogram is presented for maintenance on recorder CR. Serial outputs (RS-232) flow to writer or panel-mounted line printer L for data logging and to cassette recorder CS for storing applications programs. Results and alarm messages flow to host computer via serial link. An applications program is entered via data processor and downloaded to an analyzer RAM for execution in the analyzer. The processor controls several analyzers

and (5) analog and serial outputs to peripheral devices or process control systems and computers.

Analyzer. This equipment usually is located close to the sampling point, enclosed in a shelter for weather protection. An analyzer is typically designed to comply with NEC Class I, Groups B, C, and D, Division 1, requirements for operation in hazardous areas by combination of explosion-proof enclosures, air purging, and intrinsically safe electric circuits. Sections of the analyzer include a controlled temperature compartment (heated air bath) for the columns, sample and column switching valves, and a detector. A pneumatics section for pressure or flow controllers for the carrier gas and other auxiliary gases (such as hydrogen and combustion air for an FID), as well as service air for the heater, electronics purge, and valve actuation. The electronics compartment contains a microprocessor with a central processing unit (CPU) and RAM and ROM for program control, data acquisition and reduction, output, and all communications functions. The RAM is battery-backed to prevent loss of applications programs due to power failure.

In its most usual form, the microprocessor performs these functions:

1. Controls all sequenced analyzer functions, such as sample injection and valve switching, by means of the applications program stored in RAM.
2. Samples and digitizes the detector signal at up to 40 Hz; performs peak area integration or peak height measurement with baseline correction and deconvolution of incompletely resolved peaks.
3. Identifies components by comparing elution times with values stored in memory.
4. Calculates the composition with a choice of several calibration and calculation methods stored in ROM.
5. Controls the sequencing of sample conditioner in multistream systems.
6. Performs automatic calibrations by analyzing calibration standards at user-selected intervals and automatically updating calibration factors.

7. Performs auxiliary calculations such as determining average molecular weight, specific gravity, heating (thermal units) value, or other properties based on the calculated sample composition.

8. Monitors electromechanical sensors in the analyzer and sample conditioner system to detect abnormal conditions: oven temperature out of limits, carrier gas flow failure, sample flow failure, etc.

9. Performs software diagnostics on the detector signal and analytical results to detect abnormal conditions: change in elution time, total peak area out of limits, excessive baseline noise or drift.

10. Communicates with data processor over a serial link to transmit analytical data and calculations and receives new or modified applications programs. A digitized form of the detector signal may also be transmitted for remote reconstruction of a real-time chromatogram.

In addition, the analyzer can accept analog signals from other field-mounted analyzers or sensors such as flowmeters and pressure transducers. The signal can be scaled, digitized, and incorporated into special calculations to determine mass flow, therms per day, reactor yields, and so on.

In alternative forms of these systems, the applications program is stored in the processor; the analyzer microprocessor digitizes only the detector signal and transmits the digitized values to the data processor. Applications program event commands are received in real time from the data processor and converted at the analyzer to electrical and pneumatic signals for sample valve actuation, column switching, sample conditioner control, and so on.

Data Processor. This unit commonly is located near or in the control room in a nonhazardous environment, and as much as 2000 to 3000 feet (610 to 914 meters) from the analyzers. The data processor also has its own microprocessor with a CPU and a complement of ROM and RAM in which the operating system and user-specific applications programs are stored. Communication with the analyzers is by a serial link. In its most usual form, the processor is a special-purpose microprocessor and can control up to six or eight analyzers. In other forms the processor may be a microprocessor-based minicomputer and control as many as 32 analyzers.

The processor has several main functions:

1. Input of applications programs by means of a special-purpose keyboard, with an alphanumeric display and interactive dialogue. Prompting of the operator and screening of the input data ensures the input of all necessary parameters and prevents conflicting data inputs. The program is down-loaded into the analyzer memory and can be recalled for editing and modifications. In some versions, input may be accomplished through an ancillary video terminal with a keyboard and menu-driven operator communications.

2. Receives data from the analyzers and distributes them to the various output devices in analog or digital form as required.

3. Monitors the status of the analyzers and displays alarms and transmits them to peripherals.

4. Provides manual control of all analyzer functions during setup and maintenance and acts as a diagnostic center for troubleshooting and corrective maintenance. A real-time chromatogram is available.

Except for the initial setup and maintenance of the analyzer, all operations, including programming, manual operation, and calibration, take place at the processor location.

In some versions the applications program for all analyzers may be stored in the processor memory instead of the analyzer. Event commands are sent to the analyzer in real time over the serial link and converted to analog commands by the analyzer. Digitized data are received from each analyzer, with all data acquisition and reduction accomplished in the processor.

In yet other versions the processor is part of the analyzer and may be dedicated to, and integral to, a single analyzer at the analyzer location with all analyzer and processor functions described performed locally.

The simplest form of record is a bar graph. The record consists of a series of bars, one for each component, of height proportional to the component concentration. The output for each component is scaled to give a full-scale reading equivalent to a convenient concentration value. Each component may have a different full-scale range. A large number of components in one or many streams may be recorded on the same instrument. Different streams may be identified by the height of a flat-top bar preceding the series of bars for that stream.

Trend outputs consist of a continuous electrical signal [0 to 10 volts or 4 to 20 milli (mA)] from the processor. As many as 30 to 40 such outputs may be available from a single processor. Each output represents the concentration of a particular component in one of the sample streams on a given analyzer scaled to some convenient range. Component identity and scale factors for each output channel are user-assigned from the processor keyboard.

The output value is held constant at the last value until a new value is determined in a subsequent analysis, causing a stepwise change in the record as the signal is updated at the end of each analytical cycle. A separate recorder pen is required for each trend output recorded. Trend outputs may also be used to input analytical data to a process control system. A separate two-wire 4- to 20-mA output line is typically required for each component input.

A real-time chromatogram for any selected analyzer may be obtained at the data processor for setup and maintenance. The chromatogram is received in digital form at a rate of 10 or more data points per second and reconstructed into analog form and scaled at the processor.

Serial output ports (RS-232) are usually provided for serial transmission of data—usually in ASCII code as previously shown in Fig. 5.

Valves. Electrically or pneumatically operated valves are used for the injection of liquid or gas samples and column switching. Rotary, spool-and-O-ring, diaphragm, and sliding-plate valves with from 4 to 12 ports are used. Vaporizing liquid sampling valves mounted through the wall of the analyzer transfer the liquid sample from the cold exterior zone to the heated vaporizing zone within the oven. The sample valve meters a fixed volume of liquid or gas into the column with a repeatability of ±0.25 percent. The sample size may vary from approximately 0.1 mL to 50 mL (with an external sample loop).

Liquid-Column Chromatography

This method is particularly useful for the separation and analysis of high-molecular-weight compounds beyond the range of gas chromatography. It is generally classified according to the stationary phase or to the nature of its interaction with sample components.

1. Liquid-liquid partition chromatography, where the sample components are partitioned between a moving liquid phase and a stationary liquid phase deposited on an inert solid. The two solvent phases must be immiscible. The stationary phase may be a large molecule chemically bonded to the surface of a solid (bonded liquid phase) to prevent loss by solubility in the moving phase. This method can also be subdivided into normal-phase systems, in which the moving phase is less polar than the stationary phase, and reverse-phase systems, in which it is more polar.

2. Liquid-solid or absorption chromatography, in which the sample components are absorbed on the surface of an adsorbent such as silica gel.

3. Ion exchange, in which ionic sample components interact with functional groups on a permeable ionic resin.

4. Exclusion or gel permeation, in which compounds are separated by molecular size into a range of pore sizes in a polymeric gel. This method is useful for measuring the molecular-weight distribution of polymers.

Isocratic elution uses a solvent of constant composition throughout the analysis. Gradient elution is a modification of the technique, in which the solvent is a mixture of two solvents that differ in solvent strength. The composition or ratio of the two is changed during the analysis in accordance with a predetermined program. The change may be continuous, linear or nonlinear, stepwise, or a combination.

Apparatus Required. The need for more efficient columns and faster separations has led to the development of stationary phase packings with particles as small as 2 mm, operating at pressures as high as 10,000 psig (69 MPa), and with solvent flow rates as low as 1 mL/min or less. This has in turn led to the development of detectors with internal volumes of only a few microliters and special fittings and connectors with minimum dead volume to prevent band spreading and loss of resolution. These improvements in apparatus and technique have resulted in an ability to achieve complex separations with speeds comparable to those of gas chromatography.

The solvent is moved through the system by constant-flow or constant-pressure pumps which are driven mechanically (screw-driven syringe or reciprocating) or by gas pressure with pneumatic amplifiers. For gradient elution two pumps may be synchronized and programmed to provide a controlled, reproducible composition change.

Samples may be introduced by syringe directly into the column through a septum or by means of valves with a fixed volume that has been prefilled with the sample. Valves may be of a rotary, sliding plug, or diaphragm design and of stainless steel and fluoroplastic construction for inertness. Auto samplers are used for unattended injection of samples loaded into vials and sequentially rotated into the injection mechanism.

Major contributors to this article were: J. G. CONVERSE, Chief Chemist, Sterling Chemicals, Inc., Texas City, Texas and R. VILLALOBOS The Foxboro Company (A Siebe Company) Foxboro, Massachusetts

Additional Reading

Beesley, T.E., S. Raymond: *Chiral Chromatography,* John Wiley & Sons, Inc., New York, NY, 1998.

Berezkin, V.G. and J. de Zeeuw: *Capillary Gas Adsorption Chromatography,* John Wiley & Sons, Inc., New York, NY, 1998.

Berthod, A. et al.: *Micellar Liquid Chromatography,* Vol. 83, Marcel Dekker, Inc., New York, NY, 2000.

Bowers, M.T., et al.: "Gas-Phase Ion Chromatography: Transition Metal State Selection and Carbon Cluster Formation," *Science,* **1446** (June 4, 1993).

Brown, P.R. and E. Grushka: *Advances in Chromatography,* Vol. 40, Marcel Dekker, Inc., New York, NY, 2000.

Campbell, D. and A. Foundes: "Chromatographs Meet Environmental Needs," *Hydrocarbon Processing,* **63** (February 1991).

Converse, J.G.: *Sample Preparation for On-Site Automated Chemical Analysis,* Paper 90–458 (New Orleans Meeting), Instrument Society of America, Research Triangle Park, NC, 1990.

Converse, J.G.: "Improve On-Line Chemical Concentration Measurement," *Chem. Eng. Progress,* **73** (May 1991).

Converse, J.G.: "Process Chromatography," in D.M. Considine, Ed., *Process/Industrial Instruments & Controls Handbook,* 4th Edition, McGraw-Hill Companies, Inc., New York, NY, 1993.

Fried, B. and J. Sherma: *Thin-Layer Chromatography,* Marcel Dekker, Inc., New York, NY, 1999.

Fritz, J.S. and D.T. Gjerde: *Ion Chromatography,* John Wiley & Sons, Inc., New York, NY, 2000.

Holman, R.B., M.H. Joseph, and A.J. Cross: *High Performance Liquid Chromatography ion Neuroscience Research,* John Wiley & Sons, Inc., New York, NY, 1999.

Kastner, M.: *Protein Liquid Chromatography,* Elsevier Science, New York, NY, 1999.

Kenney, B.F.: "Applications of High-Performance Liquid Chromatography for the Flavor Research and Quality Control Laboratories in the 1990s," *Food Technology,* **76** (September 1990).

Kolb, B. and L.S. Ettre: *Static Headspace—Gas Chromatography: Theory and Practice,* John Wiley & Sons, Inc., New York, NY, 1999.

Matter, L.: *Food and Environmental Analysis of Capillary Gas Chromatography,* John Wiley & Sons, Inc., New York, NY, 1998.

McNair, H.M. and J.M. Miller: *Basic Gas Chromatography,* John Wiley & Sons, Inc., New York, NY, 1997.

Millner, P.A.: *High Resolution Chromatography: A Practical Approach,* Oxford University Press, Inc., New York, NY, 1999.

Parcher, J.F.: *Unified Chromatography,* American Chemical Society, Columbus, OH, 2000.

Poole, C.F. and S.K. Poole: *Chromatography Today,* Elsevier Science, New York, NY, 2000.

Saunders, S.J. et al.: "Modeling the Separation of Amino Acids by Ion-Exchange Chromatography," *Chem. Eng. Progress,* **47** (August 1988).

Stahly, G.P.: "Thin-Layer Chromatography," *Today's Chemist at Work,*" **28** (January 1993).

Strobel, H.A. and W.R. Heineman: *Chemical Instrumentation—A Systematic Approach in Instrumental Analysis,* John Wiley & Sons, Inc., New York, NY, 1989.

Swadesh, J.K.: *HPLC: Practical and Industry Chromatography,* CRC Press, LLC., Boca Raton, FL, 2000.

Thevenon-Emeric, G. and F.E. Regnier: "Process Monitoring by Parallel Column Gradient Elution Chromatography," *Analytical Chemistry,* **1114** (June 1, 1991).

Wahl, J.H., C.G. Eske, and V.L. McGuffin: "Solvent Modulation in Liquid Chromatography," *Analytical Chemistry,* **1117** (June 1, 1991).

CHROMITE. An important mineral in the chromite series of multiple oxides. The dominant compound is $FeCr_2O_4$, but most chromite also contains magnesium (Mg) and aluminum (Al). Crystallizes in the isometric system. Hardness, 5.5; specific gravity, 4.5–4.8; color, black. Associated with peridotite and serpentine (metamorphized peridotite). Commercial amounts occur as placer deposits in serpentine areas. In the United States chromite has been mined in Maryland, Pennsylvania, California, Montana, Oregon, Wyoming and North Carolina. Important deposits occur in Quebec, Canada. Valuable deposits occur in Asia Minor, Zimbabwe, New Caledonia, Cuba, India, and the Philippines; also in New South Wales, and in the Ural Mountains associated with platinum.

CHROMIUM. [CAS: 7440-47-3]. Chemical element, symbol Cr, at. no. 24, at. wt. 51.996, periodic table group 6, mp 1837–1877°C, bp 2672°C, density 7.2 g/cm³. Elemental chromium has a body-centered cubic crystal structure. The metal is silver-white with a slight gray-blue tinge, very hard (9.0 on the Mohs scale), capable of taking a brilliant polish, not appreciably ductile or malleable. The element is not affected by air or H_2O at ordinary temperatures, but when heated above 200°C, chromic oxide Cr_2O_3 is formed. There are four stable isotopes: ^{50}Cr, and ^{52}Cr through ^{54}Cr. Four radioactive isotopes have been identified, all with comparatively short half-lives ^{48}Cr, ^{49}Cr, ^{51}Cr, and ^{55}Cr. The element was first identified by Vauquelin in 1797.

Ionization potential 6.76 eV; second, 16.6 eV. Oxidation potentials $Cr \longrightarrow Cr^{3+} + 3e^-$, 0.71 V; $Cr^{2+} \longrightarrow Cr^{3+} + e^-$, 0.41 V; $2Cr^{3+} + 7H_2O \longrightarrow Cr_2O_7^{2-} + 14H^+ + 6e^-$, 1.33 V; $Cr + 3OH^- \longrightarrow Cr(OH)_3 + 3e^-$, 1.3 V; $Cr + 4OH^- \longrightarrow CrO_2^- + 2H_2O + 3e^-$, 1.2 V; $Cr(OH)_3 + 5OH^- \longrightarrow CrO_4^{2-} + 4H_2O + 3e^-$, 0.12 V.

Other important physical properties of chromium are given under **Chemical Elements**.

Chromium occurs chiefly as chromite (ferrous chromite), $Fe(CrO_2)_2$, in Zimbabwe, the Republic of South Africa, the former U.S.S.R., New Caledonia, India, Philippine Islands, Japan, Turkey, Greece, Cuba, and California. (1) Heating chromite in the electric furnace with carbon yields ferrochrome for alloys, and (2) when chromite is heated with sodium carbonate and nitrate, sodium chromate is formed, which is then extracted with H_2O. This is the substance from which chromium compounds are obtained. See also **Chromite**.

Chromium is used extensively for (1) decorative and wear-resistant electroplating, (2) many important alloys, and (3) the manufacture of numerous chemicals and refractory materials.

Alloys

In constructional steels, chromium imparts hardness by improving hardenability and promoting the formation of carbides. These steels have exceptional wear resistance and are relatively stable at elevated temperatures. In stainless and heat-resisting steels, chromium improves corrosion and heat resistance. As shown in Table 1, stainless steels may be grouped into three principal classes (1) austenitic, (2) martensitic, and (3) ferritic stainless steels. The fundamentals of stainless steels are further described under **Iron Metals, Alloys, and Steels**. Stainless steels, as a group of major ferrous alloys, are characterized by their high degree of resistance to chemical attack. They possess a property commonly referred to as passivity, a property eminently displayed by elemental chromium. This property is manifested in steels when the chromium content exceeds about 11%. A very marked improvement in corrosion and heat resistance is achieved when chromium is added to this or a greater extent to low-carbon steels. The addition of nickel, along with chromium, enhances these properties even more. Although a steel that contains 12% chromium will stain, it will not undergo progressive rusting in normal atmospheres. When the chromium content is increased to 18%, staining will not occur in normal atmospheres, but may occur to a limited extent in particularly bad industrial environments. However, the addition of 8% nickel, along with the 18% chromium, will make the steel stain-resistant in all but the most rigorous industrial atmospheres. Additional resistance to corrosion and heat resistance may be obtained by the presence of molybdenum and other elements in smaller quantities.

An example of the effect of chromium content on the corrosion resistance of steel is demonstrated as follows:

A low-carbon steel placed in boiling 65% HNO_3 for one month:

TABLE 1. REPRESENTATIVE STAINLESS STEELS

Type	Composition and characteristics
	Austenitic Stainless Steels
302	Basic alloy of the group: Cr, 17–19%; Ni, 6–8%
302B	Silicon (2–3%) added to increase scaling resistance
303	Sulfur (0.02–0.05%) added to improve machinability
303Se	Selenium (0.15%) added to improve machinability
304L	Extra low carbon for improved weldability
304	Lower carbon content to improve weldability and inhibit carbide formation: Cr, 18–20%; Ni, 10–12%
305	Nickel increased to lower work-hardening: Cr, 17–19%; Ni, 10–13%
308	Chromium and nickel increased to increase corrosion and scaling resistance: Cr, 19–21%; Ni, 10–12%
309S	Lower carbon content (0.08% max) for improved weldability
314	Silicon added for increased scaling resistance at high temperatures: Cr, 23–26%; Ni, 19–22%; Si, 2% max
316	Molybdenum added to improve resistance to pitting corrosion and strength at high temperatures: Cr, 16–18%; Ni, 10–14%; Mo, 2–3%
316L	Extra-low carbon content for improved weldability
317	Additional molybdenum to further improve resistance to pitting corrosion: Cr, 18–20%; Ni, 11–15%; Mo, 3–4%
321	Titanium added to prevent chromium carbide precipitation: Cr, 17–19% Ni, 9–12%; C, 0.08% max: Ti, 5×C content
347	Columbium (niobium) or tantalum added to prevent chromium carbide precipitation
	Martensitic Stainless Steels
410	Basic alloy of the group: Cr, 11.5–13.5%; Ni, 0.5% max
405	Aluminum added to prevent weld hardening
414	Nickel (2%) added to improve corrosion resistance
416	Sulfur added to improve machinability
416 Se	Selenium added to improve machinability
420	Carbon increased for higher hardness: Cr, 12–14%; Ni, 0.5% max; C, over 0.15%
431	Chromium increased to further improve corrosion resistance: Cr, 15–17%; Ni, 1.25–2.50%
440A	Carbon slightly decreased to improve toughness: Cr, 10–18%; Ni, 0.5% max; C, 0.6–0.75%
440B	Carbon decreased slightly to improve toughness even more
440C	Carbon increased to increase hardness; chromium increased to make up loss in corrosion resistance: Cr, 16–18%; Ni, 0.5% max; C, 0.95–1.2%
	Ferritic Stainless Steels
430	Basic alloy of the group: Cr, 14–18%; Ni, 0.5% max
430FSe	Selenium added to improve machinability
442	Chromium increased to improve resistance to scaling and corrosion: Cr, 23–27%; Ni, 0.5% max

4.5% chromium	corrosion rate, 12.9 in.
8.0% chromium	0.14
12.0% chromium	0.01
18.0% chromium	0.003
25.0% chromium	0.0006

The corrosion resistance of iron-chromium alloys was known in England and France in the early 1800s, but passivity was not clearly recognized and reasonably understood until 1910 as the result of studies by Borchers and Monnartz in Germany. Commercial stainless steels were introduced shortly thereafter in Germany, France, England, and a bit later in the United States. Worldwide production of stainless steels now is measured in terms of millions of tons annually.

Research programs worldwide are seeking to improve the performance of 8 to 14% chromium ferritic steels for high-temperature applications up to 650°C (1200°F), particularly with applications in advanced power plants as targets. Such alloys are needed for future coal-fired power plants as well as for fast breeder and nuclear fission reactors. In addition to traditional uses of high-temperature steels as found in steam and gas turbine and boiler tube uses, the new alloys will be intended for use as fuel cladding, wrapper and steam generator tubing for fast breeder reactors, and first-wall materials for fusion reactors as currently conceived.

In addition to ferrous alloys, chromium also is added to copper, vanadium, zirconium, and other metals to form several hundred chromium-bearing alloys. Nickel-chromium-iron alloys have high electrical resistance and are used widely as electrical heating elements. *Nichrome* and *Chromel* are examples.

Chromium Plating

Although the tonnage of chromium used for electroplating is far below its consumption for alloys, plating represents a major market for the element. In terms of special protective coatings, chromium is behind only copper, lead, and zinc in consumption. Generally, chromium plating is used for two purposes: (1) wear-resistance and (2) decorative effect, taking on polish and a much brighter surface than the other electroplated metals. The "bright work" of automotive hardware, plumbing fixtures, and electrical appliances are examples. Normally, chromium is plated over nickel where the nickel is about 100×thicker than the chromium. In decorative plating, the thickness of the chromium plate ranges only from 0.00001 to 0.00002 inch (0.00025 to 0.00051 millimeter). Hexavalent chromium also forms protective chromate coatings over aluminum, cadmium, copper, magnesium, and zinc. These chromate conversion coatings are used on hot-dipped or electrogalvanized parts, on zinc die castings, and extensively on aluminum parts for aircraft.

In 1979, scientists at the General Motors Research Laboratories (Warren, Michigan) announced some interesting findings concerning chromium plating. The conventional process uses a bath containing Cr^{6+} ions. During plating, the Cr^{6+} ions reduce to Cr^{3+} and then to metallic Cr. Investigators wondered for years why the process will not succeed if one commences with Cr^{3+} ions, thus avoiding the double step. The GM scientists found that starting with Cr^{3+} fails because it immediately forms a stable complex with water molecules from which Cr cannot be deposited. Commencing with Cr^{6+} succeeds because during reduction a chemical film forms around the cathode (part being plated). Since Cr^{3+} is bound in that film, it does not react with water, but, instead, plates out as chromium metal. See Fig. 1.

As reported in 1983 by Dr. James Hoare, Research Fellow (General Motors Research Laboratories), the electrolyte for plating is a chromic acid solution which contains various chromate ions: chromate, dichromate, and trichromate. From a series of steady-state polarization experiments, Dr. Hoare concluded that trichromate is the ion important in chromium deposition. Sulfuric acid has been recognized as essential to chromium plating and has been assumed by some to be a catalyst for the process. In this strongly acidic solution, sulfate should be mostly present as the bisulfate ion (HSO_4^-). Dr. Hoare found, contrary to expectation, that the addition of sulfuric acid to the plating bath decreased the conductivity of the solution.

Combining these findings with the results of prior investigations, Dr. Hoare concluded that the electroactive species was a trichromate-bisulfate complex. See Fig. 2. From equilibrium considerations, he theorized that the maximum concentration of this species occurred at a 100-to-1 chromic acid/sulfuric acid ratio. The observation that the maximum rate of chromium deposition also occurred at this ratio supports the conclusion that this trichromate-bisulfate complex is the electroactive species. During the plating process, the complex diffuses from the bulk solution toward the cathode. See Fig. 3. Electron transport takes place by quantum mechanical tunneling through the potential energy barrier of the Helmholtz double layer, and the unprotected chromium in the complex (Cr atom on the left

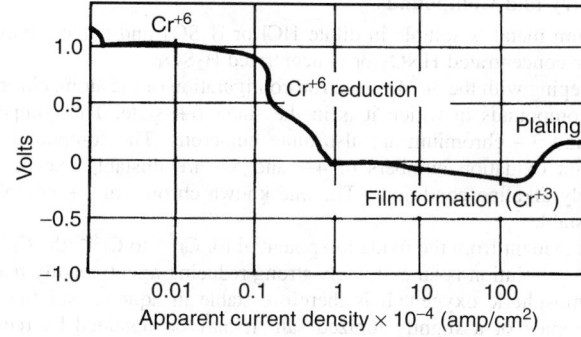

Fig. 1.   Typical polarization curve made during investigation of chromium plating process. (*General Motors Research Laboratories*)

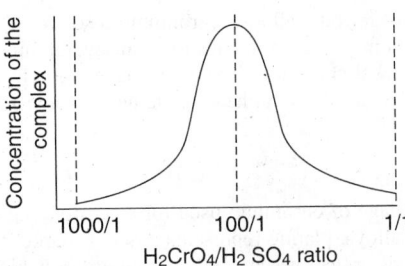

Fig. 2. The electroactive complex and a theoretical plot of its concentration as a function of chromic acid to sulfuric acid ratio. (*General Motors Research Laboratories*)

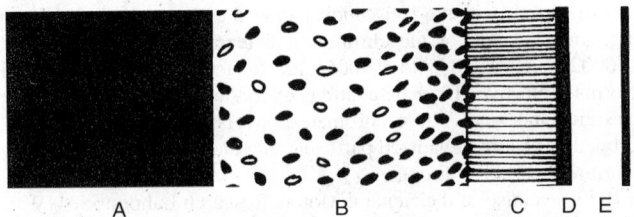

Fig. 3. The electroactive complex diffuses from the bulk electrolyte solution (A) through the diffusion layer (B) to the Helmholtz double layer (C) to be discharged as metallic chromium (D) on the cathode surface (E). (*After General Motors color sketch*)

in Fig. 1) loses electrons by successive steps, going from Cr^{+6} to Cr^{+2}. Decomposition of the resulting chromodichromate complex takes place by acid hydrolysis to form a chromous-oxybisulfate complex:

$$^+Cr\!-\!O \rightleftharpoons H \rightleftharpoons O\!-\!\overset{\overset{\displaystyle O}{\|}}{\underset{\underset{\displaystyle O}{\|}}{S}}\!-\!O^-$$

The positive end of this complex is adsorbed onto the cathode surface. Electrons are transferred from the cathode to the adsorbed chromium ion, forming metallic chromium and regenerating the HSO_4^- ion. Thus, Dr. Hoare's mechanism explains how sulfuric acid, in the form of the bisulfate ion, participates in the plating process. (Information regarding Cr plating furnished by General Motors Research Laboratories and gratefully acknowledged.)

$$^-O\!-\!\overset{\overset{\displaystyle O}{\|}}{\underset{\underset{\displaystyle O}{\|}}{Cr}}\!-\!O\!-\!\overset{\overset{\displaystyle O}{\|}}{\underset{\underset{\displaystyle O}{\|}}{Cr}}\!-\!O\!-\!\overset{\overset{\displaystyle O}{\|}}{\underset{\underset{\displaystyle O}{\|}}{Cr}}\!-\!OH \qquad \begin{matrix} O\!\leftrightarrow\!H\!\leftrightarrow\!O\!-\!\overset{\overset{\displaystyle O}{\|}}{\underset{\underset{\displaystyle O}{\|}}{S}}\!-\!O^- \\ \\ O\!\leftrightarrow\!H\!\leftrightarrow\!O\!-\!\overset{\overset{\displaystyle O}{\|}}{\underset{\underset{\displaystyle O}{\|}}{S}}\!-\!O^- \end{matrix}$$

Chemistry and Compounds

Chromium metal is soluble in dilute HCl or H_2SO_4 and made passive by dilute or concentrated HNO_3 or concentrated H_2SO_4.

In keeping with the $3d^5 4s^1$ electron configuration of the atom, chromium forms compounds in which it is in the stable 6+ state. The compounds of 2+ and 3+ chromium are also quite numerous. The compounds with chromium oxidation numbers of 4+ and 5+ are unstable except under extremely alkaline conditions. The one known chromium 1+ complex is also unstable.

As is evident from the oxidation potential for Cr^{2+} to Cr^{3+}, the Cr^{2+} ion in aqueous solution is an extremely strong reducing agent, readily reacting with atmospheric oxygen. It is therefore stable in aqueous solution only in a complex or a slightly ionized salt. It may be obtained by reducing hexavalent or trivalent chromium in acid solution with one of the active metals, such as zinc. The four halogens form halides of divalent chromium,

CrX_2, which dissolve in water with the evolution of heat, although the fluoride is less soluble than the other halides. The ammoniacal solution of chromium(II) sulfate, $CrSO_4$, is particularly reactive with gases, not only with oxygen like the other Cr^{2+} compounds, but also acetylene. Chromium(II) acetate is only slightly soluble in H_2O.

The Cr^{3+} ion readily forms complexes; it exists in aqueous solution as $Cr(H_2O)_6^{3+}$, and forms other complexes with anions, such as $Cr(H_2O)_5Cl^{2+}$, $Cr(H_2O)_4Cl_2^+$, etc. Thus chromium(III) halides may be crystallized in a number of forms differing in color and other properties due to variation in the bonding. Compounds of the trichloride have been reported with the following arrangements: $Cr(H_2O)_6Cl_3$, $[Cr(H_2O)_5Cl]Cl_2 \cdot H_2O$, and $[Cr(H_2O)_4Cl_2]Cl \cdot 2\ H_2O$. Trivalent chromium also forms double salts, notably the chromium alums, hydrated double salts of Cr(III) sulfate and the alkali metal (or thallium or ammonium) sulfates.

The three oxides of chromium, CrO, Cr_2O_3, and CrO_3, are of interest in exhibiting the transition in properties from basicity to acidity and from electrovalence to covalence, with increasing oxygen content, and the consequent increase in charge upon the Cr atom. Thus Cr_2O_3, the middle oxide, is amphiprotic, dissolving either in strong alkali hydroxide solutions or in acids.

One of the most stable of the compounds of Cr(IV) is the tetrafluoride, CrF_4, brown solid, steel-blue vapor at $150°C$, prepared by direct reaction of the elements; even this compound readily undergoes hydrolysis. Chromium tetrachloride, $CrCl_4$, may be prepared as a gas by reaction of Cl_2 and $CrCl_3$ at elevated temperature, but decomposes at room temperature. Chromium(V) occurs in CrF_5, fire-red, volatile, and in the hypochromates, M_3CrO_4, green, which may be prepared by fusion of alkali chromate and alkali hydroxide at high temperature.

CrO_3 is the anhydride of chromic and dichromic acids, H_2CrO_4 and $H_2Cr_2O_7$, which have not been isolated, but whose anions are found in many salts. pK_{A1} and $pK_{A2} = 0.745$ and 6.49, and -1.4 and 1.64 respectively. They are strong oxidants in acid solution, and are readily obtained in basic solution by oxidation of Cr(III) compounds. The oxyhalogen compounds of chromium are chiefly the CrO_2X_2 (chromyl) compounds, which are acid halides of chromic acid. Chromyl chloride, CrO_2Cl_2, deep red, liquid, is prepared by heating sodium dichromate and sodium chloride with H_2SO_4, while chromyl fluoride undergoes polymerization to a white solid. Chlorochromates, e.g., $KCrO_3Cl$, and fluorochromates, e.g., $KCrO_3F$, are also known, but are hydrolyzed in water.

Chromates and Dichromates

Sodium chromate, Na_2CrO_4, potassium chromate, K_2CrO_4, ammonium chromate, $(NH_4)_2CrO_4$, calcium chromate, $CaCrO_4$, are yellow soluble solids; barium chromate, $BaCrO_4$, pale yellow, strontium chromate, $SrCrO_4$, pale yellow, lead chromate, $PbCrO_4$, yellow (used as a pigment, "chrome yellow"), zinc chromate, $ZnCrO_4$, yellow, used as a pigment, mercurous chromate, Hg_2CrO_4, yellow to red to brown, silver chromate, Ag_2CrO_4, reddish-brown, are insoluble solids. Sodium dichromate, $Na_2Cr_2O_7$, potassium dichromate, $K_2Cr_2O_7$, readily crystallized, ammonium dichromate, $(NH_4)_2Cr_2O_7$ (forming nitrogen gas and green chromic oxide solid upon heating) are red soluble solids, of important application as oxidizing agents, e.g., sulfurous acid causes reduction to chromic; silver dichromate, $Ag_2Cr_2O_7$, red insoluble solid, changing to silver chromate, Ag_2CrO_4 upon boiling with H_2O. Solutions of chromate in the presence of acid are changed to the corresponding dichromate.

A number of peroxychromates are known, including the deep-blue, organic soluble reaction product of H_2O_2 and acid $Cr_2O_7^{2-}$ solutions, which is H_4CrO_7, i.e., $CrO_3 \cdot 2H_2O_2$.

Trivalent chromium forms many complexes. The large number is due in part to the many possible arrangements in which the same elements may be present as part of a complex anion or as the cation, as was indicated for the chloride complexes of the hydrated Cr^{3+} ion. The ligands may also be present in various proportions. As in the case with Co^{3+}, Cr^{3+} forms complexes with ammonia in the various color series, e.g., a violeo-chloride, $[Cr(NH_3)_4Cl_2]Cl$, a purpureo chloride, $[Cr(NH_3)_5Cl]$ Cl_2 and a luteo chloride, $[Cr(NH_3)_6]Cl_3$. In addition to NH_3 many amines, especially ethylenediamine and its derivatives, form stable complexes with Cr^{3+}. In the oxalato complexes, the tripositive chromium ion coordinates with oxalate groups to form such anions as $[Cr(C_2O_4)_3]^{3-}$ and $[Cr(C_2O_4)_2(H_2O)_2]^-$. There are many cyano, thiocyano, nitrato, fluoro, bromo, iodo, azido, acetylido and nitro complexes.

Organometallic Compounds

These include a large number of cyclopentadienyl compounds. In addition to dicyclopentadienyl chromium itself, $(C_5H_5)_2Cr$, there are many derivatives of it with additional radicals or molecules such as $(C_5H_5)_2CrBr$, $(C_5H_5)_2CrNH_3$, $C_5H_5Cr(NO)_2CH_2Cl$, $H(C_5H_5)Cr(CO)_3$ and $C_5H_5(CO)_3CrCr(CO)_3\ C_5H_5$. The compounds of Cr with other hydrocarbon radicals are fewer ($Cr(C_6H_6)_2$ is known), such as $(C_6H_5)_5CrOH\cdot H_2O$ and $(C_6H_5)_3Cr\cdot 3$ THF (THF =Tetrahydrofuran).

Besides $Cr(C_6H_6)_2$, chromium hexacarbonyl, $Cr(CO)_6$, white, solid, bp (extrap.) $145.7°C$, represents chromium in the zero oxidation state. This is a nonpolar compound, insoluble in H_2O, freely soluble in organic solvents. The zero-valent compound $K_6Cr(CN)_6$ has also been made by the reaction of $K_3Cr(CN)_6$ and potassium metal in liquid NH_3.

Cermets

The term *cermet* derives from the combination of ceramic and metal. Cermets are produced by powder metallurgy techniques and represent the bonding of two or more metals. They are particularly useful at high temperatures ($850-1250°C$). Chromium is used in several cermet combinations, including chromium-bonded aluminum oxide, metal-bonded chromium carbide, and metal-bonded chromium boride.

Biological Aspects

The first indication for a biological role of chromium in metabolism was derived from the enzymatic studies in 1939. The succinate-cytochrome dehydrogenase system, an important enzyme system for the production of energy, requires certain inorganic cofactors. Of various elements tested, chromium produced the greatest increase of enzyme activity. A significant stimulation of the activity of phosphoglucomutase by chromium was described by Strickland in 1949. This system which has an important function in the early steps of carbohydrate metabolism requires magnesium and one other metal for optimal activity. Chromium was outstanding as a "second metal" because it produced the highest enzyme activation and supported a measured amount of activity even when given alone.

Chromium also stimulates fatty acid and cholesterol synthesis from acetate in liver. That chromium is an essential cofactor for the action of insulin on the rat lens was shown by Farkas in 1964. In the absence of the element, no significant insulin effect on glucose utilization of lens can be demonstrated. Chromium supplementation to the donor animals results in a significant response of lens tissue to the hormone. Numerous other findings indicate that chromium may play several vital roles in biological systems.

Chromium levels in biological matter have been studied extensively. In contrast to findings with other metals, chromium concentrations in the United States population are highest at the time of birth, with a pronounced decline during the lifetime, whereas they appear to remain high in some other countries, such as Thailand and the Philippines. These findings suggest the possibility of a relative chromium deficiency in the United States. The relationship of this to disturbances of carbohydrate metabolism in humans, while under study, remains to be established.

Not all of the various chemical forms of chromium are effective in improving sugar metabolism, and the exact nature of the compound or compounds involved in activating insulin is not established. Some of the chromium in plants may not be present in nutritionally effective forms. It has not been established that chromium is essential to plants, but high concentrations of the metal are toxic. Most agricultural crops, especially their seeds, contain only low levels of chromium.

Chromium Use in Strain Gage

Strain measurements at high temperatures are required in developing designs for turbine engines and hypersonic aerospace vehicles. Operating temperature may exceed $1000°C$ ($1830°F$). Reliable strain gages for this range have not been available. Maximum operating temperature of traditional gages is $400°C$ ($750°F$). Thus, engineers were required to research a number of metal combinations that would perform reliably at such high temperatures. The ideal strain gage must change its electrical resistance only in response to strain and be unaffected by temperature changes.

Researchers at NASA (National Aeronautics and Space Administration, Cleveland, Ohio) tested a large number of alloys, such as iron-chromium-aluminum, platinum-tungsten, platinum-palladium-molybdenum, and palladium-chromium. The search was narrowed to the palladium (Pd)-chromium (Cr) combination. Ultimately, the combination of Pd-13% Cr (wt) was selected as optimal. A platinum wire that is wound around the periphery of the gage grid provides the required temperature compensation.

Additional Reading

Davis, J.R.: *Metals Handbook,* ASM International, Materials, Park, OH, 1998.

Dennis, J.K. and T.E. Such: *Nickel and Chromium Plating,* Woodhead Publishing, Ltd., Cambridge, UK, 1994.

Katz, S.A., and H. Salem: *The Biological and Environmental Chemistry of Chromium,* John Wiley & Sons, Inc., New York, NY, 1994.

Lei, Jih-Fen: "Pd-Cr Gages Measure Static Strain at High Temperatures," *Adv. Mat. & Processing,* **58** (May 1992).

Lewis, R.J. and N.I. Sax: *Sax's Dangerous Properties of Industrial Materials,* 10th Edition, John Wiley & Sons, Inc., New York, NY, 1999.

Lide, D.R.: *CRC Handbook of Chemistry and Physics,* 84th Edition, CRC Press, LLC., Boca Raton, FL, 2003.

Staff: "ASM Handbook—Properties and Selection: Nonferrous Alloys and Pure Metals," ASM International, Materials Park, OH, 1990. http://www.asm_intl.org/

CHROMIZING. Production of a high chromium content surface layer on iron and steel by heating at high temperatures in a solid packing material containing chromium powder, or in an atmosphere containing chromium chloride. The surface layer is formed by diffusion of chromium into the iron in the same manner as carbon diffuses into iron in carburizing; however, the process is much slower and requires higher temperatures than carburizing. A similar result can be obtained by high-temperature diffusion of electrodeposited chromium. Chromized coatings have corrosion resistance and elevated temperature oxidation resistance similar to the high chromium types of stainless steels.

CHROMOGENIC COUPLERS. Couplers are used in secondary color development. See also **Dyes**. The term "coupler" is applied to a large number of organic compounds that combine with a limited number of chromogenic developers to produce dye images with a wide range of color and intensity. Couplers are dye intermediates but differ from chromogenic developing agents in that they do not as a rule have the ability to develop a silver image.

Couplers may be hydroxy or amine derivatives of aromatic compounds, such as benzene, naphthalene or anthracene. Phenol, napthol, aniline, cresol, paraaminophenol and dimethylparaphenylenediamine are examples. With these compounds coupling takes place with the hydrogen atom which is in the ortho or para position to the hydroxy or amino group on the coupler.

Compounds having active methylene, $=CH_2$ groups, with strong polar groups for other valences as the cyano $-CN$, the carbonyl $=CO$, the aceto CH_3CO-, the acid ester $-COOC_2H_5$, and the phenyl, $-C_6H_5$ groups will couple. Acetoacetic ester and paranitrophenylacetonitrile are illustrations.

The methylene group may be part of a ring structure as in coumarine and indoxyl, or may be part of a heterocyclic ring attached to a phenyl group as in 1-phenyl-3-methyl-5-pyrazolone.

Compounds having an N in the ring will couple if there is a methyl $-CH_3$ group, attached to the ring in the alpha position to the nitrogen. Two illustrations for this type are picoline and 2-methylthiazole.

Because secondary color development has been particularly successful in the development of color materials, and because of the ease by which it is possible to control color by this method, the field of couplers has expanded rapidly.

CHROMOPHORE. Certain groups of atoms in an organic compound cause characteristic absorption of radiation irrespective of the nature of the rest of the compound. Such groups are called chromophores or color carriers.

CHROMOPHORIC ELECTRONS. Electrons in the double bonds of the chromophore groups. Such electrons are not bound as tightly as those of single bonds and can thus be transferred into higher energy levels with less expenditure of energy. Their electronic spectra appear at frequencies in the visible or near ultraviolet region of the spectrum.

CHROMOPLASTS. See **Plastids**.

CHRYSOBERYL. The mineral chrysoberyl, an aluminate of beryllium corresponds to the formula $BeAl_2O_4$, crystallizes in the orthorhombic system with both contact and penetration twins common, often repeated resulting in rosetted structures. Hardness, 8.5; specific gravity, 3.75; luster vitreous; color various shades of green sometimes yellow. A variety which is red by transmitted light is known as alexandrite. Streak colorless; transparent to translucent, occasionally opalescent. Chrysoberyl also is known as *cymophane* and *golden beryl*.

Chrysoberyl occurs in granitic rocks, pegmatites and mica schists; often is found in alluvial deposits. The Ural Mountains yield alexandrite. Other localities for chrysoberyl are Czechoslovakia; Ceylon; Rhodesia; Brazil; and Madagascar, where it occurs of gem quality in the pegmatites of that island. In the United States it is found in Maine, Connecticut, and New York. The word chrysoberyl is derived from the Greek words meaning golden and beryl. Cymophane has its derivation also from the Greek words meaning wave and appearance, in reference to the opalescence exhibited at times.

CHRYSOCOLLA. This mineral, a hydrous silicate of copper probably corresponding to the formula $Cu_2H_2Si_2O_5(OH)_4$, is perhaps a mineral gel, for it usually appears as an amorphous mass, in veins, or as incrustations. Common occurrence as massive cryptocrystalline character, possibly orthorhombic; extremely rare as small acicular crystals.

Chrysocolla is generally some shade of blue or green but if impure may be brown or black. It has a characteristic conchoidal fracture; hardness, 2–4; sp gr, 2.24; vitreous to dull luster; translucent to opaque.

Chrysocolla is a secondary mineral and associated commonly with other copper minerals of similar origin. It is one of the less important ores of copper and has a minor use as a gemstone. Among the localities for excellent specimens may be mentioned Cornwall and Cumberland, England; Congo; Chile; Lebanon and Berks Counties, Pennsylvania; the Clifton-Morenci Globe and Bisbee districts in Arizona; Dona Ana County, New Mexico, and the Tintic district, Utah.

The word chrysocolla is derived from the Greek words meaning gold and glue, formerly the name for gold solder.

CHRYSOTILE. A delicately fibrous variety of serpentine which separates easily into silky, flexible fibers of greenish or yellowish color, with formula $Mg_3Si_2O_5(OH)_4$. It crystallizes in the monoclinic system; hardness, 2.5; sp gr, 2.55. Its name is derived from the Greek words meaning gold and fibrous. Most of the common asbestos of commerce is chrysotile. It is mined in Thetford, Province of Quebec, and in the Republic of South Africa. See also **Asbestos**; and **Serpentine**.

CINNABAR. The mineral cinnabar, mercuric sulfide (HgS) occurs in small and often highly modified hexagonal crystals, usually of rhombohedral or tabular habit. It is found chiefly in crystalline crusts, granular or simply massive. The overture of cinnabar is subconchoidal; hardness, 2–2.5; specific gravity, 8–8.2; luster, adamantine tending toward metallic, sometimes dull. This mineral has a characteristic cochineal-red color which, however, may be brownish at times, occasionally dull lead gray. The streak is scarlet; it is transparent to opaque.

Cinnabar occurs in veins or may be in masses in shales, slates, limestones and similar rocks due to the impregnation by mineral-bearing solutions or as replacements. The former U.S.S.R., former Czechoslovakia, Bohemia, Bavaria, Italy and Spain have furnished excellent specimens. The most important of the world's mercury deposits is at Almaden in Spain. Italy, Peru, Surinam, China and Mexico have commercially valuable occurrences of cinnabar. In the United States this mineral is found in California (most important deposit), Nevada, Utah, Texas and Oregon. Cinnabar is the chief ore of mercury. Its name is supposed to be of Hindu origin.

CIS-COMPOUND. See **Isomerism**.

CITRIC ACID. [CAS: 77-92-9]. $C_3H_4(OH)(COOH)_3$, formula weight 192.12, white crystalline solid, mp 153°C, decomposes at higher temperatures, sp gr 1.542. Citric acid is soluble in H_2O or alcohol and slightly soluble in ether. The compound is a tribasic acid, forming mono-, di-, and tri- series of salts and esters. Citric acid may be obtained: (1) from some natural products, e.g., the free acid in the juice of citrus and acidic fruits, often in conjunction with malic or tartaric acid; the juice of unripe lemons

(approximately 6% citric acid) is a commercial source; (2) by fermentation of glucose (blackstrap molasses is a major source); and (3) by synthesis.

Citric acid and sodium citrate have found use as additives in effervescent beverages and medicinal salts, although excessive quantities are considered toxic. Citric acid can be used as an effective antioxidant. Because the acid is not readily soluble in fats, it is added to formulations that improve solubility, such as propylene glycol and butylated hydroxy anisol, and this can be used as a stabilizer for tallow, fats, and greases. Citric acid also is used for adjusting pH in certain electroplating baths. It finds a number of miscellaneous uses in etching, textile dyeing and printing operations.

Citrates (like tartrates) in solution change silver of ammonio-silver nitrate into metallic silver. Calcium citrate, due to its solubility characteristics, is of importance in the separation and recovery of citric acid. Calcium citrate plus dilute H_2SO_4 yields citric acid plus calcium sulfate, and the latter may be separated by filtration. Citric acid may be obtained by evaporation of the filtrate.

In most living organisms, the citric acid cycle constitutes the final common pathway in the degradation of foodstuffs and cell constituents to carbon dioxide and water. This cycle is described in the entry on **Carbohydrates**.

Derivatives

Salts. The trisodium citrate salt is made by dissolving citric acid in water at a concentration of 50% w/w or higher. A 50% solution of sodium hydroxide is carefully added to pH 8.0–8.5. The reaction is exothermic and cooling is necessary to prevent boiling. The tripotassium salt of citric acid is made in a similar manner using potassium hydroxide. The product crystallizes as the monohydrate. Ammonium salts of citric acid are made by adding either aqueous or anhydrous ammonia to citric acid dissolved in water. They are usually used in the liquid form rather than isolated as a dry product.

Esters. The significant esters of citric acid are trimethyl citrate, triethyl citrate, tributyl citrate, and acetylated triethyl- and tributyl citrate. Many other esters are available but have not been used on a commercial scale. Citric acid esters are made under azeotropic conditions with a solvent, a catalyst, and the appropriate alcohol.

Catalysts used are usually acids such as sulfuric acid, p-toluenesulfonic acid, sulfonic acid ion-exchange resins, and others. The water from the reaction of the citric acid and the alcohol is continuously removed as the azeotrope until no more water is formed. At this point, the reaction is usually complete and the solvent and any excess alcohol is distilled off under mild vacuum. The catalyst is neutralized using carbonate or sodium hydroxide, leaving a crude product. If a pure product is desired, the ester can be distilled under high vacuum.

Citric acid esters are used as plasticizers in plastics such as poly(vinyl chloride), poly(vinylidene chloride), poly(vinyl acetate), poly(vinyl butyral), polypropylene, chlorinated rubber, ethylcellulose, and cellulose nitrate. Most citrate esters are nontoxic and are acceptable by the FDA for use in food-contact packaging and for flavor in certain foods. As a plasticizer, citrate esters provide good heat and light stability and excellent flexibility at low temperatures.

CITRIC ACID CYCLE. See **Carbohydrates**.

CITRINE. The mineral citrine is a yellow variety of quartz sometimes used as a gem. It is often marketed under the name topaz and may mislead the unwary. Brazil and Madagascar have furnished material of excellent quality. See also **Quartz**.

CLARIFYING AGENTS. Chemical substances used in connection with the purification of various solutions and liquors that occur during the processing of raw materials to final end-products. These agents operate in connection with mechanical equipment to bring about the removal of suspended particles that represent product impurities. Allowing such particles to settle by gravity alone would require very long periods. Clarification is part of the total process of sedimentation, which may be defined as the removal of solid particles from a liquid stream by gravitational force. The operation is effected by slowing the velocity of a feed stream in a large-volume tank so that gravitation settling can occur. Sedimentation is divided into two functions: (1) thickening, where the primary purpose is to increase the concentration of suspended solids of the feed stream, i.e., to remove liquids (this is largely a mechanical operation, assisted sometimes by clarifying agents); (2) clarification, where the purpose is to remove fine-sized particles and produce a clear effluent, i.e., to remove solids. Some equipment does both and the dividing line between thickening and clarification is not always sharp. See also **Classifying (Process)**. A clarifying agent to assist in this operation must possess certain properties for acting on the suspended particles—chemical precipitation, attractive via ionic forces, absorption qualities (large surface areas plus weak forces).

In the sugar refining industry, for example, various soluble nonsugar compounds are present in sugar juices as the result of rupturing plant cells (either from pressing sugar cane stalks or by extraction of the sliced root of sugar beets). Lime is commonly used to precipitate impurities, followed by carbonation of the solution with carbon dioxide to remove residual lime as calcium carbonate. Where filtration is used, various filter aids, such as fuller's earth, may be used. Filtering-type centrifuges also may be used. Phosphates, frequently in the form of orthophosphoric acid, may be used to precipitate the calcium. Some authorities suggest the use of polyphosphates, such as superphosphate and pyrophosphate, along with lime in the clarification operation. Phosphates assist in regulating the pH for optimal precipitation of calcium.

Tannin is used as a clarifying agent in wine making. Polyvinylpyrrolidone (PVP) has been used as a clarifying agent in the food industry. Ion exchange processes are also used in various processes, along with or in lieu of conventional clarification. For example, in ion exchange, the demineralization of sugar solutions can be effected; iron can be removed from wine by substitution with hydrogen ions.

Generally, wherever practical and economical, food processors prefer to accomplish clarification without the aid of chemicals—because all or but a trace of these chemicals must be removed so that they do not reappear in the final food product.

CLASSIFYING (Process). An operation or series of operations designed to separate a mixture of substances of various sizes and specific gravities into two or more categories, the cuts or divisions being made both with reference to size and to specific gravity (density). These operations find wide application in mining, metallurgical, water and sewage treatment, as well as use in other fields, such as the chemical industry.

Flotation. This is a means of separating a relatively small particle from a liquid medium. The particle may have a specific gravity greater than, less than, or the same as the liquid from which it is floated. There are two fundamental requirements: (1) a gas bubble and particle must come in contract with each other; and (2) the particle should have an *affinity* for attaching itself to the bubble.

To achieve the first objective, various methods of bubble production and particle agitation have been used. Since the invention of flotation by Haynes in 1860, two basic methods have emerged:

(a) *Dissolved gas-impeller agitation*, wherein gas under pressure is sparged into the bottom of a vessel in which an impeller mixes the rising bubbles with the agitated particles.

(b) *Self-induced gas-impeller agitation*, wherein the impeller is so positioned in the liquid that it inspirates ambient gas into the liquid as bubbles. These bubbles are brought into contact with the agitated particle at the impeller's most dynamic zone. This method (see Fig. 1) has become the most accepted method.

The second requirement has been served by the development of a number of chemical reagents, which fall into five basic categories: (1) collection; (2) conditioning; (3) levitation; (4) frothing; and (5) depressant.

Frothers are chemicals whose molecules contain both a polar and a nonpolar group. The purpose of a froth is to carry mineral-laden bubbles for a period of time until the froth can be removed from the flotation machine for recovery of its mineral content. Typical frothing chemicals are alcohols, cresylic acids, eucalyptus oils, camphor oils, and pine oils, all of which are slightly soluble in water. Soluble frothers in common use include alkyl ethers and phenyl ethers of propylene and polypropylene glycols.

Collectors are chemical reagents that selectively coat the particles to be floated with a water-repellant surface that will adhere to air bubbles. Collectors generally are classified as cationic, anionic, or nonionic. Examples of collectors include the xanthates, dithiophosphates, thiocarbonilides, and thionocarbonates, all of which are anionic collectors

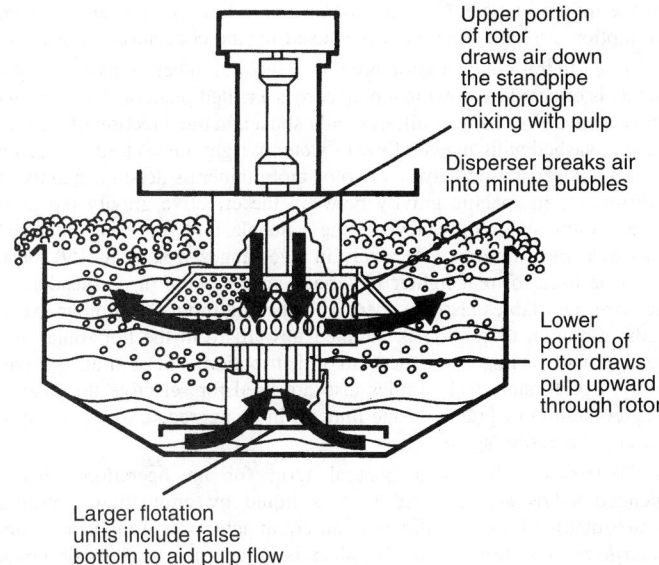

Fig. 1. Flotation cell. Upper portion of rotor draws air down the standpipe for thorough mixing with pulp. Lower portion of rotor draws pulp upward through rotor. Disperser breaks air into minute bubbles. Larger flotation units include false bottom to aid pulp flow

for sulfides, a major need in ore processing. Fatty acids and soaps are anionic collectors and serve for nonsulfides. Amine salts are cationic collectors for nonsulfides.

Depressants are mainly inorganic salts, which compete with the collector for position on the sulfide surface. This permits the separation of one sulfide mineral from another. In one case, for example, in an alkaline solution, the addition of sodium cyanide prevents flotation of sphalerite and pyrite by xanthates, but not of galena, thus producing a higher grade of galena concentrates. The cyanide solution does not permanently affect the floatability of sphalerite as it can be floated by adding cupric sulfate and xanthate.

Activators are chemical reagents, which alter the surface of a sulfide so that it can absorb a collector and float. Cupric sulfate is the most widely used activator. For example, xanthate as a collector will not readily float sphalerite, but the addition of cupric sulfate to the pulp changes the surface of the sphalerite particles to copper sulfide. Xanthate then will readily float the activated sphalerite as it behaves similarly to copper sulfide.

Although flotation was developed as a separation process for mineral processing and applies to the sulfides of copper, lead, zinc, iron, molybdenum, cobalt, nickel, and arsenic and to nonsulfides, such as phosphates, sodium chloride, potassium chloride, iron oxides, limestone, feldspar, fluorite, chromite, tungstates, silica, coal, and rhodochrosite, flotation also applies to nonmineral separations. Flotation is used in the water disposal field, particularly in connection with petroleum waste water cleanup.

Dense-media Separation. This operation is useful for the separation of solid particles of different densities. A liquid suspension of finely divided high-gravity solids is prepared. Ores of different densities, when exposed to such a suspension will tend to separate by rising or settling in the liquid suspension. Numerous types of solids have been used to obtain a high-gravity medium, but the magnetic solids (ferro-silicon and magnetite) are most frequently used. These solids, alone or in combination, can provide a suitable dense medium over a gravity range of 1.25 to 3.40. Dense-media separation is applicable to any ore in which the valuable component has an appreciable gravity difference from the gangue components. In coarse-ore heavy-media separation plants, the limiting bottom size of dense-medium feed is 10 mesh and the upper size limit is 12 inches (0.3 meter). The magnetic particles of the dense medium subsequently are removed by a magnetic separator.

Jigging. In this operation, a pulsating stream of liquid flows through a bed of materials of different specific gravities, causing the heavy material to work down to the bottom of the bed and the lighter material to rise to the top. This is a very old operation used for concentrating heavy mineral from the lighter gangue. Construction costs are low, but power and water consumption are high. The process is used for the concentration of coal.

Tabling. In this concentration process, a separation between two or more minerals is effected by flowing a pulp across a riffled plane surface inclined slightly from the horizontal, differentially shaken in the direction of the long axis, and washed with an even flow of water at right angles to the direction of motion. A separation between two or more minerals depends mainly on the difference in specific gravity between the effective gravity (sp gr of mineral minus sp gr of water) of the valuable and the waste material. Tables treat metallic ores effectively in size ranges from 6 to 150 mesh, but can be used to treat lighter materials, such as coal of a considerably larger size. Dry tables also are used. The shaking motion is similar except that the direction of motion is inclined upward from the horizontal and, instead of water acting as the medium of distribution, a blast of air is driven through a perforated deck. Tables also are used for selective flocculation or agglomeration of grains of one mineral in an aggregate by the addition of an agglomerating agent.

Sedimentation. This is a general term for an operation wherein suspended solids are removed from a liquid by gravitational settling. The two major forms of sedimentation equipment are (1) thickeners, and (2) clarifiers. The term *decanting* also is sometimes used to designate sedimentation.

Thickeners. The primary objective of thickening is to increase the concentration of the feedstream. The mechanical continuous thickener, equipped with sludge-raking arms, is the most common type. Usually the operation is performed in cylindrical tanks. The sludge collection system and the removal system are designed to move the settled material continuously across the tank floor to a discharge point. Feed enters through a central feed well designed to distribute around the periphery. Thickened sludge, raked toward the center by a slowing revolving mechanism, enters a central collecting trough or cone and is discharged through a spigot or removed by a sludge pump. See Figs. 2 and 3.

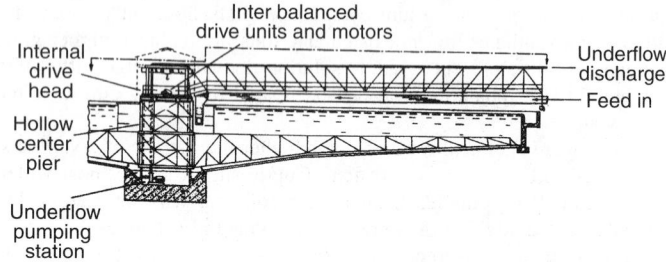

Fig. 2.   Caisson-type thickener

Fig. 3.   Center-pier type thickener, showing pumps and access

Fig. 4.   Activated sludge final clarifier

Clarifiers. The primary objective of clarifying is to free solids from a relatively dilute stream. These units operate on the basis of gravity sedimentation and utilize a raking mechanism, as in a thickener, Frequently, clarifiers are operated in conjunction with flocculation equipment, which employs chemical coagulants, such as alum, iron salts, lime, polyelectrolytes, activated silica sol, and other chemical reagents. Clarification essentially expedites the natural gravity settling process. An activated sludge final clarifier is shown in Fig. 4.

See also **Clarifying Agents**.

CLATHRATE.   See **Compound (Chemical)**.

CLAUDE PROCESS.   See **Ammonia**.

CLAYS.   The terms *clay* or *clays* commonly refer to either rocks that are consolidated or unconsolidated sediments, or a group of minerals having unique properties. Traditionally, clays (rocks) are distinctive in at least two properties that render them technologically useful: plasticity and composition. Clays are predominantly composed of hydrous phyllosilicates, referred to as clay minerals. These are hydrous silicates of Al, Mg, K, and Fe, and other less abundant elements. Clay minerals are extremely fine crystals or particles, often colloidal in size and usually plate-like in shape. The nonclay mineral portion of clays (rocks) may consist of other minerals, portions of rocks, and organic compounds.

The very fine particles yield very large specific surface areas that are physically sorptive and chemically surface reactive. Many clay mineral crystals carry an excess negative electric charge owing to internal substitution by lower valent cations, and thereby increase internal reactivity in chemical combination and ion exchange. Clays, which may have served as substrates selectively absorbing and catalyzing amino acids in the origin of life, apparently catalyze petroleum formation in rocks. See also **Petroleum**.

A clay deposit usually contains nonclay-like minerals as impurities and these impurities may actually be essential in determining the unique and specially desired properties of the clay. Both crystalline and amorphous minerals and compounds may be present in a clay deposit.

A broad definition of clays includes the following properties: *(1)* Crystalline hydrated silicates of aluminum, iron, and magnesium comprise the majority of clay minerals; however, amorphous hydrated aluminum compounds are also included. Distinctions among clay minerals are made by chemical and structural parameters. The chemical variations range from kaolinite, which is relatively uniform in chemical composition, to smectite minerals, which vary widely in chemical composition, base exchange properties, and expanding crystal lattice. Clay minerals are excellent examples of mixed layering, both random and regular, in layer-structure silicates. These mixed-layer clays are among the most ubiquitous of the various clay minerals. The structural differences among clay minerals are related to the arrangement of tetrahedral (T) and octahedral (O) layers, and the manner in which electrostatic charge imbalances, created by chemical substitution, are neutralized. Figure 1 shows several examples. *(2)* The possible content of hydrated alumina and iron. *(3)* The extreme fineness of individual clay particles, which may be of colloidal size in at least one dimension. *(4)* The property of thixotropy in various degrees of complexity. *(5)* The possible content of quartz, SiO_2, sand and silt, feldspars, mica, chlorite, opal, volcanic dust, fossil fragments, high density so-called heavy

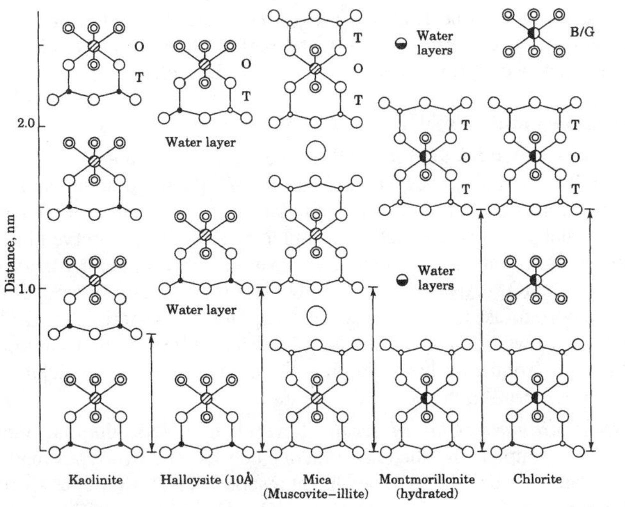

Fig. 1. Diagrammatic representation of the succession of layers in some layer lattice silicates (12) where O is oxygen; ⊚, hydroxyl; •, silicon; Si–Al; ⊘, aluminum; ◑, Al–Mg; ○, potassium; ◒, Na–Ca. Sample layers are designated as O, octahedral; T, tetrahedral; and B/G, brucite- or gibbsitelike. The distance depicted by arrows between repeating layers in nm are 0.72, kaolinite; 1.01, halloysite (10 Å); 1.00, mica; ca 1.5, montmorillonite; and 1.41, chlorite

minerals, sulfates, sulfides, carbonate minerals, zeolites, and many other rock and mineral particles ranging upward in size from colloids to pebbles.

Geology and Occurrence

Clays may originate through several processes: (1) hydrolysis and hydration of a silicate, i.e., alkali silicate + water → hydrated aluminosilicate clay + alkali hydroxide; (2) solution of a limestone or other soluble rock containing relatively insoluble clay impurities that are left behind; (3) slaking and weathering of shales (clay-rich sedimentary rocks); (4) replacement of a preexisting host rock by invading guest clay where the constituents are carried in part or wholly by solution; (5) deposition of clay in cavities or veins from solution; (6) bacterial and other organic activity, including the extraction of metal cations as nutrients by plants; (7) action of acid clays, humus, and inorganic acids on primary silicates; (8) alteration of parent material or diagenetic processes following sedimentation in marine and freshwater environments; and (9) resilication of high alumina minerals.

Clays or shales that may be utilized in the manufacture of bricks, tiles, and other heavy clay products exist in every state in the United States. Glacial clays, as unassorted glacial till or secondarily deposited melt water are abundant in the United States north of the Missouri and Ohio Rivers. Fire clays are those that resist fusion at a relatively high temperature, usually around 1600°C. Missouri, Pennsylvania, Ohio, Kentucky, Georgia, Colorado, New Jersey, Texas, Arkansas, Illinois, and Maryland are large producers of fire clays. Adobe, a calcareous, sandy to silty clay used extensively for making sundried brick, is available in the more arid southwestern states. Slip clay for glazing pottery is produced near Albany, New York.

Bentonite, widely distributed geographically and geologically, also varies widely in properties.

Each continent has clays of almost every type; however, certain deposits are outstanding.

The commercial value of a clay deposit depends on market trends, competitive materials, transportation facilities, new machinery and processes, and labor and fuel costs. Naturally exposed outcrops, geological area and structure maps, aerial photographs, hand and power auger drills, core drills, earth resistivity, and shallow seismic methods are used in exploration for clays. Clays are mined primarily by open-pit operation, including hydraulic extraction; however, underground mining is also practiced.

Mineralogy

The development of apparatus and techniques, such as x-ray diffraction, contributed greatly to research on clay minerals. Crystalline clay minerals are identified and classified primarily on the basis of crystal structure and the amount and locations of charge (deficit or excess) with respect to the basic lattice. Amorphous (to x-ray) clay minerals are poorly organized analogues of crystalline counterparts.

The structural variations among the clay minerals can be understood by considering various physical combinations of tetrahedral and octahedral sheets and the electrostatic effect chemical substitution has on the structural units. The tetrahedral sheets are composed primarily of Si^{4+} and oxygen, but minor amounts of Al^{3+} or Fe^{3+} may substitute for Si^{4+}. The substitution of M^{3+} for Si^{4+} leaves the tetrahedral sheet negatively charged. The cations of the octahedral sheet are composed primarily of Al^{3+}, Fe^{3+}, Mg^{2+}, and Fe^{2+}, but all other transition elements, except Sc, may be included. The anions of the octahedral sheet are O^{2-}, OH^-, and F^-. The smallest unit of the octahedral sheet contains three octahedral having an ideal net charge of negative six, i.e., three O^{2-}. If the negative charge is balanced by two trivalent cations, the layer is referred to as a dioctahedral layer; if balanced by three bivalent cations, the layer is referred to as a trioctahedral layer. Substitution of bivalent cations for trivalent cations, univalent cations (Li^+) for bivalent cations or unfilled octahedral sites, leaves the octahedral layer a net negative charge. The tetrahedral apical oxygen is shared with the octahedral layer to join the two types of layers.

The least complicated clay minerals are the 1:1 clay minerals composed of one tetrahedral (T) layer and one octahedral (O) layer.

Clay minerals that are composed of two tetrahedral layers and one octahedral layer are referred to as 2:1 clay minerals or TOT minerals.

The multitude of variation in clay minerals is caused by substitution in the octahedral and tetrahedral layers, resulting in charge deficits. The manner in which the charge deficit is balanced leads to many of the useful and unique properties of clay minerals.

Crystalline and Paracrystalline Groups

Kaolins. The kaolin minerals include kaolinite, dickite, and nacrite which all have composition $Al_2O_3 \cdot 2\ SiO_2 \cdot 2\ H_2O$; halloysite (7 Å), $Al_2O_3 \cdot 2\ SiO_2 \cdot 2\ H_2O$; and halloysite (10 Å), $Al_2O_3 \cdot 2\ SiO_2 \cdot 4\ H_2O$. The structural formulas for kaolinite and halloysite (10 Å), which are shown in Figure 1, are $Al_4Si_4O_{10}(OH)_8$ and $Al_4Si_4O_{10}(OH)_8 \cdot 4\ H_2O$, respectively. The so-called fire clay mineral is a b-axis disordered kaolinite; halloysite (7 Å) and halloysite (10 Å) are disordered along both the a- and b-axes. Indeed, most variations in the kaolin group originate as structural polymorphs, related to variations in layer stacks.

Halloysite, a mineral in the kaolin family, has a chemical composition similar to, but physical properties that differ greatly from, kaolinite. Halloysite differs from kaolinite in tetrahedral Al content, layer stacking sequence, and configuration of the six-fold rings. Four basic morphologies of halloysite are recognized: tubular (long and short), spheroidal, platy, and prismatic.

Kaolin most commonly originates by the alteration of feldspar or other aluminum silicates via an intermediate solution phase, usually by surface weathering or by rising warm (hydrothermal) waters.

Large deposits of relatively pure kaolinite have developed from parent, feldspar-rich pegmatites, whereas others are secondarily deposited in sedimentary beds after transportation.

The textures of kaolin (rock) include varieties similar to examples observed in igneous and metamorphic as well as sedimentary rocks. Kaolin grains and crystals may be straight or curved, sheaves, flakes, face-to-face or edge-to-edge floccules, interlocking crystals, tubes, scrolls, fibers, or spheres.

Serpentines. Substituting 3 Mg^{2+} for the 2 Al^{3+} in the kaolin structure results in the serpentine minerals, $Mg_3Si_2O_5(OH)_4$. In serpentines all three possible octahedral cation sites are filled. Most serpentine minerals are tubular to fibrous in structure presumably because of misfit between Mg octahedral and tetrahedral layers.

Talc and Pyrophyllite. Talc and pyrophyllite are 2:1 layer clay minerals having no substitution in either the tetrahedral or octahedral layer. These are electrostatically neutral particles ($x = 0$) and may be considered ideal 2:1 layer hydrous phyllosilicates. Talc and pyrophyllite are found in metamorphic rocks that are rich in Mg and Al, respectively.

Smectites (Montmorillonites). Smectites are the 2:1 clay minerals that carry a lattice charge and characteristically expand when solvated with water and alcohols, notably ethylene glycol and glycerol.

Smectites are structurally similar to pyrophyllite or talc, but differ by substitutions mainly in the octahedral layers.

The minerals of the smectite group have been formed by surface weathering, low temperature hydrothermal processes, alteration of volcanic dust in stratified beds, action of circulating water of uncertain source among fractures and in veins, and laboratory synthesis.

Illite. Illite is a general term for the clay mineral constituents of argillaceous sediments that strongly resemble mica minerals. Other names that have been used for illite include bravaisite, degraded mica, hydromica, hydromuscovite, hydrous illite, hydrous mica, K-mica, micaceous clay, and sericite. Illite and the mica minerals have a 2:1 sheet structure similar to the smectite minerals except that the maximum charge deficit in mica is typically in the tetrahedral layers and contains potassium held tenaciously in the interlayer space, which contributes to a 1.0-nm basal spacing.

The formula of illite can be expressed as $2K_2O \cdot 3MO \cdot 8R_2O_3 \cdot 24SiO_2 \cdot 12H_2O$, and the crystal structure by the formula $K_y[Al_{4-x}(Fe, Mg)_x](Si_{(8-y)+x}Al_y)O_{20}(OH)_4$ where y refers to the K^+ ions that satisfy the excess charges resulting when about 15% of the Si^{4+} positions are replaced by Al^{3+}.

Illite was defined as the most abundant clay mineral in Paleozoic shale and is widespread in many other sedimentary rocks; it is common in soils, slates, certain alteration products of igneous rocks, and recent sediments. Its origin has been attributed to alteration of silicate minerals by weathering and hydrothermal solutions, reconstitution, wetting and drying of soil clays, and diagenesis involving other three-layer minerals and potassium during geologic time and pressure under deep burial. Illitization of smectite via illite–smectite mixed-layer intermediates is a very common and an important reaction in the formation of shales during burial diagenesis.

Glauconite. Glauconite is a green, dioctahedral, micaceous clay rich in ferric iron and potassium. The generally accepted formula for glauconite is $(Na, K)_{0.78}(Fe^{3+}_{1.01}Al_{0.45}\ Mg_{0.39}Fe^{2+})_{2.05}(Si_{3.65}Al_{0.35})O_{10}(OH)_2$. Glauconite has many characteristics common to illite, but much glauconite contains random mixed expanding layers, and can be referred to as interstratified glauconite–smectite minerals. In addition, glauconite found in Late Cenozoic rocks tends to have less crystallographic order than older glauconite; therefore, the modifiers ordered (well crystalline) and disordered (poorly crystalline) are commonly used.

Glauconite occurs abundantly in sand-size or bigger pellets, or in pellets within fossils, notably foraminifera, giving it an organic connotation.

Celadonite is an iron-rich dioctahedral micaceous mineral that is similar to glauconite. Celadonite has a composition of: $(Na, K)_{0.83}(Fe^{3+}_{0.72}Al_{0.49}\ Mg_{0.63}Fe_{2+0.20})_{2.05}(Si_{3.81}Al_{0.19})O_{10}(OH)_2$ (39) and, like glauconite, has well crystalline, poorly crystalline, and interstratified varieties.

Chlorite and Vermiculite. Chlorite is a 1.4-nm (14 Å) clay mineral that cannot be expanded or collapsed by traditional laboratory procedures. Structurally, the unit layer of chloride is composed of a 2:1 layer combined with a 0.4-nm Mg or Al interlayer or hydroxide sheet.

Palygorskite and Sepiolite. Palygorskite (attapulgite) and sepiolite are clay minerals in which the 2:1 layers are linked together in chain-like or a combination of chain-sheet structures.

Palygorskite and sepiolite are different from other clay minerals in the manner in which the 2:1 layers are joined. Rather than being joined in a continuous manner, the tetrahedral sheets are joined to an adjacent inverted tetrahedral layer, making the octahedral layers noncontinuous and leaving an open channel in the mineral structure.

Palygorskite has an ideal formula that approximates $MgAl_3Si_8O_{20}(OH)_3 (OH_2)_4 \cdot x[R^{2+}(H_2O)_4]$; the ideal formula for sepiolite is $Mg_8Si_{12}O_{30} (OH)_4(OH_2)_4 \cdot x[R_{2+}(H_2O)_8]$. The chemical composition of a specific sample may vary widely because there is substitution of Na, Fe, Mn, Al, and Ni in the octahedral sheets of sepiolite, and substitution of Na, Fe, and Mn in palygorskite.

These clays have distinctive uses and properties not shown by platy clay minerals.

Mixed-Layer Minerals. In addition to polymorphism resulting from the disordering and proxying of one element for another, clay minerals exhibit ordered and random intercalation sandwiches with one another.

Mixed-layer clays, particularly illite–smectite, are very common minerals and illustrate the transitional nature of the 2:1 layered silicates. The transition from smectite to illite occurs when smectite, in the presence of potassium from another mineral such as potassium feldspar, or from thermal fluids, is heated and/or buried. With increasing temperature smectite plus potassium is converted to illite.

The physical structure of mixed-layer minerals is open to question. In the traditional view, the MacEwan crystallite is a combination of 1.0-nm (10-Å) nonexpandable units (illite) that forms as an epitaxial growth on 1.7-nm

expandable units (smectite) that yield a coherent diffraction pattern. This view is challenged by the fundamental particle hypothesis which is based on the existence of fundamental particles of different thickness.

Amorphous and Miscellaneous Groups

Allophane and Imogolite. Allophane is an amorphous clay that is essentially an amorphous solid solution of silica, alumina, and water. Allophane has been found most abundantly in soils and altered volcanic ash. It usually occurs in spherical form but has also been observed in fibers.

Imogolite is an uncommon paracrystalline clay mineral assigned the formula $1.1SiO_2 \cdot Al_2O_3 \cdot 2.3-2.8H_2O$. The morphology of imogolite has been reported as thread-shaped and as hollow spheres. Imogolite is generally viewed as an intermediate between allophane and kaolinite. In modern environments both allophane and imogolite are associated with volcanic material in areas of high rainfall.

High Alumina Clay Minerals. Several hydrated alumina minerals should be grouped with the clay minerals because the two types may occur so intimately associated as to be almost inseparable. Diaspore (α-AlO(OH)) and boehmite (γ-AlO(OH)), both $Al_2O_3 \cdot H_2O(Al_2O_3, 85\%; H_2O, 15\%)$ are the chief constituents of diaspore clay, which may contain over 75% Al_2O_3 on the raw basis. Gibbsite, $Al_2O_3 \cdot 3 H_2O (Al_2O_3, 65.4\%; H_2O, 34.6\%)$, and cliachite, the so-called amorphous alumina hydrate (much cliachite is probably cryptocrystalline), as well as the monohydrates, occur in bauxite, bauxitic kaolin, and bauxitic clays.

The hydrated alumina minerals usually occur in oolitic structures (small spherical to ellipsoidal bodies the size of BB shot, about 2 mm in diameter) and also in larger and smaller structures. High alumina minerals are found where intense weathering and leaching has dissolved the silica. It is generally believed that a very humid, subtropical climate is required for this (lateritic) stage of weathering.

T. DOMBROWSKI
Engelhard Corporation

Additional Reading

Carr, D. D. Sr. ed.: *Clays: Industrial Minerals and Rocks*, 6th Edition, AIME, pp. 229–277.
Giese, R. F.: "Hydrous Phyllosilicates (exclusive of micas)," *Miner. Soc. Am. Rev. Mineral.* **19**, 29–62 (1988).
Grim, R. E.: *Clay Mineralogy*, McGraw-Hill Book Co., Inc., New York, NY, 1968.
Weaver, C. E. and L. D. Pollard: *The Chemistry of Clay Minerals*, Elsevier, New York, NY, 1973.

CLINICAL CHEMISTRY. A subdivision of chemistry that deals with the behavior and composition of all types of body fluids, including the blood, urine, perspiration, glandular secretions, etc. It involves analysis and testing of these for content of numerous metabolic constituents, as well as foreign materials; thus it also includes toxicological factors.

CLINOZOISITE. Clinozoisite, crystallizing in the monoclinic system, is a hydrous calcium aluminum silicate, $Ca_2Al_3Si_3O_{12}(OH)$. Crystals are usually of prismatic habit with striations parallel to the b-axis. May occur as granular or columnar masses. Hardness 6.5, sp gr 3.21–3.38, vitreous luster, transparent to translucent. Color gradations from gray through green to pink.

Clinozoisite occurs in crystalline schists, which are products of metamorphism from calcic feldspar-rich dark igneous rocks. Zoisite (orthorhombic) represents its dimorphous counterpart.

CLOUD POINT. (1) The temperature at which a solution becomes cloudy as it is cooled at a specified rate. The cloud point is an important property in the specification of lacquers, oils, and other industrial solutions.

(2) In petroleum technology, the temperature at which a waxy solid material appears as a diesel fuel is cooled. This material is harmful to engine performance. See also **Petroleum**.

COACERVATION. An important equilibrium state of colloidal or macromolecular systems. It may be defined as the partial miscibility of two or more optically isotropic liquids, at least one of which is in the colloidal state. For example, gum arabic shows the phenomenon of coacervation when mixed with gelatin. It also may be defined as the production, by coagulation of a hydrophilic sol, of a liquid phase, which often appears as viscous drops, instead of forming a continuous liquid phase. See also **Colloid Systems**.

COAGULATION. 1. In its general scientific usage this term has two closely related meanings: (1) The process of complete or partial solidification of a colloidal solution to a gelatinous mass; or of the separation from a liquid system of a gelatinous mass. It involves the separation of the disperse from the continuous phase which fact distinguishes it from "gelation." (2) The result of an alteration of a disperse phase or of a dissolved solid which causes the separation of the system into a liquid phase and an insoluble mass, as the coagulation of egg albumin. See also **Colloid Systems**.

2. In cloud physics, coagulation is generally used synonymously with accretion. Less frequently, it refers to any process by which a cloud's numerous small cloud drops are converted into a smaller number of larger precipitation particles. When so used, the term is employed in analogy to the coagulation of any colloidal state. (See 1 above.)

3. In biological science, the term coagulation has two somewhat more specific meanings: (1) The clotting of blood or lymph. (2) The changes produced in tissue of the application of increased temperatures or by certain chemicals. See also **Anticoagulants**; and **Blood**.

Coagulation value, is the concentration of a coagulant that effects a given amount of coagulation of a colloidal, or other dispersed system.

COAGULATION (Hofmeister Series). A definite order of arrangement of anions and cations according to their powers of coagulation when their salts are added in quantity to lyophilic sols. Thus, the order of cations is $Mg^{2+} > Ca^{2+} > Sr^{2+} > Ba^{2+} > Li^+ > Na^+ > K^+ > Rb^+ > Cs^+$. The Hofmeister series is also called the *lyotropic series*, and the effect is called salting-out, a term applied strictly to the effect of electrolytes upon true solutions.

COAL. Containing more than 50% (weight) and 70% (volume) of carbonaceous material, including inherent moisture, coal is a readily combustible rock. Coal was formed from the compaction and induration of variously altered plant remains similar to those found in peat. Coal was formed during earlier geological periods, the process of formation acting slowly over extremely long periods of time. Coal is not a uniform substance, but reflects the conditions of its formation. These include:

1. *Differences in the kinds of plant materials* from which the coal was derived account for different types of coal.
2. *Differences in the degree of metamorphism occurring* during the formation of coal determine the different ranks of coal.
3. *Differences in the range of impurity* in coal account for the *different grades of coal.*

The fermentation of vegetable matter under conditions of no air and abundant moisture where volatiles are retained, resulting in the formation of bitumens, such as peat and coal, is known as *bituminous fermentation*. The metamorphic transformation of bituminous coal into anthracite is known as *anthracitization*. *Coalification* is the alteration or metamorphism of plant material into coal; the biochemical process of diagenesis and the geochemical process of metamorphism in the formation of coal. The peat-to-anthracite theory of coal formation is described as a process in which the progressive ranks of coal are indicative of the degree of coalification and, by inference, of the relative geologic age of the deposit. Peat, as the initial stage of coalification, is of recent geological age. Lignite, as an intermediate stage, is usually Tertiary or Mesozoic, and bituminous coal and anthracite, as the more advanced stages of coalification, are usually Carboniferous.

Status of Coal as a Major Energy Source

In 1991 coal furnished 55% of the total fuel required to generate electricity in the United States. One ton of coal consumed by a power plant generates approximately 2000 kilowatt hours of electricity. Then current consumption of 772 million tons/year for electricity generation was expected to rise to close to 1.3 billion tons by the year 2030. Clean coal technologies will be mandated. Even after installation of considerable environmental correction equipment, coal remains the least expensive of the fossil fuels by a wide margin. The cost to generate a million Btu of energy for petroleum is $2.63; for natural gas, $1.18; and for coal, $0.77. U.S. recoverable coal

reserves are approximate at 268 billion tons and, at current rates, would not be exhausted until the year 2230. That provides technology scores of years to further develop and refine renewable energy sources. Refer to **Alphabetical Index** for other energy sources.

The other principal coal-using categories are coke plants, industrial chemical and transportation applications, and residential and commercial uses.

The capital costs of environmental controls for coal-burning electric utility plants rose from 5% of total plant costs in the early 1970s to well over 35% near 1990s. The operating costs of environmental controls have followed a similar pattern. New coal-consuming generating technologies, such as fluidized-bed combustion and gasification-combined cycle, integrate emission controls so that the plant has greater flexibility in meeting emissions standards with a wide variety of coal types and may reduce the total cost of using coal. Nevertheless, such advanced technologies consume large quantities of fuel and produce significant amounts of solid wastes. The pollution problems associated with coal combustion are covered in the article on **Pollution (Air)**, the restoration of strip-mined land is covered in this article.

As reported by the World Energy Council, the United States has about 15% of the world's estimated recoverable coal, more than any other country except China. See Fig. 1. These statistics illustrate why the United States is a major exporter of coal.

Major Types of Coal
The major coals may be defined as follows:

Anthracite Coal. Coal of the highest metamorphic rank, in which the fixed carbon content is between 92 and 98%. It is hard, black, and has a semimetallic luster and semiconchoidal fracture. Anthracite ignites with difficulty and burns with a short, blue flame and without smoke. Anthracite coal is also known as hard coal, stone coal, kilkenny coal, and black coal.

Semianthracite Coal. Coal having a fixed-carbon content of between 86 and 92%. It is between bituminous coal and anthracite coal in metamorphic rank, although its physical properties more closely resemble those of anthracite.

Semibituminous Coal. Coal that ranks between bituminous coal and semianthracite. It is harder and more brittle than bituminous coal, has a high fuel ratio and burns without smoke. Semibituminous coal is also known as metabituminous coal which is defined as containing 89–91.2% carbon, analyzed on a dry, ash-free basis. The term smokeless coal also is used.

Bituminous Coal. Coal that ranks between sub-bituminous coal and semibituminous coal and that contains 15–20% volatile matter. It is dark brown-to-black in color and burns with a smoky flame. Bituminous coal is the most abundant rank of coal and is commonly Carboniferous in age. The most common synonym is soft coal.

Sub-bituminous Coal. A black coal intermediate in rank between lignite and bituminous coals, or in some classifications, the equivalent of black lignite. It is distinguished from lignite by higher carbon content and lower moisture content.

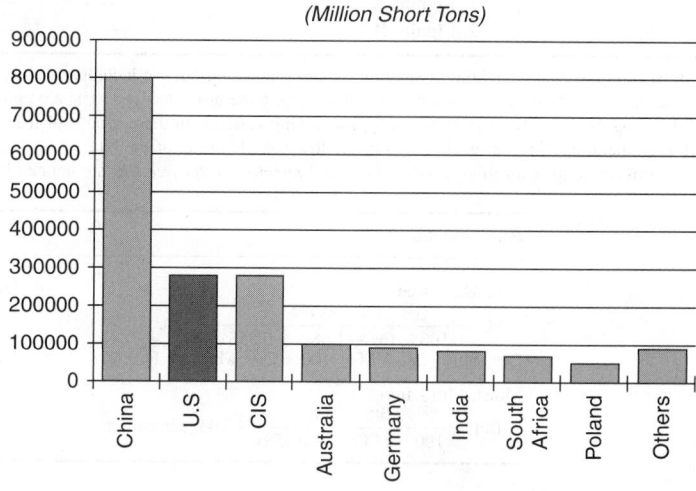

Fig. 1. World recoverable coal reserves (millions of short tons). (*World Energy Council*)

The sub-bituminous coals are further classified in terms of their calorific value:

Sub-bituminous A Coal—A type of sub-bituminous coal having 10,500 or more, but less than 13,000 Btu per pound (5838–7228 Calories/kg).
Sub-bituminous B Coal—A type of sub-bituminous coal having 9,500 or more, but less than 10,500 Btu per pound (5282–5838 Calories/kg).
Sub-bituminous C Coal—A type of sub-bituminous coal having 8,300 or more, but less than 9,500 Btu per pound (4615–5282 Calories/kg).

Lignite Coal. A brownish-black coal that is intermediate in coalification between peat and sub-bituminous coal; consolidated coal with a calorific value less than 8,300 Btu per pound (4615 Calories/kg), on a moist, mineral-matter-free basis. Synonyms include *brown lignite* and *brown coal*. Further classifications of lignite are made on the basis of calorific value:

Lignite A Coal—A lignite that contains 6,300 or more Btu per pound, but less than 8,300 Btu per pound (3503–4615 Calories/kg). Also known as *black lignite*.
Lignite B Coal—A lignite that contains less than 6,300 Btu per pound (3503 Calories/kg). Also known as brown lignite or brown coal.

Peat. This is an unconsolidated deposit of semicarbonized plant remains of a water-saturated environment, such as a bog or fen, and of persistently high moisture content (minimum of 75%). It is considered an early stage or rank in the development of coal. The carbon content is about 60%; oxygen content is about 30%. Structures of the vegetal matter can be seen. When dried, peat burns freely.

Peat Coal. This refers to two materials: (a) a coal transitional between peat and brown coal or lignite; and (b) an artificially carbonized peat that is used as a fuel.

Cannel Coal. A compact, tough *sapropelic coal* that contains spores and that is characterized by a dull-to-waxy luster, conchoidal fracture, and massiveness. It is attrital and high in volatiles. By American standards, it must contain less than 5% anthraxylon. Synonyms include *candle coal, kennel coal, cannel, cannelite, parrot coal*, and *curley cannel*. A sapropelic coal is derived from organic residues (finely divided plant material, spores, algae, etc.) in stagnant or standing bodies of water. Putrefaction is under anaerobic conditions rather than by peatification.

Ranks of Coal

Coals are classified in order to identify end-use and also to provide data useful in specifying and selecting burning and handling equipment and in the design and arrangement of heat-transfer surfaces. One classification of coal is by rank, that is, according to the degree of metamorphism, or progressive alteration, in the natural series from lignite to anthracite. Volatile matter, fixed carbon, inherent or bed moisture (equilibrated moisture at 30°C and 97% humidity), and oxygen are all indicative of rank, but no one item completely defines it. The classification of the American Society for Testing and Materials (ASTM) uses fixed carbon and calorific values, calculated on a mineral-matter-free basis, as the classifying criteria.

In establishing the rank of coals, it is necessary to use information showing an appreciable and systematic variation with age. For the older coals, a good criterion is the "dry, mineral-matter-free fixed carbon or volatile." However, this value is not suitable for designating the rank of the more recent, younger coals. A dependable means of classifying the

TABLE 1. CLASSIFICATION OF COALS BY RANK (ASTM)

Class	Group	Fixed Carbon Limits, % (Dry, Mineral-Matter-Free Basis) Equal or Greater Than	Less Than	Volatile Matter Limits, % (Dry, Mineral-Matter-Free Basis) Greater Than	Equal or Greater Than	Calorific Value Limits, Btu/Pound (Moist*, Mineral-Matter-Free Basis) Equal or Greater Than	Less Than	Agglomerating Character
I Anthracite	1. Meta-anthracite	98	—	—	2	—	—	Nonagglomerating
	2. Anthracite	92	98	2	8	—	—	Nonagglomerating
	3. Semianthracite[c]	86	92	8	14	—	—	Nonagglomerating
II Bituminous	1. Low volatile bituminous	78	86	14	22	—	—	
	2. Medium volatile bituminous	69	78	22	31	—	—	
	3. High volatile A bituminous	—	69	31	—	14,000[a]	—	Commonly Agglomerating[b]
	4. High volatile B bituminous	—	—	—	—	13,000[a]	14,000	
	5. High volatile C bituminous	—	—	—	—	11,500	13,000	
						10,500[b]	11,500	Agglomerating
III Subbituminous	1. Subbituminous A	—	—	—	—	10,500	11,500	
	2. Subbituminous B	—	—	—	—	9,500	10,500	Nonagglomerating
	3. Subbituminous C	—	—	—	—	8,300	9,500	
IV Lignitic	1. Lignite A	—	—	—	—	6,300	8,300	
	2. Lignite B	—	—	—	—	—	6,300	Nonagglomerating

* Moist refers to coal containing its natural inherent moisture, but not including visible water on the surface of the coal.
[a] Coals having 69% or more fixed carbon on the dry, mineral-matter-free basis are classified according to fixed carbob, regardless of calorific value.
[b] It is recognized that there may be nonagglomerating varieties in these groups of the bituminous class, and there are notable exceptions in high-volatile c bituminous group.
[c] If agglomerating, the coal is classified in the low-volatile group of the bituminous class.

The terms, *mineral-matter-free fixed carbon*; and *mineral-matter-free Btu* are defined by the following formulas:

Parr formulas	Approximation formulas
Dry, Mm - free $$FC = \frac{FC - 0.15S}{100 - (M + 1.08A + 0.55S)} \times 100, \%$$	Dry, Mm - free $$FC = \frac{FC}{100 - (M + 1.1A + 0.1S)} \times 100, \%$$
Dry, Mm - free $VM = 100 \times$ Dry, Mm-free FC, %	Dry, Mm-free VM = 100 − Dry, Mm-free FC, %
Moist, Mm - free $$Btu = \frac{Btu - 50S}{100 - (1.08A + 0.55S)} \times 100,\text{ per pound}$$	Moist, Mm - free $$Btu = \frac{Btu - 50S}{100 - (1.1A + 0.1S)} \times 100,\text{ per pound}$$

Symbols Used: Mm = mineral matter; Btu = heating value per pound; FC = fixed carbon,%; VM = volatile matter, %; M = bed moisture,%; A = ash, %; S = sulfur, %. all for coal on a moist basis.
Conversion Factor: 1 Btu/pound = 0.556 Calories/kg.

latter is the "moist, mineral-matter-free Btu" which varies little for the older coals, but appreciably and systematically for younger coals.

Classification of major coals according to rank or age is given in Table 1. The criteria given in the prior paragraph are used in classifying the older and younger coals. Seventeen United States coals are arranged in order of the classification of Table 1 and presented in Table 2.

Classification of coals in Europe and other parts of the world differs somewhat from the American system. European classifications include: (1) the *International Classification of Hard Coals by Type*; and (2) the *International Classification of Brown Coals*. These systems were developed by a Classification Working Party established in 1949 by the Coal Committee of the Economic Commission for Europe. The term "hard coal" is defined as a coal with a clorific value of more than 10,260 Btu per pound (5705 Calories/kg) on the moist, ash-free basis. The term "brown coal" refers to a coal containing less than 10,260 Btu per pound (5705 Calories/kg). In European terminology, the term "type" is equivalent to rank in American coal classification terminology and the term "class" approximates the ASTM rank. Space does not permit a full comparison of the various systems. Reference to various ASTM publications is suggested.

Commercial Sizes of Coal

Anthracite Coal. Standard sizes for anthracite coal are indicated in Table 3. The broken, egg, stove, nut and pea sizes are largely used for hand-fired domestic units and gas producers. Buckwheat and rice are used in mechanical types of firing equipment.

Bituminous Coal. The sizes of bituminous coal are not well standardized, but the following sizings are commonly recognized:

(*Run of Mine*)—Coal that is shipped from the mine without screening. It is used for both domestic heating and commercial steam production.

(*Run of Mine—8-inch*)—This is run-of-mine coal with oversize lumps broken up, (8 inches = 20.3 centimeters.)

(*Lump—5-inch*)—This size will not go through a 5-inch round hole. It is used for hand-firing and domestic purposes. (5 inches = 12.7 centimeters.)

(*Egg—5 by 2-inch*)—This size goes through a 5-inch hole, but is retained on 2-inch round-hole screens. It is used for hand-firing, gas producers, and domestic firing. (5 × 2 inches = 12.7 × 5.1 centimeters.)

(*Nut—2 by 1¼-inch*)—This size is used for small industrial stokers, gas producers, and hand-firing. (2 × 1¼ inches = 5.1 × 3.2 centimeters.)

(*Stoker Coal—1¼ by ¾-inch*)—This size is largely used for small industrial stokers and domestic firing. (1¼ × ¾ inches = 3.2 × 1.9 centimeters.)

TABLE 3. ANTHRACITE COAL SIZES

Name Used in the Trade	Diameter of Hole			
	Will Pass Through		Will Not Pass Through	
	Inches	~Centimeters	Inches	~Centimeters
Broken	$4\frac{3}{8}$	11.1	$3\frac{1}{4}$ to 3	8.3 to 7.6
Egg	$3\frac{1}{4}$ to 3	8.3 to 7.6	$2\frac{7}{16}$	6.2
Stove	$2\frac{7}{16}$	6.2	$1\frac{5}{8}$	4.1
Nut	$1\frac{5}{8}$	4.1	$\frac{13}{16}$	2.1
Pea	$\frac{13}{16}$	2.1	$\frac{9}{16}$	1.4
Buckwheat	$\frac{9}{16}$	1.4	$\frac{5}{16}$	0.8
Rice	$\frac{5}{16}$	0.8	$\frac{3}{16}$	0.5

(*Slack—¾-inch and under*)—This is used for pulverizers, cyclone furnaces, and industrial stokers. (¾-inch = 1.9 centimeters.)

Geology of Coal

Coal is interspersed as individual beds within other types of sedimentary rock beds, including sandstones, limestones, clays, shales, and mixtures of these materials. The plant material that ultimately became coal deposits was accumulated in upland bogs, coastal or near-coastal swamps, or delta plains. It is envisioned that the conditions were somewhat similar to the conditions existing today in the Okefenokee Swamp in Georgia or the Everglades of Florida. These areas may have varied from a few acres to several hundreds of square miles (hectares/square kilometers). Hence, the variation in the occurrence of coal as we find it today.

For the geological processes (coalification) to convert such plant materials into coal, it was necessary that the original swamps be submerged—by way of rises in sea level or land subsidence. Probably many submersive actions occurred with intermittent deposition of calcareous materials deposited from water containing muds, sands, and slimes. As the result of a series of compactions, with varying depths of burial, heat, and pressure, and length of time, the progress of coalification also varied—from peat to lignite, to sub-bituminous coal, to bituminous coal, possibly to anthracite.

TABLE 2. REPRESENTATIVE UNITED STATES COALS ARRANGED IN ORDER OF ASTM CLASSIFICATION

Coal rank				Coal analysis, bed moisture basis						Rank	Rank
Class	Group	State	County	M	VM	FC	A	S	Btu	FC	Btu
I	1	Pennsylvania	Schuylkill	4.5	1.7	84.1	9.7	0.77	12,745	99.2	14,280
I	2	Pennsylvania	Lackawanna	2.5	6.2	79.4	11.9	0.60	12,925	94.1	14,880
I	3	Virginia	Montgomery	2.0	10.6	67.2	20.2	0.62	11,925	88.7	15,340
II	1	West Virginia	McDowell	1.0	16.6	77.3	5.1	0.74	14,715	82.8	15,600
II	1	Pennsylvania	Cambria	1.3	17.5	70.9	10.3	1.68	13,800	81.3	15,595
II	2	Pennsylvania	Somerset	1.5	20.8	67.5	10.2	1.68	13,720	77.5	15,485
II	2	Pennsylvania	Indiana	1.5	23.4	64.9	10.2	2.20	13,800	74.5	15,580
II	3	Pennsylvania	Westmoreland	1.5	30.7	56.6	11.2	1.82	13,325	65.8	15,230
II	3	Kentucky	Pike	2.5	36.7	57.5	3.3	0.70	14,480	61.3	15,040
II	3	Ohio	Belmont	3.6	40.0	47.3	9.1	4.00	12,850	55.4	14,380
II	4	Illinois	Williamson	5.8	36.2	46.3	11.7	2.70	11,910	57.3	13,710
II	4	Utah	Emery	5.2	38.2	50.2	6.4	0.90	12,600	57.3	13,560
II	5	Illinois	Vermilion	12.2	38.8	40.0	9.0	3.20	11,340	51.8	12,630
III	1	Montana	Musselshell	14.1	32.2	46.7	7.0	0.43	11,140	59.0	12,075
III	2	Wyoming	Sheridan	25.0	30.5	40.8	3.7	0.30	9,345	57.5	9,745
III	3	Wyoming	Campell	31.0	31.4	32.8	4.8	0.55	8,320	51.5	8,790
IV	1	North Dakota	Mercer	37.0	26.6	32.2	4.2	0.40	7,255	55.2	7,610

Note: Definition of coal rank is given in Table 1.
M = equilibrium moisture, %; VM = volatile matter, %; FC = fixed carbon,%; A = ash, %; S = sulfur, %; Btu = high heating value, Btu per pound; Rank FC = dry, mineral-matter-free fixed carbon,%; Rank Btu = moist, mineral-matter-free Btu per pound.
All calculations are per the Parr formula defined in Table 1.
Conversion Factor: 1 Btu = 0.2520 Calorie.

Fig. 2. Main coalfields in the United States. In the United States coal is truly a national resource that is present in 38 states, stretching from the East to West coasts. About half of these coal deposits are located in the western United States, including Alaska; 28% of the deposits are in the interior of the country and 22% occur in Appalachia. The western United States includes a huge deposit called "Wyodak," which currently is the leading source of coal production. The Powder River Basin in Wyoming and southeastern Montana is part of this deposit. This western coal combines with significant low-sulfur reserves in the East to provide an important source of low-sulfur coal for electric power plants. Coal "seams" or deposits in this basin average 70 feet (21.3 meters) in thickness, although some exceed 100 feet (33.6 meters). In the East the most important deposits are in the Appalachian Basin, an area encompassing 72,990 square miles (189,044 square kilometers). The lighter shading shows a large lignite deposit. (*National Coal Association*)

Major coalfields in the United States are shown in Fig. 2.

Coal Formation in the United States. It is postulated that coal formed in the U.S. during three major geological periods.

1. During the Pennsylvanian (Carboniferous) period which dates back approximately 300 million years. Deposits include the predominately bituminous coal beds in the Appalachian Province extending from Pennsylvania (including the anthracite beds in central Pennsylvania) into northeastern Alabama. Also the contiguous Eastern Interior Region of Illinois, southwestern Indiana, and western Kentucky; in the contiguous Western Interior Region of Iowa, Kansas, Missouri, northeastern Oklahoma and northwestern Arkansas; and in the separated central portion of Texas (excluding Texas lignite).
2. During the Cretaceous period which dates back approximately 100 million years. Deposits include the predominately bituminous and sub-bituminous coal beds in the Rocky Mountain Province, extending in large, separated regions from central Montana into northeastern Arizona and northwestern New Mexico.
3. During the Tertiary period which dates back approximately 65 million years. Deposits include the sub-bituminous coal and lignite beds in the Great Plains Province, which includes northeastern Wyoming, eastern Montana, western North Dakota, and northwestern North Dakota.

Coal beds form only a very small percentage of the total thicknesses of the overall sedimentary strata comprising the so-called "Coal Measures" in coal-bearing areas. The thicknesses of individual coal beds within the United States range from a few millimeters (horizon markers) to as much as 100 feet (30 meters) or more. The number of individual coal beds of commercial significance may range from less than 10 feet (3 meters) to over 100 feet (30 meters). The coal, however, is rarely found in full vertical sequence at any one particular spot, but usually is distributed unevenly in single beds or small groups of beds around the margin or within the interior of the generally basin-shaped areas of coal-bearing strata.

Depending on the desired or feasible rate of annual production, the amounts of coal reserves required to support a new mine designed for an economic life of 20 years or more may range from a comparatively few tons to 300 million tons or more. Such amounts are dependent upon the

coal-bed thickness and the ease of mining and particularly whether surface or underground mining techniques will be required. Thus, a given mining area may range from as little as one thousand acres to as much as 50 square miles (130 square kilometers).

A variety of detrimental irregularities may accompany or interrupt an otherwise orderly accumulation of plant material either during swamp growth or shortly thereafter. While many coal beds or portions of beds are relatively low in ash content, other beds or portions of beds may contain depositional admixtures of particles of mud or silt which were washed or blown into the swamp during plant growth. Where relatively abundant, these particles serve to increase the ash content of the eventual coal bed and thus to decrease its quality correspondingly. During periods of prolonged swamp flooding, layers of mud or silt may have been deposited on the preexisting plant accumulations. Such deposition then may have been followed by additional plant accumulation. Such layers of impurities between underlying and overlying accumulations of plant materials, eventually hardening into shales or silty shales, are called *partings*. These may range from knife-edge thickness to thicknesses of up to a foot ($\frac{1}{3}$ meter) or more. Such partings within a single coal bed decrease the quality of the coal as mined and impair the mining procedures, particularly when such partings have become pyritized.

Such partings are not always evenly deposited over large portions or the entire extent of a coal-forming swamp, but may become progressively thicker toward the source of the deposited material. The partings may be wedge-shaped, with the overlying plant material occurring increasingly higher above the underlying plant material and sometimes becoming increasingly thinner to the point of disappearance. Deposition of impurities in this manner results in splitting the total thickness of the coal bed into two or more diverging *benches*. This causes mining difficulties and sometimes a bench may be so thin as not to be economically recoverable.

In some coal beds, relatively flat, lenticular masses ranging up to several feet in diameter, composed of pyrite, calcite or siderite, were formed during plant growth. Such materials (concretions) may represent the eventual immediate roof of the coal bed. These concretions impede mining operations and cause a hazard because of their tendency to drop out of the roof unexpectedly during operation of the mine.

From initial deposition and burial under overlying sedimentary materials through succeeding geological periods, coal beds are continually subject to the action of ground water. Thus, some coal beds have developed a system of essentially vertical fractures—thin cracks, often filled with coatings of pyrite, calcite, kaolinite and other minerals deposited from ground water. Impurities from these veins lower the quality of the coal.

During the long periods since formation, many coal beds have been subject to folding and sharp deformation, resulting in specific dislocations or faults. Such shifting may range from a foot or two (less than one meter) to several hundred feet (meters) and even up to thousands of feet (meters) in linear extent. The coal and lignite fields in the Great Plains Province are relatively undisturbed. The coal-bearing strata in the Appalachian Province are relatively flat along their northwestern margin, but increase in intensity of relatively mild but significant folding toward the southeast at right angles to the regional northeast-southwest trend of the component coal fields. The coal beds in the various basins comprising the Rocky Mountain Province range from comparatively gentle slopes of but a few degrees over areas of broad extent to areas of similar extent with prevailing dips of up to 20 or 30°, along with a few areas of limited extent where the coal beds are highly deformed.

Exploration Techniques. The diamond core drill historically has been the most extensively used tool in coal exploration. Cores of coal, properly recovered, enable accurate seam descriptions and measurements; also provide material for chemical analysis. Geologging or electric logging, used for several years in the oil and gas fields, is now gaining acceptance in coal-exploration technology. The system involves hoisting a sensor up the length of a drill hole while electric pulses are transmitted through the hoist cable to a console in a truck or on the surface. Here instruments record the variations in properties of the rock strata as a function of hole depth. The electric curves usually run in coal exploration are resistivity and spontaneous potential. Radiometric or nuclear curves include gamma ray, neutron, and the density or gamma-gamma log.

The impurities in coal beds, present either as distinct partings or disseminated throughout, are composed of clay materials that have a high density and a high natural radioactivity relative to coal. Consequently, the gamma ray and density curves, invaluable for bed correlation and thickness determination, also can be used as semiquantitative indices of coal quality.

Coal Mining Technology

The manner in which coal is mined depends upon several factors, including depth of the coal bed from the surface and the geological character of the terrain. Generally, coal that is 200 feet (61 meters) or more from the surface is mined *underground*, using "deep mining" techniques. Shallow deposits are extracted by surface *mining* methods. Of the approximately 3500 operating mines in the United States, the number is virtually split between the two methods. The trend in total number of mines, since 1980, has trended downward, while the production per mine has increased to 500,000 tons of coal or more. This trend has resulted from greater mechanization, the use of large-capacity equipment, and access to large blocks of coal from which to form *mining units*, especially in the western states.

Surface Mining. This method of mining permits removal of as much as 90% of the total coal from a given deposit. Very large dragline excavators are used to remove the overburden (rock and soil covering the coal). Other equipment includes power shovels for loading operations, front-end loaders, trucks, and bucketwheel excavators. Land restoration has become an intimate part of the total surface mining operation. Reclamation is required by both federal and state laws.

Since passage of the federal Surface Mining Control and Reclamation Act (1977), the U.S. coal industry has reclaimed in excess of 2.5 million acres (1+ million hectares). This is an area larger than the state of Delaware. Also, more than 100,000 additional acres of abandoned mines, remnants of neglect from prior years, have been reclaimed through funds paid by coal producers into a national land trust. Responsible coal operators are guided by the principle that the right of coal extraction carries with it the responsibility of restoring the land. Reclaimed sites are returned to productive use, depending upon location, to farms, parks, and building sites. Surface mining equipment and methodologies are shown in Figures 3 through 9.

Underground Coal Mining. In general, most underground coal is mined by the room-and-pillar system, which involves excavating a series of "rooms" into the coalbed and leaving "pillars" or columns of coal to help support the mine roof. More than half of the coal taken from underground mines is produced by *continuous mining*, which uses a specialized cutting machine that mechanizes the entire extraction process. This "continuous miner" tears the coal from a seam and automatically removes it from the area by conveyor. See Figures 10 and 11.

Another form of underground mining is known as conventional mining. This process accounts for about 11% of deep-mined coal and consists of a series of operations that involve cutting the coalbed so that it breaks easily when blasted with explosives. The broken coal then is ready to be removed from the mine. Where the geology is favorable, this is the most practical and economical underground mining method.

Shortwall Mining. This method is used in relatively few underground mines. It involves the use of a continuous mining machine and movable

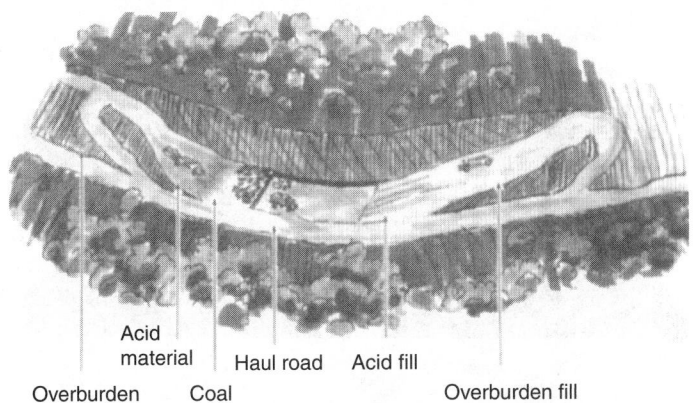

Fig. 4. Haulback method of surface mining. Using either scrapers or off-high way trucks, the major principle of this method is that all spoil except that from the initial cut is moved along the bench rather than being placed on the outslope. (*Caterpillar Inc*)

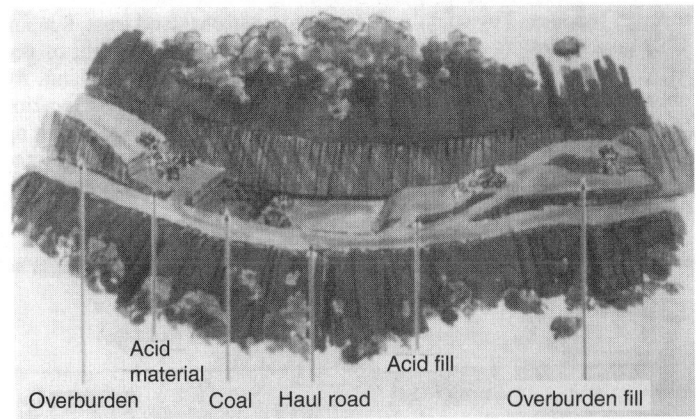

Fig. 5. Truck haulback method of surface mining is used mainly in rocky overburden, which is uneconomical for loading by scrapers. With either truck or scraper haulback operations, nonacid rock or clay overburden is placed over acidic material to facilitate revegetation. (*Caterpillar Inc*)

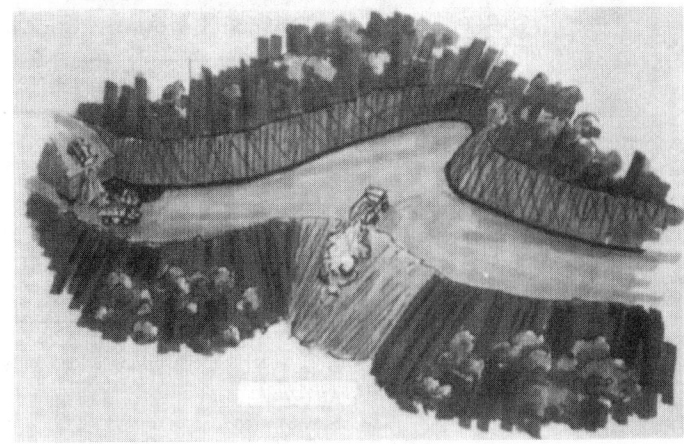

Fig. 6. Valley fill method of surface mining. In this method the miner generally hauls overburden in trucks and constructs fills over the side at bench height. When scrapers are used, the operator ejects overburden at bench height and tractors doze it over the edge. (*Caterpillar Inc*)

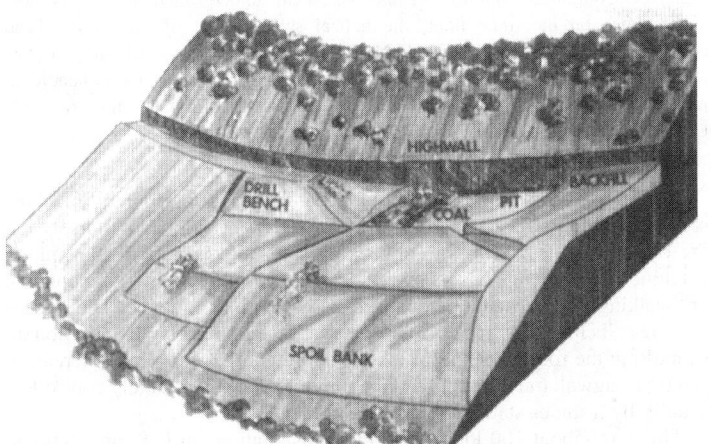

Fig. 3. Early method of contour surface mining. This was the predominant way of mining throughout Appalachia until passage of stringent legislation that ushered in a new integrated mining technique. (*Caterpillar Inc*)

roof supports to shear coal panels 150 to 200 feet (46–61 meters) wide and more than 0.5 mile (0.8 kilometer) long.

Longwall Mining. With favorable geology, an increasing amount of underground coal production is the result of *longwall mining*. This is one of the most important technological advances to impact the coal

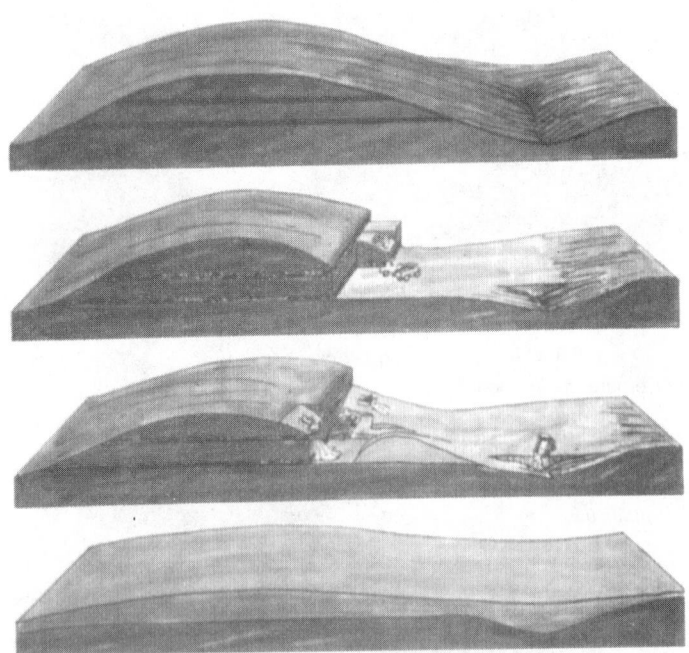

Fig. 7. In mountaintop leveling, a part of the mountain top is moved to fill an adjacent valley and build a near-level fill in the mined area. The reclaimed flat or gently rolling land that results from this type of operation can then be used for a variety of purposes which could increase the value of the land. (*Caterpillar, Inc*)

Fig. 9. An 85-ton capacity off-highway truck for use in hauling overburden in a surface mine. (*Caterpillar, Inc*)

Fig. 8. A 13.5 cubic yard (10.3 cubic meter) wheel loader working a load-and-carry operation in a surface mine. (*Caterpillar Inc*)

Fig. 10. Continuous mining machine used in underground mine. As coal accumulates on the mine floor, the helical screw effect of the cutting head constantly moves the pile toward the center of the head, contributing to fast loading and improved productivity and cleanup. Two powerful motors handle all motions of the machine. Safety provisions are incorporated in the design. (*National Coal Association*)

industry since the introduction of mechanized equipment a half-century ago. This methodology has made a significant contribution toward doubling underground mining productivity over recent decades. The production of coal per shift from longwall is, on average, more than double that of either conventional or continuous mining. Coal recovery rates of 80% are possible under favorable circumstances.

In the longwall system, which originated in Europe, two parallel entries are excavated from the main mine entry directly into the coal seam. The parallel entries, which may be as much as 750 feet (365 meters) apart or more, are then joined together at their far end by a crosscut. The coal face that is formed at the crosscut is called the "longwall."

In a longwall operation, a rotating shear or plow on a mining machine moves back and forth across this long seam, cutting and transporting the coal from the face by a conveyor system that is part of the mining machine. The machine has its own movable electrohydraulic roof supports, or "walking props," that are advanced as the seam is removed. The mined-out areas behind the roof support are allowed to cave. In addition to supporting the roof, these props also help protect the work area, increasing safety. Longwall mining units and the mining cycle are remotely controlled, usually by a miner stationed in one of the entries.

There are about 100 longwall mining installations in U.S. underground mines, most of which are located in Appalachia and the Midwest. The applicability of the methodology to a specific coalbed depends upon several factors, including the thickness and suitability of the seam and the strength of the surrounding strata. See Fig. 12.

Fig. 11. Closeup of cutting edge and helical screw, which cuts and removes coal from the mine seam. (*National Coal Association*)

High-Pressure Mining Techniques. In the early 1990s a joint effort was undertaken by the University of Missouri and the National Aeronautics and Space Administration's Jet Propulsion Laboratory, to develop a prototype of a machine known as RAPIERS (Room and Pillar In-Seam Excavator/Roof Supporter). This machine employed a pair of jet-lances to carve a horizontal slot in the center of the face of the seam. Hydraulic jets then progressively cut vertical slots, using wedge-shaped cutters to move the coal toward the center slot. Mechanical arms transferred the cut pieces of coal onto a conveyor belt. With the use of water, coal dust was kept to a minimum. The machine also could support the tunnel ceiling prior to the installation of support bolting.

Coal Preparation Plants

Raw coal from the mine must be treated (washed, sized, sorted, etc.) prior to shipment to the end user. The amount of refuse discarded may be up to 20% of the raw coal. Power plants and other coal consumers use transport systems, furnaces, heaters, reactors, and the like that demand *uniform* feedstocks. Coal preparation plants are not highly standardized because of differences in the physical and chemical properties of the raw coal from one mine location to the next. See Fig. 13.

Since most coal impurities have specific gravities greater than coal, the density of a coal particle is a direct measure of its purity. The differences in this physical parameter is the basis for mechanical separation of coal from refuse. Both gravity and centrifugal force devices are used and these may employ air or liquids as washing media. There are relatively few instances where air washing will work because of the moisture required to be on the coal to meet mining regulations. The few cases where the raw coal is dry enough for air washing require a complete dust collection system to meet air pollution standards. Thus, few plants utilize air washing. There may be a trend to air washing in the western coal fields because of the scarcity of water.

Washing processes fall into three classes: (1) hydraulic separation; (2) dense medium separation; and (3) centrifugal (cyclone) separation.

Hydraulic separation depends on a process called jigging, which creates a particle stratification from an alternate expansion and compaction of a bed of particles by a pulsating fluid flow. As originally developed, a basket filled with material was moved up and down in a tank filled with water. The more modern Baum jig process utilizes an air impulse concept in which the water is moved by air pressure from an adjacent sealed chamber. There are several refinements of the process, including the McNally Norton standard washer.

More accurate separations are made in dense medium vessels. Coal is slurried in a medium with a specific gravity close to that at which the separation is to be made. The lighter coal tends to float and the refuse to sink. The two fractions then can be mechanically separated. Theoretically, any size particle can be treated by the dense medium process. Practically,

the sizes range from about 0.5 millimeter to about 6 inches (15 centimeters). Organic liquids, salt solutions, aerated solids and water suspensions have found use as commercial media. Water suspensions meet most practical requirements and are the least costly. The bulk of coal mechanically cleaned by the dense medium process is separated in suspensions of magnetite in water.

The use of centrifugal force as an aid in coal-refuse separation is a relatively recent addition to the coal-cleaning process. As originally developed, the device employed a dense working medium. The latest units do not employ an artificial gravity suspension and are known as hydrocyclones. Design of the unit allows the formation of a hindered settling bed as the dense particles move down the side wall under the force of gravity. Less dense particles are unable to penetrate this heavy bed and move back into the main hydraulic current and are discharged out the top of the unit.

Tables are also used to wash coal. The reciprocating action of a table stratifies the high-gravity coal particles on the bottom and the low-gravity particles rise to the upper level of the bed. As the low-gravity particles rise, they are moved across riffles that separate the high- and low-gravity material by the water flowing to the low side of the table deck. The refuse is trapped in the riffle troughs and the motion of the deck moves the refuse to discharge off the end.

For separating fine coal particles from refuse particles, flotation is often used. Finely disseminated air bubbles are passed through a coal slurry. The fine coal particles adhere to the air bubbles and rise to the top where they are removed as a concentrate while the heavy refuse particles sink and are removed by the flow of water through the flotation cell. A frothing reagent, such as methylisobutylcarbinol, is added to the feed. See also **Classifying (Process)**.

Water remaining on marketable coal is a contaminant as serious as the undesirable ash. It may cause problems in handling and shipping, increase freight cost, and reduce heating value per unit weight. The difficulty of dewatering increases with increases in the surface area of the material. Several processes are used, depending upon the particle size of the coal. Vibrating-screen type centrifuges may be used. For the removal of very fine material (28-mesh and smaller), a filter process may be required. Both disk-type and drum-type filters are used. Where filters are required, filtration is usually preceded by sedimentation. Chemical flocculants sometimes are used to assist the settling process. A final reduction of moisture content frequently is accomplished by thermal drying. The use of fluidized bed coal dryers is increasing.

Transportation of Coal

Railroads handle approximately 60% of the coal mined in the United States during some part of its journey from mine to point of consumption or shipment overseas. Most American railroads offer four distinct types of service for the transportation of coal: (1) single carload; (2) multiple carload; (3) trainload volume; and (4) unit train. Each type of freight service has its own operating characteristics, which result in a distinct level of operating cost and freight rates. The first two methods are self-evident. Trainload volume is the tendering of a sufficient number of carloads of freight on one day from one origin to one destination to permit the carrier to handle the movement in a special train. The number of carloads required to form a trainload will vary from one railroad to the next. Trainload volume movements are an irregular movement on an irregular schedule. The basic operation requires simplified switching and terminal operations, resulting in economies in rail operation. Trainload volume trains use rail cars assigned to a car pool. The trains are governed by tariff provisions requiring a limited control over the loading and unloading of the railroad equipment and occasionally requiring a minimum annual volume.

A unit train movement is an integral movement of coal moving from a single origin to a single destination on a regularly scheduled train, avoiding all terminals and switching operations. Unit trains utilize specialized loading and unloading facilities and specialized railroad equipment assigned to dedicated service. The unit train movement is governed by tariff provisions requiring both controlled loading and unloading of the railroad equipment and a minimum annual tonnage. The loading of a unit train at the York Canyon Mine (New Mexico) is shown in Fig. 14. When large volumes of coal are to be loaded within a short time, some form of flood-loading is required, permitting the coal to free fall into the cars. The four basic types of flood-loading are: (1) Ground storage with loading tunnel; (2) ground storage with loading bin; (3) silo storage; and (4) silo

Fig. 12.   Example of longwall mining. (*National Coal Association*)

storage with loading bin. The first type is used in the system shown in Fig. 14. A unit train being loaded from a storage silo is shown in Fig. 15.

The unloading of coal cars can be accomplished by several methods. A unit train with conventional hopper cars would be unloaded with the cars being spotted over the storage area. The gates on the cars would be opened by laborers stationed alongside the rail cars. After the first cars over the unloading area were discharged, the train would move forward until the next loaded car is moved into place. This spotting, unloading, and spotting sequence must be repeated perhaps a hundred times until the entire train is unloaded.

A unit train with quick-opening bottom-drop cars would be unloaded by having the gates on the rail cars opened by either a mechanical tripping mechanism or by an electrical device as the cars roll over the pit or trestle. After the cars are unloaded, the gates on the cars would be closed by a similar mechanical or electrical mechanism. Motion unloading systems, although costly, represent many advantages. Proceeding across the pit area at 4 to 5 miles (6.4 to 8 kilometers) per hour, a 100-car, 10,000-ton (9000-metric ton) unit train can be unloaded in 15 minutes. Considering startup time, the total unloading time may approximate one hour. The same 100-car unit train, in an efficient two-car rotary dump facility, will require from

4 to 5 hours to unload the train. If a single-car rotary dump facility were used, the unloading time will range from 8 to 12 hours. A motion unloading facility is illustrated in Fig. 16.

After coal arrives at its destination for consumption—say an electric power generating plant—considerable handling remains prior to the coal-fired boilers. See Figs. 17 and 18.

Barge and Truck Transport of Coal. Approximately 28% of U.S. coal is transported by barges over inland waterways. Trucks account for approximately 13% and used mainly for deliveries of 100 miles (161 kilometers) or less. Because of weight limitations, trucks are not used for long hauls.

Coal Slurry Pipelines. The first patent covering the pumping of coal and water dates back to 1891. In 1914, the first commercial transport of coal in water was carried out in England, when a short 8-inch (20-centimeter) pipeline was used to carry coal from river barges to a power plant. Thereafter, several proposals were submitted for the long distance transport of coal from mine to market in the eastern United States, but failed to materialize for several reasons, not the least of which were technical problems. Intensive research into slurry transport was continued and, by 1957, technology and engineering had advanced to the point where the

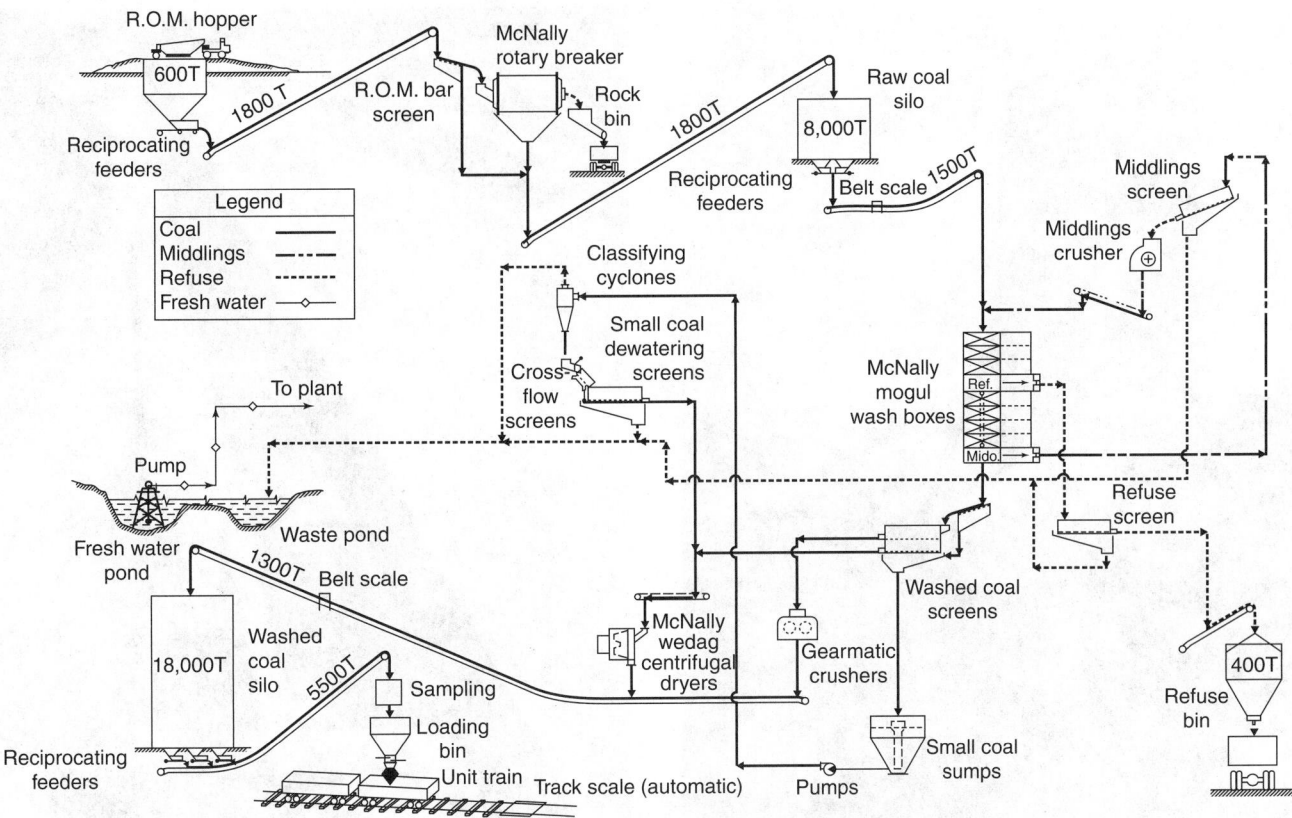

Fig. 13.   Coal preparation plant. (*McNally Pittsburgh Mfg. Corp*)

Fig. 14.   Unit train being loaded at a New Mexico mine. (*McNally Pittsburgh Mfg. Corp*)

first long distance transportation of coal in water was feasible. The result was construction and operation of the Consolidation Coal Pipeline, 10 inches (~25 centimeters) in diameter and 108 miles (174 kilometers) in length, transporting 1.25 million tons (~1.1 million metric tons) of coal per year from Cadiz, Ohio, to an electrical generating station 20 miles (32 kilometers) east of Cleveland on the shores of Lake Erie. The pipeline was powered by three pump stations, spaced about 35 miles (56 kilometers) apart, where discharge pressures reached 1,000 psi (6.9 mPa). Coal with a graded size consist, 8 mesh by 0, and a concentration of 50% solids was transported. Consist means the size makeup of the solid phase of the coal

Fig. 15.   Unit train is loaded with coal as it passes through base of storage silo

Fig. 16.   A unit train with motion unloading of hopper cars being unloaded at large power plant in Tennessee. (*TVA*)

slurry. The term 8 mesh by 0 indicates coal with a graded size makeup in the range of 8 mesh and zero (dust).

Although the Ohio line operated successfully, transporting 7 million tons (6.3 million metric tons) of coal, some unexpected operating problems had to be resolved. Much investigation was conducted with variables, such as size consist and slurry concentration and the resultant effect on slurry stability. After 7 years of operation, the line was shut down in 1963, when the unit train concept resulted in much lower freight rates on significantly higher amount of coal movement.

Economics vary in different locations, however, and slurry pipelines can be particularly, attractive where no railroad facilities exist. Thus, throughout the world today there are about ten operating coal slurry pipelines. The only one (1991) in the United States is the Black Mesa Pipeline, 273 miles (440 kilometers) long, mostly 18-inch (~46 centimeters) diameter, but with some 12-inch (7.6-centimeters) diameter sections.

Coal slurry pipelines have been constructed in several countries, including a 38-mile (61-kilometer) 12-inch (30.4-centimeter) diameter pipeline in Russia, a 51-mile (82-kilometer) pipeline in Poland, as well as others in France and other locations in Europe. The feasibility of slurry transportation depends upon the resolution of a number of variables, the most important of which from a hydraulic standpoint are: (1) Size consist; (2) velocity; and (3) concentration. The selection of a proper size consist (gradation) is important in order that homogeneous flow can be achieved at prudent operating velocities. For coal slurry, such a consist is on the order of 8 mesh by 0 (approximately 0.1-inch (2.5-millimeter) particle size to dust). Homogeneous flow (solids evenly distributed across the pipe diameter) is important if excessive wear in the bottom of the pipe is to be avoided and stable operation achieved.

Of equal importance and directly related to size consist is the proper selection of velocities for transport. The velocity cannot be excessive so as to cause abrasion of pipe wall and inordinately high pressure drops. Conversely, the velocity should not be so low as to cause heterogeneous flow, with resultant excessive wear in the pipe bottom or bed formation which will cause unstable operation.

The two prime disadvantages facing coal slurry pipelines are: (1) An adequate and assured water supply is required. In water-short areas of the western United States, this is a major consideration; (2) dewatering slurry for consumption at a power plant, or for transshipment by barge is required. Centrifuging is the primary method used to date. While reduction of coal particles to the very small size needed for movement by pipeline serves the requirements of a generating station for a finely-ground coal, the fine size makes dewatering difficult.

Testing of Coal

Proximate Analysis. This includes the determination of total moisture, volatile matter, and ash; and the calculation of fixed carbon for coals and cokes. The term "Proximate" should not be confused with the word "approximate," since all Proximate Analysis tests are performed according to rigid specifications and tolerances. Proximate Analysis results may be used to establish the rank of coals; to show the ratio of combustible to incombustible constituents, to provide the basis for buying and selling coal, and to evaluate for beneficiation, or other purposes.

Moisture in coal takes three forms: (1) free or adherent moisture, essentially surface water; (2) physically bound or inherent moisture (that moisture held by vapor pressure and other physical processes); and (3) chemically bound water (water of hydration or "combined" water). The ASTM defines *total moisture* as a loss in weight in an air atmosphere under rigidly controlled conditions of temperature, time, and air flow. Total moisture represents a measurement of all water not chemically combined. Total moisture is determined by a two-step procedure, involving air-drying for removal of surface moisture from the gross sample, division and reduction of the gross sample, and determination of residual moisture in the prepared sample. An algebraic calculation is used to obtain the total moisture value.

Ash is the noncombustible mineral matter left behind when coal is burned under rigidly controlled conditions of temperature, time, and atmosphere.

Fig. 17.   Looking down on a bucket-wheel stacker-reclaimer for handling coal. The machine is moving along a 1200-foot (366-meter) stockpile near Uniontown, Kentucky. Incoming coal from the conveyor moves to the top of the machine and is dumped into 30-foot (9-meter) trenches beside the tracks which support the machine. Later, the bucket wheel will recover coal from the trench. Excess coal is pushed back and forth from the trench by bulldozer as required. (*Dravo Corporation*)

Fig. 18.   Electric generating station in central Pennsylvania is one of several large mine-mouth electric power generating plants in the area which consume coal from nearby mines. Clearly visible are the long conveyor lines required to move the coal once it is received at the plant. (*New York State Electric and Gas Corporation*)

Total nitrogen is determined by chemical digestion (Kjeldahl-Gunning) methods.

Oxygen content is determined by calculations, subtracting total carbon, hydrogen, sulfur, nitrogen, and ash from 100%.

Chlorine is commonly included as part of the ultimate analysis.

Other important chemical and physical tests performed to characterize coal include: (1) Heating value (Btu content); (2) sulfur forms; (3) ash fusibility temperatures; (4) ash analysis; (5) trace elements; (6) free swelling index; and (7) hardgrove grindability.

Heating value is determined by burning a coal sample in an oxygen bomb and measuring the temperature rise. See also **Calorimetry**.

Three *sulfur forms* recognized by ASTM are: (1) sulfate sulfur, which may be in the form of calcium or iron sulfate; (2) pyritic sulfur, which is sulfur combined with iron in the form of minerals pyrite and/or marcasite; and (3) organic sulfur, which is bonded to the carbon structure.

Ash fusibility can be defined broadly as the melting temperature of the ash.

Ash analysis is the term used to designate analysis of the major elements commonly found in coal and coke ash. The elements, expressed as oxides, are SiO_2, Al_2O_3, Fe_2O_3, TiO_2, CaO, MgO, Na_2O, K_2O, P_2O_5, and SO_3.

Interest in *trace element analysis* has increased by environmental concerns.

Volatile matter is defined as the gaseous products, exclusive of moisture vapor, driven off during standardized test conditions. The combustible gases are carbon monoxide, hydrogen, methane, and other organic hydrocarbons. Those generally classified as noncombustible are carbon dioxide, ammonia, hydrogen sulfide, and some chlorides. Volatile matter tests are used to establish the rank of coals, to indicate coke yield upon carbonization, and to establish burning characteristics.

Fixed carbon is the solid residue, other than ash, resulting from the volatile matter test. The value is calculated by subtracting moisture, volatile matter, and ash from 100%.

Total carbon is determined by catalytic burning of the sample in oxygen to form carbon dioxide, which can be readily measured.

Total hydrogen also is determined by catalytic burning of a sample in oxygen to form water. The water is absorbed in a desiccant and weighed directly. Hydrogen results as determined include the hydrogen present in both the sample moisture and water of hydration.

Mining Safety and Health. Although steady technological progress has taken place over a number of years concerning the health and safety of coal mining personnel, a strong impetus toward further improvements has come from legislation at the state and national level over the past score of years. Studies and improvement programs continue to be well funded. Efforts have fallen into the following major categories.

Ground Control—with the objective of developing technology to prevent accidental falls of roof, rib and face, and coal bumps. Study areas include: (a) artificial support, (b) hazard detection, and (c) design of mine openings. Horizontal roof strain indicators for detection of unstable roof conditions have been tested. Other research programs have included a microseismic fracture warning system, polymeric roof bolts, and chemical impregnation techniques.

In late 1991 Australian government researchers announced the development of a remote-controlled vehicle that can scout ahead of a rescue crew to locate missing and injured miners. Three stereo-video cameras permit the vehicle to operate in murky areas of a mine. The vehicle also includes gas analysis instrumentation. A fiber-optic cable, wound on a large drum, permits surface operators to convey instructions to the vehicle and to receive the results of gas analysis data and images of what the vehicle "sees." The vehicle is named after a burrowing marsupial, *Numbat*.

Fire and Explosion Prevention—study areas have included: (a) ignition, (b) flame propagation, (c) fire detection and alarm, (d) suppression and extinguishing, and (e) methanometry. Devices and techniques tested have included explosion-proof bulkheads, coal dust and rock dust analyzers, ignition suppression devices for face equipment, and remote sealing techniques.

Industrial-type Hazards—with the objective of identifying hazard sources in electrical, mechanical, illumination, and non-emergency communication fields. Developments have included (a) advanced remote surveillance and communication systems, (b) portable-area illumination systems, (c) trolley-phone wireless systems, and (d) protective canopies for use on underground low-coal machines.

Methane Control—with the objective of developing safe methods for mining methane-laden coalbeds. Study areas have included: (a) predictions of concentrations and flow; (b) control in advance of mining; and (c) control during mining. Techniques tested have included water infusion to reduce methane in the face area, degasification through vertical boreholes, the plugging of oil and gas wells that penetrate coal beds, and the complete degasification of operational mines.

Respirable Dust—with the objective of providing improvements for protecting miners from exposure to respirable coal mine dust. Study areas have included: (a) dust formation; (b) dust control, and (c) dust measurement. Tests have included: the infusion of water into coal beds for control of respirable dust; the use of water-based, high-expansion foaming systems in conjunction with continuous mining machines, to reduce dust at the face; the use of foam systems for dust suppression on conveyors and transfer points; and the use of prototype dust meters. See also **Pneumokonioses**.

Noise—with the assessment of permissible noise levels for communication and warning signals and the development of technology for noise abatement and control. Developments have included an audio dosimeter to replace conventional sound-level meters, discriminating earmuffs, and a noise control muffler system to reduce pneumatic drill noise.

Additional Reading

Derickson, A.: *Black Lung: Anatomy of a Public Health Disaster,* Cornell University Press, Ithaca, NY, 2000.
Dunrud, C.R.: *Engineering Geology Applied to the Design and Operation of Underground Coal Mines,* DIANE Publishing Company, Collingdale, PA, 2000.
Staff: IEA, *Coal Information 1998,* Organization fro Economic Cooperation & Development, France, 1999.
Staff: *Coal Industry Annual, 1998,* United States Government Printing Office, Washington, DC, 2000.

Web References

American Coke and Coal Chemicals Institute. http://www.accci.org/
International Directory of Coal Industry Organisations. http://www.mining- technology.com/industry/
MINE-ENGINEER.COM. http://www.mine-engineer.com/
The Coal Association of Canada. http://www.coal.ca/
World Coal Institute. http://www.wci-coal.com/home.htm

COAL CONVERSION (CLEAN COAL) PROCESSES. Coal, representing a major reserve of energy in the United States and a few other countries (see also **Coal**), currently accounts for well over half of the electricity produced in the United States. Thus, although coal poses environmental threats and problems, until other essentially renewable energy sources can be developed technologically, coal obviously will be a major energy source for well into the 21st century. The problem, then, is to determine how coal can be treated or converted and how coal combustion processes can be altered so that their impact on the environment can be minimized and, of course, fall within the practical economic limitations realistically imposed on the *value of energy*. Among energy alternatives, coal probably is at the apex of the triangle made up of three interacting forces: the energy supply, the environment, and the national economy.

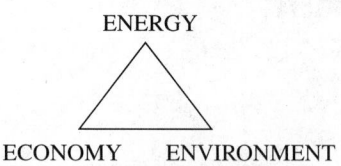

Coal conversion and utilization technology, although less spectacular, rivals the space program's engineering in terms of the numbers and difficulties of problems requiring solution. This is attested to by the scores of approaches suggested, studied, and applied in test situations. Coal has proved to be a very stubborn, scientifically unpliable substance to manipulate. Coal conversion or combustion processes are not amenable to much laboratory benchwork, but usually require fairly large and expensive pilot plants for process testing. Scale-up problems are equally difficult. For example, those coal conversion processes that have reached the large-scale testing phase pose severe materials problems and operate under high temperatures, high pressures, and with severely corrosive and abrasive materials at various stages in the process. Thus, advances in materials engineering (need for new alloys, ceramics, etc.) parallel the other processing problems of coal conversion. (These are reminiscent of materials problems encountered in the space program—for example, solving the reentry heating problems in the early days of the space shuttle.)

Assuming that coal will be required for several future decades as a major source of energy, what options are open for the near, intermediate, and far term? There is a considerable consensus that converting raw coal into essentially a new form of fuel, as through gasification, liquefaction, or treated solid forms, will provide minimum ultimate impact on the environment. Discounting the obvious conservation of energy (easy to preach; difficult to practice), there are two main avenues of approach for the immediate and short term:

1. Improving coal combustion processes to maximize electric generation efficiency, partially with the obvious target of using less coal/kilowatt and thus directly helping the environment; and
2. Designing processes that either:

   a. Create reduced air pollution; or
   b. Treat pollutants prior to their emission into the air.

Considerable progress along these lines has been made as a result of the Clean Air Act and other forces that are demanding environmental protection. Recent revisions of the Clean Air Act have placed high responsibilities on the operators of electric utilities. Emissions reduction in the absence of an entirely new coal fuel technology is extremely costly. Although the more recent amendments award and penalize operators for emissions reduction performance, the political aspects of which are not described here, the bottom line is higher electricity costs for the consumer, reemphasizing the strong inverse relationship between energy and the environment. Location of electric generating plants that fall within "allowances" program are shown in Fig. 1. This program obviously is tied in with international concerns over $FISO_2$ emissions. See also **Acid Rain**.

For those readers who are not familiar with past achievements in coal conversion, the following several paragraphs and examples are included. The use of coal is tersely reviewed prior to major environmental concerns, which belatedly did not become part of the public psyche until the mid-twentieth century.

Early Chronology of Coal and Steam

As pointed out in the entry on **Coal**, the principal ingredient of coal is carbon and, as described in the entry on **Combustion (Fuels)**, it is the combination of carbon with atmospheric oxygen to produce carbon dioxide (CO_2), an exothermic reaction that releases 14,100 Btu/pound (7840 Calories/kilogram) of carbon, that provides the heat energy derived from burning coal. Depending upon the composition of the coal, other heats of reaction will occur from the combustion of hydrogen and sulfur in the coal with air, but these are secondary factors. The fixed carbon content of coal ranges from about 98% for a Class 1 anthracite or hard coal, as may be mined in Pennsylvania, to as low as about 32% for a Class 17 brown coal or lignite, as may be mined in North Dakota.

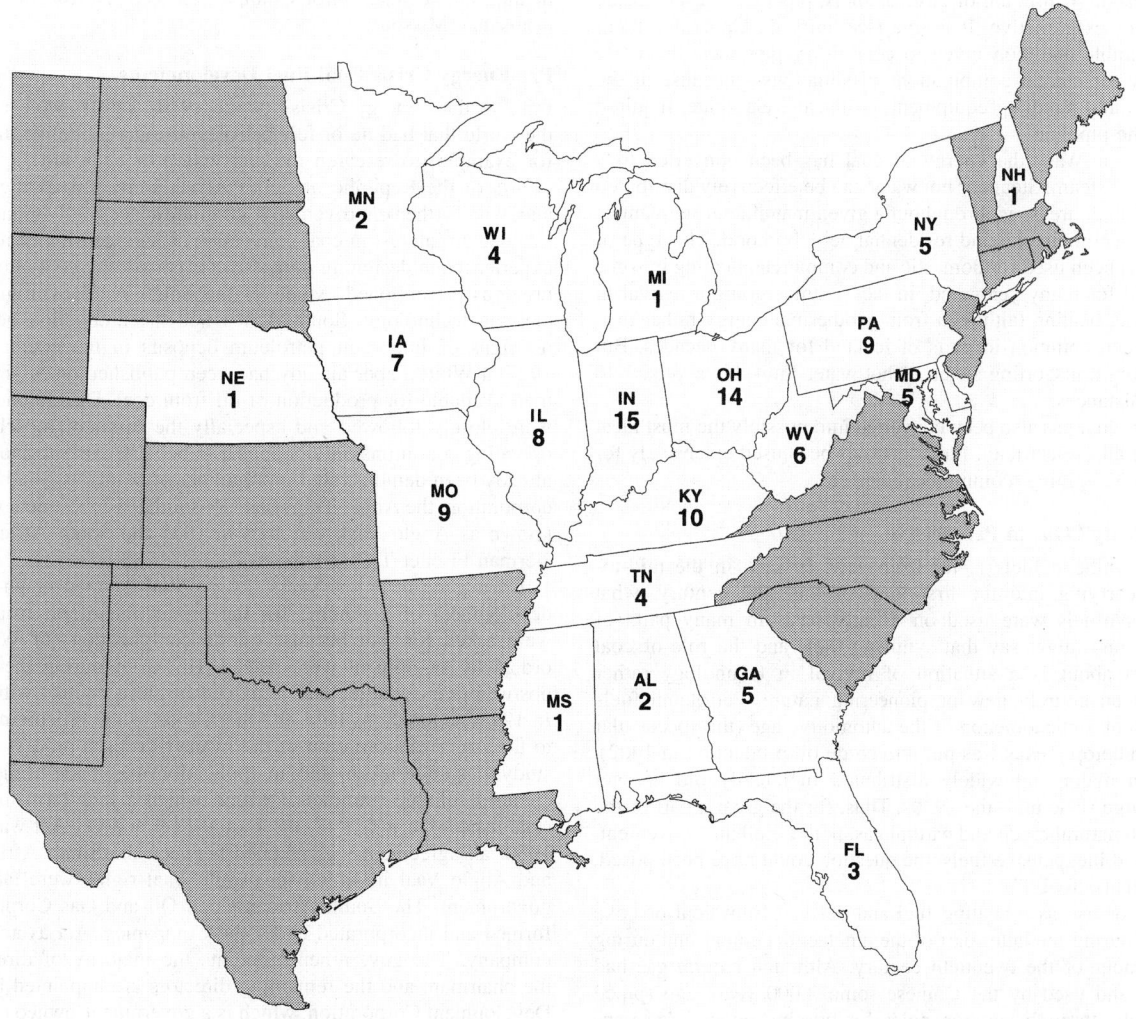

Fig. 1.   States with coal-burning electric power plants that are targeted by Clean Air Act revisions, requiring that by 1995 such plants must reduce their emissions of sulfur dioxide (SO_2). Complicated politically, the program is measured in terms of allowances, one allowance equaling 1 ton of SO_2. Depending upon the size of the power plant, allowances range from 1000 to 255,000 units of SO_2. Because of plant size variations, current state of cleanup equipment, and the fact that several plants are under the same management, the program allows for certain trade-offs. In other words, reduction at some plants may be applied to plants not performing as well environmentally. Allowances also may be purchased from other plants. In essence, this will allow some older plants to continue operating for the next few years, while others are modernizing or while new plants are being constructed. Thus, the regulatory targets are for the total 111 plants to accomplish a large regional reduction by 1995

Direct use of coal, as in pulverized form for the firing of boilers in the electric utility industry, poses a number of environmental problems that can be solved only through the use of costly and elaborate antipollution measures and equipment. But, also in considering the expanded use of solid coal as a major source of energy, there are several other limitations over and beyond the environmental.

Aside from coal-powered steam locomotives and seagoing ships, which essentially were retired from most regions of the world over the past several decades, solid coal is quite unsuited for transportation energy. The energy density of raw coal means that a significant portion of the energy obtained from combusting it is required to move it (as part of a transportation vehicle). This is further amplified by the equipment required to burn coal—massive, heavy furnaces and boilers—which also have to be moved with the vehicle.

Transportation power needs dictate high energy density with fuels that are easy and convenient to handle and that can be converted to power by much smaller, lighter-weight engines. It follows, then, that for coal to be a useful energy source for transportation, it must be preconverted in some way to overcome the aforementioned objections.

Even for stationary use, particularly where energy needs are much less concentrated than in the case of a central electric power plant, such as for commercial and residential heating and large numbers of industrial plants, the conversion of coal into liquid or gaseous forms is required to provide the needed convenience, improved cleanliness, and use efficiency. Probably of equal importance is the cost of transporting large amounts of coal from sources to many tens or hundreds of thousands of locations where it can be used. With liquid or gaseous fuels, pipeline transportation, for example, becomes attractive. It is true (see entry on **Coal**) that there has been considerable attention given to coal slurry pipelines, but these are practical only for certain combinations of situations—because of the slurry preparation and handling equipment, costly and elaborate, required at both ends of the pipeline.

Since the time of Watt, the energy of coal has been converted to a gas—in the form of steam. Steam or hot water can be effectively distributed within relatively small areas, as throughout a given manufacturing plant or complex, or even a commercial and residential neighborhood. This type of district heating has been used for domestic and commercial heating in some parts of the world for many years and, in fact, is undergoing a revival in some areas. District heating (although from geothermal energy rather than from coal) has been common in parts of Iceland for many decades. But thermal losses from transporting steam or hot water limit this approach to relatively small distances.

For many years coal has also been converted into possibly the most ideal form of energy of all—electricity. Electricity has been used extensively for powering machines, lighting, communications, etc.

"Artificial" or "City" Gas in Perspective

The concept of synthetic fuels is far from new. In fact, in the mid-to-late 1800s and carrying into the first third of the 20th century, what are now called synfuels were used on a daily basis in many parts of the world. Thus, one might say that synthetic fuels and the role of coal in bringing them about is a situation of revival in technology rather than anything of an entirely new or pioneering nature. Petroleum fuels essentially represent a phenomenon of the automotive age (the spectacular salt dome at Spindletop, Texas was put into crude oil production in 1902). Natural gas as a major and widely distributed fuel really did not get underway on a large scale until the 1930s. Thus, for the greater part of the 20th century, with natural crude and natural gas, both excellent, convenient, generally clean and inexpensive fuels, the question could have been posed, "Who needs synthetic fuels?"

Artificial gas for use as a heating fuel and derived from coal or coke was widely used during the latter part of the nineteenth century and during the first few decades of the twentieth century. Although natural gas had been discovered and used by the Chinese some 2,000 years ago (piped from shallow wells through bamboo poles for burning under large pans for evaporation of sea water to obtain salt), the first hard evidence of commercial use of natural gas dates back to 1802, when it was used for lighting the streets of Genoa, Italy. The first natural gas utility company was formed in 1858 (The Fredonia (N.Y.) Gas Light Company). The numerous advantages of a gaseous fuel that could be piped to industrial, commercial, and residential users for heating purposes were recognized long before natural gas was found on a large scale and made available to communities

hundreds and more miles from the originating wells. Thus, for many decades, artificial gas was used. The local gas utility was characterized by having one or more so-called "gas works" in which a rather poor grade of gas (on present standards) was produced essentially from coal or coke and steam. In the manufacture of producer gas, a deep hot bed of coal or coke was blasted continuously with a mixture of air and steam. The products of the process were carbon monoxide, nitrogen (from the use of air), small amounts of hydrogen, and some carbon dioxide. Because of the large percentage of nitrogen in the gas, the heating value was low [125 to 150 Btu per cubic foot (1113 to 1335 Calories per cubic meter) as compared with natural gas having a value of from 900 to 1.200 Btu per cubic foot (8,010 to 10,680 Calories per cubic meter)]. Blue water gas, carbureted water gas, and coal gas were also produced from coal or coke and, in some instances, enriched with oil and later natural gas. Because of the great availability and, at one time, apparent inexhaustible supply of natural gas in the United States (and a few other areas of the world), manufactured gases were phased out rapidly. The use of natural gas increased 730% between 1940 and 1970 in the United States, during which period the gas industry produced 313 trillion cubic feet (8.9 trillion cubic meters) of natural gas. In other areas of the world, however, where natural gas was not available locally, manufactured gas, sometimes referred to as town gas, city gas, etc., persisted. Thus, it is not surprising that the current new coal gasification technology essentially stems from Europe and the United Kingdom, where an interest in improving artificial gas manufacture continued long after such interests mainly disappeared in the United States. During the 1930s and 1940s, only a few projects for converting coal into gas were conducted in the United States—for example, the U.S. Bureau of Mines project at Louisiana, Missouri.

Pre-Energy Crisis Coal Fuel Developments

Prior to the "Energy Crisis" of the 1970s, efforts were made in parts of the world that had no or few petroleum reserves (unlike the United States, for example) to research the conversion of coal into superior fuels. The efforts of the Republic of South Africa were outstanding for that period and, with further improvements, continue successfully today. Germany also retained an interest in coal conversion as an extension of its primary major experiences in designing processes for converting coal into "artificial gas," previously mentioned. Much of the South African program relied upon German technology. South Africa had the motivation because there are no signs of important petroleum deposits in the country.[1] As early as 1927, a White Paper already had been published discussing the processes then available for production of oil from coal. Developments in Germany were closely followed and especially the Fischer-Tropsch process, as its operating conditions did not appear to be very extreme and the process had already been demonstrated in a number of plants. A South African mining corporation, the Anglo Transvaal Consolidated Investment Company, better known as Anglo Vaal, acquired in 1935 the South African rights to the German Fischer-Tropsch process.

During the next few years, Anglo Vaal devoted much attention to the development of a scheme for the production of oil from coal. Tenders were asked for, but because of the complications of World War II, no orders for equipment were placed. However, during the war and in the postwar years, Anglo Vaal remained in close contact with developments. In 1943, negotiations held in America led to the procurement of the rights to the American variation of the Fischer-Tropsch process. In 1946, a new study was undertaken, and an application was made to the government to create a suitable framework within which a long-term industry could be established. During 1947, the Liquid Fuel and Oil Act was passed and, in 1950, an agreement was reached between the South African government and Anglo Vaal in which the Anglo Vaal rights were taken over by the government. The South African Coal, Oil and Gas Corporation Ltd. was formed and incorporated under the Companies' Act as an ordinary public company. The government appoints the majority of directors, including the chairman, and the remaining directors are appointed by the Industrial Development Corporation, which is a government-owned organization with the objective, as the name implies, of stimulating industrial development in the country. Sasol operates like a normal business concern, with an autonomous board of directors, and is subject to South African company law and taxation like any other company.

---

[1] As partially related by Jan C. Hoogendoorn, Manager, Research and Development, South African Coal, Oil and Gas Corporation Ltd., Sasolburg, Republic of South Africa.

A site for the plant was selected close to the banks of the Vaal River, which is South Africa's major source of water, 50 miles (80 kilometers) south of Johannesburg and on top of a vast coal field. Sasol acquired approximately 8000 acres (3200 hectares) for the plant and its township which was to have the name of Sasolburg. The site was in the middle of an area where cattle grazing and corn (maize) production were the only activities and Sasol had to create its own infrastructure from scratch.

Although it was clear that the plant would be based on the synthesis of hydrocarbons from hydrogen and carbon monoxide as invented and developed by Fischer and Tropsch, it still had to be decided which processes to choose for the individual steps in this integrated complex. For gasification, the Lurgi pressure gasification process with steam and oxygen was selected because this process had already been demonstrated in gasifiers of a smaller size. It had the advantage of being able to work on the rather low-grade, high-ash coal available to Sasol. The fact that it operated at a pressure of approximately 350 psi (2.4 mPa), which is also the desired operating pressure for the Fischer-Tropsch plant, was an additional advantage. This avoided cumbersome compression of large volumes of gas arising from low-pressure gasification. A small part of the Sasol I facility is shown in Fig. 2. Generalized flowsheet for the Sasol II facility is given in Fig. 3.

The production of synthetic motor fuels, pipeline gas, ammonia, and chemicals, where coal gasification is the key to successful production, was first commenced in 1955. Extensive experience was gained from 20 years of operation and a large expansion (Sasol II) was commenced in 1975 and completed in the spring of 1980. The newer facility processes 40,000 metric tons of coal per day and produces the equivalent of 60,000 barrels per day of petroleum-like fuels. New mines were developed nearby to feed coal over a conveyor system into a feed preparation facility. Two huge coal piles have been built up to provide continuous operation in event of a temporary shutdown of the mines. The coal is charged to a gasification section where synthesis gas is prepared in the presence of steam and oxygen. Ash content of South African coal is relatively high (about 25%) and thus some 10,000 metric tons/day of ash are produced in the gasifiers. The gasification section contains 36 Lurgi gasifiers which produce the raw synthesis gas. These are improved versions of the gasifiers used at the original Sasol I plant. The original plant also produced town gas as well as liquids. In the new

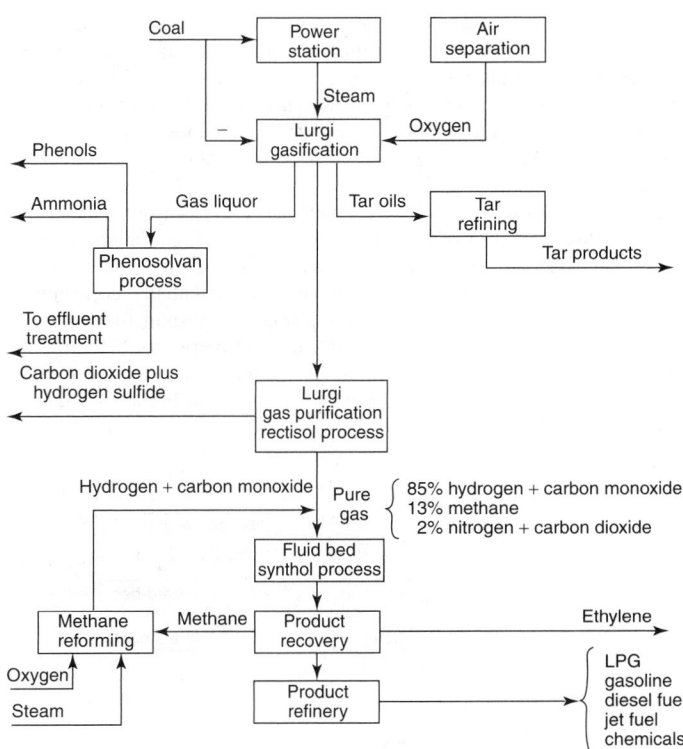

Fig. 3.   Generalized flowsheet of Sasol II project. (*South Africa Coal, Oil and Gas Corporation Limited*)

facilities, no gas will be sold. The oxygen requirements of the new facility are quite high, and six air separation plants, each with a capacity of 2500 metric tons/day, represent one of the largest oxygen facilities in the world. Liquors from the gasification section are charged to a Lurgi Phenosolvan

Fig. 2.   Gas purification portion of coal gasification complex of Sasol 1 facility at Sasolburg. (*South Africa Coal, Oil and Gas Corporation Limited*)

unit where ammonia and phenols are recovered and wastewater effluents are cleaned. The raw gas has to be further treated and this is accomplished in a Lurgi Rectisol unit, where carbon dioxide, hydrogen sulfide, and other impurities are removed. The heart of the plant is the liquefaction section. This is where the synthesis gas is liquefied in the presence of an iron-based catalyst. This Fischer-Tropsch reactor technology is proprietary to Sasol but is licensed.

Methane reforming units receive methane-rich gas from a cryogenic product recovery facility and subject the gas to partial oxidation. Some of the carbon dioxide content is removed and the gas recycled to the reactors. Once liquids are recovered, the stream goes to essentially conventional refining units. The plant's production is primarily transport fuels. Most of the gasoline production is currently sold to other refineries for blending with their stocks, but a portion of the product is marketed directly to consumers.

Other early coal conversion processes are shown in Figures 4, 5, 6.

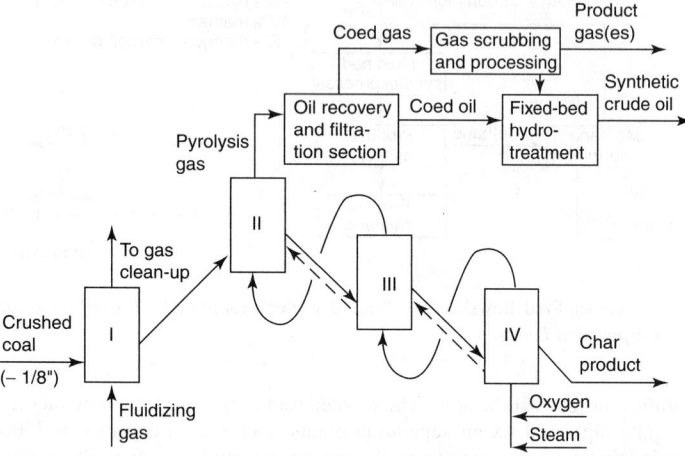

Fig. 4.   Early (1970s) process for coal pyrolysis

Synfuels Program

The Arab Oil Embargo (1973) jolted many world governments into developing programs that would make their economies immune to future energy crises. In the United States a *Synfuels Program* was legislated. The primary thrust then was to develop fuels that would substitute for crude petroleum as a starting point. The attention given to coal conversion was mainly to derive petroleum-like fuels from coal rather than simply to process control for the well-established coal burning plants of that day. However, the technology of the Synfuels Program later became the basis for converting coal adaptable to traditional coal combustion power plants—not so much in the interest of meeting an energy crisis, but rather to avoid a future environmental crisis. Thus, current programs directed toward environmental protection had somewhat of a running start because much of the technology previously described, such as the Lurgi processes, was a starting foundation. As international relations improved and oil supplies were again reassured, considerable and progressive complacency on the part of government and the public set in—with the exception of so-called "dirty coal" power plants. Thus, as of the mid-1980s, interest in coal technology had reawakened.

Clean Coal Programs of the 1990s

Traditional coal handling practices that resulted in excessive quantities of pollutants in the air with associated ill effects on the environment, including acid rain in the northeastern United States and parts of Canada are being tackled through the joint cooperation of government and private electric power interests. One of the first major goals to be achieved was set for 1995. Further background is given in Fig. 1 at the start of this article. See also Fig. 7.

Many Options to Consider. In evaluating the use of coal for electric power generation, there are many options that must be considered. Recalling the ENERGY-ENVIRONMENT-ECONOMY interdependent triangle mentioned at the beginning of this article, option preference varies—power consumers (both industrial and public), the coal mine workers, and the environment regulators, among others. This article concentrates on the environment.

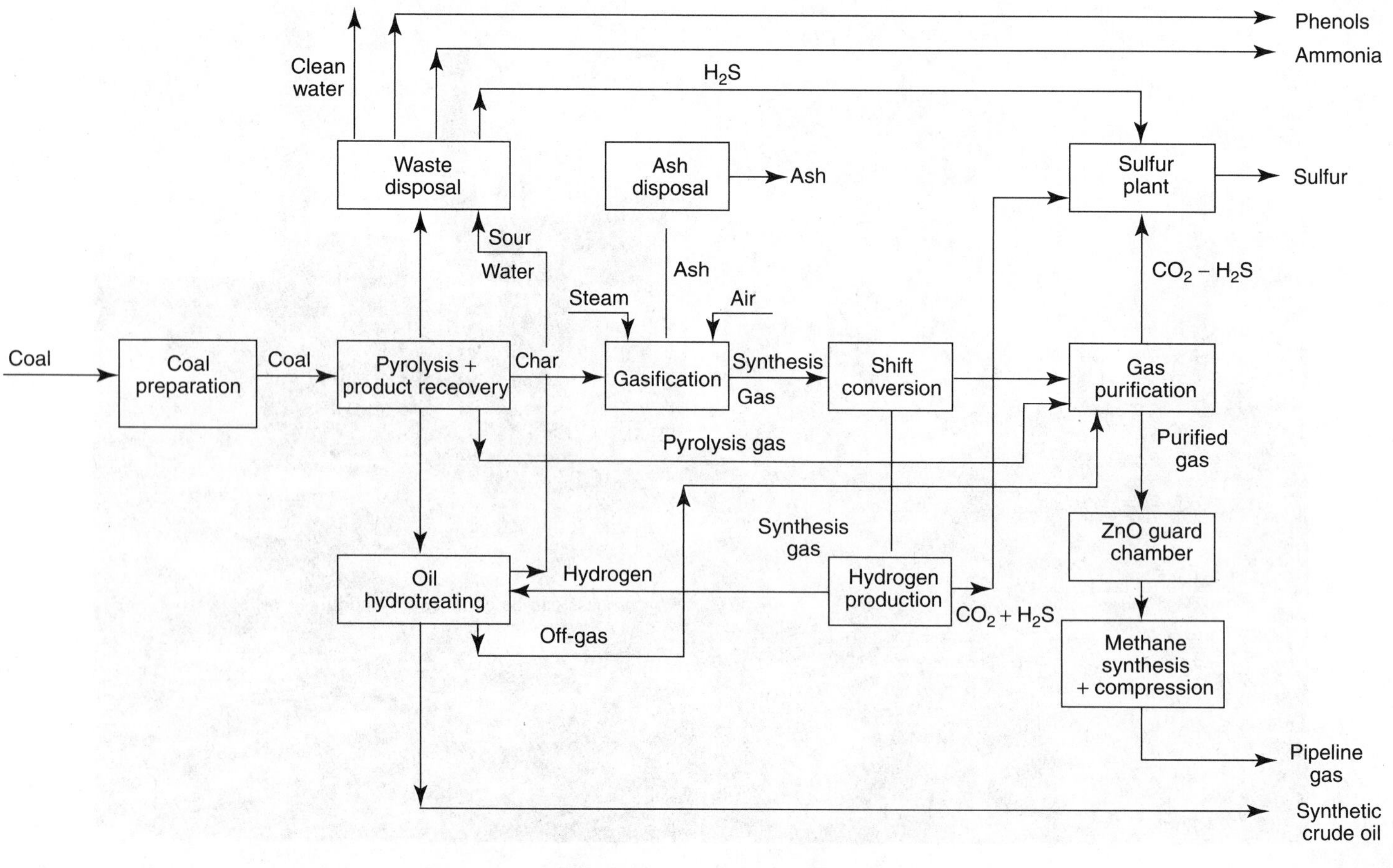

Fig. 5.   Early (1970s) process for making pipeline gas from coal

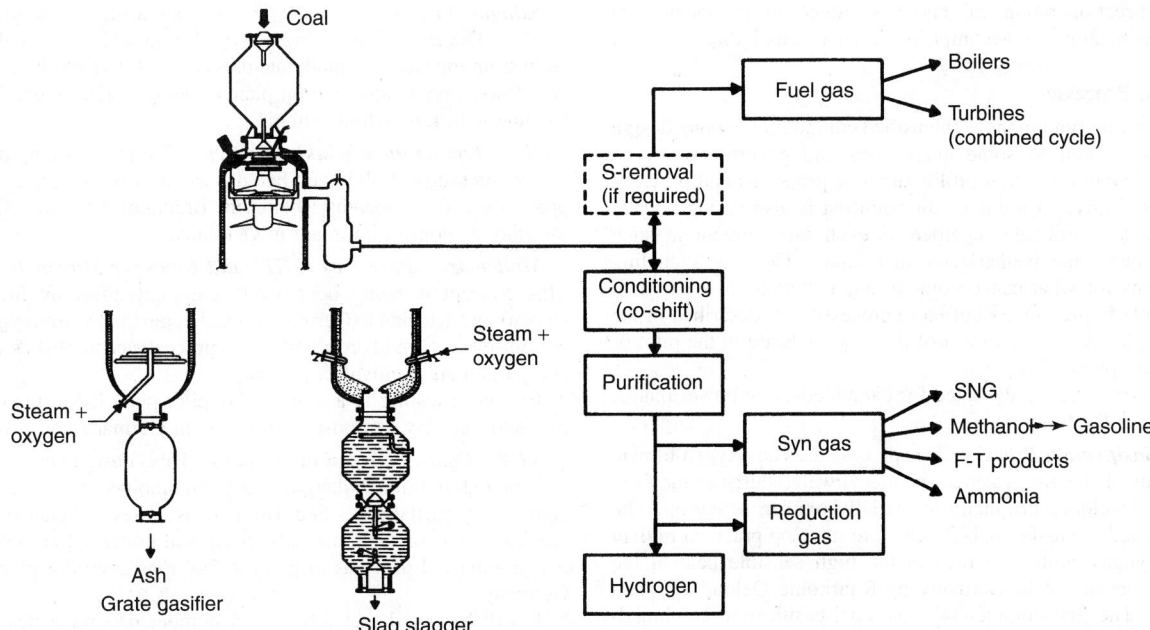

Fig. 6. Basic schemes for Lurgi gas production, the fundamental technology of which dates back to the era of "artificial" gas. The process shown here (circa early 1970s) permitted selection of various gasifiers and/or changing operating pressures to influence the final crude gas composition, as dictated by the end use and economics. (*Lurgi Kohle und Mineralotechnik GmbH, Frankfurt, Germany*)

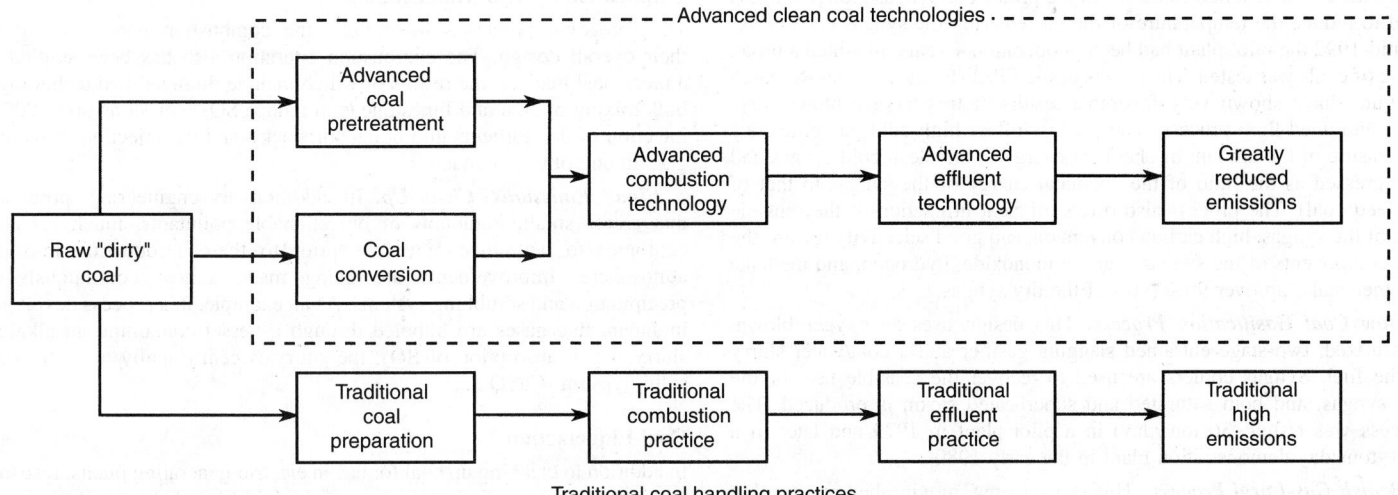

Fig. 7. Overview of coal preparation, conversion, combustion, and effluent treatment technologies that are designed to markedly reduce deleterious emissions and thus protect the environment. The line at the bottom, indicating traditional practices, now *fully* applies only to the very oldest coal-energized electric power plants. The Clean Air Act (U.S.) and the "Clean Coal" program call for progressive upgrading or modernization for achieving major goals by 1995, as described in the text. It should be mentioned that traditional practices have been improved over the years, but these have not been sufficient to meet environmental goals. Better results can be achieved only by way of marked advances in technology

Coal Preparation and Pretreatment Technology

The principal measure of the previously mentioned "allowances" program (see Fig. 1) is expressed in terms of tons of sulfur dioxide (SO_2) emitted. However, close behind the environmental effects of SO_2 are emissions of nitric oxides (NO_x). Unfortunately, the chemistry of SO_2 removal differs from that of NO_x removal.

Targeting SO_2 Emissions. As described in the article on Coal, various types of coal contain different amounts of sulfur. Thus, using low-sulfur coals provides the utility operator with a major headstart. But low-sulfur coals are not as abundant as, for example, subbituminous coal, which costs more. The relative location of a utility and the mine also is a factor because of high transportation costs. Also, when combusted, low-sulfur coals create more ash and yields fewer Btus (thermal energy). Several processes have been or are being developed to lower sulfur content by pretreatment.

Sulfur in Coal. Sulfur occurs in coal in two forms: (1) *organic* sulfur, which is chemically bonded to the coal, and (2) *pyritic* sulfur, which occurs in tiny, iron-associated and separate particles in the coal. This is a physical rather than a chemical impurity.

Pyritic sulfur can be removed by froth flotation, which takes advantage of the differences of specific gravity of the two types of sulfur. To be effective, the coal must be pulverized into particles in the micron region. The process can be enhanced by adding limestone, catalyst, and soda ash to the coal dust. After treatment, the coal is formed into briquettes for ease of handling by conventional conveyors.

By way of genetic engineering, microorganisms with metabolic capacities, unknown of as recently as 15 years ago, have been produced. Coal bioprocessing developments are directed to three objectives:

1. Reduce the sulfur content;
2. Solubilize coal and convert the soluble product to liquid or gaseous fuels; and
3. Convert coal-derived synthesis gas to liquid or gaseous fuels.

In addition to direct operations to remove or reduce sulfur content, coal liquefaction or gasification can accomplish similar results indirectly.

Coal Gasification Processes

U.S. federal and private funding have intensified competition among design engineering firms as well as some universities and government-owned laboratories. This has resulted in a proliferation of proposals and projects. Inasmuch as the objective of reducing air pollution is also strong outside the United States and probably regarded as even more urgent in such countries as Germany, the Netherlands, and Japan. There is a dueling among design firms for what can become a major market for equipment over the foreseeable future. Several of these processes are described in the following paragraphs. As of 1993 most of these projects are in the pilot or demonstration plant phase.

Most of these processes are designed for combined-cycle power plants, which are defined in the footnote.[2]

Texaco Coal Gasification Process. This process uses an oxygen-blown, pressurized, entrained slagging gasifier and a coal/water slurry as the fuel. The process was developed originally to partially oxidize heavy oil. The process was modified to gasify coal. Research to develop practical radiant and convective syngas coolers to recover the high sensible heat in the raw syngas was carried out in Germany by Ruhrkohle Oelund Gas and Ruhrchemie A.G. The first commercial-scale coal-gasification combined-cycle (IGCC) plant was built in Oberhausen, Germany. It is a 165-tons/day demonstration plant.

Shell Coal Gasification Process. This process uses an oxygen-blown, pressurized, entrained slagging gasifier and dry pulverized coal as the fuel. Syngas coolers are used to recover the sensible heat of the raw syngas, and both saturated and superheated steam are produced. Syngas recirculation is used to reduce the temperature of the gas entering the syngas coolers. As of mid-1992 the pilot plant had been in operation 4 years, in which a broad range of coals was tested. The results of the EPRI (Electric Power Research Institute) have shown very favorable results of this oxygen-blown, dry-feed, entrained-flow process. The process offers high cold gas efficiency (a measure of the amount of chemical energy in the clean cold syngas and is expressed as the ratio of the chemical energy of the syngas to that of the feed coal). The process also offers efficient utilization of the sensible heat of the syngas, high carbon conversion, and good selectivity toward the fuel components of the syngas. Carbon monoxide, hydrogen, and methane together make up over 90% (vol) of the dry syngas.

Dow Coal Gasification Process. This design uses an oxygen-blown, pressurized, two-stage entrained slagging gasifier and a coal/water slurry as the fuel. Syngas coolers are used to recover the sensible heat of the raw syngas, and both saturated and superheated steam is produced. The process was tested (36 tons/day) in a pilot plant in 1979 and later in a 1600-tons/day demonstration plant in the early 1980s.

British Gas-Lurgi Process. This is a slagging, moving-bed process that is based on the earlier Lurgi oxygen-blown, pressurized, moving-bed dry-ash gasifier. The Lurgi gasifier was modified by increasing the temperature above the melting point of the coal slag and decreasing steam consumption, thus markedly increasing efficiency. A syngas cooler is not needed because hot gases from the slagging bottom preheat the coal bed above. Tars and other organics produced contribute to difficulties in water treatment.

Kellogg/Rust and Westinghouse Process. An oxygen-blown or air-blown pressurized, agglomerating fluidized-bed gasifier is used. A 35-tons/day pilot plant has been tested. Present research is targeted to modifying the process in an effort to allow high-temperature desulfurization by injecting limestone into the fluidized bed.

Kilngas Process. This design uses an air-blown agglomerating bed gasifier. The main feature is a rotary kiln, in which the coal is gasified by adding air and steam at moderate pressures (400 to 700 kPa; 60 to 100 psi). A 600-tons/day demonstration plant is being tested by the designer, Allis-Chalmers, in East Alton, Illinois.

High-Temperature Winkler Process. The process employs an oxygen-blown, pressurized, fluidized-bed dry ash gasifier designed for lignite. The process was developed by Rheinische Braunkohle Werke AG, with an 800-tons/day demonstration plant in Germany.

Molten-Iron Pure Gas (MIP) and Klockner Molten Iron Processes. This concept is being developed in a joint effort by Sumitomo Metal (Japan) and Klockner (Germany). Coal is gasified with oxygen in a top- or bottom-blown liquid iron bath. The process accomplishes desulfurization and particulate removal in one step.

Extensive testing in a 40-tons/day pilot plant led to the construction of two 800-tons/day demonstration plants in Germany and Sweden.

VEW Coal Conversion Process. This design uses an air-blown, pressurized, entrained slagging gasifier/combined-cycle unit, in which the coal is only partially gasified. Hot exhaust gases and char from the gasifier are fed into a conventional pulverized coal boiler. After syngas is cleaned, it is combusted in a gas turbine. A 240-tons/day pilot plant is located in Germany.

It will be noted that there are a number of similarities in the various designs. Although several designs may be required to satisfy varying compositions of coal feed, it would appear that ultimately coal gasification processes will be fewer in number in the future, once the demonstration plants complete their tests.

Combustion System Modifications

The foregoing processes incorporate the combustion phase as part of their overall design. The combustion operation also has been studied in a piecemeal manner, the results of which include fluidized-bed technology, bulk mixing of coal and limestone (can reduce SO_2 emissions up to 90%), injection of dry sorbents into boiler (or stack), and the injection of natural gas, among other approaches.

Final Emissions Clean-Up. In addition to engineering processes that create smaller amounts of objectionable pollutants, much research continues to neutralize emissions prior to their introduction into the atmosphere. Improvements are being made almost continuously to precipitation and scrubbing systems. As an example, in a process developed in Japan, flue gases are bubbled through a vessel containing an alkaline slurry. Upon absorption of SO_2, the slurry is centrifugally dried to form solid gypsum ($CaSO_4$).

Coal Liquefaction

In addition to cleaning up coal for use in electric-generating plants, research continues for converting coal into useful liquid fuels that can substitute for petroleum-based products, demanded largely for transportation and for use as raw materials by the chemical industry. Present impetus in this area, however, is much less because reasonable costs of petroleum detract from business incentive.

The basic chemistry of coal in terms of creating liquid fuels is the hydrogen-to-carbon ratio. In raw coal this ratio is less than unity, whereas a ratio of 1.5 to 2 or more is needed to produce desirable liquid fuels. In hydroliquefaction processes, the first step is pyrolysis, followed by the addition of hydrogen from another source.

Underground Coal Gasification. *In situ* gasification of coal was suggested by Siemens in 1868, and the first patent for a process was granted to Betts in 1909. The first experiment was conducted in England prior to World War I and in Russia as early as 1933. Much interest was shown for this concept during the energy crisis of the 1970s, and considerable research was undertaken, for example, at the Lawrence Livermore National Laboratory. In the long term, incentives for the concept include the use of coal seams that are unsuitable for mining and less disturbance of land by comparison with present surface-mining techniques. A proposed system is shown in Fig. 8.

---

[2] Steam turbine systems cannot take full advantage of the high temperatures available from the combustion of fossil fuels because of material limitations, nor of the heat available at the low temperature end of the steam cycle because of economic considerations. The combination of two or more heat engine cycles that cover different parts of the temperature range is referred to as a *combined-cycle plant*. The combination of two different cycles represents a *binary cycle*. The addition of a second cycle at the high temperature end of the steam cycle is referred to as a *topping cycle*; if added at the low temperature end, it is a *tailing* or *bottoming* cycle. The most common binary cycle power plants are comprised of a steam turbine cycle topped with a gas turbine cycle. In this combined cycle, the hot exhaust gases from the gas turbine are used to generate steam for the steam for the steam turbine generators. Cycle efficiency near 45% can be achieved.

Additional Reading

Ackerman, B.A. and W.T. Hassler: *Clean Coal/Dirty Air,* Yale University Press, New Haven, CT, 1993.

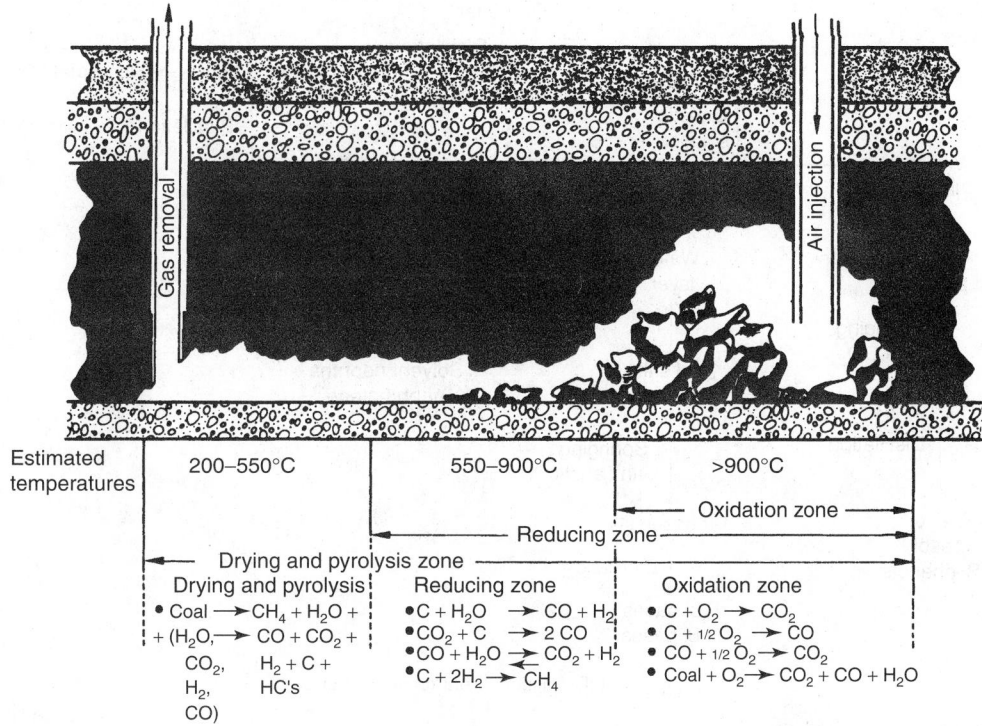

Fig. 8.   Conceptual view of channel during underground coal gasification. (*U.S. Department of Energy*)

Baumol, W.J. and W.E. Oates: *The Theory of Environmental Policy,* Cambridge University Press, New York, NY, 1990.

Borowitz, S.: *Farwell Fossil Fuels: Renewing America's Energy Policy,* Perseus Publishing Group, Boulder, CO, 1998.

Corcoran, E.: "Cleaning up Coal," *Sci. Amer.,* **106** (May 1991).

Davis, G.R.: "Energy for Plant Earth," *Sci. Amer.,* **54** (September 1990).

Hendriks, C.F.: *Carbon Dioxide Removal from Coal-Fired Power Plants,* Kluwer Academic Publishers, Norwell, MA, 1994.

Hertz, N., N. Stewart, and A. Cohn: "High-Efficiency GCC Power Plants," *EPRI J.* (July/August 1992).

Knight, E.H.: *Clean Coal Technology Demonstration Program: Program Update,* DIANE Publishing Company, Collingdale, PA, 2000.

Kubel, E.J., Jr.: "Coal Gasification: A Materials Challenge," *Adv. Materials & Processes,* **37** (March 1989).

Lumpkin, R.E.: "Recent Progress in the Direct Liquefaction of Coal," *Science,* **873** (February 19, 1988).

Miller, B.G.: "Coal-Water Slurry Fuel Utilization in Utility and Industrial Boilers," *Chem. Eng. Progress,* **29** (March 1989).

Schlosberg, R.H.: *Chemistry of Coal Conversion,* Perseus Publishing Group, Boulder, CO, 1985.

Staff: "Coal Gasification: A Worldwide Project," *Adv. Materials & Processes,* **12** (March 1989).

Staff: "Great Plains Project Starts on the Road to Profitability," *Chem. Eng. Prog.,* **7** (September 1989).

Staff: "Groundbreaking for Plant to Make Low-Sulfur Fuels," *Chem. Eng. Progress,* **11** (December 1990).

Srivastava, R.D., et al.: "Coal Bioprocessing," *Chem. Eng. Progress,* **45** (December 1989).

COAL TAR AND DERIVATIVES.

[CAS: 65996-93-2]. Coal tar constitutes the major part of the liquid condensate obtained from the "dry" distillation or carbonization of coal (mostly bituminous) to coke. The three major products of this distillation are (1) metallurgical coke, (2) gas which is suitable as a fuel after appropriate chemical treatment, and (3) condensable liquids which leave the coke oven along with the gas and which are constituted principally of ammonia liquor and coal tar. The condensable materials and gas impurities are separated from gas in the condensation and purification train of the coke oven plant. The purified coke oven gas is used as fuel to heat the coke ovens and steel producing furnaces. Prior to the widespread use of natural gas as a domestic fuel, coke oven gas was widely used for this purpose after additional purification as residential fuel.

Since metallurgical coke for use in blast furnaces is the prime product of coal carbonization, coal tar production is tied closely to the demand for metallurgical coke. Although steel production has increased progressively over many years, the demand for metallurgical coke has remained reasonably steady, for several reasons. Large improvements in blast-furnace efficiency have occurred. The amount of coke required to produce one ton of pig iron has dropped from 1760 pounds coke/ton of pig iron (878 kilograms coke/metric ton of pig iron) to 1250 pounds coke/ton of pig iron (624 kilograms coke/metric ton of pig iron) and in some modern blast furnaces, the rate is below 1000 pounds coke/ton of pig iron (500 kilograms/metric ton of pig iron). Further, coke has been partially replaced by lower-cost carbonaceous materials, such as natural gas, petroleum oils, powder coal, and tar. Fundamental changes in the production of steel are expected to further reduce the need for coke.

A number of years ago, coal tar was the primary, if not the sole, source for hundreds of important organic chemicals and derivatives, notably the phenols, cresols, naphthalene, and anthracene, as well as other important coal tar end-products, such as solvent naphtha and pitch. In recent years, synthetic processes for the production of phenol, the cresols and later the xylenols, have been developed and thus, to a large extent, have pushed coal tar into the background as a source of feedstocks for the chemical industry.

Carbonization Process

Present coking processes generally are of two types: (1) *high-temperature* (900–1200°C) carbonization for producing metallurgical coke and (2) low-temperature (500–750°C) carbonization, still practiced in some countries where there is a market for "semi-coke" as a smokeless home fuel and tars as feedstock for synthetic liquid fuels.

Currently, in the United States, slot ovens with byproduct recovery systems are used almost exclusively. These ovens are built in the shape of narrow chambers placed side by side with interspersed flues for heating. Usually, up to 90 chambers are placed together to form a battery. The chambers are charged individually with coal from the top. After carbonization is complete (14–20 hours), the chambers are discharged on one side by pushing with a ram from the opposite side. Each chamber is connected at the top to one or two collecting ducts, or mains, which carry the gas evolved and the distillate (tar) to suitable coolers and receivers. By high-temperature carbonization, one obtains generally from one ton (metric) of coal: 748 kilograms coke, 343 cubic meters of coke oven gas, and 37.9 liters of tar.

Processing of Crude Tar

With reference to Fig. 1, the crude tar, after being separated from ammonia and other gases, is subjected to an initial distillation (called *topping*) which

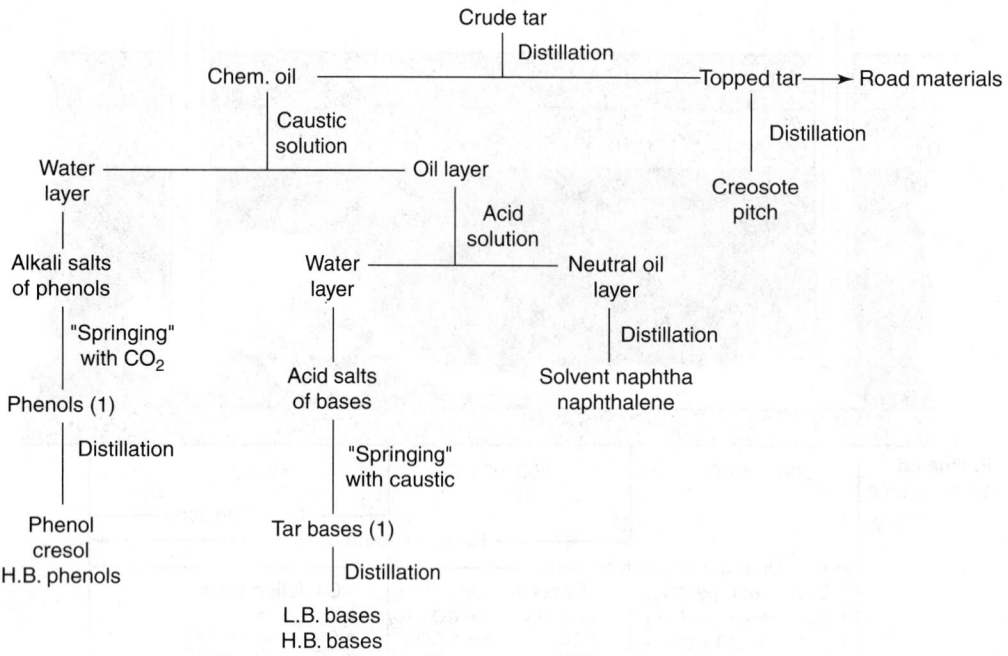

Fig. 1.   Bulk fractions from crude coal tar

separates the desired chemical constituents from the higher-boiling, more viscous tar constituents. In a typical case, the distillate from this operation (sometimes referred to as *chemical oil*) has an upper boiling point of about 250°C and contains (1) phenols (*tar acids*), (2) napthalene, which is the most prevalent single constituent of coal tar (6–10%), (3) pyridine-type bases (*tar bases*), and (4) neutral oils. The tar acids constitute about 1.5–3% of the coal tar.

Tar Acids

These materials are recovered by extraction of the chemical oil with aqueous alkali, usually caustic solution. The aqueous layer is separated from the dephenolized (*acid-free*) oil. The phenols then are recovered in crude form by acidification (*springing*) of the aqueous solution, usually by injecting carbon dioxide, followed by gravity settling. The crude phenols then are fractionated to obtain phenol, cresols, and the higher-boiling phenols (mostly xylenols). See also **Phenol**.

Tar Bases

These materials are extracted from the dephenolized oil with aqueous solutions of mineral acids. This operation may be carried out on the entire neutral oil, or it may be done on the solvent naphtha fraction. In the latter case, only the lowest-boiling bases (picolines and lutidines) are recovered, and the higher-boiling bases (mostly quinoline and isoquinoline) can be recovered from postnapthalene fractions or left in the residue for disposal. In European practice, the topping is carried out so that several fractions are obtained: *carbolic oil*, which yields the phenols and lower-boiling bases; and *naphthalene oil,* from which naphthalene is recovered by crystallization.

The tar bases form water-soluble salts with mineral acids which are separated from the oil. They are recovered from their salts by contacting with aqueous alkali (*springing*) and separating the crude bases from the salt solution. The lutidines constitute the major part of the lower-boiling bases. See also **Pyridine and Derivatives**.

Solvent Naphtha

The lower-boiling fraction of the neutral oil is a very powerful solvent, particularly for coatings containing coal tar and pitch. The material also is a source of unsaturated compounds, such as indene and, in a lesser amount, coumarone and homologues of these compounds. Resins are formed *in situ* from these compounds when solvent naphtha is treated with Friedel-Crafts type catalysts. See also **Friedel-Crafts Reaction**. These resins are useful in the manufacture of inexpensive floor tiles and coatings. The remaining solvent is recovered by distillation and used as a solvent.

Naphthalene

This compound finds a ready market principally for the production of phthalic anhydride. There is a great variety of processes to isolate napthalene from the acid-free or neutral oils. Frequently, the naphthalene is first concentrated by distillation, and the enriched oil then is worked up by crystallization. This process is prevalent in Europe. It is also possible to isolate the naphthalene by careful fractionation. Depending on the purity desired, additional chemical treatments may be required. Naphthalene usually is traded with freezing point as a measure of purity (80.3°C) for pure naphthalene. A good quality commonly used is "78°C" naphthalene, which is about 96% pure.

Topped Tar

With reference Fig. 1, it will be noted that topped tar is the residue remaining from the topping operation where the chemicals are separated as the distillate. The principal use of topped tar is in road materials. A number of standard grades (RT-1 to RT-12) are available, the grade depending on the "consistency" or viscosity of the tar. Road tar has excellent weather and skid resistance, but its use is limited by availability and price as compared with asphalt. This is borne out by the respective amounts used for road building (United States) with about 90% using asphalt.

Creosote

Chemically, creosote is a mixture of a great number of compounds, almost exclusively of cyclic structure. Individual compounds present in creosote in concentrations of 2–4% are acenaphthene, fluorene, diphenylene oxide, anthracene, and carbazole. Only one compound, phenanthrene, is present in a larger concentration (12–14%). For many years, chemists in many countries have tried to isolate individual compounds and to find profitable uses for them. Most of these attempts have failed with exception of those involving anthracene. See also **Anthracene**. The principal use of creosote is for preservation of wood. Railroad ties, poles, fence posts, marine pilings, and lumber for outdoor use are impregnated with creosote in large cylindrical vessels. If properly treated, the life of the wood is greatly extended. Materials that are competitive with creosote for wood-preservation purposes include various petroleum oils, and pentachlorophenol. Pentachlorophenol is used in solutions of creosote or of petroleum oils. Blends of creosote with petroleum oils also are used for economic reasons.

Pitch

This is the residue from the processing of coal tar. Since pitch constitutes over 50% of the crude tar, its utilization has a major effect on the economics of tar processing. Coal tar contains an estimated 5,000–10,000 compounds;

it is reasonable to assume that about half of this number is contained in the pitch. Of the roughly 300 compounds identified in coal tar, about one-half must be in the pitch on the basis of their boiling points. It is probable that none of these compounds is present in pitch in concentrations of more than a fraction of 1%. Coal tar pitch is a black, shiny material that is solid and brittle at low temperatures and liquid at high temperatures. Since it is composed of a great number of different compounds, many of which interact to form eutectic mixtures, it does not show a distinct melting or crystallizing point. Pitch usually is characterized by the *softening point*, which can be measured in several ways.

Because of the importance of pitch in various industries, many studies have been made to elucidate its composition. Solvent fractionation has been used to subdivide pitch into fractions by molecular weight, the higher-molecular-weight fractions requiring more powerful solvents. However, even the highest-molecular-weight compounds are only of moderate molecular weights, ranging up to 6 or 7 condensed rings with molecular weights in the range of 350–400. Constituents isolated from pitch have been identified and appear to be crystalline, well-defined substances. It is generally believed that the glasslike state of pitch is caused by association forces and by the mutual melting-point depression exerted by a large number of multiring compounds that tend to form eutectics. The uses of pitch fall in two general classes: (1) applications based upon the binder properties of carbonized pitch (*pitch coke* or *binder coke*); and (2) uses based upon the other physical qualities of pitch.

Carbon pitch is used for carbon electrodes in electrolytic reduction processes, such as aluminum reduction or the production of electro-steels in arc furnaces. Refractory pitch is used in the manufacture of refractory brick, usually burned magnesite or dolomite, the pores of which are filled with pitch by hot impregnation. Upon firing, the pitch in the brick is converted to carbon by carbonization. The remaining pitch coke within the refractory product retards penetration of molten metals and slags, thus prolonging the life of the brick furnace lining. Coke pitch is used in the production of foundry cores.

Roofing Pitch

A substantial amount of pitch is used as covering membrane on flat roofs on industrial plants, large office and apartment buildings, parking garages, and similar structures. Pitches of 50–60°C softening point generally are used.

GEORGE R. ROMOVACEK
Koppers Company, Inc.
Monroeville, Pennsylvania

G. G. LAUER
Pittsburgh, Pennsylvania

COATING AGENTS (Foods). Substances that are used to protect the surface (and penetration through the surface) of various materials that are being processed or in final-product format. Coating agents are widely used in the pharmaceutical and food industries.

For example, in the food field, fresh produce requires very little processing. Some fruits may be dipped in 1–2% citric acid prior to freezing. This removes any residual lye from the peeling process and any further destruction of ascorbate is prevented. However, citric acid alone is not always sufficient to prevent deteriorative effects during freezing and thus a sequestrant/antioxidant may be added. A solution of 0.5% citric acid with 0.02% D-erythroascorbic acid may be used to prevent browning of some fruits during freezing and defrosting. The procedure is also useful with some vegetables. Plain wax coatings are applied to many packaged cheeses.

Dried fruits, such as raisins, prunes, and figs, require a residual moisture content because most consumers do not like a thoroughly dry or crispy product in this category. Residual moisture, of course, encourages mold and yeast growth. Protection can be gained by dipping the fruit in solutions of 2–7% potassium sorbate, this leaving a fine coating of antimicrobial agents on the fruit pieces.

Acetostearin products (di- and triglycerides) solidify into wax-like solids and are used in some protective coatings for food products. Researchers have shown these products to be effective against moisture penetration, as well as against atmospheric gases. The coating is applied by dipping; spraying in the case of nuts.

For tablet-like confections and pharmaceuticals, sucrose is one of the most common of coating materials. It is transparent, but can be made opaque, and makes an excellent film-like coating. Gums and resins are also used. Some products such as pills and candies can be multicoated by using approved colorants in the coating agents.

The film-forming characteristics of starch have been used for many years. Numerous products can be protectively and decoratively coated with starch. Starch can be added to sugar solutions to provide a less brittle and moisture-sensitive surface. Starch coatings have the advantage of being oil and grease resistant.

Methylcellulose when incorporated into sweet dough products serves to improve adherence of the glaze.

Red meats, poultry, and fish can be coated with a protective film prior to freezing by dipping into successive solutions of 10–15% sodium alginate, 3–5% calcium chloride, and 10–20% glycerol, the latter used as a plasticizer. This coating improves retention of juices upon freezing or thawing. Sodium alginate coating for sausages, alone or with ethylcellulose, prevents salt rust and increases storage stability. Researchers also have studied the effectiveness of carrageenan as a way to prevent oxidative rancidity after freezing. In a Norwegian process, fish are block-frozen in alginate jelly, forming an air-tight coating, preventing oxidative rancidity.

Coatings.   See **Conversion Coatings**; **Paint and Finish Removers**.

COBALT.   [CAS: 7440-48-4]. Chemical element, symbol Co, at. no. 27, at. wt. 58.9332, periodic table group 9, mp 1495°C, bp 2870°C, density 8.832 g/cm^3.

There are two allotropic modifications of cobalt, a close-packed hexagonal form (ε) with space group P6$_3$/mmc, stable at temperatures below 417°C; and a face-centered cubic form (α) with space group Fm3m, stable at higher temperatures—up to the melting point. The metal is silvery gray in color. The only naturally occurring isotope, ^{59}Co, is stable, but the other twelve known isotopes are radioactive, their mass numbers ranging from 54 to 64. Half lives range from 0.2 second for ^{54}Co to 5.3 years for the industrially and medically important ^{60}Co. See also **Radioactivity**.

Cobalt was identified and described by Georg Brandt in 1735, but had to wait until the last decade of the nineteenth century before the new sources of metal supply from New Caledonia and Canada stimulated its metallurgical usage.

First and second ionization potentials are 7.86 eV and 17.05 eV, respectively; oxidation potential E° is −0.277 V. Further general specifications are given under **Chemical Elements**.

Occurrence

The cobalt content of the earth's crust is estimated to be within the range 10 to 40 ppm. Economic concentrations of the element are the exception so that supply is governed by its byproduct output from ores mined for the recovery of other elements, particularly copper and nickel. For technical reasons, the sulfides, arsenides and oxidized minerals form almost the entire economic source of the metal. Its production is restricted to a relatively few countries, the most important being the Republic of Zaire and Zambia in Africa, Canada, Finland, Morocco, and the United States. The cobaltiferous deposits in the copper belt of central southern Africa vary in content from 1 to 30 parts of cobalt to 100 parts of copper, with an estimated hundreds of millions of tons containing 2 to 15 parts cobalt to 100 parts copper. Sulfide nickel ores the world over have a cobalt content in the range 2 to 5 (very occasionally as high as 10) parts to 100 parts of nickel; oxide nickel ores vary in cobalt content from 1 to 30 parts per 100 parts nickel, with estimated thousands of millions of tons containing 5 to 15 parts cobalt per 100 nickel. The nickel lateritic deposits in the tropical and subtropical areas of the world represent a large potential future source of both nickel and cobalt.

Manheim (1986) reported that ferromanganese oxides in the open oceans are more enriched in cobalt than any other widely distributed sediments or rocks. Concentrations of cobalt in excess of 1% in ferromanganese crusts on seamounts, ocean ridges, and other raised areas of the ocean have been found. Some experts have observed that these cobalt-rich crusts are among the slowest-growing of any material source on earth, with estimates of an accumulation of only one molecular layer over a period of a few months. As observed to date, most of the cobalt-rich crusts are situated within the exclusive economic zone of the United States. This is in contrast with abyssal cobalt nodules, which occur mainly in international waters and are also difficult to recover. The U.S. Department of the Interior is preparing an environmental impact statement pertaining to the recovery of the ferromanganese crusts.

Mining and Recovery

The economically important cobalt-containing minerals exploited at present are listed in Table 1. The metal extraction processes, following the usual pretreatment of the ore, are varied and complicated because the metallurgical properties of cobalt differ insufficiently from those of the associated metals and because the cobaltiferous raw materials comprise the sulfide, arsenide, and oxide, or a mixture of these. The final refining stage invariably involves electrolysis.

Investigators at the Argonne National Laboratory (1985) reported a two-step process for extracting cobalt and manganese from low- and medium-grade ores normally mined for other metals. The new process provides an important secondary source for these strategically important metals. More detail is given in the entry on **Manganese**.

Uses of Metal and Alloys

The metallurgical applications of cobalt consume approximately 70–75% of the world production; the remainder goes into chemicals. Relatively little use has been made of the pure metal, the most important being as the radioisotope ^{60}Co for teletherapy and industrial radiation processing and gamma radiography. The increasing availability of the metal in various wrought forms will stimulate its applications in those areas where the intrinsic properties of cobalt are advantageous, e.g., magnetic devices, wear resistance and bearing properties at elevated temperatures, and the manufacture of heterogeneous welding rods for hardfacing. Major uses of the metal may be classified as follows:

High-temperature Materials. To meet the demand of the gas turbine industry, alloys capable of reliable performance at high temperature and loads have been developed based on cobalt and cobalt-containing nickel and iron-based alloys. The improvement in properties at these very high temperatures ($\sim$750–1200°C) has been brought about by the addition to the oxidation- and sulfidation-resistant cobalt-chromium matrix of varying amounts of refractory metals, mainly tungsten, molybdenum, tantalum, and columbium (niobium) to strengthen the matrix further and promote the formation of stable carbides, which makes a substantial contribution to the high-temperature strength. The application of current techniques of vacuum melting and casting insure optimum material properties and improved component reliability. A new family of alloys based on Co-Fe-Cr is being successfully applied in the exacting conditions that exist in metallurgical furnaces and petrochemical plants.

Magnetic Materials. For the past 50 years, apart from iron, cobalt has been the most important constituent of permanent magnet materials. This is because cobalt additions raise the saturation magnetization and the Curie temperature to higher values than are obtained from pure iron. Present advanced Alnico type cast alloys are the results of 40 years of development and are the most widely used permanent magnet materials. More recently, permanent magnet materials based on cobalt-rare-earth (e.g., samarium, praseodymium) compounds have emerged as the most powerful magnets available.

Hardfacing and Wear-resistant Alloys. These materials, essentially quaternary alloys of cobalt; chromium, tungsten (or molybdenum); and carbon, are widely used for industrial hardfacing purposes. They can be deposited by welding techniques, sprayed on as powders, or produced as separate castings. By using the weld deposition technique, a highly alloyed heat-, wear-, and corrosion-resistant surface can be applied to a much cheaper substrate, e.g., mild steel. Used originally as a component reclamation process, these alloys and techniques are now primary design requirements for many engineering items. The feasibility of electrode-positing composite cobalt-carbide coatings from an agitated slurry of solid carbide particles in a conventional plating bath has been demonstrated. These coatings provide excellent wear protection and are now established in the aerospace industry, while other engineering applications are being evaluated.

The demand for wear resistance allied with a low coefficient of friction has been successfully met by cobalt-molybdenum-silicon alloys. Structurally they conform to the well-proven bearing concept of hard intermetallic phases dispersed through a softer cobalt matrix. They offer outstanding bearing performance in conditions of poor lubrication.

As of the early 1990s the historical importance of cobalt in alloys for reducing wear resistance was markedly diminished by the development of *Norem* alloys, which were derived from stainless steel, particularly the Armco *Nitronic 60*, admittedly one of the few stainless steels that has excellent wear resistance. This is an outgrowth of research conducted by EPRI (Electric Power Research Institute) in an effort to lower equipment costs. Initial uses of the new cobalt-free alloys will be power plant valves and turbines. Also, in nuclear reactor applications, the *Norem* alloys will not become activated (Co does), reducing worker-protection and maintenance costs. Tests have confirmed that the new cobalt-free alloys retain their wear-resistant properties when produced as rods and powders, the most common forms used by welders.

Alloy Steels. The addition of cobalt to high-speed tool steel was one of the earliest uses of cobalt. This application is represented by the super-high-speed tool steel grades, which contain 5–12% cobalt. Two other cutting tool materials, the nonferrous cobalt-chromium tungsten-carbon cast alloy and the cemented carbides also provide a steadily growing outlet for cobalt. Hot work die steels are another group of alloy steels which benefit from the effect cobalt has on the tempering characteristics and consequent hot strength retention. The most significant development in recent years, however, is the advent of the Ni-Co-Mo miraging family of steels, which combine very high strength and toughness properties to an unusual degree. This is still an area of active development and growing engineering utilization.

Toxicity. Cobalt, like most other metals, is not entirely harmless, although it is not in any way comparable to the known toxic metals, such as mercury, cadmium, and lead. Inhalation of fine cobalt dust over long periods can cause an irritation of the respiratory organs, which may result in chronic bronchitis. Complete recovery is usually achieved upon removal from the contaminated atmosphere. Cobalt salts can cause benign dermatoses, either in people new to handling them, or after prolonged exposure, usually several years.

Chemistry of Cobalt. The metal in the massive form is not attacked by air or water at temperatures below approximately 300°C; above this temperature it is oxidized in air. The metal combines readily with the halogens to form the respective halides. It combines with most of the other metalloids when heated or in the molten state. It does not combine directly with nitrogen but decomposes ammonia at elevated temperature to form a nitride. It reacts with carbon monoxide at 225–230°C to form the carbide, Co_2C. Cobalt also forms intermetallic compounds with many metals, e.g., Al, Cr, Mo, Sn, Ti, V, W and Zr. Metallic cobalt is readily

TABLE 1. PRINCIPAL ECONOMICALLY IMPORTANT COBALT MINERALS

Group and Name	Ideal Formula	Cobalt Content %	Source Area
Sulfides			
Linnaeite	$(Co, Cu, Ni, Fe)_3S_4$	Up to $\sim$48	U.S.A., Zaire, Finland
Carrolite	$CuCo_2S_4$	Up to $\sim$38	Zambia, Zaire, U.S.A.
Pentlandite	$(Fe, Ni, Co)_9S_8$	Up to $\sim$2	Canada
Cobaltiferous Pyrite	$(Fe, Co)S_2$	Up to $\sim$2	Canada, Finland, Zambia
Arsenides			
Skutterudite	$(Co, Ni, Fe)As_3$	Up to $\sim$28	Canada, Morrocco, U.S.A.
Gersdorffite	$(Ni, Co, Fe)AsS$	Up to $\sim$12	Canada, Zaire, U.S.A.
Oxides			
Heterogenite	$Co_2O_3 \cdot H_2O$	Up to $\sim$57	Zaire, Zambia
Asbolite	$(Co, Ni)O_2MnO_24 \cdot H_2O$	Up to $\sim$27	New Caledonia, Canada

Note: See also **Pentlandite**; and **Skutterudite**.

dissolved in dilute sulfuric, hydrochloric or nitric acids to form cobaltous salts. Like iron, cobalt is passivated by strong oxidizing agents, such as the dichromates. It is slowly attacked by ammonium hydroxide and sodium hydroxide.

Cobalt Compounds. In general the chemical properties are intermediate between those of iron and nickel. The predominant oxidation states of cobalt compounds, except for a large class of organometallic compounds, are 2+ and 3+. Common usage assigns the terms *cobaltous* and *cobaltic*, respectively, to these.

In aqueous solutions and in the absence of complexing agents, cobalt compounds are stable only in the 2+ oxidation (cobaltous) state. In the complexed state the cobaltous ion is relatively unstable, being readily oxidized to the 3+ oxidation (cobaltic) state. An extremely large number of 3+ complex ions have been identified, most of which are quite stable in aqueous media.

Cobalt has an electronic configuration $1s^2 2s^2 2p^6 3s^2 3p^6 3d^7 4s^2$. The two 4_s electrons are readily removed producing the ordinary Co^{2+} ion. In principle, if the odd $3d$ electron were removed the simple cobaltic ion Co^{3+}, would be formed. This however does not occur; the ion exists only in complex ions or crystal lattices in which cases additional electron orbitals are filled.

Cobalt and oxygen form two stable oxides, cobaltous oxide, CoO, stable below 200°C and above 900°C; and cobalto-cobaltic oxide, Co_3O_4, which is stable below 900°C. Between 200 and 900°C the CoO oxidizes partially, or completely, to Co_3O_4.

Cobaltous oxide, [CAS: 1307-96-6], is usually prepared by heating the carbonate. It is insoluble in H_2O, NH_4OH, and alcohol, but dissolves in cold strong acids and in weak acids on heating. Commercial gray oxide, which may contain up to 40% Co_3O_4, is used in the ceramic, glass and enamel industries and also in the production of catalysts. It is also used in the preparation of cobalt metal powder. The Co(III) oxide can also be prepared by calcining oxides, hydroxide and salts. The oxide is insoluble in H_2O and only slightly soluble in acids. The commercial black oxide consists essentially of Co_3O_4 with possibly up to 20% of CoO.

Cobalt(II) hydroxide exists in two allotropic forms, a blue α-Co $(OH)_2$ and a pink β-Co$(OH)_2$. The hydroxide is prepared by precipitation from a cobaltous salt solution by an alkali hydroxide. When the alkali is in excess the pink β-form is produced—the blue α-form is produced when the cobalt salt is in excess. The salt slowly oxidizes in air at room temperature and changes to hydrated cobaltic oxide, $Co_2O_3 \cdot nH_2O$. The hydroxide is practically insoluble in H_2O and in bases, but highly soluble in mineral and organic acids. The commercial salt is used as the starting material in the preparation of drying agents.

Cobaltous halides are formed with the halogen group, but only fluorine forms a stable cobaltic compound. The other cobaltic halides are stable only in complex ions.

Commercial cobaltic fluoride is formed by the reaction of cobalt, its oxides or simple salts with ClF_3 or BrF_5 to yield a brown anhydrous salt. This is a powerful fluorinating agent readily replacing hydrogen in aliphatic and aromatic hydrocarbons. Cobaltous chloride is a pale-blue compound and very hygroscopic. A series of hydrates is known; the hexahydrate is pink but becomes blue on warming. It is extremely soluble in H_2O, and in numerous organic solvents. The commercial salt is used as a starting material in the manufacture of catalysts, in electroplating, agricultural chemicals, and pharmaceuticals. Cobaltous bromide is highly hygroscopic and transforms gradually to the red hexahydrate. The salt is mainly used in the catalysts industry.

A number of sulfides have been reported, the best characterized being Co_4S_3, Co_9S_8, CoS, Co_3S_4, and CoS_2. They are prepared either by metal or salt solution reaction with S or H_2S. The mixed sulfides of cobalt and molybdenum have catalytic properties of hydrogenation and isomerization.

The sulfate, one of the more important industrial salts of cobalt, is usually available as the red heptahydrate, an efflorescent substance. On heating it converts to the dark-blue monohydrate, and after prolonged heating above 250°C the red anhydrous salt. It is widely used in the electroplating and ceramic industries, and in the preparation of drying agents and agricultural pasture top-dressing.

Cobalt and nitrogen form three nitrides, Co_3N_3, Co_2N, and CoN, products of metal/ammonia reaction and compound decomposition. All are gray-black or black in color.

In its commercial form cobalt nitrate appears as the red hexahydrate, $Co(NO_3)_2 \cdot 6H_2O$ formed by dissolving the metal oxide or carbonate in dilute HNO_3 and concentrating the solution. The compound deliquesces in moist air and effloresces in dry air. The pink anhydrous cobaltous nitrate cannot be formed by dehydrating the hexahydrate but by treating the salt with nitrogen pentoxide gas (or in solution in concentrated HNO_3). The salt is used mainly in the preparation of catalysts.

Two discrete carbides have been characterized—Co_3C and Co_2C. The former, which has the same structure as Fe_3C, has been prepared by reacting cobalt with coal gas at temperatures between 500–800°C. Co_2C is formed by the reaction of carbon monoxide at atmospheric pressure on cobalt powder at 225–230°C.

Cobaltous carbonate, $CoCO_3$, is found almost pure in the mineral sphaerocobaltite in the Republic of Zaire and less extensively in Zambia. The pale-red anhydrous salt is obtained by reaction in solution of an alkaline carbonate and a cobaltous salt under a slight pressure of carbon dioxide (up to 1 atmosphere) and subsequent heating at 140°C. The commercial salt is violet-red in color, partially hydrolyzed with an indeterminate composition. It is insoluble in H_2O and alcohol, but dissolves easily in inorganic and organic acids, and is often used for the preparation of other salts. According to the thermal conditions it decomposes to the different types of oxides.

Cobaltous acetate, [CAS: 71-48-7], $Co(CH_3CO_2)_2$, is obtained as a pink salt by the dehydration of the tetrahydrate which is prepared by dissolving the hydroxide or carbonate in acetic acid. The tetrahydrate, which is the commercial form, is widely used in the preparation of catalysts, e.g., OXO synthesis, and dryers for inks and varnishes.

Cobaltous oxalate dihydrate, $CoC_2O_4 \cdot 2H_2O$ (pink), is obtained by adding oxalic acid or an alkaline oxalate to a cobaltous salt solution. This is the commercial form of the salt and is important as the starting material in the preparation of cobalt metal powders.

Cobalt(III) coordination compounds are the classics of coordination chemistry, many of the cobalt(III) amines having been prepared in the nineteenth century. They are invariably colored and undergo reaction slowly. Because of this many isomers have been isolated and studied. Much of the knowledge of isomerism, mechanisms of reaction, and general properties of octahedral species are based on cobalt(III) compounds. The important donor atoms (in order of decreasing tendency to complex) are nitrogen, carbon in the cyanides, oxygen, sulfur, and the halogens. The coordination number is invariably six. An extensive class of amines and compounds of the amines is known ranging from hexamine $[CoN_6]^{3+}$ through to monoamines $[CoNX_5]^{2-}$. The compounds are made in several stages by first oxidizing the cobalt(II) species and then various different substituted species are prepared by substitution reactions on the primary cobalt(III) product. When air is drawn through an aqueous solution containing cobalt(II) and ammonia the solution turns brown. From this mixture can be obtained a variety of products which depend on the initial concentrations of reactants, the pH of the solution, the anion present and the presence of heterogeneous catalysts, such as charcoal. The products are invariably colored ranging from blue-violets and green through various shades of red and brown to yellow. The absorption spectra are characteristic of the various structures. Complex cyanides of the types $[Co(CN)_6]^{3-}$ and $[Co(CN)_5X]^{3-}$ are stable and diamagnetic. The anion $[Co(CN)_6]^{3-}$ is pale yellow and is the ultimate product of the reaction when a solution of cobalt(III) cyanide in aqueous KCN is boiled in air. It is very unreactive, being untouched by chlorine, peroxide, alkali, aqueous HCl and H_2S, although it gives CO when treated with concentrated H_2SO_4.

The only cobalt(IV) compound appears to be $Cs_2[CoF_6]$ which is prepared as a yellow powder by the fluorination of Cs_2CoCl_4.

Evidence for cobalt(I) was first obtained from the electrolytic reduction of cyano-compounds and some of the reduced species have been isolated. There are also many cobalt(I) coordination compounds of the organometallic class carbonyl, isonitriles, and unsaturated hydrocarbon derivatives. The oxidation state cobalt(0) may be represented in the cyano-compound which has been formulated as $K_8[Co_2(CN)_8]$. It has been prepared as an air-sensitive brown-violet compound by reducing a liquid ammonia solution of $K_3[Co(CN)_6]$ with an excess of K metal. The only other known cobalt(0) species are organometallic compounds.

For the role of cobalt in biological systems, see **Cobalt (In Biological Systems)**.

Note: A large part of this article was prepared by E. WILLIAMS Cobalt Information Centre, London

Additional Reading

Davis, J.R.: *Metals Handbook*, 2th Edition, American Society for Metals International, Metals Park, OH, 1998. *http://www.asm-intl.org/*

Houston, B.: "Cobalt-Free Alloys Resist Wear," *Chem. Eng. Progress*, **9** (December 1990).

Lewis, R.J. and N.I. Sax: *Sax's Dangerous Properties of Industrial Materials*, 10th Edition, John Wiley & Sons, Inc., New York, NY, 1999.

Lide, D.R.: *CRC Handbook of Chemistry and Physics*, 84th Edition, CRC Press, LLC., Boca Raton, FL, 2003.

Manheim, F.T.: "Marine Cobalt Resources," *Science*, **232**, 600–608 (1986).

Niahizawa, T. and K. Ishida: "The Co-Fe Binary," *Metal Progress*, **129**(2), 57–58 (February 1986).

Parker, P.: *McGraw-Hill Encyclopedia of Chemistry*, 2nd Edition, McGraw-Hill Companies, Inc., New York, NY, 1993.

Perry, R.H. and D.W. Green: *Perry's Chemical Engineers' Handbook*, 7th Edition, McGraw-Hill Professional Book Group, New York, NY, 1999.

Sinfelt, J.H.: "Bimetallic Catalysts." *Sci. Amer.*, **253**(3), 90–98 (September 1985).

Staff: *ASM Handbook—Properties and Selection: Nonferrous Alloys and Pure Metals*, ASM International, Materials Park, OH, 1990.

COBALT (In Biological Systems).

Although cobalt is regarded as an essential element for animals, including humans, the element can perform its essential functions only after it has been incorporated into the vitamin B_{12} molecule. The microorganisms living in ruminants are the major producers of vitamin B_{12} in the food chain. Green plants do not synthesize the vitamin. The normal intake of this vitamin is by way of milk, cheese, meat, and eggs. Persons who follow a strictly vegetarian diet may become deficient in vitamin B_{12} unless supplementary sources are used. Single-stomach domestic and wild animals receive their vitamin B_{12} from animal flesh or from animal fecal material. See also **Vitamin B_{12} (Cobalamin)**.

Cobalt is present in vitamin B_{12} to the extent of about 4%. Lack of cobalt in the soil and feedstuffs prevents ruminants from synthesizing all of the vitamin B_{12} for their needs. Thus, cobalt can be added to feedstuffs as the chloride, sulfate, oxide, or carbonate. Excessive cobalt intakes are toxic, causing a reduction in feed intake and body weight, accompanied by emaciation, anemia, debility, and elevated levels of cobalt in the liver. It is of interest to note that clinical cobalt toxicity closely resembles clinical cobalt deficiency.

Cobalt is required by the microorganisms that live in nodules on the roots of legumes, such as bean and clover. They convert nitrogen from the air into chemical forms that can be used by higher plants. This is possibly the only well established and understood function of cobalt in plant growth. Legumes may grow normally and the microorganisms on their roots fix atmospheric nitrogen, even though the forage does not contain sufficient cobalt to meet the requirements of ruminants.

Areas of low cobalt content in the United States, where clovers and alfalfa are too low in cobalt content to meet requirements of cattle and sheep, include northeastern Maine, all of New Hampshire, Vermont, Massachusetts, Connecticut, and Rhode Island, much of New York with exception of the central portion, the northwestern portion of the lower peninsula of Michigan, a small area in Illinois with Peoria at its approximate center, all but eastern Iowa, and southwestern Minnesota. The low-cobalt soils of New England are primarily sandy and were formed from glacial deposits near and to the south of the White Mountains of New Hampshire. Along the south Atlantic Coastal Plain, legumes with very low concentrations of cobalt are primarily on the sandy soils formed in naturally wet areas. These soils, which are called spodosols, have light-colored subsurface layers overlying a dark-brown or dark-gray hard-pan layer.

Grasses and cereal grains generally contain less than the 0.07 to 0.10 parts per million of cobalt required by ruminants. Cattle and sheep that are not fed any legumes nearly always require cobalt supplementation.

Adding cobalt to soils, either as cobalt sulfate, or as cobaltized superphosphate, can be used to increase the level of cobalt is plants and prevent cobalt deficiency in cattle and sheep. Cobalt fertilization may not be effective in preventing cobalt deficiency on alkaline soils because in these soils, the added cobalt quickly reverses to forms that are not taken up by plants. Cobalt fertilization is more common in Australia than in the United States. In the United States, cobalt is usually added to mixed feeds, mineral mixes, or salt licks.

Still another method is to place heavy ceramic "bullets" containing cobalt in the animal's rumen. These bullets remain in the rumen and slowly release cobalt to meet the animal's needs for a long period. The diets of hogs and chickens are often supplemented with concentrated forms of vitamin B_{12}.

The relationship of the levels of cobalt in soils and plants to the health of ruminants is one of the striking examples of the importance of a soil and plant relationship to animal health. When some Australian scientists discovered this relationship, new areas in several parts of the world became usable for animal production. The vitamin B_{12} formed within cattle and sheep in these new areas contributed to the vitamin B_{12} nutrition of people, even though adding cobalt to soils does not directly affect human nutrition in the absence of the production of ruminants. Cobalt-deficient grazing soils are found in Australia, New Zealand, and, in the United States, mainly in Florida, although deficient regions are found elsewhere as previously mentioned. Cobalt deficiency can result in a condition known as "pinining disease," where the affected animals are quite listless. The disease is also known as "bush sickness" and "salt sickness."

COBALTITE. The mineral cobaltite is a sulfarsenide (see **Arsenic**; and **Sulfur**) of cobalt, corresponds to the formula CoAsS, crystallizing in the isometric system as cubes or pyritohedrons, also may be massive. Cobaltite has a very good cleavage parallel to the cube faces; uneven fracture; brittle; hardness, 5.5; specific gravity, 6.33; metallic luster; color, silvery-white to reddish, sometimes steel gray or violet to grayish-black; streak, grayish-black. Cobaltite is found with cobalt and nickel minerals deposited commonly by metasomatic processes. It is found in Sweden, Norway, England and the Province of Ontario. It is an ore of cobalt.

COCAINE.   See **Alkaloids**.

COCONUT OIL.   See **Vegetable Oils (Edible)**.

CODALAMIN.   See **Vitamin B_{12} (Cobalamin)**.

CODEINE.   See **Alkaloids**.

COENZYMES. A nonprotein substance that is closely associated with or bound to the protein component (*apoenzyme*) of an enzyme. Together, the coenzyme and apoenzyme form the complete enzyme known as the *holoenzyme*. The presence of a coenzyme is necessary for enzyme activity. Coenzymes are organic molecules of a size intermediate between the small-molecule intermediary metabolites, which serve as the substrates of enzymatic reactions, and the macromolecular proteins. Each coenzyme acts usually as acceptor or donor of some specific type of atom or group of atoms to be removed from or added to a small-molecule substrate in a reaction catalyzed by the holoenzyme.

Coenzyme A (CoA)

Pantothenic acid is a constituent of coenzyme A, which participates in numerous enzyme reactions. CoA (Fig. 1) was discovered as an essential cofactor for the acetylation of sulfanilamide in the liver and of choline in the brain. It has been established that CoA is involved in many biochemical reactions in the body as an "activator" of normally less reactive carbon fragments and a "transferer" of these fragments to different molecules. CoA is particularly important in the initial reaction of the citric acid cycle of carbohydrate metabolism and energy production. After oxidative decarboxylation of pyruvic acid, CoA combines with the two-carbon acetate fragment to form acetyl-CoA or "active" acetate.

$$CH_3-\overset{\overset{\displaystyle O}{\|}}{C}-COOH-CoA + CH_3 \longrightarrow CH_3-\overset{\overset{\displaystyle O}{\|}}{C}-CoA + CO_2$$

(Pyruvic acid)   Acetyl-CoA
                                  ("active acetate")

Coenzyme A is necessary for the activation, synthesis, and degradation of fatty acids. Synthesis of cholesterol and ultimately the production of steroid hormones are also coenzyme A dependent.

Nicotinic Acid Coenzymes

Nicotinic acid can be converted to nicotinamide in the body and, in this form, is found as a component of two oxidation-reduction coenzymes (Fig. 2): *nicotinamide adenine dinucleotide* (NAD); and *nicotinamide adenine dinucleotide phosphate* (NADP). The nicotinamide portion of the coenzyme transfers hydrogens by alternating between an oxidized quaternary nitrogen and a reduced tertiary nitrogen. See Fig. 3.

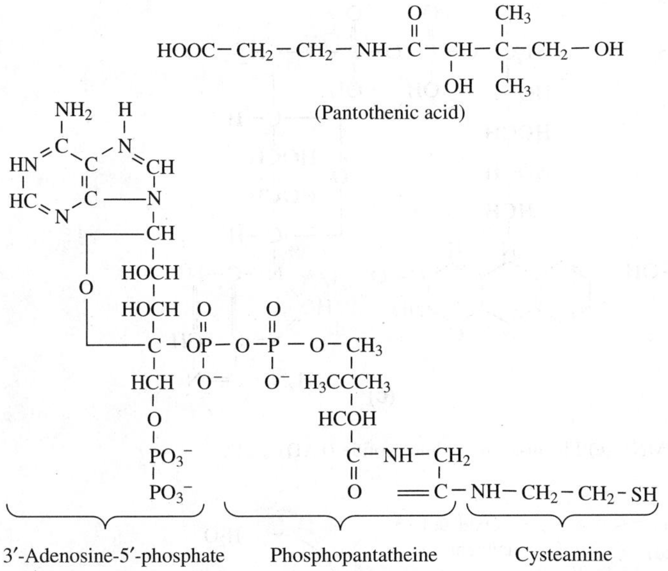

HOOC—CH₂—CH₂—NH—C—CH—C—CH₂—OH
(Pantothenic acid)

3′-Adenosine-5′-phosphate   Phosphopantatheine   Cysteamine

Fig. 1.   Structure of CoA, composed of three parts: a nucleotide part derived from 3′-adenosine-5′-phosphate, forming a phosphodiester bond with a 4-phospho derivative of pantothenic acid, and a third part derived from the amino acid, *cysteine*. The side chain SH group of the latter is free in this compound and is readily acylated, and thus able to act as a carrier for acyl groups in biochemical reactions in which it transfers that group between two substrates

Enzymes that contain NAD or NADP are usually called *dehydrogenases*. In excess of fifty NAD-dependent enzyme systems are known to exist. They participate in many biochemical reactions of lipid, carbohydrate, and protein metabolism. An example of an NAD-requiring enzyme is lactic dehydrogenase, which catalyzes the conversion of lactic acid to pyruvic acid. NADP is an essential coenzyme for glucose-6-phosphate dehydrogenase which catalyzes the oxidation of glucose-6-phosphate to 6-phosphogluconic acid. This reaction initiates metabolism of glucose by a pathway other than the citric acid cycle. The alternate route is known as the phosphogluconate oxidative pathway, or the hexose monophosphate shunt. The first step is:

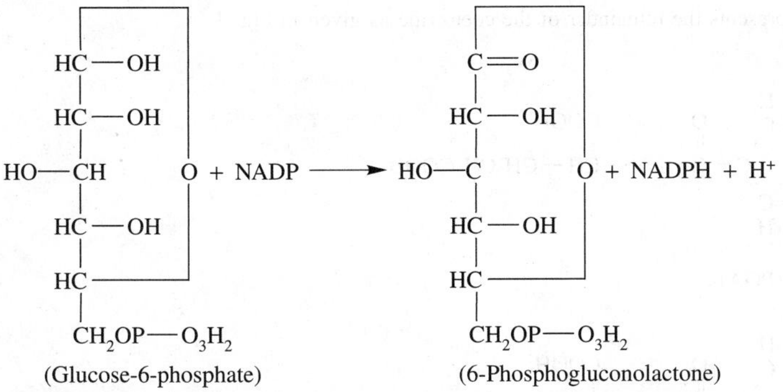

(Glucose-6-phosphate)   (6-Phosphogluconolactone)

In the biological oxidation-reduction system, reduced NAD (i.e., NADH) is reoxidized to NAD by the riboflavin-containing coenzyme FAD (*flavin-adenine dinucleotide*).

Riboflavin Coenzymes

Riboflavin has been shown to be a constituent of two coenzymes: *flavin mononucleotide* (FMN) and *flavin adenine dinucleotide* (FAD). See Fig. 4. FMN was originally discovered as the coenzyme of an enzyme system that catalyzes the oxidation of the reduced nicotinamide coenzyme, NADPH, to NADP. Most of the many other riboflavin-containing enzymes contain FAD. FAD is an integral part of the biological oxidation-reduction system, where it mediates the transfer of hydrogen ions from NADH to the oxidized cytochrome system. This is illustrated in Fig. 5. FAD can also accept hydrogen ions directly from a metabolite and transfer them, either to NAD, a metal ion, to a heme derivative, or to molecular oxygen. The various mechanisms of action of FAD are probably due to differences in the protein apoenzymes to which it is bound. The oxidized and reduced states of the flavin portion of FAD are shown in Fig. 6.

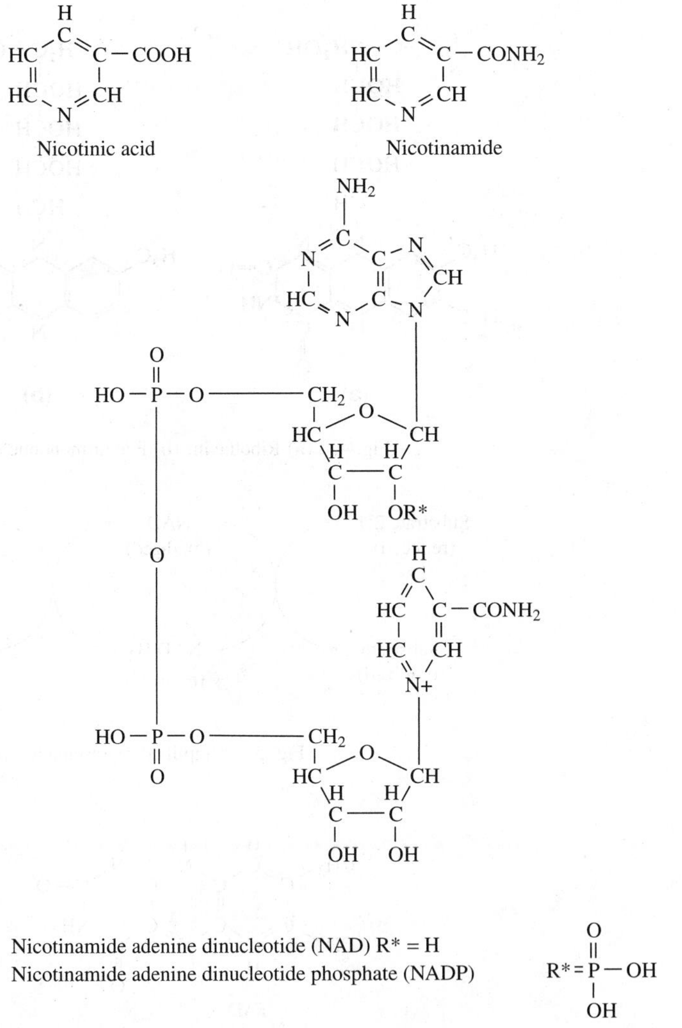

Nicotinic acid   Nicotinamide

Nicotinamide adenine dinucleotide (NAD) R* = H
Nicotinamide adenine dinucleotide phosphate (NADP)   R* = P—OH

Fig. 2.   Structures of nicotinic acid, nicotinamide, and nicotinamide coenzymes

NAD (oxidized)   NADH + H⁺ (reduced)

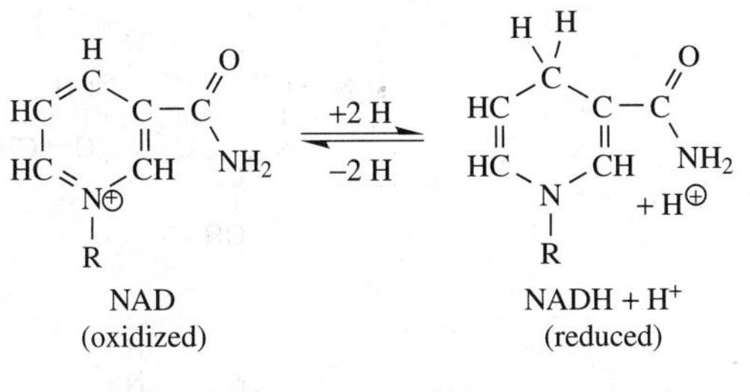

Fig. 3.   Oxidized and reduced states of nicotinamide coenzymes as shown in Fig. 2

Decarboxylation Coenzymes

Thiamine, biotin and pyridoxine (vitamin B) coenzymes are grouped together because they catalyze similar phenomena, i.e., the removal of a carboxyl group, COOH, from a metabolite. However, each requires different specific circumstances. Thiamine coenzyme decarboxylates only alpha-keto acids, is frequently accompanied by dehydrogenation, and is mainly associated with carbohydrate metabolism. Biotin enzymes do not require the alpha-keto configuration, are readily reversible, and are concerned primarily with lipid metabolism. Pyridoxine coenzymes perform nonoxidative decarboxylation and are closely allied with amino acid metabolism.

Folic Acid Coenzymes

The coenzyme forms of folic acid are derivatives of tetrahydrofolic acid, FH₄. See Fig. 7. Folic acid functions as a coenzyme in enzyme reactions which involve the transfer of one-carbon fragments at various levels of

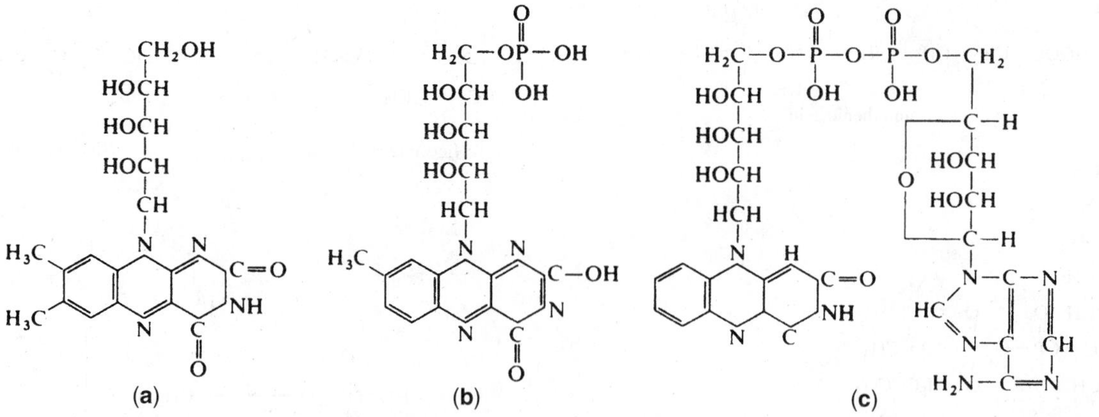

Fig. 4.   (a) Riboflavin; (b) Flavin mononucleotide (FMN); (c) Flavin-adenine dinucleotide (FAD)

Fig. 5.   Simplified representation of the biological oxidation-reduction system

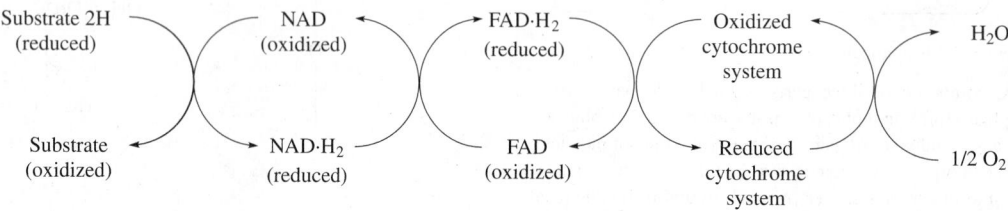

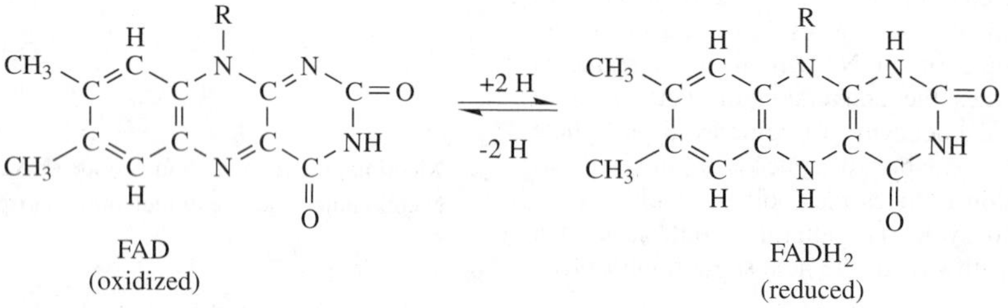

FAD
(oxidized)

$+2\,H$
$-2\,H$

FADH$_2$
(reduced)

Fig. 6.   Oxidized and reduced states of flavin coenzymes. R represents the remainder of the coenzyme as given in Fig. 4

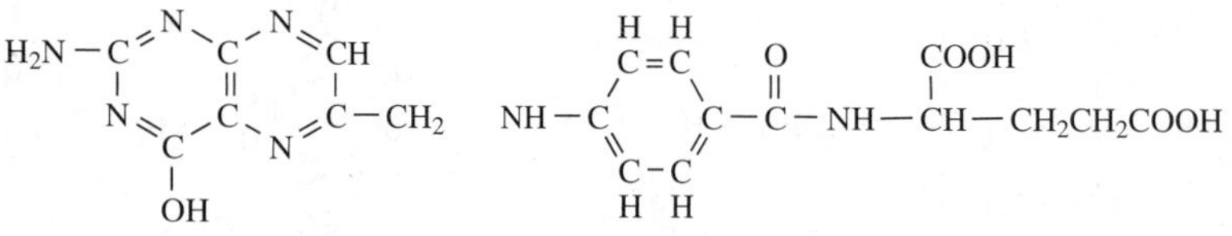

Folic acid (PGA)

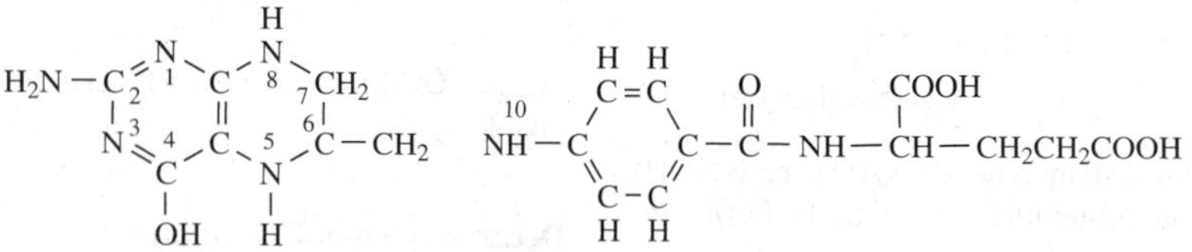

Tetrahydrofolic acid (H$_4$-PGA)

Fig. 7.   Structure of folic acid and tetrahydrofolic acid

oxidation. Vitamin B$_2$ (*cobalamin*) may be interrelated with folic acid in these reactions. Folic acid and vitamin B$_{12}$ are also considered together since certain clinical anemias can be corrected by administration of either of the two vitamins.

Coenzyme Q

A series of quinones which are widely distributed in animals, plants, and microorganisms, these quinones have been shown to function in biological electron transport systems which are responsible for energy conversion with living cells. The nature and significance of coenzyme Q was first recognized in 1957. In structure, the coenzyme Q group closely resembles the members of the vitamin K group and the tocopherylquinones, which are derived from tocopherols (vitamin E), in that they all possess a quinone ring attached to a long hydrocarbon tail. The quinones of the coenzyme Q series which are found in various biological species differ only slightly in chemical structure and form a group of related, 2,3-dimethoxy-5-methyl-benzoquinones with a polyisoprenoid side chain in the 6-position which varies in length from 30 to 50 carbon atoms. Since each isoprenoid unit in the chain contains five carbon atoms, the number of isoprenoid units in the side chain varies from 6 to 10. The different members of the group have been designated by a subscript following the Q to denote the number of isoprenoid units in the side chain, as in coenzyme Q_{10}. The members of the group known to occur naturally are Q_6 through Q_{10}.

Coenzyme Q functions as an agent for carrying out oxidation and reduction within cells. Its primary site of function is in the terminal electron transport system where it acts as an electron or hydrogen carrier between the flavoproteins (which catalyze the oxidation of succinate and reduced pyridine nucleotides) and the cytochromes. This process is carried out in the mitochondria of cells of higher organisms. Certain bacteria and other lower organisms do not contain any coenzyme Q. It has been shown that many of these organisms contain vitamin K_2 instead and that this quinone functions in electron transport in much the same way as coenzyme Q. Similarly, plant chloroplasts do not contain coenzyme Q, but do contain *plastoquinones* which are structurally related to coenzyme Q. Plastoquinone functions in the electron transport processes involved in photosynthesis. In some organisms, coenzyme Q is present together with other quinones, such as vitamin K, tocopherylquinones, and plastoquinones; and each type of quinone can carry out different parts of the electron transport functions.

Additional Reading

Bergmeyer, H.U.: *Metabolites 2: TRI- and Dicarboxylic Acids, Purines, Pyrimidines and Derivatives, Coenzymes and Inorganic Compounds,* Vol. 2, John Wiley & Sons, Inc., New York, NY, 1985.
Bugg, T.D.: *An Introduction to Enzyme and Coenzyme Chemistry,* Blackwell Science, Inc., Malden, MA, 1997.
Kagan, V.E. and P.J. Quinn: *Coenzyme Q: Molecular Mechanisms in Health and Disease,* CRC Press, LLC., Boca Raton, FL, 2000.

COLCHICINE.   See **Alkaloids**.

COLD-WORKED METAL.   When metals are plastically deformed at a relatively low fraction (frequently < 0.5) of their absolute melting temperatures, they are normally said to be *cold-worked*. An increase of *hardness* or strength is normally associated with such deformation. This hardening can be relatively stable as long as the metal is not heated high enough to cause extensive recovery and recrystallization to occur. See also **Annealing**. In metals with high melting points, such as the alloys of iron, copper, and nickel, deformation at room temperature is normally considered to be cold-working. On the other hand, room temperature deformation of a low melting point metal such as lead is more properly designated hot-working. In this regard, it is interesting to note that lead is not normally hardened by room temperature deformation.

Cold-working is frequently used for hardening of commercial metal products. Thus, piano wire and the wire used in bridge construction obtain their hardness as a result of the final wire drawing operations. Sheet stock is often supplied in different degrees of hardness obtained by cold-rolling to the desired hardness. Mechanical working is sometimes the only feasible way to harden specific metals and alloys.

COLEMANITE.   The mineral colemanite is a borate of calcium corresponding to a formula that is perhaps best represented as $Ca_2B_6O_{11}\cdot5H_2O$. It occurs either as massive deposits or in monoclinic crystals. It has a subconchoidal fracture; hardness, 4–4.5; specific gravity, 2.42; vitreous to adamantine luster, may be colorless to milky white, grayish or yellowish; transparent to translucent. Colemanite was found originally in Death Valley, Inyo County, California, and has since been found rather widely distributed in San Bernardino, Los Angeles, Kern and Ventura Counties, California, as well as in Clark, Esmeralda and Mineral Counties in Nevada.

Colemanite was, until the discovery of kernite, the chief source of borax. Kernite, $Na_2B_4O_7\cdot4H_2O$, because of its easy solubility in water, has displaced very largely other boron-bearing minerals as a source of borax. Colemanite, kernite and inyoite (probably $Ca_2B_6O_{11}\cdot5H_2O$); are lake deposits associated with other and rarer boron minerals, laid down during periods of volcanic activity or resulting from the leaching of the adjacent Tertiary sedimentary formations. Colemanite was named for Mr. William T. Coleman of San Francisco; kernite and inyoite were named from Kern and Inyo Counties, California.

COLLAGEN.   The major protein component of connective tissue. In mammals, as much as 60% of the total body protein is collagen. It comprises most of the organic matter of skin, tendons, bones, and teeth, and occurs as fibrous inclusions in most other body structures. Collagen fibers are easily identified on the basis of the following characteristic properties:

They are quite inelastic; they swell markedly when immersed in acid, alkali, or concentrated solutions of certain neutral salts and nonelectrolytes; they are quite resistant to most proteolytic enzymes, but are specifically attacked by the collagenases; they undergo thermal shrinkage to a fraction of their original length at a temperature which is characteristic of the collagen from a given animal, but this varies from one species to another; and they are converted in large part to soluble gelatin by prolonged treatment at temperatures above the thermal shrinkage level. Collagen fibers are not unique to mammals; collagen has been identified in the tissues of almost all multicellular animals, ranging from the primitive porifera and coelenterates, through the annelids and echinoderms, and up to the vertebrates.

As a protein, collagen is unusual in both chemistry and structure. Nearly one-third of its residues are glycine, and an additional 20–25% are imino acids (proline and hydroxyproline). In terms of sequence, glycine occurs regularly in essentially every third position, following as a steric requirement of the secondary-tertiary structure.

It appears that specific side-chain interactions between polar residues on adjacent collagen macromolecules are largely responsible for ordering the macromolecules into fibers. Specific cooperative interactions between functional groups on appropriately oriented macromolecules seem to be involved in the heterogeneous nucleation of hydroxyapatite crystals, and thus the initiation and control of mineralization in bones and teeth. As collagen fibers age, *in vivo*, they seem to become progressively more intermolecularly cross-linked, perhaps by the "ester-like" bonds formed. Little or no soluble collagen can be extracted from most mature connective tissue because of this extensive cross-linking, although the material can be converted into soluble gelatin by drastic thermal treatment.

Collagen and gelatin are of commercial importance. As insoluble collagen, this material may be cross-linked further by tanning and thus converted to leather. The soluble gelatins are used in the manufacture of foodstuffs, film emulsions, and glue.

COLLOID CHEMISTRY.   A subdivision of physical chemistry comprising the study of phenomena characteristic of matter when one or more of its dimensions lie in the range between 1 millimicron (nanometer) and 1 micron (micrometer). It thus includes not only finely divided particles but also films, fibers, foams, pores, and surface irregularities. Dimension, rather than the nature of the material, is characteristic. Colloidal particles may be gaseous, liquid, or solid, and occur in various types of suspensions (imprecisely called solutions), e.g., solid-gas (aerosol), solid-solid, liquid-liquid (emulsion), gas-liquid (foam). In this size range, the surface area of the particle is large with respect to its volume so that unusual phenomena occur, e.g., the particles do not settle out of the suspension by gravity and are small enough to pass through filter membranes. Macromolecules (proteins and other high polymers) are at the lower limit of this range; the upper limit is usually taken to be the point at which the particles can be resolved in an optical microscope. The first specific observations were made by Thomas Graham in approximately 1860 and were extended by Ostwald, Hatchek, and Freundlich. Though the term is often used synonymously with *surface chemistry*, in a strict sense it is limited to the size range noted in at least one dimension, whereas surface chemistry is not. Natural colloid systems include rubber latex, milk, blood, egg white, etc.

See also **Colloid Systems**; **Fibers**; **Film**, **Foam**; **Emulsions**; and **Surface Chemistry**.

COLLOID SYSTEMS.   Colloids are usually defined as disperse systems with at least one characteristic dimension in the range 10^{-7} to 10^{-4} centimeter. Examples include: *sols* (dispersions of solid in liquid); *emulsions* (dispersion of liquids in liquids); *aerosols* (dispersions of liquids or solids in gases); *foams* (dispersion of gases in liquids or solids); and *gels* (system, such as common jelly, in which one component provides a sufficient structural framework for rigidity and other components fill the space between the structural units or spaces). All forms of colloid systems are encountered in nature. Products of a colloidal nature are commonly found in industry and are notably extensive in the food field. Foams, widely used in industrial products, but also the causes of processing problems are described in entries on **Foam**; and **Foamed Plastics**.

Early Background

Thomas Graham's investigations of diffusion (1861) led him to characterize as *crystalloids* substances, such as inorganic salts which in water solutions would diffuse through a parchment membrane; and as *colloids*

(Greek word for glue) substances, such as starch and gelatin, which would not diffuse through the membrane. Sols with a given weight percent of dispersed material scatter light more strongly than a solution with the same weight percent of dissolved inorganic salt, i.e., a true solution. The Tyndall effect, in which the path of a beam of light through a turbid solution (or through dusty or smoke-filled air) is clearly defined through scattered light, is characteristic of sols. The slow diffusion and strong light scattering, together with the fact that the boiling-point elevation, freezing-point depression, and osmotic pressure caused by a given weight percent of dispersed material in sol form are much less than the corresponding magnitudes caused by the same weight percent of common inorganic salts—these observations indicated to early investigators that the particles dispersed in the sol must be larger than those resulting from dissolving inorganic salts in water.

Development of the ultramicroscope (Siedentopf and Zsigmondy, 1903) permitted particles substantially smaller than the wavelength of light to be observed in scattered light and were thus capable of counting. Invention of the ultracentrifuge by Svedberg (1924) made it possible to cause particles in sols to sediment at observable rates, to measure these rates with reasonable precision, and to infer particle sizes from these rate measurements. The ultramicroscope and ultracentrifuge permitted validation of the early conclusions that colloidal particles are much larger than ions resulting from dissolving metal salts in water.

Svedberg found that in some sols the particles were highly uniform in size. For example, he found that the gram particle weight of insulin (a protein) was 40,900 and that apparently all insulin particles had this gram particle weight. This made it extremely likely that the insulin particles were either single molecules (albeit giant ones), or aggregates of a very definite number of smaller (but still quite large by ordinary standards) molecules. The research of Staudinger (commencing about 1920) and of Carothers (1929) opened up the field of macromolecular chemistry, leading to the recognition that giant molecules were not only abundant in nature, but could be prepared by established principles of chemistry. See also **Macromolecular Science**.

The pioneering work of Svedberg and other early researchers, as continued over the years through the application of much improved instrumentation (scattered laser light, electron microscopy, among other methods for investigating microstructure) are described briefly later in this article.

It is interesting to note that Wolfgang Ostwald, in the late 1800s, stated, "There are no sharp differences between mechanical suspensions, colloidal solutions, and molecular (true) solutions. There is a gradual and continuous transition from the first through the second to the third."

Some colloidal systems are *thixotropic*, i.e., they differ in their fluid behavior from *pseudoplastic* substances in that the flow rate increases with increasing duration of agitation as well as with increased shear stress. When agitation is stopped, internal shear stress exhibits hysteresis. Upon reagitation, generally less force is required to create a given flow than is required for the first agitation. Examples of thixotropic materials include silica gel, most paints, glue, molasses, lard, fruit juice concentrates, and asphalts. By rhythmically shaking or tapping certain thixotropic suspensions, the suspensions will "set" or build up very rapidly. This type of non-Newtonian substance is said to be *rheopectic*. Bentonite sols and suspensions of gypsum in water are rheopectic. *Dilatant fluids* often are termed *inverted plastics* or inverted pseudoplastics. Initial flow under a low shear stress is at a high rate; further increases in shear stress, however, result in lower flow rate. Some liquids may change from thixotropic to dilatant or vice versa as the temperature or concentration changes. Examples of dilatant materials include quicksand, peanut (groundnut) butter, and many candy compounds. For comparison, it should be recalled that a Newtonian substance is a liquid or suspension which, when subjected to a shear stress, undergoes deformation wherein the ratio of shear rate (flow) to shear stress (force) is constant. These varying behavioral patterns of colloidal materials become important considerations in specifying pumps and other process handling equipment. See also **Gold Number**.

Role of Water Molecule—Cluster Theory

Because a very large number of colloid systems involve water, particularly in connection with biological systems, comprehension of the structure and behavior of water is paramount. Notably since the 1950s much progress has been made concerning the physical and chemical nature of water. Long established observations have shown water to be a low-molecular-weight

compound, similar in structure and molecular weight to several compounds, such as methane, ammonia, and hydrogen fluoride, among others. However, water has a number of anomalous properties when compared with other structurally similar compounds. Such comparison shows that water has unusually high melting and boiling points, dielectric constant, and density. Water also has a very high surface tension and specific heat, and a high heat of vaporization were this characteristic extrapolated from data on similar substances. That liquid water is a highly structured substance has been supported for many years through x-ray and neutron diffraction studies, as well as Raman and nuclear magnetic relaxation (NMR) techniques. That water molecules have a V-like structure with an HOH bond angle of 104.5° has long been established. Similarly it has been known that the highly electronegative oxygen atom polarizes the OH bond, thus giving the molecule an unsymmetrical charge distribution. See also **Water (Hard)**.

The apparent simplicity of structural and behavioral aspects of water disappeared with the observations of Frank and Wen (1957), the first investigators to describe water as a *mixture of monomer and polymerized molecules*. As pointed out by Busk (1984), Frank and Wen found that the formation of a single hydrogen bond enhances the ability of the other hydroxyl groups on the molecules involved to form hydrogen bonds. They described water structure as cooperative in nature. The formation of hydrogen bonds, however, restricts the molecular motion of H_2O molecules and results in a loss of entropy. Initially, the loss in entropy is more than compensated for by the gain in enthalpy of bond formation. The size that a cluster of molecules will grow to is thus a result of a balance between these two forces.

Numerous investigators have contributed to the cluster theory of water and the model proposed (Nemethey-Scheraga Flickering Cluster Model) incorporates the names of some of these investigators. The current description of the model suggests that water is composed of monomeric H_2O molecules, polymeric structures 20–90 molecules in size, and a minor amount of ions, isotopes, etc. Up to 90% of the molecules in liquid water may be in clusters at any one time. The number of clusters increases with temperature; the size decreases. The model predicts the density maximum and some of the other solid-like properties. The model forecasts the lifetime of the cluster to be very short (about 10–11 second) and the bonding forces within each cluster to be quite weak. The latter leads to accurate predictions of the liquid-like characteristics of water. Some molecules (nonbonded monomers) are present only to about 1% of the total. Computer simulations by Del Bene and Pope determined that cyclic water structures are the most stable (e.g., trimers, tetramers, pentamers, and hexamers). This study also indicated that the branching structure needed to support a continuum structure is not entropically favored.

Molecules may interact with water in at least four main ways—hydrogen bonding, ionic bonding, hydrophobic association (nonpolar molecules placed in the polar environment of water), and London dispersion or van der Waals forces.

Inasmuch as water-macromolecule interactions are complex and still not fully understood, other postulations concerning water bonding have been proposed. Fennema (1978), for example has suggested three structural states of water: (1) *constitutional water* (held in interior of folded macromolecules); (2) *interfacial water* (located at or near the surface of a macromolecule); and (3) *bulk phase water* (chemical properties identical to those of free water or water of dilute salt solutions). In making these distinctions, Fennema asserted that the categories do not imply any distinct boundaries between them. Another researcher (Labuza, 1977) proposed two additional types of forces: (1) the effects of trapping water in polymer capillaries, and (2) solute effects due to entropy considerations.

Sols

It is convenient to classify sols into three types: (1) *lyophilic* (solvent loving) colloids, for example, are solutions of gelatin or starch in water; (2) *association* colloids, of which a solution of soap in water at moderate concentration is an example; and (3) *lyophobic* (solvent repelling) colloids, for example, sulfur in water. Both lyophilic and association colloids can be prepared in thermodynamic equilibrium, so that when solvent is removed and then returned to the system, the original properties of the system are regained.

Lyophobic colloids are not (or at most, rarely) equilibrium systems. When solvent is removed and then returned to the system, the original dispersed material fails to redisperse, and it is usually convenient to regard such a system as one in which the dispersed particles are continuously aggregating. A lyophobic sol thus appears to be stable if the aggregation rate is slow; and unstable if it is fast. The terms lyophilic and lyophobic entered the literature before the characteristics of these systems were well understood and thus are somewhat anachronistic; they are nonetheless well-established.

Lyophilic sols are true solutions of large molecules in a solvent. Solutions of starch, proteins, or polyvinyl alcohol in water are representative of numerous examples. Properties of these solutions at equilibrium (for example, density and viscosity) are regular functions of concentration and temperature, independent of the method of preparation. The solvent-macromolecule compound system may consist of more than one phase, each phase in general containing both components. Thus, if a solid polymer is added to a solvent in an amount exceeding the solubility limit, the system will consist of a liquid phase (solvent with dissolved polymer) and a solid phase (polymer swollen with solvent, i.e., a polymer with dissolved solvent).

The foregoing characteristics also are found with solutions of small molecules. But properties of solutions, one of whose components is macromolecular, differ from those having only small molecular components in quite understandable ways. For example, where small molecules are involved, molecular distortion is minor. Quite generally, the shapes of small molecules are little affected by environment unless the small molecules react chemically. In contrast with small molecules, there is a considerable variation in polymer conformation with environment. See also **Molecule**; and **Polymers**.

A polymer dissolved in a good solvent will tend to stretch out, and the resulting entanglement of polymer chains and interference with solvent movement will lead to a high viscosity. If the solvent is a poor one, the polymer molecule will tend to form a small ball, and the viscosity for a given weight percent will be much less. A side group of a polymer may be ionizable. Ionization of this group distributes a charge along the backbone and charge repulsion causes the macromolecule to tend toward a rod shape. If there is a moderate salt concentration in the solution, the backbone charge will be partly shielded by ions from the salt of opposite charge. Thus, the tendency toward rod formation will be less pronounced. The tendency of oppositely charged macromolecules to aggregate is much greater than the tendency of oppositely charged small ions to pair simply because the charges involved are greater in the former case.

The foregoing special properties of solutions of large molecules are relatively easy to describe in qualitative terms, but a difference of a more subtle nature occurs when the system forms two liquid phases. In a macromolecular solution, both phases tend to be rich in the (small molecule) solvent, whereas in systems formed from two molecules of comparable size, one phase is rich in one component, while the other phase is rich in the second component. The formation of two liquid phases from a solvent-macromolecule system is sometimes call *coacervation*; and the phase with the higher percentage of macromolecule is sometimes called the *coaceryate*. See also **Coacervation**.

Association Colloids

These are generally encountered in solutions of soaps and detergents in water. These matters become important, of course, to procedures for cleaning and sterilizing equipment (food processors, biochemical manufacturers, hospitals, etc.), but the principles also apply to other association colloids also encountered industrially, notably in the food processing field. A typical soap, such as sodium stearate, $C_{17}H_{35}COONa$, or a detergent, such as sodium dodecyl benzene sulfonate, $C_{12}H_{25}C_6H_4 \cdot SO_3Na$, consists of a long hydrocarbon tail and a polar (in the examples cited, ionizable) head group. The solubility of the soap in water is largely conferred by the head group. As the soap concentration is increased, the soap molecules tend to cluster in aggregates called *micelles*, with hydrocarbon tails in the interior of the micelles and the polar groups in contact with water. The formation of micelles is favored by the interaction between hydrocarbon tails and is opposed by charge repulsion of the polar group which are placed close together at the micelle surface.

Micelle formation becomes pronounced at soap concentrations exceeding the critical micelle concentration. As hydrocarbon tail length is increased, the interaction of tails is increased, and as salt concentration is increased, the repulsion of head groups is reduced because their charges are partly shielded by ions of the salt. Both of these factors favor micelle formation, causing micelles to be larger and the critical micelle concentration to be smaller. Typically, a micelle might contain about 50 soap molecules. The micelle interior is a hydrocarbon, and as such is receptive to other molecules soluble in hydrocarbons. Hence, a soap solution can 'dissolve' such molecules (taking them up in micelle interiors) even if the molecules are quite insoluble in water. This phenomenon is called *solubilization* and is a factor in detergency.

Lyophobic Sols and Aerosols

These products can be viewed most simply and, in most cases, with sufficient as two-phase systems in which the dispersed particles are steadily and irreversibly aggregating according to a second-order rate law. Thus, where C is the number of particles per cubic centimeter (an aggregate of many primary particles being counted as one particle) at time t, and where C_0 is the number of particles per cubic centimeter at zero time, and K is a constant, C depends on t according to

$$\frac{C_0}{C} - 1 = KC_0t$$

and will be one-half its value at zero time when $KC_0t = 1$. The time required for this is longer, the smaller K and the smaller C_0. If the time required is weeks, the sol will appear quite stable over a period of days. If there is no barrier to aggregation, so that the particles aggregate as fast as diffusion brings them in contact, the rate constant K can be calculated approximately from diffusion theory and is $8kT/3\eta$, where k is Boltzmann's constant, T is the absolute temperature, and η the viscosity of the medium. Initial sol concentrations (particles per cubic centimeter) giving one-minute half-lives at room temperature are 1.4×10^9 in water; and 2.7×10^7 in air in the absence of aggregation barriers.

Although these numbers may appear large, they correspond to quite small volume percentages of dispersed particles. A particle of radius 5×10^{-5} centimeter is at the upper limit of the colloidal range; and 1.4×10^9 such particles occupy 0.07% of space. The behavior of smokes, fogs, and many dispersions of uncharged particles in water accords well with the rate equation and theoretical rate constant given. In contrast, the dispersed particles in many sols are electrically charged, manifesting this charge through electrophoresis (motion of colloidal particles under the influence of an electric field). In fact, Tiselius developed electrophoresis to a high degree, successfully fractionating and classifying proteins thereby. Evidently like charges on two colloidal particles will contribute to a repulsion between them, which will be greater, the greater the charge on each particle and the smaller the concentration of salts in solution (since ions from the salt will tend to mask the charges on the particle).

The theory of interaction between colloidal particles with a surface electrostatic potential (due to surface charges) surrounded by an electrical double layer (one layer of which is the layer of surface charges, the other a diffuse cloud of charges of opposite sign due to ions from salts in the solution) was developed by Derjaguin and Landau and independently by Verwey and Overbeek, and is generally known as the DLVO theory after the first letters in the names of these scientists. The DLVO theory shows how a barrier sufficient to reduce the rate constant K (and so to increase the half-life at a given initial concentration) by many powers of ten may arise from the interaction of charged particles in a solvent, and the magnitudes calculated agree rather well with experiment.

It is evident why the properties of lyophobic sols depend so critically on the chemistry of the interface between dispersed particle and solvent, for this chemistry establishes the means by which the surface charge can be established or altered. Particles of silver halide dispersed in water will acquire a positive charge if silver ion is in slight excess in the water, because the silver ion can readily add to the silver halide lattice. A negative charge is similarly acquired if the halide ion is in slight excess. The silver ions and halide ions are called *potential-determining ions* for the silver halide sol. They can lose their waters of hydration and adsorb on the particle side of the electrical double layer, conferring a charge on the particle. Hydrogen ion and hydroxyl ion are similarly potential-determining ions for many oxide sols, such as silica and alumina, including particles, such as carbon and many metals, which are ostensibly not oxides, but in fact usually have oxidized surfaces. Finally, charge can be conferred by the adsorption of charged macromolecules, such as gelatin, which are called *protective colloids*. Salts added to the sol form ions, which tend to mask the particle charges and so tend to promote flocculation. The ion whose charge is opposite to the particle charge (the counter ion) is of particular importance, and the greater its charge the lower the concentration at which its flocculating effect is evident.

Emulsions

These are dispersions of one liquid in another. Most commonly, one phase is an oil which is at most slightly miscible with water. The disperse phase can either be oil (an oil-in-water emulsion) or water (a water-in-oil emulsion). For apparent stability, an emulsifying agent is almost invariably required. The emulsifying agent has an oil-soluble tail and a polar head. The emulsifying agent concentrates (adsorbs) at the interface between oil and water, lowering the interfacial tension and frequently conferring a charge on the dispersed droplets. The film of emulsificant thus formed is usually only one molecule thick, but it is essential to emulsification. Mixed emulsificants, such as a mixture of sodium stearate and octadecyl alcohol, may be more effective in emulsification than either component alone, and there is a great deal of art and experience required in the formulation of emulsions. Lecithins and some proteins are effective natural emulsificants, and a mixture of lecithin and cholesterol is an effective natural mixed emulsificant. It should be noted that the difference between solubilization (described in connection with association colloids) and emulsification is not sharp, particularly insofar as large, extensively swollen micelles and ultrafine emulsions are concerned. See also **Emulsions**.

Gels

These substances involve the formation of a three-dimensional structure. A gel is a colloidal disperse system in which is contained a dispersed component and a dispersion medium, both extending continuously throughout the system. Further, the system has equilibrium-elastic (time-dependent) deformation. Thus, since they have a shear modulus of rigidity, gels are like solids, but in most other physical respects, they behave like liquids. It is conceived that the three-dimensional network is kept together by bonds or junction points, which essentially have an unlimited lifetime. Junction points may be described as primary valence bonds, attractive forces of long range, or secondary valence bonds, which maintain an association between parts of polymer chains or form submicroscopic crystalline regions. A gel may be defined as a flocculant and gelatinous precipitate. A jelly is a transparent elastic mass. Upon standing, a gel may shrink—a process known as *syneresis*.

In 1861, Thomas Graham first used the term syneresis to describe the phenomenon of exuding small quantities of liquid by gels. By definition, syneresis is the spontaneous separation of an initially homogeneous colloid system into two phases—a coherent gel and a liquid. The liquid is actually a dilute solution whose composition depends upon the original gel. When the liquid appears, the gel contracts, but there is no net volume change. Syneresis is reversible if the colloid particles do not become too coagulated immediately after their formation.

In 1937, Heller classified three types of syneresis as to cause: (1) syneresis of desorption, caused by the particle becoming less hydrophilic with time; (2) syneresis of aggregation, whereby discrete gel particles may unite into a denser gel portion; and (3) syneresis of contraction, where a gel with fibrillar structure contracts and squeezes out the intermicellar liquid. Most commonly, syneresis is the visible manifestation of further slow coagulation that follows the initial setting of the gel, the gel-forming process itself being an enmeshing of the hydrous particles into a network. It may be further explained as the exudation of liquid held by capillary forces between the heavily hydrated particles constituting the framework of the gel. Ostwald noted that the phenomenon is one of the most characteristic of the properties of gels.

A common example of syneresis is found when a mold of gelatin remains under refrigeration for a period. A general shrinkage of the body of the gel occurs and a liquid collects around the edge of the mold. The liquid is a dilute solution of the original composition. Since the total volume of the system remains the same, syneresis should not be considered simply as the opposite of imbibition (absorption). Extending the onset of syneresis in various products, notably foods, is of obvious importance.

Dispersion Processes

1. The simplest method of accomplishing dispersion is by grinding the solid (or liquid) material with the liquid medium until particles of the required size are ultimately obtained. The colloid mill (Plauson, 1921) is used for such purpose, as in mixing paints and pastes, regenerating milk from milk powder, dispersing cellulose in sodium

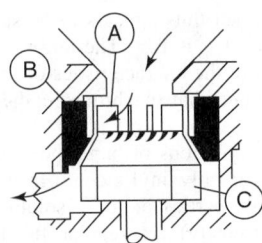

Fig. 1. One type of colloid mill. Rotor blades A break up slurry. Serrations in rotor and stator provide mechanical shear and force material into adjustable gap (0.0005–0.125 inch; 0.013–3.2 millimeters) between rotor and stator B for intense hydraulic shear. Lower part of rotor C adds further whirling action

hydroxide and carbon disulfide for the production of xanthates for viscose, and in emulsifying fats and waxes. See Fig. 1.

2. Zinc sulfide, cupric hexacyanoferrate(II), stannic acid, silver chloride are examples of precipitates which, when washed on the filter paper until the accompanying soluble electrolyte has been removed, form colloidal solutions and pass through the pores of the paper. Since this is usually to be avoided in practice, the washing is then done with an electrolyte, which does not conflict with the treatment to follow. Frequently ammonium nitrate solution is used.

3. A peptizing agent is frequently employed. Tannin is peptized by water, and by glacial acetic acid. Soaps are peptized by water. Gelatin swells in cold water but is not peptized, but is peptized in warm water. Starch, although insoluble in cold water, behaves similarly to gelatin with warm water (63 to 74°C, depending upon the kind of starch). Cellulose nitrate swells in ethyl alcohol and not in ether, but is peptized in ethyl alcohol-ether mixture. Clay is peptized by ammonium hydroxide, and it is held by some that the action of sodium hydroxide on zinc, aluminum, and chromium hydroxides is one peptization.

4. Water-peptizable colloidal substances such as gelatin, dextrin, gum arabic, and soap peptize many precipitates, and are often called protective colloids. Gelatin in the solution prevents the precipitation of silver dichromate upon mixing silver nitrate and potassium dichromate solutions. (See Condensation Processes, below.)

5. When dilute silver nitrate and dilute potassium bromide solutions are mixed so that there is a slight excess of either solution, silver bromide is peptized. Acheson's oil-dag and aqua-dag are suspensoids of graphite in oil or water containing a protective colloid, tannin. Oil-dag contains about 15% of a "deflocculated graphite," and is used in dilute solution in lubricating oil (about 0.1% graphite). Bearings gradually become coated with a thin layer of graphite.

Sonic methods also are used to create emulsions. Liquids are pumped under pressure through an orifice of special design and impinge on the edge of a blade causing it to vibrate at ultrasonic frequencies. Cavitation takes place continuously in the stream, causing violent pressure changes to be generated locally. The result is a uniform and stable emulsion and a dispersion of a very high order. See also **Cavitation**.

Condensation Processes

1. When a solution of ferric chloride is poured into a relatively large volume of boiling water, colloidal ferric hydroxide is formed. The ferric hydroxide sol does not react with hydrogen sulfide nor with potassium hexacyanoferrate(II), and like all colloidal substances does not pass readily through animal membranes or parchment.

2. When hydrogen sulfide is passed into a solution of arsenious oxide, arsenious sulfide sol is formed which in the absence of an electrolyte may be made of the high concentration of 60 grams of arsenious sulfide per 100 grams of water. Upon addition of hydrochloric acid, arsenious sulfide coagulates and is precipitated.

3. When hydrochloric acid is added to sodium silicate solution either silicic acid sol or silicic acid gel is formed.

4. When hydrogen sulfide solution is treated with an oxidizing agent, for example, the proper concentration of nitric acid, sulfur sol is formed.

5. When gold chloride very dilute solution (0.01 to 0.001% of gold chloride) is made slightly alkaline (say by the addition of magnesium oxide) and then treated with a reducing agent, for example, formaldehyde or sodium hydrosulfite $Na_2S_2O_4$, red gold sol is formed.

6. Use of a protective colloid in solution prevents the formation of the ordinary and expected precipitate in many cases, and causes the formation of the expected substance as colloidal sol. Silver nitrate (0.6 gram per liter) and potassium dichromate (0.5 gram per liter) to one of which is added 0.1 volume of hot gelatin solution (2 grams per 100 milliliters of water) are mixed with stirring silver dichromate sol is formed.

7. When an electric arc is formed under water between two metallic rods, particles of the metal of colloidal size are formed along with more or less separation of free metal. A protective colloid increases the stability. If the metal vaporizes and then condenses to the colloidal state this is strictly speaking a condensation process, if otherwise, a dispersion process.

The disappearance of the colloidal state of a substance may be accomplished in either of two directions, namely, by the colloid passing into solution or into suspension. Practically, the latter is the more important method. Coagulation, agglomeration or precipitation is readily brought about by discharge of the electric charge on the particles. Ions carrying a charge of opposite sign to that carried by the colloidal particles are active precipitants, and the higher the valency of the ion the more effective (Linder-Picton-Hardy). When the colloidal particles are made neutral the conditions are least favorable to their stability. For colloidal arsenious sulfide, which is negatively charged in water, the coagulating power of potassium iodide K^+I, calcium chloride $Ca^{2+}Cl_2$, aluminum chloride $Al^{3+}Cl_3$ is in the ratio of 1:80:1500 (Svedberg); and for colloidal ferric hydroxide, which is positively charged in water, the coagulating power of potassium chloride KCl^-, potassium sulfate $K_2SO_4{}^{2-}$ is in the ratio of 1:45. The active ion is carried down with the precipitated particles. Oppositely charged colloids, e.g., arsenious sulfide and ferric hydroxide, when mixed, precipitate each other. Other methods of coagulation are by migration of colloidal particles to and their discharge at electrodes, and by heating, as in the case of egg albumin. Coagulation is usually irreversible, especially when caused by electrolytes.

An interesting case, operating on a large scale in nature, of the precipitation of a colloidal system by an electrolyte is that of the action of sea water on the mud and silt of river water entering the ocean. When river water flows into the ocean the former, on account of its lower specific gravity, tends to flow over the latter and spread out in widening range. As the current diminishes some of the suspended mud and silt settles out, but the finer colloidal particles are coagulated by the electrolyte of the sea water and form deltas at the mouths of rivers.

Importance of Colloidal State

All living matter, whether animal or plant, are made up of many colloidal materials and are largely sustained by colloidal processes. Of similar importance is colloidal chemistry in everyday living, in almost all of our foods, such as proteins and starches, in our clothing, whether of natural or synthetic origin, and in our shelter materials, such as wood, bricks, and concrete. When there is added to these, other common things and operations of everyday life, such as pottery and porcelain, paper, rubber and leather, and cooking and washing, where colloidal matter and processes operate, it is evident how broad is the scope and how great is the importance of the field. To these there must also be added other applications in the realm of industry, such as dyeing, printing, photography, water purification, smoke prevention, ore flotation, sewage disposal and soil preparation, paints, varnishes and lacquers, plastics, adhesives, and innumerable other operations and materials.

Advanced Microstructural Analysis

In connection with gelling systems, Davis and Gordon (1984) point out that the *macrostructural* evaluation of a gel usually involves the evaluation of some chemical or physical property of the system as a whole, such as total water and protein or solids content, gel strength, turbidity changes during gelation, pH range at which gelation occurs, and degree of syneresis after gelation and storage. In contrast, in microstructural analysis, the investigator is interested in molecular isolation and characterization of

the components of the system. As reported by numerous researchers, molecular changes and interactions result in the association, aggregation, flocculation, coagulation, and gelation of the system. There are three broad areas of microstructural analytical techniques: (1) microscopic, especially electron microscopy, where water must either be removed or immobilized in the gel system prior to viewing; (2) spectroscopic; and (3) thermal analytical. The smallest resolvable size ranges from individual molecules or groups of molecules in differential thermal analysis, differential scanning calorimetry, and thermogravimetric analysis. Light microscopy and ultraviolet microscopy are limited to sizes in the range from 1000 to 3000 angstroms. Resolution of scanning electron microscopy is approximately 150 to 200 angstroms, while transmission electron microscopy has a limit of from 2 to 5 angstroms. The range of resolution of spectroscopic absorption and spectroscopic scattering techniques is from somewhat less than one angstrom to hundreds and even thousands of angstroms. Davis and Gordon report in some detail on the use of the foregoing techniques in gelatinization studies of starch, carrageenan, soy protein isolate, and other similar substances that are encountered in the food processing industries.

Additional Reading

Birdi, K.S.: *Handbook of Surface and Colloid Chemistry,* 2nd Edition, CRC Press, LLC., Boca Raton, FL 2002.

Davis, E.A. and J. Gordon: "Microstructural Analyses of Gelling Systems," *Food Technology,* **38**(5), 99–109 (1984). http://www.ift.org/

DiSalvo, F.J.: "Solid-State Chemistry: A Rediscovered Chemical Frontier," *Science,* **649** (February 2, 1990).

Dubin, P, and Y. Osada: *Polymer Gels: Fundamentals and Applications,* Okford University Press, New York, NY 2002.

Fraissard, J.P., and O. Lapina: *Magnetic Resonance in Colloid and Interface Science,* Kluwer Academic Publishers, New York, NY 2002.

Fort, T. and K.J. Mysels: *Eighteen Years of Colloid and Surface Chemistry: The Kendall Award Address 1977–1990,* American Chemical Society, Washington, DC, 1991. http://www.chemcenter.org/

Harris, P., Editor: *Food Gels,* Elsevier, New York, NY, 1990.

Hiemenz, P.C.: *Principles of Colloid and Surface Chemistry,* Marcel Dekker, New York, NY, 1974.

Holmberg, K., D.O. Shah, and M.J. Schwuger: *Handbook of Applied Surface and Colloid Chemistry,* John Wiley & Sons, Inc., New York, NY 2002.

Hunter, R.J.: *Foundations of Colloid Science,* 2nd Edition, Oxford University Press, New York, NY 2000.

Provder, T., Editor: *Particle Size Distribution: Assessment and Characterization,* American Chemical Society, Washington, DC, 1991.

Morrison, I.D., and S. Ross: *Colloidal Dispersions: Suspensions, Emulsions, and Foams,* John Wiley & Sons, Inc., New York, NY 2002.

Scheuing, D.R., Ed.: *Fourier Transform Infrared Spectroscopy in Colloid and Interface Science,* American Chemical Society, Washington, DC, 1991.

COLORANTS (Foods). Color, as a component of appearance, is important in the sensory evaluation of a food substance, including beverages. Color affects the degree of acceptability of a food product in the marketplace. Color also is a frequently useful indicator of the degree of wholesomeness of a foodstuff. Many foodstuffs are attractively colored as the result of their naturally occurring pigments. In these instances, a major objective of the food grower and processor is to protect and preserve the natural colors as long as may be required by the distribution network, considering such factors as temperature and humidity to which the food substance may be subjected before reaching the consumer.

Considering the expectations of consumers, particularly in countries that have an advanced food technology, colorants, along with flavorings and texture modifiers, are important. These factors are particularly stressed in connection with modern fabricated foods, substitute foods, and food analogues. Many years have passed since final approval was given to margarine producers to use colorants and artificial flavorings. Even though artificial flavors are used in some products, ice creams and ices, as well as soft drinks, are color matched to their fruit flavors. Yellow colorings provide a note of richness in cake mixes, eggnogs, and other products associated with their content of eggs; cheese snacks are both colored and flavored to simulate cheese; popcorn oil is usually colored; iron oxide can be added to pet foods to simulate the color of meat; the red coloration of cherries is intensified by using red colorants in the processing of maraschino cherries. Caramel colorings are widely used in both alcoholic and nonalcoholic beverages.

Particularly during the past several decades, regulatory agencies in various countries have scrutinized colorants (along with other food additives) against a backdrop of consumer health. Since the early 1900s, several countries have approved colorants for use in foods only after thorough physiological testing. In recent years, analytical instrumentation and research methodologies have become sophisticated and measurements of minute quantities are now practical. With improved analytical tools and a heightened awareness of the effects of foods upon health, a number of colorants that once were considered perfectly safe have come under question. In some countries, most or all synthetic colorants have been banned. Generally, the limitations on the use of colorants have become much more stringent.

In the United States, the first rather complete legislation involving such matters was the Food and Drug Act of 1906. As a result of that legislation, the list of colorants permitted was reduced to only seven dyes. Because the remaining seven dyes did not provide sufficient flexibility in the formulation of food products, considerable research went forth to find additional colors, not only with more desirable hues, but easier to use (solubility in oil/water, less temperature sensitivity, etc.). During the 66-year period from 1906 to 1971, several additional colors were added. In 1938, the Food, Drug, and Cosmetic Act was passed. The common names of dyes previously used were given color prefixes and numbers. For example, Amaranth became FD & C Red No. 2. Under the new act, certification became mandatory.

The color situation in the food industry was relatively without incident until the early 1950s, when, as the result of a few cases of excessive levels of usage in some candies and popcorn, two colors (FD & C Orange No. 1 and FD & C Red No. 32) were delisted. After much controversy and considerable litigation, more colors were delisted. To rectify legal complexities and unworkability of the 1938 Act, the Color Additives Amendments of 1960 were passed. Nevertheless, as pharmacological studies continued, a number of other colors were de-listed during the interim. Much more recently, FD & C Red No. 3 Lake colors were banned in the United States. The foregoing paragraphs relate largely to the use of synthetic colorants. Natural colorants include the anthocyanins, annatto colors, the betalaines, the carotenoids, cochineal, saffron, turmeric, and titanium dioxide. Caramel coloring also falls into this category. Paprika is used in some foods for its coloring attributes.

Anthocyanins are water-soluble pigments that account for many of the red, pink, purple, and blue colors found in higher plants. Most plants contain more than one of these pigments and they occur most prevalently as glycosides. Several hundred different anthocyanins are known. These compounds are most stable at a pH range of 1 to 4, thus limiting the spectrum of usage. As compared with synthetic colorants, the anthocyanins produced to date generally are less stable, have less tinctorial potency, and lack some color uniformity. They are degraded by light, heat, enzymes, and interact with ascorbic acid. They also tend to form complexes with metal ions to produce off-colors. Their main advantage stems from the fact that they are naturally derived and thus not regarded with suspicion as health deterrents as are many synthetic materials. On the other hand, attempts to alter and modify them to make up for their fundamental disadvantages could also move them toward a suspicious category—a factor which researchers on anthocyanins are taking into consideration.

Numerous commercial sources for the anthocyanins have been and are continuing to be investigated. These sources include grape anthocyanins, apparently with good potential in the carbonated beverage field. Roselle plants, native to the West Indies, have been studied and appear to have potential for use in apple and pectin jellies, but not for carbonated beverages, such as ginger ale. Cranberry pomace and blueberries have been investigated. Considerable interest has been shown in the red anthocyanin pigments of miracle fruit (*Synsepalum dulcificum*, Schum), a tropical plant that produces a red berry.

Red cabbage as a colorant source has been studied for many years. As of 1990 at least one firm has introduced San Red RC, the first commercially available food color derived from red cabbage. The color can be used alone or in combination with other colors to create strawberry, cherry, raspberry, and blueberry tones. By way of proprietary technology, the new dye is claimed to be free of flavor and odor defects, which in the past have been associated with red cabbage. The dye is pH dependent. The color tones move toward blue-red as the pH value increases. San Red RC ranks between cochineal and grape juice in percent of color retention.

Betalaines are sometimes referred to as beetroot pigments. They are made up of two main groups: (1) *betacyanins*, the principal component of which is betanin; and (2) *betaxanthins*, the principal component of

which is vulgaxanthin-I. The betacyanins contribute a red color, whereas the betaxanthins are yellowish. Another yellow pigment, betalamic acid, derives directly from cleavage of betanin and is probably the key intermediate in the biogenesis of all betalaines.

A factor of concern in connection with the betalaines is the earthy flavor associated with beets. To date, beets have been regarded as the primary source for these substances.

Carotenoids

These yellow-orange colorants are described in entry on **Carotenoids**.

The other natural colorants previously mentioned have been used for many years and are familiar to nearly everyone. Annatto colors are described in a separate entry, **Annatto Food Colors**.

Synthetic Colorants. Perkin, in 1856, synthesized the first synthetic dye, *mauve* or *mauveine*, by the oxidation of crude aniline. In that time and for about 80 years, coal tar was the principal source of aromatic compounds, which, in turn, were the sources of numerous synthesized dyes used primarily in textiles, but some of which were found to be adapted to other uses, including the coloring of food substances. This generally gave rise to the term "coal tar color" used commonly in the food and cosmetics industries for many years. Of course, with the development of more sophisticated organic syntheses and the petrochemical field, the association with coal tar no longer had a direct meaning. The term was finally eliminated from legislation in connection with the Color Additives Amendments of 1960. It is interesting to note, however, that prior to the first Act of 1906, it is estimated that some 80 such dyes were being used in a large number of food products, at a time when there were no regulations regarding the nature and purity of colorants used in foods.

Lakes. In the United States, FD & C lakes were accepted for the approved list of certified color additives for the first time in 1959. As defined by the FDA, a lake is an "Extension on a substratum of alumina, of a salt prepared from one of the water-soluble straight colors by combining such a color with the basic radical aluminum or calcium." Because the substratum of alumina hydrate or aluminum hydroxide is insoluble, the lake provides an insoluble form of the dye, i.e., a pigment. Colors from dyes result from solution in a solvent; whereas colors from pigments result from dispersion of that pigment throughout the food substance. Prior to the acceptance of lakes, insoluble colorants were formed by absorbing them on materials (insoluble), such as cellulose, flour, and starch. Generally, these forms were inadequate because of relatively low coloring power.

When utilized in solid or semisolid vehicles, dyes must be added in solution to achieve effective coloring. Thus, color migration is a problem. Dyes can migrate with the solvent during various drying or processing operations. Because lakes are insoluble, color migration is negligible in applications where distinct interfaces are required. Striped candy pieces provide an example. Where opacity is required, titanium dioxide can be added to lakes. In high-quality lakes, nearly all particles will pass through a 325-mesh screen when wet-tested. Shades of coloration can be produced by blending the various FD & C Lake Colors.

Extending Natural Colors. For some natural foods, such as meat and fish, colorants are not added, but natural colors can be extended over a longer time span by adding various approved chemicals. For example, the bright red color that consumers prefer in fresh meat is due to oxygenated myoglobin, with iron on the heme group in its ferrous state. When oxidation occurs due to exposure, the ferric form is produced which imparts a brown discoloration to the meat. A mixture of tetrasodium pyrophosphate, sodium erythorbate, and citric acid, when combined with modified-atmosphere packaging, will extend the life of the bright red coloration of the meat. Care must be exercised to avoid "masking" microbial problems that develop when meats are held too long.

Additional Reading

Ames, J.M. and T.F. Hofmann: *Chemistry and Physiology of Selected Food Colorants,* American Chemical Society, Columbus, OH, 2000.

Bigelow, S.W.: "Food Chemicals Codex: A Progress Report," *Food Technology*, **88** (May 1991).

Burdock, G.A.: *Encyclopedia of Food and Color Additives,"* CRC Press, LLC., Boca Raton, FL, 1996.

Hendry, G.A.F. and J.D. Houghton: *Natural Food Colorants,* Blackie Academic & Professional Publishers, New York, NY, 1999.

Hutchings, J.B.: *Food Color and Appearance,* Aspen Publishers, Inc., Gaithersburg, MD, 1999.

Lauro, G.J. and F.J. Francis: *Natural Food Colorants: Science and Technology,* Marcel Dekker, Inc., New York, NY, 2000.

Manu-Tawiah, W. et al.: "Extending the Color Stability and Shelf Life of Fresh Meat," *Food Technology,* **94** (March 1991).

Marmion, D.M.: *Handbook of U.S. Colorants: Foods, Drugs, Cosmetics, and Medical Devices,* John Wiley & Sons, Inc., New York, NY, 1992.

Murai, K. and D. Wilkins: "Natural Red Color Derived from Red Cabbage," *Food Technology,* **131** (June 1990).

Staff: *Food Chemicals Codex,* National Academy of Sciences, Washington, DC, 1993.

Web Reference

Institute of Food Technologists. http://www.ift.org/

COLOR CENTERS. Certain crystals, such as the alkali halides, can be colored by the introduction of excess alkali metal into the lattice, or by irradiation with x-rays, energetic electrons, etc. Thus sodium chloride acquires a yellow color and potassium chloride a blue-violet color. The absorption spectra of such crystals have definite absorption bands throughout the ultraviolet, visible and near-infrared regions. The term *color center* is applied to special electronic configurations in the solid. The simplest and best understood of these color centers is the F center. Color centers are basically lattice defects that absorb light.

COLORIMETRY. A method of chemical analysis that deals with the measurement of the light absorption by colored solutions. Since light absorption depends upon the concentration of a specific constituent in solution, colorimetry is frequently used by geologists to determine qualitatively the trace quantities of many elements.

The fundamental principle of colorimetry states that the amount of light absorbed by a given substance in solution is proportional to the intensity of incident light and to the concentration of the absorbing species. This is expressed mathematically in the *Lambert-Beer law*:

$$\log I_0/I = abc$$

where
I_0 = intensity of incident light
I_0 = intensity of transmitted light
a = absorptivity of the substance
b = light path length
c = concentration of colored substance
I/I_0 = transmittance
$\log I_0/I$ = absorbance

The term colorimetry is generally restricted to the visual comparison and matching of the color of a standard solution with that of an unknown one, whereas *spectrophotometry* involves the use of a photoelectric cell which measures a narrow band of wavelengths for transmittance.

Visual colorimetry is a simple method and is fairly precise. Essentially it requires the matching of the color of a standard solution with that of an unknown sample so that when they become identical, they must contain the same amount of colored substance in columns of equal cross-section. At this point

$$C_x b_x = C_s b_s \text{ and } C_x = C_s b_s/b_x$$

where C_x = concentration of unknown solution
b_x = length of light path of unknown solution
C_s = concentration of standard solution
b_s = length of light of standard solution

A standard series of solutions is prepared, each with a known concentration, having the same volume as the unknown, and being contained in identical flat-bottomed tubes of equal diameter (*Nessler tubes*). The solutions should be compared in daylight and examined against a white background.

A more refined method uses the Duboscq colorimeter. This instrument features a dual-matched optical system. Uniformly intense light is incident upon both colorimeter tubes and the difference in absorption of the standard and unknown solutions is compensated for by adjusting the thickness of solution through which light passes. When the two colors match $C_x = C_s b_s/b_x$.

Spectrophotometry (q.v.) is the most precise method of measuring light transmittance. It involves the use of a photoelectric cell that measures a narrow band of wavelengths for transmittance, and a prism or grating to give monochromatic light. At each wavelength, three readings are made: (1) sample cell filled with solvent, which gives the value of I_0, (2) the cell filled with standard solution, and (3) cell filled with unknown. The ratio of meter readings of the cell filled with pure solvent and sample is the I_0/I ratio. Spectrophotometry may be used analytically to measure the absorption at a single wavelength, or to determine the whole absorption spectrum of liquids and solutions in the visible and ultraviolet regions.

See also **Nephelometry**; and **Turbidimetry**.

Additional Reading

Hunter, R.S. and R.W. Harold: *The Measurement of Appearance,* John Wiley & Sons, Inc., New York, NY, 1987.

MacAdam, D.L.: *Color Measurement: Theme and Variations,* Springer-Verlag Inc., New York, NY, 1985.

MacAdam, D.L.: *Selected Papers on Colorimetry-Fundamentals,* SPIE-International Society for Optical Engineering, Bellingham, WA, 1993.

Malacara, D.: *Color Vision and Colorimetry: Theory and Applications,* SPIE-International Society for Optical Engineering, Bellingham, WA, 2002.

Saltzman, M. and A. Berger-Schunn: *Practical Color Measurement,* John Wiley & Sons, Inc., New York, NY, 1994.

Staff: *ASTM Standards on Color and Appearance Measurement,* American Society for Testing & Materials, West Conshohocken, PA, 2000.

Zollinger, H.: *Color: A Multidisciplinary Approach,* John Wiley & Sons, Inc., New York, NY, 1999.

COLUMBITE. A mineral oxide of iron, manganese, niobium (columbium), and tantalum; $(Fe,Mn)(Nb,Ta)_2O_6$. Crystallizes in the orthorhombic system. Hardness 6; sp gr 5.20; color, red to brown.

COMBUSTION (Fuels). The rapid chemical combination of oxygen with the combustible elements of a fuel. There are three combustible chemical elements of major significance—carbon, hydrogen, and sulfur. However, as a source of heat, sulfur is of minor concern. Sulfur is of particular importance in the combustion of several fuels because of the corrosion and pollution problems that its presence creates.

Carbon and hydrogen when burned to completion with oxygen unite according to:

$$C + O_2 = CO_2$$

$$+ 14,100 Btu/pound (7840 Calories/kilogram) \text{ of carbon}$$

$$2H_2 + O_2 = 2H_2O$$

$$+ 61,100 Btu/pound (33,972 Calories/kilogram) \text{ of hydrogen}$$

Air is the usual source of oxygen for boiler furnaces. These combustion reactions are exothermic as indicated by the foregoing equations.

The objective of good combustion is to release all of the indicated heat while minimizing losses from combustion imperfections and superfluous air. The combination of the combustible elements and compounds of a fuel with all the oxygen requires *temperature* high enough to ignite the constituents, mixing or *turbulence*, and sufficient *time* for complete combustion. These factors sometimes are referred to as the "three Ts" of combustion.

This description* deals with the basic chemistry necessary for understanding the phenomena of combustion in boiler furnaces. Ability to calculate the release of heat in combustion and to determine the amount and nature of the combustion products is essential for the design, for example, of a steam generating plant and determination of its performance characteristics.

Table 1 lists the chemical elements and compounds found in fuels generally used in the commercial generation of heat with their molecular weights, heats of combustion, and other combustion constants. The term "100% total air" used in Table 1 and figures and examples that appear elsewhere in this entry means 100% of the air theoretically required for combustion without excess. Higher percentages indicate the theoretical plus excess air, e.g., 125% total air means 100% theoretical air plus 25% excess air.

* Basic information for this entry from "Steam—Its Generation and Use," copyright The Babcock & Wilcox Co., New York (39th edition, 1978).

TABLE 1. COMBUSTION CONSTANTS OF CHEMICAL ELEMENTS AND COMPOUNDS GENERALLY FOUND IN FUELS

For the two right‑hand groups: the first group gives "For 100% total air — moles per mole of combustible or volume units of air per volume units of combustible" (mol); the second group gives "For 100% total air — weight units of air per weight units of combustible" (wt). Each group has "Required for combustion" (O_2, N_2, Air) and "Flue products" (CO_2, H_2O, N_2).

No	Substance	Formula	Molecular weight	Lb per Cu Ft	Cu Ft per Lb	Sp Gr Air=(1.0000)	Btu/Cu Ft Gross (High)	Btu/Cu Ft Net (Low)	Btu/Lb Gross (High)	Btu/Lb Net (Low)	Req O_2 (mol)	Req N_2 (mol)	Req Air (mol)	Flue CO_2 (mol)	Flue H_2O (mol)	Flue N_2 (mol)	Req O_2 (wt)	Req N_2 (wt)	Req Air (wt)	Flue CO_2 (wt)	Flue H_2O (wt)	Flue N_2 (wt)
1	Carbon[a]	C	12.01	—	—	—	—	—	14,093	14,093	1.0	3.76	4.76	1.0	—	3.76	2.66	8.86	11.53	3.66	—	8.80
2	Hydrogen	H_2	2.016	0.0053	187.723	0.0696	325	275	61,095	51,623	0.5	1.88	2.38	—	1.0	1.88	7.94	26.41	34.34	—	8.94	26.41
3	Oxygen	O_2	32.00	0.0846	11.819	1.1053	—	—	—	—	—	—	—	—	—	—	—	—	—	—	—	—
4	Nitrogen (atm)	N_2	28.01	0.0744	13.443	0.9718	—	—	—	—	—	—	—	—	—	—	—	—	—	—	—	—
5	Carbon monoxide	CO	28.01	0.0740	13.506	0.9672	321	321	4,347	4,347	0.5	1.88	2.38	1.0	—	1.88	0.57	1.90	2.47	1.57	—	1.90
6	Carbon dioxide	CO_2	44.01	0.1170	8.548	1.5282	—	—	—	—	—	—	—	—	—	1.88	—	—	—	—	—	1.90
	PARAFFIN SERIES																					
7	Methane	CH_4	16.04	0.0425	23.552	0.5543	1012	911	23,875	21,495	2.0	7.53	9.53	1.0	2.0	7.53	3.99	13.28	17.27	2.74	2.25	13.28
8	Ethane	C_2H_6	30.07	0.0803	12.455	1.0488	1773	1622	22,323	20,418	3.5	13.18	16.38	2.0	3.0	13.18	3.73	12.39	16.12	2.93	1.80	12.39
9	Propane	C_3H_8	44.09	0.1196	8.365	1.5617	2524	2322	21,669	19,937	5.0	18.82	23.82	3.0	4.0	18.82	3.63	12.07	15.70	2.99	1.63	12.09
10	n-Butane	C_4H_{10}	58.12	0.1582	6.321	2.0665	3271	3018	21,321	19,678	6.5	24.47	30.97	4.0	5.0	24.47	3.58	11.91	15.49	3.03	1.55	11.91
11	Isobutane	C_4H_{10}	58.12	0.1582	6.321	2.0665	3261	3009	21,271	19,628	6.5	24.47	30.97	4.0	5.0	24.47	3.58	11.91	15.49	3.03	1.55	11.91
12	n-Pentane	C_5H_{12}	72.15	0.1904	5.252	2.4872	4020	3717	21,095	19,507	8.0	30.11	38.11	5.0	6.0	30.11	3.55	11.81	15.35	3.05	1.50	11.91
13	Isopentane	C_5H_{12}	72.15	0.1904	5.252	2.4872	4011	3708	21,047	19,459	8.0	30.11	38.11	5.0	6.0	30.11	3.55	11.81	15.35	3.05	1.50	11.91
14	Neopentane	C_5H_{12}	72.15	0.1904	5.252	2.4872	3994	3692	20,978	19,390	8.0	30.11	38.11	5.0	6.0	30.11	3.55	11.81	15.35	3.05	1.50	11.91
15	n-Hexane	C_6H_{14}	86.17	0.2274	4.398	2.9704	4768	4415	20,966	19,415	9.5	35.76	45.26	6.0	7.0	35.76	3.53	11.74	15.27	3.06	1.46	11.74
	OLEFIN SERIES																					
16	Ethylene	C_2H_4	28.05	0.0742	13.475	0.9740	1604	1503	21,636	20,275	3.0	11.29	14.29	2.0	2.0	11.29	3.42	11.39	14.81	3.14	1.29	11.39
17	Propylene	C_3H_6	42.08	0.1110	9.007	1.4504	2340	2188	21,048	19,687	4.5	16.94	21.44	3.0	3.0	16.94	3.42	11.39	14.81	3.14	1.29	11.39
18	n-Butene	C_4H_8	56.10	0.1480	6.756	1.9336	3084	2885	20,854	19,493	6.0	22.59	28.59	4.0	4.0	22.59	3.42	11.39	14.81	3.14	1.29	11.39
19	Isobutene	C_4H_8	56.10	0.1480	6.756	1.9336	3069	2868	20,737	19,376	6.0	22.59	28.59	4.0	4.0	22.59	3.42	11.39	14.81	3.14	1.29	11.39
20	n-Pentene	C_5H_{10}	70.13	0.1852	5.400	2.4190	3837	3585	20,720	19,359	7.5	28.23	35.73	5.0	5.0	28.23	3.42	11.39	14.81	3.14	129	11.39
	AROMATIC SERIES																					
21	Benzene	C_6H_6	78.11	0.2060	4.852	2.6920	3752	3601	18,184	17,451	7.5	28.23	35.73	6.0	3.0	28.23	3.07	10.22	13.30	3.38	0.69	10.22
22	Toluene	C_7H_8	92.13	0.2431	4.113	3.1760	4486	4285	18,501	17,672	9.0	33.88	42.88	7.0	4.0	33.88	3.13	10.40	13.53	3.34	0.78	10.40
23	Xylene	C_8H_{10}	106.16	0.2803	3.567	3.6618	5230	4980	18,650	17,760	10.5	39.52	50.02	8.0	5.0	39.52	3.17	10.53	13.70	3.32	0.85	10.53
	MISCELLANEOUS GASES																					
24	Acetylene	C_2H_2	26.04	0.0697	14.344	0.9107	1477	1426	21,502	20,769	2.5	9.41	11.91	2.0	1.0	9.41	3.07	10.22	13.30	3.38	0.69	10.22
25	Napthalene	$C_{10}H_8$	128.16	0.3384	2.955	4.4208	5854	5654	17,303	16,708	12.0	45.17	57.17	10.0	4.0	45.17	3.00	9.97	12.96	3.43	0.56	9.97
26	Methyl alcohol	CH_3OH	32.04	0.0846	11.820	1.1052	868	767	10,258	9,066	1.5	5.65	7.15	1.0	2.0	5.65	1.50	4.98	6.48	1.37	1.13	4.98
27	Ethyl alcohol	C_2H_5OH	46.07	0.1216	8.221	1.5890	1600	1449	13,161	11,917	3.0	11.29	14.29	2.0	3.0	11.29	2.08	6.93	9.02	1.92	1.17	6.93
28	Ammonia	NH_3	17.03	0.0456	21.914	0.5961	441	364	9,667	7,985	0.75	2.82	3.57	—	1.5	2.82	1.41	4.69	6.10	—	1.59	5.51
29	Sulfur[a]	S	32.06	—	—	—	—	—	3,980	3,980	1.0	3.76	4.76	1.0 (SO_2)	—	3.76	1.00	3.29	4.29	2.00 (SO_2)	—	3.29
30	Hydrogen sulfide	H_2S	34.08	0.911	10.979	1.1898	646	595	7,097	6,537	1.5	5.65	7.15	1.0	1.0	5.65	1.41	4.69	6.10	1.88	0.53	4.69
31	Sulfur dioxide	SO_2	64.06	0.1733	5.770	2.2640	—	—	—	—	—	—	—	1.0	—	—	—	—	—	1.0	—	—
32	Water vapor	H_2O	18.02	0.0476	21.017	0.6215	—	—	—	—	—	—	—	—	—	—	—	—	—	—	—	—
33	Air	Air	28.9	0.766	13.063	1.0000	—	—	—	—	—	—	—	—	—	—	—	—	—	—	—	—

[a] Carbon and sulfur are considered as gases for molal calculations only.
All gas volumes corrected to 60 °F and 30 in. Hg dry (15.6 °C and 101.6 kilopascals).

Source: American Gas Association, Arlington, Virginia.

To convert:

lb/cu ft to kg/cu meter, multiply by	16.026
cu ft/lb to cu meters/kg, multiply by	0.0624
Btu to Calories	0.2520
Btu/ft to Calories/cu meter	8.898

Concept of the Mole

The mass of a substance in weight units equal to its molecular weight is a mole of the substance. For example, carbon (C) has a molecular weight of 12. Therefore, a pound-mole of carbon weighs 12 pounds, a gram-mole of carbon weighs 12 grams, etc. In the case of gases, the *volume* occupied by a mole is called the *molal volume* and this is constant for "ideal" gases. A pound-mole of a gas at 80°F and atmospheric pressure (1 atmosphere or 14.7 psia or 30 inches of mercury) occupies 394 cubic feet. These concepts of mass and volume are useful in combustion calculations. Pound-moles are commonly used in power plant calculations in the United States and a number of other English speaking countries. Useful metric conversions will be found in the entry on **Units and Standards**.

Fundamental Laws

Several fundamental physical laws apply to combustion calculations. These are reviewed briefly as follows:

Conservation of Matter. This is the familiar statement that "matter is neither destroyed nor created." There must be a weight balance between the sum of the weights entering a process and the sum leaving. In other words, A pounds of fuel combined with B pounds of air will always result in $A + B$ pounds of products. (It should be noted that when a pound of a typical coal is burned, releasing 13,500 Btu, the quantity of mass converted to energy amounts to only 3.5×10^{-10} pound, a loss too small to be measured or considered in conventional combustion calculations. Obviously, this conversion is of significance to nuclear reactions.)

Conservation of Energy. This is the familiar statement that "energy is neither destroyed nor created." The sum of the energy (potential, kinetic, thermal, chemical, and electrical) entering a process must equal the sum of energy leaving, although the proportionate amounts of each may change. In combustion, chemical energy is exchanged for energy in the form of heat. The parenthetical observation made in the prior paragraph also applies to this relationship.

Ideal Gas Law. The volume of an ideal gas is directly proportional to its absolute temperature and inversely proportional to its absolute pressure. The proportional constant is found to be the same for one mole of any ideal gas, so this law may be expressed as:

$$v_M = \frac{RT}{p}$$

where v_M = volume, cubic feet/mole of gas

p = absolute, pressure, pounds/square foot

T = absolute temperature, degrees Rankine = °F + 460

R = universal gas constant, 1545 foot pound/mole

The equation states that one mole of all ideal gases occupies the same volume for the same pressure and temperature conditions—394 cubic feet at 14.7 psi and 80°F. Experiments indicate that most gases approach this ideal.

Law of Combining Weights. All substances combine in accordance with simple definite weight relationships. These relationships are exactly proportional to the molecular weights of the constituents. For example, carbon (atomic weight = 12) combines with oxygen (molecular weight = 32) to form carbon dioxide (molecular weight = 44) so that 12 pounds

of carbon plus 32 pounds of oxygen unite to form 44 pounds of carbon dioxide.

Avogadro's Law. Equal volumes of different gases at the same pressure and temperature contain the same number of molecules. From the concept of the mole, a pound-mole of any substance contains a mass equal in pounds to the molecular weight of the substance. Thus the ratio of mole weight to molecular weight is a constant, and a mole of a chemically pure substance contains the same number of molecules, no matter what the substance may be. Since a mole of any ideal gas occupies the same volume at a given pressure and temperature (ideal gas law), it follows that equal volumes of different gases at the same pressure and temperature contain the same number of molecules.

Dalton's Law. The total pressure of a mixture of gases is the sum of the partial pressures that would be exerted by each of the constituents if each gas were to occupy alone the same volume as the mixture. In other words, for equal volumes, V, of three gases (A, B, and C) all at the same temperature, T, but at different pressures, P_a, P_b and P_c, when all three gases are placed in the space of the volume, V, then the resulting pressure, P, is equal to $P_a + P_b + P_c$. For gases, each gas in a mixture fills the entire volume and exerts a pressure independent of the other gases.

Amagat's Law. The total volume occupied by a mixture of gases is equal to the sum of the volumes that would be occupied by each of the constituents when at the same pressure and temperature as the mixture. This law is related to Dalton's Law, but considers the additive effects of volume instead of pressure. If all three gases are at pressure, P, and temperature, T, but at volumes V_a, V_b and V_c, then, when combined so that T and P are unchanged, the volume of the mixture, $V = V_a + V_b + V_c$.

Application of Fundamental Laws

Table 2 summarizes the molecular and weight relationships between fuel and oxygen and lists the heat of combustion for the substances commonly involved in combustion. Most of the weight and volume relationships in combustion calculations can be determined by using the information in this table and the seven fundamental laws.

The data for C and H_2 can be expressed as follows:

C	+	O_2	=	CO_2*
1 molecule	+	1 molecule	→	1 molecule
1 mole	+	1 mole	=	1 mole
		1 cubic foot	→	1 cubic foot
12 pounds	+	32 pounds	=	44 pounds
$2H_2$	+	O_2	=	$2H_2O$
2 molecules	+	1 molecule	→	2 molecules
2 moles	+	1 mole	=	2 moles
2 cubic feet	+	1 cubic foot	→	2 cubic feet
4 pounds	+	32 pounds	=	36 pounds

While there is a weight balance in these equations, there is not a molecular or volume balance; for example, 2 cubic feet of H_2 unite with

*When 1 cubic foot of oxygen combines with carbon, it forms 1 cubic foot of carbon dioxide. If carbon were an ideal gas instead of a solid, 1 cubic foot of carbon would be required.

TABLE 2. COMMON CHEMICAL REACTIONS OF COMBUSTION

Combustible	Reaction	Moles	Pounds	Heat of combustion (High) Btu pound of fuel
Carbon (to CO)	$2\,C + O_2 = 2\,CO$	$2 + 1 = 2$	$24 + 32 = 56$	4,000
Carbon (to CO_2)	$C + O_2 = CO_2$	$1 + 1 = 1$	$12 + 32 = 44$	14,100
Carbon Monoxide	$2\,CO + O_2 = 2\,CO_2$	$2 + 1 = 2$	$56 + 32 = 88$	4,345
Hydrogen	$2\,H_2 + O_2 = 2\,H_2O$	$2 + 1 = 2$	$4 + 32 = 36$	61,100
Sulfur (to SO_2)	$S + O_2 = SO_2$	$1 + 1 = 1$	$32 + 32 = 64$	3,980
Methane	$CH_4 + 2\,O_2 = CO_2 + 2\,H_2O$	$1 + 2 = 1 + 2$	$16 + 64 = 80$	23,875
Acetylene	$2\,C_2H_2 + 5\,O_2 = 4\,CO_2 + 2\,H_2O$	$2 + 5 = 4 + 2$	$52 + 160 = 212$	21,500
Ethylene	$C_2H_4 + 3\,O_2 = 2\,CO_2 + 2\,H_2O$	$1 + 3 = 2 + 2$	$28 + 96 = 124$	21,635
Ethane	$2\,C_2H_6 + 7\,O_2 = 4\,CO_2 + 6\,H_2O$	$2 + 7 = 4 + 6$	$60 + 224 = 284$	22,325
Hydrogen Sulfide	$2\,H_2S + 3\,O_2 = 2\,SO_2 + 2\,H_2O$	$2 + 3 = 2 + 2$	$68 + 96 = 164$	7,100

Source: "Steam—Its Generation and Use," 39th edition, The Babcock and Wilcox Company, New York, 1978.

1 cu ft of O_2 to form only 2 cubic feet of H_2O. This relationship is based on Avogadro's law and the law of combining weights.

The Mole in Combustion Calculations. Combustion calculations involving gaseous mixtures can be simplified by the use of the mole. Since equal volumes of gases at any given pressure and temperature contain the same number of molecules (Avogadro's law), the weights of equal volumes of gases are proportional to their molecular weights. If M is the molecular weight of the gas, 1 mole equals M pounds. Actual values are available from Table 1, e.g.:

$$1 \text{ mole of } O_2 = 32 \text{ pounds oxygen}$$

$$1 \text{ mole of } H_2 = 2 \text{ pounds hydrogen}$$

$$1 \text{ mole of } CH_4 = 16 \text{ pounds methane}$$

Data from Table 1 can be used to demonstrate that the volume of 1 mole at a given pressure and temperature is approximately fixed and independent of the kind of gas.

At 60°F and atmospheric pressure (30 inches of mercury), the specific volume of oxygen is 11.819 cubic feet per pound. Therefore, 1 mole of oxygen has a volume of $32 \times 11.819 = 378.21$ cubic feet. Similarly, at 60°F and atmospheric pressure, the specific volume of hydrogen is 187.723 cubic feet per pound, and 1 mole has a volume of $2.016 \times 187.723 = 378.45$ cubic feet. This volume, usually taken as 379 cubic feet, therefore approximates the volume of 1 mole of gas at 60°F and atmospheric pressure.

The mole fraction of a component of a mixture is the number of moles of the component divided by the sum of the number of moles of all the components of the mixture. As a mole of every ideal gas occupies the same volume, it follows by Avogadro's law that in a mixture of ideal gases the mole fraction of a component will exactly equal the volume fraction:

$$\frac{\text{Moles of component}}{\text{Total Moles}} = \frac{\text{Volume of Component}}{\text{Volume of Total Mixture}}$$

This is a valuable concept, since the volumetric analysis of a mixture of gases automatically gives the mole fractions of the different components.

In power plant practice, the practical source of oxygen is primarily air, which includes, along with the oxygen, a mixture of nitrogen, water vapor, and small amounts of inert gases, such as argon, neon, and helium. Data on the composition of air are given in Table 3.

The information in Table 2 can be used for air instead of O_2 if 3.76 moles of nitrogen (N_2) are added to both left and right side of each equation for every mole of O_2 involved. For example, the burning of CO in air becomes:

$$2 \, CO + O_2 + 3.76 \, N_2 = CO_2 + 3.76 \, N_2$$

TABLE 3. COMPOSITION OF AIR

	Composition of dry air	
	% by volume	% by weight
Oxygen, O_2	20.99	23.15
Nitrogen, N_2	78.03	76.85[a]
Inerts	0.98	—

Equivalent molecular weight of air = 29.0[a]

% Moisture = 1.3% by weight. Standard for boiler industry (ABMA)[b]

Moles air/mole oxygen

Cubic feet air/cubic feet oxygen $= \dfrac{100}{20.99} = 4.76$

Moles N_2/mole oxygen $= \dfrac{79.01}{20.99} = 3.76$

Pounds air (dry)/pound oxygen $= \dfrac{100}{23.15} = 4.32$

Pounds nitrogen/pound oxygen $= \dfrac{76.85}{23.15} = 3.32$

[a] It is convenient in combustion calculations to account for inerts as equivalent nitrogen. The equivalent weight percentage of 76.85 and the equivalent molecular weight of 29.0 have been corrected to account for the extra weight of the inerts.
[b] Air containing 0.013 pound water/pound dry air is often referred to as standard air.
Source: "Steam—Its Generation and Use," 39th edition, The Babcock and Wilcox Company, New York, 1978.

or for methane:

$$CH_4 + 2 \, O_2 + 2(3.76) \, N_2 = CO_2 + 2 \, H_2O + 7.52 \, N_2$$

As indicated by the following example for a fuel gas, molal calculations have a simple and direct application to gaseous fuels, where the analyses are usually reported in percent on a volume basis.

FUEL GAS ANALYSIS
(% by Volume)

CH_4	85.3
C_2H_6	12.6
CO_2	0.1
N_2	1.7
O_2	0.3
Total	100.0

This analysis may also be expressed as 85.3 moles of CH_4 per 100 moles of fuel; 12.6 moles of C_2H_6 per moles of fuel; and so on.

The elemental breakdown of each constituent may also be designated in moles per 100 moles of fuel, as follows:

C in CH_4	=	85.3×1	=	85.3	moles
C in C_2H_6	=	12.6×2	=	25.2	moles
C in CO_2	=	0.1×1	=	0.1	mole
Total C per 100 moles fuel			=	110.6	moles
H_2 in CH_4	=	85.3×2	=	170.6	moles
H_2 in C_2H_6	=	12.6×3	=	37.8	moles
Total H per 100 moles fuel			=	208.4	moles
O_2 in CO_2	=	0.1×1	=	0.1	mole
O_2	=	0.3×1	=	0.3	mole
Total O_2 per 100 moles fuel			=	0.4	mole
N_2 per 100 moles fuel			=	1.7	moles

An analysis of the flue gas produced by burning a gas fuel of the composition given above could be:

Constituent	% by volume
CO_2	10.4
O_2	2.8
N_2	86.8
Total	100.0

Analyses of flue gases are always reported on a volume basis, *dry*, when an Orsat or other type of gas analysis is used. Flue gases are cooled to room temperature and bubbled through water in most gas analyses, so that the gas becomes saturated with water vapor. This would occur even if no water vapor were formed during combustion. Proportionate parts of the water vapor content of the gas will be absorbed with the different constituents of the gas so that the resulting analysis may be safely assumed to be that of dry gas. These percentages may also be expressed as 10.4 moles CO_2, 2.8 moles O_2, and 86.8 moles N_2; each per 100 moles of dry flue gas.

For each mole of C burned, one mole of CO_2 is formed. From the fuel analysis used there are 110.6 moles C per 100 moles of fuel, and there are also 110.6 moles of CO_2 formed from the 110.6 moles C in the fuel. From the flue gas analysis, there are $100/10.4 = 9.62$ moles of dry flue gas per mole of CO_2. The 100 moles of fuel will then yield $110.6 \times 9.62 = 1,064$ moles of dry flue gas. By the application of the mole method, an important value has been quickly determined through knowing only the flue gas analysis and the fuel analysis.

From the flue gas analysis, the molecular weight of the dry flue gas can be easily determined, as follows:

10.4 moles of CO_2 weight			
10.4×44 pounds	=	457.6	pounds
2.8 moles of O_2 weight			
2.8×32 pounds	=	89.6	pounds
86.8 moles of N_2 weight			
86.8×28 pounds	=	2,430.4	pounds
100.0 moles of dry flue gas		2,977.6	pounds

Therefore, 1 mole equivalent of dry flue gas = 29.8 pounds, or the equivalent molecular weight of the dry flue gas = 29.8. Hence, the weight of 1,064 moles of dry flue gas is $1,064 \times 29.8 = 31,700$ pounds, or 100 moles of fuel yields 31,700 pounds of dry flue gas.

The weight of 100 moles of fuel is the sum of the products of each constituent in the fuel and its molecular weight.

CH_4	85.3	$\times$	16	=	1,365	
C_2H_6	12.6	$\times$	30	=	378	
CO_2	0.1	$\times$	44	=	4.4	
N_2	1.7	$\times$	28	=	47.6	
O_2	0.3	$\times$	32	=	9.6	
	100.0 moles			=	1804.6	pounds

Thus, 1,805 pounds of gas fuel yield 31,700 pounds of dry flue gas, and each pound of gas fuel yields $31,700/1,805 = 17.6$ pounds dry flue gas.

Heat of Combustion

In a boiler furnace (where no mechanical work is done) the heat energy evolved from the union of combustible elements with oxygen depends on the ultimate products of combustion and not on any intermediate combinations that may occur in reaching the final result.

A simple demonstration of this law is the union of 1 pound of carbon with oxygen to produce a specific amount of heat (about 14,100 Btu, Table 2). The union may be in one step to form the gaseous product of combustion, CO_2, or under certain conditions, the union may be in two steps, first to form CO, producing a much smaller amount of heat (4,345 Btu) and, second, the union of the CO so obtained to form CO_2, releasing 9,755 Btu. However, the sum of the heats released in the two steps equals the 14,100 Btu evolved when carbon is burned in one step to form CO_2 as the final product.

That carbon may enter into these two combinations with oxygen is of utmost importance in the design of combustion equipment. Firing methods must assure complete mixture of fuel and oxygen, to be certain that all of the carbon burns to CO_2 and not to CO. Failure to meet this requirement will result in appreciable losses in combustion efficiency and in the amount of heat released by the fuel, since only about 28% of the available heat in the carbon is released if CO is formed instead of CO_2.

Measurement of Heat of Combustion. In boiler practice, the heat of combustion of a fuel is the amount of heat, expressed in Btu, generated by the complete combustion (or oxidation) of a unit weight (1 pound in the United States) of fuel. *Calorific value* or *fuel Btu value* are other terms used.

The amount of heat generated by complete combustion is a constant for any given combination of combustible elements and compounds and is not affected by the manner in which the combustion takes place, provided that it is complete.

The heat of combustion of a fuel is usually determined by direct measurement in a calorimeter of the heat evolved during combustion. See entry on **Calorimetry**.

The heat of combustion of most gases encountered in boiler practice is given in Table 1. If the content of any gas mixture is known, its heat of combustion can be accurately determined by adding the products of the volume percentage of each constituent times its heat of combustion.

For accurate heat values of solid and liquid fuels calorimeter determinations are required. However, approximate heat values may be determined for most coals if the ultimate chemical analysis is known. Dulong's formula gives reasonably accurate results (within 2 to 3%) for most coals and is often used as a routine check of values determined by calorimeter:

$$Btu/pound = 14,544C + 62,028(H_2 - O_2/8) + 4050\ S \qquad (1)$$

In this formula, the symbols represent the proportionate parts by weight of the constituents of the fuel—carbon, hydrogen, oxygen and sulfur—as determined by an ultimate analysis; the coefficients represent the approximate heating values of the constituents in Btu per pound. The term $O_2/8$ is a correction applied to the hydrogen in the fuel to account for the hydrogen already combined with the oxygen in the form of moisture. This formula is not generally suitable for calculating the Btu values of gaseous fuels.

High and Low Heat Values. Water vapor is one of the products of combustion for all fuels containing hydrogen. The heat content of a fuel depends on whether this water vapor is allowed to remain in the vapor state or is condensed to liquid. In the bomb calorimeter the products of combustion are cooled to the initial temperature and all of the water vapor formed during combustion is condensed to liquid. This gives the high, or gross, heat content of the fuel with the heat of vaporization included in the reported value. For the low, or net heat of combustion, it is assumed that all products of combustion remain in the gaseous state.

While the high, or gross, heat of combustion can be accurately determined by established (ASTM) procedures, direct determination of the low heat of combustion is difficult. Therefore, it is usually calculated using the following formula:

$$Q_L = Q_H - 1040\ w \qquad (2)$$

where
Q_L = low heat of combustion of fuel, Btu/pound
Q_H = heat of combustion of fuel, Btu/pound
w = of water formed per pound of fuel
1040 = factor to reduce high heat of combustion at constant volume to low heat of combustion at constant pressure

In the United States the practice is to use the high heat of combustion in boiler combustion calculations. In Europe the low heat value is used.

Ignition Temperatures

Ignition temperature may be defined as the temperature at which more heat is generated by combustion than is lost to the surroundings so that the combustion process becomes self-sustaining. The term usually applies to rapid combustion in air at atmospheric pressure.

Ignition temperatures of combustion substances vary greatly as indicated in Table 4, which lists minimum temperatures and temperature ranges in air for fuels and for the combustible constituents of fuels commonly used in the commercial generation of heat. Many factors influence ignition temperature so that any tabulation can be used only as a guide. Pressure, velocity, enclosure configuration, catalytic materials, air-fuel-mixture uniformity, and ignition source are only a few of the variables. Ignition temperature usually decreases with rising pressure and increases with increasing moisture content in the air.

The ignition temperatures of the gases of a coal vary considerably and are appreciably higher than the ignition temperatures of the fixed carbon of the coal. However, the ignition temperature of coal may be considered as the ignition temperature of its fixed carbon content, since the gaseous constituents are usually distilled off but not ignited before this temperature is attained.

Adiabatic Flame Temperature

The adiabatic flame temperature is the maximum theoretical temperature that can be reached by the products of combustion of a specific fuel and air (or oxygen) combination assuming no loss of heat to the surroundings until combustion is complete. This theoretical temperature also assumes no dissociation, a phenomenon discussed later under this heading. The heat of combustion of the fuel is the major factor in the flame temperature, but increasing the temperature of the air or of the fuel will also have the effect of raising the flame temperature. As would be expected, this adiabatic temperature is a maximum with zero excess air (only enough air chemically required to combine with the fuel), since any excess is not

TABLE 4. IGNITION TEMPERATURES OF FUELS IN AIR (APPROXIMATE VALUES AND RANGES AT ATMOSPHERIC PRESSURE)

Combustible	Formula	Temperature, °F
Sulfur	S	470
Charcoal	C	650
Fixed carbon (bituminous coal)	C	765
Fixed carbon (semibituminous coal)	C	870
Fixed carbon (anthracite)	C	840–1115
Acetylene	C_2H_2	580–825
Ethane	C_2H_6	880–1165
Ethylene	C_2H_4	900–1020
Hydrogen	H_2	1065–1095
Methane	CH_4	1170–1380
Carbon Monoxide	CO	1130–1215
Kerosine	—	490–560
Gasoline	—	500–800

involved in the combustion process and only dilutes the temperature of the products of combustion.

The adiabatic temperature is determined from the adiabatic enthalpy of the flue gas as follows:

$$h_g = \frac{\left(\begin{array}{c}\text{heat of}\\\text{combustion}\end{array}\right) + \left(\begin{array}{c}\text{sensible heat}\\\text{in fuel}\end{array}\right) + \left(\begin{array}{c}\text{sensible heat}\\\text{in air}\end{array}\right)}{\text{weight of products of combustion}}$$

where h_g = adiabatic enthalpy (adiabatic heat content of the products of combustion), Btu/pound

Knowing the moisture content of the products of combustion and its enthalpy, the theoretical flame or gas temperature can be obtained from published graphs.*

The adiabatic temperature is a fictitiously high temperature that does not exist in fact. Actual flame temperatures are lower for two main reasons:

1. Combustion is not instantaneous. Some heat is lost to the surroundings as combustion takes place. The faster the combustion occurs the less heat is lost before combustion is complete. If combustion is slow enough, the gases may be cooled sufficiently for combustion to be incomplete with some of the fuel unburned. This is related to the time factor in the "three T's" of combustion mentioned previously.

2. At temperatures above 3,000°F (1,649°C), some of the CO_2 and H_2O in the flue gases dissociates, absorbing heat in the process. At 3,500°F (1,926°C), about 10% of the CO_2 in a typical flue gas dissociates to CO and O_2 with a heat absorption of 4,345 Btu/pound of CO formed, and about 3% of the H_2O dissociates to H_2 and O_2, with a heat absorption of 61,100 Btu/pound of H_2 formed. As the gas cools, the CO and H_2 dissociated recombine with the O_2 and liberate the heat absorbed in dissociation, so the heat is not lost. However, the effect is to lower the maximum actual flame temperature.

Combustion Calculations

The combustion calculations are the starting point for all design and performance determinations for boilers and their related component parts. They establish (a) the quantities of the constituents involved in the chemistry of combustion, (b) the quantity of heat released, and (c) the efficiency of the combustion process under both ideal and actual conditions.

Combustion Air. Since carbon, hydrogen, and sulfur are the only major combustible elements found in the fuels used for commercial steam generation, the air (pounds) theoretically required for the complete combustion of 1 pound of fuel is:

$$11.53 \text{ C} + 34.34(H_2 - O_2/8) + 4.29 \text{ S} \tag{3}$$

where C, H_2, O_2 and S represent the fraction by weight (percent/100) of carbon, hydrogen, oxygen and sulfur, and the constants are those given in Table 1. The factor $O_2/8$ in the term $(H_2 - O_2/8)$ is a correction for the hydrogen already combined with the O_2 in the fuel to form water vapor.

With gaseous fuels, instead of breaking down the hydrocarbons into their constituent elements, it is simpler to use the amount of air for the various compounds directly as given in Table 1. For instance, for a gaseous fuel containing the combustible gases indicated in the expression below, the theoretical air required for complete combustion (pounds air/pounds fuel) is:

$$2.47 \text{ CO} + 34.34 \text{ H}_2 + 17.27 \text{ CH}_4 + 13.30 \text{ C}_2\text{H}_2$$
$$+ 14.81 \text{ C}_2\text{H}_4 + 16.12 \text{ C}_2\text{H}_6 + 6.10 \text{ H}_2\text{S} - 4.32 \text{ O}_2 \tag{4}$$

Again, the molecular symbols represent the fraction by weight of the gaseous compounds and elements.

If, as is the usual custom, the analyses of gaseous fuels are given on a volumetric basis, the cubic feet of combustion air required as given in Table 1 should be used. Thus for a gaseous fuel containing the combustible gases indicated in the following expression, the number of cubic feet of theoretical air required per cubic foot of fuel for complete combustion is:

$$2.38(\text{CO} + \text{H}_2) + 9.53 \text{ CH}_4 + 11.91 \text{ C}_2\text{H}_2$$
$$+ 14.29 \text{ C}_2\text{H}_4 + 16.68 \text{ C}_2\text{H}_6 + 7.15 \text{ H}_2\text{S} - 4.76 \text{ O}_2 \tag{5}$$

where the molecular symbols represent the fraction by volume of the gaseous compounds and elements. Note that the total air requirement is reduced if oxygen is one of the constituents of the fuel.

The products of combustion can also be determined from the data given in Table 1. Assuming complete combustion with theoretical air of the fuels ordinarily used for commercial steam generation, the products of combustion in pounds (including the nitrogen carried with the combustion air) per pound of fuel are:

$$CO_2 = 3.66 \text{ C}$$

$$H_2O = 8.94 \text{ H}_2 + H_2O^{**}$$

$$SO_2 = 2.00 \text{ S}$$

$$N_2 = 8.86 \text{ C} + 26.41(H_2 - O_2/8) + 3.29 \text{ S} + N_2^{\dagger}$$

The moisture introduced with the combustion air must be added to this theoretical quantity to obtain the total weight of combustion products. The molecular symbols represent the fraction by weight of the constituents in the fuel.

Energy Losses. Not all of the Btu in the fuel are converted to heat and absorbed by the steam generation equipment. Some of the fuel may be unburned, leaving carbon in the ash or carbon may be burned incompletely to form some CO instead of all CO_2. Usually all of the H_2 in the fuel is burned. By far the greatest heat loss is the loss up the stack. Since the heat in the fuel is determined from a base of ambient temperature, all of the products of combustion must be cooled to the same temperature if all of the heat is to be utilized. Higher temperatures then represent a loss, which is the sum of four items: (1) the sensible heat in dry flue gas, (2) the sensible heat in the moisture in the air, (3) the sensible heat in the H_2O in the fuel, and (4) the latent heat of the moisture in the fuel.

It is necessary practically to use more than the theoretical air requirements to assure sufficient oxygen for complete combustion. Excess air would not be required if it were possible to have an ideally perfect union of air and fuel. It is necessary, however, to keep the excess at a minimum in order to hold down the stack loss. The excess air that is not used in the combustion of the fuel leaves the unit at stack temperature. The heat required to heat this air from room temperature to stack temperature serves no purpose and is lost heat. Table 5 gives realistic values of excess air for the fuel-burning equipment which experience has shown is required to assure complete combustion for various fuels and methods of firing.

TABLE 5. USUAL AMOUNTS OF EXCESS AIR SUPPLIED TO FUEL-BURNING EQUIPMENT

Fuel	Type of furnace or burners	Excess air % by weight
Pulverized coal	Completely water-cooled furnace for slag-tap or dry-ash-removal	15–20
	Partially water-cooled furnace for dry-ash removal	15–40
Crushed coal	Cyclone furnace—pressure or suction	10–15
Coal	Spreader stoker	30–60
	Water-cooled vibrating-grate stoker	30–60
	Chain-grate and traveling-grate stokers	15–50
	Underfeed stoker	20–50
Fuel oil	Oil burners, register-type	5–10
	Multifuel burners and flat-flame	10–20
Acid sludge	Cone and flat-flame type burners, steam-atomized	10–15
Natural, coke-oven, and refinery gas	Register-type burners	5–10
	Multifuel burners	7–12
Blast-furnace gas	Intertube nozzle-type burners	15–18
Wood	Dutch oven (10–23% through grates) and Hofft-type	20–25
Bagasse	All furnaces	25–35
Black liquor	Recovery furnaces for kraft and soda-pulping processes	5–7

Source: "Steam-Its Generation and Use," 39th edition, The Babcock and Wilcox Company, New York, 1978.

---

* Series of 8 graphs in "Steam—Its Generation and Use," pages 6–8 and 6–9, published by The Babcock and Wilcox Company, New York (39th edition, 1978).

** Fraction by weight of H_2O (percent/100) in the fuel as moisture.

† Fraction by weight of N_2 in the fuel as nitrogen.

In most furnaces operating under suction, there is also some leakage of air into the setting and, consequently, the excess air leaving the furnace and the unit is greater than that at the fuel-burning equipment. Another loss that must be considered is the radiation loss from the unit setting.

In summary, there are certain inherent heat losses over which there is no control, and others that are subject to some control. The inherent losses are the result of: (1) the discharge of the products of combustion at a temperature higher than ambient; and (2) the moisture content of the fuel plus the combination of some of the hydrogen with the oxygen in the fuel. The avoidable heat losses, or those that can be controlled by good design and careful operation, can be minimized by: (1) careful control of excess air; (2) tolerating virtually no unburned solid combustible matter in ash or refuse; (3) permitting no unburned gaseous combustibles in the exit gases; and (4) a well-insulated setting for the steam generating unit to reduce radiation loss.

The efficiency of combustion in a heat exchanger or boiler is 100 minus the sum of the heat losses expressed in percent

Additional Reading

Baukal, C.E.: *Heat Transfer in Industrial Combustion,* CRC Press, LLC., Boca Raton, FL, 2000.

Baukal, C.E., V.Y. Gershtein, and X. Li: *Computational Fluid Dynamics in Industrial Combustion,* CRC Press, LLC., Boca Raton, FL, 2000.

Chen, K., R.C. Swanekamp, and T. Elliott: *Standard Handbook of Power Plant Engineering,* 2nd Edition, The McGraw-Hill Companies, Inc., New York, NY, 1994.

Chomiak, J.: *Combustion: A Study in Theory, Fact, and Application,* Gordon and Breach, New York, NY, 1990.

Corcoran,: "Cleaning up Coal," *Sci. Amer.,* **54** (September 1990).

Culp, A.W.: *Principles of Energy Conversion,* 2nd Edition, The McGraw-Hill Companies, Inc., New York, NY, 1991.

Dry, R.J. and R.D. La Nauze: "Combustion in Fluidized Beds," *Chem. Eng. Progress,* **31** (July 1990).

Fulkerson, W., R.R. Judkins, and M.K. Sanghvi: "Energy from Fossil Fuels," *Sci. Amer.,* **128** (September 1990).

Gardiner, W.C.: *Gas-Phase Combustion Chemistry,* Springer-Verlag Inc., New York, NY, 2000.

Lide, D.R.: *CRC Handbook of Chemistry and Physics,* 84th Edition, CRC Press, LLC., Boca Raton, FL, 2003.

Perry, R.H. and D.W. Green: *Perry's Chemical Engineers' Handbook,* 7th Edition, The McGraw-Hill Companies, Inc., New York, NY, 1999.

Peters, N.: *Turbulent Combustion,* Cambridge University Press, New York, NY, 2000.

Turns, S.R.R.: *An Introduction to Combustion: Concepts and Applications,* The McGraw-Hill Companies, Inc., New York, NY, 1999.

Warnatz, J. and U. Maas: *Combustion: Physical and Chemical Fundamentals, Modeling, and Simulation, Experiments, Pollutant Formation,* Springer-Verlag Inc., New York, NY, 2001.

COMMON ION EFFECT. The reversal of ionization which occurs when a compound is added to a solution of a second compound with which it has a common ion, the volume being kept constant. The degree of ionization of the second compound then is lowered, i.e., it retrogresses. The common ion effect also can markedly affect solubility.

COMPARATIVE BIOLOGY. In any division of biological science, the comparative treatment focuses attention upon a limited subject, such as anatomy, but introduces into its treatment data drawn from many species. This method of study has been applied widely to the vertebrates; hence, comparative anatomy is likely to mean comparative anatomy of the vertebrates unless otherwise qualified.

The comparative method is valuable in determining the evolutionary development of organs and the relationship of species.

Comparative biochemistry may be defined as the study of the nature, origin, and control of biochemical diversity. This definition suggests that: (1) biochemical differences are to be found among organisms; (2) such differences arise during the evolving processes of organisms; and (3) the biochemical properties of organisms are under a variety of controls in nature, and presumably may be modified when the biochemical properties and their natural controls are understood.

In the approach to comparative biochemistry, popular before the 1950s and exemplified by the books of Baldwin and Florkin, biochemists studied the similarities and differences among higher organisms, mainly among animals. The compounds, such as the various phosphagens, blood-transport pigments, carotenoids, and the A vitamins, were correlated with the postulated phylogenetic position of the animals possessing these substances. The comparative aspects of nitrogenous excretion products and of salt and water balance have also been studied in great detail. Such types of studies have revealed metabolic differences among the *Metazoa,* differences related to their evolving nature and ecological niches. More recently, the distributions of the alkaloids and flavones of plants, organic acids of lichens, and the pterins of *Drosophila,* among other biochemical markers, have been analyzed in detail to help in establishing genetic and evolving interrelations.

Since about 1945, biochemistry has concentrated largely on cellular metabolism, and these studies tend to emphasize the uniformity of cellular biochemistry. Thus, almost all cells contain proteins built of the same 20 amino acids, RNA and DNA containing their characteristic constituents, common coenzymes ATP, NAD, coenzyme A, and so on. Also, many similarities could be detected in the metabolic events relating to energy and biosynthesis in many cell types, i.e., microorganisms, plants, and animals. This experience was soon interpreted by the school of Kluyver and van Niel and their students to signify a *uniformity of biochemistry,* a unity inferred to derive from a monophyletic process of organisms as they evolve from a primitive cellular type. In this view, biochemical differences among organisms are considered to reflect relatively late evolving divergences of which the metabolic differences among the *Metazoa,* e.g., patterns of nitrogen excretion, such as ammonotelism, ureotelism, and uricotelism, are clear examples. It is recognized that the biochemical differences found to exist between organisms are not only considered to be relatively late in their evolving processes, but are usually stated to represent minor alterations in the broad biochemical pattern common to many cells.

Thus comparative studies, while always remaining important as an investigative tool, have given ground to what appears to be the more important *uniformity* approach, with concentration on the fundamental aspects of biology which affect all life forms.

COMPLEX COMPOUND.   See **Ligand**.

COMPOSITE MATERIALS.   A mixture of mechanical combination on a macroscale of two or more materials that are solid in the finished state, are mutually insoluble and differ in chemical nature. The major types are (1) Laminates of paper, fabric, or wood (veneer) and a thermosetting material (resin, rubber, or adhesive); examples are tire carcasses, plywood, and electrical insulating structures. (2) Reinforced plastics, principally of glass fiber and a thermosetting resin; other types of fibers such as boron, aluminum silicate, and silicon carbide may be used. (3) Cermets, which are mixtures of ceramic and metal powders, heat treated and compressed. (4) Fabrics, e.g., woven combinations of wool or cotton and a synthetic fiber. (5) Filled composites in which a bonding material i.e., linseed oil, resin, or asphalt, is loaded with a filler in the form of flakes or small particles; examples are linoleum, glass flake-plastic mixtures for battery cases, and asphalt-gravel road-surfacing mixtures.

COMPOUND (Chemical).   A homogeneous, pure substance, composed of two or more essentially different chemical elements, which are present in definite proportions; compounds usually possess properties differing from those of the constituent elements.

An *addition compound* is one that is formed by the junction or union of two simpler compounds. Effectively the same as a molecular compound (see definition on the following page).

An *additive compound* is formed by an additional reaction, or by the saturation of a double bond, triple bond, or more than one of them.

An *alicyclic compound* is an organic compound containing a saturated ring of carbon atoms, such as a cycloparaffin or other hydroaromatic compound.

An *aliphatic compound* is an organic compound without ring structures, i.e., with straight chain arrangement of carbon and, possibly other, atoms. In the narrower sense, an aliphatic compound is a member of the paraffin series of hydrocarbons, or one of their derivatives.

An *aromatic compound* is an organic compound containing a ring of carbon atoms, usually unsaturated, such as a benzene, naphthalene, anthracene, and acenaphthylene ring.

An *associated compound* is a compound formed by the union of two or more molecules, usually of the same or similar chemical composition, to form a single complex molecule.

Berthollide compound

See nonstoichiometric compound in this entry.

A *binary compound* is made up of two elements in a definite molecular ratio.

A *catenation compound* has a molecular configuration resembling a linked chain, in which the atoms forming one ring pass through, but are not joined by valence forces to, the ring formed by another group of atoms. Since the two rings, while spatially interlocked, are not joined by valence forces, the application of the word compound to such aggregates may be questioned.

H.L. Frisch of AT&T Bell Laboratories has made extensive calculations of ring sizes necessary to permit the formation of various catenation compounds. He found that 20 is the minimum number of $-CH_2-$ groups in an alicyclic ring through which another ring can be catenated (threaded) without encountering excessively great repulsive forces from the alicyclic ring atoms. For threadings of two rings through a third ring, the probable minimum alicyclic ring size of the latter is 33 $-CH_2-$ groups; Borromean rings, formed by the interlocking of three rings (with no two of them locked separately), require a minimum of 30 $-CH_2-$ groups.

A laboratory preparation of the simplest of these catenation compounds, two interlocking rings, has been carried out at AT&T Bell Laboratories. They started with the dimethyl ester of a 34-carbon paraffinic dicarboxylic acid, $CH_3OOC-(CH_2)_{32}-COOCH_3$, which was reacted in a suspension of metallic sodium in xylene with acetic acid to condense the terminal ester groups to form an aceloin ring compound.

$$O=C-\overline{(CH_2)_{32}-}CHOH$$

Treatment of the latter with deuterated hydrochloric acid reduced it to a 34-carbon alicyclic (ring) compound,

$$HDC-CHD-CD_2-CHD-(CH_2)_{30}$$

containing five deuterium atoms. This hydrocarbon was added to the suspension of metallic sodium in xylene, and then more of the 34-carbon dimethyl ester was added. Ring formation of the latter compound to form the aceloin occurred as before, and a small percentage of the aceloin rings were found to be threaded through the deuterated rings in the solvent, yielding a catenation compound consisting of the 34-carbon aceloin ring and the 34-carbon deuterated alicyclic hydrocarbon threaded together, but without any atoms in one ring being joined by valence bonds to those in the other.

Separation and identification of this catenation compound was effected by chromatography and infrared spectroscopy, the latter being the reason why the hydrocarbon portion of the catenation compound was deuterated.

Chelation compound

See entry on **Chelates and Chelation**.

A *clathrate compound* means, literally, an enclosed compound, a term applied to a solid molecular compound in which a molecule of one component is physically enclosed in the crystal structure of a second compound, so that the properties of the aggregate are essentially those of the enclosing compound. Examples of such "cage compounds" are those of the small molecules of SO_2, CO_2, CO and the noble gases with ice and hydroquinone, which have very open crystal structures. Another example is the clathrate of benzene with nickel cyanide.

A *complex compound* is made up structurally of two or more compounds or ions. See also **Ligand**.

A *condensation compound* is formed by a reaction in which the largest parts, constituting the essential structural elements, of two or more molecules combine to form a new molecule, with elimination of minor elements, such as those of water.

Coordination compound

See also **Coordination Compounds**.

A *covalent compound* is formed by the sharing of electrons between atoms; as distinguished from electrovalent compounds, in which there occurs a transfer of electrons.

A *cyclic compound* has some or all of its atoms arranged in a ring structure.

An *electrovalent compound* is formed by ions, or by atoms which become ions by transfer of electrons between them. (See ionic compound below.)

An *endothermic compound* is a compound whose formation is accompanied by a positive change in heat content, i.e., by the absorption of heat.

An *epoxy compound* contains an oxygen bridge, as

$$\overline{-O-}$$
$$CH_2-CH_2-CH_2-CH_2$$

which is 1, 4 epoxy butane.

An *exothermic compound* is a compound whose formation is accompanied by a negative change in heat content, i.e., with the liberation of heat.

A *heterocyclic compound* contains one or more rings composed of atoms some of which are of dissimilar elements. A few inorganic substances fall into this classification, but by far the majority of them are carbon compounds. In organic chemistry substances of cyclic structure, as acid anhydrides, lactides, lactams, lactones, cyclic ethers, and cyclic derivatives of dicarboxylic acids which are formed by the elimination of water from aliphatic compounds, are not considered among the heterocyclic substances. Derivatives of pyridine, quinoline, thiophene, thiazole, pyrone, etc., which contain heterocyclic rings that persist in the compound through chemical reactions, are considered the true members of this class. Heterocyclic rings that contain nitrogen, sulfur, and oxygen members. The noncarbon members of the ring are termed "heteroatoms," and their number is indicated by the prefixes mono, di, tri, tetra, etc. The number of members in the ring may reach as high as sixteen, as in tetrasalicylide.

A *homocyclic compound* contains a homocyclic ring, i.e., a ring composed of atoms of the same element.

The term *inclusion compound* was once used for the clathrate compounds described in this entry.

In an *inner compound* an additional valence bond has been formed between two atoms of an already existing structure, usually by loss of the elements of water or other simple substance. Inner compound formation commonly results in the formation of a ring. The inner esters, inner anhydrides, and inner coordination compounds are well-known classes of inner compounds.

An *inorganic compound* means, in general, a compound that does not contain carbon atoms. Some very simple carbon compounds, such as carbon monoxide and dioxide, binary metallic carbon compounds (carbides) and carbonates, are also included in the group of inorganic compounds.

An *intermetallic compound* consists of metallic atoms only, which are joined by metallic bonds. Such compounds may be made semiconducting if the two metals between them contribute just sufficient electrons to fill the valence bond, e.g., InAs. See also **Alloys**.

An *interstitial compound* consists of a metal or metals and certain metalloid elements, in which the metalloid atoms occupy the interstices between the atoms of the metal lattice. Compounds of this type are, for example, TaC, TiC, ZrC, NbC, and similar compounds of carbon, nitrogen, boron, and hydrogen with metals.

An *ionic compound* is one of a class of compounds formed when atoms combine to produce molecules having stable configurations by the transfer of one or more electrons within the molecule. This type of combination is illustrated by the combination of sodium atoms and chlorine atoms to form sodium chloride. The sodium atom loses the single electron in its outer shell, and thus is left with the stable configuration of eight electrons; the chlorine atom acquires an electron to increase the number of electrons in its outer shell from seven to eight; as a result of the loss and gain of the electrons, the atoms have acquired the positive and negative charges, respectively, which constitute an electrovalent bond.

A *molecular compound* is formed by the union of two or more already saturated molecules apparently in defiance of the ordinary rules of valence. The class includes double salts, salts with water of crystallization, and metal ammonium derivatives. These salts are usually formed by Van der Waals attraction between the constituent molecules. They do not differ in any characteristic manner from compounds formed in strict accordance with the concept of valence. They are also called addition compounds.

A *nonpolar compound* is a compound in which the centers of positive and negative charge almost coincide, so that no permanent dipole moments are produced. The term nonpolar also applies to compounds in which the effect of oppositely directed dipole moments cancel. Nonpolar compounds may contain polar bonds, if their effect is canceled by opposing bonds, as may occur in a perfectly symmetrical molecule. Nonpolar compounds do not ionize or conduct electricity. Most organic compounds are to be classed as nonpolar compounds.

A *nonstoichiometric compound* has a composition not in accord with the law of definite proportions, which is therefore also called a berthollide compound. Nonstoichiometric compounds occur among the binary compounds of Group 6b, as exemplified by $TiO_{1.8}$, $Cu_{1.7}S$ and $Cu_{1.6}Se$; among the hydrides, e.g., $CeH_{2.7}$ and especially among the intermetallic compounds.

An *organic compound* is one of the great number of compounds consisting of carbon linked in chains or rings; such compounds usually also contain hydrogen and may contain elements such as oxygen, nitrogen, sulfur, chlorine, etc. Some of the simpler carbon compounds are classified as inorganic compounds.

An *organometallic* (or *metal-organic*) *compound* is an organic compound in which one or more hydrogen atoms have been replaced by a metallic atom or atoms, usually with the establishment of a valence bond between the metal atom and a carbon atom. A metallic salt of an organic acid, in which the hydrogen atoms of a COOH group is replaced by a metal atom, is not classified as an organometallic compound.

A *polar compound* is, in general, a compound that exhibits polarity, or local differences in electrical properties, and has a dipole moment associated with one or more of its interatomic valence bonds. Polar compounds have relatively high dielectric constants, associate readily in most cases, and include the substances that exhibit tautomerism. In the most general use of the term, polar compounds include all electrolytes, most inorganic substances, and many organic ones. Specifically, the term polar compound is frequently applied to the extreme type of polarity which arises in the presence of an electrovalent bond or, in wave-mechanical terms, to cases in which one ionic term dominates in the orbital function of the molecule. Such compounds are exemplified by the inorganic acids, bases, and salts which possess, to a greater or lesser degree the power to conduct electricity, associate, form double molecules and complex ions, etc.

In a *saturated compound* the valence of all the atoms is completely satisfied without linking any two atoms by more than one valence bond.

A *spiro-compound* contains two ring structures having one common carbon atom.

A *tracer compound* is a compound which, by its ease of detection, enables a reaction or process to be studied conveniently. Wide use has been made of isotopes, including radioactive isotopes of common elements, which are added in small quantities, in the form of the proper compound, to follow the course of an atom or a compound through a complicated series of reactions; or conversely, to determine the properties of a tracer—that is available only in quantities too small to handle alone—by adding it to a system containing chemically related elements, and then following its course throughout a given series of reactions. Considerable use of tracer compounds is made in the study of physiological reactions.

Unsaturated compound is a term specifically applied to a carbon compound containing one or more double bonds or triple bonds. One consequence of the presence of these double bonds or triple bonds, from which a broader concept of unsaturation stems, is the relative ease with which such bonds are split, and other constituents linked to them.

See also **Organic Chemistry**.

COMPRESSION (Gas). The compressibility of a gas is defined as the rate of volume decrease with increasing pressure, per unit volume of the gas. The compressibility depends not only on the state of the gas, but also on the conditions under which the compression is achieved. Thus, if the temperature is kept constant during compression, the compressibility so defined is called the isothermal compressibility β_T:

$$\beta_T = -\frac{1}{V}\left(\frac{\partial V}{\partial P}\right)_T = \frac{1}{\rho}\left(\frac{\partial \rho}{\partial P}\right)_T \tag{1}$$

If the compression is carried out reversibly without heat exchange with the surroundings, the *adiabatic compressibility* at constant entropy, β_S, is obtained:

$$\beta_S = -\frac{1}{V}\left(\frac{\partial V}{\partial P}\right)_S = \frac{1}{\rho}\left(\frac{\partial \rho}{\partial P}\right)_S \tag{2}$$

Here P is the pressure, V the volume, ρ the density, T the temperature, and S the entropy.

In adiabatic compression, the temperature rises, thus the pressure increases more sharply than in isothermal compression. Therefore β_S is always smaller than β_T.

The *compressibility factor* of a gas is the ratio PV/RT. This name is not well chosen since the value of the compressibility factor by itself does not indicate the compressibility of the gas.

Experimental values for the compressibility of gases can be obtained in several ways, most of which are indirect.

Since the compressibility is proportional to the pressure derivative of the volume, any experiment that establishes the P-V-T relation of a gas with sufficient accuracy also yields data for the isothermal compressibility. For obtaining the adiabatic compressibility from the P-V-T relation, some additional information is necessary [see section (c)], for instance specific heat data in the perfect gas state of the substance considered. A more direct way of determining the adiabatic compressibility is by measuring the speed of sound v, the two quantities being related by

$$v^2 = \frac{1}{\rho \beta_S} \tag{3}$$

This relation is valid only when the compressions and expansions of the sound wave are truly reversible and adiabatic. This is the case if the frequency is fairly low and the amplitude small.

Dilute gases obey the laws of Boyle and Gay-Lussac, $PV = RT$, to a good approximation. Thus, it can readily be shown that the following relations hold for the compressibility:

$$\beta_T = 1/P = V/RT$$
$$\beta_S = 1/\gamma P = V/\gamma RT \tag{4}$$

where $\gamma = c_P/c_V$, the ratio of the specific heats at constant volume and at constant pressure, respectively, and R is the gas constant.

Compressed gases show large deviations from the behavior predicted by equation (4). This is demonstrated by Fig. 1, where the isothermal compressibility of argon, divided by the corresponding value for a perfect gas at the same density, is pictured as a function of density for various temperatures. It is seen, first, that at all temperatures the compressibility at high densities falls to a small fraction of the value for a perfect gas, and second, that supercritical isotherms show a maximum in the ratio $\beta/\beta_{\text{perfect}}$ as a function of density, which maximum is the more pronounced the closer the critical temperature. It occurs roughly at the critical density ρ_c. Since at the critical point $(\partial P/\partial P)_T$ equals zero, the isothermal compressibility becomes infinite at this point. The adiabatic compressibility, however, remains finite. Qualitatively, all gases show the same behavior as pictured for argon in the diagram.

The molecular theory can explain the general features of the compressibility in its temperature and density dependence. The pressure of the gas is caused by the impact of the molecules on the wall. If the volume is decreased at constant temperature, the average molecular speed and force of impact remain constant, but the number of collisions per unit area increases and thus the pressure rises. If the gas is compressed adiabatically, the heat of compression cannot flow off, thus the average molecular speed and force of impact increase as well, giving rise to an extra increase of pressure. Therefore $\beta_S < \beta_T$. The actual magnitude of the temperature rise depends on the internal state of the molecules: the more internal degrees of freedom available, the more energy can be taken up inside the molecule and the smaller the temperature rise on adiabatic compression. Thus for gases consisting of molecules with many internal degrees of freedom, adiabatic and isothermal compressibilities differ but little.

If the gas is assumed to consist of molecules of negligible size and without interaction, then the gas can be shown to follow the laws of Boyle and Gay-Lussac; therefore its isothermal and adiabatic compressibilities must be given by Equation (4). For a perfect gas, the percentage pressure rise is proportional to the percentage volume decrease if the change is small; thus the compressibility is inversely proportional to the pressure.

To explain the very different behavior of real gases, the model must be modified. Suppose the molecular volume is small but not negligible. In states of high compression, where the total molecular volume becomes of the order of the volume available to the gas, the free space available to the molecules is only a fraction of what it would be in a perfect gas, and thus the real gas is much harder to compress than the perfect gas. This explains the low compressibility of dense gases and liquids (diagram).

Furthermore, one assumes that molecules, on approaching each other, experience a mutual attraction before they collide; this mutual attraction

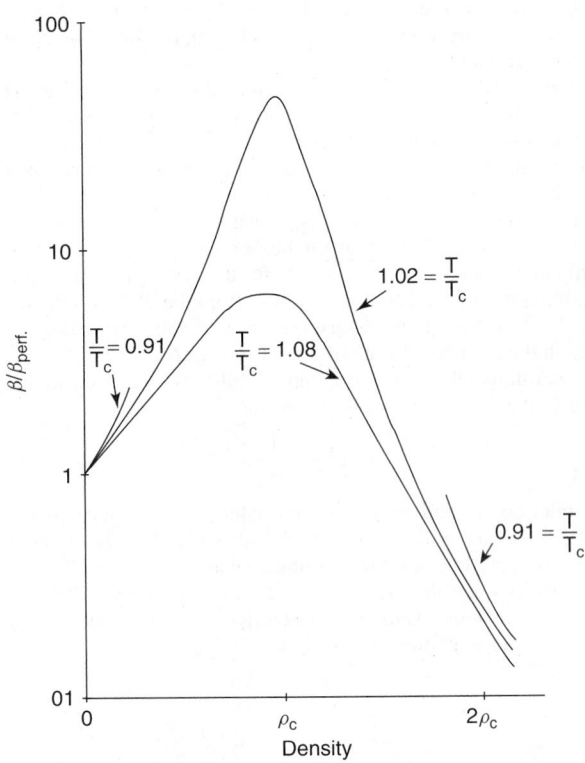

Fig. 1.   The ratio β/β_{perf} of the isothermal compressibility of argon to that of a perfect gas at the same density, as a function of the density, at 0.91, 1.02 and 1.08 times the critical temperature. The critical density is indicated by ρ_c

makes it easier to compress a real gas than a perfect gas. This explains the initial rise of the compressibility of a real gas over that of a perfect gas at temperatures not too far above the critical.

When compressed at subcritical temperatures, the gas condenses; that is, macroscopic clusters or droplets are formed under the influence of the attractive forces. During condensation, the pressure remains constant while the volume decreases, giving rise to an infinite compressibility in the two-phase region. At the critical point the system is on the verge of condensation and the compressibility is also infinite.

Theoretical predictions for the isothermal compressibility can obviously be obtained from any theory of the equation of state. If, in addition, data for the specific heat are supplied, the adiabatic compressibility can be derived in the same way. Thus the compressibility can be derived from the virtual expansion of the equation of state which expresses the ratio PV/RT in a power series in the density, the coefficients being related to the interactions of groups of two, three, etc., particles.

In the dense system the convergence of the virtual expansion is doubtful. In any case the higher coefficients are hard to calculate; here approximate theories have been developed, of which the cell model is an example.

Many semiempirical equations of state with varying degree of theoretical foundations are in use. The Van der Waals equation, a two-parameter equation that gives a qualitatively correct picture of the P-V-T relations of a gas and of the gas-liquid transition, is an example.

Modern developments are centered around the calculations of the radial distribution function $g(r)$, which is the ratio of the density of molecules at a distance r from a given molecule, to the average density in the gas. The compressibility can be expressed straightforwardly in terms of $g(r)$ as follows:

$$KT\beta_T = 1/P + \int_0^\infty [g(r) - 1]4\pi r^2 dr \qquad (5)$$

Approximate evaluations of the radial distribution function in dense systems are being obtained as solutions to integral equations derived from first principles under well-defined approximations.

COMPUTATIONAL CHEMISTRY. Use of computers in organic synthesis and in chemical engineering as a more efficient means of research than conventional laboratory experimentation. The capacity of sophisticated computers for fast mathematical calculations has made then an invaluable aid in exploring and evaluating the more likely pathways for a given organic synthesis, for which there may be innumerable possible sequences. The term *heuristic* is applied to such procedures. Computers can also handle the vast complexity of quantum-mechanical calculations and aid in the elucidation of the complicated molecular structures that occur in pharmaceutical compounds and recombinant-DNA research. The Quantum Chemistry Program Exchange at Indiana University offers many programs in this field, from subroutines to major computational systems. Chemical engineers utilize computers to develop more thermodynamically efficient procedures and to consolidate overall plant operations, especially in the areas of energy consumption, reaction rates, and hazardous waste problems.

Note: Notwithstanding the immense capability of computers to point the way to solutions of chemical and engineering problems, experimentation will remain the ultimate proof of theory. It is interesting to speculate how much time and effort such empirical scientists as Goodyear and Edison could have saved had computers been available to them.

CONANT, JAMES BRYANT (1893–1978).   An American chemist and educator, born in Boston, who received his doctorate in chemistry from Harvard in 1916 and was president of Harvard for 20 years (1933–1953). His major scientific activities included pioneering research on chlorophyll and important contributions to the Manhattan Project. Perhaps his greatest achievements lay in the educational field, in which he exerted a strong liberalizing influence at both the collegiate and secondary school levels. He also was ambassador to postwar Germany and educational adviser to Berlin. He wrote many books on science and education, including basic chemical tests, and received a number of scientific and educational awards.

CONCENTRATION (Chemical).   The quantity of matter or of a particular type of matter that exists in a unit volume, as the strength of a solution in mass of solute per unit mass of solution; or in the number of moles, hydrogen ions, etc., contained per unit volume or per unit mass.

The most commonly used method of expressing concentration is by stating the percentage, that is, parts by weight of the given substance in 100 parts by weight of the stated material. There are several exceptions to this method of expressing concentration, and the units used should be carefully recorded or observed, depending on whether the reader is operator or reader respectively. Ethyl alcohol, in water mixtures, is commonly reported as a stated percent by volume, where 50.0% by volume is equivalent to 42.47% by weight. Beverage and industrial alcohol also is rated in terms of proof, 200 proof indicating pure ethyl alcohol, that is, absolute alcohol containing no water. See also **Ethyl Alcohol.** Gases in a mixture are commonly reported by volume percent, thus nitrogen in air 78.0% by volume (equivalent to 75.5% by weight).

The concentrations of substances in solution are expressed variously: percent by weight of the actual material stated; percent by weight of a material calculated chemically from the actual material; grams of the actual material (or a material calculated chemically from this) per 100 milliliters of solution; gram moles (the formula weight taken in grams) of the actual material per liter of solution (this is molar or formal concentration and the abbreviations, M or F, respectively, are used to express it); and gram equivalents–(the equivalent weight taken in grams of the actual material per liter of solution (this is normal concentration and the abbreviation, N, is used to express it).

Since the concentration is proportional in many individual cases to an easily determined physical constant, such as specific gravity (e.g., of solutions), the index of refraction, specific rotatory power (e.g., sugar solutions, terpenes), such constants are frequently used to ascertain and express concentration data.

See also **Demal Solution; Gram-Equivalent; Gram-Molecular Weight; Molal Concentration; Molar Concentration; Mole Fraction; Mole (Stoichiometry); Mole Volume;** and **Normal Concentration.**

CONCENTRATION (Process).   In the processing of various materials, it frequently is required to increase the proportion of one material in a mixture or solution by removing all or part of one or several other components. A simple example is the retrieval of impure sodium chloride from seawater by solar evaporation of the water. Concentration is a prime reason for several of the chemical engineering unit operations described elsewhere in this volume, such as: *solid–solid* separations, as by screening, jigging, tabling, flotation, sublimation, and freeze-drying; or *liquid–solid* separations, as by filtering, centrifuging, drying, evaporating, crystallizing, leaching, expressing, and prilling; or *gas–solid* separations, as by gravity settling, electrical precipitation, cyclone and impingement settling; or *gas–gas* separations, as by absorption, adsorption, gaseous diffusion, chromatography, and electromagnetic methods; or *liquid–liquid* separations, as by distillation, dialysis, and extraction; or *liquid–gas* separations, as by drying, boiling, and condensing.

The process of concentration normally connotes an increase in the proportion of one material rather than a full separation of all other materials from the target material. Raw ores, for example, frequently will contain only a small percentage of the desired mineral. Thus, an ore may be concentrated from a fraction of 1% (wt) to 50% (wt) or more. The gaseous diffusion separation of $^{235}UF_6$ (required for nuclear fission reactors) from the natural occurring $^{238}UF_6$ (comprising 99.3%, wt) of the starting materials is an extreme example of concentration. Over 4,000 diffusion stages are required in a large atomic fuels plant to effect this concentration. Prior concentration of ores is known as *beneficiation*. This is described under **Iron**. See also **Classifying (Process)**.

CONCRETE. Concrete is a mixture of fine and coarse aggregates firmly bound into a monolithic mass by a cementing agent. The cement ordinarily employed for concrete is the standard Portland cement. The aggregates are usually sand and crushed stone or gravel. Crushed slag or cinders are used in special kinds of concrete. The formation of concrete can be thought of as a process in which the voids between the particles of coarse aggregate are filled by the fine aggregate, and the whole is cemented together by the binding action of the cement. The nature of Portland cement in this respect is described under **Cement**.

Due to its strength, permanency, and relatively low cost, concrete is one of the most important building materials employed in modern construction. It is widely used for foundations of all types, buildings, bridges, dams, retaining walls, highways, and other purposes too numerous to mention. However, the success of concrete in meeting any particular set of conditions, depends upon the proper correlation of many factors bearing on the selection and mixing of the materials, the placing of the concrete, and the original design. Concrete is strong in compression, but relatively weak in tension. Therefore structures in which the concrete is likely to be in tension must be reinforced with steel rods, which carry the tension. For strong permanent concrete, the aggregates should be clean, coarse, and well graded. River or coarse sand is better than pit sand, and should always be used where possible. Table 1 gives typical concrete mixes, with the characteristics of each.

This table is suitable for preliminary estimates only, or for small amounts of concrete. It should be remembered that research and development in the science of concrete proportioning have advanced to the point where little short of a laboratory analysis can establish the best and most economical mix for a given condition. The strength of the cementing agent is a function of the water-cement ratio, and therefore strength will vary widely depending upon the amount of water used. To obtain maximum strength, the water-cement ratio should be kept as low as possible. Type and gradation of aggregate, moisture content of the sand, and water-cement ratio are typical factors taken into account in a complete analysis for the specification of large amounts of concrete work.

TABLE 1. DATA ON CONCRETE MIXES TO YIELD 1 CUBIC YARD OF CONCRETE

Mixture	Cement, Sacks	Sand, Cu. Yd.	Stone, Cu. Yd.	Application	Weight, Tons per Cu. Yd.
1:2:3	7	0.51	0.77	Roofs, sills, tanks, Tunnels	2
1:2:4	6	0.44	0.88	R. C. floors, beams, and columns	2
$1:2\frac{1}{2}:4$	5.6	0.52	0.83	Building walls	2
1:3:5	4.7	0.52	0.86	Foundations and footings	2
			CINDERS, Cu. Yd.		
1:2:4	6.6	0.49	0.98	R. C. floors	1.5
			SLAG, Cu. Yd.		
1:2:4	6.6	0.49	0.98	R. C. floors	1.6

Concrete should be transported rapidly from the mixer to the forms, so that no initial set will have occurred before the concrete is placed in its final position. It is necessary to place concrete in the forms with care to prevent segregation of the lighter and heavier parts. This precludes dropping the concrete into place from any height. After the "green" concrete has been poured, it should be cured, or hardened, slowly over a period of about a week, during which time it should be protected from vibration, freezing, and a too-rapid rate of drying out. Strengths of concrete are usually classified on the basis of 28-day strength. Most of its strength will be acquired in this time but there is a slow increase in strength for a much longer period thereafter.

For small projects, particularly around the home, a variety of premixed dry cements and concretes are available. These simply require the addition of the requisite amount of water and adequate mixing.

Prestressed concrete elements for use in construction has increased severalfold since World War I. The concrete block industry also has expanded greatly during the same period, replacing brickwork and stonework for many applications because of economy and speed. In addition to the usual cement, sand, and gravel, concrete blocks are also available with several other ingredients to lower the weight and density of the blocks, increase thermal insulative qualities, and to add color. See also **Ceramics**.

Additional Reading

Beall, C.: *Masonry and Concrete*, The McGraw-Hill Companies, Inc., New York, NY, 2000.

Day, K.W.: *Concrete Mix Design, Quality Control and Specification*, Routledge, New York, NY, 1999.

Hover, K.C.: *Concrete Materials and Construction*, The McGraw-Hill Companies, Inc., New York, NY, 2001.

Larrard, F.De: *Concrete Mixture Proportioning*, Routledge, New York, NY, 1999.

Marchand, J. and J.P. Skalny: *Materials Science of Concrete*, The American Ceramic Society, Westerville, OH, 1999.

Nawy, E.G.: *Fundamentals of High-Performance Concrete*, John Wiley & Sons, Inc., New York, NY, 2000.

Staff: *Concrete and Aggregates*, American Society for Testing & Materials, West Conshohocken, PA, 1999.

Staff: *Annual Book of ASTM Standards 2000: Section 4: Construction: Concrete and Aggregates*, American Society for Testing & Materials, West Conshohocken, PA, 2000.

Web Reference

American Ceramics Society. http://www.acers.org/

CONDENSATE. A vapor may be reduced to liquid by removal of such portion of the latent heat of evaporation as it may contain. The act is called condensation, and the liquid is condensate. It is the property of vapors that the condensate is dense as compared to the vapor, which, in condensing, formed it; that is, there is considerable shrinkage of volume upon a reversion to the liquid state. The thermal condition of the condensate immediately upon formation is that of saturated liquid at the temperature of vapor. See also **Distillation**.

CONDUCTION (Heat). See **Heat Transfer**.

CONNATE WATER. Water that is trapped in marine sediments at the time they are laid down in the sea is commonly called *connate water*. As the term implies, connate water is produced at the same time as the rock and constitutes a sort of fossil seawater.

When marine sediments are raised above sea level and subjected to the action of circulating meteoric water, their connate water and solutes are removed by leaching and flushing and carried back to the ocean. The flushing process however is slow in fine-grained and deeply buried sediments, and water that almost certainly owes its high content of dissolved ions to remnants of marine solutions is common in such environments. The water associated with petroleum commonly is saline and occurs in formations whose porosity and structure are generally unfavorable for extensive water circulation and coincident removal of solutes. Usually the salt dissolved in brines that occur in deeply buried rocks is considered to be of connate origin. See also **Petroleum**.

Many geological processes may have modified the composition of connate brines to produce their wide range. Obviously these alterations have been extensive. Some connate waters are essentially saturated or nearly saturated solutions of sodium chloride. Others contain large proportions of calcium as well as sodium and chloride. Although the composition of connate water gives few useful clues as to the composition of the ocean in past geologic periods, the ocean has evidently been rich in chloride for a very long time, and brines in which anions other than chloride are predominant cannot logically be ascribed entirely to a connate origin.

Among the processes which might be expected to alter the concentrations of ions in connate water are: the precipitation of solids such as calcite on mineral surfaces, the solution of rock minerals and evaporites, sorption and desorption of ions on solids, and the differential movement of water molecules and ions through clay and shale strata. The latter effect is equivalent to the ultrafiltration effect of certain types of membranes used in the reverse osmosis method of removing solutes from water. Under the high pressures encountered at great depths, water and ion movements may be very different from the ones expected at low pressure. These effects may possibly explain the high concentrations of solutes in some connate brines and the difference in ion content between such brines and ordinary seawater. Another factor of considerable importance in many places appears to be the biochemical reduction of sulfur from S^{6+} as found in sulfate ions to sulfur in the more reduced forms of free sulfur, polysulfides, or sulfide ions. The brines encountered in oil and gas fields, and the gases themselves, are often high in hydrogen sulfide content as a result of sulfate reduction.

Analyses of a variety of connate brines have been published by White, Hem, and Waring (1963). White (1957) has suggested criteria for distinguishing connate water by means of ratios of concentrations of certain of the dissolved ions to one another. In most connate water the ratios of bromide and iodide to chlorine are relatively high and ratios of potassium and lithium to sodium are low.

Additional Reading

Hem, J.D. et al.: *Study and Interpretation of the Chemical Characteristics of Natural Water*, U.S. Geological Survey Water Supply Paper, 1473 (1970).

CONSERVATION LAWS AND SYMMETRY.

Basic among the natural laws are the so-called conservation laws which, in essence, state that in a given physical system under specified conditions there is a certain measurable quantity that remains changeless regardless of what actions may occur within the system. Three classical laws of this type are: (1) the law of the conservation of energy; (2) the law of the conservation of momentum; and (3) the law of the conservation of angular momentum. The concept of the conservation laws dates back to the early days of science and modifications have occurred and will continue to occur as new knowledge is gained. One of the tasks of physics is to explain the detailed rationale for laws that, on the surface, have a quality of being "self-evident."

An outstanding example of how these laws are subject to modification was Einstein's elucidation of the mass-energy equivalence ($E = mc^2$). Before that, the conservation of mass and the conservation of energy were considered to be independently valid.

General Functions of Laws of Conservation

An important function of the conservation laws is that they allow predictions about the behavior of a system without going into mechanical details of what happens during the course of a reaction. The laws provide a direct connection between the state of the system before the reaction and the state after the reaction. Also, one may conclude that any action that violates one of the conservation laws must be forbidden.

Laws of this nature are postulated as a result of many measurements of the energies and momenta involved in reactions of all kinds. It is always found that within the limit of accuracy of the experiment, for example, the amount of energy in the system after the reaction is the same as the amount of energy before the reaction. Prior to the development of elementary particle physics, there was much emphasis on transformation of energy between its various "forms," such as mechanical, electrical, and thermal energy. The present viewpoint is that the macroscopic behavior of matter is the result of interactions between elementary particles and that these elementary interactions individually obey the various conservation laws. In its elementary form, the law of conservation of energy states that when two or more particles interact, the total energy (kinetic plus potential) is always a constant. When it is said, for example, that the kinetic energy of a moving object has been transformed into the potential energy of a compressed spring, what is really meant is that, on a microscopic level, the atoms of the spring have been pressed closer together, so that there

is a greater amount of potential energy in the electric fields between the atoms of the spring. This increase of potential energy is associated with an equal decrease in the kinetic energy of the object that caused the spring to compress.

Concept of Symmetry

With the development of the Lagrangian and Hamiltonian methods of solving physical problems, and particularly with the growth of importance of quantum mechanics, it became clear that the conservation laws are closely connected with the concept of symmetry in space and time. This is based upon the ability to describe the interaction between two or more objects in terms of a potential energy function. If the potential energy of the system is known for any position of these objects in space and time, then the future motions of the objects can be predicted. Certain predictions can be made without going through a complete solution of the equations of motion. For example, it is found that if the potential energy does not depend explicitly on one of the space coordinates, then the momentum associated with that coordinate does not change, but is a constant of the motion.

Some examples of "geometrical symmetries" encountered in classical physics might include:

1. An object moves in a 3-dimensional space where its potential energy is the same at every point. The expression describing the potential does not explicitly contain the coordinates x, y, or z. That is, the system is invariant with respect to translation of the origin of the coordinate system in any direction. This symmetry is associated with conservation of linear momentum; the momentum in all three dimensions is a constant.
2. An object moves in a world that is flat so that the force of gravity is in the vertical (z) direction. The potential depends only on the height of the object above the ground, but does not depend on its location in the horizontal plane; i.e., the potential is invariant to a translation of the coordinate system in the x-y plane. There is now symmetry in two dimensions, and momentum in the x-y plane is conserved.
3. In the interaction between two spherical bodies, if the potential depends only on the distance between the bodies, then there is spherical symmetry. The system is invariant to a rotation of the coordinate system about any axis. In this case, two components of angular momentum are conserved; as the two bodies orbit around their common center of mass, the magnitude of their angular momentum is constant, while the plane of the orbit in space does not change.
4. If the interaction between two objects does not depend explicitly on the time coordinate, then the actions that take place do not depend on when one starts to measure time; i.e., the properties of the system are invariant with respect to a translation of the origin of coordinates along the time axis. This symmetry is associated with conservation of energy. Use of a 4-dimensional coordinate system allows one to associate conservation of momentum and energy in a unified manner with the geometrical symmetry of space-time.

The conservation laws and the concept of symmetry acquired importance in the area of elementary particle physics. The conservation laws act as "selection rules" to determine which reactions may take place between the many existing particles out of the very large number of otherwise conceivable reactions.

An important property of elementary particles is *parity*. Each particle has a parity number associated with it; either $+1$ or -1, depending on the type of particle. In an assembly of particles, there is a total parity, which is the sum of the individual parities. If parity is conserved, then this total parity does not change during the course of the reaction. This property of matter is associated with the so-called mirror symmetry. All the laws of nature which possess this type of symmetry are such that if the words "right" and "left" are interchanged in the statement of the law, then the behavior of the system obeying these laws is unchanged. At one time, it was thought that every natural law was of this type. As a result of conservation of parity, it was believed that it would be impossible to describe the difference between "right" and "left" by the use of words alone. Yang and Lee (1956) pointed out that in a special class of reactions involving the "weak nuclear" interaction, parity need not be conserved. As a result of this finding, it was seen that the universe does possess an asymmetry between right and left and that it is possible to describe an experiment which will definitely distinguish between the directions "right" and "left" in the universe.

Another type of symmetry of importance in elementary particle physics is that entitled *charge conjugation*. This principle states that if each particle in a given isolated system is replaced by its corresponding antiparticle, then no difference can be observed. For example, if, in a hydrogen atom, the proton is replaced by an anti-proton and the electron is replaced by a positron, then this antimatter atom will behave exactly like an ordinary atom, so long as it does not come into contact with ordinary atoms.

However, it is found that there are certain types of reactions where this rule does not hold. These are the types of reactions where conservation of parity breaks down. If one considers a piece of radioactive material emitting electrons by beta decay, the radioactive nuclei are lined up in a magnetic field which is produced by electrons traveling clockwise in a coil of wire (as seen by an observer looking down on the coil). Because of the asymmetry of the radioactive nuclei, most of the emitted electrons travel in the downward direction. If the same experiment were done with similar nuclei composed of antiparticles and the current in the magnet coil consisted of positrons instead of electrons, then the emitted positrons would be found to travel in the upward rather than in the downward direction. Thus, interchanging each particle with an antiparticle has produced a change in the experiment.

In the foregoing situation, however, the symmetry can be restored if the words "right" and "left" are interchanged in the description of the experiment at the same time each particle is exchanged with its antiparticle. In this example, this is equivalent to replacing the word "clockwise" by "counterclockwise." When this is done, the positrons are emitted in the downward direction, just like the electrons in the original experiment. The laws of nature thus have been found to be invariant to the simultaneous application of charge conjugation and mirror inversion.

Other more technical conservation laws play a role in elementary particle physics. Conservation of *baryon number* and *conservation of strangeness* are rules required to account for the fact that certain reactions involving heavy particles are forbidden. *Time reversal invariance* describes the situation that, in reactions between elementary particles, it does not make any difference if the direction of the time coordinate is reversed.

Although a symmetry idea may first suggest a conservation law, the conservation law must be tested by experiment to see if the symmetry is valid.

Conservation of Energy in Thermodynamics

The principle of conservation of energy plays a fundamental role in thermodynamics and is, therefore, also called the first law of thermodynamics. In its most general form it postulates the existence of a function of state, called the internal energy of the system U, such that its change per unit time is equal to some flow, called the energy flow from the surroundings.

This statement can be expressed symbolically by the formula

$$dU = d_e U \text{ or } d_i U = 0 \tag{1}$$

in which $d_e U$ is the energy received during the time dt from the outside, and $d_i U$ is the energy "creation" inside the system.

The explicit form of the energy flow depends on the nature of the system considered.

In closed systems and in the absence of an external field, the energy U supplied from the outside during the time interval dt is equal to the sum of the heat flow dQ expressed in units of energy and the mechanical work dW performed at the boundaries of the system. If the pressure is normal to the surface, the mechanical work is simply $-p\,dV$ and the expression of the energy conservation becomes

$$dU = dQ - p\,dV \tag{2}$$

From a purely phenomenological point of view, this expression of the conservation of energy may be considered the definition of the heat received by the system. The extension of the mechanical principle of conservation of energy to include the flow of heat is due mainly to Carnot, Joule, Helmholtz, and Clausius.

In order to express the energy conservation in continuous systems, it is useful to introduce the energy density per unit volume

$$u_v = \frac{\Delta U}{\Delta V} \tag{3}$$

In agreement with the general formulation of the principle of energy conservation

$$\frac{\partial u_c}{\partial t} = \text{div } \Phi[U] \tag{4}$$

where $\Phi[U]$ is the energy flow. This flow contains in general different contributions, among which are:

1. The convection flow corresponding to a center of mass motion ω, amounting to $u_v\omega$.
2. heat flow Q.
3. flow of energy corresponding to the pressure tensor p_{ij} in the fluid, its i component being

$$\sum_j p_{ij}\omega \tag{5}$$

4. flow of potential energy (e.g., the outward flow of electromagnetic energy).
5. flow of energy related to diffusion.

The total energy U may be split into an internal energy, a potential energy, and a macroscopic kinetic energy. Each contribution taken separately does not satisfy an equation of the simple form of Equation (4) because of possible transformation of one form of energy to another.

Conservation of Mass

This law has been put in the form that matter can be neither created nor destroyed. More accurately, the total mass of any system remains constant under all transformations. The statement, however, is subordinate to mass-energy equivalence.

Consider a homogeneous closed system containing c components ($\gamma = 1\ldots c$) among which a single chemical reaction is possible. In such a system any variation in the masses will result only from the chemical reaction. Thus, the change of the masses m_γ of component γ during the time interval dt can be written as

$$dm\gamma = v_\gamma M_\gamma d\xi \tag{6}$$

where M_γ is the molar mass of component γ and v_γ its stoichiometric coefficient in the chemical reactions.

This coefficient is generally counted positive when v appears in the right-hand member of the reaction equation, negative when it appears in the left-hand member; ξ is the degree of advancement or extent of reaction.

The total mass of the system is given by $m = \Sigma_\gamma m_\gamma$. Summing over γ, the conservation of mass for a closed system is expressed by

$$dm = \left(\sum_\gamma v_\gamma M_\gamma\right) d\xi = 0 \tag{7}$$

The equation

$$\sum_\gamma v_\gamma M_\gamma = 0 \tag{8}$$

is called the equation of the chemical reaction or, more briefly, the stoichiometric equation.

Instead of the mass of the components it is often useful to consider the mole numbers $n_1\ldots n_c$. Instead of Equation (6), this becomes

$$dn_\gamma = v_\gamma d\xi \tag{9}$$

Equations (6) and (9) are extended easily to r simultaneous reactions. The different reactions are designated by indices ($\rho = 1\ldots 3r$). Instead of Equations (6) and (9), there are

$$dm_\gamma = M_\gamma \sum_{\rho=1}^{r} v_{\gamma\rho} d\xi_\rho \tag{10}$$

$$dn_\gamma = \sum_{\rho=1}^{r} v_{\gamma\rho} d\xi_\rho \tag{11}$$

where $v_{\gamma\rho}$ denotes the stoichiometric coefficient of γ in the reaction.

The conservation of mass in a continuous system is expressed by the equation of continuity for the density, ρ,

$$\frac{\partial \rho}{\partial t} = -\text{div } \rho\omega \tag{12}$$

where ω is the macroscopic velocity. This equation expresses the fact that the local change of the density is equal to the negative divergence of the flow of matter.

Equation (12) holds also for a mixture; ω is then related to the macroscopic velocities of the different components by

$$\omega = \left(\sum_{\gamma} \rho_{\gamma} \omega_{\gamma} \right) / \rho \tag{13}$$

Thus ω is simply the velocity of the center of gravity in an element of volume.

In general, the local change of a physical quantity is due not only to the divergence of the current which is associated with it, but a "source" term has also to be taken into account. For instance, the equation of continuity for the density ρ_{γ} of a component γ participating in a chemical reaction is

$$\frac{\partial \rho_{\gamma}}{\partial t} = -\text{div } \rho_{\gamma} \omega_{\gamma} + v_{\gamma} M_{\gamma} V_{\gamma} \tag{14}$$

where V_v is the rate of the chemical reaction per unit volume.

In an open system, it is possible to split the change of mass of component γ into an external part, $d_e m_{\gamma}$ supplied from the exterior, and an internal part, $d_i m_{\gamma}$ due to changes inside the system

$$dm_{\gamma} = d_e m_{\gamma} + d_i m_{\gamma} \tag{15}$$

Taking into account the equations on conservation of mass in closed systems, this becomes

$$dm_{\gamma} = d_e m_{\gamma} + M_{\gamma} \sum_{\rho=1}^{r} v_{\gamma\rho} d\xi_{\rho} \tag{16}$$

$$dn_{\gamma} = d_e n_{\gamma} + \sum_{\gamma=1}^{r} v_{\gamma\rho} d\xi_{\rho} \tag{17}$$

Summing Equation (16) over γ and taking into account the stoichiometric equations $\Sigma_{\gamma} v_{\gamma\rho} M_{\gamma} = 0$, the total change of mass is

$$dm = d_e m \tag{18}$$

This relation expresses the conservation of mass in open systems and indicates that the change of the total mass is equal to the mass exchanged with the outside world.

Conservation of Momentum

The principle of conservation of momentum states that for a dynamical system consisting of n material particles of masses $m_1, m_2, \ldots, m_n$, respectively, and position vectors $r_1, r_2, \ldots, r_n$, respectively, if the only forces acting are the mutual interaction forces of the particles the total momentum of the system remains constant; for example

$$\sum m_i \frac{d\mathbf{r}}{dt} = \text{constant} \tag{19}$$

The law of conservation of momentum is as fundamental to physics as the law of conservation of mass energy. Like that law, it holds in quantum mechanics and relativistic mechanics as well as in classical mechanics.

Conservation of Electric Charge

Since electric charge comes in discrete quantities (there is no known way of breaking an electric charge down into bundles smaller than that contained in an electron), this law deals with the counting of objects, rather than with the measurement of continuous variables, such as momentum or energy. Conservation of electric charge means that the total number of electric charges (taking positive and negative signs into account) in a closed system does not change. This, in earlier times, meant that electric charge could not be created or destroyed. The creation of charged particles is now spoken of, but the creation of a positive charge must always be accompanied by formation of an equal negative charge (e.g., an electron-positron pair is created by a photon). Conservation of electric charge is associated with a symmetry property of Maxwell's equations known as gage invariance which states that the absolute value of the electric potential (as opposed to the relative value) plays no part in physical processes. Further developments in quantum mechanics indicate that conservation of electric charge is connected with the observation that the properties of

a system of particles do not depend on the phase of the wave function describing the system.

The following conservation principles apply particularly to interactions of elementary particles:

1. *The Total Baryon Number Remains Constant.*   A baryon is a nucleon (proton or neutron) or any particle heavier than those that can be considered to have an atomic mass number $A = 1$. Some mesons have a mass greater than the proton, but they have a mass number $A = 0$, so they are not baryons. In computing the number of baryons present in a system, each baryon counts 1; each antibaryon counts -1; and leptons and mesons count 0.
2. *The Total Lepton Number Remains Constant.*   Leptons consist of the neutrinos, electrons, muons, and their antiparticles. Here again, the basic principles applied are analogous to those for baryon number; particles count 1; antiparticles count -1; and baryons and mesons 0.
3. *Total Strangeness Quantum Number.*   This remains constant except as previously mentioned in connection with weak interactions.

Additional Reading

Bocharon, A.V., A.M. Vinogradov, and I.S. Krasil'shchik: *Symmetries and Conservation Laws for Differential Equations of Mathematical Physics,* American Mathematical Society, Providence, RI, 1999. http://www.ams.org/

Frautschi, S.C., T.M. Apostol, and D.L. Goodstein: *The Mechanical Universe,* Cambridge University Press, New York, NY, 1986.

Sato, R. and R.V. Ramachandran: *Conservation Laws and Symmetry: Applications to Economics and Finance,* Kluwer Academic Publishers, Norwell, MA, 1990.

Vinogradov, A.M.: *Symmetries of Partial Differential Equations: Conservation Laws-Applications-Algorithms,* Kluwer Academic Publishers, Norwell, MA, 1989.

Yang, C.N.: *Elementry Particles,* Princeton University Press, Princeton, NJ, 1962.

CONSISTENCY. See **Rheology**.

CONSOLUTE LIQUIDS. This term is applied to liquids when they are miscible in all proportions, i.e., mutually completely soluble, under some given conditions. It is not usually applied to gases because they are all miscible.

CONSOLUTE TEMPERATURE. The upper consolute temperature for two partially miscible liquids is the critical temperature above which the two liquids are miscible in all proportions. In some systems where the mutual solubility decreases with increasing temperature over a certain temperature range, the lower consolute temperature corresponds to the critical temperature below which the two liquids are miscible in all proportions. Some systems such as methylethyl ketone and water have both upper and lower consolute temperatures.

CONSTITUENT. 1. In general, one of the elements or parts of a compound. 2. An identifiable component in the microstructure of an alloy. It may be a phase or a characteristic configuration of several phases.

CONTRACTILITY AND CONTRACTILE PROTEINS. The fundamental property of living matter on which its power of movement depends is termed contractility. In the simplest forms of living things, it is evident in the flowing movement of the material of the cell. In more complex forms, the property is centralized in muscle tissues. Muscle cells are elongate and are so arranged that necessary movements result from their shortening when stimulated.

The study of the fibrillar proteins of muscle invites interest for two reasons. On the one hand, they form the prime ingredients of the mechanism that performs a typical vital activity, movement, in its most specialized form; thus, the elucidation of their functions, and of the physical and chemical properties basic to it, represents one of the cardinal parts of molecular biology. On the other hand, the isolated proteins display such striking physical behavior that to the macromolecular physicist they are among the most fascinating materials. Unfortunately, they are difficult to obtain and exceedingly changeable and labile; thus, working with them is somewhat difficult.

The major fibrous proteins are myosin and actin. Other proteins that may be involved in myofibrillar structure are tropomyosin and paramyosin. All four proteins share certain chemical properties; among others, they are all exceptionally rich in ionizing amino acids, thus they are highly charged molecules. Myosin is an adenosine triphosphatase. The molecular weight is not accurately known; a number of determinations cluster around 500,000. Light scattering dissymmetry suggests a mean molecular length of about 1600 micrometers, and electron microscopy suggests similar lengths.

CONTACT POTENTIAL DIFFERENCE. In his experiments with electroscopes, Volta found that when pieces of two different metals, otherwise insulated, are brought into contact, they acquire opposite charges and maintain a difference of electrical potential even while still touching. This potential difference he found to be characteristic of the given pair of metals. Thus when the metals are iron and copper, the iron has a potential about 0.15 volt higher than the copper, while for tin and iron the difference is 0.31 volt, tin being the higher. Volta listed a series of several metals, viz., zinc, lead, tin, iron, copper, silver, gold, such that when any two are put in contact, the one first named is at the higher potential. "Volta's law," which he was not in a position to demonstrate but which was established much later, states that the potential difference between any two metals in direct contact is the sum of the potential differences between intervening metals of the series. Thus for tin and copper (above) it is 0.31 volt + 0.15 volt = 0.46 volt; and it makes no difference whether the tin and copper are in direct contact or have other metals intervening between them.

The contact potential difference depends on the relative Fermi levels of the two solids. On contact, electron flow will take place until the Fermi levels of the two solids are equal. This will result in a potential difference between the two solids, called the contact potential.

A distinction must be made between the contact potentials in air and the so-called "intrinsic" contact potentials in a vacuum with all adsorbed gases removed. According to Millikan, the intrinsic potential difference between two metals A and B is expressed by $V_{AB} = h(v_A - v_B)/e$, in which h is Planck's constant, v_A and v_B are the critical frequencies of photoelectric emission for the two metals (see **Photoelectric Effect**), and e is the electronic charge. In any case, if the electronic work functions of the metals are p_A and p_B, the contact potential difference is $V_{AB} = (p_A - p_B)/e$. The work functions, and hence V_{AB}, are in general dependent upon the medium surrounding the metals. Accurate measurements of these potentials are, unfortunately, very difficult.

CONVECTION. In general, mass motions within a fluid resulting in transport and mixing of the properties of that fluid. Natural convection results from differences in density caused by temperature differences. Warn air is less dense than cool aid; the warm air rises relative to the cool air, and the cool air sinks. Forced convection involves motion caused by pumps, blowers, or other mechanical devices. See also **Heat Transfer**.

CONVERSION. (1) In its most general usage, this term denotes a change, often with the force of a directed or induced change. One specific use is a change in numerical value of a quantity resulting from the use of a different unit in the same or a different system of measurement. (2) An intramolecular rearrangement of organic substances in which the relative positions of the radicals are modified. The Beckmann rearrangement is a case in point. Radicals may be transferred from carbon, oxygen, or nitrogen to carbon; from carbon or oxygen to nitrogen; from side chains to nucleus, etc.

CONVERSION COATINGS. The industrial application of organic finishes to metals almost always requires the use of an intermediate conversion coating, particularly when the performance demands are high. Conversion coatings are formed chemically by causing the surface of the metal to be "converted" into a tightly adherent amorphous or crystalline coating, part or all of which consists of an oxidized form of the substrate metal. Conversion coatings can provide high corrosion resistance as well as strong affinity for organic coatings. They are also useful as lubricants for the drawing and forming of metals and sometimes are used for decorative purposes. The most important and widespread use of conversion coatings is on steel, zinc or galvanized steel, and aluminum alloys. The most widely used classes of conversion coatings for use on these metals are the phosphates and chromates. Depending on size, shape, volume of production, and other factors, the coating may be applied by spray, immersion, roll coat, or brush. A typical metal pretreatment sequence is: (1) cleaning; (2) rinsing; (3) conversion coating; (4) rinsing; and (5) final rinsing.

Phosphating of Steel

When a ferrous alloy is immersed in phosphoric acid, it initially forms a soluble phosphate. As the pH rises at the metal/solution interface, the phosphate becomes insoluble and crystallizes epitaxially on the substrate metal. The phosphate coating thus produced consists of a nonconductive layer of crystals that insulates the metal from any subsequently applied film and provides a topography with enhanced "tooth" for increased adhesion. The crystals insulate microanode and microcathode centers caused by stress or imperfections in the metal surface. This greatly reduces the severity of electrochemical corrosion.

Practically all phosphating processes involve patented proprietary solutions which produce superior coatings in shorter times and at lower temperatures than are obtainable with phosphoric acid alone. Treating times have been reduced from the earlier 1–2-hour to 1–5-minute periods, while temperatures have been reduced from 98 to 20–30°C.

Increasing energy costs have led to more widespread use of low-temperature phosphatizing baths. These comprise a lower free acid content which results in higher saturation of the primary zinc phosphate and thus provides a greater tendency to deposit at temperatures at or close to ambient temperature. In many cases, depending on the type of soil, it is difficult to reduce the cleaner temperature and, as a result, the cleaned parts tend to heat the phosphatizing stage to approximately 50°C even when no external heat is used.

Iron Phosphate Coatings

These coatings, weighing about 40–60 milligrams per square foot (430–646 milligrams per square meter), provide good paint adhesion, but inferior heat and corrosion resistance. They are used when coating performance is not very demanding. Iron phosphate coatings appear as a very thin blue or brown film to the naked eye.

Crystalline Zinc Phosphate Coatings

These coatings, weighing about 200 milligrams per square foot (2152 milligrams per square meter) when applied by spray, or as much as 3000 milligrams per square foot (32,280 milligrams per square meter) when applied by immersion, are of a medium-gray color and are used when higher quality is mandatory. For even greater corrosion resistance and paint adhesion, *microcrystalline zinc phosphate* coatings are used. They are dark in color and give coating weights of about 150–200 milligrams per square foot (1614–2152 milligrams per square meter) when sprayed and up to 1000 milligrams per square foot (10,760 milligrams per square meter) when applied by immersion. They are used as a paint base for products which are expected to last for years under varying environmental conditions.

Phosphating Chemistry

Iron and iron oxide react with phosphoric acid to form soluble primary iron phosphate, $Fe(H_2PO_4)_2$, liberating H_2 and H_2O, respectively. Due to the consumption of acid at the metal interface, there is a local rise in pH, which causes insoluble secondary iron phosphate, $FeHPO_4$, to coat the metal. The iron in the coating is supplied by the substrate.

In zinc phosphating, a small amount of iron phosphate is formed initially, but the bath contains primary zinc phosphate, $Zn(H_2PO_4)_2$, which crystallizes on the metal surface as secondary and tertiary zinc phosphates, $ZnHPO_4$ and $Zn_3(PO_4)_2$, respectively, when the pH rises at the metal/solution interface. The most frequently used baths contain accelerators, preferably nitrates and nitrites, which oxidize the hydrogen formed by the pickling reactions. The fundamental zinc phosphate reactions occur in three steps, all in the same bath:

(Pickling)

$$Fe^0 + 2H^+ \longrightarrow Fe^{++} + H_2$$

$$3Fe^0 + 2NO_3^- + 8H^+ \longrightarrow 3Fe^{++} + 2NO + 4H_2O$$

(Coating)

$$3Fe^{++} + 2H_2PO_4^- \longrightarrow 4H^+ + Fe_3(PO_4)_2$$

$$3Zn^{++} + 2H_2PO_4^- \longrightarrow 4H^+ + Zn_3(PO_4)_2$$

(Iron Removal)

$$4Fe(H_2PO_4)_2 + O_2 \longrightarrow 4FePO_4 + 4H_3PO_4 + 2H_2O$$

(sludge)

$$Fe(H_2PO_4)_2 + NaNO_2 \longrightarrow FePO_4 + NO + H_2O + NaH_2PO_4$$

(sludge)

In the coating reaction, each 3 moles of iron or zinc liberates 4 moles of hydrogen ion. However, in the pickling reaction, 8 moles of hydrogen ion are consumed. Thus, the pH at the metal interface rises, and insoluble tertiary ferrous phosphate and zinc phosphate crystallize on the iron surface. The coating closest to the metal interface is largely iron phosphate, while that farther away is rich in zinc phosphate.

Iron buildup in the bath is objectionable; in the iron-removal equations above, it is seen that dissolved Fe^{++} can be removed by oxidation, slowly in air or more rapidly by peroxides or nitrite, as shown in the final equation. The iron removed becomes ferric phosphate, while iron in the coating is ferrous phosphate.

Accelerators speed phosphating reactions by reacting with hydrogen liberated at the metal. Were hydrogen not removed, it would form gas bubbles which would interfere with metal/solution contact. Strong oxidizer accelerators also serve to precipitate dissolved iron and, to a degree, act as metal cleaners by oxidizing residual organic soils. In spray application, mild accelerators are usually adequate for maintaining dissolved iron at safe levels because atomization of the solution permits it to absorb from air the oxygen needed for precipitation of iron.

Zinc phosphate coatings consist of varying ratios of hopeite, $Zn_3(PO_4)_2 \cdot 2H_2O$, and phosphophyllite, $Zn_2Fe(PO_4)_2 \cdot 4H_2O$, with hopeite usually predominant. Hopeite and phosphophyllite grow epitaxially on alpha-iron crystallites. Only a slight adaptation deformation is necessary for the lattice planes of both foreign phases compared with the alpha-iron lattice of the substrate. Good adhesive strength can be expected from such a bond.

The smaller the crystals, the better the adhesion: crystal-to-metal, crystal-to-crystal, and crystal-to-final finish. Also, the smaller the crystals, the tighter the packing, the denser the coating, the less total porosity area for corrosive reactions to take place.

Crystal size depends upon such factors as growth rate, agitation, and the effects of nucleating agents and foreign atoms in the crystal lattice. Spray application provides agitation which reduces crystal size. Accelerators increase the number of nucleation sites on the substrate and result in smaller crystals. The most refined technique for producing very small crystals is the introduction of foreign elements of different atomic radii into the crystal lattice. A calcium additive in a zinc phosphating solution produces scholzite, $CaZn_2(PO_4)_2 \cdot 2H_2O$, in which the crystals may average one-twentieth the size of the finest crystals produced by other methods. It is believed that the foreign elements cause uneven growth along one crystal face, creating stresses that either stunt growth at an early stage, or rupture the crystal.

The relatively recent surge into the use of cathodic electrodeposition, especially in the automotive industry, has necessitated change in the metal pretreatment. The generation of hydroxyl ions at the cathode where the organic coating is deposited tends to dissolve zinc from the conventional zinc phosphate conversion coating with consequent deleterious coating performance. By increasing the concentration of phosphoric acid in the zinc phosphatizing stage, the ratio of iron to zinc in the conversion coating is increased. Such conversion coatings show superior performance under cathodically electrodeposited coatings.

The following sequence is typical of a production spray phosphate system: (1) cleaning for 60 seconds at $71-77°C$; (2) rinsing for 15 to 30 seconds (hot or cold); (3) phosphating for 60 seconds at $54-60°C$; (4) rinsing for $15-30$ seconds (cold); and (5) chromate-rinsing for $30-45$ seconds at $27-60°C$. The final stage, an acid chromate rinse, is extremely important to the overall adhesion and corrosion resistance of the finish. One of its functions is to seal the pores in the phosphate coating. The effect can be demonstrated by placing a drop of chromic acid in the center of a phosphatized panel and subjecting it to corrosive exposure. Corrosion in the chromated area will be strongly inhibited.

Phosphating of Zinc

The chemistry in the phosphating of zinc alloys is similar, with the exception that iron does not play an important role in the coating or in the bath. The only cation involved is Zn. Zinc phosphate coatings are used widely in the treatment of galvanized steel for refrigerators, air conditioners, kitchen cabinets, and house and building siding.

Aluminum Pretreatment

Although Al protects itself against corrosion by forming a natural oxide, the protection is not complete. In the presence of moisture and electrolytes, Al alloys, particularly the high-copper alloys, corrode much more rapidly than pure Al.

Chemically produced oxides can be formed by treatment in $2-3\%$ sodium carbonate containing 0.1% sodium dichromate for $10-20$ minutes at $66°C$, followed by "sealing" in 5% sodium dichromate at $82-88°C$ for 10 minutes. Such coatings are softer, more porous, and not as effective as those produced by chromic acid anodizing.

Electrically produced anodic coatings from chromic, sulfuric, or oxalic acid electrolytes are more dense and less porous. Their corrosion resistance is improved by hot-water sealing, which is even more effective with the inclusion of dichromate.

Generally, the higher the alloy content of the base aluminum, the heavier the oxide present and the more difficult it is to remove prior to effective chemical pretreatment. A properly formulated deoxidizer will remove the oxide only to the desired degree, with minimum attack on the base metal. The baths usually consist of fairly high concentrations $(5-10\%)$ of either nitric or sulfuric acid, along with chromates and either free or complex fluorides. A deoxidizer will also remove the smut that forms on etching in strong alkali. Smut consists of the alkali-insoluble alloying elements and their oxides.

Amorphous Phosphate Coatings for Aluminium

These coatings were introduced in 1945. Their simplicity of operation, speed, and economy have resulted in wide commercial acceptance. They provide a continuous uniform green coating with excellent paint-bonding properties and underfilm corrosion protection. The coatings consist of varying ratios of chromic phosphate and hydrated aluminum oxide. The bath contains hydrofluoric acid, which removes the natural oxide to permit contact of the coating-forming chemicals with the metal. The complexity of the reactions involved makes it difficult to present a simplified chemistry, but the results of many tests and analyses give the following coating composition: $xCrPO_4 \cdot yAl_2O_3 \cdot zH_2O$. The phosphate coatings vary from 10 to 300 milligrams per square foot (108 to 3228 milligrams per square meter), depending on the end use. The lower coating weights are used for paint bonding; the higher range is used for decorative purposes.

Gold-colored conversion coatings are formed in baths containing hydrofluoric acid to remove the natural oxide, and chromic acid. The coating composition is chromic chromate plus varying amounts of hydrated aluminum oxide. Some baths also contain ferricyanide iron, which greatly accelerates the coating action and forms some chromic ferricyanide in the coating. This constitutes one of the most widely used conversion coatings on aluminum because of its high speed, excellent corrosion resistance, and high affinity for organic finishes. The hexavalent chromium content permits these coatings to withstand somewhat more severe corrosive environments than do the amorphous phosphate coatings. The baths have a pH of about $1.2-1.9$ and can be applied by dip, brush, spray, or reverse roll coater.

Proprietary chromate rinses are frequently used over conversion coatings on aluminum for increased corrosion resistance. See also **Autodeposition**.

WILBUR S. HALL
Amchem Products, Inc.
Ambler, Pennsylvania

CONVERSION RATIO. 1. The ratio of the number of internal conversion electrons to the number of gamma rays emitted in a given time interval by a single nuclidic species during the de-excitation of one of its excited energy states. Sometimes known as the *internal-conversion coefficient.* 2. In a nuclear reactor, the number of fissionable atoms produced per fissionable atom destroyed. See also **Nuclear Power Technology**.

COOLANT. Any liquid or gas having the property of absorbing heat from its environment and transferring it effectively away from its source. Coolants are used in all types of automobiles, as well as in chemical processing and nuclear engineering equipment. One of the most effective and cheapest coolants is water, which is almost universally used in automotive and ordinary reaction equipment. Air is also used. Where intense heating requires a more efficient medium, special coolants are used; liquid sodium in nuclear reactors, liquid hydrogen in high-thrust nuclear rocket engines; carbon dioxide, propylene glycol, and "Dowtherm" in chemical-processing reactors. Methoxy propanol has been introduced for diesel engines. Some coolants provide antifreeze protection.

See also **Antifreeze Agents**.

COOLER (Thermoelectric). See **Thermoelectric Cooling**.

COORDINATION COMPOUNDS. One of a number of types of complex compounds, usually derived by addition from simpler inorganic substances. Coordination compounds are essentially compounds to which atoms or groups have been added beyond the number possible on the basis of electrovalent linkages, or the usual covalent linkages, to which each of the two atoms linked donates one electron to form the duplet. The coordinate groups are linked to the atoms of the compound usually by coordinate valences, in which both the electrons in the bond are furnished by the linked atom of the coordinated group. The amines and complex cyanides are representative of coordination compounds.

In attempting to classify coordination compounds, Sidgwick noted that the number of molecules or atoms coordinated with a metallic atom (which he called *coordination number*) is 2, 3, 4, 5, 6, or 8, and that 2, 4, and 6 are the most common. In forming such compounds, each molecule or atom donates a pair of electrons to the metallic atom, forming a semi-covalent type of bond. Thus, in the nitrocobaltates, six nitro groups each donate a pair of electrons to a cobalt(III) ion, forming the complex ion.

$$\begin{bmatrix} O_2N & & NO_2 \\ O_2N & -Co- & NO_2 \\ O_2N & & NO_2 \end{bmatrix}^{3+}$$

In the positive triammine cobalt(III) ion, three ammonia molecules donate one pair of electrons each to the cobalt(III) ion, to form the complex ion.

$$\begin{bmatrix} H_3N & & NH_3 \\ & Co & \\ & H_3N & \end{bmatrix}^{3+}$$

The covalent character of the coordination bond is evident from the fact that both the ions and the molecules that form it fail to exhibit their characteristic reactions after coordination.

Included in the coordination compounds are the double salts, the complex salts, the oxysalts, and the hydrates.

See also **Chelates and Chelation**; **Cobalt**; **Copper**; **Gold**; **Hydrate**; **Iron**; **Manganese**; and **Molybdenum (In Biological Systems)**.

COORDINATION NUMBER. The number of nearest neighbors of a given atom in a crystal structure. In covalent crystals, only those neighbors to which the atom is directly bonded are counted, and this number is usually 4 or less. In metals, the coordination number may be as high as 12, as in the close-packed structures.

COORDINATION POLYHEDRA. The arrangement of oxygen ions about the cation to which they are closely bonded in an ionic crystal, as, for example, the group SiO_4 which forms a tetrahedron. Such polyhedra pack as units in the crystal structure.

COPALITE (or Copaline). The mineral copalite or "Highgate resin" is a fossil resin found in irregular fragments in the blue clay of London, England. It resembles copal, the resin of certain modern tropical trees. Copalite is pale yellow to greenish or brownish, and emits an aromatic odor when broken. It has a hardness of 1.5; a specific gravity of 1.046; burns with a very smoky yellow flame.

COPOLYMER. See **Elastomers**; **Polymerization**.

COPPER. [CAS: 7440-50-8]. Chemical element, symbol Cu, at. no. 29, at. wt. 63.546, periodic table group 11, mp 1083°C, bp 2567°C, density 8.92 g/cm³. Elemental copper has a face-centered cubic crystal structure. The metal is yellowish-red, soft, very malleable and ductile. Very thin sheet copper is translucent and transmits greenish-blue light. The element is unattacked by dry air, but in moist air containing CO_2, a protective greenish film of basic carbonate is formed. There are two natural isotopes, ^{63}Cu and ^{65}Cu. Seven radioactive isotopes have been identified, all with comparatively short half-lives: ^{58}Cu through ^{62}Cu, ^{64}Cu, ^{66}Cu, and ^{67}Cu. Copper may have been the first metal used by humans and today ranks second, exceeded only by iron, in annual consumption.

First ionization potential 7.723 eV; second, 20.29 eV; third, 29.5 eV. Oxidation potentials: $2Cu + 2OH^- \longrightarrow Cu_2O + H_2O + 2e^-$, 0.361 V; $Cu + 2OH^- \longrightarrow Cu(OH)_2 + 2e^-$, 0.224 V; $Cu^+ \longrightarrow Cu^{2+} + e^-$, −0.153 V; $Cu \longrightarrow Cu^{2+} + 2e^-$, −0.344 V; $Cu \longrightarrow Cu^+ + e^-$, −0.522 V. Other important physical properties of copper are given under **Chemical Elements**.

Ancient Copper Metallurgy

Efforts by scientists to locate the sources of copper used in ancient Mediterranean and Near Eastern cultures through comparative chemical analyses of copper ores and archeological artifacts have largely failed for various mineralogical and metallurgical reasons. The isotopic composition of lead, an element present in a minor amount in many copper ores and bronze objects, is unchanged through metallurgical processes and may, in principle, be used to determine the sources of the copper used in Bronze Age artifacts. Results of work during the early 1980s suggest that for Late Bronze Age Crete the Laurion region in Attica, Greece, may have been a more important copper source than Cyprus.

In 1984, Lechtman (Massachusetts Institute of Technology Laboratory for Research on Archaeological Materials) reported that the metalsmiths of Andean cultures knew how to plate copper with gold or silver and how to treat alloys of copper, gold, and silver so that the surface of the metal consisted only of gold. The Spanish conquistadors, when melting gold and silver objects which had been looted from the Incas, found that the bullion was quite impure. Although appearing as pure gold or silver objects, the materials actually were alloys of those elements with copper. Andean metalsmiths had developed the alloys along with procedures for treating them so that the finished objects presented a surface of pure silver or pure gold. As further reported, the smiths also knew how to plate objects made entirely out of copper with a thin coating of gold or silver.

Readers interested in the archeology of copper will find several of the references listed at the end of this article of value. To assess the authenticity of ancient copper and bronze objects, museums such as the Cincinnati Art Museum have turned to modern scientific examination. For example, ultrahigh-voltage industrial computed tomography, coupled with x-ray inspection, has been used.

Copper Ores

Copper occurs as native copper, particularly in the region south of Lake Superior (often 99.9% Cu), as sulfides (chalcocite, copper glance, cuprous sulfide, Cu_2S; chalcopyrite, CuFeS), as oxide (cuprite, cuprous oxide, Cu_2O, red); as basic carbonates (malachite, $CuCO_3 \bullet Cu(OH)_2$, green; azurite, $2CuCO_3 \bullet Cu(OH)_2$, blue). The copper content of its ores varies from 0.3 to 8% Cu and the average is on the order of 2.5%. The value depends largely upon the content of silver and gold. The area of production is widely distributed; in the United States, Montana, Utah, New Mexico, Arizona, Michigan, and Tennessee.

Native copper ore is crushed, concentrated by washing with water, smelted, and cast into bars. Oxide and carbonate ores are treated with carbon in a smelter. Sulfide ore treatment is complex, but, in brief, consists of smelting to a matte of cuprous sulfide, ferrous sulfide, and silica, which molten matte is treated in a converter by the addition of lime and air is forced under pressure through the mass. The products are blister copper, ferrous calcium silicate slag, and SO_2. Refining is conducted by electrolysis, and the anode mud is treated to obtain the gold and silver. See Fig. 1.

See also **Azurite**; **Chalcocite**; **Chalcopyrite**; **Cuprite**; **Malachite**; and **Mineralogy**.

Leading world producers of copper include: United States, Chile, Canada, Zambia, Russia, Zaire, Peru, Philippines, South Africa, Australia, Japan, and China.

Copper is frequently detected in atmospheric particles, even in those collected at locations far removed from anthropogenic sources. Cattell and Scott (1978) reported similar enrichments found over the north Atlantic and the South Pole. It had been proposed that measured atmospheric concentrations are larger than those predicted for unenriched crustal weathering or oceanic production. It was proposed that the enrichment may result from natural processes of anomalously enriched elements in aerosol particles. These may derive from low-temperature volatilization processes, such as biological methylation, or volcanism, or direct sublimation from the earth's crust, or emissions from plants, or fractionating at the air-sea interface which enriches elements in particles produced from the oceans. Atmospheric studies conducted by Cattell and Scott near the island of Tasmania in 1977 led to the conclusion that a biogenic agent may be

Fig. 1. Beneficiation of basic minerals involves wet operations requiring large volumes of water. Here copper tailings from a mining operation are settled in a 275-foot (84-meter) diameter thickener which handles over 2000 gallons (76 hectoliters) per minute and some 13,000 tons (11,795 metric tons) of ore per day. Underflow solids are disposed of by pumping to tailing ponds. Clarified water is returned to the mill process, thus avoiding waste of water

responsible for the approximately 20,000-fold enrichment of copper during aerosol production from the ocean.

Proposed Acid Injection Mining

In the early 1990s a new copper extraction technology, known as *in situ* leach mining, was under investigation. This methodology was proposed, and demonstration plants were to be constructed to test the proposal. These were jointly sponsored projects by the U.S. Bureau of Mines and two mining companies. If successful, leaching would cut the costs and hazards of the traditional open-pit methods currently used.

In this method a known copper deposit would be injected in several locations with a solution of dilute sulfuric acid. This acid would percolate through the veins of the ore-bearing rock (usually copper oxide). The copper would go into solution as copper sulfide and be pumped to the surface by way of several recovery wells. At the surface the copper would be plated onto cathodes (electrowinning process). It is proposed that the dilute solution, when free of copper, could be returned and circulated up to several times. An automatic analytical system would maintain the required strength of the acid solution. Proponents claim that the cost of the system would be less per ton of copper produced and would increase miner safety. One factor that must be carefully tested is the possible percolation of acid solution into nearby groundwater.

Special Properties

Copper is distinguished by several properties that contribute to its extensive use: (1) a combination of mechanical workability with corrosion resistance to many substances; (2) excellent electrical conductivity, (3) superior thermal conductivity; (4) effect as an ingredient of alloys to improve their physical and chemical properties; (5) efficiency of copper and some of its compounds as catalysts for several kinds of chemical reaction; (6) nonmagnetic characteristics, advantageous in electrical and magnetic apparatus; and (7) nonsparking characteristics, mandatory for tools for use in explosive atmospheres. There are additional attractions of copper for many other applications. The metal would be used even more widely, but

for some uses, even though superior, copper cannot compete with substitute materials because of cost.

Unalloyed Copper

In the United States, the term *copper* signifies copper that contains less than 0.5% impurities or alloying elements. Copper-base alloys are those that contain no less than 40% copper. Additionally, copper appears as a minor, but important ingredient of several alloys. There are six major types of commercial, unalloyed copper. These are described briefly in Table 1.

Very-High-Copper Alloys

Although not meeting the foregoing definition of copper precisely, there is a group of copper alloys which contain only a few percent of other ingredients and commonly these are also referred to as coppers, usually with the name of the other element preceding copper in the name—as *chromium copper* or *beryllium copper*. These very-high-copper alloys are described briefly in Table 2.

Because of major changes made in the production of electronic circuits, including surface-mount technology and the use of smaller, lighter, higher-density components, interconnection design requirements have become more stringent. Higher-density electronic packaging increases the demands for improved electrical, mechanical, and thermal characteristics of connector materials. To meet some of these needs, there has been increasing demand for beryllium-copper strips in many instances to replace brass alloys and phosphor bronzes.

TABLE 1. COMMERCIAL UNALLOYED COPPERS

Electrolytic Tough-Pitch Copper

Cu, 99.90%; O, 0.04% nominal; density 8.89–8.94 g/cm³, EC, 101%
Architecture: downspouts, flashing, building fronts, gutters, screening, roofing
Automotive: radiators and gaskets
Electrical: conductive-wire contacts, terminals, switch parts, bus bars
Hardware: cotter pins, nails, rivets, soldering copper, ball floats
Other: anodes, chemical process equipment, kettles, pans, printing rolls, expansion plates, rotation bands, die-pressed forgings

Deoxidized Copper

Cu, 99.90% minimum; P, 0.025% nominal; density 8.94 g/cm³, EC, 80–90%
Industrial: condensers, evaporators, heat-exchangers, dairy tubes, fraction-ating columns, kettles, pulp and paper piping, steam and water piping, tanks
Transportation: gasoline, oil, air, and hydraulic fluid lines, oil coolers
Other: shell rotation bands, die-pressed forgings, gauge lines

Oxygen-free Copper

Cu, 99.92% minimum; no residual oxidants; density 8.89–8.94 g/cm³, EC, 101%
Electrical: conductors, electron tubes, bus bars, waveguides (for operation at high temperatures in presence of reducing gases)
Industrial: heaters, oil coolers, gasoline supply lines, radiators, refrigeration lines, water piping

Silver-bearing Copper

Cu, 99.90% minimum; 8–25 ounce (226–708 grams) Ag/ton; density 8.91 g/cm³, EC, 100–101%
Electrical: commutator bars, heavy-duty motor windings (particularly for retention of strength at elevated temperatures)
Other: brazing solders, die-pressed forgings

Arsenical Copper

Cu, 99.68%, nominal; P, 0.025% nominal; As, 0.30% nominal; density 8.94 g/cm³, EC, 90%
Industrial: heat-exchangers, boilers, radiators, condenser tubes

Free-cutting Copper

Cu, 99.4–99.5%: Te, 0.5–0.6%: density 8.94 g/cm³, EC, 90%
Industrial: electrical connectors, motor and switch parts, soldering coppers, screw-machine parts, forgings, welding-torch tips

EC, electrical conductivity (International Annealed Copper Standard)

TABLE 2. VERY-HIGH-COPPER ALLOYS

Cadmium Copper

Cu, 99.00-plus %; Cd, 0.6–1.0%
Cadmium toughens copper and increases resistance to fatigue; also increases softening temperature.
Electrical conductivity (fully annealed) is about 95% (IACS).
Essentially free of oxygen; not susceptible to gassing.
Uses: contact wires used in electrical transportation, notably long-span overhead electric transmission lines.

Chromium Copper

Cu, 99.50%; Cr, 0.5%
Chromium improves mechanical properties while retaining high thermal and electrical conductivities.
Strength and hardness depend on heat treatment and not cold working—hence alloy can be used up to temperature of about 450°C without danger of softening.

Tellurium Copper

Cu, 99.50%; Te, 0.5%
Tellurium increases softening temperature of work-hardened copper.
Alloy is excellent where combination of good machinability and electrical conductivity is required.
Uses: motor and switch parts, electrical connectors, screw-machine parts, electrical instrument parts.

Beryllium Copper

Type 1: Cu, 98%; Be, 2%
Type 2: Cu, 97%; Be, 0.4%; Co, 2.6%
Cobalt is added as a lower-cost substitute for beryllium.
Uses: instrument springs, bellows, diaphragms, bourdon tubes, nonsparking sparking tools for hazardous locations.
Alloy permits springs to be shaped while soft, followed by hardening.
Selenium copper also available for combining high electrical conductivity with free-machining and hot-working properties. Alloy makes excellent copper-to-glass seals.

Copper also is finding increased applications in composite materials, notably graphite/Cu. These composites provide both weight savings and greater strength and stiffness in advanced aerospace applications.

Growing interest also is being shown in copper-nickel-tin alloys that can be aged hardened after forming. These alloys have improved tensile strength and greater resistance to oxidation, stress relaxation, fatigue, and stress-corrosion cracking. They are finding increased usage in electronic leads, contact pins, and sockets, as well as for eyeglass frames, circuit boards, and electronic-contact clips.

In an effort to avoid penalties pertaining to lead in potable-water plumbing systems, pipe fitting and fixture manufacturers are turning away from the free-cutting copper alloys that contain lead. Copper alloyed with nontoxic bismuth and a ductility enhancer, such as phosphorus, indium, and tin, which machine as well as leaded material, are now being considered. Most possible copper alloying combinations are now under study. This program is carefully detailed in the Plewes-Loiacono reference cited.

The Brasses

There are eight principal categories of brasses, not including the leaded and alloy brasses. Brass essentially is an alloy of copper and zinc. Several of the brasses contain other ingredients, such as lead and iron. When zinc is added to copper, there is a progressive alteration of color and lowering of melting point. When the zinc content is about 10%, the metal is a bronze color; with 15% zinc, the color may be described as golden; from 20–40% zinc, there is a range of yellow colors; over 45% zinc, the color is silver-white. The melting point of a 95% copper-5% zinc brass is about 1,065°C, whereas the melting point of 50% copper-50% zinc brazing metal drops to about 880°C. The ratio of copper to zinc also progressively affects mechanical and corrosion-resistance properties. Maximum tensile strength, for example, is attained with a 55% copper content, whereas maximum ductility is attained with a 70% copper content. This exceptional range of properties accounts for the availability and demand for a wide variety of brasses. Metallurgically, brasses may be classified as (1) *alpha brass*, in which the content of zinc is less than 36% and in which the zinc is dissolved in the copper, imparting to the alloy the basic structure of copper; (2) *beta brass*, in which the content of zinc ranges between 36% and 45%. This alloy contains the CuZn as a compound and enhances the hot workability of the alloy; and (3) *gamma brass*, in which the zinc content exceeds 45% and where there are crystals in the alloy. This combination does not lend itself to either hot or cold workability. Some of the important commercial brasses are described briefly in Table 3.

The Bronzes

Classically, a bronze is defined as an alloy of copper and tin, but over the years the term has taken on a much broader meaning. The term may apply to numerous copper alloys that possess a crystalline, bronze-like structure, are of a bronze color, or may contain some tin. Further, bronze generally is considered a casting metal. In contrast, brass is generally wrought. Some alloys are commercially named bronze even though they contain no tin whatsoever.

Copper Wire and Cable

The International Annealed Copper Standard (IACS), which sets annealed copper as having 100% electrical conductivity as a basis against which to compare other metals, alloys, and materials, is accepted internationally. Using this standard for comparison, the conductivity of copper is exceeded only by silver for which the IACS figure is 108.4%. This comparison is on the basis of conductivity per gram, pound, or other mass unit. Aluminum, although widely used as an electrical conductor for selected applications, has a rating on this schedule of approximately 61%, steel a rating of 11%, and nickel-chromium alloy (valued because of its high electrical resistance rather than conductivity) has a rating of 1.5%. Thus, the value of copper for electrical conductors, considering its availability and economics, is self-evident.

Processing Equipment

In terms of thermal conductivity, assigning a value of 100 to copper, the metal is exceeded only by silver which has a value of 108. Copper is followed by gold (76), aluminum (56), magnesium (41), zinc (29), nickel (15), iron (15), steel (13–17), lead (9), and antimony (5). Thermal conductivity means good heat transfer and this is extremely important in most industrial processing equipment where heating and cooling cycles are involved. This property, when combined with corrosion resistance, makes copper attractive for the construction and lining of process vessels. Deoxidized copper, admiralty brass, and arsenical copper are effective in condenser tubes operating with fresh water. Copper is less suitable for seawater because of its inability to form a protective film. Aluminum brass and 70–30 cupronickel alloy are favored for severe seawater service. Copper and copper alloys are not suited for use in oxidizing acidic solutions, in mercury, or in the presence of free NH_3. Copper vessels are used extensively in food processing and for numerous organic materials, particularly distillation columns and hardware. The relatively high cost of copper as compared with other metals, however, is always an important factor.

Piping

Copper and copper alloys are used in a wide variety of pipes and tubes, both for industrial and domestic systems. In many areas, galvanized water pipe has been almost completely replaced by copper piping in new construction. Advantages include corrosion resistance and ease of installation, which offset higher costs. Because of excellent thermal conductivity, copper and brass fittings are used widely in hot water and steam-heating systems. However, the greater conductivity of bare copper pipe in long runs requires more attention to insulation covering.

Chemistry and Compounds

Copper is dissolved best by HNO_3; not attacked by cold dilute HCl or H_2SO_4, but in hot HCl dissolves to yield cuprous chloride, in hot concentrated H_2SO_4 to yield copper sulfate; attacked by chlorine, especially when heated, to form cuprous and cupric chlorides; only slight action by H_2S or SO_2 at ordinary temperatures in the absence of air.

In view of its $3d^{10}4s^1$ electron configuration and the relatively small energy difference between the two levels, copper forms dipositive ions as well as monopositive ones. In fact, the former are the more stable in

TABLE 3. REPRESENTATIVE BRASSES AND BRONZES

Gilding Brass

Cu, 95%; Zn. 5%; Pb, 0.03% maximum; Fe, 0.05% maximum; density 8.86 g/cm^3; mp 1066°C; AT, 427–788°C; HWT, 760–871°C
Coinage: coins, metals, tokens
Munitions: firing-pin support shells, bullet jackets, fuse caps, primers
Novelties: emblems, plaques, jewelry
Other: base for gold plate and for vitreous enamel

Commercial Bronze

Cu, 90%; Zn, 10%; Pb, 0.05% maximum; Fe, 0.05% maximum; density 8.80 g/cm^3; mp 1043°C; AT, 427–788°C; HWT, 760–871°C
Architectural: grillwork, etching bronze, screen cloth, weather stripping
Cosmetics: lipstick cases, compacts
Hardware: kickplates, line clamps, marine hardware, escutcheons, rivets, screws
Munitions: rotating bands, primer caps
Other: costume jewelry, screen wire, ornamental trim, vitreous enamel base

Red Brass

Cu, 85%; Zn, 15%; Pb, 0.06% maximum, Fe, 0.05% maximum; density 8.75 g/cm^3; mp 1027°C; AT, 427–732°C; HWT, 788–900°C
Architectural: trim, etching parts, weather stripping
Electrical: screw shells, sockets, conduit
Hardware: fasteners, fire extinguishers, eyelets
Industrial: heat-exchanger tubes, condensers, flexible hose, piping, pumps, radiator cores, pickling crates
Other: compacts, costume jewelry, dials, badges, etched articles, lipstick cases

Jewelry Bronze

Cu, 87.5%; Zn, 12.5%; Pb, 0.05% maximum; Fe, 0.10% maximum; density 8.78 g/cm^3; mp 1035°C; AT, 427–760°C; HWT, 760–900°C
Architectural: angles, channels
Hardware: chains, fasteners, slide fasteners, eyelets
Novelties: costume jewelry, emblems, compacts, etched articles, lipstick cases, plaques
Other: base for gold plate

Low Brass

Cu, 80%; Zn, 20%; Pb, 0.05% maximum; Fe, 0.05% maximum; density 8.67 g/cm^3; mp 999°C; AT, 427–704°C; HWT, 816–900°C
Architectural: medallions, spandrels, ornamental metalwork
Electrical: battery caps
Instruments: bellows and musical instruments
Hardware: flexible hose, pump lines, tokens, clock dials

Cartridge Brass

Cu, 70%; Zn, 29-plus %; P, 0.07% maximum; Fe, 0.05% maximum; density 8.53 g/cm^3; mp 954°C; AT, 427–760°C; HWT, 732–843°C
Automotive: radiator cores and tanks, reflectors
Electrical: flashlight shells, lamp fixtures, socket shells, screw shells, bead chain
Hardware: fasteners, pins, rivets, eyelets, springs, tubes, stampings
Munitions: various components. Note: Admiralty brass is similar: Cu, 71%; Zn, 28%; Sn, 1%

Yellow Brass

Cu, 65%; Zn, 34-plus %; Pb, 0.15% maximum; Fe, 0.05% maximum; density 8.47 g/cm^3; mp 932°C; AT, 427–704°C

Architectural: grillwork

Automotive: reflectors, radiator cores and tanks
Electrical: lamp fixtures, flashlight shells, screw shells, socket shells, bead chain
Hardware: kickplates, push plates, locks, hinges, grummets, fasteners, eyelets, stencils, plumbing accessories, pins, rivets, screws, springs

Muntz Metal

Cu, 60%; Zn, 39-plus %; Pb, 0.30% maximum; Fe, 0.07% maximum; density 8.39 g/cm^3; mp 904°C; AT, 427–593°C; HWT, 621–788°C
Hardware: large nuts and bolts, brazing rod, condenser plates, valve stems, hot forgings

Leaded Brasses

When lead is added to brass up to about 4%, improved machinability results. The lead has practically no effect on tensile strength or hardness. However, for cold-worked materials, lead does lower ductility and shear strength.

Phosphor Bronzes

Although tin is the primary alloying element in these alloys, their name derives from the addition of small quantities of phosphorus used as a deoxidizing agent in casting the alloys. Tensile strength ranges from moderate to very high, decreasing with amount of tin added. Tin percentage will range from 1.25 to 10%. Of the copper alloys, the phosphor bronzes are best suited for sea duty and where acid reagents may be present.

Silicon Bronzes

Most of these alloys are of proprietary compositions and are known by a variety of trade names. Silicon content ranges from 1.5 to 3.5%; usually less than 1.5% zinc content. Tin, manganese, and iron also may be added in small quantities. Because of their excellent strength, ease of welding, and corrosion resistance, the alloys have become important construction materials. As the silicon content increases, the alloys become more subject to fire cracking.

Aluminum Bronzes

The aluminum content of these alloys ranges from 4 to 10%. They are moderately hard, very ductile, and tough. The alloys resist scaling and oxidation at high temperatures because of the aluminum content. They perform well in both acids and alkalis. The alloys are good for sea duty, particularly in contact with turbulent seawater.

Nickel Silvers

Nickel essentially is added to copper-zinc alloys to enhance color. With a nickel content of about 18%, the alloy is silver-white. Also, most of the mechanical properties and corrosion resistance are improved. The alloys find wide application for operations that require ductility in the cold condition, as in stamping, spinning, deep drawing, and for articles to be plated. An alloy widely used as a spring material because of its high tensile and fatigue properties has the composition: Cu, 55%; Zn, 27%; Ni, 18%. German silver contains: Cu, 50%; Ni, 30%; Zn, 20%. It is interesting to note that the nickel silvers do not contain silver.

Cupronickels

The nickel silvers generally are classified as brasses. Cupronickels fall more into basic copper-nickel alloys. Possible minor ingredients are manganese, iron, and zinc. These alloys can be used for severe drawing, spinning, and stamping operations because they do not harden readily. They also are extensively used for condenser tubes and plates, heat exchangers, and other process equipment.

AT, annealing temperature range.
HWT, hot-working temperature range.

aqueous solution, due primarily to the larger heat of hydration of Cu^{2+} than Cu^+ Moreover, the *d*-electrons may participate in bonding, and tripositive copper, Cu(III), appears in complexes. In addition, copper forms a number of compounds essentially covalent in character, such as copper(I) oxide, Cu_2O.

This compound is less stable at room temperature than copper(II) oxide, CuO, although Cu_2O occurs in nature (as cuprite). It is the stable oxide above 1.026°C. It is prepared by fusion of copper(I) chloride, CuCl with sodium carbonate, Na_2CO_3. In its crystal, each copper atom has two colinear bonds, and each oxygen atom four tetrahedral ones; two such interpenetrating lattices constitute the structure. Copper(I) hydroxide, CuOH, is relatively stable, and is produced by electrolysis of a sodium chloride solution between copper electrodes (by action of NaOH on the cathode). Copper(II) oxide, produced by heating copper in air, is also essentially covalent (CuO, 1.95 Å); it has tetrahedral bonding of the oxygen atoms, and coplanar bonding of the copper atoms. Copper(II) hydroxide, is precipitated by alkali hydroxides from Cu^{2+} solutions. It is gelatinous, and its composition and solubility vary somewhat with the alkali concentration. It is thermodynamically unstable even in contact with liquid water with respect to dehydration to CuO, but this occurs only very slowly, except upon heating or when catalyzed by hypochlorite, hydrogen peroxide, etc.

Copper(I) halides are formed with chlorine, bromine and iodine, the chloride and bromide by reduction of the copper(II) halides with copper powder, and the iodide by reduction of copper(II) sulfate, $CuSO_4$ solution with potassium iodide. The fluoride appears never to have been made, despite reports to the contrary. All are insoluble in H_2O. Copper(II) fluoride, CuF_2 may be made from CuO and hydrofluoric acid at 400°C, copper(II) chloride, $CuCl_2$ by dissolving the oxide or carbonate in HCl, and copper(II) bromide, $CuBr_2$ from copper and bromine water; copper(II) iodide, CuI_2, is unstable at room temperature with respect to decomposition into CuI and iodine. The chloride and bromide are water-soluble, and ionic. The fluoride is only slightly water-soluble. Anhydrous copper(II) chloride, $CuCl_2$, is monoclinic and its structure contains infinite-chain molecules formed by $CuCl_4$ groups that share opposite edges. $CuBr_2$ has a similar structure.

Complex halides of both monovalent and divalent copper are known. The monovalent complexes are primarily of the composition $MCuX_2$, where X is a halogen atom and M usually an alkali metal, although $CuCl_3{}^{2-}$ ions are also known, being found in infinite chain $(CuCl_3{}^{2-})_n$ structures, as in crystals of Cs_2CuCl_3. The ion $CuCl_4{}^{3-}$ is also known. The composition of the copper(II) complex halides is primarily in terms of $CuCl_3{}^-$, $CuCl_4{}^{2-}$, or $CuBr_4{}^{2-}$, ions, although the corresponding complex fluoride has trivalent copper, as in K_3CuF_6. Its paramagnetic moment indicates two unpaired electrons.

Copper oxyhalides of a number of different compositions have been reported, but the most definitely established compositions are $Cu(OH)Cl$, $Cu_2(OH)_2Cl_2$, and $CuBr_2 \cdot 2Cu(OH)_2$. The property of forming basic salts is not limited to the halides of copper. Basic sulfates, such as $CuSO_4 \cdot 2Cu(OH)_2$, $CuSO_4 \cdot 3Cu(OH)_2$, $CuSO_4 \cdot 4Cu(OH)_2$, and $CuSO_4 \cdot 5Cu(OH)_2$, have been prepared, more or less hydrated. In copper(II) carbonate, the stable forms are oxycarbonates, $xCuCO_3 \bullet yCu(OH)_2$, (where the $x : y$ ratios may be 2:1, 1:1, 2:3, 1:9, and still other values. Many of these compositions occur in minerals, such as malachite and azurite. Copper forms complexes with larger number of ions and molecules. The halogen complexes were discussed above. In general, copper tends to be 6-coordinate, as in the complex ion $[Cu(H_2O)_2$ethylenediamine$)_2]^{2+}$. With NH_3 and many amines, stable complexes are formed, both of Cu^+, such as $[Cu(NH_3)_2]^+$, and of Cu^{2+}, such as $[Cu(NH_3)_4]^{2+}$ which add halogen, pseudohalogen, and many other anions to form compounds of the composition $Cu(NH_3)_2X_2$ and $Cu(NH_3)_4X_2$.

Among the other copper compounds, copper(II) acetate is used as a pigment and fungicide; in its basic form it is the familiar verdigris that forms on copper surfaces in the presence of moisture and organic matter. The arsenic compounds of copper are used as insecticides and wood preservatives: copper(II) arsenite is called "Scheele's green" and copper(II) acetoarsenite, $Cu(AsO_2)_2 \bullet Cu(C_2H_3O_2)_2$, is "Paris green." Copper(II) hexacyanoferrate(II), brown, is precipitated from copper(II) solutions by soluble hexacyanoferrates(II), even from very dilute solutions, and copper(II) sulfide, black, by H_2S or other soluble sulfides. Copper(II) sulfate, when hydrated, forms its characteristic blue crystals (the hydrated Cu^{2+} ion is blue).

Copper(I) cyanide dissolves in alkali cyanide solution to form cyanocuprates(I) of the general formula $M_n[Cu(CN)_{n+1}]$ where M is an alkali metal, and n ranges in value from 1 to 5. Not all of the values, of course, are found for a particular alkali or in the presence of particular anions. With sodium cyanide, NaCN, most of the complex present has an n value of 2, but if the original solute was CuCl, more of the $n = 3$ cyanocuprate(I) is present. At low temperatures, values of n of 4 and 5 are found. Copper(II) appears to coordinate four cyanide ions to form the unstable tetracyanocuprate(II) ion which decomposes at once to the copper(I) complex and cyanogen. In fact, for Cu(II) the chelated complexes

are more common than the simple ones, examples being those formed with ethylenediamine and its derivatives, oxalates, catechol, and the β-diketones.

A solution of CuCl in HCl absorbs carbon monoxide, forming copper(I) carbonyl chloride, $Cu(CO)Cl \cdot H_2O$. This reaction, which is used in gas analysis, is indicative of the ability of copper to combine with carbon monoxide. Evidence for a true carbonyl is limited to the observation that if hot carbon monoxide is passed over hot copper, a metallic mirror is produced in the hotter parts of the tube. Other organometallic compounds include the very unstable methyl copper, CH_3Cu, phenyl copper, C_6H_6Cu, and bischlorocopper acetylene $C_2H_2(CuCl)_2$.

Copper Industrial Chemicals

Copper oxides, salts, and organocopper compounds find extensive use in industry and commerce. Some of the more important compounds are summarized:

Cupric Acetate, $Cu(C_2H_3O_2)_2 \cdot H_2O$, sp gr 1.88, mp 115°C, decomposes at 240°C, dark-brown powder, slightly soluble in cold H_2O and alcohol; moderately soluble in hot H_2O and ether. Used as a fungicide, insecticide, as a catalyst, and in pigments.

Cupric Acetoarsenite (Paris Green), $(CuOAs_2O_3)_3 \cdot Cu(C_2H_3O_2)_2$, emerald-green powder, very slightly soluble in cold H_2O, soluble in alcohol and potassium cyanide. Used as a wood preservative.

Cupric Acid Orthoarsenite (Scheele's Green), $CuHAsO_2$, green powder, insoluble in H_2O, soluble in alcohol, acids, and NH_4OH. Used as a wood preservative.

Copper Carbonate (Basic), $CuCO_3 \cdot Cu(OH)_2$, dark-green monoclinic crystals, insoluble in cold H_2O decomposes in hot H_2O, soluble in potassium cyanide. Malachite, a copper ore, is of this composition. Refined compound is used as a pigment.

Cupric Hydroxide, $Cu(OH)_2$, blue, gelatinous compound, insoluble in cold H_2O decomposes in hot H_2O soluble in alcohol, NH_4OH, and potassium cyanide. Used as a pigment.

Cuprous Cyanide, $Cu_2(CN)_2$, white monoclinic crystals, insoluble in H_2O, soluble in HCl, NH_4OH, and potassium cyanide. Used in Sandmeyer's reaction to synthesize aryl cyanides.

Cuprous Iodide, Cu_2O, cubic white crystals, practically insoluble in H_2O or alcohol, soluble in NH_4OH, potassium iodide, or potassium cyanide. Used in Sandmeyer's reaction to synthesize aryl chlorides.

Cupric Oxide, CuO, black cubic crystals, insoluble in H_2O, soluble in HCl, NH_4OH, or ammonium chloride. Used as a green and blue colorant in ceramics.

Cuprous Oxide, Cu_2O, red cubic crystals, insoluble in H_2O soluble in HCl, NH_4OH, or ammonium chloride. Cuprite, a copper ore, is of this composition. Refined compound is used in electrical rectifiers.

Cupric Sulfate, $CuSO_4$ $5H_2O$, blue triclinic crystals, moderately soluble in cold H_2O, quite soluble in hot H_2O, very slightly soluble in alcohol. Used in copper plating, dyestuff manufacture, water treatment, germicides, and coppering of steels.

Cupric Chloride, $CuCl_2$, brown-yellow powder, quite soluble in cold H_2O or alcohol, very soluble in hot H_2O. Catalyst for several organic syntheses, including production of vinyl chloride monomer.

Additional Reading

Bray, W.: "Ancient American Metallurgy: Five Hundred Years of Study," in *The Art of Precolumbian Gold: The Jan Mitchell Collection* (J. Jones, Editor), Weidenfield & Nicolson, 1985.

Cattell, F.C.R. and W.D. Scott: "Copper in Aerosol Particles Produced by the Ocean," *Science*, **202**, 429–430 (1978).

Diaz, C., C. Landolt, and T. Utigard: *Copper 1999—Smelting, Technology Development, Process Modeling and Fundamentals,* The Minerals, Metals & Materials Society, Warrendale, PA, 1999.

Dutrizac, J.E. and Ji, V. Ramachandran: *Copper 1999—Electrorefining and Electrowinning,* Vol. 111, The Minerals, Metals & Materials Society, Warrendale, PA, 1999.

Eltringham, G.A., M. Sahoo, and N.L. Piret: *Copper 1999—Plenary Lectures/Movement of Copper and Industry Outlook: Copper Applications and Fabrication,* Vol. 1, The Minerals, Metals & Materials Society, Warrendale, PA, 1999.

George, D.B., P.J. Mackey, W.J. Chen, and A.J. Weddick: *Copper 1999—Smelting Operations and Advances,* The Minerals, Metals & Materials Society, Warrendale, PA, 1999.

Hancock, B.A. and M.R.L. Pan: *Copper 1999–Mineral Processing: Environment, Health and Safety,* Vol. 11, The Minerals & Materials Society, Warrendale, PA, 1999.

Herrmann, W.A.: *Copper, Silver, Gold, Zinc, Cadmium, and Mercury,* Vol. 5, Thieme Medical Publishers, Inc., New York, NY, 1999.

Joseph, G. and K.J. Kundig: *Copper: Its Trade, Manufacture, Use, and Environmental Status,* ASM International, Materials Park, OH, 1999.

Lechtman, H.: "Pre-Columbian Surface Metallurgy," *Sci. Amer.,* **250**(6), 56–63 (June 1984).

Lide, D.R.: *CRC Handbook of Chemistry and Physics,* 84th Edition, CRC Press, LLC., Boca Raton, FL, 2003.

Melnik, M.: *Heterometallic Coordination Copper (ll) Compounds: Classification and Analysis of Crystallographic and Structural Data,* Nova Science Publishers, Inc., Huntington, NY, 1999.

Mueller, D. and D.I. Groves: *Potassic Igneous Rocks and Associated Gold-Copper Mineralization,* Springer-Verlag Inc., New York, NY, 2000.

Plewes, J.T. and D.N. Loiacono: "Free-Cutting Copper Alloys Contain No Lead," *Advanced Materials & Processes,"* **23** (October 1991).

Shimada, I.: "Perception, Procurement, and Management of Resources: Archaeological Prespective," in *Andean Ecology and Civilization* (R. Maddin, Editor), MIT Press, Cambridge, MA, 1988.

Shimada, I. and J.F. Merkel: "Copper-Alloy Metallurgy in Ancient Peru," *Sci. Amer.,* **80** (July 1991).

Sousa, L.J.: *Problems and Opportunities in Metals and Materials: An Integrated Perspective,* U.S. Department of the Interior, Washington, DC, (periodically revised).

Staff: *ASM Handbook—Properties and Selection: Nonferrous Alloys and Pure Metals,* ASM International, Materials Park, OH, 1990.

Staff: *Copper and Copper Alloys,* American Society for Testing & Materials, West Conshohocken, PA, 2000.

Staff: *Nonferrous Metal Products Copper and Copper Alloys,* American Society for Testing Materials, West Conshohocken, PA, 1999.

Young, S.K., D.G. Dixon, D.B. Dreisinger, and R.P. Hackl: *Copper 1999—Hydrometallurgy of Copper,* The Minerals, Metals & Materials Society, Warrendale, PA, 1999.

COPPER (In Biological Systems). The activity of copper in plant metabolism manifests itself in two forms: (1) synthesis of chlorophyll, and (2) activity of enzymes. In leaves, most of the copper occurs in close association with chlorophyll, but little is known of its role in chlorophyll synthesis, other than the presence of copper is required. Copper is a definite constituent of several enzymes catalyzing oxidation-reduction reactions (oxidases), in which the activity is believed to be due to the shuttling of copper between the +1 and +2 oxidation states.

Traces of copper are required for the growth and reproduction of lower plant forms, such as algae and fungi, although larger amounts are toxic.

The effects of copper deficiency in plants are varied and include: dieback, inability to produce seed, chlorosis, and reduced photosynthetic activity. In contrast, excesses of copper in the soil are toxic, as in the application of soluble copper salts to foliage. For this reason, copper fungicides are formulated with a relatively insoluble copper compound. Their toxicity to fungi arises from the fact that the latter produce compounds, primarily hydroxy and amino acids, which can dissolve the copper compounds from the fungicide.

Copper is a necessary trace element in animal metabolism. The human adult requirement is 2 milligrams per day, and the adult human body contains 100–150 milligrams of copper, the greatest concentrations existing in the liver and bones. Blood contains a number of copper proteins, and copper is known to be necessary for the synthesis of hemoglobin, although there is no copper in the hemoglobin molecule.

Anemia can be induced in animals on a low copper diet, such as milk, and appears to be due to an impaired ability of the body to absorb iron. This anemia, however, is rare, because of the widespread occurrence of copper in foods. In locations, such as Australia and the Netherlands, diseases of cattle and sheep, involving diarrhea, anemia and nervous disorders, can be traced either to a lack of copper in the diet, or to excessive amounts of molybdenum, which inhibits the storage of copper in the liver.

Ingestion of copper sulfate by humans causes vomiting, cramps, convulsions, and as little as 27 grams of the compound may cause death. An important part of the toxicity of copper to both plants and animals is probably due to its combination with thiol groups of certain enzymes, thereby inactivating them. The effects of chronic exposure to copper in animals are cirrhosis of the liver, failure of growth, and jaundice.

Copper deficiency in plants is most frequent on organic soils, such as newly drained bogs, and on very sandy soils. The severe copper deficiency often found when bogs and marshes are first used for crop production is called *reclamation disease* in some parts of the world.

Ruminants are sensitive to copper deficiency. The symptoms of copper deficiency in animals vary with the species and age, but often the fading of brown or black hair is evident. On some acidic soils, the use of copper in fertilizers increases crop and pasture production, and the increases in level of copper in the plants help to prevent copper deficiency in the cattle and sheep. In parts of Australia, livestock production was impossible until copper fertilizers were used on the pastures. Application of copper fertilizers to alkaline soils generally does not increase the copper level in the crop. Farm animals are often supplied with copper in the form of dietary mineral supplements. Compounds used include copper gluconate, copper oxide, and copper sulfate.

Although copper fertilizers will sometimes increase crop yields and improve the nutritional quality of the crops, this practice must be used with caution and only on copper-deficient soils. Both plants and animals are subject to toxicity from excessive levels of copper. Ruminants, especially sheep, are sensitive to copper toxicity as well as to copper deficiency. Adding a copper fertilizer to a soil that naturally contains rather high levels of available copper may increase levels of the metal in the forage to the point of causing copper toxicity in grazing sheep. Copper toxicity from soils naturally high in copper occurs in Australia, but is uncommon in the United States. There are soils in the United States, however, that produce forage levels of copper close to toxicity limits, and if copper-bearing mineral supplements are inadvertently used with these forages, copper toxicity to sheep may result.

It is not easy to set a definite limit, in terms of the copper concentration in the diet, that will permit accurate predictions of the danger of copper deficiency or of copper toxicity in cattle and sheep. In particular, if the molybdenum concentration in the forage is high, extra amounts of copper are needed to prevent deficiency. Also, higher copper levels can be tolerated without danger of toxicity with molybdenum present.

Monogastric animals, including humans, are less sensitive than ruminants to either copper deficiency or toxicity. Copper deficiency in people has been found only when other complications, such as excessive bleeding, general starvation, and iron deficiency, are also present. Wilson's disease, an inherited disease of humans, prevents the loss of excess copper from the body and brings on copper toxicity. No direct relationships have been found between levels of available copper in the soil and the copper status of humans.

A number of copper-containing protein compounds are enzymes with an oxidase function (ascorbic acid oxidase, urease, etc.) and these play an important role in the biological oxidation-reduction system. There is a definite relationship of copper with iron in connection with utilization of iron in hemoglobin function.

Copper absorption is depressed by ascorbic acid, dietary phytates, cadmium, mercury, silver, and zinc. It appears that metals impede copper absorption through competition for metal-binding sites. Dietary copper, molybdenum, and sulfur are closely interrelated in optimum copper and molybdenum nutrition of ruminants. Increase pasture molybdenum content and low-pasture copper result in a condition known as "peat scours."

Copper toxicity tends to accumulate in the liver. The capacity to tolerate copper varies considerably with the species. Sheep are most susceptible. Swine have a much greater tolerance and copper may be added to the swine diet for pharmacological reasons (for example, use as an anthelminthic to control internal parasites).

Continuing research is providing a better understanding of the biological role of copper. The prooxidant and antioxidant effects of ascorbic acid and metal salts, including copper, in a beta-carotene-linoleate model system were studied by Israeli scientists. The interacting effects of ascorbic acid and metal ions on carotene oxidation were studied in an aqueous carotene-linoleate solution at pH of 7. Ascorbic acid at concentrations up to $10^{-3} M$ was a prooxidant. Fe^{3+} and, to a lesser extent, Co^{2+} acted synergistically with ascorbic acid, the prooxidant effect increasing with metal concentration. Cu^{2+} formed a prooxidant system with ascorbic acid only at low metal concentration, but as the copper concentration was raised, inversion of activity occurred and the copper-ascorbic acid system exerted a stabilizing action on carotene. Prooxidant effects were enhanced and antioxidant effects weakened in the presence of added linoleate hydroperoxides. The latter were unstable in the presence of ascorbic acid and especially ascorbic acid plus Cu^{2+}. Ascorbic acid itself became unstable in the presence of Cu^{2+}. Oxygen depletion, brought about by the rapid oxidation of ascorbic acid, may be partly responsible for the carotene-stabilizing effect of the Cu^{2+} couple. The investigators postulated that additional stabilization results from the radical-scavenging properties of copper or of a copper chelate formed by ascorbic and/or dehydroascorbic acid.

A team of investigators at the Department of Food Science and Human Nutrition, Michigan State University, studied the kinetics of ascorbic acid stability of tomato juice as a function of temperature, pH, and metal catalyst, including copper. The rate of copper-catalyzed destruction of ascorbic acid increased as copper concentration in tomato juice increased, and was affected by pH.

Relatively recent hypotheses concerning the effect of zinc-to-copper ratios in the diet as a determining factor of plasma cholesterol levels have been made.

A study by the Department of Foods and Nutrition, Kansas State University is exemplary of the much needed further research in determining the properties of certain elements, including copper, when contained in various food substances. Part of the study was directed at determining the effects of cooking on copper content of turkey muscle. The researchers found that copper was significantly lower in cooked than in raw breast turkey muscle, but similar in raw and cooked thigh muscle.

CORDIERITE. The mineral cordierite, composition $(Mg, Fe)_2Al_4Si_5O_{18}$ is an orthorhombic mineral frequently seen, however, in pseudo-hexagonal forms, as well as massive. It is brittle, with a subconchoidal fracture; hardness, 7–7.5; specific gravity, 2.53–2.78; luster, vitreous; color, blue of varying shades; translucent to transparent. Cordierite exhibits pleochroism (or dichroism) being dark blue, light blue and light yellow when examined by transmitted light in different directions. Hence, it is frequently called *dichroite*, and less frequently, *iolite*. It is occasionally used as a gem.

Cordierite is found as a primary mineral in the igneous rocks. It is, however, found ordinarily in gneisses, schists and in areas of contact metamorphism. Localities for good specimens are numerous in Europe, including Bavaria, Finland, Norway. It is found in Greenland, Madagascar and Sri Lanka, from which latter place come the rolled pebbles of a rich blue color known as saphir d'eau, prized as a gem. In the United States, it is found principally in Connecticut.

Named for the French geologist, Pierre Louis Antoine Cordier, this mineral has also been called iolite from the Greek word meaning violet, and stone, as well as dichroite from the Greek meaning *two-colored*.

COREY, ELIAS JAMES (1928–). An American who won the Nobel prize for chemistry in 1990 for development of novel methods for the synthesis of complex natural compounds (retrosynthetic analysis). He specialized in synthesis of terpines. He was awarded is doctorate by M.I.T. in 1951.

CORK. Cork cells are found in the outer bark of most woody-stemmed plants, but in amounts too small and with too brittle walls to be of any use to man. But in *Quercus suber*, the Cork Oak, the cork cells become a very large part of the tissue of the bark, and have been used for centuries by man. The cork oak tree is a medium-sized tree, seldom much over 50 feet (15 meters) in height, growing in nearly all countries bordering the Mediterranean Sea. The evergreen leaves are small, $1\frac{1}{2}$–3 inches long, and about an inch wide, with slightly toothed margins. The bark of the tree soon becomes rough and deeply furrowed, but is of little value except as ground cork or as a source of tannin. When the tree is about 20 years old this first formed bark is removed, care being taken not to injure the phloem and cambium layers. Within 10 days a new cork layer has formed. This layer is the first of many layers which are removed once every 10 years or so throughout the life of the tree. Removal is generally done in the early summer at a time when hot dry winds will not cause injury to the unprotected phloem and cambium.

After removal, the cork is air-dried for a time, then boiled to soften it and to remove some of the tannin. The outer part of the bark is scraped off, and the rest pressed out flat and dried. It is then ready to ship.

The physical properties of cork account for its many uses. It is very light and buoyant, more than 50% of its volume being air, and hence is used in the manufacture of floats, life-preservers, and so forth. The living protoplasm of the cork cells dries up early in their development, leaving hollow cells, each containing a small mass of air which expands after compression. Therefore, cork is very resilient, and is frequently used as a core on which to wind yarn or string in the manufacture of baseballs. In the early stages of their formation, the walls of cork cells are cellulose, but this is soon impregnated by a waterproof and nonabsorbent lipoid substance, suberin. Therefore cork is used in making handles for fishing rods, shoe-soles, and cork stoppers. Since the hollow cork cells are poor conductors both of heat and sound, cork is much used as insulating material. For this use cork is ground up and then pressed into sheets with various binding materials, giving much larger sheets than can be obtained from the tree. Ground cork is also a constituent of linoleum, gaskets, and other products.

Cork is traversed by lenticels, loose masses of porous tissue, which appear as dark spots or holes in stoppers. Usually in making stoppers the bark is cut so that these will be transverse in the stopper. In making stoppers, the forms are first punched out as cylinders, and then trimmed down by machine to the required tapering shape.

CORNFORTH, JOHN (1917–). An Australian-born chemist who won the Nobel prize for chemistry in 1975 with Vladimir Prelog for work on the chemical synthesis of organic compounds. Although deaf since childhood, he attained his doctorate from Oxford and held prestigious posts all over the world, as well as authoring many papers on organic and biochemical subjects.

CORROSION. 1. The electrochemical degradation of metals or alloys due to reaction with their environment; it is accelerated by the presence of acids or bases. In general, the corrodability of a metal or alloy depends upon its position in the activity series (electromotive force series). Corrosion products often take the form of metallic oxides; in the case of aluminum and stainless steel, this is actually beneficial, for the oxide forms a strongly adherent coating, which effectively prevents further degradation. Hence, these metals are widely used for structural purposes. Probably the most familiar kind of corrosion is that of *rusting*. This is but a special case of a general classification known as atmospheric corrosion, wherein the oxygen of the atmosphere reacts with the material in question. Most metals, with exception of the noble metals, such as gold, can be oxidized by atmospheric oxygen. In the usual case, however, water vapor must be present before any appreciable oxidation can take place. With iron, for example, about 40% relative humidity is needed at ordinary temperatures before rusting will occur.

Acidic soils are highly corrosive. Sulfur is a corrosive agent in automative fuels and in the atmosphere (SO_2) as well, and is frequently mentioned in connection with so-called acid rains. Sodium chloride in the air at locations near the sea is strongly corrosive, especially at temperatures above 70°F (21.1°C). Copper, nickel, chromium, and zinc are among the more corrosion-resistant metals and are widely used as protective coatings for other metals.

2. The term *corrosion* is also sometimes used in connection with the destruction of body tissues by strong acids and bases.

In a restricted sense, corrosion is considered to consist of the slow chemical and electrochemical reactions between metals and their environments. From a broader point of view corrosion is the slow destruction of any material by chemical agents and electrochemical reactions. This contrasts with *erosion*, which is the slow destruction of materials by mechanical agents. The character of the atmospheres to which materials are exposed may be classified as: rural, urban, industrial, urban-marine, industrial-marine, marine, tropical, and tropical-marine. In addition to these general kinds of environments, corrosion is of particular concern in the environments of chemical, petrochemical, and other processing and manufacturing environments where extremely corrosive substances may be encountered.

Metals Corrosion

The relationships between metals and hydrogen in the activity series are important because in the electrochemical processes of corrosion the discharge of hydrogen ions and the evolution of hydrogen as a gas is one of the principal cathodic reactions. The facility with which this can occur is determined by such factors as the hydrogen ion concentration (pH) of the electrolyte, the electrical potential of the corrosion cell, and the overvoltage characteristics of the cathodic surface. In a situation sometimes called *concentration cell corrosion*, two solutions of different concentrations will set up an electrical potential between them similar to that produced by a battery. If oxygen is present in the liquid and is continually replenished by contact with the air, then the oxygen concentration in the liquid will remain substantially constant. Any liquid that is contained in small holes or cracks on a metal surface will not be able to obtain oxygen from the main bulk of the solution, so when the supply in the holes and cracks is exhausted, no more oxygen can get in to replace it. Therefore, the oxygen concentration in the cracks is different from that of the main bulk of the solution and a concentration cell is set up. This minute electrical effect is sufficient to make corrosion proceed quite rapidly. A similar cell type of corrosion is that called *galvanic* or *two-metal corrosion*. Two different metals in contact will set up an electrical potential between them. If the two metals are surrounded by an electrolyte so that a closed circuit can be obtained, corrosion takes place. The magnitude of the electrical potential and, therefore, the speed and extent of the corrosion will depend upon the types of metals in each pair. In general, pairs farther apart in the activity series will corrode faster than those close together. See also **Electrochemistry**.

The electrochemical reactions in corrosion of a divalent metal may be written:

Anodic reaction:	$M^0 \rightarrow M^{++} + 2$ electrons
At the cathode:	(1) $2 H^- + 2$ electrons $\rightarrow H_2$ gas
	(2) $\frac{1}{2}O_2 + 2 H^+ + 2$ electrons $\rightarrow H_2O$
	(3) $O_2 + 2 H_2O + 2$ electrons $\rightarrow H_2O_2 + 2 OH^-$
	(4) $\frac{1}{2}O_2 + H_2O + 2$ electrons $\rightarrow 2 (OH)^-$

It is evident that oxygen as well as hydrogen plays an important part in metal corrosion. It can accelerate corrosion by participating in cathodic reactions, or it can retard corrosion by forming protective oxides or passive films. The dual effect of oxygen is one of the factors that complicates corrosion processes, including the interpretation of observations of the process and the steps to be taken to avoid corrosion damage.

Forms of Corrosion

(1) *Pitting* resulting from local action currents, as at discontinuities in protective or passive films or under or around deposits that set up concentration cells. (2) *Stress corrosion cracking* resulting from the combined effects of corrosion by a specific environment and either applied or internal static tensile stresses. Depending upon the metal and the environment, the cracks may be either intercrystalline or transcrystalline. (3) *Corrosion fatigue*, resulting from the combined effects of corrosion and cyclic stresses. Racks of this type are characteristically transcrystalline. (4) *Intergranular corrosion* resulting from preferential attack on, or around, a phase or compound that occupies grain boundaries. (5) *Corrasion-corrosion* resulting from the combined effects of corrosion and either abrasion or attrition. The mechanism usually involves local or general removal of otherwise protective corrosion product films. Particular forms are impingement attack due to the effects of high velocity or turbulence in flowing liquids, e.g., salt water in steam condensers, or other heat exchangers, in piping systems, valves, and pumps, among others. A particularly aggressive form is associated with the severe mechanical forces that are characteristic of *cavitation phenomena*. See also **Cavitation**. (6) Uniform attack or general wastage, such as may be caused by the action of strong acids as used for pickling (scale removal) or etching. This is also characteristic of the slow corrosion of durable materials in appropriate environments, such as copper roofing in suburban atmospheres, cupronickel tubes in ships, condensers, Monel-nickel copper alloy racks for pickling steel in sulfuric acid, or stainless steel columns for handling nitric acid.

Metal Corrosion Minimization

In the most recent official assessment of the economic costs of equipment damage arising from corrosion, prepared by the National Commission on Materials Policy (U.S.), it was stated that annual losses in the United States alone are on the order of many billions of dollars.

Some of the means used to combat corrosion losses include:

(1) Use of the right metal in the proper way and in the correct place. Planners tend to look too closely at first costs and not closely enough at maintenance costs; consequently, there are many applications of materials that are of lower first cost, but of severely limited life. For example, in some applications, inexpensive fasteners that will obviously corrode in a few years are used in place of stainless steel or hardened aluminum fasteners that would have cost only a few dollars more at the time of installation. When discussing applications of possible corrosion-resistant materials, it is important to define clearly the parameters of the environment of usage, such as temperature, pressure, humidity, presence of specific chemical agents, presence of living or dead organic materials, and the characteristics of associated electrical magnetic, light, and other radiation fields. Far more needs to be done on the microclimatology of environments and the specifics needed to ameliorate corrosion problems.

(2) Protective coatings—paints, enamels, other metals, oils, greases, among others. One of the more common methods used is zinc-coating, i.e., galvanizing. Galvanized iron wire with a thin layer of zinc applied to it by dipping in molten zinc or by electrical means usually will resist corrosion for an extended period. Cadmium, nickel, tin, and chromium are metals often used as protective coatings, generally applied by electroplating. See also **Electrochemistry**; **Electroplating**; **Galvanizing**; and **Paints and Coatings**.

(3) Inhibitors and neutralizers, i.e., compounds added to the environment in small concentration to form protective films which increase anodic or cathodic polarization or both, or to neutralize some corrosive constituents. For example, it is possible for corrosion to occur at many places in the piping leading to boilers or heaters, but usually it occurs in the boiler itself. The trouble is ordinarily found to be due to an acid condition of the boiler feedwater, or to dissolved oxygen contained in it. The raw water used may be acid from surface pollution or from subsurface drains. Usually this can be detected and readily remedied. A more serious factor is the oxygen dissolved in water. Under the high-temperature conditions existing in the boiler per se, this oxygen becomes extremely active in attacking metal surfaces. The operators of large, high-pressure boilers well know the necessity of removing oxygen from feedwater through the use of deactivators or de-aerators. Corrosion protection of boilers in power plants is effected by installation of ion exchange resin and other purifiers in the feedwater cycle, and by monitoring them by automatic analysis. In still other installations where acidity and oxygen are a problem, neutralizing chemicals with the dosage governed by automated pH and oxygen control systems are effective. Systems of this type are also effectively used to neutralize plant effluents to streams so that water users downstream have a reasonably neutral and clean supply of water. Clean water programs not only are desirable from an environmental standpoint, but also can help in reducing the costs of corrosion.

(4) Drying of air other gases to keep humidity below the level where corrosion becomes serious.

(5) Design of hydraulic systems to avoid excessive velocities or localized turbulence or to maintain a velocity high enough to prevent the accumulation of corrosion products or other deposits that will promote localized corrosion.

(6) Various features of design and operation of structures or equipment to favor rapid drainage and drying, prevent accumulation or concentration of corrosive chemicals in crevices or low spots, hold operating stresses and temperatures within desired limits, eliminate fabricating stresses by appropriate metallurgical treatment, avoid galvanically unfavorable combinations of different metals, and provide protection against stray electrical currents by appropriate insulation and electrical bonding.

(7) Heat-treating metals to leave them in optimum condition to resist corrosion.

(8) Applying protective electrical currents (*cathodic protection*) from sacrificial metals (galvanic anodes) such as zinc, magnesium, or aluminum or from some external source through a graphite, platinum, or other appropriate anode receiving current from a rectifier, generator, or battery. The location of the anodes, the magnitude of the current, and the applied voltage must be engineered so that without wasting current all surfaces that require protection will receive sufficient current to achieve this effect. Too much current may cause damage by the alkali generated by a cathodic reaction or by hydrogen evolved at the cathode, which can destroy protective films or embrittle metals. A current of 1 to 15 milliamperes per square foot (929 square centimeters) is usually required to protect bare steel area; for design purposes, the range is generally narrowed to 3 to 5 milliamperes. The potential required to produce this current flow depends, of course, on the resistivity of the electrolytic path between the electrodes.

Corrosion can also be suppressed by the controlled application of current to the metal as an anode. This is called *anodic protection*. Passivity is induced and preserved by maintaining the potential of the alloy at, or above, a critical potential in what is called the range of passivity in a potentiostatic diagram. Such diagrams are based on the relationship between applied anodic current density and the corresponding potential in the environment of interest.

When pipes and cables carrying an electric current are underground, they are commonly corroded by electrolytic action from unidirectional electric currents in the ground. Stray current from electric traction equipment is retarded by increasing the resistance of the ground circuit and by reducing the electric resistance of the track. Also, cathodic protection is widely used. An external source of dc voltage is applied so that the protected equipment (pipeline or cable, for example) becomes lower in potential than the soil that surrounds it. Thus, the buried material is the cathode rather than the anode. Usual forms of corrosion can be prevented when the preventive system causes the pipe or metal structure to be 0.25 to 0.30 volt negative with reference to the soil or liquid that may be surrounding it. The negative lead from a small generator, battery, or rectifier is connected to the metal structure; the positive lead to the ground at some distance. Or sacrificial magnesium or zinc rods sunk in the ground may be externally connected to the structure.

Atmospheric Corrosion of Metals and Nonmetals

Taking into consideration the relative order of corrodibility, it is preferable to describe corrosive damage as attributable to certain agents rather than to the indefinite characterization "smoke." Corrosive agents can be placed into four major groups, namely, oxygen and oxidants, acidic materials, salts, and alkalis.

Corrosion attributable to oxygen is deemed to result from the solution of oxygen by a thin film of liquid adjacent to the metallic surface, the transportation of the oxygen through the film, and the subsequent reaction at the surface of the metal. This explains why there is corrosive action even in relatively arid land. In a very dry atmosphere corrosion is, however, markedly reduced.

There are three principal categories of oxidizing agents that occur as air pollutants. These are ozone, nitrogen oxides and nitric acid, and organic peroxide. Many materials that are relatively resistant to attack by the free oxygen of the air are far less resistant to attack by such oxidants and peroxides. These dissolve in the surface film and thus convert metals to their oxides which react readily even with such relatively weak acids as carbonic acid and sulfurous acid. For instance, copper tarnishes rapidly forming the oxide, which dissolves readily in dilute acids.

The acid components given off to the air by the various processes of combustion are sulfur dioxide and sulfurous acid, sulfuric acid, hydrogen sulfide, hydrochloric acid, carbon dioxide and carbonic acid, and tar acids. There is little doubt that the material of greatest importance in respect to atmospheric corrosion in this group is sulfur dioxide. Generally, the total acidity of the atmosphere is closely related to the sulfur dioxide content.

Other soluble acidic components such as sulfuric acid, hydrogen sulfide, hydrochloric acid, nitric acid and the like are all of minor importance. Carbon dioxide and carbonic acid play a significant role in acid decomposition.

One aspect of tar and tar acids should be noted, namely, that these are sticky and cling to the surfaces with which they come in contact. This enables the acids that such tar contains to have a prolonged corrosive action. It also increases the difficulty of removal by rain or wind or other action.

It is common to consider that certain salts have a very corrosive action. This is true in the respect that the corrodibility of marine atmospheres has been shown to be greater than rural, tropical, and urban atmospheres. For example, ammonium sulfate and ammonium chloride being salts of strong acids and a weak base, that is ammonium hydroxide, hydrolyze in water to yield the respective acids. These salts then have a corrosive action, which is due actually to the acid produced in hydrolysis.

Alkalis seldom occur as air pollutants except under industrial conditions. Nevertheless, the corrosive action of alkalis should not be completely ignored. While a number of metals are relatively resistant to acid attack, they have an amphoteric action and can react with alkalis. For instance, aluminum and zinc are in this category and they are subject to corrosive attack by relatively weak alkalis.

Metals exposed to air pollution can be placed into three groups:

1. Metals that corrode rapidly because they do not form completely protective corrosion products. The major metal in this group is iron. It should be noted, however, that iron oxide Fe_2O_3 does have some protective action.
2. Metals that are initially attacked somewhat readily but subsequently become resistant to attack because of the formation of a corrosion-resisting film, which hinders further attack. Among the metals in this group are aluminum, lead, zinc, brass, copper, nickel and magnesium.
3. Metals that are almost completely corrosion resistant, as for instance stainless steel of the 18/8 type, chromium plate products, monel, and gold.

Mention has already been made of the action of oxygen and oxidants on metal. It should be noted that metals react with sulfides, such as hydrogen sulfide, and are subsequently subject to additional slow attack by oxygen and oxidants. Thus, copper reacts to form sulfide and then the basic copper sulfate.

Generally, metals are resistant to attack in dry air; even in pure humid air, corrosion is slight; when, however, air pollutants are present the rate of corrosion will increase measurably, the increase being dependent upon the humidity and the character of the pollutant. Such action may be grouped as follows:

Relative Humidity		Degree of Corrosion
Less than	60	None
More than	60	Slow but definite
	80	Decided increase
Greater than	80	Very high

A factor of note is the settling and adherence of particles on metals. Particles of carbon, ammonium sulfate, and silica cause a marked increase in corrosion, and this is accentuated in atmospheres containing sulfur dioxide. The presence of such hygroscopic particles enhances the adherence of liquid and thus provides for electrochemical attack.

Stone building materials may be placed into relatively resistant and nonresistant categories. The acids of the air, such as sulfuric acid, attack carbonate-bearing stone, such as limestone, converting the calcium carbonate to calcium sulfate. The gypsum formed is dissolved by rainwater, causing pitting. Incrustations may be formed because of the crystallization of soluble salts. These break away in time and leave pitted surfaces. Other types of damage, such as porosity of the stone, are caused by analogous reactions.

Corrosion-Resistant Metals

Some concept of the economic importance of corrosion to the production and consumption of various metals can be gleaned from Table 1. Many mineral commodities are used in more or less direct proportion to steel production. In the case of the United States and many other countries, a large number of the mineral materials important to combating corrosion come from distant sources. That is why many countries maintain a stockpile of such materials, particularly those that are regarded as critical or strategic. In the United States, 93 materials are officially classified for defense purposes as basic stockpile commodities. Seventy-nine of these are metals and minerals, including nearly every one of the metals with important corrosion-resistant properties.

The possible use of low-grade, currently noncommercial, mineral deposits, requires constant consideration. For example, chromium has been recognized as an important strategic material ever since World War I. Over the years, the U.S. government, through the U.S. Geological Survey and the Bureau of Mines, has discovered and carefully defined numerous domestic chromium deposits. The Bureau of Mines in its metallurgical laboratories has produced acceptable chrome concentrates from these deposits as well as acceptable ferrochromium chemicals. Current Bureau of Mines research includes recovering chromium, nickel, and cobalt from laterite deposits, both domestic and from other countries, and also from flue dusts, plating wastes, and other residues.

Corrosion Monitoring

Combatting corrosion in continuous processing plants, which may be scheduled for quite infrequent but thorough equipment checking and maintenance, is particularly difficult. Process downtime costs for a large unit may be several hundred thousand dollars per day in terms of lost production. To avoid excessive downtime for checking and still control the effects of corrosion (personnel and equipment safety, product quality and throughput rates, etc.) requires means to measure the status and rate of corrosion that may be taking place within the equipment. The design

TABLE 1. APPLICATIONS OF CORROSION-RESISTANT METALS

Metal	Corrosion resistance use	Consumption For corrosion applications	Consumption Percent of all uses of metal
Nickel	Alloying, 34%; high-temperature oxidation resistance, 26%; plating, 13%	144,000 MT	73
Chromium	Alloying, 57%; coatings and plating, 7%	306,000 MT	64
Titanium	Coatings, 51%; alloying, 1%	243,000 MT	52
Cadmium	Plating	2,520 MT	45
Gold	Plating and alloying	50,000 kg	35
Zinc	Galvanizing, 32%; coatings and sacrificial anodes, 3%	432,000 MT	35
Tin	Plating, tinning	19,800 MT	34
Tantalum	Alloying, 16%; cladding, 12%; high-temperature oxidation resistance, 5%	1,935 MT	33
Rare earths	Alloying	3,600 MT	30
Platinum	Resistance to chemical attack	13,375 kg	27
Silver	Alloying	1,370 kg	26
Columbium (Niobium)	Alloying; high-temperature oxidation resistance	630 MT	25
Iron oxide pigments	Coatings	30,600 MT	25
Copper	Alloying and plumbing	370,000 MT	18
Molybdenum	Alloying; coating	4,860 MT	18
Cobalt	Alloying; high-temperature oxidation resistance	1,530 MT	17
Magnesium	Alloying; sacrificial anodes	14,400 MT	15
Zirconium	Alloying; chemical resistance	340 MT	15
Thorium	Alloying	31 MT	12
Hafnium	Alloying	3 MT	11
Beryllium	Alloying	4 MT	10
Lead	Pigments and plating, 8%; cable covering, 1%	108,000 MT	9
Indium	Coatings	1,555 kg	8
Aluminum	Alloying; coatings; cladding	225,000 MT	5
Manganese	Alloying; cladding	45,000 MT	4

Source: U.S. Bureau of Mines.
MT = metric ton; kg = kilogram

of corrosion monitors is among the most recent developments in overall process instrumentation. For obvious reasons, such on-line corrosion testing must be of a nondestructive nature.

In one type of monitor, changes in electrical resistance of a measuring element or probe relate to corrosion rate. The measuring element may be a wire, tube, or strip that can be inserted in a tee or an elbow in the process piping. As the measuring element corrodes, the cross-sectional area reduces and the electrical resistance increases. The thickness of the measuring element is directly proportional to a corrosion dial reading. The difference in dial readings is plotted over a period of time and from these data, corrosion rate can be determined. Probes are available in a number of different metals for different temperature ranges and corrosion conditions. In another variation, three electrode probes are used. The corrosion rate is determined by measuring electrical current flow between the test and auxiliary electrodes. That current either cathodically protects or anodically accelerates the corrosion rate of the test electrode, depending upon the flow. The current is measured on a microammeter that has been converted to read directly the corrosion rate in mils (1 mil = 0.001 inch = 25.4 micrometers) per year of the test electrode.

In another instrumental approach, a hydrogen test probe operates on the principle that hydrogen will diffuse through the thin wall of a test probe and set up a pressure within the tube. The rate at which the pressure increases is measured by a pressure gage. The rate at which hydrogen is penetrating per unit area can then be determined, using the exposed surface area and the internal volume of the probe. Beyond a certain rate, severe hydrogen damage can be anticipated.

Measurement of wall thickness can be determined using ultrasonic nondestructive testing methodology. This method can be used to determine wall thinning, pitting, erosion, and flaws in metals, plastics, and various rubbers and polymers. Some disadvantages of the method include the need to take many readings over a period of time to determine corrosion rate and the fact that high-temperature measurements tend to be inaccurate.

Infrared thermographic techniques can be used to identify hot spots on process equipment. The camera works on the theory that the hotter the object, the higher the frequency of radiation. For off-line corrosion monitoring, borescopes for inspecting tubes, pumps, compressors, and other equipment may be used. Spot chemical testing can indicate the presence of alloy constituents of unknown materials. Television camera and holographic techniques also have been used. The monitoring of pH is an invaluable indicator of possible corrosion problems, particularly in cooling water systems. Monitoring is usually done continuously because pH shifts can take place rapidly in many systems, particularly as a result of a process leak.

Probably the weight loss coupon approach is one of the most reliable methods and is widely used. The accuracy of the data is highly dependent on good techniques and on the statistical significance of the tests. The engineering quality data produced require the efforts of many people over a period of several weeks or months, which makes this information quite costly. However, it is the technique of first choice of many processors.

Additional Reading

Bardal, E.: *Corrosion and Prevention,* Springer-Verlag New York, Inc., New York, NY, 2003.

Bierwagen, G.P.: *Organic Coatings for Corrosion Control,* Oxford University Press, Inc., New York, NY, 1998.

Cahn, R.W., P. Haasen, E.J. Kramer, and M. Schutze: *Corrosion and Environmental Degradation Set: A Comprehensive Treatment,* John Wiley & Sons, Inc., New York, NY, 2000.

Davis, J.R.: *Corrosion of Aluminum and Aluminum Alloys,* ASM International, Materials Park, OH, 1999.

Davis, J.R.: *Corrosion: Understanding the Basics,* ASM International, Materials Park, OH, 2000.

Graedel, T.E.: *Atmospheric Corrosion,* John Wiley & Sons, Inc., New York, NY, 2000.

Hou, P.Y., D.A. Shores, M.J. McHallan, et al.: *High Temperature Corrosion and Materials Chemistry,* The Electrochemical Society, Inc., Pennington, NJ, 1998.

Revie, R.W., H.H. Uhlig: *Uhlig's Corrosion Handbook,* John Wiley & Sons, Inc., New York, NY, 2000.

Roberge, P.R.: *Corrosion Engineering Handbook,* The McGraw-Hill Companies, Inc., New York, NY, 1999.

Sastri, V.S.: *Corrosion Inhibitors: Principles and Applications*, John Wiley & Sons, Inc., New York, NY, 1998.

Schweitzer, P.A.: *Encyclopedia of Corrosion Technology,* Marcel Dekker, Inc., New York, NY, 1998.

Schweitzer, P.A.: *Corrosion Resistance Tables: Metals, Nonmetals, Coatings, Mortars, Plastics, Elastomers and Lining, and Tabrics/Part A, B, C,* Marcel Dekker, Inc., New York, NY, 2000.

Stansbury, E.E. and R.A. Buchanan: *Fundamentals of Electrochemical Corrosion,* ASM International, Materials Park, OH, 2000.

Stratmann, M., and G.S. Frankel: *Encyclopedia of Electrochemistry, Corrosion and Oxide Films,* Vol. 4, John Wiley & Sons, Inc., New York, NY, 2003.

Talbot, D. and J.R. Talbot: *Corrosion Science and Technology,* CRC Press, LLC., Boca Raton, FL, 1998.

Van, O.: *Corrosion Control of Metals by Organic Coatings,* CRC Press, LLC., Boca Raton, FL, 1999.

Yeadon, M., R.F.C. Farrow, and S. Chiang: *Recent Developments in Oxide and Metal Epitaxy—Theory and Experiment: Materials Research Society Symposium Proceedings,* Materials Research Society, Warrendale, PA, 2000.

Zarras, P., Y. Wel, and J.D. Stenger-Smith: *Electroactive Polymers for Corrosion Control,* American Chemical Society (ACS), Washington, DC, 2003.

CORROSION EMBRITTLEMENT. The or loss of ductility of metals due to corrosion, usually as a result of intergranular embrittlement attack, which may not readily be visible.

CORROSION INHIBITION. See **Autodeposition**; **Conversion Coatings**; **Paint and Finish Removers**; **Petroleum**.

CORSITE. An orbicular diorite resulting from the segregation, in rounded concentric forms, of ferro-magnesian minerals. It derives its name from its occurrence on the Island of Corsica, and is also sometimes called Napoleonite.

CORTICOSTERONE. See **Hormones**.

CORTISONE. See **Steroids**.

CORUNDUM. The mineral corundum, Al_2O_3, aluminum oxide, occurs as well-developed hexagonal crystals, which may display prismatic, rhombohedral, pyramidal or tabular habits. The larger crystals are often rounded or barrel shaped. Corundum shows both basal and rhombohedral partings; the fracture is conchoidal; hardness, 9; specific gravity, 4.0–4.1; luster, vitreous to adamantine, may be pearly on base; transparent to translucent. Common corundum is gray, grayish-blue or brown, but may be red, yellow or whitish; it is sometimes called adamantine spar. Transparent corundum may be colorless or of various tints. The highly prized ruby is deep red; the sapphire, blue. Transparent yellow corundum is known as oriental topaz; if violet, oriental amethyst; if green, oriental emerald.

Emery is a mixture of granular corundum of dark color, magnetite and hematite, sometimes with spinel. Quartz may be present. For a long time emery was supposed to be an ore of iron. Until the introduction of artificial abrasives, emery was much used for such purposes.

Corundum is found as an accessory mineral in the crystalline rocks such as crystalline limestone and dolomites, gneisses, schists as well as in the igneous rock types granite and syenite. Corundum syenites are found in Canada, especially in the Province of Ontario. Rubies have long been mined in Upper Burma; both rubies and sapphires are found near Bangkok, Thailand. Numerous localities in India furnish gem stones of high quality.

In the United States, common corundum is found in New York, New Jersey, Pennsylvania, Virginia, North Carolina, South Carolina and Georgia; sapphires of gem quality near Helena, Montana, associated with alluvial gold in the Missouri River. From the crystalline limestones and schists of the islands of Naxos and Samos in the Grecian archipelago most of the emery of commerce comes. Other deposits are near Ephesus in the Middle East, and in the town of Chester in Massachusetts. The word corundum comes from the Hindu, kurand; emery is derived from the Greek name for this substance.

See also **Bauxite**.

COTTON *(Gossypium species; Malvaceae).* A natural fiber, cellulosic in composition, with the general formula $(C_6H_{10}O_5)_x$, specific gravity, 1.54; moisture regain 7–8.5% (at 70°F (21.1°C) and 65% relative humidity); and a tensile strength of $60–120 \times 10^3$ psi (same condition of temperature and humidity). Cotton is quite resistant to degradation by heat. After about 5 hours at 121°F (50°C), the material yellows, and above 300°F (149°C), the material decomposes. Cold concentrated acids and hot dilute acids disintegrate cotton. In alkalis, cotton mercerizes without damage. The fiber is quite resistant to most solvents.

Even with the large inroads of synthetic fibers into the textile industry, cotton remains as a major textile fiber, consumption still being expressed in terms of billions of pounds per day. As one of the most versatile of fibers, cotton blends well with other fibers. In addition to the cellulosic content (88–96% depending upon source), cotton contains proteins, pectin, sugar, and wax (about 0.4–0.8%). Much research is going forth to eliminate much of the loss and damage that occur in harvesting and ginning. These factors contribute to price instability.

From the standpoint of use in textiles, cotton is rated excellent for hand (general feel, softness, drape), pilling resistance, and stability to repeated launderings. Cotton is rated good for abrasion resistance, strength, wash and wear performance, and wrinkle resistance. Resistance to sunlight is rated only fair. Pressed-crease retention is rated poor. Safe-ironing temperature for most cotton fabrics is 425°F (219°C).

Important worldwide producers include Brazil, Egypt, India, and the United States.

The uses of cotton are many. First among them is the manufacture of cloth. Cloth may be woven entirely of cotton, as much is, or may be of cotton mixed with other fibers, as synthetics, linen and wool.

Cotton fibers, especially the fuzz, are frequently used to stuff mattresses, pads and upholstered furniture. Treated with chemicals that remove the thin coating of waxy substances that cover the fibers, the latter becomes absorbent cotton, which is capable of absorbing many times its weight of water.

Cotton treated in this way is almost pure cellulose, and so is in great demand by those industries using cellulose. The pure cellulose of the fiber may be dissolved and then precipitated in sheets, giving the familiar thin transparent cellophane. Or the dissolved cellulose may be pressed through fine holes and solidified, giving rayon. If treated with concentrated caustic soda, cotton fibers take on a high degree of luster. The product of this process is called mercerized cotton, after John Mercer, its discoverer.

Treated with nitric acid under various conditions, cotton yields a long series of by-products. Some of them are plastic substances. If highly nitrated, cellulose becomes gun-cotton, used in the manufacture of explosives. Collodion is one of these nitrated products. Many varnishes and lacquers are made from cotton cellulose.

Not all the derivative products of cotton come from the fibers. Some, for example, are obtained from the seeds. In preparing these, the hulls are first removed from the kernel within. These hulls are used as fuel in the ginning mill, as food for cattle, and as fertilizer. The kernels are heated and pressed to remove the oil in cotton. During this pressing the kernels are wrapped in cloth to prevent anything but oil from being expressed. The oil is purified to a soft white substance very similar to lard in appearance. Cottonseed oil is used in making salad oils, oleomargarine and soap. After the oil is expressed, the seed cake may be used as food for stock or as a fertilizer.

See also **Fibers**.

Additional Reading

Bajaj, Y.P.S.: *Cotton,* Springer-Verlag Inc., New York, NY, 1999.

Basra, A.S.: *Cotton Fibers: Developmental Biology, Quality Improvement, and Textile Processing,* Haworth Press, Inc., Binghamton, NY, 2000.

Jenkins, J.N. and S. Saha: *Genetic Improvement of Cotton: Emerging Technologies,* Science Publishers, Inc., Enfield, NH, 2001.

Smith, C.W.: *Cotton: Origin, History, Technology, and Production,* John Wiley & Sons, Inc., New York, NY, 1999.

Vandergriff, A.L.: *Ginning Cotton: An Entrepreneur's Story,* Texas Tech University Press, Lubbock, TX, 1996.

COTTONSEED OIL. See **Vegetable Oils (Edible)**.

COUMARIN. See **Anticoagulants**; **Furan and Related Compounds**.

COULOMETER. Also known as coulombmeter, a device for the measurement of electric current. Originally developed (1916) by the U.S. National Bureau of Standards, the *silver coulometer* consists of a small platinum vessel, acting as the cathode, into which a pure silver anode is immersed. An aqueous solution of silver nitrate (15% $AgNO_3$, wt) of very high purity is used as the electrolyte. In use, both the quantity of silver deposited and the time are carefully noted. These measurements permit a calculation of the average current strength.

The more practical form for laboratory use employs copper electrodes in a bath of copper sulfate. The thin copper cathode, between two heavy copper anodes, is removable for weighing; and since one coulomb deposits 0.000329 gram of copper, the weight of the copper deposit enables one to determine the quantity of electricity in coulombs. A solution recommended for this cell consists of 15 grams of crystalline copper sulfate, 5 grams of pure sulfuric acid, and 5 grams of pure alcohol, dissolved in 100 grams of distilled water.

The effect of current flow on electrolyte concentration also can be determined by titrating the electrolyte after electrolysis. A device of this type is known as a *titration coulometer*. Rarely, a coulometer may be referred to as a *voltmeter*.

COUPLING (Chemical).

Reactions for the formation of chemical compounds usually by establishing a valence bond between a carbon atom and a nitrogen atom. Phenols and several other organic substances are also said "to couple." Polyphenylene oxides, thermoplastic materials, are produced by means of oxidative-coupling technology.

COTTRELL, FREDERICK G. (1877–1948).

A native of California Cottrell obtained his doctorate from Leibig in 1902. His major contribution to industrial chemistry was the discovery of a practical method of dust elimination by electrical precipitation. Used in factory stacks and other large units, this process has contributed greatly to purifying the atmosphere of industrial areas. The principle involves charging a suspended wire with electricity. This creates a field that ionizes the surrounding air, the particles assuming the charge on contact and then moving to the wall of the stack, where they are electrically discharged and precipitated.

COVELLITE.

The mineral covellite, cupric sulfide, CuS, is hexagonal, usually in thin platey crystals, but may be massive. It has a hardness of 1.5–2; specific gravity, 4.6; luster, submetallic to resinous; color, dark indigo blue, sometimes showing a purplish tarnish, or if moistened may appear purple in color. Its streak is dark gray to black; it is opaque. Covellite is found associated with chalcopyrite, bornite, and chalcocite, and is believed to be chiefly of secondary origin. Covellite occurs in Yugoslavia, Saxony, Sardinia, Argentina, Chile, Bolivia and Peru, and in the United States at Butte, Montana, and in Colorado, Wyoming, and Utah. This mineral was named for Covelli, who discovered it in the lavas of Mt. Vesuvius.

CRACKING PROCESS.

A reaction in which a hydrocarbon molecule is broken or fractured into two or more smaller fragments. Sometimes the term *pyrolysis* is used for this reaction. Possibilities for cleavage of a molecule include: (1) a carbon-hydrogen bond; (2) a bond between an inorganic atom and a carbon or hydrogen atom; and (3) a carbon-carbon bond. Usually the objective of cracking is that of reducing the size of hydrocarbon molecules; hence the target is to fracture the carbon-carbon bonds. The main cracking processes are: (1) thermal cracking; (2) fluid catalytic cracking; and (3) hydrocracking.

Thermal Cracking

Of the thermal cracking processes, two are of major importance: (1) coking, and (2) visbreaking (viscosity breaking). Both of these processes convert nondistillable residues into more valuable products. Thermal cracking was the first of the principal cracking processes used in the petroleum industry. For increasing gasoline production and improving quality, fluid catalytic cracking has essentially replaced thermal cracking.

In *thermal coking*, heavy residual stocks are converted into gas, gasoline, distillates, and coke. Generally, the objective is that of maximizing the yield of distillates; and minimizing the production of gas, gasoline, and coke. Light distillates are used for both domestic and industrial heating oils. There are two types of thermal coking processes: (1) cyclic, semicontinuous process, sometimes referred to as *delayed coking*, decarbonizing, or low-pressure coking; and (2) a continuous fluid coking process. About 70% of the installed capacity in the United States is the delayed coking type.

As shown by Fig. 1, a delayed coking unit is comprised of three sections-a furnace, coke drums, a fractionating unit, plus coke removal and handling equipment. Usually the feedstock is charged to the lower part of the fractionator. Here the feedstock is contacted by hot vapors from the coke drum, causing any light components to be flashed from the feed before the feed joins with the recycle and charged (from the bottom of the fractionator) to the furnace. The charge in the furnace is heated to about 480°C (896°F). The heated effluent from the furnace is introduced into the bottom of one

of two or more insulated vessels (coke drums) where, as the result of its contained heat, the material cracks to form a solid coke residue. At the same time, lighter cracked products are evolved and proceed from the top of the coke drums to the fractionator. The reaction in the coke drum is endothermic and thus the temperature of the material drops to about 425°C (797°F). The cracked products leave as vapors; the coke remains in the drum. The fractionator separates the cracked vapors into several side streams as shown on the diagram. When accumulated coke reaches a certain level in the drum, that drum is temporarily taken off-stream and the flow is switched to the second drum. Prior to removal of coke, the drum is steamed to remove vapors. Water is also added to cool the coke.

The fluid coking process accomplishes the coking operation in a continuous manner. Feed is sprayed into a fluid bed of hot coke in a coking reactor. Steam introduced into the bottom of the reactor provides the fluidization energy. The cracked products are quenched in an overhead scrubber and then go to the fractionator. The coke is deposited on the particles in the reactor, which commute with a heater vessel in which a portion of the coke is burned to heat up the returning coke particles to supply the energy for the coking reaction.

Fluid Catalytic Cracking

This process is used principally to create gasoline, C_3/C_4 olefins, and light distillates by the selective decomposition of heavy distillates. The process was introduced during World War II to replace earlier thermal cracking processes. Specially prepared catalysts are used. The resulting gasoline contains substantial proportions of high-octane-number hydrocarbon components, including aromatics, branched paraffins, and olefins. The cracking proceeds in accordance with the carbonium-ion mechanisms. Consequently, there are minor amounts of fragments lighter than C_3 in the products. This is to be contrasted with thermal cracking by the free-radical mechanism, wherein large proportions of fragments lighter than C_3 result. Another product of fluid catalytic cracking is *cycle oil*, a distillate that boils above gasoline. Cycle oils are withdrawn as net products and are used as components in heating oils, feedstocks to hydrocracking units, and for blending with heavy residuals as a means of reducing viscosity. The highly aromatic clarified slurry oils have been found to be useful feeds for the manufacture of carbon black.

Indicated in Fig. 2 is a representative fluid catalytic cracking unit, comprising: (1) a reactor; (2) a regenerator; (3) the main fractionator; (4) an air blower or compressor; (5) a spent-catalyst stripper; (6) catalyst recovery equipment, including cyclones internal in the reactor and regenerator; and slurry settler, and possibly an electrostatic precipitator; and (7) a gas-recovery unit. The catalyst used is essentially a specially prepared composite of silica and alumina.

In operation, preheated feedstock meets a controlled stream of hot, regenerated catalyst. Vaporized oil and catalyst ascend in the riser, such that the catalyst particles are suspended in a dilute phase. Essentially all of the cracking occurs in the riser. The catalyst particles are separated from the cracked vapors at the end of the riser and the catalyst containing a coke deposit is returned to the regenerator. The cracked vapors pass through one or more cyclones located in the upper portion of the reactor and proceed to the fractionator (main column) that produces the side streams indicated.

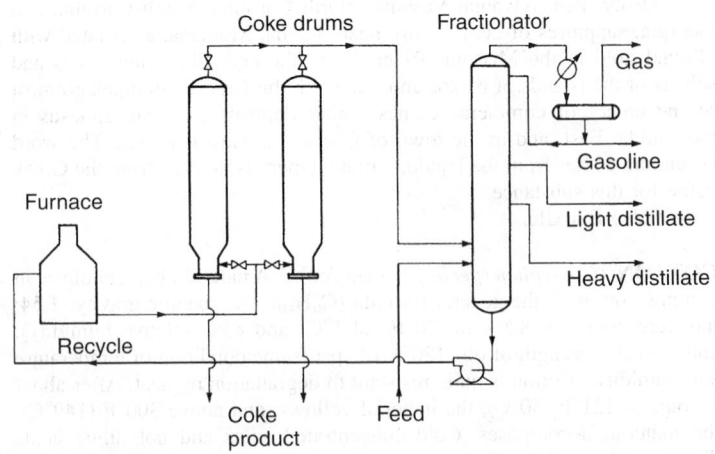

Fig. 1. Delayed coking unit shown schematically

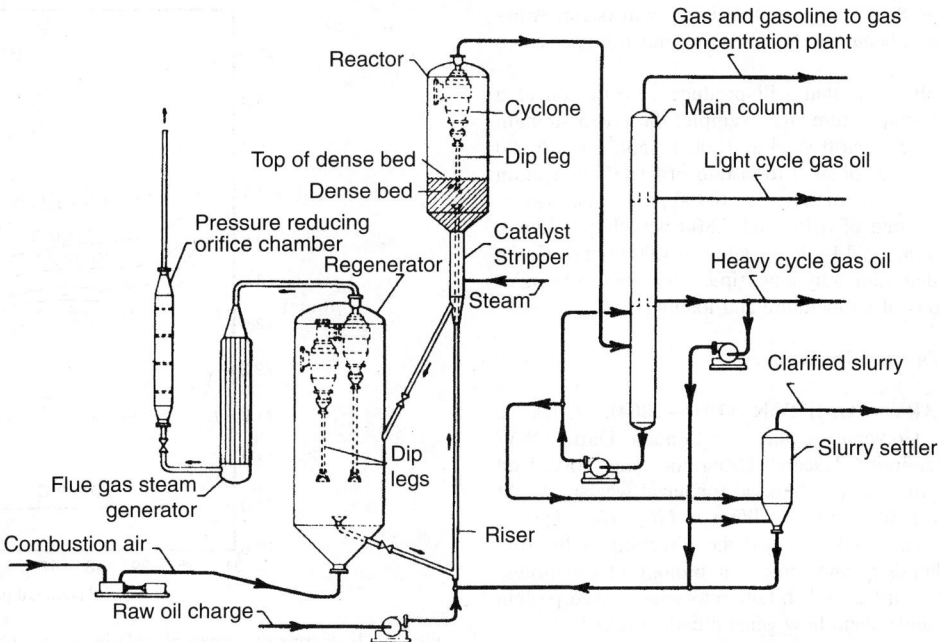

Fig. 2.   Fluid catalytic cracking process. (*UOP Inc*)

Hydrocracking

Processes in this category produce gasoline and light distillates from feed distillates that are higher-boiling than the products. Hydrocracked products are not olefinic. The light gaseous hydrocarbons produced by hydrocracking are entirely paraffinic. The processes operate at elevated pressures in the presence of hydrogen and catalysts. Temperatures are usually lower than 482°C (900°F). Pressures run from 800 to 2,500 psig. Both fixed-bed and ebullating-bed configurations are used. Because carbonaceous deposits accumulate very slowly on the catalyst, the on-line periods for these units are quite long, ranging from several months to over a year. Somewhat more costly to build than fluid catalytic cracking, the hydrocracking process has the advantage that it can handle heavier and dirtier feedstocks and also may be adapted to varying product ratios of gasoline to middle distillate.

Technical Staff, UOP Inc.
Des Plaines, Illinois

Editor's Addendum

For any given petroleum processing design, there are competing designs, including cracking processes. Each particular design will have its own specific advantages and limitations (products yielded, product specifications, operating conditions, economic factors, etc.). Once developed by a design group, a given process generally will be licensed to end users or refinery construction and engineering firms, as negotiated.

Additional Reading

Anthony, R. and M.L. Occelli: *Hydrotreating Catalysts—Preparation, Characterization and Performance,* Elsevier Science, New York, NY, 1989.

Douglas, J. *Conceptual Design of Chemical Processes,* The McGraw-Hill Companies, Inc., New York, NY, 1990.

Kerridge, A.E.: *Refining Handbook '90,* part of Hydrocarbon Processing, **83** (November 1990).

Meyers, R.A.: *Handbook of Petroleum Refining Processes,* 3rd Edition, The McGraw-Hill Companies, Inc., New York, NY, 2003.

Occelli, M.L.: *Fluid Catalytic Cracking II: Concepts in Catalyst Design,* American Chemical Society, Washington, DC, 1991.

Occelli, M.L. and J. O'Connor: *Fluid Cracking Catalysis,* Marcel Dekker, Inc., New York, NY, 1998.

Sadeghbeigi, R.: *Fluid Catalytic Cracking Handbook,* Gulf Publishing Company, Houston, TX, 2000.

Scherzer, A.L. and A.J. Gruia: *Hydrocracking Science and Technology,* Marcel Dekker, Inc., New York, NY, 1996.

Springenschmid, R.: *Prevention of Thermal Cracking in Concrete at Early Ages,* Routledge, New York, NY, 1999.

CRAM, DONALD JAMES (1919–2001). Awarded the Nobel prize for chemistry, together with Jean-Marie Lehn, and Charles J. Pedersen, in 1987 for work in elucidating mechanisms of molecular recognition, which are fundamental to enzymic catalysis, regulation, and transport. He also studied three-dimensional cyclic compounds that maintained a rigid structure, accepting substrates in a structurally preorganized cavity. Cram called these compounds cavitands, while Lehn named them cryptands. Cram received his doctorate from Harvard University in 1947. See also **Cavitand.**

CREEP (Metals).   This term usually is associated with the slow, plastic deformation of metals under constant load. Continued plastic deformation, where the applied force does not change, can result from two basic causes. (1) If a metal is deformed, as in tension, its cross section is correspondingly reduced. This raises the stress level in the material and, if the rate of this increase in stress exceeds the rate of strain hardening, creep occurs. (2) Plastic flow may be promoted by thermally activated softening processes occurring in the metal that counteract the mechanisms leading to strain hardening. Thus, thermal energy may aid dislocations in cutting through one another or in passing around inclusions. Thermal energy may also provide the means for dislocations of opposite sign to move toward each other so that they can recombine. This may eliminate dislocations entrapped in each other's force field, thereby allowing additional dislocations to move out from the sources and plastic deformation to continue.

Creep of metals at high temperatures is primarily controlled by thermally activated processes. Since there are many conceivable mechanisms, creep cannot be associated with a single activation energy. However, in a given temperature range, the creep rate may be controlled by a particular mechanism and the temperature dependence of the creep rate in this range will be related to a corresponding activation energy. Thus, several activation energies have been observed for creep in aluminum, each of which is controlling in a different temperature range. In general, the activation energy is larger the higher the temperature range in which it controls. The activation energy for creep of metals at very high temperatures (just below the melting point) often tends to be the same as that for self-diffusion. This implies that vacancy motion or dislocation climb is very important in creep at extremely high temperatures.

From a practical point of view, creep becomes an important natural phenomenon when the temperature at which a metal is loaded lies above about 0.4 to 0.5 of its melting point on an absolute scale. In some metals such as zirconium, which undergo a solid state phase change, creep becomes an important effect above about one-half of the temperature of the phase transformation. In many metals such as steel, creep is almost nonexistent at room temperatures if the metal is not loaded above its annealed yield strength. However, at 900°F (482°C) steel can creep readily at very small stresses, and equipment such as boilers and tubes for petroleum cracking stills, intended to operate at high temperatures for long periods of time, must be designed on the assumption that creep will occur. Alloy steels and other materials have been developed having much higher creep strengths than carbon steel.

Creep strength is the unit stress that will produce deformation at a specified rate at a specified temperature; for example, the creep strength of a certain 0.15% carbon open-hearth steel at 1000°F (538°C) is 6,100 pounds per square inch (415 atmospheres) for a rate of 0.01% elongation in 1,000 hours. Other values for this material are 6,900 pounds per square inch (469 atmospheres) for a rate of 0.1% and 7,800 pounds per square inch (531 atmospheres) for a rate of 1.0% elongation in 1,000 hours. Creep strength values can only be determined by long- time laboratory tests under carefully controlled conditions of temperature and loading.

CREOSOTE.   See **Coal Tar and Derivatives**.

CRICK, FRANCIS HARRY COMPTON (1916–2004).

Crick, a British physicist, studied at University College in London. During WW II, he joined the British Admiralty Research Laboratory and worked on developing radar and magnetic mines to be used for naval warfare. Right after the war, he read Erwin Schrodinger's *What is Life? The Physical Aspects of the Living Cell*. The book changed the direction of his life. Crick began studying biochemistry and molecular biology at Cambridge University's laboratory. Later, at Cavendish Laboratory he studied protein structures. Crick wanted to understand how genetic code worked.

In 1951, James Watson, an American biologist, joined Crick's research. Together they pioneered a new era in science when they proposed that the molecular structure of DNA was a double spiral helical chain in a paper published in the scientific journal, *Nature* in 1953. This astonishing work was done by Watson, just 24 years old, and Crick who had not yet even finished his PhD. In 1962, Crick and Watson shared the Nobel Prize in physiology/medicine for contributions in discovering the molecular structure of DNA. Crick continued his studies with DNA as Director of Cambridge University's Molecular Biology Laboratory.

In later years, Dr. Crick became a research professor at the Salk Institute for Biological Studies in California and at age 84, was still researching. He has written several books on the implications of the new findings in biochemistry.

See also **Genetics and Gene Science (Classical)**.

J.M.I.

CRITICAL COMPOSITION.

Systems consisting of two liquid layers that are formed by the equilibrium between two partly miscible liquids, frequently have a consolute temperature or a critical solution temperature, beyond which the two liquids are miscible in all proportions. At this temperature, the phase boundary disappears, and the two liquid layers merge into one. The composition of the mixture at that point is called the critical composition. There is, in some cases, a lower consolute temperature as well as an upper consolute temperature.

CRITICAL CONCENTRATION.

When two immiscible liquids are heated in contact with each other their mutual solubility is usually increased until, at the critical solution temperature, they become consolute. The composition of the two solutions immediately before they become consolute is termed the critical concentration. See also **Critical Solution Temperature**.

CRITICAL DENSITY.

The density of a substance which is at its critical temperature and critical pressure.

CRITICAL MASS.

The amount of concentrated fissionable material that can just support a self-sustaining fission reaction. See also **Nuclear Power Technology**.

CRITICAL POINT.

(1) A point where two phases, which are continually approximating each other, become identical and form but one phase. With a liquid in equilibrium with its vapor, the critical point is such a combination of temperature and pressure that the specific volumes of the liquid and its vapor are identical, and there is no distinction between the two states. (2) The critical solution point is such a combination of temperature and pressure that two otherwise partially miscible liquids become consolute.

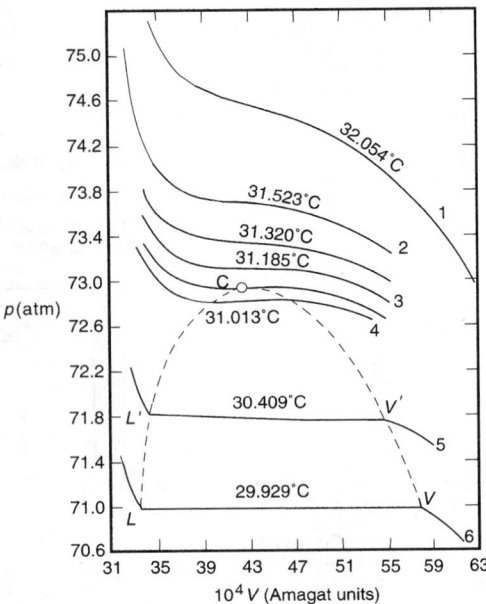

Fig. 1.   Isotherms of carbon dioxide in the neighborhood of the critical point

To consider in detail the critical point as defined in (1), examine Fig. 1, which shows the family of isotherms of a pure substance in the fluid range (liquid or gas) such for example as shown in the figure for carbon dioxide.

At sufficiently high temperatures each isotherm is a continuous curve, but at low temperatures the isotherm consists of three portions. The first section of the curve at high pressures corresponds to the liquid state, while that at low pressures refers to the gaseous state. These two curves are joined by a horizontal line, corresponding to the simultaneous presence of two phases, liquid and gas.

The isotherm between those numbered 3 and 4 in the figure represents the transition between isotherms corresponding to the gas phase only, and those including a horizontal portion corresponding to a liquid-gas equilibrium. In this isotherm the horizontal line has contracted to a single point of inflection C. This is the critical point characterized by the relations

$$\left(\frac{\partial p}{\partial V}\right)_{T_c} = 0; \left(\frac{\partial^2 p}{\partial V^2}\right)_{T_c} = 0; \left(\frac{\partial^3 p}{\partial V^3}\right)_{T_c} < 0.$$

The curve $LL'C$ gives the molar volume of the liquid. Similarly $VV'C$ gives the molar volume of the gas.

At the critical point the molar volumes of the liquid and of the gas become equal. In general a critical state is characterized by the fact that the two coexistent phases (here the liquid and the vapor) are identical.

The curve $VV'CL'L$ is called the *saturation curve*.

The experimental data do not indicate the existence of a critical point for the liquid-solid transition.

Above the critical point the substance can no longer exist in the liquid state. The critical temperature is thus the highest temperature at which the liquid and vapor can coexist.

Ternary Critical Point. The point where, upon adding a mutual solvent to two partially miscible liquids (as adding alcohol to ether and water), the two solutions become consolute and one phase results.

CRITICAL SOLUTION TEMPERATURE.

For two partially miscible liquids, the compositions of the two conjugate solutions approach each other with increasing temperature. At the critical solution temperature the two solutions have identical compositions and form one layer.

CRITICAL TEMPERATURE.

1. This term is most commonly used to denote the maximum temperature at which a gas (or vapor) may be liquefied by application of pressure alone. Above this temperature the substance exists only as a gas. 2. The critical temperature of a superconducting transition takes place in zero magnetic field.

CRITICAL VOLUME.

The volume occupied by unit mass, commonly one mole, of a substance at its critical temperature and critical pressure.

CROCIDOLITE (Blue Asbestos). The mineral crocidolite may be considered as a fibrous variety of the monoclinic amphibole, riebeckite. It is also known as a massive mineral. Its hardness is 4; specific gravity, 3.2–3.3; luster, silky to dull; color, blue or bluish-green. It is found in Austria, France, Bolivia, the Republic of South Africa (the variety known as tiger's-eye); and in the United States, in Massachusetts and Rhode Island. The name crocidolite is derived from the Greek, meaning to weave, in reference to its fibrous appearance. See also **Cat's-Eye.**

CROCOITE. The mineral crocoite, lead chromate, corresponds to the formula $PbCrO_4$, and forms prismatic monoclinic crystals, often acicular. It is also found in columnar or granular masses. It has a rather distinct cleavage parallel to the prism, and a less distinct cleavage parallel to the base. It has a conchoidal fracture, is sectile; hardness, 2.5–3; specific gravity, 5.9–6.1; luster, adamantine to vitreous; color, red; streak, orange-yellow, translucent. Crocoite is a secondary mineral believed to be formed by waters containing chromic acid acting upon lead minerals like galena, with which it is associated. It is found in the former U.S.S.R., Rumania, Tasmania, Brazil, the Philippines, and Arizona. It is not of commercial importance. The name crocoite is derived from the Greek word for saffron in reference to the color of the powdered mineral.

CROSS LINKING. See **Rubber (Natural).**

CRUCIBLE. A crucible is a vessel made of heat-resistant material and employed to hold a material that is at high temperature, or is to be subjected to high temperature. Crucibles will range in size from the small laboratory types to large ones having a capacity of several tons of molten metal. They are roughly cup- or barrel-shaped, and are made of some material such as clay. In the laboratory, platinum, iron, and porcelain crucibles are used. In the iron and steel industry, clay crucibles have been used, but at the present time most manufacturers employ the graphite crucible, which is made half and half from graphite and fireclay, well-mixed, molded, and burned to vitrification. The crucible may be used to hold a material being melted or burned, as in some processes for making steel, where the raw material is put in the crucible, which is then set in a hot furnace until the contents are melted. Or, a crucible may be used to receive molten metal, which has been produced elsewhere, as from a brass furnace or cupola, in which case it is the means for conveying it from the point of melting to the point of casting, serving thus as the intermediate reservoir between the furnace and the mold. See also **Casting.**

CRYOCHEMISTRY. That branch of chemistry devoted to the study of reactions occurring at extremely low temperatures ($-200°C$ and lower). It permits synthesis of compounds that are too unstable or too reactive to exist at normal temperatures.

CRYOGENIC MATERIALS. Those metals and alloys usable in structures operating at very low temperature, and usually possess improved strength properties at these temperatures. See also **Cryogenics.**

CRYOGENICS. The production and study of phenomena that occur at very low temperatures, i.e., below about 80 K. The first step in attaining the required temperature generally involves the liquefaction of a gas or gases. Liquids can exist over a range of temperatures limited by the critical point at the higher end and the triple point at the low-temperature end. It is thus possible to compress a gas to the liquid phase at the critical point and to cool it by boiling under reduced pressure to its triple point. A series of gases having their critical and triple points overlapping can thus be used in a cascade process each being used as the refrigerant for the next in the series. Pictet used this method to liquefy oxygen, using methyl chloride and ethylene as refrigerants. There are, however, no liquids that cover the range from 77 K to the critical point of hydrogen, or from 14 K to the critical point of helium (5.2 K). Thus, liquid hydrogen and helium cannot be produced by the cascade method.

Expansion of Gases

A gas may also be cooled by making it do work in the course of an expansion. When an ideal gas is expanded through an aperture into a constant volume, no work is done, since there are no interactions between the molecules and the molecules themselves occupy no volume. When a nonideal gas is so expanded, however, an amount of internal work. ($W = $

$(PV)_{final} - (PV)_{initial}$) is done against the intermolecular forces. This work may be positive or negative, resulting in a cooling or heating of the gas. Air is cooled by this Joule-Thomson expansion at room temperature, but hydrogen and helium must be precooled to 90 and 15 K, respectively to obtain further cooling upon expansion. Using this method, Kamerlingh Onnes first succeeded in liquefying helium in 1908. Compressed gases may also be made to do external work, for example, by expansion against a movable piston. In this case, the work is always positive and helium may be cooled and liquefied without any precooling by liquid hydrogen. See also **Ammonia**; and **Helium.**

With liquid helium readily available in the laboratory, research in the temperature range 5 to 0.8 K has become commonplace. By using the isotope of helium ^3He, it is possible to attain temperatures down to about 0.3 K since the isotope has a lower boiling point than ^4He. This is about the lowest temperature practically attainable by boiling liquids at reduced pressure. To reach lower temperatures, it is necessary to use magnetic phenomena.

Magnetic Cooling

Although this concept dates back several decades, as of the early to mid 1993, this was one of the most seriously researched areas for designing supercoolers.

Debye and Giauque pointed out that at 1 K the entropy of paramagnetic salts was still fairly large and, moreover, that it was almost all due to nonalignment of magnetic moments and that the entropy of lattice vibrations was very small. If the electron spins are aligned by application of a magnetic field, the entropy of the salt falls to a low value and the heat of magnetization can be extracted isothermally. The salt can then be thermally isolated and demagnetized adiabatically and its temperature will fall. Temperatures on the order of 0.01 K can readily be reached by this method and the lower limit for a single demagnetization would appear to be about 10^{-3} K. When demagnetization occurs, the spin system reaches equilibrium temperature in about 10^{-10} seconds. It is found that equilibrium is achieved between the spin temperature and the lattice temperature by spin-orbit coupling in times on the order of a few seconds. Paramagnetic salts have relatively high specific heats at low temperature and hence the cold salt can be used to cool other bodies. However, making good thermal contact with the cooled salt can be difficult.

Kurti, Simon, and Gorter suggested that a further reduction in temperature could be attained if adiabatic demagnetization was performed on nuclear moments rather than the electron spin. The temperature which can be reached in an adiabatic demagnetization is determined by the point at which the entropy of the system in zero external field decreases sharply with decrease in temperature due to the alignment of magnetic moments, i.e., the point at which the interaction energy μh equals kT ($h = $ internal field; $\mu = $ magnetic moment). Since the interaction energies of nuclear moments are much smaller than electron spin interactions, much lower temperatures should result. The materials used experimentally were metals cooled to 10^{-2} K by contact with a paramagnetic salt. The thermal isolation of the nuclear spins during demagnetization was achieved naturally by the nuclear-spin-conduction electron relaxation time (~ 100 seconds). Thus, while the nuclear spins cooled to between 10^{-5} and 10^{-6} K, the conduction electrons and lattice remained in thermal contact with the cooling salt at 10^{-2} K.

Magnetic Refrigerators. Magnetic cryogenic refrigerators are presently considered indispensable to the operation of future space-based defense systems, increasing their efficiency, reliability, and life span. These coolers are designed to cool space-based infrared sensors and signal processors to temperatures lower than now practicable with conventional closed, gas-cycle coolers. Magnetic refrigerators employ what is known as the *magneto-caloric* effect, which cools by alternately magnetizing and demagnetizing a solid material. For example, a magnetic solid, such as gadolinium gallium garnet, can be continuously rotated through a strong magnetic field to produce temperatures even lower than gas-cycle coolers. Other possible applications for this new technology include space astronomical observators, space power generation, high-speed rail transportation, medicine, and the commercial gas liquefaction industry.

Another method of attaining temperatures below 1 K is to take advantage of the fact that the entropy of a superconducting metal is less than that of the metal in its normal state. Quenching of a superconductor by the application of a magnetic field can cause a cooling to about 0.1 K. However, since the specific heat of metals is very small at these temperatures, they are not very suitable for cooling other bodies.

Low-Temperature Measurements

Such measurements are usually carried out using secondary thermometers calibrated at certain fixed points previously determined on the absolute scale by a standard instrument. This instrument is generally a constant-volume gas thermometer used at low pressures. When the readings of this instrument are extrapolated to zero pressure, the scale coincides with the thermodynamic scale. In the range of 0.8 to 5.2 K, the vapor pressure of ^{4}He provides the most commonly used secondary scale and it agrees with the absolute scale to within 2 millidegrees over this range. The use of ^{3}He instead of ^{4}He increases the coverage to 0.3 K.

Resistance thermometers are useful over a wide range. For example, platinum is used from 273 to 15 K, and carbon covers the range 20 to 2 K and has the advantage of being quite insensitive to magnetic fields. The foregoing are examples of secondary standards where the scales are interpolated between fixed points.

For temperatures below 0.3 K, the susceptibility of a paramagnetic salt can be measured and the temperature calculated by extrapolation from Curie's Law $\chi = C/T$. This method gives true values for the temperatures so long as Curie's Law holds. Beyond this region, it is necessary to perform a thermodynamic cycle to determine the relationship between the magnetic temperature T^* and the absolute temperature T. The method of Kurti and Simon is to demagnetize adiabatically from a known temperature on the absolute scale, using a number of different field intensities, and hence to determine the relationship between T^* and the entropy S over the required range of temperature. Measurement of the amount of heat Q necessary to raise the temperature from T_1^* to T_2^* gives the absolute value of the average temperature $T_{1,2}$ from the relationship $\Delta Q = T \Delta S$. The heat is generally supplied to the salt by gamma rays, thus ensuring even heating of the sample, a necessary precaution since the thermal conductivity is poor. The problem of nonlinearity does not arise in the case of nuclear spin demagnetization since, in this case, the susceptibility obeys Curie's Law down to 10^{-7} K.

Cryogenic Phenomena

One of the most interesting phenomena of cryogenics is that of superconductivity, which was also discovered by Onnes. When metals are cooled from room temperature, their resistances decrease and at low temperature, they attain low values, which are fairly independent of temperature. Some metals, however, have a critical temperature below which their resistance goes to zero. Such a metal is known as a superconductor.

Superfluidity

As liquid helium is cooled below 2.18 K, it undergoes a sudden discontinuity of specific heat and a second-order transition to the superfluid state. In this state, the viscosity of the helium becomes a function of the method used to measure it. Measured by an oscillating disk method, the viscosity falls from 23×10^{-6} poise just above the transition to 1×10^{-6} poise at 1.3 K. Measured by passage through very fine capillaries, the viscosity is very nearly zero in the superfluid state. Hence, it is postulated that there are two coexisting, noninteracting fluids, one having the properties of non-"superfluid" helium; and the other having virtually zero viscosity and zero entropy. It is interesting to note that the thermal conductivity of superfluid helium is about 2,000 times greater than that of copper. This is the result of the motion of the entropy-free superfluid rather than normal thermal conductivity. See also **Superfluidity**.

Magnetic Properties

There are several fundamental experiments on the magnetic properties of materials that become possible as a result of the low-temperature environment of cryogenics. The first of these was discovered by deHaas and Van Alphen in 1930. They found that at low temperatures, the susceptibility of bismuth single crystals rose and fell periodically as the magnetic field was increased. Later work shows that the periodicity occurs in all metals at low temperatures and is the result of quantization of electron motion perpendicularly to the applied field. This effect was used to determine the Fermi surface of metals.

Radioactive Decay

The method of alignment of nuclear moments has been used to study radioactive decay as a function of nuclear orientation. The aligning field in this case can be either an externally applied magnetic field or an internal crystal field. One of the more striking of these experiments has been the test of the Lee-Yang theory of nonconservation of parity in weak interactions. A third fundamental experiment of interest was the confirmation of London's concept that flux through a superconducting ring is quantized. The ring was a lead tube of 10^{-3} centimeter diameter and it was suspended from a torsion balance. The tube was made superconducting in the presence of longitudinal magnetic fields, and the frozen-in flux was measured and found to be quantized.

Several practical applications of cryogenics technology are described elsewhere in this encyclopedia. Check Alphabetical Index.

Additional Reading

Andrei, E.Y.: *Two-Dimensional Electron Systems: On Helium and Other Cryogenic Substrates,* Kluwer Academic Publishers, Norwell, MA, 1997.

Balachandran, B.B., R.P. Reed, D. Bubser, K.T. Hartwig, and D.G. Gubser: *Advances in Cryogenic Engineering (Materials),* Vol. 44, Kluwer Academic Publishers, Norwell, MA, 1998.

Barron, R.F.: *Cryogenic Systems,* Oxford University Press, Inc., New York, NY, 1992.

Barron, R.F.: *Cryogenic Heat Transfer,* Hemisphere Publishing Corporation, New York, NY, 1999.

Beardsley, T.M.: "Cold Storage (Human Organs)," *Sci. Amer.,* **20** (February 1990).

Brownell, D.L. and E.C. Guyer: *Handbook of Applied Thermal Design,* Taylor & Francis, Inc., New York, NY, 1999.

Bryson, W.E.: *Cryogenics,* Gardner Publications Inc., Cincinnati, OH, 1999.

Burt, B., J. Hull, P. Kittel, et al.: *Advances in Cryogenic Engineering,* Vol. 45, Kluwer Academic Publishers, Norwell, MA, 2000.

Edeskuty, F.J. and W.F. Stewart: *Safety in the Handling of Cryogenic Fluids,* Perseus Publishing, Boulder, CO, 1996.

Flynn, T.M.: *Cryogenic Engineering,* Marcel Dekker, Inc., New York, NY, 1996.

Giauque, W.F.: "Low Temperature, Chemical, and Magneto Thermodynamics: The Scientific Papers of William F. Giaque," *Giauque Sci. Papers F,* 1969.

Guobang, C., G. Bang, and F.W. Steimle: *Crogenics and Refrigeration,* World Scientific Publishing Company, Inc., Riveredge, NJ, 1998.

Hung, Yen-Con: "Prediction of Cooling and Freezing Times," *Food Technology,* **137** (May 1990).

Jacobsen, R.T., S.G. Penoncello, and E.W. Lemmon: *Thermodynamic Properties of Cryogenic Fluids,* Perseus Publishing, Boulder, CO, 1997.

Kakac, S., H.F. Smirnov, and M.R. Avelino: *Low Temperature and Cryogenic Refrigeration,* Kluwer Academic Publishers, Norwell, MA, 2003.

Katz, D.L. and R.L. Lee: *Natural Gas Engineering: Production and Storage,* The McGraw-Hill Companies, Inc., New York, NY, 1990.

Kurti, J.F. et al.: *Low Temperature Physics,* Pergamon Press, London, UK, 1952. (A Classic Reference.)

Poling, B.E., J.M. Prausnitz, and J.P. O'Connell: *The Properties of Gases and Liquids,* 5th Edition, The McGraw-Hill Companies, Inc., New York, NY, 2000.

Reid, R.C. and J.M. Prausnitz: *Properties of Gases and Liquids,* The McGraw-Hill Companies, Inc., New York, NY, 1987.

Staff: *Gas Processing Handbook,* Gulf Publishing Co., Houston, TX, April 1990.

Walker, G. and E.R. Bingham: *Low-Capacity Cryogenic Refrigeration,* Oxford University Press, Inc., New York, NY, 1994.

Weisend, J.G.: *Handbook of Cryogenic Engineering,* Taylor & Francis, Inc., New York, NY, 1998.

CRYOHYDRATE. An eutectic system consisting of a salt and water, having a concentration at which complete fusion or solidification occurs at a definite temperature (eutectic temperature) as if only one substance were present.

CRYOHYDRIC POINT. The eutectic point in cases in which the system contains water.

CRYOLITE. Cryolite, sodium aluminum fluoride, Na_3AlF_6, crystallizes in the monoclinic system but in forms that closely approach cubes and isometric octahedrons. It is usually found massive. Cryolite has an uneven fracture, is brittle; hardness, 2.5; specific gravity, 2.97; luster, vitreous to greasy; color, snow-white but may be colorless, reddish or brownish; translucent to transparent. The only considerable occurrence of cryolite is at Ivigtut, Greenland, where veins of this mineral are associated with granites and gneisses. Small occurrences of cryolite have been noted in the Ilmen Mountains. Russia, and at Pikes Peak, Colorado. Cryolite has its chief use in the electrolytic production of aluminum, but small amounts are employed in the manufacture of opalescent glass.

The name cryolite is derived from the Greek words meaning frost (ice) and stone in reference to its translucency.

The aluminum industry no longer uses the natural mineral for operation of electrolytic cells except infrequently for start-up operations. The synthetic cryolite used is produced by two main processes: (1) the sodium aluminate from the Bayer process (see also **Bauxite**) is reacted with hydrofluoric acid: $NaAlO_2 + 2NaOH + 6HF \longrightarrow 3NaF \cdot AlF_3 + 4H_2O$; or (2) sodium carbonate may be used instead of NaOH:

$$NaAlO_2 + Na_2CO_3 + 6HF \longrightarrow 3NaF \cdot AlF_3 + CO_2 + 3H_2O$$

The quality hydrofluoric acid required for these reactions may be prepared from fluorspar: $CaF_2 + H_2SO_4 \longrightarrow 2HF + CaSO_4$. Normally, cryolite is not considered toxic. A continued exposure to finely divided cryolite in the air may lead to *fluorosis*. However, the compound is toxic to insects.

S. J. SANSONETTI
Consultant, Metals Co. (ALCOA)
Richmond, Virginia

CRYOPUMP. (1) An exposed surface refrigerated to cryogenic temperature for the purpose of pumping gases in a vacuum chamber by condensing the gas and maintaining the condensate at a temperature such that the equilibrium vapor pressure is equal to or less than the desired ultimate pressure in the chamber.

(2) The act of removing gases from an enclosure by condensing the gases on surfaces at cryogenic temperature. Also referred to as a *cryogenic pump* and not to be confused with *cryogenic fluid pump* for *circulating cryogenic propellants.*

CRYOSCOPE. An instrument for measuring the freezing or solidification point. The Hortvet cryoscope is used for the estimation of added water in milk from the lowering of the freezing point.

CRYOSCOPIC CONSTANT. A quantity calculated to represent the molal depression of the freezing point of a solution, by the relationship

$$K = \frac{RT_0^2}{1000l_f}$$

in which K is the cryoscopic constant, R is the gas constant, T_0 is the freezing point of the pure solvent, and l_f is the latent heat of fusion per gram. The product of the cryoscopic constant and the molality of the solution gives the actual depression of the freezing point for the range of values for which this relationship applies. Unfortunately, this range is limited to very dilute solutions, usually up to molalities of $\frac{1}{100}$ and must be modified for many solutes. See also **Freezing-Point Depression**.

CRYOTRON. A device based upon the principle that superconductivity established at temperatures near absolute zero is destroyed by the application of a magnetic field.

CRYPTATE COMPLEXES. See **Chemical Elements**.

CRYPTOCRYSTALLINE. When the texture of a rock is so finely crystalline (that is, made up of such minute crystals) that its crystalline nature is but vaguely revealed even in a thin section by transmitted polarized light, the rock is said to be cryptocrystalline. Among the sedimentary rocks, chert and flint are cryptocrystalline. Lava flows, especially of the acidic type such as felsites and rhyolites, may have a cryptocrystalline ground mass as distinguished from pure obsidian (acidic), or tachylite (basic), which are natural rock glasses.

CRYPTOCRYSTALLINITY. See **Mineralogy**.

CRYSTAL. A macroscopic sample of a solid substance exhibiting some degree of geometrical regularity, or symmetry, or capable of showing these properties after suitable treatment (e.g., cleavage, etching, etc.). Almost all pure elements and compounds are capable of forming crystals.

A perfect crystal is one in which the crystal structure would be that of an ideal space lattice. No such crystals exist, all real crystals containing imperfections which have a strong influence on the physical properties of the crystal.

The mechanical and electronic properties of polycrystalline materials depend dramatically upon the boundaries between neighboring regions of different crystallographic orientation-that is the grain boundaries-and it is to these regions that minor alloying and impurity elements segregate. Their presence may be ascertained by high resolution transmission electron microscopy (HRTEM) and computer simulation.

Structure of Crystals

Early investigators suggested that the regular structure of crystals, embodied in the laws of crystallography, could be explained if they were thought of as built up by the repetition of equal polyhedral cells, fitting together to fill space, each cell representing a characteristic group of particles, perhaps the atoms and molecules of the compound. A rough calculation showed that the spacing of these units in many ionic crystals might be of the same order of magnitude as the wavelengths of x-rays, as deduced from quantum theory. Von Laue suggested, and verified, that diffraction of the x-rays occurs when they are passed through a suitably oriented crystal. He knew, from the density and atomic weights, that the number of atoms in a cubic centimeter of rock salt, for example, is about 4.488×10^{22}, and that therefore, if they are equally spaced in all three directions, their distance apart is 2.814×10^{-8} centimeters or 2.814 Å. Certain quantum theory calculations had already indicated that x-rays have wavelengths of this order. It occurred to von Laue that if a beam of x-rays were directed upon a crystal and the crystal turned into a suitable position, one might observe interference maxima analogous to those produced with light by a diffraction grating. This proved to be the case, and the result verified beyond question the existence of regular spacings between reflecting planes of some sort, presumably plane arrays of atoms or ions. The kind of pattern obtainable is demonstrated by Fig. 1.

Subsequent analysis of the problem by Bragg resulted in a formula analogous to that for interference of light reflected by thin plates. If the reflecting layers are spaced at equal distances d, and if the wavelength of the incident x-rays is λ, the angle θ between rays and layers necessary for an interference reflection maximum is given by Bragg's law, viz.,

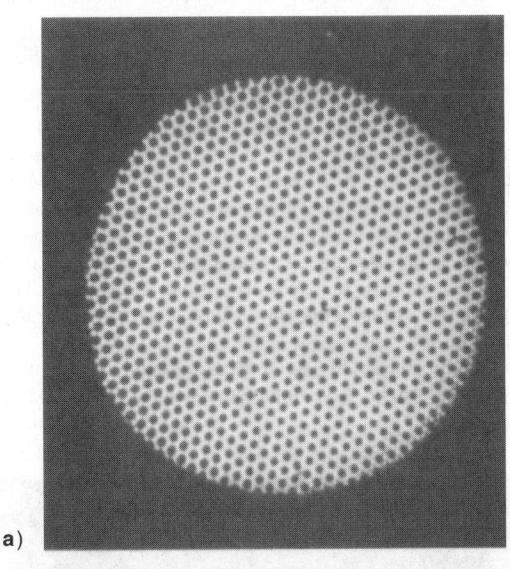

(a)

(b)

Fig. 1. Types of x-ray patterns of crystals: (a) Steel balls in a crystalline lattice; (b) Bragg reflections with shape S^2

$\sin \theta = n\lambda/2d$; in which n is an integer. By slowly turning the crystal, the various plane-families are brought into suitable orientations for the production of maxima. The result is a Laue pattern of black spots on the photographic plate placed beyond the crystal to catch the reflections. Another method, developed by Hull and by Debye and Scherrer, secures the necessary angular variation by crushing the crystal to powder and relying upon the fortuitous orientation of the fragments; the pattern in this case being a system of concentric rings, as exemplified by Fig. 2.

While it is very easy, when one knows the structure of the crystal and the wavelength of the rays, to predict the diffraction pattern, it is quite another matter to deduce the crystal structure in all its details from the observed pattern and the known wavelength. The first step is to determine the spacing of the atomic planes from the Bragg equation, and hence the dimensions of the unit cell. Any special symmetry of the space group of the structure will be apparent from space group extinction. A trial analysis may then solve the structure, or it may be necessary to measure the structure factors and try to find the phases or a Fourier synthesis. Various techniques can be used, such as the F2 series, the heavy atom, the isomorphous series, anomalous atomic scattering, expansion of the crystal and other methods.

By such methods, the structures of crystals have been determined and all of them can be shown to possess space structures corresponding to one or another of the 14 Bravais lattices. See Figs. 3 and 4.

Crystal Systems and Crystallography

The field that deals with the geometrical relations between the atomic planes within them is *crystallography*. Some minerals, if broken, will separate along given 'cleavage planes' into polyhedral fragments. Even when powdered, the minute grains will show this characteristic and measurements will indicate the planes to belong to one or another plane family, the members of any one of which are all parallel. It should be stressed, however, that all minerals do not possess cleavage planes for such experimentation.

In most crystal systems each of the more prominent crystal faces belongs to one of three plane-families intersecting along what are called the crystal axes. In the hexagonal system there are four. These may be conveniently used as coordinate axes, X, Y, Z, though they are not generally at right angles. Haüy discovered that if the ratio of the intercepts of two crystals planes on one of these axes is a simple fraction, such as $\frac{3}{5}$, the ratios of the intercepts on the other axes are likewise simple. This suggests that the two intercepts on any one axis are multiples of a common unit. The units are, however, generally different for the different axes, bearing to each other ratios called the axial ratios.

It is more convenient to use the reciprocals of the intercepts. For example, a plane might have intercepts equal to 10,000, 15,000, and 6,000 of the respective units. The reciprocals have the ratios 1/10,000: 1/15,000:1.6,000, which in lowest terms are 3:2:5. These smallest integers are the Miller indices of the family to which this plane belongs, and the family is thus designated (325). The family (201) is parallel to the Y axis but intersects the X and Z axes. (The hexagonal system has four Bravais-Miller indices for each plane-family.)

If the intercept of any plane has a negative value, that is, if it cuts the axes when extended in a direction opposite to that in the standard arrangement, the fact is shown by a bar over the Miller index, e.g., $(2\bar{2}2)$. The eight faces of an octahedron are (111), $(1\bar{1}1)$, $(\bar{1}11)$, $(\bar{1}\bar{1}1)$, $(11\bar{1})$, $(\bar{1}\bar{1}\bar{1})$, and $(\bar{1}\bar{1}\bar{1})$, that is, they all belong to the form (111).

Close study of the angles, indices, and axial ratios long since made it clear that every crystalline substance has a structure built upon a space "lattice" characteristic of the substance. It has been established that this is due to the regular arrangement of the atoms, molecules, or ions composing the substance. As shown by Table 1, the lattice structures of crystals may be classified into 32 symmetry classes (point groups), which are further divided into seven systems. This topic also is discussed under **Mineralogy**.

Practically all minerals are crystalline, although perfect natural crystals are seldom, if ever, found. Because of the laws of crystallography, however, a crystallographer can usually determine the crystal form of a known species from a fragment of the original crystal, provided that at least two of the crystal faces are visible. Crystalline aggregates are said to be cryptocrystalline when the individual particles are proved to have crystalline structure but their crystal faces are exceedingly small or indistinguishable. A mineral is pseudocrystalline if its external form does not correspond with its crystalline structure.

A new structural form has been proposed of *quasicrystals*—designated *shechtmanite*—which are neither crystalline nor completely amorphous. These materials possess a fivefold symmetry of long-range order. Despite Pauling's contention that data on these structures can be related to the common phenomenon of "twinning," increasing data indicate that this structural system does exist and has long been ignored because the icosahedral, fivefold symmetry upon which it is based cannot serve as a unit cell of a crystal. The difficulties presented in composing structures based upon icosahedra require also the concept of "tiling," that is, the filling of a three-dimensional space gap with two-dimensional structures or "tiles" while maintaining a periodicity of structure. Penrose of Oxford has laid down the essential criteria for such structures.

Determination of Crystal Structure

The object of a crystal-structure determination is to ascertain the position of all of the atoms in the unit cell, or translational building block, of a presumed completely ordered three-dimensional structure. In some cases, additional quantities of physical interest, e.g., the amplitudes of thermal motion, may also be derived from the experiment. The processes involved in such crystal-structure determinations may be divided conveniently into (1) collection of the data, (2) solution of the phase relations among the scattered x-rays (phase problem)—determination of a correct trial structure, and (3) refinement of this structure.

The data consist of intensities I (hkl), where h, k, and l (the Miller indices represent a vector triplet which conveniently identifies the beam diffracted from a single crystal. In a typical determination, there may be

Fig. 2.   Concentric ring pattern of crystalline materials

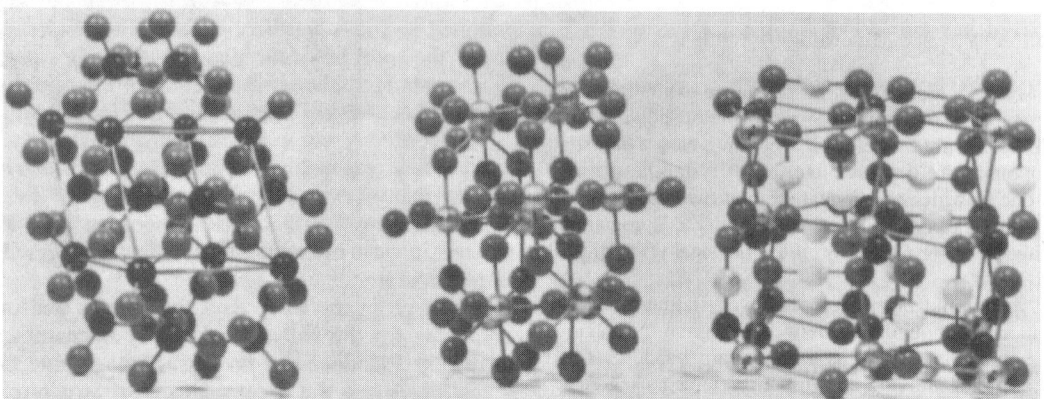

Fig. 3.   Computer-generated crystal structure models: (left to right) Beta-quartz, rutile, potassium dihydrogen phosphate (*AT&T Bell Laboratories*)

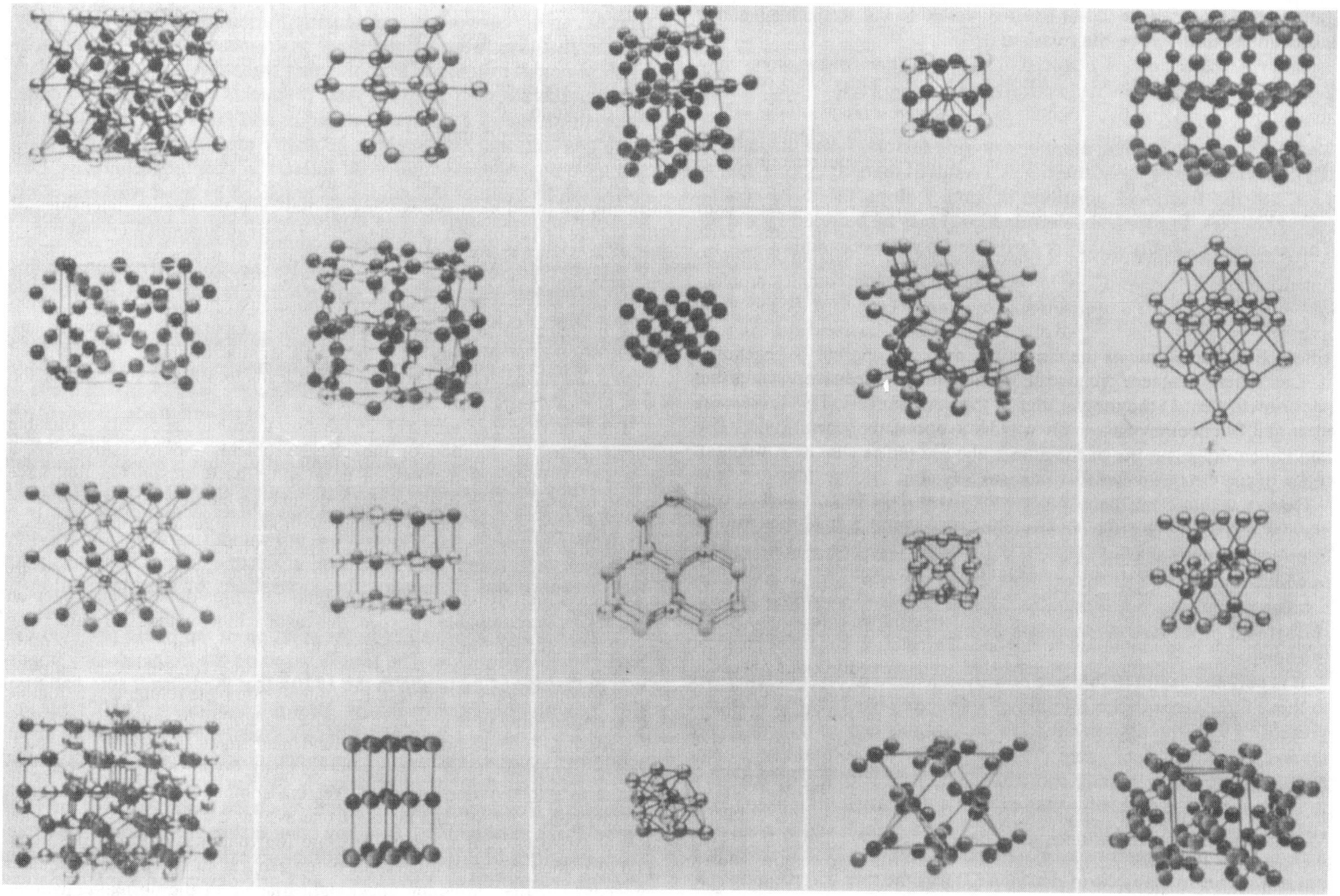

Fig. 4.   Computer-generated crystal structure models: (top row, left to right) Cuprite, zinc-blende, rutile, perovskite, tridymite; (second row) Cristobalite, potassium dihydrogen phosphate, diamond, pyrites, arsenic; (third row) Cesium chloride, sodium chloride, wurtzite, copper, niccolite; (fourth row) Spinel, graphite, beryllium, carbon dioxide, alpha quartz. (*AT&T Bell Laboratories*)

TABLE 1. ELEMENTS OF CRYSTAL SYSTEMS

System	Crystallographic elements	Essential symmetry	Number of point groups
Cubic, or regular	Three axes at right angles: all equal.	4 triad axes; 3 diad, or 3 tetrad axes	5
Tetragonal	Three axes at right angles: two equal	1 tetrad axis	7
Orthorhombic or rhombic	Three axes at right angles: unequal.	3 diad axes, or 1 diad axis and 2 perpendicular planes intersecting in a diad axis	3
Monoclinic	Three axes, one pair not at right angles: unequal.	1 diad axis, or 1 plane	3
Triclinic or anorthic	Three axes not at right angles: unequal.	No axes or planes	2
Hexagonal	Three axes co-planar at 60°: equal. Fourth axis at right angles to other three.	1 hexad axis	7
Rhombohedral or trigonal	Three axes equally inclined, not at right angles: all equal.	1 triad axis	5

one to two thousand such $I(hkl)$. The intensity is related to the structure factor $F(hkl)$ by the relation.

$$I(hkl) = K F(hkl) F^*(hkl) \qquad (1)$$

where K is a known relative factor, and where F^* is the complex conjugate of F. The structure factor itself is related to the scattering by the j atoms in the unit cell by the relation,

$$F(hkl) = \sum_j f_j T_j \exp[2\pi i(hx_j + ky_j + lz)] \qquad (2)$$

where f_j are the individual atomic scattering factors, T_j are the individual modifications of the scattering as a result of thermal motion, and x_j, y_j, z_j

are the fractional positions of atom j along the three crystallographic axes. In a typical determination, j may be between 10 and 60. The *scattering density* $\rho(xyz)$ is derivable from the relation,

$$\rho(xyz) = V^{-1} \sum_{-\infty h,k,l}^{\infty} F(hkl) \exp[-2\pi i(hx_j + ky + lz)] \qquad (3)$$

where V is the volume of the unit cell.

The "phase problem" in crystallography arises because in the usual experiment (Eq. 1) the magnitudes of the complex structure factors are obtained, but not the phases. Yet in order to obtain the scattering density, and hence the positions of the atoms, the phases as well as the magnitudes of the structure factors are necessary (Eq. 3).

Once the phase problem is solved, then the positions of the atoms may be refined by successive structure-factor calculations (Eq. 2) and Fourier summations (Eq. 3) or by a nonlinear least-squares procedure in which one minimizes, for example, $\sum w(|F_{obs}| - |F_{calc}|)$· with weights w taken in a manner appropriate to the experiment. Such a least-squares refinement procedure presupposes that a suitable calculational model is known.

It is perhaps useful to indicate how the attention of crystallographers to these three steps in the solution of a structure has changed in recent years. In 1954, the time involved in the arduous task of collecting the three-dimensional data—step (1)—needed for the solution of a complex problem was generally short in comparison with the time needed to solve the phase problem—step (2). This time involved in step (2) of course depended (and still depends) upon the complexity of the problem, and on the ingenuity, luck, and perseverance of the investigator, but it was true in many cases that step (2) was the rate-determining step in the entire process. This in part was because little attention was paid to detailed refinements—step (3); in 1954, three-dimensional least-squares refinements of complex structures were out of the question computationally, and even Fourier refinements were rare, for on computing systems advanced for those days (e.g., punched-card tabulators, sorters, and primitive electronic computers), a three-dimensional Fourier summation might require forty man-hours. In fact, in 1954 it was usual for the crystallographer to examine the unit cells of a number of related substances and to pick the problem that was crystallographically most favorable (and perhaps soluble from two-dimensional data), even though this problem might not be the one of greatest chemical or physical interest. Ten years later the situation had changed markedly, mainly because of the availability of high-speed computers. It is still true that there are classes of problems where step (2) is rate-determining, but these problems are far more complex than those attempted in 1954. Yet there is an extensive class of problems in which today the solution of the phase problem is straightforward and rapid. The crystallographer is thus often working on the problem of greatest chemical or physical interest, and is able to obtain a solution in feasible time. Relatively complete refinement of structures is now the rule, since it is a reasonably fast and effortless procedure. Thus it turns out that in many crystallographic problems the rate-determining step is data collection. For this reason, there has been a dramatic increase in interest in ways of making data collection less tedious, more rapid, and more accurate.

Although in the early days the Braggs and others used ionization chambers for the collection of x-ray intensities, these methods were gradually abandoned in favor of photographic film techniques. Until a few years ago the great majority of structure determinations were based on photographically recorded intensities, usually visually estimated. This process is a slow one: the typical time involved in the collection and estimation of a data set of two thousand intensities is perhaps six to eight weeks. Collection of intensity data from protein crystals is far more challenging and time-consuming, both because the number of data to be collected is far greater and because the crystals are unstable and rapid collection is thus desirable. For these reasons, Harker and his co-workers, particularly Furnas, then at Brooklyn Polytechnic Institute, were among those instrumental in developing scintillation-counter methods for collecting three-dimensional x-ray data. Diffractometers with single-crystal orienters, based on the so-called Eulerian geometry developed by Harker and Furnas, as well as on the more conventional Weissenberg geometry are available commercially and have engendered widespread interest in counter techniques. Data collection by counter techniques, as practiced by most workers, is still an arduous task, since the setting of a number of orientation angles is involved. Program or computer control of such setting operations is an obvious extension. Especially for neutron diffraction studies, such programmed control of diffractometers has been the rule for some time.

Nevertheless, a programmed unit will do only what it was designed to do, whereas a computer can be programmed to perform new tasks or operations as they seem necessary. There are several installations of computer-controlled diffractometers. Counter methods, particularly when semiautomatic or completely automatic, enable more rapid data collection than is possible photographically. What is equally important is that they should also enable more accurate data to be collected. The general level of accuracy of intensities obtained photographically is perhaps 15 to 20%. Such a level has proved sufficient for the solution of conformational or stereochemical problems, but not necessarily for the determination of meaningful descriptions of thermal motion or bonding.

There are two approaches to the solution of the phase problem that have remained in favor. The first is based on the tremendously important discovery of Patterson in the 1930s that the Fourier summation of Eq. 3, with the experimentally known quantities F^2 (hkl) replacing F(hkl) leads not to a map of scattering density, but to a map of all interatomic vectors. The second approach involves the use of so-called direct methods developed principally by Karle and Hauptman at the U.S. Naval Research Laboratory and which led to the award of the 1985 Nobel Prize in Chemistry. Building upon earlier proposals that the relative intensities of the spots in a diffraction pattern contain information about a crystal phase, Hauptman and Karle developed a mathematical means of extracting the information. A fundamental proposition of their direct method is that if three intense spots in the pattern have positions whose coordinates add up to zero, their relative phases will cancel out. Computations done with many triads of spots yield probable phases for a significant number of diffracted waves and further mathematical analysis leads to a likely solution for the structure of the molecule as a whole.

The *Patterson function* has been the most useful and generally applicable approach to the solution of the phase problem, and over the years a number of ingenious methods of unraveling the Patterson function have been proposed. Many of these methods involve multiple superpositions of parts of the map, or "image-seeking" with known vectors. Such processes are ideally suited to machine computation. Whereas the great increase in the power of x-ray methods of structure determination in the past few years has come simply from our ability to compute a three-dimensional Patterson function, it is reasonable to expect that, as machine methods of unraveling the Patterson function are developed, this power will increase many fold.

Step (3), the refinement of crystal structure, continues to enjoy a considerable amount of interest. Reasonably complete refinement is routine these days, owing in large measure to the availability of suitable computers. For reasons that are both practical and mathematically sound, the least-squares approach to refinement has gained favor over the successive structure-factor-Fourier approach. Yet the computational problems often tax this generation of computers. If one assigns a single isotropic thermal parameter to each atom, then there are four parameters, three positional and one thermal, to be determined for each atom. In the least-squares procedure, if one stores the upper right triangle of the normal-equations' matrix, then $\frac{1}{2}N(N + 1)$ elements are required, where N is the number of variables. In a machine with a memory of 32,000 words, a practical limit is reached at about $N = 200$, if one wishes to keep the rest of the program in core. Thus refinement of a 50-atom problem often taxes the memory capacity of the machine, and for larger problems special computational or mathematical tricks are needed. One of these tricks is to make use of known features of the structure or the thermal motion to reduce the number of parameters. But, even with increased numbers and more rapid availability of data, computerized solutions to crystallographic problems still suffer from a degree of uniqueness. Programs written to operate on one machine must often be extensively revised to be used on another. An international group of workers at the National Resource for Computation in Chemistry at the Lawrence Berkeley Laboratory is trying to overcome this problem of interchangeability by writing a program designed to run on any medium or large sized computer. When completed, the program will allow a crystallographer, armed only with experimental data and a computer, to determine the structure of any crystal without having to write special programs.

Crystal Growth

The direct growth of an ideal and perfect crystal is difficult except at very high supersaturations because of the difficulty of nucleating a new surface on a completed surface of the crystal. But, if there is a screw dislocation present, it is not necessary to start a new surface, and growth proceeds in a spiral fashion by the accretion of atoms at the edge of growth steps. The resultant growth spirals have been observed, and it is believed that most crystals grow in this manner. See Fig. 5. But spiral growth is not the only mechanism that enables crystal growth at fast rates. Gilmer, in particular, has employed computer simulation models in studies of crystal growth and has demonstrated that, among other influences, temperature and impurity levels have decisive effects upon growth rates. Again, since different crystal faces have different kinetic properties, the particular crystal plane exposed to growth will also be a partial determinant of growth rate.

In modern technology based upon solid state chemistry and physics, much emphasis is placed upon the availability of elements and compounds

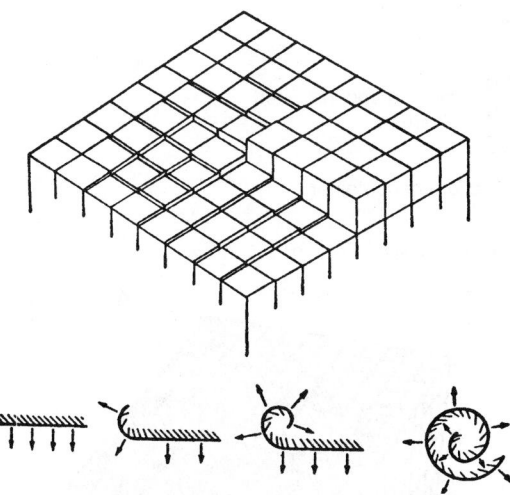

Fig. 5.   Highly schematic representation of crystal growth. (*F.C. Frank*)

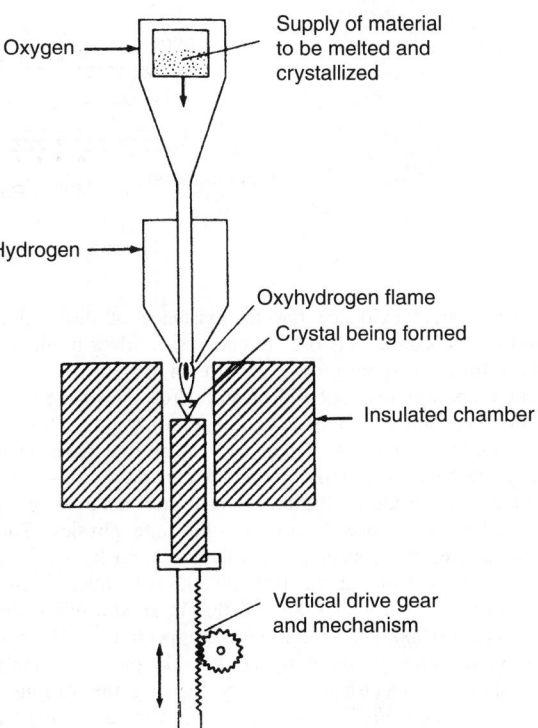

Fig. 6.   Verneuil technique for growing crystals

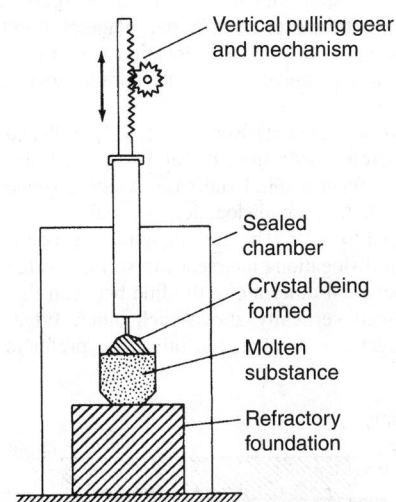

Fig. 7.   Czochralski method for growing crystals

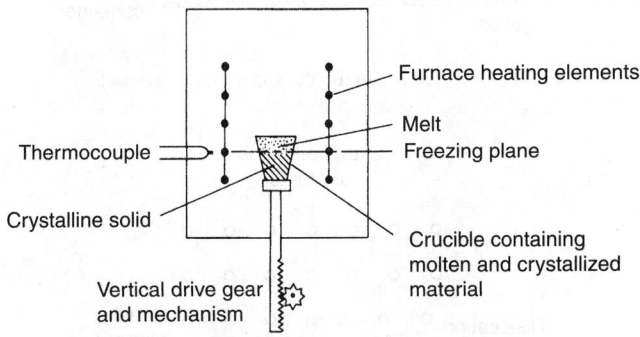

Fig. 8.   Stockbarger technique for growing crystals

in single-crystal form. Over the past twenty-five years a highly sophisticated technology has developed in this area. From the relatively simplistic growth of ammonium and potassium dihydrogen phosphates (ADP and KDP) from saturated solutions for transducer elements, a level has been obtained at which pure metals (Cu, Pb, Al, Ag, Fe, etc.), a semi-metals (Si, Ge, As, etc.), and compounds (GaAs, InAs, InSb, InP, etc.) are available and even essential in large single crystal form to the electronics industries. Synthetic gems (rubies, spinels, sapphires, emeralds, and zircons) are single crystals of aluminum, beryllium, and zirconium silicates or oxides with controlled impurity levels of transition elements.

Growth of such single crystals can follow several techniques, with thermodynamic constraints dictating the technique for any particular material: crystallization by cooling a supersaturated solution of a compound in a high-temperature flux; crystallization by dropping powder through an intense flame onto a seed pedestal, known as the Verneuil technique (see Fig. 6); crystallization by pulling a "seed" crystal from the surface of a liquid melt, known as the Czochralski method (see Fig. 7); crystallization by lowering a melt through a small, controlled thermal gradient, known as the Stockbarger technique (see Fig. 8); crystallization by zone-melting, known as the Pfann method (see Fig. 9); and crystallization by the vapor-phase approach (see Fig. 10). All of these procedures, however, require three essential ingredients: (1) A good "seed" crystal from which spiral and sometimes oriented growth can occur and develop; (2) highly precise operational conditions-movement of fractions of a millimeter, or temperature variations of 0.5°C per hour; and (3), as previously indicated, materials of a specific impurity level. Given these conditions, single crystals can be grown in large quantities-ranging from the multimillion carat operations of Linde (United States) and Djevaherdijian (Switzerland), using the Verneuil technique, to the multikilo manufacture of single crystal silicon by Texas Instruments Incorporated (United States), employing the Czochralski and zone melt methods, and including a 27-kilogram single crystal of dislocation-free silicon by the Kayex Corporation (United States) grown by a modified Czochralski approach.

See also **Semiconductors**.

Dislocation in Crystalline Solids

This type of imperfection in a crystalline solid is generated as follows: A closed curve is drawn within the solid, and a cut made along any simple surface which has this curve as boundary. The material on one side of this surface is displaced by a fixed amount called the Burgers vector relative to the other side. Any gap or overlap is made good by the addition or removal of material, and the two sides are then rejoined, leaving the strain displacement intact at the moment of rewelding, but afterwards allowing the medium to come to internal equilibrium. If the Burgers vector represents a translation vector of the lattice, the weld is invisible, and the dislocation is characterized only by the original curve, or dislocation line and by the Burgers vector.

A dislocation line may only terminate at the surface of the crystal. The energy of a dislocation is largely stored as strain in the surrounding lattice. The important property of a dislocation is its ability to move quite easily through the lattice, and hence to allow the rapid propagation of slip. The general dislocation defined above usually separates into its components, edge and screw dislocations, which may be treated as rather stable entities

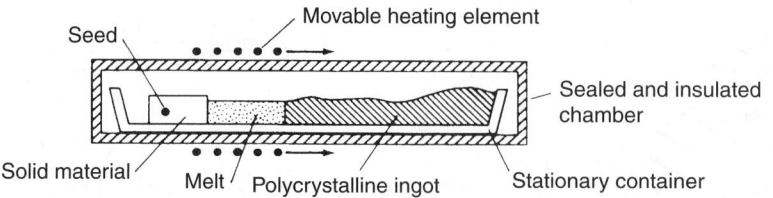

Fig. 9.   Zone-melting or Pfann method for growing crystals

in the theory. Direct evidence for the existence of dislocations is the observation of dislocation networks in crystals of silver bromide, and, for their motion, from the spiral growth patterns in crystals.

In recent years high-resolution electron microscopes have been used to study dislocations and their movements in thin crystals. This is possible in both metals and nonmetals. Other techniques that have been applied successfully are field ion microscopy and x-ray microscopy. Dislocations are important in determining the mechanical and electrical properties of solids, and play an important part in solid state physics. The density of dislocations (i.e., the concentration of dislocation lines), for example, is believed to vary from about 100,000,000 per square centimeter in good natural crystals, through 1,000,000,000 in good artificial crystals up to about 1,000,000,000,000 in cold-worked specimens. These estimates are based on the energy stored by cold work, on x-ray analysis, and on measurements of electrical resistivity. Among the various types of dislocation are the *edge dislocation*, which is defined as having its Burgers vector (line of displacement) normal to the line of the dislocation. An edge dislocation may be thought of as caused by inserting an extra plane of atoms terminating along the line of the dislocation (Fig. 11). For example, if the dislocation were along the Z-axis and its Burgers vector along the X-axis, then one might think of an extra half plane of atoms being inserted at the surface $x = 0$, $y > 0$. Such a dislocation would be of positive sign. An edge dislocation may move easily only parallel to its Burgers vector, i.e., in its slip plane.

The *screw dislocation* has its Burgers vector parallel to the line of the dislocation. In a screw dislocation, the atomic planes are joined together in such a way as to form a spiral staircase, winding round the line of the dislocation (Fig. 12). A screw dislocation is capable of easy movement in any direction normal to itself. The growth spirals formed in crystal growth appear where such dislocations intersect the surface. Edge dislocations of the same sign repel each other along the line between them, but are most stable when arranged vertically above each other. Edge dislocations of opposite signs attract one another, but otherwise prefer to lie so that the

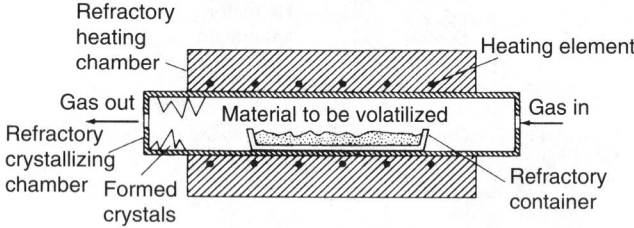

Fig. 10.   Crystallization by the vapor-phase approach

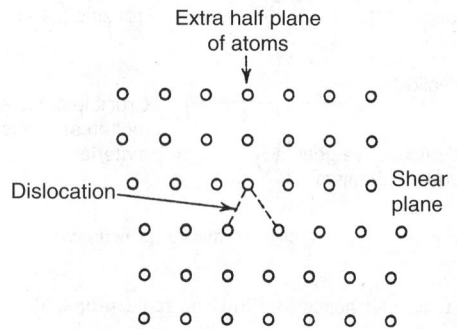

Fig. 11.   Atomic arrangement in an edge dislocation

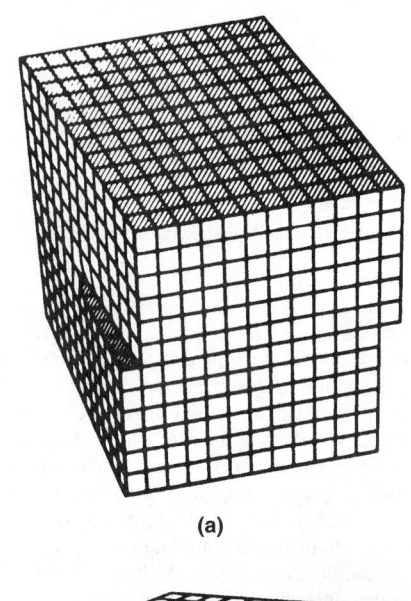

(a)

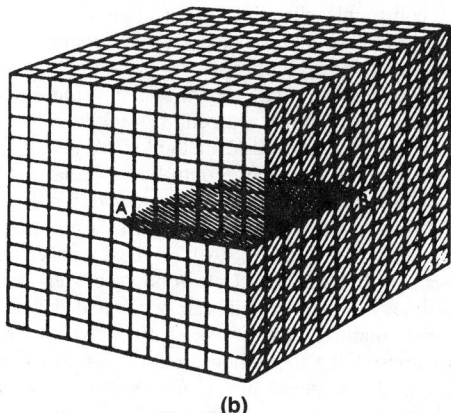

(b)

Fig. 12.   (**a**) Simple type of screw dislocation; (**b**) both screw and edge dislocation along the arc from A to B

line between them makes an angle of 45° with their slip planes. Screw dislocations of opposite sign attract, of like sign repel.

Dislocation Line

The curve separating displaced and undisplaced positions of a crystal, and thus at the center of a dislocation, is termed the dislocation line. Within a distance of one or two lattice constants of the dislocation line the atoms are displaced by an amount more than can be represented fairly as a strain. A screw location may have a substantial hole down the dislocation line, through which impurity atoms may diffuse.

Dislocation Climb

This is a type of dislocation motion, differing fundamentally from slip, associated with the edge components of dislocations. In climb, an edge dislocation moves in a direction perpendicular to its slip plane as atoms are either added to or taken away from the extra plane of the dislocation. The motion of atoms to and from the dislocation is accomplished by vacancy movements. If an atom in the plane next to the edge jumps out of its position and attaches itself to the extra plane, as indicated in the accompanying figure, a vacancy is created which can then diffuse off into the lattice. Repetition of this process over and over will cause the extra plane to grow in size and, if this occurs, the process is said to be negative climb. On the other hand, if vacancies diffuse up to the extra plane of the dislocation and remove atoms from it, the plane grows smaller in size and positive climb is said to occur.

Dislocation climb only becomes of practical significance at elevated temperatures because of its dependence upon vacancies whose number and mobility depend very strongly on the temperature. Dislocation climb is important in high temperature creep and recovery phenomena. See Fig. 13.

Crystal Slip

This is the process by which a crystal undergoes plastic deformation, as a result of which one atomic plane moves over another. Slip is believed to occur through the movement of dislocations. The total deformation of a given crystal is the sum of many small lateral displacements in parallel crystallographic planes of a given family. Moreover, each slip plane becomes more resistant to further deformation than the remaining potential slip planes.

Cross-Slip

This is slip that occurs simultaneously on several slip planes having the same slip direction. See Fig. 14. This type of plastic deformation is normally associated with the movement of screw dislocations. Screw dislocations can move on any slip plane that passes through the dislocation. This is a result of the fact that the slip plane of a dislocation is that plane which contains both the dislocation and its Burgers vector, and the fact that the Burgers vector of a screw dislocation lies parallel to the dislocation itself.

See also **Microgravity and Materials Processing; Liquid Crystals; and Solid-State Physics**.

Quasicrystals

The existence of quasicrystals was discovered by the Israeli scientist, D. Shechman, in the early 1980s. After years of experimentation, the first practical application for such materials was announced in the early 1990s in connection with the development of a nonstick, abrasion-resistant coating for cookware. They appear to be good candidates for a number of tribological uses because of their excellent wearability and abrasion resistance. Patents along these lines have been granted to the Centre National de la Recherche Scientifique in France. Researchers targeted two compounds with the nominal composition (atomic percent) of $Al_{65}Cu_{20}Fe_{15}$ (designated as alloy A) and $Al_{64}Cu_{18}Fe_8Cr_8$ (designated as alloy B). Powders were prepared by mechanically grinding ingots to a mesh size

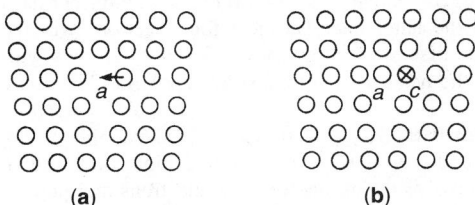

(a)   **(b)**

Fig. 13.   Negative climb of an edge dislocation

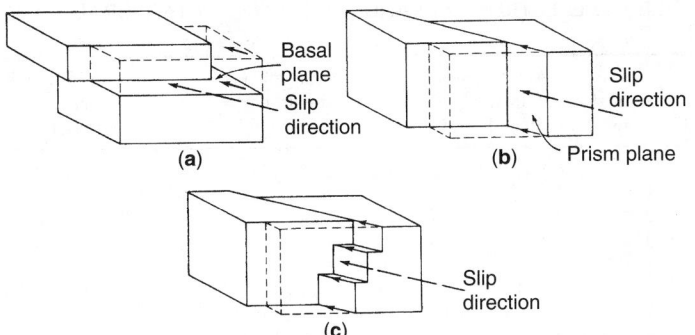

(a)   **(b)**   **(c)**

Fig. 14.   Schematic representation of cross-slip in a hexagonal metal: (**a**) slip on basal plane; (**b**) skip on prism plane; (**c**) cross-slip on basal and prism planes

ranging from 20 to 75 micrometers (0.0008 to 0.003 in.) and were thermally sprayed onto "soft" aluminum alloy, pure copper, and low-carbon steel substrates using flame, supersonic, and plasma-arc spraying. The deposited materials consist of a mixture of quasicrystals and crystalline phases. Alloy A contains about 70 percent icosahedral phase, while alloy B contains about 70 percent quasicrystalline (icosahedral plus decagonal) phases. An icosahedron is a polygon having 20 faces and a decagon is a polygon having ten angles and ten faces.

As pointed out by Stephens and Goldman (State University of New York at Stony Brook), "Quasicrystals are neither uniformly ordered like crystals nor amorphous like glasses. Many features of quasicrystals can be explained, but their atomic structure remains to be described fully." See also **Aluminum Alloys and Engineered Materials**.

Additional Reading

Amato, I.: "The High Side of Gravity," *Science,* **30** (July 5, 1991).

Amato, I.: "Atoms Do the Two-Step on Crystal Dance Floors," *Science,* **970** (August 30, 1991).

Beck, R.D. et al.: "Impact-Induced Cleaving and Melting of Alkali-Halide Nanocrystals," *Science,* **879** (August 23, 1991).

Braga, D., F. Grepioni, and A.G. Orpen: *Crystal Engineering: From Molecules and Crystals to Materials,* Kluwer Academic Publishers, Norwell, MA, 1999.

Bunning, T.J., S.H. Chen, N. Koide, et al. : *Liquid Crystal Materials and Devices: Materials Research Society Symposium Proceedings,* Materials Research Society, Warrendale, PA, 1999.

Byrappa, K., M. Yoshimura, and L. Yoshimura: *Handbook of Hydrothermal Technology: Technology for Crystal Growth and Materials Processing,* William Andrew Publishing, Park Ridge, NJ, 2000.

Cathonet, P.: *Quasicrystals at Home on the Range,* Advanced Materials & Processes, 6 (June 1991).

Chuang, I. et al.: "Cosmology in the Laboratory: Defect Dynamics in Liquid Crystals," *Science,* **1336** (March 15, 1991).

DeYoreo, J., W. Casey, and A. Malkin: *Morphology and Dynamics of Crystal Surfaces in Complex Molecular Systems,* Materials Research Society, Warrendale, PA, 2000. http://www.mrs.org/

Dierking, I.: *Textures of Liquid Crystals,* John Wiley & Sons, Inc., New York, NY, 2003.

Faust, W.L.: "Explosive Molecular Ionic Crystals," *Science,* **37** (July 7, 1989).

Flam, F.: "Liquid Crystals Meet the Cosmos at APS Meeting," *Science,* **649** (May 3, 1991).

Gavish, M. et al.: "Ice Nucleation by Alcohols Arranged in Monolayers at the Surface of Water Drops," *Science,* **973** (November 16, 1990).

Gilmer, G.H.: "Computer Model of Crystal Growth," *Science,* **208**, 355–363 (1980).

Henderson, B. and R.H. Bartram: *Crystal-Field Engineering of Solid-State Laser Materials,* Cambridge University Press, New York, NY, 2000.

Jaric, M.V. and D. Gratias: *Extended Icosahedral Structures,* Academic Press, Inc. San Diego, CA, 1989.

Kelly, A., G.W. Groves, and P. Kidd: *Crystallography and Crystal Defects,* John Wiley & Sons, Inc., New York, NY, 2000.

Langer, J.S.: "Dendrites, Viscous Fingers, and the Theory of Pattern Formation," *Science,* **1150** (March 3, 1989).

Lines, M.E., A.M. Glass: *Principles and Applications of Ferroelectrics and Telated Materials,* Oxford University Press, Inc., New York, NY, 2000.

Massa, W.: *Crystal Structure Determination,* Springer-Verlag Inc., New York, NY, 2000.

McPherson, A.: "Macromolecular Crystals," *Sci. Amer.,* **62** (March 1989).

Musevic, I., B. Zeks, and R. Blinc: *The Physics of Ferroelectric and Antiferroelectric Liquid Crystals,* World Scientific Publishing Company, Inc., River Edge, NJ, 2000.

Pool, R.: "A Stirring Tale of Crystal Growth," *Science,* **913** (November 16, 1990).

Prigogine, I., S.A. Rice: *Advances in Chemical Physics Volume 113, Advances in Liquid Crystals,* John Wiley & Sons, Inc., New York, NY, 2000.

Scheel, H., and T. Fukuda: *Crystal Growth Technology,* John Wiley & Sons, Inc., New York, NY, 2003.

Stephens, P.W. and A.I. Goldman: "The Structure of Quasicrystals," *Sci. Amer.,* **44** (April 1991).

Strauss, S.: "Impossible Matter," *Technology Review (MIT),* **10** (January 1991).

Uhrin, R.: *Selected Papers on Laser Crystal Growth,* SPIE International Society for Optical Engineering," Bellingham, WA, 2000.

Wang, Y. et al.: "Twinning in MgSio3 Perovskite," *Science,* **468** (April 27, 1990).

Warner, M., and E.M. Terentjev: *Liquid Crystal Elastomers,* Oxford University Press, New York, NY, 2003.

Weisbuch, I., Sasi, M.L., and L. Leiserowitz: "Molecular Recognition at Crystal Interface," *Science,* **637** (August 9, 1991).

Westbrook, J.H. and R.L. Fleischer: *Intermetallic Compounds: Crystal Structures of Intermetallic Compounds,* Vol. 1, John Wiley & Sons, Inc., New York, NY, 2000.

CRYSTAL FIELD THEORY.   A theory developed in the early 1930s in research on magnetism by Bethe, Van Vleck and others. It applied particularly to the transition metal ions, and is therefore conveniently treated by reference to those ions.

A transition metal ion in a complex or compound is considered to be subject to the electrostatic field of the molecules and ions in its neighborhood, particularly by those constituting its nearest neighbors. In the compounds to which the theory applies, the only nearest neighbors important to the theory are either negative ions, or molecules such as NH_3 which have unshared electron pairs and which orient themselves so that the negative end of the electron-pair dipole is directed toward the transition metal ion. The effect of these negative ions or negatively oriented dipoles (which we shall hereafter call ligands, with the understanding that ligand field theory is a later development of crystal field theory) is to produce a negative field about the positive transition metal ion.

In the absence of this negative field, the d-electrons of the central ion have orbitals of equal energy, i.e., degenerate orbitals, but the field of the ligands affects the energies of these orbitals to different degrees. To show how this effect arises, consider the example chosen by Griffith and Orgel. This is the regular octahedral complex MX_6, where M is a metal ion of the first transition series of the elements (see **Periodic Table of the Elements**), and X is a ligand such as H_2O, NH_3 or Cl^-.

The five $3d$ orbitals of M have the forms indicated in Fig. 1, in which the coordinate axes lie along the MX bond directions. It is clear from the figure that the d_{z2} and d_{x2-y2} orbitals have substantial amplitudes in the directions of the ligands but that the orbitals d_{xy}, d_{yz}, and d_{zx} tend to avoid them. Hence the energy of an electron in the d_z^2 or d_{x2-y2} orbitals will be substantially raised by the repulsive field of the ligands, whereas the energy of an electron in the d_{xy}, d_{yz}, or d_{zx} orbitals will be comparatively little affected. Furthermore, it is obvious from symmetry that the degeneracy of the last three orbitals is maintained in the octahedral complex and it can be shown by group theory that the d_{z2} and the d_{x2-y2} orbitals also remain degenerate. Consequently the five d orbitals split into a lower group of three and an upper group of two, the two groups being usually designated as t_{2g} and e_g respectively (or sometimes as γ_5 and γ_3, and as $d \in$ and $d\gamma$ respectively).

In tetrahedral complexes it can be shown similarly that the d orbitals are again split into groups of three and two, respectively, but now the doubly degenerate orbital is lower. In all other important cases the degeneracy of the d orbitals is reduced even further.

In order to understand the electronic structure of an octahedral complex and the optical transitions it can undergo, it is necessary to appreciate the principles determining the distribution of the d electrons among the t_{2g} and e_g orbitals. Let us begin by considering the ground state. Two separate tendencies are at work. The first is the tendency for the electrons to occupy, as far as possible, the orbitals of lowest energy in the ligand field. The second is for the electrons to go into different orbitals with their spins parallel, since this gives a lower electrostatic repulsive energy and a more favorable exchange energy. Let us now see how these ideas apply to an octahedral complex containing $n3d$ electrons.

The ion $[Ti(H_2O)_6]^{3+}$ has one d electron. In the ground state this will obviously occupy one of the t_{2y} orbitals. A transition is possible in which this electron is transferred to one of the e_y orbitals, and this occurs at 20,400 cm^{-1}. (The intensities of such transitions are low—ϵ_{max}; ≈ 10—as they are symmetry-forbidden.) The converse situation arises in the hydrated copper(II) ion which has nine d electrons. In this ion the vacancy in the d shell is one of the e_y orbitals. This vacancy can be filled by exciting an electron from one of the t_{2g} orbitals, giving rise to a transition at about 12,500 cm^{-1}. However, the ion $[Cu(H_2O)_6]^{2+}$ is strongly distorted in its ground state so that most of the degeneracy of the t_{2g} and e_g orbitals is removed and there is more than one transition in this region. From these two examples we see that the splitting between the t_{2g} and the e_g orbitals may be quite large, being usually in the range 20–50 kcal mole^{-1}.

It is only when we come to consider complexes with several d electrons that complications arise. If there are only two or three d electrons, both the above-mentioned tendencies can be satisfied simultaneously by placing the electrons in different t_{2g} orbitals with their spins parallel. However, when there are more than three d electrons this is no longer possible. If there are 4–$7d$ electrons we then have the choice either of putting as many as possible into the low-energy t_{2g} orbital or distributing them so as to maintain a maximum number of parallel spins. This is illustrated in Table 1.

The former choice will be favored if the orbital separation Δ is large, and the latter will be realized if Δ is small. The value of Δ depends primarily on the nature of the ligand and the charge on the ion, and the following generalizations can be made for the first transition series. (1) For hydrated bivalent ions Δ falls in the range 7,500–12,500 cm^{-1}. (2) For hydrated tervalent ions Δ falls in the range 13,500–21,000 cm^{-1}. (3) The common ligands can be arranged in a sequence so that Δ for their complexes with any given metal increases along the sequence. A shortened series is I$^-$, Br$^-$, Cl$^-$, F$^-$H$_2$O, oxalate, pyridine, NH$_3$. ethylenediamine, NO$_2^-$, CN$^-$. Finally, Δ for the compounds of the second and third series is 40–80% larger than for corresponding compounds of the first series.

With these considerations in mind let us consider in turn the two extreme possibilities, known respectively as the "strong-field" and the "weak-field" case. If Δ is very large the tendency for electrons to go into separate orbitals will be outweighed by their tendency to occupy the t_{2g} orbitals (as against the e_g orbitals) in circumstances where these two tendencies conflict. In the strong-field case, therefore, a complex with up to six d electrons will have all these in t_{2g} orbitals with the maximum number of unpaired spins consistent with the restriction to the t_{2g} orbitals. As examples we may take the hexacyanoferrates(II) and (III) which possess six and five d electrons, respectively, all in t_{2g} orbitals with the maximum number of unpaired spins consistent with the restriction to the t_{2g} orbitals. As examples we may take the hexacyanoferrates(II) and hexacyanoferrates(III) which possess six and five d electrons respectively all in t_{2g} orbitals. The former has no unpaired electrons and the latter one. The next four electrons will then enter the e_g orbitals; the first two will go into different e_g orbitals with their spins parallel as in the octahedral complexes of Ni^{2+}. In Co^{2+} there is just one e_g electron.

The complex $[Co(NH_3)_6]^{3+}$ provides a good example of the strong-field case. The ground state is $(t_{2g})^6$ and the transitions observed at the longest wavelengths involve taking one of these electrons and putting it in an e_g orbital. According to the choice of the orbitals involved the final state may be one of the two triply degenerate states, and the bands associated with

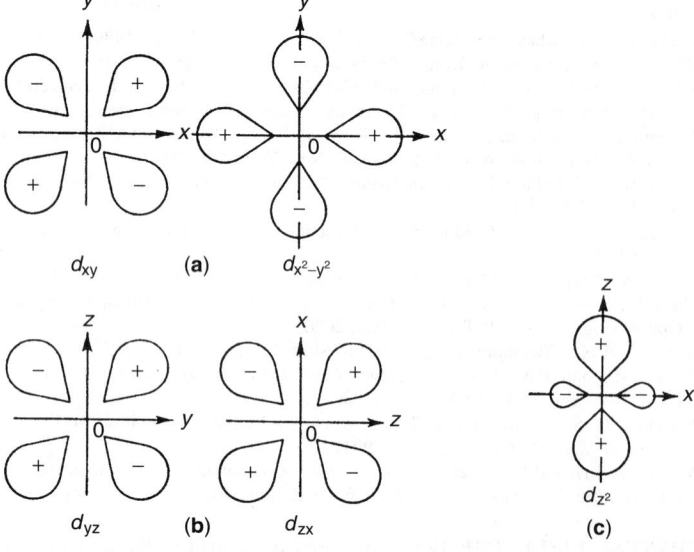

d_{xy} **(a)** $d_{x^2-y^2}$

d_{yz} **(b)** d_{zx} **(c)** d_{z^2}

Fig. 1.   Cross section of the five d orbitals, chosen in real form

TABLE 1. d-ELECTRON ARRANGEMENTS IN OCTAHEDRAL COMPLEXES

Number of d Electrons	Arrangement in weak ligand field		N	Arrangement in strong ligand field		N	Gain in orbital energy in strong field
	t_{2g}	e_g		t_{2g}	e_g		
1	↑↑	—	0	↑		0	0
2	↑↑	—	1	↑↑	—	1	0
3	↑↑↑	—	3	↑↑↑	—	3	0
4	↑↑↑	↑	6	↑↓↑↑	—	3	Δ
6	↑↓↑↑	↑↑	10	↑↓↑↓↑↓	—	6	2Δ
7	↑↓↑↓↑↑	↑↑	11	↑↓↑↓↑↓	↑	9	Δ
8	↑↓↑↓↑↓	↑↑	13	↑↓↑↓↑↓	↑↑	13	0
9	↑↓↑↓↑↓	↑↓↑	16	↑↓↑↓↑↓	↑↓↑	16	0

N Number of distinct pairs of electrons with parallel spin.

the two transitions have been observed for a number of d_6 complexes in each of the three transition series. (The separation between the bands is not in very good agreement with theory, however, so that it is difficult to obtain a reliable value of Δ for such cobalt(III) complexes. The reasons for this are not fully understood at present.)

The weak-field case is that in which the separation Δ is not large enough to overcome the tendency of the d electrons to go into different orbitals with their spins parallel. For example, in the hydrated manganese(II) ion the five d electrons each occupy one of the five d orbitals; this is because the separation Δ is insufficient to break up the highly stable half-filled shell in which all the electron spins are parallel. The same is true of $[Fe(H_2O)_6]^{3+}$ and of the hydrated iron(II) ion, which has six d electrons, the extra one being in one of the t_{2g} orbitals. It may be noted again that, only in those complexes containing four, five, six, or seven d electrons, is there an important distinction between the strong- and the weak-field cases; if there are one, two, or three d electrons, they will necessarily occupy t_{2g} orbitals, while if there are eight or nine, the vacancies in the d shell will occur in the e_g orbitals in both the strong- and the weak-field case.

Additional Reading

Ciferri, A.: *Liquid Crystallinity in Polymers: Principles and Fundamental Properties,* John Wiley & Sons, Inc., New York, NY, 1991.

Desiraju, G.R.: *Crystal Design: Structure and Function,* John Wiley & Sons, Inc., New York, NY, 2003.

Khoo, Iam-Choon C.: *Liquid Crystals: Properties and Nonlinear Optical Phenomena,* John Wiley & Sons, Inc., New York, NY, 1995.

CRYSTAL HABIT.   The external shape of a crystal, which depends on the relative development of the different faces, as well as upon the interfacial angles characteristic of the crystal.

CRYSTAL (Homometric Pairs).   Two crystal structures having the same x-ray diffraction pattern. This is possible because, basically, a diffraction pattern depends only on the relative vector distances between the atoms in the lattice, not on their absolute positions in space.

CRYSTAL (Isomorphous).   One of two or more crystals which are similar in crystalline form and in chemical properties and are related in chemical composition in that one or more of the atoms and radicals in one are of similar chemical type to the corresponding atoms or radicals in the other. Usually, one or more of their other atoms or radicals are identical.

CRYSTALLIZATION.   Crystals are formed (1) from solution, (2) from fusion, and (3) by sublimation.

The formation of crystals from solution, starting with an unsaturated solution, takes place when a solution is evaporated or cooled below the saturation point, except as retarded by supersaturation. Supersaturation is prevented by the addition of seed crystals of the substance. Since, in the case of the majority of soluble substances, solubility increases with increase of temperature cooling below the saturation point favors the formation of crystals. In very few cases, such as sodium sulfate above $32.4°C$, calcium sulfate (slightly soluble), calcium hydroxide (slightly soluble), solubility decreases with increase of temperature, and the above statement would not apply. A case of wide scope and great importance is that of crystal formation by precipitation upon mixing two solutions. Actually, this is the same as for substances of greater solubility, since the substance precipitated is first formed in solution and the excess above the saturation point separates as precipitate. As a rule the crystals are larger and more perfect the slower their growth. Conversely, when small crystals are desired, rapid stirring and quick cooling are practical. The smaller the crystals of a given substance, the purer the material generally is. Small crystals may be increased in size by allowing them to stand in the mother liquor before separation.

An industrial forced-circulation evaporative crystallizer is shown in Fig. 1. Sizes range from 18 inches ($\sim$46 centimeters) to over 42 feet (12.6 meters) in diameter, with no inherent limit to the size of the vessel. Slurry is moved by the circulation pump through the heat exchanger, where it is subjected to a temperature rise of 2 to 10°F. The heated liquor is discharged tangentially into the body at a point sufficiently far beneath the surface so that the liquor entering tangentially is just at the boiling point for the liquid depth at which it is submerged. As the liquid rotates around the body and rises toward the surface, it starts to boil, and this boiling induces a secondary circulation which creates a spinning toroid of fluid within the

body. Depending on the location of tangent inlet with respect to the cone, this toroidal circulation can result in considerable secondary circulation and agitation within the body of the vessel. When properly designed, this type of vessel is capable of producing a smooth boiling action with relatively small amounts of salt being deposited on the walls, while still maintaining a suitable suspension of product crystals within the boiling zone and in the lower part of the vessel. Crystalline materials produced in this type of equipment include sodium carbonate, sodium sulfate, and sodium chloride. Operating cycles of this equipment between washouts to remove salt growth from the walls of the body or from the heat exchangers normally range from 30 to 90 days.

To achieve some control of the number of fine particles within the crystallizer body, and thereby increase the overall particle size, it is necessary to selectively remove the fine particles so that they can be destroyed by the action of heat or dilution. A draft-tube baffle crystallizer of the type shown in Fig. 2 achieves these objectives. A body of growing crystals is suspended by the circulation flowing up the draft tube from the propeller shown close to the bottom of the vessel. From the areas surrounding this body of circulated slurry, a stream is removed at relatively low velocity so that gravitational settling will produce a separation between the product-size crystals and relatively fine crystals, which are removed with the clarified mother liquor leaving by the circulating pipe. In the draft-tube baffle crystallizer, both the velocity of the liquor in the settling zone and the quantity of liquor removed by the circulating pump are important to insure that the proper end-product size is achieved and that reasonable stability in particle size is obtained. The equipment is used to crystallize potassium chloride, ammonium sulfate, and other relatively fast-growing inorganic salts.

The foregoing descriptions cover only two of several industrial crystallizer configurations.

The formation of crystals from fusion takes place when the melted substance is cooled sufficiently slowly near and below the fusion point. If the cooling is rapid the fusion may result in the formation of an undercooled liquid of rigidity corresponding to a solid. Glasses, whether artificial, such as glass, vitreous enamels, and slags, or natural, such as vitreous rocks and minerals, e.g., obsidianite, are undercooled liquids. Rocks and minerals that have cooled sufficiently slowly from fusion form crystals, for example, granite. An important method of forming pure crystals is zone melting.

The formation of crystals by sublimation takes place when the vapor of a substance is condensed as a solid without passing through the liquid

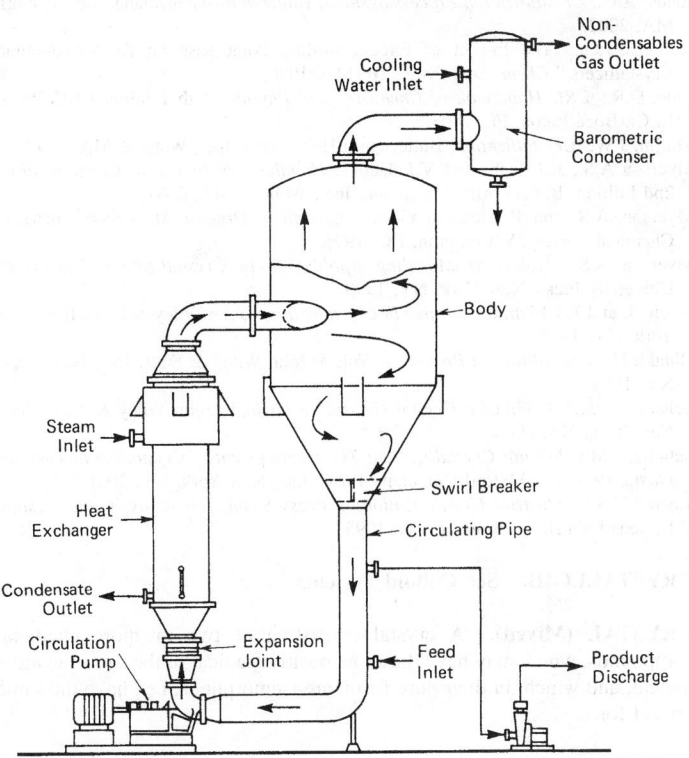

Fig. 1.   Forced-circulation evaporative crystallizer

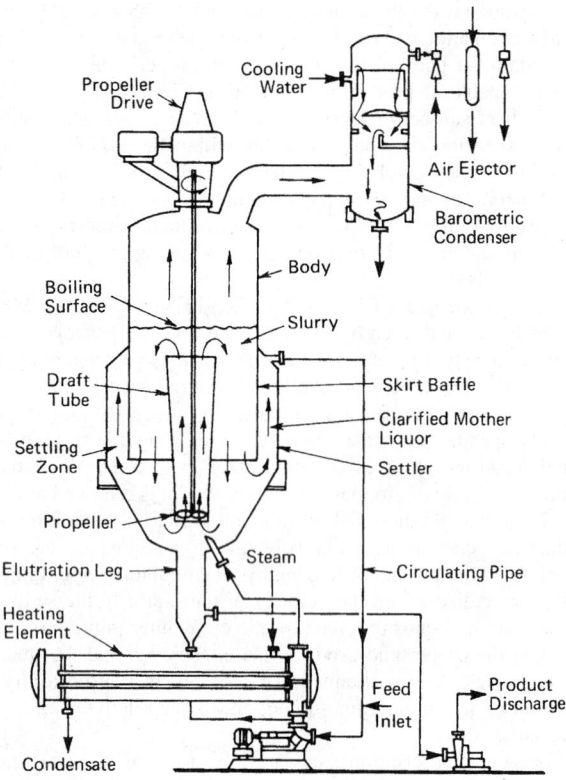

Fig. 2.   Draft-tube baffle crystallizer

phase in so doing. This occurs when the temperature of the condenser is below that of the melting point of the substance.

The heat of crystallization is in amount the same as the heat of solution of a given substance but of opposite sign.

Additional Reading

Ducruix, A. and R. Giege: *Crystallization of Nucleic Acids and Proteins: A Practical Approach,* Oxford University Press, Inc., New York, NY, 2000.

Hartel, A.: *Crystallization in Foods,* Aspen Publishers, Inc., Gaithersburg, MD, 1999.

Jones, A.G.: *Crystallization Process Systems,* Butterworth-Heinemann, Inc., Woburn MA, 2002.

Kumana, J.D.: "The Impact of Excess Boiling Point Rise on Evaporators and Crystallizers," *Chem. Eng. Prog.,* **10** (May 1990).

Lide, D.R.: *CRC Handbook of Chemistry and Physics,* 84th Edition, CRC Press, LLC., Boca Raton, FL, 2003.

Mullin, J.W.: *Crystallization,* Butterworth-Heinemann, Inc., Woburn, MA, 1997.

Myerson A.S., J. Joseph, and V.J. Jacovs: *Handbook of Industrial Crystallization,* 2nd Edition, Butterworth-Heinemann, Inc., Woburn, MA, 2001.

Myerson, A.S. and P. Meenan: *Crystal Growth of Organic Materials,* American Chemical Society, Washington, DC, 1996.

Myerson, A.S.: *Molecular Modeling Application in Crystallization,* Cambridge University Press, New York, NY, 1999.

Nyvlt, J. and J. Ulrich: *Admixtures in Crystallization,* John Wiley & Sons, Inc., New York, NY, 1995.

Ohtaki, H.: *Crystallization Processes,* Vol. 3, John Wiley & Sons, Inc., New York, NY, 1998.

Scheel, H., and T. Fukuda: *Crystal Growth Technology,* John Wiley & Sons, Inc., New York, NY, 2003.

Schultz, J.M.: *Polymer Crystallization: The Development of Crystalline in Thermoplastic Polymers,* Oxford University Press, Inc., New York, NY, 2001.

Tavare, N.S.: *Industrial Crystallization: Process Simulation Analysis and Design,* Perseus Publishing, Boulder, CO, 1995.

CRYSTALLOID.   See Colloid Systems.

CRYSTAL (Mixed).   A crystal consisting of two or more chemical compounds, which may have the same positive radical or the same negative radical, and which, in their pure form, are isomorphous, i.e., have the same crystal form.

CRYSTAL OSCILLATOR.   This device is a precise mechanical resonator and frequency generator. The need for a stable, accurate, and low-cost frequency generator for the precise control of commercial radio and other higher communication frequencies led to the development of piezoelectric resonators, notably quartz crystals, for a wide variety of applications. The quartz crystal has become highly developed as a frequency standard for timekeeping and as a time-signal generator. See also **Piezoelectric Effect**.

Piezoelectricity was discovered by the Curie brothers in 1880. The term *piezo* is derived from the Greek word meaning "to press." The effect causes a crystal to exhibit electrical polarity when the crystal is subjected to mechanical pressure. Conversely, the crystal is physically deformed when subjected to an electrical potential. Specifically, piezoelectricity is a property of nonconducting solids that have a crystal lattice that does not have a center of symmetry.

Quartz, tourmaline, and rochelle salts and such synthetic crystals as ethylene diamine tartrate (EDT), dipotassium tartrate (DKT), and ammonium dihydrogen phosphate (ADP) have varying suitability as piezoelectric elements. Tourmaline, an expensive material used mainly in hydrostatic pressure-measuring devices, is more durable than quartz and, for a given frequency, normally is more rugged than quartz. Rochelle salt has a greater piezoelectric effect than any other crystal but has the disadvantage of a greater sensitivity to temperature change than quartz. EDT has an advantage over quartz when used in frequency-modulated oscillators because of the wide gap between its resonant and antiresonant frequencies.

All piezoelectric crystals should have a good temperature coefficient, that is, should show as little change in resonant frequency as possible under large variations in temperature. Ideally, the piezoelectric constant of proportionality between the mechanical and electrical variables must be the same for both direct (pressure-to-electricity) and converse effects.

Natural quartz crystal and new synthetic quartz crystal have been found best to meet the properties required of a piezoelectric element. Quartz is hard (7 on Mohs' scale as compared with a diamond, which is 10) and is relatively abundant and stable. Because of the anisotropic structure of quartz, cutting a crystal in different orientations makes it possible to obtain crystals for the widest range of applications, including frequency control, filters, resonators, and electromechanical transducers. Each of these uses depends on the orientation of the crystal cut with respect to the crystallographic axes. As shown in Fig. 1, the principal axes in quartz are identified as the optic (Z), the mechanical (Y), and the electrical (X) axes. By use of polarized light and x-ray diffraction techniques, the various axes may be properly located, permitting a crystal plate to be cut from the quartz with the performance characteristics desired. After cutting and extensive processing, the crystal plate is mounted at the nodal points of its normal vibration. These points, which also serve as electrical connecting points, allow the crystal to vibrate freely with a minimum of damping. Finally, the mounted crystal is hermetically sealed in a dry inert atmosphere within a crystal holder.

A stability of 1 part in a billion is routinely possible by the use of proportional ovens that can regulate the crystal temperature within ±0.001°C. By employing multiplying or binary techniques, frequency-generating devices in the range of from less than 1 Hz to more than 600 MHz are available. In the temperature-compensated crystal oscillator (TCXO), stability over a wide range of temperatures is obtained by a compensating network that shifts the frequency of the crystal by an amount approximately equal and opposite to the shift of frequency caused by the

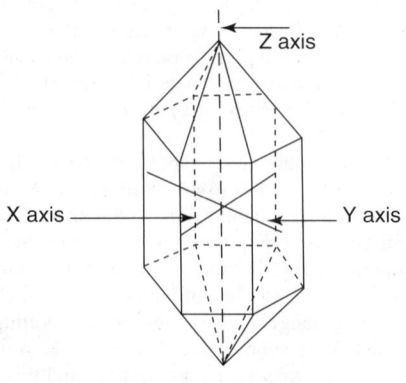

Fig. 1.   Principal axes in a quartz crystal

temperature change. This network eliminates the need for an oven. TCXOs are available with temperature stabilities of ±0.5 ppm over a temperature range of minus 40 to plus 70°C.

CRYSTAL PHASES (α^-, β^-, γ^-, ϵ^-, η^-, etc.). Certain alloy systems may form different crystal structures, according to the relative proportions of the constituents, e.g., Cu-Zn, for which no less than five different phases are known. In many cases, the same crystal structure occurs with quite different constituent metals, so that it is often possible to use one expression such, for example, as β-phase, to cover a wide variety of compounds all having the same basic structure. This effect is explained by the Hume-Rother rules. Pure substances, as well as alloys, may exhibit more than one crystal structure, depending on temperature and past history, e.g., cobalt, iron, titanium.

CRYSTAL PICKUP. Since piezoelectric crystals produce electrical voltages when subjected to mechanical stresses they offer possibilities for various electromechanical processes. One of these is the phonograph pickup, where the phonograph needle operating in the groove of the record must transmit mechanical motion to something which will convert it into electrical effects so electronic amplifiers may be used. The crystal is one of the most sensitive and at the same time one of the highest fidelity devices for doing this. While many crystals exhibit the piezoelectric effect, Rochelle salt is the most sensitive and is used for pickups, microphones, etc. Among the other crystals used, quartz is noted for its durability. Through a mechanical linkage the motion of the needle is transmitted into mechanical stresses on the crystal and hence produces electrical effects which may be amplified and then converted to sound by the loudspeaker.

CUMMINGTONITE. The mineral cummingtonite is a variety of amphibole which is essentially $(Mg, Fe, Mn)Si_8O_{22}(OH)_2$, the amounts of magnesium and iron varying as they replace one another. Cummingtonite is generally restricted to material containing from 50 to 70% $MgSiO_3$. The name *grunerite* has been applied to cummingtonite which contains more than 50% $FeSiO_3$. Cummingtonite usually occurs as a dark green to brown fibrous to lamellar mineral. It derives its name from Cummington, Massachusetts. See also **Amphibole**.

CUMULATIVE EXCITATION. An excited atom, in the metastable state, may receive a further increment of energy by collision, as with an electron, and thus be raised to a still higher energy state. This process by which an atom is raised by collision from one excited state to higher states is known as cumulative excitation. In fact, it is possible for an atom in the metastable state to receive sufficient energy by this process to be ionized and this process is designated as cumulative ionization.

CUPRITE. The mineral cuprite, cuprous oxide, Cu_2O, occurs as isometric crystals, usually octahedrons, but may be cubes, dodecahedrons or modified combinations. It also is found as a massive, earthy material. Its fracture is conchoidal to uneven; brittle; hardness, 3.5–4; specific gravity, 6.14; luster, submetallic to earthy; color, red; nearly transparent to nearly opaque. Its streak is shining brownish-red. Cuprite is a secondary mineral resulting doubtless from the oxidation of copper sulfides. It is often found associated with native copper, malachite and azurite.

Cuprite is a fairly common mineral, and of the many localities in which it occurs may be mentioned the Province of Perm, in the former U.S.S.R.; Chessy, France; Broken Hill, New South Wales; Corocoro, Bolivia; Andacollo, Chile; Bisbee, Arizona; and Del Norte County, California. Magnificent large transparent red gem crystals, some with a coating of malachite, have been found at Ojunga, S.W. Africa. The name cuprite is derived from the Latin *cuprum*, copper.

CURIE, PIERRE (1859–1906) CURIE, MARIE (1867–1934). Pierre Currie was born and raised in Paris. With his brother, Jacques, he studied crystals and in 1880 discovered piezoelectricity. Piezoelectricity is the production of an electric charge by pressure on certain crystals. Pierre became director at the School of Industrial Physics and Chemistry in Paris where he worked for 22 years. His doctoral thesis on magnetism led to his discovery, the Curie point, a temperature at which ferromagnetic substances lose their magnetism.

Pierre met a Polish born woman named Marie Sklodovska when she was studying mathematics, chemistry, and physics at the Sorbonne. They married in July 1895 and began working together. In 1897, Marie began searching for a topic for her doctoral research. She became interested in Becquerel's work and decided to investigate the rays Becquerel had discovered coming from uranium. She discovered "radioactivity" from this work. Upon further research, Marie realized that the radiation coming from uranium ore was too large to be produced by uranium alone. Pierre stopped his research and began working with Marie on radioactivity in pitchblende. As a result, they discovered polonium and radium. It took the Curies years to obtain a few grains of pure radium. Their work revolutionized the scientific world's understanding of atomic structure. The Curies also introduced the technique of radiation therapy into medicine. In 1903, the Curies and Henri Becquerel shared the physics Nobel Prize for their work on radioactivity. Marie Curie was the first female recipient of a Nobel Prize.

In 1906, Pierre was killed. Marie continued her work, was the first woman to teach at the Sorbonne, and in 1911 won an unprecedented second Nobel Prize in chemistry for her discovery and study of polonium and radium. On July 4, 1934, Marie died of leukemia from exposure to radiation.

Marie's daughter, Irene, joined in the work at the Radium Institute. With her husband, Frederic Joliot, and under the combined name Joliot-Curie, continued the work of the Curies in 1935, the pair won a Nobel Prize for their discovery of artificial radioactivity.

See also **Curie-Weiss Law**; and **Radioactivity**.

J.M.I.

CURIE POINT (or Curie Temperature). Ferromagnetic materials lose their permanent or spontaneous magnetization above a critical temperature (different for different substances). This critical temperature is called the Curie point. Similarly, ferroelectric materials lose their spontaneous polarization above a critical temperature. For some such materials, this temperature is called the "upper Curie point," for there is also a "lower Curie point," below which the ferroelectric property disappears. See also **Ferromagnetism**.

CURIE-WEISS LAW. The transition from ferromagnetic to paramagnetic properties, which occurs in iron and other ferromagnetic substances at the Curie point, is accompanied by a change in the relationship of the magnetic susceptibility to the temperature. P. Curie stated in 1895 that above this point the susceptibility varies inversely as the absolute temperature. But this was found to be not generally true, and was modified in 1907 by P. Weiss to state that the susceptibility of a paramagnetic substance above the Curie point varies inversely as the excess of the temperature above that point. At or below the Curie point, the Curie-Weiss law does not hold.

CURIUM. [CAS: 7440-51-9]. Chemical element, symbol Cm, at. no. 96, at. wt. 247 (mass number of the most stable isotope), radioactive metal of the Actinide series, also one of the Transuranium elements, mp estimated $1350 \pm 50°C$. ^{247}Cm has a half-life of 1.64×10^7 years. Other long-lived isotopes are ^{245}Cm ($t_{1/2} = 9320$ years), ^{246}Cm ($t_{1/2} = 5480$ years), ^{248}Cm ($t_{1/2} = 4.7 \times 10^5$ years), and ^{250}Cm ($t_{1/2} = 2 \times 10^4$ years). Other known isotopes are ^{238}Cm, ^{242}Cm, ^{244}Cm, and ^{249}Cm. Electronic configuration

$$1s^2 2s^2 2p^6 3s^2 3p^6 3d^{10} 4s^2 4p^6 4d^{10} 4f^{14} 5s^2 5p^6 5d^{10} 5f^7 6s^2 6p^6 6d^1 7s^2.$$

Ionic radius: Cm^{3+} 0.98 Å. See also **Chemical Elements**.

First identified in 1944 by G.T. Seaborg, R.A. James, and A. Ghiorso, who found ^{242}Cm in the product obtained by bombarding ^{239}Pu with alpha particles of resonance energies. Later L.B. Werner and I. Perlman produced and isolated the same isotope by the action of neutrons upon ^{241}Am.

In experiments the concentration of curium must be kept low in order to avoid the formation of a reducing medium due to the action of the ^{242}Cm alpha particles on H_2O. At a concentration of 10^{-5} molar in curium, and under conditions where americium(III) is oxidized to americium(VI) in the same solution, the curium is not oxidized above the (III) state with ammonium peroxydisulfate.

The solubility properties of curium(III) compounds are in every way similar to those of the other tripositive Actinide elements and the tripositive Lanthanide elements. Thus the fluoride and oxalate are insoluble in acid solution, while the nitrate, halides, sulfate, perchlorate, and sulfide are all soluble.

Solid curium trifluoride has been prepared by drying the fluoride, which precipitates from dilute HNO_3 upon the addition of hydrofluoric acid.

Curium trifluoride can be reduced to the metal by heating at 1275°C in a beryllia crucible with barium vapor. The metal is silvery in color and has the properties of an electropositive element in common with the other Actinide elements.

The ion Cm^{3+} is colorless, as are its compounds generally. CmF_3 is hexagonal, Cm_2O_3 is white and CmO_2 is black and hexagonal. $CmCl_3$ is light yellow and hexagonal. Cm^{4+} is known in solution only as the complex fluoride.

In research at the Institute of Radiochemistry, Karlsruhe, West Germany during the early 1970s, investigators prepared alloys of curium with iridium, palladium, platinum, and rhodium. These alloys were prepared by hydrogen reduction of the curium oxide or fluoride in the presence of finely divided noble metals. The reaction is called a *coupled reaction* because the reduction of the metal oxide can be done in the presence of noble metals. The hydrogen must be extremely pure, with an oxygen content of less than 10^{-25} torr.

Curium was first isolated in the form of a pure compound, the hydroxide, of curium–242 (produced by the neutron irradiation of americium–241) by Werner and Perlman at the University of California in the autumn of 1947. Much of the earlier work with curium used the isotopes ^{242}Cm and ^{244}Cm, but the heavier isotopes offer greater advantages mainly because of their longer half-lives. The isotope ^{248}Cm, obtainable in relatively high isotopic purity as the alpha particle decay daughter of ^{252}Cf, is the most practical for chemical studies. See also **Radioactivity**.

A gram of ^{242}Cm generates approximately three thermal watts of energy–very high as compared with one–half thermal watt per gram of ^{238}Pu. This property has given consideration to the possible use of curium as an isotope power source.

Additional Reading

Asprey, L.B., F.H. Ellinger, S. Fried, and W.H. Zachariasen: "Evidence for Quadrivalent Curium: X–Ray Data on Curium Oxides," *Amer. Chem. Soc. J.*, **77**, 1707–1708 (1955).
Asprey, L.B., F.H. Ellinger, S. Fried, and W.H. Zachariasen: "Evidence for Quadrivalent Curium. II. Curium Tetrafluoride," *Amer. Chem. Soc. J.*, **79**, 5825 (1957).
Lide, D.R.: *CRC Handbook of Chemistry and Physics*, 84th Edition, CRC Press, Boca Raton, FL, 2003.
Seaborg, G.T.: "The Chemical and Radioactive Properties of the Heavy Elements," *Chem. Eng. News.*, **23**, 2190–2193 (1945). (A classic reference.)
Seaborg, G.T.: *Transuranium Elements*, Dowden, Hutchinson Ross, Stroudsburg, PA, 1978. (A classic reference.)
Werner, L.B. and I. Perlman: "First Isolation of Curium," *Amer. Chem. Soc. J.*, **73**, 5215–5217 (1951). (A classic reference.)

CURL, ROBERT F., JR. (1933–). An American who won the Nobel prize for chemistry along with Sir Harold W. Kroto and Richard E. Smalley in 1996, the 100th anniversary of Alfred Nobel's death. The trio won for the discovery of the C_{60} compound called buckminsterfullerene. He graduated from Rice University and received a Ph.D. from the University of California, Berkeley in1957.

See also **Buckminsterfullerene (Buckyballs)**; and **Carbon**.

CYANAMIDES. [CAS: 420-04-2]. Cyanamide NC· NH_2 or HN:C:NH is a white solid, melting point 44°C, boiling point 140°C at 20 mm pressure, transformed at 150°C into cyanuramide, tricyantriamide (NC· $NH_2)_3$. Cyanamide reacts (1) as a base with strong acids forming salts, (2) as an acid forming metallic salts, such as calcium cyanamide $CaCN_2$. Cyanamide is formed (1) by reaction of cyanogen chloride CN· Cl plus ammonia (ammonium chloride also formed), (2) by reaction of thiourea plus lead hydroxide (lead sulfide also formed).

When calcium cyanamide is boiled with water, dicyandiamide (NC· $NH_2)_2$, melting point 207°C is formed (along with calcium hydroxide). Fusion of dicyandiamide with sodium carbonate plus carbon produces sodium cyanide plus ammonia (also some tricyantriamide). Diethylcyanamide $(C_2H_5)_2N·$ CN is a colorless liquid, boiling point 189°C at 748 mm pressure, and when hydrolyzed yields diethylamine $(C_2H_5)_2NH$ plus ammonia plus carbon dioxide. Diphenylcyanamide $C_6H_5N:C:NC_6H_5$ when hydrolyzed yields aniline plus carbon dioxide. Benzylcyanamide $C_6H_5CH_2$ NH· CN is a white solid, melting point 43°C.

CYANIC ACID AND RELATED COMPOUNDS. Cyanic acid, HCNO or HOCN, is a colorless, odorless liquid; soluble in water and in ether; volatile with decomposition when heated; passing at ordinary temperature into a mixture of cyanuric acid, $(HNCO)_3$, and cyamelide, $(CONH)_x$, white solid, which on vaporizing yields cyanic acid; when cyanic acid vapor is rapidly cooled in a freezing mixture, unstable, liquid cyanic acid is obtained, and when the vapor is condensed above 105°C, cyanuric acid

$$(HNCO)_3 \text{ or } CO \begin{array}{c} NH-CO \\ \diagup \qquad \diagdown \\ \qquad \qquad NH \\ \diagdown \qquad \diagup \\ NH-CO \end{array}$$

is obtained. Cyamelide dissolves in sulfuric acid unchanged and addition of water causes precipitation of cyamelide; passes into cyanuric acid when warmed with concentrated sulfuric acid, finally into carbon dioxide plus ammonia; dissolves in sodium hydroxide solution forming sodium cyanate. Sodium cyanate is prepared by heating sodium cyanide and an oxide such as lead monoxide, PbO, trilead tetroxide, Pb_3O_4, or lead dioxide, PbO_2, addition of water and separation of the sodium cyanate solution from the lead oxide by filtration. Sodium cyanate solution upon boiling changes into sodium carbonate plus urea, $CO(NH_2)_2$.

Ammonium cyanate, $CNONH_4$, white solid, formed by reaction of sodium cyanate and ammonium sulfate solutions, is transformed to urea upon being heated at 100°C. This reaction was carried out in 1828 by Wöhler, and is the first record of a so-called inorganic substance being transformed outside a living organism into a so-called organic substance. The following esters are known:

Methyl isocyanate, CH_3NCO, boiling point 44°C

Ethyl cyanate, C_2H_5OCN, decomposes on heating

Ethyl isocyanate, C_2H_5NCO, boiling point 60°C

Phenyl isocyanate, C_6H_5NCO, boiling point 166°C

Ethyl cyanurate $C_2H_5O·C \begin{array}{c} N-C(OC_2H_5) \\ \diagup \qquad \qquad \diagdown \\ \qquad \qquad \qquad N \\ \diagdown \qquad \qquad \diagup \\ N=C(OC_2H_5) \end{array}$

Ethyl isocyanurate $CO \begin{array}{c} N(C_2H_5)-CO \\ \diagup \qquad \qquad \diagdown \\ \qquad \qquad \qquad N(C_2H_5) \\ \diagdown \qquad \qquad \diagup \\ N(C_2H_5)-CO \end{array}$

The extensive use of organic isocyanates in various industrial processes for production of high polymers has brought about tonnage production. Toluene diisocyanate is made by nitrating toluene to the dinitro compound, which is then reduced to the diamine, and treated with phosgene to obtain the diisocyanate:

$$C_6H_5CH_3 \xrightarrow{HNO_3} C_6H_3(NO_2)_2CH_3H \xrightarrow{H}$$

$$C_6H_3(NH_2)_2CH_3 \xrightarrow{COCL_2} C_6H_3(NCO)_2CH_3$$

Toluene diisocyanate is widely used in the manufacture of urethane plastics, particularly the urethane foamed plastics. Another isocyanate, diphenylmethane 4,4′-diisocyanate, is produced by reaction of aniline and formaldehyde, followed by reaction with phosgene:

$$2C_6H_5NH_2 \xrightarrow{HCHO} CH_2(C_6H_4NH_2)_2 \xrightarrow{COCL_2} CH_2(C_6H_4NCO)_2$$

The diphenylmethane 4,4′-diisocyanate is used in the manufacture of solid urethane elastomers (primarily for heavy-duty tires) and chemically resistant coatings.

Fulminic acid, HONC, and the fulminates are violently explosive. Utilizing this property, mercuric fulminate, $Hg(ONC)_2 · \frac{1}{2}H_2O$, is used as a detonator for other explosives. Mercury fulminate is made by the reaction of ethyl alcohol and mercuric nitrate in excess of nitric acid, from which insoluble mercuric fulminate separates. Silver fulminate, Ag(ONC), is more explosive than mercuric fulminate, and is used in the manufacture of firecrackers. Free fulminic acid may be obtained by reaction of potassium fulminate and excess of ether. It volatilizes with the ether upon distilling, and changes rapidly to metafulminic acid. Related to fulminic acid is fulminuric acid, $(HONC)_3$, or $NO_2·CH(CN)·CONH_2$.

CYANOGEN. [CAS: 460-19-5]. Cyanogen $(CN)_2$ is a colorless gas of marked characteristic odor, very poisonous, density 1.8 (air equal to 1.0), melting point $-28°C$, boiling point $-20°C$, soluble. When passed into water at $0°C$, cyanogen forms hydrocyanic acid plus cyanic acid, but at ordinary temperatures the reaction is complex. With sodium hydroxide solution, there is formed with cyanogen sodium cyanide plus sodium cyanate, with dilute sulfuric acid oxamic acid $COOH·CONH_2$, oxalic acid $COOH·COOH$. By reaction with tin and hydrochloric acid, cyanogen is reduced to ethylene diamine $CH_2·NH_2·CH_2·NH_2$. Cyanogen reacts with hydrogen to form hydrocyanic acid, and with metals, e.g., zinc, copper, lead, mercury, silver, to form cyanides. Cyanogen, (1) when burned in air produces a violet flame forming carbon dioxide and nitrogen in the outer part and carbon monoxide and nitrogen in the inner part, (2) when exploded with oxygen produces carbon dioxide or carbon monoxide and nitrogen depending upon the ratio of oxygen to cyanogen (2 volumes oxygen plus 1 volume cyanogen yields 2 volumes carbon dioxide plus 1 volume nitrogen; 1 volume oxygen plus 1 volume cyanogen yields 2 volumes carbon monoxide plus 1 volume nitrogen). The flame spectrum contains characteristic bands in the blue and violet. By means of the electric spark, the electric arc or a red hot tube, cyanogen is decomposed into carbon plus nitrogen. When heated at ordinary pressure at about $300°C$, or under 300 atmospheres pressure at about $225°$, cyanogen is converted into paracyanogen, a brown powder, also formed when mercuric cyanide is heated. Cyanogen is prepared (1) by reaction of sodium cyanide and copper sulfate solutions, whereby one half the cyanogen is evolved as cyanogen gas and one half remains as cuprous cyanide. From the filtered cuprous cyanide, by treatment with ferric chloride solution, cyanogen is evolved with accompanying formation of ferrous chloride, (2) by heating ammonium oxalate $COONH_4·COONH_4$ with phosphorus pentoxide, water being abstracted. Small amounts of cyanogen are present in blast furnace gas and raw coal gas.

CYANOHYDRINS. A cyanohydrin is an organic compound that contains both a cyanide and a hydroxy group on an aliphatic section of the molecule. Cyanohydrins are usually α-hydroxy nitriles which are the products of base-catalyzed addition of hydrogen cyanide to the carbonyl group of aldehydes and ketones. The IUPAC name for cyanohydrins is based on the α-hydroxy nitrile name. Common names of cyanohydrins are derived from the aldehyde or ketone from which they are formed.

The outstanding chemical property of cyanohydrins is the ready conversion to α-hydroxy acids and derivatives, especially α-amino and α,β-unsaturated acids. Because cyanohydrins are primarily used as chemical intermediates, data on production and prices are not usually published. The industrial significance of cyanohydrins is waning as more direct and efficient routes to the desired products are developed.

Properties

Cyanohydrins are usually colorless to straw yellow liquids with an objectionable odor akin to that of hydrogen cyanide. Table 1 list physical properties of some common cyanohydrins.

Cyanohydrins can react either at the nitrile group or at the hydroxyl group.

Preparation

Cyanohydrins can be formed by the acid- or base-catalyzed reaction of hydrogen cyanide with an aldehyde or ketone, the displacement of bisulfite ion by cyanide ion on the bisulfite addition compounds of aldehydes and ketones, or the exchange of cyanide ion between a ketone cyanohydrin and an aldehyde to give the usually more stable aldehyde cyanohydrin.

Shipping, Storage, and Handling

Cyanohydrins should be stabilized with acid to pH 3–4 to prevent decomposition to hydrogen cyanide and carbonyl compound. When cyanohydrins are shipped, steel drums, carboys, tank cars, and barges are used. In general, cyanohydrins are combustible liquids and many decompose upon heating. They should be stored in a cool, dry place, preferably outside and separated from other storage. Containers should be protected against physical damage.

Health and Safety

Cyanohydrins are highly toxic by inhalation or ingestion, and moderately toxic through skin absorption. Special protective clothing should be worn and any exposure should be avoided. The area should be adequately ventilated. Immediate medical attention is essential in case of cyanohydrin poisoning.

Specific Compounds

Formaldehyde Cyanohydrin. This cyanohydrin, also known as glycolonitrile, is a colorless liquid with a cyanide odor. It is soluble in water, alcohol, and diethyl ether. Equimolar amounts of 37% formaldehyde and aqueous hydrogen cyanide mixed with a sodium hydroxide catalyst at $2°C$ for one hour give formaldehyde cyanohydrin in 79.5% yield.

Acetaldehyde Cyanohydrin. This cyanohydrin, commonly known as lactonitrile, is soluble in water and alcohol, but insoluble in diethyl ether and carbon disulfide. Lactonitrile is used chiefly to manufacture lactic acid and its derivatives, primarily ethyl lactate. Lactonitrile is manufactured from equimolar amounts of acetaldehyde and hydrogen cyanide containing 1.5% of 20% NaOH at $-10-20°C$. The product is stabilized with sulfuric acid.

Acetone Cyanohydrin. This cyanohydrin, also known as α-hydroxyisobutyronitrile and 2-methyllactonitrile, is very soluble in water, diethyl ether, and alcohol, but only slightly soluble in carbon disulfide or petroleum ether. Acetone cyanohydrin is the most important commercial cyanohydrin as it offers the principal commercial route to methacrylic acid and its derivatives, mainly methyl methacrylate. see also **Methacrylic Acid and Derivatives.** The principal U.S. manufacturers are Rohm and Haas Company, DuPont, CyRo Industries, and BP Chemicals.

Acetone cyanohydrin is manufactured by the direct reaction of hydrogen cyanide with acetone, catalyzed by base, generally in a continuous process.

Benzaldehyde Cyanohydrin. This cyanohydrin, also known as mandelonitrile, is a yellow, oily liquid, insoluble in water, but soluble in alcohol and diethyl ether. Mandelonitrile is a component of the glycoside amygdalin, a precursor of laetrile found in the leaves and seeds on most *Prunus* species (plum, peach, apricot, etc). It is commercially prepared from benzaldehyde and hydrogen cyanide.

Ethylene Cyanohydrin. This cyanohydrin, also known as hydracrylonitrile or glycocyanohydrin, is a straw-colored liquid miscible with water, acetone, methyl ethyl ketone, and ethanol, and is insoluble in benzene, carbon disulfide, and carbon tetrachloride. Ethylene cyanohydrin differs from the other cyanohydrins discussed here in that it is a β-cyanohydrin. It is formed by the reaction of ethylene oxide with hydrogen cyanide.

MICHAEL S. CHOLOD
Rohm and Haas Company

TABLE 1. PHYSICAL PROPERTIES OF SOME CYANOHYDRINS

Name	Mol wt	Mp, °C	Bp,[a] °C	Specific gravity	n_D^{20}	Flash point, °C
formaldehyde cyanohydrin	57.06	<-72	119 at 3.2 kPa[b]	1.104	1.4117	
acetaldehyde cyanohydrin	71.03	-40	182–184, dec	0.988	1.4050	77
acetone cyanohydrin	85.10	-19	85 at 3.1 kPa[b]	0.927	1.3992	74
cyclohexanone cyanohydrin	121.17	29	109–113 at 1.2 kPa[b]	1.032	1.4576	60
benzaldehyde cyanohydrin	133.15	-10	170, dec	1.117	1.5315	
ethylene cyanohydrin	71.08	-46.2	228	1.059	1.4256	129
propylene cyanohydrin	85.11		207		1.4280	

[a] At 101.1 kPa[b] unless otherwise noted.
[b] To convert kPa to mm Hg, multiply by 7.5.

Additional Reading

Migrdichian, V.: *The Chemistry of Organic Cyanogen Compounds*, Reinhold Publishing Co., New York, NY, 1947, Chapt. 9.

CYBOTAXIS. A condition in which certain liquids, under x-ray examination, give evidence of structure resembling that of crystals. By passing a beam of x-rays through various alcohols and other organic liquids, G.W. Stewart and his collaborators have obtained one, two, or even three diffraction maxima or halos, somewhat like the diffraction rings produced by powdered crystals. These suggest that molecules are temporarily arranged in rows, layers, or stacks like bricks in a pile and that they have one, two, or even three different dimensions or spacings, corresponding, in accordance with Bragg's law, to the different angles of diffraction observed.

A closely related property is exhibited by certain substances known as "liquid crystals," which appear to be intermediate between merely cybotactic liquids and true crystals. In these there appear to be large groups of molecules which, though able to move and turn about, retain their structural arrangement. Such mesomorphic substances manifest even some of the optical properties of crystals, which the former type do not.

See also **Liquid Crystals**.

CYCLOHEXANOL-CYCLOHEXANONE.

Cyclohexanol is a colorless, viscous liquid with a camphoraceous odor. It is used chiefly as a chemical intermediate, a stabilizer, and a homogenizer for various soap detergent emulsions, and as a solvent for lacquers and varnishes. Cyclohexanol was first prepared by the treatment of 4-iodocyclohexanol with zinc dust in glacial acetic acid, and later by the catalytic hydrogenation of phenol at elevated temperatures and pressures.

Cyclohexanone is a colorless, mobile liquid with an odor suggestive of peppermint and acetone. Cyclohexanone is used chiefly as a chemical intermediate and as a solvent for resins, lacquers, dyes, and insecticides. Cyclohexanone was first prepared by the dry distillation of calcium pimelate, $-OOC(CH_2)_5COO^-Ca^{2+}$, and later by Bouveault by the catalytic dehydrogenation of cyclohexanol.

Physical Properties

Important physical properties of cyclohexanol and cyclohexanone are shown in Table 1. Cyclohexanol is miscible in all proportions with most organic solvents, including those customarily used in lacquers. It dissolves many oils, waxes, gums, and resins.

Cyclohexanone is miscible with methanol, ethanol, acetone, benzene, *n*-hexane, nitrobenzene, diethyl ether, naphtha, xylene, ethylene glycol, isoamyl acetate, diethylamine, and most organic solvents. This ketone dissolves cellulose nitrate, acetate, and ethers, vinyl resins, raw rubber, waxes, fats, shellac, basic dyes, oils, latex, bitumen, kaure, elemi, and many other organic compounds.

Reactions

Cyclohexanol shows most of the typical reactions of secondary alcohols. Cyclohexanone shows most of the typical reactions of aliphatic ketones.

Economic Aspects

Estimated annual cyclohexanone production capacities were 665×10^3 t in 1992; the production is greater than 90% captive for caprolactam production. The annual cyclohexanol production is only 10 thousand

metric tons. These production figures do not include KA-oil (cyclohexanol–cyclohexanone) production for adipic acid. Worldwide annual capacity for cyclohexanone is approximately 3.0 million metric tons, also primarily for caprolactam production. Projected new capacity for caprolactam could add 0.5 million metric tons worldwide in this decade.

Health and Safety Factors

Cyclohexanol is slightly toxic by the oral route of exposure and is slightly irritating to the skin. It can cause severe eye irritation and transient corneal injury.

The ACGIH threshold limit value (TLV), time-weighted average for an 8-h workday, 40-h workweek, was set at 50 ppm (~ 200 mg/m^3) with a notation for skin absorption.

Cyclohexanone has only slight toxicity by the oral, dermal, and inhalation routes of exposure. Liquid or vapor exposures may result in transient corneal injury. Primary irritation and defatting of the skin can result from substantial or prolonged contact with cyclohexanone. Exposure to high vapor concentrations can cause central nervous system (CNS) depression, an effect which can also occur after ingestion or repeated dermal exposure.

The time-weighted average OSHA permissible exposure limit (PEL), as well as the ACGIH threshold limit value (TLV), for cyclohexanone is 25 ppm (100 mg/m^3) with a notation for skin absorption.

The precautions usually observed when handling volatile solvents should be observed as a matter of course with cyclohexanone and cyclohexanol. These include adequate and proper ventilation, avoidance of prolonged breathing of vapor or contact of the liquid with the skin, avoidance of internal consumption, and protection of the eyes against splashing liquids.

WILLIAM B. FISHER
JAN F. VANPEPPEN
Allied Signal Inc.

Additional Reading

Chemical Economics Handbook, SRI International, Menlo Park, CA, Sept. 1990; Allied Signal Co., internal data, 1992.
1991–1992 Threshold Limit Values for Chemical Substances and Physical Agents and Biological Exposure Indices, American Conference of Government Industrial Hygienists, Cincinnati, OH, 1991.
Registry of Toxic Effects of Chemical Substances (RTECS), NIOSH Database, Cincinnati, OH, 1992.
U.S. Pat. 4,092,360 (May 30, 1978), J. F. VanPeppen and W. B. Fisher (to Allied Chemical Corp.).

CYCLOTRON. A device for accelerating charged particles to high energies by giving particles traveling in a spiral path successive increments of energy from an alternating electric field between electrodes placed in a constant magnetic field. The path radius increases as energy increases. It was invented in 1929 by E. O. Lawrence (1901–1958) at the University of California at Berkeley. It is now used chiefly for basic nuclear research.

CYSTEINE. See **Amino Acids**.

CYSTINE. See **Amino Acids**.

CYTOCHROMES. The cytochrome *c*-cytochrome oxidase system represents the terminal segment of the respiratory chain common to the vast majority of organisms utilizing oxygen as the terminal oxidant in tissue respiration. The complete respiratory chain consists of a number of electron carriers, both protein and nonprotein in nature, organized in a definite sequence within the walls and internal partitions of subcellular organelles known as mitochondria. These structures carry the electrons that come from the substrates being oxidized and eventually react with oxygen. The energy released in several of the many steps of this series of reactions is utilized to make the high-energy compound adenosine triphosphate, a process known as *oxidative phosphorylation*. The high-energy compound is, in turn, employed to drive the many reactions of metabolism which require chemical energy. Every component of the terminal respiratory chain is reduced by the component immediately proceeding it and then reduces the component immediately following it in the chain, itself becoming reoxidized. The *c*-cytochrome oxidase system is common to all vertebrates and invertebrates, plants, as well as numerous microorganisms, and must

TABLE 1. PROPERTIES OF CYCLOHEXANOL AND CYCLOHEXANONE

Property	Cyclohexanol	Cyclohexanone
structure	⬡—OH	⬡=O
mp, °C	25.15	−47
bp, °C	161.1	156.7
d_4^{20}, g/mL	0.9493	0.9478
n_D^{25}	1.4648	
sp heat, 15–18°C, J/g[a]	1.75	1.81
viscosity, 25°C, mPa·s(= cP)	4.6	2.2
flash point, open cup, °C	67.2	54

[a] To convert J to cal, divide by 4.184.

be distinguished from systems having similar functions, but very different properties, which occur in numerous bacteria.

The cytochromes were first observed by MacMunn as early as 1886. He described their spectral absorption bands in a large variety of organisms and tissues. His discovery was, however, forgotten after a controversy with Hoppe-Seyler had raised doubts as to the validity of some of his conclusions, and it was not until 1925 that Keilin independently rediscovered the remarkable cytochrome spectrum in the flight muscles of a living insect.

Keilin's observations came at a time when the understanding of tissue respiration had advanced to the point of providing the foundations necessary for the unraveling of the physiological role and chemical nature of cytochromes. The first step had indeed been taken some 40 years earlier by Ehrlich when he found that a variety of animal tissues could transform a mixture of α-napththol and dimethyl-p-phenylenediamine to indophenol, in the presence of oxygen. A decade later, the enzyme responsible for this effect had been named indophenol oxidase, and it was shown that its activity was inhibited by cyanide. In the first decades of the twentieth century, Warburg, from studies of the catalysis of the oxidation of cysteine by iron-charcoal, considered a "model" of cellular respiration. He concluded that oxygen activation was the all-important process in cellular respiration and that an iron-containing enzyme, the "respiratory enzyme" or *Atmungsferment*, is solely responsible for the transport of the oxidizing equivalents of oxygen to the substrates. An opposing view was taken by Thunberg, who had detected a large variety of dehydrogenases in tissues, and by Wieland, who used palladium-hydrogen as a "model" of tissue respiration and believed that substrate-specific hydrogen activations were characteristic of all biological oxidation processes, the reaction with oxygen being nonspecific and relatively unimportant.

The controversy as to the respective roles and importance of hydrogen and oxygen activation faded into the background when, following his initial observations, Keilin demonstrated that the four-banded spectrum of cytochrome, observed in a large variety of tissues and organisms, was in fact the spectrum of the ferrous, or reduced, forms of three distinct cytochromes—cytochrome a, cytochrome b and cytochrome c. Keilin obtained a soluble preparation of cytochrome c from baker's yeast, and together with Hartree, in 1938–1939, showed that the indophenol oxidase activity of particulate tissue preparations was simply the result of a nonenzymic reduction of cytochrome c by dimethyl-p-phenylenediamine, the reduced heme protein being oxidized by indophenol oxidase in the presence of oxygen. Having established the nature of the final steps of tissue respiration, they renamed the enzyme "cytochrome oxidase," since its only function appeared to be the oxidation of cytochrome c. There had been no doubt of the overwhelming physiological importance of the system ever since 1934, when Haas found that, in a number of tissues, the rate of oxygen uptake was identical to that of cytochrome c reduction, demonstrating that nearly all of the oxidizing equivalents of oxygen were transmitted by the cytochrome c-cytochrome oxidase system.

That the material in tissues reacting directly with oxygen was in fact a heme compound had been shown by the experiments of Warburg and collaborators on the effect of carbon monoxide on tissue respiration, carried out in the late 1920s. Warburg observed that carbon monoxide inhibits the uptake of oxygen by tissues and that this inhibition is reversed in bright light. Using this phenomenon, he succeeded in measuring the absorption spectrum of the carbon monoxide complex of the respiratory enzyme, a spectrum that turned out to be clearly that of a heme compound. Thus, when Keilin and Hartree in 1939 found that in the presence of carbon monoxide, cytochrome a showed up as two spectroscopic components, they were able to demonstrate that the new cytochrome, cytochrome a_3, was the substance responsible for the photochemical action spectrum of Warburg. Cytochrome a_3 was thus identified with the respiratory enzyme reacting directly with oxygen, and the system was considered to be composed of three entities, cytochromes c, a, and a_3, reacting consecutively, like all the other components of the respiratory chain.

Cytochrome c consists of a polypeptide chain, from 104 to 108 amino acid residues in length. A single heme prosthetic group is attached by thioether bonds formed between the sulfhydryl side chains of two cysteine residues in the protein and the vinyl side chains of the porphyrin ring as shown by

ANN C. DEBALDO
University of South Florida
Tampa, Florida

Additional Reading

Arinc, E., E. Hodgson, and J.B. Schenkman: *Molecular and Applied Aspects of Oxidative Drug Metabolizing Enzymes*, Plenum Publishing Corporation, New York, NY, 1999.

Asard, H., A. Berczi, and R.J. Caubergs: *Plasma Membrane Redox Systems and Their Role in Biological Stress and Disease*, Kluwer Academic Publishers, Norwell, MA, 1999.

Ioannides, C.: *Cytochromes P450: Metabolic and Toxicological Aspects*, CRC Press, LLC., Boca Raton, FL, 1998.

Ishimura, Y., H. Shimada, and M. Suematsu: *Oxygen Homeostasis and Its Dynamics*, Springer-Verlag Inc., New York, NY, 1998.

Scott, R.A. and A.G. Mauk: *Cytochrome C: A Multidisciplinary Approach*, University Science Books, New York, NY, 1997.

CYTOTOXIC CHEMICALS. Chemical agents that damage cells to which they are applied. They are poisons, to which cells respond with injury, disease, or death. There are multitudes of cytotoxic chemicals; they act by a variety of mechanisms; and they have many different kinds of effects. See also **Carcinogens**.

A simplistic classification can start with biological alkylating agents of which a great variety exists. They all possess the ability to add alkyl groups to a wide range of electronegative groups under mild aqueous conditions. It is thought that some destroy growing cells by cross-linking the adjacent guanidine molecules on DNA. Others cause breaks in chromosomes while yet others uncouple oxidative phosphorylation, which leads to a loss of ADP and ATP and accumulation of AMP.

Antimetabolites are analogs of folic acid, the purines and the pyrimidines. In the first group, the effect is to block DNA synthesis by interfering with the deoxyuridylic acid $\rightarrow$ deoxythymidilic acid step. The second interferes with DNA synthesis by blocking conversion of inosinic acid to adenylsuccinic acid and xanthylic acid. Pyridimine analogs inhibit deoxythymidilic acid syntheses which results in failure of DNA synthesis and the death of proliferating cells.

Several antibiotics are also cytotoxic and combine with DNA, blocking its template activity in directing the synthesis of messenger RNA.

Many of these cytotoxic chemicals find application in treating various forms of cancer, but it must be remembered that they all also affect normal cells and their introduction to body organs can only be treated with great care and even apprehension.

D

DACITE. The name of a somewhat variable group of extrusive igneous rocks similar to the rhyolites but richer in plagioclase feldspar. Typical dacites are felsitic to porphyritic in texture. Dacites are the extrusive equivalents of the quartz-rich varieties of diorites and are sometimes classified as quartz-bearing andesites. The porphyritic types usually occur toward the center of the thicker dacite flows, dikes and sills, as well as the marginal zones of laccoliths. Dacites are common in the Cordilleran province of North, Central and South America. The term, dacite, was proposed by G. Stache of Austria for lavas in the old Roman province of Dacia.

DALTON, JOHN (1766–1844). Dalton was an English scientist who worked in the fields of biology, chemistry, earth science, mathematics, and physics. He became a professor of mathematics at the New College in Manchester, England. His beginning research work was on meteorology. Early in his career, he also studied color blindness. This was of special interest to him since Dalton was himself color-blind.

His teaching career required teaching chemistry and he began researching chemistry and meteorology and soon he proposed that in a mixture of gases, each component acts independently of the other gases present. This proposal is known as Dalton's Law. Continued research in this field led him to consider the nature of matter and he began to develop a systematic atomic theory based on quantitative chemical principles.

See also **Dalton Law**.

J.M.I.

DALTONIDE COMPOUNDS. See **Chemical Composition**.

DALTON LAW. The law of partial pressures in mixed gases and vapors. If several gases not reacting chemically with each other are introduced into the same container, the pressure of the resulting mixture is equal to the sum of the pressures which would be observed if each gas were separately enclosed in that container. We may, for example, regard the atmospheric pressure as the sum of a nitrogen pressure, an oxygen pressure, an argon pressure, a carbon dioxide pressure, a water-vapor pressure, etc. The same principle holds for mixtures of the saturated vapors of two or more liquids evaporating in the same closed space, provided one liquid does not dissolve the vapor from the other (as water dissolves ammonia). Like other gas laws, this law is approximately valid only within limits. See also **Combustion (Fuels)**.

DANBURITE. The mineral danburite, $CaB_2Si_2O_8$, calcium-boron silicate, crystallizes in the orthorhombic system in prismatic forms somewhat resembling the mineral topaz. Its fracture is subconchoidal; brittle; hardness, 7; specific gravity, 2.97–3.02; color, colorless, yellowish-white, yellow, dark wine yellow and brownish-yellow; luster, vitreous to greasy; translucent to transparent. It is found at Danbury, Connecticut, from whence its name was derived, Saint Lawrence County, New York, Switzerland, Japan, and Madagascar.

DARCY'S LAW. The volumetric rate of flow of water through a sand filter bed is directly proportional to the cross-sectional area of the bed and the pressure difference across the bed, and inversely proportional to the thickness of the bed.

DARMSTADIUM. See **Chemical Elements**.

DATOLITE. Datolite, basic calcium boron silicate, $CaBSiO_4(OH)$, occurs in monoclinic crystals of varied habit, mostly short stout prisms, but often in highly modified forms. Datolite reveals no cleavage, its fracture is conchoidal to uneven; brittle; hardness, 5–5.5; specific gravity, 2.9–3.0; luster, vitreous to dull; color, white to gray or may be greenish, yellowish, or brownish. It has a white streak and is transparent to translucent usually, but has been observed opaque.

Datolite is a secondary mineral, being found in veins and cavities associated with zeolites and calcite, particularly in the basic igneous rocks. It has been found in the Harz Mountains, Germany; in the Trentino district, Italy; in Norway and Tasmania. In the United States it has been found in the Triassic traps of the Connecticut River Valley in Massachusetts and Connecticut, and from similar rocks in New Jersey. In Michigan, datolite has been found associated with the copper-bearing rocks of Keweenaw County. This mineral derives its name from the Greek word meaning to divide, in reference to the granular structures of some of the massive varieties.

DAVY, SIR HUMPHRY (1778–1829). Born in Cornwell, Davy was the first to isolate the alkali metals and recognize the identity of chemical and electrical energy. A pioneer in the science of electrochemistry, he carried out basic studies of electrolysis of salts and water, and his application of electricity to the decomposition of molten caustic potash led to the isolation of metallic potassium.

DEACON PROCESS. A method of converting hydrogen chloride to chlorine by oxidation of hydrogen chloride with oxygen at 400–500°C over a copper-salt catalyst, $2HCl + O \rightarrow Cl_2 + H_2O$. It is a means of producing chlorine without caustic and of utilizing the large amounts of by-product hydrogen chloride from the chlorination of organic compounds. When conducted in the presence of an organic compound that reacts with the chlorine formed, it is known as oxychlorination, e.g., $CH_2=CH_2 + 2HCl + O \rightarrow CH_2ClCH_2Cl + H_2O$.

DEARATION. Because water dissolves, to a greater or lesser extent, many common gases, it will contain in the natural state a certain amount of dissolved gases, such as oxygen and carbon dioxide. Deaeration is the removal of this dissolved gas. Deaeration at the present is practiced where the gas that water contains would have undesirable effects. Often dissolved oxygen is objectionable because of its corrosive action. This is true in the case of the high-pressure steam boiler, where a small amount of oxygen dissolved in the feedwater may become quite active in attacking the boiler metal under the high pressure and temperature conditions there experienced. Steam boiler operators often treat their boiler feedwater in deaerators to remove this oxygen.

These deaerators are either of the deactivating or the heating type. Deactivating types employ chemical means of deaeration. Deaerating action in a heating-type deaerator is obtained by first reducing the solubility of the gas through heating the water (under pressure); second, reducing the pressure and producing explosive boiling; and third, controlling the agitation of the water subsequent to the second action in a partially evacuated region.

DE BROGLIE, LOUIS-VICTOR (1892–1987). de Broglie was a remarkable French physicist. He studied at the Sorbonne in Paris, and although he entered as a history student, he later turned to science. He became a professor of theoretical physics at the Sorbonne in 1932. He was researching at about the same time that Arthur Compton, and de Broglie's s doctoral thesis showing the electron was demonstrated to have a wave character, made Compton's finding more significant. In 1923, de Broglie published three papers on light quanta. In his dissertation he described wave-particle duality, how matter waves should behave, and suggested it should be possible to diffract a beam of electrons using crystal. His theory advanced the study of physics significantly. For his work, de Broglie received the Nobel Prize in 1929. He wrote forty-five books on physics in his lifetime.

Professor de Broglie's most important publications are:

Recherches sur la théorie des quanta (Researches on the quantum theory), Thesis Paris, 1924.

Ondes et mouvements (Waves and motions), Gauthier-Villars, Paris, 1926.

Rapport au 5e Conseil de Physique Solvay, Brussels, 1927.

La mécanique ondulatoire (Wave mechanics), Gauthier-Villars, Paris, 1928.

Une tentative d'interprétation causale et non linéaire de la mécanique ondulatoire: la théorie de la double solution, Gauthier-Villars, Paris, 1956.

English translation: *Non-linear Wave Mechanics: A Causal Interpretation*, Elsevier, Amsterdam, 1960.

Introduction á la nouvelle thèorie des particules de M.Jean-Pierre Vigier et de ses collaborateurs, Gauthier-Villars, Paris, 1961.

English translation: *Introduction to the Vigier Theory of elementary particles*, Elsevier, Amsterdam, 1963.

Étude critique des bases de l'interprétation actuelle de la mécanique ondulatoire, Gauthier-Villars, Paris, 1963.

English translation: *The Current Interpretation of Wave Mechanics: A Critical Study*, Elsevier, Amsterdam, 1964.

See also **de Broglie Wavelength**; **Electron Theory**; and **Quantum Number**

J.M.I.

DE BROGLIE WAVELENGTH.   A wavelength ascribed to any particle having momentum. For a relativistic particle, the value of this wavelength is given by the expression:

$$\lambda = \frac{h}{mv} = \frac{h}{m_0 v}\sqrt{1 - \frac{v^2}{c^2}}$$

where λ is the de Broglie wavelength, h is the Planck constant, m_0 is the rest mass of the particle, v is its velocity, and c is the velocity of light. The observed mass of the particle is m and the momentum is mv.

As an example, the wavelength of an electron moving with a kinetic energy of 1 eV (electron volt) is 1.23×10^{-7} cm, shorter than the wavelength $\lambda = c/v = hc/hv = 1.24 \times 10^{-4}$ cm for a photon with an energy of 1 eV, but longer than the wavelength of a proton moving with a kinetic energy of 1 eV, which is $\lambda = 2.86 \times 10^{-9}$ cm.

DEBYE-FALKENHAGEN EFFECT.   The variation of the conductance of an electrolytic solution with frequency. This effect, which is noted at high frequencies, is also called the dispersion of conductance.

DEBYE-HÜCKEL LIMITING LAW.   The departure from ideal behavior in a given solvent is governed by the ionic strength of the medium and the valences of the ions of the electrolyte, but is independent of their chemical nature. For dilute solutions, the logarithm of the mean activity is proportional to the product of the cation valence, anion valence, and square root of ionic strength giving the equation

$$-\log f_\pm = Az + z - \sqrt{\mu}$$

See **Electrochemistry**.

DEBYE, PETER J. W. (1884–1966).   A Dutch chemist and physicist who received the Nobel prize for chemistry in 1936 for his contributions to our knowledge of molecular structure through his investigations on dipole moments and on the diffraction of X-rays and electrons in gases. The interference patterns are still called Debye-Sherrer rings. He also made outstanding contributions to knowledge or polar molecules and to fundamental electrochemical theory.

DEBYE-SEARS EFFECT.   A piezoelectric crystal vibrating in a longitudinal mode in a liquid sets up acoustic waves consisting of regions of compression and regions of rarefaction in the liquid, which alternate at distances of half a wavelength. Hence, if a parallel beam of light shines through such a crystal tank with plate-glass walls, the regions of density and rarefaction act like a plane light diffraction-grating. If the parallel beam from the cell is focused on a single spot when no sound waves are present, first and higher order diffraction-spectra will appear on either side of the zero-order spot when sound waves are present. From the spacings of the diffraction orders, the sound wavelength can be determined, which, together with the frequency, gives the velocity of sound in the liquid. See **Piezoelectric Effect**.

DEBYE THEORY OF SPECIFIC HEAT.   The specific heat of solids is attributed to the excitation of thermal vibrations of the lattice, whose spectrum is taken to be similar to that of an elastic continuum, except that it is cut off at a maximum frequency in such a way that the total number of vibrational modes is equal to the total number of degrees of freedom of the lattice.

The Debye temperature is defined by the relation

$$\Theta = \frac{hv}{k}$$

where v is the maximum frequency of the thermal vibrations of the lattice, h is Planck's constant and k is the Boltzmann constant.

DECARBURIZATION.   Reduction in carbon content at the surface of steel or cast iron by heating in air or other oxidizing or reducing gases. In heating for hot rolling, forging, or heat treatment, decarburization is usually objectionable, and specially prepared neutral furnace atmospheres may be used to reduce or eliminate it. Molten salt or lead baths are also effective in protecting the surface during heat treatment.

In the case of heat-treated machine parts, surface decarburization is objectionable because it reduces fatigue strength and lowers the wear-resistance of bearing surfaces. Important surfaces of hardened steel parts are often finish-ground, in which case a limited amount of decarburized skin can be removed. Tool steels for cutting tools, punches, chisels, etc., are usually ground sufficiently to remove all decarburization; however, many tools and dies are machined to finish dimensions before hardening and extreme care must be taken to protect the surface.

Decarburization is intentional in the processing of low-carbon sheet steels for electrical applications. In the production of malleable cast iron by annealing white cast iron decarburization is beneficial.

DECAY PRODUCT.   A nuclide resulting from the radioactive disintegration of a radionuclide, being formed either directly or as the result of successive transformations in a radioactive series. Also called *daughter*, or *daughter element*. A decay product may be either radioactive or stable.

DECOLORIZING AGENTS.   A substance that removes color by a physical or chemical action. Charcoals, carbon blacks, clays, earths, activated alumina or bauxite, or other materials of highly adsorbent character are used to remove undesirable colors (and often odors) from sugar, vegetable and animal fats and oils, and other substances. In a broad sense, decolorizing agents also embrace bleaches, which usually remove color by chemical reaction.

Activated carbon is one of the most widely used of the adsorbants. It is an amorphous form of carbon characterized by high adsorptivity. The carbon is obtained by the destructive distillation of wood, nut shells, animal bones, or other carbonaceous material. It is "activated" by heating to 800–900°C with steam or carbon dioxide, which results in a porous internal structure. The internal surface area of activated carbon averages about 10,000 square feet (929 square meters) per gram. Numerous uses include applications in the brewing and sugar refining industries.

Diatomaceous earth also finds numerous adsorbant applications in food and chemical processing, not only in decolorizing, but as a filter aid and clarifying agent as well. This is a soft, bulky solid material (88% silica) composed of skeletons of small prehistoric aquatic plants related to algae. See also **Diatomite**. They have intricate geometric forms and expose a great deal of area per unit of weight.

Fuller's earth, also used as an adsorbant, is a porous colloidal aluminum silicate (clay) which has a high natural adsorptive power. See also **Fuller's Earth**.

Silica gel is a regenerative adsorbant consisting of amorphous silica derived from sodium silicate and sulfuric acid. In addition to color adsorbing and bleaching powers, silica gel is used as a dehumidifying and dehydrating agent and as an anticaking agent.

Prior to crystallization in the refining of sugar, bleaching of the syrup is required. This is sometimes effected through treatment of the solution with calcium hypochlorite, usually in the presence of calcium phosphate which serves as a buffer and aids in the final precipitation of calcium from the bleached solution.

The physical properties of representative adsorbents and decolorizers are given in table in entry on **Adsorption**.

DECOMPOSITION (Chemical). A chemical change in which a single chemical substance is broken up into two or more other substances, which differ from each other and from the parent substance in chemical identity. Complete decomposition refers to such a condition of the products that they are not readily decomposed further, e.g., such decomposition products as ammonia and carbon dioxide. *Degradation* refers to gradual decomposition in which the molecule is diminished in size in small steps. See also **Degradation (Chemical)**.

The *heat of decomposition* is the change of heat content when one mole of a compound is decomposed into its elements. This is equal in quantity, but opposite in sign, to the *heat of formation*.

Sensitized decomposition is a chemical decomposition that is brought about by the presence of a second substance which absorbs an exciting radiation. The essential mechanism of the reaction is the excitation of particles of the second substance by the radiation, followed by collisions between these excited particles and molecules to be decomposed. The process proceeds most effectively if the energy difference between the ground state and excited state of the sensitizer is nearly equal to the energy of the decomposition reaction.

Double decomposition is a term used to express the interaction of molecules which exchange one or more of their constituent atoms or radicals.

DEFOAMING AGENTS. Film breakers or defoaming agents are substances used to reduce foaming caused by proteins, gases, or nitrogenous materials which may interfere with processing or the desired characteristics of the end-products. Processes particularly prone to foaming conditions include the Kraft process for papermaking, where a very foamy pulp slurry is formed; phosphoric acid production from phosphate rock; beet sugar processing; several fermentation processes; and, in terms of the end-product, latex paints. See also **Paints and Coatings**.

The terms *defoamer* and *antifoam* or *antifoaming agent* are frequently used interchangeably. A *defoamer* best describes a substance that kills the foam from above, once it exists. An *antifoaming agent* stops the foam from forming in the first place. Often a defoamer will be a poor antifoaming agent, and vice versa. Defoaming agents, when used in small concentrations, can be quite effective. Where a suitable chemical substance cannot be found, physical means may be required. These may be mechanical, electrical, or thermal in nature. The mechanical devices are fundamentally simple, usually taking the form of rotating breaker arms. The presence of a hot surface near a foam tends to destroy the foam. Essentially, a portion of the foam is evaporated, causing the acceleration of its breakdown. It has also been established that electrical discharges tend to weaken or destroy films.

Mechanisms of Defoaming Agents

These agents may operate via a number of mechanisms, but the most common ones appear to be those of entry and/or spreading. The defoamer must first of all be insoluble in the foaming liquid for these mechanisms to function. Second, the surface tension of the defoamer must be as low as possible. The interfacial tension between defoamer and foamer should be low, but not so low that emulsification of the defoamer may occur. Third, the defoamer should be dispersible in the foaming liquid. It was first shown in 1948 that thermodynamically the entry of the defoamer droplet into a bubble surface occurs when the entering coefficient has a positive value. The physics of bubbles is described in entry on **Foam**.

A type of defoamer may consist of a dispersion in oil of fine particles of silica coated with silicone, the silicone surface of the particles causing them to be hydrophobic. The defoaming action of such a formulation can be explained on the basis of the entry mechanism. Hydrophobic particles can act as an emulsifying agent where the defoamer oil constitutes the continuous phase and the foam constitutes the dispersed phase.

In experimental trials, it has been found that excessively hydrophobic particles, such as powdered Teflon, do not function as well as silicone-coated particles. An emulsifier particle must be wetted to some extent by the dispersed phase in order to function as an emulsifier.

The more efficient defoaming mechanism of spreading involves transport of underlying liquid so that the liquid is replaced by a film of defoamer that does not support foam. A drop of oleic acid added to water spreads at a velocity of 30 miles (48.2 kilometers) per hour. The mechanical shock to a film by such a defoamer may be considerable. In addition to the foam-destroying aspect, spreading is also of value as a defoamer-dispersion method, particularly in viscous or poorly stirred systems.

Defoaming agents are in three principal categories, but sometimes are used in combination: (1) surfactants made soluble, (2) dispersions of hard particles, and (3) dispersions of soft particles. The fatty acid-fatty alcohol combination in hydrocarbon oil is an example of a solubilized surfactant defoaming formulation. Paraffinic waxes and fatty amides may be used in soft-particle formulations. The most common of the hard-particle formulations is silica or a mineral coated with silicone dispersed in a vehicle. A particle size as small as 0.02 micrometer may be optimal.

Choice of defoaming agents in the food field is somewhat restricted because substances used obviously must be nontoxic and not produce off odors, off colors, or off tastes. Chemical defoaming agents commonly used in food processing include decanoic acid, dimethylpolysiloxane, lauric acid, mineral oil (white), myristic acid, octanoic acid, oleic acid, oxystearin, palmitic acid, petrolatum, petroleum wax (synthetic), silicon dioxide, sorbitan monostearate, and stearic acid. See also **Foam**.

An example from the brewing industry points out the importance of defoaming agents. Advantages of their use include: (1) higher production through increased fermentation capacity—up to 20% more throughout; (2) the lid of the fermentation tank can be left on, reducing oxidation and improving sanitation; (3) lower oxidation rate, which gives a better physical and chemical stability to the beer as the denatured or partially denatured protein levels remain low, and thus reducing turbidity; and (4) less yeast build-up on the sides of the tank, thus reducing cleaning requirements. Inasmuch as foam is a consideration in the final quality of the brewed product, effective foam control during processing can later affect the "head" of the final product, providing for a stable, long-lived, creamy foam. Thus, a defoaming agent for process use must be insoluble in the beer and capable of removal so that it does not detract from the head of the final product.

DEGASIFICATION. Removal of gas, as applied particularly to the removal of the last traces of gas from wires used in vacuum apparatus, from metals to be plated, and from substances to be used in other specialized applications. Untreated glass always contains water, carbon dioxide, oxygen, and traces of other gases within it and on its surface, and these are ordinarily in a state of equilibrium with the surroundings. When the pressure is reduced, however, the equilibrium is upset, and these gases, being gradually released from solution in and adsorption on the glass, spoil the vacuum. It is usual to drive the gases out of the glass by baking the glass at a temperature of about 350°–500°C, while on the vacuum pump. Degassing of metal is necessary for the same reason as in degassing of glass, but because of the larger quantities of gas present in metals, more complex methods of degassing must be employed. These include baking at high temperature, eddy-current heating, and electron bombardment.

See also **Gettering**.

DEGENERACY. In the kinetic theory of gases, a gas that does not obey the ideal gas laws is referred to as a degenerate gas. The greater the deviation of the real gas from the ideal, the greater is its degeneracy.

A *degenerate electron gas* is an electron gas that is far below its Fermi temperature, that is, which must be described by the Fermi distribution. The essential characteristic of this state is that a very large proportion of the electrons completely fill the lower energy levels, and are unable to take part in any physical processes until excited out of these levels.

DEGRADATION (Chemical). A gradual decomposition occurring in stages with well-marked intermediate products. For example, the maltose chain loses one carbon atom under certain conditions to produce a sugar with eleven carbon atoms in its skeleton. In fact, the term degradation often means specifically a reduction of the number of carbon atoms in an organic compound, usually an aliphatic compound. Among the specific methods used for this purpose are the Hoffmann reaction, treating an amide with a hypohalite; the conversion of fatty acids to methyl ketones, followed by oxidation; and the Curtius reaction for the conversion of an acid azide to the primary amine.

Kraft Method

A method of reducing the number of carbon atoms in the molecule of an acid, especially a fatty acid by the decomposition of its calcium or barium salt in the presence of a salt of acetic acid, followed by oxidation of the resulting methyl ketone:

$$(RCH_2COO)_2Ca + (CH_3COO)_2Ca \longrightarrow 2RCH_2COCH_3 + 2CaCO_3.$$

$$\downarrow CrO_3$$

$$2RCOOH + 2CH_3COOH.$$

See also **Decomposition (Chemical)**.

DEHUMIDIFICATION.   A process, used in air conditioning and in the process industries, in which air or other gases partially saturated with water are subjected: (1) to cooling below their dew point, so that part of the water vapor is condensed and thus separated from the gas; (2) to the action of chemical desiccants, either liquid or solid, which adsorb moisture from the gas; or (3) to a combination of both actions. The most commonly used gas in industry is compressed air. This air requires drying and conditioning for trouble-free operation of pneumatic equipment, including tools, instruments, and automatic controllers. In paint manufacture, dry inert gas is used to blanket agitation operations. Dry process gases, such as nitrogen or hydrogen, are used in metal-annealing operations. The manufacture of some transistors requires blanketing with a dry gas during assembly.

Deliquescent dryers (dissolving desiccant types) and refrigeration-type dryers are adequate for most air-comfort applications and some industrial applications. Usually, these systems only partially remove moisture from air and other gases and generally require additional drying equipment. To meet tight drying specifications, a solid, regenerable desiccant that adsorbs the moisture usually is used. Desiccant regeneration usually is accomplished by the application of heat or purging, or a combination of both procedures. Normally, dual drying towers are used for continuous service, allowing one tower to be on-line, while the other tower is being regenerated. Desiccants most frequently used are silica gel, activated alumina, and molecular sieves. Desiccant systems permit efficient dew-point performance in the range of -40 to $-100°F$ (-40 to $-73°C$) and are available in a wide range of capacities and pressure ratings (from atmospheric pressure up to 340 atmospheres; 5000 psig).

DEHYDRATION (Chemical).   Removal of water from a substance or system or chemical compound; or removal of the elements of water, in correct proportion, from a chemical compound or compounds. The elements of water may be removed from a single molecule, or from more than one molecule, as in the dehydration of alcohol; this may yield ethylene, by loss of the elements of water from each molecule, or ethyl ether, by loss of the elements of water from two molecules, which then join to form a new compound:

Many reactions known in chemistry under special names, such as neutralization, esterification and etherification are dehydration reactions. See also **Esterification**.

In the food processing field, dehydration is sometimes described as the removal of 95% or more of the water from a food substance, by exposure to thermal energy by various means. The aims of dehydration are reduction in volume of the product, increase in shelf life, and lower transportation costs, among other factors. There is no clearly defined line of demarcation

between drying and dehydrating, the latter sometimes being considered as a supplement of drying. Usually, the direct use of solar energy, as in the drying of raisins, hay, etc., is not lumped in with dehydrating. The term dehydration also is not generally applied to situations where there is a loss of water as the result of evaporation. *Rehydration* or *reconstitution* is the restoration of a dehydrated food product to essentially its original edible condition by the simple addition of water, usually just prior to consumption or further processing. The distinction between the terms drying and dehydrating may be somewhat clarified by the fact that most substances can be dried beyond their capability of restoration. Important food products that are dehydrated include animal feedstuffs, hops, malt, oat, peanut (groundnut), potato, rice, and sweet potato. See also **Drying (Process)**.

DEHYDROGENATION.   A reaction resulting in the removal of hydrogen from an organic compound or compounds. This process is brought about in several ways. Simple heating of hydrocarbons to high temperature, as in thermal cracking, causes some dehydrogenation, indicated by the presence of unsaturated compounds and free hydrogen. Catalytic processes often produce commercially practicable yields of selected dehydrogenated products. The enzyme dehydrogenase is a selective catalyst of this character. There is considerable evidence to indicate that many reactions commonly classed as oxidations, e.g., the oxidation of methanol to formaldehyde are actually dehydrogenations, i.e.,

In the chemical process industries, nickel, cobalt, platinum, palladium, and mixtures containing potassium, chromium, copper, aluminum, and other metals are used in very large-scale dehydrogenation processes. For example, acetone (6 billion pounds per year) is made from isopropyl alcohol; styrene (over 2 billion pounds per year) is made from ethylbenzene. The dehydrogenation of n-paraffins yields detergent alkylates and n-olefins. The catalytic use of rhenium for selective dehydrogenation has increased in recent years. Dehydrogenation is one of the most commonly practiced of the chemical unit processes.

See also **Organic Chemistry**.

DEIONIZING.   A method for purifying water that involves two steps. First, soluble salts are converted into acids by passing through a hydrogen exchanger. Second, they are removed by an acid adsorbent or synthetic resin.

See also **Demineralization**.

DEISENHOFER, JOHANN (1943–).   Awarded the Nobel prize for chemistry in 1988, along with Robert Huber and Hartmut Michel, for work that revealed the three-dimensional structure of closely linked proteins that are essential to photosynthesis. Doctorate awarded in 1974 by Max Planck Institute of Biochemistry, Germany.

DELIQUESCENCE.   When a substance absorbs moisture upon exposure to the atmosphere, the substance is said to be *deliquescent*. At ordinary temperatures the vapor pressure of water varies as shown in Table 1. If the solution of a substance in water has a lower water vapor pressure than that of the atmosphere at the given temperature, water vapor condenses in the solution from the atmosphere until the water vapor pressure of the solution equals the water vapor pressure of the surrounding atmosphere.

TABLE 1. VARIATION OF WATER VAPOR PRESSURE WITH TEMPERATURE

Temperature °C	Water vapor pressure in mm of mercury	
	At saturation	At 50% humidity
0	4.6	2.3
10	9.2	4.6
20	17.5	8.8
30	31.8	15.9
40	55.3	27.7

Substances that are ordinarily deliquescent are sulfuric acid (concentrated), glycerol, calcium chloride crystals, sodium hydroxide (solid), and 100% ethyl alcohol. In an enclosed space, these substances deplete the water vapor present to a definite degree. Other substances are used to accomplish this end by chemical reaction, e.g., phosphorus pentoxide (forming phosphoric acid), and boron trioxide (forming boric acid). Water is absorbed from nonmiscible liquids by addition of such substances as anhydrous sodium sulfate, potassium carbonate, anhydrous calcium chloride, and solid sodium hydroxide. The converse phenomenon is known as *efflorescence*.

See also **Dehumidification**; and **Efflorescence**.

DEMAL SOLUTION.
A solution which contains one gram-equivalent of solute per cubic decimeter of solution. It is slightly weaker than a normal solution, in the ratio of the magnitude of the liter to the cubic decimeter.

DEMINERALIZATION.
Removal from water of mineral contaminants, usually present in ionized form. The methods used include ion-exchange techniques, flash distillation, or electrodialysis. Acid mine wastes may be purified in this way, thus alleviating the pollution problem.

See also **Deionizing**; and **Ion-Exchange Resins**.

DEMULSIFICATION.
The process of destroying or "breaking" and unwanted emulsion, especially water-in-oil types occurring in crude petroleum. Both chemical and physical means are used. Chemical means include addition of polyvalent ions to neutralize electrical charges or of a strong acid; physical means include heating, centrifuging, or use of high-potential alternating current.

See also **Emulsions**.

DENDRITE.
A tree-like crystal formed during solidification of metals or alloys. Dendrites generally grow inward from the surface of the mold, extending branches from a central trunk in a manner resembling a fir tree. In alloys, the central portions of a dendritic crystal are richer in higher melting point constituents, while the outer portions consist of lower melting point material, which is last to solidify. This form of segregation can usually be eliminated by diffusion during subsequent mechanical working and heat treatment.

The reasons for the branched growth of a crystal into a liquid that of the temperature of which falls ahead of, the solid, are easily understood. Whenever a small section of the interface is ahead of the surrounding surface, it will be in contact with liquid metal at a lower temperature. Its growth velocity will be increased relative to the surrounding surface, which is in contact with liquid at a higher temperature, and the formation of a spike is only to be expected. Associated with the formation of each spike is the release of a quantity of heat (latent heat of fusion). This heat raises the temperature of the liquid adjacent to any given spike and retards the formation of other similar projections on the general interface in its immediate vicinity. The net result is that a number of spikes of almost equal spacing are formed which grow parallel to each other in the fashion shown in Fig. 1. The *dendritic growth direction* depends upon the crystal structure of the metal.

The branches, or spikes, in Fig. 1 are primary in nature. Once they have formed growth at the general interface will be slow because here the supercooling is small and the latent heat of fusion associated with the formation of the spikes tends to further decrease its magnitude. At

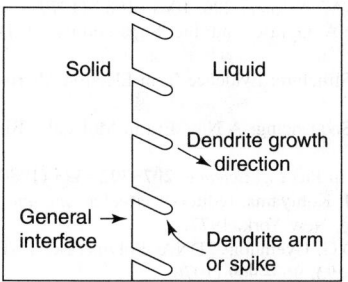

Fig. 1. Schematic representation of the first stage of dendritic growth. A temperature inversion is assumed to exist at the interface i.e., the temperature in the liquid drops in advance of the interface

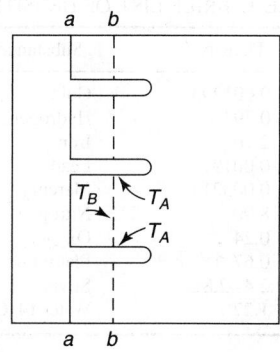

Fig. 2. Secondary dendrite arms form because there is a falling temperature gradient starting at a point close to a primary arm and moving to a point midway between the primary arms

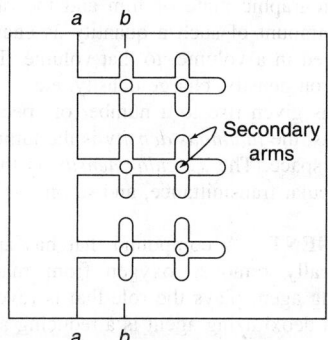

Fig. 3. In a cubic crystal, primary and secondary arms are normal to each other

section *bb* in Fig. 2 on the other hand, the average temperature of the liquid is, by definition, lower than at *aa*. However, even on this section at points in the liquid close to the spikes the temperature will be higher than midway between the spikes ($T_A > T_B$) because of the latent heat released at the spikes. There is, therefore, a decreasing temperature gradient not only in front of the primary spikes, but also in directions perpendicular to the primary branches. This temperature gradient is responsible for the formation of secondary branches, which form at more or less regular intervals along the primary branches, as shown in Fig. 3.

Additional Reading

Galenko, P.K.K., V.A. Zhuravlev: *Physics of Dendrites: Computational Experiments,* World Scientific Publishing Company, Inc., Riveredge, NJ, 1994.
Kassner, K.: *Pattern Formation in Diffusion-Limited Crystal Growth: Beyond the Single Dendrite,* World Scientific Publishing Company, Inc., Riveredge, NJ, 1995.
Stuart, G., N. Spruston and M. Hausser: *Dendrites,* Oxford University Press, Inc., New York, NY, 2000.

DENSITY.
The density of a substance is its mass per unit volume, usually expressed in grams per cubic centimeter. The specific gravity of the substance is the ratio of its density to that of water, usually at 4°C, or 20°C, or 60°F, in the same units, and is therefore an abstract number independent of units. The density of a body is the ratio of its mass to its volume.

To determine the density of a given substance, it is necessary only to ascertain the volume of a specimen whose mass is known by weighing. This may be obtained from measurements on the dimensions of the specimen, or, in the case of a liquid, by the use of a pycnometer or specific gravity bottle. For solids a more precise ethod is to measure the buoyant force, upon the specimen, of a liquid of known density in which it is immersed, or by enclosing it in a gravity bottle and determining the volume by displacement. The Mohr-Westphal balance is especially designed to give densities of liquids by the buoyant force on a solid sinker of known volume. The hydrometer may also be used for quick determinations of liquid densities. The density of a gas is best obtained by weighing a specimen of it in a large, light bulb of known capacity, concurrently observing the temperature and pressure to which the gas is subjected, much as the pycnometer is used for liquids.

TABLE 1. BRIEF LIST OF DENSITIES

Substance	Density	Substance	Density
Air	0.001293	Gold	19.3
Alcohol	0.794	Hydrogen	0.0000899
Aluminum	2.70	Iron	7.86
Carbon dioxide	0.001977	Lead	11.3
Chlorine	0.003214	Mercury	13.55
Copper	8.90	Nitrogen	0.001251
Cork	0.24	Oxygen	0.001429
Gasoline	0.67	Platinum	21.45
Glass	2.4–2.8	Silver	10.5
Glycerine	1.27	Water (4°C)	0.999973

See Table 1 for a brief list of densities, all in grams per cubic centimeter. For gases the densities are at standard temperature and pressure. The term density is also applied in length-force-time systems of units, to the weight per unit volume. Other meanings of the term density are the blackness of the image on a photographic plate or film and the ratio of the number of particles or total amount of such a quantity as energy or momentum, carried by or contained in a volume, to that volume. Thus one speaks of energy density, electron density, charge density, etc.

This last usage has given rise to a number of specific applications of the term density. Thus, the *luminous density* is the luminous energy found in a unit volume of space. The *specular density* is the logarithm of the reciprocal of the specular transmittance, and so on.

DEOXIDIZING AGENT. A compound that has an affinity for oxygen—hence, chemically removes oxygen from many substances. In essence, a deoxidizing agent plays the role that is reverse that of an oxidizing agent. Thus, a deoxidizing agent is a reducing agent. Of course, at one time, oxidation meant simply a combination with oxygen and reduction meant a loss of oxygen. In their broader interpretations, oxidation now refers to the loss of one or more electrons from the outer-shell of an atom, and the reverse for reduction. Although the broader interpretation also could apply to deoxidation, the term still is interpreted generally as removal of oxygen.

Oxygen frequently is an impurity in various metallurgical processes, particularly melting and refining processes. Deoxidizing agents are commonly used to reduce or remove oxygen from molten metals. Lithium metal, for example, will preferably absorb oxygen (by combination) from copper and copper alloys. Boron carbide also is used as a deoxidizing agent for casting copper. Silicon usually in the form of ferrosilicon, and manganese in the form of ferromanganeses are widely used in the production of steel and iron alloys. Aluminum, titanium, zirconium, and vanadium also play a deoxidizing part in the production of iron alloys. Magnesium is also a powerful deoxidizer and desulfurizer and is used in the production of such metals as beryllium, hafnium, titanium, uranium, yttrium, and zirconium. Zinc is used as a deoxidant in the refinement of silver and gold.

Normally, the process of rock formation is one of oxidation. However, a greenish or yellowish area in a red rock may be developed by reduction (deoxidation) of ferric oxide in the presence of organic materials.

Deoxidation also enters into physiological reactions. For example, ascorbic acid (vitamin C) is acclaimed as an antioxidant.

DEOXYRIBONUCLEIC ACID (DNA). A complex sugar-protein polymer of nucleoprotein which contains the genetic code for enzymes in the cell. It occurs as a major component of the genes, which are located on the chromosomes in the cell nucleus. The DNA molecule is a unique and vastly intricate structure; it is comprised of from 3000 to several million nucleotide units arranged in a double helix containing phosphoric acid, 2-deoxyribose, and the nitrogenous bases adenine, guanine, cytosine, and thymine. The spiral consists of two chains of alternating phosphate and deoxyribose units in continuous linkages. The nitrogenous bases project toward the axis of the spiral and are joined to the chains by hydrogen bonds. Adenine always unites with thymine and cytosine with guanine. The complementary of the bases on the joined chains allows each chain to act as a template for replication of the other when the chains are separated, thus producing two new strands of DNA. The sequence of the bases on the chains varies with the individual, and it is this sequence that governs the genetic code. DNA works in conjunction with ribonucleic acid (RNA).

The foregoing is a highly generalized definition. Within the last few years, considerable new knowledge has been gained concerning DNA and its genetic function. The early studies of genes concentrated on bacterial genes. In these, the bacterial genes are not spread out in pieces. More recent studies in the mid- and late 1970s concentrated on animal viruses, animals, and humans. The genes in these cases, with the possible exception of the histone genes, are found in pieces that are spread out along DNA. Thus, between gene fragments, there are long stretches of DNA, the functions of which are poorly understood as of the present. This discovery of fragmented or spaced out genes in animals raised the question as to whether such structures are exceptional or the rule. Subsequent research has indicated that they are the norm.

This discovery, while creating several new and fundamental questions, has provided at least partial answers to some former questions. For example, it has been known for some years that there are large quantities of DNA in the cells of higher organisms, that is, DNA in an amount far in excess of the DNA required if the genes were not in pieces. The spacing out of the genes into fragments accounts for all or part of the excess DNA previously noted.

As of the early 1980s, numerous hypotheses have been formulated by molecular biologists and this fundamental discovery has stimulated a whole new line of research in many laboratories throughout the world. Some scientists have observed that the extra DNA cannot be accounted for simply upon the basis of evolutionary theories. The extra DNA may play a role in controlling gene expression. The complexity and current uncertainty of these hypotheses are beyond the scope of this book at this juncture in the research program. Perhaps the topic will be better clarified at the time of the next edition. Several of the references listed shed further insights.

Studies of DNA in the human cell have suggested that the intricate DNA structure may be as much as 1.2 meters (4 feet) long—and yet it is contained within a nucleus which is less than 0.025 millimeters ($\frac{1}{1000}$ inch) in diameter. That the DNA structure is in the form of a complex tangle under these conditions is not difficult to imagine. Upon division of the cell, the DNA condenses into strands in duplicate copies, one set for each daughter cell. It follows that the copying must take place while the DNA is in the form of a tangle, but in such a manner to allow easy separation of the duplicate copies. The mechanics and time required for the copying process still remain undetermined. As early as 1974, Berezney and Coffey (Johns Hopkins University School of Medicine) estimated from electron micrographs that about 5% of the protein in a nucleus appears to be a rigid skeleton or scaffolding, a structure they termed the *nuclear matrix*. These researchers have proposed that the DNA in the nucleus is attached onto the nuclear matrix at thousands of sites, with the genetic material arrayed in thousands of loops. The loops pass through enzyme complexes located at the matrix sites and in so doing are copied. In the process, the loops are preserved and remain attached to the matrix. It is further postulated that when a cell divides, duplicate copies of loops already joined to a scaffolding are drawn apart to the daughter cells. It has been estimated that genetic material in rat liver cells consists of from 10,000 to 15,000 loops. Thus, the nuclear matrix would have a corresponding number of enzyme complexes.

See also **Genetics and Gene Science (Classical)**; and **Nucleic Acids**.

Additional Reading

Bauer, W.R., Crick, F.H.C., and J.H. White: "Supercoiled DNA," *Sci. Amer.*, **243**, 118–133 (1980).

Brown, P.O., and N.R. Cozzarelli: "A Sign Inversion Mechanism for Enzymatic Supercoiling of DNA," *Science*, **206**, 1081–1083 (1979).

Cozzarelli, N.R.: "DNA Gyrase and the Supercoiling of DNA," *Science*, **207**, 953–960 (1980).

Griffith, J.D.: "DNA Structure: Evidence from Electron Microscopy," *Science*, **201**, 525–527 (1978).

Kolata, G.B.: "DNA Sequencing: A New Era in Molecular Biology," *Science*, **192**, 645–647 (1977).

Kolata, G.B.: "Genes in Pieces," *Science*, **207**, 392–393 (1980).

Kolber, A.R., and M. Kohiyama (editors): *Mechanism and Regulation of DNA Replication*, Plenum, New York, 1977.

Lehman, I.R., and D.G. Uyemura: "DNA Polymerase I: Essential Replication Enzyme," *Science*, **193**, 963–969 (1976).

Maugh, T.H., II: "Phylogeny: Are Methanogens a Third Class of Life?" *Science*, **198**, 812 (1977).

Mitchell, R.M., et al.: "DNA Organization of *Methanobacterium thermoautotrophicum*," *Science*, **204**, 1083–1084 (1979).

Portugal, F.H., and J.S. Cohen: *A Century of DNA*, MIT Press, Cambridge, Massachusetts, 1978.

Razin, A., and A.D. Riggs: "DNA Methylation and Gene Function," *Science*, **210**, 604–610 (1980).

Smith, G.P.: "Evolution of Repeated DNA Sequences by Unequal Crossover," *Science*, **191**, 528–535 (1976).

Sparrow, A.H., and A.F. Nauman: "Evolution of Genome Size by DNA Doublings," *Science*, **192**, 524–529 (1976).

Staff: "Undogmatic Toad," in "Science and the Citizen," page 66–69, *Sci. Amer.*, **242**, 4 (1980).

Temin, H.M.: "The DNA Provirus Hypothesis," *Science*, **192**, 1075–1080 (1976).

Timberlake, W.E.: "Low Repetitive DNA Content in *Aspergillus nidulans*," *Science*, **202**, 973–975 (1978).

Yang, R.C.A., and R. Wu: "BK Virus DNA: Complete Nucleotide Sequence of a Human Tumor Virus," *Science*, **206**, 456–459 (1979).

DEPHLEGMATION.

Partial condensation of vapor from a distillation operation to produce a liquid richer in higher-boiling constituents than the original vapor. The residual vapor is richer in the lower-boiling constituents.

See also **Distillation**.

DEPOLARIZATION. See **Batteries**.

DERIVATIVE (Chemical).

A term used in organic chemistry to express the relation between certain known or hypothetical substances and the compound formed from them by simple chemical processes in which the nucleus or skeleton of the parent substance exists. Thus, phenol, aniline, and toluene are said to be derivatives of benzene; and many of the terpenes are derivatives of cymene. Or, when a paraffin such as methane is halogenated, a series of halogen-substitution products may be formed. Depending upon conditions of the reaction, CH_4 plus Cl may yield the monosubstitution product CH_3Cl (methyl chloride), or the disubstitution product CH_2Cl_2 (methylene dichloride), or the trisubstitution product $CHCl_3$ (chloroform), or the tetrasubstitution product CCl_4 (carbon tetrachloride). In essence, these compounds are derived from methane and thus are methane derivatives as the result of chlorination. Usually the term applies to those compounds where the resulting compound is formed in one step, although a chain of steps may be involved in some cases, depending essentially upon how easy it is to identify the "derivative" with the parent substance. Where a chain of steps is involved, the intervening compounds often are called intermediates rather than derivatives.

DESALINATION.

As generally used, the term describes the production of water appropriate for human consumption from seawater and brackish water. Seawater averages about 35,000 ppm of total dissolved solids (TDS). Brackish waters TDS range from 2000 ppm upwards. The maximum TDS of water considered tolerable and acceptable for continued human consumption is about 500 ppm, although water containing up to 1000 ppm TDS may be consumed for short periods.

Desalination processing techniques fall into three groups:

1. *Processes based on a change of phase*—distillation and freezing.
2. *Processes based on selective transport using membranes*—electro dialysis and reverse osmosis.
3. *Processes based on chemical bonding*—ion exchange.

Distillation and freezing process have advantages for high-salinity water, such as seawater, because the energy required is independent of salt content. Electrodialysis and ion exchange are best suited for brackish waters. Reverse osmosis is competitive for desalting both seawater and brackish water. A general desalting plant schematic of the components common to all desalting systems is shown in Fig. 1. The raw saline water is treated to prepare the water for desalination and to ensure more efficient and trouble-free performance of desalination systems. The pretreated saline water enters the desalting plant, where the product water stream is the desalting plant's primary output; a brine stream is also produced. The product water is mixed with suitable chemicals in the post-treatment step, depending upon the intended use of the water. The treated product water is then transferred to a distribution system.

Distillation

In distillation, impurities from saline water or brine are removed by boiling the saline water, collecting the water vapor and then cooling the

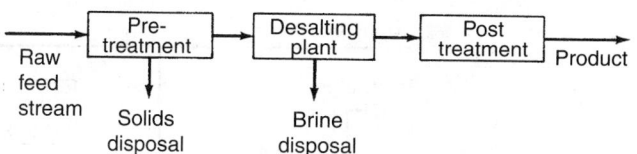

Fig. 1. Generalized schematic of a desalting plant

vapor until it condenses. The basic equipment used is an evaporator. Commercial evaporators operate under vacuum to utilize the maximum available temperature range for heat transfer. The three major classifications of distillation systems currently in use are: (1) multistage flash distillation (MSF); tube evaporation (VTE, HTE); and vapor compression (VC).

Multistage flash distillation is used widely by itself and also in conjunction with steam power plants to produce potable water from seawater. The MSF process is illustrated in Fig. 2. Multistage flash distillation incorporates flashing of heated brine as it enters a stage in which the vapor pressure is maintained so that the boiling point of the brine is below its incoming temperature. Flashing occurs in steps as the heated brine passes in series through a number of stages at successively lower pressures. The multistage distillation process is the most economic of the high-temperature processes in terms of energy and cooling water requirements. The optimum number of stages is determined by an economic and technical evaluation. Modern MSF designs typically have three or four heat rejection stages, plus twenty flash stages.

Tube evaporators may be designed as vertical tube evaporators (VTE) or horizontal tube evaporators (HTE). In the VTE, vapor produced in one effect is condensed in the next effect. The effect is an evaporator chamber receiving heat from an external source or from a higher effect and producing vapor and brine which may serve as a heat source for the next effect. To obtain high efficiency in the use of heat energy, the process is repeated in several evaporator effects arranged in series. See also **Evaporation**.

Vapor compression uses the reverse principle of multieffect distillation. As a vapor is compressed, its temperature and pressure increase. The compressed vapor can be used instead of fresh steam as a heat source at the high-temperature end of the distillation process.

All distillation processes are based on energy-consuming phase changes to convert seawater to potable water. Careful consideration of pretreatment requirement and materials of construction is necessary to provide a satisfactory system.

Electrodialysis

Electrodialysis (ED) is a membrane process based on the ability of semipermeable membranes to pass select ions in feedwater. A direct current electrical field transports the ions through the membranes. In the basic system, alternating cation- and anion-selective membranes are placed in an electrical field. An electrodialysis stack schematic is shown in Fig. 3. The cation-selective membranes permit only the transport of cations; anion-selective membranes allow only the transport of anions.

The transport of ions across the membranes results in ion depletion in some cells and ion concentration in alternate ones. Water exiting from the ion-depleted cells is the desalted product; water leaving the ion-concentrated cells is the brine. The cell pair consists of a cation-selective membrane, an ion-depleted cell, an anion-selective membrane, and an ion-concentrated cell. A commercial electrodialysis unit consists of hundreds of cell pairs stacked vertically and clamped between two huge oppositely charged electrodes. Electric energy is consumed in proportion to the quantity of salts to be removed. Economics usually limit the application of ED to feedwaters of less than 10,000 ppm total dissolved solids (TDS). The electrodialysis process does not remove organics, colloidal matter, or bacteria.

Reverse Osmosis

If a solution is placed on one side of a semipermeable membrane and water is placed on the other side, then there is a natural tendency (*osmosis*) for water to diffuse through the membrane to the solute side until an equilibrium osmotic pressure is reached.

If a pressure is applied to the solute side, substantially greater than the osmotic pressure, the water diffuses from the solution through the membrane to the fresh-water side. This phenomenon is called reverse

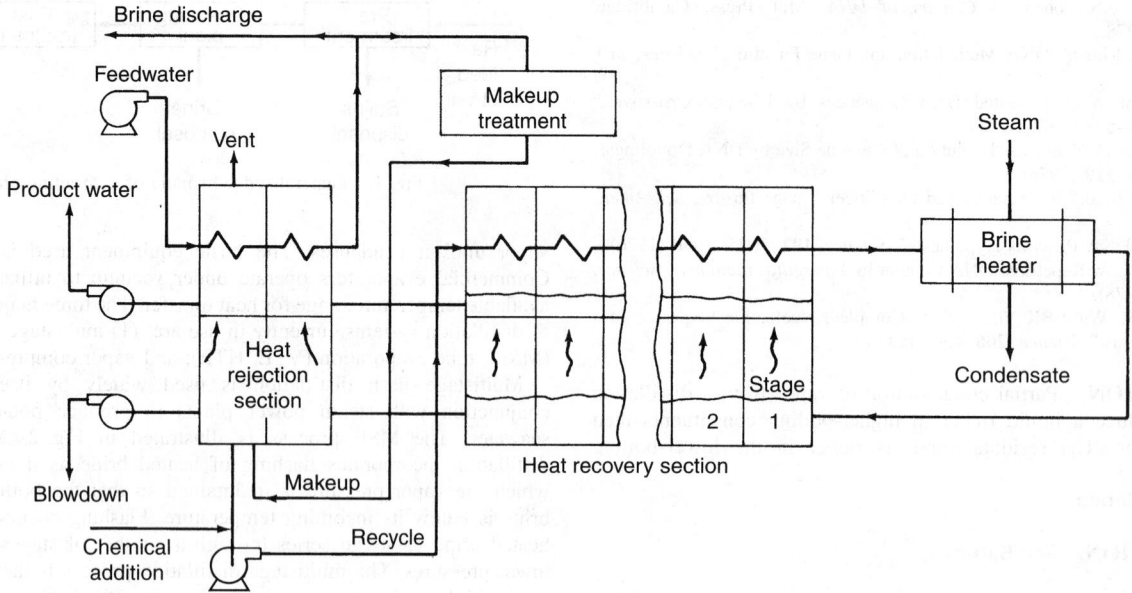

Fig. 2.   Multistage flash evaporation process used in desalination

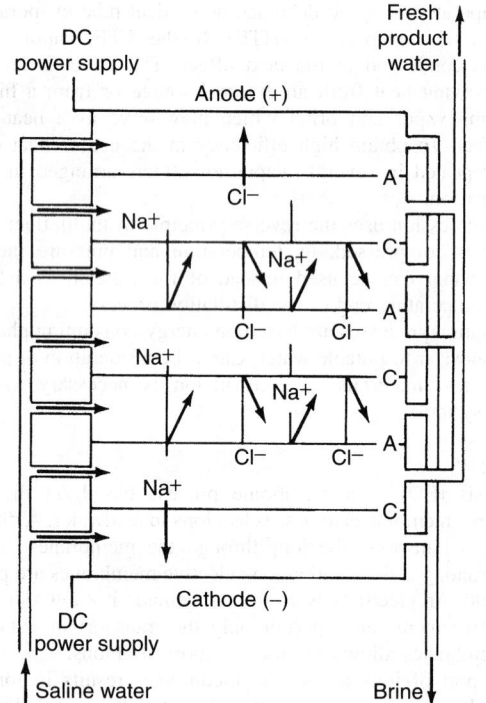

Fig. 3. Schematic of electrodialysis stack used in desalination and other separation processes

osmosis (RO). In applying this principle to desalination, hydraulic pressure is used to force pure water from saline water through a semipermeable membrane. See Fig. 4. RO membranes may be spiral wound sheets, hollow fine fibers, or tubes. In the spiral wound design, two membranes sheets are glued back to back to form an envelope. When water, under pressure, is passed over the surface of the sheet, pure water passes through the membrane, while the salts are held back on the outer surface of the envelope. The membrane envelopes are separated by mesh spacers and rolled into the spiral configuration. The hollow fiber membrane design places a large number of hairlike hollow elements in a pressure vessel. The membranes have an outside diameter of about 100 to 300 micrometers and an inside diameter of about half that dimension. Normally, the fibers are looped in a U-shape so that both ends are embedded in a plastic tube-sheet. Under pressure, desalted water passes through the fiber walls and flows down the inside of the fibers for collection. The hollow fine fiber design is shown in Fig. 5.

Reverse osmosis performs a separation without a phase change. Thus, the energy requirements are low. Typical energy consumption is 6 to 7 kWh/m^2 of product water in seawater desalination. Reverse osmosis, of course, is not only used in desalination, but also for producing high-pressure boiler feedwater, bacteria-free water, and ultrapure water for rinsing electronic components—because of its properties for rejecting colloidal matter, particle and bacteria.

Ion Exchange

Ion exchange processes use natural or synthetic granular materials that exchange one ion for another. The new (captured) ion is held for a finite period and then released to a regenerating solution. Because regeneration costs increased proportionally with the TDS reduction, the ion exchange process has not been competitive with membrane processes for desalting

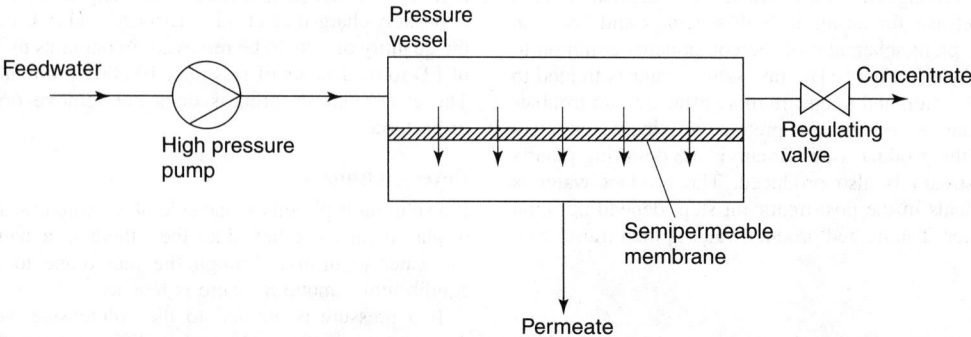

Fig. 4.   Schematic of reverse osmosis process used in desalination process. (*Toyobo Co., Ltd., Osaka, Japan*)

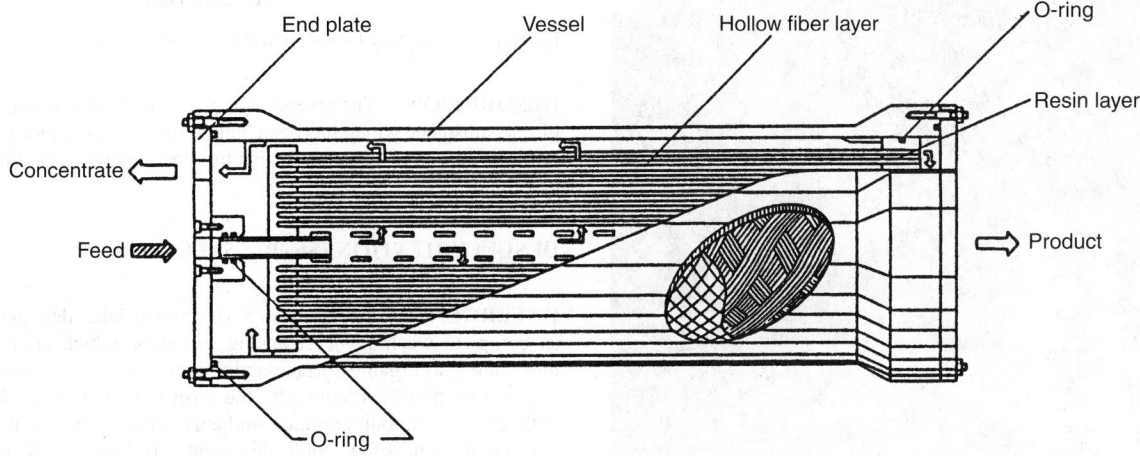

Fig. 5. Hollow fiber reverse osmosis module used in desalination and other separation processes. (*Toyobo Co., Ltd., Osaka, Japan*)

feedwaters above 1000 ppm TDS. Ion exchange technology is used worldwide in systems ranging from domestic water softeners to large municipal and industrial demineralizers. See also **Ion-Exchange Resins**.

Other Processes

A number of other desalination processes, such as freezing, membrane distillation, and solar humidification, have been used to desalt saline waters. Based on their commercial success, these processes can be considered as minor desalination processes.

In the *freezing process*, dissolved salts are naturally excluded during the formation of ice crystals. When seawater is frozen, fresh-water ice crystals are formed and the salt is concentrated in the remaining brine solution. The ice crystals can be separated from the brine, washed, and melted, to yield fresh water.

In the *membrane distillation process*, combined use of distillation and membranes is made. Salt water is warmed to produce vapor. This vapor passes through porous membranes, which are permeable to vapor but not to the liquid phase. The vapor is condensed on a cooled surface to produce fresh water. The main advantage of this process is its simplicity and the need for only small temperature differentials to operate.

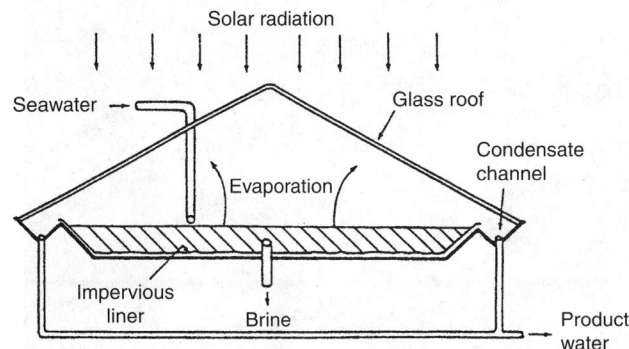

Fig. 6. Cross section of a solar still

The solar humidification process is represented schematically in Fig. 6. This process generally imitates a part of the natural hydrologic cycle in that the saline water is heated by the sun's rays to enhance vapor production.

Fig. 7. Multistage flash distillation (MSF) desalination plant in Al Jubail, Saudi Arabia. Capacity/day is 23,500 tons × 20 units. (*Toyobo Co., Ltd., Osaka, Japan*)

Fig. 8.   Portion of reverse osmosis (RO) desalination plant in a Singapore power station. Capacity is 2,000 tons/day. (*Toyobo Co., Ltd., Osaka, Japan*)

The vapor is then condensed on a cool surface and the condensate collected as product water.

Worldwide Installations

As of 1990, the worldwide inventory of total capacity of installed desalination plants was approximately 13.2 million cubic meters (nearly 3.5 billion U.S. gallons) per day.

Two representative installations are shown in Figures 7 and 8.

MASAAKI SEKINO
Toyobo Co., Ltd.
Iwakuni, Yamaguchi-Pref. Japan

Additional Reading

Balaban, M.: *Desalination Directory,* Balaban International Science Services, L'Aquila, Italy, 1998.

Ettouney, H.M., and H.T. El-Dessouky: *Fundamentals of Salt Water Desalination,* Elsevier Science, New York, NY, 2002.

Heitmann, Hans-Gunter: *Saline Water Processing: Desalination and Treatment of Seawater, Brackish Water, and Industrial Waste Water,* John Wiley & Sons, Inc., New York, NY, 1999.

Klein, E.: *Affinity Membranes: Their Chemistry and Performance in Absorptive Separation Processes,* John Wiley & Sons, Inc., New York, NY, 1991.

Matsurra, T.: *Synthetic Membranes and Membrane Separation Processes,* CRC Press, LLC., Boca Raton, FL, 1994.

Staff: Institution of Chemical Engineers: *Desalination and Water RE-Use,* Hemisphere Publishing Corporation, Rugby, UK, 1991.

Staff: International Atomic Energy Agency: *Nuclear Desalination of Sea Water: Proceedings of an International Symposium,* Bernan Associates, Lanham, MA, 1998.

Web Reference

Desalination Directory Online. http://www.desline.com/

DESORPTION.   The reverse of absorption or adsorption, as in the release of one substance which has been "taken into" another by a physical process, or the release of a substance which has been held in concentrated form upon a surface.

DESULFURIZATION (Steel).   See **Calcium**.

DETERGENTS.   For purposes of this article, detergents are defined as complete washing or cleansing products, which contain among their ingredients an organic surface-active compound (*surfactant*) that has soil-removal properties. Frequently, the term *detergent* is used synonymously with *surfactant*, but common industry practice treats the surfactant as one component of a total detergent product, as is done here. See also **Surfactants**. Additionally, this article treats primarily only the so-called synthetic detergents, excluding those products in which soap is the sole or predominant surfactant. Soaps, the alkali salts of long-chain fatty acids, differ significantly in certain important performance properties from the synthetic surfactants and are described separately in this encyclopedia. See also **Soaps**.

The synthetic detergent industry is one of the largest chemical process industries. In 1984, annual U.S. production of synthetic detergents was about 7 million tons, with approximate manufacturer's value of $9 billion. The industry differs from many other chemical process industries, however, in that the bulk of its production is sold directly to individuals for household consumption, primarily as branded products, rather than to industrial or institutional users.

Detergent Ingredients and Their Functions

Detergent formulas vary greatly, depending upon the intended end-use application. Differentiation by surfactant type as well as selection of auxiliary ingredients is involved. Some of the ingredients most commonly used in detergent formulas and their functions are as follows.

Surfactants.   By definition every detergent product contains one or more types of surfactants. Basically, every surfactant is an organic compound consisting of two parts: (1) a *hydrophobic* portion, normally including a long hydrocarbon chain, and (2) a *hydrophilic* portion, which renders the entire compound sufficiently soluble or dispersible in water or other polar solvent to serve its intended use. Together, these combined hydrophobic and hydrophilic moieties render the compound surface-active—able to concentrate at the interface between a surfactant solution and another phase, such as air, soil, and textile substrate to be cleaned.

Surfactants provide the detergent with the ability to penetrate and wet soiled surfaces, to displace, solubilize, or emulsify various soils, particularly oils and greases, and to disperse or suspend certain soils in solution to prevent their redeposition. In addition, the surfactants provide (to various degrees) whatever foaming or sudsing properties the detergent solution possesses, properties which are necessary for satisfactory use in certain industrial applications and highly desirable to many consumers in household laundry, hand-dishwashing, and personal care uses.

Surfactants in broad use may be classified into three general types: (1) *anionics*, in which the hydrophilic portion of the molecule carries a negative charge; (2) *cationics*, in which the charge of this portion is positive; and (3) nonionics, which do not dissociate but commonly derive their hydrophilic portion from polyhydroxy or polyethoxy structures. *Ampholytic* and *zwitterionic* surfactants are also known and are starting to be of commercial importance.

Of the three main types of surfactants, the anionics are by far the most commercially important class, constituting, in particular, the major surfactant type represented in laundry and hand-dishwashing detergents. Among the anionics, linear sodium alkyl benzene sulfonate (LAS), linear alkyl sulfates, and linear alkyl ethoxy sulfates are by far the most widely used compounds. (The industry converted voluntarily to linear alkyl chains in the mid-1960s to obtain improved biodegradability relative to the branched-chain alkylates formerly used.)

$$CH_3(CH_2)_x$$

(x ranges from 9 to 15)

$$SO_3Na$$

Linear alkyl benzene sulfonate (LAS)

$$CH_3(CH_2)_xOSO_3Na \quad (x \text{ ranges from 9 to 17})$$

Alkyl sulfate

$$CH_3(CH_2)_x(O-CH_2-CH_2)_y \; OSO_3Na$$

(x ranges from 7 to 15)

(y ranges from 0 to 6)

Alkyl ethoxy sulfate

Nonionic surfactants are also used in substantial amounts in laundry detergents and in automatic dishwashing detergents, both applications reflecting in particular their generally lower sudsing characteristics than the anionics. Commercially important examples of the nonionics include the alkyl ethoxylates, the ethoxylated alkyl phenols, the fatty acid ethanol amides, and complex polymers of ethylene oxide, propylene oxide, and alcohols.

$$CH_3(CH_2)_x(O-CH_2-CH_2)_yOH$$

(x ranges from 9 to 15)

(y ranges from 4 to 20)

Alkyl ethoxylates

$$CH_3(CH_2)_x$$

$$(OCH_2CH_2)_yOH$$

(x ranges from 7 to 13)

(y ranges from 3 to 12)

Ethoxylated alkyl phenols

$$CH_3(CH_2)_xCON(CH_2CH_2OH)_y(H)_z$$

(x ranges from 9 to 17)

(y ranges from 1 to 2)

Fatty acid ethanol amide

(z is 1 or 0)

Cationic surfactants are used in only limited tonnage for specialty detergent products, such as metal cleaners for electroplating, and more commonly in ancillary textile laundering products for their fabric-softening, antistatic, and germicidal properties. A typical cationic surfactant would be tallow trimethylammonium chloride:

$$CH_3(CH_2)_{13-17}N^+(CH_3)_3Cl^-$$

In contrast with soaps, virtually all the synthetic anionic and nonionic surfactants do not form readily visible insolubles in the presence of the Ca^{++} and Mg^{++} ions present in many water supplies—the familiar scum, curd, and "lime soap" of the laundry soap era. This property, in combination with the discovery that the detergency performance of many synthetic surfactants could be greatly augmented by the addition of certain phosphate chelating agents, led to the "detergent revolution" in the laundry-product field. See also **Surfactants**.

Builders. *Builder* is the term used within the industry to designate materials in the synthetic detergent which chelate (sequester) or precipitate polyvalent metal ions present in the cleaning solution, particularly Ca^{++} and Mg^{++} ions, which are present in substantial quantities in so-called hard water supplies. Nearly 2 billion pounds (0.9 billion kg) of builders are presently used in detergent products, particularly in textile-cleaning detergents, where they may constitute one-half of the total weight of product. Laundry products containing significant quantities of builders are called *heavy-duty* or *built* detergents.

Builders perform several critical functions in present-day detergents:

1. They prevent polyvalent metal ions from combining with (many) surfactant(s) to form an adduct which is less effective than the unmodified surfactant in cleansing properties.

2. They prevent polyvalent metal ions from combining with various soils, such as lipid residues and clays, to form less dispersible residues, which adhere tenaciously to the surface to be cleaned.

The aforementioned effects combine to provide greatly enhanced removal of many soils and stains by a formula with builder compared with a formula without.

3. They prevent redeposition of soils removed from a surface back onto the surface through a dispersing action associated with chelating and charge-distribution effects.

4. They provide added and buffered alkalinity to the wash solution, which is generally helpful to cleansing in most applications.

5. They provide enhanced removal and kill of microorganisms, a critical property in many applications involving sanitation of commercial facilities.

By far, the most commonly used builders are the condensed polyphosphates, particularly pentasodium tripolyphosphate (STP) and, to a lesser extent, tetrasodium pyrophosphate. These polyphosphates chelate the polyvalent metal ions to form a soluble complex (under some conditions, the pyrophosphate complexes are sparingly soluble). In response to state and local legislation aimed at reducing the phosphate levels in lakes and rivers, detergent manufacturers began large-scale utilization of synthetic zeolites (sodium aluminosilicates) in 1978. Zeolites are insoluble in water, but act as ion exchange agents; exchanging Na^+ for Ca^{++} in the wash water. Peak usage of zeolites occurred in 1982 at 300 million pounds (136 million kg) annually.

Precipitating builders, such as sodium carbonate, Na_2CO_3, are also used, but to a much lesser extent, since the precipitate formed can itself deposit on the surface to be cleaned unless special washing procedures are followed. Other chelating builders than phosphates, such as trisodium nitrilotriacetate (NTA), tetrasodium ethylenediamine tetraacetate (EDTA), and other polycarboxylates, have also been used in small to moderate quantities, but questions concerning their safety, efficacy, or economics have to date prevented major displacement of the phosphates.

Bleaches. In many detergent applications, the action of a supplementary bleaching agent is necessary or desirable. It is frequently preferable to incorporate the bleach directly into the detergent product for reasons of convenience or performance.

Two types of bleaching agents are in common use in detergent products: (1) those based on a hypochlorite bleaching species, and (2) those based on a peroxygen species. The hypochlorite, or "chlorine," bleaches are considerably more powerful in their oxidizing action than the commonly used peroxygen bleaches under most U.S. usage conditions. Commercially important examples of hypochlorite species bleaches used in detergents include: Sodium dichloroisocyanurate dihydrate, NaDCC·$2H_2O$, and chlorinated trisodium phosphate (a physical combination of NaOCl, H_2O, and Na_3PO_4,), while sodium perborate, $NaBO_3·4H_2O$, is by far the most commonly used of the peroxygen bleaches. All current detergent products containing bleach are dry solids since aqueous bleach solutions have not been found to be sufficiently stable in the presence of other desired detergent ingredients.

Types of detergent products to which chlorine bleaches are frequently added include hard-surface cleaners, such as the scouring cleansers and the automatic dishwashing detergents, and detergents for commercial sanitation, in which the disinfecting properties of the chlorine bleach are a critical performance attribute. In the cleaning of textiles, consideration of possible fiber and color damage prevents the universal use of a powerful oxidizing agent like hypochlorite. Consequently, a separate liquid solution of sodium hypochlorite is often used in conjunction with the detergent on those fabrics where its use is appropriate, while some textile detergents contain the milder sodium perborate bleach for more general application.

Corrosion Inhibitors. Early research indicated that unmodified alkaline detergents could be corrosive to certain hard surfaces such as aluminum, washing-machine porcelain, and the overglaze on fine china. It was quickly determined, however, that the addition of soluble silicates (silicates with varying ratios of SiO_2 to Na_2O are used, depending upon product processing and intended application) could essentially prevent this corrosive effect. Consequently, the soluble silicates are widely added in moderate amounts to alkaline detergents, and even contribute to detergency through their added alkalinity.

Alkalinity Boosters. It has long been known that many soils are sensitive to high pH. Fairly high levels of sodium carbonate, soluble silicates (ratio 1.6 to 3.2 SiO_2 to Na_2O) and sodium metasilicate (Na_2SiO_3) are used to supply alkalinity for improved cleaning performance.

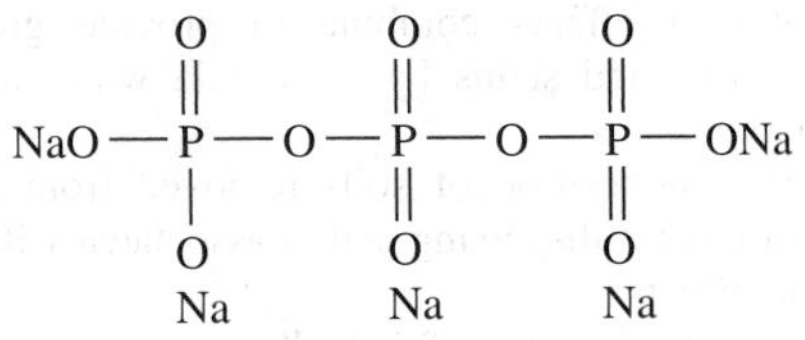

STP

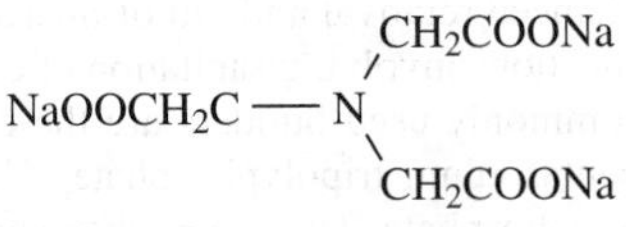

NTA

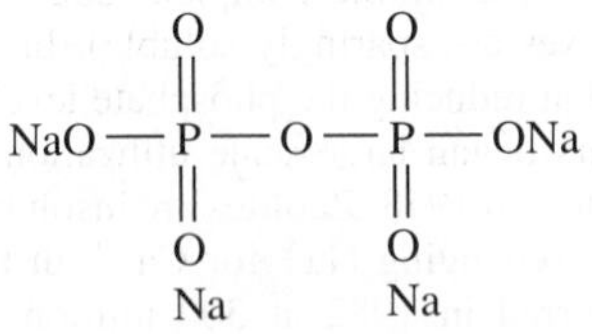

Tetrasodium pyrophosphate

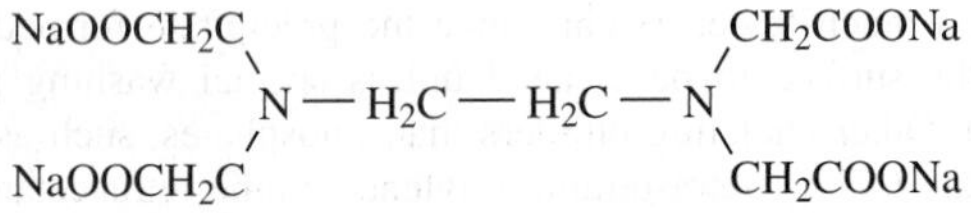

EDTA

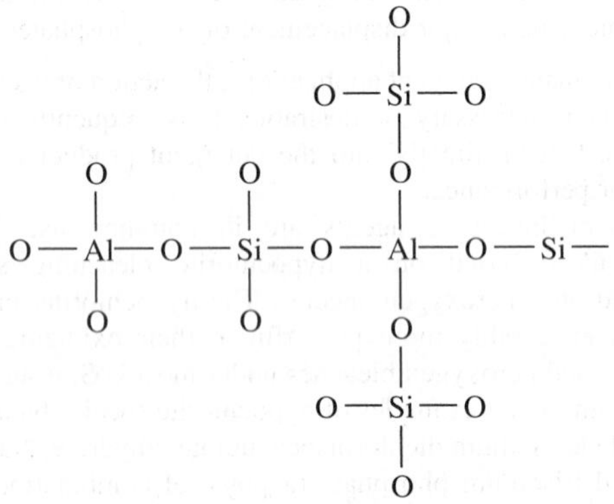

Sodium aluminosilicates
(Zeolities)

Sudsing Modifiers. In many applications, the detergent surfactants that appear to provide optimum cleansing performance do not provide satisfactory product-sudsing characteristics from a functional or aesthetic standpoint. Accordingly, materials that depress or boost the sudsing of the basic surfactant system are often added in small amounts to such products. Examples of materials which can be used to boost the sudsing of common anionic surfactant systems are the mono- and diethanol amides of C_{10-16}, fatty acids, while the long-chain, $C_{16, 22}$ fatty acids themselves and certain nonionics, such as the ethoxylated fatty alcohols, are commonly used as suds depressors.

Fluorescent Whitening Agents (FWA). Fluorescent whitening agents (also called *fluorescers, brighteners, optical bleaches*) are organic chromophores which absorb incident light in the ultraviolet region and reemit part of the absorbed energy as visible light, generally in the blue region of the visible spectrum. In addition, the chromophore is modified with organic substituents to make it substantive to one or more textile substrates from a laundry wash solution. Consequently, the brightness and whiteness of the fabrics on which FWA is deposited are enhanced. In effect, an added portion of the incident light is reflected from the fabric.

Enzymes. The most substantial advance in textile cleansing since the introduction of synthetic detergents has occurred through the introduction of low levels of enzymes into laundry detergent and presoak products. Both proteolytic and amylolytic enzymes are used by the industry to hydrolyze protein and starch so that the smaller soil fragments are easier to remove. They are effective on stains with protein and carbohydrate substituents (such as body soils, many food stains, grass stains, blood, and many others). Enzymes are catalytic and specific and thus must be used at low levels in products with excellent safety for fibers and textile colors (compared with a general chemical reactant, such as hypochlorite bleach).

Presently, all detergent enzymes are derived from fermentation cultures of specific strains of the ubiquitous bacilli *B. subtilis* and *B. licheniformis*. (The enzymes themselves are, of course, nonliving proteins, biochemical products of the bacilli which are used to hydrolyze exocellular proteins into small peptide fragments that can be absorbed by the bacteria as nutrient sources.) It had been recognized since early in this century that enzymatic action could be a most useful mechanism in textile cleansing, but available enzymes were rendered ineffective by the elevated temperatures and alkalinity required for satisfactory general detergency. The discovery and production of the *B. subtilis* and *B. licheniformis* mutants and their metabolites which remained active under laundering conditions was the essential step required to make enzymatic action available in textile detergency.

Antiredeposition Agents. In textile laundering, many soils once removed have a tendency to redeposit back onto the textile substrate during the remainder of the washing process. Certain agents, such as carboxymethyl cellulose and polyvinyl alcohol have been found to be effective in preventing or minimizing this effect and are consequently added in very small amounts to many laundry detergents.

Softeners and Anti-stats. A recent formulation trend is to incorporate fabric care benefits, such as softening and anti-static properties, into textile cleaning products in both powder and liquid form. The most common ingredient used is ditallow dimethyl ammonium chloride, the same material used in many liquid fabric-softener products:

$$(CH_3(CH_2)_{13-17})_2N^+(CH_3)_2Cl^-$$

Ditallow dimethylammonium chloride

Since this material is cationic, it is generally incompatible with anionic surfactants unless special proprietary formulation approaches are used. Therefore, fabric care detergents, especially liquids, often use nonionic surfactants.

Other Materials. In addition to the basic performance ingredients previously discussed, other materials are commonly added to facilitate product manufacture (for example, hydrotropes such as xylene sulfonate) and to enhance product acceptability among consumers (colorants and perfumes). A few materials are formed or carried along into the finished product by the manufacturing process (such as sodium sulfate and water).

Types of Detergent Products

Synthetic detergents are manufactured to perform a wide variety of household and industrial cleansing operations. Within the commercially more important (in terms of tonnage and dollar volume of sales) household-product category, the following major functional areas may be defined.

Textile Cleansing. Household laundry detergents and presoak products represent the largest single category of detergent use, currently accounting for over 2 million tons per year in U.S. consumption. These products are used in conjunction with individually owned or self-service washing machines to launder most of the clothing, bed and bath linens, curtains, and other textiles used in the typical household. Prior to the introduction of detergents over a half-century ago, soaps were almost exclusively used for these purposes.

Detergents in liquid, tablet, and granular powder forms are sold for household laundry use, the granules accounting for by far the major amount (75%) of product sold. A household laundry detergent powder typically contains 10–20% surfactant, 20–40% builder, 5–10% sodium silicates, low levels of suds builders or suppressors, anti-redeposition agents, fluorescent whitening agents, enzymes, peroxygen bleach, colorants, perfume as required to meet aesthetic and performance objectives, and the remainder being materials of processing (sodium sulfate, water). Powder bulk density varies among products, but most commonly is such as to provide a wash solution concentration of about 0.15% detergent at the product usage recommended. The presoak powders differ from the wash detergents in providing relatively higher wash water concentrations of enzymes, builders, and peroxygen bleach (if used), but a relatively lower concentration of surfactant. The liquid laundry detergents are currently the most rapidly growing segment of the market. They use the same type of materials present in powders, but typically have much higher levels of surfactants (20–30%) and lower levels of builders.

Hard-Surface Cleaning. Products used for hard-surface cleaning around the home include the hand- and automatic dishwashing products, the liquid and powder floor and wall cleaners, and the abrasive scouring cleansers.

Although the higher-sudsing laundry detergents may be used satisfactorily for hand dishwashing (and were so used extensively in the past), high-sudsing liquid products are predominantly used for this purpose. Relative to powders, the liquids offer superior ease and convenience of use, a product form into which surfactant types and levels better suited to the dishwashing task can be incorporated, and comparable economy of use since the builders required for superior laundry performance are not essential for hand dishwashing use. Dishwashing liquids typically consist of 25–45% anionic or semipolar surfactants, the remainder consisting of solvents, hydrotropes, buffers, colorants, perfumes, and water of processing. Surfactant types used in this application include LAS, alkyl sulfates, alkyl polyethoxy sulfates, di- or monoethanol fatty acid amides, alkyl glyceryl ether sulfonates, and alkyl dimethylamine oxides. The alkyl portions of these surfactants commonly average to a coconut or C_{12} chain length, as opposed to laundry detergents, where slightly longer alkyl chains are used.

Automatic dishwashing products are designed solely for use in household mechanical dishwashers, machines in which a moving high-velocity water spray is used to clean the tableware and cooking utensils. Performance requirements for an automatic dishwasher product differ substantially from those for laundry and hand dishwashing products; they include: (1) very low sudsing to prevent suds overflows with the high-velocity spray; (2) very complete rinsing to avoid residual deposits; (3) complete sequestration of Ca^{++} and Mg^{++} ions in water supplies by the use of relatively large amounts of builder to avoid deposits of sparingly soluble Ca and Mg salts; and (4) thorough removal of minute particles of food protein which can form nuclei for spot formation during drying. Consequently, automatic dishwashing products typically contain a low level of a nonionic surfactant (commonly polyoxy-ethylene/polyoxypropylene condensates), a low level of a dry bleach (KDCC or chlorinated trisodium phosphate), a high level of builder (STP and/or sodium carbonate), and moderate-to-high levels of auxiliary sources of alkalinity (sodium silicates and/or sodium carbonate). These products are typically dry powders for reasons of machine design and product bleach stability during the period between manufacture and eventual use. However, thixotropic automatic dishwashing liquid products containing bleach were recently introduced into the marketplace.

Abrasive cleaners are used to remove soils and stains from hard surfaces that are durable to the scouring action. Such surfaces include stainless steel and porcelain plumbing fixtures, metal and ceramic cooking utensils, and various stone, metal, and ceramic building surfaces. Typically, these products consist of a very high level of abrasive (commonly silica flour) with moderate to low levels of a dry chlorine bleach (KDCC or chlorinated trisodium phosphate) and low levels of surfactant (LAS) and builder (STP) for wetting action and improved stain removal.

Other hard-surface cleaners are formulated to clean larger surface areas which do not require or are less resistant to the action of an abrasive cleanser or from which the abrasive would be difficult to remove, such as floors, walls, woodwork, and large appliances. These products are sold in both liquid and powder form. The liquid products contain low levels of nonionic and anionic surfactants; moderate levels of a stable, highly soluble builder (commonly tetrapotassium pyrophosphate, $K_4P_2O_7$) for better stain removal; solvents; hydrotropes; and water of processing. The powdered products contain a very low level of surfactant (LAS), a moderate amount of builder (STP), and substantial amounts of sources of mild alkalinity (trisodium phosphate and mixtures of sodium carbonate and sodium bicarbonate) for improved soil removal. Low sudsing is a requirement for both types of products to aid in rinsing the surfaces after cleaning.

Personal Care Products. Within the broad definition of synthetic detergents, a variety of cleansing products are made for personal care. These include such products as cleansing bars, shampoos, bubble-bath products, cosmetic cleansers, and tooth pastes. Formulations of these products vary widely, depending upon their intended use.

Although essentially pure soap products continue to dominate the cleansing-bar field, a few products contain synthetic surfactants in addition to soap to act as scum and curd dispersants. Synthetic surfactants used in this application include alkyl sulfates, alkyl glyceryl ether sulfonates, alkyl esters of sodium isothionate, and alkylamides of N-methyl tauride.

Shampoos are commonly formulated in liquid, paste, and gel form and usually consist of high-sudsing anionic surfactant(s) (such as LAS and those previously listed for bars), along with specific ingredients for improved hair health or control (such as antidandruff agents and substantive collagen proteins).

Bubble baths are provided in both liquid and powder form, and commonly provide a high sudsing surfactant(s). Cosmetic cleansers vary widely in formulation, depending upon intended use. Some provide mixtures of surfactants and heavy mineral oil (cold cream), while others provide an organic solvent (for mascara removal, for example). Tooth pastes provide a low level of an anionic surfactant (several types used) and high levels of a moderate abrasive (insoluble pyrophosphates), along with other special ingredients, such as anticaries agents.

Many other products, such as rug cleaners, automobile cleaners, scouring pads, and pet care products incorporate synthetic detergents.

Industrial and Institutional Applications. In addition to household products just described, a wide variety of products based wholly or partially on synthetic detergents are made for industrial and institutional use. Included among these are products that are essentially direct analogues of the various household products. In addition, there is a wide variety of products which have no counterparts among the household detergents. Included are detergents to scour raw yarns in the textile manufacturing process, detergents to clean metals prior to painting or electroplating, surgical preparation products, detergents to clean and disinfect poultry houses, and many more. Industrial and institutional detergent products account for about 500 million pounds (227 million kg) of product annually.

Manual G. Venegas
George J. Kaminsky
The Procter & Gamble Company
Cincinnati, Ohio

Additional Reading

Ash, I. and M. Ash: *Handbook of Industrial Surfactants: An International Guide to More than 21,000 Products by Trade Name, Composition, Application, and Manufactures,* Vol. 1, Gower Publishing Ltd., Brookfield, VT, 1997.

Broze, G.: *Handbook of Detergents: Properties,* Vol. 82, Marcel Dekker, Inc., New York, NY, 1999.

Cutler, W.G. and E. Kissa: *Detergency: Theory and Technology,* Marcel Dekker, New York, NY, 1987.

Goddard E.D. and K.P. Ananthapadmanabhan (Editor): *Interactions of Surfactants with Polymers and Proteins,* CRC Press LLC., Boca Raton, FL, 1993.

Halliday, H. et al.: *Recent Advances in Surfactant Research,* S. Karger AG, Switzerland, 1999.

Hummel, D.O.: *Handbook of Surfactant Analysis: Chemical, Physicochemical and Physical Methods,* John Wiley & Sons, Inc., New York, NY, 2000.

Karsa, D.R.: *Surface Active Behaviour of Performance Surfactants,* CRC Press, LLC., Boca Raton, FL, 2000.

Lai, Kuo-Yann: *Liquid Detergents,* Marcel Dekker, Inc., New York, NY, 1996.

Rosen, M.J.: *Surfactants and Interfacial Phenomena,* 2nd Edition, John Wiley & Sons, Inc., New York, NY, 1994.

Rubingh, D.N. and P.M. Holland: *Cationic Surfactants: Physical Chemistry,* Marcel Dekker, Inc., New York, NY, 2000.

Showell, M.S.: *Powdered Detergents,* Marcel Dekker, Inc., New York, NY, 1998.

Staff: *Soaps and Other Detergents, Polishes, Leather, Resilient Floor Coverings,* American Society for Testing & Materials, West Conshohocken, PA, 1998.

Vanee, Y.H., O. Misset, and E.J. Baas: *Enzymes in Detergency,* Marcel Dekker, Inc., New York, NY, 1997.

Workman, J.: *The Handbook of Organic Compounds: NIR, IR, and UV Spectra Featuring Polymers and Surfactants,* Academic Press, Inc., San Diego, CA, 2000.

Web References

The Soap and Detergent Association. www.sdahq.org
The Surfactants Virtual Library. www.surfacants.net

DETINNING. See **Goldschmidt Detinning Process**.

DEUTERIUM. The isotope of hydrogen with mass number 2 is termed deuterium. The symbol D is sometimes used. Using ocean water as a reference, the atomic abundance of deuterium in natural hydrogen is 0.0149%. Deuterium oxide D_2O is known as heavy water and was first identified by Harold C. Urey in 1932. Urey noted a slight shift in the spectrum of deuterium and tritium as compared with protium (ordinary hydrogen). The diameter of the electron orbit for deuterium is slightly greater than for protium, and still greater for tritium. Deuterium and deuterium oxide gained prominence largely because of their excellent properties as moderators in nuclear reactors. See also **Uranium**.

DEUTERON.   The nucleus of deuterium (heavy hydrogen) is known as deuteron. A particle that contains one proton and one neutron also is termed a deuteron.

DEVITRIFICATION.   The process by which the natural rock glasses, such as obsidian and tachylyte, develop minute but definite minerals, usually quartz and feldspar.

Devitrification also applies to manufactured glasses, denoting crystallization and detected by the appearance of opaque areas. See also **Vitreous State**.

DEW POINT.   The temperature to which a given parcel of air must be cooled at constant pressure and constant watervapor content in order for saturation to occur; the temperature at which the saturation vapor pressure of the parcel is equal to the actual vapor pressure of the contained water vapor. Any further cooling usually results in the formation of dew or frost. Also called *dewpoint temperature*. When this temperature is below 0°C, it is sometimes called the frost point.

DEXTROROTATORY COMPOUND.   See **Asymmetry (Chemical)**; **Isomerism**.

DEZINCIFICATION.   A form of electrolytic corrosion observed in some brasses where the copper-zinc alloy goes into solution with subsequent redeposition of the copper. The small red copper plugs thus formed in the brass are usually porous and of low strength. In recent years, the term dezincification has also been applied in a more general sense to signify any metallic corrosion process that dissolves one of the components from an alloy.

DIAGENESIS.   A term proposed by Gumbel in 1888 for the gradual and successive chemical–physical changes that take place in sediments previous to or during their consolidation. Diagenesis may also include the numerous processes of lithification but is a useful term only when particularly applied to the more or less contemporaneous chemical alteration of sediments.

DIALLAGE.   The mineral term for a calcium-iron pyroxene, similar in chemical composition to diopside but richer in iron oxide. In addition to the typical prismatic cleavage of the pyroxene group diallage has a marked "cleavage" parallel to the vertical pinacoids, known as diallage parting. Diallage is a common constituent of gabbros. The term diallagite was proposed by Cloiseaux in 1845 for rocks particularly rich in diallage. The term diallage is derived from the Greek meaning difference, and referring to the peculiar cleavages of this variety of monoclinic pyroxene. See also **Pyroxene**.

DIALYSIS.   The process of separating compounds or materials by the difference in their rates of diffusion through a colloidal semipermeable membrane. Thus, sodium chloride diffuses eleven times as fast as tannin and twenty-one times as fast as albumin. When the process is conducted under the influence of a difference in electrical potential, as from electrodes on opposite sides of the semipermeable membrane, it is called electrodialysis.

An apparatus for carrying out a dialysis usually consists of two chambers separated by a semipermeable membrane of parchment paper latex, animal tissue, or other colloid. In one chamber the solution is placed, and in the other, the pure solvent. Crystalline substances diffuse from the solution through the membrane and into the solvent much more rapidly than amorphous substances, colloids or large molecules.

DIAMAGNETISM.   Diamagnetism is the phenomenon in which the magnetization in a substance opposes the magnetizing force that induces it. Diamagnetism is considered to exist in all substances, although in substances exhibiting paramagnetism or ferromagnetism, it is masked by the much greater opposite effect due to the orientation of the magnetic atoms or molecules.

DIAMINES AND HIGHER AMINES, ALIPHATIC.   The aliphatic diamine and polyamine family encompasses a wide range of multifunctional, multireactive compounds. This family includes ethylenediamine (EDA) and its homologues, the polyethylene polyamines (commonly referred to as ethyleneamines), the diaminopropanes and several specific alkanediamines, and analogous polyamines. The molecular structures of these compounds may be linear, branched or cyclic, or combinations of these.

The ethyleneamines have found the broadest commercial application and are the primary focus of this article. The lower molecular weight ethylenediamines, i.e., EDA, diethylenetriamine (DETA), piperazine (PIP), and N-(2-aminoethyl)-piperazine (AEP), are available commercially as industrially pure products. The tetramine (TETA), pentamine (TEPA), hexamine (PEHA), and higher polyamine products are commercially available as boiling point fractions consisting of natural mixtures of linear, branched, and cyclic compounds. Their compositions are largely determined by the chemical processes used in their production. The individual components in these higher ethyleneamines are generally not available in industrial quantities.

The predominant commercial diaminopropanes are 1,2-propylenediamine (1,2-PDA), 1,3-diaminopropane (1,3-PDA), iminobispropylamine (IBPA), and dimethylaminopropylamine (DMAPA). Other commercially important products include other higher alkylenediamines, such as hexamethylenediamine (HMDA); certain cyclic amines, such as triethylenediamine (TEDA); and various alkyl- and hydroxylalkyl-derivatives.

Physical Properties

Physical properties of some commercially available polyamines appear in Table 1. Generally, they are slightly to moderately viscous, water-soluble liquids with mild to strong ammoniacal odors. Although completely soluble in water initially, hydrates may form with time, particularly with the heavy ethyleneamines (TETA, TEPA, PEHA, and higher polyamines), to the point that gels may form or the total solution may solidify under ambient conditions. The amines are also completely miscible with alcohols, acetone, benzene, toluene and ethyl ether, but only slightly soluble in heptane. Piperazine, the lowest mol wt cyclic diamine, freezes above room temperature. As such, it is available commercially as either the anhydrous solid or an aqueous solution.

Chemical Properties

The aliphatic alkyleneamines are strong bases exhibiting behavior typical of simple aliphatic amines. Additionally, dependent on the location of the primary or secondary amino groups in the alkyleneamines, ring formation with various reactants can occur. This same feature allows for metal ion complexation or chelation. The alkyleneamines are somewhat weaker bases than aliphatic amines and much stronger bases than ammonia as the pK_b values indicate.

Alkylene Oxides and Aziridines.   Alkyleneamines react readily with epoxides, such as ethylene oxide or propylene oxide, to form mixtures of hydroxyalkyl derivatives. Aziridines react in an analogous fashion to epoxides.

Aliphatic Alcohols and Alkylene Glycols.   Simple aliphatic alcohols, such as methanol, can be used to alkylate alkyleneamines.

Organic Halides.   Alkyl halides and aryl halides, activated by electron withdrawing groups (such as NO_2) in the ortho or para positions, react with alkyleneamines to form mono- or disubstituted derivatives.

Aldehydes.   Alkyleneamines react exothermically with aliphatic aldehydes. The products depend on stoichiometry, reaction conditions, and structure of the alkyleneamine.

Organic Acids and Their Derivatives (Anhydrides, Nitriles, Ureas).   Alkyleneamines react with acids, esters, acid anhydrides or acyl halides to form amidoamines and polyamides. Various diamides of EDA are prepared from the appropriate methyl ester or acid at moderate temperatures.

Sulfur Compounds.   EDA reacts readily with two moles of CS_2 in aqueous sodium hydroxide to form the bis sodium dithiocarbamate.

Environmentally Available Reactants.   Under normal conditions ethyleneamines are considered to be thermally stable molecules. However, they are sufficiently reactive that upon exposure to adventitious water, carbon dioxide, nitrogen oxides, and oxygen, trace levels or by-products can form and increased color usually results.

TABLE 1. PROPERTIES OF COMMERCIAL DIAMINES AND HIGHER AMINES

Commercial name	Molecular weight	Freezing point, °C	Bp[a], °C	ΔH^a_{vap} kJ/mol	Refractive index, n^{20}D	Viscosity at 20°C, mPa·s(= cP)
ethylenediamine	60.1	10.8	117.0	40.7	1.4565	1.8
diethylenetriamine	103.2	−39	206.9	54.0	1.4859	7.2
triethylenetetramine	146.2[b]	−35	277.4	56.4	1.4986	26
tetraethylenepentamine	189.3[b]	>−40	315		1.5067	76
pentaethylenehexamine[c]	232.4[b]	−30	180–280[d]			100–300
aminoethylpiperazine	129.2	−17	221	41.2	1.5003	15
piperazine	86.1	109.6	144.1			
1,2-propylenediamine	74.1	−27	120–123	38.2	1.4455	1.6
1,3-diaminopropane	74.1	−12	137–140	46.4[e]	1.4555	2.0
iminobispropylamine	131.2	−16	110–120	76.2[e]	1.4791	9.6
N-(2-aminoethyl)-1,3-propylenediamine	117.2		80			
N,N'-bis-(3-aminopropyl)-ethylenediamine	174.3		170			
dimethylaminopropylamine	102.2	−56	134.9	35.6	1.4350	1.1
menthanediamine	170.3	−45	107–126[f]		1.479	17.5[g]
triethylenediamine	112.2	158	174	61.9[h]		
hexamethylenediamine[i]	116.2	41	204.0	51.0		

[a] At 101.3 kPa = 1 atm unless otherwise noted.

[b] Linear component. Commercial product consists of a mixture of linear, branched, and cyclic structures with the same number of nitrogen atoms.

[c] Commercial higher polyamine products contain up to about 40% PEHA.

[d] At 0.67 kPa, 10–60% distills in this range.

[e] At 93.3°C.

[f] At 1.3 kPa.

[g] At 25°C.

[h] Heat of sublimation, below 78°C.

[i] For manufacture of HMDA in preparation of nylon-6,6 see **Polyamides**.

Manufacture

Ethyleneamine Processes. Present industrial processes are based on ethylene and ammonia. The sixty-year-old ethylene dichloride (EDC) process is still the most widely practiced industrial route for producing ethyleneamines.

Alternative processes for the manufacture of ethyleneamines have been actively sought since the late 1960s. The catalytic reductive amination of monoethanolamine (MEA), which was the first such process to appear, produces the lighter ethyleneamines (EDA, DETA, PIP, and AEP) and coproduct aminoethylethanolamine (AEEA), and hydroxyethylpiperazine (HEP).

The condensation of MEA with EDA over heterogeneous catalysts to form primarily DETA represents the newest commercial technology for making ethyleneamines.

A process for the production of ethylenimine, a suspected carcinogen, by the vapor phase dehydration of monoethanolamine has been developed.

Diaminopropane Processes. 1,2-Propylenediamine can be produced by the reductive amination of propylene oxide, 1,2-propylene glycol, or monoisopropanolamine. 1,3-Propanediol can be used to make 1,3-diaminopropane. Various propaneamines are produced by reducing the appropriate acrylonitrile–amine adducts. Polypropaneamines can be obtained by the oligomerization of 1,3-diaminopropane.

Economic Aspects

Worldwide growth for most ethyleneamines is expected to parallel GDP. Some regional and certain applications demands will show somewhat higher growth rates.

Of the worldwide ethyleneamines capacity, over 50% is EDC-based; the balance is monoethanolamine-derived.

Storage and Handling

By virtue of their unique combination of reactivity and basicity, the polyamines react with, or catalyze the reaction of, many chemicals, sometimes rapidly and usually exothermically. Some reactions may produce derivatives that are explosive (e.g., ethylenedinitramine). The amines can catalyze a runaway reaction with other compounds (e.g., maleic anhydride, ethylene oxide, acrolein, and acrylates), sometimes resulting in an explosion.

As commercially pure materials, the ethyleneamines exhibit good temperature stability, but at elevated temperatures noticeable product breakdown may result in the formation of ammonia and lower and higher mol wt species.

Like many other combustible liquids, self-heating of ethyleneamines may occur by slow oxidation in absorbent or high surface-area media. In some cases, this may lead to spontaneous combustion; either smoldering or a flame may be observed. These media should be washed with water to remove the ethyleneamines, or thoroughly wet prior to disposal in accordance with local and Federal regulations.

Since ethyleneamines react with many other chemicals, dedicated processing equipment is usually desirable. Amines slowly absorb water, carbon dioxide, nitrogen oxides, and oxygen from the atmosphere, which may result in the formation of low concentrations of by-products and generally increase color. Storage under an inert atmosphere minimizes this sort of degradation.

Galvanized steel, copper, and copper-bearing alloys are unacceptable for all ethyleneamine service. A 300 series stainless steel or aluminum are recommended for storage of lighter amines (particularly EDA and DETA) to maintain product quality. Carbon steel generally can be used for storage of the heavier ethyleneamines without noticeable impact on product quality if the storage temperature is modest (<60°C), nitrogen blankets are maintained to exclude air, and the material is anhydrous. A 300 series stainless steel is often specified for heating coils, transfer lines, and small agitated tanks, because carbon steel can suffer enhanced corrosion as a result of the erosion of the passive film by the product velocity. Similar logic suggests cast 316 stainless steel for pumps and valves in ethyleneamine service.

Baked phenolic-lined carbon steel is acceptable for storage of many pure ethyleneamines, except EDA. Gaskets utilized in ethyleneamine service generally are made of Grafoil flexible graphite or polytetrafluoroethylene (TFE).

Most common thermal insulating materials are acceptable for ethyleneamine service. However, porous insulation may introduce the hazard of spontaneous combustion if saturated with ethyleneamines from a leak or external spill.

Certain ethyleneamines require storage above ambient temperature to keep them above their freezing points (EDA and PIP) or to lower the viscosity (the heavy amines). As a result, the vapors "breathing" from the storage tank can contain significant concentrations of the product. Water scrubbers may be used to capture these vapors.

Solid ethyleneamine carbamates, formed by the reaction of the amines with carbon dioxide, can foul tank vents and pressure relief devices. Vent fouling can be minimized by using a nitrogen blanket that prevents atmospheric CO_2 from being drawn in, or by steam-tracing the vents (>160°C) to decompose the carbamates.

Although the ethyleneamines are water soluble, solid amine hydrates may form at certain concentrations that may plug processing equipment, vent lines, and safety devices. Hydrate formation usually can be avoided by insulating and heat tracing equipment to maintain a temperature of at least 50°C. Water cleanup of ethyleneamine equipment can result in hydrate formation even in areas where routine processing is nonaqueous. Use of warm water can reduce the extent of the problem.

Health and Safety Factors

Ethyleneamine vapors are painful and irritating to the eyes, nose, throat, and respiratory system. Extremely high vapor concentration may cause lung damage. Prolonged or repeated inhalation may lead to kidney, liver, and respiratory system injury. Contact with the liquids will severely damage the eyes and may cause serious burns to the skin. When swallowed, the concentrated liquid materials may produce considerable local injury. Both vapors and liquid can cause sensitization in some individuals, resulting in contact dermatitis and/or the development of an asthmatic respiratory response. This may occur in certain susceptible individuals following exposure to extremely low concentrations of ethyleneamines, even below the irritation threshold.

The ACGIH has adopted TLVs of 10 ppm (25 mg/m^3) and 1 ppm (4 mg/m^3) for EDA and DETA, respectively. Strict precautions should be observed to prevent direct contact with ethyleneamines, including eye, skin, and respiratory protection. If contact is made, medical treatment should be obtained immediately, in addition to flushing and washing with copious amounts of water. Vomiting is not to be induced following ingestion.

Applications

Polyalkylene polyamines find use in a wide variety of applications by virtue of their unique combination of reactivity, basicity, and surface activity. With a few significant exceptions, they are used predominantly as intermediates in the production of functional products. End-use profiles for the various ethyleneamines include fungicide, oil and fuel additives, polyamides/epoxy curing, paper resins, chelating agents, fabric softeners/surfactants, petroleum production, bleach activator, and anthelmintics/pharmaceuticals.

RICHARD G. CARTER
ARTHUR R. DOUMAUX, JR.
STEVEN W. KAISER
PAMLA R. UMBERGER
Union Carbide Chemicals and Plastics Company Inc.

DIAMOND. An allotropic form of carbon, diamond over the centuries has been known best for its use by the jewelry trade as a precious stone. See also **Gemstones.** See Fig. 1. Although diamond possesses several outstanding properties as a material, its hardness largely has accounted for the industrial uses of diamond for cutting and polishing operations. Until comparatively recently, industrial diamonds were the byproducts of gem stone-mining operations. Although numerous researchers over the years have been attracted by the possibility of creating diamonds synthetically, only during the last decade or two has sufficient progress been made in diamond synthesis to forecast the production of industrial diamonds in tonnage quantities at affordable prices. In addition to the traditional use of diamond because of its hardness, other properties of this outstanding material (heat conductivity, for example) are exploited. Annual industrial diamond production, which amounted to about $200 million in 1991 is expected to exceed $1 billion by the end of the century. As of 1994, research is continuing at an aggressive pace, notably in Japan; a few firms in the United States also are quite active in the field. Over the years, Russian scientists also have been developing a position in synthetic diamond technology.

Important Properties of Diamond

Diamond has the highest atom-number density of any known material at terrestrial pressures[1]. Because of its high atom-number density and strong covalent bonding, diamond has the highest hardness and elastic modulus of any material and is the least compressible substance known.

[1] Diamond possibly may be exceeded by the superdense carbon reported by Matyusenko, et al., 1979. (See reference listed.)

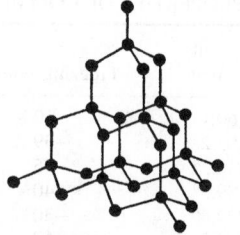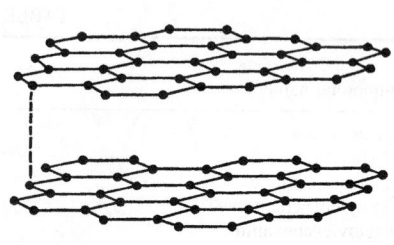

Fig. 1. A gross conceptualization of the space lattices of (**a**) diamond, and (**b**) graphite. See new spatial concept for graphite in article on **Carbon**. Diamond crystallizes in the cubic system, is the hardest of known substances (10 on the Mohs scale; 5,500–7,000 on the Knoop scale), specific gravity 3.51–3.521 (20°C), dielectric constant at 10^4 Hz, 16.5, at 10^8 Hz, 5.5, index of refraction 2.417–2.4195. Classically, a diamond crystal may be pictured as a huge polymeric molecule, very tightly packed, with a density about 1.6 times greater than the other allotropic form of carbon, graphite. The normal C—C single bond distances in the atomic lattice of diamond are all 1.54 Å, whereas in graphite the C—C bond distances are 1.42 Å. The tight packing of diamond accounts, of course, for its relatively high density and for its extreme hardness. Graphite, on the other hand, essentially is comprised of two-dimensional molecules, very laminar, and which tend to slide and thus impart lubricity to the substance

The thermal conductivity of diamond at 300 K is higher than that of any other material, and its thermal expansion coefficient at 300 K is 0.8×10^{-6}, lower than that of *Invar* (an Fe—Ni alloy). Diamond is a very wide-band gap semiconductor ($Eg = 5.5$ eV), has a high breakdown voltage (10^{7}V cm^{-1}), and its saturation velocity of 2.7×10^7 cm s^{-1} is considerably greater than that of silicon, gallium arsenide, or indium phosphide.

Because of these several outstanding qualities, as early as 1988, J.C. Angus and C.C. Hayman (Case Western Reserve University) observed, "There is great interest in using single-crystal diamond for heat sinks and as an active semiconductor element. Potential electronic applications of diamond include high-temperature devices, millimeter-wave traveling wave amplifiers, backward-wave oscillators, and picosecond high-voltage electro-optic switches. Some potential applications may require only polycrystalline films, for example, laser and x-ray windows, lenses, bearing surfaces, and tribological coatings." Several of these applications already have been demonstrated as of the mid 1990s.

Naturally Occurring Diamonds

Geologists observe that probably all of the known diamonds found prior to 1725 had been found in India. In that year, diamonds were discovered in Brazil, a source that has continued to present times. Most Brazilian stones, however, are comparatively small.

Diamonds were discovered in 1867 along the Orange River in South Africa, and since then Africa has been preeminent in the production of diamonds; in the seventies and eighties occurred a series of amazing discoveries of diamond fields and stones of extraordinary size. Diamonds have also been found in Australia, Borneo, British Guiana, and Arkansas.

Much as the matter has been studied there is no consensus as to the genesis of the diamond. It is found in alluvial deposits, both unconsolidated and consolidated, indicating the erosion of rocks containing diamonds not only during the present era but also in past geologic time. In Africa diamonds are mined in a dark basic rock of the general nature of peridotite called kimberlite from the town of Kimberley. The kimberlite occurs in vertical "pipes," resembling what once may have been volcanic necks or other types of igneous rock conduits. It is supposed that the diamonds have been formed in and brought to the surface by the magma, which was of the general nature of peridotite. Undoubtedly high pressures, and possibly high temperatures as well, are necessary for the development of crystallized carbon in the form of diamonds.

Expanded insight concerning the formation and occurrence of diamonds in southern Africa is provided by Boyd (Carnegie Institution of Washington) and Gurney (University of Cape Town) in a paper dealing with diamonds and the African lithosphere. In addition to gaining a better understanding of diamonds, the researchers point out that new knowledge on the structure and history of the Kaapvall craton in which diamonds occur has been gained. As pointed out by the investigators, high pressures are required for the crystallization of diamonds, corresponding to depths in the earth's mantle of at least 150 kilometers. Diamonds formed at those depths have been brought to the surface in volcanic eruptions of kimberlite. See also **Kimberlite**. Diamonds are found in minute concentrations (+ ppm) in kimberlite that has filled the throats of ancient volcanoes, such as those at Kimberley, where igneous occurrences of diamonds were first recognized. Erosion of kimberlite pipes and transport of diamonds in streams and rivers has dispersed them beyond the boundaries of the craton, in some cases forming secondary concentrations in stream gravels and beach deposits.

In a summary of their findings, Boyd and Gurney state that the Kaapvall craton has a root composed in large part of peridotites that are strongly depleted in basaltic components. See also **Peridotite**. The asthenosphere boundary shelves from depths of 170 to 190 kilometers beneath the craton to approximately 140 kilometers beneath the mobile belts which border the craton on the south and west. The root formed earlier than 3 billion years ago, and at that time ambient temperatures in it were 900 to 1200°C. These temperatures are near those estimated from data for xenoliths erupted in the Late Cretaceous or from present-day heat-flow measurements. Many of the diamonds in southern Africa are believed to have crystallized in this root in Archean time and were xenocrysts in the kimberlites that brought them to the surface. Information about the composition and history of this root is being gathered through study of mineral inclusions in diamonds. Multiple, separate inclusions in individual diamonds are abundant. Sulfides form the most abundant inclusions, but garnet, olivine, pyroxenes, and chromite are also relatively common. For the reader who is seriously interested in South African diamond resources, reference to the full Boyd—Gurney report is suggested.

The manner in which large diamonds are cut and the weight of famous large diamonds is given under **Gemstones**. Growing interest by the jewelry trade in the so-called *pink diamond* has been demonstrated in recent years. As pointed out by P. Proddow and M. Fasel (See reference listed), "Pink diamonds became readily available only in 1984. Now they are used to surround everything from a 79-carat peridot from Burma to a starfish broach to black and white pearls in twin heart earclips."

Synthesis of Diamond

It is reasonable to assume that interest in synthesizing diamonds, driven by the scarcity of natural diamonds[2] extends back to the days of alchemy. Contemporary research involving either high-temperature, high-pressure methods or chemical vapor deposition (CVD) processes commenced in the late 1940s. Important early work was performed at Case Western Reserve University in Cleveland, Ohio and at the Institute for Physical Chemistry, Moscow, Russia. In 1949, W.G. Eversole (Union Carbide Corporation) was the first researcher to grow diamond successfully at low pressures. Conclusive proofs and repetitions of the experiments occurred in early 1953 and thus appears to predate the diamond synthesis at high pressure by investigators at General Electric Company, which was reported as accomplished in December 1954, but which was not announced publicly until 1955. Essentially contemporaneously, Liander (Allemanna Svenska Elektriska Aktiebolaget ASEA) in Sweden reported synthesis of diamond. It is interesting to note that Eversole grew new diamond on preexisting diamond nuclei, whereas the GE and ASEA syntheses commenced with nondiamond carbons.

As reported by J.C. Angus and C.C. Hayman (Case Western Reserve University), "B. Deryagin (Russia), who started work on low-pressure diamond synthesis in 1956 has had the longest sustained research effort on metastable diamond growth of any worker." Deryagin's group initiated many approaches to the problem, starting with the growth of diamond whiskers by a metal-catalyzed vapor-liquid solid (VLS) process. Later, the group researched epitaxial growth from hydrocarbons and hydrocarbon-hydrogen mixtures, using different forms of vapor transport reactions. The Deryagin group also concentrated on theoretical investigations of the relative nucleation rates of diamond and graphite. See Fig. 2.

J.C. Angus and co-researches at Case Western Reserve University (CWRU) targeted principally the CVD process, by attempting to deposit diamond on diamond seed crystals from hydrocarbons and hydrocarbon-hydrogen mixtures. The CWRU team grew P-type semiconducting diamond from $CH_4-B_2H_6$ gas mixtures and studied the rate of diamond and graphite growth in CH_4-H_2 gas mixtures and ethylene. The group was the first to report on the preferential etching of graphite compared to diamond by atomic hydrogen and noted that boron had an unusual catalytic effect on

[2] Proof that diamond is an allotropic form of carbon was given by the English chemist, Smithson Tennant, in 1797.

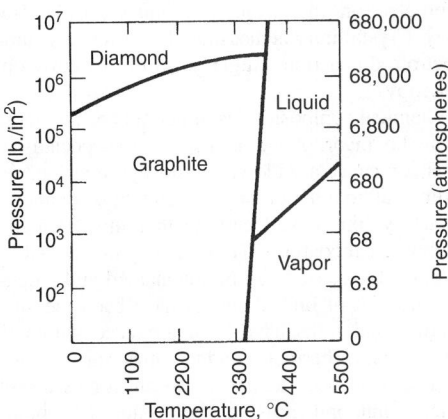

Fig. 2. Phase diagram approximation of carbon, indicating pressure-temperature parameters favoring yield of graphite and diamond. See also phase diagram in the article on **Carbon**

metastable diamond growth. The role of hydrogen in permitting metastable diamond growth was a constant thread throughout the early work.

Numerous other groups worked on the problem, and it was demonstrated that the presence of hydrogen enhanced yields of diamond. The addition of hydrogen to the hydrocarbon gas phase was shown to suppress the growth rate of graphite more than it suppressed the growth rate of diamond, resulting in higher diamond yields. However, graphitic carbons nucleated on the surface and suppressed further diamond growth. It was necessary to remove the graphitic deposits preferentially, with atomic hydrogen or oxygen, and to repeat the sequence.

Isotopically Adjusted Diamonds

The ratio of isotopes in natural diamond is about one carbon-13 atom for every 100 atoms, the remainder being carbon-12 atoms. During the early 1970s, R. Seitz (Harvard University) became convinced, it is reported, that an isotopically pure diamond (all carbon-12 atoms) would be an excellent conductor of heat and less susceptible to damage by a laser beam. In July 1990, General Electric announced that it had made an isotopically pure carbon-12 diamond. Tests and calculations have shown that a diamond of all or nearly all carbon-12 atoms had been produced. Apparently, the methane gas used in the CVD process can be enriched with either C-12 or C-13, after which thin mosaics of tiny diamond grains serve as feedstock for a week-long process involving 1000 tons of pressure and a temperature of about 1500°C. The operation may be described as a slow diffusion process. Because of the time factor and extreme conditions, costs are high. J.C. Angus (CWRU) suggests, "An alternative route for growing the gem-sized diamonds might develop from the vapor deposition methods now used for growing diamonds." Subsequent research by various groups has indicated that an all carbon-13 diamond may be even better than the pure carbon-12 structure. Quantum mechanical calculations have indicated that slightly more atoms cause the atoms to be about .015 percent closer. This could translate into practical end-use properties, including hardness, among other property changes still unknown. It is reported that a carbon-13 diamond had been produced in 1971, but that it had not been reported officially. Some scientists now note that in the future diamonds may become commercially available on a "ready to order," isotopic ratio basis.

John Angus (CWRU) has suggested that an alternative route for growing gem-size diamonds may develop from the vapor deposition methods now used for growing diamond films.

Diamond Films

Two principal classes of diamond film deposition have been developed: (1) PACVD (plasma-assisted chemical vapor deposition) and (2) IBED (ion-beam-enhanced deposition).

A.H. Deutchman and R.J. Partyka (Beam Alloy Corporation) observe, "Characterization and classification of thin diamond films depend both on advanced surface-analysis techniques capable of analyzing elemental composition and microstructure (morphology and crystallinity), and on measurement of macroscopic mechanical, electrical, optical and thermal properties. Because diamond films are very thin (1 to 2 micrometers or less) and grain and crystal sizes are very small, scanning electron microscopy and transmission electron-microscopy techniques must be used to examine film morphology. Crystallinity is measured by various techniques including x-ray and electron diffraction Auger electron spectroscopy, and laser Raman spectroscopy."

Analysis of chemical composition is important because large percentages of hydrogen can be incorporated in the films, especially with PACVD techniques. This can cause a wide variety of hydrogenated structures.

Diamond films, although not approaching bulk diamond, are harder than most refractory nitride and carbide thin films, which makes them attractive for tribological coatings. Transparency in the visible and infrared regions of the optical spectrum can be maintained and index-of-refraction values approaching that of bulk diamond have been measured. Electrical resistivities of diamond films have been produced within the full range of bulk diamond, and thermal conductivities equivalent to those of bulk diamond also have been achieved. As substrates for semiconductor electronic devices, diamond films can be produced by both the PACVD and IBED techniques.

Diamonds in Meteorites

The presence of diamond in extraterrestrial specimens first was detected in the late 1800s. Inasmuch as meteorites are believed to originate from relatively small bodies (only a few hundred km in diameter or less), the origination of meteorite diamond is believed to differ from the diamond-forming process on Earth—mainly because of the absence of high pressures as found on Earth. Researchers recently have observed, "There is considerable need to subject diamonds from Abee (enstatite chondrite) and, for that matter, diamonds from ureilites and iron meteorites to the kind of scrutiny being given to the nanometer-sized component Cδ. It may be that low-temperature, low-pressure (that is, nonshock) formation of diamond was an important process in the early inner solar system." See also **Extraterrestrial Materials**.

Additional Reading

Amato, I.: "GE Achieves Dial-an-Isotope Diamonds," *Science*, **653** (November 1, 1991).

Angus, J.C. and C.C. Hayman: "Low-Pressure, Metastable Growth of Diamond and 'Diamondlike' Phases," *Science*, **913** (August 10, 1988).

Boyd, F.R. and J.J. Gurney: "Diamonds and the African Lithosphere," *Science*, **232**, 472–476 (1986).

Davidson, J.L.: *Diamond Materials V,* Electrochemical Society, Inc., Pennington, NJ, 1998.

Deutchman, A.H. and R.J. Partyka: "Diamond Film Deposition," *Advanced Materials & Processes*, **29** (June 1989).

Galli, G., et al.: "Melting of Diamond at High Pressure," *Science*, **1547** (December 14, 1990).

Guyer, R.L. and D.E. Koshland, Jr.: "Diamond: Glittering Prize for Materials Science," *Science*, **1640** (December 21, 1990).

Marshall, E.: "GE's Cool Diamonds Prompt Warm Words," *Science*, **25** (October 5, 1990).

Pan, L.S., et al.: "Electrical Transport Properties of Undoped CVD Diamond Films," *Science*, **830** (February 14, 1992).

Prelas, M.A., G. Popovici, and L.K. Bigelow: *Handbook of Industrial Diamonds and Diamond Films,* Marcel Dekker, Inc., New York, NY, 1997.

Proddow, P. and M. Fasel: "The Pink of Perfection," *Art and Antiques*, **37** (February 1994).

Russell, S.S., et al.: "A New Type of Meteoritic Diamond in the Enstatite Chondrite Abee," *Science*, **206** (April 10, 1992).

Staff: "Diamonds Protect IR-Sensor Windows," *Advanced Materials & Processes*, **8** (May 1991).

Staff: "Diamond Thin Films: A Market Set to Soar," *Advanced Materials & Processes*, **8** (August 1991).

Staff: "Forecast—Ceramics," *Advanced Materials & Processes*, **43** (January 1991).

Stix, G.: "Muffling Unkdapp: Synthetic Diamond," *Sci. Amer.*, **169** (September 1990).

Wilson, J.I. and K. Wihelm: *Diamond Thin Films,* John Wiley & Sons, Inc., New York, NY, 1996.

Yarbrough, W.A. and R. Meissier: "Current Issues and Problems in the Chemical Vapor Deposition of Diamond," *Science*, **688** (February 9, 1990).

DIAMOND ANVIL HIGH PRESSURE CELL.   The behavior of matter under extreme pressure is of great interest to several scientific disciplines and scientists working in the field of high pressure research are continually striving to achieve higher and higher pressure, to expand our knowledge in the field. Modern high pressure research falls under two categories—*static* and *dynamic*. Static high pressure is sustained pressure acting on a sample,

while dynamic pressures last only a few millionths of a second and are generated by a shock wave. While dynamic pressures in the megabar range* (millions of atmospheres) can easily be generated, static pressures of this magnitude are hard to realize. Only recently, this has become possible through the introduction of a novel pressure generating device called the *Diamond Anvil Cell* (DAC). The DAC has practically replaced all other pressure generating devices used in high pressure research and is proving to be a versatile tool to study the high pressure behavior of matter. With the DAC, pressures of 5 megabars (5 million atmospheres) have recently been achieved.

*Pressure is usually expressed in bars, atmospheres, or pascals. One bar = 0.9868 atmosphere = 10^5 pascals. One megabar = 10^6 bars = 100 GPa (giga pascals).

Why diamond works so well is because of its two most desirable properties, namely, (1) it is the hardest substance known to science, and (2) it is very transparent to optical radiation as well as to x-rays. Compared to diamond, tungsten carbide, which was used in older pressure generating devices, has a much lower compressive strength and, further, it is opaque to radiation.

The Diamond Anvil Cell

In Fig. 1 a modern DAC capable of generating megabar pressures is shown. Quite contrary to the general conception of high pressure apparatus, the DAC is such a compact device that it fits in the palm of a hand. Figure 2 shows the basic elements of the DAC. The arrangement shown consists of two flawless (gem quality) diamonds, one-third to one-half carat in weight. The pointed ends of the diamonds are ground off and polished to produce small flats of about 0.5 millimeter in size and are set in opposition to each other. When a metal gasket is compressed between these small flat faces very high pressure is generated in the gasket. To apply this pressure on a sample, a hole is drilled in the gasket for locating the sample and the pressure-transmitting medium. Because the area over which the force is concentrated is extremely small, the pressure on the sample, which is force per unit area, can be enormous. For the same reason a force that can be applied by hand is multiplied 500 to 1000 times by the mechanism, which makes the DAC a very compact apparatus. In a practical DAC such as the one shown in Fig. 1, the two diamonds are mounted inside a hardened steel mechanism, machined to high tolerances, that imparts thrust along an axis perpendicular to the diamond faces. For the successful operation of the DAC, the two diamond flats have to be set perfectly parallel to each other and further they should be accurately centered along the common axis. The mechanism carries hardened steel or tungsten carbide rockers for diamond support and alignment. The thrust and alignment mechanisms can be designed in different ways, and hence, several types of diamond cells have evolved.

The simple operation of the DAC is the hard-won result of two decades of evolution in design. High pressure devices employing diamonds were first built in 1959 by C.E. Weir, E.R. Lippincott, A. Van Valkenburg and E.N. Bunting of the National Bureau of Standards, and J.C. Jamieson, A.W. Lawson and N.D. Nachtrieb of the University of Chicago, independently and simultaneously. Since that time, the NBS Scientists and several other groups have developed the DAC as a fine tool for ultrahigh pressure research.

In a diamond cell, the sample volume is sacrificed for the sake of higher pressures, and hence, all operations connected with the cell have to be performed under a microscope. In preparing the DAC for an experiment, the first step is to indent the metal gasket (hardened stainless steel strip or Inconel strip) with the anvil diamonds to the correct thickness (50 to 100 micrometers) and then drill a 100- to 200-micrometer hole as close to the center of the indentation as possible. The gasket is seated on the face of one of the diamonds in the same orientation as it had when the indentation was made. The sample material and a small chip of ruby for pressure calibration are then placed in the hole. Finally, to maintain hydrostatic pressure the hole is filled with a tiny drop of fluid from a syringe and then the hole is quickly sealed by the diamond faces before the fluid evaporates.

When thrust is applied, the metal gasket is compressed, and along with it the pressure medium and the sample. Because of the greater compressibility of the pressure medium, the pressure on the gasket is always greater than the pressure on the contents in the hole, and this ensures perfect sealing of the pressure chamber. The optimum hardness for the gasket is an important element in diamond cell work, for too soft a gasket would simply flow like butter, and too hard a gasket would not compress. The purpose of the

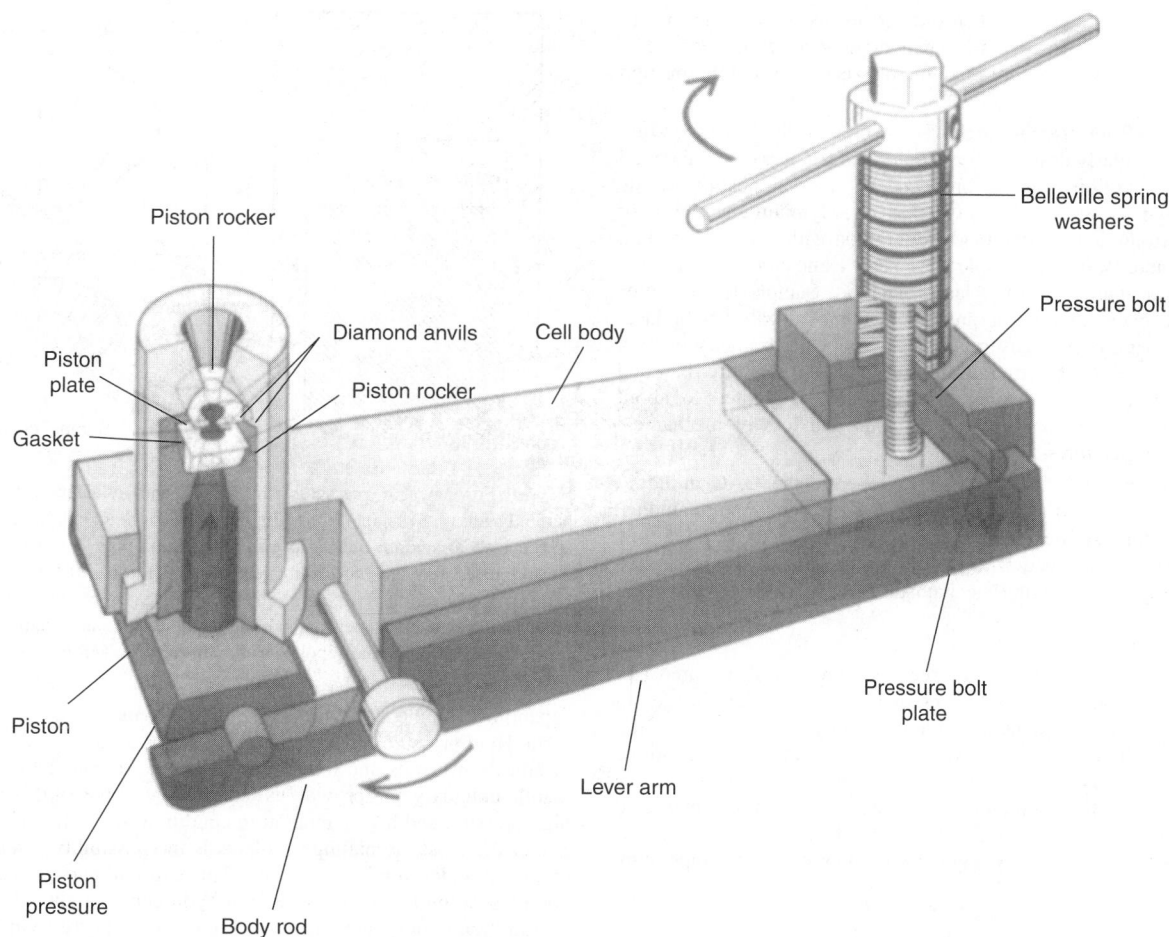

Fig. 1.   Modern diamond anvil cell of the type developed at the Geophysical Laboratory of the Carnegie Institution in Washington, DC. It is about 20 cm long and weighs about 3 kg. The piston and cylinder are machined to very high tolerances. The anvil diamond on the piston is fixed on a half-cylindrical rocker mount recessed into the piston. The other anvil diamond is inside the cylinder and sits on a similar rocker mount. The two rocker mounts are set at right angles and can be moved laterally with screws to center the diamond flats. Parallel alignment of the diamond flats is accomplished by tilting the two half-cylindrical rockers until the optical interference fringe pattern appearing in between the two flats is reduced to a uniform gray color, and then locking the rockers in position with set screws. Thrust is generated by turning the handle on the spring-loaded pressure bolt clockwise. The thrust is transmitted by the long lever arms and is delivered to the movable piston as a vertical upward force through the action of the fulcrum (the rod on the side goes into the body of the cell and acts as the fulcrum) and a swiveling piston pressure plate (not seen in the figure). The lever arm mechanism magnifies the thrust and delivers it to the piston plate which, in turn, pushes the piston diamond against the stationary diamond attached to the cylinder. One of the finest features of the diamond anvil apparatus is that pressurized samples can be viewed directly with an optical microscope. (*From "The Diamond-Anvil High-Pressure Cell" by A. Jayaraman. Copyright © April 1984 by Scientific American, Inc. All rights reserved*)

gasket is to provide containment for the pressure medium as well as to give support to the diamond flats. The higher the pressure, the harder the gasket material should be.

Megabar Pressures. To reach megabar pressures, the diamond flats have to be modified to what is known as beveled geometry. This is shown in Fig. 3. In this a part of the flat has been beveled; the bevel angle is typically 5°. The flat region B and dimension A are variable, and for megabar pressures they are usually 50 micrometers and 300 micrometers, respectively. The beveling seems to smooth out the pressure gradients, with beneficial effects on the diamond. At megabar pressures, gradients over the active area dictate the use of a very small sample and this makes experimentation at these pressures a challenge. However, these challenges are being met by using ingenious methods.

Pressure Calibration. Pressure is determined in a remarkably simple way using ruby fluorescence. Indeed, without the ruby the small size of the high-pressure cell would make the pressure quite difficult to calibrate and the scientific value of the instrument would be far less than it is. When ruby is excited by light, it fluoresces strongly with a deep red color. The fluorescent emission can be resolved by a spectrograph into two peaks, called R_1 and R_2, whose wavelengths are accurately known at atmospheric pressure. When the pressure on the ruby is increased, the fluorescence peaks shift to a higher wavelength. This shift has been calibrated for pressures known on independent grounds, and hence, the measured shift of the peaks is an indirect measure of the pressure on the ruby. Usually the

stronger of the two peaks, the R_1 peak, is used in pressure calibration. The spectral shift can be quickly and accurately measured with a simple grating spectrometer. The shift is almost exactly proportional to the pressure up to 300 thousand atmospheres (0.037 nanometer per atmosphere). For higher pressures the spectral shift is somewhat smaller for a given increment in pressure, but the shift has now been calibrated up to 2 megabars (nearly 2 million atmospheres) using gold as standard. The ruby seems to work quite satisfactorily to the highest pressures reached with the DAC, namely, 5 megabars, and an extrapolation of the above scale is used to measure these very high pressures.

Pressure Media

There are several materials currently employed as hydrostatic media in the cell. The most convenient material to load into the cell is a mixture of four parts methanol to one part ethanol, but at pressures of more than 105,000 atmospheres the mixture solidifies at room temperature and the pressure is no longer hydrostatic. For higher hydrostatic pressures, noble gases, viz. argon, xenon, or even hydrogen and helium are used. These gases can be trapped in the tiny hole in the gasket either by cooling the diamond cell to low temperatures and filling with the liquefied gas or by the high pressure gas loading technique. In the latter method, the diamond cell is enclosed in a large pressure vessel and charged with the gas at about 2000 atmospheres. The high pressure gas, filling the gasket hole, is trapped by applying pressure to the gasket with the diamonds, from outside. After the release of pressure in the large pressure vessel, the diamond cell with the trapped high pressure gas is taken out and loaded into the lever-arm mechanism for further pressurization when running an experiment. The trapped gas can serve as a pressure medium, or it can be a sample for high pressure studies.

High Pressure-High Temperature. High pressure-high temperature conditions are particularly desirable for geophysical studies. In the diamond anvil cell such conditions can be produced by internal heating of the sample with a high power laser beam (such as a neodymium glass laser) or externally by surrounding the diamond anvil region with a heater element. In the first method, only the sample gets heated and not the diamond, because of its transparency to the laser radiation. Sample temperatures of 5000°C have been attained at pressures of one megabar with laser heating. The temperature is measured either by optical pyrometry or by a radiometric technique. In the external heating method, the entire diamond anvil region is heated. The temperature attainable is limited to about 1000°C because of the damage to the diamond and loss of mechanical support at high temperatures.

Low Temperature. It is also possible to operate the DAC at liquid helium temperatures (−270°C) for high pressure-low temperature studies. For low temperature studies the diamond cell is placed inside a low temperature Cryostat provided with optical windows, and is pressurized by a suitably designed tightening mechanism.

Ultrahigh Pressure Research

The diamond anvil cell is a tool par excellence for optical, infrared and Raman spectroscopy and enables the researcher to study the changes in the electronic structure and chemical binding caused by the application of high pressure. Phase transitions, which involve changes in the atomic architecture can be determined with the diamond cell using the x-ray diffraction technique. The diamond cell has been successfully interfaced with synchrotron sources, which generate powerful x-ray beams, for x-ray diffraction studies. In geophysics, the high pressure-high temperature

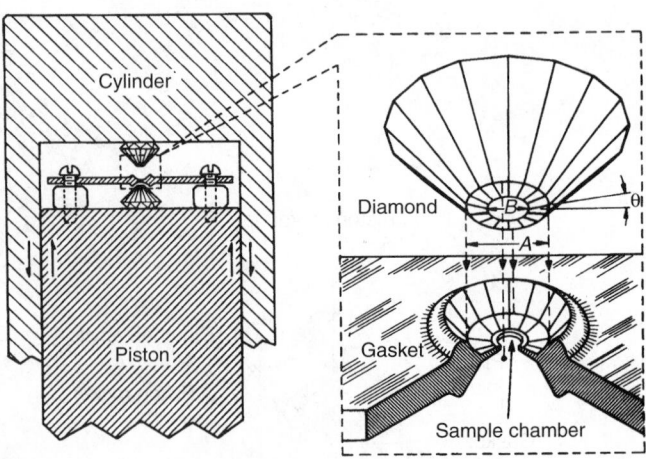

Fig. 3. For megabar pressures, the piston and cylinder assembly should be perfectly fitting, to maintain the alignment of the diamond anvils. The diamonds are beveled to smooth out the pressure gradients. The *A* dimension varies from 0.6 mm to 0.3 mm and the *B* dimension from 0.3 mm to 0.1 mm. The gasket hole varies from 0.1 mm to 0.025 mm. The bevel angle θ is usually around five degrees. After an experiment at megabar pressures, the diamonds develop ring cracks, but they can be reground and repolished for further use. (*After Mao and Bell*)

capabilities of the diamond cell is like a window to the interior of the earth. High pressure-high temperature studies give information on the state of silicate minerals and oxides in the mantle region right up to the core-mantle boundary and provide us with a view of the earth's interior, where high pressure and high temperature conditions exist. In solid state physics, one of the most fascinating problems is the possibility of making metallic hydrogen under ultrahigh pressure. This extraordinary change from a very good insulating to a metallic state in hydrogen is predicted to occur near 3 to 4 million atmospheres. High pressure scientists are trying hard to make metallic hydrogen with the DAC.

The limit of pressure attainable with the DAC could be much higher than 5 million atmospheres. Theoretical calculations reveal that diamond is stable up to 23 million atmospheres with respect to any phase transitions. Plastic deformation would limit its capability, but this limit could be much higher than the presently reported maximum pressure of 5 million atmospheres.

Since the introduction of the diamond anvil high pressure cell in the early 1980s, it has found wide usage in a variety of research fields. Possibly, the greatest pressure achieved to date was 4.16 megabars by researchers (Cornell Univ.) when "squeezing" a microscopic sample of molybdenum powder. By comparison, the pressure at the center of the Earth is estimated at about 3.6 megabars. Other extremely high pressures have been achieved at the Lawrence National Laboratory (Livermore, California).

In 1988, J.M. Brown and L.J. Slutsky (Univ. of Washington) and K.A. Nelson and L.T. Cheng (Massachusetts Institute of Technology) took advantage of the adaptability of laser-induced phonon spectroscopy and diamond anvil techniques to determine acoustic velocities and equations of state (methanol and ethanol). As pointed out by the researchers, "Acoustic velocities are directly related to interatomic force constants and these velocities of sound are a matter of considerable importance to the earth sciences." By knowing these properties, proceeding from a seismological image of the Earth's interior, a geophysical model in terms of density, temperature, and chemical composition can be created.

Godward (Univ. of California, Berkeley) and a team of researchers used a laser-heated diamond cell (modified Mao-Bel type) in their research on the ultrahigh-pressure melting of lead. This study is part of a project to *characterize* materials at ultrahigh pressures. Such data can be useful for a wide range of applications in the planetary sciences and the physics of condensed matter.

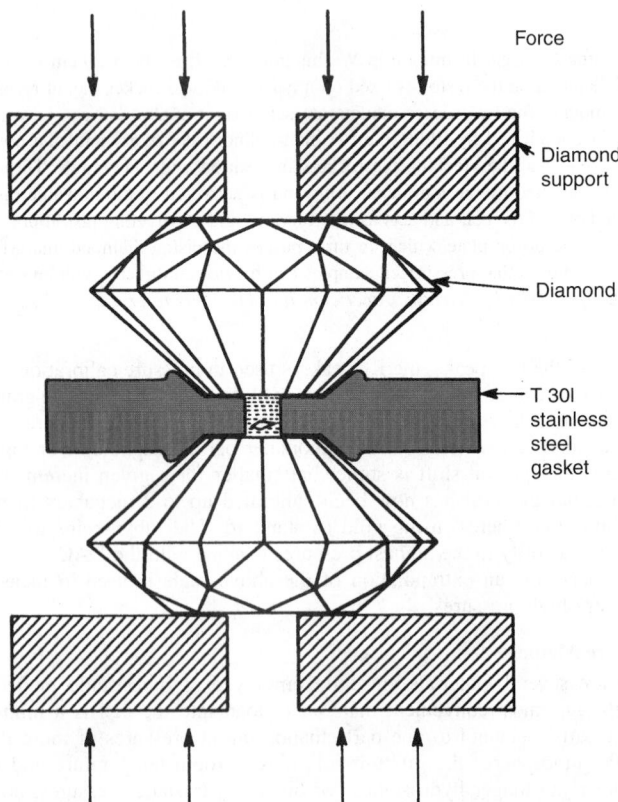

Fig. 2.   Basic principle of the diamond cell. Pressure is generated in the gasket hole when the diamonds are pushed against one another. The sample and a small chip of ruby for pressure calibration are placed in the hole and the latter is filled with a pressure-transmitting medium. The purpose of the gasket is to provide containment for the pressure medium as well as support the diamond flats. Suitable apertures in the diamond support blocks provide access to optical, x-ray, and other radiation

A. JAYARAMAN
AT&T Bell Laboratories
Murray Hill, New Jersey

Additional Reading

Amato, I.: "Unworldly Pressures," *Science News*, **72** (February 2, 1991).
Brown, J.M., et al.: "Velocity of Sound and Equations of State for Methanol and Ethanol in a Diamond-Anvil Cell," *Science*, **241**, 65–67 (1988).

Godward, B.K., et al.: "Ultrahigh-Pressure Melting of Lead: A Multidisciplinary Study," *Science*, **462** (April 27, 1990).

Jayaraman, A.: *Rev. Mod. Phys.*, **45**, 65 (1983).

Jayaraman, A.: *Sci. American*, **250**, 54 (1984).

Jayaraman, A.: *Rev. Sci. Instruments*, **57**, 1013 (1986).

Skelton, E.F.: *Physics Today*, **37**, 44 (1984).

DIAPHRAGM CELL. A type of electrolytic cell for the production of sodium hydroxide and chlorine from sodium chloride brine. The cell contains anode and cathode compartments separated by a porous diaphragm or membrane to prevent mixing of the solutions. Asbestos fibers are usually used for this diaphragm, though a recent development is a plastic material made from perfluorosulfonic acid. The brine is fed continuously to the anode compartment, where chlorine is released at the graphite anode and flows through the diaphragm to the steel cathode, where hydrogen is liberated. Sodium hydroxide accumulates in the liquid and is continuously drained from the cathode compartment. The Hooker cell and the Vorce cell are two widely used types of diaphragm cell.

See also **Chlorine.**

DIASPORE. The mineral diaspore is a hydrous oxide of aluminum corresponding to the formula AlO(OH) occurring in prismatic orthorhombic crystals, usually somewhat flattened, or massive. It displays good cleavage; conchoidal fracture; is brittle; hardness, 6.5–7; specific gravity, 3.3–3.5; luster, vitreous to pearly; color, white, grayish, greenish, yellowish, brownish or colorless; transparent to translucent. Diaspore is found associated with corundum, emery and bauxite, being probably an alteration product of the oxide. It has been made artificially. Diaspore has been found associated with emery in the Ural Mountains, in the Middle East, in the Island of Naxos, Greece, and in the United States at Chester, Massachusetts. Its name is derived from the Greek word meaning to scatter, because of its decrepitation upon heating.

DIATOMACEOUS EARTH. See **Diatomite.**

DIATOMITE. Diatomite is a naturally occurring, porous, high surface area form of hydrous silica that is used as a filter aid and as a mineral filler. Diatomite products may be classified according to manufacturing method into three categories: natural diatomite, calcined diatomite, and flux-calcined diatomite. Products from all three categories find widespread use in industrial filtration applications as a filter aid for achieving higher clarity, longer filter cycles, and removing high solids concentrations (Table 1). Products from all three categories are also used as functional fillers where diatomite properties add to the performance of paints, plastics, rubber, catalysts, agricultural chemicals, pharmaceuticals, toothpastes, polishes, and other chemicals.

Diatomite, also known as diatomaceous earth, or kieselguhr, consists mainly of accumulated shells or frustules of intricately structured amorphous hydrous silica secreted by diatoms, which are microscopic, one-celled golden brown algae of the class Bacillariophyceae. Diatoms exist in many different environments and are abundant in regions of oceanic upwelling.

Origins of Deposits

Most commercial marine diatomite deposits exploit accumulations resulting from large blooms of diatoms that occurred in the oceans during the Miocene geological epoch. Marine deposits must have been formed on the bottom of protected basins or other bodies of quiet water, undisturbed by strong currents, in an environment similar to the existing Santa Barbara Channel or Gulf of California.

The main deposits of freshwater diatomite were laid down in large lakes. Many of these deposits in the western United States formed during glacial times, when the local climate was wetter. Several tens of square kilometers in Nevada west of Tonopah are covered with diatomite as are other large areas in the Great Basin.

Location of Deposits. Deposits of diatomite are known to exist on every continent and in nearly every country. Most of the deposits are not large enough or sufficiently pure to have commercial value. Production figures show the location of the deposits that meet commercial standards in both respects.

In the United States the most extensive commercial deposits are located in California, Nevada, Oregon, and Washington. The U.S. Bureau of Mines also reports the commercial operation of diatomite deposits in Arizona.

TABLE 1. PROPERTY RANGES OF DIATOMITE PRODUCTS[a]

Property	Filter aids			Fillers, all types
	Natural	Calcined	Flux-calcined	
permeability range, μm^{2}[b]	0.06	0.5–2.0	1.0–29.6	
density, kg/m^3				
wet cake	240–350	270–350	290–380	
bulk	112	120–128	144–336	104–160
particle size distribution, μm				
10% <	1.5–3.6	2.5–4.4	7–11	2–4
50% <	7.0–13.4	10.0–16.1	25–37	6–20
90% <	25–44.5	30.0–58.9	65–97	14–30
approx. pressure differential[c], kPa[d]	36.5	2.33–4.56	1.11–0.058	
specific gravity	2.00	2.25	2.33	2.0–2.3
porosity, by vol, %	65–85	65–85	65–85	65–85
median pore size range, μm	1.5	3.5–5.0	7–22	[e]
surface area, m^2/g	10–20	4–6	1–4	0.7–30
pH	6.0–8.0	6.0–8.0	8.0–10.0	6.0–10.0
refractive index	1.42	1.44	1.48	1.40–1.49
oil absorption, %				100–210

[a] Values typical or estimated, not specifications.
[b] To convert from μm^2 to d'Arcys, multiply by 1.013.
[c] Measurement at 0.034 cm/s and 0.1 g/m^2 precoat.
[d] To convert from kPa to psi, multiply by 0.145.
[e] Hegman gauge readings, useful for paint manufacture, run from 0–55.

Physical and Chemical Properties

Chemically, diatomite consists primarily of silicon dioxide, $SiO_2 \cdot nH_2O$, and is essentially inert. It is attacked by strong alkalies and by hydrofluoric acid but is virtually unaffected by other acids. The silicon dioxide has a unique structure, resulting from the intricate form of the diatom skeletons. The chemically combined water content varies from 2 to 10%. Impurities that are often found mixed with the diatomite are other aquatic fossils such as sponge residues, Radiolaria, silicoflagellata, sand, clay, volcanic ash, mineral aerosols, calcium carbonate, magnesium carbonate, soluble salts, and organic matter.

The color of pure diatomite is white, or near white, but impurities such as carbonaceous matter, clay, iron oxide, volcanic ash, etc may darken it.

Individual diatom frustules are porous. The diatoms are highly variable in shape and size, having particles that range in effective diameter from 0.75 to 1000 μm, but most are 50 to 100 μm in diameter. Diatom shapes can range from simple cylinders and disks to complex, highly variable, but always punctate, forms.

The bulk density of powdered diatomite varies from 112 to 320 kg/m^3. The true specific gravity of diatomite is 2.1 to 2.2, the same as for opaline silica, or opal. The thermal conductivity of bulk quantities of diatomite is low but increases with higher percentages of impurities and a higher density. The fusion point depends on the purity but averages about 1430°C for pure material, which is slightly less than for pure silica. The addition of chemical agents, such as soda ash, reduces the fusion point.

Diatomite has only weak adsorption powers but shows excellent absorption because of its structure and high surface area. Acids, liquid fertilizers, alcohol, water, oils, and other fluids are absorbed by diatomite.

Mining and Processing

Diatomite deposits are usually discovered by observation of outcrop, and the value of the deposits is determined by geological prospecting and exploration.

Mining. Most diatomite is mined by open-pit methods.

Economic Aspects. Owing to the low bulk density of diatomite, freight and trucking rates (on a weight basis) are high. Domestic finished products are packed and shipped in laminated kraft-paper bags, usually containing 22.5 kg, or the product is shipped in bulk or semibulk bags. Bagged products are shipped by truck or rail boxcar.

Domestic Producers. A principal company mining diatomite and processing it into finished products is Celite Corp. (Lompoc, California), which has wholly owned mines and processing facilities in Lompoc, California; Quincy, Washington; Jalisco, Mexico; Murat, France; Alicante, Spain; and a joint venture mine in Iceland.

Production. After the United States, Romania, the former USSR, and France are the largest producers of diatomite. Combined with the United States, these countries account for more than 75% of the world's production.

KENNETH R. ENGH
Sandkuhl Clay Works

Additional Reading

ASTM D604-81; D719-86, American Society for Testing and Materials, Philadelphia, Pa., 1989.

Cain, C.W. Jr.: in J.J. McKetta, ed., *Encyclopedia of Chemical Processing and Design,* Vol. 21, Marcel Dekker, Inc., New York, NY, 1984.

Kadey, F.L.: in S.J. Lefond, ed., *Industrial Minerals and Rocks,* 5th Edition, AIME, New York, NY, 1983.

Kiefer, J.: *Brauwelt Int.* **6,** 300 (1991).

DIAZO COMPOUNDS AND DIAZONIUM SALTS. See Azo and Diazo Compounds.

DICARBOXYLIC ACIDS.

The diacids are characterized by two carboxylic acid groups attached to a linear or branched hydrocarbon chain. Aliphatic, linear dicarboxylic acids of the general formula $HOOC(CH_2)_nCOOH$ and branched dicarboxylic acids are the subject of this article. The bifunctionality of the diacids makes them versatile materials, ideally suited for a variety of condensation polymerization reactions. Several diacids are commercially important chemicals that are produced in multimillion-kg quantities and find application in a myriad of uses.

Nomenclature

Unsubstituted aliphatic dicarboxylic acids, $HOOC(CH_2)_nCOOH$, are most often referred to by their trivial names for $n = 2$ to 10 (Table 1). Higher homologues are named using the IUPAC system by adding the suffix dioic to the parent hydrocarbon.

Physical Properties

Detailed summaries of physical properties are given. The diacids are colorless, crystalline solids that melt somewhat higher than monoacids of the same molecular weight. For diacids of even carbon number, melting points decrease sharply for numbers 2–10 and remain relatively constant for numbers 12–20 (see Table 1). There is a marked alternation in melting point and other physical properties with changes in carbon number from even to odd within the series. Odd members exhibit lower melting points, and higher solubility. Theoretical treatments have been developed to correlate these physical properties. The alternating effects are the result of the inability of odd carbon number compounds to assume an in-plane orientation of both carboxyl groups with respect to the hydrocarbon chain. Other properties showing these alternations are decarboxylation temperature and index of refraction. These effects have practical consequences in the selection of material for a given preparation because acid melting point, decarboxylation temperature, and solubility are often key considerations. The effects persist in derivatives based on the diacids, particularly polyamides, polyurethanes, and polyesters.

The temperature at which decarboxylation occurs is of particular interest in manufacturing processes based on polymerization in the molten state where reaction temperatures may be near the point at which decomposition of the diacid occurs. The diacids become more heat stable at carbon number four, with even-numbered acids always more stable. Thermal decomposition is strongly influenced by trace constituents, surface effects, and other environmental factors.

Lower members of the series are water soluble; solubility falls off sharply above adipic acid. Alternating effects are again expressed with acids of odd carbon numbers being the most soluble (see Table 1). Dibasic acids are ionized in aqueous solution to varying degree depending upon the proximity of the carboxyl groups within the individual structures. The carboxyl group, being electron-withdrawing, causes the neighboring carboxyl hydrogen to be more readily dissociated.

Chemical Properties

The dibasic acids undergo the reactions typical of monocarboxylic acids.

Manufacture, Preparation, and Processes

Glutaric Acid. Until 1990–1991 glutaric acid was available commercially from DuPont as a by-product in the production of adipic acid. It is no longer available, but DuPont produces dimethyl glutarate and mixtures of dimethyl succinate and dimethyl glutarate, as well as mixtures of dimethyl glutarate and dimethyl adipate.

Several procedures for making glutaric acid have been described in *Organic Syntheses* starting with trimethylene cyanide, methylene bis (malonic acid), γ-butyrolactone, and dihydropyran.

Pimelic Acid. This acid is manufactured by Tateyama Chemical Company in Japan in quantities of about 1000–2000 kg/yr, and by Heinrich Mock Nachf in Germany. The method or process they are using has not been disclosed. Pimelic acid is available in small quantities with purities of 98% from laboratory chemical supply companies. The preparation of pimelic acid has been described in *Organic Syntheses;* cyclohexanone condenses with diethyl oxalate, followed by decarboxylation to ethyl 2-keto-hexahydrobenzoate, and then cleavage of the β-keto ester with strong alkali.

Suberic Acid. This acid is not produced commercially at this time. However, small quantities of high purity (98%) can be obtained from chemical supply houses. If a demand developed for suberic acid, the most economical method for its preparation would probably be based on one analogous to that developed for adipic and dodecanedioic acids; air oxidation of cyclooctane to a mixture of cyclooctanone and cyclooctanol. This mixture is then further oxidized with nitric acid to give suberic acid.

Azelaic Acid. This acid is produces by the Emery Group of Henkel Corporation in Cincinnati, Ohio, in multimillion kg quantities. The process that is currently used is based on the ozonolysis of oleic acid (from grease or tallow) followed by the decomposition of the ozonide with oxygen.

Sebacic Acid. This acid is produced commercially by Union Camp in Dover, Ohio, by Hokoku Oil Company in Japan, and by a state enterprise in the People's Republic of China. The process used in each case is based on the caustic oxidation of castor oil or ricinoleic acid in either a batch or continuous process.

Dodecanedioic Acid. Dodecanedioic acid (DDDA) is produced commercially by Du Pont in Victoria, Texas, and by Chemische Werke Hüls in Germany. The starting material is butadiene which is converted to cyclododecatriene using a nickel catalyst. Hydrogenation of the triene gives cyclododecane, which is air oxidized to give cyclododecanone and cyclododecanol. Oxidation of this mixture with nitric acid gives dodecanedioic acid.

Brassylic Acid. This acid is commercially available from Nippon Mining Company (Tokyo, Japan). It is made by a fermentation process.

TABLE 1. PHYSICAL PROPERTIES OF $C_2–C_{21}$ ALIPHATIC DICARBOXYLIC ACIDS

IUPAC name	Mp, °C	Bp, °C[a]	Water solubility,[b] g/100 mL	Density, g/mL
ethanedioic	187 (dec)		9.5	
propanedioic	134–136 (dec)		152	1.619
butanedioic	187.6–187.9		8.35	1.572
pentanedioic	98–99	200[c]	130	1.424
hexanedioic	153.0–153.1	265	3.08[d]	1.345
heptanedioic	105.7–105.8	272	5.0	1.287
octanedioic	143.0–143.4	279	0.16	1.270
nonanedioic	107–108	286.5	0.214	1.235
decanedioic	134.0–134.4	294.5	0.10	1.231
undecanedioic	110.5–112		0.003	
dodecanedioic	128.7–129.0	254[e]	0.004	1.16
tridecanedioic	114			
tetradecanedioic	126.5			
pentadecanedioic	114.7			
hexadecanedioic	125			
heptadecanedioic	117–118			
octadecanedioic	124.6–124.8			
nonadecanedioic	118–119.5			
eicosanedioic	124–125			
heneicosanedioic	118–120			

[a] At 13.3 kPa = 100 mm Hg unless otherwise noted.
[b] At 20–25°C unless otherwise noted.
[c] At 2.7 kPa = 20 mm Hg.
[d] At 34.1°C.
[e] At 2.0 kPa = 15 mm Hg.

C-19 Dicarboxylic Acids. The C-19 dicarboxylic acids are generally mixtures of isomers formed by the reaction of carbon monoxide on oleic acid. Because the reaction produces a mixture of isomers, no single chemical name can be used to describe them. There are currently no commercial producers of C-19 dicarboxylic acids.

C-20 Dicarboxylic Acids. These acids have been prepared from cyclohexanone via conversion to cyclohexanone peroxide followed by decomposition by ferrous ions in the presence of butadiene. Okamura Oil Mill (Japan) produces a series of commercial acids based on a modification of this reaction.

C-21 Dicarboxylic Acids. C-21 dicarboxylic acids are a mixture of predominantly 5-(6)-carboxy-4-hexyl-2-cyclohexene-1-octanoic acid, 5-isomer and 6-isomer. C-21 dicarboxylic acids are produced by Westvaco Corporation in Charleston, South Carolina in multimillion-kg quantities. The process involves reaction of tall oil fatty acids (TOFA) (containing about 50% oleic acid and 50% linoleic acid) with acrylic acid and iodine at 220–250°C for about 2 h.

Dicarboxylic Acids via Microorganisms

During the 1980s a number of patents were issued describing the preparation of dicarboxylic acids or esters using microorganisms. The α, ω-n-alkanedioic acids that have been prepared generally have 5–25 carbons. One of the first methods described the preparation of dimethyl 1,16-hexadecanedioate using a nutrient solution, n-hexadecyl bromide, and certain strains of *Torulopsis*. Other methods have been described to give dibasic acid of 8–22 carbons using n-alkanes or n-alcohols and various organisms.

Derivatives and Uses

Diacids, owing to their ready incorporation into polymers, are components in a wide variety of materials. The diacids are important industrial intermediates for the manufacture of diesters, polyesters, and polyamides. These derivatives find application as plasticizing agents, lubricants, heat transfer fluids, dielectric fluids, fibers, copolymers, inks and coatings resins, surfactants, fungicides, insecticides, hot-melt coatings, and adhesives. Of the higher diacids, azelaic, sebacic, and dodecanoic find the greatest application. Derivatives of glutaric and C-21 diacids also enjoy significant commercial applications.

Economic Aspects

In addition to azelaic, sebacic, dodecanedioic, eicosanedioic (C_{20} diacids), and C_{21} diacids, undecanedioic, brassylic, tetradecanedioic, hexadecanedioic, docosanedioic, and tetracosanedioic acids are available, expensive, and in limited quantity from research chemical supply houses.

Health and Safety Factors

In general, the higher diacids are essentially nontoxic. There are no indications that the dicarboxylic acids detailed here are carcinogenic or teratogenic in animals or humans. It is generally recognized that these diacids are ocular irritants and that the inhalation of the dust of these diacids is irritating to the mucous membranes and the respiratory tract. The water solubility of glutaric acid fosters its toxicity. Glutaric acid is a known nephrotoxin.

Environmental Effects. In general, the higher diacids do not pose substantial environmental risk; however, releases of significant quantities into surface or ground waters may be reportable under the Clean Water Act. The low biotoxicity of the higher diacids results, in part, from their limited water solubility. Glutaric acid is significantly more water soluble than the other diacids described herein, and the aquatic biotoxicity of glutaric acid and dimethyl glutarate is established.

ROBERT W. JOHNSON
CHARLES M. POLLOCK
Union Camp Corporation
ROBERT R. CANTRELL
Union University

Additional Reading

Johnson, R.W.: in R.W. Johnson and E. Fritz, eds., *Fatty Acids in Industry,* Marcel Dekker, Inc., New York, NY, 1989, Chapt. 13.
Pryde, E.H. and J.C. Cowan: in J.K. Stille and T.W. Campbell, eds., *Condensation Monomers,* Wiley-Interscience, New York, NY, 1972, pp. 1–153.

DICHLOROBENZENES. See Chlorinated Organics.

DIE. A device, usually of steel, having a specific shape or design that it imparts to such materials as metals and plastics either by impact (stamping), by the contour of a negative cavity (casting), or by passing the material through it (extrusion). Diamond dies may be used for wiredrawing. The terms "die" and "mold" are virtually synonymous in the sense of a negative cavity into which a molten metal or plastic is introduced under pressure, the former being used in reference to metals, and the latter for plastics, rubber, etc.

See also **Die Casting**; and **Extrusion**.

DIE CASTING. Die castings are produced by forcing molten metal under pressure into a steel die. The pressure is maintained until solidification is complete. The process is essentially a further development of gravity-feed casting, but the pressure function entails finer detail and better finish. While gravity-feed casting tonnage is greater than that of pressure casting, the latter has a wider field of application and is more important in the quantity production of precision parts. Zinc alloys are generally used for die castings, although aluminum alloys, brass alloys and other non-ferrous metals are used to a considerable extent.

The process of die casting is entirely automatic and requires the following elements: a die-casting machine to hold the molten metal under pressure; a metallic mold or die capable of receiving the molten metal, and designed to permit easy and economical ejection of the solidified product; and a casting alloy that will produce a satisfactory product with suitable physical characteristics.

There are two types of die-casting machines. The first, or air-operated machine, forces the material into the die by high pressure on the surface of the molten metal in a special ladle or goose; the second, or plunger type machine, forces the material into the die by means of a cylinder and piston which are submerged in the molten metal.

Die-casting dies are constructed in different styles for various production requirements. A single die contains an impression of only one part; a multiple die contains two or more impressions of any one part; a combination die contains one impression only of two or more parts; and a combination-multiple die contains a number of impressions of each of two or more parts. Single dies are comparatively cheap and are used for small-lot production, since they reduce the tool investment to a minimum for any one part. Combination dies, when properly planned, will reduce the total die cost for a given set of castings to a minimum. They are applicable to parts that will always be used in the same quantities and of the same alloy. These parts should be of the same general character and weight. Multiple dies are usually slower to operate than single dies but will give higher production rates for the same labor costs.

Die-casting dies are often vented by permitting air to escape through the clearance in the ejector and core pin bearings. The problem of venting is considerably more important than in sand casting because the mold has no porosity. Sometimes dies are vented by grinding shallow grooves on the parting surfaces of the dies; in other instances plugs with suitable vent grooves are added to the die.

In 1986, direct injection zinc die casting reached commercial status with the production of gear casings for fractional horsepower electric motors. The process features use of a heated manifold and mini-nozzles to feed molten zinc directly into the die. The Zinc Institute (New York) reports that direct injection eliminates traditional runner and gating pathways and the scrap associated with them. The process is netting a 10% saving in scrap, remelt, and processing costs. Since the molten zinc is injected directly into the die, it can fill the cavity without any chilling before contacting cavity walls. The result is a more uniform structure in the cast part and fewer rejects. Also, cast surface finish is smooth so that preplate finishing operations can often be eliminated.

DIELECTRIC HEATING. The heating of a dielectric material by molecular friction in it as a result of the application of a high-frequency, alternating electric field. Dielectric heating is applicable to nearly all nonconducting materials, such as plastics, wood, and certain liquids. The method is extensively used for the preheating of plastic materials because the materials must be heated uniformly. Prior conventional methods required hours instead of a few minutes with dielectric heating. The method also is used in the printing and dyeing industry, and in the lumber and associated industries for speeding and perfecting the drying of glued joints.

Normally the power required for dielectric heating is provided by some form of oscillator, although the power can be obtained from an amplifier driven by an oscillator. The principal engineering is involved in the design of an oscillator circuit that will provide the proper energy level and in the design of the configuration whereby the energy can be most efficiently imparted to the workpiece.

DIELECTRIC THEORY.   A dielectric is a material having electrical conductivity low in comparison to that of a metal. It is characterized by its dielectric constant and dielectric loss, both of which are functions of frequency and temperature. The dielectric constant is the ratio of the strength of an electric field in a vacuum to that in the dielectric for the same distribution of charge. It may also be defined and measured as the ratio of the capacitance C of an electrical condenser filled with the dielectric to the capacitance C_0 of the evacuated condenser:

$$\varepsilon = C/C_0$$

The increase in the capacitance of the condenser is due to the polarization of the dielectric material by the applied electric field. The terms "specific inductive capacity" and "permittivity" are occasionally used instead of *dielectric constant*. The constant ε appearing in the Coulomb law of force is called the permittivity, but it is also commonly called the dielectric constant. The relative permittivity or dielectric constant is the ratio $\varepsilon/\varepsilon_0$, where ε_0 is the permittivity or dielectric constant of free space. In the mks system of units, the dielectric constant of free space is 8.854×10^{-12} farad/m, while in the esu system the relative and the absolute dielectric constants are the same. The relative dielectric constant, which is dimensionless, is the one commonly used. When variation of the dielectric constant with frequency may occur, the symbol is commonly primed. When a condenser is charged with an alternating current, loss may occur because of dissipation of part of the energy as heat. In vector notation, the angle d between the vector for the amplitude of the charging current and that for the amplitude of the total current is the loss angle, and the loss tangent, or dissipation factor, is

$$\tan \delta = \frac{\text{Loss current}}{\text{Charging current}} = \frac{\varepsilon''}{\varepsilon'}$$

where ε'' is the loss factor, or dielectric loss, of the dielectric in the condenser and ε' is the measured dielectric constant of the material.

At low frequencies of the alternating field, the dielectric loss is normally zero and ε' is indistinguishable from the dielectric constant edc measured with a static field. Debye has shown that

$$\frac{\varepsilon_{dc} - 1}{\varepsilon_{dc} + 2} = \frac{4\pi N_1}{3}\left(\alpha_0 + \frac{\mu^2}{3kT}\right) \tag{1}$$

where N_1 is the number of molecules or ions per cubic centimeter; α_0 is the molecular or ionic polarizability, i.e., the dipole moment induced per molecule or ion by unit electric field (1 esu = 300 volts/cm); μ is the permanent dipole moment possessed by molecule; k is the molecular gas constant, 1.38×10^{-16}, and T is the absolute temperature. An electric dipole is a pair of electric charges, equal in size, opposite in sign, and very close together. The dipole moment is the product of one of the two charges by the distance between them.

In Equation (1), $\mu^2/3kT$ is the average component in the direction of the field of the permanent dipole moment of the molecule. In order that this average contribution should exist, the molecules must be able to rotate into equilibrium with the field. When the frequency of the alternating electric field used in the measurement is so high that dipolar molecules cannot respond to it, the second term on the right of the above equation decreases to zero, and we have what may be termed the optical dielectric constant ε_∞ defined by the expression

$$\frac{\varepsilon_\infty - 1}{\varepsilon_\infty + 2} = \frac{4\pi N_1}{3}\alpha_0 \tag{2}$$

ε_∞ differs from n^2, the square of the optical refractive index for visible light, only by the small amount due to infrared absorption and to the small dependence of n on frequency, as given by dispersion formulas. It is usually not a bad approximation to use $\varepsilon_\infty = n^2$. The general Maxwell relation $\varepsilon' = n^2$ holds when ε' and n are measured at the same frequency. The Debye equation may be written in the form

$$\frac{\varepsilon_{dc} - 1}{\varepsilon_{dc} + 2} - \frac{\varepsilon_\infty - 1}{\varepsilon_\infty + 2} = \frac{4\pi N_1}{9kT}\mu^2 \tag{3}$$

A much better representation of the dielectric behavior of polar liquids is given by the Onsager equation

$$\frac{\varepsilon_{dc} - 1}{\varepsilon_{dc} + 2} - \frac{\varepsilon_\infty - 1}{\varepsilon_\infty + 2} = \frac{3\varepsilon_{dc}(\varepsilon_\infty + 2)}{(2\varepsilon_{dc} + \varepsilon_\infty)(\varepsilon_{dc} + 2)}\frac{4\pi N_1 \mu^2}{9kT} \tag{4}$$

Anomalous dielectric dispersion occurs when the frequency of the field is so high that the molecules do not have time to attain equilibrium with it. One may then use a complex dielectric constant

$$\varepsilon^* = \varepsilon' - j\varepsilon'' \tag{5}$$

where $j = \sqrt{-1}$. Debye's theory of dielectric behavior gives

$$\varepsilon^* = \varepsilon_\infty + \frac{\varepsilon_{dc} - \varepsilon_\infty}{1 + j\omega\tau} \tag{6}$$

where ω is the angular frequency (2π times the number of cycles per second) and τ is the dielectric relaxation time. Dielectric relaxation is the decay with time of the polarization when the applied field is removed. The relaxation time is the time in which the polarization is reduced to $1/e$ times its value at the instant the field is removed, e being the natural logarithmic base.

Combination of the two equations for the complex dielectric constant and separation of real and imaginary parts gives

$$\varepsilon' = \varepsilon_\infty + \frac{\varepsilon_{dc} - \varepsilon_\infty}{1 + \omega^2\tau^2} \tag{7}$$

$$\varepsilon = \frac{(\varepsilon_{dc} - \varepsilon_\infty)\omega\tau}{1 + \omega^2\tau^2} \tag{8}$$

These equations require that the dielectric constant decrease from the static to the optical dielectric constant with increasing frequency, while the dielectric loss changes from zero to a maximum value ε''_m and back to zero. These changes are the phenomenon of anomalous dielectric dispersion. From the above equations, it follows that

$$\varepsilon''_m = (\varepsilon_{dc} - \varepsilon_\infty)/2 \tag{9}$$

and that the corresponding values of ω and ε' are

$$\omega_m = 1/\tau \tag{10}$$

and

$$\varepsilon''_m = (\varepsilon_{dc} + \varepsilon_\infty)/2 \tag{11}$$

The symmetrical loss-frequency curve predicted by this simple theory is commonly observed for simple substances, but its maximum is usually lower and broader because of the existence of more than one relaxation time. Various functions have been proposed to represent the distribution of relaxation times. A convenient representation of dielectric behavior is obtained, according to the method of Cole and Cole, by writing the complex dielectric constant as

$$\varepsilon^* = \varepsilon_\infty + \frac{\varepsilon_{dc} - \varepsilon_\infty}{1 + (j\omega\tau_0)^{1-\alpha}} \tag{12}$$

where τ_0 is the most probable relaxation time and α is an empirical constant with a value between 0 and 1, usually less than 0.2. When the values of ε'' are plotted as ordinates against those of ε' as abscissas, a semicircular arc is obtained intersecting the abscissa axis at $\varepsilon' = \varepsilon_\infty$ and $\varepsilon' = \varepsilon_{dc}$. The center of the circle of which this arc is a part lies below the abscissa axis, and the diameter of the circle drawn through the center from the intersection at ε_∞ makes an angle $\alpha\pi/2$ with the abscissa axis. When a is zero, the diameter lies in the abscissa axis, there is but one relaxation time, and the behavior of the material conforms to the simple Debye theory. When, as may arise from intramolecular rotation, a substance has more than one relaxation mechanism, or, when the material is a mixture, the observed loss-frequency curve is the resultant of two or more different curves and, therefore, departs from the simple Debye or Cole-Cole curve.

If the dielectric material is not a perfect dielectric, and has a specific dc conductance k'(ohms^{-1} cm^{-1}), there is an additional dielectric loss

$$\varepsilon''_{dc} = \frac{3.6 \times 10^{12}\pi k'}{\omega} \tag{13}$$

The effective specific conductance is given by

$$k' = \frac{1}{4\pi}\frac{(\varepsilon_{dc} - \varepsilon_\infty)\omega^2\tau}{1 + \omega^2\tau^2} \tag{14}$$

It is evident from this equation that k' increases with ω, approaching a limiting value, k_∞, the infinite-frequency conductivity, which is attained when 1 can be neglected in comparison with $\omega^2 \tau^2$, so that

$$k_\infty = \frac{\varepsilon_{dc} - \varepsilon_\infty}{4\pi\tau} \tag{15}$$

In a heterogeneous material, interfacial polarization may arise from the accumulation of charge at the interfaces between phases. This occurs only when two phases differ considerably from each other in dielectric constant and conductivity. It is usually observed only at very low frequencies, but, if one phase has a much higher conductivity than the other, the effect may increase the measured dielectric constant and loss at frequencies as high as those of the radio region. This so-called Maxwell-Wagner effect depends on the form and distribution of the phases as well as upon their real dielectric constants and conductances. Each type of form and distribution requires special treatment. For a commercial rubber, for example, the observed loss may be

$$\varepsilon''(\text{observed}) = \varepsilon''_{dc} + \varepsilon''(\text{Maxwell} - \text{Wagner}) + \varepsilon''(\text{Debye}) \tag{16}$$

DIELS-ALDER REACTION.

An important organic reaction for the synthesis of six-membered rings, discovered in 1928. It involves the addition of an ethylenic double bond to a conjugated diene, i.e., a compound containing two double bonds separated by one single bond, as in 1,3-butadiene ($CH_2{=}CH{-}CH{=}CH_2$) or cyclopentadiene. The ease of addition of the ethylenic compound is greatly enhanced by adjacent carbonyl groups; hence maleic anhydride reacts quantitatively with hexachlorocyclopentadiene to form chlorendic anhydride.

DIELS, OTTO P. H. (1876–1954).

A German chemist who won the Nobel prize in chemistry with Kurt Alder in 1950. He was awarded the prize for diene synthesis work, which led to improved methods of analyzing and synthesizing organic compounds. His research resulted in the discovery of carbon suboxide, methods of dehydrating cyclical hydrocarbons using selenium, and determination of the structure of steroids. A student of Fischer", he graduated from the University of Berlin.

DIESEL FUELS. See **Petroleum**.

DIFFRACTION.

In any wave disturbance, the interference pattern resulting from the rays passing through different parts of an opening, or coming from different points around an opaque object, as they unite at each point. Diffraction and interference effects are characteristic of all wave phenomena no matter of what type. They are thus found in electromagnetic waves (light, x-rays, etc.), sound waves, water waves, and in matter waves. Diffraction occurs whenever a wave passes through a restricted aperture, such as a small hole or slit, or around an edge or particle. An example in optics is the case in which light from a point source passes the edge of a postcard and falls upon a white screen; the shadow of the edge is not sharply defined, but deepens to darkness gradually on one side, and is bordered by very narrow alternate bright and dark interference fringes (*diffraction bands*) on the other. Again, the image of a minute opaque speck under magnification against a bright background is surrounded by concentric diffraction rings. The image of a bright object, such as a star, as formed in the focal plane of a converging lens, is also surrounded by diffraction rings. If two such images are close together the fringe systems will overlap and no matter how much magnification is applied it will never be possible to obtain clear, well-separated, images of the points. The resolving power of an optical instrument may be defined as a measure of the sharpness with which small images very close together may be distinguished. It is directly proportional to the diameter of the objective aperture and inversely proportional to the wavelength of the light. Diffraction thus limits the resolving power, and hence the practicable magnification, of optical instruments.

Diffraction always results in energy being carried into regions that it could not reach if the propagation of the wave were strictly rectilinear. The patterns produced are geometrically similar whenever the ratio of the wavelength to the dimensions of the aperture is the same. Thus a radio wave in the AM broadcast range will be diffracted in passing through a hole 15 meters in diameter to the same extent as blue light passing through a pin-hole of 0.001 millimeter diameter.

For a single slit of width a and light of wavelength λ, falling on the slit at normal incidence, the intensity of light at an angle θ from the normal to the slit is given by

$$I = R_0^2 \frac{\sin^2\left(\dfrac{\pi a \sin\theta}{\lambda}\right)}{\left(\dfrac{\pi a \sin\theta}{\lambda}\right)^2}$$

Fresnel diffraction. The intensity at any point is the resultant of disturbances coming directly to that point from all parts of the exposed wave front. In general, the wave front is spherical or cylindrical, resulting from a source at finite distance, and the point of observation is also at finite distance.

Fraunhofer diffraction phenomena are observed when both the source and the point of observation are effectively at infinite distance from the diffracting object, obstacle, or aperture. This condition is sometimes brought about by passing the light from the source through a collimator before it is diffracted, and then focusing the parallel diffracted rays at the point of observation.

For material particles, according to the de Broglie hypothesis and quantum mechanics, a material particle having a momentum of magnitude p behaves as though it were associated with a wave of wavelength $\lambda = h/p$, where h is the Planck constant. In any physical process in which a particle interacts simultaneously with two or more scattering centers, the waves associated with the scattering from the various centers will undergo interference with one another to produce diffraction analogous to that which would be observed with light of the same wavelength. According to the principles of quantum mechanics, the wave which is associated with the particle is described by the quantum mechanical wave function $\psi(r, t)$ which contains all of the information concerning the state of the particle which one is physically allowed to have. Born showed that $\psi^*(r, t)\psi(r, t)d\tau$ should be interpreted as the probability that the particle will be found in the volume element dt. Thus in regions where the wave functions representing the scattering from the different scattering centers give complete destructive interference, the probability of finding the particle will be zero. In regions where constructive interference occurs, the probability of finding the particle will be enhanced.

The analogy between the diffraction of light, x-rays, etc., on one hand and that of material articles on the other becomes clear if we remember that in the former phenomena it is the probability of finding a photon in a given location that is determined by the square of the absolute value of the wave amplitude. Because it is relatively easy to use electrons or neutrons having wavelengths of the order of one angstrom, electron and neutron diffraction may be used to study crystal structure in a manner very similar to x-ray diffraction. Electrons do not penetrate as deeply into matter as do x-rays, hence electron diffraction reveals structure near the surface; neutrons do penetrate easily and have the advantage that they possess an intrinsic magnetic moment that causes them to interact differently with atoms having different alignments of their magnetic moments.

Additional Reading

Born, M. and E. Wolf: *Principles of Optics: Electromagnetic Theory of Propagation, Interference and Diffraction of Light,* Cambridge University Press, Inc., New York, NY, 2000.

Chung, F.H. and D.K. Smith: *Industrial Application of X-Ray Diffraction,* Marcel Dekker Inc., New York, NY, 1999.

Halpern, A.: *3,000 Solved Problems in Physics,* McGraw-Hill, New York, NY, 1988.

Hooijmans, P.W.: *Coherent Optical System Design,* John Wiley & Sons, Inc., New York, NY, 1994.

Krawitz, A.D.: *Introduction to Diffraction on Materials, Science and Engineering,* John Wiley & Sons, Inc., New York, NY, 2001.

Lapedes, D.N.: *McGraw-Hill Dictionary of Physics and Mathematics,* The McGraw-Hill Companies, Inc., New York, NY, 1990.

Lauterborn, W. and K.M. Wiesenfeldt: *Coherent Optics Fundamentals and Applications,* Springer-Verlag Inc., New York, NY, 1995.

Lide, D.R.: *CRC Handbook of Chemistry and Physics,* 84th Edition, CRC Press, LLC., Boca Raton, FL, 2003.

Nussbaum, A.: *Optical System Design,* Prentice-Hall, Inc., New Jersey, 1997.

Parker, S.P.: *McGraw-Hill Encyclopedia of Physics,* The McGraw-Hill Companies, Inc., New York, NY, 1993.

Schwartz, A.J., M. Kumar, and B.L. Adams: *Electron Backscatter Diffraction in Materials Science,* Kluwer Academic Publishers, Norwell, MA, 2000.

Van Stryland, E.W., J.M. Enoch, and W.L. Wolfe: *Handbook of Optics,* Vol. 3, 2nd Edition, The McGraw-Hill Companies, Inc., New York, NY, 2000.

DIFFRACTION GRATING. A series of very fine, closely spaced parallel slits, or of very narrow, parallel reflecting surfaces, which, when light is incident upon it at a definite angle, produces a succession of spectra. The complete optical theory is somewhat complicated, but the action of a plane transmission grating may be explained approximately as follows.

A plane, monochromatic light wave W, incident at angle i Fig. 1, reaches the slits at different times. A lens L receives the waves emerging from any two adjacent slits, A and B (among many others), after they have traveled paths differing by $CA + AD$; that is, by $S \sin i + S \sin \delta$, in which $S = AB$. If the lens is so placed that this path difference is a whole number of wavelengths, $n\lambda$, the successive wave-trains will reach it in the same phase, so that when they are brought to the focus F, they will be in synchronism and will produce a bright image of the distant source. Therefore any angle δ for which this result is possible is subject to the condition

$$S \sin i + S \sin \delta = n\lambda$$

or

$$\sin \delta = \frac{n\lambda}{S} - \sin i$$

Bright images will be produced for those angles δ which correspond to $n = 1, 2, 3, 4, \ldots$; the numbers denote the "orders" of the images. It is easily shown that for any order the total deviation $(i + \delta)$ is least when $\delta = i$ and therefore when

$$\sin \delta = \frac{n\lambda}{2S}$$

If the incident light is composed of various wavelengths, the corresponding images of any order will appear at different points, since δ varies with λ; and the result is a spectrum. In short, the grating acts as a dispersion piece, and as such is of great value in spectroscopes.

For high dispersion the slits must be very fine and very close together (S small), and for high resolving power (sharpness of spectral lines) the total number of slits must be large. Gratings having several thousand slits to the inch of width are common. They may be made by ruling fine scratches with a diamond point on glass, or, with reflecting gratings, on polished metal. If the rulings are not spaced with absolute regularity, false lines, called ghosts appear in the spectrum.

Rowland was the first to rule reflection gratings on concave metal surfaces. Such gratings eliminate the necessity of the spectroscope collimator or focusing lenses, as they take light direct from the spectroscope slit and form the spectral-line images like a concave mirror. The echelon is another special type of grating.

For a constant angle of incidence, the angular dispersion is given by

$$\frac{d\delta}{d\lambda} = \frac{n}{S \cos \delta}$$

and for small angles from the normal to the grating, $\cos \delta$ may be replaced by unity. At the point of focus, usually a photographic plate, the so-called normal spectrum will have a constant linear dispersion, often expressed in mm per angstrom. The resolving power of a grating equals nN, where N is the total number of slits. It, too, is there a constant. For a prism spectroscope both dispersion and resolving power depend on the wavelength of the incident light. The constancy of these two quantities is thus an advantage for a grating instrument.

Typical gratings for the visible and the ultraviolet regions have 6,000–18,000 lines per centimeter, for the infrared, 700–3,000 lines per centimeter. These numbers are roughly equal to the wave number of the light to be dispersed.

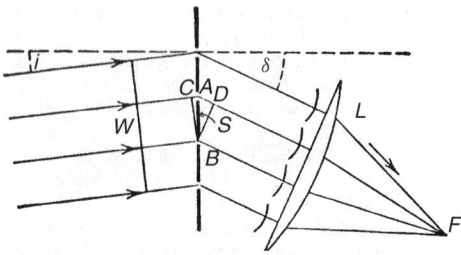

Fig. 1.   Diffraction by a plane grating

DIFFUSION. This term denotes the process by which molecules or other particles intermingle as a result of their random thermal motion. The molecules of a gas or of a liquid wander about rapidly, colliding frequently and exchanging kinetic energy, but maintaining a certain aimless progress. If an enclosure contains two gases, the lighter initially above and the heavier below, the gases at once begin to mingle because of their molecular motion. The same is true of a dense solution (as of sugar) and pure water; both the sugar and the water molecules wander across the boundary, so that in the course of time the whole body of liquid attains nearly uniform concentration. The process whereby this is effected is called diffusion. In the case of fluids of different color, its progress may be easily watched.

The rates at which different gases diffuse at a given temperature are inversely proportional to the square roots of their molecular weights. Thus, hydrogen diffuses four times as fast as oxygen. This follows, according to the kinetic theory, from the fact that the molecules of various kinds have the same mean kinetic energy and hence their mean square speeds are in the inverse ratio of their masses. In the case of a solution of non-uniform concentration, the diffusion of the solute from the more to the less concentrated regions takes place in accordance with *Fick's law*, expressed by the equation

$$\frac{dm}{dt} = -DS\frac{dc}{dx}$$

This gives the mass of solute diffused per unit time through a cross-section S, in terms of the concentration gradient dc/dx in the direction x perpendicular to the cross section. D is a constant for the given solute and solvent at a given temperature, and is called the diffusion coefficient. For any one pair of substances, D is found to be proportional to the absolute temperature. It should be stated that these statements apply only to nonelectrolytic solutions.

Diffusion in solids is a phenomenon that occurs rather slowly, but can be observed. Three basic processes may be responsible: (a) direct exchange of atoms on neighboring sites; (b) migration of interstitial atoms; (c) diffusion of vacancies. The first process requires very large energy. The energy to make an interstitial migration is rather large, but many atoms migrate easily. Vacancies are fairly readily formed, and diffuse fairly easily. From the Kirkendall effect it appears that (b) and (c) are the usual processes. The diffusion coefficient is related to the ionic mobility by the Einstein relation.

Another use of the term diffusion is to denote the passage of particles through matter in such circumstances that the probability of scattering is large compared with that of leakage or absorption. It is often limited to phenomena described by a member of the class of differential equations known as diffusion equations.

Diffusion operations are of large importance in chemical and process engineering. Both gaseous and thermal diffusion are used to separate one gas from another. In the case of *gaseous diffusion*, if a binary gaseous mixture at a high pressure is passed over a microporous barrier, a fraction of the gas will diffuse through the barrier into a low-pressure discharge chamber and will be found to be richer in the content of one gas than of the other gas. This is termed *Knudsen diffusion*. The passage of gas mixtures through the barrier is governed by the unequal collision frequency of each molecular species upon the walls of the pores. Fast, so-called light molecules separate from slower, heavier molecules within the barrier. The Oak Ridge, Tennessee plant designed for the enrichment of $^{235}UF_6$ from the naturally occurring uranium hexafluoride that contained 99.3% $^{238}UF_6$ represented the first major application of gaseous diffusion on a large scale. The molecular weight of the hexafluoride of ^{235}U is 349, whereas that of the hexafluoride of ^{238}U is 352. Inasmuch as the rate of diffusion of a gas is inversely proportional to the square root of density, the greatest separation factor for one stage of separation is the square root of 352/349, or 1.0043. Inasmuch as only part of the gas can diffuse through a given barrier, the separation factor is less. Thus, the number of diffusion stages for the Oak Ridge plant was approximately 4,000, requiring a plant that covered several acres of ground. Polymeric barriers also are under study and with scientific improvements, gaseous diffusion may become a widely used means for the recovery of carbon dioxide, helium, and nitrogen from natural gas.

In *thermal diffusion*, a thermal gradient is applied to a homogeneous solution (gas or liquid). This causes a concentration gradient and thus affords a means of separating materials. The logic of thermal diffusion is derived from the kinetic theory of gases and the cage model of liquids. If there is no marked size difference, heavier species tend to concentrate in the cold region. Where materials of identical molecular weight are involved, the larger molecules go to the cold region by virtue of their greater momentum. In the static mode, differential concentration can be established by eliminating convection currents that otherwise would tend to negate the effects of the applied thermal gradient. In the reflux method, hot and cold materials are flowed countercurrently. The reflux usually is provided using the density gradient that results from the imposition of the temperature gradient. Equipment of this latter type usually is referred to as a *thermogravitational column*, or a *Clausius-Dickel* column. Limited applications of thermal diffusion separations include those for concentrating dilute mixtures of isotopic gases. However, equipment costs tend to be high and efficiencies low.

Because chemical processing involves both the mingling and separating of fluids (gases and liquids), an understanding of diffusion processes is fundamental to process design.

Additional Reading

Chapman, S. and T.G. Cowling: *The Mathematical Theory of Non-Uniform Gases: An Account of Kinetic Theory of Viscosity, Thermal Conduction and Diffusion in Gases,* Cambridge University Press, New York, NY, 1991.

Glicksman, M.E.: *Diffusion in Solids: Field Theory, Solid-State Principles, and Applications,* John Wiley & Sons, Inc., New York, NY, 1999.

King, C.J.: "Spray Drying Food Liquids and the Retention of Volatiles (Selective Diffusion)," *Chem. Eng. Progress,* **33** (June 1990).

Lide, D.R.: *CRC Handbook of Chemistry and Physics,* 84th Edition, CRC Press, LLC., Boca Raton, FL, 2003.

Mikhailov, M.D. and M.N. Ozisik: *Unified Analysis and Solutions of Heat and Mass Diffusion,* Dover Publications, Mineola, NY, 1994.

Perry, R.H. and D.W. Green: *Perry's Chemical Engineers Handbook,* 7th Edition, The McGraw-Hill Companies, Inc., New York, NY, 1999.

Schweitzer, P.A.: *Handbook of Separation Techniques for Chemical Engineers,* 3rd Edition, The McGraw-Hill Companies, Inc., New York, NY, 1996.

Yaws, C.L.: *Handbook of Transport Property Data: Viscosity, Thermal Conductivity, and Diffusion Coefficients of Liquids and Gases,* Butterworth-Heinemann, Inc., Woburn, MA, 1995.

DIFFUSION CURRENT. The limiting current reached by electrolytic migration of the ions in a solution under the application of a potential difference to the electrodes. As the potential difference is increased the ion current to the electrodes increases rapidly at first but soon reaches a limiting value (the diffusion current value) as the potential difference is increased. If the potential difference is increased still further, a point is ultimately reached at which a new ion species begins to discharge.

The current limit is set by the rate of diffusion (of the ion being discharged) through the depleted layer surrounding the electrode. This diffusion rate is proportional to the ion concentration. For application of this effect, see **Polarographic Analyzers**.

DIFFUSION POTENTIAL. When liquid junctions exist where two electrolytic solutions are in contact, as in the case of two solutions of different concentrations of the same electrolyte, diffusion of ions occurs between the solutions, and the differences in rates of diffusion of different ions set up an electrical double layer, having a difference of potential, known as the diffusion potential or liquid junction potential.

DIGESTER (Process). In the process industries, the term *digester* is used in two principal connections: (1) the digestion of wood chips in the production of pulp prior to the manufacture of paper, and (2) the digestion of sewage sludge in waste-treatment operations. The term also appears in a number of other operations operating under varying conditions and hence a generalized definition is difficult to formulate. In chip digestion (also termed cooking), the chip digester is a large vessel provided with suitable raw-chip feed and cooked-chip discharge ports and equipped with means for heating and maintaining its contents at a specified temperature for a specific time. Batch digesters are vertical, stationary cylindrical pressure vessels into which chips and cooking liquor are charged and in which liquor is constantly moved, either by percolation within the digesters aided by direct addition of steam for heating purposes, or by continual withdrawal of liquor through screened ports and reintroduction of the liquor, after further heating. Modern batch digesters are typically 4,000 to 6,000 cubic feet (113.3 to 170 cubic meters) in volume with pulp capacities of 10 to 20 tons (9 to 18 metric tons).

By contrast in terms of operating parameters, sewage sludge is digested by aerating a lagoon or pond under normal outdoor temperatures, except that below about 40°F (4.5°C), the activity of the bio-organisms which aid in the digestion falls off considerably.

Autoclaves used in the chemical industry also are sometimes referred to as digesters.

DIGESTER (Pulp). See **Pulp (Wood) Production and Processing**.

DIGESTION. (1) The physiological processes involved in the assimilation of nutrients from ingested foods by the animal organism. Hydrochloric acid in the gastric juice plays a prominent part, aided by the saliva which initiates carbohydrate breakdown, and bile and pancreatic secretions in the intestine. Numerous types of enzymes catalyze these processes. (2) In chemical engineering, the term refers to several processes: (a) the preferential dissolution of certain mineral constituents in some ore concentrates, (b) the liquefaction of organic waste materials by microbiological action as the activated sludge, (c) the removal of lignin from wood by hot chemical solutions in the manufacture of chemical cellulose and paper pulp, (d) the separation of fabric from scrap tires by hot sodium hydroxide solution in the reclaiming of rubber. The equipment for (c) and (d) is called a digester.

DIGITALIS (*Digitalis purpurea; Foxglove, Scrophulariaceae*). The foxglove is a biennial often grown as an ornamental plant. The first year of growth produces only the long basal leaves, while in the second year the erect leafy stem 2–5 feet (0.6–1.5 meters) tall is developed. The flowers are borne in a raceme, which, through the bending of the peduncles or individual flower stalks, becomes one-sided. The purple flowers have a five-parted calyx, a tubular bell-shaped corolla obscurely five-lobed, five stamens and a single pistil. They are pollinated mainly by bees. The fruit is a two-celled capsule.

The drug digitalis is prepared mostly from leaves of the second year's growth. These are rather coarse ovate leaves covered with glandular hairs. Decoctions of the leaves have been used in Europe for many years.

Digitalis is a valued drug in medicine and is used in certain kinds of heart disease. The chief effects it has on the heart are the regulation of its rate, rhythm, tone, contraction, and conduction of impulses.

As a crop plant, digitalis is grown in England, Germany, and in the United States, especially in Michigan. Propagation is by seeds, which are sown under glass and later transplanted.

DILATANCY. The property of certain colloidal solutions of becoming solid, or setting, under pressure. Also known as "inverse plasticity" since there is an increase in the resistance to deformation with increase in the rate of shear.

DILATANT SUBSTANCES. See **Rheology**.

DILATION NUMBER. Ratio of the volume of a liquid to the volume of a solid of the same composition at the same temperature.

DILATOMETER. 1. An instrument used to measure small increments in the volume of liquids, as a solid phase separates.

2. An instrument for measuring very small length changes in a solid metal specimen such as occur during thermal expansion or phase transformations.

DILUENT. (1) An ingredient used to reduce the concentration of an active material to achieve a desirable and beneficial effect. Examples are combination of diatomaceous earth with nitroglycerin to form the much less shock-sensitive dynamite; addition of sand to cement mixes to improve workability with no serious loss of strength; addition of an organic liquid having no solvent power to a paint or lacquer to reduce viscosity and achieve suitable application properties.

(2) Low-gravity materials used primarily to reduce cost, e.g., blown asphalt, wood floc, etc., in rubber and plastic mixes. In this sense, there is no clear distinction between a diluent and an extender.

(3) An ingredient of rocket fuels, such as helium, hydrazine, or hydrogen.

DIMER ACIDS. The dimer acids, 9- and 10-carboxystearic acids, and C-21 dicarboxylic acids are products resulting from three different reactions of C-18 unsaturated fatty acids. These reactions are, respectively, self-condensation, reaction with carbon monoxide followed by oxidation of the resulting 9- or 10-formylstearic acid (or, alternatively, by hydrocarboxylation of the unsaturated fatty acid), and Diels-Alder reaction with acrylic acid. The starting materials for these reactions have been almost exclusively tall oil fatty acids or, to a lesser degree, oleic acid, although other unsaturated fatty acid feedstocks can be used.

Physical Properties

The physical properties of polymerized fatty acids are influenced by the basestock, by the dimerization conditions and catalysis, and by the degree to which monomer, dimer, and higher oligomers are separated following the dimerization.

Dimer acids are relatively high mol wt (ca 560) and yet are liquid at 25°C. This liquidity is a consequence of the many isomers present, most with branching or cyclic structures.

Most of the products listed in Tables 1 and 2 are based on manufacture from tall oil fatty acids. Dimer acids based on other feedstocks (e.g., oleic acid) may have different properties.

Chemical Properties

Structure and Mechanism of Formation. Thermal dimerization of unsaturated fatty acids has been explained both by a Diels-Alder mechanism and by a free-radical route involving hydrogen transfer. The Diels-Alder reaction appears to apply to starting materials high in linoleic acid content satisfactorily, but oleic acid oligomerization seems better rationalized by a free-radical reaction.

Chemical Reactions. The reactions of dimer acids were reviewed fully in 1975. The most important is polymerization; the greatest quantities of dimer acids are incorporated into the non-nylon polyamides. Other reactions of dimer acids that are applied commercially include polyesterification, hydrogenation, esterification, and conversion of the carboxy groups to various nitrogen-containing functional groups.

Manufacture

Clay-catalyzed oligomerization and thermal oligomerization are the two commercial processes used to manufacture dimer acids.

Process Modification. Dimer acid process modifications have fallen into three categories; those claiming higher dimer:trimer ratios, those utilizing varying types of clays, and those purporting to result in improved yields. Another aspect of process improvement is color improvement.

Other Polymerization Methods. Experimental alternatives include the use of peroxides, hydrogen fluoride, a sulfonic acid ion-exchange resin, and corona discharge.

TABLE 1. PROPERTIES OF DIMER AND DISTILLED DIMER ACID[a]

Physical property	Distilled dimer acid		Dimer acid
	Hydrogenated[b]	Unhydrogenated[c]	
composition %			
dimer acids	97	95	82–83
trimer acids (and higher)	3	4	14–16
monobasic acids	trace	1	1–5
acid number	191–197	190–196	189–197
viscosity at 25°C, mm²/s (= cSt)	~5200	7000–8000	7500–9000
specific gravity, 25/25°C	0.94	0.9	~0.95

[a] Hystrene series of dimer acids, Humko Chemical Division of Witco Corporation.
[b] Empol series of dimer acids, Henkel Corp., Emery Group (oleic-based, thus of lower viscosity).
[c] Hystrene series of dimer acids, Humko Chemical Div. of Witco Corp.

TABLE 2. PROPERTIES OF TRIMER ACIDS[a]

Physical property	Value
composition, %	
dimer	40
trimer	60
monobasic acids	trace
acid number	170–190
viscosity at 25°C, mm²/s (= cSt)	~30,000

[a] Hystrene 5460, Humko Chemical Div. of Witco Corp.

Energy Requirements. The production of dimer acids is quite energy-intensive. A standard operation sequence normally results in the expenditure of about 18.6 MJ (17,600 Btu) (equivalent to 0.67 kg coal or 0.33 kg natural gas of fuel oil) to produce each kg of crude dimer and to separate it into monomer, dimer, and trimer.

Storage and Handling

Since dimer acids, monomer acids, and trimer acids are unsaturated, they are susceptible to oxidative and thermal attack, and under certain conditions they are slightly corrosive to metals. Special precautions are necessary, therefore, to prevent product color development and equipment deterioration.

Economic and Market Aspects

According to one estimate, the current capacity for manufacturing dimer acids in the U.S. is around 55,000 t per year. Current demand is estimated at about 33,600 t per year, and is expected to grow at about 2–3% per year to 35,000 t in 1993.

The current market situation for dimer acids includes relatively high raw material costs, high energy costs, slow growth, and relatively low prices. It is generally recognized as a mature market, with hopes for future growth hinging on factors such as increased polyamide use and a resurgence of oil drilling, where dimers are used for corrosion inhibition.

Health and Safety Aspects

The acute oral toxicity and the primary skin and acute eye irritative potentials of dimer acids, distilled dimer acids, trimer acids, and monomer acids have been evaluated based on the techniques specified in the *Code of Federal Regulations* (CFR). Based on these evaluations, monomer acids, distilled dimer acids, dimer acids, and trimer acids are classified as nontoxic by ingestion, are not primary skin irritants or corrosive materials, and are not eye irritants as these terms are defined in the federal regulations.

Uses

Nonreactive Polyamide Resins. Dimer-based polyamide resin markets are divided into those for reactive polyamides and those for nonreactive polyamides. The largest-volume commercial application of dimer acids is in nonreactive polyamide resins. Dimer acids impart flexibility, corrosion resistance, chemical resistance, moisture resistance, and adhesion to nonreactive polyamides. Hot-melt adhesives, the largest commercial application of nonreactive polyamide resins, are thermoplastics that have fairly sharp melting ranges. Flexographic printing inks utilize nonreactive polyamides from dimer acids as resin binders. The most important coating application for the nonreactive polyamide resins is in producing thixotropy.

Reactive Polyamide Resins. Reactive polyamide resins are used extensively to react with epoxy or phenolic resins. The amount used in epoxy applications far exceeds the use with phenolic resins.

THOMAS E. BREUER
Humko Chemical Division of Witco Corporation

Additional Reading

Johnson, R.W.: in E.H. Pryde, ed., *Fatty Acids*, American Oil Chemists Society, Champaign, IL, 1979.
Johnson, R.W.: in R.W. Johnson and E. Fritz eds., *Fatty Acids in Industry*, Marcel Dekker, Inc., New York, NY 1989.
Leonard, E.C. ed.: *The Dimer Acids*, Humko Sheffield Chemical, Memphis, TN, 1975.
Leonard, E.C.: in E.H. Pryde, ed., *Fatty Acids*, American Oil Chemists Society, Campaign, IL, 1979.

DIOPSIDE. The mineral diopside is a monoclinic pyroxene corresponding to the chemical formula $CaMgSi_2O_6$, calcium magnesium silicate. Its crystals, like those of other pyroxenes, tend to be short stout prisms of square or octagonal cross-section. Compact, granular, lamellar and fibrous varieties are often found. The prismatic cleavage is characteristic, cleavage planes intersecting at angles of 87° and 93°. A basal parting is often noted, but should not be confused with the cleavage. The hardness of diopside is 5–6; specific gravity, 3.2–3.3; uneven fracture tending toward conchoidal; luster, vitreous to dull; sometimes pearly on the base; color, light or dark greens, but may be colorless, gray, yellow or blue, although the latter color is rare.

Diopside is a primary mineral in rocks like diorites, gabbros and the like, but is also found in schists, and, as the result of contact metamorphism, in such rocks as crystalline limestones and dolomites. Diopside is found in association with vesuvianite, garnet, spinel, scapolite, tremolite, tourmaline and similar minerals. It is a rather widespread mineral, important localities being found in the following European countries: Finland, Sweden, Switzerland, Italy; it is found in eastern Siberia near Lake Baikal. In Canada diopside localities are in Lanark and Hastings Counties, Province of Ontario, and in the United States in Lewis and St. Lawrence Counties, New York, and in Maine.

ELMER B. ROWLEY
Union College
Schenectady, New York

DIOPTASE. The mineral dioptase is a rather rare copper silicate corresponding to the formula $CuSiO_2(OH)_2$, occurring in prismatic crystals of the hexagonal system, tri-rhombohedral in form. It may be found in crystalline aggregates or simply massive. Dioptase displays a conchoidal to uneven fracture; hardness, 5; specific gravity, 3.28–3.35; luster, vitreous; color, a beautiful emerald green. It has been found in the former U.S.S.R., Congo, Central African Republic, South West Africa, Chile, and in the United States in Arizona. The name is derived from the Greek words meaning through and to see, because cleavage was observed by looking through the crystals.

DIORITE. Diorite is a deep-seated igneous rock composed dominantly of sodiaplagioclase feldspar with hornblende, biotite, and (or) augite. Orthoclase may be present in small amounts, also quartz. Any considerable proportion of the latter mineral produces a quartz-diorite. With increasing amount of orthoclase, we have granodiorite, which is generally understood to be a rock intermediate in character between quartz-diorite and granite. If quartz is absent and there are essentially equal amounts of orthoclase and plagioclase the rock is then known as a monzonite from the type locality, Monzoni, in the Tyrol. There are quartz monzonites and, where the deficiency of silica is great enough, nephelite monzonites. Rocks of the latter sort have been reported from Madagascar. A variety of quartz, diorite, containing both hornblende and biotite, is called tonalite from the Tonale Alps, although the rock found there is more nearly a granodiorite.

DIOXIN. Although 75 variations of dioxins are known, the compound considered to possess the greatest toxicity and the subject of a continuing debate among industrial and environmental scientists specifically is 2,3,7,8-tetracholor dibenzo-*p*-dioxin and generally referred to as TCDD or, in much of the literature, simply as *dioxin*.

Attention to dioxin was propelled into scientific and public scrutiny as the result of the use of the herbicide 2,3,5-T (Agent Orange) during the Vietnam conflict. Also, a major localized contamination resulted from an accident in a chemical plant in Seveso, Italy, in 1976.

Minute amounts of dioxin are created in certain combustion processes and by a few chemical manufacturing processes, including the use of chlorine in paper bleaching.

In late 1991, a group of investigators, Sutter, et al. (Chemical Industry Institute of Toxicology, Research Triangle Park, North Carolina), reported of their studies on the effects of dioxin at the genetic and molecular level. It appears that dioxin may elicit its effects by altering gene expression in susceptible cells. In brief detail, the introduction of the Sutter, et al. paper (reference listed) summarizes: "Five TCDD-responsive complementary DNA clones were isolated from a human keratinocyte cell line. One of these clones encodes plasminogen activator inhibitor-2, a factor that influences growth and differentiation by regulating proteolysis of the extracellular matrix. Another encodes the cytokine interleukin-1β. Thus, TCDD alters the expression of growth regulatory genes and has effects similar to those of other tumor-promoting agents that affect both inflammation and differentiation."

As of early to mid 1990s, studies of dioxin continued. Epidemiological studies have shown an increase in the occurrence of soft-tissue sarcoma and non-Hodgkin's lymphoma, but these findings have been challenged by other studies. Some research has indicated that a given dosage of TCCD may kill a guinea pig and leave a hamster unaffected, or that female rats will develop liver cancer, while male rats will not from the same dosage. One effect that has pretty well risen above debate is that TCCD can cause chloracne (a disfiguring skin condition) among some humans who have been exposed to dioxin.

Additional Reading

Esser, C. and E. Gleichmann: *Dioxins and the Immune System, Mechanisms and Consequences of Interference,* S. Karger Publishers, Farmmington, CT, 1994.

Holloway, M.: "A Great Poison," *Sci. Amer.,* **16** (November 1990).

Holloway, M.: "A Press Release on Dioxin Sets the Record Wrong," *Sci. Amer.,* **24** (April 1991).

Roberts, L.: "Flap Erupts Over Dioxine Meeting," *Science,* **866** (February 22, 1991).

Schecter, A.J.J.: *Dioxins and Health,* Kluwer Academic Publishers, Norwell, MA, 1994.

Shawn, J.W.: *Dioxins and Agent Orange: Index of New Information with Clinical and Research Results,* Abbe Publishing Association of Washington DC, Washington, DC, 1995.

Sutter, T.R., et al.: "Targets for Dioxin: Genes for Plasminogen Activator Inhibitor-2 and Interleukin-β," *Science,* **415** (October 18, 1991).

Wittich, Rolf-Michael: *Biodegradation of Dioxins and Furans,* Springer-Verlag Inc., New York, NY, 1998.

DIPOLE MOMENT. In the simplest case, let two electric charges $+q$ and $-q$ be separated by the distance **d**. Then the permanent electric dipole moment is the vector $\mathbf{p} = q\mathbf{d}$. More generally, if discrete charges q_i are located at points x_i, y_i, z_i the magnitude of dipole moment is given by $p_\alpha = \sum q_i \alpha_i, \alpha = x, y, z$. If the charge distribution is continuous, the summations are replaced by integrals. An induced dipole moment can be produced by an electric or magnetic field (see also **Magnetism**). Atomic or molecular dipole moments, permanent or induced, are of considerable value in the study of atomic or molecular structure. The magnitude of such moments is usually reported in Debye units. The magnetic dipole moment produced by a current i flowing in a loop area A has magnitude $m = iA$. It is a vector with directional normal to the plane of the loop and sense taken as the direction of progression of a right-handed screw rotating with the current.

DISACCHARIDES. See **Carbohydrates**.

DISINFECTANT. A substance used on inanimate objects to destroy harmful microorganisms or inhibit their activity. Disinfectants are either complete of incomplete. Complete disinfectants destroy spores as well as vegetable forms of microorganisms; incomplete disinfectants destroy vegetable forms of the organism, but do not injure spores.

Some representative disinfectants are (1) mercury compounds (mercuric chloride, phenylmercuric borate); (2) halogens and halogen compounds (chlorine, iodine, fluorine, bromine, calcium and sodium hypochlorite); (3) phenols, including cresol from coal tar, *o*-phenylphenol; (4) synthetic detergents (anionic, such as sodium alklbenzene sulfonates, and cationic, such as quaternary ammonium compounds); (5) alcohols of low molecular weight, except methanol; (6) natural products (pine oil); (7) gases (sulfur dioxide, formaldehyde, ethylene oxide). Heat and electromagnetic waves are used as disinfectants.

A number of compounds (mercurous and mercuric chlorides, copper sulfate and carbonate, and a mixture of zinc oxide and zinc hydroxide) have been employed as seed disinfectants.

Effectiveness of disinfectants is rated by the phenol coefficient.

DISLOCATION. In crystallography, a type of lattice imperfection whose existence in metals is postulated in order to account for the phenomenon of crystal growth and of slip, particularly for the low value of shear stress required to initiate slip. One section of the crystal adjacent to the slip plane is assumed to contain one more atomic plane that the section on the opposite side of the slip plane. Motion of the dislocation results in displacement of one of the sections with respect to another.

DISOLVED OXYGEN (DO). One of the most important indicators of the condition of a water supply for biological, chemical, and sanitary investigations. Adequate dissolved oxygen is necessary for the life of fish and other aquatic organisms, and is an indicator of corrosivity or water, photosynthetic activity, septicity, etc.

DISPERSANTS. Dispersants are materials that help maintain fine solid particles in a state of suspension, and inhibit their agglomeration or settling in a fluid medium. With the help of mechanical agitation, dispersants can also break up agglomerates of particles to form particle suspensions. Another use of dispersants is to inhibit the growth of crystallites in a supersaturated solution. This characteristic is also known as precipitation inhibition, threshold inhibition, or antinucleation. Overall, dispersants are useful in preventing settling, deposition, precipitation, agglomeration, flocculation, coagulation, adherence, or caking of solid particles in a fluid medium.

Physical Chemistry of Dispersants

A convenient way to understand particle dispersion is to consider the process in four successive parts: the nature of particles and surfaces, adsorption onto particles, interface properties, and forces of attraction and repulsion.

Particles and Surfaces. Dispersants are primarily used to increase stability (prevent settling) of solid particles in liquid media, whereas surfactants are used more frequently to stabilize liquid (including polymer latex) surfaces within liquids. When the surface of a liquid is increased (stressed), molecules of the liquid flow to the surface to lower its energy, "healing" it. In contrast, solids exhibit no significant flow to the surface. Any stresses applied therefore remain in the form of a higher energy surface. Thus the history of a particle is important to its surface properties. Treatments that alter particle surface properties include freshly cleaving a surface along lowest energy crystal faces, adsorption of molecules and ions, heating or cooling, friction, corrosion, and grinding or polishing. The process by which a particle is formed also affects its surface properties. Examples of this include screw and spiral dislocations, missing layers, and other defects due to contamination or stress during formation.

Adsorption onto Particles. The Gibbs Adsorption law relates how adsorption onto surfaces affects interfacial tension, $d\gamma = -RT\Gamma d\ln c$, where γ = interfacial or surface tension, in N/m (1 N/m = 1000 dyn/cm); R = gas constant; T = absolute temperature; Γ = interfacial or surface concentration, in mol/unit area (i.e., adsorption); and c = dimensionless concentration ($d\ln c = dc/c$, thus units cancel).

If adsorption occurs ($\Gamma > 0$), then increasing the concentration of dispersant in the bulk water reduces interfacial tension.

Most adsorption processes are exothermic (ΔH is negative). Adsorption processes involving nonspecific interactions are referred to as physical adsorption, a relatively weak, reversible interaction. Processes with stronger interactions (electron transfer) are termed chemisorption. Chemisorption is often irreversible and has higher heat of adsorption than physical adsorption. Most dispersants function by chemisorption, in contrast to surfactants, which tend to physically adsorb.

Interface Properties. A polymeric dispersant may have segments extended into the solution, or the segments may be coiled, depending on whether the solvent is good (polymer–solvent interactions energetically favored) or poor (polymer–polymer and solvent–solvent contacts favored). Between these two solvent–polymer interactions is a θ (theta) solvent, in which neither condition is favored. If the polymer–solvent interaction is better than θ conditions, the extending chains or segments will repel chains adsorbed on other particles, as well as making the distance between two particles greater and enhancing steric repulsion. If the interaction is worse than θ conditions, the particles may flocculate due to mutual attraction of the polymer layers. On the other hand, if the polymer–solvent interaction is too strong, the polymeric dispersant may be adsorbed only weakly or not at all. This can lead to depletion flocculation, which is due to desorbed chains that are squeezed out from between two approaching particles. The desorbed chains can cause solvent to flow from between the particles by osmotic forces, leaving a bare area so that attraction between the particles is increased.

Attractive and Repulsive Forces. The force that causes small particles to stick together after colliding is van der Waals attraction. There are three van der Waals forces: *(1)* Keesom-van der Waals, due to dipole–dipole interactions that have higher probability of attractive orientations than nonattractive; *(2)* Debye-van der Waals, due to dipole-induced dipole interactions (i.e., uneven charge distribution is induced in a nonpolar material); and *(3)* London dispersion forces, which occur between two nonpolar substances.

As the distance between two approaching particles decreases, their electrical double layers begin to overlap. As a first approximation, the potential energy of the two overlapping double layers is additive, which is a repulsive term since the process increases total energy. Electrostatic repulsion can also be considered as an osmotic force, due to the compression of ions between particles and the tendency of water to flow in to counteract the increased ion concentration.

The overall stability of a particle dispersion depends on the sum of the attractive and repulsive forces as a function of the distance separating the particles. DLVO theory, named for Derjaguin and Landau and Verwey and Overbeek, encompasses van der Waals attraction and electrostatic repulsion between particles, but does not consider steric stabilization. The net energy, ΔG_T, between two particles at a given distance is the sum of the repulsive and attractive forces: ΔG_T = (electrostatic repulsive forces) − (van der Waals attractive forces). The electrostatic repulsive forces are a function of particle kinetic energy (kT), ionic strength, zeta potential, and separation distance. The van der Waals attractive forces are a function of the Hamaker constant and separation distance.

Although some progress has been made in calculating steric repulsive forces, the theory concerning them is not as completely developed as DLVO theory. The adsorbed polymer layers of two particles (in a good solvent) begin to interpenetrate as the particles approach each other. The interaction between these polymer layers can have an osmotic effect due to an increase in the local concentration of the adsorbed polymer layers, and can have an entropic or volume restriction effect due to crowding of the interacting chains. In both cases, entropy decreases, which is unfavorable. Moreover, the osmotic effect can create unfavorable enthalpic changes due to desolvation of closely packed chains. To regain lost entropy, the particles must separate to allow the chains more freedom of movement, while the solvent moves in to resolvate the polymer layer. As with electrostatic repulsion, an energy barrier is created. A common approximation used is that the strength of the energy barrier rises steeply at slightly less than the adsorbed layer thickness. Some of the practical differences between sterically and electrostatically stabilized dispersions may be summarized as follows:

Steric stabilization	Electrostatic stabilization
insensitive to electrolyte	coagulation occurs with increased electrolyte
effective in aqueous and nonaqueous media	more effective in aqueous media
effective at high and low concentrations	more effective at low concentrations
reversible flocculation common	coagulation often irreversible
good freeze–thaw stability	freezing often induces irreversible coagulation

Dispersant Materials

Dispersant materials include condensed phosphates, organic polymeric dispersants, poly(meth)acrylates, polymaleates, condensed phosphates, polysulfonates, sulfonated polycondensates, and tannins, lignins, glucosides, and alginates.

Uses

Dispersants are used in recirculating cooling water, boiler water, geothermal fluids, seawater distillation, reverse osmosis, sugar processing, oilfields, drilling muds, cement, paints and pigments, mineral processing, caulks, sealants, roof coatings, pesticides, animal feeds, detergents, and cleaners.

Environmental Considerations

Biodegradability of Dispersants. Most reviews on biodegradable polymers suggest that, with the exception of poly(vinyl alcohol) and poly(ethylene glycol)s, most synthetic organic dispersants are recalcitrant in the environment. More recently developed dispersants displaying biodegradability are polymers containing ester linkages and ether linkages. There is currently a great deal of research activity to develop dispersants that are both effective and biodegradable. Consequently, developments in this area should be rapid.

WILLIAM M. HANN
Rohm and Haas Company

DISPERSION.   (1) A two-phase system where one phase consists of finely divided particles (often in the colloidal size range) distributed throughout a bulk substance, the particles being the disperse or internal phase, and the bulk substance the continuous or external phase. Under natural conditions, the distribution is seldom uniform; but under controlled conditions, the uniformity can be increased by addition of wetting or dispersing agents (surfactants) such as fatty acid. The various possible systems are: gas-liquid (foam), solid-gas (aerosol), gas-solid (foamed plastic), liquid-gas (fog), liquid-liquid (emulsions), solid-liquid (paint), and solid-solid (carbon black in rubber). Some types, such as milk and rubber latex, are stabilized by protective colloid that prevents agglomeration of the dispersed particles by an abherent coating. Solid-in-liquid colloidal dispersions (loosely called solutions) can be precipitated by adding electrolytes that neutralize the electrical charges on the particles. Larger particles will gradually coalesce and either rise to the top or settle out, depending upon their specific gravity.

See also **Colloid Systems**; **Surfactants**; and **Suspension**.

(2) In the field of optics, dispersion denotes the retardation of a light ray, usually resulting in a change of direction as it passes into or out of a substance, to an extent depending on the frequency. Dispersion is a critically important property of optical glass.

See also **Refraction**.

DISPOSAL (Radioactive Wastes). See **Nuclear Reactor**.

DISSOCIATION. This can be broadly defined as the separation from union or as the process of disuniting. In chemistry, dissociation is the process by which a chemical combination breaks up into simpler constituents due, for example, to added energy as in the case of the dissociation of gaseous molecules by heat, or to the effect of a solvent upon a dissolved substance, as in the action of water upon dissolved hydrogen chloride. Dissociation may occur in the gaseous, liquid or solid state or in solution.

Elementary substances, if polyatomic in the molecule, will dissociate under conditions of sufficient energy. Chlorine and iodine, which are diatomic, are half dissociated at 1700°C and 1200°C, respectively. Just above the boiling point the molecule of sulfur is S_8. Its molecular weight decreases from 250 at 450°C to 50 at 2070°C. Thus there are some monatomic sulfur molecules at 2070°C. The dissociation probably takes place in reversible steps and can be represented by the equation:

$$S_8 \rightleftharpoons 4S_2 \rightleftharpoons 8S. \tag{1}$$

Many chemical compounds dissociate readily upon heating or otherwise supplying them with energy. Acetic acid vapor consists of double molecules just above the normal boiling point, but dissociates completely into single molecules at 250°C. Nitrogen tetroxide (N_2O_4) is a pale reddish brown gas at temperatures near its normal boiling point of 21.3°C. On heating the density of the gas becomes less and the color becomes darker until it is almost black. At 140°C the molecular weight is 46 which is that of NO_2 molecules. The dissociation can be written:

$$N_2O_4 \rightleftharpoons 2NO_2. \tag{2}$$

If one mole of gas yields ν moles of gaseous products, and α is the fraction of the one mole which dissociates, then the total number of moles present is:

$$1 - \alpha + \nu\alpha = 1 + \alpha(\nu - 1). \tag{3}$$

Now the density of a given weight of gas at constant pressure is inversely proportional to the number of moles, and if d_1 is taken as the density of the undissociated gas and d_2 that of the partially dissociated gas, then:

$$\frac{d_1}{d_2} = \frac{1 + \alpha(\nu - 1)}{1} \tag{4}$$

or

$$\alpha = \frac{d_1 - d_2}{d_2(\nu - 1)} \tag{5}$$

Therefore the *degree of dissociation* of a substance can be found by measuring the densities of the undissociated and partially (or completely) dissociated substance in the gaseous state. Molecular weights may be substituted for densities giving

$$\alpha = \frac{M_1 - M_2}{M_2(\nu - 1)} \tag{6}$$

The degree of dissociation can be used to calculate the *equilibrium constant* for dissociation. The equilibrium constant may be expressed in terms of concentrations, for example, moles per liter (K_c), or in terms of partial pressures (K_p). The degree of dissociation and equilibrium constants are important theoretically and practically, e.g., the latter can be used to ascertain the extent of a chemical process.

The temperature dependence of dissociation is expressed in terms of the equilibrium constant and is

$$\frac{d \ln K_p}{dT} = \frac{\Delta H}{RT^2} \quad \text{or} \quad \frac{d \ln K_c}{dT} = \frac{\Delta H}{RT^2} \tag{7}$$

where ΔH is the heat of dissociation. Integrating between the limits T_1 and T_2 one obtains

$$\ln \frac{K_{p_2}}{K_{p_1}} = \frac{\Delta H}{R} \left(\frac{T_2 - T_1}{T_1 T_2} \right) \tag{8}$$

$$\ln \frac{K_{c_2}}{K_{c_1}} = \frac{\Delta H}{R} \left(\frac{T_2 - T_1}{T_1 T_2} \right)$$

Electrolytes, depending upon their strength, dissociate to a greater or less extent in polar solvents. The extent to which a weak electrolyte dissociates may be determined by electrical conductance, electromotive force, and freezing point depression methods. The electrical conductance method is the most used because of its accuracy and simplicity. Arrhenius proposed that the degree of dissociation, α, of a weak electrolyte at any concentration in solution could be found from the ratio of the equivalent conductance, Λ, of the electrolyte at the concentration in question to the equivalent conductance at infinite dilution Λ_0 of the electrolyte. Thus

$$\alpha = \frac{\Lambda}{\Lambda_0} \tag{9}$$

This equation involves the assumption that mobilities of the ions coming from the electrolyte are constant from infinite dilution to the concentration in question. From the degree of dissociation and the concentration, the ionization constant or protolysis constant of a weak electrolyte can be obtained.

Water is a weak electrolyte, ionizing according to the equation:

$$H_2O + H_2O \rightleftharpoons H_3O^+ + OH^- \tag{10}$$

The specific conductance L, of water at 25° is 5.5×10^{-8} mho cm^{-1}, and the equivalent conductance of water at infinite dilution is found from the equivalent conductance of its constituent ions (H_3O^+ and OH^-) to be 547.8 mhos. The equivalent conductance Λ of water at 25°C is LV, where V is the volume of water (18 ml) containing 1 gram equivalent of water. Hence $\Lambda = LV = 5.5 \times 10^{-8} \times 18 = 9.9 \times 10^{-7}$. Therefore $\alpha = \Lambda/\Lambda_0 = 9.9 \times 10^{-7}/547.8 = 1.81 \times 10^{-9}$. Now $C_{H_3O^+} = C_{OH^-} = 55.5 \times 1.81 \times 10^{-9} = 1.00 \times 10^{-7}$ and

$$K = \frac{C_{H_2O^+} \times C_{OH^-}}{C_{H_2O}^2} \tag{11}$$

but C_{H_2O} is a constant, namely 55.5 moles/l and therefore

$$K_\omega = (55.5)^2 \, K = C_{H_3O^+} \times C_{OH^-}$$
$$= 1.00 \times 10^{-7} \times 1.00 \times 10^{-7}$$
$$= 1.00 \times 10^{-14}$$

The ionization constant of pure water varies with temperature as shown below.

Temperature°C	0	10	25	40	50
$K_\omega \times 10^{14}$	0.113	0.292	1.008	2.917	5.474

Inserting corresponding values of K_ω and absolute temperature into Eq. (8) and solving for ΔH one finds the heat of ionization per mole of water to be 13.8 kilocalories.

Ionization or dissociation in general can be repressed by adding an excess of a product of the dissociation process.

The acid formed when a base accepts a proton is called the conjugate acid of the base and the base formed when an acid donates a proton is the conjugate base of the acid. Thus in the reaction

$$HA + H_2O \rightleftharpoons H_3O^+ + A^- \tag{12}$$

HA and A^- are conjugate acid and base and H_2O and H_3O^+ are conjugate base and acid, respectively.

The common ion effect then can be found as the following example shows. When using ammonium hydroxide to which the common ammonium ion in the form of ammonium chloride has been added, the ionization can be represented by the equation

$$NH_3 + HOH \rightleftharpoons NH_4^+ + OH^- \qquad (13)$$

Ammonium ion NH_4^+ is the conjugate acid of the ammonia molecule. The ionization constant can be written

$$K = \frac{C_{NH_4^+} \times C_{OH^-}}{C_{NH_3}} \qquad (14)$$

and

$$C_{OH^-} = K \frac{C_{NH_3}}{C_{NH_4^+}} \qquad (15)$$

Now the base NH_3 is such a weak base that in the presence of NH_4Cl the concentration of unionized base, C_{NH_3}, is equal to the total concentration of base represented by C_{base}, and $C_{NH_4^+}$ coming from the weak base is so small as to be negligible. Hence the NH_4^+ ions can be considered as coming exclusively from the NH_4Cl. Therefore Eq. (15) can be written

$$C_{OH^-} = K \frac{C_{base}}{C_{salt}} \qquad (16)$$

or

$$pOH = pK + \log \frac{C_{salt}}{C_{base}} \qquad (17)$$

Thus C_{OH^-} and hence the degree of ionization of the base NH_3 is decreased with increasing concentration of salt. The salt effect of adding electrolytes with no common ion to a solution of incompletely ionizable substance can be seen from the following considerations and using the equilibriums represented by Eq. (14) which in terms of activities becomes:

$$K = \frac{a_{NH_4^+} \times a_{OH^-}}{a_{NH_3}}$$
$$= \frac{C_{NH_4^+} \times C_{OH^-}}{C_{NH_3}} \cdot \frac{f_{H_3O^+} \times f_{OH^-}}{f_{NH_3}} \qquad (18)$$

This ionization constant in terms of activities is called the true or thermodynamic ionization constant. It does not differ too much from the K in Eq. (14) for sufficiently low ionic strengths. The two differ more markedly for appreciable ionic strengths. Now suppose a salt with no common ion is added to the solution. The ionic strength of the solution will be increased. This increase in ionic strength causes a decrease in the activity coefficients of the ions except in very concentrated solutions. Thus for K of Eq. (18) to stay constant the concentrations of the ions must increase to offset the decrease in their activity coefficients. The ammonia must therefore increase in ionization and K as defined by Eq. (14) must increase. This is known as a salt effect.

Ampholytes in solution give equal concentrations of a weak acid and a non-conjugate weak base. The amino acids are ampholytes which contain within their molecules equal amounts of a weak acid, the COOH group and a weak non-conjugate base, the NH_2 group.

According to Arrhenius those substances which yield the hydrogen ion in solution are acids, whereas bases produce the hydroxyl ion. As long as water was considered the only "ionizing" solvent these definitions were relatively simple. In the case of nonaqueous solvent chemistry at least three other concepts have been advanced. These are: (1) Franklin's *solvent system concept*, first limited to water and ammonia but since extended to nonprotonic media and defining an acid as a substance yielding a positive ion identical with that coming from auto-ionization of the solvent and a base as a substance yielding a negative ion identical with that coming from auto-ionization of the solvent; (2) the *protonic concept* of acids as proton donors and bases as proton acceptors advanced by Brønsted and by Lowry; and (3) Lewis' electronic theory according to which an acid is a molecule, radical or ion which can accept a pair of electrons from some other atom or group to complete its stable quota of electrons, usually an octet, and forming a covalent bond, and a base is a substance which donates a pair of electrons for the formation of such a bond.

In liquid ammonia as in water auto-ionization takes place. Ammonium and amide ions are formed by the dissociation or protolysis according to the following equation

$$2NH_3 \rightleftharpoons NH_4^+ + NH_2^- \qquad (19)$$

The acid and base analogs of ammonia as a solvent is specified by this equilibrium as NH_4^+ and NH_2^- ions. All substances which undergo ammonolysis and hence bring about an increase in the ammonium ion concentration yield acid solutions. Thus P_2S_5 dissolves in liquid ammonia to give an acid solution as follows.

$$P_2O_5 + 12NH_3 \longrightarrow 2PS(NH_2)_3 + 3(NH_4)_2S. \qquad (20)$$

The solution is acid since an ammonium salt is formed and also because a solvo acid is obtained.

Many substances dissolve in liquid sulfur dioxide to yield ionic, conducting solutions. It has been found that such conductance data extrapolated to very high dilution yield the limiting conductance of sulfur dioxide. Both the Ostwald dilution law and the law of independent mobility of ions hold for "strong" electrolytes in highly dilute solutions.

The order of increasing dissociation and conductivity of salts in liquid sulfur dioxide apparently parallel the order of increasing cationic size. Probably because of solvation effects a similar relationship does not hold with respect to anion size. The mobilities of various ions in liquid sulfur dioxide have been studied. The van't Hoff i factors or mole numbers have been obtained by the ebullioscopic method for a wide variety of solutes in liquid sulfur dioxide. For non-electrolytes the mole number is one within experimental error. In liquid sulfur dioxide, univalent electrolytes give mole numbers which indicate large effects of ion-association of some kind. As would be expected the mole numbers of these electrolytes approach two in very dilute solutions. See Jander and Mesech, Z. physik. Chem., *A183*, 277 (1939). The mole number in general can be found from the ratio of the value of a colligative property of the solute is solution to the value of the same colligative property for a normal solute such as sugar, both solutes being at the same molal concentration.

The protonic concept of acid and bases is applicable to many of these high temperatures solvent systems such as the fused ammonium salts which possess the "onium" ion or solvated proton, and the fused anionic acids which are salts possessing a metallic ion and a hydrogen containing anion. One of the most useful of the anionic acids is KHF which is used to dissolve ore minerals containing silica, titania and other refractory oxides.

In many high temperature reactions there is an absence of hydrogen-containing ions. The Lewis electron pair concept of acid and bases can be used to advantage in such systems. In such systems strong anion bases such as the $O^=$ ion coming from basic compounds such as metallic oxides, hydroxides, carbonates or sulfates react with acidic oxides such as silica through the intermediate formation of polyanionic silicate complexes. The average ionic size of these complexes depend no doubt upon the temperature and the amount of added base.

Anion bases include the sulfide and fluoride ions coming from the corresponding alkali metal compounds. Likewise, metaphosphate and metaborate melts are acid in nature. Also proton-like character can be ascribed to any positive ion. The smaller the positive particle and the higher its charge, the greater is its polarizing tendency in bringing about deformation of negative ions, and the more reasonably can such an ion be looked upon as an acid analog.

When the potential energy of a diatomic molecule is plotted versus the distance separating the nuclei in the molecule, the potential-energy curve shows a minimum of zero in the energy at the distance separating the nuclei where the molecule is most stable, that is, where the nuclei are at the equilibrium internuclear separation. Energy is required to force them closer together or to pull them farther apart. The energy required to separate the nuclei to an infinite distance is D', the dissociation energy measured from the minimum of the potential energy curve. The spectroscopic dissociation energy D is smaller than D' by the zero point energy $\frac{1}{2}h\nu_0$. This results in the relationship,

$$D' = D + \tfrac{1}{2}h\nu_0 \qquad (21)$$

The spectroscopic dissociation energy D is the dissociation energy of an ideal gas molecule at absolute zero, where all the gas molecules are in the zero potential energy level, h is Planck's constant (6.62×10^{-27} erg second), and ν_0 is the frequency of vibration of the nuclei at the lowest vibrational level, which is above the point of zero potential energy at the equilibrium internuclear separation. Thus, for the hydrogen molecule, $D = 4.476$ electron volts, $\nu_0 = 1.3185 \times 10^{14}$ sec^{-1}, and since 1 electron volt = 23.06 kilocalories per mole we calculate D' using Eq. (21) as follows:

$$D' = (4.47 \text{ eV})(23.06 \text{ kcal mole}^{-1} \text{ eV}^{-1}) +$$

$$\frac{6.023 \times 10^{23} \text{mole}^{-1} \times 6.62 \times 10^{-27} \text{ erg sec} \times 1.3185 \times 10^{14} \text{ sec}^{-1}}{(2)(4.184 \times 10^{10} \text{ ergs kcal}^{-1}).}$$

$$= 109.5 \text{ kcal mole}^{-1}$$

E. S. Amis
University of Arkansas
Fayetteville, Arkansas

DISTILLATION.

This is one of the most important and widely used of the chemical unit operations, both in the laboratory and on a large industrial scale, for separating the components of a liquid mixture. Distillation provides a means for partially vaporizing the mixture and separately recovering the vapor and residue. Consequently, the method is dependent upon the vapor pressures of the components making up the mixture. The vapor pressure of a pure substance is a constant, but varying with temperature. In distillation, the lighter, more volatile components of the original mixture (*distilland*) concentrate in the vapor when heat is applied. Advantage is taken of the fact that the ratios of the component substances in the vapor and liquid phases, except for special situations, are different. The less volatile components concentrate in the liquid residue (*bottoms*). The vapors evolved by distillation are condensed and are termed the *distillate*.

Distillation should be contrasted with evaporation wherein the vapor (frequently water) is not usually condensed (except where it is desired to conserve water). The principal product desired in evaporation is the solid material which remains in the evaporator vessel. Thus distillation would be used to separate two or more miscible liquids, such as glycol and water, whereas evaporation would be used to separate solid sodium chloride from brine.

The effectiveness of separation by distillation is largely determined by differences in the boiling points of the starting components. Where these are widely separated, as in the case of water (100°C) and ethylene glycol (197.6°C), the separation is relatively easy and redistillation is not required. Closer-boiling mixtures, such as the isomers of xylene, are much more difficult to separate by distillation. As the boiling points of the components approach each other, effective separation by distillation becomes more difficult.

Boiling-point diagrams or what also are termed boiling and condensation curves, usually experimentally determined, are useful guides in the design of distillation equipment. Fig. 1 is the phase diagram of a binary system forming a liquid and a vapor phase at constant pressure. Curve I is the boiling curve, which gives the coexistence temperature as a function of liquid composition; and curve II is the condensation curve, which gives the coexistence temperature as a function of the composition of the vapor phase. If the temperature is increased, vaporization begins when the boiling curve is crossed. Inversely, condensation begins when the temperature is decreased below the condensation curve. The use of boiling-point diagrams in still design will be discussed later.

Major Types of Distillation

Distillation can be a batch or a continuous operation. Batch distillation is frequently used in the laboratory for determining the chemical composition of mixed liquids, such as hydrocarbons. In the majority of industrial processes, distillation is continuous. In a batch operation, the charge material is boiled and vapors are removed continuously, condensed, and collected until such point is reached where there is the desired average composition. The separation is not sharp. This type of operation is also termed *simple* or *differential distillation*. In another approach, the mixture may be heated until a definite fraction of the liquid batch is vaporized, during which time the liquid and vapor are kept in intimate contact, i.e., with no vapors being removed. At the prescribed temperature and after the liquid and vapor have had opportunity to reach full equilibrium, the vapor is suddenly withdrawn and condensed. This approach is known as *equilibrium* or *flash distillation*. The method finds wider application in connection with multicomponent systems than with simple binary systems.

In the majority of industrial distillation systems, some of the distillate will be continuously returned to the distillation column. The returned condensate is contacted countercurrently with the rising vapors, thus bringing about an enrichment of the vapor in the more volatile components than otherwise would be accomplished with a single distillation and most often obviates the need for one or more redistillations to obtain the degree of purity desired. This approach is known as *rectification* or *fractional distillation*. The material returned is termed *reflux*. In most rectifying columns, the raw feed to the column is introduced at about the mid-level of the tower or column. The portion of the column above the point of feed is called the *rectifying section*; the portion below, the *stripping section*. Where the feed may be introduced at the top of the column, the entire column then is usually referred to as a *stripping column*, with no reflux used.

Dephlegmation is a means for increasing the efficiency of fractional distillation by forcing the vapors from the still to bubble through shallow layers of condensate in a column or dephelgmator. Thereby, the amount of low-boiling component in the vapor is increased and a substantial portion of the higher-boiling components is retained in the condensate.

Steam distillation is a process whereby compounds that are sparingly soluble in water may be distilled by heating with water or by blowing steam through the mixture. Compounds of relatively high boiling point may be distilled at lower temperatures by this method and thus prevent degradation.

A representative fractional distillation column of which there are thousands in use in the process industries, notably in the petroleum and petrochemical industries, is shown in Fig. 2. The material balance of the column is:

$$F = W + D$$

$$Fx_F = Wx_F + Dx_D$$

where F = feed rate, weight-moles/unit of time
  W = bottom product, weight-moles/unit of time
  D = distillate, weight-moles/unit of time
  x_F = mole fraction of low boiler in feed
  x_D = mole fraction of low boiler in distillate
  x_W = mole fraction of low boiler in bottom product

In this balance, the assumption is made that the molar heat capacities and the latent heats of vaporization of all components are identical. It is also assumed that heat losses from the column and heats of mixing are negligible. With these assumptions, the upward vapor flow and the

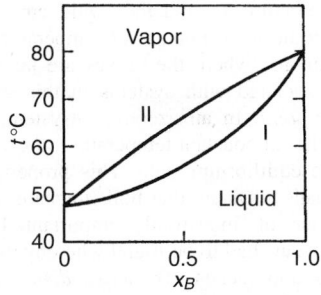

Fig. 1. Temperature-composition of a liquid-vapor system at constant pressure

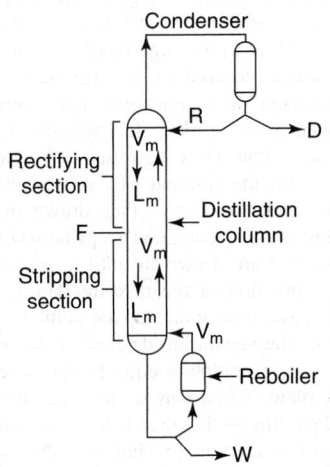

Fig. 2. Typical distillation column

downward liquid flow in both the rectifying and stripping sections will be invariant within the sections. It is also assumed that accounting for the column heat balance is independent of the compositions of the product streams. Within these qualifications, the internal material balance is:

$$L_n = (1 + b)R$$

$$V_n = D + (1 + b)R$$

$$L_m = L_n + qF$$

$$V_m = L_m - W$$

$$x_W = f\left(\frac{L_m}{V_m}\right)$$

$$x_D = g\left(\frac{L_n}{V_n}\right)$$

where V_n = vapor rate in rectifying section, weight-moles/unit of time
V_m = vapor rate in stripping section, weight-moles/unit of time
L_n = liquid rate in rectifying section, weight-moles/unit of time
D = distillate rate, weight-moles/unit of time
R = external reflux, weight-moles/unit of time
b = a numerical factor, depending upon the reflux enthalpy or temperature. (It should be noted that b is greater than zero whenever the reflux temperature is below that at the top of the column.)
q = a numerical factor, depending on the feed enthalpy whose value satisfies certain constraints
    $q < 0$, when feed temperature is below feed plate temperature.
    $q = 1$, when feed temperature and composition are identical with those of feed plate.
    $1 > q > 0$, when feed enters column partially vaporized.
    $q = 0$, when feed is fully vaporized and is at saturated temperature.
    $q < 0$, when feed is superheated vapor.

f and g are factors which account for several functional relationships which depend upon such column design criteria as the number of plates in column, location of control plates, location of feed plate, and temperature and other conditions specified for control plates.

The foregoing type of material balance is of large value in determining the best form of automatic control to apply to the column in order to maximize yields.

Distillation Calculations. In determining the type of packing or trays to be used in a fractionating column, the diameter, height, location of feed, location of reflux return, vapor and liquid rates, and all other specifications for a column to effect a given separation, equilibrium diagrams of the type shown in Fig. 3 are important. Prior to the availability of high-speed computers, distillation column designers depended heavily upon graphical solutions, notably McCabe-Thiele diagrams, named after the early developers of this concept. A typical diagram of this type for a simple binary distillation is shown in Fig. 4. Oversimplifying the method, first an equilibrium curve of the type of Fig. 3 is constructed. Next a 45° diagonal line is drawn. With knowledge of desired final composition, i.e., composition of the liquid received by the top plate from the condenser, X_p, calculate the intercept of an *operating line* with the Y-axis of the chart. This is indicated as #1 on Fig. 4. The term $X_p/O + 1$ is this Y-intercept of the operating line. O is the reflux ratio. Next, the intersection of the diagonal line with the ordinate X_p is marked. This is indicated as #2 on Fig. 4. The operating line is then drawn in by joining #1 and #2. Now, commencing at #2, rectangular steps between the operating line and the equilibrium curve are drawn in until it crosses the line $X = X_s$. X_s is the starting composition of the mixture. The number of horizontal steps counted (in this case, five) indicates the number of *theoretical plates* required to accomplish the separation desired. A theoretical plate may be defined as a plate wherein complete equilibrium is reached between the vapor rising from the plate and passing to the plate above—with the liquid leaving the plate and passing to the plate below. An actual plate, of course, will not perform with this efficiency. Thus, the designer, depending upon past experience with plates of certain designs, will include an appropriate margin in specifying the number of actual plates needed.

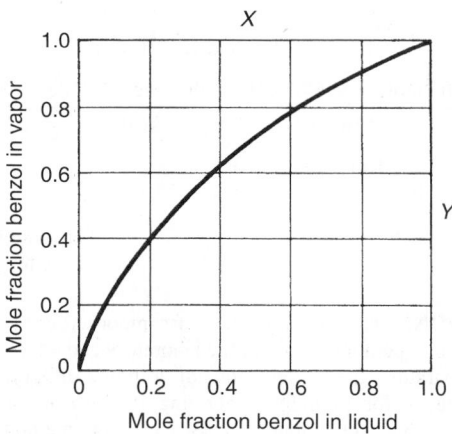

Fig. 3.   Equilibrium diagram for the benzol-toluol system

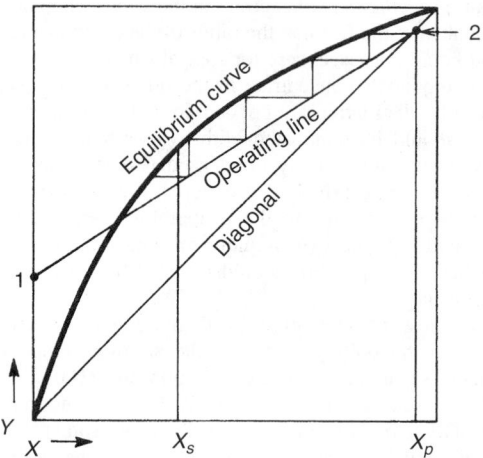

Fig. 4.   Representative solution for theoretical plates in rectifying part of a distillation column

Numerous computer software programs are available today to streamline design calculations. Of course, such software can be structured only after very careful analysis is made of the chemical dynamics of a given application. Calculations are particularly complex and difficult in the instances of azeotropic and reactive distillation. Software programs are described in some detail in the Kumana, Morris, and Venkataramkan references listed.

Azeotropic Systems. An azeotropic system is one wherein two or more components have a constant boiling point at a particular composition. Such mixtures cannot be separated by conventional distillation methods. If the constant boiling point is a minimum, the system is said to exhibit *negative azeotropy*; if it is a maximum, *positive azeotropy*. Consider a mixture of water and alcohol in the presence of the vapor. This system of two phases and two components is divariant. Now choose some fixed pressure and study the composition of the system at equilibrium as a function of temperature. The experimental results are shown schematically in Fig. 5.

The vapor curve *KLMNP* gives the composition of the vapor as a function of the temperature T, and the liquid curve *KRMSP* gives the composition of the liquid as a function of temperature. These two curves have a common point M, where the curves are tangent. Because of the special properties associated with systems in this state, the point M is called an *azeotropic point*. In an azeotropic system, one phase may be transformed to the other at constant temperature, pressure and composition without affecting the equilibrium state. This property justifies the name azeotropy, which means a system that boils unchanged.

Because a number of industrially important liquid mixtures are azeotropic systems, means had to be found whereby they may be separated by distillation. Two approaches, *extractive distillation* and *azeotropic distillation*, are used. In either case, a *separating agent* is added to the column so as to alter favorably the relative volatilities of the

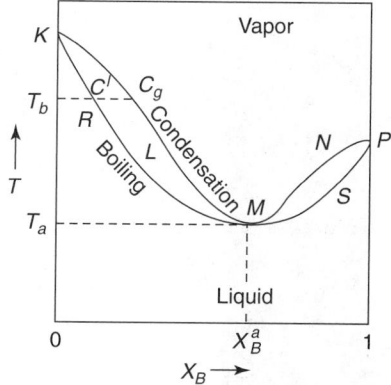

Fig. 5. Boiling-point diagram of an azeotropic system exhibiting negative azeotropy

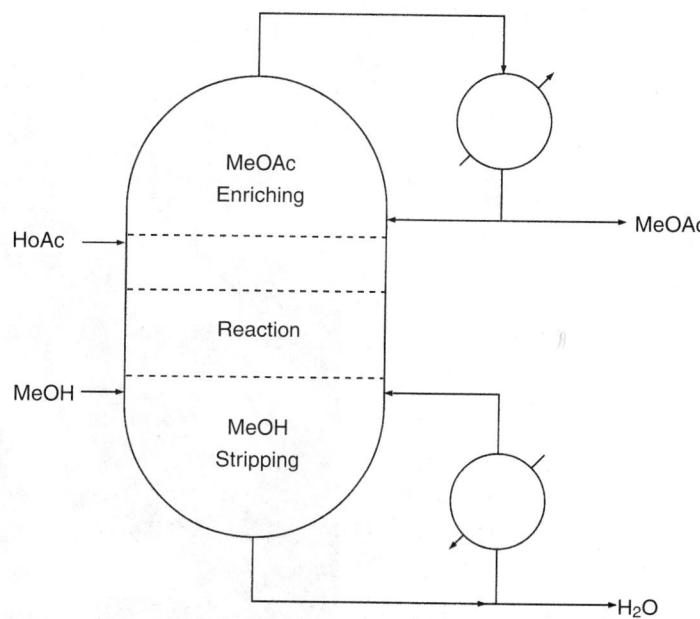

Fig. 6. General principle of reactive distillation (*After Agrada*)

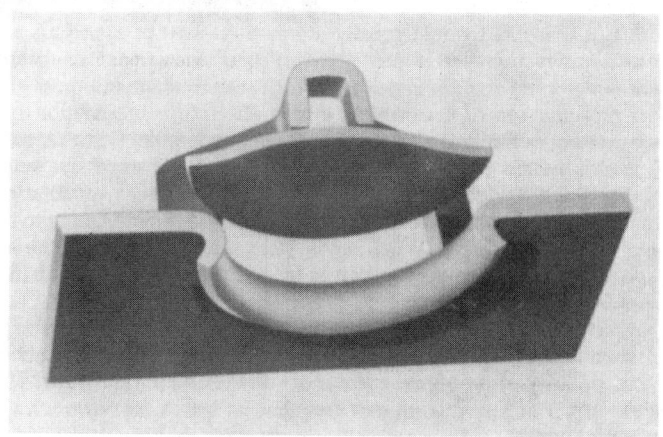

Fig. 7. Cross section of a form of bubble cap. View is from underneath plate or trap upon which cap is mounted

Fig. 8. Distillation tray incorporating several hundred shallow, tunnel-type caps

feed components. Usually, water or polar organic compounds are found most effective. They increase the liquid-phase non-ideality of one feed component more than another.

In extractive distillation, the agent (sometimes termed solvent) is significantly less volatile than the regular feed components. The agent will be added near the top of the column. The agent behaves as a *heavier-than-heavy* key component. It is also conveniently separated from the product streams. Because the agent usually must be added in fairly substantial amounts, this means that column diameters and heat loads are increased, while plate efficiencies are lowered.

In azeotropic distillation, an agent is selected that will form an azeotrope with one of the feed components. In essence, separation is accomplished between this "new" azeotrope (as an overhead product) and the other feed component as bottoms product. An agent will be selected preferably that will permit easy separation after distillation.

Agrada, et al. (see reference listed), describes the manufacture of high-purity methyl acetate via *reactive distillation*. This operation is difficult because of reaction equilibrium limitations and formation of methyl acetate-methanol and methyl acetate-water minimum-boiling azeotropes. The researchers point out, "Conventional processes use schemes with multiple reactors in which a large excess of one of the reactants is used to achieve the high conversion of the other reactant. Some use a series of vacuum and atmospheric distillation columns to change the composition of the methyl acetate-water azeotrope. The refined methyl acetate is separated from the unconverted reactants, and the methyl acetate-methanol azeotrope is then recycled to the reactors. Other schemes use several atmospheric distillation columns and a column with an extractive agent, such as ethylene glycol monomethyl ether, to act as an entrainer to separate the methyl acetate from methanol."

In the process described by Agrada, et al., the concept of a countercurrent reactive distillation column (Fig. 6) is used.

Trays and Packing. Trays with bubble caps or other suitable configurations for enhancing a maximum intermingling of rising vapors with falling liquid in a column are usually used where efficiency and close separations are major considerations. Packed columns, filled with ceramic shapes of various types, such as Berl saddles and Raschig rings, are used primarily where cost and acid-resistance are factors. The sectional view of a form of bubble cap is shown in Fig. 7. A tunnel cap is much more shallow and replete with peripheral perforations through which vapor and liquid flow. One column tray, incorporating several hundred tunnel caps, is shown in Fig. 8. In some very tight separations, several trays separated a few feet apart up the vertical length of the column may be required.

A petroleum crude atmospheric distillation column is illustrated in Fig. 9. This unit separates the crude into three major fractions which then are subjected to later separations: (1) a light straight-run fraction, consisting primarily of C_5 and C_6 hydrocarbons, but also containing any C_4 and lighter gaseous hydrocarbons dissolved in the crude; (2) a naphtha fraction having a nominal boiling range of 200–400°F (93–204°C); and (3) a light distillate with boiling range of 400–650°F (204–343°C).

Parastillation. This relatively new approach to distillation, as proposed by A.E.O. Jenkins in 1983, is a method for multistage, counter-current contact between vapor and liquid and is reputed to provide 33% more ideal stages than traditional factional distillation at the same tray spacing. The

differences between the old and the proposed methodologies are shown in Fig. 10. This is a schematic of the vapor and liquid flows in the distillation column. In parastillation, the vapor is divided into two parts by a partition running the full height of the column; liquid entering the top of the column flows from one vapor side to the other. As pointed out by Canfield, a secondary advantage of the parastillation process over traditional distillation is that the liquid on a given vapor side of the column always flows in the same direction. Analysis of other advantages of the Jenkins system is outlined in the Canfield reference.

Fig. 9.   View from base upward of atmospheric crude distillation tower

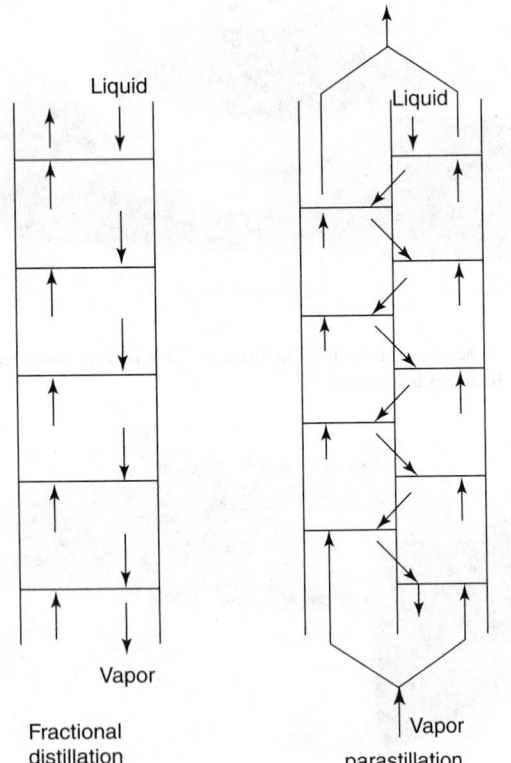

Fractional
distillation

parastillation

Fig. 10.   Comparison of parastillation with fractional distillation

The literature is rich in descriptions of numerous improvements that have been made in distillation equipment over the past few years. For those readers desiring more detail, refer to the articles listed.

Additional Reading

Agrada, V.H., L.R. Partin, and W.H. Heise: "High-Purity Methyl Acetate Via Reactive Distillation," *Chem. Eng. Progress*, **40** (February 1990).

Balchen, J.G.: *Dynamics and Control of Chemical Reactors, Distillation Columns, and Batch Processes: Selected Papers from the 3rd IFAC Symposium, Maryland, U.S.A.*, Elsevier Science, New York, NY, 1993.

Bowman, J.D.: "Use Column Scanning for Predictive Maintenance," *Chem. Eng. Progress*, **25** (February 1991).

Canfield, F.B.: "Computer Simulation of the Parastillation Process," *Chem. Eng. Progress*, **58** (February 1984).

Coker, A.K.: "Understand the Basics of Packed-Column Design," *Chem. Eng. Progress*, **93** (November 1991).

Doherty, M.F. and M.F. Malone: *Conceptual Design of Distillation System*, The McGraw-Hill Companies, Inc., New York, NY, 2001.

Fair, J.R. and J.L. Bravo: "Distillation Columns Containing Structured Packing," *Chem. Eng. Progress*, **19** (January 1990).

Harrison, M.E.: "Consider Three-Phase Distillation in Packed Columns," *Chem. Eng. Progress*, **80** (November 1990).

Kiarwe, H.Z. and D.R. Gill: "Predict Flood Point and Pressure Drop for Modern Random Packings," *Chem. Eng. Progress*, **32** (February 1991).

Kister, H.Z. and J.R. Haas: "Predict Entrainment Flooding on Sieves and Valve Trays," *Chem. Eng. Prog.*, **63** (September 1990).

Kister, H.Z.: *Distillation Design*, The McGraw-Hill Companies, Inc., New York, NY, 1991.

Kumana, J.D.: "Run Batch Distillation Processes with Spreadsheet Software," *Chem. Eng. Prog.*, **53** (December 1990).

Kurtz, D.P., K.J. McNulty, and R.D. Morgan: "Stretch the Capacity of High-Pressure Distillation Columns," *Chem. Eng. Progress*, **43** (February 1991).

Lide, D.R.: *CRC Handbook of Chemistry and Physics*, 84th Edition, CRC Press, LLC., Boca Raton, FL, 2003.

Luyben, W.L.: *Practical Distillation Control*, Chapman & Hall, New York, NY, 1992.

Meilli, A.: "Heat Pump for Distillation Columns," *Chem. Eng. Progress*, **60** (June 1990).

Morris, C.G., et al.: "Crude Tower Simulation on a Personal Computer," *Chem. Eng. Progress*, **63** (November 1988).

Perry, R.H. and D.W. Green: *Perry's Chemical Engineers' Handbook*, 7th Edition, The McGraw-Hill Companies, Inc., New York, NY, 1999.

Rawlings, J.B.: *Dynamics and Control of Chemical Reactors, Distillation Columns and Batch Processes (Dycord '95): A Postprint Volume From the 4th IFAC Symposium, Helsingor, Denmark, 1995*, Elsevier Science, New York, NY, 1995.

Stichlmair, J. and J.R. Fair: *Distillation: Principles and Practice*, John Wiley & Sons, Inc., New York, NY, 1998.

Venkataramkan, S., W.K. Chan, and J.F. Boston: "Reactive Distillation Using ASPEN PLUS," *Chem. Eng. Progress*, **45** (August 1990).

DIURETIC AGENTS.   Diuretics are agents that increase urine output or flow. The term is generally used to describe all drugs that act on the kidney to increase the production of urine. More specifically, the terms saliuretic or natriuretic are used to describe those agents that exert diuretic effects by primarily increasing the excretion of sodium chloride. Aquaretics are agents that increase urine output by producing a water diuresis but do not promote the urinary excretion of electrolytes. Both the use of diuretics in the treatment of hypertension, and the effects of diuretics on the kidney to promote the excretion of urine to normalize derangements in body fluid distribution leading to edematous states are discussed herein.

Disturbances in body fluid distribution may occur at three principal sites: *(1)* within the interstitial space, as occurs in peritonitis or cirrhosis with ascites; *(2)* between the interstitial space and the vascular tree, as in the nephrotic syndrome; or *(3)* within the vascular tree, as in congestive heart failure. Thus, the underlying disease may be of cardiac, hepatic, or renal origin. These derangements provide what amounts to a low blood volume signal to the kidneys that activates renal mechanisms to retain salt and water. If the retained salt and water do not terminate the low volume signal, the kidneys continuously retain salt and water, resulting in edema. The clinical outcome of excessive accumulation of salt and water depends on the particular sector of the extracellular space to which the retained fluid is relegated. Clinical signs of edema appear when the volume of the extracellular space is exceeded by several liters. Diuretics are used in the treatment of edematous states because, in most cases, they produce satisfactory mobilization and subsequent prevention of fluid accumulation in the interstitial space, the abdominal cavity, the lungs, and/or thoracic cavity. However, diuretic therapy is symptomatic in nature, and unless the underlying pathology is corrected, the kidneys continue to retain salt and water and the retained fluid and electrolytes are redistributed to the various compartments described. The principal indications of diuretics in the treatment of edema are in congestive heart failure; renal disease; hepatic cirrhosis with ascites; obesity, where salt and water retention are prominent; premenstrual tension; edema of pregnancy, including toxemia; and steroid administration. Edema may also be associated with other clinical conditions such as inflammation or hypersensitivity reactions.

Pharmacology and Mechanism of Action

Low Ceiling Diuretics. The designation of low ceiling diuretics denotes that the total excretion of the filtered sodium ion load is less than 10% compared to about 30% for the high ceiling diuretics. There are many chemical classes in this category, i.e., thiazides, quinazoline sulfonamides, chlorthalidone, indapamide, etc, but their site of action in the kidney is similar, and they are grouped as thiazide-type diuretics for general discussion.

The most popular diuretics in this class are hydrochlorothiazide and chlorthalidone; there are more potent low ceiling diuretics available. The long duration of action of chlorthalidone, 24 to 72 h, makes once a day dosing possible, and achieves good patient compliance. Cyclothiazide, polythiazide, and trichlormethiazide are about 15 to 30 times more potent than hydrochlorothiazide, and about 500 to 1000 times more potent than chlorothiazide, the first member of the thiazide family marketed.

The low ceiling diuretics increase urinary sodium excretion by acting directly on the $Na^+ - Cl^-$ transport mechanism in the convoluted distal tubules of the kidney.

Indapamide has been shown to possess diuretic and independent vasodilatory effects. It lowers the elevated blood pressure and reduces total peripheral resistance without an increase in heart rate. Indapamide antagonizes the vasoconstricting effects of the catecholamines and angiotensin II, a property not shared by other thiazide-type diuretics. Tripamide is also reported to have direct vasodilatory effects.

Weight loss is a good indicator of fluid loss and excretion. The first wave of fluid mobilized is from the periphery. The excretion of chloride and water is considered passive, and the excretion of potassium and magnesium is increased. In long-term use, the excretion of calcium is decreased.

In long-term treatment, the thiazides may produce hypokalemia, hyperglycemia, hyperuricemia, and a 5% increase in plasma cholesterol; indapamide has been shown not to increase plasma cholesterol or lipids at therapeutic doses. Thiazides can cause hyponatremia in patients with large water intake while on the drug; hyponatremia may be associated with nausea, vomiting, and headaches.

Paradoxically, the thiazides are efficacious, especially if combined with a prostaglandin synthetase inhibitor such as indomethacin or aspirin, in the treatment of nephrogenic diabetes insidipus, in which the patient's renal tubules fail to reabsorb water despite the excessive production of ADH. Thiazides can decrease the urine volume up to 50% in these patients.

High Ceiling (Loop) Diuretics. The principal action of the loop diuretics is inhibition of sodium and chloride reabsorption in the thick ascending limb of the loop of Henle. They can produce an excretion of 20 to 30% of the filtered sodium ion load. Most loop diuretics have a rapid onset of action, steep diuretic dose-response curves, and usually are short acting.

The most commonly used loop diuretics are furosemide, bumetanide, and ethacrynic acid. Newer agents available in some countries include torasemide and piretanide. The potency ratio for furosemide:ethacrynic acid:torasemide:piretanide:bumetanide is 1:1:2:4:40.

After long-term use of a loop diuretic, the extracellular fluid volume contracts and water reabsorption in the proximal and distal tubules increases, thus overriding and diminishing the diuretic's effects. A second high ceiling diuretic may sometimes induce diuresis again. This may be due to additional mechanisms. The loop diuretics increase urinary excretion of potassium and magnesium, as do the thiazides, but loop diuretics also increase urinary excretion of calcium; the excretion of chloride ion is greater than that of sodium ion, suggesting the inhibition of active chloride transport by the high ceiling diuretics in the loop of Henle. Alkalosis may develop in patients treated with the loop diuretics, and plasma renin activity (PRA) is markedly elevated. The high ceiling diuretics inhibit the ability of the kidney to concentrate urine even in the presence of high concentrations of vasopressin.

Ototoxicity, as evidenced by transient or permanent hearing loss, is a serious side effect of ethacrynic acid, and occurs less frequently with furosemide. Bumetanide is claimed to have only 20% of the ototoxic potential of furosemide. It has been reported that patients treated with torasemide at high doses for four weeks did not suffer hearing loss.

Potassium-Sparing Diuretics. Potassium-sparing diuretics act on the aldosterone-sensitive portion of cortical collecting tubules, and partially in the distal convoluted tubules of the nephron. The commonly used potassium-sparing diuretics are triamterene, amiloride, and spironolactone. Spironolactone is a competitive aldosterone receptor antagonist, whereas triamterene and amiloride are not.

Amiloride is far more soluble than triamterene, and is the most widely studied potassium-sparing diuretic. Its natriuretic effect is minimal because only 2 to 3% of the filtered sodium ion load reaches the collecting tubules of the nephron. Etozolin is a newer, long-lasting agent that has a gradual onset of action.

Spironolactone antagonizes the effects of aldosterone by binding at the aldosterone receptor in the cytosol of the late distal tubules and renal collecting ducts. Side effects of spironolactone are gynecomastia, decreased libido, and impotency.

Potassium-sparing by diuretic agents, particularly spironolactone, enhances the effectiveness of other diuretics because the secondary hyperaldosteronism is blocked. This class of diuretics decreases magnesium excretion. The most important and dangerous adverse effect of all potassium-sparing diuretics is hyperkalemia, which can be potentially fatal; the incidence is about 0.5%. Therefore, blood potassium concentrations should be monitored carefully.

Natriuretic Peptide Diuretics. Atrial natriuretic peptide (ANP), an endogenous diuretic, natriuretic, and vasodilator, is a peptide hormone primarily synthesized and stored by atrial cardiocytes, and secreted by the atria in response to mechanical stretch of the atria. ANP is also known as anaritide, $C_{112}H_{175}N_{39}O_{35}S_3$; atrial natriuretic factor (ANF); auriculin; cardionatrin; and atriopeptide. Its primary action is in the kidney and the vascular system.

It has been suggested that both the increased glomerular filtration rate (GFR) caused by ANP and the direct epithelial action in the collecting ducts by ANP are necessary to explain the diuretic effects of ANP. It appears that ANP may increase GFR by relaxing the glomerular mesangial cells resulting in increased surface area for filtration. Due to the augmented GFR, ANP increases the delivery of sodium and water to the renal tubules beyond the distal convoluted tubule. In the collecting ducts, ANP reduces sodium and free-water reabsorption by antagonizing the action of vasopressin. Therefore, the increased loads of sodium and water passing through the collecting ducts without the increased compensatory reabsorption result in profound diuresis and natriuresis.

Atrial Natriuretic Peptide Potentiator Diuretics. Neural endopeptidase inhibitors or atrial peptidase inhibitors are compounds that inhibit the enzyme that degrades ANP, resulting in higher plasma concentrations, and longer duration of action, of ANP. The diuretic effects of this class of compounds resemble those resulting from administration of ANP. Compounds such as thiorphan, candoxatril, SCH-34826, and SCH-39370 have been studied in hypertension and congestive heart failure in humans with only limited success.

Osmotic Diuretics. An effective osmotic diuretic is a nonionic compound, freely filterable at the glomerulus, not reabsorbed by the tubules of the nephron, and biologically inert except for its osmotic properties. One of the best examples of an osmotic diuretic is mannitol, used to prevent acute renal failure in many major surgeries and traumatic injuries. An osmotic diuretic increases urine flow, rather than the excretion of sodium, by maintaining a high osmotic gradient resulting from the presence of large amounts of nonreabsorbable solutes in the luminal side of the proximal tubules. Under such conditions, the reabsorption of water is impaired along the descending loop of Henle as well as the collecting ducts, and urine flow increases. At high concentrations of an osmotic diuretic, the urinary excretion of sodium is also increased. This is due to the reduced reabsorption of sodium in the proximal tubules.

Carbonic Anhydrase Inhibitor Diuretics. Carbonic anhydrase accelerates the hydration of carbon dioxide to carbonic acid in aqueous solution, up to 7500-fold as compared to the nonenzymatic reaction. The hydrogen ions liberated from carbonic acid in the epithelial cells of the proximal tubules of the nephron are exchanged for sodium ions in the renal tubular lumen. When the generation of hydrogen ions is inhibited by a carbonic anhydrase inhibitor, the exchange of hydrogen for sodium ions is greatly diminished and the diuretic effect ensues. The site of action of carbonic anhydrase inhibitors is in the proximal tubules. In addition to sodium, the excretion of bicarbonate and potassium is also increased. Owing to the increased urinary bicarbonate excretion, the urine becomes alkaline and the blood becomes acidotic. The diuretic effect ceases once metabolic acidosis occurs.

Acetazolamide, the best example of this class of diuretics, is rarely used as a diuretic since the introduction of the thiazides. Its main use is for the treatment of glaucoma and some minor uses, e.g., for the alkalinization of the urine to accelerate the renal excretion of some weak acidic drugs, and for the prevention of acute high altitude mountain sickness.

Methylxanthine Diuretics. The methylxanthines are of very limited efficacy when used as diuretics. The excretion of sodium and chloride ions are increased, but the potassium excretion is normal. Even though the methylxanthines have been demonstrated to have minor direct effects in the renal tubules, it is believed that they exert their diuretic effects through increased renal blood flow and GFR.

Organomercurial Diuretics. Before the advent of the thiazide diuretics, mercurial and organomercurial diuretics were the mainstay therapy for the treatment of edema. They have become obsolete and are of historical value only.

Water Diuretics (Aquaretics). A water diuretic, i.e., aquaretic, decreases urinary osmolality by influencing the kidney to excrete water selectively without a concomitant proportionally increased excretion of sodium ions. A water diuretic should be efficacious for the treatment of hyponatremia, i.e., low plasma sodium concentration, and the syndrome of inappropriate antidiuretic hormone secretion (SIADH). In many diseases and conditions, when water is retained to a greater extent as related to sodium ions, hyponatremia results. This is seen in many edema cases arising from congestive heart failure (CHF), hepatic cirrhosis, renal failure, and nephrotic syndrome. In the treatment of these conditions, the conventional diuretics will lose their effectiveness once hyponatremia occurs. Diseases of the brain and the lung, certain surgeries, and some tumors also will cause hyponatremia; in these conditions, with any given plasma osmolality, the plasma antidiuretic hormone (ADH) concentrations are inappropriately high. When this occurs, the patient is inferred to have SIADH.

There is no specific water diuretic marketed as of this writing. Demeclocycline has been used clinically with only limited success.

Economic Aspects

The sales of oral diuretics are declining, and are forecast to continue their decline in constant dollars during the 1990s. Several possible explanations can be offered for these trends. The patents of market leaders are expiring, leading to the introduction of generic brands at ca 40% below the cost of the branded market leaders; physicians are switching to newer treatments for hypertension, e.g., calcium channel blockers and angiotensin-converting enzyme inhibitors; and concerns are growing about the possible adverse effects of diuretics, e.g., hypokalemia, the progression of atherosclerosis, and the increase in mortality, serum cholesterol, glucose tolerance, and diabetes.

PETER CERVONI
PETER S. CHAN
American Cyanamid Company

Additional Reading

Andreucci, V.E. and A. Dal Canton (Editors): *Diuretics: Basic, Pharmacological, and Clinical Aspects,* Kluwer Academic Publishers, Norwell, MA, 1987.

Cragoe, E.J. Jr.: *Diuretics, Chemistry, Pharmacology, and Medicine,* John Wiley & Sons, Inc., New York, NY, 1983.

Dirks, J.H.: *Diuretics: Physiology, Pharmacology and Clinical Use,* W.B. Saunders Company, Philadelphia, PA, 1995.

Greger, R. (Editor), et al.: *Diuretics, Vol. 117,* Springer-Verlag New York, Inc., New York, NY, 1996.

Hook, J.B. and R.Z. Gussin: in M. Antonaccio, ed., *Cardiovascular Pharmacology,* 2nd Edition, Raven Press, New York, NY, 1984.

Larkin, N.D. and D.D. Fanestil: in J.B. West, ed., *Physiological Basis of Medical Practice,* 12th Edition, Williams and Wilkins Co., Baltimore, MD, 1991.

Puschett, J.B. and A. Greenberg: *Diuretics IV: Chemistry, Pharmacology, and Clinical Applications, Proceedings of the Fourth International Conference on Diuretics,* Elsevier Science, New York, NY, 1993.

Seldin, D.W. and G.H. Giebisch (Editors): *Diuretic Agents: Clinical Physiology and Pharmacology,* Morgan Kaufmann Publishers, Orlando, FL, 1997.

Taylor, S.H. (Editor): *New Advances in Diuretic Treatment,* S. Karger Publishers, Inc., Farmington, CT, 1994.

Weiner, I.M.: in A.G. Gilman and co-workers, eds., *The Pharmacological Basis of Therapeutics,* 8th Edition, Pergamon Press, Inc., Elmsford, NY, 1990.

DNA (Recombinant). See **RECOMBINANT DNA**.

DOLERITE. The term dolerite, derived from the Greek meaning deceitful, was originally applied to all dark, heavy, fine-grained igneous rocks of doubtful character. It is now used to indicate gabbroid or basaltic types occurring as dikes or sills whose mineralogical composition is plagioclase, feldspar, hornblende or pyroxene or both, olivine and perhaps biotite, magnetite or ilmenite and pyrite. Included in the dolerites are the diabases, which display plagioclase laths in a somewhat radial arrangement, and from this circumstance we have the textural term diabasic which is synonymous with ophitic.

DOLOMITE. The mineral dolomite, the carbonate of calcium and magnesium, corresponds to the formula $CaMg(CO_3)_2$ and closely resembles calcite. Its crystals, rhombohedral in habit, fall in the hexagonal system. Like calcite, it may be massive or granular, some marbles being dolomite rather than calcite. It displays a perfect cleavage parallel to the rhombohedron; subconchoidal fracture, brittle; hardness, 3.5–4; specific gravity, 2.85; luster vitreous to pearly; color varies widely, white, reds, greens, black, browns, yellows or colorless; transparent to translucent. Unlike calcite, dolomite dissolves very slowly if at all in dilute cold hydrochloric acid; powdered dolomite will dissolve in warm acid. This is the common test for the two minerals.

Much dolomite occurs as stratified rocks where it is believed to have been formed by a secondary process, probably by the action of waters charged with magnesium compounds. Dolomite also is found as a vein mineral, as is calcite. Iron or manganese, rarely zinc or cobalt, may replace some of the magnesium. Ankerite is the name given to a mineral whose composition is essentially a calcium-magnesium-iron carbonate. Among the many noted localities for dolomite are Saxony, Switzerland, Italy, France, Spain, Brazil, Mexico; in the United States, Roxbury, Vermont; Lockport, New York, Phoenixville, Pennsylvania; Alexander County, North Carolina; Hancock County, Illinois, and the Joplin District, Missouri. Dolomite was named for Deodat deDolomieu, who first described its characteristics. See also **Lime and Limestone**.

DOMAIN STRUCTURE. The theory of the macroscopic behavior of ferromagnetic and ferroelectric crystals depends on their consisting of large numbers of domains, each polarized to saturation but pointing in different directions so as to minimize the energy. An applied field tends to make those domains grow which are already favorably oriented, at the expense of those opposed to the field. Due to anisotropy energy, domains tend to be oriented along certain directions of easy magnetization, but the detailed arrangement of domains in a crystal is a complicated compromise between the tendency of each domain to be as large as possible and the necessity of creating closed loops of magnetic flux.

DOUBLE SALT. A hydrated compound resulting from crystallization of a mixture of ions in aqueous solution. Common examples are the alums, made by crystallizing from solution either potassium or ammonium sulfate and aluminum sulfate; Rochelle salt (potassium sodium tartrate), made from a water solution of potassium acid tartrate treated with sodium carbonate; and Mohr's salt (ferrous ammonium sulfate), crystallized from mixed solutions of ferrous sulfate and ammonium sulfate.

DOUBLET. 1. Two elements that are shared by two atoms so as to form a nonpolar valence bond.

2. A pair of spectral lines resulting from transitions between a common state and two states which differ only in total angular momentum (J), i.e., have identical values of orbital (L) and spin (S) angular momenta.

3. Two stationary states having common values of (L) and (S), but different values of (J).

DRYING OILS. Drying oils oxidize upon exposure to air from a liquid film to a solid, dry film. Consumption of drying oils in the United States peaked in the late 1940s or early 1950s; in the early 1990s much smaller, but still significant, amounts are used. The use of synthetic drying oil-based resins has exceeded the use of natural drying oils; however, their consumption is declining owing to discoloration and embrittlement caused by continued oxidation after film formation.

Natural Oils

Occurrence and Isolation. Most drying oils are derived from plant seeds. The largest volume drying oil, linseed oil, is obtained from flaxseed, *Linum usitatissimum*. In the United States flax is grown in North Dakota, Minnesota, and South Dakota. Flax for oil is also raised in Canada, Argentina, India, and parts of the former USSR. Soybean oil, the second most important oil, is obtained from soybeans, the seed of *Glycine max* (L) Merrill. It is produced in the United States, Brazil, Argentina, and China. Without modification, it is semidrying oil, not a drying oil. Perilla, safflower, sunflower, and walnut oils have limited uses as drying oils. Tung oil, also called wood oil or chinawood oil, is obtained from the seed kernels of the tung tree, *Aleurites fordii*. The principal source of the oil is China. Limited quantities of another conjugated oil, oiticica oil from the oiticica tree, *Licania rigida*, also are used. The only animal oils used as drying oils on a significant scale are marine fish oils, primarily from anchovy, menhaden, pilchard, and sardines. Fish oil is isolated by steam treatment of the fish.

Castor oil, derived from the beans of *Ricinus communis*, is converted to a drying oil by heating with catalysts to yield dehydrated castor oil.

Trees, especially conifers, contain tall oils. Tall oil is not isolated directly; tall oil fatty acids are isolated from the soaps generated as by-products of the sulfate pulping process for making paper.

A wide variety of other oils have been investigated and, in many cases, used commercially over the years. More complete listings are available.

Composition and Analysis. Naturally occurring drying oils are triglycerides. The reactivity of the oils results from the presence of esters of fatty acids with two or more nonconjugated double bonds separated by single methylene groups, $-CH=CHCH_2CH=CH$ hose with two or more conjugated double bonds. Typical compositions of some of the more important oils are listed in Table 1.

Autoxidation. Oils are classified as drying oils, which form solid films on exposure to air; semidrying oils, which form tacky, sticky films; and nondrying oils, which do not undergo marked increase in viscosity on exposure to air. Drying oils are further classified as nonconjugated and conjugated oils, depending on whether the double bonds in the predominant fatty acids are separated by one methylene group or are conjugated.

A general statement, useful in considering synthetic drying oils as well as natural oils, is that if the average number of methylene groups between two double bonds per molecule is greater than 2.2, the oil is a drying oil; if less than 2.2, the oil is a drying oil. There is no sharp dividing line between semidrying and nondrying oils.

The reactivity of nonconjugated drying oils is related to the average number of methylene groups between double bonds per molecule. Such

TABLE 1. TYPICAL FATTY ACID COMPOSITION OF DRYING OILS FROM SEEDS,[a]%

Oil	Saturated[b]	Oleic	Linoleic	Linolenic
linseed	10	22	16	52
perilla	7	14	16	63
safflower	10	13	77	
soybean	16	24	51	9
sunflower[c]	14	14	72	
sunflower[c]	9	72	19	
tung[d]	6	4	8	
walnut	8	16	72	

[a] Proportions shown are approximate; actual compositions can vary greatly.
[b] Palmitic and stearic acids.
[c] Examples of the especially large variations in composition of available sunflower oils.
[d] Also 82% α-eleostearic acid.

methylene groups are allylic to two double bonds, and show much greater reactivity than methylene groups allylic to only one double bond.

The reactions taking place during drying are complex, with many side reactions. Films form from a drying oil, such as linseed oil, in the following steps: an induction period during which naturally present antioxidants, mainly tocopherols, are consumed; a period of rapid oxygen uptake with a weight gain of about 10% (ftir shows an increase in hydroperoxides and appearance of conjugated dienes during this stage); a complex sequence of autocatalytic reactions in which hydroperoxides are consumed and the cross-linked film is formed; and cleavage reactions to form low mol wt by-products.

The rates at which nonconjugated drying oils dry are slow. Metal salts (driers) are known to catalyze the drying rate. The most widely used are the oil-soluble cobalt, manganese, lead, zirconium, and calcium salts of 2-ethylhexanoic acid or naphthenic acids. See also **Paints and Coatings**.

Combinations of metal salts are almost always used. Although mixtures of lead with cobalt and/or manganese are particularly effective, toxicity regulations ban the use of lead driers in consumer paints sold in interstate commerce in the United States. Combinations of cobalt and/or manganese with zirconium, and frequently also with calcium, are commonly used. The amounts of driers needed are very system specific. Their use should be kept to the minimum possible level since they not only catalyze drying but also catalyze the post-drying embrittlement and discoloration reactions.

Oils containing conjugated double bonds, such as tung oil, dry more rapidly than nonconjugated drying oils. Free-radical polymerization of the conjugated diene systems can lead to chain-growth polymerization rather than just combination of free radicals to form cross-links. In general, the water and alkali resistance of films formed using conjugated oils are superior, presumably because more of the cross-links are stable carbon-to-carbon bonds. However, because α-eleostearic acid in tung oil has three double bonds, discoloration on baking or aging is severe.

Both nonconjugated and conjugated drying oils can be polymerized by heating under an inert atmosphere to form so-called bodied oils. Bodied oils have higher viscosities and are often used in oil paints to improve application and performance characteristics.

Synthetic and Modified Drying Oils

Varnishes. The drying rate of drying oils can be increased by dissolving a solid resin in the oil and diluting with a hydrocarbon solvent. Such a solution is called a varnish. The solid resin serves to increase the glass-transition temperature, T_g, of the solvent-free film so that film hardness is achieved more rapidly.

In varnish manufacture, the drying oil, i.e., linseed oil, tung oil, or mixtures of the two, and the resin are cooked together to high temperatures to yield a homogeneous solution for the proper viscosity. The varnish is then thinned with hydrocarbon solvents to application viscosity. Varnishes were widely used in the nineteenth and early twentieth centuries. They have been replace almost completely by a wide variety of other products, especially alkyds, epoxy esters, and urethane oils.

Synthetic Conjugated Oils. Tung oil dries rapidly, but is expensive, and its films discolor rapidly owing to the presence of three double bonds. These defects led to efforts to synthesize conjugated oils, especially those containing esters of fatty acids with two conjugated double bonds.

Esters of Higher Functionality Polyols. When oil-derived fatty acids react with polyols having more than three hydroxyl groups per molecule, the average number of cross-linking sites per molecule increases proportionally to the functionality of the polyol. Because the number of reactive sites in soybean oil with the composition listed in Table 1 is 2.07, soybean oil is a semidrying oil. However, the pentaerythritol (2,2-bis(hydroxymethyl)-1,3-propanediol) ester of soybean fatty acids with 2.76 reactive sites per molecule is a drying oil.

Oxidizing alkyds can be considered as still-higher functionality drying oils. Their drying speed is faster owing to both the higher functionality and the higher T_g of the rigid aromatic rings in the phthalate esters.

Oils Modified with Maleic Anhydride (Maleated Oils). Oils, with either conjugated or nonconjugated double bonds, react with maleic anhydride (2,5-furandione) to form adducts. The products of these reactions with maleic anhydride, termed maleated oils, react with polyols to give moderate mol wt derivatives that dry faster than the unmodified oils. Such products have not found significant commercial use, but similar reactions with alkyds and epoxy esters are used on a large scale to make water-dilutable derivatives.

Vinyl-Modified Oils. Both conjugated and nonconjugated drying oils react in the presence of free-radical initiators with such vinyl monomers as styrene, vinyltoluene, acrylic esters, and cyclopentadiene. High degrees of chain transfer cause the formation of wide varieties of products, including low mol wt homopolymers of the vinyl monomer, short-chain graft copolymers, and dimerized drying oil molecules. The reaction products with drying oils, except cyclopentadiene, are not commercially important, but the same principle is widely used in making modified alkyds.

Uses

Although some drying oils continue to be used as drying oils, the largest use of drying and semidrying oils in the early 1990s is as raw materials, either directly or as the fatty acids obtained by saponification, in the manufacture of oxidizing alkyds, epoxy esters, urethane oils, and synthetic drying oils. Tall oil fatty acids also are used to manufacture dimer acids.

Since drying oils are considered a renewable resource, they may again become important in paints and printing inks, depending on the cost of drying oils compared with petroleum-derived raw materials. However, in many cases, the properties that can be obtained with synthetic binders, especially retention of flexibility, gloss, and color, are superior to those that can be obtained with a drying oil or drying oil-derived binder.

Paints. Although most drying oils have been replaced as paint vehicles by latexes and other synthetic resins, oils are still being used to a degree in paint and allied products. In exterior house paints, linseed oil or oxidizing alkyds are used when paint must be applied at temperatures as low as 4 to 5°C, i.e., temperatures at which latexes do not coalesce satisfactorily. They also are used in primers over chalky surfaces where latex paints do not provide adequate adhesion.

Most stains used for finishing shingles and other natural wood exterior products are made with pigmented linseed oil, diluted with hydrocarbon solvents for penetration.

Red lead-in-oil primers have been used for many years for corrosion protection of steel where it is not practical to remove oily rust particles from the surface of the steel; concern about lead content has resulted in use reduction.

Printing Inks. The use of drying oils in printing inks has decreased. Some inks based on drying oils are still used for sheet-fed letterpress printing. The principal remaining use is in lithographic printing, particularly sheet-fed lithographic printing.

ZENO W. WICKS, JR.
Consultant

Additional Reading

Formo, M.W.: in D. Swern, ed. *Bailey's Industrial Oil and Fat Products,* John Wiley & Sons, Inc., New York, NY, Vol. I, 1979, pp. 177–232 and 687–816; Vol. II, 1982, pp. 343–406.

Rheineck, A.E. and R.O. Austin: in R.R. Myers and J.S. Long, eds., *Treatise on Coatings,* Vol. I, No. 2, Marcel Dekker, New York, 1968, pp. 181–248.

Wicks, Z.W. Jr., F.N. Jones, and S.P. Pappas: in *Organic Coatings. Science and Technology,* Wiley-Interscience, New York, NY, Vol. I, 1993, pp. 133–143.

DRYING (Process). Frequently in the process industries, materials must be dried, i.e., liquid must be removed from a solid or gaseous phase. In most instances, the liquid to be removed is water, although in solvent recovery systems, for example, the liquid may be an organic solvent. See **Dehumidification** for a discussion of the drying of gases. Some materials which must be dried include: (1) solutions, colloidal suspensions, and emulsions, such as extracts, milk, blood, waste liquors, rubber latex, and inorganic salt solutions; (2) slurries, which are pumpable suspensions, as found in calcium carbonate, bentonite, clay slip, and lead concentrates; (3) sludges and pastes, such as centrifuged solids, starch, filter-press cakes, and sedimentation sludges; (4) powders that may be relatively free-flowing when wet, but very dusty when dry, including pigments, cement, clay, and centrifuged precipitates; (5) fibrous solids and granular and crystalline solids, such as sand, ores, rayons staple, salt crystals, and synthetic rubber; (6) formed and shapes solids, such as pottery, rayon cakes, shotgun shells, brick, rayon skeins, lumber, and objects that have been painted or otherwise coated; (7) sheeted materials, such as impregnated fabrics, paper, plastic, and fiber-board—in the continuous form; or veneers, wallboards, foam-rubber sheets, and photographic prints—in the individual-piece configuration. Because of this very wide variety of drying requirements, it is obvious that there does not exist what might be termed a universal dryer. Further, universal dryer design criteria are difficult to develop and summarize. See Table 1.

Basic Concepts of Drying

Two criteria hold for a large number of drying situations, and they break the drying process down into two main periods: (1) the constant-rate period, the rate of removal of liquid per unit of drying surface is essentially steady, but to qualify the surface of the material must remain fully wet (saturated) during this period; and (2) the falling-off period, during which time the rate of drying decreases as it becomes increasingly difficult to move moisture from the capillaries and interstices of the material to the surface where the moisture can be taken up and moved away. With temperature and air flow (if air is the absorbing and moisture conveying medium to be used) well established, the constant-rate period is relatively easy to forecast and to design because this situation is somewhat analogous to that of drying a shallow container of water. Knowing the foregoing conditions as well as the wetted area involved, drying time periods can be predicted. But, in the case of the falling-off period, the dryer designer must be intimately familiar with the mechanism whereby moisture is held to the material and the mechanics involved in moving the moisture to the surface where it can be picked up.

Thus, the falling-off period has been broken down into two phases: (1) the unsaturated-surface drying period; and (2) the internal-moisture movement period. This first period does not impose unusually difficult analysis because essentially the condition is analogous to a partially-filled shallow vessel whose surface is decreasing as drying proceeds. Essentially, this is an extension of the analysis of the constant-rate period, still leaving the internal-movement situation to be predicted. Formulas are available to assist the designer, based upon past experience with specific drying situations and materials. This is a process that is difficult to assess theoretically, and almost always requires experimental runs or comparisons with other at least somewhat similar materials that have been dried successfully. One can generally state that where materials are to be dried to a low-moisture content, the internal-moisture movement period will require the largest portion of the total drying time.

Classification of Dryers

In reviewing the wide variety of drying equipment available, there are a few major classifications that are helpful. There is the distinction between continuous and batch operations. If a continuous drying operation is desired because it will fit into the overall manufacturing operations best, then this decision will rule out those drying concepts that can be applied only in a batch manner. In some instances, it may turn out that in an otherwise fully continuous manufacturing process, the drying portion will have to be handled by batches simply because, for the particular product, batch drying offers the greatest efficiency. There is also the distinction between direct heating and indirect heating methods used in dryers. In direct heating, the heat needed is applied by way of immediate contact between the wet material and hot gases. In indirect heating, the heat needed is transferred to the wet material through an intervening medium, commonly pipes or a retaining wall. There are numerous instances, for example, where a product may be too sensitive to withstand exposure to a moving hot gas.

Since a large majority of drying equipment involves the use of hot gases (usually air), other means of heating can be overlooked. Dielectric heating and freeze-drying, for example, are other means where practical.

TABLE 1. MAJOR TYPES OF PROCESS DRYERS

Direct heating	Indirect heating

CONTINUOUS DRYERS

Tunnel Dryers. Material to be dried is placed on trucks or carts which are moved through a tunnel in which there is a flow of hot gases. Temperature of the tunnel may be zone controlled.

Through-circulation Dryers. Material is supported on a conveying screen that moves continuously. Hot gases from below or above conveyor pass through the material and pick up moisture.

Rotary Dryers. Material (liquid) is pumped to and showered within a rotating cylinder through which hot gases flow.

Tray Dryers. Material is placed on vibrating trays over or under which hot gases flow.

Sheeting Dryers. Material in sheet form passes continuously through a hot chamber. Depending upon product, sheet may be taut (as pinned to a frame). or it may pass through dryer in festoon manner.

Pneumatic Conveyor Dryers. Material is moved in a stream of gas at high-velocity and high temperature and finally collected by a cyclone separator.

Drum Dryers. For materials in liquid and slurry form. One or several drums dip into the liquid or slurry and thus coat the heated drums. The drum temperature is controlled to effect drying during a part of the rotation of the drum from which the dried material is removed by knife prior to dipping one again into the liquid material.

Cylinder Dryers. Material in the form of a continuous sheet passes over and around cylinders which rotate and which are heated, usually by steam or hot water.

Screw-conveyor Dryers. A conveyor is housed within a closed, heated housing. This operation may proceed at atmospheric pressure or under vacuum.

Vibrating-tray Dryers. Similar to the directly heated tray dryer except that the heat is conducted to the trays indirectly (as by electrical heating) rather than by hot gases.

Steam-tube Rotary Dryers. Material is passed through a long rotating cylinder. A shell around the cylinder contains steam, hot water, or other heating medium.

BATCH DRYERS

Through-circulation Dryers. Material is placed on trays with screen bottoms. Hot gases are blown from below through material.

Vacuum Rotary Dryers. Material is subjected to agitation within a stationary, horizontal shell under vacuum. The agitation may be heated to increase drying effectiveness.

Tray and Compartment Dryers. Material is placed on trays which then may be placed on trucks or on permanent shelves within are blown dryer. Hot gases across the trays.

Agitated-pan Dryers. Material is placed in covered shallow pans. Pans are jacketed for heating. An agitator stirs the material constantly. This design may be operated at atmospheric pressure or under vacuum.

Vacuum Tray Dryers. Trays in which material is placed are heated by conduction from supporting shelves. The whole compartment may be under a relatively high vacuum. The material is not agitated.

Drying equipment also can be made available with means to accelerate the drying process if these means are acceptable to product quality. For example, agitation, stirring, and otherwise keeping the material to be dried in constant motion naturally assists the amount of exposure of surfaces to the drying medium. Numbers of materials, however, cannot be handled in this fashion and require relatively conservative, still conditions.

Principal design configurations of drum-type dryers are illustrated and described in Figs. 1, 2, 3, and 4; conveyor-type dryers in Figs. 5, 6, and 7.

Dehydration

This operation is sometimes described as the removal of 95% or more of the water from a substance, by exposure to thermal energy by various means. Frequently used in food processing, the aims of dehydration are reduction in volume of the product, increase in shelf life, and lower transportation costs, among other factors. There is no clearly defined line of demarcation

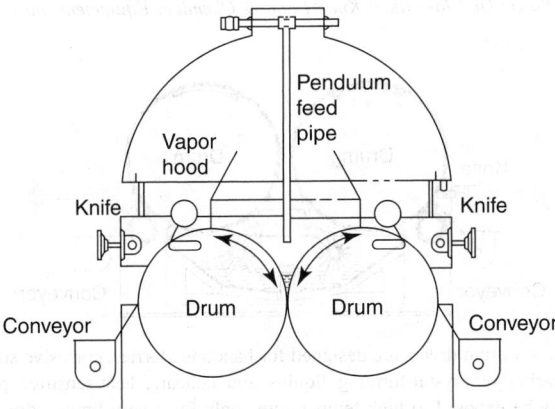

Fig. 2. Double-drum dryer (atmospheric). Dryers of this type handle a variety of food products of widely varying densities and viscosities—dilute solutions, heavy liquids, or pasty materials. A number of products can be dried successfully with this kind of configuration, inasmuch as exposure to temperature above the boiling point is restricted to just a few seconds. The movable drum permits effective control over product film thickness. Feed may be by perforated tube trough, pendulum, or various special configurations. (*Buflovak Division, Blaw-Knox Food & Chemical Equipment, Inc*)

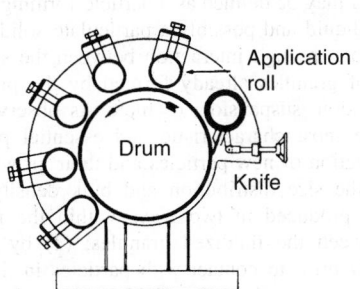

Fig. 1. Single-drum dryer (atmospheric). Dryers of this type may be dip or splash fed (not shown), or, as shown, equipped with applicator rolls. The latter is particularly effective for drying high-viscosity liquids or pasty materials, such as mashed potatoes, applesauce, fruit-starch mixtures, gelatin, dextrine-type adhesives, and various star ches. The applicator rolls eliminate void areas, permit drying between successive layers of fresh material and form the product sheet gradually. While single applications may dry to a lacy sheet or flakes, the multiple layers generally result in a product of uniform thickness and density with minimum dusting tendencies. (*Buflovak Division, Blaw-Knox Food & Chemical Equipment, Inc*)

between drying and dehydrating, the latter sometimes being considered as a subelement of drying. Usually, the direct use of solar energy, as in the drying of raisins, hay, etc., is not lumped in with dehydrating. The term dehydration also is not generally applied to situations where there is loss of water as the result of evaporation. Rehydration or reconstitution of a dehydrated product is the act of restoring a product to essentially its original condition by the simple addition of water, usually just prior to use (as consumption of a food product). The distinction between the terms drying and dehydrating may be somewhat clarified by the fact that most substances can be dried beyond their capability of restoration.

In the case of drying foodstuffs, this is a complex combination of coupled heat and mass transfer through natural tissues. The need to take

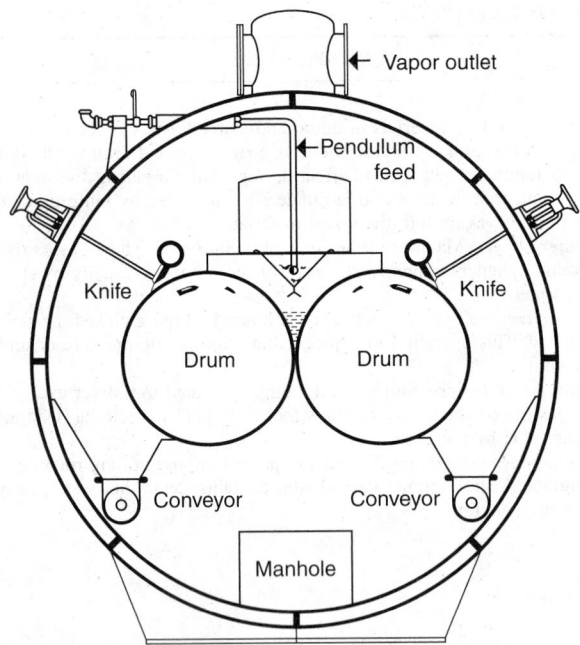

Fig. 3. Single- and double-drum dryer (vacuum). Design configurations like this are applicable whenever products must be dried without exposure to high temperatures or reactive atmospheres. Vacuum operation is particularly suited for vitamin extracts, protein hydrolyzates, soluble coffee, and malt products. Continuous drying without breaking vacuum is achieved by a special conveyance system that utilizes two receivers and air locks. Single-drum feed is usually the pan type with pump and spreading device, or a spray film for materials that are repelled by contact with heated surfaces. Double-drum utilize pendulum or perforated tube feed. Specially designed dispersion devices for feed also can be used. (*Buflovak Division, Blaw-Knox Food & Chemical Equipment, Inc*)

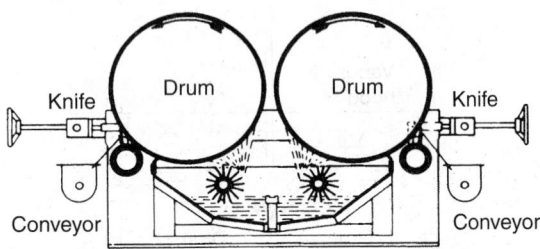

Fig. 4. Twin-drum dryers are designed for handling slurries, corrosive solutions, crystal-bearing or crystal-forming liquids and delicate, heat-sensitive products which may be exposed to high temperatures only for a very limited time. These drums may utilize pendulum or perforated tube for top feed, or splash or spray for bottom feed. Heat-sensitive materials are not in direct contact with the drums. The temperature remains fairly constant and preconcentration is minimized. Cooling and agitation of the material in the pan may also be provided when necessary. (*Buflovak Division, Blaw-Knox Food & Chemical Equipment, Inc*)

into account the structure of the food, as opposed to considering it as an homogenous solid, is important. In general, the main source of water in a tissue is the cell. Thus, the transport of water to the outside involves migration through the cell and its enveloping structure, through the porous structure of the tissue, and then through the outside boundary layer. To

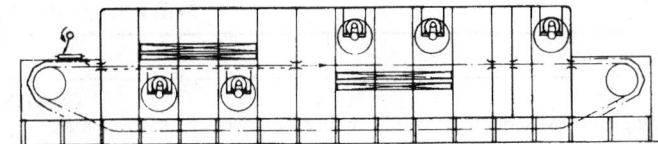

Fig. 5. Single-stage, single-pass conveyor dryer of convection type commonly used to dry, cool, toast, roast, bake, and heat materials. Such units can be zoned for various product drying temperatures, cooling, humidifying, and conditioning requirements. The air flow through the product can be upward or downward and the heat source can be steam, gas, oil, electric, or waste heat. Dryers with this design configuration are applicable to the processing of breakfast cereals, pet foods, fresh vegetables, fruits, and nuts, among numerous other food substances. (*The National Drying Machinery Company*)

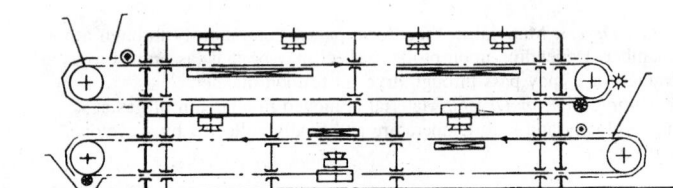

Fig. 6. A multi-tier, multi-pass conveyor dryer. Due to space limitations it is sometimes necessary to tier dryers and use gravity or transfer conveyors to carry the product from one tier to the other. The location of the discharge end will vary with the number of tiers. (*The National Drying Machinery Company*)

be able to predict drying behavior, it is necessary to establish the extent to which the foregoing transport steps are controlling. There is no general answer because both the cellular structure and the characteristics of the porous-like tissue structure are involved and the roles they play in the transport process may differ widely—simply because tissue properties change from one foodstuff to the next.

The energy requirements for drying have been studied rather intensively in recent years because the cost of a product can be markedly affected by this energy intense operation. Some researchers have observed that drying costs are much lower for air drying and drum drying than for freeze-drying.

A simplified flow diagram for the preparation of dehydrated fruits and vegetables is given in Fig. 8.

Combination Drying Operations

For some products, it is sometimes advantageous to combine drying with other associated operations. For example, in a fluidized-bed spray granulation process (FBSG), drying, cooling, granulating, and coating of final products, such as various organic and inorganic salts, minerals, dyestuffs, herbicides, pharmaceuticals, and pulping waste liquors, among others, can be accomplished with a reduction of materials-handling and energy costs. FBSG may be defined as a particle-forming process by which a solid containing liquid and possibly a particulate solid is converted into a granular solid state through interaction between the sprayed liquid and a fluidized layer of granules already formed by the process. The liquid feed can be a solution, suspension, or melt. As observed by Mortensen and Hovmand, the most characteristic and essential part of the FBSG process is the formation of new particles and their growth in the fluidized bed, determining the size distribution and bulk density of the product. New particles are produced in two ways within the fluidized bed: (1) from attrition between the fluidized granules; (2) by spray drying the droplets sufficiently prior to contact with particles in the fluidized layer. The process involves a number of variables, most of which are subject

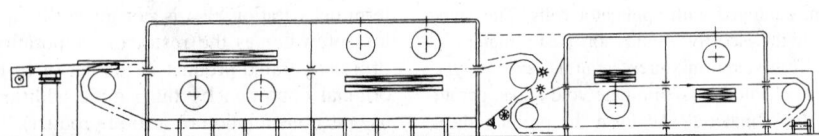

Fig. 7. A multi-stage, 4-zone drying range consisting of a series of single-stage conveyor dryers designed so that the individual stages present a new set of conditions to the product being processed. These stages, in turn, can be zoned to give maximum processing flexibility. Transfer devices between stages reorient the product to provide effective processing uniformity. (*The National Drying Machinery Company*)

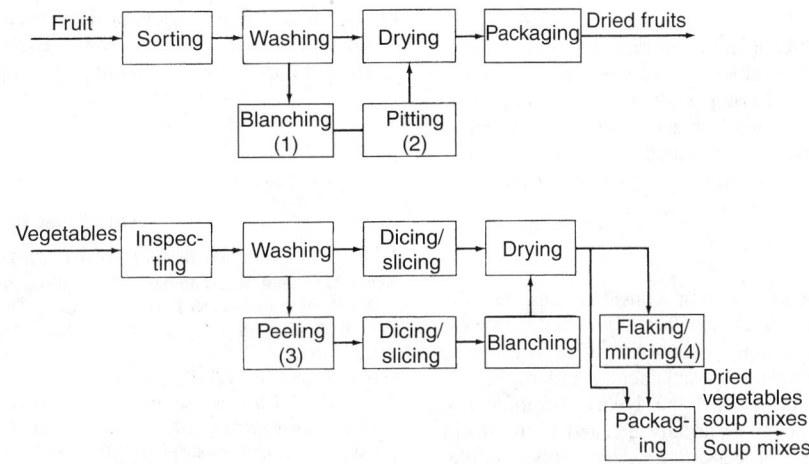

Fig. 8. Simplified flowsheets of operations required in preparation of (**a**) dried fruit and (**b**) dried vegetable products. Operations (1) and (2) are required only for certain fruits, such as prunes. Operation (3) is required only for certain vegetables, such as potatoes. Operation (4) is required only for certain end products

to control: (1) feed concentration, (2) ambient air temperature, (3) dryer inlet gas temperature, (4) dryer outlet gas temperature, (5) heating surface temperature, and (6) fluidizing gas velocity.

Spiral Drying Technique

Also sometimes referred to as the DRT (*Drallrohr Trocknung*) technique, this is a relatively recent, innovative process targeted at reducing energy costs. As pointed out by Hess and Rossi, the dryer is designed for continuous moisture removal from wetwater products via fresh ambient air or of solvents via recirculated inert gas. The product's residence time is a matter of seconds, thus precluding thermal degradation of the product. A jacketed outer cylinder is designed for pressurized heat transfer media. The cylinder rests on a base which also incorporates the product inlet. Dry powder is discharged at the top of the cylinder after the product film has moved spirally up the inner wall. Placed within the cylinder is a steam-heated concentric displacement body, which rotates slowly via an external geared motor. The body's outer surface has numerous segmented air-guide plates that are arranged at an appropriate angle. The distance between these plates and the inner wall is greater than the product film thickness. A blower moves the conveying gas tangentially into the tube base, entering opposite a wet-feed metering screw. Thus, the gas disperses the wet feed via intense mixing. This mixture is prevented from bypassing up the column by the displacer's segmented guide plates and is forced into a corkscrew-shaped path along the jacketed surface inner face. Inertial force creates a product film that threads its way upward at constant speed, but in a spiral path until it reaches the top of the dryer. Claims for energy conservation, protection of product, and corrosion resistance, among other factors, are well detailed in the Hess/Rossi reference.

Molecular Sieve Dehydration

This is also a relatively new technology that extends the use of molecular sieve adsorptive dehydration to the removal of over 20% water from organic admixtures.

Numerous other improvements in drying technology are described in the references given. See also **Spray Drying**.

See also **Dielectric Heating**; and **Freeze-Drying**.

Additional Reading

Cary, J.D. and E.B. Gutoff: "Analyze the Drying of Aqueous Coatings," *Chem. Eng. Progress* **73** (February 1991).

Cook, E.M. and H.D. DuMont: *Process Drying Practice,* The McGraw-Hill Companies, Inc., New York, NY, 1991.

Etzel, M.R. and K. Waananen: *Drying of Foods and Biological Materials,* Amer. Inst. of Chem. Engrs., New York, NY, 1992.

Hess, D.: "The DRT Spiral Drying Technique," *Chem. Eng. Progress*, 43–50 (April 1983).

Irudayaraj, J.: *Microwave Processing of Foods,* Amer. Inst. of Chem. Engrs., New York, NY, 1992.

Pabis, S., D.S. Jayas, and S. Cenkowski: *Grain Drying: Theory and Practice,* John Wiley & Sons, Inc., New York, NY, 1998.

Parlmutter, B.A.: "Combine Filtration and Drying," *Chem. Eng. Progress*, **29** (July 1991).

Perry, R.H. and D.W. Green, Editors: *Perry's Chemical Engineers' Handbook,* 7th Edition, The McGraw-Hill Companies, Inc., New York, NY, 1999.

Staff: *Drying Process Software,* Energy Saving Consultants, Boynton Beach, FL, 1992.

DUBNIUM. See **Chemical Elements**.

DULONG AND PETIT LAW OF SPECIFIC HEATS. It has long been known that the atomic heats of the great majority of elements have nearly the same value at room temperature; in fact, the thermal capacity of a gram-atom of most elements is not far from 6 calories per degree. Dulong and Petit expressed this by stating that the specific heats of elements are in inverse proportion to their atomic weights.

That this should be the case for gases, easily follows from the kinetic theory and the principle of equipartition of energy. For example, if the same mean energy per molecule is necessary to raise the temperature of oxygen and of hydrogen 1°, the same is true per atom, and weights of these gases having the same number of atoms will have equal thermal capacities.

For solids the matter is not quite so simple, and the more exacting theories of Einstein, Debye, and others show that the atomic heat should be expected to vary with the temperature. According to Debye, there is a certain characteristic temperature for each crystalline solid at which its atomic heat should equal 5.67 calories per degree. Einstein's theory expresses this temperature as hv_m/k, in which h is Planck's constant, k is Boltzmann's constant, and v_m is a frequency characteristic of the atom in question vibrating in the crystal lattice.

DUMORTIERITE. This mineral found within schists and gneisses is valuable for use in the manufacture of high-grade porcelain. It is a borosilicate of aluminum $Al_7(BO)_3(SiO_4)O_3$, crystallizing in the orthorhombic system. Crystals are rare; commonly occurs as fibrous aggregates of blue to pink color, with vitreous luster. Transparent to translucent with hardness of 8.5, and specific gravity of 3.41.

World occurrences include France, Madagascar, Brazil, and Mexico, with California and Nevada the principal United States localities.

DU VIGNEAUD, VINCENT (1901–1978). An American biochemist who won the Nobel prize for chemistry in 1955. His work involved the study of the metabolism of biologically significant sulfur compounds, which led to the finding of transmethylation in mammalian metabolism. He isolated and proved the structure of the vitamin biotin, and synthesized penicillin, oxytocin, and the vasopressin hormone of the posterior pituitary. His education was at Rochester, Yale, St. Louis, and George Washington Universities.

DYE CARRIERS. Dye carriers are needed for complete dye penetration of polyester fibers. Carriers cause the glass-transition temperature, T_g, of the polyester polymer to become lower and allow the penetration of water-insoluble dyes into the fiber.

Dye carriers, occasionally called dyeing accelerants, are used on cellulose triacetate fibers, but have found their greatest use in the dyeing of polyester.

Carrier Properties

Most carriers are aromatic compounds, and have similar solubility parameters to the poly(ethylene terephthalate) fibers and to some disperse dyes.

Table 1 lists the four main groups of compounds most commonly used as dye carriers. In order for these compounds to act effectively as carriers, they must be homogeneously dispersed in the dyebath. Because the carrier-active compounds have little or no solubility in water, emulsifiers are needed to disperse these compounds in the dyebath.

Carrier Formulation

The formulation of a carrier depends on four considerations: *(1)* the carrier-active chemical compound; *(2)* the emulsifier; *(3)* special additives; and *(4)* environmental concerns. Additional parameters to be considered in the formulation of a carrier product with satisfactory and repeatable performance arise from the equipment in which the dyeing operation is to be carried out. The choice of equipment is usually dictated by the form in which the fiber substrate is to be processed, e.g., loose fiber, staple, continuous or texturized filament, woven or knot fabric, yarn on packages or in skeins.

Carrier Selection

A carrier is selected by the dyer according to various criteria. The type of equipment and conditions under which it is to be used have already been mentioned. Other considerations include color yield, dye migration, and product and emulsion stability.

Health and Safety Factors

Most carrier-active compounds are based on aromatic chemicals with characteristic odor. An exception is the phthalate esters, which are often preferred when ambient odor is objectionable or residual odor on the fabric cannot be tolerated. The toxicity of carrier-active compounds and of their ultimate compositions varies with the chemical or chemicals involved. The environment surrounding the dyeing equipment where carriers are used should always be well-ventilated, and operators should wear protective clothing (e.g., rubber gloves, aprons, and safety glasses or face shields, and possibly an appropriate respirator). Specific handling information can be obtained from the supplier or manufacturer.

OSHA and EPA have established exposure limits that must be carefully considered in relation to the waste disposal method available and the environment in which dye carriers are to be used.

The increasingly stringent government regulations and the introduction of carrierless–dyeable polyester have not substantially affected the use of carriers. The factor with the greatest impact on their use has been provided by the spectacular technological advances in dyeing equipment that have taken place since the early 1980s. High temperature dyeing has greatly reduced the time element (dyeing cycle) and the need for carriers.

ERNESTO DE GUZMAN
BOYCE SUTTON, JR.
Sybron Chemicals Inc.

Additional Reading

Hansen, C.M.: *J. Paint Technol* **39**, 104 (1967).
Keen, M.C. and R.J. Thomas: "Absorption Properties of Latyl Disperse Dyes on Application to Dacron Polyester Fibers," *Dyes and Chemicals Technical Bulletin*, E. I. du Pont de Nemours & Co., Inc., Organic Chemicals Dept., Wilmington, Del., 1992.
Salvin, S. and co-workers: *Am. Dyest. Rep* (22) (Nov. 2, 1959).
Tandy, K. Jr.: Characteristics of Polyester Homopolymer Fiber Which Affect Dyeing Properties, paper presented at the *14th AATCC New England Regional Technical Conference*, New Hamsphire, May 19–21, 1977.

DYE AND DYE INTERMEDIATES

Classification Systems for Dyes

Dyes may be classified according to chemical structure or by their usage or application method. The former approach is adopted by practicing dye chemists who use terms such as azo dyes, anthraquinone dyes, and phthalocyanine dyes. The latter approach is used predominantly by the dye user, the dye technologist, who speaks of reactive dyes for cotton and disperse dyes for polyester. Very often, both terminologies are used, for example, an azo disperse dye for polyester and a phthalocyanine reactive dye for cotton.

Classification of Dyes by Use or Application Method

The classification of dyes according to their usage is summarized in Table 1, which is arranged according to the *Colour Index* (CI) application classification. It shows the principal substrates, the methods of application, and the representative chemical types for each application class.

Although not shown in Table 1, dyes are also being used in high technology applications, such as in the medical, electronics, and especially the reprographics industries.

Nomenclature of Dyes

Dyes are named either by their commercial trade name or by their *Colour Index* (CI) name. In the *Colour Index* these are cross-referenced.

The commercial names of dyes are usually made up of three parts. The first is a trademark used by the particular manufacturer to designate both the manufacturer and the class of dye, the second is the color, and the third is a series of letters and numbers used as a code by the manufacturer to

TABLE 1. COMPOUNDS MOST COMMONLY USED AS DYE CARRIERS

Compounds	Mol wt	Bp, °C
phenolics		
o-phenylphenol	170.2	280–284
p-phenylphenol	170.2	305–308
methyl cresotinate	166.0	240
chlorinated aromatics		
o-dichlorobenzene	147.0	172–178
1,3,5-trichlorobenzene	181.45	214–219
aromatic hydrocarbons and ethers		
biphenyl	154.2	255.9
methylbiphenyl	168.24	255.3
diphenyl oxide	170.0	259.0
1-methylnaphthalene	142.2	244.6
2-methylnaphthalene	142.2	241
aromatic esters		
methyl benzoate	136.14	198–200
butyl benzoate	178.22	250
benzyl benzoate	212.24	323–324
phthalates		
dimethyl phthalate	194.18	298
diethyl phthalate	212.18	298
diallyl phthalate	246.25	290
dimethyl terephthalate	194.18	284

TABLE 1. USAGE CLASSIFICATION OF DYES

Class	Principal substrates
acid	nylon, wool, silk, paper, inks, and leather
azoic components and compositions	cotton, rayon, cellulose acetate, and polyester
basic	paper, polyacrylonitrile-modified nylon, polyester, and inks
direct	cotton, rayon, paper, leather, and nylon
disperse	polyester, polyamide, acetate, acrylic, and plastics
fluorescent brighteners	soaps and detergents, all fibers, oils, paints, and plastics[c]
food, drug, and cosmetic	foods, drugs, and cosmetics
mordant	wool, leather, and anodized aluminum
natural	food
oxidation bases	hair, fur, and cotton
pigments	paints, inks, plastics, and textiles
reactive	cotton, wool, silk, and nylon
solvent	plastics, gasoline, varnish, lacquer, stains, inks, fats, oils, and waxes
sulfur	cotton and rayon
vat	cotton, rayon, and wool

define more precisely the hue, and also to indicate important properties the dye possesses.

Classification of Dyes by Chemical Structure

The two overriding trends in dyestuffs research for many years have been improved cost-effectiveness and increased technical excellence. Improved cost-effectiveness usually means replacing tinctorially weak dyes such as anthraquinone, the second largest class after the azo dyes, with tinctorially stronger dyes such as heterocyclic azos, triphendioxazines, and benzodifuranones. This theme will be pursued throughout this section discussing dyes by chemical structure.

Azo Dyes. These dyes are by far the most important class, accounting for over 50% of all commercial dyes, and having been studied more than any other class. Azo dyes contain at least one azo group ($-N=N-$) but can contain two (disazo), three (trisazo), or, more rarely, four or more (polyazo) azo groups. The azo group is attached to two radicals of which at least one, but, more usually, both are aromatic.

In monoazo dyes, the most important type, the A radical often contains electron-accepting groups, and the E radical contains electron-donating groups, particularly hydroxy and amino groups.

Almost without exception, azo dyes are made by diazotization of a primary aromatic amine followed by coupling of the resultant diazonium salt with an electron-rich nucleophile.

In theory, azo dyes can undergo tautomerism: azo/hydrazone for hydroxyazo dyes; azo/imino for aminoazo dyes, and azonium/ammonium for protonated azo dyes. A more detailed account of azo dye tautomerism can be found elsewhere.

The three metals of importance in azo dyes are copper, chromium, and cobalt. The most important copper dyes are the 1:1 copper(II): azo dye complexes of formula; they have a planar structure.

In contrast, chromium(III) and cobalt(III) form 2:1 dye:metal complexes that have nonplanar structures. Geometrical isomerism exists.

Premetallized dyes are now used widely in various outlets to improve the properties of the dye, particularly lightfastness. However, this is at the expense of brightness, because metallized azo dyes are duller than nonmetallized dyes.

Carbocyclic azo dyes are the backbone of most commercial dye ranges. Based totally on benzene and naphthalene derivatives, they provide yellow, red, blue, and green colors for all the major substrates such as polyester, cellulose, nylon, polyacrylonitrile, and leather.

The carbocyclic azo dye class provides dyes having high cost-effectiveness combined with good all-around fastness properties. However, they lack brightness, and consequently, they cannot compete with anthraquinone dyes for brightness. This shortcoming of carbocyclic azo dyes is overcome by heterocyclic azo dyes.

One long-term aim of dyestuffs research has been to combine the brightness and high fastness properties of anthraquinone dyes with the strength and economy of azo dyes. This aim is now being realized with heterocyclic azo dyes, which fall into two main groups: those derived from heterocyclic coupling components, and those derived from heterocyclic diazo components.

All the heterocyclic coupling components that provide commercially important azo dyes contain only nitrogen as the hetero atom. They are indoles, pyrazolones, and especially pyridones; they provide yellow to orange dyes for various substrates.

In contrast to the heterocyclic coupling components, virtually all the heterocyclic diazo components that provide commercially important azo dyes contain sulfur, either alone or in combination with nitrogen (the one notable exception is the triazole system). These S or S/N heterocyclic azo dyes provide bright, strong shades that range from red through blue to green, and therefore complement the yellow-orange colors of the nitrogen heterocyclic azo dyes in providing a complete coverage of the entire shade gamut.

Anthraquinone Dyes. Anthraquinone dyes are based on 9,10-anthraquinone, which is essentially colorless. To produce commercially useful dyes, powerful electron-donor groups such as amino or hydroxy are introduced into one or more of the four alpha positions (1,4,5, and 8). The most common substitution patterns are 1,4-, 1,2,4-, and 1,4,5,8-. To maximize the properties, primary and secondary amino groups (not tertiary) and hydroxy groups are employed.

Anthraquinone dyes are prepared by the stepwise introduction of substituents on to the performed anthraquinone skeleton or ring closure of appropriately substituted precursors.

The principal advantages of anthraquinone dyes are brightness and good fastness properties, but they are both expensive and tinctorially weak. However, they are still used extensively, particularly in the red and blue shade areas, because other dyes cannot provide the combination of properties offered by anthraquinone dyes, albeit at a price.

Benzodifuranone Dyes. BDFs are unusual in that they span the whole color spectrum from yellow through red to blue, depending on the electron-donating power of the R group on the phenyl ring of the aryl acetic acid, i.e., Ar $= =C_6H_4R$ (R = H, yellow–orange; R = alkoxy, red; R = amino, blue). The first commercial BDF, Dispersol Red C-BN, a red disperse dye for polyester, is already making a tremendous impact. Its brightness even surpasses that of the anthraquinone reds, while its high tinctorial strength (ca 3–4 times that of anthraquinones) makes it cost-effective.

Polycyclic Aromatic Carbonyl Dyes. Structurally, these dyes contain one or more carbonyl groups linked by a quinonoid system. They tend to be relatively large molecules built up from smaller units, typically anthraquinones. Since they are applied to the substrate (usually cellulose) by a vatting process, the polycyclic aromatic carbonyl dyes are often called the anthraquinonoid vat dyes.

Although the colors of the polycyclic aromatic carbonyl dyes cover the entire shade gamut, only the blue dyes and the tertiary shade dyes, namely, browns, greens, and blacks, are important commercially. As a class, the polycyclic aromatic carbonyl dyes exhibit the highest order of lightfastness and wetfastness.

Indigoid Dyes. Like the anthraquinone, benzodifuranone, and polycyclic aromatic carbonyl dyes, the indigoid dyes also contain carbonyl groups. They are also vat dyes.

Indigoid dyes represent one of the oldest known classes of dyes. Although many indigoid dyes have been synthesized, only indigo itself is of any importance today. Indigo is the blue used almost exclusively for dyeing denim jeans and jackets and is held in high esteem because it fades in tone to give progressively paler blue shades.

Polymethine and Related Dyes. Cyanine dyes are the best known polymethine dyes. Nowadays, their commercial use is limited to sensitizing dyes for silver halide photography. However, derivatives of cyanine dyes provide important dyes for polyacrylonitrile. They include azacarbocyanines, hemicyanines, and diazahemicyanines.

Styryl Dyes. The styryl dyes are uncharged molecules containing a styryl group $C_6H_5-CH=C$ usually in conjugation with an *N,N*-dialkylaminoaryl group. Styryl dyes were once a fairly important group of yellow dyes for a variety of substrates. They are synthesized by condensation of an active methylene compound, especially malononitrile with a carbonyl group, especially an aldehyde. As such, styryl dyes have small molecular structures and are ideal for dyeing densely packed hydrophobic substrates such as polyester.

Yellow styryl dyes have now been largely superseded by superior dyes such as azopyridones, but there has been a resurgence of interest in red and blue styryl dyes. The addition of a third cyano group to produce a tricyanovinyl group causes a large bathochromic shift: the resulting dyes are bright red rather than the greenish yellow color of the dicyanovinyl dyes. These tricyanovinyl dyes have been patented by Mitsubishi for the transfer printing of polyester substrates. Two synthetic routes to the dyes are shown: one is by the replacement of a cyano group in tetracyanoethylene, and the second is by oxidative cyanation of a dicyanovinyl dye with cyanide. The use of such toxic reagents could hinder the commercialization of the tricyanovinyl dyes.

Di- and Triaryl Carbonium and Related Dyes. As a class, these dyes are bright and strong, but are generally deficient in lightfastness. Consequently, they are used in outlets where brightness and cost-effectiveness, rather than permanence, are paramount, for example, the coloration of paper. Many dyes of this class, especially derivatives of pyronines (xanthenes), are among the most fluorescent dyes known.

Resurgence of interest in triphendioxazine dyes arose through successful modification of the intrinsically strong and bright triphendioxazine chromogen to produce blue reactive dyes for cotton. These blue reactive dyes combine the advantages of azo dyes and anthraquinone dyes. Thus they are bright, strong dyes with good fastness properties.

Phthalocyanines. Apart from the recent discoveries of benzodifuranone dyes and diketopyrrolopyrrole pigments, phthalocyanine is the only novel chromogen of commercial importance discovered since the nineteenth century.

Phthalocyanines are analogues of the natural pigments chlorophyll and heme. However, unlike these natural pigments, which have extremely poor stability, phthalocyanines are probably the most stable of all the colorants in use today. Substituents can extend the absorption to longer wavelengths, into the near infrared, but not to shorter wavelengths, and so their hues are restricted to blue and green.

Of all the metal complexes evaluated, copper phthalocyanines give the best combination of color and properties and consequently the majority of phthalocyanine dyes are based on copper phthalocyanine.

Besides being extremely stable, phthalocyanines are bright and tinctorially strong ($\varepsilon_{max} \sim 100,000$); this renders them cost-effective. Consequently, phthalocyanines are used extensively in printing inks and paints.

Quinophthalones. Like the hydroxy azo dyes, quinophthalone dyes can, in theory, exhibit tautomerism. The dyes are synthesized by the condensation of quinaldine derivatives with phthalic anhydride. Quinophthalones provide important yellow dyes for the coloration of plastics and for the coloration of polyester.

Sulfur Dyes. These dyes are synthesized by heating aromatic amines, phenols, or nitro compounds with sulfur or, more usually, alkali polysulfides. Sulfur dyes are used for dyeing cellulosic fibers. They are insoluble in water and are reduced to the water-soluble leuco form for application to the substrate by using sodium sulfide solution. The sulfur dye proper is then formed within the fiber pores by atmospheric oxidation. Sulfur dyes constitute an important class of dye for producing cost-effective tertiary shades, especially black, on cellulosic fibers.

Nitro and Nitroso Dyes. These dyes are now of only minor commercial importance, but are of interest for their small molecular structures. The most important nitro dyes are the nitrodiphenylamines. Their small molecules are ideal for penetrating dense fibers such as polyester, and are therefore used as disperse dyes for polyester. All the important dyes are yellow. Although the dyes are not terribly strong ($\varepsilon_{max} \sim 20,000$), they are cost-effective because of their easy synthesis from inexpensive intermediates.

Nitroso dyes are metal-complex derivatives of *o*-nitrosophenols or naphthols. Tautomerism is possible in the metal-free precursor between the nitrosohydroxy tautomer and the quinoneoxime tautomer. The only nitroso dyes important commercially are the iron complexes of sulfonated 1-nitroso-2-naphthol. These inexpensive colorants are used mainly for coloring paper.

Miscellaneous Dyes. Other classes of dyes that still have some importance are the stilbene dyes and the formazan dyes. Stilbene dyes are in most cases mixtures of dyes of indeterminate constitution that are formed from the condensation of sulfonated nitroaromatic compounds in aqueous caustic alkali either alone or with other aromatic compounds, typically arylamines. The sulfonated nitrostilbene is the most important nitroaromatic, and the aminoazobenzenes are the most important arylamines.

Formazan dyes bear a formal resemblance to azo dyes, since they contain an azo group. The most important formazan dyes are the metal complexes, particularly copper complexes, of tetradentate formazans. They are used as reactive dyes for cotton.

Dye Intermediates

The precursors of dyes are called dye intermediates. They are obtained from simple raw materials, such as benzene and naphthalene, by a variety of chemical reactions. Usually, the raw materials are cyclic aromatic compounds, but acyclic precursors are used to synthesize heterocyclic intermediates. The intermediates are derived from two principal sources, coal tar and petroleum.

Intermediates Classification. Intermediates may be conveniently divided into primary intermediates (primaries) and dye intermediates. Large amounts of inorganic materials are consumed in both intermediates and dyes manufacture.

Inorganic materials include acids (sulfuric, nitric, hydrochloric, and phosphoric), bases (caustic soda, caustic potash, soda ash, sodium carbonate, ammonia, and lime), salts (sodium chloride, sodium nitrite, and sodium sulfide) and other substances such as chloride, bromine, phosphorus chlorides, and sulfur chlorides. The important point is that there is a significant usage of at least one inorganic material in all processes, and the overall tonnage used by, and therefore the cost to, the dye industry is high.

Dye intermediates are defined as those precursors to colorants that are manufactured within the dyes industry, and they are nearly always colorless. Colored precursors are conveniently termed color bases. As distinct from primaries they are only rarely manufactured in single-product units because of the comparatively low tonnages required. Fluorescent brightening agents (FBAs) are neither intermediates nor true colorants.

There are at least 3000 different intermediates in current manufacture (over half that number are specifically mentioned in the *Colour Index*), and in addition there is a comparatively small number of products manufactured by individual companies for their own specialties.

The Chemistry of Dye Intermediates

The chemistry of dye intermediates may be conveniently divided into the chemistry of carbocycles, such as benzene and naphthalene, and the chemistry of heterocycles, such as pyridenes and thiophenes.

Chemistry of Aromatic Carbocycles. Benzene and naphthalene are by far the most important aromatic carbocycles used in the dyes industry. The hundreds of benzene and naphthalene intermediates used can be prepared from these parent compounds by the sequential introduction of a variety of substituents e.g., NO_2, NR^1R^2, Cl, SO_3H, etc. Introduction of these groups are known as unit processes. The substituents are introduced into the aromatic ring by either electrophilic or nucleophilic substitution. In general, aromatic rings, because of their inherently high electron density, are much more susceptible to electrophilic attack than to nucleophilic attack. Nucleophilic attack only occurs under forcing conditions unless the aromatic ring already contains a powerful electron-withdrawing group, e.g., NO_2. In this case, nucleophilic attack is greatly facilitated because of the reduced electron density at the ring carbon atoms.

Unit Processes. The unit processes encountered in intermediate and dye chemistry are summarized in Table 2.

Chemistry of Aromatic Heterocycles. In contrast to the benzenoid intermediates, it is unusual to find a heterocyclic intermediate that

TABLE 2. UNIT PROCESSES IN DYES MANUFACTURE

Process	Primaries[a]	Intermediates (common usage)	Colorants (common usage)
nitration	6	√	
reduction	8	√	
sulfonation	4	√[b]	√
oxidation	5	√	
fusion/hydroxylation	3	√	
amination	3	√[c]	
alkylation	2	√	√
halogenation	2	√	√
hydrolysis	2	√	
condensation	1	√	√
alkoxylation	1	√	
esterification	1	√	
carboxylation	1	√	
acylation	1	√	√
phosgenation	1	√	√
diazotization	1	√	√
coupling (azo)	1	√	√

[a] Number of occurrences within 30 identified product manufactures.

[b] Includes chlorosulfonation.

[c] Includes the Bucherer reaction.

is synthesized via the parent heterocycle. They are synthesized from acyclic precursors.

The most important heterocycles are those with five- or six-membered rings; these rings may be fused to other rings, especially a benzene ring. Nitrogen, sulfur, and to a lesser extent oxygen, are the most frequently encountered heteroatoms. They are often considered in two groups: those containing only nitrogen, such as pyrazolones, indoles, pyridones, and triazoles which, except for triazoles, are used as coupling components in azo dyes, and those containing sulfur (and also optionally nitrogen), such as thiazoles, thiophenes, and isothiazoles, that are used as diazo components in azo dyes. Triazines are treated separately since they are used as the reactive system in many reactive dyes.

Equipment and Manufacture

The basic types of dye (and intermediate) manufacture are shown in Figure 1. There are usually several reaction steps or unit processes.

The reactions for the production of intermediates and dyes are carried out in bomb-shaped reaction vessels made from cast iron, stainless steel, or steel lined with rubber, glass (enamel), brick, or carbon blocks. Wooden vats are also still used in some countries, e.g., India. These vessels have capacities of 2–40 m^3 (ca 500–10,000 gal) and are equipped with mechanical agitators, thermometers or temperature recorders, condensers, pH-probes, etc, depending on the nature of the operation. Jackets or coils are used for heating and cooling by circulating through them high boiling fluids (e.g., hot oil, or Dowtherm), steam, or hot water to raise the temperature, and air, cold water, or chilled brine to lower it. Unjacketed vessels are often used for aqueous reactions, where heating is affected by direct introduction of steam, and cooling by addition of ice or by heat exchangers. The reaction vessels normally span two or more floors in a plant to facilitate ease of operation.

Products are transferred from one piece of equipment to another by gravity flow, pumping, or by blowing with air or inert gas. Solid products are separated from liquids in centrifuges, on filter boxes, on continuous belt filters, and perhaps most frequently, in various designs of plate-and-frame or recessed plate filter presses. The presses are dressed with cloths of cotton, Dynel, polypropylene, etc. Some provide separate channels for efficient washing, others have membranes for increasing the solids content of the presscake by pneumatic or hydraulic squeezing.

The plates and frames are made of wood, cast iron, or now usually hard rubber, polyethylene, and polyester.

When possible, the intermediates are taken for the subsequent manufacture of other intermediates or dyes without drying because of savings in energy costs and handling losses. There are, however, many cases where products, usually in the form of pastes discharged from a filter, must be dried. Where drying is required, air or vacuum ovens (in which the product is spread on trays), rotary dryers, spray dryers, or less frequently, drum dryers (flakers) are used. Spray dryers have become increasingly important.

The final stage in dye manufacture is grinding or milling. Dry grinding is usually carried out in impact mills (Atritor, KEK, or ST); considerable amounts of dust are generated, and well-established methods are available to control this problem. Dry grinding is an inevitable consequence of oven drying, but more modern methods of drying, especially continuous drying, allow the production of materials that do not require a final comminution stage. Wet milling has become increasingly important for pigments and disperse dyes.

In the past the successful operation of batch processes depended mainly on the skill and accumulated experience of the operator. This operating experience was difficult to codify in a form that enabled full use to be made of it in developing new designs. The gradual evolution of better instrumentation, followed by the installation of sequence control systems, has enabled much more process data to be recorded, permitting maintenance of process variations within the minimum possible limits.

Full computerization of multiproduct batch plants is much more difficult than with single-product continuous units because the control parameters vary fundamentally with respect to time. The first computerized azo and intermediates plants were brought on stream by ICI Organics Division (now Zeneca Specialties) in the early 1970s, and have now been followed by many others. The additional cost (ca 10%) of computerization has been estimated to give a saving of 30 to 45% in labor costs. However, highly trained process operators and instrument engineers are required.

Economic Aspects

With the state of the world economy as it stands in 1993, the growth rate for the industry is likely to be as low as 2–3%, but even this low figure represents something like an additional 20,000 t/yr.

Health and Safety Aspects

Toxicology and Registration. The toxic nature of some dyes and intermediates has long been recognized. Acute, or short-term, effects are generally well known. They are controlled by keeping the concentration of the chemicals in the workplace atmosphere below prescribed limits and avoiding physical contact with the material. Chronic effects, on the other hand, frequently do not become apparent until after many years of exposure.

The positive links between benzidine derivatives and 2-naphthylamine with bladder cancer prompted the introduction of stringent government regulations to minimize such occurrences in the future. Currently, the three principal regulatory agencies worldwide are European Core Inventory (ECOIN) and European Inventory of Existing Commercial Substances (EINECS) in Europe, Toxic Substances Control Act (TOSCA) in the United States, and Ministry of Technology and Industry (MITI) in Japan. Each of these has its own set of data and testing protocols for registration of a new chemical substance.

Environmental Concerns. Dyes, because they are intensely colored, present special problems in effluent discharge; even a very small amount is noticeable. However, the effect is more aesthetically displeasing rather than hazardous. Of greater concern is the discharge of toxic heavy metals such as mercury and chromium.

Effluents from both dye works and dyehouses are treated both before leaving the plant, e.g., neutralization of acidic and alkaline liquors and heavy metal removal, and in municipal sewage works. Various treatments are used.

Biological treatment is the most common and most widespread technique used in effluent treatment, having been employed for over 140 years. There are two types of treatment, aerobic and anaerobic.

Removal of color by adsorption using activated carbon is also employed. Activated carbon is very good at removing low levels of soluble chemicals, including dyes. Its main drawback is its limited capacity. Consequently, activated carbon is best for removing color from dilute effluent.

Chemical treatment of the effluent with a flocculating agent is the most robust and generally most efficient way to remove color.

Chemical oxidation is a more recent method of effluent treatment, especially chemical effluent. This procedure uses strong oxidizing agents like ozone, hydrogen peroxide, chlorine, and potassium permanganate in order to force degradation of even some of the more resilient organic molecules. As of this writing (ca 1993), these treatments remain very expensive and are of limited size, thought they may have some promise in the future.

Additional strategies being implemented to minimize dye and related chemical effluent include designing more environmentally friendly chemicals, more efficient (higher yielding) manufacturing processes, and more effective dyes, e.g., reactive dyes having higher fixation.

Peter Gregory
Zeneca Specialties

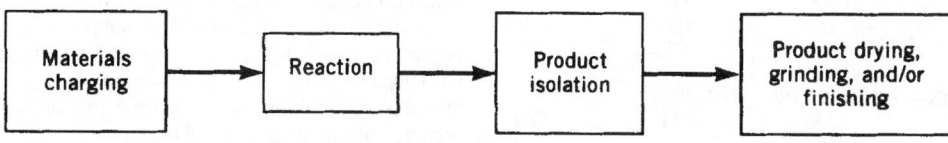

Fig. 1.   Operation sequence in dye and intermediate manufacture.

Additional Reading

Colour Index, 3rd Edition, The Society of Dyers and Colorists, Bradford, U.K., 1971.

Gordon, P.F. and P. Gregory, *Organic Chemistry in Color*, Springer-Verlag, Berlin, 1983.

Venkataraman, K. *The Chemistry of Synthetic Dyes*, Vols. I–VIII, Academic Press, Inc., New York, 1952–1974.

Zollinger, H. *Color Chemistry: Synthesis, Properties and Applications of Organic Dyes and Pigments*, 2nd Edition, VCH, 1991.

DYES: ANTHRAQUINONE.

The synthesis of an anthraquinone dye generally involves a large number of steps. Highly toxic metals such as mercury or chromium(VI) are sometimes required. Some processes need to employ a large amount of organic solvent, and others involve a great quantity of waste acids. With the increasing demand for environmental protection, the regulation of pollutant effluents has become more stringent year after year, which has caused a sharp increase in the costs for wastewater treatment. This situation has led to intensive improvement of conventional methods and the development of new synthetic routes as well.

Efforts have also been made to overcome complicated processes. Methods to reduce the number of steps or to use new starting materials have been studied extensively.

Because of their small extinction coefficients anthraquinone dyes have less tinctorial strength than azo dyes; that is the intrinsic disadvantage of anthraquinones. This fact and the complexity of preparation have made their production costs higher than those of azo dyes. However, the anthraquinone dyes have excellent properties that are not attainable by azo dyes, such as brilliancy of color, fastness, and excellent dyeing properties (leveling and dye bath stability). Thus the anthraquinone dyes have been widely used in the areas where these properties are required. Cotton or polyester–cotton blend fibers for military wear and working wear that require extreme fastness are dyed mainly with anthraquinone vat dyes. Most polyester fabrics for automobile seats are dyed with anthraquinone disperse dyes, since the requirement for lightfastness is extremely high and, simultaneously, bright shades are needed.

World dye manufacturers have already begun to develop new types of dyes that can replace the anthraquinones technically and economically. Some successful examples can be found in azo disperse red and blue dyes. In the reactive dye area intensive studies have continued to develop triphenodioxazine compounds to replace anthraquinone blues.

Color and Structure

The uv–vis spectrum of anthraquinone shows an absorption maximum at 323 nm ($\varepsilon = 4500$) due to a $\pi - \pi^*$ transition and very weak absorption in the visible range, 405 nm ($\varepsilon = 60$) due to a $n-\pi^*$ transition. Thus anthraquinone is almost colorless. Introduction of electron-donating substituents causes a bathochromic shift. This is due to the charge-transfer band from the lone pair of amino or hydroxyl groups to the oxygen atom of the carbonyl group. By increasing the electron-donating ability of substituents, the bathochromic shifts are enhanced (Table 1). In the case of the same substituent, the bathochromic shift is larger when the substituent is in the 1-position rather than in the 2-position. The introduction of an electron-withdrawing group has little effect on the absorption maximum of the spectrum.

The absorption maximum of a disubstituted anthraquinone greatly depends on the substituents and their positions. The 1,4-disubstituted compound shows a remarkable bathochromic shift. Larger bathochromic shifts are observed with increasing electron-withdrawing ability of β-substituents.

1,4,5,8-Tetrasubstituted anthraquinones give a slightly reddish blue tint to greenish blue color, depending on the substituents and their positions.

In addition to the color and the tinctorial strength, which are very important factors for the molecular design of anthraquinone dyes, affinity for fibers, various kinds of fastness (light, wet, sublimation, nitrogen oxides (NO_x) gas, washing, etc), and application properties (sensitivity for dyeing temperature, pH, etc) must be considered thoroughly as well.

Method of Synthesis

Anthraquinone dyes are derived from several key compounds called dye intermediates, and the methods for preparing these key intermediates can be divided into two types: *(1)* introduction of substituent(s) onto the anthraquinone nucleus, and *(2)* synthesis of an anthraquinone nucleus having the desired substituents, starting from benzene or naphthalene derivatives (nucleus synthesis). The principal reactions are nitration and sulfonation, which are very important in preparing α-substituted anthraquinones by electrophilic substitution. Nucleus synthesis is important for the production of β-substituted anthraquinones such as 2-methylanthraquinone and 2-chloroanthraquinone. Friedel-Crafts acylation using aluminum chloride is applied for this purpose. Synthesis of quinizarin (1,4-dihydroxyanthraquinone) is also important.

Key Intermediates

1-Aminoanthraquinone and Related Compounds. 1-Aminoanthraquinone is the most important intermediate for manufacturing acid, reactive, disperse, and vat dyes. It has been manufactured from anthraquinone-1-sulfonic acid by ammonolysis of the sulfo group with aqueous ammonia in the presence of an oxidizing agent such as nitrobenzene-3-sulfonic acid. In this process the starting material can only be obtained by mercury-catalyzed sulfonation of anthraquinone with oleum. For improved ecology, the alternative route based on 1-nitroanthraquinone was established. 1-Nitroanthraquinone is prepared from anthraquinone by nitration in sulfuric acid or organic solvent. 1-Aminoanthraquinone is prepared from 1-nitroanthraquinone by reduction with sodium hydrogen sulfide or by catalytic hydrogenation. Highly purified product is manufactured by continuous vacuum distillation.

1-Amino-4-bromoanthraquinone-2-sulfonic acid (bromamine acid) is the most important intermediate for manufacturing reactive and acid dyes. Bromamine acid is manufactured from 1-aminoanthraquinone-2-sulfonic acid by bromination in aqueous medium, or in concentrated sulfuric acid.

1-Amino-2-bromo-4-hydroxyanthraquinone is one of the most important intermediates for manufacturing red disperse dyes. It is prepared by dibrominating 1-aminoanthraquinone in concentrated sulfuric acid and subsequent hydrolysis in the presence of boric acid.

1-Amino-2-chloro-4-hydroxyanthraquinone is another important intermediate in red disperse dye manufacture. 1-Amino-2-chloro-4-hydroxyanthraquinone is prepared via a route from chlorobenzene and phthalic anhydride as the raw materials.

1,4-Dihydroxyanthraquinone. This anthraquinone, also known as quinizarin, is of great importance in manufacturing disperse, acid, and vat dyes. It is manufactured by condensation of phthalic anhydride with 4-chlorophenol in oleum in the presence of boric acid.

1,4-Diaminoanthraquinone and Related Compounds. Leuco-1,4-diaminoanthraquinone (leucamine) is an important precursor for 1,4-diaminoanthraquinone and is prepared by heating 1,4-dihydroxyanthraquinone with sodium dithionite in aqueous ammonia under pressure. 1,4-Diaminoanthraquinone is an important intermediate for vat dyes and disperse dyes, and is prepared by oxidizing leuco-1,4-diaminoanthraquinone with nitrobenzene in the presence of piperidine.

1,4-Diamino-2,3-dichloroanthraquinone (CI Disperse Violet 28) is an important compound as an intermediate for CI Disperse Blue 60 and CI Disperse Violet 26, and is prepared by chlorination of leuco-1,4-diaminoanthraquinone with chlorine gas or sulfuryl chloride in an inert organic solvent such as nitrobenzene.

1,4-Diamino-2,3-dicyanoanthraquinone is the key intermediate for manufacturing CI Disperse Blue 60. 1,4-Diamino-2,3-dicyanoanthraquinone is

TABLE 1. SPECTRAL DATA FOR SOME MONOSUBSTITUTED ANTHRAQUINONES[a] IN METHANOL

Substituent	1-position		2-position	
	λ_{max}, nm	ε	λ_{max}, nm	ε
ELECTRON-DONATING GROUPS				
OCH_3	378	5200	363	3950
OH	402	5500	368	3900
$NHCOCH_3$	400	5600	367	4200
NH_2	475	6300	440	4500
$NHCH_3$	503	7100	462	5700
$N(CH_3)_2$	503	4900	472	5900
ELECTRON-WITHDRAWING GROUPS				
NO_2	325	4300	323	5200
Cl	333	5000	325	3900

[a] Unsubstituted anthraquinone $\lambda_{max} = 323$ nm; $\varepsilon = 4500$.

manufactured by reaction of 1,4-diaminoanthraquinone-2,3-disulfonic acid with alkali metal cyanide.

1,4,-Diaminoanthraquinone-2,3-dicarboxyimide is the intermediate for CI Disperse Blue 60 and is prepared by hydrolysis of 1,4-diamino-2,3-dicyanoanthraquinone in concentrated sulfuric acid.

Anthraquinone-1-Sulfonic Acid and Its Derivatives. Anthraquinone-1-sulfonic acid has become less competitive than 1-nitroanthraquinone as the intermediate for 1-aminoanthraquinone. However, it still has a great importance as an intermediate for manufacturing vat dyes via 1-chloroanthraquinone.

Anthraquinone-1-sulfonic acid is prepared from anthraquinone by sulfonation with 20% oleum in the presence of mercury catalyst, a Hg(II) salt such as $HgSO_4$ or HgO, at 120°C.

1-Chloroanthraquinone is an intermediate for manufacturing vat dyes such as CI Vat Brown 1. 1-Chloroanthraquinone is prepared by chlorination of anthraquinone-1-sulfonic acid with sodium chlorate in hydrochloric acid at elevated temperature.

1-Methylaminoanthraquinone is an important intermediate for manufacturing solvent dyes and acid dyes, and is prepared from anthraquinone-1-sulfonic acid by replacing the SO_3H group with methylamine.

Anthraquinone-α,α′-Disulfonic Acids and Related Compounds. Anthraquinone-α,α′-disulfonic acids and their derivatives are important intermediates for manufacturing disperse blue dyes (via 1,5- or 1,8-dihydroxyanthraquinone, or 1,5-dichloroanthraquinone) and vat dyes (via 1,5-dichloroanthraquinone).

Anthraquinone-1,5-disulfonic acid and anthraquinone-1,8-disulfonic acid are produced from anthraquinone by disulfonation in oleum.

1,5-Dichloroanthraquinone is an important intermediate for vat dyes and disperse blue dyes. 1,5-Dichloroanthraquinone is prepared by the reaction of anthraquinone-1,5-disulfonic acid with $NaClO_3$ in hot hydrochloric acid solution.

1,5-Dihydroxyanthraquinone (anthrarufin) is an important intermediate for manufacturing disperse blue dyes and is prepared from anthraquinone-1,5-disulfonic acid by heating with an aqueous suspension of calcium oxide and magnesium chloride under pressure at 200–250°C.

α,α′-Dinitroanthraquinones and Related Compounds. 1,5- and 1,8-Dinitroanthraquinone are the key intermediates for manufacturing disperse blue dyes via dinitrodihydroxyanthraquinone and vat dyes via diaminoanthraquinones. 1,5-Dinitroanthraquinone and 1,8-dinitroanthraquinone are prepared by nitration of anthraquinone with nitric acid in sulfuric acid. α,β′-Dinitroanthraquinones are also formed in the reaction.

1,5-Diaminoanthraquinone is prepared from 1,5-dinitroanthraquinone by ammonolysis, by catalytic hydrogenation, or by reduction with sodium sulfide. It is also prepared from anthraquinone-1,5-disulfonic acid by ammonolysis. 1,5-Diaminoanthraquinone is an important intermediate for manufacturing vat dyes.

1,5-Dihydroxy-4, 8-dinitroanthraquinone is an important dye precursor for CI Disperse Blue 56, and is prepared from 1,5-diphenoxyanthraquinone by hexanitration in sulfuric acid and subsequent hydrolysis with aqueous alkali. 1,5-Dinitro-4,8-dihydroxyanthraquinone is also prepared from dimethoxyanthraquinone. High purity of 1,5-dimethoxyanthraquinone is required for manufacturing disperse blue dyes.

2-Methylanthraquinone and Related Compounds. 2-Methylanthraquinone and its derivatives are important as intermediates for manufacturing various kinds of vat dyes and brilliant blue (turquoise blue) disperse dyes. 2-Methylanthraquinone is prepared from phthalic anhydride and toluene via a benzoylbenzoic acid. 1-Nitroanthraquinone-2-carboxylic acid is of great importance as an intermediate for manufacture of vat dyes as well as disperse dyes. It is conventionally prepared from 2-methyl-1-nitroanthraquinone, by oxidation in sulfuric acid with sodium dichromate. 1-Aminoanthraquinone-2-carboxylic acid is also an important intermediate for vat dyes and disperse dyes and is prepared from 1-nitroanthraquinone-2-carboxylic acid by reaction with ammonia.

2-Chloroanthraquinone and Its Derivatives. 2-Chloroanthraquinone and its derivatives are the most important intermediates for vat dyes and high performance organic pigments. 2-Chloroanthraquinone is prepared by Friedel-Crafts reaction of chlorobenzene and phthalic anhydride in the presence of aluminum chloride, followed by ring closure in concentrated sulfuric acid. 2-Amino-3-hydroxyanthraquinone is prepared by heating 5-benzoylbenzoxazolone-2′-carboxylic acid in sulfuric acid. This compound is an intermediate for CI Vat Red 10.

Benzanthrone and Related Compounds. Benzanthrone is prepared by the reaction of anthraquinone with glycerol, sulfuric acid, and a reducing agent such as iron. Benzanthrone is an important intermediate for manufacturing vat dyes.

N-Methylanthrapyridone and Its Derivatives. 6-Bromo-3-methylanthrapyridone is an important intermediate for manufacturing dyes soluble in organic solvents. These solvent dyes are prepared by replacing the bromine atom with various kinds of aromatic amines. 6-Bromo-3-methylanthrapyridone is prepared from 1-methylamino-4-bromoanthraquinone by acetylation with acetic anhydride followed by ring closure in alkali. The starting material of this route is anthraquinone-1-sulfonic acid.

Reactive Dyes

Most of the anthraquinone reactive dyes are derived from bromamine acid. These dyes give a bright blue shade and excellent lightfastness. A great number of reactive groups have been proposed; typical examples include sulfatoethylsulfone, dichlorotriazine, monochlorotriazine, monofluorotriazine, and other heterocyclic groups (see **Dyes: Reactive**).

Disperse Dyes

Disperse dyes are water-insoluble, aqueous dispersed materials that are used for dyeing hydrophobic synthetic fibers, including polyester, acetate, and polyamide.

By introducing amino, hydroxy, or methyl groups onto the anthraquinone moiety as the principal auxochromes, dyes that have yellow through greenish blue shades are obtained. Among these dyes many that have brilliant red, violet, blue, and greenish blue shades have great industrial importance in view of their affinity for polyester or cellulose acetate fibers and lightfastness and sublimation resistance. On the contrary, yellow or orange dyes are not satisfactory because of the rather simple molecular structure. Therefore these shades are obtained from other chromophores.

On the basis of the kind and the position of their substituents and their color range, the anthraquinoid disperse dyes may be classified as follows:

Color range	Chemical description
red	1-amino-4-hydroxyanthraquinones
blue, greenish blue	1,1,4,5,8-substituted anthraquinones
greenish blue	1,4-diaminoanthraquinone-2,3-dicarboxyimides
violet, blue	1,4-diaminoanthraquinone derivatives
violet, blue	N-substituted 1-amino-4-hydroxyanthraquinones

Acid Dyes

Acid dyes are used for dyeing wool, synthetic polyamides, and silk in aqueous media. Anthraquinone acid dyes give brilliant reds, violets, blues, and greens and exhibit excellent lightfastness. Because of their relatively high cost, they are used to dye high grade textiles in pale and moderate shades. Various kinds of anthraquinone acid dyes have been developed so far mainly by IG-Farbenindustrie in Germany applying chemical reactions that were studied in developing vat dyes. However, the number of commercial products has declined because of poor properties or unavailable raw materials. Anthraquinone acid dyes may be classified into two groups: bromamine acid derivatives and quinizarin derivatives.

Vat Dyes

Anthraquinone vat dyes have been used to dye cotton and other cellulose fibers for many decades. Despite their high cost, relatively muted colors, and difficulty in application, anthraquinone vat dyes still form one of the most important dye classes of synthetic dyes because of their all-around superior fastness.

Anthraquinone vat dyes are water-insoluble dyes. They are converted to leuco compounds (anthrahydroquinones) by reducing agents such as sodium hydrosulfite in alkaline conditions. These water-soluble leuco compounds have an affinity to cellulose fibers and penetrate them. After reoxidation by means of air or other oxidizing agents, the dye becomes water-insoluble again and fixes firmly on the fiber.

The anthraquinone vat dyes can be classified into several groups on the basis of their chemical structures: (1) benzanthrone dyes, (2) indanthrones, (3) anthrimides, (4) anthrimidocarbazoles, (5) acylaminoanthraquinones, (6) anthraquinoneazoles, (7) anthraquinone acridones, (8) anthrapyrimidines, and (9) highly condensed ring systems. Recently, research and development efforts have focused on improved manufacturing of traditional vat dyes.

Mordant Dyes

Mordant dyes have hydroxy groups in their molecular structure that are capable of forming complexes with metals. Although a variety of metals such as iron, copper, aluminum, and cobalt have been used, chromium is most preferable as a mordant. Alizarin or CI Mordant Red 11 (CI 58000), the principal component of the natural dye obtained from madder root, is the most typical mordant dye. Many mordant dyes have given way to the vat or the azoic dyes, which are applied by much simpler dyeing procedures.

Acid–mordant dyes have characteristics similar to those of acid dyes which have a relatively low molecular weight, anionic substituents, and an affinity to polyamide fibers and mordant dyes. In general, brilliant shades cannot be obtained by acid–mordant dyes because they are used as their chromium mordant by treatment with dichromate in the course of the dyeing procedure. However, because of their excellent fastness for light and wet treatment, they are predominantly used to dye wool in heavy shades (navy blue, brown, and black). In terms of chemical constitution, most of the acid–mordant dyes are azo dyes; some are triphenylmethane dyes; and very few anthraquinone dyes are used in this area. CI Mordant Black 13 is one of the few examples of currently produced anthraquinone acid–mordant dyes.

Functional Dyes

The investigation of new dyes has always been focused on the development of fast, brilliant, inexpensive, and easy applicable dyes. Because a great emphasis has been placed especially on fastness, the dyes with poorer fastness have been ignored in the past. However, in recent years new needs for dyes that change color in response to low energy stimuli including light, electricity, or heat have arisen in the electronics industry. This new application includes information recording, information display, and energy conversion. The term *functional dye* has been applied to dyes that are used in advanced technologies based on optoelectronics since 1981 when the book entitled *The Chemistry of Functional Dyes* was published in Japan.

In order to develop the dyes for these fields, characteristics of known dyes have been re-examined, and some anthraquinone dyes have been found usable. One example of use is in thermal-transfer recording where the sublimation properties of disperse dyes are applied. Anthraquinone compounds have also been found to be useful dichroic dyes for guest-host liquid crystal displays when the substituents are properly selected to have high order parameters. These dichroic dyes can be used for polarizer films of LCD systems as well. Anthraquinone derivatives that absorb in the near-infrared region have also been discovered, which may be applicable in semiconductor laser recording.

Economic Aspects

Production Capacity and Demand. The production capacity for each dye or dye intermediate has rarely been announced officially by the individual manufacturers. Principal manufacturers of anthraquinone dyes and their intermediates are as follows: Bayer, BASF, and Hoechst (Germany); Hoechst Mitsubishi Kasei, Sumitomo Chemical, Mitsui Badische Dyes, and Nippon Kayaku (Japan); Ciba-Geigy and Sandoz (Switzerland); Zeneca and Holliday (U.K.); Crompton & Knowles (U.S.); and IDI (India).

Recent increases in environmental costs have become a serious problem, and future prospects for the anthraquinone dye industry are not optimistic. Some traditional manufacturers have stopped the production of a certain dye class or dye intermediates that were especially burdened by environmental costs, e.g., vat dyes and their intermediates derived from anthraquinone-1-sulfonic acid and 1,5-disulfonic acid. However, several manufacturers have succeeded in process improvement and continue production, even expanding their capacity. In the forthcoming century the worldwide framework of production will change drastically.

Consumption

Anthraquinone dyes are the most important dye class after azo dyes. The consumption of each dye class or set of classes is approximately parallel to the consumption of fibers to which they are applied.

Among these dye classes, anthraquinone dyes are in an important position in reactive dyes and vat dyes for cellulose fibers, disperse dyes for polyester, and acid dyes for polyamide. Applications for high performance organic pigments for plastics and paints are also important areas.

Health and Safety Information

In general, anthraquinone dyes and their intermediates have not been reported as strongly toxic substances, but for many compounds safety data have not been evaluated. 1-Nitroanthraquinone, 1-chloroanthraquinone, and benzanthrone are reported to cause mild skin irritation in a test with rabbits, 500 mg/24 h. Some eye irritation data have been reported.

There are some tumorigenic data for anthraquinone dyes and intermediates which have been evaluated thoroughly. Data for 2-aminoanthraquinone and 2-methyl-1-nitroanthraquinone are available. 2-Aminoanthraquinone has been assessed by the United Nations International Agency for Research on Cancer (IARC) from studies in animals, and is judged to fall into the *Animal: Limited Evidence* group. 2-Aminoanthraquinone has been evaluated by EPA (Genetic Toxicology program) and a positive carcinogenic effect for rat and mouse is designated. 2-Methyl-1-nitroanthraquinone has been assessed by IARC and judged as belonging in the *Animal: Sufficient Evidence* group. 2-Methyl-1-nitroanthraquinone has been evaluated by the National Cancer Institute (NCI) and clear evidence of carcinogenicity for rat and mouse is demonstrated.

Most anthraquinone dyes and their intermediates are handled in a powder form. Their dust poses the threat of contact to eyes and skin or contamination of surroundings. Attention must be paid to avoid these hazards. Special attention should be paid to avoid contact with compounds that are recognized to have probable carcinogenicity.

In the case of handling in relatively small quantities, i.e., for laboratory use, normal personal equipment, i.e., dust masks, safety glasses, and gloves, and hoods with local exhaust ventilation should be used. In plant operations, special technical handling measures should be taken because the possibility of contact is extremely high, especially when charging the raw materials or isolating or packaging the intermediates or final products.

MAKOTO HATTORI
Sumitomo Chemical Company

Additional Reading

Hallas, G.: in J. Shore, ed., *Colorants and Auxiliaries,* Society of Dyers and Colourists, Bradford, UK, 1990, pp. 230–267.

Schweizer, H.R.: *Künstriche Organische Farbstoffe und ihre Zwischenprodukte,* Springer-Verlag, Berlin, Germany, 1964, pp. 301–320.

Stilmar, F.B. et al: in H.A. Lubs, ed., *The Chemistry of Synthetic Dyes and Pigments,* Reinhold Publishing Corp., New York, NY, 1955, pp. 335–550.

Venkataraman, K.: *The Chemistry of Synthetic Dyes,* Vol. 2, Academic Press, Inc., New York, NY, 1952.

DYES: APPLICATION AND EVALUATION. The global consumption of textiles is estimated at around 30 million t, and this is expected to grow at 3% per year. The coloration of this amount needs ca 700,000 t of dye, with a value of $4400 million. The principal reasons for coloring textiles are for aesthetic appearance and decoration or for utilitarian purposes, and unless there is an unpredicted change in human behavior the majority of textiles will continue to be dyed to produce colored apparel, home furnishings, carpets, etc. Among the aesthetic uses are fashion garments and household articles such as drapes, towels, and carpets. In the utilitarian group are uniforms (military and civil), and work wear.

In order for a colored substance to be regarded as a dyestuff, a number of requirements must be satisfied. A dyestuff must be substantive for a textile and exhaust from an aqueous solution into the fiber; have a high exhaustion; exhaust at a rate allowing economic processing; give a uniform level dyeing; and have satisfactory fastness for the particular end use the textile is intended for. The process of dyeing is therefore a combination of chemistry, application technology, economics, and customer needs.

Classification of Dyestuffs According to Application

The *Colour Index* categorizes all coloring matters according to application characteristics. The following are the main types currently of interest as dyes.

Acid Dyes. These are anionic dyes, usually containing sulfonic acid groups, that are substantive to wool, other protein fibers, and polyamides when dyed from an acidic dyebath. The lower the pH the more rapid the dyeing, and exhaustion efficiency is enhanced by increased acidity.

Mordant Dyes. This group includes many natural as well as synthetic dyes. They have no or low substantivity for textile fibers and are therefore applied to cellulosic or protein fibers that have been treated (mordanted) with metallic oxides to give points of attraction for the dye. The dye forms a complex with the metal and depending on the metal and fiber can simply form a large macromolecule incapable of desorbing, or a dye molecule bound to the fiber resulting from chelation with the metal. An important subgroup is chrome dyes where wool is treated with Cr^{3+} with which it reacts; dye is then applied which in turn complexes with the chromium.

Metals such as chromium and cobalt can be introduced into dye molecules to give larger molecules. They can be regarded as being a special form of mordant dye. The complexes can be formed by chelating one or two molecules of dye with metal. They are applied in a similar manner to acid dyes.

Direct Dyes. These are defined as anionic dyes, again containing sulfonic acid groups, with substantivity for cellulosic fibers. They are usually azo dyes and can be mono-, dis-, or polyazo, and are in general planar structures. They are applied to cellulosic fibers from neutral dyebaths, i.e., they have direct substantivity without the need of other agents. Salt is used to enhance dyebath exhaustion. Some direct dyes can be applied to wool and polyamides under acidic conditions, but these are the exception.

Fiber-Reactive Dyes. These dyes can enter into chemical reaction with the fiber and form a covalent bond to become an integral part of the fiber polymer. They therefore have exceptional wetfastness. Their main use is on cellulosic fibers where they are applied neutral and then chemical reaction is initiated by the addition of alkali. Reaction with the cellulose can be by either nucleophilic substitution, using, for example, dyes containing activated halogen substituents, or by addition to the double bond in, for example, vinyl sulfone, $-SO_2CH=CH_2$, groups.

Basic Dyes. These are usually the salts of organic bases where the colored portion of the molecule is the cation. They are therefore sometimes referred to as cationic dyes. They are applied from mild acid, to induce solubility, and applied to fibers containing anionic groups. Their main outlet is for dyeing fibers based on polyacrylonitrile.

Vat Dyes. The basic mechanism of vat dye application is the conversion of an insoluble complex polycyclic molecule based on the quinone structure into a soluble leuco form by treatment with alkaline-reducing agents. This leuco form is then absorbed onto cellulose. Once the dye has been exhausted into the cellulose it is reconverted *in situ* to the insoluble pigment form which is trapped within the fiber. These dyes have high wet- and lightfastness. A subgroup of vat dyes is the solubilized vat dyes which are temporarily solubilized to allow easy application without reducing agents followed by regeneration of the insoluble dye after dyeing. These dyes are no longer of commercial importance.

Sulfur Dyes. These are complex molecules containing sulfur obtained from the reaction between selected organic intermediates such as 4-aminophenol, or *p*-phenylenediamine and molten sulfur or polysulfide. The actual structures of sulfur dyes are largely unknown although it is considered that they possess sulfur-containing heterocyclic rings. They are applied like vat dyes with the leuco form being generated by using a reducing agent such as sodium sulfide.

Disperse Dyes. These are substantially water-insoluble dyes applied from aqueous dyebath in a finely dispersed form. They are the most important class of dye for dyeing hydrophobic synthetic fibers such as polyester and acetates.

Ingrain Dyes/Azoic Dyes. These are dyes that are formed in the fiber by applying precursors. An example of this class are the azoic dyes. With these dyes a coupling component is applied to the fabric followed by a diazonium compound to form the insoluble dyes. Alternatively, a stabilized mix of the two can be applied and the insoluble azoic dye created on the fiber in a separate treatment, e.g., acid steaming. This is a traditional method for obtaining bright heavy shades cheaply but has lost some popularity as a result of poor rubbing fastness and the decreasing availability of the amines and diazonium salts. Other dyes in this group are phthalocyanine compounds which still have commercial importance, particularly in textile printing.

Other Dyes. Other dye classes listed in the *Colour Index* include dyes for leather, solvents, paper, and food. Leather dyes are those acid, direct, mordant, and basic dyes that show substantivity for leather, good diffusion into it, and acceptable fastness. They are essentially applied in an analogous manner to acid or basic dyes. Paper is colored by both inorganic pigments and natural and synthetic organic colorants. The main dyes used are basic, acid, and direct dyes. Solvent dyes can be regarded as similar to disperse dyes. They are small, unsulfonated molecules, plus a few basic dyes that show high solubility in solvents. Finally food dyes, nontoxic colored substances that can be added to food, are included. See also **Colorants (Food)**.

Fluorescent Whitening Agents. These are fluorescent substances that transform invisible ultraviolet light into visible blue light. Fluorescent whitening agents change the appearance of substrates in two ways: by emitting light and therefore increasing the luminosity (brightness); and by changing the shade from yellowish white to bluish white. They are unfavorably influenced by factors such as low uv content light and high uv light absorption of the substrate or other chemicals present on the substrate. Like dyes, fluorescent whitening agents are available in classes analogous to acid, basic, direct, and disperse dyes for application to all substrates, including paper.

Physical and Organic Chemistry of Dyes and the Dyeing Process

The practical characteristic of a dyestuff is that when a textile is immersed in a solution containing a dye, the dye preferentially adsorbs onto and diffuses into the textile. The thermodynamic equations defining this process have been reviewed in detail. The driving force for this adsorption process is the difference in chemical potential between the dye in the solution phase and the dye in the fiber phase. In practice it is only necessary to consider changes in chemical potential and to understand that the driving force is the reduction in free energy associated with the dye molecule moving from one phase to the other, as the molecule always moves to the state of lowest chemical potential.

Influence of the Fiber. In order for a dye to move from the aqueous dyebath to the fiber phase the combination of dye and fiber must be at a lower energy level than dye and water. This in turn implies that there is a more efficient, lower energy sharing of electrons or intramolecular energy forces, and there are a number of mechanisms that allow this to happen.

Fibers exist as natural, or synthetic, hydrophilic, hydrophobic, nonionic, and ionic. Natural fibers have complex chemical structures with a multitude of possible points of attraction for a dyestuff and are difficult to characterize because of the structure being strongly influenced by regional, climatic variations and the species of plant or animal. Dyeing of natural fibers is therefore much more complex than dyeing synthetic fibers where structures can be characterized and the availability of points of attraction can be deliberately engineered into the fiber's molecular chain. The various types of fiber are summarized in Table 1. The fiber type dictates the type of dye needed.

Modes of Attraction

The force of attraction between a dye and fiber results from the usual electronic interactions. They include ionic forces (coulombic attraction), ion-dipole forces, hydrogen bonds, charge-transfer forces, van der Waals forces, hydrophobic interaction, and covalent bonds.

Dyestuff Organic Chemistry

Dyestuffs impart color to textiles because of their ability to absorb electromagnetic radiation in the wavelengths visible to the human eye (400–650 nm). When white light strikes a dyestuff molecule certain wavelengths are absorbed, depending on the molecular construction, and others are reflected. The wavelengths of the reflected light give the specific color of the dyestuff.

Dyestuff organic chemistry is concerned with designing molecules that can selectively absorb visible electromagnetic radiation and have affinity for the specified fiber, and balancing these requirements to achieve optimum performance. To be colored the dyestuff molecule must contain unsaturated chromophore groups, such as azo, nitro, nitroso, carbonyl, etc. In addition, the molecule can contain auxochromes, groups that supplement the chromophore. Typical auxochromes are amino, substituted amino, hydroxyl, sulfonic, and carboxyl groups.

There is little correlation between classifications according to chemical type and application properties. Application classifications are of most practical usefulness to the dyer.

The Dyeing Process

The physical chemistry associated with dyeing has been described elsewhere both in detail and in summary. The purpose of this treatment is to outline those basic concepts that have a direct impact on dyeing in order to appreciate the fundamental processes taking place.

Zeta Potential. When a textile is immersed in water a negative charge is developed on its surface. This is called the zeta potential. This happens even with ionic fibers in neutral dyebaths. Negatively charged dyes therefore are coulombically repelled.

Internal and External Phases. When dyeing hydrated fibers, for example, hydrophilic fibers in aqueous dyebaths, two distinct solvent phases exist, the external and the internal. The external solvent phase consists of the mobile molecules that are in the external dyebath so far away from the fiber that they are not influenced by it. The internal phase comprises the water that is within the fiber infrastructure in a bound or static state and is an integral part of the internal structure in terms of defining the physical chemistry and thermodynamics of the system. Thus dye molecules have different chemical potentials when in the internal solvent phase than when in the external phase. Further, the effects of hydrogen ions (H^+) or hydroxyl ions (OH^-) have a different impact. In the external phase acids or bases are completely dissociated and give an external or dyebath pH. In the internal phase these ions can interact with the fiber polymer chain and cause ionization of functional groups. This results in the pH of the internal phase being different from the external phase and the theoretical concept of internal pH.

Isotherms. When a fiber is immersed in a dyebath, dye moves from the external phase into the fiber. Initially the rate is quick but with time this slows and eventually an equilibrium is reached between the concentration of dye in the fiber and the concentration of dye in the dyebath. For a given initial dyebath concentration of a dye under given dyebath conditions, e.g., temperature, pH, and conductivity, there is an equilibrium concentration of dye in fiber, D_f, and dye in the dyebath external solution, D_s. Three models describe this relationship: simple partition isotherm, Freundlich isotherm, and Langmuir isotherm.

With simple partition the situation is comparable to the partition of a solute between two solvents. The bonding forces involved between uncharged dye and uncharged fiber, and uncharged dye and uncharged solvent are considered to be the same. The dye is sometimes referred to as in solid solution in the fiber. The simple partition isotherm is found in practice with disperse dyes on cellulose acetate and polyester. It represents the dyeing situation with the minimum restrictions for the dye to enter the fiber; the only restriction is when the fiber solution becomes saturated.

The Freundlich isotherm, where the dye in fiber D_f is directly proportional to $(D_s)^x$ and a plot of $\log D_f$ against $\log D_s$ gives a straight line, is generally found with cellulosic and other ionic hydrophobic fibers.

In a Langmuir isotherm, the reciprocal of dye in fiber $1/DF_f$ is directly proportional to the reciprocal of dye in the dyebath $1/DF_s$. A plot of $1/D_f$ against $1/D_s$ therefore gives a straight line. Langmuir isotherms are typically found with ionic synthetic fibers and ionic dyes, e.g., dyeing polyacrylonitrile with modified basic dyes, and on hydrophilic fibers in situations when the number of sites becomes very low. This may arise when the internal pH is such that only a small number of sites ionize.

Aggregation, Activity, and Solution. Theoretical treatments use the term activity which assumes that the dyestuff is present in a monomolecular dissociated state in solution. Dyes are not generally in this state except at very dilute concentrations; some molecular interaction is more likely. Thus practical situations are not fully described or characterized by theoretical treatments assuming monomolecularity, but the errors involved are usually of no practical consequence. However, when gross aggregation takes place there is significant interference as the dyestuff available for absorption is removed and previous theoretical considerations become invalid. Where aggregation takes place less dye is absorbed than predicted by theory. In general, aggregation is therefore to be avoided, except in exceptional situations where it is introduced to deliberately slow down the rate of dyeing. Dyebath conditions are usually adopted to minimize the potential for aggregation.

Rate of Diffusion. Diffusion is the process by which molecules are transported from one part of a system to another as a result of random molecular motion. This eventually leads to an equalization of chemical potential and concentration throughout the system, and in the case of dyeing an equilibrium between dye in the fiber and dye in the dyebath. In dyeing there are three stages to diffusion: diffusion of dye through the bulk solution of the dyebath to the fiber surface, diffusion through this surface, and diffusion of dye from the surface into the body of the fiber to

allow for more dye to diffuse through the surface layer. These processes have been summarized elsewhere.

Level Dyeing. The concept of obtaining a level dyeing in reasonable time is fundamental in practical processes. Dyes that have a low affinity are likely to diffuse to the fiber surface slowly (a low rate of strike), and once in the fiber quickly diffuse through the fiber to give a uniform distribution of dye molecules. Because of the low affinity of these dyes there are no strong forces of attraction restricting movement. These dyes exhibit good migration and are likely to give level dyeings.

Level dyeings can be obtained by applying dyes of low affinity that dye quickly if unevenly, and allow them to migrate to give uniform level dyeings by extending the time of dyeing. This simple approach has the disadvantage that such low affinity dyes produce dyeings of relatively low wetfastness. Good wetfastness requires high affinity dyes. Because these exhibit low migration it is necessary to ensure that a level dyeing is obtained from the start and maintained throughout the dyeing process.

In order to ensure level dyeings from high affinity dyes it is necessary to make them behave as if they were of low affinity. This is done by reducing the difference between their chemical potential in solution and fiber. The techniques used include: higher temperature for more diffusion; change in pH to control the number of sites available; addition of an electrolyte to either compete with the dye for sites, or neutralize the sites preventing ionic attraction; and addition of auxiliary agents that either compete with the dye for the fiber by lowering the chemical potential in the dyebath phase thus making it a more attractive environment for the dye, or by removing dye from the equilibrium by forming a temporary complex or aggregate in the dyebath. In essence, in order to obtain level dyeings with higher affinity dyes the objective is to slow down the rate of dyeing.

Compatibility. To produce a desired shade more than one dye is usually needed. Often combinations of three dyes are used, e.g., yellow, red, and blue, in order to obtain the maximum number of shades available from the minimum number of dyestuffs. In order to give uniform coloration it is necessary to apply such mixtures of dyes from the same dyebath. As the dyeing proceeds the textile takes up more of the dyes. If the hue of the fabric is the same throughout all stages of dyeings, simply becoming stronger with time, and the hue of the final textile is reproducibly uniform both on its surface and within its interior, e.g., for a wound package of yarn, then the dyes are said to be compatible under the dyeing conditions used. Compatibility is mainly a function of exhaustion rate, but can also be influenced by migration.

Dyeing of Cellulosic Fibers

Preparation for Dyeing. Cotton fibers are coated with natural waxes and pectins. These can be removed by aqueous alkalies at 80°C or above or by solvent treatment to improve the absorbancy and dyeability of the fibers. Cotton may be made suitable for dyeing in a variety of forms, such as raw stock, yarn, or piece goods. Raw stock is normally dyed without thorough dewaxing, since the natural waxes aid in subsequent spinning operations. Careful preparation of cotton piece goods is essential to achieve suitable dye penetration, fastness, and general appearance. Fabric construction dictates whether the fabrics will be processed in rope or open-width forms. Heavy piece goods, and those which are subject to rubs and crease marks, are handled in open width.

Before dyeing in light or bright shades, the goods should be bleached with hydrogen peroxide and caustic soda to bleach the motes. This operation also helps in the removal of trace impurities that remain after boil-off.

Mercerizing is accomplished by passing the cotton fabric through 15–30% caustic soda under tension. Improved luster and increased dye affinity result. With knitted fabrics it is necessary to remove the knitting oils by either alkali treatment or solvents. Viscose rayon, because of its low wet strength, must be processed under minimum tension at all stages of preparation. Skeins contain few impurities and require only light scouring.

The Dyeing Process. When cotton fiber is immersed in water it develops a negative charge. In order for dyes to show good buildup on cotton the dyes must be soluble, planar, aromatic structures. Solubility is obtained by incorporating negatively charged sulfonic acid groups, i.e., the anions. Therefore the dyes show long-range, coulombic (ionic) repulsion, but very strong short-range van der Waals forces of attraction. Thus there is a potential barrier that the dyestuff molecule has to overcome. Natural and introduced thermal agitation tend to equalize the distribution of ions.

TABLE 1. FIBER–DYE PROPERTY REQUIREMENTS

Fiber name	Type/general classification	Chemical constitution	Ionic nature in dyebath
cotton, linen, and other vegetable fibers	natural, hydrophilic	cellulose	anionic
viscose rayon	synthetic,[a] hydrophilic	regenerated cellulose	anionic
wool, silk, hair	natural, hydrophilic	complex proteins	cationic
nylon	synthetic, somewhat hydrophobic	polyamide	usually cationic
acrylics	synthetic, hydrophobic	modified polyacrylonitriles	anionic
acetate	synthetic,[a] hydrophobic	acetylated cellulose	nonionic
triacetate	synthetic,[a] hydrophobic	acetylated cellulose	nonionic
polyester	synthetic, hydrophobic	polyester	usually nonionic
polypropylene	synthetic, hydrophobic	polyolefin	nonionic

[a] Some references distinguish between synthetic fibers made from synthetic polymers and those made by modification of cellulose (man-made fibers).

The use of salt or similar electrolyte is critical in the dyeing of cellulose. When sodium chloride is added to the dyebath, sodium ions (Na^+) diffuse to the negative charges on the cellulose and neutralize them. With the introduction of the sodium ions (Na^+) the coulombic repulsion force between fiber and negatively charged dye is removed and only the strong attraction forces exist. It is then possible for negatively charged dye to diffuse unhindered to the fiber surface.

Direct Dyes

The simplest way of coloring cellulosic fibers is with direct dyes. The dyeing mechanism follows exactly the outline just described where the addition of salt is used to allow dyestuff to be absorbed on the fiber. This is done carefully to ensure that level dyeing is achieved, especially during the early stages of dyeing.

Leveling Power. Direct dyes are classified according to their leveling characteristics. Class A direct dyes migrate well and have high leveling power, i.e., they have low affinity and high diffusion. Class B direct dyes have poor leveling power and exhaustion must be brought about by controlled salt addition. Class C direct dyes are dyes of poor leveling power which exhaust well in the absence of salt and the only way of controlling the rate of exhaustion is by temperature control.

In all these application methods the same procedure is adopted. At the beginning the rate of dyeing must be as slow as needed to give levelness; once this has been achieved the rate of dyeing is systematically increased to give complete exhaustion.

Wetfastness. Class A direct dyes offer the most trouble-free process for dyeing cellulose. However, they do not always provide sufficient wetfastness.

The Class B and C dyes show better resistance to desorption, i.e., they show higher wetfastness, but they do not overcome it fully, and even the Class C direct dyes show inadequate wetfastness and poor staining of adjacents in fastness tests as a result of the reversible nature of the dyeing process. Attempts to overcome this problem have concentrated on chemical treatment of the direct dyes after they were applied. These treatments essentially make the direct dye molecules already on the fiber much bigger and thereby increase the nonionic forces of attraction or reduce solubility in order to reduce desorption and give good wetfastness. Methods used include applying dyes containing free amino groups, diazotizing them and coupling with a base; aftertreatment with formaldehyde; applying dyes containing hydroxyl groups ortho to the azo group and then aftertreating with metallic salts, e.g., Cu; and applying 1–4% of a cationic surface-active agent (over 15–30 min at 25–60°C) to form a sparingly soluble complex with the dye. Today only the latter two processes are used.

Fiber-Reactive Dyes

Because of the limitations of direct dyes and the ability to use simple acid dye chromophores to give bright washfast dyeings, fiber-reactive dyes have become a well-established, popular way of dyeing cellulose. The growth rate of reactive dye consumption of 3.9% per annum is four times the growth rate of other dyes for cellulosic fibers.

A reactive dye for cellulose contains a chemical group that reacts with ionized hydroxyl ions in the cellulose to form a covalent bond. When alkali is added to a dyebath containing cellulose and a reactive dye, ionization of cellulose and the reaction between dye and fiber is initiated. As this destroys the equilibrium more dye is then absorbed by the fiber in order to re-establish the equilibrium between active dye in the dyebath and fiber phases. At the same time the addition of extra cations, e.g., Na^+ from using Na_2CO_3 as alkali, has the same effect as adding extra salt to a direct dye. Thus the addition of alkali produces a secondary exhaustion.

At the end of the dyeing process there is fixed dye on the fiber, and hydrolyzed dye in both the dyebath and fiber. All active dye disappears. In order to take advantage of the fastness offered by covalently bonding the dye to the cellulose it is necessary to remove all the hydrolyzed dye from the fiber. Unfortunately, the hydrolyzed dye does exhibit some affinity for the cellulose and removing it is a desorption process rather than a rapid physical removal. Thus the removal is a relatively difficult procedure that is a critical part of the total dyeing process. Once a dye is fixed it cannot migrate and therefore dyeings achieved using fiber-reactive dyes must be level before they are fixed.

The Ideal Fiber-Reactive Dye Profile. Figure 1 shows the general profile for the application of a reactive dye. In addition to showing the rate profile of fixation between dye and fiber, three other practical parameters (A–C) are noted.

The overall objective is to make the fixation (C) as high as possible for economic reasons. The closer the fixation (C) is to the total dye on the fiber (B) then the smaller the concentration of [dye–OH] and the less that needs to be removed in the washing-off process after dyeing.

Application Methods. There are many detailed application methods used for applying reactive dyes, and all have been described in detail. Examples of the main methods include cold exhaust dyeing fiber-reactive dyes, warm, hot exhaust dyeing dyes, migration exhaust technique for less than 0.5% depth of shade, all-in method, continuous dyeing, and cold pad-batch dyeing.

Chemical Types. A wide range of reactive groups have been investigated, with 20–30 used commercially and over 200 patented. These have been described in detail elsewhere. Because these reactive groups differ chemically the activation of the reactive systems is different as are the rates of reaction with cellulose, from one reactive system to another. This rate of reaction with cellulose, or reactivity, dictates the temperature and pH needed for dyeing.

The most important reactive groups are those based on halotriazine or halopyrimidine systems, where an activated halogen substituent undergoes a nucleophilic substitution reaction with ionized cellulose, or dyes based on sulfatoethylsulfonyl groups.

Bifunctional fiber-reactive dyes have been developed. The concept behind bifunctional dyes is that if two distinct reactive groups are used, the probability of obtaining dye covalently bonded to the cellulose at the end of dyeing instead of being hydrolyzed is increased because each molecule must react twice with OH to be fully hydrolyzed. The claimed benefits of bifunctional reactive dyes are a generally higher level of fixation, and in the case of mixed reactive groups, suitability for application over a range of temperatures and methods.

Correlation of Application, Affinity, and Reactivity. Figure 2 correlates fiber-reactive dye application suitability to reactivity and affinity.

Vat Dyes on Cellulose

Most are based on the quinone structure and are solubilized by reduction with alkaline reducing agents such as sodium dithionite. Conversion back to the insoluble pigment is achieved by oxidation. The dyes are applied by either exhaust or continuous dyeing techniques. In both cases the process is comprised of five stages: preparation of the dispersion, reduction, dye exhaustion, oxidation, and soaping.

Uses of Vat Dyes. The main characteristic of vat dyes is their excellent fastness to light, water, and other agents, e.g., chlorine. Vat dyes are

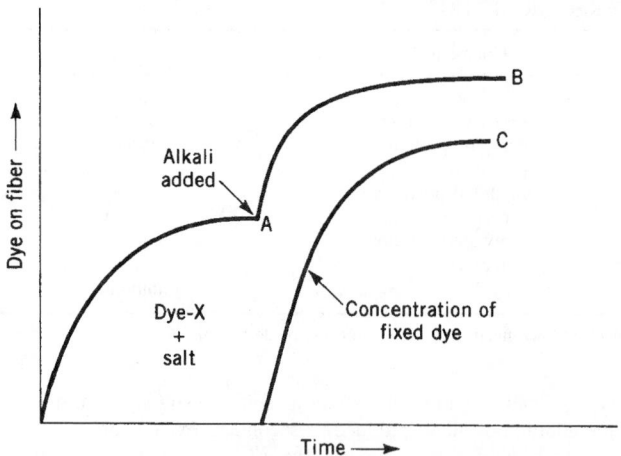

Fig. 1.   Amounts and forms of fiber-reactive dye on the fiber as a function of time for a low affinity dye, where X represents the reactive group. Point A represents the amount of dye exhausted in neutral conditions; B is the total amount of dye exhausted at the end of the dyeing process, i.e., [dye-OH] + [dye-X] + [dye-O-cell]; and C is the amount of dye fixed [dye-O-cell].

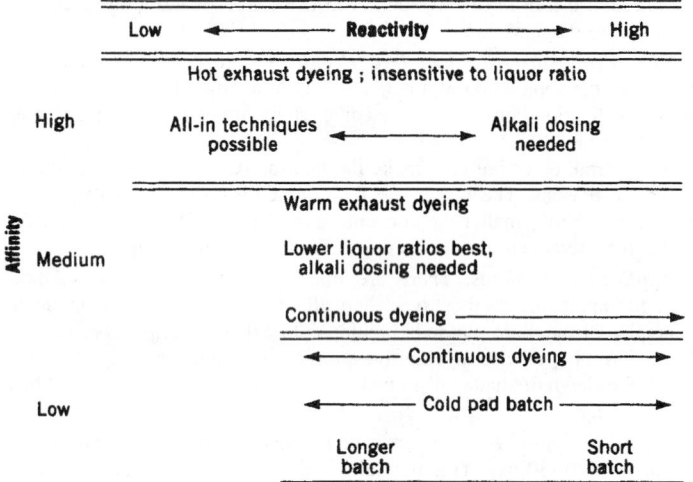

Fig. 2.   Summary of dyeing techniques related to dye reactivity and affinity characteristics.

therefore widely used in outlets demanding high lightfastness such as outerwear, furnishings, drapes, etc.; high wetfastness and fastness to repeated washing such as workwear; high chlorine fastness such as institutional articles; or where general excellent fastness is required as in the case of sewing threads where it is impossible to know the use of final garment they will be used to construct. The majority of vat dyes used worldwide are applied by continuous dyeing; polyester–cotton blends are the most important substrate.

Sulfur Dyes. These are a special case of vat dyes and behave in an analogous manner except that the reducing agent used is sodium sulfide. In order to obtain rapid oxidation chemical oxidizing agents are used. The main outlet for these dyes is in the economic production of navy and black shades on woven fabrics by continuous dyeing, often applying the pre-reduced form of the sulfur dye.

Indigo. Indigo is similar to vat dyes in its application, however, it is not based on a quinone structure but on indigotin. In the presence of alkaline-reducing agents the C=O group is reduced to CH—OH and the dye rendered soluble. As with vat dyes the reaction is reversible, oxidation being achieved with atmospheric oxygen. The principal use for indigo is in denim.

Dyeing of Wool

Preparation for Dyeing. Raw wool must be cleaned before it can be efficiently carded, combed, otherwise processed, or dyed.

Mechanism of Dyeing. Wool has a polymeric structure based on amino acids. It is dyed either in its neutral or in its net positively charged form. As with cellulose, wool, being a hydrophilic fiber, is dyed with water-soluble dyes that contain sulfonic acid groups to impart solubility.

The dyeing of wool is carried out by applying a negatively charged dye to a neutral (or slightly negatively charged substrate) or to a strongly positively charged fiber, depending on pH. Strong ionic attraction exists that results in high affinity and rapid uptake of dye, so rapid that it is essential to control this rate of uptake if level dyeing is to be achieved. The wool dyeing processes are therefore designed around methods of obtaining level dyeings under practical application conditions.

Acid Dyes

Classes. There are three classes of acid dyes: acid leveling, acid milling, and super milling. Acid leveling dyes are molecular dispersions at low temperatures (true solutions) and are simple molecules. Acid milling dyes are colloidal dispersions at low temperatures and true solutions at high temperatures. Super milling dyes are colloidal dispersions at both low and high temperatures and are complex molecules containing low alkyl hydrophobic chains to enhance fastness.

Controlling Dyeing Behavior. As with all other dyes the dyeing process concentrates on obtaining level dyeings within an economic time period, and once again slower dyeing means better control and level dyeing is enhanced. When dyeing wool with acid dyes four factors control the dyeing behavior: pH of the dyebath, presence and concentration of electrolyte, temperature of the dyebath, and choice of dyestuff class.

Practical Processes. With acid leveling dyes no real problems exist because the dyes show good migration, electrolyte is added from the beginning, and rather like Class A direct dyes on cotton, level dyeing is achieved by prolonging the times at the boil.

The other extreme is found with super milling dyes when at the start ammonium acetate, sulfate, or an organic ester is present without any electrolyte. Dyeing is carried out more slowly taking some 60 min to reach the boil, and often the dye is applied with a cationic leveling agent.

Acid milling dyes are intermediate in behavior, being applied with acetic or formic acid in the presence of sodium sulfate. A disadvantage of acid dyes is that their wetfastness depends on the formation and maintenance of a salt linkage between the charged wool and dye. This requires an acidic internal pH to be maintained in the wool.

Mordant Dyes/Metal Complex Dyes

Certain acid dyes can have their fastness properties improved by combining the dye with a metal atom (chelation). The most common metal is chromium, although cobalt is sometimes used, and this can be introduced in a number of ways. The basic mechanism is donation of electron pairs by groups in the dye (ligands) to a metal ion.

Methods of Introducing Metal. Methods of introducing metal include chroming and dyeing together, afterchroming, and metal introduced into the dyestuff molecule in manufacture.

Dyeing Wool with Fiber-Reactive Dyes

Fiber-reactive dyes are by no means as popular for dyeing wool as they are for cotton because the fastness of fiber-reactive dyes on wool is not that much better than other dyes. They are difficult to apply and the severe washing treatments needed to remove unfixed dye can damage the wool itself. They are used on specially treated wools that have been made suitable for washing in automatic washing machines by treating with a polymer. The dye reacts with cations in the polymer. These dyes are not very commercially important, as the need for bright high fast shades on wool is not as high as on cotton. The dyeing mechanism has been described in detail elsewhere.

Silk

Because it is also a protein, silk can be dyed as wool, but in practice the dyes used are generally acid dyes in view of the fiber not being treated to any severe washing in its life. The main difference between wool and silk is in the preparation of the fiber for dyeing.

Silk in its raw state is coated with sericin. It is necessary to remove this gum in order to develop the silk luster and dyeability. Synthetic detergent systems, such as higher alcohol sulfates, and soda ash and boric acid have replaced soap to a large extent for degumming.

Dyeing of Synthetic Polyamides

Dyeing Mechanism. Nylon is similar in its general chemical structure to the natural fiber wool, and therefore all the previously described processes for wool are applicable to dyeing nylon with acid, metallized, and other dyes. There are, however, significant differences. Nylon is synthetic, it has defined chemical structure depending on the manufacturing process, and it is hydrophobic.

Chemically there are important differences. There are no side chains and unlike wool the number of amino and carboxylic groups differs; there is an excess of carboxylic groups. The numbers of amino groups can be changed by chemical modification, e.g., in deep dyeing nylon, but for the most part nylon fibers can be considered to have a limited number of sites, which can differ from one chemical type to another.

Physically there are differences. Like all polymeric fibers nylon contains crystalline and noncrystalline areas. Only amino groups in the noncrystalline regions are accessible.

Finally, because the fiber is synthetic, polymer formation followed by drawing into a yarn presents the likelihood of chemical and physical variations in the yarn. It is usual to stabilize the fibers by a heat-setting process before dyeing, and again further physical variation can be introduced at this stage. The manufacturing history of the polymer therefore plays a role in determining the dyeing performance. In order to obtain level dyeings it is necessary to consider not only the basic chemical reactions taking place but also the relative sensitivities of the dyes to physical and chemical variation in the fiber.

Acid Dyes. The majority of acid dyes is applied to nylon rather than to wool. There are three groups of dyes: Group 1 includes dyes with little affinity at neutral or acidic pH but which exhaust under strongly acidic conditions; Group 2, the largest group of dyes which exhaust onto nylon in the pH range 3.0–5.0; and Group 3, dyes with a high affinity for nylon under neutral or weakly acidic pH. Only dyes within one group should be used together, and dyestuff manufacturers assist in this by having different nomenclature for dyes in each group.

Tanning Agents. It is possible to improve the wetfastness of acid dyes by aftertreatment. The original method was to apply tannic acid, tartar emetic, and formic acid.

Metal Complex Dyes. The 1:2 metal–dye complexes are of commercial interest because of their excellent lightfastness in pale shades. These macromolecules are difficult to apply level and are sensitive to both chemical and physical variations. In their application they are treated as the Group 3 acid dyes.

Other Soluble Hydrophilic Dyes. Some direct dyes have profiles on nylon very similar to Group 3 dyes and therefore, to supplement the range of shades available, they are sometimes applied with Group 3 dyes.

Disperse Dyes. The insoluble, hydrophobic disperse dyes readily dye nylon, and because their mode of attraction is completely nonionic they are completely insensitive to chemical variations and pH. Small molecular-sized disperse dyes (ca mol wt 400) show very high rates of diffusion and excellent migration properties and they are insensitive to physical variations in the nylon. As the molecular size of disperse dyes increases they show increasing sensitivity to physical variation.

Although when using disperse dyes on nylon they are readily absorbed at temperatures up to the boil, they are also readily desorbed when the dyed fabric is immersed in wash liquors.

The main use for disperse dyes is where excellent coverage of fibers likely to have physical and chemical variations is needed, and where wetfastness is not critical. The small molecular-weight dyes are therefore widely used for pale shades on continuous filament yarns used in hosiery. There is also some use made in exhaust dyeing of carpets made from continuous bulk filament nylon to give good coverage.

Dyeing is relatively simple. The disperse dye is added to a dyebath containing a nonionic dispersing agent, sodium hexametaphosphate, and sometimes acetic acid is added to give pH 5.5 to prevent decomposition of some disperse dyes. Dyeing is carried out by bringing the dyebath to the boil, and continuing until exhaustion is completed.

Dyeing of Acrylic Fibers

In order to make fibers of commercial interest acrylonitrile is copolymerized with other monomers such as methacrylic acid, methyl methacrylate, vinyl compounds, etc., to improve mechanical, structural, and dyeing properties. Fibers based on at least 85% of acrylonitrile monomer are termed acrylic fibers; those containing between 35–85% acrylonitrile monomer, modacrylic fibers. The two types are in general dyed the same, although the type and number of dye sites generated by the fiber manufacturing process have an influence.

Basic dyes are the most popular class applied to acrylic fibers. Like nylon, acrylic can be dyed with disperse dyes, but with the same reservations of fastness. Disperse dyes are therefore only used for pale shades where excellent levelness is needed or difficult to obtain by any other method owing to variations in the fiber.

Preparation for Dyeing. Fabrics are scoured with a synthetic detergent at 45–65°C and are rinsed before further processing to remove tints, size, wax, grease, spinning oils, or other impurities that were applied or picked up during the manufacturing operation. Bleaching, when required, is usually accomplished by means of a sodium chlorite bleach, a selected optical brightener, or a suitable combination of the two. Acrylic-blend fabrics may require other bleaching agents if chlorine-sensitive fibers are present. Most acrylic fibers require a presetting in open-width in boiling water to avoid dimensional stability problems during subsequent wet-processing steps.

Dyeing Mechanism. As the importance of acrylic fibers grew, basic dyes were developed having localized charge in one specific part of the molecule, allowing stronger salt links to be formed than with the delocalized type. These newer dyes are often referred to as modified basic dyes. Essentially, their structure is that of a disperse dye that has been protonated. These dyes therefore have high rates of diffusion into the fiber, and their mode of attraction is almost entirely ionic.

The effect of pH depends on the fiber type. The SO_3^- groups on fibers are so strong that they are deprotonated even in neutral dyebaths. The dyebath pH therefore has no influence on the availability of these sites in the fiber and therefore pH cannot be used to control the uptake and level dyeing behavior of the dye. For carboxylic acid groups the pK_a is about 5.5. At lower pH values there are considerably fewer sites available for the dye and at higher pH values considerably more COO^-.

As with wool and nylon when applying dyes to the fiber where dye and site are oppositely charged the need is to control the rate of exhaustion to promote level dyeing. With acrylic this need is made all the more important by two additional factors: first, the modified basic dyes show poor migration because they form very strong salt bonds especially between dyes with delocalized charges and fibers with strongly negative SO_3^- sites; secondly acrylic fibers do not readily dye below their glass-transition temperature, T_g, which is usually around 80°C.

Compatibility Values. The need to apply dyes in admixture to give more shades necessitates a way of measuring the compatibility of dyes. Depending on charge, the degree of localization, and molecular shape and size, dyes have different affinities and behavior and hence different dyeing rates. A qualitative testing procedure has been defined where dyes are applied with a range of known dyes and the unknown dye is ascribed the same compatibility value as that already given to the known dye with which it dyes compatibly under all practical exhaust-dyeing conditions except in the presence of anionic dyes or auxiliaries. There are five values and in combinations the dye with the lowest value exhausts most rapidly. For best results dyes should be mixed with dyes having the same compatibility value, or at least no more than one value different.

Level Dyeing Techniques. It is exceptionally difficult to obtain level dyeings on acrylic, and temperature and pH control depend on fiber type and are not always adequate. Sodium sulfate in limited amounts can be used to some effect.

The more popular method to control leveling is to use cationic products that act as colorless dyes competing with the colored cationic dyes for the fiber sites. If amounts of colored modified basic dye and colorless modified basic dye equal to the saturation value of the fiber are uniformly dissolved in the dyebath, then level dyeing behavior is promoted.

Dyeing of Polyester

Polyester fibers are based on poly(ethylene terephthalate) (PET); some modified versions are formed by copolymerization, e.g., basic dyeable polyester. The modified forms dye in analogous manner to other fibers of similar charge.

Preparation for Dyeing. A hot alkaline scour with a synthetic surfactant and with 1% soda ash or caustic soda is used to remove size, lubricants, and oils. Sodium hypochlorite is sometimes included in the alkaline scouring bath when bleaching is required. After bleaching, the polyester fabric is given a bisulfite rinse and, when required, a further scouring in a formulated oxalic acid bath to remove rust stains and mill dirt which is resistant to alkaline scouring.

Dyeing Mechanism. Unmodified polyester fibers are very hydrophobic and absorb only minimal amounts of water, and are therefore only dyeable with hydrophobic disperse dyes. The mechanism of dyeing is by simple partition, the so-called solid solution mechanism. Disperse dyes are only sparingly soluble and therefore high temperatures are needed to increase the amount of soluble dye in the system. Disperse dyes on polyester generally have good fastness as a result of the fiber being below its glass-transition temperature in wash treatments and the slow rate of desorption. The degree of crystallinity, the drawing, and heat-setting temperatures of polyester all play a role in determining the rate and amount of dye uptake.

Disperse Dyes. There is a general correlation between heat fastness, the propensity to desorb under conditions of dry heat onto a white piece of polyester, and the dyeing properties of disperse dyes. Low energy dyes are not usually used in thermofixation, as their low heat fastness at the thermofixation temperatures used (200–210°C) results in the subliming of them from the hot fabric.

Medium energy dyes are based on larger sized molecules than the low energy dyes. They have slower rates of dyeing, better heat fastness, and generally higher wetfastness. They are not suitable for carrier dyeing. Their main application methods are exhaust dyeing at temperatures of 125–135°C, and for continuous dyeing by thermofixation at around 30–60 s at 190–210°C. Because of their medium molecular size these dyes dye rapidly (15–30 min) at 125°C.

High energy dyes are based on large molecules with polar groups. These dyes have excellent heat fastness resulting from extremely low rates of sublimation. Their main use is in dyeing fabrics that are to be given a subsequent high temperature heat treatment. Dyeing with these dyes requires either longer times or temperatures than with medium energy dyes to achieve full exhaustion, e.g., 45–60 min at 125–135°C in exhaust dyeing.

Dyeing Processes. Polyester yarns and fabrics are usually dyed by exhaust techniques; continuous dyeing is largely used only for blends with cellulose. The basic dyeing process is relatively simple. The dyebath is set with disperse dye and dispersing agent (a nonionic or anionic surface-active agent) at pH 5.5 obtained with, for example, acetic acid. The temperature is slowly raised up to the dyeing temperature (125–135°C) and kept there to complete exhaustion and promote migration followed by cooling to below the boil for removal of the dyed material.

During the cooling process after dyeing, the solubility of the disperse dye remaining in the dyebath decreases rapidly and it can precipitate on the surface of the polyester fibers. If it is not removed, the resulting dyeing will exhibit both poor fastness to rubbing and poor wetfastness. This precipitated dye is removed by a combined chemical decomposition and stripping. The dye is destroyed by reduction using hot (70°C) caustic soda and sodium hydrosulfite, optionally in the presence of a detergent.

This process, based on strong reducing agents, can be avoided by the use of disperse dyes that are removed by aqueous alkali alone. Two types of dye are used: dyes containing diesters of carboxylic acid and dyes destroyed by mild alkali.

Thermal Migration. In any subsequent heat treatment of the polyester such as heat-setting to stabilize the fiber or fabric, or in the application of a finishing agent such as a softener or antistat, the polyester is again taken above its glass-transition temperature, and dyestuff molecules again have mobility within the fiber.

Some general observations are that low energy dyes thermally migrate more than medium or high energy dyes, presumably because of their tendency to sublime out of the fiber at high temperature; the behavior of medium and high energy dyes has no correlation to their heat fastness; the more polar the dye the greater the likelihood of thermal migration; the higher the temperature the greater the risk; and the presence of hydrophobic finishing agents increases the likelihood of thermal migration.

Dyeing of Cellulose Esters

Acetate fibers are dyed usually with disperse dyes specially synthesized for these fibers. They tend to have lower molecular size (low and medium energy dyes) and contain polar groups presumably to enhance the forces of attraction by hydrogen bonding with the numerous potential sites in the cellulose acetate polymer. Other dyes can be applied to acetates such as

acid dyes with selected solvents, and azoic or ingrain dyes can be applied especially for black colorants. However their use is very limited.

Cellulose Diacetate. When preparing cellulose diacetate for dyeing, strong alkalies must be avoided in the scouring of acetate because the surface of the cellulose acetate would be saponified by such treatment.

Very small quantities of acetate staple are dyed; however, large quantities of acetate filament are found in satin, taffeta, and tricot fabrics. These are usually dyed open-width on a jig owing to their inclination to crease or crack easily.

Cellulose Triacetate. Cellulose acetate having 92% or more of the hydroxyl groups acetylated is referred to as triacetate. This fiber is characteristically more resistant to alkali than the usual acetate and may be scoured, generally, in open-width, with aqueous solutions of a synthetic surfactant and soda ash. Triacetate is a hydrophobic fiber, as compared to secondary acetate, and consequently does not dye rapidly. It is necessary to increase the rate of diffusion of the disperse dye into the fiber by increasing the dyeing temperature to 110–130°C or using a dye accelerant or carrier, or both.

Dyeing of Fiber Blends

Fiber blends combine the advantageous properties of two or more fibers into one fabric. They are available as blends of natural fibers, synthetic fibers, or natural fibers blended with synthetic. The differences in dyeability between the many fibers on the market open a wide field of multicolored yarns and fabrics to the stylist.

Fiber blends can be dyed into union shades (tone-on-tone) or multicolor effects can be obtained by coloring the individual components in different shades or by maintaining one fiber in an undyed state (reserving). A complete reserving of a fiber is not possible in all cases.

When dyeing fiber blends it must be decided whether the fibers can be dyed simultaneously from the same dyebath, or separately and in what order from different dyebaths.

With respect to fiber components that are dyed with completely different dye classes, the ability to use single-bath techniques (exhaust and continuous) depends on the interaction between the dyes and the compatibility of their dyeing procedures.

Cellulosic Fiber Blends

Cellulosic–Polyester Fibers. One of the most important fiber blends on the market is the mix of 35/65 or 50/50 cotton–polyester. High tenacity viscose fibers are sometimes used instead of cotton. Although the knitgoods are dyed in exhaust dyeing procedures, most of the woven fabrics are dyed according to one of the continuous dyeing processes. The choice of dyes and hence dyeing method is determined by the fastness properties required.

Cellulosic–Acrylic Fibers. Commonly this blend is used in knitgoods, woven fabrics for slacks, drapery, and upholstery fabrics. Since anionic direct dyes are used for the cellulosic fiber and cationic dyes for the acrylics, a one-bath dyeing process is only suitable for light to medium shades. Auxiliaries are needed to prevent precipitation of any dye complexes.

In two-bath processes either the cotton or the acrylic can be dyed first. Heavy shades are best dyed by first dyeing the acrylic and then dyeing the cotton under alkaline conditions. In order to prevent desorption of the cationic dye the dyeing temperature for the cotton dyeing must be below the glass-transition temperature for the acrylic of 80°C.

Cotton–acrylic fiber blends are also used for high quality upholstery pile fabrics. Besides the one-bath exhaust dyeing procedure involving a very high ratio of liquor to bath, a continuous pad-steam process is used to dye these fabrics.

Cellulosic Fiber–Nylon Blends. These blends are used in fabrics for apparel, corduroy, and swimwear. If wetfastness requirements are relatively low, the nylon portion can be dyed with disperse dyes and the cellulosic fiber with direct dyes and a one-bath procedure can be employed. For better wetfastness, the nylon portion is dyed with level dyeing acid colors together with the direct dyes in one bath at 95°C using a reserving agent to prevent the direct dyes from dyeing the nylon. An aftertreatment with a cationic fixative improves the wetfastness properties. For swimwear, the cotton portion is dyed with fiber-reactive dyes. After rinsing hot and cold and soaping at the boil, the nylon portion is dyed with a phosphate buffer system. Selected acid and/or acid milling colors are applied. An aftertreatment with a phenolsulfonic acid condensation product results in best wetfastness properties.

Wool Blends

Wool–Cellulosic Fibers. One of the oldest fiber blends in the textile market is the combination of wool and cotton or wool and viscose. In a one-bath process, selected direct and acid dyes are applied at pH 4.5–5.0 at 98–100°C. A phenolsulfonic acid condensation product is added as a reserving agent, to prevent the direct dyes from dyeing the wool under acid conditions. If optimum wetfastness properties are required, fiber-reactive dyes can be applied to both fibers by use of a two-bath process.

Wool–Nylon. Nylon has been blended with wool in order to give additional strength to the yarn or fabric. It is used mainly in the woollen industry for coats and jackets and, to a lesser extent, for socks and carpet yarns. Both fibers are dyed with the same products, however the fibers have different affinity to them. Generally level dyeing acid dyes are applied.

Wool–nylon upholstery fabrics and carpet yarns require higher light–and wetfastness properties. Neutral premetallized dyes are used in these cases. However, they have a much higher affinity to the nylon than the wool. Therefore, stronger retarding agents have to be employed, e.g., phenolsulfonic acid condensation products.

Wool–Acrylic Fibers. This blend is being used for industrial and hand knitting yarns. Special precautions are necessary because the two fibers are colored with dyes of opposite ionic type. Usually, level dyeing acid dyes are used for the wool portion in combination with the cationic dyes for acrylic fiber.

Wool–Polyester Fibers. The wool–polyester blend is the most common fiber combination in the worsted industry. Disperse dyes for polyester and acid or neutral premetallized dyes for wool are employed in a one-bath process.

Blends of Synthetic Fibers

Polyester Fiber Blends. Disperse dyeable and cationic dyeable polyester fibers are frequently combined in apparel fabrics for styling purposes. Whereas the disperse dyes dye both fibers, but in different depths, selected cationic dyes reserve the disperse dyeable fiber completely, resulting in color/white effects.

Polyester Fiber–Nylon Blends. This fiber blend is used in apparel fabrics as well as in carpets. Disperse dyes dye both fibers, however they possess only marginal fastness properties on nylon. Therefore it is important to select those disperse dyes that dye nylon least under the given circumstances. The nylon is dyed with acid dyes, selected according to the fastness requirements.

Polyester Fiber–Acrylic Fiber Blends. This fiber blend is dyed in a similar fashion to that of the blends of the different polyester fibers. The selection of cationic dyes is substantially larger for the acrylic blend.

Nylon Blends. Differential dyeing nylon types and cationic dyeable nylon blends are used primarily in the carpet industry. The selection of cationic dyes for nylon is rather limited; most products have very poor fastness to light. These blends are dyed in a one-bath procedure at 95–100°C. Selected acid dyes are used for differential dyeing. Disperse dyes will dye all different types in the same depth.

Elastomeric Fibers. Elastomeric fibers are polyurethanes combined with other nonelastic fibers to produce fabrics with controlled elasticity. See also **Fibers: Elastomeric**. Processing chemicals must be carefully selected to protect all fibers present in the blend.

Dyeing is carried out by the method best suited to the fiber used as the outer sheath, e.g., acid or premetallized dyes for nylon-based, reactive or direct dyes for cotton-based.

Other Application Procedures

Pigment Dyeing. Many dyers do not look upon this form of coloration as dyeing; nevertheless millions of meters of fabric are dyed by this system each year. A finely dispersed (0.5–5.0 μ diameter) organic pigment is applied by padding together with organic binders and, depending on the binder system, a catalyst. After drying, the fabric is cured at 170–175°C when polymerization and optionally cross-linking of the binder takes place. The typical binder systems used are acrylic and butadiene resins. The pigments used cover the range of azoics, carbon black, phthalocyanines, triphenylmethanes, and dioxazine derivatives. They give excellent lightfastness and good washfastness in pale to medium shades. Fabrics being used range from lightweight poplins and sheetings to corduroy of cellulosic or fabrics of polyester–cellulosic blends. In addition

to polyester–cellulosic and cellulosic fabrics, pigments may also be applied to 100% synthetic fibers of special construction for unique uses.

Solvent Dyeing. Solvent dyeing generally refers to dyeing in nonaqueous media. In the early 1970s, solvent dyeing was expected to become the dyeing process of the future and was discussed and researched extensively. This interest did not materialize into practical acceptance and the technique has not achieved importance.

Dyeing Machinery

In the application of dyes three techniques are used: the dye liquor is moved as the material is held stationary, the textile material is moved without mechanical movement of the liquor, or both move.

Transportation of Dye Liquor through the Textiles. Regardless of the form of the textile, raw stock, sliver, yarn, or cloth, the principle is generally the same. A large stainless steel kier, capable of withstanding sufficient pressure to reach a maximum operating temperature of 145°C, has one or more perforated spindles through which the dye liquor is pumped. Around this spindle the textile is packed tightly in the form of a cake in a perforated basket, as yarn, or in a package as a sliver or tow in a can, or as cloth around a beam barrel. The dye liquor is pumped through the textile, then flows to the bottom of the machine and into the return side of the pump.

Transportation of the Textile Material with No Mechanical Movement of the Liquor. The chain warp dyeing procedure is widely used in the dyeing of indigo on warp yarns which are in the form of ropes or chains. In this procedure, several warp ropes are pulled through a series of tubs containing the dye liquor and gradually are dyed to the desired shade. Piece goods are dyed by means of a batch process on a dye jig. Winch or beck dyeing is one of the oldest forms for dyeing mechanized piece goods (fabric) (Fig. 3). In continuous dyeing, the equipment at hand may be simple padder or a complete dye range. Most dye padders consist of a medium density rubber roller across the width of which pressure can be applied. This roller presses against a stainless steel or a hard rubber roller. These rollers may be mounted either vertically or horizontally. The cloth is passed through a stainless steel pad box, down under a rod or a roller which is below the dye liquor level, between the squeeze rollers.

Dye ranges can be of different configuration depending on the composition of the fabrics being processed and dye system used. All contain similar units; differences are mainly in the method of heating.

The following sequence is typical for polyester–cotton blended fabric. The infrared units reduce moisture content 20–30% and greatly minimize uneven migration of the dye on the wet goods when they go onto the drying cylinders. The dried cloth progresses into the thermosol unit where the dyestuff for the synthetic portion of the fabric is fixed. Goods then continue through the chemical pad for immersion into an alkaline (or reducing) solution, depending on whether fiber-reactive or vat dyes are applied to the cellulosic fibers. They then pass through a steamer and finally through 8–10 wash boxes, which contain various chemicals depending on the class of dyestuff applied.

Machines Based on Movement of Both Dye Liquor and Material. One example of a machine in which both yarn and dye liquor are moved is the

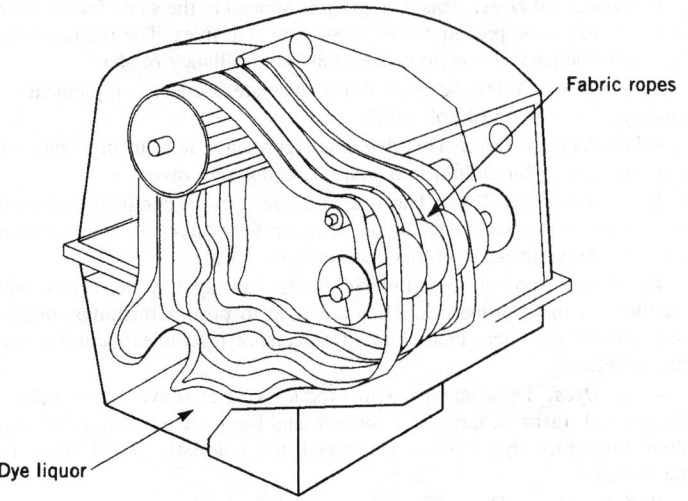

Fig. 3. Winch or beck dyeing machine.

Klauder-Weldon skein dye machine; not only do the skeins turn, but the liquor is pumped in small streams over the yarn as the threads pass over the spindles. This process assures maximum uniformity and levelness.

Another extremely popular machine of this type is the jet dyeing machine which conserves energy by reducing the cloth-to-liquor ratio to 1:10 or lower as compared to 1:20 for the winch. In this machine, the fabric which is in a rope form is transported by movement of the dye liquor through a Venturi jet. This method provides intimate contact between the dye liquor and each meter of material. The machine operates at 40–135°C.

Overflow machines may be thought of as hybrids of jets and winches. They usually feature a winch reel which, unlike the cloth-guiding roller in a jet, is normally driven and provides motive power to the fabric. There is also some driving force on the cloth from the circulation of the liquor through the overflow tube down which they both pass. Both pressurized and nonpressurized versions exist and are available from a large number of different makers. The principal advantage that overflow machines possess over jets is a gentler action on the fabric.

Control of Dyeing Equipment. Over the years, the dyer and machinery manufacturer have applied any mechanical or electrical equipment that would enable them, day after day, to produce repeatable dyeings of top quality. First, thermometers were installed in dye lines; these soon evolved into thermocouples with remote recording. Other improvements were soon developed, such as automatic four-way valves with variable-interval controls, flow controls, pressure recorders, hydraulic and air pressure sets on rollers, pH controls, etc.

Textile Printing

The term textile printing is used to describe the production of colored designs or patterns on textile substrates through a combination of various mechanical and chemical means. In printing on textiles, a localized dyeing process takes place, whereby in general the chemical and physical parameters of dyeing apply.

The process of textile print coloration can be divided into three steps. First, the colorant is applied as pigment dispersion, dye dispersion, or dye solution from a vehicle called print paste or printing ink, containing in addition to the colorant such solutions or dispersions of chemicals as may be required by the colorant or textile substrate to improve and assist in dye solubility, dispersion stability, pH, lubricity, hygroscopicity, rate of dye fixation to the substrate, and colorant-fiber bonding. The second step is the fixation process. During the afterscouring, the third step, the prints are rinsed and scoured in a detergent solution.

Colorants for Textile Printing

Pigments. Pigment-printed textiles represent the highest percentage of all printed textiles, accounting for between 40 and 50% of all cellulose and over 90% of polyester–cotton blend prints.

Disperse Dyes. Disperse dyes are used in powder or pasted form, or ready-to-prepare aqueous dispersions for incorporation into a thickener solution.

Acid Dyes. These dyes have their greatest importance in printing of polyamide.

Premetallized Dyes. This dye group is applied to the same textile fibers and with the same procedures as those with acid dyes. The premetallized dyes offer better fastness properties, but lack brilliancy of shade.

Direct Dyes. A few selected direct dyes are used to complement the acid dyes in printing of polyamide.

Fiber-Reactive Dyes. This dye class represents, next to pigments, the main dye group for cellulosic fibers, i.e., cotton and rayon.

Basic (Cationic) Dyes. The use of basic dyes is confined mainly to acrylic textile fibers, acetate, and as complementary dyes for acid-modified polyester fibers that accept this class of dyes.

Vat Dyes. Applied to cellulosic fibers, vat dyes yield prints with excellent fastness properties. They are used to print furnishings, drapes, and camouflage where their infrared reflectance resembles natural terrain and foliage.

Azoic Dyes. These are used to produce cost-effective heavy yellow, orange, red, maroon, navy blue, brown, and black shades and are printed alongside other dye classes to extend the coloristic possibilities for the designer.

Phthalocyanine Dyes. These dyes are synthesized as the metal complex on the textile fiber from, e.g., phthalonitrile and metal salts.

Dye Combinations. In certain cases it is desirable to print fiber blends with combinations of the appropriate dye classes, rather than with pigments. Only polyester–cellulose blends are of commercial importance and the following dye systems have been developed for them. The dyes of the different classes are contained in the same print paste and, therefore, are applied simultaneously in one print operation. They include disperse–reactive combinations, disperse–pigment combinations, and disperse–vat combinations.

Styles of Printing

Styles of printing include direct printing, discharge printing, resist printing, and wax printing.

Printing Machinery

Textile materials can be printed at different steps of the textile manufacturing process. Woven fabrics comprise the largest percentage of printed goods. In recent years, knitted textile fabrics have considerably increased in importance. However, printing can also be done on yarns in skein form, or on warps being passed from a warp beam to another beam, or as yarn strands. Space printing is a process where a yarn, temporarily knitted into a loose fabric, is printed and then deknitted. Carpets can be printed in woven or tufted constructions. Vigoureux printing is the printing of woolen slubbing. Regardless of the state of the textile material, any printing process makes use of one of the following methods: screen printing or roller printing.

Paper Coloring

Colorants for Paper. Among the colorants that have been and are being used for the dyeing of paper are natural inorganic pigments (ochre, sienna, etc.); synthetic inorganic pigments (chromium oxides, iron oxides, carbon blacks, etc.); natural organic colorants (indigo, alizarin, etc.); and synthetic organic colorants. The last is the largest and most important group.

Basic Dyestuffs. Basic dyestuffs are usually used for dyeing of unbleached pulp in mechanical pulp such as wrapping paper, kraft paper, box board, news, and other inexpensive packaging papers. Their strong and brilliant shades also make them suitable for calendar staining and surface coloring where lightfastness is not critical.

Acid Dyestuffs. Because of poor affinity and good solubility, acid dyestuffs have poor bleedfastness and form colored backwater and are therefore suitable for paper that does not require wetfastness, such as construction grades. Acid dyestuffs are most suitable for calendar staining or surface coloring because of their solubility and brightness of shade.

Direct Dyestuffs. Direct dyestuffs' bonding ability to nonligneous pulps and excellent fastness properties to light and bleeding make them useful for all fine papers. The shades of direct dyestuffs are not as bright as those of acid or basic dyestuffs and in blended furnishes (bleached–ligneous pulps) mottling or graniting may occur.

Pigments. Synthetic organic pigments are replacing the use of some inorganic pigments for ecological reasons. Pigments do not react chemically with the fiber, but are fixed physically and are dependent on filtration, absorption, occlusion, and flocculation. Paper dyeings with pigments have outstanding fastness properties, but poor affinity, low tinctorial strength, and two-sidedness problems limit their application to paper.

Fluorescent Whitening Agents. Fluorescent whitening agents change the appearance of paper in two ways: by emitting light and therefore increasing the luminosity (brightness); and by changing the shade from yellowish white to bluish white.

Dyeing Processes

Paper may be colored by dyeing the fibers in a water suspension by batch or continuous methods. The classic process is by batch dyeing in the beater, pulper, or stock chest. Continuous dyeing of the fibers in a water suspension is adaptive to modern paper machine processes with high production speeds in modern mills. Solutions of dyestuffs can be metered into the high density or low density pulp suspensions in continuous operation.

Nonimpact Printing. Interest is growing in the use of nonimpact styles because of the quickness of color changeover and the ability to interface these machines to computer-aided design systems. Two basic types exist: drop on demand and constant drop techniques.

Dyeing of Leather

Not only may the compound used to convert hide substance into leather vary chemically over a wide range, but the quantities used, the method of application, and the physical condition of the hide prior to tanning or dyeing may vary, with each factor in turn affecting the dyeing properties of the resultant leather. Also, leather retains many of the properties originally associated with the parent substance, and these affect profoundly and, in many ways, limit the dyeing properties of the final product. Chief among these properties are sensitivity to extremes of pH, thermolability, and the tendency to combine with acidic or basic compounds.

Leather Dyes. The main classes of dyes employed in the coloring of leather are the acid, acid/direct, direct, and basic types. On chrome leather, the direct dyes usually have greater affinity and produce fuller or heavier shades than do acid or chrome dyes. Acid/direct dyes as well as the metallized-type dyestuffs may be classified for the purpose of leather dyeing as the main types in use. Basic dyes color chrome leather weakly and unevenly, unless the leather is first mordanted or retanned with suitable materials, such as vegetable tannin, syntans, or previously applied acid and/or direct dyes. They may be used alone on vegetable tanned leather to produce full shades or, as is done more frequently, following a preliminary coloring with acid or acid/direct dyes. In the latter case, basic dyes are used to impart fullness of shade with minimum coloring matter and cost.

Leather Dyeing Methods and Equipment. The methods used in dyeing leather are quite simple and they obtain their names from the equipment employed, such as drum, wheel, paddle, brush, tray, or spray dyeing. Most leather is dyed in drums.

Fastness Tests for Textiles

The principal active bodies in the field of colorfastness testing have been the American Association of Textile Chemists and Colorists (AATCC) and the Europaische-Convention für Echtheitprüfung/Groupement d'Etudes des Commissions Européenes pour la Soliditié (ECC). The ISO subcommittee concerned with colorfastness tests is ISO TC 38/SCI. This meets every two or three years to coordinate developments in standard testing methods and to seek international agreement on proposed new tests and modifications to existing tests. The purpose of ISO is solely to produce useful standard test methods. The setting of specifications and levels of acceptance on the basis of such test methods is a matter to be resolved between buyer and seller.

Testing of Dyes. At the 1989 meeting of ISO/TC38/SC1 in Williamsburg, Va., a new work group (WG11), Characterization of Dyestuffs, was established. The following tests are significant, together with alternative techniques currently being considered for introduction as ISO standards: evaluation of dyestuff migration, thermal fixation properties of disperse dyes on polyester–cotton, transfer of disperse dye on polyester, transfer of basic dyes on acrylics, transfer of acid and premetallized dyes in nylon, dispersion stability of disperse dyes at high temperature, foaming propensity of disperse dyes, and evaluation of the dusting properties of powder (or other solid dyes).

Other New Methods. Because the values obtained are dependent on the conditions of measurement, standard test procedures are under review by ISO for: determination of cold-water solubility of water-soluble dyes; determination of the solubility and solution stability of water-soluble dyes; and determination of the electrolyte stability of reactive dyes.

Safe Handling of Dyes

The Ecological and Toxicological Association of Dyes and Organic Pigments Manufacturers (ETAD), an international body of all primary manufacturers based in Europe but also with standing committees in the United States, Brazil, and Japan, issues clear guidelines for the safe handling of dyes. In December 1991 the United States Operating Committee of ETAD joined with the United States Environmental Protection Agency in publishing a pollution prevention guidance manual for the dye manufacturing industry. See also **Dyes: Environmental Chemistry**.

Colour Index Generic Names

The *Colour Index* assigns CI generic names to commercial dyes. This CI name is defined as "a classification name and serial number which when allocated to a commercial product allows that product to be uniquely identified within any *Colour Index* Application Class." This enables the particular commercial products to be classified along with other products whose essential colorant has the same chemical constitution.

BRIAN GLOVER
Zeneca Colours

Additional Reading

Aspland, J.R.: *Textile Chemist Colourist, A Series on Dyeing*, AATCC, Research Triangle Park, N.C., Oct. 1991–Nov. 1993, Chapts. 1–15.

Colour Index, 3rd Edition. (4th edition in preparation), Society of Dyers and Colourists (SDC), U.K., in collaboration with American Association of Textile Chemists and Colourists (AATCC), Research Triangle Park N.C., 1971.

Shore, J. ed.: *Colourants and Auxiliaries, Organic Chemistry and Application Processes*, Vol. 1, Colourants, Society of Dyers and Colourists, U.K., 1990.

Textiles—Tests for Colour Fastness, ISO 105, AATCC, Research Triangle Park, N.C., 1990.

DYES: ENVIRONMENTAL CHEMISTRY

Effluent Treatment Methods

Methods of effluent treatment for dyes may be classified broadly into three main categories: physical, chemical, and biological.

Physical	Chemical	Biological
adsorption	neuralization	stabilization ponds
sedimentation	reduction	aerated lagoons
flotation	oxidation	trickling filters
flocculation	electrolysis	activated sludge
coagulation	ion exchange	anaerobic digestion
foam fractionation	wet-air oxidation	bioaugmentation
polymer flocculation		
reverse osmosis/ultrafiltration		
ionization radiation		
incineration		

There are four stages: preliminary, primary, secondary, and tertiary treatment processes, which differ mainly by the number of operations performed on the waste steams.

Preliminary treatment processes of dye waste include equalization, neutralization, and possibly disinfection. Primary stages are mainly physical and include screening, sedimentation, flotation, and flocculation. The objective is to remove debris, undissolved chemicals, and particulate matter. Secondary stages are used to reduce the organic load, which essentially is a combination of physical/chemical separation and biological oxidation. Tertiary stages are important because they serve as a polishing of effluent treatment. These methods are adsorption, ion exchange, chemical oxidation, hyperfiltration (reverse osmosis), electrochemical, etc.

Fate of Dyes

Recent estimates indicate 12% of the synthetic textile dyes used yearly are lost to waste streams during dyestuff manufacturing and textile processing operations. Approximately 20% of these losses enter the environment through effluents from wastewater treatment plants.

With few exceptions, the normal use of organic colorants poses few problems in terms of acute ecological effects. On the other hand, certain dyestuffs exhibit toxic effects toward microbial populations and can be toxic and/or carcinogenic to animals. Also, the possible contamination of drinking water supplies is of concern because certain classes of dyes are known to be enzymatically degraded in the human digestive system, producing carcinogenic substances.

Until recently, few papers appeared on the fate of dyes in the environment. But because of the importance of this subject, work is being done primarily by the U.S. Environmental Protection Agency (U.S. EPA) and the Ecological and Toxicological Association of the Dyestuff Manufacturing Industry (ETAD).

One of the reasons for lack of literature was probably because environmental analysis depends heavily on gas chromatography/mass spectrometry, which is not suitable for most dyes because of their lack of volatility. However, significant progress is being made in analyzing nonvolatile dyes by newer mass spectral methods such as fast atom bombardment (FAB), desorption chemical ionization, thermospray ionization, etc.

Dyestuffs in general, and azo dyes in particular, are likely to undergo substantial primary biodegradation in an anaerobic environment through reductive cleavage of azo bonds into aromatic amines. Lipophilic aromatic primary amines are aerobically degradable, but depending on their precise structure, some sulfonated aromatic amines may not be degradable.

The fate of one dye that has been thoroughly studied is the azo dye Disperse Blue 79 which may be designated 6-bromo-2, 4-dinitroaniline → 3-(*N,N*-diacetoxyethylamino)-4-ethoxyacetanilide.

Disperse Blue 79 is the largest volume dye on the market today. It has been estimated that during the manufacture of Disperse Blue 79 there should be released the following amounts of dye: 4.5–14 t per year at a total of nine sites with an estimated 3–20 kg per day.

Of particular importance is the degradation, cleavage, or reduction of the dye into aromatic amines, one of which is 6-bromo-2,4-dinitroaniline. This amine is toxic, mutagenic, and was selected for carcinogenic study. The precursor for preparation of 6-bromo-2,4-dinitroaniline, is 2,4-dinitroaniline, which has been extensively evaluated and found to be highly toxic, mutagenic, and was selected for carcinogenic study.

Small amounts of the following aromatic compounds were found to be present in the effluent after manufacture of Disperse Blue 79: 6-bromo-2,4-dinitroaniline, 6-bromo-2,4-dinitrophenol, 3-diacetoxyethylamino-4-ethoxy acetanilide, 3-dihydroxyethylamino-4-ethoxyacetanilide, 3-[(*N*-hydroxyethyl, *N*-acetoxyethyl)amino]-4-ethoxyacetanilide, and 3-acetoxyethylamino-4-ethoxy acetanilide. However, after a neutralization and heat stabilization step is added after coupling in manufacture, no diazotizable amine or phenol and only traces of the coupling component and its impurities were found.

The fate study of Disperse Blue 79 in anaerobic sediment–water systems shows the following degradation products:

where X = NH$_2$, Y = NO$_2$; X = NO$_2$, Y = NH$_2$; or X = Y = NH$_2$

These products suggest that this dye may undergo reduction in bottom sediments in the environment, resulting in the subsequent release of potentially hazardous aromatic amines into water.

A large study was done to determine the fate of Disperse Blue 79 in a conventionally operated activated sludge process and in an anaerobic sludge digestion system. The results showed no degradation in the activated sludge system, but did show degradation in the anaerobic digester of which no positive identification of compounds were made.

Pollution Prevention

Cooperation between industry, government, academia, and private environmental groups to implement these pollution prevention acts has begun in earnest.

The dye and dyeing industries have begun to give pollution prevention and its other forms of lessening or eliminating waste generation, such as waste minimization and source reduction, a high priority. The USEPA and the Ecological and Toxicological Association of the Dyestuffs Manufacturing Industry (ETAD) have jointly set up a program for pollution prevention in the dyestuff industry. The original goals were to develop a pollution prevention guidance manual and conduct a baseline survey of industry prevention practices for dye manufacture. Recently the manual was published and the data for the baseline survey collected by USEPA and ETAD.

There was a number of papers and patents on recycling dye and textile industry wastewater for reuse of dye, textile auxiliaries, and water. Recycling is considered a part of pollution prevention.

Heavy Metals

The heavy metals, copper, chromium, mercury, nickel, and zinc, which are used as catalysts and complexing agents for the synthesis of dyes and dye intermediates, are considered priority pollutants.

A number of papers have appeared on the removal of heavy metals in the effluents of dyestuff and textile mill plants. The methods used were coagulation, polymeric adsorption, ultrafiltration, carbon adsorption, electrochemical, and incineration and landfill. Of interest is the removal of these heavy metals, especially copper by chelation using trimercaptotriazine and reactive dyed jute or sawdust.

Toxicity

The past experience of the dyestuff industry in its use of dye intermediates such as β-naphthylamine and benzidine, known human bladder carcinogens, have led to studies as to whether or not handlers of dyes are exposed cancer, dermatitis, and other disorders.

The National Institute of Occupational Safety and Health (NIOSH) and the Occupational Safety and Health Administration (OSHA) reported in 1978 that the three primary benzidine-based azo dyes, namely Direct Black 38, Direct Blue 6, and Direct Brown 95, were carcinogenic in animals as a result of being converted to benzidine. These dyes are characterized by having a biphenyl diazo linkage:

This has led to concern about possible carcinogenicity from these dyestuffs, and therefore benzidine-based dyes and pigments are no longer produced by the large dyestuff manufacturers.

Two large studies were done for the selection of azo, nitro, and anthraquinone dyes for carcinogen bioassay. Based on previous information or testing, a total of 30 dyes were selected based on chemical structure, potential exposure, and suspicion of carcinogenicity.

Because of the large number of dyestuffs and the fact that most of these colorants have not been tested for carcinogenicity, structure–activity theory may help predict possible candidates for study.

In order to minimize the possible toxicity and damage of humans and the environment arising from the production and applications of colorants, an international association, the Ecological and Toxicological Association of the Dyestuff Manufacturing Industry (ETAD), was established in 1974. ETAD coordinates the ecological and toxicological efforts of synthetic organic dyes and pigment manufacturers. To date, ETAD consists of 32 members in Western Europe, North America, Japan, and India. The purpose of ETAD's toxicological work is to identify and assess risks caused by colorants and their intermediates with respect to their potential acute toxicity and their chronic effects on human health. This is accomplished by recommended methods to member firms following appraisal and development by appropriate ETAD committees, lectures, and publications. One of the projects of ETAD was a survey of acute oral toxicity, as measured by LD$_{50}$, the 50% lethal dose, which showed that of 4461 colorants tested, only 44 had a LD$_{50}$ < 250 mg/kg, but 3669 exhibited practically no toxicity (LD$_{50}$ > 5 g/kg). The evaluation of these colorants by chemical and coloristic classification showed that the most toxic ones are found among diazo and cationic dyes. Pigments and vat dyes, on the other hand, have a low acute toxicity, presumably because of their low solubility in water and in lipophilic systems.

Legislation

There has been a tremendous increase in regulatory activities worldwide aimed at achieving safer manufacture, use, and disposal of chemicals, including colorants. Table 1 is a summary of important United States, European, and Japanese Environmental Legislation affecting workers, consumers, and the public and environment.

The two most important pieces of chemical control legislation enacted affecting the dye and pigment industries are the United States' toxic Substance Control Act (TSCA) and EEC's Classification, Packaging, and Labeling of Dangerous Substances and its amendments.

Besides the federal laws, all 50 U.S. states have also passed environmental laws. As previously mentioned, both the United States and the state of New Jersey have passed Pollution Prevention Acts.

There is a difference between the United States, European, and Japanese environmental laws in introducing a new chemical compound. The EEC and Japan require animal and other toxicological and ecological testing, whereas the United States encourages but does not require these tests. This could result in rejection of a U.S. chemical exported product. Subsequently, greater harmonization is needed in many federal versus state versus overseas regulatory issues impacting the dye industries.

ABRAHAM REIFE
CIBA-GEIGY Corporation

Additional Reading

Anliker, R. and E.A. Clarke: *J. Soc. Dyer. Color.* **98**, 42 (1982).

Clarke, E.A. and R. Anliker: *Environmental Chemistry, Anthropogenic Compounds,* Vol. 3, Part A, Springer-Verlag, New York, NY 1980, p. 181.

Helmes, C.T. et al.: *J. Env. Sci. Health* **A19**(2), 97 (1984).

Houk, J., M.J. Doa, M. Dezube, and J.M. Rovinski: "Evaluation of Dyes Submitted Under the Toxic Substance Control Act New Chemicals Programme," *Colour Chemistry,* Elsevier Applied Science, London and New York, 1991.

TABLE 1. IMPORTANT ENVIRONMENTAL LEGISLATION

Law	Worker	Consumer	Public and environment
UNITED STATES			
Toxic Substances Control Act (TSCA)	X	X	X
Occupational Safety and Health Act (OSHA)	X		
Food, Drug, and Cosmetic Law		X	
Consumer Product Safety Act		X	
Labeling of Hazardous Materials		X	X
Federal Water Pollution Control Act			X
Clean Air Act			X
Resource Conservation and Recovery Act (RCRA)			X
Superfund Amendments and Reauthorization Act (SARA) Plus Title III	X		X
Superfund (CERCLA)			X
Hazardous Materials Transportation (DOT)			X
CONEG (Heavy Metals)			X
USDA			X
California Proposition 65			X
EEC			
Control of Certain Industrial Activities	X	X	X
Classification, Packaging, and Labeling of Dangerous Substances *UK*		X	X
Health and Safety at Work Act	X		
Carcinogenic Substances Regulations	X		
Toy (Safety) Regulations		X	
Pencil and Graphic Instruments (Safety) Regulations		X	
Poison Act		X	
Clean Air Act			X
SWITZERLAND			
Environment Protection Law		X	X
Poison Law		X	
FRANCE			
Control of Chemical Products		X	X
Consumer Protection and Information		X	
Pollution Control Law			X
GERMANY			
Environmental Chemicals Law	X	X	X
JAPAN			
Chemical Control Law			X

DYES: NATURAL. Natural dyes were replaced by synthetic dyes, although lately there has been a revival of the use of natural dyes for coloring foods, and some textile manufacturers are using natural dyes for dyeing their products. This article discusses those natural dyes formerly manufactured.

Anthraquinones

The anthraquinone structure occurs in both the plant and animal kingdom. Those natural dyes having this structure surpass all other natural dyes in fastness properties. See also **Dyes: Anthraquinone.**

Alizarin. There is only one significant plant anthraquinone dye, alizarin (CI Natural Red 6, 8, 9, 10, 11, and 12; CI 75330).

Alizarin is a mordant dye forming various colored coordination complexes with different metallic salts. Based on analytical results, a structural formula has been proposed for the alizarin complex.

Anthracene was oxidized to anthraquinone, dibrominated, and the dibromo derivative subjected to a caustic fusion. Alizarin was obtained in an impure form and in low yield. This represented the first synthesis of a natural dye.

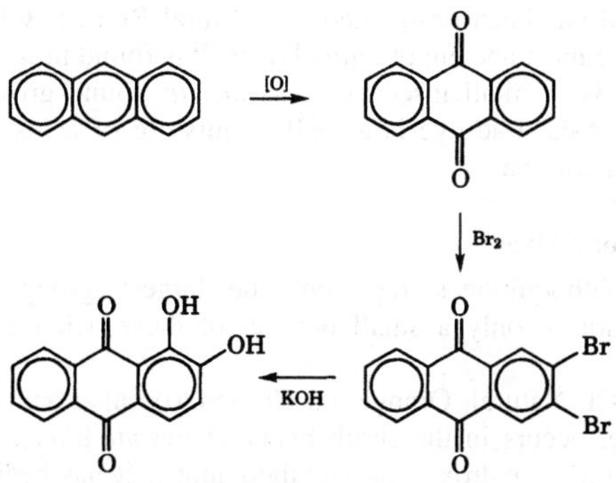

(1)

Later, at BASF, a process was developed for the manufacture of alizarin by the caustic fusion of anthraquinone-2-sulfonic acid (so-called silver salt) which was patented in England on the 25th of June, 1869. One day later, W. Perkin applied for a patent for the manufacture of alizarin by a process almost identical to the German process except that the "silver salt" was prepared as follows:

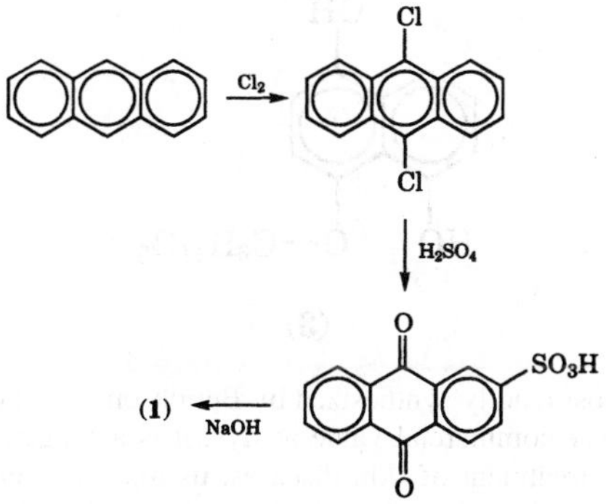

Later, improvements were made in the process: use of oleum for sulfonating anthraquinone and the addition of an oxidizing agent to the caustic melt.

For years this was the process used to manufacture alizarin, although it was claimed that a more economical process would result if 2-chloroanthraquinone were to be used instead of silver salt.

Just as synthetic alizarin forced natural alizarin out of the market, synthetic alizarin has been replaced by azoic dyes because they are easier to apply.

Animal Anthraquinone Dyes. Kermisic Acid. Many accounts claim that kermisic acid (CI Natural Red 3; CI 75460) is the oldest dyestuff ever recorded. The dye produces a brilliant scarlet color with an alum mordant. Although expensive, it was cheaper than its rival Tyrian Purple. It was in great demand until the sixteenth century, when it was displaced by carminic acid.

The structure of kermisic acid is 1,3,4,5-tetrahydroxy-7-carboxy-8-methylanthraquinone. Carminic acid (CI Natural Red 4; CI 75470), is a red dye occurring as a glycoside in the body of the cochineal insect *Dactylopius coccus* of the order Homoptera, family Coccidae. Until the advent of synthetic dyes, the principal use for carminic acid was for dyeing tin-mordanted wool or silk. Its aluminum lake, carmine, finds use in the coloring of foods. The structural formula of carminic acid is (2).

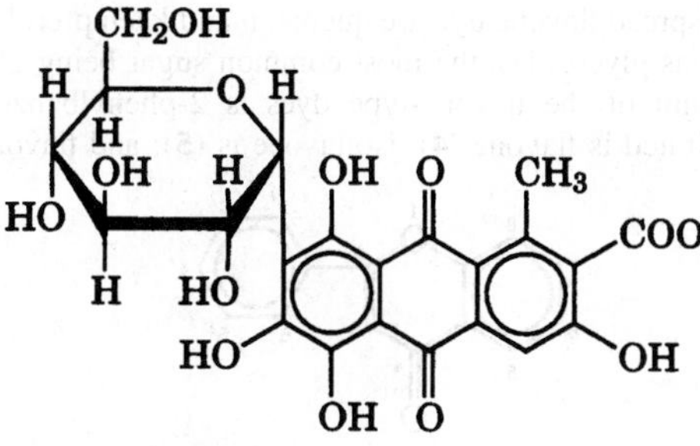

(2)

Laccaic acid has been designated (CI Natural Red 25; CI 75450). Lac dye ranks as the most ancient of animal dyes. It is found in lac, the resinous secretion of a very small insect, *Coccus laccae*, found growing in India and Southeast Asia. Lac dye is actually a mixture of acids derived from 2-phenylanthraquinone

Naphthoquinone Dyes

Although naphthoquinones represent the largest group of naturally occurring quinones, only a small number of these achieved importance as dyestuffs.

Lawsone (CI Natural Orange 6; CI 75420), also known as henna and isojuglone, occurs in the shrub henna (*Lawsone alba*). Lawsone has been identified as 2-hydroxy-1,4-naphthoquinone. It has been synthesized by the Thiele acetylation of 1,4-naphthoquinone followed by hydrolysis and oxidation.

Lapacol (CI Natural Yellow 16; CI 75490) (lapachic acid, taiguie acid, tecomin) is a yellow pigment occurring in the wood of trees of the genus Tecoma, native to the West Indies and tropical South America.

Juglone (CI Natural Brown 7; CI 75500) was isolated from the husks of walnuts in 1856. Juglone occurs in walnuts as a glycoside of its reduced form, 1,4,5-trihydroxynaphthalene. Its structure is (3).

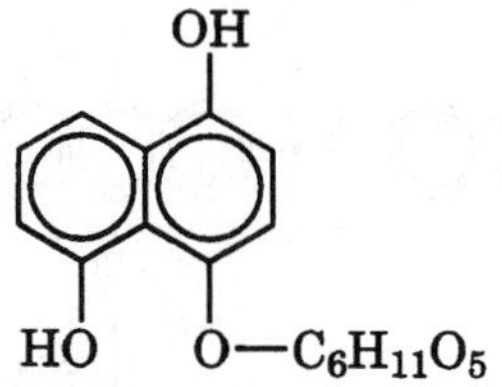

(3)

Juglone is most readily synthesized by Bernthsen's method. Although it no longer has any commercial value as dye, it is a fungicide and as such finds use in the treatment of skin diseases. Its toxic properties have been made use of in catching fish. Juglone has been used to detect very small amount of nickel salts since it gives a deep violet color with such salts.

Alkannin, shikonin, and shikalkin are grouped together because the first two are enantiomers and the last one is their racemate. Alkannin (CI Natural Red 20; CI 75530) (*Anchusa tinctoria* or *alkanna tinctoria*) is a member of the Boraginaceae family. It is found in the roots of alkanet, a perennial shrub native to Southern Europe.

Alkannin occurs in the roots of the plant as the alkali-sensitive ester of angelic acid. It may be extracted from the roots by using boiling light petroleum ether. Treatment of this extract with dilute sodium hydroxide gives a blue solution from which the dye is precipitated by the addition of acid. The crude product is purified by vacuum sublimation. Its structure is a hydroxylated naphthoquinone with a long, unsaturated side chain; it has the (*S*)-configuration.

Shikonin (CI 75535) occurs as an acetyl derivative in the Japanese shikone, *Lithospermum erythrorhizon*, another member of the Boraginaceae family. It is the (*R*)-optical isomer of alkannin. Tissue cultures of *L. erythrorhizon* are used in Japan to manufacture shikonin mainly for cosmetic use. Both alkannin and shikonin are mordant dyes producing violet to gray colors on fabrics. Shikalkin the racemate, has been synthesized.

Flavones. These compounds are the most widely distributed natural coloring matter formerly used as dyestuffs. Flavone-type dyes occur in all the higher plants: in the leaves, roots, bark, fruits, pollen, and flower petals. The most widespread flavone dye are quercetin and kaempferol. In general, the dyes occur as glycosides, the most common sugar being glucose.

The basic unit of the flavone-type dyes is 2-phenylbenzopyrone (4) which unsubstituted is flavone (4); isoflavone is (5); and flavonol (6) is

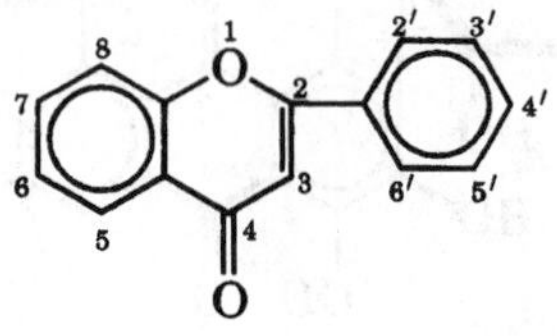

(4)

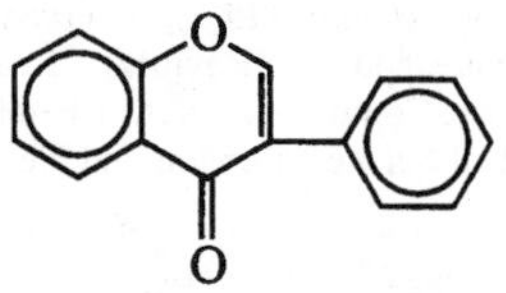

(5)

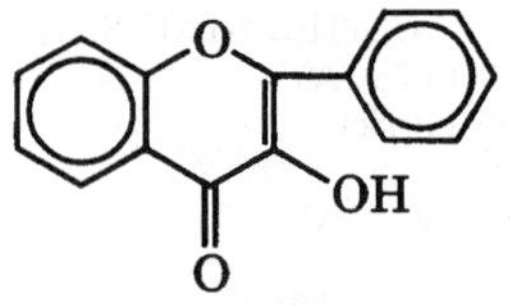

(6)

Flavone dyes having these structures are hydroxylated and methoxylated derivatives. Those dyes containing not more than three hydroxyls are generally termed flavones, whereas those containing up to and including six are flavonols.

The flavone, isoflavone, and flavonol-type dyes owe their importance to the presence of an *o*-hydroxy carbonyl structure within the molecule. Positions 4 and 5 can chelate with different metallic salts to give colored insoluble complexes. In other words, these dyes require a mordant in order to fix them onto the fiber.

Anthocyanins. Like the flavones, the anthocyanins are found throughout nature. This class of polyphenolic compounds is responsible for the pink, red, violet, and blue colors found in plants. Like many other natural phenolic substances, anthocyanins occur in plants as glycosides; the sugar-free anthocyanins are known as anthocyanidins.

All anthocyanidins have the 2-phenylbenzopyrylium or flavylium cation structure, a resonance hybrid of oxonium forms and carbenium forms. There are three fundamental groups of anthocyanidins to which all the other anthocyanidins could be referred. In the following structure R = R′ = H designates pelargonidin R = OH, R′ = H is cyanidin; and R = R′ = OH is delphinidin.

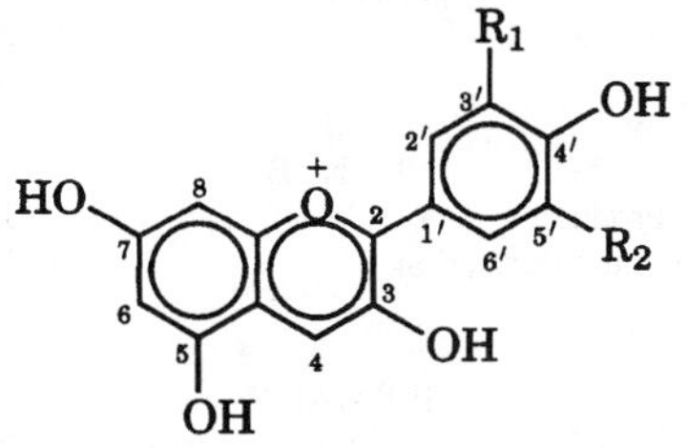

The anthocyanins are pH sensitive. Their color, in part, is determined by the pH of the sap.

A convenient method for synthesizing anthocyanidins involves the condensation of an *o*-hydroxybenzaldehyde with an acetophenone.

Indigoid Dyes

Tyrian Purple. The ancient kingdom of Tyre owed its fame and fortune to the purple dye produced from the lowly mollusks found on its shores. The dye became known as Tyrian Purple (CI 75800) reflecting its place of origin. These mollusks belong to the Muricidae family and the genera *Murex* and *Purpura* which include *M. brandaris* and *M. trunculus*, the principal sources of the dye. The dye is 6,6′-bromoindigotin.

Indigotin. The blue dye of the ancient world was derived from indigo and woad (CI Natural Blue; CI 75780). Indigo belongs to the legume family. The two most important species are *Indigo tinctoria* and *I. suffruticosa*, found in India and the Americas, respectively. The leaves of the indigo plant do not contain the dye as such, but in the form of its precursor, a glycoside known as indican.

Woad, *Isatis tinctoris*, belongs to a genus that comprises some thirty species.

In 1890, it was observed that treatment of ω-bromoacetanilide with alkali produced oxindole. Based on this observation, K. Heumann treated *N*-phenylglycine with alkali and obtained indoxyl (keto form), which on aerial oxidation converted to indigotin. Later, a variation of the original Heumann process was made: aniline, formaldehyde, and hydrogen cyanide react to form phenylglycinonitrile, which is hydrolyzed to phenylglycine. This is the most widely used process for manufacturing indigotin.

The greatest improvement in the manufacture of indigotin came when sodamide was used with alkali in the conversion of phenylglycine to indoxyl.

Although there is still demand for indigotin for dyeing blue jeans, it has lost a good part of the market to other blue dyes with better dyeing properties. At present, practically all the indigotin consumed in the United States comes from abroad.

Natural Food Colors

In the 1970s, decertification of the important food colors FD&C Reds 2 and 4 caused much concern among manufacturers of food dyes. With the possibility that other synthetic dyes would be banned present, attention was turned to the use of natural dyes as food colorants. Many such dyes had been in use for hundreds of years until they were replaced by synthetic dyes.

The yellow dye curcumin (CI Natural Yellow 3; CI 75300), also known as tumeric, occurs in the roots of the plant *Curcuma tinctoria*, found growing wild in Asia. The dye is an oil-soluble bright yellow material, and is the only natural yellow dye that requires no mordant. It finds use as a colorant for baked goods such as cakes. Carmine is the trade name for the aluminum lake of the red anthraquinone dye carminic acid obtained from the cochineal bug. The dye is obtained from the powdery form of cochineal by extraction with hot water, the extracts treated with aluminum salts, and the dye precipitated from the solution by the addition of ethanol. This water-soluble bright red dye is used for coloring shrimp, pork sausages, pharmaceuticals, and cosmetics. It is the only animal-derived dye approved as a colorant for foods and other products.

Carotenoids. The carotenoids are a group of widely distributed, highly colored, fat-insoluble, naturally occurring organic compounds. They owe their color to the four repeating isopyrene units found in the molecule and may, therefore, be classified as tetraterpenoids. The carotenoids may be divided into two principal groups, the carotenes, which are strictly hydrocarbons, and the xanthophylls, which contain oxygen. Carotenoids are found in almost all fruits and vegetables, egg yolk, dairy products, and sea foods. The chemistry of the carotenoids has been described in a number of reviews.

Although the carotenoids can be obtained from natural sources, it is far more economical to manufacture them for commercial use. Three have been manufactured for many years: β-carotene, canthaxanthin and β-apo-8'-carotenal.

In general, the low solubility of the carotenoids creates a problem when they are applied as food dyes. To overcome this, a microcrystalline powder form has been prepared. This is then mixed into an edible fat. In this form it finds use for coloring margarine, butter, cake mixtures, and other fact-containing foods. β-Carotene is available also as an emulsion, in a water-dispersed form, and as a liquid suspension. Canthaxanthin is commercially available as a 10% water-dispersable beadlet or spray-dried powder. Because of its exceptionally good tomato color-enhancing properties, it finds use in tomato-based products such as pizza and spaghetti sauce. It is useful in water-based foods such as peach ice cream and pink grapefruit beverages. β-Apo-8'-carotenal has high tinctorial strength; because of this, it is marketed in several different strengths, usually as a dispersion in vegetable oils. Its main use is for coloring process cheese and French dressing.

Carotenoids have two general characteristics of importance to the food industry: they are not pH sensitive in the normal 2–7 range found in foods, and they are not affected by vitamin C, making them especially important for beverages.

Bixin (CI Natural Orange 4; CI 75120) is found in the seed of the plant *Bixa orellana*, native to India. Later it was found growing in South America, where the Indians used the red dye from the seeds as a body paint. An extract of the seeds appears on the market as annatto. This extract is used in coloring butter, margarine, and cheese such as Leicester cheese. In Mexican and South American cuisine, it finds special use as a flavor and coloring matter. Annato is available as an aqueous solution, as an oleaginous dispersion, and a spray-dried powder.

Crocetin (CI Natural Yellow 6; CI 75100) occurs in saffron as crocin, the digentiobiose ester of crocetin. Saffron is found in the pistils of the plant *Crocus sativus*.

Betalaines. In 1968, the term betalaines was used to describe collectively two groups of plant pigments: the red betacyanins and the yellow betaxanthins. The red and yellow dyes found in beets, *Beta vulgaris*, fall into this category.

The color of betalaines is barely affected by the pH range normally found in foods. However, the dyes are heat-sensitive, which places some limitations on their use as food dyes.

Beet juice contains about 80% of fermentable carbohydrates and nitrogenous compounds. To remove these compounds, a yeast fermentation utilizing *Candida utillis* has been suggested. By so doing, a more concentrated form of the dye becomes available. The red dye from beets is sold as beet juice concentrate, as dehydrated beet root, and as a dried powder.

Chlorophyll. Chemically pure chlorophyll is difficult to prepare, since it occurs mixed with other colored substances such as carotenoids. Commercially it is solvent extracted from the dried leaves of various plants such as broccoli or spinach. Chlorophyll is water-insoluble. It has none of the characteristics of a dye in that it has no affinity for the usual fibers such as cotton or wool. Chlorophyll is properly classified as a pigment (CI Natural Green 3; CI 75810). As such, it finds use for coloring soaps, waxes, inks, fats, or oils. Chlorophyll is an ester composed of an acidic part, chlorophyllin, esterified by an aliphatic alcohol known as phytol. Hydrolysis of chlorophyll using sodium hydroxide produces the moderately water-soluble sodium salts of chlorophyllin, phytol, and methanol. The magnesium in chlorophyllin may be replaced by copper. The sodium copper chlorophyllin salt is heat-stable, and is ideal for coloring foods where heat is involved, such as in canning.

Health, Safety, and Environmental Factors of Natural Dyes

Natural dyes comprise those colors derived from plant or animal matter without chemical processing. Modern tests have verified the safety of natural dyes as food colorants; many of these dyes are on the FDA's list of approved food dyes.

Natural dyes processed for the market do not undergo any chemical operations. Those operations involved are purely physical, such as grinding, spray or vacuum drying, and water or solvent extractions. None of these operations create any great environmental problems.

A. J. COFRANCESCO
Consultant

Additional Reading

Britton, G., et al.: *Carotenoids: Biosynthesis and Metabolism,* Birkhauser Verlag, Cambridge, MA, 1998.

Fiechter, A.: *Downstream Processing Biosurfactants Carotenoids,* Springer-Verlag, Inc., New York, NY, 1995.

Frank, H.A.: *The Photochemistry of Carotenoids,* Kluwer Academic Publishers, Norwell, MA, 2000.

Leggett, W.F.: *Ancient and Medieval Dyes,* Chemical Publishing Co., New York, NY, 1964.

Passwater, R.A. and S. Davis: *Beta-Carotene and Other Carotenoids: The Antioxidant Family That Protects Against Cancer and Heart Disease and Strengthens the Immune System,* Keats Publishing, Chicago, IL, 1996.

Perkin, A.G. and A.E. Everest: *The Natural Organic Colouring Matters,* Longmans, Green and Co., New York, NY 1981.

Staff: *Iarc Handbooks of Cancer Prevention, Vol. 2: Carotenoids,* International Agency for Research on Cancer World Health Organization, Oxford University Press, New York, NY, 1998.

Thomson, R.H.: *Naturally Occurring Quinones,* 3rd Edition, Academic Press, Inc., New York, NY 1987

DYES: REACTIVE. Reactive dyes are those dyes containing electrophilic functional groups capable of reacting with a nucleophile to form a covalent bond either through addition or displacement. Nucleophiles within fibers that typically react with dyes are hydroxyl groups in cellulose, amino, hydroxyl, and thiol groups in wool, and amino groups in polyamide. The outstanding characteristic feature of reactive dyes is their high wetfastness properties attributed to covalent bonding, an advantage over those dyes fixed through adsorption or mechanical entrapment. Unlike large bulky direct dyes, since reactive dyes are chemically bonded to the substrate, smaller molecules giving brighter colors are possible. The principal use of reactive dyes is far and away greatest for cellulose (cotton and rayon), followed by wool, with polyamide (nylon) a very distant third. Applications to silk and leather represent only very minor uses. See also **Dyes: Application and Evaluation**.

Reactive Dye Structure

Reactive dyes consist basically of three components: a dye, a bridging group (B), and the reactive group (R), dye—B—R. The reactive group may be considered in two parts as a carrying group and the reactive component.

Most commonly used chromophores parallel those of other dye classes. Azo dyes represent the largest number with anthraquinone and phthalocyanine making up most of the difference. Metallized azo and formazan dyes are important and have gained in importance as a chromophore for blue dyes during recent years. See also **Dye and Dye Intermediates**.

Yellow dyes are generally monoazo, and most are pyrazolone or pyridone couplings. Orange dyes are generally monoazo derived from couplings to pyrazolones or of slightly substituted phenyl and naphthyl groups.

Many red dyes are based on H-acid (**1**), e.g., Reactive Reds 2, 24, and 218. Others are substituted phenyl and naphthyl or metallized systems. Violet dyes are also metallized monoazo dyes.

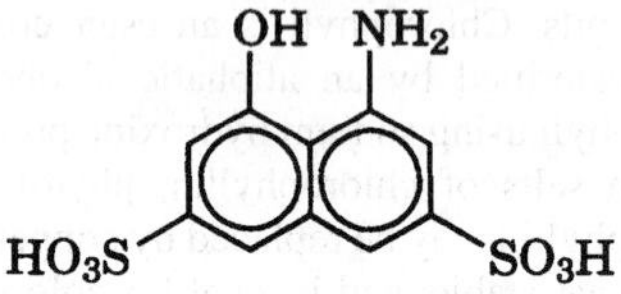

(1)

Blue dyes are derived from anthraquinone, phthalocyanine, or metallized formazan. See also **Anthraquinone** (Figs. 1 and 2). There are also oxazine and thiazine dyes reported (Fig. 2).

Brown and black dyes are generally disazo with exceptions for metallized or polycyclic structures. Two disazo dyes are Reactive Brown 11 (**9**) and Reactive Black 5 (CI 20505) (**10**).

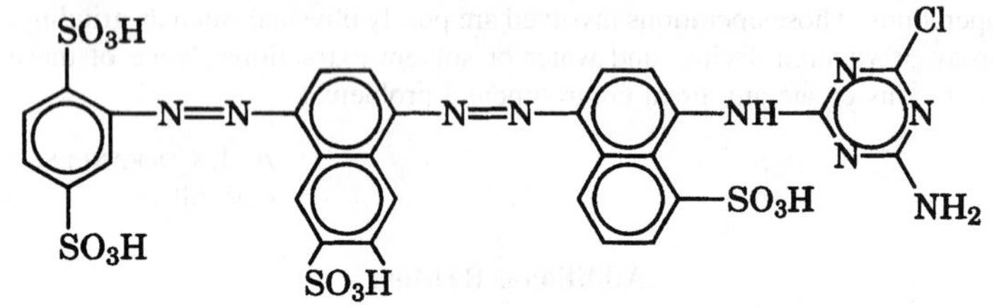

(9)

Green dyes are obtained by bridging an anthraquinone blue chromogen with a yellow chromogen, as in the following reactive green (**11**) or from phthalocyanine.

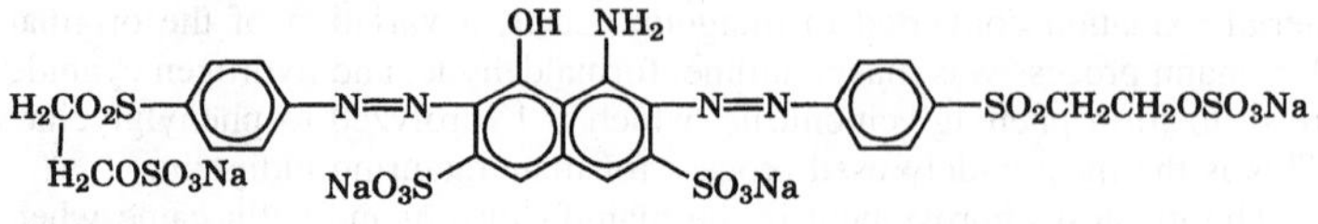

(10)

Properties

Reactive Groups. Although fastness of reactive dyes to washing is good, once fixed on the substrate, hydrolysis during application leads to low dye fixation rate, i.e., dye that hydrolyzes in the application process is not fixed on the fiber. Reactive dyes are highly sulfonated, and as such are very water-soluble. Fairly large amounts of salt are required to force the dye into the substrate during application. The hydrolyzed dye is even more soluble than the dye, and presents concerns for the dyer because of effluent color and low biodegradability as well as economic losses. A very large part of the development effort spent on reactive dyes during recent years has been toward improving fixation on the substrate. Approaches taken have been to find different reactive groups as well as to include more than one reactive group in the dye structure and select reactive groups less sensitive to hydrolysis or more reactive to the substrate under application conditions.

During the 1980s, there was a revived interest in bireactive dyes. Every principal dye manufacturer has introduced dyes with more than

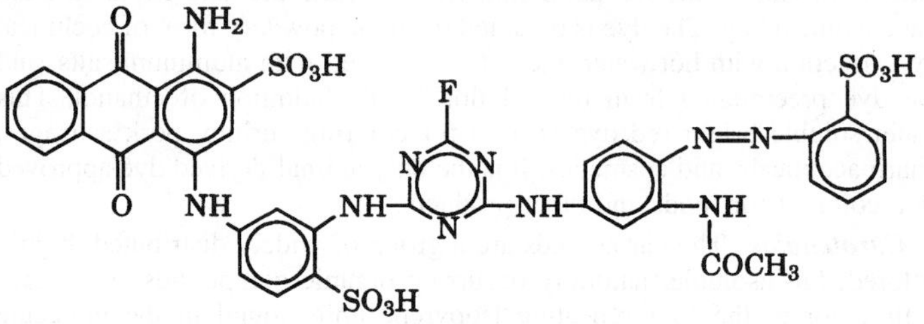

(11)

one reactive group. The ones offering highest fixation have two or more reactive groups with different rates of reaction. Different rates of reaction may be due to selection of different groups, e.g., dichlorotriazine and trichloropyrimidine, or by varying substituents on the triazine ring; electron-donating groups decrease reactivity, and electron-withdrawing groups increase reactivity.

Reactive groups have minimal auxochrome effect on color intensity, and color yield per molecular weight decreases with increasing numbers of reactive groups. Increased dye fixation and reduced environmental impact of hydrolyzed dye more than compensate for color reduction of additional reactive groups.

The fluorotriazine reactive group was reported in the mid-1970s, and has been the subject of many literature references since that time.

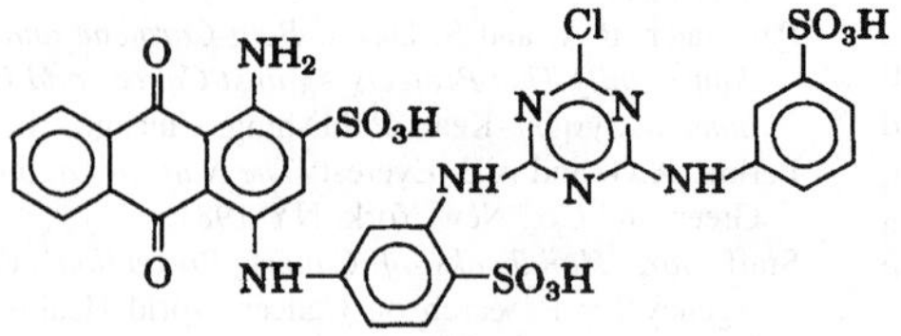

(2)

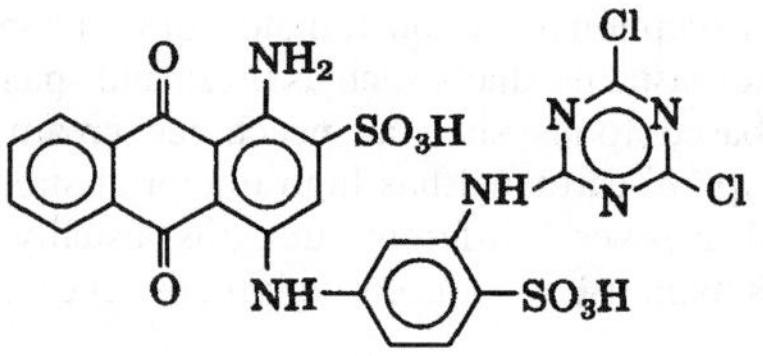

(3)

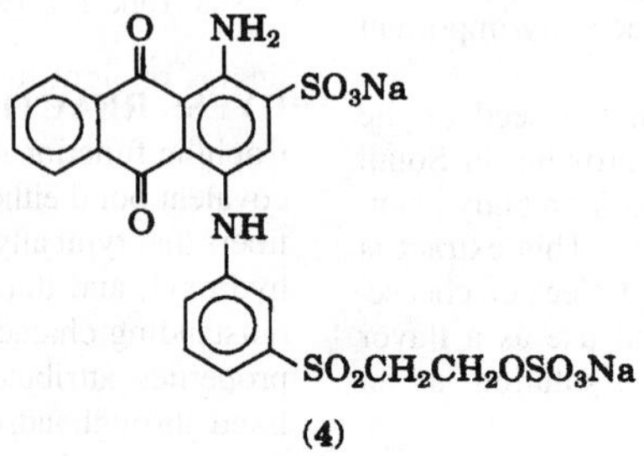

(4)

Fig. 1.   Blue reactive dyes from anthraquinone. Reactive Blue 5 (CI 61210) (**2**), Reactive Blue 4 (CI 61205) (**3**), e.g., Procion Blue MXR, Reactive Blue 19 (CI 61200) (**4**), e.g., Remazol Brilliant Blue R.

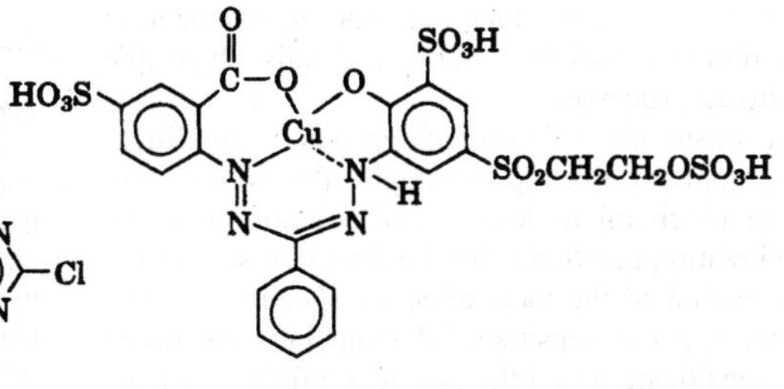

Fig. 2.   Phthalocyanine (5); and metallized formazan (6,7) and azine (8) blue reactive dyes. Reactive Blue 15 (CI 74459) (5); Reactive Blues (6) and (7); Reactive Blue 204 (8). Other blue oxazine dyes are other fluorotriazines and a dichlorotriazine.

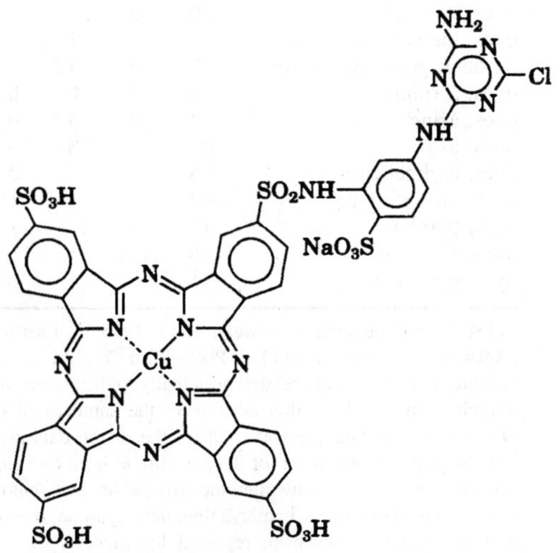

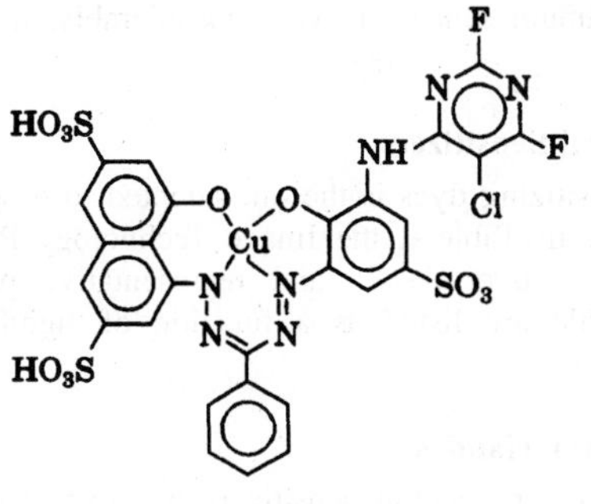

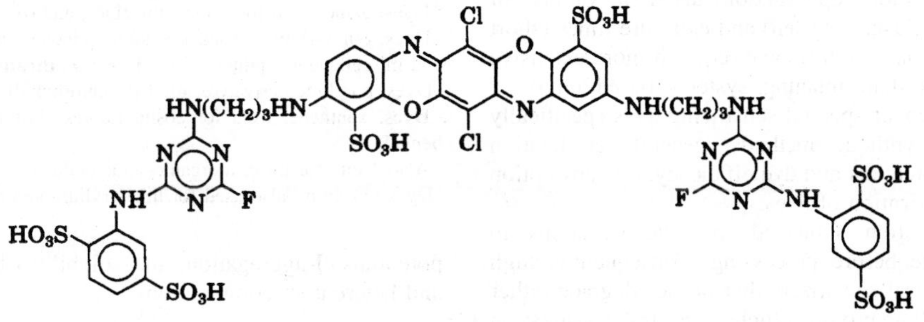

Monofluorotriazine dyes are especially attractive because of their high color yield at low temperatures (below 40°C).

Reactivity in the Dyebath

The most important discovery in dyeing cellulose with reactive dyes was the application of Schotten-Baumaun principles. Reaction of alcohols proceeds more readily and completely in the presence of dilute alkali, and the cellulose anion (cell−O⁻) is considerably more nucleophilic than is the hydroxide ion. Thus the fixation reaction (eq. 1) competes favorably with hydrolysis of the dye (eq. 2).

$$\text{dye}-\text{B}-\text{R} + \text{cell} - \text{O}^- \rightarrow \text{dye}-\text{B}-\text{O}-\text{cell} \qquad (1)$$

$$\text{dye}-\text{B}-\text{R} + \text{OH}^- \rightarrow \text{dye}-\text{B}-\text{OH} \qquad (2)$$

The most important reactive groups today are monochlorotriazine (high energy), vinyl sulfone (medium energy), and monofluorotriazine (low to medium energy), although dichlorotriazine, 2,3-dichloroquinoxaline and 2,4-difluoro-5-chloropyrimidine have a significant presence. α-Bromoacrylamide and vinyl sulfone from *N*-methyltaurine are the most important reactive dyes for wool.

Bifunctional Dyes. There are many examples of dyes with two or more reactive groups, including many mixed reactive systems. Dye fixation is increased significantly with increasing number of reactive groups. Some multiple reactive dyes are claimed to have as high as 95% fixation.

Methods of Synthesis

Reactive dyes are synthesized by *(1)* condensation of an amine function in the chromogen molecule with a reactive group; *(2)* by coupling a diazonium salt with a coupling component that has a reactive group, or by coupling a diazonium salt containing a reactive group with a coupler; or *(3)*, in the case of copper phthalocyanine (CPC), by condensing CPC sulfonyl chloride with an amino-containing bridging group attached to a reactive group.

Economic Aspects

Development in reactive dyes over the past two decades has clearly been the most active of any class of dyes. Every significant dye manufacturer is now offering reactive dyes. Reactive dyes are offered commercially both as dry powders and buffered liquid forms.

Roy E. Smith
CIBA-GEIGY Corporation

Additional Reading

Bradbury, M.J., P.S. Collishaw, and D.A.S. Phillips: *J. Soc. Dyers Colourists*, 108, (1992).
Renfrew, A.H.M. and J.A. Taylor: *Rev. Prog. Color. Realt. Top.*, **20** (1990).
Zollinger, H.: *Text. Chem. Color.* **23**, 12 (1991).

DYES: SENSITIZING.

Spectral sensitizing dyes extend the wavelengths of light to which inorganic semiconductors, organic semiconductors, and chemical (biological) reactions can be photosensitized. Spectral sensitizers are needed for the blue, green, and red portions of the visible spectrum. For infrared photographic, electrophotographic, and biological applications, sensitizing dyes are also needed to match output wavelengths of solid-state lasers (optical data storage, laser printing, range finding, and data transmission), to provide color-selective infrared photography (an effective environmental survey method), or to match transmission wavelengths of body tissue.

Spectral sensitizing dyes are considered "functional" dyes to distinguish them from conventional colorants. The absorption of radiation by a functional dye causes some additional function(s) to occur, and in many cases, the sensitizing dyes can exhibit more than one type of sensitization reaction. Many texts and reviews provide extensive details about the diversity of spectrally sensitized processes.

The detection of spectral sensitizing action often depends on amplification methods such as photographic or electrophotographic development or, alternatively, on chemical or biochemical detection of reaction products. Separation of the photosensitization reaction from the detection step or the chemical reaction allows selection of the most effective spectral sensitizers. Prime considerations for spectral sensitizing dyes include the range of wavelengths needed for sensitization and the absolute efficiency of the spectrally sensitized process. Because both sensitization wavelength and efficiency are important, optimum sensitizers vary considerably in their structures and properties.

Structural Classes of Spectral Sensitizers

A useful classification of sensitizing dyes is the one adopted to describe patents in image technology. In Table 1, the Image Technology Patent Information System (ITPAIS), dye classes and representative patent citations from the ITPAIS file are listed as a function of significant dye class.

Spectral Sensitization of Silver Halides

The large number of patents for spectral sensitizers of silver halides indicates their extensive use in photographic plates, films, and papers. Color films that give good color reproduction under a variety of illuminants (daylight, electronic flash, tungsten) and exposure times (short for electronic flash, long for available light) have required more extensive tailoring of spectral sensitizers than imaging systems based on other semiconductors. Early reviews cover spectral sensitizing dyes specifically for silver halides, including synthetic methods, general sensitization mechanisms, electrochemical potentials and dye efficiency, dye absorption and aggregation, and supersensitization of dyes.

Photographic products are often subjected to wide variations in temperature, humidity, and time before processing. Consequently, high priorities are also given to spectral sensitizers that do not degrade either the film or image quality, as well as provide efficient spectral sensitization at the desired wavelengths. Although there are many available dyes and pigments, commercial silver halide films, papers, and plates are efficiently sensitized by just a few types of chromophores in the cyanine and merocyanine dye classes (Fig. 1).

The suitability of these chromophores rests in large measure in being able to simultaneously optimize three properties (electrochemical

TABLE 1. DYES AND SEMICONDUCTORS, PATENT CITATIONS[a]

ITPAIS classification	Silver halide	ZnO	TiO$_2$	ZnS	Se	CdS, CdSe	Misc. organic semiconductor
dyes, cyanine[b]	420	33	27	10	45	16[c]	11
dyes, merocyanine[b]	118	2	1	0	27	2	0
all other polymethine dyes[b]	277	5	4	1	34	3	4
dyes, acridine	0	2	0	0	1	0	0
dyes, azine[d]	4	4	3	0	7	5	0
dyes, azo	107	18	28	3	28	10	4
dyes, arylmethane	3	3	1	2	7	2	0
dyes, quinone type[e]	48	10	16	1	21	2	1
dyes, porphine[f]	12	8	17	1	24	2	1
dyes, xanthene[g]	6	14	11	3	17	3[h]	3
dyes, pyrylium[i]	1	0	1	0	42	3[h]	4

[a] ITPAIS was developed between 1975–1985 by Eastman Kodak Co., Agfa-Gevaert (Antwerp/Leverkusen), and Fuji Photo Film Co., Ltd., and encompasses selected patents and literature references related principally to the chemical aspects of image technology. Search terms used for this table were the same as in the previous edition, and the Derwent patent database was used for the search data presented here.

[b] Dyes, polymethine: used for dyes having at least one electron donor and one electron acceptor group linked by methine groups or aza analogues; allopolar cyanine, dye bases, complex cyanine, hemicyanine, merocyanine, oxonol, streptocyanine, and styryl. Supersensitization has been reported for these types—18 cites for cyanines, 3 for merocyanine, and 6 for all other polymethine types.

[c] Also 3 citations for misc. inorganic semiconductors.

[d] Dyes, azine: used for azines, thiazines, and oxazines.

[e] Dyes, quinone type: used for anthraquinones, indamines, indoanilines, indophenols, and miscellaneous quinones (see **Dyes: Anthraquinone**).

[f] Dyes, porphine derivative: used for chlorophyll, phthalocyanines, and hemin.

[g] Dyes, xanthene: used for eosin, fluorescein-type phthaleins, rhodamines, and rose bengal.

[h] Also 1 cite for misc. inorganic semiconductors.

[i] Dyes pyrylium: also used for thiapyrylium and benzothiapyrylium.

potentials, J-aggregation, and solubility) by choosing various substituent and heteroatom combinations.

Spectral Sensitization of Inorganic and Organic Solids

Relatively few spectral sensitizing dyes continue to serve the large-volume color (imaging and printing) photographic markets, where silver halide is the dominant semiconductor. Technology to serve both color printing and office copy markets will employ not only silver halides but other inorganic and organic semiconductors as well. Inorganic semiconductors

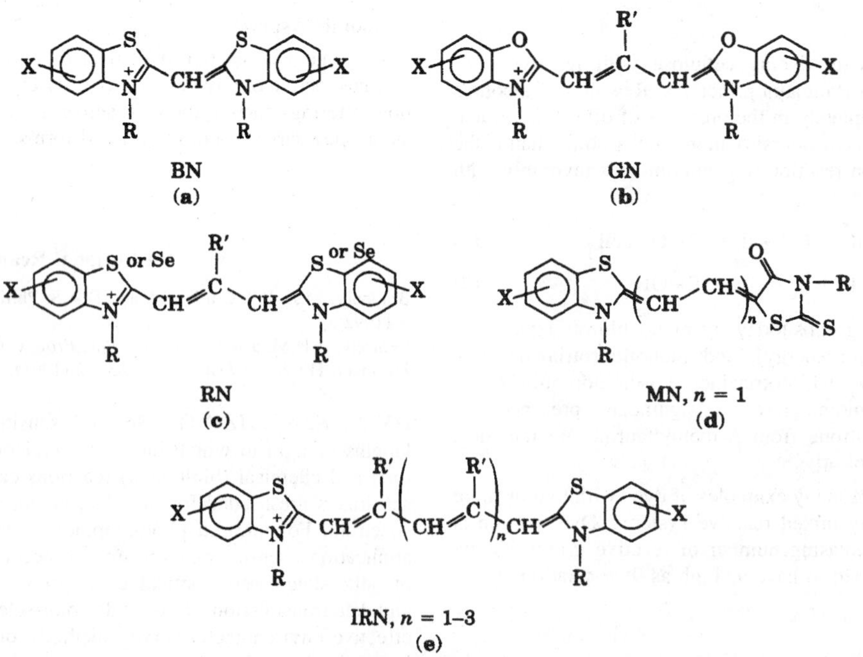

Fig. 1.   Spectral sensitizing dyes for silver halides (**a**) Blue sensitizers (400–500 nm) are designated BN; (**b**) green sensitizers (500–600 nm) are designated GN (the ring oxygen may be replaced by N(R)); (**c**) red sensitizers (600–700 nm) are designated RN; (**d**) MN designates a merocyanine dye; and (**e**), infrared sensitizers (>700 nm) are designated IRN.

include selenium, germanium, CdS, HgO, HgI_2, ZnO, PbO, Cu_2O, thallium halides, and TiO_2. As noted by patent citations and the published literature, both zinc oxide and titanium dioxide have been extensively investigated. Spectrally sensitized photoconduction has also been observed for organic semiconductors, i.e., anthracene, poly(N-vinylcarbazole), polyacetylene, copper phenylacetylide, phthalocyanines, and other solid dyes.

Continued commercial development of nonsilver office copying into color printing and optical data storage to higher densities provides many opportunities for spectral sensitizer improvement among competing technologies. Spectral sensitizers for these other materials are already quite varied (Table 2). Recent handbooks review both sensitizing dye technology and imaging systems that utilize dyes.

Spectral Sensitization in Photochemical Technology

Functional dyes of many types are important photochemical sensitizers for chemical reactions involving oxidation, polymerization, (polymer) degradation, isomerization, and photodynamic therapy. Often, dye structures from several classes of materials can fulfill a similar technological need, particularly for laboratory or small-scale reactions where production efficiency may be of secondary importance. Commercial photochemical technology, however, is more selective and requires photochemical efficiency, ease of product separation, and lack of unwanted side reactions to an extent similar to that required by imaging processes. In addition, reusability of the spectral sensitizer is also preferred in commercial photochemical reactions.

Uses and Suppliers

Sensitizing dyes are used primarily for specialty purposes: photographic sensitizers, electrophotographic sensitizers, laser dyes, infrared (optical disk, etc.) imaging, and certain medicinal applications. Because of this, their manufacture is limited to significantly smaller quantities than for fabric dyes or other widely used coloring agents. However, the photographic, laser, and medicinal uses place high demands on the degree of purity required, and the reproducibility of synthetic methods and purification steps is very important. Suppliers of cyanine dyes include manufacturers of other specialty organic and photographic chemicals: Aldrich Chemical Co. (Milwaukee, Wisconsin), Eastman Chemical Company (Kingsport, Tennessee), Japanese Institute for Photosensitizing Dyes (Okayama, Japan), Molecular Probes (Eugene, Oregon), NK Dyes (Japan), Pfaltz and Bauer (Stamford, Connecticut), Riedel deHaen (Karlsruhe, Germany), and H. W. Sands. More importantly, these firms provide sources of generally useful reagents that, in two or three synthetic steps, lead to many of the commonly used sensitizers.

Sensitizing Dye Toxicology

Nearly every significant class of dyes and pigments has some members that function as sensitizers. Toxicological data are often included in surveys of

TABLE 2. SPECTRAL SENSITIZERS FOR INORGANIC AND ORGANIC SEMICONDUCTORS

Semiconductor	Sensitizer
CdS (Cu doped)	thiacarbocyanine
	benzothiazolylrhodanines
	rhodamine B
ZnO	thiacarbocyanine
	erythrosin
	eosin
	phthalocyanine
	rhodamine B
	rose bengal
	methylene blue
	fluorescein
TiO_2	benzothiazolylstyryl dye
	thiacarbocyanine
poly(N-vinylcarbazole)	rose bengal
	2,4,7-trinitrofluorenone
	thiapyrylium dyes
solid dye particles	phthalocyanines
photoelectrophoresis	thioindigo
	flavanthrone
	quinacridone
dye electrodes	benzothiazolylrhodanines
	phthalocyanines

dyes, reviews of toxic substance identification programs, and in material safety data sheets provided by manufacturers of dyes.

DAVID M. STURMER
Eastman Kodak Company

Additional Reading

Braun, A.M., M.-T Maurette, and E. Oliveros: trans. by D.F. Ollis and N. Serpone, *Photochemical Technology,* John Wiley & Sons, Inc., New York, NY, 1991.
Diamond, A.S.: ed., *Handbook of Imaging Materials,* Marcel Dekker, Inc., New York, NY, 1991.
Okawara, M., T. Kitao, T. Hirashima, and M. Matsuoka: *Organic Colorants, Handbook of Selected Dyes for Electro-optical Applications,* Elsevier, Amsterdam, The Netherlands, 1988.
Sturmer, D.M.: in E.C. Taylor and A. Weissberger, eds., *Special Topics in Heterocyclic Chemistry, The Chemistry of Heterocyclic Compounds,* Vol. 30, Wiley-Interscience, New York, NY, 1977, p. 441.

DYSPROSIUM. [CAS: 7429-91-6]. Chemical element, symbol Dy, at. no. 66, at. wt. 162.50, ninth in the Lanthanide Series in the periodic table, mp 1412°C, bp 2567°C, density 8.551 g/cm^3 (20°C). Elemental dysprosium has a close-packed hexagonal crystal structure at 25°C. The pure metallic dysprosium is silver-gray in color and retains its luster at room temperature. Although stable up to approximately 400°C, the metal then oxidizes at a slow rate up to 600°C. Because of its comparative softness, the metal can be worked by conventional equipment to form rod, foil, and ribbon configurations. There are seven natural isotopes ^{156}Dy, ^{158}Dy, and ^{160}Dy through ^{164}Dy. Twelve artificial isotopes have been identified. In terms of abundance, dysprosium is present on the average of 3 ppm in the earth's crust and is ranked forty-second in terms of total abundance, thus making it potentially more available than tin or beryllium. First identified the element by Lecoq de Boisbaudran in 1886. The metal has a low acute-toxicity rating. Electronic configuration

$$1s^2 2s^2 p^6 3s^2 3p^6 3d^{10} 4s^2 4p^6 4d^9 4f^{10} 5s^2 5p^6 5d^1 6s^2$$

Ionic radius Dy^{3+} 0.908 Å. Metallic radius 1.775 Å. First ionization potential 5.93 eV; second 11.67 eV. Dysprosium appears to be exclusively trivalent. Other important physical properties of dysprosium are given under **Rare-Earth Elements and Metals**.

Dysprosium occurs in apatite, euxenite, gadolinite, and xenotime. All of these minerals also are processed for their yttrium content. With liquid-liquid organic and solid-resin organic ion-exchange techniques, the separation of dysprosium from yttrium is favorable.

Because dysprosium has a high neutron-absorbing capability, the metal, usually in the form of foil, is used for detecting and measuring nuclear reactions and exposures. Dy_2O_3 also is used in dosimeters. Dysprosium does not emit harmful decay radiations when it is used as a neutron sponge. Further, helium-generated alpha particles, which may ultimately crack structural parts for containing nuclear fuel, are not generated. Thus, stainless steel, containing about 3% Dy_2O_3 sometimes is used in control rod hardware for high-flux-beam reactors. Although Dy_2O_3 catalyzes the polymerization of ethylene and other synthetic reactions, the cost to date has limited these uses. Since dysprosium oxide fluoresces yellow in glass under ultraviolet radiation, it has been considered as an activator for the yellow component of the phosphors used in black-and-white television picture tubes. Investigations are continuing in connection with future uses of dysprosium in thermoelectric, semiconducting, photoelectric materials, and garnet microwave devices.

K. A. GSCHNEIDNER, JR.
B. EVANS
Iowa State University, Iowa

Additional Reading

Bochkarev, M. et al.: *Organoderivatives of Rare Earth Elements,* Kluwer Academic Publishers, New York, NY, 1995.
Elliott, J.: *Structure and Chemistry of the Apatites and Other Calcium Orthophosphates,* Elsevier Science, New York, NY, 1994.
Henderson, P.: *Rare Earth Element Geochemistry,* Elsevier Science, New York, NY, 1984.
Stwertka, A. and E. Stwertka: *A Guide to the Elements,* Oxford University Press, Inc., New York, NY, 1998.

E

EBULLIOMETER. An instrument, sometimes referred to as an ebullioscope, that measures the property of a substance by noting a deviation from a normal known boiling point. The term applies to apparatus for estimating the percentage of alcohol in a mixture through observation of the boiling point. Beckmann's apparatus for molecular weight determination is an ebullioscope. The ebullioscopic constant is a quantity calculated to represent the molal elevation of the boiling point of a solution by the relationship:

$$K = \frac{RT_0^2}{1000l_e}$$

in which K is the ebullioscopic constant, R is the gas constant, T_0 is the boiling point of the pure solvent, and l_e is the latent heat of evaporation per gram. The product of the ebullioscopic constant and the molality of the solution gives the actual elevation of the boiling point for the range of values for which this relationship applies. Unfortunately, this range is limited to very dilute solutions, not extending to solutions of unit molality.

See also **Beckmann Method**.

EBULLISM. The formation of bubbles, with particular reference to water vapor bubbles in biological fluids caused by reduced ambient pressure; the boiling of body fluids. See also **Ebulliometer**.

ECLOGITE. This is a coarse, granular rock composed chiefly of garnet and pyroxene with subordinate amounts of various minerals such as rutile, magnetite, and apatite. Hornblende sometimes is present, replacing the pyroxene, often to the extent that a garnet amphibolite is produced. The origin of eclogites is obscure, they may result in part from the deep-seated metamorphism of gabbronic rocks, but some may have resulted from crystallization of a primary basic magma under conditions of great pressure. They may represent segregations in a highly basic magma analogous to segregations of basic minerals in granites and other common igneous rocks. Seemingly confirmatory evidence of this idea is found in the chunks of ecologite-like material found in kimberlite in the Republic of South Africa.

ECOLOGY. A branch of biology that deals with the relations of organisms and their environment, including their relations with other organisms. It is an interdisciplinary field, cutting across the life and geophysical sciences. Ecological investigations look into two directions: (1) The nature of environments and the demands which these environments make upon the organisms that inhabit them; and (2) the characteristics of organisms (plant or animal), species, and groups that permit or promote their tolerance of specific environmental conditions. In recent years, particular emphasis has been placed on the studies of groups rather than single species and this has given rise to the term *ecosystem*. Odum defines an ecosystem as "any entity or natural unit that includes living and nonliving parts interacting to produce a stable system in which the exchange of materials between the living and nonliving parts follows circular paths." However, primarily a branch of biology, ecology does involve chemistry in respect to plant and animal nutrients, metabolism, photosynthesis, etc., especially interference's that may occur in connection with these. Thus, insecticides, chemical-waste disposal, air and water pollution, oil spills, and radioactive contaminants have a direct bearing on the ecology of a given area.

See also **Environmental Chemistry**.

EDIBLE OIL. As commonly used, the term refers to any fatty oil obtained from the flesh or seeds of plants that is used primarily in foodstuffs (margarine, salad dressing, shortening, etc.). Among these are olive, safflower, cottonseed, coconut, peanut, soybean, and corn oils, some of which may be hydrogenated to solid form. They vary in degree of

unsaturation, ranging from 78% for safflower to about 10% for coconut. Castor oil, though technically edible, is not usually considered in this classification, nor are medicinal oils derived from animal sources (cod liver, mineral oil, etc.).

See also **Vegetable Oils (Edible)**.

EFFLORESCENCE. When a substance evolves moisture upon exposure to the atmosphere, the substance is said to be efflorescent, and the phenomenon is known as efflorescence. At ordinary temperatures, the vapor pressure of water is shown by Table 1. If the substance has a higher water vapor pressure than corresponds to that of the atmosphere at the given temperature, water vapor is evolved from the substance until the water vapor pressure of the substance equals the water vapor pressure of the surrounding atmosphere.

TABLE 1. VAPOR PRESSURE OF WATER

Temperature, °C	Water vapor pressure in mm mercury	
	At saturation	At 50% humidity
0	4.6	2.3
10	9.2	4.6
20	17.5	8.8
30	31.8	15.9
40	55.3	27.7

Substances that are ordinarily efflorescent are sodium sulfate decahydrate, sodium carbonate decahydrate, magnesium sulfate heptahydrate, and ferrous sulfate heptahydrate. When the saturated solution of a substance in water has a water vapor pressure greater than that of the surrounding atmosphere, evaporation of the water from solution takes place.

See **Deliquescence** for the converse phenomenon.

EFFUSION. Effusion is a general term denoting a process of discharge; it is also used specifically to denote the passage of gas under pressure through a small orifice.

EHRLICH, PAUL (1854–1915). A native of Silesia, Enrlich is considered the founder of the science of chemotherapy, or the treatment of diseases by chemical agents. He did fundamental work on immunity, which earned him the Nobel prize in medicine in 1908, and also developed the famous neoarsphenamine (salvarsan or 606) treatment for syphilis (1910), which was not improved upon until the discovery of penicillin.

EICOSANOID. Any of a number of biochemically active compounds resulting from enzymic oxidation of arachidonic acid, e.g., prostaglandins, thromboxanes, prostacyclin, and leukotrienes. As a group they compose what is known arachidonic acid cascade. They have many pharmacological and medical possibilities.

See also **Prostaglandins**.

EIGEN, MANFRED (1927–). A German physicist who shared the Nobel prize for chemistry with Ronald George Wreyford Norrish and George Porter in 1967, for their studies of extremely fast chemical reactions, effected by disturbing the equilibrium by means of very short pulses of energy. His research concerned the rate of hydrogen-ion formation through disassociation of water. He also was concerned with enzyme control. He received his degree at the University of Gottingen.

EINSTENIUM. [CAS: 7429-92-7]. Chemical element symbol Es, at. no. 99, at. wt. 254 (mass number of the most stable isotope), radioactive metal of the *Actinide* series, also one of the *Transuranium* elements. Both einsteinium and fermium were formed in a thermonuclear explosion that occurred in the South Pacific in 1952. The elements were identified by scientists from the University of California's Radiation Laboratory, the Argonne National Laboratory, and the Los Alamos Scientific Laboratory. It was observed that very heavy uranium isotopes which resulted from the action of the instantaneous neutron flux on uranium (contained in the explosive device) decayed to form Es and Fm. The probable electronic configuration of Es is

$$1s^2 2s^2 2p^6 3s^2 3p^6 3d^{10} 4s^2 4p^6 4d^{10} 4f^{14} 5s^2 5p^6 5d^{10} 5f^{11} 6s^2 6p^6 7s^2.$$

Ionic radius Es^{3+} 0.97 Å. See also **Chemical Elements**.

All known isotopes of einsteinium are radioactive. The first evidence of their existence was obtained by ion-exchange methods applied to coral rocks obtained from Eniwetok Atoll after the thermonuclear explosion. The first pure isotope found was ^{253}Es produced by prolonged treatment of plutonium-239 with neutrons in the Arco, Idaho, Testing Reactor. The most stable is ^{254}Es, half-life 270 days, and therefore the mass number 254 is carried in the atomic weight table. Others include ^{245}Es–^{246}Es, ^{248}Es–^{252}Es, and ^{253}Es, ^{255}Es.

The ion Es^{3+} is stable. The isotopes of mass numbers 245, 252, 253 and 254 decay by alpha-particle emission; that of mass number 250 by electron capture, those of mass numbers 246, 248, 249, and 251 by both of these processes, while those of mass numbers 255 and 256 emit electrons to form the corresponding fermium isotopes.

Sufficient einsteinium, produced through intense neutron bombardments of plutonium-239 in a reactor, was not available until 1961 to allow separation of a macroscopic amount. Cunningham, Wallmann, L. Phillips, and Gatti worked with submicrogram quantities to separate a small fraction of pure einsteinium-235 compound and measure its magnetic susceptibility. Only a few hundredths of a microgram were available at that time.

Additional Reading

Armbruster, P. and G. Münzenberg: "Creating Super-heavy Elements," *Sci. Amer.*, 66–72 (May 1989).

Cunningham, B.B., J.R. Peterson, Baybarz, R.D., and T.C. Parsons: "The Absorption Spectrum of Es^{3+} in Hydrochloric Acid Solutions," *Inorg. Nucl. Chem. Lett.*, **5**, 519–523 (1967).

Cunningham, B.B. and T.C. Parsons: "Preparation and Determination of the Crystal Structure of Californium and Einsteinium Metals," reportnoLawrence Berkeley Laboratory Nuclear Chemistry Annual Report, UCRL-20426, University of California, Berkeley, California, 1970.

Fisk, Z., et al. : "Heavy-Electron Metals: New Highly Correlated States of Matter," *Science*, **239**, 33–41 (1988).

Fuger, J. and L.R. Morss: *Transuranium Elements: A Half Century*, American Chemical Society, Washington, DC, 1992.

Lide, D.R.: *CRC Handbook of Chemistry and Physics*, 84th Edition, CRC Press, LLC., Boca Raton, FL, 2003.

Seaborg, G.T. and W.D. Loveland: *The Elements beyond Uranium*, John Wiley & Sons, Inc., New York, NY, 1990.

ELAIDINIZATION. Originally, the reaction by which oleic acid is converted into elaidic acid. However, it is now used in a more general sense to indicate the conversion of any unsaturated fatty acid or related compound from the geometric *cis* to the corresponding *trans* form. Nitrous acid and selenium compounds are commonly used as catalysts for this reaction. The resulting *trans* acids are more stable to oxidation.

ELASTICITY. The property whereby a body, when deformed, automatically recovers its normal configuration as the deforming forces are removed. Each of its several types is probably due to the action of intermolecular forces that are in equilibrium only for certain configurations.

Deformation or, more briefly, strain is of various kinds; in each case its measure is a certain abstract ratio. For example, the elongation of a rod under tension is expressed as the ratio of the increase in length to the unstretched length. Linear compression is the reverse of elongation. They are both accompanied by a fractional change in diameter, the ratio of which to the elongation is called the Poisson ratio. Shear is a strain involving change of shape, such that an imaginary cube traced in the unstrained material becomes a rhombic prism. The measure of shear is the tangent of the angle through which the oblique edges have been made to depart from their original perpendicular direction. Volume strain is the ratio of a decrease in volume to the normal volume. Flexure, or bending, and torsion, or twisting, are combinations of these more elementary strains. A straight rod bent into a plane curve undergoes elongation on the convex side and linear compression on the concave side, while there is an intermediate neutral layer, which suffers neither.

In tests such as the tensile test it is necessary to differentiate between two basic ways of measuring strain. "Conventional strain" is the increase in length of the gage section divided by the original gage length, whereas the "true strain" is the natural logarithm

$$\ln \frac{l}{l_0}$$

where l_0 is the original value and l the instantaneous value of the length of the gage section.

For every strain, there arises, in an elastic substance, a corresponding stress, which represents the tendency of the substance to recover its normal condition. Stress is expressed in units of force per unit area. Tensile stress, for example, is the ratio of the force of tension to the normal cross section of the rod subjected to it. Shearing stress is the force tending to push one layer of the material past the adjacent layer, per unit area of the layers. Pressure, expressed in like units, is the stress corresponding to volume compression, etc.

For each type of strain and stress there is a modulus, which is the ratio of the stress to the corresponding strain. In the case of elongation or linear compression, it is commonly called Young's modulus; we also have the bulk modulus and the shear modulus of rigidity. See Table 1.

In engineering design, Young's modulus is used for tension and compression and the rigidity modulus for shear, as in torsion springs. According to Hooke's Law, the stress set up within an elastic body is proportional to the strain to which the body is subjected by the applied load.

Theory of Elasticity

The fundamental quantities in elasticity are second-order tensors, or dyadics: the deformation is represented by the *strain dyadic*, and the internal forces are represented by the *stress dyadic*. The physical constitution of the deformable body determines the relation between the strain dyadic and the stress dyadic, which relation is, in the infinitesimal theory, assumed to be linear and homogeneous. While for anisotropic bodies this relation may involve as much as 21 independent constants, in the case of isotropic bodies, the number of elastic constants is reduced to two.

Let **s**(**r**) be the displacement vector, due to the deformation, of a particle that before the deformation was situated at point P having **r** as position vector with respect to some arbitrary origin. A neighboring point Q, whose position vector was $\mathbf{r} + d\mathbf{r}$ before the deformation, will suffer a displacement $\mathbf{s}(\mathbf{r} + d\mathbf{r})$ which will differ from **s**(**r**) by the quantity

$$d\mathbf{s} = d\mathbf{r} \cdot \nabla s$$

The hypothesis of small deformations means that ds, the change in the displacement vector when we go from P to the neighboring point Q, is very small compared to $d\mathbf{r}$, the position vector of Q relative to P. Consequently, the scalar components of the dyadic ∇s are all very small compared to unity. The geometrical meaning of the dyadic ∇s is obtained by separating it into its symmetric part $\mathbf{S} = \frac{1}{2}(\nabla s + s\nabla)$ and its antisymmetric part $\mathbf{R} = -\frac{1}{2}\mathbf{1} \times (\nabla \times \mathbf{s})$, where **1** is the unity dyadic. The antisymmetric part is interpreted as follows: if at some point M the symmetric part vanishes, then we have for the neighborhood of M the relation

$$d\mathbf{s} = d\mathbf{r} \cdot R_M = \omega_M \times d\mathbf{r}$$

TABLE 1. NOMINAL MODULUS VALUES OF REPRESENTATIVE METALS (values in pounds per square inch)

Material	Young's modulus	Rigidity of shear modulus
Magnesium	6.5×10^6	2.4×10^6
Aluminum	10.2×10^6	3.6×10^6
Copper	14.5×10^6	6.1×10^{6s}
Steel	30×10^6	11.6×10^6

where $\omega_M = \frac{1}{2}(\nabla \times s)_M$ is an infinitesimal vector. This means that the neighborhood of point M undergoes an infinitesimal rigid rotation, without any change in shape or size. Consequently, the deformation is represented by the symmetric part S, which is called the *strain dyadic*.

In a Cartesian orthonormal basis, in which we have $r = \sum_{i=1}^{3} x_i a_i$, we write $s = \sum_{i=1}^{3} s_i a_i$, and obtain

$$S = \sum_{i,j=1}^{3} a_i a_j S_{ij}$$

where

$$S_{ij} = \frac{1}{2}\left[\frac{\partial}{\partial x_i}s_j + \frac{\partial}{\partial x_j}s_i\right].$$

The diagonal components S_{11}, S_{12}, and S_{33} are the coefficients of linear extension in the directions a_1, a_2, and a_3, respectively, while the nondiagonal components $S_{12} = S_{21}$, $S_{13} = S_{31}$, and $S_{23} = S_{32}$ are called shear strains. For instance, $2S_{12}$ is the change in the angle of the dihedron formed by the planes that before the deformation were respectively normal to the directions a_1 and a_2. The shear strains are not essential for the complete representation of a deformation since they can be made to vanish by expressing S in the basis of its principal axes.

If an infinitesimal element of the body occupies the volume dV before the deformation and the volume dV' after, the relative increase of volume, or volumetric dilatation, is given by

$$\frac{dV' - dV}{dV} = S_{11} + S_{22} + S_{33} = |S| = \nabla \cdot s$$

The forces applied to a finite deformable body are either body forces acting on every volume element dV and represented by the notation $F = dV\rho K$, where F is the force per unit volume, K is the force per unit mass, and ρ is the density, or surface forces acting on every element dS of the bounding surface and represented by $dS\, T$, where T is the surface stress, or surface force per unit area. The effect of these applied forces is transmitted throughout the body, so that through any surface element inside the body, there is a force exerted by the matter on one side of the element upon the matter on the other side. Such forces are called internal stresses and are defined as follows: let dS be a surface element completely inside the body, and let us choose arbitrarily the positive sense of the normal n to this surface element; this defines for dS a positive side, the one containing n, and a negative side. Then T_n, the stress vector on the positive side of dS is defined as a vector such that $dS\, T_n$ is the surface force on the positive side of dS—i.e., the resultant of all the forces exerted through dS by the matter on the positive side of dS upon the matter on the negative side. In general there is a normal component $T_n \cdot nn$, which is a pressure or a traction depending upon whether the sign of $T_n \cdot n$ is negative or positive, and a tangent component $n \times T_n \times n$ called the shear stress. The value of the stress vector T_n depends upon the orientation of the normal n, so that we can characterize the state of stress at a point by defining the *stress dyadic* T through the relation

$$T_n = n \cdot T$$

The mechanical equilibrium conditions applied to an arbitrary volume V, bounded by the closed surface S, and completely inside the deformable body give

$$\int_V dV F + \int_S dS n \cdot T = 0$$

and

$$\int_V dV r \times F + \int_S dS r \times (n \cdot T) = 0$$

By the use of the divergence theorem, the first condition gives the equation

$$\nabla \cdot T + F = 0$$

at any point inside the body, and the second condition implies that T is a symmetric dyadic. On the external surface of the body, we have usually to fulfill the boundary condition

$$n \cdot T = T$$

where T is the applied external force per unit area. Other boundary conditions can also be met, such that the value of the displacement be prescribed.

For infinitesimal deformations, we assume that the relation between strain and stress is expressed by Hooke's law: the deformation is proportional to the applied force. For isotropic bodies, this linear relation is

$$S = \frac{1}{E}[(1 + v)T - v|T|1]$$

where E is Young's modulus and v is Poisson's ratio. These two elastic constants can be defined by considering the stretching of a cylindrical bar by normal traction forces uniformly distributed on the end sections; then we have

$$\text{Young's modulus} = \frac{\text{Normal traction force per unit cross sectional area}}{\text{Relative longitudinal extention}}$$

and

$$\text{Poisson's ratio} = \frac{\text{Relative lateral contraction}}{\text{Relative longitudinal extension}}$$

We can also write

$$T = 2\mu S + \lambda|S|1$$

where $\mu = E/2(1 + v)$ and $\lambda = vE/(1 + v)(1 - 2v)$ are Lamé's constants. μ is the rigidity modulus, the only constant necessary when the volumetric dilatation vanishes everywhere.

Substituting the preceding relation into the equilibrium equations, we transform them into

$$2\mu\nabla \cdot S + \lambda\nabla|S| + F = 0 \text{ in side the body}$$

and

$$2\mu n \cdot S + \lambda n|S| = T \text{ on the bounding surface.}$$

These vector relations are not sufficient for the complete determination of the symmetric dyadic S. To insure that a solution of the above equations corresponds to a possible displacement vector s, we must be able to integrate the relation

$$S = \frac{1}{2}(\nabla s + s\nabla)$$

i.e., from a given expression for S obtain the value of s. From the vanishing of the curl of a gradient, it is easily seen that this integrability condition, also called the compatibility equation, is

$$\nabla \times S \times \nabla = 0$$

By elimination of the vector products, we obtain the equivalent form

$$\nabla\nabla \cdot S + \nabla \cdot S\nabla - \nabla\nabla|S| - \nabla \cdot \nabla S = 0$$

Using the stress-strain relation and the equilibrium conditions, we obtain the Beltrami-Michell form of the compatibility equation:

$$\nabla \cdot \nabla T + \frac{1}{1 + v}\nabla\nabla|T| = -\frac{v}{1 - v}\nabla \cdot F1 - (\nabla F + F\nabla)$$

Finally, by expressing the strain dyadic in terms of the displacement vector, we obtain Navier's form of the equilibrium equations:

$$\mu\nabla \cdot \nabla s + (\lambda + \mu)\nabla\nabla \cdot s + F = 0$$

inside the body and

$$\lambda n\nabla \cdot s + 2\mu n \cdot \nabla s + \mu n \times (\nabla \times s) = T$$

on the bounding surface. Dealing here directly with the displacement vector, there is no need of considering the compatibility equation.

The propagation equation for elastic disturbances is obtained by adding the inertia force to the body force. We get then

$$\mu\nabla \cdot \nabla s + (\lambda + \mu)\nabla\nabla \cdot s + \rho K = \rho\frac{\partial^2}{\partial t^2}s$$

inside the body. The stress-strain relation and the boundary conditions are not affected, but we generally have to take into account initial conditions.

The energy density u, or energy per unit volume, is given by

$$u = \frac{1}{2}SS : TT + \frac{1}{2}\rho\frac{\partial s}{\partial t} \cdot \frac{\partial s}{\partial t}$$

where the first term is potential, or strain energy, and the second term is kinetic energy. The energy flux density vector

$$S = -\frac{\partial s}{\partial t}.$$

is a vector such that $dS\,\mathbf{n}\,\mathbf{S}$ gives the quantity of energy that flows per unit time through the surface element dS in the positive direction of $\mathbf{n}$, the normal to dS. At any point the energy continuity equation

$$\frac{\partial u}{\partial t} + \nabla \cdot \mathbf{S} - \rho\frac{\partial \mathbf{s}}{\partial t} \cdot \mathbf{K} = 0$$

expresses the conservation of mechanical energy.

ELASTOMERS.

Of natural or synthetic origin, an elastomer is a polymer possessing elastic (rubbery) properties. A polymer is a substance consisting of molecules which are, in the most part, multiples of low-molecular-weight units, or monomers. As an example, isoprene (2-methylbutadiene-1,3) is C_5H_8, whereas polyisoprene is $(C_5H_8)_x$, where $x \geq 2$ and normally is from 1,000 to 10,000 for rubbers. Although they differ in composition from natural rubber, many of these high-molecular-weight materials are termed *synthetic rubbers*. See also **Rubber (Natural)**.

The serious development of synthetic rubbers commenced in the late 1930s and early 1940s, accelerated by a cutoff of supplies of natural rubber because of political turmoil and war. Synthetic rubbers fall into two major classifications: (1) general-purpose rubbers, the major volume of which is nevertheless used for tire production; and (2) specialty rubbers that essentially find little use in tires, but that are important for a number of other categories. Synthetic rubbers have not replaced natural rubber for numerous uses. For large, heavy-duty truck and bus tires, natural rubber tends to run considerably cooler and wears better than a blend of natural and synthetic rubbers. On the other hand, a tire tread made of a blend of styrene-butadiene (SBR) and butadiene rubber (polybutadiene) wears longer than natural rubber in conventional automobile, usage, where lower temperatures can be maintained.

Styrene-Butadiene (SBR) Rubbers

This series of rubbers includes monomer ratios up to about 50% styrene. The addition of more than 50% styrene makes the materials more like plastic than rubber. The most commonly used SBR rubbers contain about 25% styrene, which is polymerized in emulsion systems at 5–10°C. Most SBR goes into tires, but the type for the tread differs from that of the sidewall or carcass. SBRs for adhesives, shoe soles, and other products also differ. The formulation permits vast varieties of end products. Among the processing variables that can be manipulated to provide different end characteristics are temperature, viscosity, use of different emulsifiers and solvents, use of different antioxidants for stabilization, different oils, carbon blacks, and coagulation techniques.

Initial processes for emulsion (E-SBR) called for polymerization at 50°C. However, it was found later that cold processing produced a rubber particularly good for tires. This is sometimes referred to as *cold*

SBR. The overall process for making SBR is shown in Fig. 1. Typical formulations are:

	Parts by weight	
	Hot SBR	Cold SBR
Deionized water	180	200
Fatty acid soap	5	4.5
Styrene	25	28
t-Dodecyl-mercaptan	0.35	0.2
Butadiene	75	72
Potassium persulfate	0.3	0
Redox initiator	0	small quantity
Temperature	50°C	5°C
Time	12 hours	8 hours
Conversion	75%	60%

Polymerization of emulsion SBR is started by free radicals generated by the redox system in cold SBR and by persulfate or other initiator in hot SBR. The initiators are not involved in the molecular structure of the polymers. Almost all molecules are terminated by fragments of the chain transfer agent (a mercaptan). Schematically, the molecules are RSM_n H, where RS is the $C_{12}H_{25}S$ part of a dodecyl mercaptan molecule; M is the monomer involved; n is the degree of polymerization, and H is a hydrogen atom formerly attached to the sulfur of a mercaptan. In the case of free-radical-initiated polymerization of butadiene, by itself to form homopolymers or with other monomers for form copolymers, the butadiene will be about 18%; 16% *cis*-1,4; and 66% *trans*-1,4.

Considerable quantities of SBR latex are used in the manufacture of foam rubber, adhesives, fabric treating, and paints. The solid content of lattices runs from 50% to 65–70%.

Solution (S-SBR) consists of styrene butadiene copolymers prepared in solution. A wide range of styrene-butadiene ratios and molecular structures is possible. Copolymers with no chemically detectable blocks of polystyrene constitute a distinct class of solution SBRs and are most like styrene-butadiene copolymers made by emulsion processes. Solution SBRs with terminal blocks of polystyrene (S-B-S) have the properties of self-cured elastomers. They are processed like thermoplastics and do not require vulcanization. Lithium alkyls are used as the catalyst.

Stereospecific solution polymerization has been emphasized since the discovery of the complex coordination catalysts that yield polymers of butadiene and isoprene having highly ordered microstructures. The catalysts used are usually mixtures of organometallic and transition metal compounds. An example of one of these polymers is *cis*-1,4-polybutadiene,

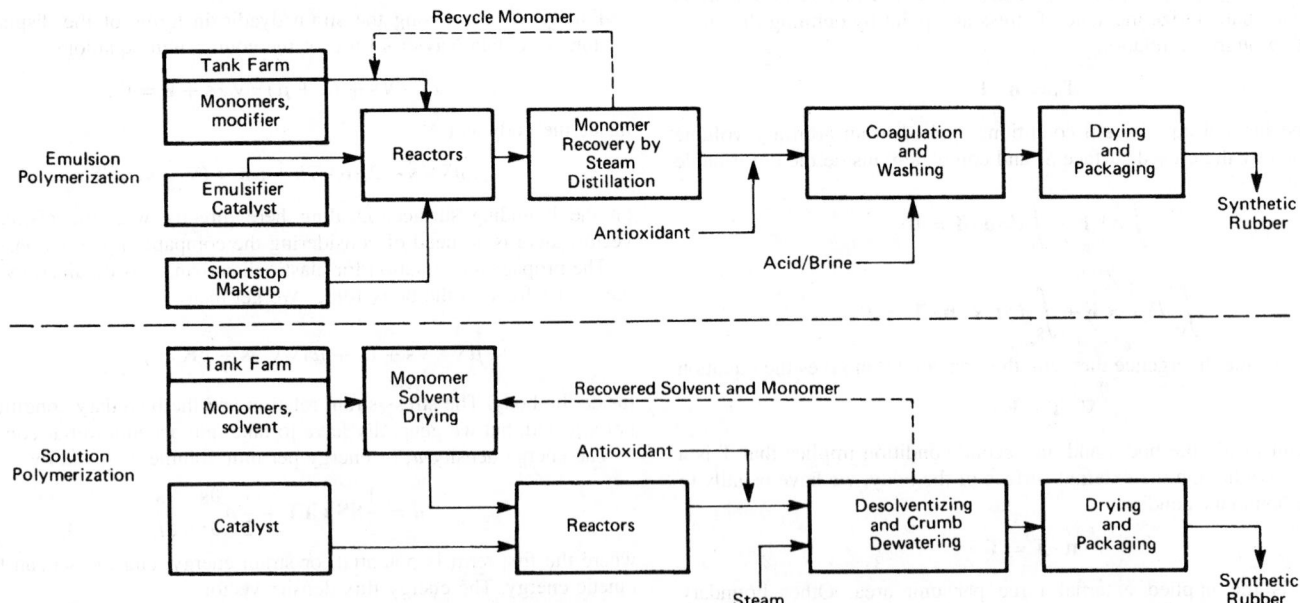

Fig. 1.   Synthetic rubber production process

the bulk of which is used in tires. However, it must be blended with other materials because of its poor processability and traction.

Butyl Rubber

Known as IIR, butyl rubber is a copolymer of isobutylene and isoprene. The elastomers contain only 0.5–2.5 mole % of isoprene. This is introduced to effect sufficient unsaturation to make the rubber vulcanizable. Polymerizations are usually carried out at low temperature (−80 to −100°C) with methyl chloride as solvent. Anhydrous aluminum chloride and a trace of water serve as catalyst.

Butyl is one of the earlier synthetic rubbers. However, it lost favor after the development of SBR. Butyl rubber is incompatible with natural rubber and is difficult to cure. Chlorobutyl rubbers, a much more recent development and containing up to 1.3% chlorine, apparently do not exhibit these former problems. Major uses for butyl rubber have been inner tubes for tires. The appearance of tubeless tires, however, depressed this market. Butyl rubber makes excellent motor mounts because of its high energy absorption and low rebound. Essentially free of double bonds, butyl rubber has a high resistance to aging, attractive properties for use in curing bags for tire production and in outside coating materials.

Acrylonitrile-Butadiene Rubbers (NBR)

Except for the monomers used, the production of NBRs is quite similar to that described for the SBRs. The NBR family is sometimes referred to as the *nitrile rubbers*. The acrylonitrile-butadiene ratios cover a wide range from 15:85 to 50:50. NBRs are noted for their solvent resistance, increasing with the acrylonitrile content. Thus, they are used for gaskets and oil and gasoline hoses, solvent-resistant electrical insulation, and food-wrapping films. Nitrile lattices also are used in treating fabrics for dry-cleaning durability. Because the NBRs become quite inflexible (stiff) at low temperatures (actually brittle at about −20°C), they are blended with polyvinyl chloride for some applications.

Neoprene

This family of dry rubbers and lattices was introduced in 1932, called *Duprene* by Du Pont at that time. The material is made by the free-radical-initiated polymerization of chloroprene in emulsion systems. As with most synthetic rubbers, a variety of neoprenes is made possible by variation of the polymerization conditions and ingredients. Neoprene is particularly good for its fire-retardant, solvent-resistant, and high-temperature stability properties. The chlorine in each segment deactivates the adjoining carbon-carbon double-bond, thus making it less sensitive to oxidative attack. Metal oxides, such as zinc oxide and magnesium, serve as curatives rather than sulfur.

Polyurethanes

Polyethylene in solution is treated with chlorine and sulfur dioxide to introduce approximately 1.3% sulfur and 29% chlorine into the polymer. Most of the chlorine is attached directly to the carbon atoms in the backbone of the polymer. The remainder is in the form of sulfuryl chloride groups, · SO_2Cl, through which crosslinking occurs in the curing step with metal oxides. The material has good oxidation and ozone resistance and thus overall excellent weather resistance. Calendered stocks are used for lining ditches and ponds, for example.

Thiokol® Rubbers

These are polysulfide rubbers and are prepared by the condensation polymerization of sodium polysulfides with a dichloro (sometimes blended with a trichloro) organic compound. Type A, the first family of rubbers, was made from Na_2S_4 and ethylene dichloride. Thiokols are known for high resistance to organic solvents.

Polyacrylate Elastomers

These are made in emulsion or suspension systems involving the copolymerization of ethyl acrylate with the acrylate esters of higher-molecular-weight alcohols. These materials have excellent solvent-resistant properties and stability at elevated temperatures. A major use is for automatic-transmission gaskets for automobiles.

Silicone Elastomers

These materials have alternating Si and O atoms for a backbone and the members differ mainly in the nature of the organic substituents on the Si atoms and the degree of polymerization. Because of the absence of double bonds in the backbone, the numerous forms of stereoisomers found in unsaturated hydrocarbon rubbers do not have counterparts in the silicone rubbers. The chemical combination of organic and inorganic materials gives the silicone rubbers useful properties over a wide temperature range (−70 to +225°C). These rubbers are well known for excellent dielectric stability and high resistance to weathering, oils and chemicals.

Fluoroelastomers

These materials are prepared by emulsion copolymerization of perfluoropropylene and vinylidene fluoride; or of chlorotrifluoroethylene and vinylidene fluoride. Also there are fluorosilicones in this family. Useful at temperatures up to and over 300°C, the fluoroelastomers have excellent resistance to aromatic solvents, acids, and alkalies. They are also among the more costly of the available commercial elastomers.

Ethylene-Propylene Elastomers

Known as EPR, this material is of limited use because it cannot be vulcanized in readily available systems. However, the rubbers are made from low-cost monomers, have good mechanical and elastic properties, and outstanding resistance to ozone, heat, and chemical attack. They remain flexible to very low temperatures (brittle point about −95°C). They are superior to butyl rubber in dynamic resilience.

Additional Reading

Harper, C.: *Handbook of Plastics, Elastomers, and Composites,* McGraw-Hill Companies, New York, NY, 1996.

Holden, G. (Editor), et al.: *Thermoplastic Elastomers,* Hanser-Gardner Publications, Cincinnati, OH, 1996.

Mark, J. (Editor), et al.: *Science and Technology of Rubber,* Academic Press, Inc., San Diego, CA, 1994.

Schweitzer, A.: *Corrosion Resistance Tables: Metals, Nonmetals, Coatings, Mortars, Plastics, Elastomers and Linings,* (4th Edition), Marcel Dekker, Inc., New York, NY, 1995.

ELECTRET. A permanently polarized piece of dielectric material; the analog of a magnet. Barium titanate ceramics, preferably containing a small percentage of lead titanate, can be polarized by cooling from a temperature above the Curie point in an applied electric field. Electrets are also produced by solidification of mixtures of certain organic waxes in a strong electric field.

ELECTRIDE. An experimental compound composed of an alkali-metal cation and an electron in which the electron functions as a chemical element (e.g., a halogen) in salt formation. Several such compounds have been made in the U.S. and abroad. The phenomenon is reported to be one that challenges accepted concepts of compound formation.

ELECTROCAPILLARITY. The surface tension between two conducting liquids in contact, such as mercury and a dilute acid, is sensibly altered when an electric current passes across the interface. As a result, when the contact is in a capillary tube, the pressure difference on the opposite sides of the meniscus is affected by a current traversing the capillary column, to an extent dependent upon the direction of the current across the boundary.

ELECTROCHEMICAL MACHINING (ECM). In this method of metalworking, electrical and chemical energy become the cutting edges of the tool. Electrical energy causes a chemical reaction which, in turn, dissolves metal from a workpiece into an electrolytic solution. The ECM tool (cathode) is brought very close to the workpiece (anode). The distance will range from less than 0.001 to 0.010 inch (0.025 to 0.25 millimeter). A low-voltage, high-density direct current passes between them through the electrically conductive electrolyte solution. This solution is pumped through the gap at pressures ranging up to 300 psi (20.4 atmospheres). The solution normally is maintained at about 100–120°F (38–49°C). The current that is passed through the electrolyte solution ranges widely. Machines using up to 20,000 amperes have been used. As the current passes from the workpiece to the tool, metallic particles on the surface of the workpiece (ions) are caused to go into solution. These particles are then swept away by the rapidly flowing electrolyte. Some of the advantages claimed for ECM include: (1) there is virtually no tool wear; (2) no burrs are produced; (3) both hard and soft metals can be machined at same rate; (4) usually, additional finishing operations are not required;

(5) no mechanical stresses are produced in workpiece surface; (6) no thermal effects are produced in the workpiece because of the relatively low temperature of the operation; (7) tough metals often can be removed faster by ECM than conventional methods; (8) good tolerances and repeatability are produced in complex as well as simple shapes; and (9) the process is easily automated. The basics of this process are also applied in deburring, grinding, and polishing operations.

ELECTROCHEMISTRY.

That branch of science that deals with the interconversion of chemical and electrical energies, i.e., with chemical changes produced by electricity as in electrolysis or with the production of electricity by chemical action as in electric cells or batteries. The science of electrochemistry began about the turn of the eighteenth century.

Background

In 1796, Alessandro Volta observed that an electric current was produced if unlike metals separated by paper or hide moistened with water or a salt solution were brought into contact. Volta used the sensation of pain to detect the electric current. His observation was similar to that observed ten years earlier by Luigi Galvani who noted that a frog's leg could be made to twitch if copper and iron, attached respectively to a nerve and a muscle, were brought into contact.

In his original design Volta stacked couples of unlike metals one upon another in order to increase the intensity of the current. This arrangement became known as the "voltaic pile." He studied many metallic combinations and was able to arrange the metals in an "electromotive series" in which each metal was positive when connected to the one below it in the series. Volta's pile was the precursor of modern batteries.

In 1800, William Nicholson and Anthony Carlisle decomposed water into hydrogen and oxygen by an electric current supplied by a voltaic pile. Whereas Volta had produced electricity from chemical action these experimenters reversed the process and utilized electricity to produce chemical changes. In 1807, Sir Humphry Davy discovered two new elements, potassium and sodium, by the electrolysis of the respective solid hydroxides, utilizing a voltaic pile as the source of electric power. These electrolytic processes were the forerunners of the many industrial electrolytic processes used today to obtain aluminum, chlorine, hydrogen, or oxygen, for example, or in the electroplating of metals such as silver or chromium.

Since in the interconversion of electrical and chemical energies, electrical energy flows to or from the system in which chemical changes take place, it is essential that the system be, in large part, conducting or consist of electrical conductors. These are of two general types—electronic and electrolytic—though some materials exhibit both types of conduction. Metals are the most common electronic conductors. Typical electrolytic conductors are molten salts and solutions of acids, bases, and salts.

A current of electricity in an electronic conductor is due to a stream of electrons, particles of subatomic size, and the current causes no net transfer of matter. The flow is, therefore, in a direction contrary to what is conventionally known as the "direction of the current." In electrolytic conductors, the carriers are charged particles of atomic or molecular size called *ions*, and under a potential gradient, a transfer of matter occurs.

An electrolytic solution contains an equivalent quantity of positively and negatively charged ions whereby electroneutrality prevails. Under a potential gradient, the positive and negative ions move in opposite directions with their own characteristic velocities; each, accordingly, carries a different fraction of the total current through any one solution. Each fraction is referred to as the ionic transference number. Furthermore, the velocity increases with temperature causing a corresponding increase in electrolytic conductivity. This characteristic is opposite to that observed for most electronic conductors which show less conductivity as their temperature is increased.

The concept that charged particles are responsible for the transport of electric charges through electrolytic solutions was accepted early in the history of electrochemistry. The existence of ions was first postulated by Michael Faraday in 1834; he called negative ions "anions" and positive ones "cations." In 1853, Hittorf showed that ions move with different velocities and exist as separate entities and not momentarily as believed by Faraday. In 1887, Svante Arrhenius postulated that solute molecules dissociated spontaneously into *free ions* having no influence on each other. However, it is known that ions are subject to coulombic forces, and only at infinite dilution do ions behave ideally, i.e., independently of other ions

in the solution. Ionization is influenced by the nature of the solvent and solute, the ion size, and solute-solvent interaction. The dielectric constant and viscosity of the solvent play dominant roles in conductivity. The higher the dielectric constant, the less are the electrostatic forces between ions and the greater is the conductivity. The higher the viscosity of the solvent, the greater are the frictional forces between ions and solvent molecules and the lower is the electrolytic conductivity.

In 1923, Debye and Hückel presented a theory that took into account the effect of coulombic forces between ions. They introduced the concept of the ion atmosphere, in which at some radial distance r from a central ion, there is, on a time average, an ionic cloud of opposite charge which sets up a potential field whose magnitude depends on the magnitude of r. This interionic attraction leads to two effects on the electrolytic conductivity. Under a potential gradient, an ion moves in a certain direction. However, the ion cloud, being of opposite sign, will tend to move in the opposite direction, and because of its attraction for the central ion, will have a retarding effect, or *drag effect*, on the ion velocity, thereby leading to a lowering in the electrolytic conductivity. On the other hand, the central ion will tend to pull the ion cloud with it to a new location. The ion atmosphere will adjust to its new location in time, but not instantaneously, and the delay results in a dissymmetry in the potential field around the ion. This also causes a lowering in the conductance of the solution. These effects become more pronounced as the concentration of the solution is increased; for dilute solutions, below about 0.1 molal, the equivalent conductance decreases with the square root of the concentration. For more concentrated solutions, the relation between conductivity and concentration is much more complex and depends more specifically on individual solute properties.

Interionic attraction in dilute solutions also leads to an effective ionic concentration or activity that is less than the stoichiometric value. The *activity* of an ion species is its thermodynamic concentration, i.e., the ion concentration corrected for the deviation from ideal behavior. For dilute solutions the activity of ions is less than one, for concentrated solutions it may be greater than one. It is the ionic activity that is used in expressing the variation of electrode potentials, and other electrochemical phenomena, with composition.

When electricity passes through a circuit consisting of both types of electrical conductors, a chemical reaction always occurs at their interface. These reactions are electrochemical. When electrons flow from the electrolytic conductor, oxidation occurs at the interface while reduction occurs if electrons flow in the opposite direction. These electronic-electrolytic interfaces are referred to as *electrodes*; interfaces where oxidation occurs are known as *anodes*; and those at which reduction occurs, as *cathodes*. An anode is also defined as that electrode by which "conventional" current enters an electrolytic solution, a cathode as that electrode by which "conventional" current leaves. Positive ions, for example, ions of hydrogen and the metals, are called *cations* while negative ions, for example, acid radicals and ions of nonmetals, are called *anions*.

Laws of Electrolysis

In 1833, Faraday enunciated two laws of electrolysis which give the relation between chemical changes and the product of the current and time, i.e., the total charge (coulombs) passed through a solution. These laws are: (1) the amount of chemical change. e.g., chemical decomposition, dissolution, deposition, oxidation, or reduction, produced by an electric current is directly proportional to the quantity of electricity passed through the solution; (2) the amounts of different substances decomposed, dissolved, deposited, oxidized, or reduced are proportional to their chemical equivalent weights. A chemical equivalent weight of an element or a radical is given by the atomic or molecular weight of the element or radical divided by its valence; the valence used depends on the electrochemical reaction involved. The electric charge on an ion is equal to the electronic charge or some integral multiple of it. Accordingly, a univalent negative ion has a charge equal in magnitude and of the same sign as a single electron, and its chemical equivalent weight is equal to its atomic weight, if an element, or to its molecular weight, if a radical. A trivalent ion has +3 or −3 electronic charges, depending on whether it is a positive or a negative trivalent ion. For trivalent ions, then, the equivalent weight would be equal to its atomic weight, if an element, or to its molecular weight, if a radical, divided by three.

The quantity of electricity required to produce a gram-equivalent weight of chemical change is known as the *faraday*. A faraday corresponds, then, to an *Avogadro number of charges*. The most accurate determination of the faraday has been made by a silver-perchloric acid coulometer in which the amount of silver electrolytically dissolved in an aqueous solution of perchloric acid is measured. This method gives 96,487 coulombs (or ampere-seconds) per gram-equivalent for the faraday on the unified ^{12}C scale of atomic weights adopted in 1961 by the International Commission on Atomic Weights.

Electrochemical Equivalent

Preferably termed *coulomb equivalent* of an element or radical, this is the weight in grams which is equivalent to 1 coulomb of electricity and is given by the gram-equivalent weight divided by the faraday (96,487 coulombs per gram-equivalent); for example, the electrochemical equivalent of silver is given by 107.870/96,487 or 0.00111797 grams/coulomb, where 107.870 is the atomic weight of silver based on the unified ^{12}C scale adopted in 1961. The electrochemical equivalents of other elements may be calculated in like fashion.

In electrolysis and in any electric cell or battery, there is an electromotive force (emf) or voltage across the terminals. This emf is expressed in the practical unit, the volt, which is equal to the electromagnetic unit in the meter-kilogram-second system. In any one cell, the emf is the sum of the potentials of the two electrodes and of any liquid-junction potentials that may be present. Neither of the individual electrode potentials can be evaluated without reference to a chosen reference electrode of assigned value. For this purpose, the hydrogen electrode has been universally adopted and is arbitrarily assigned a zero potential for all temperatures when the hydrogen ion is at unit activity and the hydrogen gas is at atmospheric pressure. A hydrogen electrode consists of a stream of hydrogen gas bubbling over platinized platinum or gold foil and immersed in a solution containing hydrogen ions; the electrochemical reaction is: $\frac{1}{2}$H2 (gas) = H + (solution) + ε, where ε represents the electron. The potential of the hydrogen electrode, E_H, as a function of hydrogen ion concentration and hydrogen-gas pressure is given by

$$E_H = E_H^0 - (RT/nF) \ln(a_{H+}/p_{H2}^{1/2})$$
$$= E_H^0 - (RT/nF) \ln(c_{H+} f_{H+}/p_{H2}^{1/2})$$

where E_H^0 is the standard quantity assigned a value of zero, R is the gas constant, T the absolute temperature, n the number of equivalents, F the faraday, p_{H2} the pressure of hydrogen, and a_{H+}, c_{H+}, and f_{H+}, respectively, the activity, concentration, and activity coefficient of hydrogen ions. When a_{H+}, and $p_{H2}^{1/2}$, equal one, $E_H = E_H^0$. For very dilute solutions below 0.01 molal f_{H+} may be taken as unity without appreciable error.

The standard potentials, E^0, of other electrodes are obtained by direct or indirect comparison with the hydrogen electrode. Values are determined at 25°C. The values for several metals and other elements are given in entry on **Activity Series**. The reducing power of the elements decreased on going down the column from those elements with negative standard electrode potentials to those with positive potentials. These values are for the ions at unit activity, and reversible or thermodynamic values as a function of metal or radical concentration are given by equations similar to the one above. For the general reaction: $M = M^{n+} + n\varepsilon$, the potential is given by $E_m = E_H^0 - (RT/nF)1 \, na_{Mn+}$.

In electrolysis, at very low current densities, the potentials of the electrodes approximate in magnitude their reversible values and deviate somewhat from these values because of an *IR* drop in the solution and possible concentration polarization (the concentration at the electrode surface may differ from that in the bulk of the solution). Also for high current densities, especially for the generation of gases such as hydrogen, oxygen or chlorine, the voltage required exceeds the reversible voltage; the excess voltage is known as overvoltage, or overpotential for a single electrode, and arises from energy barriers at the electrode. Overpotential, in general, increases logarithmically with an increase in current density.

Scope of Electrochemistry

In addition to the foregoing, it is customary to include under electrochemistry: (1) processes for which the net reaction is physical transfer, e.g., concentration cells; (2) electrokinetic phenomena, e.g., electrophoresis, electroosmosis, and streaming potential; (3) properties of electrolytic solutions, if they are determined by electrochemical or other means, e.g., activity coefficients and hydrogen ion concentration; (4) processes in which electrical energy is first converted to heat, which in turn causes a chemical reaction that would not occur spontaneously at ordinary temperature. The first three are frequently considered a portion of physical chemistry, and the last one is a part of electrothermics or electrometallurgy.

The passage of electricity through gases is sometimes included under electrochemistry. However, in electrical discharges in gases, the principles are entirely different from what they are in the electrolysis of electrolytic solutions. Whereas in the latter, ionic dissociation occurs spontaneously as a result of forces between solvent and solute and without the application of an external field, for gases, relatively high voltages must be applied to accelerate the electrons from the electrode to a velocity at which they can ionize the gas molecules they strike. In this case, the resulting chemical reaction taking place between ions, free radicals, and molecules occurs in the gas phase and not at the electrodes as in the electrolysis of solutions. Studies of the electrical conduction of gases, accordingly, are generally considered under the physics of gases.

Electrochemistry finds wide application. In addition to industrial electrolytic processes, electroplating, and the manufacture and use of batteries already mentioned, the principles of electrochemistry are used: in chemical analysis, e.g., polarography, and electrometric or conductometric titrations; in chemical synthesis, e.g., dyestuffs, fertilizers, plastics, insecticides; in biology and medicine, e.g., electrophoretic separation of proteins, membrane potentials; in metallurgy, e.g., corrosion prevention, electrorefining; and in electricity, e.g., electrolytic rectifiers, electrolytic capacitors.

Additional Reading

Adams, R.N.: *Electrochemistry at Solid Electrodes*, Marcel Dekker, Inc., New York, NY, 1969. A classic reference.

Bard, A.J., et al.: *Standard Potentials in Aqueous Solution*, Marcel Dekker, Inc., New York, NY, 1985.

Bard, A.J.: *Encyclopedia of Electrochemistry of the Elements*, Vol. 16, Marcel Dekker, Inc., New York, NY, 1999.

Bruce, P.G.: *Solid State Electrochemistry*, Vol. 5, Cambridge University Press, New York, NY, 1997.

Conway, B.E.: *Modern Aspects of Electrochemistry*, Vol. 33, Kluwer Academic Publishers, Norwell, MA, 1999.

Hamann, C.H. and W. Vielstich: *Electrochemistry*, John Wiley & Sons, Inc., New York, NY, 1998.

Kaifer, A.E. and M. Gomez-Kaifer: *Supramolecular Electrochemistry*, John Wiley & Sons, Inc., New York, NY, 1999.

Kerridge, D.H. and E.G. Polyakov: *Refactory Metals in Molten Salts: The Chemistry, Electrochemistry and Technology*, Kluwer Academic Publishers, Norwell, MA, 1998.

Kissinger, P.T. and W.K. Heineman: *Laboratory Techniques in Electroanalytical Chemistry*, 2nd Edition, Marcel Dekker, Inc., New York, NY, 1996.

Lund, H. and M.M. Baizer: *Organic Electrochemistry: An Introduction and Guide*, 3rd Edition, Marcel Dekker, Inc., New York, NY, 1991.

Mach, H., D. Krause, and F.G. Baucke: *Electrochemistry of Glasses and Glass Melts: Including Glass Electrodes*, Springer-Verlag, Inc., New York, NY, 2000.

Memming, R.: *Semiconductor Electrochemistry*, John Wiley & Sons, Inc., New York, NY, 2001.

Sato, N.: *Electrochemistry at Metal and Semiconductor Electrodes*, Elsevier Science, New York, NY, 1998.

Sawyer, D.T., J.L. Roberts, Jr., and A. Sobkowiak: *Electrochemistry for Chemists*, John Wiley & Sons, Inc., New York, NY, 1995.

Stratmann, M.: *Encyclopedia of Electrochemistry, Corrosion and Oxide Films*, Vol. 4, John Wiley & Sons, Inc., New York, NY, 2003.

Trescott, M.M.: *The Rise of the American Electrochemicals Industry, 1880–1910*, Greenwood, Westport, CT, 1981.

Wang, J.: *Analytical Electrochemistry*, John Wiley & Sons, Inc., New York, NY, 2000.

ELECTRODE. Either of two substances having different electromotive activity that enables an electric current to flow in the presence of an electrolyte. See also **Electrolyte**. Electrodes are sometimes called plates or terminals. Commercial electrodes are made of a number of materials that vary widely in electrical conductivity, i.e., lead, lead dioxide, zinc, aluminum, copper, iron, manganese dioxide, nickel, cadmium, mercury, titanium, and graphite; research electrodes may be calomel (mercurous chloride), platinum, glass or hydrogen.

In an electric circuit, part of which is composed of other than the usual conductor of copper, or other metal, the terminal connecting the conventional conductor and the conducting substances is an electrode. Examples of electrodes are: the electric cell, where they dip in the electrolyte; the electric furnace, where the electrodes connect the external circuit with the heating arc; the metallic elements in thermionic tubes and gas-discharge devices; and in semiconductor devices, where electrodes perform one or more of the functions of emitting, collecting or controlling by an electric field the movements of electrons and ions. See also **Graphite**.

ELECTRODEPOSITION. The precipitation of a material at an electrode as the result of the passage of an electric current through a solution or suspension of the material, for example, alkaline-earth carbonates, rubber from latex, paint films on metal. A technique for electrodepositing refractory carbide coatings on metal has been reported. The electrode is in the shape of the desired article. An important advantage of electrodeposition is its ability to coat complex shapes having small and irregular cavities with exact thickness control.

See also **Electrophoresis**.

ELECTRODE POTENTIALS. See **Activity Series**.

ELECTRODIALYSIS. Electrodialysis (ED) is a process for moving ions across a membrane from one solution to another under the influence of a direct electric current. In 1940, a multicompartment ED was invented by K. Meyer and W. Strauss in Zürich. Such process and apparatus is what is now meant by ED. In such an apparatus, many membranes *A* selective to anions alternate with membranes *C* selective to cations between a single pair of electrodes (Fig. 1). When a d-c potential is applied, cations M^+ move toward the negatively charged cathode and are able to permeate the cation-selective membranes, but not the anion-selective membranes, if the latter are perfectly selective. Similarly, anions X^- move toward the positively charged anode and are able to permeate the anion-selective membranes, but not the cation-selective ones. As a result, the odd-numbered compartments in the figure become depleted in the electrolyte, the even-numbered compartments enriched. It is shown in Figure 1 that the choice of which electrode is positive and which negative is arbitrary. If the left hand electrode is negative and the right hand one positive, then the odd numbered compartments become enriched in electrolyte, the even-numbered ones are depleted. This is the basis of reversing type ED (EDR) invented by W. McRae in 1956, in which the polarities of the electrodes are reversed regularly at intervals of 15 minutes to 1 week depending on the details of the application.

The depletion compartments of the figure (and in some applications, the enrichment compartments as well) may be filled with ion-exchange beads, fibers, or fabric (see **Ion-Exchange Resins**), leaving interstices through which fluid may flow. This type of filled cell ED was invented by W. Walters, D. Weiser, and L. Marek in 1955. They filled both depletion and enrichment compartments with a mixture of anion exchange (AX) and cation exchange (CX) beads. This process is called electrodeionization (EDI).

Ion-Exchange Membranes

In 1948, ion-selective membranes having high selectivity, low electrical resistance, good mechanical strength, and good chemical stability were invented by W. McRae. They were essentially insoluble, synthetic, polymeric, organic ion-exchange resins in sheet form. Typical chemical structures for modern membranes of this type are shown schematically in Figure 2. The cation-selective membranes (Fig. 2a) consist of cation-exchange (CX) resin composed of polystyrene having negatively charged sulfonate groups chemically bonded to most of its phenyl groups. The charges of the sulfonate groups are electrically balanced by positively

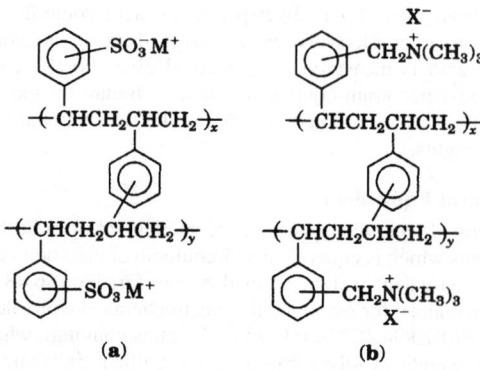

Fig. 2. Schematic representation of (**a**) cation-exchange resin, and (**b**), anion-exchange resin

charged cations (counterions). Sulfonated polystyrene swells greatly in water. The amount of swelling is typically controlled by including cross-linking agents in the polymer, e.g., divinyl benzene; by incorporating electrically neutral polymers; or by having extensive regions (blocks) in the polymer which lead to substantial microcrystallinity. The positively charged counter-ions, e.g., Na^+, Ca^{2+}, or Mg^{2+}, are appreciably dissociated from the chemically bound negatively charged groups once the membrane is exposed to water. Thus the counter-ions are mobile and may be exchanged for other cations from an ambient solution, maintaining the electrical neutrality of the membrane. This high (typically > 1 meq/cm³ of membrane) concentration of counter-ions in ion-exchange resins is responsible for the low electrical resistance of the membrane. The high concentration of bound negatively charged groups tends to exclude mobile, negatively charged ions (co-ions) from an ambient solution and is responsible for the high ion selectivity of the membranes.

The anion-selective (AX) membranes (Fig. 2b) may also consist of cross-linked polystyrene, but have positively charged quaternary ammonium groups chemically bonded to most of the phenyl groups in the polystyrene instead of the negatively charged sulfonates. In this case, the counter-ions are negatively charged, e.g., Cl^-, HCO_3^-, NO_3^-, or SO_4^{2-}.

The membranes described above called homogeneous ion-exchange membranes, represent about half the world production of ion-exchange membranes and are made in the United States and in Japan. The other half (made principally in the CIS and in the People's Republic of China) are made by bonding about 75 parts of AX or CX powder with 25 parts of, e.g., polyethylene and are called heterogeneous. Whether homogeneous or heterogeneous, the ion-exchange membranes tend to transfer divalent ions (such as Ca^{2+}, Mg^{2+}, SO_4^{2-}) relatively more rapidly than univalent ions. Skinned homogeneous membranes are made in Japan and on the contrary tend to transfer univalent ions much more rapidly than divalent. CX membranes based on perfluorinated polymers and having carboxylate CX groups on one surface and sulfonate groups on the other are used in membrane chloralkali cells. Such membranes are made principally in the United States and Japan.

A large quantity of homogeneous ion-exchange membranes are made annually for separators in alkaline batteries. See also **Batteries**.

Commercial membranes are usually reinforced with woven, synthetic fabrics to improve the mechanical properties. Several hundred thousand

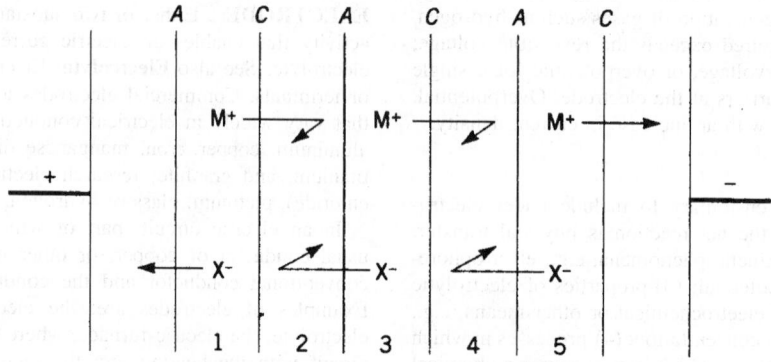

Fig. 1.   Principle of multicompartment electrodialysis. See text

square meters of ion-exchange membranes are now produced annually. The mechanical and electrochemical properties are varied by the manufacturers to suit the proposed applications. The electrochemical properties of most importance for ED are (1) the electrical resistance per unit area of membrane; (2) the ion transport number, related to current efficiency; (3) the electrical water transport, related to process efficiency; and (4) the back-diffusion, also related to process efficiency.

Apparatus

The apparatus for ED is fundamentally an array of alternating AX and CX membranes terminated by electrodes. The membranes are separated from each other by gaskets which form the fluid compartments between the membranes. The enrichment and depletion compartments also alternate through the array. Holes in the gaskets and membranes register with each other to provide two pairs of internal hydraulic manifolds to carry fluid into and out of the compartments. One pair communicates with the depletion compartments and the other with the enrichment compartments. Much effort has been expended on the design of the entrance and exit channels from the manifolds to the compartments to prevent unwanted cross-leak of fluid intended for one class of compartment into the other class. This effort is made increasingly difficult by the continuing trend to thinner membranes and gaskets, the latter determining membrane spacing and the thickness of the fluid compartments; such trends are intended to reduce energy consumption. A contiguous group of two membranes and the associated two fluid compartments is called a cell pair. A group of cell pairs and the associated end electrodes is called a stack or pack. Generally 100–600 cell pairs are arranged in a single stack, the choice being made on the basis of ED capacity desired, the uniformity of flow distribution achieved among the several compartments of the same class in a stack, and the maximum total direct current potential desired. One or more stacks may be arranged in a press, designed to compress the membranes and gaskets against the force of fluid flowing through the compartments thereby preventing fluid leaks to the outside and internal cross-leaks between compartments. For small presses such compression is usually provided by tie-rods; for larger presses, hydraulic rams are frequently used.

Commercial membranes have typical thicknesses of ca 0.15–0.5 mm; the compartments between the membranes have typical thicknesses of ca 0.3–2 mm. The thickness of a cell pair is therefore in the 1.3–5.0 mm range, commonly about 3.0 mm. One hundred cell pairs have a combined thickness of about 300 mm. The effective area of a cell pair for current conduction is generally on the order of 0.2–2 m^2.

Polarization. In concentrated electrolytes the electric current applied to a stack is limited by economic conditions, the higher the current I the greater the power consumption W in accordance with the equation $W = I^2 R_s$ where R_s is the electrical resistance of the stack. In relatively dilute electrolytes, the electric current which can be applied is limited by the ability of ions to diffuse to the membranes across a convection free, laminar flow, diffusion layer contiguous with the depletion faces of the membranes. The need for such diffusion is due to the fact that almost all the current passing through a membrane is carried by its counter-ions, whereas, in the ambient solution roughly only half the current is carried by such counter-ions, the remainder by the co-ions. Therefore, there is a deficit of counter-ions at the membrane-solution interface. When the concentration c of electrolyte at the interface between an AX membrane and the solution in the depletion compartment is negligible compared to the concentration c_b in the bulk solution (i.e., outside the laminar flow layer), then the rate at which ions can diffuse to the interface is a maximum and the AX membrane and the apparatus are said to be concentration polarized or simply polarized. The applied current density corresponding to such maximum rate of diffusion is the limiting current density i_{lim}. As the applied current density approaches i_{lim}, the electrical resistance of the ED stack increases substantially even though the bulk solution in the depletion compartments may still contain an appreciable concentration of electrolyte. This is a sign of polarization; i_{lim} is not a wall. At applied current densities above i_{lim} much of the increase in current through the AX membrane is carried by hydroxide ions (OH^-) resulting from dissociation of water at the interface. As a result, the enrichment compartments become somewhat alkaline and the depletion compartments somewhat acidic. This is also a sign of polarization.

The rate at which ions can diffuse to the interface, and therefore, i_{lim}, is inversely proportional to δ (the thickness of the convection free, laminar flow, diffusion layer). The latter is typically reduced by including turbulence promoting structures (such as screens) in the depletion compartments, and generally also in the enrichment compartments. In such case, i_{lim} in A/cm^2 is, in order of magnitude, equal numerically to the concentration c_b in gram-equivalents/L. i_{lim}, measured in terms of the AX membrane area, can also be increased (by a factor of 10 or more) by including ion-exchange beads, fibers, or fabrics in the depletion compartments as in EDI.

Applications

ED is used principally in the production of high purity water, in demineralization of brackish water (about 1/10th the concentration of seawater), in concentration of seawater (to 18–20%), and in deashing cheese whey. For the production of high purity water it is typical to use EDI. When EDI is operated at currents well above i_{lim} for the filled cells, then substantial water dissociation occurs and EDI behaves in many ways as a continuously electrically regenerated mixed-bed ion exchanger producing ultrapure water equivalent to that produced by chemically regenerated mixed beds. EDI is rapidly replacing chemically regenerated ion-exchange for the production of high purity water.

EDR is typically used for demineralization of brackish water, which often contains poorly soluble minerals such as calcium bicarbonate and calcium sulfate, as well as colloids such as humic and fulvic acids and iron hydroxides. The periodic reversal of the direction of the electric current avoids scaling and fouling of the membranes by such substances.

Univalent ion selective membranes are used in nonreversing ED for seawater concentration. Both EDR and nonreversing ED are used for whey deashing.

Bipolar ED. Bipolar ion-exchange membranes have one surface consisting of CX resin and the opposite surface of AX resin. The interface between the CX and AX resins may be regarded as a zero gap ED compartment. When a direct current is passed through such a membrane in a direction to pull anions out of such interface and through the AX resin, the interface rapidly becomes depleted of all ions other than those resulting from the dissociation of water. Dilute alkali can therefore be produced at the outer surface of the AX region and dilute acid at the outer surface of the CX layer.

The bipolar membranes are used in a more or less conventional ED stack together with conventional unipolar membranes. Such a stack has many acid–alkali producing membranes between a single pair of end electrodes. The advantages of the process compared to direct electrolysis seem to be that because only end electrodes are required, the cost of the electrodes used in direct electrolysis is avoided, and the energy consumption at such electrodes is also avoided.

The disadvantages appear to be that the bipolar membranes are comparatively expensive, and the economic life is limited to about one year. Such short lifetime appears to result from the very high ($\sim 10^6$ V/cm) potential gradients at the interface between the AX and CX regions. Additionally, practical current densities are limited to about 1000 A/m^2 available area.

Economic Aspects

About 5000 ED plants of all types have been installed worldwide for the demineralization of brackish and potable water. These range in capacity from a few to more than 10,000 m^3/d. The total installed capacity is greater than 1.5×10^6 m^3/d and uses more than 3×10^6 m^2 of ion-exchange membranes. Ionics, Incorporated (Watertown, Massachusetts) the leading supplier outside the CIS and the People's Republic China has sold more than 2000 ED plants all using homogeneous membranes. Most of their plants are of the EDR type. Their total installed capacity is greater than 600,000 m^3/d and uses more than 1.2 million m^2 ion-exchange membrane. Most of the remaining 3000 ED plants were built by CIS and the People's Republic of China entities using heterogeneous membranes and are used almost exclusively in the countries in which they were made.

About 10 ED plants for concentrating seawater have been built, exclusively using nonreversing ED stacks and univalent ion-selective membranes made by the Japanese companies Asahi Chemical Industry Company, Asahi Glass Company, and Tokuyama Corporation. The combined capacity is about 1,700,000 metric tons of salt per year using about 850,000 m^2 of ion-exchange membranes. Obviously each plant is very large. Seven of the plants are in Japan as a result of the decision by the Japanese government that all domestic salt had to be made in Japan. Salt produced by ED is not economic compared to imported solar salt. A recent decision by the government removing the above monopoly probably means such ED plants will be phased out.

More than 150,000 metric tons per year of 90% demineralized, dry, whey solids are made by ED or by ED followed by chemically regenerated ion-exchange. See also **Desalination**.

Wayne A. McRae
Consultant

Additional Reading

Rautenbach, R. and R. Albrecht: *Membrane Processes,* John Wiley & Sons, Inc., New York, NY, 1989.
Shaffer, L.H. and M.S. Mintz: in K.S. Spiegler, ed., *Principles of Desalination,* Academic Press, Inc., New York, NY, 1966, pp. 199–289.
Strathmann, H.: *Membrane Separation Systems—A Research Needs Assessment,* U.S. Dept. of Energy, Washington, D.C., p. 8–1.
Wilson, J.R. ed.: *Demineralization by Electrodialysis,* Butterworths, London, 1960.

ELECTROFORMING. The electrolytic depositing of metal upon a conducting mold to make a desired metal object, such as precision tubing or medals. The mold is often of graphite-coated wax, so that it can be removed by melting. See also **Electroplating**.

ELECTROLUMINESCENCE. Luminescence generated in crystals by electric fields or currents in the absence of bombardment or other means of excitation. It is a solid-state phenomenon involving *p*- and *n*-type semiconductors, and is observed in many crystalline substances, especially silicon carbide, zinc sulfide, and gallium arsenide, as well as in silicon, germanium, and diamond.

See also **Luminescence**; and **Phosphors and Phosphorescence**.

ELECTROLYSIS. Decomposition of water and other inorganic compounds in aqueous solution by means of and electric current, the extent being proportional to the quantity of electricity passing through the solution. The positive and negative ions formed are carried by the current to the oppositely charged electrodes, where they are collected (if wanted) or released (if unwanted). Metallic ions deposited on the electrode form a coating. A simple electrolysis is the separation of water into oxygen and hydrogen. Somewhat more complicated is electrolysis of brine to chlorine and sodium hydroxide; this is carried out in electrolytic cells of the diaphragm or mercury type, with water taking part in the reaction. In electroplating, metal salts dissociate into their constituent ions, the positively charge metal ions coating the cathode. There are a number of variations of this process (electrodeposition, electrocoating, electroforming).

See also **Electrochemistry**; **Electrolytic Cell**; and **Ionization**.

ELECTROLYSIS-TYPE CHEMICAL ANALYZER. Sometimes referred to as electroplating analyzers, these devices can be used for determining metals and other materials that will plate out on an electrode which is part of an electrolytic cell. Alloys, such as stainless steel, brass, and bronze that contain chromium, copper, lead, iron, nickel, and tin-bearing metals containing bismuth, can be analyzed in this fashion. The sample must be relatively easy to dissolve so that an electrolyte can be formed and it must be sufficiently large to permit plating out a quantity of material that can be accurately weighted. The potential required to effect plating of a specific material should be known in advance. Complex materials usually are identified on a cumulative basis by making stepwise increases in plating potential, with intermittent weighing of the plated electrode.

ELECTROLYTE. A substance that will provide ionic conductivity when dissolved in water or when in contact with it; such compounds may be either solid or liquid. Familiar types are sulfuric acid and sodium chloride, which ionize in solution. One solid electrolyte, used originally in fuel cells, is a polymer of perfluorinated sulfonic acid used as the core of a water electrolysis cell for production of hydrogen and oxygen. When saturated with water it has high conductivity. Another solid type is a ceramic mixture of sodium, aluminum, lithium, and magnesium used as a separating medium in the liquid-sodium-sulfur (β) battery under continuing development. The most common application of electrolytes is in electroplating of metals in which dissolved (ionized) metal salts are the electrolytes.

See also **Electrolysis**; and **Electroplating**.

ELECTROLYTIC CELL. An electrochemical device in which electrolysis occurs when an electric current is passed through it. Ionizable compounds dissociate in the aqueous solution with which the electrodes are in contact. Such cells are of two types: (1) the diaphragm cell, which has two compartments separated by a porous membrane; and (2) the mercury cell, in which mercury is the cathode. The anodes of both types have long been made of graphite. Because this decomposes rapidly as electrolysis progresses, they are being replaced with dimensionally stable types consisting of titanium coated with oxides of ruthenium and other rare metals, which are also much more efficient. In electrolysis of sodium chloride, the current causes the chloride ion to the anode, where it is collected as chlorine gas; sodium hydroxide and hydrogen are also formed, the hydrogen being discharged. The overall cell reaction is: $2HaCl + 2H_2O \rightarrow H_2 + Cl_2 + 2NaOH$. This principle is applied in the electroplating of metals, electrodeposition of colloids, and similar processes.

See also **Diaphragm Cell**; and **Electroplating**.

ELECTROLYTIC CONDUCTIVITY AND RESISTIVITY MEASUREMENTS. Industrial interest in the measurement of electrolytic conductivity (of which electrolytic resistivity is the reciprocal) arises chiefly from its usefulness as a measure of ion concentrations in water solutions. Also, by comparison with other analytical methods, this is relatively simple and inexpensive.

Pure water is a very poor conductor. Water, such as may be produced by passage through a mixed-bed ion exchanger, has a conductivity approaching very closely the theoretical minimum of approximately 0.05 μS/cm (18 MΩ•cm) at 25°C, which is due to the dissociation products of water itself (S/cm = siemens/centimeter). The conductivity of a water solution, as encountered in industrial practice, is almost exclusively due to a dissolved electrolyte rather than to the solvent (water) ions, and thus a criterion for electrolyte concentration can be established. Solutions of strong electrolytes follow a rather uniform pattern of change in conductivity with concentration, which is almost linear at low concentrations, rising more gradually to a maximum (usually about 20–30% wt) and then falling as the concentration rises further. A series of such conductivity-concentration curves is given in Fig. 1.

Practical applications of these measurements include: (1) Gauging the quality of pure water, such as distilled or demineralized water and condensed boiler steam; (2) measuring the extent of reactions, such as neutralization, precipitation, and washing of soluble electrolytes from insoluble materials; (3) detecting contamination, such as leaks in heat exchangers and resultant contamination of heating or cooling media as in acid coolers, condenser coils, and steam coils; (4) checking possible saltwater intrusions in streams and wells; (5) checking on process interface levels where, for example, oil-in-water and water-in-oil emulsions may be distinguished by the fact that the former are conductive and the latter are

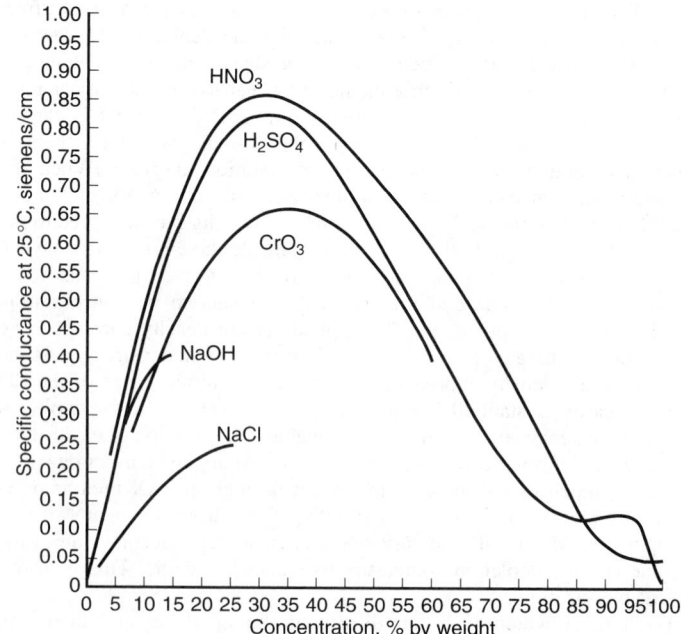

Fig. 1.   Conductivity-concentration curves for certain electrolytes

essentially nonconductive; and (6) the enhancement of the conductivity of certain ions of interest. For example, very conductive hydroxyl ions may be removed from boiler water by the addition of a weakly ionized organic acid to reveal a conductivity which is more nearly proportional to that of the remaining dissolved salts. Or, conductive ammonium hydroxide may be removed from steam condensate by passage through a hydrogen cation exchanger to reveal the conductivity of the remaining dissolved salts, which are converted to their corresponding mineral acids. Similarly, samples may be boiled or sparged to remove conductive dissolved gases.

Measurement Fundamentals

The flow of electricity through matter is accomplished by movement of electric charges, which, in metallic conductors, are electrons, and in electrolytic conductors are positive and negative ions. Conducting solutions in general are electrolytic conductors. In electrolytic conductors, current is usually introduced and leaves the system through metallic electrodes on the surface of which chemical reactions occur. It is possible when using alternating current to cause current to flow by inductive coupling as well as by direct contact between electrode and electrolyte. Positive ions or cations move toward the cathode, where reduction takes place, and negative ions or anions move toward the anode, where oxidation takes place. The conductivity of a solution depends upon the concentration and mobility of all ions present. The ion mobility, in turn, depends upon ion size and charge, as well as the dielectric constant of the solvent and the solution temperature and viscosity.

The determination of electrolytic conductivity presently consists of measuring the A.C. electrical conductance of a column of solution. Although A.C. measurement methods greatly reduce errors associated with electrolysis, when electrodes are used, they introduce other errors associated with series and shunting capacitance which must be compensated for in the design of the measuring instrument. A precision of 0.01% can be obtained in laboratory measurements by following the bridge techniques first discussed by Grinnell Jones and his coworkers. Industrial conductivity meters are capable of providing accuracies of 1% of the actual conductivity under ideal conditions.

Electrolytic conductivity is most often measured by placing electrodes in contact with the electrolytic solution which is contained in such a way that the measured electrical conductance between the electrodes can be related to the conductivity of the solution. The conductivity cell most commonly comprises an enclosure made of electrically insulating material, such as glass or plastic, which serves to hold or isolate a portion of the electrolytic solution and to accommodate the two electrodes. The cell constant of such a device is then used to relate the measured electrical conductance between the electrodes to the actual electrolytic conductivity.

Two electrodes, 1 centimeter square, located on opposite interior faces of a hollow cube, 1 centimeter on an edge, would have a cell constant of 1/cm; a measured conductance of 100 microsiemens at 25°C would indicate a conductivity of 100 microsiemens/cm (10 millisiemens/m) at 25°C.

Definitions and Units. Electrolytic conductivity is often defined as the electrical conductance of a unit cube of solution as measured between opposite faces. It is expressed in the same units as electrical conductivity, i.e., reciprocal ohms per unit length. Most commonly we find Mho/centimeter ($\Omega^{-1}cm^{-1}$), siemens/centimeter (S cm^{-1}), and siemens/meter (S cm^{-1}):

$$1\Omega^{-1}cm^{-1} = S\ cm^{-1} = 100\ S\ m^{-1}$$

Few solutions exhibit conductivities as great as 1 siemens/cm. The most commonly used decimal submultiples are microohm/centimeter ($\mu\Omega^{-1}cm^{-1}$), microsiemens/centimeter (μS cm^{-1}), and millisiemens/meter (mS m^{-1}):

$$1\Omega^{-1}cm^{-1} = 1\ S\ cm^{-1} = 0.1\ m\ S\ m^{-1}$$

Electrolytic resistivity (the reciprocal of conductivity) is similarly defined as the electrical resistance of a unit cube of solution. It is expressed in the same units as electrical resistivity, i.e., ohms times a unit of length. Most commonly we find: ohm-cm (Ω-cm) and ohm-meter (Ω-m):

$$100\Omega - cm = 1\Omega - m$$

Again, decimal multiples commonly encountered are megohm-centimeter and megohm-meter:

$$100\ M\Omega - cm = 1\ M\Omega - m$$

Resistivity units are used almost exclusively to describe ultrapure water in the 10 megohm-cm to 18 megohm-cm (0.1 microsiemens/cm to 0.055 microsiemens/cm) range generated by mixed bed ion exchange and used as boiler feed water and in certain critical washing applications.

The cell constant of a conductivity cell is defined as a factor that relates the measured conductance between the cell terminals to the conductivity of the electrolyte being measured. It is generally expressed in reciprocal units of length (although occasionally in units of length for certain European manufacturers). Most commonly we find 1/centimeters (cm^{-1}) and 1/meters (m^{-1}):

$$1\ cm^{-1} = 100\ m^{-1}$$

The conductance measured between the cell terminals is multiplied by the cell constant given in reciprocal units of length to calculate the conductivity. To calculate the resistivity, the measured resistance between the cell terminals is divided by the cell constant. Although the cell constant (in reciprocal units of length) can be calculated from the dimensions of the conductivity cell by dividing the length of the electrical path through the solution by the cross-sectional area of the path, in practice, these measurements are difficult to make and are only used to approximate the cell constant, which is determined by use of standard solutions of known conductivity or by comparison with other conductivity cells which have been so standardized.

Measuring Circuits

Although there are several circuits used for measuring electrolytic conductivity, the ac Wheatstone bridge is widely applied and is potentially the most stable and accurate.

AC Wheatstone Bridge. A typical system is shown in Fig. 2 and comprises the bridge, including the voltage source, the null indicator, and the conductivity cell. In Fig. 2, *D* represents an ac voltage-sensitive device called the *detector*. The ac source may be the low-voltage tap on a line-frequency operated transformer or battery or line-powered electronic oscillator for higher frequencies. The magnitude of the bridge voltage necessarily is related to the sensitivity of the detector and also to the general characteristics of the electrolytes to be tested.

The usual industrial measuring and control equipment is supplied with bridge voltages of 1–10 V. The frequency of this ac source in commercial units is commonly 60 Hz and more rarely 1000 Hz. Where measurements are to be made on high-resistance electrolytes, such as distilled water or steam condensate, the lower bridge source frequency is preferable. For measurements in high-conductivity solutions, the higher bridge frequencies are of advantage.

R_S is the so-called *standard arm* of the bridge; it is generally made variable, as a device either to change the range of the instrument by selecting one of a number of resistors differing in resistance by powers of 10, or to correct for the temperature coefficient of resistance of the electrolyte. R_3 and R_4 are end resistors whose function is to establish the limits of the bridge calibration. R_5 is the calibrated slidewire potentiometer. With R_3 and R_4 short-circuited, the range of the bridge would be zero to infinity in resistance or conductance. Increasing values of R_3 and R_4

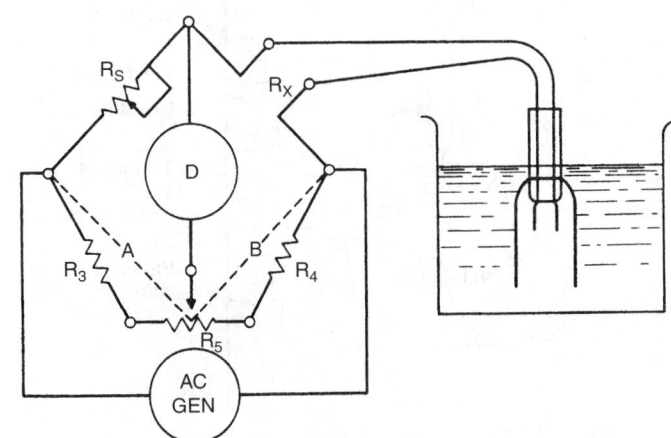

Fig. 2. Alternating current Wheatstone bridge used in electrolytic conductivity measurements

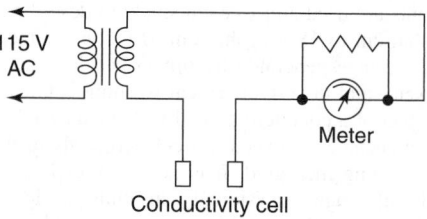

Fig. 3.   Conductivity measurement system utilizing a simple ohmmeter circuit

compared with the value of R_5 will reduce the range covered. It should be noted that the slidewire contact resistance is in series with the detector, and thus variable values of this resistance cause no error in bridge readings. R_X is effectively the resistance of the electrolyte measured between the two electrodes of the conductivity cell immersed in the liquid under test. The condition for balance of the Wheatstone bridge is that $A/B = R_S/R_X$, and this condition is indicated by no current flow through the detector D.

While most laboratory conductivity bridges are manually balanced, the Wheatstone bridge circuit also finds use in a variety of conductivity monitors, controllers, and recorders where it is mechanically rebalanced by a servomechanism operated by the detector. Generally in these devices, advantage is taken of the phase shift, which occurs in the detected signal as the bridge is driven through balance by the servo motor.

Conductivity Meter. A second system utilizes a simple ohmmeter circuit, shown in Fig. 3. A meter, transformer secondary winding, and conductivity cell are connected in series so that the current is a function of the cell conductance. The meter may be calibrated in resistivity or conductivity units.

While early circuits of this type suffered from inaccuracies due to line voltage variations, the addition of a regulated power supply to drive the transformer has brought this relatively simple and inexpensive circuit into wide use. Complete isolation may be achieved by interposing a second transformer between the cell and meter. Generally, a stage of amplification is added to increase sensitivity and to reduce nonlinearity caused by meter resistance. This, combined with gated detection, reduces those polarization errors associated with series capacitance at the electrodes. Driven shields are employed to reduce the errors associated with the shunt capacitance of long cell leads. The addition of automatic temperature compensation, alarm contacts, and electrical outputs make the conductivity meter the most widely used instrument for industrial measurement and control applications.

Electrodeless Circuit. An electric current may be caused to flow in an electrolyte by means of induction without the use of contacting electrodes. In such electrodeless systems, the electrolyte is contained in an electrically insulating tube which passes through the cores of two transformers (Fig. 4) in such a way that the electrolyte forms a closed loop linking the flux in both cores. In the first transformer, this loop of electrolyte serves as a single-turn secondary winding in which an alternating voltage is induced. In the second transformer, the loop forms a single-turn primary winding, providing a means for measuring the resulting current which is directly proportional to the specific conductance of the electrolyte comprising the loop. Alternatively, both transformers may be located about an insulated tube immersed in the electrolytic solution.

Variations of these systems employing glass tubes were in use before 1907. However, more recently, the introduction of chemically resistant, high-temperature electrical insulators, such as the fluorocarbons, has simplified the design of the insulated tube comprising the electrodeless conductivity cells. Since no contacting electrodes are used, all the problems associated with electrodes, such as polarization and electrode surface maintenance, simply disappear. Wide application of these systems is found in highly conductive electrolytes, such as the strong mineral acids and bases—often in conjunction with abrasive slurries or materials containing entangling fibers.

Four-Electrode Circuit. This method avoids errors caused by polarization and fouling by using a set of measuring electrodes located between a set of current-producing electrodes. See Fig. 5. The measuring electrodes are used to determine the voltage drop in the electrolyte caused by the current. The conductivity of the electrolyte between the measuring electrodes is proportional to the current divided by the potential. Laboratory measurements may be made with either AC or DC. However, process instrumentation almost always utilizes AC. In practice, the alternating current through the entire cell is varied to maintain a constant potential between the measuring electrodes. When this is done, the conductivity of the electrolyte between the measuring electrodes is proportional to the cell current. Changes external to the measuring electrodes, such as may be caused by current electrode polarization or fouling, will not cause a change in the cell current, which will be maintained at such a value that the potential between the measuring electrodes is constant. Systems have been designed that can accommodate a tenfold increase in impedance at the current electrodes due to polarization and fouling. Fouling and polarization errors do not occur at the measuring electrodes since the measurement there is essentially potentiometric with no current flowing through the measuring circuit. Size and orientation of the measuring electrodes are also chosen to avoid such errors.

Conductivity Cells

These are simple in basic structure, consisting typically of two metal plates or electrodes spaced within an insulating chamber. Examples are shown in Figs. 6 and 7. This arrangement permits isolation and measurement of a portion of the solution and serves to make the measured resistance independent of sample volume and proximity to conductive and nonconductive surfaces. In laboratory cells, platinum electrodes mounted in a glass are commonly employed for their excellent chemical resistance.

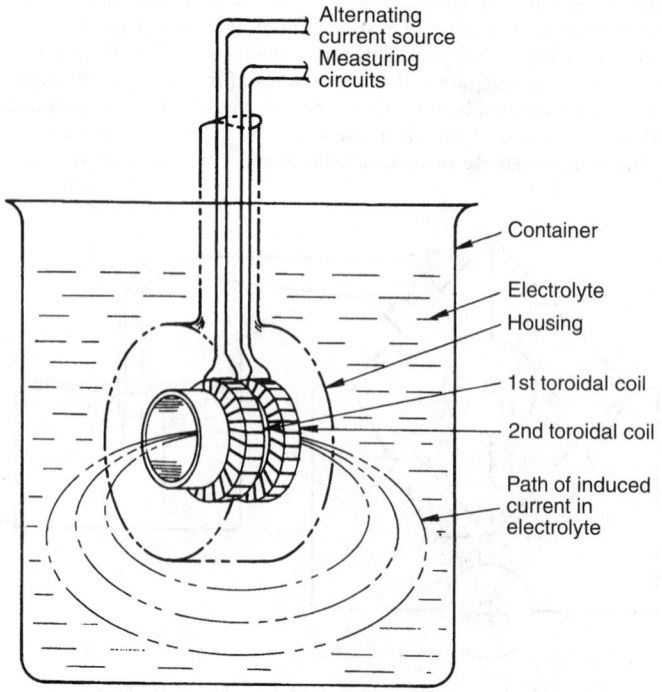

Fig. 4.   Inductive electrolytic conductivity measuring circuit

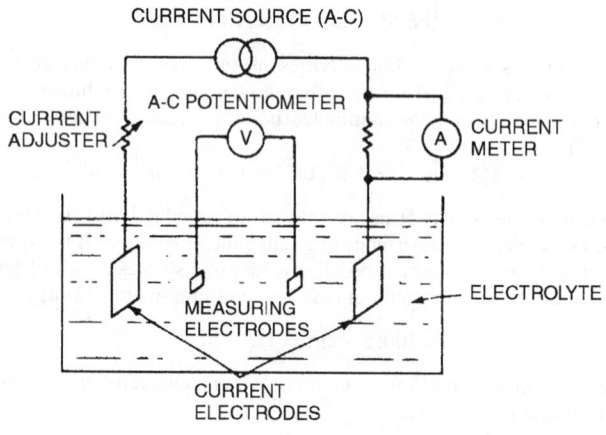

Fig. 5.   Four-electrode conductivity circuit

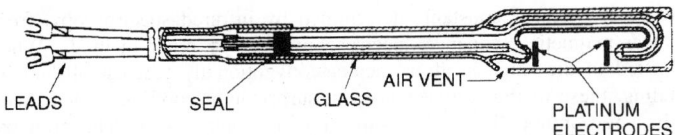

Fig. 6. Dip-type conductivity cell

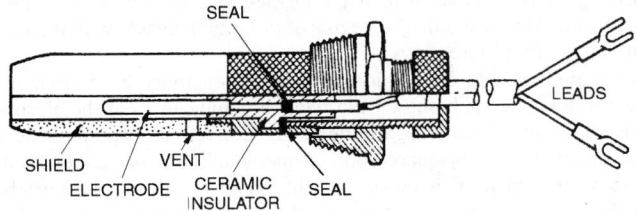

Fig. 7. Screw-in conductivity cell for high-pressure service

Dip Cell. This is designed for dipping or immersing into open vessels. See Fig. 6.

Screw-in Cell. This is designed for permanent installation in pipelines and tanks and is equipped with threaded fittings. See Fig. 7.

Insertion Cell with Removal Device. This is configured to permit removal of the element without closing down or depressurizing the line in which it is installed.

Flow Cell. This type of cell is built in sections of plastic or glass tubing with bores from several millimeters to one inch (25.4 mm) and more. The cell has internal electrodes, usually metallic or carbon rings, mounted flush with the wall to offer little resistance to flow.

ELMER SPERRY and JOHN NAGY
Beckman Industrial Corporation
Cedar Grove, New Jersey

ELECTROMAGNETIC RADIATION. Energy propagated through space, or through material media, in the form of an advancing disturbance in electric and magnetic fields existing in space or in the media. The term *radiation*, alone, is used commonly for this type of energy, although it actually has a broader meaning. Also called *electromagnetic energy* or simply *radiation*.

See also **Electromagnetic Spectrum**.

ELECTROMAGNETIC SEPARATION. Separation of isotopes, especially those of uranium, by first accelerating them by means of an electrostatic field and then passing them through a magnetic field. The effect of this is to cause all the particles to take a curved path; the heavier ones, having higher kinetic energy, describe a wider arc than the lighter ones. Thus, two isotopes of closely similar masses can be separated and collected.

See also **Magnetic Separation** and **Mass Spectrometry**.

ELECTROMAGNETIC SPECTRUM. The ordered array of known electromagnetic radiations, extending from the shortest cosmic rays, through gamma rays, X-rays, ultraviolet radiation, visible radiation, infrared radiation, and including microwave and all other wavelengths of radio energy.

The division of this continuum of wavelengths (or frequencies) into a number of named subportions is rather arbitrary and, with one or two exceptions, the boundaries of the several subportions are only vaguely defined. Nevertheless, to each of the commonly identified subportions there correspond characteristic types of physical systems capable of emitting radiation of those wavelengths. Thus, gamma rays are emitted from the nuclei of atoms as they undergo any of several types of nuclear rearrangements; visible light is emitted, for the most part, by atoms whose planetary electrons are undergoing transitions to lower energy states; infrared radiations are associated with characteristic molecular vibrations and rotations; and radio waves, broadly speaking, are emitted by virtue of the accelerations of free electrons as, for example, the moving electrons in a radio antenna wire.

See also **Absorption Spectrum**.

ELECTROMOTIVE SERIES. See **Activity Series**.

ELECTRON. Discovered by J. J. Thompson in 1896, the electron is an elementary particle of rest mass

$$m = 9.107 \times 10^{-31}$$

kilogram, a charge of 1.602×10^{-19} coulomb, and a spin quantum number $\frac{1}{2}$. Its charge may be positive or negative, although the term electron is commonly used for the negative particle, which is also called the negatron. The positive electron is called the positron. The electron is a constituent of all matter, thus the normal atom consists of a positively charged nucleus surrounded by a sufficient number of electrons so that their total charge is equal to the positive charge on the nucleus. The electron also has wave characteristics, with a frequency and a wavelength. According to wave mechanics, an electron traveling with speed v is associated with a "de Broglie wave" train of wavelength $\lambda = h/mv$ (in which m is the electronic mass and h is Planck's constant).

Bonding Electron

An electron in a molecule that holds two adjacent nuclei together.

Conduction Electron

An electron which plays an important part in electrical or thermal conduction by solids, i.e., by metals or semiconductors, e.g., the electrons in the conduction band, which are free to move under the influence of an electric field.

Electron Donor

1. When a valence bond between two atoms is that type of covalent linkage in which both the electrons of the duplet are supplied by one atom, then that atom, or portion of the molecule of which it forms a part, is called the electron donor. The other atom in the linkage is called the electron acceptor. 2. A donor is also an impurity added to a pure semiconductor to increase the number of free electrons.

Electron Duplet

A pair of electrons shared by two atoms; it is equivalent to a single, nonpolar chemical bond.

Electron Octet

A group of eight valence electrons which constitutes the most stable configuration of the outermost, or valency, electron-shell of the atom, and hence the form which frequently results from electron transfer or sharing between two atoms in the course of a chemical reaction.

Electron Pair

The negatron and positron that result from the pair-production process or interact to initiate an annihilation process.

Electron Shell

The structure of a neutral atom consists of a positively charged nucleus with a number of electrons moving about it—the number being such that their total negative charge is equal to the positive charge on the nucleus. These electrons may be assigned to various shells, characterized by different principal quantum numbers.

Electron Spin

The intrinsic angular momentum of an electron, independent of any orbital motion. Spin ($= h/2$) contributes to the total angular momentum of the electron and is quantized. It gives rise to multiplicity in line spectra, which may be characterized by introduction of the spin quantum number.

Electron Transfer

The process of the shifting of an electron from one electrical field to another, as in the formation of an electrovalent bond, in which an electron moving in an orbit about one atom shifts to move in an orbit around the two bonded atoms.

Free Electron

An electron that is not held in the immediate neighborhood of an atom or molecule.

Orbital Electron

An electron remaining with a high degree of probability in the immediate neighborhood of a nucleus, where it occupies a quantized orbital.

Photo Electron

An electron ejected from a substance by the action of a single photon of light or other electromagnetic radiation.

Secondary Electron

An electron deriving its motion from a transfer of momentum from primary radiation, which may be either particulate or electromagnetic.

Valence Electron

The electrons in the outermost shell of the structure of an atom. Since these electrons are commonly the means by which the atom enters into chemical combinations—either by giving them up, or by adding others to their shell, or by sharing electrons in this shell—these outermost electrons are called valence electrons.

Additional Reading

Warwick, A. and J.Z. Buchwald: *Histories of the Electron,* The Birth of Microphysics, The McGraw-Hill Companies, Inc., New York, NY, 2001.

ELECTRON AFFINITY.   1. Degree of electronegativity, or the extent to which an atom holds valence electrons in its immediate neighborhood, compared to other atoms of the molecule.

2. The work required to remove an electron from a negative ion, and hence to restore the neutrality of an atom or molecule, is called the electron affinity of the atom or molecule.

ELECTRON DIFFRACTION.   Beams of high-speed electrons exhibit diffraction phenomena analogous to those obtained with light, thus showing the wave-like character of electron beams. Such patterns are useful in the interpretation of the structure of matter.

See also **Electron Microscope**.

ELECTRONEGATIVITY.   This term refers to the relative tendency of an atom to acquire negative charge. This tendency is not precisely defined because no exact theoretical or experimental method of evaluation has been devised. Electronegativity exists because the electronic cloud that surrounds each atomic nucleus is inadequate to block off the nuclear charge completely at the periphery. In other words, although every complete atom is electrically neutral as observable from a distance, a small fraction of the total nuclear positive charge can be detected at any point near the surface of the atom. This "effective nuclear charge" at the surface is relatively insignificant in the absence of low energy vacancies capable of accommodating a foreign electron, as in the atoms of M 8 elements (commonly called the *inert* or *noble gases*). However, wherever a low energy electron vacancy occurs in the outer shell of an atom's electronic cloud, the effective nuclear charge as manifest within that vacancy is of major importance. Indeed, it constitutes the cause and means of chemical bond formation, and largely determines both polarity and strength of the bond. Electronegativity is a measure of the force of this effective nuclear charge within an orbital vacancy at the distance of the atomic radius.

Evaluation of relative electronegativity was first accomplished by Pauling. He considered the energy of a heteronuclear single covalent bond to consist of the geometric mean of the homonuclear single bond energies of the two elements, *supplemented* by an electrostatic or ionic energy resulting from uneven electron sharing in the bond. This he attributed to an electronegativity difference between the two elements. He used the difference between the observed bond energy and the average of the homonuclear energies as a measure of the ionic energy. From such differences for various pairs of atoms he established a relative scale in which the most electronegative element, fluorine, has the value 4.0, and within the same period are O 3.5, N 3.0, C 2.5, B 2.0, Be 1.5, and Li 1.0.

Mulliken defined electronegativity as the average of the "valence state" ionization energy and electron affinity. Gordy suggested that electronegativity is a measure of the electrostatic potential at the surface of an atom, expressed as the effective nuclear charge, $Z_{eff}e$ divided by the radius. Allred and Rochow modified this concept by considering the electrostatic force, $Z_{eff}e^2/r^2$, as the measure of electronegativity. Electronegativities have also been estimated from work functions of

metals, from force constants determined by infrared spectroscopy, from nuclear magnetic resonance spectroscopy, and by other methods. When adjusted to the same arbitrary scale, conventionally that established by Pauling, these methods give values in surprisingly good agreement, with only a few minor discrepancies that remain controversial. The principal application of *Pauling scale electronegativities*, which are those almost universally given in textbooks, has been qualitative prediction of bond polarity. That is, a given bond between atoms initially differing in electronegativity is polar with a partial negative charge on the initially more electronegative atom. The degree of polarity increases with increasing electronegativity difference.

Far more valuable applications of electronegativity have been made using a different scale based on the relative compactness of the electronic clouds of atoms. Electronegativities thus derived are approximately a linear function of the square root of the Pauling scale values, and in this sense in generally good agreement. They (see Table 1) have been used for quantitative estimation of the partial charges on combined atoms, which in themselves permit correlation of a vast quantity of chemical data and interpretations of many chemical phenomena. The partial charges in turn have been applied to the quantitative calculation of bond energy. Furthermore, more recently, a simple quantitative relationship between homonuclear covalent bond energy and electronegativity has been demonstrated. Experimental homonuclear bond energy can be used to calculate electronegativity, or vice versa.

Space does not permit a detailed description of the concepts and methods mentioned here, but an example of the results obtainable may illustrate the principles involved. Silica, SiO_2, consists of silicon atoms initially of 2.84 electronegativity, tetrahedrally surrounded by oxygen atoms, initially of 5.21 electronegativity, each of which bridges two silicon atoms. The principle of electronegativity equalization states that when two or more atoms initially different in electronegativity combine, their electronegativities become equalized to the geometric mean. For SiO_2 the electronegativity of the compound is $(2.84 \times 5.21^2)^{1/3} = 4.26$. The equalization is brought about through the uneven sharing of the bonding electrons. The oxygen being initially more electronegative, attracts more than a half share of the bonding electrons. By spending more than half time more closely associated with the oxygen, the bonding electrons impart a partial negative charge on the oxygen, expanding the cloud through increased repulsions and decreasing the electronegativity. Simultaneously the silicon atoms shrink because of reduced repulsions, and increase in electronegativity because of reduced shielding between nucleus and bonding electrons. The decrease in oxygen electronegativity from 5.21 to 4.26 is 0.95. If oxygen had acquired an electron completely the electronegativity would have dropped by 4.75. The partial charge on oxygen is defined as the ratio $0.95/4.75 = -0.20$ (minus because the electronegativity decreased). The silicon is left with a partial positive charge of 0.40.

The silicon-oxygen bond, like all heteronuclear bonds, can be treated *as if* its energy were partly covalent and partly ionic. Instead of ionic

TABLE 1. RELATIVE ELECTRONEGATIVITIES OF SOME ELEMENTS (relative compactness scale)

H	3.55	K	0.42	Rb	0.36	Cs	0.28
Li	0.74	Ca	1.22	Sr	1.06	Ba	0.78
Be	2.39	Zn	3.00	Cd	2.59	Hg	2.93
B	2.93	Ga	3.28	In	2.84	Tl(I)	1.89
				Sn(II)	2.31		
C	3.79	Ge	3.59	Sn(IV)	3.09	Tl(III)	3.02
N	4.49	As	3.90	Sb	3.34	Pb(II)	2.38
O	5.21	Se	4.21	Te	3.59	Pb(IV)	3.08
F	5.75	Br	4.53	I	3.84	Bi	3.16
Na	0.70						

Mg	1.56	Sc	1.30	Y	1.05	La	0.88
Al	2.22	Ti	1.40	Zr	1.10	Hf	1.05
Si	2.84	V	1.60	Nb	1.36	Ta	1.21
P	3.43	Cr	1.88	Mo	1.62	W	1.39
S	4.12	Mn	2.07	Tc	1.80	Re	1.53
Cl	4.93	Fe	2.10	Ru	1.95	Os	1.67
		Co	2.10	Rh	2.10	Ir	1.78
		Ni	2.10	Pd	2.29	Pt	1.91
		Cu	2.60	Ag	2.57	Au	2.57

[a] Values for the transitional elements are tentative estimates only.

energy supplementing the covalent energy, as suggested by Pauling, the ionic energy *substitutes* for a part of the covalent energy. The ionic weighting coefficient is half the charge difference: $(0.40 + 0.20)/2 = 0.30$. The covalent weighting coefficient, $1.00 - 0.30$, is 0.70. For the covalent energy one takes the geometric mean, 59.7, of the homonuclear bond energies of silicon (53.4) and oxygen (66.7, as calculated from the O_2 molecule), multiplies by 0.70, and corrects for bond length by the factor (covalent radius sum)/(observed bond length) $= 1.90/1.61$.

This calculation gives 49.1 kcal/mole of bonds. The ionic energy is simply the conversion factor (to kcal/mole) 332, times the weighting coefficient 0.30, divided by the bond length 1.61, or 61.8 kcal/mole of bonds. The sum, 110.9 kcal, is the Si–O bond energy. Atomization of SiO_2 requires the rupture of four SiO bonds per formula unit; $4 \times 110.9 = 443.6$ kcal/mole for the atomization energy of $SiO_2(c)$. Subtraction of the atomization energies 108.9 for Si and 119.2 for two O gives -215.5 kcal/mole for the calculated standard heat of formation of $SiO_2(c)$. The experimental value is -217.7.

Electronegativity thus permits a quantitative interpretation of the bonding in SiO_2 or any other compound for which appropriate data are known. The same principles allow a superior alternative to the "ionic" model of nonmolecular solids, and offer high hope of eventually elucidating the thermochemistry of mineral substances in general.

ELECTRON EMISSION. The liberation of electrons from an electrode into the surrounding space. Quantitatively, it is the rate at which electrons are emitted from an electrode.

ELECTRONEUTRALITY. If one describes the properties of electrolytic solutions in terms of ionic species, one has to take account of the fact that the concentrations of all species are not independent because the solution as a whole is neutral.

One generally uses the symbol z_i to denote the charge on an ion measured in units of the charge of a proton (for example, for Na^+, $z = 1$; for La^{3+}, $z = 3$; for PO_4^{3-}, $z = -3$); z is also called the *charge number* of the ion.

If n_i is the number of moles of the ionic species i, the condition of electrical neutrality is

$$\sum_i z_i n_i = 0 \tag{1}$$

Alternatively if one uses the subscript $+$ to denote positively charged ions or *cations* and $-$ to denote negatively charged ions or *anions*, then one may write (1) in the form

$$\sum_+ z_+ n_+ = \sum_- z_- n_- \tag{2}$$

ELECTRON GAS. The term electron gas is used to denote a system of mobile electrons, as, for example, the electrons in a metal that are free to move. In the free electron theory of metals, these electrons move through the metal in the region of nearly uniform positive potential created by the ions of the crystal lattice. This theory when modified by the Pauli exclusion principle, serves to explain many properties of metals, especially the alkali metals. For metals with more complex electronic structure, and semiconductors, the band theory of solids gives a better picture.

ELECTRON MICROSCOPE. The concepts that eventually led to the development of electron microscopes came out of the discovery of the wave nature of the electron in 1924. The effective wavelength of the electrons varies with accelerating voltage and is less than 1 Å: $\lambda = \sqrt{(150/V)}$ Å. This short wavelength makes possible far better resolution and higher magnification in the electron microscope as compared with the optical microscope.

The lenses used in electron microscopes act on the beams of electrons in much the same way that ordinary glass lenses act on beams of light. Most electron microscopes have electromagnetic lenses, although electrostatic lenses also can be used. Application of a uniform axial magnetic field causes the electrons to travel in a spiral path and return to the axis as shown schematically in Fig. 1. Except for the spiral part of the motion, this behavior is just like that of a simple glass lens, and the equations for

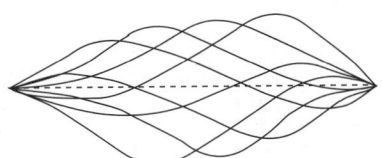

Fig. 1.   Courses of electron beams in a homogeneous magnetic field

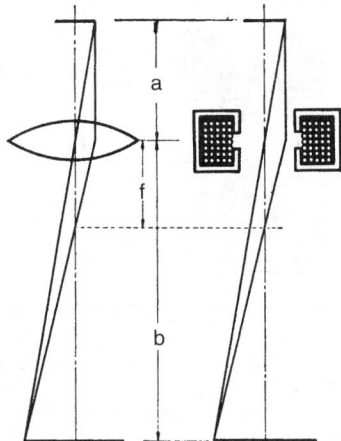

Fig. 2.   Lens system used in transmission-type electron microscope. (*Left*) optical convex lens; (*Right*) electromagnetic lens

optical lenses apply, as shown in Fig. 2:

$$\frac{1}{a} + \frac{1}{b} = \frac{1}{f}$$

$$\text{Magnification} = \frac{b}{a}$$

However, in the electromagnetic lens, the angle of deflection and the focal length depend upon the strength of the magnetic field. This can be controlled by adjusting the current in the coils so that the magnification of the lens can be continuously varied over a broad range. See Fig. 3.

A simple electron microscope, as illustrated in Fig. 4, is operated by producing a beam of electrons from a heated filament, accelerating the beam with a high voltage applied to an anode, and then directing this beam of illuminating electrons onto the specimen. The specimen is in some respects similar to that used for optical microscopy, but because electrons are not so penetrating as light, the specimen must be much thinner (on the order of 1,000 Å or less for most materials). For biological materials, these thin sections are produced by ultramicrotomes. Many materials, especially metals, can be thinned chemically or electrochemically. Rough surfaces

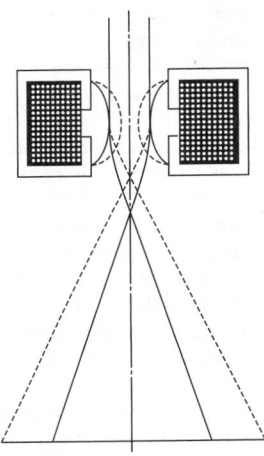

Fig. 3.   Relation between magnetic field intensity and magnification in case of large magnetic field intensity (i.e., a large current flowing through the coil); or small magnetic field intensity (i.e., a small current flowing through the coil)

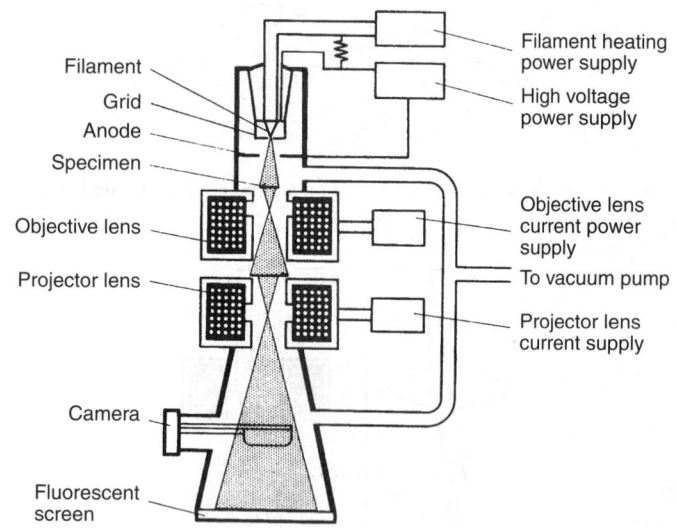

Fig. 4.   Construction of a simple electron microscope

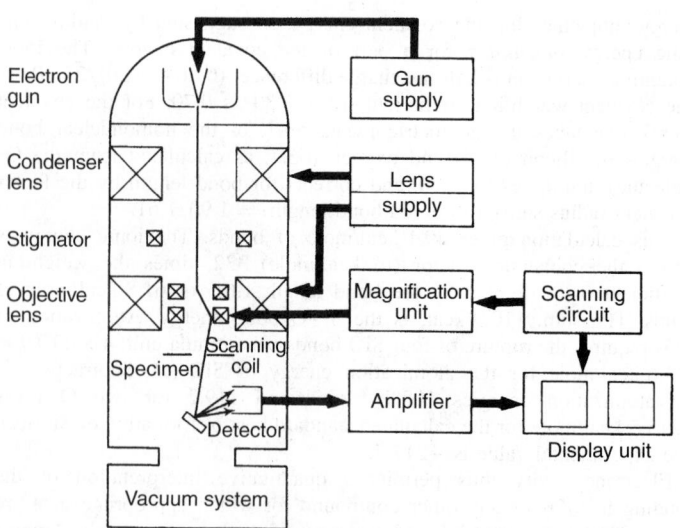

Fig. 5.   Simplified diagram of scanning-type electron microscope

can be examined by evaporating carbon on the sample and then removing it and using the carbon replica in the microscope.

The beam of electrons that passes through the specimen is then magnified by an objective lens and a projector lens and finally strikes either film held in a camera, or a fluorescent screen. The image seen on the fluorescent screen has varying shades of gray that depend upon the distribution of density and thickness of the specimen. This is most important for biological samples. For crystalline materials, the regular arrays of atoms aligned in critical directions can act like a mirror, diffracting the beam to another direction. This makes it possible to observe imperfections in the atomic arrangement of a material.

More complex electron microscopes use additional lenses, both above and below the specimen. The condenser lenses above the specimen concentrate the electron beam and increase the illumination. The addition of intermediate lenses below the specimen make it possible to go to higher magnification in the final image. Various alignment controls, apertures for the lenses, specimen handling devices, and suitable airlocks and anticontamination traps also are provided.

The central column of the instrument must be maintained as a vacuum because the electrons would be absorbed if any atmospheric gases were present. Typical resolutions obtainable by commercial instruments are on the order of 2 to 5 Å. Accelerating voltages from 20,000 to 1 million volts have been used. The higher accelerating voltages are useful for penetrating thick specimens (in some cases, up to 1 micrometer or more).

Scanning-Type Electron Microscope

This type of electron microscope is completely different in principle and application from the conventional transmission-type electron microscope. In the scanning instrument, the surface of a solid sample is bombarded with a fine probe of electrons, generally less than 100 Å in diameter. The sample emits secondary electrons that are generated by the action of the primary beam. These secondary electrons are collected and amplified by the instrument. Since the beam strikes only one point on the sample at a time, the beam must be scanned over the sample surface in a raster pattern to generate a picture of the surface sample. The picture is displayed on a cathode ray tube from which it can be photographed.

A block diagram of a scanning-type electron microscope is given in Fig. 5. Major elements of the instrument include the electromagnetic lenses that are used to form the electron probe, the scan coils that sweep the beam over the sample, the detector that collects the secondary electrons, and the amplifying means where the secondary electrons are amplified and fed to the cathode ray tube for display. Since the cathode ray tube is scanned in synchronization with the electron beam, the resulting picture corresponds to the area of the sample being examined.

The advantages of the scanning microscope as compared with conventional optical microscopes include superior resolution and depth of field. Resolution of 200 Å is obtained with a depth of field several hundred times that of a conventional optical microscope. The scanning microscope also makes possible the display of other kinds of data obtained from the specimen, notably cathodoluminescent photons emitted by fluorescing samples,

electrical voltages generated in semiconducting samples by the passage of the electron probe, and characteristic x-rays that can be used to determine sample composition. Also see **X-Ray Analysis**.

The range of magnification of scanning electron microscopes generally is from less than $30\times$ to about $40,000\times$, limited by the resolution of the instrument. Most specimens can be examined without any special preparation, but nonconducting specimens usually are coated with 100 Å to 500 Å of metal to conduct away the beam current. A conventional diffusion pump vacuum system is used since a vacuum is required for operation of the electron beam. The image that is formed is easily interpreted as surface topography inasmuch as the illuminating and shadowing effects on the sample are similar to the appearance of large objects as they normally are seen by the unaided eye. The scanning electron microscope has found broad application in the transistor industry to show voltage distributions in such devices. Other uses include industrial quality control and a broad range of industrial and biological research.

High-Resolution Scanning Transmission Electron Microscope (STEM)

The concept for constructing a STEM dates back to 1963, growing out of the techniques and practices of research in nuclear and high-energy physics, with minimal consideration given to the large record of experience gained with the STEM. The operating principles of the STEM reflect those of accelerator physics. As described by A.V. Crewe, a leading authority in the field, electrons from a field emission source are accelerated to a final potential V_0 and then focused on the specimen. Scattered electrons leaving the specimen normally are refocused by the magnetic field of the lens at some point farther down the column and then diverge. The elastically scattered electrons strike an annular detector; the inelastic and unscattered electrons are separated by a spectrometer. The beam is scanned across the specimen by using deflection coils; below the specimen, the scanning action is removed with additional deflection coils. The maximum scattering angle for the electrons is only $2°$ or $3°$. Present resolution of the STEM is from 2 to 2.5 angstroms, still insufficient to resolve distances between atoms in most solids. A point resolution of 0.5 angstrom is needed to obtain images that will resolve such distances. See also **Scanning Tunneling Microscope**.

Additional Reading

Amelinckx, S.: *Electron Microscopy: Principles and Fundamentals,* John Wiley & Sons, Inc., New York, NY, 1997.

Crewe, A.V.: "High-Resolution Scanning Transmission Electron Microscopy," *Science,* **221,** 325–350 (1983).

Hayat, M.A.: *Principles and Techniques of Electron Microscopy: Biological Applications,* Cambridge University Press, New York, NY, 1999.

Reimer, L., et al.: *Scanning Electron Microscopy: Physics of Image Formation and Microanalysis,* Springer-Verlag, Inc., New York, NY, 1998.

Shindo, D. and H. Kenji: *High-Resolution Electron Microscopy for Materials Science,* Springer-Verlag, Inc., New York, NY, 1998.

Watt, I.M.: *The Principles and Practice of Electron Microscopy,* Cambridge University Press, New York, NY, 1996.

ELECTRON (Photoelectron) SPECTROSCOPY. A valuable research tool for conducting basic research in molecular chemistry and surface physics, in chemical characterization of atmospheric particulate matter in environmental studies, and in the analysis of the performance of catalysts in industrial processes. Electron spectroscopy is based on the high-resolution analysis of the kinetic energy distributions of electrons emitted from solid, liquid, or gaseous substances after they are irradiated with a beam of monoenergetic x-rays or ultraviolet radiation. Monochromatized synchrotron radiation has also become important as a tunable photo source. The physical quantity measured is the electron binding energy B, which is given by Einstein's relation $h\nu = B + K$, where $h\nu$ is the known photon energy and K is the measured photoelectron kinetic energy. Chemical information is obtained from chemically induced changes in the binding energies. In principle, all electron orbitals from the K shell out to the valence levels can be studied. When x-ray excitation is used, the technique is called x-ray photoelectron spectroscopy (XPS); when ultraviolet light is used, the term ultraviolet photoelectron spectroscopy (UPS) is used. The photoelectric effect was discovered by Hertz in 1883. K.M. Siegbahn, University of Uppsala, was awarded half of the 1981 Nobel Prize in Physics for his work in this field.

ELECTRON THEORY. In the 1830s, Faraday had tentatively suggested that his experiments in electrochemistry could be interpreted in terms of a small unit of charge attached to ions. This notion of individual "atoms of charge" was somewhat eclipsed, however, by the enormous success of Maxwell's theory of electromagnetism, which was generally interpreted, by 1880, as favoring a view that electrical phenomena were due to continuous charge distributions and motions. G. Johnstone Stoney, in 1874, and Helmholtz, in 1881, had suggested again an atomic interpretation of electricity, but it was not until the brilliant experiments of Perrin, J.J. Thomson, Zeeman, and others in the 1890s that the concept of the electron received firm experimental foundation. Later experiments and theory (Millikan, Bohr, etc.) established the constancy of the electronic charge and interwove the concept of an electron of definite charge and mass into the basic structure of the atom.

The Cathode Ray Controversy

After the discovery of the cathode ray in high-vacuum discharge tubes by Plücker in 1858, there developed, with the experiments of Goldstein, Crookes, Hertz, Lenard, and Schuster, a controversy over the nature of the rays. A predominately German school held that the rays were a peculiar form of electromagnetic rays. The British physicists thought they were negatively charged particles. The controversy provides a classic "case history" of the typical scientific controversy in which two quite different models both explain most, but not all, of the observable facts.

Thomson's Determination of e/m

In 1897 Thomson devised an apparatus in which he could deflect a beam of cathode rays with a magnetic field of induction B and also with an electric field of strength E. If the fields are perpendicular both to each other and to the original path of the beam, and if they occupy the same region, then (with proper polarities and magnitudes of fields) the electric force on the beam can equal the magnetic force. Thus, the beam hits the same point on a fluorescent screen as when no fields are applied. If e is the charge of a given particle, m its mass, and v its velocity, $v = E/B$. Thus, velocities of typical cathode ray beams could be measured. If the magnetic field is used alone, and the curvature R of the beam is measured, then one can equate centripetal and magnetic field forces $mv^2/R = Bev$, and then deduce $e/m = v/BR$. With v known from the previous experiment, e/m can be calculated. Thomson's early values were not very precise, but later experiments of a similar type gave values close to 1.76×10^{11} coulombs/kg. More recent experiments, drawing on measurements of many kinds, give $e/m = (1.75890 \pm 0.00002) \times 10^{11}$ coulombs/kg.

The Zeeman Effect

In 1896 Zeeman discovered the broadening of spectral lines when a light source was in a strong magnetic field. Experimental refinements by Zeeman and others, and theoretical work by Lorentz and Zeeman, permitted the interpretation of this effect as due to the influence of the magnetic field on oscillating or orbiting negatively charged particles within the light-emitting or absorbing atoms. From the spectroscopic data, the ratio of charge to mass of these hypothetical particles could be shown to be equal to that of cathode rays. The Zeeman effect thus provided the first experimental evidence that the negative particles emitted by atoms when heated (Edison effect) or subjected to high fields and/or ionic bombardment (cathode rays) or bombarded by short-wavelength light (photoelectric effect) were, indeed, actual constituents of the atoms and were probably responsible for the emission and absorption of light.

The Charge on the Electron

In the decade following 1897, many different methods were evolved for determination of ionic charges. Some methods depended upon measuring the total charge of a number of ions used as nuclei for cloud droplet formation. Other methods were more indirect experiments, for example, which, combined with the kinetic theory of gases, could give crude values for Avogadro's number, N. By dividing the Faraday constant (the charge carried in electrolysis by ions formed from one gram-atom of a univalent element) by N, one could determine the average charge per ion. Similarly, the constants in Planck's theory of blackbody radiation, when evaluated experimentally, could provide a numerical value for N, as could certain experiments in radioactivity. All such methods gave values of N of the order of 6×10^{23}, and hence 1.6×10^{-19} coulomb for the ionic charge. None of these methods measured individual charges; strictly speaking, the value for the ionic charge could be thought of only as an average value.

Millikan's experiments with single oil drops, beginning in 1906, provided a method for measuring extremely small charges with precision. He was able to show that the charge on his drops was *always ne*, with $e = 1.60 \times 10^{-19}$ coulomb (modern value) and n a positive or negative integer.

He observed the motions of very small charged oil drops in uniform vertical electric fields. The drops were so small that they moved with constant velocity (except for Brownian fluctuations) for a given force. The force in each case was due to gravity acting on the mass of the drop and to the electric field (if any) acting on the charge, q, on the drop. The charge on a given drop could be changed by shining x-rays upon it. Using Stokes' law (in a form modified to correct for the fact that the drops were *not* large in comparison to the inhomogeneities of the surrounding air) and the velocity of a drop in free (gravitational) fall, Millikan could infer the diameter and mass of a given drop, and then calculate its charge. The charge q always equaled ne. A few other physicists, in similar experiments, thought they had detected electric charges smaller than Millikan's e, but their experimental techniques were probably faulty.

Millikan's experiment did not prove, of course, that the charge on the cathode ray, beta ray, photoelectric, or Zeeman particle was e. But if we call all such particles electrons, and assume that they have $e/m = 1.76 \times 10^{11}$ coulombs/kg, and $e = 1.60 \times 10^{-19}$ coulomb (and hence $m = 9.1 \times 10^{-31}$ kg), we find that they fit very well into Bohr's theory of the hydrogen atom and successive, more comprehensive atomic theories, into Richardson's equations for thermionic emission, into Fermi's theory of beta decay, and so on. In other words, a whole web of modern theory and experiment defines the electron. The best current value of $e = (1.60206 \pm 0.00003) \times 10^{-19}$ coulomb.

The Wave Nature of the Electron

De Broglie had suggested in 1924 that electrons have in some respects the characteristics of waves, and deduced, for the wavelength equivalent to a moving electron, the expression $\lambda = h/mv$ in which m and v are the mass and speed of the electron and h is Planck's constant. If the electron is moving, for example, with a speed corresponding to 65 eV of energy, the corresponding "De Broglie wavelength" is 1.52 angstroms, which is in the X-ray range. This led Davisson and Germer to see whether electrons might be reflected from crystals after the manner of X-rays. They used a single crystal of nickel cut parallel to the (111) planes, and upon varying the electron speed at a fixed angle of incidence, they found not only a distinct "regular" reflection, but also a series of diffraction maxima strikingly similar to those obtained with the same crystal for X-rays of varying wavelength. The differences observed were satisfactorily explained as due to the refraction of the nickel for the electron waves.

G.P. Thomson independently reached the same conclusions in 1927. The hypothesis that matter exhibits both corpuscular and wavelike characteristics served as a stimulus for the formal development of quantum mechanics by E. Schrödinger, M. Born, W. Heisenberg, and others. Following the discovery, which eventually led to a Nobel Prize to Davisson and Thomson, electron diffraction was immediately utilized as a tool for the study of the structure of matter.

Other Characteristics of Electrons

In applying quantum mechanics to certain problems in atomic spectroscopy, in 1925 and 1926, Pauli, and Goudsmit and Uhlenbeck found that electrons must possess angular momentum of amount $\pm\frac{1}{2}(h/2\pi)$. Dirac's work on a generalized quantum theory of the electron showed that it possessed a related magnetic dipole moment of magnitude $eh/4\pi\ mc$. The ratio of the dipole moment to the angular momentum (e/mc) is larger than can be accounted for in classical terms with any homogeneous wholly negative model. The concept of electronic dipole magnetic moment is essential not only in spectroscopy but in theories of ferromagnetism. See also **Magnetism**.

One may speak of the "classical radius of the electron," $a = e^2/mc^2$, derived by setting the self-energy of the coulomb field of a charge e contained at a radius a equal to the relativistic rest energy, mc^2 of the electron. This $a = 2.82 \times 10^{-13}$ cm, comfortably smaller than any atom, but larger than the usual estimates of sizes of protons and neutrons.

Positive Electrons

Dirac's paper in 1928 could be interpreted as predicting the existence of electrons that are positive. But until such particles were found experimentally by C.D. Anderson in 1932 in cloud chamber pictures of cosmic ray particle tracks, most physicists preferred other interpretations of Dirac's paper. Positive electrons, or positrons are now known (1) to occur as decay products from certain radioactive isotopes, (2) to be produced (paired with a negative electron) in certain interactions of high-energy gamma rays with intense electric fields near nuclei, and (3) to be the product of certain decays of certain mesons. In principle, positrons could form anti-atoms with nuclei made from anti-protons and anti-neutrons, but in practice almost all positrons produced in the observable universe quickly meet their end by annihilating themselves together with some negative electron. The end product of a positron-electron annihilation is a pair of gamma rays. See also **Electron**.

Additional Reading

Kubler, J.: *Theory of Itinerant Electron Magnetism,* Oxford University Press, Inc., New York, NY, 2000.

Kuzne, A.M., T. Sov, and J. Ulstrup: *Electron Transfer in Chemistry and Biology: An Introduction to the Theory,* John Wiley & Sons, Inc., New York, NY, 1998.

March, N.H.: *Electron Density Theory of Atoms and Molecules,* Academic Press, Inc., San Diego, CA, 1997.

Ppauli, W. and C.P. Enz: *Optics and the Theory of Electrons,* Dover Publications, Inc., Mineola, NY, 2000.

Tonomura, A.: *The Quantum World Unveiled by Electron Waves,* World Scientific Publishing Company, Inc., River Edge, NJ, 1999.

ELECTRON VOLT.   This is a convenient unit of energy for calculations in electronics and in connection with ionization or excitation of atoms or molecules. When an electric charge e is transferred from a region where the electric potential is V_1 to one where the potential is V_2, its potential energy changes by an amount equal to $e(V_1 - V_2)$. If the charge e is the electronic charge 1.602×10^{-19} coulomb (as it is when the transferred particle is an electron or a proton), and if the potential difference $V_1 - V_2$ is one volt, then the corresponding change in energy is equal to 1.602×10^{-19} joule or 1.602×10^{-12} erg, and this is called an electron volt. Thus if a doubly ionized positive oxygen molecule moves in an electric field through a potential drop of 500 volts, it receives $2 \times 500 = 1000$ electron volts or 1.602×10^{-9} erg of energy; and since the mass of the oxygen molecule is about 5.31×10^{-23} grams, this energy would give the molecule, if unimpeded, a speed of about 7.77×10^{6} centimeters per second or 48.1 miles per second. The abbreviation for electron volt is eV. In x-rays, nuclear physics, and elementary particle physics, the use of higher energies is encountered and additional abbreviations in common use are keV for 10^3 eV, MeV for 10^6 eV, and GeV (sometimes BeV in the United States) for 10^9 eV.

ELECTROPHILIC REACTION.   The reaction in which an electrophilic reagent attacks a nucleophilic compound. The reagent is taken to be the inorganic substance (in the case of reactions of inorganic and organic substances) or the simpler of two reacting organic compounds. The electron pair for the bond formed is furnished by the nucleophilic compound. The term *electrophilic* connotes *electron-seeking* and is applied, for example, to positively charged cations, or to reactions brought about by them.

ELECTROPHORESIS.   Electrophoresis is a separation technique most often applied to the analysis of biological or other polymeric samples. It has frequent application to analysis of proteins and DNA fragment mixtures. See also **Nucleic Acids**. The high resolution of electrophoresis has made it a key tool in the advancement of biotechnology. Variations of this methodology are being used for DNA sequencing. See also **Genetic Engineering**. Isolating active biological factors associated with diseases such as cystic fibrosis, sickle-cell anemia, myelomas, and leukemia, and establishing immunological reactions between samples on the basis of individual compounds. Electrophoresis is an extremely effective analytical tool because it does not affect a molecule's structure, and it is highly sensitive to small differences in molecular charge and mass.

The term electrophoresis refers to the movement of a solid particle through a stationary fluid under the influence of an electric field. The study of electrophoresis has included the movement of large molecules, colloids, fibers, clay particles, latex spheres. See also **Latex Technology**. Basically anything that can be said to be distinct from the fluid in which the substance is suspended. This diversity in particle size makes electrophoresis theory very general.

The fundamental principle behind electrophoresis is the existence of charge separation between the surface of a particle and the fluid immediately surrounding it. An applied electric field acts on the resulting charge density, causing the particle to move, the fluid around the particle to move, or both. An applied electric field also generates heat, through resistive heating, and gases, through electrolysis reactions. Each is important in understanding and designing working electrophoresis equipment.

There are three distinct modes of electrophoresis: zone electrophoresis, isoelectric focusing, and isotachophoresis. These three methods may be used alone or in combination to separate molecules on both an analytical (μL of a mixture separated) and preparative (mL of a mixture separated) scale. Separations in these three modes are based on different physical properties of the molecules in the mixture, making at least three different analyses possible on the same mixture.

Distinction is also made among electrophoretic techniques in terms of the type of matrix employed for analysis. Matrices include polymer gels such as agarose and polyacrylamide, paper, capillaries, and flowing buffers. Each matrix is used for different types of mixtures, and each has unique advantages.

There are a variety of techniques for detecting separated sample compounds using chemical stains, photographic media, and immunochemistry. Each detection technique also gives different information about the identity, quantity, and physical properties of the molecules in the mixture. Detection is often the focus of electrophoresis, and usually yields basic information about the mixture being studied.

Electrolysis Reactions.   The electrodes in electrophoresis equipment are typically constructed from platinum wire, and sodium chloride generally carries the bulk of the current in any electrophoretic medium. This results in the reactions at the cathode of $2\ H_2O + 2e^- \rightarrow 2\ OH^- + H_2$, and $H_2O + OH^- \rightleftharpoons A^- + H_2O$; at the anode: $H_2O \rightarrow 2\ H^+ + 20.5\ O_2 + 2e^-$, $H^+ + A^- \rightleftharpoons HA$. That is, water is electrolyzed. The hydrogen gas produced at the cathode can be hazardous, especially because it is in the vicinity of an electrode that is also producing heat. For this reason, electrode chambers are usually open to the atmosphere so that gases can vent.

The other reactions at the electrodes produce acid (anode) and base (cathode) so that there is a possibility of a pH gradient throughout the electrophoresis medium unless the system is well buffered. See also **Hydrogen-Ion Activity**. Buffering must take the current load into account because the electrolysis reactions proceed at the rate of the current. Electrophoresis systems sometimes mix and recirculate the buffers from the individual electrode reservoirs to equalize the pH.

Modes of Electrophoretic Separations

Electrophoresis Equipment.   Most electrophoresis equipment shares a basic design, a diagram of which is shown in Figure 1.

Zone Electrophoresis.   In zone electrophoresis multicomponent samples are applied to an electrophoretic medium, most commonly a gel, an electric field is applied, and after a predetermined length of time or after a certain level of power, current, or voltage has been applied, the electrophoretic medium is inspected for resolution of the sample components.

Disc Electrophoresis.   Resolution in zone electrophoresis depends critically on getting sample components to migrate in a focused band, thus some techniques are employed to concentrate the sample as it migrates

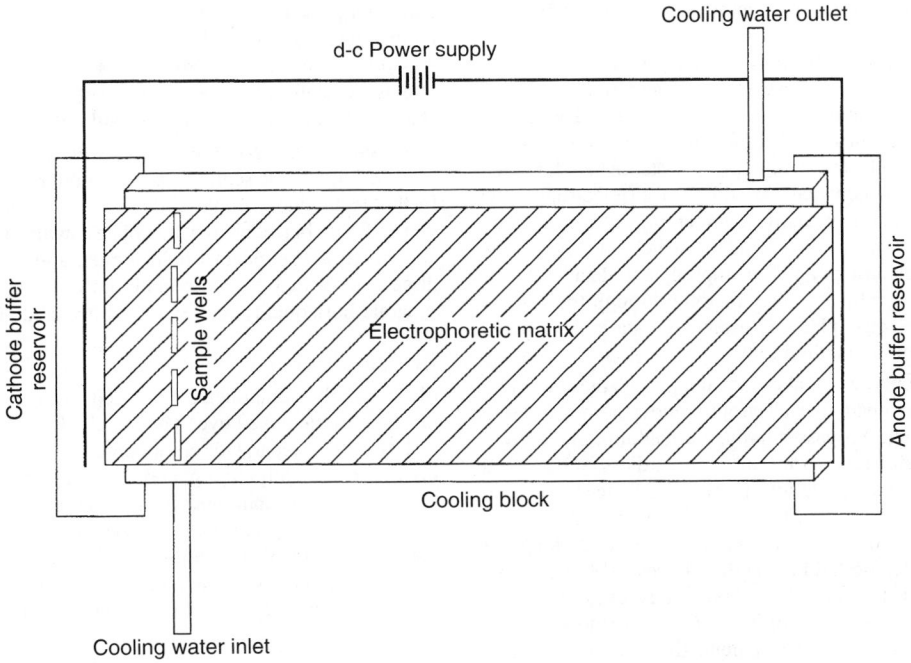

Fig. 1.   Electrophoresis equipment

through the gel. The most common technique is referred to as discontinuous pH or disc electrophoresis. Disc electrophoresis employs a two-gel system, where the properties of the two gels are different.

Two-dimensional electrophoresis, introduced in 1976, combines polyacrylamide (sodium dodecyl sulfate-polyacrylamide) gel electrophoresis with isoelectric (electrofocusing) electrophoresis. The method is attractive because it can separate large numbers of proteins. A more recent two-dimensional technique is called *electrophoretic titration curve analysis*. As reported by Maugh (1983) a pH gradient from 3 to 9 is established in a horizontal slab. A trough is then cut into the slab at a right angle to the gradient. The sample is applied to the trough and conventional electrophoresis is performed, perpendicular to the gradient to produce a classical titration curve. The latter provides useful information about the protein, including its stability and the binding of ligands. The data also predict behavior of the protein in ion-exchange chromatography.

Another recent technique, preparative electrophoresis, is *recycling isoelectric focusing* developed by Bier and colleagues (University of Arizona). See Fig. 2.

A general trend in electrophoresis is to use thinner gels to provide better retention of nucleotides for DNA sequencing and, because of more efficient cooling, higher voltages can be used for faster separations. Prior gels typically have been 2 mm in thickness; current gels are typically less than 0.3 mm thick. Numerous equipment and procedural advancements are continuing apace.

Capillary Electrophoresis. This technique, which appeared in the late 1980s, makes possible rapid and automated analysis of small volumes of complex mixtures with excellent resolution and sensitivity. The procedure is well described by M.J. Gordon et al. (see reference listed).

Native Zone Electrophoresis. In some cases, good resolution between sample species can be obtained with little or no sample pretreatment. In these cases, the gels are said to be native gels. In this method, the charge on the individual sample component is primarily responsible for its differential migration.

Pulsed Field Gel Electrophoresis. In pulsed field electrophoresis, the direction of the field is intermittently changed, either forward and backward or from side to side. A small molecule notices no real difference in its electrophoresis because it can completely reorient in each field direction. However, the redirectioning of the electric field causes larger molecules to travel in a zigzag pattern, putting kinks in to the length of the molecule. The longer the molecule, the more kinks in its length, and the slower it travels down the length of the gel. The larger molecule finds itself traveling backward along some sections of its length with respect to the direction of the electric field. This has allowed resolution of megabase size strands of DNA, and has made the Human Genome Project feasible.

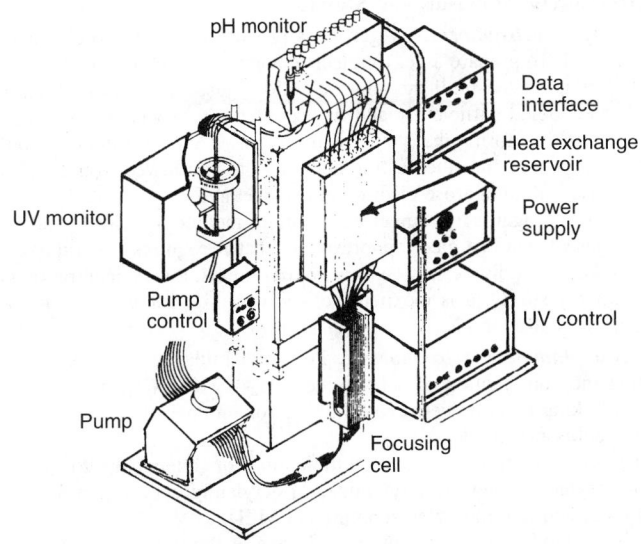

Fig. 2.   Schematic representation of the recycling isoelectric focusing (RIEF) apparatus developed by Milan Bier (University of Arizona). In this preparative electrophoresis technique, no gel is used; instead the solution is recycled to be fractionated through a focusing cell and heat-exchange reservoir. The focusing cell is a series of parallel-flow chambers which are separated by monofilament nylon screen elements. The screens streamline the flow through the apparatus and avoid loss due to convection. The proteins are allowed to migrate from chamber to chamber exclusively under the influence of the electric field. Each chamber is connected to a separate glass channel in the heat-exchange reservoir. Capacity of the apparatus is determined by the volume of the heat exchanger and the cross-sectional area of the focusing chambers. A modular design permits scale-up to larger volumes. The pH gradient is established by the electric field itself, in the same manner as in analytical gel focusing. The apparatus has been used to process milligram quantities of antibodies to single-band purity within a period of 4 hours. (*M. Bier*)

Isoelectric Focusing. Isoelectric focusing is a technique used for protein separation, by driving proteins to a pH where they have no mobility. Resolution depends on the slope of a pH gradient that can be achieved in a gel.

Ampholytes or Zwitterions. An ampholyte is a molecule that can be either positively or negatively charged, depending on the pH. These molecules are also called zwitterions. All amino acids and proteins are ampholytes, or amphoteric.

A special class of ampholytes has been synthesized for the purpose of isoelectric focusing. These ampholytes have an amino end and a carboxyl end that are separated by varying numbers of methylene groups. The further apart the amino and carboxyl groups, the less one affects the ionization of the other; thus a different isoelectric point (pI) is established for each molecule. These ampholytes, which may be added to an electrophoretic medium, migrate in one direction or another, under the influence of an applied electric field, until they reach a zone in which the pH is the same as that ampholyte's isoelectric point. The ampholyte molecules buffer themselves and establish the local pH as they migrate through the gel. As the ampholytes reach an isoelectric pH, they establish a stationary spatial pH gradient in the electrophoretic medium.

Isotachophoresis. Isotachophoresis takes advantage of the fact that electroneutrality must be maintained in an electrophoretic system in order to support an electric field. If a current passes through a medium, that current must be constant from one electrode to the other, regardless of the local ion concentration or mobility; i.e., dilute ions must move faster to keep up with a zone of more concentrated ions. Electric fields compensate for this because the electric field strength does not have to be constant along the length of the medium. The electric field strength is lowest where the ions are most concentrated and most mobile. Isotachophoresis takes advantage of this phenomenon by lining up the ions of interest, fastest (most mobile) to slowest. This is a highly specialized technique that requires detailed knowledge of the properties of the sample to be separated, and is generally not applicable to analytical separations.

Electrophoretic Materials and Matrices

Agarose Electrophoresis. Agarose is produced from the processing of red seaweed. To prepare a gel for electrophoresis a combination of agarose and buffer is heated until the agarose solid is dissolved and boiling. The solution is cooled sufficiently and then poured into a warmed gel casting apparatus which forms the shape of the gel as it cools. After the cooled solution is poured into the casting apparatus, it is allowed to gel and can then be used in an agarose electrophoresis method.

The use of agarose as an electrophoretic method is widespread. The advantages of agarose electrophoresis are that it requires no additives of cross-linkers for polymerization, it is not hazardous, low concentration gels are relatively sturdy, it is inexpensive, and it can be combined with many other analytical methods.

Polyacrylamide Electrophoresis. Polyacrylamide gels are synthesized through the combination of acrylamide, $CH_2=CHCONH_2$, monomer and a cross-linking comonomer. Cross-linking comonomer of choice is N, N'-methylenebisacrylamide.

The most commonly used combination of chemicals to produce a polyacrylamide gel is acrylamide, bisacrylamide, buffer, ammonium persulfate, and tetramethylenediamine (TEMED).

Polyacrylamide gel electrophoresis is one of the most commonly used electrophoretic methods. Analytical uses of this technique center around protein characterization, for example, purity, size, or molecular weight, and composition of a protein. Polyacrylamide gels can be used in both reduced and nonreduced systems as well as in combination with discontinuous and isoelectric focusing (ief) systems.

Paper Electrophoresis. Besides being easy to obtain, paper is a good medium because it does not contain many of the charges that interfere with the separation of different compounds. Two types of paper employed in this type of electrophoresis are Whatman 3 MM (0.3 mm) and Whatman No. 1 (0.17 mm).

In paper electrophoresis, the sample is placed directly onto chromatographic or filter paper and then exposed to a buffer solution at each end and an electric field is applied. As in most electrophoretic techniques, charged dyes are combined with samples and standards to see the progress of the electrophoresis.

Capillary Electrophoresis. The glass capillaries used are typically 20 to 200 μm in diameter, may be filled with buffer or gel, and are frequently coated on the inside. Capillaries are used because of the high surface-to-volume ratio which allows high voltages without heating effects. The only limitations associated with capillaries are limits of detection and clearance of sample components.

Capillary electrophoresis is a commercially available technique, and has been integrated with most automated lab equipment such as autosamplers,

computer peak analysis (the charts generated are called electropherograms), temperature control, and recirculating buffers. The use of capillaries as a support medium for electrophoresis is advantageous because it avoids the effects of heating that occur in a gel, can be very rapid, and produces a chart recording rather than a stained gel for archiving.

Free-Flow Electrophoresis. Free-flow electrophoresis is the most common technique for scaling up electrophoresis for commercial application. In this technique, sample compounds are injected into a curtain of buffer which flows between two flat plates, with electrodes parallel to the flow at each end. The electric field is then applied perpendicularly to the flow direction, so that as compounds flow down between the electrodes they separate horizontally and exit the flow field at different locations.

Detection Techniques

Most sample components analyzed with electrophoretic techniques are invisible to the naked eye. Thus methods have been developed to visualize and quantify separated compounds. These techniques most commonly involve chemically fixing and then staining the compounds in the gel. Other detection techniques can sometimes yield more information, such as detection using antibodies to specific compounds, which gives positive identification of a sample component either by immunoelectrophoretic or blotting techniques, or enhanced detection by combining two different electrophoresis methods in two-dimensional electrophoretic techniques.

SCOTT RUDGE
KATHLEEN MARKEY
Synergen, Inc.

Additional Reading

Allen, R.C., C.A. Saravis, and H.R. Maurer: *Gel Electrophoresis and Isoelectric Focusing of Proteins,* Walter de Gruyter, New York, NY, 1984.

Baldo, B.A. and E.R. Tovey: *Protein Blotting: Methodology, Research, and Diagnostic Applications,* Karger, Switzerland, 1989.

Camilleri, P.: *Capillary Electrophoresis: Theory and Practice,* 2nd Edition, CRC Press, LLC., Boca Raton, FL, 1997.

Chrambach, A.: *The Practice of Quantitative Gel Electrophoresis,* VCH Publishers, New York, NY, 1985.

Dose, E.V. and G.A. Guiochon: Internal Standardization Techniques for Capillary Zone Electrophoresis, *Analytical Chemistry,* **1154** (June 1991).

Gersten, D.M.: *Gel Electrophoresis: Proteins,* John Wiley & Sons, Inc., New York, NY, 1996.

Gordon, M.J. et al.: Capillary Electrophoresis, *Science,* **224** (October 14, 1988).

Hames, B.D. and D. Rickwood: *Gel Electrophoresis of Proteins: A Practical Approach,* Oxford University Press, Inc., New York, NY, 1998.

Khaledi, M.G.: *High-Performance Capillary Electrophoresis: Theory, Techniques, and Applications,* John Wiley & Sons, Inc., New York, NY, 1998.

Lunn, G.: *Capillary Electrophoresis Methods for Pharmaceutical Analysis,* John Wiley & Sons, Inc., New York, NY, 1999.

Mitchelson, K.R. and J. Cheng: *Capillary Electrophoresis of Nucleic Acids,* Humana Press, Totowa, NJ, 2000.

Peterson, J.R. and A.A. Mohammad: *Clinical and Forensic Applications of Capillary Electrophoresis,* Humana Press, Totowa, NJ, 2000.

Rabilloud, T.H.: *Proteome Research: Two-Dimensional Gel Electrophoresis and Identification Methods,* Springer-Verlag, Inc., New York, NY, 1999.

Robyt, J.F. and B.J. White: *Biochemical Techniques: Theory and Practice,* Waveland Press, Inc., Prospect Heights, IL, 1987.

Smisek, D.L. and D.A. Hoagland: Electrophoresis of Flexible Macromolecules, *Science,* **1221** (June 8, 1990).

Srensen, H. and C. Bjergegaard: *Chromatography and Capillary Electrophoresis in Food Analysis,* Springer-Verlag, Inc., New York, NY, 1998.

Weinberger, R.: *Practical Capillary Electrophoresis,* Academic Press, Inc., San Diego, CA, 2000.

ELECTROPLATING. The coating of an object with a thin layer of some metal through electrolytic deposition. The process is widely used, either for the purpose of rendering a lustrous or noncorrosive finish on some article. In electroplating, the general object is to employ the article to be plated as the cathode in an electrolytic bath composed of a solution of the salt of the metal being plated. The other terminal, the anode, may be made of the same metal, or it may be some chemically unaffected conductor. A low-voltage current is passed through the solution, which electrolyzes and plates the cathodic articles with the metal to the desired thickness. In this way, table utensils are silver plated, various parts are made weatherproof by cadmium or chromium plating, and a high finish may be imparted through nickel plating. Copper, zinc, and gold are also plated. As the plating proceeds, the strength of the solution must be kept up by the addition of crystals of the plating salt, or a renewal of the anode if it is of the plating metal. A firm bond between the anode and the deposited metal is to be secured when the two metals are of a type that tends to alloy. If this is not the case, some intermediate metal, which will alloy between the base and the plate, is first deposited. For example, in silver plating on steel, the iron would otherwise form a poor bond with the silver, so a thin layer of copper is first deposited on it.

Because of the excellent conducting properties of a metallic salt solution, only a low voltage is required. As this must be D.C. the process of electroplating calls for a supply of current from special low-voltage D.C. generators or rectifiers. The voltage will be of the order of 6 volts or less between anode and cathode.

Some of the solutions used to place various metals are as follows: for silver or gold plating, double cyanide of the metal and potassium; copper plating, copper sulfate; nickel plating, nickel ammonium sulfate. The articles to be plated must be thoroughly and effectively cleaned of all grease and dirt by washing in caustic or acid solutions. While the above is a brief outline of the process of electroplating, in commercial operations there are many troublesome angles which would not be suspected from the foregoing. Irregularity of the plate, poor surface graining, "trees," insufficient bond, and other troubles develop. The overcoming of these requires the use of various expedients, such as careful control of temperature, or the addition of certain colloids and other compounds which have been found effective in preventing formation of defects on the plated articles. See Table 1.

As will be noted from the table, in order to plate out the metals above hydrogen from an aqueous solution, the concentration of the hydrogen in the cathode film must be low. Hence a basic solution, such as provided by a cyanide bath, may be required. A cyanide bath also may be used to provide a smooth adherent plate.

To plate out an alloy, that is, to codeposit at least two kinds of metals, the plating bath must be so designed that the electrodeposition potentials of the two metals in the cathode film are equal or nearly so.

Four typical electroplating baths are:

Copper: CuCN 26 g/l, NaCN 35 g/l, Na_2CO_3 30 g/l
          $KNaC_4H_4O_6 \cdot 4H_2O$ 45 g/l, NaOH to give pH of 12.6
Copper: $CuSO_4 \cdot 5H_2O$ 188 g/l, H_2SO_4 74 g/l
Tin: Tin (as tin fluoborate concentrate) 60 g/l, free fluoboric acid 100 g/l, free boric acid 15 g/l.
Zinc: $Zn(CN)_2$ 60 g/l, NaCN 23 g/l, NaOH 53 g/l

The bright, hard, ornamental chrome so popular on automobiles and household and office articles is produced by electroplating. In this process, the chromium is not present in the bath as a positive metal ion, but rather as part of the anion of chromic acid, H_2CrO_4. The object being plated is made the cathode. Usually the article will have previously been plated with copper and then with nickel. Sulfuric acid serves as a catalyst. The final chromium plate ranges from 0.00001 to 0.0005 inch (0.025–0.127 mm) in thickness. Although improvements have been made in recent years for certain chromium-plating applications, the traditional bath is an aqueous solution of chromic trioxide, CrO_3, and sulfuric acid, with a ratio of approximately 100 to 1 (wt). Fluosilicate catalysts are also added to some chromium-plating baths. See also **Chromium**.

The voltage for electrodeposition under ideal conditions would be:

$$E = E^0 + \frac{RT}{VF} \ln A$$

where E is the required voltage relative to the solution as measured by a hydrogen electrode in a unit molal activity hydrogen ion solution. E^0 is the

TABLE 1. METALLIC IONS COMMONLY ELECTROPLATED

Ion	E^0	Ion	E^0
Zn^{+2}	−0.762	Cu^{+2}	+0.345
Cr^{+3}	−0.71	Cu^{+1}	+0.522
Cd^{+2}	−0.402	Ag^{+1}	+0.800
Ni^{+2}	−0.250	Au^{+3}	+1.42
Sn^{+2}	−0.136	Au^{+1}	+1.68
$(H^{+1}$	0.000)		

electrolytic potential, in volts, of the metal being plated when immersed in a solution containing its ions at unit molal activity (approximately unit molal concentration). $R = 8.31$ joules per degree mole; T is the Kelvin temperature; F 96,500 joules per gram-equivalent; V is the valence of the ions which are depositing out; $\ln A$ is the natural logarithm of the activity of these ions (approximately the natural logarithm of their molality).

In actual practice the concentration and therefore the activity of the ions soon after the electroplating process starts is different in the solution just next to the cathode, called the cathode "film," than in the main body of the bath. The foregoing equation must in practice be modified to read:

$$E = E^0 + \frac{RT}{VF} \ln A - P,$$

where A is the molal activity in the cathode film of the ions being electrodeposited and P is the extra potential required to keep the plating going. A and P depend on temperature, current, density, concentration, valence, pH, and ion mobility.

P. S. ALBRIGHT
Wichita, Kansas

Additional Reading

Paunovic, M.: *Modern Electroplating*, John Wiley & Sons, Inc., New York, NY, 2000.

ELECTROPOLISHING. Production of a smooth surface on metals by electrochemical means.

In all electroplating processes, metals (and hydrogen) are deposited on the cathode and dissolved from the anode, except when insoluble anodes are used in which case oxygen is liberated at the anode. Electropolishing is the reverse of electroplating. The work is made the anode and tends to be dissolved. The operating conditions are controlled so that atomic oxygen forms continuously and reacts with the metal surface. Part of this oxygen may be liberated as gas. According to one theory, the high points of the metal surface are most readily oxidized and this oxidized material is thereupon dissolved in the electrolyte or otherwise removed. In any case, selective solution of the high points of the surface tends to give a very smooth finish comparable or superior to a mechanically buffed surface. A wide variety of electrolytes is used. A typical one for stainless steels contains phosphoric acid and butyl alcohol.

All mechanical methods of polishing, including those used for metallographic samples, produce a thin surface layer of work-hardened metal. Electropolishing produces a strain-free surface that is especially suitable for microscopic examination.

An important commercial application of the process is the polishing of stainless steel parts of irregular contour which would be difficult or impossible to buff. Copper and its alloys, Monel metal, aluminum, and many other alloys can be electropolished.

ELECTROSOL. A colloidal solution produced by electrical means, as by passing a spark between metal electrodes in a liquid.

ELECTROSTATIC PRECIPITATOR. The concept of the electrostatic precipitator for the removal of particulates from smoke and industrial emissions dates back to the late–1800s and the pioneering work of Frederick Gardner Cottrell. Some of the earliest work in connection with electrical precipitators was directed to copper smelting in the early 1900s. Quoting briefly from reports, "In the first decade of the century there were three well-publicized and classic examples of smelter smoke injury in the United States—at Ducktown, Tennessee, at Salt Lake City, and at Anaconda, Montana. Some indication of the scope of the problem can be seen from a study that was eventually made at the latter and which revealed some startling figures. In a normal day's operation, up and out the stack of that copper smelter went the amazing total of 3,200 metric tons of sulfur dioxide, 200 metric tons of sulfur trioxide, 30 metric tons of arsenic trioxide, 3 metric tons of zinc, and over 2 metric tons each of copper, lead, and antimony trioxide. The marvel is that anything remained, but nothing could more clearly demonstrate that here in full bloom was a cardinal essence of successful invention: a need existed. . . . The emissions from the low stacks of an old plant operated at a neighboring location had killed all vegetation, and losses of livestock by arsenical poisoning had been heavy over the near-lying area. Years after the plant was dismantled, the topsoil of a large area centering at the old site was stripped off, sent through

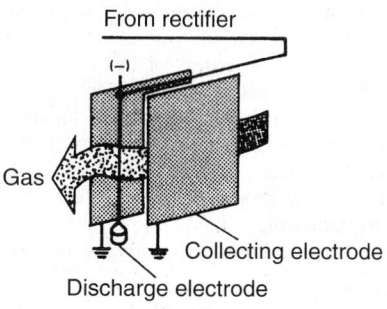

From rectifier

(−)

Gas

Collecting electrode

Discharge electrode

Fig. 1.   Arrangement of electrodes in electrostatic precipitator

concentrators, and smelted at the new plant with a reported recovery of over $1 million in copper and other metals."

But despite the need and the invention, Cottrell had considerable difficulty over a number of years in gaining acceptance of the electrical precipitator by industry. Today, and for a number of decades past, the electrostatic precipitator has been a major device for combatting air pollution. Since the precipitator functions only against particulates, numerous other items of air pollution control equipment, such as absorbers, scrubbers, and filters, are required and are described elsewhere in this volume.

Operating Principle

The basic operating principle of the electrostatic precipitator is demonstrated by the familiar experiment in which a glass rod is rubbed with a silk cloth; the action gives the rod an electrostatic charge, making it capable of attracting uncharged bits of paper, lint, or cork. In the electrostatic precipitator, it is the collecting surfaces that are grounded, while the charge is created on the particulates to be collected.

The power supply is a transformer-rectifier set that steps up ordinary 220-V AC supply to the high level necessary for precipitator operation, and rectifies it to direct current. The DC voltage is applied to discharge electrode wires suspended in the gas flow path. See Fig. 1. In the most common industrial type of precipitator, the discharge electrodes hang between rows of collecting electrode plates, which form a series of parallel gas flow ducts. The high potential on the discharge electrodes causes a corona discharge, from which electrons migrate out into the gas. These create gas ions, which attach themselves to particulates in the gas and give the particles a charge.

The collecting electrodes are grounded, so that the high potential difference between them and the discharge electrodes creates a powerful electric field through which the gas must flow. According to Coulomb's law, such a field exerts a force on charged particles in the field. In the precipitator, this force moves particles out of the gas stream to the collecting electrodes. In a typical precipitator, the force on a particle 0.5 micrometer in diameter is several thousand times the force of gravity on such a particle.

At the grounded collecting electrodes, the particulates lose their charge. They drain off, if liquid, accumulate until washed off or, more commonly in the case of dry dust, are dislodged by mechanical agitation of the electrodes. In a few applications, the collecting electrodes are vertical pipes, instead of parallel plates, each pipe with a discharge electrode wire hanging down its axis.

ELECTRUM. Electrum is a native alloy of gold and silver in which the latter metal may be present in quantities up to 40%. Electrum from the Urals is said to carry 20% copper. The color of electrum is a pale yellow or yellowish-white and the name is derived from the Greek word mentioned in the "Odyssey," meaning a metallic substance consisting of gold alloyed with silver. This same word was also used for the substance amber, doubtless because of the pale yellow color of certain varieties.

ELEMENTARY PARTICLES. See **Particles (Subatomic).**

ELEMENTS (Chemical). See **Chemical Elements.**

ELUTION. In general, a process for extracting a solid substance from a mixture of solids by means of a liquid; as in the recovery of a vitamin

adsorbed on an adsorbent by means of a solution. Specifically, elution is a process for the recovery of sucrose from molasses. Quicklime in the proportion of 25% of the weight of the molasses is added, the resulting mass is freed from much impurity by percolating (in "elutors") with 35% alcohol and is then decomposed by carbon dioxide, which liberates the sucrose. See also **Chromatography.**

ELUTRIATION. The separation of solids by the action of water or other liquids: hence, also the washing of a solid by decantation or a related process.

EMBDEN-MEYERHOF PATHWAY.   See **Carbohydrates.**

EMBRITTLEMENT.   A lowering of the ductility of a metal as a result of physical or chemical changes. Metals may be embrittled under many different conditions. Ordinary steel, wrought iron, and body-centered cubic metals generally, as well as zinc alloys and magnesium alloys suffer a reduction in impact toughness at subnormal temperatures. The effect is only temporary, full recovery of toughness occurring upon return to normal temperatures. Austenitic stainless steels, brasses and bronzes, nickel alloys, aluminum alloys, and lead alloys are not subject to severe embrittlement at low temperatures. Nickel additions to ordinary steels have a favorable effect.

Hydrogen embrittlement of iron and steel may be caused by absorption of atomic hydrogen in electroplating processes or in pickling baths. After such exposure the normal toughness can usually be restored by prolonged aging or a short period of heating at a slightly elevated temperature, as in a steam bath.

Season cracking of high zinc brasses is a severe form of embrittlement resulting in cracking or disintegration. Somewhat similar forms of stress-corrosion cracking occur in many other metals and alloys. Embrittlement of boiler plate, discussed below, may be considered a special case.

Steels and ingot iron may be embrittled by any heat treatment that deposits films of either oxides or carbides in the grain boundaries.

Caustic embrittlement is the development of brittleness in metals such as steel or ferrous alloys, upon prolonged exposure to alkaline substances, like caustic soda, in solution. Failures and explosions in boilers and evaporators have been caused by this action. Effective water treatment essentially has eliminated this condition in boilers. See also **Corrosion Embrittlement.**

EMDE DEGRADATION. Modification of the Hofmann degradation method for reductive cleavage of the carbon-nitrogen bond by treatment of an alcoholic or aqueous solution of a quaternary ammonium halide with sodium amalgam. Also used as a catalytic method with palladium and platinum catalysts. The method succeeds with ring compounds not degraded by the Hofmann procedure.

EMERALD. This beautiful green variety of the mineral beryl has been known since ancient times and always prized as a gem, both because of its color and relative rarity. It is frequently cloudy or flawed, hence the expression "rare as an emerald without a flaw." The original source of emeralds seems to be the so-called Cleopatra's mines in Egypt, where in a range of low mountains about 15 miles (24 kilometers) from the Red Sea, they are found in schists. The quality of these emeralds is not high, but there is much evidence of considerable workings in a former period. See also **Beryl.**

EMISSION SPECTROSCOPY. Study of the composition of substances and identification of elements by observation of the wavelengths of radiation they emit as they return to a normal state after excitation by an external energy source. When atoms or molecules are excited by energy input from an arc, spark, or flame, they respond in a characteristic manner; their identity and composition are signaled by the wavelengths of incident light they emit. The spectra of elements are in the form of lines of distinctive color, such as the yellow sodium D line of sodium; those of molecules are groups of lines called bands. The number of lines present in an emission spectrum depends on the number and position of the outermost electrons and the degree of excitation of the atoms. The first application of emission spectra was identification of sodium in the solar spectrum (1814).

See also **Spectroscopy (Instrumental Analysis).**

EMULSIONS

Simple Emulsions

A (macro)emulsion is formed when two immiscible liquids, usually water and a hydrophobic organic solvent, an oil, are mechanically agitated so that one liquid forms droplets in the other one. A microemulsion, on the other hand, forms spontaneously because of the self-association of added amphiphilic molecules. During the emulsification agitation both liquids form droplets, and with no stabilization, two emulsion layers are formed, one with oil droplets in water (o/w) and one of water in oil (w/o). However, if not stabilized the droplets separate into two phases when the agitation ceases. If an emulsifier (a stabilizing compound) is added to the two immiscible liquids, one of them becomes continuous and the other one remains in droplet form.

During emulsification new surfaces are created between the two phases. Such a process requires energy; the surface free energy, numerically identical to the easily measure surface tension, reflects this amount.

There are two fairly common misconceptions about the conclusions that may be drawn from the value of the surface free energy. First, the energy input to enlarge the interface during emulsification is not a significant part of the total energy needed for this process. The viscous resistance during the agitation absorbs most of the energy, giving a small temperature rise to the system. Secondly, the emulsion is stabilized by the addition of a compound adsorbed to the interface, termed an emulsifier. Emulsifiers are molecules with nonpolar and polar parts that reside at the interface. Their presence causes a reduction of the surface tension and a stabilization of the emulsion, but these two features are not directly related. The interfacial tension reflects the amount of emulsifier at the interface.

Any conclusion that a low interfacial tension per se is an indication of enhanced emulsion stability is not reliable. In fact, very low interfacial tensions lead to instability. The stability of an emulsion is influenced by the charge at the interface and by the packing of the emulsifier molecules, but the interfacial tension at the levels found in the common emulsion has no influence on stability.

Emulsification

Much commercial equipment is available for emulsification (Fig. 1) and has been well described. The following discusses the relations between energy input and emulsion droplet size.

The emulsification process in principle consists of the break-up of large droplets into smaller ones due to shear forces. The simplest form of shear is experienced in lamellar flow, and the droplet break-up may be visualized according to Figure 2. The phenomenon is governed by two forces, i.e., the Laplace pressure, which preserves the droplet, and the stress from the velocity gradient, which causes the deformation. The ratio between the two is called the Weber number, We, where η_c is the viscosity of the continuous phase, G the velocity gradient, r the droplet radius, and γ the interfacial tension: $We = \eta_c \cdot G \cdot r / \gamma$.

As an approximate rule, break-up of droplets occurs for a Weber number in excess of one, a rule of thumb that is actually valid for the range of viscosity ratios of the dispersed phase to the continuous phase of less than approximately five. Higher viscosities of the disperse phase lead to serious

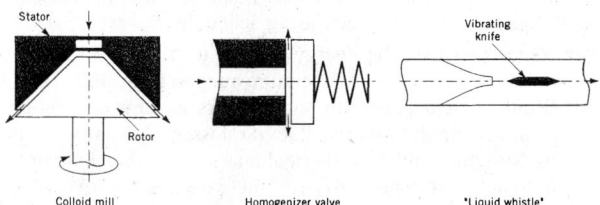

Stator
Vibrating knife

Rotor

Colloid mill   Homogenizer valve   'Liquid whistle'

Fig. 1.   Commercial emulsification equipment is built on different principles for mixing one liquid into another

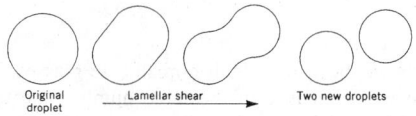

Original droplet   Lamellar shear   Two new droplets

Fig. 2.   A droplet is broken into two droplets in lamellar shear

difficulties with emulsification because the shear energy is then dispersed in rotation of the droplets.

As may be expected, turbulent flow is more efficient for droplet formation in low viscosity liquids.

Finally, some general rules for the amount of surfactant appear to be valid. For anionic surfactants the average size of droplets is reduced for an increase of surfactant concentration up to the critical micellization concentration, whereas for nonionic surfactants a reduction occurs also for concentrations in excess of this value. The latter case may reflect the solubility of the nonionic surfactant in both phases, causing a reduction of interfacial tension at higher concentrations, or may reflect the stabilizing action of the micelles *per se*.

Stability of Two-Phase Emulsions

Before determining the degree of stability of an emulsion and the reason for this stability, the mechanisms of this destabilization should be considered. When an emulsion starts to separate, an oil layer appears on top, and an aqueous layer appears on the bottom. This separation is the final state of the destabilization of the emulsion; the initial two processes are called flocculation and coalescence. In flocculation, two droplets become attached to each other but are still separated by a thin film of the liquid. When more droplets are added, an aggregate is formed, in which the individual droplets cluster but retain the thin liquid films between them. The emulsifier molecules remain at the surface of the individual droplets during this process.

In the coalescence step, the thin liquid film between the droplets is destabilized, and a large droplet is formed. Hence, the coalescing emulsion is characterized by a wide size distribution of the droplets, but no clusters are present. Finally, the droplets achieve such a size that they are recognized by the naked eye as a separate phase. A fully separated emulsion consists of an oil layer and an aqueous layer.

The sequence, flocculation → coalescence → separation, is complicated by the fact that creaming or sedimentation occurs and that this process is determined by the droplet size.

The definition of stability and the appropriate measure of the rate of destabilization of an emulsion depend on the application. For an application such as the use of a fluorocarbon as a blood substitute the destabilization stage of importance is obviously the flocculation; aggregates of droplets would clog blood vessels. At the other extreme are beverages such as soda or reconstituted fruit juices in which aggregation of flavor droplets is not noticed by the consumer and is not an essential disadvantage. The consumer instead would notice any separated oil forming a "greasy"ring in the bottle neck. In this case creaming must be prevented, a distinctly different problem from stabilization against flocculation.

Stabilization Mechanisms. The stabilization of an emulsion involves slowing the destabilization, primarily the flocculation process. This may be achieved in two principal manners: by reducing the mobility of droplets through enhanced viscosity or by inserting an energy barrier between them.

Two kinds of barriers are important for two-phase emulsions: the electric double layer and steric repulsion from adsorbed polymers. The repulsion from the electric double layer is famous because it played a decisive role in the theory for colloidal stability that is called DLVO, after its originators Derjaguin, Landau, Vervey, and Overbeek. The theory provided substantial progress in the understanding of colloidal stability, and its treatment dominated the colloid science literature for several decades.

Polymer Stabilization. Polymers have so far been used comparatively less than the common surfactants to stabilize emulsions in spite of the fact that excellent stabilization by them can be achieved. Application probably has been limited because the adsorption of polymers to emulsion droplets has displayed some intricate phenomena; small changes in polymer structure or in solvent properties may lead to drastic changes in adsorption.

A polymer is adsorbed in the form of loops, tails, and trains. Sufficiently long tails or loops provide stabilization. The action is extremely efficient; a single loop or tail gives a barrier of approximately 20 kT.

However, polymer stabilization is sensitive to the properties of the environment. Adsorption energies lower than the optimal range give no adsorption of the polymer and no stabilization. At a higher adsorption energy, the polymer adsorbs flat at the interface. Such an adsorption is also without stabilization effect because at short distances the van der Waals potential has already reached such large negative values that the potential well is too deep for the droplets to be deaggregated. Only in a limited range of adsorption energies, in which loops and tails are formed, does the polymer serve. In addition, the same phenomenon means that a minimum molecular weight is necessary to obtain stability.

These problems may be overcome by using block copolymers. The polymer blocks are chosen to be selectively soluble in the aqueous and oil phase, respectively.

Stability of Three-Phase Emulsions

In the simplest emulsions just described, the final separation is into two liquid phases upon destabilization. The majority of emulsions are of this kind, but in some cases the emulsion is divided into more than two phases. One obvious reason for such a behavior is the presence of a material that does not dissolve in the oil or the water. One such case is the presence of solid particles, which is common in emulsions for food, pharmaceuticals, and cosmetics. Another less trivial reason is that the surfactant associates with the water and/or the oil to form a colloidal structure that spontaneously separates from the two liquid phases. This colloidal structure may be an isotropic liquid or may be a semisolid phase, a liquid crystal, with long-range order.

In the case of emulsions with three liquids the presence of the third phase results in a reduction of the energy input for the emulsification process, whereas the emulsion with a liquid crystal as the third phase shows interesting stabilization mechanisms. Finally, the emulsion with added particles illustrates the importance of liquid–solid wetting for stability.

See also **Colloid Systems**.

STIG E. FRIBERG
STEVEN JONES
Clarkson University

Additional Reading

Becher, P.: *Emulsions: Theory and Practice,* 3rd Edition, American Chemical Society, Washington, DC, 2001.

International Union of Pure and Applied Chemistry, *Manual on Colloid and Surface Science,* Butterworths, London, 1972.

Klose, A., V. Schmitt, and P. Poulin: *Emulsion Science: Basic Principles: An Overview,* Springer-Verlag New York, LLC., New York, NY, 2003.

Mittal, K.L. and P. Kumar: *Handbook of Microemulsion Science and Technology,* Marcel Dekker, Inc., New York, NY, 1999.

Mollet, H., and A. Grubenmann: *Formulation Technology: Emulsions, Suspensions, Solid Forms,* John Wiley & Sons, Inc., New York, NY, 2001.

Napper, D.E.: *Polymeric Stabilization of Colloidal Dispersion,* Academic Press, Inc., New York, 1983.

Ross, S., and I.D. Morrison: *Colloidal Dispersions: Suspensions, Emulsions, and Foams,* 2nd Edition, John Wiley & Sons, Inc., New York, NY, 2002.

Sherman, P.: *Emulsion Science,* Academic Press, Inc., New York, NY, 1969.

Sjoblom, J.: *Emulsions and Emulsion Stability,* Marcel Dekker, Inc., New York, NY, 1996.

ENAMELS, PORCELAIN OR VITREOUS.

Porcelain enamel is a glassy coating applied to various metal substrates through a high (usually >425°C) temperature fusion process. The resultant chemically bonded glass–metal composite exhibits properties which reflect the chemical, physical, electrical, and aesthetic properties of the glass while combining the strength, ease of fabrication, and durability of the metal. Porcelain enamels are formulated to develop specific properties on metals or alloys.

The porcelain enamel is composed of various inorganic metal oxides fused between 1100 and 1400°C to form an alkali borosilicate glass. The glass is rapidly quenched to produce either flakes by rapid chilling using water-cooled rollers, or glass granules by water quenching a narrow stream of molten glass. The resultant product is known as frit. An enameled article is produced by applying either wet or dry finely ground frit particles to the metal substrate. The enamel is fired to form one or more glassy layers to achieve a desired surface finish and property group for the article.

Modern appliances were developed utilizing the chemical and physical properties of the glass-on-metal coatings on cast iron, sheet steel, and, more recently, aluminum. Enamels have been used on a wide array of kitchen utensils, cooking devices such as ovens and range tops, sanitary ware such as bathtubs and lavatories, water heater tanks, specially built industrial chemical vessels, cast-iron piping, storage silos for agricultural and industrial products, as well as architectural interior and exterior surfaces.

Porcelain enamels are used in modern mass transit facilities because of inherent fire resistance as well as ease of maintenance and durability in the face of highly corrosive atmospheres resulting from vehicular traffic. Newer applications of porcelain enamels include electronic substrates, pyrolytic self-cleaning oven interiors, microwave oven interiors, outdoor cookers, and fireplace liners as well as wood-stove exteriors, writing boards for erasable markers, institutional surfaces such as bathroom stalls and hospital operating room walls that need to be easily disinfected, and subway car interiors and elevator walls that are durable and easy to clean. Porcelain enamel surfaces are frequently used in food contact applications. The glassy nature of porcelain enamels has a significant effect on encapsulating and minimizing any solubility of constituents. Enamels can also be formulated to have specific acid, alkali, and abrasion-resistant properties.

Metals for Enameling. Sheet steel is usually bought in precut sheets or coils for subsequent stamping and pressing into shapes. The steels are chemically formulated to make them suitable for the fabrication and porcelain enameling operations. The ASTM has classified enameling-grade steels into Type I and Type II (A or B) as well as specifying various qualities such as commercial, drawing, and drawing quality special killed. Type I has an extremely low carbon level, commonly produced by decarburizing in an open-coil process. This material is suitable for direct cover coat enameling practice. The less expensive Type II has moderately low carbon and is suitable for ground coat enameling operations.

The basic steel types are undergoing gradual modifications to adapt the steels to the continuous casting process. This has led to changes in the minor constituents of steel such as boron, nitrogen, titanium, and other alloying elements.

Cast Iron. Cast irons for enameling contain between 2.8 and 3.7 wt % carbon; the more usual content is between 3.25 and 3.6 wt %. The carbon is usually found in two forms: graphitic carbon and combined carbon. Some additional elements in the cast iron are silicon, manganese, sulfur, and phosphorus. The cast iron known as gray cast iron is the most widely used for enameling purposes. Before enameling, a casting must be cleaned, usually by abrasively blasting the surface with sand or steel shot.

Aluminum. The most commonly used aluminum alloys for enameling are 3003 P.E. grade and 6061 P.E. grade. It is important that the magnesium and copper contents of the alloys be kept to a minimum; otherwise, spalling of the enamel can occur.

Other Metals. Metals such as the austenitic series, Types 301–347, and the ferritic series, Types 409–446, of stainless steels may be enameled, as well as a number of other alloys. The metal preparation usually consists of degreasing and grit blasting. Copper, gold, and silver are also enameled. These metals are usually prepared for application by degreasing.

Metal Preparation. Enameling cannot be successful unless the metal is thoroughly cleaned and kept clean until the final coat is fired. Simply touching the surface with a hand can cause defects. Cast iron, thick steel parts, and aluminum castings may be sand-blasted. The thickness precludes the danger of deformation resulting from metal loss. Sand, silicon carbide, and steel grit are satisfactory abrasives. Products made from thin sheet material are most satisfactorily and most economically cleaned by chemical methods that require alkali and soap solutions to remove grease and dirt, and acid solutions to remove oxidized metal.

Firing of Enamels. Firing can be carried out in intermittent box-type furnaces or continuous furnaces. The dryer and the furnace form one continuous unit or function as separate units in the continuous firing process. Most industrial furnaces are fiber-lined (low thermal mass), which lowers cost and downtime between firing schedules.

Energy Requirements. The energy needed to heat 0.45 kg of enamel ware and tooling to the firing temperature ranges from 400–1000 kJ/kg (300–450 Btu/lb) and depends on such factors as furnace loading, tool weight, and, in gas-fired furnaces, flue-gas losses. Advances in furnace design, firing systems, and low thermal mass insulating materials have reduced the total energy needed over the years. Thus overall energy requirements for enameling have dropped as much as 50% since the 1970s. Technologies such as electrostatic frit powder application, various two-coat–one-fire processes (e.g., dry-over-dry, dry-over-wet, and wet-over-wet), as well as reductions in the metal pretreatment requirements have also contributed.

Composition

Porcelain enamels are basically alkali borosilicate glasses. These enamels are complex, however, because of the large number and types of oxides which are needed to develop proper adherence and functional properties. Network-forming ingredients and modifiers are used as in normal glass making practice. The principal network formers are SiO_2, B_2O_3, and P_2O_5. Modifiers include the alkali metal oxides (Na_2O, K_2O, and Li_2O) and alkaline-earth metal oxides (CaO, BaO, and SrO). Other common oxides include Al_2O_3, MgO, ZrO_2, ZnO, TiO_2, Sb_2O_3, and the halide F^-. Transition-metal oxides such as Fe_2O_3, CoO, NiO, CuO, and MnO_2 are used for adherence to the sheet-steel substrate and color development in ground coats. Continuous-cleaning (catalytic) oven enamels have high percentages of the transition-metal oxides to achieve cleaning effectiveness. Less frequently used oxides of cerium, cadmium, lead, and tin are used in sheet-steel, cast-iron, and aluminum enamels.

Enamels used on cast iron and aluminum have traditionally been composed of SiO_2, B_2O_3, P_2O_5, and PbO. The lead oxide produces good surface quality, fusibility, and acid resistance when properly formulated with other oxides. More recently some nonlead-bearing compositions have been developed for both cast-iron and aluminum metals. Glasses containing lead oxide are not recommended for food contact surfaces.

Porcelain enamels meet a variety of performance characteristics required for different applications. The common characteristics of all enamels include good adherence to the substrate and good thermal expansion fit to the metal. Specific properties depend on usage; for example, acid and alkali resistance, hot water resistance, abrasion resistance, thermal shock resistance, high gloss, high reflectance, specific color, heat resistance, and cleanability.

Titanium Dioxide. The recrystallization of titanium dioxide in a cover-coat glass is very important to the development of thin, highly opaque finish coats. Titania, TiO_2, is the primary opacifying agent for white finish coats. Two polymorphic forms of titania, anatase and rutile, may be present in the enamel. See also **Titanium**. Anatase is preferred because anatase crystals are present in the size range (0.1–0.2 μm) for maximum reflectance, and therefore generate the most desirable bluish white color.

Properties

Thermal Fit and Residual Stresses. Thermal expansion measurements are typically carried out on dilatometric equipment consisting of a fused quartz pusher rod inside a tube of the same material.

A porcelain enamel glass becomes less viscous as the temperature is increased during firing. Above the softening point, the enamel is relatively fluid and conforms to the metal surface. As the porcelain enamel is cooled from the firing temperature of 750–800°C to the softening point, the fluid glass does not retain stress. However, as cooling proceeds below the softening point, the expansion (or contraction) or the coating exceeds that of the steel, and tensile stresses begin to develop in the coating. On further cooling, the stress increases until the temperature at which the expansion of the glass equals that of the metal. With still further cooling, the coefficient of expansion of the glass is less than that of the metal, coating tensile stresses decrease, and compressive stresses develop and are retained at room temperature.

Thermal expansion comparisons of the coating and metal have often been used to determine residual stresses in the coatings.

Composite Modulus of Elasticity. The modulus of elasticity of the enamel glass–steel composite system has been shown to lie between the modulus of the glass and that of the metal. The composite modulus can be calculated by $E_c = (E_m - E_e)Q^3 + E_e$, where E_c, E_m, and E_e are the modulus of elasticity of the composite, the metal, and the enamel, respectively; and $Q =$ thickness of the metal divided by total thickness of the composite.

Residual Compressive Stress. Residual compressive stress in commercial ground coat enamels varies with enamel thickness.

Thinner coatings have higher compressive stresses, other factors being equal. Higher residual compressive stress in the coating also can be obtained by using enamel glass having a lower thermal expansion coefficient, or a metal having a higher expansion coefficient or a higher modulus of elasticity.

Maximum Strain. Strain in enamels that leads to failure is on the order of 0.002–0.003 cm/cm. Thinner enamels having higher residual compressive stresses are more flexible and can be strained to a greater degree.

Some other physical properties of enamel glass are density, from 2.5–3.5 g/mL; Mohs' hardness, 5–6; tensile strength, 34–103 MPa (4,900–15,000 psi); compressive strength, 1380–2760 MPa ($2–4 \times 10^5$ psi); modulus of elasticity, 55–83 GPa ($8–12 \times 10^6$ psi); and dielectric constant, 5–10.

Appearance and Color. Porcelain enamel allows the designer great variety with regard to color, texture, and aesthetic appeal. The enamels exhibit exceptional color stability whether used in domestic interiors or architectural exteriors. Colors should be grouped according to the type and the usage of enamel. Categories are (1) ground coats, which can generally be used by themselves as a finish coat or a base coat; (2) cover coats that are self-opacified; and (3) clear and semiopaque enamels, used for developing a wide range of colors. Transition-metal oxides used principally to develop a bond between the glass and the substrate metal also produce specific colors. Those range in shades of grays from blue to yellow and red to green. Combinations of each are also possible. Cobalt oxide is the principal colorant for blue, nickel oxide, and iron oxide, Fe_2O_3, for green, manganese dioxide for red, and nickel oxide with other oxides for yellow.

Cover coats such as self-opacified titanium enamels derive their color from titanium dioxide crystals nucleated in the glass during firing of the coating.

Textures in enamels are developed by the use of semicrystalline glasses or by the addition of refractory materials such as quartz, alumina, zirconia and zircon, feldspar, various clays, and titania. More refractory glasses are also added to impart a lower gloss and a texture.

Colored enamels are produced by tinting the titania enamels during the smelting operation through the addition of colorant oxides. Control of enamel color is most popularly done using computer-controlled color measuring systems consisting of spectrophotometric detectors (380–720 nm) and hardware to measure, integrate, evaluate, report, and store the color data. The most common systems in use are the Commission Internationale de l'Eclairage (CIE) $L^*a^*b^*$ and Rd, a,b color scales and various illuminates.

Decorating. The decoration of enamels is primarily done by silk screening. Other methods include indirect printing such as use of decals (decalcomania), and indirect lithographic, thermoplastic, and total transfer printing. Photographic printing is also done as a specialty.

Microstructure and Thickness. The microstructure of enamels is of importance to understanding and thus being able to control the macroproperties of the enamel. The microstructure is also related to thickness of the enamel and its firing history. Porcelain enamels typically have a bubble structure which is a result of gas evolution during firing.

The thickness of the enamel layer varies with the type of use and metal. However, typical thicknesses are from 75 to 150 μm (3 to 6 mils) for sheet steel, 175 to 359 μm (7 to 14 mils) for hot-rolled steel, 100 to 125 μm (4 to 5 mils) for each coat of wet process cast iron (760 to 788°C fire), 15 to 25 μm (0.5 to 1 mil) for dry process cast-iron base coats and 750 μm (average) to nearly 2250 μm (30 mils to 90 mils) for the dusted cover coats (898 to 955°C fires), and 25 to 50 μm (1 to 2 mils) for aluminum alloys. Stainless steel and copper may have enamel coatings from 40 μm to 175 μm (1.5 to 7 mils) thick.

The study of cross sections is extremely valuable as a way of tracking defects and determining sources of contaminant once the enameled article has been produced.

Enamel Testing

Standards. The development of standards for porcelain enamel coatings is shared by several national and international organizations. The American Society for Testing and Materials (ASTM), the Porcelain Enamel Institute (PEI), and the American National Standards Institute (ANSI), as well as the Association of Home Appliance Manufacturers (AHAM) are active in developing, collecting, and disseminating information to interested organizations. Cooperation with the International Standards Organization (ISO) is also fostering the development of internationally unified standards for vitreous enamel coatings. Enamel is tested for abrasion resistance, adherence, impact resistance, thermal shock resistance, resistance to chemical attack, and enamel defects.

WILLIAM D. FAUST
Ferro Corporation

Additional Reading

Andrews, A.I.: *Porcelain Enamels*, 2nd ed., Garrard Press, Champaign, IL, 1961.
Dietzel, A.H.: *Emaillierung: Wissenschaftliche Grundlagen und Grundzuge der Technologie*, Springer-Verlag, Berlin, Germany, 1981.
Watril, J.: *Vitreous Enamels*, Borax Holdings Ltd., London, 1984.

Wright, J.F., C.G. Bergeron, and J.C. Oliver: *Porcelain Enamel*, in S. J. Schneider, Jr., vol. chrmn., "Engineering Materials Handbook," Vol. 4, ASM International, Materials Park, OH, 1991, pp. 937–942.

ENANTIOTROPY.   The property possessed by a substance of existing in two crystal forms, one stable below, and the other stable above, a certain temperature called the transition point.

ENARGITE.   A grayish-black or iron-black orthorhombic mineral Cu_2AsS_4. It is an important copper ore, occurring in veins of small crystals or granular masses. Often contains antimony up to about 6% and sometimes, small amounts of iron and zinc.

ENDORPHINS.   See **Enkephalins and Endorphins**.

ENERGY.   In most contemporary texts and those of the last several decades, energy generally has been defined simply as "the ability or capacity to do work." This is a broadening of the earlier definition in terms of Newtonian mechanics, which was "a property of moving masses."

The concept of energy is central to thermodynamics, quantitative chemistry, and electromagnetism. Consider Einstein's mass-energy equation, $E = mc^2$ for the interconversion of mass and energy, where E = energy in ergs; m = mass in grams; and c is the velocity of light in centimeters per second. Or, Planck's equation, which expresses the fundamental law of quantum theory, stating that the energy transfers associated with radiation are made up of definite quanta of energy proportional to the frequency of the radiation: $E = hv$, where E = the value of the quantum units of energy; v = the frequency of radiation; and h is the elementary quantum of action, more commonly known as Planck's constant (6.6256×10^{-27} erg-second—the proportionality factor that, when multiplied by the frequency of a photon, gives the energy of the photon).

Although the fundamental definition of energy can be brief, it immediately calls for an explanation of work, and of power. In the strict physical sense, work is performed only when a force is exerted on a body while the body moves at the same time in such a way that the force has a component in the direction of motion. The amount of work done during motion from point "a" to point "b" can be expressed by

$$W = \int_a^b F \cos\theta \, ds$$

where F is the total force exerted and θ is the angle between the direction of F and the direction of the elemental displacement, ds. In the cgs system, the unit of work is the *dyne-centimeter* or *erg*; in the mks system, the *newton-meter* or *joule*; and in the English system, the *foot-pound*.

In rotational motion, the definition just given can be exactly applied, but it is often convenient to express the force as a torque and the motion as an angular displacement. The work done will be

$$W = \int_a^b \tau \cos\theta \, d\omega$$

where in this case, θ is always the angle between the torque τ, expressed as a vector quantity and the elemental angular motion $d\omega$, also expressed as a vector. The units of work performed in angular motion will, of course, be the same as in the case of linear motion. Notice that the definition of work involves no time element.

Power is defined as the rate at which work is performed. The average power accomplished by an agent during a given period of time is equal to the total work performed by the agent during the period, divided by the length of the time interval. The instantaneous power can be expressed simply as

$$P = dW/dt$$

In the cgs system, power has the units of *ergs per second*; in the mks system, units of *joules per second* (or *watts*); and in the English system, units of *foot-pounds per second*. A common engineering unit is the *horsepower*, defined as 550 foot-pounds per second; or 33,000 foot-pounds per minute. The SI unit of power is the watt. 1 watt = 1 joule per second. (1 joule is the work done by 1 newton acting through a distance of 1 meter.) 1 joule = 1 watt-second = 10^7 ergs = 10^7 dyne-centimeters. The SI unit of force is the newton. (1 newton = 10^5 dynes). See also entry on **Units and Standards**.

Now, returning to the basic definition of energy as the capacity for performing work. This definition may be better understood when stated as:

"The energy is that which diminishes when work is done by an amount equal to the work so done." The units of energy are identical with the units of work previously given.

Energy can exist in a variety of forms, some more recognizable as being capable of performing work than others. Forms in which the energy is not dependent upon mechanical motion are generally referred to as forms of *potential energy*. The most common example in this category is gravitational potential energy. A body near the earth's surface undergoes a change in potential energy when it is changed in elevation, the amount being equal to the product of the weight of the body and the change in elevation.

Potential energy also may be stored in an elastic body, such as a spring or a container of compressed gas. It may exist in the form of chemical potential energy, as measured by the amount of energy made available when given substances react chemically. Potential energy also exists in the nuclei of atoms and can be released by certain nuclear rearrangements.

Kinetic energy is the energy associated with mechanical motion of bodies. It is quantitatively equal to $\frac{1}{2}mv^2$, where m is the mass of a body moving with velocity v. In the case of rotational motion, the kinetic energy is more easily calculated, using the expression $\frac{1}{2}I\omega^2$, where I is the moment of inertia of the body about its axis of rotation and w is the angular velocity. Kinetic energy, like all forms of energy, is a scalar quantity (having magnitude but not direction). In a system made up of an assembly of particles, such as a given volume of gas, the total kinetic energy is equal to the sum of the kinetic energies of all the molecules contained in the volume. Calculation of the energy of such systems is very successfully treated theoretically on the basis of statistical averages.

Within a given system, energy may be transformed back and forth from one form to another, without changing the total energy of the system. A simple example is the pendulum, in which the energy is periodically converted from gravitational potential energy to kinetic energy and then back to gravitational potential energy. A similar situation, but on a submicroscopic scale, occurs in solid materials where the atoms are vibrating under the effect of interatomic rather than gravitational forces. As the temperature of a solid increases, the energy associated with the vibration of the atoms increases.

The example just given illustrates how, on a macroscopic scale, heat can be considered a form of energy. Regardless of the material involved, any amount of heat absorbed or released may be quantitatively expressed as an amount of energy. A *gram-calorie* of heat is equivalent to 4.19 joules, and in the English system, a *British thermal unit* (Btu) is equivalent to 778 foot-pounds.

Potential energy is also present in electric and magnetic fields. The energy available in a region of electric field is equal to $E^2/8\pi$ per unit volume, where E is the electric field strength. Within a given volume, the total energy represented by the electric field is the integral of $E^2/8\pi$ over the volume. Similarly, the energy represented by a magnetic field may be independently calculated by integrating $H^2/8\pi$ over any given volume, where H represents the magnetic field strength. In the case of an electrically charged capacitor, the total energy in the electric field, and hence in the capacitor, can be shown to be $\frac{1}{2}CV^2$. Here C is the capacitance and V the electric potential to which the capacitor is charged. Similarly the total energy in the magnetic field associated with an inductor carrying an electric current is $\frac{1}{2}LI^2$, where L is the inductance and I is the current.

Electromagnetic radiation is a combination of rapidly alternating electric and magnetic fields. Energy is associated with these fields and is exchanged between the electric and magnetic forms. This energy in a quantum of electromagnetic radiation, such as light or gamma radiation, can be expressed in different ways, but is commonly expressed as $E = hv$, as previously mentioned.

For particulate radiation or any very rapidly moving mass, the expression previously given for the kinetic energy, $\frac{1}{2}mv^2$, is not accurate when the velocity approaches that of the velocity of light. The theory of relativity requires a correction be made, and the exact kinetic energy, T, may be calculated in terms of the mass, m_0, of light in vacuum, c, as follows:

$$T = m_0c^2\left[\left(1 - \frac{v^2}{c^2}\right)^{-1/2} - 1\right]$$

Notice that this formula may also be written:

$$T = (m - m_0)c^2$$

where m is the variable quantity $m_0(1 - (v^2/c^2))^{-1/2}$. This quantity represents the mass of the body, reducing to m_0 when v is zero, and approaching infinity as v approaches the speed of light.

This example illustrates another result of the theory of relativity, namely, the equivalence of mass and energy. Rewriting the last equation,

$$m = m_0 + \frac{T}{c^2}$$

The mass is seen to increase linearly with the kinetic energy of the body, the proportionality factor being c^2. It should be noted that even the rest mass, m_0, represents an amount of energy equal to $m_0 c^2$. The total energy of a body of mass, m, can be generally given as:

$$E = mc^2 \quad \text{or} \quad E = m_0 c^2 + T$$

In dealing with radiation, whether particulate or electromagnetic, it is customary to express energy in terms of electron volts. An electron volt is equal to the amount of work done when an electron moves through an electric field produced by a potential difference of one volt. One electron volt is equivalent to 1.60×10^{-12} erg. When charged particles, such as electrons or protons, are given kinetic energy by an accelerator, their kinetic energy is stated in terms of electron volts (eV), million electron volts (MeV), or billion electron volts (BeV).

A basic principle of physics known as the conservation of energy requires that within any closed system, the total energy must remain constant. Energy can be changed from one form to another; but the total, so long as no energy is added to or lost from the system, must be constant. In the case of the swinging pendulum, decreases in kinetic energy reappear as increases in potential energy and vice versa. Eventually, of course, the pendulum will stop due to the effect of frictional forces. At that time, all of the kinetic energy and gravitational potential energy will have been converted to heat.

In another example involving a radioactive atom, the total energy represented by the atom and the emitted radiation must be constant. If a gamma ray is emitted, the rest mass of the atom will be decreased by an amount equivalent to the sum of the energy of the gamma ray and the recoil kinetic energy of the atom, which will be very small. If a beta ray is emitted, the rest mass of the atom will be decreased by an amount equivalent to the sum of the rest mass of the emitted electron, the kinetic energy of the electron, and the recoil kinetic energy of the atom.

Entropy

In the mathematical treatment of thermodynamic processes there occurs very often a quantity, now relating energy to absolute temperature, now associated with the probability of a given distribution of momentum among molecules, and again expressing the degree in which the energy of a system has ceased to be *available energy*. Its mathematical form suggests that these are all aspects of a single physical magnitude. Application of the second law of thermodynamics leads to this conclusion: if any physical system is left to itself and allowed to distribute its energy in its own way, it always does so in a manner such that this quantity, called entropy, increases; while at the same time, the available energy of the system diminishes. This has led to the observation of a so-called "order of merit" for the various forms of energy.

With reference to Table 1, the energy usually flows from higher levels to lower levels—in a direction such that the entropy increases. Thus, cosmic microwave background radiation is defined as the ultimate heat sink, i.e., it represents the ultimate in energy degradation with no lower form in which to be converted.

TABLE 1. ENERGY FLOW AND ENTROPY

Form of energy	Entropy per unit energy
Gravitation	0
Energy of rotation	0
Energy of orbital motion	0
Nuclear reactions	10^{-6}
Internal heat of stars	10^{-3}
Sunlight	1
Chemical reactions	1–10
Terrestrial waste heat	10–100
Cosmic microwave radiation	10^4

The universe evolved by the gravitational contraction of objects of all sizes, from clusters of galaxies to planets. In considering that thermodynamics appears to favor the degradation of gravitational energy to other forms, why is it that after an estimated 10 billion years since cosmic evolution, gravitational energy remains the predominant form? This is explained in terms of a series of phenomena that can be termed "hangups." These are, in essence: (1) The cosmos is large to the extreme; distances between objects are tremendously long; the average density is extremely low. Thus, matter cannot collapse gravitationally in a time shorter than the "free fall time." In relating free fall time (t) with density (d) with the formula, $Gdt^2 = 1$, where G is the constant in Newton's law of gravitation, it is apparent that the free fall time is extremely long. It is estimated that the time is about 100 billion years. But, since the density of our own galaxy is estimated at one million times that of the universe, more than the size hangup is required to preserve the galaxy. (2) An extended object cannot collapse gravitationally if it is spinning rapidly. The object assumes a stationary orbit revolving about the inner parts instead of collapsing. Thus, the earth has not collapsed into the sun. Other examples of the spin hangup at work include galaxies, planetary systems, double stars, and the rings of Saturn. (3) Hydrogen "burns" to form helium when it is heated and compressed. But, this thermonuclear burning releases energy, which opposes any further compression. Thus, a star with a lot of hydrogen cannot collapse gravitationally beyond a certain point until the hydrogen is burned up. It is estimated that the sun has been "stuck" on this thermonuclear hangup for 4.5 billion years and will need another 5 billion years to burn hydrogen before its gravitational contraction can be resumed. (4) Whereas a thermonuclear bomb is made mainly of heavy hydrogen, the sun contains ordinary hydrogen with only a trace of the heavy hydrogen isotopes. Whereas heavy hydrogen can burn explosively by strong nuclear reactions, ordinary hydrogen can react with itself only by the weak-interaction process. This proton-proton reaction proceeds about 10^{18} times more slowly than a strong nuclear reaction at same density and temperature. At least three fortunate circumstances contribute to the weak-interaction hangup: (a) without it, there would not have been a long-lived and stable sun; (b) the ocean would constitute an excellent thermonuclear high explosive; (c) hydrogen has survived rather than having been consumed in the initial, hot, dense phase of the evolution of the universe. (4) Because the transport of energy from the hot interior of the earth to the surface requires billions of years, the earth remains geologically active, these processes deriving their energy from the original gravitational condensation of the earth estimated as some 4 billion years ago. (5) There also is a special surface tension hangup, accounting for the survival of fissionable uranium and thorium nuclei in the earth's crust. They contain a high positive charge and excessive electrostatic energy such that they are ready to explode when triggered. However, before this can happen their surface must be stretched into a nonspherical shape. This process is opposed by an extremely powerful force of surface tension, estimated at about 10^{18} times stronger than that of a drop of water. Thus, it is estimated that fewer than one in a million of the earth's uranium nuclei fission spontaneously.

Energy Technology

Breadth in the "Packaging" of Energy. In Fig. 1 a packet of energy of 1 joule (1 newton-meter) is represented by the box in the upper center. Various energy packets, ranging from 1 electron volt (10^{-19} joule) to the daily energy output of the sun (total—in all directions) of 10^{32} joules are indicated.

Perfecting contemporary energy resources and power generation and consumption and developing presently nonconventional energy resources and systems possibly pose the greatest challenge to scientists and technologists in the last quarter of this century. In particular, scientists and technologists must be encouraged to work better together in an effective and realistic fashion to create constructive solutions to the problem of the energy/environment interface.

There are numerous entries in this *Encyclopedia* on various energy topics. Consult the alphabetical index for such energy sources as coal, electric power, fuel cells, geothermal energy, hydroelectric power, hydrogen as a fuel, natural gas, nuclear power, oil shale, petroleum, solar energy, substitute natural gas and other synthetic fuels, tar sands, tidal energy, and waste materials. Also, a number of energy-converting and generating processes are described, including boilers and combustion, as well as energy-utilizing systems, such as diesel engines, gas and expansion turbines, internal combustion engines, steam engines, and steam turbines.

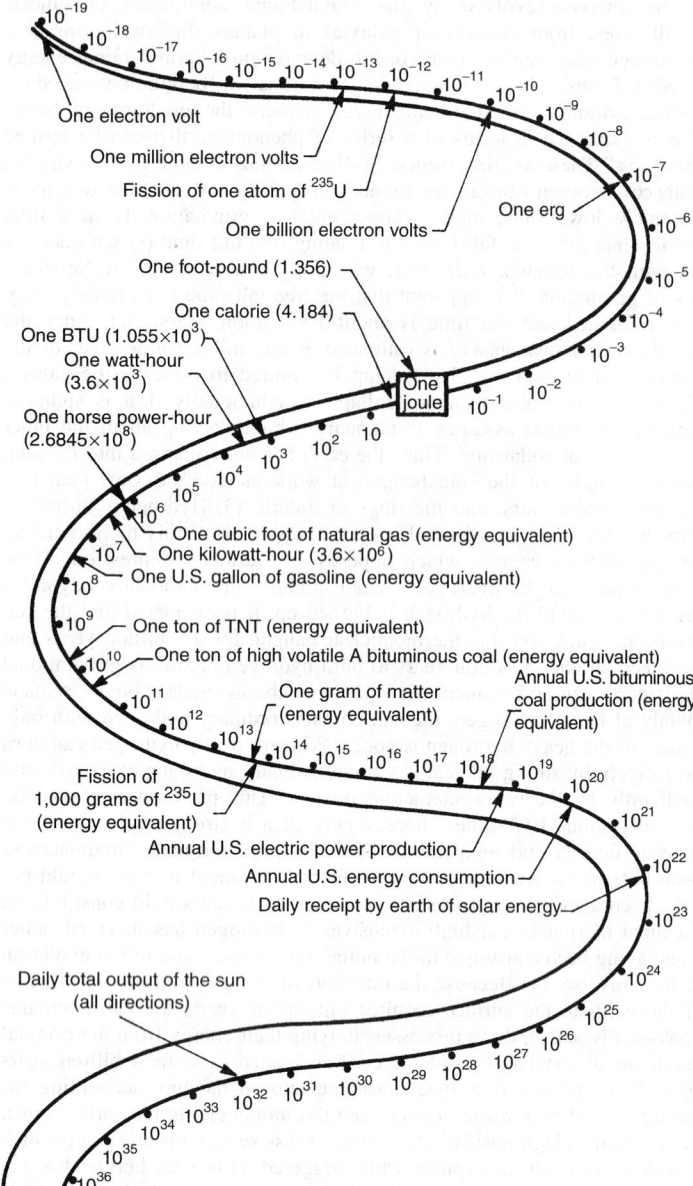

One electron volt
One million electron volts
Fission of one atom of ^{235}U
One billion electron volts
One erg
One foot-pound (1.356)
One calorie (4.184)
One BTU (1.055×10^3)
One watt-hour (3.6×10^3)
One horse power-hour (2.6845×10^6)
One cubic foot of natural gas (energy equivalent)
One kilowatt-hour (3.6×10^6)
One U.S. gallon of gasoline (energy equivalent)
One ton of TNT (energy equivalent)
One ton of high volatile A bituminous coal (energy equivalent)
One gram of matter (energy equivalent)
Annual U.S. bituminous coal production (energy equivalent)
Fission of 1,000 grams of ^{235}U (energy equivalent)
Annual U.S. electric power production
Annual U.S. energy consumption
Daily receipt by earth of solar energy
Daily total output of the sun (all directions)

Fig. 1.   Spectrum of various energy quantities. (*Source: Omnibix U.S.A.*)

Additional Reading

Beggs, P.: *Energy Management and Conservation,* Butterworth-Heinemann, Inc., Woburn, MA, 2001.

Bose, J., I. Fischer: *Geothermal Heat Pumps: Introductory Guide,* Ground Source Heat Pump Publications, Stillwater, OK, 1997.

Considine, D.M. (editor): *Energy Technology Handbook,* McGraw-Hill, New York, NY, 1976. (A classic reference.)

Laird, F.: *Solar Energy, Technology Policy and Institutional Values,* Cambridge University Press, New York, NY, 2001.

Leondes, C.T.: *Energy and Power Systems,* Gordon & Breach Publishing Group, Newark, NJ, 2000.

Mori, Y.H. and K. Ohnishi: *Energy and Environment: Technological Challenges for the Future,* Springer-Verlag, Inc., New York, NY, 2000.

Perlmutter, A., S.L. Mintz, and B.N. Kursunoglu: *Global Energy Demand in Transition: The New Role of Electricity,* Kluwer Academic Publishers, Norwell, MA, 1995.

Shepherd, D.W.: *Energy Studies,* World Scientific Publishing Company, Inc., River Edge, NJ, 2000.

Sieniutycz, S. and A. de Vos: *Thermodynamics of Energy Conversion and Transport,* Springer-Verlag, Inc., New York, NY, 2000.

Sorensen, B: *Renewable Energy,* Academic Press, Inc., San Diego, CA, 2000.

Staff: *U.S. Energy Outlook—Fuels for Electricity,* National Petroleum Council, Washington, DC. (Revised periodically.)

Staff: *Statistical Year Book on the Electric Utility Industry,* Edison Electric Institute, New York (Published annually). *http://www.eei.org/*

Staff: *Semi-Annual Electric Power Survey,* Edison Electric Institute, New York (Published semi-annually).

Thumann, A. and D.P. Mehta: *Handbook of Energy Engineering,* Prentice-Hall, Inc., New Jersey, 1997.

Turner, W.C.: *Energy Management Handbook,* Prentice-Hall, Inc., New Jersey, 1997.

Wheeler, R.A.: *Bioenergetics: Simulations of Electron, Proton, and Energy Transfer,* Oxford University Press, Inc., New York, NY, 2001.

Wulfinghoff, D.R.: *Energy Efficiency Manual,* Energy Institute Press, Edison, NJ, 2000.

Zumerchik, J.: *MacMillan Encyclopedia of Energy,* Vol. 2, Macmillan Publishing Company, Inc., New York, NY, 2000.

Web References

Association of Energy Engineers. *http://www.aeecenter.org/*
Atomic Energy Council. *http://www.aec.gov.tw/meco/e0.htm*
Department of Infrastructure, Energy and Resources. *http://www.dier.tas.gov.au/*
Energy Information Administration. *http://www.eia.doe.gov/*
EPRI—The Electric Power Research Institute. *http://www.epri.com/*
United States Geological Survey (USGS). *http://www.usgs.gov/*

ENERGY (Fuel Cell).   See **Fuel Cells.**

ENERGY LEVEL.   A stationary state of energy of any physical system. The existence of many stable, or quasi-stable, states, in which the energy of the system stays constant for some reasonable length of time, is an essential characteristic of quantum-mechanical systems, and is the basis of large areas of modern physics.

ENERGY STATE TERMS.   Terms designating the discrete energy states of a particle in a system. Thus the energy states of an atom are called S, P, D, F, ... terms, respectively, corresponding to the values 0, 1, 2, 3, ... of L, the resultant angular momentum quantum number of the atom. The energy states of a molecule are called Σ, Π, Δ, Φ, ... terms, respectively, corresponding to the values 0, 2, 3, ... of λ, the electronic orbital angular momentum (about internuclear axis) quantum number.

The letters indicating the value of L are usually preceded by a superscript denoting the multiplicity and followed by a subscript denoting the total angular momentum quantum number J. In addition, the principal quantum number is often written as a coefficient. Energy state terms and their transitions are shown in energy level diagrams.

Magnetic Energy State

A magnetic dipole of a moment μ in a magnetic field of flux density B has an energy that depends on orientation, $E = -\mu B \cos\theta$, or in vector notation $E = -\mu \cdot \mathbf{B}$, in mksa units ($-\mu \cdot \mathbf{H}$ in emu). In atomic and nuclear systems the orientation of μ relative to $\mathbf{B}$ is quantized, only certain values of $\cos\theta$ being allowed. Transitions between these allowed magnetic energy states may take place with the emission or absorption of electromagnetic (magnetic dipole) radiation of frequency given by the Bohr condition:

$$v = \Delta E/h, \quad \text{or} \quad \omega = \Delta E/h$$

Particles such as electrons, protons, nuclei, etc., have intrinsic magnetic moments $\mu = eghl/2M$, where $\mathbf{l}h$ is the spin angular momentum, g the g-factor of the particle, and M is the mass of the electron or the mass of the proton. The magnetic energy states are thus given by $E = -heg\mathbf{l} \cdot \mathbf{B}/2M = -hegmB/2M$. Here m is the magnetic quantum number, which can take on the values $-l$, $-(l-1)$, ... $(l-1)$, l where l is the spin quantum number ($\frac{1}{2}$ for electrons and protons). The energy is also often written $E = -\mu_B gmB$ or $-\mu_N gmB$ where μ_B and μ_N are the Bohr magneton and nuclear magneton respectively. The magnetic quantum number can only change by ± 1 as a result of the emission of radiation, so that there is only one emission or absorption frequency $\omega = egB/2M$.

Negative Energy State

1. Any bound state, in which the sum of the kinetic energy and the potential energy, the latter reckoned relative to zero at infinity, is less than zero. The existence of such states is essential for the stability of any system that is not surrounded by a region of positive potential energy, such as the Coulomb barrier.

2. A consequence of the Dirac electron theory is that there exist electron states of negative total energy (including both rest mass energy and kinetic energy). Electrons in such states of negative energy are unobservable, only electrons of positive total energy being observable. The allowed

positive and negative states are shown in the diagram (only $E > m_0c^2$ and $E < -m_0c^2$ are allowed in a field-free region). If a γ-ray photon of energy greater than $2m_0c^2$ (where m_0 is the rest mass energy of the electron) is absorbed by an electron of negative energy, it will be lifted into a positive energy state and will become observable. The positron is identified with the hole that is left behind.

ENKEPHALINS AND ENDORPHINS. During the past decade of brain research, the number of chemical-messenger systems identified has increased dramatically. This advance is highlighted by the discovery of a large family of brain chemicals known as *neuropeptides*. As shown by Fig. 1, these molecules are made up of long chains of amino acids, ranging from a few to as many as 39 amino acids. Research indicates that these substances are resident within the neutrons. A few of these substances, such as corticotropin (ACTH) and vasopressin, have been known for many years and identified with the hypothalamus and pituitary gland. Probably of greatest interest to researchers have been the *enkephalins* and *endorphins*. These chemicals are strikingly similar to the opiate morphine.

Identification of these substances followed the finding that specific regions of the brain possess receptor sites that bind opiates with a high affinity. These sites were revealed through the use of radioactively labeled opiate compounds. Further research showed that opiate receptors are located in those regions of the brain and spinal cord which are known to be associated with emotion and pain. Pioneering research in the field included the work of Snyder and Pert (Johns Hopkins University School of Medicine and Terenius (University of Uppsala). The enkephalins (Met- and Leu-), as shown in Fig. 1, were first isolated by Hughes and Kosterlitz (University of Aberdeen) in 1975. Each enkephalin contains five amino acids, one of these being methionine in one case, and leucine in the other

case. Shortly after the isolation of these compounds, the endorphins, also morphinelike compounds, were isolated from the pituitary gland. Some researchers have suggested that some of the non-traditional methods used for relieving chronic pain, such as acupuncture, direct electrical stimulation of the brain, and possibly hypnosis, may be effective because these procedures may cause enkephalins or endorphins or both to be released to the brain and spinal cord. The drug naloxone (*Narcan®*) blocks the binding of morphine and experiments have shown that naloxone also blocks the effects of the aforementioned pain-relieving procedures; hence the tentative hypothesis.

Early research findings indicate that neuropeptides are released from axon terminals through the presence of calcium ions known to be the releasing mechanism in connection with established transmitters. Thus, some investigators believe that the neuropeptides also may be transmitters. This is particularly true of a compound simply identified as *substance P*. See Fig. 1. Some investigators have found that substance P is associated with spinal neurons involved in pain stimuli.

One of the surprising findings of recent brain and central nervous system research is that chemical substances previously considered to be exclusive to the province of the brain have been found in organs outside the nervous system (e.g., somatostatin, neurotensin, and enkephalins have been found in the gut), and conversely, that other substances active in other organs, but not previously associated with the central nervous system, have been found in the latter (e.g., gastrin, vasoactive intestinal polypeptide (VIP), and cholecystokinin, traditionally associated with the gastrointestinal tract, but now found in the central nervous system).

Also among recent discoveries in brain chemistry are the so-called *trophic substances*, which are believed to be secreted from nerve terminals. One of these is *nerve growth factor* (NGF). It has been established that

Tyr-Gly-Gly-Phe-Met
(Met-Enkephalin)

Tyr-Gly-Gly-Phe-Leu
(Leu-Enkephalin)

Arg-Pro-Lys-Pro-Gln-Gln-Phe-Phe-Gly-Leu-Met-NH$_2$
(Substance P)

p-Glu-Leu-Tyr-Glu-Asn-Lys-Pro-Arg-Arg-Pro-Tyr-Ile-Leu
(Neurotensin)

Asp-Arg-Val-Tyr-Ile-His-Pro-Phe-NH$_2$
(Angiotensin II)

Tyr-Gly-Gly-Phe-Met-Thr-Ser-Glu-Lys-Ser-Gln-Thr-Pro-Leu-Val-Thr-Leu-Phe-Lys-Asn-Ala-Ile-Val-Lys-Asn-Ala-His-Lys-Lys-Gly-Gln
(Beta-Endorphin)

Ser-Tyr-Ser-Met-Glu-His-Phe-Arg-Tyr-Gly-Lys-Pro-Val-Gly-Lys-Lys-Arg-Arg-Pro-Val-Lys-Val-Tyr-
Pro-Asp-Gly-Ala-Glu-Asp-Glu-Leu-Ala-Glu-Ala-Phe-Pro-Leu-Glu-Phe
(ACTH, Corticotropin)

Asp-Tyr-Met-Gly-Trp-Met-Asp-Phe-NH$_2$
(Cholecystokinin-Like Peptide)

His-Ser-Asp-Ala-Val-Phe-Thr-Asp-Asn-Tyr-Thr-Arg-Leu-Arg-Lys-Gln-Met-Ala-Val-Lys-Lys-Tyr-Leu-Asn-Ser-Ile-Leu-Asn-NH$_2$
(Vasoactive Intestinal Polypeptide, VIP)

p-Glu-His-Pro-NH$_2$
(Thyrotropin Releasing Hormone, TRH)

p-Glu-His-Trp-Ser-Tyr-Gly-Leu-Arg-Pro-Gly-NH$_2$
(Luteinizing-Hormone Releasing Hormone, LHRH)

Ala-His
(Carnosine)

p-Glu-Gln-Arg-Leu-Gly-Asn-Gln-Trp-Ala-Val-Gly-His-Leu-Met-NH$_2$
(Bombesin)

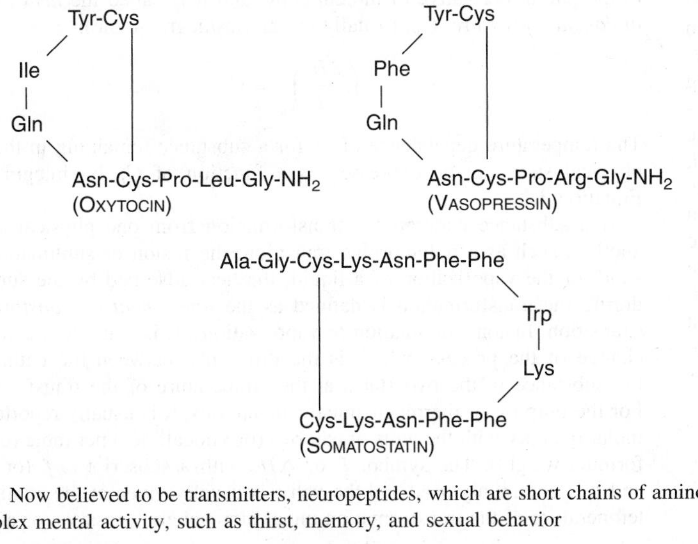

Tyr-Cys
/
Ile
|
Gln
\
Asn-Cys-Pro-Leu-Gly-NH$_2$
(Oxytocin)

Tyr-Cys
/
Phe
|
Gln
\
Asn-Cys-Pro-Arg-Gly-NH$_2$
(Vasopressin)

Ala-Gly-Cys-Lys-Asn-Phe-Phe
|
Trp
|
Lys
Cys-Lys-Asn-Phe-Phe
(Somatostatin)

Abbreviations of Amino Acids

Ala	Alanine	Leu	Leucine
Arg	Arginine	Lys	Lysine
Asn	Asparagine	Met	Methionine
Asp	Aspartic Acid	Phe	Phenylalanine
Cys	Cysteine	Pro	Proline
Gln	Glutamine	Ser	Serine
Glu	Glutamic Acid	Thr	Threonine
Gly	Glycine	Trp	Tryptophan
His	Histidine	Tyr	Tyrosine
Ile	Isoleucine	Val	Valine

Fig. 1.   Now believed to be transmitters, neuropeptides, which are short chains of amino acids found in brain tissue and notably localized in axon terminals, participate in complex mental activity, such as thirst, memory, and sexual behavior

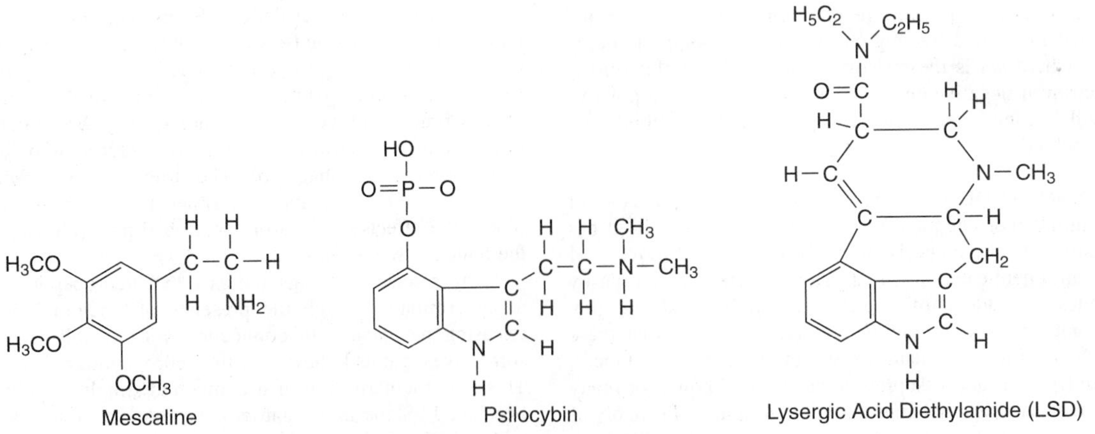

Fig. 2. Representative hallucinogenic drugs which bear structural similarities to some of the monoamine transmitters. It has been hypothesized that these similarities may cause the hallucinogens to mimic natural transmitters at synaptic receptors in the brain. Note presence of benzene-ring structure in these substances, a structure that is present in four out of the five monoamines previously shown. Also note presence of indole ring in psilocybin and lysergic acid diethylamide, a structure that is also present in the monoamines serotonin and histamine

this protein is required for the differentiation and survival of peripheral sensory and sympathetic neurons. Another benefit of recent research is a better understanding of how psychoactive drugs interact with the brain and central nervous system. See Fig. 2.

Additional Reading

Barchas, J.D., et al.: "Behavioral Neurochemistry: Neuroregulators and Behavioral States," *Science*, **200**, 964–973 (1978).
Costa, E., and M. Trabucchi, Editors: *The Endorphins*, Raven, New York, NY, 1978.
Iversen, L.L.: "The Chemistry of the Brain," *Sci. Amer.*, **241**(3), 134–149 (1979).
Snyder, S.H.: "Brain Peptides as Neurotransmitters," *Science*, **209**, 976–983 (1980).

ENRICHMENT. 1. Also "secondary enrichment." The term applied by students of ore deposits to the natural processes by which the lower levels of an ore deposit are enriched at the expense of the upper levels, or the original protore. Particularly applied to lodes in which the sulfide ores have been concentrated by the leaching of the upper levels of the vein and redeposition below the groundwater table. Important ore minerals belonging to this type are chalcocite and argentite.

2. Any process which changes the isotopic ratio; in reference to uranium, it is a process that increases the ratio of ^{235}U to ^{238}U in uranium by separation of isotopes.

3. In food technology, the addition to a foodstuff of various nutrient substances during manufacture to increase the dietary value of the food, e.g., addition to wheat flour of vitamins B_1, B_2, niacin, and iron. In this way, the food is brought up to specific nutritional standard.

4. Addition of oxygen to air to increase it combustion-supporting ability.

ENSTATITE. The mineral enstatite is an orthorhombic pyroxene, rarely in distinct crystals, usually found as fibrous or lamellar masses or perhaps compact. It has one easy cleavage parallel to the prism; brittle with uneven fracture; hardness 5–6; specific gravity 3.2–3.4; luster pearly to vitreous, sometimes somewhat metallic in bronzite, a variety of enstatite carrying up to 15% ferrous oxide, FeO. Color grayish to greenish or yellowish-white, green and brown. Chemically, enstatite is a silicate of magnesium, $MgSiO_3$. It occurs in igneous rocks which are high in magnesium content, like gabbros, diorites, and pyroxenites, and less commonly in metamorphic rocks. Meteorites of both the stony and metallic types have been shown to contain enstatite. It has been found at many places in Europe, (the former Czechoslovakia, Austria, Bavaria, Germany, Norway), and the Republic of South Africa. In the United States it occurs in Putnam and St. Lawrence Counties, New York; Lancaster County, Pennsylvania; Jackson County, North Carolina, and near Baltimore, Maryland. The name enstatite is derived from the Greek word meaning *opponent*, in reference to its refractory nature; it is almost infusible. See also **Pyroxene**.

ENTER-DOUDOROFF PATHWAY. See **Carbohydrates**.

ENTEROVIRUSES. See **Virus**.

ENTHALPY. The *enthalpy, H* or *heat content*, of a substance is a thermodynamic property defined as the *internal energy, E*, plus the product of the pressure, *P*, times the *volume, V*, of the substance

$$H = E + PV \qquad (1)$$

The enthalpy is an extensive state function; its value depends only on the state and the amount of the substance and not on its previous history. It has the units of energy and it is usually expressed in calories (or kilocalories).

For a process at *constant pressure* ($\Delta P = 0$), in which the only work performed is the mechanical pressure-volume work ($P\Delta V$), the *change in enthalpy*, ΔH, is equal to the heat adsorbed by the system, q (hence the name heat content):

$$\Delta H = \Delta E + P\Delta V = q \qquad (2)$$

This relation is a direct consequence of the definition of enthalpy by Equation (1) and of the mathematical statement of the first law of thermodynamics, namely that the change in internal energy, ΔE, is equal to the heat adsorbed minus the work done ($q - P\Delta V$). It is clear that this thermodynamic relation does not define absolute values of enthalpy or internal energy. Changes in enthalpy, however, are readily measured by calorimetric techniques, and the relative enthalpy values are sufficient for all thermochemical calculations.

Enthalpy-Temperature Relation and Heat Capacity

When heat is adsorbed by a substance, under conditions such that no chemical reaction or state transition occur and only pressure-volume work is done, the temperature, T, rises and the ratio of the heat adsorbed, over the differential temperature increase, is by definition the heat capacity. For a process at constant pressure (following Equation (2)), this ratio is equal to the partial derivative of the enthalpy, and it is called the *heat capacity at constant pressure*, C_p, (usually in calories/degree-mole):

$$\left(\frac{\partial H}{\partial T}\right)_p = C_p \qquad (3)$$

The temperature dependence of H for a substance remaining in the same physical state can be expressed as a function of C_p by integration of Equation 3.

If a substance undergoes a transformation from one physical state to another, such as a polymorphic transition, the fusion or sublimation of a solid, or the vaporization of a liquid, the heat adsorbed by the substance during the transformation is defined as the *latent heat of transformation* (transition, fusion, sublimation or vaporization). It is equal to the enthalpy change of the process, which is the difference between the enthalpy of the substance in the two states at the temperature of the transformation. For the purpose of thermochemical calculations, it is usually reported as a molar quantity with the units of calories (or kilocalories) per mole (or gram formula weight). The symbol L or ΔH, with a subscript t, f (or m), s, and v is commonly used and the value is usually given at the equilibrium temperature of the transformation under atmospheric pressure, or at 25°C.

For a substance undergoing one phase transformation, with a latent heat Δh_t, at a temperature T_t, the enthalpy change between two temperatures, T_1 and T_2, such that $T_1 < T_t < T_2$, is given by

$$H_T - H_T = \int_{T_1}^{T_1} C'_p\,dT + \Delta h_t + \int_{T_1}^{T_2} C'_p\,dT \qquad (4)$$

where C'_p and C''_p are the heat capacities of the substance in the two different physical states. For several successive transformations, additional terms are added. Fig. 1 illustrates the temperature dependence of enthalpy and heat capacity.

Very precise measurements of the heat capacity of liquids and solids can be obtained by calorimetric techniques at relatively low temperatures (below 200°C) and they can be extrapolated down to the absolute zero of temperature (-273.15°C) by reliable theoretical expressions. In that temperature range, heat capacity data are usually very accurate and enthalpy values are obtained by integration (Equation (4)). The most reliable method for determining high-temperature enthalpies and heat capacities is the dropping method (or method of mixtures) which consists of dropping the substance under investigation from a furnace at a known temperature into a calorimeter at room temperature. This method determines directly the change in enthalpy (or heat content) of the substance between the temperature of the furnace and that of the calorimeter. Heat capacities are obtained by differentiation (Equation (3)). The measurement of heat capacity of gases is usually more difficult, and their thermodynamic properties can be more accurately calculated by methods of statistical mechanics based upon energy level of gas molecules obtained from spectroscopic data, or upon the knowledge of the molecular configuration and the vibration frequencies of the molecules.

Molar enthalpy data for elements and inorganic compounds above room temperature are usually tabulated in the form of the heat content above a reference temperature, usually 298.15°K = 25°C. They are represented by: $H_T - H_{298.15}$ in calories/mole. The data are correlated over a range of temperature by empirical equations such as a series of powers of the absolute temperature or such as the following expression adopted by K.K. Kelley (1960) for his extensive compilation of data on inorganic compounds:

$$H_T - H_{298.15} = aT + bT^2 + c/T + d \qquad (5)$$

where T is the absolute temperature (°K) and a, b, c, d are constants determined from experimental data. The corresponding equation for heat capacity is:

$$C_p = a + 2bT - c/T^2 \qquad (6)$$

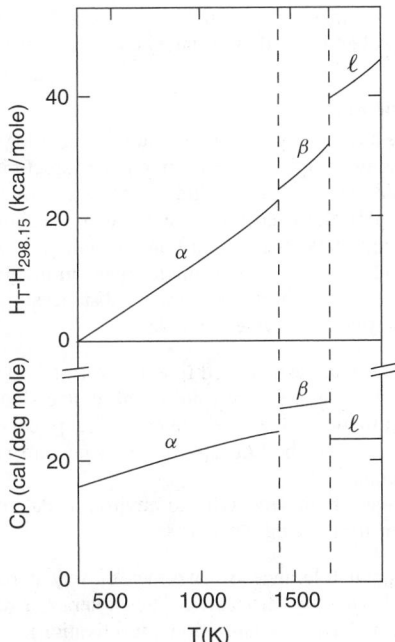

Fig. 1.   Example of the temperature dependence of enthalpy relative to 25°C, $H_T - H_{298.15}$, and heat capacity, C_p. The data are for fluorite, CaF_2 (*K.K. Kelly, 1960*). The discontinuities in the lines correspond to the α to β transition (1424 K) and the fusion (1691 K)

Standard Enthalpy of Formation

For the convenience of tabulation and computation of thermodynamic data, it is essential to present them in a commonly accepted form relative to a single standard state of reference. At all temperatures, the *standard state* for a *pure liquid or solid* is the *condensed phase under a pressure of 1 atmosphere*. The standard state for a *gas* is the *hypothetical ideal gas at unit fugacity* (equivalent to a "perfect gas" state), in which state the enthalpy is that of the real gas at the same temperature when the pressure approaches zero. Values of thermodynamic quantities for standard-state conditions are identified by a superscript ^{0}, and H^0, for instance, is the enthalpy change of a reaction when reactants and products are in the standard state.

The *standard enthalpy of formation*, ΔH_f^0 (also represented by ΔH_f^0 or simply H_f^0), of a substance at a given temperature is by definition, the enthalpy change when 1 mole of the substance in its standard state is formed, isothermally, at the indicated temperature from the elements, each in its standard state. Usual units are kilocalories/mole. *For all elements* in their *stable form at* 25°C (298.15 K), the *enthalpy of formation is zero*. If solid substances have more than one crystalline form, the most stable one is taken as the standard state, and the others have slightly different enthalpies. This convention about zero enthalpy is arbitrary but universally accepted, and it may be compared to the arbitrary choice of zero for terrestrial altitudes. The combination of enthalpies of formation, enthalpies of transition, and heat capacities makes possible the calculation of the enthalpy of a substance, in a given state at a given temperature, relative to a commonly accepted reference.

Enthalpy calculation for mixtures is more complex than for pure substances and its discussion is beyond the scope of the present article. *Aqueous solutions* (q.v.), however, are very important from a geochemical point of view, and reliable data are usually available. The *enthalpy of solution* is the enthalpy change resulting from the dissolution of a substance; it is a function of the solute concentration and its values are reported accordingly in the literature. For a *solute in aqueous solution*, a standard state is defined as the *hypothetical ideal solution of unit molality* (1 mole of solute per 1000 grams of water). In this state, the partial molal enthalpy and heat capacity of the solute are the same as in the infinitely dilute real solution. Although it is impossible to prepare a solution of only one ionic species (since the system must remain neutral), it is convenient to apportion the enthalpy (and other thermodynamic properties) between the various ions. This appointment is not unique and an additional convention has to be made, namely that the *standard enthalpy of hydrogen* ion in aqueous solution (aq) at unit activity, ΔH_f^0 for H^+ (aq), is *zero*. The properties of a neutral electrolyte in the standard state are equal to the algebraic sum of the values corresponding to the individual ions.

Heat of Reaction and Gibbs Free Energy Change

For a chemical reaction, at constant pressure with only pressure-volume work performed, the heat adsorbed by the process, q, is equal to the enthalpy change, ΔH, or the sum of enthalpies of the products of the reaction minus the sum of enthalpies of the reactants (taking into account the amount of each).

$$q = \Delta H = \Sigma H_{\text{products}} - \Sigma H_{\text{reactants}} \qquad (7)$$

Those heat effects can be easily calculated when the enthalpies of formation and the enthalpy-temperature relations are available for the substances considered. Usually, the *heat of reaction* is defined as the heat evolved by the process, and it is equal to the enthalpy change but opposite in sign, while heats of fusion or vaporization always refer to the heat adsorbed, and for heats of solution the usage varies. In order to avoid any confusion, it is recommended to express heat effects of chemical process by reporting the enthalpy change, ΔH.

Early chemists thought that the heat of reaction, $-\Delta H$, should be a measure of the "*chemical affinity*" of a reaction. With the introduction of the concept of *netropy* (q.v.) and the application of the second law of thermodynamics to chemical equilibria, it is easily shown that the true measure of chemical affinity and the driving force for a reaction occurring at constant temperature and pressure is $-\Delta G$, where ΔG represents the change in thermodynamic state function, G, called *Gibbs free energy* or *free enthalpy*, and defined as the enthalpy, H, minus the entropy, S, times the temperature, T ($G = H - TS$). For a chemical reaction at constant pressure and temperature:

$$\Delta G = \Delta H - T\Delta S \qquad (8)$$

and the Gibbs free energy change can be obtained by calculating the enthalpy and entropy change and applying Equation (8). The criterion for *a spontaneous chemical reaction* is that ΔG be *negative*, and a chemical equilibrium corresponds to the condition $\Delta G = 0$.

Conversely, if the Gibbs free energy change is known as a function of temperature at constant pressure, the enthalpy change can be obtained by a relation which is an alternate form of the *Gibbs-Helmholtz equation*, and which can be derived from Equation (8).

$$\left(\frac{\partial(\Delta g/T)}{\partial(1/T)} \right)_P \Delta H \qquad (9)$$

This means that ΔH is the slope of the line representing $\Delta G/T$ versus $1/T$ at constant pressure.

See also **Thermochemistry**.

For references, see entries on **Heat Transfer**; and **Thermodynamics**.

ENTROPY.

1. In the mathematical treatment of thermodynamic processes there occurs very often a quantity, now relating energy to absolute temperature, now associated with the probability of a given distribution of momentum among molecules, and again expressing the degree in which the energy of a system has ceased to be available energy. Its mathematical form suggests that these are all aspects of a single physical magnitude. Application of the second law of thermodynamics leads to the conclusion that if any physical system is left to itself and allowed to distribute its energy in its own way, it always does so in a manner such that this quantity, called "entropy," increases. At the same time the available energy of the system diminishes. This law applies to the universe as a whole, hence the proposition that the total entropy increases as time goes on. An interesting conclusion as to entropy in the vicinity of absolute zero is expressed by the Nernst heat theorem; viz, that all physical and chemical changes in this region take place at constant entropy. Any process during which there is no change of entropy is said to be "isentropic." This is true, for example, of an adiabatic process in which there is no dissipation of energy, i.e., one which is also a reversible process. In thermodynamics discussions, entropy is commonly classed, along with temperature, pressure, and volume, as one of the variables defining the state of a body, and is often graphed as such on thermodynamic diagrams.

2. In information theory, entropy is a measure of the uncertainty of our knowledge.

3. In thermodynamics, entropy is defined by the equation

$$dS = dQ/T$$

where dS is an infinitesimal change in the entropy of a system, dQ is the infinitesimal amount of heat that enters the system, and T is the absolute temperature.

In statistical mechanics, entropy is

$$k \log_e P + \text{constant}$$

where k is Boltzmann's constant, and P is the statistical probability of the state considered.

Standard Entropy

The total entropy of a substance in a state defined as standard. Thus, the standard states of a solid or a liquid are regarded as those of the pure solid or the pure liquid, respectively, and at a stated temperature. The standard state of a gas is at 1 atmosphere pressure and specified temperature, and its standard entropy is the change of entropy accompanying its expansion to zero pressure, or its compression from zero pressure to 1 atmosphere. The standard entropy of an ion is defined in a solution of unit activity, by assuming that the standard entropy of the hydrogen ion is zero.

Entropy of Disorder

That part of the entropy of a substance that is due to a disordered arrangement of the particles as opposed to a similar but ordered arrangement. The most clear-cut example is the order-disorder transition in binary alloys, in which virtually the whole entropy change is of this kind. The entropy change on fusion of a solid is largely due to entropy of disorder.

See also **Energy**.

ENVIRONMENTAL CHEMISTRY. That aspect of chemistry concerned with air and water pollution, pesticides, and chemical and radioactive waste disposal. A random selection of specific areas of research includes. (1) Lead and other toxic chemicals in the air. (2) Effects of increased burning of coal, biological modification of wastes. (3) Detoxification methods, pesticide content of fish, environmental analytical and monitoring techniques. (4) Utilization of biomass, drinking water quality, organic contaminants in lakes and rivers, and effect of deforestation on carbon dioxide and oxygen content of air.

ENZYME. An enzyme is a protein that serves as a catalyst for a particular biological transformation—as, for example, the conversion of sugar into alcohol and water. Because of the make-up of the genetic material, most enzymes are highly specific. As discussed in the article on **Industrial Biotechnology**, this specificity is very advantageous in bioprocessing.

Fermentation, one of the most common transformations to be accomplished with the aid of an enzyme as catalyst, has been known for about 4000 years, mainly in connection with brewing, winemaking, and dairy products, such as cheese and yogurt. It was not until the early 1600s that the concept of an enzyme was recognized. Well over 300 years passed by, however, before the first enzyme to be isolated, *urease*, was produced in crystalline form. Shortly thereafter, numerous other enzymes were isolated in pure form, including amylase, carboxy-peptidase, chymopapain, papain, pepsin, and starch phosphorylase. Today, there are many hundreds of known enzymes with many specific purposes. Enzymes, once created from microorganisms right in the fermenting vessel, can in some instances be purchased in pure form for addition to the fermentation batch vessel. As mentioned in the article on **Enzyme Preparations**, purified enzymes are widely used in the food industry for enhancing flavor and stabilization of food quality and, among other uses, are compounded in packaged detergents. Major classes of enzymes include the oxidoreductases, transferases, hydrolases, lyases, isomerases, and ligases or synthetases, their names indicative of their functions.

Sources of Enzymes

Enzyme complexes are generated by living cells, notably yeasts, molds, bacteria, and actinomycetes. Enzymes are involved in numerous biological transformations, as in the metabolism of living organisms, and thus play a vital role at practically all levels of food involvement—production, processing, and consumption, whether by fish, bird, insect, or primate. Enzymes and the transformations which they promote are ever present during the entirety of the food chain. Investigations in botany, pursuits of agronomy, studies of nutrition, inquiries into plant and animal pathology, and the numerous other aspects of science that are involved in life processes, when probed in depth, ultimately encounter the vital roles played by enzymes.

Characteristics of Enzymes

Although with the advent of gene recombination technology the knowledge of enzymes is gaining rapidly, their principal characteristics have been established for decades. These include sensitivity to the environment (temperature and pH), the need for a clean watery medium (solvent), the requirements for nutrients, such as carbon (for energy), oxygen (absence or presence depending upon whether microorganism involved is anaerobic or aerobic), nitrogen, phosphorus, and trace substances.

Common properties of enzymes include:

1. Their predominant, established role as catalysts, often providing the means of effecting chemical (biological) conversions that otherwise would be difficult and at lower rates of energy expenditure.
2. Their structure which suggests that enzymes are simple or conjugated proteins.
3. Their relatively high sensitivity to environmental conditions.
4. Their origin from living cells.

The environmental tolerance of enzymes closely parallels other substances associated with live processes. They tolerate a relatively narrow temperature span and with denaturation (deactivation) occurring at temperatures generally above 50 C (122 F). Greatly reduced activity usually occurs well above the freezing point of water. Enzymes have a low tolerance to a pH below 4.0, and a minimal to no tolerance of certain organic solvents (alcohol, acetone, etc.), and destruction by numerous organic and inorganic substances.

Unlike most inorganic catalysts, enzymes are very specific for the transformations they catalyze. An acid catalyst, for example, will yield glucose, fructose, and galactose in the hydrolysis of raffinose (a trisaccharide). But, the enzyme diastase will yield melibose and fructose; emulsin will yield sucrose and galactose. The glucosidic linkages are hydrolyzed at about equal rates with an acid catalyst, whereas the enzyme catalysts act on just one kind of linkage even though the difference in linkages is small. Whereas acids may catalyze numerous compounds, including amides, acetals, and esters, a given enzyme will confine its actions to a very specific compound or closely related group. This characteristics adds very much to the efficiency of the enzymes when used in bioprocesses.

Another advantage of enzymes relates back to their microorganism precursors. Through the application of genetic recombination technology, enzyme-source microorganisms can be customized, that is, a wild bacterium or fungus can be manipulated to call for more desirable properties and the elimination of undesirable characteristics. Mutation is one way to bring this about. As pointed out by Hopwood (see reference), in point mutation, one can change one base pair (example: adenine-thymine to guanine-cytosine; or a base pair on a short stretch of DNA may be deleted from a sequence. Such (spontaneous) changes occur naturally, but they happen rarely (one in a million). This frequency of mutation can be multiplied hundreds of time by exposing microorganisms to mutagenic x-rays, gamma rays, or neutrons. With this technique, one can hope to find the desired mutant by examining only hundreds or a few thousands of samples instead of millions.

Suitability for Bioprocesses

In addition to the very large role that enzymes play in life processes and medicine and in industrial fermentation and related processes, enzymes are finding a growing role in industrial products, such as detergents, where enzymes tend to break down proteins to water-soluble proteoses or peptones. Enzymes for such use must remain active at relatively high pH values (8.5 to 9.5) and remain stable for a long product shelf life. See also **Detergents**.

The great number of reactions catalyzed by the enzymes in living organisms can be indicated by mentioning some of the major types. They include: all the oxidation processes by which the organism obtains its energy—mechanical and thermal; the hydrolysis processes by which food carbohydrates, proteins, and fats are broken down into simpler molecules capable of direct oxidation or of use by the organism in constructing its own structure; and all the detoxification reactions by which many harmful substances that may be absorbed by the organism, as well as its normal waste products, are converted into forms suitable for excretion.

Activators

Most enzymes can function only with the assistance of certain other substances. These are broadly designated as *activators*, and are commonly grouped into two classes. The first is that of the nonspecific activators, which take no part in the conversion and appear to act by their effect upon the enzyme itself. The most important of these are the metallic ions K^+, Na^+, Rb^+, Cs^+, Mg^{2+}, Ca^{2+}, Al^{3+}, Zn^{2+}, Cd^{2+}, Cr^{2+}, Mn^{2+}, Fe^{2+}, Co^{2+}, Ni^{2+}, Cu^{2+}. The second class of activators, mentioned earlier in this entry, are organic molecules, which enter into the conversion itself, often as carriers of a particular group. These substances are the group discussed in the entry on **Coenzymes**. In general, they are regenerated in their original form by other processes, so that they are not strictly substrates. On the other hand, the nicotinamideadenine nucleotides, which act as hydrogen carriers for various oxidoreductase reactions, may well be regarded as substrates.

A *substrate* may be defined as a substance modified by the action of an enzyme, or by the growing upon it of microorganisms. A *coenzyme* is a low-molecular-weight organic substance that can attach itself and thus supplement specific proteins to form active enzyme systems.

Inhibitors and Primers. Two other adjunct substances are the *inhibitors*, which retard or block enzyme action; and the *primers*, which enhance, or in some cases, are essential to it. An example is the priming of polyribonucleotide phosphotransferase by short ribonucleotide polymers. See also **Enzyme Inhibitors**.

Structure of Enzymes. The knowledge of the structure of enzymes is growing at a rapid rate, but much research remains before a high confidence level can be established pertaining to even the fundamentals of certain basic conversions.

Molecular Biology of Enzymes

One of the early examples of molecular biology which occurred after the 1950s when the DNA molecular configuration was established, but before restriction enzymes were discovered (early 1970s) was the work of researchers during the late 1950s and early 1960s on investigating the enzyme *ribonuclease*. The systematic name of this enzyme is polyribonucleoside-2-oligonucleotide-transferase. This enzyme transfers a phosphate group from one position to another within a polynucleotide, forming a cyclic compound, and the pancreatic form of this enzyme can also catalyze the transfer of the phosphate group to water, which is a step in the depolymerization of RNA.

The molecular weight of this enzyme was found (by analysis of its constituent amino acids) to be 13,700. It was found to consist of a single polypeptide chain, internally cross-linked by four cystine residues, as evidenced by the lack of any drop in molecular weight to accompany the oxidation of all four cystines to cysteic acid, and also by the occurrence of only one terminal $-NH_2$ group, and one terminal $-COOH$ group.

From this point, the primary structure was fully determined. The *primary structure* of an enzyme, or other protein, is the number, length, and composition of the polypeptide chains, the linear arrangement of their amino acids, and the number and position of the cross-links between chains. (The geometrical configuration of the molecule, which is usually a three-dimensional coiled and folded structure, and the side chains and their interactions were not determined.)

To determine the primary structure, after oxidation of the disulfide bridges between the cysteine molecules, the enzyme was cleaved into a series of linear polypeptides by the enzymatic action of trypsin. The fragments were separated by chromatography, and their individual amino acids were split off by acid hydrolysis. Then by repeating this process with another enzyme, a different series of polypeptide fragments were obtained, because the chains split at different points. By studying the overlaps of the two series, the fragments could be arranged in linear order. Combining the peptide structure so determined with the amino acids found gave the provisional primary structure of the enzyme shown in Fig. 1.

The final step in complete elucidation of the three-dimensional structure of ribonuclease was made by researchers at the Roswell Park Memorial Institute, Buffalo, New York. This group, headed by Dr. David Harker, employed x-ray diffraction techniques. Roughly 500,000 diffraction points were recorded and the data so obtained was fed to a computer.

The elucidation of the structure of ribonuclease follows that of lycozyme by a group at London's Royal Institute headed by Dr. David C. Phillips, and that of the other protein, myoglobin, for which Dr. Max F. Perutz and Dr. John C. Kendrew of Cambridge University received the Nobel Prize in chemistry in 1962.

Note in this structure of enzyme that specific points are marked as those at which other enzymes act. This feature of "active sites" is characteristic of enzyme behavior. In the case of the enzymes trypsin or chymotrypsin, the active sites for peptide or ester hydrolysis contain the functional groups of two histidine residues and of a serine residue. The ester enters the active region, forming temporary bonds with the enzyme at that point, the $-OR$ group of the ester becoming bonded to a hydrogen atom of the enzyme. Then the bond between the hydrogen atom and the enzyme breaks, releasing the alcohol of the ester. As a next step, H_2O adds from the solution to the complex, and by another bond rupture, the acid part of the ester is released, leaving the enzyme in its original condition. The overall reaction is a simple hydrolysis of the ester,

$$R'COOR + H_2O \longrightarrow ROH + R'COOH$$

but a large number of steps may be involved.

The determination of sites of active centers is effected not only by splitting of enzymes, but also by treating them with temporary or permanent inhibitors, and determining their points of attachment. Other methods of studying enzymes are by means of *enzyme induction* and *enzyme repression*.

Enzyme Induction

An example of *enzyme induction* is the growth of the bacteria *Escherichia coli* in a suitable culture medium. If no beta-galactoside is added to the medium, the bacteria form scarcely any of the enzyme that hydrolyzes that sugar. The addition of the sugar to the medium increases the production of the enzyme by the cell by as much as 10,000 times. On the other hand, the same bacteria will produce the enzyme tryptophan synthase only if tryptophan is absent from the culture medium. These observations are useful, not only

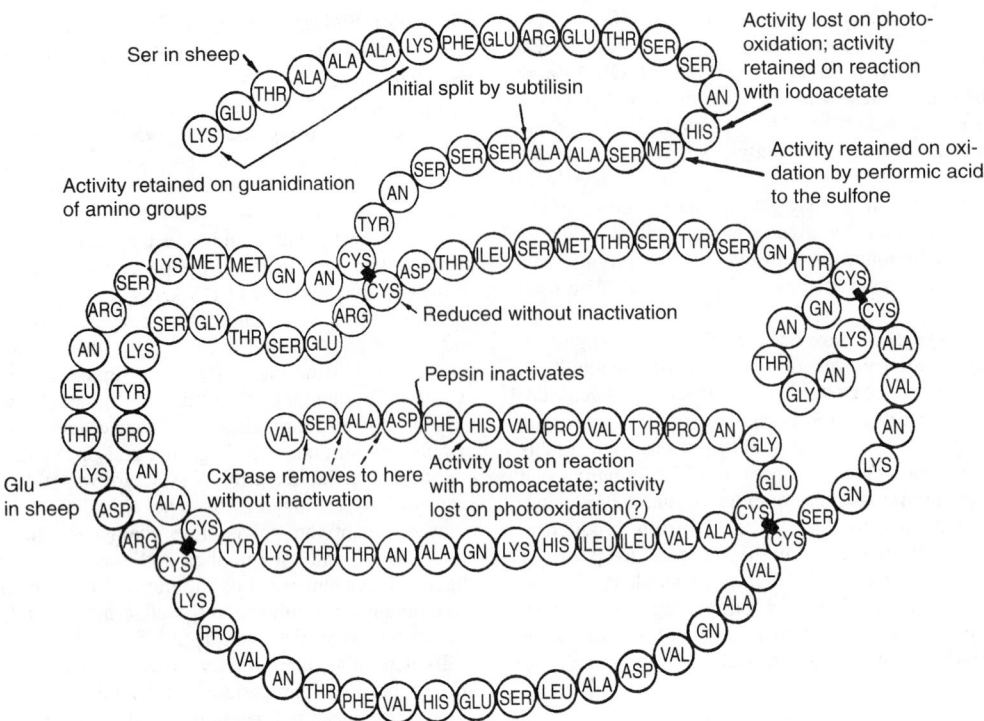

Fig. 1.   Early representation of ribonuclease structure. Legend: Alanine (ALA), adenine (AN), arginine (ARG), aspartic acid (ASP), cysteine (CYS), cytosine (CN), glutamic acid (GLU), glycine (GLY), guanine (GN), histidine (HIS), isoleucine (ILEU), leucine (LEU), lysine (LYS), methionine (MET), phenylalanine (PHE), proline (PRO), serine (SER), threonine (THR), thymine (TN), tryptophan (TRY), tyrosine (TYR), uracil (UN), and valine (VAL)

in interpreting enzyme action, but also in determining its relationship to genetics, for in this case, the genes determining the ability to synthesize both enzymes are present on the chromosome map of the organism.

There are methods used to study enzymes other than those of chemical instrumental analysis, such as chromatography, that have already been mentioned. Many enzymes can be crystallized, and their structure investigated by x-ray or electron diffraction methods. Studies of the kinetics of enzyme-catalyzed reactions often yield useful data, much of this work being based on the Michaelis-Menten treatment. Basic to this approach is the concept that the action of enzymes depends upon the formation by the enzyme and substrate molecules of a complex, which has a definite, though transient, existence, and then decomposes into the products of the reaction. Note that this point of view was the basis of the discussion of the specificity of the active sites discussed above.

A simple enzyme reaction may thus be written as

$$E + S \Longleftrightarrow ES \longrightarrow \text{products}$$

In the Michaelis-Menten treatment, this equation can be regarded as the result of the three processes:

rate of formation of $ES = k_1[E][S]$
rate of decomposition of ES into products $= k_2[ES]$
rate of decomposition of ES into original reactants $= k_3[ES]$

where the terms in square brackets denote concentrations of E, S, and ES, and the k's are rate constants. By representing the ratio $(k_2 + k_3)/k_1$ by K_m, the *Michaelis constant*, we can obtain a form of the Michaelis equation

$$\frac{[E_t]}{v} = \frac{K_m}{k_3 + [S]} + \frac{1}{k_3}$$

where E_t is the total concentration of enzyme (as distinguished from $[E]$, the concentration of free enzyme), and v is the velocity of the reaction. By plotting $[E_t]/v$ against $1/[S]$, we can obtain, at the axis-intercepts, values of $1/K_m$ and $1/v$.

This approach has been successfully extended to the more complex enzymatic conversions involving inhibitors, activators, and even multiple reactions in which the successive action of more than one enzyme is involved.

The observation follows that in addition to the specificity of the enzymes with respect to reaction type and to structure of the substrate, the action is also confined to a single configuration of the substrate. If the molecular structure of the substrate is unsymmetrical (asymmetric) and therefore

two compounds exist—one the mirror image of the other—a specific enzyme will act upon only one of the stereoisomers. This specificity is undoubtedly due to the fact that the region of interaction on the enzyme is also asymmetric and exists in only one form or configuration. Racemic mixtures or certain substrates may sometimes be separated by making use of the fact that enzymic action will affect only one of the two forms.

Enzyme Repression

Repression as applied to biochemical reactions is a process of feedback control whereby a cell limits its production of the substances produced within it. An example that has been investigated shows the nature and mechanism by which this limitation is effected. It has been found that production of the amino acid L-isoleucine by cells of the bacterium *Escherichia coli* is repressed in the presence of an excess of the product. This excess is obtained experimentally by adding the substance to the culture medium in which the bacterium is grown. A form of the L-isoleucine is used that has been labeled with a radioactive isotope, so that the mechanism of the repression can be followed.

By this means, it has been found that the excess of L-isoleucine has two distinct effects—one that is relatively slow, and another that is rapid. The slower effect is to repress production by the cell of all the enzymes required to catalyze the series of biochemical reactions in the metabolic pathway by which the cell synthesizes L-isoleucine. The fast effect is to inhibit production of the enzyme for the first reaction in the series. This enzyme is L-threonine deaminase, which removes the amino group from L-threonine, as a preliminary step to its oxidation and reintroduction of the amino group, in order to produce L-isoleucine from it.

The independent existence of these two effects was demonstrated by the discovery among mutations of *E. coli*, a mutation that exhibited only one of the effects, and of another mutation that exhibited only the other, leading to the conclusion that two distinct genes are involved in the control system.

An even more striking instance of feedback control is found in the synthesis of DNA (see also **Nucleoproteins and Nucleic Acids**). As pointed out in that entry, normal DNA is composed of the nucleotides deoxyguanosine, deoxycytidine, deoxyadenosine, and thymidine, and the amounts of the first and second of these are the same, as are those of the third and fourth. Obviously, close control is required of the amounts of these nucleotides that are synthesized by the cell, if they are to be made in the quantities required for DNA synthesis. Evidence has been found that the enzyme carbamoylphosphate: L-aspartate carbamoyl transferase, which

TABLE 1. CLASSIFICATION OF ENZYMES

1. Oxidoreductases
  1.1 *Acting on the CH−OH group of donors*
    1.1.1 With NAD or NADP as acceptor
    1.1.2 With cytochrome as an acceptor
    1.1.3 With O_2 as acceptor
    1.1.99 With other acceptors
  1.2 *Acting on the aldehyde or keto group of donors*
    1.2.1 With NAD or NADP as acceptor
    1.2.2 With a cytochrome as an acceptor
    1.2.3 With O_2 as acceptor
    1.2.4 With lipoate as acceptor
    1.2.99 With other acceptors
  1.3 *Acting on the CH−H group of donors*
    1.3.1 With NAD or NADP as acceptor
    1.3.2 With a cytochrome as an acceptor
    1.3.3 With O_2 as acceptor
    1.3.99 With other acceptors
  1.4 *Acting on the CH−H_2 groups of donors*
    1.4.1 With NAD or NADP as acceptor
    1.4.3 With O_2 as acceptor
  1.5 *Acting on the C−H group of donors*
    1.5.1 With NAD or NADP as acceptor
    1.5.3 With O_2 as acceptor
  1.6 *Acting on reduced NAD or NADP as donor*
    1.6.1 With NAD or NADP as acceptor
    1.6.2 With a cytochrome as an acceptor
    1.6.4 With a disulfide compound as acceptor
    1.6.5 With a quinone or related compound as acceptor
    1.6.6 With a nitrogenous group as acceptor
    1.6.99 With other acceptors
  1.7 *Acting on other nitrogens compounds as donors*
    1.7.3 With O_2 as acceptors
    1.7.99 With other acceptors
  1.8 *Acting on sulfur groups of donors*
    1.8.1 With NAD or NADP as acceptor
    1.8.3 With O_2 as acceptor
    1.8.4 With a disulfide compound as acceptor
    1.8.5 With a quinone or related compound as acceptor
    1.8.6 With a nitrogenous group as acceptor
  1.9 *Acting on heme groups of donors*
    1.9.3 With O_2 as acceptor
    1.9.6 With a nitrogenous group as acceptor
  1.10 *Acting on diphenols and related substances as donors*
    1.10.3 With O_2 as acceptor
  1.11 *Acting on H_2O_2 as acceptor*
  1.12 Acting *on hydrogen as donor*
  1.13 Acting *on single donors with incorporation of oxygen (oxygenases)*
  1.14 *Acting on paired donors with incorporation of oxygen into one donor (hydroxylases)*
    1.14.1 Using reduced NAD or NADP as one donor
    1.14.2 Using ascorbate as one donor
    1.14.3 Using reduced pteridine as one donor
2. Transferases
  2.1 *Transferring one-carbon groups*
    2.1.1 Methyltransferases
    2.1.2 Hydroxymethyl-, formyl-, and related transferases
    2.1.3 Carboxyl- and carbamoyltransferases
    2.1.4 Amidinotransferases
  2.2 *Transferring aldehydic or ketonic residues*
  2.3 *Acyltransferases*
    2.3.1 Acyltransferases
    2.3.2 Aminoacyltransferases
  2.4 *Glycosyltransferases*
    2.4.1 Hexosyltransferases
    2.4.2 Pentosyltransferases
  2.5 *Transferring alkyl or related groups*
  2.6 Transferring *nitrogenous groups*
    2.6.1 Aminotransferases
    2.6.3 Oximinotransferases
  2.7 *Transferring phosphorus-containing groups*
    2.7.1 Phosphotransferases with an alcohol group as acceptor
    2.7.2 Phosphotransferases with a carboxyl group as acceptor
    2.7.3 Phosphotransferases with a nitrogenous group as acceptor
    2.7.4 Phosphotransferases with a phospho-group as acceptor
    2.7.5 Phosphotransferases, apparently intramolecular
    2.7.6 Pyrophosphotransferases
    2.7.7 Nucleotidyltransferases
    2.7.8 Transferases for other substituted phospho-groups
  2.8 *Transferring sulfur-containing groups*
    2.8.1 Sulfurtransferases

    2.8.2 Sulfotransferases
    2.8.3 CoA-transferases
3. Hydrolases
  3.1 *Acting on ester bonds*
    3.1.1 Carboxylic ester hydrolases
    3.1.2 Thiolester hydrolases
    3.1.3 Phosphoric monoester hydrolases
    3.1.4 Phosphoric diester hydrolases
    3.1.5 Triphosphoric monoester hydrolases
    3.1.6 Sulfuric ester hydrolases
  3.2 *Acting on glycosyl compounds*
    3.2.1 Glycoside hydrolases
    3.2.2 Hydrolyzing N-glycosyl compounds
    3.2.3 Hydrolyzing S-glycosyl compounds
  3.3 *Acting on ether bonds*
    3.3.1 Thioether hydrolases
  3.4 *Acting on peptide bonds (peptide hydrolases)*
    3.4.1 α-Aminoacyl-peptide hydrolases
    3.4.2 Peptidyl-amino acid hydrolases
    3.4.3 Dipeptide hydrolases
    3.4.4 Peptidyl-peptide hydrolases
  3.5 *Acting on C−bonds other than peptide bonds*
    3.5.1 In linear amides
    3.5.2 In cyclic amides
    3.5.3 In linear amidines
    3.5.4 In cyclic amidines
    3.5.5 In cyanides
    3.5.99 In other compounds
  3.6 *Acting on acid-anhydride bonds*
    3.6.1 In phosphoryl-containing anhydrides
  3.7 Acting on C−C bonds
    3.7.1 In ketonic substances
  3.8 *Acting on halide bonds*
    3.8.1 In C-halide compounds
    3.8.2 In P-halide compounds
  3.9 *Acting on P−bonds*
4. Lyases
  4.1 *Carbon-carbon lyases*
    4.1.1 Carboxyl-lyases
    4.1.2 Aldehyde-lyases
    4.1.3 Ketoacid-lyases
  4.2 *Carbon-oxygen lyases*
    4.2.1 Hydro-lyases
    4.2.99 Other carbon-oxygen lyases
  4.3 *Carbon-nitrogen lyases*
    4.3.1 Ammonia-lyases
    4.3.2 Amidine-lyases
  4.4 *Carbon-sulfur lyases*

  4.5 *Carbon-halide lyases*
  4.99 *Other lyases*
5. Isomerases
  5.1 *Racemases and epimerases*
    5.1.1 Acting on amino acids and derivatives
    5.1.2 Acting on hydroxyacids and derivatives
    5.1.3 Acting on carbohydrates and derivatives
    5.1.99 Acting on other compounds
  5.2 *Cis-trans isomerases*
  5.3 *Intramolecular oxidoreductases*
    5.3.1 Interconverting aldoses and ketoses
    5.3.2 Interconverting keto- and enol-groups
    5.3.3 Transposing C−bonds
  5.4 *Intramolecular transferases*
    5.4.1 Transferring acyl groups
    5.4.2 Transferring phosphoryl groups
    5.4.99 Transferring other groups
  5.5 Intramolecular *lyases*
  5.99 *Other isomerases*
6. Ligases or Synthetases
  6.1 *Forming C−bonds*
    6.1.1 Aminoacid-RNA ligases
  6.2 *Forming C−bonds*
    6.2.1 Acid-thiol ligases
  6.3 *Forming C−bonds*
    6.3.1 Acid-ammonia ligases (amide synthetases)
    6.3.2 Acid-amino acid ligases (peptide synthetases)
    6.3.3 Cyclo-ligases
    6.3.4 Other C−ligases
    6.3.5 C−ligases with glutamine as N-donor
  6.4 *Forming C−bonds*

catalyzes the conversion between aspartic acid and carbamoyl phosphate (which has deoxycytidine triphosphate, CTP, as its final product), is inhibited by an excess of the CTP, and is also initiated (or activated) by an excess of deoxyadenosine triphosphate, which requires an equal amount of CTP to react with it in forming DNA. There are thus both positive and negative feedback controls on the synthesis of the enzymes that catalyze the synthesis of the nucleotides. Since all enzymes are proteins, the mechanism of this control is believed to be that suggested for protein synthesis in the entry on **Nucleoproteins and Nucleic Acids**. See also **Genetics and Gene Science (Classical)**; and **Molecular Biology**.

An aspect of the control of enzymatic action that is related to the effect of initiation (or activation) just discussed is the effect of *induction*, which can readily be illustrated experimentally. Many years ago, it was found that the yeast, *Saccharomyces ludwigii*, although able to ferment many sugars, was ineffective on lactose (milk sugar), because it did not synthesize the necessary enzyme, lactase (ζ D-galactoside galactohydrolase). However, if this yeast was grown for several generations on a medium containing lactose, it acquired the ability to make lactase, and its subsequent generations retained that ability. In the years since this discovery, so many instances of induction have been discovered that they are regularly cited in discussions of the properties of those enzymes for which they are known, as are also the repressing and blocking substances.

Classification of Enzymes by Function

As shown in Table 1, enzymes are classified into six groups: (1) oxidoreductases, (2) transferases, (3) hydrolases, (4) lyases, (5) isomerases, and (6) ligases or synthetases. The main group to which an enzyme belongs is indicated by the first figure of the code number. The second figure indicates the subclass. For the oxidoreductases, it shows the type of group in the *donors* which undergoes oxidation. For the transferases, it indicates the nature of the group which is transferred. For the hydrolases, it shows the type of bond hydrolyzed; for the lyases, the type of link that is broken between the group removed and the remainder; for the isomerases, the type of isomerization involved; and for ligases, the type of bond formed.

The third figure of the code number, indicating the sub-sub class, shows for the oxidoreductases the type of acceptor involved; for the transferases and hydrolases, it shows more precisely the type of group transferred or bond hydrolyzed; for the lyases, it shows the nature of the group removed; for the isomerases, it indicates in more detail the nature of the isomerization; and for the ligases, it shows the nature of the substance formed. Thus, an enzyme number, commonly indicated by the prefix EC, provides fairly detailed information about a specific enzyme.

Categories of Conversions

The comparative simplicity of the classification scheme bears testimony to the underlying unity of enzymatic catalysis.

Oxidoreductases. The overall conversion catalyzed by the oxidoreductases can be written as hydrogen transfer, and these enzymes might be considered to be merely one section of the transferases. The oxidoreductases are classified separately because of their large number and because of their great biological importance in bringing about the main energy-yielding conversions of living tissues.

Transferases. The main groups of transferases are concerned with the transfer of *one-carbon* groups, acyl groups, glycosyl residues, amino- and other nitrogen-containing groups, phosphate, and sulfate. Oxidoreductases and transferases together represent about half or more of the enzymes presently recognized. A general conversion for both oxidoreductases and transferases can be written:

$$AX + B \rightleftharpoons A + BX$$

Hydrolases. These enzymes include esterases, glycosidases, peptidases, deaminases, and enzymes which hydrolyze acid anhydrides (such as the pyrophosphate group in adenosinetriphosphate). Many hydrolases have been shown to be able, under appropriate conditions, to catalyze transfer conversions; a high concentration of acceptor is usually necessary, since there is competition between the added acceptor and water for the group transferred. The detailed mechanism in these cases probably involves transfer of a part of the substrate onto a group on the enzyme, with subsequent transfer to an acceptor or hydrolysis, e.g., for a hydrolase acting on a substrate AB to produce AOH and BH:

$$EH + AB \longrightarrow E - A + BH$$

and

$$E - A + X \longrightarrow E + AX$$

or

$$E - A + H_2O \longrightarrow EH + AOH$$

These hydrolases, if not all, can therefore be regarded as transferases which include H_2O among their possible acceptors. Under normal conditions, in aqueous solution, hydrolysis will be the dominant conversion.

Lyases. Enzymes in this grouping catalyze conversions of the type:

$$AX - BY \rightleftharpoons A = B + X - Y$$

Molecules, such as H_2O, H_2S, NH_3, or aldehydes, are added across the double bond of a second unsaturated molecule. Decarboxylases, such as those acting on amino acids can be regarded as lyases (carboxylases), assuming CO_2 and not H_2CO_3 to be the immediate product of decarboxylation. Over one hundred lyases are known.

Isomerases. These include enzymes that bring about conversions similar to those in several other groups, but distinguished in that the reaction takes place entirely within one molecule, which is not cleaved, so that the overall reaction is

$$A \rightleftharpoons B$$

Thus, there are intramolecular oxidoreductases (e.g., ketolisomerases), intramolecular transferases (e.g., phosphomutases), and intramolecular lyases. About fifty isomerases are known.

Ligases. These enzymes catalyze conversions, which are more complex than those of the other groups, and must involve at least two separate stages in the reaction. The overall result is the synthesis of a molecule from two components with a coupled breakdown of adenosine triphosphate, or some other nucleoside triphosphate. In general, this may be written:

$$X + Y + ATP \longrightarrow XY + AMP$$

$$+ \text{Pyrophosphate (or ADP + Phosphate)}$$

These enzymes, of which many are known, are of great importance in the conservation of chemical energy within the cell and in the coupling of synthetic processes with energy-yielding breakdown conversions. See also **Enzymes in Organic Synthesis**.

Additional Reading

Beynon, R.J. and J.S. Bond: *Proteolytic Enzymes: A Practical Approach,* Oxford University Press, Inc., New York, NY, 2001.

Hopwood, D.A.: "The Genetic Programming of Industrial Microorganisms," *Sci. Amer.,* **245**(3), 90–102 (September 1981).

Hui, Y.H.: *Data Source Book for Food Scientists and Technologists,* John Wiley & Sons, Inc., New York, NY, 1991.

Kornberg, A.: *For the Love of Enzymes: The Odyssey of a Biochemist,* Harvard University Press, Cambridge, MA, 1991.

Laskin, A.I., et al.: *Enzyme Engineering XIV,* Vol. 864, New York Academy of Sciences, New York, NY, 1998.

Lauwers, A. and S. Scharpe: *Pharmaceutical Enzymes,* Vol. 84, Marcel Dekker, Inc., New York, NY, 1997.

Liebman, J.F.: *Molecular Structure and Energetics-Mechanistic Principles of Enzyme Activity,* Vol. 9, John Wiley & Sons, Inc., New York, NY, 1989.

Nagodawithana, T.W.: *Enzymes in Food Processing,* Academic Press, Inc., San Diego, CA, 1997.

Phillips, D.C.: *Protein Engineering,* Science & Technology Review, 46–51, The University of Wales, Cardiff, UK, (March 1987).

Schomburg, D. and D. Stephan: *Enzyme Handbook,* Vol. 14, Springer-Verlag, Inc., Upper Saddle River, NJ, NY, 1999.

Stauffer, C.E.: *Enzyme Assays for Food Scientists,* Van Nostrand Reinhold, New York, NY, 1989.

Stephen J.: *Source Book of Enzymes,* CRC Press, LLC., Boca Raton, FL, 1997.

Suckling, C.J., C.G. Gibson, and A.R. Pitt: *Enzyme Chemistry,* Kluwer Academic Publishers, Norwell, MA, 2000.

Tamanoi, F. and D.S. Sigman: *The Enzymes,* Vol. 21, Academic Press, Inc., San Diego, CA, 2000.

Uhlig, H: *Industrial Enzymes and Their Applications,* John Wiley & Sons, Inc., New York, NY, 1998.

Whistler, R.L. and J.N. BeMiller: *Carbohydrate Chemistry for Food Scientists,* American Association of Cereal Chemists, St. Paul, MN, 1997. http://www.scisoc.org/aacc/

Wong, D.W.S.: *Food Enzymes: Structure and Mechanism,* Chapman & Hall, New York, NY, 1999.

ENZYME INHIBITORS.

The development of potent enzyme inhibitors has led to an increased understanding of enzyme mechanisms and has provided effective therapeutic agents for the treatment of diseases. Enzymes are natural biocatalysts that promote specific reactions essential for the viability of living organisms. They have unique recognition sites that allow them to select their substrates out of the vast pool of biologically important compounds in living cells. The substrate binds to the active site, that portion of the enzyme responsible for promoting the chemistry involved in converting the substrate to product. Inhibitors of enzymes prevent this chemical reaction from occurring by altering the active site and thereby rendering the enzyme at least temporarily inactive.

Applications for Enzyme Inhibitors

Enzyme inhibitors are often used to further the understanding of enzyme mechanisms. Inhibitors serve as probes for kinetic and chemical processes during catalysis. Inhibitors are also used for *in vivo* studies to localize and quantify enzymes in organs or to mimic certain genetic diseases that involve the absence of an enzyme in a given biosynthetic pathway. In pharmacological research, enzyme inhibitors are used to inactivate specific enzymes or groups of enzymes, leading to the treatment of many diseases.

An important area of drug design is the development of drugs against microorganisms and parasites in humans. A powerful strategy consists of choosing a target enzyme that is essential for the existence of the invader, but not the host. Therefore, these drugs usually exhibit low toxicity toward the host. An alternative approach is the choice of a target enzyme that exists as different species, called isozymes, in the invader and the host.

Another class of therapeutic agents is used for the treatment of certain genetic diseases or other enzymatic disorders caused by the dysfunction or absence of one particular enzyme. This often leads to an unwanted accumulation or imbalance of metabolites in the organism.

A third class of enzyme inhibitors consists of antitumor agents. Their design is quite challenging because tumor cells do not appear to contain enzymes that are very different from those in normal cells. Most antitumor drugs, called antiproliferative agents, take advantage of the fact that tumor cells generally grow and divide much faster than normal cells.

Classification of Enzyme Inhibitors

Enzyme inhibitors may be classified as follows:

Noncovalent inhibitors	Covalent inhibitors
rapid reversible inhibitors tight, slow, slow-tight binding inhibitors transition-state analogues multisubstrate analogues	mechanism-based inhibitors affinity labels pseudoirreversible inhibitors

All of these enzyme inhibitors are active site-directed.

Computer-Aided Inhibitor Design

Computer-aided inhibitor design is a relatively new and powerful approach for the development of novel, potentially potent, nonsubstrate-analogue enzyme inhibitors. Computer-aided methods and biological screening can each lead to new classes of novel inhibitors. However, computer-aided design methods can focus the search for inhibitors, thereby circumventing much of the time-consuming synthetic and natural product purification procedures for those compounds they find unlikely to function as inhibitors.

Design of Inhibitors from an Enzyme Structure. As high quality, three-dimensional structures of enzymes have become increasingly available, they have been used for the structure-based design of inhibitors. The enzyme structure provides a map showing the shape of available spaces, the locations of potential hydrogen-bonding groups, and the locations of potentially charged groups.

Computers are integral to structure-based design processes. Graphics workstations and a variety of computer algorithms (procedures) are used to help visualization of three-dimensional structures. Molecular graphics programs can provide tools for displaying, building, and energy-minimizing three-dimensional structures, in addition to docking structures, bringing them together in three-dimensional space like an inhibitor into the active-site of an enzyme. Visual display requires molecular surface contours, color codes, depth perception cues, and tools for three-dimensional manipulation. Currently available program packages that contain molecular graphics capabilities include BIOGRAPH (from BioDesign), CHEM-X (from Chemical Design, Ltd.), FRODO, INSIGHT (from Biosym Technologies, Inc.) MACROMODEL, MIDAS, MOGLI (from Evans and Sutherland Computer Corp.) QUANTA (from Polygen Corp.), and SYBYL (from Tripos Association).

In addition to assisting with structure visualization, computer algorithms are used to automate the search for inhibitors by performing tasks such as screening databases and analyzing enzyme active sites. DOCK, a computer program package, brings molecules together in space to score the fit of a single potential inhibitor in an enzyme active site, or to screen small molecule databases for compounds that score well. Another program, GRID, finds regions in the active site that should have favorable interactions with a specific chemical group. A third program, HINT, identifies and maps potential hydrophobic and polar interactions.

Design of Inhibitors from a Series of Structurally Related Compounds. Traditionally, medicinal chemists have used trial-and-error methods to study the relationship between structure and function for the development of inhibitors in a fashion more rational than biological screening. After measuring the inhibitory effect of an original compound, a small chemical modification is made to create a structural analogue, which is tested for a change in inhibition. It is assumed that the structural change is small enough so that the mechanism of action is not drastically altered. By testing several structural analogues, the spatial limits of the enzyme's binding site are explored without knowing the structure of the enzyme. Modern, quantitative approaches use the speed and computational power of computer algorithms to determine specific structure–function relationships for the design of more potent analogues of known inhibitors. Methods include quantitative structure–activity relationship (QSAR) methods, comparative molecular field analysis methods (CoMFA), and perturbation free-energy calculations.

Design of Inhibitors from a Pharmacophore. Pharmacophore-based inhibitor design is a method for developing inhibitors that combines the structural elements known to be important for binding with new structural templates. From a series of compounds that bind to the enzyme, the pharmacophore, the structural unit containing the chemical elements in the orientation required for binding, must first be identified. Once the pharmacophore has been defined, it can be used in many ways to develop inhibitors. For example, structural databases can be searched for existing molecules that contain the proposed pharmacophore. Alternatively, potential inhibitors can be built from structural pieces that connect the elements of the pharmacophore in the proper fashion.

ANGELIKA MUSCATE
CYNTHIA L. LEVINSON
GEORGE L. KENYON
University of California, San Francisco

Additional Reading

Fersht, A.: *Enzyme Structure and Mechanism,* 2nd ed., W. H. Freeman and Co., New York, NY 1985.

Kuntz, I.D. *Science* **257**, 1078–1082 (1992).

Martin, Y.C.: *Methods Enzymol.* **203**, 587–613 (1991).

Santi, D.V. and G.L. Kenyon: in M.E. Wolff, ed., *Burger's Medicinal Chemistry,* 4th ed., Wiley-Interscience, New York, NY 1980, pp. 349–391.

ENZYME PREPARATIONS.

During the past several years, a number of commercially prepared enzyme preparations have been available to processors, notably for use in the food industry. These preparations fall into three basic categories: (1) Animal-derived preparations; (2) plant-derived preparations; and (3) microbially derived preparations.

Fruit juices, jams, and jellies, corn (maize) syrups and sweeteners, structured protein foods, and tenderized meats are exemplary of products, the quality of which has been improved through the use of enzyme preparations. Principal areas of development in food-grade enzyme research have been toward upgrading quality and byproduct utilization, higher rates and levels of extractions, synthetic food development, sweetener development, improving flavor of foods, and the stabilization of food quality and nutrition. Enzyme preparations also are used in the detergent field.

Animal-derived enzyme preparations include catalase (bovine liver), lipase, pepsin, rennet, and trypsin. Plant-derived preparations include bromelain, cellulase, ficin, malt, papain, and pectinase. Microbially derived preparations include amylases, carbohydrase, catalase, glucose oxidase, lipase, protease, and zymase.

ENZYME THERAPEUTIC

Physical and Chemical Properties

Enzymes are protein catalysts of remarkable efficiency and specificity. Lipid, carbohydrate, nucleotide, or metal-containing prosthetic groups may be attached to these enzymes and serve as essential components of their catalyses by enhancing specificity and/or stability. Each enzyme has a specific temperature and pH range where it functions to its optimal capacity; the optima for these proteins usually lie between 37–47°C, and pH optima range from acidic, i.e., 1.0 in the case of gastric pepsin, to alkaline, i.e., 10.5 in the case of alkaline phosphatase. However, enzymes from extremely thermotolerant bacteria have become available; these can function at or near the boiling point of water, and therapeutic use of these ultrastable proteins can be anticipated.

Hydrolases represent a significant classes of therapeutic enzymes (Table 1). Another group of enzymes with pharmacological uses has built-in cofactors, e.g., in the form of pyridoxal phosphate, flavin nucleotides, or zinc.

Commercial enzymes are available in oral form, sometimes formulated with appropriate stabilizers and excipients. However, such preparations are seldom suitable for parenteral use. Therefore, dry preparations devoid of high salts, excipients, or reducing agents have been adopted for the final formulation of therapeutic enzymes involves lyophilization in the presence of mannitol and a physiological buffer. Since many enzymes can be denatured by heat even in the dry state, refrigeration or freezing during transit or storage is customary.

The therapeutic utility of an enzyme preparation is largely dependent on its stability as finally formulated. A number of chemical modifications have been employed and include binding to inert surfaces and encapsulation to increase the resistance of these intrinsically labile macromolecules to decomposition. Unfortunately, some of these modifications can interfere with the optimal kinetic performance of the enzyme. The development and deployment of recombinant DNA technologies have made significant contributions toward meeting the goal of mass production of human enzymes for human use.

Uses

Therapeutic enzymes have a broad variety of specific uses, i.e., as oncolytics, thrombolytics, or replacements for inherited deficiencies. Additionally, there is a growing group of miscellaneous enzymes of diverse function.

Immobilized, Derivatized, and Entrapped Enzymes

Immobilized Enzymes. With the development of techniques for binding enzymes to insoluble supports, immobilized enzymes have been used not only in clinical analysis but also for therapeutic purposes.

Enzyme Conjugates. One approach used to prolong residence time of a given enzyme in the circulation is to conjugate that enzyme with albumin, a natural plasma protein.

Erythrocyte Entrapment of Enzymes. Erythrocytes have been used as carriers for therapeutic enzymes in the treatment of inborn errors. Exogenous enzymes encapsulated in erythrocytes may be useful both for delivery of a given enzyme to the site of its intended function and for the degradation of pathologically elevated, diffusible substances in the plasma.

Chemically or enzymatically modified preparations of enzymes present another potential immunologic problem since modifications designed to stabilized enzyme activity may produce new antigenic determinants.

As of this writing (ca 1993) there are eight methods of erythrocyte entrapment. Six methods depend on loading via hypotonic exchange, one method depends on chlorpromazine-induced endocytosis, and one depends on voltage-step induced transitory permeation.

Health and Safety Factors

Repetitive doses of foreign proteins may produce severe immunologic reactions, ranging from mild allergy to anaphylactic shock and death.

TABLE 1. THERAPEUTIC ENZYMES

Enzyme	Catalysis	Use
neuraminidase	hydrolysis of terminal acylneuraminyl residues	antineoplastic
ribonuclease	RNA → oligoribonucleotides	antineoplastic
L-α-arabino-furanosidase	L-α-arabinofuranoside → alcohol + L-arabinose	antineoplastic
brinase	fibrinogen → fibrin	fibrinolytic
α-glucosidase	D-1, 4α-glucoside → α-D-glucose	metachromatic leukodystrophy
β-glucosidase	D-1, 4β-glucoside → β-D-glucose	type A glycogenosis
arylsulfatase	phenolsulfate → phenol + sulfate	metachromatic leukodystrophy
α-galactosidase	α-D-galactoside → α-D-galactose	Fabry's disease
β-galactosidase	β-D-galactoside → β-D-galactose	Fabry's disease
bromelain	protein → amino acids and peptides	antiinflammatory
collagenase	collagen → amino acids and peptides	dermal ulcers
papain	protein → amino acids and peptides	reduction of edema after dental surgery
L-asparaginase	L-asparagine → L-aspartic acid + NH_3	antineoplastic
streptokinase	plasminogen → plasmin	thrombolytic
arvin	fibrinogen → fibrin	fibrinolytic
urokinase or tissue plasminogen activator	plasminogen → plasmin	thrombolytic
coagulation factor VIII	prothrombin → thrombin	hemophilia
glucocerebrosidase	glycolipid glucocerebroside → α-D-glucose + ceramide	Gaucher's disease
lipase	carboxylic ester → alcohol + carboxylic acid	pancreatic deficiency
L-glutaminase	L-glutamine → L-glutamic acid + NH_3	antineoplastic
L-arginase	L-arginine → L-ornithine + urea	antineoplastic
L-tyrosinase	L-tyrosine + O_2 → dihydrophenylalanine + H_2O	antineoplastic
L-serine dehydratase	L-serine → pyruvate + NH_3	antineoplastic
L-threonine deaminase	L-threonine → 2-ketobutyric acid + NH_3	antineoplastic
L-tryptophanase	L-tryptophan → indole + pyruvate + NH_3	antineoplastic
deoxyribonuclease	DNA → oligodeoxyribo-nucleotides	chronic bronchitis
trypsin	protein → peptides	athletic injuries
chymotrypsin	protein → peptides	athletic injuries
superoxide dismutase	$O_2^{[[mindo]]} + O_2^{[[mindo]]} + 2 H^+ \rightarrow O_2 + H_2O_2$	antiinflammatory

Parenteral administration of metabolically active enzymes can be acutely toxic on account of their biochemical effects. However, large metabolically effective doses of these products may be less toxic than small doses, due to immune paralysis, wherein large doses of antigen repress the expression of the complementary antibody. It is essential to circumvent such allergic responses in the enzyme therapy of genetic diseases, because replacement enzyme therapy may be required over a lifetime. By contrast, treatment of cancer with enzymes may be somewhat less problematic because of the attenuation for the immunologic reactivity of patients in many cases. A significant effort is being directed toward the use of human sources of enzymes, or of human-type enzymes produced in cultures of prokaryotic or eukaryotic organisms, with a goal of reducing or eliminating the problems of antigenicity.

For enzymes intended for parenteral use, the manufacturer must assure that the enzyme preparation is essentially pure and free of endotoxins. All preparations of enzymes intended for parenteral use are tested for safety in lower animals under the conditions anticipated in clinical trials.

HIREMAGULAR N. JAYARAM
Indiana University School of Medicine
GURPREET S. AHLUWALIA
Gillette Research Institute
DAVID A. COONEY
National Cancer Institute

Additional Reading

Cowan, M.J.: *Clin. Biochem.* **24**, 375–381 (1991).
Goldberg, D.M.: *Clin. Chim. Acta* **206**, 45–76 (1992).
Holcenberg, J.S. and J. Roberts, eds.: *Enzymes as Drugs,* Wiley-Interscience, New York, 1981.
Kohn, D.B. and et al.: *Human Gene Ther.* **2**, 101–105 (1991).

ENZYMES IN ORGANIC SYNTHESIS.

The application of enzymes in organic synthesis is a modern and rapidly growing area in synthetic organic chemistry. The rapid expansion of this area in recent years has been brought about by a number of factors, most importantly that a large number of enzymes have become commercially available. Of about 2500 enzymes identified thus far about 300 are available in a partly purified form. Moreover, because of advances in molecular biology, fermentation, and purification techniques, the cost of some enzymes has been reduced to less than $500/kg.

The synthetic utility of enzymes for the preparation of organic molecules is tremendous. Enzymes catalyze virtually all types of chemical reactions with the exception of Diels-Alder condensation. They possess remarkably high catalytic power (up to 10^{12} rate acceleration compared to the nonenzymatic reactions) and unsurpassed stereo- and regioselectivity. From a technological standpoint they also offer a number of advantages: the mild temperatures, neutral pH, and atmospheric pressure under which most enzymes operate result in processes that are environmentally acceptable, low energy consuming, and usually do not require high capital investments.

Biotransformations are carried out by either whole cells (microbial, plant, or animal) or by isolated enzymes. Both methods have advantages and disadvantages. In general, multistep transformations, such as hydroxylations of steroids, or the synthesis of amino acids, riboflavin, vitamins, and alkaloids that require the presence of several enzymes and cofactors are carried out by whole cells. Simple one- or two-step transformations, on the other hand, are usually carried out by isolated enzymes. Compared to fermentations, enzymatic reactions have a number of advantages including simple instrumentation; reduced side reactions, easy control, and product isolation.

The principal emphasis of this article is on reactions carried out by isolated enzymes that have the broadest synthetic utility, i.e., hydrolases, oxidoreductases, and lyases. Biotransformations catalyzed by living cells are considered to a lesser extent. For more detailed information on biotransformations the reader may consult several books and numerous reviews.

Hydrolases

Quantitative Analysis of Selectivity. One of the principal synthetic values of enzymes stems from their unique enantioselectivity, i.e., ability to discriminate between enantiomers of a racemic pair. Detailed quantitative analysis of kinetic resolutions of enantiomers relating the extent of conversion of racemic substrate (c), enantiomeric excess (ee), and the enantiomeric ratio (E) has been described elsewhere.

Enzyme-Catalyzed Asymmetric Synthesis. The extent of kinetic resolution of racemates is determined by differences in the reaction rates for the two enantiomers. At the end of the reaction the faster reacting enantiomer is transformed, leaving the slower reacting enantiomer unchanged. It is apparent that the maximum product yield of any kinetic resolution cannot exceed 50%.

The situation is different if the substrate is a prochiral or meso compound. Since these molecules have a center or plane of symmetry the binding of pro-S or pro-R forms is equivalent. The chirality appears only as a result of the transformation. Hence, at least theoretically, the compound can be converted to one enantiomer quantitatively.

Hydrolytic enzymes such as esterases and lipases have proven particularly useful for asymmetric synthesis because of their abilities to discriminate between enantiotopic ester and hydroxyl groups. A large number of esterases and lipases are commercially available in large quantities; many are inexpensive and accept a broad range of substrates.

Dicarboxylic Acid Monoesters. Enzymatic synthesis of monoesters of dicarboxylic acids by hydrolysis of the corresponding diesters is a widely used and thoroughly studied reaction. It is catalyzed by a number of esterases, lipases, and proteases and is usually carried out in an aqueous buffer, pH 6–8 at room temperature. Organic cosolvents may be added to increase solubility of the substrates. The pH is maintained at a constant level by the addition of aqueous hydroxide. After one equivalent of base is consumed the monoesters are isolated by conventional means. Recent developments in the area of enzymatic catalysis in nonaqueous media have significantly broadened the repertoire of hydrolytic enzymes.

Monoacyl Diols. Enzymatic synthesis of chiral monoacyl diols can be carried out either by direct enzymatic acylation of prochiral diols or by hydrolysis of chemically synthesized dicarboxylates.

$$
\underset{\substack{\text{HO} \qquad \text{OH}}}{\overset{\substack{\text{R} \\ | \\ \text{CH} \\ \diagup \quad \diagdown \\ \text{CH}_2 \quad \text{CH}_2}}{}} \quad \xrightarrow{\text{acylation}} \quad \underset{\substack{\text{HO} \qquad \text{OOCR}'}}{\overset{\substack{\text{R} \\ | \\ \overset{*}{\text{C}}\text{H} \\ \diagup \quad \diagdown \\ \text{CH}_2 \quad \text{CH}_2}}{}} \quad \xleftarrow{\text{hydrolysis}} \quad \underset{\substack{\text{R}'\text{COO} \qquad \text{OOCR}'}}{\overset{\substack{\text{R} \\ | \\ \text{CH} \\ \diagup \quad \diagdown \\ \text{CH}_2 \quad \text{CH}_2}}{}}
$$

Generally, these two methods complement each other. With some rare exceptions an enzyme that produces an S ester in the hydrolysis reaction produces an R isomer in acylation reaction and vice versa.

Kinetic Resolutions. From a practical standpoint the principal difference between formation of a chiral molecule by kinetic resolution of a racemate and formation by asymmetric synthesis is that in the former case the maximum theoretical yield of the chiral product is 50% based on a racemic starting material. In the latter case a maximum yield of 100% is possible. If the reactivity of two enantiomers is substantially different the reaction virtually stops at 50% conversion, and enantiomerically pure substrate and product may be obtained in close to 50% yield. Conveniently, the enantiomeric purity of the substrate and the product depends strongly on the degree of conversion so that even in those instances where reactivity of enantiomers is not substantially different, a high purity material may be obtained by sacrificing the overall yield.

The variety of enzyme-catalyzed kinetic resolutions of enantiomers reported in recent years is enormous. Similar to asymmetric synthesis, enantioselective resolutions are carried out in either hydrolytic or esterification–transesterification modes. Both modes have advantages and disadvantages. Hydrolytic resolutions that are carried out in a predominantly aqueous medium are usually faster and, as a consequence, require smaller quantities of enzymes. On the other hand, esterifications in organic solvents are experimentally simpler procedures, allowing easy product isolation and reuse of the enzyme without immobilization.

Optically Active Acids and Esters. Enantioselective hydrolysis of esters of simple alcohols is a common method for the production of pure enantiomers of esters or the corresponding acids. Lipases, esterases, and proteases accept a wide variety of esters and convert them to the corresponding acids, often in a highly enantioselective manner. Lipase-catalyzed kinetic resolutions are often practical for the preparation of optically active pharmaceuticals.

Optically Active Alcohols and Esters. In addition to the hydrolysis of esters formed by simple alcohols described above, lipases and esterases also catalyze the hydrolysis of a wide range of esters based on more complex and synthetically useful cyclic and acyclic alcohols. Although the hydrolysis of acetates often gives the desirable resolution, to achieve maximum selectivity and reaction efficiency, comparison of various esters is recommended. Both saturated and unsaturated derivatives are easily accepted by lipases and esterases.

Hydrolysis of Enol Esters. Enzyme-mediated enantioface-differentiating hydrolysis of enol esters is an original method for generating optically active α-substituted ketones.

Chiral Lactones and Polyesters. Similar to intermolecular reactions described previously, lipases also catalyze intramolecular acylations of hydroxy acids; the reaction results in the formation of lactones.

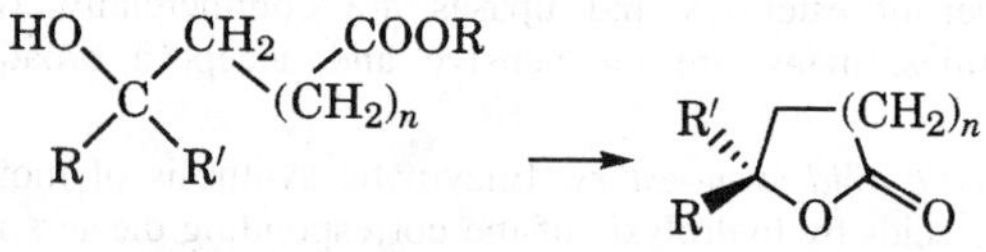

$$n = 1,2$$

Regioselective Acylation of Hydroxy Compounds. Aliphatic diols can be selectively acylated at the primary and secondary positions by a number of lipases in nonaqueous solvents.

Resolution of Racemic Amines and Amino Acids. Acylases (EC 3.5.1.14) are the most commonly used enzymes for the resolution of amino acids. Porcine kidney acylase (PKA) and the fungal *Aspergillus* acylase (AA) are commercially available, inexpensive, and stable. Amino alcohols can be resolved by a number of pathways, including hydrolysis, esterification, and transesterification.

Unprotected racemic amines can be resolved by enantioselective acylations with activated esters. This approach is based on the discovery that enantioselectivity of some enzymes strongly depends on the nature of the reaction medium.

Two interesting approaches for resolution of racemic amino acids have been reported. Both are based on enantioselective ring opening with *in situ* racemization of the nonreactive isomer (Fig. 1).

Hydrolysis of Nitriles. The chemical hydrolysis of nitriles to acids takes place only under strong acidic or basic conditions and may be accompanied by formation of unwanted and sometimes toxic by-products. Enzymatic hydrolysis of nitriles by nitrile hydratases, nitrilases, and amidases is often advantageous since amides or acids can be produced under very mild conditions and in a stereo- or regioselective manner.

There are two distinct classes of enzymes that hydrolyze nitriles. Nitrilases (EC 3.5.5.1) hydrolyze nitriles directly to corresponding acids and ammonia without forming the amide. In fact, amides are not substrates for these enzymes. Nitriles also may be first hydrated by nitrile hydratases to yield amides which are then converted to carboxylic acid with amidases. This is a two-enzyme process, in which enantioselectivity is generally exhibited by the amidase, rather than the hydratase.

Peptide Synthesis. Despite significant progress in the chemical synthesis of peptides and proteins by both liquid- and solid-phase methodologies, a number of shortcomings still exist. The principal limitation of chemical methods stems from the formation of by-products at each individual condensation step that accumulate during the course of the repeated reactions. Moreover, even the small amount of racemization that often occurs in the presence of highly activated coupling reagents reduces the purity of final product and complicates purification dramatically. In this respect the use of enzymes for amino acid or peptide coupling has a number of advantages. Mild reaction conditions and the excellent stereo- and regioselectivity of enzymes require only minimal protection, precludes racemization, and guarantees the structural fidelity of the product.

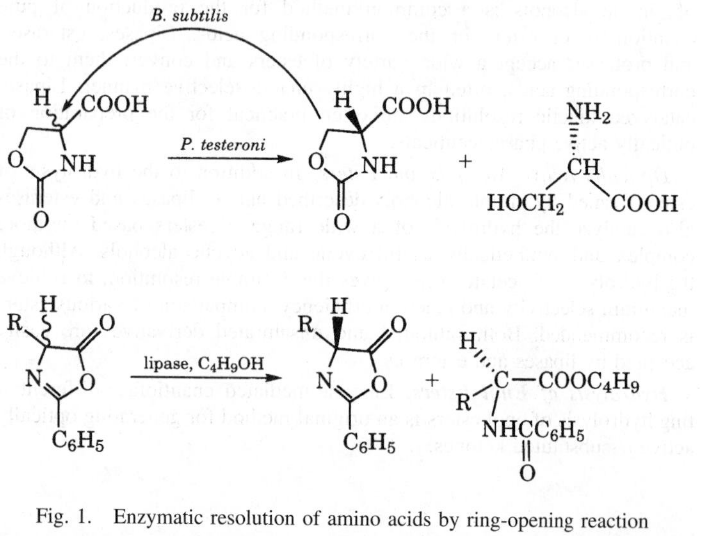

There are two basic strategies for enzyme-catalyzed peptide synthesis: equilibrium- and kinetically controlled synthesis. The former is the direct reversal of proteolysis and involves the condensation of an amino component with unactivated carboxyl component. The latter proceeds by the aminolysis of an activated peptide ester.

Lyases

Aldol Additions. These reactions catalyzed by lyases are perhaps the most synthetically useful enzymatic reactions for carbon–carbon bond formation.

There are two distinct groups of aldolases. Type I aldolases, found in higher plants and animals, require no metal cofactor and catalyze aldol addition via Schiff base formation between the lysine ε-amino group of the enzyme and a carbonyl group of the substrate. Class II aldolases are found primarily in microorganisms and utilize a divalent zinc to activate the electrophilic component of the reaction. A great variety of aldol additions have been carried out, allowing the synthesis of numerous nitrogen-containing sugars, deoxysugars, and fluorosugars, based, among others, on D-threose, L-threose, D-ribose, L-arabinose, D-xylose, D-glucose-6-P, and D-mannose-6-P, where P = phosphate.

Cyanohydrin Synthesis. Another synthetically useful enzyme that catalyzes carbon–carbon bond formation is oxynitrilase (EC 4.1.2.10). This enzyme catalyzes the addition of cyanides to various aldehydes that may come either in the form of hydrogen cyanide or acetone cyanohydrin (Fig. 2). The reaction constitutes a convenient route for the preparation of α-hydroxy acids and β-amino alcohols. Acetone cyanohydrin can also be used as the cyanide carrier, and is considered to be superior since it does not involve hazardous gaseous HCN and also virtually eliminates the spontaneous nonenzymatic reaction.

Oxidoreductases

Biocatalytic redox reactions offer great synthetic utility to organic chemists. The majority of oxidase-catalyzed preparative bioconversions are still performed using a whole-cell technique, despite the fact that the presence of more than one oxidoreductase in cells often leads to product degradation and lower selectivity. Fortunately, several efficient cofactor regeneration systems have been developed, making some cell-free enzymatic bioconversions economically feasible.

The two oxidoreductase systems most frequently used for preparation of chiral synthons include baker's yeast and horse liver alcohol dehydrogenase (HLAD). The use of baker's yeast has been recently reviewed in great detail and therefore will not be covered here. The emphasis here is on dehydrogenase-catalyzed oxidation and reduction of alcohols, ketones, and keto acid, oxidations at unsaturated carbon, and Bayer-Villiger oxidations.

Chiral Alcohols and Lactones. HLAD has been widely used for stereoselective oxidations of a variety of prochiral diols to lactones on a preparative scale. In most cases pro-(S) hydroxyl is oxidized irrespective of the substituents. The method is applicable to, among others, *cis*-1,2-bis(hydroxymethyl) derivatives of cyclopropane, cyclobutane, cyclohexane, and cyclohexene. Resulting γ-lactones are isolated in 68–90% yields and of 100% ee.

Although alcohol dehydrogenases (ADH) also catalyze the oxidation of aldehydes to the corresponding acids, the rate of this reaction is significantly lower. The systems that combine ADH and aldehyde dehydrogenases (EC 1.2.1.5) (AldDH) are much more efficient.

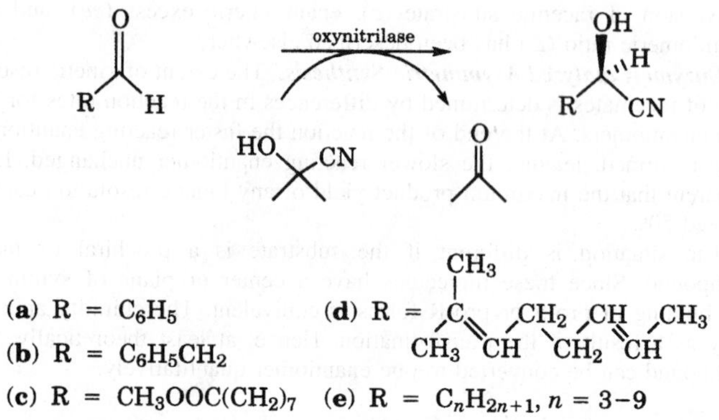

Fig. 1.   Enzymatic resolution of amino acids by ring-opening reaction

(a) R = C_6H_5

(b) R = $C_6H_5CH_2$

(c) R = $CH_3OOC(CH_2)_7$

(d) R = $\begin{array}{c} CH_3 \\ C \\ CH_3 \end{array} CH \begin{array}{c} CH_2 \\ CH_2 \end{array} CH \begin{array}{c} CH_3 \\ CH \end{array}$

(e) R = C_nH_{2n+1}, n = 3–9

Fig. 2.   Cyanohydrin formation

Alcohol dehydrogenase-catalyzed reduction of ketones is a convenient method for the production of chiral alcohols. HLAD, the most thoroughly studied enzyme, has a broad substrate specificity and accommodates a variety of substrates.

Hydroxy and Amino Acids. Reduction of 2-oxoacids with NADH, catalyzed by lactate dehydrogenase, is an established method for the preparation of homochiral α-hydroxy acids of both S and R configurations. The enzyme is found in all higher organisms and can be easily isolated from a variety of mammalian and bacterial sources.

Baeyer-Villiger Oxidations. The biological equivalent of the Baeyer-Villiger reaction is a useful transformation in providing a mild technique for converting ketones into esters or lactones.

Lipoxygenase-Catalyzed Oxidations. Lipoxygenase-1 catalyzes the incorporation of dioxygen into polyunsaturated fatty acids possessing a $1(Z),4(Z)$-pentadienyl moiety to yield $(E),(Z)$-conjugated hydroperoxides. A highly active preparation of the enzyme from soybean is commercially available in purified form.

Summary

The use of enzymatic catalysis inorganic synthesis has grown tremendously in recent years and continues to grow. Greater availability of enzymes, development of new methodologies for their utilization, use of nonconventional environments, and the design and synthesis of new biocatalysts with altered selectivity and increased stability were essential for the successful development of this field. As more is learned about selectivity of enzymes toward unnatural substrates, the choice of an enzyme for a particular transformation will become easier to predict. It will simplify a search for an appropriate catalyst and help to establish biocatalytic procedures as a useful supplement to classical organic synthesis.

ALEKSEY ZAKS
Schering-Plough Research Institute

Additional Reading

Davis, H.G., R.H. Green, D.R. Kelly, and S.M. Roberts, eds.: *Biotransformations in Preparative Organic Chemistry,* Academic Press Ltd. London, UK, 1989.
Holland, H.L.: *Organic Synthesis with Oxidative Enzymes,* VCH Publishers, Inc., New York, 1992.
Santaniello, E., P. Ferraboschi, P. Grisenti, and A. Manzocchi: *Chem. Rev.* **92,** 1071 (1992).
Wong, C.-H. and G.M. Whitesides: *Enzymes in Synthetic Organic Chemistry,* Tetrahedron Organic Chemistry Series, Vol. 12, Elsevier Science Inc., Tarrytown, NY, 1994.

EOSINOPHILS. See **Blood**.

EPHEDRINE. See **Alkaloids**.

EPICHLOROHYDRIN. A highly reactive and industrial important compound with the structural formula

It is also called 1-chloro-2,3-epoxypropane and is classified as an organic epoxide. The compound is a colorless, clear, mobile liquid with an odor something like chloroform. Molecular weight, 92.53; freezing point, $-57.1°C$; boiling point, $116.07°C$; density 1.1750 g/cm^3 at 25°C with reference to water at 4°C. Solubility is 6.53 g/100 g of water. Epichlorohydrin is made by the chlorohydrination of allyl chloride, in which 1,2-dichlorohydrin and 1,3-dichlorohydrin are produced as intermediates.

One of the most common epoxy resins is produced by the reaction between epichlorohydrin and bisphenol A. See also **Epoxy Resins**. The compound also is used in the production of epichlorohydrin-based rubbers which have good aging, high resiliency, and flexibility at low temperatures, advantage of which is taken in automotive and aircraft parts, seals, gaskets, hose, belting, wire, and cable jackets. These rubbers also have good resistance to solvents, fuels, oils, and ozone. A number of wet-strength resins for use in the paper industry also are derived from epichlorohydrin,

including (a) epichlorohydrin-modified polyamides; and (b) the addition of epichlorohydrin to high-molecular-weight polyalkylene polyamines. The advantages of these resins is that no alum or acid medium is required for incorporating the resin into the cellulose pulp. During the drying process, the resin cross-links and thus yields a paper with permanent wet-strength properties. Ion-exchange resins also can be prepared by reacting epichlorohydrin with ethylene diamine or a similar amine. The resulting material is a stable, water-insoluble anion-exchange resin.

In addition to its use in the production of epoxy resins epichlorohydrin is used in large quantities in the manufacture of glycerin. Other uses include textile applications where it is used to modify the carboxy groups of wool, thus increasing durability and improving moth resistance; in the synthesis of antistatic agents, wrinkle-resistant agents, and coating sizings. Effective against the larvae of certain insects, the compound is used in control chemicals for agriculture where permitted.

EPIDIORITE. A term applied to gabbros, dolerites, and diabases, the augite of which has been partly altered to hornblende, thus approaching a diorite in mineral composition. The term is derived from the Greek, meaning upon, plus diorite.

EPIDOTE. This mineral is a hydrous silicate of calcium, aluminum, and iron with the formula, $Ca_2(Al, Fe)_3Si_3O_{12}(OH)$. The ratio of aluminum to iron ranges from 6:1 to 3:2. Epidote is found in prismatic monoclinic crystals, which may be acicular to fibrous. Fine granular and compact masses are common. The mineral displays one good cleavage, an uneven fracture; is brittle; hardness, 6–7; specific gravity, 3.25–3.5; luster, vitreous to resinous; typical color, pistachio green, but may be yellowish- to brownish-green, sometimes red, yellow, gray, white or colorless. Colorless to grayish streak; transparent to opaque. The characteristic color of ordinary epidote makes it usually an easily identified mineral.

It occurs commonly in metamorphic rocks as gneisses and schists; however, it seems probable that under certain conditions it may appear as a primary mineral, for example in granitic rocks. The Urals, Austria, Switzerland, Italy, France and Norway are known for their occurrences of fine epidote crystals. In the United States epidote has been found in excellent specimens at Franconia and Warren, New Hampshire; Huntington, Massachusetts; Willimantic and Haddam, Connecticut; Chaffee County, Colorado, and Riverside County, California. The word epidote is derived from the Greek. The name pistacite, from the Greek word meaning pistachio nut, has been occasionally applied to this mineral. It has been used as a gemstone but is in little demand for this purpose.

EPINEPHRINE. See **Alkaloids; Hormones**.

EPITAXY. An oriented crystalline growth between two crystalline solid surfaces of different chemical composition, in which the surface of one crystal provides, through its lattice structure, preferred positions for the deposition of the second crystal. This behavior is characteristic of some types of high polymers.

EPOXY RESINS. A family of thermosetting resins known for their excellent mechanical and electrical properties, dimensional stability, resistance to high temperatures and numerous chemicals, and for their strong adhesion to glass, metal, fibers, and numerous other materials. Structurally, epoxy resins are characterized by the presence of a three-membered ring known as the epoxy, epoxide, oxirane, or ethoxylline group. Commercial epoxy resins contain aliphatic, cycloaliphatic, or aromatic backbones. The capability of the epoxy ring to react with a variety of substrates imparts versatility to the resins. Treatment with curing agents gives insoluble and intractable thermoset polymers. In order to facilitate processing and modify cured resin properties, other constituents may be included in the compositions: fillers, solvents, diluents, plasticizers, accelerators, curatives, and tougheners.

Resin Properties

Epichlorohydrin and Bisphenol A-Derived Resins. The most widely used epoxy resins are diglycidyl ethers of bisphenol A (1) derived from bisphenol A and epichlorohydrin.

(1)

The outstanding performance characteristics of the resins are conveyed by the bisphenol A moiety (toughness, rigidity, and elevated temperature performance), the ether linkages (chemical resistance), and the hydroxyl and epoxy groups (adhesive properties and formulation latitude, or reactivity with a wide variety of chemical curing agents). See also **Phenolic Resins**.

The bisphenol A-derived epoxy resins are most frequently cured with anhydrides, aliphatic amines, or polyamides.

Diluents are commonly used to reduce the viscosity of epoxy systems to aid handling, improve ease of application, and to facilitate higher filler loading to reduce formulation cost. This, however, is achieved at the expense of other properties. To achieve a balance of properties, careful selection of diluent is needed.

Specialty Epoxy Resins. In addition to bisphenol, other polyols such as aliphatic glycols and novolaks are used to produce specialty resins. Epoxy resins may also include compounds based on aliphatic, cycloaliphatic, aromatic, and heterocyclic backbones. Glycidylation of active hydrogen-containing structures with epichlorohydrin and epoxidation of olefins with peracetic acid remain the important commercial procedures for introducing the oxirane group into various precursors of epoxy resins.

Epoxy Cresol–Novolak Resins (ECN). The cresol–novolak epoxy resins (**2**) are multifunctional, solid polymers characterized by low ionic and hydrolyzable chlorine impurities, high chemical resistance, and good thermal performance. ECN resins are widely used as base components in high performance electronic and structural molding compounds, high temperature adhesives, castings and laminating systems, tooling applications, and powder coatings.

(2)

The epoxy cresol–novolak resins (**2**) are prepared by glycidylation of *o*-cresol–formaldehyde condensates in the same manner as the phenol–novolak resins.

Bisphenol F Resin. Bisphenol F epoxy resin is of the same general structure as the epoxy phenol novolaks. Bisphenol F is 2,2'-methylene bisphenol.

Owing to relatively low viscosity, these resins offer advantages for 100% solids (solvent-free) systems. Higher filler levels are possible because of the low viscosity. Faster bubble release is also achieved. Higher epoxy content and functionality of bisphenol F epoxy resins can provide improved chemical resistance compared to conventional epoxies.

Bisphenol F epoxy resins are used in high-solids-high-build systems such as tank and pipe linings, industrial floors, road and bridge deck toppings, structural adhesives, grouts, coatings, and electrical varnishes. Bisphenol F epoxy resins are manufactured in Europe and Japan.

Epoxy Phenol–Novolak Resins. Epoxy phenol–novolak resins are represented by the general idealized structure (**3**) whereby multifunctional products are formed containing a phenolic hydroxyl group per phenyl ring in random para–para', ortho–para', and ortho–ortho' combinations.

Subsequent epoxidation with epichlorohydrin yields the highly functional epoxy novolak. The product can range from a high viscosity liquid of *n* = 0.2 to a solid of *n* value greater than 3.

(3)

The thermal stability of epoxy phenol–novolak resins is useful in adhesives, structural and electrical laminates, coatings, castings, and encapsulations for elevated temperature service. Filament-wound pipe and storage tanks, liners for pumps and other chemical process equipment, and corrosion-resistant coatings are typical applications using the chemically resistant properties of epoxy novolak resins.

Curing agents that give the optimum in elevated temperature properties for epoxy novolaks are those with good high temperature performance such as aromatic amines, catalytic curing agents, phenolics, and some anhydrides.

Polynuclear Phenol–Glycidyl Ether-Derived Resins. This is one of the first commercially available polyfunctional products. Its polyfunctionality permits upgrading of thermal stability, chemical resistance, and electrical and mechanical properties of bisphenol A–epoxy systems. It is used in molding compounds and adhesives.

Cycloaliphatic Epoxy Resins. This family of aliphatic, low viscosity epoxy resins consists of two principal varieties, cycloolefins epoxidized with peracetic acid and diglycidyl esters of cyclic dicarboxylic acids.

The nonaromatic nature of these materials provides for improved uv resistance and arc-track resistance compared to conventional epoxies. The best properties are generally achieved with anhydride and phenolic curing agents.

Recommended applications include transformers, insulators, bushings, wire and cable coatings, generators, motors and switchgear, additives for adhesives, vinyl stabilization, and as viscosity depressants.

Aromatic and Heterocyclic Glycidyl Amine Resins. Among the specialty epoxy resins containing an aromatic amine backbone, the following are commercially significant.

Tetraglycidylmethylenedianiline-Derived Resins. Resins from aromatic glycidyl amines can be formulated into hot-melt or solution-binder systems with various reinforcements, e.g., glass, graphite, boron, or aramid. They are utilized for graphite-reinforced composites in aerospace and leisure products, structural adhesives, laminates, tooling and casting applications, and structures such as wings and fuselages.

Triglycidyl p-Aminophenol-Derived Resins. Resins derived from triglycidyl *p*-aminophenol, originally developed by Union Carbide Corp., are currently marketed by CIBA-GEIGY. Synthesis is conducted by reaction of epichlorohydrin with the phenolic and amino groups followed by dehydrohalogenation. The product is a viscous liquid (15–5 Pa · s (15–50 P) at 25°C) which is considerably more reactive toward amines than standard bisphenol A-derived resins.

Used to increase heat resistance and cure speed of bisphenol A epoxy resins, it has utility in such diverse applications as adhesives, tooling compounds, and laminating systems.

Triazine-Based Resin. Triglycidyl isocyanurate is a solid resin that provides superior thermal, electrical, and mechanical properties and is recommended for laminates, insulating varnishes, coatings, and adhesives. Widely used as a curing agent for special polyester-based weatherable powder coatings, it is also used in electronic applications owing to its retention of optical transparency after aging at temperatures up to 150°C and minimal smoke evolution on thermal decomposition (see EMBEDDING).

The triazine ring-containing product 1,3,5-triglycidyl isocyanurate is synthesized by glycidylation of cyanuric acid with epichlorohydrin.

Resin Synthesis and Manufacture

Epichlorohydrin and Bisphenol A-Derived Resins. Liquid epoxy resins may be synthesized by a two-step reaction of an excess of epichlorohydrin to bisphenol A in the presence of an alkaline catalyst. The reaction consists initially in the formation of the dichlorohydrin of bisphenol A and further reaction by dehydrohalogenation of the intermediate product with a stoichiometric quantity of alkali.

In recent years, production of liquid resins of higher purity, i.e., higher monomer content and fewer side-reactions, has been accomplished. This is in response to more stringent product quality requirements.

Aliphatic Glycidyl Ethers. Aliphatic epoxy resins have been synthesized by glycidylation of difunctional or polyfunctional polyols such as a 1,4-butanediol, 2,2-dimethyl-1,3-propanediol (neopentyl glycol), polypropylene glycols, glycerol, trimethylolpropane, and pentaerythritol.

The epoxidation is generally conducted in two steps: (1) the polyol is added to epichlorohydrin in the presence of a Lewis acid catalyst (stannic chloride, boron trifluoride) to produce the chlorohydrin intermediate, and (2) the intermediate is dehydrohalogenated with sodium hydroxide to yield the aliphatic glycidyl ether. Solid epoxy resins are prepared by the Taffy or Advancement processes.

Taffy Process. Bisphenol A reacts directly with epichlorohydrin in the presence of a stoichiometric amount of caustic. The molecular weight of the product is governed by the ratio of epichlorohydrin–bisphenol

A. In practice, the taffy process is generally employed for only medium molecular-weight resins ($n = 1 - 4$).

Advancement Process. In the advancement process, sometimes referred to as the fusion method, liquid epoxy resin (crude diglycidyl ether of bisphenol A) is chain-extended with bisphenol A in the presence of a catalyst to yield higher polymerized products. The advancement process is more widely used in commercial practice.

In recent years, proprietary catalysts for advancement have been incorporated in precatalyzed liquid resins. Thus only the addition of bisphenol A is needed to produce solid epoxy resins. Use of the catalysts is claimed to provide resins free from branching which can occur in conventional fusion processes. Additionally, use of the catalysts results in rapid chain-extension reactions because of the high amount of heat generated in the processing.

The preparation of flame-retardant epoxy resins is accompanied by inclusion of tetrabromobisphenol A in the advancement process See also **Flame-Retarding Agents**. Products containing ca 20 wt % Br are extensively employed in the printed circuit board industry.

Liquid resins containing bromine (ca 49 wt %) can also be prepared directly from tetrabromobisphenol A and epichlorohydrin and are used for critical applications where a high degree of flame retardancy is required.

Curing Reactions

A variety of reagents has been described for converting the liquid and solid epoxy resins to the cured state, which is necessary for the development of the inherent properties of the resins. Liquid epoxy resins contain mainly epoxy groups and solid resins are composed of both epoxy and hydroxyl curing sites. The curing agents or hardeners are categorized as either catalytic or coreactive and the functional groups of the resins are terminal epoxy together with a pendent hydroxyl per repeat unit of the polymer chain.

Economic Aspects

Epoxy resin sales increased rapidly in the 1970s and continued the increase into the 1980s as new applications were developed.

High performance multifunctional epoxy resins have achieved a higher compounded growth rate than the significantly larger volume DGEBPA conventional resins. Although epoxy resins cost more than competitive products, their longer service life and high performance capabilities provide a better cost-performance ratio.

The fastest growth over the past decade has been in laminates and composites. Tremendous growth in the electronics market has markedly increased the demand for epoxy laminating resins for the manufacture of printed wiring boards and epoxy-molding compounds for semiconductor encapsulation. Use of epoxy composites in the transportation industry, including the military and commercial aircraft and automotive fields, where a high strength-to-weight ratio is required, is growing at a steady rate that is expected to increase through the end of the century.

As with other petrochemical-based products, prices of epoxies have risen rapidly during the 1980s and 1990s as oil prices have escalated.

Trademarks of the principal epoxy resin producers are as follows: Shell Chemical Co. (EPON, EPONOL, Epikote); Dow Chemical Co. (D.E.R., D.E.N., D.E.H., Tactix, Quatrex); CIBA-GEIGY Corp. (Araldite); Dainippon Ink and Chemicals (Epotuf, Kelpoxy); and Union Carbide Co. (Unox).

Health and Safety Factors

There have been many investigations of the toxicity of various classes of epoxy-containing materials (glycidyloxy compounds). The use and interpretation of the vast amount of data available has been obscured by two factors: (1) proper identification of the epoxy systems in question and (2) lack of meaningful classification of the epoxy materials. In general, the toxicity of many of the glycidyloxy derivatives is low but the diversity of compounds found within this group does not permit broad generalizations for the class.

Applications

Epoxy resins are used in protective coatings, e.g., waterborne coatings, high solids coatings, and powder coatings, structural composites, electrical laminates, electrical, electronics, and structural applications (molding components and casting and encapsulation), and adhesives.

JOHN GANNON
Consultant

Additional Reading

Ballauf, M.: *Polymer Latexes/Expoxide Resins/Polyampholytes,* Springer-Verlag, Inc., New York, NY, 1999.

Bauer, R. (Editor): *Epoxy Resin Chemistry II,* American Chemical Society, Washington, DC, 1983.

Flick, E.W.: *Epoxy Resins, Curing Agents, Compounds, and Modifiers: An Industrial Guide,* Noyes Publications, Park Ridge, NJ, 1993.

Lee, H. and K. Neville: *Handbook of Epoxy Resins,* McGraw-Hill, Inc., New York, NY, 1967, reprinted 1982.

May, C.A. ed.: *Epoxy Resins* Chemistry and Technology, 2nd ed., Marcel Dekker, Inc., New York, NY, 1988.

McAdams, L.V. and J.A. Gannon: in J.I. Kroschwitz, ed., *Encyclopedia of Polymer Science and Engineering,* 2nd ed., Vol. 6, Wiley-Interscience, New York, NY, 1986, pp. 322–382.

Potter, W.G.: *Epoxide Resins* Springer-Verlag, New York, NY, 1970.

Web Reference

University of Southern Mississippi, Department of Polymer Science. http://www.psrc.usm.edu/macrog/epoxy.htm

EPSOMITE. Epsomite is normally found as an efflorescence on mine and cave walls. It belongs to the orthorhombic crystal system, being a hydrous sulfate of magnesium, $MgSO_4 \cdot 7H_2O$, with vitreous to earthy luster and colorless to white, of transparent/translucent quality. Hardness of 2–2.5, and specific gravity of 1.68. It is very bitter to the taste. Found with the soluble salt lake deposits in Stassfurt, Germany, and in limestone caves in Kentucky, Tennessee and Indiana; also in several California and Colorado abandoned mines.

EPSOM SALTS. See **Magnesium**.

EPSTEIN-BARR VIRUSES. See **Virus**.

EQUATION OF STATE.   Also called *characteristic equation,* a relation, empirical or derived, between thermodynamic properties of a substance or system. The equation of state must be single-valued in terms of its variables. This is a direct consequence of the concept of state.

There exist systems, namely systems which undergo processes involving hysteresis (plastic deformation or ferromagnetism, for example) for which no equation of state can be indicated. Although the laws of thermodynamics may apply to such systems, the rigorous results of classical thermodynamics are not applicable because the science of thermodynamics is developed on the assumption of the existence of the single-valued function.

In the realm of classical thermodynamics, equations of state are assumed given. They can be derived from first principles only by the methods of statistical mechanics and quantum mechanics. These rely on the adoption of suitable molecular models for substances, and so far no universal, generally applicable model has been discovered even for narrow classes of substances such as gases.

It is an experimental fact that every thermodynamic system possesses a definite number n of independent properties that determine its state. Consequently, an equation of state is a relation between n properties (mutually independent) chosen (otherwise arbitrarily) as the independent properties ($x_1, x_2 \ldots x_n$ of the system and one more property, the dependent property y. Hence the equation of state is a function of the form

$$F(y, x_1, x_2, \ldots x_n) = 0$$

The simplest thermodynamic systems possess two independent properties, consequently the simplest equation of state is written in terms of three variables. When it is written in terms of pressure p, volume V, and absolute temperature T, it is called the p-V-T relation for the system or the thermal equation of state. When one of the caloric-thermodynamic properties (better called caloric properties, because p, V, T are thermodynamic properties also), such as enthalpy, entropy, Gibbs function, or work function (Helmholtz function) are given, the equation is called a thermodynamic equation, or better, a caloric equation of state, although the latter is not a commonly accepted designation.

Even in the case of a simple system, one equation of state, e.g., the equation $f(p, V, T) = 0$, does not necessarily determine the form of all the other equations of state. This is connected with the fact that the derivation of the other equations of state may involve the integration of partial derivatives which leads to the appearance of whole functions in the integration "constant." An equation from which other equations of state can be derived by differentiation only, is called a *fundamental equation* (of state). In the case of a pure substance in a specified phase, the p, V, T relation does not constitute a fundamental equation with respect to the properties U, H, S, G, A, or their derived properties C_p, C_v, γ, etc. Consequently, it is possible to have two or more substances whose p, V, T relations are identical but whose specific heats, for example, are different.

In the case of continuous systems, for which the state changes from point to point(for example, a flow field of a viscous fluid), it is assumed that at every point, the equation of state is the same as for a homogeneous system and does not involve the gradients of the thermodynamic properties. Hence, such systems can only be studied with the aid of thermodynamics if local departures from equilibrium are small (near-equilibrium processes), i.e., if the gradients of the thermodynamic properties are not too great.

An equation of state must necessarily involve a finite (even if very large) number of independent variables. The particular variables which are chosen as independent is immaterial, on condition that they are mutually independent, and that their number is appropriate to the physical nature of the system.

Equations of states of various types of systems are numerous. The Curie equation is the equation of state of a paramagnetic solid. The Beattie and Bridgeman equation, Berthelot equation, Clausius equation, Dieterice equation, Keyes equation, and Van der Waals equation are other examples in this category.

It should be noted that equations of state for systems which consist of several components, rather than a single substance, can be written by introducing the variables N_1, N_2, ... N_c, which are the respective mole numbers of the components present.)

EQUILIBRIUM.

In the elementary sense of the macroscopic (visible to the naked eye) system, equilibrium is obtained if the system does not tend to undergo any further change of its own accord.

Mechanical and Electromagnetic Systems

Equilibrium in mechanical and/or electromagnetic systems is reached when the vectorial summation of generalized forces applied to the system is equal to zero. In any potential field, that is, gravitational or electric vector potential, force can be expressed as gradient of potential (magnetic force however, is a curl of a vector potential). The potential energy therefore has an extremum at the equilibrium configuration. For example, a system such as a mass suspended by a string against the gravitational force (or its weight) is at mechanical equilibrium if the tensile force in the string is equal to the weight of the mass it supports. The d'Alembert principle further states that the condition for equilibrium of a system is that the virtual work of the applied forces vanishes.

Thermodynamic Systems

When a hot body and a cold body are brought into physical contact, they tend to achieve the same warmth after a long time. These two bodies are then said to be at thermal equilibrium with each other. The zeroth law of thermodynamics (R.H. Fowler) states that two bodies individually at equilibrium with a third are at equilibrium with each other. This led to the comparison of the states of thermal equilibrium of two bodies in terms of a third body called a thermometer. The temperature scale is a measure of state of thermal equilibrium, and two systems at thermal equilibrium must have the same temperature.

Generalization of equilibrium consideration by the second law of thermodynamics specifies that the state of thermodynamic equilibrium of a system is characterized by the attainment of the maximum of its entropy. Thermodynamic coordinates are defined in terms of equilibrium states.

Equilibrium between two phases of a system is reached when there is no net transfer of mass or energy between the phases. Phase equilibrium is determined by the equality of the Gibbs functions (also called free enthalpy, free energy, or chemical potential) of the phases in addition to equality of their temperatures and stresses (such as pressure and/or field intensities—intensive properties). Equilibrium of first-order phase change requires continuity of slope or first derivative of the Gibbs function with respect to an intensive property and is generalized as the Clapeyron relation. Second- and higher-order phase changes are given by the condition of continuity of curvature or second derivative of the Gibbs function and so on.

Chemical or nuclear equilibrium of a reactive system is reached when there is no net transfer of mass and/or energy between the components

of a system. At chemical or nuclear equilibrium, the Gibbs function of the reactants and the products must be equal according to stoichiometric proportions, in addition to uniformity in temperature and stresses. Chemical equilibrium is summarized in the form of the Law of Mass Action. The trend for the displacement from an equilibrium state is specified by LeChâtelier's principle.

Thermodynamic equilibrium is reached when the condition of mechanical, electromagnetic, thermal, phase, and chemical and nuclear equilibrium is reached.

Stability of Equilibrium

A process or change of state carried out on a system such that it is always near a state of equilibrium is called a quasi-stationary equilibrium. This requires that the process be carried out slowly. If a mechanical system is initially at the equilibrium position with zero initial velocity, then the system will continue at equilibrium indefinitely. An equilibrium position is said to be stable if a small disturbance of the system from equilibrium results only in small, bounded motion about the rest position. The equilibrium is unstable if an infinitesimal displacement produces unbounded motion. In the gravitational field, a marble at rest in the bottom of a bowl is in stable equilibrium, but an egg standing on its end is in unstable equilibrium. When motion can occur about an equilibrium position without disturbing the equilibrium, the system is in neutral (or labile, or indifferent) equilibrium, an example being a marble resting on a perfectly flat plane normal to the direction of gravity. It is readily seen that stable equilibrium is the case when the extremum of potential is a minimum.

When dealing with general thermodynamic systems, the fact that entropy tends to a maximum in the trend toward equilibrium of a natural process generalizes the above mechanical consideration with respect to stability. An equilibrium state can be characterized as a stable equilibrium when the entropy is a maximum; neutral equilibrium when displacement from one equilibrium state to another does not involve changing entropy; and unstable equilibrium when entropy is a minimum. Any slight disturbance from an unstable equilibrium state of a system will lead to transition to another state of equilibrium.

Statistical Equilibrium

In the microscopic sense, that is, treating systems in terms of elemental particles such as molecules, atoms, and other material or quasi-particles (such as photons in radiation, phonons in solids and liquids), equilibrium states are recognized as the most probable states. An equilibrium state of a system is therefore defined in terms of most probable distributions of its elements among microscopic states which may be defined in terms of energy states. In this sense, statistical equilibrium is a condition for macroscopic equilibrium and an equilibrium state of a system is one of its extremal states. In the methods of statistical mechanics, the probability of distribution is expressed in terms of the density of distributions in the phase space. Based on the Liouville theorem, if a system is in statistical equilibrium, the number of the elements in a given state must be constant in time; which is to say that the density of distribution at a given location in phase space does not change with time. For an isolated system, the distribution is represented by a microcanonical ensemble. At equilibrium, no phase point can cross over a surface of constant energy, and the density of distribution is preserved. In this case individual molecules of a system can be represented by phase points. Any part of an isolated system in statistical equilibrium can be represented by a canonical ensemble. A subsystem of a large system in thermal equilibrium also behaves like the average system of a canonical ensemble. A system and a constant temperature bath together can be considered as an isolated system. A phase point in a canonical ensemble can represent a large number of molecules, thus accounting for strong interactions. A canonical ensemble is characterized by its temperature and is therefore pertinent to the concept of thermal equilibrium. When applied to equilibrium of systems involving mass exchange, such as a chemical system, we have a "particle bath" in addition to a constant temperature bath. The pertinent representation for equilibrium including mass exchange as well as energy exchange is known as a grand canonical ensemble, which accounts for the chemical potentials of its elements.

When applied to a system with a large number of elements, the distributions are measured by thermodynamic probability (W); the most probable distribution is such that W is a maximum. This optimal principle is consistent with the condition of maximum entropy (S) cited under **Entropy**. The Boltzmann hypothesis states that $S = k \ln W$, where k is the Boltzmann constant.

Depending on the specifications of W, namely, those of Maxwell-Boltzmann (for low concentration of distinguishable particles, weak interaction and high temperature, such as a dilute perfect gas), Fermi-Dirac (for elemental particles with antisymmetric wave functions at high concentrations of indistinguishable particles and low temperatures, such as electrons in metal), or Einstein-Bose (for elemental particles with symmetric wave functions, such as He^4 at high concentration of indistinguishable particles and low temperature), equilibrium distributions take different forms. The Maxwellian speed distribution in a dilute perfect gas is a distribution based on Maxwell-Boltzmann statistics.

As a consequence of molecular considerations, when two systems are connected for transfer of mass without significant transfer of energy, such as two containers at different temperatures connected by a capillary tube, we have the relation of thermal transpiration.

Trend toward Equilibrium

The mechanism by which equilibrium is attained can only be visualized in terms of microscopic theories. In the kinetic sense, equilibrium is reached in a gas when collisions among molecules redistribute the velocities (or kinetic energies) of each molecule until a Maxwellian distribution is reached for the whole bulk. In the case of the trend toward equilibrium for two solid bodies brought into physical contact, we visualize the transfer of energy by means of free electrons and phonons (lattice vibrations).

The Boltzmann H-theorem generalizes the condition that with a state of a system represented by its distribution function f, a quantity H, defined as the statistical average of $\ln f$, approaches a minimum when equilibrium is reached. This conforms to the Boltzmann hypothesis of distribution in the above in that $S = -kH$ accounts for equilibrium as a consequence of collisions which change the distribution toward that of equilibrium conditions.

Consideration of perturbation from an equilibrium state leads to methods for dealing with rate processes and methods of irreversible thermodynamics in general.

Fluctuation from Equilibrium

A necessary consequence of the random nature of elemental particles in a body is that the property of such a body is not at every instant equal to its average value but fluctuates about this average. A precise meaning of equilibrium can only be attained from consideration of the nature of such fluctuations. In the above, we have repeatedly considered a "large" number of particles. It is important to know how large a number is "large." When considering fluctuation of energy from an average value in an isolated system, the ratio of the two is given to be proportional to $1/\sqrt{N}$, where N is the total number of elements in the system. This is also the magnitude of the fluctuation of number of particles in a system involving transformation of phases and chemical and nuclear species. An equilibrium state is one at which the longtime mean magnitude of fluctuation from the average state is independent of time and this magnitude has reached a minimum value.

Large perturbation from a given state of fluctuation leads to a relaxation process toward a state of equilibrium. The relaxation time, for instance, measures the deviation from quasistationary equilibrium of a process carried out at a finite rate.

Additional Reading

Balescu, R.: *Statistical Dynamics: Matter out of Equilibrium,* World Scientific Publishing Company, Inc., River Edge, NJ, 1997.

Denbigh, K.: *The Principles of Chemical Equilibrium,* Cambridge University Press, New York, NY, 1994.

Jones, W. and N. March: *Theoretical Solid State Physics: Perfect Lattices in Equilibrium, Vol. 1,* Dover Publications, Inc., Mineola, NY, 1990.

Mazenko, G.F.: *Equilibrium Statistical Mechanics,* John Wiley & Sons, Inc., New York, NY, 2000.

Ozkaya, N. and M. Nordin: *Fundamentals of Biomechanics: Equilibrium, Motion and Deformation,* 2nd Edition, Springer-Verlag, Inc., New York, NY, 1999.

Prausnitz, J.M., R.N. Lichtenthaler, and E.G. Azevedo: *Molecular Thermodynamics of Fluid-Phase Equilibria,* Prentice-Hall, Inc., Upper Saddle River, NJ, 1998.

Stauffer, D. and D. Chowdhury: *Principals of Equilibrium Statistical Mechanics,* John Wiley & Sons, Inc., New York, NY, 2000.

Zwanzig, R.: *Non-Equilibrium Statistical Mechanics,* Oxford University Press, Inc., New York, NY, 2001.

EQUILIBRIUM DIAGRAM. A diagram showing the phase fields of an alloy system under the conditions of complete equilibrium using as coordinates the temperature, the compositions in terms of the components, and the pressure. The most frequently used equilibrium diagrams in metallurgy are drawn with the pressure considered constant. See iron-carbon diagram under **Iron Metals, Alloys, and Steels**. See also **Distillation**.

EQUIVALENT ELECTRONS. For an atom, electrons in the same orbital (whereby they have the same principal quantum number and the same azimuthal quantum number). For a molecule, electrons having the same quantum numbers, apart from spin, and the same symmetry g or u.

ERBIUM[1]. [CAS: 7440-52-0]. Chemical element symbol Er, at. no. 68, at. wt. 167.26, eleventh in the Lanthanide Series in the periodic table, mp 1529°C, bp 2868°C, density 9.066 g/cm³ (20°C). Elemental erbium has a close-packed hexagonal crystal structure at 25°C. The pure metallic erbium is silver-gray in color and retains its luster at room temperature, not affected by moisture or normal atmospheric gases. Large pieces of the metal do not oxidize readily even when heated. Fine chips and powder, however, will ignite and burn. Because of its comparative softness, the metal can be worked by conventional equipment. The metal should be annealed after size-reduction. There are six natural isotopes ^{162}Er, ^{164}Er, ^{166}Er through ^{168}Er and ^{170}Er. Twelve artificial isotopes have been prepared. The natural isotopes are not radioactive. In terms of abundance, erbium is present on the average of 2.8 ppm in the earth's crust, making its potential availability about equal with uranium. The element was first identified by C.G. Mosander in 1843. The thermal-neutron-absorption cross section of erbium is 160 borns per atom, relatively high and tenth among the natural elements. The metal has a low acute-toxicity rating. Electronic configuration

$$1s^2 2s^2 2p^6 3s^2 3p^6 3d^{10} 4s^2 4p^6 4d^{10} 4f^{11} 5s^2 5p^6 5d^1 6s^2.$$

First ionization potential 6.10 eV; second 11.93 eV. Ionic radius Er^{3+} 0.881 Å. Metallic radius 1.758 Å. Other important physical properties of erbium are given under **Rare-Earth Elements and Metals**.

Erbium occurs in certain types of apatites, xenotime, and gadolinite. These minerals also are processed for their yttrium content as well as for other heavy *Lanthanide* elements. With liquid-liquid organic and solid-resin organic ion-exchange techniques, the separation of erbium from the other elements is favorable.

Because of the metal's high thermal-neutron-absorption cross section, it has been of much interest in terms of use in nuclear reactor hardware. When an erbium-activated phosphor is coated onto a gallium-arsenide diode, the latter emits infrared radiation, which is converted to visible light by the phosphor. Through variation of the energizing power and by use of a combination of rare-earth-activated phosphors, the primary colors of light can be produced. Thus, erbium holds promise for use in display panels and color-television picture tubes. An erbium hydride-hydrogen system at a fixed temperature creates an extreme vacuum and when used for comparative purposes makes it possible to measure vacuums in the range of 10^{-4} to 10^{-11} torr with much precision. The system has been used for the calibration of ionization gauges used for very high vacuums as found in outer space. Erbium is in an early stage of investigation for application to lasers, semiconductor devices, garnet microwave devices, ferrite bubble devices, and catalysts.

See references listed at ends of entries on **Chemical Elements**; and **Rare-Earth Elements and Metals**.

Erbium in Lightwave Communication Amplifier

In January 1992, E. Desurvire (Columbia University Center for Telecommunications Research) reported that optical fibers made from silica glass and traces of erbium can amplify light signals when they are energized by infrared radiation. Desurvire developed an efficient radiation source (referred to as a laser diode chip) that, when integrated into a fiber optic communication system, can increase transmission capacity by a factor of 100.

[1]The main portion of this article was revised and updated by K.A. Gschneidner, Jr., Director, and B. Evans, Assistant Chemist, Rare-Earth Information Center, Energy and Mineral Resources Institute, Iowa State Univ., Ames, Iowa.

The device can be effective in very long stretches of communication cable, such as used in transoceanic service. Each cable will be capable of carrying 500,000 messages simultaneously, a factor of 12 times greater than present cables.

ERGOSTEROL. See Photochemistry and Photolysis.

ERNST, RICHARD R. (1933–). A native of Switzerland who won the Nobel prize in chemistry in 1991 for important methodological developments in NMR spectroscopy. He invented Fourier-transform NMR (FT-NMR), which multiplied sensitivity 10 to 100 times compared to dispersive instruments. He also devised two-dimensional NMR techniques, increasing resolution and enabling structure determinations of biologically important macromolecules. Ernst received his Ph.D from the Federal Technical Institute (ETH) in Zurich, Switzerland.

ERYTHRITE. A mineral of the composition, $Co_3(AsO_4)_2 \cdot 8H_2O$, isomorphous with annabergite. The color ranges from rose to crimson. The mineral sometimes contains nickel and occurs in monoclinic crystals, in earthy forms (as a weathering product of cobalt ores) in the oxidized portions of the veins, or in globular and reniform masses. The mineral sometimes is referred to as *erythrine, cobalt bloom,* and *peachblossom ore.*

ERYTHROCYTES.   See Blood.

ESSENTIAL AMINO ACIDS.   See Amino Acids.

ESTERIFICATION.   This article describes methods for the production of carboxylic esters:

$$\begin{array}{c} O \\ \parallel \\ R-C-OR' \end{array}$$

For the properties of some of these compounds, see **Esters, Organic.**

Esters are most commonly prepared by the reaction of a carboxylic acid and an alcohol with the elimination of water. Esters are also formed by a number of other reactions utilizing acid anhydrides, acid chlorides, amides, nitriles, unsaturated hydrocarbons, ethers, aldehydes, ketones, alcohols, and esters (via ester interchange).

On the basis of bulk production, poly(ethylene terephthalate) manufacture is the most important ester producing process. This polymer is produced by either the direct esterification of terephthalic acid and ethylene glycol, or by the transesterification of dimethyl terephthalate with ethylene glycol. Dimethyl terephthalate is produced by the direct esterification of terephthalic acid and methanol.

Other large-volume esters are vinyl acetate (VAM), methyl methacrylate (MMA), and dioctyl phthalate (DOP). VAM is produced for the most part by the vapor-phase oxidative acetoxylation of ethylene. MMA and DOP are produced by direct esterification techniques involving methacrylic acid and phthalic anhydride, respectively.

The acetates of most alcohols are also commercially available and have diverse uses. Because of their high solvent power, ethyl, isopropyl, butyl, isobutyl, amyl, and isoamyl acetates are used in cellulose nitrate and other lacquer-type coatings. Butyl and hexyl acetates are excellent solvents for polyurethane coating systems, see also **Urethane Polymers.** Ethyl, isobutyl, amyl, and isoamyl acetates are frequently used as components in flavoring, see also **Flavors and Essences,** and isopropyl, benzyl, octyl, geranyl, linalyl, and methyl acetates are important additives in perfumes.

Effect of Structure. The rate at which different alcohols and acids are esterified as well as the extent of the equilibrium reaction are dependent on the structure of the molecule and types of functional substituents of the alcohols and acids.

In making acetate esters, the primary alcohols are esterified most rapidly and completely, i.e., methanol gives the highest yield and the most rapid reaction. Under the same conditions, the secondary alcohols react much more slowly and afford lower conversions to ester products; however, wide variations are observed among the different members of this series. The tertiary alcohols react slowly and the conversions are generally low (1–10% conversion at equilibrium).

The introduction of a nitrile group on an aliphatic acid has a pronounced inhibiting effect on the rate of esterification.

Substitutions that displace electrons toward the carboxyl group of aromatic acids diminish the rate of the reaction.

Kinetic Considerations. Extensive kinetic and mechanistic studies have been made on the esterification of carboxylic acids since Berthelot and Saint-Gilles first studied the esterification of acetic acid. A number of mechanisms for acid- and base-catalyzed esterification have been proposed. One possible mechanism for the bimolecular acid-catalyzed ester hydrolysis and esterification is shown below.

This mechanism leads to the rate equation for hydrolysis and to an analogous expression for the esterification:

$$-\frac{d[E]}{dt} = \frac{k_1 K_1 [E][H_2O][H^+]}{1+\alpha} - \frac{k_2 K_2 [A][R'OH][H^+]}{1+1/\alpha}$$

In this expression, α depends on those rate coefficients in the above mechanism whose values are assumed to be high. Other mechanisms for the acid hydrolysis and esterification differ mainly with respect to the number of participating water molecules and possible intermediates.

Applications of kinetic principles to industrial reactions are often useful. Initial kinetic studies of the esterification reaction are usually conducted on a small scale in a well stirred batch reactor. In many cases, results from batch studies can be used in the evaluation of the esterification reaction in a continuous operating configuration.

Equilibrium Constants. The reaction between an organic acid and an alcohol to produce an ester and water is expressed as:

$$\begin{array}{ccc} O & & O \\ \parallel & & \parallel \\ RCOH + R'OH & \rightleftharpoons & RCOR' + H_2O \end{array}$$

This was first demonstrated in 1862 by Berthelot and Saint-Gilles, who found that when equivalent quantities of ethyl alcohol and acetic acid were allowed to react, the esterification stopped when two-thirds of the acid had reacted. Similarly, when equal molar proportions of ethyl acetate and water were heated together, hydrolysis of the ester stopped when about one-third of the ester was hydrolyzed. By varying the molar ratios of alcohol to acid, yields of ester >66% were obtained by displacement of the equilibrium. The results of these tests were in accordance with the mass action law shown. $K = [ester][water]/[acid][alcohol]$. However, in many cases the equilibrium constant is affected by the proportion of reactants. The temperature as well as the presence of salts may also affect the value of the equilibrium constant.

Completion of Esterification. Because the esterification of an alcohol and an organic acid involves a reversible equilibrium, these reactions usually do not go to completion. Conversions approaching 100% can often be achieved by removing one of the products formed, either the ester or the water, provided the esterification reaction is equilibrium limited and not rate limited. A variety of distillation methods can be applied to afford ester and water product removal from the esterification reaction. See also **Distillation.** Other methods such as reactive extraction and reverse osmosis can be used to remove the esterification products to maximize the reaction conversion. In general, esterifications are divided into three broad classes, depending on the volatility of the esters: (1) Esters of high volatility, such as methyl formate, methyl acetate, and ethyl formate, have lower boiling points than those of the corresponding alcohols, and therefore can be readily removed from the reaction mixture by distillation. (2) Esters of medium volatility are capable of removing the water formed by distillation. Examples are propyl, butyl, and amyl formates, ethyl, propyl, butyl, and amyl acetates, and the methyl and ethyl esters of propionic, butyric, and valeric acids. (3) Esters of low volatility are accessible via several types of esterification.

Use of Azeotropes to Remove Water. With the aliphatic alcohols and esters of medium volatility, a variety of azeotropes is encountered on distillation. Removal of these azeotropes from the esterification reaction mixture drives the equilibrium in favor of the ester product.

Use of Desiccants and Chemical Means to Remove Water. Another means to remove the water of esterification is calcium carbide supported in a thimble of a continuous extractor through which the condensed vapor from the esterification mixture is percolated. A column of activated bauxite (Florite) mounted over the reaction vessel has been used to remove the water of reaction from the vapor by adsorption.

Catalysts. The choice of the proper catalyst for an esterification reaction is dependent on several factors. The most common catalysts used are strong mineral acids such as sulfuric and hydrochloric acids. Lewis acids such as boron trifluoride, tin and zinc salts, aluminum halides, and organo–titanates have been used. Cation-exchange resins and zeolites are often employed also.

Acid-Regenerated Cation Exchangers. The use of acid-regenerated cation resin exchangers as catalysts for effecting esterification offers distinct advantages over conventional methods. Several types of cation-exchange resins can be used as solid catalysts for esterification. In general, the strongly acidic sulfonated resins comprised of copolymers of styrene, ethylvinylbenzene, and divinylbenzene are used most widely. With the continued improvement of ion-exchange resins, such as the macroporous sulfonated resins, esterification has become one of the most fertile areas for use of these solid catalysts.

Despite the higher cost compared with ordinary catalysts such as sulfuric or hydrochloric acid, the cation exchangers present several features that make their use economical. The ability to use these agents in a fixed-bed reactor operation makes them attractive for a continuous process. Cation-exchange catalysts can be used also in continuous stirred tank reaction (CSTR) operation.

Batch Esterification

Batch esterification is used to produce ethyl acetate (Fig. 1) and *n*-butyl acetate.

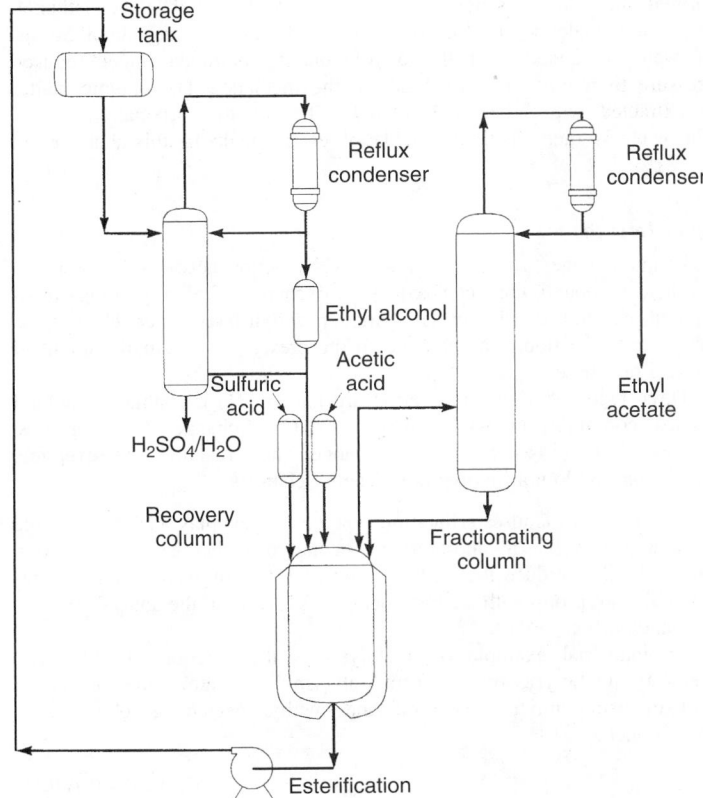

Fig. 1.   Batch ethyl acetate process

Continuous Esterification

The law of mass action, the laws of kinetics, and the laws of distillation all operate simultaneously in a process of this type. Esterification can occur only when the concentrations of the acid and alcohol are in excess of equilibrium values; otherwise, hydrolysis must occur. The equations governing the rate of the reaction and the variation of the rate constant (as a function of such variables as temperature, catalyst strength, and proportion of reactants) describe the kinetics of the liquid-phase reaction. The usual distillation laws must be modified, since most esterifications are somewhat exothermic and reaction is occurring on each plate. Since these kinetic considerations are superimposed on distillation operations, each plate must be treated separately by successive calculations after the extent of conversion has been determined. See also **Distillation**.

Continuous esterification of acetic acid in an excess of *n*-butyl alcohol with sulfuric acid catalyst using a four-plate single bubblecap column with reboiler has been studied. The rate constant and the theoretical extent of reaction were calculated for each plate, based on plate composition and on the total incoming material to the plate. Good agreement with the analytical data was obtained.

A continuous distillation process has been studied for the production of high boiling esters from intermediate boiling polyhydric alcohols and low boiling monocarboxylic aliphatic or aromatic acids. The water of reaction and some of the organic acid were continuously removed from the base of the column.

Continuous esterification is used to produce methyl acetate (Fig. 2).

Vapor-Phase Esterification

Catalytic esterification of alcohols and acids in the vapor phase has received attention because the conversions obtained are generally higher than in the corresponding liquid-phase reactions.

Physicochemical Considerations. The determination of the equilibrium constant K_G for the reaction $C_2H_5OH + CH_3COOH \rightarrow C_2H_5OOCH_3 + H_2O$ has been the subject of a number of investigations over the temperature range of 40–300°C. The values of the equilibrium constant range from 6–559 with 71–95% ester as the equilibrium concentration from an equimolar mixture of ethyl alcohol and acetic acid, depending on the technique used. A study of the reaction mechanism indicates that adsorption of acetic acid is the rate-controlling step; the molecularly adsorbed acetic acid then reacts with alcohol in the vapor phase.

Ethyl Acetate. Catalysts proposed for the vapor-phase production of ethyl acetate include silica gel, zirconium dioxide, activated charcoal, and potassium hydrogen sulfate. More recently, phosphoric-acid-treated coal and calcium phosphate catalysts have been described.

Esterification of Other Compounds

Acid Anhydrides. Acid anhydrides react with alcohols to form esters (in high yields in many cases) with a carboxylic acid formed as by-product:

$$\underset{RC-O-CR}{\overset{O \quad\quad O}{\parallel \quad\quad \parallel}} + R'OH \longrightarrow \underset{RC-OR'}{\overset{O}{\parallel}} + \underset{RC-OH}{\overset{O}{\parallel}}$$

However, this method is applied only when esterification cannot be effected by the usual acid–alcohol reaction because of the higher cost of the anhydrides. The production of cellulose acetate. See also **Fibers: Cellulose Esters**, phenyl acetate (used in acetaminophen production), and aspirin (acetylsalicylic acid) (see also **Salicylic Acid and Related Compounds** are examples of the large-scale use of acetic anhydride.

Formic anhydride is not stable. However, formate esters of alcohols and phenolics can be prepared using formic–acetic anhydride. Dibasic acid anhydrides such as phthalic anhydride and maleic anhydride readily react with alcohols to form the monoalkyl ester. Ketene, like acid anhydrides, reacts with alcohols to form (acetate) esters.

Acid Chlorides. Acid chlorides react with alcohols to form esters.

Amides. Alcoholysis of amides provides another method for synthesizing esters.

Other methods of converting amides to esters have been described. Alkyl halides can be treated with amides to give esters. Also, esters can be synthesized from *N*-alkyl-*N*-nitrosoamides, which are derived from the corresponding amides.

Nitriles. Alcoholysis of nitriles offers a convenient way to produce esters without isolating the acid. Catalysts such as hydrogen chloride,

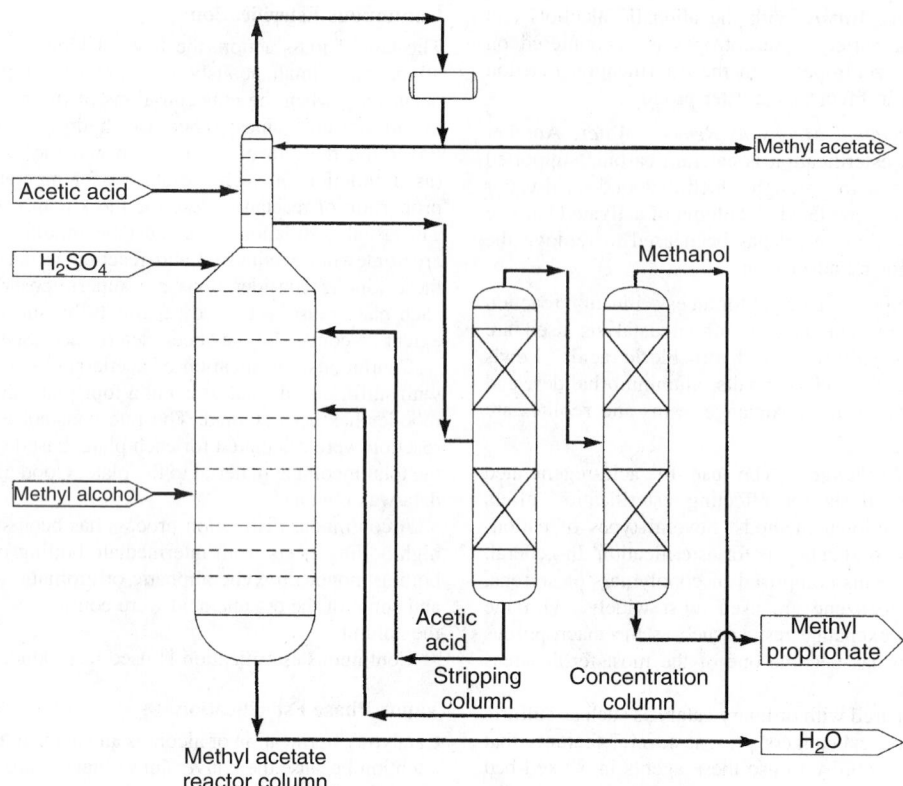

Fig. 2.   Continuous methyl acetate process

hydrogen bromide, and sulfuric acid have been employed. One of the most important applications of this process is that of methyl methacrylate manufacture.

Unsaturated Hydrocarbons. Olefins from ethylene through octene have been converted into esters via acid-catalyzed nucleophilic addition.

Most of the vinyl acetate produced in the United States is made by the vapor-phase ethylene process.

Ethers. In the presence of anhydrous agents such as ferric chloride, hydrogen bromide, and acid chlorides, ethers react to form esters. Esters can also be prepared from ethers by an oxidative process.

Unsaturated esters can be prepared from the corresponding acetylenic ethers with yields in most cases of >50%. β-Hydroxyethyl esters can be prepared from carboxylic acids and ethylene oxide.

Aldehydes and Ketones. Esters are obtained readily by condensation of aldehydes in the presence of alcoholate catalysts such as aluminum ethylate, $Al(OC_2H_5)_3$, by the Tishchenko reaction.

Alcohols. The direct synthesis of esters by dehydrogenation or oxidative hydrogenation of alcohols offers a simple method for the preparation of certain types of esters, such as ethyl acetate.

Technical Preparation of Esters

Esterification is generally carried out by refluxing the reaction mixture until the carboxylic acid has reacted with the alcohol and the water has been split off. The water of the ester is removed from the equilibrium by distillation. The choice of the esterification process to obtain a maximum yield is dependent on many factors, i.e., no single process has universal applicability.

Methyl Esters. Methyl esters are obtained in good yield using methylene dichloride or ethylene dichloride as solvent.

Medium Boiling Esters. Esterification of ethyl and propyl alcohols, ethylene glycol, and glycerol with various acids, e.g., chloro- or bromoacetic, or pyruvic, by the use of a third component such as benzene, toluene, hexane, cyclohexane, or carbon tetrachloride to remove the water produced is quite common.

High Boiling Esters. The following procedure can be used for making diethyl phthalate and other high boiling esters. Phthalic anhydride (1 equiv) and 2.5 equivalents of ethanol are refluxed for 2 h in the presence of 1% of concentrated H_2SO_4.

Difficulty Esterifiable Acids. The sterically hindered acids, such as 2,6-disubstituted benzoic acids, cannot usually be esterified by conventional means. Several esters of sterically hindered acids such as 2,4,6-triisopropylbenzoic acid have been prepared by dissolving 2 g of the acid in 14–20 mL of 100% H_2SO_4. After standing a few minutes at room temperature, when presumably the acylium cation is formed, the solution is poured into an excess of cold absolute methanol. Most of the alcohol is removed under reduced pressure, about 50 mL of water is added, and the distillation is continued under reduced pressure to remove the remainder of the methanol. The organic matter is extracted with ether and treated with sodium carbonate solution. The ester is then distilled. Yields of esters made in this manner are 57–81%.

Ester Interchange

Ester interchange (transesterification) is a reaction between an ester and another compound, characterized by an exchange of alkoxy groups or of acyl groups, and resulting in the formation of a different ester. The process of transesterification is accelerated in the presence of a small amount of an acid or a base.

Three types of transesterification are known: (1) exchange of alcohol groups, commonly known as alcoholysis, (2) exchange of acid groups, acidolysis, and (3) ester–ester interchange. These reactions are reversible and ordinarily do not involve large energy changes.

Applications. Transesterifications via alcoholysis play a significant role in industry as well as in laboratory and in analytical chemistry. The reaction can be used to reduce the boiling point of esters by exchanging a long-chain alcohol group with a short one, eg, methanol, in the analysis of fats, oils, and waxes.

An industrial example of acidolysis is the reaction of poly(vinyl acetate) with butyric acid to form poly(vinyl butyrate). Often a butyric acid–methanol mixture is used and methyl acetate is obtained as a coproduct.

MOHAMMAD ASLAM
G. PAULL TORRENCE
EDWARD G. ZEY
Hoechst Celanese Corporation

Additional Reading

Larock, R.C.: *Comprehensive Organic Transformations,* VCH Publishers, Inc., New York, NY, 1989.

Markley, K.S. in Markley, K.S. ed.: *Fatty Acids,* part 2, Wiley-Interscience, New York, NY, 1961, p. 757.

Otera, J.: *Esterification: Methods, Reactions, and Applications,* John Wiley & Sons, Inc., New York, NY, 2003.

Patai, S.: *The Chemistry of Carboxylic Acids and Esters,* Wiley-Interscience, New York, NY, 1969.

Reid, E.E. in P. Grotggins: *Unit Processes in Organic Synthesis,* 5th ed., McGraw-Hill Book Co., Inc., New York, NY, 1958.

ESTERS, ORGANIC. Esters are compounds that, on hydrolysis, yield alcohols or phenols and acids according to the equation: $RA + H_2O \leftrightarrows ROH + HA$, where R is a hydrocarbon fragment and A is the anion portion of an organic acid.

Nomenclature

The names of esters consist of two words that reflect their formation from an alcohol and a carboxylic acid. According to the IUPAC rule, the alkyl or aryl group of the alcohol is cited first followed by the carboxylate group of the acid with the ending -ate replacing the -ic of the acid. For example, $CH_3CH_2COOCH_3$, the methyl ester of propanoic acid, is called methyl propanoate (or methyl propionate, if the trivial name, propionic acid, is used for the carboxylic acid).

Physical Properties

The physical properties of organic esters vary according to the molecular weight of each component. Lower molecular weight esters are colorless, mobile, and highly volatile liquids that usually have pleasant odors. As the molecular weight increases, volatility decreases and the consistency becomes waxy, then solid, and eventually even brittle, often with formation of lustrous crystals. The melting point of an ester is generally lower than that of the corresponding carboxylic acid. However, the boiling point depends on the chain length of the alcohol component and eventually exceeds that of the acid. Lower molecular weight esters are relatively stable when dry and can be distilled without decomposition. Organic esters are generally insoluble in water, but soluble in various organic liquids. Lower esters are themselves good solvents for many organic compounds. The physical properties of commercially important aliphatic and aromatic organic esters are listed in Table 1.

Chemical Properties

The reactions of esters have been reviewed. Because of the large number of possible acid and alcohol moieties, the chemical properties of esters may differ considerably. Only typical reactions applicable to the majority of esters are described in the following sections.

Hydrolysis. Esters are cleaved (hydrolyzed) into an acid and an alcohol through the action of water. This hydrolysis is catalyzed by acids or bases. The mechanistic aspects of ester hydrolysis have received considerable attention and have been reviewed.

Enzymatic Hydrolysis. Enzymatic hydrolysis has received enormous attention. The enzymes generally employed are lipases from microorganisms, plants, or mammalian liver. The great advantage of the enzymatic process is its high chemo- and stereoselectivity.

Transesterification. When esters are heated with alcohols, acids, or other esters in the presence of a catalyst, the alcohol or acid groups are exchanged. This process is called transesterification. It is accelerated by the presence of a small amount of acid or alkali. Three types of transesterification are known: (*1*) exchange of alcohol groups (alcoholysis), (*2*) exchange of acid groups (acidolysis), and (*3*) ester–ester interchange. See also **Esterification**.

Ammonolysis and Aminolysis. Esters and ammonia react to form amides and alcohols.

If primary or secondary amines are used, *N*-substituted amides are formed. This reaction is called aminolysis.

Reduction. Esters can be reduced to alcohols by catalytic hydrogenation using molecular hydrogen or by chemical reduction.

Reaction of Enolate Anions. In the presence of certain bases, eg, sodium alkoxide, an ester having a hydrogen on the α-carbon atom undergoes a wide variety of characteristic enolate reactions.

Mechanistically, the base removes a proton from the α-carbon, giving an enolate that then can react with an electrophile.

Grignard and Related Reactions. Esters react with alkyl magnesium halides in a two-stage process to give alcohols. The reaction involves nucleophilic substitution of R^3 or OR^2 and addition of R^3MgX to the carbonyl group. With 1,4-dimagnesium compounds, esters are converted to cyclopentanols. Lactones react with Grignard reagents and give diols as products. Many other organometallic compounds also react with carbonyl.

α-Halo esters react with aldehydes or ketones in the presence of zinc to form β-hydroxy esters. This is known as the Reformatsky reaction.

TABLE 1. PHYSICAL PROPERTIES OF SOME COMMON ESTERS

Ester	Mol wt	n_D^{20}	Bp, °C[a]	Freezing point, °C
methyl formate	60.05	1.344	32	−99.8
ethyl formate	74.08	1.3598	54.3	−80
butyl formate	102.13	1.3889	106	−91.9
methyl acetate	74.08	1.3594	57	−98.1
ethyl acetate	88.1	1.3723	77.1	−83.6
vinyl acetate	86.1	1.3959	72.2	−93.2
propyl acetate	102.13	1.3844	101.6	−92.5
isopropyl acetate	102.13	1.3773	90	−73.4
butyl acetate	116.16	1.3951	126	−73.5
isobutyl acetate	116.16	1.3902	117.2	−98.6
sec-butyl acetate	116.16	1.3877	112	
t-butyl acetate	116.16	1.3855	97	
pentyl acetate	130.18	1.4023	149.3	−70.8
isoamyl acetate	130.18	1.4000	142	−78
sec-hexyl acetate	144.22	1.4014	157	0
2-ethylhexyl acetate	172.26	1.4204	199.3	−93
ethylene glycol diacetate	146.14	1.415	191	−31
2-methoxyethyl acetate	118.13	1.4019	145	−65.1
2-ethoxyethyl acetate	132.16	1.4058	156.4	−61.7
2-butoxyethyl acetate	160.12	1.42	187.8	−32
2-(2-ethoxyethoxy)ethyl acetate	176.21	1.423	217.4	−25
2-(2-butoxyethoxy)ethyl acetate	204.27	1.4265	247	−32.2
benzyl acetate	150.18	1.5232	215.5	−51.5
glyceryl triacetate	218.23	1.4296	258	−78
ethyl 3-ethoxypropionate	146.19		165–172	−50
glyceryl tripropionate	260.3	1.4318	176	−58
methyl acrylate	86.09	1.4040	80.5	<−75
ethyl acrylate	100.11	1.4068	99.8	<−72
butyl acrylate	128.17	1.4185	69	−64.6
2-ethylhexyl acrylate	184.28		130[i]	−90
methyl methacrylate	100.12	1.4119	100	−48
methyl butyrate	102.13	1.3878	102.3	−84.8
ethyl butyrate	116.16	1.4000	121.6	−100.8
butyl butyrate	144.22	1.4075	166.6	−91.5
methyl isobutyrate	102.13	1.3840	92.6	−84.7
ethyl isobutyrate	116.16	1.3870	110	−88
isobutyl isobutyrate	144.22	1.3999	148.7	−80.7
methyl stearate	298.5	1.457	215	40
ethyl stearate	312.52	1.429	213–215	33.7
butyl stearate	340.58		343	27.5
dodecyl stearate	440.8	1.433		28
hexadecyl stearate	496.91	1.441		57
dimethyl maleate	144.13	1.4409	204	
dimethyl oxalate	111.09	1.4096	185	−41
dimethyl adipate	174.2	1.4283	115	10.3
diethyl adipate	202.25	1.4372	245	−19.8
di(2-ethylhexyl) adipate	370.58	1.4472	214	−60
methyl benzoate	136.15	1.517	199.5	−12.5
ethyl benzoate	150.18	1.505	212.9	−34.2
methyl salicylate	152.15	1.536	223.3	−8.6
ethyl salicylate	166.18	1.522	231.5	1.3
dimethyl phthalate	194.19	1.515	282	−2
diethyl phthalate	222.24	1.499	295	−33
dibutyl phthalate	278.35	1.4911	340	−35
di(2-ethylhexyl) phthalate	390.56	1.486	231	−50
dimethyl isophthalate	194.19	1.5168	124	67
dimethyl terephthalate	194.19		288	140
methyl anthranilate	151.17	1.584	132	24
benzyl cinnamate	238.29		244	39
dimethyl carbonate	90.08	1.3682	90	3
diethyl carbonate	118.13	1.3854	127	−43

[a] At 101.3 kPa = 760 mm Hg unless otherwise stated.

Preparation of Acyloins. When aliphatic esters are allowed to react with metallic sodium, potassium, or sodium–potassium alloy in inert solvents, acyloins (α-hydroxyketones) are formed.

Pyrolysis. The pyrolysis of simple esters of the formula $RCOOCR^1R^2CHR^3_2$ to form the free acid and an alkene is a general reaction that is used for producing olefins.

Carbonylation Reaction. The carbonylation of methyl acetate is an important industrial reaction for producing acetic anhydride.

Substitution, Alkylation, and Rearrangement. The reaction of alkaline phenoxides with alkyl *S*-2-(chloro)- or *S*-2-(mesyloxy)propionate gives optically active *R*-2-aryloxyalkanoic acid esters in good chemical and optical yields (>97%*ee*). The reaction is utilized in the synthesis of several phenoxy herbicides.

Optically active 2-arylalkanoic acid esters have been prepared by Friedel-Crafts alkylation of arenes with optically active esters, such as methyl *S*-2-(chlorosulfonoxy)- or *S*-2-(mesyloxy)propionate, in the presence of aluminum chloride.

The Fries rearrangement of phenol esters gives a mixture of 2- and 4-acylphenols. Similarly, enol esters undergo rearrangement to give the corresponding 1,3-diketones.

Occurrence and Preparation

Currently, most of the simple esters used commercially are of synthetic origin, although esters occur naturally in large quantities in fats, oils, and waxes. Microorganisms produce a complex array of compounds containing the ester linkage, ranging from simple esters to macrocyclic lactones, such as erythromycin, which are important because of their antibacterial properties.

Recovery of naturally occurring esters is accomplished by steam distillation, extraction, pressing, or by a combination of these processes. Synthetic esters are generally prepared by reaction of an alcohol with an organic acid in the presence of a catalyst such as sulfuric acid, *p*-toluenesulfonic acid, or methanesulfonic acid.

Stability and Storage

All organic esters are unstable in the presence of acid or base and nucleophiles such as water or alcohols. However, if stored anhydrous, they are stable. Storage vessels can be constructed of steel, aluminum, or other metallic materials, but plastic storage tanks are unsuitable because the highly lipophilic esters can sometimes permeate into the container boundary and soften or even dissolve it.

The properties of flash point, autoignition temperature, and flammable limit should be considered when an ester is to be handled in any fashion.

Health and Safety Factors

Toxicity. The degree of toxicity of organic esters covers a wide range. These toxicities are usually described in terms of threshold limiting values (TLV), or permissible exposure limits (PEL). Both the PEL and the TLV describe the average concentration over an 8-h period to which a worker may be exposed without adverse effects. The lethal dosages for 50% of the exposed animals, LD_{50}s, are also used as an indicator of the relative toxicity. The LD_{50}s of organic esters for small mammals range between 0.4 and 16 g/kg. The TLVs of organic esters range between 5 and 400 ppm.

When ingested or absorbed, organic esters are likely to be hydrolyzed to the corresponding alcohols and carboxylic acids. Therefore the toxicities of the hydrolysis products should also be considered. Some organic esters are highly volatile and can act as asphixiant or narcotic. Also, skin absorption and inhalation are among the hazards associated with esters that are volatile or have good solvent action. Because of the high solubility of fats and oils in organic esters, prolonged or repeated exposure to skin can cause drying and irritation.

Acetates generally do not cause any physiological effects unless high exposure occurs since they are usually converted into or occur naturally as metabolites. However, large enough exposure to acetate esters can cause narcotic effects.

Propionates and higher aliphatic esters generally become less toxic as the size of the alkyl carboxylate increases.

The acrylate esters are more physiologically hazardous than their saturated homologues. They are usually lachrymators and irritants, and their toxicities decrease with increasing molecular weight.

Among adipates, oxalates, malonates, and succinates, the adipates are the least toxic.

Benzoate esters, like most organic esters, are not very toxic. They are not absorbed through the skin as rapidly as alkyl esters but are more potent physiologically. They are also moderate skin irritants.

The phthalate esters are one of the most widely used classes of organic esters, and fortunately they exhibit low toxicity.

More information on the toxicities of a range of organic esters is available in the literature.

Exposure Limits. The Occupational Safety and Health Act (OSHA) of 1990 lists a multitude of acetates, phthalates, formates, and acrylates along with the corresponding permissible exposure limits and threshold limit values. If there is potential for exposure to an organic ester for which PEL or TLV data has been identified, then an exposure limit lower than that listed.

Regulation and Waste

Waste from production of organic esters is usually not a problem since the method of synthesis often involves a carboxylic acid condensation with an alcohol and the only by-product is water. Any organic remnants lost to the process water can usually be biologically degraded. The biochemical oxygen demand (BOD) or chemical oxygen demand (COD) should be measured if biological treatment is used on the process waste from ester production. Organic ester vapor emitted in processing usually can be burned.

Extensive federal environmental regulations exist that govern organic esters as well as many other substances. These regulations must always be consulted for complete information before using large amounts of organic esters. State and local regulations must also be met, which in some cases are more stringent than federal regulations.

Uses

Organic esters are used as solvents, plasticizers, in resins, plastics, and coatings, as lubricants, in perfumes, flavors, cosmetics, and soap, as surface-active agents, as medicinals, and as herbicides and pesticides.

See also **Organic Chemistry**

KWOLIANG D. TAU
VARADARAJ ELANGO
JOSEPH A. MCDONOUGH
Hoechst Celanese Corporation

Additional Reading

Johnson, R.W. and E. Fritz eds.: *Fatty Acids in Industry*, Marcel Dekker, Inc., New York, 1989.

March, J. *Advanced Organic Chemistry*, 4th ed., John Wiley & Sons, Inc., New York, NY, 1992.

Patai, S. ed.: *The Chemistry of Acid Derivatives*, Suppl. B, Parts 1 and 2, John Wiley & Sons, Inc., New York, NY, 1979.

Trost, B.M. and I. Fleming, eds.: *Comprehensive Organic Synthesis: Selectivity, Strategy and Efficiency in Modern Organic Chemistry*, Vol. 1–9, Pergamon Press, Inc., Elmsford, NY, 1991.

ESTROGEN. See **Hormones; Steroids**.

ETHANE. (Dimethyl; Methylmethane). [CAS: 74-84-0]. C_2H_6, formula weight 30.07, colorless, odorless gas, mp $-172°C$, bp $88.6°C$, sp gr 1.05 (air = 1.0), practically insoluble in H_2O, moderately soluble in alcohol. The compound burns when ignited in air with a pale faintly luminous flame; forms an explosive mixture with air over a moderate range. With excess air, products of combustion are CO_2 and H_2O. Ethane is among the chemically less reactive organic substances. However, ethane reacts with chlorine and bromine to form substitution compounds. Ethane occurs, usually in small amounts, in natural gas. The fuel value of ethane is high, 1,730 Btu per cubic foot. Ethane may be prepared by reaction of magnesium ethyl iodide in anhydrous ether (Grignard's reagent) with H_2O or alcohols. Ethyl iodide, ethyl bromide, or ethyl chloride, are preferably made by reaction with ethyl alcohol and the appropriate phosphorus halide. Important ethane derivatives, by successive oxidation, are ethyl alcohol, acetaldehyde, and acetic acid.

ETHANOL. See **Ethyl Alcohol**.

ETHANOLAMINES. [CAS: 141-43-5]. There are three ethanolamines, all hydroxy-amines, and all high-tonnage industrial chemicals. Production is about 300 mil pounds annually.

Monoethanolamine, $NH_2CH_2CH_2OH$, industrial symbol (MEA) Formula weight 61.08, mp 10.5°C, bp 171°C, sp gr 1.018

Diethanolamine, $NH(CH_2CH_2OH)_2$, industrial symbol (DEA) Formula weight 105.14, mp 28.0°C, bp 270°C, sp gr 1.019

Triethanolamine, $N(CH_2CH_2OH)_3$, industrial symbol (TEA) Formula weight 149.19, mp 21.2°C, bp 360°C, sp gr 1.126

Mono- and triethanolamine are miscible with H_2O or alcohol in all proportions and are only slightly soluble in ether. Dietanolamine will dissolve in H_2O up to 96.4% at 20°C, is very soluble in alcohol, and only slightly soluble in ether.

Wurtz first reported the ethanolamines in 1860, but they were not used commercially on any scale until the late 1920s. All of the compounds are clear, viscous liquids at standard conditions and white crystalline solids when frozen. They have a relatively low toxicity. Industrially, the compounds are important (1) because they form numerous derivatives, notably with fatty acids, soaps, esters, amides, and esteramides; and (2) for their exceptional ability for scrubbing acidic compounds out of gases. Monoethanolamine, for example, will effectively remove H_2S from hydrocarbon gases. The compounds also remove CO_2 from process streams and, where desired, the CO_2 may easily be recovered by heating the absorptive solutions. The soaps of the ethanolamines are extensively used in textile treating agents, in shampoos, and emulsifiers. The fatty acid amides of diethanolamine are applied as builders in heavy-duty detergents, particularly those in which alkylaryl sulfonates are the surfactant ingredients. The use of triethanolamine in photographic developing baths promotes fine grain structure in the film when developed. Ethanolamine also is used as a humectant and plasticizing agent for textiles, glues, and leather coatings; and as a softening agent for numerous materials. Morpholine is an important derivative.

In early processes, the ethanolamines were prepared by reacting ethylene chlorohydrin $ClCH_2 \cdot CH_2OH$ with NH_3. Current processes react ethylene oxide $\langle(CH_2)_2\rangle O$ with NH_3, usually in aqueous solution. The ratio of mono-, di-, and triethanolamines varies in accordance with the amount of NH_3 present. This is controlled by the quantities of MEA and DEA recycled. Higher NH_3-ethylene oxide ratios favor high DEA and TEA yields, whereas lower ratios are used where maximum production of MEA is desired. The reaction is noncatalytic. The pressure is moderate, just sufficient to prevent vaporization of components in the reactor. The bulk of the H_2O produced in the reaction is removed by subsequent evaporation. The dehydrated ethanolamines then proceed to a further drying column, after which they are separated in a series of fractionating columns, not difficult because of the comparatively wide separation of their boiling points.

e (The Number)

A transcendental number, used as the base of the system of natural or Napierian logarithms. It is defined by

$$e = \lim_{n \to \infty} (1 + 1/n)^n$$

or by

$$e = \lim_{x \to 0} (1 + x)^{1/x}$$

It is represented by the infinite series

$$e = 1 + \frac{1}{1!} + \frac{1}{2!} + \frac{1}{3!} + \frac{1}{4!} + \cdots + \frac{1}{n!} + \cdots$$

and it equals 2.71828, approximately. In this book, the number e is written in italic form, except in equations involving electrical quantities, where the Roman e is used to avoid confusion.

ETHER (Chlorinated).   See Chlorinated Organics.

ETHERS.
In general, ethers are neutral, pleasant-smelling compounds that have little or no solubility in water, but are easily soluble in organic liquids. Their boiling points approximate those of hydrocarbons having comparable molecular weights and geometries. Detailed physical properties for ethers most commonly used as solvents are listed in Table 1.

The homologous series of ethers has the formula $C_nH_{2n+2}O$. Structurally, the ethers have an oxygen linkage between two radicals ($R-O-R'$). R and R' may be the same as in dimethyl ether CH_3-O-CH_3; or they may differ as in ethylisopropyl ether $C_2H_5-O-C_3H_7$. The latter may be referred

TABLE 1. TYPICAL PHYSICAL AND CHEMICAL PROPERTIES OF COMMONLY USED ETHERS

Property	Ethyl	MTBE	THF	Isopropyl	n-Butyl
chemical formula	$C_4H_{10}O$	$C_5H_{12}O$	C_4H_8O	$C_6H_{14}O$	$C_8H_{18}O$
molecular weight	74.12	88.15	72.10	102.17	139.22
boiling point at 101.3 kPa,[a] °C	34.5	55.0	66.0	68.4	142.0
vapor pressure at 20°C, kPa[a]	56	27	17	16	0.67
evaporation rate[b]	11.8	8.5	5.7	8.0	0.66
viscosity at 20°C, mPa·s (= cP)	0.23	0.35	0.48	0.38	0.65
flash point	−40	−30	−17	−13	25
autoignition temp, °C	160	426	321	440	185
flammability limits in air, vol %					
lower	1.9	1.6	1.8	1.0	0.9
higher	48.0	8.4	11.8	21.0	8.5

[a] To convert kPa to mm Hg, multiply by 7.5.
[b] n-Butyl acetate = 1.

to as a *mixed ether*. Mixed ethers frequently are made from mixed alcohols. Where R and R' are alkyls, the ether may be called an *alkyl ether* or an *alphyl oxide*. They may be considered to be derivatives of the monohydric alcohols. Each ether is isomeric with a saturated alcohol. Both diethyl ether and butyl alcohol are $C_4H_{10}O$. Also, there are many isomeric ethers, starting with $C_4H_{10}O$. Methylpropyl ether and diethyl ether are isomeric. Where compounds such as these have the same general formula, are members of the same family, and differ only by the alkyl group present, they are termed *metameric*.

Since they are similar structurally to the alcohols, phenols also form ethers. An example of an aromatic ether is methylphenyl ether (anisole) $C_6H_5-O-CH_3$. There are few ethers where both R and R' are aryls. The structure of thioethers is similar to the other ethers, but with a sulfur atom in the link instead of an oxygen atom, as $R-S-R'$. Examples of thioethers include diethyl sulfide $C_2H_5-S-C_2H_5$ and methylethyl sulfide $CH_3-S-C_2H_5$, which is a mixed thioether.

The properties of the ethers may be summarized by: (1) with the exception of dimethyl ether which is a gas, the ethers are volatile, mobile, inflammable liquids that are lighter than H_2O; (2) they are relatively inert chemically, not being acted on by alkali metals or alkalis and not reacting with dilute acids; (3) they form substitution products when reacted with chlorine and bromine; and (4) they decompose when heated with strong acids, yielding esters.

Ether

$(C_2H_5)_2O$, formula weight 74.12, mp −116.3°C, bp 34.6°C, sp gr 0.708. Probably the best known of the ethers, diethyl ether, commonly called simply *ether*, is slightly soluble in H_2O (1 volume in 10 volumes H_2O) and is miscible with alcohol in all proportions. Ether dissolves iodine and many organic substances, e.g., oils and fats, waxes, resins, and alkaloids and hence is widely used as a solvent for these substances in the preparation of numerous products, including explosives and collodion. Ether explodes in oxygen in the presence of a flame or spark, yielding H_2O and CO_2. When heated with an acid, such as H_2SO_4, ether yields ethyl alcohol. With phosphorus halides, ethyl halide (2 moles) is formed. Ether reacts with HNO_3 to form ethyl oxide.

Although still used medically, at one time ether was the major anesthetic, for which it must be scrupulously pure. In addition to various side effects which may result from the use of ether as an anesthetic, it is a definite hazard in the operating room because of its explosive properties, particularly in enriched oxygen atmospheres.

Methyl Tertiary Butyl Ether

MTBE is easily made by the selective reaction of isobutylene and methanol over an acidic ion-exchange resin catalyst, in the liquid phase and at temperatures below 100°C. To be economically competitive, MTBE's use as an octane enhancer in gasoline has been dependent on low cost isobutylene. There are a number of possible isobutylene sources for making MTBE.

MTBE production capacity has grown steadily, usually at an annual rate of 10 to 20% per year. MTBE may be the second largest organic chemical produced in the United States, second only to ethylene.

MTBE and related ethers are used to add octane to gasoline. MTBE also adds oxygen to the gasoline, which allows for more efficient combustion, and therefore less carbon monoxide and unburned hydrocarbon in the exhaust emissions. A refined grade of MTBE is used in the solvents and pharmaceutical industries. Its higher autoignition temperature and narrower flammability range also make it relatively safer to use compared to other ethers.

One other unique use of MTBE is a medical procedure for the removal of gallstones. This alternative to gallbladder surgery was developed at the Mayo Clinic, and takes advantage of MTBE's capability to quickly dissolve cholesterol.

See also **Organic Chemistry**.

ETHYL ALCOHOL. [CAS: 64-17-5]. C_2H_5OH, formula weight 46.07, is a colorless liquid with mild characteristic odor, mp $-114.1°C$, bp $78.32°C$, sp gr 0.789. Also known as *ethanol*, the compound is miscible in all proportions with H_2O or ether. When ignited, ethyl alcohol burns in air with a pale blue, transparent flame, producing H_2O and CO_2. The vapor forms an explosive mixture with air and it is used in some internal combustion engines under compression as a fuel. See also **Fuel**. Such mixtures are frequently referred to as *gasohol*.

Anhydrous ethyl alcohol is made from the constant boiling mixture with H_2O (95.6% ethyl alcohol by weight)—(1) by heating with a substance such as calcium oxide, which reacts with H_2O and not with alcohol, and then distilling, or (2) by distilling with a volatile liquid, such as benzene (bp $79.6°C$), which forms a constant low-boiling mixture with H_2O and alcohol (bp $64.9°C$), so that H_2O is removed from the main portion of the alcohol; after which alcohol plus benzene distills over (bp $78.5°C$). Anhydrous ethyl alcohol is required for certain purposes as a solvent and reagent and fuel applications.

Commercially, ethyl alcohol is marketed by the proof gallon, 200 proof on the scale representing pure alcohol (100%). When the term *alcohol* alone is used, it refers to a liquid that ranges from 188 to 192 proof (94% to 96% ethyl alcohol). When the terms *grain alcohol, high-purity* alcohol, or *pure ethyl* alcohol are used, these usually refer to a liquid that is 190 proof. In most countries, beverage alcohol is highly taxed and to make the product available for nonbeverage purposes, denaturants will be added. Denaturants include methyl alcohol, pyridine, benzene, kerosene, pine oil, mixtures of primary and secondary aliphatic higher alcohols, and hydrogenated organic compounds. Thousands of nonbeverage industrial and commercial products, notably food extracts, toiletries, pharmaceuticals, solvents, and cleaning products, contain denatured ethyl alcohol.

Worldwide, ethyl alcohol is the basis for a huge alcoholic beverage industry, offering a wide range of products wherein the alcoholic content varies from a few to over 50% (100 proof). Industrially, ethyl alcohol is very important high-tonnage raw and intermediate material for numerous processes, and is used extensively in solvents, antiseptics, antifreeze compounds, and fuels.

Production

Natural fermentation is the oldest process for making ethyl alcohol and still constitutes the principal means for creating the alcoholic content of beverages. Except in connection with other alcohol-containing products, industrial producers of ethyl alcohol use processes other than fermentation. For fermentation, almost any agricultural raw material with a carbohydrate content in the form of sugars or starches that are easily converted to sugars can be used. Once the raw materials are in the form of sugars, yeast enzymes are added to commence natural fermentation. Traditionally, in the United States, industrial alcohol prepared by fermentation has used blackstrap molasses, which contains up to 50% sugars and can be easily fermented. The starting mash is prepared by diluting the molasses with H_2O to bring the sugar content down to about 15% (weight). The mash is slightly acidified, after which invertase (enzyme to convert sucrose) and zymase (enzyme to convert glucose and fructose) are added. The products are ethyl alcohol and CO_2. Yeast activity is sustained by the addition of nutrients. With careful control of temperature and acidity, the fermentation process can be completed in about two days. The resulting mash (beer) usually contains about 12% ethyl alcohol which is recovered from the beer by distillation. See also **Fermentation**.

In modern industrial ethyl alcohol plants, the compound is produced in two principal ways: (1) by *direct hydration of ethylene*, or (2) by *indirect hydration of ethylene*. In the direct hydration process, H_2O is added to ethylene in the vapor phase in the presence of a catalyst: $CH_2:CH_2 + H_2O \rightleftharpoons CH_3CH_2OH$. A supported acid catalyst usually is used. Important factors affecting the conversion include temperature, pressure, the $H_2O/CH_2:CH_2$ ratio, and the purity of the ethylene. Further, some byproducts are formed by other reactions taking place, a primary side reaction being the dehydration of ethyl alcohol into diethyl ether: $2C_2H_5OH \rightleftharpoons (C_2H_5)_2O + H_2O$. To overcome these problems, a large recycle volume of unconverted ethylene usually is required. The process usually consists of a reaction section in which crude ethyl alcohol is formed, a purification section with a product of 95% (volume) ethyl alcohol, and a dehydration section which produces high-purity ethyl alcohol free of H_2O. For many industrial uses, the 95%-purity product from the purification section suffices.

In the indirect hydration process, ethylene first is absorbed in concentrated H_2SO_4 to form mono- and diethyl sulfates: $CH_2:CH_2 + H_2SO_4 \rightleftharpoons CH_3CH_2OSO_3H$; and

$$2CH_2:CH_2 + H_2SO_4 \longrightarrow (CH_3CH_2)_2SO_4.$$

The ethyl sulfates then are hydrolyzed to ethyl alcohol: $CH_3CH_2OSO_3H + H_2O \rightleftharpoons CH_3CH_2OH + H_2SO_4$; and

$$(CH_3CH_2)_2SO_4 + 2H_2O \longrightarrow 2CH_3CH_2OH + H_2SO_4.$$

Remaining steps in the process include recovery and purification of the crude ethyl alcohol and reconcentration of the dilute H_2SO_4. The crude ethyl alcohol is steam-stripped from the dilute acid solution, followed by distillation for purification.

Azeotropes

The physical properties of ethyl alcohol are influenced by the hydroxyl group that imparts hydrogen-bonding characteristics and polarity to the substance that are analogous to water. Ethyl alcohol displays a highly nonideal behavior in numerous solutions, forming several azeotropes. The list of binary azeotropes of ethanol is long, including acetonitrile, benzene, carbon disulfide, chloroform, ethyl acetate, hexane, toluene, and water. See also **Azeotropic System**.

Chemistry

Ethyl alcohol reacts (1) with sodium metal, forming sodium ethoxide C_2H_5ONa plus hydrogen gas, (2) with phosphorus chloride, bromide, iodide, forming ethyl chloride, bromide, iodide, respectively, (3) with H_2SO_4 concentrated, forming at $100°C$ ethyl hydrogen sulfate $C_2H_5OSO_2OH$, at $140°C$ diethyl ether $(C_2H_5)_2O$, at $200°C$ ethylene $CH_2:CH_2$, (4) with organic acids, warmed in the presence of H_2SO_4, forming esters, e.g., ethyl acetate $CH_3COOC_2H_5$, ethyl benzoate $C_2H_5COOC_2H_5$ (see various individual acids), (5) with magnesium methyl iodide in anhydrous ether (Grignard's solution), forming methane as in the case of primary alcohols, (6) with calcium chloride to form a solid addition compound $4C_2H_5OH·CaCl_2$, which is decomposed by H_2O, (7) with oxygen, using sodium dichromate solution and H_2SO_4, to form acetaldehyde (and acetic acid), using air, in the presence of acetic bacteria, to form vinegar (dilute acetic acid along with the substances present in the alcohol used, e.g., wine, cider), (8) with HNO_3 (a) concentrated, free from nitrogen tetroxide, to form ethyl nitrate, (b) dilute to form glycollic acid, (c) concentrated acid containing nitrogen tetroxide (fuming HNO_3) explosive reaction, (9) with chlorine (or bromine) to form chloral CCl_3CHO (or bromal).

See also **Organic Chemistry**.

Additional Reading

Erwin, V. and R. Deitrich: *Pharmacological Effects of Ethanol on the Nervous System*, CRC Press, LLC., Boca Raton, FL, 1995.
Heyns, M.: *The Influence of Ethyl Alcohol on the Development of the Chondrocranium of Gallus Gallus*, Vol. 134, Springer-Verlag, Inc., New York, NY, 1997.
Lide, D.R.: *CRC Handbook of Chemistry and Physics*, 84th Edition, CRC Press, LLC., Boca Raton, FL, 2003.
Milleer, N. and M. Gold: *Alcohol*, Vol. 2, Kluwer Academic Publishers, Norwell, MA, 1991.

ETHYL CELLULOSE. A versatile, thermoplastic cellulose ether that is compatible with a wide variety of solvent systems, resins, oils, and plasticizers. This versatility permits a wide diversity of end-product properties. It is an excellent film former as from a wide range of neat, lacquer, or dispersion formulations. Molded ethyl cellulose has excellent toughness, flexibility, and shock resistance. Useful temperature range is from about −40 to +100°C. In the preparation of ethyl cellulose, wood pulp or cotton linters with a high alpha-cellulose content are reacted with ethyl chloride and sodium hydroxide. Structural formulas of cellulose and ethyl cellulose with complete (54.9%) ethoxyl substitution are:

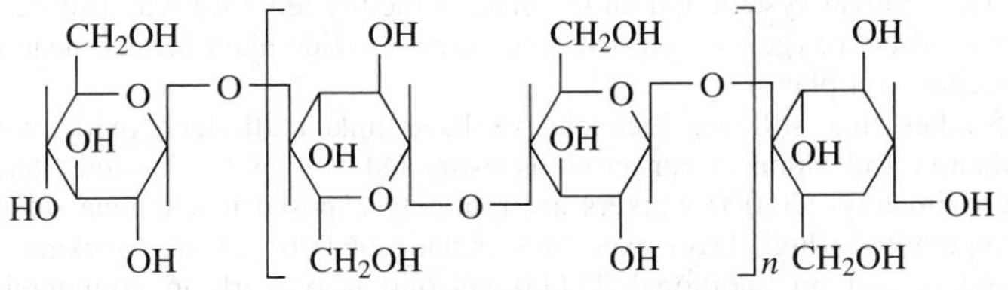

Cellulose, $n > 500$

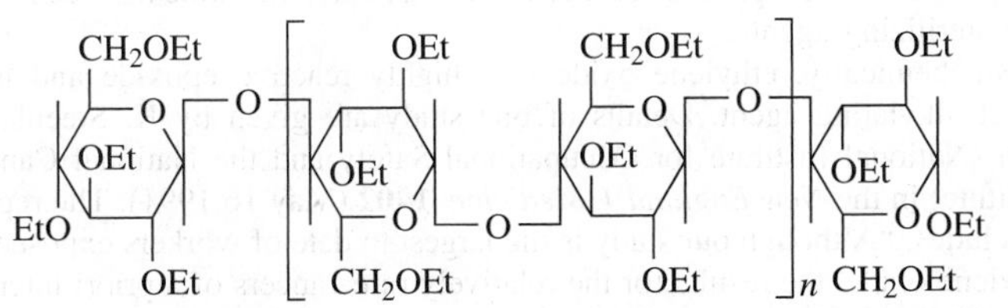

Tri-O-ethyl cellulose, $n = 50$–150; Et = CH_3CH_2

The natural color of ethyl cellulose is colorless to light amber, but it can be formulated into a wide range of transparent, translucent, and opaque colors. The material should be dried before molding because it is slightly hygroscopic. Compression molding temperatures range from 121–200°C and pressures from 500–5,000 psi. Injection molding temperatures range from 175–260°C and pressures from 8,000–32,000 psi. Strong acids decompose the material, but weak acids and strong alkalis have only a slight effect. Weak alkalis do not attack the material. Ethyl cellulose is soluble in a large number of organic solvents.

Among the commercial uses for ethyl cellulose are: strippable coating for metal parts, paper coatings, and in medicinal tablets. An interesting application is coating for bowling pins. Ethyl cellulose sheeting is tough, flexible, and transparent, yet sufficiently rigid to withstand rough handling.

ETHYL CHLORIDE.   See Chlorinated Organics.

ETHYLENE. [CAS: 74-85-1]. C_2H_4, formula weight 28.03, colorless gas with slight odor, normal bp −103.7°C, critical pressure of 49.98 atmospheres, and critical temperature of 9.5°C, density 1.26 grams per liter (0°C and 760 mm), sp gr 0.97 (air = 1.0), very slightly soluble in H_2O, slightly soluble in alcohol. Ethylene burns when ignited in air with a luminous flame. The presence of ethylene in coal gas is chiefly responsible for the luminosity of the latter gas. Ethylene forms an explosive mixture with air and has a high fuel value, 1,615 Btu per cubic foot.

Even though there are few direct end-uses foe ethylene, it is probably the most important petrochemical feedstock, both in terms of quantities used and economic value. Ethylene is the feedstock for ethylene oxide, ethylbenzene, ethyl chloride, ethylene dichloride, ethyl alcohol, and polyethylene, most of which, in turn, are used to produce hundreds of other end-products. Most ethylene is produced by steam cracking of ethane or propane.

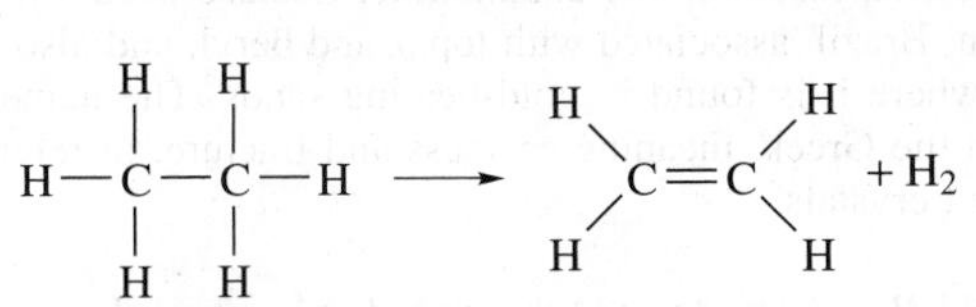

Ethane   Ethylene

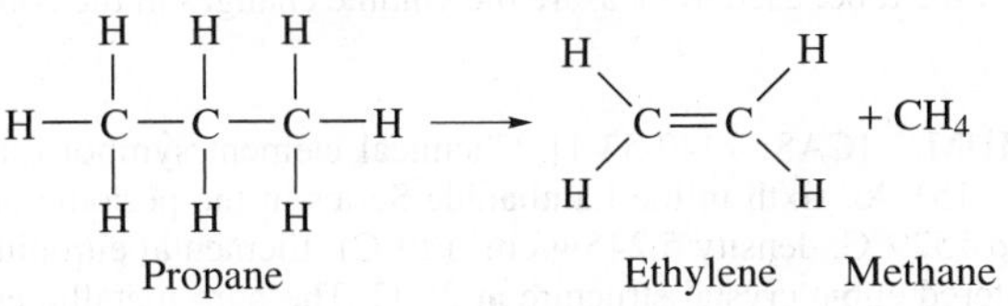

Propane   Ethylene   Methane

Ethylene also may be produced from other paraffinic or naphthenic hydrocarbons. The reactions are highly endothermic (34,400 kcal/kg mole of ethane cracked at approximately 900°C) and proceed in the direction indicated at temperature exceeding approximately 620°C without a catalyst.

Ethylene is of importance as a petrochemical feedstock because of its great versatility in reacting to form several chemical intermediates. The double bond provides reactivity; the compound also has the ability to homopolymerize and copolymerize with other monomers. Some of the important reactions involving ethylene include:

Chlorination

$$CH_2 = CH_2 + HCl \xrightarrow[\text{cat.}]{\text{acidic}} CH_3 - CH_2Cl$$
Ethyl chloride

Oxidation

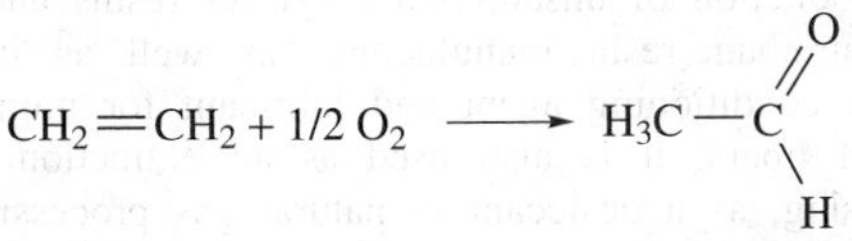

Acetaldehyde

Hydration

$$CH_2 = CH_2 + H_2O \longrightarrow C_2H_5OH$$
Ethyl alcohol

Oxychlorination

$$CH_2 = CH_2 + 2\,HCl + 1/2\,O_2 \longrightarrow C_2H_4Cl_2 + H_2O$$
Ethyl dichloride

Ethylene dichloride is used for the production of vinyl chloride.

Alkylation

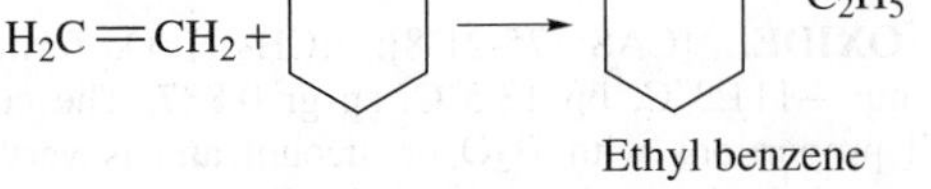

Ethyl benzene

Ethyl benzene is used for the production of styrene.

Polymerization

$$n\,CH_2 = CH_2 \longrightarrow (-CH_2-CH_2)_n$$
High- and low-density polyethylene

Ethylene is also oxidized in large quantities to ethylene oxide:

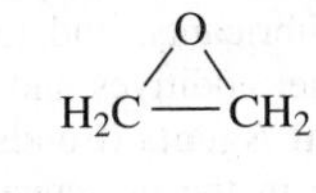

At one time, ethylene was produced by the dehydration of ethyl alcohol over alumina.

Almost any naphthenic or paraffinic hydrocarbon heavier then methane can be steam-cracked to yield ethylene. The preferred feedstock in the United States has been ethane and/or propane recovered from natural gas, or from the volatile fractions of petroleum. However, because of long-term uncertainties pertaining to natural gas, many producers have been turning to heavier petroleum fractions, such as gas oils, as feedstock. The consumption of ethylene throughout the free world is estimated to be about 40×10^9 pounds per year.

Ethylene reacts: (1) with the halogens to form substitution halides; (2) with hypochlorous and hypobromous acid to form ethylene chlorohydrin or ethylene bromohydrin, respectively; (3) with hydrogen iodide or bromide (not chloride) to form ethyl iodide or ethyl bromide; (4) with hydrogen, in the presence of a catalyst, e.g., finely divided nickel at 150°C, to form ethane; (5) with concentrated sulfuric acid at 160°C to form ethyl hydrogen sulfate; and (6) with potassium permanganate to form ethylene glycol, although glycol is preferably made from ethylene dichloride or chlorohydrin.

In addition to its uses in the preparation of intermediates for a large variety of petrochemical reactions, ethylene is used as an anesthetic, as a fuel with oxygen for high-temperature flames, and as a coloring and ripening agent for citrus fruits and tomatoes. Ethylene chlorohydrin is used as an agent for decreasing the dormant period of seeds. See also **Polyethylene**.

ETHYLENE DICHLORIDE.   See Chlorinated Organics.

ETHYLENE GLYCOL. [CAS: 107-21-1]. $HOCH_2CH_2OH$, this compound, traditionally associated with its use as a type of permanent antifreeze for internal-combustion engine cooling systems. However, since the early 1960s, large tonnages of ethylene glycol have been used in the production of polyesters for fibers, films, and coatings. The compound also finds important uses in hydraulic fluids, in the manufacture of low freezing-point explosives, glycol ethers, and deicing solutions. Di-and triethylene glycols are important coproducts usually produced in the manufacture of ethylene glycol. Diethylene glycol, $HOCH_2CH_2OCH_2CH_2OH$, is used in the production of unsaturated polyester resins and polyester polyols for polyurethane-resin manufacture, as well as in the textile industry as a conditioning agent and lubricant for numerous synthetic and natural fibers. It is also used as an extraction solvent in petroleum processing, as a desiccant in natural gas processing, and in the manufacture of some plasticizers and surfactants. Triethylene glycol,

$$HOCH_2CH_2OCH_2CH_2OCH_2CH_2OH,$$

finds principal use in the dehydration of natural gas and as a humectant.

In one process for the manufacture of the aforementioned glycols, ethylene oxide is formed by direct oxidation of ethylene with oxygen over a silver catalyst. After purification, the stabilized ethylene oxide is mixed with a large excess of water, preheated, and fed to an ethylene oxide reactor. Here the ethylene oxide and water react under high temperature and high pressure to form principally ethylene glycol, with the other aforementioned glycols as coproducts. The crude glycols are dehydrated and then recovered individually as highly pure overhead streams from a series of vacuum-operated purification columns.

Principal properties of the glycols are summarized in Table 1.

ETHYLENE OXIDE. [CAS: 75-21-8]. $\langle(CH_2)_2\rangle$ O, formula weight 44.05, liquid, mp $-111.3°C$, bp $13.5°C$, sp gr 0.887. The compound is miscible in all proportions with H_2O or alcohol and is very soluble in ether. Ethylene oxide is slowly decomposed by H_2O at standard conditions, converting into glycol $CH_2OH·CH_2OH$. Ethylene oxide is a very high-tonnage chemical, approaching nearly 4 billion pounds annually. In terms of consumption, it is used as follows: (1) 60% for manufacture of ethylene glycol, an antifreeze compound, as well as a raw material for production of polyethylene terephthalate (used in the manufacture of polyester fibers); (2) 12% for preparation of surfactants; (3) 8% for the manufacture of ethanolamines; (4) 10% for production of ethylene glycols which are used in plasticizers, solvents, and lubricants; and (5) 10% for making glycol ethers, which are used as jet-fuel additives and solvents. See also **Rocket Propellants**; and **Antimicrobial Agents (Foods)**.

Direct oxidation of ethylene in the presence of a silver catalyst is the predominant large-scale process used: $CH_2 : CH_2 + \frac{1}{2}O_2 \rightarrow \langle(CH_2)_2O\rangle$. The yield is approximately 70% of the theoretical. For maximum yield, very careful temperature control is required, the yield dropping as the temperature climbs. The side reaction: $CH_2 : CH_2 \rightarrow 3CO_2 + 2H_2O$ is the main factor for reducing yield. Thus far, silver has proved to be the most effective catalyst. Several compounds have been investigated that can inhibit the side reaction and also be compatible with the catalyst. These compounds have included ethylene dichloride, ethylene dibromide, alcohol, amines, and organometallic compounds, but their success has been limited. Plants have been designed to use either air or pure oxygen for oxidation. Selection presents an interesting study in economics because (1) where air is used, a purge reactor and associated purge absorber are required (not required by the O_2 process), and (2) where O_2 is used, both a CO_2 removal system and an O_2-making facility are required. The trend is toward the oxygen system with the ethylene oxide plant located near an air-separation plant.

Studies that still are inconclusive have linked ethylene oxide with leukemia and stomach cancer. It is estimated that in the United States approximately 270,000 workers are routinely exposed to ethylene oxide. Comparatively high level exposures include 96,000 persons working in hospitals and an additional 21,000 persons who work in commercial medical supply sterilization facilities, as well as in the production of spices and pharmaceutical products. Since the 1950s, ethylene oxide has been used as a sterilizing agent.

Biochemically, ethylene oxide is a highly reactive epoxide and is a direct alkylating agent. Details of one study are given by K. Steenland, et al (National Institute for Occupational Safety and the National Cancer Institute) in the *New England J. Medicine*, **1402** (May 16,1991). The report concludes: "Although our study is the largest to date of workers exposed to ethylene oxide, the results for the relatively rare cancers of a priori interest are still limited by the small numbers of cases and perhaps limited by the short follow-up. Our findings are therefore not conclusive."

It has been established, however, that ethylene oxide is a potent mutagen and animal carcinogen.

ETHYLENE-PROPLENE ELASTOMERS.   See Elastomers.

ETHYLENE-VINYL ACETATE COPOLYMERS. Known as EVA copolymers, these materials are polyolefins that can be processed like other thermoplastics, but they approach rubbery materials in softness and elasticity. The resins meet regulatory requirements for use in direct contact with food in food-processing machinery and in packaging applications. They are used in a number of applications to replace plasticized polyvinyl chloride and rubber. EVA copolymers require no curing or plasticizer. Parts made from EVA have little or no odor. Their elasticity is permanent. The copolymers can be injection-, blow-, compression-, transfer- and rotationally molded or extruded into film, sheeting, pipe, and profiles. EVA copolymers offer advantages over polyvinyl chloride and rubber in that they have good clarity and gloss, stress-crack resistance, good barrier properties, low-temperature flexibility and toughness, good adhesive properties, and good resistance to ultraviolet radiation. Their main limitation is a comparatively low resistance to heat and solvents. The resins soften at a temperature of about $70°C$. EVA copolymers are not attacked by alcohols, glycols, or weak organic acids. However, to a varying degree, the materials are attacked by chlorinated hydrocarbons, straight-chain paraffinic solvents, and benzene and its derivatives.

EUCLASE. The mineral euclase is a silicate of beryllium and aluminum corresponding to the formula $BeAlSiO_4(OH)$, which crystallizes in the monoclinic system. It has a perfect prismatic cleavage; hardness, 7.5; specific gravity, 3.1; luster, vitreous; is colorless to sea-green or blue. It has been used to a very slight extent for jewelry as its transparent crystals somewhat resemble the aquamarine. Euclase occurs in the Minas Gerais region, Brazil, associated with topaz and beryl, and also in the Ural Mountains, where it is found in gold-bearing sands. The name euclase is derived from the Greek, meaning easiness and fracture, in reference to its easily cleaved crystals.

EUDIOMETER. A graduated tube closed at one end. In one form, two platinum wires are sealed so that a spark may be passed through the contents of the tube; used to measure the volume changes in the combustion of gases.

EUROPIUM. [CAS: 7440-53-1]. Chemical element symbol Eu, at. no. 63, at. wt. 151.96, sixth in the Lanthanide Series in the periodic table, mp $822°C$, bp $1529°C$, density 5.245 g/cm^3 $(20°C)$. Elemental europium has a body-centered cubic crystal structure at $25°C$. The pure metallic europium

TABLE 1. PROPERTIES OF ETHYLENE, DIETHYLENE, AND TRIETHYLENE GLYCOLS

	Ethylene glycol	Diethylene glycol	Triethylene glycol
Molecular weight	62.07	106.12	150.17
Boiling point (760 mm Hg)	197.6°C	245°C	287.4°C
Vapor pressure (20°C)	0.06 mm Hg	<0.01 mm Hg	<0.01 mm Hg
Specific gravity (20°C/20°C)	1.1155	1.1184	1.1254
Freezing point in air (760 mm Hg)	−13°C	83°C	99°C
Water solubility	- - - - - - - complete - - - - - - -		

Source: Glycols, Shell Chemical Company Bull. IC: 67–58.

is silver-gray in color under vacuum, but oxidizes readily in air and must be handled in an inert atmosphere. Europium is very soft as compared with the other rare-earth metals. Two stable isotopes of the element occur naturally ^{151}Eu and ^{153}Eu. Upon absorption of thermal neutrons, ^{151}Eu forms ^{152}Eu with a half-life of 13 years; ^{153}Eu forms ^{154}Eu with a half-life of 16 years. The latter further decays to ^{155}Eu with a half-life of 1.7 years. In terms of abundance, europium is present on the average of 1.2 ppm in the earth's crust, making its potential availability greater than antimony, bismuth, or cadmium. The element was first identified by Sir William Crookes in 1889. Europium dissolves readily in dilute mineral acids and reacts with H_2O at room temperature. The metal is not known to be toxic but because of its high reactivity in air, great care in handling is mandatory. Electronic configuration

$$1s^2 2s^2 2p^6 3s^2 2p^6 3d^{10} 4s^2 4p^6 4d^{10} 4f^6 5fs^2 5p^6 5d^1 6s^1$$

Ionic radius Eu^{2+} 1.09 Å, Eu^{3+} 0.950 Å. Metallic radius 1.995 Å. First ionization potential 5.67 eV; second 11.25 eV. Oxidation potential $Eu^{2+} \rightarrow Eu^{3+} + e^-$, 0.43 V. Other important physical properties of europium are given under **Rare-Earth Elements and Metals**.

Europium occurs in the rare-earth fluocarbonate mineral bastnasite, mainly found in southern California. The mineral contains between 0.09 and 0.11% Eu_2O_3. Other minerals, such as xenotime and monazite, also contain europium compounds, and sometimes they are used as sources of the element.

Because of the desirable nuclear properties of the element, europium has received serious consideration for the construction of nuclear reactor hardware. Earlier commercial unavailability of the element, however, favored the use of other materials. Some small reactors have been constructed in which europium molybdate has been the major control-rod component. With much increased availability of the metal in recent years, the prospect of further usage of europium in reactor design are good. A europium-activated yttrium orthovanadate, $Eu:YVO_4$, has shown promise as a red phosphor for commercial television. An increase of 40% in light output has been claimed. With this system, the average color television set would require about $\frac{1}{2}$ g of Eu_2O_3 and 6 g of Y_2O_3. The stimulus resulting from this discovery resulted in the development of other new phosphors involving europium in various host matrices. These new materials have been used in high-intensity mercury-vapor lamps, general-purpose fluorescent lamps, x-ray screens, charged-particle detectors, and neutron scintillators. In some optically read memory systems, ferromagnetic europium chalcogenides (sulfides, selenides, and tellurides) have been used. Other electronic and semiconductor uses of europium are under serious investigation.

K. A. GSCHNEIDNER, JR.
Iowa State University
Ames, Iowa

EUTECTIC. An eutectic reaction is a reversible isothermal transformation in which, during cooling, a single liquid phase is transformed into two or more solid phases, the number of solid phases being equal to the number of components. In a given alloy system, at a fixed pressure, all phases will have fixed compositions during the isothermal transformation. The temperatures at which the freezing occurs is known as the eutectic temperature, while the composition of the liquid phase is called the eutectic composition. On a temperature-composition binary phase diagram, the eutectic point is determined by the eutectic composition and the eutectic temperature. In general, an alloy of the eutectic composition freezes at a minimum temperature. For this reason, eutectic compositions, or compositions close to the eutectic, are frequently used in low melting point solders.

By a similar usage in petrology, a eutectic is a discrete mixture of two or more minerals, in definite proportions, which have simultaneously crystallized from the mutual solution of their constituents. The eutectic point is the lowest temperature at any given pressure at which the above physical-chemical process may take place. The eutectic ratio is the ratio by weight of two minerals that originate by the above process.

EUTECTOID. This is a phase transformation analogous to an eutectic where a single solid phase, instead of a liquid phase, is transformed into two or more different solid phases. The number of solid phases in the

resulting eutectoid structure is equal to the number of components in the system. Under very slow cooling, the eutectoid transformation should occur at the eutectoid temperature. However, due to the sluggishness of solid state transformations, there is usually some hysteresis with the transformation temperature depressed on cooling and raised on heating. Under equilibrium conditions, the compositions of the various phases are fixed in an eutectoid reaction just as they are in an eutectic transformation.

The best-known eutectoid reaction is that which occurs in steel where the austenite phase, stable at high temperatures, transforms into the eutectoid structure known as pearlite. In this transformation, the austenite phase, containing 0.8% carbon in solid solution, transforms to a mixture of ferrite (nearly pure body-centered cubic iron) and iron-carbide (Fe_3C). At atmospheric pressure, the equilibrium temperature for this reaction is 723°C. This temperature is the eutectoid temperature.

In binary alloy systems, a eutectoid alloy is a mechanical mixture of two phases which form simultaneously from a solid solution when it cools through the eutectoid temperature. Alloys leaner or richer in one of the metals undergo transformation from the solid solution phase over a range of temperatures beginning above and ending at the eutectoid temperature. The structure of such alloys will consist of primary particles of one of the stable phases in addition to the eutectoid, for example ferrite and pearlite in low-carbon steel. See also **Iron Metals, Alloys, and Steels**.

EVAPORATION. The evaporation of a liquid consists in the escape from the main body of the liquid of those molecules which, in their thermal agitation, are moving with a sufficient speed to break through the surface tension. That is, evaporation is the escape of molecules whose kinetic energy exceeds the work function of cohesion at the liquid surface. Since only a small proportion of the molecules are at any instant located near enough to the surface and are moving in the proper direction to escape, the rate of the evaporation is limited. It is easy to see why it proceeds more rapidly with higher temperature, and why liquids of low surface tension are relatively volatile. Also, as the faster moving molecules emerge, those left behind have less average energy, and the temperature of the liquid is thereby lowered. If the evaporation takes place in a closed vessel, the escaping molecules accumulate as a vapor above the liquid. Many of them return to the liquid, such returns being more frequent, the greater the density and pressure of the vapor. Presently the processes of escape and return come to equilibrium; the vapor is then said to be "saturated," its density and pressure no longer increase, and the cooling effect ceases. Even a warm breeze cools the skin because it removes the evaporating perspiration and prevents saturation.

Evaporation is a major chemical engineering unit operation for bringing about separations of liquids and solids and, in particular, to recover the solute (such as a dissolved salt) from the solvent (frequently water). Usually, the main object of the separation is the solute. The pulp and paper industry is a large user of evaporation equipment. In pulp mills, after the digestion system, the pulp is leached with water and the chemical solids are dissolved out almost completely by a pulp-washing system. The recovered liquid from these operations is fed to an evaporator, generally at about 15% total dissolved solids content. The evaporator removes much of the water and in so doing concentrates the liquid to 55–65% total dissolved solids, whereupon the solution then can be further processed in a chemical recovery furnace. Other types of pulp processing also involve chemical-containing solutions that must be evaporated for recovery of valuable chemicals. Evaporation also is used extensively in the production of table and industrial salt (sodium chloride) as well as other salts, in caustic-chlorine production, in the phosphate industry, and in food processing. Evaporators can be large structures as the illustrations indicate.

Evaporation is, in principle, the same operation as plain distillation, with three modifications in practice. First, the vapor may or may not be recovered. Second, the residue in the evaporator may or may not contain solids. Third, vacuum evaporation is frequently used in a single compartment or in multiple stages, with each successive stage operated at an increasing vacuum, utilizing the heat of condensation of the vapor from the preceding stage. In multiple stage evaporators, there is a saving in the cost of heat and an increased expenditure for apparatus. Vacuum evaporation is frequently utilized to lower the temperature to which a substance is subjected and thus avoid decomposition by passing a current of warm dry air over the substance. Combined high-vacuum and very low temperature evaporation or drying is practiced in the final removal of water vapor from frozen penicillin, due to the heat-sensitive nature of this material. Water vapor passes from the place of higher concentration, that is, the substance, to the place of lower concentration, that is, the air. Water vapor is thus removed from the substance. If oxygen of the air reacts with the substance, an inert gas such as nitrogen may be substituted for air.

An evaporator system may be single effect (Fig. 1), in which the steam is produced from one evaporator, or multiple effect, in which the steam is produced from several evaporators in series. In a multiple effect system, the vapor from one evaporator becomes the heating steam in the succeeding. Unusual conditions met in industrial or steam heating plants may require so large a fraction of make-up as to warrant double, triple, or quadruple effect evaporators. The central generating station ordinarily employs single effect and rarely requires more than a double effect system. The ratio vapor produced/steam used is about 0.8 for the single effect, 1.5 for the double effect, and 2.5 for the triple effect system.

Fig. 2.   The sixth-effect evaporator body of a multiple-effect evaporation system

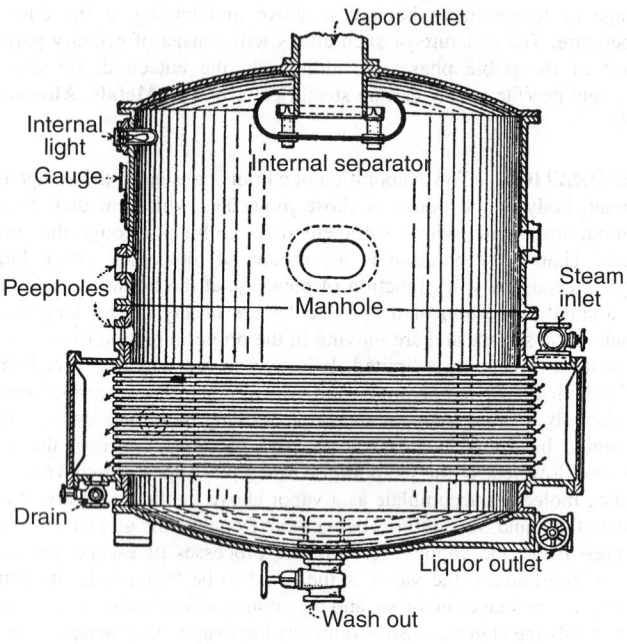

Fig. 1.   Horizontal-tube evaporator

See Figs. 2 and 3. Evaporator feed is sometimes preheated to increase evaporator capacity.

Evaporators are classed as film, flash, or submerged-tube types. The first and last are steam-tube types; in the former the raw water trickles over the hot tubes, in the latter the tubes are entirely surrounded by the water being evaporated. The flash type produces steam by dropping the pressure on water at the saturation temperature. The excess heat flashes part of the water into steam, then the remainder is drawn off, reheated, and again flashed.

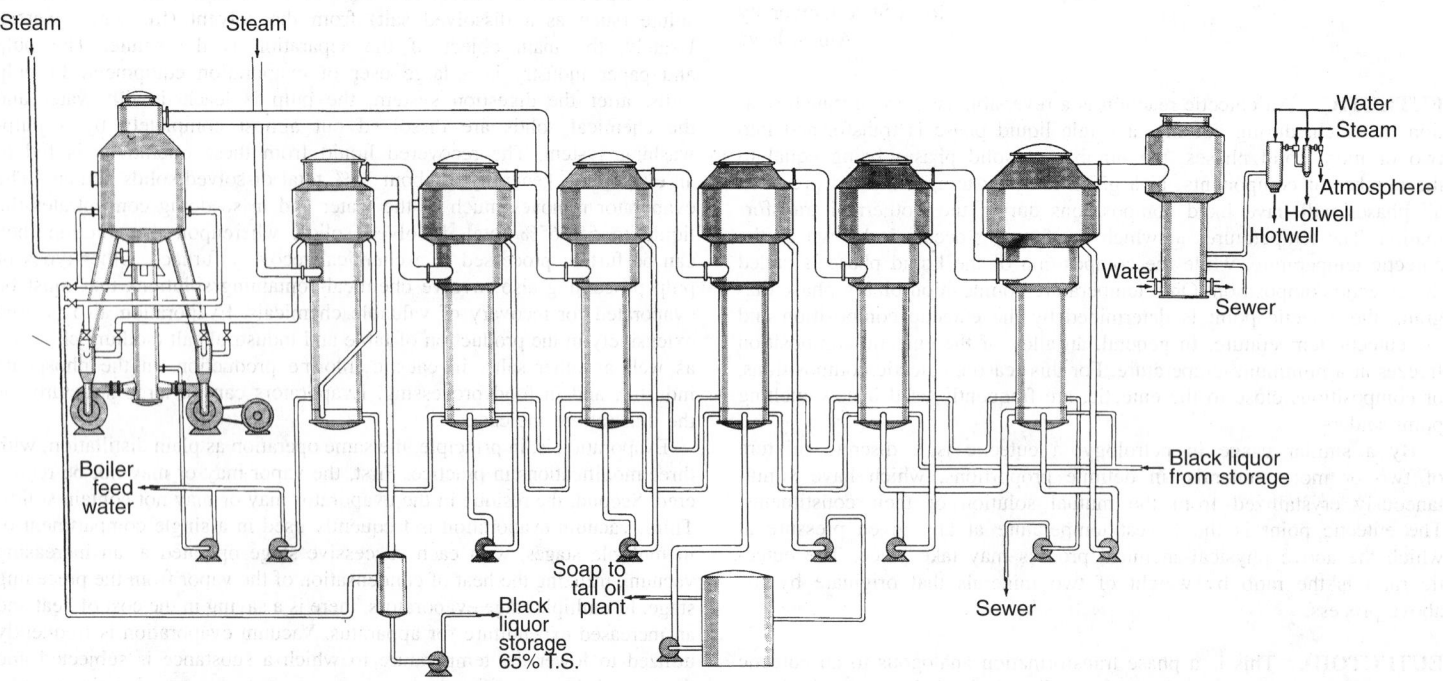

Fig. 3.   Compound multiple-effect kraft mill evaporator

Additional Reading

Billet, R.: *Evaporation Technology and Its Technical Applications,* John Wiley & Sons, Inc., New York, NY, 1989.

Jones, F.E.: *Evaporation of Water: With Emphasis on Applications and Measurements,* CRC Press, LLC, Boca Raton, FL, 1992.

Minton, P.E.: *Handbook of Evaporation Technology,* Noyes Publications, Park Ridge, NJ, 1986.

To, L.C.: *Robust Nonlinear Control of Industrial Evaporation Systems,* World Scientific Publishing Company, Inc., River Edge, NJ, 1998.

EVAPORATION (Desalination).   See **Desalination**.

EVAPORITE.   A sedimentary rock formed by precipitation from waters at the earth's surface. As described by Lowenstein (*Science,* **1090**, September 8, 1989), ancient evaporites have been used to track the chemistry of ancient surface waters, particularly seawater. Study of marine evaporites has led to the general (not unanimous) conclusion that the major elemental chemistry of seawater has not changed significantly during the last 600 million years.

EVEN-EVEN NUCLEI.   Atomic nuclei that contain an even number of protons and an even number of neutrons.

EXCHANGE DEGENERACY.   An exchange process that does not entail a change in value or configuration. For example, by the Heitler-London theory, the essential reason for the strong attraction (or repulsion), of the two H-atoms in the H_2 molecule is the exchange degeneracy, i.e., the fact that, for very large internuclear distance, by exchange of the two electrons of the two atoms, a configuration results that is indistinguishable from the original configuration. Therefore, as they approach, there arises an interaction between them, which may be treated mathematically as electron exchange.

EXCITED STATE.   A higher than normal energy level of the electrons of an atom, radical, or molecule, typically resulting from absorption of photons (quanta) from a radiation source (arc, flame, spark, etc.) in any wavelength of the electromagnetic spectrum, X-ray, UV, visible, infrared, microwave, and radio frequencies are used for excitation in various types of spectroscopy. When the energizing source is removed or discontinued, the atom or molecule returns to its normal or stable state either by emitting the absorbed photons or by transferring the energy to other atoms or molecules. The emission by the atom or molecule yields line or band spectra characteristic of its structure, thus permitting identification. Photochemical reactions are induced by excited chemical entities, which are also responsible for the phenomena of luminescence (phosphorescence and fluorescence).

EXOSMOSIS.   An osmotic process by which a diffusible substance passes from the inner or closed, to the outer parts of a system, as in the loss of substances from a portion of a plant root to water in the surrounding soil.

EXPLOSIVES.   In the conventional sense, a solid, gas, or liquid material which, when triggered, will release a great amount of heat and pressure by way of a very rapid, self-sustaining exothermic decomposition. This entry does not describe nuclear explosives.

There are two principal classes of explosives: (1) *deflagrating explosives,* whose burning processes are rather slow—with progressive reaction rates and buildup of pressure that create a heaving action; and (2) *detonating explosives,* which are characterized by very rapid chemical reactions, thus causing tremendously high pressure and brisance (shattering action). In the latter, detonation waves may obtain a velocity in excess of 20,000 feet per second. The decomposition of cellulose nitrate used in propellants typifies the deflagrating type: $C_{24}H_{30}N_{10}O_{40} \rightarrow 5N_2 + 10H_2 + 5H_2O + 11CO_2 + 13CO$. The decomposition of nitroglycerine typifies the detonating type: $4C_3H_5(ONO_2)_3 \rightarrow 12CO_2 + 10H_2O + 6N_2 + O_2$.

Black powder, using KNO_3 or $NaNO_3$, charcoal and sulfur was probably the first explosive developed and is attributed either to Chinese or Egyptian ingenuity. The time of first use occurred before the birth of Christ. This is a deflagrating explosive and was adapted for blasting purposes as early as the 1600s.

Black powder (gunpowder) consists of an intimate mixture of finely divided solids, 75% potassium nitrate, 15% carbon, 10% sulfur. Powders for sporting guns contain a slightly larger percentage of potassium nitrate

(75 to 78%, smaller percentage of carbon (15 to 12%), and a variation in sulfur from 9 to 12%. Mining or blasting powders, where large volumes of gas are desired, may have 14 to 21% carbon and 13 to 18% sulfur. When ignited, potassium nitrate supplies oxygen for the combustion of explosives, of carbon to carbon dioxide and of sulfur to sulfur dioxide. One gram of powder yields 250 to 300 milliliters of gas measured at $0°C$ and 760 mm pressure. The heat evolved per gram is 500 to 700 calories, and the temperature of the explosion is estimated at $2,700°C$.

Among the earliest high explosives were *mercury fulminate*, $HgC_2N_2O_2$, developed late in the seventeenth century, and *nitrostarch*, $C_{12}H_5(ONO_2)_{30}$, discovered by Braconnot in 1832 and still used as a sensitizing ingredient in modern commercial explosives. *Nitrocotton* was produced in 1838 by Dumas and Pelouse by treating cotton and paper with nitric acid, in the same way that Braconnot treated starch with HNO_3. *Nitroglycerin*, $C_3H_5(ONO_2)_3$, was first made by Sobrero. Early nitroglycerin formulations were highly dangerous and caused numerous accidents. In 1867, Nobel found that nitroglycerin could be rendered safe by absorbing it in a porous material, such as kieselguhr, or diatomaceous earth. After formulation of this first *dynamite*, Nobel introduced $NaNO_3$ and later NH_4NO_3 into his dynamite formulations. Nobel, in 1875, while experimenting with cellulose tetranitrate, mixed collodion with nitroglycerine, resulting in the development of blasting gelatin. The development of the blasting cap used with a safety fuse allowed for safe, positive initiation of dynamite.

Wilbrand, in 1863, first prepared *trinitrotoluene* (TNT), $C_6H_2(CH_3)(NO_2)_3$. The material was not manufactured in production quantities until about 1900. The German military recognized the advantages of TNT as a replacement for *picric acid*, which they had used earlier. TNT was used extensively during World War I and became a standard military explosive.

Tollens, in 1891, prepared *pentaerythritol tetranitrate* (PETN), $C(CH_2NO_3)_4$, but this compound was not commercially available until after World War I. Commercial production had to await a lowering in the cost of formaldehyde and acetaldehyde used in its production.

Henning, in 1899, discovered *cyclotrimethylenetrinitramine* (cyclonite-RDX), but its potential was not realized until about 1920. RDX was used extensively during World War II as a component of numerous cyclotols, plastic explosives, and bursting charges.

$$
\begin{array}{c}
NO_2 \\
| \\
N \\
H_2C \diagup \quad \diagdown CH_2 \\
| \qquad \qquad | \\
O_2N - N \qquad N - NO_2 \\
\diagdown \; C \; \diagup \\
H_2
\end{array}
$$

(RDX)

Although ammonium nitrate, NH_4NO_3, was known to have explosive qualities as early as 1659, it was not used much in explosive formulations until about 1867. At that time, it was used by Nobel to replace a portion of nitroglycerin in dynamite. Because of critical toluene shortages during World War I, NH_4NO_3 was used as a way of conserving TNT. Mixtures containing 80% NH_4NO_3 and 20% TNT; or 50–50 mixes were used. These became known as 80:20 or 50:50 *amatols* and were used as military explosives for shells and bombs. In World War II, particularly by the Axis powers, amatols were used, again to conserve TNT. The Texas City, Texas disaster of 1947 (explosion of ship loaded with NH_4NO_3) reemphasized the potential of this substance for explosives. Later, Robert Akre used a combination of prilled NH_4NO_3 and carbon black (94:6 mix) in making an explosive for blasting in open-pit strip mines. The substance was patented under the name *Akremite*. Later experiments included mixing liquid hydrocarbons to replace the carbon black. This resulted in ANFO explosives. It was found that diesel fuel oil can be mixed with NH_4NO_3 with a consistent quality. Commencing in the late 1950s, ANFO explosives became widely accepted.

Unfortunately, the water resistance of ANFO is low and numerous experiments in attempting to dry-package it were not markedly successful. This shortcoming of ANFO led Cook, Farnum, and others to develop *slurry explosives*. These materials are comprised of: oxidizers, such as NH_4NO_3 and $NaNO_3$ fuels, such as coals, oils, aluminum, and other carbonaceous materials; sensitizers, such as TNT, nitrostarch and smokeless powder; and water—all mixed with a gelling agent to form a thick, viscous explosive having excellent water-resistant properties. These explosives are made as cartridged units or are mixed at the site in bulk and then pumped into place.

The largest consumers of commercial explosives are the mining, construction, and seismic-prospecting industries. Explosives are also used in agriculture for blasting stumps, setting posts, breaking up hardpan, clearing land, digging wells, and blasting drainage and irrigation ditches. Explosive technology also is applied by industry in the forming, cladding, bonding, hardening, and welding of metals. On a small scale, explosive-actuated devices are used in valves, switches, and relays, as well as for cutting, punching, riveting, and fastening metals.

See Fig. 1, a typical quarry blast using a millisecond delay electric blasting cap.

Fig. 1. Quarry blast utilizing high explosives initiated with electrical delay devices (millisecond delay electric blasting caps)

Additional Reading

Akhavan, J.: *The Chemistry of Explosives,* Springer-Verlag Inc., New York, NY, 1998.

Brown, G.I.: *The Big Bang: A History of Explosives,* Sutton Publishing, Ltd., UK, 2000.

Cook, J.R.: *Chemistry and Characteristics of Explosive Materials,* Vantage Press, Inc., New York, NY, 2000.

Cooper, P.W.: *Explosives Engineering,* John Wiley & Sons, Inc., New York, NY, 1997.

Cooper, P.W. and S.R. Kurowski: *Introduction to the Technology of Explosives,* John Wiley & Sons, Inc., New York, NY, 1997.

Holmberg, R.: *Explosives and Blasting Technique,* A.A. Balkema Publishers, Rotterdam, 2000.

Lipanov, A.: *Theory of Combustion Powder and Explosives,* Nova Science Publishers, Inc., Huntington, NY, 1996.

Mader, C.L.: *Numerical Modelling of Explosives and Propellants,* CRC Press, LLC., Boca Raton, FL, 1997.

Spani, J.C. and Knackmuss Hans-Joachim: *Biodegradation of Nitroaromatic Compounds and Explosives,* Lewis Publishers, Boca Raton, FL, 2000.

EXTENDER. A low-gravity material used in paint, ink, plastic, and rubber formulations chiefly to reduce cost per unit volume by increasing bulk. Extenders include diatomaceous earth, wood flock, mineral rubber, liquid asphalt, etc. Microscopic droplets of water fixed permanently in a plastic matrix are an efficient extender for polyester resins. In the food industry, the term refers to certain extruded proteins, especially those derived from soybeans, which are used in meat products to provide equivalent nutrient values at lower cost. Made from defatted soy flour, they are often called *textured proteins*.

EXTRACTION (Liquid–Liquid). The physical process of liquid–liquid extraction separates a dissolved component from its solvent by transfer to a second solvent, immiscible with the first but having a higher affinity for the transferred component. The latter is sometimes called the consolute component. Liquid–liquid extraction can purify a consolute component with respect to dissolved components which are not soluble in the second solvent, and often the extract solution contains a higher concentration of the consolute component than the initial solution. In the process of fractional extraction, two or more consolute components can be extracted and also separated if these have different distribution ratios between the two solvents.

The principle of liquid–liquid extraction, and some of the special terminology, are illustrated in Figure 1 which shows a single contacting stage. If equilibrium is fully established after contact, the stage is defined as an ideal or theoretical stage. The two resulting liquid phases are the raffinate from which most of solute C has been removed, and the extract, consisting mainly of solvent B and C.

In the simplest case, the feed solution consists of a solvent A containing a consolute component C, which is brought into contact with a second solvent B. For efficient contact there must be a large interfacial area across which component C can transfer until equilibrium is reached or closely approached. On the laboratory scale this can be achieved in a few minutes simply by hand agitation of the two liquid phases in a stoppered flask or separatory funnel. Under continuous flow conditions it is usually necessary to use mechanical agitation to promote coalescence of the phases. After sufficient time and agitation, the system approaches equilibrium which can be expressed in terms of the extraction factor ε for component C:

$$\varepsilon = \frac{\text{quantity of C in B-rich phase}}{\text{quantity of C in A-rich phase}} = m \frac{B}{A} \tag{1}$$

where B and A refer to the quantities of the two solvents and m is the distribution coefficient.

The component C in the separated extract from the stage contact shown in Figure 1 may be separated from the solvent B by distillation, evaporation, or other means, allowing solvent B to be reused for further extraction. Alternatively, the extract can be subjected to back-extraction (stripping) with solvent A under different conditions, eg, a different temperature; again, the stripped solvent B can be reused for further extraction. Solvent recovery is an important factor in the economics of industrial extraction processes.

Whereas Figure 1 assumes a physical extraction based on different solubilities as expressed by the distribution coefficient, many extractions depend on chemical changes. In such cases the component C in the feed solvent may not itself have any solubility in the extracting solvent B, but can be made to react with an extractant to produce a compound or species which is soluble in B. Many metals can be extracted from aqueous solutions of their salts into organic carrier solvents by using organic extractants which can form organometallic compounds or complexes. Stripping of the metals from the organic to an aqueous phase can be effected by changing a chemical condition such as pH.

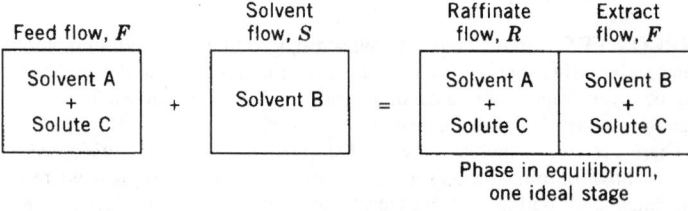

Fig. 1. Single contacting stage

Principles

Physical Equilibria and Solvent Selection. In order for two separate liquid phases to exist in equilibrium, there must be a considerable degree of thermodynamically nonideal behavior. If the Gibbs free energy, G, of a mixture of two solutions exceeds the energies of the initial solutions, mixing does not occur and the system remains in two phases. For the binary system containing only components A and B, the condition for the formation of two phases is

$$\frac{d^2 G}{d x_A^2} > \tag{2}$$

The selection of solvents for a given separation depends largely 0 on equilibrium considerations. Other important factors include cost, ease of solvent recovery by distillation or other means, safety and environmental impact, and physical properties which must permit easy phase dispersion and separation. Solvent selection is therefore a broad-based exercise which is hard to quantify. However a useful quantitative approach has been proposed for comparing simplified equilibrium estimations on the basis of regular solution theory.

Chemical Equilibria. In many cases, mass transfer between two liquid phases is accompanied by a chemical change. The transferring species can dissociate or polymerize depending on the nature of the solvent, or a reaction may occur between the transferring species and an extractant present in one phase.

In addition to the liquid–liquid reaction processes, there are many cases in both analytical and industrial chemistry where the main objective of separation is achieved by extraction using a chemical extractant. The technique of dissociation extraction is very valuable for separating mixtures of weakly acidic or basic organic compounds.

In hydrometallurgical separations, a metal ion in aqueous solution can be selectively converted to an organometallic compound or complex which is soluble in an organic carrier solvent.

Chelating extractants owe effectiveness to the attraction of adjacent groups on the molecule for the metal. Anionic extractants are commonly based on high molecular weight amines. Solvating extractants contain one or more electron donor atoms, usually oxygen, which can supplant or partially supplant the water which is attached to the metal ions.

Interfacial Mass-Transfer Coefficients. Whereas equilibrium relationships are important in determining the ultimate degree of extraction attainable, in practice the rate of extraction is of equal importance. Equilibrium is approached asymptotically with increasing contact time in a batch extraction. In continuous extractors the approach to equilibrium is determined primarily by the residence time, defined as the volume of the phase contact region divided by the volume flow rate of the phases.

The rate of mass transfer depends on the interfacial contact area and on the rate of mass transfer per unit interfacial area, i.e., the mass flux. The mass flux very close to the liquid–liquid interface is determined by molecular diffusion in accordance with Fick's first law:

$$N = -D \frac{\partial c}{\partial z} \tag{3}$$

where N refers to the flux in the z direction, c is the concentration of the consolute component, and D is its molecular diffusivity in the solvent.

Mass-Transfer Coefficients with Chemical Reaction. Chemical reaction can occur in any of the five regions shown in Figure 2, i.e., the bulk of each phase, the film in each phase adjacent to the interface, and at the interface itself. Irreversible homogeneous reaction between the consolute component C and a reactant D in phase B can be described as

$$C + zD \rightarrow products \tag{4}$$

The equations of combined diffusion and reaction, and their solutions, are analogous to those for gas absorption. It has been shown how the concentration profiles and rate-controlling steps change as the rate constant increases. When the reaction is very slow and the B-rich phase is essentially saturated with C, the mass-transfer rate is governed by the kinetics within the bulk of the B-rich phase. This is defined as regime 1. For a slow reaction defined as regime 2, the consolute component C is almost entirely depleted in the bulk of the B-rich phase and the mass transfer of C between the phases controls the rate of the reaction. For a very fast reaction the depletion of C affects the concentration profile in the diffusion film. The steepening of the concentration profile for regime 3 leads to an enhancement in the film mass-transfer coefficient in the B-rich phase. Finally, the case of an instantaneous reaction (regime 4) leads to the formation of a thin reaction zone to which components C and D diffuse in stoichiometric amounts.

Interfacial Contact Area and Approach to Equilibrium. Experimental extraction cells such as the original Lewis stirred cell are often operated with a flat liquid–liquid interface the area of which can easily be measured. In the single-drop apparatus, a regular sequence of drops of known diameter is released through the continuous phase. These units are useful for the direct calculation of the mass flux N and hence the mass-transfer coefficient for a given system. In industrial equipment, however, it is usually necessary to create a dispersion of drops in order to achieve a large specific interfacial area, a, defined as the interfacial contact area per unit volume of two-phase dispersion. Thus the mass-transfer rate obtainable per unit volume

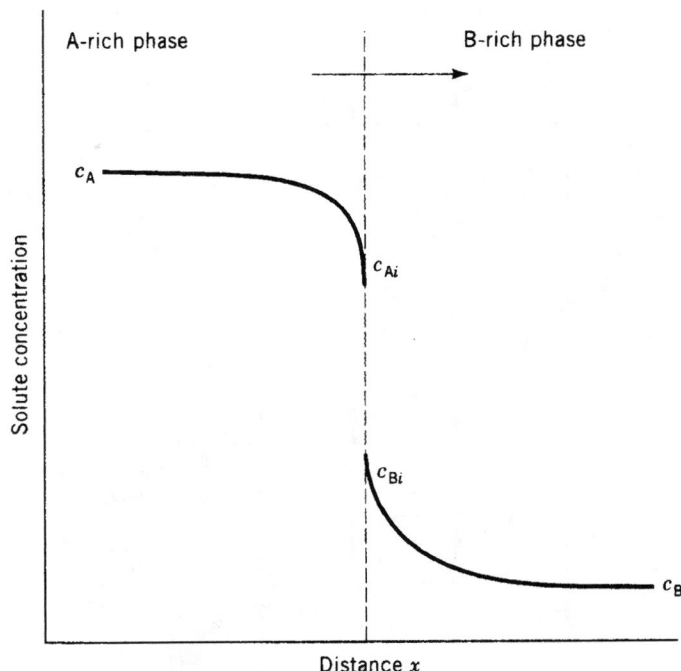

Fig. 2. Concentration profiles near an interface where the arrow represents the direction of mass transfer, c_A = concentration of C in A-rich phase, c_B = concentration of C in B-rich phase, and the subscript i denotes the interface

is given as

$$(N \cdot a) = K_A a (c_A - c_A^*) \tag{5}$$

Calculation of Equilibrium Stages. Multistage contacting can be arranged in a concurrent, crosscurrent, or countercurrent manner. The sequence of stages is sometimes referred to as a cascade, referring to the early use of gravity overflow from stage to stage. The countercurrent arrangement represents the best compromise between the objectives of high extract concentration and a high degree of extraction of the solute, for a given solvent-to-feed ratio. For the case of a partially miscible ternary system, the number of ideal stages in a countercurrent cascade can be estimated graphically on a triangular diagram, using the Hunter-Nash method.

Fractional Extraction. Fractional extraction is the separation of two or more consolute components by solvent extraction. Single-solvent fractional extraction has been known for many years, but the range of solvents available is limited because of the requirement that the solvents must be sparingly miscible with each of the feed components.

Dual solvent fractional extraction makes use of the selectivity of two solvents (A and D) with respect to consolute components.

Differential Contacting. Although the equilibrium stage concept has proved extremely useful in describing the performance of mixer-settlers and plate columns having discrete stages, it is not appropriate for spray towers, packed columns, etc, in which no discrete stages can be identified. In such differential types of contactors, equilibrium between phases is never reached and therefore the mass-transfer rate is important in the design procedure.

A differential countercurrent contactor operating with a dilute solution of the consolute component C and immiscible components A and B is shown in Figure 3. Under these conditions, the superficial velocities of the A-rich and B-rich streams can be assumed not to vary significantly with position in the contactor, and are taken to be U_A and U_B, respectively. The concentration of C in the A-rich stream is c_A and that in the B-rich stream is c_B.

A steady-state material balance can be carried out on a small section of length dz and volume dz (on the basis of unit cross-sectional area) in the contactor:

$$U_B \, dc_B = U_A \, dc_A = K_A a (c_A - c_A) \, dz \tag{6}$$

Axial Dispersion. Elementary texts assume that all the fluid in each phase has the same resident time in a countercurrent extractor. In practice,

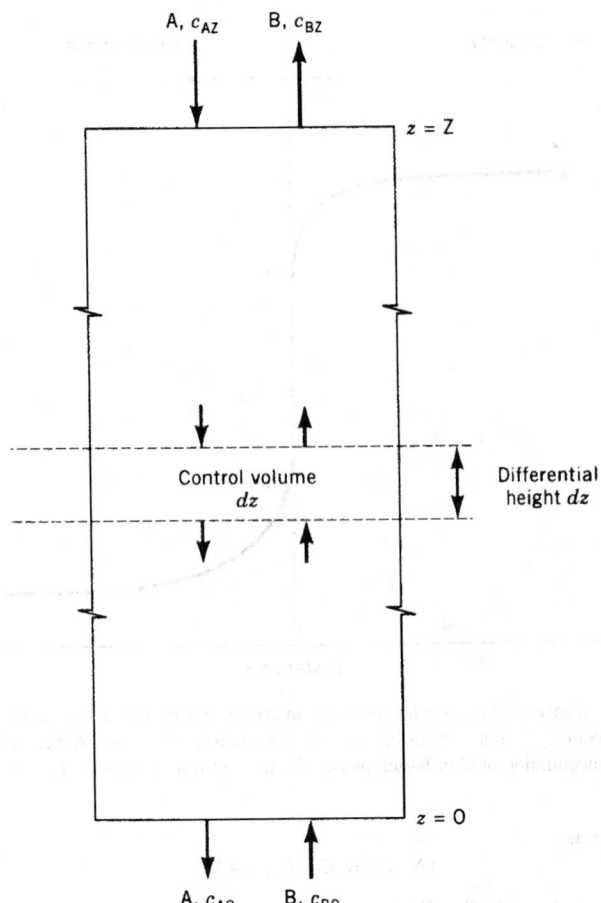

Fig. 3.   Mass transfer in a differential contactor. Terms are defined in the text

the two phases rarely move countercurrently in plug flow because of axial mixing which arises from the action of turbulent eddies, circulation currents, or the effects of drop wakes. The effect is to flatten the axial concentration profiles within each phase. Axial mixing can lead to a reduction in the effective driving force for mass transfer which in turn reduces the NTU below that expected for the plug flow case. An important feature of the profile is the discontinuity or "jump" in concentration which occurs at entry to the contactor when the liquid in the feed line enters the mixed region of the column.

Two alternative approaches are used in axial mixing calculations. For differential contactors, the axial on model is used:

$$N = -E \frac{\partial c}{\partial z} \qquad (7)$$

For contactors in which discrete well-mixed compartments can be identified, for example sieve-plate columns, axial mixing effects are incorporated into the stagewise model by means of the backflow ratio α which is defined as the fraction of the net interstage flow of one phase which is considered to flow in the reverse direction. For a contactor in which there are many compartments, the axial dispersion coefficient and the backflow ratio, α, are interrelated as follows:

$$E = \frac{U H}{\ln((1+\alpha)/\alpha)} \qquad (8)$$

where H is the height of one compartment and U is the superficial velocity. The detailed calculations of concentration profiles and mass-transfer rates with axial mixing require the solution of a fourth-order differential equation (dispersion model) or the equivalent difference equation (backflow model) along with appropriate boundary conditions.

Drop Diameter. In extraction equipment, drops are initially formed at distributor nozzles; in some types of plate column the drops are repeatedly formed at the perforations on each plate. Under such conditions, the diameter is determined primarily by the balance between interfacial forces and buoyancy forces at the orifice or perforation.

In many types of contactors, such as stirred tanks, rotary agitated columns, and pulsed columns, mechanical energy is applied externally in order to reduce the drop size and thereby increase the rate of mass transfer.

Holdup and Flooding. The volume fraction of the dispersed phase, commonly known as the holdup h, can be adjusted in a batch extractor by means of the relative volumes of each liquid phase added. However, in a countercurrent column contactor, the holdup of the dispersed phase is considerably less than this, because the dispersed drops travel quite fast through the continuous phase and therefore have a relatively short residence time in the equipment. The holdup is related to the superficial velocities U of each phase, defined as the flow rate per unit cross section of the contactor, and to a slip velocity U_s:

$$U_s = U_d/h + U_c/(1-h) \qquad (9)$$

As the throughput in a contactor represented by the superficial velocities U_c and U_d is increased, the holdup h increases in a nonlinear fashion. A flooding point is reached at which the countercurrent flow of the two liquid phases cannot be maintained. The flow rates at which flooding occurs depend on system properties, in particular density difference and interfacial tension, and on the equipment design and the amount of agitation supplied.

The nonuniformity of drop dispersions can often be important in extraction. This nonuniformity can lead to axial variation of holdup in a column even though the flow rates and other conditions are held constant.

Coalescence and Phase Separation. Coalescence between adjacent drops and between drops and contactor internals is important for two reasons. It usually plays a part, in combination with breakup, in determining the equilibrium drop size in a dispersion, and it can therefore affect holdup and flooding in a countercurrent extraction column. Secondly, it is an essential step in the disengagement of the phases and the control of entrainment after extraction has been completed.

Membrane Extraction. An extraction technique which uses a thin liquid membrane or film has been introduced. The principal advantages of liquid-membrane extraction are that the inventory of solvent and extractant is extremely small and the specific interfacial area can be increased without the problems which accompany fine drop dispersions.

Supercritical Extraction. The use of a supercritical fluid such as carbon dioxide as extractant is growing in industrial importance, particularly in the food-related industries. The advantages of supercritical fluids as extractants include favorable solubility and transport properties, and the ability to complete an extraction rapidly at moderate temperature.

Two-Phase Aqueous Extraction. Liquid–liquid extraction usually involves an aqueous phase and an organic phase, but systems having two or more aqueous phases can also be formed from solutions of mutually incompatible polymers such as poly(ethylene glycol) (PEG) or dextran.

Because of the growth in biotechnology, two-phase aqueous extraction is becoming more important industrially. Two-phase aqueous systems have low interfacial tension, low interphase density difference, and high viscosity in comparison with most aqueous–organic systems. Although interfacial contact is very efficient, the separation of the phases after contact can be slow, requiring centrifugation. The performance of a spray column for two-phase aqueous extraction has also been reported.

Equipment and Processing

Laboratory Extractors, Pilot-Scale Testing, and Scale-Up. Several laboratory units are useful in analysis, process control, and process studies. The AKUFVE contactor incorporates a separate mixer and centrifugal separator. It is an efficient instrument for rapid and accurate measurement of partition coefficients, as well as for obtaining reaction kinetic data. Miniature mixer–settler assemblies set up as continuous, bench-scale, multistage, countercurrent, liquid–liquid contactors are particularly useful for the preliminary laboratory work associated with flow-sheet development and optimization because these give a known number of theoretical stages.

Because the factors relating to mass transfer and fluid dynamics of the systems in an extractor are extremely complex, particularly for mixed solvents and feedstocks of commercial interest, pilot-scale testing remains an almost inevitable preliminary to a full-scale contactor design. These tests provide *(1)* total throughput and agitation speed; *(2)* HETS or HTU; *(3)* stage efficiency; *(4)* hydrodynamic conditions, such as droplet dispersion, phase separation, flooding, emulsion layer formation, etc; *(5)* selection of direction of mass transfer; *(6)* solvent-to-feed ratio; *(7)* material of

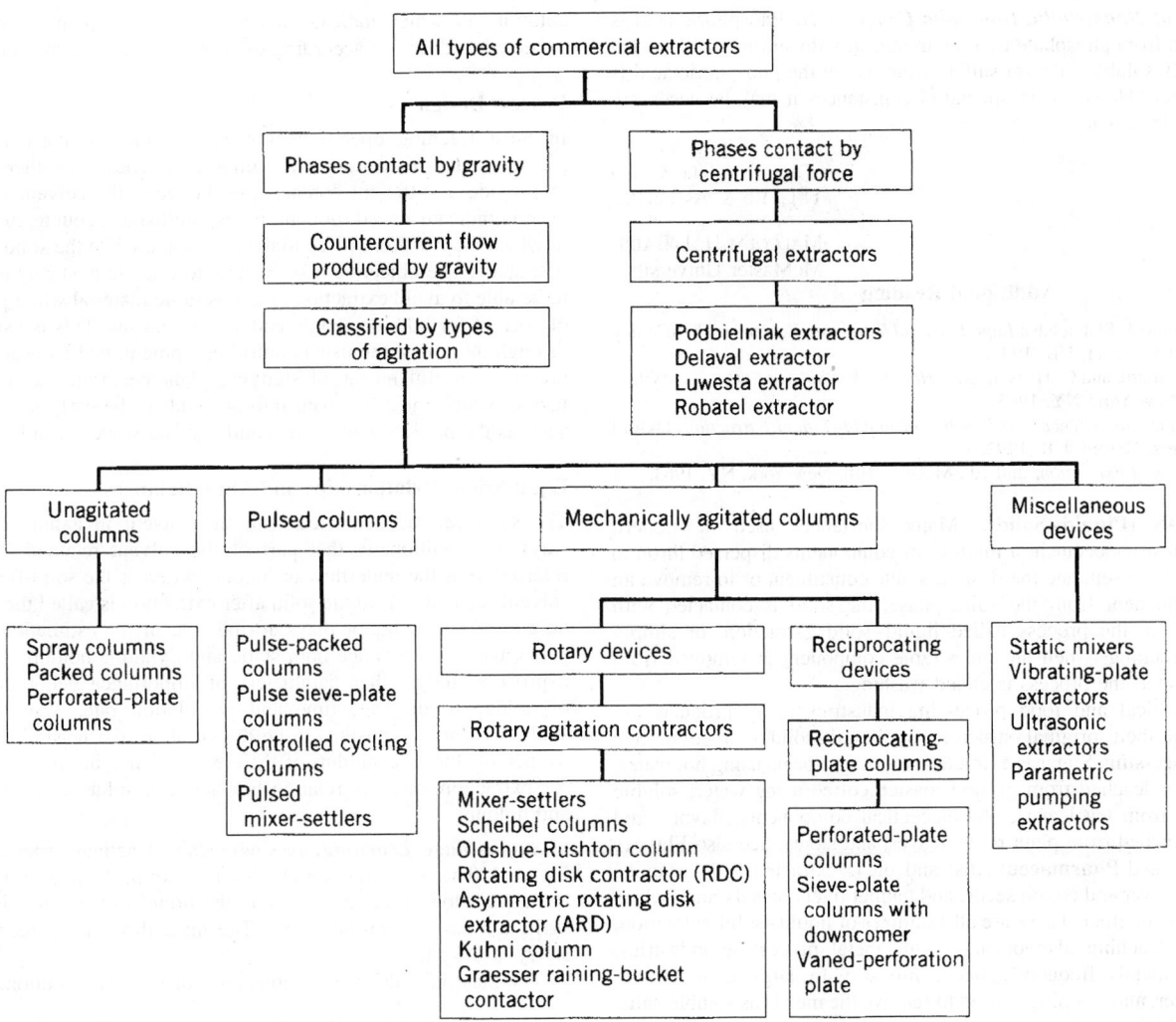

Fig. 4.   Classification of commercial extractors

construction and its wetting characteristics; and (8) confirmation of the desired separation in cases where equilibrium data are not available.

For design of a large-scale commercial extractor, the pilot-scale extractor should be of the same type as that to be used on the large scale. Reliable scaleup for industrial-scale extractors still depends on correlations based on extensive performance data collected from both pilot-scale and large-scale extractors covering a wide range of liquid systems. Only limited data for a few types of large commercial extractors are available in the literature.

Commercial Extractors. Extractors can be classified according to the methods applied for interdispersing the phases and producing the countercurrent flow pattern. Figure 4 summarizes the classification of the principal types of commercial extractors.

Organic Processes

Petroleum and Petrochemical Processes. The first large-scale application of extraction was the removal of aromatics from kerosene to improve its burning properties. Solvent extraction is also extensively used to meet the growing demand for the high purity aromatics such as benzene, toluene, and xylene (BTX) as feedstocks for the petrochemical industry. Additionally, the separation of aromatics from aliphatics is one of the largest applications of solvent extraction.

Pharmaceutical Processes. The pharmaceutical industry is a principal user of extraction because many pharmaceutical intermediates and products are heat-sensitive and cannot be processed by methods such as distillation. A useful broad review can be found in the literature. Extraction is used in the production of antibiotics, vitamins, sulfa drugs, methaqualone, phenobarbital, antihistamines, cortisone, estrogens and other hormones (qv), and reserpine and alkaloids.

Food Processing. Food processing makes use of solvent extraction in several ways. Industrial refining of fats and oils using propane is known

as the Solexol process. Solvent extraction is used in many protein refining processes, for example the extraction of fish protein from ground fish using *i*-propyl alcohol. Recovery of lactic acid by an extractive fermentation has recently been reported. The applications of extraction in the food industry have been reviewed.

Other Organic Processes. Solvent extraction has found application in the coal-tar industry for many years, as for example in the recovery of phenols from coal-tar distillates by washing with caustic soda solution. Solvent extraction of fatty and resimic acid from tall oil has been reported. Dissociation extraction is used to separate *m*-cresol from *p*-cresol and 2,4-xylenol from 2,5-xylenol. Solvent extraction can play a role in the direct manufacture of chemicals from coal, treatment of industrial effluents, biopolymer extraction, and difficult separations.

Inorganic Processes

Nuclear Fuel Reprocessing. Spent fuel from a nuclear reactor contains ^{238}U, ^{235}U, ^{239}Pu, ^{232}Th, and many other radioactive isotopes (fission products). Reprocessing involves the treatment of the spent fuel to separate plutonium and unconsumed uranium from other isotopes so that these can be recycled or safely stored.

Copper. The recovery of copper, Cu, from ore leach liquors as a stage in the hydrometallurgical route to the pure metal is one of the largest applications of liquid–liquid extraction.

Nickel and Cobalt. Often present with copper in sulfuric acid leach liquors are nickel and cobalt. In the case of chloride leach liquors, separation of cobalt from nickel is inherently simpler because cobalt, unlike nickel, has a strong tendency to form anionic chloro-complexes. Thus cobalt can be separated by amine extractants, provided the chloride content of the aqueous phase is carefully controlled. A successful example of this approach is the Falconbridge process developed in Norway.

Extraction of Nonmetallic Inorganic Compounds. Phosphoric acid is usually formed from phosphate rock by treatment with sulfuric acid, which forms sparingly soluble calcium sulfate from which the phosphoric acid is readily separated. However, in special circumstances it may be necessary to use hydrochloric acid.

TEH C. LO
T. C. Lo & Associates

MALCOLM H. I. BAIRD
McMaster University

Additional Reading

Godfrey, J.C. and M.J. Slater, eds.: *Liquid–Liquid Extraction Equipment,* John Wiley & Sons, Ltd. Chichester, UK, 1994.

Lo, T.C., M.H.I. Baird, and C. Hanson, eds.: *Handbook of Solvent Extraction,* Wiley-Interscience, New York, NY, 1983.

Thornton, J.D. ed.: *The Science and Practice of Liquid–Liquid Extraction,* Oxford University Press, Oxford, UK, 1992.

Treybal, R.E.: *Liquid Extraction,* 2nd ed., McGraw-Hill, New York, NY, 1963.

EXTRACTION (Liquid–Solid).

Many substances used in modern processing industries occur in a mixture of components dispersed through a solid material. To separate the desired solute constituent or to remove an unwanted component from the solid phase, the solid is contacted with a liquid phase in the process called liquid–solid extraction, or simply leaching. In leaching, when an undesirable component is removed from a solid with water, the process is called washing.

In the biological and food processing industries many products are extracted from their original structure by liquid–solid extraction. See also **Food Processing**. Sugar is extracted from sugar beets using hot water; instant coffee is leached from ground roasted coffee using water; soluble tea is leached from tea leaves; pharmaceutical components, flavors, and essences are leached from plant roots, leaves, and stems. See also **Flavors and Essences**; and **Pharmaceuticals**; and oil is extracted from peanuts, soybeans, sunflower and cotton seeds, and halibut livers by solvents such as hexane, acetone, or ether. These are all examples of liquid–solid extraction.

Large-scale leaching also occurs in the metal processing industries, where useful metals frequently occur mixed with large quantities of unwanted matter, and leaching is used to remove the metals as soluble salts.

Mechanisms of Extraction

If the solute is uniformly distributed through the solid phase the material near the surface dissolves first to leave a porous structure in the solid residue. In order to reach further solute the solvent has to penetrate this outer porous region; the process becomes progressively more difficult and the rate of extraction decreases. If the solute forms a large proportion of the volume of the original particle, its removal can destroy the structure of the particle which may crumble away, and further solute may be easily accessed by solvent. In such cases the extraction rate does not fall as rapidly.

In general, the following steps can occur in an overall liquid–solid extraction process: solvent transfer from the bulk of the solution to the surface of the solid; penetration or diffusion of the solvent into the pores of the solid; dissolution of the solvent into the solute; solute diffusion to the surface of the particle; and solute transfer to the bulk of the solution. Any one of the five basic processes may be responsible for limiting the extraction rate.

The overall extraction process is sometimes subdivided into two general categories according to the main mechanisms responsible for the dissolution stage: *(1)* those operations that occur because of the solubility of the solute in or its miscibility with the solvent, e.g., oilseed extraction, and *(2)* extractions where the solvent must react with a constituent of the solid material in order to produce a compound soluble in the solvent, eg, the extraction of metals from metalliferous ores. In the former case the rate of extraction is most likely to be controlled by diffusion phenomena, but in the latter the kinetics of the reaction producing the solute may play a dominant role.

Diffusion and Mass Transfer During Leaching. Rates of extraction from individual particles are difficult to assess because it is impossible to define the shapes of the pores or channels through which mass transfer has to take place. However, the nature of the diffusional process in a porous solid could be illustrated by considering the diffusion of solute through a pore. This is described mathematically by the diffusion equation, the solutions of which indicate that the concentration in the pore would be expected to decrease according to an exponential decay function.

Process Design

In most leaching operations the maintenance of constant fluid flows, pressures, and temperatures are important. These, together with the need to provide a sufficient contact time between the solvent and the solids, usually indicate a need for continuous, multistage, countercurrent processes in which fresh solvent is fed to the final stage while the solids are fed to the first stage. The objective is to be able to operate at steady conditions, and to be able to avoid extraction of undesirable material while preventing loss of solvent for both economic and safety reasons. This is usually achieved through the use of the usual control equipment, and recording instruments provide a useful means of studying plant performance. There are other factors which must be taken into account in the early stages of a design such as the particle size of the solid and the solvent employed.

Equilibrium Relationships and Mass Balances

The solid can be contacted with the solvent in a number of different ways but traditionally that part of the solvent retained by the solid is referred to as the underflow or holdup, whereas the solid-free solute-laden solvent separated from the solid after extraction is called the overflow. The holdup of bound liquor plays a vital role in the estimation of separation performance. In practice both static and dynamic holdup are measured in a process study, other parameters of importance being the relationship of holdup to drainage time and percolation rate. The results of such studies permit conclusions to be drawn about the feasibility of extraction by percolation, the holdup of different bed heights of material prepared for extraction, and the relationship between solute content of the liquor and holdup.

Single-Stage Leaching. A single-stage leaching process is shown in Figure 1. The solution overflow rate is V kg/h; the mass fraction of solute in the overflow solution is x_A; and the liquid in the slurry is flowing at L kg/h, and has a composition y_A. The mass flow of dry inert solids in the slurry is B kg/h.

The material balance equations are, for the total solution:

$$L_0 + V_2 = L_1 + V_1 = M \qquad (1)$$

where M is the total input flow rate of solution to the unit; for the solute component A:

$$L_0 y_{A0} + V_2 x_{A2} = L_1 y_{A1} + V_1 x_{A1} = M x_{Am} \qquad (2)$$

where $x_1 = y_1 = x_{Am}$; and for the solids:

$$B = L_0 N_0 = L_1 N_1 \qquad (3)$$

where N_i is the mass concentration of inert solids in the ith stream, i.e., kg of inert solid per kg solution. From these balances the concentration of the discharged solution can be estimated.

Countercurrent Multistage Leaching. Countercurrent extraction offers the most economical use of solvent, permitting high concentrations in the final extract and high recovery from the initial solid but utilizing the least amount of solvent. When the amount of solvent removed with the insoluble solid in the underflow is constant, it is convenient to define the ratio

$$R = \frac{\text{amount of solvent in overflow}}{\text{amount of solvent in underflow}} = \frac{V_n}{L_n} \qquad (4)$$

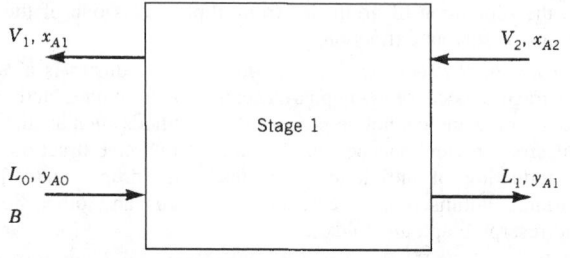

Fig. 1.   Flow diagram for single-stage leaching

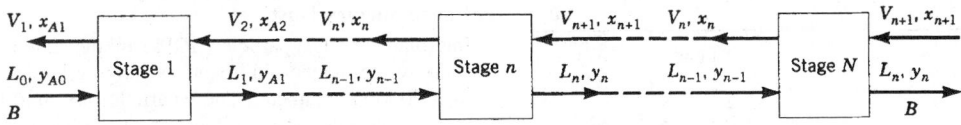

Fig. 2. Flow diagram for countercurrent multistage leaching

If perfect mixing occurs in each stage and the solute is not adsorbed preferentially at the surface of the solid, then the concentration of the solution in the underflow is the same as that in the overflow and

$$R = \frac{\text{amount of solute in overflow}}{\text{amount of solute in underflow}} = \frac{V_n x_{An}}{L_n y_{An}} \quad (5)$$

Referring to Figure 2, by considering solute mass balances over n, $(n-1)$, ...2, 1 units in turn and eliminating intermediate solute mass fractions and flow rates, the amount of solute associated with the leached solid may be calculated in terms of the composition of the solid and solvent streams fed to the system. The resulting equation is

$$L_0 y_{A0} = \frac{R^{n+1} - 1}{R - 1} L_n y_{An} - \frac{R^n - 1}{R - 1} V_{n+1} x_{An+1} \quad (6)$$

Countercurrent Leaching with Variable Underflow. In practice most applications have a variable underflow, which is normally greatest at that point in the process where the solute concentration in the solvent is highest.

Extractors

The variety of extractors used in liquid–solid extraction is diverse, ranging from batchwise dump or heap leaching for the extraction of low grade ores to continuous countercurrent extractors to extract materials such as oilseeds and sugar beets where problems of solids transport have dominated equipment and development.

Safety and Environmental Considerations

Solvent flammability, the solvent, and dust loading in the atmosphere of the working environment and of the products in the case of edible materials are the main factors that constitute health and safety hazards in extraction plants. General safety and environmental standards must therefore be applied and due recognition taken of the most recently published national regulations relating to acceptable threshold limit values (TLVs) for solvents and dusts.

Disposal of exhausted solids can be easily overlooked at the plant design stage, particularly when these have no intrinsic value; alternative disposal methods might include landfill of inert material or incineration, hydrolysis, or pyrolysis of organic materials. Liquid, solid, and gaseous emissions are all subject to the usual environmental considerations.

RICHARD J. WAKEMAN
University of Exeter

Additional Reading

Burkin, A.R.: *The Chemistry of Hydrometallurgical Processes*, E. & F. N. Spon, London, 1966.

Coulson, J.M., J.F. Richardson, J.R. Backhurst, and J.H. Harker: *Chemical Engineering*, 4th ed., Vol. 2, Pergamon Press, Oxford, UK, 1991.

McCabe, W.L. and J.C. Smith: *Unit Operations in Chemical Engineering*, 3rd ed., McGraw-Hill Book Co., Inc., London, UK, 1976.

Perry, R.H. and D. Green, eds., *Perry's Chemical Engineers Handbook*, 50th ed., McGraw-Hill Book Co., Inc., New York, NY, 1984.

EXTRACTIVE METALLURGY. That phase of metallurgy dealing with the removal of metals from minerals. Methods are discussed under individual metals.

EXTRATERRESTRIAL MATERIALS. Extraterrestrial materials are samples from other bodies in the solar system that can be studied in earthbound laboratories. Sensitive and ever-improving analytical techniques are used to provide information at levels of detail and sophistication that cannot be matched by telescopic or spacecraft investigations. Much of the knowledge of early solar system bodies, processes, environments, and chronology has come from the study of these samples. Extraterrestrial materials that are available for laboratory study include meteoritic materials that fall naturally to the earth, some meteoritic material that has been captured in space, and lunar samples that were recovered by the Apollo and Luna sample-return missions flown to the moon during the years 1969 to 1972. The meteoritic materials in existing collections include samples from asteroids, comets, and the moon, and probably Mars. The comet and asteroid samples are the best preserved solids from the early solar system and are the oldest and most cosmochemically primitive samples available for direct study. Because of their primitive and unfractionated nature, these samples provide the best estimate of the composition of the sun and the solar system as a whole. It has been shown that many meteorites contain preserved interstellar grains, particles older than the Sun that formed around other stars and served as the initial building blocks of the solar system.

Meteorites

Meteorites by definition are extraterrestrial materials that fall from the sky and actually hit the surface of the earth. Meteorites fall randomly to the Earth but are not found randomly distributed on the Earth's surface. The highest general concentrations of meteorites occur in Antarctica, where long exposure time and the combined effects of ice movement and sublimation concentrate meteorites on top of blue ice fields. In Antarctica and elsewhere, meteorites are often found in clusters created by the breakup of a larger body during hypervelocity entry into the atmosphere. When a meteor breaks up at high altitude, the resulting fragments impact over an elliptical region several kilometers across the ground, forming a strewn field where sometimes thousands of individual specimens are found. Because of atmospheric breakup, the number of individual meteorite specimens that are collected is much larger than the actual number of meteoroids that produced them. Meteoroids are themselves fragments of bodies that broke up in space, and the actual lineage of meteoritic samples may trace back to a relatively limited number of initial parent bodies.

Types. Most meteorites can be classified into definite groups distinguished by elemental, mineralogical, petrographic, and isotopic composition. The general groups are the chondrites, achondrites, irons, and stony irons. Although fragments of one meteorite class are often found inside another as a result of collisional mixing, in general the bulk properties of meteorites fall into quantified groups without a continuum of compositions between established groups. It is likely that some groups are samples of single asteroids, the apparent source of most meteorites. The majority of asteroids are located in the asteroid belt between Jupiter and Mars, and are believed to be relic solar nebula planetismals that escaped incorporation into planets.

Chondrites. Over 90% of the meteorites that are observed to fall out of the sky are classified as chondrites, samples that are distinguished from terrestrial rocks in many ways. One of the most fundamental is age. Like most meteorites, chondrites have formation ages close to 4.55 Gyr. Chondrites also have basically undifferentiated elemental compositions for most nonvolatile elements and match solar abundances except for moderately volatile elements. The most compositionally primitive chondrites are members for the type 1 carbonaceous (CI) class.

Another unique property of chondrites is the presence of chondrules, objects found in nearly all chondrites except those of the CI class. Chondrules are millimeter-sized, spheroidal bodies composed predominantly of olivine, $(Mg, Fe)_2 SiO_4$, pyroxene, and a glass of approximate feldspathic composition. It is believed that these were objects individually orbiting the Sun that formed by rapid melting and cooling of millimeter-sized precursors. The processes must have been highly efficient, as chondrules comprise $> 75\%$ of the mass of many meteorites. It is possible that chondrules were primary building blocks of the earth and the terrestrial planets.

Chondrites are divided into eight subclasses distinguished by elemental, isotopic, and mineralogical composition. A characteristic distinguishing different chondrite groups is the abundance and oxidation state of iron, Fe. Chondrite classes are also distinguished by their abundances of both volatile and refractory elements and by the oxygen isotope compositions. This remarkable distinction, illustrated in the three-isotope plot shown in Figure 1, correlates with compositional and mineralogical classification.

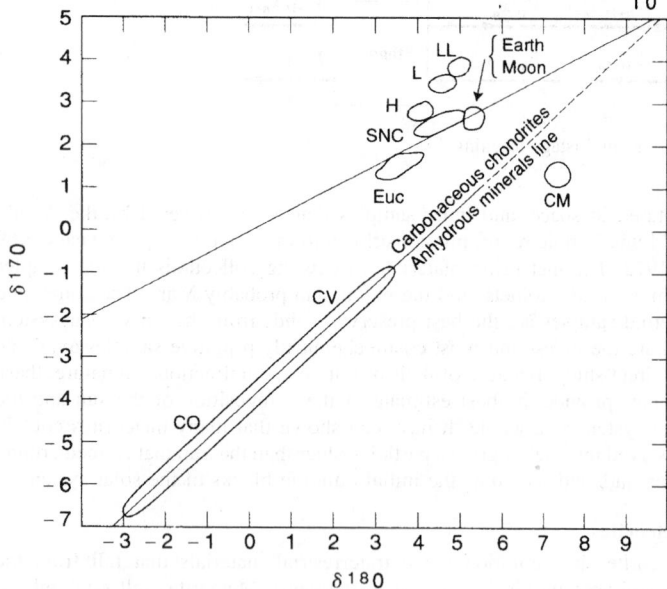

Fig. 1. The bulk oxygen isotopic composition of different meteorite classes where (—) is the terrestrial fractionation line. The δ notation refers to the normalized difference between $^{17}O : ^{16}O$ or $^{18}O : ^{16}O$ ratios to those in standard mean ocean water (SMOW) in relative units of parts per thousand. The meteorites formed from materials that were enriched or depleted in ^{16}O and their bulk compositions plot off of the terrestrial line which has a slope of 1/2 owing to mass dependent fractionation. Some of the anhydrous minerals from carbonaceous chondrites fall on a line having a slope of 1. These anomalous compositions may be produced by mixing with an ^{16}O-rich component that has a different nucleosynthetic history than mean solar system material

Within each chondrite class there are petro graphic grades that relate to alteration processes that occurred within the meteorite parent body. The grades range from 1 to 6, although no class has examples in more than four grades. Grades 3 to 6 represent the effects of thermal metamorphism where the higher number is the more strongly altered. Grades 1 and 2 occur only for the CI and CM chondrites, respectively. CI and CM chondrites have been extensively altered by aqueous alteration in their parent bodies probably as a result of the melting of ice followed by reactions of preexisting phases.

Achondrites. The achondrites are differentiated stony meteorites that are apparently derived from parent bodies that were heated to at least partial melting temperatures. Achondrites do not contain chondrules and do not have elemental compositions that match solar abundances for condensable elements. These materials are old but their properties were more determined by planetary processes such as melting and differentiation than by primary nebular processes, such as condensation and accretion. Many of the achondrite subclasses can be combined into three basic groups: HED, SNC, and lunar groups. The HED group comprises the howardite, eucrite, and diogenite subclasses, which together are responsible for more than 6% of all meteorite falls. The SNC achondrites are composed of the Shergotty, Nakhla, and Chassigny subgroups.

Irons. Approximately 4% of meteorite falls are irons. Because they are distinctive rocks and weather relatively slowly, most meteorites that were not seen to fall, but were found accidentally, are irons. Iron meteorites are composed of metallic iron and siderophile elements that fractionated from molten parent bodies. They may have been cores of asteroids or they may have only been localized metal accumulations.

Origin. Typical meteorites have formation ages of 4.55 Gyr and exposure ages of only 10^7 years, during which time they existed as meter-sized bodies unshielded to the effects of cosmic rays. With the exception of the SNC (Martian) and lunar meteorites, it is widely believed that most conventional meteorites are asteroid fragments liberated relatively recently by collisions and transferred to the earth by gravitational perturbations. The principal source location is thought to be the zone 2.5 AU (one AU is equal to the mean distance between the Earth and the sun) from the sun where the orbital frequency about the sun is exactly three times that of Jupiter.

Interplanetary Dust

Interplanetary dust particles (IDPs) are the submillimeter-size regime of the solar system meteoroid inventory ranging in size from tens of nanometers to 1000 km in diameter. These particles are short lived in the interplanetary medium because of the effects of self-collisions and orbital decay caused by the drag component of sunlight (the Poynting-Robertson effect). Most of the IDPs that now reach the earth were liberated from comets and asteroids within the last 10^6 years. Over 40,000 tons of IDPs impinge on the earth annually and cumulatively they are the dominant meteoritic mass input on time scales shorter than 10^7 years. Interplanetary dust in the size range of a few micrometers to a millimeter can be collected in and below the atmosphere, and recovered particles provide an important sampling of asteroids and comets. These samples complement conventional meteorites because they include specimens of objects that for a variety of reasons do not either reach the earth or survive atmospheric entry in greater than cm-sized pieces to become conventional meteorites. Dust provides a broader and less biased sampling of interplanetary materials because small samples of fragile materials can survive atmospheric entry without crushing. Additionally, sunlight pressure effects cause all dust beyond the earth's orbit to evolve toward the sun where collisions with earth are possible.

Collection. IDPs can be collected in space although the high relative velocity makes nondestructive capture difficult. Below 80 km altitude, IDPs have decelerated from cosmic velocity and collection is not a problem; however, particles that are large or enter a very high velocity are modified by heating.

The flux of 10-μm particles is $1/(m^2 \cdot d)$, a value high enough that these IDPs can be collected directly using high altitude aircraft. The spatial density of 10-μm IDPs at 20-km altitude is $10^{-3}/m^3$. Particles >100 μm fall at a rate of only $1/(m^2 \cdot yr)$ and can only effectively be collected after they have fallen to the ground and concentrated in a surface deposit. The small particles are collected primarily using stratospheric aircraft although they have also been recovered by melting pristine Antarctic ice. The larger IDPs have been collected from deep sediments, Greenland ice, and Antarctic ice, and a few other selected terrestrial environments that allow the extraterrestrial to be efficiently isolated and distinguished from terrestrial particles. Many of the >100-μm cosmic particles are spherules and their shape assists in making a distinction from other materials.

The most common IDPs are black objects having approximately solar elemental composition except for very volatile elements such as the noble gases. There are particles that deviate strongly from this pattern but they are rare and are usually dominated by a single mineral such as FeS, olivine, or FeNi metal. Most of the particles can be grouped into two classes: one contains hydrated minerals such as serpentine and smectite; the other, ones that are anhydrous.

Origin. Individual IDPs are short-lived and the long-term presence of dust in the solar system implies that there must be sources capable of generating the approximately 10^7 kg/s of new dust required to balance losses by collisions, ejection by radiation pressure, and spiraling into the sun owing to the Poynting-Robertson effect. For sizes >10 μm, it has long been known that comets and asteroids are the principal source. Particles are liberated from asteroids by collisions of both asteroids and asteroid debris. Dust from comets is released when solar heating sublimes the ice matrix in comets. Dust is ejected, and because of the effects of light-pressure drag and ejection velocity, it forms the dust tail that can extend to lengths of over 10^8 km. Most of the comet dust that is collectable on the earth is believed to have been derived from the Kuiper belt comets that reside in a flattened distribution extending from the region of the outer planets to distances of a few 100 AU from the sun. The dust from both comets and asteroids is believed to be samples of early solar system materials that have been relatively well preserved over the age of the solar system inside moderately small bodies.

DONALD E. BROWNLEE
University of Washington

Additional Reading

Hutchinson, R.: *The Search for Our Beginning,* Oxford University Press, New York, NY, 1983.

Kerridge J.F. and M.S. Matthews, eds.: *Meteorites and the Early Solar System,* University of Arizona Press, Tucson, AZ, 1988.

McSween, H.Y. *Meteorites and Their Parent Planets,* Cambridge University Press, New York, NY, 1987.

EXTRUSION.   A majority of stock plastic shapes (bars, cylinders, special cross sections) are made in this way. Thermoplastic materials are heated in a plasticizing cylinder and by means of a rotating screw are forced through a die to provide the desired cross section. A variation of the process is used for extruding coatings of soft plastic materials over other materials. Almost any profile can be imparted to the product, but of course, variations in profile are limited to two dimensions. The tooling costs for extrusion are low compared with injection molding. Thickness of the material can be controlled quite precisely. Production rates are high.

Certain metals, including aluminum, and various rubbers are also extruded.

Extrusion is also combined with mixing in some applications. In the type of device shown in the Fig. 1. The material is fed as a dry solid, is fluxed in the barrel to form a paste, and then resolidified at the discharge. Action in the barrel is one of shearing, rubbing, and kneading. One continuous screw or two screws rotate in the closely fitting barrel. The work of the screw is augmented by forcing the material through breaker screens and around breaker disks just before the material is forced out through the exit nozzle. Such machines are often used for extruding soft chemical and food mixes which do not require fluxing, as well as for the extrusion of hard plastics, some of which must be fluxed at temperatures above 400°F (204°C). Wires can be covered and shapes of intricate cross section can be produced. Plastic resins also are blended in extruders to form pellets for later press and injection molding.

An excellent and detailed summary of the use of extrusion in the plastics and polymer industries is given by J.A. Gibbons, et al. in the "Modern Plastics Encyclopedia," pp. 219–234, McGraw-Hill, New York, 1986.

Extrusion in the Food Industry

Numerous food formulations, as exemplified by spaghetti, macaroni, and other pasta products, are characterized by a uniform cross-sectional shape

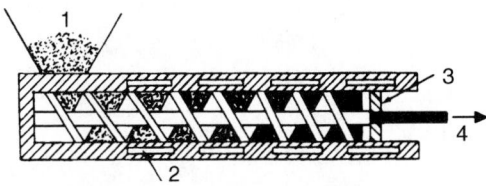

Fig. 1.   Mixed-type extruder: (1) Charge stock; (2) heating or cooling chambers in extruder jacket; (3) die; and (4) extruded product

(round, rectangular, etc.) and length (rod-like, tube-like, ribbon-like, etc.). These characteristics are imparted by forcing an initial pasty mass through the dies of an extruder. Temperature control of extruding is extremely important. For more detail, reference is suggested to the "Foods and Food Production Encyclopedia," (D.M. and G.D. Considine, Eds.), Van Nostrand Reinhold, New York, 1982.

Additional Reading

Berghaus, U. and F. Hensen: *Plastics Extrusion Technology,* Hanser Gardner Publications, Cincinnati, OH, 1997.

Chung, C.I.: *Extrusion of Polymers: Theory and Practice,* Hanser Gardner Publications, Cincinnati, OH, 2000.

Rauwendaal, C.: *Understanding Extrusion,* Hanser Gardner Publications, Cincinnati, OH, 1998.

Rauwendaal, C.: *SPC Statistical Process Control in Injection Molding and Extrusion,* Hanser Gardner Publications, Cincinnati, OH, 2000.

EXTRUSIVE.   A property of an igneous rock that has been ejected onto the earth's surface. Lava flows and detrital material, such as volcanic ash, are extrusives. See also **Mineralogy**.

F

FACE-CENTERED CUBIC STRUCTURE. An internal crystal structure, determined by X-rays, in which the equivalent points are at the corners of the unit cell and at the centers of the six faces of a cube.

FAHRENHEIT TEMPERATURE SCALE (abbr F). A temperature scale with the ice point at 32° and the boiling point of water at 212°. Conversion with the Celsius "centigrade" temperature scale (abbr C) is by the formula

$$F = 9/5C + 32$$

The scale was invented by a German physicist G. D. Fahrenheit (1686–1736), who introduced the use of mercury instead of alcohol in thermometers.

FALLOUT (Radioactive). The term *fallout* generally has been used to refer to particulate matter that is thrown into the atmosphere by a nuclear process of short time duration. Primary examples are nuclear weapon debris and effluents from a nuclear reactor excursion. The name fallout is applied both to matter that is aloft and to matter that has been deposited on the surface of the earth. Depending on the conditions of formation, this material ranges in texture from an aerosol to granules of considerable size. The aerodynamic principles governing its deposition are the same as for any other material of comparable physical nature that is thrown into the air, such as volcanic ash or particles from chimneys. Therefore, many of the principles learned in studies of fallout from nuclear weapons can be applied to studies of other particulate pollution in the atmosphere.

The topographic distribution of fallout is divided into three categories called: (1) local (or close-in); (2) tropospheric (or intermediate); and (3) stratospheric (or worldwide) fallout. No distinct boundaries exist between these categories. The distinction between local and tropospheric fallout is a function of distance from source to point of deposit. The primary distinction between tropospheric and stratospheric fallout is the place of injection of the debris into the atmosphere, above or below the tropopause. Whether the radioactive debris from a nuclear weapon becomes tropospheric or stratospheric fallout depends on yield, height, and latitude of burst (the height of the tropopause is a function of latitude).

Because air acts as a viscous medium, a drag force is developed to oppose the gravitational force that acts on airborne particulate matter. This makes the velocity of fall dependent on particle size. The larger particles (diameters greater than about 20 micrometers) have a higher rate of settling and create local fallout. Smaller particles injected below the tropopause are carried by prevailing winds over large regions of the surface of the earth, and they create the tropospheric fallout. Tropospheric fallout particles larger than about 0.1 micrometer diameter continually mix through the circulating air mass that is in contact with the surface of the earth and gradually settle to the ground, or are washed down by rain or snow. Many smaller particles form nuclei for raindrops. Parts of the tropospheric fallout may remain in the atmosphere for a month or more, long enough to circle the earth several times. The mean residence time above the tropopause of stratospheric fallout is from 5 to 30 months, during which time it completely encircles the earth. It gradually returns through the tropopause, primarily in certain regions where mixing between the two layers is more probable.

The exact characteristics of the radiation associated with fallout depend on the nature of the nuclear processes from which its radioactivity originates. Generally, these radioactive nuclides are products formed from the fission of uranium or plutonium but, under appropriate circumstances, considerable quantities of radioactivity can be formed through nuclear reactions induced by neutrons that are produced by the weapon or reactor. The radiation problems associated with local fallout are usually those of high-intensity gamma-ray radiation fields resulting from the relatively large quantities of radioactive material that fall back to earth within a few tens of miles from the point of origin. The important radioactive materials consist in this case of short-lived fission products and neutron-induced radioactive nuclides. The hazards of worldwide fallout come more from the problems of the long-lived radionuclides, such as ^{134}Cs, ^{137}Cs, and ^{90}Sr, that can enter the human food chain and ultimately be absorbed by the body.

For a nuclear weapon burst in air, all materials in the fireball are vaporized. Condensation of fission products and other bomb materials is then governed by the saturation vapor pressures of the most abundant constituents. Primary debris can combine with naturally-occurring aerosols, and almost all of the fallout becomes tropospheric or stratospheric. If the weapon detonation takes place within a few hundred feet of (either above or below) a land or water surface, large quantities of surface materials are drawn up or thrown into the air above the place of detonation. Condensation of radioactive nuclides in this material then leads to considerable quantities of local fallout, but some of the radioactivity still goes into tropospheric and stratospheric fallout. If the burst occurs sufficiently far underground, the surface is not broken and no fallout results.

See also **Nuclear Power Technology**.

C. Sharp Cook
The University of Texas
El Paso, Texas

FARADAY, MICHAEL (1791–1867). A native of England, Faraday did more to advance the science of electrochemistry than any other scientist. After hearing Sir Humphrey Davy, professor of chemistry, at the Royal Institution speak, Faraday sought a job with Davy as his assistant, and thus his scientific career began in chemistry and electricity.

A profound thinker and accurate experimentalist and observer, he was the first to propound correct ideas as to the nature of electrical phenomena, not only in chemistry but in other fields. His contributions to chemistry include the basic laws of electrolysis, electrochemical decomposition (the basis of corrosion of metals) of battery science, and electrometallurgy.

Faraday is credited with discovering electrical rotation caused by magnetism and making in 1821 the first electric motor. Later, Faraday also discovered electromagnetic induction and invented the transformer. His work led to the growth of electrical technology.

See also **Electrochemistry**.

J.M.I.

FAT. A glyceryl ester of higher fatty acids such as stearic and palmitic. Such esters and their mixtures are solids at room temperature and exhibit crystalline structure. Lard and tallow are examples. There is no chemical difference between a fat and an oil, the only distinction being that fats are solid at room temperature and oils are liquid. The term fat usually refers to triglycerides specifically, whereas lipid is all-inclusive.

See also **Lipids**.

FATIGUE (Metals). Failure of metal parts by progressive cracking caused by repeated application of stress. Most fatigue failures start at the surface where discontinuities in section such as square shoulders, screw threads, or even tool marks cause a high concentration of stress.

Internal discontinuities may also start a fatigue crack. The most notable example is "transverse fissures" in rails, which are believed to originate in areas within the rail section known as "flakes," a defect originating during the cooling period after hot rolling. Once a minute crack is started anywhere in the section, the root of the crack becomes the seat of high stress concentration upon subsequent applications of tensile stress. Thus, the crack will spread until the section is too weak to carry the load and the remaining portion will fracture suddenly.

The portion of the section that failed progressively will be worn quite smooth due to the rubbing action of successive stress applications (e.g., alternate tension and compression in a rotating member loaded as a beam). The suddenly fractured portion will have the usual crystalline appearance characteristic of fractures in heat-treated steels. For this or other reasons, fatigue failures have wrongly been blamed on "crystallization" of the metal.

All metals are crystalline and no alteration in the size or shape of the grains or crystals takes place in service during or before fatigue failure. (Exceptions might be made in the case of lead and other alloys which recrystallize when cold worked at room temperature.)

The fatigue strength, also called endurance limit, is the maximum stress that can be applied repeatedly without failure. In the case of steel, tests are run at a given maximum stress to 10,000,000 reversals or cycles of stress unless failure occurs earlier. It has been found that failures do not occur in steels after a successful run of this duration (4 days at 1700 rpm or less than 1 day at 10,000 rpm). In the case of aluminum alloys and certain other non-ferrous metals, fatigue, failures have occurred after much longer runs, hence tests are sometimes made to 500,000,000 cycles. The materials do not have a true endurance limit; and the number of reversals of stress is stated in reporting the fatigue strength.

The most common test is a rotating beam type in which a carefully machined and polished sample is loaded as a beam while rotating in anti-friction bearings. As any point in the periphery rotates from top to bottom to top position the stress changes from maximum compression to maximum tension and back to maximum compression. From 4 to 8 or more individual tests at various maximum stress levels may be required to determine the endurance limit.

The endurance limit for smooth test specimens run in normal atmospheres at room temperature is an ideal or limiting value. In the case of steels not hardened, or heat-treated to moderate hardness, the smooth specimen endurance limit is approximately one-half of the tensile strength. The endurance limit-tensile strength ratio is less than one-half for many other materials.

The presence of stress raisers, particularly at the surface, will lower the endurance limit. Notch sensitivity can be evaluated as the ratio of the endurance limit of a standardized notched specimen to that of a smooth specimen. Fatigue tests run in a corrosive medium, either gaseous or liquid, generally give much lower endurance limits than tests run in normal atmosphere. Corrosion-fatigue is responsible for many service failures of shafts and other stressed parts of pumps, engines, or processing equipment operating in corrosive media. Protective coatings are sometimes used to improve service life. Nitriding of alloy steel parts has proved very effective.

In order to guard against fatigue failures in critical parts such as connecting rods, they should be fabricated from high-quality steels using designs that avoid regions of stress concentration such as sharp fillets and engraved part numbers. They should be finished over all, avoiding tool or grinding marks. As a further aid in obtaining high fatigue strength such parts are being surface peened by a shot blasting process which work-hardens the surface and sets up compressive stresses in the surface layers. This raises the endurance limit, apparently by reducing the maximum tensile stresses that can be developed at the surface in normal operation.

Additional Reading

Carlson, R.W.: *An Introduction to Fatigue Metals and Composite,* Chapman & Hall, New York, NY, 1995.

Conway, J.B. and L.H. Sjodahl: *Analysis and Representation of Fatigue Data,* ASM International, Materials Park, OH, 1991.

Davis, J.R.: *Metals Handbook,* ASM International, Materials Park, OH, 1998.

Dixon, J.I., Editor: *Failure Analysis: Techniques and Applications,* ASM International, Materials Park, OH, 1992.

Frost, N.E., K.J. Marsh, and L.P. Pook: *Metal Fatigue,* Dover Publications, Inc., Mineola, NY, 1999.

Henry, S.D. and F. Reidenbach: *Fatigue Data Book: Light Structural Alloys,* ASM International, Materials Park, OH, 1995.

Kuhn, H. and D. Medlin: *ASM Handbook: Mechanical Testing and Evaluation,* Vol. 8, ASM International, Materials Park, OH, 2000.

Shiozawa, K. and T. Sakai: *Databook on Fatigue Strength of Metallic Materials,* Elsevier Science, New York, NY, 1996.

Staff: ASM: *Failure Analysis and Prevention,* Vol. 11 of ASM Handbook, ASM International, Materials Park, OH, 1986.

Stephens, R.I., H.O. Fuchs, and A. Faterni: *Metal Fatigue in Engineering,* John Wiley & Sons, Inc., New York, NY, 2000.

Weronski, A. and T. Hejwowski: *Thermal Fatigue of Metals,* Marcel Dekker Inc., New York, NY, 2000.

FATTY ACIDS. A carboxylic acid derived from or contained in an animal or vegetable fat or oil. They are non-toxic and readily biodegradable.

All fatty acids are composed of a chain of alkyl groups containing from 4 to 22 carbon atoms (usually an even number) and characterized by a terminal carboxyl group $-COOH$. The generic formula for mentioned acetic is $CH_3(CH_2)_xCOOH$ (the carbon atom count includes the carboxyl group). Fatty acids may be saturated or unsaturated (olefinic), and solid, semisolid, or liquid. They are classed among the lipids, together with soap and waxes. See also **Lipids**.

Saturated

A fatty acid in which the carbon atoms of the alkyl chain are connected by single bonds. The most important of these are butyric (C_4), lauric (C_{12}), palmitic (C_{16}), and stearic (C_{18}). Stearic acid leads all other fatty acids in industrial use, primarily as a dispersing agent and accelerator activator in rubber products and soaps. See also **Stearic Acid and Stearates**.

Unsaturated

A fatty acid in which there are one or more double bonds between the carbon atoms in the alkyl chain. These acids are usually vegetable derived and consist of alkyl chains containing 18 or more carbon atoms with the characteristic end group $-COOH$. Most vegetable oils are mixtures of several fatty acids or their glycerides; the unsaturation accounts for the broad chemical utility of these substances, especially of drying oils. See also **Vegetable Oils (Edible)**. The most common unsaturated acids are oleic, linoleic, and linolenic (all C_{18}). Safflower oil is high in linoleic acid, peanut oil contains 21% linoleic acid, olive oil is 38% oleic acid, palmitoleic acid is abundant in fish oils. Aromatic fatty acids are now available. See also **Linoleic Acid**; **Linolenic Acid**; and **Oleic Acid**.

Note: Linoleic, linolenic, and arachidonic acids are called essential fatty acids by biochemists because such acids are necessary nutrients that are not synthesized in the animal body.

Uses

Special soaps, heavy-metal soap, lubricants, paints and lacquers (drying oils), candles, salad oil, shortening, synthetic detergents, cosmetics, and emulsifiers.

For further details, refer to Fatty Acid Producers Council, 475 Park Ave. S., New York, NY, 10016.

See also **Carboxylic Acids**; and **Chlorinated Organics**.

FATTY ALCOHOL. A primary alcohol (from C_8 to C_{20}), usually straight chain. High molecular weight alcohols are produced synthetically by the Oxo and Ziegler processes. See also **Oxo Process**. Those from C_8 to C_{11} are oily liquids; those greater than C_{11} are solids. Other methods of production are (1) reduction of vegetable seed oils and their fatty acids with sodium, (2) catalytic hydrogenation at elevated temperatures and pressures, and (3) hydrolysis of spermaceti and sperm oil by saponification and vacuum fractional distillation. See also **Saponification**; and **Soaps**. The more important commercial saturated alcohols are octyl, decyl, lauryl, myristyl, cetyl, and stearyl. The commercially important unsaturated alcohols, such as oleyl, linoleyl, and lynolenyl, are also normally included in this group. The odor tends to disappear as the chain length increases.

Uses

Solvent for fats, waxes, gums, and resins; pharmaceutical salves and lotions; lubricating-oil additives; detergents and emulsifiers; textile anti-static and finishing agents; plasticizers; nonionic surfactants and cosmetics.

See also **Detergents**; and **Plasticizers**.

FATTY AMINE. A normal aliphatic amine derived from fats and oils. May be saturated or unsaturated, and primary, secondary, or tetiary, but the alkyl groups are straight chain and have an even number of carbons in each. The length varies from 8 to 22 carbon atoms.

FATTY ESTER. A fatty acid with the active hydrogen replaced by the alkyl group of a monohydric alcohol. The esterification of a fatty acid, RCOOH, by and alcohol, R′OH, yields the fatty ester RCOOR′. The most common alcohol used is methanol, yielding the methyl ester $RCOOCH_3$. The methyl esters of fatty acids have higher vapor pressures than the corresponding acids and are distilled more easily.

FEEDER (Gravimetric). Frequently in the form of belt scales, the function of a gravimetric feeder is to continuously weigh material as it leaves a hopper, chute, or bin while in transport to a process. The gravimetric feeder delivers a desired amount of material within a given time period and, in this sense, are solid mass flowmeters, the setpoint (desired rate) of which can be adjusted by the operator. Some situations require adding a small quantity of additive material to an uncontrolled flow of bulk material. A belt scale may be used to measure the "wild flow" stream whose rate signal feeds the ratio input circuit of a setpoint controller which incorporates a ratio adjustment that establishes the ratio of additive to wild flow. An increase or decrease in the wild flow rate produces a corresponding change in the additive flow rate of the feeder, thereby maintaining the correct proportion. More than one additive may be involved in such a system. The wild flow may be of a solid or a liquid; the additive materials may be solids or liquids.

The use of gravimetric feeders in a multiunit proportioning system involving the simultaneous blending of three bulk materials is shown in Fig. 1. The system incorporates a master control that permits increasing or decreasing the total flow rate of the system. This may be accomplished in several ways: (1) A manually adjusted power supply that provides a master reference signal to feed the setpoint controllers for each feeder; (2) a demand signal generated by a primary control loop designed to maintain a process variable that is affected by the total feed from the system; or (3) a total flow control loop that maintains a uniform total flow by summing the individual feed rates. The last mentioned provides feedback to a total flow setpoint controller which, in turn, provides the demand signal to the individual feeder controllers.

Digital blending systems provide the most accurate means for automatic, continuous-control flow of bulk materials. They are dependent entirely on the accuracy of the digital pulses transmitted from the individual weigh feeders—with a control accuracy of plus or minus one pulse.

Automatic pacing provides the solution to systems where one or more materials may lag the remainder of the system at start-up because the material characteristics may produce a resistance to flow from the storage bins.

Belt scales are applied and installed on practically any size, length, and shape of conveyor—from 1-foot (0.3-meter) wide belts, handling as low as 1 pound (0.45 kilogram) per minute at a minimum belt speed of 1 foot (0.3 meter) per minute, to a 10-foot (3-meter) wide belt handling as high as 20,000 short tons (18,000 metric tons) per hour at speeds up to 1000 feet (300 meters) per minute.

The basic design and applications of a conventional belt conveyor scale include a weigh platform (or weigh bridge), on which are mounted standard conveyor idlers used to sense the weight of the material passing over it. A belt-speed pickup system measures the speed of the conveyor belt. Weight and speed signals are transmitted to a multiplier whose product or true rate output signal is integrated and displayed on a continuous totalizer. The rate, of course, is equal to weight times speed. The basic components used to sense the weight and speed may be mechanical, electrical, pneumatic, or hydraulic.

Of a different design principle is the loss-in-weight feeder. The decreasing weight of the material in the weigh hopper is continuously sensed and compared with a diminishing programmed setpoint which is preestablished to satisfy the flow rate desired to process. The deviation between these two signals is measured and converted for corrective action to the feeder whose speed is adjusted accordingly to maintain the

programmed set point—hence the desired rate of flow. When the material in the weigh hopper reaches a selected low-level point, the hopper is quickly refilled to a selected higher level, thereby permitting uninterrupted flow rate to process.

FEEDER (Vibratory). A mechanical device used to provide controlled bulk material flow from storage to processes, or to and from various processes. Generally, these feeders can be driven hydraulically, electromechanically, pneumatically, or electromagnetically.

The electromagnetic vibratory feeder is basically a two-mass spring-connected system. One of the masses is the trough in which the bulk material flows; the other mass is the base or reaction mass. The two masses are connected through a set of leaf springs, which allow the two masses to vibrate relative to each other. The electromagnetic feeder creates vibratory feeding at a steady 3,600 vibrations per minute from a half-wave rectified 60-cycle alternating current. The feeders have no rotating parts, bearings, eccentrics, or sliding joints, which eliminate the need for lubrication and minimize wear.

In order for the two-mass driven system to take advantage of the natural magnification factor, the system must have its natural vibration frequency close to the operating frequency. This is referred to as subresonant tuning, that is, operating below the resonant frequency. A subresonant feeder can operate favorably regardless of high headloads on the feeder that may be caused by large hopper openings, wide heavy troughs, or the need to convey material a considerable distance from the hopper to the end of the trough. In addition to headload, the damping effect of the bulk material being handled must be considered in feeder design and selection. The damping effect of the material is a direct measure of the energy that is absorbed by the material moving from the hopper along the vibrating trough. Headloads, when considered alone, tend to cause a feeder to increase in amplitude under the effects of this load. The damping effect tends to decrease the amplitude. In practice, as proved by tests, the two effects tend to offset each other in a subresonant-tuned feeder. The general result—a feeder system that operates at reasonably constant amplitude regardless of material effects. Operating details are shown in Fig. 1.

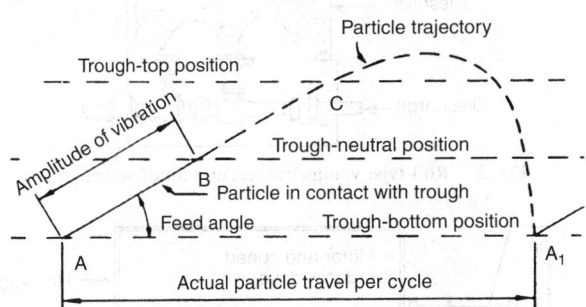

Fig. 1. Action of a vibratory feeder in moving a single particle from point A to point A_1. The trough motion approximates sinusoidal motion and the particle is in contact with the trough surface from the lowest point to approximately the mid-point of the stroke (about one-fourth of a cycle from point A to point B). At this point, the particle has been accelerated to its maximum velocity and leaves the trough surface on a free flight trajectory (trough is decelerating from point B to point C) rejoining the trough surface at point A_1, the lowest point of the stroke—which completes one cycle

FEEDER (Volumetric). A device for metering a particulate solid by volume. Functionally, volumetric feeders for solids are analogous to volumetric flowmeters for fluids and are important elements of many process control installations. Basic characteristics of a volumetric feeding device are: (1) a determinable geometry that, in essence, becomes the basic unit of flow measurement, and (2) a controlled rate of transfer of material from the feeder to the process. These characteristics are illustrated by the rotary lock feeder shown in Fig. 1. Here, the geometry is fixed for a given size unit by cross-sectional area A. This is the area bounded by two adjacent vanes or lobes of the rotor, the housing, and the width of the rotor. The rate of transfer is determined by the rotor speed, which may be fixed or variable.

The following factors are important in specifying a feeder for a particular application: particle size, bulk density, adhesion, cohesion, abrasion, moisture, segregation, degradation, and various control factors, such that the required ease and sensitivity of control may be provided. Other factors

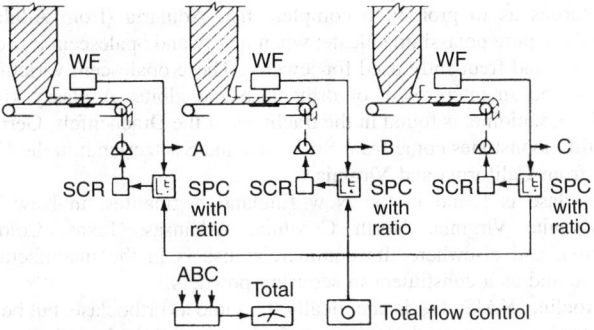

Fig. 1. Multiunit proportioning system involving the simultaneous blending of three bulk materials. WF = weigh feeder; SPC = set point controller; SCR = silicon-controlled rectifier drive

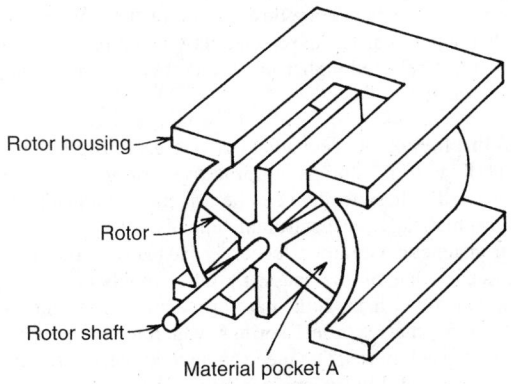

Fig. 1.   Rotary lock feeder

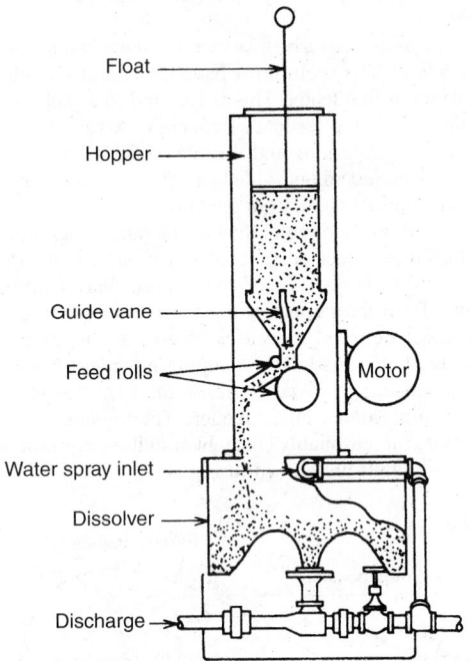

Fig. 2.   Roll-type volumetric feeder with dissolver

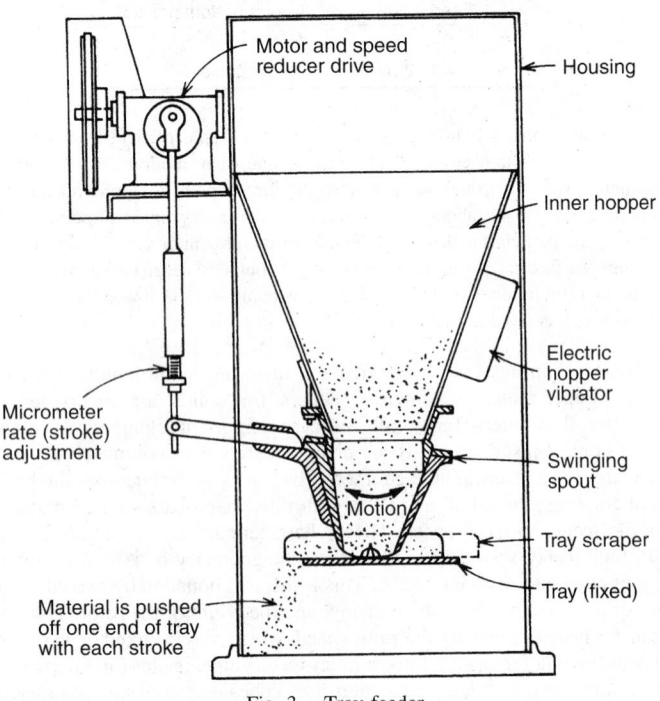

Fig. 3.   Tray feeder

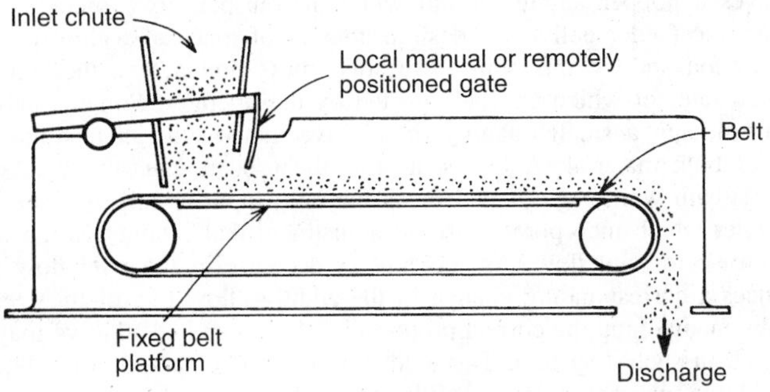

Inlet chute

Local manual or remotely positioned gate

Belt

Fixed belt platform

Discharge

Fig. 4.   Volumetric belt feeder

include fire and explosion prevention, operator and maintenance skills needed, and both upstream and downstream conditions that may affect flow of material to and from the feeder.

There are several types of volumetric feeders in addition to the rotary lock feeder. These include a roll-type feeder (Fig. 2), a tray feeder (Fig. 3), and a belt feeder (Fig. 4).

FEEDSTOCKS.   Gaseous or liquid petroleum-derived hydrocarbons or mixture of hydrocarbons from which gasoline, fuel oil, and petrochemicals are produced by thermal or catalytic cracking. It is also called charging stock. Feedstocks commonly used include ethane, propane, butane, butene, benzene, toluene, xylene, maphtha, and gas oils.

See also **Petrochemicals**; and **Petroleum**.

FEHLING'S SOLUTION.   A reagent used as a test for sugars, aldehydes, etc. It consists of two solutions, copper sulfate and alkaline tartrate, which are mixed just before use. Benedict's modification is a one-solution preparation.

Additional details are available in the Book of Methods, Association of Official Analytical Chemists.

FELDSPAR.   Feldspar is the name of a group which includes the most important of the rock-forming minerals, making up perhaps as much as 60% of the earth's crust.

This group of minerals consists of three silicates: a potassium-aluminum silicate, a sodium-aluminum silicate, and a calcium-aluminum silicate ($KAlSi_3O_8$, $NaAlSi_3O_8$, and $CaAl_2Si_2O_8$) and their isomorphous mixtures.

The various members of the feldspar group show many characteristics in common. Crystallizing in the monoclinic and triclinic systems, they show similarity of crystal habit, cleavage and other physical properties as well as similar chemical relationships.

Orthoclase, $KAlSi_3O_8$, derives its name from the Greek words meaning right or straight, and fracture, because its two cleavages are at right angles to each other. It crystallizes in the monoclinic system and its crystals are usually prismatic; it occurs also in coarsely cleavable masses. Hardness, 6; specific gravity, 2.56–2.58; luster, vitreous to pearly; colorless to white, gray, yellow or red, rarely green. Twin crystals are not uncommon.

Orthoclase is a common constituent of many igneous rocks and is often found in huge masses in pegmatite veins. Localities for orthoclase are so numerous as to prohibit a complete list. Adularia (from Adular) is essentially a pure potassium silicate; when pearly and opalescent it is called moonstone and frequently used for jewelry. These opalescent varieties are known to be an intergrowth of orthoclase and albite. A glassy kind of orthoclase, sanidine, is found in the trachytes of the Drachenfels, Germany. Beautiful moonstones come from Sri Lanka and Switzerland; in the United States, from California and Virginia.

Orthoclase is found in the New England pegmatites, in New York, Pennsylvania, Virginia, North Carolina, Arkansas, Texas, Colorado, California, and elsewhere. Its commercial use is in the manufacture of porcelain and as a constituent in scouring powders.

Microcline, $KAlSi_3O_8$, is chemically the same as orthoclase, but belongs to the triclinic system, the prism angle being slightly less than a right angle (89°30′), hence the name microcline from the Greek meaning small, and to slope. Microcline is like orthoclase in all physical properties and can be distinguished from it surely only by optical examination.

Under the polarizing microscope microcline displays a minute multiple twinning which results in a grating-like structure that is unmistakable. It is probable that much orthoclase would, upon proper examination, prove to be microcline. Amazon stone or amazonite is a beautiful green microcline occurring in the Ilmen Mountains in the Urals, Italy, Norway, Madagascar. It also occurs in the United States, in the Pikes Peak region, Colorado, Virginia, North Carolina, and sparingly in the pegmatites of New England.

The name amazon stone is derived from the application of this term to some green mineral found by the Spaniards among the aborigines of the Amazon Valley in South America. As no microcline is known to occur in the region there must have been some confusion with another green-colored substance.

A soda microcline, anorthoclase, is known, which is probably an isomorphous mixture of $KAlSi_3O_8$ and $NaAlSi_3O_8$, the sodium-aluminum silicate being in the greater proportion. The soda feldspar albite, $NaAlSi_3O_8$ and the calcium feldspar anorthite, $CaAl_2Si_2O_8$ form an isomorphous series from pure albite at one end to pure anorthite at the other, the two molecules appearing to be completely miscible one with the other. The members of this series are spoken of as the soda-lime (or lime-soda) feldspars. As a group, they are called the plagioclase feldspars (from the Greek meaning *oblique* and *fracture*, referring to the two cleavages at an angle that differs slightly from a right angle). Nearly always present are the striations, fine parallel lines, resulting from minute multiple twinning, which, never seen on orthoclase or microcline, are therefore an important diagnostic feature.

More or less arbitrarily, four intermediate plagioclase feldspars are recognized between albite and anorthite; these are listed below together with the approximate percentage of each molecule present.

	% of $NaAlSi_3O_8$	% of $CaAl_2Si_2O_8$
Albite	100 to 90	0 to 10
Oligoclase	90 to 70	10 to 30
Andesine	70 to 50	30 to 50
Labradorite	50 to 30	50 to 70
Bytownite	30 to 10	70 to 90
Anorthite	10 to 0	90 to 100

Albite is so called from the Latin, *albus*, in reference to its usual pure white color. It is a sodium aluminum silicate corresponding to the formula $NaAlSi_3O_8$. It crystallizes in the triclinic system commonly in tabular crystals. Twinning is very common, thin twinning lamellae producing a series of fine striations on certain crystal faces. There are two good cleavages at an angle of $86° 24'$ to each other. Hardness, 6; specific gravity, 2.62; luster, vitreous to pearly. It may be colorless to white or gray and transparent to opaque.

Albite is a relatively common and important rock-making mineral associated with the more acid rock types and in pegmatite dikes, often with rarer minerals like tourmaline and beryl. There are many famous localities in Europe in the Swiss and Austrian Alps, the Urals, the Harz Mountains, in Italy, France, and Norway. Brazil has yielded fine specimens. In the United States, notable localities are Paris and Auburn, Maine; Chesterfield, Mass.; Haddam, Connecticut; Amelia County, Virginia; and the Pikes Peak region of Colorado. It is used in the ceramic industries and also in the manufacture of artificial teeth.

Anorthite was named by Rose in 1823 from the Greek meaning oblique, referring to its triclinic crystallization. The physical properties are essentially the same as for albite, except that the specific gravity of anorthite is somewhat greater, 2.74–2.76. Anorthite is characteristic of the basic igneous rocks such as gabbro and basalt. Anorthite is found in the lavas of Vesuvius and Monte Somma, Italy; in Finland; Japan; and in the United States, in Sussex County, New Jersey.

The intermediate members of the plagioclase group are all very similar and with the exception of certain labradorites, cannot be distinguished from each other ordinarily save by optical means. Oligoclase is a common mineral in such rocks as granites, syenites, diorites, their extrusive equivalents and many gneisses. It is a frequent associate of orthoclase. The word oligoclase is derived from the Greek meaning little, and fracture, in reference to the fact that its cleavage angle differs slightly from $90°$. Sunstone is mainly oligoclase (sometimes albite) spangled with flakes of hematite.

Andesine is a characteristic mineral of rocks such as diorites, which contain a moderate amount of silica and related extrusives, such as andesites. Because of its occurrence in these latter, andesine derives its name from them as well as from the Andes Mountains.

Labradorite is the characteristic feldspar of the more basic rock types like diorite, gabbro, andesite or basalt and it is usually associated with some one of the pyroxenes or amphiboles. Labradorite frequently shows a beautiful play of iridescent colors due to minute inclusions of another mineral. However, the labradorescent phenomenon has not been fully determined. The classic location for this mineral is, of course, Labrador, whence its name. It is a constituent there of the rock anorthosite and is found in the anorthosites of the Provinces of Quebec and Ontario, and in the Adirondack region in New York State.

Bytownite, named from Bytown, the former name for Ottawa, Canada, is a rare mineral occasionally found in the more basic rocks.

The feldspars crystallize from the magma in both extrusive and intrusive rocks; they occur as contact minerals in veins, and they are developed in many sorts of metamorphic rocks, e.g., albite schists. They may also be found as mechanical deposits in various sedimentary rocks.

ELMER B. ROWLEY
Union College
Schenectady, New York

FELSITE.   Felsites are defined by American geologists as dense, fine-grained, light-colored rocks rich in silica, hence classified with the rhyolites, from which some of them have been formed by devitrification. Felsites may occur as intrusive dikes but, in general, they are found as extrusive rocks. They frequently occur interbedded with volcanic ash, tuff or breccia. According to American usage, any light-colored lava whose ground mass or matrix is so fine-grained that the individual minerals cannot be distinguished by the naked eye (macroscopically), may be roughly classified as felsite, hence the prevalence of the term felsitic texture. When felsites show phenocrysts they are called felsite porphyries. The term felsite was first applied by Gerhard in 1814 to the fine ground mass (matrix) of porphyries, and is therefore one of the oldest, commonly used, petrological terms.

FEMIC.   This term is used by petrologists to designate the more common ferromagnesian minerals such as pyroxene and olivine. Rocks relatively rich in femic minerals are said to be urafic.

FERGUSONITE.   A mineral multiple oxide containing niobium (columbium), tantalum, and titanium, corresponding to formula $Y(Nb, Ta)O_4$. Essentially an oxide, or niobate-tantalate of yttrium with varying amounts of erbium, cerium, and iron. Crystallizes in the tetragonal system. Hardness, 5.5–6.5; specific gravity, 5.6–5.8; color, variable. Named after Robert Ferguson (1799–1865), of Scotland.

FERMENTATION.   A form of respiration that requires no oxygen. There is an incomplete breakdown of food; carbon dioxide and other products, such as alcohol, are formed. The word is commonly used to refer to the conversion of sugars (and sugars derived from starch) into ethyl alcohol by the enzymes of yeast.

The process of fermentation has been used from prehistoric times in the preparation of foods and beverages, but the causative agents of fermentation were not recognized until the middle of the nineteenth century. The end-products resulting from the natural fermentation of glucose, namely, alcohol and carbon dioxide, were identified by Gay-Lussac in 1810, but it was thought that this process resulted from contact catalysis and the decay of animal or vegetable materials. This explanation was refuted by the work of Pasteur (1857) on the lactic acid fermentation. In the course of this investigation, Pasteur determined that fermentation was caused by living cells, that different microbial species caused different fermentations, that the nitrogenous materials present served only to support the growth of the cells, that lactic acid was produced when cells (removed from the fermentation mixture) were added to a sugar solution, and that the natural fermentation yielded both alcohol and lactic acid, but that the amount of each could be altered by changes in pH. In later studies, Pasteur showed that the conversion of glucose to alcohol, $C_6H_{12}O_6 \rightarrow 2CO_2 + 2C_2H_5OH$, was caused by yeast cells growing under anaerobic conditions, thus leading to the definition that fermentation was "life without air." A more modern definition of fermentation would be those energy-yielding reactions in which organic compounds act as both oxidizable substrates and oxidizing agents. Anaerobic reactions in which inorganic compounds are utilized as electron acceptors may be termed "anaerobic respirations," whereas reactions in which oxygen serves as a terminal electron acceptor are respirations.

Almost any organic compound may be fermented provided it is neither too oxidized nor too reduced, since it must function as both electron donor and electron acceptor. In some fermentations, a compound is degraded via a series of reactions in which intermediates in the sequence act as electron donors and acceptors. In others, one molecule of the substrate may be oxidized while another molecule is reduced, or two different organic compounds may be degraded after a coupled oxidation-reduction reaction. These fermentations provide energy required for the growth of a variety of cells. In addition, many microorganisms can carry out, in appropriate conditions, a number of fermentative reactions (e.g., oxidations, reduction, cleavages) which do not yield useful energy, or do not yield sufficient energy for growth.

In view of the great variety of different compounds that may be fermented and the enzymatic capabilities of different microorganisms, it is not surprising that numerous compounds important in industry (e.g., ethyl alcohol, butyl alcohol, acetone, 2,3-butylene glycol), in the production, preservation, and seasoning of food (e.g., lactic, citric, and glutamic acids), and in medicine (e.g., vitamins are extracted from the yeast carrying out the alcoholic fermentation) may be produced most cheaply through microbial fermentations. In addition, fermentations continue to be important in the production of foods (e.g., the lactic and propionic acid fermentations in the making of cheeses), beverages (e.g., the alcoholic fermentations in the making of wine and beer), and in the leavening of breads (by the carbon dioxide produced in the equation previously given). It should be pointed out that fermentation, while sometimes requiring the least expensive processing equipment to handle, and often fairly low-cost raw materials, is not always the most economic route. At one time, nearly all industrial alcohol (ethyl) was produced via fermentation. Currently, most ethanol is prepared by the direct hydration (vapor phase, catalytic); addition of water to ethylene; or by the indirect hydration of ethanol via the sulfation-hydrolysis process. See also **Ethyl Alcohol**. However, with conservation measures possibly affecting the availability of hydrocarbons for chemicals (ethylene is produced by thermally cracking hydrocarbon feedstocks), interest in fermentation of agricultural feedstocks may return.

In addition to alcoholic fermentation, there are hundreds of other types of fermentative processes. Some of these include:

Amolytic fermentation—the fermentation of starch, but specifically it is an incomplete fermentation of starch in which simple sugars are not produced.

Butyric Fermentation—in which butyric acid is produced. The organisms producing this type of fermentation are mainly anaerobic like *Clostridium butyricum*. Some organisms, such as *Clostridium tetani*, the organism causing tetanus, and *Clostridium botulinum*, the organism causing botulism, also produce this type of fermentation.

Lactic Fermentation—in which lactic acid is produced. This is an important fermentation for the preservation of food. *Lactobacillus bulgaricus*, *L. casei*, and *Streptococcus lactis* are used for the manufacture of dairy products, such as sour cream. *Lactobacillus plantarum* is used in the preservation of certain vegetables, such as the production of pickles and kraut.

Controlled Oxidative Fermentations—by which a number of industrial chemicals are produced. *Citromyces*, for example, can be used for the production of citric acid from sugar. *Aspergillus niger* will yield oxalic acid by partial oxidative fermentation, but if the mold is permitted to remain in contact with the acid, it will convert it to carbon dioxide.

Some sugars such as glucose may be completely oxidized to carbon dioxide by certain bacteria, most molds, and some yeasts. Such microorganisms produce complete oxidation by fermentation. Many bacteria and yeasts are able to produce a gassy fermentation. The gaseous end product in the fermentation of vegetable products with *Leuconostoc mesenteroides* and *Lactobacillus brevis* is carbon dioxide. The gaseous end products of the coliform group are carbon dioxide and hydrogen.

Ropy Fermentation—which causes the spoilage of foods. Ropy milk is caused by *Aerobacter aerogenes*, *Lactobacillus bulgaricus*, *L. casei*, and *Alcaligenes viscosus*. Ropy bread is caused by members of the *Bacillus mesentericus* group which are identical with or are strains of *B. subtilis* or *B. pumilus*. Rope in maple syrup is produced by *A. aerogenes*.

A characteristic cultural reaction of *Clostridium perfringens* and many other clostridia is known as a *stormy fermentation*. When the organism is inoculated into milk the lactose is fermented and the casein is coagulated.

The anaerobic respiration that takes place in the muscles of higher animals, when insufficient oxygen is available for a complete breakdown of the food, is also called fermentation. Lactic acid and carbon dioxide are the products of this type of fermentation.

Fermentation is discussed in numerous other entries in this encyclopedia. In particular, see also **Enzyme**; **Ethyl Alcohol**; **Genetics and Gene Science (Classical)**; **Industrial Biotechnology**; **Molecular Biology**; and **Yeasts and Molds**.

Additional Reading

Austin, G.T. and R.N. Shreve: *Shreve's Chemical Process Industries,* 5th Edition, McGraw-Hill Companies, Inc., New York, NY, 1984.

Boulton, C. and D. Quain: *Brewing Yeast and Fermentation,* Blackwell Science, Inc., 2000.

Cave, S. and D.R. Harper: *Introduction to Fermentation Practices,* Springer-Verlag, Inc., New York, NY, 2000.

Erickson, L.E. and D. Yee-Chak Fung: *Handbook on Anaerobic Fermentation,* Marcel Dekker, New York, NY, 1988.

Fickinger, M.C. and S. Drew: *The Encyclopedia of Bioprocess Technology: Fermentation, Biocatalysis, and Bioseparation,* John Wiley & Sons, Inc., New York, NY, 1998.

Kristiansen, B.: *Integrated Design of a Fermentation Plant: The Production of Baker's Yeast,* VCH Publishers, Inc., New York, NY, 1994.

Murray, R.K. et al.: *Harper's Biochemistry,* Prentice-Hall, Inc., New Jersey, 1996.

Neway, J.O.: *Fermentation Process Development of Industrial Organisms,* Marcel Dekker, New York, NY, 1989.

Russell, I. and G.G. Stewart: "Contribution of Yeast and Immobilization Technology to Flavor Development in Fermented Beverages," *Food Technology,* 146 (November 1992).

Stanbury, P.F., S.J. Hall, and A. Whitaker: *Principles of Fermentation Technology,* 2nd Edition, Butterworth-Heinemann, Inc., Woburn, MA, 1995.

Vogel, H.C. and C.L. Todaro: *Fermentation and Biochemical Engineering Handbook: Principles, Process Design, and Equipment,* Noyes Publications Inc., New York, NY, 1996.

FERMI.   The name used by some scientists to designate a length of 10^{-15} meter $= 10^{-13}$ centimeter, a length equal approximately to the radius of a nucleon.

FERMI, ENRICO (1901–1954).   Fermi was an American physicist born in Rome. It should be noted that he won a fellowship to the University of Pisa for a paper on sound waves and after only four years at Pisa, he received a Ph.D. in physics.

Fermi is remembered for first producing nuclear fission in 1934 while bombarding uranium with neutrons. He received the Nobel Prize for physics in 1938 for his pioneering work on the production of induced radioactivity by neutron irradiation. He also built, under the football bleachers at University of Chicago, the first nuclear reactor in the United States in 1942. Although it only produced one-half watt of power in one-half minute, it was a success, which gave hope for the development of the atomic bomb. Element number 100 was named fermium in his honor.

See also **Fermions**; **Fermi Resonance**; **Fermi Selection Rules**; **Fermi Surface**; and **Fermium**.

J.M.I.

FERMIONS.   Those elementary particles for which there is antisymmetry under intrapair production. They obey Fermi-Dirac statistics. Fermionic hadrons are called *baryons*; other hadrons are *mesons*. See also **Baryons**; and **Mesons**. The experimentally observed fermions are the leptons and baryons, each of which has a spin angular momentum of $\sqrt{3}h/4\pi$, where h is Planck's constant, which is equivalent to stating that each fermion has a spin quantum number of magnitude $\frac{1}{2}$.

See also **Particles (Subatomic)**.

FERMI RESONANCE.   In polyatomic molecules, two vibrational levels belonging to different vibrations (or combinations of vibrations) may happen to have nearly the same energy, and therefore be accidentally degenerate. As was recognized by Fermi in the case of CO_2 such a "resonance" leads to a perturbation of the energy levels that is very similar to the vibrational perturbations of diatomic molecules.

FERMI SELECTION RULES.   A set of selection rules for beta decay which state that an allowed transition between parent and daughter states must have no change of spin quantum number and no change of parity.

FERMI SURFACE.   Of a metal, semi-metal, or semiconductor that surface in momentum space which separates the energy states which are filled with free or quasi-free electrons from those which are unfilled. Such a surface exists simply because the electrons obey Fermi-Dirac statistics. It is a surface of constant energy and is sometimes called the Fermi level.

If one considers an elementary model of a metal consisting of a lattice of fixed positive ions immersed in a sea of conduction electrons that are free to move through the lattice, every direction of electron motion will be equally probable. Since the electrons fill the available quantized energy states starting with the lowest, a three-dimensional picture in momentum coordinates will show a spherical distribution of electron momenta and, hence, will yield a spherical Fermi surface. In this model, no account has been taken of the interaction between the fixed positive ions and the electrons. The only restriction on the movement or "freedom" of the electrons is the physical confines of the metal itself.

A short derivation, starting with the Schrödinger equation, shows that the total energy of an electron (and thus also its kinetic energy) is given by

$$E = \hbar^2 k^2 / 2m = \rho^2 / 2m$$

where h is Planck's constant divided by 2π, k is the magnitude of the electron wave vector, m is the mass of the electron, and ρ is its momentum. A plot of E against k is then a parabola, as shown in Fig. 1(**a**). The Cartesian components of those values of k which are possible solutions to the Schrödinger equation are $k_i = 2\pi \, n_i / L$, where the n_i's are integers and L is a physical dimension of the metal. Since for each energy value so defined there are actually two states (one for an electron with spin up; one with spin down), it can be shown that the density of energy states available to the electrons is

$$g(E) = \frac{(2m)^{3/2}}{2\pi^2 \hbar^3} E^{1/2}$$

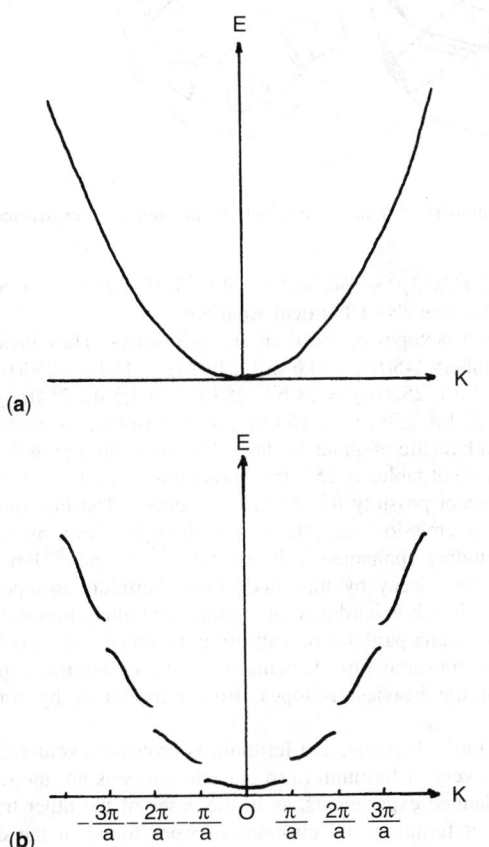

(a)

(b)

$$-\frac{3\pi}{a} \quad -\frac{2\pi}{a} \quad -\frac{\pi}{a} \quad O \quad \frac{\pi}{a} \quad \frac{2\pi}{a} \quad \frac{3\pi}{a}$$

Fig. 1.   (**a**) Energy plotted against wave number for the free electron model. (**b**) Energy plotted against wave number for the "quasi-free" electron model, showing energy discontinuities at Brillouin zone boundaries

where $g(E) \, dE$ is the number of states in the energy range E to $E + dE$. Then $n(E)$, the number of electrons per unit volume occupying energy states in this energy range, is

$$n(E)dE = g(E)f(E)dE$$

where $f(E) = \{\exp [(E - E_f)/bT] + 1\}^{-1}$, a function characteristic of particles which obey Fermi-Dirac statistics. In this expression, T is the absolute temperature, b is Boltzmann's constant, and E_f is a parameter depending on the number of electrons involved. This turns out to be the Fermi energy. E_f can be evaluated by integrating $n(E) \, dE$ from $E = 0$ to $E = \infty$ and recognizing that the integral is equal to N, the total number of electrons per unit volume. The result (at $T = 0$ K) is

$$E_f = \frac{\pi^2 \hbar^2}{2m} \left(\frac{3N}{\pi} \right)^{2/3}$$

At $T = 0$ K, for $E < E_f$, $f(E) = 1$, while for $E > E_f$, $f(E) = 0$. Physically, this means that the probability of a state below the Fermi level being occupied is one; whereas for states with $E > E_f$, the occupancy probability drops abruptly to zero. For temperatures greater than absolute zero, the occupancy probability drops smoothly from 1 to 0 in a range of energy of width approximately equal to bT. This shell of partially filled states gives rise to the following definition: *The Fermi level is the energy level at which the probability of a state being filled is just equal to one half.*

A numerical evaluation of the Fermi energy for a simple metal having one or two conduction electrons per atom yields a value of approximately 10^{-11} erg, or a few electron volts. The equivalent temperature, E_f/b, is several tens of thousands of degrees Kelvin. Thus, except in extraordinary circumstances, when dealing with metals, $bT \ll E_f$; i.e., the energy range of partially filled states is small, and the Fermi surface is well defined by the foregoing statement. It must be noted, however, that this is not necessarily true for semiconductors where the number of free electrons per unit volume may be very much smaller.

The foregoing description provides a qualitative look at the physics of metals and, under some circumstances, semi-metals and semiconductors. A more detailed analysis requires that the effects of the ions in the lattice be recognized. This can be accomplished by introducing the periodic potential due to the lattice through which the electrons must move. Then the electrons are no longer "free," but, depending on the strength and character of the potentials and the approximations used in solving the Schrödinger equation, act as "quasi-free" particles. Another approach is the "tight-binding approximation"; occasionally a combination of the two approaches is used. In any case, introduction of lattice effects changes the characteristics of the model; the total energy and kinetic energy of an electron are no longer equivalent. The periodic lattice can be described in terms of Brillouin zones, each of which is large enough (in momentum space) to accommodate two electrons per atom. The Brillouin zone boundaries appear to the electrons as Bragg reflection planes or energy discontinuities, resulting in an energy versus wave number plot as shown in Fig. 1(**b**).

For many metals, the "nearly free" electron description corresponds quite closely to the physical situation. The Fermi surface remains nearly spherical in shape. However, it may now be intersected by several Brillouin zone boundaries which break the surface into a number of separate sheets. It becomes useful to describe the Fermi surface in terms not only of zones or sheets filled with electrons, but also of zones or sheets of holes, that is, momentum space volumes which are empty of electrons. A conceptually simple method of constructing these successive sheets, often also referred to as "first zone," "second zone," and so on was demonstrated by Harrison. An example of such construction is shown in Fig. 2.

This construction works out quite well, for example, for aluminum, which has three valence electrons per atom. Experiments and more elegant theoretical calculations show that the fourth zone is totally unoccupied and that the third zone is not multiple-connected in the manner shown.

See also **Semiconductors**; and **Solid-State Physics**.

In a scholarly paper, W.E. Pickett and D.J. Singh (Naval Research Laboratory), R.E. Cohen (Carnegie Institution of Washington), and H. Krakauer (College of William and Mary) point out that, "Recent experimental results are beginning to limit seriously the theories that can be considered to explain high-temperature superconductivity. The unmistakable observation of a Fermi surface, by several groups and methods, make it the focus of realistic theories of the metallic phases. Data from angle-resolved photoemission, positron annihilation, and deHaas-van Alphen experiments are in

1st band   2nd band   3rd band   4th band

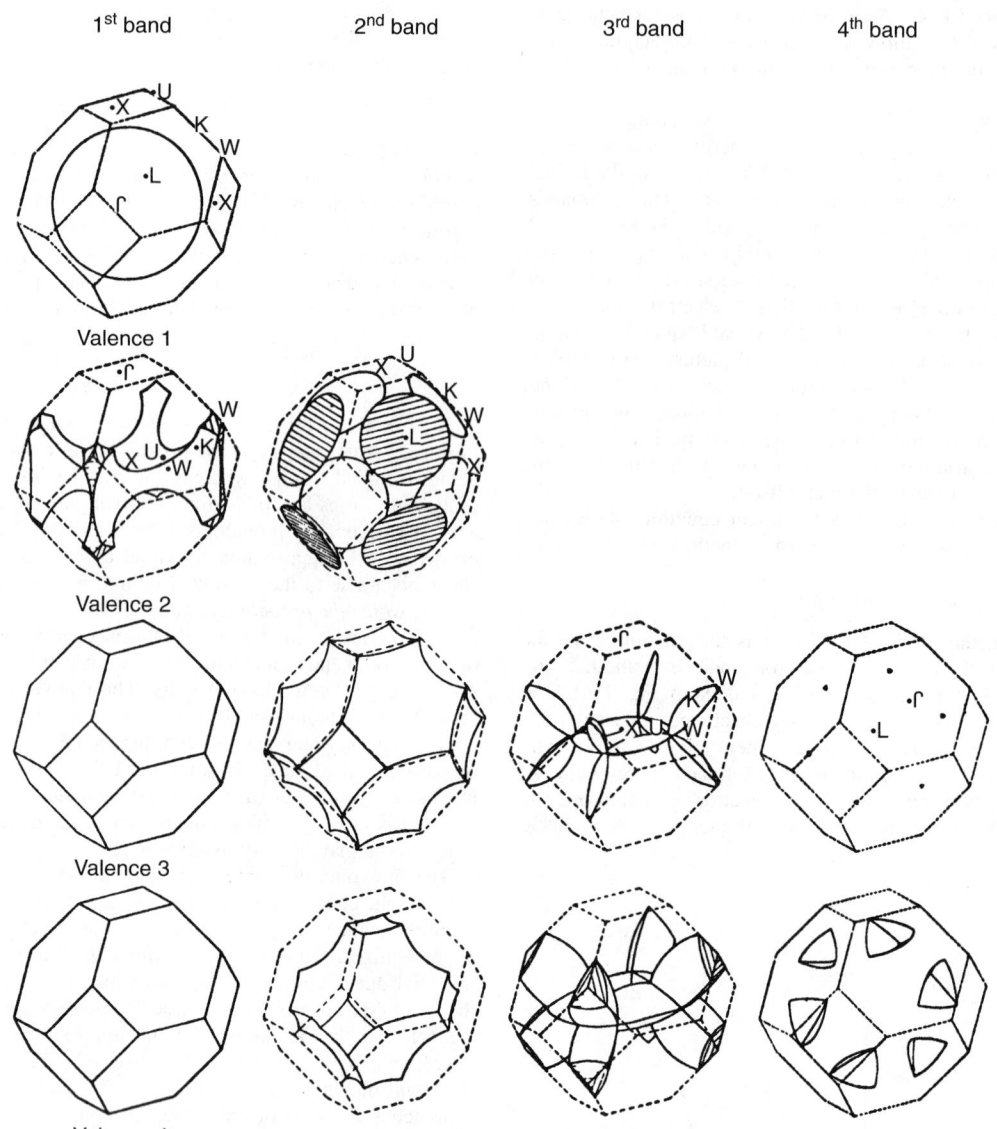

Valence 1

Valence 2

Valence 3

Valence 4

Fig. 2.   Fermi surfaces in several zones or bands, for face-centered cubic metals having various numbers of "quasi-free" electrons per atom, as constructed by Harrison

agreement with band theory predictions, implying that the metallic phases cannot be pictured as doped insulators. The character of the low energy excitations ("quasiparticles"), which interact strongly with atomic motions, with magnetic fluctuations, and possibly with charge fluctuations, must be sorted out before the superconducting pairing mechanism can be given a microscopic basis." In their paper, the authors describe three primary experimental methods of measuring Fermi surfaces. Reference to "Fermi Surfaces, Fermi Liquids, and High-Temperature Superconductors," *Science*, **46** (January 3,1992) is suggested.

FERMIUM.   [CAS: 7440-72-4]. Chemical element symbol Fm, at. no. 100, at. wt. 257 (mass number of the most stable isotope), radioactive metal of the Actinide series, also one of the Transuranium elements. During the period 1953–1954, a group of scientists at the Nobel Institute of Physics (Stockholm) bombarded ^{238}U with ^{16}O ions, producing and isolating a 30-min alpha emitter. This was called $^{250}100$. However, discovery of element 100 was not claimed at that time. Subsequently, the isotope was identified and the 30-min half-life confirmed. Both fermium and einsteinium were formed in a thermonuclear explosion that occurred in the South Pacific in 1952. The elements were identified by scientists from the University of California's Radiation Laboratory, the Argonne National Laboratory, and the Los Alamos Scientific Laboratory. It was observed that very heavy uranium isotopes that resulted from the action of the instantaneous neutron flux on uranium (contained in the explosive device) decayed to form Es and Fm. The probable electronic configuration of

Fm $1s^2 2s^2 2p^6 3s^2 3p^6 3d^{10} 4s^2 4p^6 4d^{10} 4f^{14} 5s^2 5p^6 d^{10} 5f^{12} 6s^2 6p^6 7s^2$. Ionic radius 0.97A. See also **Chemical Elements**.

All known isotopes of fermium are radioactive. They include isotopes of mass numbers 248($t_{1/2}$ = 0.6 m.), 249($t_{1/2}$ = 150 s.), 250($t_{1/2}$ = 0.5 h.), 251($t_{1/2}$ = 7 h.), 252($t_{1/2}$ = 23 h.), 253($t_{1/2}$ = 4.5 d.), 254($t_{1/2}$ = 3.24 h.), 255($t_{1/2}$ = 22 h.), 256($t_{1/2}$ = 160 m.), and an isotope of mass number 257 which has a half-life of about 10 days. The mass number of fermium given in atomic weight tables is 253, the mass number of the isotope of longest half-life (except possibly for the last one cited). The first three decay by alpha-particle emission only; the others, through ^{255}Fm, by that and other modes, including spontaneous fission for ^{254}Fm and ^{255}Fm. Isotopes of 256 and higher decay by this mode only. Fermium isotopes are made: (1) by heavy iron bombardment of uranium and plutonium isotopes; (2) by the action of alpha particles on californium isotopes; and (3) from several of the other transuranium elements by multiple neutron capture—or in the case of the heaviest isotopes, from einsteinium by single neutron capture.

The ion Fm^{3+} is stable, and fermium is probably exclusively trivalent.

The discovery of fermium (also einsteinium) was not the result of very carefully planned experiments, as in the cases of the other trans-uranium elements, but fermium and einsteinium were found in the debris of an atomic weapon test in the Pacific in November 1952. Researchers, using the Oak Ridge High Flux Isotope Reactor (HFIR) which produced 3.2-hour ^{254}Fm, determined the magnetic moment of the atomic ground state of the neutral fermium atom with a modified atomic beam magnetic resonance apparatus. In essence, this observation represented that of a macroscopic property of the metallic 0-valent state of fermium.

Additional Reading

Greenwood, N.N. and A. Earnshaw: *Chemistry of the Elements,* 2nd Edition, Butterworth-Heinemann, Inc., Woburn, MA, 1997.

Krebs, R.E.: *The History and Use of Our Earth's Chemical Elements: A Reference Guide,* Greenwood Publishing Group, Inc.

Lide, D.R.: *CRC Handbook of Chemistry and Physics,* 8th Edition, CRC Press, LLC., Boca Raton, FL, 2003.

Parker, P.: *McGraw-Hill Encyclopedia of Chemistry,* 2nd Edition, The McGraw-Hill Companies, Inc., New York, NY, 1993.

Seaborg, G.T. and W.D. Loveland: *The Elements beyond Uranium,* John Wiley & Sons, Inc., New York, NY, 1990.

FERRIMAGNETISM. A type of magnetism, macroscopically similar to ferromagnetism, but microscopically more like antiferromagnetism in that the magnetic moments of neighboring ions tend to align antiparallel. These moments are, however, of different magnitudes, and hence may still have quite a large resultant magnetization.

FERRITES. The term *ferrite* is commonly used generically to describe a class of magnetic oxide compounds which contain iron oxide as a principal component. In metallurgy (qv), however, the term ferrite is often used as a metallographic indication of the α-iron crystalline phase.

Ferrites can be classified according to crystal structure, i.e., cubic vs hexagonal, or magnetic behavior, i.e., soft vs hard ferrites.

The existence of ceramic magnetic materials capable of combining the resistivity of a good insulator (10^{12} ohm-cm) with high permeability was announced by Snoek (1946). Shortly thereafter, Nèel (1948) introduced the term ferrimagnetism to describe the novel magnetic properties of these materials. A simple ferrite is composed of two interpenetrating ferromagnetic sublattices with magnetizations $M_a(T)$ and $M_o(T)$ which decrease with increasing temperature and vanish at the Curie point, T_c. In a ferromagnetic material, the resulting saturation magnetization M would be $M_a + M_b$. See also **Ferromagnetism**. However in a ferrite, strong antiferromagnetic interaction between sublattices results in antiparallel alignment, and $M = M_a - M_b$. In general, $M_a(T) \neq M_b(T)$, and the material behaves in most respects like a ferromagnet, exhibiting domains, a hysteresis loop, and saturation of the magnetization at relatively low applied magnetic fields. Practical values for saturation magnetization and Curie temperature range from 250–5,000 oersteds (19,984–397,887 ampere turns/meter) and from 100 to 600°C.

Ferrites resemble ceramic materials in production processes and physical properties. The high resistance of ferrites makes eddy-current losses extremely low at high frequencies. The direct current resistivities correspond to those of semiconductors, being at least one million times those of metals. Magnetic permeabilities may be as high as 5,000 and dielectric constants in excess of 100,000. Ferrites provide design advantages over strip and powder cores for filter cores, deflection transformers, and yokes and in antenna rods, pulse transformers, delay lines, waveguide elements, and a number of other electronic components.

Ferrimagnetic materials have spinel, garnet, and hexagonal structures. A typical spinel ferrite is $NiFe_2O_4$. Other ferrites may be obtained by substituting magnetic (cobalt, nickel, manganese) or nonmagnetic (aluminum, zinc, copper) ions for some of the nickel or iron ions. That is, $Ni_{1-y}Co_yAl_xFe_{2-x}O_4$, where x and y may be varied to modify M and T_c. Yttrium iron garnet (YIG), $Y_3Fe_5O_{12}$, is the classical ferrimagnetic garnet which combines very low magnetic loss with high resistivity. Substitution of magnetic rare-earth ions (gadolinium, ytterbium, holmium, etc.) for Y and of nonmagnetic ions (gallium, aluminum) for some of the Fe ions leads to many different compositions with a wide range of M and magnetic loss. The rare-earth ions form a third magnetic sublattice with attendant magnetization M_c antiparallel to the resultant magnetization $M_{a,b}$ of the two Fe sublattices. Since M_c and $M_{a,b}$ exhibit different variations with temperature, the net magnetization may vanish twice, at F_c and at an intermediate temperature called the compensation point, T_{comp}, where $M_c = M_{a,b}$.

A typical hexagonal ferrite is $BaFe_{12}O_{19}$. Again, other magnetic ions, such as manganese, cobalt, and nickel may be introduced to produce wide variations in M and T_c. Hexagonal ferrites are characterized by large

anisotropy fields with an axis of symmetry that may be either a direction of hard (planar ferrites) or easy (uniaxial ferrites) magnetization.

To distinguish among major fields of applications, ferrites can be separated into five groups: Soft, square-loop, hard, microwave, and single-crystal ferrites.

Soft ferrites have a slender S-shaped hysteresis loop with low remanence and low coercive force permitting easy magnetization and demagnetization with little magnetic loss. These ferrites are uniquely suited to low-loss inductor and transformer cores for radio, television, and carrier telephony.

Square-loop ferrites are materials exhibiting an almost rectangular hysteresis loop with two distinct states of remanence and with a coercive force of a few oersteds. All practical square-loop ferrites have a spinel structure. The Mg-Mn(Zn) system has retained an important position in computer memory applications. Lithium-nickel ferrites and more complex systems containing Li, Mn, and Al have shown stability and fast switching over a wide range of temperatures.

Hard ferrites are characterized by hexagonal structure, a hysteresis loop enclosing a large area, and a coercive force of several thousand oersteds. These ferrites can store a significant amount of magnetic energy. They have been used as permanent magnets in loudspeakers, small motors, generators, and measuring instruments.

Microwave ferrites have garnet, spinel, or hexagonal crystal structure and very low electric and magnetic loss factors. In general, the required M increases with the frequency f of applications. Substituted and pure garnets, Mg-Mn-Al ferrites and Mg-Mn ferrites are used at the lower part of the microwave spectrum where $M = 200$ to 3000 gauss is adequate. In the millimeter wave region, $f = 30$ to 100 GHz, Ni-Zn ferrites ($M = 5000$ gauss) and hexagonal ferrites of various compositions may be used. Microwave ferrite devices, such as isolators, circulators, switches, phase shifters, limiters, parametric amplifiers, and harmonic generators are based upon interactions of rf signals with the ferrite magnetization.

Single-crystal ferrites of practical importance are rare-earth garnets grown in a flux of molten lead oxide. Some of these are optically transparent, permitting direct observation of magnetic domains. Interaction of infrared and visible light with the electron spins is called the magneto-optic effect. It permits electronic modulation of a beam of light that propagates through a single-crystal garnet. These devices are also of interest in laser technology.

Single-crystal, rare-earth garnet sheets have been grown on a substrate with a preferred direction of magnetization perpendicular to the plate. In these plates, tiny round magnetic domains called "bubbles" can be formed by an applied magnetic field. These bubbles can be propagated, erased, and manipulated to perform binary functions in computers, including logic, memory, counting, and switching. See also **Rare-Earth Elements and Metals**.

Additional Reading

Goldman, J.: *Handbook of Modern Ferromagnetic Materials,* Kluwer Academic Publishers, Norwell, MA, 1999.

Gurevich, A.G. and G.A. Melkov: *Magnetization Oscillations and Waves,* CRC Press, LLC., Boca Raton, FL, 1996.

Koul, S. and B. Bhat: *Microwave and Millimeter Wave Phase Shifters: Dielectric and Ferrite Phase Shifters,* Vol. 1, Artech House, Inc., Norwood, MA, 1991.

Valenzuela, R.: *Magnetic Ceramics,* Cambridge University Press, New York, NY, 1994.

Whicker, L.: *Ferrite Phasers and Ferrite MIC Components,* Vol. 2, Artech House, Inc., Norwood, MA, 1974.

FERROCHROMIUM. An alloy, composed principally of iron and chromium, used as a means of adding chromium to steels (low, medium, and high-carbon) and cast iron. Available in several classifications and grades, generally containing between 60 to 70% chromium, in crushed sizes and lumps up to 75 pounds that readily dissolve in molten steel.

See also **Chromium**.

FERROELECTRIC EFFECT. The phenomenon whereby certain crystals may exhibit a spontaneous dipole moment (which is called ferroelectric by analogy with ferromagnetic—exhibiting a permanent magnetic moment). The effect in the most typical case, barium titanate, seems to be due to a polarization catastrophe, in which the local electric fields due to the polarization itself increase faster than the elastic restoring forces on the ions in the crystal, thereby leading to an asymmetrical shift in ionic positions, and hence to a permanent dipole moment. Ferroelectric crystals often show several Curie points, domain structure and hysteresis, much as do ferromagnetic crystals.

FERROELECTRIC MATERIALS. The dielectric analogs of ferromagnetic materials. Their uses parallel those of ferromagnetic materials in such applications as magnetostrictive transducers, magnetic amplifiers, and magnetic information storage devices. Rochelle salt was the first ferroelectric material to be discovered and the barium titanate ceramics are materials of this type.

FERROMAGNETISM. The property of certain materials that gives them relative permeabilities noticeably exceeding unity, in practice from 1.1 to 10^6. Such materials generally exhibit hysteresis, hence can be used for permanent magnets. Ferromagnetism is an extreme case of paramagnetism, and results from the spontaneous alignment of the electron magnetic moments associated with spin even in the absence of an externally applied field.

Ferromagnetism exemplifies cooperative phenomena in solids. It is characterized by a spontaneous macroscopic magnetization M (magnetic moment per unit volume) in the absence of an applied magnetic field at temperatures below a critical value, known as the Curie temperature, T_C. This property is exhibited by the transition metals, iron, cobalt, and nickel; by the rare-earth metals, such as gadolinium, terbium, dysprosium, holmium, erbium, and thulium; and by a variety of alloys, compounds, and solid solutions involving the transition, rare-earth, and actinide elements. See also **Rare-Earth Elements and Metals**. Curie temperatures range from a fraction of a degree to hundreds of Kelvin.

The apparent permeability of a magnetic material at microwave frequencies is affected (in the presence of a transverse, steady field) by the precession of electron orbits in the atoms. If the microwave frequency equals the precession frequency, resonance occurs and the apparent permeability reaches a sharp maximum. The resonance frequency depends upon the strength of the transverse field. Thus, a thin film of a ferromagnetic substance placed in a static magnetic field H is found to be capable of absorbing from an oscillating field whose magnetic vector is perpendicular to H at a frequency given by

$$\omega = \left(\frac{ge}{2mc}\right)(BH)^{1/2}$$

where b is the magnetic induction associated with H, e and m are the charge and mass of the electron, c is the velocity of light, and g is very near to 2, the LandÈ factor for free electrons.

The exchange interaction between electrons in neighboring atoms can be shown to depend on the relative orientations of the electronic spins. If it should turn out that parallel spins are favored, there is a strong tendency for all the spins in the lattice to become aligned, the transition to the ordered state corresponding to the Curie point. The concept of localized spins (e.g., d-electrons in the transition metals) is confirmed by neutron diffraction, but the theory is incomplete at the stage of calculating the actual magnitude and sign of the interaction.

FERROMANGANESE. An alloy consisting of manganese (approximately 48%) plus iron and carbon.
See also **Manganese**.

FERROMOLYBDENUM. An alloy, composed largely of iron and molybdenum, used as a means of adding molybdenum to steel. Engineering steels rarely contain more than 1% molybdenum, stainless steels may contain 3%, and tool steels as much as 10%. Ferromolybdenum is available in several grades in which molybdenum ranges from 55 to 75% and the maximum carbon content is 1.10%, 0.60%, or 2.50%. It is generally added to the furnace since it does not oxidize under steelmaking conditions. Mp approximately 1630C. Available in crushed sizes up to one inch.
See also **Molybdenum**.

FERRONICKEL. See **Nickel**.

FERRONIOBIUM. An alloy of iron and niobium made by reducing the ore columbite with silicon.
Uses: Stainless steels and other alloys for welding rods.
See also **Niobium**.

FERROPHOSPHORUS. An alloy of iron and phosphorus used in the steel industry for adjustments of phosphorus content of special steels.
See also **Phosphorus**.

FERROSILICON. An alloy of iron and silicon used to add silicon to steel and iron, d 5.4, insoluble in water. Small quantities of silicon deoxidize the iron, and larger amounts impart special properties.
See also Silicon.

FERROTITANIUM. An alloy composed principally of iron and titanium, used to add titanium to steel. It is often made from titanium scrap. Three classifications are available: low, high, and medium carbon content. Furnished in various lump, crushed, and ground sizes.
See also **Titanium**.

FERROTUNGSTEN. An alloy of iron and tungsten used as a means of adding tungsten to steel. Contains 70 to 80% tungsten and no more than 0.6% carbon. Melting range 1648–2750°C, dissolves readily in molten steel. Furnished in ground and crushed sizes up to one inch.
See also **Tungsten**.

FERROVANADIUM. CAS: 12604-58-9. An iron-vanadium alloy used to add vanadium to steel. Vanadium is used in engineering steels to the extent of 0.1–0.25% and in high-speed steels to the extent of 1–2.5% or higher. Melting range 1482–1521°C. Furnished in a variety of lump, crushed, and ground sizes; formed by reduction of the oxide with aluminum or silicon in the presence of iron in an electric furnace.
See also **Vanadium**.

FERROZIRCONIUM. Alloys used in the manufacture of steel. (1) 12–15%, silicon 39–43%, iron 40–45%; application: steel of high silicon content. (2) 35–40% zirconium alloy. Approximate analysis: zirconium 35–$0%, silicon 47–52%, iron 8–12%; application: steel of low silicon content.
See also **Zirconium**.

FERTILE MATERIAL. Material that is not in itself fissionable, but that, in a nuclear reactor, may be transformed into fissionable material through a nuclear transformation; e.g.:

$$^{232}\text{Th} + n \longrightarrow {}^{233}\text{Th} \xrightarrow[23\text{ min}]{\beta^-} {}^{233}\text{Pa} \xrightarrow[27.4\text{ days}]{\beta^-} {}^{233}\text{U}$$

the ^{233}U is fissionable.

FERTILIZER. A substance, but often a combination of substances, of organic composition, natural and/or manufactured, in solid or liquid slurry forms (in some cases in the gaseous phase) made available to plants to promote normal, healthy, and often vigorous growth. Most frequently added to soils, fertilizers also are applied directly to plant parts above ground (foliar sprays), in nutrient fluids as furnished in hydroponic systems, and by irrigation systems.

Unless poisoned or severely leached, some soils still may retain some of the nutrients required by growing plants and may support plants of a weak, straggly nature, with submarginal yields and poor quality for a number of years. Some soils may be generally poor, that is, they originally did not contain any of the primary plant nutrients in adequate concentration; or they may be poor soils because they have an imbalance of nutrients. It should be stressed that the proper chemical nutrients must be present along with suitable physical properties of the soil if highest yields and quality are to be achieved. A given soil may be classified as generally rich and yet lack the needed concentration of only one or two micronutrients. Further, considering the soil-plant system, soils should be customized to the needs of specific crops; or, in reverse, certain crops should not be planted on soils that are severely out of balance with their needs.

Because of the intensity of cultivation in many regions of the world, notably among the larger producers of food and fiber crops (where there is repetitive use of the same tracts of land year after year), the uniformity of nutrient content, as may have been present in the virgin soil, cannot be assumed. In fact, experience has shown that the same soil class may vary widely in nutrient content from one field to the next. The ability of soils to provide plant nutrition is a reflection of how the soils have been artificially treated (fertilized and conditioned) over the remote and recent past and what kinds of crops have been grown on them. With increasing use of control chemicals, soils require increasing frequency of testing and analysis, not only for required plant nutrients, but also for the possible presence of chemicals that tend to accumulate rather than dissipate downward to deeper ground levels or to the groundwater. Some substances may not be biodegradable and hence they may be present from one growing season to the next.

Particularly during the last few decades, astute growers have learned to depend upon reliable scientific analysis of their soils, augmented by tissue analyses of plant parts, and sometimes total plant analysis, as a basis for planning an annual fertilizing program. Perhaps not sufficiently stressed are problems that can arise from overfertilization as well as from underfertilization. Crops vary immensely in their ability to tolerate deficiencies and excesses of nutrients. Overfertilization also can cause serious pollution problems. Overfertilization also adds a needless element to the cost of food production.

Basic Fertilizer Functions. Major reasons for adding fertilizer to soils and plants include: (1) replenishment of chemical elements that have been reduced or exhausted by the soils to the crops previously grown or leached from the soils (as the result of poor tillage practices, overirrigation, natural flooding, and, in some cases, adding nutrients that are naturally deficient in a given type of soil); and (2) customizing the nutrient content of soils to particular growing objectives.

Fertilization, in combination with irrigation, has been responsible for converting vast semiarid and arid lands with lean soils into land useful for production of a number of important food and fiber crops. This effective combination is particularly important to many of the developing countries that have large holdings of land in these categories.

Although essentially self-evident, the fact that crops exhaust the soil of key chemical ingredients is made even clearer by examination of Table 1. In the analysis, nitrogen is lost and does not appear in the ash, but other methods for analyzing for nitrogen confirm its significant content in crops. Carbon, oxygen, and hydrogen, of course, are made available to crops by way of water and the atmosphere. See Table 1.

Principal Categories of Nutrients. The nutrients required by plants fall into three categories: (1) *primary nutrients or elements*—nitrogen, phosphorus, and potassium, because they are generally required by plants in larger amounts and are often present in more limited amounts in soil; (2) *secondary nutrients or elements*—calcium, magnesium, and sulfur, because they generally are not so limited in soils and they are required in smaller amounts; and (3) *micronutrient elements*, of which there are several and which are required in very small amounts. See Table 2.

Nitrogen Requirements

Although the requirements for nitrogen are well understood today, this was far from self-evident just a few centuries ago. See Fig. 1. It seems reasonably safe to postulate that the early growers of plants practiced rudimentary forms of fertilization without understanding what they were doing beyond "something taken from the soil must be returned." The latter observation in itself was rather profound for ancient peoples. It remained for the awakenings of chemistry in the 1600s and notably by the early scientists of the late 1700s and early 1800s to commence investigations of the technical links between plants and their nutrients.

Chilean saltpeter, $NaNO_3$, was the first of the chemical nitrogenous fertilizers. Ammonium sulfate, $(NH_4)_2SO_4$, made available as a byproduct of coal-gas produced in large quantities prior to wide use of natural gas,

TABLE 2. PRINCIPAL CATEGORIES OF PLANT NUTRIENTS

Primary nutrients or elements	Secondary nutrients or elements
Nitrogen (N)	Calcium (Ca)
Phosphorus (P)	Magnesium (Mg)
Potassium (K)	Sulfur (S)

	Micronutrients	
		General range in soils
Element	Pounds/Acre	Kilograms/Hectare
Boron (B)	20–200	22.4–224
Manganese (Mn)	100–10,000	112–11,200
Zinc (Zn)	10–600	11.2–672
Copper (Cu)	2–400	2.2–448
Iron (Fe)	10,000–200,000	11,200–224,000
Molybdenum (Mo)	1–7	1.1–7.8

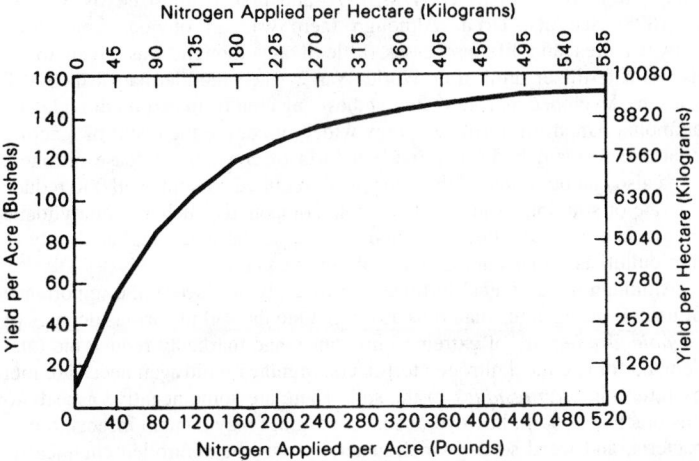

Fig. 1. Effect of nitrogen fertilizer application on crop of irrigated corn (maize) in state of Washington. It should be pointed out that the very high yields were obtained under experimental, carefully controlled conditions. Although much higher yields are obtained in some regions, the average yield in the United States is 87.3 bushels/acre (5489 kilograms/hectare). The world average is 45 bushels/acre (2829 kilograms/hectare)

was also an early fertilizer, but followed Chilean saltpeter by several years. Ammonium sulfate is still an important source of nitrogen, but no longer holds the lead role in most parts of the world.

In the early 1900s, attempts to fix atmospheric nitrogen in a compound that could be applied directly to soil were made. During that period, nitrogen fixation was a much-discussed aspect of industrial chemistry, much as synthetic fuels are a timely topic today. For some years, calcium cyanamide, $CaCN_2$, was produced in the United States, largely as the result, at that time, of the newfound hydroelectric energy of the Tennessee Valley region. A large facility was located at Muscle Shoals, Alabama. The last plant for making $CaCN_2$ in the United States was closed in

TABLE 1. CONSTITUENTS OF ASH OF NORMAL CROPS

	Pounds of constituent removed per acre of ground (Kilograms per hectare given in parentheses)								
Crop and part	Silica	Potash	Soda	Magnesia	Lime	Ferric oxide	Chloride	Sulfate	Phosphate
Grain	15	14	7	2	8	1	0	0	36
	(16.8)	(15.7)	(7.8)	(2.2)	(9)	(1.1)	(0)	(0)	(40.3)
Straw	233	33	1	28	12	6	4	13	11
	(261)	(37)	(1.1)	(31.4)	(13.4)	(6.7)	(4.5)	(14.6)	(12.3)
Roots	27	143	17	46	18	4	12	46	26
	(30.2)	(160)	(19)	(51.5)	(20.2)	(4.5)	(13.4)	(5.5)	(29.1)
Tops	3	89	17	72	10	3	50	39	29
	(3.4)	(99.7)	(19)	(80.6)	(11.2)	(3.4)	(56)	(43.7)	(32.8)
Hay	78	38	12	45	7	1	4	9	15
	(87.4)	(42.6)	(13.4)	(50.4)	(7.8)	(1.1)	(4.5)	(10.1)	(16.8)

June 1971. In Norway, also because of low-cost hydroelectric energy once available, production of calcium nitrate by way of first producing nitric acid was pioneered.

The first breakthrough in the large-scale synthesis of ammonia resulted from the work of Fritz Haber (Germany, 1913), who found that ammonia could be produced by the direct combination of nitrogen and hydrogen in the presence of a catalyst at a relatively high temperature and very high pressure. See also **Ammonia**. During the interim, much-improved processes for producing synthetic ammonia have been designed.

For many years, the bulk of nitrogen fertilizers has been based upon ammonia, as synthesized. There has been an increasing trend in some major agricultural regions and countries to favor the use of liquid nitrogen for some crops. These products include anhydrous ammonia, aqua ammonia, ammonium salts in solution, and numerous combinations, such as formulations also containing phosphate salts. Because of the relatively easy solubility of these materials, some ecologists have expressed concern over so-called *nitrogen runoff*. Fortunately, however, most soils will fix ammonium ions quite rapidly, thus reducing a pollution threat.

It is interesting to observe that the compound urea, NH_2CONH_2, was discovered (in urine) as early as 1773 and first synthesized by Wöhler in 1828. See also **Urea**. Although there was an obvious connection between urea and life processes, little if any thought was given to its use as a fertilizer until after World War I, when the German firm BASF (Bosch) developed a process for synthesizing urea from carbon dioxide and ammonia. Ureaform fertilizers, now widely used, are the result of reacting urea and formaldehyde to provide a form of controlled-release nitrogen. Urea also can be coated with sulfur (itself required by some soils) to reduce the rate of solution. And slowly soluble compounds, such as isobutylidene diurea, can be made, but because of high cost, these are marketed only to horticulturists rather than growers of commercial crops.

Although not chemical fertilizers (artificially produced), the importance of natural nitrogenous materials as returned to the soil in various degrees of *organic farming* are of extreme importance and markedly reduce the total demand for chemical nitrogen fertilizers. Significant nitrogen needs are met by returning *crop residues* to the soil. There are some negative aspects to this practice as well, however, because residues may contain insects, fungi, bacteria, and weed seeds, which generally must be controlled chemically. *Green manure* is a crop purposely grown for plowing under to enrich the soil. Animal manure also is an important contributor to the soil, not only in terms of nitrogen, but phosphate and potash as well. Also, certain crops, notably legumes, create more nitrogen for the soil than they use—this as a result of bacteria that inhabit the root zones of these plants. Collectively, these kinds of materials are sometimes called *organic fertilizer*. See Fig. 2.

Other chemical fertilizers used as sources of nitrogen include ammonium nitrate, which contains approximately 35% nitrogen by weight. Calcium ammonium nitrate contains between 20.5 and 28% nitrogen by weight,

depending upon the amount of calcium carbonate added. Ammonium sulfate nitrate contains from 26 to 28% nitrogen by weight. See also **Nitrogen** and specific nitrogen compounds described in this book.

Phosphorus Requirements

Deficiencies of available phosphorus in soils are a major cause of limited crop production. Phosphorus deficiency is regarded by some authorities as the most critical mineral deficiency in grazing livestock.

Liebig and other investigators in the mid-1800s indicated that the much earlier noted fertilizing properties of bones were derived mainly from their phosphate content and that the treatment of bones with sulfuric acid increased their effectiveness in soils. With these advantages proclaimed, a large market for what was then called "chemical manure" resulted in a shortage of bones throughout the European agricultural community. Fortunately, a number of years later, several guano deposits were located in Peru and quite a bit later, phosphate minerals were located in Florida and other parts of the world. Guano fertilizers today are used mainly for special horticultural products, whereas commercial chemical phosphate fertilizers are derived from phosphate rock.

In formulating phosphate fertilizers, solubility is of particular concern. Good solubility assures availability of phosphorus to the plants. For example, in water and in alkaline and neutral soils, the apatites, which are $Ca_5(PO_4)_3R$ (where R is usually but not always fluorine), are quite insoluble and hence of little value to the soil. Tricalcium phosphate is also quite insoluble. On the other hand, these compounds are moderately soluble in acid soils and thus can be used with discretion. Somewhat in contrast, dicalcium phosphate, $CaHPO_4$, is quite soluble in acid soils and only moderately soluble in water and alkaline and neutral soils. Fortunately, monocalcium phosphate $Ca(H_2PO_4)_2$, is soluble in water and all moist soils. The presence of iron and aluminum phosphates also has an effect on total solubility, these compounds being insoluble in water, but soluble in weak acids. Some countries require total water solubility and thus monocalcium and ammonium phosphates must be used. In other areas, slight solubility in water or appreciable solubility in weak acids is adequate to meet regulations. Some humic soils can assimilate ground phosphate rock even without chemical treatment. Commonly in phosphate fertilizer manufacture, phosphorus is recovered from phosphate rock. See also **Phosphorus**; and **Phosphoric Acid**.

Nomenclature. Because the ammonium phosphate fertilizers contain two of the primary plant nutrients, it is pertinent at this point to briefly comment on the manner in which these fertilizers are named. A series of three numbers, separated by dashes, is used to indicate the primary nutrient content of fertilizer mixtures. In order, from left to right, the numbers show the percentage of nitrogen, phosphoric oxide, and potash:

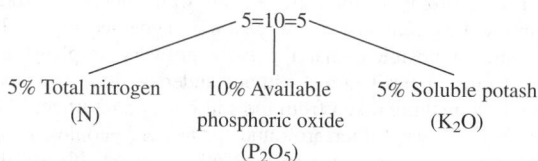

| 5% Total nitrogen (N) | 10% Available phosphoric oxide (P_2O_5) | 5% Soluble potash (K_2O) |

Single Superphosphate is produced in large quantities and is the oldest of the water-soluble phosphates. The material contains about 20% P_2O_5 by weight. Single superphosphate is made by reacting ground phosphate rock with 70% sulfuric acid. This reaction results in a solid mass of monocalcium phosphate and gypsum. The fluorine and silicon evolved are removed by water scrubbing.

Wet-Process Orthophosphoric Acid contains 30–54% P_2O_5 by weight. When sulfuric acid is added to phosphate rocks in a proportion greater than needed to make single superphosphate, orthophosphoric acid, H_3PO_4, is produced. This acid is used as an intermediate in preparing other phosphates.

Triple Superphosphate is made by acidulating phosphate rock with phosphoric acid. The concentrated triple superphosphate produced is essentially monocalcium phosphate containing very little gypsum. The principle use of triple superphosphate is in mixed fertilizers to make P_2O_5 available in water-soluble form.

Ammonium Phosphates. Although several ammonium phosphates can be prepared, only the mono- and the di-compounds are produced for fertilizer use. In some processes, anhydrous ammonia is reacted with phosphoric acid, with the resultant slurry converted to solid form by drying.

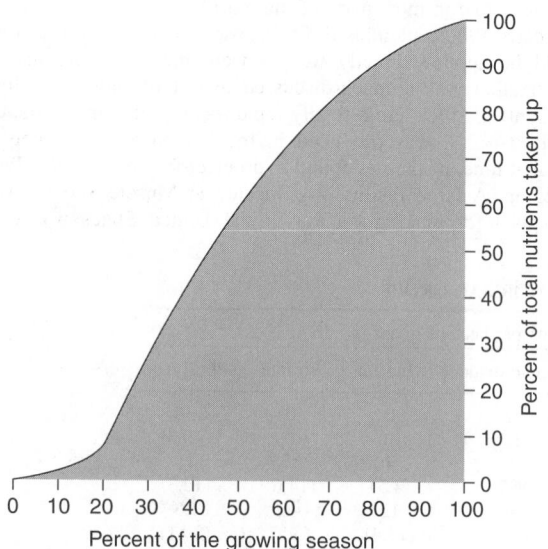

Fig. 2. Composite nutrient uptake chart representing an average of many plants. Each plant has its own specific nutrient uptake curves. (*Source: Leitch reference listed*)

The ratio of ammonia to phosphoric acid can be varied between 1 and 2, and consequently several product grades can be made.

Nitrophosphates. Phosphate rock is readily dissolved in nitric acid to yield a mixture of calcium nitrate, phosphoric acid, and monocalcium phosphate. When calcium nitrate is converted to solid form, the material is highly hygroscopic and thus not desirable for packaging and storing. This is overcome by forming calcium nitrate tetrahydrate crystals through chilling and later removal by filtering or centrifuging. Ammoniation of the mother liquor produces a mixture of ammonium phosphate, dicalcium phosphate, and ammonium nitrate. These can be concentrated and prilled or granulated. Thus, in terms of the traditional fertilizer nomenclature, that is, %N = %P_2O_5 = %K_2O, numerous product grades are possible according to raw-material ratios and process conditions. By adding potash to 20-20-0, a formulation of 15-15-15 can be obtained, as one of numerous examples.

Nonorthophosphates. If 54% orthophosphoric acid is dehydrated to remove the remaining free water, a pyro acid is yielded. Continued heating removes more waters and various insoluble compounds are formed. Evaporation by submerged combustion or under vacuum yields a eutectic between ortho and pyro acids with a P_2O_5 content of about 72% by weight. This superphosphoric acid can be ammoniated to yield liquid ammonium polyphosphate (APP) fertilizers, such as 10-34-0. Because of their strong sequestering properties, these liquids keep impurities in solution, as well as the salts of micronutrient metals, which are insoluble derivatives of orthophosphoric acid.

APP fertilizers also are produced by reacting ammonia directly with wet-process phosphoric acid and dissolving the melt in ammonia solution to produce 10-34-0. By adding potash and a bit of clay, nitrogen-phosphorus-potassium suspensions, such as 13-13-13, are possible. If the melt is granulated, a solid with proportions 12-57-0 will result. Thus, by adding urea and potash, a wide variety of mixes (28-28-0, 19-19-19, and so on) are possible.

Potassium Requirements

Researchers in the early 1800s found that potassium is an essential plant nutrient although the details of its function were unknown. The potassium in wood ashes was first used in Europe, later to be replaced by sylvite, KCl, and carnalite, $KCl \cdot MgCl_2 \cdot 6H_2O$, found in deposits in Germany. During the interim, billions of tons of these materials have been located. A number of sites are continuously mined. Large reserves are found in the former U.S.S.R., Canada, and east and west Germany. Significant deposits also are found in Israel and Jordan (Dead Sea region), Spain, France, the United Kingdom, and the United States. Deposits are as water-soluble salts, such as sylvite, and minerals with varying content of magnesium.

The physiological role of potassium in life processes is described in entry on **Potassium and Sodium (In Biological Systems)**. Potassium is a usual, but variable constituent of most soils. However, available potassium may be depleted through loss arising from leaching by rain water, flooding, and over-irrigation, or through loss to crops continuously planted on a given tract. Thus, potassium is categorized as a primary nutrient. Potassium differs from most other essential constituents of plant cells in that it is not built into the cell as part of an organic compound, but rather it is an ion from a soluble inorganic or organic salt. Potassium ions may chelate with cellular constituents, such as polyphosphates. The ion is of the correct size to fit into the water lattice adsorbed to the proteins in the cell. In general, potassium ions are attracted to protein or other colloidal or structural units having a negative charge. Mucopolysaccharides within the cell, on the cell surfaces and of the intercellular structures, are of particular importance in holding potassium. Active centers or other configurational features of the proteins in the cell may be affected or altered by the potassium held by electrostatic or covalent binding. There are several enzyme systems activated by potassium.

In plants, the meristematic tissues in general are particularly rich in potassium, as are other metabolically active regions, such as buds, young leaves, and root tips. Potassium deficiency may produce both gross and microscopic changes in the structure of plants. Effects of deficiency include leaf damage, high or low water content of leaves, decreased photosynthesis, disturbed carbohydrate metabolism, and low protein content, among other abnormalities. The importance of potassium is also reflected by livestock that consumes plants and feedstuffs prepared from plant materials.

Potassium fertilizers are prepared from the potassium minerals previously mentioned. The ores are crushed, beneficiated, crystallized, and dried to commercial *potash* or *muriate* (KCl), in various grades and particles containing from 60 to 62% KCl. Relatively small amounts of other potassium salts are used as fertilizers. Some vegetables and tobacco are adversely affected by high chloride concentrations and some growers prefer to use potassium sulfate, K_2SO_4, or potassium nitrate, KNO_3.

Potash is frequently applied to the soil along with salts containing nitrogen and/or phosphoric oxide, P_2O_5, in amounts varied for different soil and crop requirements. One method used is that of combining crushed muriate with moist nitrogen- and P_2O_5- containing compounds, followed by granulation of the mixture. Another method is that of dry-blending materials, such as urea, diammonium phosphate, and potash, and applying the mixture to the soil. The total water solubility of potash results in full initial K_2O availability in moist soils. However, in clays, when rainfall or irrigation is limited, excessive chloride build-up can be harmful.

Calcium Requirements

The role of calcium in biosystems is described in entry on **Calcium (In Biological Systems)**. Fortunately, as the fifth most abundant element in the earth's crust, calcium is usually available to plants in abundance. Nevertheless, large amounts of calcium are added to soils by virtue of the use of lime (calcium oxide, CaO) as means to adjust the pH of soils. Among many reasons why soil pH is so important are the effects of acidic soils on the availability of manganese, a micronutrient, and also of aluminum, more recently identified as toxic to some crops. The use of dolomitic lime, $CaMg(CO_3)_2$, also is an effective way to correct magnesium deficiencies. Agricultural lime is not water soluble and, therefore, it will not correct soil acidity immediately after application.

Sulfur Requirements

The role of sulfur in biosystems is described in entry on **Sulfur (In Biological Systems)**. Sulfur in some form is required by all living organisms. Among important sulfur-containing compounds are the amino acids cysteine, cystine, and methionine; the vitamins thiamine and biotin; and certain complex lipids, such as sulfatides, among others. In the chain from soils to plants to animals, including humans, inorganic sulfur (sulfate ion, SO_4^{-2}) is taken up by plants and converted within the plant to organic compounds (sulfur amino acids). The most important feature of sulfur in the food chain is that plants use inorganic sulfur compounds to make the aforementioned amino acids, whereas animals use the sulfur amino acids for their own processes and excrete inorganic sulfur compounds.

Although sulfur is a widespread element in the earth's crust, ranking as the 14th element in abundance, it is obviously less abundant than calcium and, further, it is not so evenly distributed. Thus some soils show sulfur deficiency. The trend toward high-analysis fertilizers without sulfur can create a need for more deliberate use of sulfur fertilizers.

Sulfur is present in ammonium sulfate, used as a fertilizer, but ammonium sulfate is only one of many fertilizers that may be used. Elemental sulfur and sulfur-containing compounds are also used for controlling various plant pests. The soils also pick up sulfur from air pollution. Nevertheless, soil analysis should be made to provide an accurate diagnosis of whether or not a particular soil may require additional sulfur.

Magnesium Requirements

The role of magnesium in biosystems is described in entry on **Magnesium (In Biological Systems)**. Magnesium is generally abundant in the earth's crust, ranking 8th in abundance. Nevertheless, magnesium deficiencies are quite common. Magnesium deficiency is a fairly common cause of poor crop yields, especially among crops produced on sandy soils. Accumulation of magnesium from the soil by plants is strongly affected by the species of plant. Legumes usually contain more magnesium than grasses, tomatoes, corn (maize), regardless of the level of magnesium in the soil. A high level of available potassium in the soil interferes with the uptake of magnesium by plants, and thus magnesium deficiency can occur even if there are adequate amounts of the element in the soil. The role of magnesium highlights the systems aspects of agricultural management, exemplified by need to maintain a proper magnesium intake from pasture and from feedstuffs. When animals are fed diets primarily of grains, a proper balance among magnesium, calcium, and phosphorus must be maintained to minimize danger of urinary calculi. Magnesium deficiency among cattle (grass tetany or grass staggers) is observed most frequently when animals are first grazed on lush grass or wheat pastures, indirectly indicating the relative low uptake of magnesium by certain crops.

TABLE 3. STATUS OF MICRONUTRIENT DEFICIENCIES IN SOILS OF THE UNITED STATES

State	Boron ND	Mod	Ser	Copper ND	Mod	Ser	Iron ND	Mod	Ser	Manganese ND	Mod	Ser	Molybdenum ND	Mod	Ser	Zinc ND	Mod	Ser
Alabama		x		x			x				x		x				x	
Arizona	x				x			x			x		x				x	
Arkansas		x		x			x			x			x			x		
California			x			x	x				x			x				x
Colorado	x			x				x		x			x				x	
Connecticut			x	x				x		x				x			x	
Delaware			x	x				x			x				x	x		
Florida			x			x		x			x		x					x
Georgia		x		x			x				x			x				x
Idaho		x		x				x		x				x			x	
Illinois		x		x			x				x			x		x		
Indiana		x			x			x			x			x		x		
Iowa		x		x				x		x			x			x		
Kansas		x		x					x	x				x			x	
Kentucky		x		x			x			x			x				x	
Louisiana		x		x				x		x			x				x	
Maine			x	x			x			x			x			x		
Maryland			x	x			x			x				x		x		
Massachusetts			x	x			x			x				x		x		
Michigan		x				x	x				x			x		x		
Minnesota		x			x		x			x			x				x	
Mississippi		x		x					x	x			x				x	
Montana			x	x					x	x			x				x	
Nebraska		x		x					x	x				x				x
Nevada	x			x					x	x			x				x	
New Hampshire			x		x				x		x			x			x	
New Jersey			x		x				x			x		x			x	
New Mexico	x			x					x	x				x			x	
New York			x		x		x					x		x			x	
North Carolina			x		x		x					x		x			x	
North Dakota	x			x			x			x			x				x	
Ohio			x	x			x					x	x			x		
Oklahoma		x			x				x	x			x				x	
Oregon			x		x			x			x			x			x	
Pennsylvania			x	x			x				x			x		x		
Rhode Island		x		x			x				x		x			x		
South Carolina			x	x			x				x		x				x	
South Dakota	x			x					x	x			x				x	
Texas		x		x					x	x			x				x	
Utah	x			x					x			x	x					x
Vermont		x		x			x			x			x			x		
Virginia			x		x				x			x	x				x	
Washington			x	x			x			x				x				x
Wisconsin		x				x	x				x			x				x
Wyoming	x			x			x			x			x				x	

Source: University of Wisconsin.
ND = no deficiency.
Mod = moderate deficiency.
Sev = severe deficiency.

Magnesium deficiencies are easily corrected by applying magnesium minerals, such as kieserite or dolomite.

Micronutrients

The principal micronutrients and their deficiencies in soils of the United States are shown in Table 3. Even though the traditional micronutrients may be required only in minute quantities, deficiencies can lead to diseased crops and stunted livestock. See also entries on **Boron**; **Copper (In Biological Systems)**; **Iron**; **Manganese**; **Molybdenum (In Biological Systems)**; and **Zinc (In Biological Systems)**.

Much more detail on all aspects of fertilizers, including worldwide consumption, methods of application, etc., can be found in the "Foods and Food Production Encyclopedia," (D.M. Considine, editor), Van Nostrand Reinhold, New York, 1981.

Additional Reading

Anac, D., P. Martin-Prevel: *Improved Crop Quality by Nutrient Management,* Kluwer Academic Publishers, Norwell, MA, 1999.

Bacon, P.E.: *Nitrogen Fertilization in the Environment,* Marcel Dekker Inc., New York, NY, 1995.

Barrueco-Rodriguez, C.: *Fertilizers and Environment,* Kluwer Academic Publishers, Norwell, MA, 1995.

Bergstrom, L. and H. Kirchmann: *Carbon and Nutrient Dynamics in Natural and Agricultural Tropical Ecosystems,* CAB International, New York, NY, 1998.

Borges, R.J.: "Choose the Right Alloy for Fertilizer Acids," *Chem. Eng. Progress,* **82** (November 1992).

Bumb, B.L.: *Global Fertilizer Perspective, 1980–2000: The Challenges in Structural Transformation,* International Fertilizer Development Center, Muscle Shoals, AL, 1995. *http://www.ifdc.org/*

Foth, H.D. and B.G. Ellis: *Soil Fertility,* CRC Press, LLC., Boca Raton, FL, 1996.

Francis, M.: *Background Report on Fertilizer Use, Contaminants and Regulators,* DIANE Publishing Company, Collingdale, PA, 2000.

Havlin, J.L., S.T. Tisdale, and W. Helson: *Soil Fertility and Fertilizers: An Introduction to Nutrient Management,* Prentice-Hall, Inc., Upper Saddle River, NJ, 1998.

Hedrick, J.L.: *The Molecular and Cellular Biology of Fertilization,* Perseus Publishing, Boulder, CO, 1986.

Laegreid, M. and O. Bockman: *Agriculture, Fertilizers and the Environment,* Oxford University Press, Inc., New York, NY, 2000.

Leitch, D.: *Matching Fertilizer Application to Crop Requirements,* Chevron Chemical Company, San Francisco, CA, (Revised periodically).

Lemaire, G.: *Diagnosis of the Nitrogen Status in Crops,* Springer-Verlag, Inc., New York, NY, 1997.

Mortvedt, J.J., R.H. Follett, and L.S. Murphy: *Meister Publishing Company,* Willoughby, OH, 1999.

Prasad, R. and J.F. Power: *Soil Fertility Management for Sustainable Agriculture,* Lewis Publishers, Boca Raton, FL, 1997.

Raymond, J.R. and J.R. Postgate: *Nitrogen Fixation,* Cambridge University Press, New York, NY, 1998.

Rengel, Z.: *Nutrient Use in Crop Production,* The Haworth Press, Inc., Binghamton, NY, 2000.

Smil, V.: *Enriching the Earth: Fritz Haber, Carl Bosch, and the Transformation of Word Food Production,* MIT Press, Cambridge, MA, 2000.

Srivastava, H.S.S. and R.P. Singh: *Nitrogen Nutrition and Plant Growth,* Science Publishers, Inc., Gainesville, FL, 1998.

Staff: *Annual Fertilizer Review,* Food and Agriculture Organization (United Nations), Rome (Issued annually).

Staff: *Fertilizer Situation,* U.S. Department of Agriculture, Washington, DC, (Issued several times each year).

Staff: *Production Yearbook,* Food and Agriculture Organization (United Nations), Rome (Issued annually).

Staff: United Nations Industrial Development Organization: *Fertilizer Manual,* Kluwer Academic Publishers, Norwell, MA, 1998.

Stevenson, F.J. and M.A. Cole: *Cycles of Soils: Carbon, Nitrogen, Phosphorus, Sulfur Micronutrients,* 2nd Edition, John Wiley & Sons, Inc., New York, NY, 1999.

Tiessen, H.: *Phosphorus in the Global Environment: Transfers, Cycles, and Management,* Vol. 54, John Wiley & Sons, Inc., New York, NY, 1995.

FIBER (Dietary).   Cell-wall dietary fiber is a complex system composed of variable amounts of cellulose, other polysaccharides such as hemicellulose and pectin, and lignin. The precise composition and proportion of polysaccharide types is related to the plant source, stage of maturity, and growing conditions. The composition and physical properties of dietary fiber also may be affected by both postharvest physiological changes and food processing. Cereal grains, legumes, vegetables, and fruits are primary sources of dietary fiber. A smaller proportion of total dietary fiber comes from polysaccharides (gums and mucilages) added for their functionality in processed foods.

The variation in water solubility among polysaccharides results in varied physiological roles. Plant cell-wall polysaccharides and lignin provide insoluble dietary fiber (IDF); nondigestible storage polysaccharides, some pectic polysaccharides, and most of the functional additives contribute soluble dietary fiber (SDF).

A resurgence of interest in dietary fiber has been stimulated by epidemiological evidence of differences in colonic disease patterns between cultures with diets containing large quantities of fiber, and Western cultures having more highly refined diets. Many African countries, for example, are relatively free of diverticular disease, ulcerative colitis, hemorrhoids, polyps, and cancer of the colon. Whereas most interest has focused on the beneficial role of dietary fiber, there is also concern that high fiber diets may cause disturbances in the absorption of nutrients such as minerals (see **Mineral Nutrients**) and vitamins.

Terminology. Dietary fiber is the accepted terminology in the United States for nutritional labeling. Total dietary fiber (TDF) and its subfractions, insoluble dietary fiber (IDF) and soluble dietary fiber (SDF), are defined analytically by official methods.

Sources, Composition, and Structure of Dietary Fiber

Natural sources of fiber in the diet include fruits, vegetables, legumes, and cereal grain products. The insoluble fiber content of some processed foods and breads is supplemented by incorporation of purified cellulose, cereal brans, or other plant fiber preparations. Cellulose and its chemically modified derivatives; seaweed polysaccharides, alginates and carrageenans; seed mucilaginous polysaccharides, guar and locust bean galactomannans; highly complex plant exudate polysaccharides, gum arabic, tragacanth, and others; microbially synthesized polysaccharides, xanthan and gellan gum; pectins; and other plant polysaccharides are added to foods for a variety of purposes. Because these materials are also nondigestible, they contribute to the total effect of dietary fiber and are encompassed by its definition.

Dietary fiber is a mixture of simple and complex polysaccharides and lignin. In intact plant tissue these components are organized into a complex matrix, which is not completely understood. The physical and chemical interactions that sustain this matrix affect its physicochemical properties and probably its physiological effects. Several of the polysaccharides classified as soluble fiber are soluble only after they have been extracted under fairly rigorous conditions.

Any starch escaping digestion in the upper gastrointestinal tract also contributes to dietary fiber effects. Some food starches, and the amylose fraction in particular, are readily converted into a nondigestible or slowly digestible physical form under certain food processing conditions. These resistant starches are readily fermented by colonic bacteria. Small amounts of waxes, cutin, and minerals in fruits and vegetables contribute to total dietary fiber values but may be physiologically inert.

Physiological Properties

The beneficial effects of dietary fiber, including both soluble and insoluble fiber, are generally recognized. Current recommendations are for daily intakes of 20–35 g in a balanced diet of cereal products, fruits, vegetables, and legumes. However, the specific preventive role of dietary fiber in certain diseases has been difficult to establish, in part because dietary risk factors such as high saturated fat and high protein levels are reduced as fiber levels increase.

Dietary fiber is important in the functioning of the entire gastrointestinal (GI) tract and affects the structure and morphology of the intestine. The process of chewing insoluble fiber-rich foods increases salivation and the flow of gastric secretions. Fiber components that increase viscosity or gel, for example, guar gum and pectin, delay gastric emptying. Fibers exert buffering action and may alter gastric pH by their effect on gastrointestinal hormones. A reduced glycemic response, probably resulting from delayed absorption of glucose, is associated with various fiber fractions, particularly viscosity-enhancing fibers.

Dietary fiber has a pronounced effect on the characteristics of the fecal mass and on the rate of passage of digest through the GI tract. High fiber diets also play a role in the excretion of bile acids and cholesterol.

Physicochemical Properties

Several physicochemical properties of dietary fiber contribute to its physiological role. Water-holding capacity, ion-exchange capacity, solution viscosity, density, and molecular interactions are characteristics determined by the chemical structure of the component polysaccharides, their crystallinity, and surface area.

Sources of Dietary Fiber for Processed Foods

An increasing number of fiber sources are available for food processing, representing a diversity of sources and technological advances. Commercial food-grade purified cellulose products and cereal brans are used in food products to enhance the fiber content or for other functional purposes. Many purified or partially purified nondigestible polysaccharides are used in food systems for their physicochemical properties, for example, for viscosity or as suspending agents. These polysaccharides contribute to the dietary fiber content even though they are not used for that purpose. Fiber sources include commercial cellulose products (powdered cellulose, microcrystalline cellulose, cellulose derivatives, and bacterial cellulose), bran, sugar-beet pulp, soybean cotyledons, and pea fiber. Sources of functional polysaccharides contributing to SDF include seaweed or algal, microbial, plant exudate, legume seeds, and tubers.

B. A. Lewis
Cornell University

Additional Reading

Anderson, J.W., D.A. Deakins, T.L. Floore, B.M. Smith, and S.E. White: *Crit. Rev. Food Sci. Nutr.* **29,** 95 (1990).

Dreher, M.L.: *Handbook of Dietary Fiber, An Applied Approach,* Marcel Dekker, New York, NY, 1987.

Furda, I. and C.J. Brine, eds.: *New Developments in Dietary Fiber,* Plenum Press, New York, NY, 1990.

Kritchevsky, D., C. Bonfield, and J.W. Anderson, eds.: *Dietary Fiber, Chemistry, Physiology and Health Effects,* Plenum Press, New York, NY, 1990.

McCleary, B.V., and L. Prosky: *Advanced Dietary Fiber Technology,* Blackwell Publishers, Malden, MA, 2001.

Spiller, G.A.: *CRC Handbook of Dietary Fiber in Human Nutrition,* 3rd Edition, CRC Press, LLC., Boca Raton, FL, 2003.

FIBER GLASS.   Glass in fibrous form. The material generally has properties similar to the glass from which it is made except that the tensile strength may be increased up to over 100 times that of the base glass. History records the use of strands of glass for decorating vases by the early Egyptians. The famed Venetian craftsmen had a limited knowledge of drawing glass fibers, but it was not until the 1930s and 1940s that glass producers perfected a way to make fibers commercially.

Two forms of glass fibers are produced; a staple or short-length fiber or monofilament, and continuous strand composed of many-monofilaments bonded together in a threadlike form. The continuous strands are often chopped into short lengths, ranging from 1/8-inch to 2 inches (3 milli meters to 5 centimeters) or longer, and this product is referred to as chopped strand. Staple fibers are used for thermal and acoustical insulation. Continuous strands are used for yarn (as in fabrics), tire cord, and plastic reinforcement. Both thermoset and thermoplastic resins are reinforced by chopped strand. Some varieties of chopped strand are also converted to monofilament paper, which is utilized in roofing shingles, flooring materials, and other products. Products using continuous filaments have excellent tensile strength (as high as 400,000 pounds per square inch; 2759 megapascals) compared with organic fiber strengths of less than 150,000 psi (1034 megapascals).

The base glass is made by heating raw materials, such as silica sand, limestone, dolomite, clay, boric acid, soda ash, and other minor ingredients, in a high-temperature furnace. See also Glass. Typical glass-fiber compositions are given in Table 1. Fiber made from electrical-grade glass E is used most commonly for yarn, tire cord, and plastic reinforcement because of its high strength and electrical properties. Specialty glasses, although low in volume of production, fill important needs. S glass is a superior-strength glass primarily for defense applications, such as missile cases. C glass is more chemically resistant than E and is used for battery separator plates and chemical filters. Alkali glass A is used to some extent in the production of plastic reinforcement products.

Continuous-Fiber Products

In the "direct-melt" process, shown in Fig. 1, raw materials are fed to a tank furnace to convert the mixture to glass. The glass flows to forehearths, which have platinum-alloy bushings or spinnerettes in the bottom. The bushings contain many holes, or orifices, each of which supplies a small stream of molten glass from which monofilaments are drawn. Mechanical attenuation, which produces a forming package, is accomplished by attaching the fibers to a rotating drum which turns up to 20,000 peripheral feet (6,000 meters) per minute.

The "marble-melt" process consists of producing 1-inch (2.5-centimeter) marbles by a separate tank furnace. The marbles are then fed to a bushing unit, which is heated by electrical resistance. From this point, the process is identical to the direct-melt process.

Sizing. Because of the basic character of glass, the filaments are somewhat fragile and tend to abrade each other in close contact. A protective coating or sizing is necessary for the production, processing, and end use of all continuous-fiber glass products. Generally, a fiberglass sizing or binder for textile or reinforcement products may contain: (1) a film former, generally resinous in nature, that forms a strand or thread from grouped monofilaments; (2) a lubricant, to aid in processing and end use of the fiber-glass product; and (3) additives to accomplish specified purposes, e.g., providing antistatic characteristics.

Sizings for plastic reinforcement also will have a coupling agent, such as a chrome complex, a silane, or combination of these two, to assure an interfacial bond between the glass surface and the resin matrix. Yarns for weaving normally have an oil-starch sizing. These coatings are applied before winding of the forming package.

Filaments ranging in number from 20 to 2000 are then gathered together as a thread or strand before winding. As shown by Table 2, the filaments are available in many diameters and are letter-designated. In continuous-fiber products, filaments range from designation B to U.

Continuous-filament products are designated by a letter-number system, which specifies properties important to end users. For example, listing a strand as ECK67.5 (200) 630 indicates that the material is made from E glass and is a C continuous fiber of K diameter; the strand contains 200 monofilaments and has a yield of 67.5 × 100, or 6,750 yards per pound. (In the metric system, yield is expressed in Tex, or grams per kilometer. The number 486,235 divided by yards per pound is equal to Tex. Hence the yield in this case would be 72 Tex). This product is coated at the bushing with 630 binder, an oil-starch type making it suitable for weaving into fabric.

Forming packages composed of wound strands normally are not supplied to industrial users without further processing. Strands are twisted and plied before being woven into fabric. A plied yarn, for example, is coated with a latex binder before being used as a tire-cord reinforcement.

End products made from continuous strand include fire-resistant curtains, reinforced tires and transmission belts, and many reinforced plastic items, such as boats, auto bodies, corrosion-proof pipe, roofing panels, and missile cases.

Staple-Fiber Products. Monofilament, short-length fibers are used for thermal and acoustical insulation, filtration, and cushioning. These products are made in basically three ways. In the high-temperature blast-jet process, 30-mil-diameter fibers or rods first are produced by a bushingtype process. The coarse primary rods then are filamentized by a high-temperature, high-velocity blast burner. The blown mass of filaments is collected on a conveyor belt and bonded together by an inert thermosetting resin in a manner which creates many tiny air spaces throughout the material. The bonding process may be modified to produce flexible blankets, rigid board, or special molded shapes, such as pipe insulation. Coatings, facings, or jackets usually are applied for reflective, vaporbarrier, or decorative purposes.

In another process (replacing the high-temperature blast-jet process in some areas), a stream of molten glass is directed onto a rapidly rotating wheel which contains holes in its periphery. Centrifugal force directs glass through each hole to create fibers. A third process involves conversion of 2-to-6-inch (5-to 15-centimeter) chopped strand and textile-type yarns into separate and random monofilaments by a garnetting machine.

Properties of Staple Fibers. Fiber diameters range from AAA to G (Table 2), with the largest production volume in the range C−G. Thermal conductivity of glass-fiber products is influenced by fiber diameter, density or compactness of the fiber mass, and temperature conditions. Generally, thermal conductivity ranges between 0.20 and 0.80 (Btu)(in.)/(hr)(ft^2)(°F). In metric units, this is: 0.029 and 0.115 watt/meter-Kelvin W/m · K.

Temperature of applications ranges between near absolute zero to 593°C (1100°F), or to the softening point of the glass. Unbonded mat is used at extreme temperature conditions, whereas standard bonded insulation covers the range of −40 to 232°C (−40 to 450°F). Fiberglass acoustical products are particularly good energy absorbers at the frequency levels of 500−2000 Hz. Fiber diameter, density, and method of mounting control the absorbing characteristics.

TABLE 1. FORMULATIONS FOR TYPICAL FIBER GLASS TYPES

Ingredient	Type of glass. wt%				
	E	Insulating	A	S	C
SiO_2	54	63	73	64	65
Al_2O_3	14	5	1	24	4
MgO	4	2	2	10	3
CaO	19	6	10		14
R_2O	0.5	16	14	—	8
B_2O_3	8	7	—	0–2	6
Fe_2O_3	0.3				
F_2	0.2	1			

Note: R = rare-earth element.

TABLE 2. CODING SYSTEM FOR FIBER GLASS DIAMETERS

Designation	Diameter $\times 10^{-5}$ in.
AAA	<3.0
AA	3.0–5.9
A	6.0–9.9
B	10.0–14.9
C	15.0–19.9
DE	23.0–27.9
G	35.0–39.9
H	40.0–44.9
K	50.0–54.9
P	70.0–74.9
Q	75.0–80.0
R	81.0–85.0
S	86.0–90.0
T	91.0–95.0
U	96.0–100.0

Note: To convert to microns (micrometers), multiply by 25.4 × 10^{-2}.

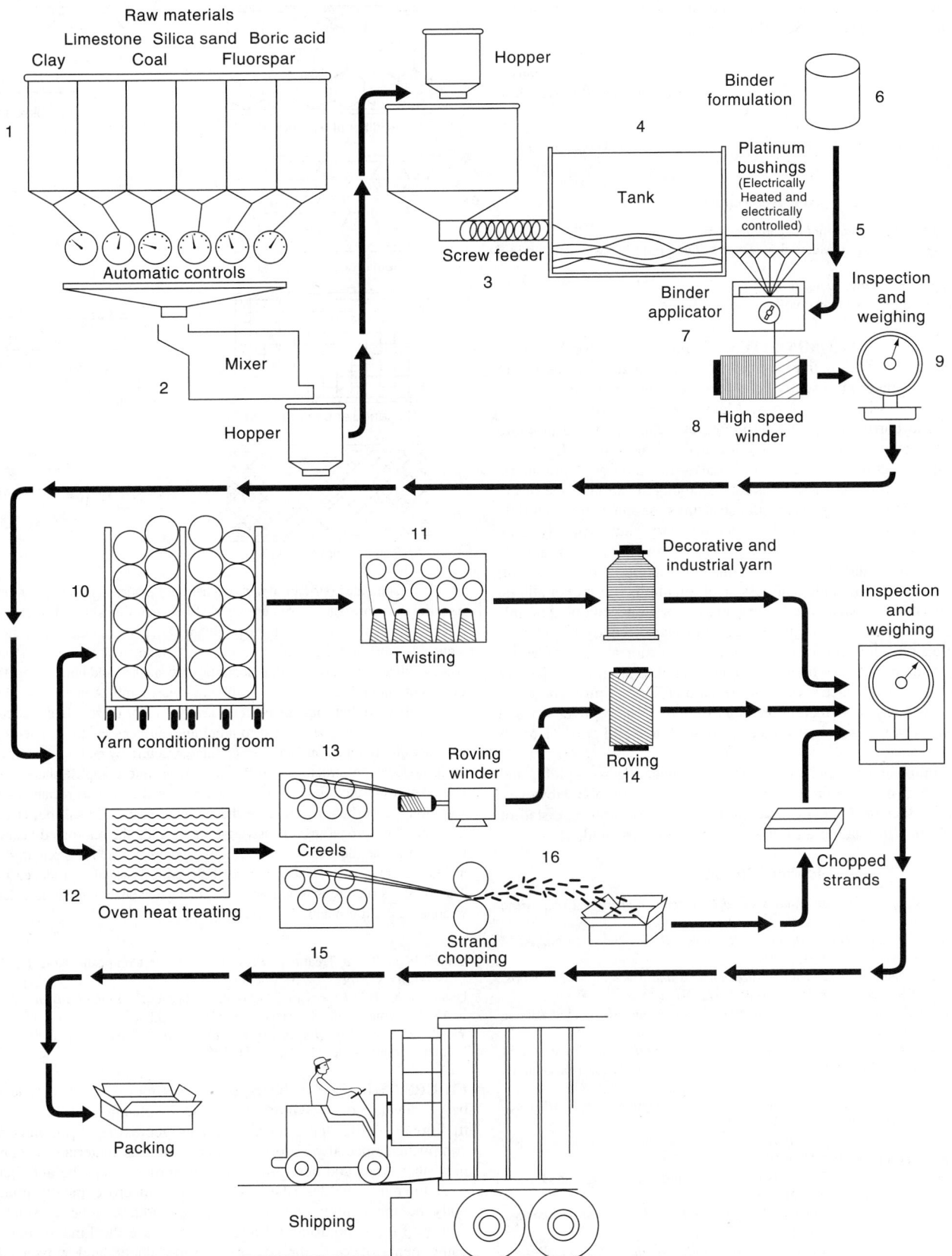

Fig. 1. Direct-melt process for producing fiber glass. Raw materials (1) are automatically weighed and batched to mixer (2) prior to passing through screw feeder (3) to the glass melting tank (4). The molten glass flows to forehearths (5), at the bottom of which are platinum-alloy bushings or spinners. The latter are electrically heated and carefully temperature-controlled. Formulated binder material (6) is applied to the newly formed filaments (7) prior to high-speed winding (8). After weighing and inspecting (9), the wound multi-filament (in the form of a strand) follows one of three paths in accordance with desired end product. In making decorative and industrial yarn, the package are placed in a conditioning room (10) prior to twisting (11). For the production of roving and chopped strand, the material from inspection operation (9) passes to an oven (12), where the filaments are heat-treated. This is followed by creeling (13) and roving winding (14) for production of roving. Following creeling (15), chopped strands (16) may also be made. There are several additional weighing and inspecting stations. (*PPG Industries*)

Glass fibers as used in lightwave communications are described under **Optical Fiber Systems**. See also Alphabetical Index.

L. Dow Moore
PPG Industries
Pittsburgh, Pennsylvania

Additional Reading

Loewenstein, K.L.: *The Manufacturing Technology of Continuous Glass Fibers,* 3rd Edition, Elsevier Science, New York, NY, 1993.

Web References

Johns Manville Inc. http://www.johnsmanville.com/
Owens Corning (Composite Materials), Toledo, OH. http://www.owenscorning.com/composites/
Owens Corning. http://www.owenscorning.com/
PPG Fiber Glass. http://www.ppg.com/frames/fibergls.htm

FIBER-REINFORCED COMPOSITES. Many advances were made in the 1980s toward the design and application of metal and other composite structures. These materials usually offer large advantages in strength-to-weight and stiffness-to-weight ratios. Many composites have excellent temperature- and corrosion-resistance properties. Aluminum and graphite are among the favored materials for constructing composites. However, the technology has applied scores of materials—metals, fibers, polymers, etc.—in the search for better materials for specific applications. One of the earliest and still the most popular cellular structures for composite materials is the hexagonal honeycomb. Cell structures vary, but typically they measure from 1.5 to 25.5 mm (0.06 to 1 in.) across. As shown in Figure 1, several other shapes and contours are made. In addition to aluminum, metals commonly used include stainless steel, titanium, and nickel alloys. Reinforced plastics are also used, particularly for honeycomb. Aramid-reinforced epoxy, phenolic, and polyimide cores were recently introduced. These materials offer low coefficient of thermal expansion and good electric and thermal insulation. Glass-fiber reinforced phenolics are now frequently specified for heat resistance, while some carbon-fiber reinforced resins also match the strength of metal-core honeycomb. Density of composite materials ranges rather widely—typically from 16 to 880 kg/m³ (1 to 55 lb/ft³).

Because composite materials technology is advancing so rapidly, and hence changing, it is not in order here to make a detailed probe. However, the interested reader of this encyclopedia will find the references listed of considerable value in making an initial assessment of the field.

Additional Reading

Agarwal, B.D. and L.J. Broutman: *Analysis and Performance of Fiber Composites,* John Wiley & Sons, Inc., New York, NY, 1990.

Bauccio, M.L.: *ASM Engineered Materials Reference Book,* 2nd Edition, ASM International, Materials Park, OH, 1994.

Bhagat, R., R. Arsenault, and S. Fishman: *Mechanisms and Mechanics of Composites Fracture,* ASM International, Materials Park, OH, 1993.

Carter, G.F. and D.E. Paul: *Materials Science and Engineering,* ASM International, Materials Park, OH, 1991.

CCM: "Challenges for Plastics/Composites," *Advanced Materials & Processes,* **31** (Report of Center for Composite Materials, University of Delaware), January, 1992.

Fujine, M.: "Alternate Materials Reduce Weight in Automobiles," *Advanced Materials & Processes,* **20** (June 1993).

Jang, B.Z.: *Advanced Polymer Composites: Principles and Applications,* ASM International, Materials Park, OH, 1993.

Karbhari, V.M. and D.S. Kukich: "Polymer Composites Technology in Japan," *Advanced Materials & Processes,* **26** (August 1993).

Mai, Y.W. and K. Jang-Kyo: *Engineered Interfaces in Fiber Reinforced Composites,* Elsevier Science, New York, NY, 1998.

Mallick, P.K.: *Fiber-Reinforced Composites: Materials, Manufacturing, and Design,* Marcel Dekker, Inc., New York, NY, 1993.

Mallick, P.K.: *Composites Engineering Handbook,* Marcel Dekker, Inc., New York, NY, 1997.

Rohatgi, P., Ed.: *Friction, Lubrication and Wear Technology for Advanced Composite Materials,* ASM International, Materials Park, OH, 1993.

Staff: *Advanced Composites: Design, Materials and Processing Technologies,* ASM International, Materials Park, OH, 1992.

Staff: *Advanced Synthesis of Engineered Structural Materials,* ASM International, Materials Park, OH, 1993.

Summerscales, J. *Microstructural Characterization of Fiber-Reinforced Composites,* CRC Press, LLC., Boca Raton, FL, 1998.

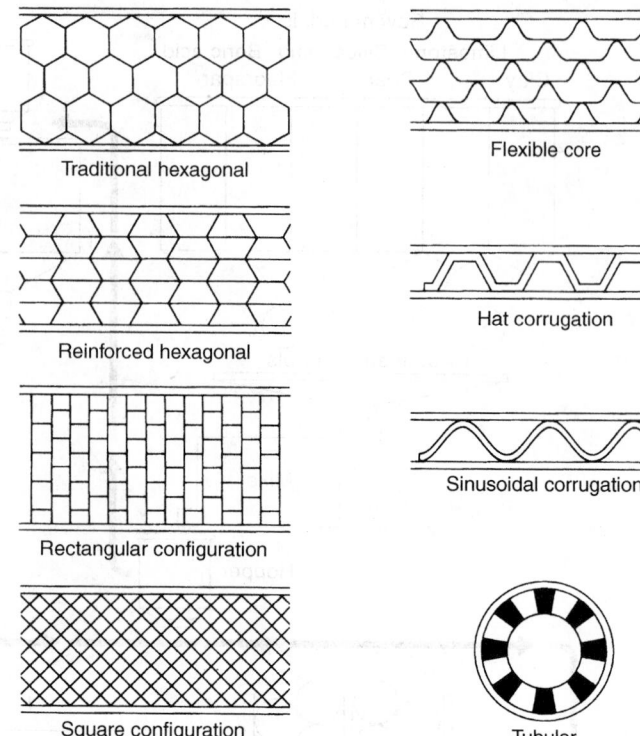

Fig. 1. Laminates (structural sandwiches) are made up of a core of either a solid material or a corrugated structure bonded between thin skins. The corrugated core often is filled with another lightweight material to provide added strength. The sinusoidal core and the "hat" core are two fundamental corrugated structures. For some applications, layers of core materials are stacked and bonded together to form even more complex structures, such as a honeycomb. Adhesive bonding is most commonly used, but some metal cores are welded or brazed. In structures from flat sheet, adhesive may be applied to adjoining surfaces in strips. Then the stack of laminates is stretched into an open cellular structure by pulling the plies apart. The basic honeycomb structure is difficult to form into complex shapes or contours. Overexpanding the plies to stretch the hexagonal cells into rectangles produces a variation on the honeycomb core that is more easily formed, but at the sacrifice of some shear properties. Other variations include the reinforced-hexagonal core, in which a flat sheet is inserted in the structure for added strength, but with an added weight penalty. The square-cell structure is usually produced by welding. The tubular structure is produced from corrugated sheet bonded to a flat sheet and wrapped around a mandrel. (*After Bittence*)

Unger, W.J.: "Heat Treatment Prevents Polymer Composite Cracking," *Advanced Materials & Processes,* **33** (October 1993).

Upadhya, K., Ed.: *Processing, Fabrication and Application of Advanced Composites,* ASM International, Materials Park, Ohio, 1993.

Woishnis, W.A., Ed.: *Engineering Plastics and Composites,* 2nd Edition, ASM International, Materials Park, OH, 1993.

FIBERS. The field of fibers is an evolving one, with new technologies being developed constantly. With the increasing use of fibers in non-traditional textile applications, such as geotextiles (qv), fiber-reinforced composites, specialty absorption media, and as materials of construction, new fiber types and new processing technologies can be anticipated.

A fiber may be described as a flexible, macroscopically homogeneous body having a high ratio of length to width and being small in cross section. Fibers, both natural and synthetic, are the fundamental structural components of yarn, thread, string, rope, paper, and woven and matted goods. Fibers usually exhibit considerable elasticity, the ability to return to their original dimension without permanent stretching. Most fibers tend to interlock or mechanically bond with other fibers, forming fiber matrices. Natural fibers are usually quite un-uniform; for example, cotton staple ranges from $\frac{1}{2}$ to $2\frac{1}{2}$ inches (1.3 to 6.4 centimeters) in length, with a diameter of about 1/1,000 inch (0.025 millimeter). Some synthetic fibers are made in the form of very thin filaments and thus may be quite uniform, with fiber length controlled and tailored for specific applications. Many materials required for the production of synthetic fibers are derived from petroleum and, along with synthetic plastics and resins, the synthetic fiber industry contributed importantly to the great growth of the chemical and petrochemical industries. To some extent, the essential raw materials for synthetic fibers are threatened because of competition for the same raw materials that are consumed for power in terms of fuels—both petroleum and natural gas.

The use of fibers dates back to antiquity. Very early uses (and still found among primitive peoples) are *tying* applications, using easily obtainable fibers from plants. Fibers have been used for centuries for making rough cordage, huts, and rope suspension bridges. Broomcorn and broomroot fibers have long been used for what might be termed *brush* applications. Straw, bamboo, rattan, and palm leaves were among the early *plaiting* and *rough-weaving fibers* for use in furniture making and basketry. Of course, they are still extensively used in various parts of the world for these purposes. The use of various reeds, husks, and grasses as *filling fibers* is very old, but large tonnages of such fibers still are used for packing materials, upholstery padding, and like applications.

The use of fibers in nonwoven sheet-like products is quite old, although there has been a resurgence of interest in so-called nonwovens during recent years. Felts, paper, and, more recently, some of the nonwoven disposable garments and products, notably for hospital use, are representative of the use of fibers for what might be termed *matting* and webbing applications. But by far the most important use of fibers, and the application most often visualized in connection with their use, is for the manufacture of knitted and woven textile fabrics and products. Spinning of the fibers to make yarn increases the utility of fibers for uses that far exceeded the imagination of those persons who first applied fibers in their cruder forms.

Although fibers can be classified in numerous ways, in terms of present-day technology, they are fundamentally classified as (1) natural fibers, and (2) synthetic fibers. The principal natural fibers are cotton, wool, and, to a much lesser extent, silk, flax, and mohair. Synthetic fibers have made inroads into the use of all natural fibers, but the greatest impact has occurred in connection with the latter three fibers. Cotton continues to be a major textile fiber, measured in terms of billions of pounds used per year. Cotton is one of the most versatile of all fibers and blends well with synthetics. This is also true of wool, but to a somewhat lesser extent. *Synthetic Fibers.* Introduced in 1910 as a substitute for silk, rayon was the first artificial or synthetic fiber. Rayon, of course, differs completely in chemical constitution from silk. Rayon typifies most reconstituted or synthetic fibers, which perform almost as well and, in a number of respects, far better than their natural "counterparts." Some of the more recently developed synthetic fibers have little if any resemblance to naturally available fibers and thus entirely new types of end-products with previously unobtainable end-qualities are available.

It is interesting to note that some authorities define a synthetic fiber as a "noncellulosic fiber of synthetic origin," a definition which excludes rayon and acetate. Other authorities, however, include rayon and acetate, along with nylons, polyesters, acrylics, and others, in the full spectrum of synthetic fibers. The reasoning behind the fine distinction is that, with cellulose-derived synthetics, one commences with a naturally fibrous material and grossly modifies it, whereas with most other synthetics, the starting materials are strictly chemicals that bear no relationship whatever to a fibrous structure, many of the starting ingredients actually being in the gaseous or liquid phase.

In classifying synthetic fibers, there is also a narrow, twilight zone between fibers and elastomers. There are elastomers with fiber-like qualities; and vice versa. For example, spandex is a fiber with rubber-like qualities. See also **Elastomers**.

Particularly in the period between the late 1930s and late 1950s, there was a vigorous development of new and different synthetic fibers, largely stemming from the research efforts of competing firms. Much of the technology was well guarded and there was an overwhelming tendency to give all new fibers tradenames rather than generic designations. This created much confusion, particularly in the marketplace. Some of the former tradenames were lost because they were widely and variously used and ultimately became the generic term for a class of fibers. As a tool for obtaining clarification, and for protecting buyers of various synthetic fibers, the U.S. Congress passed the Textile Fiber Products Identification Act, which is administered by the U.S. Federal Trade Commission. The principal synthetic fibers as so defined are listed in Table 1. For more complete definitions, refer to ASTM Standards on Textile Materials.

Many textile products combine the advantageous physical and chemical properties of two or more fibers (synthetic or natural). Careful preblending operations are usually required.

TABLE 1. GENERIC DESIGNATIONS AND DEFINITIONS OF SYNTHETIC FIBERS

ACETATE FIBERS

A manufactured fiber in which the fiber-forming substance is cellulose acetate. Where not less than 92% of the hydroxyl groups are acetylated, the term *triacetate* may be used as a generic description of the fiber. A portion of the molecule may appear as:

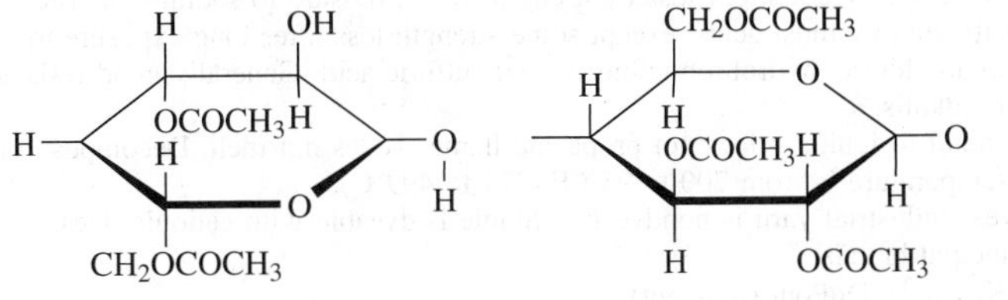

(acetate) (triacetate)

Specific gravity: 1.3–1.32
Moisture regain: 3.2% (triacetate); 6.3–6.5% (acetate)
Tensile strength: $18-22 \times 10^3$ psi (124–152 MPa) (triacetate)
  $20-24 \times 10^3$ psi (138–166 MPa) (acetate)
Excellent to impervious to aging. Good resistance to mildew discoloration and sunlight (acetate), although there may be some loss of strength from long exposure to sunlight. Triacetate has poor resistance to sunlight. Fair resistance to abrasion.
Attacked by strong oxidizing agents. Resists common solvents, normal hypochlorite, peroxide bleaching. Dissolves or swells in acetone, ketones, trichloroethylene, concentrated and glacial acetic acid, and methylene chloride.
Triacetate does not stick when ironed at cotton setting temperature, 450°F (232°C). Melts at 572°F (301°C). Acetate sticks at 350–375°F (177–191°C). Softens at 400–445°F (204–230°C). Melts at 500°F (260°C).
Dyes: Dispersed and developed dyes are commonly used. Acid dyes are used for printing. Solution dyed available.
Types available:
  Triacetate (filament)
  Acetate (filament and staple)
  See also **Fibers: Acetate**.

ACRYLIC FIBERS

A manufactured fiber in which the fiber-forming substance is any long-chain synthetic polymer composed of at least 85% by weight of acrylonitrile units:

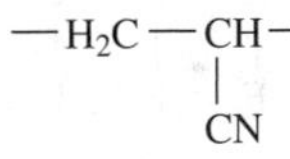

$$-H_2C-CH-$$
$$|$$
$$CN$$

A portion of the molecule may appear as:

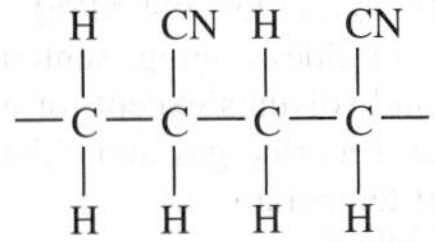

Specific gravity: 1.16–1.18
Moisture regain: 1.0–2.5%
Tensile strength: $30-54 \times 10^3$ psi (207–373 MPa)
Excellent resistance to mildew and aging. Good resistance to sunlight and abrasion.
Good resistance to bleaches and common solvents.
Generally good resistance to mineral acids and weak alkalis.*
Safe ironing temperature up to 300°F (150°C).*
Does not support combustion.
Dyes: Disperse and cationic.*
Principal brands:
  Acrilan®, Monsanto (staple)
  Creslan®, American Cyanamid (staple and tow)
  Mannacryl®, Mann Industries (staple, tow, and pulp)
  See also **Fibers: Acrylic**.

ARAMID FIBERS

A manufactured fiber in which the fiber-forming substance is a long-chain synthetic polyamide in which at least 85% of the amide linkages are attached directly to two aromatic rings. Amide linkage:

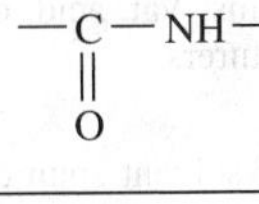

$$-C-NH-$$
$$\|$$
$$O$$

(continued overleaf)

TABLE 1. (*continued*)

Specific gravity: 1.38–1.44
Moisture regain: 4.5–7% (at 55% RH)
Tensile strength: $90-400 \times 10^3$ psi (621–2760 MPa)
Excellent resistance to mildew and aging. Prolonged exposure to sunlight causes deterioration, but fibers are self-screening. Good abrasion resistance.*
Some are degraded by bleaching; others are not affected. No degradation in solvents, except slight loss of strength from exposure to sodium chlorite.*
Unaffected by most acids, except some strength loss after long exposure to hydrochloric, hydrobromic, nitric, and sulfuric acid. Generally good resistance to alkalis.*
Difficult to ignite—does not propagate flame—does not melt. Decomposition temperature is from 700 to 930°F (371 to 499°C).*
Dyes: Industrial yarn is nondyeable. Staple is dyeable with cationic dyes.
Principal brands:
  Kevlar®, DuPont (filament)
  Nomex®, DuPont (staple, tow and filament)

CARBON FIBERS

A manufactured fiber made by pyrolysis of an organic precursor—rayon, polyacrylonitrile, or pitch in an inert atmosphere.
Specific gravity: 1.77 (1.96 for high-modulus)
Moisture regain: 0
Tensile strength: 515×10^3 psi (3554 MPa)
  360×10^3 psi (2484 MPa)
  270×10^3 psi (1863 MPa)
Does not melt. Oxidizes slowly in air at temperatures above 600°F (316°C).
Cannot be dyed.
Excellent resistance to acids and alkalies, even at high concentration and temperature.
Strong oxidizers will degrade fiber.
Inert to all known solvents, but poor resistance to hypochlorite.
Excellent resistance to mildew, aging, and sunlight. Poor resistance to abrasion.
Supplier:
  Celion^R, BASF Structural Materials (high-strength, high-modulus, and ultrahigh-modulus fibers).

FLUOROCARBON FIBERS

Fiber formed of long-chain carbon molecules whose available bonds are saturated with fluorine. A portion of the molecule may appear as:

```
     F   F   F   F   F
     |   |   |   |   |
  —C—C—C—C—C—
     |   |   |   |   |
     F   F   F   F   F
```

Specific gravity: 0.8–2.2
Moisture regain: 0
Tensile strength: $25-115 \times 10^3$ psi (173–794 MPa)
Good to excellent resistance to mildew, aging, sunlight, and abrasion.
Essentially inert to bleaches and solvents except for alkali metals at high temperature and/or pressure. Fluorine gas and chlorine trifluoride react with fibers at high pressures and temperatures.*
Essentially inert to acids and alkalis.
Very heat resistant. Usually can be safely handled from −350 to +550°F (−212 to +288°C).* Melts between 550 and 620°F (288 and 327°C).*
Dyes: Some cannot be dyed; others can be pigmented and dyed with selected solvent system.
Principal brands:
  Gore-Tex®, W.L. Gore (expanded PTFE staple, filament, tow, and slit film-RT)
  Teflon® DuPont (TFE multifilament, staple, tow and flock; FEP monofilament)

GLASS FIBERS

A manufactured fiber in which the fiber-forming substance is glass.
Specific gravity: 2.48–2.69
Moisture regain: None
Tensile strength: $313-700 \times 10^3$ psi (2160–4830 MPa)
Not attacked by mildew, although binder may be affected by it. Excellent resistance to aging and sunlight.
Unaffected by bleaches and solvents.
Resists most acids and alkalis.
Nonburning. Generally holds 75% tensility up to 650°F (343°C). Softens between 1560 and 1778°F (843 and 970°C). Melts at 2720°F (1493°C).
Dyes: Resin-bonded pigment systems. Vat, acid, or chrome dyes will tint.
Available from numerous manufacturers.

LYOCELL FIBERS

A manufactured fiber composed of solvent spun cellulose.

Specific gravity: 1.56
Moisture regain: 11.5%
Tensile strength: Not available
Attacked by mildew, but has good resistance to aging, sunlight, and abrasion.
Attacked by strong oxidizing agents. Not damaged by bleaches.
Generally insoluble in common organic solvents.
Hot dilute or cold concentrated acids disintegrate the fiber, similar to cotton. Strong alkaline solutions cause swelling and reduce strength. Can be mercerized.
Does not melt. Loses strength at about 300°F (572°F) and begins to decompose at about 350°F (171°C) under extended periods of exposure.
May be dyed with all classes of dyes normally used for dyeing cellulosic fibers.
Supplied by:
  Tencel®, Courtaulds

MODACRYLIC FIBERS

A manufactured fiber in which, when not qualified as rubber or anidex, the fiber-forming substance is any long-chain synthetic polymer composed of less than 85%, but at least 35% by weight of acrylonitrile units. A portion of the molecule may appear as:

```
     H   H   H   H
     |   |   |   |
  —C—C—C—C—
     |   |   |   |
     H   Cl  H   CN
```

Specific gravity: 1.35–1.37
Moisture regain: 2.5–3.0%
Tensile strength: $29-47 \times 10^3$ psi (200–324 MPa)
Good to excellent resistance to mildew, aging, and sunlight. Good resistance to abrasion.
Good resistance to bleaches, dry-cleaning fluids, and most common solvents. Dissolves in warm acetone and acrylic-type solvents.
Resistant to most acids and good resistance to weak alkalis; some discoloration may result.*
Boiling-water shrinkage, about 1%. Good resistance to shrinkage in dry heat, with about 5% shrinkage at 390°F (200°C). Pressure and heat at 300°F + (150°C+) may cause stiffening and discoloration.*
Does not support combustion.
Dyes: Neutral-premetalized, cationic (basic), and disperse.*
Principal brands:
  SEF, Monsanto (staple)

NYLON FIBERS

A manufactured fiber in which the fiber-forming substance is a long-chain synthetic polyamide in which less than 85% of the amide linkages are attached to two aromatic rings. Amide linkage:

```
     —C—NH—
       ‖
       O
```

A portion of the Nylon 6,6 molecule, based upon hexamethylene diamine and adipic acid, may appear as:

```
    H  H  H  H  H  H      O  H  H  H  H  O
    |  |  |  |  |  |      ‖  |  |  |  |  ‖
  —N—C—C—C—C—C—C—N—C—C—C—C—C—C—
    |  |  |  |  |  |  |  |     |  |  |  |
    H  H  H  H  H  H  H  H     H  H  H  H
```

A portion of the Nylon 6 molecule, based upon caprolactam, may appear as:

```
    H  H  H  H  H  O      H  H  H  H  H  O
    |  |  |  |  |  ‖      |  |  |  |  |  ‖
  —N—C—C—C—C—C—C—N—C—C—C—C—C—C—
    |  |  |  |  |        |  |  |  |  |
    H  H  H  H  H        H  H  H  H  H
```

Specific gravity: 1.03–1.14 (most = 1.14)
Moisture regain: 2.8–5%*
Tensile strength: $40-134 \times 10^3$ psi (276–925 MPa)*
Excellent resistance to mildew, aging, and good-to-excellent resistance to abrasion. Prolonged exposure to sunlight causes some deterioration.

TABLE 1. (continued)

Excellent resistance to bleaches and other oxidizing agents. Generally insoluble in most organic solvents except some phenolic compounds.

Strong oxidizing agents and mineral acids may cause degradation of some brands. However, generally unaffected by most mineral acids, except when hot. Dissolves with partial decomposition in concentrated solutions of hydrochloric, sulfuric, and nitric acids.* Substantially inert in alkalis.

Sticking temperature is about 445°F (229°C).* Melts between 480 and 525°F (249 and 274°C).* Some yellow slightly if held at 300°F (150°C) for several hours. Decomposes between 600 and 730°F (316 and 388°C).*

Dyes: Has marked affinity for all types of dyestuffs, including pigment, direct, acid, premetalized acid, disperse, chrome, and vat colors, including complex types.*

Principal brands:

Nylon 6, DuPont and others (staple, monofilament and filament-RT and -HT, staple and tow)

Nylon 6,6, DuPont and others (staple and tow; monofilament and filament-RT; and filament-HT)

OLEFIN FIBERS

A manufactured fiber in which the fiber-forming substance is any long-chain synthetic polymer composed of at least 85% by weight of ethylene, propylene, or other olefin units. A portion of the molecule may appear as:

$$-\overset{\underset{\displaystyle H}{|}}{\underset{\underset{\displaystyle H}{|}}{C}}-\overset{\underset{\displaystyle CH_3}{|}}{\underset{\underset{\displaystyle H}{|}}{C}}-\overset{\underset{\displaystyle H}{|}}{\underset{\underset{\displaystyle H}{|}}{C}}-\overset{\underset{\displaystyle CH_3}{|}}{\underset{\underset{\displaystyle H}{|}}{C}}-$$

(Polypropylene)

Specific gravity: 0.9–0.96.

Moisture regain: Negligible (polyethylene); 0.01–0.1% (polypropylene)

Tensile strength: $11-90 \times 10^3$ psi (76–621 MPa)*

Not attacked by mildew. Good to excellent resistance to sunlight, abrasion, and aging.

Resistance to bleaches and most solvents, but some swelling in chlorinated hydrocarbons at room temperature and dissolves at 160°F (71°C) and higher.*

Excellent resistance to acids and alkalis, with exception of oxidizing agents, such as chlorosulfonic acid and concentrated nitric acid.

Softens at 225–235°F (107–113°C) (polyethylene): at 285–330°F (141–166°C) (polypropylene).

Melts at 230–250°F (110–121°C) (polyethylene); at 320–350°F (160–177°C) (polypropylene).

Dyes: Traditionally fibers are pigmented during manufacture, but some can be dyed with disperse, acid, and chelating dyes and certain vats, sulfurs, and azoics.

Types available:

Polyethylene, Hercules (monofilament)

Herculon®, Hercules (staple, bulk filament)

Marves®, Alpha®, Phillips (staple, tow, multifilament)

Essera®, Amoco XXV®, Marquesa®, Lana®, Palton III, Amoco (staple, tow, multifilament)

Spectra 900 and 1000®, Allied (various types)

Fibrilon®, Synthetic Industries (fibrillated and monofilament)

See also **Olefin Fibers**.

POLYBENZIMIDAZOLE FIBERS

A manufactured fiber in which the polymer is a sulfonated poly(2,2'-m-phenylene-5,5'-dibenzimidazole).

Specific gravity: 1.43

Moisture regain: 15–20%

Tensile strength: 50×10^3 psi (345 MPa)

Will not ignite. Does not melt. Decomposes in air at 860°F (460°C). Retains fiber integrity and suppleness upon flame exposure. High char yield.

Excellent resistance to most acids and alkalies. Some loss of strength in strong alkalis at elevated temperature. Excellent resistance to organic solvents. Unaffected by most bleaches and solvents.

Natural fiber color is gold. Dyeable to dark shades with basic dyes following caustic treatment.

Good resistance to mildew and aging. Prolonged exposure to sunlight causes darkening and some loss of tensile strength. Good abrasion resistance.

Principal brands:

PBI®, Hoechst Celanese (staple)

POLYESTER FIBERS

A manufactured fiber in which the fiber-forming substance is any long-chain synthetic polymer composed of at least 85% by weight of an ester of a substituted aromatic carboxylic acid, including but not restricted to substituted terephthalate units:

$$p(-R-O-\overset{\underset{\displaystyle \|}{\displaystyle O}}{C}-C_6H_4-C-O-)$$

and parasubstituted hydroxybenzoate units:

$$p(-R-O-C-C_6H_4-C-O-)$$

A portion of the molecule may appear as

$$-O-\overset{\underset{\displaystyle H}{|}}{\underset{\underset{\displaystyle H}{|}}{C}}-\overset{\underset{\displaystyle H}{|}}{\underset{\underset{\displaystyle H}{|}}{C}}-O-\overset{\underset{\displaystyle O}{\|}}{C}-C_6H_4-\overset{\underset{\displaystyle O}{\|}}{C}$$

Specific gravity: 1.34–1.39 (most = 1.38)

Moisture regain: 0.4%

Tensile strength: $33-165 \times 10^3$ psi (228–1139 MPa)*

Good to excellent resistance to mildew and sunlight, although prolonged exposure to full sunlight degrades some brands (strength loss). Abrasion resistance ranges from good to excellent.

Good to excellent resistance to bleaches, soaps, synthetic detergents, dry-cleaning agents, sea water, and perspiration. May be soluble in some phenolic compounds.

Sticking temperature is 440–445°F (227–230°C). Melts between 480–500°F (249–260°C).*

Good resistance to most mineral acids. Dissolves with partial decomposition in concentrated sulfuric acid. Good resistance to weak alkalis. Moderate resistance to strong alkalis.*

Dyes: Disperse, azoic, and cationic dyes.*

Principal brands:

A.C.E.®, Compet®, Allied (filament)

Dacron®, DuPont (staple and tow; partially oriented filament; filament-RT; filament-HT)

Fortrel®, Fiber Industries Div. Celanese (staple, RT AND HT; filament, RT and HT)

Kodel®, Eastman (staple)

Polyester, BASF (filament)

Trevira®, Hoechst Celanese (staple and filament)

See also **Polyester Fibers**.

RAYON FIBERS

A manufactured fiber composed of regenerated cellulose, as well as manufactured fibers composed of regenerated cellulose in which substituents have replaced not more than 15% of the hydrogens of the hydroxyl groups. A portion of the molecule may appear as:

$$\begin{array}{c} O \\ | \\ HC- \\ | \\ HCOH \quad\quad O \\ | \\ HOCH \\ | \\ HCOH \\ | \\ HC- \\ | \\ HCH \end{array}$$

Specific gravity: 1.46–1.54

Moisture regain: 11–13%

Tensile strength: $28-66 \times 10^3$ psi (193–455 MPa)*

Attacked by mildew. Resistant or stable to aging.* Mostly good resistance to sunlight and abrasion, but long exposure may yellow some intermediate rayons.*

Not affected by solvents. Insoluble in common organic solvents. Some attacked by strong oxidizing agents, but generally not damaged by hypochlorite or peroxide.*

Most rayons behave to acids much as cotton. Hot dilute or cold concentrated acids cause disintegration of fibers. Strong alkaline solutions cause swelling and reduce strength.*

TABLE 1. (*continued*)

Fibers do not melt, but may lose strength above 300°F (150°C) and decompose between 350 and 464°F (177 and 240°C).*

Dyes: Direct, vat, fiber-reactive, sulfur and pigment.*

Principal types and brands: Cuprammonium, available from several sources (filament).

Fibro®, Courtaulds (filament)

Rayon, North American Rayon Corp. (filament)

Saran®, Pittsfield Weaving (monofilament)

SPANDEX FIBERS

A manufactured fiber in which the fiber-forming substance is a long-chain synthetic polymer comprised of at least 85% of a segmented polyurethane. A portion of the molecule may appear as:

$$— O(CH_2)_4OCONH \underset{CH_3}{\bigcirc} NHCOO(CH_2)_4O —$$

Specific gravity: 1.2

Moisture regain: <1.0–1.3

Tensile strength: $11–15 \times 10^3$ psi (76–104 MPa)

Good to excellent resistance to mildew, aging, sunlight, and abrasion.

Good resistance to deterioration by bleaches, but some discolored slightly by hypochlorite bleaches. Resistant to solvents, including dry-cleaning fluids, and oils, except glycols.*

Good resistance to mild acids and alkalis, but may be degraded by strong acids and alkalis at high temperatures. Some are slightly yellowed by dilute hydrochloric and sulfuric acids.*

Sticking point from 347 to 420°F (75 to 216°C).* Melts at 511–518°F (267–269°C).*

Dyes: Good affinity for most classes of dyes, but disperse, acid, and premetalized dyes are generally preferred.*

Principal brands: Glospan/Cleerspan®, Globe (multifilament) Lycra®, DuPont (coalesced monofilament)

SULFUR FIBERS

Fiber-forming substance is a long-chain synthetic polysulfide with at least 85% of the sulfide linkages attached directly to two aromatic rings.

Specific gravity: 1.37

Moisture regain: 0.6%

Tensile strength: $35–40 \times 10^3$ psi (242–276 MPa)

Outstanding resistance to heat. Retains more than 70% of original strength after exposure to air at 400°F (204°C) for 5000 hours.

Excellent resistance to acids and alkalies, except hot, concentrated sulfuric acid and concentrated nitric acid.

Resistant to bleaches and solvents.

Nondyeable.

Excellent resistance to mildew, sunlight, aging, and abrasion.

Principal brands:

Ryton® Phillips (staple)

Characteristics of Major Natural Fibers

COTTON

General formula, $(C_6H_{10}O_5)_x$, Chemical composition is cellulose.

Specific gravity: 1.54

Moisture regain: 7.0–8.5%

Cotton is attacked by cold concentrated acids and by hot dilute acids. Alkalis cause mercerization, but without damage. Cotton is quite resistant to most solvents.

Cotton fabrics have an excellent hand, good abrasion resistance, excellent pilling resistance, excellent stability to repeated launderings (if preshrunk), fair sunlight resistance, excellent colorfastness, good wash and wear performance (if resin-treated), and good wrinkle resistance (if resin-treated).

Safe ironing temperature: 425°F (219°C)

Dyes used: Direct, vat, azoic, basic, mordant, pigment, sulfur, and fiber-reactive.

WOOL

General formula $(C_{42}H_{157}O_{15}N_5S)_x$. Chemical composition is keratin.

Specific gravity: 1.32

Moisture regain: 11–17%

Tensile strength: $17–29 \times 10^3$ psi (117–200 MPa)

Wool shows marked effects of thermal degradation above 212°F (100°C). Scorches at 400°F (204°C): chars at 570°F (299°C).

Wool is destroyed by hot sulfuric acid; otherwise it is quite resistant to acids. Strong alkalis destroy the material; it is attacked by weak alkalis. Wool is quite resistant to most solvents.

Tensile strength: $60–120 \times 10^3$ psi (414–828 MPa)

Cotton fibers are quite resistant to thermal degradation. After about 5 hours at 250°F (121°C), material yellows. Decomposes above 300°F (150°C).

Wool fabrics have an excellent hand, fair abrasion resistance (but good in carpets), good pilling resistance (pills form, but tend to break off), poor stability to repeated launderings, good sunlight resistance, good colorfastness, poor wash and wear performance, and good wrinkle resistance.

Safe ironing temperature: 300°F (149°C)

Dyes used: Acid, milling, chrome, mordant, vat, and indigo.

SILK

Silk is comprised essentially of the protein fibroin.

Silk fabrics have an excellent hand, fair abrasion resistance, good pilling resistance, good stability to repeated launderings, poor sunlight resistance, good colorfastness, poor wash and wear performance, and good wrinkle resistance.

Safe iron temperature: 300°F (149°C)

FLAX

A bast fiber.

Flax fabrics (linen) have an excellent hand, fair abrasion resistance, fair pilling resistance, good stability to repeated launderings, fair sunlight resistance, excellent colorfastness, very poor wash and wear performance, and poor wrinkle resistance.

Safe ironing temperature: 450°F (232°C).

---

* indicates variation of this property with particular brand of fiber.

Note: Unless otherwise noted, moisture regain is stated for a relative humidity of 65% at 70°F (21.1°C). RT = regular tenacity; HT = high tenacity; IT = intermediate tenacity.

See also **Acrylic Plastics**; **Elastomers**; **Fiber-Reinforced Composites**; **Polyester Fibers**; and **Polymers**.

Additional Reading

Bates, F.S.: "Polymer-Polymer Behavior," *Science*, **898** (February 22, 1991).

Brauman, J.I.: "Polymers," *Science*, **853** (February 22, 1991).

Chawla, K.K.: *Fibrous Materials,* Cambridge University Press, New York, NY, 1998.

Lewin, M., H.F. Mark, and J. Preston: *Handbook of Fiber Science and Technology: High Technology Fibers,* Vol. 3, Marcel Dekker, Inc., New York, NY, 1996.

Lewin, M.: *Handbook of Fiber Chemistry,* Marcel Dekker, Inc., New York, NY, 1998.

McAllister, I.: *Manmade Fiber Chart,* Textile World, Atlanta, Georgia, 1992.

Nakamura A.: *Fiber Science and Technology,* Science Publishers, Inc., Enfield, NH, 2000.

FIBERS: ACETATE. Cellulose acetate fiber, or *acetate*, is a chemical derivative of the naturally occurring polymer *cellulose*. Two types of acetate fibers are produced:

1. Fibers made from partially hydrolyzed cellulose triacetate, called *secondary acetate* (or simply acetate).
2. Fibers that are fully acetylated cellulose, called *triacetate*.

Chemistry

Highly purified cellulose wood pulp (greater than 95% alpha cellulose) is the basic raw material for making cellulose acetate. The natural polymer, cellulose, in wood pulp has a degree of polymerization of 500 to 1000, the basic repeating unit of which is cellubiose:

Cellubiose contains two anhydroglucose units connected by a B-1, 4-glucoside linkage. Each anhydroglucose unit contains two secondary alcohols in the 2 and 3 positions and one primary alcohol in the 6 position. Esterification of the three hydroxyl groups produces cellulose triacetate, but in practice, the commercial triacetate has a degree of substitution (DS) of 2.80–2.95, and secondary acetate has a DS of 2.35–2.40. The DS number is the average number of hydroxyl groups esterified per anhydroglucose unit. Since cellulose and acetic acid do not react directly to an appreciable extent, the ester is prepared by reacting cellulose with acetic anhydride in a solvent of glacial acetic acid, using sulfuric acid as a catalyst. The chemical reaction for converting cellulose to the triacetate thus is:

$$C_6H_7O_2(HO)_3 + (CH_3CO_2)_2O \longrightarrow$$

$$C_6H_7O_2(OCOCH_3)_3 + 3\ CH_3COOH$$

where $C_6H_7O_2(OH)_3$ is one anhydroglucose unit in the cellulose molecule.

The solution of triacetate is converted to secondary acetate through the addition of aqueous acetic acid, which results in hydrolysis of about 20% of the ester groups. Addition of water precipitates the secondary acetate and stops the deesterification. This procedure results in a random distribution of the acetyl and hydroxyl groups on the polymer. This, in turn, enhances the solubility of the secondary acetate in acetone. Acetone is commonly used as the solvent for spinning operations. The triacetate is produced by precipitating the product without hydrolysis.

Properties

Commercial acetate and triacetate yarns are produced in a range of 45–900 denier. Denier is a measure of the fineness of a yarn and is the weight in grams of 9000 meters of yarn. The largest portion is in the range of 55–150 denier. The filament count for these yarns ranges from 14 to 100 and the denier per filament ranges from 2.5 to 9. The yarns are produced in both bright and dull lusters, the latter type containing 1–2% titanium dioxide (TiO_2). Color-pigmented or "solution dyed" acetate yarns are also produced. The cross-sectional shape of acetate fibers is typically crenellated and circularly symmetric. The crenellated cross section results from the formation of a rigid skin during the initial filament formation when the fiber is largely fluid. As the acetone volatilizes through the skin, the fiber solidifies and shrinks. This results in skin folds or crenellations. Shaped fiber cross sections, such as Y shapes, are also produced and provide the capability for increasing the bulk of yarns and fabrics.

Key physical properties are given in the accompanying table.

PHYSICAL PROPERTIES OF ACETATE YARN

Property	Secondary acetate	Triacetate
Tenacity, g/denier:		
Conditioned	1.2–1.4	1.1–1.3
Wet	0.8–1.0	0.8–1.0
Breaking elongation, %:		
Conditioned	25–45	26–35
Wet	35–50	30–40
Density, g/cm³	1.32	1.30
Percent moisture regain (65% relative humidity at 22°C)	6.3–6.5	3.2

Chemical Properties and Dyeing. Acetate and triacetate fibers are readily dyed with disperse and azoic dyestuffs. Acid dyes may be used for print dyeing the triacetate fibers. Acetate and triacetate fibers, being organic esters, are susceptible to hydrolysis in strong alkaline solutions. Weakly basic or acid solutions have little effect on these fibers. Severe degradation occurs through hydrolysis when these fibers are subjected to strong mineral acids. Both the secondary acetate and triacetate show strong resistance to hypochlorite and peroxide bleaches. The fibers are not affected by normal dry-cleaning solvents.

Thermal Properties. Secondary acetate, having a random substitution of the hydroxyl groups, shows very little crystallinity and cannot be induced to crystallize or heat set. On the other hand, triacetate is a stereospecific molecule capable of crystallizing when heated above 205°C, the glass transition temperature, for a short period. Hence, triacetate offers heat-setting advantages in fabric finishing processes. Triacetate fabrics, heated at 240°C for 30 seconds, in a state of tension, exhibit good retention of the flat geometry and resist wrinkling at lower temperatures. Pleats and creases applied to triacetate garments, which are then heated above the glass transition temperature, will be retained over the normal period of usage. The melting points for secondary acetate and triacetate are 260 and 300°C respectively. Secondary acetate softens in the range of 205–230°C, which necessitates a moderate ironing temperature. Triacetate does not soften to the sticking temperature of 232°C and hence can accept ironing temperatures usually used for cotton.

Manufacture of Acetate Fibers

Acetic anhydride is produced by some of the acetate fiber producers because cellulose acetate is the major end user for this chemical. The anhydride is the product of reaction with ketene and acetic acid. Ketene is made by the catalytic pyrolysis of either acetic acid or acetone. The manufacturer of acetate fibers proceeds through a number of steps as delineated below.

(1) Wood pulp (high alpha type) in roll or sheet form is shredded to permit rapid penetration of the pulp by the reactants.

(2) The shredded pulp is treated with glacial acetic acid which may contain part or all of the sulfuric acid catalyst. This "pretreatment" swells the cellulose fibers and increases the accessibility of the acetylating agent. Ratios of acetic acid to cellulose may be 1:1 to 3:1 and the treatment time may vary from one-half to several hours, depending upon temperature.

(3) The activated pulp is then treated with the acetylation solution, a mixture of acetic anhydride and acetic acid in about 2:3 ratio and containing the remainder of the sulfuric acid catalyst (5–20% based on cellulose). Since the reaction is exothermic and it is necessary to keep the reaction temperature low (<50°C), precooling of the acetylation mixture and/or external cooling of the reaction vessel are employed. Reaction times of 5 to 10 hours, under controlled temperature, result in the pulp mass dissolving in the acetylating mixture. At this point, the cellulose is completely acetylated in the triacetate form. If triacetate is the desired product, aqueous acetic acid is applied to destroy the anhydride and the product is precipitated, washed, and dried.

(4) In producing the secondary acetate, the excess anhydride is reacted with aqueous acetic acid under careful temperature control. When all the anhydride is reacted, excess water is added to the extent of 10–30% of the mixture. The acetate is allowed to hydrolyze until a DS of 2.35–2.40 is obtained. Modern practice permits hydrolysis to occur in 4 to 8 hours at temperatures of 70–80°C. Sulfur esters are also hydrolyzed and sodium or magnesium acetate may be introduced with the hydrolysis water to neutralize the acid produced. Additional water precipitates the secondary acetate and control is exercised to obtain an open "flake" rather than a pelletized material. The open flake makes more efficient the removal of acetic acid in subsequent washing. The flake is dried and conveyed to storage bins.

Cellulose acetate is converted into fibers by solvating the polymer in an acetone solution which is then spun by a process called *dry spinning*. In the spinning process, the polymer solution is metered through a spinneret into a column containing warm air which evaporates the acetone. The solidified filaments are oiled and package collected at the base of the spinning tube.

(5) Preparation of the polymer spinning solution or "dope" begins with blending of flake batches to ensure uniformity of chemical composition, particularly acetyl content, which is important for uniform dyeing properties of the spun yarn. Either batch or continuous mixers are used for dissolving the blended flake in acetone-water solution. The final composition of the "dope" is about 26% acetate, 72% acetone, and 2% water. The water serves as a cosolvent with the acetone, leading to a marked reduction in the viscosity of the solution, which is approximately 900–1000 P (poises). A delustrant, such as titanium dioxide (TiO_2) may be added to the dope in concentrations of 1–2%, based on the acetate content. Colored pigments may also be added to the dope to produce mass-dyed fibers.

(6) The viscous "dope" solution is filtered several times to eliminate insoluble fiber and debris. Plate-and-frame as well as continuous filtration may be used. After filtration, the dope is transported to storage tanks where it is permitted to deaerate just prior to spinning.

(7) The deaerated dope is pumped to the spinning machine headers which distribute the dope to the individual spinning positions. Each machine has a hundred or more spinning positions where a single yarn of acetate yarn is produced. A metering pump provides accurate control of the quantity of dope pumped through a small line filter and heat exchanger which raises the temperature to 50–70°C just prior to reaching the spinneret. Spinnerets are made of stainless steel and incorporate 14–100 or more holes. The holes are 0.0015–0.0025 inch (0.04–0.06 millimeter) in diameter, having a depth slightly less than their diameter.

(8) The nascent acetate filaments leaving the jet are carried through the vertical spinning tube which may be 10–20 feet (3–6 meters) long and from 6 to 12 inches (15 to 30 centimeters) in diameter. Hot air moving concurrent or countercurrent to the path of the filaments evaporates most of the acetone and solidifies the liquid stream into the acetate filament.

(9) The acetone-laden air removed from the spinning tubes is stripped of acetone as it passes through activated carbon absorbers. The absorbers are steam-stripped and the acetone recovered by distillation. An acetone recovery efficiency of 96% overall is commonly experienced in the industry.

(10) As the yarn leaves the spinning tube, it is passed over an oiling device which applies 2–3% of lubricant of the weight of the yarn. The lubricant serves to reduce friction in subsequent textile processing and to diminish static electrification. From the oiler, the yarn passes over a rotating *godet* or wheel which supplies the force to transport the yarn from the jet to the take-up mechanism which packages the product.

Spinning speeds for acetate yarn range from 492 to 2297 feet (150 to 700 meters) per minute. Since acetone must be removed from the yarn in the spinning tube, the spinning speed depends upon the denier of the yarn, air velocity and temperature, dope temperature, and composition and spinning tube length.

Spinning of cellulose triacetate yarns follows the procedures used for the secondary acetate except that a different solvent is used. This is generally a 90:10 mixture of methylene chloride and methanol.

The manufacture of acetate staple and tow follow the same scheme as employed for continuous filament yarns, except that the yarns are combined as they leave the spinning tube to form a tow which is mechanically crimped and cut into staple lengths. Acetate and triacetate are packaged in bales of about 400 pounds (181 kilograms).

JOSEPH W. SCHAPPEL
Avtex Fibers, Inc.
Front Royal, Virginia

FIBERS: ACRYLIC.

During the 1970s there was rapid growth of acrylic fiber production in Japan, Eastern Europe, and developing countries. By 1981 an estimated overcapacity of approximately 21% had developed. This overcapacity is expected to decrease through the 1990s with continued increases in world production balanced by markets opening in China, Eastern Europe, Russia, and the Americas. Acrylics retain their traditional market, in knitted goods, like sweaters, and men's half-nose, and carpets. In addition, small market shares have developed in many new areas, such as carbon fiber precursors, asbestos replacement fibers, and conductive/metallized fibers.

Physical Properties

Acrylic and modacrylic fibers are sold mainly as staple and tow products with small amounts of continuous filament fiber sold in Europe and Japan.

Staple lengths may vary from 25 to 150 mm, depending on the end use. Fiber deniers may vary from 1.3 to 17 dtex (1.2 to 15 den); 3.2 dtex (3.0 den) is the standard form. The appearance of acrylics under microscopical examination may differ from that of modacrylics in two respects. First, the cross sections of acrylics are generally round, bean-shaped, or dogbone-shaped. The modacrylics, on the other hand, vary from irregularly round to ribbon-like. The modacrylics may also contain pigment-like particles of antimony oxide to enhance their flame-retardant properties.

The physical properties of these fibers are compared with those of natural fibers and other synthetic fibers in Table 1.

Chemical Properties

Among the outstanding properties of acrylic fibers is the very strong resistance to sunlight. Acrylic fibers are also resistant to all biological and most chemical agents. In terms of resistance of acrylic fibers to oxidizing agents, Orlon acrylic has been compared to nylon, cotton, and acetate yarns. Acrylic fibers, when heated, discolor and decompose rather than melt, but they have very good color and heat stability at temperatures less than 120°C. The excellent chemical resistance of acrylic fibers may stem from its unique laterally bonded structure.

Flammability

A most important property of acrylic and modacrylic fibers is their flammability and ignition behavior. Fibers used in textiles must not ignite readily when placed in contact with a flame. In this respect acrylic fibers compare favorably to the other natural and synthetic fibers currently on the market. There are, however, significant differences in the burning characteristics of the various fiber types. Cotton and rayon, for example, burn with the formation of a char. Nylon, polyester, olefin, wool, and acrylics, on the other hand, burn and melt simultaneously. More rigorous standards are required for end uses such as carpets, sleepwear, draperies, and bedding. Fibers for these applications must also be self-extinguishing after removal from the ignition source. The modacrylics, e.g., SEF and Kanekalon, melt and self-extinguish. This is generally achieved in acrylonitrile-based fibers by incorporating halogen comonomers such as vinylidene chloride, vinyl chloride, and vinyl bromide.

Fiber Identification

Although visual and microscopical examination, together with simple manual tests, are still the primary methods of identification, there are many new sophisticated instrumental methods available, based on chemical and

TABLE 1. PHYSICAL PROPERTIES OF STAPLE FIBERS

Property	Acrylic	Modacrylic	Nylon-6,6	Polyester	Polyolefin	Cotton	Wool
sp gr	1.14–1.19	1.28–1.37	1.14	1.38	0.90–1.0	1.54	1.28–1.32
tenacity, N/tex[a]							
dry	0.09–0.33	0.13–0.25	0.26–0.64	0.31–0.53	0.31–0.40	0.18–0.44	0.09–0.15
wet	0.14–0.24	0.11–0.23	0.22–0.54	0.31–0.53	0.31–0.40	0.21–0.53	0.07–0.14
loop/knot tenacity	0.09–0.3	0.11–0.19	0.33–0.52	0.11–0.50	0.27–0.35		
breaking elongation, %							
dry	35–55	45–60	16–75	18–60	30–150	<10	25–35
wet	40–60	45–65	18–78	18–60	30–150	25–50	
average modulus, N/tex[a]							
dry	0.44–0.62	0.34	0.88–0.40	0.62–2.75	1.8–2.65		
elastic recovery, %							
2% stretch	99	99–100		67–86		74	99
10% stretch		95	99	57–74	96		
20% stretch							65
electrical resistance	high	high	very high	high	high	low	low
static buildup	moderate	moderate	very high	high	high	low	low
flammability	moderate	low	self-extinguishing	moderate	moderate	spontaneous ignition at 360°C	self-extinguishing
limiting oxygen index	0.18	0.27	0.20	0.21		0.18	0.25
char/melt	melts	melts	melts, drips	melts, drips	melts	chars	chars
resistance to sunlight	excellent	excellent	poor; must be stabilized	good	poor; must be stabilized	fair; degrades	fair; degrades
resistance to chemical attack	excellent	excellent	good	good	excellent	attacked by acids	attacked by alkalies, oxidizing, and reducing agents moderate
abrasion resistance	moderate	moderate	very good	very good	excellent	good	
index of birefringence	0.1		0.6	0.16			0.01
moisture regain, std %	1.5–2.5	1.5–3.5	4–5	0.1–0.2	0	7–8	13–15

[a] To convert N/tex to gf/den, multiply by 11.3.

physical properties. These methods are able to distinguish between closely related fibers which differ only in chemical composition or morphology.

Instrumental Analysis. It is difficult to distinguish between the various acrylics and modacrylics. Elemental analysis may be the most effective method of identification. Specific compositional data can be gained by determining the percentages of C, N, O, H, S, Br, Cl, Na, and K. In addition the levels of many comonomers can be established using ir and uv spectroscopy.

Fiber Characterization

In addition to characterizing the many properties introduced by the choice of monomers and the polymerization process itself, considerable further characterization is required to quantitatively describe the properties imparted by spinning and subsequent downstream processing. These important properties relate to the crystalline order and microstructure of the fibers, and the resultant performance characteristics, such as crimp retention, abrasion resistance, mechanical properties, etc.

Acrylonitrile Polymerization

Except for fibers designed for industrial applications where resistance to chemical attack is of prime importance, all acrylic fibers are made from acrylonitrile combined with at least one other monomer. The comonomers most commonly used are neutral comonomers, such as methyl acrylate and vinyl acetate to increase the solubility of the polymer in spinning solvents, modify the fiber morphology, and improve the rate of diffusion of dyes into the fiber. Sulfonated monomers, such as sodium styrenesulfonate (SSS), sodium methallyl sulfonate (SMAS), and sodium sulfophenyl methallyl ether (SPME) are used to provide dyesites or to provide a hydrophilic component in water reversible crimp bicomponent fibers. Halogenated monomers, usually vinylidene chloride, vinyl bromide, and vinyl chloride, impart flame resistance to fibers used in the home furnishings, awning, and sleep-wear markets. Modacrylic compositions are used when the end use requires high flame resistance. Almost all of the modacrylics are flame-resistant fibers with high levels of halogen monomers.

Copolymerization

Homogeneous Copolymerization. Nearly all acrylic fibers are made from acrylonitrile copolymers containing one or more additional monomers that modify the properties of the fiber. Thus, copolymerization kinetics is a key technical area in the acrylic fiber industry. When carried out in a homogeneous solution, the copolymerization of acrylonitrile follows the normal kinetic rate laws of copolymerization. Comprehensive treatments of this general subject have been published. The more specific subject of acrylonitrile copolymerization has been reviewed. The general subject of the reactivity of polymer radicals has been treated in depth.

Heterogeneous Copolymerization. When copolymer is prepared in a homogeneous solution, kinetic expressions can be used to predict copolymer composition. Bulk and dispersion polymerization are somewhat different since the reaction medium is heterogeneous and polymerization occurs simultaneously in separate loci. In bulk polymerization, for example, the monomer swollen polymer particles support polymerization within the particle core as well as on the particle surface. In aqueous dispersion or emulsion polymerization the monomer is actually dispersed in two or three distinct phases: a continuous aqueous phase, a monomer droplet phase, and a phase consisting of polymer particles swollen at the surface with monomer. This affects the ultimate polymer composition because the monomers are partitioned such that the monomer mixture in the aqueous phase is richer in the more water-soluble monomers than the two organic phases.

Polymerization Methods. Acrylonitrile and its comonomers can be polymerized by any of the well-known free-radical methods. Bulk polymerization is the most fundamental of these, but its commercial use is limited by its autocatalytic nature. Aqueous dispersion polymerization is the most common commercial method, whereas solution polymerization is used in cases where the spinning dope can be prepared directly from the polymerization reaction product. Emulsion polymerization is used primarily for modacrylic compositions where a high level of a water-insoluble monomer is used or where the monomer mixture is relatively slow reacting.

Commercial Polymerization Methods. Aqueous media, such as emulsion, suspension, and dispersion polymerization, are by far the most widely used in the acrylic fiber industry. Water acts as a convenient heat-transfer and cooling medium and the polymer is easily recovered by filtration or centrifugation.

Processes using a continuous stirred tank reactor have replaced the semibatch process except where low volume specialty products are made. For startup, the reactor is charged with a certain amount of the reaction medium, usually solvent or pH adjusted water. In more sophisticated processes the start-up period may be minimized by filling the reactor with overflow from a reactor already operating at steady state. The reactor feeds are metered in at a constant rate for the entire course of the production run, which normally continues until equipment maintenance is needed. A steady state is established by taking an overflow stream at the same mass flow rate as the combined feed streams. The main advantage of this process over the semibatch process is that control of molecular weight, dye site level, and polymer composition is greatly improved.

The only other commercial polymerization process used for acrylic fibers is solution polymerization. This type of process can be implemented by feeding the monomers to a continuous mixing tank along with a solvent for the polymer. The overflow stream from this tank is then routed to a form of continuous reactor where the polymerization is carried out in a homogeneous solution. Monomer is removed from the product stream and the resulting polymeric solution is used directly for spinning. An obvious advantage of this process is that considerable cost savings can be achieved by eliminating the filtration, drying, and dope making steps required in the aqueous dispersion process. There are two drawbacks associated with solution polymerization. First, it is difficult to produce dopes of high solids, particularly with organic solvents. Second, most of the effective solvents have very high chain-transfer constants, making it difficult to produce polymer of high molecular weight. Solvents suitable for this type of commercial polymerization are DMF and DMSO.

Because of the highly exothermic nature of acrylonitrile polymerization, bulk processes are not normally used commercially. However, a commercially feasible process for bulk polymerization in a continuous stirred tank reactor has been developed. The heat of reaction is controlled by operating at relatively low conversion levels and supplementing the normal jacket cooling with reflux condensation of unreacted monomer.

The problems of monomer recovery, reaction medium viscosity, and control of reaction heat are effectively dealt with by the process design of Montedison Fibre. This process produces polymer of exceptionally high density. Thus, though the polymer is still swollen with monomer, the medium viscosity remains low because the amount of monomer absorbed in the porous areas of the polymer particles is greatly reduced. The process is carried out in a CSTR with a residence time, Q, such that the product k/d $\times$ Q is greater than or equal to $1.k_d$ is the initiator decomposition rate constant. This condition controls the autocatalytic nature of the reaction because the catalyst and residence time combination assures that the catalyst is almost totally expended in the reactor.

Solution Spinning

One of the principal problems in early commercialization of acrylic fibers was the lack of a suitable spinning method. The polymer cannot be melt spun, except possibly at high pressure in the presence of water. Solution spinning was the only feasible commercial route. The first real breakthrough occurred at DuPont in 1948 when dimethylformamide (DMF) was used to commercialize Orlon. DMF is still the most important spinning solvent for acrylic fibers, but other effective solvents are in wide use.

Dry Spinning. This is the process first employed commercially by DuPont in 1948. The process is similar to early acetate fiber spinning. The dope is pumped through spinnerettes with 200–900 holes placed at the top of a solvent drying tower. The solvent is removed by circulating an inert gas through the tower at 150–300°C. However, unlike acetate spinning which employs a low boiling solvent (acetone), acrylic solvents are all high boiling and hard to extract completely in the drying tower. Consequently, the fiber from the bottom of the drying tower contains 10–25% solvent. This is removed in a second step by passing the threadline through a hot water bath and possibly by a series of wash rolls. Vapor from the drying tower and wash water from the secondary solvent extraction are routed to a solvent recovery plant where the solvent and wash water are recovered and purified for reuse. This is essential for economic reasons, but environmental protection also requires recovery rather than discharge.

Wet Spinning. Wet spinning differs from dry spinning primarily in the way solvent is removed from the extruded filaments. Instead of evaporating the solvent in a drying tower, the fiber is spun into a liquid bath containing a solvent/nonsolvent mixture called the coagulant. The solvent is almost always the same as the solvent used in the dope and the nonsolvent is usually water.

Tow Processing. After the spinbath step the tow processing is similar for both wet and dry spun yarns. Wet spun yarns, however, may contain 100 to 300% of solvent/nonsolvent mixture per dry pound of fiber whereas dry spun tows generally hold only 10–30% solvent. Therefore, the initial washing steps differ in their details. The key tow processing steps are washing, stretching, finish application, collapse, drying, crimping, and relaxing.

Modification of Properties

Handle. For a given fabric construction the denier, compliance, cross-sectional configuration, degree of crimp, moisture absorption, and surface smoothness of the fibers all influence the softness of the final product. Very fine filament deniers are effective where good draping, anticrease properties, and softness of handle are desired. Softness of handle can also be achieved through modifying the fiber cross section. Flattened cross sections reduce the bending modulus. Sheath-core spinning, conjugate spinning, and drawing into modified spinbath compositions also yield modified handle.

Improved Comfort Properties. Wear comfort generally means cotton-like properties. The ability to absorb moisture from the skin and the softness of cotton fabrics are considered to be the two key properties for comfort. The extremely fine denier of cotton fibers accounts for its softness. Both properties can be achieved in acrylic fibers. Improved moisture retention can be achieved by incorporating hydrophilic comonomers that decrease ultimate fiber density, by modifying the fiber spinning process, or by using after-treatments such as modified finishes.

Reduced Pilling. Pilling can be reduced by increasing the likelihood that the pills will break or wear off. Thus the most effective approaches include reducing fiber strength, incorporating defects in the fiber, increasing fiber brittleness, and reducing shear strength.

Improved Hot—Wet Properties. Acrylic fibers tend to lose modulus under hot—wet conditions. Knits and woven fabrics tend to lose their bulk and shape in dyeing and, to a more limited extent, in washing and drying cycles as well as in high humidity weather. Moisture lowers the glass-transition temperature, T_g, of acrylonitrile copolymers and, therefore, crimp is lost when the yarn is exposed to conditions required for dyeing and laundering. A number of polymer and fiber modifications have been devised to overcome this problem, though none has been successful enough to allow acrylics to compete successfully in easy care apparel markets.

Improved Abrasion Resistance. Abrasion resistance is generally improved by reducing the microvoid size and increasing the fiber density. Abrasion-resistant fibers have been produced by incorporating hydrophilic comonomers or comonomers with small molar volumes. Sulfonated monomers, acrylamide derivatives, and N-vinylpyrrolidinone are some of the hydrophilic comonomers that can be used to reduce void content; vinylidene chloride, with its relatively small molar volume, is effective in increasing fiber density. The spinning process itself has a significant effect on fiber density and abrasion resistance. Dry spinning, for example, is known to produce a denser fiber structure than conventional wet spinning.

Fiber Whiteness and Thermal Stability. The most effective route to improved whiteness and thermal stability is modification of the polymerization. The polymer must be linear and free of conjugated or unstable chemical structures formed by side reactions during the polymerization process.

Commercial Products

The majority of acrylic fiber production is 3.3–5.6 dtex (3 to 5 den) staple and tow furnished, undyed, in either bright or semidull luster. The principal markets are in apparel and home furnishings. Within the apparel sector these fibers find extensive use in sweaters and in single jersey, double-knit, and warp-knit fabrics for a variety of knitted outerwear garments such as dresses, suits, and children's wear. Other large markets for acrylics in the knit goods area are hand knitting yarns, deep pile fabrics, circular knit, fleece fabrics, half-hose, coarse-cut knitwear, and deep pile fabrics for blankets. Acrylics also hold a strong position in most broadwoven fabric categories including apparel and home furnishings for area rugs, carpets, curtains, and upholstery. Much of the growth in acrylics fibers usage has come from the replacement of wool.

Acrylics and modacrylics are also useful industrial fibers. Fibers low in comonomer content, such as Dolan 10 and DuPont's PAN Type A, have exceptional resistance to chemicals and very good dimensional stability under hot—wet conditions. These fibers are useful in industrial filters, battery separators, asbestos fiber replacement, hospital cubical curtains, office room dividers, uniform fabrics, and carbon fiber precursors. The excellent resistance of acrylic fibers to sunlight also makes them highly suitable for outdoor use. Typical applications include modacrylics, awnings, sandbags, tents, tarpaulins, covers for boats and swimming pools, cabanas, and duck for outdoor furniture.

Besides the standard staple and tow products, acrylic and modacrylic fibers are offered in many forms for specialized applications. Fibers with enhanced properties are in great demand. Yarn bulk is enhanced by using bicomponent or biconstituent fibers. Pilling is reduced by producing fibers that are more brittle, and yarns with exceptionally soft handle are produced by using fine denier fibers or fibers treated with special friction reducing finishes. There have been many efforts to develop premium products based on proprietary technology. Specialty fibers, such as ultrafine denier fibers, acid-dyeable, producer dyed, and pigmented fibers, ion-exchange fibers, metallized and semiconducting fibers, hollow fibers for apparel, coarse denier fibers for wigs, and simulated animal hair, are available. Other examples of specialty applications are the functional fabrics, e.g., antimicrobial fabrics. Other functional fabrics are made from moisture-repellent fibers, moisture-absorbent sheath-core fibers, antistatic fibers, and flame-resistant fibers.

Economic Aspects

Because of the rapid capital investment in acrylics that occurred in the early 1970s, there is a large excess capacity. Prices have consequently been soft since 1977. Since that time there has been only minimal investment in plants or equipment and a curtailment in research and development work.

Two U.S. producers remain. Monsanto, best known for its Acrilan acrylic fiber, is by far the largest and most diversified. With new spinning technology, originally introduced as Acrilan II, Monsanto now produces Duraspun, a fiber with enhanced abrasion resistance for the sock market. Other new products utilizing new spinning technology are Softlon and Ultrette, part of Monsanto's HP apparel line. Cytec (formerly American Cyanamid) continues to do well with Creslan acrylic and other premium products like MicroSupreme, a microfiber comparable to Monsanto's Fi-Lana. A third U.S. producer, Mann Industries, Inc. recently withdraw from the acrylic business and sold much of its technology to other producers, like Monsanto. The other U.S. producers are making carbon fiber precursors and carbon fiber. European producers like Courtaulch and Hoescht, are now attempting to sell more fiber in the profitable U.S. market.

Industrial use of acrylic fiber broadwoven goods has increased in recent years. The advantages are uv stability and its wide range of options for dyeing. The primary products are marine fabrics, awnings, and outdoor furniture. The worldwide picture for acrylics is summarized in Table 2.

In general, production is forecast to decline in the most developed countries and in regions that depend heavily on acrylic exports. Production appears to be shifting to the next generation of low cost producers, namely South and East Asia, the Middle East, Africa, and Eastern Europe. Countries such as Taiwan and South Korea are losing their competitive advantage to rising wages and prices.

The principal thrust of Japan's joint research programs is the development of process technologies to reduce conversion costs and development of high value-added products. Energy cost reduction is a prime concern. Opportunities exist in low energy consuming polymerization and spinning

TABLE 2. WORLDWIDE SYNTHETIC FIBER CAPACITY AND PRODUCTION, 1991–2001, 10^3 T

Year	Acrylic	Polyester	Nylon	Polypropylene
CAPACITY				
1991	2,870	11,200	4,700	3,170
1995	3,360	14,710	5,310	4,000
1998	3,540	15,650	5,410	4,310
2001	3,550	15,980	5,470	4,550
PRODUCTION				
1991	2,350	9,170	3,670	2,580
1995	2,620	11,250	4,100	3,230
1998	2,730	13,280	4,390	3,540
2001	2,790	15,050	4,710	3,850

technology. High productivity can be achieved by high speed polymerization and spinning, robotization, automation, multi-end spinning, and high speed crimping. Other strategies are withdrawal from unprofitable market sectors and consolidation of production and research and development effort into areas where special cost or property advantages seem apparent. New high volume markets are also being studied. These include acrylics for asbestos replacement, cement reinforcement, and geotextile applications. Products that have limited volume potential but have potential as high value-added premium products are carbon and graphite fiber precursors, high strength fibers, fibers for reverse osmosis, ion-exchange fibers, and functional fibers such as antimicrobial, antistatic, water-repellant, and highly reflectant fibers. Chief among the new products being sought are high strength, high toughness and high modulus fibers, and functional fabrics. Recent developments in man-made fiber technology have been reviewed.

RAYMOND S. KNORR
Monsanto Company

Additional Reading

Cernia, E.: "Acrylic Fibers," in H.F. Mark, S.M. Atlas, and E. Cernia, eds., *Man-Made Fibers*, Vol. III, Wiley-Interscience, New York, NY, 1968.

Frushour, B.G. and R.S. Knorr: "Acrylic Fibers," in M. Lewin and Eli M. Pearce, eds. *Handbook of Fiber Science and Technology*, Vol. IV, Fiber Chemistry, Marcel Dekker, Inc., New York, NY, 1985.

Kennedy, R.K.: "Modacrylic Fibers," in H.F. Mark, S.M. Atlas, and E. Cernia, eds., *Man-Made Fibers*, Vol. III, Wiley-Interscience, New York, NY, 1968.

Masson, J.C.: *Acrylic Fiber Technology and Applications*, Marcel Dekker, Inc., New York, NY, 1995.

Moncrieff, R.W.: *Man-Made Fibers*, 6th Edition, John Wiley & Sons, Inc., New York, NY, 1975.

FIBERS: CELLULOSE ESTERS. The predominant cellulose ester fiber is cellulose acetate, a partially acetylated cellulose, also called acetate or secondary acetate. It is widely used in textiles because of its attractive economics, bright color, styling versatility, and other favorable aesthetic properties. However, its largest commercial application is as the fibrous material in cigarette filters, where its smoke removal properties and contribution to taste make it the standard for the cigarette industry. Cellulose triacetate fiber, also known as primary cellulose acetate, is an almost completely acetylated cellulose. Although it has fiber properties that are different, and in many ways better than cellulose acetate, it is of lower commercial significance primarily because of environmental considerations in fiber preparation.

Polymer Characteristics

Cellulose triacetate is obtained by the esterification of cellulose with acetic anhydride. Commercial triacetate is not quite the precise chemical entity depicted as (1) because acetylation does not quite reach the maximum 3.0 acetyl groups per glucose unit. Secondary cellulose acetate is obtained by hydrolysis of the triacetate to an average degree of substitution (DS) of 2.4 acetyl groups per glucose unit. There is no satisfactory commercial means to acetylate directly to the 2.4 acetyl level and obtain a secondary acetate that has the desired solubility needed for fiber preparation.

$$OOCCH_3 \quad CH_2OOCCH_3$$
$$O_2CCH_3 \qquad O_2CCH_3$$
$$CH_2OOCCH_3 \qquad OOCCH_3$$

(1)

The degree of acetylation is specified by two separate terms: acetyl value (%) and combined acetic acid (%). The ratio of these two values is always 43:60, reflective of the molecular weight ratio of the acetyl group to acetic acid. Commercial cellulose triacetate has a combined acetic acid content of 61.5%, corresponding to 2.92 acetyl groups per glucose unit. Cellulose acetate, with 2.4 acetyl groups per glucose unit, has a combined acetic acid content of approximately 55%.

Fiber Properties

Mechanical Properties. Acetate and triacetate have a tenacity in the range of 0.10–0.12 N/tex (1.1–1.4 gf/den) with a breaking elongation of about 25–30%. Compared to other common textile fibers, acetate and triacetate are relatively weak, e.g., 20–25% the tenacity of polyester. This is not necessarily a disadvantage, because fabric construction can be used to obtain the desired fabric performance targets. Pilling, the accumulation of fuzz balls on the fabric with wear, is not a problem as it is with the higher tenacity fibers.

This modulus of elasticity, or Young's modulus, is related to many of the mechanical performance characteristics of textile products. The modulus of elasticity can be affected by drawing, i.e., elongating the fiber; environment, i.e., wet or dry, temperature; or other procedures. Values for commercial acetate and triacetate fibers are generally in the 2.2–4.0 N/tex (25–45 gf/den) range.

The wet modulus of fibers at various temperatures influences the creasing and mussiness caused by laundering. Acetate, triacetate, and rayon behave quite similarly, with a lower sensitivity than acrylic.

For acetate and triacetate the work of rupture is essentially the same at 0.022 N/tex (0.25 gf/den). This is higher than for cotton (0.010 N/tex = 0.113 gf/den), similar to rayon and wool, but less than for nylon (0.076 N/tex = 0.86 gf/den) and silk (0.072 N/tex = 0.81 gf/den).

The elongation of a stretched fiber is best described as a combination of instantaneous extension and a time-dependent extension or creep. This viscoelastic behavior is common to many textile fibers, including acetate. Conversely, recovery of viscoelastic fibers is typically described as a combination of immediate elastic recovery, delayed recovery, and permanent set or secondary creep. The permanent set is the residual extension that is not recoverable. These three components of recovery for acetate are given in Table 1. The elastic recovery of acetate fibers alone and in blends has also been reported. In textile processing strains of more than 10% are avoided in order to produce a fabric of acceptable dimensional or shape stability.

Absorption and Swelling Behavior. The absorption of moisture by acetate and triacetate fibers generally depends on the relative humidity and whether equilibrium is approached from the dry or wet side. The percentage of moisture regain of commercial fibers (ASTM D1909-68), taken at 65% relative humidity for the absorption cycle, is 6.5 for acetate fiber and 3.5 for triacetate. Heat treatment can lower the moisture regain of triacetate fiber, and values of 2.5–3.2% have been observed.

Percentage of water imbibition is an important property of ease-of-care and quick-drying fabrics. This value is determined by measuring the moisture remaining in a fiber in equilibrium with air at 100% rh while the fiber is being centrifuged at forces up to 1000 g. The average recorded value for acetate is 24%; triacetate not heat-treated, 16%; and heat-treated triacetate, 10%.

Specific Gravity. The values of 1.32 for acetate and 1.30 for triacetate are accepted for fibers of combined acetic acid contents of 55 and 61.5%, respectively.

Refractive Index. The refractive index parallel to the fiber axis (ε) is 1.478 for acetate and 1.472 for triacetate. The index perpendicular to the axis (ω) is 1.473 for acetate and 1.471 for triacetate.

Thermal Behavior. Acetate softens and sticks in the 190–205°C range, and fuses at ca 260°C. The apparent shining or glazing temperature is usually lower than the sticking temperature. The sole-plate temperature of an iron should not exceed 170–180°C when used on acetate fabrics. The sticking and glazing temperatures of untreated triacetate fiber are in the same range as those of acetate, whereas those exhibited by heat-treated triacetate fibers are considerably higher. Fabrics made of the latter can be

TABLE 1. ELONGATION RECOVERY OF ACETATE FIBERS

Fiber	Immediate elastic recovery, %	Delayed recovery, %	Permanent set, %
acetate multifilament			
at 50% of breaking tenacity	74	26	0
at breaking point	14	16	70
acetate staple yarn			
at 50% of breaking tenacity	58	42	0
at breaking point	12	18	70

ironed at temperatures as high as 240°C. The melting point of triacetate is ca 300°C.

Acetate and triacetate exhibit moderate changes in mechanical properties as a function of temperature. As the temperature is raised, the tensile modulus of acetate and triacetate fibers is reduced, and the fibers extend more readily under stress. Acetate and triacetate are weakened by prolonged exposure to elevated temperatures in air.

Light Stability. The resistance of textile fibers to sunlight degradation depends on the wavelength of the incident light, relative humidity, and atmospheric fumes. Acetate and triacetate fibers have essentially the same light-absorption characteristics in the visible spectrum; absorption in the uv region is slightly higher. Both fibers, when exposed under glass, behave similarly to cotton and rayon, i.e., they are somewhat more resistant than unstabilized pigmented nylon and silk and appreciably less resistant than acrylic and polyester fibers.

Electrical Behavior. Because of their high resistivity both acetate and triacetate yarns readily develop static charges, and an antistatic finish is usually applied to aid in fiber processing. Both yarns have also been used for electrical insulation after lubricants and other finishing agents are removed.

Dyeing Characteristics. Disperse dyes, high melting crystalline compounds with low solubility in the dye bath, are most frequently used for cellulose acetate and triacetate fibers.

Colored acetate and triacetate yarns are produced by incorporating colored pigments (inorganic or organic), soluble dyes, or carbon in the polymer solution before extrusion. Solution-dyed acetate and triacetate yarns are extremely colorfast to washing, dry cleaning, sunlight, perspiration, seawater, and crocking, and usually surpass the performance of vat-dyed yarns. In addition, acetate and triacetate dyed by conventional methods are susceptible to gases or fumes and fade; such fading is absent in solution-spun, pigment-dyed yarn.

Chemical Properties. Under slightly acidic or basic conditions at room temperature, acetate and triacetate fibers are resistant to chlorine bleach at the concentrations normally used in laundering.

Triacetate fiber is significantly more resistant than acetate to alkalies encountered in normal textile operations.

Acetate and triacetate are essentially unaffected by dilute solutions of weak acids, but strong mineral acids cause serious degradation.

Resistance to Microorganisms and Insects. Resistance of triacetate to microorganisms, based on soil-burial tests, is high, approaching that of polyester, acrylic, and nylon fibers.

Manufacture

Cellulose Acetate and Triacetate Polymer. The production of acetate and triacetate polymer is accomplished by the esterification of high purity chemical cellulose, except for special plastic-grade acetates requiring low color and high clarity, where cotton linters are used

Cellulose Acetate and Triacetate Fibers. Polymer solutions are converted into fibers by extrusion. The dry-extrusion process, also called dry spinning, is primarily used for acetate and triacetate.

The dry-extrusion process consists of four operations; dissolution of the polymer in a volatile solvent; filtration of the solution to remove insoluble matter; extrusion of the solution to form fibers; and lubrication, yarn formation, and packaging.

Anisotropic Solutions

Many cellulosic derivatives form anisotropic, i.e., liquid crystalline, solutions, and cellulose acetate and triacetate are no exception. Various cellulose acetate anisotropic solutions have been made using a variety of solvents. The nature of the polymer–solvent interaction determines the concentration at which liquid crystalline behavior is initiated. The better the interaction, the lower the concentration needed to form the anisotropic, birefringent polymer solution. Strong organic acids, e.g., trifluoroacetic acid, are most effective and can produce an anisotropic phase with concentrations as low as 28%.

Products

Yarns and Fibers. Many different acetate and triacetate continuous filament yarns, staples, and tows are manufactured. The variable properties are tex (wt in g of a 1000-m filament) or denier (wt in g of a 9000-m filament), cross-sectional shape, and number of filaments. Individual filament fineness (tex per filament or denier per filament, dpf) is usually in the range of 0.2–0.4 tex per filament (2–4 dpf). Common continuous filament yarns have 6.1, 6.7, 8.3, and 16.7 tex (55, 60, 75, and 150 den, respectively). However, different fabric properties can be obtained by varying the filament count (tex per filament or dpf) to reach the total tex (denier).

Yarn Packages. The principal package types used by the textile industry are tubes, cones, and beams.

Staple and Tow. The same extrusion technology that produces continuous filament yarn also produces staple and tow. The principal difference is that spinnerets with more holes are used, and instead of winding the output of each spinneret on an individual package, the filaments from a number of spinnerettes are gathered together into a ribbon-like strand, or tow. A mechanical device uniformly plaits the tow into a carton, from which it can be continuously withdrawn without tangling. Staple is produced by cutting a crimped tow into short (usually 4–5-cm) lengths resembling short, natural fibers.

Fibrillated Fibers. Instead of extruding cellulose acetate into a continuous fiber, discrete, pulp-like agglomerates of fine, individual fibrils, called fibrets or fibrids, can be produced by rapid precipitation with an attenuating coagulation fluid.

Economic Aspects

Cellulose acetate, the second oldest synthetic fiber, is an important factor in the textile and tobacco industries. Acetate belongs to the group of less expensive fibers; triacetate is slightly more expensive. An annual listing of worldwide fiber producers, locations, and fiber types is published by the Fiber Economics Bureau, Inc.

The principal textile applications of both acetate and triacetate fibers are in women's apparel and home-furnishing fabrics. Although the use of acetate fiber for textile applications has generally declined, the total worldwide production of cellulose acetate increased owing to tow for cigarette filters.

The combined annual world acetate production (filament, staple, and tow) peaked in 1980 with 672,000 t, dropped to 574,000 t in 1984, and rose to 713,000 t in 1993. The United States accounted for ca 45% of the world total. Other principal acetate producing countries include the U.K., Japan, Canada, Italy, and the former USSR.

GEORGE A. SERAD
Hoechst Celanese Corporation

Additional Reading

"Data of Fiber Economics Bureau," as published in *Text. Organon*, 112–113, and 115 (June 1986); *Fiber Organon*, 138–139 and 144 (July 1992); and author information.

Maim C.J. and G.D. Hiatt: in E. Ott, H.M. Spurlin, and M.W. Graffin, eds., *Cellulose and Cellulose Derivatives,* High Polymers Series, 2nd Edition, Vol. V, Wiley-Interscience, New York, NY, 1954, Pt. II.

Segal, L.: in N.M. Bikales and L. Segal, eds., *Cellulose and Cellulose Derivatives,* High Polymers Series, Vol. V, Wiley-Interscience, New York, NY, 1971, Chapt, XVII-A.

Serad, G.A.: in J.I. Kroschwitz, ed., *Encyclopedia of Polymer Science and Engineering,* Vol. 3, 2nd Edition Wiley-Interscience, New York, NY, 1985, pp. 200–226.

FIBERS: ELASTOMERIC. Elastomeric fibers can be made from natural or synthetic polymeric materials that provide a product with high elongation, low modulus, and good recovery from stretching. Currently, these fibers are made primarily from polyisoprenes (natural rubber) or segmented polyurethanes and to a lesser extent from segmented polyesters. In the United States the generic designation spandex has been given to a manufactured fiber in which the fiber-forming substance is a long-chain synthetic polymer comprised of at least 85% of a segmented polyurethane; in Europe the equivalent term elastane is commonly used.

Thermoplastic, inelastic fibers, such as nylon and polyester, may be processed to provide spring-like, helical, or zigzag structures. These fibers can exhibit high elongations as the helical or zigzag structure is stretched, but the recovery force is very low. This apparent elasticity results from the geometric form of the filaments as opposed to elastomeric fibers whose elastic properties depend primarily on entropy changes inherent within their polymer structure. Thus processed inelastic fibers must comprise a significant portion of a stretch fabric whereas an elastomeric fiber provides the necessary stretch properties at 5–20% of fabric weight.

Other elastomeric-type fibers include the biconstituents, which usually combine a polyamide or polyester with a segmented polyurethane-based fiber. These two constituents are melt-extruded simultaneously through the same spinneret hole and may be arranged either side by side or in an eccentric sheath–core configuration. As these fibers are drawn, a differential shrinkage of the two components develops to produce a helical fiber configuration with elastic properties. An applied tensile force pulls out the helix and is resisted by the elastomeric component. Kanebo Ltd. has introduced a nylon–spandex sheath–core biconstituent fiber for hosiery with the trade name Sideria.

Nonspandex elastomeric fibers based on segmented polyesters and polyethers are currently being developed that can be melt-spun into threads. Teijin Ltd. produces an elastomeric fiber of this type with the trade name Rexe.

Mechanical Properties

In both rubber thread and spandex fibers, mechanical properties may be varied over a relatively broad range. In rubber, variations are made in the degree of cross-linking or vulcanization by changing the amount of vulcanizing agent, usually sulfur, and the accelerants used. In spandex fibers, many more possibilities for variation are available. By definition spandex fibers contain urethane linkages with the following repeat structure:

$$-\!\!\left(\!R\!-\!O\!-\!\overset{\overset{\displaystyle O}{\|}}{C}\!-\!\overset{\overset{\displaystyle H}{|}}{N}\!-\!R'\!-\!\overset{\overset{\displaystyle H}{|}}{N}\!-\!\overset{\overset{\displaystyle O}{\|}}{C}\!-\!O\!\right)_{\!n}$$

The number of polymers in the classification is obviously very large. Most urethane polymers in current use for the manufacture of spandex fibers are made by the reaction of 1000–4000 molecular weight hydroxy terminated polyethers or polyesters with a diisocyanate at a molar ratio of ca 1:1.4 to 1:2.5, followed by reaction of the resulting isocyanate-terminated prepolymer with one or more diamines to produce a high molecular weight urethane polymer. Small amounts of monofunctional amines may also be included to control polymer molecular weight. Mechanical properties may be affected by changing the particular polyester or polyether glycol, diisocyanate, diamine(s), and monoamine used; they can be further modified by changing the molecular weight of the glycol and by changing the glycol–diisocyanate molar ratio.

The physical characteristics of current commercial rubber and spandex fibers are summarized in Table 1.

Manufacture

Cut Rubber. To produce cut rubber thread, smoked rubber sheet or crepe rubber is milled with vulcanizing agents, stabilizers, and pigments. This milled stock is calendered into sheets 0.3–1.3 mm thickness, depending on the final size of the rubber thread desired. Multiple sheets are layered, heat-treated to vulcanize, then slit into threads for textile uses. Individual threads have either square or rectangular cross-sections.

Extruded Latex Thread. In the manufacture of extruded latex thread, a concentrated (up to ca 50% solids) natural rubber latex is blended with aqueous dispersions of vulcanizing agents, stabilizers, and white pigments. This compounded latex is held under controlled temperature conditions until partial vulcanization occurs. This has the effect of increasing wet strength and thus the processability of the extruded threads. The matured latex is extruded at constant pressure through precision-bore glass capillaries into a 15–55% acetic acid bath where coagulation into thread form occurs. Threads are removed from the coagulation bath by transfer rollers, washed free of excess acid with water, and conducted through a dryer, after which a silicone oil-based finish is applied and the threads are formed into multiend ribbons. The ribbons are then vulcanized by multiple passes on a conveyer belt through an oven that can increase curing temperature in stages up to about 150°C. After vulcanization the multiend ribbons are packed without support in boxes for shipment to the customer.

Spandex Fibers. Four different processes are currently used to produce spandex fibers commercially: melt extrusion, reaction spinning, solution dry spinning, and solution wet spinning. These processes involve different practical applications of basically similar chemistry. An isocyanate terminated prepolymer is formed by the reaction of a 1000–4000 molecular weight macroglycol with a diisocyanate at a glycol–diisocyanate ratio that may range from 1:1.4 up to about 1:2.5. The soft segment macroglycol can be either a polyether, a polyester, a polycarbonate, hydroxyl-terminated polycaprolactone, or a combination of these. The prepolymer subsequently reacts with either a glycol or diamine(s) at near stoichiometry; a small amount of monofunctional amine may be included to control final polymer molecular weight. If the diol or diamine(s) reaction with the prepolymer is carried out in a solvent, the resulting block copolymer solution may be wet or dry spun into fiber. Alternatively, the prepolymer may be reaction spun by extrusion into a bath containing diamine to form a fiber, or the prepolymer may be permitted to react in bulk with a diol and the resulting polymer melt extruded in fiber form.

Chemical Properties

Stabilization. Both rubber and spandex fibers are subject to oxidative attack by heat, light, atmospheric contaminants such as NO_x, and active chlorine. Both rubber and spandex fibers are likely to contain antioxidants; the spandex fibers may also be stabilized to uv light and to atmospheric contaminants that cause discoloration.

Solvent Resistance. Elastomeric fibers tend to swell in certain organic solvents; rubber fibers swell in hydrocarbon solvents such as hexane. Spandex fibers become highly swollen in chlorinated solvents such as tetrachloroethylene.

Dyeing. Spandex fibers have an affinity for dispersed or acid dyes; rubber fibers normally cannot be dyed.

Economic Aspects

Most process developments have occurred in the United States, Germany, Japan, and Korea. A large proportion of worldwide capacity is controlled by DuPont, either directly or through subsidiaries and joint ventures.

Commercially, elastomeric fibers are almost always used in combination with hard fibers such as nylon, polyester, or cotton. Prices of spandex fibers are highly dependent on thread size; selling price generally increases as fiber tex decreases. Factors that contribute to the relatively high cost of spandex fibers include *(1)* the relatively high cost of raw materials; *(2)* the small size of the spandex market compared to that of hard fibers, which limits scale and thus efficiency of production units; and *(3)* the technical problems associated with stretch fibers that limit productivity rates and conversion efficiencies.

Uses

Elastomeric fibers are used in cut rubber and extruded latex and spandex fibers.

JOHN E. BOLIEK
ARNOLD W. JENSEN
E. I. du Pont de Nemours & Co., Inc.

TABLE 1. PHYSICAL PROPERTIES OF ELASTOMERIC FIBERS

Property	Spandex	Extruded rubber	Cut rubber
sizes available[a]	1.1–250[b]	16–610 tex[b]	2.5–21 μm dia[c]
tenacity, N/tex[d]	0.05–0.13	0.02–0.03	0.01–0.02
elongation, %	400–800	600–700	600–700
modulus[e], N/tex[d]	0.013–0.045	0.004–0.005	0.002–0.004
stability[f]			
uv light	good	fair	fair
ozone	good	poor	poor
NO_x	fair, yellows	poor	poor
active Cl	fair, yellows	poor	poor
body oils	fair	poor	poor
cosmetics	good	fair	fair
dyeability	dyeable	not dyeable	not dyeable
abrasion resistance	very good	poor	poor

[a] Spandex size is usually expressed in denier which is weight in g/9000 m length. However, the SI unit is tex, the weight in g/1000 m. Rubber size is expressed as gauge, which is the reciprocal of diameter or size in inches. [b] To convert tex to den, multiply by 9.09. [c] 1,200–10,000 gauge. [d] To convert N/tex to gf/den, multiply by 11.33. [e] First cycle stress at 300% elongation. [f] Both spandex fibers and rubber threads normally contain antioxidants and other stabilizers.

Additional Reading

Kotani, T. et al.: *J. Macromol. Sci-Phys.* **831**, 65 (1992).
U.S. Pat. 2,957,852 (Oct. 25, 1960), P. Frankenburg and A. Frazier (to DuPont Co.).
U.S. Pat. 3,296,063 (Jan 3, 1967), C. Chandler (to DuPont Co.).
U.S. Pat. 5,028,642 (July 2, 1991), C. W. Goodrich & W. Evans (to DuPont Co.).

FIBERS: VEGETABLE. Vegetable fibers, as the name implies, are derived from plants. The principal chemical component in plants is cellulose, and therefore they are also referred to as cellulosic fibers. The fibers are usually bound by a natural phenolic polymer, lignin, which also is frequently present in the cell wall of the fiber; thus vegetable fibers are also often referred to as lignocellulosic fibers, except for cotton, which does not contain lignin.

Vegetable fibers are classified according to their source in plants as follows: (1) the bast or stem fibers, which form the fibrous bundles in the inner bark (phloem or bast) of the plant stems, are often referred to as soft fibers for textile use; (2) the leaf fibers, which run lengthwise through the leaves of monocotyledonous plants, are also referred to as hard fibers; and (3) the seed-hair fibers, the source of cotton (qv), are the most important vegetable fiber. There are over 250,000 species of higher plants; however, only a very limited number of species have been exploited for commercial uses (less than 0.1%). The commercially important fibers are given in Table 1.

World markets for vegetable fibers have been steadily declining in recent years, mainly as a result of substitution with synthetic materials.

General Properties

Chemical Composition. Chemically, cotton is the purest vegetable fiber, containing over 90% cellulose with little or no lignin. The other fibers contain 70–75% cellulose, depending on processing. Boiled and bleached flax and degummed ramie may contain over 95% cellulose. Kenaf and jute contain higher contents of lignin, which contributes to their stiffness. Although the cellulose contents are fairly uniform, the other components, e.g., hemicelluloses, pectins, extractives, and lignin, vary widely without obvious pattern. These differences may characterize specific fibers.

Fiber Dimensions. Except for the seed-hair fibers, the vegetable fibers of bast or leaf origins are multicelled and are used as strands. However, for papermaking the strands are broken down to the ultimate cells.

Physical Properties. Bast and leaf fibers are stronger (higher tensile strength and modulus of elasticity) but lower in elongation (extensibility) than cotton. Vegetable fibers are stiffer but less tough than synthetic fibers. Kapok and coir are relatively low in strength; kapok is known for its buoyancy.

Among the bast textile fibers, the density is close to 1.5 g/cm^3, or that of cellulose itself, and they are denser than polyester. Moisture regain (absorbency) is highest in jute at 14%, whereas that of polyester is below 1%. The bast fibers are typically low in elongation and recovery from stretch. Ramie fiber has a particularly high fiber length/width ratio.

The microfibrils in vegetable fibers are spiral and parallel to one another in the cell wall. The spiral angles in flax, hemp, ramie, and other bast fibers are lower than cotton, which accounts for the low extensibility of bast fibers.

Processing and Fiber Characteristics

Bast Fibers. Bast fibers occur in the phloem or bark of certain plants. The bast fibers are in the form of bundles or strands that act as reinforcing elements and help the plant to remain erect. The plants are harvested and the strands of bast fibers are released from the rest of the tissue by retting, common for isolation of most bast fibers. Retting involves the biological (bacterial) breakdown of pectins in the plant through immersion of the stalks in ponds or streams for periods of time sufficient to release the desired outer bast fibers from the woody inner core. The retted material is then further processed by breaking, scutching, and hackling.

Leaf (Hard) Fibers. Hard or cordage fibers are found in the fibrovascular systems of the leaves of perennial, monocotyledonous plants growing in Central America, East Africa, Indonesia, Mexico, and the Philippines. They are generally of the *Agave* and *Musa* genera. The leaf elements are harvested by cutting at the base with a sickle-like tool, and bundled for processing by hand or by machine decortication. In the latter case, the leaves are crushed, scraped, and washed. The fibers are generally coarser than the bast fibers and are graded for export according to national rules for fineness, luster, cleanliness, color, and strength.

Seed- and Fruit-Hair Fibers. The seeds and fruits of plants are often attached to hairs or fibers or encased in a husk that may be fibrous. These fibers are cellulosic based and of commercial importance, especially cotton (qv), the most important natural textile fiber.

TABLE 1. VEGETABLE FIBERS OF COMMERCIAL INTEREST

Commercial name	Source	Botanical name of plant	Growing area
Bast or soft fibers			
China jute	Abutilon	*Abutilon theophrasti*	China
flax		*Linum usitatissimum*	north and south temperate zones
hemp		*Cannabis sativa*	all temperate zones
jute		*Corchorus capsularis; C. olitorius*	India
kenaf		*Hibiscus cannabinus*	India, Iran, CIS, South America
ramie		*Boehmeria nivea*	China, Japan, United States
roselle		*Hibiscus sabdariffa*	Brazil, Indonesia (Java)
sunn		*Crotalaria juncea*	India
urena	cadillo	*Urena lobata*	Zaire, Brazil
Leaf or hard fibers			
abaca		*Musa textilis*	Borneo, Philippines, Sumatra
cantala	Manila maguey	*Agave cantala*	Philippines, Indonesia
caroa		*Neoglaziovia variegata*	Brazil
henequen		*Agave fourcroydes*	Australia, Cuba, Mexico
istle		*Agave* (various species)	Mexico
mauritius		*Furcraea gigantea*	Brazil, Mauritius, Venezuela, tropics
phormium		*Phormium tenax*	Argentina, Chile, New Zealand
pineapple	piña	*Ananas comasus*	Hawaii, Philippines, Indonesia, India, West Indies
sansevieria	bowstring hemp	*Sansevieria* (entire genus)	Africa, Asia, South America
sisal		*Agave sisalana*	Haiti, Java, Mexico, South Africa
Seed-hair fibers			
coir	coconut husk fiber	*Cocos nucifera*	tropics, India, Mexico
cotton		*Gossypium* sp.	United States, Asia, Africa
kapok		*Ceiba pentandra*	tropics
milkweed floss		*Chorisia* sp.	North America
Other fibers			
broom root	roots	*Muhlenbergia macroura*	Mexico
broom corn	flower head	*Sorghum vulgare technicum*	United States
crin vegetal	palm leaf segments	*Chamaerops humilis*	North Africa
palmyra palm	palm leaf stem	*Brossus flabellifera*	India
pissava	palm leaf base fibers	*Attalea funifera*	Brazil
raffia	palm leaf segments	*Raphia raffia*	East Africa

Economic Aspects

The principal bast and leaf fibers are produced in yields of 2–5%, with some exceptions such as flax (15%) and kapok (17%), on a green plant basis. Vegetable fiber production on the world market has dropped 25–35% since 1970 because of periods of economic recession and synthetic fiber replacements. Imports of vegetable fibers have dropped 70–90% since

1970 (with the exception of flax wastes). These market trends reflect the recessions and substitution with synthetic fibers. Although most vegetable fibers are converted to lower cost commodity products, some of the fibers are converted into the most expensive products in their respective industries, e.g., U.S. currency (paper). The value of most of the fibers has dropped since 1981.

Uses

Vegetable fibers have application in a broad range of fibrous products, including textiles and woven goods, cordage and twines, stuffing and upholstery materials, brushes, and paper. The traditional uses for the vegetable fibers have been eroded by substitution with synthetics on the world market. The declining uses include cordage, mats, filling material, brushes, etc. However, the unique properties of the bast fibers have allowed continued use in such specialty papers as bank notes, some writing papers, and cigarette papers.

Recent work by the USDA and Kenaf International (Texas) has demonstrated the potential of both growing and processing kenaf fibers for newsprint and other paper products in the United States. Another promising potential use for vegetable fibers is in the new lignocellulosic-based composites under development in various parts of the industrialized world. Such products are already utilized in the automotive industry for automobile interior door and head liners and as trunk liners.

RAYMOND A. YOUNG
University of Wisconsin, Madison

Additional Reading

Cook, J. G.: *Handbook of Textile Fibers I. Natural Fibers,* 5th ed., Merrow Publishing, Durham, UK, 1984.
Morton, W. E. and J. W. S. Hearle: *Physical Properties of Textile Fibers,* 2nd Edition, John Wiley & Sons, Inc., New York, NY, 1975, p. 284.
Nelson, E. E.: in S. P. Parker, ed., *Encyclopedia of Science and Technology,* Vol. 9, McGrawHill, Inc., New York, NY, 1982, pp. 8–12.
Rebenfeld, L.: in J. I. Kroschwitz, ed., *Encyclopedia of Polymer Science and Engineering,* Vol. 6, 2nd ed., Wiley-Interscience, New York, NY, 1986, p. 647.

FICTILE. Descriptive of certain molecules which have no permanent structure but are constantly changing their shapes and arrangements. An example is the metal carbonyl $Fe_3(CO)_{12}$, in which, according to Dr. F. Albert Cotton, originator of the term, "carbonyl groups readily move from one iron atom to another through the rapid formation and dissolution of carbonyl bridges between iron atoms."

FIESER, LOUIS F. (1899–1977). A distinguished American chemist, Fieser became professor of organic chemistry at Harvard in 1930 after teaching for several years at Bryn Mawr. He achieved the synthesis of vitamin K_1 and did fundamental research on cortisone, the chemistry of steroids, and aromatic carcinogens. His achievements as a chemist and educator are recognized throughout the world. Unique in his facility in laboratory demonstration and as a lecturer and author, he exemplified that rare combination of a great teacher and a profound scholar.

FILAMENT. The resistive element (1) in a common electric lamp and (2) in a thermionic tube through which current is passed to provide the temperature required for thermionic emission. The surface of the filament may supply the emission, or the filament may be employed as a heater for an indirectly heated cathode.

The term also is used in astronomy in connection with prominences. In the textile field, particularly in connection with synthetic fibers, individual strands of extruded nylon, rayon, and so on are often referred to as filaments. Often, many filaments are twisted together to form yarn.

FILLER. (1) An inert mineral powder of rather high specific gravity (2.00–4.50) used in plastic products and rubber mix to provide a certain degree of stiffness and hardness and to decrease cost. Examples are calcium carbonate (whiting), barytes, blanc fixe, silicates, glass spheres and bubbles, slate flour, soft clays, etc. Fillers have neither reinforcing nor coloring properties, and the term should not be applied to materials that do, i.e., reinforcing agents or pigments. Fillers are similar to extenders and diluents in their cost-reducing function; exact lines of distinction between these terms are difficult, if not impossible, to draw. Use of fillers and extenders in plastics has increased in recent years due to shortages of basic material.

(2) The cross or transverse thread in a fabric or other textile structure.

(3) A metal or alloy used in brazing and soldering to effect union of the metals being joined.

See also **Diluent**; **Extender**; and **Reinforcing Agent**.

FILM. In its most general usage, film means any thin sheet of material used for covering, coating or wrapping, or any thin layer that enters into the structure, usually on or near the surface, of a substance or object. The term film also denotes the monomolecular layer that is formed on the surface of a solution or at an interface between two immiscible liquids. The adsorption is of such a nature that the free surface energy is a minimum. Insoluble and non-volatile substances placed on the surface of a liquid such as water may also under certain conditions spread out on the water surface to give a monomolecular film. Adsorbed films of gases or liquids are also formed on solids such as mica, sodium chloride, glass and metals. In some cases, such as the adsorption of vapors on solids at relatively high pressures, it seems that the films may be thicker than one molecular layer and may attain thickness of three or four molecules.

The term film is also applied to sheets of cellophane, polyethylene, polyvinylidene chloride, etc., used for wrapping and packaging of food products, meats, and poultry (especially shrink films that are stretched before application). These function as a moisture vapor barrier. Plastic films are also used as slip surfaces in concrete structures such as airstrips, ice rinks, and highways. Photographic film is made from cellulose acetate.

Condensed Film

A surface film in which the molecules are closely packed and steeply oriented to the surface. The molecular packing approaches that observed in the crystalline state.

Expanded Film

A state of film intermediate in area and other properties between gaseous and condensed films.

Gaseous Film

A film in which molecules move about independently on the surface and their lateral adhesion for each other is very small. At low surface pressures (π) and large area (A), a gaseous film obeys the relation $\pi A = kT$. At higher pressures an equation of the form $(\pi A - A_0) = xkT$ holds, where x is a constant.

Liquid-Expanded Film

This film occupies a much larger area than a condensed film, but is still a coherent film. It can form a separate phase from a gaseous film with which it is in equilibrium, and obeys the relation

$$(\pi - \pi_0)(A - A_0) = C$$

where π is the surface pressure, A the surface area, and A_0 the co-area of the molecule.

See also **Fibers: Acetate**; and **Thin Films**.

FILM (Bubble). See **Foam**.

FILTER MEDIUM. Almost any water-insoluble, porous material having a reasonable degree of rigidity can serve as a filter. Sand is used in simple large-scale water filtration, the voids between the grains providing the porosity. In industrial operation, cotton duck, woven wire cloth, nylon cloth, and glass cloth are used. For laboratory work, Whatman filter paper, diatomaceous earth, and closely packed glass fibers are standard materials. Plastics membranes containing more than a million pores per square inch are used in bacteriological filtration.

See also **Filtration**.

FILTRATION. Filtration is the separation of two phases, particulate form, i.e., solid particles or liquid droplets, and continuous, i.e., liquid or gas, from a mixture by passing the mixture through a porous medium. This article discusses the more predominant separation of solids from liquids.

Filtration is often referred to as mechanical separation because the separation is accomplished by physical means. This does not preclude chemical or thermal pretreatment used to enhance filtration. Although some slurries separate well without chemical conditioning, most pulps of a widely varying nature can benefit from pretreatment. (See also **Flocculating Agents**).

In a filtration system (Fig. 1) the porous filtration medium is housed in a housing, with flow of liquid in and out. A driving force, usually in the form of a static pressure difference, must be applied to achieve flow through the filter medium. It is immaterial from the fundamental point of view how the pressure difference is generated but there are four main types of driving force, i.e., gravity, vacuum, pressure, and centrifugal. The two-types of filtration used most often in practice are cake, or surface, filtration (Fig. 2) and deep bed filtration (Fig. 3). This division usually is unambiguous but in some cases, such as cartridge filters, there is no sharp dividing line.

Separation Efficiency. Similarly to other unit operations in chemical engineering, filtration is never complete. Some solids may leave in the liquid stream, and some liquid will be entrained with the separated solids. As emphasis on the separation efficiency of solids or liquid varies with application, the two are usually measured separately. Separation of solids is measured by total or fractional recovery, i.e., how much of the incoming solids is collected by the filter. Separation of liquid usually is measured in how much of it has been left in the filtration cake for a surface filter, i.e., moisture content, or in the concentrated slurry for a filter-thickener, i.e., solids concentration.

Filtration-Related Processes

Several processes are used to enhance the filtration process itself. They may also be related processes in their own right. They include washing of solids, cake dewatering, pretreatment of suspensions (addition of inert filter aids), mechanical squeezing of cakes, electro-kinetic effects (Table 1), and magnetic separation.

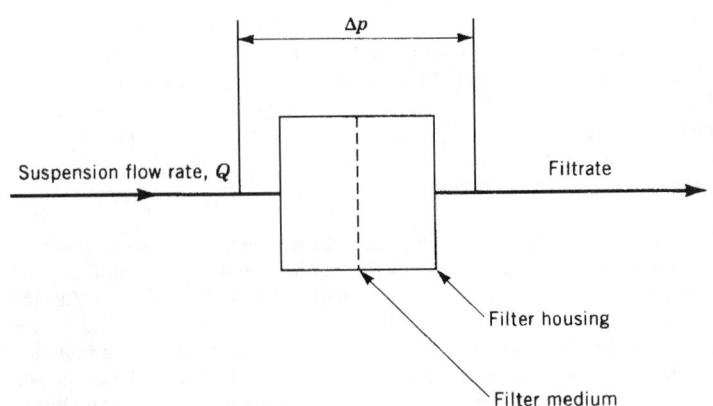

Fig. 1.   Schematic diagram of a filter

TABLE 1. SUMMARY OF ELECTRO-KINETIC EFFECTS

Effect	Fluid	Particles	Electric field
electrophoresis	still	moving	applied
sedimentation (or migration) potential	still	moving	measured
electroosmosis	moving	still	applied
streaming potential	moving	still	measured
electroosmotic pressure	still	still	applied

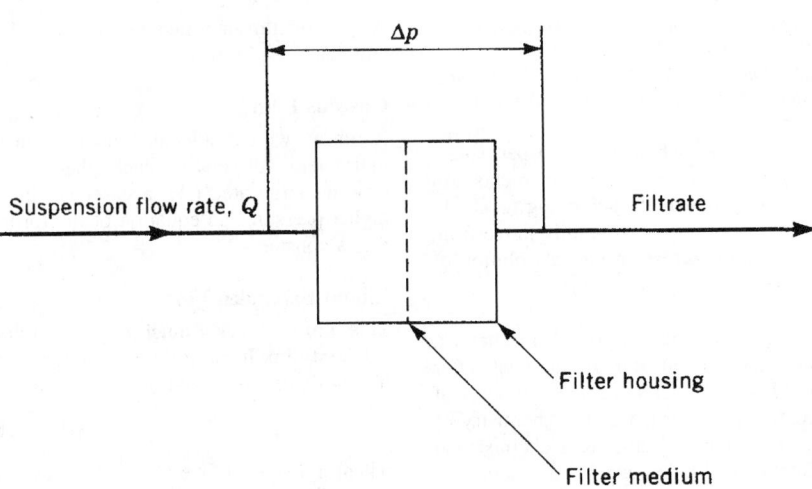

Fig. 2.   Schematic diagram of a surface filter, i.e., the cake filtration mechanism

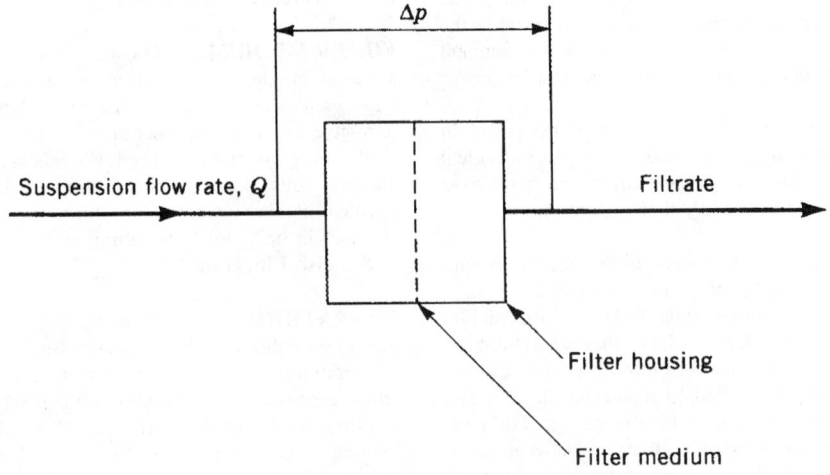

Fig. 3.   Mechanism of deep bed filtration

Cake Filtration Theory

It does not matter, from the fundamental point of view, how the pressure drop is generated in the filter. In the case of the centrifugal filters there is an additional phenomenon of the mass forces acting on the liquid within the cake. The conventional filtration theory must be amended to include this effect.

Carman-Kozeny Equation. Flow through packed beds under laminar conditions can be described by the Carman-Kozeny equation in the form

$$\frac{Q}{A} = \frac{\Delta p}{\mu L} \frac{\varepsilon^3}{5(1-\varepsilon)^2 S_0^2} \quad (1)$$

where Q is volumetric flow rate, A is face area of the bed, L is depth of the bed, Δp is applied pressure drop, ε is voidage of the bed (porosity), S_0 is volume specific surface of the bed, and μ is liquid viscosity.

Darcy's Law and the Basic Filtration Equation. Darcy's law combines the constants in the last term of equation 1 into one factor K known as the permeability of the bed, i.e.,

$$K = \frac{\varepsilon^3}{5(1-\varepsilon)^2 S_0^2} \quad (2)$$

where K is a constant for incompressible solids. For compressible cakes, K depends on applied pressure, approach velocity, and concentration, and therefore presents serious problems in cake filtration testing and scale-up.

Modern filtration theory tends to prefer the Ruth form of Darcy's law, i.e.,

$$\frac{Q}{A} = \frac{\Delta p}{\mu R} \quad (3)$$

where R is the bed resistance. In cake filtration the bed resistance consists of the medium resistance in series with the resistance of the deposited cake, assuming no penetration of solids into the filtration medium; the general filtration equation is then written as

$$\frac{Q}{A} = \frac{\Delta p}{\mu R + \sigma \mu c V/A} \quad (4)$$

where σ is specific cake resistance, μ is liquid viscosity, c is solids concentration in the feed, V is filtrate volume collected since commencement of filtration, and R is medium resistance.

There is hidden assumption in equation 4 that the volume of the solids and liquid retained in the cake is negligible. This is reasonable at low concentrations but can lead to errors at high solids concentration and moisture contents of cakes. A corrected value of the concentration c can be used in equation 4 to reduce the errors, i.e.,

$$c(\text{corrected}) = \frac{1}{(1/c) - (1/\rho_s) - [(m-1)/\rho]} \quad (5)$$

where m is the mass ratio of wet to dry filter cake, ρ_s is solids density, and ρ is liquid density. This correction is necessary only at feed concentrations of greater than 200 g/L.

The scale-up of conventional cake filtration uses the basic filtration equation (eq. 4).

Benefits of Prethickening. The feed solids concentration has a profound effect on the performance of any cake filtration equipment. It affects the capacity and the cake resistance, as well as the penetration of the solids into the cloth which influences filtrate clarity and medium resistance. Thicker feeds lead to improved performance of most filters through higher capacity and lower cake resistance.

Prethickening of filter feeds can be done with a variety of equipment such as gravity thickeners, hydrocyclones, or sedimenting centrifuges. Even cake filters can be designed to limit or completely eliminate cake formation and therefore act as thickening filters and be used in this thickening duty.

Pressure Filtration. High pressure drops have a twofold effect, i.e., on capacity and on displacement dewatering which often follows.

The most important feature of the pressure filters which use hydraulic pressure to drive the process is that they can generate a pressure drop across the medium of more than 1×10^5 Pa which is the theoretical limit of vacuum filters. While the use of a high pressure drop is often advantageous, leading to higher outputs, drier cakes, or greater clarity of the overflow, this is not necessarily the case. For compressible cakes, an increase in pressure drop leads to a decrease in permeability of the cake and hence to a lower filtration rate relative to a given pressure drop.

Optimization of Cycle Times. In batch filters, one of the important decisions is how much time is allocated to the different operations such as filtration, displacement dewatering, cake washing, and cake discharge, which may involve opening of the pressure vessel. All of this has to happen within a cycle time t_c which itself is not fixed, though some of the times involved may be defined, such as the cake discharge time.

If all of the nonfiltration operations are grouped together into a downtime, t_d, assumed to be fixed and known, an optimum filtration time t_{opt} in relation to t_d can be derived by optimizing the average dry cake production obtained from the cycle. For constant pressure filtration and where the medium resistance R and the specific cake resistance σ are constant, the following equation applies:

$$t_{opt} = t_d \left[1 + \sqrt{\frac{2\mu R}{\sigma c \Delta p t_d}} \right] \quad (6)$$

where Δp and c are the operating pressure drop and the feed solids concentration, respectively.

When the medium resistance R is small compared with the specific cake resistance σ, the second term in the above equation becomes negligible and the optimum filtration time t_{opt} becomes equal to downtime t_d. For any other case, t_{opt} is always greater than t_d. It follows, therefore, that the filtration time should be at least equal to the sum of the other nonfiltration periods involved in the cycle.

Vacuum Filters

In vacuum filters, the driving force for filtration results from the application of a suction on the filtrate side of the medium. Although the theoretical pressure drop available for vacuum filtration is 100 kPa, in practice it is often limited to 70 or 80 kPa.

In applications where the fraction of fine particles in the solids of the feed slurry is low, a simple and relatively cheap vacuum filter can yield cakes with moisture contents comparable to those discharged by pressure filters. Vacuum filters include the only truly continuous filters built in large sizes that can provide for washing, drying, and other process requirements.

Vacuum filters are available in a variety of types, and are usually classified as either batch operated or continuous. An important distinguishing feature is the position of the filtration area with respect to gravity, i.e., horizontal or nonhorizontal filtering surface.

Vacuum filters include the nutsche filter, enclosed agitated vacuum filters, the vacuum leaf filter, the tipping pan filter, horizontal rotating pan filters, horizontal belt vacuum filters, rotary vacuum drum filters, and rotary vacuum disk filters.

Batch Pressure Filters

Excluding variable chamber presses, which rely on mechanical squeezing of the cake and are discussed in a separate section, pressure filters may be grouped into two categories, i.e., plate-and-frame filter presses, and pressure vessels containing filter elements. The latter group also includes cartridge filters; these are discussed separately. All of the above pressure filters are suited to handling different types of cake. Pressure vessel filters (leaf-type) handle incompressible or slightly compressible cakes. Filter presses handle both compressible and incompressible cakes, especially with the flexibility potential of membranes. Variable chamber presses cannot be used on incompressible materials. Cylindrical element filters, i.e., candle filters, are used for clarification applications, using precoat and often body feed, resulting in cakes that are slightly compressible. Cartridge filters are for clarification only, with little if any cake formed.

Mechanical Batch Compression Filters. In conventional cake filtration the liquid is expelled from the slurry by fluid pressure in a fixed-volume filtration chamber; in mechanical compression this is achieved by reduction of the volume of the retaining chamber. This compression of either a slurry or a cake, which might have been formed by conventional filtration, offers advantages to industries handling a variety of different materials. Such materials include highly compressible, sponge-like solids; very fine particles such as clays; fibrous pulps; gelatinous mixtures like starch residues or some pharmaceuticals; and flocculated wastewater sludges.

Continuous Pressure Filters

A continuous pressure filter may be defined as a filter that operates at pressure drops greater than 100 kPa and does not require interruption of its operation to discharge the cake; the cake discharge itself, however, does not have to be continuous. There is little or no downtime involved, and the dry solids rates can sometimes be as high as 1750 kg/m²h with continuous pressure filters.

Most continuous pressure filters available (ca 1993) have their roots in vacuum filtration technology. A rotary drum or rotary disk vacuum filter can be adapted to pressure by enclosing it in a pressure cover; however, the disadvantages of this measure are evident. The enclosure is a pressure vessel which is heavy and expensive, the progress of filtration cannot be watched, and the removal of the cake from the vessel is difficult. Other complications of this method are caused by the necessity of arranging for two or more differential pressures between the inside and outside of the filter, which requires a trouble-some system of pressure regulating valves.

Despite the disadvantages, the advantages of high throughputs and low moisture contents in the filtration cakes have justified the vigorous development of continuous pressure filters.

Horizontal or vertical vessel filters, especially those with vertical rotating elements, have undergone rapid development with the aim of making truly continuous pressure filters, particularly but not exclusively for the filtration of fine coal. There are basically three categories of continuous pressure filters available, i.e., disk filters, drum filters, and belt filters including both hydraulic and compression varieties.

Disk filters include the McGaskell and Gaudfrin disk filters, the KDF filter, and the KHD pressure filter.

Drum filters include the TDF drum filter and the BHS-Fest filter.

Continuous compression filters include belt presses and screw presses.

Thickening Pressure Filters. The most important disadvantage of conventional cake filtration is the declining rate due to the increased pressure drop caused by the growth of the cake on the filter medium. A high flow rate of liquid through the medium can be maintained if little or no cake is allowed to form on the medium. This leads to thickening of the slurry on the upstream part of the medium; filters based on this principle are sometimes called filter thickeners.

The methods of limiting cake growth are classified into five groups, i.e., removal of cake by mass forces (gravity or centrifugal), or by electrophoretic forces tangential to or away from the filter medium; mechanical removal of the cake by brushes, liquid jets, or scrapers; dislodging of the cake by intermittent reverse flow; prevention of cake deposition by vibration; and cross-flow filtration by moving the slurry tangentially to the filter medium so that the cake is continuously sheared off. The extent of the commercial exploitation of these principles in the available equipment varies, but cross-flow filtration is exploited most often.

Centrifugal Filters

The driving force for filtration in centrifugal filters is centrifugal forces acting on the fluid. Such filters essentially consist of a rotating basket equipped with a filter medium. Similar to other filters, centrifugal filtration does not require a density difference between the solids and the suspending liquid. If such density difference exists sedimentation takes place in the liquid head above the cake. This may lead to particle size stratification in the cake, with coarser particles being closer to the filter medium and acting as a precoat for the fines to follow.

In centrifuges, in addition to the pressure due to the centrifugal head due to the layer of the liquid on top of the cake, the liquid flowing through the cake is also subjected to centrifugal forces that tend to pull it out of the cake. This makes filtering centrifuges excellent for dewatering applications. From the fundamental point of view, there are two important consequences of these additional dewatering forces. First, Darcy's law and all of the theory based on it is incomplete because it does not take into account the effect of mass forces. Second, pressures below atmospheric can occur in the cake in the same way as in gravity fed deep bed filters. The conventional filtration theory has been modified to make it applicable to centrifugal filters.

Due to good performance and high cost, centrifuges are often referred to as the Rolls-Royces of solid–liquid separation. They have parts rotating at high speeds and require high engineering standards of manufacture, high maintenance costs, and special foundations or suspensions to absorb vibrations. Another feature distinguishing the filtering centrifuges from other cake filters is that the particle size range they are applicable to is generally coarser, from 10 μm to 10 mm. In particular, cake filters that move the cake across the filter medium are restricted to using metal screens, which by their very nature are coarse. No cloth can withstand the abrasion due to the cake forced on the cloth and pushed over its surface. Only the

fixed bed, batch-operated centrifuges can use cloth as the filtration medium and be used, therefore, with fine suspensions.

See also **Centrifugation**; **Classifying (Process)**; **Electrostatic Precipitator**; and **Membrane Separations Technology**.

LADISLAV SVAROVSKY
Consultant Engineers and Fine Particle Software

Additional Reading

ASM: *Filtration Tests and Practices,* D 1889, F 660, 661, 662, 795, 796, 838, and 901 American Society for Testing and Materials, Philadelphia, PA, (Varying dates of publication).

Cheremisinoff, N.P.: *Liquid Filtration,* Butterworth-Heinemann, Inc., Woburn, MA, 1998.

Considine, D.M. and G.D. Considine, Eds.: *Foods and Food Technology Encyclopedia,* Van Nostrand Reinhold, New York, NY, 1982.

Dickenson, T.C.: *Filters and Filtration Handbook,* 4th Edition, Elsevier Science, New York, NY, 1997.

Ernst, M. et al.: "Tackle Solid–Liquid Separation Problems," *Chem. Eng. Progress,* **22** (June 1991).

Johnston, P.R.: "Liquid Filtration," *Chem. Eng. Progress,* **18** (November 1988).

Johnston, P.R.: "Misleading Practice of Using Filtration Date to Deduce Pore Size in Filter Media," Proceedings of International Filtration Conference, Ocean City, Maryland, American Institute of Chemical Engineers, New York, NY, 1988.

Meltzer, T.H. and M.W. Jornitz: *Filtration in the Biopharmaceutical Industry,* Marcel Dekker, Inc., New York, NY, 1998.

Perlmutter, B.A.: "Combine Filtration and Drying," *Chem. Eng. Progress,* **29** (July 1991).

Rushton A., A.S. Ward, and R.G. Holdich: *Solid–Liquid Filtration and Separation Technology,* 2nd Edition, John Wiley & Sons, Inc., New York, NY, 2000.

Svarovsky, L. *Solid–Liquid Separation Processes and Technology,* Vol. 5, "Handbook of Powder Technology," Elsevier, Amsterdam, the Netherlands, 1985.

Svarovsky, L. ed.: *Solid–Liquid Separation,* 3rd Edition, Butterworths, London, 1990.

Tiller, F.M. "Tutorial: Interpretation of Filtration Data, I," *Fluid/Part. Sept. J.* **3**(2) (June 1990).

Tiller, F.M. "Tutorial: Interpretation of Filtration Data, II," *Fluid/Part. Sep. J.* **3**(3) (Sept. 1990).

Wnek, W.: "Electrokinetic and Chemical Aspects of Water Filtration," *Filt. Separ.,* **11**, 237 (1974).

Yada, K.: "Study of Chrysotile Asbestos by a High Resolution Electron Microscope," *Acta Cryst,* **23**, 704 (1967).

FINE STRUCTURE. In atomic spectra, the occurrence of a spectral line as a doublet, triplet, etc., due to the interaction or coupling between the orbital angular momentum and the spin angular momentum of the electrons in the emitting atoms. Fine structure is also exhibited by spectra of particles. See also **Atomic Spectra**.

FIRE-BRICK. Fire-brick is a type of brick capable of withstanding high temperatures and is used to line flues, stacks, furnaces, etc. Good resistance to heat flow is not to be secured simultaneously with refractoriness—indeed, the most refractory bricks generally have the highest thermal conductivities. Where necessary, insulation is added to minimize heat leaks. It is important for the refractory brick to be satisfactory on a number of points in addition to refractoriness, for resistance to melting is only one of several requirements to be met. Among these might be cited resistance to erosion by ash-laden gases, and to the fluxing action of molten slag. A good refractory should not spall badly under rapid temperature changes. The structural strength of fire-brick should hold up well as its temperature approaches the fusion temperature.

Modern installations often impose furnace conditions so severe that refractories other than fire-clay are needed. High aluminum and silicon carbide refractories are typical of these. The heat conductivities of the super-refractories are larger than those of fire-clay brick, and such construction should be backed up with high temperature insulation. Silicon carbide blocks are the most refractory and have the quality of resisting clinker adhesion better than ordinary fire-brick. Their fusion temperature is about 4000°F (2204°C).

Clay fuses at from 2800 to 3200°F, (1538 to 1760°C) the upper limit being for flint clay and the lower for the plastic form which, due to its cementing qualities, is especially valuable in fire-brick manufacture. Red brick is not suitable for refractory service, nor is insulating brick. There are several fire-clay furnace cements on the market that are adaptable to monolithic lining. The standard size of fire-brick and insulating brick is 9 inches by 4½ inches by 2½ inches (22.9 by 11.4 by 6.4 centimeters).

FIRECLAY. This term is chiefly used by British geologists to designate the leached clays, rich in silica and alumina and low in alkalies and lime, which lie directly beneath coal beds. These clays are of economic importance because they are refractory, and do not melt when heated to high temperatures.

FIRE-RETARDANT PAINTS. See **Paint and Finish Removers**.

FIRMING AGENTS (Foods). Certain foodstuffs, such as apples, potatoes, and beans, tend to be rather fragile when subjected to processing operations prior to packaging (canning, freezing, etc.) and, if not treated in some way to retain their natural firmness to a relatively high degree, the texture of the final product will be disappointing (mushiness versus slight chewiness or crispness). Certain chemical substances can be added prior to or during processing to protect and retain natural firmness. These substances include: aluminum potassium sulfate, aluminum sodium sulfate, aluminum sulfate, calcium carbonate, calcium chloride, calcium citrate, calcium gluconate, calcium hydroxide, calcium lactobionate, calcium phosphate (monobasic), calcium sulfate, and magnesium chloride, among others.

Researchers have found that calcium lactate, in particular, can be an effective agent for preserving the firmness of apple slices during processing and prior to canning or freezing. Studies have shown that calcium salts participate in firming the tissues of various fruits and vegetables by forming calcium pectates. Calcium citrate is quite useful for firming peppers, potatoes, tomatoes, lima beans, and snap beans. Manufacturers of pet foods have found that from 1 to 2.5% monoglyceride contributes to the firming of pet foods, as well as aiding in the prevention of fat separation.

FISCHER, EDMOND H. (1920–1992). Fischer was born in Shanghai, China and spent his early childhood studying in Geneva. He received degrees in biology and chemistry from the University of Geneva. He came to America in 1953 to work at Caltech but soon accepted a position at the University of Washington in their biochemistry department. Here he met Edwin Krebs and the two men started researching together. They discovered cellular enzymes could be activated and inactivated through phosphorylation-dephosphorylation by protein kinases.

Fischer and Krebs were awarded the Nobel Prize in Physiology or Medicine for their work on biological regulation mechanisms in 1992.

See also **Carbohydrates**.

J.M.I.

FISCHER, EMIL (1852–1919). A German organic chemist, recipient of the Nobel prize in chemistry (1902) for his original research in the chemistry of purines and sugars. He was professor of Chemistry at the University of Berlin (1882), succeeding Hofmann. He synthesized fructose and glucose and elucidated their stereochemical configurations; he also established the nature of uric acid and its derivatives. Additional work included enzyme chemistry, proteins, synthetic nitric acid, and ammonia production.

FISCHER, ERNST OTTO (1918–). A German inorganic chemist who won the Nobel prize for chemistry in 1973 with Geoffrey Wilkinson for their independent work on the chemistry of organometallic "sandwich compounds." He was the contributor to many publications on organometallic chemistry. His education and work were primarily in Munich.

FISCHER, HANS (1881–1945). A German biochemist who studied under Emil Fischer. He was awarded the Nobel prize in chemistry in 1930 for his synthesis of the blood pigment hemin. He also did important fundamental research on chlorophyll, and porphyrins, and carotene.

FISCHER-TROPSCH PROCESS. Synthesis of liquid or gaseous hydrocarbons or their oxygenated derivatives from the carbon monoxide and hydrogen mixture (synthesis gas) obtained by passing steam over hot coal. The synthesis is carried out with metallic catalysts such as iron, cobalt, or nickel at high temperature and pressure. The process was developed in Germany in 1923 by F. Fischer and H. Tropsch and was used there for making synthetic fuels before and during World War II. It has never been used for this purpose in the U.S., the only coal-to-gasoline conversion plant using this process is Sasol in South Africa, though the closely related Lurgi process is being used rather extensively in a number of locations. Easing of the petroleum crisis has tended to diminish conversion activity in the U.S.

FISCHER-TROPSCH SYNTHESIS. (Synthol process; Oxo synthesis). Synthesis of hydrocarbons, aliphatic alcohols, aldehydes, and ketones by the catalytic hydrogenation of carbon monoxide using enriched synthesis gas from passage of steam over heated coke. The ratio of products varies with conditions. The high-pressure Synthol process gives mainly oxygenated products, and addition of olefins in the presence of cobalt catalyst (Oxo synthesis) produces aldehydes. Normal-pressure synthesis leads mainly to petroleum-like hydrocarbons.

FISSION (Nuclear). See **Nuclear Fission**.

FITTING REACTION. The formation of aromatic hydrocarbons from aryl or aryl and alkyl bromides by the use of sodium, e.g., bromobenzene plus ethyl bromide plus sodium forms ethylbenzene plus sodium bromide, $C_6H_5Br + C_2H_5Br + 2Na \rightarrow C_6H_5 \cdot C_2H_5 + 2NaBr$.

FIXED BED. In processing terminology, a fixed-bed installation (usually a reactor) requires that materials in the solid phase that are to be reacted with gases and vapors remain in a fixed location. In other words, the flow in such equipment is that of the materials in the gaseous or vapor phase. The solid materials require careful preparation to permit a maximum of surface to be exposed to the gases that pass through them and to avoid the occurrence of channels, through which the gases would pass without contacting the bulk of the solids. Beds of this type are often used in connection with various catalytic operations and are used, for example, in the production of benzene, in catalytic reforming, hydrocracking, hydrotreating, vinyl chloride monomer production, and in ion-exchange operations. A later reaction development, but also one that is several decades old, is the fluid-bed reactor, in which the solids to be reacted are essentially "fluidized" and intermix with other solids or with gases in a rapidly moving turbulent stream—in contrast with the solids remaining in a fixed bed. Whether or not a fixed bed is used instead of a fluidized bed is determined by numerous factors, notably time and cost. Fixed-bed reactors tend to be simpler and less costly, but often require cyclic operation because of the need to replenish the solids (catalysts, etc.). Thus to achieve continuous production, two or more beds are required (one on stream; the other regenerating).

See also **Fluidization**.

FIXED OILS. These are fats, compounds of glycerin and various complex fatty acids. Fixed oils are often called the nonvolatile oils, in distinction to the essential or volatile oils, which are readily vaporized by heat. It is characteristic of the fixed oils to leave a spot when dropped on paper. Many of them remain liquid at common room temperatures, others are solid at such temperatures. Solid forms are usually called fats, a purely arbitrary distinction since slight changes in temperature will cause many of them to change from liquid to solid or vice versa.

Fixed oils, especially those of economic importance, are largely obtained from the seeds of plants. They have a high energy value, so form a valuable food if they prove palatable.

Various methods are employed to obtain the oil from the vegetable tissues. Quite commonly, the seeds containing the oil are subjected to great pressures in hydraulic presses. This may be done without heating, but is frequently facilitated by heating the seeds, the oil being then hot-pressed instead of cold-pressed. More recently, the screw press has come into use. This press has a rotating screw, which presses the ground seeds under very high pressure through a cage-like cylinder. The oil is squeezed through openings in the cylinder walls, while the seed meal is discharged through an opening in the end of the cylinder. This procedure is advantageous in that it is continuous, and the machine need not be stopped for loading. A third method of obtaining the oil is by means of solvents. Following expression of the oil from the plant tissues, various methods of refining, decolorizing and deodorizing are employed.

Fixed oils are usually classified into three groups, drying, semidrying, and nondrying oils. Often a fourth group is made of those which are usually seen in solid form, the vegetable fats, although they differ but little otherwise from the other groups.

Drying oils are those which on exposure to air form a tough elastic film. Linseed oil from flax seeds is one of the most important and is largely used in making paints and varnishes. Tung oil, obtained from the fruits of *Aleurites fordii*, is a valuable oil much used in the manufacture of waterproof varnishes and quick-drying enamels. The tree, a native of China and Japan, has been introduced into Florida. Other drying oils are nut oil, from walnut seeds, poppy seed oil, hemp seed oil, and sunflower oil, the latter largely a product of the former U.S.S.R.

Nondrying oils are those which remain permanently greasy or sticky, becoming rancid after a time. Among these oils the most important are olive oil, castor oil from the seeds of the castor bean plant, rape seed oil, peanut oil, almond oil, used medicinally, and tea seed oil.

Semidrying oils are intermediate in nature. The principal semidrying oils are cotton-seed oil, Soyabean oil, corn or maize oil and sesame oil. The latter is obtained from the seeds of *Sesamun indicum*, a member of the *Pedaliaceae*, cultivated in India, China and Japan, where the oil is much used as a food oil and for cooking.

Nondrying oils that are ordinarily solid are palm and palm-kernel oil, coconut oil, and cocoa butter. Another interesting oil of this group is macassar oil, obtained from the seeds of *Schleichera trijuga*, one of the *Sapindaceae*, occurring in tropical Asia. The oil was formerly much used as a potential "hair restorer," necessitating the use of removable covers, or antimacassars, on the backs of upholstered chairs. The same tree also yields a useful timber.

FLAME HARDENING.

Surface hardening of steel or cast iron by heating a thin surface layer to the hardening temperature with an oxy-acetylene flame, followed by rapid cooling. Depending on the nature of the part to be hardened, either the torch system or the work itself may be moved. Cylindrical parts are rotated before a stationary flame. An air jet or liquid spray following the torch is used to quench-harden the surface. The relatively cool metal in the interior hastens cooling of the surface by conduction. The depth of flame hardening may be less than $\frac{1}{16}$ inch to about $\frac{1}{4}$ inch (1.6–6 millimeters), depending on the thickness of the section and the service requirements. Distortion is generally less than in parts hardened by general heating and quenching.

Since no hardening agent such as carbon or nitrogen is added to the surface of the steel by this process, only steels having sufficient carbon to harden readily upon quenching are used for flame hardening. The most desirable range is 0.35–0.70% carbon. The hardening treatment is followed by a low-temperature tempering treatment to relieve quenching strains. Typical applications of flame hardening are gear teeth, cams, bearing surfaces, rail ends, crankshafts, and many other machine parts and tools.

FLAME PHOTOMETRY AND SPECTROMETRY.

The basic principle of flame emission spectrometry rests on the fact that salts of metals, when introduced under carefully controlled conditions into a suitable flame, are vaporized and excited to emit radiations that are characteristic for each element. Correlation of the emission intensity with the concentration of that element forms the basis of quantitative evaluation.

The determinations of sodium and potassium constitute the majority of published applications. However, the flame is a suitable emission source for at least 45 elements, which may be grouped as follows:

1. *Elements determined*: aluminum, barium, boron, calcium, cesium, chromium, copper, iron, lead, lithium, magnesium, manganese, potassium, rubidium, sodium, strontium.
2. *Elements determined but sometimes overlooked*: antimony, arsenic, bismuth, cadmium, cobalt, gallium, indium, lanthanum, nickel, palladium, rare earths (except cerium), rhodium, ruthenium, scandium, silver, tellurium, thallium, tin, and yttrium.
3. *Elements with distinctive but less sensitive flame spectra*: beryllium, germanium, gold, mercury, molybdenum, niobium, rhenium, selenium, silicon, titanium, tungsten.
4. *Elements determined by indirect means*: bromine, chlorine, fluorine, iodine (although bromine, chlorine and fluorine can be determined by their metallic halide spectra), phosphorus, and silicon.

Materials in which these elements are determined by flame spectrometry include water, glasses, cement, soils, fertilizers, plant materials, biological fluids and tissues, petroleum products and metallurgical products.

The *flame spectrometer*, used in emission spectrometry, consists of: (1) the pressure regulators and flow meters for the fuel gases; (2) the atomizing device; (3) the flame source; (4) the optical system;

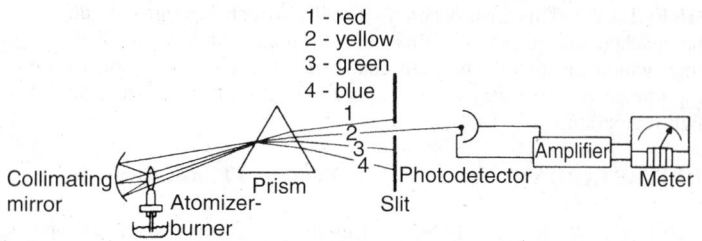

Fig. 1.   Flame spectrometer

(5) appropriate photosensitive detectors; and (6) the electrical circuit for measuring or recording the intensity of the radiation. Depending upon the use intended, the instrument may be a relatively simple assemblage of interference filters and a photodetector, i.e., a flame photometer, or it may be an elaborate prism or grating monochromator, i.e., a flame spectrometer such as the instrument illustrated in Fig. 1.

Virtually all flame spectrometers rely on atomization to deliver a steady flow of solution to the flame. The solution is drawn through a capillary positioned either concentric with or at right angles to the annulus or capillary from which the aspirating gas (oxygen or air under pressure) enters. At the tip of the solution capillary, the liquid is sheared off and dispersed into droplets by the blast of oxygen or air.

The best isolation of radiant energy can be achieved with flame spectrometers that incorporate either a prism or grating monochromator, those with prisms having variable gauged entrance and exit slits. Both these spectrometers provide a continuous selection of wavelengths with resolving power sufficient to separate completely most of the easily excited emission lines, and afford freedom from scattered radiation sufficient to minimize interferences. Fused silica or quartz optical components are necessary to permit measurements in the ultraviolet portion of the spectrum below 350 nanometers. See also **Analysis (Chemical)**; **Atomic Spectroscopy**; **Photometers**; and **Spectro Instruments**.

FLAME-RETARDING AGENTS.

Some pertinent definitions include *fire retardant* (flame retardant), used to describe polymers in which basic flammability has been reduced by some modification as measured by one of the accepted test methods: *fire-retardant chemical*, used to denote a compound or mixture of compounds that when added to or incorporated chemically into a polymer serves to slow or hinder the ignition or growth of fire, the foregoing effect occurring primarily in the vapor phase; *materials*, single substances of which things are constructed that may be composed of single or blended polymers, may be layered or fiber-reinforced, and might contain a variety of additives; and *products*, consumer items made of one or more materials.

Measuring Fire Performance of Products

Laws have been promulgated to improve the fire performance of everyday fuels. Most of the fire test methods in regulations have been developed by consensus standards organizations in response to a particular fire hazard. The two leading entities are the American Society for Testing and Materials (ASTM) and the National Fire Protection Association (NFPA). Methods are then referenced in the model building codes, such as the Standard Building Code (Southern Building Code), Basic Building Code (Building Code Officials Administration International), and the Uniform Building Code (International Conference of Building Officials), as well as NFPA's National Fire Codes, National Electrical Code, and Life Safety Code. Selected portions of these structures are in turn incorporated into laws by a governmental jurisdiction. In addition, there are a number of voluntary practices. For example, Underwriters Laboratories (UL) allows the use of its endorsement on products that meet their test criteria, and the upholstered furniture industry has adopted voluntary cigarette ignition-resistance standards.

Fire test methods attempt to provide correct information on the fire contribution of a product by exposing a small sample to conditions expected in a fire scenario.

The assessment of the contribution of a product to the fire severity and the resulting hazard to people and property combines appropriate product flammability data, descriptions of the building and occupants, and computer software that includes the dynamics and chemistry of fires.

Methods for Improved Performance

The materials of attention in promoting fire safety are generally organic polymers, both natural, such as wood (qv) and wool (qv), and synthetic, nylon (see **Polyamides**), vinyl, and rubber (qv). Less fire-prone products generally have either inherently more stable polymeric structures or fire-retardant additives. The former are usually higher priced engineering plastics (qv) which achieve increased stability at elevated temperatures by incorporating stronger (often aromatic) chemical bonds in the backbone of the polymer. Examples are the polyimides, polybenzimidazoles, and polyetherketones. There are also some advanced polymers, such as the polyphosphazenes and the polysiloxanes, which have strong inorganic backbones. Thermally stable pendent groups are also necessary. Strongly bonded polymers may, however, be brittle or difficult to process.

Fire-retardant additives are most often used to improve fire performance of low-to-moderate cost commodity polymers. These additives may be physically blended with or chemically bonded to the host polymer. They generally effect either lower ignition susceptibility or, once ignited, lower flammability. Ignition resistance can be improved solely from the thermal behavior of the additive in the condensed phase. Retardants such as hydrated alumina add to the heat capacity of the product, thus increasing the enthalpy needed to bring the polymer to a temperature at which fracture of the chemical bonds occurs. The endothermic volatilization of bound water can be a significant component of the effectiveness of this family of retardants. Other additives, such as the organophosphates, change polymer decomposition chemistry. These materials can induce the formation of a cross-linked, more stable solid and can also lead to the formation of a surface char layer. This layer both insulates the product from further thermal degradation and impedes the flow of potentially flammable decomposition products from the interior of the product to the gas phase where combustion would occur.

RICHARD G. GANN
National Institute of Standards and Technology

ANTIMONY AND OTHER INORGANIC FLAME RETARDANTS

Antimony Compounds

Toxicity. Antimony has been found not to be a carcinogen or to present any undue risk to the environment. However, because antimony compounds also contain minor amounts of arsenic which is a poison and a carcinogen, warning labels are placed on all packages of antimony trioxide.

Mixed Metal Antimony Synergists. Worldwide scarcities of antimony have prompted manufacturers to develop synergists that contain less antimony. Other metals have been found to work in concert with antimony to form a synergist that is as effective as antimony alone. Thermoguard CPA from Elf Atochem NA, which contains zinc in addition to antimony, can be used instead of antimony oxide in flexible poly(vinyl chloride) (PVC) as well as some polyolefin applications. The Oncor and AZ products which contain silicon, zinc, and phosphorus from Anzon Inc. can be used in a similar manner. The mixed metal synergists are 10 to 20% less expensive than antimony trioxide.

Boron Mechanism. Boron functions as a flame retardant in both the condensed and vapor phases. Under flaming conditions boron and halogens form the corresponding trihalide. Because boron trihalides are effective Lewis acids, they promote cross-linking, minimizing decomposition of the polymer into volatile flammable gases. These trihalides are also volatile; thus they vaporize into the flame and release halogen which then functions as a flame inhibitor.

Boron also reacts with hydroxyl-containing polymers such as cellulose. When exposed to a flame the boron and hydroxyl groups form a glassy ester that coats the substrate and reduces polymer degradation. A similar type of action has been observed in the boron–alumina trihydrate system.

Mechanism. Alumina trihydrate functions as a flame retardant in both the condensed and vapor phases. When activated, it decomposes endothermically, eliminating water.

$$2\,Al\,(OH)_3 \longrightarrow Al_2O_3 + 3\,H_2O$$

In the flame phase the water vapor forms an envelope around the flame, which tends to exclude air and dilute the flammable gases.

Other Inorganic Materials

Other inorganic flame retardants include magnesium hydroxide, molybdenum oxides, and tin.

Applications

Poly(vinyl chloride). PVC is a hard, brittle polymer that is self-extinguishing. In order to make PVC useful and more pliable, plasticizers (qv) are added. More often than not the plasticizers are flammable and make the formulation less flame-resistant. The flame resistance of the poly(vinyl chloride) can be increased by the addition of an inorganic flame-retardant synergist, e.g., antimony oxide, mixed metal antimony synergists, zinc borate, molybdenum oxide, zinc stannates, and alumina trihydrate.

Unsaturated Polyesters. There are two approaches used to provide flame retardancy to unsaturated polyesters. These materials can be made flame-resistant by incorporating halogen when made, or by adding some organic halogen compound when cured. In either case a synergist is needed.

Olefin Polymers. The flame resistance of polyethylene can be increased by the addition of either a halogen synergist system or hydrated fillers. Similar flame-retarder packages are used for polypropylene.

IRVING TOUVAL
Touval Associates

HALOGENATED FLAME RETARDANTS

Fundamentals of Flammability

In order for a solid to burn it must be volatilized, because combustion is almost exclusively a gas-phase phenomenon. In the case of a polymer, this means that decomposition must occur. The decomposition begins in the solid phase and may continue in the liquid (melt) and gas phases. Decomposition produces low molecular weight chemical compounds that eventually enter the gas phase. Heat from combustion causes further decomposition and volatilization and, therefore, further combustion. Thus the burning of a solid is like a chain reaction. For a compound to function as a flame retardant it must interrupt this cycle in some way. There are several mechanistic descriptions by which flame retardants modify flammability: inert gas dilution, thermal quenching, protective coatings, physical dilution, and chemical interaction.

Flammability Testing

One problem associated with discussing flame retardants is the lack of a clear, uniform definition of flammability. Hence, no clear, uniform definition of decreased flammability exists. The latest American Society for Testing and Materials (ASTM) compilation of fire tests lists over one hundred methods for assessing the flammability of materials. Several of the most common tests used on plastics include ASTM E162-87, ASTM E119-88, MVSS 302, ASTM D2863-87, ASTM E662-83, ASTM E84, UL 723, UL 910, UL 94, UL 790, ASTM E108-90, UL 1715, CAL 133, CAL 117, ASTM E1353-90, ASTM E1354-90, ASTM E1354-90, and ASTM E906.

Flame Retardants

Compounds of chlorine and bromine are the halogen compounds having commercial significance as flame-retardant chemicals. Halogenated flame retardants can be broken down into three classes: brominated aliphatic, chlorinated aliphatic, and brominated aromatic. As a general rule, the thermal stability increases as brominated aliphatic < chlorinated aliphatic < brominated aromatic. The thermal stability of the aliphatic compounds is such that with few exceptions, thermal stabilizers such as a tin compound must be used. Brominated aromatic compounds are much more stable and may be used in thermoplastics at fairly high temperatures without the use of stabilizers and at very high temperatures with stabilizers.

Antimony–Halogen Synergism. Antimony oxide is commonly employed as a fire-retardant supplement for halogen-containing polymer systems as a means of reducing the halogen levels required to obtain a given degree of flame retardancy. This reduction is desirable because the required halogen content may be so high that it affects the physical properties of the final polymer. In many cases, the antimony oxide is used simply to give a more cost-effective system.

Brominated Additive Flame Retardants. Additive flame retardants are those that do not react in the application designated. There are a few compounds that can be used as an additive in one application and as a reactive in another. Tetrabromobisphenol A (TBBPA) is the most notable example. Table 1 lists the properties of most commercially available bromine-containing additive flame retardants.

Chlorinated Additive Flame Retardants. Bis(hexachlorocyclopentadieno)cyclooctane and pentabromochlorocyclohexane are chlorinated compounds used as additive flame retardants.

Oligomeric Flame Retardants. There are several oligomeric flame retardants. The principal advantage claimed for these materials is their

TABLE 1. ADDITIVE FLAME RETARDANTS

Common name	Mol formula	Bromine, %	Specific gravity	Mp, °C
		BROMINATED		
ethylenebisdi-bromonorbor-nanedicarbox-imide	$C_{20}H_{20}Br_4N_2O_4$	45	2.07	294
tetrabromo-bisphenol A	$C_{15}H_{12}O_2Br_4$	58.4	2.17	180
tris-dibromopro-pylisocyanurate	$C_{12}H_{15}Br_6N_3O_3$	65.8		106–108
ethylenebistetra-bromophthal-imide	$C_{18}H_4N_2O_4Br_8$	68	2.66	445
tetrabromo-bisphenol S-bis(2,3 dibromo-propylether)	$C_{18}H_{14}Br_8O_4S$	70.8		52–55
tetrabromocy-clooctane	$C_8H_{12}Br_4$	74.7	2.27	73
dibromo-ethyldibromo-cyclohexane	$C_8H_{12}Br_4$	74.7	2.38	70–76
tetradecabromo-diphenoxy-benzene	$C_{18}Br_{14}O_2$	82	3.25	370
decabromo-diphenyl oxide	$C_{12}Br_{10}O$	83	3.00	305
	POLYMERIC AND OLIGOMERIC BROMINATED			
tetrabromo-bisphenol A carbonate oligomer, phenoxy end capped	$(C_{16}H_{12}O_3Br_4)_n$	52	2.2	210–230
epoxy oligomers of tetrabromo-bisphenol A	$(C_{18}H_{16}O_3Br_4)_n$	52–54		
poly(dibromostyrene)	$(C_8H_6Br_2)_n$	59	1.9	155–165[a]
brominated polystyrene, low molecular weight	$(C_8H_{5.3}Br_{2.7})_n$	66	2.1	130–140
poly(pentabromo-benzylacrylate)	$(C_{10}H_6Br_5O_2)_n$	70	2.05	210

[a] A higher molecular weight version has a mp of 210–230°C.

resistance to bloom and plate-out. All of the available oligomeric flame retardants are brominated (Table 1).

Reactive Flame Retardants. Table 2 lists the commercially available reactive flame retardants and intermediates.

Economic Aspects

There are a relatively small number of producers of halogenated flame retardants, especially for brominated flame retardants, where three producers account for greater than 80% of world production.

Health and Safety

In general, the acute toxicity of halogenated flame retardants is quite low. Continual use of decabromidiphenyl oxide has been placed in question based on the discovery that under certain laboratory conditions brominated dibenzo-*p*-dioxins are generated.

Research sponsored by the Brominated Flame Retardants Industry Panel regarding the use of brominated flame retardants shows that there is no evidence that the use of decabromodiphenyl oxide leads to any unusual risk. In addition, a study by the National Bureau of Standards (now National Institute of Science and Technology) showed that the use of flame retardants significantly decreased the hazards associated with burning of common materials under realistic fire conditions. Work in Japan confirms this finding.

TABLE 2. BROMINATED REACTIVE FLAME RETARDANTS AND INTERMEDIATES

Compound	Molecular formula	Bromine, %	Specific gravity	Mp, °C
diester/ether diol of tetrabromoph-thalic anhydride	$C_{15}H_{16}O_7Br_4$	46	1.80	liquid
tetrabromobisphe-nol A-bis(2-hydroxyethyl ether)	$C_{19}H_{20}O_4Br_4$	51.6	1.8	116
tetrabromobisphe-nol A	$C_{15}H_{12}O_2Br_4$	58.4	2.17	180
disodium salt of tetrabromo-phthalate	$C_8O_4Br_4Na_2$	61	2.8	>500
tetrabromodipen-taerythritol	$C_9H_{20}O_2Br_4$	63.2	1.98	81.5–82.5
tribromophenyl allyl ether	$C_9H_7Br_3O$	64.6	2.20	74
tribromostyrene	$C_8H_5Br_3$	68		65–67
2,4,6-tribromo-phenol	$C_6H_3OBr_3$	72.5	2.22–2.55	95.5
tribomoneopentyl alcohol	$C_5H_9OBr_3$	73.6	2.28	62–67
pentabromobenzyl bromide	$C_7H_2Br_6$	84.8		
hexachlorocyclopen-tadiene	C_5Cl_6	78.0[a]	1.710	11[b]
chlorendic acid	$C_9H_2O_4Cl_6$	55.0[a]		[c]

[a] Chlorine, %.
[b] 239°C bp.
[c] Decomposes to the anhydride.

ALEX PETTIGREW
Ethyl Technical Center

PHOSPHORUS FLAME RETARDANTS

Mechanisms of Action

Condensed-Phase Mechanisms. The mode of action of phosphorus-based flame retardants in cellulosic systems is probably best understood. Cellulose decomposes by a noncatalyzed route to tarry depolymerization products, notably levoglucosan, which then decomposes to volatile combustible fragments such as alcohols, aldehydes, ketones, and hydrocarbons. However, when catalyzed by acids, the decomposition of cellulose proceeds primarily as an endothermic dehydration of the carbohydrate to water vapor and char. Phosphoric acid is particularly efficaceous in this catalytic role because of its low volatility (see **Phosphoric Acids and Phosphates**). Also, when strongly heated, phosphoric acid yields polyphosphoric acid which is even more effective in catalyzing the cellulose dehydration reaction. The flame-retardant action is believed to proceed by way of initial phosphorylation of the cellulose.

Vapor-Phase Mechanisms. Phosphorus flame retardants can also exert vapor-phase flame-retardant action. Both physical and chemical vapor-phase mechanisms have been proposed for the flame-retardant action of certain phosphorus compounds, such as triphenyl phosphate.

Interaction with Other Flame Retardants. Some claims have been made for a phosphorus–halogen synergism. A few cases are well established; however, phosphorus–halogen interactions are often merely additive, and in some cases slightly less than additive.

Antagonism between antimony oxide and phosphorus flame retardants has been reported in several polymer systems, and has been explained on the basis of phosphorus interfering with the formation or volatilization of antimony halides, perhaps by forming antimony phosphate.

Commercial Phosphorus-Based Flame Retardants

Many thousands of phosphorus compounds have been described as having flame-retardant utility. The compounds demonstrating commercial utility are much more limited in number. They include inorganic phosphorus compounds (red phosphorus, ammonium phosphates, insoluble ammonium polyphosphate, phosphoric acid-based systems for cellulosics), additive organic phosphorus flame retardants (melamine phosphates and other amine phosphates, trialkyl phosphates, dimethyl methylphosphonate, diethyl ethylphosphonate), halogenated alkyl phosphates and phosphonates (2-chloroethanol phosphate (3:1), 1-chloro-2-propanol phosphate (3:1), 1,3-dichloro-2-propanol phosphate (3:1), bis(2-chloroethyl) 2-chloroethylphosphonate, diphosphates, oligomeric 2-chloroethyl phosphate, 2-chloroethyl 2-bromoethyl 3-bromoneopentyl phosphate, oligomeric cyclic phosphonates, pentaerythritol phosphates, cyclic neopentyl thiophosphoric anhydride, aryl phosphates, phosphine oxides), reactive organic phosphorus compounds (organophosphorus monomers, phosphorus-containing diols and polyols, oligomeric phosphate–phosphonate), reactive organophosphorus compounds in textile finishing (tetrakis(hydroxymethyl)phosphonium salts), dimethyl 3-[(hydroxymethyl)amino]-3-oxopropylphosphonate, and phosphorus-containing polymers (polyester fibers containing phosphorus).

Health, Safety, and Environmental Factors

Toxicology. The structure–toxicity relationships of organophosphorus compounds have been extensively researched and are relatively well understood. The phosphorus-based flame retardants as a class exhibit only moderate-to-low toxicity.

A particular mode of neurotoxicity was discovered for tricresyl phosphate that correlated with the presence of the *o*-cresyl isomer (or certain other specific alkylphenyl isomers) in the triaryl phosphates. The use of low ortho-content cresols has become the accepted practice in industrial production of tricresyl phosphate.

Mutagenic and later carcinogenic properties were found for tris(2,3-dibromopropyl) phosphate, a flame retardant used on polyester fabric in the 1970s. This product is no longer on the market. The chemically somewhat-related tris(dichloroisopropyl) phosphate has been intensively studied and found not to display mutagenic activity. Tris(2-chloroethyl) phosphate appears to be a weak tumor-inducer in a susceptible rodent strain.

There appears to be no documented case of any type of fire retardant contributing to human fire casualties. Most smoke inhalation casualties appear to be caused by carbon monoxide.

Effects on Visible Smoke. Smoke is a main impediment to egress from a burning building. Although some examples are known where specific phosphorus flame retardants increased smoke in small-scale tests, other instances are reported where the presence of the retardant reduced smoke. The effect appears to be a complex function of burning conditions and of other ingredients in the formulation.

Environmental Considerations. The phosphate flame retardants, plasticizers, and functional fluids have come under intense environmental scrutiny. Results published to date on acute toxicity to aquatic algae, invertebrates, and fish indicate substantial differences between the various aryl phosphates. The EPA has summarized this data as well as the apparent need for additional testing.

Tests in pure water, river water, and activated sludge showed that commercial triaryl phosphates and alkyl diphenyl phosphates undergo reasonably facile degradation by hydrolysis and biodegradation. The phosphonates can undergo biodegradation of the carbon-to-phosphorus bond by certain microorganisms.

Economic Aspects

The largest volume use of phosphorus-based flame retardants may be in plasticized vinyl. Other use areas for phosphorus flame retardants are flexible urethane foams, polyester resins and other thermoset resins, adhesives, textiles, polycarbonate–ABS blends, and some other thermoplastics. Development efforts are well advanced to find applications for phosphorus flame retardants, especially ammonium polyphosphate combinations, in polyolefins, and red phosphorus in nylons. Interest is strong in finding phosphorus-based alternatives to those halogen-containing systems which have encountered environmental opposition, especially in Europe.

EDWARD D. WEIL
Polytechnic University

FLAME RETARDANTS FOR TEXTILES

Flame Resistance

Factors Affecting Performance. The flame resistance of a textile fiber is affected by the chemical nature of the fiber, its ease of combustion, the fabric weight and construction, the efficiency of the flame retardant, the environment, and laundering conditions.

The weight and construction of the fabric affect its burning rate and ease of ignition. Lightweight, loose-weave fabrics burn much faster than heavier weight fabrics; therefore, a higher weight add-on of fire retardant is needed to impart adequate flame resistance.

Mechanism of Flame Retardants. The burning process of cellulose depends on both a source of ignition and the presence of oxygen. A low temperature degradation of cellulose proceeds by the formation of levoglucosan, which in turn undergoes dehydration and polymerization, leading to tars, flammable gases, liquids, and other solids. The flammable gases thus produced ignite, causing the liquids and tars to volatize to some extent. This produces additional volatile fractions which ignite and produce a carbonized residue that does not burn readily. The process continues until only carbonaceous material remains. After the flame has subsided, the carbonized residue slowly oxidizes and glowing continues until the carbonaceous char is consumed.

In general, cotton treated with an effective flame retardant provides the same decomposition products upon burning as does untreated cotton; however, the amount of tar is greatly reduced, with a corresponding increase in the solid char. Consequently, as decomposition takes place, smaller amounts of flammable gases are available from the tar, and greater amounts of nonflammable gases from the decomposition of the char fraction. Char is essentially carbon. Its oxidation causes afterglow. Phosphorus-containing compounds, in some cases polymers, are particularly effective in inhibiting char oxidation. Numerous studies have been made on burning of untreated and flame-retardant-treated cellulose.

Several theories have been postulated to explain the various types of flame retardants for cotton. These theories include coating, gas, thermal, and dehydration or chemical.

Durability of Retardant Finishes

Fire resistance of a treated cellulosic fabric is reduced when the retardant contains acid groups and the treated fabric is soaked or laundered in water containing calcium, magnesium, or alkali metal ions. Phosphate- and carbonate-based detergents affect durability of fire retardants. Soap-based detergents can result in a substantial loss of fire resistance because of the deposit of fatty acid salts. Phosphorus-based flame retardants are adversely affected by water hardness and laundry bleach, sodium hypochlorite. Exposure to sunlight and weathering can lead to sufficient loss of flame retardant so that the fabric is no longer flame resistant. Similarly, a combination of sunlight followed by laundering or autoclaving can also lead to loss of flame resistance in a cellulosic fabric.

Nondurable Finishes. Flame-retardant finishes that are not durable to laundering and bleaching are, in general, relatively inexpensive and efficient. In some cases, a mixture of two or more salts is more effective than either of the components alone.

The water-soluble flame retardants are most easily applied by impregnating the fabric with a water solution of a retardant, followed by drying. The water-soluble flame retardants used most widely for textiles are listed in Table 3. Less commonly used retardants include sulfamates of urea or other amides and amines; aliphatic amine phosphates, such as triethanolamine phosphate, phosphamic acid (amido phosphoric acid, $H_2PO_3NH_2$), and its salts; and alkylamine bromides, phosphates, and borates.

Semidurable Finishes. Semidurable fire retardants resist removal from 1 to approximately 15 launderings. Such retardants are adequate for applications such as drapes, upholstery, and mattress ticking. If they are sufficiently resistant to sunlight or can be easily protected from actinic degradation, they can also be applied to outdoor textile products. The principal disadvantage of water-soluble flame retardants is their lack of durability. This undesirable property can be overcome by precipitating their inorganic oxides on the fabric, e.g., $WO_3 \cdot xH_2O$ and $SnO_2 \cdot yH_2O$:

$$2\,Na_2WO_4 + SnCl_4 + (2x + y)H_2O \longrightarrow 4\,NaCl$$
$$+ 2\,WO_3 \cdot xH_2O + SnO_2 \cdot yH_2O$$

There are several methods for introducing the insoluble deposits into the fabric structure. The multiple bath method, in which the fabric is first impregnated with a water-soluble salt or salts in one bath and is then passed into a second bath which contains the precipitant, is used most often. Most semidurable retardants used on cotton are based on a combination of phosphorus and nitrogen compounds.

Early Durable Finishes. Early studies to produce durable flame retardants for cellulose were based on treatment with inorganic compounds containing antimony and titanium.

Outdoor Finishes. Excellent fire-resistant fabric has been obtained by treating fabric with a suspension or emulsion of insoluble fire-retardant

TABLE 3. WATER-SOLUBLE FLAME-RETARDANT FORMULATIONS,[a] % COMPOSITION

Formulation	Borax	Boric acid	Diammonium phosphate	Sodium phosphate dodecahydrate	Other
	$Na_2B_4O_7 \cdot 10\ H_2O$	H_3BO_3	$(NH_4)_2HPO_4$	$Na_3PO_4 \cdot 12\ H_2O$	
1	70	30			
2	47	20	33		
3		50		50	
4		50	50		
5	50	35		15	
6			25		75[b]
7	15	47			38[c]

[a] 100% Ammonium bromide, NH_4Br, is also used.
[b] Ammonium sulfamate, $NH_4OSO_2NH_2$.
[c] 18% Sodium phosphate, Na_3PO_4, and 20% sodium tungstate dihydrate, $Na_2WO \cdot 2H_2O$.

salts or oxides, e.g., antimony(III) oxide, along with a chlorinated organic vehicle such as chlorinated paraffin.

In the 1990s, two types of flame retardants are preferred for outdoor fabrics, i.e., a system based on phosphorus and nitrogen such as the precondensate–NH_3 finish and an antimony–bromine system based on decabromodiphenyl oxide and antimony(III) oxide.

FWWMR Finish. The abbreviation for fire, water, weather, and mildew resistance, FWWMR, has been used to describe treatment with a chlorinated organic metal oxide. Plasticizers, coloring pigments, fillers, stabilizers, or fungicides usually are added. However, hand, drape, flexibility, and color of the fabric are more affected by this type of finish than by other flame retardants. Add-ons of up to 60% are required in many cases to obtain adequate flame resistance. Durability of this finish is good and fabric processed properly retains its flame resistance after four to five years of outdoor exposure. This type of finish is suited for very heavy fabrics, e.g., tents, tarpaulins, or awnings.

Test Methods

Numerous tests covering flame retardancy and related matters are available. The requirements most often specified for fire resistance of a textile material are that it must pass either Federal Specification Method 5903 or NFPA 701.

Types of Retardants

Fire Retardants for Cellulosics. Phosphorus-containing materials are by far the most important class of compounds used to impart durable flame resistance to cellulose. Flame-retardant finishes containing phosphorus compounds usually also contain nitrogen or bromine or sometimes both.

Flame retardant fabrics and finishes include mesylated and tosylated celluloses, urea–phosphate type, phosphonomethylated ethers, amide-based systems, cyanamide, dialkyl phosphite and related retardants (Pyrovatex CP "New", dialkylphosphonopropionamides, triazines), THPC-based retardants (THPC–amide process, THPC–urea–disodium phosphate, THPOH–amide process, THP–amide process), ammonia–gas-cured flame retardants (THPOH–NH_3 process, precondensate–NH_3 process), and pentamethylphosphorotriamide.

Application techniques include radiation and incorporation of flame retardants in fiber.

Textile-Specific Uses of Flame Retardants

Flame retardants are used in smolder-resistant upholstery fabric, combination flame retardant–durable press performance, flame-retardant treatments for wool, thermoplastic fibers (Tris, decabromodiphenyl oxide–polyacrylate finishes, Antiblaze 19, nylon finishes), polyester–cotton fiber blends (THPOH–ammonia–Tris finish, decabromodiphenyl oxide–polyacrylate finish, THPC–amide–poly(vinyl bromide) finish, THPOH–NH_3 and Fyrol 76, LRC-100 finish, phosphonium salt–urea precondensate), cotton–wool blends, and core-yarn fabric.

Economic Aspects

The identification of Tris as a potential carcinogen dealt a resounding blow to the flame-retardant finishing industry. From 1977 to 1984, several principal suppliers of flame-retardant chemicals either reduced the size of their operations or abandoned the market completely. However, Albright and Wilson Corporation (U.K.) continues to produce THPC–urea

precondensate and market it worldwide, and Westex Corporation (Chicago) continues to apply precondensate–NH_3 finish to millions of yards of goods for various end uses. American Cyanamid reentered the market with a precondensate-type flame retardant based on THPS.

The largest commission finishers of fire-resistant textiles in the United states (ca 1993) are Westex and MF&H Textiles, Inc. (Butler, Georgia). Specialized flame-retardant applications to cotton-wrapped polyester, Kevlar, nylon, and glass core yarns are beginning to attract the interest of the industry for special-purpose fabrics.

Health and Safety

Because Tris polyester flame-retardant chemical has been demonstrated to be a potential carcinogen, workers in this field have tested a number of commonly used chemicals for potential mutagenicity. Neither the THPOH–NH_3 finish nor its extracts caused a significant systematic increase in mutations when tested by the Ames mutagenicity test. The Hooker Chemical Co. has reported results of tests conducted by an independent laboratory which indicate no significant mutagenic potential from any of the company's proprietary textile flame retardants. Although Fyrol 76 was reported to be nontoxic, results from its mutagenic screening are not known. Stauffer's substitute for Tris, Fyrol FR2, was accused of mutagenic activity by the Environmental Defense Fund, and has been withdrawn from the market by the company. A study has been made by the National Toxicology Program Study on the carcinogenicity of THPC and THPS which concluded that there is no evidence of carcinogenic activity for either compound in rats or mice.

Regulatory Legislation. In February 1978, the Consumer Products Safety Commission approved changes in the FF-3 and FF-5 standards for children's sleepwear. It eliminated the melt–drip time limit and coverage for sizes below 1 and revised the method of testing the trim. This permits the use of untreated 100% nylon and 100% polyester for children's sleepwear.

TIMOTHY A. CALAMARI, JR.
ROBERT J. HARPER, JR.
United States Department of Agriculture

Additional Reading

Ainsworth, S.J. *Chem. Eng. News* **70**, 34 (Aug. 31, 1992).

Barnard, J.A. and J.N. Bradley: *Flame and Combustion,* Chapman and Hall, London, 1985.

Drew, M.J., C.W. Jarves, and G.C. Lickfield: in Nelson, G.L. ed., *Fire and Polymers,* ACS Symposium Series 425, Washington, DC.

Gann, R.G., R.A. Dipert, and M.J. Drews: in J.I. Kroschwitz, ed., *Encyclopedia of Polymer Science and Engineering,* 2nd Edition, John Wiley & Sons, Inc., New York, NY, 1986, pp. 154–210.

Granzow, A.: *Accounts Chem. Res.* **11**(5), 177–183 (1978).

Hastie, J.W.: *High Temperature Vapors,* Academic Press, Inc., New York, NY, 1975.

Horrocks, A.R.: *Rev. Prog. Coloration* **16**, 62–101 (1986).

Kuryla, W.C. and A.J. Papa, eds.: (1973–1975), *Flame Retardancy of Polymeric Materials,* Vol. 5, Marcel Dekker, Inc., New York, NY, 1979.

Lewin, M., S.M. Atlas, and E.M. Pearce, eds.: *Flame Retardant Polymeric Materials,* Plenum Press, New York, 1975.

Lyons, J.W.: *The Chemistry and Uses of Fire Retardants,* Wiley-Interscience, New York, NY, 1970, Chapt. 5.

Lyons, J.W.: *The Chemistry and Uses of Fire Retardants,* Wiley-Interscience, New York, NY, 1970.

Nelson, G.L. ed.: *Fire and Polymers,* ACS Symposium Series 425, American Chemical Society, Washington, DC, 1990.

Price, D., B. Iddon, and B.J. Wakefield, eds.: *Bromine Compounds Chemistry and Applications,* Elsevier, Amsterdam, the Netherlands, 1988.

Reeves, W.A., G.L. Drake, Jr., and R.M. Perkins: *Fire-Resistant Textiles Handbook,* Technomic Publishing Co., Inc., Westport, CT, 1974.

Touval, I.: *J. Fire Flam.* **3**, 130 (1972).

Troitzsch, J.: *International Plastics Flammability Handbook,* Hanser Publishers, Munich, Germany, 1990.

Textile Flammability, A Handbook of Regulations, Standards and Test Methods, American Association of Textile Chemists and Colorists, Research Triangle Park, NC, 1975.

Weil, E.D.: in R.E. Engel, ed., *Handbook of Organophosphorus Chemistry,* Marcel Dekker, Inc., New York, NY, 1992, pp. 683–738.

Weil, E.D.: "Additivity, Synergism and Antagonism in Flame Retardancy—Recent Developments," paper presented at *3rd Annual BCC Conference on Recent Advances in Flame Retardancy of Polymeric Materials,* Stamford, Conn., May 19–21, 1992.

FLASH DISTILLIATION.   See **Distillation; Desalination**.

FLASHING (Thermal).   Liquids may exist with thermal stability at high temperatures provided they are subjected to sufficiently high pressure. Water, for example, may be heated to about 700°F (371°C) without boiling if under a pressure of 3,200 psi (217.7 atmospheres). It is true of liquids in general that the lower the pressure on them the lower the boiling temperature and the lower the heat contained in the "saturated" liquid. Thus high-temperature liquids when passed from a region of pressure sufficient for stability into a low-pressure region are not able to contain all the heat originally possessed as heat of fluid, and will be spontaneously partially evaporated by the surplus. This violent readjustment to thermal equilibrium is called "flashing," and is a common occurrence, having many uses and occasionally creating hazards. For example, the destructiveness of a boiler explosion arises mainly from the violence of flashing action, since the water originally contained in a ruptured boiler drum at 600 psi (40.8 atmospheres) pressure suffers an almost instantaneous *four-hundred fold* expansion in volume.

FLASH POINT.   The lowest temperature at which an oil will volatilize to yield sufficient vapor to form with air an inflammable gaseous mixture, demonstrable through the production of a flash on contact with a small open flame. The flash point occurs at a temperature lower than the burning point, which is the lowest temperature at which the production of combustible gas occurs rapidly enough to support a steady flame. It is also to be noted that the flash point is the temperature of formation, under the test conditions, of the lower explosive mixture of the substance tested, with air. (The higher explosive mixture is the *maximum* concentration of vapor, with air, which will sustain combustion.)

The flash point is an important characteristic of oils used for various purposes, such as lubrication, because a low flash point indicates the presence or absence of undesirable lower-boiling compounds. On the other hand, the flash temperature is less important then the burning temperature in determining the fire risk of an oil. The flash point is tested experimentally by heating the oil under certain specified conditions in a cup. A thermometer is suspended in the oil so that the temperature may be read during the test. Periodically an open test flame is introduced through an opening in the cover to detect the slight explosive puff which follows when the flash point has been reached.

See also **Petroleum**.

FLAVIN ADENINE DINUCLEOTIDE (FAD).   See **Coenzymes**; **Vitamin**.

FLAVONOIDS.   A group of aromatic, oxygen-containing heterocyclic pigments widely distributed among higher plants. They constitute most of the yellow, red, and blue colors in flowers and fruits. (Most of the other pigments are carotenoids. See also **Carotenoids**.) The flavonoids include the catechins, leucoanthocyanidins and flavonones, flavanols, flavones, the anthocyanins, and the flavonols. See also **Anthocyanins**.

FLAVOR ENHANCERS AND POTENTIATORS.   A *flavor enhancer* is a substance which when present in a food accentuates the taste of the food without contributing any flavor of its own. This is reminiscent of the role of a catalyst in a chemical reaction which promotes a reaction without chemically participating in the reaction. Although not usually regarded as a flavor enhancer, common salt, if not used excessively, enhances the taste of food substances. Salt does not fully meet the definition of an enhancer, however, because the salt is detectable as such.

Monosodium glutamate for many years has been the best known and most widely used of the flavor enhancers. MSG is normally effective in terms of a relatively few parts per thousand, but far less powerful than the newer flavor potentiators. Like enhancers, potentiators do not add any taste of their own to food substances, but intensify the taste response to the flavorings already present in the food. Because a potentiator is more powerful, smaller quantities of the substances are required than in the case of the enhancers. Generally, the available potentiators are from about 15 to nearly 100 times more effective than the enhancer.

Explanations for the actions of enhancers and potentiators, as of the early 1980s, remain qualitative and rather vague. It is not likely that the actions of these substances will be well understood until there are new theories or refinement of existing theories pertaining to the sensations and

perception of taste and odor. Experience does indicate that enhancers and potentiators act more in terms of taste than odor.

Monosodium Glutamate. The chronology of MSG commenced centuries ago when certain seaweeds were used in the Far East to improve the flavor of soups and certain other foods. It was not until 1908, however, when the curiosity of K. Ikeda (university of Tokyo) caused him to study the seaweed *Laminaria japonica*, traditionally used by Japanese cooks to enhance food flavoring. After much research on the seaweed, MSG was isolated and identified as an excellent flavor enhancer, particularly for high-protein foods. As an aside, it is interesting to note that Ritthausen in Germany had isolated glutamic acid as early as 1866 and his associates had prepared the sodium salt of the acid, namely monosodium glutamate. But the path of research in Germany was targeted in other directions and the flavor enhancing qualities of MSG were left to Ikeda to determine.

The Japanese throughout the first half of the century produced glutamic acid by extraction from natural materials, a slow and costly method. Nevertheless, the demand for MSG grew rapidly and cost tended to be a secondary factor. It was not until 1956 that Japanese microbiologists succeeded in developing the first industrial production of L-glutamic acid by means of fermentation. See Fig. 1. The problem of producing glutamic acid, as well as a number of other important amino acids, by fermentation was the lack of suitable strains of microorganisms for starting the cultures. Initially, the Japanese researchers were successful in isolating microbial strains from natural sources that possessed good abilities to excrete and accumulate a large amount of the amino acid in the cultural broth, but only under very carefully controlled conditions. For example, S. Kinoshia found that a high yield of glutamate could be attained only when the level of biotin (a vitamin required by glutamate-producing bacteria) was held within certain limits. An excess of biotin killed the microorganisms. Both antibiotics and detergents were used to control the biotin levels. Later, work was conducted with an artificial mutant by way of investigating genetic techniques. Ultimately, large-scale MSG production was achieved by the fermentation route, sugar beets commonly used as a raw material. More detail on the fermentation process can be found in the Oeda (1974) reference listed. MSG also can be produced by chemical synthesis, as shown in Fig. 2.

Although listed as a GRAS substance (generally regarded as safe) for many years, questions concerning its safe usage have arisen from time to time and, as of the mid-1980s, MSG still remains somewhat controversial. It is known that overconsumption of MSG can produce an illness, usually

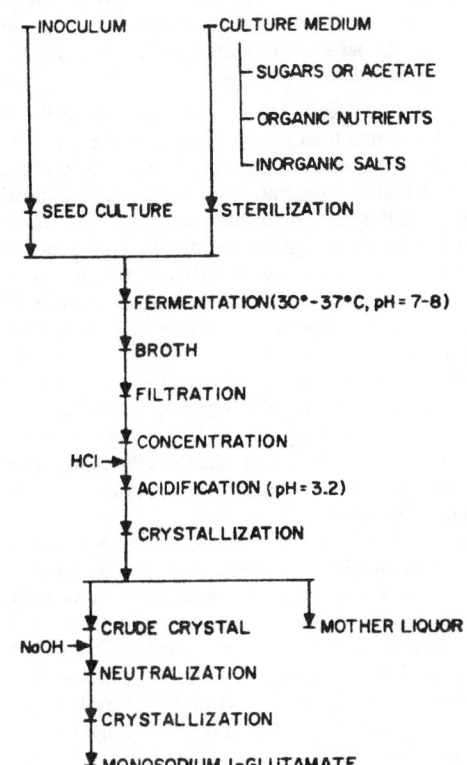

Fig. 1.   Preparation of monosodium L-glutamate by fermentation. Sugar beets, corn (maize), and wheat gluten have been used in the process.

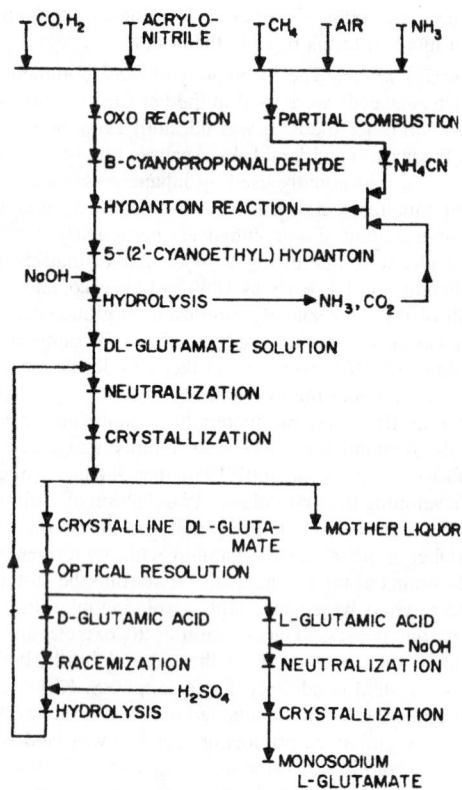

Fig. 2. Preparation of monosodium L-glutamate by chemical synthesis.

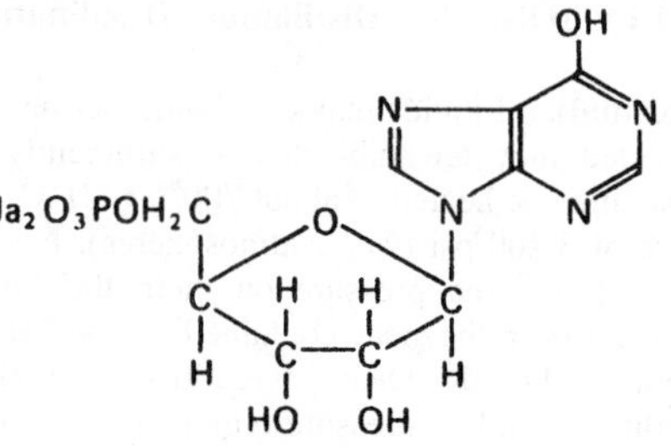

Disodium 5′-inosinate
$C_{10}H_{11}N_4Na_2O_8P \cdot H_2O$
Molecular Weight (anhydrous): 392.17

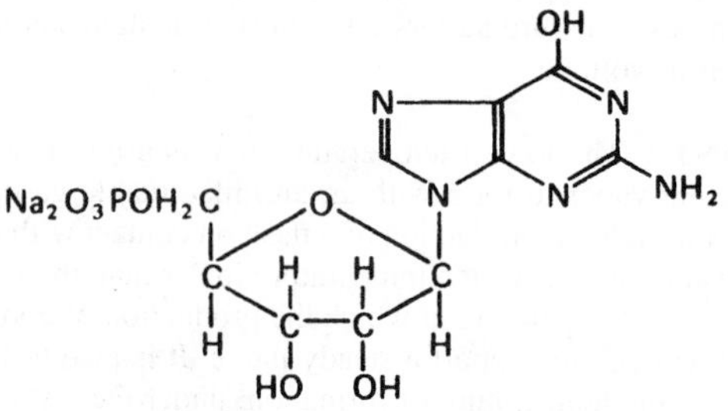

Disodium 5′-guanylate
$C_{10}H_{12}N_5Na_2O_8P \cdot xH_2O$
Molecular Weight (anhydrous): 407.19

Fig. 3. Salts of ribonucleotides. (*Takeda, Tokyo.*)

of just a few hours duration, in some persons. This is commonly referred to as the "Chinese Restaurant Syndrome." Apparently even when usage is somewhat excessive, only a relatively few people exhibit the symptoms of the syndrome. The possible seriousness of any deleterious effects of MSG tend to be countered by the many years the substance has been used by literally thousands of food processors and many millions of chefs and household food preparers. Some authorities also have a comfortable regard for MSG because of occurrence of the substance in many natural foods, notably mushrooms, tomatoes, and human milk. The topic is summarized by Krueger (1979).

The 5′-Nucleotides. Also dating back many years in the Far East was the knowledge that bonita tuna possesses a substance that very effectively enhances the flavor of foods. However, it was not until 1913 that S. Kodama (Tokyo University) commenced a serious investigation directed toward identifying and isolating the substance from tuna. Initially, Kodama believed that the substance was the histidine salt of 5′-inosinic acid, but later found that the substance was actually 5′-inosinic acid itself. This nucleotide was found to be many more times as effective as MSG. Further research by Kodama and others has shown that these nucleotides are present in many natural foods.

The nucleotides, in addition to their effectiveness at much lower concentrations, have been found to be superior to MSG for certain types of foods (in addition to high-protein foods). It also has been observed that the nucleotides tend to create a sense of increased viscosity, providing more body, for example, to soups. One manufacturer (Takeda) produces a series of the nucleotides by the enzymatic hydrolysis of ribonucleic acid. See formulas of Fig. 3. As of the early 1980s, these compounds are enjoying a high volume of production and usage. It should be pointed out that the nucelotides are commonly used together with MSG. Some researchers point out that while the nucelotides and MSG have a lot in common, there is a considerable difference in their use. The nucelotides are up to 100 times more effective than MSG on a weight basis, and whereas MSG has been a favorite of processors for the enhancement of "meaty flavor," the range of the nucleotides is broader, modifying salty or sweet flavors and suppressing many undesirable flavors. The nucleotides are not a replacement for MSG. The substances do have a synergistic effect when used together. Generally, 1 kilogram of nucelotide used with 50 kilograms of MSG will have the same flavor intensifying result as 100 kilograms of MSG alone.

Other Potentiators. Because the market is so large, research continues at a good pace in seeking other potentiators. Established since the early 1940s, *maltol* is effectively used in foods that are high in carbohydrates, such as beverages, jams, and gelatins. Claims of reducing sugar content by 15% in products using maltol have been made. Other potentiators used or proposed include dioctyl sodium sulfosuccinate, N,N'-di-*o*-tolyethylenediamine, and cyclamic acid.

Additional Reading

Furia, T.E., and N. Bellanca: *Fenaroli's Handbook of Flavor Ingredients,* CRS Press, Cleveland, Ohio (1971).

Furia, T.E.: *Handbook of Food Additives,* 2nd Edition, CRC Press, Cleveland, Ohio (1972).

Krueger, J.: "MSG: One of the Food Industry's Most Studied Ingredients," *Processed Prepared Foods,* **148,** 1, 128–140 (1979).

Oeda, H.: "Amino Acids," in *Chemical and Process Technology Encyclopedia,* (D.M. Considine, Editor), McGraw-Hill, New York (1974).

Staff: *An Introduction to Nucleotide Seasonings,* Ajinomoto, Tokyo (1980).

Staff: *Ribonucleotide Flavor Enhancers,* Takeda, Tokyo (1980).

FLAVORS AND ESSENCES. Normal, healthy people have keen senses of taste and odor, as well as of some of the collateral senses that interact with and contribute to the overall sensations of taste and odor. Much progress has been made over the last few decades in gaining an improved understanding of the physiology and chemistry of these perceptions, but research in the field is mainly of a qualitative and statistical nature. Numerous theories have been developed, but no single theory has been universally accepted. Flavor and essence scientists face tremendous challenges in their attempts to convert an art that dates back centuries into a body of science that can illuminate producers and consumers alike in terms of their odor and taste preferences and dislikes.

The importance of flavor and odor to product successes and failures in the food field alone accounts for an estimated $15–$20 billion loss in profitability by food processors just in the United States.[1] These costs entail the development and premarketing costs that are classified as *nonproductive* when a product does not achieve a close–to–break-even status. This

[1] See van Osnabrugge, W., reference listed.

becomes serious when one realizes that, in an average year, 80% of food introductions fail in the marketplace. Other factors, of course, make a product unacceptable, such as texture, dietary content, and appearance, but most experts agree that taste and odor account for a high portion of these failures. As of the early 1990s, it was estimated that approximately 3500 new foods and beverages were introduced into the retail food chain per year.

Well over 5500 compounds have been identified as flavor components of foods.

General Principles of Taste and Odor

The *flavor* of an edible substance is the combined sensation of taste and odor as perceived by the eater/drinker of that substance. Although the components (*flavorings*) are present in food substances, the full aspects of flavor require intimate contact between substance and consumer. The odors emanating from a bakery tend to be richer and more pleasant than the bread itself; the flavor of coffee seldom attains the richness of aroma that one perceives in the vicinity of a coffee roasting plant. Flavor is a unique combination of nerve impulses on the brain centers as the result of actions upon receptors located on the tongue and in the lining of the nose. It is thus the result of interaction between the food substance and the consumer.

Very broadly, tastes can be divided into three, possibly four, categories—*sweet, sour* (or acid), and *bitter* (alkaline) and *salty*. Salty, which for practical purposes is a major component of taste perception, involves physiological differences; in some ways it may be classified as a flavor potentiator.

The manner in which people acquire taste preferences is poorly understood. Sinki (see reference listed) suggests that flavor preference may be explained by:

1. *Genetic factors* (race, sex, individual characteristics, such as acuity, health, and aging);
2. *Physiological factors* (caloric, nutrient, and general health needs); and
3. *Psychological factors*, with three subclassifications:

   a. *Environmental factors* (economic, geographic, legislative, and individual or group habits),
   b. *Personal factors* (nostalgia, religion, attitude, intellect, and belief), and
   c. *Association factors* (as with positive, pleasurable situations or, by contrast, with negative, unpleasant situations, such as illness). As with other organoleptic senses of humans, the brain's ability to sense a very large variety of tastes, odors, and colors upon exposure is acute, but the ability to recall these sensations is quite limited. For example, the human eye can distinguish several thousand shades of color, but attempts to "remember" or match color samples from sample charts in a paint store is highly inexact.

Practical experience abetted by statistical surveys reveals a wide diversity of taste and odor preferences among peoples worldwide. For example, in studying preferences for flavored yogurt, strawberry is the clear winner in a majority of countries surveyed. Exceptions were a preference for cherry (Germany), citrus (Japan), coffee (Switzerland), and blueberry (Austria). The least preferred flavors were orange, tropical, peach, and banana flavors.

Tastes are also acquired over a period of time. For example, When many North Americans first taste cola drinks or coffee as children or youths, the tastes may be repugnant, but over a period of time these tastes become personal favorites, or they may be shunned for a lifetime. Similarly, many Europeans upon their first exposure to the taste of popular American cola beverages, peanut butter, root beer, and so on, react negatively and may never acquire a real taste for such products. Likewise, flavors such as cassis- or black-current-flavored drinks, which are popular in many European countries, have not enjoyed acceptance in North America.

Over the centuries, humans as well as other animals have learned to associate tastes and odors, in particular, with life-threatening situations. Human excreta, for example, is a foul odor. In some fashion, humans have learned to associate such odors with unpleasant, unhealthy living conditions. The odor of smoke produces dual associations—a log burning on the family hearth, or a burning house or building. The "associative" aspects of odors and tastes remain very poorly understood.

Flavor Characteristic Terminology. The flavorist uses terms (reminiscent of sound harmonics) to describe the roles of certain ingredients in a flavor compound. *Top note*—indicates the flavor first perceived by the food monitor or consumer—and is usually a flavoring substance with a volatility relatively higher than the other flavor components present. *Main note* (sometimes called *middle note*)—is the predominating flavor of the food substance, coming on strong so to speak immediately after sensing the top note. *Bottom note* (sometimes called *undertone*)—flavors that are perceived slightly later in the cycle of tasting or smelling a food substance. An optimal blend of the various notes results in what is sometimes called a *full-bodied flavor*, important, for example, in coffees. Although some of the flavors listed in Table 1 appear immensely undesirable (and often are), sometimes, in the proper combination, they contribute to the desired full-bodied flavor effect. For example, in a garlic flavoring, the characterizing flavor, of course, is garlic. However, an expert panel may detect in a satisfactory product a number of flavor notes, some in addition to or desired in garlic flavor. An expert panel has reported the following flavor notes in such a product: acid, astringent, biting, bitter, boiled, brown, cabbage, earthy, green, heat, iodine, leek, metallic, musty, onion, plastic, potato, rubbery, scallion, sharp, sour, skunky, sulfitic, and toasted.

Role of Odor in Flavor Perception. In terms of total flavor sensation, many authorities agree that odor is usually more important than taste. Experience, of course, demonstrates the marked reduction of flavor sensation when the nasal passages are partially blocked, as in the case of a common cold. In such instances, the layperson may refer to the "flat taste" of the food. In actuality, the taste buds are functioning normally; it is the odor component of flavor that is missing.

The odor component of flavor is made up of at least two vectors. Sniffing a substance without contact with the tongue provides a partial indication of odor, that is, molecular vapors or gases pass directly to the olfactory sensors in the nose via the nasal cavities. This vector might be called the absolute external odor or fundamental odor of a substance. This vector is dependent upon the vapor pressure (volatility) of the food substance itself. The other vector of odor is what researchers call internal odor because the molecules reach the olfactory sensors by way of the pharynx, a flattened tubular passage that connects the back of the mouth with the nasal cavities. In the mouth, the food substance is wetted by saliva, altering not only the vapor pressure of the flavoring agents present, but sometimes exposing more and different flavorings, thus affecting flavor intensity and quality. It is well known, of course, that exceedingly dry substances tend to be odorless or nearly so. The odor of a polished metallic surface, for example, is difficult for most persons to detect. The addition of only modest amounts of moisture to most dry substances significantly increases their fundamental odor, as the result of increasing vapor pressure and by activating all flavoring substances present. The effect of moisture on odor is dramatically illustrated by the dog at the fireside and the dog that has just come in out of the rain.

Classification of Flavoring Substance Sources

There are two fundamental classifications of flavors: (1) *natural*, and (2) *synthetic* or *artificial*. Prior to the early beginnings of organic chemistry (circa 1828), all flavors, essences, aromas, and like substances were derived from naturally occurring materials.

Natural Flavors and Essences. Natural food flavors either occur in nature or are generated during heating or processing by enzymatic reactions or by fermentation.

It is interesting to note that very early civilizations (B.C. and A.D.) gave greater attention to essences than to flavors. In those early times, perfume was considered one of life's few luxuries and, as observed by Tyrrell, ranked higher than learning and many forms of tangible wealth. Early sources of essential oils were derived from resinous gums that exuded from cuts in the bark of certain trees and shrubs. Other odorous substances of antiquity were fruit juices and extracts derived from flowers.[2] Ancient China, Egypt, and India accumulated considerable knowledge for the preparation of pleasant odorous formulations. Such concoctions were

---

[2] Additional sources included arils, balsams, barks, beans, berries, branches, buds, bulbs, calyxes, capsules, catkins, cones, flowering tops, fronds, gums, hips, husks, kernels, needles, nuts, peels, pits, pulps, rhizomes, rind, roots, seeds, shoots, stalks, stigmas, stolons thalli, twigs, wood, and wood sawdust, as well as some entire plants.

TABLE 1. WORDS AND PHRASES SOMETIMES USED TO DESCRIBE FLAVOR (KEY FLAVORS AND UNDERTONES)

Animal-associated	Terpeny	Perfumey
Animalic	Tobaccolike	Rosy (many more flower
Balcony		sterms like this)
Barbequelike	**DAIRY ASSOCIATED**	
Barnyardy	Baby biblike	**USUALLY HOTNESS ASSOCIATED**
Beefy	Buttery	Acrid
Birdy	Butyric	Astringent
Bird-cagey	Cheesy	Biting
Bloody	Cowy	Brown
Chickeny	Creamy	Burnt
Clammy	Curdled milklike	Burnt tirelike
Cured	Fatty	Burnt coffeelike
Eggy	Goaty	Charcoaly
Fatty	Oily	Chililike
Feathery	Sour	(Cooling)
Fishy	Waxy	Garlicy
Gluelike		Hickorylike
Guanolike	**DAMP-EARTH ASSOCIATED**	Horseradishy
Hammy	Claylike	Hot
Lion cagey	Damp cellarlike	Mexican
Meaty	Dank	(Minty)
Muttonlike	Dirtlike	Mustardy
Oily	Earthy	Peppery
Porky	Funguslike	Piquant
Salt-airy	Mildewy	Pungent
Sea-misty	Moldy	Red peppery
Sulfitic	Mushroomlike	Sharp
	Musky	Smokey
BOTANY-ASSOCIATED	Musty	Spicy (many specific
Barky	Potato skinlike	terms in this
Beanlike	Rooty	category)
Grassy	Wet haylike	Tangy
Green	Woodsy	Tartlike
Green applelike		Tart
Haylike	**DEGRADATION ASSOCIATED**	Toasty
Herby	Boiled egglike	Warm
Leafy	Burnt rubbery	
Melonish	Dead	**NEUTRAL**
Mown-haylike	Decayed	Bland
Piney	Fried onionlike	Cardboardy
Seedy	Gym lockerlike	Cerealike
Stalky	Limburgy	Chalky
Tomato-y (many more	Mercaptanlike	Characterless
vegetative terms	Natural gaslike	Crackerless
like this)	Overripe	Flat
Twiggy	Putrid	Fluory
Unripe	Rancid	Light
Vegetabaly	Rottens	Matzolike
Viney	Skunky	Mellow
Woody	Sulfitic	Mild
	Sweat socklike	Nondescript
CHEMICAL-ASSOCIATED	Sweaty	Overcooked
Acetic	Wet doglike	Papery
Aminelike		Pasty
Ammoniacal	**FERMENTATION ASSOCIATED**	Raw
Alcoholy	Beery	Stale
Aldehydic	Bread doughy	Starchy
Balsamic	Cidery	Strawlike
Camphoraceous	Moonshiny	Tasteless
Chloriney	Rummy	Wallpaper pasty
Cough mediciny	Vinegary	Waxy
Cresylic	Whiskey breathlike	Weak
Furniture polishlike	Winey	
Fusel oily	Yeasty	**RESINOUS**
Hydrocarbonlike		Balsamic
Iodiny	**FLOWER-FRUIT ASSOCIATED**	Leathery
Medicinal	Berrylike	Pruny
Metallic	Citrusy	Resinous
Mothbally	Estery	Woody

TABLE 1 (continued)

Animal-associated	Terpeny	Perfumey
Nicotiny	Floral	
Phenolic	Flowery	**SWEETNESS ASSOCIATED**
Plasticlike	Fragrant	Butterscotchy
Salty	Fruity	Candylike
Shoe polishlike	Lemony	Caramellike
Soapy	Orangey	Chocolaty
Jammy	Bitter	Peculiar
Jellylike	Boiled	Persistent
Malty	Broiled	Refreshing
Mapley	Characteristic	Rich
Marshmallowy	Delicate	Roasted
Sacchariny	Fresh	Soft
Sugary	Fried	Strong
Sweet	Full bodied	Tenacious
Vanillic	Grilled	
	Intense	
MISCELLANEOUS	Lingering	
Acidy	Overpowering	
Baked	Overwhelming	

Note: Obviously some of the foregoing terms connote dimensions beyond those strictly conveyed by taste and odor, providing credence to the concept that the total experience of eating or drinking goes well beyond two vectors.

used in religious and perfunctory political ceremonies. Also, they were used simply for personal gratification by those who could afford them, as typified today by the large annual expenditures made for perfume and aromatic cosmetics. A major breakthrough in the preparation of aromatics occurred during the Middle Ages when the Moslems discovered the process of distillation for purifying and concentrating odorous substances.

Enfleurage. This is an ancient process for capturing aromatic essential oils from flowers, such as jasmine and tuberose. In this now essentially obsolete process, freshly gathered flower petals are carefully spread on a sheet, usually glass, upon which is spread a very thin film of highly purified fat. The petals remain in contact with the fat film for 24 hours, after which the petals are removed and replaced with a fresh batch. The process requires from 30 to 40 repetitions before the fat becomes saturated with the essential oil. The fat at this point is called *pomade*, which is extracted with pure alcohol. Prior to the availability of more advanced technology, some essence manufacturers would have as many as a thousand petal frames in operation at one time.

Steam Distillation. Although in use for well over a century, steam distillation is still used in some essence plants. The raw aromatic substances are placed in a vessel with water. The steam produced carries the odorous substances and, when condensed, yields essential oil and scented waters. Most installations of this type are found in the underdeveloped countries.

Vacuum distillation is preferred by many flavoring producers because it is more rapid than steam distillation. In this method, the drying and resinification of flavoring constituents is avoided because of the absence of air. The separation of closer-boiling-point materials is more effective under vacuum. Undesirable side reactions between constituents are avoided because of the relatively low temperatures required. Thus, many delicate substances can be handled without thermal degradation.

Rectification (fractional distillation) is used generally in those situations where a higher degree of separation and purity are desired. The products are known as *rectified essential oils*. Traces of water, solid and resinous materials, color bodies, and, depending upon the degree of rectification, terpenes and sesquiterpenes may be removed. Properly designed, a fractional distillation system can effect high purity of final product with only a loss of from 1.5 to 2.5% of the quantity of essential oil input. In deterpenization, the terpenes come off as head fractions. Manufacturers of high-quality flavorings generally prefer the well-defined chemical and physical constants obtainable with a terpeneless essential oil.

In rectification, a part of the vapor is condensed and the resulting liquid contacted with more vapor, usually in a column with plates or packing, by returning (refluxing) some of the condensate back to the column. Greater rectification is accomplished where the reflux ratio is high. In a practical way, this arrangement accomplishes what would otherwise require a series of stills in series. Where several fractions (different boiling points) are desired, these can be taken from different plates or height locations of the column. Heavier components, of course, collect in the lower portions of the column.

Molecular distillation is used for certain flavoring substances, such as fruit juice concentrates. This technique is used where the other methods do not effect the separations desired or cause thermal degradation. In this procedure, distillation is carried out at very low pressures (of the order of 0.001 millimeters). A molecular still is distinguished by the fact that the distance from the surface of the liquid being vaporized and the condenser is less than the mean free path (the average distance traveled by a molecule between collisions) of the vapor at the operating temperature and pressure. This distance is usually of the order of magnitude of a few inches (several centimeters). Close separations are often required in the preparation of fractions and isolates of essential oils that are to be used in the preparation of synthetic flavorings.

Extraction. This process was introduced in the early 1900s. The natural raw materials are soaked with an organic solvent. The temperature, time of extraction, and number of extractions required depend upon the solvent used and the characteristics of the raw material. These operating parameters usually are carefully guarded trade secrets. As described by Tyrrell, after maceration, the saturated solvent is filtered and then pumped into the still where 90% of the solvent is evaporated. The final concentration is carried out under vacuum to yield a *concrete* or, where alcohol is used, a *resoinoid*. The concrete normally is a solid waxy mass containing all of the hydrocarbon soluble and odorous matter of the plant. By way of low-temperature washing with alcohol, a waxy substance precipitates out. Subsequent removal of the alcohol at low pressure yields a highly concentrated product known as the *absolute*.

Tinctures, extracts, and absolutes are produced by extraction rather than distillation. A tincture is more dilute than a fluid extract and usually is less volatile. Some food processors prefer these qualities. In other cases, a much more concentrated fluid product (*extract*) or a solid or semisolid (*concrete*) extract is desired. Thus, after filtering, part of the alcohol will be removed by vaporization, leaving a much more concentrated fluid. There are differing degrees of concentration. Most of the familiar extracts found at retailers (almond, lemon, vanilla, etc.) are essential oils dissolved in an alcohol-water mixture.

Oleoresins. To the food processor, an oleoresin represents a flavoring substance midway between the spices and the essential oils. Oleoresins are usually solid and tacky substances at room temperature and soft and sticky at elevated temperatures. Using various spices as starting ingredients, a volatile solvent is percolated through the ground mass, followed by vacuum removal and recovery of the solvent. Generally, oleoresins are uniform and provide a concentrated flavoring power as contrasted with the spices. Oleoresins represent less bulk for the processor to handle (no celluloses, as in spices), but they tend to impart less color to a food product (small quantities required because of flavoring concentration) and not all flavor notes may be extracted during their manufacture. Sometimes the manufacturer will add small amounts of essential oils to their oleoresins to provide a more full-bodied flavoring. Oleoresins are also prepared from some spices, such as turmeric and paprika, not for their flavoring, but more for their coloring. Oleoresins of these two spices are commonly used, for example, in French-type salad dressings.

In numerous instances, additional processing may be required to remove stubborn impurities, waxes, colors, and notably terpene fractions. Solvent extraction is also used in processing herbs and spices to yield oleoresins.

Because the processing of natural substances often involves multi-step operations, costs are generally considered to be higher than a majority of synthesized compounds. A number of natural flavors and essences are given in Table 2, which illustrates the very wide range of ingredient costs.

Animal Sources of Flavors and Essences. At one time, animal glandular secretions were widely used in scents, particularly as fixatives. With so much success in synthesis today, coupled with growing empathy for animal welfare, the amounts used of these substances have been significantly reduced. The musk deer (*Moschus moschiferus* L), found in the Himalayan highlands, was at one time a traditional source. The reddish-brown secretion of the male is the odorous principle identified as 3-methylcyclopentadecanonone-one, the principal use of which is in perfumery as a fixative, but which has been reported as an additive in certain food products. There is the civet, *Viverra civetta* Schreber, a cat that

TABLE 2. COST RANGE OF FLAVORS AND ESSENCES (SCALE OF 1 (LOW COST) TO 1500 (HIGH COST))

Concentrated form	Index
Orange oil	1
Grapefruit oil	1
Cedarwood oil	2
Camphor oil	3
Eucalyptus oil	3
Sassafras oil	4
Citronella oil	4
Bitter almond oil	4
Clove oil	5
Cinnamon oil	6
Lemongrass oil	6
Cornmint oil	9
Pettigrain oil	9
Rosemary oil	10
Lignaloe (Boise de rose oil)	10
Anise oil	12
Pine oil	13
Lemon oil	13
Pineneedle oil	14
Onion and garlic oil	16
Nutmeg oil	17
Patchouli oil	17
Lime oil	20
Thyme oil	22
Lavender oil (spike oil)	24
Origanum oil	24
Palmarose oil	25
Caraway oil	26
Bergamot oil	31
Cedar leaf oil	31
Peppermint oil (mentha piperita)	45
Ylang ylang (canaga oil)	53
Geranium oil	60
Cassia oil	67
Vetiver oil	67
Sandalwood oil	110
Orris oil	942
Neroli oil (orange flower oil)	987
Rose oil (attar of rose)	1500

lives in Africa and southeastern Asia. The glandular secretion is of main interest in perfumery as a fixative, but it has been reported in foodstuffs at low levels (about 4 parts per million). There is the beaver of genus *Castor* of the northern climes of Alaska, Canada, and Siberia, the dried and ground glandular secretions of which are also used in perfumery, but also in chewing gum up to concentrations of 400 ppm. The flavoring additive is known as castoreum. And there is beeswax, a crude yellow wax that represents a secondary secretion of the honeybee. In addition to use as a modifier in perfumery, the substance is used up to levels of 5 ppm to enhance the flavor and textural qualities of honey. There always has been a close link between the technology of flavorings for the food field and of fragrances used in perfumes, cosmetics, and related products.

Synthetic (Artificial) Flavors and Essences. Much as the synthesis of aniline (Perkin, England, 1856) revolutionized the textile dyeing industry, the later synthesis of flavors and essences caused a major turn of the aromatics industry. Because of the almost limitless number of organic compounds, artificial equivalents of the natural substances became possible, equally important, entirely new flavors and essences could be created and combined in innumerable ways. Table 3 lists a number of synthetic equivalents or analogues of several natural flavors. Usually the cost of research in finding equivalents and the large number of steps (chemical reactions and separations) required to isolate a given compound are high. Thus, although synthetic or artificial, these compounds are not necessarily less expensive in many instances. Although there are tight regulatory controls to approve the safety of synthetic aromatics and colors, there is a strong preference in the marketplace today for natural flavors where they are in good supply.

Regardless of origin, a large majority of flavor substances are volatile, with a boiling point ranging between 20°C (68°F) to 300°C (572°F). Most food flavors are lipophilic (affinity for fats) and with molecular weights

TABLE 3. SYNTHETIC EQUIVALENTS (APPROXIMATIONS OR ANALOGUES) OF VARIOUS NATURAL FLAVORS

Almond	Tolualdehyde (*o, m, p*)
Apple	Allyl butyrate, cyclohexylvalerate, isovalerate, propionate; Benzyl isovalerate; Butyl isovalerate, valerate; Cinnamyl formate, isobutyrate, isovalerate: Citronellyl isovalerate; Cyclohexyl acetate, butyrate, isovalerate: Ethyl isovalerate, valerate; Isopropyl acetate, valerate; 2-Methylallyl butyrate; Methyl butyrate; Terpenyl isovalerate.
Apricot	Allyl butyrate, cyclohexylcaproate, cyclohexylvalerate, propionate; Amyl phenylacetate; Benzyl formate, propionate; Butyl propionate; Cinnamic acid; Citronellyl acetate: gamma-Decalactone; gamma-Dodecalactone; Ethyl cinnamate; Geranyl butyrate, isobutyrate, isovalerate. Heptyl acetate, propionate; Methyl ionone; Phenylethyl dimethyl carbinol; Phenylpropyl alcohol; Propyl cinnamate; Santalyl acetate; Tetrahydrofurfuryl propionate; gamma-Undecalactone.
Banana	Cyclohexyl acetate, butyrate, propionate; Ethyl valerate.
Butter	Diacetyl.
Caramel-	
Butterscotch	Maltol.
Caraway	D-carvone.
Cheese	Hexanoic acid; Isovaleric acid (rancid).
Cherry	Allyl benzoate; Anisyl butyrate, propionate; Cyclohexyl cinnamate, formate; Methyl anthranilate; Rhodinyl formate; Tetrahydrogeraniol; Tolualdehyde (*o, m, p*).
Chocolate	Tetrahydrofurfuryl propionate.
Cinnamon	Cinnamaldehyde; Alpha-methylcinnamaldehyde.
Citrus	Decanal dimethyl acetal.
Cloves	Methyl cinnamate; isoeugenol.
Cocoa	Neryl butyrate; Phenylpropyl cinnamate.
Coconut	Allyl undecylate; Ethyl undecylate; Methyl undecyl ketone; gamma-Nonalactone; gamma-Octalactone.
Cognac	Allyl pelargonate; Cyclohexyl caproate.
Cola	2-Ethyl-3-furylacrolein.
Currant	Cyclohexyl butyrate; Guaiol acetate (black currant); Linalyl acetate, isobutyrate, propionate (black currant); Methyl ionone; Methyl propionate (black currant).
Fatty	Decanal; Ethyl nonanoate; Heptyl alcohol; Lauryl alcohol, aldehyde; Nonanal; Octanal; 1-Octanol; Undecanal; 10-Undecenal.
Flowery	Anisyl alcohol; Benzyl acetate, phenylacetate; Cinnamic acid; Cinnamyl acetate; Citronellyl formate; Cresyl acetate; Decanal; Dimethyl benzyl carbinol; Dimethyl benzyl carbinyl acetate; Ethyl anthranilate; Geranyl acetate; Hydroxycitronellal dimethyl acetate; Linalool; Linalyl acetate; Methyl benzoate; Penethyl acetate; 2-Phenylpropionaldehyde; 3-Phenylpropionaldehyde.
Flowery/Fruity	Anisyl acetate; Cinnamyl isovalerate; Citronellyl; Ethyl laurate, octanoate; Geranyl butyrate, propionate; Nonyl acetate.
Fruity	Benzyl propionate; Butyl acetate; Cinnamyl anthranilate, formate; Citronellyl acetate, butyrate, isobutyrate; propionate; Delta-decalactone; Diethyl malonate; Dimethylbenzyl carbinyl acetate; Delta-dodecalactone; Ethyl *p*-anisate, benzoate, butyrate, heptanoate, hexanoate, maltol, nonanoate; Isoamyl butyrate, hexanoate, isovalerate, cinnamate; Linalyl isobutyrate, propionate; Maltol; Methyl benzoate, cinnamate; 2-Methyl-undecanal; Nerolidol; Octanol; Octyl formate; Phenethyl isobutyrate, isovalerate; gamma-Indecalactone.
Grape	Allyl salicylate; Cinnamyl anthranilate; Guaiol acetate; Isobutyl anthranilate; Isovalerophenone; Octyl isobutyrate; Phenylpropyl acetate, ether.
Grapefruit	Styralyl acetate.
Green Leaves	Allyl anthranilate
Hawthorne	Acetanisole; *p*-Methoxybenzaldehyde.
Heliotrope	Piperonal
Honey	Allyl phenoxyacetate, phenylacetate; Benzyl cinnamate; Carvacryl acetate; Cinnamyl butyrate; *p*-Cresyl acetate; *p*-Cresyl ethyl ether; *m*-Cresyl phenylacetate; *p*-Cresyl phenylacetate; Cyclohexyl phenylacetate; Ethyl phenoxyacetate, phenylacetate; Guaiol phenylacetate; Isobutyl phenylacetate; Linalyl butyrate; Methyl phenylacetate; Phenethyl acetate, butyrate, phenylacetate; Phenylacetic acid; Propyl phenylacetate; Santalyl phenylacetate.
Lemon	Citral; Citronellal.
Licorice	Methylcyclopenteneolone.
Maple	Methylcyclopenteneolone.
Melon	Cinnamaldehyde; Ethyl hexadienoate; Methyl amyl ketone; Octyl butyrate.
Menthol	3-*p*-Methanol.
Mushroom	Hexyl furan carboxylate.
Mustard	Allyl formate.
Orange	Linalyl anthranilate.
Peach	Allyl cyclohexylcaproate, cyclohexylvalerate, undecylate; Amyl phenylacetate; Anisyl alcohol, butyrate; l-Citronellol; Cyclohexyl caproate, cinnamate; gamma-Dodecalactone; Ethyl cinnamate; Isopropyl benzyl carbinol; Methyl methylanthranilate; Methyl nonyl ketone; Methyl octine carbonate; gamma-Octalactone; Oxyl acetate; Phenethyl alcohol, isovalerate, salicylate; Phenylallyl alcohol; Phenylpropyl isobutyrate; Propyl cinnamate; Rhodinyl formate; gamma-Undecalactone.
Pear	Benzyl butyrate; 2-Ethylbutyl acetate; Ethyl heptylate; Hexyl acetate; Hexyl furan carboxylate; Isoamyl acetate; 2-Methylallyl caproate; Methylheptenone; Propyl acetate.
Pineapple	Allyl caproate, cyclohexylacetate, cyclohexylbutyrate, cyclohexylpropionate, 2-nonylenate, phenoxyacetate; Benzyl formate; Bornyl acetate; *n*-Butyl acetate; Butyl isobutyrate; Cinnamyl acetate; Decanal dimethyl acetal; Ethyl butyrate, hexadienoate, phenoxyacetate; Hexyl butyrate; 2-Methylallyl caproate; Methyl beta-methylpropionate; Methyl undecylate; Propyl isobutyrate.
Plum	Butyl formate: Citronellyl butyrate, formate, propionate; gamma-Decalactone; Guaiol butyrate; heptyl formate; Hexyl formate; Isoamyl formate; Isopropyl formate, propionate; Linalool; Neryl propionate; Phenethyl formate (green plum); Phenethyl isobutyrate (green plum); Phenylallyl alcohol; Phenylpropyl butyrate; Propyl formate; Terpenyl butyrate.
Raspberry	Benzyl salicylate; alpha-Ionone; Isobutyl cinnamate; Methyl ionone; Neryl acetate; Santalol.
Rose	Phenethyl alcohol; Phenethyl dimethyl carbinyl isovalerate; Rhodinol.
Rum	Ethyl formate; isobutyl formate.
Sassafrass	*p*-Propyl anisole.
Spearmint	1-Carvone.
"Spice"	Eugenol; Isoeugenyl acetate; 3-Phenylpropyl acetate.
Strawberry	Anisyl formate; benzyl isobutyrate; Cuminic alcohol; Ethyl methylphenylglycidate; Ethyl phenylglycidate; Isoamyl salicylate; Isobutyl anthranilate; Methylacetophenone; Methyl cinnamate; Methyl naphthyl ketone; Nerolin; Neryl isobutyrate; Phenylglycidate
Vanilla	Propenyl guaiethol; Vanillydene acetone.
Violet	alpha-Ionone; beta-Ionone; Methyl-2-octynoate.
Walnut	gamma-Octalactone.
Wine	Ethyl acetate; heptylate.
Wintergreen	Allyl salicylate; Methyl salicylate.

between 50 and 250. Many foods contain hundreds of different compounds, not all of which, of course, contribute to odor and flavor. For example, over 700 compounds may be found in coffee, but with only a few having significant impact on the flavor profile of coffee.

Compounds that impact directly upon odor or taste are sometimes called "key" compounds. Some of these are listed below:

Apple	ethyl-2-methyl butyrate
Bell pepper	3-isobutyl pyrazine
Butter	diacetyl
Coffee	furfuryl mercaptan
Cucumber	*trans*-2-cis-6-nonadienal
Grape (Concord)	methyl N-methyl-anthrailate
Grapefruit	nootkatone
Mushroom	1-octene-3-ol
Popcorn	methyl-2-pyridyl ketone
Rice (basmatic)	2-acetyl-pyrroline

With the precision characteristic of chromatographic separations, it is self-evident that this instrument has become an indispensable tool for the flavor and essence chemist.

In most European countries, flavors that occur naturally or are generated during heating or processing by enzymatic reactions or modification generally are considered "natural flavors." Flavors that are often referred to in the United States as "synthetic" are usually termed "artificial" in Europe. These would include such compounds as ethyl vanillin, allyl-α-ionone, and ethyl maltol. However, substances that are synthesized but chemically identical to the naturally occurring substances are classified as "natural-identical." This class would include diacetyl, benzaldehyde, anisyl acetate, and benzophenone.

Traditional strawberry flavoring for soft drinks may include amyl acetate, amyl butyrate, butyl isovalerate, ethyl acetoacetate, ethyl butyrate, ethyl caproate, and ethylfuran carbonate—where the beverage is to be sweetened with natural sugar. On the other hand, if synthetic sweetener is used, the formulation will be quite different: aldehyde C_{18}, diethyl acetal, geranil, beta-ionone, maltol, neroli essential oil, octanyl dimethyl acetal, phenethyl alcohol, terpineol, and vanillin. The multiplicity of ingredients to obtain a given flavor is also illustrated by the flavoring substances used in an imitation rose flavor: aldehyde C_8, citral, citronellol, geraniol, linalool, phenethyl alcohol, and rhodinol. Many scores of additional examples of this nature are given and excellently described in the Fischetti reference listed.

Flavor and Essence Research. Successful food processors realize how important the selection of flavoring ingredients is to the ultimate success of a product in the marketplace and consequently support flavor research and extensive premarketing testing prior to introducing a new product nationwide or worldwide. Much of the basic flavor research is carried out by flavor manufacturers, as exemplified in the United States by member firms of FEMA, as previously mentioned. The costs for such research, of course, become part of the price for flavor ingredients. These costs, coupled with the cost of chemically processing many flavor and essence ingredients, whether obtained from natural sources or synthesized, can become quite high and measured in terms of $/ounce or $/gram. On the other hand, many ingredients can be quite inexpensive, as previously shown in Table 2.

Some flavor and essence research also is conducted at the university level. For example, the Monell Chemical Senses Center (affiliated with the University of Pennsylvania) is renowned worldwide for its scientific approach to a better understanding of the human senses of taste and odor.

Although odor (essence) is an important part of flavor sensitivity, there are numerous inedible consumer products where olfactory sensitivity is the principal concern. In addition to perfumes, soaps, hair sprays, and other cosmetics, odor is important to the acceptability of laundry and cleaning products, polishes and waxes, and decorative materials. Deodorizing agents also fall within this general sphere of interest.

The general steps involved in flavor and essence research, as suggested in a "Scientific Status Summary" prepared by the Institute of Food Technologists' (IFT) Expert Panel on Food Safety and Nutrition, include the following:

1. Selection of sample containing target flavor.
2. Isolation, concentration, and preliminary fractionation.
3. Final separation.
4. Synthesis of authentic compound.
5. Confirmation of identification.
6. Sensory evaluation.
7. Data interpretation.

The ultimate objective of flavor research is to understand the biological pathways leading to the formation of the compound, as well as the chemical mechanisms responsible for the development of objectionable flavors in agricultural and ocean produce.

An interesting example is given in the IFT Report. In the flavor analysis of packaged potato chips, 2,5-dimethylpyrazine, along with a number of other pyrazine compounds, was identified. For certain foods, the pyrazines are a major class of flavor ingredients. Thus, in seeking aroma and flavor improvement, numerous combinations and proportions of various pyrazine compounds were tested in an effort to find the most effectively pleasing formulation.

Flavor research owes much to the science of chromatography, which made its appearance a relatively few decades ago. Stofberg observes that more than 5000 compounds have been identified as flavor components of foods. The Fischetti reference listed contains exhaustive listings of these compounds and their properties.

New Paths in Flavor and Essence Research. Although much research continues in the search for new, more promising, and sometimes less expensive substitutes, a considerable portion of research today considers a number of other factors that shape the taste and odor profile of a product. For example, studies are directed toward better understanding the interactions of flavor compounds with other chemical compounds in the final product or during processing, such as during preserving, baking, or sterilizing. Interactions with carbohydrates, proteins, fats, and enzymes are among these objectives. Chang also mentions two additional areas of technology that will contribute to advances in flavor technology. These include biotechnology, which is bringing such techniques as genetic engineering (recombinant DNA) and plant tissue culture to flavor research. The other major development is supercritical fluid extraction (SFE), which will impact on fermentation and enzymatic modification as flavor-creating sources. Chang notes that the use of SFE will produce essential oils with a higher yield and improved quality as compared with traditional extraction with organic compounds. However, SFE equipment is more costly.

Carbohydrates and Flavor Development. In considering the flavor of food products, the researcher must be concerned not only with achieving flavor by adding natural or synthetic flavoring substances, but also with biochemical processes that may occur during processing, storing, and later heating or cooling of the product just prior to serving the product. Inasmuch as carbohydrates predominate in many food products, this is a class of chemical substances that has received much attention. The flavor scientist must distinguish the simple sugars from the complex polysaccharides. These types of compounds differ widely in their functional properties. While the simple carbohydrates consist of mono- and disaccharides, the complex polysaccharides include, for example, gums and hydrocolloids.

As pointed out by Godshall (Sugar Processing Research, Inc.), sweetness is a primary functional property of the common simple carbohydrates, including sucrose, glucose, fructose, and lactose. Sometimes this flavor is referred to as "pure sweet," with a relative sweetness value arbitrarily set at 100 for comparing the sweetening power of other carbohydrates. This value usually is fixed at 100 when comparing the high-intensity sweeteners with sucrose. See also **Sweeteners**. With the exception of fructose, xylitol, and invert syrup, sucrose is the sweetest of the common carbohydrates. Some sweeteners exhibit surprising performance. For example, glucose has a property of "self-potentiation"—that is, at concentrations of 2–10%, it has a sweetness index of 50 to 60, whereas at concentrations of 50–60%, the sweetness index rises to 90 to 100. Fructose, although sweeter than sucrose in solution, is essentially no sweeter than sucrose in baked goods. The molecular structure of closely related sweeteners also affects taste. For example, beta-glucopyranose is only about two-thirds as sweet as alpha-glucopyranose.

Advantage can be taken of what may be termed sweetness synergism. For example, Godshall cites glucose-sucrose mixtures that can be 20–30% sweeter than either constituent alone of similar concentration. A number of substances have the property of suppressing sweetness. Carbohydrates, such as guar gum, cornstarch, and carboxymethylcellulose, suppress sweetness, particularly of sucrose.

The food flavorist also must be aware of metal complexes, which carbohydrates are capable of forming. Iron salts not only form complexes with dietary fiber, but with nearly all of the known natural sugars. Fructose, maltitol, sorbitol, and xylitol can easily form complexes with the ferric ion. In the same manner, formation of complexes can help to minimize unpleasant metallic tastes.

Carbohydrates characteristically absorb volatile substances from the environment, and they also retain their volatility, even under drying cycles. Flavorists have taken advantage of these properties, purposely using sugars and polysaccharides as flavor "carriers," which is of particular interest in making stable, dry, and flowable flavor powders.

Flavor chemists also are aware of the Maillard or browning reaction, which can occur in food products as caused by the condensation of an amino group by a reducing sugar. This is a non-enzymatic reaction, the effects of which have been studied, both on flavor and color, each of which can be desirable or harmful, depending upon the food product. For example, the browning reaction is favorable in the flavor development of aroma in baked bread, roast meat, coffee, chocolate, toasted nuts, baked potatoes, and maple and cane syrups. However, browning in such products as milk powders and dried orange juice are highly undesirable.

Encapsulated Flavors. Modified procedures during the past decade have permitted the preparation of encapsulated flavors with flavor levels over twice that of prior available products. Spray drying has been the principal key to this success. First, an oil flavor is emulsified into an aqueous solution or is dispersed in an edible carrier material, after which the emulsion is pumped through an atomizer into a high-temperature chamber. The water evaporates rapidly, and particles of carrier material are formed around the flavor. However, some of the flavor component reaches the surface of the product. This requires the addition of antioxidants to suppress oxidative changes in the flavor ingredient.

Even more recent has been the introduction of encapsulation/extrusion, which also permits conversion of flavorants, such as essential oils, into solid form. Spray drying is not required. In the encapsulation process, the flavor substance is "enrobed." A viscous carbohydrate, with less than 10% water, is created by heating, after which an emulsifier and acid flavoring ingredients are added. The ingredients are reacted under pressure in a cool alcohol bath, and then the product is extruded to form filaments. Thus, the final easy-to-handle product contains the flavor within a small capsule.

Effects of Microwaves on Flavors. When microwave cookery was introduced, it added another dimension to the food flavorist's concern. When one first uses microwave heating for foods, the rate at which different food substances increase in their temperature for a given intensity and time of exposure is readily noticeable. This phenomenon, of course, is not present when conductive or convective heating is used. Like water, most flavor components are polar molecules and tend to be vaporized and thus lost; the flavors that are not lost may be somewhat decomposed and thus distort the flavor effect intended. Among the common flavorings used, there is a considerable difference in their performance when subjected to microwave radiation. Generally, a microwavable food product enjoys a distinct advantage in the marketplace. Thus, flavoring formulations must be adjusted so that flavor will be retained, regardless of the manner in which the product is heated.

In researching this problem, one flavor supplier tested more than 500 chemicals, solvents, essential oils, and natural raw materials (acids, aldehydes, ketones, esters, lactones, amines, thio compounds, and hydrocarbons), which may be used in varying amounts in compounding complex flavors. By way of testing these materials at various powers in the microwave oven, values representing the ratio of the temperature increase of the sample to the temperature increase of a standard (water) were established. These data have been quite useful in compounding flavors for microwave-able foods.

The research indicated that comparatively few common flavorings absorb heat (thus, temperature rise/unit of time) as fast as water. Among the more volatile under microwave radiation are fenugreek and onion oleoresin, whereas, in decreasing order of volatility, are sage oleoresin, ginger oleoresin, carrot seed oil, anise oil, basil sweet oil, oleoresin celery, and oleoresin black pepper.

As part of the testing program, two lemon flavors were considered for addition to a microwave-able cake mix. The test showed that one flavor (highly concentrated in citral) became extremely hot and produced over ten chemical by-products, with consequent serious change in the flavor profile. By contrast, a flavor precisely engineered for microwave heating retained its integrity and consistency throughout the baking process. Further details are given in Pszcaola reference listed.

Spices and Seasonings

These products may be defined as dried aromatic substances (natural) used to season food products. They play an important role in seasoning, but also

are frequently used for their food coloring power as well as their effects on other properties, such as texture. Spice science traditionally is treated as a subscience of that pertaining to flavors and essences.

Regulations and Labeling of Food Flavors. Government regulation of flavor additives for foods range considerably from one country to the next. In the United States, the Food and Drug Administration (FDA) prevails. In addition to testing new flavors, the FDA is assisted by an independent expert panel, made up of expert toxicologists, pharmacologists, and biochemists. Industry self-policing is provided by the Flavor and Extract Manufacturers Association (FEMA). Flavors fall within the province of the 1938 Federal Food, Drug, and Cosmetics Act. Premarketing safety evaluations and clearance by the FDA were not required by law until 1958, when an amendment to the Act was passed by the U.S. Congress.

At that time the GRAS (Generally Regarded as Safe) classification was also provided. The criterion for GRAS: "Chemicals normally present in food consumed by man through the ages without any apparent adverse effects can be presumed to be safe at the concentrations found in those foods." The sound basis for the GRAS classification is reflected by the fact that only ten substances, among many hundreds, have been removed from the GRAS classification over a period of nearly 50 years.

The primary concern of regulation has been and continues to be the banning of substances that may induce cancer in people and animals. Labeling regulations for food flavors require that one or more designations should appear on the food container: Spice, Natural Flavor, and Artificial Flavor.

The aforementioned regulations also apply to other food additives, such as colors, texturizing agents, and emulsifiers.

In food products, the concentration of flavor substances ranges from a few parts per million (ppm) to about 100 ppm, with the concentration of individual compounds as low as parts per billion (ppb) or even parts per trillion (ppt). Such concentrations can be determined today by chromatographic techniques. Of course, sugar (sucrose and others) are used in much higher concentrations, particularly in baked goods, jellies, jams, fruit drinks, and cured meats, among others. In many foods, sugars contribute in a major way to other properties, such as texture, moisture retention, and resistance to spoiling over extended periods.

Physiological Aspects of Flavors and Essences

Receptor cells especially sensitive to chemicals are found in virtually all animals. By convention, those receptors normally excited by contact with chemicals in liquid phase at relatively high concentrations are termed taste or *gustatory receptors*, although the distinctions between taste and smell are not critical at cellular or molecular levels.

Physiologists use the term *papillae* to identify so-called taste buds on the surface of the tongue. There are several thousand papillae on the human tongue. Because the tongue takes so much abuse when a person is eating a variety, often very coarse and rough foods, some authorities believe that the papillae are constantly being renewed. Small fibers, almost like tiny hairs, extend from these cells to the surface of the tongue. Some investigators have described the papillae as being of three general shapes: those that look like tiny mushrooms; those that appear like miniature hills with moats around them; and the tiny threads and cones. Observable differences in shape tend to reinforce earlier theories, which embrace the use of differently structured organisms to sense different taste categories, notably the traditional four basic taste sensations—sweet, sour, bitter, salty. These concepts have been difficult to refine or confirm.

Chemical aspects of taste receptor functions can be studied by recording the patterns of electrical potentials in receptor cells while the cells are being stimulated with pure chemicals of known structures and properties. Since the mid-1950s, when this method was first successfully applied to single taste receptor cells, using receptors on the mouth parts of a fly, many earlier theories of taste stimulation have been revised.

Intracellular recordings from taste cells of rat and hamster show that even primary receptor cells are sensitive to three or four of the so-called basic taste modalities. Consequently, it is generally held that a variety of different receptor *sites* commonly exist on the receptor membrane of any one receptor cell. Biochemical characterization of events at receptor sites has progressed in analyzing electrolyte and carbohydrate stimulation.

Electrolyte stimulation is chiefly a function of monovalent cations in all animals that have been studied. Consequently, the receptor sites are thought to be anionic. The pH relationships of stimulation also indicate that strongly acidic (e.g., PO_4^{2-} or SO_4^{2-}) receptor groups are involved. Calculations of free energy changes of the reaction between salt and receptor site give values between 0 and -1 kcal/mole; and low ΔF values suggest that the reaction involves only weak physical forces. The reaction occurs extremely rapidly, since typical nerve impulses can be recorded within 1 millisecond after stimulating electrolytes are applied. In blow-flies, 0.004 M NaCl, which produces 1 impulse per second, represents the threshold for behavior response. These thresholds appear to be somewhat higher in humans.

No one type of receptor site or reaction can account for the extreme structural specificities observed. A curious assortment of molecules can elicit "sweet" sensations. Early studies on a variety of organisms demonstrated that ring structures and D-isomers were more stimulating in polyol compounds than straight-chain and L-isomers. Thus, inositol (Fig. 1), with its ring structure, was found to stimulate. The straight-chain polyhydric alcohols sorbitol, dulcitol, (Fig. 2), and mannitol did not stimulate. Possession of an alpha-D-glucopyrano side linkage was found to generally increase the stimulating capacity of sugars. Maltose, with a 1,4-linkage; turanose, with a 1,3-linkage; and the nonreducing sugars stimulate. Lactose, with its 1,4-linkage, and melibiose, with a 1,6-linkage, both lack the alpha link and are relatively nonstimulating.

Conformation, as well as configuration, is important in determining the stimulating power of sugar molecules. Glucose, which exists in solution almost entirely in an aldopyranose "chair" formation has derivatives of both 1C and C1 conformations (Figs. 3 and 4). Those of the C1 type are considerably the more stimulating. The hydroxyl groups attached to C3 and C4, inclined 19 degrees above and 19 degrees below the adjacent plane of the molecule, appear to be necessary for the critical linkage at the receptor site. Lack of effects by metabolic inhibitors (azide, fluoride, iodoacetate, etc.) or of temperature effects upon the initial excitatory process, suggests that this step depends upon specific physical rather than chemical reactions.

The nature of other polyol receptor sites and the molecular basis for genetic and species differences in taste capabilities remain largely unknown. Saccharin (o-sulfobenzime, Fig. 5) exemplifies both puzzles, since its molecule does not fit any known sugar receptor site, yet it is confused with sugar stimuli by humans and other primates, but probably not by nonprimate animals. The substitution of other groups for one hydrogen (dotted lines in Fig. 5) renders saccharin tasteless. The genetic basis of taste has been studied with phenylthiocarbamide (PTC). A strong bitter taste of PTC depends upon the chemical components indicated by dotted lines in Fig. 6, and upon possession of a dominant "taster" gene in humans. Curiously, a small change in the molecule (Fig. 7) yields a product 250 to 300 times sweeter than sugar.

Taste receptors for water have been reported to occur on mouth parts of mammals and invertebrates. Specialized amino acid and amine receptors are found on the legs of many anthropoids. The mechanisms by which adequate stimuli initiate nerve impulses in these cells offer a rich field for further investigation. Some stimuli, especially long-chain hydrocarbons, are known to act in an opposing manner, i.e., by decreasing, rather than increasing, the output of receptor impulses. Their effects resemble the actions of narcotics. Some authorities suggest that taste sensation, as ultimately perceived, probably results from a complex coded pattern of augmented or depressed frequencies of nerve impulses, originating in the different cells of a heterogeneous population of taste receptors.

Sense of Smell. For many years, physiologists have explained that the sense of smell is located in the mucus lining inside the nose. Traditionally, it has been observed that a person with a "dry" nose has little if any sense of smell. Molecules characteristic of certain flavorings must be moistened by the mucus before they can be detected. Traditionally, it also has been observed that these nerve cells (estimated to be a million or more) tend to become blocked (refuse further transmission of signals) upon prolonged exposure to any given odor. This blocking phenomenon can occur within just a few minutes after exposure to certain, usually powerful odors. When all is sorted out concerning the mechanics of tasting and smelling, it is highly likely that operationally the cells in the nose lining and the papillae of the tongue will be highly similar, if not identical—simply because many other interrelationships have been shown to be similar. Of course, over the years, some researchers have approached the phenomena of taste and odor separately. Some of the odor detection theories proposed have included: (1) the *vibrational theory* (Demerdach; Dyson; Wright); (2) the *stereochemical theory* (Amoore, Johnston, Naves); (3) the *theory of interfacial adsorption* (Beck; Davies); and the *profile functional group theory* (Beets).

The aforementioned theories are concerned with the size and shape of odorant molecules, but differ in certain underlying concepts. For example, accommodating for functional groups, electron donor-acceptor characteristics, as well as the sorptive nature of odorants on sensor sites. The vibration theory largely concentrates on the far-infrared and Raman spectral characteristics of odoriferous substances. The remaining theories concentrate on structural and behavior characteristics of odorant molecules, stressing direct interactions physically, chemically, and biologically with the olfactory sensor system.

The vibrational theory stresses so-called osmic frequencies (Wright) as setting up resonances in the sensory organs. In commenting on this concept, Dravnieks observed that spectra are codes describing molecular structures and shapes with emphasis on the distribution of atomic masses, distances, and bond polarities. Questions regarding corresponding intra-molecular vibrations as direct factors in odor discrimination are independent of the validity of the spectral code.

In connection with the stereochemical theory, Amoore, although emphasizing the importance of molecular shape and size, also gives important emphasis to such chemical factors as electrophilic-nucleophilic characteristics, rotational moment, and the presence of functional groups. In the theory, the functional groups play a more important role in small molecules than in larger ones. Size and shape similarities were analyzed and demonstrated by preparing silhouettes patterned from three-dimensional molecular models.

Critics of the Amoore theory point out that electrophysiological data do not support the concept that sensors are sensitive to size and shape. Other authorities, however, point out that molecular size and shape, operating with a reactive character, most likely are odor relevant, but that this is not explained satisfactorily by the Amoore theory. Another objection to the theory is that equal weight is given to positive and negative differences in odorant molecules. Further, it is observed that the theory does not account for odor blocking or fatiguing (*anosmia*).

Fig. 1 through 7.   Molecules illustrating relationships of seven different structures and their effects on taste receptors. (*After Hodgson.*)

The profile functional group theory is a relatively complex two-step concept. First, it is visualized that the functional groups of the odorant molecule interact with the receptor site, thus causing a given orientation of the molecule. This, in turn, determines the final odor-relevant profile. Any similarity in profile at the receptor site causes similar signal transmissions to the brain. Because some adorant molecules will react more strongly at receptor sites than others, stronger signals will be transmitted.

The interfacial adsorption theory proposes that the orientation of odorant molecules is dependent upon their behavior at the hydrophilic-hydrophobic interface, taking into account interaction with the mucus and adjacent olfactory membrane.

Additional Reading

Amoore, J.E., Johnston, J.W., Jr., and M. Rubin: "The Sereochemical Theory of Odor," *Sci. Amer.* (February 1964).

Beck, L.H.: "A Quantitative Theory of the Olfactory Threshold Based upon the Amount of the Sensory Cell Covered by an Adsorbed Film," *Proc. New York Academy of Sciences*, **116**, 448 (1964).

Beets, M.G.J.: *Structure and Odor,* Molecular Structure and Organoleptic Quality, Society of Chemical Industry, London, 1961.

Chang, S.S.: "Food Flavors," *Food Technology*, **99** (December 1989).

Contis, E.T. et al.: *Food Flavors: Formation, Analysis, and Packaging Influences,* Elsevier Science, New York, NY, 1998.

Davies, J.T. and F.H. Taylor: "Olfactory Thresholds: A Text of a New Theory," *Perfume, Essential Oil Record*, **46** (1955).

Demerdache, A. and R.H. Wright: "Low-Frequency Molecular Vibration in Relation to Odor," in *Olfaction and Taste* (T. Hayaski, editor), Vol. 2, Pergamon, Elmsford, New York, NY, 1979.

Dravnieks, A.: *Comparison of Theories on Relations between Odor Parameters and Other Properties of Odorants,* NATO Advanced Study Institute on Odor Theories and Odor Measurements, Robert College, Blebes, Istanbul, Turkey, 1966.

Dyson, G.M.: "Raman Effect and Concept of Odor," *Perfume, Essential Oil Record*, **28** (1937).

Dziezak, J.D.: "Spices," *Food Technology*, **102** (January 1988).

Dziezak, J.D.: "Microencapsulation and Encapsulated Ingredients," *Food Technology*, **136** (April 1988).

Dziezak, J.D.: "New Spice Alternative Maximizes Flavor and Stability," *Food Technology*, **104** (September 1988).

Fischetti, F., Jr.: "Natural and Artificial Flavors," in *CRC Handbook of Food Additives*, 2nd Edition, T.E. Furia, Edition, Vol. **2**, CRS Press, Boca Raton, FL, 1991.

Giese, J.: "FEMA Expert Panel: 30 Years of Safety Evaluations for the Flavor Industry," *Food Technology*, **84** (November 1991).

Godshall, M.: "The Role of Carbohydrates in Flavor Development," *Food Technology*, **71** (November 1988).

Johnston, J.W. and A. Sandoval: "Organoleptic Qualities and the Stereochemical Theory of Olfaction," *Proc. Sci. Sect. Toilet Goods Association*, **34** (1960).

Jutka, J.R. and D.B. Nelson: "Preparation of Encapsulated Flavors with High Flavor Level," *Food Technology*, **154** (April 1988).

Moody, W.G.: "Beef Flavor, A Review," *Food Technology*, **227–232** (May 1983).

Mussinan, C.J. and M.J. Morello: *Flavor Analysis: Developments in Isolation and Characterization*, Vol. 705, Oxford University Press, Inc., New York, NY, 1998.

Naves, Y.R.: *The Relationship between the Stereochemistry and Odorous Properties of Organic Substances,* Molecular Structure and Organoleptic Quality, Society of Chemical Industry, London, 1957.

Oser, B.L. and R.A. Ford: "FSMA Flavor and Extract Manufacturers Association Expert Panel: 30 Years of Safety Evaluation for the Flavor Industry," *Food Technology*, **84** (November 1991).

Piggott, J.R. and A. Paterson: *Understanding Natural Flavors,* Blackie Academic & Professional, London, UK, 1999.

Pszcaola, D.E.: "Application of the Delta T Theory in the Design of Microwavable Flavors," *Food Technology*, **102** (September 1988).

Riley, K.A. and D.H. Kleyn: "Fundamental Principles of Vanilla/Vanilla Extract Processing and Methods of Detecting Adulteration in Vanilla Extracts," *Food Technology*, **64** (October 1989).

Risch, S.J. and Chi-Tang, Ho: *"Spices," Flavor Chemistry and Antioxidant Properties*, Vol. 660, American Chemical Society, Washington, DC, 1997.

Roberts, D.D. and A. Taylor: *Flavor Release,* American Chemical Society, Washington, DC, 2000.

Shahidi, F. and K.R. Cadwallader: *Flavor and Lipid Chemistry of Seafoods,* American Chemical Society, Washington, DC, 1997.

Shell, E.R.: "Chemists Whip Up a Tasty Mess of Artificial Flavors," *Smithsonian*, **78** (May 1986).

Sinki, G.S.: "Finding the Universally Acceptable Taste," *Food Technology*, **90** (July 1988).

Staff: *Food Chemicals Codex,* National Academy of Sciences, Washington, DC, (Revised periodically).

Staff: "Food Additives, Who Needs Them?" *Food Technology*, **55–57** (January 1985).

Teranishi, R. and P.J. Williams: *Biotechnology for Improved Foods and Flavors,* Vol. 637, American Chemical Society, Washington, DC, 1996.

Tyrrell, M.H.: "Evolution of Natural Flavor Development with the Assistance of Modern Technologies," *Food Technology*, **68** (January 1990).

van Osnabrugge, W.: "How to Flavor Baked Goods and Snacks Effectively," *Food Technology*, **74** (January 1989).

Wright, R.H.: "Odor and Molecular Vibration," *J. of Applied Chemistry*, **4**, London (1954).

Young, G.: "Chocolate—Food for the Gods," *National Geographic*, **664** (November 1984).

Early Classical References Dealing with Odor Sensing

Web References

Institute of Food Technologists. *http://www.ift.org/*

The Institute of Food Science and Technology. *http://ifst.org/*

FLINT. Flint is a rock composed essentially of a crypto-crystalline form of silica. It is very dense and tough, breaking with a conchoidal fracture; colors, usually dark grays, blues, or browns, often black. It occurs chiefly as nodules and masses in chalks and limestones. Flint is particularly interesting because it was used by primitive man for making instruments (artifacts) for thousands of years before he learned to use bone and metal. Flint remained an essential mineral resource for making fire, including the flint locks on guns, until the close of the eighteenth century. From the dawn of civilization, the best flint has come from Belgium and the coastal chalks of the English Channel and the Paris Basin. See also **Chert**.

FLOCCULATING AGENTS. Flocculation is defined as the process by which fine particles, suspended in a liquid medium, form stable aggregates called flocs. The degree of flocculation can be defined mathematically as the number of particles in a system before flocculation divided by the number of particles (flocs) after flocculation. Flocculation makes the suspension nonhomogeneous on a macroscopic scale. A complete or partial separation of the solid from the liquid phase can then be made by using a number of different mechanical devices. Flocculating agents are chemical additives which, at relatively low levels compared to the weight of the solid phase, increase the degree of flocculation of a suspension. They act on a molecular level on the surfaces of the particles to reduce repulsive forces and increase attractive forces.

Applications

The principal use of flocculating agents is to aid in making solid–liquid separations. These applications include:

1. Removing small amounts of suspended inorganic or organic particles from surface water prior to its use as drinking water or industrial process water.
2. Concentrating the organic solids in municipal or industrial wastewater to produce a sludge with a minimum volume and water content for incineration or other means of disposal, and a clarified (very low suspended solids) water that can be discharged or recycled. This operation is often called dewatering.
3. Removing suspended inorganic material from waste streams generated in the beneficiation of ores or nonmetallic minerals, to form a concentrated slurry that can be used for reclamation of mined out areas or other uses and a clarified water that can be discharged or recycled.
4. Separating the solid and liquid phases in leaching operations, where a valuable material is contained in the liquid phase, so its recovery is to be maximized.
5. Binding fine cellulose fibers and solid inorganic additives to long cellulose fibers as the paper pulp is being formed into sheets on a paper machine (see **Papermaking and Finishing**).

Chemical Composition

Flocculants can be classified as inorganic or organic. The inorganic group as well as some highly charged cationic organic flocculants are sometimes referred to as coagulants; however, no such distinction is made in this article.

Inorganic Flocculating Agents. The inorganic flocculating agents are water-soluble salts of divalent or trivalent metals. For all practical purposes these metals are aluminum, iron, and calcium. The principal materials currently in use are: aluminum sulfate, aluminum chloride hydroxide, sodium aluminate, ferric chloride, ferric sulfate, ferrous sulfate, calcium hydroxide, and lime.

Organic Flocculants. The organic flocculants are all water-soluble natural or synthetic polymers. Since the 1950s the use of natural products as flocculating agents as steadily declined as more effective synthetics have taken their place. The only natural polymers used to a significant degree as flocculants are starch and guar gum. Examples of synthetic polymers include acrylamide–acrylic polymers and their derivatives, polyamines and their derivatives, poly(ethylene oxide), and allylamine polymers.

Mechanism of Flocculation

In order to form flocs the individual particles must move and collide. Flocculation can be classified as either orthokinetic or perikinetic. In the first case particle motion results from turbulence in the suspension, and in the latter from Brownian motion. Orthokinetic motion is almost always the case in industrial applications. At very close distances, polar materials are attracted by dipole-induced dipole interactions commonly called van der Waals forces. In most aqueous suspensions, ionization of surface groups gives the particle an overall negative charge. The charged particles in suspension are surrounded by a group of positive ions referred to as the double layer. As particles approach each other the resulting electrostatic repulsion of the double layers prevents flocculation. Increasing the ionic strength of the liquid medium reduces the repulsion until the particles start to aggregate at the critical flocculation concentration. As the charge of these positive ions forming the double layer is increased by adding higher charged ions to the system, the double layer gets nearer to the surface allowing the particles to become closer and be attracted by the van der Waals forces. This is the explanation for the empirically derived Schulze-Hardy rule that the critical flocculation concentration of positive ions for a particular system decreases proportionally with the sixth power of the charge. This mechanism is called double-layer compression and is often cited for the inorganic flocculating agents, such as alum and ferric salts, which add trivalent ions to the system. However, this explanation of the action of aluminum and ferric salts does not take into account the fact that they are present at least partially as polymeric species when added to many systems, and that polymeric precipitates may be formed at the usual concentrations and the pH range that they are used.

The second flocculation mechanism is referred to as the charge patch or electrostatic mechanism. A highly cationic polymer is adsorbed on a negative particle surface in a flat conformation. This promotes flocculation by first reducing the overall negative charge on the particle thus reducing interparticle repulsion. A third mechanism is called bridging. Some individual segments of a very high molecular weight polymer, usually a high molecular weight anionic polyacrylamide, adsorb on a surface. Large segments of the polymer extend into the liquid phase where other segments are adsorbed on other particles, effectively linking the particles together with polymer bridges. In contrast to the first two mechanisms, bridging is strongly affected by molecular weight and the ionic content of the solution.

A fourth mechanism is called sweep flocculation. It is used primarily in very low solids systems such as raw water clarification. Addition of an inorganic salt produces a metal hydroxide precipitate which entrains fine particles of other suspended solids as it settles.

Flocculant Performance and Selection

There is no comprehensive quantitative theory for predicting flocculation behavior that can be used for flocculant selection. This must ultimately be determined experimentally. There are three variables that affect the results obtained in any particular flocculation system. These are the type of flocculant, type of substrate, and type of mechanical treatment of the flocculated substrate. The size and physical properties of the flocs that form, rather than the degree of flocculation, are the key elements in determining the practical effectiveness of a flocculant in any specific application. The effect of mechanical treatment can be viewed in terms of the type of force applied to the flocs. In thickeners and settling basins the flocs are acted on by gravity and by the weight of material added on top of them. In vacuum filters the flocs are subjected to atmospheric pressure. In belt presses and plate-and-frame filters the flocs are subjected to mechanical pressure and in centrifuges they are subject to centrifugal forces. In a flowing system, such as a continuous paper machine, they are subjected to shear and elongational forces on the same scale as the particle size. In addition to the type of force that is applied to the flocs, the kinetics of floc formation also plays an important role in the results obtained in their application.

There are some general principles that can serve as guidelines for initial screening in terms of both flocculant chemistry and molecular weight. In general the large flocs formed by high molecular weight polymers tend to settle faster than smaller ones.

In the case of thickeners, the process of compaction of the flocculated material is important. The flocs settle to the bottom and gradually coalesce under the weight of the material on top of them. As the bed of flocculated material compacts, water is released. Usually the bed is slowly stirred with a rotating rake to release trapped water. The concentrated slurry, called the underflow, is pumped out the bottom. Compaction can often be promoted by mixing coarse material with the substrate because it creates channels for the upward flow of water as it falls through the bed of flocculated material. The amount of compaction is critical in terms of calculating the size of the thickener needed for a particular operation. The process of compaction has been extensively reviewed in the literature.

For most substrates the operating dosage of flocculant necessary to give the settling rate necessary to operate a thickener is well below the maximum amount that can be adsorbed on the substrate. As more and more polymer is added above this operating dosage the flocs can become larger and somewhat sticky. The bed of flocculated material then becomes very viscous. The rake mechanism may become overloaded and the flocculated material may not flow into the underflow pump. The dosage response and the sensitivity to overdosing may affect the selection of flocculating agent.

For filter belt presses and centrifuges, resistance to shear and mechanical pressure is the most important parameter. In general, flocs produced by charge patch neutralization are stronger than those produced by inorganic salts alone.

For vacuum filters, both the rate of filtration and the dryness of the cake may be important. The filter cake can be modeled as a porous solid, and the best flocculants are the ones that can keep the pores open.

Retention aid polymers are used in a very high shear environment, so floc strength and the ability for flocs to reform after being sheared is important. The optimum floc size is a compromise. Larger flocs give better free drainage, but tend to produce an uneven sheet owing to air breakthrough in the suction portions of the paper machine. In some cases the type of floc needed for retention can be seen as similar to that needed for vacuum filtration. Floc size can be controlled by both the type of flocculant and the addition point.

General guidelines concerning the initial selection of flocculant chemistry are (1) suspensions of organic materials, such as municipal waste, are usually treated with a cationic flocculant, either inorganic or organic; and (2) suspensions of inorganic materials such as clay are usually treated with an anionic polymer or a combination of an anionic polymer with a cationic flocculating agent.

Laboratory Flocculant Testing. The objective of laboratory testing of flocculants is to determine which chemical composition and molecular weight will give the best cost performance. The usual method is to simulate on a laboratory scale the formation of flocs and then subject them to the same or similar types of forces as would be encountered in a full-scale dewatering device.

Operating Parameters and Control

Flocculating agents differ from other materials used in the chemical process industries in that their effect not only depends on the amount added, but also on the concentration of the solution and the point at which it is added. The process streams to which flocculants are added often vary in composition over relatively short time periods. This presents special problems in process control.

Dilution. In many applications, dilution of the flocculant solution before it is mixed with the substrate stream can improve performance. The mechanism probably involves getting a more uniform distribution of the polymer molecules.

Addition Point. The flocculant addition point in a continuous system can also have a significant effect on flocculant performance.

Automatic Control. In some industries, the waste streams can vary in composition over a relatively short time period. When the solids level of a slurry changes, the entire dosage response may change. Automatic systems are available for thickeners that adjust the dosage according to the incoming solids level, overflow turbidity, and streaming current potential.

Analysis

Inorganic flocculants are analyzed by the usual methods for compounds of this type. Residual metal ions in the effluent are measured by spectroscopic techniques such as atomic absorption.

The detection of organic polymers in solution represents a more difficult problem, especially in industrial water and wastewater. In theory, charged polymers react with polymers of the opposite charge in solution and such reactions can be used to titrate the concentration of polymer present. There are a number of techniques using this method.

The molecular weights and molecular weight distributions of lower molecular weight polymeric flocculants are determined by viscosity measurements. High molecular weight acrylamide-based polymers are characterized by light scattering techniques.

Toxicology and Environmental Issues

Based on animal studies and mutagenicity studies, trace amounts of organic polymers do not appear to present a toxicity problem in drinking water. The reaction products with both chlorine and ozone also appear to have low toxicity. The principal concern is the presence of unreacted monomer and other toxic and potentially carcinogenic nonpolymeric organic compounds in commercial polymeric flocculants. The principal compounds are acrylamide in acrylamide based polymers, dimethyldiallyammonium chloride in allylic polymers, and epichlorohydrin and chlorinated propanols in polyamines, as well as the reaction products of these compounds with ozone and chlorine.

Until 1990 the EPA maintained a list of chemicals suitable for potable water treatment in the United States. Since then the entire question of certification and standards has been turned over to a group of organizations headed by the National Sanitation Foundation, which has issued voluntary standards. As of January 1992, standards had been issued for most of the principal inorganic products, but only for three polymers, poly(DADMAC), Epi-DMA (epichlorohydrindimethylamine) polymers and polyacrylamide. Certifications for commercial products meeting specified standards are issued by the National Sanitation Foundation, Underwriter Laboratories, and Safe Water Additives Institute (SWAI).

The same questions about the safety of organic flocculants have been raised in other countries. The most drastic response has occurred in Japan and Switzerland where the use of any synthetic polymers for drinking water treatment is not permitted.

HOWARD I. HEITNER
Cytec Industries

Additional Reading

Gregory, J.: in B.M. Moudgil and P. Somasundaran, eds., *Flocculation, Sedimentation and Consolidation,* American Institute of Chemical Engineers, New York, NY, 1985, pp. 125–138.
Halverson, F.: in K.J. Hipolit, ed., *Chemical Processing Aids in Papermaking: A Practical Guide,* TAPPI Press, Atlanta, Ga., 1992, pp. 103–127.

FLORY, PAUL J. (1910–1986). An American chemist who won the Nobel prize in 1974 for his work in polymer chemistry. He published extensive work on the physical chemistry of polymers and macromolecules. He held many medals and awards. Flory received his doctorate from Ohio State University in 1934. He was the C. J. Wood professor of chemistry at Stanford University.

FLOTATION. A method of separating minerals from waste rock or solids of different kinds by agitating the pulverized mixture of solids with water, oil, and special chemicals that cause preferential wetting of solid particles of certain types by the oil, while other kinds are not wet. The unwetted particles are carried to the surface by the air bubbles and thus separated from the wetted particles. A frothing agent is also used to stabilize the bubbles in the form of a froth that can be easily separated from the body of the liquid (froth flotation). Do not confuse with floatation.

FLUID AND FLUID FLOW. The word fluid refers to a state of matter in which only a uniform isotropic pressure can be supported without indefinite distortion; so, a gas or a liquid. The distinction between highly viscous liquids and solids is a difficult one, the same material acting as an ordinary liquid under some circumstances and as a solid under others. Fluids may be described in various ways. A *perfect fluid* is

frictionless offering no resistance to flow except through inertial reaction. A *homogeneous fluid* has the same properties at all points. An *isotropic fluid* has local properties that are independent of rotation of the axis of reference along which those properties are measured. An *incompressible fluid* is a fluid whose density is substantially unaffected by change of pressure. The behavior of a real fluid is similar to that of an incompressible fluid only if the pressure variations in the flow are small compared with the bulk modulus of elasticity. For a fluid in motion in a gravitational field with velocities of order v, it is necessary that both v and $\sqrt{gh}$ should be small compared with the velocity of sound in the fluid. (h is the depth of the fluid and g the acceleration due to gravity.) An *elastic fluid* is a fluid for which elastic stresses and hydrostatic pressures are large compared with viscous stresses. A *viscous fluid* has an appreciable fluid friction. A *Newtonian fluid* is a viscous fluid in which the viscous stresses are a multiple of the rate of strain. The contact of proportionality is the fluid *viscosity*. A *Maxwellian fluid* is a viscous fluid in which the stress-strain relationship includes the relaxation effect (which takes a measurable time) of the relaxation of the elastic stresses set up by a sudden deformation. A *thixotropic fluid* is a fluid whose viscosity is a function not only of the shearing stress, but also of the previous history of motion within the fluid. The viscosity usually decreases with the length of time the fluid has been in motion. Such systems commonly are concentrated solutions of substances of high molecular weight, or colloidal suspensions. See also **Viscosity.**

Fluidity is the property of a substance that expresses its ability to *flow,* as contrasted with viscosity, which is the resistance to flow. Fluidity is a measure of the rate at which a fluid is deformed by a shearing stress, and is mathematically the reciprocal of the viscosity.

Fluid Dynamics

The study of the motion of matter in the gas, liquid, plastic, or plasma state is *fluid dynamics.* When restricted to the flow of incompressible (i.e., constant density) fluids, the term *hydrodynamics* is used. When dealing with electrically conducting fluids with magnet fields present, the term *magneto-fluid dynamics* is used. When dealing with practical problems of air flow past airplane wings, through ventilating equipment, etc., the term *aerodynamics* is used. See also **Aerodynamics.**

Basically, two fundamental approaches are used: (1) continuum or field dynamics and (2) kinetic theory and nonequilibrium statistical mechanics. The study of fluids tends to be quite complex.

Continuum Dynamics. In this approach, fluid properties, such as velocity, density, pressure, temperature, viscosity, and conductivity, among others, are assumed to be physically meaningful functions of three spatial variables x_1, x_2, and x_3, and time t. Nonlinear partial differential equations are set up to relate these variables. Such equations have no general solutions even for the most restrictive boundary conditions. But solutions are carried out for very idealized flows. Couette flow is one of these. See Fig. 1.

The flow is between parallel plates, lower plate at $y = 0$ at rest. upper plate y_B moving with constant speed u_B in the x direction. Stress throughout the fluid is constant, given by $P_{xy} = \mu\,(du/dy) = \mu\,(u_B/y_B)$. This is pure shear flow and experimentally is often considered to define and measure the viscosity coefficient μ, assumed constant for the homogeneous fluid. The velocity profile appearing at the right in Fig. 1 shows by velocity arrows of different length at the various positions of y how the velocity varies with position. Steady flow (no dependence of any quantity on time), constant pressure, constant density, and laminar flow are additional assumptions for Couette flow. The flow is realized experimentally by confining the fluid in the narrow annulus between rotating concentric cylinders of nearly equal radius; the cylinders rotate at different speeds.

In Fig. 2, the special flow (Poiseuille flow) is in a pipe of uniform cross section, pressure is assumed to be constant across each cross section but to vary linearly with distance x along the axis of the pipe so that $dp/dx = (p_1 - p_2)L$. Pistons driving the flow are assumed to be infinitely far away, so that the flow velocity, parallel to pipe axis, has the same

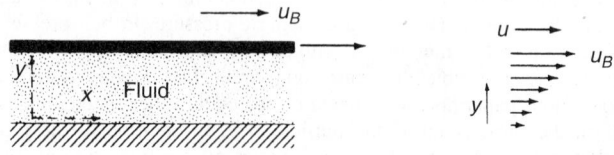

Fig. 1. Couette flow

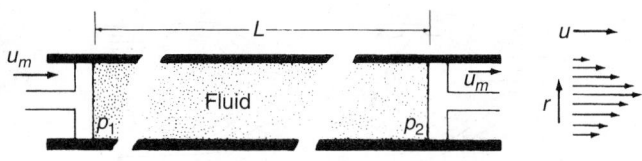

Fig. 2.   Poiseuille flow

dependence upon y and z for all x. The velocity profile is parabolic in both the two-dimensional case (infinite parallel plates) and in the circular cross-section case. Mean flow velocity u_m and viscosity coefficient are assumed constant: the flow is assumed steady and laminar. For a circular cross-section pipe of radius α, at any distance r from the center. $u = 2u_m(1 - r^2/a^2)$, and the volume passing a cross section per second is $Q = \pi\,a^2 u_m = \pi\,a^4\,(p_1 - p_2)/8\mu\,L$. Since these formulas do not apply near pipe entrances, discretion in applying them to pipes of finite length is necessary even when the flow is steady and laminar. See also **Laminar Flow**; and **Turbulent Flow**.

Other examples of idealized solutions are: one-dimensional flow of an ideal gas through a normal shock wave; flow of an ideal gas without viscosity through a pipe of slowly changing cross section (wind tunnel); and one-dimensional finite waves in an ideal gas. Numerous other solutions involve making whatever approximations and assumptions necessary to obtain descriptions of observed flows.

Kinetic Theory. In the kinetic theory and nonequilibrium statistical mechanics, fluid properties are associated with averages of properties of microscopic entities. Density, for example, is the average number of molecules per unit volume, times the mass per molecule. While much of the molecular theory in fluid dynamics aims to interpret processes already adequately described by the continuum approach, additional properties and processes are presented. The distribution of molecular velocities (i.e., how many molecules have each particular velocity), time-dependent adjustments of internal molecular motions, and momentum and energy transfer processes at boundaries are examples.

When motion of the fluid consists of only small fluctuations about a state of near-rest, the continuum equations are linearized by neglecting nonlinear terms; and they become the equations of acoustics. A large variety of fluid motions are described as sound waves; when the small-motion or acoustic description can be used, the principle of superposition is valid. This powerful principle allows addition of simple simultaneous motions to represent a more complex motion, such as the sound reaching the audience from the instruments of a symphony orchestra. The superposition principle does not apply to large-scale (nonacoustical) motions, and the subject of fluid dynamics (in distinction from acoustics) treats nonlinear flows, i.e., those that cannot be described as superpositions of other flows.

Since sound waves travel with a speed relative to the fluid, waves moving in a moving field can sometimes be carried off in a direction opposite to the direction of sound travel. The flow where this occurs is called *supersonic*; the flow speed is greater than the sound speed at the spot where the flow is supersonic. Supersonic flow occurs around high-speed vehicles and missiles, and in pipes when high pressure gas escapes through a nozzle into a region of sufficiently lower pressure. A steady supersonic flow always must pass through a *shock front* to slow down to subsonic flow again.

The continuum description of flow fails to describe nearly all actual flow because actual flows when looked at carefully are *turbulent*. Turbulent flows have violent and erratic fluctuations of velocity and pressure not associated with any corresponding fluctuations of the boundaries containing or driving the fluid. Turbulence is generally considered to be the manifestation of the nonlinear nature of the fundamental equations. Under certain conditions, nonturbulent or *laminar* flow exists. A common example is cigarette smoke rising from a cigarette held at rest; near the cigarette, the stream is smooth and straight, or laminar, and further up the flow breaks into turbulence. See also **Reynolds Number**.

Flowing Fluids

Fluids exhibit a number of specific characteristics when they are caused to flow, as contrasted with remaining at rest.

Conservation Laws. The fundamental conservation laws of physics can be used to obtain the basic equations of fluid motion, the equations of continuity (mass conservation), of flow (momentum conservation), of energy (first law of thermodynamics). In addition, conservation over the whole flow system imposes constraints on the flow that can be very useful. The best known of these is the *Kármán momentum integral* for nearly unidirectional mean flow,

$$\frac{d}{dx}\int \rho v(v_1 - v)\,dy - \frac{dP}{dx}\int \frac{(v_1 - v)}{v_1}\,dy = [\tau]$$

equating the momentum flux by flow to the total forces applied by pressure gradients and boundary stresses. See also **Bernoulli Law**.

Fluid Dynamic Pressure. The pressure necessary to accelerate a fluid from rest to a speed of V is the dynamic pressure equivalent to that speed. If p is taken as mass density of the fluid,

$$\text{dynamic pressure, } q = \frac{\rho V^2}{2}$$

This conception is useful in aerodynamics as the dynamic pressure represents the unit air pressure acting on a surface increment in atmospheric air moving with velocity V over the surface. By Bernoulli's theorem,

$$p = p_0 - \frac{\rho v^2}{2} p_0 - q$$

The vacuum caused by air in motion over a surface is greatest when this imaginary dynamic pressure is greatest since q represents the vacuum.

Fluid Dynamical Similarity. Two geometrically similar fluid flows are dynamically similar if the flow field of one may be transformed into the flow field of the other by the same change of length and velocity scales that was necessary to make the boundary conditions identical. If the equations of motion of the flow are made nondimensional by expressing velocities and lengths as fractions of these scales, these equations contain a number of nondimensional coefficients that determine the character of the flow. The general condition for dynamical similarity is that all these coefficients should be the same for the two flows. For geometrically similar flows, the conditions to be observed are generally expressed as the Reynolds number, the Prandtl number, the Grashof (or Rayleigh) number, the Mach number, and the Froude number.

Boundary Layer in Flowing Fluids. Motion of a fluid of low viscosity, such as air or water, around a stationary body or through a stationary conduit, possesses the free velocity of an ideal fluid everywhere except in an extremely thin layer immediately next to the stationary body.

Many of the phenomena of fluid flow may be studied and analyzed without consideration of this boundary layer. But, thin as this layer may be [usually a few thousandths of an inch (less than a millimeter)], its internal mechanics must be understood and evaluated in certain of the phenomena of fluid motion. Some of the more important of these considerations are:

1. The magnitude of the maximum lift coefficient of the airfoils.
2. Profile drag of airfoils.
3. The drag of bluff bodies.
4. The large variations of drag coefficient at critical Reynolds number for laminar-turbulent transition.
5. The transfer of heat through surface films.

Many of the phenomena of the boundary layer are explainable on the basis of the theory advanced by Prandtl at the University of Göttingen laboratory nearly half a century ago. In the same flow-research group were others, like Blasius, who broadened and experimentally confirmed the original hypotheses.

An elementary understanding of the effect of fluid viscosity will be had by considering a two-dimensional flow along the upper surface of a very thin flat-plate, as shown in Fig. 3. The thickness of the boundary layer, greatly exaggerated, is y_v; the free stream velocity is V, the variable velocity in the boundary layer is u. A basic assumption of the theory is that a fluid layer of infinitesimal thickness resting against the plate "sticks" to it, so shearing force of the next fluid layer on the stationary layer determines the skin friction. Assuming that the boundary layer consists of lamina of fluid sliding on each other, the velocities of these lamina increase with y until, at the edge of the boundary layer, $u = V$. A series of boundary layer velocity profiles for stations x_1, x_2, x_3 are drawn, to enable the reader to

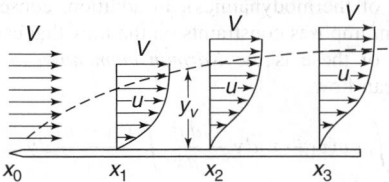

Fig. 3.   Velocity profiles in the boundary layer. Dashed line is hypothetical upper edge of the boundary layer

visualize the effect of friction on the momentum in the boundary layer, and the thickening of it due to lower average u.

Note also the variation of the profile near the surface of the plate. Skin friction has steadily decelerated the individual fluid particles. The profile at x_3 indicates that the lower portion has come to rest. This is known as the stagnation point. Air-flow phenomena in this region are important in many ways, especially when there is an intended rising downstream pressure gradient, as in diffuser tubes or over the surface of airfoils.

Fluid flow in a divergent tube is illustrated in Fig. 4. On the lower profile, greatly enlarged velocity profiles are shown for the boundary layer at stations $x_1, x_2, \ldots x_6$. The x_4 profile indicates a stagnation point. Since the pressure gradient of the diffuser has a pressure at x_5 above that at x_4, a reverse flow is produced toward the stagnation point. Streamlines, drawn in the upper half of the tube, show what happens to the fluid flow. The region of the surface of separation between the reverse flow along the wall and the forward flow is unstable and breaks up into random vortices. Kinetic energy is irreversibly transferred to heat and the diffusion fails to produce the expected pressure gradient. Had the divergence of the conduit been sufficiently small, the pressure gradient would have been lowered per unit length and turbulence in free stream or boundary layer would have delayed the stagnation point.

For the airfoil (Fig. 5) with burbled or partly burbled air flow, a similar explanation exists. Streamline flow over the upper surface of the airfoil increases in velocity as angle of attack (and circulation Γ) increases. However, the surface friction in the boundary layer is decelerating the air particles next to the surface and, with high adverse pressure gradient (existing at high angle of attack) opposing the kinetic energy of the boundary layer, the velocity profiles ultimately show a stagnation point toward the rear of the airfoil. The flow separates from the surface and vortices form a turbulent wake in place of the streamline wake previously

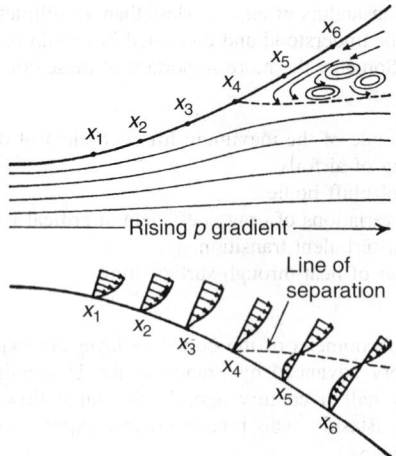

Fig. 4.   Effect of boundary layer viscosity of flow on a diffuser

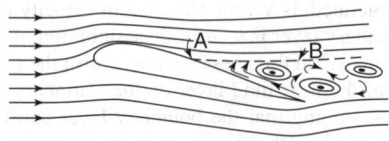

Fig. 5.   Partial burble on airfoil at high angle of attack: (A) stagnation point; (B) line of separation

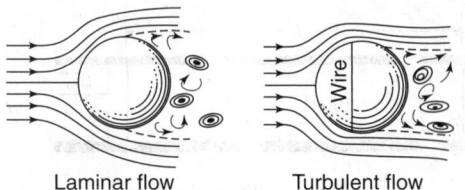

Fig. 6.   Boundary layer turbulence reduces width of low-pressure wake

existing. Once the separation surface moves onto an airfoil, minor increases of angle of attack bring it rapidly forward. High circulation strength is no longer needed to fulfill the requirement of unity of upper and lower flows at the trailing edge (Kutta's hypothesis) so Γ decreases sharply, and with it the lift. Boundary layer theory accounts in this manner for the maximum lift coefficient. At the same time the extended turbulent wake sharply increases the profile drag coefficient.

Now return to a view of the nature of flow in the boundary layer. It has been called laminar, and so it is for values of the Reynolds number below a *critical* value. But for years, beginning about the time of Osborne Reynolds' experiments and revelations in the field of fluid flow, it has been known that the laminar property disappears, and the flow suddenly becomes turbulent, when the critical Vl/v is reached. Usually flow starts over a surface as laminar but after passing over a suitable length the boundary layer becomes turbulent, with a thin laminar sublayer thought to exist because of damping of normal turbulent components at the surface. See Fig. 6.

This transition has profound effects in all fluid dynamics, and certainly so in aerodynamics. The velocity profile in the boundary layer becomes fuller near the surface on account of the higher average kinetic energy of the layer created by turbulent energy exchange from layer to layer. The effective viscosity is therefore larger in turbulent than laminar flow, the turbulent boundary layer thickens more rapidly downstream, the skin friction increases.

The importance to aerodynamics is the beneficial effect of turbulence on the wake existing in the rear of bluff bodies. Even thin airfoils become bluff bodies at high angles of attack. A turbulent boundary layer has more kinetic energy than a laminar one. This carries the air farther toward the rear of the surface before a stagnation point is reached and so reduced the width of the wake and that part of the profile drag. This reduction of drag can be, and usually is, much larger than the increase of skin friction so turbulence has a good effect on both profile drag and maximum lift coefficient of an airfoil. The critical value of Reynolds number lies between 200,000 and 2,000,000, being affected by the initial turbulence existing in the air prior to meeting the surface. Decreasing initial turbulence increases the critical Reynolds number. The effect of turbulence on the boundary layer is strikingly displayed by observing the drag wake behind a smooth sphere mounted in an airstream whose velocity is just under that required to produce breakdown of the laminar boundary layer. An artificial roughness is provided in the form of a fine wire or thread encircling the sphere in the laminar flow. The boundary layer, of course, becomes turbulent downstream from this irregularity resulting in a delayed stagnation point and narrower wake.

Constant-Stress Layer in Flowing Fluids. In the boundary layer of a fluid flowing over a solid wall, the shear stress varies with distance from the wall but it may be considered nearly constant within a small fraction of the layer thickness. The concept is of particular importance in turbulent flow where it leads to a theoretical derivation of the "law of the wall," the logarithmic distribution of mean velocity. The constant stress layer is the best-known example of the equilibrium flows near a wall.

Classes of Flowing Fluids

Just as there are many types of fluids, so there are, partly as a result, many types of fluid flow. *Uniform flow* is steady in time, or the same at all points in space. Steady flow is flow of which the velocity at a point fixed with respect to a fixed system of coordinates is independent of time. Many common types of flow can be made steady by a suitable choice of coordinates. *Rotational flows* have appreciable vorticity, and they cannot be described mathematically by a velocity potential function. *Turbulent flow* is flow in which the fluid velocity at a fixed point fluctuates with time in a nearly random way. The motion is essentially rotational, and is characterized by rates of momentum and mass transfer considerably larger than in the corresponding laminar flow. *Laminar flow* is flow in which the mass of fluid may be considered as advancing in separate laminae (sheets), with simple shear existing at the surface of contact of laminae, should there be any difference in mean speed of the separate laminae. If turbulence exists, its effect is confined to a lamina, and there is not exchange of momentum between laminae. *Streamline flow* is flow in which fluid particles move along the streamlines. This motion is characteristic of viscous flow at low Reynolds numbers or of inviscid, irrotational flow. *Secondary flow* is a less rigorously defined term than many of the foregoing types of flow. The flow in pipes and channels is frequently found to possess components at right angles to the axis. These components which take the form of diffuse vortices with axes parallel to the main flow form the secondary flow. Three types may be mentioned: 1. Secondary flow in curved pipes or channels, being a motion outward near the flow center and inwards near the walls. 2. Secondary flow in straight pipes and channels of noncircular section, being a motion along the walls toward corners or places of large curvature and from there to the center of the flow. This only occurs in turbulent flow. 3. Secondary flow in pulsating flow. This is due to second-order effects and is particularly striking with ultrasonic waves.

For many types of flow, calculations are complex. Where they can be made at all, they require the methods of tensor analysis of the stresses and strains involved. One important relationship that is widely useful in the systematic study of fluid flow in the equation of continuity.

A great simplification of the fluid calculations can be effected by assuming that the fluid is perfect, homogeneous, totally incompressible, not viscous, and that therefore its properties are not affected by changes in temperature or pressure. While such an ideal fluid does not exist, its properties are often approached closely enough by real fluids so that calculations based upon it are often useful in practice.

Thus, elementary hydraulics always includes Bernoulli's law, and it is repeated here due to its great importance to the subject of fluid flow.

Bernoulli's law is

$$\frac{p}{w} + \frac{v^2}{2g} + z = \text{a constant}$$

and the symbols are defined as follows:

p = the static pressure in pounds per square feet
w = the specific weight of fluid in pounds per cubic feet
v = the velocity in feet per second
g = the gravitational acceleration in feet per second
z = the potential, or "elevation," head in feet

Bernoulli's law states that in steady flow, the total head is a constant at any point and equal to the sum of the pressure head, p/w, the velocity head, $v^2/2g$, and the potential head (z). Since there is actually a loss of head between any two points due to friction, the difference between the total heads at any two points must equal the friction head when the flow is steady.

In this equation, $v^2/2g$ represents the velocity head, a pressure that could be recovered by the efficient reduction of the velocity in a conduit of expanding cross section. When this velocity head is multiplied by the specific weight of the fluid, it is reduced dimensionally to the unit of pressure, pounds per square feet. It may be designated the dynamic pressure, in counter distinction to the static pressure p. In the case of an expansionable fluid, such as gas, the total energy at two points in the flow must be the same, that is, the heat energy, plus the kinetic energy of motion, must be constant. It is necessary to invoke this law of continuity of energy in dealing with the flow of gases or vapors through nozzles. See also **Viscosity**.

Additional Reading

Brower, W.B.: *A Primer in Fluid Mechanics: Dynamics of Flows in One Space,* CRC Press, LLC., Boca Raton, FL, 1998.

Chopey, N.P., T.G. Hicks: *Handbook of Chemical Engineering Calculation,* 2nd Edition, The McGraw-Hill Companies, Inc., New York, NY, 1993.

Datta-Barua, L.: *Natural Gas Measurement and Control,* The McGraw-Hill Companies, Inc., New York, NY, 1992.

Heinrich, J.C., D.W. Pepper: *Intermediate Finite Element Method: Fluid Flow and Heat Transfer,* Taylor & Francis, Inc., Philadelphia, PA, 1997.

Perry, R.H., D.W. Green, Editors: *Perry's Chemical Engineers' Handbook,* 7th Edition, The McGraw-Hill Companies, Inc., New York, NY, 1997.

Poling, B.: *Properties of Gases and Liquids,* 4th Edition. McGraw-Hill Companies, Inc., New York, NY, 2000.

Rahman, M.A.: *Potential Flow of Fluids,* WIT Press, Southampton, UK, 1995.

Reid, R.C., J.M. Prausnitz: *Properties of Gases and Liquids,* 4th Edition, The McGraw-Hill Companies, Inc., New York, NY, 1987.

Sabersky, R.H., A.J. Acosta: *Flid Flow: A First Course in Fluid Mechanics,* 4th Edition, Prentice-Hall, Inc., Upper Saddle River, NJ, 1998.

Sell, G.R., R. Temam, and C. Foias: *Turbulence in Fluid Flows: A Dynamical Systems Approach,* Springer-Verlag, Inc., New York, NY, 1998.

FLUID FRICTION. The flow of any actual fluid must of necessity be attended by the presence of friction, due to the physical nature of fluids, none of which meets the requirements of the ideal fluid, as mentioned in fluid flow. A great deal of time and attention have been devoted to the study of the properties of a flowing fluid. The frictional effects present in the flow of liquids have been rationalized much more thoroughly than for vapors and gases. However, for all three the friction is found to depend upon the nature of the fluid itself, its viscosity, and upon the conduit which contains it. Due to the different molecular arrangement of liquids, vapors, and gases, the study of friction of fluids has become a specialized study of each of these three.

Fluid flow rarely follows the commonly accepted idea of streamlines, since the velocities necessary for viscous flow of this nature are almost always lower than those found expedient to employ. Most flows are turbulent in nature. They become turbulent at a definite velocity, the value of which was studied by Reynolds; and this value is incorporated in the well-known Reynolds Number. A general thermodynamic equation of energy of a fluid under flow conditions would be as follows:

Gain in kinetic energy + gain in potential energy + network

received + energy liberated by any chemical change

= change in heat content between two states.

In the case of a liquid, this equation can be considerably simplified—in fact, it becomes Bernoulli's well-known equation—but in the case of compressible fluids, which may also undergo some change of form, such as condensation or compression, the longer equation applies. Most practical problems in fluid friction arise in connection with the flow of fluid through pipes.

FLUIDIZATION. This term is used in engineering to denote the preparation of a solid material having a particle size and other properties such that it may be handled, in many respects, like a fluid. It also refers to the technology of handling the material under such conditions. One of the early developments in this field was pulverized coal; its development received great impetus in its use in catalytic processes in the petroleum industry, and it has since been extended far more widely. One great advantage of having a solid catalyst that flows like a liquid is that its particle size can be much smaller than possible in the old type of solid-bed catalyst, where entrainment in the passing fluid was objectionable. This smaller particle size greatly increases the surface area (which increases as the inverse square of the particle radius) and therefore increases the effectiveness of the catalyst. Another advantage of fluidization is greater ease and hence lower cost of handling, which has permitted the use of catalysts in processes that require their frequent reactivation.

Other applications of fluidization have been made to such materials as sodium chloride (table salt), soda ash, sodium phosphate, sodium sulfate, starch, talc, magnesium oxide, dry clay, boric acid, hydrated lime, and various high polymers in powdered or "bead" form. Fluidization is especially effective in loading and unloading materials from railroad cars and trucks, as well as in moving them about within the plant.

Additional Reading

Chen, J.C.: *Fluidization and Fluid Particle Systems: Recent Research and Development,* American Institute of Chemical Engineers, New York, NY, 1998.

Gidaspow, D.: *Multiphase Flow and Fluidization,* Academic Press, Inc., San Diego, CA, 1993.

King, D.: *Advances in Fluidization and Fluid Particle Systems,* American Institute of Chemical Engineers, New York, NY, 1997.

Perry, R.W., D.W. Green: *Perry's Chemical Engineers' Handbook,* 7th Edition, The McGraw-Hill Companies, Inc., New York, NY, 1997.

FLUID PARCEL. In any fluid, an imaginary portion of that fluid which for theoretical studies may be considered to have all the basic dynamic and thermodynamic properties of the fluid but which is small enough so that its motion with respect to the surrounding fluid does not induce marked compensatory movements. Also called *parcel*. The size of the fluid parcel cannot be given precise numerical definition but it must be large enough to contain a great number of molecules and small enough so that the properties assigned to it are approximately uniform within it.

See also **Fluid and Fluid Flow**.

FLUORESCENCE. This term has three common usages: 1. The process of emission of electromagnetic radiation by a substance resulting from of the absorption of energy from radiation. This energy may be either electromagnetic or particulate, provided that the emission continues only as long as the stimulus producing it is maintained. That is, fluorescence is a luminescence which ceases within about 10^{-8} second after excitation stops; this period of time being the lifetime of an atomic state for a normal allowed transition.

2. The term fluorescence may also be applied to the radiation emitted, as well as to the emission process.

3. In X-ray terminology, the term fluorescence may be used in the more specific sense (than given in the general definition above) to denote the characteristic X-rays emitted as a result of the absorption of X-rays of higher frequency.

FLUORIDATION. See **Fluorine (In Biological Systems)**.

FLUORINE. [CAS: 7782-41-4]. Chemical element symbol F, at. no. 9, at. wt. 18.9984, periodic table group 17 (halogens), mp $-219.62°C$, bp $-188.1°C$, density 1.696 g/l (gas at $0°C$), 1.108 g/cm^3 (liquid at bp). Fluorine is a pale yellow gas, poisonous, very reactive, combines with most other elements in the dark, except it does not combine readily with oxygen. Critical pressure is 55 atm; critical temperature is $-129.2°C$. First identified by Scheele in 1771, but not isolated until 1886 by Moissan who electrolyzed fused potassium hydrogen fluoride in a platinum apparatus. Fluorine is a high-tonnage chemical, used mainly in the production of fluorides, in the synthesis of fluorocarbons, and as an oxidizer for rocket fuel.

First ionization potential 17.42 eV; second, 34.6 eV; third, 58.02 eV; fourth, 84.88 eV; fifth, 113.0 eV; sixth, 152.9 eV. Oxidation potential $F^- \rightarrow \frac{1}{2}F_2 + e^-$, -2.85 V; $2F^- + H_2O \rightarrow F_2O + 2H^+ + 4e^-$, 2.1 V; $HF \rightarrow \frac{1}{2}F_2 + H^+ + e^-$, 3.03 V. Other important physical characteristics of fluorine are given under **Chemical Elements**.

Production

Because fluorine is the most reactive element and one of the strongest oxidizing agents known, its preparation caused difficulties for many years. The requirements for fluorine, for separating ^{235}U from ^{238}U during the development of the atomic bomb, accelerated research on finding improved production methods. Much research went into the development of compounds and materials that would resist the actions of fluorine for use in diffusion plants. Materials finally selected and in use in modern fluorine production plants include (1) a cathode integral with a mild-steel cell body, (2) a carbon anode, (3) a steel cell head, including a Monel skirt, and (4) a Monel screen diaphragm. The skirt is required to prevent admixture of the fluorine gas formed with the hydrogen gas also formed. The electrolyte consists of a fused mixture of potassium fluoride and hydrofluoric acid. The overall reaction: $2HF \rightarrow H_2 + F_2$. Hydrogen is liberated at the cathode; fluorine at the anode. The potassium fluoride is required because hydrofluoric acid of the purity required does not conduct an electric current.

Fluorine causes both chemical and thermal burns and, unfortunately, they may not be detected immediately, depending upon the concentration. Further, upon contact with the skin, fluorine gas reacts with water in the skin to form hydrofluoric acid, an excellent solvent for protein. Personnel directly involved in the handling of fluorine must be equipped with gauntlet gloves, neoprene rubber apron, chemical goggles and when in atmospheres above the TLV, gas masks with cannisters must be used. In emergency situations where there is a very high concentration of fluorine, the area must be evacuated, directing personnel upwind.

All containers, processing equipment, and piping to be used in fluorine service first must be passivated before use and thereafter designated for fluorine service. These requirements result from the severe oxidizing characteristics of fluorine gas. Passivation removes any easily oxidized materials, such as paint, pipe dopes, metal oxides, grease, and metal filings.

During the procedure, a metal fluoride film will form on metal surfaces, thus minimizing further corrosion of the metal by fluorine.

Special permits are required to ship fluorine. Generally, fluorine is transported as a nonliquefied compressed gas in seamless steel or nickel cylinders. Upon receipt, multijacketed dewars frequently are used to contain the product.

Chemistry and Compounds

Fluorine exhibits in common with the other halogen elements a marked readiness to form singly charged negative ions, as would be expected from the fact that these atoms need only one electron to acquire an inert gas configuration. However, the electron affinity of fluorine (3.74 eV) is not the highest of the four common halogens, but is less than that of chlorine and bromine (4.02 eV and 3.78 eV respectively). The greater reactivity of fluorine in aqueous solution is due to the fact that its lower electron affinity is more than offset by its lower energy of dissociation (38 kcal against 58 kcal for Cl_2 and 46 kcal for Br_2) and the higher energy of hydration (122 kcal for F^- against 89 kcal for Cl^- and 81 kcal for Br^-). The overall result is to give the system $F^- \rightarrow \frac{1}{2}F_2 + e^-$, the largest negative oxidation potential (-2.85 V) of any simple ion to its element.

The reactions of fluorine have, in general, high temperature coefficients. At low temperatures its reactivity with hydrogen is very slight, but becomes rapid and even violent at higher temperatures and in the presence of impurities. Fluorine reacts with all metals, the vigor of the reaction and the composition of the resulting fluoride depending upon the temperature and the reactivity of the metal. Sulfur, silicon, carbon, and antimony ignite in fluorine; cesium, rubidium, and potassium form trifluorides, which, however, do not contain trivalent cations, while the noble metals react only at very high temperatures. Unlike the three alkali metals mentioned, however, most metals do not form fluorides in exceptional oxidation states, and, in fact, many elements form oxyanions of higher valence than they do fluorocomplexes. Thus manganese forms two fluorides, MnF_2 and MnF_3, and its fluorocomplex ions of highest valence are MnF_6^{2-} and MnF_5^-; chromium forms four stable fluorides, CrF_2, CrF_3, and CrF_4, and CrF_5 (and possibly CrF_6).

Four binary compounds of fluorine and oxygen have been reported, O_4F_2, O_3F_2, O_2F_2, and OF_2. The polyoxygen difluorides are produced from the elements by action of the silent electric discharge at low pressures, lower temperature favoring higher oxygen content. Dioxygen difluoride is a yellow to orange solid, melting at about $-160°C$. It decomposes rapidly at temperatures above $-25°C$. The others are even less stable. Oxygen difluoride, OF_2, is prepared by passing fluorine rapidly through weak NaOH solutions or by electrolysis of liquid HF solutions of H_2O or other oxygen compounds. It has an $O-F$ bond distance of 1.4 Å, $F-F$, 2.22 Å, and FOF angle, about $105°$. It reacts with metals to form fluorides and oxygen, with other halides to form fluorides, oxygen, and the other halogens, and with other compounds usually to yield fluorides.

Hydrogen fluoride is the most stable of the hydrogen halides (heat of formation 64 kcal). Its bond moment shows the compound to have a marked covalent character. In the pure state, liquid hydrogen fluoride is slightly more conducting than pure water, and in the anhydrous state it reacts only with the alkali metals, alkaline earth metals (excluding beryllium and magnesium) and with thallium. Liquid HF is an extremely strong acid, having a Hammett acidity function of 10.2 (compared to H_2SO_4 11.3), HCl, hydrobromic and hydriodic acids being essentially unionized in it, although the dielectric constant is comparable to that of water. However, in 0.1 N aqueous solution, hydrogen fluoride is only about 15% ionized. A correlative property is the extensive polymerization through hydrogen bonding, various polymers being present. In the vapor state, the degree of polymerization depends upon temperature and pressure, varying from mostly monomer to linear hexamer or even higher polymers, with the ring hexamer being particularly favored. The liquid likewise contains monomer and polymers, the average degree of polymerization being three or four. The units of the polymers undergo very rapid exchange. In aqueous solution there is a strong tendency for fluoride ion to associate with hydrogen fluoride molecules, forming the symmetrical HF_2^- ion. It reacts with metal oxides and hydroxides to produce the fluorides, and with metal ions to produce complex ions, or their salts. It reacts with phosphorus pentoxide to produce complex fluorides, oxyfluorides, and, on continued action, mono-, di- or hexafluorophosphates. See also **Hydrofluoric Acid**.

Due to the high oxidation potential of fluorine, and the small size of the fluoride ion, the element enters into many compounds with the other halogens. Diatomic compounds of this type include CIF, BrF, and IF (the latter two having been identified but not isolated in a pure state); tetratomic compounds include ClF_3, BrF_3, and IF_3. Hexatomic compounds include BrF_5, and IF_5, while the octatomic type is limited to one member, IF_7. These compounds are discussed under the entry for the halogen forming the donor atom in the compound.

Fluorine forms polyhalide anionic complexes including $IFBr^-$, $IFCl_3^-$, BrF_4^-, and IF_6^-, which occur in salts of the higher alkalies, as well as cations such as BrF_2^+, which occurs in such salts as BrF_2SbF_6, BrF_2AuF_4, BrF_2SO_3F, etc. The difluoriodate ion, $IO_2F_2^-$ is also known in such salts as KIO_2F_2.

Organic compounds: Organic fluorine compounds are made by reaction of the corresponding alkane chloro-compounds with silver fluoride, mercurous fluoride, antimony trifluoride, titanium tetrafluoride, and the arene fluoro-compounds by the diazo-reaction using hydrogen fluoride, and otherwise. The effect of the continued replacement of hydrogen atoms by fluorine atoms is an initial increase in reactivity, followed by a reversal of this effect, so that the highly substituted compounds are relatively inert. See also **Fluorocarbon**.

Additional Reading

Banks, R.E.: *Fluorine Chemistry at the Millennium: Fascinated by Fluorine,* Elsevier Science, New York, NY, 2000.
Howe-Grant, M.: *Fluorine Chemistry: A Comprehensive Treatment,* John Wiley & Sons, Inc., New York, NY, 1995.
Hudlicky, M.: *Fluorine Chemistry for Organic Chemists: Problems and Solutions,* Oxford University Press, Inc., New York, NY, 2000.
Kitazume, T., T. Yamazaki: *Experimental Methods in Organic Fluorine Chemistry,* Gordon & Breach Publishing Group, Newark, NJ, 1999.
Krebs, R.E.: *The History and Use of Our Earth's Chemical Elements: A Reference Guide,* Greenwood Publishing Group, Inc., Westport, CT, 1998.
Lewis, R.J., N.I. Sax:. *Sax's Dangerous Properties of Industrial Materials,* 10th Edition, John Wiley & Sons, Inc., New York, NY, 1999.
Lide, D.R.: *CRC Handbook of Chemistry and Physics,* 84th Edition, CRC Press LLC., Boca Raton, FL, 2003.
Parker, P.: *McGraw-Hill Encyclopedia of Chemistry,* 2nd Edition, The McGraw-Hill Companies, Inc., New York, NY, 1993.

FLUORINE (In Biological Systems). Fluorides are not required for plant growth, but in animals, including humans, low levels of fluorides have been shown to have beneficial effects on teeth and on bone structure. Growth increases in experimental animals have been reported, when low levels of fluorides have been added to purified diets. However, fluoride substances show toxicity in both animals and plants when encountered in fumes and dusts from industrial facilities as well as natural emissions from the eruption of volcanoes. Abnormally high levels of fluoride in water also have caused fluorine toxicity in animals and mottled teeth in humans.

Fluorides do not usually move from the soil to plants and on to livestock feedstuffs and human foodstuffs in amounts that are toxic. Injury to plants from fluoride in the soil has been noted on soils that are too acid for the satisfactory growth of most plants. On limed soils or soils with sufficient calcium for optimum growth, any fluorine added to the soil reacts with the calcium and other soil constituents to form insoluble compounds, which are not taken up by the plants. Rock phosphate and some kinds of superphosphate fertilizers contain large amounts of calcium fluoride, but the fluorine content of the plants grown on soils that have been heavily fertilized with these phosphates is not appreciably increased. Tea and some other members of the *Theaceae* family are the only plants that take up very much fluorine from the soil.

While the soil-to-plant segment of the food chain contains some built-in safeguards against fluorine toxicity, this toxicity occurs as a result of the deposition of airborne fumes and dusts on the above-ground parts of plants, followed by the consumption of these contaminated plants by animals, including humans. Also, fluorine toxicity has been caused by direct inhalation of the fumes and dusts, or by drinking water with abnormally high fluorine levels. If the fumes and dusts are mixed into the soil, they will be inactivated, and they will not find their way into the food chain in toxic amounts.

The safeguards against toxicity provided by the chemistry of fluorine in soils make it unlikely that applying fluorine-containing compounds to soils will be a useful way to insure that plants will contain sufficient fluorine to prevent dental caries. However, tea and mechanically deboned meats may contribute to these needs. When increased fluoride intake is desirable, *carefully controlled* direct additions to drinking water, to dentrifices, or

to specific foods are more promising than adding fluorides to soils that produce food crops.

Fluoridation

In a broad sense, this term would signify the addition of fluorine to a substance much as chlorination means the addition of chlorine. In a more specific, but commonly used sense, fluoridation means the addition of very small amounts of a fluoride-containing compound to water supplies for the purpose of preventing dental caries. It has been shown over a number of years that the introduction of about 1 part per million (ppm) of fluoride to drinking water will reduce the incidence of tooth decay in children by as much as 60%, as compared with similar groups of children who consume nonfluoridated water. Because of the striking nature of these findings, a few cities in the United States commenced experimental treatment of water supplies during the mid-1940s. As of the early 1980s, it was estimated that nearly 100 million persons in the United States were supplied with fluoridated water.

The commonly used compound is sodium fluoride or sodium silico-fluoride in a dry crystalline or powdered form. Hydrofluosilicic (flusilicic) acid is also used in liquid form. Inasmuch as concentrations of fluoride in excess of 1.5 ppm may cause mottling of tooth enamel, it is mandatory to exercise very careful control to maintain the desired 1.0 ppm dosage. Water supply samples are frequently tested by municipal authorities. Fluoride concentrations may be determined by colorimetric or electrometric methods, and the latter can be adapted to continuous reporting and controlling.

The concept of fluoridation has created numerous controversies among the populace, a situation that occurs frequently when decisions to install fluoridation systems for the first time are under consideration. Until the mid-70s, it was believed that such practice was unquestionably safe and that arguments against fluoridation were essentially emotionally motivated. However, some second thoughts are now being taken, particularly with reference to possible reactions of fluorine with certain pollutants now found in raw water supplies that once were not present.

Feedstuffs

Excessive amounts of fluoride in the soil can cause tooth and bone damage in livestock. Parts of Arkansas, California, South Carolina, and Texas have soils abnormally high in fluorine content. In serious situations, diarrhea and emaciation will be exhibited by the livestock. The effects depend upon the fluorine source and species of livestock. Exceptionally high fluoride levels can be encountered near smelters where pollution safeguards have not been installed or are ineffectively maintained. As compared with other livestock, pigs can tolerate much more fluorine (up to nearly 300 ppm of fluorine derived from rock phosphates).

Fluorine in Tea

The majority of foods found in the average diet contain 0.2–0.3 ppm or less fluorine in the food as consumed. Tea and seafoods are notable exceptions (McClure, 1949). Different values are reported for fluorine content of various teas by different investigators (Wang et al., 1949; Fabre and de Campos, 1950; de Campos, 1950; Zimmerman et al., 1957; Quentin et al., 1960; Okada and Furuya, 1969; Cook, 1970; Venkateswarlu and Sita, 1971). The fluorine content of tea depends upon the origin of the plant, the type of soil and fertilizer, the age of the leaves, and the time of harvesting (Garber, 1962).

In 1978, investigators at the University of Teheran (Iran) undertook a study to find out the fluorine content of teas consumed in Iran and to evaluate the potentiality of tea as a contributor of fluorine. Tea is an important item in the Iranian diet and drunk mostly by labourers and peasants; furthermore, diluted infused tea is used as a supplement in between breast feedings of infants. The investigators concluded that, considering the optimal intake of fluorine of 1 milligram per day suggested for protection from dental caries, the drinking of tea in Iran provides about half of this amount without considering the fluorine content of water and other sources.

Fluorine Content of Mechanically Deboned Beef and Pork

One question that has arisen from time to time in connection with mechanically deboned meat (MDM) is its possible fluoride content because some microscopic bone particles may be present in the product. Investigators Kruggel and Field, Division of Biochemistry and Division of Animal Science, University of Wyoming, made a study of this in 1977. Samples were collected from regions where high levels of fluoride occurring in the water and vegetation have been reported. Higher magnesium, iron, and fluoride contents were found in beef MDM from the western and midwestern regions of the United States when compared with the southern region. Higher iron and fluoride contents were found in beef MDM than in pork MDM. One conclusion drawn was that the consumption of fluoride from MDM and other foods combined would be far below the 20–80 milligrams or more of fluoride that must be consumed daily to produce toxicity (Food and Nutrition Board, 1974). Mottling of teeth in children has been observed at fluoride concentrations in the diet and drinking water of 2–8 ppm. A frankfurter containing 10% MDM would contain about 1.7 ppm fluoride. Since the daily fluoride intake in many areas of the United States is not sufficient to afford optimal protection against dental cavities (Food and Nutrition Board, 1974), products which contain MDM may be of value in furnishing needed fluoride and in reducing the incidence of tooth decay (Kruggel/Field, 1977).

Fluorine in Marine Sponge *Halichondria moorei*

It is well known that many marine organisms accumulate the halogens iodine, bromine, and chlorine. However, reports of fluorine accumulation have been rare. Thus the report of findings by Gregson et al. (1979) that the marine sponge *Halichondria moorei* has a fluorine content of 10% of the total dry weight is of interest. In this species, the fluorine occurs as potassium fluorosilicate, which is known to be a powerful anti-inflammatory agent. It is of interest that closely related sponge varieties of the same habitat contain little if any fluorine—and, further, that the habitat is free of fluorine except for the small amount naturally present in seawater.

Additional Reading

Banks, R.E.: *Fluorine Chemistry at the Millennium: Fascinated by Fluorine,* Elsevier Science, New York, NY, 2000.

Cook, H.A.: "Fluoride Intake through Tea by British Children," *Fluoride Quart. Rept.* **3**, 12 (1970).

de Campos, P.: "Fluorine Content of Tea Cultivated in Sao Paulo," in *Chem. Abstr.* **44**, 22, 11498, 1952.

Fabre R., P. de Campos: "Distribution of Fluorine in Plants—Tea Leaves," *Ann. Pharm. Franc.* **8**, 391 (1950).

Food and Nutrition Board: *Effects of Fluoride in Animals,* National Academy of Sciences, Washington, DC, 1974.

Garber, K.: "Plants and Fluorine" (in German), *Qualitas Plant. Mater. Vegetabiles,* **9**, 33 (in *Chem. Abstr.,* **52**, 2, 2589, 1962).

Gregson, R.P. et al.: "Fluorine is a Major Constituent of the Marine Sponge Halichondria moorei," *Science* **206**, 1108–1109 (1979).

Holden, C.: "Rat Fluoride Study 'Equivocal," *Science,* **681** (May 11, 1990).

Howe-Grant, M.: *Fluorine Chemistry: A Comprehensive Treatment,* John Wiley & Sons, Inc., New York, NY, 1995.

Hudlicky, M.: *Fluorine Chemistry for Organic Chemists: Problems and Solutions,* Oxford University Press, Inc., New York, NY, 2000.

Kitazume, T., T. Yamazaki: *Experimental Methods in Organic Fluorine Chemistry,* Gordon & Breach Publishing Group, Newark, NJ, 1999.

Krebs, R.E.: *The History and Use of Our Earth's Chemical Elements: A Reference Guide,* Greenwood Publishing Group, Inc., Westport, CT, 1998.

Kruggel, W.G. and R.A. Field: "Fluoride Content of Mechanically Deboned Beef and Pork from Commercial Sources in Different Geographical Areas," *J. Food Sci.,* **42**, 1, 190–192 (1977).

Lewis, R.J., N.I. Sax:. *Sax's Dangerous Properties of Industrial Materials,* 10th Edition, John Wiley & Sons, Inc., New York, NY, 1999.

Lide, D.R.: *CRC Handbook of Chemistry and Physics,* 84th Edition, CRC Press LLC., Boca Raton, FL, 2003.

McClure, F.J.: "Fluoride in Foods," *U.S. Public Health Report,* **64**, 1061 (1949).

Okada, F., K. Furuya: "Fluoride Content in Tea" (in Japanese), *Chago Gijutsu Kenkyu* **37**, 32 (in *Chem. Abstr.* **71**, 23, 245, 1966).

Parker, P.: *McGraw-Hill Encyclopedia of Chemistry,* 2nd Edition, The McGraw-Hill Companies, Inc., New York, NY, 1993.

Quentin, K.E. et al.: "Analytical Determination of Small Amounts of Fluorine in Foods and Waters, 4. Fluoride Studies in Foods" (in German), *Chem. Abstr.* **54**, 6, 5967, 1960.

Venkateswarlu, A., P. Sita: "A New Approach to the Microdetermination of Fluoride Adsorption-Diffusion Technique," *Anal. Chemi.* **43**, 78 (1971).

Wang, T.H. et al.: "Fluorine Content of Fukein Teas," *Food Research* **14**, 98 (1949).

Zimmerman, P.W. et al.: "Fluorinein Food with Special Reference to Tea" (in German), *Chem. Abstr.* **52**, 1, 610, 1958.

FLUORITE. Fluorite is a calcium fluoride mineral CaF_2 crystallizing in the isometric system, often in superb cubic crystals. Twinned crystals are common, usually as cubic penetration twins. It is found in many diverse geological environments, from vein material associated with metallic ores, especially lead and silver, to sedimentary formations associated with celestite, gypsum, dolomite, and calcite. It is also found as a component mineral in high-temperature pneumatolytic deposits with cassiterite, topaz and tourmaline, and in pegmatites. Exceptional crystals are found in Alpine type veins on quartz crystals from Switzerland. Also occurs as massive compact to granular aggregates. Possesses perfect 4-directional cleavage planes, with uneven to splintery fracture. It is a brittle mineral with a hardness of 4 and a specific gravity of 3.180. Vitreous luster when crystallized, dull to glimmering in massive material. Colorless when pure, but shades of blue, green, yellow, brown, white, rarely rose-red and pink, are known, including intermediate color graduations of each type. Certain colored crystals appear blue by reflected light, green by transmitted light. This phenomenon may be a product of heat, ultraviolet light, pressure, or exposure to radiation, as from x-rays. Varying color zones are commonly observed in areas parallel to the crystal faces. Massive varieties may also exhibit parallel zones of varying color.

Phosphorescence is not uncommon when certain fluorites are exposed to sunlight, ultraviolet rays, or are heated. Vivid fluorescence is a common attribute of many fluorites, with blue to violet fluorescence predominant. The word fluorescence is derived from the mineral name, fluorite, owing to its strong fluorescent character.

Certain dark blue fluorite from Bavaria known as antozonite contains free fluorine and calcium, which when released either by grinding or exposure to cathode rays produce a distinctive odor, caused by the reaction of the fluorine with water.

Fluorite is a ubiquitous mineral and is so widespread in its occurrence that only the most noteworthy can be mentioned. The English localities at Cumberland, Durham, and Weardale are world famous. Exceptionally beautiful banded material of blue fibrous character from Derbyshire, known as Blue-John, has been much-used for decorative carved pieces, such as vases and other ornamental objects. Norway has produced exceptional specimens from the famous Kongsberg silver veins, as well as yttrium-rich fluorite from northern Norway associated with rare-earth minerals. Fine material has been obtained from the Transvaal in the Republic of South Africa, Tasmania, and Australia. Large quantities of fluorite are mined in Mexico at Guadalcazar and Guanajuato.

Notable United States localities include Hardin and Pope Counties in Illinois, and also adjacent Kentucky areas where it is intimately associated with calcite, barite, quartz, with minor galena and sphalerite in sedimentary rock veins. Large deep green masses yielding exceptional cleavage octahedrons were obtained from Westmoreland, New Hampshire. Macomb, New York produced large sea-green crystallized cubes. Various sedimentary formations in Ohio have yielded fine brown crystals associated with celestite. Many occurrences are known throughout Colorado and Idaho. Optical-quality crystals have been obtained from Madoc, Ontario, Canada in association with barite and calcite; also in British Columbia near Grand Forks, and at several localities in Mexico.

Fluorite is highly valued as a flux in the manufacture of steel; also as a raw material for hydrofluoric acid. When of optical quality, the mineral is used for lens and prisms in scientific instruments.

ELMER B. ROWLEY, F.M.S.A.
Formerly Mineral Curator, Department of Civil Engineering
Union College
Schenectady, New York

Additional Reading

Carr, D.D. and N. Herz: *Concise Encyclopedia of Mineral Resources,* Elsevier Science, New York, NY, 1989.

Gallant, R.A.: *Minerals,* Benchmark Books, Tarrytown, NY, 2000.

Perkins, D. and K. Henke: *Minerals in Thin Section,* Prentice-Hall, Inc, Upper Saddle River, NJ, 1999.

FLUOROCARBON. A number of organic compounds analogous to hydrocarbons, in which the hydrogen atoms have been replaced by fluorine. The term is loosely used to include fluorocarbons that contain chlorine; these should properly be called chlorofluorocarbons or fluorocarbon chlorides, since it is these which are thought to deplete the ozone layer of the upper atmosphere. Fluorocarbons are chemically inert, nonflammable, and stable to heat up to 260–316°C. They are denser and more volatile than the corresponding hydrocarbons, and have low refractive indices, low dielectric constants, low solubilities, low surface tensions, and viscosities comparable to hydrocarbons. Some are compressed gases; others are liquids. These compounds were once used extensively in aerosol packages. They are used as refrigerants, solvents, blowing agents, fire extinguishers, lubricants and hydraulic fluids—as components of complete systems.

Fluorocarbon polymers include polytetrafluoroethylene, polymers of chlorotrifluoroethylene, fluorinated ethylene-propylene polymers, polyvinylidene fluoride, and hexafluoropropylene, among others. These are thermoplastic substances, resistant to chemicals and oxidation; noncombustible; with broad useful temperature range (up to 285°C); with high dielectric constant; resistant to moisture, weathering, ozone, and ultraviolet radiation. Their structure comprises a straight backbone of carbon atoms symmetrically surrounded by fluorine atoms. These materials are available as powders and dispersions for further processing, as films, sheets, tubes, rods, tapes, and fibers. They find use in high-temperature wire and cable insulation, other electrical equipment, chemical processing equipment, coatings for cooking utensils, piping, gaskets. Among the fluorocarbon polymers are a number of fluoroelastomers. These polymers are amorphous, thermally stable, noncombustible, have low glass transition temperature (−77°C), and are generally resistant to attack by solvents and chemicals.

FLUOROELASTOMER. Any elastomeric high polymer containing fluorine; they may be homopolymers or copolymers. Fluorocarbon polymers include a large group of fluoroelastomers, including a copolymer in which the molecular skeleton is a −P = N− chain containing approximately equal numbers of tri- and heptafluoroethoxy side groups. Such polymers are amorphous, thermally stable, noncombustible, generally resistant to attack by solvents and chemicals, and have low glass transition temperature (−77°C).

FLUOROMETERS. In fluorescence analysis, the amount of light emitted characteristically under suitable excitation is used as a measure of the concentration of the responsible material under observation. Thus, the method is closely related to colorimetric or spectrophotometric analysis, in which the amount of light absorbed characteristically is used to measure the concentration of the dissolved species.

The main advantage of fluorescence methods is their high sensitivity, about one part in 10^8, in many determinations both inorganic and organic. This is two or three orders of magnitude better than absorption methods, where the sensitivity is limited by the necessity of detecting a very small fractional decrease in the light transmitted by the solution.

In fluorescence, the situation is inherently more favorable. Inasmuch as zero concentration corresponds to darkness (neglecting reagent blanks), and the sensitivity depends on detecting the first faint emission of light as the concentration is increased, advantage can be taken of highly sensitive detectors, such as photomultipliers, and high-intensity ultraviolet sources for excitation. Combining these with sophisticated electronic and optical techniques has led to a remarkable achievement; under favorable conditions, it is possible to detect Rhodamine 5DGN down to the extremely low concentration of one part in 10^{12} in an instrument designed for tracing ocean currents with a fluorescent dye marker.

The use of fluorescent methods requires that the substance to be determined is fluorescent under suitable irradiation, or can be made so by a chemical reaction. Among organic substances, fluorescence is shown mainly by aromatic compounds (including such hydrocarbons as benzene, naphthalene, anthracene, and their derivatives) rather than the aliphatic series. Among the metal ions, only a few, such as uranium and thallium, show intrinsic fluorescence; but many others can be determined fluorometrically by adding a specific reagent that reacts with the metal to form a fluorescent complex. Thus, aluminum is complexed with the dye Pontachrome BBR, beryllium with morin, and zirconium with flavanol.

Various sources are used for exciting fluorescence, including proprietary lamps, mercury vapor lamps, and the xenon arc lamp. A tungsten lamp may be used for substances having a strong excitation band above 450 mµ. Both the desired excitation band and emission band may be isolated by means of interference filters. However, since the desired excitation band is usually in the ultraviolet, the tungsten lamp is not in general use, but may have specific applications. It gives a band spectrum and does not have the sharp line limitation of the mercury vapor lamps.

One proprietary lamp is similar to the ordinary fluorescent lighting tube, but contains a phosphor that emits an abundance of radiation in the 350- to 360-mµ region of the spectrum. These lamps are usually of 5 or 15 watts and operate with a simple starter and ballast. The phosphor emits visible light, which must be excluded by a filter. Mercury vapor lamps provide the

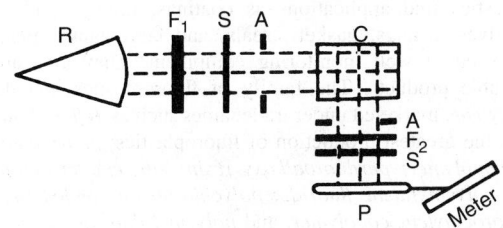

Fig. 1.   Simple 90°-axis fluorometer as viewed from above: *R*, radiation source; F_1 and F_2, filters; *S*, shutters; *A*, apertures or slits; *C*, sample container

only practical type of metallic arc used in fluorometry. These are designed to operate at high pressure or low pressure. The high-pressure type was made with a mercury arc at a pressure of about 8 atmospheres surrounded by a protective envelope.

If the emission is in the visible spectrum it may be estimated by visual comparison with standards. In any range of the spectrum, the intensity of the emission may be measured with a phototube, a barrier layer cell, or with a photographic plate and densitometer. By far the most common procedure is to use a phototube or an electron multiplier phototube attached to a microphotometer or a recorder.

Fluorometers are made, containing the lamps and measuring devices just discussed, along with filters and other components. Fig. 1 shows the functional elements of a fluorometer.

A fluorometer constructed with two monochromators is called a *spectrofluorometer*. With the spectrofluorometer, two types of information can be obtained easily: the wavelength of best excitation and the wavelength of the strongest emission. Two curves are generally plotted on the recorder for each fluorescing material: an excitation curve and an emission curve. The *excitation spectrum* is a plot of the wavelength of the exciting source, against the intensity of the emission. The excitation wavelength producing the greatest intensity of emission would seem to be best exciting wavelength. However, this statement is true only for the particular light source and grating.

See also **Analysis (Chemical)**.

Additional Reading

Ichinose, N. et al.: *Fluorometric Analysis in Biomedical Chemistry: Trends and Techniques Including HPLC Applications*, Vol. 109, John Wiley and Sons, Inc., New York, NY, 1991.

Lakowicz, J.: *Principles of Fluorescence Spectroscopy*, Kluwer Academic Publishers, Norwell, MA, 1999.

Wilkinson, M., F. Schut: *Digital Image Analysis of Microbes: Imaging Morphometry Fluorometry Techniques and Applications*, John Wiley and Sons, Inc., New York, NY, 1998.

FLUOROPLASTICS. These plastic materials may be placed into two convenient categories: (1) fluorocarbon plastics, and (2) other fluoroplastics. Fluoroplastics are produced by free radical initiated polymerization or copolymerization of the monomers. Fluorocarbon plastics contain no C−H bonds; other fluoroplastics contain some C−H and/or C−Cl bonds in the basic structure.

These plastics generally are considered as tough but relatively soft materials, with high elongation. Their useful characteristics are maintained over a wide temperature range—just above absolute zero to as high as 260°C (500°F). They withstand chemical environments well and essentially are unaffected by all organic solvents and reactive organic and inorganic compounds. Insulation resistance is quite high; dielectric constant and dissipation factor are low; arc resistance is high; surface energy is low (good antistick performance). Fluorocarbon plastics do not support combustion in air. When exposed to fire, they resist ignition and do not promote flame spread. Other fluoroplastics are somewhat less rugged than the fluorocarbons, with somewhat less resistance to certain organic materials and extreme environmental conditions. The other fluoroplastics are regarded as stiffer and stronger, and they display less deformation under load and creep.

All fluoroplastics except polytetrafluoroethylene can be processed using melt techniques commonly applicable to thermoplastics. Processing temperatures are somewhat higher than normally used with other thermoplastics and corrosion-resistant equipment is required to resist the corrosive effects of the molten polymer.

Fluoroplastics find applications as coatings, linings, and as components of valves, fittings, gaskets, seals, and large tanks. Fluoroplastics have been used in well monitoring equipment. They also are used in wire and cable products. The family of fluoroplastics includes *polytetrafluoroethylene*, marketed under tradenames such as *Teflon, Fluon, Halon*, constituting the greatest production of fluoroplastics; *fluorinated ethylene-propylene copolymer; perfluoroalkoxy resin; ethylene-tetrafluoroethylene copolymer, polyvinylidene fluoride; polychlorotrifluoroethylene; ethylene-chlorotrifluoroethylene copolymer*, and *polyvinyl fluoride*.

Additional Reading

Harper, C.A.: *Modern Plastics Handbook*, The McGraw-Hill Companies, Inc., New York, NY, 1999.

Rosato, D.V., M.G. Rosato: *Concise Encyclopedia of Plastilcs*, Kluwer Academic Publishers, Norwell, MA, 1999.

Salamone, J.C.: *Concise Polymeric Materials Encyclopedia*, CRC Press, LLC., Boca Raton, FL, 1998.

Strong, A.B.: *Plastics: Materials and Processes*, 2nd Edition, Prentice-Hall, Inc., Upper Saddle River, NJ, 1999.

FLUX (Slag).   A material added to the contents of a smelting furnace or a cupola for the purpose of purging the metal of impurities, and of rendering the slag more liquid. The flux most commonly used in iron and steel furnaces is limestone, which is charged in the proper proportions with the iron and fuel. The slag is a liquid mixture of ash, flux, and other impurities.

FLUX (Solder).   A material which by its chemical action facilitates the soldering and brazing of metals. Such a flux applied to a metallic surface cleans it and renders it receptive to amalgamation with the solder or brazing metal. Some fluxes are rosin, for soldering tin; muriatic acid, for galvanized iron and other zinc surface; and borax for brazing.

FOAM.   Foam is a nonequilibrium dispersion of gas bubbles in a relatively smaller volume of liquid. An essential ingredient in a liquid-based foam is surface-active molecules. These reside at the interfaces and are responsible for both the tendency of a liquid to foam and the stability of the resulting dispersion of bubbles. Important uses for custom-designed foams vary widely from familiar examples of detergents, cosmetics, and foods, to fire extinguishing, oil recovery, and a host of physical and chemical separation techniques. Unwanted generation of foam, on the other hand, is a common problem affecting the efficiency and speed of a vast number of industrial processes involving the mixing or agitation of multicomponent liquids. In all cases, control of foam rheology and stability is desired. These physical properties, in turn, are determined by both the physical chemistry of their liquid—vapor interfaces and by the structure formed from the collection of gas bubbles.

Physical Chemistry of Interfaces

The chemical composition, physical structure, and key physical properties of a foam, namely its stability and rheology, are all closely interrelated. Since there is a large interfacial area of contact between liquid and vapor inside a foam, the physical chemistry of liquid—vapor interfaces and their modification by surface-active molecules plays a primary role underlying these interrelationships.

For aqueous solutions, the chemical constituents most commonly responsible for foaming are surfactants, i.e., surface-active agents. Such molecules find wide use in other settings, and are distinguished by having both hydrophilic and hydrophobic regions.

Reduced Surface Tension. Just as surfactants self-organize in the bulk solution as a result of their hydrophilic and hydrophobic segments, they also preferentially adsorb and organize at the solution—vapor interface. In the case of aqueous surfactant solutions, the hydrophobic tails protrude into the vapor and leave only the hydrophilic head groups in contact with the solution. The favorable energetics of the arrangement can be seen by the reduction in the interfacial free energy per unit area, or surface tension, σ.

Gibbs Elasticity and Marangoni Flows. The reduction of surface tension with increasing surfactant adsorption gives rise to a nonequilibrium effect which can, in some cases, promote foaming. A sudden increase in the interfacial area by mechanical perturbation or thermal fluctuation results in a locally higher surface tension because the number of surfactant molecules per unit area simultaneously decreases. The Gibbs elasticity, E, is often used to quantify the instantaneous change in surface tension σ with area

A, i.e., $E = d\sigma/d \ln A$. If the film of liquid separating two neighboring bubbles in a foam develops a thin spot, the surface tension gradient in the vicinity of the thin spot will induce a Marangoni flow of liquid toward the direction of higher σ. This flow of liquid toward the thin spot helps heal the fluctuation and thus keeps the neighboring bubbles from coalescing.

Interfacial Forces. Neighboring bubbles in a foam interact through a variety of forces which depend on the composition and thickness of liquid between them, and on the physical chemistry of their liquid—vapor interfaces. For a foam to be relatively stable, the net interaction must be sufficiently repulsive at short distances to maintain a significant layer of liquid in between neighboring bubbles. Interfacial forces include the van der Waals interaction, the electrostatic double layer interaction, and disjoining pressure.

Physical Properties of Foam

Based on the underlying physical chemistry of surfactants at interfaces, important features of foam structure, stability, rheology, and their interrelationships can be considered as ultimately originating in the molecular composition of the base liquid.

Structure. Foam structure is characterized by the "wetness" of the system. Foams with arbitrarily large liquid to gas ratios can be generated by excessive agitation or by intentionally bubbling gas through a fluid. If the liquid content is sufficiently great, the foam consists of well-separated spherical bubbles that rapidly rise upwards displacing the heavier liquid. Such a system is usually called a froth, or bubbly liquid, rather than a foam.

If there are sufficiently strong repulsive interactions, such as from the electric double-layer force, then the gas bubbles at the top of a froth collect together without bursting. Furthermore, their interfaces approach as closely as these repulsive forces allow; typically on the order of 100 nm. Thus bubbles on top of a froth can pack together very closely and still allow most of the liquid to escape downward under the influence of gravity while maintaining their spherical shape. Given sufficient liquid, such a foam can resemble the random close-packed structure formed by hard spheres.

A dry foam, by contrast, is one with so little liquid that the bubbles are severely distorted into approximately polyhedral shapes. Typically this occurs for foams with less than 1% liquid by volume.

A complete characterization of the structure of a foam requires a characterization of the structure of the bubbles that comprise the foam. The total liquid content can be readily found from the mass densities of the foam and the liquid from which it was made. However, a more detailed determination of the bubble structure, including their average size, their shape, their structure and their size distribution is much more difficult, and is typically impeded by the problems in visualizing the interior of a foam. Even in the absence of any intrinsic optical absorption of the liquid, the strong mismatch in the indexes of refraction between the gas and the fluid results in a large scattering of light, usually precluding direct visualization of the interior structure of a foam. As a result, other, less direct, methods have been developed, and must be used, except in exceptional cases where the foam structure has been optimized for visualization.

One optical imaging technique that circumvents the problem of multiple light scattering is to estimate the bubble size distribution from the area individual foam bubbles occupy at a glass surface. Such experiments, and the systematic differences between bulk and surface bubble distributions, have been reviewed. Another technique that also directly measures the bubble size distribution is the use of a Coulter counter, where individual bubbles are drawn through a small tube and counted. This yields a direct measure of the bubble size distribution, but it is invasive and cannot probe the structure of the foam.

One technique that does probe the foam structure directly is cryomicroscopy. The foam is rapidly frozen, and the solid structure is cut open and imaged with an optical or electron microscope. Such methods are widely applicable and provide a direct image of the foam structure; however, they destroy the sample and may also perturb the foam structure in an uncontrolled manner during the freezing.

Stability. Control of foam stability is important in all applications, whether degradation of a custom foam is to be minimized or whether excessive foaming is to be prevented. In all cases, the time evolution of the foam structure provides a natural means of quantifying foam stability. There are three basic mechanisms whereby the structure may change: by the gravitational segregation of liquid and bubbles, by the coalescence of neighboring bubbles via film rupture, and by the diffusion of gas across the liquid between neighboring bubbles.

Any means of characterizing foam structure can be used to study foam evolution, provided that the measurement can be made noninvasively and sufficiently rapidly. One technique that has been applied successfully is the measurement of the change in the pressure head over an evolving foam. Multiple light scattering can also be used to follow the time evolution of a foam. A common engineering technique for determining foam stability entails measuring the amount of foam produced. For defoaming applications, this is often a more important measure of stability than the foam structure.

Rheology. The rheology of foam is striking; it simultaneously shares the hallmark rheological properties of solids, liquids, and gases, and their mechanical response to external forces can be very complex.

One simple rheological model that is often used to describe the behavior of foams is that of a Bingham plastic. This applies for flows over length scales sufficiently large that the foam can be reasonably considered as a continuous medium. The Bingham plastic model combines the properties of a yield stress like that of a solid with the viscous flow of a liquid.

While the Bingham plastic model is an adequate approximate description of foam rheology, it is by no means exact, especially at low strain rates. More detailed models attempt to relate the rheological properties of foams to the structure and behavior of the bubbles.

To determine rheological parameters such as the yield stress and effective viscosity of a foam, commercial rheometers are available; rotational and continuous-flow-tube viscometry are most commonly employed (See also **Rheology**). However, obtaining reproducible results independent of the sample geometry is a difficult goal which arguably has not been achieved in most of the experiments reported in the scientific literature.

Production

Several techniques are available for the generation of special-purpose foam with the desired properties. The simplest method is to disperse compressed gas directly into an aqueous surfactant solution by means of a glass frit. A variation of this method that allows for control of liquid content is to simultaneously pump gas and surfactant solution through a bead pack or steel wool, for example, at fixed rates. Less reproducible mechanical means of foam generation include brute force shaking and blending. For highly reproducible foams composed of small bubbles, such as shaving creams, the aerosol technique is especially suitable.

Applications

Foams have a wide variety of applications that exploit their different physical properties. The low density, or high volume fraction of gas, enable foams to float on top of other fluids and to fill large volumes with relatively little fluid material. These features are of particular importance in their use for fire fighting. The very high internal surface area of foams makes them useful in many separation processes. The unique rheology of foams also results in a wide variety of uses, as a foam can behave as a solid, while still being able to flow once its yield stress is exceeded. Foams are used in food, oil recovery, detergents, textiles, and cosmetics.

Safety, Health, and Environment

Foams play important roles in environmental issues, both beneficial and detrimental.

Natural Waters. Many water systems have a natural tendency to produce foam upon agitation. The presence of pollutants exacerbates this problem. This was particularly severe when detergents contained surfactants that were resistant to biodegradation. Then, water near industrial sites or sewage disposal plants could be covered with a blanket of stable, standing foam. However, surfactant use has switched to biodegradable molecules, which has greatly reduced the incidence of these problems.

Wastewater Treatment. The treatment of wastewater, either from sewage or from industrial processes, typically entails a preliminary filtration to remove the large volumes of solids, and then a slower settling to remove the sand and gravel. See also **Water (Hard)**. The water is then treated by an activated sludge process to remove the remaining dissolved solids and organic colloidal particles. Activated sludge is a biomass that assists in the degradation of the organic waste in the water. The process entails a mixing and aeration of the wastewater with the activated sludge, which can lead to problems of foaming. The foams produced can be quite stable, resulting in additional problems for waste disposal. The foams

produced in this process differ from those normally encountered in that the foam producing and stabilizing agents are microbial, primarily including *Nocardia*, *Microthrix parvicella*, and *Rhodococcus*. These foams are more difficult to treat with defoaming agents. Moreover, it is very difficult to predict the degree of foamability of the waste being treated. In other, more specialized wastewater treatments, these problems do not arise, and defoaming agents can be used effectively.

Chlorofluorocarbon Alternatives. There still is no completely satisfactory propellant for use in the aerosol method of foam production. Chlorofluorocarbons, still widely used, are harmful to atmospheric ozone and low molecular weight hydrocarbons, now popular, e.g., in producing shaving cream, are explosive and promote the greenhouse effect. See also **Fluorine**.

DOUGLAS J. DURIAN
UCLA
DAVID A. WEITZ
Exxon Research & Engineering Company

Additional Reading

Adamson, A.W., A.P. Gast: *Physical Chemistry of Surfaces,* 6th Edition, John Wiley & Sons, Inc., New York, NY, 1997.
Aubert, J. H., A. M. Kraynik, and P. B. Rand: *Sci. Am.* **254**, 74 (May 1986).
Aubert, J.H., A.M. Kraynik, and P.B. Rand: "Aqueous Foams," *Scientific Amer.*, **254**(5), 74–82 (May 1986).
Bikerman, J. J.: *Foams,* Springer-Verlag, New York, NY, 1973.
Cheng H. C. and T. E. Natan, in N. P. Cheremisinoff, ed.: *Encyclopedia of Fluid Mechanics,* Vol. 3, Gulf Publishing Co., Houston, TX, 1986, p. 3.
Ekserova, D.R., P.M. Krugliakov: *Foam and Foam Films: Theory, Experiment, Application,* Elsevier Science, New York, NY, 1997.
Khan, S.A., R.K. Prud'homme: *Foams,* Marcel Dekker, Inc., New York, NY, 1996.
Kraynik, A. M.: *Ann. Rev. Fluid Mech.* **20**, 325 (1988).
Landrock, A.H.: *Handbook of Plastics Foams: Types, Properties, Manufacture, and Applications,* Noyes Publications, Park Ridge, NJ, 1995.
Sadoc, J.F., N. River: *Foams and Emulsion,* Kluwer Academic Publishers, Norwell, MA, 1999.
Weaire, D.L., S. Hutzler: *The Physics of Foams,* Oxford University Press, Inc., New York, NY, 1999.

FOAMED PLASTICS. Foamed polymers, otherwise known as cellular polymers or polymeric foams, or expanded plastics have been important to human life since primitive people began to use wood, a cellular form of the polymer cellulose. Cellular polymers have been commercially accepted in a wide variety of applications since the 1940s. The total usage of foamed plastics in the United States has risen from 441×10^3 t in 1967 to a projected 2.8×10^6 t in 1995.

Classification

A cellular plastic has been defined as a plastic the apparent density of which is decreased substantially by the presence of numerous cells disposed throughout its mass. In this article the terms cellular plastic, foamed plastic, expanded plastic, and plastic foam are used interchangeably to denote all two-phase gas-solid systems in which the solid is continuous and composed of a synthetic polymer or rubber.

Theory of the Expansion Process

Foamed plastics can be prepared by a variety of methods. The most important process, by far, consists of expanding a fluid polymer phase to a low density cellular state and then preserving this state. This is the foaming or expanding process. Other methods of producing the cellular state include leaching out solid or liquid materials that have been dispersed in a polymer, sintering small particles, and dispersing small cellular particles in a polymer. The latter processes are relatively straightforward processing techniques but are of minor importance.

The expansion process consists of three steps: creating small discontinuities or cells in a fluid or plastic phase; causing these cells to grow to a desired volume; and stabilizing this cellular structure by physical or chemical means.

Initiation and Growth of Cells. The initiation or nucleation of cells is the formation of cells of such size that they are capable of growth under the given conditions of foam expansion. The growth of a hole or cell in a fluid medium at equilibrium is controlled by the pressure difference (ΔP) between the inside and the outside of the cell, the surface tension of the fluid phase γ, and the radius r of the cell:

$$\Delta P = 2\gamma/r \qquad (1)$$

Stabilization of the Cellular State. The increase in surface area corresponding to the formation of many cells in the plastic phase is accompanied by an increase in the free energy of the system; hence the foamed state is inherently unstable. Methods of stabilizing this foamed state can be classified as chemical, e.g., the polymerization of a fluid resin into a three-dimensional thermoset polymer, or physical, e.g., the cooling of an expanded thermoplastic polymer to a temperature below its second-order transition temperature or its crystalline melting point to prevent polymer flow.

Manufacturing Processes

A summary of the methods for commercially producing cellular polymers is presented in Table 1.

Expandable Formulations

Physical Stabilization Process. Cellular polystyrene, the outstanding example; poly(vinyl chloride); copolymers of styrene and acrylonitrile (SAN copolymers); and polyethylene can be manufactured by this process.

Chemical Stabilization Processes. This method is more versatile and thus has been used successfully for more materials than the physical stabilization process. Chemical stabilization is more adaptable for condensation polymers than for vinyl polymers because of the fast yet controllable curing reactions and the absence of atmospheric inhibition. Foamed plastics produced by these processes include polyurethane foams, polyisocyanurates, and polyphenols.

Decompression Expansion Processes

Physical Stabilization Process. Cellular polystyrene, cellulose acetate, polyolefins, and poly(vinyl chloride) can be manufactured by this process.

Chemical Stabilization Processes. Cellular rubber and ebonite are produced by chemical stabilization processes.

Dispersion Processes

Frothing. The frothing process for producing cellular polymers is the same process used for making meringue topping for pies. A gas is dispersed in a fluid that has surface properties suitable for producing a foam of transient stability. The foam is then permanently stabilized by chemical reaction. The fluid may be a homogeneous material, a solution, or a heterogeneous material. Foamed plastics produced by frothing include latex foam rubber, urea–formaldehyde resins, and polyurethanes.

Syntactic Cellular Polymers. Syntactic cellular polymer is produced by dispersing rigid, foamed, microscopic particles in a fluid polymer and then stabilizing the system. The particles are generally spheres or microballoons of phenolic resin, urea–formaldehyde resin, glass, or silica. The fluid polymers used are the usual coating resins, e.g., epoxy resin, polyesters, and urea–formaldehyde resin.

Properties of Cellular Polymers

Mechanical Properties of Commercial Foamed Plastics. The properties of commercial rigid foamed plastics are presented in Table 2. The properties of commercial flexible foamed plastics are presented in Table 3.

TABLE 1. METHODS FOR PRODUCTION OF CELLULAR POLYMERS

Type of polymer	Extrusion	Expandable formulation	Froth foam	Compression mold	Injection mold	Sintering
cellulose acetate[a]	+					
epoxy resin[b]		+	+			
phenolic resin		+				
polyethylene[a]	+	+		+	+	+
polystyrene	+	+			+	+
silicones		+				
urea-formaldehyde resin			+			
urethane polymers[b]		+	+		+	
latex foam rubber			+			
natural rubber	+	+		+		
synthetic elastomers	+	+		+		
poly(vinyl chloride)[a]	+	+	+	+	+	
ebonite			+			
polytetrafluoroethylene						+

[a] Also by leaching.
[b] Also by spray.

TABLE 2. PHYSICAL PROPERTIES OF COMMERCIAL RIGID FOAMED PLASTICS

Property	Cellulose acetate	Polystyrene, extruded sheet		PVC		Polyurethane Polyether		Isocyanurate, laminate
density, kg/m³[a]	96	96	160	32	64	32	64	32
mechanical properties compressive strength, kPa[b] at 10%	862	290	469	345	1035	138	482	117–206
tensile strength, kPa[b]	1172	2070–3450	4137–6900	551	1207	138	620	248–290
flexural strength, kPa[b]	1014			586	1620	413	1380	
shear strength, kPa[b]	965			241	793	138	413	117
compression modulus, MPa[c]	38–90			13.1	35	2.0	10.3	
flexural modulus, MPa[c]	38			10.3	36	5.5	5.5	
shear modulus, MPa[c]				6.2	21	1.2	3.4	1.7
thermal properties thermal conductivity, W/(m·I)	0.045	0.035	0.035	0.023		0.016–0.025	0.022–0.030	0.019
coefficient of linear expansion, 10^{-5}/°C						5.4	7.2	
max service temperature, °C	177	77–80	80			93	121	149
specific heat, kJ/(kg·K)[d]						ca 0.9	ca 0.9	
electrical properties dielectric constant dissipation factor	1.12	1.27		1.28		1.05	1.1	
	20	0.00011		0.00014		13	18	
moisture resistance water absorption, vol %	4.5							
moisture vapor transmission, g/(m·s·GPa)[e]		86	56	15		35	50	230

[a] To convert kg/m³ to lb/ft³, multiply by 0.0624.
[b] To convert kPa to psi, divide by 6.895.
[c] To convert MPa to psi, multiply by 145.
[d] To convert kJ/(kg·K) to Btu/(lb·F), divide by 4.184.
[e] To convert GPa to psi, multiply by 145,000.

TABLE 3. PHYSICAL PROPERTIES OF COMMERCIAL FLEXIBLE FOAMED PLASTICS

Property	Expanded NR[a]	Expanded CR[a]	Expanded SBR	Latex foam rubber		PE extruded plank			PE sheet, extruded	Polypropylene, sheet	Polyurethane, standard cushioning		PVC		Silicon, Sheet
density, kg/m^3[b]	56	320	192	72	80	35	96	144	43	10	16	24	112	96	160
cell structure[c]	C	C	C	C	O	C	C	C	C		O	O	C	O	O
tensile strength, kPa[d]		206	758	551	103	138	413	690	41		88	118	24	3.4	
tensile elongation, %		500			310	60	60	60	276	138–275	160	205		220	310
rebound resilience, %					73				50			40			
tear strength, $(N/m)[e] \times 10^2$						10.5	26	51	26		3.3	4.4			
max service temp., °C	70	70	70	70		82	82	82	82	121					260
thermal conductivity, W/(m·K)	0.036	0.043	0.065	0.030		0.053	0.058	0.058	0.040–0.049	0.039			0.040		0.086

[a] NR = natural rubber; CR = chloroprene rubber.
[b] To convert kg/m^3 to lb/ft^3, multiply by 0.0624.
[c] C = closed; O = open.
[d] To convert kPa to psi, multiply by 0.145.
[e] To convert N/m to lbf/in., divide by 1.75.

TABLE 4. TYPICAL PHYSICAL PROPERTIES OF COMMERCIAL STRUCTURAL FOAMS

Property	ABS	Nylon[a]	PC[b]	Polyester[c]		HDPE	Polypropylene		High impact polystyrene		Polyurethane	
glass-reinforced	no	yes	yes	no	30%	no	no	20%	no	20%	no	no
density, g/cm^3	0.80	0.85	0.97	0.80	1.10	0.60	0.60	0.73	0.70	0.84	0.40	0.60
tensile strength, kPa[d]	18,600	48,000	101,000	37,900	76,000	8,900	13,800	20,700	12,400	34,500	11,000	23,400
compression strength, kPa[d] at 10% compression	6,900		51,700		76,000	8,900					5,500	19,300
flexural strength, kPa[d]	25,500	82,700	172,000	68,900	137,900	18,800	22,000	41,400	31,000	58,600	22,000	41,400
flexural modulus, GPa[e]	0.86	5.2	5.2	2.1	6.6	0.83	0.83	2.8	1.4	5.2	0.7	1.1
max use temperature, °C	82	203	132	193		110	115					

[a] Nylon-6,6 glass-reinforced.
[b] Polycarbonate.
[c] Thermoplastic polyester.
[d] To convert kPa to psi, divide by 6,895.
[e] To convert GPa to psi, multiply by 145,000.

The properties that are achieved in commercial structural foams (density >0.3 g/cm^3) are shown in Table 4.

Structural Variables

The properties of a foamed plastic can be related to several variables of composition and geometry often referred to as structural variables. These variables include polymer composition, density, cell structure (i.e., cell size, cell geometry, and the fraction of open cells), and gas composition.

Rigid Cellular Polymers. A separate class of high density, rigid cellular polymers has grown continually since the 1970s to become significant commercially. These are the structural foams with a density >300 kg/m^3. They are treated here as a separate category of rigid foams.

Compressive strength and modulus are widely used as general criteria to characterize the mechanical properties of rigid plastic foams. Rigid cellular polymers generally do not exhibit a definite yield point when compressed but instead show an increased deviation from Hooke's law as the compressive load is increased. Structural variables that affect the compressive strength and modulus of a rigid plastic foam are, in order of decreasing importance: plastic-phase composition, density, cell structure, and plastic state.

The creep characteristic of plastic foams must be considered when they are used in structural applications. Data on the deformation of polystyrene foam under various static loads have been compiled. There are two types of creep in this material: short-term and long-term. The minimum load required to cause long-term creep in molded polystyrene foam varies with density ranging from 50 kPa (7.3 psi) for foam density 16 kg/m^3 (1 lb/ft^3) to 455 kPa (66 psi) at foam density 160 kg/m^3 (10 lb/ft^3).

The successful application of time–temperature superposition for polystyrene foam is particularly significant in that it allows prediction of long-term behavior from short-term measurements. This is of interest in building and construction applications.

Structural Foams. Structural foams are usually produced as fabricated articles in injection molding or extrusion processes. The optimum product and process match differs for each fabricated article, so there are no standard commercial products for one to characterize. Rather there are a number of foams with varying properties. The properties of typical structural foams of different compositions are reported in Table 4. The most important structural variables are again polymer composition, density, and cell size and shape.

Flexible Cellular Polymers. The application of flexible foams has been predominantly in comfort cushioning, packaging, and wearing apparel, resulting in emphasis on a different set of mechanical properties than for rigid foams. The compressive nature of flexible foams (both static and dynamic) is their most significant mechanical property for most uses (Table 3). Other important properties are tensile strength and elongation, tear strength, and compression set. These properties can be related to the same set of structural variables as those for rigid foams.

Other Properties

The thermal, electrical, acoustical, and chemical properties of all cellular polymers are of such a similar nature that the discussions of these properties are not separated into rigid and flexible groups.

Thermal Properties. More information is available relating thermal conductivity to structural variables of cellular polymers than for any other property.

The following separation of the total heat transfer into its component parts, even if not completely rigorous, proves valuable to understanding the total thermal conductivity, k, of foams:

$$k = k_s + k_g + k_r + k_c \qquad (2)$$

where k_s, k_g, k_r, and k_c are the components of thermal conductivity attributable to solid conduction, gaseous conduction, radiation, and convection, respectively.

As a good first approximation, the heat conduction of low density foams through the solid and gas phases can be expressed as the product of the thermal conductivity of each phase times its volume fraction. Most rigid polymers have thermal conductivities of 0.07–0.28 W/(m · K) and the corresponding conduction through the solid phase of a 32 kg/m³ (2 lbs/ft³) foam (3 vol %) ranges 0.003–0.009 W/(m · K). In most cellular polymers this value is determined primarily by the density of the foam and the polymer-phase composition. Smaller variations can result from changes in cell structure.

Although conductivity through gases is much lower than that through solids, the amount of heat transferred through the gas phase in a foam is generally the largest contribution to the total heat transfer because the gas phase is the principal part of the total value (ca 97 vol % in a 32 kg/m³ foam). The thermal conductivities of the halocarbon gases are considerably less than those of oxygen and nitrogen. It has, therefore, proved advantageous to prepare cellular polymers using such gases that measurably lower the k of the polymer foam.

The variation in total thermal conductivity with density has the same general nature for all cellular polymers. The increase in k at low densities is owing to an increased radiant heat transfer, the rise at high densities to an increasing contribution of k_s.

The thermal conductivity of most materials decreases with temperature and can change upon aging under ambient conditions if the gas composition is influenced by such aging. Thermal conductivity of foamed plastics has been shown to vary with thickness. This has been attributed to the boundary effects of the radiant contribution to heat-transfer.

The specific heat of a cellular polymer is simply the sum of the specific heats of each of its components.

The coefficients of linear thermal expansion of polymers are higher than those for most rigid materials at ambient temperatures because of the supercooled-liquid nature of the polymeric state, and this applies to the cellular state as well. When cellular polymers are used as components of large structures, the coefficient of thermal expansion must be considered carefully because of its magnitude compared with those of most nonpolymeric structural materials.

Because the cellular materials, like their parent polymers, gradually decrease in modulus as the temperature rises rather than undergoing a sharp change in properties, it is difficult to precisely define the maximum service temperature of cellular polymers. The upper temperature limit of use for most cellular polymers is governed predominantly by the plastic phase.

Work aimed at developing tests to evaluate the performance of plastic foams in actual fire situations continues. All plastic foams are combustible, some burning more readily than others when exposed to fire. Some additives, when added in small quantities to the polymer, markedly improve the behavior of the foam in the presence of small fire sources.

Plastic foams are advantageous compared to other thermal insulations in several applications where they are exposed to moisture pickup, particularly when subjected to a combination of thermal and moisture gradients.

Electrical Properties. Cellular polymers have two important electrical applications. One takes advantage of the combination of inherent toughness and moisture resistance of polymers along with the decreased dielectric constant and dissipation factor of the foamed state to use cellular polymers as electrical-wire insulation. The other combines the low dissipation factor and the rigidity of plastic foams in the construction of radar domes. Polyurethane foams have been used as high voltage electrical insulation.

Environmental Aging. All cellular polymers are subject to a deterioration of properties under the combined effects of light or heat and oxygen. The response of cellular materials to the action of light and oxygen is governed almost entirely by the composition and state of the polymer phase.

Comfort cushioning is the largest single application of cellular polymers; flexible foams are the principal contributors to this field. However, the rapid growth rate of structural, packaging, and insulation applications has brought their volume over that of flexible foams during the past few years. Table 5 shows U.S. consumption of foamed plastics by resin and market.

TABLE 5. MARKET FOR CELLULAR POLYMERS, 10³ T

Item	1967	1982	1995[a]
BY MARKET			
insulation	58	261	472
flooring	20	98	154
other construction	9	136	288
cushioning	52	195	336
other furniture	40	103	175
packaging	43	177	311
transportation	76	140	238
consumer	44	136	225
bedding	18	57	113
appliances	14	40	61
other	68	225	408
Total	*441*	*1567*	*2781*
BY RESIN			
flexible urethane	181	511	844
rigid urethane	68	248	449
styrene	125	410	699
vinyl	61	232	413
others	6	165	376
Total	*441*	*1567*	*2781*

[a] Projected as of 1994.

Commercial Products and Processes

Flexible Polyurethane. These foams are produced from long-chain, lightly branched polyols reacting with a diisocyanate, usually toluene diisocyanate (TDI), to form an open-celled structure with free air flow during flexure. During manufacture these foams are closely controlled for proper density, ranging from 13 to 80 kg/m³ (0.8–5 lbs/ft³), to achieve the desired physical properties and cost.

In flexible polyurethane foams, the primary blowing agent is carbon dioxide, which is formed by the reaction of water and toluene diisocyanate. Softer foams with lower densities require an auxiliary blowing agent such as CFC-11, HCFC-141b, or methylene chloride. Since the load bearing characteristics of the foam are of great importance to the ultimate consumer this property is also closely controlled during manufacture.

Applications. Carpet underlayment is a substantial market. Most furniture cushioning is made from blocks of slab-produced polyurethane foam in the density range of 16 to 29 kg/m³ (1.0–1.8 lbs/ft³). The furniture market for polyurethane foams tends to reflect the current economic trends.

For passenger car seating about 90% is made by the molded foam process. The transportation market has experienced a decline since 1979 due to decreased automotive production and also because U.S. cars have been downsized, resulting in the use of less polyurethane foam per ca.

Consumption of polyurethane foam in bedding reached a maximum in 1978 and has since declined. Textile uses, however, are a relatively stable area and consist of the lamination of polyester foams to textile products, usually by flame lamination or electronic heat sealing techniques.

Rigid Polyurethane. These foams are characterized by closed-celled structure and very high compressive strength. They are produced by using a highly branched, short-chain polyol reacted with an aromatic isocyanate of two or more functionality which is often polymeric. Pour-in-place and free rise rigid polyurethane foams usually have a density in the region of 32.0 kg/m³ (2.0 lbs/ft³), although molded rigid foams have densities ranging up to 640 kg/m³ (40 lbs/ft³) in structural foams. Insulation effectiveness is one of the outstanding characteristics of rigid polyurethane foams which display thermal conductivities as low as 0.017 W/(m·K).

Process and Equipment. Rigid polyurethane foam processes use the same high or low pressure pumping, metering, and mixing equipment for flexible foams. Subsequent handling of the mixture is determined by the end product desired. Processes include lamination, pour-in-place, molding, bun stock, box foams, and spray.

Applications. The principal use for rigid polyurethane foams is for insulation in various forms utilized by a variety of industries. Packaging constitutes another significant use and is often a foam-in-place operation to protect industrial equipment such as pumps or motors.

Polystyrene. There are five basic types of polystyrene foams produced in a wide range of densities and employed in a wide variety of applications:

(1) extruded polystyrene board; (2) extruded polystyrene sheet; (3) expanded bead molding; (4) injection molded structural foam; and (5) expanded polystyrene loose-fill packaging.

Expanded polystyrene (EPS) beadboard insulation is produced with expandable polystyrene beads. These beads are produced by impregnating with 5 to 8% pentane and sometimes with flame retardants such as hexabromocyclododecane, pentabromomonochlorocyclohexane, or a synergistic mixture of antimony trioxide and dicumyl peroxide during suspension polymerization. The beads are preexpanded by fabricators with steam or vacuum and then allowed to age. The preexpanded beads are fed to the steam heated block molds where further expansion and fusion of beads take place. The molded blocks are then sliced into various sizes needed for specific applications after curing.

Expanded polystyrene bead molding products account for the largest portion of the drinking cup market and are used in fabricating a variety of other products including packaging materials, insulation board, and ice chests. The insulation value, the moisture resistance, and physical properties are inferior to extruded boardstock, but the material cost is much less.

Expanded polystyrene loose-fill packaging materials are produced normally by extrusion process followed by multiple steam expansions to give low density foam shapes that resemble "S", "8", and hollow shells. Expandable polystyrene loose-fill packaging material is also produced by suspension polymerization process with blowing agent incorporated into the polymer during the polymerization. These products are used as dunnage or space filling materials for cushion packaging. They have good shock absorbency, excellent resiliency, and are odorless.

Extruded polystyrene board was first introduced in the early 1940s by Dow Chemical Co. with the tradename Styrofoam. The Styrofoam process consists of the extrusion of a mixture of polystyrene and volatile liquid blowing agent expanded through a die to form boards in various sizes. The continuous boards are then passed through the finishing equipment for further sizing.

In residential sheathing insulation, fiberboard is still the most widely used product, although the use of extruded and molded polystyrene foam and of foilfaced isocyanurate foam is increasing depending on the cost, the amount of insulation required, and compatibility of insulation with other construction systems. In cavity-wall insulation, mineral wool, polyurethane, urea–formaldehyde, and fiber glass are widely used, although fiber glass batt is the most economical insulation for stud-wall construction. In mobile and modular homes, cellular plastics are used widely because of their light weight and more efficient insulation value.

Extruded polystyrene foam sheet is primarily produced in a single-screw tandem extrusion line. Primary application of foam sheet is as a packaging material in items such as disposable dishes and food containers, trays for meat, poultry and produce products, and egg cartons.

Injection molded structural foam is used widely for high density items such as picture frames, furniture, appliances, housewares, utensils, toys, pipes, and fittings. Most of these products are produced by injection molding or profile extrusion methods from impact modified polystyrene. Almost all high density foam products are produced with a chemical blowing agent that releases either nitrogen or carbon dioxide, typically sodium bicarbonate or azodicarbonamides. Medium density products can be produced with either a physical or chemical blowing agent, or a combination of both.

Poly(vinyl chloride). Cellular poly(vinyl chloride) (PVC) foam is available in both flexible and rigid foams. Flexible PVC foams are primarily produced by spread coating and calendering of fluid plastisols by means of a chemical blowing agent or mechanical frothing with air. Flexible PVC foams also are made by the extrusion process. Rigid PVC foams are produced by the extrusion or injection molding processes. Blowing is achieved by a chemical blowing agent or gas injection into the extruder.

Raw Materials. PVC is inherently a hard and brittle material and very sensitive to heat; it thus must be modified with a variety of plasticizers, stabilizers, and other processing aids to form heat-stable flexible or semiflexible products or with lesser amounts of these processing aids for the manufacture of rigid products. See also **Vinyl Chloride Polymers**.

Applications. Furniture and motor vehicle upholstery is the largest market for flexible vinyl foams. Because of better aesthetics (leather-like plastics), comfort, and favorable pricing, they are expected to show good growth in upholstery, carpet backing, resilient floor coverings, outerwear, footwear, luggage, and handbags. The only application for flexible vinyl foams in protective packaging applications is for stretch pallet wraps. These wraps are produced by extrusion.

Rigid vinyl foams in construction markets have grown substantially due to improved techniques to manufacture articles with controlled densities and smooth outer surfaces. Wood molding substitute for door frames and other wood products is an area that has grown. Rigid vinyl foams are also used in the manufacture of pipes and wires as resin extenders and in sidings and windows as the replacement of wood or wood substitutes.

Polyethylene. There are three basic types of polyethylene foams of importance: (1) extruded foams from low density polyethylene (LDPE); (2) foam products from high density polyethylene (HDPE); and (3) cross-linked polyethylene foams. Other polyolefin foams have an insignificant volume as compared to polyethylene foams and most of their uses are as resin extenders.

Extruded low density foam produced from LDPE is a tough, flexible, and resilient closed-celled foam used in a wide variety of applications such as cushion packaging and safety components.

HDPE foam is primarily used as a high density rigid product. Shipping pallets are a rapidly growing market at a projected growth rate of about 26% per year through the mid-1990s. Most of these products are produced by thermoforming sheet and injection molding.

Cross-linked polyethylene foams are produced by either radiation or chemical cross-linking of an extruded expandable sheet containing a chemical blowing agent. These products have finer texture and a softer, more resilient feel than extruded low density polyethylene foams and are used in comfort cushioning and cushion packaging applications.

Kanegafuchi Chemical of Japan has introduced a chemical cross-linking process for producing PE foams by the bead technique similar to EPS. Their Eperan beads have been used to produce molded articles as cushioning materials, sound insulating panels, etc. Asahi-Dow and BASF have also been reported to have developed similar products.

Health and Safety

Flammability. Plastic foams are organic in nature and, therefore, are combustible. They vary in their response to small sources of ignition because of composition and/or additives. All plastic foams should be handled, transported, and used according to manufacturers' recommendations as well as applicable local and national codes and regulations.

Virtually all plastic foams are blown with inert gases (CO_2, N_2, H_2O). Among these blowing agents, hydrocarbons and some of the HCFs and HFCs are flammable and pose a fire hazard in handing at the manufacturing plants.

Atmospheric Emissions. Certain organic compounds are found to be smog generating substances because of their high photochemical reactivity at ambient conditions. Since fully or partially halogenated hydrocarbons are considered to have low reactivity in the lower atmosphere (troposphere), substitution of photochemically reactive compounds for the current blowing agents may reduce ozone depletion in the stratosphere, but has adverse impact on the indoor ambient air quality. Therefore, ozone/oxidant interaction with the total environment needs to be considered in developing environmentally acceptable alternative blowing agents.

Toxicity. The presence of additives or unreacted monomers in certain plastic foams can limit their use where food or human contact is anticipated. Heavy metals can also be found in various additives. The manufacturers' recommendations or existing regulations again should be followed for such applications.

KYUNG W. SUH
The Dow Chemical Company

Additional Reading

Benning, C.J.: *Plastic Foams,* Vols. 1 and 2, Wiley-Interscience, New York, NY, 1969.

Frisch, K.C. and J.H. Saunders: *Plastic Foams,* Vol. 1, Pts. 1 and 2, Marcel Dekker, Inc., New York, NY, 1972 and 1973.

Hilyard, N.C. et al: *Mechanics of Cellular Plastics,* Macmillan Publishing Co., Inc., New York, NY, 1982.

Klempner, D., and V. Sendijarevi'c: *Handbook of Polymeric Foams and Foam Technology,* 2nd Edition, Hanser-Gardner Publications, Inc., Cincinnati, OH, 2004.

FOG. Fog is a hydrometeor, a visible aggregate of minute droplets or ice crystals suspended in the atmosphere near the earth's surface, the result of condensation and consequent formation of water droplets or ice crystals in the atmosphere; and aerosol. The size of the droplets ranges from colloidal to macroscopic. "Synthetic" fogs can be produced on a laboratory scale by ultrasonic vibrations, and natural fogs can be precipitated by the same means. Mists or fogs composed of atomized particles of oil are used as military concealment screens and for insecticidal purposes in orchards and truck gardens. See also **Aerosols**; and **Smog**.

Supercooled fog is fog having a temperature less than 0°C. It consists of small droplets at temperatures less than freezing which can exist as liquid down to approximately −40°C. Below this temperature, ice crystals tend to form automatically and the fog changes to ice fog.

Fog Seeding for Thinning and Dispersion

Supercooled fog, when "seeded" with dry ice (solid carbon dioxide) particles or with vaporizing liquid propane gas, will change to ice crystals, which fall to the ground and clear the fog. Dry ice particles falling through super-cooled fog leave a trail of very small ice crystals which, become nuclei upon which the water vapor of the supercooled droplets is deposited. The ice nuclei grow and the droplets disappear. Many of the ice crystal particles fall to the ground, thus clearing the fog. Vaporizing liquid propane causes excessive cooling which, in turn, stimulates the creation and growth of ice nuclei. These ice particles act in the same manner as in the case of dry ice seeding, and the fog is cleared.

Commercial airports have been cleared of supercooled fog for many years by dry ice seeding as a standard practice. Military airfields have been cleared by use of both propane and dry ice methods.

Warm fog is fog wherein the air temperature is above freezing. The aforementioned methods of seeding do not function in warm fog. Unfortunately, 95% of all fogs are warm fogs that blanket airports, highways, and harbors. Warm fogs have been seeded with a variety of chemicals and agents in an attempt to clear, or at least to thin them. None has proven more than slightly successful. Ordinary table salt has proven to be the most useful, but it is corrosive and ecologically unacceptable. A variety of organic and inorganic chemicals in various combinations, with and without electrical charges, has been tried, but with only slight success. A very fine spray of water droplets that are highly charged electrically when injected into warm fog has some promise.

Helicopter downwash has been successful in a very limited number of cases. To be successful, the fog must be very shallow, and the air above the fog, both dry and warm. Exhaust plumes from jet engines have had some success in a number of European airports, notably at Orly (Paris).

Basically, however, the problem of warm fog clearing remains unsolved.

FOG TRACKS. Linear regions of condensation, produced in air or other gases that are supersaturated with water vapor, by the passage of electrified particles. Fog tracks are useful in following the courses and collisions of such particles.

FOLIC ACID. Frequently identified with the other B vitamins, folic acid plays a number of important roles in human and animal biological systems. The substance, *pteroylmonoglutamic acid* or *folacin*, is involved in the synthesis of nucleic acid, in purine-pyrimidine metabolism, in serine-glycine conversion, in the differentiation of embryonic nervous systems, in one-carbon transfer mechanisms, in the metabolism of tyrosine and histidine, and in the synthesis of choline, among other biological processes. Folic acid is closely related to vitamin B_{12} (cobalamine) because of their interdependence in biological processes.

Dietary deficiencies of folic acid are most frequently associated with anemias (macrocytic, megaloblastic, and pernicious), glossitis, diarrhea, gastrointestinal lesions, intestinal malabsorption, and sprue.

Sources of Folic Acid

Neither folic acid nor vitamin B_{12} is produced by humans in adequate amounts; the substances must be absorbed from food. Natural sources of folic acid include:

High folic acid content (90–300 micrograms/100 grams)
Asparagus, dry beans (lentils, limas, navy), liver (beef, chicken, lamb, pork), spinach, wheat bran, yeast

Medium folic acid content
Beef kidney
Low folic acid content
Most fruits, nuts, vegetables, grains, and dairy products.

The substance is synthesized by bacteria in some vertebrates, including human, rat, dog, pig, and rabbit. Exogenous sources are required by most other vertebrates and invertebrates. In ruminants, synthesis of folic acid occurs in the rumen, but some researchers believe that newborn lambs require a dietary supplement. The most common manifestation of a deficiency in livestock is development of a characteristic macrocytic, hyperchromic anemia (also called megaloblastic anemia). Bone marrow changes, red cells are large and immature, usually with an accompanying reduction of white cell numbers. Folic acid deficiency in poultry retards growth.

Causes of Folic Acid Deficiency

These may be placed in four principal categories: (1) *inadequate intake* as the result of general *nutritional deficiencies* caused by food faddism or *alcoholism*; (2) *increased demand* of the body for folic acid in combination with inadequate intake, which may be the result of pregnancy, severe hemolysis (destruction of red blood cells), or chronic hemodialysis or peritoneal dialysis; (3) *inadequate absorption*, which may result from the presence of certain diseases, such as tropical sprue, gluten-sensitive enteropathy (nontropical sprue), Crohn's disease, lymphoma or amyloidosis of small bowel, diabetic enteropathy, and various intestinal procedures, such as resections or diversions; (4) *interference with folic acid metabolism*, as may result from blocking and interfering reactions of certain drugs, such as methotrexate, trimethoprim, pyrimethamine, phenytoin, ethanol (as in alcoholism), antituberculosis drugs, and possibly oral contraceptives (debatable among experts).

Replacement of Folic Acid

With so many causes of deficiency as just delineated, obviously replacement measures must be matched against the causes. A primary objective, where feasible, of course, is to remove or alleviate the causative factors. Also, where there are no contraindication factors, folic acid can be administered in therapeutic dosages. Very small dosages (1 mg orally per day) can often be effective.

Biochemistry

Folic acid coenzymes are derivatives of tetrahydrofolic acid. See also **Coenzymes**. Structurally, these are:

Pteridine   p-Amino-benzoic acid   Glutamic acid

Folic acid

Tetrahydrofolic acid

One-carbon fragments in various oxidation states are: (1) formyl (—CHO); (2) hydroxymethyl (−CH₂OH); and (3) methyl (−CH₃). The coenzyme forms of folic acid have one of these groups attached to either the 5-N or 10-N of tetrahydrofolic acid. One folic acid coenzyme, methyltetrahydrofolate (CH_3-FH_4) transfers in methyl group to homocysteine to yield methionine, in a reaction which also requires a vitamin B_{12} coenzyme:

$$HS-CH_2CH_2\overset{\overset{\displaystyle NH_2}{|}}{C}HCOOH + CH_3-FH_4$$

Homocysteine

$$\xrightarrow{B_{12}\ coenzyme} CH_3-S-CH_2CH_2\overset{\overset{\displaystyle NH_2}{|}}{C}HCOOH+FH_4$$

Methionine

As pointed out by Chen and Cooper (Division of Food Science and Nutrition, California State University, Northridge, California), the group of compounds denoted by the term *folacin* is a heterogeneous group of derivatives with a similar basic structure and biological function. Folic acid is the basic structural unit in these compounds. Other monoglutamate folates are formed when the pteridine moiety of this basic molecule is reduced or substituted in the 5-N or 10-N position. In addition, all of these monoglutamate folates may be transformed into polyglutamates of various length by the addition of glutamic acid residues to the basic molecule.

Antagonists of folic acid include aminopterin (4-amino-pteroylglutamic acid), methotrexate (amethopterin), pyrimethamine, and 4-amino-pteroylaspartic acid. Synergists include biotin, pantothenic acid, niacin, vitamins B_1, B_2, B_6, B_{12}, C, and E, somatotrophin (growth hormone), and testosterone.

Bioavailability of Folic Acid

Factors which cause a decrease in bioavailability include: (1) high urinary excretion; (2) destruction by certain intestinal bacteria; (3) increased urinary excretion caused by vitamin C; (4) presence of sulfonamides which block intestinal synthesis; and (5) a decrease in absorption mechanisms. Increase in bioavailability can be provided by stimulating intestinal bacterial synthesis in certain species. No toxicity due to folic acid has been reported in humans.

Some of the unusual features of folic acid noted by investigators include: (1) folic acid antagonists used in cancer therapy with temporary remissions; (2) folic acid occurs in chromosomes; (3) folic acid is distributed throughout cells; (4) needed for mitotic step metaphase to anaphase; (5) antibody formation decreased in folic acid deficiency; (6) choline-sparing effects; (7) analgesic in humans—pain threshold is increased; (8) antisulfonamide effects; (9) enterohepatic circulation of folate; (10) synthesized by psittacosis virus; (11) concentrated in spinal fluid.

Historical Perspective

Over the years, folic acid has been variously referred to as vitamin B_c, vitamin M, and the *L. casei* factor.

In 1931, Wills demonstrated a factor from yeast active in treating anemia. In 1938, Day et al. found yeast or liver extracts active in treating anemia in monkeys. Hogan and Parrot, in 1939, showed how anemia in chicks could be prevented by using liver extract. The *L. casei* growth factor was isolated from liver and yeast by Snell and Peterson in 1940. Hutchings et al., in 1941, found the *L. casei* factor also essential for chicks. Also, in 1941, Mitchell, Snell, and Williams isolated bacterial (*S. lactis* R.) growth factor similar to *L. casei* factor from yeast and named the substance folic acid. Stokstad, in 1943, reported *L. casei* factor from liver more active than from yeast; and provided evidence of multiple factors. Pteroylmonoglutamic acid was finally isolated, the structure proved, and the substance synthesized by Angier et al. in 1946. Commercial production of folic acid is either by extraction from yeast or liver, or by synthesis wherein 2,3-dibromopropanol, 2,4,5-triamino-6-hydroxypyrimidine, and para-aminobenzoyl glutamic acid are reacted.

Folic Acid Assay

Deficiencies of folic acid and vitamin B_{12} are relatively common. Whenever macrocytic anemia is present, evaluation of these two vitamins is necessary to determine the cause of the condition. The standard method of measuring folic acid has been the microbiological assay (Bailey et al., 1982), which can be used to measure folic acid in serum, blood, tissues, and foods. Improved high performance liquid chromatography (HPLC) methods have simplified differential analysis of the metabolites of folic acid (Shane, 1982), but low percent recovery is compounded by the degree of glutamate conjugation.

The evaluation of folic acid status must often also include evaluation of vitamin B_{12} because of its effect on folate metabolism. A vitamin B_{12}-dependent reaction is necessary for an enzyme involved in the catabolism of branched-chain amino acids (methylmalonyl CoA to succinyl CoA). This reaction may provide the basis for a functional assessment method for vitamin B_{12} status. See also **Hormones**; and **Vitamin**.

Additional Reading

Ayling, J.E., M.G. Nair, and C.M. Baugh: *Chemistry and Biology of Pteridines and Folates,* Kluwer Academic Publishers, Norwell, MA, 1993.

Bailey, L.B. et al.: "Folacin and Iron Status and Hematological Findings in Black and Spanish American Adolescents," *Am. Geriatr. Soc.,* **27**, 444 (1982).

Ball, G.F.M.: *Fat Soluble Vitamin Assays in Food Analysis: A Comprehensive Review,* Elsevier, New York, NY, 1989.

Bradshaw, S.D.: *Folic Acid (Pteroylglutamc Acid) in Health, Deficiency and Therapy,* Abbe Publishing Association, Washington, DC, 1998.

Chen, T.S., R.G. Cooper: "Thermal Destruction of Folacin: Effect of Ascorbic Acid, Oxygen and Temperature," *J. Food Sci.,* **44**(3), 713–716 (1979).

Cooper, R.G., T.S. Chen, and M.A. King: "Thermal Destruction of Folacin in Microwave and Conventional Heating," *J. Amer. Dietet. Ass.,* **73**, 406 (1978).

Czeizel, A.E., I. Dudas: "Prevention of the First Occurrence of Neural-Tube Defects by Periconceptional Vitamin Supplementation," *N. Eng. J. Med.,* **1832** (December 24, 1992).

Gaby, S.K. et al.: *Vitamin Intake and Health: A Scientific Review,* Marcel Dekker, Inc., New York, NY, 1991.

Gaull, G.E. et al.: *Nutrition in the '90s,* Marcel Dekker, Inc., New York, NY, 1991.

Kotsonis, F.N., M.A. MacKey: *Nutrition in the 90s,* Vol. 2, Marcel Dekker, Inc., New York, NY, 1994.

Machlin, L.J.: *Handbook of Vitamins,* 2nd Edition, Marcel Dekker, New York, NY, 1991.

Picciano, M.F., J.F. Gregory, and E.L. Stokstad: *Folic Acid Metabolism in Health and Disease,* John Wiley & Sons Inc, New York, NY, 1990.

Rosenberg, L.H.: "Folic Acid and Neural-Tube Defects," *N. Eng. J. Med.* **1875** (December 24, 1992).

Rothfeld, G.S., Suzanne Levert: *Folic Acid and the Amazing B Vitamins,* Berkley Publishing Group, New York, NY, 2000.

Shane, B.: "High Performance Liquid Chromatography of Folates," *Am. J. Clin. Nutr.* **35**, 599 (1982).

Spreen, A.N.: *Folic Acid: The Essential B Vitamin That Prevents Birth Defects and Promotes Optimal Health,* Woodland Publishing, Inc., Pleasant Grove, UT, 2000.

Walji, H.: *Folilc Acid,* Thorsons Guide, Tulsa, OK, 1997.

FOOD ADDITIVES.

According to the *U.S. Code of Federal Regulations,* food additives may be defined as "substances ... the intended use of which results or may reasonably be expected to result, directly or indirectly, either in their becoming a component of food or otherwise affecting the characteristics of food" (*Title 21, Part 170.3,* Apr. 1, 1990). Canada and the European Community have adopted similar definitions. According to this broad definition, a food additive is synonymous to a food ingredient. In practice, however, the word additive is limited to substances that are used in small quantities.

In the United States, substances permitted in food and beverages are regulated by the U.S. Food and Drug Administration (FDA), an agency of the Department of Health and Human Services. Additives used in meat and poultry are regulated by the U.S. Department of Agriculture, and additives for alcoholic beverages are regulated by the Bureau of Alcohol, Tobacco and Firearms of the U.S. Department of Treasury. Premarketing approval is required.

In the United States, additional ramifications may be expected from FDA's announcement of final regulations for new food labeling requirements under the directive of the Nutrition Labeling and Education Act of 1990. Among other things, these regulations limit health claims that can be made on food labels. They also require new information on nutrient content, and limit the use of descriptors such as low and free in association with calories, fat levels, and other food product characteristics.

In Europe, the formation of the European Economic Community has created a requirement to bring food additive approvals of the member nations into alignment, so as to eliminate differences in laws that hinder the movement of foodstuffs among these nations.

Classes of Food Additives

Acidulants. Acidulants, the most versatile and widely used ingredients in the food industry, function well as flavoring agents. They include adipic acid, citric acid, fumaric acid, gluconolactone, lactic acid, malic acid, phosphoric acid, and tartaric acid. See also **Acidulants and Alkalizers (Foods)**.

Anticaking Agents. Anticaking agents function by absorbing excess moisture, or by coating particles and making them water repellent. They include calcium silicate, $CaSiO_3$, and calcium and magnesium salts of long-chain fatty acids, such as calcium stearate, $C_{36}H_{70}CaO_4$. See also **Anticaking Agents**.

Antifoaming Agents. Polydimethylsiloxane, or silicone, is used at a level of approximately 10 parts per million to control foam in food products.

Antioxidants. Antioxidants work by donating a hydrogen atom to the reactive peroxide radical, ending the chain reaction. Both synthetic and natural antioxidants exist. The most commonly used synthetic antioxidants include butylated hydroxyanisole (BHA), $C_{11}H_{16}O_2$, butylated hydroxytoluene (BHT), $C_{15}H_{24}O$, propyl gallate (PG), $C_{10}H_{12}O_5$, and *tert*-butylhydroquinone (TBHQ), $C_{14}H_{22}O_2$. Although BHT was never removed from the GRAS list, continuing concern over its safety has resulted in decreased usage.

The most popular natural antioxidants on the market are rosemary extracts and tocopherols. Certain compounds, known as chelating agents, react synergistically with many antioxidants. Citric acid and ethylenediaminetetraacetic acid (EDTA), $C_{10}H_{16}N_2O_8$, are the most common chelating agents used.

Another group of compounds called oxygen scavengers retard oxidation by reducing the available molecular oxygen. Products in this group are water soluble and include erythorbic acid, $C_6H_8O_6$, and its salt sodium erythorbate, $C_6H_8O_6Na$, ascorbyl palmitate, $C_{22}H_{38}O_7$, ascorbic acid, $C_6H_8O_6$, glucose oxidase, and sulfites. See also **Antioxidants**.

Bulking Agents and Bulking Sweeteners. Bulking agents are substances that add bulk to food products while contributing fewer calories than the ingredients they replace.

Polydextrose, a polymer of glucose that contains traces of sorbitol and citric acid, is the most widely used soluble bulking agent in the United States.

Low calorie bulking agents represent an ingredient category having a great deal of potential, and several companies are developing products. The most common are naturally derived polymers of glucose and other sugars (polydextrose falls into this category), enantiomers of natural sugars, or synthetic polymers. Bulking sweeteners provide a bulking effect, along with some of the sweetness and functional properties of sugar. Products that fall into this category include mannitol, $C_6H_{14}O_6$, isomaltitol, some L-sugars, and fructooligosaccharides. See also **Bodying and Bulking Agents (Foods)**.

Colorants. According to U.S. regulations, colorants are divided into two classes: certified and exempt. The FD&C certified colors are all water-soluble dyes, but can be transformed into insoluble pigments known as lakes by precipitating the dyes with aluminum, calcium, or magnesium salts on a substrate of aluminum hydroxide.

Exempt colors do not have to undergo formal FDA certification requirements, but are monitored for purity. The colorants exempt from FD&C certification are annatto extract, β-carotene, beet powder, β-apo$-8'$-carotenol, canthaxanthin, caramel, carmine, carrot oil, cochineal extract, cottonseed flour, ferrous gluconate, fruit juices, grape skin extract, paprika, paprika oleoresin, riboflavin, saffron, titanium dioxide, turmeric, turmeric oleoresin, ultramarine blue, and vegetable juices. See also **Colorants (Foods)**.

Dietary Fiber. Dietary fiber is a broad term that encompasses the indigestible carbohydrate and carbohydrate-like components of foods that are found predominantly in plant cell walls (see **Carbohydrates**). It includes cellulose lignin, hemicelluloses, pentosans, gums, and pectins.

Emulsifiers. The chemical structures of emulsifiers, or surfactants, enable these materials to reduce the surface tension at the interface of two immiscible surfaces, thus allowing the surfaces to mix and form an emulsion. An emulsifier consists of a polar group, which is attracted to aqueous substances, and a hydrocarbon chain, which is attracted to lipids. Emulsifiers include mono- and diglycerides, lecithin, propylene glycol esters, lactylated esters, sorbitan and sorbitol esters, polysorbates, and sucrose esters.

Enzymes. In the food industry, the largest use of enzymes is in starch processing, cheese production, fruit and vegetable juice processing, baking, and brewing. Commercial enzyme preparations are obtained from animals and plants via extraction, or through cultivation of select microorganisms. Enzymes are divided into six main classes: hydrolases, isomerases, ligases, lyases, oxidoreductases, and transferases. See also **Enzyme**.

Fat Replacers. Two classes of fat replacers exist: mimetics, which are compounds that help replace the mouthfeel of fats but cannot substitute for fat on a weight for weight basis; and substitutes, compounds having physical and thermal properties similar to those of fat, that can theoretically replace fat in all applications. Because fats play a complex role in so many food applications, one fat replacer is often not a satisfactory substitute. Thus a systems approach to fat replacement, which relies on a combination of emulsifiers, gums, and thickeners, is often used.

Existing fat mimetics are either carbohydrate-, cellulosic (fiber)-, protein-, or gum-based.

As of this writing, only one fat substitute, caprenin, a triglyceride composed of capric acid, $C_{10}H_{20}O_2$, caprylic acid, $C_8H_{16}O_2$, and behenic acid, $C_{22}H_{44}O_2$, has had any commercial application.

Firming Agents. During thermal processing and freezing, the bonds of pectic substances in plant walls that help to stabilize structure are modified, resulting in an unacceptably soft product. The cell wall structure of fruits and vegetables can be strengthened by adding polyvalent cations that promote the cross-linking of the free-carboxyl groups of pectic substances. Fruits such as tomatoes, berries, and apple slices are commonly firmed by added calcium salts prior to processing. Acidic aluminum salts are added during the preparation of pickles and relishes to provide the same effect. See also **Firming Agents (Foods)**.

Flavors. Flavorings are used in the food industry to replace or enhance flavors that are lost during processing, to create flavor combinations that do not exist in nature, and to mask objectionable flavors. Over 6000 flavor ingredients exist. They include essential oils, oleoresins, fruit juices and concentrates, botanical and animal extracts, aroma chemicals, and compounded flavors. See also **Flavors and Essences**.

Flavor Enhancers. Flavor enhancers have the ability to enhance flavors at a level below which they contribute any flavor of their own. Worldwide, the most popular flavor enhancers are monosodium L-glutamate (MSG), $NaC_5H_8NO_4$, and the 5'-ribonucleotides: disodium 5'-inosinate (IMP), $C_{10}H_{11}N_4O_8P \cdot 2\,Na$, and disodium 5'-guanylate (GMP), $C_{10}H_{12}-N_5O_8P \cdot 2\,Na$.

Ammonium glycyrrhizinate (AG), $C_{42}H_{65}NO_{16}$, is a flavor enhancer derived from licorice root. Maltol, $C_6H_6O_3$, and ethyl maltol, $C_7H_8O_3$, are used as flavor enhancers in products such as cake mixes, confections, cookies, ice cream, fruit juices, puddings, and beverages. See also **Flavor Enhancers and Potentiators**.

Flour Bleaching Agents and Bread Improvers. Benzoyl peroxide, $C_{14}H_{10}O_4$, is a bleaching agent that is typically added at the flour mill at a level between 0.015 and 0.075%. This additive oxidizes the carotenoid pigments, resulting in a white flour. Gases that exert an effect on the flour upon immediate contact include chlorine gas, chlorine dioxide, ClO_2, nitrosyl chloride, $NOCl$, nitrogen oxides, N_xO_x, and nitrogen tetroxide, N_2O_4. Others that exert their effect when the flour is made into dough include potassium bromate, $KBrO_3$, potassium iodate, KIO_3, calcium iodate, $Ca(IO_2)_3$, and calcium peroxide, CaO_2.

Formulation Aids. Formulation acids, which include carriers, binders, fillers, plasticizers, and film-formers, are ingredients used in processing to impart a particular physical state or textural characteristic. Table 1 gives an overview of the formulation aids used in the food industry.

Fumigants. Fumigants are volatile substances used for controlling insects or pests. They include ethylene oxide, C_2H_4O, propylene oxide, C_3H_6O, and methyl bromide, CH_3Br.

Gases. Gases provide three basic functions as food ingredients: preservation, carbonation, and aeration.

Humectants. In certain foods, it is necessary to control the amount of water that enters or exits the product. It is for this purpose that humectants are employed. Polyhydric alcohols (polyols), which include propylene glycol, $C_3H_8O_2$, glycerol, $C_3H_8O_3$, sorbitol, $C_6H_{14}O_6$, and mannitol $C_6H_{14}O_6$, contain numerous hyroxyl groups. Their structure makes them hydrophilic and enables them to bind water in foods. See also **Humecants and Moisture-Retaining Agents**.

TABLE 1. FORMULATION AIDS USED IN FOOD PROCESSING

Category	Function	Typical applications
CARRIERS		
starches	allow addition of incompatible substances to food product	cheese, dry mixes, flour, flavor compounds
dextrins		
cellulose compounds		
silicas		
BINDERS		
starches	hold food together	prepared meat, fish, poultry, chewing gum, confections
salts		
dextrins		
oils		
gums		
FILLERS		
maltodextrin	add bulk to food products	confections, dietary products, chewing gum, cereal mixes
polydextrose		
starches		
PLASTICIZERS		
oils	maintain soft texture of food products	chewing gum, confections, margarine, cheese products
waxes		
resins		
humectants		
FILM FORMERS		
carnauba wax	increase palatability, preserve gloss, inhibit discoloration, protect food surfaces	confections, snack foods, nuts, fresh and dried fruits and vegetables
paraffin		
sodium caseinate		
mineral oil		

Leavening Agents. Sodium bicarbonate, $NaHCO_3$, is the most commonly used product, but ammonium bicarbonate, NH_4HCO_3, and potassium bicarbonate, $KHCO_3$, are used as well. When used alone, sodium bicarbonate reacts to give products a bitter, soapy flavor. Thus it is always combined with a leavening acid.

Leavening acids are classified according to the rate at which they release carbon dioxide from sodium bicarbonate. Most products use both slow- and fast-acting leavening acids to obtain appropriate volume. The leavening acids most frequently used include potassium acid tartrate, sodium aluminum sulfate, δ-gluconolactone, and ortho- and pyrophosphates. See also **Leavening Agents**.

Lubricants and Release Agents. Ingredients that fall into this category include oils, lecithin, starch, distilled acetylated monoglycerides, and magnesium silicate, $MgSiO_3$. See also **Lubricating Agents**.

Nonnutritive Sweeteners. As of this writing there are only three nonnutritive sweeteners approved for used in the United States: aspartame, $C_{14}H_{18}N_2O_5$, saccharin, $C_7H_5NO_3S$, and acesulfame K, $C_4H_5NO_4S \cdot K$. See also **Sweeteners**.

Nutrients. In the United States, foods are either restored, enriched, or fortified with nutrients. The enrichment program followed in the United States is *(1)* the enrichment of flour, bread, and degerminated and white rice using thiamin, $C_{12}H_{17}N_5O_4S$, riboflavin, $C_{17}H_{20}N_4NaO_9P$, niacin, $C_6H_5NO_2$, and iron; *(2)* the retention or restoration of thiamin, riboflavin, niacin, and iron in processed food cereals; *(3)* the addition of vitamin D to milk, fluid skimmed milk, and nonfat dry milk; *(4)* the addition of vitamin A, $C_{20}H_{30}O$, to margarine, fluid skimmed milk, and nonfat dry milk; *(5)* the addition of iodine to table salt; and *(6)* the addition of fluoride to areas in which the water supply has a low fluoride content.

Preservatives. Most preservatives do not kill microorganisms present in food. Rather, they prevent further growth and proliferation of anything that is present by either lowering the water activity or increasing the pH of the foods in which they are used. Preservatives include benzoates, sorbates, propionates, organic acids, sulfur dioxide and sulfites, parabens, sodium nitrate and sodium nitrites, and natamycin and nisin. See also **Preservative**.

Processing Aids. Manufacturing aids used to improve the appearance or performance of food products include clarifying agents (flocculants), clouding agents, catalysts, and filter aids.

Solvents. Solvents are generally used to either extract particular compounds, such as an essential oil from a plant, or to carry additives into a food system, such as a flavor into a powdered mix. Common solvents include ethanol, C_2H_6O, glycerine $C_3H_8O_3$, propylene glycol, $C_3H_8O_2$, triethyl citrate, $C_{12}H_{20}O_7$, polyhydric alcohols, carbon dioxide, acetylated monoglycerides, hexane, C_6H_{14}, methylene chloride, CH_2Cl_2, acetone, C_3H_6O, and trichloroethylene, C_2HCl_3. See also **Solvent**.

Stabilizers and Thickeners. Many food products receive their textural properties from a group of compounds known as hydrocolloids. Hydrocolloids fall into two classes: polysaccharides and proteins. They include locust bean gum, guar gum, gum arabic, carrageenan, xanthan gum, cellulose, agar, starch, pectin, alginates, and gelatin. See also **Stablizer**.

Leslie J. Friedman
C. Gail Greenwald
Arthur D. Little, Inc.

Additional Reading

Code of Federal Regulations, Title 21 Part 170.3, U.S. Government Printing Office, Washington, DC, Apr. 1, 1990.

Lindsay, R. C.: in O. R. Fennema, ed., *Food Chemistry*, 2nd Edition, Marcel Dekker, Inc., New York, NY, 1985, pp. 665–666.

Otles, S.: *Methods of Analysis of Food Components and Additives*, CRC Press, LLC., Boca Raton, FL, 2004.

Russell, N. J., and G. W. Gould: *Food Preservatives*, 2nd Edition, Kluwer Academic Publishers, Norwell, MA, 2003.

Smith, J., and L. Hong: *Food Additives Databook*, Iowa State Press, Ames, IA, 2002.

Yannai, S.: *Dictionary of Food Compounds with CD-ROM: Additives, Flavors, and Ingredients*, CRC Press, LLC., Boca Raton, FL, 2003.

FOOD PROCESSING. Food processing operations can be grouped into three categories: preparation, assembly, and preservation of foods. Preparation processes are used to convert raw plant or animal tissue into edible ingredients. This may include separation of inedible and hazardous components, extraction or concentration of nutrients, flavors, colors, and other useful components, and removal of water. Assembly processes are used to combine and form ingredients into consumer products. Preservation processes are used to prevent the spoilage of foods. Five sources of food spoilage must be addressed in order to deliver fresh, safe foods and ingredients: microbial contamination, including viruses; enzyme activity from enzymes in the food itself and from external enzymes such as from microbial activity; chemical deterioration such as oxidation and nonenzymatic browning; contamination from animals, insects, and parasites; and losses owing to mechanical damage such as bruising. Preservation processes can be used to extend the shelf life of fresh foods, such as produce, or to manufacture products for long-term storage where shelf lives are measured in years. The processing of foods is regulated by federal food laws that cover good manufacturing practices, nutritional content of foods, and food and ingredient standards. See also **Food Additives**.

Plants and animals are the primary sources of food. The food processing industry devotes considerable research to the selection and improvement of plants and animals for raw materials. Genetic engineering, as well as conventional breeding methods, are being used to improve the yield, color, flavor, texture, nutrient content, and resistance to diseases, insect loss, and climatic stress. However, product quality can vary owing to weather, soil, growing practices, harvest methods, and post-harvest handling. Thus food processing unit operations must be designed to accept raw materials having a wide range of qualities. In addition, provision often must be made for profitable use of by-products and waste streams.

Regulations

Food processing operations are usually regulated and mandated by national and international laws, regulations, and standards which define nutritional requirements, certain ingredients, process conditions, and even the composition of some products. Food safety and toxicology regulations include standards for toxic and carcinogenic substances in foods; pathogenic microbes; and physical hazards.

Process Optimization

Food processing operations can be optimized according to the principles used for other chemical processes if the composition, thermo-physical properties, and structure of the food is known. However, the complex chemical composition and physical structures of most foods can make process optimization difficult. Moreover, the quality of a processed product may depend more on consumer sensory responses than on measurable chemical or physical attributes. Retention levels of ascorbic acid, $C_6H_8O_6$, or thiamine can often be used as an indicator of process conditions.

Theoretical Basis. Food preservation theory has yielded mathematical models for predicting the heating times and temperatures needed to produce foods free of pathogenic or spoilage microbes. Mild heat treatments used to inactivate viruses, vegetative pathogenic bacteria such as *Salmonella* sp., and certain yeasts and molds, are referred to as pasteurization operations.

Spore-forming bacteria are among the most heat-resistant organisms known. Research since the early 1920s has been directed toward the development of mathematical models to predict the rate of heat inactivation of *Clostridium botulinum* spores as a function of heating time and temperature, and the composition of the suspending media. Spore germination can be inhibited by antibiotic substances produced by several types of lactic acid producing bacteria. These substances, called bacteriosins, are finding increased use in preventing the growth of gram-positive bacteria.

Two other broad areas of food preservation have been studied with the objective of developing predictive models. Enzyme inactivation by heat has been subjected to mathematical modeling in a manner similar to microbial inactivation. Chemical deterioration mechanisms have been studied to allow the prediction of shelf life, particularly the shelf life of foods susceptible to nonenzymatic browning and lipid oxidation.

Water Activity. The rates of chemical reactions as well as microbial and enzyme activities related to food deterioration have been linked to the activity of water in food. Water activity, at any selected temperature, can be measured by determining the equilibrium relative humidity surrounding the food. Water activity can be related to the moisture content of the food as measured by standard moisture tests.

Preservation of Foods

Preservation operations to reduce or eliminate food spoilage can be grouped into five categories: heat treatments; storage near or below the freezing point of water; dehydration and control of water activity; chemical preservation; and use of mechanical operations such as washing, peeling, filtration, centrifugation, grinding, ultrahigh hydrostatic pressure, and most importantly, the packaging. Most food preservation technologies use two or more preservation operations because virtually all processed foods are packaged.

Short-Term Storage. Short-term storage operations include refrigeration, heat treatment, and preservatives.

Long-Term Storage. Inactivation of microbes and enzymes in foods and food ingredients is necessary to ensure a long useful packaged shelf life. This can be achieved by using one or more preservation operations such as applying heat; using storage temperatures below −18°C; drying to water activities below 0.65; and by adding chemical preservatives such as organic acids (acetic or lactic) or table salt.

Thermal Preservation Technology. The heat preservation of foods can be accomplished by various combinations of heating times and temperatures, depending on the number and type of heat-resistant spores present, the composition of the food, and the physical characteristics of the food and package.

The inactivation of heat-resistant spores appears to follow first-order kinetics. Thus if the rate of inactivation of a spore population is known at several temperatures, and the rate of heating of the slowest point in a package can be determined or calculated from heat-transfer principles, then the time needed to sterilize the package can be calculated for any external heating condition.

The establishment of safe thermal processes for preserving food in hermetically sealed containers depends on the slowest heating volume of the containers. Heat-treated foods are called commercially sterile.

Chemical changes in foods resulting from heating, such as the loss of pigments, flavors, and vitamins, can also be approximated by first-order kinetics.

Rapid heating and cooling of liquid foods, such as milk, can be performed in a heat exchanger and is known as high temperature–short time (HTST) processing. HTST processing can yield heat-preserved foods of superior quality because heat-induced flavor, color, and nutrient losses are minimized.

Equipment and processes for thermal preservation depend on the physical form of the food and its pH. Foods having a pH < 4.5 often can be sterilized, for commercial purposes, at or near a temperature of 100°C. Commercial sterility for these products means that the product will not spoil owing to microbial growth as long as the pH remains at or below 4.5 The spores of *Bacillus coagulans* are an important exception. This latter microbe is found in tomato products, and these products are often adjusted to a pH of 4.0 or lower, or given an additional heat treatment.

Acid foods generally require the simplest equipment for heat preservation. The food can be heated to 100°C and filled hot into suitable containers. The containers are sealed, inverted to sterilize the closure, held at the filling temperature for a short time to ensure that the package is thoroughly heated, and then cooled. Tomato sauces, jellies, fruits, fruit juices, and pickles are routinely preserved in this fashion.

Low acid foods have a pH > 4.5, require sterilization at temperatures above 100°C, and thus require treatment in pressure vessels. Heat preservation processes above 100°C can be carried out in batch or continuous heat-exchange equipment.

Freezing Preservation. The rate of loss of color, flavor, texture, and nutrients, the growth of microbes, and the activity of enzymes and other life forms are all functions of temperature. Thus lower storage temperatures prolong the useful life of foods.

Equipment for food freezing is designed to maximize the rate at which foods are cooled to −18°C to ensure as brief a time as possible in the temperature zone of maximum ice crystal formation. This rapid cooling favors the formation of small ice crystals which minimize the disruption of cells and may reduce the effects of solute concentration damage. Rapid freezing requires equipment that can deliver large temperature differences and/or high heat-transfer rates.

Many formulated foods and certain animal products tolerate freezing and thawing well because their structures can accommodate ice crystallization, movement of water, and related changes in solute concentrations. Starches can be modified for freeze–thaw stability against gel breakdown through several cycles. By contrast, most fruits and vegetables lose significant structural quality on freezing and during storage because their rigid cell structures fail to accommodate to ice crystal formation. Frozen food storage equipment must be designed to minimize temperature fluctuations.

Most frozen foods have a useful storage life of one year at −18°C. However, foods high in fat such as sausage products may become rancid after two weeks in frozen storage if not protected from oxygen by special packaging and antioxidants.

Food freezing equipment can be classified by the method and medium of heat transfer used. High velocity air is the most common medium used for direct contact freezing of nonpackaged foods. However, rapid freezing requires high air velocities and low operating temperatures. For these and other reasons many foods are frozen in equipment using conduction or liquid heat-transfer methods. Capital and energy savings, reduced moisture loss, and elimination of defrosting are some advantages of these methods.

Liquid heat-transfer media for immersion freezing include solutions of edible salts, sugars, alcohols, and esters. These heat-transfer agents offer high heat-transfer rates, reduced pumping costs, and allow operating at higher refrigerant temperatures.

Conduction freezing between chilled plates is a very cost-effective method of heat removal for products that can be packaged in a geometry to fit between refrigerated plates. Cryogenic freezing equipment uses liquid nitrogen or carbon dioxide snow. These units have the advantage of portability and simplicity and can produce extremely fast freezing rates. The refrigerant can be sprayed directly on the product to ensure rapid heat transfer.

The quality of a frozen food may be determined more by the temperature at which it is stored than by the method or rate of freezing. Storage temperatures may fluctuate as products move from manufacturing through distribution channels to the consumer's home freezer. The useful shelf life of a frozen food may be severely limited by exposure to storage temperatures above −18°C, even for a few hours.

Dehydration Processing. Dehydration is one of the oldest means of preserving food. Microbes generally do not grow below a minimum water activity, A_w, of 0.65.

Foods dried to water activities in the range of 0.65 to 0.85 are often referred to as intermediate moisture foods. These partially dried foods tend to be soft and to rehydrate easily. The remaining water acts as a plasticizer. Because molds and yeast may be able to grow in these partially dried products, they must be preserved by heat, vacuum, or modified atmosphere packaging, refrigeration, or chemical means.

Foods high in sucrose, protein, or starch tend to bind water less firmly and must be dried to a low moisture content to obtain microbial stability.

Fresh plant and animal tissue when dried to a water activity much below 0.97 show irreversible disruption of metabolic processes. Products susceptible to oxidation and oxidative rancidity can be treated with antioxidants and vacuum or inert gas packed to minimize exposure to oxygen. Low temperature storage can further reduce the rate of chemical deterioration.

Continuous hot air driers are used to prepare most of the high quality, dried, piece-form fruits and vegetables produced in the United States. Liquids and pastes are commonly dried in spray, drum, or freeze dryers. Particulate foods can be dried in batch or continuous air-fluidized beds or freeze dryers. Many agricultural commodities are sun-dried when weather conditions at harvest provide low humidity, warm temperatures, and good air circulation.

Chemical Preservation. Food additives can enhance the effectiveness of food preservation by heat, refrigeration, and drying methods. The addition of a food-grade acid to a low acid food to shift the pH to a value below 4.5 allows heat preservation at a temperature of 100°C instead of in the range of 121°C. Antioxidants such as butylated hydroxyanisole (BHA) can be added to potato chips to reduce the need for expensive oxygen-impermeable flexible packaging. Sulfur dioxide is used in wine and in dry fruit and vegetable products to preserve colors and flavors and prevent nonenzymatic browning.

Food can be preserved by fermentation using selected strains of yeast, lactic acid-producing bacteria, or molds. The production of ethanol, lactic and other organic acids, and antimicrobial agents in the food, along with the removal of fermentable sugars, can yield a product having an extended shelf life.

Lactic acid-producing bacteria associated with fermented dairy products have been found to produce antibiotic-like compounds called bacteriocins. Concentrations of these natural antibiotics can be added to refrigerated foods in the form of an extract of the fermentation process to help prevent microbial spoilage. Other natural antibiotics are produced by *Penicillium roqueforti*, the mold associated with Roquefort and blue cheese, and by *Propionibacterium* sp., which produce propionic acid and are associated with Swiss-type cheeses.

Ionizing radiation is considered to be a chemical preservation method and applications must be cleared by the Food and Drug Administration for use, not only on a product-by-product basis, but also on a dose basis.

Other Technologies. Several technologies for the preservation of foods using a minimum of heat are being explored. The application of ultrahigh pressure to the preservation of foods has been investigated. Capacitance discharge has also been investigated as a means to pasteurize or commercially sterilize foods which can pass between plates sufficiently close together to allow an electric field of approximately 25,000 V/cm. Very high intensity flashes of visible light can be used to pasteurize fruit juices using a minimum of heating in a manner which appears to be similar to capacitance discharge.

Computer Integrated Manufacturing, Instrumentation, and Controls. Large food processing firms are exploring the use of computer integrated manufacturing. Thermal processing controls have been developed to the point where time and temperature process deviations can be corrected on line. Freezer, dryer, and vacuum evaporator operating conditions can be controlled and optimized using systems already available to the process industry.

An important aspect of food processing, common with other processing industries, is yield of finished product from starting raw materials for any shift and for specific unit operations. Computer-integrated manufacturing can start with the measurement of material flows and build upon this information. Instrumentation for the on-line measurement of specific food qualities of importance to the consumer such as food flavor, aroma, texture, and microbial content are under development. These quality factors are monitored using statistical quality control procedures using standard sampling plans and control strategies.

DANIEL F. FARKAS
Oregon State University

Additional Reading

Canned Foods, Principles of Thermal Process Control, Acidification, and Container Closure Evaluation, 5th Edition, The Food Processors Institute, Washington, DC, 1988.

Heldman, D.R. and D.B. Lund, eds.: *Handbook of Food Engineering,* Marcel Dekker, Inc., New York, NY, 1992.

Potter, N.N.: *Food Science,* 5th Edition, Van Nostrand Reinhold, New York, NY, 1995.

Tressler, D.K., W.B. Van Arsdel, and M.J. Copley, eds.: *The Freezing Preservation of Foods,* 4th Edition, Vols. 1–4, Avi Publishing Co., Westport, CT, 1979.

FOOD TOXICANTS, NATURALLY OCCURRING.

Toxicants are substances which, upon ingestion, product changes in homeostasis that are threatening to the normal function of the organism. There are substantial differences in the toxicity thresholds of individuals to specific agents. Factors affecting toxicity include body weight, sex, age, general state of health, and the presence of potentiating or inhibitory substances.

Toxic Proteins, Peptides, Amides, and Amino Acids

Nitrogenous compounds are the most frequently implicated natural toxicants in foods. These compounds may be grouped either according to gross manifestations or specific structural characteristics. Accordingly, vitamin-destroying enzymes, hemagglutenins, enzyme inhibitors, and many hepatotoxins, are of protein, peptide, or amino acid composition. Many of the hepatotoxins are also carcinogens.

Enzyme inhibitors of a protein nature are of significant concern because of widespread occurrence. The most common of these affect the pancreatic enzymes, trypsin and chymotrypsin, and are found in legumes, as well as in egg whites and potatoes.

Many protein inhibitors cause little nutritional difficulty because these compounds are heat labile under ordinary processing and cooking procedures including microwaving, and significant numbers are water soluble. Many are found in highest concentrations in the outer portions of plants, e.g., wheat bran; thus normal peeling and milling operations also give some protection. It has been demonstrated that treatment with compounds such as sodium sulfite, ascorbic acid, and cupric sulfate, as well as fermentation with *Rhizopus oligosporus*, such as in the production of tempeh, is also an effective means of reducing trypsin inhibitor activity in both fresh and hardened common beans.

Reports of specific amino acid toxicities from normal eating patterns are rare. Although all amino acids except alanine have been shown to be toxic, the probability of intoxication is very remote. Humans seem able to tolerate all amino acids in excess of 10 times the recommended intake.

Lathyrus sativus (khesari), which constitutes a principal food crop for many in India, has been known for decades to cause neolathyrism. The causative agent, *N*-oxalyl-L-α-diaminopropionic acid (ODAP), produces partial or total loss of control of the lower limbs with associated neurological symptoms. Common methods of preparation, including soaking in lime water and boiling, are effective in destroying this amino acid.

The seeds of legumes may contain hemagglutenins and lectins that may cause destruction of the epithelia of the gastrointestinal tract; interfere with cell mitosis; cause hemorrhage; impede renal, cardiac, and hepatic function; and produce red blood cell agglutination. Many of these compounds are rendered inactive by moist heat, and the toxicity may be further reduced or neutralized by digestive enzymes, making them poorly absorbed. Because lectins reach the colon mostly in an inactive state, they appear to protect humans from colon cancer by causing hypersecretion of intestinal mucus, or by direct toxic effect on tumor cells.

Saponins disrupt red blood cells and may produce diarrhea and vomiting. They may also have a beneficial effect by complexing with cholesterol and thus lowering serum cholesterol levels. In humans, intestinal microflora seem to either destroy saponins or inactivate them in small concentrations.

Acute toxicoses resulting from consumption of toxic mushrooms is infrequent, yet of increasing concern because of the practice of gathering fungi in the wild. The most serious of these toxicoses result from the

Amanita family of mushrooms which contains several toxic peptides belonging to the amatoxin and phallotoxin groups.

Both the common cultivated mushroom as well as the Shitake mushroom contain hydrazines that have been shown to be carcinogenic precursors in experimental animals. Hydrazine levels, however, vary considerably as a function of variety, processing, storage, and preparation. One week of refrigerated storage reduces levels significantly, and all hydrazines are lost during canning and/or cooking.

Phytoalexins

Phytoalexins are low molecular weight compounds produced in plants as a defense mechanism against microorganisms. They do, however, exhibit toxicity to humans and other animals in addition to microbes. Coumarins, glycoalkaloids, isocoumarins, isoflavonoids, linear furanocoumarins, stilbenes, and terpenes all fall into the category of phytoalexins. Because phytoalexins are natural components of plants, and because their concentration may increase as a response to production and management stimuli, it is useful to recognize the possible effects of phytoalexins in the human diet.

Linear furanocoumarins are potent photosensitizing agents in celery, parsley, parsnips, limes, and figs. The most commonly reported symptoms include contact dermatitis and photodermatitis, particularly on the hands and forearms.

Enumerable phytoalexins, including furanosesquiterpene, ipomeamarone, eudesmanes, and others, have been isolated from mold-infected sweet potatoes. The clinical symptoms seem to revolve around lung edema. Whereas high concentrations of these chemicals can occur in damaged sweet potatoes, the occurrence is much less (by as much as 20-fold) in nondamaged sweet potatoes. Of possible concern to human health is the fact that blemishes sufficient to result in large increases in concentration of lung-edema toxins are not always easily detected by the naked eye. Additionally, these compounds are heat stable.

Oligosaccharides

Oligosaccharides, specifically the α-galactosides raffinose, stachyose, and verbascose, are widely present in legumes and are indigestible by humans because of a lack of α-galactosidase. As a result, these compounds undergo fermentation in the colon with the concomitant production of CO_2, H_2, and CH_4, commonly referred to as flatulence. Reports have shown germination to be effective in reducing α-galactoside content of cowpeas and other legumes.

Goitrogens are compounds that produce goiter by interfering with thyroxine synthesis in the thyroid gland. Foodborne goitrogens are often characterized by the presence of sulfur and most are thiocyanates or closely related compounds. Because of their widespread occurrence in Cruciferae, goitrogens are among the most common and longest recognized substances of toxic nature in the human food supply.

Oxalates, Phytates, and Other Chelates

Of nutrient chelates in the human diet, oxalates and phytates are the most common. Oxalic acid, found principally in spinach, rhubarb leaves, beet leaves, some fruits, and mushrooms, is a primary chelator of calcium. Oxalate present in pineapple, kiwifruit, and possibly in other foods, occurs as calcium oxalate, CaC_2O_4. This compound is in the form of needle-like crystals, known as raphides, which can produce painful sensations in the mouth when eaten raw. The effects of oxalic acid in the diet may be twofold. First, it forms strong chelates with dietary calcium, rendering the calcium unavailable for absorption and assimilation. Secondly, absorbed oxalic acid causes assimilated Ca to be precipitated as insoluble salts that accumulate in the renal glomeruli and contribute to the formation of renal calculi.

Phytic acid, although restricted to a more narrow range of food products, mainly grains, complexes a broader spectrum of minerals than does oxalic acid. Decreased availability of P is probably the most widely recognized result of excessive intakes of phytic acid, yet Ca, Cu, Zn, Fe, and Mn are also complexed and rendered unavailable by this compound. High intakes of both calcium and vitamin D help to offset the deleterious effects of oxalates.

Vasoactive and Psychoactive Amines and Alkaloids

Most compounds producing hypertensive episodes are classified as amines and are found in greatest concentration in banana, plantain, tomato, avocado, pineapple, broad beans, and various cheeses. Amines that are vasoactive include dopamine, $C_8H_{11}NO_2$; tyramine; histamine, $C_5H_9N_3$; tryptamine, $C_{10}H_{12}N_2$; noradrenaline, $C_8H_{11}NO_3$; and dihydroxyphenylalanine (DOPA), $C_9H_{11}NO_4$.

Patients receiving monoamine oxidase inhibitors (MAOI) as antidepressant therapy have been especially subject to the hypertensive effects of vasoactive amines. These dietary amines have also been implicated as causative agents in migraine. Other naturally occurring alkaloids have been recognized for centuries as possessing neurological stimulant and depressant properties.

Caffeine, a xanthine derivative, has been consumed for thousands of years and is present in over 60 plant species including coffee beans, tea leaves, cacao beans and cola nuts. In addition to naturally occurring sources, caffeine has been used as a food ingredient (flavoring) for over 100 years. Caffeine produces stimulatory effects by facilitating mental and muscular effort and diminishes drowsiness and fatigue. Individual thresholds of toxicity vary considerably, but symptoms such as restlessness, increased respiration, muscular tension and twitching, and tachycardia may imply acute toxicity.

Depressant symptoms, which include burning abdominal pain, decreased excitability, convulsions, nausea, and coma, become the general syndrome for all oral alkaloid poisoning. Myristicin, found in both nutmeg and mace, is a psychoactive agent that may be fatal in infants who consume as little as two whole nutmegs. Its toxicity resemble alcohol intoxications.

Several glycoalkaloids present in food are of toxicological interest. Solanine, found in potatoes, tomatoes, apples, eggplant, and sugar beets, has been responsible for several cases of moderate to severe poisoning. Solanine is a cholinesterase inhibitor and toxic doses are probably ca 200 mg. Market potatoes contain about 1–5 mg of solanine per 100 g fresh weight. The USDA establishes solanine levels of 20 mg/100 g as the limit for safe consumption.

Antinutrients

The presence of antivitamins in certain foods means that merely assuring an adequate intake of a vitamin is no guarantee that a deficiency state cannot exist physiologically. The enzyme thiaminase acts by either specific splitting of the thiamine molecule or nonspecific hydrolysis. Niacin inhibitors, acting through nicotinamide mononucleotide (NMN) depression in erythrocytes, have been studied in corn and millet, and a biotin antagonist, avidin, has long been recognized in raw egg white. Linatine, found in flaxseed, is the only pyridoxine antagonist known and seems to function by the formation of a stable complex. Yeast and pea seedlings contain specific pantothenic acid antagonists, although the structure and mode of action are unexplained. Riboflavin antagonism, found only in the Akee plum of Jamaica, is rather rare, but is of interest because it can be fatal.

The antagonisms that exist between unsaturated fatty acids, and carotene and vitamin E are complicated and largely undefined. Linoleic acid acts as an antivitamin to dl-α-tocopherol (vitamin E) by reducing availability through direct intestinal destruction. Various lipoxidases destroy carotenes and vitamin A.

Investigations have focused on the content of polyphenolics, tannins, and related compounds in various foods and the influence on nutrient availability and protein digestibility. It has been established that naturally occurring concentrations of polyphenoloxidase and polyphenols in products such as mushrooms can result in reduced iron bioavailability. Likewise, several studies have focused on decreased protein digestibility caused by the tannins of common beans and rapeseed (canola).

Vitamin Toxicity

Reported cases of vitamin toxicity owing to overdose are usually associated with increased over-the-counter availability of supplemental vitamins and indiscriminate supplementation. Fat-soluble vitamins tend to accumulate in the body with relatively inactive mechanism for excretion and cause greater toxicological difficulties than do water-soluble vitamins.

Infants may be sensitive to doses of vitamin A in the range of 75,000–200,000 IU (90.6–1.5 g). Dangerous doses of vitamin D seem to lie in the range of 1000–3000 IU/kg body wt (25–75 µg/kg body wt). Cases of toxicity of both vitamins E and K have been reported, but under ordinary circumstances these vitamins are considered relatively innocuous.

Of the water-soluble vitamins, intakes of nicotinic acid on the order of 10 to 30 times the recommended daily allowance (RDA) have been shown to cause flushing, headache, nausea, and moderate lowering of serum cholesterol with concurrent increases in serum glucose. Toxic levels of folic acid are ca 20 mg/d in infants, and probably approach 400 mg/d in adults. The body seems able to tolerate very large intakes of ascorbic acid (vitamin C) without ill effect, but levels in excess of 9 g/d have been reported to cause increases in urinary oxalic acid excretion.

Essential Minerals and Heavy Trace Elements

Ingestion of at least 10 times normal levels of essential minerals would be required to approach toxic proportions. See also **Mineral Nutrients**. The only exceptions occur in cases of plant foods grown on soils unusually high in Mo, Se, and Cu. Levels can reach toxic quantities in these cases, but these are rare occurrences.

Cases involving human toxicity from heavy trace elements, such as Pb, Hg, As, and Cd, are much more common but are almost exclusively traced to accidental contamination rather than true natural occurrences.

Cyanogenic Glycosides

Complex glycosides, which upon hydrolysis yield hydrogen cyanide, are commonly found among plant materials. The toxicity of this class of compounds, found in the bitter almond, pits of stone fruits, sorghum, and lima beans, is directly related to HCN liberation upon digestive hydrolysis.

Nitrates, Nitrites, and Nitrosamines

The carcinogenicity of nitrosamines has created widespread concern over the safety of food products that are significant sources of nitrates and nitrites. Nitrosamines are readily formed by reaction of secondary amines with nitrites at acid pH, conditions which may occur in the gastrointestinal tract.

Nitrates are found in fairly high concentrations in beets, spinach, kale, collards, eggplant, celery, and lettuce. Additionally, nitrates and nitrites are commonly used in the curing solutions of bacon, ham, and other cured meats. In cured meats, nitrates and nitrites control the growth of microorganisms, particularly *Clostridium botulinum*, and also serve as color preservatives.

Although the potentially carcinogenic nitrosamines may be present in foods, particularly cured meats, occurrence is infrequent and at low levels. USDA regulations stipulate that ascorbic acid be added to cured meats at five times the level of nitrates and nitrites to prevent the formation of carcinogenic *N*-nitroso compounds. See also **Food Additives**.

Sodium Chloride

Excessive intake of NaCl contributes to increased fluid retention, and in some individuals there may be a relationship between NaCl intake and hypertension. Both consumers and food processors have reduced use of NaCl.

Toxins

Mycotoxins. The condition produced by the consumption of moldy foods containing toxic material is referred to as mycotoxicosis. Molds and fungi fall into this category and several derive their toxicity from the production of oxalic acid, although the majority of mycotoxins are much more complex.

Mycotoxins find their way into the human diet by way of mold-contaminated cereal and legume crops, meat, and milk products. Corn and peanuts probably represent the most common sources of mycotoxins in the human diet. Many mycotoxins are acutely toxic as well as being potent carcinogens.

Many parasitic fungi have been shown to produce toxins; however, the toxins of *Aspergillus* and *Penicillium* have perhaps the greatest potency against humans.

Seafood Toxins. Virtually scores of fish and shellfish species have been reported to have toxic manifestations. Most of these toxicities have been shown to be microbiological in origin. There are a few, however, that are natural components of seafoods.

Several species of the moray eel (*Gymnothorax*) have caused toxic reactions, especially in Japan. The toxic principle appears to be proteinaceous and is found predominately in the blood but it may occur in the flesh as well. Its exact structure remains somewhat uncertain.

Amnesic shellfish poisoning resulted in four deaths in northeastern Canada in the early 1990s, and domoic acid, the causative agent, was first documented on the Washington and Oregon coasts in 1991. The toxin is produced by a single-celled phytoplankton that constitutes part of the food chain of some shellfish, including Dungeness crab.

Pufferfish toxin, isolated from a dozen or more species, has been identified as having the empirical formula $C_{11}H_{17}N_3O_8$, but the structure is not well-established, nor is it certain that the same structure is universally responsible for poisoning, although this is assumed to be the case. The so-called paralytic shellfish poisoning reported in many areas of the world has a microbiological etiology, and is thus more accurately a contamination rather than a natural toxicosis.

The liver of sharks and other oily fishes sometimes accumulate toxic levels of vitamin A, and cases of acute poisoning have been reported both among Eskimos and the Japanese.

Legislation and Regulatory Considerations

There exists little specific legislation dealing with natural toxicants in foods. The *1958 Food Additives Amendment to the Federal Food, Drug, and Cosmetic Act* stipulates that no substance that has been shown to be carcinogenic to either humans or animals may be added to the food supply. Accordingly, those foods that contain added carcinogens are subject to the Delaney Clause. Maximum tolerances of heavy metals, such as Pb and Hg, have been established by FDA at 0.5 ppm in the food product. For aflatoxins, there is presently a zero tolerance in effect (based on the Delaney Clause), and screening is generally on a qualitative basis. With these exceptions, natural toxicants in food products are generally not treated by specific food legislation. Naturally occurring toxicants have been reviewed in greater detail elsewhere. See also **Toxicology**.

FRED H. HOSKINS
Washington State University

Additional Reading

Hall, R.L. and S.L. Taylor, *Food Tech.* **43**(9), 270 (1989).
Jones, J.M. *Food Safety*, Eagan Press, St. Paul, MN, 1992.
Morgan, M.R.A. and D.T. Coxon, *Natural Toxicants in Food*, DCH Publishers, New York, NY, 1987, pp. 221–230.
Sharma, R.P. and D.K. Salunke, *Toxicants of Plant Origin*, CRC Press, Boca Raton, FL, 1989, pp. 179–236.

FORENSIC CHEMISTRY. Forensic science is an applied science having a focus on practical scientific issues that come up during criminal investigations or at trial. Some components are unique to the field because it is conducted within the legal arena.

Physical Evidence

Forensic scientists work with physical evidence, i.e., "data presented to a court or jury in proof of the facts in issue and which may include the testimony of witnesses, records, documents or objects." Physical evidence is real or tangible and can literally include almost anything, e.g., the transient scent of perfume on the clothing of an assault victim; the metabolite of a drug detected in the urine of an individual in a driving-under-the-influence-of-drugs case; the scene of an explosion; or bullets removed from a murder victim's body.

Examination of physical evidence provides two subtle and different types of conclusion. All members of a class or group have identical characteristics. Types of physical evidence which exhibit class characteristics are paint, glass, fibers, fabric, building material, etc. This type of physical evidence is said to be identified. The best that chemical and physical examinations can ever do is to place items into groups of similarly manufactured items. It is not possible to differentiate one item of evidence as being uniquely distinguishable from another.

Some types of physical evidence, because of the manner in which the material is made, are unique; such evidence can be individualized. Examination can show an item of individualized evidence is unique and comes from one, and only one source. The classic example is fingerprints. Other categories of evidence exhibiting individualization are handwriting, markings on bullets fired from the same gun, broken pieces of glass or plastic which can be physically fit together again, and forensic deoxyribonucleic acid (DNA) evidence.

Physical evidence serves two purposes. In some cases it is used to prove a component or element of a crime. The other purpose for which physical evidence is used is to develop associative evidence in a case. Physical evidence may help to prove a victim or suspect was at a specific location, or that the two came in contact with one another.

Most of the forensic science or crime laboratories located in North America are associated with law enforcement agencies, medical examiner–coroner departments, or prosecutors' offices. There are a large number of independent consultants, also. Laboratories exist at the municipal, county, state, and federal levels of government. There are approximately 300 government-operated forensic science laboratories in the United States.

Forensic science laboratories are generally divided into separate specialty areas. These typically include forensic toxicology, solid-dose drug testing, forensic serology, trace evidence analysis, firearms and tool mark examination, questioned documents examination, and latent fingerprint examination. Laboratories principally employ chemists, biochemists, and biologists at various degree levels.

The bulk of the scientific testing in crime laboratories involves the analysis and characterization of either synthetic or biochemical organic substances or both. Additionally there are a number of evidence categories classified as inorganic.

Forensic Testing

Toxicology. Psychoactive substances, illicit and ethical (licit) drugs and alcohol (ethanol), are the greatest source of physical evidence analyzed in most crime laboratories. Drug testing falls into two categories: solid dose samples and toxicology related cases, e.g., blood, urine, or tissue specimens in post-mortem cases or cases involving driving under the influence of alcohol or drugs, as well as workplace or employee drug testing.

Blood and urine are most often analyzed for alcohol by headspace gas chromatography using an internal standard, e.g., 1-propanol. Breath alcohol testing is accomplished by a number of techniques. The oldest reliable procedure involves bubbling a measured volume of deep-lung air containing alcohol through an acidic solution of potassium dichromate, $K_2Cr_2O_7$. Newer instruments rely on infrared spectroscopy to measure the blood alcohol concentration in breath.

Driving under the influence of alcohol cases are complicated because people sometimes consume alcohol with other substances. The most common illicit substances taken with alcohol are marijuana and cocaine. Forensic toxicology laboratories having large caseloads rely on immunoassay techniques to screen specimens. Immunoassay technology involves the manufacture of antibodies that are specific to particular drugs or to a class of drugs.

There are several immunological techniques in use. Enzyme multiplied immunological technique (EMIT) employs an enzymatic reaction to determine concentration, whereas radioimmunoassay (RIA) uses radioactively tagged reagents such as ^{125}I to measure concentration.

Thin-layer chromatography (tlc) is frequently used. A drawback to tlc, however, is that the technique is not especially sensitive and low levels of drugs may be missed.

Gas chromatography (gc) and gas chromatography-mass spectroscopy (gc/ms) are the most common analytical procedures used in modern forensic toxicology laboratories. Drugs are separated from their biological matrices, i.e., blood, urine, liver, etc., by liquid–liquid or solid-phase extraction utilizing the solubility of the suspect drug in acid or alkaline aqueous solution relative to the organic solution containing the specimen.

Solid-Dose Narcotics and Dangerous Drugs. Solid-dose drug testing differs from forensic toxicology in that the solid form of the drug is tested, rather than a biological specimen containing the drug and its metabolite. The typical drugs of abuse in North America are heroin; cocaine, i.e., freebase, crack, and the HCl salt; marijuana; hashish, a concentrated form of marijuana; amphetamine; methamphetamine; phencyclidine; and LSD.

Trace Evidence. Trace evidence refers to minute, sometimes microscopic material found during the examination of a crime scene or a victim's or suspect's clothing. Trace evidence often helps police investigators develop connections between suspect and victim and the crime scene. The challenge to the forensic scientist is to locate, collect, preserve, and characterize the trace evidence.

Trace evidence in criminal investigations typically consists of hairs; both natural and synthetic fibers, fabrics; glass; plastics; soil; plant material; building material such as cement, paint, stucco, wood, etc, flammable fluid residues, e.g., in arson investigations; explosive residues, e.g., from bombings, and so on. Perhaps the simplest examination done is the physical match. Other examinations result only in demonstrating class characteristics.

Microscopy plays a key role in examining trace evidence owing to the small size of the evidence and a desire to use nondestructive testing techniques whenever possible. Polarizing light microscopy is a method of choice for crystalline materials. Other microscopic procedures involving infrared, visible, and ultraviolet spectroscopy also are used to examine many types of trace evidence.

More traditional analytical techniques also are used. Capillary column gas chromatography is the method of choice for characterizing flammable fluid residues in arson cases. Scanning electron microscopy (sem) and energy dispersive x-ray analysis (edx) are used frequently in gunshot residue examination and to characterize evidence of an inorganic origin. Pattern recognition examinations are important in footwear and tire impression cases. Lasers and other high intensity or alternative light sources are useful in crime laboratories to visualize latent fingerprints, seminal fluid stains, obliterated writings, and erasures, and to aid in specialized photographic work. Infrared and ultraviolet light sources are also used to view items of evidence.

Forensic Serology. Blood, often associated with crimes of violence, is powerful physical evidence. Its presence suggests association with the criminal act and blood can be used to associate suspects and locations with the bleeder. Blood is a complex mixture of cellular material, proteins, and enzymes and several tests are available for suspected bloody evidence. A typical test protocol involves *(1)* determining whether blood is present, *(2)* determining if it is human blood, *(3)* typing the blood, and *(4)* when applicable, performing DNA typing.

Many of the chemical tests of the presence of blood rely on the catalytic peroxidase activity of heme. Species origin tests, used to determine whether the specimen is human or from another source, are immunological in nature.

Blood collected as evidence in criminal acts is usually dried and deposited on a variety of substrates. Sample size is usually on the order of a 2 or 3 mm diameter stain. Traditional typing involves ABO blood grouping, and characterizing stable polymorphic proteins or enzymes present in blood by means of electrophoresis. See also **Electrophoresis**.

More recently, the forensic application of DNA testing has dramatically enhanced the ability to determine the source of a blood sample. Two procedures are in forensic use: restriction fragment length polymorphism (RFLP) and polymerase chain reaction (PCR).

BARRY A. J. FISHER
Scientific Services Bureau
Los Angeles County Sheriff's Department

Additional Reading

Brenner, J.C.: *Forensic Science: An Illustrated Dictionary,* CRC Press, LLC., Boca Raton, FL, 2003.

Fisher, B.A.J.: *Techniques of Crime Scene Investigation*, 5th Edition, CRC Press, Boca Raton, FL, 1992.

Kirk, P.L.: *Crime Investigation*, 2nd Edition, John Wiley & Sons, Inc., New York, NY 1974.

Moenssens, A., F.E. Imbau, and J.E. Starrs: *Scientific Evidence in Criminal Cases*, 3rd Edition Foundation Press, Mineola, NY, 1986.

Saferstein, R. ed.: *Forensic Science Handbook*. Prentice-Hall, Englewood Cliffs, NJ, 1982 and "Forensic Science Handbook", Vol. II, 1988.

Saferstein, R., R.E. James, and C.E. Meloan: *Criminalistics: An Introduction to Forensic Science,* 8th Edition, Pearson Custom Publishing, Boston, MA, 2003.

FORGING. Historical records do not note when the first humans discovered that certain malleable metals (such as gold, copper, and zinc) and (probably at a somewhat later time) some formulations of iron could be hammered and pounded into approximate shapes. Artifacts prove that the art of forging dates back to antiquity.

Forging, as an effective industrial metal-shaping process, made its first appearance during the industrial revolution of the late 18th century, when the first steam-powered hammers were developed. Over the years, larger forging presses have been developed to replace these earliest tools; they have been automated to an impressive extent.

See also **Iron Metals, Alloys, and Steels**.

FORMALDEHYDE. [CAS: 50-00-0]. HCHO, formula weight 30.03, colorless gas with pungent odor, mp $-92°C$, bp $-21°C$, sp gr 0.815 (at $-20°C$). The gas is very soluble in H_2O, alcohol, and ether. Formaldehyde usually is produced and marketed as a 37% (weight) solution in water. From 3 to 15% methyl alcohol normally is added as a stabilizer to prevent paraformaldehyde formation. The commercial trend is to furnish a more concentrated product (up to 50% HCHO by weight) which contain as little as 0.5 to 1% methyl alcohol. The addition of special stabilizing agents and storage at elevated temperatures reduces the formation of paraformaldehyde.

Polymerized formaldehyde (trioxane) is a ring compound of anhydrous formaldehyde with the formula $(HCHO)_3$. See also Acetal Resins. Trioxane is a colorless crystalline solid with a pleasant odor, mp 62°C, bp 115°C, sp gr 1.17. This compound is used as a tanning agent and solvent and as a source of dry HCHO gas. Because trioxane ignites readily at 113°C and burns with an odorless, hot flame, it has been furnished in tablet form as a replacement for solidified alcohol in portable heating applications.

Paraformaldehyde $(CH_2O)_x$, sometimes called paraform, is an amorphous white powder and may be used for applications where an aqueous solution of HCHO may not be desirable. It finds use as an antiseptic and as a catalyst and hardener for certain synthetic resins. Formaldehyde gas also may be dissolved in methyl or butyl alcohol for applications where H_2O is undesirable.

Formaldehyde is a high-tonnage chemical and, in addition to the uses already mentioned, finds wide application in the manufacture of urea-formaldehyde resins (growing annually at a rate of about 6%), melamineformaldehyde resins (growth rate of about 5%), and acetal resins (growth rate of about 10%). Formaldehyde also is used in the production of pentaerythritol which, in turn, is used in the manufacture of lubricant additives, resin esters, pentaerythritol tetranitrate, and alkyd resins. Formaldehyde also is required in the manufacture of hexamethylene tetramine, a compound important in explosives manufacture and as a resin-curing agent. Other uses for formaldehyde include the production of ethylene glycol, acrylic esters, urea-formaldehyde fertilizers, textile-treating agents, tetrahydrofuran for elastomeric fibers, trimethylol propane for urethanes, and as a solvent for synthetic and natural resins. The growing need for nitrilotriacetic acid (NTA) and isoprene, for which formaldehyde is required, will account for additional tonnage production in the near future.

Production

All major commercial processes for making formaldehyde initially yield an aqueous solution of HCHO. In over 90% of the installations, methyl alcohol is the chargestock. Other feedstocks uncommonly used include methane, hydrocarbon gases, and dimethyl ether. See Fig. 1. Those processes commencing with methyl alcohol are of two types: (1) the silver-catalyzed process in which formaldehyde results from a combination dehydrogenation-oxidation reaction: $CH_3OH \rightarrow HCHO + H_2 + \frac{1}{2}O_2 \rightarrow H_2O$. The first part of the two-step reaction is endothermic, the second part is exothermic; and (2) the oxide-catalyzed process in which methyl alcohol is directly oxidized: $CH_3OH + \frac{1}{2}O_2 \rightarrow HCHO + H_2O$. The latter is an exothermic reaction. A process representing the silver technology is shown in the accompanying flowsheet. The process essentially consists of

two sections: (1) the synthesis portion, which yields products containing from 3 to 15% methyl alcohol, and (2) the distillation portion, which is required only where the methyl alcohol content of the final product must be low. In the oxide-catalyzed process, catalysts used generally are mixtures of oxides of molybdenum, iron, and vanadium.

Formaldehyde reacts with many chemicals in a marked manner: (1) with ammonio-silver nitrate (Tollen's solution), to form metallic silver, either as a black precipitate or as an adherent mirror film on glass; (2) with alkaline cupric solution (Fehling's solution), to form cuprous oxide, red to yellow precipitate; (3) with rosaniline (fuchsine, magenta) which has been decolorized by sulfurous acid (Schiff's solution), the pink color of rosaniline is restored; (4) with NaOH, yields methyl alcohol plus sodium formate; (5) with NH_4OH, when evaporated, yields hexamethylene tetramine "urotropine" $(CH_2)_6N_4$, white solid, mp 263°C; (6) with sodium or hydrogen peroxide in sodium hydroxide, yields sodium formate; and (7) with manganese dioxide and H_2SO_4, forms methyl, dimethoxymethane $CH_2 (OCH_3)_2$, colorless liquid, bp 42°C.

Formaldehyde gas, when cooled under certain conditions, yields trioxymethylene, metaformaldehyde $(CH_2O)_3$; formaldehyde solution, when evaporated, upon standing, or upon being subjected to low temperatures, yields paraformaldehyde $(CH_2O)_x$, white solid, from which formaldehyde is regenerated upon heating; dilute formaldehyde, in the presence of calcium hydroxide solution, yields a mixture of sugars called formose from which fructose $C_6H_{12}O_6$ has been prepared, suggesting the intermediate formation in nature of formaldehyde in the photosynthetic process of the conversion of carbon dioxide to sugars. Formaldehyde stands chemically between methyl alcohol on the one hand—to which it can be reduced—and formic acid on the other hand—to which it can be oxidized.

Formaldehyde is commonly detected by the Schiff test (above), and confirmed by the formation of a dimethyl derivative with a mp of 189°C.

J. R. MASSON
DAVY McKEE
Oil & Chemicals Ltd.
London, England

Additional Reading

Davis, R.: *Formaldehyde Toxicology,* Gordon & Breach Publishing Group, Newark, NJ, 1993.

Graham, J.D., M.J. Roberts, and L.C. Green: *In Search of Safety: Chemicals and Cancer Risk,* Harvard University Press, Cambridge, MA, 1991.

Lewis, R.J. and N.I. Sax: *Sax's Dangerous Properties of Industrial Materials,* 10th Edition, John Wiley & Sons, Inc., New York, NY, 1999.

Staff: *Wood Dust and Formaldehyde,* World Health Organization, New York, NY, 1995.

Zimmerman, N.: *Hazardous Material: Formaldehyde,* Gordon & Breach Publishing Group, Newark, NJ, 1991.

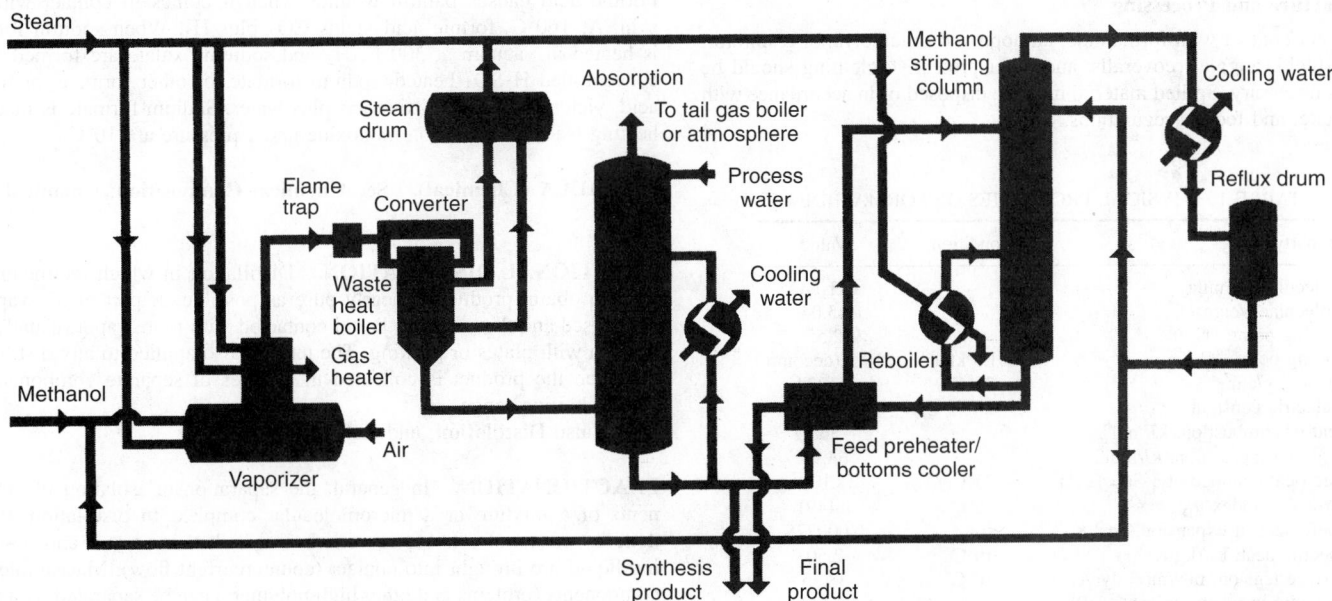

Fig. 1. Process for making formaldehyde, using a silver catalyst. Developed and operated by Imperial Chemical Industries Limited and available through Davy McKee (Oil) and Chemicals Ltd. London, England

FORMAMIDE

FORMAMIDE. Formamide (methanamide), $HCONH_2$, is the first member of the primary amide series and is the only one liquid at room temperature. It is hygroscopic and has a faint odor of ammonia. Formamide is a colorless to pale yellowish liquid, freely miscible with water, lower alcohols and glycols, and lower esters and acetone. It is virtually immiscible in almost all aliphatic and aromatic hydrocarbons, chlorinated hydrocarbons, and ethers. By virtue of its high dielectric constant, close to that of water and unusual for an organic compound, formamide has a high solvent capacity for many heavy-metal salts and for salts of alkali and alkaline-earth metals. It is an important solvent, in particular for resins and plasticizers. As a chemical intermediate, formamide is especially useful in the synthesis of heterocyclic compounds, pharmaceuticals, crop protection agents, pesticides, and for the manufacture of hydrocyanic acid.

In the 1990s, formamide is mainly manufactured either by direct synthesis from carbon monoxide and ammonia, or more importantly in a two-stage process by reaction of methyl formate (from carbon monoxide and methanol) with ammonia.

Properties

Table 1 lists the important physical properties of formamide.

Reactions

As a result of its bifunctionality, formamide is a highly reactive intermediate that is useful in a wide variety of synthetic applications.

Shipment

Formamide is a registered substance, e.g., in TSCA, EINECS, and MITI, and can, therefore, be produced in and imported into the United States, EEC, and Japan in compliance with the above-mentioned acts.

Formamide is best shipped in containers made of stainless steel or in drums made of, or coated with, polyethylene.

Economic Aspects

The estimated capacity of formamide was approximately 100,000 t/yr worldwide in 1990. In 1994, there are only three significant producers; BASF in Germany is the leading manufacturer. Most of the formamide produced in utilized directly by the manufacturers.

Storage

The shelf life is unlimited in sealed containers. The product is neither explosive nor spontaneously flammable in air. It is stable to the effects of light and air below ca 100°C. However, the product is combustible and should accordingly be stored with adequate precautions.

Manufacture and Processing

To prevent contact with formamide, an approved organic vapor respirator, a face shield, goggles, coveralls, and other protective clothing should be worn as necessary. Spilled material must be disposed of in accordance with local, state, and federal regulations.

TABLE 1. PHYSICAL PROPERTIES OF FORMAMIDE

Property	Condition	Value
molecular formula	–	CH_3NO
molecular weight	–	45.04
melting point, °C	–	2.55
boiling point, °C	101.3 kPa[a]	210.5 (decomp.)
density, g/cm³	20°C	1.1334
dielectric constant	20°C	109 ± 1.5
heat of combustion, kJ/mol[b]	–	−568.2
heat of vaporization, kJ/mol[b]	–	64.98
electrical conductivity, $S(= \Omega^{-1})$	25°C	2×10^{-7}
refractive index n_D^{15}	15°C	1.4491
coefficient of expansion, cm³/K	–	0.000775
specific heat, kJ/(kg·K)[b]	19°C	2.30
surface tension, mN/m(= dyn/cm)	20°C	58.35
dynamic viscosity, mPa·s(= cP)	15°C	4.320

[a] To convert kPa to mm Hg, multiply by 7.5.

[b] To convert J to cal, divide by 4.184.

Health and Safety

Formamide exhibits no particular acute toxicity with oral, dermal, and other applications in rats and other species. Precautions that should be observed when handling formamide include avoidance of prolonged inhalation of vapors or contact of the liquid with skin or eyes. When handling the chemical its teratogenic property has to be taken into account.

Small quantities of spilled formamide can be washed away with plenty of water. Larger amounts should be absorbed appropriately or pumped into containers for proper disposal by incineration or biological degradation in a sewage water treatment plant.

A. Hohn
BASF AG

Additional Reading

Bipp, H.: in B. Elvers, S. Hawkins, M. Ravenscroft, J. Rounsaville, and G. Schuls, eds., *Ullmann's Encyclopedia of Industrial Chemistry,* 5th Edition, Vol. A12, VCH Publishers, Weinheim, Germany, pp. 1–5.

"Formamide," BASF Technical Leaflet 1983; BASF data sheet, BASF AG, Operating Division Zwischenprodukte, Ludwigshafen, Germany, 1989.

FORMIC ACID. [CAS: 64-18-6]. HCOOH, formula weight 46.03, colorless liquid, mp 8.6°C, bp 100.8°C, sp gr 1.220. Sometimes referred to as methanoic acid or hydrogen carboxylic acid, this compound is miscible with H_2O, alcohol, or ether in all proportions. Formic acid occurs in some living organisms, such as nettles, ants, and caterpillars. Commercially, the compound is obtained from the black liquor of sulfite paper mills where it is present as sodium formate. In the laboratory, it may be prepared by reacting oxalic acid and glycerol at about 110°C, or by reacting solid lead formate with H_2S gas at about 100°C, yielding anhydrous formic acid. In the textile and leather industries, formic acid is used as a reducing agent, particularly in connection with the chrome dyeing of wool. Formic acid, in a reaction with glycerol at 220°C, is a source of allyl alcohol. Miscellaneous uses for formic acid have included brewing (acts as a fermentation assistant), electroplating for pH and redox adjustment, and use as a food preservative, germicide, and coagulant for rubber latex. The compound also is useful in the preparation of metallic formates and esters. The most important ester is methyl formate $HCOOCH_3$, a solvent for acrylic resins and cellulose esters as well as an intermediate for certain organic syntheses.

Formic acid solution reacts as follows: (1) with hydroxides, oxides, carbonates, to form formates, e.g., sodium formate, calcium formate, and with alcohols to form esters; (2) with silver of ammonio-silver nitrate to form metallic silver; (3) with ferric formate solution, upon heating, to form red precipitate of basic ferric formate; (4) with mercuric chloride solution to form mercurous chloride, white precipitate; and (5) with permanganate (in the presence of dilute H_2SO_4) to form CO_2 and manganous salt solution. Formic acid causes painful wounds when it comes in contact with the skin. At 160°C, formic acid yields CO_2 plus H_2. When sodium formate is heated in vacuum at 300°C, H_2 and sodium oxalate are formed. With concentrated H_2SO_4 heated, sodium formate, or other formate, or formic acid, yields carbon monoxide gas plus water. Sodium formate is made by heating NaOH and carbon monoxide under pressure at 210°C.

FORMULA (Chemical). See **Chemical Composition; Chemical Formula**.

FRACTIONAL DISTILLATION. Distillation in which rectification is used to obtain product as nearly pure as possible. A part of the vapor is condensed and the resulting liquid contacted with more vapor, usually in a column with plates or packing. The term is also applied to any distillation in which the product is collected in a series of separate components of similar boiling range.

See also **Distillation**; and **Reflux**.

FRACTIONATION. In general, the separation or isolation of components of a mixture or a micromolecular complex. In distillation, this is done by means of a tower or column in which rising vapor and descending liquid are brought into contact (countercurrent flow). Macromolecular components (proteins and other high polymers) can be separated by a number of methods, including electrophoresis, gel filtration, chromatography, centrifugation, foam fractionation, and partition.

See also **Distillation**; and **Reflux**.

FRANCIUM. [CAS: 7440-73-5]. Chemical element symbol Fr, at. no. 87, at. wt. 223 (mass number of the most stable isotope), periodic table group 1, mp 26–28°C, bp 676–678°C, density 2.4 g/cm³. To date, 22 isotopes of francium, with mass numbers ranging from 203 to 224, have been identified. All are radioactive. See also **Radioactivity**. Although Mendeleev visualized that element 87 would occupy the bottom position among the alkali metals in his periodic classification, the discovery of francium did not occur until 1939 when it was confirmed by Marguerite Perey, a collaborator of Marie Curie. Many earlier attempts had been made, including the observations of J.A. Cranston in 1913, the search for the element in radioactive ores by O. Hahn and G. Hevesy in 1926, and efforts by M.C. GuÉben in 1932. Also, earlier investigations by S.J. Meyer, V. Hess, and F. Paneth in 1914, when they were studying the emissions by ^{227}Ac, contributed to the network of information that finally led to the firm identification of francium. See also **Chemical Elements**.

Studies indicate that no further isotopes of francium with half-lives longer than those already known should exist. Thus, among the first 101 elements of the periodic chart, francium is the most unstable. The short half-lives of the isotopes explain the difficulties in isolating and confirming the element and of learning more of the detailed chemistry of the element.

The isotopes of francium that occur in nature are found in thorium and uranium ores, in which they are continually formed by disintegration chains. These start with ^{232}Th ($4n$ family), ^{237}Np ($4n + 1$ family), and ^{235}U ($4n + 3$ family). It is estimated that 1 ton of natural uranium contains about 3.8×10^{-3} g of ^{223}Fr and 10^{-17} g of ^{221}Fr. The separation of these isotopes from natural sources involves long and complex chemical procedures. Like the other alkaline elements, francium remains in solution when other elements are precipitated as carbonates, hydroxides, fluorides, chromates, sulfates, and sulfides. However, at a pH of 9, francium can be extracted from solution by nitrobenzene in the presence of sodium tetraphenylborate. Extraction from a very dilute sodium solution can be effected with dipicrylamine in nitrobenzene solution, the separation being much easier to effect than in the case of cesium. Francium can be separated from rubidium and cesium by chromatography on cation-exchange resins or on mineral exchangers.

The heavy isotopes of francium can be formed by irradiation of uranium or thorium by protons of high energy; the lighter isotopes can be obtained by nuclear reactions induced in gold, tellurium, or lead targets by heavy ions.

Because of the difficulties in obtaining any significant quantities of francium, use of the element is confined to scientific investigations. ^{223}Fr is used for the measurement of ^{227}Ac. Studies have shown that francium fixes itself in induced sarcomas in rats. Because of the short half-lives of ^{223}Fr and ^{212}Fr, which would cause no radiation risk to organisms, the property could become useful for the early diagnosis of certain kinds of cancers.

See list of references at end of entry on **Chemical Elements**.

FRANKLINITE. The mineral franklinite is a zinc-iron-manganese mineral whose formula may be written $(Zn, Mn^{2+}, Fe^{2+}) (Fe^{3+}, Mn^{3+})_2 O_4$, but the composition varies considerably, in respect to the amounts of the several metals that may be present. Its isometric crystals have an octahedral habit; it may be coarse or finely granular or compact. It shows a parting parallel to the octahedron; fracture, uneven; brittle; hardness, 5.5–6.5; specific gravity, 5–5.2; luster, usually metallic, occasionally dull; color, black; streak, brown to black; opaque; may be slightly magnetic. Only in one place in the world does franklinite occur in quantity: at Franklin Furnace, New Jersey, from whence it was named. Here there are two bodies of this mineral, which is used as a zinc ore, about 3 miles distant from each other. The franklinite is found in pre-Cambrian limestones that are associated with gneisses believed to be of igneous origin and responsible for the mineralization. Associated minerals are willemite, zinc, silicate, and zincite, zinc oxide, manganoan calcite.

FRASCH PROCESS. A process by which much of the world's sulfur is obtained. Developed about 1900 by Herman Frasch, the process involves melting sulfur underground by introducing superheated water through a pipe under pressure and forcing the molten sulfur to the surface by compressed air.

FREEDOM (Degrees of). 1. The number of variables which must be fixed before the state of a system may be defined according to the phase rule. See also **Phase Rule**. The relationship between the number of degrees of freedom (F), the components (C), and the phases (P) of a system is expressed by the formula,

$$F = C + 2 - P$$

Thus a pure gas has two degrees of freedom. At any temperature its volume and pressure are variable, but if one of these is fixed then the other is automatically determined for, at any given temperature and pressure, each pure gas assumes one, and only one, volume at equilibrium.

2. The number of independent coordinates necessary for the unique determination of the position of every particle in a dynamical system is the number of degrees of freedom. Each degree of freedom is represented by a coordinate that can vary with time independently of all the rest. Thus a single particle which may move anywhere in three-dimensional space has three degrees of freedom. A particle constrained to move on a surface has two degrees of freedom. A system composed of three particles has 9 degrees of freedom; for it takes nine independent coordinates to specify the positions of the particles in space, and their arrangement may therefore be changed in nine different ways. A single rigid body, on the other hand, has 6 degrees of freedom, since it may have motions of translation in three coordinate directions, and it may also rotate about any one of the three coordinate axes through its center of mass. Any actual motion of the body is in general made up of all six, its linear motion being the resultant of three linear components and its rotation the resultant of three angular components. Each molecule of a diatomic gas has 7 degrees of freedom; viz., the six just mentioned for the molecule as a whole (regarded as a rigid body), and, in addition, one corresponding to the possible vibration of the two atoms toward and away from each other. If the body is not rigid, the number of degrees of freedom may be virtually infinite. It should, however, be added that, because of restrictions imposed by the quantum theory, not all of the possible degrees of freedom can in general be expected to participate in changes of molecular energy.

3. The number of degrees of freedom in a statistical quantity is the number of independent values necessary to determine it. A sample of n values $x_1, x_2, \ldots, x_n$ has n degrees of freedom but the sum

$$\sum_{i=1}^{n} (x_i - \overline{x})^2$$

is regarded as having $n - 1$ because, for given $\overline{x}$ only $n - 1$ values are assignable at will.

By extension, the number of degrees of freedom of a statistical distribution relates to the degrees of freedom of the distributed statistic. For example, χ^2, being the distribution of the sum

$$\sum_{i=1}^{n} (x_i - \overline{x})^2/\sigma^2$$

has $n - 1$ degrees; and the F-distribution, which concerns the ratio of two independent such quantities has a pair of degrees of freedom, one relating to the numerator and the other to the denominator of the ratio.

FREE ELECTRON THEORY OF METALS. The most characteristic property of a metal is its electrical conductivity. It was early recognized (by Drude) that this could be explained if the electrons in the metal were relatively free to move. A model of a metal as a gas of free electrons, moving in the region of nearly uniform positive potential created by the ions of the crystal lattice, although satisfactory in some respects, led to serious difficulties, until it was pointed out by Sommerfeld that the electrons must obey the Pauli exclusion principle and hence constitute a highly degenerate Fermi-Dirac gas. On this basis, one can calculate the electronic specific heat, magnetic susceptibility, Hall coefficient, Wiedemann-Franz ratio, thermionic emission, and other physical properties.

The reason for the success of this simple theory is that in certain elements, notably the alkali metals, the outer electrons are only very loosely bound to the remainder of the atom. When the atoms are brought together to make up the crystal lattice these electrons can jump from ion to ion as if they were free. To understand the behavior of the more complex metals, and of semiconductors, it was necessary to introduce the further complications of the band theory of solids.

See also **Electron**; and **Solid-State Physics**.

FREE ENERGY. There are two quantities to which this term has been applied.

1. The Gibbs free energy, which is also called the Gibbs function and the thermodynamic potential, is most generally understood when the term free energy is used without qualification. It is defined by the equation.

$$G = U - TS + pV$$

where U is the internal energy, T, the absolute temperature, S, the entropy, p, the pressure, V, the volume, and G, the Gibbs free energy (the letter F is also used to denote this quantity).

2. The Helmholtz free energy, which was called the psi function, and which is perhaps most commonly known as the work function. It is defined by the equation

$$A = U - TS$$

where U is the internal energy, T, the absolute temperature, S, the entropy and A, the Helmholtz free energy. (The letter ψ or the letter F are sometimes used instead of A for this quantity.) The decrease in A is equal to the maximum work done on the system in a constant-temperature, reversible change. In terms of the partition function, $A = -RT \, Z$. Like the Gibbs free energy, the Helmholtz free energy is a thermodynamic potential, although the latter term is commonly used to refer specifically to the Gibbs free energy.

FREE ENERGY CHANGE.

The change in the Gibbs free energy for a chemical reaction, defined as

$$\Delta G = \sum_{r=1}^{n} v_r g_r$$

where g_r is the molar Gibbs free energy of the rth component in the pure state, under the same conditions of temperature and pressure as those in which the reaction takes place. v_r is the stoichiometric coefficient of the rth component.

This quantity ΔG is equal to the *maximum net* work available (i.e., work, other than work of expansion, in a reversible process) for a given change in state under constant temperature and pressure.

FREE MACHINING.

Many alloys are prepared in free machining variants or forms that require less power for machining, give better surface finishes, and are less wearing on the tools. These alloys have incorporated in them small particles of another metal or some compound that act as stress raisers, which cause chips to break off easily during machining. Sulfur, selenium, lead, and bismuth are among the elements added or left in the alloys for this purpose.

FREE RADICAL.

Free radicals have been defined as highly reactive groups of atoms containing unpaired electrons. This definition is imprecise, because it would include ions, such as those of the lanthanide and actinide series, which not only possess such unpaired electrons, but also—because of this—exhibit the color, magnetic and other characteristics of "free radicals". The term is in general reserved for short-lived alkyl radicals possessing a magnetically noncompensated electron, or somewhat longer-lived, larger organic molecules of the aryl-alkyl type, similarly possessing unpaired electrons in their valence shell.

A few very reactive inorganic radicals, such as NO, ClO$_2$; or NO$_2$, may also be construed as "free," but knowledge of the chemistry of free radicals is best exemplified by unsaturated organic fragments, such as CH_3, C_6H_5, among others. These free radical fragments are formed by breaking one or more bonds in a stable molecule by photolysis, electrolysis, pyrolysis, and some other processes. The first free radical to be synthesized was that of triphenylmethyl by Moses Gomberg in 1900. Paneth and Hofeditz, in 1926, found free radicals in the pyrolysis of lead tetramethyl, and Rice subsequently demonstrated their existence in the breakdown products of many organic compounds.

Because of the affinity of their unpaired electrons, free radicals have short lives, tend to dimerize and thus lose their reactivity. Because of their generally short half-lives (1–100 milliseconds), detection and identification of these entities is essentially through spectrophotometric methods. However, in solid systems, free radicals can be trapped for appreciable lengths of time and at least one of these, 2,2-diphenyl-1-picrylhydrazyl, has such a long half-life that it is sold as such for the photometric determination of tocopherol.

Free radicals formed by photolysis play significant roles in the chemistry of the earth's atmosphere, not the least of which is the destruction of ultraviolet-protective ozone by fluorocarbons from anthropogenic sources. But the most important reactions involving free radicals as intermediates in organic chemistry are polymerizations, such as may develop from a free radical and ethylene. These may also be ensured through use of substances known to produce free radicals, e.g., benzoyl peroxide, of which the benzoyl free radical can initiate a chain reaction that may continue indefinitely. Or, it may lose CO_2 itself to give a new free radical, phenyl, which is itself capable of polymerization.

A free radical chain reaction proceeds through a succession of free radicals. In the photochemical chlorination of an alkane, the initiating step is the homolytic fission of chlorine molecules to produce chloroalkane molecules and chlorine free radicals. These two reactions constitute the propagating step. However, the chlorine free radicals may also combine to form chlorine molecules or react with the alkane free radicals to form chloroalkane molecules. Both of these reactions constitute terminating steps of the chain reaction. It should be noted, however, that the foregoing sequence cannot take place in the dark. Exposure to light allows the series of reactions then to proceed rather violently.

Hydrocarbon oxidation may also be considered a free radical chain-type reaction. At elevated temperatures, hydrocarbon free radicals (R) are formed which react with oxygen to form peroxy radicals (ROO). These, in turn, take up a hydrogen atom from the hydrocarbon to form a hydroperoxide (ROOH) and another hydrocarbon free radical. The cycle repeats itself with the addition of oxygen. The unstable hydroperoxides remaining are the major points for degradation and lead to rancidity and color development in oils, fats, and waxes; decomposition and gum formation in gasolines; sludging in lubricants; and breakdown of plastics and rubber products. Antioxidants, such as amines and phenols, are often introduced into hydrocarbon systems in order to prevent this free radical oxidation sequence.

See also **Antioxidants**; and **Carcinogens**.

R. C. VICKERY
Blanton/Dade City, Florida

Additional Reading

Haliwell, B. and J. Gutteridge: *Free Radicals in Biology and Medicine,* 5th Edition, Oxford University Press, New York, NY, 1999.

Kaslner, S.: *Nitric Oxide and Free Radicals in Peripheral Neurotransmission,* Vol. 2, Birkhauser Verlag, Cambridge, MA, 2000.

Packer, Lester, E. Cadenas, and G. Poli: *Free Radicals in Brain Pathophysiology,* Marcel Dekker, Inc., New York, NY, 2000.

Parson, A.W.: *An Introduction to Free Radical Chemistry,* Blackwell Science, Inc., Malden, MA, 2000.

Rosen, G.M. et al.: *Free Radicals: Biology and Detection by Spin Trapping,* Oxford University Press, Inc., New York, NY, 1999.

Walling, C.: *Fifty Years of Free Radicals,* American Chemical Society, Washington, DC, 1994.

Wiseman, H., P. Goldfarb, and T. Ridgway: *Biomolecular Free Radical Toxicity: Causes and Prevention,* John Wiley & Sons, Inc., New York, NY, 2000.

FREE ROTATION.

The power of two structural entities joined by a linkage to occupy any position relative to each other and relative to their planes of symmetry. For example, in diphenyl $C_6H_5-C_6H_5$ the two planar benzene rings may occupy any position relative to each other from parallelism to perpendicularity of the planes of their rings. However, the o-, o'-dinitrodiphenic acids can be separated into optical isomers, showing that the position of the substituents cannot be planar, i.e., that the free rotation of the rings has undergone steric hindrance.

FREE VOLUME.

A liquid differs from a solid having the same type of packing of the molecule in having a certain additional volume, the free volume, which provides the necessary looseness in the structure to permit free movement of the molecules. The concept of free volume is used in several theories of the liquid state.

FREEZE-CONCENTRATING.

In lieu of evaporation and other means for concentrating liquid substances, notably foods, freeze-concentrating may be used, particularly where it is desirable to retain volatile constituents, as in the instances of increasing the alcohol content of wines, and to prepare and preserve flavor, as in the case of orange juice or coffee extract concentrates. From an energy standpoint, the energy required by freeze-concentrating is considerably less than that used by evaporative systems.

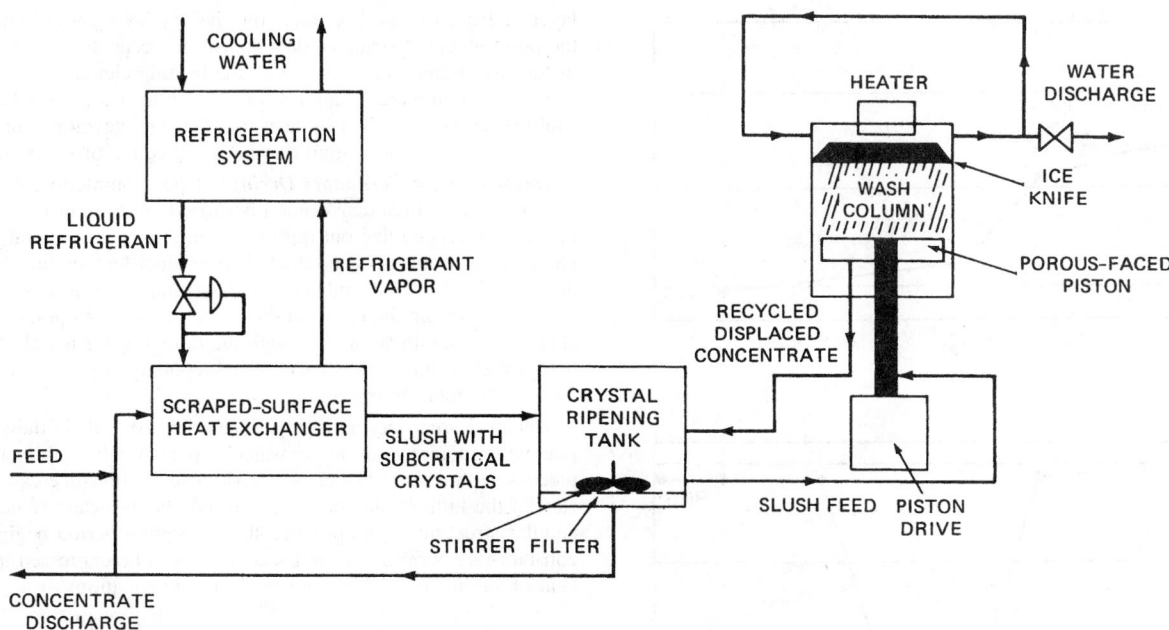

Fig. 1. Proprietary freeze-concentrating system

There are difficulties associated with the achievement of high concentrations through freezing. The liquid viscosity may increase so much as concentration increases and freezing point drops that difficulty in handling the ice-concentrate mixture and of separating the concentrate from the ice will arise. Improvements in recent years have been made by the use of ripening-induced growth of large ice crystals through sacrificial melting of small, easily formed subcritical ice crystals. Improvements also came from the development of wash columns, which provide efficient solute recovery from the ice-concentrate mixture. Even then, it is estimated that 35–50% dissolved solids represents the maximum concentration that can be practically achieved by freeze-concentration. By comparison, the liquid food concentrations achievable using two other attractive concentrating methods are even less: 20–35% by reverse osmosis; and 20–30% by ultrafiltration. However, these latter processes, when considered in conjunction with (prior to) evaporation may be attractive from an energy-expenditure standpoint. A proprietary freeze-concentration system is depicted in Fig. 1.

Additional Reading

Note: See references at end of article on **Freeze-Preserving**.

FREEZE-DRYING. A process for removing moisture from a wet material by bringing the material to the solid state and subsequently subliming it. This process is used for drying and preserving a number of products, notably food products, including instant coffee, vegetables, fruit juices, and meats. The needs of the food industry, coupled with those of pharmaceutical manufacturers, accelerated research into this process several years ago and commercial applications of freeze-drying are now commonplace in these industries.

The wet material in the form of a wet solid or in the form of a suspension or solution is frozen under vacuum or at atmospheric pressure, followed by transforming the ice into vapor and removing it. In the usual case, the dried material remaining will be a spongy mass of about the same size and shape as the original frozen mass. It frequently will be found to have excellent stability, convenient reconstitution when placed in cold water, and will maintain flavor and texture sometimes indistinguishable from the original materials. These properties differ markedly with various materials. Some products are much better adapted to the process than others.

Usually materials to be freeze-dried are complex mixtures of water and several other substances. When such materials are cooled below 32°F (0°C), pure ice crystals will separate out first. With further cooling, the mass will become rigid as the result of formation of eutectics. (A eutectic is that particular mixture out of a possible combination of two or more mixtures of materials that has the lowest melting point.) See Fig. 1. Most food products and biologicals solidify completely at a temperature in the range of −5 to −100°F (−15 to −73°C). At solidification of the entire

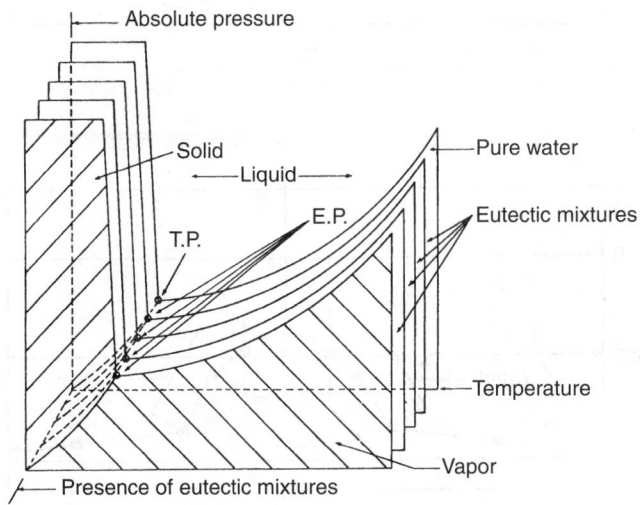

Fig. 1. Eutectic phase diagram in freeze-drying process

mass, all of the free water has been transformed into ice. Only a small quantity of the original water, the bound water, remains fixed in the internal structure of the material.

The quality of the finished product as well as the rate of drying will be affected by the size, shape, and size distribution of the ice crystals which form during freezing. These properties also will be affected by the homogeneity of the frozen mass. Thus, freezing must be effected under carefully controlled conditions (time, pressure, and temperature). Large ice crystals result from slow freezing rates. These may be injurious to certain substances. On the other hand, too-rapid freezing results in small ice crystals, which may cause undesirable color and texture changes.

Sublimation (or Primary Drying). For the sublimation phase of the process, the frozen material usually is subjected to a vacuum of about 4.6 millimeters of mercury. The ice-crystal sublimation process can be regarded as comprised of two basic processes: (1) Heat transfer, and (2) mass transfer. In essence, heat is furnished to the ice crystals to sublime them; the generated water vapor resulting is transferred out of the sublimation interface. Thus, it is evident that sublimation will be rate-limited by both resistances to heat and mass transfer as they occur within the material.

As the sublimation interface recedes in the material (Fig. 2), the dry layer presents a resistance to the flow of water vapor and a pressure difference must exist between the ice interface and the surface of the dry

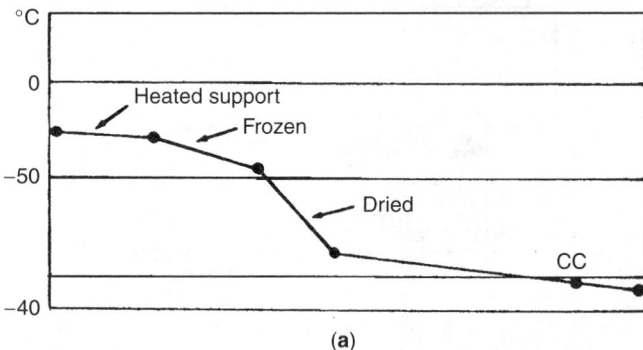

(a)

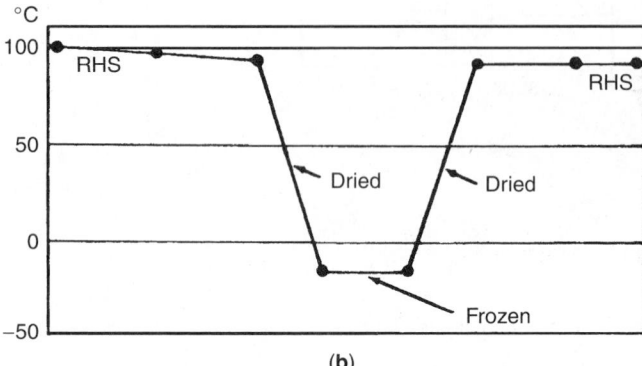

(b)

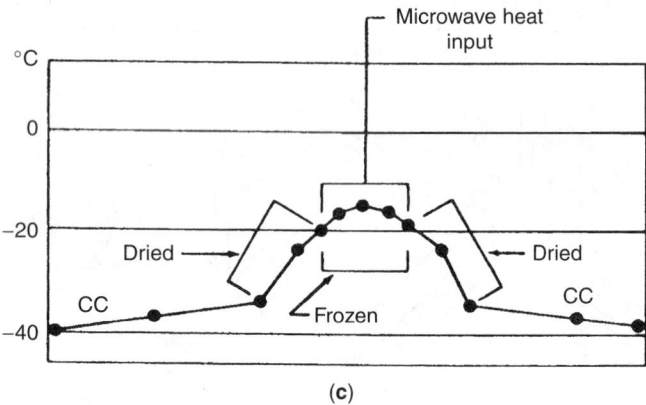

(c)

Fig. 2. Heat-input methods for freeze-drying processes: (a) conduction, (b) radiation, (c) microwave. CC = cold condenser; RHS = radiant-heat device

layer. A large pressure difference will facilitate high mass-transfer rates; however, the maximum allowable sublimation temperature at which no melting will occur and the cost of the vacuum equipment restrict this driving force to a limited range. In practice, the maximum allowable temperature and corresponding pressure at the sublimation interface is in the range of +15° to −40°F (−9.4° to −40°C) and 2000 to 100 micrometers of mercury, respectively.

The rate of heat input to the frozen material is a function of the operating-vacuum method of heat transfer and the properties of the dried product. The operating vacuum determines the pressure difference and, in turn, the rate of mass transfer, which must be in balance with the rate of heat input. Otherwise, either melting will occur at the sublimation interface and the purpose of freeze-drying will be defeated; or the sublimation temperature will decrease and the cost of processing will increase.

The heat required for sublimation (1200 Btu per pound of ice; 664 kilogram-calories per kilogram of ice) can be supplied by conduction, radiation, electric resistance, microwave, or infrared heating. Three methods of heat input that have been investigated extensively are shown in Fig. 2. Depending on the method of heat transfer, the temperature gradient between the sublimation interface and the heat source is limited by the maximum temperature which can be tolerated on the surface of the dry

layer or frozen mass. For radiation, the dry layer should not be heated to the point where charring or decomposition occur. For conduction, melting of the frozen mass in contact with the heating element should be avoided.

In most commercial applications, conditions are such that the rate of sublimation is controlled by heat transfer. The development of techniques for improving the heat-input rate is the objective of many investigations.

Desorption (or Secondary Drying). Upon completion of sublimation of the ice crystals, final dehydration is carried out to remove the bound water that did not crystallize out during freezing; and it is bound by adsorption phenomena to the dried product. The product temperature is increased to 80–120°F (27–49°C), and under high vacuum, the bound water and oxygen are removed from the dried product. The rate of desorption is considerably slower than sublimation. Although the bound water is only 5–10% of the total water in many substances, the secondary drying may require up to 35% of the total drying time.

Drying Rates. Drying a frozen material proceeds initially at a constant rate with rapid evolution of water vapor. As the sublimation interface recedes within the product, water-vapor evolution decreases. This is the start of the falling-rate period. When only bound water remains within the cellular structure of the product, the desorption period begins. During the constant-rate period, the sublimation rate can be expressed in terms of the heat of sublimation of ice and the heat-rate equation:

$$\text{Rate of sublimination} = \frac{U A \Delta T}{\Delta H_{\text{ice}}}$$

The overall heat-transfer coefficient U depends upon the properties of the dry product and the method of heat transfer. The heat-transfer rate A is influenced by the mechanical design of the heating elements and the conditioning of the frozen mass. The temperature gradient ΔT is limited by the maximum allowable temperatures at the sublimation interface and dry-layer surface. In the constant-rate period, the first one-half to two-thirds of the drying cycle, about 80% of the water is removed.

Processes and Equipment

In addition to the three fundamental operations just described, the freeze-drying process involves several other operations necessary to achieve an economically feasible system for large-scale production. The general commercial process comprises: (1) Preparation of the material; (2) freezing; (3) conditioning of the frozen mass; (4) drying, that is, sublimation and desorption; and (5) conditioning the product. See Fig. 3.

Preparation of the Material. It is not always economically practical to subject a product in its original state to freeze-drying. One or more operations may be required to prepare the product. Wet solids, such as fruits and meats, are usually ground or sliced to facilitate drying by increasing the surface and reducing the thickness. Coffee extract and fruit juices are preconcentrated in order to minimize the water to be removed by sublimation and, in turn, to reduce the processing cost or to ensure that the final product will not be fragile.

Freezing. (1) *Vacuum cooling*: The material freezes itself by the evaporation of water as it is subjected quickly to high vacuum. (2) *Direct contact*: The material is immersed in a cold liquid or in a stream of cold air or inert gases. (3) *Indirect contact*: The material is frozen on cold surfaces.

Freezing the material is accomplished in the vacuum chamber where drying takes place in a separate piece of equipment, or if the frozen mass requires some conditioning prior to drying.

Vacuum cooling is normally accomplished in the drying chamber. Its advantage is that large quantities of water are removed rapidly and no prior refrigeration is required. On the other hand, volatile flavor components are removed which may affect the quality of the final product. Also, removing water from the outer layer prior to complete freezing makes the cellular structure of certain materials collapse, and subsequent drying and reconstitution may be inhibited. Therefore, this freezing technique has limited application.

In direct-contact freezing, wet solid materials are placed in cold chambers and sprayed directly or immersed in cold air, inert gases, or liquid refrigerants. Solutions or slurries may be frozen by spraying them in a cold stream of gas or liquid. Poor control of the rate of freezing limits these techniques to cases where quick freezing is desirable. Indirect-contact freezing is generally carried out in trays placed on refrigerated shelves inside the vacuum chamber.

A combination of indirect- and direct-contact equipment is commonly used. Trays containing the material to be dried may be placed on

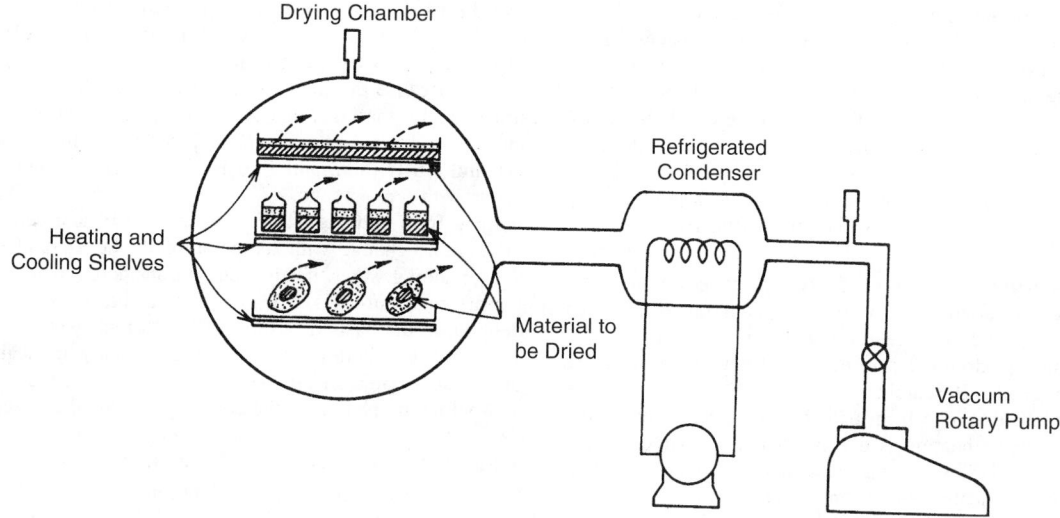

Fig. 3.   Schematic diagram of freeze-drying process

refrigerated shelves in a cold chamber and blasted with cold air, or inert gas or freezing belts may be used. Freezing belts in cold rooms are excellent for continuous operations. By controlling the temperature of the surrounding air and the refrigerant temperature along the length of the belt, good control of the freezing rate can be maintained.

Conditioning of the Frozen Mass. Materials frozen in a bulky or block form, meat, and solutions frozen on a belt or trays require further processing before drying. Granulation or slicing of such materials increases the available surface and minimizes the resistance to heat and mass transfer during drying. Standard size-reduction devices operating in cold chambers at about $-50°F$ ($-46°C$) are sometimes used.

Substances that do not freeze into a rigid solid at low temperatures, such as fruit juices, may be subjected to a devitrification treatment to avoid the soft-glass structure detrimental to optimum drying.

Drying. The conditioned frozen material to be dried is placed in a vacuum chamber, where sublimation and desorption of water occur. As soon as the chamber has been evacuated and the optimum vacuum has been reached (0.5 to 0.05 millimeters of mercury), heating is applied so that the ice sublimes. For large-scale production of food products, the combination of conduction and radiation that results from circulating a hot fluid through coils or plates has proved quite satisfactory for heating.

A number of vacuum-chamber designs have been used. For large installations, custom engineering and fabrication are dictated to assure optimum performance for a given product. The three vacuum chambers commonly used may be classified as batch, semicontinuous, and continuous.

Batch units are frequently cylindrical shelf-type dryers equipped with heating and cooling coils or plates. Designs for better heat input, that is, spikes or expanded metal sheet that penetrate the frozen mass or movable heated shelves compressing the frozen materials, have met with some success.

Semicontinuous units are usually long, cylindrical tunnels. Trays with the frozen material are continuously conveyed through a series of heated zones. Interlocks are used in both ends for proper vacuum control. The frozen material is heated along the length of the tunnel, with each zone maintained at a different temperature, and is removed as a fully dried product.

Continuous units are designed to move the frozen material continuously on the heated surface and transfer it through a series of zones in which the temperature and vacuum are maintained at different levels. The frozen material is fed via interlocks in the chamber. Vibrators or other mechanical means are used to maintain the product in continuous motion.

During the constant-rate drying period, the temperature of the heat source (radiation) is from $200-300°F$ ($93-149°C$) for many food products. Thus, high heat-input rates are achieved. This temperature is reduced to $125°$ to $150°F$ ($51°$ to $66°C$) in the falling-rate and desorption periods to avoid charring and decomposition of the dried products.

The drying process is discontinued when the residual moisture content is sufficiently low to ensure good preservation of the specific product. This may range from 1 to 3%.

Water Removal. The three common methods for removing the water vapor are condensers, direct- and indirect-contact; desiccants, such as calcium chloride and zeolites; and vacuum pumps. The indirect-contact refrigerated condenser offers a good arrangement for large-scale operations. The condensing surface can be located in the drying chamber, or in a separate chamber. The water vapor condenses and forms an ice layer on the cold surface and subsequently is removed by intermittent melting or scraping.

Vacuum Pumps. The function of these pumps is to evacuate the drying chamber quickly without allowing the prefrozen material to melt—and thereafter to reduce the pressure progressively to the desired vacuum and maintain it at this level by removing the noncondensable gases. The vacuum equipment can be either an oil-sealed rotary vacuum pump, or a multistage stream-ejector system.

Conditioning of the Product. The high porosity and low moisture content of the freeze-dried product require that the vacuum be broken and packaging be done under a dried inert-gas blanket, in many cases, to prevent oxidation during storage and maintain the low moisture content. Carbon dioxide or nitrogen are commonly used.

Variety of Problems. Each type of food product presents specific problems that affect the design and operation of freeze-drying processes. For some products, such as freeze-dried coffee (annual production of 50 million pounds; 22.5 million kilograms), the process has reached a mature and sophisticated stage. In this process, coffee extract with 20–25% solids is the raw material. Major steps in the process include: (1) clarification of the extract; (2) freeze-concentration of the extract to 30–40% solids; (3) freezing extract to a completely frozen mass at -13 to $-45°F$ (-25 to $-43°C$); (4) granulation of the frozen mass; (5) sublimation of the ice at a vacuum of approximately 200 micrometers of mercury absolute; and (6) drying the final product to a moisture content of 1–3%. For batch dryers, the overall drying cycle is on the order of 6–8 hours.

Intensive research continues in the interest of improving the freeze-drying process for certain foods and for expanding the use of this process to a broader spectrum of food products. The Sharma reference (listed at end of this entry) provides a good summary of work in this area being done in connection with freeze-dried meat patties. In another reference, Schmidt describes a comparison of vacuum and atmospheric freeze-drying of carrots. As pointed out by James M. Flink (Massachusetts Institute of Technology), "Specific information on the costs of producing processed foods is generally not available in the scientific literature, it presumably being considered proprietary information by those having the best data, the food processing industry." In the Flink reference, based upon available data, a cost comparison is made of three food-preserving processes—canned, frozen, freeze-dried, and freeze-dried compressed.

Increasing attention in the late 1970s was given to leafy vegetables, such as spinach. The large surface area of spinach seems well adapted for freeze-drying and compression. Foods produced by freeze-drying have less weight and a preserved flavor and structure, but the volume, in terms of packaging, transportation, and storage, is not changed. In an effort to alleviate this problem, different methods of compressing the freeze-dried products have been developed to eliminate most of the void spaces. Many fruits and vegetables have been compressed and then reconstituted to a normal appearance and texture. Most of the research has been directed toward vegetables. When properly preconditioned, freeze-dried foods can be compressed with little or no fragmentation. Freeze-dried foods have an average bulk density of 0.3 gram per cubic centimeter. With existing technology, it may be possible to compress most foods to a bulk density of 0.9 gram per cubic centimeter, without interfering with reconstitution. This concept has been studied in some depth at Texas A& M University.

Microwave Freeze-drying. Because microwave energy penetrates very well into ice, it would appear to offer an excellent solution to the heat-transfer problems of conventional freeze-drying. Because the microwave process has a much shorter drying cycle, decreased equipment capacity can be an economic result. Researchers have found that microwave energy utilization efficiencies range from 65% to 90% over much of the drying cycle. But microwave heating is dependent upon relatively high-cost electrical energy. Regardless of the economics, a major problem remaining with microwave heating is the melt-back of the frozen core and/or overheating in the dried layer. This situation occurs when microwave energy is put into the food faster than the sublimation-diffusion process can remove it, causing the pressure at the ice interface to rise above the triple point, usually resulting in melting of the ice. When melt-back occurs, the microwaves couple selectively into the water rather than the ice, thus causing intense local heating, accelerated melting, and a "runaway" condition (first reported by Gunn in 1967). Also for certain products, such as meat, a maximum allowable dried layer temperature of 140°F (60°C) should be observed during the process so that thermal degradation of the dried product can be prevented. This requires a matching of the electric heating rate with the mass flow rate in order to optimize the process. In early 1977, researchers at the University of Waterloo (Ontario, Canada) prepared a mathematical analysis of this problem.

Additional Reading

Day, J.G. and M.R. McLellan: *Cryopreservation and Freeze-Drying Protocols,* Vol. 38, Humana Press, Totowa, NJ, 1995.
Fink, J.M.: "A Simplified Cost Comparison of Freeze-Dried Food with Its Canned and Frozen Counterparts," *Food Technology,* **31,** 4, 50–56 (1977).
Fink, J.M.: "Energy Analysis in Dehydration Processes," *Food Technology* **37,** 3, 77–84 (1977).
Gunn, R.D.: Ph.D. Thesis, University of California, Berkeley, CA, 1967.
Oetjen, Georg-Wilhelm: *Freeze-Drying,* John Wiley & Sons, Inc., New York, NY, 1999.
Rey, L. and J.C. May: *Freeze-Drying/Lyophilization of Pharmaceutical and Biological Products,* Vol. 96, Marcel Dekker, Inc., New York, NY, 1999.
Sharma, S.C. and E. Seltzer: "Development of Procedures to Minimize Mechanical Damage in Freeze Dried Meat Patties," *J. Food Sci.,* **42,** 5, 1336–1343 (1977).

FREEZE-PRESERVING. Knowledge of the fact that food substances remain edible for longer periods of time when cooled probably dates back to antiquity, centuries before the process for making ice was developed. The latter led to cold storage, a practice that persisted for several decades and which, of course, remains useful for a number of products in current times and in certain regions. Cold storage was first limited by the minimum achievable temperature dictated by the melting point of ice. Chemicals to depress the freezing point were an additional step toward cold-preservation. Generally accredited with the initial breakthrough from cold-storage practices to present freezing technology was the step to quick-frozen foods, pioneered by Clarence Birdseye, among others, in the late 1920s. Consumers began to accept the fact that fresh, high-quality food when frozen quickly and when retained at a temperature of about −17.8°C (0°F) was a good substitute for fresh produce out of season. Frozen and stored in this way, these foods represented a marked improvement over the earlier available, slowly cooled food products.

Freeze-preserving is an across-the-board operation in the food industry of countries with advanced technology. There are relatively few foods—vegetables, fruits, fish, poultry, and meat—that cannot be frozen with reasonable success.

Fundamentals of Freezing. In food materials, water is the major component. Thus, when foods are cooled below 0°C, ice formation occurs, starting at a temperature between 0 and −3°C (32 and 26.6°F), which depends upon the molar concentration of soluble cell components. As the temperature is progressively reduced, more and more water is turned into

ice; the latent heat of ice formation adds to the sensible heat involved in cooling both ice and the unfrozen portion. This leads to large variations in heat capacities while thermal conductivities also change considerably, mainly because the thermal conductivity coefficient of ice is nearly four times greater than that of water. For most biological materials, the largest part of the freezing process takes place in a temperature interval between −1 and −8°C (30.2 and 17.6°F). The largest variations of heat capacity occur between −1 and −3°C (30.2 and 26.6°F). Only at temperatures ranging from −20 to −40°C (−4 to −40°F) and below, there is no more measurable change with temperature in the amount of ice present, and the remaining water, if any, can be considered as nonfreezable. However, for practical purposes, a lower limit to the phase-change interval can be defined on the basis of a ratio of ice to total water content of, say, 90%. This choice, in addition to providing an easily applicable criterion, allows one to approximate heat capacity and thermal conductivity curves, above and below the phase-change zone, by means of constant values.

Wide Range of Freezing Configurations

The food processor has several options available when selecting the best freezing format for a given set of product characteristics and marketing objectives. Methods can be classified in several ways, as for example the media used to contact and extract heat from the food substance—air or other gases, liquids, or mechanical contact.

Air-blast systems commonly take the form of large rooms, tunnels, or cells. In a room, the air velocity may be low or range up to 1500 feet (457 meters) per minute. The temperature for air-blast freezing usually ranges from −29 to −40°C (−20 to −40°F). Blast freezing requires longer than other available methods and product quality cannot be assured unless efficiently insulated. Insulation can be improved by using cold storage doors and air curtains. Where insulation is inadequate, frosting will occur on the coils, lowering the refrigerating capabilities. Air-blast freezing also can be effected in a tunnel on a fluidized bed, or on a belt. For individually quick frozen (IQF) products, fluidized-bed, air-blast systems are frequently used. An advantage of the fluidized bed is that it keeps the product in motion and separated during the freezing process.

Plate freezers generally are limited in application to prepackaged products. In this system, the food substance is placed in direct contact with refrigerated metal plates (usually steel or aluminum). Cooling coils are located within the interior of the metal plates. The required contact refrigeration time ranges from about 30 to 90 minutes, depending upon size and nature of food substance.

Liquid-immersion freezing has grown in acceptance during the past few years. This system requires placing the product in a bath of cooling liquid, which must be nontoxic, noncorrosive, have a low freezing point, low viscosity, and high thermal conductivity. Wrapping of the product is required in many cases. Salt solutions and propylene glycol are frequently used. Advantages over air-blast freezing include operational energy savings that result from high heat-transfer coefficients and high heat capacities.

Aqueous freezants, composed of a single solute species, such as an inorganic salt, acid, or base, or an organic compound, such as sucrose, have been proposed and used for freezing fruits, vegetables, and fish. Freezants containing more than one solute also have been suggested, including sodium chloride with minor amounts of calcium chloride or potassium chloride added to water or sea water for freezing fish.

In considering a ternary system (15% sodium chloride, 15% ethanol, and 70% water), among other observations, Cipolletti et al. noted that: (1) freezing times to 0°F (−17.8°C) as fast as 2 minutes were achievable for carrots and peas; (2) photomicrographs showed greatly reduced cell damage as compared with air-blast frozen samples; (3) sodium chloride uptake in products after freezing varied from a minimum of 0.89% for peas to a maximum of 2.06% for carrots; (4) ethanol uptake in products after freezing was small (0.05–0.27%); and (5) no significant organoleptic preference differences were indicated for peas, snap beans, and corn, frozen either by immersion with the salt-ethanol-water medium (with or without blotting) or by air blast. Preference for air-blast frozen carrot samples was indicated by a panel because of the absence of salt. Panels evaluating mixed vegetable samples containing carrots, peas, beans, and corn indicated no exclusive preference between immersion-frozen and air-blast frozen vegetables.

Cryogenic freezing has gained wide acceptance during recent years. Some of the advantages of this method over blast freezing are immediately obvious from Table 1. Because cryogenic freezing is very fast and

TABLE 1. COMPARATIVE PERFORMANCE OF FREEZING METHODS FOR SELECTED SUBSTANCES

| Commodity | Cryogenic freezing | | Blast freezing |
	Liquid freon	Liquid nitrogen	
Strawberry			
Freezing time	3 minutes	5 minutes	900 minutes
Temperature after freeze	−25°C (−13°F)	−28°C (−18°F)	−20°C (−4°F)
Percent weight loss	0.0	1.4	2.7
Mushroom			
Freezing time	3 minutes	5 minutes	180 minutes
Temperature after freeze	−30°C (−22°F)	−26°C (−15°F)	−20°C (−4°F)
Percent weight loss	0.0	1.9	2.5
Beef			
Freezing time	4 minutes	8 minutes	180 minutes
Temperature after freeze	−28°C (−18°F)	−50°C (−58°F)	−20°C (−4°F)
Percent weight loss	0.1	1.4	1.3

Source: "Freezing Equipment Influence on Weight Losses," Sture Astrom, Helsingborg, Sweden.

accomplished at extremely low temperatures (down to −196°C; −320°F), less dehydration occurs. Problems of cell damage, caused by sharp ice crystals formed during slower freezing processes, are largely overcome with short freezing times. Also, the sooner a product is deeply frozen, the sooner will be the halting of bacterial and enzyme degradation.

For cryogenic freezing, nitrogen is used in several forms—as a shower of liquid droplets, as a liquid bath for direct immersion, or as a cold gas. Carbon dioxide is used as a liquid or in solid "snow" form. When used in a tunnel for IQF applications, liquid carbon dioxide can freeze products at a temperature from −62 to −78°C (−80 to 109°F). Fluorocarbons and halocarbons also have been used in conjunction with tunnel and spiral-type freezers that are used in IQF methods.

Additional Reading

Bryan, F.L.: "Application of HACCP to Ready-to-Eat Chilled Foods," *Food Technology*, **70** (July 1990).

Cipolletti, J.C., Robertson, G.H., and D.F. Farkas: "Freezing of Vegetables by Direct Contact with Aqueous Solutions of Ethanol and Sodium Chloride," *J. Food Sci.*, **42**, 4, 911–916 (1977).

Dougherty, R.H.: "Future Prospects for Processed Fruit and Vegetable Products," *Food Technology*, **124** (May 1990).

Lechowich, R.V.: "Microbiological Challenges of Refrigerated Foods," *Food Technology*, **84** (December 1988).

Lund, D.: "Food Processing: From Art to Engineering," *Food Technology*, **242** (September 1989).

Reid, D.S.: "Optimizing the Quality of Frozen Foods," *Food Technology*, **78** (July 1990).

Reineccius, G.A.: "Flavor and Nutritional Concerns Relating to the Quality of Refrigerated Foods," *Food Technology*, **84** (January 1989).

Schwartzberg, H.G. and M.A. Rao: *Biotechnology and Food Process Engineering*, Marcel Dekker, Inc., New York, NY, 1990.

Staff: *Developments in Biological Standardization: Biological Freeze-Drying and Formulation*, Vol. 74, S.Karger Publishers, Inc., Farmington, CT, 1992.

Staff: "Freeze Concentration," *EPRI J.*, **6** (April/May 1992).

Thorne, S.: *Developments in Food Preservation*, Elsevier Science, New York, NY, 1989.

FREEZING POINT. The temperature at which a liquid solidifies under any given set of conditions. It may or may not be the same as one of the following: (a) the melting point, or the temperature at which a solid substance changes from solid to liquid form; (b) the "true freezing point," or the temperature at which the liquid and solid forms of a substance exist in equilibrium at a given pressure, usually one standard atmosphere; or (c) the *ice point*, or the temperature at which a mixture of air-saturated pure water and pure ice may exist in equilibrium at a pressure of one standard atmosphere. The freezing point is not an "equilibrium" property of a substance; it applies to the liquid phase only. It is somewhat dependent upon the "purity" of the liquid, the volume and shape of the liquid mass, the availability of freezing nuclei and the pressure acting upon the liquid.

The freezing point of a solution becomes proportionally lower with an increasing amount of dissolved matter. Therefore, since natural water almost invariably contains some solutes, its freezing point usually is found to be slightly below 0°C. For example, bulk samples of normal seawater freeze at about −1.9°C, or 28.6°F. The *maximum freezing point* is that temperature for a particular composition of a two-component or multicomponent liquid system at which the freezing point is higher than that for any other composition or for the pure components.

The *minimum freezing point* is that temperature for a particular composition of a two-component or multicomponent liquid system at which the freezing point is lower than that for any other composition or for the pure components.

FREEZING-POINT DEPRESSION. The freezing point of a solution is, in general, lower than that of the pure solvent. The depression is proportional to the active mass of the solute. For dilute (ideal) solutions

$$\Delta T = K m$$

where ΔT is the lowering of the freezing point, K, the *cryoscopic constant* for the given solvent, and m, the molality of the solution. This is known as the *Blagden law*.

There are several methods for measuring this depression: In the *Beckmann method* the freezing point of pure solvent and that of solution is measured by a special type of thermometer, the "Beckmann thermometer." The solvent or solution is contained in a double-walled glass apparatus and placed in a freezing mixture not more than 5°C below the freezing point of the solvent. By rapid stirring when the liquid has supercooled about $\frac{1}{2}°$, crystallization is induced and the temperature rises to the freezing point.

In the *equilibrium method* a relatively large amount of solvent crystals are allowed to form and the system allowed to come to equilibrium. The temperature is recorded and some of the solution withdrawn and analyzed.

See also **Cryoscopic Constant**.

FRICTION (Mechanical). Friction is the force between surfaces that opposes their relative motion. The chief causes of friction are the interlocking of the minute irregularities on the rubbing surfaces, adhesion between the surfaces, and the indentation of the softer by the harder body. Friction between solid bodies may be classified as sliding and rolling. The laws of sliding friction were investigated by Coulomb, who found that, approximately and within limits: (1) the friction between two surfaces is slightly greater just before motion begins than when the surfaces are in steady relative motion; (2) the friction is proportional to the force pressing the surfaces together; and (3) it is independent of the area of contact, and (except at start) of the speed of relative motion. The constant ratio of the friction to the force pressing the surfaces together is called the coefficient of friction, some typical values of which are as follows:

Dry wood on dry wood	0.35
Leather on metal	0.55
Iron on stone	0.50
Wood on stone	0.40
Stone on stone or brick	0.65
Well-oiled metals	0.05

By means of such coefficients, it is possible to calculate what the friction will be between two bodies, as a wooden sill on a stone foundation, when the force pressing them together is given.

The angle at which a plane surface must be inclined for a solid block to slide steadily down it is the angle of friction; its tangent is the coefficient of friction between plane and block. Lubrication greatly reduces the coefficient by separating the solid surfaces. Rolling friction, due to the indention of the surfaces in rolling contact, is much less than sliding friction, as illustrated by the use of ball-bearings. The viscosity of liquids and gases is sometimes called "internal friction."

FRIEDEL-CRAFTS REACTION. Aluminum chloride anhydrous, introduced by Friedel and Crafts, is used as reagent, generally in CS_2 solution to avoid rise in temperature, for the preparation of (1) aryl-alkyl hydrocarbons, (2) di- and triphenylmethane and derivatives, and (3) aryl-alkyl and diaryl ketones. Other chlorides, such as those of zinc, iron(III), and tin(IV), are often effective in certain cases.

1. Aryl-alkyl hydrocarbons. The reaction takes place between benzene or its homologues and the alkyl haloid, thus:

$$C_6H_5 \cdot H + Cl \cdot CH_3 \longrightarrow C_6H_5 \cdot CH_3 + HCl$$

| Benzene | Methyl chloride | Toluene | Hydrogen chloride gas evolved |

$$C_6H_4 \genfrac{}{}{0pt}{}{H}{H} \genfrac{}{}{0pt}{}{Cl \cdot CH_3}{Cl \cdot CH_3} \longrightarrow C_6H_4(CH_3)_2 + HCl$$

| Benzene | Methyl chloride | Benzene |

2. Di- and triphenylmethane, derivatives. The reaction takes place between benzene and benzyl haloid or methylene haloid in the case of diphenylmethane, and between benzene and benzal haloid or chloroform in the case of triphenylmethane, thus:

$$C_6H_5CH_2Cl + HC_6H_5 \longrightarrow C_6H_5CH_2C_6H_5 + HCl$$

| Benzyl chloride | Benzene | Diphenylmethane |

$$H_2CCl_2 + \genfrac{}{}{0pt}{}{HC_6H_5}{HC_6H_5} \longrightarrow C_6H_5CH_2C_6H_5 + \genfrac{}{}{0pt}{}{HCl}{HCl}$$

| Methylene chloride | Benzene | Diphenylmethane |

$$C_6H_5CHCl_2 + \genfrac{}{}{0pt}{}{HC_6H_5}{HC_6H_5} \longrightarrow C_6H_5CH \genfrac{}{}{0pt}{}{C_6H_5}{C_6H_5} + \genfrac{}{}{0pt}{}{HCl}{HCl}$$

| Benzyl chloride | Benzene | Triphenylmethane |

$$HCCl_3 + \genfrac{}{}{0pt}{}{HC_6H_5}{HC_6H_5}{HC_6H_5} \longrightarrow C_6H_5CH \genfrac{}{}{0pt}{}{C_6H_5}{C_6H_5} + \genfrac{}{}{0pt}{}{HCl}{HCl}{HCl}$$

| Chloroform | Benzene | Triphenylmethane |

3. Ketones. The reaction takes place between benzene and paraffin or benzenoid acyl haloid thus:

$$CH_5COCl + HC_6H_5 \longrightarrow C_6H_5COCH_3 + HCl$$

| acetyl chloride | Benzene | Acetophenone |

$$C_6H_5COCL + HC_6H_5 \longrightarrow C_6H_5COC_6H_5 + HCl$$

| Benzoylchloride | Benzene | Benzophenone |

The keto-group occupies the position para to alkyl already present. Two acyl groups have been placed in mesitylene to form diacetylmesitylene:

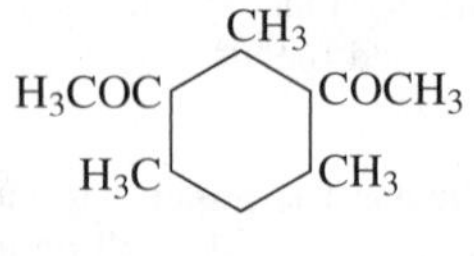

Summarizing: benzene or its homologues plus paraffin-substituted haloid in the presence of aluminum chloride anhydrous react with the elimination of hydrogen chloride. In several cases an intermediate compound of the reactants with aluminum chloride has been identified.

1. Xylene plus benzene to yield toluene, and the reverse, namely, toluene to yield xylene plus benzene. Boiling temperature.
2. Benzene, toluene and homologues chlorinated by reaction with chlorine gas.
3. Benzene sulfinated by reaction with SO_2. Benzene sulfinic acid $C_6H_5 \cdot SOOH$ formed.

FROTHING AGENTS.   See **Classifying (Process).**

FRUCTOSE.   See **Carbohydrates; Sweeteners.**

FUEL. In the conventional sense, a fuel is a material or combination of materials which, when burned with air, produces heat. This heat, in turn, can be used in numerous ways—as in the conversion of water to steam. The steam, in turn, can be used in many ways—as in a steam turbine to produce electricity. Fuels also are burned to obtain explosive or mechanical energy—as in an internal combustion engine where heat per se is an inevitable, but undesired byproduct. The term *fuel* is also used in connection with nuclear reactions—as the material, such as uranium and plutonium isotopes, which undergoes fission and, in so doing, yields heat energy. Fuel also appears in the term *fuel cell*, in which chemical reactions other than what may be considered as conventional combustion are carried out to yield electrical energy.

Conventional fuels may be solids (coal, coke, wood, etc.); liquids (fuel oil, gasoline, alcohol, etc.); or gases (natural gas, synthetic gases, hydrogen, etc.). Where natural fuels are derived from geochemical processes in the earth over long periods of time, as coalification in the production of peat and various grades of coal from prehistoric vegetation, the term *fossil fuel* is applied. The principal fuels in this very large and important class of fuels are coal, petroleum, and natural gas. Where fuels are produced by chemical means, often by synthesis from other materials, the term chemical fuel may be applied, as in the instances, for example, of alcohol (either from natural fermentation or synthesis), hydrogen (from chemical or electrochemical reactions, including the electrolysis of water), and various synthesis gases (water gas, producer gas, town gas, etc.). Rocket fuels generally are chemical-type fuels.

Some of the desired properties of fuels, depending upon particular applications, include heat content, that is, Btus or calories released upon combustion per unit weight or volume of the fuel. Energy density is particularly important where a fuel is used in some form of vehicle where the fuel must be carried and thus part of the fuel is expended simply to transport itself. Cleanliness of burning is of major concern, not only in terms of pollutants that may be produced as the result of combustion, but

TABLE 1. HEATING VALUE OF VARIOUS FUEL SUBSTANCES[A]

Gases	Btu cubic foot at 60°F and 30 inches mercury pressure	Kilogram-calories/gram-molecular weight
Acetylene	1,455	312
Butane	3,200	680
Carbon monoxide	317	67.1
Ethane	1,730	368
Ethylene	1,615	332
Hydrogen	319	68.4
Methane	995	211
Natural gas	975–1,180	207–253
Substitute natural gas (SNG)		
Pipeline quality (high-Btu)	950–1,050	202–224
Low-Btu quality	400–600	85–128

Liquids	Kilogram-calories/gram	Kilogram-calories/gram-molecular weight
Benzene	10.0	782
Ethyl alcohol	7.13	328
n-Heptane	11.5	1,150
n-Hexane	11.5	990
Methyl alcohol	5.34	171
n-Octane	11.4	1,303
n-Pentane	11.6	883
n-Propyl alcohol	8.00	481
Toluene	10.2	934

Solids		Kilogram-calories/gram
Carbon (Amorphous to CO_2)	8.08	(97.0 kg-cal per gram-molecular molecular weight CO_2)
Carbon (Amorphous to CO)	2.49	(29.9 kg-cal per gram-molecular weight CO)
Cellulose	4.21	

[a] Where applicable, higher heating value (water condensed as formed in combustion) figures are given.

also in terms of additional costs of equipment that may be required to process and handle the byproducts, such as flue gas and ash.

When selecting a fuel, the power engineer will consider the heating value of a fuel in terms of cost effectiveness and, in particular, the polluting byproducts of a given fuel. The heating values of several fuels are given in Table 1. Much more detail pertaining to specific fuels and energy sources is given in several separate articles in this encyclopedia. See also **Batteries; Coal; Coal Conversion (Clean Coal) Processes; Combustion (Fuels); Energy; Fuel Cells; Hydrogen (Fuel); Natural Gas; Nuclear Power Technology; Petroleum; Tar Sands**.

FUEL CELLS. The fuel cell, approximately 150 years old, was invented by Sir William Grove in England in 1839. The inventor called it a "gaseous battery" at that time to distinguish the fuel cell from another invention of his, the electric storage battery. The fuel cell is an electrochemical device that directly combines hydrogen and oxygen from air to produce electricity and water. With prior processing, a wide range of fuels, including natural gas and coal-derived synthetic fuels, can be converted to electric power.

The basic process is attractively efficient, basically pollution-free, and inasmuch as single fuel cells can be assembled into stacks of varying sizes, systems can be designed to produce a wide range of output levels and thus accommodate numerous types of applications—large and small.

The initial practical uses of fuel cell power plants date back to the early U.S. space missions—*Gemini* (1965) and *Apollo* (1969). Spurred by successful applications in the space program and by the energy embargo crisis of the early 1970s, fuel cells were among several innovative energy sources researched. Considerable government funding during that period was made available to fuel cell research. When the oil crisis waned, the scale of effort directed toward new energy technologies slowed. During the early fuel cell research period, most concentration was placed on the phosphoric acid fuel cell (PAFC).

When serious concern was expressed over increasing air pollution caused by the combustion of fossil fuels, most of the early attention was given to the generation of sulfur oxides (SO_x) and nitrogen oxides (NO_x). The major thrust was given to ways of reducing and cleaning up these emissions and not a great deal of emphasis was placed on fuel cells because of their essentially unproven practical nature at that time. There were a few exceptions; they are described later in this article.

Initially, most of the research effort in connection with reducing air pollution by fossil fuels was targeted toward reducing NO_x and SO_2 emissions, either by switching to less-polluting fuels, such as natural gas, or by pretreating coal, as well as using various means for cleaning up emissions prior to release to the atmosphere. The serious problem of carbon dioxide emissions was not an initial target, thus awaiting a fuller development of the global warming hypothesis. The latter concern refocused attention on essentially nonpolluting sources, such as the fuel cell. The prospects of ultimate CO_2 regulations and taxes instilled interest in the electric power producers, envisioning that perhaps the fuel cell (also battery and photovoltaic power) could be cost-effective, particularly as improvements in the technology came along.

Also, for several years, the electric power industry has shown increasing interest in improving the efficiency of power distribution. Long transmission lines are a source of power loss. It was envisioned that a practical solution could be that of locating smaller generating facilities at strategic locations. For this type of application, the modular construction of fuel cell-battery-, or photovoltaic-energy conversion units offered considerable attraction. Diesel-operated supplementary generators have been used for many years and also enter into the mix of possibilities, but diesel engines are polluting. It is interesting to note that despite their relatively high generating costs per kWh, a few photovoltaic installations have been made during the early 1990s.

With practical, low- to medium-capacity generating schemes, local generating stations for isolated and sparse geographical areas can conserve major costs in installing and maintaining long-power transmission lines.

In the EPRI Study, a bottom-up approach was developed to define sites and applications where distributed generation may provide high value. Methods first were developed with the cooperation of the Los Angeles Dept. of Water& Power (LADWP). These were tested and refined with the assistance of South West Corporation (CSW) and Oglethorpe Power Corporation (OPC). In the OPC study, the methodology used applied not only to fuel cells, but also to distributed diesels and batteries. More detail is given in the D. Rastler reference listed.

TABLE 1. EMERGING ELECTRIC POWER DISTRIBUTION-GENERATION TECHNOLOGIES

Characteristic	Batteries	Fuel cells[2]	Photovoltaics
Size	500–10,000 kW	500–5000 kW	1–1000 kW
Area	2–4 kWh/ft²	0.44 kW/ft²	100 kW/acre
Timing of Entry	1994–2000	1997–2000	1995–2000
Cost per kW[1]	$600 (1 hr storage) $900 (3 hr)	$1500 down to $1000	$5000 down to $2500
Fuel/Energy	Off-peak electricity at Natural gas, LPG, propane, incremental cost	Solar energy landfill gas, shut-in gas	

Source: Adapted from EPRI data.

[1] Molten carbonate fuel cells at 52–60% electrical efficiency.

[2] For batteries, learned-out costs for first-generation 500 kW plants. Later-generation technologies and/or larger plants are expected to be less expensive. For fuel cells and photovoltaics, cost will decrease as manufacturing production increases.

TABLE 2. DISTRIBUTED BENEFITS FOR 2–MW FUEL CELLS ($/MWH IN 1991)

Benefit	Case studies		
	LADWP	CSW	OPC
Spinning Reserve	1,1	1,8	2.0
Peak Operation	0.6	0.7	1.4
Reserve Margin	na	0.9–1.7	1.7–3.4
Transmission and Distribution Deferral	1.1–7.1	1.7–13	1.5–4.8
Energy Loss Savings	1.9–16	4.1–17.1	3.6
Improved Reliability	0–1.3	2.7–13	0
Low Emissions[1]	8.1–21	0.2–58	0.1–38
Thermal Waste Heat	0–5.8	0–8.4	0–12
Fuel Diversity	0–8.4	8–20	0

Source: Adapted from EPRI data.

[1] The CSW and OPC cases considered NO_x, SO_x, and CO_2. The LADWP study considered NO_x only.

TABLE 3. BENEFIT/COST RATIOS FOR 2-MW FUEL CELLS

Factor	LADWP	CSW	OPC
Gross Levilized Cost ($/MWh)			
Market-entry Unit[1]	73	87	109
Commercial Unit[2]	49	64	85
Range of Distributed by Site ($/MWh)	14–46	22–85	9–64
Cost of Deferrable Resources ($/MWh)	52–60	54–73	92
Benefit/Cost Ratio			
Market-entry Unit	0.9–2.3	0.8–3.4	0.9–1.6
Commercial Unit	1.7–15	1.6–7	1.2–4.3

Source: Adapted from EPRI data. Dollar amounts are in 1990 $.

[1] $ 1500/kW, available in 1997.

[2] $ 970/kW, available in 2000.

As pointed out by D. Rastler (EPRI), "Distributed generation is unlikely to replace future needs for large central station generation. However, if cost performance and reliability targets can be achieved—through volume production—distributed generation can have far-reaching implications with respect to siting future generating resources, ratemaking, and competition." As of 1994, EPRI is aggressively continuing to research, including test site comparisons, fuel cells, batteries, and photovoltaics (solar power). Some tentative findings are reflected in Tables 1, 2, and 3.

Early Experience with the Phosphoric Acid Fuel Cell (PAFC)

As an energy conversion device, the fuel cell is distinguished from a conventional battery by the fact that the electrodes are invariable and catalytically active. Current is generated by reaction on the electrode surfaces, which are in contact with an electrolyte. As a rule, fuel and oxidant are supplied as required by the current load; water is continuously removed.

Single Cell. Under load, the voltage of one individual fuel cell element is less than one volt. Therefore, the assembly of many cells, connected in series as a stack, is required. Each individual cell contains the elements needed for feeding reactants to the electrode surface, and removal of water from the cell, as shown in Fig. 1 for a hydrogen air-cell with acid electrolyte.

The electrode reactions are comprised of the oxidation of hydrogen on the anode (the negative electrode) to hydrated protons with the release of electrons; and on the cathode (the positive electrode) the reaction of oxygen with protons to form water vapor with the consumption of electrons. Electrons flow from the anode through the external load to the cathode and the circuit is closed by ionic current transport through the electrolyte. In an acid cell, the current is carried by protons.

Reactants in this cell need not be pure. Hydrogen may be extracted from fuel mixtures and oxygen from air. Since product moisture is formed in an acid cell on the cathode, the air depleted in oxygen can be used for water removal if the cell is operated at a sufficiently high temperature to vaporize the water as it is formed.

The electrode has a central function in cell operation. In its catalyzed layer, it provides a large number of sites where gases and electrolyte can react. By virtue of a porous configuration, fast reactant transport and removal of inerts and products moisture is possible. The electrode also provides a path for current to flow to the terminals and serves to contain the electrolyte. The latter not only provides ionic conduction, but also assures separation of the reactants.

Cell Voltage. The cell voltage and the free energy of the underlying reaction are defined by

$$U = \frac{-\Delta F}{nF}$$

where U = theoretical cell voltage; n = number of electrons transferred in the reaction; and F = Faraday constant.

Since $\Delta F = \Delta H - T \Delta S$, it follows that, depending upon the value of ΔS the electrical energy to be derived from the cell, can be larger or smaller than the energy ΔH obtained by direct combustion of the fuel. ΔH = reaction enthalpy for the current-generating reaction; ΔS = entropy change; and T = absolute temperature.

For the hydrogen-oxygen couple, the corresponding theoretical cell voltage at 25°C is 1.23 volt if liquid coater is formed; or 1.18 volt if the product water is vaporized.

The thermodynamically possible conversion efficiency, however, is only partly realized in a practical fuel cell. Two basic losses are encountered: (1) the ohmic loss and (2) the electrode polarization, that is, the deviation of

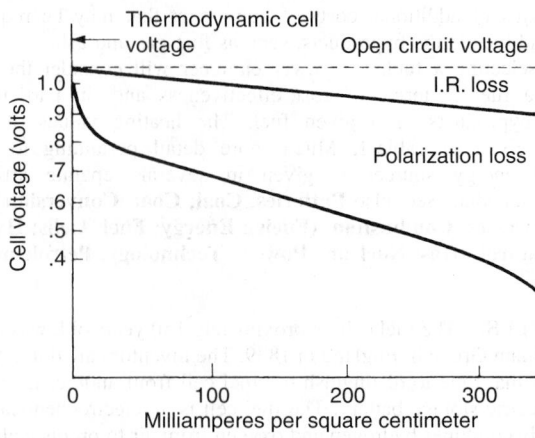

Fig. 2.   Current voltage characteristics of hydrogen-air cell with phosphoric acid electrolyte. Operating temperature is 125°C (257°F)

the actual from the thermodynamic electrode potential. The polarization is the result of the irreversibility of the electrode process, that is, the activation polarization and the voltage loss, which develops from concentration gradients of the reactants. This leads to the current-voltage characteristics as shown in Fig. 2.

Phosphoric Acid Matrix Cell. In the matrix-type cell construction, a limited amount of electrolyte is trapped in a microporous structure by capillary forces. As a result, thin, highly porous, and comparatively low-cost electrodes can be used inasmuch as the electrodes are not required to contain the electrolyte. In practice, for electrolyte absorption, a thin layer of plastic-bonded silicon carbide powder (0.3 mm thick) is applied to the electrode surface. See Fig. 3. Single cells are sandwiched between carbon plates with a suitable pattern for current collection and grooved for reactant distribution. The directions of flow channels for air and hydrogen fuel are perpendicular to each other. Air is used to remove the product water as it is generated and air may serve to remove reject heat. In large cells, in order to minimize temperature gradients, heat is removed through cooling plates or coils located in the stack, either by recirculation of liquid coolants, or by generation of steam.

Diversity of Fuels. The reactants converted in the fuel cell to electric power are hydrogen and oxygen from air. Hydrogen, however, is not the primary fuel source in many modern cells. It is obtained from selected fossil fuels to be substituted eventually by coal-derived synthetic fuels. These fuels are converted by suitable processing, such as steam reforming, into a hydrogen-containing gas stream from which hydrogen is extracted in the fuel cell for power generation.

Although natural gas feedstock currently is preferred by some demonstration plants, alternative fuels, such as light distillates, coal gas, and fuel-grade methanol may be used. Methanol can be steam reformed at relatively low temperatures and, for this reason, can be adapted to smaller, transportable fuel-cell power plants of the type desired for certain military and commercial gear.

4.5 MW Demonstration Plant

A process flow schematic of the demonstration unit[1] installed at the Tokyo Electric Power facility at Goi, Japan and in operation since 1983 is shown in Fig. 4. Fuel cell generators have three unique major subsystems that are unfamiliar to electric utilities: (1) A fuel processing subsystem, (2) a fuel cell power section, and (3) a power conditioning subsystem. See Fig. 5.

Natural gas feedstock enters the fuel processing subsystem at about 65 psig (4.5 atm). The fuel is first processed in hydrodesulfurizing unit (HDS) and zinc-oxide (ZnO) beds to remove any sulfur compounds. The desulfurized fuel is mixed with process steam and preheated to about 850°F (454°C) before entering the reformer, which consists of reactor tubes containing a

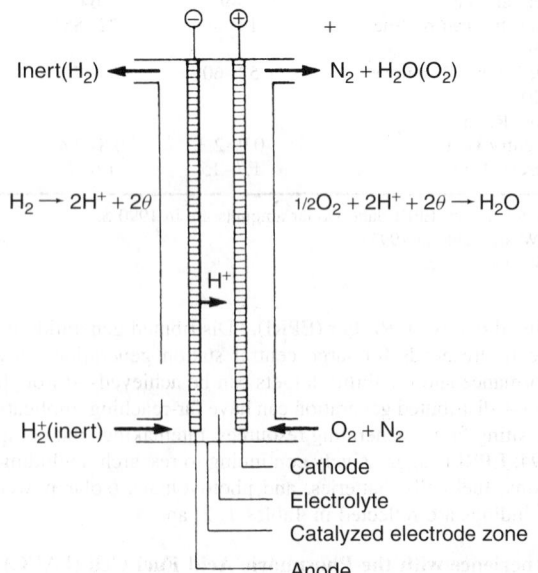

Fig. 1.   Principles of operation of the hydrogen-air cell with acid electrolyte. Product water is removed by the flowing air

---

[1] Consolidated Edison Company of New York installed a similar 4.8 MW unit in New York City, but after a few years, the project was closed down. The demonstration did not see the production of power as planned because the acid electrolyte of the fuel cells became depleted due to extended program delays. An attempt to replenish the aid was unsuccessful. However, must was learned during the engineering, licensing, construction and start-up phases of the program.

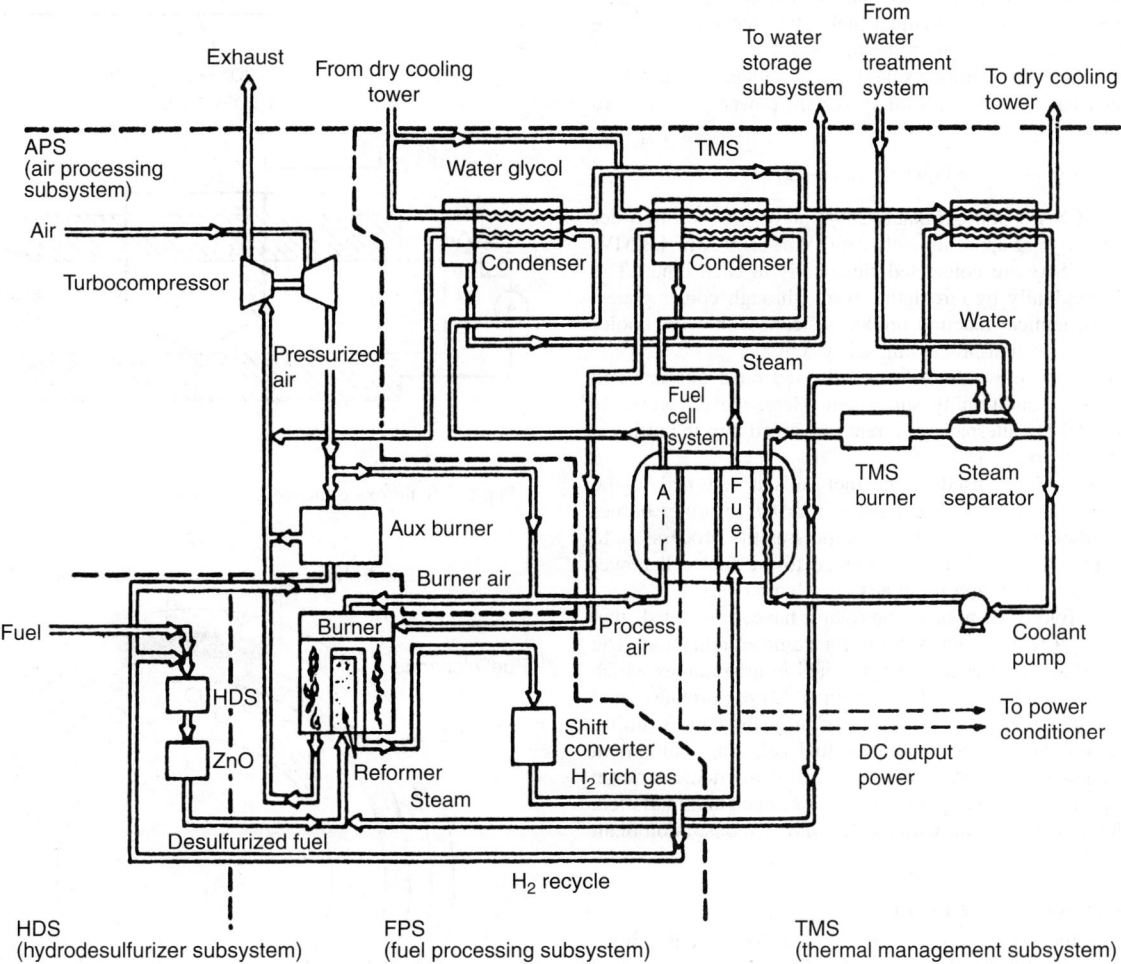

Fig. 4.   Process flow diagram of 4.5 MW fuel cell power demonstration plant installed at the Tokyo Electric Power Company, Goi, Japan

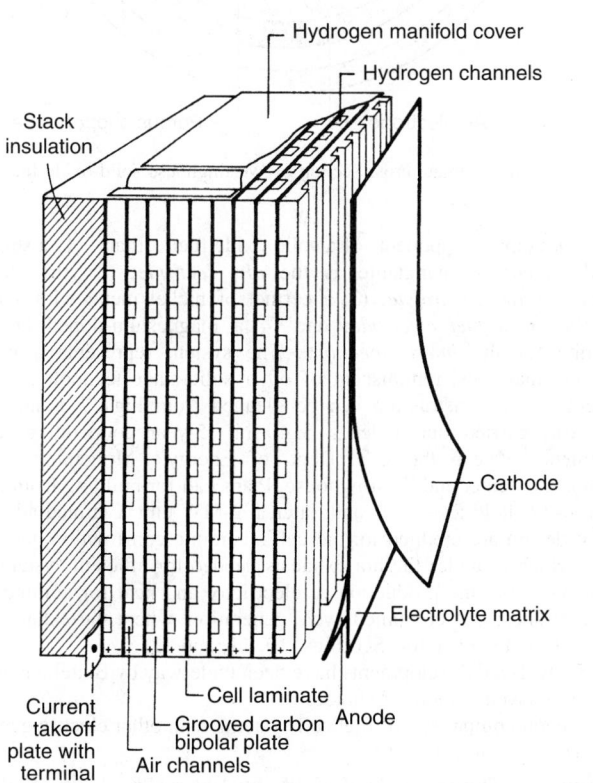

Fig. 3.   Stack section of phosphoric acid matrix cell. Operation and design of cell are exceedingly simple. Product water is removed with air

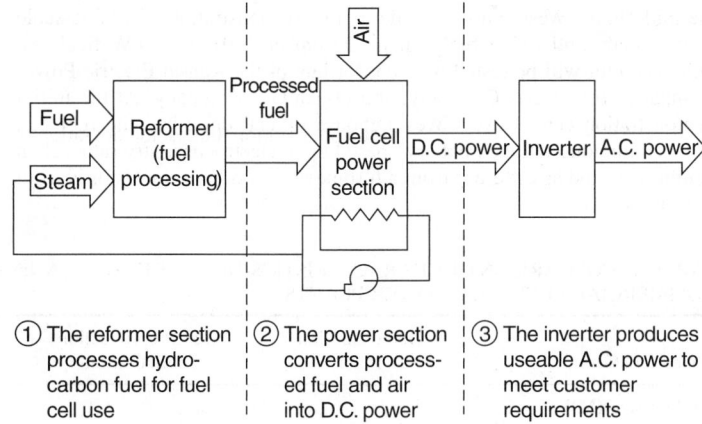

① The reformer section processes hydrocarbon fuel for fuel cell use

② The power section converts processed fuel and air into D.C. power

③ The inverter produces useable A.C. power to meet customer requirements

Fig. 5.   Three principal subsystems required to convert hydrocarbon fuel to electric power

nickel catalyst. This converts the natural gas and steam mixture into a synthesis fuel gas consisting of hydrogen, carbon monoxide, carbon dioxide, and water. The heat for this reaction is provided by burning the unused hydrogen in the exhaust from the fuel cell power section anode. The processed fuel exits the reformer at about $1000°F$ ($538°C$) and enters the shift converter, which further enriches the fuel's hydrogen content. In the two-stage shift converter, a catalyst converts the fuel's carbon monoxide and steam into hydrogen and carbon dioxide. The processed fuel, which now contains about 70% hydrogen, is delivered to the fuel cell power section.

The power section consists of twenty cell-stack assemblies (CSAs). Each CSA contains approximately 450 individual fuel cells, stacked one above the other. Manifolds on the CSAs direct the fuel gas to the anode side of the fuel cells and the air to the cathode side. Electrochemically, the CSAs convert the hydrogen and oxygen into direct-current power according to the following reaction:

$$2H_2 + O_2 \longrightarrow dcPower + Heat + 2H_2O$$

The output of each CSA is approximately 280 V at 850 A (dc). In order to deliver a desired voltage of approximately 2800 V at 1700 A of 4.8 MW dc at full power, the CSAs are connected electrically to each other. The CSAs are cooled individually by circulating water through cooler plates. Heat from the electrochemical reaction produces steam within the cooler plates, which is used in the fuel-processing subsystem.

Direct current from the power section is collected on a bus bar and connected to the power conditioning subsystem. Here, the dc power is converted to 3-phase, 50 Hz alternating current that is fed into the utility's 66 kV transmission network.

Any hydrogen-rich fuel not used in the fuel cells is returned to the reformer burner. The pressurized exhaust gases from the reformer burner are expanded in a turbocompressor, which compresses the process air to about 37 psig (2.4 atm). Process air is distributed to the fuel cell power section, to the reformer burner, and to an auxiliary burner.

Operators of the Tokyo demonstration plant have concluded that phosphoric acid fuel cell technology is ready for commercialization. The project demonstrated that: (1) fuel cells can be sited in urban areas which are regulated by strict environmental constraints; (2) performance and operational characteristics were very close to design goals; and (3) utility personnel can efficiently operate and maintain fuel cell plant equipment with minimal additional training. As a consequence of the demonstration plant success, a new 11-MW power plant will be developed and marketed. A comparison of the new PC23 Unit with the 4.5 MW demonstration plant is given in Table 4.

Shift of Emphasis to Solid Oxide Fuel Cells

Fluid electrolytes in fuel cells (and batteries as well) present unique handling problems that tend to be alien to the usual problems encountered in managing an electric utility. Consequently, a solid electrolyte is an attractive concept. Westinghouse has developed a solid-oxide fuel cell, as shown schematically in Fig. 6. Under development continuously since the mid-1980s, Westinghouse made its first demonstration of a large-scale commercial solid oxide fuel cell in December 1991, a 25-kW field test unit. The unit will be tested by a consortium of the Kansai Electric Power Company, Tokyo Gas Company, and Osaka Gas Company. At the initial demonstration ceremony, a Westinghouse official described the fuel cell as a "continuously-fueled battery" based upon electrochemistry rather than combustion and as different from a turbine generator as a transistor is from a vacuum tube.

TABLE 4. COMPARISON OF CHARACTERISTICS-DEMONSTRATION AND COMMERCIAL FUEL CELL POWER PLANTS

Characteristics	Demonstration Plant	PC23 Preproduction Configuration
Module size (MW)	4.5	11
Heat rate (Btu/kWh) HHV[1]	9300	8300
Power rate (%)	25–100	30–100
Plant operating pressure (psia)	50	120
CSA[2] component size (ft^2)	3.7	10
Plant area (acres)	0.8	0.8–1.0
Start-up time (hours)	4	6
Emissions (lb/10^6 Btu)		
NO$_x$	0.02	0.003
SO$_x$	0.00003	trace
Smoke	none	none
Fuel	natural gas, naphtha	Natural gas, synthetic natural gas, light petroleum distillates; medium-Btu gas and methanol with additional equipment

[1] High heating value.
[2] Cell-stack assembly.

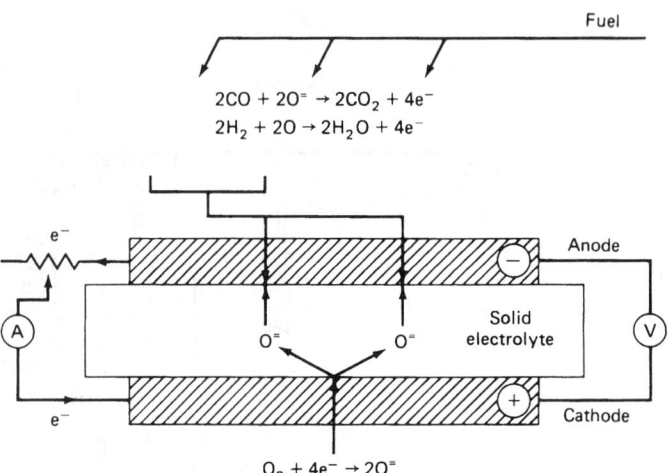

Fig. 6.   Solid-oxide fuel cell. Not to scale. (*Westinghouse Electric Corp*)

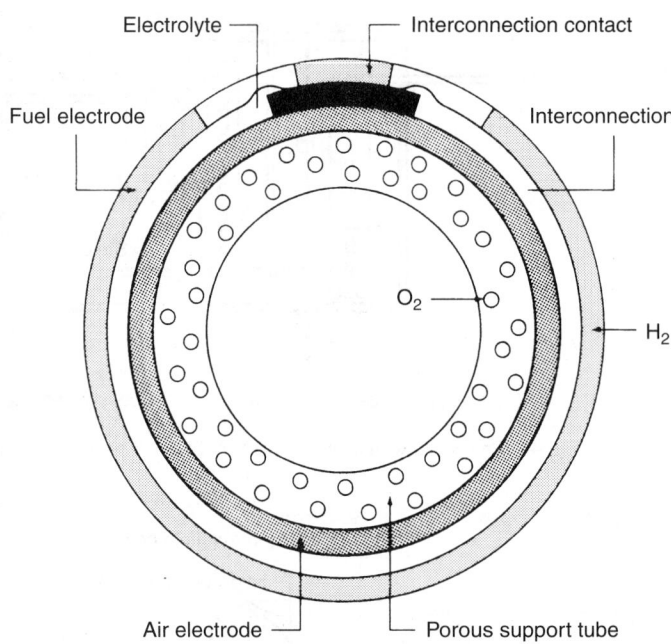

Fig. 7.   Approximate cross section of Westinghouse solid-oxide fuel cell

The principal components of a solid oxide fuel cell are: (1) a strontium-doped lanthanum manganite as the *air electrode*; (2) yttria-stabilized zirconia as the *electrolyte*; (3) a cermet of nickel metal with stabilized zirconia for a *fuel electrode*; and (4) a magnesium-doped lanthanum chromite for the *interconnections*. The system represents a grouping of exotic materials, reminiscent of other solid-stage technologies. Some concept of the construction can be gleaned from Figs. 7, 8, and 9. The work was carried out under a $140 mil, 5-year cooperative research agreement between the U.S. Dept. of Energy's Morgantown Energy Technology Center and Westinghouse. Plans call for building from three to five 100-kW field tests units and, later, a 2-MW unit. Claimed advantages for the design are credited to: the cell's ceramic construction and tubular shape, which provides the hot exhausts needed for efficient systems, such as cogeneration; the production of electricity and heat for commercial or industrial sites; and combined-cycle generation, where the exhaust drives a steam turbine-generator. See Figs. 10, 11, and 12.

Other fuel-cell developments have been underway by Battelle Northwest and Brookhaven National Laboratory.

A general comparison of fuel cell systems with other electric generating systems is shown in Fig. 13.

Miniature Thin-Film Fuel Cells. In 1990, C.K. Dyer (Bell Communications Research, Morristown, New Jersey) reported the successful construction of a tiny electrochemical device (unconventional fuel cell)

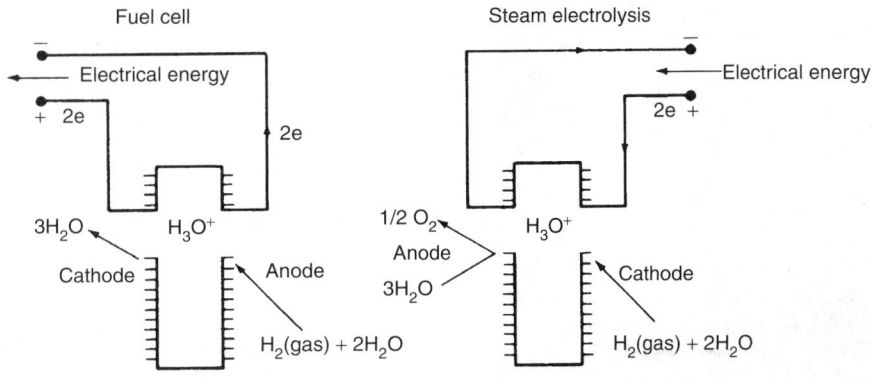

Fig. 9. Schematic representation of principle utilized in a H_3O^+-β/β-Al_2O_3 electrolyte fuel cell and steam electrolysis cell. As proposed by researchers at McMaster University (Canada), oxygen and wet hydrogen are supplied and power is produced by the migration of H_3O^+ through the electrolyte. This is the reserve of steam electrolysis where steam and power are supplied to both sides of the electrolyte. The passage of current via H_3O^+ ions results in oxygenated and hydrogenated steam

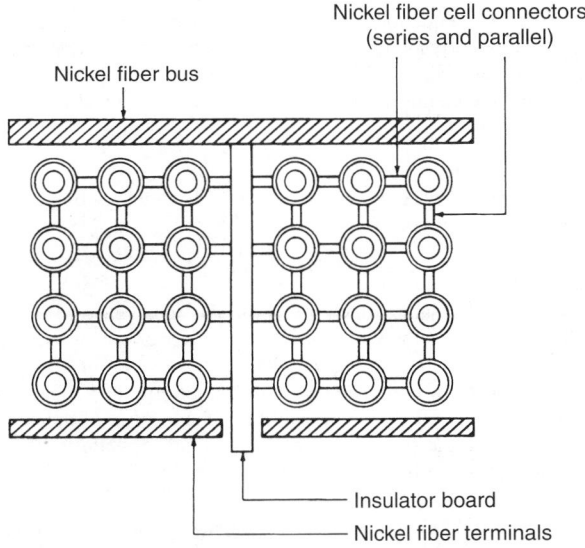

Fig. 8. Cross section of interconnection arrangement to form a 24-cell bundle solid-oxide fuel cell generator. Three cells are connected in parallel and eight in series. (*Westinghouse design approximation*)

Fig. 10. Researcher, Joe Makiel, monitors operation of a 20 kW solid oxide fuel cell generator while it is tested on a variety of hydrocarbon fuels. (*Westinghouse Electric Corporation*)

Fig. 11. Assemblers Dionne Davis and John Sige clean excess nickel anode material from ceramic tubes which are the heart of the Westinghouse solid-oxide fuel cell. (*Westinghouse Electric Corporation*)

power supply, comparable to a small battery where the cost of the fuel is not important." The device consists of a porous aluminum-oxide membrane (2,000 to 5,000 angstroms thick) that is "sandwiched" between two thin platinum films that serve as electrodes. The device develops approximately 1V between its electrodes (a few milliwatts of power per sq. cm) when it is exposed to a mixture of air and hydrogen at room temperature. In an early state of development, the miniature fuel cell remains to be proved as commercially viable. The relative simplicity of the design, coupled with its small size, may have considerable potential in term of high-speed mass production.

The cooperation of J. R. Benke, Westinghouse Electric Corporation, Pittsburgh, Pennsylvania and D. Rastler, Electric Power Research Institute, Palo Alto, California is gratefully acknowledged

Additional Reading

Abelson, P.H.: "Applications of Fuel Cells," *Science*, **1469** (June 22, 1990).

Caruana, C.: "Electrical Vehicle Research," *Chem. Eng. Progress*, **11** (February 1993).

Dicks, A. and J. Larminie: *Fuel Cell Systems Explained,* John Wiley & Sons, Inc., New York, NY, 2000.

Doughty, D.H., H. Brack, K. Naoi, and L.F. Nazar: *New Materials for Batteries and Fuel Cells,* Materials Research Society, Warrendale, PA, 2000.

Houston, B.: "Molten Carbonate Fuel Cell Project," *Chem. Eng. Progress*, **12** (September 1990).

Koppel, T.: *Powering the Future,* John Wiley & Sons, Inc., New York, NY, 1999.

Kordesch, K. and G. Simader: *Fuel Cells and Their Applications,* VCH Publications, Inc., New York, NY, 1996.

Mugerwa, M.N., J. Leo, and M.J. Blomen: *Fuel Cell Systems,* Kluwer Academic Publishers, Norwell, MA, 1999.

that may find application for furnishing power to microscopic electronic components used in portable power supplies and information processing equipment. The researcher observes. "The device is purely a convenience

Fig. 12.   Technician Terry Sickeler adjusts nickel felt between bundles of the tubular solid-oxide cells that are central to the generating unit. The conducting felt interconnects the tubes electrically while allowing room for expansion when heated. (*Westinghouse Electric Corporation*)

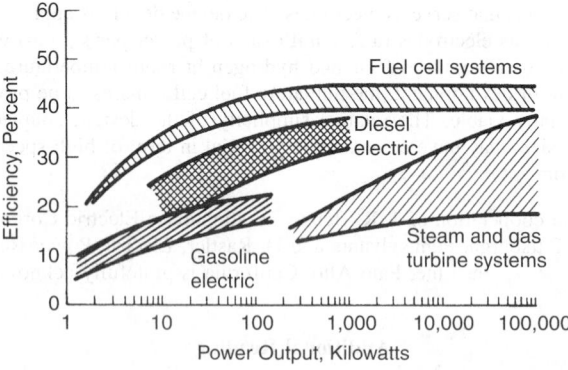

Fig. 13.   Thermal efficiency of fossil-fuel operated fuel-cell power plants compares favorably with conventional means of energy conversion. The efficiency is reduced in small units mainly because of losses in the fuel processing

Nguyen, Q.M. and T. Takahashi: *Science and Technology of Ceramic Fuel Cells*, Elsevier Science, New York, NY, 2000.

Peterson, J.: "Microchip Power from a Shrunken Fuel Cell," *Sci. News*, **85** (February 10, 1990).

Rastler, D.: "Distributed Generation," *Electric Power Research Institute J.*, **28** (April/May 1992).

Savage, D.: *Fuel Cells: New Materials and Concepts Open Major Markets*, John Wiley & Sons, Inc., New York, NY, 1998.

Staff: "Fuel Cells—New Concepts," *Westinghouse Technology*, **21** (April 1989).

Staff: "Energy for a Cleaner Environment (Fuel Cell)," *Westinghouse Technology*, **8** (Summer 1990).

FUEL CONVERSION FACTOR.   See **Nuclear Reactor**.

FUEL CYCLE.   See **Nuclear Reactor**.

FUEL (Hydrogen).   See **Hydrogen (Fuel)**.

FUEL (Nuclear).   See **Nuclear Reactor**.

FUELS.   See **Coal**; **Natural Gas**; **Petroleum**.

FUGACITY.   Only perfect gases obey exactly the ideal gas law, which is the basis for the derivation of many other equations for the properties of gases. Therefore we cannot substitute the measured pressure of real gases for the p term in such equations without more or less inaccuracy. Since, however, calculations are simplified by using ideal equations for real gases, the quantity fugacity is defined as the equivalent pressure of a real gas for which the ideal gas equations are valid, so that by tabulating calculated values of fugacity corresponding to measured pressures for real gases, we can use the relatively simple equations derived for real gases.

For example, the chemical potential for a mixture of ideal gases can be written in the form

$$\mu_i = u_i^*(T) + RT \ln p_i$$

where p_i is the partial pressure of component i. By analogy one may write for a mixture of real gases

$$\mu_i = \mu_i^*(T) + RT \ln p_i^*$$

where $\mu_i^*(T)$ is the same function as for the ideal gas, while all the effects of molecular interactions (that is, of the departure of the real gas from ideality) are included in the p_i^*. This function $p_i^*(T, p, n_1, \ldots n_c)$ is called the fugacity of component i. This definition, due to G.N. Lewis, permits the preservation for real gases of the general form of the equations for ideal gases, with the fugacities replacing partial pressures.

In the lower pressure limit, p_i^* reduces to p_i.

FUKUI, KENICHI (1918–1998).   A Japanese professor who was corecipient of the Nobel prize for chemistry along with Roald Hoffmann ins 1981. His work involved quantum mechanical studies of chemical reactivity. Fukui's entire career has been at Kyoto University.

FULLER'S EARTH.   A fine-grained earthy substance similar to clay, both in appearance and composition, but lacking the usual plasticity, possessing a higher water content, and usually high in magnesia. The material consists mainly of hydrated aluminum silicates, such as the clay minerals, montmorillonite and palygorskite. Generally, it is believed that fuller's earth was formed as a residual deposit as the result of decomposition of rock in place, perhaps by the devitrification of volcanic glass. The color of the material ranges from light brown through yellow and white to light and dark green. Fuller's earth is used for decolorizing oils, degreasing raw wool, and as a natural bleaching agent.

FUMARIC ACID.   See **Isomerism**.

FUME.   A suspension of fine solid or liquid particles (0.2 to 1 micrometer in diameter) in a gas. Technically, fumes are colloidal systems formed from chemical reactions, such as combustion, distillation, sublimation, calcination, and condensation.

FUNGICIDE (Pyridine).   See **Pyridine and Derivatives**.

FUNGICIDES.   The fungicides are a heterogeneous group of chemicals that mitigate, inhibit or destroy fungi. Most chemicals, except the very inert ones, are fungitoxic if present in sufficient quantity, but they usually are not designated as fungicides unless they are effective at nominal dosages of 1000 ppm or less in aqueous suspensions. Those chemicals that inhibit spore germination or mycelial growth without destroying the fungous body are properly known as fungistats, although in common usage they are referred to as fungicides.

The commercial fungicides are indispensible to the welfare of people in preventing or curing the diseases of plants, man and animals and in suppressing the deterioration of stored agricultural produce, material and structures made of cellulosic, lignified or plastic materials. Such diseases as athlete's foot, skin mycosis and ringworm of the human scalp, pulmonary infection of fowl, and moist eczema of dogs are amenable to control by fungicides. The major use of fungicides, however, is in preventing plant diseases such as the leaf blights, powdery mildews, downy mildews, rusts, anthracnoses, fruit rots, fruit scab and stem cankers by spray or dust application of protective or eradicant fungicides. They are also used as soil fumigants and ground sprays to destroy spores and mycelium in their natural habitat before they can attack plants, to disinfest seed known to bear spores or mycelium of smut and other types of fungi, and to protect seed from decay and damping-off organisms in the soil. Other uses include impregnation of fabrics to prevent mildewing and decomposition when in contact with moist substrates, impregnation of wallpaper, paints and leather goods to suppress mildew, treatment of structural timbers, piles, fenceposts, etc. to prevent dry rot and decay, and incorporation into bread to suppress mold growth.

About 100 fungicides are required for these various uses in the United States. The principal ones are sulfur; lime-sulfur (polysulfides of calcium); copper sulfate (or its equivalent in the oxides, basic sulfates, oxychloride and other relatively insoluble copper compounds); creosote products and zinc chloride, both used as wood preservatives; and a wide variety of organic compounds. Among the latter are several dithiocarbamates, such as ferbam and zineb, and other thio compounds, like N-(trichloromethylthio)-phthalimide (folpet); cis-N((trichloromethyl)thio)-4-cyclohexane-1,2-dicarboxyimide (captan); and 8-hydroxyquinoline.

Prior to 1939, the inorganic sulfur and copper compounds were used almost exclusively as spray and dust materials and the copper and organic mercury compounds as seed treatments. Sulfur had been used from before the time of the Romans in various plant prescriptions and had been used for powdery mildews since the beginning of the 19th century. Copper sprays were introduced in 1882 as Bordeaux mixture for control of downy mildew on grapes, and as a copper carbonate seed treatment in 1917. Lime-sulfur was developed as an apple spray in 1906. A new era in fungicides was initiated in the period 1934–1939 with the announcement of the dithiocarbamate and quinone fungicides which indicated the potentialities of organic compounds.

Mercurial fungicides were abandoned in 1971 because of environmental pollution and hazard of conversion into poisonous methyl mercury.

The intensive search for new organic fungicides has continued for over 30 years. The technique employed in most laboratories is to make an extensive survey of various structures by empirical methods to locate materials that suppress spore germination on glass slides or prevent mycelial growth on nutrient agar plates or rolled tubes. Those materials having an ED_{50} (effective dose for 50% inhibition) in the order of 10 ppm are further tested in use applications.

The mechanism of action of the fungicides has been only partially solved. Sulfur deposits volatilize and within two to three minutes after coming into contact with a spore the sulfur begins to be released as hydrogen sulfide. It was formerly thought that hydrogen sulfide was the lethal agent, but it is now evident that it is the product rather than the cause of fungus destruction, probably because the hydrogen transport system of the sulfur-sensitive spore becomes overtaxed after about 10,000 ppm of sulfur has been reduced. In any event, sulfur appears unique in that its lethal effect does not depend upon its accumulation inside or on the spore.

The organic fungicides have tremendous affinity for spores. For example, within 30 seconds after spores are placed in suspension containing 2 ppm of 2-heptadecyl-2-imidazoline they will remove up to 6,000 ppm of their body weight. There is reason to believe that this attribute may very well depend upon the compound's lipoid solubility induced by the 17-carbon chain in the 2-position. Homologs containing longer or shorter chains are less fungitoxic and fungitoxicity in this series is directly correlated with ability to fractionate into the lipoid phase of an aqueous-lipoid system.

Once the fungicides penetrate to the cell membrane or into the cytoplasm they may operate by devious means to disrupt vital functions. There is substantial evidence that the quinones immobilize the sulfhydryl and imino prosthetic group of enzymes. The 8-hydroxyquinoline and dithiocarbamate compounds are active against copper and other metallic members of an enzyme system, presumably by their ability to chelate metals. Heavy metals such as mercury affect certain enzymes such as amylases and may serve as general protein precipitants.

The best evidence available indicates that most, if not all, fungicides are not particularly specific in their action on vital cell systems and may react with nonvital molecules; thus a large percentage of them is detoxified before an essential biochemical process can be effected. The actual dosage required on a spore weight or mycelial basis for a lethal effect has been determined for very few compounds but most of those investigated by Miller and McCallan, who used radioactively labelled molecules, were not lethal until 5,000 to 20,000 ppm were accumulated by the spores.

There is appreciable specificity in the action of fungitoxicants on different types of fungi. Ferric dimethyldithiocarbamate, for example, is much more effective than sulfur against the cedar-apple rust fungus (*Gymnosporangium juniperi-virginian-nae*) but has no particular advantage in control of apple scab (*Venturia inaqualis*). Ferbam is also superior to copper sprays against anthracnose of tomato (*Colletotrichum phomoides*) but is totally inadequate against *Alternaria* and *Phytophthora* leaf blights of this crop. Insofar as is known at present, specificity is only qualitative and not absolute so fungicides may be detected in a general group of candidate compounds with reasonable accuracy by measuring their effects on two or three indicator species of fungi.

See also **Pyridine and Derivatives**.

GEORGE L. McNEW
Boyce Thompson Laboratories
Yonkers, New York

FUNK, CASIMIR (1884–1948). Born in Poland and later becoming an American citizen, Funk in 1911 isolated a food factor, extracted from rice hulls, that he found to be a cure for a disease caused by malnutrition (beri-beri). Believing this to be an amine compound essential to life, he coined the name *vitamine*, from which the final *e* was later dropped. The various types and functions of vitamins were not differentiated until some years later as a result of the work of McCollum, Szent-Gyorgi, R.J. Williams, and others.

FURAN AND RELATED COMPOUNDS. Furan [CAS: 110-00-9]. C_4H_4O, contains a ring of 1 oxygen and 4 carbons, with 1 hydrogen attached to each carbon:

Beta, prime HC———CH Beta
Alpha, prime HC   CH Alpha
O

Furan is a colorless liquid, boiling point 32°C, insoluble in water, soluble in alcohol or ether. Furan vapor produces a green coloration on pine wood moistened with hydrochloric acid. Furan may be made from mucic acid, COOH(CHOH)$_4$COOH, by dry distillation into pyromucic acid, $C_4H_3O \cdot COOH$, and then heating the latter under pressure at 270°C. Furan derivatives are known, namely, methyl, primary alcohol, aldehyde, carboxylic acid, in which the group attachment is at carbon number 2:

HC——CH   HC——CH
HC   C·CH$_3$ HC   C·CH$_2$OH
O   O

Sylvane   Furfuryl alcohol
Alpha-methyl   Alpha-furyl
furan;   carbinol;
boiling point   boiling point
65°C   170°C (750 mm)

HC——CH   HC——CH
HC   C·CHO   HC   C·COOH
O   O

"Furfural,"   Pyromucic acid,
Alpha furfur-   furoic acid,
aldehyde;   furane-alpha-
boiling point   carboxylic acid;
160°C (740 mm)   melting point 133°C
                    boiling point 230°C

See also **Furfuraldehyde**.

Coumarone is benzofuran, C_8H_6O or $C_6H_4CH{:}CH \cdot O$, a colorless liquid,

boiling point 173°C, and diphenylene oxide is dibenzofuran, $C_{12}H_8O$ or $C_6H_4 \cdot C_6H_4 \cdot O$, a white solid,

melting point 81°C, boiling point 288°C.

Coumarin is 1,2-benzopyrone, $C_9H_6O_2$ or $C_6H_4OCOCH{:}CH$, a white solid,

melting point 67–68°C, boiling point 301°C.

Gamma-pyrone, $C_5H_4O{:}O(4)$, is a gamma-ketone (4) containing a ring of 1 oxygen and 5 carbons with 1 hydrogen attached to each of 4 carbons, namely, 2,3,5,6.

Gamma-pyrone is a colorless liquid, melting point 32°C, boiling point 218°C.

Pyrone derivatives are known, e.g.,

Alpha, alpha prime dimethyl-gamma-pyrone

Chelidonic acid gamma-pyrone-alpha, alpha-prime-dicarboxylic acid

Chromone is benzo-pyrone, $C_9H_6O_2$, a

white solid, melting point 59°C, and chromane is a colorless liquid, boiling point

214°C, 750 mm.

Flavone is phenyl chromone:

white solid, melting point 97°C.

Xanthone is dibenzon-pyrone, $C_{13}H_8O_2$ or

or

while solid, melting point 174°C, boiling point 351°C, and xanthene is

white solid, melting point 100°C, boiling point 315°C. From chromone and xanthone a number of yellow dyes are made, which dyes also occur in nature. Such dyes are chrysin, fisetin, buteolin, morin, quercetin, rhamnetin.

FURFURALDEHYDE. [CAS: 98-01-1]. $2\text{-}C_4H_3O \cdot CHO$, formula weight 192.16, colorless, odorous (pungent, almond-like) liquid aldehyde, mp -38.7°C, bp 161.7°C, sp gr 1.159. Also known as 2-furaldehyde or 2-furancarboxalde hyde, this compound becomes brown in color when in contact with air. Furfural is modestly soluble in H_2O (up to 8% by weight at 20°C) and is miscible in all proportions with alcohol and ether. At atmospheric pressure, a mixture of furfural and H_2O (65%) forms a minimum-boiling azeotrope when a distillation temperature of 97.9°C is reached.

Aside from a darkening in color, furfural is relatively stable thermally and does not exhibit changes in physical properties after prolonged heating up to 230°C. The reactions of furfural are typical of those of the aromatic aldehydes, although some complex side reactions occur because of the reactive ring. Furfural yields acetals, condenses with active methylene compounds, reacts with Grignard reagents, and provides a bisulfite complex. Upon reduction, furfural yields furfural alcohol; upon oxidation, it yields furoic acid. It can be decarbonylated to furan.

Furfural is obtained commercially by treating pentosan-rich agricultural residues (corncobs, oat hulls, cottonseed hulls, bagasse, rice hulls) with a dilute acid and removing the furfural by steam distillation. Major industrial uses of furfuraldehyde include: (1) the production of furans and tetrahydrofurans where the compound is an intermediate; (2) the solvent refining of petroleum and rosin products; (3) the solvent binding of bonded phenolic products; and (4) the extractive distillation of butadiene from other C_4 hydrocarbons.

When pentoses, e.g., arabinose, xylose, are heated with dilute HCl, furfuraldehyde is formed, recognizable by deep red coloration with phloroglucinol, or by the formation, with phenylhydrazine, of furfuraldehyde phenyl-hydrazone $C_4H_3O \cdot CH{:}NNHC_6H_5$, solid, mp 97°C.

FUSION ENERGY. The ultimate probability of creating on Earth sources of essentially unlimited quantities of usable energy in safe nuclear fusion reactors at comparatively low cost and free of pollutive byproducts is so overwhelmingly attractive that the concept has supported a large research effort of international scope over a period approaching a half-century. Considering the magnitude of the technical problems facing fusion power scientists, the progress made to date is impressive.

The achievement of energy independence through nuclear fusion would provide earthly cultures with advantages and benefits of a magnitude unrivaled by but few past technological breakthroughs. The point—the goal of energy independence is not one of simply satisfying the curiosity of theorists and academicians, but rather the success of this endeavor would affect the very fabric of society. *This is the true driving force behind fusion energy research.*

Philosophers, over the years, have stressed that great returns almost always require great investments, not only of a monetary-materials nature, but also of brain power and perhaps, above all, patience. Successful fusion reactors, as of 1994, may be within the reach of a comparatively few more years of effort, or additional decades may be required.

Stated simplistically, capturing energy from nuclear fusion is like that of building miniature suns on Earth, where temperatures and other physical parameters of dimensions that stretch the imagination are involved. Superbly durable materials are required and remain to be developed. Scaled-down models for experimental purposes are large and consequently costly to construct. Thus, it is clear that fusion power research is no abode for the short-term pragmatist, the irresolute, or the impatient.[1]

The fusion of deuterium and tritium nuclei into a helium nucleus plus a neutron results in the release of approximately 1000 times more energy than that required to cause the reaction. The deuterium isotope is readily abundant, and the tritium isotope can be produced as a part of the fusion fuel cycle, so that the deuterium-tritium (DT) fuel for fusion appears to be virtually inexhaustible. In addition, making fusion work is so difficult that, if anything goes wrong with the reaction, the process simply stops. Inertial confinement fusion (ICF) has so little fuel mass available at any one time in the reaction chamber that there is no possibility of meltdown in case of an accident. The surrounding materials can be chosen so that the problem of long-lived nuclear waste is greatly reduced. However, the fusion reaction is so difficult to ignite that the only sustained (beyond a few seconds) working examples are those of the stars and perhaps a few thermonuclear weapons.

Fusion Theory and Chronology

The character of the atomic nucleus is such that the individual nuclear particles are most tightly bound in elements of intermediate atomic number. When energy is sought, attention is focused on the more loosely assembled elements, releasing energy by splitting (fissioning) the heavy isotopes or by joining (fusing) the lighter ones. There is less energy release per fusion reaction than there is per fission reaction, but the reactants are more plentiful and easier to handle. A particular fusion reaction is of interest if the power produced can be sufficiently large to offset the power consumed in generating and maintaining the reacting medium, and if the relevant rates can be large enough so that economically interesting regimes are accessible to modern technology. There are over thirty such reactions possible. The most appealing of the fusion reactions as possible routes to fusion energy are (1) those which involve the heavy hydrogen isotopes, deuterium (H_1^2) or D; and (2) those which involve tritium (H_1^3) or T. These tend to have the largest fusion reaction probability (cross section) at the lowest energies. Deuterium is abundant, naturally occurring and in wide use now as D_2O in heavy-water-moderated reactors. Tritium is a radioactive isotope with a 12.3-year half-life and does not occur in nature. Tritium emits an electron and decays to stable helium-3.

The deuterium (D-D) reaction chain may be represented by:

$$D + D \longrightarrow {}^3He + n + 3.2 \text{ MeV}$$

$$D + D \longrightarrow T + p + 4.0 \text{ MeV}$$

$$D + T \longrightarrow {}^4He + n + 17.6 \text{ MeV}$$

$$D + {}^3He \longrightarrow {}^4He + p + 18.3 \text{ MeV}$$

$$\cdots\cdots\cdots\cdots\cdots\cdots\cdots\cdots$$

$$6D \longrightarrow 2{}^4He + 2p + 2n + 43.1 \text{ MeV}$$

The first two equations represent the fact that the D-D reaction can follow either of two paths, producing tritium and one proton; or helium-3 and one neutron, with equal probability. The products of the first two reactions form the fuel for the third and fourth reactions, and they are burned with additional deuterium. The net reaction consists of the conversion of six deuterium nuclei into two helium nuclei, two hydrogen nuclei, and two neutrons along with a net energy release of 43.1 MeV. The reaction products—helium, hydrogen, and neutrons—are harmless as

contrasted with the myriad fission products obtained in a fission reactor. The neutrons produced may be absorbed in sodium to produce an additional 0.25 MeV per cycle. Therefore, the D-D reaction produces at least 7 MeV per deuterium atom (deuteron) and, with absorption in sodium, more than 10 meV per fuel atom.

The peak reaction rate coefficient of the D-D reaction is considerably less than that of the deuterium-tritium (D-T) reaction occurring within the (D-D) cycle. Thus, attention tends to focus on the latter. Because tritium does not occur naturally, the reaction must be supplemented by one using lithium to reproduce the tritium fuel:

$$D + T \longrightarrow {}^4He + n + 17.6 \text{ MeV}$$

$$n + {}^6Li \longrightarrow {}^4He + T + 4.8 \text{ MeV}$$

$$\cdots\cdots\cdots\cdots\cdots\cdots\cdots\cdots$$

$$D + {}^6Li \longrightarrow 2{}^4He + 22.4 \text{ MeV}$$

This reaction is tritium-regenerating and produces only helium as a reaction product.

The D-T reactor is technologically more complex than the D-D reactor because of the need to facilitate the second reaction (which takes place outside the plasma) and because very energetic neutrons must be slowed down to allow the reaction with lithium to take place. However, the conditions needed to achieve net power output are less demanding than for the D-D fuel reactor. The D-T reaction will probably be exploited first, but its ultimate, very long term use may be limited by the availability of lithium.

Fusion reactions can take place only when the nuclei of the fuel atoms are brought into close enough conjunction. The nuclei are positively charged and so they repel each other. This repulsion is equivalent to an energy barrier, which can be penetrated with reasonable efficiency only if the reacting nuclei have kinetic energy comparable to the barrier height. The level of kinetic energy required depends upon the particular reaction and the desired reaction rate, but in general, plasmas of interest have average energy per particle in excess of 5 keV. A collection of particles with average energy 5 keV has an effective temperature of at least 10^8 degrees Kelvin. At these temperatures, the gas is completely dissociated into its constituent positively charged nuclei and free electrons. The density ranges between 10^{13} to 10^{14} cm^{-3}. The electrical charge density is such that the behavior of the collection of particles is completely dominated by electrostatic and electromagnetic phenomena. Such a charge-dominated collection of ionized matter is known as *plasma*. This plasma at such extremely high temperatures cannot be confined by walls made of materials, known or imagined. But confinement, even for a nanosecond or less, is required if fusion reactions are to occur.

Early Research on Fusion Fundamentals

Means to confine the plasma have been the major objective of fusion power research in several countries for a number of years. Methods researched for confining the plasma include the use of strong magnetic fields and inertial confinement methods in which the fuel is pelletized in a special way and fusion reactions are initiated either by laser beams, beams of particles, and heating the plasma with high-power microwave radiation.[2] Other means are being researched.

Two principal problems face fusion power scientists in their quest to achieve energy breakeven, that is, a condition where the plasma produces as much fusion energy as is required to heat it up. First, the fuel must be heated to a temperature of 10^8 degrees Celsius in order for the nuclei to overcome strong electrostatic repulsion and get close enough to fuse. The scattering cross section is much greater than the fusion cross section. Consequently, the particles must collide a large number of times before fusion can occur. Reacting nuclei must be confined for a long period of time, or be at very high densities. The second requirement, closely related to the first, is confinement time. The Lawson number is commonly used to indicate the efficiency of plasma confinement. Plasma temperature and the Lawson number are key parameters that determine if a fusion reaction will produce net positive energy.

---

[1] When the concept of fusion power was new, there was genuine worldwide interest in the subject and this continued for at least a few decades. It is interesting to contrast the views of most planners at the political and scientific level in Europe and the United Kingdom with those planners in the United States during the last few years. Some authorities attribute this to the budgetary process, which in the United States is subject to annual reviews and, in particular, places emphasis on short-term failures and successes, whereas at the international level, budgetary planning generally is in terms of longer time spans, such as 5-year intervals. Short-cycle financing makes long-range planning difficult, including the ability of U.S. scientists to cooperate with others worldwide.

[2] On Earth, fusion ignition, that is, the efficient burn-up of deuterium and tritium was demonstrated by the thermonuclear (hydrogen) bomb. In that case, obviously tremendously greater amounts of energy were released than that required to trigger fusion reactions. In that event, an atomic bomb was used to generate the extremely high temperature and the degree of confinement required for hydrogen nuclei to fuse.

An early record was achieved in the mid-1980s by the Princeton Plasma Physics Laboratory's *Tokamak Fusion Test Reactor* (TFTR). For magnetically confined plasma, a temperature of 200×10^6 K and a confinement parameter of 1.5×10^{14} seconds per cubic centimeter were set. The prior record had been held by the *Alcator C* tokamak at the Massachusetts Institute of Technology (1983). Further progress is mentioned later in this article.

Toroidal Magnetic Confinement. The original and one of the most thoroughly researched methods of plasma containment was suggested by Russian scientists several decades ago. In this scheme, the plasma is retained at a pressure of a few atmospheres. As the plasma expands in the magnetic field, currents that retard this expansion are excited. The plasma is surrounded by a vacuum insulation in order to sustain the required high temperature at which thermonuclear reactions occur. The method of confinement is time-limited and thus requires that the reactor be operated in a pulsed mode. Theoretically, the amount of energy required for heating the plasma initially is *small* when compared with the thermonuclear energy generated. Devices of this configuration are known as *tokamaks*. In the magnetic-confinement approach, the fuel must be maintained at a fairly low density because of practical limits posed by the magnetic field strengths available. Thus, confinement times of seconds (or minutes) must be achieved in order to get a substantial burn-up of the fuel.

It was found that a tokamak can be started as a bevatron by discharging condensers through the coils of the transformer yoke. See also **Particles (Subatomic)**. However, considerable difficulties arise in stabilizing the plasma ring in the magnetic field. The ratio of R to A (Fig. 1) is quite critical to stability. In early experiments, the time of confinement of the plasma was found to be but a fraction of a second. This led to the thinking that as the tokamak is scaled up in size, the time of confinement may be proportional to the square of the size of the machine.

Another problem of considerable magnitude was learned, namely, that of the D and T ions (see previous equations). Since the plasma is heated by an electric field, the energy is initially and principally transferred to the electrons (of small mass). thus lengthening the time to heat the ions (of much greater mass). Transfer of energy from electrons to ions is essentially by collision. The problem of transferring energy in a more efficient manner was the subject of much early research. Two methods were approached: (1) injection of atoms of deuterium or tritium, which were already accelerated to temperatures required for the thermonuclear reaction, into the plasma ring; and (2) exciting radial AlfÉn magnetoacoustic waves in the external magnetic field by the circulating high-frequency current. This concept was based upon the knowledge that energy dissipated by magnetoacoustic waves is directly passed onto the ions and the transmitted power may be sufficient to heat the ions and sustain their temperature for a sufficiently long time.

Another early problem found in connection with tokamaks is the fact that the plasma attracts and absorbs impurities form the walls of the container, thus lowering the reaction rates. This apparently arises from emissions of neutral atoms by the plasma that impacts and erodes the wall.

The early T-3 machine (Kurchatov Institute of Technology, Russia) was (as early as 1962) the prototypical example of toroidal confinement of plasma. It was found that the magnetic field lines in such closed geometry are constrained to follow toroidal surfaces and the plasma particles spiral along the field lines. Thus, it was learned that simple toroidal fields cannot confine a plasma in equilibrium. Some stabilizing scheme must be used. The T-3 supplied equilibrium by means of a large circulating current induced around the torus. This current also served to heat the plasma by resistive ($I^2 R$) heating.

Early experience also showed that the induced plasma current in a tokamak generates a magnetic field that loops the minor axis of the torus. The field lines form helices along the toroidal surface; the plasma must cross the lines to escape. It does so through the cumulative action of many random displacements caused by interparticle collisions. (in effect, diffusing across the field lines and out of the system). Thermal energy is transported by much the same process.

In other early, commonly explored toroidal devices, the required equilibrium was supplied by means of externally imposed twisted multiple magnetic fields. The *stellerator* design developed by the Princeton Plasma Physics Laboratory was representative of this approach.

Particle orbits in toroidal fields are exceedingly complex. Because of the spatially varying field and the acceleration experienced in moving along the curved field lines, the particles drift away from and return to the original magnetic field lines. These excursions are quite large and the particle and energy diffusion are enhanced by large factors. Thus, it was learned that a principal goal of tokamak designs is that of measuring particle and energy confinement times under numerous conditions, and thus to be able to predict how these relationships may change as the size of the torus is increased. It also became apparent that the simple resistive heating of T-3 could not be scaled up to allow ignition of a toroidal reactor and this led to the development of another heating scheme.

Several candidate heating schemes, including plasma compression, the induction of moderate turbulence, the absorption of high-power oscillations at various characteristic frequencies, and heating to neutral beam injection—all have been subjected to research.

The latter scheme was one of the most favored methods for consideration. It was reasoned that beams of neutral atoms easily can penetrate the magnetic field surrounding the plasma only to be ionized with very high efficiency upon entering the plasma itself. The resulting trapped ions become a significant energy source and also provide additional particles to offset plasma losses. Thus, neutral beam injectors became the subject of intense developmental effort.

Scientists at the Lawrence Radiation Laboratory reasoned that, if some way could be found to stop the charged particles of a plasma in their motion along the magnetic field lines, the magnetic field lines need not be closed within the plasma region and that many problems associated with the toroidal system could be avoided. Thus, the "magnetic mirror" was developed, and it was based on the simple phenomenon that charged particles will reflect from a region of increasing magnetic field if their encounter with the field lines is sufficiently close to perpendicular—and, further, that they will not be reflected if they are moving nearly parallel to the lines of force. Mirror devices take advantage of this by trapping the plasma in a bulge of the field, where the lines are spread. When the field increases, the lines reapproach each other and a reflecting region is formed. A prototype, *Mirror 2X*, was built and tested by the Lawrence Radiation Laboratory.

Theta pinch experiments also were explored in still another regime of high density parameters. The object of these experiments was the rapid compression of an existing low-density, low-temperature plasma. It was reasoned that if the heating pulse is sufficiently rapid, there can follow a two-step process: (1) the plasma is heated first by the shock wave generated by the rapidly rising field, and (2) heated further by adiabatic compression as the field continues to grow more slowly in time. It was soon found, of course, that the technology required to serve these targets indeed was impressive. Capacitors storing nearly a megajoule of energy at 50 kV would be discharged through hundreds of parallel paths (to reduce inductance) into a massive single-turn compression coil to generate fields in excess of 100 kilogauss. In the *Scylla IV* (Los Alamos Scientific Laboratory) experiment, the current rose to 8.6 million amperes in only 3.7 microseconds. The resulting plasma had a density $n = \times 10^{16}/cm^3$ and ion temperature of

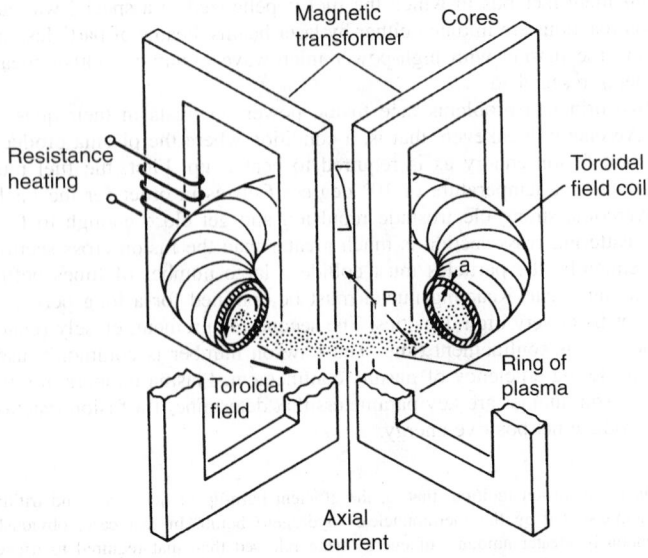

Fig. 1.   The basics of a tokamak. (*After Kapitza*)

$T_i = 3.2$ keV. The plasma lifetime was very short, because the plasma particles simply stream out of the ends of the device, although the plasma is in stable equilibrium, occupying a thin cylinder in the center of the coil during its residence there. It was found that plasmas achieved in the theta pinch process are so hot and dense that the kinetic pressure is several hundreds of atmospheres and the plasma energy density very nearly equals the magnetic field energy density. Plasma behavior under such conditions continues to be investigated. The large toroidal, theta-pinch apparatus (*Scyllac*) which was installed at the Los Alamos Laboratory is shown in Fig. 2.

Lawson Criteria. The fusion process is such that there is no sharp dividing line in a fusion reactor, as there is in a fission reactor, that is, between "go" and "no go." Instead, there simply is the question of achieving a sufficiently high steady state fusion reaction rate that will provide a margin of excess energy for power generation. This condition is expressed approximately by the Lawson criterion:

Plasma density n/cm^3	5×10^{14}
Maximum plasma temperature T_i (keV)	10
Confinement time (microseconds)	200

Over the years, there has been a lingering concern over the confinement conditions that could be achieved when a device is scaled up to represent the conditions that can be expected in future commercial models. These concerns were alleviated partially in 1979 when the Princeton Plasma Physics Laboratory demonstrated a scale-up tokamak (the Princeton Large Torus, PLT), at which time plasma conditions were close to those required for a fusion reactor. In 1989, it was reported by Princeton scientists that the TFTR achieved temperatures of 200×10^6 K at a Lawson parameter of 10^{13}. However, the Lawson parameter was estimated to be about a factor of 20 too low. The prior record of 80×10^6 K had been set by the Princeton Large Torus in 1980. Also noted in 1986 tests was preliminary evidence of a "bootstrap current," which had been predicted theoretically as early as 1971. This hypothetical current arises spontaneously in hot, high-density tokamak plasma and flows in such a manner that it sustains the confining magnetic field with a minimum of input from external transformers. This was considered fortuitous in the quest to produce a steady level of power as contrasted with pulsing on and off. Early steps to improve the Lawson parameter as well as temperature included heating the plasma by neutral beam (electrically neutral deuterium atoms) injection as well as by a technique referred to as rf heating.

Fig. 2. View of apparatus for testing toroidal theta pinch concept. Shown is the Scyllac torus. (*Los Alamos Scientific Laboratory*.)

Inertial Confinement. In the inertial approach, the fuel is heated as it is compressed to a very high density, estimated at about 1000 times that of the normal density of the solid fuel. An intense energy source is focused onto the outer surface of a specially formed spherical pellet. This produces ablation on the outer surface somewhat similar to the ablation of a rocket as it is exposed to extremely high temperatures. The energy also causes an implosion (an inward bursting) of the deuterium-tritium fuel mixture in the inner portion of the pellet. The compression process heats the fuel to ignition temperature and also contributes to the quantity of fuel that can be burned. Inasmuch as the compressed fuel is restrained by its own inertia, the fuel burns before it can fly apart. This is a time span of a billionth of a second or less.[3]

Laser Beam Fusion Technology. Research in this methodology was commenced in the 1960s and by the early 1980s had achieved a higher performance than the particle beam fusion technology. However, it was concluded later that although the laser beam radiation of fuel pellets had numerous advantages, the scheme was inefficient and costly.

Particle Beam Fusion Technology. Research on particle beam fuel technology has continued. It is reported that researchers at the Kurchatov Institute decided in 1971 to pursue the electron beam approach. This approach was made possible largely because of advancements in pulsed-power technology, which had commenced in the mid-1960s. The technology was supported by the U.S. Atomic Energy Commission and the Department of Defense during that period. The technology was regarded as a means of providing radiation sources for testing the survivability of intercontinental ballistic missiles against bursts of radiation from antiballistic missile warheads. This testing required powerful x-ray bursts. Researchers found that this condition can be created when a multimillion-volt electron beam hits a thin foil of metal, such as tantalum. Work on apparatus to create intense x- and gamma-radiation was undertaken at Sandia as well as by the British Atomic Weapons Research Establishment. In 1967, the British developed an electronbeam accelerator capable of discharging a maximum energy of 100,000 joules into an x-ray tube at an electric potential of 10 million volts. The trillion-watt accelerator was named *Hermes*. A similar device was developed by DOD. Many authorities claim that the required power for pellet ignition is 100 trillion watts. An operating current of between 10 and 100 million amperes must be attained. An accelerating potential of between 1 and 10 million volts is required. See also **Particles (Subatomic)**.

In as much as electrons are charged particles, they repel each other, adding to the difficulties of focusing a beam, particularly as compared with focusing photons from a laser. As pointed out by one scientist, the problem can be overcome by utilizing the "self-pinching" effect of the magnetic field generated by a high-current beam when the electric-repulsion force is neutralized. The phenomenon is illustrated in Fig. 3. Spence and Yonas investigated this effect when at Physics International in the late 1960s. Although beyond the scope of this book, the principles have since been fully explained by Poukey of Sandia Laboratories, using a computer model that simulates the motion of trillions of electrons as they swarm within a diode. Similar modeling has been conducted by Goldstein of the Naval Research Laboratory. Yonas also observed that, in addition to improvements in beam focusing, the technological questions of pulse formation were largely answered during the late 1970s through the development of high-power switches, as well as an understanding of electric breakdown.

During earlier work at Physics International and the Kurchatov Institute and Sandia Laboratories, a solution to the problem of power concentration was demonstrated, again beyond the scope of this article. See also Fig. 4.

Light-Ion Beams. In 1986, VanDevender and Cook (Sandia National Laboratories) observed that the Particle Beam Fusion Accelerator II (PBFA II) was under development at the laboratory and had the potential of igniting thermonuclear fuel. See Fig. 5. The accelerator was designed to generate up to 5 megamperes of Li ions at 30 million electron volts. The light-ion approach was found to combine relative low cost and high payoff, but also featured a number of risks pertaining to practicality. One factor of concern was that the beams of very high density generate electric and magnetic

---

[3] In view of the ultimate use of a fusion reactor for the generation of power, possible shortcomings of the methodology must be considered along with its advantages. The complexity of the system, the materials requirements, and the overall costs must be such that fusion power, in its early commercial applications, must compete well with other sources of electric power generation.

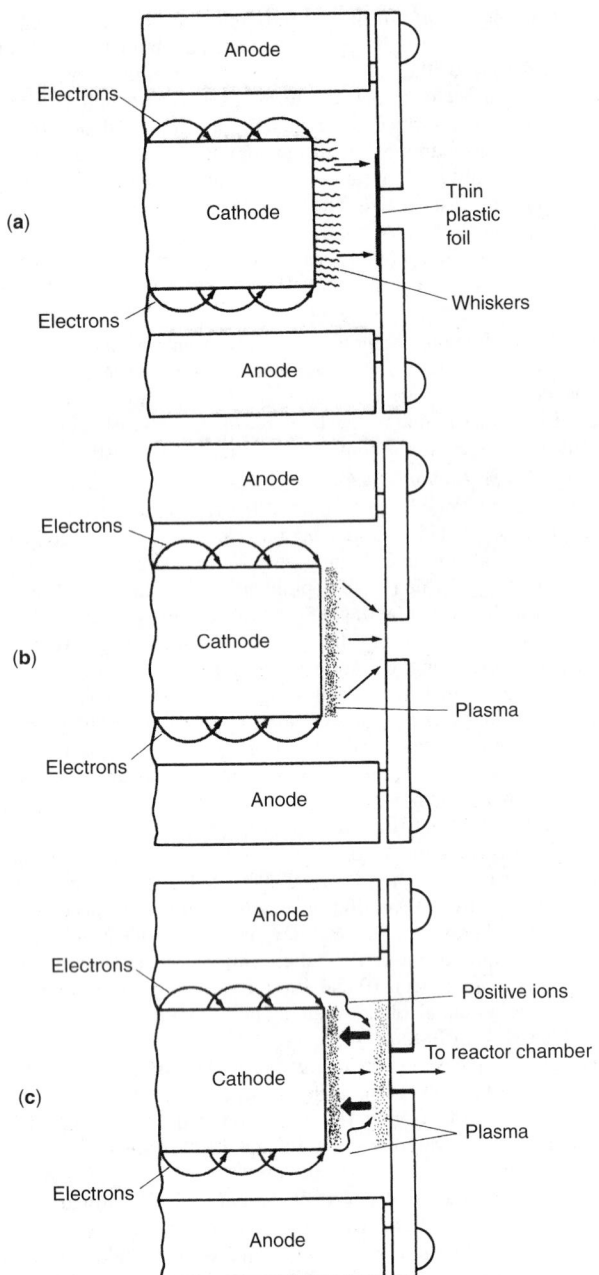

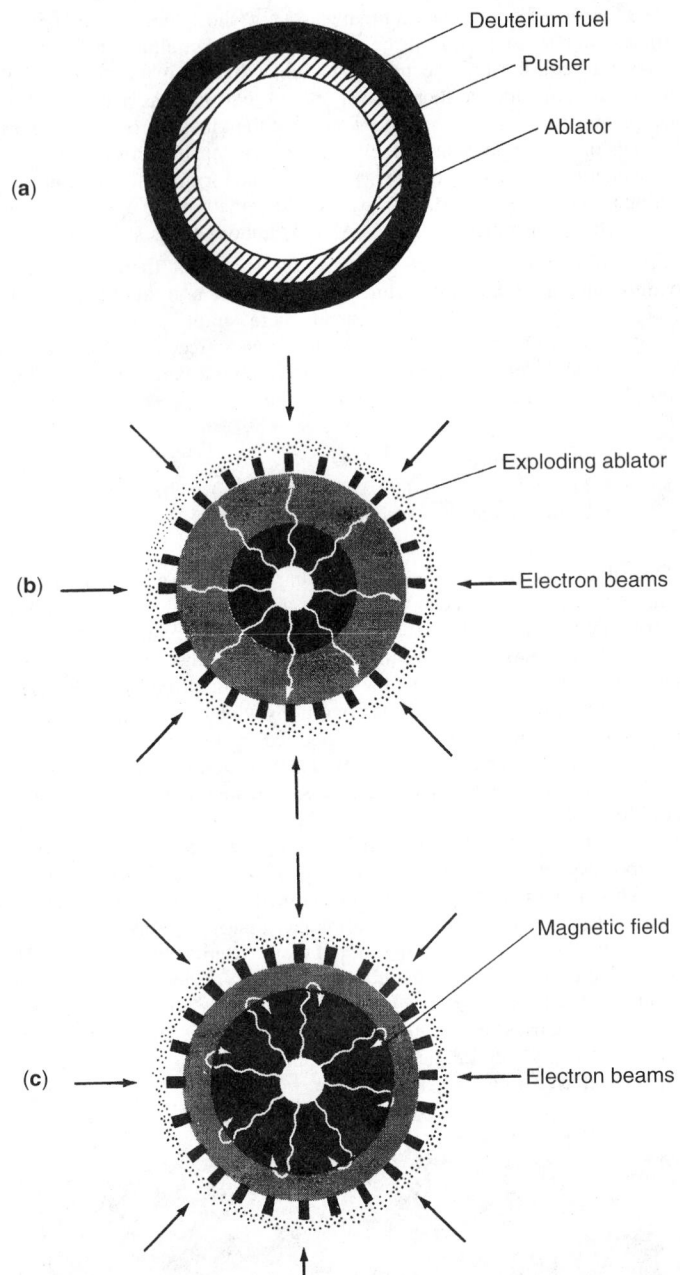

Fig. 3. Illustration of self-focusing effect designed to permit electrons flowing along cathode of magnetically insulated transmission line to focus on the anode at axis where magnetic field drops to zero. A plasma layer, shown as bands of black dots in (b) and (c), is important to the focusing effect. The initial emission of the electrons is from microscopic protrusions on the surface of the cathode referred to as "whiskers." The electrons are accelerated to the anode in essentially an unfocused array. The heated whiskers shown in (a) explode to form a plasma. This intensifies the flow of electrons across gap between cathode and anode. An additional plasma layer is formed on the surface of the anode (b). As shown by the prominent leftward-pointing arrows, positively charged ions flow from the anode plasma to the cathode plasma. These ions partially neutralize the self-repulsive forces of the electrons, thus sharpening the focus of the electron beams as it proceeds to the reactor chamber (c). (*After Yonas*)

Fig. 4. The design and configuration of the heavy-hydrogen fuel pellets are critical to obtaining an effective and efficient implosion. As shown by (a), the concept of the hollow, multiple shell is made up of three elements: (1) An outside layer or shell of an explosive "ablator" material: (2) a middle layer or shell, known as a "pusher," and (3) the fuel per se. This construction drives the inner layers toward the center of the pellet, thus igniting fusion reactions between deuterium nuclei in the fuel, as indicated by (b). Scientists at Sandia Laboratories have successfully produced a trapped magnetic field, as shown in (c), which assists in insulating the imploding fuel from the surrounding high-density pusher shell. This conserves energy that otherwise would be lost to the pusher material by way of heat conduction, as well as by confining slow electrons to the implosion region and, of course, adding to the final high temperature of the fuel required. With this scheme, scientists have been able to obtain up to a million neutrons from a deuterium fuel pellet at a slow implosion velocity (4 centimeters/microsecond). (*After Yonas*)

fields, once thought to be prohibitive to achieving high focal intensities. Subsequent advances in beam production and focusing have shown that such self-generating forces can be controlled, at least to the extent required for ignition, breakeven, and high gain experiments.

Recent Research on Fusion Fundamentals

During intervening years to the present, research has continued essentially along the lines of the prior investigations just described. Studies have involved the use of several machines (about 70% tokamaks) to achieve a self-sustaining energy process. Several of these machines either have

been retired, or are facing a period of diminishing returns. Exemplary of the experimental machines are Princeton's Tokamak Fusion Test Reactor (TFTR) and the Joint European Torus (JET) located in Culham, England, each of which has created plasmas hotter than the interior of the sun, but only for two seconds. In November 1991, for example, the JET ignited a blend of 14% tritium and 86% deuterium to deliver nearly 2 million watts

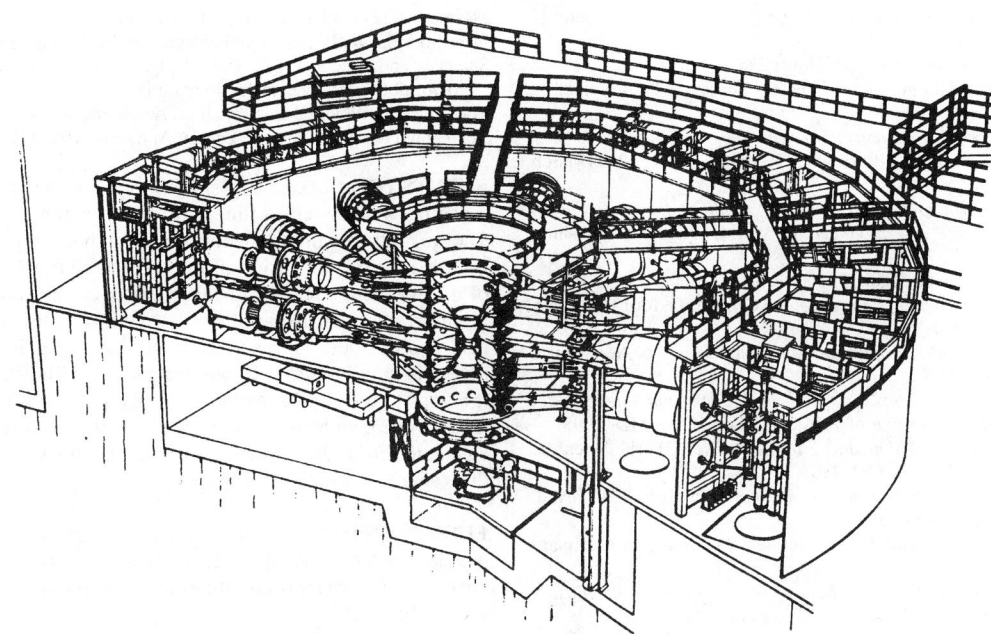

Fig. 5. Conceptual diagram of the PFBA II. Capacitors in the outer annulus produce energy that is delivered to the target (center of apparatus) through a multimodule power-conditioning network. (*Sandia National Laboratories.*)

of power in a two-second burn. During the burn, the JET's confinement chamber was rendered radioactive.

The planned International Thermonuclear Experimental Reactor (ITER), as of 1994, is the center-stage attraction for the fusion power community. As observed by Paul-Henri Rebut, director of ITER, "If ITER fails, fusion will be delayed half a century—or more." Rebut also directs fusion research at the present JET facility.

The ITER will be an extension of the conventional tokamak design. The new design will be 40 feet (12.2 m) in diameter, approximately double that of the JET reactor, which currently is the largest in the world. The design will improve heat retention, a major requirement for achieving ignition and break-even energy performance and better. The fuel to be used will be a combination of tritium and deuterium, which will burn more efficiently and cleanly than deuterium alone.

ITER is being designed to run in a "steady-state" for a period of two weeks or longer. This time span compares with a few seconds in contemporary tokamaks. Design plans now underway and awaiting approval by the nations sponsoring the unit require the solution to numerous problems, including: (1) obtaining materials that will withstand bombardment of neutrons produced during the fusion reaction (currently used stainless steel absorbs radiation and, over time, becomes fatigued); and (2) a much more durable and tough diverter material than that currently available must be developed. The diverter may become the principal problem and limit the performance of the machine. Some researchers have expressed doubt that existing carbon composite materials or even tungsten will be able to withstand the heat during extended reaction runs. (3) Superior superconducting magnets are required for handling most of the work of compressing the plasma. Such magnets are vital in reducing electric power consumption so that the reactor can operate at a break even or better efficiency. There is also the problem of positioning the magnets sufficiently close to the vessel so that the most powerful magnetic lines possible can be achieved, while at the same time shielding the magnets from heat produced by the fusion-heat generated. (4) The density of the swirling plasma must be increased in an effort to improve the chances of achieving a sustainable fusion reaction. One approach under consideration is the use of high-intensity microwaves for replacing particle beams used in the past to compress the plasma. (5) All design elements and the system as a whole must be environmentally safe.

The Parties to ITER have now assessed the details, and it is expected that an optimised design will be approved in mid-2000. This should allow the Final Design Report of ITER-FEAT to be completed by mid-2001 as originally planned.

With good fortune, commercial versions of the ITER could be available by 2030.

Heavy-Ion Fusion. As of the late 1980s and early 1990s, some special attention was paid to the use of heavy-ion fusion. In this technology, miniature thermonuclear explosions are created by accelerating charged particles of lead or other massive elements into capsules of hydrogen isotopes. Some researchers contend that ultimately this may prove to be the secret required to make fusion power a viable energy alternative. Experimentation has been carried out in a Multiple-Beam Experiment (MBE-4) at the Lawrence Berkeley Laboratory. Unlike most accelerators, which boost particles by means of powerful radio waves, the MBE-4 uses induction. Thus far heavy-ion research has received little funding. The accelerators can fire rapidly—up to hundreds of times per second. It has been estimated that a maximum of ten shots per second would be sufficient to create a commercial power generator. Considerable attention to the technology has been given in Europe (Heavy-Ion Research, Darmstadt), in Japan and at the School of Plasma Physics (Varenna).

Cold Fusion. As reported throughout this article, known nuclear fission occurs only at very high temperatures. During the 1989–1990 period, there was considerable discussion of the findings of chemists at a western university in the United States—to the effect that they had observed nuclear fission at laboratory temperatures. Although, initially, this was a rather startling revelation, several attempts to duplicate the results of the experiment failed. However, to a dwindling degree, the topic remains controversial.

Additional Reading

Aldhous, P.: "JET Strike Hits Brussels," *Science*, **1755** (June 26, 1992).

Baker, D.: "Advanced Diagnostics and Instrumentation Progress in Nuclear Fusion Research," *InTech*, **48** (April 1990).

Cherfas, J.: "Europe: Betting Heavily on Fusion," *Science*, **1500** (December 14, 1990).

Conkin, W.F.: *Fusion: Power Unlimited,* Xlibris Corporation, New York, NY, 2000.

Crawford, M.: "Fusion Panel Drafts a Wish List for the '90s," *Science*, **110** (July 13, 1990).

Crawford, M.: "Hot Fusion: A Meltdown in Political Support," *Science*, **1534** (March 10, 1990).

Crawford, M.: "U.S. Fusion Program Struggling to Stay in the Game," *Science*, **1561** (December 14, 1990).

Craxton, R.S., R.L. McCrory, and J.M. Coures: "Progress in Laser Fusion," *Sci. Amer.*, **68** (August 1986).

Fillo, J.A. and P. Lindenfeld: *Introduction to Nuclear Fusion Power and the Design of Fusion Reactors,* American Association of Physics Teachers, College Park, MD, 1984.

Furth, H.P.: "Magnetic Confinement Fusion," *Science*, **1522** (September 28, 1990).

Graham, D.: "Quest for Fusion," *Technology Review (MIT)*, **14** (July 1992).

Hamilton, D.P.: "Energy Science Takes a Heavy Budget Hit," *Science*, **501** (October 26, 1990).

Hamilton, D.P.: "A Fusion First," *Science*, **927** (November 15, 1991).

Hamilton, D.P.: "Allocating the Pain in Energy Science," *Science*, **1482** (September 27, 1991).

Hamilton, D.P.: "Fusion Megabucks," *Science*, **507** (February 1, 1991).

Hamilton, D.P.: "The Fusion Community Picks up the Pieces," *Science*, **1203** (March 6, 1992).

Harms, A.A. and G.H. Miley: *Principles of Fusion Energy: An Introduction to Fusion Energy for Students of Science and Engineering*, World Scientific Publishing Company, Inc., River Edge, NJ, 2000.

Holden, C.: "Fusion Panel Lowers Its Sights," *Science*, **193** (October 11, 1991).

Huizenga, J.R.: *Cold Fusion: The Scientific Fiasco of the Century*, Oxford University Press, Inc., New York, NY, 1994.

Kammash, T.: *Fusion Energy in Space Propulsion*, American Institute of Aeronautics & Astronautics, Reston, VA, 1995.

Lindl, J.D.: *Inertial Confinement Fusion: The Quest for Ignition and Energy Gain Using Indirect Drive*, American Institute of Physics, College Park, MD, 1998.

Liu, C.S. and F. Wagner: *Nuclear Fusion and Plasma Physics*, World Scientific Publishing Company, Inc., River Edge, NJ, 1995.

Mizuno, T. and J. Rothwell: *Nuclear Transmutation: The Reality of Cold Fusion*, Infinite Energy Press, Concord, NH, 1998.

Musso, B.: "Ansaldo: The Italian Art of Re-Structuring," *Sci. Amer.*, **10** (January 1990).

Nishikawa, K. and M. Wakatani: *Plasma Physics: Basic Theory with Fusion Applications*, Springer-Verlag, Inc., New York, NY, 2000.

Panarella, E.: *Current Trends in International Fusion Research*, Kluwer Academic Publishers, Norwell, MA, 1997.

Ress, D. et al.: "Neutron Imaging of Laser Fusion Targets," *Science*, **956** (August 19, 1988).

Scott, P.E., E. Sindoni, and G. Gorini: *Diagnostics for Experimental Thermonuclear Fusion Reactors*, Plenum Publishing Corporation, New York, NY, 1996.

Staff: IAEA: *Plasma Physics & Controlled Nuclear Fusion Research 1994*, Vol. 4, Bernan Associates, Lanham, MD, 1993.

Staff: *Fusion Energy Program: The Role of TPX and Alternate Concepts*, DIANE Publishing Company, Collingdale, PA, 1995.

Staff: IAEA: *Fusion Energy 1996: Proceedings of the Sixteenth International Conference on Fusion Energy, Organized by the International Atomic Energy Agency*, Bernan Associates, Lanham, MD, 1997.

Staff: National Research Council: *Review of the Department of Energy's Inertial Confinement Fusion Program*, National Academy Press, Washington, DC, 1997.

Stefanini, A.M., G. Hebbia, and S. Lunardi: *Heavy-Ion Fusion: Exploring the Variety of Nuclear Properties*, World Scientific Publishing Company, Inc., River Edge, NJ, 1994.

Stone, R.: "A Tritium Boost for JET," *Science*, **841** (August 23, 1991).

Surko, C.M. and R.E. Slusher: "Waves and Turbulence in a Tokamak Fusion Plasma," *Science*, **221**, 817–818 (1983).

Szoke, A. and R.W. Moir: "A Practical Route to Fusion Power," *Technology Review (MIT)*, **20** (July 1991).

VanDevender, J.P. and D.L. Cook: "Inertial Confinement Fusion with Light Ion Beams," *Science*, **232**, 831–836 (1986).

Velarde, G., Y. Ronen, et al.: *Nuclear Fusion by Inertial Confinement*, CRC Press, LLC., Boca Raton, FL, 1993.

Waldrop, M.M.: "Tokamak Sets Records in Temperature and Confinement," *Science*, **233**, 937 (1986).

Web References

Davis Diverted Tokamak(DDT) Facility. http://www.llnl.gov/das/das_re- search/plasma.html

DIII-D Fusion Home Page, General Atomic, San Diego. http://fusioned.gat.com/

European Fusion Development Agreement–JET. http://www.jet.efda.org/

International Thermonuclear Experimental Reactor (ITER). http://www.iter.org/

International Atomic Energy Agency: http://www.iaea.org/worldatom/

Los Alamos National Laboratory: http://wsx.lanl.gov/

MIT Plasma Science and Fusion Centre Atomic, San Diego. http://www.psfc.mit.edu/

National Fusion Energy Science Web Site: http://www.fusionscience.org/

Office of Fusion Energy, U.S. Fusion Energy Sciences Program: http://wwwofe.er.doe.gov/

Plasma Science and Technology: http://www.plasmas.org/

Princeton Plasma Physics Laboratory: http://www.pppl.gov/

Science and Technology, Oak Ridge National Laboratory: http://www.ornl.gov/ornlhome/science_technology.htm

The American Physical Society: http://www.aps.org/

The United Kingdom Atomic Energy Authority. UKAEA: http://www.fusion.org.uk/

FUSION (Heat of). Very simple experiments show that the fusion of a given mass of any crystalline substance requires a definite quantity of heat. The quantity required per unit mass, without any change of temperature, is called the heat of fusion of the substance. It may be measured by means of a calorimeter. The fused substance is introduced into the calorimeter at a temperature somewhat above its melting point and allowed to cool, the heat evolved being measured. At the melting point it ceases to cool for a time, but continues to give out heat as it solidifies; and when all congealed, it begins to cool again. At this stage the process is terminated; and the total heat evolved, with corrections for the cooling before and after solidification calculated from the known specific heats, gives the heat of fusion. For ice the value is about 79.71 calories per gram.

FUSION (Nuclear). The combination of two light nuclei to form a heavier nucleus, with the release of the difference of the nuclear binding energy of the products and the sum of the binding energies of the two light nuclei. Examples are:

$$^2\text{H} + {}^2\text{H} \longrightarrow {}^3\text{He} + n + 3.27 \text{ MeV}$$

$$^2\text{H} + {}^3\text{H} \longrightarrow {}^4\text{He} + n + 17.59 \text{ MeV}$$

$$^2\text{H} + {}^6\text{Li} \longrightarrow {}^8\text{Be} \longrightarrow 2{}^4\text{He} + 22.37 \text{ MeV}$$

Fusion reactions can take place only if the reacting nuclei possess sufficiently high energies to overcome their mutual Coulomb repulsion and to approach within the range of nuclear forces, hence they are favored by high temperatures. See also **Nuclear Power Technology**.

FUSION (Phase Change). A change from the solid to the liquid phase of matter. In crystalline bodies, and, as has now become understood, also in many other solids not exhibiting well-defined crystal structure, the atoms are held in positions of stable equilibrium by intermolecular forces. They of course move with thermal agitation, but their movements are oscillatory and do not carry them outside a limited range of distance from their equilibrium positions. Stable equilibrium may, however, become unstable when the system is disturbed beyond a certain limit. Thus if a solid body is sufficiently heated, the molecules break loose from their stable configuration and wander about or diffuse among each other. When this condition has become general, the body exhibits the characteristics of a liquid, and we say it has undergone fusion. In some cases, such as ice, the change is quite abrupt, the substance having a well-defined melting point; in others, like glass or pitch, it is gradual. The difference is probably due to the more uniform potential energy of the atoms in the former case, so that they all "break loose" at the same stage of thermal agitation. In the latter case, some atoms require more energy to dislodge them than others. In any case, the process requires a supply of energy, which is recognized as the heat of fusion. With most substances, fusion is accompanied by an increase in volume; but with some, like ice, the volume becomes definitely less.

Fusion, as an order-disorder transition, is the concept that fusion of a crystalline solid is essentially a change from the almost perfectly ordered solid state to a disordered liquid state. The vacant spaces in the crystal lattice correspond to the other component in the binary alloys, which undergo order-disorder transition in the pure form. Evidence from x-ray diffraction measurements indicates that short-range order is retained during fusion but long-range order is lost.

G

GABBRO. Gabbro is a deep-seated and often very coarse-grained igneous rock composed of plagioclase feldspar, usually labradorite or bytownite and monoclinic pyroxene, with occasionally as accessories olivine (when it is then called olivine gabbro), biotite, magnetite, ilmenite, and hornblende. Norite is a variety of gabbro, carrying orthorhombic pyroxene, usually hypersthene instead of the monoclinic sort. Troctolite is essentially olivine and plagioclase. Quartz gabbros are known and have probably been derived from magmas somewhat oversaturated with silica. On the other hand, essexites represent gabbros whose parent magma doubtless had an insufficiency of silica resulting in the formation of nephelite. Gabbros are frequently rich in sulfides that may be of commercial value, a notable occurrence of which is at Sudbury, Canada. Here a norite carrying chalcopyrite and nickeliferous pyrrhotite forms the most important deposits of nickel known. Gold, silver and platinum are also recovered from this ore.

GADOLINIUM. [CAS: 7440-54-2]. Chemical element symbol Gd, at. no. 64, at. wt. 157.25, seventh in the Lanthanide series in the periodic table, mp. 1,313°C, bp 3,273°C, density 7.901 g/cm³ (20°C). Elemental gadolinium has a close-packed hexagonal crystal structure at 25°C. The pure metallic gadolinium is silver-gray in color, slow to tarnish in normal atmospheres. The metal is soft, malleable, and easy to fabricate with normal tools provided that processing temperatures are maintained below 150°C. The turnings and chips of gadolinium are mildly pyrophoric and care must be exercised in their handling. There are seven natural isotopes of gadolinium: ^{152}Gd, ^{154}Gd through ^{158}Gd, and ^{160}Gd. Eleven artificial isotopes have been prepared. The natural isotopes are not radioactive. In terms of abundance, gadolinium is present on the average of 5.4 ppm in the earth's crust, making it potentially more available than tantalum, tin, or tungsten. The element was first identified by J.C.G. Marignac in 1880. The natural isotopic mixture of gadolinium has the greatest thermal-neutron-absorption cross section of all elements, 40,000 barns. This is approximately 10 times greater than the next two elements, samarium (5,800 barns) and europium (4,300 barns). However, gadolinium is limited to nuclear applications mainly as a start-up and shutdown material because only two of the natural isotopes ^{155}Gd and ^{157}Gd behave in this manner. ^{155}Gd and ^{157}Gd make up 31% of the total weight of elemental gadolinium. The metal has a low acute-toxicity rating. Electronic configuration

$$1s^2 2s^2 2p^6 3s^2 3d^{10} 4s^2 4p^6 4d^{10} 4f^7 5s^2 5p^6 5d^1 6s^2.$$

Ionic radius Gd³⁺ 0.938 Å. Metallic radius 1.801 Å. First ionization potential 6.1 eV; second 12.09 eV.

Other important physical properties of gadolinium are given under **Rare-Earth Elements and Metals**.

Gadolinium reacts vigorously with dilute mineral acids, but is practically inert to strong bases and boiling H_2O. Gadolinium is an active reducing agent for metals, including iron, chromium, manganese, tin, lead, and zinc. The major sources of gadolinium are xenotime, monazite, gadolinite, residues from uranium mining, and ion-exchange clays found in Southern China.

Although the nuclear properties of the element are attractive, gadolinium has enjoyed rather limited applications in reactor technology. A major use of gadolinium is in amorphous Gd-Co(Fe) alloys for magnetic recording and information storage. An important discovery in the 1960s showed that gadolinium iron garnets (called GIGs) $Gd_6Fe_5O_{12}$ possess a crystalline structure which finds useful application in microwave frequency control, circulators, isolators, and bandpass filters in electronic circuitry. Gadolinium oxide also is used as the host matrix in the red phosphor for color television picture tubes, where it is activated by europium. Gadolinium oxysulfide Gd_2O_2S is used as an x-ray image intensifier making it possible to reduce the exposure of patients to x-rays. Along with yttrium and lanthanum activated by cerium, gadolinium is used in a phosphor for single-gun beam-indexing flying-spot scanning cathode ray tubes. Gadolinium complexes are used as MRI (Magnetic Resonance Imaging) contrasting agents to improve the images obtained in MRI scans of various organs of the human body. Gadolinium also provides magnetic properties when alloyed with cobalt, cerium, iron, and copper ($Co_{3.5}CuFe_{0.5}Ce$) in permanent magnets, imparting a desirable negative temperature coefficient of magnetic saturation. Gadolinium metal and compounds are under consideration for use in a variety of magnetic refrigeration and cooling applications ranging from the liquifaction of hydrogen and natural gases, to refrigerator/freezers, supermarket chillers, and air conditioners. See also **Refrigeration**.

See references listed at ends of entries on **Chemical Elements**; and **Rare-Earth Elements and Metals**.

K. A. GSCHNEIDNER, JR.
B. EVANS
Iowa State University
Ames, Iowa

GAHNITE. The mineral gahnite, also known as *zinc spinel*, is isometric with an octahedral habit but may appear as dodecahedrons or modified cubes. Chemically it is zinc aluminate corresponding to the formula $ZnAl_2O_4$. There is a tendency for cleavage parallel to the octahedron, fracture varies from conchoidal to uneven; brittle; hardness 7.5–8; specific gravity 4.6; luster, vitreous; color ranges from dark green through various shades of greenish- or bluish-black, yellowish-black or grayish, subtransparent to almost opaque. Gahnite is found in association with other zinc minerals at several European localities, notably in Bavaria and Sweden. In the United States it is found at Franklin and Sterling Hill, New Jersey; at Rowe, Massachusetts and in Maryland, North Carolina, Georgia and Colorado. Gahnite was named in honor of the Swedish chemist, J.G. Gahn.

GALENA. [CAS: 1314-87-0]. The mineral galena, lead sulfide, PbS, crystallizes in the isometric system, usually in cubes or cube-octahedron combinations, less frequently in octahedrons. It is often found in cleavable masses, but may be granular or fibrous. The highly perfect cubic cleavage is an important characteristic of this mineral: it may, however, sometimes show an octahedral parting. Its hardness is 2.5; specific gravity, 7.58; luster, metallic; color, lead gray; streak, grayish-black; opaque. Galena is the most important ore of lead and in addition often carries values of silver; it is then known as argentiferous galena. It occasionally is actually mined as a silver ore. Sometimes galena contains small amounts of zinc, cadmium, antimony, bismuth, and copper as sulfides.

Galena is a very common and widely spread mineral, it occurs in veins and beds in various rocks, both crystalline and sedimentary. Some of these deposits are doubtless replacements, others seem to show a close connection with intrusive igneous rocks. Of the many European localities, the classics are Freiberg, Saxony, and the silver mines of the Harz Mountains. This mineral has been found in the lavas of Vesuvius, in Italy, and fine specimens came from Cornwall and Cumberland, England. Australia, South America, Chile, and Peru produce galena. In the United States, Missouri, Illinois, Iowa, and Wisconsin contain large and important galena deposits. In Colorado and Idaho it has been mined for its silver content. Galena is usually associated with sphalerite, smithsonite, and at Phoenixville, Pennsylvania, with beautiful pyromorphite crystals. The name is derived from the Latin galena, a term which was applied both to the lead ore and slag from refining.

ELMER B. ROWLEY
Union College
Schenectady, New York

GALLIUM. [CAS: 7440-55-3]. Chemical element symbol Ga, at. no. 31, at. wt. 69.72, periodic table group 13, mp 29.78°C, bp 2403 ± 0.5°C, density 5.90 (solid at 20°C), 6.095 (liquid at 29.8°C), 5.445 (liquid at 1100°C). Elemental gallium has a one-face-centered orthorhombic crystal structure. Among the elements, gallium (like mercury) is liquid at ordinary temperatures. Gallium is a white, tough metal, but so soft that it can be cut with a knife. A freshly exposed surface soon oxidizes superficially to a bluish-gray color. When heated about 500 C, the metal burns in air. Gallium is only slightly affected by H_2O at room temperature, but reacts vigorously in boiling H_2O. The metal is only slowly attacked by concentrated acids, but does dissolve readily in aqua regia. The two stable isotopes of gallium are ^{69}Ga and ^{71}Ga. The eight radioactive isotopes include ^{64}Ga through ^{68}Ga, ^{70}Ga, ^{72}Ga, and ^{73}Ga. All have a relatively short half-life, the longest, ^{67}Ga with a half-life of 78 hours. See also **Radioactivity**. Gallium was one of the elements predicted by Mendeleev from his early periodic arrangement of the chemical elements. The element first was identified by Francois Lecoz de Boisbaudran in 1875 from observations in a spectroscopic study of zinc blende. In terms of abundance, gallium ranks 31st among the elements, with about 15 ppm in the earth's crust.

First ionization potential 6.00 eV; second, 20.43 eV; third, 30.6 eV. Oxidation potentials $Ga \rightarrow Ga^{3+} + 3e^-$, -0.52 V; $Ga + 4OH^- \rightarrow H_2GaO_3^- + H_2O + 3e^-$, 1.22 V.

Other important physical characteristics of gallium are given under **Chemical Elements**.

Gallium's renown as a valuable chemical element stems from its increasing use over the past decade in electronic devices. See also **Semiconductors**; and **Solid-State Physics**.

Gallium occurs in very small amount in zinc blende, magnetite, pyrite, bauxite, and kaolin of certain localities. A few parts per million is present in Oklahoma zinc ores. The recovery of gallium from zinc flue dust is effected by solution of the dust in excess of HCl, addition of potassium chlorate, and distillation to remove germanium. When the residue is converted into sulfate, fractional electrolysis of the slightly acid solution removes zinc, and the gallium is obtained almost free from indium. The only known deposit of gallite, $CuGaS_2$, is in southwest Africa. The mineral contains about 1% gallium. The most important commercial source of gallium is bauxite, which contains up to 0.01% gallium. The metal is recovered from the sodium aluminate used in the extraction of aluminum from bauxite. In one process, calcium hydroxide is mixed with the sodium aluminate solution. At this juncture the ratio of gallium to aluminum is about 1 to 3,000. By precipitating and filtering out calcium aluminate, a gallium-rich solution remains. The filtrate then is agitated with CO_2 which precipitates more aluminum out as aluminum hydroxide. At this point, the enriched gallate-in-caustic solution contains approximately 0.2 grams of gallium per liter. This solution is used as an electrolyte in a mercury cathode cell. The gallium amalgamates with the mercury. It is dissolved out of the mercury with boiling NaOH in the presence of iron, which serves as a catalyst. At this point, the concentration is approximately 80 g of gallium per liter. The process is repeated several times, after which the gallium concentrate is electrolyzed, using a stainless steel cathode on which the gallium plates out. The gallium is easily removed from the cathode by raising the temperature above the melting point. For highly-pure metal, subsequent purification processes are required, including (1) crystallization as monocrystals, (2) chemical treatment with acids or oxygen at high temperatures, or (3) repeat resolution in pure boiling NaOH and reelectrolyzing. A metal of 99.99999% purity thus can be obtained.

Uses

The availability of gallium in very high purity is important to its use as a semiconductor in various electronic devices, such as diodes, laser diodes, and electroluminescent diodes. The compound usually used in these applications is gallium arsenide GaAs which is prepared by reacting hydrogen and arsenic vapor with gallium oxide Ga_2O_3 (prepared from very pure metal) at a temperature of about 600°C. Properties of the GaAs so produced include: intrinsic electron concentration, 10^7; energy gap, 1.38 eV at 20°C; electron mobility, 8,800 cm^2/V-s.

Gallium arsenide [CAS: 1303-00-0] also is used in solar batteries. See also **Solar Energy**. Gallium metal is used as an activator in luminous paints and phosphors, as well as in arc rectifiers, dental amalgams, as a sealant in vacuum systems, in transistors, and in some organic syntheses. Because the metal expands upon solidifying (3.1%), it should not be stored in fragile containers. Although potentially useful in high-temperature thermometers because of its liquidity over a wide temperature range, these applications have been limited, partially because of the high cost of the element.

Chemistry and Compounds

Gallium metal is quite corrosive to most other metals because of the rapidity with which it diffuses into the crystal lattices of metals. For example, only a very small amount of gallium in contact with an aluminum plate or sheet will result in immediate embrittlement as the result of the diffusion of gallium through the grain boundaries separating them. Gallium readily forms alloys with most metals over 600°C, including barium, copper, gold, iron, lead, lithium, magnesium, manganese, nickel, platinum, silver, sodium, titanium, vanadium, zirconium, and zinc. The few metals that tend to resist attack by gallium are molybdenum, niobium, tantalum, and tungsten.

Gallium trihalides include the trifluoride, tribromide, triiodide, and the trichloride. The trichloride is readily formed by heating the metal with chlorine or HCl, is soluble in ether, and, like aluminum chloride, is effective as a catalyst in various organic reactions. Both the trichloride and the tribromide are dimeric in the vapor state. Other known trivalent gallium compounds are the sesquisulfide, sesquisulfate (which forms double salts analogous to the alums), trinitrate, nitride, sesquioxide (which is polymorphic like alumina), and trihydroxide, which is, however, of variable composition, and which forms salts, the gallates, in alkaline solution.

Known gallium(II) compounds include the sulfide, selenide, telluride, dichloride, and dibromide. The last two are unstable, reacting vigorously with water to give hydrogen, and also undergoing oxidation, or disproportionation to the metal and the gallium(III) compound. They also are diamagnetic and their structure is $Ga^+ [GaX_4]^-$.

Simple gallium(I) compounds are also unstable, but Ga^+ may be stabilized in the presence of large anions, e.g., in $Ga[AlCl_4]$. The sulfur and selenium compounds Ga_2S and Ga_2Se have been shown to exist, but the oxide is uncertain.

Triethylgallium and trimethylgallium have been prepared, but are extremely reactive, even with air and H_2O. Like aluminum and indium, gallium forms a number of chelated oxy compounds, almost all of which are of 6-coordinate type. They include the stable crystalline inner complexes of which the β-diketones coordinate in the proportion of 3 molecules of diketone per atom of gallium. Trioxalato as well as dioxalato salts are known, and compounds such as 8-quinolinol and substituted 8-quinolinols form trimolecular chelate rings involving nitrogen and donor oxygen.

Gallium, like boron, forms a dimeric hydride, Ga_2H_6, from which a series of tetrahydrogallates, containing the GaH_4^- ion, is derived.

Gallium and most of its compounds are not highly toxic. For rats and rabbits, the LD_{100} has been established at approximately 100 mg of gallium per kilogram.

Toxicology

The toxicity of metallic gallium or gallium salts is very low. The corrosive, poisonous, or irritating nature of some gallium compounds is attributable to the anions or radicals with which it is associated. Gallium metal-organics, such as $Ga(CH_3)_3$, react vigorously with air, and can be explosive. The gallium halides, except the fluoride, hydrolyze in water to form corresponding halogen acids. Gallium phosphide, arsenide, selenide, and telluride react slowly with water, and more vigorously with acids and bases, to liberate toxic compounds.

Additional Reading

Davis, J.R.: *Metals Handbook,* 2nd Edition, ASM International, Materials Park, OH, 1998.

Greenwood, N.N. and A. Earnshaw: *Chemistry of the Elements,* 2nd Edition, Butterworth-Heinemann, Inc., Woburn, MA, 1997.

Krebs, R.E.: *The History and Use of Our Earth's Chemical Elements, A Reference Guide,* Greenwood Publishing Group, Inc., Westport, CT, 1998.

Lide, D.R.: *CRC Handbook of Chemistry and Physics,* 84th Edition, CRC Press, LLC., Boca Raton, FL, 2003.

Mahajan, S. and L.C. Kimerling: *Concise Encyclopedia of Semiconducting Materials and Related Technologies,* Elsevier Science, New York, NY, 1992.

Parker, P.: *McGraw-Hill Encyclopedia of Chemistry,* 2nd Edition, The McGraw-Hill Companies, Inc., New York, NY, 1993.

Sandroff, C.J. et al.: "Gas Clusters in the Quantum Size Regime: Growth on High Surface Area Silica by Molecular Beam Epitaxy," *Science,* **391** (July 28, 1989).

Vander Veen, M.R.: "Gallium Arsenide Sandwich Lasers," *Advanced Materials 7 Processes,* **39** (May 1988).

Westbrook, J.H.: "Electrical Materials" in *Encyclopedia of Materials Science and Engineering,* MIT Press, Cambridge, MA, 1986.

Willardson, R.K.: "Advances in Gallium Arsenide Crystal Growth," *Advanced Materials & Processes,* **24** (June 1986).

Wolsky, A.M., R.F. Giese, and E.J. Daniels: "The New Superconductors: Prospects for Applications," *Sci. Amer.,* **61** (February 1989).

Yablonovitch, E.: "The Chemistry of Solid-State Electronics," *Science,* **347** (October 20, 1989).

GALVANIZING. A process for rustproofing, and otherwise protecting, iron and steel, by applying a metallic zinc coating. The process can be used with nearly any size or shape of product, including large structural assemblies and steel sheet in coils and cut lengths. Millions of tons of new steel are galvanized each year, much of which is used prior to the application of other coatings, such as paint. Metallic zinc is applied to iron and steel by three processes: (1) hot dip galvanizing, (2) electrogalvanizing, and (3) zinc spraying. Most galvanized sheet steel is coated by the hot dip process. See also **Zinc.**

Hot-Dip Galvanizing

In the hot dip process, the sheets or other articles to be coated must be free from scale, dirt, grease, etc., and are usually prepared by pickling and washing before immersion in molten zinc commercially known as spelter. Articles fabricated from iron and steel sheets and wire are hand-dipped. Sheets and wire are handled mechanically.

An increasing proportion of sheet-metal products is being coated as sheet or strip before fabrication. This requires a tightly adhering coating to prevent peeling during stamping or forming operations. In order to obtain good adherence in hot-dipped coatings special processing is necessary, especially with the heavier weights of coating which give longer protection. For lighter weight coatings a duplex bath consisting of a layer of molten lead under the molten zinc is often used. The steel sheet passes through the lead, which does not adhere, and up through the zinc. The time in which the steel is in contact with the spelter is greatly reduced and consequently less zinc is deposited.

Some galvanized sheets are annealed after dipping in order to form a coating consisting entirely of iron-zinc compounds, a process which tends to increase resistance to peeling.

Electrogalvanizing

Electrodeposited zinc coatings are simpler in structure than hot dip coatings. They are composed of pure zinc, have a homogeneous structure, and are highly adherent. These coatings generally are not as thick as those produced by hot dipping. Coatings range in thickness up to 13.7 micrometer (0.065 mil). This process is particularly suitable for very thin, formable products. The electrogalvanized surface is smooth and fine and can readily be prepared for painting by phosphatizing. The coating is free of the characteristic spangled pattern of hot-dipped surfaces.

Electrogalvanizing can be done essentially at room temperature, thus the process does not alter the mechanical properties that could result from the higher temperatures encountered in dipping.

Zinc Spraying

This process involves the projection of atomized particles of molten zinc onto a prepared surface. Three types of spraying pistols are currently in use: (1) the molten metal pistol, (2) the powder pistol, and (3) the wire pistol. Sprayed coatings are slightly rough and porous. The slight porosity, however, does not adversely affect the protective value of the coating because zinc is anodic to steel. The zinc corrosion products that form during service fill the pores of the coating, giving a solid appearance. The slight roughness of the surface makes it an ideal basis for paint when properly pretreated. Spraying can be applied to nearly any shape or size of product—at the factory or at the site of final use. Spraying is the only satisfactory method of deposition available for applying very heavy zinc coatings up to 0.25 mm (0.01 in.) and greater in thickness.

GAMMA GLOBULIN. The fraction of the protein globulins of the blood plasma in which are found the antibodies.

See also **Blood.**

GAMMA RADIATION. A photon, or quantum of electromagnetic radiation, that is emitted when an atomic nucleus undergoes a transition from one of its excited energy levels to a lower level. The name *gamma* ray was applied in the earlier years of radioactivity investigations, while the exact nature of these radiations was still a mystery. Gamma-ray energies range from 10^4 to 10^7 eV. They are often emitted as a part of a nuclear reaction, when an atomic nucleus is left in an excited state, or during an isomeric transition. Gamma rays also can be emitted following alpha-particle decay, beta-particle decay, or orbital electron capture, if the daughter nuclide is left in an excited state.

In the strictest sense, the term gamma ray is applicable only to photons produced as a result of transitions in atomic nuclei. However, the term is also sometimes used to denote bremsstrahlung radiation produced when the high energy electrons in the beam of an electron accelerator, such as an electrostatic generator, a betatron, a synchrotron, or a linear accelerator, strike the target of that accelerator.

Gamma rays carry away the full energy of the transition with which they are associated. As a result, if detecting systems are used that are capable of absorbing the full energy of the gamma ray, a spectrum of gamma-ray numbers as a function of energy shows a series of distinct peaks, each associated with an individual gamma-ray transition. On the other hand, the discrete energy characteristics of gamma rays are more difficult to observe if the detecting system separates the effects of different types of gamma-ray interactions with matter, such as the Compton, photoelectric, and pair-production interactions. Under certain circumstances, a transition that would normally be expected to emit a gamma ray may sometimes release its energy through an internal conversion process.

See also **Particles (Subatomic);** and **Radioactivity.**

GAMMA-RAY SPECTROSCOPY. Gamma rays of concern here originate in the nucleus of radioactive isotopes, i.e., chemical elements whose nuclei are unstable and emit radiation as they decay to stable states. Such radioactive isotope disintegration follows rules that are always the same for the nucleus. These rules can be set down in a so-called decay scheme. An example is shown in Fig. 1 for the case of the radioisotope ^{137}Cs (cesium-137). The basic decay scheme shown indicates that cesium-137 decays into ^{137}Ba (barium-137) by emitting beta particles (electrons). Eight percent of the cesium nuclei decay directly into barium-137 nuclei; then about 2.5 minutes later, the excited nuclei decay to the lowest energy or ground state by emitting gamma rays having an energy level of 662 keV. Some heavy nuclei emit alpha particles. An alpha particle is a ^{4}He (helium-4) nucleus (two protons and two neutrons). The cesium-137 isotope, with a nucleus containing a total of 137 neutrons and protons, disintegrates with a half-life of 30 years. Since the number of nuclei is halved, the amount of radiation (intensity) is halved. With existing electronic systems, half-lives between 10^{-10} second and 10^{10} years can be measured.

Like most natural events, radioactive decay is not a uniform function. Consequently, the term *half-life* is meant to describe the value that would result if an infinite number of half-life measurements were made and the average calculated. Individual decays, however, follow a Poisson distribution, i.e., the standard deviation is equal to the square root of the number of observed decay events. This fact enables the experimenter to calculate the probable accuracy of his result, assuming no instrumentation inaccuracy.

Gamma Ray Detection

Gamma rays are high energy electromagnetic radiation with very short wavelengths (10^{-18} to 10^{-11} cm). They penetrate matter deeply—on the average much more deeply than do alpha and beta rays, which are charged

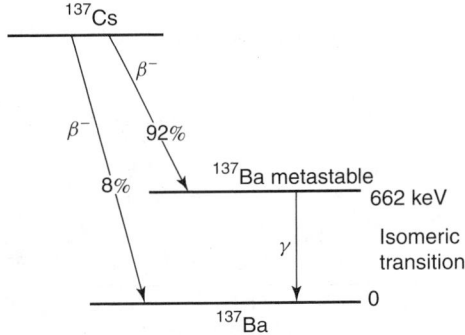

Fig. 1.   Decay scheme for ^{137}Cs

particles. It is their deep penetration that makes gamma rays useful in the laboratory and industry, in much the same way as X-rays. X-rays originate from shell transitions by orbital electrons, whereas gamma rays originate in the nucleus. Gamma rays usually are detected by observing effects that they produce in matter and when they encounter an atom. Important among these effects are: (1) the photoelectric effect; and (2) the Compton effect. The photoelectric effect occurs when the gamma ray strikes one of the orbital electrons of the atom, transferring its energy to the electron. This process produces a free electron and an ionized atom. The Compton effect arises in the case where the gamma ray strikes an orbital electron without imparting all of its energy to the electron. The electron is detached from the atom but receives only part of the gamma energy. The remaining energy persists as a scattered gamma ray with lower energy than the initial ray. This scattered ray may further collide with one or more other atoms, freeing other electrons. These types of interactions occur variously in nuclear radiation detectors. In each detector type, some observable reaction results, and in one manner or another produces an electrical output charge suitable as input for an electronic measuring system.

Gamma Ray Spectra

Measurements of gamma radiation are chiefly made in two ways: (1) a record is made of the number of counts as a function of energy, in which case a gamma ray spectrum is obtained; and (2) time relations are observed, in which case several types of information may be desired. A gamma spectrum, as measured by an ideal system, might appear as in Fig. 2. This is the ideal spectrum of the cesium-137 gamma radiation phenomena discussed earlier. In this spectrum, a large peak appears at 662 keV—caused by the gamma energy radiated when the metastable barium-137 nucleus returns to its ground state. There is also a continuum representing the energies imparted to Compton-scattered electrons. In practice, the spectra measured are not so well defined. See Fig. 3. Most noticeable is that the peaks of the spectrum are broadened to a greater or lesser extent by the characteristics of the devices used to detect gamma rays. Relating to this broadening as a measure of system quality, is its "resolution." This is a function both of the detector and of the associated circuitry. Resolution commonly is defined as: the ratio of the full width at half the maximum height of peak (FWHM) to the energy of the center of the peak. Thus, resolution indicates how well the detector can separate or resolve two different energy peaks. Typical resolutions for common gamma ray detectors range from about 10% to a few tenths of a percent. Also evident in Fig. 3 is a backscatter peak, which results because a large number of gamma rays squarely strike matter between the source and the detector, losing much of their energy before detection.

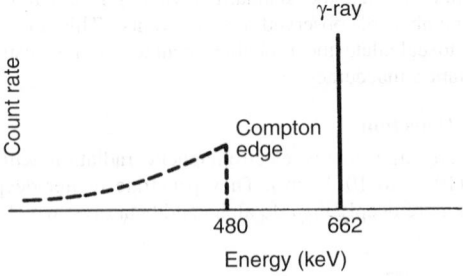

Fig. 2.   Ideal gamma spectrum of ^{137}Cs

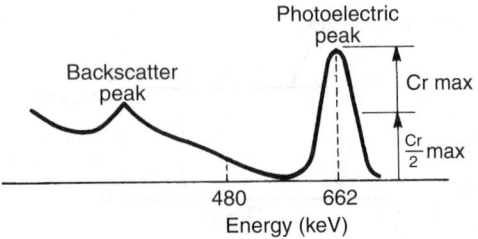

Fig. 3.   Typical gamma spectrum of ^{137}Cs

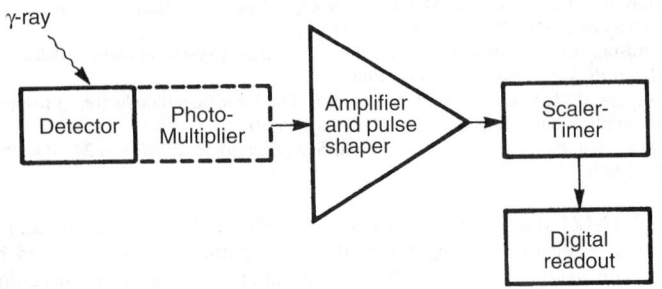

Fig. 4.   Gross counting measures radiation intensity of gamma ray regardless of energy

Energy Measurements

The measurements usually made in gamma ray work fall into two broad groups: (1) those made of the energy of the radiation; and (2) those made of its timing relative to another event. In addition, counting without regard to energy (often called gross counting) is also done to measure the intensity of the radiation. See Fig. 4. Intensity is measured in terms of counts/minute (or second).

Time Measurements

The second general class of measurements is one in which the time of occurrence of the gamma ray relative to a reference event is of interest to the experimenter. Such situations occur when gamma radiation is known to occur a specific interval of time after a trigger event.

Detectors

Commonly used detectors include scintillation, semiconductor, and gas proportional detectors. The scintillation detector often is preferred where high efficiency is more important than resolution—efficiency defined as a measure of the probability that an incident gamma ray will interact with the material in the detector. Semiconductor types are used increasingly, particularly where high resolution is required.

Signal Processing

The signal from the detector is a relatively short current pulse; the time integral of this current impulse is a charge proportional to the energy of the absorbed radiation. The preamplifiers and amplifiers which follow these detectors convert this impulse of a charge into a voltage pulse whose height (peak amplitude) is proportional to energy. Thus, signal processing prepares the charge from the detector for the final step, pulse height analysis. In the case of a timing measurement, signal processing prepares the charge signal for use with a timing pickoff (time discriminator). See also **Radioactivity**.

Additional Reading

Debertin, K. and R. Helmer: *Gamma and X-Ray Spectrometry Semiconductor Detectors,* Elsevier Science, New York, NY, 1999.
Hoff, R.: *Capture Gamma-ray Spectroscopy,* American Institute of Physics, College Park, MD, 1991.
Kern, J.: *Proceedings of the 8th International Symposium on Capture Gamma-Ray Spectroscopy and Related Topics,* World Scientific Publishing Company, Inc., Riveredge, NJ, 1994.
Raman, S.: *Capture Gamma-Ray Spectroscopy and Related Topics, 1984: International Symposium, Knoxville, Tennessee,* American Institute of Physics, College Park, MD, 1985.
Wender, S.: *Capture Gamma-Ray Spectroscopy and Related Topics,* American Institute of Physics, College Park, MD, 2000.

GANGLIOSIDES.   Identified by Kleng in 1935, the gangliosides are a family of acidic glycolipids that are characterized by the presence of sialic acid. The compounds bear a strong negative charge and are unusual in that they contain both hydrophobic and hydrophilic regions. These compounds are membrane components. Plasma cell membranes are rich with gangliosides. It has been suggested that gangliosides participate in the transmission of membrane-mediated information in living systems. As described by Fishman and Brady, "the carbohydrate portion of gangliosides is made up of molecules of sialic acid, hexoses, and *N*-acetylated hexosamines. The hydrophobic moiety is called ceramide, and it consists of a long-chain fatty acid linked through an amide bond to the nitrogen atom

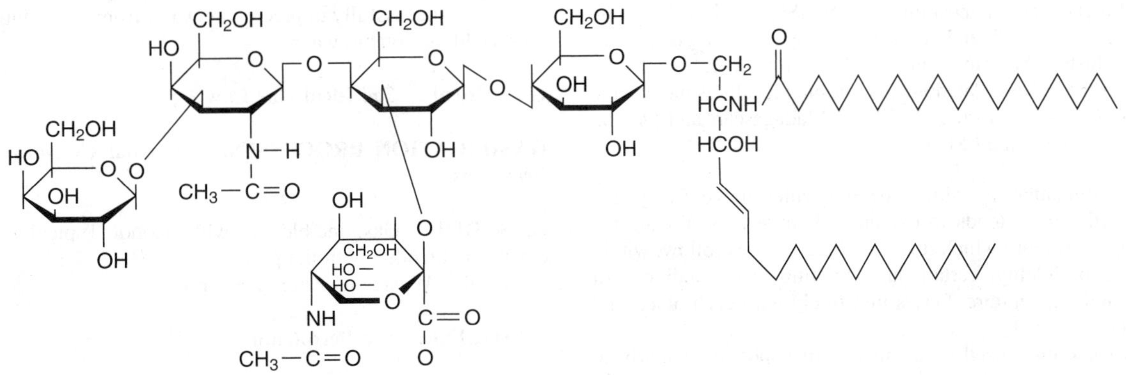

Fig. 1. Configuration of monosialoganglioside G_{M1} as suggested by Svennerholm

on carbon 2 (C-2) of the amino alcohol, sphingosine. Oligosaccharides are linked through a glycosidic bond to C-1 of the sphingosine portion of ceramide." Svennerholm (1963) suggested the configuration given in Fig. 1. The role of gangliosides is still rather discrete, but Fishman and Brady (1976) studied in some detail the interaction of cholera toxin with ganglioside-deficient cells. They also studied, the interaction of cholera toxin with ganglioside-deficient cells, as well as the interaction with glycoprotein hormones and their effect on the action of these hormones.

Additional Reading

Fishman, P.H. and R.O. Brady: "Biosynthesis and Function of Gangliosides," *Science*, **194**, 906–915 (1976).
Kleng, E.: *Z. Physiol. Chem.*, **235**, 24 (1935).
Ledeen, R.W., G. Tettamanti, E.L. Hogan, and A.J. Yates: *New Trends in Ganglioside Research: Neurochemical and Neuroregenerative Aspects,* Springer-Verlag New York, LLC., New York, NY, 1988.
Rahmann, H.: *Gangliosides and Modulation of Neuronal Functions,* Springer-Verlag New York, LLC., New York, NY, 1987.

GANGUE. The minerals and rock mined with a metallic ore but valueless in themselves or used only as a by-product. They are separated from the ore in the milling and extraction processes, often as slag. Common gangue materials are quartz, calcite, limonite, feldspar, pyrite, etc.

GARNET. The name garnet is now applied to a group of very important minerals crystallizing in the isometric system and showing the same habitat of dodecahedrons and trapezohedrons. Garnets belong to the nesosilicate group of silicate minerals and conform to the general formula $A_3B_2(SiO_4)_3$. The elements represented by A and B, respectively, may include calcium, magnesium, manganese, and ferrous iron; aluminum, ferric iron, chromium or titanium. While garnets show no cleavage, a dodecahedral parting is rarely noted; fracture conchoidal to uneven; some varieties very tough and valuable for abrasive purposes and for polishing eyeglass lenses. The hardness of garnet varies between the different varieties from 6.5 to 7.5, and the specific gravity from 3.4 to 4.3. Luster, vitreous to resinous; colors, red, yellow, brown, black, green, or colorless; transparent to opaque. The word garnet is derived from the Latin granatus, a grain.

In general, six varieties of garnet are recognized, based on their chemical composition: grossularite (which is also called hessonite and cinnamon-stone); pyrope; almandine or carbuncle; spessartine; uvarovite; and andradite. Grossularite is a calcium-aluminum garnet which corresponds to the formula $Ca_3Al_2(SiO_4)_3$; the calcium may, however, be in part replaced by ferrous iron and the aluminum by ferric iron. The name grossularite is derived from the botanical name for the gooseberry, grossularia, in reference to the green garnet of this composition found in Siberia. Other shades are the well-known cinnamon brown, reds, and yellows. Because of its inferior hardness to zircon, which mineral the yellow crystals resemble, they have been termed hessonite, from the Greek meaning inferior. Curiously, in the gem-bearing gravels of Ceylon, both zircon and hessonite are found and indiscriminately called hyacinth. This term, from the Greek, was apparently a general term used by Pliny for the transparent varieties of corundum; later it was used for yellow zircons.

Grossularite is found in crystalline limestones with vesuvianite, diopside, wollastonite and wernerite. Among the many localities are the Urals, Italy, Switzerland, Mexico, and, in the United States, Maine and New

Hampshire. Fine specimens are obtained from the Jeffrey Mine, Asbestos, Quebec, Canada.

Pyrope, sometimes called Cape ruby, is ruby-red in color and chemically a magnesium aluminum silicate with the formula $(Mg,Fe)_3Al_2(SiO)_3$; the magnesium may be replaced in part by calcium and ferrous iron. The color of pyrope varies from deep red to almost black. The transparent pyropes are used as gems, but some have a slight tinge of yellow. The name pyrope is derived from the Greek word meaning *fire-like*. A subvariety of pyrope from Macon County, North Carolina, is of a violet-red shade and has been called rhodolite, from the Greek meaning *a rose*. In chemical composition it may be considered as essentially an isomorphous mixture of pyrope and almandine, in the proportion of two molecules of pyrope to one molecule of almandine. Pyrope is found at Teplitz and Aussig, Bohemia; in the Kimberley diamond mines in the Republic of South Africa; in Australia and elsewhere. In the United States, important localities are in Arizona, New Mexico, and Utah.

Almandine is the modern gem the carbuncle, although in Pliny's time this term was used for almost any red stone. The term carbuncle is derived from the Latin *carbunculus*, meaning a little spark. The name almandine is a corruption of Alabanda, a locality in the Middle East where, in ancient times, these red stones were cut. Chemically almandine is an iron-aluminum garnet corresponding to the formula $Fe_3Al_2(SiO_4)_3$. The deep red transparent stones are often called precious garnet and used for gems. Almandine occurs in metamorphic rocks like mica schists usually associated with typically metamorphic minerals such as staurolite, kyanite, and andalusite. Good gem material comes from India and Brazil. Almandine is also found in Australia, Alaska, Africa, Norway, Sweden, Madagascar, and Japan. In the United States almandine with 11.48% MgO pyrope content is found in the gneisses of the Adirondack region of New York, sometimes of very large size, in New England, and elsewhere.

Spessartine is manganese aluminum garnet, $Mn_3Al_2(SiO_4)_3$. The name of this mineral is derived from Spessart in Bavaria, a well-known European locality. Spessartine of a beautiful orange-yellow comes from Madagascar. Violet-red spessartine has occurred in rhyolites in Colorado and Maine. Uvarovite is a calcium chromium silicate the formula being $Ca_3Cr_2(SiO_4)_3$. It is a rather rare garnet, bright green in color, usually in small crystals associated with chromite in serpentines, sometimes in crystalline limestones or schists. It is found in the Urals, the Republic of South Africa, Canada, and, in the United States, in California and Pennsylvania. Andradite, calcium-iron garnet, $Ca_3Fe_2(SiO_4)_3$, is of variable composition and may be red, yellow, brown, green, or black, or of intermediate shades. The subvarieties topazolite, yellow or green, demantoid, green, and melanite, a black sort, are recognized. Andradite is found both in deep-seated igneous rocks like syenite as well as in serpentines, schists, and crystalline limestones. Demantoid has been called the "emerald of the Urals" from its occurrence there. Varieties of andradite are found in many localities in Europe: Italy, Switzerland, Norway, and Saxony. In the United States it is found at Franklin, New Jersey; Magnet Cove, Arkansas; and elsewhere.

ELMER B. ROWLEY
Union College
Schenectady, New York

GARNET (Synthetic). See **Yag and Yig**.

GARNIERITE. This mineral occurs as amorphous masses, presumably as a product of secondary alteration of nickel-bearing peridotites. It is a hydrous silicate of nickel and magnesium, $(Ni, Mg)_3Si_2O_5(OH)_4$. Hardness is 2–3; specific gravity 2.2–2.8, and characterized by its apple green color with dull-to-earthy luster. An important nickel-ore mineral is found with chromite and serpentine in New Caledonia. Additional localities include the Republic of South Africa, the former U.S.S.R., Madagascar, and Oregon and North Carolina in the United States.

GAS. 1. A state of matter, in which the molecules move freely and consequently the entire mass tends to expand indefinitely, occupying the total volume of any vessel into which it is introduced. Gases follow, within considerable degree of fidelity, certain laws relating their conditions of pressure, volume, and temperature. Gases mix freely with each other, and they can be liquefied.

2. The term is sometimes used as distinct from vapor, particularly to indicate a substance having a critical temperature below room temperature.

The fundamental gas laws are described elsewhere in this volume. In particular, see also **Equation of State**.

An inert gas is a gas that does not react chemically. The rare gases of the atmosphere were long considered to be completely inert. Also known as noble gases, these included argon, helium, krypton, neon, radon, and xenon. Definite compounds of radon and xenon, for example, have been identified in recent years, but generally their identification as being inert is well justified. Some gases are termed permanent gases, including oxygen, nitrogen, and hydrogen, which require low temperatures and, in practice, high pressures for their liquefaction. The term arises from the fact that in the early years of scientific investigation of these materials, long before the conditions of liquefaction were obtainable, it was believed that these gases could not be liquefied under any circumstances, and hence termed permanent gases.

The laws pertaining to the forces of gas pressure and to the flow of gases are based ultimately upon the kinetic theory, but certain principles can be stated without analyzing their origin to that extent. To a first approximation, the ideal gas law, or the Boyle-Charles law, represents the dynamics of gases at rest. At a given temperature, the pressure of a body of gas varies inversely as its volume, and hence directly as its density (Boyle law). And at a fixed volume, the pressure is a linear function of the temperature, varying at the same rate ($\frac{1}{273}$ per centigrade degree) for all gases (Charles law). But dynamic processes in a gas are complicated by the fact that change in volume is, in general, accompanied by change in temperature, so that simple dynamics is overshadowed by thermodynamics. It was for this reason, for example, that the correct formula for the speed of sound in air proved, for a time, elusive. A gas is highly compressible, and this property affords ready opportunity for the energy of mechanical impulses, which would be merely transmitted by a noncompressible fluid, to be transformed into heat, or for the gas to use its thermal energy to create impulses of its own. The same circumstance complicates the effect of gravity. The atmosphere is not an ocean of uniform density and definite depth; its pressure and density are logarithmic functions of the altitude. The forces associated with moving gases form the subject matter of aerodynamics.

GAS CHROMATOGRAPHY. See **Chromatography**.

GAS-COOLED REACTOR. See **Nuclear Reactor**.

GAS CONSTANT. The constant of proportionality R in the equation of state of a perfect gas $pv = RT$, when referring to one gram-molecule of gas. R has the value of 1.985 calories per mole degree (°C).

GASEOUS DIFFUSION. See **Diffusion**; **Graham Law**.

GAS HYDRATE. A clathrate compound formed by gas (either noble or reactive) and water. The compounds are crystalline solids and are insoluble in water. They usually form (only at relatively low temperatures and high pressures) directly by contact of gas and liquid water. From 6 to 18 molecules of water may combine with each molecule of gas, depending on the nature of the gas.

The best-known gas hydrates are those of ethane, ethylene, propane, and isobutane. Others include methane and 1-butene, most of the fluorocarbon refrigerant gases, nitrous oxide, acetylene, vinyl chloride, carbon dioxide, methyl and ethyl chloride, methyl and ethyl bromide, cyclopropane, hydrogen sulfide, methyl mercaptan, and sulfur dioxide.

Interest in the gas hydrates originated mainly because of the nuisance of such compound formation in gas pipelines. In recent years, propane has been used successfully to precipitate water from salt solution (or seawater), thus yielding potable water.

GAS (Ideal). See **Ideal Gas Law**.

GASIFICATION PROCESSES. See **Coal Conversion (Clean Coal) Processes**.

GASOHOL. Gasoline blended with alcohol. Typically 10% methanol or ethanol is blended with the gasoline.

See also **Wastes as Energy Sources**.

GASOLINE. See **Petroleum**

GAS (Perfect). See **Perfect Gas**.

GATTERMANN ALDEHYDE SYNTHESIS. Preparation of aldehydes of phenols, phenol ethers, or heterocyclic compounds by treatment of the aromatic substrate with hydrogen cyanide and hydrochloric acid in the presence of Lewis acid catalysts.

GATTERMANN-KOCH REACTION. Formulation of benzene, alkylbenzenes, or polycyclic aromatic hydrocarbons with carbon monoxide and hydrochloric acid in the presence of aluminum chloride at high pressure. Addition of cuprous chloride allows the reaction to proceed at atmospheric pressure.

GAY-LUSSAC, JOSEPH LOUIS (1778–1850). A French chemist and physicist noted for the brilliance and accuracy of his reasoning and experimental work. He contributed greatly to the knowledge of gases in his discovery (1808) of the law of combining volumes and his independent discovery (1802) of the law of Charles, the relationship of temperature to the volume of gases. He graduated from and taught at the Ecole Polytechnique, becoming a full professor in 1810. His work in chemistry was extensive, resulting in the discovery of boron, which he named, with Louis-Jacque Thenard, and a variety of compounds such as boron trifluoride, chloric acid, and dithionic acid ($H_2S_2O_6$). He identified iodine as an element, named it, and studied its properties. He investigated the relationship of acids and bases and introduced many analytical techniques (such as the use of litmus as an indicator). Among his many contributions to industrial chemistry were improvements in the production of sulfuric acid. Much of the progress of chemistry in the early 19th century is associated with his career.

GAY-LUSSAC'S LAW. A modification of Charles' law to state the following: At constant pressure the volume of a confined gas is proportional to its absolute temperature. The volumes of gases involved in a chemical change can always be represented by the ratio of small whole numbers.

GEL. A colloid in which the disperse phase has combined with the continuous phase to produce a viscous jellylike product. Only 2% gelatin in water forms a stiff gel. A gel is made by cooling a solution, whereupon certain kinds of solutes (gelatin) form submicroscopic crystalline particle groups that retain much solvent in the interstices (so-called "brush-heap" structure). Gels are usually transparent but may become opalescent.

See also **Colloid Systems**; and **Pectins**.

GELATIN. Gelatin is a protein obtained by partial hydrolysis of collagen, the chief protein component in skin, bones, hides, and white connective tissues of the animal body. Type A gelatin is produced by acid processing of collagenous raw material; type B is produced by alkaline or lime processing. Because it is obtained from collagen by a controlled partial hydrolysis and does not exist in nature, gelatin is classified as a derived protein.

Uses of gelatin are based on its combination of properties; reversible gel-to-sol transition of aqueous solution; viscosity of warm aqueous solutions; ability to act as a protective colloid; water permeability; and insolubility in cold water, but complete solubility in hot water. It is also nutritious. These properties are utilized in the food, pharmaceutical, and photographic industries. In addition, gelatin forms strong, uniform, clear, moderately flexible coatings which readily swell and absorb water and are ideal for the manufacture of photographic films and pharmaceutical capsules.

Chemical Composition and Structure

Gelatin is not a single chemical substance. The main constituents of gelatin are large and complex polypeptide molecules of the same amino acid composition as the parent collagen, covering a broad molecular weight distribution range. In the parent collagen, the 18 different amino acids are arranged in ordered, long chains, each having ~95,000 mol wt. These chains are arranged in a rod-like, triple-helix structure consisting of two identical chains, called α_1, and one slightly different chain called α_2. These chains are partially separated and broken, i.e., hydrolyzed, in the gelatin manufacturing process. Different grades of gelatin have average molecular weight ranging from ~20,000 to 250,000.

Analysis shows the presence of amino acids from 0.2% tyrosine to 30.5% glycine. The five most common amino acids are glycine, 26.4–30.5%; proline, 14.8–18%; hydroxyproline, 13.3–14.5%; glutamic acid, 11.1–11.7%; and alanine, 8.6–11.3%. The remaining amino acids in decreasing order are arginine, aspartic acid, lysine, serine, leucine, valine, phenylalanine, threonine, isoleucine, hydroxylysine, histidine, methionine, and tyrosine.

Stability. Dry gelatin stored in airtight containers at room temperature has a shelf life of many years. However, it decomposes above 100°C. Aqueous solutions or gels of gelatin are highly susceptible to microbial growth and breakdown by proteolytic enzymes. Stability is a function of pH and electrolytes and decreases with increasing temperature because of hydrolysis.

Physical and Chemical Properties

Commercial gelatin is produced in mesh sizes ranging from coarse granules to fine powder. In Europe, gelatin is also produced in thin sheets for use in cooking. It is a vitreous, brittle solid, faintly yellow in color. Dry commercial gelatin contains about 9–13% moisture and is essentially tasteless and odorless with specific gravity between 1.3 and 1.4. Most physical and chemical properties of gelatin are measured on aqueous solutions and are functions of the source of collagen, method of manufacture, conditions during extraction and concentration, thermal history, pH, and chemical nature of impurities or additives.

Economic Aspects

Of the gelatin produced in the United States, 55% is acid processed, i.e., type A. The U.S. food industry consumes about 20,000 t/yr, with an annual growth rate of 0.5%; the pharmaceutical industry consumes about 10,000 t/yr; and the photographic industry about 7,000 t/yr. In the United States, the pharmaceutical gelatin market is expected to grow on the average of 2.5% per year. The photographic gelatin market has been stable or growing slightly. Color paper and x-ray products use over 55% of the photographic gelatin in the United States, with graphic arts and instant films using an additional 30%.

See also **Colloid Systems**.

THOMAS R. KEENAN
Kind & Knox Gelatine, Inc.

Additional Reading

Band, S.J. ed.: "Photographic Gelatin," *Proceedings of the Fifth RPS Symposium,* Oxford, UK, 1985, The Imaging Science and Technology Group of the Royal Photographic Society, 1987.
J. Photogr. Sci. **40**(5,6), 122–251 (1992).
Ridgway, K. ed.: *Hard Capsules Development and Technology,* The Pharmaceutical Press, London, UK, 1987.
Ward, A.G. and A. Courts, eds.: *The Science and Technology of Gelatin,* Academic Press, Inc., New York, NY, 1977.

GEMSTONES.

A gemstone is a mineral substance which because of its beauty or rarity is in demand for ornamental purposes, chiefly personal adornment. The origin of such use for what we now call gem minerals is lost in the dim vistas of early human history. Ancient records describe the various gemstones, and archaeologists find them in their investigations of bygone peoples. When we look at a collection of minerals with their bright colors and varying degrees of transparency or light-reflecting power, we cannot doubt that primitive man was much attracted by them and valued them greatly. We may imagine, too, that the occasionally found crystals with their regular geometric forms were more highly prized than broken fragments of the same minerals. Later they learned to polish them. Apparently, the oldest form into which stones were shaped is that known as *en cabochon*, a French term derived from the Latin word for head and referring to its rounded shape. The forms were either hemispherical or hemi-ellipsoidal. The Emperor Nero is supposed to have had a large emerald cut en cabochon, and, indeed, for several centuries after his time this seems to have been the only sort of cutting employed. The supposedly accidental discovery in 1475 that diamonds would mutually scratch each other began the era of modern gem cutting. Previously it had been believed that diamonds were so hard that they could not be artificially shaped. At first, however, little progress was made in fashioning gems other than polishing a number of facets without any definite arrangement.

We owe to Vicenzio Peruzzi, a Venetian, the credit for devising the so-called "brilliant cut," the style of the modern diamond cutting. Diamond cutting except for certain refinements due to a more thorough understanding of the behavior of minerals toward light, remains the same as in Peruzzi's day. At the present time, transparent stones of all sorts are usually "brilliant cut," whereas translucent or opaque are cut en cabochon.

Since time immemorial, dealers in gems have used as the unit of weight the carat, undoubtedly introduced from the east. The word is derived from the Greek meaning a small horn. This refers to the pods of the locust tree, *Ceratonia siliqua*, a common Mediterranean tree whose seeds were said to have been taken as the unit of weight in buying and selling gems. In the nineteenth century, the actual weight of the carat differed slightly in different countries of Europe, from a little under to somewhat over $\frac{1}{5}$ of a gram. The metric carat is exactly $\frac{1}{5}$ of a gram.

Gemstone Materials

There are three types of gemstone materials as defined by the U.S. Federal Trade Commission: *(1)* natural gemstones are found in nature and at most are enhanced; *(2)* imitation or simulated, fake, faux, etc., material resembles the natural material in appearance only and is frequently only colored glass or even plastic; and *(3)* synthetic material is the exact duplicate of the natural material, having the same chemical composition, optical properties, etc., as the natural, but made in the laboratory. Moreover, the word gem cannot be used for synthetic gemstone material. The synthetic equivalent of a natural material may be used as an imitation of another, e.g., synthetic cubic zirconia is widely used as a diamond imitation.

Synthetic gemstone materials often have multiple uses. Synthetic ruby and colorless sapphire are used for watch bearings, unscratchable watch crystals, and bar-code reader windows. Synthetic quartz oscillators are used for precision timekeeping, citizen's band radio (CB) crystals, and filters. Synthetic ruby, emerald, and garnets are used for masers and lasers.

In the gemstone jewelry market, synthetics provide a less expensive alternative to natural gemstones, but of a better quality than that available in costume jewelry. In general, a synthetic should be available for no more than 10% of the cost of equivalent-quality natural gemstone to be commercially viable. Synthetics are frequently divided into three groups: *(1)* luxury synthetics, involving slow and difficult growth processes, produced in small quantities for a price-restricted market; *(2)* intermediates; and *(3)* low cost synthetics, produced on a large scale.

Properties

The important properties are those of importance in natural gemstones. First is hardness, H. A value of 7 or greater on Mohs' scale is desirable to avoid scratches from the quartz (H = 7) sand present in dust. Next is color or a total lack of color, as in diamond and its simulants. A high refractive index (RI) permits the return by total internal reflection of most of the light falling onto a well-cut gemstone, giving brilliance, and a high dispersion (DISP) spreads the internally reflected light into spectral colors, resulting in fire.

Several gemstone species occur in various colors, depending on the presence of impurities or irradiation-induced color centers. Any material can have its color modified by the addition of various impurities: synthetic ruby, sapphires, and spinel are produced commercially in over 100 colors.

Manufacture

The most frequently used techniques for the commercial manufacture of synthetic gemstone materials are summarized in Table 1. Only rarely used for synthetics are such alternative growth techniques as the Bridgman technique of solidification in a crucible and the float zone technique, both involving growth from the melt.

TABLE 1. TECHNIQUES FOR COMMERCIAL GEMSTONE MATERIAL SYNTHESIS

Technique	Material
CRYSTAL GROWTH FROM THE MELT	
Verneuil (flame fusion)	ruby, sapphires, and stars; spinel; rutile; strontium titanate
Czochralski (pulling)	ruby and sapphire; alexandrite; garnets: YAG and GGG
float zone	ruby and sapphire; alexandrite
skull melting	cubic zirconia
CRYSTAL GROWTH FROM SOLUTION	
flux	alexandrite; emerald; ruby and sapphire; spinel
hydrothermal	colorless, amethyst, citrine, and smoky quartz; emerald; ruby and sapphire
high pressure	diamond;[a] jadeite[a]
OTHER TECHNIQUES	
complex chemical	opal

[a] Grown for other purposes or experimental production.

Materials

Alexandrite. Alexandrite, which is a colorless chrysoberyl, $BeAl_2O_4$, when pure, has a color change derived from Cr.

Beryl. Beryl, $Be_3Al_2Si_6O_{18}$, is called aquamarine when pale green or blue from the presence of Fe, emerald when dark green from Cr or at times V, and morganite or red beryl when pink or red, respectively, from Mn. Only synthetic emerald is in commercial production.

Corundum. Crystalline Al_2O_3, corundum, is called ruby when colored red by about 1% Cr, and sapphire for colorless and other colors particularly when blue from charge transfer between about 0.01% each of Fe^{2+} and Ti^{4+}.

Cubic Zirconia. As of this writing, cubic zirconia, ZrO_2, is the best diamond imitation available. This material can also be made in almost any color.

Diamond. The synthesis by a high pressure process of single-crystal diamond large enough for gemstone use was revealed by the General Electric Company in 1971. The yellow color (containing N) is grown much more easily than colorless (pure) and blue (B). None of these is likely to be viable for use in jewelry in the near future.

Garnets. Both YAG, yttrium aluminum garnet, $Y_3Al_5O_{12}$, and GGG, gadolinium gallium garnet, $Gd_3Ga_5O_{12}$, have the garnet structure and were used at one time as diamond imitations. These have been supplanted by cubic zirconia.

Opal. Opal is the only commercial synthetic gemstone material that is not a single crystal. It consists of a three-dimensional diffraction grating of geometrically aligned spheres of $SiO_2 \cdot xH_2O$, where x is usually <10%.

Quartz. When colorless, quartz is also known as rock crystal; irradiation of this produces smoky quartz. The name citrine is used when quartz is colored by Fe, and irradiation of this can produce purple-colored amethyst under certain circumstances.

Rutile. Rutile, a form of TiO_2, was at one time used as a rather poor diamond imitation. Related is strontium titanate, $SrTiO_3$, now more properly called synthetic tausonite.

Spinel. Colorless (pure), blue (Co), and other colored synthetic spinels made by the Verneuil process are widely seen in class rings and in other jewelry uses, where the blue is often mislabeled as synthetic sapphire.

Other Synthetic Materials. Many other natural gemstone materials have been duplicated in the laboratory on an experimental basis, often only in small sizes. Examples include tourmaline, topaz, and zircon. Of some potential is synthetic jadeite, one of the two forms of jade.

Discredited Synthetics. There are several materials that have in the past been considered to be synthetics, but were found on closer examination not to deserve such a designation, being merely imitations. Examples include imitation coral, lapis lazuli, and turquoise, all made by ceramic processes.

Gemstone Treatment

Color and clarity are two of the attributes that give gemstones used in jewelry value. Gemstones deficient in either color or clarity can be enhanced. Almost worthless material can at times be converted into valuable-appearing gemstones. An estimated two-thirds of all colored gemstones used in jewelry have been treated. Accordingly, the identification of the use of treatments and the disclosure of enhancements to the purchaser are important.

Some treatments are practiced so widely that untreated material is essentially unknown in the jewelry trade. The heating of pale Fe-containing chalcedony to produce red-brown carnelian is one of these.

The stability of a particular treatment is also important. The enhancement should survive during normal wear or display conditions.

Heat Treatments

The most commonly seen of the gemstones that have been enhanced by heat treatment are listed in Table 2. Parameters for specifying the conditions for heat treatment of a gemstone material include the maximum temperature reached and the time for which the maximum temperature is sustained; the rate of heating to temperature, the rate of cooling down from temperature, and any holding stages while heating and cooling; the chemistry and pressure of the atmosphere; and any material in contact with the gemstone. Exact conditions for heat treatments vary widely according to the natural materials used.

Irradiation Treatments

The process of irradiation involves the exposure of a specimen to one of a variety of radiations. A summary is given in Table 3.

When radiation interacts with matter, a displacement of the outermost electrons in atoms occurs. This displacement can lead to the formation of color centers or to valence state changes. The most commonly seen gemstones enhanced by irradiation are summarized in Table 4. When properly performed, there is no significant residual radioactivity.

TABLE 2. GEMSTONES ENHANCED BY HEATING

Material	Change[a]	Product	Use[b]
amber	clarified, sun-spangled	amber	F
amber	reconstructed, aged	amber	R
beryl	green to blue	aquamarine	W
chalcedony	pale to red-brown or red	carnelian, agate, tiger's eye, etc.	W
corundum	develop, intensify, or lighten blue	blue sapphire	W
corundum	develop or intensify yellow	yellow sapphire	W
corundum (ruby)	remove off-shades	ruby	F
corundum (ruby, sapphires)	remove silk, remove or develop asterism	starting material	W
corundum	diffuse in color or asterism	ruby, sapphires	R
diamond	change color after irradiation	starting material	R
quartz	amethyst to yellow citrine	starting material	W
quartz	crackled and dyed	various colors	R
zircon	brown to colorless or blue	starting material	W
zoisite	brown to deep purple-blue	starting material	W

[a] All product colors listed are stable.
[b] Prevalence of treatment occurring in product: R = rare to occasional; F = frequent; W = widespread or near-total.

TABLE 3. RAYS AND PARTICLES COMMONLY USED FOR THE IRRADIATION OF GEMSTONES

Irradiation type	Average energy, eV	Coloration uniformity	Induced radioactivity	Localized heating
x-rays	1×10^4	poor	none	none
γ-rays from Co-60 or Cs-137	1×10^6	good	none	none
neutron beam	1×10^6	good	strong	none
electron beam	1×10^6	poor	none	strong
electron beam	2×10^7	good	some	strong

TABLE 4. GEMSTONES ENHANCED BY IRRADIATION

Material	Change or product	Comments[a]	Use[b]
corundum	colorless to yellow	S,U,R	R
diamond	near colorless to black, blue, green, yellow, or red	S,R	F
pearl	darken to black	S	F
quartz	colorless to smoky	S,R	W
quartz	amethyst to amethyst–citrine	S,R	F
spodumene	pink kunzite to deep green	U,R	R
topaz	colorless or pale to blue	S,R	W
topaz	colorless or pale to brown	S,U,R	R
tourmaline	colorless or pale to red or multicolor	S,R	F

[a] S = stable; U = unstable, may fade; R = can be reversed by another treatment.
[b] Prevalence of treatment occurring in product: R = rare or occasional; F = frequent; W = widespread to near-total.

Other Treatments

Other treatments fall into three groups: impregnations, surface modifications and composite gemstones.

Identification of Treated Gems

A trained gemologist, taught by the Gemological Institute of America of Santa Monica, California, and New York, the Gemmological Association of Great Britain of London, or elsewhere, is needed for identification of treated gems. This topic is also discussed in textbooks. In some materials the induced change is the exact equivalent of a process that also occurs naturally, so that such treatments cannot be identified.

KURT NASSAU
Nassau Consultants

Additional Reading

Anderson, B.W. and E.A. Jobbins: *Gem Testing,* 10th Edition, Butterworths, London, UK, 1990.
Guides for the Jewelry Industry, U.S. Federal Trade Commission, Washington, DC, Feb. 27, 1979 (under revision in 1993).
Hurlbut, C.S. Jr. and R.C. Kammerling: *Gemology,* 2nd Edition, John Wiley & Sons, Inc., New York, NY, 1991.
Liddicoat, R.T. Jr.: *Handbook of Gem Identification,* 12th Edition, Gemological Institute of America, Santa Monica, CA, 1989.
Nassau, K. *Gemstone Enhancement,* 2nd Edition, Butterworths, Boston, MA, 1994.
Nassau, K.: *The Physics and Chemistry of Color,* John Wiley & Sons, Inc., New York, NY, 1983.
Nassau, K.: *Gems Made by Man,* Gemological Institute of America, Santa Monica, CA, 1980.

GENETIC CODE. See **Genetics and Gene Science (Classical)**.

GENETIC ENGINEERING. Biotechnological methods of genetic engineering are relatively new techniques that plant breeders have to make direct modifications of DNA, a living thing's genetic materials. Scientists make copies to genes for desired traits and introduce the gene copy into an organism such as a food crop. The new gene is usually a single gene whose function is well understood. These new techniques avoid one of the major problems encountered by plant breeders who use cross hybridization, no unwanted or undesirable genes are introduced with the desired gene. In addition, scientists can make copies of genes from any organism, plant, animal, or microbe, that may yield a desired trait and introduce that gene into a food crop. This greatly expands the pool of potentially useful traits available to plant breeders to improve food crops.

Once a desired gene has been introduced into a crop via genetic engineering, the gene is usually crossed into other crop lines that have desired commercial traits. Such crossing also permits the breeder to evaluate the genetic stability of the new gene. Overall, genetic engineering allows breeders to develop new varieties more rapidly, but at this stage of the technology, the new methods are used in conjunction with other methods of plant breeding, such as cross-hybridization.

The time required to evaluate new varieties and the number of field trials will vary depending on the need to confirm performance, to evaluate characteristics of the food, to evaluate environmental effects, and to produce the required amount of seed before the new plant variety can be grown commercially by farmers.

Genetic engineering is used to achieve the same goals of agronomic and quality characteristics as traditional techniques and allows the breeder to make some modifications that would not be possible through other methods of plant breeding.

Genetically engineered food crops have been developed to: resist pests and disease and to tolerate chemical herbicides; exhibit improved food processing traits; exhibit improved nutritional content; resist adverse soil and weather conditions; and to exhibit improved fruit ripening or softening, texture, or flavor.

In 1992, the FDA published a policy statement that explains how foods, fruits, vegetables, grains, and their by products such as vegetable oils, are regulated under the FD&C Act. This policy applies to foods and food ingredients, including animal feeds, derived from plants modified through all methods of plant breeding, including genetic engineering.

FDA's guidance to industry is the following:

Genetic Modification

The introduced genetic materials should be well-characterized to ensure that any introduced genes do not encode harmful substances and should be inserted stably in the plant genome to minimize the chance for subsequent undesired genetic rearrangements.

Toxicants

Plants are known to produce toxicants and antinutritional factors, such as protease inhibitors, hemolytic agents, and alkaloids, which often protect the plant against pests and disease. Many of these toxicants are present in today's crops at levels that do not cause acute toxicity or do not affect humans or animals when the food is properly prepared. New plant varieties should not contain levels of such toxicants that are above the range that exists in today's crops.

Nutrients

Another unintended consequence of genetic modification of the plant may be an alteration (relative to the total diet) in levels of important nutrients and bioavailability of a nutrient due to changes in the form of the nutrient or of other constituents that effect absorption or metabolism of nutrients.

New Substance

In some cases using genetic engineering, plant breeders may introduce genes into food crops that encode substances that differ substantially in structure and function from substances currently found in food. Based on current developments, such substances would be expected to be proteins or protein enzymes that modify carbohydrates and fatty acids in the food. In some cases, such substances will require premarket approval as food additives; in other cases, the food may require new labeling to properly inform consumers of the new attributes of the food. However, in most cases to date, the substances that occur in food as a result of gene transfer have been safely consumed as food previously or are substantially similar to food substances and would not require premarket review by FDA.

Allergenicity

There are thousands of different proteins in our food supply, and only a few cause food allergy reactions. However, because genetic engineering can result in the introduction of genetic material from essentially any source (plant, animal, or microbe) into food, there is a possibility that a protein encoded by the newly introduced genetic material will be an allergen and produce an allergic response in some members of the population. FDA has raised this issue in its guidance to industry, especially in cases where the transferred genetic material is derived from a source that is known to be commonly allergenic. Examples of such foods that affect the U.S. population include milk, eggs, fish crustacea, mollusks, tree nuts, wheat, and legumes (particularly peanuts and soybeans). FDA believes that proteins derived from commonly allergenic sources should be presumed to be allergens and special labeling would be required, unless scientific evidence demonstrates otherwise.

In cases where a protein is derived from a source that is not known to be allergenic, it is not possible to predict definitively allergenic potential. While it is unlikely that a new protein that occurs in very low concentrations in food will be an allergen (as is the case for most proteins introduced via genetic engineering at this time), developers have taken steps to minimize the likelihood that a new protein will be an allergen by evaluating whether the new proteins exhibit characteristics typical of allergenic proteins (such as stability to heat, acid, and enzyme degradation).

The FDA encourages developers to discuss questions regarding allergenicity with agency scientists.

Antibiotic Resistance Markers

In experiments involving genetic engineering, only a few plants cells take up the desired new gene. Developers use selectable marker genes during gene transfer experiments to improve their chances of selecting plants that have successfully incorporated the desired gene. The most widely used marker is kanamycin resistance gene that produces the enzyme, aminoglycoside 3′-phosphotransferase II (also referred to as APH(3′)II and neomycin phosphotransferase II). Plant cells are normally killed by antibiotics. APH3′II inactivates the antibiotics kanamycin and neomycin and permits plant cells to grow in culture that have incorporated gene and express the APH(3′)II enzyme.

Once the desired plant variety has been selected, the marker gene serves no useful purpose in the new plant, but it does continue to produce the gene product, APH(3′)II in the case of kanamycin resistance. This enzyme is present at very low concentrations in food.

The use of marker genes that encode resistance to clinically important antibiotics raises questions regarding whether the enzyme in the food could inactivate oral doses of the antibiotic or whether the gene present in the plant DNA could be transferred to pathogenic microbes in the GI tract or in soil rendering them resistant to treatment with the antibiotic. FDA evaluated these questions for the use of kanamycin resistance in tomato, cotton, canola.

The FDA found that kanamycin and neomycin are very toxic antibiotics and as such have very limited oral clinical use and are used only in situations where patients are not consuming food. There is also too little of the essential cofactor, ATP, present in food for the enzyme to degrade a significant amount of antibiotic.

There is no known mechanism by which a gene can be transferred from a plant chromosome to a microbe. Thus, the possibility of that such transfer would generate new resistant organisms is very small, especially when compared to the high rate of spread of resistance through known mechanisms of microbe to microbe transfer to antibiotic resistance. FDA believes that the use of marker genes that encode resistance to other clinically useful antibiotics can be evaluated by similar criteria that were used for kanamycin resistance.

Animal Feeds

Feeds developed for animals raised as food sources must be meet the same safety standards as human food under the FD&C Act. In contrast to the human diet, an animal feed derived from a single plant may make up over half of the animal's diet. Further, animals consume plants and plant parts that are not part of the human diet. Nutrient composition and availability of nutrients are important considerations for animal health.

Labeling

The FD&C Act defines the information that must be disclosed in labeling (including the food label). The Act requires that all labeling be truthful and not misleading. The Act does not require disclosure in labeling of information solely on the basis of consumer desire to know. The Act does require that a food be given a common or usual name, and that the label disclose information that is material to representations made or suggested about the product and consequences that may arise from the use of the product.

The FDA will require special labeling if the composition of a food developed through genetic engineering differs significantly from its conventional counterpart. For example, if a food contained a major new sweetener as a result of genetic modification, new common or usual name or other labeling may be required. Similarly, if a new food contains an allergen that consumers would not expect in that food, labeling would be necessary to alert sensitive consumer. However, if a protein commonly produces very serious allergic reactions (e.g., peanut protein) and is transferred to another food, FDA would need to evaluate whether labeling would provide sufficient consumer protection.

To date, the FDA is not aware of information that would distinguish genetically engineered foods as a class from foods developed through other methods of plant breeding and, thus, require such foods to be specially labeled to disclose the method of development. The agency has not required labeling for other methods of plant breeding such as chemical, or radiation-induced, not required to be labeled "hybrid sweet corn" because it was developed through cross-hybridization.

The FDA is reviewing public comments on labeling issues. One issue that is particularly difficult is the question of whether special labeling should be required for a food derived from a plant that has been modified to express a gene derived from an animal and whether the presence of such a gene or its product affects certain ethical or religious beliefs. Currently, no foods are approaching the market that raise this issue. However, the issue is very complex, and FDA believes that further discussion is warranted.

Summary

The FDA has provided guidance for developers, which establishes a standard to care to ensure that foods derived from new plant varieties are safe and wholesome. Irrespective of the method by which a food is produced, all foods must meet the same stringent safety standards and be properly labeled in accordance with the FD&C Act.

See also **Biotechnology (Bioprocess Engineering)**; **Genetics and Gene Science (Classical)**; and **Industrial Biotechnology**.

Additional Reading

Shannon, T.A.: *Genetic Engineering: A Documentary History,* Greenwood Publishing Group, Inc., Westport, CT, 1999.
Steinberg, M.L. and S.D. Cosloy: *The Facts on File Dictionary of Biotechnology and Genetic Engineering,* Revised Edition, Facts on File, Inc., New York, NY, 2000.
Yount, L.: *Biotechnology and Genetic Engineering,* Facts on File, Inc., New York, NY, 2000.

Web References

Food and Agriculture Organization of the United Nations: http://www.fao.org/
Food and Drug Administration: http://www.fda.gov/
Glossary of Biotechnology and Genetic Engineering: http://www.fao.org/DOCREP/003/X3910E/X3910E00.HTM
One World Guide to Biotechnology and Genetic Engineering: http://www.oneworld.org/guides/biotech/index.html
The Institute of Science, Technology and Public Policy: http://www.istpp.org/genetic_engineering.htm

GENETICS AND GENE SCIENCE (Classical).

Early biologists essentially targeted the hereditary aspects of life (animals and plants) and depended on empiricism and statistics for their knowledge. Toward the late 1880s, with the availability of improved microscopy and an increased interest in biochemistry, researchers turned much of their attention to the structure and performance of the individual cell. Soon cytology became an important biological discipline and from this, over a further time span, cells and their components were reduced to the molecular level. Although even in modern times hereditary processes remain very important research objectives, particularly as they relate to diseases, the new gene-based sciences also encompass such diverse fields as crop improvement and criminology. (See Table 1 for Chronology of Genetic Science.)

Fundamentals of Genetics

Gregor Johann Mendel (1866) sometimes is referred to as the father of genetics. By studying the crosses of garden peas in his garden, Mendel worked out the basic principles of inheritance. Over the years, genetics and the gene sciences have proceeded along six major pathways:

1. *Experimental Breeding*, a procedure dating back several centuries, requires considerable time and patience because the animals or plants studied must experience a number of lifetimes (generations). Statistical methods typify this kind of genetic research.

2. *Pedigree Analysis*, an approach widely used where experimental breeding is not practical. Pedigrees show the inheritance of specific traits, which can be traced, in all of the members of a family line. Human pedigrees have been very useful in terms of tracing the familial aspects of certain diseases. One of the first diseases so traced was hemophilia. Stock

TABLE 1. AN ABRIDGED CHRONOLOGY OF PROGRESS IN GENETIC SCIENCE. (Early Years to Commencement of the Human Genome Project)

1543	Andreas Vesalius, Belgian anatomist, produced the first anatomical map of humans in the paper, "De Humani Corporis Fabrica." This publication is recognized as a first step toward what may be termed, *intellectual medicine*. For many decades thereafter, chromosomes and genes (prior to their identification and naming) were simply considered the *minutia* of life and beyond research—prior to the introduction of improved microscopes.
1665	Cytology (science of cells) had its beginnings when Robert Hooke, English physicist, described the nature of cork cells.
1820	Robert Brown, Scottish botanist, postulated the "nucleus" of individual cells.
1838	Mathias Jacob Schleiden, German botanist, adopted Brown's views of the nucleus and proposed the general concept that living organisms are made up of cells and that the nucleus is essential to the formation of new cells.
1839	Theodore Schwann, German physiologist, published a paper, "Microscopic Investigations on the Accordance in the Structure and Growth of Plants and Animals," this leading to the first acceptance of the cellular origin and structure of animals and plants. Schwann observed, "The entire animal or plant is composed either of cells or of substances thrown off by cells; cells have a life that is somewhat independent, and this individual life of all the cells is subject to that of the organism as a whole."
1866	Gregor Johann Mendel, Austrian naturalist and botanist, published a paper entitled, "Experiments in Plant Hybridization," which was based upon his personal experimentation with garden plants (mostly peas) for tracing the dominance of traits from one generation to the next—and, in this sense, was the father of traditional genetic science. His work, however, was not received with acclaim, but rather was considered of little importance by Karl Näageli, a revered botanist during that period. Ironically, Mendel's principles were "rediscovered" independently many years later (1900) by Hugo De Vries, a Dutch botanist, by Karl Correns, a German biologist, and by Von S. Tschermak, an Austrian naturalist. This group put to final rest the prior concept that "heredity is transmitted by fusable parental *bloods*."
1876	Johann Friedrich Horner, Swiss ophthalmologist, was the first researcher to establish a connection between a cellular deformity (gene abnormality) and the familial aspects of color blindness.
1900	Hugo De Vries observed that heredity is a conservative force and that, if heredity were perfect, all organisms would carry the same genotype and evolution would not occur. De Vries pointed out, however, that this conservatism is opposed by a factor of change, that is, *mutation*. He suggested that mutational changes must be drastic and sudden, whereas it was soon to be learned that mutations range widely in their cause and effect.
1903	Camillio Golgi, Italian pathologist, while researching malarial parasites, demonstrated the nervous system as being interlaced rather than connected in a complete network. Golgi developed a method for staining nerve cells. Previously, Golgi had described the Golgi complex (apparatus) of the cell and considered that to be a cytoplasmic organelle occurring in almost every type of vertebrate cell. Golgi also described the importance of membranes in cells.
1903	Wilhelm Ludwig Johannsen, Dutch geneticist, introduced the words *gene, genotype*, and *phenotype* to the literature of genetics.
1909	A.E. Garrod, British physician, pioneered the field of developmental genetics and visualized development as a network of chemical reactions, many of which are facilitated by specific catalysts or enzymes. By extrapolation, he surmised that each enzyme is produced by just one gene and that each gene produces just one enzyme (Garrod-Beadle concept).
1910	Thomas Hunt Morgan and E.B. Wilson (Johns Hopkins University) became interested in using the fruit fly as an experimental model for heredity studies after having discovered a fly with white eyes, as contrasted with the normal red coloration. Subsequently, because of its very short reproductive span allowing many generations to be studied over a brief time period, the fruit fly (*Drosophila melanogaster*) became the focus of thousands of genetic studies continuing to the present. The genome of the fruit fly is approaching completion as of 1993. As a somewhat later date, mice became a model for geneticists and its genome is nearing completion as of 1994.
1911	E.B. Wilson (Columbia University) confirmed link of color blindness with the X-chromosome.
1946	Frederick Sanger, British biochemist, determined the complete amino acid sequence in the protein insulin. In prior years, Sanger had developed the use of 2,4-dinitrofluorobenzene (Sanger's reagent) which became an important tool for protein analysis. Sanger was first researcher to show that proteins are polypeptides in which alpha amino and imino acids are bound together by peptide bonds between their alpha-amino and alpha-carboxyl groups. Sanger was awarded the Nobel Prize (chemistry) in 1948.
1950s	Linus Carl Pauling, American physical chemist and 1954 Nobelist (chemistry), contributed new knowledge to the understanding of proteins, enzymes, and nucleic acids. Pauling also proposed the gene structure of hemoglobin, particularly as it relates to sickle cell anemia. Pauling and others also pioneered procedures for sequencing amino acids.
1952	Alexander Robertus Todd (Lord), British biochemist first researcher to synthesize adenosine diphosphate (ADP) and adenosine triphosphate (ATP). Todd was awarded the Nobel Prize (chemistry) in 1957.
1953	J.D. Watson, American chemist and Nobelist (1962), and Francis Harry Compton Crick, American scientist and Nobelist (1962), proposed that the molecular structure of DNA is composed of deoxyribonucleic acid and proteins (histones and high-molecular-weight proteins). These researchers proposed that the molecular structure of DNA is a double spiral helical chain. James H. White, American mathematician, shared the 1962 Nobel Prize.
1968	It was reported that 68 human genes had been mapped to the X-chromosome.
1970	Restriction enzymes, which cut DNA in specific places, were discovered and when coupled with recombinant DNA technology, made it possible to identify a specific stretch of genetic material.
1970	The concept of Recombinant DNA was proposed by several geneticists. Thus, new DNA structures could be created. Both positive results and negative concerns were expressed. For example, the addition of new genes to bacteria and viruses could confer qualities that could be harmful to other forms of life, including humans, with possibly epidemic, even catastrophic proportions. Researchers attending 1973 Gordon Research Council proposed that the National Academy of Sciences address these concerns. Guidelines and regulatory actions were initiated, some of which continue to the present. Regulations vary somewhat between one country and the next.
1970s	Torbjorn Caspersson and Lore Zech (Karolinska Institute, Stockholm) developed a staining technique (using quinacrine mustard) that fluoresces under ultraviolet light, revealing that each chromosome has a unique banding pattern.
1976	A.M. McKusick (then at University of Washington) published a catalog of 1,487 genetic disorders. This was revised in 1990 to include nearly 5,000 inherited characteristics. About a decade later, McKusick became the first head of the International Genome Organization.
1988	The National Academy of Sciences (U.S.) endorsed a massive national effort to map and sequence the human genome. The project target—to produce genetic and physical maps of increasing resolution, with a fully detailed map of the chromosomes—the project to be completed within a decade and at a cost estimated to be $3 billion.

Further details are given within text of article.

breeders keep careful pedigree records as breeding guides. Horses and other high-performing animals are bought and sold based upon their pedigrees.

3. *Cytogenetics* (cytology) is a study of the chromosomes and cellular infrastructure that are keys to heredity. This field now embraces the study of individual genes.

4. *Biochemistry* and, in particular, molecular biology is a study of the genes—what they are, how they perform, and how they reproduce. Through an analysis of gene action, biochemical geneticists—working with such diverse organisms as molds, bacteria, viruses, fruit-flies, mice, and human cells—have been able to trace the course of the breakdown of particular amino acids in the cells and to learn of abnormalities that arise when a gene fails to produce a particular enzyme.

5. Population genetics deals with the distribution of genes in various populations. Human population geneticists have traced population migration and the intermixing of races through an analysis of the frequency of the various blood antigens. Within recent years, some geneticists have

turned to analyzing fossil genetic material to trace the process of heredity over many thousands of years.

6. *Genetic Recombination*, made possible by the discovery of the recombinant DNA procedure in the 1970s, makes it possible to develop extensive and detailed maps of the nucleotide sequences of gene molecules—to the point where, in 1990, plans were outlined for mapping the complete human genome, a program that was well underway as of 1994.

Defining the Gene

Genes are the physical units of heredity. The precise definition for gene has changed over the years as more has been learned about the chemical nature of genetic material and function. In modern terms, a gene may be defined as a segment of genetic material that determines the sequence of amino acids in specific polypeptides. In lieu of additional findings, geneticists have noted a one-to-one relation between gene and polypeptide. It appears that this definition applies at least to those genes called *structural* genes because they determine the primary structure of proteins.

Structural Genes. So far as known, structural genes in all organisms are composed of nucleic acids. In the RNA viruses, the genes are RNA (ribonucleic acid) only, but in all other organisms, the DNA viruses and the cellular forms, which all possess both DNA (deoxyribonucleic acid) and RNA, the gene material is either known to be DNA, or assumed to be for good reason.

The genes of viruses and bacteria appear to consist of nucleic acid unaccompanied by closely bound protein. Ordinarily this naked nucleic acid is in the two-stranded condition; exceptions are known among both the RNA and DNA viruses, some of which possess single-stranded genetic material. In those organisms with true nuclei, the genetic material is always double-stranded DNA associated with protein ordinarily of the histone type. The function of the protein is not considered to be genetic. It probably controls DNA in its role of determining protein structure. Also it may serve to hold genes together and attached to the chromosomes of which they are a part.

Structural genes carry out their role of dictating protein structure by producing a messenger RNA (mRNA) which is a single strand of RNA containing nucleotide bases complementary to one of the strands of the double-stranded DNA of the gene from which it is copied or "transcribed." The evidence is that the same DNA strand of a gene is always transcribed into mRNA. In this way, only one kind of mRNA is made for each gene. In the transcription process the C, T, A and G bases of the DNA determine G, A, U and C, respectively, in the mRNA strand. Transcription effectively constitutes *gene action*. By definition, if a gene is not actively forming mRNA, it is inactive or "turned off."

Each kind of gene is different from every other gene in its DNA sequence. Hence, as many different kinds of mRNA are formed as there are different genes in the organism.

Genes in eukaryotic cells are often not collinear with their products. Instead genes contain intervening sequences of DNA (*introns*) which result in a gene that is much longer than required for the simple coding of amino acid sequence. An enzymatic reaction, gene splicing, is required for the expression of the genes. That is, the entire gene, including introns, is transcribed as a long mRNA precursor. The intervening sequences are clipped out and the ends rejoined to yield the mRNA with the correct coding sequence for the gene product. After their formation, the mRNA strands attach to ribosomes in the cytoplasm, and the process of protein biosynthesis commences. The significant point to be emphasized here is that the sequence of nucleotide bases, of the "genetic code," in a particular gene is reflected in a specific sequence of amino acids in the polypeptide produced through the protein synthetic mechanism.

The one-to-one relation between gene and polypeptide is a more accurate statement of the situation than the earlier one gene-one enzyme hypothesis. It is now known that a number of proteins are constituted in their functional state of subunits, which are polypeptides. When subunits are all identical, the one gene-one protein statement holds with certain exceptions. However, proteins such as vertebrate lactic acid dehydrogenase (LDH) and hemoglobin are made up of different subunits. For example, the dominant adult hemoglobin in man contains both α and β polypeptides as subunits. These have somewhat different amino acid sequences, and each has been shown to be under the control of a different gene. The genes are not even on the same chromosome. A similar situation has been found for LDH which may be made up of at least two different subunits, each one again under the control of a separate gene.

A term currently used by many synonymously with structural gene is *cistron*. Its original definition was based on complementation tests. If two chromosomes bearing the same kinds of genes (homologous chromosomes) are introduced into the same cell, "product interactions" may be observed between the genes of the same type, *i.e.*, genes which control the same kind of polypeptide. If two genes of the same type are mutant, but mutant at different sites, they may *complement* and produce a protein that has an activity comparable to the nonmutant, even though each mutation alone, or together on the same chromosome, can produce only a mutant, inactive protein. Those mutants that do not complement with the production of an active protein are said to have mutational sites within the same cistron.

Controlling Genes. Genes which do not carry codes for the synthesis of proteins which constitute the enzymes, structural components, etc., of the cell almost certainly exist. These genes may produce proteins, but the proteins presumably act by the regulation of the activity of the structural genes, turning them on and off according to circumstances within the cell.

Examples of such genes are found in *Escherichia coli*. These, termed *regulator genes*, presumably produce substances, possibly proteins, which prevent or repress structural genes from synthesizing mRNA unless other substances, the inducers, are present to inhibit the repressor substances. Alternatively, repressor substances from other types of regulator genes are active in repression only when certain substances activate the repressor substances. The reason for the existence of these genes would seem to be for the regulation of metabolism by preventing the overproduction of enzymes when their substrates are not present, or of end products such as amino acids. In the latter case, the end product is usually considered to be the substance which activates the repressor produced by the regulator.

Genetic Code. Genetic information stored in the genes, as a linear sequence of the bases (A, C, G, and T) in deoxyribonucleic acid molecules, is transcribed into a complementary base sequence (U, G, C, and A, respectively) in the messenger RNA molecules. This "coded message" contained in the mRNA, as a linear sequence or 4-letter "language," is "translated" in the process of protein biosynthesis into a linear sequence of the 20 amino acids within the protein polypeptide chain synthesized. Each nucleotide triplet or "code word", consisting of one of the 64 possible triplet combinations of U, G, C, and A nucleotides in a messenger RNA molecule, may specify one particular amino acid for incorporation into the polypeptide chain. It appears that certain amino acids may be specified by more than one of the 64 nucleotide triplets; in this respect, the genetic code is said to be "degenerate." A few particular triplet "words" may have special functions, such as to signal polypeptide-chain initiation, or chain termination. The first identification of a particular triplet as the code word for a particular amino acid was the discovery that the sequence UUU (in the form of polyuridylate) appears to be the "code word" specifying incorporation of phenylalanine into a polypeptide, in a cell-free, *in vitro* system containing ribosomes and other required components.

Evidence that a nucleotide *triplet* (and not some smaller or larger run of nucleotides) is the "code word" for incorporation of a specific amino acid has come from studies of the fine structure of genes or DNA of a bacteriophage (virus). Many tentative formulations of a "code dictionary" of messenger RNA triplets, with the corresponding amino acid specified by each triplet, have been proposed, on the basis of both experimental results (primarily those of the Nirenberg group and of the Ochoa group) and theoretical considerations. The exact determination of the genetic code, or pattern of correspondence between each possible nucleotide triplet of mRNA and the amino acid specified by that triplet for incorporation into proteins, has been an active field.

Deoxyribonucleic Acid (DNA)

DNA is a complex sugar-protein polymer of nucleoprotein, which contains the genetic code for enzymes in the cell. It occurs as a major component of the genes, which are located on the chromosomes in the cell nucleus. The DNA molecule is a unique and vastly intricate structure. It is comprised of from 3000 to several million nucleotide units arranged in a double helix containing phosphoric acid, 2-deoxyribose, and the nitrogenous bases adenine, guanine, cytosine, and thymine. The spiral (see Fig. 1) consists of two chains of alternating phosphate and deoxyribose units in continuous linkages. See Fig. 2. The nitrogenous bases project toward the axis of the spiral; they are joined to the chains by hydrogen bonds. Adenine units pair with thymine, and cytosine units with guanine. The complementarity of the bases on the joined chains allows each chain to act as a template for replication of the other when the chains are separated, thus producing two new strands of DNA. See Fig. 3. The sequence of the bases on the chains varies with the individual, and it is this sequence that governs the genetic code. DNA works in conjunction with ribonucleic acid (RNA). Genes are found in pieces that are spread out along DNA. Between gene fragments, there are long stretches of DNA, the functions of which are only recently being clarified.

DNA in Perspective. As early as 1838, Schleiden and Schwann proposed that large organisms, as represented by the complete animal, are constructed from large numbers of very small cells, all of which are derived from a single original cell by the repeated process of cell division. The nature of the molecular processes underlying cell division did not emerge until the early 1950s. The chemical nature of DNA and RNA was not established until 1952 by Brown and Todd. In that same year, through a detailed analysis of insulin, Sanger showed that proteins are polypeptides in which alpha-amino and imino acids are bound together by peptide bonds between their alpha-amino and alpha-carboxyl groups. These molecules were shown to be polymers in which limited numbers of monomers linked together to form molecules having complex properties.

For several years, the biological roles of these substances were controversial topics in the scientific community. In 1944, investigators Avery, McLeod, and McCarty suggested an essential distinction between DNA and RNA; they were joined in 1952 by Hershey and Chase in this opinion. It was concluded at that time that DNA is the fundamental storehouse of genetic information.

Phillips suggests that during the early 1950s, molecular biologists were seeking the answers to three fundamental questions: (1) How is the information-embedded in the DNA of the genes copied for transmission to successive generations of cells? (2) How does this information direct the synthesis of proteins? and (3) How do proteins, essentially having simple structures, acquire their diverse and subtle chemical properties?

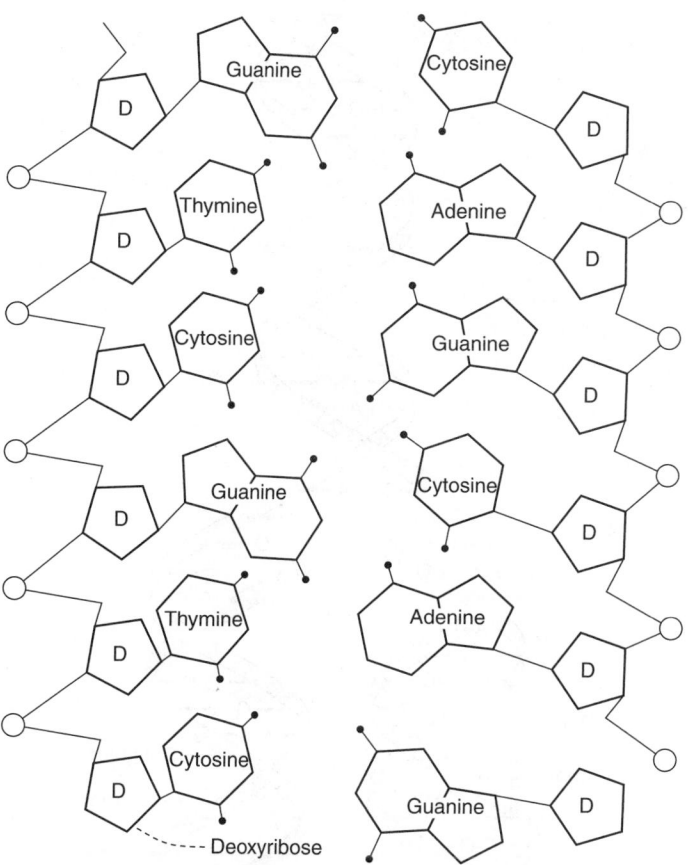

Fig. 2. Schematic of DNA molecule showing repeating sequences of deoxyribose (white pentagons) and phosphodiester units that provide structural support. The varying sequences of pyrimidine and purine bases encode genetic information. The purines are guanine and adenine; the pyrimidines are thymine and cysine. Note that guanine pairs with cytosine; adenine pairs with thymine

Very shortly, the first question was answered in principle by Watson and Crick who proposed the three-dimensional structure of DNA in 1953. Their proposal that DNA is composed of two polynucleotide chains forming a double helix was based upon studies of x-ray diffraction patterns of DNA fibers.

Ribonucleic Acid (RNA)

Ribonucleic acids comprise a group of natural polymers consisting of long chains of alternating phosphate and D-ribose units, with the bases adenine, guanine, cytosine, and uracil bonded to the 1-position of the ribose. Ribonucleic acid is universally present in living cells and has a functional genetic specificity due to the sequence of bases along the polyribonucleotide chain.

Types of RNA include the following. (1) *Messenger RNA*, synthesized in the living cell by the action of an enzyme that carries out the polymerization of ribonucleotides on a DNA template region which carries the information for the primary sequence of amino acids in a structural protein. It is a ribonucleotide copy of the deoxynucleotide sequences in the primary genetic material. (2) *Ribosomal RNA*, which exists as a part of a functional unit within living cells called the ribosome, a particle containing protein and ribosomal RNA in roughly 1:2 parts by weight, having a particle weight of about 3 million. Messenger RNA combines with ribosomes to form polysomes containing several ribosome units, usually five (e.g., during hemoglobin synthesis), complexed to the messenger RNA molecule. This aggregate structure is the active template for protein biosynthesis. (3) *Transfer RNA*, the smallest and best-characterized RNA class. Its molecules contain only about 80 nucleotides per chain. Within the class of transfer RNA molecules, there must be at least 20 separate kinds, correspondingly related to each of the 20 amino acids naturally occurring in proteins. Transfer RNA must have at least two kinds of specificity. (1) It must recognize (or be recognized by) the proper amino acid activating enzyme so that the proper amino acid will be transferred to its free 2' or 3'

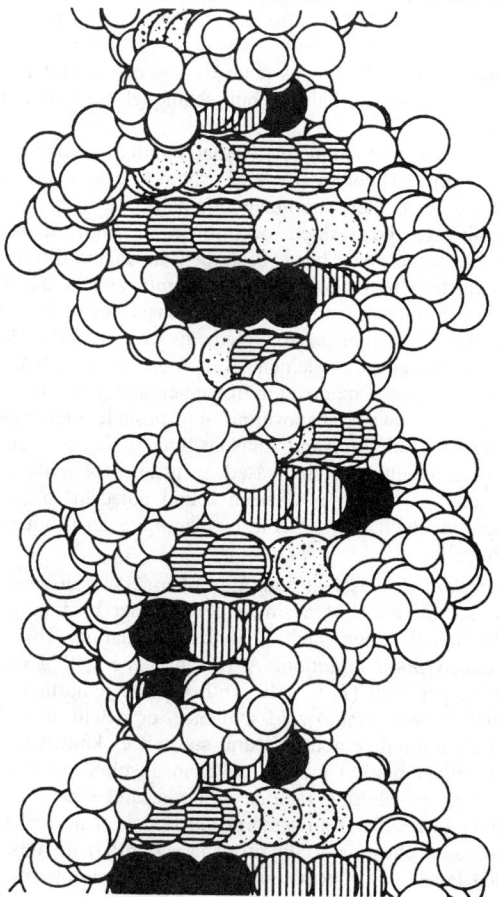

Fig. 1. Consisting of two helically intertwined strands, the DNA molecule is composed of deoxyribose and phosphate. As shown here, at periodic intervals the sugar-phosphate backbones are joined together by the complementary purine and pyrimidine bases. A single base linked to a deoxyribose-phosphate moiety constitutes a deoxyribonucleotide. Legend: Solid black circles = Thymine; Vertical bars = Adenine; Horizontal bars = Guanine; Dotted circles = Cytosine

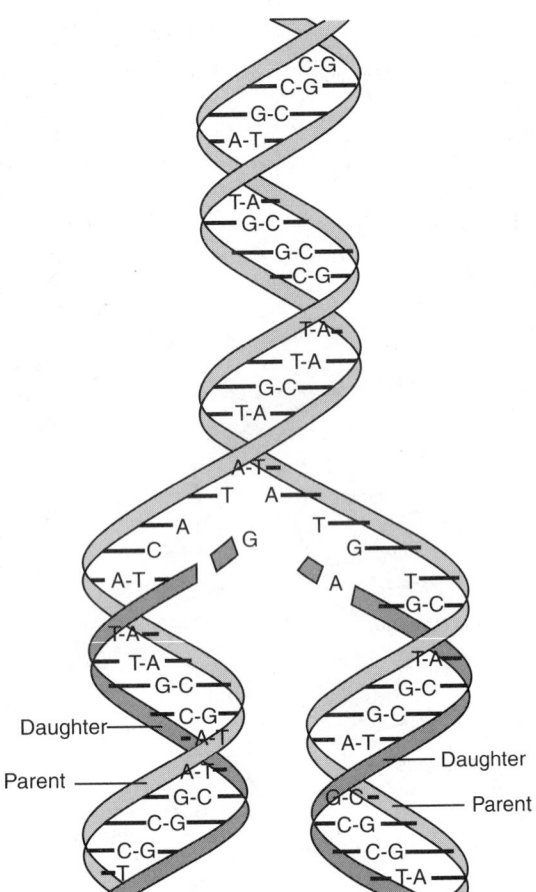

Fig. 3. For replication, the two strands of the parent DNA molecule (light gray) separate as the base pairs detach. The replicated (daughter) strands (dark gray) form as guanine (G) pairs with cytosine (C) and adenine (A) pairs with thymine (T)

OH group. (2) it must recognize the proper triplet on the messenger RNA-ribosome aggregate. Having these properties, the transfer RNA accepts or forms an intermediate transfer RNA-amino acid that finds its way to the polysome, complexes at a triplet coding for the activated amino acid, and allows transfer of the amino acid into peptide linkage.

Mutations of Genes

New organisms in nature normally are formed by very slow processes. A change in the base sequence of the DNA constituting a gene results in an inherited alteration in the code and is called a *gene mutation*. Mutations are genetic changes that occur suddenly and are thereafter heritable.

Mutations arise through three general mechanisms: (1) chemical modification of preformed DNA, such as breakage and aberrant reunion of molecules or the changes elicited by ultraviolet light, for example; (2) errors in incorporation of the purine and pyrimidine bases, or additions and subtractions of bases, during DNA replication; and (3) unequal exchange between two identical or similar DNA molecules ("unequal crossing over") during recombination. These chemical changes normally occur with low frequency (spontaneous mutations), but the frequency can be increased by means of various chemical and physical treatments (induced mutations). Even when so induced, the frequency of bacterial mutants for a particular trait, for example, is low, e.g., one mutant in 10^4 to 10^{10} bacteria. Thus, any biological evolutionary alterations brought about by the mechanism of mutation represent a very slow pathway. Such procedures do not comprise effective tools for what has been referred to as *genetic engineering* (genetic manipulation) wherein gene structures can be willfully directed under laboratory conditions.

A change in the base sequence of the DNA constituting a gene results in an inherited alteration in the code and is called a gene mutation. Changes in base sequence may conceivably result from: (1) the deletion or addition of one or more nucleotide pairs in the DNA chain; (2) changes in one or more bases along the chain; or (3) inversion of a segment of the chain.

Good evidence for the occurrence of the first type of mutation exists at least in bacteriophage of the T series, which infect *Escherichia coli*. The deletion or addition of a single base pair into a DNA chain of a gene should be expected to cause considerable difficulties in the translation of the code in the derived mRNA, into an amino acid sequence. For example, if the mRNA of the nonmutant strain has the sequence:

$$\overline{GCU}\ \overline{AAU}\ \overline{GAA}\ \overline{UUU}\ \overline{AAA}\ \overline{CAU}\ \overline{\cdots}$$

which is read in triplets from left to right to give a particular sequence of amino acids, say ala · asp NH² · glu · phe · lys · his, a deletion of a base in the mutant would change the "reading frame" starting at the point of deletion. Thus, the sequence

$$\overline{GCU}\ \overline{AAU}\ G\!\downarrow\!\overline{AU}\ \overline{UUA}\ \overline{AAC}\ \overline{AU}\cdots$$
$$_A$$

would produce the sequence ala · asp NH_2 · asp · leu · aspNH$_2$ as one possibility. A similar result would be expected from a duplication, by shifting the reading frame. Such mutations as these should be expected to produce "nonsense" sequences of amino acids after the point of change, and they are referred to as frame shift mutations.

Mutations which are the result of the simple changing of bases, say G $\Longleftrightarrow$ A, C $\Longleftrightarrow$ T, or G $\Longleftrightarrow$ C, or A $\Longleftrightarrow$ T should obviously cause changes in a single triplet rather than a whole sequence. As a result only a single amino acid change should occur in the polypeptide, if but a single base is changed. Many mutant proteins from a variety of organisms are now known which have but a single amino acid change from the nonmutant, and are therefore presumably the result of a single, or adjacent changes within a single triplet. The nonoccurrence of single mutations causing the substitution of two adjacent amino acids within a chain is evidence that the genetic code is not overlapping. As might be expected, a number of different amino acids may be substituted for the nonmutant acid, but the number of substitutions has been found to be limited for any particular amino acid. This also has connotations for the nature of the genetic code.

Gene mutations of other types such as inversions probably occur in addition to the two discussed above, but techniques have yet to be devised to analyze them.

For the present it is enough to say that a mutation may occur at any point within a gene. Theoretically there should be as many "mutational sites" within a gene as there are nucleotide pairs.

Gene instability, the sudden occurrence of high mutability of a normally stable gene, has been described at many different loci in maize, the galactose region of *Escherichia coli*, and in the white locus of *Drosophila*. Studies of the molecular basis for this instability in bacteria have identified transposable elements (*transposons*) as the agents responsible. A transposable element is a segment of DNA capable of transposition intact from one position in the genome to another. In addition to promoting their own transposition, transposable elements can also promote inversion, deletion, and transposition of adjacent chromosomal DNA sequences, resulting in increased occurrence of mutations. Indeed transposable elements have now been found adjacent to many mutant genes in bacteria. Such transposable elements are also thought to occur in eukaryotic cells.

Point Mutation. A classical definition of a genetic disease is one that results from the mutation of a single gene, either by inheritance or by some environmental factor, such as ionizing radiation. This situation is sometimes called *point mutation*. A mutant gene will generally cause one of two happenings: (1) it will synthesize an abnormal protein that has an altered primary amino acid sequence, or it will alter the level of production of a normal protein. Natural substances known to be affected by inherited point mutation include collagen, insulin, myoglobin, a large number of enzymes, clotting factors, albumin, and others. For example, hemolytic anemia results from underproduction of a normal form of the enzyme, glucose-6-phosphate dehydrogenase (G6PD). In this case, there are decreased levels of a normal stable enzyme. It is interesting to note that many genetic defects are not observed in utero because the mother may generate sufficient required enzymes. After birth, several weeks may elapse before the newborn indicates a lack of a given enzyme. There are other situations where years may be required for the abnormality to be detected. This may be true, for example, in the case of a degrading enzyme that very slowly causes the accumulation of metabolic waste products. In the case of Gaucher's disease, undegraded macromolecules in the liver and spleen will cause these organs to enlarge over a period of time. The time span of detection may range from the development of gross mental deficiency, blindness, and death in a child's first year of life; or, much later in life, to the detection of an enlarged spleen during surgery for some other condition.

Thus, it has been found that genetic diseases run the gamut of time and of severity. In Pompe's disease, a deficient enzyme (alpha-1,4-glucosidase) may range from death (total deficiency) to the progressive manifestation of cardiac or peripheral myopathy in later life (mild deficiency).

Recombinant DNA Technology

In the early 1970s, there was an interesting observation of great significance, that is, the discovery of certain enzymes that have the ability to cut and splice hereditary material. The cut pieces are about the order of a gene in length. Also, some of these enzymes have the further ability to cut a few bases further down than the others, so that what sometimes are known as "sticky ends" are produced. Thus, any species of DNA, if cut by the same enzyme, will possess the same type of sticky ends, and fragments of differing DNAs, through a form of biological "scissors and paste" process, can cause the lower part of one DNA molecule to stick well onto the upper part of another molecule. The result is a hybrid molecule. Theoretically, the technique can cross the boundaries of species by selecting DNA material from fully different sources. The ability to cut and recombine is the basis for the term *recombinant DNA*.

A useful modification of the basic clip-and-paste process involves inserting the DNA fragments into a DNA molecule, which has the power of self-replication. Many bacteria contain small circular cytoplasmic DNA molecules called *plasmids*, which are capable of self-replication inside the bacterial cell. The characteristics of rapid bacterial growth and multiplication allow quantity replication of the recombinant plasmids in short periods of time. This technique thus offers an obvious advantage over the slow and laborious chemical methods.

However, obtaining sufficient quantities of a specific gene in purified form, for insertion into a plasmid, is difficult when one considers the genetic complexity of living organisms. An approach to the problem has been the use of an enzyme known as *reverse transcriptase*. This enzyme synthesizes DNA from RNA. The primary product of genes is mRNA, which possesses base sequences complementary to the genes. The large quantities of specific mRNA available, coded for by the single gene, allow biochemical purification of the mRNA. Thus, if one can isolate the mRNA coded from a particular gene, the corresponding DNA sequence, identical to the gene, can be reconstructed using reverse transcriptase. This synthesized DNA then can be inserted into a plasmid by standard recombinant DNA methods and amplified by growing the plasmid in bacteria.

The advantages of recombinant techniques for increasing knowledge of the genetic construction of any organism are immediately recognized. A number of practical findings from such investigations can be envisaged. These include the incorporation of nitrogen-fixing genes in agricultural plants, to eliminate the need for nitrogen fertilizers; and the bacterial manufacture of large quantities of polypeptide hormones, such as insulin; and the bacterial production of vaccines and enzymes, as well as the treatment of genetic diseases. Possible production of fermentation products (alcohol, methane, etc.) as fossil fuel substitutes may be aided by this technique.

It should be stressed that recombinant DNA methodology is *not* a way of constructing new forms of life in vitro. Even the simplest organisms are extremely complex and the maximum alteration of the simplest genome would be of the order of 1%. Also, the genomes of the simplest organisms are highly ordered and the random insertion of a few genes from an unrelated organism is unlikely to create a whole new organism.

In the initial stages of recombinant DNA research, there was considerable concern regarding possible serious consequences of producing biologically hazardous DNA molecules. Both self-policing and governmental guidelines, which are under continuous review, were established, and they continue in most countries where recombinant DNA research is being conducted. The concept of *biological containment* was developed. By this means, safety factors may be built into the genetic structure of the organism to be studied. For example, as bases for recombinant experiments, EK2 derivatives of *E. coli* cells have been used. These are 100 million times less able to survive in nature outside an artificial laboratory environment and thus present no biohazards to the community. These mutant cell lines are usually constructed by causing a deletion of a portion of DNA in a gene responsible for critical cell characteristics, such as

ability to metabolize a certain substrate or to construct a rigid cell wall. Alternatively, defective mutant genes may be inserted into the genome replacing normal genes responsible for properties critical to the survival of the cell.

When two homologous (i.e., bearing the same kinds of genes) chromosomes are paired in synapsis (as in early meiosis in the nucleated organisms) recombination may occur. Recombination is the exchange, usually equal, of segments of chromosomes. Thus, if a chromosome marked:

$$A b C D e f g H i J$$

recombines with one marked:

$$A b c d E f G h I j$$

between d and e, the recombinant products will be A b c D E f G h I j and A b c d e f g H i J. This natural process presumably occurs in all organisms both *between* or *within* genes. First, it provides a powerful tool for establishing that the genes are ordered linearly on the chromosome, and in what order. Second, it allows one to establish that there exists a collinearity between the genetic material of a gene and the polypeptide it produces. This has been done by mapping a number of mutational sites for a gene that determines one of the polypeptides (protein A) forming the enzyme tryptophan synthetase in *Escherichia coli*. Each mutant produces a modified protein A, which can be shown to differ from the wild type by a single amino acid substitution. When the order of the mutant sites on the *coli* chromosome was compared to the order of amino acids within protein A affected by the mutations, it was found that they were the same. This fundamental finding could only have been possible with the use of a recombination analysis.

Laboratory equipment and reagents for accelerating the manipulation of genetic material have improved markedly in recent years, but the details are beyond the scope of this encyclopedia.

Genes and Diseases

A considerable burden of human disease is attributable to an individual's genetic inheritance. Advances have enabled the detection of an increasing variety of diseases in fetal development and, in some cases, provide a basis for successful treatment.

Studies on human genetics have long been confined to observations of pedigrees and populations with respect to phenotypic traits. Most recently, however, advances in cell biology, biochemistry, cytogenetics and immunology have enabled geneticists to study the human genome more directly and techniques utilizing recombinant DNA have revolutionized these studies. Additionally, technologies employing monoclonal antibodies, hybrid cells, sophisticated protein chemistry, and prophase chromosome banding are all being brought to bear on a variety of problems in human genetics.

At the cytological level, the power and resolution of a variety of chromosome staining and banding techniques has been increased by their application to prophase chromosomes and the genetic map now locates over one thousand bands. At the nucleosomal level, the association of DNA with histone proteins is reasonably well understood. However, knowledge of the higher order structure and the nature of the association between DNA and the acidic structural scaffold, or core proteins, of the chromosome, remains unresolved.

A number of recent surprises have been: the discovery of the split nature of the gene, with its intervening introns; and the later findings of nonfunctioning gene copies or *pseudogenes* and, particularly, the finding of scattered pseudogenes representing DNA copies of processed mRNAs that had become incorporated into the genome. Large and clinically important gene clusters, such as those of the major histocompatibility complex, beta-globulins and the immunoglobulins, have been the subjects of much recent study.

The mechanisms involved in gene activation and inactivation are major problems in biology, so that transient, or permanent structures associated with such phenomena will continue to attract much attention. There is now evidence for changes in chromatin structure at chromosome sites prior to their becoming transcriptionally active; nuclease sensitive sites, enhancers, and promoters have also been identified at various loci.

Defining the location and association of genes and gene clusters in the genome is essential for the understanding of genome organization and in order that genetic techniques may identify and enlighten inherited diseases. Both family (meiotic) and somatic (mitotic) approaches have been dramatically extended, not only by introduction of recombinant DNA technology, but also through use of restriction fragment length polymorphisms and the isolation and cloning of DNA sequences of known and unknown function.

Many genes coding for proteins involved in the disease process have been isolated and cloned. A direct comparison between genomic DNAs of individuals with and without a specific inherited disease is, however, not at present practical because of the large size of the genome and the multitude of nonrandom base changes in on-coding DNA. If the disease is a consequence of lack of expression of a given gene in a specific tissue, then tissue-specific cDNA libraries can be made from mRNAs from the tissues of normal and affected individuals and the libraries compared by crosshybridization to identify a missing sequence.

Specific DNA probes exist for a number of chromosomes so that diagnosis of fetal sex, sex chromosome anomalies, trisomes, and other aneuploidies will shortly be available. Diagnosis of hemoglobinopathies by fetal blood sampling has already been superceded by DNA analysis. Recombinant DNA technology is obviously going to play a major role in antenatal diagnoses.

Although most human cancers are acquired diseases, all types may occur in heritable or nonheritable forms, and heritability may be associated with a dominant or recessive expression at a single locus, or with a constitutional chromosome anomaly. The changes associated with inherited predisposition to cancer must involve genetic alterations or mutational events at the sites of chromosome anomalies. There is now evidence for this in retinoblastomas.

In acquired malignancies, oncogene activity appears to occur in association with chromosomal rearrangement. There is some evidence that the cooperation of two or more oncogenes, acting in concert, or in sequence, may effect transformation of a normal state to a malignant one. However, further studies are needed to clarify this situation.

Diseases arising from genetic causes may be metabolic, endocrinologic, neurologic, or may develop as the result of mutation, organ implantation, and other factors.

Metabolic Disorders. These fall into four general categories:

- Lipid—hyperlipoproteinemias.
- Purine—gout and Lesch-Nyhan syndrome.
- Metal—Wilson's disease (hepatolenticular degeneration), and hemo chromatosis.
- Porphyrin—porphyrias and idiopathic hyperbilirubinemia.

Generally, metabolic disorders result from:

- *Carbohydrate abnormalities,* such as renal glycosuria (a transport defect), pentosuria (enzyme deficiency, xylitol dehydrogenase), lactase deficiencies, fructose intolerance, galactosemia, galactokinase deficiency, oxalosis, and several glycogenoses (von Gierke's, Forbes', Andersen's, Hers's, and Tarui's diseases).
- *Lysosomal storage abnormalities,* such as glycogenosis (Pompe's disease), Tay-Sachs, Krabbe's, Gaucher's, and Fabry's diseases, as well as metachromatic leukodystrophy, aspartylglycosaminuria, and Niemann-Pick disease. Also included in this category are mucopolysaccharidoses, Hunter's, Schele's, and Hurler's syndromes.
- *Amino acid abnormalities,* such as phenylketonuria, tyrosinemia, alkaptonuria, albinism, histidinemia, hyperprolinemia, homocystinuria, cystinuria, and ketoaciduria. Note that these names, in general, imply the germane amino acid.
- *Urea cycle abnormalities* including hyperammonemia, cirtullinemia, argininosuccinicaciduria, and argininemia.
- *Collagen abnormalities,* such as Ehlers-Damlos syndrome, Marfan's syndrome, pseudoxanthoma elasticum, and osteogenesis imperfecta.

Endocrinologic Disorders. These fall into two general categories:

- *Polypeptide hormonal dysfunctions,* such as diabetes mellitus, familial goiter, pseudohypoparathyroidism, and congenital adrenal hyperplasia.

- *Steroid hormonal dysfunctions,* including male pseudobermaphro ditism and testicular feminization.

Neurological Disorders. Although there are other disorders that are suspect, but fully connected to genetic causes, the principal connections already positively made are the muscular dystrophies.

Hematological Disorders. Blood related diseases include hereditary spherocytosis, pyruvate kinase deficiency, glucose-6-phosphate dehydrogenase deficiency, and hemoglobinopathies, such as thalassemias.

Renal Disorders. Kidney and urinary tract diseases include hypophosphatemic and vitamin D-resistant rickets, renal tubular acidosis, and Fanconi's syndrome.

Immunological Disorders. There are several kinds, for example, amyloidosis.

Genetic diagnosis and therapy are discussed in several articles on specific diseases throughout this encyclopedia.

New Advances in Understanding DNA Replication

Many important details have emerged concerning the mechanisms of DNA replication in both bacteria (Prokaryotes) and higher cells (Eukaryotes). These mechanisms are vital in understanding how a cell duplicates its genetic material (DNA), and how this duplication is related to cell division. For these reasons, cells have evolved elaborate mechanisms to ensure that the process of duplication (DNA replication) is error free. This level of control is so important that cells will actually cease cell division if errors become too frequent and wait until the DNA is repaired.

Two relatively recent developments have added to our knowledge significantly concerning how DNA replication occurs with fidelity or in what molecular biologists and biochemists call a processive polymerase activity. DNA polymerase is the enzyme which actually polymerizes (adds DNA precursors or building blocks) DNA. There are many such DNA polymerases in pro- and eukaryotic cells that have different functions but the main enzyme in prokaryotes is DNA polymerase III and in Eukaryotes, DNA polymerases alpha, delta, and epsilon. All four of these DNA polymerases are made of subunits.

The most important of these subunits for processive activities are termed the "brace" and the "clamp". They are both remarkable structures, themselves composed of protein subunits. The brace is a complex of proteins which binds to DNA at a site where DNA replication will begin (a double stranded primed DNA template), and in the presence of the high energy compound ATP (adenosine triphosphate) will help load the clamp. The clamp is a ring-shaped DNA polymerase accessory protein complex that looks like a "doughnut" and slips over the double stranded primed template like a curtain rod sing. Once this occurs, the rest of the DNA polymerase will associate to form the processive polymerase. The clamp-DNA polymerase complex will then slide down the template DNA strand (with the clamp behind the polymerase pushing it) easily, picking up DNA precursors to be polymerized into newly synthesized DNA, until the entire DNA molecule has been duplicated faithfully.

This remarkable mechanism relates structure to function in DNA replication in ways not easily noted previously.

New Advances in Understanding DNA Replication written by William Firshein, Professor, Department of Molecular Biology and Biochemistry, Wesleyan University.

The Human Genome Project (HGP)

The Human Genome Initiative is a worldwide research effort that has the goal of analyzing the structure of human DNA and determining the location of all human genes. In parallel with this effort, the DNA of a set of model organisms will be studied to provide the comparative information necessary for understanding the functioning of the human genome. The information generated by the human genome project is expected to be the source book for biomedical science in the 21st century. See Fig. 4. It will have a profound impact on and expedite progress in a variety of biological fields, including those such as developmental biology and neurobiology, where scientists are just beginning to understand the underlying molecular mechanisms. The analysis and interpretation of the information will occupy scientists for many years to come. Thus, the maximal benefit of the human genome project will only be achieved if it is surrounded by research efforts

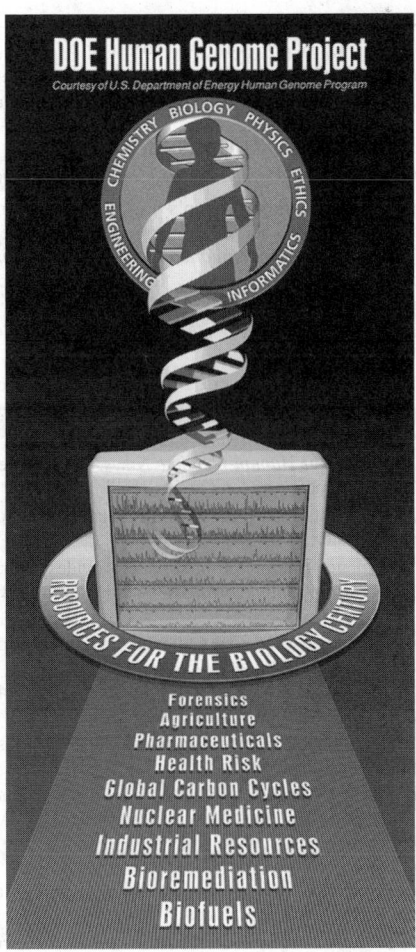

Fig. 4. Illustration of the DOE Human Genome Project. (Courtesy of U.S. Department of Energy Human Genome Project)

that are focussed on understanding and taking advantage of the human genetic information.

To Know Ourselves: The Human Genome Project 1983–2001

The biosciences research community is now embarked on a program whose boldness, even audacity, has prompted comparisons with such visionary efforts as the Apollo space program and the Manhattan project. That life scientists should conceive such an ambitious project is not remarkable; what is surprising—at least at first blush—is that the project should trace its roots to the Department of Energy.

For close to a half-century, the DOE and its governmental predecessors have been charged with pursuing a deeper understanding of the potential health risks posed by energy use and by energy-production technologies with special interest focused on the effects of radiation on humans. Indeed, it is fair to say that most of what we know today about radiological health hazards stems from studies supported by these government agencies. Among these investigations are long-standing studies of the survivors of the atomic bombings of Hiroshima and Nagasaki, as well as any number of experimental studies using animals, cells in culture, and nonliving systems. Much has been learned, especially about the consequences of exposure to high doses of radiation. On the other hand, many questions remain unanswered; in particular, we have much to learn about how low doses produce their insidious effects. When present merely in low but significant amounts, toxic agents such as radiation or mutagenic chemicals work their mischief in the most subtle ways, altering only slightly the genetic instructions in our cells. The consequences can be heritable mutations too slight to produce discernible effects in a generation or two but in their persistence and irreversibility, deeply troublesome nonetheless.

Until recently, science offered little hope for detecting at first hand these tiny changes to the DNA that encodes our genetic program. Needed was a tool that could detect a change in one "word" of the program, among

perhaps a hundred million. Then, in 1984, at a meeting convened jointly by the DOE and the International Commission for Protection Against Environmental Mutagens and Carcinogens, the question was first seriously asked: Can and should we sequence the human genome? That is, can we develop the technology to obtain a word-by-word copy of the entire genetic script for an "average" human being, and thus to establish a benchmark for detecting the elusive mutagenic effects of radiation and cancer-causing toxins? Answering such a question was not simple. Workshops were convened in 1985 and 1986; the issue was studied by a DOE advisory group, by the Congressional Office of Technology Assessment, and by the National Academy of Sciences; and the matter was debated publicly and privately among biologists themselves. In the end, however, a consensus emerged that we should make a start. For a timeline of major events in the U.S. Human Genome Project see Table 2.

Adding impetus to the DOE's earliest interest in the human genome was the Department's stewardship of the national laboratories, with their demonstrated ability to conduct large multidisciplinary projects—just the sort of effort that would be needed to develop and implement the technological know-how needed for the Human Genome Project. Biological research programs already in place at the national labs benefited from the contributions of engineers, physicists, chemists, computer scientists, and mathematicians, working together in teams. Thus, with the infrastructure in place and with a particular interest in the ultimate results, the Department of Energy, in 1986, was the first federal agency to announce and to fund an initiative to pursue a detailed understanding of the human genome. See Table 2.

Of course, interest was not restricted to the DOE. The National Institutes of Health, the Cold Spring Harbor Laboratory and the Howard Hughes Medical Institute had also sponsored workshops. In 1988 the NIH joined in the pursuit, and in the fall of that year, the DOE and the NIH signed a memorandum of understanding that laid the foundation for a concerted interagency effort. See Table 2. The basis for this community-wide excitement is not hard to comprehend. The first impulse behind the DOE's commitment was only one of many reasons for coveting a deeper insight into the human genetic script. Defective genes directly account for an estimated 4000 hereditary human diseases; maladies such as Huntington disease and cystic fibrosis. In some such cases, a single misplaced letter among three billion can have lethal consequences. For most of us, though, even greater interest focuses on the far more common ailments in, which altered genes influence but do not prescribe. Heart disease, many cancers, and some psychiatric disorders, for example, can emerge from complicated interplays of environmental factors and genetic misinformation.

The first steps in the Human Genome Project are to develop the needed technologies, then to "map" and "sequence" the genome. But in a sense, these well-publicized efforts aim only to provide the raw material for the next, longer strides. The ultimate goal is to exploit those resources for a truly profound molecular-level understanding of how we develop from embryo to adult, what makes us work, and what causes things to go wrong. The benefits to be reaped stretch the imagination. In the offing is a new era of molecular medicine characterized not by treating symptoms, but rather by looking to the deepest causes of disease. Rapid and more accurate diagnostic tests will make possible earlier treatment for countless maladies. Even more promising, insights into genetic susceptibilities to disease and to environmental insults, coupled with preventive therapies, will thwart some diseases altogether. New, highly targeted pharmaceuticals, not just for heritable diseases, but for communicable ailments as well, will attack diseases at their molecular foundations. And even gene therapy will become possible, in some cases actually "fixing" genetic errors. All of this in addition to a new intellectual perspective on who we are and where we came from.

For all the diversity of the world's five and a half billion people, full of creativity and contradictions, the machinery of every human mind and body is built and run with fewer than 100,000 kinds of protein molecules. And for each of these proteins, we can imagine a single corresponding gene (though there is sometimes some redundancy) whose job it is to ensure an adequate and timely supply. In a material sense, then, all of the subtlety of our species, all of our art and science, is ultimately accounted for by a surprisingly small set of discrete genetic instructions. More surprising still, the differences between two unrelated individuals, between the man next door and Mozart, may reflect a mere handful of differences in their genomic recipes—perhaps one altered word in five hundred. We are far more alike than we are different. At the same time, there is room for near-infinite variety.

TABLE 2. MAJOR EVENTS IN THE U.S. HUMAN GENOME PROJECT AND RELATED PROJECTS[a]

Date	Event
1983	• The Los Alamos National Laboratory, a Department of Energy Laboratory (LANL) and Lawrence Berkeley National Laboratory, a Department of Energy Laboratory (LLNL) begin production of DNA clone (cosmid) libraries representing single chromosomes.
1984	• DOE OHER and ICPEMC cosponsor Alta, Utah, conference highlighting the growing role of recombinant DNA technologies. See *http://www.ornl.gov/hgmis/project/alta.html* for more information on this conference. The Office of Technology Assessment (OTA) incorporates Alta proceedings into report acknowledging value of human genome reference sequence.
1985	• Robert Sinsheimer holds meeting on human genome sequencing at University of California, Santa Cruz. • At OHER Charles DeLisi and David A. Smith (Director of the DOE Human Genome Program) commission the first Santa Fe conference to assess the feasibility of a Human Genome Initiative. See *http://www.ornl.gov/hgmis/publicat/hgn/v11n3/05delisi.html* for a perspective by Charles DeLisi, HGP Pioneer. See *http://www.ornl.gov/TechResources/Human. Genome/publicat/hgn/v7n3/02smithr.html* Evolution of a Vision: Genome Project Origins, Present and Future Challenges, and Far-Reaching Benefits by David A. Smith.
1986	• Following the Santa Fe conference, DOE OHER announces Human Genome Initiative. With $5.3 million, pilot projects begin at DOE national laboratories to develop critical resources and technologies.
1987	• Congressionally chartered DOE advisory committee, HERAC, recommends a 15-year, multidisciplinary, scientific, and technological undertaking to map and sequence the human genome. DOE designates multidisciplinary human genome centers. See *http://www.ornl.gov/hgmis/project/herac2.html* for the Human Genome Initiative Office of Health and Environmental Research report. • NIH NIGMS begins funding of genome projects.
1988	• Reports by congressional OTA and NAS NRC committees recommend concerted genome research program. • HUGO founded by scientists to coordinate efforts internationally. See *http://www.gene.ucl.ac.uk/hugo/* • First annual Cold Spring Harbor Laboratory meeting on human genome mapping and sequencing. • DOE and NIH sign MOU outlining plans for cooperation on genome research. See *http://www.ornl.gov/hgmis/publicat/hgn/v2n1/03memo.html* for more information on this memorandum of understanding. • Telomere (chromosome end) sequence having implications for aging and cancer research is identified at LANL.
1989	• DNA STSs recommended to correlate diverse types of DNA clones. See *http://www.ornl.gov/hgmis/publicat/hgn/v2n3/01stsnew.html Human Genome News*, **2**(3) (September 1990); "STS-New Strategy May Provide Common Link for Mapping." • DOE and NIH establish Joint ELSI Working Group. See *http://www.ornl.gov/hgmis/publicat/hgn/v2n1/05elsi.html Human Genome News*, **2**(1) (May 1990); "NIH-DOE Joint Working Group on Ethical, Legal, and Social Issues Established."
1990	• DOE and NIH present joint 5-year U.S. HGP plan to Congress. The 15-year project formally begins. See *http://www.ornl.gov/hgmis/publicat/hgn/v2n1/04five.html Human Genome News*, **2**(1) (May 1990); "Five-year Plan Goes to Capitol Hill." Projects begun to mark gene sites on chromosome maps as sites of mRNA expression. • Research and development begun for efficient production of more stable, large-insert BACs.
1991	• Human chromosome mapping data repository, GDB, established. See *http://www.gdb.org/* The Genome Database. Established at Johns Hopkins University in Baltimore, Maryland, USA in 1990, the Genome Database (GDB) is the official central repository for genomic mapping data resulting from the Human Genome Initiative.
1992	• Low-resolution genetic linkage map of entire human genome published. • Guidelines for data release and resource sharing announced by DOE and NIH. See *http://www.ornl.gov/hgmis/publicat/hgn/v4n5/04share.html Human Genome News*, **4**(5) (January 1993); "NIH, DOE Guidelines Encourage Sharing of Data, Resources."
1993	• International IMAGE Consortium established to coordinate efficient mapping and sequencing of gene-representing cDNAs. See *http://www.ornl.gov/hgmis/publicat/hgn/v6n6/3image.html Human Genome News*, **6**(6) (Mar.–Apr. 1995); 3 "IMAGE Characterizes cDNA Clones."
	The Scientist **13**(4); 17 (Feb. 15, 1999) Hot Papers In Genomics: G. Lennon, C. Auffray, M. Polymeropoulos, M.B. Soares, "The I.M.A.G.E. Consortium: An Integrated Molecular Analysis of Genomes and Their Expression," *Genomics*, **33**, 151–152, 1996. (Cited in more than 290 papers since publication). See *http://www.ornl.gov/meetings/wccs/hot1_990215.html* • DOE-NIH ELSI Working Group's Task Force on Genetic and Insurance Information releases recommendations. See *http://www.ornl.gov/hgmis/publicat/hgn/v5n2/01task.html Human Genome News*, **5**(2) (July 1993); "Insurance Task Force Makes Recommendations." • DOE and NIH revise 5-year goals [*Science* **262**, 43–46 (Oct. 1, 1993)]. • French Généthon provides mega-YACs to the genome community. • IOM releases U.S. HGP-funded report, "Assessing Genetic Risks." See *http://www.ornl.gov/hgmis/publicat/hgn/v5n4/iomreprt.html Human Genome News*, **5**(4) (November 1993): "IOM Issues Report on Genetic Testing." • LBNL implements novel transposon-mediated chromosome sequencing system. • GRAIL sequence-interpretation service provides Internet access at ORNL. See *http://genome.ornl.gov/* Computational Biology at ORNL.
1994	• Genetic-mapping 5-year goal achieved 1 year ahead of schedule. See *http://www.ornl.gov/hgmis/publicat/hgn/V6N4/MAPGOALS. HTML Human Genome News*, **6**(4), 1 (Nov. 1994); "Genetic Map Goal Met Ahead of Schedule." • Completion of second-generation DNA clone libraries representing each human chromosome by LLNL and LBNL. • Genetic Privacy Act, first U.S. HGP legislative product, proposed to regulate collection, analysis, storage, and use of DNA samples and genetic information obtained from them; endorsed by ELSI Working Group. See *http://www.ornl.gov/hgmis/publicat/hgn/v6n6/4genetic.html Human Genome News*, **6**(6), 4 (Mar.–Apr. 1995); "Genetic Privacy Act Introduced." • DOE MGP launched; spin-off of HGP. See *http://www.ornl.gov/hgmis/publicat/hgn/V6N3/OHER.HTML Human Genome News*, **6**(3), 7 (September 1994); "OHER Launches Microbial Genome Initiative." • LLNL chromosome paints commercialized. See *http://www.ornl.gov/hgmis/publicat/hgn/v7n5/01collab.html Human Genome News*, **7**(5), 1 (January–March 1996); "Collaborations Multiply Research, Commercial Benefits." • SBH technologies from ANL commercialized. • DOE HGP Information Web site activated for public and researchers. See *http://www.ornl.gov/hgmis/* Human Genome Project Information.
1995	• LANL and LLNL announce high-resolution physical maps of chromosome 16 and chromosome 19, respectively. See *http://www.ornl.gov/hgmis/publicat/hgn/v6n5/2safchrm.html Human Genome News*, **6**(5), 2 (Jan.–Feb. 1995); "High-Resolution Physical Maps of Chromosomes 16 and 19 Completed." • Moderate-resolution maps of chromosomes 3, 11, 12, and 22 maps published. See *http://www.ornl.gov/hgmis/publicat/hgn/v6n5/4chrom2.html Human Genome News*, **6**(5), 14 (Jan.–Feb. 1995); "Groups Publish Detailed Chromosome 22 Map." • Physical map with over 15,000 STS markers published. See *http://www.ornl.gov/hgmis/publicat/hgn/v7n5/05detail.html Human Genome News*, **7**(5) (January–March 1996); "Detailed Human Physical Map Published by Whitehead-MIT." • First (nonviral) whole genome sequenced (for the bacterium Haemophilus influenzae). See *http://www.ornl.gov/hgmis/publicat/hgn/v7n1/05microb.html Human Genome News*, **7**(1) (May–June 1995); "Two Bacterial Genomes Sequenced." • Sequence of smallest bacterium, *Mycoplasma genitalium*, completed; provides a model of the minimum number of genes needed for independent existence. See *http://www.ornl.gov/hgmis/publicat/hgn/v7n1/05microb.html Human Genome News*, **7**(1) (May–June 1995). • EEOC guidelines extend ADA employment protection to cover discrimination based on genetic information related to illness, disease, or other conditions. See *http://www.ornl.gov/hgmis/publicat/hgn/v7n2/4eeocada.html Human Genome News*, **7**(2), 4 (July–August 1995); "New EEOC Guidelines Clarify Disability."
1996	• *Methanococcus jannaschii* genome sequenced; confirms existence of third major branch of life on earth. See *http://www.ornl.gov/hgmis/publicat/hgn/v8n1/01archae.html Human Genome News*, **8**(1) (July–September 1996); "Third Branch of Life Confirmed."

TABLE 2. *(continued)*

Date	Event

- DOE initiates 6 pilot projects on BAC end sequencing. See *http://www.ornl.gov/hgmis/publicat/hgn/v8n1/08bacend.html Human Genome News*, **8**(1) (July–September 1996); "BAC End-Sequencing Projects Initiated."
- Health Care Portability and Accountability Act prohibits use of genetic information in certain health-insurance eligibility decisions, requires DHHS to enforce health-information privacy provisions. See *http://www.ornl.gov/hgmis/publicat/hgn/ v8n3/01fear.html Human Genome News*, **8**(3 & 4) (January–June 1997); "Fear of Genetic Discrimination Drives Legislative Interest."
- HGP Participants Agree on Sequencing Data Release Policies Bermuda Conference. See *http://www.ornl.gov/hgmis/ research/bermuda.html*
  Summary of the Report of the Second International Strategy Meeting on Human Genome Sequencing (Bermuda, 27th February–2nd March, 1997) as reported by HUGO.
- DOE and NCHGR issue guidelines on use of human subjects for large-scale sequencing projects. See *http://www.ornl.gov/hgmis/ publicat/hgn/v8n1/08humans.html Human Genome News*, **8**(1) (July–September 1996); DOE, NCHGR Issue Human Subject Guidelines."
- *Saccharomyces cerevisiae* (yeast) genome sequence completed by international consortium. See *http://www.ornl.gov/hgmis/ publicat/hgn/v8n3/08yeast.html Human Genome News*, **8**(3 & 4) (January–June 1997); "Yeast Genome Directory."
- Sequence of the human T-cell receptor region completed. See *http://www.ornl.gov/hgmis/publicat/hgn/v8n1/09immune.html Human Genome News*, **8**(1) (July–September 1996); Immune System Genes Reveal Surprises."
- Wellcome Trust sponsors large-scale sequencing strategy meeting for international coordination of human genome sequencing. See *http://www.ornl.gov/hgmis/publicat/hgn/v7n6/19intern.html Human Genome News*, **7**(6) (April–June 1996); International Large-scale Sequencing Meeting."

1997
- NIH NCHGR becomes National Human Genome Research Institute (NHGRI). See *http://www.ornl.gov/hgmis/publicat/ hgn/v8n3/08nchgr.html Human Genome News*, **8**(3 & 4) (January–June 1997); "NCHGR Becomes NIH Institute."
- *Escherichia coli* genome sequence completed. See *http://www.ornl.gov/hgmis/publicat/hgn/v9n1/07ecoli.html Human Genome News*, **9**(1–2) (January 1998); "Complete *E. coli* Genome Sequence in Public Database."
- Second large-scale sequencing strategy meeting held in Bermuda. See *http://www.gene.ucl.ac.uk/hugo/bermuda2.htm* "Summary of the Report of the Second International Strategy Meeting on Human Genome Sequencing Bermuda, 27th February–2nd March 1997."
- High-resolution physical maps of chromosomes X and 7 completed. See *http://www.ornl.gov/hgmis/publicat/hgn/ v8n3/08xmap.html Human Genome News*, **8**(3 & 4) (January–June 1997); "Chromosome X Map Completed," and *http://www.ornl.gov/hgmis/publicat/hgn/v9n1/05mile.html Human Genome News*, **9**(1–2) (January 1998); "Team Completes HGP Milestone: Human Chromosome 7 Map."
- DOE-NIH Task Force on Genetic Testing releases final report and recommendations. See *http://www.med.jhu.edu/tfgtelsi/promoting/* "Promoting Safe and Effective Genetic Testing in the United States: Principles and Recommendations, Task Force on Genetic Testing."
- DOE forms Joint Genome Institute for implementing high-throughput activities at DOE human genome centers, initially in sequencing and functional genomics. See *http://www.ornl.gov/hgmis/publicat/ hgn/v8n2/01doe.html Human Genome News*, **8**(2) (October–December 1996); "DOE Merges Genome Center Sequencing Efforts."
- UNESCO See *http://www.unesco.org/* adopts Universal Declaration on the Human Genome and Human Rights. See *http://www.unesco.org/human_rights/hrbc.htm*

1998
- Hospital for Sick Children, Toronto, Ontario, to continue GDB data collection, curation. See *http://www.ornl.gov/hgmis/publicat/ hgn/v10n1/21gdb.html*
- *Caenorhabditis elegans* genome sequence completed. See *http://www.ornl.gov/hgmis/publicat/hgn/v10n1/08celeg.html*
- DOE and NIH reveal new five-year plan for HGP, predict project completion by 2003. See *http://www.ornl.gov/hgmis/publicat/hgn/ v10n1/03goal.html Human Genome News*, **10**(1–2) (February 1999); "Five-Year Research Goals of the U.S. Human Genome Project: October 1, 1998 to September 30, 2003."

- JGI exceeds sequencing goal, achieves 20 Mb for FY 1998. See *http://www.ornl.gov/hgmis/publicat/hgn/v10n1/01jgi.html Human Genome News*, **10**(1–2) (February 1999); "DOE Joint Genome Institute Exceeds DNA Sequencing Goal."
- GeneMap'98 containing 30,000 markers released. See *http://www.ornl.gov/hgmis/publicat/hgn/v10n1/07genmap.html Human Genome News*, **10**(1–2) (February 1999); "GeneMap 98 Doubles Density of 1996 Map."
- Incyte Pharmaceuticals announces plans to sequence human genome in 2 years. See *http://www.ornl.gov/hgmis/publicat/hgn/ v10n1/07incyte.html Human Genome News*, **10**(1–2) (February 1999); "Second Private Human Genome Sequencing Project Under Way."
- *Mycobacterium tuberculosis* bacterium sequenced. See *http://www.ornl.gov/hgmis/publicat/hgn/v9n3/16tigr.html#tb Human Genome News*, **9**(3) (July 1998); "Microbial Genome News."
- Celera Genomics formed to sequence much of human genome in 3 years using HGP-generated resources. See *http://www.ornl.gov/ hgmis/publicat/hgn/v9n3/01venter.html Human Genome News*, **9**(3) (July 1998); "Private-Sector Sequencing Planned." In May, J. Craig Venter [The Institute for Genomic Research (TIGR)] announced plans to form a new company with Perkin-Elmer's Applied Biosystems Division (PE-ABD) to sequence a large portion of the human genome in 3 years for $300 million.
- DOE funds production BAC end sequencing projects. See *http://www.ornl.gov/hgmis/publicat/hgn/v10n1/04bacend.html Human Genome News*, **10**(1–2) (February 1999); "BAC End Sequencing Speeds Large and Small Projects."
- Largest-ever ELSI meeting attended by over 800 from diverse disciplines and sponsored by DOE; Whitehead Institute; and the American Society of Law, Medicine, and Ethics. See *http://www.ornl.gov/hgmis/publicat/hgn/v10n1/13white.html Human Genome News*, **10**(1–2) (February 1999); "The Human Genome Project: Science, Law, and Social Change in the 21st Century: Reports from Cambridge Symposium."
- Human Genome Project passes midpoint. See *http://www.ornl. gov/hgmis/project/midpt.html*

1999
- First Human Chromosome Completely Sequenced! On December 1, researchers in the Human Genome Project announced the complete sequencing of the DNA making up human chromosome 22. See *http://www.ornl.gov/hgmis/project/chr22.html*
- Joint Genome Institute sequencing facility opens in Walnut Creek, CA. See *http://www.ornl.gov/hgmis/publicat/hgn/ v10n3/03richard.html Human Genome News*, **10**(3–4) (October 1999); "Richardson Attends Facility Opening."
- Major Drug Firms Create Public SNP Consortium. See *http://www.ornl.gov/hgmis/publicat/hgn/v10n3/13snp.html Human Genome News*, **10**(3–4) (October 1999); "Major Drug Firms Create Public SNP Resource."
- The Billion Base Pair Celebration November 23, 1999. Bruce Alberts, President, National Academy of Sciences and early planner of the Genome Project; Francis Collins, Director, NHGRI; Secretary of HHS, Donna Shalala; Secretary of DOE, Bill Richardson. See *http://www.ornl.gov/hgmis/graphics/ video/nhgri112399.ram* (Total Running Time: 01:09:45; Bandwidth: 146 Kbps).
- HGP advances goal for obtaining a draft sequence of the entire human genome from 2001 to 2000. See *http://www.ornl.gov/ hgmis/project/update.html*

2000
- HGP leaders and President Clinton announce the completion of a "working draft" DNA sequence of the human genome. See *http://www.ornl.gov/hgmis/project/clinton1.html* White House Press Conference: The Human Genome Project, June 26, 2000 (Total Running Time: 00:41:23; Bandwidth: 33 Kbps) See *http://www.ornl.gov/hgmis/graphics/video/whpc062600.ram*
- International research consortium publishes chromosome 21 genome, the smallest human chromosome and the fifth to be completed. See *http://hgp.gsc.riken.go.jp/chr21/*
- DOE researchers announce completion of chromosomes 5, 16, and 19 draft sequence. See *http://www.ornl.gov/hgmis/project/ 51619jgi.html* "Researchers Decode Three Human Chromosomes: Information May Lead to Treatments for Kidney Disease, Diabetes, and Prostate and Colorectal Cancer."
- International collaborators publish genome of fruit fly *Drosophila melanogaster*, the largest organism sequenced to date. See *http://www.ornl.gov/hgmis/archive/articles/drosophila.html*

(continued overleaf)

TABLE 2. (continued)

Date	Event
	• President Clinton signs executive order prohibiting federal departments and agencies from using genetic information in hiring or promoting workers. See *http://www.ornl.gov/hgmis/elsi/ legislat.html#clinton*
2001	• **Publication of Initial Working Draft Sequence February 12, 2001** Special issues of *Science* (Feb. 16, 2001) and *Nature* (Feb. 15, 2001) contain the working draft of the human genome sequence. Nature papers include initial analysis of the descriptions of the sequence generated by the publicly sponsored Human Genome Project, while Science publications focus on the draft sequence reported by the private company, Celera Genomics. Links for more information are: *Nature: http://www.nature.com/genomics/Science: http://www.sciencemag.org/content/vol291/issue5507/index. shtml* **Press releases on First Analysis of Genome Sequence:** *http://www.ornl.gov/hgmis/resource/media.html#releases*
	• Pieter de Jong's team (now at the Oakland Children's Hospital, Oakland, CA) was a major provider of the BAC libraries used in the sequencing of the human and several other genomes. See *http://www.ornl.gov/meetings/bacpac/*

[a] Acronyms

ADA—Americans with Disabilities Act.
ANL—Argonne National Laboratory, a Department of Energy Laboratory.
BAC—bacterial artificial chromosome.
cDNA—complementary deoxyribonucleic acid.
DHHS—Department of Health and Human Services at National Institutes of Health (NIH).
DNA—deoxyribonucleic acid.
DOE—Department of Energy.
EEOC—Equal Employment Opportunity Commission.
ELSI—ethical, legal, and social issues.
FY—federal fiscal year (October 1 to September 30).
GDB—Genome Database.
GRAIL—Gene Recognition and Analysis Internet Link.
HERAC—Health and Environmental Research Advisory Committee.
HGI—Human Genome Initiative.
HGP—Human Genome Project, Human Genome Program.
HUGO—Human Genome Organisation.
ICPEMC—International Commission for Protection Against Environmental Mutagens and Carcinogens.
IMAGE—Integrated Molecular Analysis of Gene Expression.
IOM—Institute of Medicine.
JGI—the Department of Energy's Joint Genome Institute in Walnut Creek, California. The JGI houses the DOE's production sequencing facility.
LANL—Los Alamos National Laboratory, a Department of Energy Laboratory
LBNL—Lawrence Berkeley National Laboratory, a Department of Energy Laboratory.
LLNL—Lawrence Livermore National Laboratory, a Department of Energy Laboratory.
MGP—Microbial Genome Project.
MOU—memorandum of understanding.
mRNA—messenger ribonucleic acid.
NAS—National Academy of Sciences.
NCHGR—National Center for Human Genome Research at National Institutes of Health (NIH).
NHGRI—National Human Genome Research Institute at National Institutes of Health (NIH).
NIGMS—National Institute of General Medical Sciences at National Institutes of Health (NIH).
NIH—National Institutes of Health.
NRC—National Research Council.
OBER—Office of Biological and Environmental Research, U.S. Department of Energy (formerly Office of Health and Environmental Research).
OHER—Office of Health and Environmental Research, U.S. Department of Energy (now Office of Biological and Environmental Research).
ORNL—Oak Ridge National Laboratory, a Department of Energy Laboratory.
OTA—Office of Technology Assessment.
R&D—research and development.
SBH—Sequencing by hybridization.
STS—sequence tagged site.
UNESCO—United Nations Educational, Scientific, and Cultural Organization.
YAC—yeast artificial chromosome.

It is no overstatement to say that to decode our 30,000 genes in some fundamental way would be an epochal step toward unraveling the manifold mysteries of life.

See also **Biotechnology (Bioprocess Engineering)**; **Genetic Engineering**; **Industrial Biotechnology**; and **Proteins**.

Major portions of this article were prepared by
A. C. DeBaldo, Ph.D.
University of South Florida
Tampa, Florida

Additional Reading

Adler, R.G.: "Genome Research: Fulfilling the Public's Expectations for Knowledge and Commercialization," *Science*, 908 (August 14, 1992).

Adolph, K.W.: *Genome Research in Molecular Medicine and Virology*, Academic Press, Orlando, FL, 1993.

Aldhous, P.: "Managing the Genome Data Deluge," *Science*, 502 (October 22, 1993).

Anderson W.F.: "Human Gene Therapy," *Science*, 808 (May 8, 1992).

Bauer, W.R., F.H.C. Crick, and J.H. White: "Supercoiled DNA" (*A Classic Nobel Laureate Paper*), *Sci. Amer.*, (July 1980).

Beardsley, T.: "From Mice to Men," *Sci. Amer.*, 18 (December 1993).

Bodmer, Sir Walter: "Genome Research in Europe," *Science*, 480 (April 24, 1992).

Bull, J.J., I.J. Molineux, and J.H. Werren: "Selfish Genes," *Science*, 65 (April 3, 1992).

Calladine, C. and H. Drew: *Understanding DNA*, Academic Press, Orlando, FL, 1992.

Collins, F. and D. Galas: "A New Five-Year Plan for the U.S. Human Genome Project," *Science*, 43 (October 19, 1993).

Copeland, N.G. et al.: "A Genetic Linkage Map of the Mouse: Current Applications and Future Prospects," *Science*, 57 (October 1, 1993).

Considine, G.D., and P.H. Kulik: *Van Nostrand's Scientific Encyclopedia*, 9th Edition, John Wiley & Sons, Inc., New York, NY, 2002.

Culliton, B.J.: "Mapping Terra Incognita (Humani Corporis)," *Science*, 210 (October 12, 1990).

Cuticchia, A.J. et al.: "Managing All Those Bytes: The Human Genome Project," *Science*, 47 (October 1, 1993).

Eigen, M., W. Gardiner, Schuster, P., and R. Winkler-Oswatitsch: "The Origin of Genetic Information" (*A Classic Nobel Laureate Paper*), *Sci. Amer.*, (April 1981).

Eisenberg, R.S.: "Genes, Patents, and Product Development," *Science*, 903 (August 14, 1992).

Erickson, D.: "Hacking the Genome," *Sci. Amer.*, 128 (April 1992).

Erickson, D.: "Diagnosis by DNA," *Sci. Amer.*, 116 (October 1992).

Farkas, D.H.: *Molecular Biology and Pathology*, Academic Press, Orlando, FL, 1993.

Farr, C.J. and P.N. Goodfellow: "Hidden Messages in Genetic Maps," *Science*, 49 (October 2, 1992).

Farrell, R.E., Jr.: *RNA Methodologies*, Academic Press, Orlando, FL, 1993.

Fischman, J.: "Going for the Old: Ancient DNA Draws a Crowd," *Science*, 655 (October 29, 1993).

Fox, S.: "Applications for Synthesizing and Sequencing DNA Beyond the Genome Project," *Genetic Eng. News*, 6 (June 1991).

Friedmann, T., Ed.: *Molecular Genetic Medicine*, Academic Press, Orlando, FL, 1992.

Grunstein, M.: "Histones as Regulators of Genes," *Sci. Amer.*, 68 (October 1992).

Jasny, B.R.: "Genome Delight," *Science*, 11 (October 2, 1992).

Jürgens, G.: "Genes to Greens: Embryonic Pattern Formation in Plants," *Science*, 487 (April 24, 1992).

Karlin, S. and V. Brendel: "Chance and Statistical Significance in Protein and DNA Sequence Analysis," *Science*, 39 (July 3, 1992).

Kessler, D.A. et al.: "The Safety of Foods Developed by Biotechnology," *Science*, (June 26, 1992).

Kevles, D.J. and L. Hood, Eds.: *The Code of Codes: Scientific and Social Issues in the Human Genome Project*, Harvard University Press, Cambridge, MA, 1992.

Kiley, T.D.: "Patents on Random Complementary DNA Fragments?" *Science*, 915 (August 14, 1992).

Klug, A. and R.D. Kornberg: "The Nucleosome" (*A Classic Nobel Laureate Paper*), *Sci. Amer.*, (February 1981).

Marx, J.: "Genome Project Plans Described," *Science*, 152 (April 9, 1993).

Morell, V.: "30-Million-Year-Old DNA Boosts an Emerging Field," *Science*, 1860 (September 25, 1992).

Pääbo, S.: "Ancient DNA," *Sci. Amer.*, 86 (November 1993).

Pearson, P.L. et al.: "The Human Genome Initiative—Do Databases Reflect Current Progress?" *Science*, 214 (October 11, 1991).

Rajewsky, K.: "A Phenotype or Not: Targeting Genes in the Immune System," *Science*, 483 (April 24, 1992).

Rhodes, D. and A. Klug: "Zinc Fingers," *Sci. Amer.*, 56 (February 1993).

Risch, N.L.: "Genetic Linkage: Interpreting Lod Scores," *Science*, 803 (February 14, 1992).

Roberts, L.: "Academy Backs Genome Project," *Science*, 725 (February 12, 1988).

Roberts, L.: "Taking Stock of the Genome Project," *Science*, 20 (October 1, 1993).

Roberts, L.: "NIH, DOE Battle for Custody of DNA Sequence Data," *Science*, 504 (October 22, 1993).

Selvin, P.R. et al.: "Torsional Rigidity of Positively and Negatively Supercoiled DNA," *Science*, 82 (January 3, 1992).

Singer, M. and P. Berg: *Genes and Genomes,* University Science Books, Mill Valley, CA, 1990.

Stephens, J.C. et al.: "Mapping the Human Genome: Current Status," *Science,* 237 (October 12, 1990).

Suzuki, D.T. et al.: *An Introduction to Genetic Analysis,* Fourth Edition, Freeman, Salt Lake City, UT, 1989.

Thompson, L.: "At Age 2, Gene Therapy Enters a Growth Phase," *Science,* 744 (October 30, 1992).

Varmus, H.: "Reverse Transcription" (*A Classic Nobel Laureate Paper*), *Sci. Amer.,* (November 1978).

von Hippel, P.H. and T.D. Yager: "The Elongation-Termination Decision in Transcription," *Science,* 809 (February 14, 1992).

Withka, J.M.: "Toward a Dynamical Structure of DNA," *Science,* 597 (January 31, 1992).

Wolffe, A.: *Chromatin,* Academic Press, Orlando, FL, 1992.

Zyskind, J. and S.I. Bernstein: *Recombinant DNA Laboratory Manual,* Academic Press, Orlando, FL, 1992.

Web References

Brookhaven National Laboratory: http://www.bnl.gov/world/

Cold Spring Harbor Laboratory: http://www.cshl.org/

DOE Human Genome Program: http://www.er.doe.gov/production/ober/hug_top.html

"Evolution of a Vision—Part I" by David Smith: http://www.ornl.gov/break hgmis/publicat/hgn/v7n3/02smithr.html

"Evolution of a Vision—Part II" by Francis S. Collins: http://www.ornl.gov/hgmis/publicat/hgn/v7n3/03collin.html

Genome Therapeutics Corporation: http://www.genomecorp.com/

Human Genome Management Information System at Oak Ridge National Laboratory: http://www.ornl.gov/TechResources/Human_Genome/home.html

Human Genome News: Full text of all issues is available online: http://www.ornl.gov/hgmis/publicat/hgn/hgn.html

"Introducing the Human Genome Project: Its Relevance, Triumphs, and Challenges " by Ari Patrinos and Daniel W. Drell: http://www.ornl.gov/TechResources/Human_Genome/publicat/judges/drell.html

Lawrence Berkeley National Laboratory Human Genome Center: http://www-hgc.lbl.gov/GenomeHome.html

Lawrence Livermore National Laboratory: http://www.llnl.gov/

Livermore Human Genome Center: http://www-bio.llnl.gov/bbrp/genome/genome.html

Los Alamos National Laboratory Center for Human Genome http://www-ls.lanl.gov/ index.html

National Academy of Sciences: http://www.nas.edu/

National Institutes of Health: http://www.nih.gov/

NIH National Center for Human Genome Research: http://www.nhgri.nih.gov/

The Genome Database at Johns Hopkins University School of Medicine: http://gdbwww.gdb.org/

The Howard Hughes Medical Institute: http://www.hhmi.org/intro/

The Institute for Genomic Research—TIGR: http://www.tigr.org/cet/

The National Center for Genome Resources: http://www.ncgr.org/

GEOCHEMISTRY. The study of the chemical composition of the earth in terms of the physicochemical and geological processes and principles that produce and modify minerals and rocks. Of practical importance in discovering and establishing the limits of ore deposits, petroleum, tar sands, salt, sulfur, and other valuable resources.

GEOCRONITE. A mineral sulfide of lead, antimony and arsenic, Pb_5SbAsS_8. Crystallizes in the monoclinic system. Hardness, 2.5; specific gravity, 6.4±; color, gray to blue with metallic luster; opaque.

GERMANIUM. [CAS: 7440-56-4]. Chemical element symbol Ge, at. no. 32, at. wt. 72.59, periodic table group 14, mp 937°C, bp 2830°C, density 5.36 g/cm³(20°C). Elemental germanium has a diamond cubic crystal structure. Germanium is a silver-white, lustrous, hard, brittle metal. When heated in oxygen to 730°C, the metal is partially oxidized to dioxide. The element is unaffected by solutions of acids and bases, but is soluble in fused NaOH. In the form of powder (dull gray), combines readily with chlorine to form the volatile tetrachloride. Although predicted by Mendeleev as early as 1871, the element was not fully identified until 1886 by Winkler. Mendeleev had previously termed the missing element *eka-silicon*. There are five natural isotopes ^{70}Ge, ^{72}Ge through ^{74}Ge, and ^{76}Ge. Seven radioactive isotopes include ^{67}Ge through ^{69}Ge, ^{71}Ge, ^{75}Ge, ^{77}Ge, and ^{78}Ge. All have a relatively short half-life, the longest, ^{68}Ge with a half-life of 275 days. In terms of abundance, germanium ranks 32nd among the element and thus is about as abundant as gallium, selenium, arsenic, and bromine. First ionization potential 8.13 eV; second, 15.86 eV;

third, 31.97 eV; fourth, 45.5 eV. Other important physical characteristics of germanium are given under **Chemical Elements**.

Germanium occurs in very small amounts in many sulfide ores, such as American zinc ores (0.25% GeO_2), and the rare mineral argyrodite (silver germanium sulfide) of Saxony and Bolivia. The primary source is flue dust from the zinc industry. Also, it may be obtained from the reduction of oxide and sulfide ores. A major ore is germanite, a copper ore found in southwest Africa. The ore is quite complex, containing some 20 different elements. The copper content ranges as high as 45%, sulfur up to 30%, whereas the germanium content is from 6 to 9%. The ore also contains up to 1% gallium. A major sulfide ore is renierite, which contains up to about 8% germanium. Small quantities of germanium are found in lepidolite, sphalerite, and spodumene. Some English coals contain as much as 1.6% germanium oxide. The germanium metal of 99.99+% purity is obtained by zone melting. In this system, electric heating coils are moved slowly along the length of an ingot. Impurities in the metal tend to raise or lower the freezing point of the molten alloy. By progressively melting the metal along the length of the ingot, the impurities that tend to lower the melting point will be swept to the last portion of the ingot to freeze, whereas the impurities that tend to raise the melting point will concentrate in the first region to freeze.

Uses

The principal uses of germanium have been in solid-state electronic devices, notably transistors, which can be used as amplifiers and oscillators. The electrical properties of germanium metal, which have brought about its wide use in semiconductors, are its high specific resistance at ordinary temperatures and the narrow gap between its filled energy band and its conduction band. Thus, germanium is an intrinsic semiconductor, wherein an increase of temperature or the addition of very small amounts of group 3 or group 5 elements can cause electrons to move readily to the conduction band to form "holes," thus making the material conductive. A key to the manufacture of semiconductor devices is making materials of high purity, great uniformity, and in sufficient quantity. See also **Semiconductors**.

The addition of as little as 0.35% germanium to tin doubles the hardness of tin. Similarly, germanium improves the strength and hardness of aluminum and magnesium alloys. These applications are limited, however, because of the current high costs of germanium. Germanium-silicon alloys are under intensive study for use in thermoelectric generators. Advantages claimed for these metals include better thermoelectric qualities above 600°C, an improved efficiency per unit weight factor, and virtually no corrosion or decomposition.

Chemistry and Compounds

Germanium forms compounds in which the oxidation states are (II) and (IV). The divalent ones are unstable. Thus, the monoxide is readily oxidized by air when hydrated. However, when completely dehydrated it resists the action of H_2SO_4 and potassium hydroxide, and reacts only slowly with fuming HNO_3. On heating in an inert atmosphere, it disproportionates to the elements and germanium dioxide, GeO_2. The latter resembles silicon dioxide in existing in more than one form, with a difference in chemical properties. The stable form at room temperature has the rutile structure, but just below the melting point the stable form has the cristobalite structure. Germanium(IV) oxide, GeO_2, prepared by hydrolysis of germanium(IV) chloride, $GeCl_4$, is somewhat soluble in water, acids, and alkalis, but GeO_2 from heating of germanic acid is insoluble. Like silicon dioxide, GeO_2 forms gels readily.

Germanium(II) hydroxide, $Ge(OH)_2$, is obtained by action of alkali hydroxides upon germanium(II) chloride, $GeCl_2$, solutions; it is amphiprotic, dissolving in excess of the alkali. Moreover, the acid form, sometimes called germanous acid, is obtained upon heating the hydroxide: $Ge(OH)_2 \rightarrow HGe(O)H$. GeO_2 is slightly acid in solution and when freshly precipitated ($pK_A = 9.4$). There is no experimental evidence for the existence of a definite hydrate, although melting point diagrams of germanate salts have indicated the existence of ortho($\equiv GeO_4$), meta($= GeO_3$), and tetra($\equiv Ge_4O_9$) compounds.

Germanium forms dihalides and tetrahalides with all four of the common halogens. In general, the dihalides readily react with halogens or other oxidizing agents to form tetravalent germanium compounds, and some, e.g., the iodide, disproportionate to the metal and tetravalent compound.

Suggestive of carbon and silicon is the existence of hydrides of germanium, though they are much fewer in number. The compound GeH_4

is called germane (mp $-165°C$, bp $-90°C$). Compounds having the general formula Ge_nH_{2n+2} ($n = 2$, 3, etc.) are called digermane, trigermane, etc., according to the number of germanium atoms present. The first three compounds in this series have been obtained by treatment of magnesium germanide with ammonium bromide in liquid ammonia. Compounds such as $GeHCl_3$ and alkylgermanes are also known. Germane and the alkyl- and aryl-substituted germanes retaining at least one hydrogen atom are somewhat more acidic than the corresponding silanes in nonaqueous media, easily forming alkali salts, R_3GeM and even dialkali salts R_2GeM_2 under some circumstances. Germane, GeH_4, appears to be thermodynamically stable, although no quantitative data are available on its heat of formation. It decomposes at about $285°C$.

Germanium also forms organometallic compounds. Over two hundred have been reported, from chloromethyl trichlorogermane, $ClCH_2GeCl_3$ to cyclotetrakis (diphenyl germanoxane), $[(C_6H_5)_2GeO]_4$.

Additional Reading

Anderson, D.L.: "Composition of the Earth," *Science,* **367** (January 20, 1989).

Avallone, E.A. and T. Baumeister: *Mark's Standard Handbook for Mechanical Engineers,* 10th Edition, The McGraw-Hill Companies, Inc., New York, NY, 1996.

Belz, L.H.: "Special Metals in Electronics," *Advanced Materials & Processes,* 65 (November 1987).

Dahmen, U. and K.H. Westmacott: Observations of Pentagonally Twinned Precipitate Needles of Germanium in Aluminum." *Science,* **233,** 875–876 (1986).

Davis, J.R.: *Metals Handbook,* 2nd Edition, ASM International, Materials Park, OH, 1998.

DiSalvo, F.J.: "Solid-State Chemistry: A Rediscovered Chemical Frontier," *Science,* 649 (February 9, 1990).

Greenwood, N.N. and A. Earnshaw: *Chemistry of the Elements,* 2nd Edition, Butterworth-Heinemann, Inc., Woburn, MA, 1997.

Hull, R. and J.C. Bean: *Germanium Silicon: Physics and Materials,* Vol. 5, Academic Press, Inc., San Diego, CA, 1999.

Krebs, R.E.: *The History and Use of Our Earth's Chemical Elements: A Reference Guide,* Greenwood Publishing Group, Inc., Westport, CT, 1998.

Lide, D.R.: *CRC Handbook of Chemistry and Physics,* 84th Edition, CRC Press, LLC, Boca Raton, FL, 2003.

Parker, P.: *McGraw-Hill Encyclopedia of Chemistry,* 2nd Edition, The McGraw-Hill Companies, Inc., New York, NY, 1993.

Patai, S.E.: *The Chemistry of Organic Germanium, Tin and Lead Compounds,* John Wiley & Sons, Inc., New York, NY, 1995.

Staff: *ASM Handbook: Properties and Selection of Nonferrous Alloys and Pure Metals,* ASM International, Materials Park, OH, 1990.

Westbrook, J.H.: *Electrical Properties,* in Encyclopedia of Materials Science and Engineering, MIT Press, Cambridge, MA, 1986.

GERSDORFFITE. A mineral related to cobaltite and ullmannite in the cobaltite group. A sulfide-arsenide of nickel, $NiAsS$. Crystallizes in the isometric system. Hardness, 5.5; specific gravity, 5.9; color, white to gray with metallic luster; opaque.

GETTERING. The absorption of gas by a getter film. When this process occurs during the dispersal of the getter through an evacuated system (such as an electron tube), it is called dispersal gettering; when by action of the already dispersed film, it is called contact gettering. In electric-discharge gettering, the process is accelerated by passing an ionizing electron discharge through the gas. The gas is ionized, and the ions are neutralized when they impinge on an electrode, so that the final product is neutral gas atoms. These are then easily absorbed by the getter.

A getter film is a metallic deposit in a vacuum system with the function of absorbing residual gas. Electropositive metals, such as sodium, potassium, magnesium, calcium, strontium, and barium have been used as getters. The process of depositing a getter film upon a surface may be done in various ways. In the distillation method, the metal to be deposited is volatilized into the vacuum system from a side tube provided with constructions for sealing-off when the process is completed. The electrolytic method is applicable where the metal to be deposited is sodium, and where the system is made of soda-lime glass. It is well known that sodium may be electrolyzed through soda-lime glass. If, therefore, a thermionic source of electrons is provided inside an evacuated sealed-off vessel, part of which is dipped into a suitable liquid kept at a high potential relative to the source of electrons, a current will pass, carried by electrons between the thermionic cathode and the inner surface of the glass, and by ions within the glass. The only ions in the glass that are mobile are sodium ions, and thus pure sodium is released at the inner surface of the envelope.

Other modern getter materials include cesium-rubidium alloys, tantalum, titanium, zirconium, and several of the rare-earth elements, such as hafnium.

GIAUQUE, WILLIAM F. (1895–1982). An American chemist who achieved distinction for his studies of the properties of matter at temperatures approaching absolute zero ($-273°C$). This research established the science of cryogenics. Giauque received the Novel prize in chemistry in 1949. He was professor and research director at the University of California at Berkeley. One of his most significant contributions was the invention of a magnetic cooling device that made it possible to attain cryogenic temperatures. An important property of matter discovered as a result of his work in superconductivity.

See also **Cryogenics**; and **Superconductivity**.

GIBBERELLIC ACID AND GIBBERELLIN PLANT GROWTH HORMONES. These organic chemical compounds, first isolated from the parasitic fungus *Gibberella fujikuori* in Japan in the late 1930s, produce unusual results when applied to plants, including various food crops. The results can be advantageous or disadvantageous. The phenomena of the gibberellins were uncovered as the result of studying the excessive leaf elongation in rice plants. This fungus disease of rice is sometimes referred to as the "foolish seedling" disease in rice. When infected with this fungus, the rice plants grow ridiculously tall and the stems break before the plants can flower and produce seed. When experimentally applied to higher plants, the gibberellins have varied effects. The most common reaction is the rapid lengthening of the stems. The stems of citrus trees, for example, have been stimulated to grow at a rate six times greater than normal. When applied to the young fruit of seedless grapes, the gibberellins cause the fruit to grow much larger and to stay on the vine longer. Although some results can be predicted from experience with other species, generally results must be observed through long trial-and-error experimentation with many plants and many different concentrations and forms of the chemical growth hormones. The gibberellins are but one category of several kinds of plant hormones that affect food crop production. See also **Plant Growth Modification and Regulation**.

Since the 1960s, commercial gibberellin formulations have been available. These take several forms, ranging from liquid concentrates through tablets and powders. In some countries, registration is required of these compounds. The following practical results, among others, have been achieved when gibberellins are used properly on certain food plants:

Artichoke: prolongs picking period
Barley: enzyme content increased
Bean: more rapid emergence of plant
Blueberry: better fruit set
Celery: extends winter crop
Cherry (sour): combats cherry yellow virus
Cucumber: produces staminate flowers
Grape: loosens and elongates clusters; increases grape size
Hops: increases yields; aids harvesting
Lemon: delays yellow color development
Lettuce: increases seed production; effects uniform bolting
Oats: promotes more rapid emergence of plant
Orange (navel): retards aging of rind
Potato: stimulates sprouting
Prune (Italian): increases yield; reduces internal browning
Rhubarb: for forced crops, increases yield
Rye: promotes more rapid emergence of plant
Soyabean: promotes more rapid emergence of plant
Sugarcane: increases sucrose yield
Tangerine: increases yield and fruit set
Wheat: promotes more rapid emergence of plant

The gibberellins are actually a family of closely related substances. To date, structures have been determined for well over a dozen of these and a number have been isolated from higher plants. See Structures 1 and 2. The structure of three fused saturated or nearly saturated rings, with two additional rings perpendicular to them, suggests relationship to the diterpens for which there is strong isotopic evidence. For example, C^{14}-kaurene is readily converted to gibberellic acid (GA_3) by *Gibberella* cultures. The biosynthesis is apparently inhibited by chlorocholine, which is suspected as the basis for the dwarfing action of this compound. GA_7 to date has had the highest activity in most tests.

Gibberellic acid (GA$_3$)

Gibberellic acid (GA$_7$)

Gibberellins cause rapid elongation of shoots; many of the dwarf forms of maize (corn), bean, pea, and morning glory (closely allied to sweet potato) are caused to grow into tall forms indistinguishable from their tall genetic relatives. Many long-day plants are brought into flower in short days by gibberellin, and some biennials, including *Hyoscyamus* (henbane), are made to flower in one year. This process depends on the activation of cell divisions in the shoot apex. Like auxins (other plant hormones), gibberellins produce parthenocarpic fruits, especially on tomato, but unlike auxins, they do not inhibit lateral bud development, but they inhibit rooting of cuttings and promote the germination of many seeds. Their transport shows no polarity. They are active at concentrations comparable to those of the auxins. There is good evidence that the gibberellins act only when auxin is present.

In their biological function, it is believed that the gibberellins destroy or bypass naturally occurring inhibitors that normally prevent premature germination. However, high concentrations of the gibberellins and like substances actually prevent germination in certain varieties of seed.

An excellent example of the performance of gibberellins is given by S.B. Ross, et al. in "Gibberellins: A Phytohormonal Basis for Heterosis in Maize," *Science*, **1216** (September 2, 1988).

GIBBS-DUHEM EQUATION.

In a system of two or more components at constant temperature and pressure, the sum of the changes for the various components, of any partial molar quantity, each multiplied by the number of moles of the component present, is zero. The special case of two components is the basis of the Gibbs-Duhem equation of the form:

$$n_1 d_{\overline{X}_1} = -n_2 d_{\overline{X}_2}$$

in which n_1 and n_2 are the number of moles of the respective components and $\overline{X}_1$ and $\overline{X}_2$ are the partial molar values of any extensive property of the components.

GIBBS, JOSIAH WILLARD (1839–1903).

The father of modern thermodynamics. During his lifelong post as professor of mathematical physics at Yale, he stated the fundamental concepts embraced by the three laws of thermodynamics, especially the nature of entropy. A theorist rather than an experimenter, Gibbs was the first to expound with mathematical rigor the "relation between chemical, electrical, and thermal energy and capacity for work." It has been said that throughout his adult life Gibbs did nothing but think. The results established him as a great creative scientist.

See also **Thermodynamics**.

GIBBS-KONOVALOV THEOREMS.

Consider a binary system containing two phases (e.g., liquid and vapor). Both components can pass from one phase to another. The Gibbs-Konovalov theorems refer to the properties of the phase diagrams of such systems (see also **Azeotropic System**). The first theorem is: *At constant pressure, the temperature of coexistence passes through an extreme value (maximum, minimum or inflexion with a horizontal value), if the composition of the two phases is the same. Conversely, at a point at which the temperature passes through an extreme value, the phases have the same composition.* The second theorem is similar. It refers to the coexistence pressure at constant temperature.

GIBBS PARADOX.

When two samples of the same gas at a given temperature and pressure are allowed to mingle by the removal of a separating partition, the entropy of the resulting system is equal to the sum of the entropies of the two original parts. And, there is no extra term which arises when the two original systems are composed of different gases. This paradoxical absence is called the Gibbs paradox; it can be explained by using the theory of grand canonical ensembles.

GIBBS PHASE RULE.

See **Phase Rule**.

GILBERT, WALTER (1932–).

An American molecular biochemist who won the Nobel prize for chemistry in 1980 along with Paul Berg and Frederick Sanger for their studies of the chemical structure of nucleic acid. Author of many papers on theoretical physics and molecular biology. He has been at Harvard since 1972.

GILSONITE.

The mineral Gilsonite, named for S.H. Gilson of Salt Lake City, is a variety of asphaltum that occurs in Uinta County, Utah. It is found in black lustrous masses that ignite easily. A less frequently used name for it is uintaite.

GIRBOTOL ADSORPTION.

A process for the removal of hydrogen sulfide or carbon dioxide from a gaseous mixture. An organic amine (ethanolamine or diethanolamine, which are basic) is allowed to flow down a tortuous path through a tower where it is contacted by and absorbs (acidic) hydrogen sulfide or carbon dioxide from the gas to be purified as it moves up the tower. The amine, contaminated with these products, is then sent from the bottom of the tower to a steam stripper where it flows countercurrent to steam, which strips the hydrogen sulfide or carbon dioxide from it. The amine is then returned to the top of the tower. The process is widely used in the petroleum industry for purifying refinery and natural gases and for recovery of hydrogen sulfide for sulfur manufacture. Removal of carbon dioxide from gases is usually done with nonoethanolamine.

GLASS.

Traditional glass is an inorganic product of fusion that has cooled to a rigid solid without undergoing crystallization. Within the last few years, sol-gel glass has been introduced to the commercial market. Sol-gel processing is a chemically based method for producing glass at temperatures much lower than the traditional melting methods. Sol-gel glasses are described later in this article.

Glass may be transparent, translucent, or opaque, and it may be colored. The chemical composition and corresponding properties may vary over a wide range. Glass will support a load, and it may be shaped, broken, or cut. It is much like other solid materials, and yet it is unique.

Its uniqueness becomes obvious when it is examined on a submicroscopic level. Most solids have regular, orderly patterns for the arrangement of atoms, molecules, and ions, but glassy materials are highly disordered. There is some short-range order in glass, but beyond one or two atoms or ions the ordering may be described as random. Thus, on a submicroscopic level, glassy solids look more like liquids than solids.

Since glasses do not have ordered structures with correspondingly specific bonding energies between rows, stacks, planes, or discrete ions, they do not have definite melting points. When a glassy material is heated, it softens slowly and transforms to the liquid state. Crystalline solids generally transform from a solid to a liquid at a single specific temperature, the melting point. On cooling, a material that has a tendency to crystallize to solid will do so at the same temperature at which it transformed to a liquid. When a glass is cooled from a high temperature, it becomes increasingly viscous in a manner related to the inverse of the temperature until it becomes a rigid solid again. Thus, a specific temperature where melting or freezing takes place cannot be found for glass; i.e., glass does not have a melting point.

Most glasses can be made to crystallize if they are subjected to the right conditions of temperature and rate of cooling, which suggests that the glassy state is like a supercooled liquid. This is not borne out by measurements of density and other volume properties, which do not decrease in a linear manner as glass is cooled below its crystallization temperature.

Why is it that some melts, when cooled through a crystallization temperature, form glasses while others do not? It is simply a question of whether the melt can be cooled through the temperature range of maximum crystal growth rate faster than the crystals can grow. Thus table salt cannot be formed as a glass, but sand, or SiO_2 can be. The maximum crystal growth rate is normally just below the melting point of the material, but materials that tend to form glasses easily are much more viscous at these temperatures. For example, in the extreme cases of salt and sand, the differences in viscosities at their respective melting points is about eight orders!

The two-dimensional drawing in Fig. 1 shows SiO_2 in the ordered, or crystalline, and in the random, or glassy, state to illustrate the difference on a submicroscopic scale. Figure 2 shows how the volume properties of a material would respond to temperature if they could be prepared as a glass, a supercooled liquid, or crystalline material.[1]

Most glasses are composed of inorganic oxides, and most commercial glasses contain SiO_2 as their major constituent, but there are organic glasses and elemental metallic glasses. Glass is typically hard and brittle, and exhibits a conchoidal fracture. Most commercial glasses are transparent or translucent in the visible portion of the spectrum.

The continuous and smooth relationship of the viscosity of glass with its temperature is an important property. Figure 3 shows a typical viscosity versus temperature curve for a commercial glass. The working range is the viscosity in which most commercial glasses are formed. Glassware formed by automatic forming equipment would be made from glass at a temperature such that the glass will have a viscosity in the lower portion of this range (10^3 to 10^5). Some other operations, such as hand working, might be done at higher viscosities.

Generally, freshly formed glasses are in danger of deforming under their own weight when they are at viscosity below the softening point. At the annealing point, the glass is rigid and at this viscosity (temperature) the internal strains caused by the forming and nonuniform cooling would be

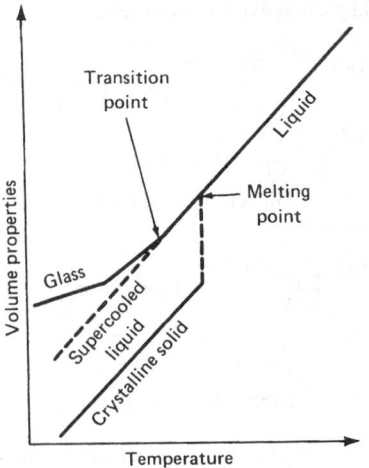

Fig. 2.   Volume properties of glass in contrast with crystalline solids as a function of temperature

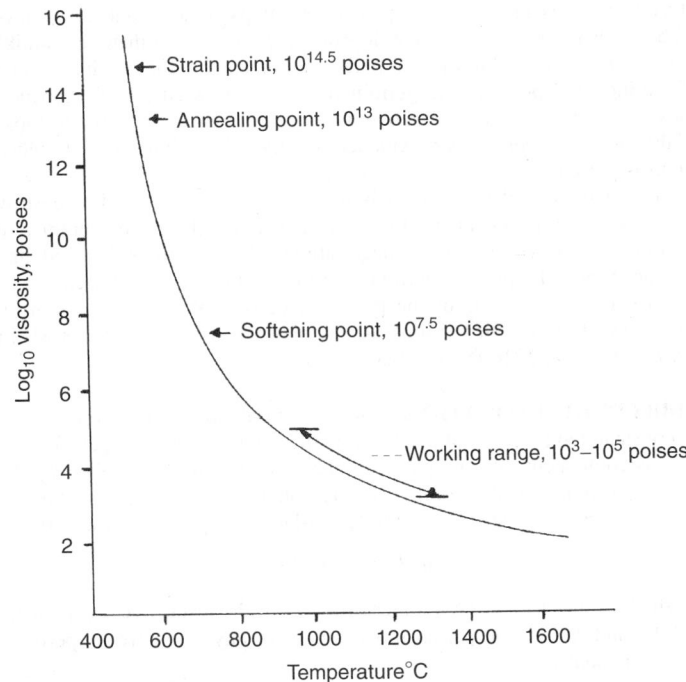

Fig. 3.   Viscosity-temperature relationship of a typical commercial soda-lime glass

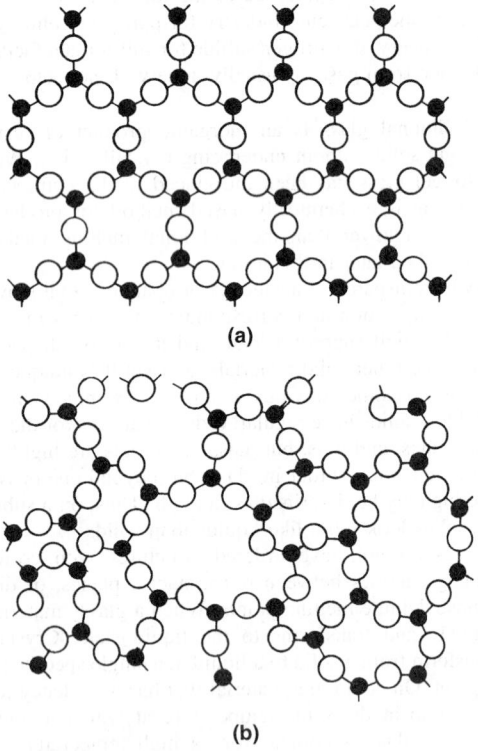

Fig. 1.   Silicon dioxide (SiO_2): (**a**) crystalline, and (**b**) glassy state. (Course structure is shown. Some authorities have recently suggested that, when studied at a much finer structure (such as by neutron scattering techniques), glass shows a much more orderly structure)

[1] *Note*: Traditionally, the structure of glass has been determined by means of x-ray crystallography. Which reveals a random network of disorderly structure. Neutron scattering of glass, however, makes it possible to examine the much finer structure of the material. It has been found that in glasses the angles between bonds that link atomic or molecular building blocks vary, whereas in crystals, of course, the links are orderly—that is, an endless repetition of a regular atomic or molecular geometry. In recent experiments at Grenoble, neutrons were beamed at samples of silicate glass. From these studies at this much finer scale, researchers now believe that the molecular structure of glass is far from random. As pointed out in 1991 by Nicholas Borrelli (Corning), "Normally glass is considered a random network, but that really is a misnomer." See Amato reference listed.

decreased to an acceptable commercial level in 15 minutes. At the strain point, the glass is substantially rigid, and at the temperature equivalent to this viscosity, the internal stresses would be reduced to very low values if the temperature were maintained for four hours.

Types of Traditional Glass

A wide range of glass products exists, each type having special properties. The properties of glass are determined primarily by chemical composition, and since the composition may be varied almost infinitely, there are many thousands of different glasses. However, they may be generally classified into soda-lime-silica glasses; lead glasses; borosilicate glasses; and a number of special glasses, including solder glasses, laser glasses, silica glass, glass-ceramics, and colored glasses. These types essentially bracket the commercial glasses.

Soda-Lime-Silica Glasses. This is the most important group in terms of tonnage melted and variety of use. The combination of silica sand, soda ash, and limestone produces a glass that is easily melted and shaped and has good chemical durability. The raw materials are indigenous to most areas of the world and inexpensive. Soda-lime glasses are particularly suited to automatic-machine-forming methods and are the basis for most of the bottle-, sheet-, and window-glass industry. Very small amounts (often less than 3% of the total batch) of alumina, magnesia, boric oxide, and other chemicals are added to act as stabilizers and to increase durability.

Lead Glasses

The glasses of this group, composed basically of silica sand and lead oxide, have a high refractive index and high electrical resistivity. Potash is present as a significant constituent in most of these glasses. The slow rate of increase in viscosity with decrease in temperature makes lead glass particularly suitable to hand fabrication. The amount of lead may vary considerably, even up to 92% lead oxide; it is a more expensive glass, as the raw materials are relatively expensive and special care is needed in melting to avoid bubbles and seeds. Glasses of this type are used in high-quality art and tableware and for special electrical applications.

Borosilicate Glasses

This group of glasses is basically a combination of silica sand with boric oxide and soda ash. The glasses have excellent chemical durability and electrical properties, and their low thermal expansion yields a glass with a high resistance to thermal shock. High durability makes them ideal for demanding industrial and domestic use, such as chemical laboratory ware, cook ware, and pharmaceutical ware. These glasses were developed in the early part of this century to cope with the problem of cold rain on hot railway-signal lights.

Special-Purpose Traditional Glasses

Solder Glasses. These glasses have low softening and annealing temperatures together with expansion characteristics which permit them to be used as intermediate glasses in making seals between two glass surfaces, between a glass and a metal, or between two ceramic surfaces. In fact, solder glass might be described as a high-grade glass glue. Normally, sealing temperatures are well below the annealing temperature of the glass being sealed, and there is little permanent effect on the glass parts being joined. The major constituents of these glasses include lead oxide, boric oxide, and zinc oxide.

Laser Glasses. Glass has various characteristics that make it an ideal laser host material. Its random structure permits broad emission and absorption bands, which provide higher efficiency, more energy storage, and greater energy per pulse than any other material. In addition, most lasing ions are easily soluble in the glass, and rods, fibers, or disks of any size and of high optical quality are easily fabricated. Of the several rare-earth ions that have been made to lase in a glass host, only neodymium has received commercial application. When a neodymium glass lases, it emits light at a rather fixed wavelength of 1.06 nm. Neodymium-doped silicate, phosphate, and fluoride have been used to provide the energy source for laser fusion research throughout the world.

Silica Glass. A glass composed of silicon dioxide as the only constituent has a very high softening temperature and a very low thermal expansion.

It is costly to make and fabricate because temperature in excess of $1800°C$ is required to manufacture it. However, its refractory character coupled with its very high resistance to thermal shock makes it ideal for special laboratory equipment, windows in high-temperature environments, and instruments.

Glass-Ceramics. These materials are formed in the same manner as conventional glasses and then subjected to heat treatments which caused controlled nucleation and crystallization. Although nearly completely crystalline, their properties can range from transparent to opaque; electrically insulating to weakly conducting; hard to machineable; and with positive, zero, or negative thermal expansions depending upon the composition and heat treatment. This family of materials is based on glasses whose major constituents are magnesium oxide, lithium oxide, aluminum oxide, and silicon dioxide. The crystalline phase or phases and their morphology control the properties of the materials, but the starting chemical composition and the heat treatment determine which crystalline phases will result. Glass-ceramics, which are the result of recent research efforts, have found applications as household cooking ware, reflective optics and laser gyro substrates, chemical processing components, and cooking-stove tops.

Colored Glasses. Nearly all glasses can be colored by adding one or more colorants to the batch in correct amounts. Production of some colors requires, or is enhanced by, the state of oxidation of the coloring agents and by the atmospheres in which the glasses are melted. Table 1 indicates the colors obtainable, colorants used, and chemical states required or utilized.

While the preceding paragraphs describe several classes of glass, within each class there can be infinite composition variations to fit the exact requirements of the user. Table 2 shows typical composition ranges for commercial glasses.

TABLE 1. COMMONLY USED INGREDIENTS FOR COLORING GLASS

Glass color	Coloring agent	State
Red........	Cadmium sulfide, cadmium selenide	Reduced
	Cuprous oxide	Reduced
	Gold (metal)	
Yellow........	Cerium oxide with titanium oxide	
Yellow-green......	Chromic oxide	Oxidized
Blue-green........	Iron chromite	Reduced
Blue........	Cobalt oxide	
Purple........	Neodymium oxide	
Gray........	Nickel oxide with titanium oxide	
Black........	Copper, cobalt, nickel, and iron oxides in combinations of two or more	
Amber........	Iron sulfide	Reduced
Flint (or colorless)	Selenium and cobalt oxide*	Oxidized

* Selenium and cobalt are used in flint glass to add red and blue hues in amounts only sufficient to balance the green hue resulting from iron oxide present as impurity in most naturally occurring raw materials. The intended result is an even light transmission over the whole visible spectrum.

TABLE 2. COMPOSITION OF COMMERCIAL GLASSES (Weight Percent)

	Soda-lime silica glass				Borosilicate glass	Laser glass	Solder glass	Lead glass	Glass-ceramics
	Containers	Plate and window glass	Tableware	Fibre glass fabrics and insulation					
SiO_2	70–74	71–74	71–74	65–74	70–82	61–69	0.5–16	35–70	62–70
Al_2O_3	1.5–2.5	1–2	0.5–2	2–4.5	2–7.5	0–5	0.1–4	0.5–2.0	17–22
B_2O_3				3–5.5	9–14		7–20		
Li_2O									3–5
Na_2O	13–16	12–15	13–15	8–16	3–8	12–24		4–8	
K_2O				0–1				5–10	
CaO	10–14	8–12	5.5–7.5	5–16	0.1–1.2	3–10			0–5
MgO			4.0–6.5	3–5.5					0–7
BaO					0–2.5		0–4		
ZnO							7–62		
PbO							4–77	12–60	
CuO							0–10		
Nd_2O_3						1–6			
CeO_2						0.1–1			
F_2							0–2		
ZrO_2									3–10
TiO_2									

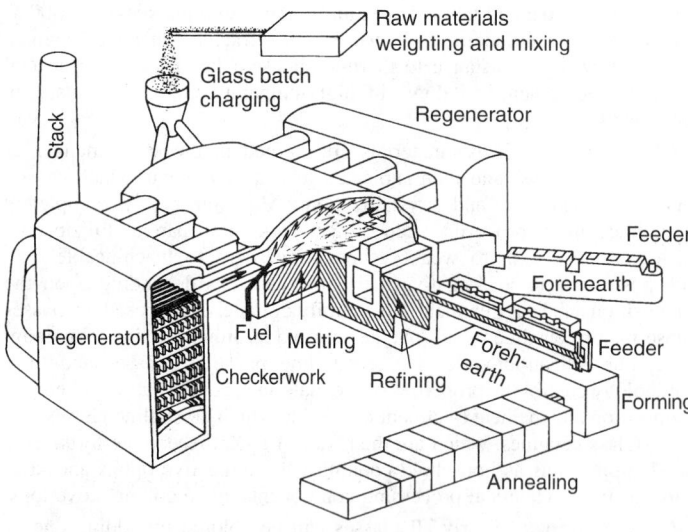

Fig. 4.   Representative glass-producing facility

Traditional Manufacturing Process

Glass products are many and varied, and glass compositions range rather widely, depending on the desired products. Figure 4 shows a typical cross section of a glass manufacturing facility. Raw materials weighing, mixing, charging, and melting are common requirements regardless of the forming operation that is to follow. Most melting furnaces have a primary melting area, followed by a refining or homogenizing section, which is connected to the forming operation by channels called feeders. Although fiber glass is not passed through an annealing furnace after it is formed, most other glass products are annealed to relieve stresses caused by uneven cooling during and immediately after forming.

It is apparent that although there are several similar steps in all glass-manufacturing processes, the forming operations are the most diverse.

Batch Preparation. This begins with the selection, procurement, and storage of an adequate quantity of the raw materials. Selection is made on the basis of the oxides that each material contains and will provide to a glass and on the basis of purity and grain size. Naturally occurring raw materials are used wherever possible for economy, e.g., silica sand, limestones, feldspars, borates, soda ash, boric acid, potash, and barium carbonate. The prescribed quantities of these raw materials, depending on their chemical composition, are measured carefully and mixed together to provide a homogeneous batch. Such mixing is done on an intermittent or a continuous basis, depending on the volume of batch needed to charge the furnaces. The batch is conveyed by a variety of means to the furnaces but always in such a way that segregation is avoided. The importance of grain size of the various raw materials becomes evident in preventing dusting and/or segregation.

Furnaces. A variety of furnaces is used in the industry to melt the batch to produce glass. They must all accomplish the two purposes of confining the heat to the necessary area and containing the melted glass within the furnace. Crucibles or pots are sometimes used to contain the batch and the melted glass, in which cases the furnace merely retains heat; however, tank furnaces (Fig. 3) are far more common. They are so constructed that the lower portion contains the glass and the superstructure retains the heat and provides combustion space for the fuels used. "Day" tanks are used in some instances where the operation is intermittent and the quantity of glass is small. The great majority of glass produced is melted in continuous furnaces, which are charged initially with batch and cullet (broken-up pieces of previously melted glass) which are melted, filling the tank to a specified depth. Thereafter, batch and cullet are charged continuously at a rate equal to that at which the molten glass is withdrawn from the working end.

Continuous tank furnaces are designed to provide for a separate melter section and a refiner or conditioning section. The melting end is maintained at the necessary high temperatures to accomplish the melting and chemical reactions of the batch materials. The refining, or conditioning, section retains the glass long enough for it to cool to the necessary lower working temperatures.

Glass-melting furnaces are built of refractory materials of various types, which will withstand the severe conditions to which they are exposed. The lower portion of the melter section, for instance, must be of the highest quality to withstand the corrosive action of the glass as well as the high temperatures used. Some sections may use lower-quality refractories because the temperature or corrosion conditions are not as severe.

Fuels used in today's furnaces in the United States are natural gas or oil. The fuel is fed to burners that project flames over the surface of the glass. Nearly all continuous furnaces utilize regenerators, which reclaim a portion of the heat from the exhausting combustion gases. Although some glass is melted entirely by the use of electric power, it is generally too expensive to use as a sole source of energy. When electric power is used to augment the fossil fuels, it is called electric boosting.

For the areas that do have sufficiently low-cost electric power, the furnaces are constructed with conventional bottoms but with superstructure only adequate for initial heat-up. They depend on a blanket of batch floating on the surface of the glass to retain the heat within the tank that is provided by the submerged electrodes. Fresh batch is added to the blanket at a rate equal to the rate of melted glass withdrawn.

Melting. This provides the mutual solution of the oxide material high temperatures to yield a homogeneous liquid. Temperatures may range from 1427°C to over 1593°C, depending on the glass composition. Water vapor, entrapped air, and CO_2 are given off, some of which become entrapped in the glass, resulting, initially, in a foamy mass. As the melt moves to the higher-temperature regions, the viscosity is lowered and the gases escape. Deliberate hot spots enhance the natural convection currents, promoting homogeneity. More modern furnaces utilize bubblers, which introduce controlled pulses of air through furnace bottom, further enhancing convection. This is particularly valuable for increasing temperatures near the tank bottom in melting those glasses that are more opaque to infrared radiation.

The glass is essentially free from bubbles (or seeds) when it reaches the end of the melting chamber. It then passes under floaters in some furnaces, or through submerged throats in most, to the so-called refining section (more properly, the conditioning section). Here the refining conditioning consists of allowing the glass to increase to a more useable viscosity level by uniformly lowering the temperature, which also allows the remaining tiny seeds or gaseous inclusions to dissolve.

Furnaces supply glass to up to eight forming machines. Forehearths or alcoves serve to channel the glass to the individual machines or machine locations and to further change the temperature and viscosity.

Forming Operations. These are many and varied, involving two three, or four major steps. The first is a further temperature conditioning to place the glass in the exact viscosity range, sometimes wide but often quite narrow, suitable for the selected primary forming operation. The second step is the primary forming itself, followed usually, but not always, by an annealing step. Single or multiple secondary operations may ensue. Only the major forming processes of drawing, pressing, blowing, and casting will be discussed.

Drawing is one of the simpler forming methods by which thousands of tons of window glass and millions of feet of rod and tubing are produced annually. Drawing window glass frequently utilizes a rectangular refractory frame, called a debiteuse, placed on the surface of the conditioned glass. It has a slot roughly 4–8 in. (10–20 cm) wide and 8 ft (2.4 m) or more long through which the glass is pulled vertically. The width and length of the slot in the debiteuse, together with the drawing speed, aid materially in controlling the width and thickness of the sheet. The upward draw may continue until the sheet is nearly cold, when it can be stored and cracked off in suitable lengths. Or, it may be bent over a large roller at nearly the last moment it will withstand bending and conveyed horizontally into the annealing lehr. This method of making window glass has been largely replaced by the float glass process described below.

Glass tubing may be drawn vertically in a manner similar to that for window glass. Another common method is the Danner process, in which a suitable stream of glass is flowed onto a conical rotating mandrel supported with its small end downward and its axis at a suitable angle to the horizontal. The tubing is drawn from the small end, through which sufficient air is blown to retain the desired cross section of the tubing. Drawing continues horizontally over rollers until the tubing can be cracked off in lengths at the cold end. Glass tubing is also made by the downdraw process, where air is blown into the tube as it is drawn from the bottom of a refractory bowl of molten glass.

Plate glass may be formed by flowing the molten glass over the lip of the discharge end of the furnace between a set of large water-cooled rollers and then pulling it away by means of driven rollers. The resulting sheet is up to 1 in. (2.5 cm) or more thick and 10–12 ft. (3–3.7 m) wide. However, most flat glass made throughout the world today is made by the recently developed float-glass process. In this process the molten glass is formed into a sheet by floating it on a bath of molten metal such as tin. The glass flowing onto the bath of tin is pulled across the surface and cooled to the temperature at which it is rigid while still on the molten metal. The outstanding advantage of this process is that it produces a plate of glass both surfaces of which require no further polishing.

Modern methods of pressing, blowing, and casting usually involve an intermediate step, the formation of a suitable charge of glass, or gob, for the ensuing operation. The most common method involves a gob feeder located at the end of the forehearth. This consists of a bowl, or spout, kept full of glass by flow from the forehearth and having an orifice in its bottom and a refractory tube suspended in the bowl over the spout. The tube may be lowered to shut off the flow of glass or raised to permit flow at a selected rate. A refractory plunger operates vertically inside the tube. It provides a pumping action on its upstroke, momentarily restraining the flow of the glass. Its downstroke forces the accumulated glass out of the orifice, where it is sheared off. The result is a charge of glass, called a gob, of controlled size, which is delivered to the forming machine by gravity.

Pressing, or press-forming, operations normally are used for relatively shallow, heavy-walled products. Pressing is accomplished by means of a metal mold (usually iron or steel), a ring, which is centered on top of the mold, and a plunger, which is forced into the mold through the ring. The mold shapes the exterior of the product, the ring the sides, and the plunger the interior. A pressing machine may have many molds mounted on its circular, rotating table, a ring for each mold or, more commonly, a single ring mounted on the same mechanism as the plunger, and a single plunger. After a gob is charged into the mold, the machine indexes one station under the plunger and the plunger moves down into the mold, dwells momentarily, then retracts. It is noteworthy that the plunger action flows the glass into the mold cavity rather than stamping out the product by a quick movement. Since considerable heat is removed from the glass by the plunger, it is cooled with water internally. The product remains in the mold for about half the revolution of the press table before removal to allow it to cool below its deformation temperature. The molds may be cooled by forced air.

Blowing methods work best for deep products and it frequently must be used for thin-walled items. A common procedure, called the blow and blow, involves two steps, of which the first is shaping the glass charge into a form called a blank or parison. Gob-fed machines receive the gob in the parison mold, where it is shaped into a cylinder about two-thirds the

height of the bottle. The finish, or top, of the bottle is formed in the same operation at the bottom of the mold by action of a small plunger entering the mold from below and delivering a puff of air. A transfer mechanism holding the parison by the completed finish then swings and inverts it into a second mold for the second step, blowing the glass into its final shape. A cross section of the molds shows this process in Fig. 5. The most modern machinery for rapidly forming containers and bottles commercially are individual section (IS) machines. Each section is capable of forming up to four gobs at the same time and there are as many as ten sections per machine. The individual sections can be sequenced electronically to produce more than 400 bottles per minute on a 10-section machine. See also Fig. 6.

The Owens process employs vacuum to charge the glass into the blank or parison mold. Here, a blank mold dips into a shallow pot of molten glass, a vacuum is applied, and a charge of viscous glass is pulled into the blank mold. The finish is formed simultaneously at the top of the blank. This blank or parison is subsequently transferred into the blow mold, where the bottle is blown into its final form. See Fig. 7.

In another modern machine, the glass flows downward from an orifice in a continuous stream that passes between rollers that flatten it into a ribbon with alternate thick and thin spots. The ribbon is picked up by a horizontally moving support in which voids coincide with the thick portions

Fig. 6.   Three white-hot bottles immediately after being formed on a section of an IS machine. The bottles will be transferred immediately to an annealing lehr for cooling and annealing. (*Owens-Illinois, Inc.*)

Fig. 5.   A high-productivity IS machine manufacturing three bottles on each section at the same time. (*Owens-Illinois, Inc.*)

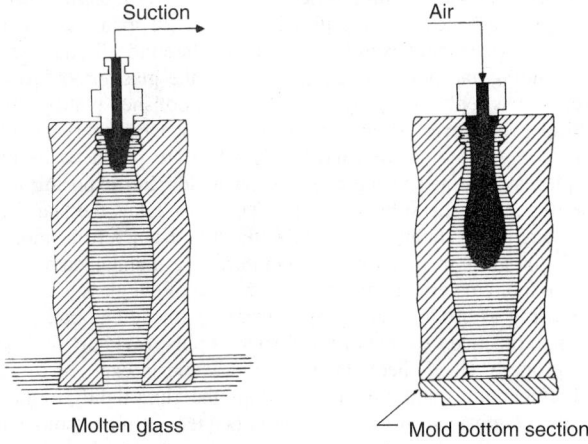

Fig. 7.   Owens process. *Left*: Blank mold is dipped into the surface of molten glass, where it is filled by a vacuum suction. As the mold is lifted from the glass, a knife cuts off the glass and closes the mold. *Right*: The blank mold opens, and a puff of air is introduced to shape the parison before transferring it to the blow mold, where it is blown to its final shape

of the ribbon. Blow heads on an endless belt operating from above the ribbon provide puffs of air to aid in producing a bulbous sagging in the thick portion of the ribbon. After sufficient sagging, molds on an endless belt close around the sagging glass from below, and air from the blow heads blows the glass into the shape of the mold. After the molds open, the product, frequently light bulbs or Christmas ornaments, can be cracked off the ribbon.

Casting is usually restricted to two types of operations. The first involves the simple pouring of molten glass into molds. Examples include such massive shapes as the borosilicate mirror blank for the Mt. Palomar telescope and the large glass-ceramic mirror blanks for observatories in Australia and South America. The molds are specially constructed for refractory materials.

The second type of casting is spin casting, in which a gob from a gob feeder is fed into the bottom of a metal mold supported so that it can be rotated rapidly or spun on its vertical axis. The centrifugal force thus generated causes the glass to flow up the inclined sides of the mold, producing a conical shape. The initial movement of the glass is aided by insertion of a conical plunger into the glass at the bottom of the mold when spinning is begun. Mold speeds of up to 1,600 rpm are attained within one second. The funnel portion of television tubes is sometimes produced by this method.

Annealing. As with most substances on cooling, the temperature differential between the surface and interior layers of a piece of glass establishes temporary stresses, and the higher this differential the greater the stresses. Fracturing can occur when the stresses exceed the tensile strength of the glass. Permanent stresses can be avoided by carefully controlled cooling from a little below the annealing point to the strain point. This is the annealing range. Thereafter, the rate of cooling need only be such that the temporary stresses do not exceed the tensile strength of the glass. Glass manufacturers have learned to take advantage of these phenomena.

Annealing immediately follows glass-forming operations. In continuous processes, the ware is placed on an endless belt, which carries it through the lehr, a tunnel in which the temperature is carefully controlled. Temperature of the ware is raised initially to near the softening point, then lowered slowly through the annealing range and thereafter at a more rapid rate to the point where it can be packed or stored. The process is designed to result in the degree of permanent stresses desired. Optical glass must be annealed very thoroughly to produce an essentially distortion- and strain-free lens; however, some stresses can be tolerated or become beneficial to most other products. Small rods and tubing, for instance, are strong enough because of their regular cross section to require no annealing, while tempered glass has uniformly controlled stresses to increase its mechanical performance.

Secondary Operations. Lampworking is one of the many and varied operations utilized to produce glassware following the initial forming. The materials used are rod and tubing, which are softened in the flame of burners and shaped or blown as desired.

Grinding and polishing are important steps in many glass-manufacturing processes. Use of a sequence of increasingly finer gradations of abrasives, usually ending with jewelers' rouge or cerium oxide powder for polishing, produces the desired results. Optical lenses, prisms, and reflective optics parts are prominent examples. The plate-glass industry has used long lines of grinding and polishing equipment, but the glass produced by the float process has replaced nearly all ground and polished plate glass.

Bending procedures are utilized to produce shapes otherwise difficult to fabricate, e.g., automotive windshields. They are produced by placing the flat pieces of proper shape and size on molds and exposing them to temperatures above the softening point. The glass takes the shape of the mold by sagging or slumping with or without assistance from mold parts contacting the glass from above. Temperatures are maintained sufficiently low and the mold material is such that the surface of the glass is unaffected.

Laminating to produce safety-glass parts, as for automotive windows, is a common practice. A sheet of resin, such as polyvinyl butyral, is placed between properly sized sheets of glass; the whole is exposed to slightly elevated temperatures and pressures, to bond the glass tightly to the resin.

Coating of glass products such as containers is quite common, the objective being to protect the container from abuse to which it is subjected in handling during filling and shipping. A coating which is not visible, can be labeled, protects the surface, and provides lubricity is required and usually calls for a two-layer coating such as tin or titanium oxide, followed by a lubricious coating such as polyethylene. The oxide coatings are obtained by subjecting the hot container to a vapor of chloride, which

oxidizes to the oxide. Thick opaque or translucent oxide and metallic coatings are sometimes used to provide attractive color effects or light protection. Many precision optical lenses are coated with thin, vapor-deposited layers, which reduce the light lost by reflection from the surface, and some architectural glass is coated to provide attractive colors and reflect undesirable infrared radiation.

Decorating glass or glassware is an old art that takes many and varied forms. Cutting, grinding, and mechanical or chemical polishing or etching, are well known. Opaque, translucent, and transparent enamels can be applied by silk screens or other means in multiple colors and in almost any pattern. Low-melting vitreous enamels have been used for many years, and when properly fired, they provide good durability. More recently, organic polymers have been substituted for the vitreous enamel. They are not quite as durable as vitreous enamels, but they do not require high curing temperatures.

Tempering is the direct reverse of annealing; i.e., high permanent stress is induced in the glass. Rapid cooling or quenching is applied to the glass surfaces at a temperature slightly below the softening point, placing the surfaces in a high degree of compression while the balancing tensile forces are confined to the interior. Since glass always breaks in tension, very considerable strength is incorporated. Typical products are glass doors, automotive glass, windows, goggles, spectacles, and table ware. Tempering must be the final step in the production line. Other products can be strengthened by judicious control of the degree of annealing if their shapes permit it.

Sealing glasses to each other or to other materials must take into account the thermal expansion-and-contraction characteristics. Many glasses have thermal-expansion properties that allow them to be sealed to metals, but each metal usually requires a different glass composition. Solder glasses are used to seal two pieces of glass to each other, two pieces of metal, or a piece of metal and a piece of glass. The glass seals on light bulbs and vacuum tubes are examples of commercial glass-metal seals, while color TV tubes are sealed together with solder glass at a temperature at which the phosphors are not degraded.

Sealing glasses used for color television tubes are devitrifying or crystallizing sealing glasses. They crystallize during the sealing process to produce a seal that will not soften during the processing of the bulb—because the crystallized glass has a higher melting temperature than the starting sealing glass.

See also **Ceramics**.

EARL D. DIETZ
Toledo, Ohio

GLASS BLOCKS

Introduced during the art deco period (1920s–1930s), glass blocks for structural and decorative purposes were quite popular. Interest faded, but has returned within the last few years.

In addition to their decorative appeal, glass blocks are claimed to provide better energy conservation (solar reflective blocks are available), lower sound transmission, aesthetic flexibility, minimal maintenance, and enhanced security. In addition to plain blocks, they are available with various decorative designs. See Fig. 8. An effective use of glass blocks

Fig. 8. Decorative glass block with pattern Decora®. (*Pittsburgh Corning Corporation*)

Fig. 9. Use of glass blocks for entrance to a high-rise building. (*Pittsburgh Corning Corporation*)

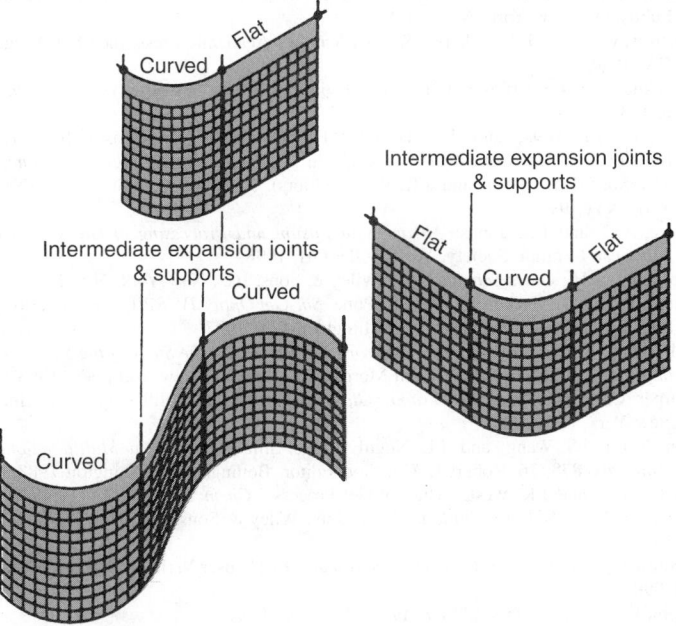

Fig. 10. Various ways to arrange glass block walls for interior or exterior. (*Pittsburgh Corning Corporation*)

for an external wall is shown in Fig. 9. Design schemes for obtaining architectural effects are shown in Fig. 10.

In making glass blocks, molten glass is extruded in "gobs" that are poured into an open half-block mold. A plunger, which creates the pattern on the inner surface of the block, presses the glass into the mold to produce

a half-block. Using direct heat, the two halves are fused together to form a complete hollow block unit. This process thus creates an insulating air space, which makes the blocks energy efficient.

Sol-Gel Glass

Sol-gel processing is a new, chemically based method for producing glass at much lower temperatures than traditional melting methods (see above). Due to the low temperatures there are many advantages of sol-gel glass processing, such as casting of net shapes and net surfaces, improved physical properties, and the production of a new type of material, transparent porous glass matrices. See Table 3.

Three methods can be used to make sol-gel glasses:

1. Gelation of colloidal powders.
2. Hypercritical drying.
3. Controlled hydrolysis and condensation of metal alkoxide precursors, followed by drying at ambient pressure and temperature.

Definitions. Colloids are solid particles with diameters of 1 □ 100 nanometers. A sol is a dispersion of colloidal particles in a liquid. A gel is an interconnected rigid network of sub-micrometer dimensions. A gel can be formed from an array of discrete colloidal particles (Method 1) or the 3-D network can be formed from the hydrolysis and condensation of liquid metal alkoxide precursors (Methods 2 and 3), shown in Fig. 11. The metal alkoxide precursors used in Methods 2 and 3 are usually $Si(OR)_4$ where R is CH_3, C_2H_5, or C_3H_7. The metal ions can be Si, Ti, Sn, Al, and so on.

Processing Steps. Seven steps are involved in making glass by the sol-gel method. See Fig. 12. A low-viscosity sol is formed by mixing (Step 1). The viscosity of the sol increases greatly as a gel begins to form. Prior to gelation the sol is applied as a coating, pulled into a fiber, or cast into a mold with a precise shape and surface features (Step 2). Gelation (Step 3) occurs in the mold, forming a solid object with the desired shape and surface. Low-cost polymer molds can be used, but interfacial bubbles must be prevented and contamination must be avoided, since it can nucleate cracks in the weak gel.

After gelation, the interconnected 3-D gel network is completely filled with pore liquid. Holding the gel in its pore liquid for several hours at 25–80°C leads to localized solution and precipitation of the solid network, called aging (Step 4). The thickness of the interparticle necks increases during aging, as does density and strength of the gel. Aging must continue until the gel is strong enough to resist cracking during drying.

The pore liquid is removed during drying (Step 5). Drying of colloidal gels (Method 1) is relatively easy because the pores are large (100 nm). Alkoxide-based gels have very small pores (1–10 nm), and thus large

TABLE 3. ADVANTAGES OF SOL-GEL GLASS

Net-Shape/Surface Casting
Complex geometries
Lightweight optics
Aspheric optics
Surface replication (e.g., fresnel lenses)
Binary/diffractive optics
Internal structures
Reduced grinding
Reduced polishing

Improved Physical Properties (Type V Silica)[*]
Lower coefficient of thermal expansion
Lower vacuum ultraviolet cutoff wavelength
Higher optical transmission
No absorption due to H_2O or OH bands
Lower solarization
Higher homogeneity
Fewer defects

Transparent Porous Structures (Type VI Silica)[*]
Impregnated with optically active organics, such as laser dyes, NLO molecules
Graded refractive index (GRIN) lenses
Laser-enhanced densification
Laser-written microoptical arrays and wavelengths
Controlled chemical doping
Control of variable oxidation states of dopants

[*] Types I–IV silicas are discussed in Bruckner reference listed.

CH₃
|
O
|
H₃C—O—Si—O—CH₃ + 4(H₂O) $\longrightarrow$
|
O
|
CH₃
(TMOS)

OH
|
HO—Si—OH + 4(CH₃OH)
|
OH

Hydrolysis
The hydrated silica tetrahedra immediately interact in a condensation reaction, forming ≡Si—O—Si≡ bonds.

OH   OH   OH   OH
| | | |
HO—Si—OH + HO—Si—OH $\longrightarrow$ HO—Si—O—Si—OH + (H₂O)
| | | |
OH   OH   OH   OH

Condensation
Linkage of additional ≡Si—OH tetrahedra occurs as a polycondensation reaction and eventually results in a SiO₂ network.

HO—Si—O—Si—OH + 6(Si(OH)₄) $\longrightarrow$

HO   OH
| H H |
HO—Si—O O—Si—OH
| |
HO   OH
| |
O   O
| |
HO—Si—O—Si—O—Si—O—Si—OH + 6(H₂O)
| | | |
HO   O   O   OH
       | |
       O   O
       | |
HO—Si—O O—Si—OH
       H H
HO   OH

Polycondensation

Fig. 11. Chemical reactions involved in sol-gel alkoxide processing of silica gel-glass

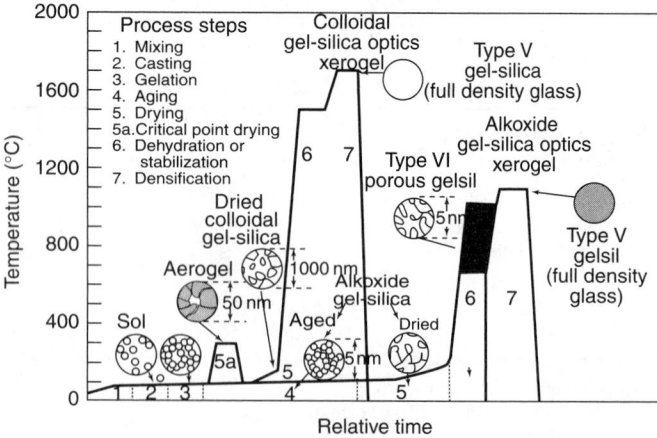

Fig. 12. Processing sequence for sol-gel silica optics

capillary stresses can arise during drying. Hypercritical evaporation at elevated temperature and pressure (Method 2) avoids the solid–liquid interface and eliminates drying stresses. A gel produced in this method is called an aerogel. Aerogels have very low densities—as low as 80 kg/m³—and very large void volumes (95–99%).

Careful control of hydrolysis and condensation rates by use of acid catalysts in Method 3 results in very narrow pore size distributions, which minimizes stress gradients during drying by thermal evaporation under ambient pressure and low temperatures. Gels dried in this manner are

termed xerogels. The generic term gel usually applies to a xerogel. A gel is defined to be dried when the physically adsorbed water is completely gone, between 120–180°C (248–356°F) (Stage 5 in Fig. 12). The surface area of gels made by Method 3 is very large (200–900 m²/g), depending upon pore size, which can vary from 1.2 to 10 nm.

Chemical stabilization of a dried gel, Step 6, is necessary to use the material as a transparent porous matrix. Thermal treatment in the range of 800–1000°C (1472–1832°F) (Fig. 12) desorbs silanols and eliminates three-membered silica rings from the gel, which can interact with atmospheric water and cause cracking. Stabilization increases density, strength, and hardness of the gel and converts the network to a glass with network properties similar to fully dense amorphous silica. A stabilized optically transparent porous matrix is designated as Type VI gel-silica. Applications of this new type of optical glass are indicated in Table 3.

Densification of an alkoxide-derived silica gel-glass is completed around 1150°C (2101°F), where all the pores are eliminated (Stage 7, Fig. 12). Removal of hydroxyls and water from the pores of the gel-glass prior to densification results in fully dense Type V gel-silica, which has a purity and homogeneity superior to silica glass made by traditional methods. The density becomes equivalent to that of (Types I and II) fused quartz or (Types VI and IV) fused silica (i.e., 2.2 g/cc).

The ability to make optics without grinding or polishing and to replicate surface features from a master mold with high accuracy (1 part in 10⁴) is an important advance in optical glass technology offered by sol-gel processing of Type V gel-silica. The new Type VI porous optical matrices made by sol-gel processing make it possible to achieve multifunctional optical components, also an important advance in the field.

L. L. HENCH
University of Florida
Gainesville, Florida

Additional Reading

Amato, I.: "A New Order in Glass," *Science,* **1377** (June 7, 1991).

Amstock, J.S.: *Handbook of Glass in Construction,* The McGraw-Hill Companies, Inc., New York, NY, 1997.

Bach, H. and D. Krause: *Analysis of the Composition and Structure of Glass and Glass Ceramics,* Springer-Verlag, Inc., New York, NY, 1999.

Bach, H. and D. Krause: *Thin Films on Glass,* Springer-Verlag, Inc., New York, NY, 2000.

Behling, S. and S. Behling: *Glass: Structure and Technology in Architecture,* Prestel Publishing, New York, NY, 2000.

Brinker, C.J. and G.W. Scherer: *Sol-Gel Science,* Academic Press, Inc., San Diego, CA, 1990.

Bruckner, R.: "Properties and Structure of Vitreous Silica," *J. Non-Crystalline Solids,* **5**, 123 (1990).

Chia, T., J.K. West, and L.L. Hench: "Fabrication of Microlenses by Laser Densification on Gel-Silica Glasses," in *Chemical Processing of Advanced Materials* (L.L. Hench and J.K. West, Editors), John Wiley & Sons, Inc., New York, NY, 1992.

Clare, A.G. and L.E. Jones: *Advances in Fusion and Processing of Glass II,* The American Ceramic Society, Westerville, OH, 1998.

Doremus, R.H.: *Glass Science,* John Wiley & Sons, Inc., New York, NY, 1994.

Dunn, B.S., J.D. Mackenzie, and J.E. Pope: *Sol-Gel Optics IV,* SPIE—International Society for Optical Engineering, Bellingham WA, 1997.

Ellis, W.S.: *Glass: From the First Mirror to Fiber Optics, the Story of the Substance that Changed the World,* William Morrow & Company, New York, NY, 1998.

Harper, C.A.: *Handbook of Ceramic Glasses,* The McGraw-Hill Companies, Inc., New York, NY, 2000.

Hench, L.L., S. Wang, and J.L. NoguÈs: "Gel-Silica Optics," in *Multifunctional Materials,* **878**, 76, Robert L. Gunshor, Editor, Bellingham, Washington 1988.

Hench, L.L. and J.K. West: "The Sol-Gel Process," *Chem. Rev.,* **90**, 33 (1990).

Pfaender, H.C.: *Schott's Guide to Glass,* John Wiley & Sons, Inc., New York, NY, 1983.

Schittich, C. et al.: *Glass Construction Manual,* Birkhauser Verlag, Cambridge, MA, 1999.

Stein, D.L.: "Spin Glasses," *Sci. Amer.,* 52 (July 1989).

Tonucci, R.J. et al.: "Nanochannel Array Glass," *Science,* 783 (October 30, 1992).

Tse, J.S. and D.D. Klug: "Structural Memory in Pressure-Amorphized AlPO4," *Science,* 1559 (March 20, 1992).

Varshneya, A.K.: *Fundamentals of Inorganic Glasses,* Academic Press, Inc., San Diego, CA, 1997.

GLASSY METALS. Metallic glasses do not occur naturally but are produced by various techniques, the oldest of which is by cooling molten metals so rapidly that the atoms do not get to form regular crystalline structures. The atoms are frozen in random or nonrepeating atomic patterns similar to those found in organic glasses. By metallic it is meant that the amorphous material is composed primarily, but not necessarily exclusively, of metallic elements that exhibit the properties of metals, such as electrical or magnetic behavior.

Primarily through modification of the chemistries, glasses have been produced that can truly be considered bulk, even exceeding 1 cm in all dimensions. Warm extrusion and consolidation of metallic glass powders produce blocks of glassy metals that can later be machined into components. Besides rapid solidification into thin ribbons or flakes, there exists a wide range of techniques to produce metallic glass. Methods available include chemical means, mechanical alloying, vaporization, and solid-state reactions.

Interest is maintained in these materials because of the combination of mechanical, corrosion, electric, and magnetic properties. However, it is their ferromagnetic properties that lead to the principal application of glassy metals. The soft magnetic properties and remarkably low coercivity offer tremendous opportunities for this application.

A limitation of metallic glass is that it exists in metastable form, which means that it tends to crystallize if heated with sufficient thermal energy to allow the kinetics of crystallization, i.e., both nucleation and growth, to occur. If glassy metal alloys were all intrinsically unstable, however, they would be much less promising as an engineering material. Understanding the solid-state structure, the unusual mechanical properties, the liquid-like electrical properties, and the ferromagnetic properties of metallic glasses has been the focus of research and development efforts since the late 1950s.

Examples of the various categories of metallic glasses are given in Table 1.

Formability

To form an amorphous, i.e., glassy metal alloy from the liquid state means that the crystallization step must be avoided during solidification. This can be understood by considering a time–temperature–transformation (TTT) diagram. Nucleating phases require an incubation time to assemble atoms through a statistical process into the correct crystal structure which is capable of surmounting an activation barrier ΔG^*. Incubation times can vary from fractions of a second to many seconds the shape of the TTT curve is in the form of a C because of competing phenomena. As temperature is lowered, the free energy available to nucleate and grow a crystalline phase increases but the kinetic ability to do so through atomic diffusion decreases, resulting in a nose at T'. The glass-forming ability of a material is determined by the kinetics of this process, followed by the initial stages of crystal growth. If the liquid alloy is cooled from above the melting temperature to a temperature below the nose of the TTT curve at T' within a time less than the time where crystallization begins, the alloy's liquid-like structure becomes frozen in when the temperature drops below the glass-forming temperature, T_g, whereby an amorphous solid is obtained. Once the alloy is below T_g, diffusion processes are such that growth of a crystalline embryo is essentially halted, and the liquid structure is preserved.

A reduced glass temperature can be defined as $T_{rg} = T_g/T_l$, which represents a measure of glass-forming ability. The higher the T_g and the lower the liquidus temperature, T_l, the easier it is to supercool the metal melt to a glassy state. Conventional theory predicts $T_{rg} = 0.65-0.7$ for good glass formers.

TABLE 1. METALLIC GLASSES

Alloy	Useful properties/applications
$Fe_{78}B_{13}Si_9$	Metglas 2605S2,[a] good magnetic properties
$Fe_{80}B_{20}$	Metglas 2605[a]
$Pd_{80}Si_{20}$	easy to form glass, thick samples produced
$Pt_{60}Ni_{15}P_{25}$	
$Cu_{84.3}P_{15.7}$ eutectic	brazing foil
$Al_{85}Ni_5Fe_2Gd_8$	high strength low density thick ribbons
$Mg_{80}Cu_{15}Sn_5$	low density cast 4-mm dia rods
$Co_{83}Gd_{17}$	sputtered sample can support magnetic bubbles
$Mo_{64}Re_{16}P_{10}B_{10}$	superconducting below 8.7 K
$W_{60}Ir_{20}B_{20}$	crystallization temperature above 1200 K
$Zr_{41.2}Ti_{13.8}Cu_{12.5}Ni_{10}Be_{22.5}$	14-mm rod produced

[a] Produced by AlliedSignal.

Processing

Traditionally, production of metallic glasses requires rapid heat removal from the material which normally involves a combination of a cooling process that has a high heat-transfer coefficient at the interface of the liquid and quenching medium, and a thin cross section in at least one-dimension. Besides rapid cooling, a variety of techniques are available to produce metallic glasses. Processes not dependent on rapid solidification include plastic deformation, mechanical alloying, and diffusional transformations.

Splat quenching or gun techniques involve rapid solidification through atomizing molten metal by blowing it out a tube. The resulting liquid vapor is quenched by impingement upon a metal substrate having a high thermal coefficient. Melt-spinning can produce large quantities of very uniform ribbons, filaments, or tapes.

Lasers can be used to obtain very fast quench rates up to 10^8 K/s.

Metallic glass powders can be made in various sizes through atomization and comminution processes.

Ion implantation has a large (10^{14} K/s) effective quench rate. This surface treatment technique allows a wide variety of atomic species to be introduced into the surface. Sputtering and evaporation methods are other very slow approaches to making amorphous films, atom by atom.

Investigations have been carried out on the formation of amorphous phase through solid-state reactions, typically by the use of mechanical alloying and interdiffusion between thin films.

Crystallization

Metallic glasses are metastable. A combination of thermal energy and time leads to crystallization. At room temperature, this may require a very long time but at moderate temperatures of 373 K or more (depending on the alloy) devitrification can occur in minutes. Usually the crystallization temperature is given as the temperature at which crystallization begins as the alloy is heated at a constant rate. Crystallization leads to an increase in density of about 1%. In general, T_x, the crystallization temperature, is somewhere between 0.4 and $0.65T_m$ where T_m is the melting temperature.

Crystallization, by definition, implies that the initial structure be a glass, followed by the nucleation and growth of a crystalline phase, be it the equilibrium one or a metastable phase. The process is a first-order transformation and involves atomic diffusion, or at least atomic shuffles. Types of crystallization reactions that occur include polymorphous crystallization, which is a composition invariant transformation such as that in Fe–B, and eutectic crystallization, T_e, in FeNiPB glass, where fine lamellae of iron–nickel austenite and metastable $(FeNi)_3$ PB phases grow cooperatively.

Crystallization need not always be deleterious to properties such as strength. In Al-based glasses, partially crystallized material actually increased the fracture strength by 30%.

Mechanical Properties

Of the various physical properties, it is the mechanical properties that make metallic glasses so unique when compared to their crystalline counterparts. A metallic glass obtains its mechanical strength in quite a different way from crystalline alloys. The disordered atomic structure increases the resistance to flow in metallic glasses so that these materials approach their theoretical strength. An attractive feature is that metallic glasses are equally strong in all directions because of the random order of their atomic structure.

The ductility of glassy metals varies according to the kind of stress applied. Owing to the large intrinsic ductility, metallic glasses can be plastically deformed into useful shapes at no loss of mechanical strength.

The fracture of metallic glasses occurs by the formation of shear bands having a 45° orientation relative to the tensile axis. Typical veining on the fracture surface is a characteristic of almost all metallic glasses.

The deleterious embrittlement of a glassy metal during annealing is also accompanied by a change in fracture mode. Almost all metallic glasses containing Fe, Co, and Al show this behavior.

At low applied stress levels crystalline alloys display fatigue properties superior to metallic glasses. At large loads approaching the fracture strength, metallic glasses can survive many more cycles than comparable strength crystalline alloys. This is because the metallic glass deforms elastically up to the high stress levels, whereas crystalline alloys generate localized regions of plastic deformation which lead to nucleation of fatigue cracks.

TABLE 2. MAGNETIC PROPERTIES

Alloy	Coercive field, H_c, A/m[a]	Curie temp, θ_c, °C
$Fe_{80}B_{20}$[b]	3.1	374
$Fe_{80}P_{16}B_1C_3$[c]	4.0	292
$Fe_3Co_{72}P_{16}B_6Al_3$	1.2	260
$Fe_{96.8}Si_{3.2}$[d]	20–40	730

[a] To convert A/m to oersted, multiply by 0.0126.
[b] Metglas 2605.
[c] Metglas 2615.
[d] Material is crystalline.

Properties

Chemical. Along with magnetic and mechanical behavior, high corrosion resistance is one of the most desirable properties of metallic glasses.

Magnetic. More experimental research and development work has been done on magnetic effects in metallic glasses than on any other property (Table 2). It is their soft magnetism and microstructural homogeneity that has led to the most significant applications.

Electrical. Unlike their crystalline counterparts, amorphous metals generally have high electrical resistivity not only at room temperature, but also, because of a very small temperature coefficient, near absolute zero. Certain metallic glasses, e.g., $La_{80}Au_{20}$, do show superconductivity.

Uses

The magnetic properties of glassy metals provide the only commercial use in bulk quantities, although brazing foils provide another niche for metallic glasses. Metallic glasses have yet to find their way into commercial products in structural applications, in spite of the great strength of some glasses. This is related to their shearing instability and size limitations of the glassy product which requires that it employ a fabrication step to obtain bulk material. This situation is changing, however; currently, major breakthroughs are being made in producing bulk metallic glasses by conventional techniques.

GARY J. SHIFLET
University of Virginia

Additional Reading

Greer, A.L.: *Science,* **267**, 1947–1953 (Mar. 31, 1995).
Mercier, J.P., G. Zambelli, W. Kurz, and A. Gay: *Introduction To Materials Science (Series in Applied Chemistry and Materials Science),* Elsevier Science, New York, NY, 2003.

GLAUBERITE. This anhydrous sulfate of sodium and calcium mineral, $Na_2Ca(SO_4)_2$, crystallizes in the monoclinic system. Hardness of 2.5–3, specific gravity of 2.8, with vitreous luster and pale yellow to gray in color. Grades from transparent to translucent. Perfect basal pinacoidal cleavage with conchoidal fracture. Glauberite is a product of salt lake evaporation. World occurrences include the Stassfurt, Germany saline deposits, and Borax Lake in San Bernardino County, California.

GLAUCONITE. Glauconite is a hydrous silicate of potassium, iron, and aluminum with considerable ionic substitution, crystallizing in the monoclinic system. A general formula is $(K, Na)(Al, Fe^{3+}, Mg)_2(Al, Si)_4O_{10}(OH)_2$. It possesses perfect basal cleavage; hardness, 2; specific gravity, 2.4–2.95; color dull green to blue-green; and is often a constituent of marine deposits, forming "green sands." It is believed to have been produced through the alteration of iron-bearing silicates, chiefly biotite and possibly augite and hornblende. It occurs along the Atlantic Coastal Plain of the United States. Frequently found filling the interiors of the shells of *Globigerina,* a common genus of the foraminifera (*Protozoa*). Since *Globigerina* occurs as a deep-sea deposit, some European geologists have claimed that glauconite is only found in deep water. On the other hand, typical green sands occur associated with sand and clays which are certainly of shallow marine origin. Glauconite derives its name from the Greek word meaning *bluish-gray.*

Glauconite, being of sedimentary origin, can be used to determine the age of those sediments by evaluating its $^{40}K/^{40}Ar$ ratio (potassium-argon isotope ratio). See also **Ocean Resources (Mineral).**

GLAUCOPHANE. Glaucophane, essentially a complex silicate of sodium, and iron or aluminum, $Na_2(Mg, Fe)_3Al_2Si_8O_{22}(OH)_2$, is a rather rare mineral although it has been noted from widely separated occurrences. It is monoclinic and ordinarily is fibrous or granular. It is brittle; hardness, 6; specific gravity, 3–3.1; color, azure blue, blackish-blue or gray; luster, vitreous to pearly; translucent to opaque. Glaucophane is found only in the metamorphic rocks sometimes forming glaucophane schists. It is found in Switzerland, Italy, Siberia, Japan and in the United States chiefly in the rocks of the Coast Ranges in California and Oregon. The name glaucophane is derived from the Greek words meaning *bluish-gray,* and *appear.*

GLOBULINS. Proteins that are insoluble in water, but that dissolve readily in aqueous salt solutions. The term globulins is applied to certain subgroups of the plasma proteins. See also **Antibody**; and **Blood.**

GLUCOSE. See **Carbohydrates**; **Starch**; **Sweeteners.**

GLUTAMINE. See **Amino Acids**; **Starch.**

GLYCEROL. [CAS: 56-81-5]. Glycerol, propanetriol, glycyl alcohol, "glycerine," $CH_2OH \cdot CHOH \cdot CH_2OH$, is a colorless, viscous liquid, of sweetish taste, odorless, boiling point 290°C. Glycerol reacts (1) with phosphorus pentachloride to form glyceryl trichloride, $CH_2Cl \cdot CHCl \cdot CH_2Cl$, (2) with acids to form esters, e.g., glycerol monoacetate $CH_2OH \cdot CHOH \cdot CH_2OOCCH_3$, glycerol diacetate $C_3H_5(OH)(OCOCH_3)_2$, glycerol triacetate (triacetin), $CH_2OOCCH_3 \cdot CHOOCCH_3 \cdot CH_2OOCCH_3$, glycerol mononitrates (alpha, $CH_2OH \cdot CHOH \cdot CH_2ONO_2$; beta, $CH_2OH \cdot CHONO_2 \cdot CH_2OH$), glycerol dinitrates (1, 2, $CH_2OH \cdot CHONO_2 \cdot CH_2ONO_2$; 1, 3, $CH_2ONO_2 \cdot CHOH \cdot CH_2ONO_2$), glyceryl trinitrate ("nitroglycerine"), $CH_2ONO_2 \cdot CHONO_2 \cdot CH_2ONO_2$, glyceryl tristearate (tristearin), $CH_2OOCC_{17}H_{35} \cdot CHOO\text{-}CC_{17}H_{35} \cdot CH_2OOCC_{17}H_{35}$, indirectly, glycerol monophosphates (alpha, $CH_2OH \cdot CHOH \cdot CH_2OPO(OH)_2$, beta, $CH_2OH \cdot CHOPO(OH)_2 \cdot CH_2OH$, (3) with oxidizing agents, e.g., dilute nitric acid, to form glyceric acid, $CH_2OH \cdot CHOH \cdot COOH$, tartaric acid, $COOH \cdot CHOH \cdot COOH$, mesoxalic acid, $COOH \cdot CO \cdot COOH$, (4) with phosphorus plus iodine, to form allyl iodide, $CH_2: CHCH_2I$, which with hydrogen iodide yields propylene, $CH_2:CHCH_3$, and then isopropyl iodide, CH_3CHICH_3, (5) with sodium or sodium hydroxide to form alcoholates, (6) with sodium hydrogen sulfate or phosphorus pentoxide heated, to form acrolein, $CH_2:CHCHO$. Glycide alcohol is obtained by treatment of glycerol alphamonochlorohydrin

$$CH_2OH \cdot \underset{\underset{O}{\rule{1.2em}{0.4pt}}}{CH \cdot CH_2}$$

$CH_2OH \cdot CHOH \cdot CH_2Cl$, which is made by reaction of hypochlorous acid and allyl alcohol with barium hydroxide. With hydrogen chloride, glycide alcohol yields epichlorohydrin

$$CH_2Cl \cdot \underset{\underset{O}{\rule{1.2em}{0.4pt}}}{CH \cdot CH_2}$$

Glycerol may be detected by the characteristic odor of acrolein, found on heating with potassium bisulfate.

Glycerol is used (1) in the manufacture of high explosives, e.g., glyceryl trinitrate ("nitroglycerin"), which is the main component of dynamite, (2) in antifreeze solutions, especially for automobile radiators, (3) to maintain a moist condition in fruits and tobacco, (4) in cosmetics and skin preparations, and (5) to prepare glycerol phosphoric acid, used in medicine, and "boroglyceride" used as a preservative. See Table 1.

TABLE 1. CHARACTERISTICS OF WATER SOLUTIONS OF GLYCEROL

% Glycerol by weight	Specific gravity (15.6°C/60°F)	Freezing point, °C
20	1.049	−5.0
40	1.103	−15.6
60	1.158	−34.0

GLYCOGEN. See **Carbohydrates.**

GLYCOL. A dihydric alcohol (i.e., a compound containing two alcoholic hydroxyl groups). The chemical properties are represented by those of the simplest members of the class, ethylene glycol, 1,2-ethanediol, $CH_2OH \cdot CH_2OH$, which is a colorless, viscous liquid, of sweetish taste, odorless, boiling point 197°C, miscible in all proportions with water or alcohol, slightly soluble in ether. Like ethyl alcohol, ethylene glycol is often called by the class name.

Glycol reacts (1) with sodium to form sodium glycol, $CH_2OH \cdot CH_2ONa$, and disodium glycol, $CH_2ONa \cdot CH_2ONa$; (2) with phosphorus pentachloride to form ethylene dichloride, $CH_2Cl \cdot CH_2Cl$ (3) with carboxy acids to form mono- and disubstituted esters, e.g., glycol monoacetate, $CH_2OH \cdot CH_2OOCCH_3$, glycol diacetate, $CH_3COOCH_2 \cdot CH_2OOCCH_3$; (4) with nitric acid (with sulfuric acid), to form glycol mononitrate, $CH_2OH \cdot CH_2ONO_2$, glycol dinitrate, $CH_2ONO_2 \cdot CH_2ONO_2$; (5) with hydrogen chloride, heated, to form glycol chlorohydrin (ethylene chlorohydrin, $CH_2OH \cdot CHCl$); (6) upon regulated oxidation to form glycollic aldehyde, $CH_2OH \cdot CHO$, glyoxal, $CHO \cdot CHO$, glycollic acid, $CH_2OH \cdot COOH$, glyoxalic acid, $CHO \cdot COOH$, oxalic acid, $COOH \cdot COOH$.

Glycol is made by reaction of ethylene and chlorine or hypochlorous acid to form ethylene dichloride or ethylene chlorohydrin, respectively, followed by treatment of either of these with sodium carbonate solution heated under pressure. Glycol is also formed when ethylene is treated with potassium permanganate.

Glycol is used (1) in antifreeze solutions, especially for automobile radiators; (2) in the preparation of ethers and esters, especially nitrate for explosive; (3) as a solvent substitute for glycerol.

See Table 1.

TABLE 1. CHARACTERISTICS OF WATER SOLUTIONS OF GLYCOL

% Glycerol by weight	Specific gravity (15.6°C/60°F)	Freezing point, °C
17	1.026	−6.7
32.5	1.048	−17.8
44	1.063	−28.9

GLYCOLYSIS. A series of about 10 enzyme-stimulated reactions in which glucose is broken down into pyruvic acid in cell respiration. No oxygen is needed for glycolysis and it is used as the sole energy source for anaerobic organisms. In aerobic metabolism, however, the pyruvic acid is then taken through the tricarboxylic acid (TCA) cycle, and the balance of the energy is extracted. It appears that glycolysis can take place free in the cytoplasm, but that the tricarboxylic acid cycle must take place within the mitochondria of the cell.

Glycolysis was defined in the late 1920s by Otto Warburg as "the splitting of carbohydrate into lactic acid." This type of lactic acid fermentation was well known to Berzelius, Liebig, Pasteur, and Claude Bernard in the mid-1800s, as was also alcoholic fermentation. Various kinds of carbohydrates may serve as substrates for glycolysis. It is remarkable that although glycolysis is the sum of a very large number of consecutive intermediate compounds, enzymes, and coenzymes, knowledge of these components and their sequences was acquired many years ago. For most animal cells studied, the biochemical sequence from glucose may be summarized as shown in Table 1.

The splitting of sugar to lactic acid is thus, briefly, the shifting of hydrogen by means of the nicotinamide moiety of diphosphopyridine nucleotide (also termed nicotinamide adenine dinucleotide). Nicotinamide in DPN takes away two atoms of hydrogen from phosphorylated carbohydrate, and after dephosphorylation gives back two hydrogens (in $DPNH_2$) to pyruvic acid.

The biochemical importance of the foregoing sequence in glycolysis is at least twofold. (1) Each one of the intermediate compounds formed leads to one or more important possible side reactions. These, in turn, lead to innumerable reactions indispensable to life processes, including respiration. (2) In the entire sequence, and also in some of its parts, comparatively large amounts of free energy are made available—up to a maximum of 28,000 cal/mole lactate formed under common *in vivo* conditions from one-half mole of glucose. This free energy available is considerably larger than the approximately 9,000 calories free energy available from hydrolysis

TABLE 1. EMBDEN-MEYERHOF PATHWAY

Step	Product	By way of
	Glucose (start)	
1	D-Glucose	Glucokinase, ATP, Mg^{2+}, insulin: anti-insulin regulators
2	D-Glucose-6-phosphate	Phosphoglucoisomerase
3	D-Fructose-6-phosphate	Phosphofructokinase, ATP, Mg^{2+}
4	D-Fructose-1,6-diphosphate	Fructaldose
5	D-Glyceraldehyde-3-phosphate	Glyceraldehyde-3-phosphate dehydrogenase, DPN, $HOPO_3{}^{2-}$
6	1,3-Diphospho-D-glycerate	3-Phosphoglycerate kinase, ADP, Mg^{2+}
7	3-Phospho-D-glycerate	Phosphoglycerate mutase, Mg^{2+}
8	2-Phospho-D-glycerate	Enolase, Mg^{2+}
9	Phosphoenolpyruvate	Pyruvate kinase, ADP, Mg^{2+}
10	Pyruvate	Pyruvate reductase = lactate dehydrogenase, $DPNH_2$
11	L-Lactate	

of the high-energy ATP to ADP and inorganic phosphate, although much smaller than the free energy of combustion of a mole of lactate to carbon dioxide and water, some 332,000 calories. Whereas the free and heat energies of combustion of lactate are nearly equal, lactic acid fermentation from glucose represents an instance of the relatively rare situation in which the free energy liberated is considerably greater (about 50%) than the heat energy liberated, owing to the large entropy change involved in the formation of the additional carbonyl ($=C=O$) bond in two lactates derived from one glucose molecule.

The foregoing reaction sequence, commonly called the Embden-Meyerhof pathway after its initial investigators, was in due course worked out in greater detail by Warburg. This pathway is also common to ethyl alcohol fermentation down to the pyruvate stage, which then branches off (via carboxylase) to form acetaldehyde and finally (via alcohol dehydrogenase, $DPNH_2$) to ethanol. Alcoholic fermentation is sometimes erroneously referred to as glycolysis. Ordinary respiration, by this same reasoning, could be called glycolysis, since it too shares the common pathway down to pyruvate. Just as lactate fermentation is the most common fermentation met with in animal cells, so alcoholic fermentation is the most common fermentation met with in plant cells, a distinction most easily observed under anaerobic conditions.

See also **Carbohydrates.**

GLYCOSIDES. Substances that by reaction with water, either in the presence of certain enzymes or of dilute acids or alkalis, yield a sugar (see also **Carbohydrates**) as one of the products, plus a *principle* (see also Table 1) characteristic of the individual glycoside. When the sugar is glucose, the parent compound is called a glucoside, and further called an alpha- or beta-glucoside according to the type of glucose produced. Analogous terms are the *alpha* and *beta glycosides*, applied generally to this class of compounds yielding sugars on hydrolysis. Most glycosides are soluble in cold or hot water, and in alcohol (95% C_2H_5OH), and insoluble or slightly soluble in ether (used to separate from alcohol solution). Most optically active glycosides are levorotatory. The di- and polysaccharides are to be considered glycosides. Glycosides occur in plants, especially in leaves, buds, young shoots where metabolism is active, and in the bark and seeds. Anthocyanins, the plant colors of flowers, are glycosides, as are also some tannins.

GLYCOSIDES (Steroid). In plants, steroids occur as glycosides, as acyl glycosides, as esters, and in the free form. Many of the steroid glycosides are important drugs or starting materials for the partial synthesis of drugs.

Sterolins

Although the sterols are largely in the free and esterified form, plants contain significant quantities of sterolins (steryl glycosides and acyl glycosides). As a rule, the 3-hydroxyl group of the sterol is linked to glucose or some other common sugar to form a heteroside, but other hydroxyl groups in the sterol molecule and higher saccharides may be involved. The two most common sterol aglycones in higher plants are sitosterol (previously called β-sitosterol) and stigmasterol, shown in Fig. 1. Cholesterol is usually present in very small amounts. (See **Cholesterol.**) The biosynthesis of sitosterol differs from that of cholesterol in the

TABLE 1. SELECTED REPRESENTATIVE GLYCOSIDES

Glycoside	Formula	Melting point °C	Hydrolysis	
			Sugar	Principle
1. Aesculin in horsechestnut bark	$C_{15}H_{16}O_9 \cdot 1\frac{1}{2}H_2O$	205	glucose	aesculetin
2. Amygdalin in peach kernels, cherry laurel leaves, bitter almonds	$C_{12}H_{16}O_7 \cdot 3H_2O$	200 (anhyd.)	glucose	mandelocyanides
			glucose	benzaldehyde + hydrocyanic acid
3. Arbutin in bearberry leaves	$C_{12}H_{16}O_7 \cdot \frac{1}{2}H_2O$	165	glucose	hydroquinone
4. Coniferin in sap of coniferous trees	$C_{16}H_{22}O_8$	185	glucose	coniferyl alcohol
5. Dhurrin in sorghum seedlings, millet	$C_{14}H_{17}O_7N$	—	glucose	para-hydroxy-benzaldehyde + hydrocyanic acid
6. Digitalin in digitalis	$C_{35}H_{56}O_{14}$	217	glucose	digitaligenin, digitalose
7. Digitonin in digitalis	$C_{55}H_{90}O_{29}$	235 approx. decom.	glucose galactose	digitogenin
8. Digitoxin in digitalis	$C_{34}H_{54}O_{11}$	240 (anhyd.)	digotoxose	digitoxigenin
9. Helleborein	$C_{37}H_{56}O_{18}$	200–230 decom.	glucose	helleboretin
10. Hesperidin in unripe oranges	$C_{50}H_{60}O_{27}$	251	glucose rhamnose	hesperetin
11. Indican in natural indigo	$C_{14}H_{17}O_6N \cdot 3H_2O$	100 (anhyd.)	glucose	indigo
12. Phloridzin in bark of fruit trees	$C_{21}H_{24}O_{10} \cdot 2H_2O$	108 Remelts 170 decom.	glucose	phloretin
13. Quercitrin	$C_{21}H_{22}O_{12} \cdot 2H_2O$	168 decom. (anhyd.)	glucose rhamnose	quercitin
14. Saponin in soapwort root, forms foam with water. toxic to cold blooded animals	$C_{32}H_{52}O_{17}$	195 decom.	sugar	sapogenin
Tannins in nut galls	—	—	glucose	gallic acid
Anthocyanins	—	—	—	anthocyanidins
Red (with acids), violet (free), blue (with alkalis) pigments of flowers				
Cyanin	$C_{15}H_{10}O_6$	—	glucose	cyanidin
Idaein	—	—	galactose	cyanidin
Pelargonin	$C_{15}H_{10}O_5$	—	—	pelargonidin
Delphinin	$C_{15}H_{10}O_7$	—	glucose	delphinidin + para-hydroxy-benzoic acid

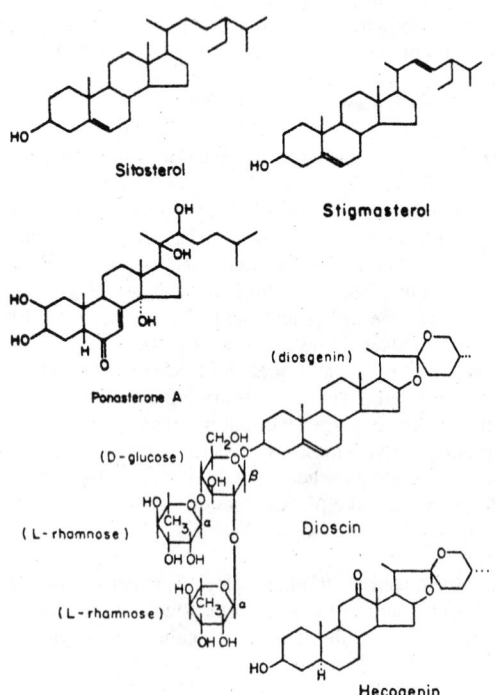

Fig. 1. Steroidal saponins and related compounds.

alkylation of a precursor at C-24 and involves the successive introduction of two methyl groups. Stigmasterol is formed by the dehydrogenation of sitosterol.

Cholesterol and the C_{29} sterols are used by plants as the starting materials for the biosynthesis of other steroids with the same number of carbons, such as the insect-molting hormones and the steroidal sapogenins and alkaloids or they are converted to progesterone and other steroids with a lower number of carbons. Fig. 1 shows the structure of one representative of the insect-molting hormones, ponasterone A, which has been isolated from plants in the form of its 3-glucoside, ponasteroside A.

Steroidal Saponins and Glycoalkaloids

Certain plants have the ability to hydroxylate cholesterol stereospecifically at C-26 or C-27. A glycoside of cholesterol with a glucose residue attached to this terminal hydroxyl group and chacotriose (2L-rhamnoses + D-glucose) attached to its 3-hydroxyl group is a biogenetic precursor of the saponin, dioscin, shown in Fig. 1. Steroidal saponins are glycosides of spiroketals, which form spontaneously, when the terminal sugar in an analogous 16-hydroxy-22-keto-steroid is enzymatically removed. While the configuration at C-22 is the same in all natural sapogenins, the orientation of the methyl group at C-25 depends on the position of the terminal hydroxyl group in the sterol precursor. In the D- or isosapogenin series the methyl group is α-oriented (equatorial), as in diosgenin (see Fig. 1), whereas in the L-, normal, or neosapogenin series it is β-oriented (axial).

Sapogenins are widely distributed in monocots belonging to the genera *Yucca, Trillium, Chlorogalum, Smilax, Nolina, Agapanthus, Agave, Manfreda,* and *Dioscorea,* and in dicots belonging to the genera *Digitalis, Solanum, Lycopersicon,* and *Cestrum.* Diosgenin (Fig. 1) from Mexican barbasco root (*Dioscorea* tubers) and hecogenin (Fig. 1) from wastes of African sisal fibers (*Agave sisalana*) are important starting materials for the commercial preparation of synthetic hormones.

The glycoalkaloids are nitrogen analogs of the saponins, occurring in the Solanaceae. Two of their aglycones are shown in Fig. 2. Tomatidine occurs in the form of a glycoside, tomatine, in tomato vines and may also be used for the partial synthesis of steroid hormones. Solanidine is found in potatoes and other *Solanum* species in the form of various glycosides, the solanines and chaconines. For instance α-chaconine differs from dioscin (Fig. 1) only in the nature of the aglycone. The glycoalkaloids are likewise synthesized by plants from cholesterol.

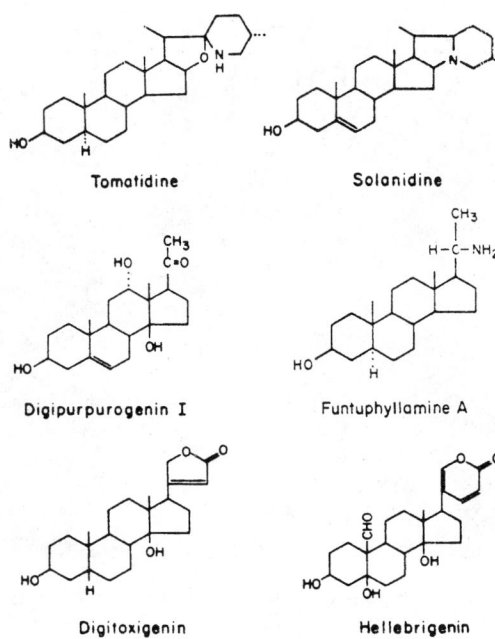

Tomatidine   Solanidine

Digipurpurogenin I   Funtuphyllamine A

Digitoxigenin   Hellebrigenin

Fig. 2.   Representative glycoalkaloids occurring in Solanaceae.

Saponins and glycoalkaloids are characterized by their surface activity and hemolytic effect as well as their ability to form complexes with cholesterol and similar sterols. The best-known cholesterol-precipitating agent is digitonin, a saponin in *Digitalis*. While ingested saponins are nontoxic to warmblooded animals, the glycoalkaloids are toxic. Tomatine and other glycoalkaloids have antifungal and cytostatic activity.

Pregnane Derivatives

The degradation of cholesterol in plants and animals produces pregnenolone (Δ^5-pregnen-3β-ol-20-one), which is oxidized to progesterone (Δ^4-pregnene-3,20-dione). (See **Steroids**.) Various plants contain neutral pregnane derivatives with C and D rings in cis-fusion in the form of glycosides. They have been called digitanol glycosides, because they were first isolated from *Digitalis* plants, and contain the rare hexoses otherwise only found in the cardiac glycosides. Fig. 2 shows an example of the digitanols, digipurpurogenin I, which occurs in *Digitalis purpurea* as digipurpurin, a glycoside in which three molecules of D-digitoxose are attached to the 3-hydroxyl group.

Plants belonging to the Apocynaceae and Buxaceae aminate the pregnane derivatives to produce alkaloids, which may be present as either glycosides or esters. Only one example of Apocynaceae alkaloid is shown in Fig. 2, funtuphyllamine A, which was isolated from *Funtumia africana*. The alkaloids of kurchi bark (*Holarrhena antidysenterica*) have many interesting pharmacological properties and one of them, conessine, is used as an amebicide and as a starting material for the partial synthesis of aldosterone.

Cardiac Glycosides

About eleven plant families are known to elaborate cardiac glycosides. Their genins have either 23 (cardenolides) or 24 (bufadienolides) carbon atoms and their sugars are not found elsewhere in nature. Cardenolides and bufadienolides have not been found together in the same genus. Both types of genins are synthesized in plants from a C_{21} steroid, usually progesterone, and contain a 14β-hydroxyl group.

Fig. 2 gives one example of a cardenolide, digitoxigenin, and one representative of the bufadienolides, hellebrigenin. *Digitalis* plants contain three cardenolides: digitoxigenin, gitoxigenin (16β-hydroxy digitoxigenin), and digoxigenin (12β-hydroxy-digitoxigenin). These genins are combined with 2 molecules of digitoxose, 1 molecule of acetyldigitoxose, and 1 molecule of glucose to form the lanatosides (digilanides) A, B, and C, respectively. When the acetyl group is removed by mild alkaline hydrolysis, one obtains the corresponding purpurea glycosides (desacetyllanatosides or desacetyldigilanides). Enzymatic removal of the glucose unit, on the other hand, gives the acetyl derivatives of the digitoxose triosides. Combination

of both hydrolytic procedures yields the digitoxose triosides digitoxin, gitoxin, and digoxin.

The bufadienolides occur in plants as well as animals, but only in plants are they in the form of glycosides. Their 3-hydroxyl group is attached to glucose, rhamnose, or thevetose. Hellebrigenin (Fig. 2), which is also known as bufotalidin, occurs in the rhizomes of the Christmas rose and other *Helleborus* species in the form of a rhamnoside. Bufadienolides have so far been found in plants in only two families, the buttercup and the lily family.

Crude leaf preparations of *Digitalis* have been in medical use since 1785. Pure cardiac glycosides are now available. These preparations in injectable tinctures or powdered leaf tablets are used extensively for the treatment of congestive heart failure. They increase the force of the heart muscle and the power of systolic contraction, apparently by inhibiting the active transport of K^+ and Na^+ ions through cell membranes.

GLYCYRRHIZINS.   See **Sweeteners**.

GLYOXYLATE SHUNT PATHWAY.   See **Carbohydrates**.

GOETHITE.   The mineral goethite is a hydroxide of iron corresponding to the formula FeO(OH) crystallizing in the orthorhombic system. It occurs in prisms, but is often found in foliated or other massive forms. When observable it shows one good cleavage parallel to the prism; fracture, uneven; hardness, 5–5.5; specific gravity, 3.3–4.3; luster, adamantine to dull; color; yellowish, reddish, brownish to nearly black; translucent to opaque. It is found associated with hematite and limonite, being perhaps in part an alteration product of the latter mineral. Goethite is used as an ore of iron. There are many European localities, including Bohemia, Saxony, Westphalia, and Cornwall. In the United States it is found in the hematite mines of the Lake Superior region and in Colorado. This mineral was named in honor of the German poet Johannes Wolfgang von Goethe.

GOLD.   [CAS: 7440-57-5]. Chemical element symbol Au (from Latin aurum), at. no. 79, at. wt. 196.967, periodic table group 11 (transition metals), mp 1,064.43°C, bp approximately 3080°C, density 19.32 g/cm³ (20°C). Elemental gold has a face-centered cubic crystal structure.

Gold is a yellow metal, soft, and extremely malleable. The purity of gold (sometimes referred to as "fineness") is expressed in karats. Pure gold is 24 karat. See also **Radioactivity**. In terms of cosmic abundance, in the estimate of Harold C. Urey (1952), using silicon as a base with a figure of 10,000, gold was ranked number 79 among the elements, with an abundance figure of 0.0015. In terms of abundance in seawater, gold is ranked number 59 among the elements, with an estimated content of 38 pounds per cubic mile (4 kilograms per cubic kilometer) of seawater.

Electronic configuration is

$$1s^2 2s^2 2p^6 3s^2 3p^6 3d^{10} 4s^2 4p^6 4d^{10} 4f^{14} 5s^2 5p^6 5d^{10} 6s^1.$$

First ionization potential is 9.223 eV; second 19.95 eV. Oxidation potentials: Au → Au^{1+}, $E°$ = −1.68 V: Au → Au^{3+}, $E°$ = −1.50 V. Other important physical properties of gold are given under Chemical Elements.

Gold is one of the most ancient metals. Gold jewelry and ornaments made as early as 3500 B.C. have been discovered at Ur in Mesopotamia. During the period from 3000 to 2000 B.C., lead cupellation was used to purify gold and most modern jewelry techniques were developed during that time.

Occurrence and Processing

Gold is found chiefly as the free metal scattered through gravel (placer gold) or disseminated in veins of quartz (vein gold). Small quantities also are found in lead and copper sulfide ores. Nuggets of native gold, varying in size from that of a tiny pebble to a mass weighing as much as 248 pounds (112.5 kilograms), have been found. In a combined state, gold occurs in sylvanite, a telluride of gold and silver, (Au, Ag)Te₂, a rich ore found in Colorado. The bulk of the gold ores contain very little gold (about 5 to 15 grams/metric ton). Some of the richest ores found in Africa contain from 20 to 30 grams/metric ton. Almost all countries produce some gold. The leader, by far, is the Republic of South Africa, followed by Russia and Canada. Far behind, other producers include the United States, Australia, Ghana, and Zimbabwe. See also **Mineralogy**.

The treatment of gold ores involves: (1) grinding, amalgamation, and/or cyanidation of those ores containing coarse free gold, and (2) the very fine grinding, flotation, roasting, and amalgamation and/or cyanidation of those ores containing gold telluride or sulfide. These processes produce an impure gold metal containing considerable silver and some copper plus other base metals. The impure gold is purified by melting and oxidizing the base metals or by melting and chlorinating (Miller process) which removes the base metals and silver. The silver-containing oxidized gold is purified by the electrolysis of gold chloride solutions containing an HCl solution (Wohlwill process). In the latter process, the anode is the alloy (gold-silver) and the cathode is pure gold. The gold deposits then on the cathode and the silver forms silver chloride and remains as a deposit about the anode.

Throughout early mining history, it was believed that ores, such as placer gold, resulted from mechanical weathering, wind, and water erosion of the veins of ore. However, since the early 1800s, geologists have found that biological processes also play a role in shaping some mineral deposits. Watterson (U.S. Geological Survey), in the early 1980s, made a serendipitous observation that gold solutions are lethal to many soil bacteria. Thin coats of gold tend to condense around the bacterial spores, clogging the narrow pores in their cell walls, through which nutrients enter. Watterson's findings were confirmed by inspection of placer gold particles in an Alaskan stream. Masses of gilded cells were found as the result of this biological process in connection with Pedomicrobia and related bacteria. Stephen Mann (Univ. of Bath), who specializes in biomineralization, observes that many bacteria can become encased in mineral coatings under favorable conditions. It should be noted that some mining firms are using bacteria to assist in extracting metals from low-grade ores. Further detail is given in the Rennie reference listed.

Uses of Gold

The monetary aspects of gold have long dominated commercial interest in the metal. Gold through history has provided a common base from which the value of materials and services can be measured. Gold probably became a medium of exchange as early as 3400 B.C.

Jewelry is the largest commercial user of gold, accounting for nearly 65% of the total consumption. Most jewelry is made by the "lost wax process," a casting method that dates to 3000 B.C. or earlier. Usually these jewelry products employ karat golds which contain 10 and 14 karats, and less commonly 18 karat, of gold (41.7, 58.3 and 75.0 weight percent of gold, respectively). These gold alloys are of two general types. Red, yellow, and green golds are basically alloys of gold, copper, and silver. A wide variety of color shades can be produced by varying composition within this ternary alloy system, with reddish hues provided by high copper to silver ratios, and pale green tint when silver is predominant. These alloys almost always contain minor amounts of zinc and deoxidizers or grain refiners to facilitate fabrication. The second widely used class is the white karat golds, which are produced in two basic alloy types. These are the original gold-nickel-zinc-copper (18 karat) and the gold-copper-nickel-zinc (10 and 14 karats) alloys, and the more recent gold-palladium-silver-copper, and gold-copper-nickel-palladium-silver alloys which are usually 14- and 10-karat alloys. The pink golds are derived from the system gold-silver-copper-nickel-zinc. These are essentially red golds, which are "whitened" by the addition of silver, nickel, and zinc.

Considerable brazing is done by jewelry manufacturers and the solders that are used may be of a lower karat content than the alloy being brazed. Usually they contain much more silver and zinc than the alloys themselves.

The use of gold in the electrical, electronic, and other industrial fields has grown considerably in recent years, estimated at about 25%. The electrical and thermal conductivity, resistance to oxidation, and ease of being electroplated make gold an excellent coating for electrical contacts. See Fig. 1. This has been particularly true in metallized ceramics for use in microelectronics and other electronic components. Here gold does not migrate into the ceramic as does silver. Gold is widely used as a conductor in thin and thick film circuitry. It is also useful as bonding wire for integrated circuit electrical connections and mechanical packaging of semiconductor chips (die bonding).

Gold is used extensively in many industrial solders and brazing alloys. These range from the low-melting eutectics of gold with germanium, silicon, and tin to gold-copper, gold-nickel, and gold-palladium-nickel alloys. The latter brazing materials have the ability to withstand long use at high temperatures and are particularly applicable to jet engine fabrication.

Gold is also used in dentistry. This application has declined in recent years; however, it still accounts for about 7% of gold consumption. Gold alloys, such as gold-silver-copper with varying amounts of platinum and

Fig. 1.   Electrolytically deposited gold crystals. (*Bausch & Lomb*)

palladium, are used for restorations and for bridges, inlays, and partial dentures. These are cast with much more precision than jewelry, and have, in fact, replaced wrought gold wire in many of these dental appliances. Gold wire is now used principally in orthodontic and prosthetic appliances. These are complex alloys containing gold, platinum, palladium, silver, copper, nickel, and zinc.

Some of the minor commercial uses of gold are among the most interesting. Gold is used to produce a very beautiful ruby glass. When an oxidizing glass is melted with a gold salt, the gold dissolves, forming colorless ions. If reducing agents like Sn, Sb, Bi, Pb, Se, or Te are present, the glass will become red after heating at temperatures between 600–700°C, as a result of the precipitation of minute particles of gold. Gold films deposited on glass by evaporation are superior to other metals for reflectivity in the infrared. Mirrors thus coated have application in spectroscopy and space science. Thin films applied to plate glass give adequate transmission of light combined with good infrared reflectivity, reducing the overheating of office windows during hot weather. Gold is extremely malleable. It can be rolled and beaten into foil less than 5 millionths of an inch (0.00013 millimeter) thick. Such foil has been used for indoor and outdoor decoration for centuries. One of the most conspicuous examples is the gold leaf dome, an architectural highlight in many important structures.

Chemistry of Gold

Gold has a $5d^{10}6s^1$ electron configuration, like the similar ones at lower levels of copper and silver, and thus the d electrons can take part in bonding. However, for gold the +3 oxidation state is the most stable, and the +1 state next to it in stability, so that Au^{3+} as well as Au^+ are found both in simple compounds and in complexes. As with copper and silver, the bonds in most gold compounds, including the oxides, are largely covalent. In most of its compounds gold is univalent or trivalent. While a few compounds are known in which it is divalent, some of these are considered to consist of Au(I) and Au(III), rather than Au(II). Thus, the compound with cesium and chlorine, $CsAuCl_3$, is black and diamagnetic, and so contains both Au(I) and Au(III). A similar compound with cesium, silver, and chlorine, $Cs_2AuAgCl_6$, yields $[AuCl_4]$ and $[AgCl_2]^-$ ions on hydrolysis. However, the sulfide, AuS, probably contains divalent gold.

Gold does not combine directly with oxygen. Gold(I) oxide, Au_2O, formed by heating AuOH to 200°C, is very easily reduced to gold. It is essentially covalent. Gold(I) hydroxide, AuOH, is prepared from a gold(I) solution by the addition of potassium hydroxide solution in theoretical amounts. It forms a deep-blue "solution" believed to be a colloidal sol. It dissolves in excess alkali to form aurates(I), such as $KAu(OH)_2$. Gold(III) oxide, Au_2O_3, is formed by heating $Au(OH)_3$ at 100°C in the presence of a dehydrating agent. Like Au_2O, it is easily reduced to gold. It dissolves in hydrochloric, hydrobromic, and hydriodic acids, forming the haloauric acids, $HAuX_4$. It also dissolves in excess of alkali hydroxide, forming an aurate, containing the ion $[Au(OH)_4]^-$.

Gold(III) hydroxide, Au(OH)$_3$, is precipitated by the addition of potassium hydroxide solution in equivalent amount, to a solution of chloroauric acid (obtained by dissolution of gold in aqua regia). It is insoluble in H$_2$O, gives many of the reactions of Au$_2$O$_3$, and may be a hydrous form of that compound. Gold(II) oxide, AuO, formed by the action of potassium bicarbonate upon solutions of chloroauric acid, is believed, as stated above, to consist of gold(I) and gold(III), based on properties of other divalent gold compounds.

Gold does not react directly with fluorine, but dissolves in bromine trifluoride, BrF$_3$, to form BrF$_2$AuF$_4$, which loses BrF$_3$ at 120°C to give gold(III) trifluoride, AuF$_3$ which decomposes into the elements at about 500°C. Water decomposes AuF$_3$, into hydrogen fluoride and Au(OH)$_3$. The chlorides, on the other hand, are the most important of the gold salts. Gold(I) chloride, AuCl, may be produced by heating gold(III) chloride, AuCl$_3$, in air at 170°C; it is hydrolyzed by H$_2$O, to AuCl$_3$ and gold. Gold(III) chloride, AuCl$_3$, is formed directly from the elements at 200°C; unlike AuCl, it is soluble in H$_2$O forming initially H[AuCl$_3$(OH)], which then undergoes further hydrolysis. With hydrochloric acid, AuCl$_3$ forms tetrachloroauric(III) acid. [AuCl$_4$], of which many salts are known. Gold(I) bromide, AuBr, is formed by continued heating of bromoauric(III) acid above 100°C. Like the AuCl, it readily undergoes hydrolysis. Gold(III) bromide is formed by the action of bromine water upon gold. The equivalence of its three Au-Br bonds have been proved by a tracer technique with radioactive bromine. With hydrobromic acid it forms H[AuBr$_4$]. Gold(I) iodide is prepared from the elements at 50°C, or by the slow decomposition of AuI$_3$ at room temperature. It decomposes on heating above 120°C. It dissolves in potassium iodide, KI, solution, forming KAuI$_2$, which then decomposes to gold and KAuI$_4$. Gold(III) iodide, obtained by evaporation of a 1:1 hydriodic acid solution of AuCl$_3$, is unstable, decomposing, when dry or when heated with H$_2$O, into the elements. It dissolves in hydriodic acid as H[AuI$_4$]. The gold(I) halides are the least soluble of the univalent halides except for silver iodide. The solubility product constants are AuI, 1.6×10^{-23}; AuBr, 5.0×10^{-17}; AgI, 8.30×10^{17}; AuCl, 2.0×10^{-13}; and AgBr, 4.27×10^{-13}.

There are many gold complexes. The gold(I) and gold(III) halocomplexes, involving the groups [AuX$_2$]$^-$ and [AuX$_4$]$^-$ have already been discussed. Apparently, there are no other gold(I) halocomplexes than the chloro-compound. There are also fluorocomplexes of the form M[AuF$_4$] formed by fluorination of M[AuCl$_4$] where M is an alkali metal or ammonium. Due to the polar character of the AuF bond, they are readily hydrolyzed. Many hexachloroaurates, such as Cs$_2$M[AuCl$_6$], are known.

Gold forms complexes with ammonia much less readily than do copper and silver. A few ammonia complexes of gold(III), such as KAuCl$_4 \cdot 3$NH$_3$, have been prepared. Gold(I) halides react more readily. AuCl forms [Au(NH$_3$)$_2$]Cl, while AuBr and AuI react, but only with anhydrous ammonia, to form [Au(NH$_3$)$_2$]Br and [Au(NH$_3$)$_6$]. Gold(I) cyanide dissolves in excess cyanide to form the very stable ion [Au(CN)$_2$]$^-$, $K_{inst} = 10^{-38.3}$. This complex is so stable that gold metal dissolves in potassium cyanide solution in the presence of air. This is of importance in the separation of gold from its ores; while Au(CN)$_3$ reacts to form [Au(CN)$_4$]$^-$, $K_{inst} = 10^{-56}$. Treatment of salts of this ion with sulfites, gold(III) forms such complexes as K$_5$[Au(SO$_3$)$_4$] $\cdot$ 5H$_2$O and Na$_5$[Au(SO$_3$)$_4$] $\cdot$ 14H$_2$O. In these complexes, the sulfite group is monodentate and is attached to the gold atom through the sulfur atom (really an aurisulfonate ion); however, a bidentate compound is also known.

Gold(III) chloride or tetrachloroaurates(III) also form thiosulfate complexes, especially in the presence of NaI, of the form Na$_3$[Au(S$_2$O$_3$)$_2$], in which the gold is monovalent.

Gold forms thiocyanate complexes M[Au(SCN)$_2$] and M[Au(SCN)$_4$].

A striking difference between gold and copper or silver is the fact that its oxyacid compounds do not exist in stable form, and few have been isolated. Among the few that are known are the gold(III) orthoarsenite, AuAsO$_3 \cdot$ H$_2$O, the gold(III) selenate, Au$_2$(SeO$_4$)$_3$, and the gold(III) iodate, Au(IO$_3$)$_3$. Nevertheless a number of complexes of oxyacids are known, including M[Au(NO$_3$)$_4$] $\cdot$ 2H$_2$O (M = H$_3$O$^+$, NH$_4^+$, K$^+$, Rb$^+$), Mg[Au(CH$_3$CO$_2$)$_4$].

Gold is unique among the coinage metals in forming true (i.e., sigma-bonded) stable organometallics. The action of methyllithium on AuBr$_3$ in ether at −65°C produces a solution of (CH$_3$)$_3$Au, which begins to decompose at −35°C into gold, ethane, and methane. The presence of benzylamine or ethylenediamine, however, stabilizes the solution up to room temperature. Triethylgold is less stable than trimethylgold. The action of a hydrogen halide on a trialkylgold or the action of an alkyl Grignard

reagent in pyridine on gold(III) halides produces dialkylgold halides, which are much more stable. Appropriate methathetical reactions of these produce the corresponding cyanides, sulfates, etc. These are all covalent compounds, as attested by the solubility of the sulfates, (R$_2$Au)$_2$SO$_4$, in benzene and chloroform. The melting points of a few dialkylgold compounds are: (CH$_3$)$_2$AuBr, 68°C; (C$_2$H$_5$)$_2$AuCl, 48°C; (C$_2$H$_5$)$_2$AuBr, 58°C; (C$_2$H$_5$)$_2$AuCN, 103–105°C; (n-C$_3$H$_7$)$_2$AuCN. 94–95°C; (i-C$_3$H$_7$)$_2$ AuCN, 88–90°C; (i-C$_5$H$_{11}$)$_2$AuCN. 70°C; (C$_6$H$_5$CH$_2$)$_2$AuCl, 100°C decomposes; (C$_6$H$_5$CH$_2$CH$_2$)$_2$AuBr, 112.5°C. The n-propyl chloride and bromide, and the n-butyl, i-butyl, and i-amyl bromides are liquid at room temperature.

The dialkylgold halides are dimeric, having the planar structure:

$$\begin{array}{ccccc} R & & X & & R \\ & \diagdown & & \diagup & \\ & Au & & Au & \\ & \diagup & & \diagdown & \\ R & & X & & R \end{array}$$

The cyanides, on the other hand are tetrameric, having the structure shown below.

$$\begin{array}{ccc} R & & R \\ | & & | \\ R-Au-CN-Au-R \\ | & & | \\ N & & N \\ C & & C \\ | & & | \\ R-Au-NC-Au-R \\ | & & | \\ R & & R \end{array}$$

Donald A. Corrigan
Handy & Harman
Fairfield, Connecticut

Additional Reading

Brady, G.S.S., H.R. Clauser, and J.A. Vaccari: *Materials Handbook,* 14th Edition, The McGraw-Hill Companies, Inc., New York, NY, 1996.

Davis, J.R.: *Metals Handbook,* 2nd Edition, ASM International, Materials Park, OH, 1998.

Gasparrini, C.: *Gold and Other Precious Metals: From Ore to Market,* Springer-Verlag, Inc., New York, NY, 1993.

Greener, E.H.: *Dental Materials,* Encyclopedia of Materials Science and Engineering, MIT Press, Cambridge, MA, 1986.

Krebs, R.E.: *The History and Use of Our Earth's Chemical Elements: A Reference Guide,* Greenwood Publishing Group, Inc., Westport, CT, 1998.

Lagowski, J.J.: *MacMillan Encyclopedia of Chemistry,* Vol. 1, Macmillan Library Reference, New York, NY, 1997.

Lechtman, H.: "Pre-Columbian Surface Metallurgy," *Sci. Amer.,* **53** (June 1984).

Lide, D.R.: *CRC Handbook of Chemistry and Physics,* 84th Edition, CRC Press, LLC., Boca Raton, Fl, 2003.

Meyer, C.: "Ore Metals Through Geologic History," *Science,* **227,** 1421–1428 (1985).

Parker, P.: *McGraw-Hill Encyclopedia of Chemistry,* 2nd Edition, The McGraw-Hill Companies, Inc., New York, NY, 1993.

Rennie, J.: "Bug in a Gilded Cage: All That Glitters is Sometimes Bacterial," *Sci. Amer.,* **27** (September 1992).

Schmidbauer, H.: *Gold: Progress in Chemistry, Biochemistry, and Technology,* John Wiley & Sons, Inc., New York, NY, 1998.

Stwertka, A.: *A Guide to the Elements,* Oxford University Press, Inc., New York, NY, 1998.

GOLD NUMBER.   When certain colloids (hydrophilic), such as gelatine, are added to a gold sol, the gold sol is strongly protected against the flocculating action of electrolytes. This protective action on red gold sols may be measured by utilizing the color change red to blue which indicates the first stage of coagulation. The "gold number" as defined by Zsigmondy is the weight in milligrams of protective colloid which is just sufficient to prevent the change from red to blue in 10 cm^3 of a standard gold sol (0.0053 to 0.0058 percent Au) after the addition of 1 cm^3 of a 10 percent sodium chloride solution.

GOLDSCHMIDT DETINNING PROCESS.   A method for the recovery of tin, based upon the action of dry chlorine gas on scrap tinplate. The tin reacts readily to form stannic chloride, and the iron reacts only slightly. By fractional distillation, the stannic chloride is separated from the small amount of ferric chloride that is formed.

GOLDSCHMIDT REDUCTION PROCESS.   Reaction of oxides of various metals with aluminum to yield aluminum oxide and the free metal. This reaction has been used to produce certain metals, e.g., chromium and zirconium, from oxide ores; and it is also used in welding (iron oxide plus aluminum giving metallic iron and aluminum oxide, plus considerable heat). (Thermite process.)

A method of producing formates by heating sodium hydroxide with carbon monoxide under pressure.

A process for recovery of tin, by treatment of scrap tinplate with dry chlorine, better known as the **Goldschmidt Detinning Process**.

GOODYEAR, CHARLES (1800–1860).   Born in Woburn, MA. Goodyear was the first to realize the potentialities of natural rubber. Frustrated by its lack of stability to temperature and other weaknesses in the uncured state, he experimented with additives such as magnesium and sulfur. The discovery of vulcanization was not accidental, as is often stated, but the result of intelligent trials and correct evaluation of their results. Though Goodyear's patents were contested by Hancock in England, he well merits the credit for making rubber usable in countless ways and helping to make the automobile possible.

See also **Vulcanization**.

GRAHAM LAW.   The rates of diffusion of two gases are inversely proportional to the square roots of their densities.

GRAHAM, THOMAS (1805–1869).   Born in Scotland, Graham is famous for his basic studies in diffusion that led to the development of colloid chemistry. He was the first to observe a marked difference in the rate of passage of certain types of substances through a parchment membrane. Those that readily crystallize, like sugar, pass rapidly through the membrane, but gelatinous types are "slow in the extreme." Graham designated the later, which comprise albumin, starch, gums, etc., as colloids and their solutions as colloidal solutions. The former, which he called crystalloids, form "true" or molecularly, dispersed solutions.

See also **Colloid Chemistry**; and **Colloid Systems**.

GRAIN.   (1) The smallest unit of mass in the avoidupois system; 1 grain = 0.0648 gram; one ounce contains 437.5 grains. (2) Any cereal plant, as wheat, corn, barley, etc., (3) Crystalline particles of metals. (4) The dehaired side of a skin of hide.

GRAIN BOUNDARY.   The surface separating two regions of a solid in which the crystal axes are differently oriented. It has been shown that such a boundary may be thought of as built up of an array, or network of dislocations, whose spacing depends on the tilt θ of the axes across the surface. The energy (per unit area) of a grain boundary is given by

$$E/E_m = (\theta/\theta_m)\{1 - \ln(\theta/\theta_m)\}$$

where E_m and θ_m are parameters depending on the material.

Grain boundary relaxation is a source of internal friction in solids due to the motion of grain boundaries under stress.

GRAIN REFINER.   An additive agent used to obtain finer grains in a casting. The material is added to the molten metal prior to casting.

GRAIN SIZE.   In metallurgy, it is common practice to call the crystals of a polycrystalline metal its grains. The grain or crystal size of metals is determined by microscopic examination of a suitably prepared section. There are two principal standards of grain size in use in the United States. Both are standards of the American Society for Testing and Materials.

For most nonferrous alloys, particularly brass and bronze and other alloys having homogeneous grain structures with twin bands, a set of ten photomicrographs having average grain diameters ranging from 0.010 to 0.200 millimeter are used for direct comparison with microstructures at a magnification of 75 times.

The A.S.T.M. standard grain size chart for steels covers about the same range of average grain diameters but the comparison is made at 100 times magnification and the grain size is expressed by numbers from 1 to 8. The following single equation relates the grain size number to the grain sizes:

$$n = 2^{N-1}$$

where N is the grain size number and n the number of grains per square inch. In general, grain sizes 1 to 3 are considered coarse, 4 to 6 intermediate, and 7 to 8 fine. The grain size of steel can also be judged from a clean fracture if the steel can be fractured without appreciable plastic deformation because the fracture surface mirrors the grain structure. This is possible with most heat-treated machine steels and tool steels, but low-carbon steels are often too tough to break with a crystalline fracture. A series of standard fractures is available for direct visual comparison, and the numbering system for these standards coincides with that of the charts used for microscopic determination of grain size.

The grain size of metals is related to many important properties. In general, fine grain size is an indication of relatively high strength, hardness, and toughness while coarse grain indicates softness and plasticity. However, the hardenability of steels by heat treatment is highest for coarse grain steel. Coarse grain size is usually desirable for creep strength at elevated temperatures.

In the case of sheet and strip for drawing or stamping, coarse grain may give a rough surface. On the other hand, metal with too fine a grain size may lack plasticity and crack in the dies; therefore, a compromise must be reached.

The grain size of castings is generally much coarser than that of wrought products such as rod or sheet. In the case of steel castings the original coarse structure may be refined by heat treatment. This is not possible in the case of most nonferrous alloys because they do not undergo a change in type of crystal structure on heating or cooling.

In the case of hot-rolled or forged metals, the finishing temperature has an important influence on grain size. A high finish-forging temperature, for example, will permit grain growth after recrystallization. In the case of metals finished by cold-working processes, the final annealing temperature establishes the grain size. A high annealing temperature results in coarse grain size.

GRAM-ATOM.   That quantity of an element having a mass in grams numerically equal to the atomic weight. One gram-atom contains the Avogadro number of atoms.

GRAM-EQUIVALENT.   The gram-atomic weight of an element (or formula weight of a radical) divided by its valence. In the case of multivalent substances there will be more than one value for the gram-equivalent, viz., Fe(II) = 27.92 grams, Fe(III) = 18.61 grams, and the proper value for the particular reaction must be chosen.

GRAM-MOLECULAR WEIGHT.   That amount of a pure substance having a weight in grams numerically equal to the molecular weight. One gram-molecular weight contains the Avogadro number of molecules. It is also designated as the mole or mol.

GRAM STAIN.   A method of staining microorganisms which enables such organisms to be classified into two main groups, those which retain the stain being described as *Gram-positive*; and those from which the stain is decolorized being described as *Gram-negative*. The organisms are first stained with either gentian violet or its analogue, crystal violet, and then treated with the solution of iodine. An organic solvent, usually alcohol, is then applied which washes out the stain from the Gram-negative organisms, leaving Gram-positive organisms with the violet stain unaffected. A counterstain of some contrasting color is then applied to demonstrate the Gram-negative organisms. Gram-positive organisms include staphylococci, streptococci, pneumococci; among the Gram-negative organisms are gonococci, meningococci, *Bacillus coli*, and the salmonellae.

GRANITE.   This name is applied to a common and widely occurring group of deep-seated igneous rocks consisting of orthoclase, plagioclase, quartz, hornblende, biotite, muscovite and minor accessories such as magnetite, garnet, zircon and apatite. Rarely, a pyroxene is present. Ordinary granite always carries a small amount of plagioclase, but when this is absent the rock is then referred to as an alkali-granite. An increasing proportion of plagioclase feldspar causes granite to pass into granodiorite. A rock consisting of equal proportions of orthoclase and plagioclase plus quartz may be considered a quartz monzonite. A granite containing both muscovite and biotite micas is called a binary granite.

The word granite comes from the Latin *granum*, a grain, in reference to the grained structure of such a crystalline rock.

Granite occurs as stock-like masses and as batholiths often associated with mountain ranges and frequently of great extent. Granite has been intruded into the crust of the earth during all geologic periods, except perhaps the most recent; much of it is of pre-Cambrian age. Granite is widely distributed throughout the earth.

Graphic granite is a coarsely crystalline variety of granite or pegmatite composed almost entirely of quartz and feldspar that have intergrown in such a manner as to simulate Semitic or cuneiform characters.

GRANITOID. A textural term derived from granite and signifying the relatively uniform and coarse grain of batholithic rocks, such as granite, syenite, anorthosite, etc. In a typical granitoid rock, each species of mineral occurs as a single generation; the silicates crystallize first, and any surplus of free silica crystallizes last in the form of quartz, or is finally driven off with the surplus water to form quartz veins.

GRAPHITE. [CAS: 7440-44-0]. An allotropic form of carbon, graphite occurs in nature and also is produced artificially. Graphite crystallizes in the hexagonal system, often in the form of scales or plates, or in large foliated masses. Graphite has a perfect basal cleavage, is soft (hardness between 0.5–1 on the Mohs scale—similar to talc), and feels greasy to the touch. Specific gravity 2–2.2, black to steel gray, lustrous metallic appearance, very opaque. Graphite finds many uses: (1) in the manufacture of "lead" pencils, graphite (the marking medium) is mixed with clay as a bonder, the amount of clay used determining the hardness of the pencil lead; (2) in the manufacture of self-lubricative metals in which graphite is mixed with copper, lead, and tin, after which the mix is sintered and subjected to powder metallurgy techniques to form alloys which will hold relatively large volumes of lubricating oil over long periods of use; (3) in the construction of heat-resistance structures, such as rocket casings and chemical process equipment, allowing operating temperatures up to 3,000°C and greater; (4) in the manufacture of corrosion-resistant apparatus for chemical processing; (5) in the manufacture of packings where the lubricative and corrosion-resistant characteristics of graphite are advantageous; (6) in the production of electrodes for electric furnaces and electrolysis equipment; and (7) a special pyrolytic graphite, with excellent electrical and thermal conductivity properties, good tensile strength at temperatures up to about 2,800°C, and impervious to gases and liquids, finds use in various electrical apparatus and, when mixed with boron, makes an effective nuclear radiation shield. Graphite slows the flow of neutrons without capturing them.

Graphite in Composites

Graphite has been used in composite materials of construction for a number of years, notably pioneered in structures for aircraft. The use of composite materials based upon graphite (carbon) fibers, fiber glass, numerous plastics (including epoxies), and ceramic fibers, among others, has received zealous attention in the materials community in the last half of the 1980s. Carbon-carbon (C/C) composites emerged from requirements of the aerospace field and their numerous advantages are now being extended to a variety of industrial and transportation equipment applications, including the automotive field. As observed by Klein (Nov. 1986 reference), not only can C/C withstand the heat generated at the nose cone and leading edge of space vehicles, C/C has endured such conditions mission after mission. The temperature capabilities of C/C extend to over 3300°C (5972°F), and C/C composites are twenty times stronger than conventional graphite, yet are 30% lighter, with a density of about 85 lb/ft3 (1.38 g/cc). C/C can endure higher temperatures for longer periods of time than other ablative materials. It also resists thermal shock, permitting rapid transition from −158°C (−250°F) in the cold of space to nearly 1650°C (3002°F) during reentry, well beyond the capabilities of metals and ceramics.

The C/C nose cone is made by a two-dimensional layup. In a first step, graphite cloth, preimpregnated with phenolic resin, is laid in a mold and cured. The part is trimmed, then pyrolyzed, driving off gases and moisture as the phenolic resin converts to graphite. At this point, the relatively soft composite is impregnated with furfuryl alcohol and pyrolyzed three additional times, each step increasing the density, strength, and modulus. A ceramic coating of silica and alumina is applied in the form of a powder that is finer than the pores in a human hand. To prevent the C/C from oxidizing, a coating of silicon carbide is caused to form on the top two layers of the laminate. Because the SiC is brittle and susceptible to craze-cracking, additional protection is provided by impregnating the surface with

tetraethylorthosilicate, which is cured, leaving a silicon dioxide residue throughout the coating, further reducing the area of exposed carbon. C/C is stiff and resists buckling, maintaining its aerodynamic shape over a wide temperature range. The composite has long fatigue life when subjected to thermal cycling. Numerous other, similar techniques are used to make C/C composites for a variety of applications, an excellent example of which is racing car brake disks. See Figs. 1 and 2.

Sources of Graphite

Graphite is formed during the metallurgical operations of producing pig iron, cast iron, malleable cast iron, and some special die steels and has a marked effect upon the characteristics of these materials. See also **Iron Metals, Alloys, and Steels**. The effects may be positive or negative. When present in cast iron in excessive amounts, or in the form of large interlocking flakes or films, graphite reduces the tensile strength.

Graphite is a rather widely distributed mineral and is found in a variety of rocks. It occurs in marbles, gneisses or schists; granites and other igneous rocks often carry graphite. It has been noted in pegmatites. It is likely that graphite has been formed by different processes, by magmatic separation of the graphite as an original constituent or as the result of assimilation of carbonaceous rocks, by pneumatolytic action, or by the metamorphism of sedimentary rocks that contained original carbonaceous

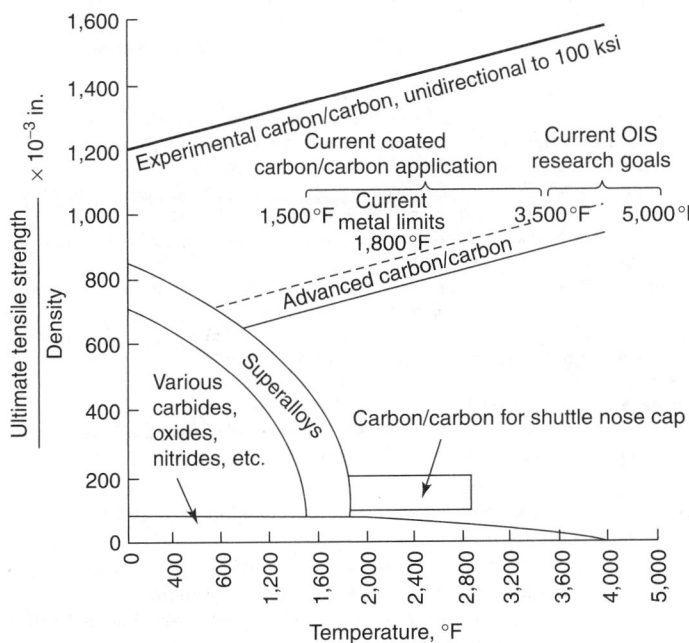

Fig. 1. Carbon-carbon retention of strength at high temperatures. (*LTV Aerospace and Defense*)

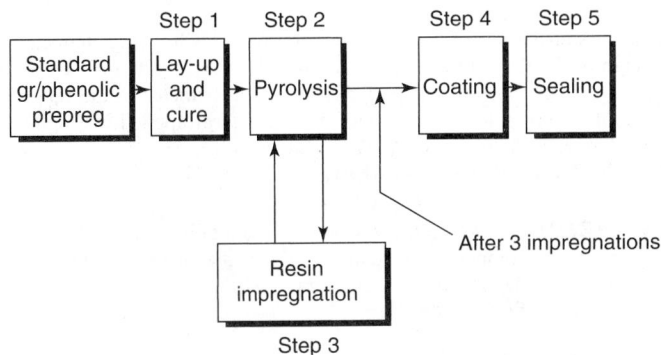

Fig. 2. Processing C/C composites. Graphite cloth is impregnated with furfuryl alcohol and pyrolyzed three or more times, each time increasing part density, strength, and modulus. Next, the part is packed with ceramic powder and fired at 1650°C to form a silicon carbide coating on the top two layers of the laminate, to prevent oxidation. (*LTV Aerospace and Defense*)

matter. Well-known localities are in Siberia, on the Island of Ceylon, which is the chief producing district at present; England, Madagascar, Mexico, and Canada. In the United States it is found in the Adirondack region of New York State, in Massachusetts, Rhode Island, Pennsylvania, Alabama, New Mexico, and Montana. Natural graphite sometimes is referred to as plumbago, black lead, and Flanders stone.

Graphite is made artificially by heating coke to a very high temperature, usually in an electric furnace. To prevent oxidation, the coke is covered with a layer of sand.

The German mineralogist, A.G. Werner, devised the name graphite from the Greek meaning *to write*, with reference to its use in pencils.

For a comparison of the characteristics and crystalline structure of graphite and diamond, see also **Carbon**; and **Diamond**.

Additional Reading

Arsenault, R., R. Bhaget, and S. Fishman: *Mechanisms and Mechanics of Composites Fracture,* ASM International, Materials Park, OH, 1993.

Clemmer, C.R. and T.P. Beebe, Jr.: "Graphite: A Mimic for DNA and Other Biomolecules in Scanning Tunneling Microscope Studies," *Science,* 640 (February 8, 1991).

Gutowski, T.G.: *Advanced Composites Manufacturing,* John Wiley & Sons, Inc., New York, NY, 1997.

Harper, C.A.: *Handbook of Plastics, Elastomers, and Composites,* The McGraw-Hill Companies, Inc., New York, NY, 1996.

Inagaki, M.: *New Carbons: Control of Structure and Functions,* Elsevier Science, New York, NY, 2000.

Klein, A.J.: "Composites that Fight Fatigue," *Adv. Mat. And Processes,* 33–36 (February 1986).

Pierson, H.O.: *Handbook of Carbon, Graphite, Diamond, and Fullerenes: Properties, Processing, and Applications,* Noyes Publications, New York, NY, 1994.

Rabe, J.P. and S. Buchholz: "Commensurability and Mobility in Two-Dimensional Molecular Patterns on Graphite," *Science,* 424 (July 26, 1991).

Rohatgi, P., Editor: *Friction, Lubrication, and Wear Technologies for Advanced Composite Materials,* ASM International, Materials Park, OH, 1993.

Staff: *Advanced Synthesis of Engineered Structural Materials,* ASM International, Materials Park, OH, 1993.

Upadhya, K., Editor: *Processing, Fabrication and Application of Advanced Composites,* ASM International, Materials Park, OH, 1993.

Utsumi, W. and T. Yagi: "Light-Transparent Phase Formed by Room-Temperature Compression of Graphite," *Science,* 1542 (June 14, 1991).

Woishnis, W.A.: *Engineering Plastics and Composites,* 2nd Edition, ASM International, Materials Park, OH, 1993.

Yoshimura, S. and R.P. Chang: *Supercarbon: Synthesis, Properties and Applications,* Springer-Verlag, Inc., New York, NY, 2000.

GRAVEL. An unconsolidated, natural accumulation of rounded rock fragments resulting from erosion, consisting predominantly of particles larger than sand (diameter greater than 2 millimeters; $\frac{1}{12}$ inch), such as boulders, cobbles, pebbles, granules, or any combination of these fragments; the unconsolidated equivalent of conglomerate. In the United Kingdom, the range of 2–10 millimeters has been specified.

Gravel is also a popularly used term for loose accumulation of rock fragments, such as detrital sediment associated especially with streams or beaches, composed predominantly of more or less rounded pebbles and small stones, and mixed with sand that may compose 50–70% of the total mass.

Gravel is also a term for rock or mineral particles having a diameter in the range of 2–50 millimeters. In the United States, the term is used for rounded rock or mineral soil particles having a diameter in the range of 2–75 millimeters $\frac{1}{6}$ to 3 inches); formerly the term applied to fragments having diameters ranging from 1–2 millimeters.

See also **Ocean Resources (Mineral)**.

GRAVIMETRIC ANALYSIS. A type of quantitative analysis involving precipitation of a compound that can be weighed and analyzed after drying. It is also used in determining specific gravity.

See also **Specific Gravity**.

GRAVITY AND MICROGRAVITY. Gravity is such an accepted part of our lives that we rarely think about it, even though it affects everything we do. Any time we drop or throw something and watch it fall to the ground, we see gravity in action. Although gravity is a universal force, there are times when it is not desirable to conduct scientific research under its full influence. In these cases, scientists perform their experiments in microgravity—a condition in which the effects of gravity are greatly reduced, sometimes described as "weightlessness."

Any object in free-fall experiences microgravity conditions, which occur when the object falls toward the Earth with an acceleration equal to that due to gravity alone [approximately 9.8 meters per second squared (m/s^2), or 1 g at Earth's surface].

Brief periods of microgravity can be achieved on Earth by dropping objects from tall structures. Longer periods are created through the use of airplanes, rockets, and spacecraft. The microgravity environment associated with the space shuttle is a result of the spacecraft being in orbit, which is a state of continuous freefall around the Earth.

Newton offered a thought experiment to explain how an object could stay in orbit while falling toward the Earth. He imagined a cannon at the top of a tall mountain that fired cannonballs. Each cannonball was acted upon by two forces: the force from the explosion and the force of gravity. The combination of the two forces would cause the cannonballs to travel in an arc. If the cannonballs were fired with more and more energy, they would hit the ground farther and farther away from the cannon. If the cannonball was fired with enough energy, it would fall entirely around the Earth and return to its starting point, completing an orbit.

Instead of being fired from a cannon atop a mountain, a spacecraft is launched in a trajectory that arcs above the Earth. When a particular speed and altitude are attained, the craft's falling path will be parallel to the curvature of the Earth, and a microgravity environment is established.

This microgravity environment gives researchers a unique opportunity to study the fundamental states of matter, solids, liquids, and gases, and the forces that affect them. In microgravity, researchers can isolate and study the influence of gravity on physical processes, as well as phenomena that are normally masked by gravity and thus difficult, if not impossible, to study on Earth.

Microgravity Research

Working in partnership with the scientific community and commercial industry, NASA's Microgravity Research Program strives to increase understanding of the effects of gravity on biological, chemical and physical systems.

Using both space flight and ground-based experiments, researchers throughout the nation, as well as international partners, are working together to benefit economic, social and industrial aspects of life for the United States and the entire Earth. Ten U.S. universities, designated by NASA as "Commercial Space Centers," share these space advancements with U.S. industry to create new commercial products, applications and processes.

Under the NASA Headquarters' Office of Life and Microgravity Sciences and Application, the Microgravity Research Program supports NASA's strategic plan in the Human Exploration and Development of Space Enterprise.

Microgravity research has been performed by NASA for more than 25 years. The term "microgravity" literally means a state of very little gravity. The prefix "micro" comes from the Greek word mikros, meaning "small." In metric terms, the prefix means "one part in a million" (0.000001). Gravity dominates everything on Earth, from the way life has developed to the way materials interact. But aboard a spacecraft orbiting the Earth, the effects of gravity are barely felt. In this "microgravity environment," scientists can conduct experiments that are all but impossible to perform on Earth. In this virtual absence of gravity as we know it, space flight gives scientists a unique opportunity to study the states of matter (solids, liquids and gases), and the forces and processes that affect them.

Marshall Space Flight Center in Huntsville, Ala. is the lead center for NASA's Microgravity Research Program. The program manages Microgravity Science and Applications Project Offices at the Lewis Research Center in Cleveland, Ohio, the Jet Propulsion Laboratory in Pasadena, Calif., and also project offices at the Marshall Center.

Under the project offices, the Microgravity Research Program is divided into nine major areas: five science disciplines, three research infrastructure programs and the Space Product Development Office.

The science disciplines include Biotechnology, Fluid Physics, Materials Science, Combustion Science and Fundamental Physics. The infrastructure activities include Acceleration Measurement, Advanced Technology and the Glovebox Flight Programs.

Marshall Center manages the Biotechnology program and Material Science program as well as the Glovebox Flight program and the Space

Products Development office. Lewis Research Center manages the Fluid Physics, Combustion Science and Acceleration Measurement programs. The Jet Propulsion Laboratory manages the Fundamental Physics and the Advanced Technology Development program.

Microgravity Biotechnology

Biotechnology is the application of engineering and technology to life sciences research. NASA partnerships with private industry and academia in space-based, biotechnological research are helping ensure that scientific advances in the field continue to produce technical innovations for improved health care on Earth. Biotechnology research in space focuses on protein crystal growth, (1) growing organic crystals with thousands of atoms and (2) on cell/tissue culturing, the study of how cells interact in a low-gravity or low-shear environment.

Pure, precisely ordered protein crystals of sufficient size and uniformity for X-ray analysis are in demand by the pharmaceutical industry as tools for research. Structural information gained from protein crystals can provide a better understanding of the role of a given protein in the body's immune system. Protein crystal research could ultimately aid in the development of more effective drugs and life-saving treatments for many diseases.

Since the mid-1980s, NASA has sponsored protein crystal growth experiments to learn about the effects of space on the growth process and to refine techniques for obtaining the highest quality crystals in space and on the ground. The result is that generally, protein crystals produced in space are larger and more precisely ordered than those produced on Earth. These improvements are important to scientists who analyze a crystal's three-dimensional structure (the key to understanding a protein's activity) and possibly develop new and more effective medicines. Knowledge of the molecular structure of the antibody may lead to development of treatments and vaccines to cure the disease that causes pneumonia and severe upper respiratory infection in nearly four million children, ages 1 to 5 in the United States annually.

The other focus of biotechnology in microgravity is cell and tissue culturing experiments. Located at Johnson Space Flight Center in Houston, Texas, the goal of this research is to grow cells on a tissue in near-weightlessness, that otherwise is unachievable on Earth.

The medical benefit of microgravity tissue and culture engineering may lead to new research models in cellular and molecular biology. These studies also are developing new tissues for potential transplant operations.

Biotechnology research results have provided significant advances in the understanding of many diseases including AIDS, heart disease, cancer, diabetes, and hepatitis. See also **Biotechnology (Bioprocess Engineering)**.

Everyone has practical experience with fluids, i.e., liquids and gases, and knows how a fluid will behave under "normal" circumstances. Steam rises from the surface of a hot spring or a boiling pot, and water spilled on a tabletop runs over, then off, the surface. Gravity drives much of the fluid behavior we are accustomed to on Earth.

Many of our intuitive expectations do not hold up in microgravity, though, because other forces such as surface tension control fluid behavior. Surface tension causes drops of any liquid to form almost perfect spheres when the influence of gravity is absent. On Earth, gravity distorts the shape when liquid is resting on or attached to a surface. Although these differences in fluid behavior often present engineers and astronauts with practical problems, they also offer scientists unique opportunities to explore different aspects of the physics of fluids.

Research conducted in microgravity is increasing our understanding of fluid physics to provide a foundation for predicting, controlling and improving a vast range of technological processes. The behavior of fluids is at the heart of many phenomena in materials science, biotechnology and combustion science. Surface tension-driven flows, for example, affect some techniques of semiconductor crystal growth, welding, and the spread of flames on liquids. The dynamics of liquid drops are an important aspect of chemical process technologies and meteorology.

Results from microgravity Fluid physics research will lead to better understanding of the effects of miniaturization of electronic materials. Advances in the field will lead to even smaller and more efficient electronic devices with reduced costs for the consumer.

Microgravity Materials Science

Materials science investigates the relationships between the structure, properties, and processing of materials. Structure is the arrangement of the atoms in the material. Properties include physical, chemical, electronic, thermal and magnetic characteristics. Processing is the method by which materials are formed. They can be solidified, evaporated and condensed, or dissolved and then separated from a solution. NASA's materials science microgravity program uses the unique characteristics of the microgravity space environment to study these fundamental relationships in materials solidification and crystal growth.

In the production of electronic materials, crystals have achieved far greater value as conductors than they ever had as gemstones. Pioneering research is leading to next-generation commercial crystal products.

Material science also has a focus on the production of alloys and composites. High-strength metals are needed in the aviation, aerospace, power generation and propulsion industries. Processing these materials in space helps researchers understand how to make better materials on Earth and is allowing scientists to create new metal alloys. Alloys are mixtures of metals or metals and nonmetals. When combined, they can produce materials with improved strength or better resistance to corrosion.

On Earth, when a melted alloy solidifies, it forms pine-tree-shaped crystals called dendrites. These dendrites play a very important role in determining the properties of the alloy and its subsequent usefulness. Gravity causes fluid flows in the alloy, leading to the formation of irregular dendrites that weaken the alloy or metal structure. This type of processing is so complex that it is difficult to measure and predict, and even more difficult to control. In space, gravity-related phenomena such as convection are reduced, thus simplifying the process for study. See also **Dendrite**.

Ceramics and glass experiments also are part of the Material Science program. Optical engineering is being revolutionized by new glasses, crystals and other materials that surpass conventional substances in quality. However, production of these superior materials is difficult. Some glasses have chemical mixes that react with their containers. Others are extremely sensitive to contamination levels from impurities of even a few parts per billion. For example, certain fluoride glasses are of great interest for their infrared transmission properties. These glasses can be made on Earth, but trace contaminants from processing containers have prevented them from reaching their highest potential.

Containerless processing, in which a sample is suspended and manipulated without touching contaminating containers, is an attractive solution to these problems. Containerless processing on massive samples can only be done in the microgravity environment of space where the forces used for suspending and manipulating the samples are not overwhelmed by gravity.

Microgravity Materials results will contribute to future models of industrial and manufacturing processes. This will lead to new, stronger, lighter alloys with never-seen-before properties.

Microgravity Combustion Science

NASA's combustion research program focuses on understanding the important processes of ignition, flame spreading and flame extinction during combustion in low gravity. Research is directed at gaining basic knowledge of combustion processes, as well as addressing issues of fire safety in space.

Scientists are interested in the physical characteristics of flame, such as size and shape, and the role of soot formation in combustion. Investigations also study air flows and the transfer of heat and mass in fuel vapors, liquid pools, paper and metal solids.

Since the physical and chemical mechanisms that cause flames to spread on Earth are strongly influenced by gravity, researchers are finding out flames behave very differently in the low-gravity of an orbiting spacecraft. It is well known that material flammability and flame growth are strongly affected by the environment, including oxygen content, pressure and air flow. However, the effects of these conditions in the microgravity environment are largely unknown. Scientists want to understand combustion to improve efficiency of our fuel-driven machines and to evaluate potential fire hazards aboard spacecraft. Combustion research will lead to more efficient fuels, better fire safety and a cleaner environment.

Microgravity Fundamental Physics

Fundamental physics researchers use the low-gravity environment of space to test basic scientific theories not possible in the gravity environment on Earth in fields such as thermophysical measurements, atomic physics and relativistic physics.

This research is important because it seeks to uncover principles that govern the behavior of the physical world, such as the influence of heat energy, new forms of matter and low-temperature physics.

Fundamental physics research in microgravity is driving the development of new technologies that will advance scientific knowledge and improve life on Earth. The benefits of this research can be seen in improvements in ultra- sensitive detectors of temperature and magnetic fields, as well as valves that can function at temperatures close to absolute zero. Also, scientists have discovered the transition between different forms of matter, whether magnetic or nonmagnetic, solid or liquid, and the similarities between the different systems. Theories resulting from studies of superfluid helium in microgravity can help to understand many other systems. Scientists can use these theories to better understand the formation of weather systems such as tornadoes and hurricanes, how water seeps through soil, and how cracks propagate in metals.

New understanding of nature's processes have made much exploratory surgery unnecessary with the advent of magnetic resonance imagers (MRIs). Another application of this research is liquefied gases. Liquid oxygen is used to supply breathing gas in hospitals, and helps fuel the powerful rockets that have made human exploration of space possible.

Acceleration Measurement Program

The Acceleration Measurement Program is an operational space flight infrastructure program to measure and track accelerations with a recording system. Acceleration is the force that pushes the passengers in a car against the side opposite of a turn. Acceleration measurement systems serve a wide variety of microgravity science and technology experiments. These systems can measure and record low-gravity accelerations at as many as three experiment sites simultaneously. They can be mounted on or near an experiment to measure the accelerations experienced by the experiment. Understanding the interactions of accelerations contributes to improved microgravity research.

Units mounted on spacecraft exteriors historically have had the capability for remote commanding from the ground and downlinking mission data. Units mounted inside recently have been modified to incorporate this capability. The data are displayed at NASA's Lewis Telescience Support Center in Cleveland, Ohio or NASA's Marshall Payload Operations Control Center (POCC) in Huntsville, Ala.

Advanced Technology Development Program

The Advanced Technology Development Program was developed by NASA's Microgravity Science and Applications Division in response to the challenges researchers face when defining experiment requirements and designing associated hardware. The program provides efficient, cost-effective, state-of-the-art technological support for microgravity science investigations. The Advanced Technology infrastructure program enables new types of scientific investigations and gives researchers capable, high-quality experimental hardware to overcome existing technology-based constraints. The goal is to investigate and develop high-risk microgravity research technologies before they are needed on the critical development path for actual flight hardware.

Historically, Advanced Technology projects have encompassed a broad range of activities. Project funding includes the development of diagnostic instrumentation and measurement techniques, observational instrumentation and data recording methods, acceleration characterization and control techniques, and advancements in methodologies associated with hardware design technology.

Glovebox Flight Program

The Glovebox Flight Program is a microgravity infrastructure program that provides facilities for performing investigations not requiring large, specialized equipment. The glovebox offers scientists the capability to conduct experiments, test science procedures, and develop new technologies in microgravity. The facility enables crew members to handle, transfer and manipulate experiment hardware and materials that are not approved for use in the open Spacelab.

By providing hardware, development and investigation integration services, the program allows researchers to concentrate on scientific objectives, rather than facility development and vehicle and mission issues. The Glovebox Flight Program lowers cost and allows for quicker and easier access to space for proof-of-concept demonstrations and new investigations. The program office is developing laboratory support equipment items for the International Space Station.

Space Product Development Office

The Space Product Development office examines the opportunities for space commerce, offering a full range of support capabilities to demonstrate the commercial value of space. To ensure continued growth of U.S. industry, the office initiates and guides pilot projects in an effort to eliminate barriers to viable space commercialization.

The Space Product Development Office manages projects such as a new advance in insulation called "Aerogel," a NASA flight project aimed at revolutionizing window insulation and light emitting diodes, a project originally aimed at developing plant growth facilities for space flight and today is funding state-of-the-art cancer-fighting treatments.

Ground-Based Research

A major challenge facing NASA's space-based microgravity program is to conduct scientifically significant and productive research through the wisest possible use of space. To achieve this, NASA uses a ground-based research program to assess whether scientific investigations are worthy of a space flight opportunity. These studies are then refined for the ultimate experimental test during a space mission.

To create low-gravity environments on Earth for research, free-fall facilities are used in a variety of ways. Releasing experiment samples from tall drop towers provides about four seconds of microgravity. Research aircraft can expose experiments to about 30 seconds of low-gravity while the aircraft approaches the top of a steep climb and begins a sharp descent. This parabolic curve is generally repeated about 40 times during each flight, which is used primarily to perform experiments requiring short times for experimental equipment tests or for crew training.

Low-cost sounding rockets, such as Space Processing Applications Rockets, also have parabolic flight paths. They ascend and then descend, rather than proceeding into orbit around the Earth. Sounding rocket flights provide five to seven minutes of low-gravity. Although these periods of microgravity are brief, the test facilities are beneficial both for space flight preparation and for some actual microgravity research.

Shuttle-Mir Microgravity

The Microgravity Research Program Office manages the development and integration of microgravity science experiments of the Shuttle-Mir program. Both United States and international microgravity science partners used the facilities aboard Mir to conduct investigations in fluid physics, combustion, biotechnology and materials science. The microgravity facilities aboard the Mir space station included furnaces, glovebox and a system to isolate experiments from the station's vibration environment.

The Future of Microgravity Science; The International Space Station

The NASA Microgravity Research Program is evolving to take maximum advantage of the upcoming International Space Station. The Space Station will permit long-duration microgravity experiments in an environment otherwise more similar to Earth-based laboratories minus the gravity.

Rather than experiments being limited to a week or two, as they are aboard the Shuttle, Space Station microgravity experiments will stretch over long periods of time. This longer duration will greatly increase the number and types of materials that can be processed to full term. This will be a great advantage to experiments in areas such as solution and vapor crystal growth, which require 15 to 30 days of continuous growth to produce crystals of the desired size.

On the Space Station, with crew members to observe experiments and with equipment for analyzing samples in orbit, it will not be necessary to return all specimens to Earth for analysis before running the next experiment. This will allow researchers to conduct experiments in a series which builds on prior results without waiting years for another flight opportunity.

Future space research will stress both scientific and commercial goals. Products will include crystals, metals, ceramics, glasses, and biological materials. Processes will include solidification of metals and alloys, as well as transporting fluids and chemicals in microgravity. As research in these areas develops, the benefits will become increasingly apparent on Earth: new materials, more efficient use of fuel resources, new medicines, advanced computers and lasers and better communications. Like space, opportunities offered by microgravity science are vast, and only beginning to be explored.

NASA's Microgravity Research Program Office at Marshall Center is responsible for the definition and development of microgravity science and space product development projects planned for the International Space Station.

See also **Microgravity and Materials Processing**.

Additional Reading

Barlow, P.W. and D. Moore: *Life Sciences: Microgravity,* Elsevier Science, New York, NY, 1999.

El-Genk, M.S.: *Space Technology and Applications International Forum 2000: Conference on International Space Station Utilization; Conference on Thermophysics in Microgravity; Conference on Enabling Technology and Required Science,* American Institute of Physics, College Park, MD, 2000.

El-Genk, M.S. *Space Technology and Applications International Forum: Conference on International Space Station Utilization; Conference on Global Virtual Presence; Conference on Applications of Thermophysics in Microgravity and Breakthrough Propulsion Physics,* American Institute of Physics, College Park, MD, 2000.

Moore, D., P. Bie, and H. Oser: *Biological and Medical Research in Space: An Overview of Life Sciences Research in Microgravity,* Springer-Verlag Inc., New York, NY, 1996.

Staff: National Research Council; Committee on Physical Science; Space Studies Board: *Microgravity Research in Support of Technologies for the Human Exploration and Development of Space and Planetary Bodies,* National Academy Press, New York, NY, 2000.

Zee, A.: *Einstein's Universe: Gravity at Work and Play,* Oxford University Press, Inc., New York, NY, 2001.

Web References

Jet Propulsion Laboratory. *http://www.jpl.nasa.gov/*
Marshall Space Flight Center. *http://www.msfc.nasa.gov/*

GREASE. A lubricating agent of higher viscosity than oils, consisting originally of a calcium or sodium soap jelly emulsified with mineral oil. Greases are employed where heavy pressures exist, where oil drip from the bearings is undesirable, and where the motion of the contacting surfaces is discontinuous so that it is difficult to maintain a separating film in the bearing. Grease-lubricated bearings have greater frictional characteristics at the beginning of operation, causing a temperature rise which tends to melt the grease and give the effect of an oil-lubricated bearing.

The principal categories of greases are: (1) calcium soap greases; (2) sodium soap greases; (3) complex soap greases—combinations of soaps and fatty acids used to impart high-temperature properties and moisture resistance. A low-molecular-weight soap can be used as a binding agent between the oil and soap in place of water; (4) lithium soap greases—excellent as multipurpose greases; (5) extreme-pressure greases, usually containing some form of sulfur, phosphorus, or other reactive agent—particularly suited to uses where there are sudden shock loads or continuous high pressures, as in steel rollingmill bearings; (6) nonsoap greases—exemplified by organically modified clays which hold the lubricating oil both by absorption and adsorption. Such greases are often used in high-temperature applications because they actually have no melting point; (7) asphalt-base greases—blends of asphaltic materials with lubricating oil, enabling a wide range of consistencies; and (8) filler-type greases—frequently calcium-base greases that contain solid materials having unctuous properties. The filler essentially serves as a cushion for absorbing impacts. Calcium and sodium base greases are most commonly used; sodium base greases have higher melting point than calcium base greases but are not resistant to the action of water. Graphite, either by itself or mixed with grease, is also employed as a lubricant. Gear greases consist of rosin oil, thickened with lime and mixed with mineral oil, with some percentage of water. The special-purpose greases often contain glycerol and sorbitan esters. They are used, for example, for low temperature conditions. See also **Lubricant**.

Standard methods for testing greases are published by the American Society for Testing and Materials, Philadelphia, Pennsylvania.

GREENHOUSE EFFECT. The water vapor and carbon dioxide found naturally in the atmosphere keep the Earth warmer than it would otherwise be. The clear atmosphere allows sunlight to penetrate to the Earth's surface and warm it. The surface releases this energy as infrared radiation, which is absorbed by water vapor and CO_2 in the atmosphere. This mechanism is commonly known as the "greenhouse effect." Without the greenhouse effect, the earth would be about 33°C (60°F) colder than it is currently.

Humanity is altering the energy balance of the planet by adding gases that absorb infrared radiation to the atmosphere, and thereby strengthening the greenhouse effect. The chief "greenhouse gases" are CO_2, methane, and nitrous oxide. Whenever oil, coal, gas, or wood are burned, carbon dioxide is released into the atmosphere. Approximately half of the CO_2 that is released is soon absorbed by the oceans or by increased plant photosynthesis. The other half remains in the atmosphere for many decades. As a result, the atmospheric concentration of CO_2 is increasing. The average concentration of carbon dioxide has increased from around 275 parts per million before the industrial revolution, to 315 ppm when precise monitoring stations were set up in 1958, to 368 ppm in 1999. This change has increased the amount of energy striking the earth's surface by about 1.5 watts for every square meter of the earth's surface. This increased energy is equal to about 1% of the energy in the sunlight that reaches the earth's surface. See also **Greenhouse Gases**.

About two thirds of the current emissions of methane into the atmosphere result from cattle farming, rice paddies, landfills, coal mining, oil and gas production, and several other human activities. The other third comes from natural sources, particularly wetlands and termites. The total greenhouse effect from methane has increased by about 0.5 watts (0.3%) the energy striking each square meter of the earth's surface.

Several other gases collectively may have as much of a greenhouse effect as methane. Nitrous oxide (also known as "laughing gas") is released by the use of nitrogen fertilizers, the burning of wood, and some industrial processes. Higher levels of ozone, an urban pollutant regulated by EPA NAAQS, also add to the greenhouse effect. (The loss of ozone in the upper atmosphere tends to reduce the greenhouse effect.) Other gases with a greenhouse effect include chlorofluorocarbons (CFCs), hydrofluorocarbons (HFCs), perfluorocarbons (PFCs), hydrochlorofluorocarbons (HCFCs), and sulfur hexafluoride (SF_6).

Additional Reading

Dobson, A., A. Jolly, and D. Rubenstein: "The Greenhouse Effect and Biological Diversity," *Trends Ecol. Evol.* **4**(3), 64–68 (1989).

Hardy, J.T.: *Climate Change: Causes, Effects, and Solutions,* John Wiley & Sons, Inc., New York, NY, 2003.

Hocking, C., C. Sneider, and L. Bergman: *Global Warming and the Greenhouse Effect,* University of California, Berkeley, Hall of Science, Berkeley, CA, 1999.

Long, D.: *Global Warming,* Facts on File, Inc., New York, NY, 2003.

GREENHOUSE GASES. Greenhouse gases are global in their effect upon the atmosphere. The primary greenhouse gases, unlike many local air pollutants like carbon monoxide, oxides of nitrogen, and volatile organic compounds, are considered stock pollutants. A stock air pollutant is one that has a long lifetime in the atmosphere, and therefore can accumulate over time. Stock air pollutants are also generally well mixed in the atmosphere. As a consequence of this mixing, the impact a greenhouse gas has on the atmosphere is mostly independent of where it was emitted. These characteristics of greenhouse gases imply that they should be addressed on a global (i.e., international) scale. See also **Pollution (Air)**.

Anthropogenic emissions of greenhouse gases occur in every country of the world. These emissions result from many of the industrial, transportation, agricultural, and other activities that take place in each country. Countries that are signatories to the United Nations Framework Convention on Climate Change (UNFCCC) are committed to reporting their anthropogenic emissions of greenhouse gases to the Secretariat of the convention.

What are Greenhouse Gases?

Some greenhouse gases occur naturally in the atmosphere, while others result from human activities. Naturally occurring greenhouse gases include water vapor, carbon dioxide, methane, nitrous oxide, and ozone. Certain human activities, however, add to the levels of most of these naturally occurring gases:

Carbon dioxide is released to the atmosphere when solid waste, fossil fuels (oil, natural gas, and coal), and wood and wood products are burned.

Methane is emitted during the production and transport of coal, natural gas, and oil. Methane emissions also result from the decomposition of organic wastes in municipal solid waste landfills, and the raising of livestock.

Nitrous oxide is emitted during agricultural and industrial activities, as well as during combustion of solid waste and fossil fuels.

Very powerful greenhouse gases that are not naturally occurring include *hydrofluorocarbons* (HFCs), *perfluorocarbons* (PFCs), and *sulfur hexafluoride* (SF_6), which are generated in a variety of industrial processes.

Each greenhouse gas differs in its ability to absorb heat in the atmosphere. HFCs and PFCs are the most heat-absorbent. Methane traps over 21 times more heat per molecule than carbon dioxide, and nitrous oxide absorbs 270 times more heat per molecule than carbon dioxide. Often, estimates of greenhouse gas emissions are presented in units of millions of metric tons of carbon equivalents (MMTCE), which weights each gas by its GWP value, or *Global Warming Potential*.

See also **Greenhouse Effect**.

GREENOCKITE. The mineral greenockite is cadmium sulfide, CdS, and is used as an ore of that metal. It is found rarely in hexagonal crystals, sometimes as earthy coatings on other minerals. Its hardness is 3–3.5; specific gravity, 4.9–5.0; luster, adamantine to earthy; color, yellow to yellowish-orange; subtransparent. It is found in Scotland, Bohemia, and France; also, in the United States, at Franklin Furnace, New Jersey; and Marion County, Arkansas, where it occurs as a yellow coloring matter in smithsonite; and in Mono County, California. It was named for Lord Greenock.

GREENSTONE. Greenstone is an old field term for more or less altered basalts and dolerites, which, because of the development of chlorite, or perhaps hornblende or epidote, develop a characteristic green color. Many diabases and epidiorites have been called greenstones.

GREISEN. An old German petrological term originally proposed by Werner for an igneous rock of granitic or aplitic texture composed principally of quartz, alkali feldspar, the fluorine-rich micas, and sometimes containing topaz. Greisens are pneumatolytically altered granites which are closely associated with the development of the tin ore mineral cassiterite.

GRIGNARD REACTIONS. Very important to the synthesis of numerous organic compounds, both in the laboratory and on a large scale in industry, is a two-step reaction involving the use of organo-magnesium halides. These reactions were studied intensively by Victor Grignard during the early 1900s and for this work he was awarded the Nobel Prize in Chemistry in 1912. The reactions are universally referred to as Grignard reactions and the many magnesium compounds required by the reactions are known as Grignard reagents. Grignard's work stemmed from a discovery by Barbier in 1899 that dimethylheptenol could be prepared by reacting methyl iodide, dimethylheptenone, and magnesium in ethyl ether. In studying the mechanics of Barbier's reaction, Grignard found that the reaction proceeds in two steps: (1) the reaction of magnesium and an alkyl halide to form the corresponding alkyl magnesium halides; and (2) the reaction of the alkyl magnesium halide with a compound containing a carbonyl group to form a new carbon-carbon bond. Through subsequent years of experience, researchers have learned that nearly all alkyl and aryl halides react with magnesium to form Grignard reagents. However, the aryl and vinyl derivatives are, with more difficulty, achieved. In the mid-1950s, Normant and Ramsden showed that some of the less reactive halides, such as vinyl chloride and chlorobenzene, will form a Grignard reagent with comparative ease if tetrahydrofuran is used as the solvent. See Table 1.

Because of the importance of the Grignard reaction techniques, they have received much study and numerous proposals have been made concerning the detailed mechanics involved. Originally, Grignard represented a Grignard reagent by RMgX, where R is the alkyl or aryl radical and X is the halide. Thus, magnesium ethyl bromide, a Grignard reagent, would appear in Grignard's symbolism as C_2H_5MgBr. Two of the main factors which make Grignard reagents so important are: (1) the many kinds of reagents that can be formulated, considering the substitution possibilities of the R and the X in the formula; and (2) the variety of reactions in which the Grignard reagents participate to yield numerous kinds of compounds. This versatility is demonstrated partially by Table 1.

The general sequence of the reactions is now embodied in the following generic forms, where RX = an organic halide (most typically a chloride or bromide, although fluorides can be induced to react); S = a coordinating solvent (such as an ether or an amine); and AZ = a substrate with an electronegative group, Z:

$$RX + Mg + nS \longrightarrow RMgX \cdot S_n$$

TABLE 1. REACTIONS OF GRIGNARD REAGENTS

Grignard reagents react with	To yield
H₂O, alcohols, primary or secondary amines	Hydrocarbons
Oxygen	Alcohols and phenols
CO₂	Carboxylic acids
Nitriles	Ketones
Metal halides	Organometallic compounds
NH₃	Hydrocarbons
γ-Lactones	Glycols
Acid esters	Tertiary alcohols (except formic acid which yields secondary alcohols or aldehydes)
Aldehydes	Secondary alcohols (except formaldehyde which yields primary alcohols)
Carboxylic acids	Tertiary alcohols
Acid halides	Tertiary alcohols or ketones
Ketones	Tertiary alcohols
Hydrogen halides	Hydrocarbons
Sulfur	Mercaptans

$$RMgX \cdot S_n + AZ \longrightarrow RAZMgX \cdot S_n$$

$$RAZMgX \cdot S_n \longrightarrow RA + ZMgX \cdot S_n$$

The heterolysis of AZ is dependent on the substrate and does not always occur. The final isolation of the product usually involves a hydrolysis step.

The development of improved industrial procedures, including the substitution of tetrahydrofuran (THF) for diethyl ether and the demonstration that the less reactive, but significantly less expensive, vinyl and aryl chlorides could be successfully used, has greatly expanded the commercial possibilities of this reaction. In the flavor, fragrance, pharmaceutical, and fine chemical industries, its use can generally be regarded as routine. Tens of thousands of metric tons of Grignard reagents are produced annually for captive use or merchant sale.

The great value of the Grignard reaction to the synthetic chemist is its general applicability as a building block for an impressive range of structures and functional groups. The Grignard reagent can act both as a prototypical carbon nucleophile that can undergo addition and substitution reactions and as a strong base that can deprotonate acidic substrates, resulting in the conjugate base or in some cases elimination reactions. Grignard reagents react with most functional groups containing polar multiple bonds (e.g., ketones, nitriles, sulfones, and imines), highly strained rings (epoxides), acidic hydrogens (e.g., alkynes), and certain highly polar single bonds (e.g., carbon–halogen and metal–halogen).

Preparation of Grignard Reagents

A Grignard reagent is prepared by first adding magnesium and a partial charge of solvent to the reactor, followed by the addition of RX, in the remaining solvent, to the reaction flask.

Solvent Preparation. The most critical aspect of the solvent is that is must be dry (less than 0.02 wt % of H_2O) and free of O_2.

Other considerations for the solvent are the solubility of the Grignard reagent and the temperatures required for initiation and adventitious reactions of the Grignard with the solvent. Based on these three considerations, the best general solvent for the preparation of a Grignard reagent is THF. However, other solvents that are commonly used are diethyl ether, methyl *t*-butyl ether, di-*n*-butyl ether, glycol diethers, toluene, dioxane (R_2Mg), and hexane.

Magnesium Preparation. A surface coating resulting from the oxidation or hydration of the metal surface is the principal problem encountered for the magnesium reaction surface. Fortunately, there are dozens of methods to remove the inert coating, thus activating the magnesium. For industrial use, the best method is using freshly chipped Mg turnings with a small quantity of the desired Grignard added to the reactor before addition of RX.

The Organohalogen Component. Just as for Mg and the solvent, the organic halide must be dry (less than 0.02 wt % of H_2O) and free of O_2. The relative reactivity of the halogens is reflected in the rate of disappearance of Mg, which follows the general order I > Br > Cl >> F. Unfortunately, the rate of disappearance Mg of does not always correlate with the formation of active Grignard. Typically, the more reactive the RX is, the higher the probability of forming a homocoupled product. Therefore, when choosing X, the rate of reactivity, product selectivity, and cost must be taken into account.

Other Methods. There are several common alternative methods for making Grignard reagents. Metal-exchange reactions are straightforward and MgR_2 can easily be prepared by this route.

Hydromagnesation reactions allow for the economical preparation of a Grignard reagent from an olefin.

Industrial Manufacturing Process

In spite of its industrial use for many years, the commercial-scale production of Grignard reagents has not been extensively described. The only practically important method is the batch method described by Grignard in 1900, namely formation of the Grignard reagent, reaction with a substrate, followed by hydrolysis of the reaction mixture.

The equipment can usually be constructed of carbon steel except for the hydrolysis vessel, which is usually glass-lined to avoid corrosion by aqueous acids. All vessels must be supplied with an inert gas (nitrogen or argon) for purging and blanketing and are vented to release off-gases. It is imperative that the reaction vessel be protected with a rupture disk.

Analysis of Grignard Reagents

There are three potential problems that may occur during Grignard reagent preparation: oxidation by O_2, hydrolysis by H_2O, or homocoupling during the addition of alkyl or aryl halide. All three of these reactions decrease the active Grignard reagent while maintaining the same equivalents of base. Consequently, the concentration of a Grignard reagent should not be assumed, based on the reactants. The disadvantages of not analyzing the Grignard reagent are improper stoichiometry, potentially deleterious side reactions, highly exothermic quenching processes, phase splits, waste disposal, and cost problems. The analytical technique must be able to differentiate between active Grignard and total basicity. Many methods are available to measure the active Grignard, ranging from titration to electrophilic quenching followed by gc analysis.

Economic Aspects

The Grignard reaction has been commercially important for a number of years, and for certain industrial processes it remains the favored (or only) practical route to construct various element-to-carbon bonds.

There are five components to the cost of using a Grignard reagent: *(1)* magnesium metal, *(2)* the halide, *(3)* the solvent, *(4)* the substrate, and *(5)* disposal of the by-products. Prices for tetrahydrofuran and diethyl ether, the two most commonly used solvents, have increased. The cost of the halide depends on its structure, but as a general rule the order of cost is chloride < bromide < iodide.

Health and Safety Factors

Fire Hazards. The hazards associated with the manufacture, transport, and use of Grignard reagents are related to the flammability of the solvents employed and the exothermic reactions involved in their preparation and use.

Toxicology. Because of their high reactivity, there is little meaningful information on the health hazards of Grignard reagents *per se*. Rather, consideration needs to be given to the reagents employed, including the solvents and the products (or by-products) of the reaction. Some starting materials, such as organic halides (notably methyl bromide and vinyl chloride), are particularly toxic.

Regulatory Considerations. Commercial use of a Grignard reagent in the United States requires that it appear on the Environmental Protection Agency (EPA) list of Chemical Substances in Commerce. A corresponding registration exists for the European community and for Japan.

Because they are classified as flammable liquids, Grignard reagents in the United States must be packaged in drums or other suitable containers bearing a red U.S. Department of Transportation label.

Reactions and Applications of Grignard Reagents

Reactions and applications of Grignard reagents include asymmetric syntheses using Grignard reagents, Grignard reactions with inorganic chlorides, Grignard reagents as bases, metal-assisted modified Grignard reactions, intramolecular Grignard reactions, Grignards as methacrylate polymerization catalysts, and Grignard reagents as supports for the Ziegler-Natta process.

GARY S. SILVERMAN
Elf Atochem North America
PHILIP E. RAKITA
Elf Atochem Japan

Additional Reading

Kharasch, M. and O. Reinmuth: *Grignard Reactions of Nonmetallic Substances,* Prentice-Hall, Inc., New York, NY, 1954.
Okubo, M. and K. Matsuo: *Rev. Heteroatom Chem.* **10**, 213 (1994).
Raston, C. and G. Salem: *Chem. Met.–Carbon Bond* **4**, 159 (1987).
Silverman, G.S. and P.E. Rakita, eds.: *The Grignard Reagent Handbook,* Marcel Dekker, Inc., New York, NY, 1996.

GRINDING AND POLISHING AGENTS. These materials are comprised of abrasives in some form. An abrasive is a hard substance that, in particulate form, is capable of effecting a physical change in a surface, ranging from the removal of a thin film of tarnish to the cutting of heavy metal cross sections and cutting stone. Abrasive action can be negative, as in the case of grit in lubrication oil that will cause engine wear; or it can be positive, as used in scores of different abrasive products, ranging from sandpaper to grinding wheels.

The two principal categories of abrasives are: (1) natural abrasives, such as quartz, emery, corundum, garnet, tripoli, diatomaceous earth (diatomite), pumice, and diamond; and (2) synthetic abrasives, such as fused alumina, silicon carbide, boron nitride, metallic abrasives, and synthetic diamond. Quartz, emery, garnet, and corundum were used in prehistoric times. Natural diamond was first used in India as an abrasive in about 800 B.C. The first grinding machines were developed in France in about 1300 A.D. Early shellac-bonded abrasives were developed in India about 1825. The first cylindrical grinding machine was made in the United States in 1860. Vitrified bonded abrasives were developed in the United States in 1872. Fused alumina and silicon were developed in 1901. Resinoid bonded abrasives were developed by L. Baekeland in Belgium in 1923. Metal bonded diamond wheels appeared in 1936, and synthetic diamond abrasives and cubic boron nitride were developed in the United States during 1955 and 1957, respectively.

TABLE 1. GENERAL PROPERTIES OF ABRASIVES

Abrasive	Composition	Trade names and synonyms	Knoop hardness	Melting point C	Specific gravity
Quartz	SiO_2	Sand, flint	820	1,700	2.65
Emery	imp. Al_2O_3		2000	1,900	4.00
Corundum	imp. Al_2O_3		2000	2,050	3.95
Garnet			1360	1,200[a]	4.25
Tripoli	98% SiO_2	Rottenstone	820	1,700	2.50
Diatomite	89% SiO_2	Diatomaceous earth Kieselguhr	820	1,700	2.50
Pumice	70% SiO_2 + Oxides	Pumicite, black ash			2.50
Diamond	C'		6500	1,000[a]	3.51
Fused alumina	93–97% Al_2O_3	Alundum, Aloxite, Lionite	2000	2,050	3.95
Silicon carbide	SiC	Carborundum, Crystolon	2450	2,400?	3.20
Cubic boron nitride	BN	Borazon	4700	2,000[a]	

[a] decomposes

Source: Norton Company.

TABLE 2. AVERAGE PARTICLE SIZE OF ABRASIVE GRAIN USED IN GRINDING WHEELS

Grit size	Inches	Micrometers
8	.1817	4,620
10	.1366	3,460
12	.1003	2,550
14	.0830	2,100
16	.0655	1,660
20	.0528	1,340
24	.0408	1,035
30	.0365	930
36	.0280	710
46	.0200	508
54	.0170	430
60	.0160	406
70	.0131	328
80	.0105	266
90	.0085	216
100	.0068	173
120	.0056	142
150	.0048	122
180	.0034	86
220	.0026	66
240	.00248	63
280	.00175	44
320	.00128	32
400	.00090	23
500	.00065	16
600	.00033	8

The general properties of natural and synthetic abrasives are given in Table 1. Both natural and synthetic abrasives must be crushed to small particle size before bonding to cloth or paper for mounting on various tools. See Table 2.

The Nature of the Grinding Process has been the object of much research. It appears to be a mixture of chemical and physical processes. Chips are cut from the metal by the sharp abrasive points which are heated in the process to the melting temperature of the metal. At this temperature chemical reactions take place which involve the abrasive, the metal, and the surrounding atmosphere. These reactions cause a dulling of the abrasive points necessitating a wearing of the wheel structure to expose new ones. The reactions also have a beneficial effect in preventing the rewelding of the chips to the base metal and their adhesion to the wheel structure. The latter is termed "loading." The detrimental reactions involving the abrasive are minimized by the control of its purity. The beneficial reactions can be augmented by the incorporation of chemical aids either into the wheel structure or into a fluid applied to the point of contact. Substances commonly used for this purpose are organic and inorganic sulfides and chlorides.

Polishing appears to be a quasi-chemical process in which the metal (or other material) is removed in particles approaching molecular size.

GROUNDWATER. At varying depths below the surface of the earth, depending upon wet or dry seasons, underground structures, and other natural and unnatural factors, is a zone which is saturated with water most of which comes from rain which has penetrated the ground. The upper surface of this saturated zone is called the water table, and the water itself, the groundwater or the subsurface water. The region above the upper surface of the water table is called the zone of aeration or vadose zone.

There is a lower limit to the saturated zone as well as an upper limit. Little groundwater exists at depths below 2,000–3,000 feet (610–914 meters). Deep down in the earth's crust the pressure must be so great that all pores in the rocks are completely closed; thus at depths of several miles below the surface there could exist no zone of saturation.

The groundwater moves through the rocks and unconsolidated materials of the earth near the surface, constantly seeping into streams and lakes to maintain these bodies of water between rains. If this seepage is sufficiently strong on hillsides or elsewhere springs may result. A well is simply an opening dug deep enough to encounter the zone of saturation.

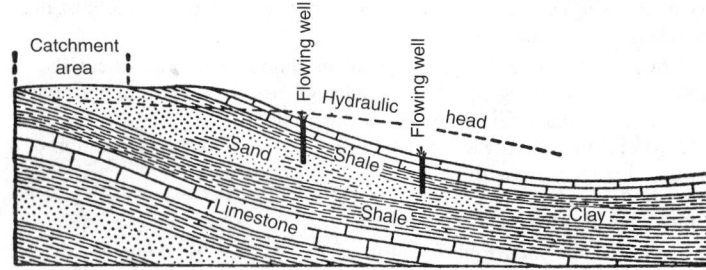

Fig. 1.   Ground cross section showing flowing artesian wells in a monocline.

In certain cases, the groundwater will flow through porous tilted beds called aquifers, from higher to lower localities, establishing a "head." This is sometimes sufficiently great to cause the water to flow out under pressure and rise above the surface of the ground, when the aquifer is penetrated by a drill. Such a source of water is called an artesian well (see Fig. 1), from Artois, France, a classic locality for such waters. Artesian conditions exist along much of the Atlantic Coastal Plain of the United States and in North and South Dakota, Nebraska, Kansas, Illinois, Indiana, Missouri, and Arkansas. Since the supply of underground water is largely dependent upon structure, the geology of water supply is one of the most important economic phases of the earth sciences. From the point of view of their origin, groundwaters are classified as juvenile, connate, and meteoric. Juvenile waters are of volcanic or magmatic origin, hence original. Connate waters are those in which the sediments were originally deposited. Meteoric waters are those of atmospheric origin.

All pure water, and most of all of the underground waters are of meteoric or surface-water origin. See also **Wastes and Pollution**.

GROUP. A set of elements, finite or infinite in number, satisfying the following conditions: (1) There is a defined operation by which to each ordered pair of elements A and B in the group G there is associated an element C of G, denoted by $C = AB$, and called the product of A and B. (2) For this operation the associative law holds: $(AB)C = A(BC) = ABC$ for any three elements A, B, C of G. There exists: (3) a unit element E in G such that $EA = A$ for every element A of G, and (4) to each element A of G a reciprocal (or inverse) element A^{-1} of G such that $A^{-1}A = E$.

It must be understood that product, as defined in (1), is a convenient word to use for the result of combining two or more elements in a group but the law of combination is not confined to multiplication. For example, let the group elements be the integers $0, \pm1, \pm2, \ldots$ and let the combination law be addition, then the product of any two elements is their algebraic sum. These integers, regarded as elements of a group, will be seen to satisfy the requirements (1)–(4).

Infinite groups are discrete if the elements are denumerable; continuous, if they contain a nondenumerable infinity of elements. A finite group containing n elements is of order n. If $m < n$ elements satisfy the requirements of (1)–(4), they form a subgroup. Every group contains at least two subgroups: the unit element and the group itself.

The elements of a group may be symbols only, with no meaning attached to them and one then speaks of an abstract group. However, the elements may be numbers, matrices, geometrical operations, etc., and these are special groups.

If X is an element of a group G not contained in one of its subgroups H, then the set of elements HX is called a right coset and XH is a left coset. Cosets are not groups because they do not contain E, the unit element. Nevertheless, they are called "Nebengruppen" in German. If A, B, X are three elements of a group, then $B = X^{-1}AX$ is the transform of A by X and A, B are conjugate to each other. The complete set of group elements conjugate among themselves is a class of the group.

If H is a subgroup of the group G and X is an element of G, but not necessarily contained in H, then $X^{-1}HX$ is also a subgroup of G and a conjugate subgroup to H. If H and $H' = X^{-1}HX$ are conjugate then these two subgroups are invariant if $H = H'$. It is also called a normal subgroup or a normal divisor.

Suppose H is an invariant subgroup of a group G and that HX, HY, $\ldots$ are its cosets. The elements of H can be considered collectively as the unit element of another group and the various cosets as the remaining elements.

It is called the quotient or factor group and is often designated by **G/H**. The multiplication properties of this group are similar to those of **G**.

Given a group **G′** of order m with elements $A_1, A_2, \ldots, A_m$ and a second group **G″** of order n with elements $B_1, B_2, \ldots, B_n$ such that every element of **G′** commutes with every element of **G″**, then the mn element $A_i B_j$ for a group $\mathbf{G} = \mathbf{G}' \times \mathbf{G}''$ is of order mn and is called the direct product of **G′** and **G″**.

Many other types of groups have been studied. They are of interest in geometry, differential equations, topology, and other branches of mathematics. In physics and chemistry, groups are used in the study of quantum mechanics; molecular, crystal, and nuclear structure; electrical circuits, etc.

GROWTH REGULATOR (Plant).   See **Plant Growth Modification and Regulation**.

GRUNDMANN ALDEHYDE SYNTHESIS.

Transformation of an acid into an aldehyde of the same chain length by conversion of the acid chloride via the diazo ketone to the acetoxy ketone, reduction with aluminum isopropoxide and hydrolysis to the glycol, and cleavage with lead tetraacetate.

GUANIDINE.

[CAS: 113-00-8]. Guanidine, or carbamidine or iminourea, $(NH_2)C{=}NH$ is formed (1) by heating ammonium thiocyanate to 180°C, (2) by ammonolysis of orthocarbonates, $C(OC_2H_5)_4 + 3NH_3 \rightarrow (NH_2)_2C{=}NH + 4C_2H_5OH$, (3) by ammonolysis of chloropicrin, $Cl_3CNO_2 + 7NH_3 \rightarrow (NH_2)_2C{=}NH + 3NH_4Cl + N_2 + 3H_2O$, (4) by ammonolysis of cyanogen chloride, $ClCN + NH_3 \rightarrow ClC(NH_2){=}NH \rightarrow HN{=}C{=}NH \rightarrow (NH_2)_2 C{=}NH$.

Guanidine forms salts with acids, e.g., guanidine nitrate, $HNC(NH_2)_2 \cdot HNO_3$. By heating at 120°C for several hours, a mixture of ammonium thiocyanate and dicyanodiamide, guanidine thiocyanate solution is obtained by extracting with water. Treating guanidine with a mixture of nitric and sulfuric acids forms nitroguanidine

$$\left(\begin{array}{c} \quad\quad NH \cdot NO_2 \\ HN{:}C \\ \quad\quad NH_2 \end{array} \right)$$

which is reduced by zinc and acetic acid to aminoguanidine

$$\left(\begin{array}{c} \quad\quad NH \cdot NH_2 \\ HN{:}C \\ \quad\quad NH_2 \end{array} \right)$$

By treating aminoguanidine (1) with dilute acid or alkali, there is obtained, first, semicarbazide, finally hydrazine; (2) with nitrous acid, diazoguanidine

$$\left(\begin{array}{c} \quad\quad NHN{:}NOH \\ HN{:}C \\ \quad\quad NH_2 \end{array} \right)$$

which is decomposed by alkali into alkali azide (e.g., NaN_3) plus cyanamide ($H_2N \cdot CN$) plus water.

In the Pauling theory of its structure, guanidine is a resonance compound of the molecular structure cited $[(NH_2)_2C{=}NH]$ and two ionic structures in which the nitrogen of the imino group gains an electron lost by one of the amino groups.

The monoalkyl- and N,N-dialkyl guanidines are somewhat weaker bases than guanidine, because resonance of the double bond to the substituted $-NH_2$ group is restricted by the fact that carbon is more electronegative than hydrogen, and renders more difficult the acquisition of a positive charge by an adjacent nitrogen atom. This effect is still more marked with the N,N′-dialkyl guanidines, while, in contrast, the N,N′,N″-trialkyl guanidines are essentially as strong bases as guanidine.

Table 1 lists seven representative substituted guanidines.

GUMS AND MUCILAGES.

Natural gums and mucilages are carbohydrate polymers of high molecular weight obtained from plants. They can be dispersed in cold water to give viscous or mucilaginous solutions

TABLE 1. GUANIDINES

Guanidine	Formula	Melting point °C
1. Guanidine	$HN{:}C{\displaystyle <}^{NH_2}_{NH_2}$	
2. 1,3-diphenylguanidine	$HN{:}C{\displaystyle <}^{NHC_6H_5}_{NHC_6H_5}$	147
3. 1,1,3,3-tetraphenylguanidine	$HN{:}C{\displaystyle <}^{N(C_6H_5)_2}_{N(C_6H_5)_2}$	130
4. 1,2,3-triphenylguanidine	$C_6H_5N{:}C{\displaystyle <}^{NHC_6H_5}_{NHC_6H_5}$	144
5. 1,1,3-triphenylguanidine	$HN{:}C{\displaystyle <}^{NHC_6H_5}_{N(C_6H_5)_2}$	131
6. Guanylurea	$HN{:}C{\displaystyle <}^{NH_2}_{NHCONH_2}$	105
7. Aminoguanidine decomposes	$HN{:}C{\displaystyle <}^{NHNH_2}_{NH_2}$	

that normally do not gel. They are composed of acidic and/or neutral monosaccharide building units joined by glycosidic bonds. The acid groups ($-CO_2H$, $-SO_3H$) are usually present as salts of calcium, magnesium, sodium, and potassium; in certain cases substituents such as acetyl (karaya gum) and methyl groups (mesquite gum) may be present as well. Pyruvic acid residues, linked as ketals, are present in several cases (such as agar). The properties of several gums are described in Table 1.

Gums are of particular importance in the food processing field where they perform at least three functions—emulsifying, stabilizing, and thickening. A few also function as gelling agents, bodying agents, foam enhancers, and suspension agents. Gum guiac also serves as an antioxidant and preservative.

See separate entry on **Xanthan Gum**.

Sources of Gums

Gums and mucilages may be found either in the *intracellular parts* of plants or as *extracellular exudates*. Those found within plant cells represent storage material in seeds and roots. They also serve as a water reservoir and as protection for germinating seed. The polysaccharides found as extracellular exudates of higher plants appear to be produced as a result of injury caused by mechanical means or by insects. It has not been well established whether the exudates are formed at the site of the injury,

TABLE 1. GUMS AND MUCILAGES—PROPERTIES AND APPLICATIONS

ACACIA GUM (ARABIC GUM)

The dried water-soluble exudate from stems of *Acacia senegal* or related species. Thin flakes, powder, granules, or angular fragments; color white to yellowish white; almost odorless, mucilaginous taste. Completely soluble in hot and cold water, yielding a viscous solution of mucilage; insoluble in alcohol. Aqueous solution is acid to litmus. Produced in the Sudan, Nigeria, and other parts of west Africa. Used in adhesives, inks, textile printing, cosmetics; as a thickening agent and colloidal stabilizer in confectionery and other food products.

ALGINIC ACID $(C_6H_8O_6)_n$

White to yellow powder, possessing marked hydrophilic colloidal properties for suspending, thickening, emulsifying, and stabilizing. Insoluble in organic solvents; slowly soluble in alkaline solutions. Used in food industry as thickener and emulsifier; as a protective colloid; in tooth paste, cosmetics, pharmaceuticals, textile sizing, coatings; as a waterproofing agent for concrete; in boiler water treatment; in oil-well drilling muds; in storage of gasoline as a solid.

AGAR

Thin, translucent, membranous pieces or pale bluff powder. Strongly hydrophilic—absorbs 20 times its weight of cold water with swelling; forms strong gels at about 40°C. Agar (sometimes called agar-agar) is a phycocolloid derived from red algae, such as *Gelidium* and *Gracilaria*. It is a polysaccharide mixture of agarose and agaropectin. Agar is used as a culture medium in microbiology and bacteriology; as an antistaling agent in bakery products; in confectionery; in meats and poultry; as a gelation agent; in desserts and beverages; as a protective colloid in ice cream; in pet foods, health foods; as a laxative, in pharmaceuticals; for making dental impressions; as a laboratory reagent; in photographic emulsions.

CALCIUM ALGINATE

White or cream-colored powder, or filaments, grains, or granules. Slight odor and taste. Insoluble in water; insoluble in acids, but soluble in alkaline solutions. It is used in pharmaceutical products; as a food additive; as a thickening agent and stabilizer in ice cream, cheese products, canned fruits, and sausage casings also used in synthetic fibers.

CARRAGEENAN

A yellowish to colorless, coarse to fine powder, practically odorless, but with a mucilaginous taste. Moderately soluble (1 gram in 100 milliliters of water at 27°C), forming a viscous, clear, or slightly opalescent solution which flows readily. Carrageenan disperses in water more readily if first moistened with alcohol, glycerin, or a saturated solution of sucrose in water. Carrageenan is a hydrocolloid consisting mainly of a sulfated polysaccharide, the dominant hexose units of which are galactose and anhydrogalactose. It is a two-component, polyanionic colloid. The *kappa* and *lambda* components occur in varying proportions and degrees of polymerization and are associated with ammonium, calcium, potassium, or sodium ions, or with a combination of these four. Varying proportions alter the physical qualities of the substance. Carrageenan is obtained by extraction with water of members of the *Gigartinaceae* and *Solieriaceae* families of the class *Rhodophyceae* (red seaweed). The seaweed is also called Irish Moss and is prevalent off the coasts of Canada, New England, and New Jersey, but is found in other parts of the world. Carrageenan is used as an emulsifier in food products, especially chocolate milk; in toothpastes, cosmetics, pharmaceuticals; as a protective colloid; and as a stabilizing aid in ice cream (0.02%).

GUAR GUM

Yellowish-white powder. Dispersible in hot or cold water. It possesses 5–8 times the thickening power of starch. Reduces friction drag of water on metals. Guar gum is obtained from the ground endosperms of *Cyanopsis tetragonoloba*, which is cultivated in Pakistan and used there as a livestock feed. The water-soluble portion of the flour (85%) is called *guaran* and consists of 35% galactose, 63% mannose, probably combined in a polysaccharide, and $5\frac{7}{8}\%$ protein. Guar gum is used in paper manufacture; cosmetics; pharmaceuticals; as an interior coating of fire-bose nozzles; as a fracturing aid in oil wells, in textiles, printing, polishing; as a thickener and emulsifier in food products.

GUIAC GUM

Moderate yellow-brown powder, becoming olive brown upon exposure to air. Odor is balsamic. Taste is slightly acrid. Dissolves incompletely but readily in alcohol, ether, chloroform, and in solutions of alkalies. Slightly soluble in carbon disulfide and benzene. Occurs as irregular masses enclosing fragments of vegetable tissues, or in large, nearly homogenous masses. Source is resin of the wood of *Guajacum officinale*, principally found in Central America.

KARAYA GUM

A pale yellow to pinkish brown, translucent, and horny gum with a slightly acetous odor and a mucilaginous and slightly acetous taste. In powdered form it is light gray to pinkish gray. Karaya gum is insoluble in alcohol, but swells in water to form a gel. Karaya gum is obtained as a dried gummy exudate from *Sterculia urens* and other species of *Sterculiaceae* family, or from *Cochlospermum gossypium*. It occurs in tears of variable size or in broken irregular pieces having a somewhat crystalline appearance. The properties depend upon freshness and time of storage. Viscosity greatly decreases over a 6-month period. The gum is used in pharmaceuticals, textile coatings, ice cream and other food products, adhesives; as a protective colloid, stabilizer, thickener, and emulsifier.

LOCUST BEAN GUM (CAROB-BEAN GUM)

White to yellowish-white, nearly odorless powder. It is dispersible in either hot or cold water, forming a sol, having a pH between 5.4 and 7.0, which may be converted to a gel by the addition of small amounts of sodium borate. It has a molecular weight of about 310,000. The gum swells in water, but viscosity increases when heated. Insoluble in organic solvents. The gum is extracted from the ground endosperms of *Ceratonia siliqua* of the *Leguminosae* family. The gum is used in foods as a stabilizer, thickener, and emulsifier; in packaging material, cosmetics, sizing and finishes for textiles, pharmaceuticals, paints.

POTASSIUM ALGINATE

Occurs in filamentous, grainy, granular, and powdered forms. It is colorless or slightly yellow and may have a slight characteristic odor and taste. Slowly soluble in water, forming a viscous solution; insoluble in alcohol. The gum is used as a thickening agent and stabilizer in dairy products, canned fruits, and sausage casings. It is variously used as an emulsifier.

SODIUM ALGINATE

A colorless or slightly yellow solid occurring in filamentous, granular, and powdered form. Forms a viscous colloidal solution with water, insoluble in alcohol, ether, and chloroform. It is extracted from brown seaweeds. The gum is used as a thickener, stabilizer, and emulsifier in foods, especially ice cream. Also used in boiler compounds, pharmaceuticals, textile printing, cement compositions, paper coatings, and in some water-base paints.

TRAGACANTH GUM

Dull white, translucent plates or yellowish powder. Soluble in alkaline solutions, aqueous hydrogen peroxide solution; strongly hydrophilic; insoluble in alcohol. One gram in 50 milliliters of water swells to form a smooth, stiff, opalescent mucilage free from cellular fragments. It is obtained as a dried gummy exudate from *Astragalus gummifer*, or other Asiatic species of *Astragalus* (*Leguminosae* family). The gum is used in pharmaceutical emulsions, adhesives, leather dressings, textile printing and sizing, dyes, food products (notably ice cream and desserts), toothpastes; for coating soap chips and powders; and in hair wave preparations.

or whether they are generated elsewhere and then transported to the injured area.

The true exudates, such as gum arabic and the East African and Indian gums are picked by hand. Seldom are commercial samples pure. This is a serious disadvantage in product control. They are classified according to grade, which, in turn, depends upon color and contamination with foreign bodies, such as wood and bark. The exudates are processed simply by grinding, their only prior treatment being sorting and sometimes bleaching under the sun. In some cases, they are purified by extraction with water and precipitated by alcohol.

Gums and mucilages present in roots, tubers and seaweeds are usually extracted with hot water, dried, and marketed as a powder. Those gums found on the inner side of the seed coat as vitreous layers (e.g., locust bean, guar bean, etc.) are best obtained by a suitable milling process. This first removes the seed coat and then makes use of the fact that the gum layer is very hard and tough as compared with the seed endosperm. The intracellular gums and mucilages can be purified by precipitation with alcohol from aqueous solution as in the case of the plant gum exudates, or by a process such as acetylation. In a similar way, the bacterial polysaccharides can be precipitated from the cell-free culture fluid with alcohol, or as the salt of a quaternary ammonium compound where acidic groups are present.

Characteristics of Gums

The extracellular plant gums and mucilages (gum arabic, karaya gum, and tragacanth, for example) generally have a more complex structure than the intracellular types. They are made up of a number of different sugar-building units linked together by a variety of glycosidic bonds. They possess a central core or nucleus composed mainly of D-galactose and D-glucuronic acid units joined by glycosidic bonds, which are relatively stable to hydrolysis by acids. To this central nucleus are attached as side chains those sugar units, which are removed by mild acid hydrolysis. Thus, in the case of gum arabic, the acid-resistant portion of the molecule is composed of D-glucuronic acid and D-galactose; to this nucleus are attached units of L-arabinose, L-rhamnose, and D-galactopyranosyl $(1 \rightarrow 3)$ L-arabinose.

The neutral mucilages and gums, such as mannans, glactomannans, and glucomannans extracted from seed and roots, have a relatively simple structure. The kinds of building units are fewer and the molecules are much less branched. The galactomannans are usually composed of a backbone of linear chains of D-mannose units jointed by 1,6-glycosidic bonds, to which are attached at regular intervals side chains of D-galactose residues. The glucomannans are essentially linear polymers united by 1,4-linkages.

The algal polysaccharides resembled the relatively simplified structures of the neutral mucilages, as in the case of carrageenan. A wider spectrum of structures is found in the bacterial gums, which are generally of the highly branched type exuded by higher plants.

Food processing and other industrial applications of gums and mucilages take advantage of their physical properties, especially the viscosity and colloidal nature. They are substances of high molecular weight. For example, gum arabic has a molecular weight of 250,000 to 300,000. The gums and mucilages that possess relatively linear molecules, such as gum tragacanth, form more viscous solutions than the more spherically shaped gums, such as gum arabic, when at the same concentration. Consequently, for some applications, the gums with linear molecules are more economic to use. Due also to the elongated molecular shape of the seed gums and mucilages, the viscosity of their aqueous solutions varies widely with concentration. They exhibit structure viscosity. In contrast, the gums and mucilages of more spherical shape, i.e., the exudates, give solutions whose viscosities do not depend so much upon concentration.

Gums and mucilages influence each other. Mixing of two gums of the same viscosity may result in a mixture with a different viscosity. The viscosity of solutions of gums and the mucilages is dependent upon the pH, especially for those containing acid groups. In certain cases, the viscosity decreases upon standing as the result of enzymatic breakdown of the molecules. The molecules can undergo large changes in shape and size under the osmotic influence of opposing ions. Some of them, such as carrageenan from Irish Moss, can be fractionated by dilute salt solutions (potassium chloride) and the poly-β-glucosan from barley grain may be precipitated with ammonium sulfate. Gum arabic shows the phenomenon of coacervation when mixed with gelatin. See also **Coacervation**.

The specific uses of gums are wide and diverse. By way of a few examples, seaweed gums (e.g., carrageenan) and seed mucilages (guar gum) are used as stabilizers in dairy products, such as ice cream and certain cheeses. They are used in confectionery, in making jams, jellies, and in stabilizing citrus oil emulsions and salad dressings. They have been used as fixatives for 2,3-butanedione in the baking industry. Outside the food field gums and mucilages find scores of applications.

In 1974, the Northern Regional Research Center (Peoria, Illinois) of the U.S. Department of Agriculture and the Kelco Company were joint recipients of the Institute of Food Technologists award for the development and commercialization of xanthan gum. See also **Xanthan Gum**. This gum differs by virtue of its production by pure-culture fermentation of a carbohydrate as contrasted with refining a naturally occurring substance.

See also list of references under **Colloid Systems**. A particularly good reference covering the physical properties and procedures for testing various gums and mucilages is 'Food Chemicals Codex,' published by the National Academy of Sciences, Washington, DC, (revised periodically).

Additional Reading

Nussinovitch, A.: *Hydrocolloid Applications: Gum Technology in the Food and Other Industries,* Chapman and Hall, New York, NY, 1999.

Satterlee, D.:   www.herbaldave.com/Herbs/Phytochemicals/Classes/Gums+Muculage.htm, 2000.

Williams, P. and G. Phillips: *Gums and Stabilizers for the Food Industry,* Springer-Verlag, Inc., New York, NY, 1998.

GUTZEIT TEST.   A test for arsenic. Zinc and dilute sulfuric acid are added to the substance, which is then covered with a filter paper moistened with mercuric chloride solution. A yellow spot forms on the paper if arsenic is in the sample.

See also **Bettendorf's Reagent**.

GYPSUM.   [CAS: 10101-41-4]. The mineral gypsum is hydrous calcium sulfate, $CaSO_4 \cdot 2H_2O$. It occurs as flattened monoclinic crystals, often twinned, transparent cleavable masses, called selenite, or silky and fibrous, called satin spar; it may also be granular or quite compact. It is a soft mineral, hardness 2; has two good cleavages, which yield rhombic plates whose angles are 66° and 114°. Its specific gravity is 2.31–2.33; luster, vitreous to silky or pearly; color, colorless to white and gray, may be tinted red, yellow, blue, brown, etc., by impurities; transparent to opaque. A very fine-grained white or lightly tinted variety of gypsum is called alabaster, and prized for ornamental work of various sorts.

Gypsum is a very common mineral, thick and extensive beds of which are associated with sedimentary rocks. The largest deposits known occur in strata of Permian age. Besides being a result of deposition in sea and lake waters, gypsum has been deposited by hot springs, from volcanic vapors, and by sulfate solutions in veins. Notable localities for gypsum are in Greece, the Czech Republic and Slovakia, Austria, Saxony, Bavaria, Italy, France, Spain, England and Mexico. In the United States, well-known localities are at Lockport, New York; the Mammoth Cave, Kentucky; Ellsworth, Ohio; Grand Rapids, Michigan; Hermosa, South Dakota; Wayne County, Utah; and San Bernardino County, California. In Canada, the Provinces of New Brunswick and Nova Scotia have large gypsum deposits. Because the gypsum from the quarries of the Montmartre district of Paris has long furnished burnt gypsum used for various purposes, this material has been called plaster of Paris.

Often, there is confusion between the mineral gypsum, $CaSO_4.2H_2O$, and the useful product of partial dehydration, $CaSO_4.1/2H_2O$. See Table 1. There are numerous commercial products based upon gypsum. *Plaster*, made from gypsum, is widely used for the economical fabrication of building products. Importantly, the setting time of gypsum plaster can be carefully controlled through the addition of fractional percentages of *accelerators* (typically water-soluble salts, such as K_2SO_4, or finely-ground

TABLE 1. TERMINOLOGY AND PROPERTIES OF CALCIUM SULFATE-WATER COMPOUNDS

Chemical formula	Designations commonly used	Properties
$CaSO_4 \cdot 2H_2O$	Calcium sulfate dihydrate; rock gypsum; chemical gypsum; alabaster (white fine-grained); selenite (translucent platey); satin spar (fibrous); land plaster (pulverized gypsum)	All forms (natural, synthetic, and recrystallized) are thermodynamically and crystallographically equivalent. Habit may be needles, plates, or prisms.
$CaSO_4 \cdot 1/2 H_2O$	Calcium sulfate hemihydrate; calcined gypsum; stucco; plaster of Paris; molding plaster; gypsum plaster; chemical hemihydrate.	Alpha and beta types exist, depending upon conditions of calcination. Alpha type is more stable, crystalline, of lower energy. Beta type is less stable, disordered, of higher energy.
$CaSO_4$	Anhydrite	
I	Anhydrite 1: high-temperature anhydrite.	Produced by high-temperature (>1,000°C) calcining. Contains free CaO.
II	Anhydrite II: insoluble anhydrite; inactive anhydrite; dead-burned gypsum; chemical anhydrite; mineral anhydrite	Produced by calcining at 250–1,000°C. Relatively inert. Reactivity depends upon calcining-time-temperature relationship and particle size.
III	Anhydrite III: soluble anhydrite; active anhydrite; dehydrated hemihydrate.	Produced by low-temperature (175–250°C) dehydration of hemihydrate. Reacts vigorously with water and moist air to form hemihydrate.

Source: United States Gypsum Company, Des Plaines, Illinois.

gypsum) and *retarders*, which frequently are modified organic substances, such as glue, casein, blood, hair, and hoof meal; or citric, boric, and phosphoric acids and their salts. Accelerators are believed to function by providing additional nuclei for crystallization, whereas retarders are believed to provide protective colloids or insoluble salts which block water access to the plaster particle. A controlled rate of reaction can be obtained by incorporating a combination of retarders and accelerators in the gypsum plaster mix.

Wallboard (Sheetrock) is a large single user of gypsum. The product usually consists of a core of gypsum sandwiched between two layers of paper. Characteristics of the product include fire resistance, dimensional stability, low cost, and easy workability. Wallboard conventionally measures $\frac{1}{2}$ inch (1.3 centimeters) thick, 48 inches (1.2 meters) wide, and 8 to 20 feet (2.4 to 6 meters) in length. In manufacture, foamed plaster slurry is mixed and discharged on a moving web of paper. The edges of the bottom paper are scored and folded so that the slurry is completely contained between that sheet and the top paper, which is laid on the slurry. The paper surfaces not only provide strength and paintability to the finished board, but also form a continuous mold within which the gypsum is cast. The board machine operates continuously. Within five minutes after forming, the gypsum is sufficiently hard to be cut, after which the sheets are dried further before storage and shipment. Fibers may be added to provide crack resistance and additional fire resistance. Water-repellent chemicals may be added to the board core or to the paper surface. Also, decorative and functional finishes may be factory-applied.

Industrial plasters of a gypsum base include dental plasters, used in making tooth impressions, orthopedic plasters for immobilizing broken bones, pottery plasters, oil-well cements, permeable plasters for casting nonferrous metals, art and statuary casting, lamp bases, patching and grouting compounds, insulating-brick production, and pattern and model making for the aircraft and automotive industries. Water-reducing additives and reinforcing resins and cements may be added to achieve a compressive strength of over 15,000 pounds per square inch (1021 atmospheres).

Portland cement also consumes large quantities of gypsum. About 5% of gypsum is added to the cement clinker before grinding. Addition of gypsum aids in increasing the early strength of the cement and prevents undesirable false set.

Agriculturally, gypsum serves as a soil conditioner, providing a source of available calcium and sulfate, assisting the retention of organic nitrogen, without the addition of acidity or alkalinity to the soil. Gypsum is widely used in areas where the soils are deficient in sulfur. Gypsum also has been used in mixed fertilizers and animal feeds.

Terra alba or dead-burned, fine white gypsum is used as a paper filler, in plastics, and as an extender for titanium dioxide. Pharmaceutically pure gypsum can be added to bread and other bakery products, finds use in beer production, and as a pharmaceutical-tablet diluent. In Japan, calcium sulfate is used in making *tofu*, a soyabean curd.

Gypsum may be a potential source of sulfur and sulfuric acid. Some European plants make Portland cement and sulfuric acid from gypsum or anhydrite. In the Muller-Kuhne process, gypsum is mixed with clay and silica in quantities necessary to make cement, along with coke to reduce $CaSO_4$ to CaO. In equipment similar to that for portland-cement manufacture, the SO_2 is driven off and converted to sulfuric acid by the contact process.

Additional Reading

Coburn, A. et al.: *Gypsum Plaster: Its Manufacture & Use,* Intermediate Technology Publications, London, UK, 1989.

Gerhartz, W. (Editor): *Benzyl Alcohol to Calcium Sulfate, Vol. 4,* John Wiley & Sons, Inc., New York, NY, 1985.

Staff: *Cement, Lime, Gypsum,* American Society for Testing & Materials, West Conshohocken, PA, 2000.

Web References

Natural Resources Canada, Minerals and Metals Sector: http://nrcan.gc.ca/mms/efab/mmsd/minerals/gypsum.htm, 1999.

Technology Information, Forecasting & Assessment Council (TIFAC): http://www.tifac.org.in/offer/tlbo/rep/TMS149.htm

H

HABER-BOSCH PROCESS. See **Ammonia**.

HABER, FRITZ (1868–1934). Born in Breslau, Germany, Haber's great contribution to chemistry for which he was awarded the Nobel prize in 1918, was his development (with Bosch) of a workable method for synthesizing ammonia by the water-gas reaction from hot coke, air, and steam; the gas mixture obtained includes nitrogen from the air, as well as hydrogen from the steam. It was the first successful attempt to "fix" atmospheric nitrogen in an industrial process. This discovery was developed to production scale in approximately 1912; it enabled Germany to manufacture an independent supply of explosives for World War I.

HABIT. The type of geometric structure that a given crystalline material invariably forms, e.g., cubic, orthorhombic, monoclinic, tetragonal, hexagonal, etc. Each of these types has several subclasses. Thus, crystals may have the form of thin sheets or plates, cubes, rhomboids, and even more complicated geometric structures. For example, the crystalline habit of mica is monoclinic, with formation of extremely thin sheets.

See also **Crystal**.

HADRONS. These are subatomic particles, the strong interactions of which are manifested by the forces that hold neutrons and protons together in the atomic nucleus. Hadrons include the proton, the neutron, and pion, among others. These particles show signs of an inner structure, i.e., they are made up of other particles, which has led over a period of the last several years to consider the hadrons as combinations of constituents known as *quarks*. See also **Quarks**; and **Particles (Subatomic)**.

HAFNIUM. [CAS: 7440-58-6]. Chemical element symbol Hf, at. no. 72, at. wt. 178.49, periodic table group 4, mp 2207–2247°C, bp 4601–4603°C, density 13.3 g/cm^3. The alpha form of elemental hafnium has a close-packed hexagonal crystal structure; the beta form, a body-centered cubic structure. Metallic hafnium, like zirconium, exhibits passivity in air due to formation of adherent coatings of oxide or nitride. Urbain reported evidence of the element in 1911, but hafnium was not fully identified until 1923 by D. Coster and G.C. de Hevesy. The remarkable similarity between hafnium and zirconium accounts mainly for its late isolation, as compared with the majority of elements. In terms of abundance, there is an average of about 4 ppm hafnium in the earth's crust. The element occurs with zirconium in certain varieties of zircon, including malacon, cyrtolite, and alvite. One mineral found in Scandinavia, thortveitite, contains more hafnium than zirconium. Pegmatite, monazite, baddeleyite, and zerkelite also contain hafnium. First ionization potential 5.5 eV. Oxidation potentials Hf + $H_2O \longrightarrow HfO^{2+}$ + 2H + 4e$^-$ 1.68 V; Hf + 4OH$^- \longrightarrow$ HfO(OH)$_2$ + H_2O + 4e$^-$, 2.60 V. Electron configuration $1s^2 2s^2 2p^6 3s^2 4d^{10} 4s^{24} p^6 4d^{10} 4f^{14} 5s^2 5p^6 5d^2 6s^2$. Ionic radius Hf^{+4}, 0.75 Å. Other important physical properties of hafnium are given under **Chemical Elements**.

Hafnium usually is extracted from ores along with zirconium. In one process, zircon sand is broken down by carbiding or carbonitriding, followed by chlorination. The mixture formed is dissolved with a complexing agent, after which it is introduced into a liquid-liquid extraction process. The final product is HfCl$_4$. Fractional crystallization of the fluorides of hafnium and zirconium also is practiced. Metallic hafnium is made by the Kroll process in which the HfCl$_4$ is reduced in an inert atmosphere by magnesium. The hafnium sponge and magnesium chloride resulting is vacuum-distilled to accomplish the final separation. In a modified Kroll process, sodium or sodium amalgam may be used. The latter requires less rigid temperature and pressure control during processing, costs less, and introduces fewer impurities into the process. For further purification of hafnium metal, a number of methods have been used,

including electrorefining, arc and induction melting, zone refining, and the hot-wire or van Arkelde Boer process.

Uses

Compared with most metals, the annual production of hafnium is low. Mainly produced in the United States, France, and Russia, the combined production is in the range of 100 metric tons annually, or less. Several uses have been found for hafnium: (1) as a control material in water-cooled nuclear reactors. Also hafnium is an effective flux-depressor in a reactor for absorbing neutrons to decrease the peaks in neutron flux; (2) as a filament in gas-filled incandescent light bulbs; (3) as an alloying ingredient to add strength to tungsten and molybdenum filaments and electrodes used in high-pressure discharge tubes; (4) as a cathode in x-rays tubes; (5) as a getter material in vacuum tubes and systems; (6) as a minor alloying ingredient in nichrome heating elements where hafnium appears to significantly increase the lifespan of the elements; and (7) usually with zirconium, as an ingredient of several alloys.

Chemistry and Compounds

Hafnium metal dissolves in HCl (warm) and slowly in H_2SO_4, more rapidly if fluoride ion F$^-$ is present, forming compounds of HfO^{2+}, or fluoro complexes in the latter case. The metal resists the attack of weak acids and their salts.

Due to its $5d^2 6s^2$ electron configuration, hafnium forms tetravalent compounds readily, although the Hf^{4+} ion does not exist as such in aqueous solution except at very low pH values, the common cation being HfO^{2+} (or Hf(OH)$_2^{2+}$) and many of the tetravalent compounds are partly covalent. There are also less stable Hf(III) compounds. There is close similarity in chemical properties to those of zirconium due to the similar outer electron configuration ($4d^2 5s^2$ for zirconium) and the almost identical ionic radii (Zr^{4+} is 0.80 Å) the relatively low value for Hf^{4+} being due to the Lanthanide contraction.

With improved means to separate the compounds of these two elements, future research will yield more details of specific hafnium compounds. The methods of separation used effectively include ion exchange techniques, a particularly effective one using a column of silica gel, with a solution of the tetrachlorides in methanol as feed and a 1.9 N HCl solution as eluant for zirconium. Separations also have been accomplished through the distillation of the phosphorus oxychloride addition products.

See list of references at end of entry on **Chemical Elements**.

HALF-LIFE (Elements). See **Chemical Elements**.

HALIDES. A compound made up of a halogen (astatine, bromine, chlorine, fluorine, or iodine) and another element or radical may be termed a *halide*. Fundamentally, there are three classes: (1) the *ionic* (saline) halides, (2) the *covalent* (acid) halides, and (3) the *complex* halides. The ionic halides are most sharply characterized by the halides of the alkali and alkaline earth metals, plus those of certain Lanthanide and Actinide metals. They form ionic or semi-ionic crystals in the solid state, have high boiling points and melting points, and are soluble in polar solvents. Their bonding is electrovalent, varying in degree with the difference between the electronegativities of the halogen and the metal. Potassium iodide and silver fluoride are ionic, but silver iodide is essentially covalent. The fluorides exhibit a primarily ionic character for most of the metals, but the other halogens form fewer ionic compounds. The degree of ionicity varies down as well as across the periodic table.

The covalent (acid) halides have low boiling and melting points, are soluble in nonpolar solvents and insoluble in polar solvents, although they often react with the latter. The degree of covalence generally is greatest for the nonmetals. For a given nonmetal, the boiling point depends upon both the number of atoms of the halogen with which it is combined and the symmetry of the molecule. For example, the boiling points of bromine(I) fluoride, bromine(III) trifluoride, and bromine(V) pentafluoride, BrF, BrF$_3$ and BrF$_5$, are 20,135, and 40.5°C, respectively.

The complex halides are very numerous, because of the readiness with which halide ions form coordination compounds with metals. In general, stability of these complexes depends upon the size and electronic structure of the metal ion—the smaller cations form their more stable compounds with the smaller halide ions, notably with fluoride. With larger cations the order of stability is that of the ability to be polarized of the halide, i.e., decreasing from iodide to fluoride. The more electronegative transition elements form especially stable complexes; e.g., those of palladium, platinum, etc., PdCl$_4{}^{2-}$, PtF$_6{}^{2-}$, etc. The most common halo complexes have four or six halogen ions coordinated with the cation, although such complexes as those of copper, gold and mercury, e.g., CuI$_2{}^-$, AuCl$_2{}^-$, HgCl$_3{}^-$, etc., are notable exceptions.

See also **Bromine**; **Carbon**; **Chlorine**; **Chlorinated Organics**; **Fluorine**; **Halogen Group**; and **Iodine**.

HALITE (Rock Salt).

The mineral halite (rock salt) is naturally occurring sodium chloride, NaCl, common salt. It is isometric with cubic habit and cleavage. It is brittle; hardness, 2.5; specific gravity, 2.168; luster, vitreous; colorless when pure, but usually white, yellow, red, or blue. It is soluble in water. Halite occurs interbedded with sedimentary rocks in all parts of the world and in all but the very oldest rocks. It frequently occurs in association with anhydrite and gypsum. In the United States this type of "salt beds" has been exploited in Michigan, New York, Ohio, and Pennsylvania. Louisiana produces salt from great subsurface dome-shaped masses, often 2,000–4,000 feet thick. The salt domes of the Gulf Coastal Plain are particularly important as subsurface structures, on the flanks of which are apt to occur large and important pools of petroleum. Poland, Saxony, Austria, and France possess well-known deposits of salt, as well as the former U.S.S.R., England, Algeria, India, and China. Salt is chiefly used in cooking and as a preservative; in the manufacture of soda ash for the glass industry; and as a source of many sodium compounds. It derives its name from the halogen group of elements to which chlorine belongs.

See also **Sodium Chloride**.

HALL, CHARLES MARTIN (1863–1914).

A native of Ohio, Hall invented a method of reducing aluminum oxide in molten cryolite by electrochemical means. This discovery made possible the large-scale production of metallic aluminum and resulted in formation of the Aluminum Company of America. The process requires high electric power input. Hall is generally considered the founder of the aluminum industry.

See also **Hall Process**.

HALL EFFECT AND QUANTIZED HALL EFFECT.

In 1879, Edwin H. Hall (Johns Hopkins University), discovered that if a strip of gold leaf, carrying an electric current longitudinally, was placed in a magnetic field with the plane of the strip perpendicular to the direction of the field, the points directly opposite each other on the edges of the strip acquired different electric potentials; and that if such points were joined through a sensitive galvanometer, a feeble current would be indicated. In other words, the equipotential lines, ordinarily running across at right angles to the edges, were skewed into an oblique position, and the electric lines of flow in the plane of the strip were deflected to one side.

If one looks along the strip in the direction of the current, with the magnetic field directed downward, then, with strips of antimony, cobalt, zinc, or iron, the electric potential drop is toward the right and the effect is said to be positive. With gold, silver, platinum, nickel, bismuth, copper, and aluminum, it is toward the left, and the effect is called negative. The transverse electric potential gradient per unit magnetic field intensity per unit current density is called the "Hall coefficient" for the metal in question. Thus, the Hall coefficient R_H is defined as

$$R_H = \frac{E_y}{j_x H_z}$$

where E_y is the electric field developed in the y direction when a current of current density j_x flows in the x direction through a magnetic field H_z in the z direction. According to the free electron theory of metals, the Hall

coefficient should be given by

$$R_H = \frac{B}{ne}$$

where N is the number of free electrons per unit volume, of charge e (in esu), and c is the velocity of light. The observed result that for some metals the carriers would seem to have positive charges is explained by the band theory of solids. In a nearly filled band, the wave functions of the electrons near the top of the band are so modified that it is the holes in the band that behave like particles. Since a hole represents the absence of negative charge, it behaves as if positively charged. The Hall angle is the ratio of E_y (defined above) to the field E_x, generating the current in the magnetic field H_z. The Hall mobility is the mobility of the electrons or holes in a semiconductor as measured by the Hall effect.

A number of transducers utilize the Hall effect. Shown in Fig. 1 is a direct-current oscilloscope probe based on the effect. A steady direct current I_c is applied to one axis of the Hall generator and a magnetic field B, proportional to the current through the conductor, is applied to a second axis. An output voltage V_c is taken across the third axis of the Hall generator. The output voltage can be calculated from:

$$V_c = \frac{10^{-5} R_H}{t} I_c B$$

where V_c = Hall voltage, volts
R_H = Hall coefficient, cm^3 coulomb
t = thickness, cm
B = magnetic field density, kilogauss

Exploring the Complexities of the Hall Effect

Over the intervening century since Hall's discovery and notably since the advent of semiconductor technology, the Hall effect has inspired research. A number of related effects have been observed. One of these, for example, is the widely studied galvanomagnetic effect, referred to as *transfer magnetoresistance*. By shorting the Hall field or by choosing a disk geometry so that such a field does not exist, one obtains a "magnetoresistance" (more strictly, a *magnetoconductivity*) which does not saturate. This is called the Corbino magnetoresistance or Corbino effect. There are several thermal effects in a magnetic field that can produce transverse voltages or temperature gradients. These result from the velocity separation of charge carriers by the Lorentz force—the energetic ones going to one side, the slower ones going to the other. Temperature gradients are produced, and also electric fields. In the Righi-Leduc effect, a longitudinal temperature gradient produces a transverse temperature gradient (thermal analog of the Hall effect). In the Nernst effect, it produces a transverse electric field. In the Ettingshausen effect, a longitudinal electric current produces a transverse temperature gradient. This latter effect, if large, can disturb the Hall field, since the potential probes and leads are seldom made of the same material as the specimen. Therefore, the

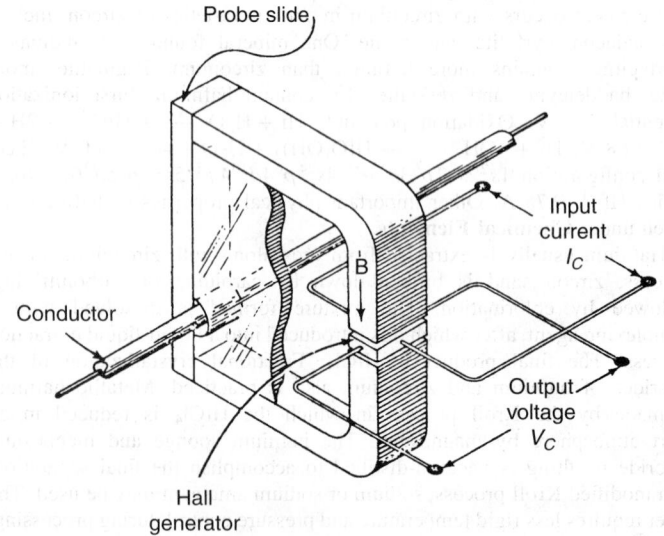

Fig. 1.   Direct-current oscilloscope probe based on Hall effect

Ettingshausen temperature gradient can produce a thermoelectric voltage, which adds to the Hall voltage.

Analysis of Hall-effect data has been one of the most widely used techniques for studying conduction mechanisms in solids, especially semiconductors. For the single-carrier case, one readily obtains carrier concentrations and mobilities, and it is usually of interest to study these as functions of temperature. This can supply information on the predominant charge-carrier scattering mechanisms and on activation energies, i.e., the energies necessary to excite carriers from impurity levels into the conduction band. Where two or more carriers are present, the analysis becomes more complex, but much more information can be obtained from studies of the temperature and magnetic field dependencies.

Unlike, for example, the magnetoresistance, the Hall effect is a first-order phenomenon. A weak magnetic field, it depends linearly on the magnetic field intensity and it does not vanish in isotropic solids if all the carriers have essentially the same velocity or if the scattering is characterized by a relaxation time which is independent of the carrier energy. As previously indicated, the Hall effect forms the basis of a number of devices used in isolating circuits, transducers, multipliers, converters, rectifiers, and gaussmeters (for measurement of magnetic fields). The fundamental component of such devices is a slab of material (often called a "Hall generator") possessing favorable Hall characteristics.

The Quantized Hall Effect

In 1980, Klaus von Klitzing[1] (High Magnetic Field Laboratory, Max Planck Institute), made some unusual findings while studying the Hall effect in devices in which the electrons free to carry current are confined within a *thin layer* of material. The researcher found that by cooling an experimental device to within a degree of absolute zero and by placing the device in a *very strong* magnetic field, the behavior of the ordinary resistance and the Hall resistance differed dramatically from that expected of a traditional Hall-effect device. Instead of increasing steadily and linearly as the strength of the magnetic field was increased, the Hall resistance increased in a *series of plateaus*. There were intervals observed in which the Hall resistance did not vary at all when the strength of the magnetic field was varied. Between the plateaus, the Hall resistance increased smoothly with increasing magnetic field. It was also found that during the same intervals of magnetic field strength during which the Hall resistance exhibited plateaus, the voltage drop parallel to the current was noted to disappear completely (no electrical resistance in sample and current flows without dissipating any energy).

The vanishing electrical resistance and the plateaus in the Hall resistance are remarkable phenomena. It is even more remarkable, as pointed out by Halperin (See reference), that, on each plateau, the value of the Hall resistance satisfies a remarkably simple condition. That is, the reciprocal of the Hall resistance is equal to an integer multiplied by the square of the charge on the electron and divided by Planck's constant (the fundamental constant of quantum mechanics). Each plateau is characterized by a different integer. Essentially, in such a system, the Hall resistance is reduced to the formula

$$R_H = \frac{1}{Nce}$$

where n = density of electrons (per square meter) in the sample. If the two-dimensional system is connected to an external reservoir of electrons and the magnetic field B is allowed to vary, then the density of electrons in the layer will vary with B in such a way as to minimize the combined energy of the layer and reservoir.

The degree of precision of the quantized Hall effect has amazed even the experts. Measured values of the Hall resistance at various integer plateaus are accurate to about one part in six million. The effect can be used to construct a laboratory standard of electrical resistance that is much more accurate than the standard resistors currently in use. Authorities also observe that, if the quantized Hall effect is combined with a new calibration of an absolute resistance standard, it should be able to yield an improved measurement of the fundamental dimensionless constant of quantum electrodynamics, the fine-structure constant α.

In his original experiment, von Klitzing used a silicon field-effect transistor (MOSFET) of exceptional quality and of the type used on integrated circuit chips. In the device, electrons are trapped in a so-called inversion layer near the surface of a silicon crystal that is covered with

[1] Nobel Prize winner for Physics, 1985.

a film of insulating silicon oxide, on top of which is deposited a metal "gate electrode," used to control the density of conduction electrons in the inversion layer.

A somewhat similar Hall effect phenomenon, known as the *fractional quantized Hall effect*, was observed at the National Magnet Laboratory (Cambridge, Massachusetts) by Tsui, Störmer, and Gossard (AT&T Bell Laboratories) a couple of years after von Klitzing's finding. The fractional quantized Hall effect was first noted in a heterojunction (an interface of crystals made of two different semiconducting materials). As pointed out by Halperin, in a heterojunction, electrons from one semiconductor are attracted to more energetically favorable locations in the other semiconductor. The positive charge thereby created in the "donor" semiconductor provides a force attracting the electrons back, however, and they become trapped in a thin layer at the interface of the two crystals.

Additional Reading

Abrikosov, A. and R.A. Silverman: *Quantum Field Theory,* Dover Publications, Inc., Mineola, NY, 1975.

Ando, T. et al.: *Mesoscopic Physics and Electronics,* Springer-Verlag, Inc., New York, NY, 1998.

Cerdeira, H.A., B. Kramer, and G. Schon: *Quantum Dynamics of Submicron Structures,* Kluwer Academic Publishers, Norwell, MA, 1995.

Chakraborty, T. and P. Pietilainen: *The Quantum Hall Effects: Integral and Fractional,* Vol. 85, Springer-Verlag, Inc., New York, NY, 1995.

Eisenstein, J.F. and H.L. Stormer: "The Fractional Quantum Hall Effect, *Science,* **1510** (June 22, 1990).

Ellis, P.J. and Y.C. Tang, Editors: *Trends in Theoretical Physics,* Addison-Wesley Longman, Inc., Reading, MA, 1990.

Falomir, H., R.E. Gomboa, and F.A. Schaposnki: *Trends in Theoretical Physics,* American Institute of Physics, College Park, MD, 1998.

Fisk, Z. et al.: *Physical Phenomena at High Magnetic Fields II,* World Scientific Publishing Company, Inc., Riveredge, NJ, 1996.

Halperin, B.I.: "Theory of the Quantized Hall Conductance," *Helvetica Physica Acta,* **56**(4603), 1241–1246 (June 17, 1983).

Halperin, B.I.: "The Quantized Hall Effect," *Sci. Amer.*, 52–60 (April 1986).

Halperin, B.I.: "The 1985 Nobel Prize in Physics," *Science,* **231**, 820–822 (1986).

Janssen, M. and J. Hajdu: *Introduction to the Theory of the Integer Quantum Hall Effect,* John Wiley & Sons, Inc., New York, NY, 1994.

Nicholas, R.J., Ed.: "Proceedings of the Fifth Int. Conf. on Electrical Properties of Two-Dimensional Systems," in *Surface Science,* 142 (1984).

Pinczuk, A. and S. Das Sarma: *Perspectives in Quantum Hall Effects,* John Wiley & Sons, Inc., New York, NY, 1996.

Schwartzschild, B.: "Von Klitzing Wins Nobel Prize for Quantum Hall Effect," *Physics Today,* **38**(12), 17–20 (December 1985).

Semenoff, G.W. and L. Vinet: *Particles and Fields,* Springer-Verlag, Inc., New York, NY, 2000.

von Klitzing, K., G. Dorada, and M. Pepper: *Phys. Rev. Lett.*, **45**, 494 (1980).

HÄLLEFLINTA. A Swedish term for hard, dense, metamorphic rocks composed chiefly of microscopic crystals of quartz and feldspar with occasional phenocrysts. Accessory minerals may be hornblende, chlorite, hematite or magnetite. The texture and composition of hÄlleflinta suggests that it is the metamorphosed equivalent of acid lava flows or tuffs.

HALL PROCESS. The electrolytic recovery of aluminum from bauxite or, more specifically, from the alumina extracted from it. See also Bayer process. A typical cell for this process consists of a rectangular steel shell lined with insulating brick and block carbon. The cell holds a molten cryolite-alumina electrolyte, commonly called the "bath." The carbon bottom is covered by a pad of molten aluminum and serves as the cathode. The anodes are pre-backed carbon blocks suspended in the electrolyte. The cathodic current is collected from the carbon bottom by embedded steel bars that protrude through the shell to connect with the cathode bus. During electrolysis, aluminum is deposited on the metal pad, and the oxygen, liberated at the anode, reacts with the carbon to form carbon dioxide, some of which is reduced to carbon monoxide by secondary reactions. At 24–48-h intervals, aluminum is tapped from the cell by a siphon. The process requires large amounts of electric power (from 4–5% of total U.S. production). Disposition of the toxic fluoride waste is a problem.

HALLUCINOGENS. There are many substances that will, if taken in appropriate quantities, produce distortion of perception, vivid images, or hallucinations. Most of these substances will produce powerful peripheral as well as the central effects. Some few agents are characterized by the predominance of their actions on mental and psychic functions. This group of drugs has been called hallucinogens, psychotomimetics, psycholytics, and psychodelics, among several ambiguous terms. None of these names is adequately descriptive of these compounds.

Hallucinogens may be classified into five groups of chemically distinct compounds: (1) lysergic acid derivatives, of which lysergic acid diethylamide (LSD-25) is the prototype; (2) phenylethylamines, such as mescaline; (3) indolealkylamines, which include psilocybin, psilocin, and bufotenin; (4) piperidyl benzilate esters, typified by ditran (a 70:30 mixture of N-ethyl-2-pyrrolidymethyl phenylcyclopentylglycolate and N-ethyl-3-piperidyl phenylcyclopentylglycolate); and (5) phenylcyclohexyl piperidines (sernyl). The chemical structures of these compounds are shown in Fig. 1.

Drugs from the first three groups have been isolated from naturally occurring sources. LSD-25 is a molecular component of ergot, a fungus that infects cereal grains. Mescaline, historically the oldest hallucinogen, was isolated from a Mexican peyote cactus. Psilocybin and psilocin were isolated from the Mexican mushroom, *Psilocybe mexicana*. Bufotenin is found in some varieties of toadstools. The indole derivatives are chemically closely related to serotonin (5-hydroxytryptamine), a compound which plays an important, yet unknown role in the central nervous system.

The piperidyl benzilate esters and phenylcyclohexyl piperidines are synthetic compounds, and have not been shown to occur naturally. Some authorities do not consider them to be hallucinogens, but active researchers in the field include them among the most active psychotomimetics.

Clinical syndromes from LSD-25, mescaline, and the indoleamines are similar. Somatic symptoms are nausea, dizziness, loss of appetite, blurred vision, paresthesia, weakness, drowsiness, and trembling. These result frequently and are usually associated with sympathomimetic effects, such as increased pulse rate and slight temperature elevation. Perceptual and psychic changes are marked. Visual illusions and vivid hallucinations, decreased concentration, slow thinking, depersonalization, dreamy states, changes in mood, and often anxiety are commonly found.

The clinical syndromes from ditran are different from those produced by the aforementioned drugs in some respects. Disorganization of thought, disorientation, confusion, mood changes, and visual and auditory hallucinations are observed. The piperidyl benzilate esters are central anticholinergics, and mental states produced by them are reminiscent of those from other anticholinergics, such as scopolamine.

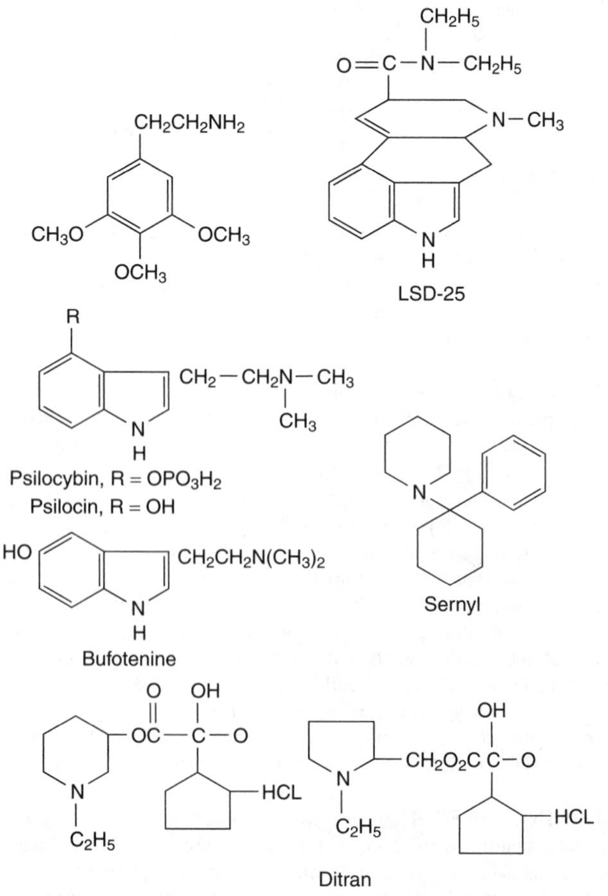

The effects of phenylcyclohexyl derivatives are also distinctive. Comparatively minor somatic symptoms are evoked. Psychic effects predominate, being typically characterized by feelings of unreality, depression, anxiety, and delusional or illusional experiences. The effects of these drugs are said to be more analogous to natural psychoses than those of the other drugs; however, the same claim has been made for ditran.

When LSD-25 was discovered, it was believed that the drug would provide an extremely useful tool in the investigation of psychoses and mental illness. However, therapeutically, the hallucinogens, including LSD-25, have been of little value to psychiatrists.

Additional Reading

Laing, R. and J.A. Siegel: *Hallucinogens: A Forensic Handbook,* Elsevier Science, New York, NY, 2001.
Robbins, P.R.: *Hallucinogens,* Enslow Publishers, Inc., Berkeley Heights, NJ, 2001.

HALOCARBON. A compound containing carbon, one or more halogens, and sometimes hydrogen. The lower members of the various homologous series are used as refrigerants, propellant gases, fire-extinguishing agents, and blowing agents for urethane foams. When polymerized, they yield plastics characterized by extreme chemical resistance, high electrical resistivity, and good heat resistance.

See also **Fluorocarbon.**

HALOGENATED COMPOUNDS. See **Chlorinated Organics.**

HALOGEN GROUP. The elements of group 17 (formerly 7a) of the periodic classification sometimes are referred to as the Halogen Group. The individual elements commonly are called *halogens*. In order of increasing atomic number, they are fluorine, chlorine, bromine, iodine, and astatine. The elements of this group are characterized by the presence of seven electrons in an outer shell, and hence have the ability to gain an electron to form negative ions with a completed octet of valence electrons. The halogens present striking similarities of chemical behavior, all being very reactive and, in particular, readily form substitution compounds with numerous organic compounds. Although these elements also have other valences all have a −1 valence in common.

HAMILTON, ALICE (1869–1970). The first American physician to devote her life to the practice of industrial medicine. In studying the lead industries in Illinois, she discovered and ameliorated lead poisoning among bathtub enamelers in Chicago. She wrote about phossy jaw, which occurred among American matchmakers who used white or yellow phosphorus. She studied the effects of carbon monoxide among steel workers, the toxicity of nitroglycerin among munitions makers during World War I, the symptoms of hatters exposed to mercury in Danbury, Connecticut, and the "dead fingers" syndrome of workers utilizing the early jackhammers. She also described the toxic effects to the blood-forming cells from benzol, and the neurologic and psychological responses of workers in the viscose rayon industry. In 1919, Dr. Hamilton was appointed assistant professor of industrial medicine at Harvard Medical School. The first woman on the Harvard faculty, she gave occupational medicine respectability as an academic pursuit.

HANN, OTTO (1879–1968). A German physical chemist who won the Nobel prize for chemistry in 1944 for his discovery of the fission of heavy nuclei and the principle of the chain reaction. Well-known for work on nuclear fission he discovered protactinium and transuranium elements with atomic numbers 94, 95, and 96. After receiving his doctorate at the University of Munich, he worked in Canada before returning to Europe.

HARDENABILITY OF STEEL. The hardenability of steel refers to the ease with which it can be hardened rather than the maximum hardness value attainable. For example, a 1-inch diameter bar of a certain 0.20% carbon alloy steel can be hardened to 50 Rockwell "C" in the center by quenching in oil. A similar bar of plain carbon steel requires a drastic quench in brine to attain the same hardness, and therefore, has a lower hardenability. Neither bar can be quenched to a greater hardness because 50 Rockwell "C" is the maximum attainable for a 0.20% carbon steel. A

Fig. 1. Structures of some hallucinogenic drugs

0.40% carbon steel can be hardened to a maximum of about 60 Rockwell "C" and the maximum for high-carbon steel is about 65 Rockwell "C."

Of the several methods for determining the relative hardenability of steels, the Jominy test is the most widely used. A cylindrical specimen 1 inch (2.5 centimeters) in diameter and about 3 inches (7.6 centimeters) long is heated to the hardening temperature and quenched in a special fixture which holds the specimen in a vertical position and directs a stream of water on the bottom surface. The stream takes an "umbrella" shape and does not wet the sides. Cooling occurs progressively from the bottom to the top of the cylinder and the cooling rate at any distance from the bottom is known and reproducible from one sample to another. The hardness along the length of a quenched Jominy bar decreases from bottom to top. The distance from the bottom, expressed in sixteenths of an inch, to the point where the hardness is 50 Rockwell "C" is one method of reporting the hardenability.

One of the principal functions of alloying elements in steel, such as manganese, chromium, nickel, molybdenum, etc., is to increase the hardenability. Whereas prodigious amounts of expensive alloys were formerly used to insure full hardening, especially in medium and heavy sections, wartime shortages focused attention on the use of as little alloy as possible within the hardenability requirements. A large number of steels were developed containing relatively small additions of a number of elements, and a number of these steels have continued in use.

HARDEN, SIR ARTHUR (1865–1940). An English chemist who won the Nobel prize in chemistry in 1929 along with Hans von-Euler-Chelpin. He discovered fermentation enzymes and demonstrated the structure of zymase. His fermentation work proved how inorganic phosphates speeded the process. Born in England, he received his doctorate in Germany.

HARDENING OF METALS. There are three principal methods of hardening metals and alloys: cold working by plastic deformation, precipitation hardening, and quench hardening as applied to steel. The last two methods involve heating and cooling operations. A pure metal may also be hardened through the addition of alloying elements. When a solid solution is formed it is normally harder than the pure metal. If additional phases are formed by alloying, these may also be harder than the pure metal and contribute to the hardness of the metal.

HARDNESS. The significance of this term as applied to solids has various interpretations. Commonly, it refers to the resistance of the substance to surface abrasion, so that of two solids, the one that will scratch the other, as diamond scratches glass, is the harder. Again, it may denote rigidity, or lack of plasticity, or even strength; in some cases a combination of several such properties. The original Mohs' Scale of Hardness is delineated in Table 1 and further described under **Mineralogy**.

In metallurgy and engineering, hardness is determined by methods based on resistance to penetration by an indenter of greater hardness than the material being tested. Aluminum, copper, lead, magnesium, tin, and their alloys, as well as plastics are generally indented by hardened steel balls ranging in size in the various tests from $\frac{1}{16}$ inch to 10 millimeters in diameter. The same methods may be used for soft steels and irons, but for heat-treated steels and all other alloys that develop high hardness special diamond indenters, or in some cases sintered tungsten carbide balls, are used. In all of the technological tests, the indenters are impressed into the test material under carefully regulated loads; thus, the relative size of the resulting indentation becomes a measure of hardness. (See Table 2.) The operating principles of the instruments most widely used in this country follow:

Brinell

The indenter is a 10-millimeter diameter hardened steel ball. A sintered tungsten-carbide ball is also coming into use, especially for testing hard metals. The load applied is generally 500 kilograms for soft metals and 3,000 kilograms for steels and hard metals. Brinell hardness is equal to the load (kilogram) divided by the surface area (square millimeter) of the impression made in the test material. Tables are available for direct conversion to hardness from the diameter of the indentation as measured with a calibrated magnifier after removal of the piece from the testing machine.

TABLE 1. HARDNESS SCALES

Moh's scale	Ridgway's extension of Mohs' scale	Metal equivalent	Others
1. Talc			
2. Gypsum			
			2.5. Finger Nail
3. Calcite			
4. Fluorite			
5. Apatite			
			5.5. Window Glass
6. Feldspar (Orthoclase)	6. Orthoclase or Periclase		
			6.5. Steel (Knife Blade; File)
7. Quartz	7. Vitreous Pure Silica		
8. Topaz	8. Quartz	8. Stellite	
9. Corundum or Sapphire	9. Garnet		
	10. Topaz		
	11. Fused Zirconia	11. Tantalum Carbide	
	12. Fused Alumina	12. Tungsten Carbide	
	13. Silicon Carbide		
	14. Boron Carbide		
10. Diamond	15. Diamond		

1. In the above scales each abrasive is capable of scratching all others above it in each scale and may be scratched by all abrasives below it.
2. The gap between 9 and 10 in the original Mohs' scale is much greater than that between 1 and 9 in the same scale.
3. Various Additional hardness scales have been devised by different investigators; in general, different materials maintain the same order of hardness in all these scales.

TABLE 2. TYPICAL HARDNESS VALUES

Material	Brinell		Rockwell	Vickers 50 kg
	500 kg	3000 kg		
Aluminum, annealed	23		H 45	25
Magnesium alloy	63		B 21	63
Armco iron	66	73	B 31	71
Yellow brass, annealed	72	82	B 40	77
Copper, cold rolled	99	83	B 55	110
Mild steel, annealed	107	117	B 70	123
Aluminum alloy, 24st	130	144	B 78	146
Stainless steel, annealed	121	145	B 80	153
Yellow brass, cold rolled	174	178	B 91	189
Ni-Moly steel, quenched in water, tempered at 1200°F (649°C)		241	C 23	255
Same, 1000°F (538°C)		293	C 31	310
Same, 800°F (427°C)		363	C 38	380
High-speed tool steel		684	C 62	740

Rockwell

Indenter is $\frac{1}{16}$-, $\frac{1}{8}$-, or $\frac{1}{4}$- inch-diameter (1.6, 3.2, or 6.4 millimeter) steel ball or a conical diamond having an apex angle of 120° and a slightly rounded point. The various scales used are designated by letters. Rockwell "B," for example, indicates a 100-kilogram load on a $\frac{1}{16}$-inch (1.6 millimeter) diameter ball. Rockwell "C" indicates a 150-kilogram load on the diamond indenter. Rockwell "30T" designates a load of 30 kilograms on a $\frac{1}{16}$-inch (1.6 millimeter) diameter ball. (An instrument of higher sensitivity known as the Rockwell Superficial Tester is used for loads of 15, 30, and 45 kilograms.) The size of the indentation is measured by a dial gauge as the final depth minus a small preliminary penetration produced by a minor preload of 10-kilograms. The Rockwell hardness values are arbitrary numbers having an inverse relationship to the depth of the indentation.

Vickers

Also known as Diamond Pyramid Hardness. Indenter is a square-based diamond pyramid with included angle between faces of 136°. Loads may vary from 1 to 120 kilograms with 10, 30, and 50 kilograms in common use. Hardness is equal to load (kilograms) divided by surface area (square millimeter) of the permanent indentation. It is determined directly from optical measurements of the diagonals of the indentation, which appear square at the surface of the metal.

Tukon

A highly sensitive instrument for determining hardness under very light loads down to 25 grams. The small indentations are measured at high magnifications up to 1,000 times. The indenter is a diamond pyramid that makes an elongated impression, one diagonal being 7 times the other in length.

Eberbach

Also used for very light loads. Consists of a spring-loaded, Vickers-type diamond pyramid indenter arranged for use on a metallurgical microscope.

Scleroscope

Depends on the height of rebound of a diamond-tipped body falling under the force of gravity from a fixed height. The instrument is relatively small and is portable. One type reads directly on a graduated dial.

While there is overlapping in the field of useful application of the various hardness tests, each has certain special qualifications. The Brinell test makes a large indentation, giving an average hardness value for several grains even in rather coarse-grained metals; however, it cannot be used on small or thin specimens. The various Rockwell tests are widely used, especially for rapid production inspection of parts. The Vickers test, which originated in England, is less rapid than the Rockwell but has the advantage of a single scale covering the hardness of all metals from lead to the hardest tool materials. The Tukon test makes it possible to determine the hardness of very thin sheets and of thin metallic coatings such as chromium plate, or zinc on galvanized steel. The Scleroscope test is used principally on heavy forgings or castings that cannot be placed in an indentation-type instrument, or for field tests where a portable instrument is required.

HARDNESS (Mineral). See **Mineralogy**.

HARGREAVES PROCESS. The manufacture of sodium sulfate (salt cake) from sodium chloride and sulfur dioxide. A mixture of sulfur dioxide and air is passed over briquettes of sodium chloride in a countercurrent manner to produce sodium sulfate and hydrogen chloride. This process accounts for only a small amount of the salt cake produced in the U.S.

HARMOTOME. The mineral harmotome is a zeolite, composition approximately (Ba, K) (Al, Si)$_2$Si$_6$O$_{16}$·6H$_2$O; it is monoclinic but often forms double twins giving the effect of a square prism. It is a brittle mineral; hardness, 4.5; specific gravity, 2.41–2.50; luster, vitreous; color, white to gray or perhaps yellow, red or brown; white streak; translucent. Harmotome like other zeolites is found in cavities in basalts and similar rocks, sometimes in trachytes or in gneisses, occasionally as a gangue mineral in veins of metallic minerals. Some well-known localities are in Bavaria; the Harz Mountains; Norway; and Scotland. Harmotome occurs in the United States with stilbite, near Port Arthur, Lake Superior. The name harmotome comes from the Greek meaning joint and to cut, referring to the division of the pyramid formed by the prismatic faces of the mineral when in the twinned position.

HASSEL, ODD (1897–1981). A Norwegian chemist who won the Nobel prize for chemistry in 1969 with Derek Barton for their contributions to the development of the concept of conformation and its application in chemistry. A great deal of his work was concerned with using X-ray and electron differentiation methods of crystal and molecular structures. He also researched stereochemistry and conformational analysis. His education and teaching career were in his homeland.

HASSIUM. See **Chemical Elements**.

HASTELLOY. See **Nickel**.

HAUPTMAN, HERBERT A. (1917–). An American biophysicist who won the Nobel prize for chemistry in 1985 along with Jerome Karle for their outstanding achievements in the development of direct methods for the determination of crystal structures. Hauptman's work involved developing equations that allow determination of phase information from X-ray crystallography intensity patterns. The use of computers permitted use of the equations to determine the conformation of thousands of chemicals. Hauptman was director of research and vice president of the Medical Foundation of Buffalo and a professor of biophysics in Buffalo at the State University of New York.

HAWORTH, SIR WALTER N. (1883–1950). An English chemist who received the Nobel prize in chemistry in 1937 along with Paul Karrer. He recommended the name *ascorbic acid* and synthesized vitamin C. He accomplished much work on carbohydrate structure and developed a substitute for blood plasma using carbohydrates. During World War II, he developed gaseous diffusion separation on uranium isotopes. He received his Ph.D. in Manchester, England.

HEAT. The agency whose addition to or removal from a physical system is the cause of thermal changes of various types. These include rise and fall of temperature, changes in length and volume, changes of physical states, such as melting, evaporations, etc.

During the eighteenth century heat was assumed to be a subtle fluid called *caloric*, filling the interstices between the ultimate particles of matter and, under conditions of isolation from the surroundings, known to satisfy a conservation law. The production of heat by friction as well as its disappearance during the performance of external mechanical work established its essential physical nature as another form of *energy* and led to the overthrow of the caloric theory. Nevertheless, we still speak of the *flow* of heat as though it were a fluid and have retained the methods of measuring the *quantity of heat* originally devised by the upholders of the caloric view.

Our direct knowledge of heat is provided by the sensation of hotness and coldness when we come in contact with various physical bodies. It is possible to arrange a set of bodies in a sequence such that A feels hotter than B, B hotter than C, etc. We say that A has a higher *temperature* than B, B a higher one than C, and so on. Of course our sensations are qualitative and are considerably influenced by the thermal conductivity of the body we touch. Thus, on a frosty morning, the head of an ax being metal feels considerably colder than the wooden handle though the two are presumably at the same temperature. To obtain a continuous and reproducible physical scale of temperature, various types of thermometers have been devised of which the mercury-in-glass or colored-alcohol-in-glass are familiar examples. The two temperature scales in common use are the Fahrenheit scale and the Celsius scale. The first assigns values of 32° and 212° to the normal freezing and boiling points of pure water, respectively, and divides this interval into 180 equal subintervals or degrees. The Celsius, formerly called the Centigrade scale assigns the respective values of 0° and 100° to the above fixed points; the standard interval is then divided into 100 equal degrees.

Temperature changes are produced by the addition or subtraction of heat from a body. Thus, temperature may be regarded as a measure of the concentration or *intensity* of heat. In general, the more heat we add to a given body the more its temperature rises.

Measurement of Heat

Since heat is imponderable and not directly observable, it is necessary to measure the size of a given quantity of heat by its effect on another body. If this effect is the production of a rise in temperature from some initial temperature, t_1, to a final temperature, t, then the rise $(t - t_1)$ is found to vary inversely with the mass of the test body. It is thus natural, following the calorists, to regard the quantity of heat, say Q, as determined by the product of m and $(t - t_1)$. Thus we say

$$Q \text{ is proportional to } m \times (t - t_1)$$

To make this statement into an equation we write

$$Q = \text{constant} \times m \times (t - t_1) \qquad (1)$$

where the constant of proportionality depends on the substance, being large for some materials and small for others. This constant for water, for example, is about 33 times as great as for lead; water is said therefore to have a greater *heat capacity* than lead. Notice that the constant in Equation (1) actually gives the numerical value of Q, which is required to warm a unit mass of the substance through a temperature interval of exactly 1°. This constant is accordingly called the *specific heat capacity* (usually abbreviated to *specific heat*) and is indicated by c. Since it is found that the value of the specific heat, particularly for gases, but in principle for all materials, depends on the conditions under which the heat is absorbed, this must be indicated. We thus have c_p and c_r, for example, for the two important cases of absorption at constant pressure and constant volume, respectively. Since the former characterizes the common laboratory case of working under atmospheric pressure, we accordingly rewrite Equation (1) as

$$Q_p = c_p m(t - t_1) \qquad (2)$$

Q_p now measures the heat absorbed under constant pressure, and c_p is the constant pressure specific heat. Since the right side of Equation (2) contains *three* quantities, a mere choice of a mass unit and a degree unit is insufficient to establish a unit of heat. It is necessary to select some substance as a standard reference body and assign an arbitrary value of, say c_p equal to unity for it. Water is the universal choice for this standard body due not only to its cheapness and ease of purification, but also to its large heat capacity.

With the selection of water as the standard with $c_p = 1$, the left side of Equation (2) clearly becomes of unit value when m and $(t - t_1)$ are each of unit value. In the English system, we accordingly have the *British thermal-unit* (or Btu) as the heat required to warm 1 pound of pure water through an interval of 1°F. In the metric system, the corresponding unit is the *calorie*, the heat required to warm 1 gram of water 1°C. A large unit or *kilocalorie* corresponding to 1,000 ordinary calories is also frequently used in scientific work.

Specific Heats

Use of Equation (2) reveals that the values of c_p obtained experimentally depend on the temperature interval used, indicating a dependence of c_p on temperature. Thus, if c_p for water were actually uniform throughout the 0 to 100°C range, a mass of water at 100°C mixed with an equal mass at 10°C would give a final mixture at exactly 50°C. The actual value is near 50.05°; this difference although small, indicates the need to specify the calorie at some particular temperature. For this purpose, we suppose a system of mass m is warmed from t to $t + \Delta t$ by the addition at constant pressure of an increment of heat ΔQ_p. Then Equation (2) becomes

$$\Delta Q_p = m \bar{c}_p \Delta t \qquad (3)$$

where now $\bar{c}_p$ is an average value of c_p over this interval. Then we define the *instantaneous* heat capacity, c_p at t by the following relation

$$c_p \frac{1}{m} \lim_{\Delta t \to 0} \frac{\Delta Q_p}{\Delta t} = \frac{1}{m} \frac{dQ_p}{dt}$$

i.e., the heat absorbed per unit mass per degree as the interval becomes smaller and smaller without limit. This leads to the differential form of Equation (3)

$$dQ_p = m c_p dt \qquad (4)$$

where dQ_p is the differential heat absorption which produces a differential temperature rise dt in a body of mass m and specific heat c_p.

The standard or 15° calorie is now defined as the rate of absorption of heat per gram per degree at 15°C and in practice is essentially the same as the average calorie over the 1° interval from 14.5 to 15.5°C.

If a mass m of water is warmed from t_1 to t, the integral of Equation (4) gives for the total heat absorbed in 15° calories

$$Q_p = \int_{t_1}^t dQ_p = m \int_{t_1}^t c_p dt = m \left[\int_{0°}^t c_p dt - \int_{0°}^t c_p dt \right] \qquad (5)$$

where the integral of c_p over the range t_1 to t has been written as the difference of two integrals from a common lower limit of 0°C. If, therefore, we evaluate an integral of the type $\int_0^t c_p dt$ with t varying in 1° steps and arrange these in a table, the right side of Equation (5) may be evaluated by merely subtracting appropriate entries.

In Fig. 1, the value of c_p in 15° calories per gram per degree is plotted graphically from 0 to 100°C, and the integrals on the right of Equation (5) are represented by appropriate areas under the c_p curve. Thus the integral from 0° to t is hatched with lines sloping up to the right, while that from

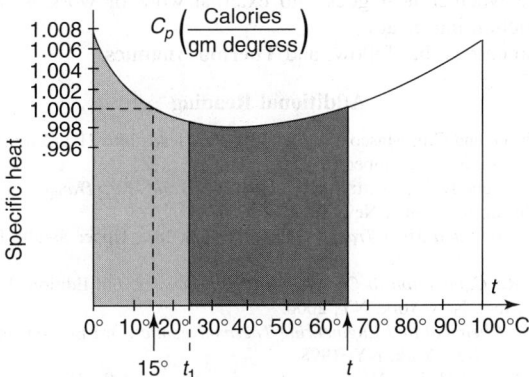

Fig. 1. Specific heat of water versus temperature

0° to t_1 has the lines sloping up to the left. The value of Q_p is then the singly hatched area.

With heat quantities measured in 15° calories, from the observed rise or fall of temperature in known masses of water, the specific heats of various substances—the heats absorbed on melting solids to liquids (heats of fusion), the heats absorbed on passage from the liquid to the vapor state (heats of vaporization), the heats evolved on combination of various substances, and the heats absorbed or evolved in chemical changes—are at once determinable (see also **Calorimetry**). For the present purpose, Table 1 gives the values of the constant pressure heat capacities of a few typical substances, variations with temperature being disregarded. Notice that c_p, although expressed in terms of calories per gram per degree, is in fact independent of the system of units since water is the reference body in all systems. Thus the specific heat of water in the English system would be 1 Btu per pound per degree Fahrenheit.

The Mechanical Nature of Heat

The conservation of heat *per se* is observed only for systems involving the performance of no mechanical or electrical work. Count Rumford (ca. 1800) was the first to establish this fact in his famous cannon-boring experiments carried out in the arsenal of the Duchy of Bavaria in Munich. He observed that when his drills became dull, heat was produced in great quantities limited only by the amount of work done against friction. He concluded that the large scale mechanical energy used in overcoming friction could only be converted into the motions of the ultimate particles of matter, a motion not directly observable but detected by our senses as heat. His results were confirmed and extended by the later work of Joule and Helmholtz, in particular, and also provided a more reliable value for the so-called *mechanical equivalent of heat*. This is taken as the amount of mechanical (or electrical) energy which when converted into heat is equivalent to exactly 1 calorie. The presently accepted value for this important constant is 4.185 joules per 15° calorie. Here the joule is the work performed when power is expended at the rate of 1 watt for 1 second. Thus an ordinary 100-watt lamp bulb converts 100 joules of electrical energy to thermal each second; this amounts to 100/4.185 or about 24 calories.

As a result of experiments such as these and a host of others, we are forced to recognize that heat is merely another form of the universal

TABLE 1. APPROXIMATE CONSTANT-PRESSURE SPECIFIC HEAT OF SELECTED MATERIALS

Substance	State	c_p (cal/g deg)
Water	Vapor	0.48
Water	Liquid	1.00
Water	Solid	0.50
Ethyl alcohol	Liquid	.54
Hydrogen	Gas	3.44
Air	Gas	.24
Aluminum	Solid	.22
Iron	Solid	.11
Lead	Solid	.03

quantity *energy*. Its transformation always occurs at the rate of 4.185 joules per calorie whether heat goes into external work or work is dissipated through friction into heat.

See also entries that follow; and **Thermodynamics**.

Additional Reading

Butterworth, D. and C.F. Mascone: "Heat Transfer Heads Into 21st Century," *Chem. Eng. Progress,* **30** (September 1991).

Incropera, F.P. and D.P. Dewitt: *Introduction to Heat and Mass Transfer,* 3rd Edition, John Wiley & Sons, Inc., New York, NY, 2000.

Rolle, K.C.: *Heat and Mass Transfer,* Prentice-Hall, Inc., Upper Saddle River, NJ, 1999.

Rosenberg, R.: *Companion to Chemical Thermodynamics,* 6th Edition, John Wiley and Sons, Inc., New York, NY, 2000.

Sandler, S.I.: *Chemical and Engineering Thermodynamics,* 3rd Edition, John Wiley & Sons, Inc., New York, NY, 1998.

Smith, J.M.M. and H. Van Ness: *Introduction to Chemical Engineering Thermodynamics,* The McGraw-Hill Companies, Inc., New York, NY, 2000.

Welty, J.R. et al.: *Fundamentals of Momentum, Heat, and Mass Transfer,* 4th Edition, John Wiley & Sons, Inc., New York, NY, 2000.

HEAT CAPACITY. The amount of heat necessary, to raise the temperature of a system, entity, or substance by one degree of temperature. It is most frequently expressed in calories per degree centigrade or Btu per degree Fahrenheit. If the mass of a substance is specified, then certain derived values of the heat capacity can be obtained, such as the atomic heat, molar heat, or specific heat.

HEAT CAPACITY EQUATION (Einstein). A quantum relationship for the heat capacity at constant volume of an element of the form:

$$C_v = 3R \left(\frac{h\upsilon}{kT} \right)^2 \left(\frac{e^{h\upsilon/kT}}{(e^{h\upsilon/kT} - 1)^2} \right)$$

in which C_v is the heat capacity at constant volume for one gram-atom of an element, R is the gas constant, h is Planck's constant, k is the Boltzmann constant, υ is the characteristic frequency of oscillation of the atoms of the element, T is the absolute temperature, and e is the natural logarithmic base.

The Einstein equation was the first approximation to a quantum theoretical explanation of the variation of specific heat with temperature. It was later replaced by the Debye theory of specific heat and its modifications.

HEAT CONSERVATION. See **Insulation (Thermal)**.

HEAT CONTENT. See **Enthalpy**.

HEAT EXCHANGER. A vessel in which an outgoing hot liquid or vapor transfers a large part of its heat to an incoming cool liquid; in the case of vapors, the latent heat of condensation is thus utilized to heat the entering liquid. The shell-and—tube type is widely used; here the hot liquid or vapor is contained in the shell while the cool liquid passes through the tubes, which are usually arranged in coils for maximum contact with the heat source. Heat exchangers are used in many chemical operations, e.g., evaporation and pulp manufacture, as well as to produce steam from the heat developed in nuclear reactors for power generation.

See also **Evaporation**; and **Heat Transfer**.

HEAT SINK. 1. In thermodynamic theory, a means by which heat is stored, or is dissipated or transferred from the system under consideration.

2. A place toward which the heat moves in a system.

3. A material capable of absorbing heat; a device utilizing such a material and used as a thermal protection device on a spacecraft or reentry vehicle.

4. In nuclear propulsion, any thermodynamic device, such as a radiator or condenser, that is designed to absorb the excess heat energy of the working fluid. Also called *heat dump*.

HEAT TRANSFER. Although there are three generally accepted methods for transferring heat from one medium to another, or from one locale to another within a given medium, it is uncommon for one method to act unilaterally. Particularly where convection may predominate, some conduction of heat will be involved. In conduction, heat must diffuse through material substances; in convection, heat is essentially carried from one locale to another by actual movement of the transport medium; in radiation, heat transfer involves radiant wave energy.

Conduction

From a microscopic standpoint, thermal conduction refers to energy being handed down from one atom or molecule to the next one. In a liquid or gas, these particles change their position continuously even without visible movement and they transport energy also in this way. From a macroscopic or continuum viewpoint, thermal conduction is quantitatively described by Fourier's equation, which states that the heat flux q per unit time and unit area through an area element arbitrarily located in the medium is proportional to the drop in temperature, -grad T, per unit length in the direction normal to the area and to a transport property k characteristic of the medium and called *thermal conductivity*:

$$q = -k \text{ grad } T \tag{1}$$

Predictions for the value of the thermal conductivity k can be made from considerations of the atomic structure. Accurate values, however, require experimentation in which the heat flux q and the temperature gradient, grad T, are measured and these values are inserted into Fourier's equation. Thermal conductivity values for a number of media over a large temperature range are shown in Fig. 1. Metals have the largest conductivities and, among these, pure metals have larger values than alloys. Gases, in contrast, have very low heat conductivity values. Electrically nonconducting solids and liquids are arranged in between. The low thermal conductivity of air is utilized in the development of thermally insulating materials. Such materials, like cork or glass fiber, consist of a solid substance with a very large number of small spaces filled by air. The thermal transport occurs then essentially through the air spaces, and the solid structure only supplies the framework that prevents convective currents. It will be noted that the thermal conductivities indicated in Fig. 1 (at ambient temperature) extend through five powers of 10. This range is still small when compared with the range for the electric conductivity of various substances, where electric conductors have values that are larger by 25 powers of 10 than electric insulators. As a consequence, it is much easier to channel electricity along a desired path than to do so with heat, a fact that accounts for the difficulty in accurate experimentation in the field of heat transfer.

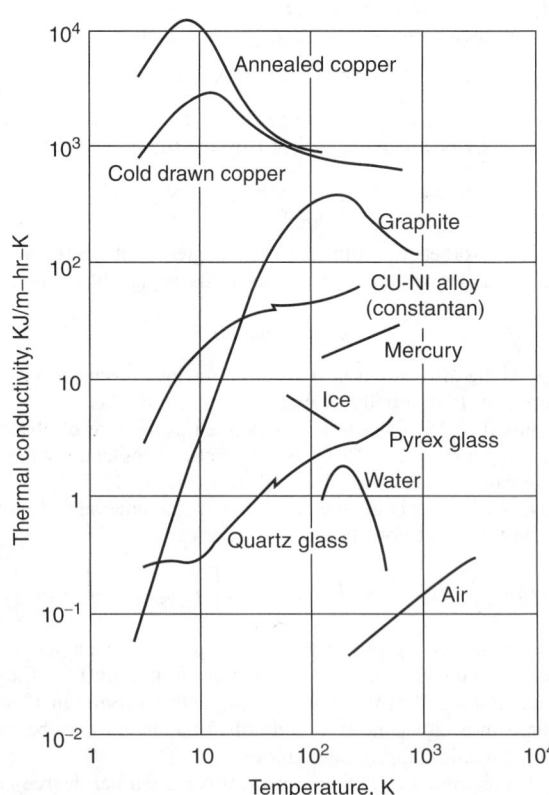

Fig. 1.   Thermal conductivity values for a wide range of substances and over a temperature range of 1 to 10^4 K

Fourier's equation can be used together with a statement on energy conservation to derive a differential equation describing the temperature field in a medium. Fourier was the first person to develop this equation and to devise means for its solution. In vector notation, this equation is:

$$\rho c = \frac{\partial \mathbf{T}}{\partial t} = \nabla(k\nabla \mathbf{T}) \qquad (2)$$

where ρ is the density, c is the specific heat, t is time, and ∇ is the Nabla (vector differential) operator. The temperature field in a substance can either change in time (unsteady state), or it can be independent of time (steady state, $\partial \mathbf{T}/\partial t = 0$). For a steady-state situation, the temperature field depends primarily on the geometry of the body involved and on the boundary conditions. The simplest case of a steady-state temperature field is a plane wall with temperatures that are uniform on each surface, but different at the two surfaces. The temperature in the wall then changes linearly in the direction of the surface normal as long as the variation of the thermal conductivity in the temperature range involved can be neglected. For an unsteady process, the capacity of the medium to store energy enters the energy conservation equation; correspondingly, the specific heat of the material and its density become factors for the conduction process, as well as the thermal conductivity. A combination of these properties, defined as the ratio of the thermal conductivity to the product of specific heat and density, called *thermal diffusivity* ($k/\rho\, c$), then determines how fast existing temperature differences in a medium equalizes in time. It is found that metals and gases have thermal diffusivity values approximately equal in magnitude and are considerably higher than thermal diffusivities of liquid and solid nonconductors. This means that temperature differences equalize much faster in metals and gases than in other substances.

Various other physical processes lead in their mathematical description to equations of the same form as Eq. (2), especially in its steady-state form. Such processes include the conduction of electricity in a conductor, or the shape of a thin membrane stretched over a curved boundary. This situation has led to the development of analogies (electric analogy, soap film analogy) to heat conduction processes, which are useful because they often offer the advantages of simpler experimentation.

Convection

When energy is transported by convection in fluids, conduction usually takes care of the transport of heat from one stream tube to another and is the dominating mode of transfer near solid walls. Convection transports heat along the stream lines and is dominating in the main body of the fluid where the velocities are large. In many situations, the flow is turbulent; this means that unsteady mixing motions are superimposed on the mean flow. These mixing motions contribute also to a transport of heat between stream tubes, a process which can be described by an "effective" conductivity which often has values by several powers of ten larger than the actual conductivity of the fluid.

Movement of the fluid may be generated by means external to the heat transfer process, as by fans, blowers, or pumps. It may also be created by density differences connected with the heat transfer process itself. The first mode is called *forced convection*; the second one *natural* or *free convection*. Convection heat transfer may also be classified as heat transfer in *duct flow*, or in *internal flow* (over cylinders, spheres, air foils, and similar objects). In the case of external flow, the heat transfer process is essentially concentrated in a thin fluid layer surrounding the object (boundary layer).

Of special interest in such heat transfer processes is the knowledge of the heat flux from the surface of a solid object exposed to the flow. This heat flux q_w per unit area and time is conventionally described by Newton's equation:

$$q_w = h(T_w - T_f) \qquad (3)$$

where T_w is the surface temperature and T_f is a characteristic temperature in the fluid. This equation defining the heat transfer coefficient h is convenient, because, in many situations, the heat flux is at least approximately proportional to the temperature difference $T_w - T_f$. Information on the heat transfer coefficients can be obtained by a solution of the Navier-Stokes equation describing the flow of a viscous fluid and the related energy equation, or they are found by experimentation. Computers enhance the ability to study heat transfer analytically at least for laminar flow, whereas in turbulent flow the bulk of the information is determined experimentally. Experimentation is difficult because of the large number of parameters involved. Dimensional analysis has been applied to reduce the number of influencing parameters, and relations for convective heat transfer are correspondingly presented in many handbooks as relations between dimensionless parameters. Such an analysis demonstrates that heat transfer in forced flow can be described by a relation of the form

$$Nu = f(Re, Pr) \qquad (4)$$

in which the Nusselt number Nu is a dimensionless parameter hL/k, containing the heat transfer coefficient h, the Reynolds number describes essentially the nature of the flow, and the Prandtl number $Pr = c_p\mu/k$ can be considered a dimensionless transport property characterizing the fluid involved. L and V are an arbitrarily selected characteristic length and velocity, respectively; ρ denotes the density, μ the viscosity, and C_p the specific heat of the fluid at constant pressure. See also **Reynolds Number**.

Convection is frequently thought of in terms of space heating and industrial heat-exchange processes. It should be pointed out that convection plays a cosmic role (in the sun's photosphere, for example), and a very large role in connection with the atmosphere of the earth and some other planetary bodies. For example, when normal convective transport is inadequate, temperature inversions occur and create smog hazards over large cities.

Attempts to develop a theory for convection date back at least to the 1790s when Thompson (Count Rumford) introduced the concept of heat convection. Very little theoretical work was undertaken, however, until the early 1900s, when BÉnard (France) undertook experimental investigations. Modern convection physics stems from the work of Lord Rayleigh, who first published on the subject in 1916. In current times, advanced convection research studies have been undertaken by Velarde and Normand (1980), among others. See reference listed. See also **Heat**.

Radiation

In the transfer of energy from one location to another in the form of photons (electromagnetic waves), usually a multiplicity of wavelengths is involved. In vacuum, all waves regardless of their wavelength move with the same speed (2.9977×10^8 meters per second). In various substances, the wave velocity c changes somewhat with wavelength, and the ratio of the wave velocity in vacuum to the velocity in a substance is equal to the optical refraction index. Air and generally all gases have refractive indices that differ from one only in the fourth decimal. Their wave velocity is therefore practically equal to that in vacuum.

Prévost's principle states that the amount of energy emitted by a volume element within a radiating substance is completely independent of its surroundings. Whether the volume element increases or decreases its temperature by the process of radiation depends upon whether it absorbs more foreign radiation than it emits or vice versa. One refers to thermal radiation when the emission of photons is thermally excited, i.e., when the substance within the volume element is nearly in thermodynamic equilibrium. For such radiation, Kirchhoff was able to derive a number of relations by consideration of a system of media in thermodynamic equilibrium. If jv indicates the coefficient of emission, i.e., the radiative flux at the frequency v^* emitted per unit volume into a unit solid angle, and η is the coefficient of absorption at the same frequency, i.e., the fraction of the intensity of a radiant beam that is absorbed per unit path length, then one of these relations states

$$c^2 \frac{jv}{\eta_v} = f(T, v) \qquad (5)$$

with c denoting the wave velocity. According to this relation, the combination of parameters on the left-hand side of Eq. (5) is a function of temperature T and frequency v of the radiation only, but does not depend upon the substance under consideration. Kirchhoff's law can also be expressed in parameters that refer to the interface of two media (1 and 2). It then takes the form

$$c^2 \frac{i_v}{\alpha_v} = f(T, v) \qquad (6)$$

in which i_v is the monochromatic intensity of the radiative flux at frequency v originating in medium 2 and traveling through the interface into medium 1 per unit solid angle and area normal to the direction of the radiant beam. α_v is the monochromatic absorptance or absorptivity, i.e., that fraction of a radiant beam approaching the interface in the medium 1 in the opposite direction that is absorbed in medium 2. The wave velocity in medium 1 is *c.* Kirchhoff's law states that the combination of the parameters on the left-hand side of Eq. (6) is again a function of temperature and frequency only, but does not depend upon the nature of the medium. A medium which absorbs all the radiation traveling into it through an interface ($\alpha_v = 1$) is called a *blackbody*. The intensity of radiation emitted by an arbitrary medium is, according to Eq. (6), in the following way related to the intensity of radiation i_{bv} emitted by a black body at the same temperature and frequency:

$$\frac{i_v}{v} i_{bv} \qquad (7)$$

The amount of heat transferred by radiation can be determined by use of the *Stefan-Boltzmann Law*

$$Q = bA(T_1^4 - T_2^4) \qquad (8)$$

where Q is the amount of heat transferred per unit time, b is a constant, A is the area of the radiating surface, T_1 is the absolute temperature of the radiating body and T_2 is the absolute temperature of the receiving body. Various correction factors are introduced into the formula to account for the shape of the bodies, their thermal radiation characteristics and the properties of the media through which the radiant rays must pass while traveling from radiator to absorber. The thermal radiation characteristics are its emissivity, a measure of its ability to radiate at a given temperature, its absorptivity, a measure of its ability to absorb heat and its reflectivity, which measures its ability to reflect without absorbing.

Radiant energy travels in a straight line. Therefore to transmit it to an object out of sight of the radiator requires a reflector, such as a furnace wall, to deflect the rays to their objective.

It is possible to set up controlled laboratory radiation between simple plane surfaces and determine therefrom accurate coefficients to incorporate into radiation equations. However, the radiation of heat from furnace gases, consisting of non-luminous gases, luminous carbon particles in flame, ash globules, etc., to the walls and tubes of a steam generator in commercial operation at variable load, is another matter. Here, empirical data which are gathered and interpreted from field tests on similar equipment, must still be resorted to however great the designer's urge to go back to basic laws of heat transfer.

Radiant heat transfer in furnaces is roughly proportioned to the difference in the fourth power of the absolute temperatures of the radiating and receiving surfaces. The water wall surface is approximately at boiler saturation temperature, while the superheater surface varies from this to somewhat above the temperature of the steam at the superheater outlet. However, the mean radiating temperature of the furnace gases is usually over 1204°C. The fourth power of the receiving surface temperature is thus seen to be small compared to the fourth power of the transmitting surface temperature; consequently the latter controls the transmittance, and boiler tube temperature does not need to be considered a variable to be accounted for.

Figure 2 shows some of the arrangements in which radiant heat-absorbing surface is disposed. It may be used to illustrate another of the difficulties that beset the designer in following a rational or semi-rational form of radiation analysis. Projected radiant surface is one thing; actual radiant energy receiving surface may be quite a different area. For example, suppose the tubes of case (a) to be separated and spaced l_1 inches on centers. The *projected* areas of cases (a) and (c) would then be the same, but it seems obvious that re-radiation from the wall causes more of a (c) tube to receive radiant energy than is the case with an (a) tube. Also, if δ is a factor correcting projected area to *equivalent* absorbing surface, what value should be assigned to it in the case of a bank of tubes which may receive by re-radiation some radiant energy deep in the tube bank? Here δ has a minimum value of 1, but some investigators have derived expressions which indicate that δ may have a magnitude of 3 or more.

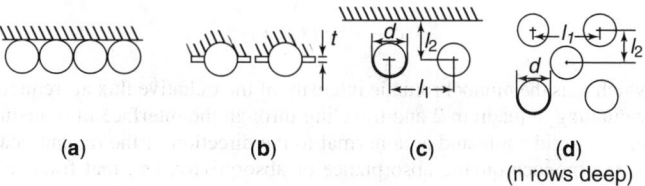

(a)   (b)   (c)   (d)
(n rows deep)

Fig. 2.   Arrangements of radiant heat-absorbing surface

Industrial Heat Transfer Equipment

Some of the more common cases of industrial heat transfer are:

1. Radiation from fuel beds and luminous gases to absorptive surfaces such as boilers, cylinder walls, etc.
2. Radiation from heat generators such as drying lamps.
3. Convection of heat out of combustion regions.
4. Convection of heat from hot surfaces under either free or forced convection.
5. Conduction of heat through the tubes of boilers, heaters, heat exchangers, condensers, etc.
6. Conduction in walls, pipe covering, and other so-called "heat insulators."
7. Conduction of heat through the plates of plate-type heat exchangers and regenerators.

Heat exchangers perform many functions within a manufacturing facility. Often they are given special names, even though they remain fundamentally heat exchangers. These include:

Chiller—a device which cools fluids to temperature below those obtainable with ordinary cooling water by using the vaporization of a refrigerant. The fluid to be cooled is routed through the tubes while the low-boiling refrigerant vaporizes from a pool of liquid in the shell.

Partial Condensers—Many overhead vapors from distillation columns in petroleum-refinery services are a mixture of light and heavy hydrocarbons and noncondensable gases, i.e., gases that are not condensed at the outlet temperature and pressure of the condenser (air, hydrogen sulfide, methane, and other light ends). These vapors are routed through the shell side while water is used as the cooling medium on the tube side of the unit. Condensation on the shell side begins at the saturation temperature of the heavy components and continues over a decreasing temperature range until part of the lighter components are condensed. Part of the existing liquid is sent back to the tower as reflux, while the remainder is further refined or passes to the trim cooler and storage.

Trim Cooler—This unit condenses the last remaining light-end vapors and cools the liquid to the ultimate storage temperature (often about 100°F: 38°C) by using cooling water. This cooling usually is not conducted in the main condenser because it would reduce column pressure.

Thermosiphon Reboiler—Flow of the vaporizing fluid depends upon the difference in static head between the column of liquid flowing from the tower to the reboiler and the partially vaporized column of liquid returning from the exchanger to the tower.

Reboilers—These exchangers operate in conjunction with a distillation tower to vaporize enough liquid to assure vaporization of the overhead product. A hot process stream of steam may be used as the heating medium. Most reboilers are shell-and-tube exchangers located at the base of the tower. The vaporizing fluid is routed through the shell side of the exchanger.

Forced-circulation Reboiler—A pump is used to provide more positive circulation than available with the thermosiphon effect, e.g., in the vaporization of viscous fluids.

Vapor Heat Exchanger—Units of this type preheat a cool stream of process fluid by using heat from partially condensing vapor. The objective is to conserve heat and eliminate the requirement for a separate preheater.

Air-cooled Exchanger—As used in the petroleum industry, air-cooled exchangers normally comprise two headers joined by a horizontal bank of finned tubes. Usually two motor-driven fans located above (induced draft) or below (forced draft) the tubes are used to circulate the air over the finned surface.

Superheater—A unit of this type heats vapor above the saturation temperature.

Waste-heat Boiler—A unit of this type generates steam and is similar to a regular steam generator except that hot gas or liquid produced by a chemical reaction (often combustion) is the heating medium.

Types of Heat Exchangers. In terms of heat exchange for recovering and recycling thermal energy, the shell-and-tube heat exchanger of the

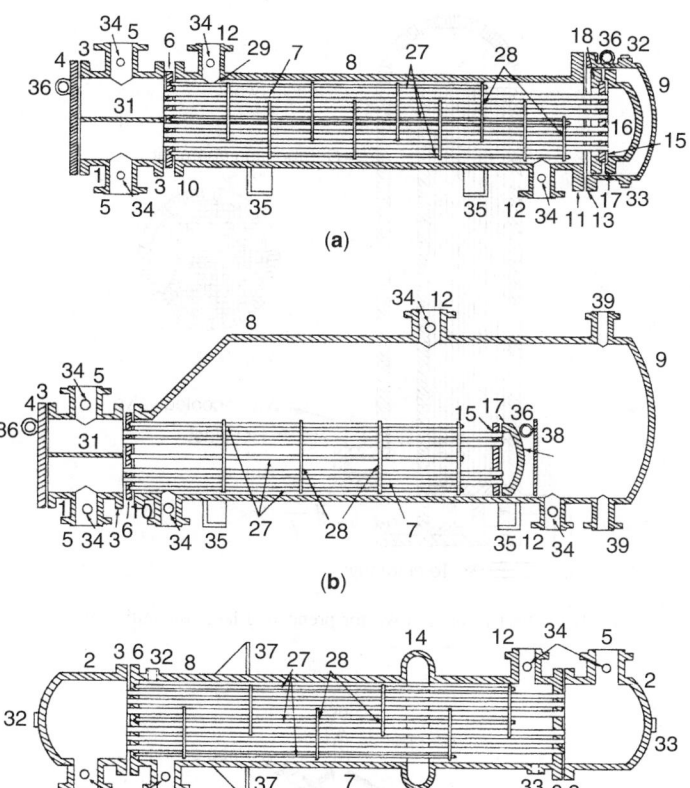

(a)

(b)

(c)

Fig. 3.   Three common types of shell-and-tube heat exchangers: (**a**) Type AES, internal-floating-head exchanger (with floating-head backing device); (**b**) Type AKT, kettle-type floating-head reboiler; and (**c**) Type BEM, fixed-tube sheet exchanger. (*Sketches adapted from specification diagrams from the Standards of Tubular Exchange Manufacturers Association.*) Legend: (1) Stationary head, channel, (2) stationary head bonnet, (3) stationary-head flange, channel, or bonnet, (4) channel cover, (5) stationary-head nozzle, (11) stationary tube sheet, (12) tubes, (13) shell, (9) shell cover, (10) shell flange, stationary-head end, (11) shell flange, rear-head end, (12) shell nozzle, (13) shell cover flange, (14) expansion joint, (15) floating tube sheet, (16) floating-head cover, (17) floating-head flange, (18) floating-head backing device, (19) split shear ring, (20) slip-on backing flange, (21) floating-head cover, external, (22) floating tube sheet skirt, (23) packing-box flange, (24) packing, (25) packing follower ring, (26) lantern ring, (27) tie rods and spacers, (28) transverse baffle, (29) impingement baffle, (30) longitudinal baffle, (31) pass partition, (32) vent connection, (33) drain connection, (34) instrument connection, (35) support saddle, (36) lifting lug, (37) support bracket, (38) weir, and (39) liquid-level connection

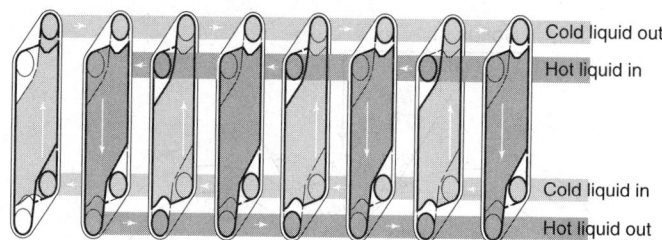

Fig. 4.   Principle of plate-and-frame heat exchanger. (*After Carlson*)

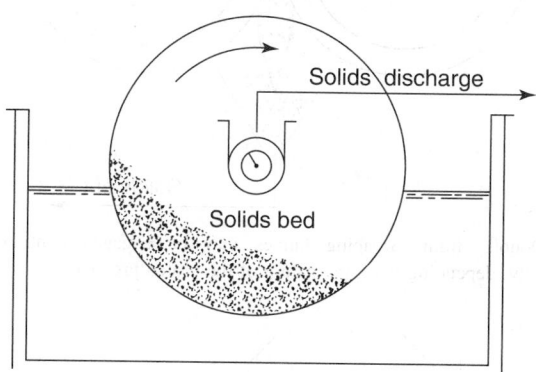

Fig. 5.   Plain rotating shell used for both heating and cooling. For high-range heating, tempered combustion gases may be used instead of water

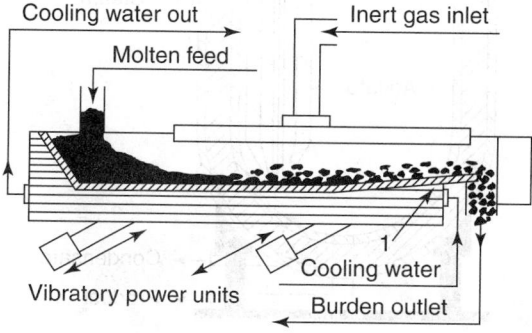

Fig. 6.   Vibrating-type heat-transfer equipment for batch solidification. Sometimes referred to as a *caster*, the machine is used widely in a number of industries. After cooling and solidification, intense vibratory action shatters cake into lumps

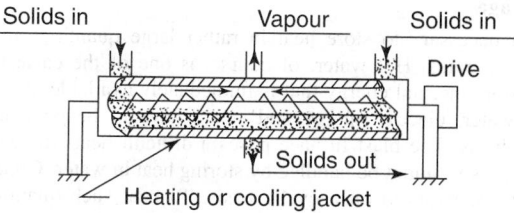

Fig. 7.   Tank equipped with mixing ribbon spirals provides considerable agitation and is useful for melting or cooking dry powdered solids. Heat-transfer efficiency is only moderate because of the relatively deep beds of solid particles

type shown in Fig. 3 for many decades has been the most common type. It can be used with liquid on both sides, gas on both sides, or liquid on one side and gas on the other side. The most common requirement is for liquid-liquid exchangers. Heat exchanges may be used strictly for processing purposes—that is, materials need to be heated (or cooled) prior to entering some processing application, such as reacting, distilling, vaporizing, and the like. Or heat exchangers may be used simply for recovering the energy from hot fluids for use elsewhere.

Although immensely improved over the years from the standpoint of design efficiency, resistance to corrosion, and ease of maintenance, among other objectives, the fundamental design has remained unchanged. However, within the past few years, the plate-and-frame heat exchanger has been introduced. See Fig. 4. This type of exchanger consists of a frame that carries a series of closely spaced metal plates that have been pressed, with a corrugated trough pattern. The plates, which are clamped between a fixed head and movable follower, have corner ports to permit the passage of process and service liquids. There are elastomeric gaskets around the ports and plate edges to avoid leakage. The plates are grouped into passes within the heat exchanger. The product and service fluids

flow countercurrent to each other between the parallel passages in each pass. Initially, plate-and-frame exchangers were used for liquid-liquid thermal exchange purposes. Increasingly, they are finding use in condensing and boiling applications, where their compact size and thinner material requirements for wetted parts offer advantages over other types of heat-exchange designs.

Less commonly used are the heat-transfer configurations shown in Figs. 5 through 9.

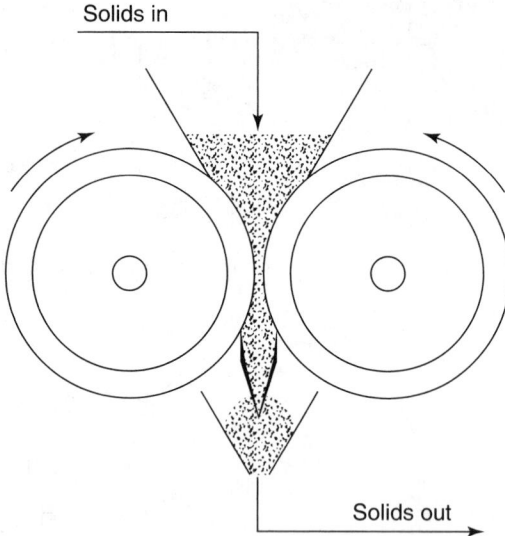

Fig. 8. Double drum. Scraping knives may be engaged continuously or intermittently, depending upon the nature of the heated product

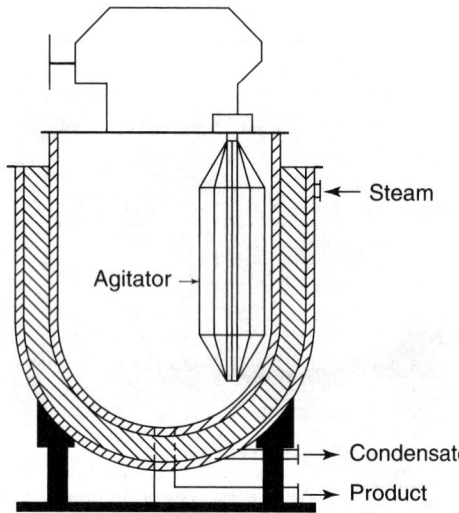

Fig. 9. Vertical agitated kettle. Although heat transfer through the jacket normally is quite poor, a kettle offers convenience in handling and cleaning and is particularly useful where batches of different materials must be processed frequently

Heat Storage

It is often necessary to store heat in rather large quantities in specially designed apparatus. Hot water, of course, is one of the easiest forms in which to store thermal energy that is immediately available. As contrasted with hot water, electric energy and steam have to be generated on an as-needed basis. The blast furnace poses a difficult heat storage problem, which obviously cannot be handled by storing heat in water. Great amounts of hot gas are required on a cyclic basis. To heat such quantities of air on a continuous, as-required basis would be quite impractical with the present stage of the art. The solution used involves several stoves that are quite large, often over 100 feet (30 meters) in height and about 25 feet (7.5 meters) in diameter. The blast temperature of approximately 1,000°F (538°C) is accomplished by preheating the stove checkerwork to a much higher temperature. Checkerwork is comprised of refractory material forms constructed in high walls in checkerboard fashion to permit free passage of air through the interstices when under pressure. The gas passing through the stove exhausts initially at 2000°F (1093°C). Mixing this with unheated air produces the required blast temperature for the blast furnace. The stoves usually are heated for a period of three hours and exhaust (termed "on wind") for a period of about one hour. See Fig. 10.

A similar system of checkerwork regenerators is used in connection with glass-tank heat-storage systems.

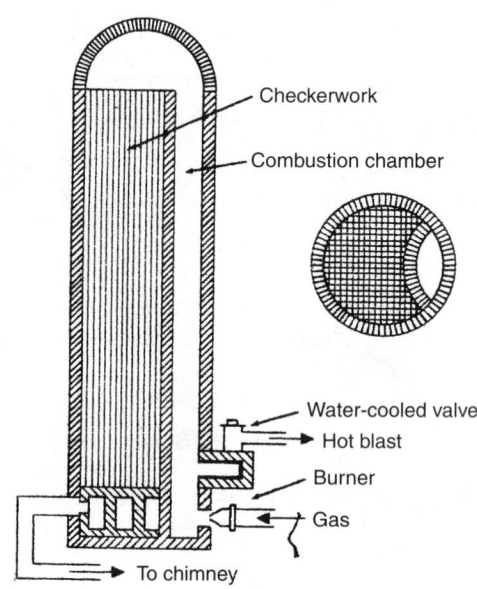

Fig. 10. Blast furnace stove for preheating large quantities of air

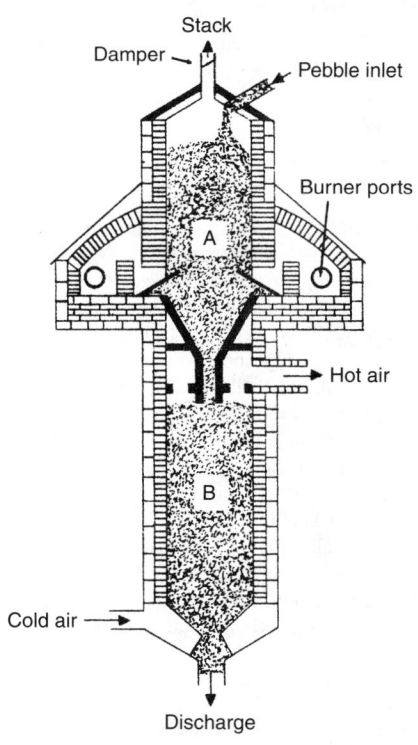

Fig. 11. Pebble heater for heating steam to temperatures impractical in metallic units. Also used for heating air, hydrogen, methane, and other gases for processing purposes. In reverse, a pebble heater may be used to recover heat from hot gases. The pebbles are heated in top chamber *A* by direct contact with combustion gases and passed through a throat to lower chamber *B*, where heat is transferred to cool gases. The two chambers are maintained at the same temperature so that there will be no gas flow between them. An average cycle on the pebbles is 30–50 minutes

Flowing streams of pebbles also have been used in the chemical industry for removing heat from gases. Pebbles and stones are also used in some solar energy storage systems. See Fig. 11. See also **Solar Energy**.

Additional Reading

Arpaci, V.S. et al.: *Introduction to Heat Transfer*, Prentice-Hall, Inc., Upper Saddle River, NJ, 2000.

Bejan, A. and J.S. Jones: *Heat Transfer*, John Wiley & Sons, Inc., New York, NY, 1999.

Butterworth, D. and C.F. Mascone: "Heat Transfer Heads Into the 21st Century," *Chem. Eng. Progress,* **30** (September 1991).

Carlson, J.A.: "Understand the Capabilities of Plate-and-Frame Heat Exchangers," *Chem. Eng. Progress,* **26** (July 1992).

Corsi, R.: "Specify Bayonet Heat Exchangers Properly," *Chem. Eng. Progress,* **32** (July 1992).

Ganapathy, V.: "Heat-Recovery Boilers: The Options," *Chem. Eng. Progress,* **59** (February 1992).

Incropera, F.P. and D.P. Dewitt: *Introduction to Heat and Mass Transfer,* 3rd Edition, John Wiley & Sons, Inc., New York, NY, 2000.

Incropera, F.P. and D.P. Dewitt: *Fundamentals of Heat and Mass Transfer,* 4th Edition, John Wiley & Sons, Inc., New York, NY, 2000.

Janna, W.S.: *Engineering Heat Transfer,* 2nd Edition, John Wiley & Sons, Inc., New York, NY, 1999.

Kakapc, S. and H. Liu: *Heat Exchanges: Selection, Rating, and Thermal Design,* CRC Press, LLC., Boca Raton, FL, 1997.

Kakapc, S., A.E. Bergles, F. Mayinger, and H. Yuncu: *Heat Transfer Enhancement of Heat Exchangers,* Kluwer Academic Publishers, Norwell, MA, 1999.

McKetta, J.J.: *Heat Transfer Design Methods,* Marcel Dekker, Inc., New York, NY, 1992.

Mukherjee, R.: "Use Double Segmental Baffles in Shell-and-Tube Heat Exchangers," *Chem. Eng. Progress,* **47** (November 1992).

Perry, R.H., D.W. Green, and J.O. Maloney: *Perry's Chemical Engineers' Handbook,* 7th Edition, McGraw-Hill Companies, Inc., New York, NY, 1997.

Reddy, J.N. and D.K. Gartling: *The Finite Element Method in Heat Transfer and Fluid Dynamics,* CRC Press, LLC., Boca Raton, FL, 2000.

Someah, K.: "On-Line Tube Cleaning: The Basics," *Chem. Eng. Progress,* **39** (July 1992).

Velarde, M.G. and C. Normand: "Convection," *Sci. Amer.,* **243**, 1, 92–108 (1980).

Welty, J.R. et al.: *Fundamentals of Momentum, Heat, and Mass Transfer,* 4th Edition, John Wiley & Sons, Inc., New York, NY, 2000.

Wood, R.M. et al.: "A New Option for Heat Exchanger Network Design," *Chem. Eng. Progress,* **38** (September 1991).

Yokell, S.A.: *A Working Guide to Shell-and-Tube Heat Exchangers,* The McGraw-Hill Companies, Inc., New York, NY, 1990.

HEAT TREATING. Heating and cooling of metals to effect changes in properties. Annealing and normalizing are generally for the purpose of softening or improving the grain structure. Patenting is also a softening process in which cold drawn carbon-steel wire is heated above its critical temperature range followed by cooling to below this range in a molten lead or molten salt bath, with subsequent cooling to room temperature.

While heat treating includes the softening treatments, it most often implies hardening and strengthening. In the case of steels this requires heating to above the critical temperature range followed by rapid cooling (quenching) in oil, water, or brine, except in the case of special grades which harden on cooling in air. This is followed by tempering, a low-temperature reheating treatment which reduces the internal stresses caused by the hardening treatment. Tempering may be carried to a high enough temperature to reduce somewhat the extreme hardness of the as-quenched steel and increase the toughness and ductility, depending on the requirements of the part. See also **Iron Metals, Alloys, and Steels.**

Another important form of heat treatment for hardening is precipitation hardening. See also **Annealing; Carbonitriding; Carburizing; Case Hardening;** and **Nitriding.** See Table 1.

Thermal or heat treating has been an inherent part of metalworking and fabricating for well over a century, and, in fact, some aspects of the topic date back to ancient times. Within the last few decades, heat treating has become quite sophisticated through the incorporation of modern computing and modeling techniques. Process modeling can be used as a scheduling tool to optimize throughput of a continuous furnace, for example. Much more instrumentation has been added to heat-treating processes, allowing better control over a larger number of variables that affect final product quality.

As of the early 1990s, one of the most interesting new processes is the use of solar energy as a direct heat source for surface hardening, cladding, and other surface modifications. An impressive demonstration project has been established at the Solar Energy Research Institute (SERI) in Golden, Colorado. Its heliostat has an area of 31.8 square meters (342 sq ft) and features an ultraviolet (UV)–enhanced aluminum coating on its front surface. The primary concentrator consists of 23 hexagonal facets, each of which is a spherical mirror ground to a 14.6-meter (48-ft) radius of curvature and is aluminum coated. At the target, 94% of the energy falls inside a 100 millimeter (4-in–diameter) circle. The beam has a Gaussian shape, with a peak flux of 2.5 MW/m^2, without a secondary concentrator.

TABLE 1. PRINCIPAL HEAT-TREATING PROCESSES

CARBURIZING
Pack and gas carburizing create a diffused carbon case. Base metals are low-carbon steels and low-carbon alloy steels. Process temperature range is 815–980°C (1500–2000°F).

Liquid carburizing creates a diffused carbon (possibly nitrogen) case. Base metals are low-carbon steels and low-carbon alloy steels. Process temperature range is 815–980°C (1500–1800°F).

Vacuum carburizing creates a diffused carbon case. Base metals are low-carbon steels and low-carbon alloy steels. Process temperature range is 815–1090°C (1500–2000°F).

NITRIDING
Gas nitriding creates a diffused nitrogen (nitrogen compounds) case. Base metals are alloy steels, nitriding steels, and stainless steels. Process temperature range is 480–590°C (900–1100°F).

Salt nitriding creates a diffused nitrogen (nitrogen compounds) case. Base metals are ferrous metals, including cast irons. Process temperature range is 510–565°C (950–1050°F).

Ion nitriding creates a diffused nitrogen (nitrogen compounds) case. Base metals are alloy steels, nitriding steels, and stainless steels. Process temperature range is 340–565°C (650–1050°F).

CARBONITRIDING
Gas carbonitriding creates a diffused carbon and nitrogen case. Base metals are low-carbon steels, low-carbon alloy steels, and stainless steels. Process temperature range is 760–870°C (1400–1600°F).

Liquid (cyaniding) creates a diffused carbon and nitrogen case. Base metals are low-carbon steels. Process temperature range is 760–870°C (1400–1600°F).

Ferric nitrocarburizing creates a diffused carbon and nitrogen case. Base metals are low-carbon steels. Process temperature range is 565–675°C (1050–1250°F).

ALUMINIZING
Creates a diffused aluminum case. Base metals are low-carbon steels. Process temperature range is 870–980°C (1600–1800°F).

SILICONIZING (chemical vapor deposition)
Creates a diffused silicon case. Base metals are low-carbon steels. Process temperature range is 925–1040°C (1700–1900°F).

CHROMIZING (chemical vapor deposition)
Creates a diffused chromium case. Base metals are low- and high-carbon steels. Process temperature range is 980–1090°C (1800–2000°F).

TITANIUM CARBIDE
Creates a diffused carbon, titanium, and TIC case. Base metals are alloy and tool steels. Process temperature range is 900–1010°C (1650–1850°F).

BORIDING
Creates a diffused born (boron compounds) case. Base metals are alloy and tool steels; cobalt- and nickel-base alloys. Process temperature range is 400–1150°C (750–2100°F).

TABLE 2. HIGH-FLUX SOLAR FACILITIES WORLDWIDE

Location	Total power kW	Peak flux MW/m^2
Alburquerque, New Mexico		
(Central receiver test facility)	5000	2.4
Furnace	22	3.0
Atlanta, Georgia		
Furnace	1.3	9.5
Golden, Colorado	10	2.5
White Stands, New Mexico	30	3.6
Odello, France		
Horizontal furnace	1000	16.0
Vertical furnace	6.5	15.0
Rehovot, Israel		
(Central receiver test facility)	2900	—
Furnace	16	11.0
Uzbek, Russia	1000	17.0

Source: Solar Energy Research Institute, Golden, Colorado.

A solar facility requires a major resource of direct normal radiation. Thus, in the United States, a facility of this type is limited to locations in Arizona, Colorado, Nevada, New Mexico, and Utah, or in a nearby region of one of these bordering states. Obviously, the facility cannot operate at night or during periods of dense cloud cover. Even with these limitations, design calculations show that a solar furnace can compete economically with laser and arc-lamp sources. The case for solar-furnace technology becomes even more attractive for materials-processing applications in space.

Currently, high-flux solar facilities are comparatively few, as shown in Table 2.

Additional Reading

Coffey, J.A.: "Supercell Carburizing," *Adv. Mat. & Proc.*, **81** (September 1989).

Conybear, J.G.: "Advanced Controls Offer Heat-Treating Flexibility," *Adv. Mat. & Proc.*, **38** (October 1989).

Conybear, J.G.: "Gas-Fired Vacuum Furnace Speeds Ion Nitriding," *Adv. Mat. & Proc.*, **87** (September 1991).

Dekumbis, R.: "Surface Treatment of Materials by Lasers," *Chem. Eng. Progress*, **23** (December 1987).

Doak, K.W.: "Furnaces Focus on New Processes, New Materials," *Adv. Mat & Proc.*, **84** (September 1989).

Holm, T.: "Synthetic Heat-Treating Atmospheres," *Adv. Mat. & Proc.*, **45** (October 1989).

Jones, L.E. and E.D. Jamieson: "Computers Tackle Heat-Treating Problems," *Adv. Mat. & Proc.*, **33** (March 1990).

Krauss, G.: "Thermal Processing of Steel," *Adv. Mat & Proc.*, **57** (January 1990).

Moerdijk, I.W.: "Polymer Quenchants," *Adv. Mat. & Proc.*, **19** (March 1990).

Persampieri, D., A. San Roman, and P.D. Hilton: "Process Modeling for Improved Heat Treating," *Adv. Mat. & Proc.*, **19** (March 1991).

Smidt, F.A.: "Surface Modification," *Adv. Mat. & Proc.*, **61** (January 1990).

Staff: *ASM Handbook: Heat Treating*, Vol. 4, ASM International, Materials Park, OH, 1991.

Staff: *Metal Casting and Heat Treating Industry*, DIANE Publishing Company, Collingdale, PA, 1996.

Stanley, J.T., C.L. Fields, and J.R. Pitts: "Surface Treating (Solar)," *Adv. Mat. & Proc.*, **16** (December 1990).

Totten, G.E.: "Polymer Quenchants: The Basics," *Adv. Mat. & Proc.*, **51** (March 1990).

HEAVY HYDROGEN. See **Deuteron**.

HEAVY WATER. Water in which the hydrogen of the water molecule consists entirely of the heavy hydrogen isotope of mass 2 (deuterium). Written D_2O. Density, 1.1076 at 20 degrees C. It is used as a moderator in certain types of nuclear reactors. The term is sometimes applied to water whose deuterium content is greater than natural water. See also **Nuclear Power Technology**.

HEISENBERG, WERNER KARL (1901–1976). Heisenberg was a German physicist who earned his Ph.D. from the University of Munich, and then became Assistant to Max Born at the University of Göttingen, and in 1924 he gained the *venia legendi* at that University. From 1924 until 1925 he worked, with a Rockefeller Grant, with Niels Bohr, at the University of Copenhagen, returning for the summer of 1925 to Göttingen.

In 1926 he was appointed Lecturer in Theoretical Physics at the University of Copenhagen under Niels Bohr and in 1927, when he was only 26, he was appointed Professor of Theoretical Physics at the University of Leipzig, and worked in the field of quantum mechanics.

In 1941 he was appointed Professor of Physics at the University of Berlin and Director of the Kaiser Wilhelm.

At the end of the Second World War he, and other German physicists, were taken prisoner by American troops and sent to England, but in 1946 he returned to Germany and reorganized, with his colleagues, the Institute for Physics at Göttingen. This Institute was, in 1948, renamed the Max Planck Institute for Physics.

Heisenberg's name will always be associated with his theory of quantum mechanics, published in 1925, when he was only 23 years old. For this theory and the applications of it, which resulted especially in the discovery of allotropic forms of hydrogen, Heisenberg was awarded the Nobel Prize for Physics for 1932.

His new theory was based only on what can be observed, that is to say, on the radiation emitted by the atom. We cannot, he said, always assign to an electron a position in space at a given time, nor follow it in its orbit, so that we cannot assume that the planetary orbits postulated by Niels Bohr

actually exist. Mechanical quantities, such as position, velocity, etc., should be represented, not by ordinary numbers, but by abstract mathematical structures called "matrices" and he formulated his new theory in terms of matrix equations.

See also **Field Theory**; and **Quantum Mechanics**.

J.M.I.

HELIUM. [CAS: 7440-59-7]. Chemical element symbol He, at. no. 2, at. wt. 4.0026, periodic table group 18 (inert or noble gases), mp $-272.2°C$ (20 atmospheres), bp $-268.93°C$ (4.2144 K), specific gravity 0.124 at 4.2144 K. The element has no triple point and can be solidified only by applying high pressure to the liquid phase. Described later, liquid helium undergoes a change in its physical properties at 2.178 K, known as the *lambda point*. Solid helium has a close-packed hexagonal crystal structure (subject to further study and confirmation). At standard conditions, helium is a colorless, tasteless, odorless gas. There are two natural isotopes ^3He and ^4He, with ^4He being slightly less than 100% abundant. The boiling point is 3.2 K for ^3He. Radioactive ^5He and ^6He have extremely short half-lives. See also **Radioactivity**. The first ionization potential for helium is 24.58 eV; second, 54.14 eV. Other physical properties of helium are described under **Chemical Elements**.

Like the other rare gases, helium exhibits negative chemical properties with ordinary materials under normal conditions. Under the influence of electric glow discharge or electron bombardment, helium forms compounds with tungsten and other metals, as well as with iodine, sulfur, and phosphorus. In a vacuum electric discharge tube helium shows green to canary-yellow glow. Discovered first in the vapors surrounding the sun by Lockyer in 1868, through the yellow spectral line near the two yellow lines of sodium, then by Ramsay in 1895 in the mineral clevite.

Helium occurs (1) in minerals of uranium and thorium, such as clevites, pitchblende, carnotite, monazite, and also in beryl, (2) in mineral waters (1 part He per thousand of water, in some Iceland waters), (3) in volcanic gases, (4) especially in certain natural gases of the United States. The first discovery of this kind was made in Kansas.

Uses

Industrially, helium is used: to provide an inert gaseous shield for arc welding, for growing transistor crystals, in the production of titanium and zirconium, to fill the space between optical lenses in instruments, as the carrier gas in some chromatographic apparatus, as a liquid bath for masers and cryotrons, as a refrigerant for furnishing the low temperature required for superconducting electrical equipment, in lasers, as a diluent gas in deep-sea diving applications, as a heat-transfer medium in gas-cooled nuclear reactors, and as a leak-detecting medium for testing pressure and vacuum equipment. Now, to a rather limited extent, helium is used as a lifting gas for airships and for balloons used in meteorological investigations. Helium is used in aerospace programs in several ways, including its use in propellant tanks as a compressed gas which expands and takes the place of fuel as the fuel is consumed, in ground-support equipment, and in communication satellites for providing the low temperature required for sensitive electronic systems. In medicine, helium sometimes is mixed with oxygen for patients with certain respiratory ailments and also it is mixed with certain anesthetics to reduce the hazards of forming an explosive mixture with air.

Future possible uses of helium have a direct influence on the conservation of helium resources. The known resources are not large in comparison with most other raw materials, and most authorities are of the opinion that conservation measures should be continued. However, most helium demand projections have proved overly optimistic and consequently have lessened the pressure for conservation. Natural gas streams, of which He is but a minor constituent, are produced commercially for sale and consumption as fuel. To separate helium by stripping from natural gas is an expense that is not attractive to the producers of natural gas. Thus, it is evident that helium conservation, under current supply/demand conditions, must stem from government regulation. Even though stripping helium from natural gas is costly, later costs to recover helium lost to the atmosphere from burning He-bearing natural gas would be many times greater.

The history of helium conservation measures dates back to 1925 when the U.S. Congress passed the Helium Act of 1925. Congress amended the act in 1960 to provide for stripping natural gas of its helium, for purchase of the separated helium by the government, and for its long-term storage. In 1971, after about 28 billion cubic feet had been stored (in a federally owned gas field called Cliffside near Amarillo, Texas), the purchase program was terminated by the government, an action that, as reported by Hammel et al., unleashed several lawsuits and not a little acrimony. As of the present, most of the litigation has been concluded, much of the He that could have been saved has been wasted to the atmosphere, and the gas fields supplying the He are almost depleted. However, in the meantime, a new and rich source of He has been discovered in southwestern Wyoming that could ensure adequate supplies for many decades if an appropriate new federal policy on He were developed and implemented. The new field, first explored by Mobil in 1960, led to an initial estimate of 3 to 15 billion cubic feet of He in the Tip Top drilling unit of that field. However, the natural gas from that unit was initially judged unfit for sale as a natural gas fuel. The borehole was cemented shut and the well abandoned. In the early 1980s, it was established by Additional drilling that the amount of He recoverable from the Wyoming field is at least 200 billion cubic feet. Left untouched this would represent an excellent long-term helium reserve. More than 90% of this field lies within federal land boundaries.

With increasing incentives because of natural gas pricing (Natural Gas Policy Act of 1978), several firms have plans to drill and develop nearly 250 deep wells, from which 2.8 billion cubic feet of acidic gas per day would be produced. If private developers elect not to conserve the helium from this project (Riley Ridge Natural Gas Project), it is estimated that about 5 billion cubic feet of He would be vented to the atmosphere each year.

Unexpected Uses for Helium

Hammel (1984) points out that new uses for He continue to emerge. These include a 49-meter-diameter He-filled sphere that has been proposed as a lighter-than-air hoist capable of moving loads in excess of 90 tons. Another use is in superconducting magnets for imaging with nuclear magnetic resonance. Proposals for the use of large amounts of He continue to be made in the national security area.

Origin of Helium

Helium has a geologic occurrence and distribution unique among the elements. It is a product of radioactive disintegration of uranium and thorium within the earth's mantle and crust. But it flows to the surface at a rate less than that of its generation, because most of it is driven into crystal structures of rock minerals until released by alpha radiation damage near radioactive concentrations. Mobile helium rising through the crust may then be trapped, along with other gases, beneath relatively impermeable barriers. Nitrogen is almost always associated with helium in natural gases, although this has not been fully explained. Also, carbon dioxide is abundant in some helium-rich gas mixtures.

Liquefaction of Helium

This was accomplished by Onnes in 1908 in Leiden, and Keosom in 1926 succeeded in solidifying helium in the same laboratory. Relatively recently, helium has been solidified at room temperature. The melting pressure at 24°C is 115 kilobars, in complete agreement with the Simon equation. Besson and Pinceaux (1979) developed an original apparatus for the experiment, which allowed loading of the cell at room temperature. Diamond anvil cells were used in the procedure.

Liquid Helium II

Upon cooling, ^{4}He liquefies at atmospheric pressure at 4.216 K to form an essentially normal liquid, liquid helium I. On further cooling to the lambda-point, 2.178 K at one atmosphere, a change occurs to liquid helium II. The latter has a very low viscosity (hence the name "superfluid") and a very high thermal conductivity, which produce such phenomena as the creeping of a film over the edge of the container, and the fountain effect, in which the liquid sprays out of a capillary. Superfluidity is commonly explained in terms of a two-fluid theory. Thus, London and Tirza attribute the properties of helium II to a mathematical peculiarity in the distribution function of Bose-Einstein statistics, whereby below the λ-point, a finite fraction of the atoms fall into a ground state of zero thermal energy. In this state they would have the properties of a superfluid. However, this theory has not yielded good quantitative predictions of the properties of the aggregate liquid helium. ^{3}He, which follows Fermi-Dirac statistics, does not have a superfluid state.

Landau treats liquid helium by an approach similar to that of the Debye theory of solids. The longitudinal and transverse sound waves, which are the elementary excitations of that theory of solids, correspond in the case of liquid helium to phonons and rotons. The *phonons* are the longitudinal sound waves, while the *rotons* are another type of elementary excitation postulated by Landau to represent the rotational motion of the liquid, because a liquid cannot support transverse waves. The specific heat can be expressed as the sum of contributions from phonons and rotons. Landau derived expressions for these that fit the data and experiments quite closely up to 1.6 K.

Feynman developed wave functions to provide an atomistic interpretation of Landau's spectrum of elementary excitations.

The complexity of the helium II problem is apparent at once when one attempts to extend the equations of classical hydrodynamics to this two-component system, in which each component has its own density and velocity. Khalatnikov derived such equations by ignoring terms of second order.

Still another area of investigation has been that of the properties of ^{3}He-^{4}He mixtures. As stated above, ^{3}He exhibits no λ-transition and no superfluidity. It has a critical temperature of 3.35 K and a boiling point of 3.2 K, against values of 5.2 K and 4.216 K for ^{4}He.

The most abundant helium atoms, ^{4}He, are bosons, but the ^{3}He atoms are fermions. This has a consequence that liquid ^{3}He does not show superfluidity—a property very probably connected with the Bose-Einstein statistics obeyed by the ^{4}He atoms.

Donnelly and associated researchers (University of Oregon) has observed that in the future liquid helium rather than air may be used in a much down-scaled wind tunnel, perhaps with experiments conducted within the space of an average room versus current, very large wind tunnels. It is envisioned that a tunnel could be filled with superfluid liquid helium, taking advantage of the liquid's absence of viscosity and friction as previously described. Based upon quantum-mechanical factors, a source of heat in the tunnel could cause extremely fast currents to flow in the liquid.

Peterson (reference listed) reported in early 1991 that researchers at Harvard University made what is considered a remarkable prediction regarding the energy-level transitions that occur in a helium atom. The agreement between theoretical calculations and experimental results show that computational methods for constructing a model of a two-electron atom can work, thus bridging the gap between theory and practice.

Chemistry

The most striking properties of helium are: its emission as the positively charged (+2) alpha particles in radioactive changes, its formation in radioactive change by uranium-radium and thorium-containing substances, emitting alpha particles, later losing the charge to become helium, and its production artificially by bombardment of lithium or boron with high-velocity protons or alpha rays.

Unlike the other inert gases, helium gives little evidence of compound formation with organic substances. Like neon, but unlike the others, it forms no hydrate. However, it forms compounds much more readily under excitation, due apparently to unpairing of its $1s$ electrons and promoting of one of them to the $2s$ state. The 460 kcal/g-atom of energy is readily obtained by electric discharge or electron bombardment. Under such conditions the helium molecule-ion, He_2^+, with a pair of bonding electrons ($1s$) and a single antibonding electron ($1s$), is formed, as are combinations of the type of HeH^+ and HeH_2^+. In a mercury discharge tube, the compound $HgHe_{10}$ has been found, and with various metallic electrodes corresponding helides, such as the compounds of tungsten, platinum, iron, palladium, bismuth, etc., e.g., WHe_2, Pt_3He, $FeHe$, $PdHe$, $BiHe$, etc., have been formed.

Additional Reading

Andronikashvili, E.L. and A.K. Ishkhneli: *Reflections on Liquid Helium,* American Institute of Physics, College Park, MD, 1998.

Besson, J.M. and J.P. Pinceaux: "Melting of Helium at Room Temperature and High Pressure," *Science,* **206**, 1073–1075 (1979).

Donnelly, R.J.: "Superfluid Turbulence (Helium)," *Sci. Amer.,* **100** (November 1988).

Epstein, A.W.: "Cool Breeze: A Helium Superwind for Wind-Tunnel Experiments," *Sci. Amer.,* **30** (May 1990).

Greenwood, N.N. and A. Earnshaw: *Chemistry of the Elements,* 2nd Edition, Butterworth-Heinemann, Inc., Woburn, MA, 1997.

Hammel, E.F., M.C. Krupka, and K.D. Williamson, Jr.: "The Continuing U.S. Helium Saga," *Science,* **223**, 789–792 (1984).

Krebs, R.E.: *The History and Use of Our Earth's Chemical Elements: A Reference,* Greenwood Publishing Group, Inc., Westport, CT, 1998.

Lide, D.R.: *CRC Handbook of Chemistry and Physics,* 84th Edition, CRC Press, LLC., Boca Raton, Florida, 2003.

Parker, P.: *McGraw-Hill Encyclopedia of Chemistry,* 2nd Edition, The McGraw-Hill Companies, Inc., New York, NY, 1993.

Peterson, I.: "Helium Theory Gets High-Precision Test," *Science News,* **86** (February 9, 1991).

Vollhardt, D. and P. Wolfle: *The Superfluid Phases of Helium 3.* Taylor and Francis, Inc., New York, NY, 1990.

Volovik, G.E.: *Exotic Properties of Superfluid Helium Three,* World Scientific Publishing Company, Inc., Riveredge, NJ, 1992.

HEMATITE.

The mineral hematite, ferric oxide, Fe_2O_3, occurs as thick or thin tabular rhombohedral forms, sometimes in pyramids but rarely in hexagonal prisms. It also assumes botryoidal, columnar and lamellar shapes, and may be granular or compact. Its hardness is 5.6; specific gravity, 5.26; luster, metallic to earthy or dull; color, dark gray to black; earthy forms may be different shades of red; streak, red to red-brown; translucent (in very thin flakes) to opaque. Hematite with a metallic luster is called specular iron.

It is a widely distributed and common mineral, found in igneous, sedimentary and metamorphic rocks as beds and veins, having probably been formed in many different ways under very different conditions. Beautifully crystallized hematite has been found in the Urals of the former U.S.S.R.; Rumania; Switzerland; the Island of Elba; Alsace, France; Cumberland, England. Extremely rich, large hematite ore bodies have been found and are being worked in Minas Gerais, Brazil; Cerro de Mercado, Durango, Mexico; Quebec and Labrador in Canada. The hematite ore deposits that lie along the southern and northwestern sides of Lake Superior in Michigan, Wisconsin and Minnesota have been worked to near depletion. Extensive beds of hematite are found throughout the Appalachian region from New York to Alabama, being mined near Birmingham in the latter state. Hematite occurs in quantity in Nova Scotia and Newfoundland. It is the most important ore of iron, and has other industrial uses in paint manufacture and polishing compounds. The name hematite is derived from the Greek word meaning blood.

HEMICELLULOSE.

Hemicellulose is the least utilized component of the biomass triad comprising cellulose, lignin, and hemicellulose.

Pure hemicellulose components are seldom extracted directly from their source. Extracts are a mixture of polysaccharides, lignin, and lignin—hemicellulose complexes (by chemical linkages and possibly physical interactions) characteristic of their origin and the solvent employed. Hemicellulose has a lower degree of polymerization (DP) than cellulose (about 200 vs more than 10,000) and its lower limits have not been clearly defined. The extract may contain two or more polymers of similar composition but different structures (polydiversity) or of different distributions and amounts of branching or bonding in otherwise similar molecules (polydispersity). If a single polymer is present, it may exhibit a spectrum of molecular weights (polymolecularity) which may exhibit a Gaussian or biased distribution. A pure hemicellulose component is one where polydiversity has been avoided and a degree of heterogeneity has been attained compatible with end use application.

The most common hemicellulose in angiosperms is composed of D-xylose arranged in a linear manner. D-Mannose is derived from a glucomannan which is the most common hemicellulose in most gymnosperms. Both contain other sugars and exist in a variety of configurations and molecular weights. This article concentrates primarily on the components of tracheids and fibers of arborescent plants.

The common hemicellulose components of arborescent plants are listed in Table 1.

Isolation and Analysis

Techniques for the isolation of hemicellulose depend on the intended end use and whether it occurs in soluble waste material or is part of a solid matrix. Isolation is more difficult from solids as diminution of particle size and removal of undesired encrustants such as lignin is necessary to increase accessibility and destroy lignin hemicellulose bonds. Delignification techniques, except for those using ethanolamine, employ oxidants. Peroxides, peroxyacetic acid, and chlorine dioxide have been used but the most common reagents are chlorine and acidified sodium chlorite.

Delignification extracts varying amounts of hemicellulose. Low reaction temperatures and (where possible) high salt concentrations minimize losses and concomitant chemical degradations such as oxidation and the effects of pH. Pectic substances and easily soluble arabinans, arabinogalactans, galactoglucomannans, xylans, and compression wood galactans are found

TABLE 1. HEMICELLULOSE COMPONENTS OF ARBORESCENT PLANTS

Gymnosperms	Dicotyledons	Monocotyledon (bamboo)
arabino-(4-*O*-methyl-glucurono)xylan	*O*-acetyl-(4-*O*-methyl-glucurono)xylan	arabino-(4-*O*-methyl-glucurono)xylan
O-acetylgalacto-glucomannan (0.1:1:3)	glucomannan	heteroxylans
O-acetylgalacto-glucomannan (1:1:3)	arabinogalactan	D-glucans
arabinogalactan	pectic substances	arabinogalactans
pectic substances	tension wood components	pectic substances
compression wood components		

in waste chlorite liquors. Carbonyl groups (excepting carboxyls) are frequently reduced with a suitable reagent before alkaline extractions are attempted to minimize β-elimination reactions. The use of an inert atmosphere during alkaline extraction prevents oxidation by oxygen.

Extraction of hemicellulose is a complex process that alters or degrades hemicellulose in some manner. Alkaline reagents that break hydrogen bonds are the most effective solvents but they de-esterify and initiate β-elimination reactions. Polar solvents such as DMSO and dimethylformamide are more specific and are used to extract partially acetylated polymers from milled wood or holocellulose.

The separation of the polysaccharide components utilizes their different solubilities, polar groups, extents of branching, molecular weights, and molecular flexibilities and may be accomplished batchwise or with easily automated column techniques such as column or high performance liquid chromatography. These procedures have been summarized in several reviews.

The increasing sophistication of analytical techniques coupled with suitable fractionation procedures has made the heterogeneity of hemicellulose components increasingly apparent. These techniques common to polymer chemistry include gas chromatography—mass spectroscopy, and proton and ^{13}C-nuclear magnetic resonance spectroscopy. Molecular studies employ viscometry, osmometry, x-ray techniques, light scattering, and chromatographic and centrifugal techniques, as well as the use of optical rotatory dispersion and circular dichroism.

Pulping

The complex behavior of hemicellulose during pulping has been reviewed. When hemicellulose and lignin dissolve with the help of chemical transformations, fresh cellulosic surfaces are created and competition for deposition in these spaces arises between the dissolved components. Hemicellulose degradations also occur and are related to pH (acid hydrolysis, β-elimination reactions, redox reactions, etc.), and pyrolytic effects to an extent dependent upon the time, temperature, and liquor composition of the cook. These reactions are rendered more complex because the cell wall controls the diffusion of the reactants and products into and out of the fiber so that hemicellulose may not be able to react or diffuse out of the fiber before the cook is completed. Under pulping conditions, the formation and cleavage of lignin-hemicellulose bonds and possible carbohydrate—carbohydrate bonds occurs complicating the nature of the product. The extent to which these competing reactions is accomplished is reflected in product composition and end use quality and is the subject of much empirical research.

Those pulps with about 15% hemicellulose are usually used for paper manufacture, whereas those with 5% or less are used where a high cellulose content is required. The proportion of glucomannan is slightly greater in sulfite pulps, whereas the quantity of xylan is somewhat greater in kraft and soda pulps. These proportions can be altered slightly by changes in the cooking schedule.

Suitable pretreatment of wood before pulping alters the behavior of hemicellulose significantly. Saponification of the acetyl groups of softwood before sulfite cooking results in glucomannan retention in the final product. Those treatments that limit the peeling reaction during alkaline pulping processes (reductions with $NaBH_4$, H_2S, or oxidations with chlorite, polysulfide, anthraquinone, etc), can result in polysaccharide retention. The pretreatment of wood with mineral acid or liberated acids of wood at elevated temperatures (i.e., 170°C for 30 min) diminishes the DP of hemicellulose components sufficiently that they will be consumed mostly during a subsequent alkaline cook. The resulting pulp behaves more like cotton cellulose in many industrial applications.

The Effect of Hemicellulose in Commercial Products

Hemicellulose components have an effect on the properties of products in which they are present. In the case of viscose manufacture much of the hemicellulose which remains in the product has only a marginal effect on strength properties and brightness and no effect on heat stability. It does contribute to swelling in yarn, and that remaining in the spent viscose liquor is harmful to filter presses. Increased clogging of spinnerets, low color index, and decreased yarn strength can also result when resin, cations, and hemicellulose are together in the steeping liquor.

When uronic acid and L-arabinofuranosyl branches are lost as a result of processing, the poor solubility of xylan acetate in organic solvents contributes adversely to cellulose acetate processing. The haze density (opaqueness) of solutions and solid products increases as hemicellulose components become less branched. Glucomannan in an acetate product contributes less haze than xylans but more than arabinoxylans. It is responsible for the false viscosity of cellulose acetate solutions derived from wood pulp. Glucomannan also causes filtration difficulties during processing, but the effects of xylans are unpredictable.

An understanding of the effect of hemicellulose on paper products is less clear because the formation of paper webs largely depends on their structure, which is influenced by many factors besides hemicellulose. Hemicellulose and related gums and mucilages help maintain a random dispersion of fibers in the furnish which results in more uniform and mechanically stronger paper webs. It also increases the rate at which pulp fibers respond to mechanical action (beating) and increases fiber bonding. The strong interfiber bonds formed after drying can alter paper structure and lead to losses in tearing strength and decreases in opacity if they are too extensive.

Applications

Hemicellulose and hemicellulose-like polysaccharides are beneficial components of foodstuffs because of their interactions between water and water-insoluble components. Endogenous polysaccharides are responsible for processing characteristics as well as texture and mouthfeel. Other properties such as gel formation, swelling of dough, fermentation, and optical properties are achieved using suitable treatments with enzymes and chemicals.

Besides inherent usefulness as a naturally occurring component of some manufactured products, hemicellulose can be utilized either as a polymer or as the source of chemical intermediates. The former use is complicated since the mixture of polysaccharides and lignin in extracts requires special treatment if one component is to be isolated. As a result, the naturally occurring gums and mucilages are at a competitive advantage in the marketplace.

Larch arabinogalactan is easy to isolate and requires limited purification for many uses. The mixture of sugars, oligosaccharides, degraded hemicellulose, and lignin found in the liquors and condensates from Asplund-like and prehydrolysis pulping processes is used for binding and extending animal fodder. The sugars and oligosaccharides in waste sulfite liquor can be used for furfural production and the growth of yeast. Pectin is used in the food industry, and apart from specialty uses finds limited application in the paper industry.

Derivatives of hemicellulose components have properties similar to the cellulosic equivalents but modified by the effects of their lower molecular weight, more extensive branching, labile constituents, and more heterogeneous nature. Acetates, ethers, carboxymethylxylan, and xylan–poly(sodium acrylate) have been prepared.

NORMAN S. THOMPSON
Consultant

Additional Reading

Clayton, D. et al.: "Chemistry of Alkaline Pulping, in *Pulp and Paper Manufacture,* 3rd ed., Vol. 5, Alkaline Pulping, The Joint Textbook Committee of the Paper Industry, TAPPI, CPPA, Technology Park, Atlanta, GA, 1989.

Gatenholm, P. and M. Tenkanen: *Hemicelluloses Science and Technology,* American Chemical Society, Washington, DC, 2003.

Shimizu, K.: in N.S. Hon and N. Shiraishi, eds., *Wood and Cellulosic Chemistry,* Marcel Dekker, Inc., New York, NY, 1991, Chapt. 5.

Thompson, N.S.: in I.S. Goldstein ed., *Organic Chemicals from Biomass,* CRC press, Boca Raton, Fla., 1981.

Whistler, R.L. and C.-C. Chen: in M. Lewin and I.S. Goldstein, eds., *Wood Structure and Composition,* Vol. 11, International Fiber Science and Technology Series, Marcel Dekker, Inc., New York, 1991, Chapt. 7.

HEMIMORPHITE. This mineral is zinc silicate, $Zn_4Si_2O_7(OH)_2 \cdot 2H_2O$, occurring in tabular and prismatic orthorhombic crystals, although often in massive and fibrous forms. There is a perfect cleavage parallel to the prism; it is brittle with a subconchoidal fracture; hardness, 4.5–5; specific gravity, 3.40–3.50; luster, vitreous; color, white, tending to translucent. Hemimorphite differs from willemite, also a zinc silicate, in that the former contains considerable water which may be driven off when heated to a high temperature.

There are many localities for hemimorphite in Europe, fine specimens having come from Saxony; Sardinia; Cumberland, Alston Moor and Derbyshire, England. It is found in Siberia, Algeria, and Mexico. In the United States, hemimorphite has been found at Sterling Hill, New Jersey; in Lehigh County, Pennsylvania, and in Virginia, Missouri, Montana, Colorado, Utah, New Mexico, and Nevada.

The mineral is so named because of the tendency to form doubly terminated crystals showing a different grouping of faces at either end. The name is derived from the Greek words meaning half and form.

HEMOGLOBIN. The main function of the hemoglobin molecule is oxygen transport. The hemoglobin molecules from each species of organism that has been examined differ in the sequence of amino acids in their polypeptide chains unless they are very closely related. Chimpanzee and human hemoglobins are apparently identical. Sometimes two or more different kinds of hemoglobin are found simultaneously in the same organism. These structural variations may give rise to differences in the physiological properties that help to determine the efficiency of oxygen transport by the blood from lungs or gills to the tissues. Hemoglobin also plays an important role in carbon dioxide transport. See also **Blood**.

Vertebrate hemoglobins are usually composed of four polypeptide chains of two types, called α and β. The molecules can, therefore, be described as $\alpha_2\beta_2$. An iron porphyrin moiety, *heme*, is associated with each chain. Evidence indicates that combination of the heme with oxygen results in structural changes in the protein to which it is bound. Studies of single crystals of horse and human hemoglobins by x-ray diffraction show that removal of oxygen from the iron atoms of the four hemes results in a separation of the β-chains from one another; the relative positions of the α-chains do not appear to change. Although the molecular basis is not fully understood, the consequences are important. It is certain that any change in the mutual relationships of the polypeptide chains will alter the environment of many amino acid residues. These environmental changes are probably responsible for the degree of oxygenation to the oxygen pressure; and the dependence of the oxygenation upon pH and upon carbon dioxide concentration.

Mutations which alter the amino acid sequence can occur in either the α or the β-chain of the adult. However, most mutations are deleterious and changes in the α-chain would be more severely selected against in a process of natural selection because any change in the α-chain would affect the sensitive fetus, whereas changes in the β-chain would affect only the adult. This means that evolution tends to favor changes in the β-chain over changes in the α-chain. These considerations indicate that molecular adaptation of hemoglobin, at least in mammals, may involve changes more in the β-chain than in the α-chain.

Hemoglobins can be dissociated into their α- and β-subunits. Not only are hemoglobins capable of dissociating into their polypeptide subunits, but certain hemoglobins are also capable of polymerization. Many reptiles and amphibians and certain mice possess hemoglobins which polymerize to form double molecules $(\alpha_2\beta_2)_2$ and sometimes triple or quadruple molecules. Many hemoglobins from invertebrate animals have very large molecular weights and are composed of a large number of subunits—as many as 180 in some species. The nature of the forces holding these large aggregates together is under study.

The amino acid sequences of hemoglobins have been extensively altered by mutation during evolution. Data on the amino acid sequences of the chains from a variety of mammalian and other vertebrate hemoglobins show that the sequence can be varied extensively without drastic change

TABLE 1. DIVERGENCE OF HEMOGLOBIN CHAINS WITH TIME

Type of chain divergence	Number of differences	Estimated time since divergence
β-δ	10	35 million years
β-γ	37	150 million years
β-α	76	380 million years
(α-β)-myoglobin	~135	650 million years

in function. There appears to exist a hierarchy in the functional importance of different parts of a protein. Substitutions in different segments of a polypeptide chain may, according to the type and position of the substitution, exhibit a spectrum of effects, ranging from detectable to catastrophic. For example, the single substitution of valine for glutamic acid in the 6th position of the β-chain in human sickle cell hemoglobin results in a large decrease in the solubility of deoxygenated hemoglobin within the red cells. The hemoglobin, by forming a gel, distorts the red cell shape ("sickle") in such a way that flow through the capillaries is retarded. Such drastic consequences do not result if the substitution is lysine rather than glutamic acid (hemoglobin C). Histidine in position 63 of the human β-chain has an essential role stabilizing the ferrous state of the heme iron. Substitution by tyrosine (in hemoglobins "M") results in the loss of this stability because the ferric iron can form a strong linkage with the $-OH$ group of tyrosine. Such a substitution results in a complete loss of capacity to combine reversibly with oxygen.

The foregoing are radical substitutions. Most effective substitutions appear to be relatively conservative and do not drastically affect the oxygen transport function. Therefore, the number of differences between homologous chains appears to be related not to functional differences, but to the time which has elapsed since the chains diverged from a hypothetical polypeptide ancestor. The mean number of differences between the hemoglobin chains of man, horse, pig, rabbit, and cattle is approximately 11. The common ancestor of these mammals may have existed some 80 million years ago. Thus, approximately 11 effective mutations per chain occurred in 80 million years, or 1 substitution per chain in 7 million years. Zuckerkandl and Pauling, using standard probability theory, have used this figure to estimate the time at which the different human hemoglobin chains (α, β, γ, and δ) are believed to have arisen by gene duplication. These estimates are shown in Table 1.

Estimates like these indicate that hemoglobins are very old and that it may be possible to find relatives of vertebrate hemoglobins in invertebrate animals. They also suggest that the gene duplication believed to be responsible for the divergence of the α- and β-chains took place in the Devonian period at the time of the appearance of early amphibians and the dominance of fish.

The suggested relationship between numbers of differences and evolutionary time is not wholly secure. It assumes uniformity in the rate of effective amino acid substitution, but this rate may be neither uniform with time, nor uniform in different parts of the polypeptide chain. Differences in the rate of effective substitution along the polypeptide chain may be due not only to restrictions imposed by the required tertiary structure, but also to differences in the rate at which various parts of the DNA or the gene mutate. The evolution of hemoglobin may be contrasted with that of cytochrome c in which approximately 50% of the molecule appears to have remained invariant during the time yeast and man have evolved.

Additional Reading

Chang, T.: *Blood Substitutes and Oxygen Carriers,* Marcel Dekker, Inc., New York, NY, 1992.
Honig, G. and J. Adams: *Human Hemoglobin Genetics,* Springer-Verlag, New York, Inc., New York, NY, 1986.
Perutz, M.: *Science Is Not a Quiet Life: Unravelling the Atomic Mechanism of Haemoglobin,* Imperial College Press, London, UK, 1999.

HEMP. The fibers of the hemp plant, *Cannabis sativa,* of the family *Cannabinaceae* (hemp family) are coarse and rather harsh and much less pliable than flax fibers. They are dark colored and not easily bleached without damage. The fibers are used mainly for making rope and coarse twine, warp of carpet, belt and upholstery webbing, and wherever strength and durability without concern for appearance are of importance. Short fibers of hemp, called tow, are used in packing joints in pipes, for pump

packing, and for stuffing upholstery. The woody waste from hemp fiber is sometimes used in the manufacture of certain papers.

Hemp is obtained from the stem pericycle of a tall hollow-stemmed annual, which is a native of central and western Asia. In cultivation, the slight branching, which characterizes the plant, is considerably reduced by planting thickly. The plants grow from 5 to 16 feet high. They have digitately compound dark green leaves and small inconspicuous flowers, which are of two kinds, occurring on different plants. The staminate flowers appear in small axillary clusters on male plants, and the pistillate flowers are borne in leafy spikes on female plants. The fruit, an achene, is a hard ovoid structure, often called hemp seed. Hemp grows best in regions having a warm humid growing season of about 5 months. The plants grow rapidly, soon shading the ground so effectively as to suppress other plants, and thus plantings of hemp have been used as a means to eradicate weeds. When the staminate flowers are mature, the plants are ready for harvest. To delay after that is not desirable, since the male plants die soon after flowering. After flowering, the fibers become coarser. Harvesting and the treatment of the plants after harvesting are similar to the procedures used with flax plants.

The hemp plants are cut off or pulled up, denuded of leaves, roots and tops, and tied in bunches and left to dry for about 2 weeks. They are then immersed in water to ret. In retting, the intercellular substance of the stems is acted upon by bacteria and softened so that the fibers are readily cleaned of surrounding tissues. Scutching removes the woody tissue, after which the rough hemp fibers are hackled, or drawn over coarse combs that pull out the fibers.

In recent years, the cultivation of hemp has been subject to controls because marijuana is prepared from the dried leaves and flowers of the plant, which then are smoked in the form of cigarettes as a narcotic. The species *Cannabis indica* is usually used in this connection.

Additional Reading

Bosca, I. and M. Karus: *The Cultivation of Hemp: Botany, Varieties, Cultivation and Harvesting,* Hemptech, Sebastopol, CA, 1998.
Ranalli, P. (Editor): *Advances in Hemp Research,* Haworth Press, Inc., Binghamton, NY, 1998.
Robinson, R. and R. Nelson: *The Great Book of Hemp: The Complete Guide to the Environmental, Commercial, & Medicinal Uses of the World's Most Extraordinary Plant,* Inner Traditions International, Ltd., Rochester, VT, 1995.

Web References

The Industrial Hemp Information Network. http://hemptech.com/
North American Industrial Hemp Council. www.NAIHC.org

HENRY'S LAW. When a liquid and a gas are in contact, the mass of the gas that dissolves in a given quantity of liquid is proportional to the pressure of the gas above the liquid. Thus, if air is kept in contact with water at standard atmospheric pressure, each kg of water dissolves 0.017 g oxygen at 20°C; if this pressure is halved (by doing the experiment at high altitude where the pressure is only 0.5 atm), the water dissolves only 0.0085 g oxygen. The law holds true only for equilibrium conditions, i.e., when enough time has elapsed so that the quantity of gas dissolved is no longer changing.

HEPARIN. [CAS: 9005-49-6]. A complex organic acid (mucopolysaccharide) present in mammalian tissue; a strong inhibitor of blood coagulation. Precise chemical formula has not been fully established, but the formula $(C_{12}H_{16}NS_2Na_3)_{20}$, with a molecular weight of 12,000, has been suggested for sodium heparinate. The drug is derived from animal livers or lungs. Heparin is used in deep venous thrombosis therapy. It is also used in rodenticides which cause internal hemorrhaging. Pets exposed to such poisons must receive immediate treatment with the administration of vitamin K, also sometimes called the antihemorrhagic vitamin. See also **Anticoagulants**; and **Vitamin K**.

HERBICIDES

Herbicide Classes and Databases

Herbicides can be classified as selective and nonselective. Selective herbicides, like 2,4-D (2,4-dichlorophenoxyacetic acid), metolachlor, and EPTC, are more effective against some types of plants than others, e.g., broadleaved plants vs grasses. Glyphosate is representative of the nonselective herbicides used for total vegetable control.

The classes of herbicidally active toxophores are limited in number. Arbitrary classification by toxophore reveals eight generic herbicide groupings, i.e., triazines, amides (haloacetanilides), carbamates, toluidines (dinitroanilines), ureas, plant growth hormones (phenoxy acids), diphenyl ethers, and miscellaneous unrelated compounds. Classification of commercial herbicides by chemical structure yields 10 related groupings with subgroups, i.e., phenoxy alkanoic acids; bipyridiniums; benzonitriles with phthalic compounds; dinitroanilines; acid amides; carbamates; thiocarbamates; heterocyclic nitrogen compounds including triazines, pyridines, pyridazinones, sulfonylureas, and imidazoles; substituted ureas; and miscellaneous groupings that include halogenated aliphatic carboxylic acids, inorganics and organometallics, and derivatives of biologically important amino acids.

Herbicides are also sometimes classified according to mode of action, selectivity, registered uses, and toxicity. The ever-increasing importance of herbicides and other pesticides and agrochemicals to a wide range of users, regulators, and researchers has led to the development of multiple and extensive computer databases. The primary database resources contain collected information relevant to herbicides, and numerous resource publications are available to those needing information on the various aspects of herbicides.

Database	Type of information
Agribusiness	agricultural chemicals and finance; agribusiness companies, product development history, and government policies
Agricola	National Agricultural Library database; general coverage of U.S. agriculture
Agrochemicals handbook	active components of agrochemicals
BIOSIS previews (Biological Abstracts)	research literature in biological, biomedical, and life sciences
Chemical Abstracts search	chemical literature and applications
CAB Abstracts	general agricultural and biological information, including weed science, from U.S. sources
Claims/U.S. patent abstracts	patents issued by the U.S. Patent Office; chemical patent records from 1950 to present
Derwent world patents index	patent data from 30 patent-issuing authorities around the world; agricultural chemical patents from 1965 to present
EMMI	U.S. EPA Environmental Monitoring Methods Index, including regulatory lists, analytical methods, detection and regulatory limits
Enviroline	coverage of worldwide environmental information
Pollution abstracts	references to environmental literature including pollutant sources and control
Toxline	National Library of Medicine toxicological information; references to pesticides, herbicides, environmental pollution, carcinogenic chemicals, food contamination, toxicological analyses
Toxnet	National Library of Medicine Toxicology Data Network, including Hazardous Substances Data Bank (HSDB), Registry of Toxic Effects on Chemical Substances (RTECS), Toxic Chemical Release Inventory (TRI), plus several other toxicological/carcinogenesis-related files

Herbicide Development

Examination of the various classified listings of herbicides provides insight into the processes and approaches that lead to the discovery of new pesticides. The four principal development approaches are random screening, imitative chemistry, testing natural products, and biorational development.

Quantitative Structure–Activity Relationship Design.

Increasing economic pressures toward more, better, and cheaper pesticides have led to the development and application of the Quantitative Structure–Activity Relationship (QSAR) paradigm and related experimental design principles for pesticides. Theoretically, quantitative determination of the relationships between chemical structure and biological and environmental properties of

a molecule should permit the design of a novel molecule with exactly those properties considered ideal for the intended application.

Modes of Herbicide Action

Modes of herbicide action include Photosystem I inhibitors [acifluorfen ($C_{14}H_{17}ClF_3NO_5$), nitrofen ($C_{12}H_7Cl_2NO_3$), oxyfluorofen ($C_{15}H_{11}ClF_3NO_4$)], electron transport between Photosystem I and Photosystem II inhibitors, Photosystem II inhibitors [atrazine ($C_8H_{14}ClN_5$), metribuzin ($C_8H_{14}N_4O_5$), diuron ($C_9H_{10}Cl_2N_2O$), bromacil ($C_9H_{13}BrN_2O_2$), ioxynil ($C_7H_3I_2NO$), dinoseb ($C_{10}H_{12}N_2O_5$), bromoxynil ($C_7H_3Br_2NO$), dinitrocresol ($C_7H_6N_2O_5$)], bleaching herbicides [fluridone ($C_{19}H_{14}F_3NO$), flurochloridone ($C_{12}H_{10}Cl_2F_3NO$), flurtamone, S3442 ($C_{17}H_{19}NO_2$), diflufenican ($C_{19}H_{11}F_5N_2O_2$), difunon ($C_{14}H_{12}N_2O_2$), norflurazon ($C_{12}H_9ClF_3N_3O$), amitrole ($C_2H_4N_4$), fluometuron ($C_{10}H_{11}F_3N_2O$), fomesafen ($C_{15}H_{10}ClF_3N_2O_6S$)], chlorophyll biosynthesis inhibitors [oxadiazon ($C_{15}H_{18}Cl_2N_2O_3$), DTP, MK−616 ($C_{14}H_{12}ClNO_2$), clethodim ($C_{17}H_{26}ClNO_3S$), sethoxydim ($C_{17}H_{29}NO_3S$), haloxyfop, methyl ($C_{16}H_{13}ClF_3NO_4$), tralkoxydim, fenoxaprop, ethyl ($C_{18}H_{16}ClNO_5$), fluazifop, butyl ($C_{19}H_{20}F_3NO_4$), alachlor ($C_{14}H_{20}ClNO_2$), metolachlor ($C_{15}H_{22}ClNO_2$), diclofop, methyl ($C_{16}H_{14}Cl_2O_4$), CDEC ($C_8H_{14}ClNS_2$), diallate ($C_{10}H_{17}Cl_2NOS$), EPTC ($C_9H_{19}NOS$), triallate ($C_{10}H_{16}Cl_3NOS$), metflurazon ($C_{13}H_{11}ClF_3N_3O$)], lipid and wax synthesis inhibitors, radical damage to antioxidative systems and cellular components inducers [paraquat ($C_{12}H_{14}N_2$), diquatop ($C_{12}H_{12}N_2$), tridiphane ($C_{10}H_7Cl_5O$)], herbicidal inhibition of enzymes [MAA (CH_5AsO_3), MSMA ($CH_5AsO_3 \cdot Na$), DSMA ($CH_5AsO_3 \cdot 2Na$), AMA, cacodylic acid ($C_2H_7AsO_2$), glufosinate ($C_5H_{12}NO_4P$), ammonium glufosinate ($C_5H_{12}NO_4P \cdot H_3N$)], amino acid and nucleotide biosynthesis inhibitors [phaseolotoxin ($C_{12}H_{33}N_8O_9PS$), glyphosate ($C_3H_8NO_5P$), rhizobitoxine ($C_7H_{14}N_2O_4$), chlorsulfuron ($C_{12}H_{12}ClN_5O_4S$), chlorimuron, ethyl ($C_{15}H_{15}ClN_4O_6S$), sulfometuron ($C_{15}H_{16}N_4O_5S$), bensulfuron, methyl ($C_{16}H_{18}N_4O_7S$), imazaquin ($C_{17}H_{17}N_3O_3$), imazapyr ($C_{13}H_{15}N_3O_3$), imazethapyr ($C_{15}H_{19}N_3O_3$), imazamethabenz ($C_6H_{20}N_2O_3$)], cell division inhibitors [trifluralin ($C_{13}H_{16}F_3N_3O_4$), oryzalin ($C_{12}H_{18}N_4O_6S$), pendimethalin ($C_{13}H_{19}N_3O_4$), nitralin, dinitramine ($C_{11}H_{13}F_3N_4O_4$), asulam ($C_8H_{10}N_2O_4S$), propham ($C_{10}H_{13}NO_2$), chloropropham ($C_{10}H_{12}ClNO_2$), barban ($C_{11}H_9Cl_2NO_2$), butylate ($C_{11}H_{23}NOS$), cycloate ($C_{11}H_{21}NOS$), propachlor ($C_{11}H_{14}ClNO$), DCPA ($C_9H_9Cl_2NO$), pronamide ($C_{12}H_{11}Cl_2NO$), bensulfide ($C_{11}H_{24}NO_4PS_3$), cinmethylin ($C_{18}H_{26}O_2$)], and plant growth regulator synthesis and function inhibitors [(naphthalene acetic acid) ($C_{12}H_{10}O_2$), indolebutyric acid ($C_{12}H_{13}NO_2$), 2,4-D ($C_8H_6Cl_2O_3$), 2,4,5-T ($C_8H_5Cl_3O_3$), MCPA ($C_9H_9ClO_3$), dicamba ($C_8H_6Cl_2O_3$), chloramben ($C_7H_5Cl_2NO_2$), picloram ($C_6H_3Cl_3N_2O_2$), naptalam ($C_{18}H_{13}NO_3$), TIBA ($C_7H_3I_3O_2$), diclofop ($C_{15}H_{12}Cl_2O_4$), ethephon ($C_2H_{16}ClO_3P$), tetcyclacis ($C_{13}H_{12}ClN_5$), AMO−1618 ($C_{19}H_{31}N_2O_2 \cdot Cl$), chlormequat chloride ($C_5H_{13}ClN \cdot Cl$), mepiquat chloride ($C_7H_{16}N \cdot Cl$), ancymidol ($C_{15}H_{16}N_2O_2$), uniconazole ($C_{15}H_{18}ClN_3O$), paclobutrazol ($C_{15}H_{20}ClN_3O$), BAS 11100W ($C_{16}H_{23}N_3O_2$)].

Environmental Fate of Herbicides

Herbicide Fates in Plants. Beyond modes of action and structure–activity relationships, developers of new herbicides must also consider uptake by plants, translocation within the plant, and possible deactivation of herbicides by contact with soil. Some of these problematic factors can be addressed as part of the QSAR studies and during the screening process. Considerable attention is also being paid to the use of safeners which protect the crop from herbicides that specifically target the weeds usually associated with that crop. Environmental protection and pesticide regulation concerns are the driving forces in the current efforts toward minimizing application rates, optimizing delivery through improved formulations and application equipment, and increasing target specificity. These research and development efforts include other important and related areas of interest to chemists, e.g., the fate and detection of herbicides in the soil and ground and surface water.

Factors Affecting Environmental Fate. The fate of herbicides in the environment is influenced by many chemical, biological, and physical factors. The principal transport and dissipation pathways include sorption to organic and mineral soil and sediment constituents; transport to groundwater in the solution phase by mass flow and/or diffusion; transport to surface water in either the solution or sorbed phases; loss to the atmosphere through volatilization, with redeposition at a later time and location; transformation or mineralization by biological, chemical, or photochemical processes; and uptake by plant or animal species. These processes do not operate as isolated systems, but occur simultaneously and involve significant interaction and feedback. Although the environmental fates of most herbicides are controlled primarily by one or two of the outlined processes, all of these factors influence the fate to some extent.

Measurement of Environmental Fate. Continued concern is expressed over the potential contamination of surface and groundwaters by agricultural chemicals. Herbicides have received much of this attention, due to their widespread use and the large total volume applied. However, this perceived threat to groundwater resources appears to be largely unfounded. A survey of private wells and public water well supplies in the United States has revealed that <1% contain herbicides at levels that would affect human or animal health. In addition, contaminated sources can usually be attributed to point rather than nonpoint sources. Nonpoint sources are generally treated by modifications in agricultural management practices. Typical modifications would include the use of alternative herbicide formulations, the splitting of the herbicide application in time, or the installation of vegetative buffer strips to trap runoff.

A re-evaluation of the water quality problem has revealed that surface water resources, rather than groundwater resources, are at higher risk of contamination from agricultural chemicals.

The public health implications of drinking water contamination by herbicides are unclear. The levels that have been detected in groundwater are generally in the part per billion (ppb) or part per trillion (ppt) range and are below estimated acute toxicity levels. However, the long-term health effects of this exposure are generally unknown. Several studies have demonstrated that the mortality from some types of cancer is significantly higher in rural residents of many corn belt states. The U.S. Environmental Protection Agency (EPA) developed (ca 1993) a classification scheme in an attempt to further evaluate the carcinogenic potential of herbicides and pesticides. In this system, chemicals are placed in one of five groups, A–E, according to their carcinogenic potential, ranging from definite (A) human carcinogens to no evidence of carcinogenicity for humans (E). The principal difference between these groups is the amount of accumulated evidence demonstrating carcinogenic potential.

This classification scheme is used in part in the determination and calculation of health advisory (HA) drinking water levels or carcinogenic risk estimates. The majority of herbicides in use in the United States for which HAs have been issued fall into Group D, with a smaller percentage falling into Group C. This would indicate that there are insufficient data to classify the carcinogenic potential of many herbicides. The lack of data does indicate however, that further testing will be required before the carcinogenic potential of many herbicides is known. Based on available HAs and the U.S. EPA classification scheme, acifluorfen, alachlor, amitrole, haloxyfop–methyl, lactofen, and oxadiazon have been listed as B2 carcinogens. Further information on carcinogenic risk assessment is available.

Since 1984, dramatic technical advances have been made in the analysis of trace organic chemicals in the environment. Indeed, these advances have been largely responsible for the increased public and governmental awareness of the wide distribution of herbicides in the environment. The ability to detect herbicides at ppb and ppt levels has resulted in the discovery of trace herbicide residues in many unexpected and unwanted areas. The realization that herbicides are being transported throughout the environment, albeit at extremely low levels, has caused much public and governmental concern. However, the public health implications remain unclear.

Traditionally, herbicides have been analyzed by gas chromatography (gc) or spectrophotometric methods. The method of choice when accuracy and sensitivity are of the utmost importance is gc, especially when combined with mass spectrometry. However, several other methods are used for routine monitoring or screening purposes. High pressure liquid chromatography (hplc) provides detection limits that nearly rival gc and require significantly less sample preparation and cleanup. Advances in the 1980s have made thin-layer chromatography (tlc) a valuable tool in herbicide analysis. Another analytical tool that has received much attention and shows great promise for routine analysis is enzyme immunoassay (eia). This technique offers the advantages of a low cost analysis, few interferences, high specificity and sensitivity, and a minimal amount of sample preparation.

A mobility ranking based on soil thin-layer chromatography (stlc) is used to classify the herbicide leaching potential of various herbicides. The rankings range from I (immobile) to V (very mobile) with intermediate categories of II (low mobility), III (intermediate), and IV (mobile). This method is widely used and has been accepted for submission of leaching data for herbicide registration purposes by the U.S. EPA. A comprehensive search of the STORET water quality database, maintained by the U.S. EPA Office of Water, is used to evaluate the potential water quality implications of various herbicides.

Herbicide Groups

Herbicides can be grouped according to common structural features. Sometimes the assignment is arbitrary when there are a multitude of functional groups, e.g., acifluorfen which is a diphenyl ether (phenoxy compound) as well as a trifluoromethyl compound.

Phenoxyalkanoics. The phenoxyalkanoic herbicide grouping is composed of two subgroups, the phenoxyacetic acids and the phenoxypropionic acids. They are widely used for foliar control of broadleaf weeds. The more heavily functionalized phenoxypropionic acid herbicides are relatively new herbicides compared to the phenoxyacetic acids and are used primarily for selective control of grassy weeds in broadleaf crops.

Considerable concern has been raised over the carcinogenic potential of the phenoxyacetic acid herbicide 2,4-D. However, the World Health Organization (WHO) has evaluated the environmental health aspects of this chemical and concluded that 2,4-D posed an insignificant threat to the environment. They did indicate, however, that only limited data on toxicology in humans are available. An HA has been issued for the phenoxyacetic acid herbicide MCPA. It was found in 4 of 18 SW samples analyzed and in none of 118 GW samples, and has been placed in group D for carcinogenic potential. EPA has published two gc methods for the analysis of the phenoxyalkanoic herbicides.

Bipyridiniums. The bipyridinium herbicides (Table 1), paraquat and diquat, are nonselective contact herbicides and crop desiccants. Diquat is also used as a general aquatic herbicide. Paraquat and diquat are much more toxic than most herbicides, and ingestion of sufficient quantities can result in death if prompt medical treatment is not obtained.

Benzonitrile, Acetic Acid, and Phthalic Compounds. Benzonitrile herbicides (Table 1) are generally used for pre-emergence and post-emergence control of broadleaf weeds. Dichlobenil also controls grass weeds and dichlobenil, endothall, and fenac are used as aquatic herbicides. Most benzonitriles are selective in their control. Benzonitrile herbicides are acidic in nature, thus their environmental fate is influenced by changes in soil pH. Sorption of these herbicides is expected to increase with decreasing pH.

Dinitroanilines and Derivatives. Dinitroaniline herbicides are used principally for the selective, pre-emergence control of annual grasses and broadleaved weeds.

Acid Amides. The principal use of acid amide herbicides is the selective control of seedling grass and certain broadleaved weeds. The majority of acid amide herbicides are applied pre-emergence or pre-plant incorporated, except for propanil which is applied post-emergence.

Phenylcarbamates. Phenylcarbamate herbicides represent one of two subgroups of carbamate herbicides, the phenylcarbamates and the thiocarbamates. The carbamate herbicides are used, in general, for the selective pre-emergence control of grass and broadleaved weeds. Exceptions would include barban, desmedipham, and phenmedipham, which are applied post-emergence.

Thiocarbamates. Thiocarbamate herbicides are nonionic. Diallate and triallate were strongly sorbed to both cation- and anion-exchange resins but minimally to kaolinite or montmorillonite. This behavior suggests a physical, rather than ionic mechanism of attraction.

Triazines. Triazine herbicides are one of several herbicide groups that are heterocyclic nitrogen derivatives. Triazine herbicides include the chloro-, methylthio-, and methoxytriazines. They are used for the selective pre-emergence control and early post-emergence control of seedling grass and broadleaved weeds in cropland. In addition, some of the triazines, particularly atrazine, prometon, and simazine, are used for the nonselective control of vegetation in noncropland. Simazine may be used for selective control of aquatic weeds.

Pyridines and Pyridazinones. Pyridine herbicides are auxin-type herbicides generally used for selective control of broadleaved weeds in cropland, rangelands, and noncroplands. The pyridazinones are used primarily for the

TABLE 1. ENVIRONMENTAL HEALTH ADVISORIES FOR HERBICIDES

Herbicide	Health advisories[a] SW	GW	Mobility[b]	Carcinogenic potential group[c]	Analytical methods[d]
BIPYRIDINIUM COMPOUNDS					
diquatop			immobile		hplc
paraquat		0/843	immobile	E	hplc
BENZONITRILE, ACETIC ACID, AND PHTHALIC COMPOUNDS					
chloramben	13/34	1/566	very mobile	D	gc[e]
DCPA	386/1995	12/982		D	gc[e]
dicamba	262/806	2/230	very mobile	D	gc[e]
dichlobenil			low		
endothall	0/3	0/604		D	gc
naptalam					uv
DINITROANILINE AND DERIVATIVES					
benefin					gc
dinitramine					gc
dinoseb	1/89	0/1270		D	
fluchloralin					gc
oryzalin					uv
pendimethalin					gc
trifluralin	172/2047	1/507	immobile	C	ir
ACID AMIDES					
alachlor					gc
bensulide			immobile		hplc
diphenamide	0/3	0/678	intermediate	D	gc
metolachlor	2091/4161	13/596		C	gc
napropamide					
pronamide	20/391			C	gc
propachlor	34/1690	2/99	intermediate	D	gc
propanil			low		
PHENYL CARBAMATES					
chloropropham			low		
karbutilate					hplc
propham	1/392	0/583	intermediate	D	hplc
THIOCARBAMATES					
asulam					uv
butylate	91/836	2/152		D	gc, glc
EPTC					hplc
thiobencarb			relatively immobile		gc, glc
triallate					gc, glc
vernolate					hplc
TRIAZINES					
ametryn	2/1190	24/560	intermediate	D	general[f]
atrazine	4123/10,942	343/3208	intermediate	C	gc
cyanazine	1708/5297	21/1821	intermediate	D	ir
hexazinone			relatively immobile	D	gc
metribuzin	938/4651	0/416		D	general[f]
prometon	386/1419	36/746	intermediate	D	gc
prometryn			low		
propazine	33/1097	15/906	intermediate	C	general[f]
simazine	922/5873	202/2654	intermediate	C	gc
terbutryn					general[f]
PYRIDINES					
clopyralid			minimal[g]		
fluroxypyr			varied		
picloram	420/744	3/64	mobile[g]	D	general[h]
triclopyr			intermediate		hplc
PYRIDAZINONES					
norflurazon			low		
pyrazon					uv
SULFONYLUREAS					
chlorimuron, ethyl			mobile		
chlorsulfuron			intermediate to very mobile		hplc, gc
metsulfuron, methyl					gc
sulfometuron			mobile to very mobile		
IMIDAZOLE COMPOUNDS					
buthidazole					eia
imazamethabenz					eia
imazapyr					eia
imazaquin			mobile to very mobile[i]		eia
imazethapyr			immobile to mobile[i]		eia
OTHER HETEROCYCLIC NITROGEN DERIVATIVES					
amitrole			mobile		vis
bentazon			very mobile	D	hplc
isoxaben			immobile		
UREAS AND URACILS					
bromacil	0/3	0/841	mobile	C	glc
chloroxuron			immobile		glc
diuron	0/25	0/1337	low	D	ir
fluometuron	0/14	0/156	intermediate	D	uv
linuron					uv
tebuthiuron			intermediate to very mobile	D	uv
terbacil				E	uv
ALIPHATIC-CARBOXYLIC					
dalapon	0/14	0/14	very mobile	D	ir
TCA			very mobile		
INORGANICS AND METAL ORGANICS					
AMS				D	titration
MISCELLANEOUS TRIFLUOROMETHYL COMPOUNDS					
acifluorfen				B$_2$	hplc
fluridone					gc
lactofen					hplc
AMINO ACID ANALOGUES					
glufosinate, glyphosate			intermediate		
	0/6	0/98	immobile to low mobility	D	hplc
OTHER MISCELLANEOUS COMPOUNDS					
cinmethylin					gc
ethofumesate					gc
tridiphane					

[a] SW = surface water; GW = ground water. Positive results/number of tests.

[b] Mobility ranking based on soil thin-layer chromatography (stlc).

[c] Group A, human carcinogen; Group B, probable human carcinogen; Group C, possible human carcinogen; Group D, not classifiable; Group E, no evidence of carcinogenicity for humans.

[d] gc = gas chromatography; hplc = high pressure liquid chromatography; ir = infrared spectroscopy; uv = ultraviolet spectroscopy; glc = gas-liquid chromatography; eia = enzyme immunoassay; vis = visible spectroscopy.

[e] Gc for chlorinated pesticides can be used.

[f] General draft method for nitrogen- and phosphorus-containing pesticides.

[g] Mobility has been reported to be mobile and minimal in different studies.

[h] General draft method for determination of chlorinated acids in water.

[i] Mobility is a function of soil pH.

selective pre- and post-emergence control of seedling grass and broadleaved weeds in cotton and sugarbeets.

Sulfonylureas. Sulfonylurea herbicides are a relatively new class of herbicides generally used for selective pre- and post-emergence control of broadleaved weeds in croplands. Sulfometuron—methyl is used for broad-spectrum selective or nonselective weed control in noncroplands.

Imidazoles. Imidazole herbicides are generally used for selective pre- and post-emergence control of grass and broadleaved weeds in croplands. Buthidazole and imazapyr are used for broad-spectrum, nonselective weed control in noncroplands.

Other Heterocyclic Nitrogen Derivative Herbicides. The herbicides in this group are heterocyclic nitrogen derivatives that do not readily fall into one of the previously discussed groups. They have a wide range of uses and properties. Most of these herbicides are used for selective, pre- and/or post-emergence weed control. Amitrole is used for post-emergence, nonselective weed control in noncroplands and also as an aquatic herbicide.

Ureas and Uracils. Urea herbicides are generally used for selective pre-emergence and early post-emergence control of seedling grass and broadleaved weeds. Uracil herbicides are generally used for selective control of annual and perennial weed control in certain crops and for general weed control in noncrop areas. Bromacil, linuron, and tebuthiuron are used for the nonselective control of weeds in noncropland.

Aliphatic–Carboxylics. These are used primarily for the selective control of annual and perennial grass weeds in cropland and noncropland. Dalapon is also used as a selective aquatic herbicides.

Metal Organics and Inorganics. The metal organic herbicides are arsenicals used for the selective, post-emergence control of grass and broadleaved weeds in cropland and noncroplands.

Miscellaneous Trifluoromethyl Compounds. The herbicides in this group are used for a wide variety of weed-control purposes. Acifluorfen, lactofen, and oxyfluorfen are used for selective, pre-, and post-emergence weed control in croplands. Fluorochloridone is used for selective, pre-emergence weed control in cropland, and fluridone, fomesafen, and mefluidide are used for post-emergence control. Fluridone is also used as an aquatic herbicide.

Amino Acid Analogues. Amino acid analogue herbicides also control a large variety of weeds. Glyphosate and glufosinate are used for the broad-spectrum, nonselective control of grass and broadleaved weeds. Diethatyl is used for selective, pre-emergence control of grass and broadleaved weeds. Flamprop is used to control the growth of wild oats in wheat.

Miscellaneous Other Herbicides. The herbicides in this group are not readily included in any of the preceding groups. Acrolein (2-propenal) is used as a contact, aquatic herbicide. Sethoxydim, clethodim, and tridiphane are used for selective, post-emergence weed control. Cinmethylin and clomazone are used for selective pre-emergence control and ethofumesate for selective pre- and post-emergence weed control.

Economic Aspects

During the period from 1979 through 1991, the estimated U.S. total annual volume of herbicide usage increased somewhat from 254 million kg active ingredient (AI) to 285 million kg. Peak herbicide usage of 306 million kg occurred in 1984. During the years between 1979 and 1992, agricultural uses accounted for 76 to 81% of the herbicide applied in the United States. Combined U.S. government, industrial, and commercial herbicide usage during those years ranged from 14% in 1979, to 19% in 1986, and 17% in 1991. Home gardens and lawns received the remaining 4 to 5%.

Based on 1990–1991 estimates, the most used herbicides in the United States are, in descending order of usage, atrazine, alachlor, meto-lachlor, 2,4-D, trifluralin, cyanazine, EPTC, metham—sodium, glyphosate, and butylate.

Although the ratios have varied from year to year since 1979, the selective herbicides used in corn production have accounted for approximately 21% of herbicide use on a per crop basis. Herbicide use in soybean and cotton production combined account for ca 23% of the selective herbicide market. Graminicides, which selectively kill grasses, constitute 40% of the total market, leaving a market share of approximately 16% for the nonselective herbicides.

Innovative Weed Management Agents

Adoption by the agricultural community requires that an innovative weed management agent must be an effective control of the target species, be cost-effective, and be practical to employ. It must not interfere with crop production practices such as crop rotation or the use of other pesticides. Additionally, new weed-control agents cannot pose a significant threat to human health or the environment. Considerable costs are incurred in the development, registration, production, and marketing of weed control agents. These costs require that an herbicide have sufficient long-term market viability and market niche potential to justify these costs in time and money. The need for safe and effective methods of crop production in an environment that contains competitive weeds is becoming increasingly critical.

Weed Management Strategies. Managers of agroecosystems are being encouraged to manage weed populations at levels that are below their economic optimum thresholds, rather than attempting to eliminate or control all noncrop plants, regardless of their actual impact. Decisions concerning management of weed populations should be governed by both agroecological principles and site-specific considerations in the context of an overall integrated pest management program. However, the practical implementation of integrated pest management (IPM) programs can be difficult.

Nonchemical or traditional practices, such as weed seed removal, optimal crop seeding rates, crop selection, enhanced crop competitiveness, crop rotation, and mechanical weed control are all important components of an effective weed management program. In the context of modern intensive chemical herbicide application, nonchemical practices may represent an innovative approach to weed management and should receive careful consideration.

Natural Products and Allelopathic Compounds as Herbicides. There is growing concern that compounds that do no occur in nature may produce unanticipated health and environmental problems. However, plants, fungi, marine organisms, and certain bacteria produce a vast array of organic compounds, and many of these natural products exhibit biological activity. In nature, these compounds are produced in minute quantities and present interesting chemical problems in detection, identification, quantification, and production of active and stable analogues. Although these compounds appear to be ecologically safe in naturally occurring amounts, the large quantities required for agricultural applications may cause environmental problems similar to those associated with chemical herbicides.

Investigations of natural product chemistries have aided in the development of bialaphos, cinmethylin, picloram, glufosinate, and other important herbicides. Additional compounds may be found through investigations of natural products that cause plants and other organisms to undergo rapid physiological change, such as plant hormones and phytotoxins. Many plant hormones and phytotoxins are also produced by microorganisms. Additionally, microorganisms have been reported to contain novel natural products that could provide basic structural templates for the development of new herbicides.

Plant Pathogens and Insects as Control Agents. Concerns about accumulations of chemical control agents in the environmental and food resources have also increased interest in microbial weed control agents. Controlling weeds with carefully screened plant pathogens offers several benefits, including a high degree of specificity for a given target weed, low potential for negative human health and environmental impact, inability to accumulate in the food chain, and other advantages. The high degree of host specificity may limit the market size for some biological control agents, but these bio-control agents can be combined with chemical herbicides and other pathogens to increase the spectrum of weeds controlled. The marketing of biological control agents may also be constrained by slow expression of phytotoxicity, pathogen dependence on optimum environmental conditions, potential resistance of the weed towards the pathogen, and lack of formulation stability under field conditions and during preuse storage. These constraints can be addressed by genetic manipulation of selected pathogenic strains to produce more effective control agents and by the investigation of the mechanisms of disease resistance in plants.

There are two principal approaches to the biological control of weeds. The first approach is referred to as classical or inoculative biological weed control. The intent of classical biological weed control approaches is the management of introduced weed populations by introduction of host-specific pathogens from the weed's native range, thus moderating the growth of weed populations by the reestablishment of an old association between host and pathogen populations in the expanded range.

An additional approach to biological weed control is referred to as the inundative or augment approach to biological weed management. This approach utilizes pathogenic propagules formulated as a weed control agent, e.g., mycoherbicides. The mass-inoculation of pathogenic propagules in an effective formulation can enhance the dissemination and survival of the pathogens, overwhelm target weed resistance, and produce results similar to those achieved with chemical herbicides. Mycoherbicides often contain native pathogens that are active against native weeds and are thus highly selective against the target weed species.

Research concerning plant pathogen control agents has resulted in two commercially available mycoherbicides. The mycoherbicide Collego is a formulated product consisting of propagules of the fungus *Colletotrichum gloeosporioides*, and Devine is a formulated product containing the sexual spores of the oomycete, *Phytophthora palmivora*. Devine is used to control stranglevine (*Morrenia odorata*) in citrus, and Collego is used in northern jointvetch (*Aeschynomene virginica*) control in rice and soybean.

Control of Weed Seeds. If agents that control weed seed germination could be applied prior to planting, interference from weeds would be prevented until reintroduction of weed propagules. Additionally, if a very large portion of the weed seed bank could be stimulated to germinate prior to planting, weeds could be controlled by a single cultivation or application of nonselective herbicide.

Development of agents that stimulate weed seed germination and/or attack weed seeds would have a profound impact on weed management. However, herbicide development programs in the early 1990s do not focus on identifying agents that are effective on weed propagules. A systematic search for compounds that render weed seeds nonviable or cause them to germinate simultaneously could provide important new weed management tools.

JUDITH M. BRADOW
CHRISTOPHER P. DIONIGI
RICHARD M. JOHNSON
SUHAD WOJKOWSKI
U.S. Department of Agriculture

Additional Reading

Cobb, A.H., and R.C. Kirkwood: *Herbicides and Their Mechanisms of Action,* CRC Press LLC., Boca Raton, FL, 2000.

Draber, W. and T. Fujita, eds.: *Rational Approaches to Structure, Activity, and Ecotoxicology of Agrochemicals,* CRC Press, Boca Raton, FL, 1992.

Grover, R. and A.J. Cessna, eds.: *Environmental Chemistry of Herbicides,* Vols. 1 and 2, CRC Press, Boca Raton, FL, 1991.

Hakansson, S.: *Weeds and Weed Management on Arable Land: An Ecological Approach,* CAB International, New York, NY, 2003.

Herbicide Handbook, 7th Edition, Weed Science Society of America, Champaign, IL, 1994.

Liebman, M., C.L. Mohler, and C.P. Staver: *Ecological Management of Agricultural Weeds,* Cambridge University Press, New York, NY, 2001.

Marrs, T.C., and B. Ballantyne: *Pesticide Toxicology and International Regulation,* John Wiley & Sons, Inc., New York, NY, 2003.

Monaco, T.J., S.C. Weller, S.C. Weller, and F.M. Ashton: *Weed Science: Principles and Practices,* 4th Edition, John Wiley & Sons, Inc., New York, NY, 2002.

U.S. EPA, *Manual of Chemical Methods for Pesticides and Devices,* 2nd Edition, Assoc. Off. Anal. Chem., Arlington, VA, 1992.

Whitehead, R. ed.: *U.K. Pesticide Guide* 1995, CAB International, Wallingford, Oxon, UK.

HEROIN. See **Alkaloids**.

HERSCHBACH, DUDLEY R. (1932–). Awarded the Nobel prize in chemistry in 1986 for work reporting that the energies of reactions of crossed molecular beams of isolated alkali metal atoms and alkyl halide molecules appeared mostly as vibrational excited states of products. This method of studying all types of chemical reactions led to a more detailed knowledge of reaction processes. Doctorate awarded from Harvard in 1958.

HERZBERG, GERHARD (1904–1999). A German-born physicist who won the Nobel prize for chemistry in 1971, for his work on the composition of molecules. His research involved the spectroscopy of atoms and molecules and their excitation behavior. He became a Canadian citizen and was the director of the Division of Pure Physics of the National Research Council of Canada.

HESSITE. A mineral telluride of silver, Ag_2Te, with some gold, crystallizing in the monoclinic system at normal temperatures; isometric system above 149.5°F (65.3°C). Crystalline form not obvious at normal temperatures. Hardness, 2–3; specific gravity, 8.24–8.45; color, gray with metallic luster; opaque. Named after G.H. Hess (1802–1850).

HETEROCYCLIC COMPOUNDS. See **Compound (Chemical)**; **Organic Chemistry**.

HETEROGENEOUS. (Latin "different kinds"). Any mixture or solution comprised of two or more substances regardless of whether they are uniformly dispersed. Common examples are such diverse materials as air (a mixture of 20% oxygen and 80% nitrogen), milk, marble, paint, gasoline, blood and mayonnaise. In all such cases, the mixtures can be separated mechanically into their components. "Homogenized" milk is a heterogeneous as regular milk and the term is, strictly speaking, a misnomer.

See also **Homogeneous**.

HETEROPOLYACIDS. Acids derived from two or more other acids, under such conditions that the negative radicals of the individual acids retain their structural identity within the complex radical or molecule formed. The term heteropolyacids is usually restricted to complex acids in which both radicals are derived from oxides, such as phosphomolybdic acid.

HETROMOLYBDATES. A large group of complex molybdenum salts and acids in which the anion contains oxygen atoms and from 2 to 18 hexavalent molybdenum atoms, as well as one or more other metal or nonmetal atoms (phosphorus, arsenic, iron, and tellurium). The latter are referred to as hetero atoms, and any of approximately 35 elements may be present in this manner. Example: $Na_3PMo_{12}O_{40}$, sodium phospho-12-molybdate. The molecular weights of these compounds range up to 3000. The acids and most of the salts are very soluble in water, and the acids and some salts are soluble in organic solvents.

HEULANDITE. The mineral heulandite is a monoclinic zeolite whose crystals are often quite suggestive of orthorhombic forms. Its chemical composition is probably $(Na, Ca)_{4-6}Al_6(Al, Si)_4Si_{26}O_{72} \cdot 24H_2O$; strontium may be present. Heulandite has one good cleavage; is brittle with a conchoidal fracture; hardness, 3.4–4; specific gravity, 2.18–2.22; luster, vitreous to pearly; color, white to gray, red or brown; streak, white; transparent to translucent. Occurs chiefly in cavities in basaltic rocks with other zeolites, but may be found in granites, pegmatites, gneisses, and schists. Famous localities are in Iceland, India, the Harz Mountains, Italy, Switzerland, Scotland, Nova Scotia; and in the United States at Bergen Hill and West Paterson, New Jersey. This mineral was named for the English mineralogist Heuland.

HEVESY, GEORG de (1885–1966). A Hungarian chemist who won the Nobel prize in chemistry in 1943, for his work on the use of isotopes as tracers in the study of chemical processes. He discovered the element hafnium in 1923. One of his interesting projects involved the calculation of the percentages of chemical elements in the universe. He also was involved in research using radioactive lead and phosphorus traces. His work included the separation of isotopes by physical means. His Ph.D. was granted at Freiburg in 1908.

HEXAMINE. [CAS: 100-97-0]. $(CH_2)_6N_4$, formula weight 140.19, white crystalline solid, mp 280°C, decomposes at higher temperatures. Also known as hexamethylenetetramine, methenamine, and urotropine, the compound is soluble in H_2O and only very slightly soluble in alcohol or ether. Although used to some extent in medicine as an internal antiseptic, the primary use of hexamine is in the manufacture of synthetic resins where the compound is a substitute for formalin (aqueous solution of paraformaldehyde) and its NaOH catalyst. Hexamine also is used as an accelerator for rubber.

On a commercial scale, hexamine is manufactured from anhydrous NH_3 and a 45% solution of methanol-free formaldehyde. These raw materials, plus recycle mother liquor, are charged continuously at carefully controlled rates to a high-velocity reactor. The reaction is exothermic. The reactor effluent is discharged into a vacuum evaporator which also serves as a crystallizer. The hexamine crystals then are washed, dried, and screened. Average yield of the process is about 96% conversion of ingredients to produce hexamine.

HEXOSE MONOPHOSPHATE OXIDATIVE PATHWAY.   See Carbohydrates.

HEYROVSKY, JAROSLAV (1890–1967).   A Czechoslovakian physiochemist who won the Nobel prize for chemistry in 1959, for his discovery and development of the polarographic and oscillo-polarographic methods of analysis. Although his Ph.D. was from the University of Prague, he later studied in London.

HIGH TEMPERATURE ALLOYS.   High temperature alloys are those combinations of metals that are used specifically for their heat resisting properties. Physical properties such as melting temperatures, elastic modulus, density, and thermal conductivity of the elemental metals that serve as the bases for most high temperature alloys are listed in Table 1. Contrary to expectations, melting point is not the primary indicator of adequate high temperature strength. For example, nickel, which has the lowest melting point of any element in Table 1, is the choice for the most severe high temperature structural applications in air. There are several reasons for this, including the lack of an allotropic phase transformation in nickel below its melting point, its high tolerance for alloying elements without causing a phase change from the close packed face-centered cubic (fcc) crystal structure, and the ability to produce a very stable precipitate, γ'-(Ni_3Al), that is the primary source of high temperature strengthening in nickel-base superalloys.

Density is a particularly important characteristic of alloys used in rotating machinery, because centrifugal stresses increase with density. Alloys which contain the heavier elements, i.e., molybdenum, tantalum, or tungsten, have correspondingly high densities.

Thermal expansion coefficients are important for mating components, as well as in the development of thermal stresses, which vary directly with the expansion coefficients. In general, thermal expansion coefficients are inversely proportional to melting point.

Mechanical Behavior

Creep Rupture.   Metals and their alloys lose appreciable strength at elevated temperatures. For most materials, the ultimate tensile and yield strengths fall off regularly as the temperature increases. The exceptions are some intermetallics, e.g., nickel aluminide(3:1), Ni_3Al, and alloys containing a large volume fraction of Ni_3Al, e.g., single-crystal alloys PWA 1480. Not only is temperature important, but time also influences mechanical behavior at elevated temperatures. Stresses well below the yield strength can cause gradual deformation and eventual fracture if sustained long enough. For this reason, the standard tensile test is inadequate for providing design data at elevated temperatures, and creep-rupture tests must be conducted, in which time-dependent deformation and fracture are determined from periodic measurements under a fixed stress or load.

Relaxation.   Relaxation also is associated with creep at high temperatures. If a specimen is stretched or compressed and is then held over a period of time at a high temperature with its ends in fixed positions, the stresses within the specimen gradually diminish. After sufficient time has

elapsed, the tensile or compressive stresses may relax to only a fraction of the original values. Creep and relaxation are thus complementary processes, and an alloy having high creep strength generally resists relaxation.

Relaxation is an important example of a creep phenomenon encountered in practice. Bolts, studs, flanges, and springs of all kinds are subject to relaxation when used at high temperatures. Bolts can become loose so that bolted joints develop leaks after operation at elevated temperatures.

Fatigue.   Engineering components often experience repeated cycles of load or deflection during their service lives. Under repetitive loading most metallic materials fracture at stresses well below their ultimate tensile strengths, by a process known as fatigue. The actual lifetime of the part depends on service conditions, e.g., magnitude of stress or strain, temperature, environment, surface condition of the part, as well as on the microstructure.

Resistance to fatigue fracture is an important consideration in selecting materials for many high temperature applications, most notably in rotating machinery such as gas or steam turbines. Generally, two classifications of fatigue behavior are made, depending on the parameter being reversed, stress, or strain.

High temperature materials which exhibit the greatest resistance to high cycle fatigue on a strength basis, i.e., fatigue limit/tensile strength vs N_f, are composite materials and dispersion strengthened alloys. Precipitation hardened alloys, on the other hand, usually demonstrate very poor fatigue resistance relative to the tensile strengths, perhaps because the precipitates become unstable as a result of repeated cycling.

The rate of crack growth is often a more useful parameter than is fatigue life. Fracture mechanics techniques have been widely applied to the crack growth behavior of high temperature alloys.

Summary of Strengthening Methods.   In practice, few alloys are strengthened by only one or merely a few of the mechanisms described, as shown in Table 2.

Surface Stability

Oxidation.   Immense progress in technology has imposed everincreasing demands on the mechanical and chemical properties, in particular the oxidation and scaling resistance, of metallic materials.

The scale morphology is dependent on the conditions of reaction, the time of oxidation, the composition of the corrosive medium, and the type and composition of the particular alloy involved. Complex alloys may form two or more layers differing in either composition or microstructure or both. In order to maintain good oxidation resistance at least one of the layers must be compact and preferably be a slow growing oxide.

The oxidation of most modern alloys is dependent on the formation of a compact protective film of a slow growing chemically stable oxide such as chromium(III) oxide, C_2O_3, alumina, Al_2O_3, or silica, SiO_2. The oxidation behavior of multicomponent γ'-strengthened alloys can be estimated by considering the Ni, C_2, and Al, content of the alloy.

Hot Corrosion.   Hot corrosion is an accelerated form of oxidation that arises from the presence not only of an oxidizing gas, but also of a molten salt on the component surface. The molten salt interacts with the protective oxide so as to render the oxide nonprotective. Most commonly, hot corrosion is associated with the condensation of a thin molten film of sodium sulfate, Na_2SO_4, on superalloys commonly used in components for gas turbines, particularly first-stage turbine blades and vanes.

The deposition of molten Na_2SO_4 in gas turbines is believed to be related to the reaction between the residual sulfur in fuel and sodium, which may

TABLE 1. PHYSICAL PROPERTIES OF HIGH TEMPERATURE METALS

Metal	Melting point, °C	Density, g/cm^3	Thermal expansion coefficient at RT, 10^6/°C	Thermal conductivity at RT, W/(m · K)^a
Co	1495	8.85	13.8	69.0
Ni	1453	8.90	13.3	92
Fe	1537	7.87	11.76	75
Cr	1890	7.2	6.2	66.9
Nb	2468	8.6	7.1	52.3
W	3410	19.3	4.5	20.1
Ta	2996	16.6	6.6	54
V	1900	6.1	9.7	31
Mo	2610	10.22	5.4	146

^{a} To convert W/(m·K) to cal/(cm·s·°C), multiply by 2.39×10^{-3}.

TABLE 2. STRENGTHENING MECHANISMS IN HIGH TEMPERATURE ALLOYS

Alloys	Primary strengthening	Secondary strengthening
superalloys		
Ni-base	precipitation of γ'	solid soln
Co-base	solid soln	precipitation of carbides
bcc refractory metals	cold working	solid soln, precipitation
directionally solidified eutectics	composite	solid soln, precipitation
dispersion-strengthened alloys	dispersion	solid soln, grain size
intermetallics	long-range order	solid soln, precipitation

be contained either in the fuel or the intake air. The sodium in the air is normally present as an aerosol of sea salt.

For most alloys, the corrosion rate displays a maximum at 850–900°C, and decreases very rapidly at temperatures up to 1000°C, again strongly suggesting that a molten salt is necessary in order to initiate hot corrosion.

It is generally conceded that the chromium content is the most important factor in hot corrosion resistance. For this reason cobalt alloys, which generally contain 20% or more chromium, display better hot corrosion resistance than nickel alloys, which typically contain 8–15% chromium.

Coatings. It is common practice to apply some type of protective coating to extend the surface stability of superalloy or refractory metal components. Superalloy coatings provide an aluminum reservoir for growth of protective Al_2O_3 scales and inhibit further oxidation. Coating microstructures can be varied from the relatively brittle high aluminum content intermetallic matrix phase, e.g., CoAl or NiAl, to the relatively ductile low aluminum content metal plus intermetallic-type structures, e.g., Co + CoAl. During cyclic oxidation and hot corrosion, excessive spallation occurs. Coatings have been developed, particularly Co–25%Cr–2%Al–0.1%Y and Ni–15%Cr–6%Al–0.1%Y, that exhibit excellent resistance to thermal cycling. These overlay coating compositions are based on the knowledge that aluminum is required to form protective Al_2O_3, and chromium is necessary to enhance Al_2O_3 formation and to improve hot corrosion resistance further. Excellent resistance to spallation is generally attributed to the addition of yttrium. A compact adhesive $Al_2O_3/CoAl_2O_4$ scale is formed having a low parabolic growth rate, thereby protecting the underlying base alloy. Both $CoAl_2O_4$ and $NiAl_2O_4$ spinels grow rather slowly when compared with most oxides. Such coatings degrade after a period of time, because in addition to attack of the protective oxide by spallation, erosion, and chemical means, inward diffusion of aluminum and chromium occurs. Therefore, although coatings can improve the oxidation resistance, degradation can occur and oxidation of the alloy ensues. NiCoAlY compositions offer superior elevated temperature oxidation resistance and diffusional stability on nickel-base superalloys. Additions of cobalt to NiCrAlY enhance hot corrosion resistance and improve coating ductility. CoCrAlY coatings provide superior protection in a hot corrosion environment.

In the case of refractory metals, coatings generally are silicides, applied by pack cementation or slurry processes. Typical silicide compositions are Si–20Cr–20Fe for niobium alloys and $MoSi_2$ for molybdenum alloys.

Specific Alloy Systems

Plain Carbon and Low Alloy Steels. For the purposes herein plain carbon and low alloy steels include those containing up to 10% chromium and 1.5% molybdenum, plus small amounts of other alloying elements. These steels are generally cheaper and easier to fabricate than the more highly alloyed steels, and are the most widely used class of alloys within their serviceable temperature range.

Of the common alloying elements in steel, molybdenum is the most effective in increasing creep—rupture strength, and the carbon—molybdenum steels generally have more than twice the creep—rupture strength of plain carbon steel at the same temperature. The most commonly used steels for high temperature service contain from 0.5 to 1.5% molybdenum.

Chromium is the most effective addition to improve the resistance of steels to corrosion and oxidation at elevated temperatures, and the chromium—molybdenum steels are an important class of alloys for use in steam power plants, petroleum refineries, and chemical-process equipment. The chromium content in these steels varies from 0.5 to 10%. As a group, the low carbon chromium—molybdenum steels have similar creep—rupture strengths, regardless of the chromium content, but corrosion and oxidation resistance increase progressively with chromium content. Most of the chromium—molybdenum steels are used in the annealed or in the normalized and tempered condition; some of the modified grades have better properties in the quench and tempered condition.

Stainless Steels. Steels containing 11% and more of chromium are classed as stainless steels. The prime characteristics are corrosion and oxidation resistance, which increase as the chromium content is increased. Three groups of wrought stainless steels, series 200, 300, and 400, have composition limits that have been standardized by the American Iron and Steel Institute (AISI).

Although the stainless steels usually are specified in the wrought condition, a number of iron—chromium, iron—chromium—nickel, and nickel or cobalt-base alloys are produced as castings. Castings are classified as heat resistant when utilized in applications at 650°C or higher. Examples of Fe—Ni—Cr heat resisting castings are HC (equivalent to wrought AISI type 446), HH (AISI type 309), and HD (AISI type 327). These alloys are used in metallurgical furnaces, oil-refinery furnaces, power plant equipment, gas turbines, and in the manufacture of glass and synthetic rubber. Iron—chromium castings containing 10–30% Cr are useful chiefly for oxidation resistance, whereas iron alloys containing more than 18% Cr and more than 7% Ni have superior strength and ductility. Other iron-base alloys containing more than 10% Cr and more than 25% Ni are used in both reducing and oxidizing atmospheres.

The 12%-chromium ferritic superalloys are a group of proprietary steels that are essentially modifications of AISI 403 stainless steel. Examples are Crucible 422, Lapelloy (AISI 619), and Jessop-H46. The modifications include adding up to several percent of molybdenum and/or tungsten to stiffen the matrix, and up to 0.5% of niobium and vanadium to improve the dispersion and stability of the carbides. Up to 2% nickel, copper, and aluminum also may be present in these steels. The modified steels have a substantially greater creep—rupture strength than the standard AISI 403 stainless steel, and about the same level of corrosion and oxidation resistance. Typical applications include high temperature bolts, blades for jet-engine compressors and for high temperature steam turbines, compressor and turbine disks for jet engines, boiler, superheater, and reheater tubes and valve parts. These steels are available in most wrought forms.

The precipitation-hardening stainless steels are proprietary grades hardened by both the martensitic transformation and precipitation hardening. These contain higher amounts of chromium (16–17%) and nickel (4–7%) than the 12% chromium ferritic alloys. These steels are normally used at lower temperatures than the 12% chromium ferritic superalloys.

The highly alloyed austenitic stainless steels are proprietary modifications of the standard AISI 316 stainless steel. These have higher creep-rupture strengths than the standard steels, yet retain the good corrosion resistance and forming characteristics of the standard austenitic stainless steels.

Nickel-Base Superalloys. The nickel-base superalloys are the most complex in composition and microstructures and, in most respects, the most successful high temperature alloys. The earliest superalloys were wrought, ie fabricated to final size by a mechanical working operation. Later alloys have incorporated higher aluminum plus titanium contents, as well as molybdenum for solid-solution strengthening (Nimonics 115 and 120).

Alloys developed by processing through the investment casting process had higher strength and design flexibility, which led to many further advances through air cooling. The cast alloys tended to contain less chromium, which was replaced by molybdenum, tungsten, and tantalum, while retaining high volume fractions (to 60%) of γ'.

Apart from γ' and solid-solution strengthening, many alloys benefit from the presence of carbides, carbonitrides, and borides.

An undesirable feature of the most highly alloyed superalloys is the tendency to develop unwanted phases such as σ and μ. Sigma (σ) phase, a platelike intermetallic compound of two or more transition metals, e.g., Cr_xFe_y or $(CrMo)_x(NiCo)_y$ where x and y can vary from 1 to 7, may precipitate from alloys containing a high refractory metal content, e.g., IN-100. There is a critical temperature range, centered around 800°C, for the precipitation of σ, and precipitation leads to a decrease in rupture properties. Low temperature ductility also is adversely affected. The recognition that σ and other topologically close packed (tcp) phases are electron compounds where precipitation from solution can be predicted by knowledge of the average electron vacancy number of the alloy matrix was the basis of the Phacomp system for predicting safe alloy compositions. In brief, the total concentration of elements having high electron vacancy number (Cr, Mo, W, Mn) must be limited to avoid σ-phase precipitation, either during alloy processing or in service. Computer programs derived from the principles of electron vacancy numbers and phase stability are used in engineering specifications for superalloys utilized in aircraft engines and industrial gas turbines.

The temperature capability of nickel-base alloys has been improved markedly by processing techniques such as vacuum arc melting.

Cleanliness is critical in modern superalloys because of the role of inclusions in initiating fatigue cracks and fracture. Several refining processes to produce cleaner superalloys have been introduced, including the use of ceramic-foam filters in conjunction with vacuum induction melting, the reintroduction of electroslag remelting, and the development of electron-beam cold-hearth refining. Directional solidification (DS), in which heat withdrawal is made to occur parallel to the ingot axis, has been introduced to produce large columnar grains parallel to that axis. The technique, which resembles the Bridgeman method for growing single crystals, produces increased ductility at intermediate temperatures (760°C), improved rupture strength in thin sections, and improved low cycle fatigue life. The best elevated temperature properties of nickel-base alloys are obtained by an adaptation of the DS process to produce single crystals.

The coarse grains developed by conventional casting processes usually are deleterious to fatigue life. For parts such as turbine disks that are life-limited by fatigue rather than creep, fine grains are produced by powder metallurgical techniques.

Some of the superalloys can be welded by arc melting processes, as well as by resistance and electron-beam techniques. Alloys having low contents are readily weldable.

Binary Fe−Ni alloys as well as several alloys of the type Fe−Ni−X, where X = Cr or Co, are utilized for their low thermal expansion coefficients over a limited temperature range. Other elements also may be added to provide altered mechanical or physical properties. Common trade names include Invar (64%Fe−36%Ni), Elinvar (52%Fe−36%Ni−12%Cr) and super Invar (63%Fe−32%Ni−5%Co). These alloys, which have many commercial applications, are typically used at low (25−500°C) temperatures.

The latest class of ODS alloys to be developed, based on Ni, Cr, and Al, relies on an Al_2O_3 protective scale for dynamic oxidation resistance; Y_2O_3 is the dispersoid in each of these alloys.

The inadequate low temperature strength displayed by many dispersion strengthened alloys led to attempted combinations of dispersion, and γ'-hardening through the mechanical alloying technique, i.e., simultaneous melding of all constituents: master alloy, solutes, and oxide dispersoids, in a special high energy ball mill. At low temperatures γ'-hardening is achieved and above 1000°C strength is retained to a greater extent than for conventional alloys.

The principal applications of nickel-base superalloys are in gas turbines, where they are utilized as blades, disks, and sheet metal parts.

Iron−Nickel Base Superalloys

Iron−nickel base superalloys were developed primarily from the stainless steels. In the United States, these alloys included 19-9 DL, 16-25-6, and A-286. Later, higher nickel contents were employed to take advantage of the superior oxidation resistance of nickel and the beneficial effects of γ'-forming elements. All iron−nickel base superalloys rely on solid solution hardening to some extent.

Because iron−nickel alloys tend to contain large amounts of ferrite stabilizers such as chromium and molybdenum, the minimum nickel content required to maintain a fcc matrix is about 25 wt%. High iron contents lower cost, increase fabricability, and tend to raise the melting point, at the expense of poorer oxidation resistance than nickel-base alloys. Chromium is added for surface protection and solid-solution strengthening of gamma. Molybdenum also is added for solid-solution strengthening, but is present also in carbides and γ'. Small quantities of boron or zirconium are added to improve workability and stress−rupture properties, and carbon is useful as a deoxidant and to provide MC carbides to help refine grain size during hot working. Finally, ductilizing effects may be realized with small addition of magnesium, calcium, and certain rare-earth elements. Iron−nickel alloys are used extensively in aircraft gas turbines and in the space shuttle main engine.

Cobalt-Base Superalloys. Cobalt-base superalloys are used principally where operating metal temperatures range from 650 to 1000°C and stresses are relatively low. Strengthened primarily by carbide precipitation and solid-solution effects, these alloys are widely used as forgings and castings for nozzle vanes in gas turbine engines, because of good thermal shock and hot corrosion resistance, and in sheet metal assemblies, such as combustion chamber liners, tail pipes, and afterburners. The cobalt alloys generally are inferior in strength to the strongest cast nickel-base superalloys. Cobalt-base alloys generally rely on chromium for high temperature corrosion resistance, and most contain at least 20−25% Cr to form protective Cr_2O_3.

Refractory Metals and Their Alloys. Many elements which could be called refractory are found in the Periodic Table, but those which have received the most attention for potential structural applications are the

TABLE 3. COMMERCIALLY AVAILABLE REFRACTORY METAL ALLOYS

Alloy designation	Nominal compositions, wt %
unalloyed niobium	Nb−0.030O−0.01C−0.03N
Nb−1Zr	Nb−1Zr
WC-103	Nb−10Hf−1Ti
FS-85	Nb−27Ta−10W−1Zr
SCB-291	Nb−10W−10Ta
B-88	Nb−28W−2Hf−0.07C
WC-129Y	Nb−10W−10Hf−0.24Y
D-43	Nb−10W−1Zr−0.1C
unalloyed tantalum	Ta−0.0150−0.01C−0.01N
Ta−10W	Ta−10W
T-111	Ta−8W−2Hf
T-222	Ta−10W−2.5Hf−0.01C
Astar 811C	Ta−8W−1Re−1Hf−0.025C
unalloyed molybdenum	Mo−0.04C−0.0030−0.001N
Mo-TZM	Mo−0.5Ti−0.1Zr−0.03C
Mo−42Re	Mo−42Re
Mo−50Re	Mo−50Re
unalloyed tungsten	W−0.01C−0.006O−0.005N
W−3Re	W−3Re
W−5Re	W−5Re
W−25Re	W−25Re
W−0.3Hf−0.025C	W−0.3Hf−0.025C
W−4Re−0.3Hf−0.025C	W−4Re−0.3Hf−0.025C
W−24Re−0.3Hf−0.025C	W−24Re−0.3Hf−0.025C

bcc metals, tantalum, molybdenum, vanadium, niobium, and tungsten, all of which melt at or above about 2000°C, (see Table 1). Commercially available alloys are as shown in Table 3.

Since about 1950, arc melting of refractory metals in vacuum or inert atmospheres by the consumable electrode technique has been used commercially to produce large ingots and billets. Extrusion, forging, and sheet rolling technologies have advanced rapidly so that many of the refractory metal alloys are now available in various mill forms. Electron-beam melting, plasma-arc spraying, fused-salt electroplating, and vapor deposition are among the specialized techniques being used to produce and fabricate the refractory metals.

In many high temperature applications in the electrical and electronics industry, the refractory metals are protected by a vacuum or an inert gas, so that oxidation is not a problem. However, for most other high temperature applications, poor oxidation resistance has limited use. The oxides of the refractory metals, rather than existing as tight, protective barriers, suffer from porosity at moderately elevated temperatures, volatility at higher temperature, and spalling of the oxide scales away from the substrate, especially at corners and edges.

In addition to oxidation itself, gas diffusion into the base metal can be more damaging than the actual loss of metal from the surface.

Efforts in the 1960s concentrated on developing oxidation-resistant coatings. The most successful coatings are disilicides of the base metal, 2−5 mm thick, usually applied by a high temperature pack cementation process. Aluminide coatings also have been applied successfully. Other coating systems have been based on noble metals and Ni−Cr alloys.

Joining is another problem. Fusion welding in inert atmospheres often develops recrystallized structures in the heat-affected zones, so that welded parts lose strength and are embrittled. Special joining techniques being used to help overcome these deficiencies include electron-beam and solid-phase welding, and the development of special brazing materials.

New Materials and Processes. New materials and processes include aligned eutectics, oxide and fiber-reinforced superalloys, intermetallic compounds and other ordered phases including titanium aluminides, nickel aluminides, and iron aluminides.

NORMAN S. STOLOFF
Rensselaer Polytechnic Institute

Additional Reading

Buckman, R.W. Jr.: in J.L. Walter, M.R. Jackson, and C.T. Sims, eds., *Alloying*, ASM, Materials Park, OH, 1988, pp. 419−445.

Gessinger, G.H. *Powder Metallurgy of Superalloys*, Butterworths, London, UK, 1984, p. 282.

Smialek, J.L. and G.H. Meier: in C.T. Sims, N.S. Stoloff, and W.C. Hagel, eds., *Superalloys II,* John Wiley & Sons, Inc., New York, NY, 1987, pp. 293–320.

Vasudevan, A.K. and J.J. Petrovic: *Mat. Sci. and Eng.* 155A, 1–17 (1992).

Wasielewski, G.E. and R.A. Rapp, in C.T. Sims and W.C. Hagel, eds., *The Superalloys,* John Wiley & Sons, Inc., New York, NY, 1972, pp. 287–316.

HILDEBRAND, JOEL (1891–1983). One of the most distinguished American chemists and teacher. Born in New Jersey, he obtained his doctorate in chemistry and physics from the University of Pennsylvania. After studying abroad under Nernst and van't Hoff, he became professor of Chemistry at the University of California, Berkeley, in 1913 where he remained until retirement. He made many important contributions to physical chemistry, particularly in the area of non-electrolyte solutions; his treatise on this subject is a recognized classic and his textbook on the principles of chemistry established a new standard of excellence. He also made important contributions to the thermodynamics of vaporization of liquids. He proposed the use of helium in deep-sea diving equipment, which has become accepted practice. A gifted teacher and lecturer, he continued his constructive research to the end of his long life. He was unusually active in outdoor sports such as swimming, skiing, and hiking. Among his numerous awards were the Nichols and William Gibbs medals and the Priestley medal.

Additional Reading

Hildebrand, J., R.L. Scott, and J.M. Prausnitz: *Regular and Related Solutions: The Solubility of Gases, Liquids and Solids,* John Wiley & Sons, Inc., New York, NY, 1979.

Hildebrand, J.: *Principles of Chemistry,* The Macmillan Company, New York, NY, 1949.

HILDEBRAND RULE. The entropy of vaporization, i.e., the ratio of the heat of vaporization to the temperature at which it occurs, is a constant for many substances if it is determined at the same molal concentration of vapor for each substance.

HINSBERG REACTION. Reaction of primary and secondary amines with sulfonyl halides to give sulfonamides; because the products from primary amines are soluble in alkali and those from secondary amines are not, and since tertiary amines do not react, this method is useful for the separation and identification of amines.

HINSHELWOOD, SIR CYRIL N. (1897–1968). An English chemist who won the Nobel prize for chemistry in 1956 along with Nikolay Semenov, a Russian. He authored "The Kinetics of Chemical Change," "The Structure of Physical Chemistry," and many other journal articles. His work clarified inorganic and organic reactions. He was educated at Oxford before he began lecturing and research.

HISTAMINE AND HISTAMINE ANTAGONISTS. The history of histamine, $C_5H_9N_3$, and the development of antihistamines have been reviewed.

Histamine Synthesis, Metabolism, and Distribution

The synthesis and disposition of histamine is well described both in allergy textbooks and in review articles.

Synthesis. Histamine, 2-(4-imidazolyl)ethylamine, is formed by decarboxylation of histidine by the enzyme L-histidine decarboxylase. Most histamine is stored preformed in cytoplasmic granules of mast cells and basophils.

Histamine Release. Histamine release is mainly caused by cross-linking of immunoglobulin E on the mast cell surface by antigens. Basophil degranulation is caused mainly by histamine-releasing factors produced by inflammatory cells, such as neutrophils, platelets, and eosinophils. After its release, histamine diffuses rapidly into the blood stream and surrounding tissues.

Metabolism. Metabolism of histamine occurs via two principal enzymatic pathways. Most (50 to 70%) histamine is metabolized to *N*-methylhistamine by *N*-methyltransferase, and some is metabolized further by monoamine oxidase to *N*-methylimidazoleacetic acid and excreted in the urine. The remaining 30 to 40% of histamine is metabolized to imidazoleacetic acid by diamine oxidase, also called histaminase. Only 2 to 3% of histamine is excreted unchanged in the urine.

Histamine in the Brain. There is evidence that histamine functions as a neurotransmitter or a neuromodulator in the brain. In the brain, histamine is related to functions such as the regulation of neuroendocrine and cardiovascular systems, thermoregulation, the circadian rhythm of sleep-wakefulness, behavior, vestibular function, cerebral vascular regulation, and antinociception and analgesia.

Histamine in the Cardiovascular System. Histamine is present in sympathetic nerves and has a distribution within the heart that parallels that of norepinephrine. A physiological role for cardiac histamine as a modulator of sympathetic responses is highly plausible.

Histamine Receptors

The actions of histamine are mediated through at least three distinct receptors.

The H_1 Receptor and its Ligands. The H_1 receptor mediates most of the important histamine effects in allergic diseases. These include smooth muscle contraction, increased vascular permeability, pruritus, prostaglandin generation, decreased atrioventricular node conduction time with resultant tachycardia, activation of vagal reflexes, and increased cyclic guanosine monophosphate (cGMP) production.

In the histamine molecule there are two principal structural elements: an imidazole moiety and an ethylamine side chain. Only the N_π;-position is absolutely necessary for H_1 agonism. The imidazole ring can be replaced, e.g., 2-pyridylethylamine, 2-thiazolylethylamine, or substituted at the 2-position. 2-Methylhistamine is often used as a selective H_1 agonist; however, larger substituents are not allowed unless a phenyl ring is used. 2-Phenylhistamine analogues appear to be very selective H_1 receptor agonists.

The classical H_1 receptor antagonists are reversible, competitive, dose-dependent inhibitors of the action of histamine on H_1 receptors. Histamine H_1 antagonists are usually divided into two classes: the first-generation or classical H_1 antagonists and the second-generation H_1 antagonists. The main distinction between the first- and second-generation drugs is the absence of sedative and anticholinergic side effects in the latter.

The classical histamine H_1 receptor antagonists are structurally very similar, all being substituted ethylamines. The classical H_1 receptor antagonists can be subdivided into six classes: aminoalkylethers (diphenhydramine, $C_{17}H_{21}NO$), ethylenediamines (tripelennamine, $C_{16}H_{18}N_3$), alkylamines (chlorpheniramine, $C_{16}H_{19}ClN_2$), piperazines (hydroxyzine, $C_{21}H_{27}ClN_2O_2$), phenothiazines (promethazine, $C_{17}H_{20}N_2S$), piperidines (cyproheptadine, $C_{21}H_{21}N$).

Several antihistamines have been derived from classical H_1-receptor antagonists, but do not penetrate into the brain. They include acrivastine ($C_{22}H_{24}N_2O_2$), cetirizine ($C_{21}H_{25}ClN_2O_3$), ebastine ($C_{32}H_{39}NO_2$), epinastine ($C_{16}H_{15}N_3$), loratadine ($C_{22}H_{23}N_2O_2Cl$), pibaxizine ($C_{24}H_{29}NO_4$), and terfenadine ($C_{32}H_{41}NO_2$).

Some of the second-generation H_1 antagonists have nonclassical structures, e.g., astemizole ($C_{28}H_{31}FN_4O$), azelastine ($C_{22}H_{24}ClN_3O$), emedastine ($C_{17}H_{26}N_4O$), levocabastine ($C_{26}H_{29}FN_2O_2$), and mizolastine ($C_{24}H_{25}FN_6O$).

The H_2 Receptor and its Ligands. The H_2 receptor mediates effects, through an increase in cyclic adenosine monophosphate (cAMP), such as gastric acid secretion; relaxation of airway smooth muscle and of pulmonary vessels; increased lower airway mucus secretion; esophageal contraction; inhibition of basophil, but not mast cell histamine release; inhibition of neutrophil activation; and induction of suppressor T cells. There is no evidence that the H_2 receptor causes significant modulation of lung function in the healthy human subject or in the asthmatic.

Combined H_1/H_2 receptor stimulation by histamine is responsible for vasodilation-related symptoms, such as hypotension, flushing, and headache, as well as for tachycardia stimulated indirectly through vasodilation and catecholamine secretion.

Structural requirements of histamine as an H_2 agonist are considered to be the protonated side-chain nitrogen atom and the ability of the imidazole amidine system to undergo a tautomeric shift. 4-Methylhistamine is often used as a selective H_2 agonist. Larger substituents are not allowed. Methylation of the amine group is allowed, but leads to nonselective analogues. H_2 agonists are divided in three chemical classes, i.e., analogues of histamine, dimaprit, and impromidine.

The prototype examples of H_2 antagonists are cimetidine ($C_{10}H_{16}N_6S$), famotidine ($C_8H_{15}N_7O_2S_3$), and ranitidine ($C_{13}H_{22}N_4O_3S$).

The H_3 Receptor and its Ligands. The H_3 receptor has been reported to modulate the release of a variety of neurotransmitters and can be regarded as a general regulatory mechanism.

In contrast to the development of selective agonists for the H_1 and H_2 receptor, potent agonists for the H_3 receptor can be obtained by simple modification of the histamine molecule. Histamine itself is already a rather potent agonist of the H_3 receptor, although it is of course not very selective.

H_3 receptor antagonists include betahistine ($C_8H_{12}N_2$), clobenpropit ($C_{14}H_{17}N_4SCl$), and thioperamide ($C_{15}H_{24}N_4S$).

Uses of Histamine Receptor Ligands

H_1 Antihistamine Treatment in Allergic Diseases. H_1-receptor antagonists are used for the symptomatic treatment of several allergic diseases where histamine release form mast cells is induced via immunological or nonimmunological mechanisms. H_1-receptor antagonists are used mainly in allergic seasonal or perennial rhinoconjunctivitis and urticaria.

Clinical Efficacy and Side-Effects of H_1 Antihistamines. It is evident from the mechanism of action of antihistamines and the etiology of allergic diseases that antihistamines in no sense achieve a cure of the patient's allergy. After the administration of a therapeutic dose, a temporal blockade of the effects of histamine is obtained. Whereas classical antihistamines needed at least twice daily administration, for most of the more recently introduced agents administration once daily is sufficient.

Nevertheless, although the nonsedating H_1 antihistamines have substantially improved the acceptability and clinical efficacy of this class of compounds, these do not provide complete relief; eye disease responds less well than nasal disease, of the rhinitis symptoms nasal congestion responds poorly, breakthrough symptoms occur at high pollen counts, and only some 70% of patients report excellent to good treatment responses. Considerable research therefore still continues in the H_1 antihistamine field. New antihistamines are continually being introduced.

The classical H_1 antihistamines are not very specific, and several compounds have some degree of anticholinergic activity. Anticholinergic effects can present side effects such as dry mouth, blurred vision, and urine retention, whereas the anticholinergic action of some antihistamines is probably the reason for effectiveness in motion sickness. Interactions in the brain with noradrenergic, serotonergic, and dopaminergic uptake systems may play a role in behavioral effects of H_1 antihistamines.

At therapeutic doses, the classical H_1-receptor antagonists generally produce sedation. This usually unwanted effect is probably caused by the H_1-receptor blockade in the CNS.

The second-generation H_1 antihistamines generally present few side effects and, in particular, are considered not to cause sedation, mainly because of reduced ability to penetrate the CNS. However, terfenadine and astemizole have been associated with prolongation of the QT-interval in the electrocardiogram (ECG) and ventricular arrhythmias, generally at higher than therapeutic plasma levels.

Azelastine and levocabastine have been developed for topical application. In contrast to earlier reports of sensitization with older antihistamines locally applied to the skin, sensitization has not been reported with local application to the nose or eyes.

H_2 Agonists and Antagonists. H_2-antagonists inhibit histamine-induced gastric acid secretion. They are used widely in the treatment of peptic ulcer disease and esophageal reflux.

H_3 Agonists and Antagonists. No clear therapeutic indications have been reported for H_3 receptor ligands, yet with their use insights in the role of histamine H_3 receptors in various (patho)physiological processes have been obtained. Interesting options for therapeutic application of H_3 agonists could be in asthmatic diseases, gastrointestinal disorders, and in the regulation of sleep/wakefulness patterns.

Economic Aspects

The sales of antagonists of H_1 receptors, used in the treatment of allergic diseases, represent 1% of the overall pharmaceutical market, i.e., $1.7 billion (U.S.). Sales of H_2 antagonists, used mainly in peptic ulcer disease and esophageal reflux, represent 3.5% of the world market, ie \$6 billion (U.S.). H_3 agonists or antagonists have not yet found a clear indication.

MONIQUE M.-L. JANSSENS
Janssen Research Foundation
HENDRIK TIMMERMAN
ROBERT LEURS
Leiden/Amsterdam Center for Drug Research

Additional Reading

Leurs, R. and H. Timmerman: *Histamine H3 Receptor: A Target for New Drugs,* Elsevier Science, New York, NY, 1998.

Middleton, E. Jr. et al.: *Principles and Practice,* 3rd Edition, The C. V. Mosby Co., St. Louis, Mo., 1988.

Uvnäs, B. ed.: *Histamine and Histamine Antagonists,* Springer-Verlag, Berlin, 1991.

HISTONES. Basic proteins which occur in the nuclei of both plant and animal cells. They are less basic than the protamines, having isoelectric points at about pH11. Some investigators restrict the term *histone* to only those basic proteins anatomically and chemically associated with DNA (deoxyribonucleic acids). The close associations of histones with DNA led to the hypothesis that histones might play a role in the control of genetic expression at the cellular level. Advances in molecular biology have permitted more detailed mechanisms for such control to be proposed. Histones, by blocking some areas of the DNA molecule, may permit only part of the DNA base sequences to act as templates for the formation of messenger RNA. Thus histones, by controlling messenger RNA formation, may ultimately control protein biosynthesis within the cell. Or, the primary role of histone may be structural, histone being essential for stabilizing the DNA helix, for the integration of DNA strands into more complex chromosomal structures, and for fixing and maintaining during cell division chromosomal changes occurring during differentiation and development. The foregoing two concepts are not mutually exclusive, i.e., histone may fix chromosomal structure in a specific configuration in which the position of the histone molecules also limit RNA formation. The possibility that histones play a role in genetic mechanisms suggests the possibility that histone changes may initiate or accompany early cellular changes, leading to the formation of tumors. Further investigation is needed to ascertain whether or not tumor histones differ from those of corresponding normal tissues.

HODGKIN, DOROTHY C. (1910–1994). An Egyptian-born chemist who was recipient of the Nobel prize for chemistry in 1964, for her determinations by X-ray techniques of the structures of important biochemical substances. Her work involved determining the structure of vitamin B_{12}, cholesterol iodide, and the antibiotic penicillin by using X-ray crystallographic analysis. She was educated at Oxford and Cambridge.

HOFFMAN, ROALD (1937–). A Polish-born chemist who won the Nobel prize for chemistry with Kenichi Fukui in 1981. His work involved applying the theories of quantum mechanics to predict the course of chemical reactions.

HOFMANN, AUGUST WHILHELM (1818–1892). A German organic chemist who studied under Liebig. While professor of chemistry at the Royal College of Chemistry in London, he did original research on coaltar derivatives that later led him into a study of organic dyes. Perkin, who first synthesized the dye mauveine in England, was a student of Hofmann. When the latter returned to Germany he continued his work in the field of dyes, which became the basis of German leadership in synthetic dye manufacture that continued until World War I.

HOFMANN DEGRADATION. Formation of an olefin and a tertiary amine by pyrolysis of a quaternary ammonium hydroxide; useful for the preparation of some cyclic olefins and for opening nitrogen-containing ring compounds.

HOFMANN ISONITRILE SYNTHESIS. Formation of isonitriles by the reaction of primary amines with chloroform in the presence of an alkali; the odor of the isocyanide is a test for a primary amine.

HOFMANN-LOFFLER-FREYTAG REACTION. Formation of pyrrolidines or piperidines by thermal of photochemical decomposition of protonated *N*-haloamines.

HOFMANN RULE. When a quaternary ammonium hydroxide containing different primary alkyl radicals is decomposed, the least-substituted olefin is formed preferentially.

HOFMANN'S REACTION. Reaction used for preparation of a primary amine from an amide by treatment with a halogen (usually bromine) and caustic soda. The resulting amine has one fewer carbon atom than the amide used.

HOLLOW-FIBER MEMBRANES. A hollow-fiber membrane is a capillary having an inside diameter of >25 μm and an outside diameter <1 mm and whose wall functions as a semipermeable membrane. The fibers can be employed singly or grouped into a bundle which may contain tens of thousands of fibers and up to several million fibers as in reverse osmosis (Fig. 1). In most cases, hollow fibers are used as cylindrical membranes that permit selective exchange of materials across their walls. However, they can also be used as containers to effect the controlled release of a specific material, or as reactors to chemically modify a permeate as it diffuses through a chemically activated hollow-fiber wall, e.g., loaded with immobilized enzyme.

The excellent mass-transfer properties conferred by the hollow-fiber configuration led to numerous applications. Commercial applications have been established in the medical field in gas separations and pervaporation; other applications are in various stages of development.

Hollow-fiber membranes may be divided into two categories: open hollow fibers where a gas or liquid permeates across the fiber wall, while flow of the lumen medium gas or liquid is not restricted, and loaded fibers where the lumen is filled with an immobilized solid, liquid, or gas. The open hollow fiber has two basic geometries: the first is a loop of fiber or a closed bundle contained in a pressurized vessel. Gas or liquid passes through the small diameter fiber wall and exits via the open fiber ends. In the second type, fibers are open at both ends. The feed fluid can be circulated on the inside or outside of the relatively large diameter fibers. These so-called large capillary (spaghetti) fibers are used in microfiltration, ultrafiltration, pervaporation, and some low pressure (<1035 kPa = 10 atm) gas applications.

In open fibers the fiber wall may be a permselective membrane, and uses include dialysis, ultrafiltration, reverse osmosis, Donnan exchange (dialysis), osmotic pumping, pervaporation, gaseous separation, and stream filtration. Alternatively, the fiber wall may act as a catalytic reactor and immobilization of catalyst and enzyme in the wall entity may occur. Loaded fibers are used as sorbents, and in ion exchange and controlled release. Special uses of hollow fibers include tissue-culture growth, heat exchangers, and others.

Hollow fibers offer three primary advantages over flat-sheet or tubular membranes. First, hollow fibers exhibit higher productivity per unit volume; second, they are self-supporting; and third, high recovery in individual units can be tolerated.

Properties

Morphology. The desired fiber-wall morphology frequently dictates the spinning method. The basic morphologies are isotropic, dense, or porous; and asymmetric (anisotropic), having a tight surface (interior or exterior) extending from a highly porous wall structure. The tight surface can be a dense, selective skin, permitting only diffusive transport, or a porous skin, allowing viscous flow of the permeate as in conventional ultrafiltration or reverse osmosis. Membrane-separation technology is achieved by use of these basic morphologies.

Mechanical Considerations and Fiber Dimensions. The hollow fiber is self-supporting, and is actually a thick wall cylinder. The ratio of outside to inside diameter in some reverse-osmosis applications is about 2 to 1 thus providing the strength to withstand high operating pressures, commercially up to 10,000 kPa (96 atm), without collapsing. A hollow fiber that is exposed to external pressure would exhibit a collapse pressure P_c that depends on the inner and outer fiber radii (IR, OR) and the Young's modulus E and Poisson ratio v of the material. The approximate relationship is given by the expression

$$P_c = \frac{2E}{(1 - v^2)}[(OR - IR)/(OR + IR)]^3 \qquad (1)$$

When the operation of the hollow-fiber membrane is to be reversed, and permeation from the bore to outer zone is required, circumferential stress and pressure drop along the fiber capillary (bore) must be considered in the design of the fiber unit.

Spinning

In preparation of permselective hollow-fiber membranes, morphology must be controlled to obtain desired mechanical and transport properties. Fiber fabrication is performed without a casting surface. Therefore, in the moving unsupported thread line, the nascent hollow-fiber membrane must establish mechanical integrity in a very short time.

There are three conventional synthetic fiber spinning methods that can be applied to the production of hollow-fiber membranes: melt spinning, solution (wet) spinning, and a combination of these first two methods, dry-jet wet-spinning.

Spinnerettes. In all methods, a tubular cross section is formed by delivering the spinning dope through an extrusion orifice. Four schemes of spinnerette nozzle cross sections are *(1)* the segmented-arc design; *(2)* the plug-in-orifice design; *(3)* the tube-in-orifice jet design; *(4)* the multiannular design.

Macrovoids. Hollow-fiber membranes that are solution-spun by the foregoing methods can exhibit large voids in conical, droplet, or lobe configurations. These voids may extend through the entire fiber cross section. The voids, in general, result from fast coagulation of a spinning solution that is relatively low in either polymer concentration or viscosity. The use of a less severe quenching medium, on the other hand, yields a macrovoid-free hollow fiber.

The presence of macrovoids in hollow-fiber membranes is a serious drawback since it increases the fragility of the fiber and limits its ability to withstand hydraulic pressures. Such fibers have lower elongation and tensile strength.

Fiber Treatment

Treatment In-line. The coagulated fiber on the moving thread-line may be subjected to cooling (for melt spinning, or for dry-jet wet spinning conducted at high temperatures), washing to remove trace solvents and dope additives, swelling with diluted solvent and/or plasticizers, stretching between godets, and heat-treating (annealing) to consolidate its morphology and impose transport properties (such as closing the skin pores of an asymmetric hollow-fiber membrane for reverse-osmosis applications). In the continuous processing of hollow fibers, these steps add little to costs but are fundamental to achieving the desired functionality of the product.

Post-Treatment of Hollow Fibers. End use of the hollow-fiber membrane dictates the type of post-treatment, if any. There are three main categories: fibers that are spun, fibers that will be chemically or physically modified, and fibers that will serve as a porous matrix for support of another (active) polymer deposited (or entrapped) upon (or within) its walls. There is no theoretical impediment to the inclusion of all conventional treatments in the spinning line: photochemical cross-linking, fluorination, and antiplasticizers have been successful.

Fiber Modification. Chemical modification of the fiber is usually a separate operation. The largest such commercial processing is the deacetylation of cellulose acetate hollow fibers, which converts them into regenerated cellulose hollow fibers employed in hemodialysis.

Fig. 1. A, hollow-fiber spool; B, hollow-fiber cartridge employed in hemodialysis; C, cartridge identical to item B demonstrating high packing density; D, hollow-fiber assembly employed for tissue cell growth; E, hollow-fiber bundle potted at its ends to be inserted into a cartridge or employed in a situation that requires mechanical flexibility.

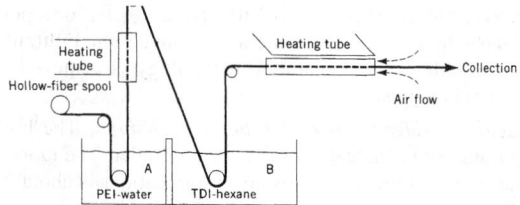

Fig. 2.   Composite hollow-fiber production scheme (PEI = polyethyleneimine; TDI = toluene 2,4-diisocyanate). Anisotropic (porous skin) polysulfone hollow fiber is rolled into bath A and is lifted vertically (to avoid droplet formation) into a heating tube. The fiber is then passed through bath B and is annealed in a ventilated heating tube (110°C).

Composite Hollow-Fiber Membrane. Composite membranes consist of highly porous substrates, having minimum resistance to the permeates, which support ultrathin semipermeable membranes. The production scheme for a composite hollow-fiber membrane, consisting of polysulfone coated by polyethyleneimine (PEI) that is cross-linked *in situ* (on the exterior surface of the fiber), is shown in Figure 2.

Although these composite fibers were developed for reverse osmosis their acceptance in the desalination industry has been limited due to insufficient selectivity and oxidative stability. The concept, however, is extremely viable; composite membrane flat films made from interfacial polymerization have gained wide industry approval. Hollow fibers using this technique to give equivalent properties and life, yet to be developed, should be market tested during the 1990s.

Interpenetrated Wall Matrix. Ion-exchange hollow fibers can be produced by polymerizing an ionic monomer within the porous wall matrix of a hollow fiber. Requirements of such a fabrication are *(1)* the monomers should not dissolve or plasticize the polymer from which the fibers are made; *(2)* the heat generated during the polymerization and contraction prior to the formation of new interpenetrating polymer should be minimized; and *(3)* the polymerization should not occur within the lumen (and hence cause plugging of the fiber). One drawback of such fibers is brittleness.

Fiber Handling and Unit Assembly. Most hollow fibers can be collected on spools by winding machines analogous to those used in the textile industry. Individual or multifilaments can be crosswound, or may be wound in a simple parallel arrangement (for highly plasticized, or large ID fibers, where cross-winding intersections may weaken the structure). Subsequent handling of the filament depends on the intended use of the hollow fiber.

Assembling and potting (cementing together) of hollow-fiber bundles, as shown in Figure 1, require great care and precision technology. The potting agent must be compatible with the function assigned to the fiber, as well as with the fiber material. Another factor important in the selection of a potting agent is its surface tension (ability to wet fibers yet not excessively wick). Commonly employed potting agents include epoxy resins, polyurethanes, and silicone rubbers.

In general there are three main types of hollow-fiber flow configurations. In the most common, for reverse osmosis and ultrafiltration, the feed enters outside the fiber; permeate is inside the fibers and flow is countercurrent. In the second, for large diameter fibers, where the feed has a high loading of particulates, the feed is through the fiber bore; permeate is outside the fiber; flow is usually countercurrent. In cross flow, shell side feed is prevalent as with microfiltration. Gas permeation uses all three flow patterns.

Materials

The components employed in spinning-dope formulations must be consistent in every batch preparation, because numerous parameters are involved in the spinning process. Thus stringent criteria are imposed on the selection of components to be used in each spinning operation. The components are rigorously tested for purity, molecular weight, molecular weight distribution, chemical composition, viscoelastic properties, and other specific parameters that might influence hollow-fiber production and final membrane properties. This often requires close cooperation between the producers of the polymer and the hollow fiber manufacturers. Materials include cellulose, cellulose ester (cellulose acetate and cellulose triacetate), polysulfone, poly(methyl methacrylate), polyamide, other nitrogen-containing polymers (polybenzimidazole, polyacrylonitrile (PAN)), and glass and inorganic hollow-fiber membranes.

Sorbent Fibers

Filled Fibers. Interest in the encapsulation of specific active materials (e.g., activated charcoal, enzymes, drugs) led to the development of encapsulation spinning, usually employing a wet- or dry wet-spinning process.

The rationale for the development of such fibers is demonstrated by their application in the medical field, notably hemoperfusion, where cartridges loaded with activated charcoal-filled hollow fiber contact blood. Low molecular weight body wastes diffuse through the fiber walls and are absorbed in the fiber core. In such processes, the blood does not contact the active sorbent directly, but faces the nontoxic, blood compatible membrane. Other uses include waste industrial applications as general as chromates and phosphates and as specific as radioactive/nuclear materials.

Hollow Fiber with Sorbent Walls. A cellulose sorbent and dialyzing membrane hollow fiber was reported in 1977 by Enka Glanzstoff AG. This hollow fiber, with an inside diameter of about 300 μm, has a double-layer wall. The inner wall consists of Cuprophan cellulose and is very thin, approximately 8 μm. The outer wall, which is ca 40-μm thick, consists mainly of sorbent substance bonded by cellulose. The advantage of such a fiber is that it combines the principles of hemodialysis with those of hemoperfusion. Two such fibers have been made: one with activated carbon in the fiber wall, and one with aluminum oxide, which is a phosphate binder.

Future Prospects

Hollow-fiber membranes are subjected to extensive studies for gaseous separation (e.g., CO_2, H_2, O_2, N_2, H_2S, CO, CH_4), where the capillary configuration has an advantage over the spiral-wound flat film and plate-and-frame devices. Another significant area of development and commercialization is pervaporation. These membranes are dense, rather than porous structures. Generally asymmetric composite constructions are employed with the ultrathin membranes on an open support.

Considerable research and development effort is being placed on a chlorine-resistant membrane that will maintain permeability and selectivity over considerable time periods (years).

When pressurized liquid is used to separate micrometer-size particles from fluids, the process is called microfiltration. Generally particle sizes are from 0.02 to 10 μm. Thus compared to ultrafiltration and reverse osmosis, fluxes and pore sizes are large, osmotic pressure low, and pressures moderate. Two types of microfiltration processes exist, crossflow and deadend. Commercially, the former is growing at the expense of the latter.

Sorbent fibers were developed in the late 1970s, in particular by California Institute of Technology and Gulf South Research Institute. The concept of encapsulation within a hollow fiber, gas, liquid, suspended solid, catalyst, or others, has potential.

IRVING MOCH, JR.
E. I. du Pont de Nemours & Co., Inc.

Additional Reading

Baker, R.W. et al.: *Membrane Separation Systems, Recent Developments and Future Directions,* Noyes Data Corp., Park Ridge, NJ, 1991.
Loeb, S. and S. Sourirajan: *Adv. Chem. Ser.* **38,** 117 (1962).
Scott, J. ed.: *Hollow Fibers Manufacture and Applications, Chemical Technology Review No. 194,* Noyes Data Corp., Park Ridge, NJ, 1981.
Torrey, S. ed.: *Membrane and Ultrafiltration Technology, Developments Since 1981,* Noyes Data Corp., Park Ridge, NJ, 1984.

HOLMIUM.   [CAS: 7440-60-0]. Chemical element symbol Ho, at. no. 67, at. wt. 164.93, tenth in the Lanthanide Series in the periodic table, mp 1,474°C, bp 2700°C, density 8.795 g/cm³ (20°C). Elemental holmium has a close-packed hexagonal crystal structure at 25°C. The pure holmium is silver-gray in color, slow to tarnish or oxidize at room temperature in normal atmospheres. Even at relatively high temperatures, the metal is slow to oxidize. Under a vacuum of about 10 torr, holmium will react when hot with water vapor, CO_2, NH_3, and hydrocarbons. Holmium is soft and can be worked by conventional equipment. There is one natural isotope of holmium, [165]Ho, and 18 artificial isotopes have been produced. The natural isotope is not radioactive. In terms of abundance, holmium is present on the average of 1.2 ppm in the earth's crust, ranking ahead of bismuth, antimony, cadmium, and mercury in potential availability. The element was first identified by P.T. Cleve and J.L. Soret in 1879. The metal has a low acute-toxicity rating. Electronic configuration

$1s^2 2s^2 2p^6 3s^2 3p^6 3d^{10} 4s^2 4p^6 4d^{10} 4f^{10} 5s^2 5p^6 5d^1 6s^2$. Ionic radius Ho^{3+} 0.901 Å. Metallic radius 1.766 Å. First ionization potential is 5.43 eV; second 13.9 eV. Other important physical properties of holmium are given under **Rare-Earth Elements and Metals**.

Holmium occurs in apatite, xenotime, and yttrium and heavy rare-earth minerals. The element of a purity of 99.9% can be obtained through organic ion-exchange techniques. Supplies of holmium are available commercially as the result of yttrium production. To date, the applications for holmium have been very limited. When added to orthoferrites, it has shown promise for use in electronic circuits. Uses in semiconductors, lasers, thermoelectric devices and phosphors currently are being studied.

See references listed at ends of entries on **Chemical Elements**; and **Rare-Earth Elements and Metals**.

K. A. GSCHNEIDNER, JR.
B. EVANS
Iowa State University
Ames, Iowa

HOLOGRAPHY. The technique of holography is similar to photography in many respects, yet it is fundamentally different. With photography, one generally records, by means of lens and film, the two-dimensional irradiance distribution in the image of an object. With holography, one records not the optically formed image of an object, but the object wave itself. This wave is recorded (frequently on photographic film) in such a way that a subsequent illumination of this record, called a *hologram*, reconstructs the original object wave. A visual observation of this reconstructed wavefront then yields a view of the object that is practically indiscernible from the original, including three-dimensional parallax effects. The process was discovered by Gabor[1] (England) in 1948. It was then identified as a two-step method of optical imagery. During the past couple of decades, holography has become widely known and a limited number of practical uses for it have been developed. This later progress is attributed to the general availability of the laser, with the outstanding temporal and spatial coherence of its light. Much of the work in the laser to was carried out by Upatnieks and Leith at the University of Michigan during the early 1960s.

With reference to Fig. 1, one starts with a single, monochromatic beam of light that has originated from a very small source. This single beam is split into two components, one of which is directed toward the

[1] For which he received the Nobel Prize in physics (1971).

object and the other to a suitable recording medium, most commonly a photographic emulsion. The component that is incident on the object is scattered by it, and this scattered radiation, now called the object wave, impinges on the recording medium. The wave that proceeds directly to the recording medium is called the *reference wave*. Since the object and reference waves originate from the same source, they are mutually coherent and form a stable interference pattern when they meet at the recording medium. The detailed record of this interference pattern constitutes the hologram.

Types of Holograms

When the hologram is illuminated with a beam similar to the original reference wave, it modulates the phase and/or amplitude of the illuminating wave in such a way that the transmitted wave divided into three separate components, one of which exactly duplicates the original object wave.

If the two interfering beams are traveling in substantially the same direction, the recording of the interference pattern is said to be a *Gabor hologram* or *in-line hologram*. If the two interfering beams arrive at the recording medium from substantially different directions, the recording is a *Leith-Upatnieks* or *off-axis hologram*. If the two interfering beams are traveling in essentially opposite directions, the recorded hologram is said to be a *Lippmann* or *reflection hologram*, first invented by Denisyuk.

Electromagnetic radiation is most commonly used, although acoustic radiation can be used. The most common electromagnetic radiation employed is light, but holograms have also been recorded successfully with electron beams, x-radiation, and microwaves.

Holograms can be classified by the way they diffract light. In an *amplitude hologram*, the varying irradiance distribution of the interference pattern is recorded as a density variation of the recording medium. In this type of hologram, the illuminating wave is always partially absorbed, i.e., the illuminating wave is *amplitude-modulated*. In the *phase hologram*, a *phase modulation* is imposed on the illuminating beam which, in turn, results in diffraction of the light. Phase modulation occurs when the optical path (thickness×index) varies with position. A phase hologram results from either relief-image or index variation, or both.

Either phase or amplitude holograms can be classified further as *Fresnel holograms* or as *Fraunhofer holograms*. Generally speaking, if the object is reasonably close to the recording medium, say just a few hologram or object diameters distant, the field at the hologram plane is the Fresnel diffraction pattern of the object. A hologram recorded in this manner is termed a *Fresnel hologram*.

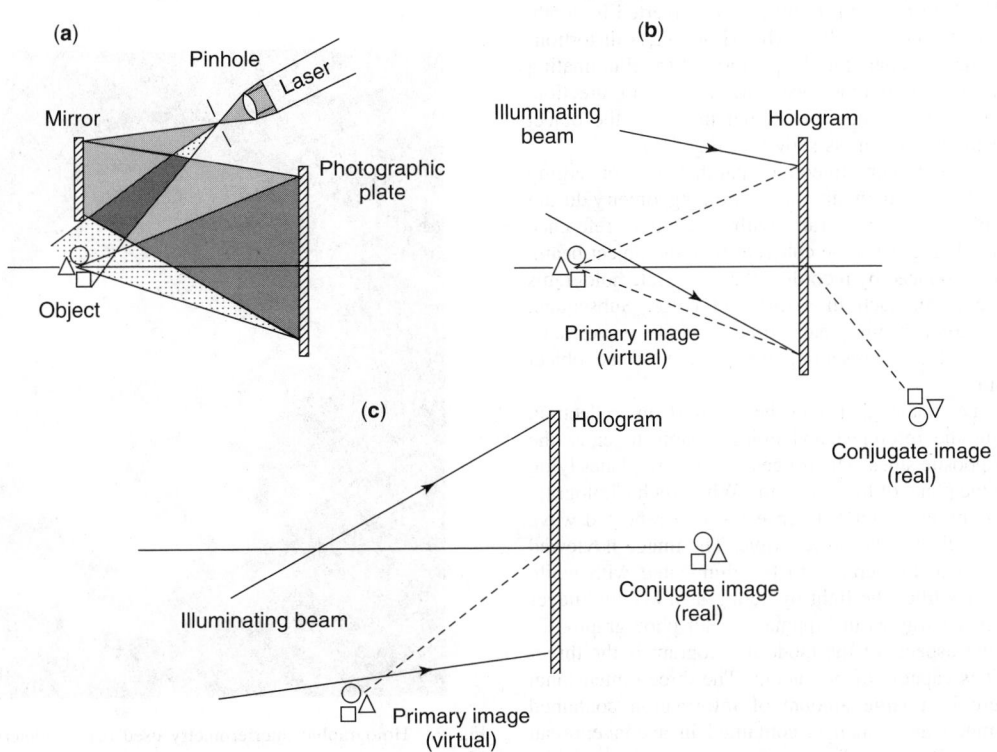

Fig. 1.   A typical holographic arrangement: (**a**) Recording the hologram; (**b**) reconstructing the primary object wave; (**c**) reconstructing an undistorted conjugate wave

If the object and hologram are separated by many object or hologram diameters, the field at the hologram due to the object alone is the Fraunhofer diffraction pattern of the object. A hologram recorded in this manner is termed a *Fraunhofer hologram*.

Any of these hologram types may be recorded as either a *thick* or a *thin* hologram. A thin hologram is one for which the thickness of the recording medium is thin compared to the space between the recorded interference fringes. A thick or volume hologram is one in which the thickness of the recording medium is of the order of or greater than the spacing of the recorded fringes.

Conceptually, the simplest form of an off-axis hologram is one for which the object is just a single, infinitely distant point so that the object wave at the recording medium is a plane wave. If the reference wave is also plane, and incident on the recording medium at an angle to the object wave, the hologram will consist of a series of Young's interference fringes. These recorded fringes are equally spaced straight lines running perpendicular to the plane of incidence. Since the hologram consists of a series of alternating clear and opaque strips, it is in the form of a diffraction grating. When the hologram is illuminated with a plane wave, the transmitted light consists of a zero-order wave traveling in the direction of the illuminating wave, plus two first-order waves. The higher diffracted orders are generally missing or very weak, inasmuch as the irradiance distribution of a two-beam interference pattern is sinusoidal. As long as the recording is essentially linear (irradiance proportional to final amplitude transmittance), the hologram will be a diffraction grating varying sinusoidally in amplitude transmittance, and only the first diffracted orders will be observed. One of these first-order waves will be traveling in the same direction as the object wave. This is the reconstructed wave.

Holographic Recording

The recording of a hologram and the subsequent reconstruction is shown in Fig. 1. In Fig. 1(**a**), the laser beam is first expanded and then divided by a mirror, which directs part of the beam directly onto the photographic plate; the rest of the light is reflected from the object. After processing, the hologram plate may be replaced in its original position (Fig. 1(**b**)), and the object removed. The light diffracted by the hologram forms, in part, the same wavefront that was originally scattered by the object. A viewer looking through the hologram will See an undistorted view of the object, just as if it were still present.

In addition to the *virtual* or *primary image*, a real, or *conjugate image* will be formed on the observer's side of the hologram. This image will appear unsharp and highly distorted, and it will also be inverted in depth, i.e., reversed front to back, as shown in Fig. 1(**b**). However, a distortion-free real image can be formed by changing the position of the illuminating beam so that all of the rays of the reference beam are reversed in direction. In this way, an undistorted, real, three-dimensional image of the object scene appears in front of the hologram, as shown in Fig. 1(**c**).

Holograms may be recorded with diverging, parallel, or converging reference beams. If care is taken to maintain the recording geometry during reconstruction, it is possible to form holograms with an arbitrary reference beam, the only requirement being that it be coherent with the object beam.

Color holograms can be produced by recording three separate holograms on a single photographic plate, each in a different color. Subsequent illumination with a three-color beam yields three separate wavefronts, one in each of the three colors representing the portion of the object corresponding to that color.

Holograms also can be made that can be viewed in reflection. This is done by allowing the reference and object beams to enter the recording medium from opposite sides. The fringes formed are planes lying approximately parallel to the plane of the hologram. When such a hologram is illuminated by a beam similar to the reference wave, a reflected wave is formed which exactly duplicates the object wave. The image is viewed in reflected light. This type of hologram can be illuminated with white light. The interference planes filter the light by acting as a $\lambda/2$ multilayer interference filter, in the same way as in Lippmann color photography.

One of the most striking aspects of the modern hologram is the three-dimensional image that it is capable of producing. The three-dimensional image indicates that there is a large amount of information contained in a single hologram—much more than is contained in a conventional photograph of the same size. Because of the many perspectives available, the hologram is well suited to display purposes. With a hologram, one can present all of the observable characteristics of a three-dimensional object clearly and concisely. Complex molecular or anatomical structure can be simply presented with a single holographic image, with little chance of error or misinterpretation on the part of the viewer. Thus holograms may reduce the number of conventional drawings or photographs to illustrate a single object. It has been proposed that the use of holograms in textbooks would be an aid to readers, particularly in fields where three dimensions are important. Holograms can be made to be viewed with a small penlight and a colored filter.

Applications of Holography

Early applications of holography, essentially prior to the early 1980s, were more of a novel than scientific nature. In 1984, Chang pointed out that holography is more than simply a curiosity related to three-dimensional photography. It is a technology involving the precise structure of light waves, with advanced implications for solutions to engineering problems. Engineering applications of holography utilize the interference patterns created by superimposing holographic images from a target made under slightly different conditions. The patterns can be examined visually or the data can be digitized for computer analysis. In realtime holography, the object is viewed through a hologram of itself. Double-exposure holography involves recording the interference patterns obtained from the same target before and after distortion. Time-average holograms, employed for vibration analysis for example, are made by exposing the plate while the object is driven in resonance.

As described by Chang, some of the more recently developed applications for holography include: (1) Analysis of dimensional instability when an object is stressed. Small distortions resulting from the applications of forces, changes in environment, and other factors yield interference fringes in the superimposed images equivalent to strain contour lines. The approach has been used in connection with thermally induced changes in large-dish antennas at very low temperatures, for observing the performance of miniature gyroscopes, and for examining deformations due to pressure in a vessel or pipe. (2) Checking for cracks in welding by seeking fringe discontinuity across a seam. (3) Determining voids in layered objects, including the inspection of composite aircraft components, clutch plate facings, multiple-layer circuit boards, tires, O-rings, antique paintings, nuclear fuel rods, and detecting the delamination in the composite blades of a helicopter, among others. See Fig. 2. (4) Vibration analysis of such components and subsystems as turbine blades, loudspeakers, rocket castings, and automobile engines. See Fig. 3. (5) Studies for flow visualization in connection with air foils, plasmas,

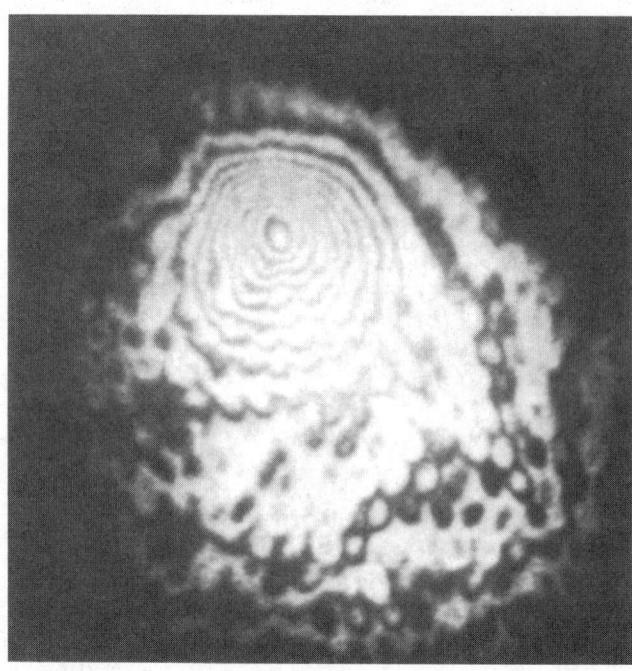

Fig. 2. Holographic interferometry used here to determine delamination in the bonding between skin and honeycomb structure of a composite helicopter rotor blade. (*After M. Chang, Newport Corp*)

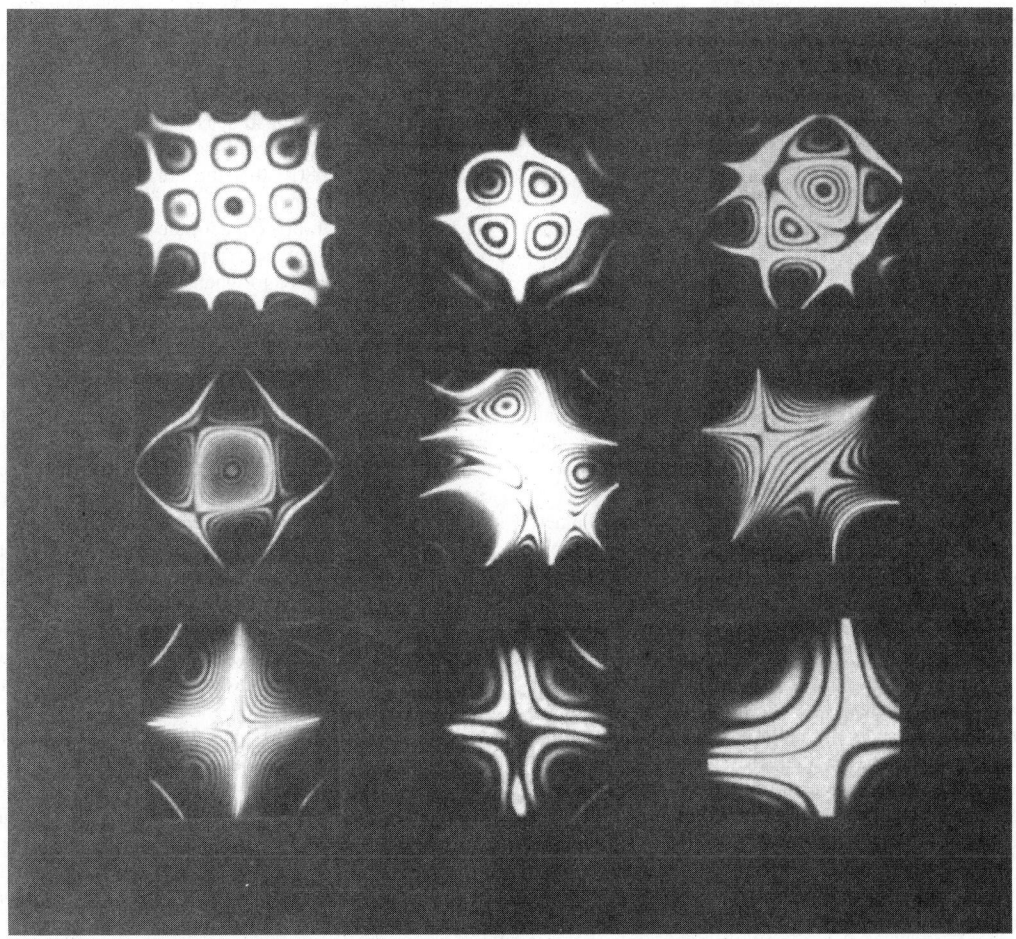

Fig. 3.   Examples of resonant modes of plate as obtained through time-averaged holographic interferometry. (*After M. Chang, Newport Corp*)

and combustion flames—as an alternate to Schlieren photography. (6) Studies of biological and crystal growth. (7) Image processing in connection with pattern recognition in robotics and in finding defects in parts, such as integrated circuits. (8) Monitoring materials properties, such as index of refraction. (9) Analysis of particle size distribution and movement, of interest in improving combustion efficiency and for checking particulate contamination of food and drug products. (10) Use of holograms instead of lens systems for transforming light beams and images and thus use in optical scanners, diffraction gratings, and pattern generators, among others. (11) Microscopic and interferometric studies where objects may be only several microns in diameter—without depth-of-field limitations.

As the potential for holographic techniques become better known, more sophisticated uses are being uncovered. One such application is picosecond holographic-grating spectroscopy. As reported by Wiersma and Duppen (1987 reference listed), interfering light waves produce an optical interference pattern in any medium that interacts with light. This modulation of some physical parameter of the system acts as a classical holographic grating for optical radiation. When such a grating is produced through interaction of pulsed light waves with an optical transition, a transient grating is formed whose decay is a measure of the relaxation time of the excited state. Transient gratings can be formed in real space or in frequency space, depending on the time ordering of the interfering light waves. The two gratings are related by a space-time transformation and contain complementary information on the optical dynamics of the system. The status of a grating can be probed by a delayed third pulse, which diffracts off this grating in a direction determined by the wave vector difference of the interfering light beams. This generalized concept of a transient grating can be used to interpret many picosecond-pulse optical experiments on condensed-phase systems. In their paper, Wiersma and Duppen illustrate some low-temperature experiments. The impact of nonlinear photon-interference spectroscopy on the field of transient-grating and more generally on four-wave mixing spectroscopy is currently significant and is expanding.

Additional Reading

Brcic, V.: *Application of Holography and Hologram Interferometry to Photoelasticity,* Springer-Verlag Inc., New York, NY, 1975.

Chang, M.: "Holography," *Instrumentation Technology,* 39–41 (March 1984).

Fournier, J.M., H. Weber, T. Asakura, et al.: *Holography: The First 50 Years,* Springer-Verlag, Inc., New York, NY, 2000.

Hariharan, P.: *Optical Holography: Principles, Techniques, and Applications,* 2nd Edition, Cambridge University Press, New York, NY, 1996.

Mehta, P.C. and V.V. Rampal: *Lasers and Holography,* World Scientific Publishing Company, Inc., Riveredge, NJ, 1993.

Staff: Optical Society of America *1996 Technical Digest of Holography,* Optical Society of America, Washington, DC, 1996.

Tonomura, A.: *Electron Holography,* Springer-Verlag, Inc., New York, NY, 1999.

Vacca, J.: *Holograms and Holography: Design Techniques and Commercial Applications,* Charles River Media, Rockland, MA, 2000.

Voelkl, E., L.F. Allard, D.C. Joy: *Introduction to Electron Holography,* Perseus Publishing Company, Boulder, CO, 1999.

Wiersma, D.A. and K. Duppen: "Piosecond Holographic-Grating Spectroscopy," *Science,* **237,** 1147–1153 (1987).

Web Reference

http://www.holo.com/

HOLOHEDRAL CRYSTAL. A crystal in which the full number of faces are developed, corresponding to the maximum and complete symmetry of the system. See also **Mineralogy**.

HOMOGENEOUS. (Latin, "the same kind"). This term, in its strict sense, describes the chemical constitution of a compound or element. A compound is homogeneous since it is composed of one and only one group of atoms represented by a formula. For example, pure water is homogeneous because it contains no other substance than is indicated by its formula, H_2O. Homogeneity is a characteristic property of compounds and elements (collectively called substances) as opposed to mixtures. The term is often loosely used to describe a mixture or solution composed of two or more compounds or elements that are uniformly dispersed in each other. Actually, no solution or mixture can be homogeneous; the situation is more accurately described by the phrase "uniformly dispersed." Thus so-called homogenized milk is not truly homogeneous; it is a mixture in which the fat particles have been mechanically reduced to a size that permits uniform dispersion and consequent stability.

See also **Heterogeneous.**

HOMOGENIZING.   A process for reducing the size of particles in a liquid and useful in the preparation of numerous food substances, including milk, ice cream, salad dressings, various fruit juices, flavor concentrates, infant foods, among others.

A reduction of particle or globule size in a mixture of two immiscible liquids makes an emulsion possible. If an emulsifying agent is present, a more stable emulsion can be produced and coalescence of the dispersed phase is prevented. The homogenizer is also used to produce dispersions by reducing the particle size in solid-in-liquid mixtures. As in the preparation of an emulsion, a dispersing agent is needed to maintain a homogeneous mixture.

Typically, a homogenizer consists of a high-pressure, positive-displacement pump and an adjustable orifice. The pump is a piston or plunger type, usually consisting of three plungers, although some homogenizers are made with five or even seven plungers. The cylinder for each plunger has an inlet and discharge valve. The plunger pump must push the product through the homogenizing valve (adjustable orifice). For two-stage homogenization, two valves are arranged in series.

A typical homogenizing valve consists of a seat and plug of very hard abrasion-resistant materials (alloys such as Stellite are used). The seating surfaces must be lapped smooth and be parallel. In operation, the plug is spring-loaded against the seat. Spring compression is adjusted so that when the product flows, energy in the form of pressure is required to lift the plug. Although many products can be homogenized at pressures below 3000 pounds per square inch (204 atmospheres), machines are made to develop pressures in excess of 8000 pounds per square inch (544 atmospheres). In another design, a valve uses a compressed cone of stainless-steel wire inserted into a socket, the product being homogenized by flowing between the wires.

A number of theories have been proposed as to what actually breaks up the particles in the homogenizer. (1) As the product enters the area between the lapped surfaces, it is suddenly accelerated to velocities as high as 30,000 feet per minute (9,144 meters per minute) at a pressure of 5,000 pounds per square inch (340 atmospheres). When acceleration is this sudden, the particle (especially the liquid particle) is stretched or elongated to the point of breaking. (2) At this high velocity, there are shear forces between layers of liquids under flow that break up particles. (3) Cavitation may be the major cause of homogenization. When the pressure energy is converted into velocity energy, the vapor pressure of the product exceeds product pressure, resulting in the formation of vapor cavities, which collapse upon leaving the valve at higher pressures. This collapsing, or implosion, of cavitation exerts tremendous force, breaking up the particles. Most homogenizers are designed to incorporate one or more of the foregoing principles.

HOMOLOGOUS SERIES.   Two organic compounds are said to be homologous if their molecular formulas differ by CH_2, or a multiple of CH_2. For example, the alkane series has the general formula, C_nH_{2n+2}, its first three members being methane, CH_4, ethane, C_2H_6, and propane, C_3H_8.

HOMOPOLYMER.   A natural or synthetic high polymer derived from a single monomer. An example of a natural homopolymer is rubber hydrocarbon, whose monomer is isoprene; a synthetic homopolymer is typified by polychloroprene or polystyrene, whose monomers are, respectively, chloroprene and styrene.

HOOKE'S LAW.   When a load is applied to any elastic body so that the body is deformed or strained, then the resulting stress (the tendency of the body to resume its normal condition) is proportional to the strain. Stress is measured in units of force per unit area; strain is the extent of the deformation. For example, when a bar of metal is subjected to a stretching load, the extent of the increase in length of the bar is directly proportional to the force per unit area, i.e., to the stretching load or stress. In general, Hooke's law applies only up to a certain stress called the *yield strength.*

HORMONES.   During the past few years, the classical concept of hormone has been undergoing revision and expansion. A half-century ago, when only a few hormones were reasonably well understood, it was generally believed that there were a comparatively few very important complex biological substances generated by a few glands that stimulate and inhibit principal body functions. Investigators have learned, particularly since the early 1980s, that the creation of hormones is not an exclusive process for the endocrine glands, that receptors for any given hormone are not usually simply concentrated in relatively limited locations of the body, but are found sometimes where they were least suspected. For example, a few years ago when J. Roth and J. Kova (National Institutes of Health) confirmed insulin receptors in the human brain, they sought and found receptors in the testes and liver. In the case of insulin, this had led investigators to explore the possibility of insulin synthesis occurring, not just in the pancreas, but elsewhere in the body.

Researchers have also commenced studies of other life forms, insects and annelida for example, in a search for hormone receptors and, indeed, have found material similar to insulin and receptors that are reminiscent of human insulin receptors. That hormone production is not limited to the endocrine glands has been suspected for a number of years, but such cases were considered the exception rather than the rule. Modern investigators no longer accept this hypothesis. Suspected earlier, but not well confirmed until the mid-1980s, the human heart, once simply considered as a pump, is now known to create at least one hormone, *atrial natriuretic factor* (ANF). A few years earlier, researchers found that the human brain synthesizes important substances, such as endorphins and enkephalins.

Thus, the study and indeed the definition of hormones is departing rapidly from the former exclusive association with endocrinology. The field is becoming broader and more complex and the number of previously unidentified substances playing some form of hormonal role is increasing. More hormones are being found and in more locations in the living process, both human and other life forms. The field no longer is limited to a comparatively exclusive few sophisticated substances.

In connection with a new hypothesis for hormones, Roth has suggested that cell hormones and neurotransmitters began as what cell biologists term *tissue factor*—substances that stimulate cells to grow or come together or otherwise react biochemically. Only when animals evolved to have extreme cell differentiation and cellular organization did glands evolve to overproduce these hormones so the animals could use them in more clever and sophisticated ways. This theory would explain why many mammalian hormones are also tissue factors. As examples, insulin and glucagon, in addition to playing roles as hormones, also act locally as tissue factors on cells within the pancreas. Exocrine and endocrine functions overlap—there is no difference between exocrine and endocrine functions at the level of unicellular organisms. Roth points out that such a hypothesis would explain the finding that many classical messenger molecules, such as prostaglandins, nerve growth factor, and the hormonal substances are found in exocrine fluids such as saliva, intestinal secretions, milk, and semen.

Investigations are also being conducted apace in what might be called hormone genetics. It has been found, for example, that guinea pigs produce two different insulins, one type made in the pancreas, the other type synthesized in the brain and other organs. Thus, there are differences in gene expression. New findings may explain why cancer cells sometimes secrete hormones that cause severe metabolic disturbances. Lung cancers, for example, are prone to produce vasopressin, a cause of water retention. Perhaps the tumor-generated vasopressin is not normal vasopressin, but a slightly different hormone that has escaped detection by radioimmunoassays. There may be numerous other instances where investigators are seeking one of the better understood hormones, when a somewhat different, uncataloged hormone should be the target. Admittedly, the thoughtful speculation involved in the reevaluation of hormone science will require considerably more research and proof.

In studies of the immune system, investigators have reported on the finding of a heretofore unknown immunoregulatory hormone, 1,25-dihydroxyvitamin D_3. The substance was found to be effective in suppressing interleukin-2.

By applying recombinant DNA techniques, a group of researchers has produced two human fertility hormones, human chorionic gonadotropin (hCG) and human luteinizing hormone (hLH). This is one of the first examples in which recombinant DNA techniques have been used to produce molecules that are a combination of proteins and carbohydrates in mammalian cells. The two hormones are similarly structured, consisting of two polypeptide chains that are put together inside cells and processed. It has been suggested that the hormones will be useful in the treatment of infertility because they can induce both ovulation and sperm production. Past hormone treatments have involved extracts from pituitaries, urine, or placentas which do not yield a pure product.

Hormone Science in the Traditional Sense

In animals, hormones are organic compounds (usually of considerable complexity and even after years of research not fully understood) that are secreted by endocrine (ductless) glands, such as the adrenal gland, the thyroid and parathyroid glands, the pituitary gland, and the gonads, among others Hormones are sometimes commonly called by the names of the glands that secrete them. Thus, there are adrenal cortical hormones, thyroid and parathyroid hormones, etc. Hormones are regulators of physiological processes within the body, exerting control over such processes as metabolism, growth, reproduction, molting, pigmentation, and electrolytic and osmotic balance, among other processes. Apparently, hormones achieve these objectives chemically and electrically, although the mechanisms are not fully understood and, in fact, the mechanisms may vary from one situation to the next. At one time, hormones were loosely called "chemical messengers" because they are transported from point to point within the organism and thus effect actions at distances from the region where they are made. If one visualizes secreting glands as sensors of a type detecting need for correction of some physiological process, then the hormones might be visualized as both the transmitters or carriers of this information and the initiators of actions as well. The conventional concept is that cells have receptors on their surfaces, which sense the presence of specific hormones. At one time, it was firmly believed that hormones, particularly polypeptide hormones, such as insulin, prolactin, and growth hormone, all of which are large charged molecules, could not penetrate through the cell's membrane and actually enter the cell. This belief has since been altered because researchers have shown that insulin, for example, can enter into the cell. Referring to this process as "internalization," one investigator in 1978 suggested that the internalization of polypeptide hormones will be one of the most active topics in cell biology for a number of years.

Research has shown that hormones and/or their receptors may be degraded. As gross examples of this type of situation, it is known that many obese people with high concentrations of insulin in their blood also have normal concentrations of blood sugar. Why doesn't the insulin decrease the blood sugar concentration in these cases? It is well known that pregnant women produce much angiotensin II, which normally increases blood pressure, but these women usually do not have hypertension. What alterations in the hormone-cell mechanism provide this result? There are also instances where males have tumors that secrete large quantities of a hormone that stimulates the production of testosterone, and yet there is no evidence of abnormal amounts of testosterone. At least two questions can be posed: Do certain hormones lose their effectiveness with time? Or, are there changes in target receptor cells? Research has indicated that there may be a relationship between concentration of hormones and the surface receptors which bind them, such receptors being inactive for a time or possibly disappearing from the cell surface altogether. A number of investigators have observed that a better understanding of the manner in which hormones affect their own and other receptors possibly may result in new ways to treat certain diseases, including insulin-resistant diabetes.

In recent years, it has been shown that a wide variety of receptors are regulated by hormones. Some receptors are sensitive to only one hormone; this appears to be the case with insulin. Others appear to be regulated not only by one hormone, but others as well. For example, it has been shown that the receptors for TRH (thyrotropin-releasing hormone) are not exclusively regulated by TRH, but also by other hormones. Receptors for gonadotropins (pituitary hormones that act on the gonads) appear to be regulated by hormones in addition to gonadotropin. There are numerous other instances of this kind.

Hormones display not only great variations in function, but also in their chemical nature, of which there is a great diversity. Some are steroids, such as estrogen, progesterone, cortisone, etc., while others are amino acids (thyroxine), polypeptides (vasopressin), low-molecular-weight proteins, and conjugated proteins. Amino acid and steroid hormones have been isolated and many, including insulin, have been synthesized. Other types are prepared directly from the endocrine organs of animals.

Hormones produced by one species usually show similar activity in other species. The hormones showing greatest species specificity are proteins or conjugated proteins.

Hormones are markedly affected by deficiencies or excesses of the various vitamins and other dietary essentials.

Because of the great complexity of a number of the natural hormones, conventional approaches of organic synthesis which have been used so successfully over the years in connection with many drugs have not proved viable to date with some of the hormones. Insulin is an example. Presently, millions of diabetics still depend upon animal insulin as extracted from the pancreatic glands of slaughtered pigs. If diabetes mellitus continues to become more prevalent, as it has over the past several years, natural sources may not be sufficient.

Recombinant DNA technology was applied to the problem the first time a few decades ago. One group has been successful in inducing the bacterium *Escherichia coli* to manufacture and secrete rat proinsulin, an immediate precursor of rat insulin that incorporates insulin itself. Research like this is an important step toward the objective of developing bacterium-based industrial systems that can replace animal and human tissues as the source of medically useful proteins, such as insulin, growth hormone, and clotting factor.

Classes of Hormones

Hormones may be grouped into two distinct types: (1) *direct-acting*; and (2) *stimulating*-substances that stimulate other organs to produce their own characteristic hormones. The latter group is sometimes called the *tropic hormones*. See summary of hormones in Table 1.

Thyroid Hormones. These are compounds of the amino acid *thyronine*. They are present in the free form only to a slight extent, existing chiefly as constituents of the protein thyroglobulin. The most important of these acids in terms of hormone action are the 3,5,3′-tri, and the 3,5,3′,5′-tetraiodocompounds, *triiodothyronine* and *thyroxin*, the structures of which are given in Part 1 and Part 2 of Table 1. The action of thyroid hormones is to accelerate cellular reactions and to increase the metabolic rate and oxygen consumption of tissues. They effect this action by stimulating many of the enzyme systems, not only the glucose oxidation system and the cytochrome chain for dehydrogenating the coenzyme NADPH, but other processes, such as the synthesis of proteins from amino acids. Their effects are clearly apparent in the pathological changes in the organism caused by their excess or deficiency. The thyrotrophic hormone and other biochemical interactions with the thyroid gland are discussed later in this entry.

Parathyroid Hormones. The influence of the parathyroid glands on the regulation of calcium concentrations in the blood of mammals was first recognized by MacCullum and Voegtlin in 1909.

More recently, several groups of investigators have succeeded in purifying and partially identifying the structure of the hormone, variously called *parathormone* and *parathyroid hormone*. This is a single chain peptide hormone with a molecular weight of about 8,000. A second parathyroid hormone, *calcitonin*, was postulated by Copp (1961). Subsequent research has indicated that this hormone is actually the hormone produced by the thyroid gland. However, a parathyroid calcitonin may exist in certain species.

The more classical function of parathyroid hormone is concerned with its control of the maintenance of constant circulating calcium levels. Its action is on (1) the kidney, where it increases the phosphate in the urine, (2) the skeletal system, where it causes calcium resorption from bone, and (3) the digestive system, where it accelerates (stimulates) calcium absorption into the blood. The hormone and gland exhibit characteristics of feedback control; when the concentration of calcium ions in the blood falls, the secretion of the hormone increases, and when their concentration rises, the secretion of hormone decreases.

Adrenal Cortical Hormones. The adrenal gland is made up of two parts, the medulla and the cortex, each of which secretes characteristic hormones. The hormones of the adrenal medulla are the catecholamines, epinephrine (adrenalin) and norepinephrine (noradrenalin), which are closely related chemically, differing only in that epinephrine has an added methyl group. See Table 1. In fact, animal experiments have established a metabolic pathway for the biosynthesis of both compounds from the amino acid phenylalanine, which involves enzymatic oxidation and decarboxylation reactions. It is also to be noted that the isomeric form of norepinephrine is most important; the natural D-form (which incidentally, is levorotatory) has many times the activity of the synthetic isomer. Epinephrine has a pronounced action upon the circulatory system, increasing both blood

TABLE 1. REPRESENTATIVE HUMAN HORMONES

Hormone Common names, (synonyms), Structure and production site	Principal physiological functions	Interrelationships with vitamins
Adrenocorticotropic Hormone (ACTH) (Adrenocorticotropin; corticotropic hormone) Straight-chain, simple, polypeptide, 39 amino acids, no S—S bridges. (See text, Fig. 1.) Molecular Weight ~ 4500 Production Site: Anterior pituitary	Maintenance of adrenal cortex Promotes secretion of steroids, oxidative phosphorylation in adrenal cortex Mobilizes and increases oxidation of free fatty acid in adipose tissue Increases gluconeogenesis in liver; increases cyclic adenosine monophosphate (AMP) in adrenal cortex Decreases urea formation in liver	Ascorbic acid: depleted in adrenal cortex on stimulation by ACTH Biotin and vitamin A: adrenocortical insufficiency noted in biotin and vitamin A deficiency Niacin: production of reduced nicotinamide adenine dinucleotide (phosphate) (NADPH) by ACTH via cyclic adenosine monophosphate (AMP) Niacin and pantothenic acid: synergistic with ACTH in steroid hormone synthesis Vitamin D: antagonized directly by ACTH via cortisol action
Aldosterone (Aldocortin; electrocortin; mineralocorticoid; 18-oxocorticosterone) Molecular Weight 360.4 Production Site: Adrenal cortex	Maintenance of normal electrolyte blood balances Prolongs survival of adrenalectomized animals Accelerates gluconeogenesis Regulates kidney function	Ascorbic acid: adrenal cortex depleted of ascorbic acid on production of aldosterone Biotin: prolongs life in adrenalectomized rats Niacin: nicotinamide adenine dinucleotide (phosphate) (NADPH) involved in synthesis of aldosterone
Cortisol (Hydrocortisone, 17-hydroxycorticosterone) Molecular Weight 362.5 Production Site: Adrenal cortex	Increases (1) protein catabolism (excepting liver) gluconeogenesis; (2) carbohydrate anabolism (liver); (3) blood sugar; (4) glucose absorption; (5) brain excitation; (6) spread of infections; (7) urinary glucose and nitrogen; (8) stress tolerance; (9) lactation; (10) water diuresis Regulates general adaptation syndrome, water balance, blood pressure, and hormone release. Decreases (1) fat anabolism; (2) growth rate; (3) inflammation; (4) eosinophils; (5) lymphocytes; (6) antigen sensitivity; (7) respiratory quotient; (8) ketosis; (9) wound healing; (10) skin pigmentation; (11) RBC hemolysis	Ascorbic acid: may be required for steroid hormone biosynthesis; depleted from adrenal cortex on cortical secretion Biotin: adrenocortical insufficiency noted in biotin deficiency Folic acid and pantothenic acids maintain secretions of steroids by adrenal cortex Niacin: nicotinamide adenine dinucleotide (phosphate) (NADPH) required for steroid hormone biosynthesis Vitamin A: deficiency causes cortical necrosis Vitamin D: action antagonized by cortisol by reducing calcium absorption in intestine
Epinephrine (Adrenaline, adrenin, suprarenin, vasotonin, vasoconstrictine, adrenamine, levorenine) Molecular Weight 183.2	Blood circulation: increases blood pressure; peripheral vasodilator; increases heart output and rate; flow increases in brain, liver, and skeletal muscle Central nervous system: causes restlessness, anxiety Kidney: reduces glomerular filtration rate Lung, intestine, genital system: inhibited motility	Ascorbic acid: maintains reduced state of epinephrine Ascorbic acid, folic acid, and vitamins B_6, and B_{12} are cofactors in synthesis of epinephrine from phenylalanine

TABLE 1. (*continued*)

Hormone Common names, (synonyms), Structure and production site	Principal physiological functions	Interrelationships with vitamins
Production Site: Adrenal medulla and chromaffin cells in gut	Metabolic effects: increases oxygen consumption, temperature, basal metabolic rate, gluconeogenesis Pituitary effects: stimulates production and release of ACTH and corticoids	
Estradiol (Female hormone; dihydrotheelin; dihydrofollicular hormone dihydrofolliculin) 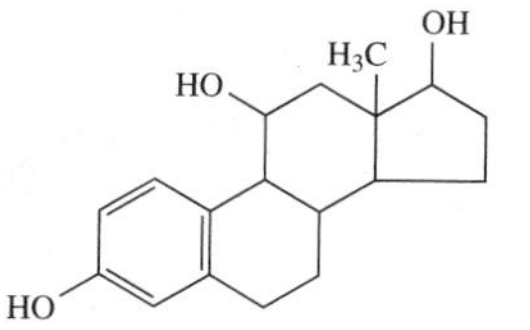 Molecular Weight 272.4 Production Sites: Ovarian follicles; tests; corpus luteum; adrenal cortex; placenta	Regulates menstrual cycle, female sex behavior Maintains secondary sex characteristics Affects antibody properties Induces estrus, uterine hypertrophy, vaginal cornification; potentiate sand stimulates calcitonin secretion	Folic acid: involved in mitotic effect of estradiol Niacin, diphosphopyridine nucleotide (DPN), triphosphopyridine nucleotide (TPN): involved in increased respiration and in cholesterol precursor synthesis Pyridixine: competes as cofactor with estrogen sulfate in kynurenine aminotransferase activity Vitamin D: synergistic in calcium metabolism with estradiol Vitamin E: involved in follotropin production or release
Follicle-Stimulating Hormone (FSH) (Follotropin, luteoantine, thylakentrin, Prolan A, gonadotropin 1, gametogenic hormone, follicle ripening hormone, gametokinetic hormone) Structure: Not fully definitized. Production Site: Anterior pituitary.	Female: stimulates ovarian follicles to grow and to develop, forming multiple layers and antra Male: stimulates seminferous tubules; stimulates spermatogenesis	Ascorbic acid: depletion in ovary due to follicle-stimulating hormone and luteinizing hormone action Vitamin E: required to maintenance of membranes in sex organs
Glucagon (HGF) (Hyperglycemic-glycogenolyltic factor; glucagon; HG-factor) Structure: Polypeptide, 29 amino acids (structure determined). No S—S bridges Molecular Weight ~3500 Production Site: Alpha cells in pancreas.	Increases: blood sugar; blood K^+, oxygen consumption, liver glycogenolysis, gluconeogenesis, nitrogen and salt excretion Decreases: liver glycogen, protein formation, gastric juice, fatty acid synthesis	Ascorbic acid: depletion of adrenal ascorbic acid by glucagon
Insulin (no synonyms) Structure: 51 amino acids. Known and synthesized. 3 S—S bridges. (See text, Fig. 4) Molecular Weight 5,734 (monomer); 12,000–48,000 (polymer), depending upon pH. Production Site: Beta cells of islets of pancreas.	Regulates carbohydrate and fat metabolism, especially glucose and fat oxidations Stimulates amino acid and glucose transport into cells and protein synthesis Female: promotes estrogen and progesterone secretion, ovulation; maintains ovarian tissues Male: stimulates Leydig cells to secrete testosterone; gametogenic with follotropin (FSH)	Ascorbic acid: acts similarly to alloxan (i.e., antagonist) Ascorbic acid: ovarian depletion on LH stimulation Vitamin E: involved in spermatogenesis
Luteinizing Hormone (LH) (Luteotropin, ISCH) Structure: Globular glycoprotein with S—S bridges. Molecular Weight 26,000 Production Site: Anterior pituitary.		
Melanocyte-stimulating Hormone (MSH) (Melanotropin, chromatophorotropic hormone; pigmentation hormone) Structure: Polypeptide; purified, synthesized; alpha and beta forms; straight chains. Molecular Weight: 1500 (alpha) 2100–2600 (beta) Production Site: Intermediate lobe of pituitary.	Mammals: exerts small effect on skin pigmentation (protection from sunlight not fully proved) Expands or contracts pigments in various chromatophores Expands melanophore pigments with color changes in amphibia (adaptation to environment) Lower vertebrates: increases sensitivity to light; decreases dark adaptation time.	Ascorbic acid: adrenal cortex depleted on ACTH and MSH activity Vitamin A: MSH decreases dark adaptation time.
Norepinephrine	Blood circulation: increase blood pressure; peripheral vasoconstrictor without change or slight decrease in output and heart rate. No flow increase in brain, liver, or muscle	Ascorbic acid: protects against oxidation of norepinephrine

(continued overleaf)

TABLE 1. (*continued*)

Hormone Common names, (synonyms), Structure and production site	Principal physiological functions	Interrelationships with vitamins
(Arterenol; noradrenaline; levarterenol) HO—⬡—CH(OH)—CH₂—NH₂ (with HO groups) Molecular Weight 169.2 Production Site: Adrenal medulla; adrenergic nerve endings; chromaffin cells.	Central nervous system effects: adrenergic transmitter agent at synapses; no brain excitation Kidney: decreases glomerular filtration rate Lung, intestine, genital system: inhibited Metabolic effects: weak epinephrine effect	Ascorbic acid, folic acid, and vitamin B_6 are cofactors in synthesis of norepinephrine from phenylalanine
Oxytocin (Oxytocic hormone; pitocin; uteracon; α-hypophamine) Cys—Tyr—Ile—Gln—Asn—Cys—Pro—Leu—Gly NH₂ Molecular Weight 1007 Production Site: Hypothalamus.	Uterine contraction, milk ejection, facilitates sperm ascent in female tract Decreases membrane potential of myometrium, basic metabolic rate, and liver glycogen Stimulates oviposition in hen, releases luteinizing hormone (LH) Increases blood sugar and urinary sodium and potassium	Findings on interrelationships with vitamins are not extensive
Parathyroid Hormone (PTH) (Parathormone) Structure: Simple polypeptide (83 amino acids), sequence determined; straight chain; No S—S bridges. Production Site: Parathyroid glands.	Increases blood calcium, kidney calcium reabsorption, phosphate excretion, and blood citrate level Mobilizes calcium and phosphate from bone Activates calcium and phosphate absorption from the gastrointestinal tract (for which vitamin D is required) Increases osteoclast formation	Vitamin D: synergistic with PTH in maintenance of serum calcium
Progesterone (Progestin, luteosterone) (steroid structure) Molecular Weight 314.5 Production Sites: Ovary (follicles, corpus luteum); testicles; adrenal cortex; placenta	In low concentrations: prepares uterus for blastocyst implantation; promotes ovulation and mammary gland development; regulates female sex accessory organs; weak corticosteroid properties; precursor to sex hormones In high concentrations: maintains pregnancy; represses ovulation and sex activity; inhibits vaginal cornification and parturition; decreases myometrial excitation	Ascorbic acid: depleted from adrenal cortex or ovary on progesterone formation Niacin: diphosphopyridine nucleotide (DPN) involved in progesterone synthesis
Prolactin LTH (Lactogenic hormone; lactogen; galactin; mammotropin) Structure: Single-chain protein, 205 amino acids Molecular Weight 23,000–25,000 Production Site: Anterior pituitary	Initiates lactation Develops mammary glands in female Increases weight and growth (similar to somatotrophin in some species) Protein anabolism (some species) Growth and secretion of crop gland (birds) Luteotropic (only in mouse, rat) Promotes maternal behavior	Not fully determined. Generally participates with other substances having growth action Participates in nidation of zygote
Relaxin (Releasin, cervilaxin) Structure: Polypeptide (4 peptides with activity have been isolated); about 30–40 amino acids in each peptide Molecular Weight 4000–5000 Production Site: Corpus luteum in pregnancy	Enlarges birth canal in preparation for parturition Separation of symphysis pubis, loss of rigidity in pelvic bones Decreases uterine motility Maintains pregnancy Increases sensitivity to oxytocin; release oxytocin Stimulates mammary gland	Ascorbic acid: maintains mucoprotein ground substance in connective tissue, affected by relaxin

TABLE 1. (*continued*)

Hormone Common names, (synonyms), Structure and production site	Principal physiological functions	Interrelationships with vitamins
	Stimulates inhibition of water in uterus Inhibits uterine contraction	

Somatotropin (STH)
(Growth hormone, GH; somatotrophic hormone;
  hypophyseal growth hormone)
Structure: Known and synthesized; coiled, unbranched
protein; 188 amino acid residues; 2 S—S bridges
 	
Molecular Weight 21,500
 	
Production Site: Anterior pituitary
	Promotes general growth of organism
	Promotes skeletal growth, protein
	  anabolism, fat metabolism, carbohydrate
	  metabolism, water, and salt metabolism
		Relates with all vitamins in connection
		  with growth actions

Testosterone
(17 beta-hydroxy-4-androsten-3-one)
 	
Molecular Weight 288.4
 	
Production Sites: Interstitial cells of ovary and testis;
  adrenal cortex; embryonic placenta
	Controls secondary male sex characteristics
	Maintains functional competence of male
	  reproductive ducts and glands
	Increases protein anabolism; maintains
	  spermatogenesis; inhibits follotropin
	Increases male sex behavior; increases
	  closure of epiphyseal plates
		Ascorbic acid, folic acid, vitamins A
		  and E are synergists with testosterone
		  for maturation of germ cells and
		  increased anabolic activity

Thyroid-stimulating Hormone (TSH)
(Thyrotrophic hormone, thyrotropin)
Structure: Glycoprotein (300 amino acids)
Molecular Weight 26,000–30,000
Production Site: S^2 type cell, anterior pituitary
	Regulates body temperature via thyroxine
	Maintains thyroid gland and its secretory
	  activity (colloid discharge)
	Maintains iodine uptake by thyroid gland
	Promotes differentiation in embryo during
	  development via thyroxine
	Stimulates coupling of diodotyrosine to
	  form thyroxine
		Ascorbic acid, thiamine, riboflavin, and
		  vitamin B_{12}: requirements increase in
		  hyperthyroidism; tissue
		  concentrations reduced
		Vitamin A: massive doses of vitamin A
		  inhibit secretion of TSH; thyroid
		  hormones required for carotene and
		  retimene conversions
		Vitamins A, D, E, and K: requirements
		  increased in hyperthyroidism; tissue
		  concentrations reduced in
		Vitamin B_6, niacin: conversion to
		  phosphorylated reactive forms
		  impaired in hyperthyroidism

Thyroxine (T_4)
(3,5,3′,5′ tetraiodothyronine)
 	
Molecular Weight 776.9
 	
Production Site: Thyroid gland
	Regulates growth, differentiation, oxidative
	  metabolism, electrolytic balance
	Increases carbohydrate metabolism,
	  calorigenesis, protein anabolism, basal
	  metabolic rate, oxygen consumption, fat
	  catabolism, fertility
	Sensitizes nervous system
		Ascorbic acid: synergist in cold survival
		Niacin: synergist in mitochondrial
		  metabolism
		Vitamin A: T_4 is required for vitamin A
		  synthesis in liver
		Vitamin B_{12}: T_4 aids in B_{12} absorption
		B complex vitamins: deficiencies
		  develop in hyperthyroidism

Vasopressin
 	
(Arginine vasopressin; antidiuretic hormone; ADH;
  pitressin; tonephin; vasophysin)
 	
Cys – Tyr – Phe – Glu NH_2 – Asp NH_2 – Cys – Pro – Leu – Gly NH_2
                        Vasopressin
 	
Molecular Weight 1084 (arginine-vasopressin)
Production Site: Hypothalamus
	Elevates blood pressure (mammals)
	  (reverse effect in birds)
	
	Decreases kidney blood flow
	Antidiuretic, releases ACTH
	
	Increases sodium chloride and urea
	  excretion
	Regulates water balance
	Stimulates contraction of smooth muscles
	Increases renal tubular water reabsorption
	Releases anterior pituitary hormones
		Not fully determined

pressure and pulse rate, and hence the cardiac output by its direct action upon the heart muscle, and especially because it causes constriction of the arterioles. However, its effects upon smooth muscles vary; it relaxes the muscles of the digestive system, but contracts the pyloric sphincter.

Norepinephrine does not affect the cardiac output, although it does raise the blood pressure by constricting the arterioles. Its muscular effects are less pronounced. Both epinephrine and norepinephrine release free fatty acids from adipose tissue, so raising its level in the blood. This effect is due to the action of the hormones in accelerating enzymatic reactions whereby the esters of the fatty acids are hydrolyzed. The third type of action of epinephrine is its effect upon the carbohydrate metabolism, notably the acceleration of the hydrolysis of glycogen in muscular tissue and the liver. It raises the glucose level in the blood, and the rate of glucose oxidation, with resulting increase in oxygen utilization, carbon dioxide production, and body temperature. See also See also **Adrenal Cortical Hormones**; and **Adrenal Medulla Hormones**. The hormones of the adrenal cortex are steroids. See also **Steroids**. Among them there are a number of hormones with androgenic activity, such as adrenosterone and 17α-hydroxyprogesterone, which are discussed under the sex hormones later in this entry. In all, over ten steroids have been identified in the adrenal cortex, including seven of characteristic cortical activity. These are corticosterone, from which the others are named, 17α-hydroxyl-11-dehydrocorticosterone (cortisone), 17α-hydroxycorticosterone (cortisol or hydrocortisone), and 18-oxocorticosterone (aldosterone). Only two hormones, cortisol and corticosterone, are normally released in fairly large quantities, and another, aldosterone, deserves mention because of its somewhat different effects, even though it is released to a far lesser extent.

All of these hormones are synthesized from cholesterol in the adrenal cortex, by an extended series of reactions, which include many related compounds. Although these hormones have widespread effects throughout the organism, their primary mechanism is not known, so that many of the effects may be indirect. Much of the knowledge of their action arises from studies of insufficiency or hyperactivity of the adrenal cortex, which produces a wide variety of pathological conditions. See Table 1.

It is generally considered that aldosterone, and to some extent the other hormones, have a regulatory effect upon the metabolism of electrolytes and water, particularly upon the concentration of the ions of the alkali metals in intracellular fluids. Administration of steroids also increases the concentration of calcium ions in those fluids. However, all three of these hormones have a number of other effects, roughly in the order of potency—cortisol, corticosterone, aldosterone. They produce changes in the metabolism of carbohydrates, proteins, and fats.

For the carbohydrates alone, three major effects are evident—increase in the rate of formation of glucose, increase in the rate of release of glucose from the liver, and increase in the rate of utilization of glucose. These hormones affect the digestive system, increasing the secretion of hydrochloric acid, pepsinogen, and trypsinogen. They prevent inflammatory responses to bacterial or even chemical stimuli; they counteract anaphylactic shock, and other effects of hypersensitivity. Obviously, these properties have led to their widespread therapeutic use.

There are relationships between the adrenal cortical hormones and the thyroid and pituitary glands. Depression of the function of the adrenals produces thyroid deficiency, whereas administration of thyroxine stimulates the ACTH-adrenal cortical mechanism.

Pituitary Hormones. The hormones of the hypophysis (pituitary gland) are quite numerous, being secreted variously in three parts of the gland—the neurohypophysis (posterior lobe), the adenohypophysis (anterior lobe), and the *pars intermedia*, which connects the other two.

The chief hormones of the neurohypophysis are the polypeptides oxytocin and vasopressin. The hormone characteristic of the *pars intermedia* is the melanocyte-stimulating hormone. It is usually spoken of in the plural, since in most mammals both alpha and beta forms are known. The structures of the first two are shown in the Table 1. The most prominent effect of oxytocin is the contraction of smooth muscle, especially of the uterus. It also has a major effect upon the muscles about the breast, and so stimulates the ejection of milk in lactating animals. It has a definite stimulating effect upon the muscles of the ureter, urinary bladder, intestine, and gall bladder.

The most prominent effect of vasopressin is upon the kidneys, where it stimulates the resorption of water in the tubules (which by repeated release and absorption concentrate the urine). It also constricts the coronary

arteries, raises the blood pressure, and exhibits the effect of oxytocin upon smooth muscles, but generally to a lesser degree.

The action of the melanocyte-stimulating hormones has been established by studies of animals, in which they cause dispersal of certain black pigments from the cells that contain them, with resulting darkening of the skin.

The adenohypophysis is the part of the gland in which the tropic hormones are secreted. They include the adrenocorticotropic hormone (ACTH), the thyrotropic hormone (TSH), and somatotropin, as well as three hormones with pronounced effects upon the gonads: the hormone prolactin, the follicle-stimulating hormone (FSH) and the luteinizing or interstitial cell stimulating hormone (LH or ISCH).

ACTH. Adrenocorticotropin (ACTH) in humans is a polypeptide containing a sequence of 39 amino acids, although work with animal forms of it and with degradation products of the human form have shown that not all of them are essential to the activity of the hormone. This sequence for the human ACTH is shown in Fig. 1.

The primary function of ACTH is the stimulation of the adrenal cortex to produce its hormones, which have already been discussed. This is evident from the therapeutic effect of administration of ACTH, which is closely similar to that of these hormones, so that if the action of only one of them is sought, its administration is preferable. Moreover, ACTH stimulates secretion of the androgenic substances mentioned as produced by the adrenal cortex.

Thyrotropic Hormone. This hormone (TSH) stimulates the development of the thyroid and controls its secretion. Although purified preparations of it have been obtained, they consist of a mixture of proteins of high mean molecular weight (about 30,000). Some of their amino acids have been determined, as well as their carbohydrates, but the structures have not been elucidated.

Growth Hormone. Somatotropin is the growth hormone. Purified preparations of extracts of it from the human adenohypophysis have been crystalized. They are known to be proteins, of mean molecular weight 21,000, and containing a single polypeptide chain. This hormone differs from the others of its group in not acting primarily upon the other endocrine glands, but in controlling the gain in body weight and the rate of skeletal growth. The growth abnormalities, such as dwarfism and giantism, have been shown to result from its hypo- and hypersecretion. In addition to its effect upon growth and anabolism generally, it has been found to affect the kidneys and pancreas, and to influence glucose, galactose, and lipid metabolism.

Gonadotropic Hormones. These include follicle stimulating hormones (FSH), luteinizing or interstitial cell stimulating hormone (LH or ISCH), and prolactin. Their structures are not known; the molecular weight of human LH is about 26,000, that of human FSH is about 30,000, and that of human prolactin is uncertain. They are proteins, with variable amounts of carbohydrates. FSH induces the growth of Graafian follicles in the ovary and the production of spermatozoa in the testis. LH stimulates the final development of the ovarian follicles, the appearance of estrus, and the change of the follicles to corpora lutea. In the male, it stimulates the secretion of testosterone. Since these effects are due to the effect of this hormone upon interstitial cells, it is also called ISCH. Prolactin stimulates lactation after birth, acts with estrogen to promote the growth of the mammary gland, and influences the activity of the corpora lutea.

Male Hormones. The androgenic hormones produced in the testes (and adrenal gland) have a widespread effect upon the development of secondary sexual characteristics (musculature, facial hair, larynx, etc.), as well as upon the sexual organs and responses themselves. They also promote anabolism to a marked degree by their effect upon nitrogen and calcium metabolism. The structure of testosterone is shown in the Table 1.

Ser—Tyr—Ser—Met—Glu—His—Phe—Arg—Tyr—Gly—Lys—Pro—
Val—Gly—Lys—Lys—Arg—Arg—Pro—Val—Lys—Val—Tyr—Pro—
                                                    NH₂
                                                     |
Asp—Ala—Gly—Glu—Asp—Glu—Ser—Ala—Glu—Ala—Phe—Pro—
                                                    Leu—Glu—Phe

Fig. 1. Amino acid sequence of human adrenocorticotropin (ACTH). Abbreviations of amino acids will be found in entry on **Amino Acids**

Female Hormones. Closely related to the male androgenic hormones, and probably synthesized from them in the female organism, are the estrogenic hormones which are produced principally in the ovary. Although β-estradiol is the normally secreted ovarian hormone, a number of other estrogenic substances have been isolated from urine and from animal studies. They include α-estradiol, estriol, and estrone. The structures of these hormones are given in Fig. 2.

These hormones are important in both the menstrual cycle and the reproductive cycle, and of course play an important role in oral contraceptives (the "pill"). They induce growth of the vaginal epithelium, secretion of mucus by the glands of the cervix, and initiate the growth of the endometrium, which is taken over by progesterone (from the corpus luteum) later in the cycle. They activate the proliferation of the mammary gland during pregnancy. As the androgens do for the male, the estrogens bring about the secondary sexual characteristics of the female. They have a number of effects upon metabolism, notably that of calcium and phosphorus, and of lipids and proteins. A number of other estrogens, some made synthetically and others obtained from animals, are known.

The corpus luteum produces two hormones, progesterone and relaxin. The structures of these hormones are shown in Part 1 and Part 2 of table.

Progesterone acts to complete the proliferation of the endometrium, which was initiated by the estrogenic hormones, and to prepare it for the ovum. In pregnancy the continued action of progesterone is necessary. It aids the growth of the breasts and has a definite effect against ovulation. It is also the biosynthetic precursor of some of the estrogenic hormones. Relaxin has been shown to have a relaxing effect on the cartilaginous junction of the public bones in preparation for parturition.

Feedback in Hormone Control Systems

Not only do the hormones initiate or stimulate biological processes, both directly and by bringing about production of other hormones in other glands, but they also act to maintain the organism in a steady state, or *homeostasis*. Thus the gonadotropic hormones from the hypophysis stimulate the testes, but the resulting production there of androgens like testosterone, inhibits the action of the hypophysis in producing the gonadotropic hormones. The complicated cycle of adjustment in the human female is shown in the cycle illustrated in Fig. 3.

As shown in the figure, the regulation of the ovarian hormones in the human female involves both positive and negative feedback. The follicle-stimulating hormone (FSH) from the adenohypophysis stimulates the Graafian follicles, which thus produce estrogens. These not only inhibit FSH production through negative feedback, but also stimulate the adenohypophysis to increase its production of luteinizing hormone (LH) through positive feedback. This hormone in turn brings about ovulation from the Graafian follicle. After the ova are discharged, the LH stimulates the empty follicle, now the corpus luteum, to produce progesterone.

This hormone brings about the changes in the reproductive organs required for the development of the embryo. Then the progesterone partly inhibits the adenohypophysis from producing further LH, an example of negative feedback; as a result, there is no further ovulation. The

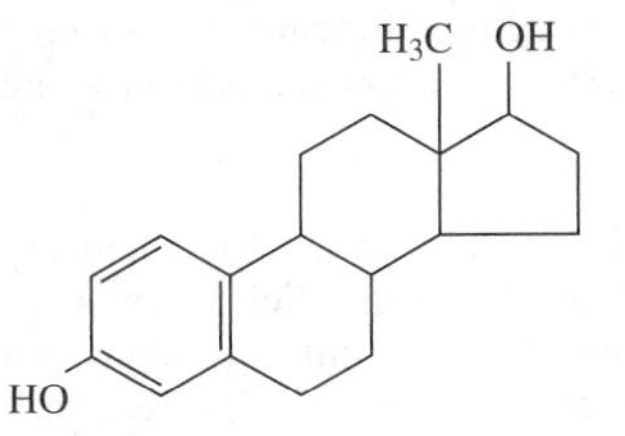

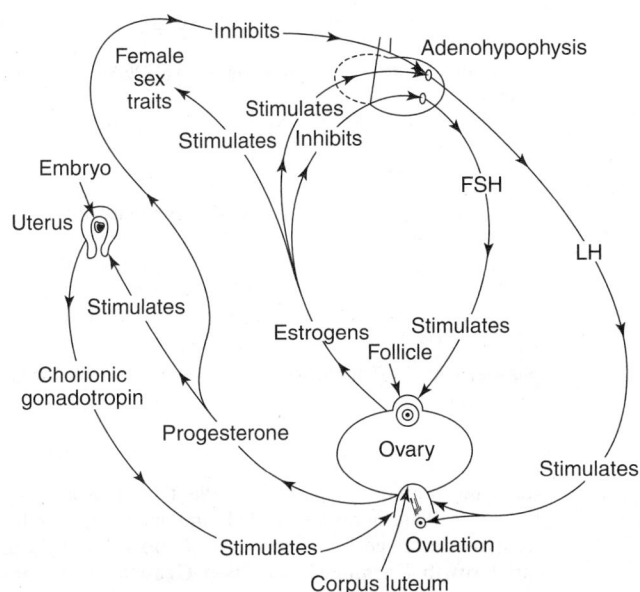

Fig. 3. Cycle of hormone adjustment in human female

progesterone also acts as a positive feedback and stimulates the production of FSH.

When pregnancy intervenes, a new feedback mechanism must be introduced, or the embryo would be expelled, by the shedding of the lining of the uterus in menstruation. Here the placenta (chorion) of the embryo itself produces hormones, as already noted. Its LH stimulates continuing production of progesterone from the corpus luteum, thus preventing menstruation and stimulating the continuing development of the uterus as needed by the growing embryo. The extra progesterone also inhibits further ovulation in spite of the presence of the gonadotropin from the placenta (chorion).

Pancreas and Nonendocrine Hormone Sources. In addition to producing hormones, the pancreas also generates digestive fluids (*pancreatic juice*). It is the hormone function that makes the pancreas a part of the endocrine system. The pancreas secretes *insulin* and *glucagon*, both hormones. The structure of glucagon consists of a single chain of amino acids. See Figs. 4 and 5.

These two hormones have two opposing effects. That of insulin is *hypoglycemic*, i.e., it increases the rate of utilization of glucose, the probable process being an effect of insulin to increase the penetration of glucose through the cell walls as well as increased phosphorylation. The overall result of action of insulin in its relation to glucose is to increase the rate of the reactions by which glucose is oxidized, but also its transformation to glycogen. The enzyme glucagon raises blood glucose levels by increasing the rate of hydrolysis of glycogen (in the liver) to increase the formation ultimately of glucose. Insulin increases the rate of entry of amino acids into cells and their rate of protein biosynthesis. Insulin also accelerates the formation of lipids from carbohydrates, whereas glucagon stimulates the formulation of keto compounds from lipids, inhibits the synthesis of fatty acids, and accelerates the breakdown of various phosphorus and nitrogen compounds. The primary result of insulin deficiency is diabetes mellitus.

Hormones may be produced by organs other than the endocrine glands. Conspicuous among such organs is the placenta, the organ on the wall of the uterus to which the umbilical cord is attached. It has been found to produce the same estrogenic hormones as the ovary, the same hormones (progesterone and relaxin) as does the corpus luteum, and gonadotropic hormones (and luteinizing hormones) similar to, but not identical with, those produced by the adenohypophysis.

Other hormones which do not originate in endocrine glands are the cholecystokinin of the intestine, and the enterogastrone and gastrin of the stomach. The first is produced by the upper intestinal mucosa and causes the gall bladder to contract; the enterogastrone is produced in the same tissue and inhibits gastric motility and secretion; it also excites secretion of digestive fluids, principally hydrochloric acid.

In plants, a *plant hormone* or "phytohormone" is an organic compound produced by the plant, controlling growth and other functions at sites

α-Estradiol

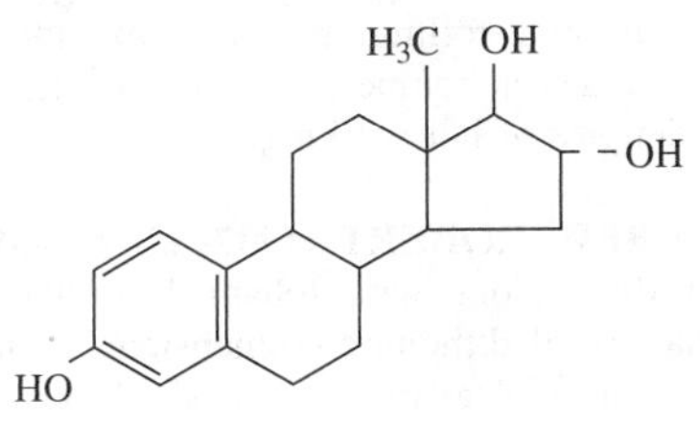

Estriol

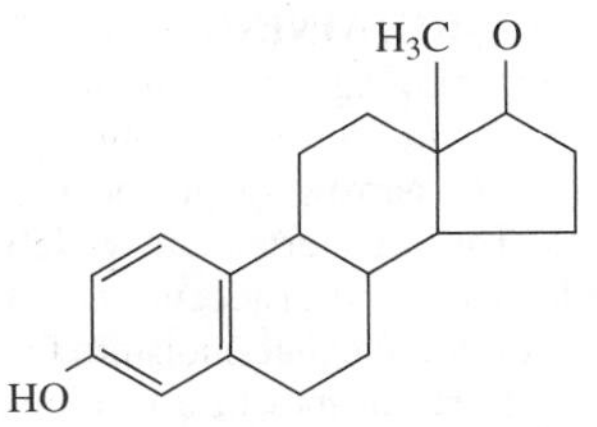

Estrone

Fig. 2. Major ovarian hormones

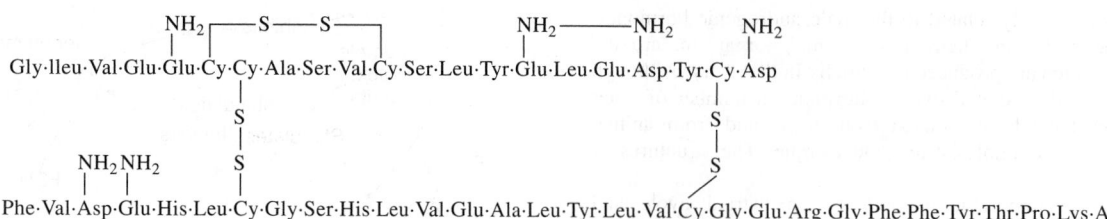

Fig. 4.   Primary structure of bovine insulin

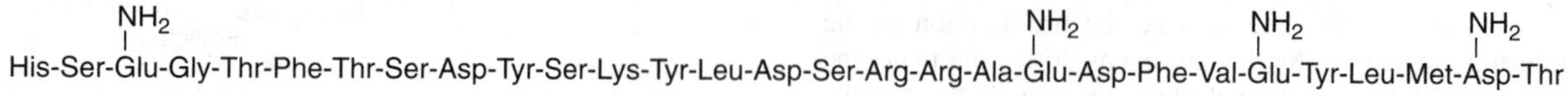

NH₂
|
His-Ser-Glu-Gly-Thr-Phe-Thr-Ser-Asp-Tyr-Ser-Lys-Tyr-Leu-Asp-Ser-Arg-Arg-Ala-Glu-Asp-Phe-Val-Glu-Tyr-Leu-Met-Asp-Thr

Fig. 5.   Glucagon

remote from where the hormone is produced. Plant hormones also act in very minute amounts. Plant hormones include the auxins, gibberellins, and kinetins. These are described in the entries on **Gibberellic Acid and Gibberellin Plant Growth Hormones**; and **Plant Growth Modification and Regulation**.

Additional Reading

Bengtsson, Bengt-Ake: *Growth Hormone,* Kluwer Academic Publishers, Norwell, MA, 1999.

Buckingham, J.C., G. Gillies, and A.M. Cowell: *Stress, Stress Hormones and the Immune System,* John Wiley & Sons, Inc., New York, NY, 1997.

Conn, P.M. and W.F. Crowley, Jr.: "Gonadotropin-Releasing Hormone and Its Analogues," *N. Eng. J. Med.,* **93** (January 10, 1991).

Copp, D.H.: "Parathyroids, Calcitonin, and Control of Plasma Calcium," *Recent Progr. Hormone Res.,* **20**, 59–77 (1964).

Erickson, D.: "Human Growth Hormone," *Sci. Amer.,* **164** (September 1990).

Erickson, D.: "Hormone Derivatives May Combat PMS and Epilepsy," *Sci. Amer.,* **124** (May 1991).

Evans, R.M.: "The Steroid and Thyroid Hormone Receptor Superfamily," *Science,* **889** (May 13, 1988).

Fackelmann, K.A.: "High-Pressure Hormone," *Sci. News,* **344** (December 1, 1990).

Hadley, M.E.: *Endocrinology,* Prentice-Hall, Inc., Upper Saddle River, NJ, 2000.

Harvey, P.W., A. Cockburn, and K.C. Rush: *Endocrine and Hormonal Toxicology,* John Wiley & Sons, Inc., New York, NY, 1999.

Kanellis, A.K. et al.: *Biology and Biotechnology of the Plant Hormone Ethylene II,* Kluwer Academic Publishers, Norwell, MA, 1999.

Kritchevsky, D. and D. Heber: *Dietary Fats, Lipids, Hormones, and Tumorigenesis: New Horizons in Basic Research,* Vol. 299, Kluwer Academic Publishers, Norwell, MA, 1996.

Kutsky, J.F.: *Handbook of Vitamins and Hormones,* Van Nostrand Reinhold, New York, NY, 1973.

Litwack, G. and T. Begley: *Vitamins and Hormones,* Vol. 61, Academic Press, Inc., San Diego, CA, 2000.

Marx, J.: "How Peptide Hormones Get Ready for Work," *Science,* **779** (May 10, 1991).

Norman, A.W. and G. Litwack: *Hormones,* Academic Press, Inc., San Diego, CA, 1997.

Seifer, D.B. and E.A. Kennard: *Menopause: Endocrinology and Management,* Vol. 18, Humana Press, Totowa, NJ, 1999.

Smith, R.G. and M.O. Thorner: *Human Growth Hormone: Research and Clinical Practice,* Vol. 19, Humana Press, Totowa, NJ, 1999.

Strauss, J. et al.: *Molecular Biology in Reproductive Medicine,* Parthenon Publishing Group, New York, NY, 1999.

Timiras, P.S., A. Vernadakis, and W.B. Quay: *Hormones and Aging,* CRC Press, LLC, Boca Raton, FL, 1996.

HORNBLENDE.   The mineral hornblende is a complex silicate. It is probably an isomorphous mixture of three molecules, a calcium-iron-magnesium silicate, an aluminum-iron-magnesium silicate and an iron-magnesium silicate. A general formula is

$$(Ca, Na, K)_{2-3}(Mg, Fe^{2+}, Fe^{3+}, Al)_5(Al, Si)_8O_{22}(OH)_2.$$

Manganese and alkalies are sometimes present as is also titanium. It is monoclinic, with prismatic crystals, often pseudo-hexagonal. Bladed, fibrous, columnar, granular and compact massive varieties also are common. It has a perfect prismatic cleavage; hardness, 5–6; specific gravity, 3.02–3.27; color, green, greenish-brown, brown and black; luster, vitreous to silky; transparent to opaque.

Hornblende is a common constituent of many of the igneous rocks such as granite, syenite, diorite, or gabbro, of gneisses and schists and is the principal mineral of the amphibolites. Hornblende alters easily to chlorite and epidote. A variety of hornblende that contains little (less than 5%) of iron oxides is gray to white in color and named edenite, from its locality in Edenville, New York. Very dark brown to black hornblendes, which contain titanium, ordinarily are called basaltic hornblende from the fact that they are usually a constituent of basalts and similar rocks.

Well-known localities for hornblende are in The Czech Republic and Slovakia, Mount Vesuvius, Italy, Norway, Sweden, and, in the United States, in Massachusetts, New Hampshire, and New York. Black hornblende is found in Renfrew County, Canada. The word hornblende is derived from the German horn, and blende, to blind or dazzle. The term blende was often used to refer to a brilliant nonmetallic luster, i.e., zinc blende.

See also terms listed under **Mineralogy**.

ELMER B. ROWLEY
Union College
Schenectady, New York

HORNBLENDITE.   A coarse-grained rock related to gabbro that consists almost wholly of hornblende. Olivine being present, this rock may grade into a hornblende-peridotite (cortlandtite). Hornblendite is a rare rock type and of relatively little importance. See also **Gabbro**.

HOT WORKING.   Plastic deformation of metals at temperatures sufficiently elevated so that the effects of the working are nullified by concurrent softening processes. Thus, when steel is hot worked by rolling at a white heat, the metal recrystallizes and softens almost immediately after it is deformed. Similarly, the deformation of lead at room temperature is also hot working and accounts for the fact that it is not possible to work-harden this material at this temperature. An empirical rule states that the lower limit of the hot-working temperature range is the recrystallization temperature.

Forging, rolling, pressing, extruding, swaging, drawing, or forming of metals at temperatures above their recrystallization temperatures are examples of hot working.

HUBER, ROBERT (1937–).   Awarded the Nobel prize for chemistry in 1988, along with Johann Deisenhofer and Hartmut Michel, for work that revealed the three-dimensional structure of closely linked proteins that are essential to photosynthesis. Doctorate awarded in 1963 by Technical University of Munich, Germany.

HUMECANTS AND MOISTURE-RETAINING AGENTS.   Substances that have affinity for water, with stabilizing action on the water content of a material, are called *humectants* or moisture-retaining agents. Ideally, a humectant maintains within a rather narrow range the moisture content caused by humidity fluctuations. These materials are widely used in certain food products, as well as tobacco, and in recent years have taken on increasing importance in the case of intermediate-moisture foods. Traditionally, humectants have been used to retain moisture in foods like coconut and marshmallows which otherwise would quickly dry and become tasteless. For example, flaked coconut is kept moist in the container by adding glycerine and glyceryl monostearate.

Among the most commonly used humectant are glycerine, potassium polymetaphosphate, propylene glycol, sodium chloride, sorbitol, sucrose, and triacetin. Also, phosphates are added to the pickling solutions used to treat cured meats, such as ham, bacon, corned beef, etc., by soaking or injection. Their principal purpose is for moisture binding to reduce the loss of fluids during curing and cooking.

During the last few years, important research has gone into the addition of multiple humectants and water to food systems. Studies have shown that a hysteresis effect may occur with certain humectants, i.e., a different rate of moisture absorption than the rate for moisture desorption. Multiple humectants tend to compensate these hysteresis effects, giving uniform rates in both directions.

HUME-ROTHERY RULES. When alloy systems form distinct phases, it is found that the ratio of the number of valence electrons to the number of atoms is characteristic of the phase (e.g., β, γ-, $\in$-) whatever the actual elements making up the alloy. Thus, both $Na_{31}Pb_8$ and Ni_5Zn_{21} are γ-structures, with the electron-atom ratio 21:13. The rules are explained by the tendency to form a structure in which all the Brillouin zones are nearly full, or else entirely empty.

HUMIDITY. Generally, some measure of water-vapor content of air. *Absolute humidity* is the ratio of the mass of water vapor present to the volume occupied by the mixture; that is, the density of the water vapor component. The percentage of water vapor in the total composition of the air may be determined by passing a measured quantity of air through a tube containing an absorbing substance that removes all the vapor, and which can be weighed before and after the absorption.

Absolute humidity is usually expressed in grams of water vapor per cubic meter or, in engineering practice, in grains per cubic foot. Because this measure of atmospheric humidity is not conservative with respect to adiabatic expansion or compression, it is not commonly used by meteorologists. As occasionally used in air-conditioning practice, absolute humidity refers to the number of grains of water vapor per pound of moist air, which is dimensionally identical with the specific humidity (defined below).

Critical humidity is the point at which the partial pressure of water vapor in the atmosphere is equal to the saturation vapor pressure. Condensation on suitable nuclei will occur when the humidity reaches or exceeds this value.

Relative humidity is the ratio of the actual vapor pressure of the air, at any temperature, to the maximum of saturation vapor pressure at the same temperature. It expresses the vapor content as a fraction or percentage of the concentration necessary to render the vapor saturated at the given temperature. At the dew point, the relative humidity is 100%. A rise of temperature without the addition of more vapor reduces the relative humidity (but not the absolute humidity), while a fall of temperature increases it and may bring about saturation. Relative humidity is measured by the hygrometer.

Specific humidity is the (dimensionless) ratio of the mass of water vapor to the total mass of the system. It may be approximated by the mixing ratio for many purposes:

$$q = \frac{w}{1 + w}$$

where q is the specific humidity and w the mixing ratio.

See also **Hygrometry and Psychrometry.**

HYATT, JOHN WESLEY (1837–1920). Hyatt is generally credited as being the father of the plastics industry. In 1869, he and his brother patented a mixture of cellulose nitrate and camphor which could be molded and hardened. Its first commercial use was for billiard balls. The TM "Celluloid" was the first ever applied to a synthetic plastic product; it flammability hazard limits its use.

HYDANTOIN AND ITS DERIVATIVES. Hydantoin is an accepted name for 2,4-imidazolidinedione. This ring system rarely occurs in nature, although some natural products with hydantoin substructures are known. A huge number of derivatives have been prepared.

Physical Properties

Hydantoins are crystalline solids with high melting points, particularly those compounds in which nitrogen is unsubstituted, because this allows intermolecular association by hydrogen bonds. Hydantoins are weak acids which dissociate at the imidic $N-3-H$ atom because this allows more efficient delocalization of the negative charge than ionization at N-1.

Several structure—acidity relationships have been established for hydantoin derivatives. Thus, ionization is known to be unaffected by alkyl substituents at N-1 and at C-5. However, aryl and other electron withdrawing groups can considerably enhance the acidity of hydantoins. Introduction of an arylmethylene side chain at C-5 increases the acidity of the N-1 hydrogen, making it measurable. This is due to delocalization of the negative charge at N-1 into the C-5 substituent.

Solvent variation can greatly affect the acidity of hydantoins. Water provides a better stabilization for the hydantoin anion and hence an increased acidity when compared to DMSO. 2-Thiohydantoin (pK_a 8.5) is a slightly stronger acid than hydantoin (pK_a 9.0). 4-Thiohydantoins appear to be weaker acids.

Spectral Properties. Hydantoin derivatives show weak absorption in the uv-visible region, unless a part of the molecule other than the imidazolidinedione ring behaves as a chromophore; however, pK_a values have been determined by spectrophotometry in favorable cases. Absorption of uv by thiohydantoins is more intense. Several pK_a values of thiohydantoins have been determined by uv-visible spectrophotometry.

Chemical Properties. Hydantoins can react with electrophiles at both nitrogen atoms and at C-5. The electrophilic carbonyl groups can be attacked by nucleophiles, leading to hydrolysis of the ring or to partial or total reduction of the carbonyl system. Other reactions are possible, including photochemical cleavage of the ring.

Synthesis

Synthesis From α-Amino Acids and Related Compounds. Addition of cyanates, isocyanates, and urea derivatives to α-amino acids yields hydantoin precursors. This method is called the Read synthesis, and can be considered as the reverse of hydantoin hydrolysis. Thus the reaction of α-amino acids with alkaline cyanates affords hydantoic acids, which cyclize to hydantoins in an acidic medium.

In a modification of the original method, Read replaced α-amino acids with α-amino nitriles. Chlorosulfonyl isocyanate is an excellent alternative to alkaline cyanates in the preparation of hydantoins from sterically hindered or labile amino nitriles.

Substitution of alkaline cyanates by isocyanates allows the preparation of 3-substituted hydantoins, both from amino acids and amino nitriles.

A variety of α-amino acid derivatives, including the acids themselves, halides, esters, and amides can be transformed into hydantoins by condensation with urea. α-Hydroxy acids and their nitriles give a similar reaction.

Synthesis From Aldehydes and Ketones. Treatment of aldehydes and ketones with potassium cyanide and ammonium carbonate gives hydantoins in a one-pot procedure (Bucherer-Bergs reaction) that proceeds through a complex mechanism. Some derivatives, like oximes, semicarbazones, thiosemicarbazones, and others, are also suitable starting materials.

Synthesis From Thiohydantoins. A modification of the Bucherer-Bergs reaction consisting of treatment of an aldehyde or ketone with carbon disulfide, ammonium chloride, and sodium cyanide affords 2,4-dithiohydantoins. 4-Thiohydantoins are available from reaction of amino nitriles with carbon oxysulfide. Both thiocompounds can be transformed into hydantoins.

Health and Safety Factors (Toxicology)

The acute toxicity of hydantoin derivatives seems to be low. Most studies on long-term toxicity of hydantoins deal with phenytoin, due to the wide use of this compound as an anticonvulsant. Long-term toxic effects of phenytoin include folate deficiency due to impaired folate absorption, hypocalcemia and osteomalacia, alterations of carbohydrate metabolism, gingival hyperplasia, and teratogenic effects. These are grouped under the term fetal hydantoin syndrome, which is associated with continued use of hydantoins during the early stages of pregnancy, and consists of mild retardation of physical and mental indexes, dysmorphic faces and, occasionally, cleft palate, cleft lip, and cardiac defects. Formation of cyanide by degradation of hydantoin derivatives used as antiseptics for water treatment has been described, and this fact might have toxicological relevance.

Applications

Halogenated Hydantoins. Halogenation has been achieved by use of a variety of halogenating reagents. These derivatives are employed as reagents in synthesis and analysis and also as disinfectants and biocides in water treatment.

N-Methylolhydantoins. 1,3-Bis(hydroxymethyl)-5,5-dimethylhydantoin is used extensively as a preservative in cosmetic and industrial applications, and carries EPA registration for the industrial segment. 1-Hydroxymethyl-5,5-dimethylhydantoin is used as an odorless donor of formaldehyde for adhesive applications.

Epoxy Resins. Urethane and ester-extended hydantoin epoxy resins cured with several compounds seem to have better properties than the previous ones.

5-Substituted Hydantoins. 5-Methylhydantoin has been elected from several structures as a formaldehyde scavenger for color photosensitive materials and water-thinned inks and coatings. Although the hydantoin ring itself does not present any medicinal activity, many 5,5-disubstituted hydantoins have shown interesting biological properties, and some of them are used in medicine, particularly as anticonvulsants. Several types of hydantoin derivatives find use as herbicides.

CARMEN AVENDAÑO
J. CARLOS MENÉNDEZ
Universidad Complutense

Additional Reading

Avendaño, C. and G.G. Trigo: *Adv. Heterocycl. Chem.* **38**, 177 (1985).
Philip, J., I.J. Holcomb, and S.A. Furasi: in K. Florey, ed., *Analytical Profiles of Drug Substances,* Vol. 13, Academic Press, Orlando, FL, 1984.
Schipper, E.S. and A.R. Day: in R.C. Elderfield, ed., *Heterocyclic Compounds,* Vol. 5, John Wiley & Sons, Inc., New York, NY, 1957.
Ware, E. *Chem. Rev.* **46**, 403 (1950).

HYDRATE. Excluding the loose usages in which the term hydrate indicates merely the presence of water or of its elements in 2:1 ratio, as in carbohydrate, the term hydrate denotes the appearance of water in compounds. There are a number of ways in which water may appear in stoichiometric proportions in compounds. Moreover, these ways may be described from more than one point of view. A somewhat systematic approach is to view these compounds from the point of view of the extent of integration of the water, or its elements, into the compound.

The term "water of constitution" is a somewhat old usage, applied to compounds in which no H_2O groupings appear in the structure of the compound, but the compound may undergo reaction, usually reversible, in which water is one of the products. Magnesium hydroxide and sulfuric acid could thus be said to have "water of constitution," even though it appears in their structure as hydroxyl groups, or hydroxyl groups and hydrogen atoms (protons).

The term "cationic water" may be used to describe the situation in which water appears in coordination compounds apparently joined to cations by covalent bonds. However, the fact that a number of such compounds exhibit "hydrate isomerism" is evidence for cationic bonding, as well as it is for the existence of other forms of these compounds in which the presence of water is due to electrostatic attractions or crystal stability requirements.

The term "anionic water" describes the situation in which water is joined to anions through covalent bonds, or more frequently, through hydrogen bonds. The type case is copper(II) sulfate pentahydrate, where the cation has a coordination number of four and presumably the fifth molecule of H_2O is bound to the sulfate ion (as well as to other H_2O molecules) by hydrogen bonds.

The term "lattice water" is commonly applied to cases in which the water molecules are occupying definite positions in the crystal lattice but are apparently not coordinated with either cations or anions. Again, clear-cut cases are those in which the compound is so highly hydrated that both lattice water and "ion water" are present.

The water in crystals may, however, be present in other than definite lattice positions. For example, the water molecules may be found in holes in the lattices, or they may occupy random positions in the lattices. The latter situation is often found in ion exchange resins where loss of water, up to a certain point, does not materially change the lattice structure.

Finally, in essentially noncrystalline materials, such as hydrous precipitates and colloidal gels, the water present is at the limiting case of being a hydrate, in which virtually no bonding, in the chemical sense, exists.

HYDRAULIC FLUID. A liquid or mixture of liquids designed to transfer pressure from one point to another in a system on the basis of Pascal's law, i.e., pressure on a confined liquid transmitted equally in all directions. For industrial use, such fluids are based on paraffinic and cycloparaffinic petroleum fractions, usually with added antioxidant and viscosity index improvers. Flame-resistant types include additives such as phosphate esters or emulsions of water and ethylene glycol. The brake fluids used in auto are composed of (1) a lubricant (polypropylene glycol of 1000–2000 mw, a castor oil derivative or a synthetic polymeric mixture of monobutyl ethers of oxyethylene and oxypropylene glycols); (2) a solvent blend (mixture of glycol ethers); and (3) additives for corrosive resistance buffering, etc., bp 375–550°F. The composition and performance characteristics are specified by the Society of Automotive Engineers: http://www.sae.org/.

HYDRAZINE. [CAS: 302-01-2]. $H_2N \cdot NH_2$, formula weight 32.04, colorless, fuming liquid, mp 1°C, bp 113°C, sp gr 1.011, decomposes when heated above 350°C at atmospheric pressure into N_2 and NH_2, also decomposes in presence of a catalyst (e.g., platinum) into N_2 and NH_3. Hydrazine burns when ignited in air with a violet-colored flame. The compound is soluble in all proportions with H_2O and is soluble in alcohol. Hydrazine forms a hydrate with one molecule of H_2O. Upon moderate heating or in a vacuum, the hydrate yields hydrazine and H_2O. Hydrazine is a base slightly weaker than NH_4OH.

Hydrazine is a tonnage chemical with numerous uses, including that of a propellant for rockets, yielding exhaust products at a high temperature and of a low molecular weight; use as a strong reducing agent in the manufacture of various chemicals; and as a blowing agent for foamed rubber. The compound reacts with citric acid to form *Continazin*, an antituberculan drug.

Although the earlier processes for the commercial production of hydrazine used urea as a raw material, modern processes employ direct ammonia oxidation. In one such process, reactions occur in two steps:

$$NH_3 + NaOCl \longrightarrow NH_2Cl + NaOH \qquad (1)$$

$$NH_3 + NH_2Cl + NaOH \longrightarrow H_2N \cdot NH_2 + NaCl + H_2O. \qquad (2)$$

High-grade hypochlorite is required for Step 1. Special agents, such as gelatin, ethylenediamine tetracetic acid, glue, high alcohols, and formaldehyde, are required to inhibit undesirable side reactions that would reduce the hydrazine yield through formation of ammonium chloride and N_2. In another hydrazine process, chlorine, NH_3, and H_2SO_4, along with methylethyl ketone, are used as the charge. The products of this process include hydrazine hydrate, hydrazine sulfate, ketazine, and dialkyldiazacyclopropane. Hydrazine also is used as a start-up ingredient in the preparation of cooling water for nuclear reactors where it is desired to keep the oxygen content of the water to an absolute minimum and thus decrease corrosion. Oxygen reacts with hydrazine. $H_2N \cdot NH_2 + O_2 \longrightarrow N_2 + 2H_2O$. When no oxygen is present in the water, the hydrazine acts as a sink for dissolved oxygen that may enter later, by maintaining metal oxides at their lower oxidation states.

Hydrazine forms two series of salts: (1) hydrazinium (1+) chloride, $H_2NNH_3^+ Cl^-$, nitrate, $H_2NNH_3^+ NO_3^-$, hemisulfate, $(H_2NNH_3^+)_2SO_4^{2-}$, (2) hydrazinium (2+) chloride, $H_3NNH_3^{2+} (Cl^-)_2$, dinitrate, $H_3NNH_3^{2+} (NO_3^-)_2$, hydrogen sulfate, $H_3NNH_3^{2+} (HSO_4^-)_2$, all soluble in H_2O. This last is produced when hydrogen azide reacts with concentrated H_2SO_4. It is very hygroscopic and decomposes in aqueous solution to give the slightly soluble monosulfate and H_2SO_4. The monosulfate and difluoride, which have been thought to have the structures $N_2H_5^+ HSO_4^-$ and $N_2H_5^+ HF_2^-$ in the solids, have been shown in fact to be $N_2H_6^{2+} SO_4^{2-}$ and $N_2H_6^{2+} (F^-)_2$. Hydrazinium azide, $N_2H_5^+ N_3^-$, is a soluble solid.

In the laboratory, hydrazine can be prepared by converting one-half of a given amount of NH_3 into chloramine, NH_2Cl, by sodium hypochlorite solution in the presence of a colloid and heating. The remaining one-half of the NH_3 reacts with chloramine to form hydrazine. The product is then cooled to 0°C and H_2SO_4 added in amount to react with the hydrazine to form hydrazine sulfate, $N_2H_6SO_4$, insoluble solid. Hydrazine hemisulfate, $(N_2H_5)_2SO_4$, is soluble in H_2O. It can also be made by the reaction of NH_3 and hydroxylamine-O-sulfonic acid.

Phenylhydrazine is a colorless liquid, slightly soluble in H_2O, miscible in all proportions with alcohol or ether, forms salts with acids, e.g., phenylhydrazine hydrochloride or phenylhydrazinium chloride, $C_6H_5NHNH_3Cl$, is a powerful reducing agent, with alkaline copper(II) salt solution (Fehling's solution) yields copper(I) oxide precipitate, reacts with carbonyl group of aldehydes or ketones yielding phenylhydrazones, white solids, of definite melting point and utilized in identification of aldehydes and ketones, e.g., acetaldehyde phenylhydrazone, $CH_3CH:NNHC_6H_5$.

Phenylhydrazine, as hydrochloride solution plus sodium acetate, reacts with polyhydroxy aldehydes or ketones yielding *osazones* or diphenyl-hydrazones, yellow solids, of definite melting point and utilized in identification of sugars, e.g., phenyl-d-glucosazone, CH_2OH $(CHOH)_3C$: $(NNHC_6H_5)CH:(NNHC_6H_5)$ plus aniline $C_6H_5NH_2$ plus NH_3.

Attention should be given to the difference between osazones and osones. An *osone* is formed by reaction of an osazone with HCl, e.g., glucosone, $CH_2OH(CHOH)_3CO \cdot CHO$.

1,1-Diphenylhydrazine is made by reduction of diphenylnitrosamine,, by zinc plus acetic acid, the nitrosamine being formed by reaction of diphenylamine,, and nitrous acid.

Tetraphenylhydrazine is a white solid, soluble in chloroform, acetone, benzene, or toluene, and upon standing is changed into triphenylamine plus azobenzene. In solution, tetraphenylhydrazine dissociates into nitrogen diphenyl, $(C_6H_5)_2N\cdot$, free radical, which in toluene at 90°C reacts with nitric oxide, NO. Tetraphenylhydrazine is formed by oxidation of diphenylamine,, by lead dioxide.

Hydrazine reacts with ketones to form *azines*.

Toxicology

Hydrazine is toxic and readily absorbed by oral, dermal, or inhalation routes of exposure. Contact with hydrazine irritates the skin, eyes, and respiratory tract. Liquid splashed into the eyes may cause permanent damage to the cornea. At high doses it can cause convulsions, but even low doses may result in central nervous system depression. Death from acute exposure results from convulsions, respiratory arrest, and cardiovascular collapse. Repeated exposure may affect the lungs, liver, and kidneys. Evidence is limited as to the effect of hydrazine on reproduction and/or development; however, animal studies demonstrate that only doses that produce toxicity in pregnant rats result in embryo-toxicity.

The TLV is set at 0.1 ppm (hydrazine); 0.2 ppm (MMH); and 0.5 ppm (UDMH). The International Agency for Research on Cancer (IARC) classifies hydrazine as a 2B or possible human carcinogen. The American Conference of Governmental Industrial Hygienists (ACGIH) classifies hydrazine as an A2 or suspect human carcinogen.

HYDRAZOIC ACID. [CAS: 7782-79-8]. HN_3, formula weight 43.03, colorless, odorous, poisonous liquid, mp −80°C, bp 37°C, explodes with marked violence. Also known as azoimide and hydronitric acid, the compound is miscible in all proportions with H_2O, alcohol, and ether. Hydrazoic acid reacts (1) with metals, e.g., magnesium, aluminum, zinc, iron, to form azides or hydrazoates (or trinitrides), (2) with heavy metal salt solutions to form insoluble azides, e.g., silver azide AgN_3, mercury(I) azide HgN_3, lead azide PbN_6. Silver, mercury(I), and copper(I) azides decompose in the light to form nitrogen plus the metal. (3) It reacts with NH_4OH to form ammonium azide $NH_4 \cdot N_3$, (4) with hydrazine to form hydrazine azide $N_2H_4 \cdot HN_3$, (5) with sodium hypochlorite plus acetic acid to form chlorazide ClN_3, explosive, (6) with sodium amalgam to form NH_3 with some hydrazine, (7) with potassium permanganate to form nitrogen and H_2O.

Hydrazoic acid is formed (1) by reaction of sodium nitrate with molten sodamide, (2) by reaction of nitrous oxide with molten sodamide, (3) by reaction of nitrous acid and hydrazinium ion ($N_2H_5^+$), (4) by oxidation of hydrazinium salts, (5) by reaction of ethyl nitrite with NaOH solution and acidifying. See also **Azides**.

HYDRAZONES. The products of the reaction between an aldehyde or a ketone with phenylhydrazine are termed *hydrazones*. Sometimes the compounds are referred to as phenylhydrazones.

$$CH_3 \cdot CHO + C_6H_5 \cdot NH \cdot NH_2 \longrightarrow CH_3 \cdot CH:N \cdot NH \cdot C_6H_5 + H_2O$$
(acetaldehyde) (phenylhydrazine) (acetaldehyde hydrazone)

$$C_6H_5CHO + C_6H_5 \cdot NH \cdot NH_2 \longrightarrow C_6H_5 \cdot CH:N \cdot NH \cdot C_6H_5 + H_2O$$
(benzaldehyde) (benzylidenehydrazone)

$$(CH_3)_2CO + C_6H_5 \cdot NH \cdot NH_2 \longrightarrow (CH_3)_2C:N \cdot NH \cdot C_6H_5 + H_2O$$
(acetone) (acetonehydrazone)

$$C_6H_5 \cdot CO \cdot CH_3 + C_6H_5 \cdot NH \cdot NH_2 \longrightarrow (C_6H_5)(CH_3)C:N \cdot NH \cdot C_6H_5 + H_2O$$
(acetophenone) (acetophenonehydrazone)

Several of the hydrazones may be decomposed by strong acids whereupon the original aldehyde or ketone is regenerated, along with the formation of a phenylhydrazine salt. When reduced, hydrazones yield primary amines.

HYDRIDES. Hydrides are compounds that contain hydrogen in a reduced or electron-rich state. Hydrides may be either simple binary compounds or complex ones. In the former, the negative hydrogen is bonded ionically or covalently to a metal, or is present as a solid solution in the metal lattice. In the latter, which comprise a large group of chemical compounds, complex hydridic anions such as BH_4^-, AlH_4^-, and derivatives of these, exist.

Commercial applications of hydrides have become important and some of these compounds have become industrial chemicals manufactured and used on a large scale.

Simple (Binary) Hydrides

Ionic Hydrides. The ionic or saline hydrides contain metal cations and negatively charged hydrogen ions. They crystallize in the cubic lattice similar to the corresponding metal halide, and when pure, are white solids. When dissolved in molten salts or hydroxides and electrolyzed, hydrogen gas is liberated at the anode. Their densities are greater than those of the parent metal, and their formation is exothermic. All are strong bases.

Physical properties of the alkali metal hydrides are given in Table 1. Sodium hydride finds commercial usage in organic synthesis in condensation and alkylation reactions.

Table 2 gives thermochemical data of alkaline-earth metal hydrides. All form orthorhombic crystals. Calcium hydride is a convenient portable source of hydrogen gas, which results from its reaction with water.

TABLE 1. PHYSICAL PROPERTIES OF ALKALI METAL HYDRIDES

Hydride	Mp, °C	$\Delta H_{(298)}$, kJ/mol[a]	$\Delta F_{(298)}$, kJ/mol[a]	S, J/(mol · K)[a]	Lattice energy, kJ/mol[a]	Density, g/cm³
LiH	688	−90.7	−70	25	916	0.77
NaH	420 dec	−56.5	−37.7	48	791	1.36
KH	dec	−57.9	−37.3	61	720	1.43
RbH	300 dec					2.60
CsH	dec					3.4

[a] To convert J to cal, divide by 4.184.

TABLE 2. PHYSICAL PROPERTIES OF ALKALINE-EARTH METAL HYDRIDES

Hydride	$\Delta H_{(298)}$, kJ/mol[a]	$\Delta F_{(298)}$, kJ/mol[a]	S, J/(mol · K)[a]	Density, g/cm³
CaH₂	−186.3	−147.4	42	1.90
SrH₂	−180.5	−138.6	54	3.27
BaH₂	−171.2	−132.3	67	4.16

[a] To convert J to cal, divide by 4.184.

TABLE 3. PROPERTIES OF COVALENT HYDRIDES

Hydride	Formula	Mp, °C	Bp, °C	Density[a] g/cm³	Density[a] g/L
beryllium hydride[b]	BeH₂	125 dec	220[c]		
magnesium hydride	MgH₂	280 dec		1.45	
aluminum hydride	AlH₃				(−185)
silane[d]	SiH₄	−185[e]	−119.9	0.68[f]	1.44[g] (20)
germane	GeH₄	−165	−90	1.523[f] (−142)	3.43[g] (0)
stannane	SnH₄	−150	52		
arsine	AsH₃	−116.9[e]	−62	1.604[f] (64)	2.695[g]

[a] Temperature in °C is given in parentheses.
[b] ΔH_{298} = 19.3 kJ/mol (4.6 kcal/mol).
[c] Begins to dissociate.
[d] ΔH_{298} = 30.55 kJ/mol (7.30 kcal/mol).
[e] Freezing point.
[f] Liquid.
[g] Gas at atmospheric pressure.

TABLE 4. GROUP 5 (VB) HYDRIDES

Compound	Formula	Density, g/cm^3
vanadium hydride	VH	5.4
niobium hydride	NbH	6.6
niobium dihydride[a]	NbH$_2$	
tantalum hydride	TaH	15.1

[a] Decomposes slowly.

Covalent Hydrides. Table 3 gives some properties of these compounds. Transition-metal hydrides, i.e., interstitial metal hydrides, have metallic properties, conduct electricity, and are less dense than the parent metal. These hydrides are much harder and more brittle than the parent metal, and most have catalytic activity. They include titanium hydride, TiH$_2$, zirconium hydride, ZrH$_2$, rare earth hydrides (lanthanum dihydride, lanthanum trihydride, cerium hydride, CeH$_2$), group 5 (VB) hydrides (Table 4), and hydrogen storage alloys (FeTi, LaNi$_5$, Mg$_2$TiH$_6$, and MgTi$_2$H$_6$).

Complex Hydrides

The complex hydrides are a large group of compounds in which hydrogen is combined in fixed proportions with two other constituents, generally metallic elements. These compounds have the general formula M(M'H$_4$)$_n$, where n is the valence of M, and M' is a trivalent Group 3 (IIIA) element such as boron, aluminum, or gallium. The most important complex hydrides are listed in Table 5.

Borohydrides. The alkali metal borohydrides are the most important complex hydrides. They are ionic, white, crystalline, high melting solids that are sensitive to moisture but not to oxygen. They include lithium borohydride, LiBH$_4$, and sodium borohydride, NaBH$_4$.

Complete hydrolysis of NaBH$_4$ produces 2.37 L hydrogen (STP) per gram of borohydride; similarly, addition of acid to a cold aqueous solution liberates the theoretical amount of hydrogen. The inorganic reductions of NaBH$_4$ are numerous and varied. Sodium borohydride reacts with boron halides to form diborane, B$_2$H$_6$, which is more conveniently handled as the monomer BH$_3$ complexed with an ether, sulfide, or amine. Sodium borohydride is used extensively for the reduction of organic compounds. Sodium borohydride is manufactured from sodium hydride and trimethyl borate in a mineral oil medium at about 275°C. Sodium borohydride is classified as a flammable solid. It is available as powder, caplets, and

TABLE 5. COMPLEX HYDRIDES

Formula	Density, g/cm^3	Mp, °C
LiBH$_4$	0.66	278
NaBH$_4$	1.074	505
KBH$_4$	1.177	585
Be(BH$_4$)$_2$	0.702	123 dec
Mg(BH$_4$)$_2$		320 dec
Ca(BH$_4$)$_2$		260 dec
Zn(BH$_4$)$_2$		> 50 dec
Al(BH$_4$)$_3$	0.549	−64.5[a]
Zr(BH$_4$)$_4$	1.13	28.7
Th(BH$_4$)$_4$	2.59	204 dec
U(BH$_4$)$_4$	2.67	100 dec
(CH$_3$)$_4$NBH$_4$	0.84	>310
(C$_2$H$_5$)$_4$NBH$_4$	0.926	225 dec
(C$_4$H$_9$)$_4$NBH$_4$		>300
(C$_8$H$_{17}$)$_3$CH$_3$NBH$_4$	0.9	ca 30
C$_{16}$H$_{33}$(CH$_3$)$_3$NBH$_4$	0.9	ca 160
NaBH$_3$CN	1.20	240 dec
NaBH(OCH$_3$)$_3$	1.24	230 dec
LiAlH$_4$	0.917	190 dec
NaAlH$_4$	1.28	178
Mg(AlH$_4$)$_2$		140 dec
Ca(AlH$_4$)$_2$		>230 dec
LiAlH(OCH$_3$)$_3$		
LiAlH(OC$_2$H$_5$)$_3$		
LiAlH(OC$_4$H$_9$)$_3$	1.03	>400
NaAlH$_2$(OC$_2$H$_4$OCH$_3$)$_2$	1.122	205 dec
NaAlH$_2$(C$_2$H$_5$)$_2$		85

[a] Bp, 44.5°C.

granules and as a 12% solution in caustic soda. The principal uses of NaBH$_4$ are in synthesis of pharmaceuticals and fine organic chemicals; removal of trace impurities from bulk organic chemicals; wood-pulp bleaching, clay leaching, and vat-dye reductions; and removal and recovery of trace metals from plant effluents.

Potassium borohydride was formerly used in color reversal development of photographic film and was preferred over sodium borohydride because of its much lower hygroscopicity. Because other borohydrides are made from sodium borohydride, they are correspondingly more expensive. Generally their reducing properties are not sufficiently different to warrant the added cost Zinc borohydride, Zn(BH$_4$)$_2$, however, has found many applications in stereoselective reductions.

Borohydride Derivatives. Modification of the BH$_4^-$ anion has provided derivatives of widely differing reducing properties. Alkoxyborohydrides, such as sodium trimethyoxyborohydride, NaBH(OCH$_3$)$_3$, exhibit enhanced reducing power but are less selective and more sensitive to decomposition by water. Sodium cyanoborohydride, NaBH$_3$CN, on the other hand, shows weakened reducing properties and is unique among the complex hydrides because it is stable in acidic aqueous solutions to a pH of about 3.

Sodium or tetramethylammonium triacetoxyborohydride has become the reagent of choice for diastereoselective reduction of β-hydroxyketones to antidiols. Trialkylborohydrides, e.g., alkali metal tri-sec-butylborohydrides, show outstanding stereoselectivity in ketone reductions.

Aluminohydrides. In general, the aluminohydrides are more active and powerful reducing agents than the corresponding borohydrides. They decompose vigorously with water. Reaction also occurs with alcohols, although more moderately, providing a route to substituted derivatives.

Freshly prepared lithium aluminum hydride is a white crystalline solid that tends to become gray during storage, although very little loss in purity occurs. Although lithium aluminum hydride is best known as a nucleophilic reagent for organic reductions, it converts many metal halides to the corresponding hydride, e.g., Ge, As, Sn, Sb, and Si. Commercial manufacture of LiAlH$_4$ uses the original synthetic method, i.e., addition of a diethyl ether solution of aluminum chloride to a slurry of lithium hydride.

Sodium aluminum hydride can be prepared from NaH, but direct synthesis from the elements is more economical.

Aluminohydride Derivatives. The few known derivatives of the aluminohydrides are principally alkoxy substitutions, including the trimethoxy, LiAlH(OCH$_3$)$_3$, triethoxy. LiAlH(OC$_2$H$_5$)$_3$, and tri-t-butoxy aluminohydrides, LiAlH(O-t-C$_4$H$_9$)$_3$.

Health and Safety Factors

In general, hydrides react exothermically with water, resulting in the generation of hydrogen. This hydrolysis reaction is accelerated by acids or heat and, in some instances, by catalysts. Because the flammable gas hydrogen is formed, a potential fire hazard may result unless adequate ventilation is provided. Ingestion of hydrides must be avoided because hydrolysis to form hydrogen could result in gas embolism.

Another aspect of the hydrolysis of hydrides is the alkalinity that results, especially from alkali metal and alkaline-earth hydrides. This alkalinity can cause chemical burns in skin and other tissues. Hydrolysis considerations obviously demand that hydrides be kept away from contact with acids.

Although there is little toxicity information published on hydrides, a threshold limit value (TLV) for lithium hydride in air of 25 μg/m^3 has been established. More extensive data are available for sodium borohydride in the powder and solution forms. The acute oral LD$_{50}$ of NaBH$_4$ is 50–100 mg/kg for NaBH$_4$ and 500–1000 mg/kg for the solution. The acute dermal LD$_{50}$ (on dry skin) is 4–8 g/kg for NaBH$_4$ and 100–500 mg/kg for the solution. The reaction or decomposition by-product sodium metaborate is slightly toxic orally (LD$_{50}$ is 2000–4000 mg/kg) and nontoxic dermally.

EDWARD A. SULLIVAN
Morton International

Additional Reading

Adams, R. M. and A. R. Siedle: *Boron, Metallo-Boron Compounds and Boranes,* Wiley-Interscience, New York, NY, 1964, Chapt. 6, pp. 373–506.

Dedieu, A.: *Transition Metal Hydrides,* John Wiley and Sons, Inc., New York, NY, 1991.

Dedina, J. and D. Tsalev: *Hydride Generation Atomic Absorption Spectrometry,* John Wiley and Sons, Inc., New York, NY, 1995.

James, B. D. and M. G. H. Wallbridge: *Progr. in Inorganic Chemistry,* Vol. 11, Wiley-Interscience, New York, NY, 1970, pp. 99–231.

Mueller, W. M., J. P. Blackledge, and G. G. Libovitz: *Metal Hydrides,* Academic Press, Inc., New York, NY, 1968, Chapt. 12, pp. 546–674.

Sastri, M.V.C., B. Viswanathan, and S.S. Murthy: *Metal Hydrides: Fundamentals and Applications,* Springer-Verlag, Inc., New York, NY, 1998.

Shrilrain, E. and S. Amoretty: *Thermophysical Properties of Lithium Hydride, Deuteride, and Tritide and of Their Solutions with Lithium,* Springer Verlag-New York, New York, NY, 1987.

Walker, E. R. H.: *Chem. Soc. Rev.* **5**, 23 (1976).

HYDROBORATION. The reaction of diboranes either with alkenes (olefins) to form trialkylboron compounds or with acetylene to yield alkenylboranes. Much research has been devoted to developing these reactions, the products of which are called organoboranes. They are useful in many complex organic syntheses, including prostaglandins and insect pheeromones.

See also **Borane**; **Carborane**; and **Organoborane**.

HYDROCARBONS. See **Organic Chemistry**.

HYDROCHLORIC ACID. [CAS: 7647-01-0]. HCl (hydrogen chloride gas) in aqueous solution, colorless when pure. Commercial grades of HCl (also known as muriatic acid) generally are marketed in three concentrations: (1) 18° Bé (sp gr 1.1417 at 15.6°C, 27.92% HCl); (2) 20° Bé (sp gr 1.160, 31.45% HCl); and (3) 22° Bé (sp gr 1.1789, 35.21% HCl). Frequently the commercial grades are slightly yellow because of impurities, notably dissolved iron. Fuming hydrochloric acid contains about 37% HCl, with a sp gr 1.194. Reagent grade hydrochloric acid usually is of this latter high strength, and is perfectly clear and colorless. The maximum limits set on impurities commonly are: NH_4 0.003%; arsenic 0.000001%; free chlorine 0.0001%; heavy metals, such as lead 0.001%; iron 0.00002%; sulfates 0.0001%; sulfites 0.0001%; and residue after ignition 0.0005%. A mixture of three parts HCl and one part is HNO_3 known as *aqua regia*, a powerful solvent and oxidizing agent which will dissolve materials that may be unaffected by either acid alone. Gold and platinum are soluble in aqua regia.

Hydrochloric acid is a very-high-tonnage chemical, finding major uses in (1) the cleaning and preparation of metals prior to application of coatings, (2) the recovery of zinc from galvanized iron scrap, (3) the production of numerous chlorides, and (4) production of chlorine. At one time, HCl was extensively used as a source of both hydrogen and chlorine by way of electrolysis. This process was made obsolete many years ago when the chlor-alkali process (electrolysis of sodium chloride brines) was introduced for the production of chlorine. In recent years, however, the production of byproduct HCl, resulting from chlorination of numerous organic compounds, has increased. In some of these instances, the installation of a HCl electrolysis plant may be economically feasible. For industrial consumption anhydrous HCl gas also is available in steel cylinders under a pressure of 1,000 psi (68 atmospheres). Hydrochloric acid forms a constant-boiling solution with H_2O (20.22% HCl) which has a bp 108.58°C (760 mm Hg).

Dilute HCl reacts (1) with many hydroxides, e.g., NaOH, to yield the corresponding chloride, e.g., sodium chloride, solution, (2) with many ordinary oxides, e.g., magnesium oxide, to yield the corresponding chloride, e.g., magnesium chloride, solution, (3) with many carbonates, e.g., calcium carbonate, to yield the corresponding chloride, e.g., calcium chloride solution plus CO_2, (4) with many sulfides, e.g., ferrous sulfide, to yield the corresponding chloride, e.g., ferrous chloride, solution plus H_2S, (5) with many metals, e.g., zinc (but not copper) to yield the corresponding chloride, e.g., zinc chloride, solution plus hydrogen gas, (6) with some special oxides, e.g., lead or manganese dioxide, to yield lead or manganese chloride plus chlorine gas, (7) with solution of some salts, e.g., silver nitrate, to yield the corresponding chloride, silver chloride, precipitate. Higher strengths of hydrochloric acid usually react similarly to the dilute. Hydrochloric acid sometimes reacts as a reducing acid, e.g., (6) above.

All metallic chlorides, except silver chloride and mercurous chloride, are soluble in H_2O, but lead chloride, cuprous chloride and thallium chloride are only slightly soluble. Metallic chlorides when heated melt, and volatilize or decompose, e.g., sodium chloride, mp 804°C; calcium, strontium, barium chloride volatilize at red heat; magnesium chloride crystals yield magnesium oxide residue and hydrogen chloride; cupric chloride yields cuprous chloride and chlorine. See also **Chlorine**; **Chlorinated Organics**; **Halides**; **Hypochlorites**; and **Sodium Chloride**.

Hydrogen Chloride

This is a colorless gas, heavier than air, density 1.639 g/l at standard conditions. The gas is poisonous and quickly causes suffocation. Formula weight 36.47, mp −111°C, bp −85°C, critical pressure 83 atm, critical temperature 51.3°C. The gas is very soluble in H_2O, accounting for the high concentrations of hydrochloric acid obtainable. Although hydrogen chloride gas may be used directly in some industrial operations, normally it is generated for the purpose of dissolving in H_2O to form hydrochloric acid. The most common route to HCl is by reacting sodium chloride with H_2SO_4. This is a two-step, exothermic reaction: (1) $NaCl + H_2SO_4 \longrightarrow NaHSO_4 + HCl$, and (2) $NaCl + NaHSO_4 \longrightarrow Na_2SO_4 + HCl$. Preparation of hydrochloric acid from the gas involves an absorption tower where the gas meets a fine spray of H_2O. Ratio controllers are used to assure maximum yield of the acid of desired concentration. These controls are easily adjusted for obtaining different concentrations. In most chlorinations of organic compounds, only half of the chlorine is used to substitute for hydrogen atoms, the remaining chlorine forming HCl. Frequently, this byproduct HCl is recycled or recovered.

Additional Reading

Behrens, D.: *DECHEMA Corrosion Handbook: Corrosive Agents and Their Interaction with Materials, Vol. 5, Aliphatic Amines, Alkaline Earth Chlorides, Alkaline Earth Hydroxides, Fluorine, Hydrogen Fluoride and Hydrofluoric Acid Hydrochloric Acid,* John Wiley and Sons, Inc., New York, NY, 1989.

Lewis, R.J. and N.I. Sax: *Sax's Dangerous Properties of Industrial Materials,* 10th Edition, John Wiley & Sons, Inc., New York, NY, 1999.

Wu, T. and T. Young: *Enthalpies of Dilution of Aqueous Electrolytes: Sulfuric Acid, Hydrochloric Acid, and Lithium Chloride,* National Bureau of Standards, National Engineering Lab, Washington, DC, 1979.

Web Reference

Clinical Center of the United States Government National Institutes of Health. http://www.cc.nih.gov/cp/about_clin_path/hcl.html

HYDROCOLLOID. A hydrophilic colloidal material used largely in food products as emulsifying, thickening, and gelling agents. They readily absorb water, thus increasing viscosity and imparting smoothness and body texture to the product, even in concentrations of less than 1%. Natural types are plant exudates (gum arabic), seaweed extracts (agar), plant seed gums or mucilages (guar gum), cereal gums (starches), fermentation gums (dextran), and animal products (gelatin). Semisynthetic types are modified celluloses and modified starches. Completely synthetic types are also available, e.g., polyvinylpyrolidone. Most are carbohydrate polymers, but a few such as gelatin and casein are proteins.

HYDRODEALKYLATION (HDA). A type of hydrogenation used in petroleum refining in which heat and pressure in the presence of hydrogen are used to remove methyl or larger alkyl groups from hydrocarbon molecules, or to change the position of such groups. The process is used to upgrade products of low value, such as heavy reformate fractions, naphthenic crudes, or recycle stocks from catalytic cracking. Also toluene and pyrolysis gasoline are converted to benzene, and methyl naphthalenes to naphthalene, by this process.

See also **Hydrogenation**.

HYDROFORMING. The use of hydrogen in the presence of heat, pressure, and catalysts (usually platinum) to convert olefinic hydrocarbons to branched chain paraffins (isomerization) to yield high-octane gasoline. Catforming and similar terms are often used in the same sense.

HYDROGEN. [CAS: 1333-74-0]. Chemical element symbol H, at. no. 1, at. wt. 1,008, periodic table group 1, mp −259.14°C, bp −252.87°C, density 0.089 (solid at 4.2 K), 0.071 (liquid at 20.4 K), sp gr 0.0696 (air = 10,000). Solid hydrogen has a hexagonal crystal structure. Hydrogen at standard conditions is a colorless, odorless, tasteless gas, suffocating, but not toxic. Hydrogen occurs chiefly combined with oxygen in H_2O, with carbon in hydrocarbons, with carbon and oxygen, and with carbon and several other elements, including oxygen, nitrogen, sulfur, phosphorus, and most metals in a vast variety of hundreds of thousands of organic compounds. See also **Organic Chemistry**. Hydrogen is considered by some scientists as the primordial substance from which all other elements in the universe were developed. In terms of cosmic abundance, with a rating of silicon = 10,000, it has been estimated that the figure for hydrogen is about 3.5×10^8, this figure compared with that of carbon = 80,000, nitrogen = 160,000, and oxygen = 220,000. For further comparison, the figure for gold is 0.0015 and for uranium it is 0.0002. In terms of abundance of the chemical elements in seawater, hydrogen ranks second (behind oxygen) with an estimated 510 million tons per cubic mile (~109 million metric tons per cubic kilometer). Hydrogen ranks eleventh in terms of content in igneous rocks in the earth's crust, the estimate of average content being 0.13%. Although free hydrogen escaped from the earth's lower atmosphere, some of the planets appear to have significant amounts, including the atmospheres of Jupiter, Saturn, and Uranus. At an altitude of 1,000 miles (1609 kilometers) above the surface of the earth, there is a greater abundance of hydrogen atoms than of nitrogen or oxygen atoms.

Hydrogen was first identified by Cavendish in 1766. The element was named by Lavoisier in 1783. However, it was not until 1931 that a second isotope of hydrogen (deuterium) with a mass number 2 was discovered by Urey. In 1934, Rutherford, Oliphant, and Harteck prepared a third isotope (tritium) with a mass number 3. Normal hydrogen (protium) and deuterium are stable, whereas tritium is radioactive, with a half-life of 12.26 years. Tritium emits a negative electron to form $^{-3}$H. It is estimated that the isotopic abundance of ^{1}H (protium) in natural occurring hydrogen is 99.9851% and on the basis of carbon = 12 (atomic weight scale), protium has a mass of 1.007825 amu. The isotopic abundance of ^{2}H (deuterium) is estimated at 0.0149% with a mass of 2.014101 amu. The artificially prepared ^{3}H (tritium), $^9Be + ^2H \rightarrow 2\ ^4He + ^3H$, has a mass of 3.01605 amu. Heavy water is deuterium oxide, ^2H_2O, usually written D_2O. Deuterium and deuterium oxide gained prominence largely because of their excellent properties as moderators in nuclear reactors. The ionization potential of hydrogen is 13.59765 ± 0.00022 eV. Other physical properties of hydrogen are given under **Chemical Elements**. See also **Deuteron**; and **Deuterium**.

When ignited, hydrogen burns in air with a pale blue to colorless, nonluminous flame, yielding H_2O. When mixed with air, the flammability limit is 4–74% hydrogen. When mixed with oxygen, the flammability limit is 4–94% hydrogen. Care always must be exercised where there may be hydrogen mixtures with air or oxygen because violent explosions may occur. In sunlight or magnesium light, hydrogen combines with chlorine with violent release of energy, forming hydrogen chloride HCl. When hydrogen is heated with sodium, calcium, and several other metals, the corresponding hydride is formed. In the presence of a catalyst, hydrogen reacts with nitrogen to form ammonia NH_3. Upon heating sulfur in the presence of hydrogen, hydrogen sulfide, H_2S, is formed. At elevated temperatures, hydrogen will reduce many of the metal oxides to the metal, notably copper, iron, nickel, tin, and lead. The oxides of zinc, aluminum, and magnesium are not so reduced. Hydrogen reacts with unsaturated organic compounds in most cases to form saturated compounds. For example, in the presence of a catalyst, hydrogen will add to oleic acid $C_{17}H_{33}COOH$ to form stearic acid $C_{17}H_{35}COOH$. See also **Hydrogenation**.

Production of Hydrogen

For chemical and petroleum processes, hydrogen is an extremely high-tonnage and one of the most fundamental raw materials. Sources of hydrogen and processes for producing it are described in entry on **Hydrogen (Fuel)**.

Uses

In terms of consumption, NH_3 is by far the largest user of hydrogen. Petroleum refining processes and methanol synthesis are the next largest consumers. Hydrogen needs for these uses are almost always fulfilled by hydrogen-generation capacity on the premises. What might be termed commodity hydrogen is shipped from hydrogen plants to various users. Some of the more important uses include the hydrogenation of numerous organic compounds, such as vegetable and animal oils, the oxyhydrogen and atomic-hydrogen welding applications, the reduction of several metallic oxides, such as iron, copper, nickel, cobalt, tungsten, and molybdenum, and the use of liquid hydrogen as a rocket fuel. See also **Ammonia**; **Hydrogenation**; **Methyl Alcohol**; **Petrochemicals**; and **Synthesis Gas**. For the potential role as a fuel. See also **Hydrogen (Fuel)**.

Ortho- and Para-Hydrogen

On the basis of nuclear spin, two forms of hydrogen are known: *ortho-hydrogen*, in which the two nuclei in the H_2 molecule have parallel spins, and *para-hydrogen*, in which the nuclear spins are antiparallel. At ordinary temperatures (and above) ortho-hydrogen is present to the extent of about 75%; at lower temperatures, the ortho changes to para-hydrogen, until at very low temperatures, as that of liquid hydrogen, the para form is present to the extent of 99.7%. There is some difference in properties between the two, notably in thermal conductivity.

The transition from ortho- to para-hydrogen releases heat in amount of 168 cal/g. The heat of vaporization of liquid hydrogen is 107 cal/g. Thus, more than ample heat is released to revaporize liquid hydrogen. Knowledge of the existence of the ortho-para transition and the development of catalysts to equilibrate the liquid during liquefaction essentially have made possible the very large-scale manufacture, use, and storage of liquid hydrogen.

Below $-220°C$ the specific heat of hydrogen is that of a monoatomic gas like helium (He). Practically pure para-hydrogen may be obtained by adsorption of ordinary hydrogen, which is three-fourths ortho and one-fourth para, on charcoal at about $-225°C$. The mp of para-hydrogen is $0.13°C$ lower (ortho-hydrogen $0.04°C$ higher) than ordinary hydrogen, and the bp at 60 mm pressure is $0.13°C$ lower (ortho-hydrogen $0.04°C$ higher) than ordinary hydrogen. Para-hydrogen reverts slowly to ordinary hydrogen, but immediately in the presence of platinized asbestos.

Atomic Hydrogen

At high temperatures, the loss of heat from a glowing wire in hydrogen is larger than expected on regular assumptions. This is believed to be due to dissociation of ordinary hydrogen into atomic hydrogen (H). See Table 1.

When hydrogen is passed through an electric arc between tungsten poles, a considerable transformation into atomic hydrogen occurs, and when a stream of this gas strikes a surface a large evolution of heat takes place through recombination to ordinary hydrogen. This atomic hydrogen flame is of temperature sufficiently high to melt tungsten (mp 3, 370°C). The half-life of the hydrogen atom is one-third second at 0.5 mm pressure. This reaction is endothermic, values of 98–105 kcal per mole having been reported for it. It is an active reducing agent, reducing many metallic oxides and halides to the free metals, and forming hydrides with many nonmetals. The energy of its exothermic recombination is utilized, in combination with the energy released by the oxidation of the H_2 formed, by atmospheric oxygen, in the oxyhydrogen welding process.

Ionization

The ionization potential of hydrogen is $13.59765 \pm .00022$ eV, and the ionization process (in the case of protium) yields an electron and a free proton. The electric field of the proton is strong, due to its small radius, so that it readily combines with polarizable atoms. Thus, in aqueous solution, it shares an unshared pair of electrons of the oxygen atom of H_2O to form H_3O^+, the hydronium ion; with NH_3 it forms NH_4^+, the ammonium ion; with phosphine it forms the phosphonium ion, PH_4^+, etc. The hydrogen atom can also add an electron, to form the hydride anion, H^-, this potential (electron affinity) being only about 0.7 eV. Hydride ions have been shown (by electrolysis, crystal structure, etc.) to exist in the hydrides which hydrogen forms with the alkali metals and some of the other metals on the left side of the periodic table. While most other hydrogen compounds are essentially covalent, the binary compounds with the halogens and some of the other elements on the right side of the periodic table exhibit a considerable degree of ionicity, varying considerably in the same group.

The hydrogen atoms in many compounds tend to be shared between the electronegative atom or group to which they are attached and similar groups on other molecules. These hydrogen bonds increase the intermolecular forces and boiling points of hydrogen fluoride, water, organic acids and alcohols, etc. A descriptive explanation of the process is the positive polarity of the H atom that is attached to the electronegative atom or group, which gives it an effective coordination number of 2, so that it can attract an unshared electron pair of a fluorine, oxygen, nitrogen, atom of

TABLE 1. DISSOCIATION OF HYDROGEN

Temperature, °C	Pressure	
	At 760 mm	At 1 mm
1730	0.33%	8.7%
2230	3.1	57.5
2730	34	99.3

another molecule. The atom having the unshared pair must be negatively polarized or easily polarizable. For example, tertiary arsines form stronger hydrogen bonds with phenols than do tertiary phosphines.

A number of hydrogen compounds ionize to yield solvated protons, i.e., $2H_2O \Longleftrightarrow OH_3^+ + OH^-$, and $2NH_3(liq.) \Longleftrightarrow NH_4^+ + NH_2^-$. Moreover, many hydrogen compounds, when dissolved in such solvents, ionize more or less completely to give solvated protons and anions. In the case of polybasic acids, ionization constants are reported for each step in this dissociation.

Hydrides

See section on hydrides in entry on **Hydrogen (Fuel)**; and separate entry on **Hydrides**.

Water and Acids

The properties of the most prevailing hydrogen-bearing compound, water, are given under **Water**. The characteristics of acids are attributed essentially to the presence of hydrogen ions. These topics are treated under **Acids and Bases**; and **pH (Hydrogen Ion Concentration)**.

Hydrogen Under Extreme Pressure

Interest in the possible existence of hydrogen as a metal is spurred by the prospect that hydrogen may be able to conduct electricity with zero resistance near room temperature and that, because of the tremendous concentration of energy, as contrasted with liquid hydrogen, it could serve as a rocket fuel and high explosive.

In 1989, Mao and Hemley (Geophysical Laboratory, Carnegie Institution of Washington) reported on an investigation of the insulator-metal transformation in solid hydrogen at high pressure. Much earlier, theoretical calculations made by Wigner and Huntington (1935) revealed that the transition may occur in the 250-to-400 GPa (2.5-to-40 megabar) range. With the high-pressure research tools (diamond anvil cell) available today, this transition point of hydrogen has become a primary target for some researchers.

Mao and Hemley (see reference listed) reported that direct optical observations of solid hydrogen at the aforementioned pressure range and at 77 K indicated that the hydrogen sample appeared nearly opaque and that optical data were consistent with a band-overlap mechanism of metallization. These findings were later challenged by Silvera (reference listed) to the effect that "Visual darkening of a sample is not sufficient evidence of metallization, just as a lustery metallic reflection is not. Good examples are the semiconductors germanium and silicon, which as thin films, are metallic in appearance. Darkening of a sample can arise from any number of physical mechanisms that cause absorption throughout the visible spectrum." Further justification, however, was given by Mao and Hemley.

In mid-1991, Badding, Hemley, and Mao reported on studies of the high-pressure chemistry of hydrogen in metals and made specific studies of the reaction between iron and hydrogen at sudden pressure-induced expansion at 3.5 gigapascals of iron samples immersed in fluid hydrogen. The investigators mention numerous specific areas of interest that may be addressed with a better understanding of the behavior of hydrogen with metallic environments, including the hydrogen degradation of ferrous metals.

Additional Reading

Badding, J.V., R.J. Hemley, and H.K. Mao: "High-Pressure Chemistry of Hydrogen in Metals: In Situ Study of Iron Hydride," *Science*, 421 (July 26, 1991).

Crawford, M.: "Accelerator Eyes for Warhead Tritium," *Science*, 469 (January 27, 1989).

Giacobbe, F.G., Iaquaniello, and O. Loiacono: "Increase Hydrogen Production," *Hydrocarbon Processing*, 69 (March 1992).

Greenwood, N.N. and A. Earnshaw: *Chemistry of the Elements*, 2nd Edition, Butterworth-Heinemann, Inc., Woburn MA, 1997.

Krebs, R.E.: *The History and Use of Our Earth's Chemical Elements: A Reference Guide*, Greenwood Publishing Group, Inc., Westport, CT, 1998.

Lide, D.R.: *CRC Handbook of Chemistry and Physics*, 84th Edition, CRC Press, LLC., Boca Raton, FL, 2003.

Mao, H.K. and R.J. Hemley: "Optical Studies of Hydrogen Atoms Above 200 Gigapascals: Evidence for Metallization by Band Overlap," *Science*, 1462 (June 23, 1989).

Parker, P.: *McGraw-Hill Encyclopedia of Chemistry*, 2nd Edition, The McGraw-Hill Companies, Inc., New York, NY, 1993.

Peterson, I.: "Squeezing Hydrogen to Molecular Metal," *Science News*, 164 (March 17, 1990).

Pool, R.: "The Chase Continues for Metallic Hydrogen," *Science*, 1545 (March 30, 1990).

Ross, P. and R. Ruthen: "Hard Pressed: Squeezed Hydrogen Forms Metal with Superconducting Potential," *Sci. Amer.*, 26 (November 1989).

Silvera, I.F.: "Evidence for Band Overlap Metallization of Hydrogen—(Technical Comments)," *Science*, 863 (February 16, 1990).

HYDROGENATION.   In its simplest interpretation, to hydrogenate is to add hydrogen. There are scores of examples where hydrogenation is used as a unit process throughout the chemical and process industries. Generally, the process is associated with relatively high pressure, elevated temperature, and the presence of a catalyst.

Nickel, prepared in finely divided form by reduction of nickel oxide in a stream of hydrogen gas at about 300°C, was introduced by Sabatier (1897) as a catalyst for the reaction of hydrogen with unsaturated organic substances to be conducted at about 175°C. Nickel proved to be one of the most successful catalysts for such reactions. The unsaturated organic substances that are hydrogenated are usually those containing a double bond, but those containing a triple bond also may be hydrogenated. Platinum black, palladium black, copper metal, copper oxide (Adkin catalyst), nickel oxide, aluminum, and other materials have subsequently been developed as hydrogenation catalysts. Temperatures and pressures have been increased in many instances to improve yields of desired product. The hydrogenation of methyl ester to fatty alcohol and methanol for example, occurs at about 3,000 psig (204 atmospheres) and 290–315°C. In the hydrotreating of liquid hydrocarbon fuels to improve quality, the reaction may take place in fixed-bed reactors at pressures ranging from 100 to 3,000 (7 to 204 atmospheres) psig. Many hydrogenation processes are of a proprietary nature, with numerous combinations of catalysts, temperature, and pressure possible.

Among the better known products of hydrogenation are hydrogenated vegetable and fish oils, which may be hardened or solidified by catalytic hydrogenation. Some of these oils can be partially hydrogenated to clarify and deodorize them. Fatty oils, such as oleic acid, may be converted into stearic acid by hydrogenation. Through hydrogenation, peanut oil, cottonseed oil, and coconut oil can be converted to materials that taste, appear, and smell like lard; or by varying the process, they can be made to resemble tallow. Most synthetic shortenings are comprised of hydrogenated oils. Usually, hydrogenated oils will have higher melting points and lower iodine values than the natural untreated oils.

Hydrogenation of Coal and Crudes

The interest in hydrogenation has been greatly intensified since the early- and mid-1970s in connection with the synthesis of new types of fuels to augment the world energy supplies. Basically, however, the hydrogenation of coal is not a new concept, but dates back at least a half-century to the time when manufactured gas (artificial, illuminating, producer, water gas, etc.) was used prior to the more general availability of low-cost, cleaner natural gas. In 1927, a White Paper was published discussing the processes then available for production of oil from coal. One of the first large-scale applications of the Fischer-Tropsch process for the production of oil from coal was that of the South African Coal, Oil and Gas Corporation's plant in Sasolburg, Republic of South Africa, constructed in the mid-1950s and expanded and improved several times during the interim.

Similarly, sour crudes, heavy residuums, and other petroleum-base starting materials can be hydrogenated, sometimes coupled with other processes, to sweeten, reduce viscosity, and otherwise improve the materials for better use as fuels. See also **Coal**; **Hydrotreating**; and **Petroleum**.

HYDROGENATION (Vegetable Oils).   See **Vegetable Oils (Edible)**.

HYDROGEN FLUORIDE.   See **Fluorine**.

HYDROGEN (Fuel).   Because of the wide use of hydrogen in the processing industries and for the hydrogenation of various oils and fats in the food and related industries, hydrogen has become much better understood during the past several decades. For many years, hydrogen has served as a specialized fuel for certain applications, such as oxyhydrogen cutting and welding torches. But generally, until the late 1960s, the possible role of hydrogen as a major energy source fuel was rarely discussed. The word *hydrogen* took on a negative connotation with the development of the hydrogen bomb, as it also did some years ago when the hydrogen-filled dirigible Hindenburg exploded as it moved toward its mooring mast in Lakewood, New Jersey in 1937.

The probable future of hydrogen in the world's energy system was the subject of prophecy over one hundred years ago. In 1874, Jules Verne wrote: "I believe that water will one day be employed as a fuel; that hydrogen or oxygen, which constitute it, used singly or together, will furnish an inexhaustible source of heat and light." And, in the early 1900s, Britain's Lord Haldane said: "It is axiomatic that the exhaustion of our coal and oil fields is a matter of centuries only. ... As it has often been assumed that their exhaustion would lead to the collapse of industrial civilization, I may perhaps be pardoned if I give some of the reasons which led me to doubt this proposition." Haldane envisioned networks of windmills generating the electricity needed to separate hydrogen from water. The hydrogen would then be liquefied and stored underground.

Some readily apparent advantages of hydrogen, both as a direct and an indirect fuel (discussed later) have been extrapolated into terms of a future hydrogen economy. As the result of continuing and concentrated research and development in the energy field, many experts see a hydrogen energy economy gradually emerging.

Like other energy proposals (and there have been many in the past decade or so) three factors will likely determine the pace of hydrogen energy technology: (1) the manner in which, step-by-step, hydrogen-oriented systems and subsystems will compete economically and environmentally with other energy source, conservation, and utilization proposals; (2) the pace of technological advancement in related fields, such as nuclear engineering, upon which hydrogen systems may depend; and (3) the pace of unilateral efforts on behalf of hydrogen-oriented systems, including the refinement of current planning-purpose data and opinions into actual operating information relating to hydrogen generation, transportation, conversion and/or end-utilization, and safety. Without the funding of a series of "crash programs," unilateral developments probably will be relatively slow. Most likely, the information bank for hydrogen systems will stem from an increasing awareness of the energy characteristics of hydrogen and the progressive use of hydrogen subsystems in situations where they are eminently superior.

The present concept of a hydrogen fuel economy includes a primary energy source, such as a nuclear fission or fusion reactor, a geothermal source, or a solar-powered source, with hydrogen being produced as the portable energy carrier. See Fig. 1. Thermal energy from nuclear sources would be used to generate electricity that would then be used to electrolyze water for the production of hydrogen and oxygen. The hydrogen would be distributed by pipeline to distant points of use, with storage provided by underground gas storage, or by liquefaction and refrigerated storage.

Fuel-related Background of Hydrogen. Although the abundant hydrogen isotope *protium* is the simplest known atom, it forms two diatomic molecules, namely, *ortho-hydrogen*, in which the two atomic nuclei spin in the same direction; and *para-hydrogen*, in which the nuclei spin in opposite directions. While the equilibrium composition of hydrogen gas is 75% ortho at ambient temperature, it changes to 99.8% para in the liquid state. The transition from ortho- to para-hydrogen is exothermic (168 cal/gram), so that the heat released is more than enough to revaporize liquid hydrogen (heat of vaporization 107 cal/gram). Recognition of the existence of the ortho-para transition and the development of catalysts to equilibrate the liquid during liquefaction have made possible the large-scale production, use, and storage of liquid hydrogen.

Hydrogen molecules dissociate to atoms endothermally at high temperatures (heat of dissociation about 103 cal/gram mole), in an electric arc, or by irradiation. This property is used to effect atomic-hydrogen arc welding, in which hydrogen gas is dissociated by an electric arc between two tungsten electrodes, the hydrogen atoms recombining at the metal surface to provide the heat required for welding.

Pertinent properties of hydrogen are given in Table 1.

Actual and potential uses for hydrogen can be predicted by inspection of its properties. Its low density, 7% that of air, plus its high thermal conductivity, 6.7 times that of air, have led to its use as a coolant in large rotating electrical equipment. The low density reduces windage friction losses to less than 10% those with air, while its high thermal conductivity and heat capacity permit more efficient heat transfer, the result being an overall increase in generator efficiency of as much as 1%.

The high heats of reaction of hydrogen with oxygen or fluorine, plus the low molecular weights of the product gases, have made hydrogen a prime fuel for rocket propulsion, since rocket thrust increases directly with the temperature and inversely with the molecular weight of the exhaust gases. Liquid hydrogen and oxygen were used in the second- and third-stage Saturn engines in the Apollo moon flights. The low atomic weight of hydrogen has made it the preferred propellant for nuclear rockets, in which nuclear emission provides heat for exhausting hydrogen gas at high temperatures.

Some studies have indicated that the cost of transporting and distributing hydrogen by pipeline may be less than the cost of transporting and distributing electric power. Presumably existing natural gas pipelines and distribution systems can be adapted to the use of hydrogen. Although hydrogen has a net heating value of only 275 Btus per cubic foot (2448 Calories per cubic meter) as compared with 913 Btus per cubic foot (8126 Calories per cubic meter) for methane, the lower density and viscosity of hydrogen make it possible for a pipeline to deliver about the same amount of thermal energy as with methane, at a somewhat greater compression cost. The thermal energy in hydrogen can be utilized more efficiently in home

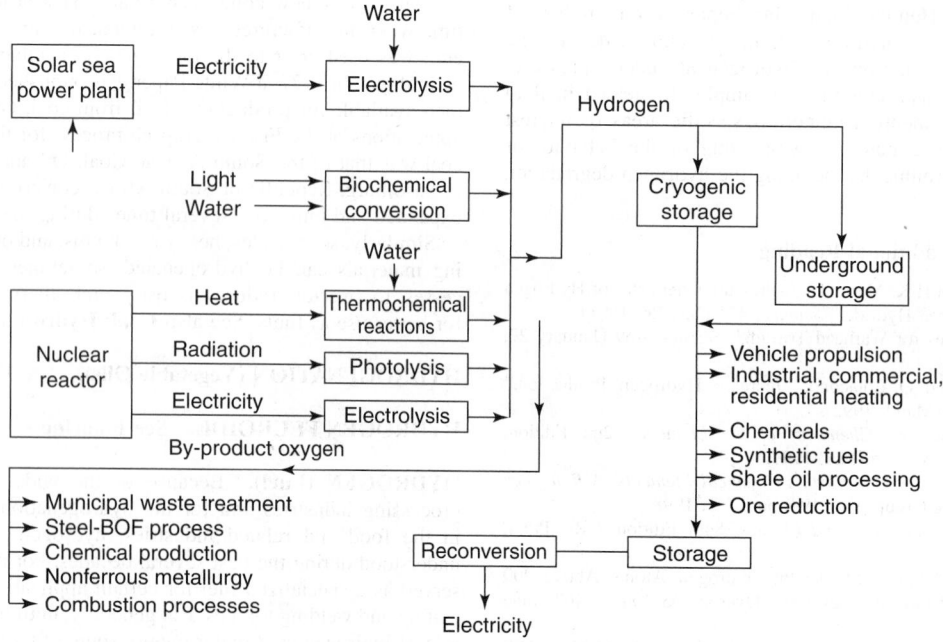

Fig. 1.   Major elements of a hydrogen fuel economy

TABLE 1. FUEL PROPERTIES OF HYDROGEN

Melting point, K	13.96
Heat of fusion at 14.0 K, calories/gram	14.0
Boiling point at 1 atmosphere, K	20.39
Heat of vaporization at 20.4 K, calories/gram	107
Density, grams/cubic centimeter	
Solid at 4.2 K	0.089
Liquid at 20.4 K	0.071
Critical temperature, K	33.3
Critical pressure, atmosphere absolute	12.8
Critical volume, cubic centimeters/mole	65.0
Critical density, grams/cubic centimeter	0.031
Heat of transition, ortho to para at 20.4 K, calories/gram	168
Specific heat (At constant pressure C_p, calories/gram)	
Liquid at 17.2 K	1.93
Solid at 13.4 K	0.63
0–200°C	3.44
Specific heat (At constant volume C_p (0–200°C) calories/gram)	2.46
Specific heat: Ratio C_p/C_r (0–200°C)	1.40
Gas density, 0°C and 1 atmosphere, grams/liter	0.0899
Gas specific gravity (Air = 1.0)	0.0695
Gas thermal conductivity, 25°C, (cal)(cm)/(s)(cm²)(°C)	0.00044
Gas viscosity, 25°C and 1 atmosphere, centipoise	0.0089
Coefficient of thermal expansion per °C	0.00356
Heat of combustion at 25°C, kcal/gram mole	
Gross	63.3174
Net	57.7976
Energy release upon combustion, calories/gram	29.000
calories/cubic centimeter	2.050
joule/gram	1.21×10^5
Flame temperature, K	2.483
Autoignition temperature, K	858
Heat of formation of HF at 25°C, kcal/gram mole ΔH	−64.2
Flammability limit, percent	
In oxygen	4 to 94
In air	4 to 74

heating than natural gas, because hydrogen can be burned in nonvented heaters, with no loss of heat, since its only primary combustion product is water. By using flameless catalytic heaters, nitrogen oxide formation can be eliminated. However, oxygen depletion of closed spaces will still present a hazard.

One advantage of hydrogen as a source of thermal energy, as compared with electricity, is that it can be stored for later use—it is a commodity with weight and volume. Electricity, although it can be converted into chemical energy in batteries, essentially is a form of energy that must be used as it is generated. Hydrogen, like natural gas or substitute natural gases, may be stored and transported as a refrigerated liquid, or stored as a gas under pressure in underground systems. Hydrogen also may be stored as a metallic hydride.

Categories of Energy-related Hydrogen Uses. The probable functions of hydrogen in future energy technology may be put into two major categories: (1) *direct functions,* in which hydrogen serves as a fuel, that is, as the source of heat, power, and light without prior conversion to some other energy form; and (2) *indirect functions,* in which hydrogen is an important component of the total energy system, but before the end-use of that energy, the hydrogen is involved in some conversion, possibly chemically, used in the creation of a synthetic fuel, such as substitute natural gas, or possibly converted into electrical energy which becomes the final end-energy used. One of the major indirect or secondary roles proposed for hydrogen is that of an energy transporter, wherein in one scheme, other forms of energy would be consumed to generate hydrogen which then would be pipelined and stored at distant points available for another conversion step—for example, converted into electrical energy as needed.

Hydrogen as Energy Source for Motive Power

Aside from their relatively low costs until the mid-1970s and continuing into the 1980s, the hydrocarbon fuels, notably gasoline and kerosene, have offered convenience in handling and transportability for use in connection with powered vehicles. And, during the past decade, the political factors that arise from the striking geographic imbalance between petroleum resources and petroleum consumption in most regions of the world have provided ample incentives to strike out for alternative sources of vehicular power.

Hydrogen, when cost competitive, can provide many of the advantages of petroleum liquids and offer the additional attraction of decreasing air pollution. Because of its low density, the net storage volume required would be at least as much as for gasoline. The storage tank must be maintained at a temperature of −423°F (−253°C), which is the boiling point of hydrogen at atmospheric pressure. This would require insulation that would increase the overall size of the storage container. Vaporization losses from the storage tank, amounting to perhaps 2% or more per day, must be vented so that no ignition of the vented hydrogen gas can occur, and no accumulation of explosive hydrogen-air mixtures are possible. The lower explosive limit of hydrogen in air is 4%, so adequate ventilation must be provided. Fortunately, hydrogen gas, being the lightest gas with a specific gravity of 0.07 referred to air, will rise and diffuse rapidly and thus can be easily dispersed. Service stations for dispensing liquid hydrogen will require more expensive storage and pumping facilities than required for gasoline.

The estimated weights and volumes expressed in Table 2 are relative to the same energy content of gasoline. Relative weight includes that of containers. The data indicate that magnesium hydride would be at a 4.6 weight disadvantage and thus require four times the tankage in comparison with the use of gasoline in a conventional automobile. New hydrides, as described later, may change this. This would be equivalent to 450 pounds (204 kilograms) of added vehicle weight and 60 more gallons (227 liters) (2 × 2 × 2 feet storage; 0.2 cubic meter) over that required for a vehicle with a 20-gallon (76 liters) gasoline tank. Bursting upon collision for liquid storage can be overcome by using containers capable of withstanding 30 Gs, which are presently available.

If a designer were to elect the option of using hydrogen in the gaseous phase at 2000 psi (136 atmospheres), this would require a metal container weighing some 30 times and requiring a volume of some 24 times that required for an energy equivalent volume of a hydrocarbon fuel. Also important in the total energy equation is the additional energy required to compress hydrogen (gaseous phase) or to liquefy it.

It is most likely that the first major use of liquid hydrogen as an energy source for motive power will be jet aircraft, largely because of the excellent weight advantage and the less serious nature of the boil-off loss and distribution problems as compared with other forms of transportation. City buses and long-haul motor trucks, already equipped mainly with hydride hydrogen power, have been tested in the United States and Germany, among other countries. These may follow as candidates wherein refueling may be effected through replacement of entire storage tanks (dewars). Because the private motorcar presents the most crucial logistics problems, including the small-capacity fuel system, concern with safety, boil-off loss of fuel even when vehicle is not in use, and the education and acceptance involving millions of users, it probably will follow rather than lead the use of hydrogen in other modes of transportation. However, during the last few years, a few firms have offered hydrogen-powered private motor vehicles, set up to switch from hydrocarbon fuel to hydrogen and vice versa, but at a cost that is not competitive with mass-marketed vehicles.

Metal Hydrides. For a number of years many scientists and advanced planners have considered the possible use of metal hydrides to store hydrogen at atmospheric or reasonable pressures and at relatively low temperatures (comparable to current metal temperatures in some conventional engines). The fact that hydrogen will form hydrides with most metals has been known for many years, during which time, a number of these hydrides have been formed and tested. Until comparatively recently, magnesium hydride and a hydride of a rare-earth metal plus nickel, such as $LaNi_5$, appeared to be best suited as hydrogen storage media.

TABLE 2. SOME HYDROGEN STORAGE OPTIONS FOR VEHICLES

Storage system	Relative system weight[a]	Relative contained volume[a]
Gaseous phase, 2,000 psi (136 atmospheres)	~30.0	~24.0
Solid (as magnesium hydride with 40% porosity)	4.6	4.0
Liquid phase at 37°R	2.4	3.8

[a] Relative to gasoline, as unity for same energy content.

The hydride-forming reaction is exothermic and reversible: Metal + Hydrogen $\rightleftharpoons$ Metal Hydride + Heat. Thus, when it is desired to call for the separation of hydrogen from the hydride, heat (of decomposition) is required. As may be expected, the heat of decomposition is roughly proportional to the stability of the particular hydride. It is thus evident that for a metal hydride to serve as an efficient and viable means for storing hydrogen, it should be capable of decomposition at a relatively low temperature—say 300°C or lower. At the same time, the hydride must be reasonably stable and, of course, not require a high hydrogen pressure to manufacture it. It is further evident that the metal portion of the hydride must be comparatively inexpensive and thus common and readily available in the quantities that may be required. Metal hydride storage, operating as it does in terms of hydrogen as a battery operates in terms of electricity, must be capable of easy and efficient replenishing or "recharging" cycles. In every respect, the hydride storage element must be as safe as current vehicular fuel systems.

Researchers have investigated a large number of known binary hydrides, i.e., compounds that contain one metal and hydrogen. Investigators now regard magnesium hydride (MgH_2) as a borderline possibility. This binary hydride evolves hydrogen at a pressure of one atmosphere and requires a decomposition temperature of 289°C.

In comparatively recent research, much has been learned concerning the manner in which hydride compounds hold hydrogen. It has been known for a long time, of course, that the metal portion of the hydride should be comprised of tiny particles so that there is a large surface area available for reaction. In searching for reasons why hydrides permit such a high density of hydrogen, Reilly/Sandrock (1980) have observed that it is possible to pack more hydrogen into a metal hydride than into the same volume of liquid hydrogen. When the subject metal is first exposed to diatomic hydrogen (H_2), the hydrogen atoms are adsorbed onto the surface of the metal. Immediately, some of the hydrogen is dissociated into monoatomic hydrogen (H). This permits the monoatomic hydrogen to penetrate deeply into the crystal lattice of the metal and to occupy what are known as *interstitial sites*. Investigators have found that these sites must have a critical minimum volume if they are to easily receive the hydrogen atom. Upon increasing the pressure of the hydrogen applied, the metal reaches a saturated phase—the metal hydride phase. It has been found that under certain conditions and with certain metals, the number of hydrogen atoms contained in the crystal will range from 2 to 3 times the number of metal atoms.

The most recent experimentation with metal hydrides has involved multiple-metal hydrides. It has been known for some time that hydrogen reacts with alloy metal combinations. Considerable research has gone forth in connection with ternary hydrides (2 metals + hydrogen), and one of the most promising of these compounds as of the early 1980s is iron-titanium hydride ($FeTiH_x$), where x may range from 1 to 2. Reilly/Sandrock report that the hydrogen storage capacity by weight percent of this ternary hydride is 1.75 and by volume (grams per milliliter) is 0.096. The energy density by weight is 593 calories per gram; and the energy density by volume is 3254 calories per milliliter. Thus, this ternary hydride has a higher hydrogen-storage capacity than an equal volume of liquid or gaseous hydrogen (at 100 atmospheres). Another promising intermetallic hydride is lanthanum-pentanickel hydride ($LaNi_5H_x$), although it is more costly to produce. In this hydride, x may range from 1 to 6. Both the iron-titanium hydride and the lanthanum-pentanickel hydride have low temperatures of formation and decomposition, contributing to easy charging and discharging at ambient temperature.

In connection with hydrogen engine design, it is assumed that the heat of decomposition required by the hydride can be furnished from the inevitable waste heat generated by any engine. See also **Hydrides**.

Hydrogen as a Heating Fuel

The routine use of hydrogen as a heating fuel for industry and commercial-residential installations entails even greater complications and would appear to be much more dependent upon the overall economic and technical aspects of a so-called hydrogen fuel economy. From many standpoints, assuming availability, hydrogen can be an excellent fuel for almost any heating application. Hydrogen can be used in the home for cooking and heating (and even lighting) and likewise in commerce and industry. Compared with natural gas, hydrogen burns with a faster, hotter flame. Hydrogen-air mixtures are flammable over wider limits of mixtures.

Hydrogen burns without producing noxious exhaust products, allowing unvented appliances except where water vapor and resulting increased humidity may be objectionable. In winter, the additional humidity can, in fact, be highly desirable. But, in humid locations in summer, the water vapor produced could be objectionable. Adequate ventilation must be provided to prevent depletion of oxygen in closed spaces.

But, generally because of the absence of hazards from carbon monoxide and other fumes, large savings could be achieved from the elimination or at least simplification of flues. Some experts suggest that not only construction costs could be lowered as the result of clean burning, but that an increase of some 30% in the efficiency of a gas-fired home heating system could be achieved. The concept of peripherally placed unflued devices, particularly through the use of catalytic "flameless" heaters, could ultimately lead to a serious revision of the widely accepted central heating concept. By maintaining the temperature of a catalytic bed as low as 100°C, the production of nitrogen oxides would be virtually eliminated.

Because hydrogen burns with a hotter flame, some design features of heating apparatus would require change. The energy content per unit mass of liquid hydrogen is about 2.75 times greater than that of hydrocarbon fuels. On the other hand, there are only 325 Btus per standard cubic foot (2893 Calories per cubic meter) of hydrogen as compared with about 1,000 Btus per standard cubic foot (8900 Calories per cubic meter) of natural gas, thus dictating further design changes. The ignition energy of hydrogen is about 0.02 millijoules, which is less than 7% that of natural gas, a major factor in making low-temperature catalytic burners possible; also a major factor in designing for safe operation.

Despite the numerous advantages of hydrogen as a direct heating fuel, particularly in the home, the application of hydrogen must be viewed in terms of the total energy concept of an exclusively hydrogen-supplied (all-hydrogen home) installation. Where the direct use of hydrogen for heating is large, the economy will be most favorable. If a substantial amount of the hydrogen must be converted into electrical energy, as by a fuel cell, then economic justification becomes more difficult.

Lighting in the all-hydrogen home may be accomplished by condoluminescence, a cold process. A phosphor is spread on the inside of a tube similar to the conventional fluorescent lamp. Upon coming in contact with the phosphor, small amounts of hydrogen combine with the oxygen in the air to excite bright luminescence in the phosphor.

Conversion of burners and other design aspects of heating systems and appliances to pure hydrogen, or to a hydrogen-enriched natural or substitute natural gas supply, while costly and inconvenient, is certainly not in the economically insurmountable category. Similar alterations over the years were made in the United States when communities switched from manufactured gas (about 50% hydrogen) to natural gas. Such switchovers are even more recent in European communities.

As more hydrogen becomes available for transportation use and as more hydrogen is pipelined regionally or transcontinentally (depending largely on the demand placed upon the supply of hydrogen for industrial uses), it may be that the hydrogen content of community gas supplies will be progressively enriched (in a periodic, stepwise manner because of switchover problems) and thus contribute in a gradual manner to less pollution and to the conservation of natural gas.

Hydrogen as an Energy Transporter

With the possible use of hydrogen as a source of motive power in the transportation field, the *non*chemical interest in hydrogen in the total energy picture is directed to the use of hydrogen as a means or mode of storing and transporting energy. It is in this area that hydrogen directly confronts the past ever-increasing trend toward a fully electrical energy economy. Undeniably, hydrogen energy has a major starting advantage over electrical energy, namely, hydrogen is a storable energy form. Investigations are showing that hydrogen in pipelines may cost less to transport than electricity flowing over long power lines. Thus, hydrogen may play an important future role simply as a mode of storage and transport, even though source and terminal energy conversions may be required.

Electrical power plants are most efficient when operated at constant output at full-rate load. Because of wide fluctuations in consumer load (daily and seasonally), generating rates require constant adjustment. Communication systems and some emergency systems employ batteries for interim storage of electrical energy, but these applications are minuscule when compared with the total electrical generating and distribution system. The principal means of large-scale storage is the use of pumped storage, i.e., in essence a reversible hydroelectric station wherein electrical energy is temporarily converted to a hydraulic head by pumping water to an elevated reservoir. Unfortunately, the topography has to be suitable for such an installation and thus this approach is limited to comparatively few power-generating sites.

The high-voltage cables required to transmit electricity from generating stations to load centers are costly. The cost of going to underground cables for transmitting bulk current ranges from 9 to 20 times that of overhead configurations. The effective use of cryogenic superconducting cables may lower underground costs considerably, but much research remains to be completed before this is possible.

Because of the tremendous volumes of fuel that can be moved in pipelines, the construction, maintenance, and operating costs of a buried pipeline are much less in terms of a percentage of the total product moved. Pipeline operations have been profitable even at the relatively low price ranges for liquid and gaseous fuels prevailing prior to the price rises of the 1970s and 1980s. Pipeline technology, of course, is well established—with several hundred thousand miles of trunklines installed and operating in the United States. These lines transport nearly 23 trillion cubic feet (0.65 trillion cubic meters) of gas. Typical pipelines range from 600 to 1,000 miles (965–1609 kilometers) in length and are up to 48 inches (1.2 meters) in diameter. Line pressures may range from 600 to 800 psi (41–54 atmospheres) but go up to 1,000 psi (68 atmospheres). A representative 36-inch (0.9 meter) pipeline will carry a gaseous fuel with the equivalent of 37,500 billion Btus (9450 billion Calories) per hour. The electrical energy equivalent would be 11,000 megawatts. By comparison, this is ten times the energy-carrying capacity of a single-circuit 500-kilovolt overhead transmission line.

The figures for pipeline transportation of pure hydrogen are not quite so attractive, but nevertheless the comparison with electric transmission costs remains highly significant.

One study shows that the pipeline transmission costs for hydrogen will range from 30% to 50% more than for natural gas. Conversion of an existing natural gas line to hydrogen service is estimated to require a rise of compressor capacity by a factor of 3.8 and compressor horsepower by 5.5.

Obviously, hydrogen transmission costs represent but one part of a total system. Should the costs of generating hydrogen in the first place, and the subsequent conversion of hydrogen into electricity at the terminal end of the system remain excessively high, then the savings in energy transportation costs, of course, become academic.

Sources of Hydrogen

The major source of chemical hydrogen over the past several decades has been natural gas. In strictly terms of chemical needs, where economic factors are favorable, natural gas has served this need well. Obviously, in terms of total energy conservation, where hydrogen is looked to as a means of conserving fossil-fuel sources, a much less costly and much more abundant hydrogen-containing raw material must be sought. The logical candidate is water. Particularly in areas of the world where hydrocarbons are not readily available, reasonably large water electrolysis installations have been made, notably in locations with low electricity costs.

In addition to electrolysis, the principal means under consideration for deriving hydrogen from water is that of thermochemical splitting. The waste heat and high temperature available from certain types of nuclear reactors would effect a series of chemical reactions, still much in the research phase, to free hydrogen and oxygen from water. Additional proposals have included the use of ultraviolet radiation from the plasma of a fusion reactor for the direct photolysis of water vapor (Department of Energy) and the use of some forms of algae, under the stimulation of light, to convert hydrogen ions to hydrogen gas by a complex chain of biochemical reactions (Case Western Reserve University).

Electrolysis. Because of years of operating experience, electrolysis is possibly an order of magnitude ahead of other proposals from a technological standpoint. Although simple in concept, electrolysis is costly—hence the research efforts to find other ways of splitting water carry a high incentive. Nevertheless, this side of one or more breakthroughs in other areas, most likely electrolysis operations will continue to serve as the basis for costs in extending the use of hydrogen in the relatively near term.

As of the early 1980s, industrial electrolyzers ranged in size from 500 standard cubic feet (14.2 cubic meters) of hydrogen production per day,

consuming 3 kilowatts of electricity, to more than 40 million standard cubic feet (~1.1 million cubic meters) of hydrogen per day, consuming 240,000 kilowatts. Most common installations are from 10,000 to 500,000 standard cubic feet (283–14,160 cubic meters) of hydrogen per day. Two factors generally characterize an electrolyzer installation: (1) access to comparatively low-cost electricity, as found in some areas served by hydroelectric installations; and (2) need for the oxygen which accompanies the production of the hydrogen. Industrial electrolyzers usually operate at efficiencies of about 60% to 70%. Some high-pressure prototype models have reached 85%. It has been pointed out (D.P. Gregory, Institute of Gas Technology) that, in theory, electrolyzers can approach a maximum electrical efficiency of nearly 120% as the result of the ideal unit absorbing ambient heat and also converting this energy into hydrogen. A reasonable, practical target for an improved electrolyzer appears to be around 100%. Thus, the production of electrolytic hydrogen would be limited only by the efficiency of electric current generation, namely, between 35% and 45%. An estimate has been made (E.C. Tanner, Princeton University; R. Huse, Public Service Electric & Gas Co.) that the overall conversion efficiency of electricity-to-hydrogen-to-electricity will approximate 38%. The theoretical power required to produce hydrogen from water is 79 kilowatts per 1,000 cubic feet (~28 cubic meters) of hydrogen gas. One of the largest electrolyzers operating commercially is that of Cominco, Limited (British Columbia). This is a 90-megawatt installation that produces approximately 36 tons (32.4 metric tons) of hydrogen gas per day for use in ammonia synthesis. Other large plants are located in Norway and Egypt.

Two main types of electrolyzers are in commercial use: (1) Tank cells with monopolar electrodes. Porous diaphragms separate the alternate cathodes and anodes to prevent gas mixing. The anodes and cathodes are connected in parallel to keep the required voltage at approximately 2 volts and to permit high current densities. This arrangement requires a large floor area; (2) bipolar electrodes, connected in series and suitably insulated. The electrodes are cathodic on one side; anodic on the other side. This arrangement requires less floor space, is more complex, and requires high voltages.

High pressure can increase efficiency and this concept has been under development for many years. A commercial electrolyzer (Lurgi) is available which operates at a pressure of 30 atmospheres and 90°C, requiring 300 amperes of electric current at 217 volts. In the mid-1960s, bipolar cells of porous nickel electrodes were developed which operate at current densities of 800 and 1600 amperes per square foot (0.09 square meter).

In the mid-1960s, electric-high-temperature, vapor-phase electrolysis (General Electric Co.) was developed. In this process, the electrolyte is solid, porous zirconia, which contains dopants. Operating temperature ranges from 500° to 800°C. A modification of the process is under development which will produce only hydrogen by consuming byproduct oxygen.

Among electrolyzer design improvements that may occur are better electrodes which may result as a spinoff from fuel-cell work. There are indications that electrode improvement could cut the costs of electrolytic hydrogen by about 20% to 25%. Electrolysis looms high in consideration of utilization of ocean thermal gradients and thus these two technologies are closely interacting.

Thermochemical Splitting. The major objective is to find one or more series of chemical reactions that will result in the satisfactory separation of hydrogen (and oxygen) from water. Considerable work has been going forth at the Nuclear Research Center, Julich, Federal Republic of Germany, where much attention has been given to sulfur- and chlorine-base thermochemical cycles. Other researchers (Institute of Gas Technology; General Electric Co.; European Atomic Energy Community) have been probing various combinations of at least 56 chemical elements, including over 700 different compounds that may show promise in various schemes for a closed-water-splitting cycle. It is understood that approximately 20 promising schemes have emerged, mainly centered in chlorine compounds. The most frequent flaw encountered among prospective reactions is the large amount of free energy required to force one or possibly two of the series of reactions; and the appearance of reactions that produce stable compounds incapable of regeneration.

Some of these reactions would rely upon a nuclear reactor as a heat source and would not have to await the emergence of a practical, operating fusion reactor. One sequence of reactions, in particular, is of interest:

$$CaBr_2 + 2H_2O \longrightarrow Ca(OH)_2 + 2HBr$$

$$Hg + 2HBr \longrightarrow HgBr_2 + H_2$$

$$HgBr_2 + Ca(OH)_2 \longrightarrow CaBr_2 + HgO + H_2O$$

$$HgO \longrightarrow Hg + \frac{1}{2}O_2$$

A drawback of this sequence is its use of highly corrosive hydrogen bromide. The scheme also requires a large inventory of mercury.

Of major concern to investigators in the thermochemical splitting schemes is the availability of appropriate materials of construction. Heat exchangers between the nuclear side and the chemical side must withstand both corrosion and radioactive contamination. The conventional nickel-chromium alloys are capable up to about 1050 K; exotic, but available alloys, up to about 1400 K. Above these temperatures, ceramics and new alloys may have to be used. Considerable materials research along these lines is going forth at the Los Alamos Scientific Laboratory.

Conventional Hydrogen Uses. Even before its serious consideration in the fuel economy, the demand for hydrogen grew at a rate of about 15% annually since World War II. About 3 trillion standard cubic feet (~85 million cubic meters) of hydrogen (8 million tons; 7.2 million metric tons) were produced in the United States in 1970. Not including energy applications, the chemical requirements for hydrogen are expected to increase by about 7% per year through the year 2000. Among demands for hydrogen include petroleum refining, plastics, elastomers, increased desulfurization of fuel oils, increased use in iron ore reduction, aerospace uses, and hydrogen/air fuel cells. About 42% of the hydrogen produced now is consumed in ammonia production; about 38% is used in petroleum refining. The other large consumers are metallurgical and food processing.

In terms of presently nonconventional fuels that will require increasing quantities of hydrogen as new processes develop, it is estimated that (1) synthetic crude oil from coal will require 6,500 standard cubic feet (184 cubic meters) of hydrogen per barrel of oil; (2) 1,300 standard cubic feet (37 cubic meters) of hydrogen will be required per barrel of oil from shale; and (3) 1,500 standard cubic feet (42 cubic meters) of hydrogen will be required for every 1,000 standard cubic feet (~28 cubic meters) of synthetic pipeline gas produced from the gasification of coal. Petroleum refining use of hydrogen is expected to increase to 610 standard cubic feet (~17.3 cubic meters) per barrel of crude refined. Direct iron ore reduction use of hydrogen is expected to increase to 20,000 standard cubic feet (566 cubic meters) per ton (0.9 metric ton) of iron. If there were not other hydrogen sources available, the hydrogen needs could be met by using approximately 10% of the natural gas production.

Additional Reading

Considine, D.M.: *Energy Technology Handbook,* McGraw-Hill Companies, Inc., New York, NY, 1973. A classic reference.

Gregory, D.P.: "The Hydrogen Economy," *Sci. Amer.,* **228**, 1, 13–21 (1973).

Norbeck, J.: *Hydrogen Fuel for Surface Transportation,* Society of Automotive Engineers, Warrendale, PA, 1996.

Peavey, M.: *Fuel from Water: Energy Independence with Hydrogen,* Merit Products, Bloomington, CA, 1988.

Peschka, W.: *Liquid Hydrogen-Fuel of the Future,* Springer Verlag-New York, New York, NY, 1992.

Reilly, J.J. and G.D. Sandrock: "Hydrogen Storage in Metal Hydrides," *Sci. Amer.,* 118–129 (February 1980).

HYDROGEN CYANIDE. [CAS: 74-90-8]. HCN, formula weight 27.03, colorless gas with characteristic odor, very poisonous, mp $-14°C$, bp $26°C$, critical temperature $183.5°C$, critical pressure 50 atmospheres, density 0.20 g/cm^3, sp gr 0.697 ($18°C$). There are two isomeric forms: (1) HCN which forms cyanides, (2) HNC (inferred from its derivatives) which forms isocyanides. Hydrogen cyanide is soluble in H_2O, or alcohol, or ether in all proportions. The compound usually is marketed as an aqueous solution containing 2–10% (weight) HCN. For many process uses, it is frequently more convenient to generate HCN as needed and thus avoid storage and handling problems. HCN burns with a red-blue flame, yielding CO_2, nitrogen, and H_2O. Aqueous solutions of HCN decompose slowly, yielding ammonium formate: $HCN + 2H_2O \longrightarrow HCOONH_4$. Decomposition is slowed by storage in dark locations. Peaches, apricots, bitter almonds, cherries, and plums contain some HCN derivatives in their kernels, frequently in combination with glucose and benzaldehyde as a glucoside (amygdalin). The bitter almond fragrance of HCN and its derivatives sometimes can be detected in such kernels.

Production

Hydrogen cyanide can be prepared from a mixture of NH_3, methane, and air by partial combustion in the presence of a platinum catalyst:

$$HN_3 + CH_4 + 1.5\ O_2 + 6\ N_2 \longrightarrow HCN + 3\ H_2O + 6N_2$$

The process is carried out at about $900–1,000°C$; yield ranges from 55–60%. In another process, methane (contained in natural gas) is reacted with NH_3 over a platinum catalyst at from $1,200–1,300°C$, the reaction requiring considerable heat input. In still another process, a mixture of methane and propane is reacted with $NH_3 : C_3H_8 + 3NH_3 \longrightarrow 3HCN + 7H_2$; or $CH_4 + NH_3 \longrightarrow HCN + 3H_2$. An electrically heated fluidized bed reactor is used. Reaction temperature is approximately $1,510°C$.

The high-tonnage uses of HCN are in the preparation of numerous chemical products and intermediates for organic syntheses. As a gas, HCN sometimes is applied as a disinfectant; or cellulosic disks impregnated with HCN may be used. In ore processing and metal treating, cyanides are widely used.

Hydrogen cyanide reacts with hydrogen at $140°C$ in the presence of a catalyst, e.g., platinum black, to form methyl amine CH_3NH_2. When burned in air, it produces a pale violet flame; when heated with dilute sulfuric acid, it forms formamide $HCONH_2$ and ammonium formate $HCOONH_4$; when exposed to sunlight with chlorine it forms cyanogen chloride CNCl, plus hydrogen chloride. An important reaction of hydrogen cyanide is that with aldehydes or ketones, whereby cyanhydrins are formed, e.g., acetaldehyde cyanhydrin $CH_3CHOH \cdot CH$, and the resulting cyanhydrins are readily converted into alpha-hydroxy acids, e.g., alpha-hydroxypropionic acid $CH_3 \cdot CHOH \cdot COOH$.

Metallic cyanides are (1) soluble, e.g., sodium cyanide NaCN, potassium cyanide KCN, calcium cyanide $Ca(CN)_2$, mercuric cyanide $Hg(CN)_2$, aurous cyanide AuCN, (2) insoluble, e.g., silver cyanide AgCN, cuprous cyanide CuCN, (3) complex, (a) decomposed by dilute H_2SO_4 and not affected by dilute NaOH, e.g., sodium silver cyanide $NaAg(CN)_2$ solution, sodium cuprous cyanide $NaCu(CN)_2$ colorless solution, (b) changed only to acid by dilute H_2SO_4 and reactive with dilute NaOH, e.g., potassium hexacyanoferrate(II) $K_4Fe(CN)_6$ yields, with dilute H_2SO_4, hexacyanoferric(II) acid, cupric hexacyanoferrate(II) $Cu_2Fe(CN)_6$ yields, with dilute NaOH, cupric hydroxide.

Sodium cyanide solution dissolves certain metals (1) with absorption of oxygen, e.g., gold, silver, mercury, lead, and (2) with evolution of hydrogen, e.g., copper, nickel, iron, zinc, aluminum, magnesium; and solid sodium cyanide, when heated with certain oxides, e.g., lead monoxide PbO, stannic oxide SnO_2, yields the metal of the oxide, e.g., lead, tin, respectively, and sodium cyanate NaCNO. Two classes of esters are known, cyanides or nitriles, and isocyanides, isonitriles or carbylamines, the latter being very poisonous and of marked nauseating odor.

Methyl cyanide CH_3CN, bp $82°C$, formed by reaction of (1) methyl iodide and potassium cyanide, (2) acetamide and phosphorus pentoxide. Methyl isocyanide CH_3NC, bp $60°C$, formed by reaction (1) of methyl iodide and silver cyanide, (2) of methylamine, chloroform and NaOH solution warmed. Ethyl isocyanide C_2H_5NC, bp $78°C$. Phenyl isocyanide C_6H_5NC, bp $78°C$ at 40 torr pressure.

Oxidation of cyanide ion (e.g., by copper(II) gives cyanogen or oxaloni-trile NCCN, poisonous colorless gas, bp $-21°C$. This reacts with organic compounds and bases like a halogen, for example, disproportionating in aqueous alkali to cyanide and cyanate. In aqueous acid, hydrolysis to oxalamide and ultimately oxalic acid takes place. Oxidation of cyanides by oxygen donors (e.g., lead monoxide or dioxide, manganese dioxide or dichromate) a little below red heat produces cyanates.

HYDROGEN-ION ACTIVITY. Hydrogen ions are involved in a wide variety of natural and industrial reactions, and the equilibrium positions as well as the rates of these reactions are therefore dependent on hydrogen-ion concentration. The hydrogen ion is more correctly termed hydronium ion. The unhydrated proton does not exist in aqueous solution but rather is bound to several molecules of water. This ion, sometimes represented as $H(H_2O)_n^+$, is usually written simply as H^+. More important is the distinction between the hydrogen ion concentration and its activity. The hydrogen ion concentration, or total acidity, is obtained by titration and corresponds to the total concentration of hydrogen ions available in a solution, i.e., free, unbound hydrogen ions as well as hydrogen ions associated with weak acids. The hydrogen ion activity refers to the effective concentration of unbound hydrogen ions, i.e., the form which affects physicochemical reaction rates and equilibria. The effective concentration of hydrogen ion in solution is expressed in terms of pH, which is the negative logarithm of the hydrogen-ion activity, a_{H^+}

$$pH = -\log_{10} a_{H^+} \tag{1}$$

The relationship between activity, a, and concentration, c, is

$$a = \gamma c \tag{2}$$

where the activity coefficient γ is a function of the ionic strength of the solution and approaches unity as the ionic strength decreases; i.e., the difference between the activity and the concentration of free hydrogen ions diminishes as the solution becomes more dilute. The pH of a solution may have little relationship to the titratable acidity of a solution that contains weak acids or buffering substances; the pH of a solution indicates only the free hydrogen-ion activity. If total acid concentration is to be determined, an acid–base titration must be performed.

pH Determination

Two methods are used to measure pH: electrometric and chemical indicator. The most common is electrometric and uses the commercial pH meter with a glass electrode. This procedure is based on the measurement of the difference between the pH of an unknown or test solution and that of a standard solution.

More recently, two different types of nonglass pH electrodes have been described which have shown excellent pH-response behavior. In the neutral-carrier, ion-selective electrode type of potentiometric sensor, synthetic organic ionophores, selective for hydrogen ions, are immobilized in polymeric membranes. Another type of pH sensor is based on an integrated ion-selective electrode and insulated-gate field-effect transistor. These sensors, usually termed ion-selective field-effect transistors (ISFETs), are based on the modulation of the transistor source-drain current by a potential (or charge) applied to the transistor gate region.

The second method for measuring pH, the optical indicator method, has more limited applications. The success of this procedure depends on matching the color that is produced by the addition of a suitable indicator dye to a portion of the unknown solution with the color produced by adding the same quantity of the same dye to a series of standard solutions of known pH. The indicator dyes can also be immobilized onto paper strips (e.g., litmus paper) or, more recently, have been placed onto the distal end of fiber-optic probes which, when combined with photometric readout, provide more quantitative indicator-dye pH determinations.

Accuracy and Interpretation of Measured pH Values. To define the pH scale and permit the calibration of pH measurement systems, a series of reference buffer solutions have been certified by the U.S. National Institute of Standards and Technology (NIST). The acidity function which is the experimental basis for the assignment of pH, is reproducible within about 0.003 pH unit from 10 to 40°C. However, errors in the standard potential of the cell, in the composition of the buffer materials, and in the preparation of the solutions may raise the uncertainty to 0.005 pH unit. The accuracy of the practical scale may be further reduced to 0.008–0.01 pH unit as a result of variations in the liquid-junction potential.

Sources of Error. Several common causes of measurement problems are electrode interferences and/or fouling of the pH sensor, sample matrix effects, reference electrode instability, and improper calibration of the measurement system.

pH Measurement Systems

Glass Electrodes. The glass electrode is the hydrogen-ion sensor in most pH-measurement systems. The pH-responsive surface of the glass electrode consists of a thin membrane formed from a special glass that, after suitable conditioning, develops a surface potential that is an accurate index of the acidity of the solution in which the electrode is immersed. To permit changes in the potential of the active surface of the glass membrane to be measured, an inner reference electrode of constant potential is placed in the internal compartment of the glass membrane. The inner cell commonly consists of a silver–silver chloride electrode or calomel electrode in a buffered chloride solution. Immersion electrodes are the most common glass electrodes. Miniature and microelectrodes are also used widely, particularly in physiological studies. Capillary electrodes permit the use of small samples and provide protection from exposure to air during the measurements. The composition of the glass has a profound effect on the

electrical resistance, the chemical durability of the pH-sensitive surface, and the accuracy of the pH response in alkaline solutions.

Reference Electrodes and Liquid Junctions. The electrical circuit of the pH cell is completed through a salt bridge that usually consists of a concentrated solution of potassium chloride. The solution makes contact at one end with the test solution and at the other with a reference electrode of constant potential. The liquid junction is formed at the area of contact between the salt bridge and the test solution.

The commercially used reference electrode–salt bridge combination usually is of the immersion type. Some provision is made to allow a slow leakage of the bridge solution out of the tip of the electrode to establish the liquid junction with the test solution.

Combination electrodes have increased in use and are a consolidation of the glass and reference electrodes in a single probe, usually in a concentric arrangement, with the reference electrode compartment surrounding the pH sensor. The advantages of combination electrodes include the convenience of using a single probe and the ability to measure small volumes of sample solution or in restricted-access containers.

Theoretical considerations favor liquid junctions by which cylindrical symmetry and a steady state of ionic diffusion are achieved.

Samples that contain suspended matter are among the most difficult types from which to obtain accurate pH readings because of the so-called suspension effect, i.e., the suspended particles produce abnormal liquid-junction potentials at the reference electrode. Internal consistency is achieved by pH measurement using carefully prescribed measurement protocols, as has been used in the determination of soil pH.

Another effect that may result in spurious pH readings is caused by streaming potentials. Presumably, these are attributable to changes in the reference electrode liquid junction that are caused by variations in the flow rate of the sample solution. This problem may be avoided by maintaining constant flow and geometry characteristics and calibrating the system under operating conditions that are identical to those of the sample measurement.

pH Instrumentation. The pH meter is an electronic voltmeter that provides a direct conversion of voltage differences to differences of pH at the measurement temperature.

Because of the very large resistance of the glass membrane in a conventional pH electrode, an input amplifier of high impedance (usually 10^{12}–10^{14} Ω) is required to avoid errors in the pH (or mV) readings.

In addition, most devices provide operator control of settings for temperature and/or response slope, isopotential point, zero or standardization, and function (pH, mV, or monovalent–bivalent cation–anion). Microprocessors are incorporated in advanced-design meters to facilitate calibration, calculation of measurement parameters, and automatic temperature compensation.

Temperature Effects. The emf, E, of a pH cell may be written

$$E = E_g^{o'} - k\text{pH} \tag{3}$$

where k is the Nernst factor $(2.303\,RT)/F$, and $E_g^{o'}$ includes the liquid-junction potential and the half-cell emf on the reference side of the glass membrane. Changes of temperature alter the scale slope because k is proportional to T. The scale position also is changed because the standard potential is temperature dependent: $E_g^{o'}$ is usually a quadratic function of the temperature.

The objective of temperature compensation in a pH meter is to nullify changes in emf from any source except changes in the true pH of the test solution. Nearly all pH meters provide automatic or manual adjustment for the change of k with T.

Nonaqueous Solvents

The activity of the hydrogen ion is affected by the properties of the solvent in which it is measured. Scales of pH only apply to the medium, i.e., the solvent or mixed solvents, e.g., water–alcohol, for which the scales are developed. The comparison of the pH values of a buffer in aqueous solution to one in a nonaqueous solvent has neither direct quantitative nor thermodynamic significance. Consequently, operational pH scales must be developed for the individual solvent systems.

Other difficulties of measuring pH in nonaqueous solvents are the complications that result from dehydration of the glass pH membrane, increased sample resistance, and large liquid-junction potentials. These effects are complex and highly dependent on the type of solvent or mixture used.

Indicator pH Measurements

The indicator method is especially convenient when the pH of a well-buffered colorless solution must be measured at room temperature with an accuracy no greater than 0.5 pH unit. Under optimum conditions an accuracy of 0.2 pH unit is obtainable.

Because they are weak acids or bases, the indicators may affect the pH of the sample, especially in the case of a poorly buffered solution. Variations in the ionic strength or solvent composition, or both, also can produce large uncertainties in pH measurements, presumably caused by changes in the equilibria of the indicator species. Specific chemical reactions also may occur between solutes in the sample and the indicator species to produce appreciable pH errors.

Industrial Process Control

The pH meters and electrodes for process control do not differ materially from those used for measurements in the laboratory, but the emphasis in industrial applications is on rugged construction to withstand mechanical stresses and extremes in ambient conditions.

The pH meter usually is coupled to a data recording device and often to a pneumatic or electric controller. The controller governs the addition of reagent so that the pH of the process stream is maintained at the desired level.

RICHARD A. DURST
Cornell University
ROGER G. BATES
University of Florida

Additional Reading

Bates, R.G.: *Determination of pH, Theory and Practice,* 2nd Edition, Wiley-Interscience, New York, NY, 1973.

Eisenman, G. ed.: *Glass Electrodes for Hydrogen and Other Cations,* Marcel Dekker, New York, NY, 1967.

Galster, H.: *pH Measurement: Fundamentals, Methods, Applications, Instrumentation,* VCH, New York, NY, 1991.

Wu, Y.C., W.F. Koch, and R.A. Durst: *Standardization of pH Measurements,* National Bureau of Standards Special Publication 260–53, U.S. Government Printing Office, Washington, DC, 1988.

HYDROGEN PEROXIDE. [CAS: 7722-84-1]. H_2O_2, formula weight 34.02, in pure, anhydrous form is a viscous, colorless liquid, sp gr 1.44, mp $-0.89°C$, bp $151.4°C$. Hydrogen peroxide is soluble in H_2O in all proportions, soluble in alcohol, or ether, but not in hydrocarbons. Reagent, chemically pure (CP) grade H_2O_2 is a solution of 90% H_2O_2 and 10% H_2O, sp gr 1.39. This concentration contains 42% active oxygen by weight. One volume yields 410 volumes of oxygen. Hydrogen peroxide solutions are high-tonnage chemicals and are supplied commercially in several strengths, ranging from 3–35% H_2O_2 by weight. Commercial grades for oxidation and bleaching normally contain 27.5–35% H_2O_2.

To reduce the tendency of H_2O_2 solutions to decompose, storage must be at comparatively low temperatures and in light-tight containers. Often, an organic material, such as acetanalide, will retard degradation. H_2O_2 has been used as an oxidizer in liquid bipropellant systems, or as a monopropellant through controlled catalytic decomposition, in supplying oxygen to various fuel mixtures for rockets and torpedoes. Low-concentration (normally 3% H_2O_2) solutions have been used for many years as antiseptics in medical applications. Bleaching is a primary outlet for H_2O_2, particularly in connection with cotton, wool, groundwood pulp—as well as hair-bleaching formulations. The compound is used as a source of gas in foaming rubber plastics. The highly reactive H_2O_2 molecule readily participates in oxidation, epoxidation, and hydroxylation reactions and is frequently used in an intermediate capacity in chemical syntheses. In restoring old paintings, H_2O_2 has been used to convert black PbS tarnish into the original white lead sulfate.

Use in Food Processing

Within recent years, there has been increased interest in the use of H_2O_2 as a bactericidal and sporicidal agent in aseptic systems used for sterilizing food processing equipment and packaging materials. Several factors affect the success of such use. At low concentration, H_2O_2 may be regarded as bactericidal, but not highly sporicidal. The latter requires concentrations of up to 35% H_2O_2. Elevated solution temperature also increases effectiveness. Hydrogen peroxide solutions can be applied at a temperature up to 95°C because of their excellent thermal stability. Such treatment must be followed by hot-air heating at about 125°C in order to dissipate H_2O_2 residuals, which must be ≤ 0.1 ppm H_2O_2. Some inorganic salts, notably cupric salts, increase bactericidal activity. Treatment with H_2O_2 and ultrasonic radiation has been shown to produce a synergistic effect on the destruction of bacterial spores. Similarly, the combination of ultraviolet radiation and H_2O_2 appears to be synergistic. In one system, a UV-irradiated solution of H_2O_2 is used. The resistance of spores varies with species. In general, the resistance of clostridial spores to H_2O_2 is lower than that of spores of bacilli. Further details are given in the Stevenson/Shafer reference listed.

Researchers have found that alkaline hydrogen peroxide renders plant fibers more digestible by ruminants and thus suggests that a number of alternative feed sources, including crop residues and other cellulosic plant biomass, may be used in animal production. Researchers at the University of Illinois and the U.S. Department of Agriculture have treated lignocellulosic residues (wheat straw, corncobs, and cornstalks) with dilute alkaline H_2O_2 solutions and have found that the fermentability of such substances increases and that the byproducts produced may be considered an acceptable energy source for the ruminant animal. Details are given in the Kerley, et al. reference listed.

Industrial Production of Hydrogen Peroxide

The traditional process for manufacturing H_2O_2 has been the electrolysis of aqueous solutions of $KHSO_4$, H_2SO_4, or NH_4HSO_4. In recent years, chemical auto-oxidation processes have grown in favor, largely because of energy costs. In these processes, the feedstock may be an alkylated quinone, alkylated anthraquinone, and hydroquinone solvents, together with hydrogen, air or oxygen, H_2O, and a nickel, palladium, or platinum catalyst. The process yields a 15–75% solution of H_2O_2 in H_2O, depending upon adjustment of process concentrations and conditions to provide desired concentration. The yield for this type of process is about 90% of theoretical. The process proceeds essentially in two steps. In the first step, anthraquinone contained in a solvent is hydrogenated at a temperature of about 40°C and a pressure of 1–3 atmospheres. The anthraquinone is reduced to hydroquinone (*p*-dihydroxybenzene):

$$C_6H_4 : (CO)_2 : C_6H_2 \longrightarrow C_6H_4 : (COH)_2 : C_6H_3R$$

R is a radical such as ethyl or tertiary butyl. In the second step, the hydroquinone solution is oxidized with air or oxygen: $C_6H_4 : (COH)_2 : C_6H_3R + O_2 \longrightarrow C_6H_4 : (CO)_2 : C_6H_3R + H_2O_2$. In theory, the process consumes only hydrogen, atmospheric oxygen, and H_2O. A solvent must be used that will minimize side reactions during hydrogenation while also dissolving both the hydrogenated and oxidized forms of the organic compound. Solvents referred to in this connection are benzene-methyl-cyclohexanol mixtures and primary and secondary nonyl alcohols. Very tight purity precautions are required because any impurities in the H_2O_2 cause spontaneous catalytic decomposition of the product. As the result of these necessary precautions, the resulting H_2O_2 is one of the purest of commercial chemicals.

The process is highly corrosive. At one time, enameled steel vessels were standard for H_2O_2 processing. Aluminum, once properly passified through pickling and treatment after fabrication, has been found satisfactory.

Hydrogen peroxide reacts (1) with alkalis to form peroxides, (2) with potassium iodide solution, in presence of ferrous sulfate, to liberate iodine. This reaction serves to indicate the presence of as small an amount as 1 part by weight of hydrogen peroxide in 25,000,000 parts of H_2O, (3) with lead sulfide PbS, brown solid, to form lead sulfate $PbSO_4$, white solid, and sometimes used to brighten the lead pigment of darkened oil paintings, (4) with lead dioxide to form lead oxide, (5) with sulfites, especially in alkaline solution, to form sulfates, (6) with nitrites to form nitrates, (7) with arsenites for form arsenates, (8) with ferrous compounds to form ferric, (9) with chromic compounds to form chromates (See also **Chromium**), (10) with permanganates in acid solution to form manganous compounds plus oxygen of twice the volume available from the hydrogen peroxide, (11) with dichromates in acid solution cold to form perchromic acid, blue solution, more soluble in ether than in acid, (12) with titanic salt solutions to form pertitanic acid, yellow solution, (13) with colored organic materials, e.g., litmus, indigo, to destroy the color, and thus used for bleaching hair, silk, feathers, straw, ivory, teeth, bones, gelatin, flour. When hydrogen peroxide solution is treated with finely divided platinum or other substances, or comes in contact with rough surfaces, e.g., ground glass, oxygen is evolved (water also formed).

In the laboratory, hydrogen peroxide is prepared from barium peroxide by treatment with ice-cold dilute acid; when H_2SO_4 is used barium sulfate insoluble may be separated by filtration. Other peroxides, e.g., sodium peroxide, react similarly with acids to form hydrogen peroxide plus the salt corresponding to the peroxide and acid used. Hydrogen peroxide is formed when ether is exposed to sunlight, when a hydrogen-oxygen flame impinges on ice, and when H_2O in a quartz vessel is exposed to ultraviolet light.

Additional Reading

Kent, J.A.: *Reigel's Handbook of Industrial Chemistry,* Chapman & Hall, New York, NY, 1992.

Kerley, M.S. et al.: "Alkaline Hydrogen Peroxide Treatment Unlocks Energy in Agricultural By-Products," *Science,* **230**, 820–822 (1985).

Lide, D.R.: *CRC Handbook of Chemistry and Physics,* 84th Edition, CRC Press, LLC., Boca Raton, FL, 2003.

Stevenson, K.E. and B.D. Shafer: "Bacterial Spore Resistance to Hydrogen Peroxide," *Food Technology,* **37**(11), 111–114 (1983).

HYDROGEN SCALE.
1. A thermometric scale. (See also **Temperature Scales and Standards**.)

2. Since there is no reliable method for determining the absolute potential of a single electrode, electrode potentials are measured against a reference electrode whose potential is arbitrarily taken as zero. The arbitrary zero in general use is the potential of a reversible hydrogen electrode, with gas at 1 atmosphere pressure, in a solution of hydrogen ions of unit activity, or other electrodes calibrated against the hydrogen electrode.

HYDROGEN SULFIDE.
[CAS: 7783-06-4]. H_2S, formula weight 34.08, colorless, odorous gas, mp $-82.9°C$, bp $-59.6°C$, sp gr 1.1895 (air = 1). The gas must be handled carefully because of (1) its toxic properties (particularly dangerous because it may paralyze the olfactory nerves), and (2) its explosive tendencies (low ignition temperature of $260°C$ and wide flammability range from 4.3 to 44% by volume in air). Hydrogen sulfide liberates considerable heat upon burning (6,230 calories/liter at $15.6°C$). The gas is produced by acid hydrolysis of many sulfides and by water hydrolysis of those elements higher in the hydrogen scale.

An aqueous solution of hydrogen sulfide is termed hydrosulfuric acid, which undergoes slow atmospheric oxidation to sulfur. The acid is a strong reducing agent, usually with the separation of sulfur, e.g., with nitric acid (nitric oxide formed), with concentrated H_2SO_4(SO_2 is formed), with permanganate (manganous ion formed in the presence of acid), dichromate (chromic ion formed in the presence of acid).

Fluorine, chlorine, bromine, and iodine react with H_2S to form the corresponding halogen acid. Metal sulfides are formed when H_2S is passed into solutions of the heavy metals, such as Ag, Pb, Cu, and Mn. This reaction is responsible for the tarnishing of Ag and is the basis for the separation of these metals in classical wet qualitative analytical methods. Hydrogen sulfide reacts with many organic compounds.

The gas results from the decomposition of metal sulfides and albuminous matter and is found in the areas of mineral springs, sewers, and in some mines where it is referred to as "stink damp." H_2S also is a byproduct of several industrial processes, including synthetic rubber, viscose rayon, petroleum refining, dyeing, and leather-treating operations. In the laboratory, H_2S usually is prepared by treating a sulfide with an acid, such as iron pyrites and HCl, or by heating thioacetamide $CH_3C(:S)NH_2$. Three processes are used industrially to produce H_2S in large quantities: (1) treating a sulfide with an acid, $2NaHS + H_2SO_4 \longrightarrow 2H_2S + Na_2SO_4$, (2) reacting sulfur with an alkali, $4S + 2NaOH + 2H_2O \longrightarrow 2H_2S + Na_2S_2O_3$, and (3) directly reacting sulfur with hydrogen, $S + H_2 \longrightarrow H_2S$. Large quantities of byproduct H_2S usually are converted into elemental sulfur or H_2SO_4.

Industrial uses for H_2S include: (1) the preparation of sulfides, such as sodium sulfide and sodium hydrosulfide; (2) the production of sulfur-bearing organic compounds, such as thiophenes, mercaptans, and organic sulfides; (3) the removal of Cu, Cd, and Ti from spent catalysts where the gas acts as a precipitant; (4) the formulation of extreme-pressure lubricants; and (5) the preparation of rare-earth phosphors used in color TV tubes.

See also **Coal**.

HYDROLYSIS.
A chemical reaction in which water reacts with another substance to form two or more substances. This involves ionization of

the water molecules as well as splitting of the compound hydrolyzed, e.g., $CH_3COOC_2H_5 + H \cdot OH \longrightarrow CH_3COOH + C_2H_5OH$. Examples are: conversion of starch to glucose by water in the presence of suitable catalysts; or the conversion of sucrose (cane sugar) to glucose and fructose by reaction with water in the presence of an enzyme or acid catalyst; or conversion of natural fats into fatty acids and glycerin by reaction with water, as occurs in one stage of soap manufacturing; or the reaction of the ions of a dissolved salt to form various products, such as acids, complex ions, etc. See also **Cellulose Ester Plastics (Organic)**; **Organic Chemistry**; and **Starch**.

HYDRONIUM ION.
An ion found in water and all its solutions, which has the formula H_3O^+ and which consists of a proton combined with a water molecule. It has been established that hydrogen ions do not exist free in aqueous solution, but are present as hydronium ions. Formation of such ions is statistically rare, resulting from the interaction of water molecules in a ratio of 1 to 556 million.

HYDROPHILIC.
Having a strong tendency to bind or absorb water, which results in swelling and formation of reversible gels. This property is characteristic of carbohydrates, such as algin, vegetable gums, pectins, starches, and of complex proteins, such as gelatin and collagen. See also **Colloid Systems**; and **Detergents**.

HYDROPHOBIC.
Antagonistic to water; incapable of dissolving in water. This property is characteristic of oils, fats, waxes, and many resins, as well as of finely divided powders, such as carbon black and magnesium carbonate. Some interesting concepts are explored in "The Hydrophobic Effect and the Organization of Living Matter," by C. Tanford, *Science,* **200**, 1012–1018 (1978). See also **Colloid Systems**.

HYDROQUINONES.
These are dihydroxy aromatic compounds with the two groups in positions corresponding to *ortho* or *para* substitution in the benzene ring. They are closely related to the quinones from which they can be obtained by reduction. Thus *o*-dihydroxybenzene (catechol) can be obtained from *o*-benzoquinone, and hydroquinone (*p*-dihydroxybenzene or quinol) from *p*-benzoquinone. Resorcinol (*m*-dihydroxybenzene) is not properly a hydroquinone since the corresponding *meta* quinone is not known to exist. Homologs of hydroquinone are usually named after the parent hydrocarbon. Thus toluhydroquinone is 2,5-dihydroxy-1- methylbenzene and naphthohydroquinone is 1,4-dihydroxy-naphthalene. Unlike many of the quinones, the ring systems are fully aromatic and undergo substitution reactions common to phenols and other benzene derivatives. However they are easily oxidized by some reagents to the less stable quinones and degradation products frequently result. Thus treatment of hydroquinone with nitric acid yields oxalic acid while halogenation with sulfuryl chloride results in a mixture of chlorohydroquinones, quinone, quinone chlorides, and tetrachloro-*p*-benzoquinone. The formation of side and degradation products can be minimized if the molecule is protected against oxidation by acetylating or benzoylating at least one of the hydroxyl groups. For example 2-nitro-hydroquinone can be prepared in good yield by the nitration of monobenzoyl hydroquinone followed by hydrolysis. Concentrated sulfuric acid gives hydroquinone-2.5-disulfonic acid directly, and tertiary amyl groups can be introduced into the ring in the 2 and 5 positions by treatment with amylene in the presence of sulfuric acid. The hydroxyl groups are weakly acidic and can readily be converted to ethers by treatment with alkyl halides or sulfates in the presence of alkali. A diacetate is formed on treatment with acetic anhydride. The most characteristic reaction of hydroquinones is their reversible oxidation to quinones.

Catechol, or 1,2-dihydroxybenzene, was first prepared by the dry distillation of catechin obtained from *Mimosa catechu*. It can also be formed by the hydrolysis of its methyl ether, guaiacol, which is a constituent of beechwood tar. It is prepared synthetically by fusing phenol-*o*-sulfonic acid with sodium hydroxide, or treating *o*-chlorophenol with aqueous alkali in the presence of copper at a high temperature and pressure. It crystallizes from benzene in colorless monoclinic plates which melt at $105°C$. The lead salt can be oxidized to *o*-benzoquinone by a solution of iodine in chloroform. The ethers of catechol are of considerable importance and can be derived from a number of naturally occurring substances. The methylene ether of protocatechualdehyde is known as piperonal, and is closely related to various natural products including piperine, safrole, and isosafrole, from which it can be derived. These compounds have been used for the synthesis of pyrethrin synergists. *Vanillin*, the principal flavoring constituent of vanilla, is the 3-methyl ether of protocatechualdehyde.

Hydroquinone is found in nature combined in the glycoside arbutin, from which it can be released by hydrolysis with emulsin or dilute sulfuric acid. It is prepared commercially from *p*-benzoquinone by reduction with sulfur dioxide. It is a dimorphic solid with the stable form melting at 170.5°C. Hydroquinone is one of a number of compounds that possess the property of forming molecular compounds with gases such as hydrogen sulfide, sulfur dioxide, krypton, xenon, etc. These are known as clathrate compounds, and their existence is due to the entrapment of atoms or molecules of the gas in the crystal lattice of the hydroquinone. Three moles of hydroquinone can entrap one mole of gas, which is firmly held but which is liberated when the clathrate is dissolved in water. The most important commercial use of hydroquinone is for the development of photographic film. Its effectiveness is dependent on its ability to reduce the silver subhalide formed on exposure of the film to light to metallic silver. It gives films of high density and it is often necessary to reduce the harshness of contrast by using it in combination with other developers such as metol or paramidophenol. Hydroquinone and its derivatives are effective antioxidants for the preservation of fats, oils, and rubber. It has also been used as a short-stopping agent for controlling polymerization in the production of synthetic rubber of the butadienestyrene type.

H. P. BURCHFIELD
Gulf South Research Institute
New Iberia, Louisiana

HYDROTREATING.

A specialized kind of hydrogenation in which the quality of liquid hydrocarbon streams is improved by subjecting them to mild or severe conditions of hydrogen pressure in the presence of a catalyst. The objective is to convert undesirable material in the feedstock to either desired materials or easily disposed byproducts, on a highly selective basis. As of the early 1980s about 45% of the crude oil refined in the United States is hydrotreated. Some applications of hydrotreating include: (1) improvement of the burning quality of jet fuels, kerosines, and diesel fuels; (2) purification of light aromatic byproducts from pyrolysis operations; (3) pretreatment of naphtha feeds for catalytic reforming units; (4) reduction in sulfur content of residual fuel oils; (5) pretreatment of catalytic cracking feeds and cycle oils by removal of metals, sulfur, nitrogen, and reduction of polycyclic aromatics; (6) desulfurization of distillate fuels; (7) upgrading of lubricating oil quality; and (8) improvement of color, odor, and storage stability of various fuels.

Some of the specific reactions involved include: (1) hydrogenation of monoaromatics to naphthenes to improve burning quality of certain fuels; (2) removal of nitrogen as ammonia from its organic combinations; (3) removal of oxygen from its organic combinations as water; (4) hydrogenation of polycyclic aromatics so that only one aromatic ring remains in the molecule; (5) hydrogenation of diolefins and olefins to paraffins or naphthenes; (6) removal of sulfur from its organic combinations in various types of sulfur compounds by hydrodesulfurization to form hydrogen sulfide; and (7) decomposition and removal of organometals, such as arsenic compounds in naphthas, by retention of these metals on the catalyst. Vanadium and nickel also can be removed.

In the hydrotreating process shown by the Fig. 1, the liquid feed is preheated by exchange with the reactor effluent. It is then heated to the desired reactor-inlet temperature in a fired heater. At this point, recycle hydrogen joins the feedstock. An excess of hydrogen is used to suppress accumulation of deactivating carbonaceous deposits on the catalyst. Fresh makeup hydrogen enters the process to maintain a sufficient supply and also pressure on the system. Cooled effluent from the reactor goes to a separator vessel at which point the recycle or net hydrogen is removed. The liquid then goes to a stripper or stabilizer where hydrogen, hydrogen sulfide, ammonia, water, and light hydrocarbons dissolved in the separator liquid are removed. The stabilized hydro-treated liquid, free of dissolved, unwanted contaminants, is routed to subsequent processing or to product fuel blending.

It is interesting to note that there are over 25 proprietary versions of this basic process. Numerous modifications are required, depending upon the nature of the feedstock and desired end products.

Technical Staff, UOP Inc.
Des Plaines, Illinois

Additional Reading

Kabe, T., A. Ishihara, and W. Qian: *Hydrodesulfurization and Hydrodenitrogenation: Chemistry and Engineering,* John Wiley & Sons, Inc., New York, NY, 2000.
Occelli, M.L. and R. Chianelli: *Hydrotreating Technology for Pollution Control: Catalysts, Catalysis, and Processes,* Marcel Dekker, Inc., New York, NY, 1996.

HYDROXPROLINE.   See **Amino Acids**.

HYDROXYCARBOXYLIC ACIDS

Lactic Acid

Lactic acid (2-hydroxypropanoic acid), $CH_3CHOHCOOH$, is the most widely occurring hydroxycarboxylic acid and thus is the principal topic of this article. Lactic acid is a naturally occurring organic acid that can be produced by fermentation or chemical synthesis. It is present in many foods both naturally or as a product of *in situ* microbial fermentation, as in sauerkraut, yogurt, buttermilk, sourdough breads, and many other fermented foods. Lactic acid is also a principal metabolic intermediate in most living organisms, from anaerobic prokaryotes to humans.

Two significant producers of lactic acids are CCA Biochem by of the Netherlands, with subsidiaries in Brazil and Spain, and Sterling Chemicals, Inc. in Texas City, Tex. CCA uses carbohydrate feedstocks and fermentation technology, and Sterling uses a chemical technology. Lactic acid has been considered a relatively mature fine chemical in that only its use in new applications, e.g., as a monomer in plastics or as an intermediate in the synthesis of high volume oxygenated chemicals, would cause a significant increase in its anticipated demand.

Physical Properties. Pure, anhydrous lactic acid is a white, crystalline solid with a low melting point. However, it is difficult to prepare the pure anhydrous form of lactic acid; generally, it is available as a dilute or concentrated aqueous solution. The properties of lactic acid and its derivatives have been reviewed. A few important physical and thermodynamic properties from this reference are summarized in Table 1.

Lactic acid is also the simplest hydroxy acid that is optically active. $L(+)$-Lactic acid (**1**) occurs naturally in blood and in many fermentation products. The chemically produced lactic acid is a racemic mixture and

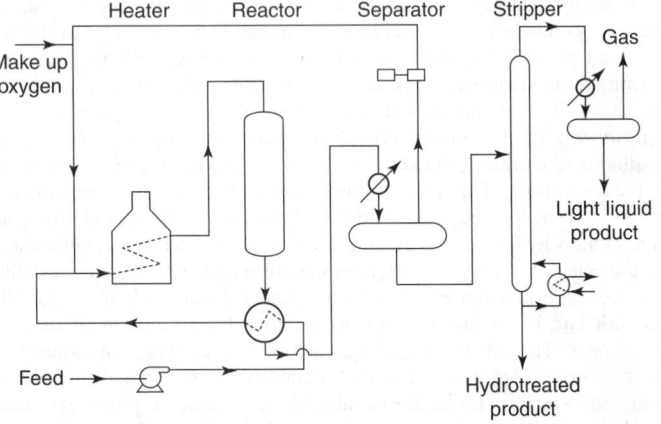

Fig. 1.   Representative hydrotreating unit used in petroleum industry. (*UOP Inc.*)

Make up oxygen · Feed · Heater · Reactor · Separator · Stripper · Gas · Light liquid product · Hydrotreated product

TABLE 1. PHYSICAL AND THERMODYNAMIC PROPERTIES OF LACTIC ACID

Property	Value
density, g/mL at 20°C	1.2243
viscosity,[a] mPa · s(= cP)	36.9
heat of solution, $L(+)$ at 25°C, kJ/mol[b]	7.79
heat of fusion, kJ/mol[b]	
racemic	11.33
$L(+)$	16.86
heat of combustion, MJ/mol[b]	
racemic	−1.355
$L(+)$	−1.343

[a] 88.6 wt% solution at 25°C.

[b] To convert J to cal, divide by 4.184.

some fermentations also produce the racemic mixture or an enantiomeric excess of $D(-)$-lactic acid (**2**).

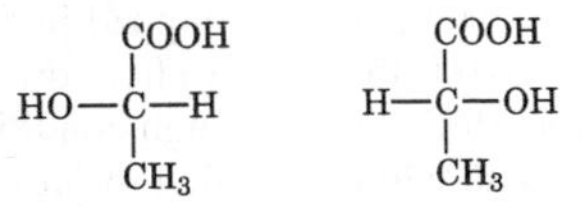

$$
\begin{array}{cc}
\text{COOH} & \text{COOH} \\
\text{HO—C—H} & \text{H—C—OH} \\
\text{CH}_3 & \text{CH}_3 \\
(\mathbf{1}) & (\mathbf{2})
\end{array}
$$

Many of the physical properties are not affected by the optical composition, with the important exception of the melting point of the crystalline acid, which is estimated to be 52.7–52.8°C for either optically pure isomer, whereas the reported melting point of the racemic mixture ranges from 17 to 33°C.

Chemical Properties. Its two functional groups permit a wide variety of chemical reactions for lactic acid. The primary classes of these reactions are oxidation, reduction, condensation, and substitution at the alcohol group.

Economic Aspects. The incentive for economical production of lactic acid is coming from the development of new, large-volume uses of lactic acid, particularly as feedstocks for biodegradable polymers and oxygenated chemicals. The advent and deployment of highly efficient, membrane-based separation processes and chemical and catalytic conversion technologies, together with commercial interest by several agriprocessing and chemical companies, will lead to the production of low cost lactic acid, which in turn will result in new opportunities for large-scale use of lactic acid.

Specifications, Quality Control, and Analytical Methods. Lactic acid is generally sold under four general product categories: synthetic, fermentation, heat-stable fermentation, and technical.

Lactic acid is generally recognized as safe (GRAS) for multipurpose food use. Lactate salts such as calcium and sodium lactates and esters such as ethyl lactate used in pharmaceutical preparations are also considered safe and nontoxic.

Uses. Currently, the principal use of lactic acid is in food and food-related applications, which in the United States accounts for approximately 85% of the demand. The rest (~15%) of the uses are for nonfood industrial applications.

Hydroxyacetic Acid

Hydroxyacetic acid (glycolic acid), $HOCH_2COOH$, is the first and simplest member of the family of hydroxycarboxylic acids. It occurs naturally as the chief acidic constituent of sugar-cane juice and also occurs in sugar beets and unripe grape juice. It is widely used as a cleaning agent for a variety of industrial applications, and also as a specialty chemical and biodegradable copolymer feedstock.

Properties. Glycolic acid is a colorless, translucent solid; mp = 10°C; bp = 112°C; d at 25°C = 1.26 g/mL; K_a at 25°C = 1.5×10^{-4} pH at 25°C = 0.5; heat of combustion = 697.1 kJ/mol (166.6 kcal/mol); heat of solution = −11.55 kJ/mol; and flash point >300°C.

Glycolic acid is soluble in water, methanol, ethanol, acetone, acetic acid, and ethyl acetate. It is slightly soluble in ethyl ether and sparingly soluble in hydrocarbon solvents.

Reactions. Because it contains both a carboxyl and a primary hydroxyl group, glycolic acid can react as an acid or an alcohol or both. Thus some of the important reactions it can undergo are esterification, amidation, salt formation, and complexation with metal ions, which lead to many of its uses. As a fairly strong acid it can liberate gases (often toxic) when it reacts with the corresponding salts.

Manufacture, Processing, and Economic Aspects. Hydroxyacetic acid is produced commercially in the U.S. by the reaction of formaldehyde with carbon monoxide and water.

Other Hydroxy Acids

Apart from lactic and hydroxyacetic acids, other α- and β-hydroxy acids have been small-volume specialty products produced in a variety of methods for specialized uses.

Preparation. The general preparation of α-hydroxy acids is by the hydrolysis of an α-halo acid or by the acid hydrolysis of the cyanohydrins of an aldehyde or a ketone. β-Hydroxy acids may be made by catalytic reduction of β-keto esters followed by hydrolysis. β-Hydroxy acids can also be prepared by the Reformatsky reaction. γ-Hydroxy acids are seldom obtained in the free state because of the ease with which they form

monomeric inner esters, which form stable five-membered rings. Thus the lactones of these acids are the common chemical forms and among these lactones γ-butyrolactone is one of the larger volume specialty chemicals derived from dehydrogenation of 1,4-butanediol.

Reactions and Uses. The common reactions that α-hydroxy acids undergo such as self- or bimolecular esterification to oligomers or cyclic esters, hydrogenation, oxidation, etc, have been discussed in connection with lactic and hydroxyacetic acid. A reaction that is of value for the synthesis of higher aldehydes is decarbonylation under boiling sulfuric acid with loss of water.

β-Hydroxy acids lose water, especially in the presence of an acid catalyst, to give α, β-unsaturated acids, and frequently β, γ-unsaturated acids. γ-Hydroxybutyric acid and its derivatives, particularly its sodium salt, have been studied and used as anesthetics, tranquilizers, sedatives, and hypnotics in surgery and general obstetrics.

Certain bacterial species produce polymers of γ-hydroxybutyric acid and other hydroxyalkanoic acids as storage polymers. These are biodegradable polymers with some desirable properties for manufacture of biodegradable packaging materials, and considerable effort is being devoted by ICI Ltd. and others to the development of bacterial fermentation processes to produce these polymers at a high molecular weight.

γ-Butyrolactone undergoes amination reactions with methylamine or ammonia to produce N-methyl-2-pyrrolidinone (NMP) or 2-pyrrolidinone (PDO) respectively, both of which are commercially important derivatives.

Other multifunctional hydroxycarboxylic acids are mevalonic and aldonic acids which can be prepared for specialized uses as aldol reaction products (mevalonic acid) and mild oxidation of aldoses (aldonic acids).

RATHIN DATTA
Consultant

Additional Reading

Datta, R. et al.: *FEMS Microbiol. Revs.* **16**, 221–231 (1995).

Glycolic. (Hydroxyacetic) Acid: Properties, Uses, Storage, and Handling; Glycolide S.G.: Properties, Uses, Storage, and Handling, bulletins, Du Pont Chemicals, Wilmington, DE, 1992.

Holten, C.H., A. Muller, and D. Rehbinder: *Lactic Acid,* International Research Association, Verlag Chemie, Copenhagen, Denmark, 1971.

Lactic Acid and Lactates, product bulletin, Purac Inc., Arlington Heights, IL, 1989.

HYDROXYLAMINE. [CAS: 7803-49-8]. H_2NOH, formula weight 33.02, white, odorless solid, mp 33°C, bp 56°C (22 mm pressure), explosive, soluble in all proportions in H_2O or alcohol. Hydroxylamine is: (1) A weak base forming with acids soluble salts that decompose more or less violently when heated, e.g., hydroxylamine hydrochloride (hydroxylammonium chloride, $H_2NOH \cdot HCl$), mp 151°C, nitrate $H_2NOH \cdot HNO_3$, hemisulfate $H_2NOH \cdot \frac{1}{2}H_2SO_4$. Dihydroxylamine oxalate and trihydroxylamine phosphate are insoluble in H_2O. Hydroxylamine hydrochloride is soluble in alcohol. (2) A weak acid forming with bases soluble salts, e.g., sodium hydroxylamite H_2NONa. Hydroxylamine salt solution is a powerful reducing agent, more especially in alkaline than in acid solution, for example, cupric salt solutions changed to cuprous oxide, silver salt solutions to silver, mercuric chloride solution to mercurous chloride, ferric salt solutions (in acid) to ferrous. Ferrous hydroxide in sodium hydroxide is, however, oxidized by hydroxylamine to ferric hydroxide plus NH_3.

Hydroxylamine reacts with carbonyl group =CO of aldehydes, ketones or quinones, yielding *oximes*, white solids, of definite melting point and used in identification of aldehydes and ketones, e.g., acetaldehyde oxime $CH_3CH:NOH$:

Beta-phenylhydroxylamine, N-phenylhydroxylamine, is a white solid, slightly soluble in water, very soluble in alcohol or ether, forms salts with acids, e.g., beta-phenylhydroxylamine hydrochloride $C_6H_5 NHOH \cdot HCl$, upon exposure to air the water solution forms azobenzene $C_6H_5N:NC_6H_5$. Beta-phenylhydroxylamine reacts (1) with oxidizing agents, such as chromic acid or ferric chloride, to form nitrosobenzene C_6H_5NO, (2) with reducing agents, such as tin plus hydrochloric acid, to form aniline $C_6H_5NH_2$, (3) with alkaline cupric salt solution (Fehling's solution) at room temperature to form cuprous oxide, (4) with ammonio-silver salt solution (Tollen's solution) at room temperature to form silver, (5) in the presence of hydrochloric acid to form paraminophenol HO·$C_6H_4 \cdot NH_2(1,4)$.

Beta-phenylhydroxylamine is formed by reduction of nitrobenzene (1) by zinc and calcium chloride or ammonium chloride solution, (2) by electrolysis in acetic acid plus sodium acetate solution.

Diphenylhydroxylamine is prepared by reaction of nitrosobenzene and phenylmagnesium bromide in anhydrous ether, followed by treatment with H_2O (magnesium hydroxybromide also formed).

When hydroxylamine reacts with aldehydes, the resulting compounds are termed *aldoximes* as, for example, acetaldoxime. $CH_3 \cdot CHO + H_2NOH \longrightarrow CH_3 \cdot CH:N \cdot OH$(acetaldoxime) $+ H_2O$. Hydroxylamine reactions with ketones produce *ketoximes*. $(CH_3)_2CO + H_2NOH \longrightarrow (CH_3)_2C:N \cdot OH$(dimethylketoxime) $+ H_2O$.

The lower aldoximes are essentially odorless, volatile liquids, and miscible with H_2O in all proportions. The higher members are only slightly soluble. Ketoximes have similar properties.

HYDROXY DICARBOXYLIC ACIDS.

Many natural and synthetic organic compounds are hydroxy dicarboxylic acids. This article discusses mainly malic and tartaric acids; thiomalic acid is included because of its structural similarity to malic acid.

Malic Acid

Malic acid (hydroxysuccinic acid, hydroxybutanedioic acid, or 1-hydroxy-1,2-ethanedicarboxylic acid), $C_4H_6O_5$, is a white, crystalline material. The levorotatory isomer, $S(-)$-malic acid (L-malic acid), is a natural constituent and common metabolite of plants and animals. The racemic compound, R,S-malic acid (DL-malic acid), is a widely used food acidulant. This material is also used in some industrial applications as a sequestrant and as a buffer for pH control. $R(+)$-Malic acid (D-malic acid) is available only as a laboratory chemical. Following the introduction of a modern, continuous manufacturing process in the early 1960s, malic acid gradually became a large-volume industrial organic acid.

Physical Properties. Malic acid crystallizes from aqueous solutions as white, translucent, anhydrous crystals. The $S(-)$ isomer melts at $100-103°C$ and the $R(+)$ isomer at $98-99°C$. On heating, D,L-malic acid decomposes at ca $180°C$, by forming fumaric acid and maleic anhydride. Under normal conditions, malic acid is stable; under conditions of high humidity, it is hygroscopic. Malic acid is a relatively strong acid. Its dissociation constants are given in Table 1.

Chemical Properties. Because of its chiral center, malic acid is optically active.

Reactions. Malic acid undergoes many of the characteristic reactions of dibasic acids, monohydric alcohols, and α-hydroxycarboxylic acids.

Manufacture. In the United States, Canada, and Europe, only the synthetic R,S-malic acid is produced commercially, whereas both the S and R,S forms are produced in Japan.

Aqueous fumaric acid is converted to levorotatory malic acid by the intracellular enzyme, fumarase, which is produced by various microorganisms.

The commercial synthesis of R,S-malic acid involves hydration of maleic acid or fumaric acid at elevated temperature and pressure.

Energy And Environmental Considerations. The energy requirements to produce malic acid via conventional processes are fairly moderate. Malic acid production generates low levels of solid, airborne, and liquid waste. Solid waste is primarily nontoxic malic acid salts resulting form regenerating carbon cells and ion-exchange resins. Airborne emissions are primarily particulates. A 1% malic acid solution is readily biodegradable, with BOD of 5300 mg/L.

Shipping and Storage. Malic acid is shipped in 50-lb, 100-lb, and 25-kg, multiwall paper bags or 100-lb (45.5 kg) fiber drums. Malic acid can be stored in dry form without difficulty, although conditions of high humidity and elevated temperatures should be avoided to prevent caking.

Economic Aspects. Malic acid is manufactured in over 10 countries. The production is primarily used for food (26.6%) and beverages (54.7%); however, some industrial applications (18.7%) exist, e.g., coatings, polymers, and resins. (Historical patterns of use in the United States have been stable and are as noted in parentheses).

Health and Safety. The U.S. FDA has affirmed R,S- and $S(-)$-malic acid as substances that are generally recognized as safe (GRAS) as flavor enhancers, flavoring agents and adjuvants, and as pH control agents. R,S- and $S(-)$-malic acid may not be used in baby foods. Malic acid is also cleared to correct natural acid deficiencies in juice or wine.

Uses. R,S-Malic acid is utilized in a variety of food and beverage and some industrial applications because of its unique combination of properties. These include having unusual taste-blending characteristics, flavor-fixing qualities, the ability to retain sour taste longer, high water solubility, and chelating and buffering properties. Malic acid is also a reactive intermediate in chemical synthesis.

Thiomalic Acid

Thiomalic acid (mercaptosuccinic acid), $C_4H_6O_4S$, mol wt = 150.2, is a sulfur analogue of malic acid. The properties of the crystalline, solid thiomalic acids are given in Table 2. The racemic acid has the following acid dissociation constants at $25°C$: $pK_{a1} = 3.30$; $pK_{a2} = 4.94$.

R,S-Thiomalic acid can be prepared from bromosuccinic acid by reaction with K_2S. The two enantiomers can be obtained from the corresponding optically active potassium bromosuccinates.

Thiomalic acid is a skin sensitizer and an antidote in heavy-metal poisoning. Traditionally, it was a component of cold permanent hair-waving solutions and of rust-removing and corrosion-inhibiting compositions. Sodium aurothiomalate (Myochrisin) and other gold thiomalate complexes have antiarthritic properties. The well-known insecticide, malathion, is the thiomalate S-ethyl ester of O,O-dimethylphosphonodithioic acid.

$$CH_3O \diagdown \overset{\overset{S}{\|}}{P}-SCHCOOC_2H_5$$
$$CH_3O \diagup \qquad \underset{CH_2COOC_2H_5}{|}$$

Tartaric Acid

Tartaric acid (2,3-dihydroxybutanedioic acid, 2,3-dihydroxysuccinic acid), $C_4H_6O_6$, is a dihydroxy dicarboxylic acid with two chiral centers. It exists as the dextro- and levorotatory acid: the meso form (which is inactive owing to internal compensation), and the racemic mixture (which is commonly known as racemic acid). The commercial product in the United States is the natural, dextrorotatory form, $(R-R^*, R^*)$-tartaric acid (L(+)-tartaric acid). This enantiomer occurs in grapes as its acid potassium salt (cream of tartar). In the fermentation of wine, this salt forms deposits in the vats.

Physical Properties. When crystallized from aqueous solutions above $5°C$, natural $(R-R^*,R^*;)$-tartaric acid is obtained in the anhydrous form. Below $5°C$, tartaric acid forms a monohydrate which is unstable at room temperature. Some of the physical properties of $(R-R^*,R^*;)$-tartaric acid are listed in Table 3.

The solubility of $(R-R^*,R^*;)$-tartaric acid in water varies from 115 g/100 g H_2O at $0°C$ to 343 g/100 g H_2O at $100°C$. One hundred grams of absolute ethanol dissolves 20.4 g of tartaric acid at $18°C$, and 100 g of ethyl ether dissolves 0.3 g at $18°C$.

TABLE 1. PHYSICAL PROPERTIES OF R,S-MALIC ACID

Property	Value
mol wt	134.09
melting point, °C	ca 130
d_4^{20}	1.601
dissociation constant	
K_1	4×10^{-4}
K_2	9×10^{-6}
viscosity (50% aqueous solution at 25°C), mPa·s(s = cP)	6.5
solubility in nonaqueous solvents, % wt/wt	
ethanol	45.5
acetone	17.8
methanol	82.7

TABLE 2. PROPERTIES OF THIOMALIC ACIDS

Acid	Mp, °C	Solubility		$[\alpha]_D^{17 17a}$
		Water	Ethanol	
R,S-	151	very sol	very sol	
R	154	sol	sol	$+64.4°$
S	152–153	sol	slightly sol	$-64.8°$

[a] 5% acid in ethanol.

TABLE 3. PHYSICAL PROPERTIES OF $(R\text{-}R^*,R^*;)$-TARTARIC ACID

Property	Value
mol wt	150.086
mp, °C (anhydrous)	169–170
d^{20}, g/cm^3	1.76
heat of solution, kJ/mol[a]	−13.8

[a] To convert kJ to kcal, divide by 4.184.

Chemical Properties. The notation used by *Chemical Abstracts* to reflect the configuration of tartaric acid is as follows: $(R\text{-}R^*,R^*;)$-tartaric acid, $(S\text{-}R^*,R^*;)$-tartaric acid, and *meso*-tartaric acid. Racemic acid is an equimolar mixture of the two optically active enantiomers and, hence, like the meso acid, is optically inactive.

When free $(R\text{-}R^*,R^*;)$-tartaric acid is heated above its melting point, amorphous anhydrides are formed which, on boiling with water, regenerate the acid. Further heating causes simultaneous formation of pyruvic acid, $CH_3COCOOH$; pyrotartaric acid, $HOOCCH_2CH(CH_3)COOH$; and, finally, a black, charred residue. In common with other hydroxy organic acids, tartaric acid complexes many metal ions.

Occurrence. $(R\text{-}R^*,R^*;)$-Tartaric acid occurs in the juice of the grape and in a few other fruits and plants. It is not as widely distributed as citric acid or $S(-)$-malic acid. The only commercial source is the residues from the wine industry. The racemic acid is not a primary product of plant processes but is formed readily from the dextrorotatory acid by heating alone or with strong alkali or strong acid. *meso*-Tartaric acid is not found in nature. It is obtained from the other isomers by prolonged boiling with caustic alkali.

Manufacture. The chemical reactions involved in tartaric acid production are formation of calcium tartrate from crude potassium acid tartrate, formation of tartaric acid from calcium tartrate, formation of Rochelle salt from argols, and formation of cream of tartar from tartaric acid and Rochelle salt (RS) liquors.

Economic Aspects. The estimated total worldwide market for tartaric acid is 58,000 t and potassium bitartrate (acid basis) is 20,000 t.

Health and Safety. The FDA affirmed $(R\text{-}R^*,R^*;)$-tartaric acid as a generally- recognized-as-safe (GRAS) food substance.

Uses. Tartaric acid is used in carbonated beverages, wine making, and other foods. It is also used to produce emulsifiers, in the manufacture of pharmaceuticals, and in many industrial uses.

Salts. Rochelle salt is used in the silvering of mirrors. Its properties of piezoelectricity make it valuable in electric oscillators. Medicinally, it is an ingredient of mild saline cathartic preparations, e.g., compound effervescing powder. In food, it can be used as an emulsifying agent in the manufacture of process cheese. Cream of tartar is used in baking powder and in prepared baking mixes.

GARY T. BLAIR
JEFFREY J. DEFRATIES
Haarmann & Reimer Corporation

Additional Reading

Chem. Mktg. Rep., 35 (Jan. 25, 1993).
Code of Federal Regulations, 21 CFR 184.1069, Office of the Federal Register, U.S. Government Printing Office, Washington, DC, 1993.
"Food Chemicals Codex," 3rd Edition, 3rd Suppl., *National Academy of Sciences*, National Research Council, Washington, DC, 1992.
Ockerman, H.W., *Source Book for Food Scientists*, The Avi Publishing Co., Inc., Westport, CT, 1978, p. 276.

HYGROMETRY AND PSYCHROMETRY.

These are instrumental methods for measuring humidity. Humidity can be expressed in a variety of different forms: wet bulb temperature; percent relative humidity (% RH); vapor pressure; mixing ratio; dew/frost point; grains per pound; grams per kilogram; and parts per million, among others. These parameters can be measured by a number of different instruments, each capable of accurate measurement under certain conditions and within specific limitations.

Definition of Humidity

Unless one is routinely working with humidity measurements, there is a tendency to overlook the fact that humidity is water gas, behaving in accordance with the ideal gas laws. One of the easiest ways to put humidity in its proper perspective is through application of Dalton's law of partial pressures to the most commonly encountered gas–*air*.

Dalton's law states that the total pressure P_m exerted by a mixture of gases or vapors is the sum of the pressure of each gas if it were to occupy the same volume by itself. The pressure of each individual gas is called its *partial pressure*. The total pressure of an air-water gas mixture, containing oxygen, nitrogen, and water, is equal to the sum of the partial pressures of each gas:

$$P_m = P_{N_2} + P_{O_2} + P_{H_2O} + \cdots$$

Therefore, the partial pressure of water vapor in air is directly related to the measurement of humidity. This vapor pressure varies from 1.22×10^{-3} mb (millibar) of mercury (0.122 Pascal) at the −75°C frost point of "bone dry" arctic or industrial dry air—to 1.013×10^3 mb of mercury (0.1013×10^6 Pascal) at the 100°C dew point of saturated hot air in a product drier. This is a change of almost a million to one over the span of interest in industrial humidity measurement.

The ideal humidity instrument would be a linear, wide-range pressure gage, specific to water vapor and employing a primary or fundamental measuring method. Such an instrument, although physically possible, would be cumbersome. Most humidity measurements are made by some secondary instrument which is responsive to humidity-related phenomena.

Humidity Parameters

The humidity parameters most often encountered in scientific and industrial applications are given in Table 1. In addition to these common parameters, numerous other formats exist for use in narrow applications or specific technologies. However, most of these are variations of the parameters listed.

The psychrometric chart provides a quick means for converting from one humidity format to another because dew point, relative humidity, ambient temperature, and wet bulb temperature can be conveniently related to each other on a single sheet of paper. The psychrometric chart has

TABLE 1. HUMIDITY MEASUREMENT METHODS

Parameter	Description	Units	Typical applications
Wet bulb temperature	Minimum temperature reached by a wetted thermometer in an airstream	°F or °C	High temperature driers, air conditioning, meteorology, test chambers
Percent relative humidity	The ratio of the actual vapor pressure to the saturation vapor pressure, with respect to water, at the prevailing dry bulb temperature	0–100%	Monitoring conditioning rooms, test chambers, pharmaceutical and foodpackaging
Dew/frost point	Dew point is the temperature to which the air must be cooled to achieve saturation. If the temperature is below 32°F, it is called the frost point	°F or °C	Heat treating, annealing atmospheres, drier control, instrument air monitoring, meteorological/environmental measurements
Volume or mass ratio	Parts per million (ppm) by volume is the ratio of the partial pressure of the water vapor to the partial pressure of the dry carrier gas. PPM by weight is identical to ppm by volume, but the ratio changes according to the molecular weight of the carrier gas.	ppm$_v$, ppm$_w$	Used primarily to insure dryness of industrial process gases such as air, nitrogen, oxygen, methane, hydrogen, etc.

long been the basic tool of air conditioning engineers. A chart of this type is given in the entry entitled **Psychrometric Chart**. Psychrometric charts are available for higher temperatures and humidities and are quite useful in drier and condensation system design. Charts are also available for lower temperatures, but these tend to be less useful because wet bulb measurements are difficult to make with any accuracy at temperatures below −7°C.

Wet Bulb/Dry Bulb Measurements

Psychrometry has long been a popular method for monitoring humidity, primarily due to its simplicity and inherent low cost. A typical industrial psychrometer consists of a pair of matched electrical thermometers, one of which is maintained in a wetted condition. Water evaporation cools the wetted thermometer, resulting in a measurable difference between it and the ambient, or dry bulb measurement. When the wet bulb reaches its maximum temperature depression, the humidity is determined by comparing the wet bulb/dry bulb temperatures on a psychrometric chart. In a properly designed psychrometer, both sensors are aspirated at an airstream rate between 4 and 10 meters per second for proper cooling of the wet bulb, and both are thermally shielded to minimize errors from radiation.

A properly designed and utilized psychrometer, such as the Assman laboratory type, is capable of providing accurate data. However, very few industrial psychrometers meet these criteria and are limited to applications where low cost and moderate accuracy are the underlying requirements. The psychrometer does have certain inherent advantages: (1) The psychrometer is capable of highest accuracy near 100% RH. From an accuracy standpoint, it is superior to most other humidity sensors near saturation. Since the dry bulb and wet bulb sensors can be connected differentially, this allows the wet bulb depression (which approaches zero as the relative humidity approaches 100%) to be measured with a minimum of error. (2) Although large errors can occur if the wet bulb becomes contaminated or improperly fitted, the simplicity of the device affords easy repair at minimum cost. (3) The psychrometer can be used at ambient temperature above 100°C, and the wet bulb measurement is usable up to 100°C.

Major limitations of the psychrometer include: (1) As relative humidity drops below about 20% RH, the problem of cooling the wet bulb to its full depression becomes difficult. The result is impaired accuracy below 20% RH, and few psychrometers work at all below 10% RH. (2) Wet bulb measurement at temperatures below 0°C are difficult to obtain with any high degree of confidence. Automatic water feeds are not feasible, because of freezing. (3) Because a wet bulb psychrometer is a source of moisture, it can only be used in environments where added water vapor from the psychrometer exhaust is not a significant component of the total volume. (4) Generally speaking, psychrometers cannot be used in small, closed volumes.

Percent Relative Humidity

Percent relative humidity is the best known and perhaps the most widely used method for expressing the water vapor content of air. Percent relative humidity is defined as the ratio of the prevailing water vapor pressure e_a to the water vapor pressure if the air were saturated, e_s, multiplied by 100:

$$\% \text{ RH} = (e_a/e_s) \times 100$$

The term "percent relative humidity" appears to be derived from the invention of the hair hygrometer in the 17th Century. The hair hygrometer operates on the principle that many organic filaments, such as hair, goldbeater's skin, and even nylon, change length as a nearly linear function of the *ratio* of *prevailing water vapor pressure* to the *saturation vapor pressure*.

Basically, percent relative humidity is an indicator of the water vapor saturation deficit of the gas mixture, rather than an indicator of sorption, desorption, comfort, or evaporation. A measurement of % RH without a corresponding measurement of dry bulb temperature is not of particular value, since the water vapor content cannot be determined from % RH alone.

Sensors for Measuring % RH. Over the years, devices other than the simple hair hygrometer have evolved which permit a direct measurement of % RH. These devices are, for the most part, electrochemical sensors that offer a degree of ruggedness, compactness, and remote electronic readout ability not afforded by hair devices.

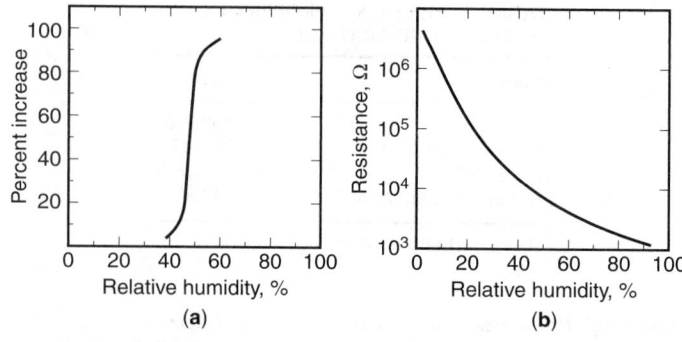

Fig. 1. Resistance characteristics of typical Dunmore and Pope sensors. (**a**) Dunmore sensors are limited to a narrow range of humidity. This sensor operates between 40 and 60% RH. (**b**) Pope sensors operate over a wide humidity range, but output impedance of the sensor varies from 1000 ohms (100% RH) to several megohms (10% RH), which complicates readout circuitry

Two widely used electronic % RH sensors are the Dunmore element and the Pope cell. The Dunmore sensor employees a bifilar-wound inert wire grid on an insulative substrate which is coated with a lithium chloride solution of a controlled concentration. The hygroscopic nature of this salt causes it to take up water vapor from the surrounding atmosphere. The ac resistance of the sensor is an indication of the prevailing % RH. Dunmore cells are excellent RH sensors, but, because of the characteristics of lithium chloride, are usually designed to cover a narrow range of interest. For example, a single sensor may cover from 40 to 60% RH and the sensor output is usable only in that range. See Fig. 1(a).

Wide-range Dunmore sensors can be made with a cluster of narrow range sensors in a common housing, mated with an electrical matching network. This arrangement, however, usually results in a rather bulky sensor.

The Pope cell employs a similar bifilar conductive grid on an insulative substrate. In this sensor, the substrate is made from polystyrene, which has been treated in a prescribed fashion with sulfuric acid. This results in sulfonation of the longer-chain polystyrene molecules. Because the sulfate radical (SO_4) is highly mobile in the presence of hydrogen ions (available from the water molecule in vapor form), the $(SO_4)^{2-}$ ions can detach and take on H^+ ions, thereby altering the surface resistivity of the sensor as a function of humidity.

In both the Dunmore and Pope sensors, the element is arranged in an ac-excited Wheatstone bridge so that only alternating current flows through the grid. Direct current excitation of either the Dunmore or Pope elements polarizes the sensor, eventually causing loss of calibration.

The Pope sensor has one significant advantage over the Dunmore sensor in that the Pope unit is a wide-range sensor, typically covering 15% RH to 99% RH in a single element. See Fig. 1(b). Considerable attention must be given to readout circuitry for the Pope sensor because the resistance varies in a nonlinear fashion from 1000 ohms to several megohms.

Dew Point Hygrometry

Dew point measurements are widely used in scientific and industrial applications when precise measurement of water vapor pressure is needed. Dew point, the temperature at which water condensate begins to form on a surface, can be accurately measured from −75°C to +100°C across the entire range of humidity with a condensation (chilled mirror) hygrometer.

Three types of instruments have received wide acceptance in dew point measurements: (1) the saturated salt dew point sensor; (2) the condensation-type hygrometer; and (3) the aluminum oxide sensor. Many other instruments are used in specialized applications, including pressure ratio devices, dewcups, and fog chambers. The latter are manually operated.

Saturated Salt Dew Point Sensors. The saturated salt (lithium chloride) dew point sensor is widely used because of its inherent simplicity, ruggedness, and low cost. Both the United States and Canadian government weather services use this type of sensor for most official groundbased humidity measurements. However, some of these are being converted to the more accurate condensation hygrometers.

The principle of the saturated salt dew point sensor is based on the relationship that the vapor pressure of water is reduced in the presence of a salt. When water vapor in the air condenses on a soluble salt, it forms a saturated layer on the surface of the salt. This saturated layer has a lower vapor pressure than water vapor in the surrounding air. If the salt is heated, its vapor pressure increases to a point where it matches the water vapor pressure in the surrounding air and the evaporation/condensation process reaches equilibrium. The temperature at which equilibrium is reached is directly related to the dew point.

A saturated salt sensor is constructed with an absorbent fabric bobbin covered with a bifilar winding of inert electrodes and coated with a dilute solution of lithium chloride. Lithium chloride (LiCl) is often used as the saturating salt because of its hygroscopic nature, which permits application in relative humidities between 11 and 100%.

An alternating current is passed through the winding and salt solution, causing resistive heating. As the bobbin heats, water evaporates into the surrounding air from the diluted LiCl solution. The rate of evaporation is determined by the vapor pressure of water in the surrounding air. When the bobbin begins to dry out, due to evaporation of water, resistance of the salt solution increases. With less current through the winding, because of increased resistance, the bobbin cools and water begins to condense, forming a saturated solution on the bobbin surface. Eventually, equilibrium is reached and the bobbin neither takes on nor loses any water.

Properly used, a saturated salt sensor is accurate to $\pm 1°C$ between dew point temperatures of -12 and $+38°C$. Outside these limits, small errors may occur as a result of the multiple hydration characteristics of lithium chloride, which may produce ambiguous results at $41°C$, $-12°C$, and $-34°C$ dew points. Maximum errors at these ambiguity points are 1.4, 1.6, and 3.4°C, respectively, but actual errors encountered in typical applications are usually less.

Applications. The saturated salt sensor has certain advantages over other electrical humidity sensors, such as % RH instruments. Because the salt sensor operates as a current carrier saturated with Li and Cl ions, addition of contaminating ions has little effect on its behavior compared to a typical RH sensor, which operates "starved" of ions and is easily contaminated. A properly designed saturated salt sensor is not easily contaminated since, from an ionic standpoint, it can be considered precontaminated.

If a saturated salt sensor does become contaminated, it can be washed with an ordinary sudsy ammonia solution, rinsed and recharged with lithium chloride. It is seldom necessary to discard a saturated salt sensor if proper maintenance procedures are observed.

Limitations of saturated salt sensors include: (1) relatively slow response time; and (2) a lower limit to the measurement range imposed by the nature of lithium chloride. The sensor cannot be used to measure dew points when the vapor pressure of water is below the saturation vapor pressure of lithium chloride, which occurs at about 11% RH. In certain gases, ambient temperatures can be reduced, increasing the RH to above 11%; but the extra effort needed to cool the gas usually warrants selection of a different type of sensor. Fortunately, a large number of scientific and industrial measurements fall above this limitation and are readily handled by the sensor.

Condensation-Type Hygrometers. The condensation-type dew point hygrometer is one of the most accurate and reliable of sensors for humidity measurements, and has the widest range. These features are achieved, however, through increased complexity and cost. In the condensation-type hygrometer, a surface is cooled (either thermoelectrically, mechanically, or chemically) until dew or frost begins to condense out. The condensate surface is maintained electronically in vapor pressure equilibrium with the surrounding gas, while surface condensation is detected by optical, electrical, or nuclear techniques. See Fig. 2. The surface temperature is then the dew point temperature, by definition.

The largest source of error in a condensation hygrometer stems from the difficulty in measuring condensate surface temperature accurately. Typical industrial versions of the instrument are accurate to $\pm 0.2°C$ over very wide temperature spans. Laboratory models offer accuracies up to $\pm 0.1°C$.

Wide span and minimal errors are two main features. A properly designed condensation hygrometer can measure dew points from $100°C$ down to frost points of $-75°C$.

Response time of a condensation dew point hygrometer is usually specified in terms of its cooling/heating rate, typically $1.5°$/second, making it considerably faster than a saturated salt dew point sensor and nearly as fast as most electrical % RH sensors. Perhaps the most significant feature of the condensation hygrometer is its fundamental measuring technique, which essentially renders the instrument self-calibrating. For calibration, it is only necessary to manually override the surface-cooling control loop, causing the surface to heat, and witness that the instrument recools to

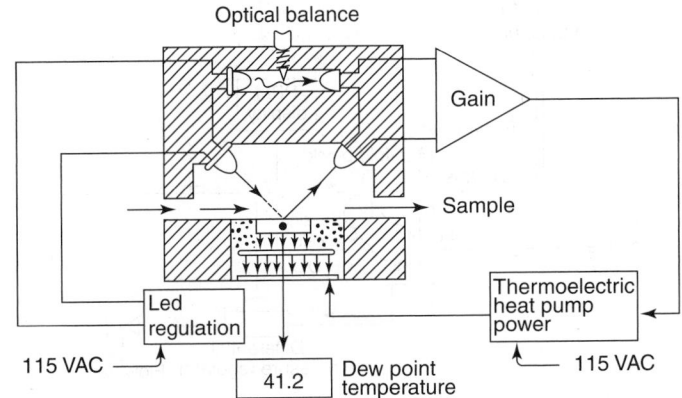

Fig. 2.   Dew is detected in a condensation hygrometer by cooling a surface until water begins to condense. Condensation is detected optically or electronically. The signal is fed into a control circuit which maintains the surface temperature at the precise dew point

the same dew point when the loop is closed. Assuming that the surface temperature measuring system is calibrated, this is a reasonable and valid check on the instrument's performance.

Because of its fundamental nature and superior accuracy and repeatability, this kind of instrument is widely used as a secondary standard (National Bureau of Standards) for calibrating other lower level humidity instruments.

The inert construction of the condensation hygrometer makes it virtually indestructible. Although the instrument can become contaminated, it is easy to wash and return to service without impairment of performance or calibration.

The condensation (chilled mirror) hygrometer measures dew/frost temperature. Unfortunately, many applications require measurement of % RH, water vapor in parts per million, or some other humidity parameter. In such cases, the user must decide whether to employ the fundamental, high accuracy condensation hygrometer and convert the dew/frost point measurement to the desired parameter, or use lower level instrumentation to measure these parameters directly. In recent years, microprocessors have been developed which can be incorporated in the design of a condensation hygrometer, resulting in instrumentation that can offer accurate measurements of humidity in terms of almost any humidity parameter.

Electrolytic Hygrometer. A typical electrolytic hygrometer utilizes a cell coated with a thin film of phosphorous pentoxide (P_2O_5), which absorbs water from the sample gas. See Fig. 3. The cell has a bifilar winding of inert electrodes on a fluorinated hydrocarbon capillary. Direct current applied to the electrodes dissociates the water, which is absorbed by the P_2O_5, into hydrogen and oxygen. Two electrons are required to electrolyze each water molecule and thus the current in the cell represents the number of molecules dissociated. A further calculation, based on flow rate, temperature and current, yields the parts per million concentration of water vapor.

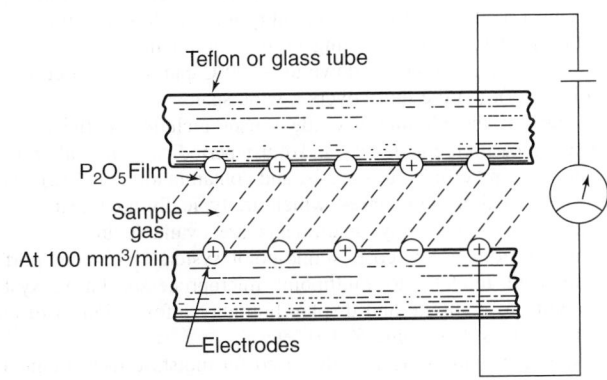

Fig. 3.   An electrolytic hygrometer dissociates water, absorbed by P_2O_5, into hydrogen and oxygen by electrolysis. Since two electrons are required to electrolyze a molecule of water, the amount of current used by the hygrometer relates to parts per million of water vapor

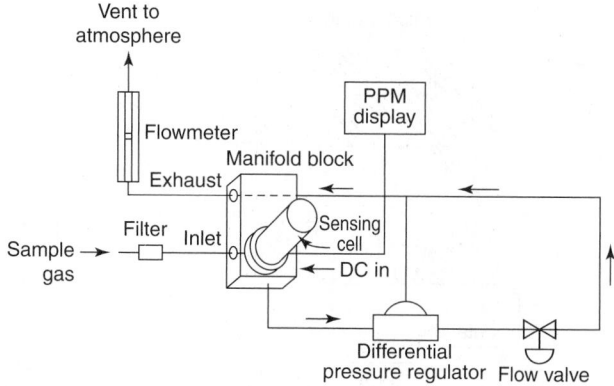

Fig. 4. Calculation of the water vapor content in an electrolytic hygrometer is dependent on precise control of the flow rate. This arrangement controls the sample pressure across the cell, ensuring correct flow regardless of input pressure fluctuations

In order to obtain accurate data, the flow rate of the sample gas through the cell must be known and constant. Since the ppm calculation is partially based on flow, an error in the flow rate causes a direct error in measurement.

A typical sampling system for insuring constant flow is shown in Fig. 4. Constant pressure is maintained within the cell. Sample gas enters the inlet, passes through a stainless steel filter, and enters a stainless steel manifold block. It is very important that all components prior to the sensor be made of an inert material, such as stainless steel, to minimize contamination. After passing through the sensor, the sample gas pressure is controlled by a differential pressure regulator which compares pressure of the gas leaving the sensor with the pressure of the gas venting to atmosphere through a preset valve and flowmeter. In this way, constant flow is maintained even though there may be nominal pressure fluctuations at the inlet port.

A typical electrolytic hygrometer can cover a span from 0 to 2000 ppm with an accuracy of $\pm 5\%$ of the reading, more than adequate for most industrial applications. The sensor is suitable for most inert elemental gases and organic and inorganic gas compounds that do not react with P_2O_5.

Electrolytic hygrometers cannot be exposed to high water vapor levels for any long period of time because this results in a high usage rate for the P_2O_5 and high cell currents.

Aluminum Oxide Moisture Sensor. This type of sensor is a capacitor, formed by depositing a layer of porous aluminum oxide onto a conductive substrate, and then coating the oxide with a thin film of gold. The conductive base and the gold layer become the capacitor's electrodes. Water vapor penetrates the gold layer and is absorbed by the porous oxidation layer. The number of water molecules absorbed determines the electrical impedance of the capacity, which is, in turn, a measure of water vapor pressure.

Advantages of the aluminum oxide sensor are: (1) small size and suitability for *in situ* use; (2) it can be used very economically in multiple sensor arrangements; (3) suitability for very low dew point levels without the need for sensor cooling () as required in condensation-type sensors—(typically, dew points down to $-100°C$ can be measured without serious difficulty); (4) the unit covers a wide span.

Limitations of the aluminum oxide sensor include: (1) the sensor is a secondary measurement device and must periodically be calibrated to accommodate aging effects, hysteresis, and contamination; and (2) sensors require separate calibration curves, which are typically nonlinear.

Aluminum oxide humidity instruments are available in a variety of types, ranging from a low-cost, single-point system, including portable battery operated models, to multipoint microprocessor based systems with capability to compute and display humidity information in different parameters, such as dew point, % RH, etc.

The aluminum oxide sensor is also used for moisture measurements in liquids (hydrocarbons). Because of its low power usage, it is suitable for use in explosion proof installations. These sensors are frequently used in petrochemical applications where low dew points are to be monitored on line and where the reduced accuracies and other limitations are acceptable. The advantages of the sensor must be weighted against the fact that

accuracy is lower than with any of the fundamental measurement sensor types. As a secondary measurement device, it can provide reliable data only if kept in calibration and if damage due to incompatible contaminants is avoided.

PIETER R. WIEDERHOLD
General Eastern Instruments Corp.
Watertown, Massachusetts

HYGROSCOPIC. 1. Pertaining to a marked ability to accelerate the condensation of water vapor. In meteorology, this term is applied principally to those condensation nuclei composed of salts that yield aqueous solutions of a very low equilibrium vapor pressure compared with that of pure water at the same temperature. Condensation on hygroscopic nuclei may begin at a relative humidity much lower than 100% (about 75% for sodium chloride); while on so-called nonhygroscopic nuclei, which merely furnish sufficiently large (by molecular standards) wettable surfaces, relative humidities of nearly 100% are required.

2. Descriptive of a substance, the physical characteristics of which are appreciably altered by effects of water vapor. The hygroscopicity of certain materials has been advantageously utilized in humidity measurement and control devices; for example, the hair element of a hair hygrometer.

HYPERCONJUGATION. The description of the properties of a molecule in terms of resonance structures in which an atom or group is not joined by any sort of bond to the atom to which it is ordinarily considered linked. Also called no-bond resonance. The hypothesis of hyperconjugation has been advanced to interpret some properties of substances containing but 1 double bond by analogy with those of substances containing conjugated double bonds. Consider a substance with a terminal structure $H_3C-CH=CH....$ One of the possible resonating structures of this group is

$$H_3 \equiv C - CH = CH —— H_3 = C = CH - CH —$$

the dotted line indicating two unpaired electrons with opposite spins.

HYPEREUTECTIC ALLOY. An alloy with a composition falling on the right of the eutectic point of a binary phase diagram that freezes with a structure containing some eutectic.

HYPERFINE STRUCTURE. In general, a set of very closely spaced lines in atomic spectra or other kinds of spectra. There may be many causes of hyperfine structure: (1) for a single atomic species or nuclide, the occurrence of spectral lines as doublets, triplets, etc., due to the interaction, or coupling, of the total angular momentum of the orbital electrons with the nuclear spin and associated magnetic moment; (2) for an element consisting of several isotopes, the occurrence of components for each spectral line that is observable under high resolution, each isotope contributing one or more components. This type of hyperfine structure is often called isotope structure to differentiate it from the first type of hyperfine structure discussed above. See also **Atomic Spectra**.

HYPERONS. These are subatomic particles that are more massive than nucleons (protons and neutrons). *Strangeness* is a property of elementary particles found useful in classifying hyperons. Each particle is assigned a strangeness quantum number S which is related to the electric charge Q, the isospin number T, and the baryon number B by the formula $Q = T + (S + B)/2$. ($T = \frac{1}{2}$ for a proton and $-\frac{1}{2}$ for a neutron; other particles may have $T = 0$ or $T = 1$, depending on the type.) Strangeness is conserved in reactions involving the strong interaction. The selection rules resulting from strangeness conservation are important in understanding why some reactions take place much more slowly than others. See also **Particles (Subatomic)**.

HYPERSORPTION. Process in which activated carbon selectively absorbs the less-volatile components from a gaseous mix, while the more-volatile components pass on unaffected. Particularly applicable to separations of low-boiling mixtures such as hydrogen and methane, ethane from natural gas, ethylene from refinery gas, etc.

HYPERSTHENE. The mineral hypersthene is an orthorhombic pyroxene, chemically a ferromagnesian silicate, differing from enstatite in that the iron content is considerable (FeO being greater than 15%). A general formula is $(Mg, Fe)SiO_3$. It is usually found as a massive mineral, whose crystals tend to be prismatic or tabular in habit. It has a distinct prismatic cleavage; fracture, uneven; brittle; hardness, 5–6; specific gravity, 3.42–3.84; luster, pearly to somewhat metallic; color, brownish-green, brown, greenish-black to grayish-black; streak, grayish-brown; translucent to opaque. Hypersthene is often associated with labradorite in gabbro and norite and in extrusive rocks like andesite. It is occasionally encountered in meteorites. Hypersthene is associated with pyrrhotite in Bavaria, with labradorite on the Isle St. Paul, Labrador. It is also found in Montmorency County, Quebec; and in the United States in the rocks of the Cortlandt series in the Hudson River Valley, and the andesites of Colorado and northern California. Superb crystals of exceptional size and quality have been found growing into and within the almandinepyrope garnets at Gore Mountain, North River, New York. The rarity of hypersthene in crystal form makes this occurrence noteworthy. The word hypersthene comes from the Greek words meaning *strong* or *tough*.

See also **Pyroxene**.

HYPOBARIC (CONTROLLED-ATMOSPHERE) SYSTEMS.

Sensitive materials, notably fresh foods, normally cannot withstand long periods of transportation and storage prior to consumption. Over the years, much of the effort extended toward offering produce in marketplaces far distant from the source was concentrated on reducing the time for delivery. Thus, the extensive use of air express and air freight. Conventional refrigeration systems for trucks and railway cars also were especially adapted for use during transport. But even with all of these improvements in technology, certain transporting feats (as, for example, shipping midwestern pork to the California and even Hawaii markets) were difficult to achieve. The controlled-atmosphere hypobaric concept has greatly extended the potential for distant shipping of delicate, perishable materials (not necessarily limited to foodstuffs).

In 1964, the Institute of Food Technologists annual award was given in recognition of the development of a controlled-atmosphere storage system. Essentially, the process was designed to reduce the rate of deterioration of certain fruits and vegetables in refrigerated storage by reducing the oxygen level, increasing the carbon dioxide level, and maintaining the relative humidity close to 100%. In an initial design, the conditions were created by using a home-furnace size catalytic generator that burned natural gas or propane gas to create the atmosphere that essentially halts the natural respiration of the stored products. Later, the gas generator was replaced by cryogenic liquefied gases, allowing additional flexibility and the creation of any desired gas mixture. Atmospheres can be tailored to particular perishables. For example, an atmosphere of 15–20% carbon dioxide and 80–85% nitrogen is optimal for strawberries. For iceberg lettuce, an atmosphere of 8–10% oxygen, less than 10% carbon dioxide, with the remainder nitrogen is used. As of the early 1980s, the system has been installed on over ten thousand rail cars and 7000 sea vans.

In 1979, another IFT award was given in recognition of a hypobaric transport and storage system for fresh meats and meat products. Hypobarics is defined as a precisely controlled combination of low pressure, low temperature, high humidity, and ventilation which, when properly applied, extends up to six times the length of time a perishable commodity remains fresh. This makes possible the shipment of perishable items by way of relatively low-cost surface transportation to distant points. In developing the concept, it was observed that refrigerated storage of fruits in closed containers will result in accumulation of gases generated by the fruit, i.e., ethylene and carbon dioxide, an atmosphere which hastens ripening and spoilage. Although ventilation of fruit containers can prevent accumulation of the gases, the gases are not removed from within the product itself—with no prevention of accumulation of gases within the cells of the fruit. The researchers made the supposition that by drawing a partial vacuum on a closed vessel containing the fruit, the low pressure would increase the diffusivity of the gases, thus promoting release and removal of the gases. At the same time, a reduction of pressure would reduce the oxygen concentration, thus retarding respiration and attendant spoilage. Combined with refrigeration, this would decelerate the metabolic processes, not only of the fruit, but also of any bacteria present. Humidification of the chamber would prevent any drying of the fruit. After testing the concept on bananas and other perishables, the system was patented.

Generally the storage temperature for meats is about $-1°C$, and up to 10 or 12°C for various fruits and vegetables. In all cases, the relative humidity is controlled at about 95%. Pressure ranges between 10 and 80 millimeters of mercury. Lower pressures are maintained for meats and seafoods; somewhat higher pressures for fruits and vegetables.

HYPOCHLORITES.

When chlorine is reacted with an alkali, a hypochlorite is formed. These compounds are very high-tonnage chemicals for sanitizing and bleaching purposes. Commercial sodium hypochlorite NaClO, CAS: 7681-52-9, usually is available in two strengths (1) the familiar household liquid bleach which contains about 5.25% (weight) NaClO, and (2) commercial bleach which contains about 13% (weight) NaClO. The latter compound sometimes is referred to as 15% bleach because the chlorine content is approximately 150 grams/liter of available chlorine. The term "liquid chlorine" usually refers to a solution of NaClO (up to 10%) used in the swimming-pool trade. "Dry chlorine" is part of the registered trademark of a proprietary calcium hypochlorite product containing 70% available chlorine. See also **Bleaching Agents**.

Sodium hypochlorite normally is manufactured in batches by diluting caustic soda to the proper starting concentration. This is approximately 6.8% NaOH for the 5.25% bleach; and about 18.5% NaOH for the 15% bleach. After cooling the caustic soda solution, chlorine gas is added through a sparger pipe until the desired concentration is reached. This usually is determined by making a series of titration analyses. Bleaching powder $CaOCl_2$ is made by passing chlorine gas over slaked lime. This was the first type of chlorine bleaching agent made and dates back to 1799. The product usually contains about 30% available chlorine. Over the years, it was used extensively in the bleaching of textiles and for sanitizing even though the compound is unstable and difficult to use. The original bleaching powder largely has been replaced by an improved calcium hypochlorite product, which contains about 70% available chlorine. The compound essentially is a calcium hypochlorite dihydrate and, in one process, is made by chlorinating a slurry of lime and caustic soda. The crystals that precipitate out are mixed with calcium chloride and chlorinated lime. When warmed, the calcium hypochlorite dihydrate precipitates, with sodium chloride remaining in solution. After filtering, the cake is dried, granulated, sized, and packaged. In addition to use in swimming pools, products of this type are used widely for water purification, algae control, and sanitation. On a very high-tonnage basis, calcium hypochlorite $Ca(ClO)_2 \cdot 4H_2O$ is used for pulp bleaching in the paper industry. Bleach liquor containing from 20–40% available chlorine may be produced in batches or continuously. In a continuous system, the flow of chlorine is controlled by making frequent (or continuous) measurements of oxidation-reduction potential.

A common means of detecting hypochlorites is the production of a blue color (caused by free iodine) with starch iodide paper by hypochlorites in weakly alkaline solution. Silver nitrate also precipitates part of the hypochlorite in solutions as white silver chloride.

Hypochlorous Acid. [CAS: 7790-92-3]. This compound, HOCl, is prepared by the reaction of (1) chlorine monoxide Cl_2O with H_2O, (2) sodium hypochlorite and an acid, excess acid yielding chlorine and oxygen, and (3) chlorine with mercuric oxide suspended in water, mercuric chloride being formed simultaneously. Hypochlorous acid is a yellow solution of characteristic odor. It decomposes upon standing, the rate depending upon (1) concentration, (2) exposure to light, (3) presence of a catalyst (cobaltous hydroxide, for example, promotes the evolution of oxygen), and (4) acidity or alkalinity. Hypochlorous acid is a powerful oxidizing agent and sometimes used as a bleaching agent for organic colors.

Perchloric Acid. [CAS: 7601-90-3]. This compound, $HClO_4$, is a colorless, fuming, oily liquid, miscible with H_2O, volatile under diminished pressure. A maximum constant-boiling solution (203°C, 760 millimeters Hg) results when the concentration of $HClO_4$ reaches 73% in H_2O. Cold dilute perchloric acid reacts with such metals as zinc and iron, yielding hydrogen gas and the corresponding perchlorate in solution. It is stable from the point of view of oxidation and reduction (except that iodine is oxidized to periodic acid, with liberation of chlorine, ferrous salt solutions to ferric, titanous salt solutions to titanic). Concentrated hot perchloric acid, on the other hand, is a powerful oxidizing agent, exploding violently in contact with charcoal, paper, or alcohol; causes serious wounds in contact with the skin. Prepared by distilling ammonium perchlorate with HNO_3 and HCl.

Metallic perchlorates are soluble in water, except that potassium perchlorate is slightly soluble. Potassium perchlorate is, however, insoluble in alcohol containing perchloric acid, a property made use of in the qualitative recognition and quantitative estimation of potassium in salt solutions. Perchlorates, when heated, evolve oxygen and leave the chloride as a residue. Potassium perchlorate decomposes at 400°C.

HYPOEUTECTIC ALLOY.

An alloy to the left of the eutectic point in a binary phase diagram that freezes with a structure containing some eutectic.

HYPOFLUORITE.

Any compound containing the group—OF. The simple anion FO^- is unknown. A number of covalent hypofluorites are known, including such compounds with carbon, oxygen, nitrogen, sulfur, chlorine and arsenic (uncertain), CF_3OF, CF_3COOF, C_2F_5COOF, NO_2OF, OF_2, O_2F_2, O_3F_2, SF_5OF, FSO_2OF, ClO_3OF and possibly AsF_4OF. These are all powerful fluorinating agents. They react violently with water yielding OF_2 as one product. The oxygen fluorides O_3F_2 and O_2F_2 decompose about $-158°C$ and $-100°C$, respectively, the former into the latter and the latter into the elements. Nitryl and perchloryl hypofluorites (fluorine nitrate and fluorine perchlorate) easily detonate. The perfluoracyl hypofluorites are much more stable but may also decompose violently. The others appear to be stable.

HYPOIODOUS ACID AND HYPOIODITES.

Hypoiodous acid (HOI) is a greenish-yellow solution, of characteristic odor. It is unstable, and cannot be distilled unchanged.

Prepared by reaction (1) of iodine and mercuric oxide (see also **Mercury**) suspension in water, mercuric iodide being simultaneously formed, (2) of sodium hypoiodite and an acid, excess acid yielding iodine.

Sodium hydroxide solution reacts with iodine to form iodide and hypoiodite, the latter decomposing in a few hours at ordinary temperatures to form iodide and iodate.

HYPONITROUS ACID AND HYPONITRITES.

Hyponitrous acid $H_2N_2O_2$ is a white solid, explosive even at as low a temperature as 0°C, soluble in water, more soluble in ether, can thus be extracted from water solution by ether and the latter evaporated, water solution decomposes quickly into nitrous oxide plus water. Hyponitrous acid is nonreactive with hydriodic acid (a strong reducing agent), but reactive with permanganic acid (a strong oxidizing agent) to form nitrous or nitric acid.

Prepared (1) by reaction of silver hyponitrite $Ag_2N_2O_2$ and hydrogen chloride in anhydrous ether, an evaporation of the resulting solution, (2) by reaction of hydroxylamine H_2NOH plus nitrous acid HONO.

Sodium hyponitrite $Na_2N_2O_2$ is formed (1) by reaction of sodium nitrate or nitrite solution with sodium amalgam (sodium dissolved in mercury), after which acetic acid is added to neutralize the alkali. Sodium stannite ferrous hydroxide, or electrolytic reduction with mercury cathode may also be utilized, (2) by reaction of hydroxylamine sulfonic acid and sodium hydroxide. Silver hyponitrite is formed by reaction of silver nitrate solution and sodium hyponitrite.

HYPOPHOSPHORIC ACID AND HYPOPHOSPHATES.

Hypo-phosphoric acid (H_2PO_3 or $H_4P_2O_6$) is a solid, melting point 55°C, decomposing in solution to form phosphorous plus phosphoric acids. Hypophosphoric acid is used in solution and is a reducing agent, but only with strong oxidizing agents, such as potassium permanganate; and the acid is unaffected by zinc and dilute sulfuric acid (distinction from phosphorous acid). Dehydration of hypophosphoric acid does not yield phosphorus tetroxide; hydration of phosphorus tetroxide does not yield hypophosphoric acid but phosphorous plus phosphoric acids.

Hypophosphoric acid is formed by reaction (1) of yellow phosphorous and potassium permanganate in sodium hydroxide medium, (2) of red phosphorus and calcium hypochlorite solution, (3) also one of the products of slow oxidation at ordinary temperatures of phosphorus in moist air.

There are recorded the following sodium hypophosphates: Na_2PO_3 (or $Na_4P_2O_6$), $NaHPO_3$ (or $Na_2H_2P_2O_6$), $Na_3H(PO_3)_2$ (or $Na_3HP_2O_6$), and $(NaH_3PO_3)_2$ (or $NaH_3P_2O_6$). There is evidence in support of each of the formulas H_2PO_3, $H_4P_2O_6$ for hypophosphoric acid.

Ester: Dimethyl hypophosphate $(CH_3)_2PO_3$ or $(CH_3O)_2PO$. See also **Phosphorus**.

HYPOPHOSPHOROUS ACID AND HYPOPHOSPHITES.

Hypo phosphorous acid (H_3PO_2, or $H \cdot PO_2H_2$) is a colorless liquid, melting point 26.5°C, density 1.493.

Hypophosphorous acid is miscible with water in all proportions and a commercial strength is 30% H_3PO_2. Hypophosphites are used in medicine.

Hypophosphorous acid is a powerful reducing agent, e.g., with copper sulfate forms cuprous hydride Cu_2H_2, brown precipitate, which evolves hydrogen gas and leaves copper on warming; with silver nitrate yields finely divided silver; with sulfurous acid yields sulfur and some hydrogen sulfide; with sulfuric acid yields sulfurous acid, which reacts as above; forms manganous immediately with permanganate.

Hypophosphorous acid is formed by reaction of barium hypophosphite and sulfuric acid, and filtering off barium sulfate. By evaporation of the solution in vacuum at 80°C, and then cooling to 0°C, hypophosphorous acid crystallizes.

Sodium hypophosphite $NaPO_2H_2$, the only sodium hypophosphite, is formed (1) by reaction of yellow phosphorus and sodium hydroxide solution (phosphine simultaneously formed), (2) by reaction of hypophosphorous acid and sodium hydroxide, and evaporating. Sodium hypophosphite, upon heating, yields sodium phosphate and sodium phosphide. Common tests for the hypophosphites are as follows:

1. Zinc reduces dilute sulfuric acid solution of hypophosphites to phosphine recognizable by odor (difference from phosphates).
2. Barium chloride produces no precipitate (difference from phosphites). See also **Phosphorus**.

HYPOSULFUROUS ACID AND HYPOSULFITES.

Hyposulfurous acid $H_2S_2O_4$ is a yellow solution rapidly oxidized in air to sulfurous acid and then to sulfuric acid. Commercially known as hydrosulfurous acid and its salts as hydrosulfites (but not to be confused with "hypo" which is sodium thiosulfate).

Hyposulfurous acid is a powerful reducing agent, e.g., with copper sulfate forms cuprous hydride Cu_2H_2, brown precipitate, which evolves hydrogen gas and leaves copper on warning, with silver nitrate yields finely divided silver, with permanganate yields manganous compounds. Hyposulfurous acid is formed by reaction of sodium hyposulfite and an acid.

Sodium hyposulfite, sodium hydrosulfite $Na_2S_2O_4 \cdot 2H_2O$ is formed (1) by reaction of zinc and sulfurous acid (or sodium hydrogen sulfite), yielding zinc hyposulfite and then converted by sodium chloride into sodium hyposulfite, (2) by electrolysis of sodium hydrogen sulfite and then addition of sodium chloride.

Sodium hyposulfite is used to bleach sugar, indigo, wood pulp. With moist hydrogen sulfide, sulfur is precipitated and sodium thiosulfate simultaneously formed.

HYPOTHESIS.

A tentative assumption, usually based upon some reasonable concept, made in order to generate interest in obtaining proof and to consider the consequences of the assumption.

HYSTERESIS.

(Derived from the Greek word meaning "to lag behind.") In general, the phenomenon exhibited by a system whose state depends

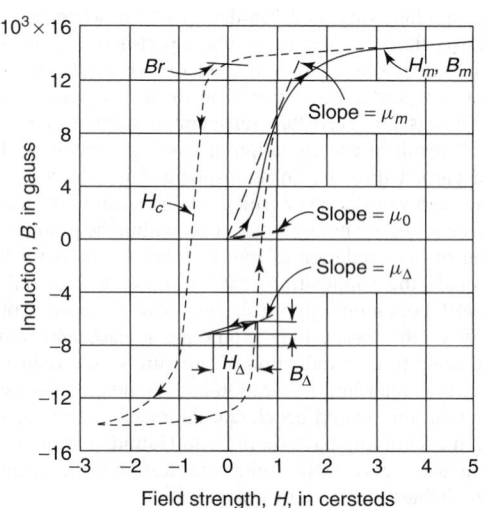

Fig. 1.   Hysteresis loop (dotted). Some important magnetic quantities are shown

on its previous history. This term usually refers to magnetic hysteresis, of importance in alternating-current machinery. When a ferromagnetic material such as iron is placed in a magnetic field, a certain amount of energy is involved in bringing about its magnetization. If the field is a rapidly alternating one, the material may become noticeably warm. It appears that the repeated changes of orientation in whatever it is within the substance that responds to the reversals of field are opposed by something like viscous friction.

A quantitative study of the process indicates that, as the field intensity H increases, the magnetic induction B also increases in a manner characteristic of the substance. This is conveniently represented by a graph, which is called the magnetization curve. See Fig. 1. Its initial slope is the initial permeability (μ_0). If H is carried to some maximum value H_m and then reduced (to $-H_m$), B follows the dotted hysteresis curve. B does not fall off as it was built up (solid line); the residual induction B_r is the induction remaining when H has been reduced to zero; the reverse H needed to reduce B to zero is called the coercive force (H_c). From this point the cycle proceeds to describe the closed curve shown by the dotted lines, which is called the hysteresis loop. The initial portion (solid line) is not retraced. The amount of energy converted into heat is proportional to the area of the cycle.

Electric hysteresis is a somewhat analogous phenomenon exhibited by dielectrics in the electric field and gives rise to heating in capacitors.

Some solids exhibit what is called elastic hysteresis, in which the variables corresponding to H and B in the magnetic case are the stress and the strain or deformation. Elastic bodies such as metals operating at stresses below the proportional limit also undergo hysteresis.

Hysteresis energy is that energy used per cycle of operation to overcome the effect of hysteresis.

I

-IC. A suffix, used in naming inorganic compounds, that indicates that the central element is present in its highest oxidation state. Thus in ferric chloride ($FeCl_3$) the iron has an oxidation number of $+3$, equivalent to its valence: in an ionized state it would have three positive charges (Fe^{3+}). (A recommended change in this system of nomenclature is to use the common name of the element (iron) together with a Roman numeral showing the oxidation number; thus, ferric chloride would be iron (III) chloride.)

ICE. The solid form of water. All commonly occurring forms of ice are crystalline, although large single crystals are relatively rare except in glaciers. At a pressure of one atmosphere, ice melts at $0°C$ by definition of the centigrade temperature scale. On the other hand, ice does not invariably form in liquid water cooled to $0°C$, because of supercooling and the absence of ice nuclei. Ice is found in the atmosphere in such forms as ice crystals, snow, hail, and ice pellets; and on the earth's surface in such forms as hoarfrost, rime, and glaze, and the following:

Anchor ice is ice attached to the bed of streams, lakes, and shallow seas, irrespective of its nature of formation. On clear, cold nights in relatively still water, it may form directly on submerged objects. It also develops in supercooled water of turbulence sufficient to maintain uniform temperature at all depths. When the water temperature increases to above $0°C$, the ice rises to the surface, often carrying with it the object on which it had accumulated. (Cf. ground ice, defined below.)

Droxtal is a tiny ice particle, about 10 to 20 microns in diameter, formed by direct freezing of supercooled water droplets at temperatures below $-30°C$. The term is coined by combining the words "drop" and "crystal."

Fossil ice is ice that was formed in the geologic past, found in regions of permafrost or where present-day temperatures are not low enough to have formed it.

Frazil crystals are ice crystals that form in supercooled water too turbulent to permit coagulation into smooth sheet ice. This is most common in swiftly flowing streams, but is also found in a turbulent sea (cf. lolly ice, defined below). It may accumulate as anchor ice on submerged objects obstructing the water flow.

Glacier ice is any ice that is or was once a part of a glacier. It has been consolidated from firn (i.e., old snow that has become granular and compacted) by further melting and refreezing, and by static pressure. It may be found in the sea as icebergs.

Ground ice is a body of clear ice in frozen ground. It is most commonly found in more-or-less permanently frozen ground, and may be of sufficient age to be termed fossil ice.

Lolly ice is salt water frazil.

Pack ice is ice covering more than half the visible sea surface; no open water whatever is visible in unbroken pack ice, such as that which sometimes covers the central Arctic Ocean.

Sludge is a dense accumulation of frazil or lolly ice; an early stage in the freezing of a body of water. The sea surface becomes thick and soupy and sometimes greasy in appearance. Sludge depth seldom exceeds 1 foot (0.3 meter).

Molecular Forms of Ice

The H_2O molecules in an ice crystal are much further apart than the molecules in liquid water and thus the density of ice is less than that of liquid water, permitting ice to float atop liquid water. In the ice crystal, the molecules are joined by highly directional, obtuse-angled hydrogen bonds, forming a regular hexagonal design. As ice is warmed and passes through the freezing point, the characteristic *rigid* but open structure of the ice crystal gives way, thus allowing H_2O molecules to crowd into the former "open spaces."

Through the application of pressure (2000 atmospheres and higher), water molecules can be forced to assume various deformed patterns

(as compared with ordinary ice). For many years, eight solid forms of water have been so produced, designated ice II through ice IX (water is designated ice I). These high-pressure forms of solid water exist at specific temperature-pressure domains in the extended phase diagram of water. Upon release of the applied high pressure, these ice structures revert to common ice or water depending upon the temperature.

Still another form of solid water was proposed several years ago by Holzapfel (a German physicist). This form, designated ice X, was described by Holzapfel as a very dense crystal structure that would not be made up of well-defined molecules linked by hydrogen bonds. Rather, each oxygen atom would be surrounded by a tight cubic array of nearest-neighbor oxygen atoms, and a hydrogen atom would be located halfway between each pair of oxygen atoms such that a hydrogen atom would be associated no more with one oxygen atom than with the other. It was predicted that this form of solid water would exist at a pressure greater than 350,000 atmospheres. At the time of this prediction, equipment for producing pressure that high was not available.

With the development of the diamond-anvil high-pressure cell, researchers at the Argonne National Laboratory, in a series of experiments, subjected water to high pressures ranging between 300,000 and 670,000 atmospheres. Through the use of Brillouin-scattering spectroscopy, an anomaly was noted at a pressure of 440,000 atmospheres. The researchers tentatively observe that this "is the tenth known solid phase of H_2O and . . . is probably the predicted symmetric ice. As such it would be the first non-molecular structure for H_2O." (In Brillouin-scattering spectroscopy, the compressibility of a sample is ascertained indirectly by measuring the reflection of laser light from highly directional sound waves in the sample.)

See also **Water (Hard)**.

Additional Reading

Fairbridge, R.W., ed.: *Encyclopedia of Geochemistry and Environmental Sciences*, John Wiley & Sons, Inc., New York, NY, 1972.

Hochheimer, H.D. and R.D. Etters: *Frontiers of High Pressure Research*, Kluwer Academic Publishers, Norwell, MA, 1991.

Jayaraman, A.: "Diamond Anvil Cell and High-Pressure Physical Investigations," *Rev. of Modern Physics*, **55**(1), 65–108 (January 1983).

Jayaraman, A.: "The Diamond-Anvil High-Pressure Cell," *Sci. Amer.*, **250**(4), 54–62 (April 1984).

Kukla, G. and J. Gavin: "Summer Ice and Carbon Dioxide," *Science*, **214**, 497–503 (1981).

Lide, D.R.: *CRC Handbook of Chemistry and Physics*, 84th Edition, CRC Press, LLC., Boca Raton, FL, 2003.

Oerlemans, J. and C.J. Van Der Veen: *Ice Sheets and Climate*, Kluwer Academic Publishers, Norwell, MA, 1984.

Peltier, W.R.: *Ice in the Climate System*, Springer-Verlag, Inc., New York, NY, 1994.

ICE POINT. The temperature at which, a mixture of air-saturated pure water and pure ice may exist in equilibrium at a pressure of one standard atmosphere.

-IDE. A suffix, used in naming compounds composed of two elements; in such names the first (electropositive) element retains its name without change, while the second (electronegative) bears the suffix—*ide* as a modification of the elemental name. Examples: sodium hydroxide, magnesium chloride, hydrogen sulfide, etc. Similarly, oxygen is modified to oxide, fluorine to fluoride, phosphorus to phosphide, and carbon to carbide.

IDEAL GAS LAW. An "ideal gas" would, if kept at a constant temperature, behave as respects volume and pressure in strict accord with Boyle's law. If now the temperature is also allowed to vary, we must combine the law of Charles (or of Gay Lussac) with Boyle's law, yielding the Boyle-Charles law:

$$pv = p_0 v_0 (1 + at), \qquad (1)$$

in which $p_0 v_0$ is the value of the pressure-volume product pv when the temperature t is zero, a is the coefficient of expansion of the gas, practically the same for all gases, and in the ideal case equal to the reciprocal of the absolute temperature of the scale zero. If the centigrade scale is used, the value of a is approximately 1/273.2 per degree. Substituting this, Equation (1) may be written

$$pv = \frac{p_0 v_0 (t + 273.2°)}{273.2°} \qquad (2)$$

which is one expression for the ideal gas law.

The factor $t + 273.2°$ will be recognized as the absolute temperature T of the gas. And since the ideal gas obeys Boyle's law, the product $p_0 v_0$ is constant however p_0 and v_0 may vary between themselves. We may thus denote the coefficient $p_0 v_0 / 273.2°$ by a single constant symbol, say R, and the ideal gas equation then takes the usual form

$$pv = RT \qquad (3)$$

The value of R depends, of course, upon the quantity of gas used, since at any pressure p_0 it is proportional to the volume v_0. For 1 gram of air, R equals about 2,868,000 gcm^2/sec^2 deg. At the zero of temperature and at any given pressure p_0, the gram molecular weights, or moles, of all pure gases have equal volumes. (This follows from Avogadro's law.) Hence if one mole of any pure gas is used, R will always have the same value, in c.g.s. units about 8.316×10^7 g cm^2/sec^2 deg; which is called the "ideal gas constant." Many physical formulas involve a quantity which may be regarded as the ideal gas constant per molecule, that is, the above molar gas constant divided by the number of molecules in a mole, 6.025×10^{23}, giving 1.3803×10^{-16} g cm^2/sec^2 deg. This is the "Boltzmann constant."

Since actual gases, even those with the smallest molecules, hydrogen and helium, do not obey the ideal gas law exactly, various empirical characteristic equations have been devised to represent their behavior.

See also **Avogadro Law**; **Boyle-Charles Law**; **Boyle's Law**; **Characteristic Equation**; and **Combustion (Fuels)**.

IDEAL SYSTEM.   A thermodynamic system is called an ideal system when the chemical potentials of all the components are of the form

$$\mu_i = \mu_i(T, p) + RT \ln x_i \qquad (1)$$

where $\mu_i(T, p)$ is a function only of the variables absolute temperature, T, and pressure, p. The x_i are the mole fractions of the components.

Systems for which μ_i has this form possess remarkably simple properties. Moreover, mixtures of perfect gases (i.e., gases under conditions which can be approximated with sufficient accuracy by the ideal gas law) and very dilute solutions have these properties.

According to Equation (2), a system is called ideal if the chemical potential of component i varies linearly with the logarithm of the mole fraction of i, with a slope RT. This linear relation need not necessarily extend over the whole concentration range, so that the quantity $\mu_i(T, p)$ is, in general, the value of μ_i extrapolated to $x_i = 1$ at constant T, p. If the system is ideal in a concentration range which extends to $x_i = 1$, then

$$\mu_i(T, p) = \mu_i^0(T, p) \qquad (2)$$

where μ_i^0 is the chemical potential of the pure component i. They are two important cases: (1) The mixture is ideal for all values of x_i and for all i. It is then called a *perfect mixture* and Equation (2) is verified for all i. (2) The mixture is ideal when all components but one (index 1) are present in very small amount. Such systems are called *ideal dilute solutions*. Then Equation (2) is only valid for component 1.

Different kinds of ideal systems are distinguished by the form of $\mu_i(T, p)$. In a mixture of perfect gases, $\mu_i(T, p)$ varies logarithmically with pressure, while for a liquid or solid solution, one can, to a first approximation, regard μ_i as independent of pressure.

See also **Perfect Gas**.

ILMENITE.   A mineral oxide of iron and titanium, FeTiO$_3$. Magnesium and manganous manganese may replace ferrous iron to form a complete isomorphous series between ilmenite, and its magnesium-manganese end members, geikielite and pyrophanite. It crystallizes in the rhombohedral

division of the hexagonal system; hardness, 5–6; specific gravity, 4.72; brittle, with uneven to conchoidal fracture. Crystals tabular, rarely rhombohedral, also massive, lamellar, granular. Color, iron black; opaque, with metallic to dull luster.

Ilmenite occurs as a common accessory mineral in both igneous and metamorphic rocks, and as heavy concentrations in certain black beach sands with magnetite, rutile, and zircon. Also found in pegmatites and as vein deposits. Valuable deposits are found in Norway; Sweden; Mexico; Finland; Ilmen Mountains, former U.S.S.R.; Canada; England; Brazil; and Italy. Brazil and India are rich in beach sand deposits. United States localities include California, Idaho, Colorado, Wyoming, Arkansas, Kentucky, Pennsylvania, Massachusetts, Connecticut, Orange County and the Adirondack Mountain Deposits in New York, and as beach sands in Florida north of St. Augustine.

Named after the Ilmen Mountains, former U.S.S.R.

IMIDES.   An imide may be defined as a compound that has the divalent radical NH combined with two acid radicals. The definition implies that the acid from which an imide is derived must be a dibasic acid, such as oxalic acid, HOOC · COOH, or succinic acid, HOOC · CH$_2$ · CH$_2$ · COOH. The derivatives of these two acids illustrate the relationship between amides and imides.

COOH	CO·NH$_2$	CO·NH$_2$	CO $\backslash$
			NH
COOH	CO·NH$_2$	CO·OH	CO $\diagup$
(oxalic acid)	(oxamide)	(oxamic acid)	(oximide)

CH$_2$CO·OH	CH$_2$CO·NH$_2$	CH$_2$CO·NH$_2$	CO $\backslash$
			NH
CH$_2$CO·OH	CH$_2$CO·NH$_2$	CH$_2$CO·OH	CO $\diagup$
(succinic acid)	(succinamic)	(succinamic acid)	(succinamide)

Phthalimide, C$_6$H$_4$ · (CO)$_2$ · NH, is an imide of commercial and industrial importance, forming a number of interesting derivatives. With alcoholic potash, phthalimide forms a potassium derivative, C$_6$H$_4$ · (CO)$_2$ · NK, which, when reacted with ethyl iodide (or other alkyl halides), yields ethylphthalimide, C$_6$H$_4$ · (CO)$_2$ · N · C$_2$H$_5$. The latter product, when hydrolyzed with an acid or alkali, further yields ethylamine. Such reaction chains are useful in the preparation of certain primary amines and their derivatives.

IMINES, CYCLIC.   Ethyleneimine (aziridine, azacyclopropane) is the smallest cyclic imine consisting of a three-membered N-heterocyclic ring ($n = 2$):

$$(CH_2)_n \quad NH$$

This article describes ethyleneimine and the most important aziridine derivatives. Unsubstituted ethyleneimine is industrially the most important representative of the aziridine class. The BASF group is by far the largest manufacturer of ethyleneimine and has production plants in Germany and the U.S. Another important producer is the Nippon Shokubai Co. Ltd. of Japan.

Physical Properties

Ethyleneimine (EI) and its two most important derivatives, 2-methylaziridine (propyleneimine), and 1-(2-hydroxyethyl)aziridine (HEA) are colorless liquids. They are miscible in all proportions with water and the majority of organic solvents. Ethyleneimine is not miscible with concentrated aqueous NaOH solutions (>17% by weight). Ethyleneimine has an odor similar to ammonia. The physical properties of ethyleneimine and the derivatives mentioned are given in Table 1.

Chemical Properties

Ethyleneimine is the only C$_2$H$_5$N isomer stable at room temperature provided CO$_2$ is excluded from the air. Unexpectedly, ethyleneimine has the highest calculated relative heat of formation of the C$_2$H$_5$N isomers. Ethyleneimine shows a lower basicity than noncyclic aliphatic amines such as dimethylamine.

Reactions

Depending on the experimental conditions used, the basicity or the ring strain can be the driving force in reactions involving ethyleneimine. With

TABLE 1. PHYSICAL PROPERTIES OF ETHYLENEIMINE AND DERIVATIVES

Property	EI	PI[a]	HEA[b]
solidification point, °C	−74	−65	
boiling point, °C	57	66	156
density, g/mL	0.837[c]	0.8017[d]	1.088
refractive index n_D	1.4130[c]	1.4084[d]	1.453[d]
flashpoint, °C	−13	−10	67
viscosity at 25°C, mPa·s(= cP)	0.418	0.491	
dielectric constant at 25°C	18.3		

[a] Propyleneimine.
[b] Hydroxyethylaziridine.
[c] At 20°C.
[d] At 25°C.

catalysis by Brönsted or Lewis acids, the aziridine ring can be opened by a large number of nucleophiles to give β-substituted ethylamines. In the absence of strong nucleophiles and at elevated temperatures, preparation of polyethyleneimines from aziridines is possible by acid-catalyzed reaction of the aziridine with itself. On the other hand, ethyleneimine and other aziridines substituted only on carbon show the typical reactions of a secondary amine, such as addition onto unsaturated systems, complex formation with metals, and reaction with halogen compounds. At low temperatures and alkaline pH the N-substituted aziridines are generally formed in these reactions. High temperatures and catalysis by acids or nucleophiles promote secondary reactions with opening of the three-membered ring, and these can be used for synthesis of heterocyclic compounds.

Nucleophilic Ring Opening. Opening of the ethyleneimine ring with acid catalysis can generally be accomplished by the formation of an intermediate aziridinium salt, with subsequent nucleophilic substitution on the carbon atom which loses the amino group. In the following, R represents a Lewis acid, usually H^+; A^- = the nucleophile.

Electrophilic Reactions on the Aziridine Nitrogen. The generalized reaction of aziridines with an electrophile (R^+) is as follows.

Reactions with Transition-Metal Compounds. The numerous published products of reactions of transition-metal compounds with aziridines can be divided into complexes in which the aziridine ring is intact, compounds formed by reaction of aziridine with the ligands of a complex, and complexes in which the aziridine molecule is fragmented (imido complexes).

Other reactions include reductive ring opening, oxidative ring opening, thermal and photochemical reactions, and polymerization.

Preparation

The Wenker process has been carried out by BASF and various other companies since the end of the 1960s. In this process the hemisulfate of monoethanolamine, a nonvolatile, crystalline substance, is used for the alkaline cyclization. The reaction can be carried out under pressure.

A production plant for salt-free ethyleneimine synthesis by catalytic dehydration of monoethanolamine in the gas phase has started operation at the Japanese company Nippon Shokubai.

G. SCHERR
U. STEUERLE
R. FIKENTSCHER
BASF AG

Additional Reading

Dermer, O.C. and G.E. Ham: *Ethylenimine and Other Aziridines,* Academic Press, Inc., New York, NY, 1969.
Goethals, E.J.: in K.J. Ivin and T. Saegusa, eds., *Ring-Opening Polymerization,* Vol. 2, Elsevier Applied Science Publishers, New York, NY, 1984.
Horn, D. and F. Linhart: in J.C. Roberts, ed., *Paper Chemistry,* Blackie, London, 1991.
Tomalia, D.A. and G.R. Killat: in J.I. Kroschwitz, ed., *Encyclopedia of Polymer Science and Engineering,* Vol. 1, John Wiley & Sons, Inc., New York, NY, 1985.

IMINO COMPOUNDS. Imino compounds are organic compounds containing the imino group $\diagdown NH \diagup$ e.g., dimethylamine, $(CH_3)_2NH$, dibenzamide, $(C_6H_5CO)_2NH$, succinimide,

$$\begin{array}{c} CH_2 \cdot CO \\ | \quad\quad\quad NH \\ CH_2 \cdot CO \end{array}$$

NH, purrole (C_4H_4NH), and uric acid,

$$\begin{array}{c} NH-CO-C-NH \\ CO \quad\quad\quad \| \quad\quad CO \\ NH-----C-NH \end{array}$$

IMMUNE SYSTEM AND IMMUNOCHEMISTRY. The word *immunity* is derived from the Latin *immunis* (free of). The term originally referred to the ability of the body to resist invasion by pathogenic organisms, but has now been expanded to include specific reactions to antigens (Ags) in general, and to include reactions observed in the emerging field of tumor immunology. See also **Antigen.**

Immunity is derived from the *immune system* which, when functioning properly, protects the organism from infection. Failures of the immune system produce some of the most challenging and serious diseases that a physician can meet in the patient population.

The immune system consists of a number of lymphoid organs, including the thymus, lymph nodes, spleen, and tonsils. It also includes aggregates of lymphoid tissue in nonlymphoid organs, such as Peyer's patches in the intestines and clusters of lymphoid tissue dispersed throughout the connective and epithelial tissues of the body. The immunologically active cells of the immune system comprise the various classes of lymphocytes. A number of cells, however, including monocytes (macrophages) and polymorphonuclear leucocytes play important accessory roles. The stem cells from which the lymphocytes arise are derived from the yolk sac and the fetal liver, later some stem cells originate from the bone marrow and differentiate into lymphocytes in the primary lymphoid organs.

The function of the immune system is the preservation of the body's integrity against antigens recognized by the lymphocytes as foreign, e.g., surface structures of microorganisms, tissue transplants, or a wide variety of chemicals. Specifically, antigens include such structurally diverse substances as proteins, polysaccharides, nucleic acids, and lipids. Large, rigid proteins are the most antigenic, and the more insoluble the foreign material, the more antigenic it appears.

The various antigens are recognized by lymphocytes, which have a memory and specificity, an ability to increase the number of antigen-specific lymphocytes following the antigenic stimulus, and an ability to distinguish between self and nonself. The production of immunoglobulins (Igs—see later) by the immune response is under the control of genes which are located in the same chromosome as, and very close to, another group of genes which control the production of the histocompatibility antigens (HLA). These are antigens that identify as self the cells they are on and differentiate from cells of other individuals. These two groups of genes form the major histocompatibility complex (MHC) which plays a crucial role in the immune system. An intriguing aspect of the genetic control of immunoglobulin synthesis is the diversity of the product; plasma cells can make antibodies that react with more than a million antigens.

As previously indicated, the primary cells involved in the immune response are lymphocytes which have a centrally located round nucleus, lack specific granules, and have a basophilic cytoplasm containing free ribosomes. The (thymus-dependent) T-lymphocytes are involved in cell mediated reactions and also interact with B-lymphocytes (see later) to regulate the production of antibody. The B cells differentiate into the antibody-producing plasma cells. There is growing evidence that neither T nor B cells constitute a homogeneous population, but actually consist of a number of subgroups which can be differentiated from each other by their surface markers and by their function.

Thymus-dependent antigens are those in which antibody production requires thymus-derived (T) cell participation, i.e., serum proteins. Thymus-independent antigens do not require this participation, i.e., polysaccharides, such as endotoxins.

Antigen is any substance capable of generating an immune response that is reacting with T and B cells to induce the formation of antibodies and sensitized lymphocytes and then reacting with those antibodies and cells once they are formed. The basis for the general immunogenicity of proteins is not known, but is probably related to their unique and stable configuration. Antigens invoke immune responses by the host which include the production of *antibodies* (Abs) possessing a specificity for the antigen which is determined by the latter's structure. See also **Antibody.** Antibodies belong to a group of *immunoglobulins* (Igs) which bind with the antigens to form complexes in which the two components are held together by weak hydrogen bonds, van der Waals forces, or ionic bonds, but not by covalent bonding. An antibody that binds a given antigen will also bind antigens having similar structural configurations. This is referred to as *cross-reactivity*. The extent to which this binding occurs indicates the measure of similarity of the two antigens. Most antigens have several antigenic determinants or Ab binding sites and the antibody response to any antigen is thus the sum of responses to each individual determinant.

In addition to antigens per se, two other types of substance are recognized by the immune system: (1) *haptens*, which are molecules capable of reacting with antibodies, but which are unable to stimulate their production unless coupled to a carrier—an immunogenic substance that is usually a protein or a synthetic polypeptide; and (2) *adjuvants*, which enhance the immune response to an antigen. These include, but are not limited to, aluminum salts, bacterial endotoxins, *Bacillus Calmette-GuÉrin* (BCG), *Bordetella pertussis*, and mycobacteria.

The immune response to antigens can proceed through several paths, the two major ones being *humoral* and *cellular*. Most immune responses involve both pathways. Humoral immunity is mediated by antigen-specific antibodies, which circulate through the body; they act at a site distant from that of their production; it can be transferred from one person to another by serum transfusion. Cell-mediated immunity is mediated by specifically sensitized cells, which release mediators in the vicinity of the antigen. This form of immunity can be transferred from one person to another by cell transfer. Cell-mediated immune responses usually take longer to develop and are responsible for resistance to many infectious agents and tumors as well as for some drug allergies, rejection of foreign organ grafts, and some autoimmune diseases.

The level of immune response to antigens varies. A *primary response* is seen to antigens the body has never before encountered. In this, the first antibody to be produced is IgM (described later), the serum concentration of which peaks after a lag of several days and then decreases. IgG production shows a longer lag time, but its concentration remains elevated for a more extended period. *Secondary responses* are more rapid and of greater intensity; they differ markedly from primary responses in having a shorter lag period, higher serum antibody levels with earlier and more pronounced emphasis upon IgG production. These secondary responses occur when the immune system faces a previously encountered antigen and illustrates the basic characteristic of immune response—*memory*. For this reason, secondary responses are also known as *anamnestic*. These memorized responses are derived from two of the major cell types which work together to produce immune responses—T and B lymphocytes. The former are thymus-dependent, are responsible for cell-mediated immune responses and for providing help for most antibody responses. B cells depend upon another central lymphoid organ (in birds, this has been identified as the Bursa of Fabricus). There does not appear to be a bursa equivalent in mammals, however, even though it was originally thought that such animals had some gut-associated primary lymphoid organ. Current evidence suggests that stem cells can differentiate into B cells in the bone marrow and in the peripheral lymphoid organs themselves. B lymphocytes have immunoglobulin on their surface and it can be shown that the cell itself has produced this. When stimulated by antigens, B cells become *plasma cells* and produce antibodies specific for that antigen. T and B cells cannot be distinguished morphologically, but may be differentiated, such as by theta antigen on T cells and surface immunoglobulin on B cells. Unlike T cells, B cells tend to migrate.

The spleen, lymph nodes, tonsils, and gut-associated lymphoid tissue comprise the secondary or peripheral system wherein T and B lymphocytes undergo terminal differentiation in response to antigen stimulation.

The third major type of cell involved in the immune process is the *macrophage*. The B and T lymphocytes are rarely phagocytic—this is the function of the macrophage. After specific recognition of the invading antigen by the lymphocytes, the macrophage acts nonspecifically, and it moves by *chemotaxis* (movement along a chemical concentration of increasing gradient) to the site of the immune response. The macrophage ingests and eventually digests the antigenic inert particle or the living or dead microorganism responsible for engendering the immune response.

These major cell types may be subdivided into populations that interact by means of soluble mediators called *lymphokines* or *monokines*. Lymphokines are products of activated lymphocytes that exert regulatory effects upon other cells of the immune system. Monokines are products of activated microphages. The lymphokines or lymphocyte mediators are soluble substances produced by lymphocytes that help to amplify and regulate a variety of immune responses. They are not immunoglobulins. There are clearly many different lymphokines and they are classified on the basis of the target cells they affect. It is likely that they have more than one biological function. Lymphokines are generally synthesized and secreted by sensitized lymphocytes stimulated by a specific antigen; they are not stored in a preformed state. Most are proteins or glycoproteins, and while it has been shown that protein synthesis inhibitors prevent their formation, their exact structures are undetermined. Some play a role in humoral reactions and others in cell-mediated responses. The soluble mediators allow such activities as help, suppression, or cytotoxicity to be manifested by the target cells.

A given antigen induces proliferation and differentiation of clones of cells capable of producing antibodies in response to that antigen; a process called *clonal selection*. Each stimulated cell produces antibodies of only one specificity. Thus, the immense heterogenicity of antibodies to a given antigen results from a great diversity of responsive cells.

As previously indicated, antibodies are immunoglobulins (Igs), produced by B-cell-derived plasma cells during a humoral response to antigen. They are synthesized on polyribosomes attached to the rough endoplasmic reticulum of plasma cells upon inoculation of antigen into the host, and they are specific for that antigen. They have two functions—specific recognition of the antigen and effector functions, such as agglutination or lysis of bacteria, complement fixation, or opsonization.

Immunoglobulins are high-molecular-weight glycoproteins having symmetrical four-polypeptide chain structures composed of two heavy and two light chains held in configuration by disulfide linkages. Based upon serological characteristics of the heavy chains, five distinct classes or isotypes have been found—IgG, IgA, IgM, IgD, and IgE. Subclasses of these isotypes have also been found. Bonds between chains—heavy-heavy or heavy-light—are by disulfide bridges with the number and positions of these being characteristic of different classes and subclasses of immunoglobulins. See Fig. 1. In a given immunoglobulin molecule, all of the heavy chains

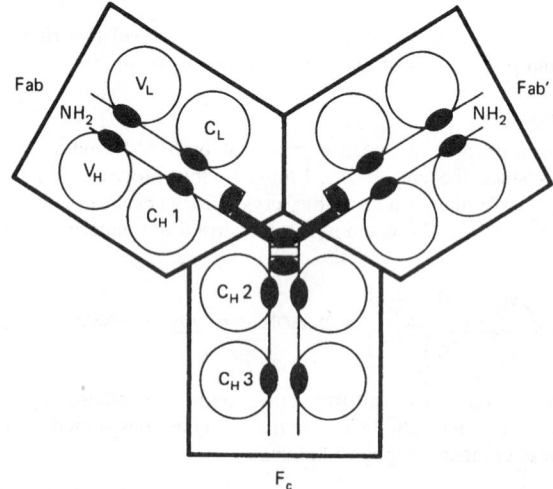

Fig. 1.   Hypothetical model of human IgG along the lines proposed by Poljak and others. Note heavy solid bars, which indicate hinge region (at center of configuration). Solid ellipses indicate S—S linkages

are identical and, although there are two types of light chains—kappa and lambda—each of which may be associated with any heavy chain, both of the light chains in a given Ig are identical. Polymerization of the basic Ig configuration can occur, but IgG, IgD, and IgE occur only as monomers. IgA may be either monomeric or dimeric; IgM is pentameric. See Fig. 2.

In addition to the basic configurations of the immunoglobulin molecule, the light and heavy chains have variable and constant regions where some variability in amino acid sequences occurs. The variable regions of the antibody molecule contain the structures responsible for the antigenic specificity of the Ig. These are found in the amino terminal of the heavy and light polypeptide chains—the Fab portion obtained as a cleavage fragment following treatment with pepsin or papain. Since each antigen-binding region is composed of the variable region of one light and one heavy chain, an immunoglobulin molecule has two identical antigen binding sites. The antigen-binding site of every Ig molecule is structurally unique. Therefore, each Ig has unique antigenic determinants not shared by any other Ig molecule. These unique determinants are called the *idiotypes* of the particular immunoglobulin. On the other hand, an *allotype* is an antigenic specificity representing any one of a number of possible allelic markers present on a significant percentage of Ig molecules in an individual which are recognizable as foreign by another individual lacking that allotype. Allotypic determinants distinguish between the Igs of a particular isotype and are genetically determined by Mendelian laws in a manner similar to those that determine the ABO blood group. *Isotypic* determinants are found in the sera of normal individuals and differentiate the various classes and subclasses of heavy and light chains.

The F_c portion of the immunoglobulin molecule identified by papain cleavage is responsible for biological activity other than antigen binding, e.g., complement fixation, transplacental transfer, binding to cells such as macrophages and granulocytes, and the rate of synthesis and catabolism of the Ig molecule.

IgG and IgM are present in the highest concentrations throughout the body and compose the major portion of a systemic humoral response to antigenic challenge. IgM exists as a pentamer of five immunoglobulin molecules bound together by a protein molecule called a J chain. IgM has a molecular weight of approximately 900,000 and a sedimentation coefficient of 19S, and is thus referred to as 19S Ig. It is the main immunoglobulin of early humoral response, will fix complement and is not cytophilic. IgG is the major component of secondary humoral response, and its ability to diffuse through tissues makes it indispensable for host defense. It has a molecular weight of 150,000 and a sedimentation coefficient of 7S. Eighty percent of serum immunoglobulin is IgG and several subclasses of this immunoglobulin exist which have different functions, i.e., complement fixation, binding to macrophages or passage across the placenta.

IgA is the primary secretory immunoglobulin and occurs in tears, nasal and intestinal secretions, saliva, and bile. It is present as a 7S IgA, it is an 11S dimer with the monomeric structures joined by a J chain and linked to a glycoprotein called secretory component. Secretory IgA plays a role in initial protection against external pathogens, aggregates bacteria, and neutralizes viruses.

IgE, with a molecular weight of 190,000, has a short half-life and occurs at low concentrations in the serum. It is cytophilic and binds strongly to most cells and basophils. Plasma cells producing IgE are primarily found in gastrointestinal mucosa and along the respiratory tract. This immunoglobulin is responsible for immediate *hypersensitivity reactions*, since crosslinking by antigens of two IgE molecules causes the release of mediators such as histamine, SRS-A (Slow Reacting Substance of Anaphylaxis), and ECF-A (Eosinophil Chemotactic Factor) which are responsible for the immediate hypersensitivity reactions. Hence, IgE plays a definitive role in allergies and may have a part in providing resistance to parasites and in protection of mucosal surfaces.

IgD occurs in extremely low concentrations in serum and has a short serum half-life. It has a molecular weight of 170,000–200,000, is monomeric, and fixes complement. Otherwise its role is unknown; it may act as a cell surface receptor on B lymphocytes.

Associated with humoral responses involving antibodies is *complement*. The *complement system* is a complex system of 18 plasma proteins circulating in inactive form in the extracellular liquid. It is the principal humoral effector of immunologically induced inflammation. It plays a crucial role, both in immunologically induced and nonspecific resistance to infection and in the pathogenesis of tissue injury. The products of complement activity regulate a number of biological events including the release of mediators from mast cells, which increases vascular permeability. Activation of complement involves a series of steps, each one generating a new active component that, in turn, activates the next component and so on. There are two major pathways of complement activation—the classical pathway and the alternate or properdin pathway. The classical pathway is initiated by the binding of antigen to antibody (two molecules of IgG or one molecule of IgM). The alternate pathway does not require the presence of antigen-specific antibody, but can be activated by endotoxin, bacterial cell-wall polysaccharides, and aggregated immunoglobulins. The ability to be initiated before an immune response can occur makes the alternate pathway a first line of defense against microorganisms. The ultimate result of complement activation is lysis of the invading cell. Mediators produced during the activation sequence play major roles by allowing *immunoadherence, opsonization*, or *chemotaxis* to occur, but others, such as *anaphylatoxins* may be harmful to the host. The importance of complement to normal host defense is illustrated by many pathologic conditions and high susceptibility to infections seen in persons with congenital complement deficiencies.

ANN C. DEBALDO
College of Public Health, University of South Florida
Tampa, Florida

Additional Reading

Abbas, A.K. and A.H. Lichtman: *Basic Immunology: The Function of the Immune System,* W.B. Saunders Company, Philadelphia, PA, 2001.

Angell, M.: "A Dual Approach to the AIDS Epidemic," *N. Eng. J. Med.*, 1498 (May 23, 1991).

Austin, K.F., S.J. Burakoff, and T.B. Storm: *Therapeutic Immunology,* 2nd Edition, Blackwell Science, Inc., Malden, MA, 2000.

Balter, M.: "East Europe: A Chance to Stop HIV," *Science*, 1964 (December 24, 1993).

Bayer, R.: "Public Health Policy and the AIDS Epidemic," *N. Eng. J. Med.*, 1500 (May 23, 1991).

Chapel, H. and M. Haeney: *Essentials of Clinical Immunology,* Blackwell Science, Inc., Malden, MA, 1993.

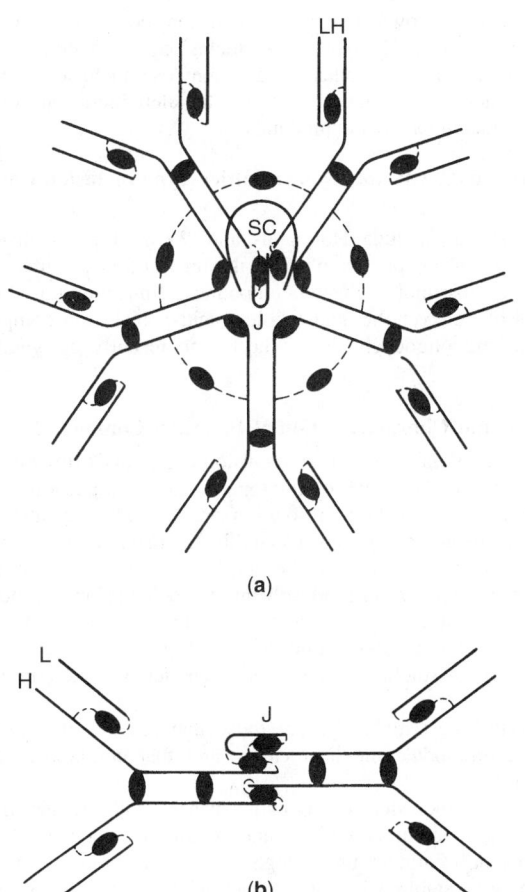

Fig. 2. Hypothetical models utilizing the clasp configuration proposed by Koshland, et al. (1975). (a) Pentameric IgM; (b) dimeric IgA. L = light chain; H = heavy chain; J = J chain. Solid ellipses indicate S—S linkages

Coe-Clough, R. and J.A. Roth: *Understanding Immunology,* Mosby-Year Book, Inc., St. Louis, MO, 1998.

Cohen, J.: "Aids Vaccine Research," *Science,* 1820 (December 17, 1993).

Cohen, J.: "T Cell Shift: Key to AIDS Therapy?" *Science,* 175 (October 8, 1993).

Cohen, I. and L.A. Segal: *Design Principles for the Immune System and Other Distributed Autonomous Systems,* Oxford University Press, Inc., New York, NY, 2001.

DasGupta, D.: *Artificial Immune Systems and Their Applications,* Springer-Verlag, Inc., New York, NY, 1998.

Descotes, J. and C. Bernard: *An Introduction to Immunotoxicology,* Taylor & Francis, Inc., Philadelphia, PA, 1999.

Diamond, J.: "The Mysterious Origin of AIDS," *Natural History,* 24 (September 1992).

Fauci, A.S.: "Multifactorial Nature of Human Immunodeficiency Virus Disease: Implications for Therapy," *Science,* 1011 (November 12, 1993).

Frank, M.M. and J.E. Volanakis: *The Human Complement System in Health Disease,* Marvel Dekker, Inc., New York, NY, 1998.

Greene, W.C.: "AIDS and the Immune System," *Sci. Amer.,* 98 (September 1993).

Janeway, C.A., Jr.: "How the Immune System Recognizes Invaders," *Sci. Amer.,* 772 (September 1993).

Jenkins, M.K.: "The Role of Cell Division in the Induction of Clonal Anergy," *Immunology Today,* 69 (February 1992).

Kuby, J.: *Immunology,* 3rd Edition, W.H. Freeman Company, New York, NY, 1999.

Leffell, M.S., M.R. Rose, and A.D. Donnenberg: *Handbook of Human Immunology,* CRC Press, LLC., Boca Raton, FL, 1997.

Lichtenstein, L.M.: "Allergy and the Immune System," *Sci. Amer.,* 116 (September 1993).

Marrack, P. and J.W. Kappler: "How the Immune System Recognizes the Body," *Sci. Amer.,* 80 (September 1993).

Marx, J.: "Cell Communication Failure Leads to Immune Disorder," *Science,* 896 (February 12, 1993).

Nossal, G.J.V.: "Life, Death and the Immune System," *Sci. Amer.,* 52 (September 1993).

Passwater, R.A. and S. Davis: *Beta-Carotene and Other Carotenoids: The Antioxidant Family That Protects against Cancer and Heart Disease and Strengthens the Immune System,* Keats Publishing, Inc., Chicago, IL, 1999.

Patterson, P.H., C. Kordon and Y. Christen: *Neuro-Immune Interactions in Neurologic and Psychiatric Disorders,* Springer-Verlag, Inc., New York, NY, 1999.

Paul, W.E.: "Infectious Diseases and the Immune System," *Sci. Amer.,* 90 (September 1993).

Paul, W. (Editor): *Fundamental Immunology,* Lippincott-Raven Publishers, Philadelphia, PA, 1998.

Roitt, I.M.: *Roitt's Essential Immunology,* Blackwell Science, Inc. Malden, MA, 1997.

Rothwell, N.J.: *Immune Responses in the Nervous System,* Oxford University Press, Inc., New York, NY, 1998.

Schwartz, R.H.: "A Cell Culture Model for T Lymphocyte Clonal Anergy," *Science,* 1349 (June 15, 1990).

Schwartz, R.H.: "Costimulation of T Lymphocytes: The Role of CD28, CTLA-4 and B7/BBi in Interleukin-2 Production and Immunotherapy," *Cell,* 1065 (December 24, 1992).

Schwartz, R.H.: *Immunologic Tolerance,* in Fundamental Immunology, 3rd Edition, Raven Press, New York, NY, 1993.

Schwartz, R.H.: "T Cell Anergy," *Sci. Amer.,* 62 (August 1993).

Sheehan, C.: *Clinical Immunology: Principles and Laboratory Diagnosis,* 2nd Edition, Lippincott Williams & Wilkins, Philadelphia, PA, 1997.

Staff: *Signaling and Gene Expression in the Immune System,* Vol. 64, Cold Spring Harbor Laboratory, Cold Spring, NY, 2000.

Steinman, L.: "Autoimmune Disease," *Sci. Amer.,* 106 (September 1993).

Stine, G.: *AIDS: Update 2000,* Prentice Hall, Inc., Upper Saddle River, NJ, 1999.

Storad, C.: *Inside AIDS: HIV Attacks the Immune System,* The Lerner Publishing Group, North Minneapolis, MN, 1998.

Toufu, Z., et al.: "Genotypic and Phenotypic Characterization of HIV-1 in Patients with Primary Infection," *Science,* 1179 (August 27, 1993).

Weissan, I.L. and M.D. Cooper: "How the Immune System Develops," *Sci. Amer.,* 64 (September 1993).

Wigzell, H.: "The Immune System as a Therapeutic Agent," *Sci. Amer.,* 126 (September 1993).

INCINERATION. Disposal of solid and liquid organic waste materials by burning at temperatures 1200 to 1500°C. This method is approved by the EPA for use on very toxic organic chemicals and chemical wastes. Use of specially equipped incinerator ships for burning chemical wastes at sea has become common place.

See also **Wastes and Pollution**.

INCLUSION COMPOUNDS. Notwithstanding the immense number and great variety of inclusion compounds, all of them may be classified into three main categories being either a complex, a cavitate, or a clathrate

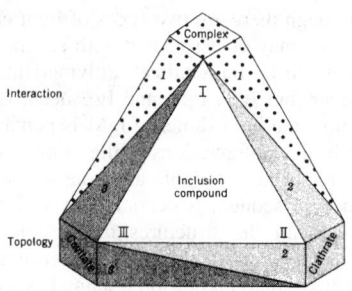

Fig. 1.   Classification/nomenclature of host–guest type inclusion compounds, definitions and relations: (*1*) coordinative interaction, (*2*) lattice barrier interaction, (*3*) monomolecular shielding interaction; (I) coordination-type inclusion compound (inclusion complex), (II) lattice-type inclusion compound (multimolecular/extramolecular inclusion compound, clathrate), (III) cavitate-type inclusion compound (monomolecular/intramolecular inclusion compound)

according to the criteria given in Figure 1. Typical examples for each class of inclusion compounds are the crown complexes, the calix-cavitates, and the hydroquinone clathrates, but in many of the recently known inclusion situations there are borderline cases treated as complex-clathrate hybrids (coordinatatoclathrates or clathratocomplexes depending on the dominant inclusion character.) By way of contrast, the description addition compound (adduct) may be used to the best advantage if a cavity does not exist either at the host molecule or in the lattice buildup. Inclusion compound, therefore, is the generic term of choice which refers to the presence of any not precisely defined cavity. In a more detailed topological characterization, there are two-dimensional open intercalates (layer- or sandwich-type inclusions), one-dimensional open channel inclusions (tubulates), and totally enclosed cage inclusions (cryptates).

Intramolecular Cavity Inclusions: Cavitates

Cavitates include crown macroring inclusion compounds (coronates), cryptates, podates, cyclophane host inclusion compounds, calixarene inclusion compounds, cyclodextrin and amylose inclusion compounds, cucurbituril inclusion compounds, molecular cleft inclusion compounds, and anionic guest inclusion compounds.

Extramolecular Cavity Inclusions: Lattice-Type Inclusion Compounds Clathrates

These compounds include Hofmann- and Werner-type inclusion compounds, inclusion compounds of urea, thiourea and selenourea, inclusion compounds of gossypol, inclusion compounds of phenolic hosts, inclusion compounds of deoxycholic acid (choleic acids), inclusion compounds of macrocyclic and oligocyclic lattice hosts and recently designed organic host lattices.

Preparation and Characterization of Inclusion Compounds

There are several ways to prepare inclusion compounds. In solution, they may simply be formed by dissolving together host and guest in a common solvent. Inclusion formation in solution applies only for intramolecular cavity inclusions and complexes. Crystalline inclusion compounds may be prepared by crystallization from the guest solvent or by cocrystallization of host and guest from an inert solvent. Solid inclusion compounds are also formed by direct exposure of the host to the vapor or liquid guest or, sometimes, by grinding solid host and guest together. Moreover, replacement of an included guest has been demonstrated in particular cases.

Appropriate guest molecules are those that have a suitable size and shape to accommodate the host cavity and that complement the host cavity chemically.

Stabilities of inclusion compounds span a wide range. Some are very stable at ambient conditions and require heating to considerable temperatures or treatment under high vacuum to cause decomposition. Others are only stable when in contact with mother liquor or excess guest solvent from which the inclusion compound was grown. A simple yet informative way for estimation of inclusion stabilities is to relate the decomposition point of the inclusion compound to the usual boiling point of the respective guest liquid.

Uses

Inclusion compounds open up a wide area of applications. An important aspect in this connection is the specific microenvironment created by the host enclosure of the guest which exerts an influence on the physical spectroscopic, chemical, and other properties of the guest.

Retardation and Control. This influence may manifest itself in a reduced volatility, and therefore, lower possible storage and handling problems of a compound when included; toxic and hazardous substances become safer.

Shielding and Stabilization. Inclusion compounds may be used as sources and reservoirs of unstable species. The inner phases of inclusion compounds uniquely constrain guest movements, provide a medium for reactions, and shelter molecules that self-destruct in the bulk phase or transform and react under atmospheric conditions.

Solubilization and Activation. Compounds included in a host take solubility properties of the host shell, and thus, become more soluble when trapped in polar or apolar media, depending on the nature of the host. This leads to important uses in chemical synthesis known as the phase-transfer principle.

Organized Media Effects. Another general reason for using host-guest inclusion chemistry in synthesis is controlled selectivity and artificial enzyme mimicry.

Sensing. Crown compounds modified by responsible chromogenic groups (chromoionophores) proved valuable tools for measuring metal ions and even enantiomeric guest concentrations in solution. Ion selective electrodes based on crown compounds and podands as the sensitive component have broad analytical applications from industrial wastewater control to clinical bedside monitoring of blood.

EDWIN WEBER
Technische Universität Bergakademie Freiberg

Additional Reading

Atwood, J.L., and J.E. Davies: *Inclusion Phenomena in Inorganic, Organic and Organometallic Hosts,* Kluwer Academic Publishers, Norwell, MA, 2002.

Atwood, J.L., J.E.D. Davies, and D.D. MacNicol, eds.: *Inclusion Compounds,* Vols. 1–3, Academic Press, Inc., London, 1984; Vols. 4–5, Oxford University Press, Oxford, UK, 1991.

Lehn, J.-M.: *Supramolecular Chemistry—Concepts and Perspectives,* VCH Verlagsgesellschaft, Weinheim, Germany, 1995.

Vögtle, F.: *Supramolecular Chemistry—An Introduction,* John Wiley & Sons, Ltd., Chichester, UK, 1991.

Weber, E. ed.: *Molecular Inclusion and Molecular Recognition—Clathrates I and II,* Springer, Berlin-Heidelberg, 1987 and 1988.

INDICATOR (Chemical).

A substance which shows by a color change, or other visible manifestation, some change in, or particular condition of, the chemical nature of a system. Thus acid-base indicators may be used to indicate the end point of a particular neutralization reaction, or they may also be used to indicate the pH value of a system. For example, there are over fifty useful indicators for determining pH, covering the range from 0 to 14. Although indicators still are used in connection with colorimetric pH determinations and are extensively applied, sometimes in the form of dye-impregnated paper tape, for ascertaining the approximate pH of soils, swimming pools, and fish tanks where convenience and cost are predominating factors, indicators for pH and other chemical determinations are not nearly so important as they were before the advent of improved electrometric instrumental analytical techniques. Several pH (hydrogen ion indicators) are listed in Table 1.

Indicators also are useful in following oxidation-reduction reactions, precipitation reactions, and, in general, throughout all volumetric analysis, and in many other chemical control operations.

INDICATOR, pH. See **pH (Hydrogen Ion Concentration)**.

INDIUM.

[CAS: 7440-74-6]. Chemical element symbol In, at. no. 49, at. wt. 114.82, periodic table group 13, mp 156.6°C, bp 2,078–2,082°C, density 7.31 g/cm^3 (20°C). Elemental indium has a face-centered tetragonal crystal structure. Indium is a silver-white metal, softer than lead, malleable, ductile, and crystalline. It is stable in dry air, but upon heating in air burns with a blue flame to form indium trioxide In_2O_3. Up to a temperature of 100°C, the element does not decompose H_2O. Indium becomes a superconductor at 3.37 K. The element dissolves in HCl, H_2SO_4, or HNO_3, but not in $NaOH$. Metallic indium combines readily with chlorine and sulfur. 113In is the only nonradioactive isotope and is in isotopic abundance of 4.23%. ^{115}In, with an extremely long half-life of 6×10^{14} years, accounts for the other 95.77% of naturally-occurring indium. Other radioactive isotopes include ^{107}In through ^{112}In, ^{114}In, ^{116}In, and ^{117}In. The half-lives of these isotopes are relatively short, measured in minutes, hours, or days. First ionization potential 5.785 eV; second, 18.86 eV; third, 28.03 eV. Oxidation potentials In → $In^{3+} + 3e^-$, −0.34 V; In → $In^+ + e^-$, −0.25 V.

Other important physical properties of indium are given under **Chemical Elements**.

Indium occurs in very small amounts in zinc blende, tungsten, tin and iron ores of certain localities. The recovery of indium from zinc flue dust (sometimes, 1 part per thousand) is effected by treating with a slight deficiency of HCl and allowing to stand. The residue is subjected to a series of treatments until finally pure indium sulfate is obtained, a solution of which when electrolyzed yields compact indium metal. A thin surface layer of indium is used on some bearings.

As the result of spectroscopic studies, the element was discovered by Reich and Richter in 1863.

On the scale of nonferrous metals, the production of indium is very limited, annual production probably not exceeding 1.3 million troy ounces (40.4 million grams). The availability of the element is affected by zinc production because it is a minor coproduct in the refining of zinc ores. The metal, in the form of an electroplate over lead and silver, has been used in aircraft bearings, the primary benefit being improved corrosion resistance. Indium also has been used as a dopant for germanium diodes and transistors. Several significant semiconductor compounds have been formulated, including InAs, InSb, and InP. The oxide has been used in electroluminescent panels. Indium alloys readily with several metals and it has been found particularly effective as a low-melting point fusible alloy when alloyed with bismuth, lead, tin, and cadmium. The eutectic alloy of indium-tin is an effective solder for glass-to-glass or glass-to-metal seals. With a melting range of 700–800°C, copper-gold-indium and copper-silver-indium alloys are used as brazing materials. The eutectic alloy of mercury, thallium, and indium has a solidifying temperature of −63°C, considerably below the mp of mercury, a feature that makes the alloy attractive for seals, switches, and thermometers for low-temperature applications. Control rods for nuclear reactors sometimes are produced from an alloy of silver, indium, and cadmium. Indium is also used in the manufacture of low-pressure sodium lamps. Indium for electroplating generally is furnished as the normal sulfate $In_2(SO_4)_3 \cdot 9H_2O$, the acid salt $In_2(SO_4)_3 \cdot H_2SO_4 \cdot 7H_2O$, or the basic salt $In_2O(SO_4)_2 \cdot 6H_2O$.

Chemistry and Compounds

Since indium has only three electrons in its valence shell, it is an electron acceptor. Indium trihalides include the trifluoride, trichloride, tribromide, and triiodide. They can be prepared by heating the metal or oxide in the halogen acid, or in the case of the trichloride and

TABLE 1. pH RANGES AND COLOR CHANGES OF SELECTED INDICATORS

Indicator	pH range of color change	Color change with Increasing pH
Alpha naphtholbenzein	0–0.8	Colorless to yellow
Methyl violet	0.2–1.9	Yellow to blue-violet
Para methyl red	1.0–3.0	Red to yellow
Thymolsulfonphthalein (Thymol Blue)	1.2–2.8	Red to yellow
	8.0–9.6	Yellow to blue
Methyl orange	3.3–4.5	Red to yellow
Methyl red	4.2–6.2	Red to yellow
Aurin (rosolic acid)	6.2–7.2	Amber to pink
Phenolsulfonphthalein (Phenol Red)	6.8–8.5	Yellow to red
Phenolphthalein	8.3–10.2	Colorless to purple
Thymolphthalein	9.4–10.7	Colorless to blue
Sodium nitrobenzeneazo-salicylate (Alizarin Yellow R)	10.1–12.0	Yellow to red
Malachite green	11.4–13.0	Blue-green to colorless
1,3,5-Trinitrobenzene	12.0–14.0	Colorless to orange

tribromide, by use of the halogen acid, or in the case of the trichloride and tribromide, by use of the halogen itself. Indium sesquisulfate forms double salts like the alums with alkali metal sulfates. A monohydrogen sulfate, $HIn(SO_4)_2 3\frac{1}{2}H_2O$, is known. Other compounds of indium(III) include the oxide (and its gelatinous hydrate), nitride, the nitrate, and the sulfide, selenide, and telluride.

Indium(II) compounds include the oxide, sulfide, fluoride, and chloride. They are prepared either by reduction of the corresponding trivalent compounds or, in the case of the chloride, by heating the metal in hydrogen chloride. They disproportionate, under suitable conditions, to give as end products the metal and the stable trivalent compound. Like gallium, indium(II) chloride is diamagnetic, having the structure $In^+[InCl_4]^-$.

Indium(I) compounds are formed by reduction of the corresponding In(III) compounds with hydrogen (on heating) or with indium metal, as in the case of the chloride. They are reactive compounds, the chloride disproportionating with water to form the metal and $InCl_3$; the oxide being oxidized on heating in air to the sesquioxide; and the sulfide reacting in dilute acids to form the sesquisulfide.

A number of indium trialkyls have been prepared, starting from the trimethyl, and some diaryl compounds are known, such as the diphenyl bromide. The lower trialkyls are tetramers. Like aluminum, indium forms a polymeric hydride, $(InH_3)_n$, from which tetrahydroindates, such as the lithium compound, $LiInH_4$, can be derived.

Like aluminum and gallium, indium forms a number of chelated oxycompounds, almost all of which are of 6-coordinate type. They include the stable crystalline inner complexes of which the β-diketones coordinate in the proportion of 3 molecules of diketone per atom of indium. Trioxalato as well as dioxalato salts are known, and compounds such as 8-quinolinol and substituted 8-quinolinols form trimolecular chelate rings involving nitrogen and donor oxygen.

Additional Reading

Davis, J.R.: *Metals Handbook,* 2nd Edition, ASM International, Materials Park, OH, 1998.
Greenwood, N.N. and A. Earnshaw: *Chemistry of the Elements,* 2nd Edition, Butterworth-Heinemann, Inc., Woburn, MA, 1997.
Krebs, R.E.: *The History and Use of Our Earth's Chemical Elements: A Reference Guide,* Greenwood Publishing Group, Inc., Westport, CT, 1998.
Lide, D.R.: *CRC Handbook of Chemistry and Physics,* 84th Edition, CRC Press, LLC., Boca Raton, FL, 2003.

Web Reference

Indium Corporation of America: http://www.indium.com/

INDOLES.

Indole is a heteroaromatic compound consisting of a fused benzene and pyrrole ring, specifically benzo[*b*]pyrrole. The indole ring is incorporated into the structure of the amino acid tryptophan and occurs in proteins and in a wide variety of plant and animal metabolites.

Properties

Indole is a colorless solid, mp 53–53°C, which is easily soluble in most organic solvents but sparingly soluble in water. Indole has a musty odor which is very persistent and its derivatives have some applications in the formulation of fragrances.

The industrial source of indole has been isolation from coal-tar distillate. Several patents for the manufacture of indole have been issued with aniline and ethylene glycol, aniline and ethylene oxide, 2-ethylaniline, and *N*-ethylaniline as the starting materials.

Reactivity

Indole is a heterocyclic analogue of naphthalene. The basic reactivity patterns of indole can be understood as resulting from the fusion of an electron-rich protein pyrrole ring with a benzene ring.

Reactions include electrophilic aromatic substitution (e.g., halogenation, nitration, *C*-acylation, and alkylation), *N*-alkylation, arylation, lithiation and subsequent transformations, and oxidation.

Syntheses

Although there are a wide variety of indole ring syntheses, most of the more useful examples fall within a small number of groups. Indole syntheses usually start with an aromatic compound, either monosubstituted or ortho-disubstituted.

Processes include the Fischer indole synthesis from arylhydrazones and related sigmatropic syntheses, reductive cyclizations of nitro compounds, the Madelung synthesis from anilides and related base-catalyzed condensations, and transition-metal catalyzed cyclizations.

Biologically Active Indole Derivatives

Synthetic Derivatives of Indoles as Pharmaceuticals. Thousands of indole derivatives have been prepared and evaluated as potential pharmaceuticals. Of those which have been put into use perhaps the most important are the nonsteroidal antiinflammatory agent, indomethacin, and the β-adrenergic blocker, pindolol.

Naturally Occurring Compounds. Many derivatives of indole are found in plants and animals where they are derived from the amino acid tryptophan. Several of these have important biological function or activity. Serotonin functions as a neurotransmitter and vasoconstrictor. Melatonin production is controlled by the circadian cycle and its physiological level influences daily and seasonal rhythms in humans and other species. Indole-3-acetic acid is a plant growth stimulant used in several horticultural applications.

The largest single class of naturally occurring indoles are the plant alkaloids. These occur with a wide range of structural diversity and are typically derived from tryptophan and terpenoid structural units. Several of these compounds are pharmacologically significant. Reserpine acts as a tranquilizer and hypotensive agent. The dimeric vinca alkaloids, vincristine and vinblastine, are used in the treatment of Hodgkin's disease, leukemia, and other forms of cancer. Derivatives of the ergot alkaloid lysergic acid are used in the treatment of migraine and the diethylamide is lysergic acid diethylamide.

Toxic Indole Derivatives. There are several documented cases where indole derivatives, both natural and of synthetic origin, have been linked to pathological effects in humans. 3-Methylindole, which is produced by bacterial fermentation in cattle, can lead to pulmonary edema. The pyridoindoles Trp-P-1 and Trp-P-2 are genotoxic substances which originate from pyrolysis of tryptophan and have been identified in foods cooked at excessively high temperatures. 4-Chloro-6-methoxyindole, which can be extracted from fava beans, yields a potent mutagen on interaction with nitrate ion.

RICHARD J. SUNDBERG
University of Virginia

Additional Reading

Atta-ur-Rahman, and A. Basha: *Indole Alkaloids,* Taylor & Francis, Inc., Philadelphia PA, 1998.
Bird, C.W. and G.W.H. Cheeseman, eds.: *Comprehensive Heterocyclic Chemistry,* Vol. 4, Pergamon Press, Oxford, 1984, Chapts. 3.04, 3.05, and 3.06.
Houlihan, W.J. eds.: *The Chemistry of Heterocyclic Compounds,* Vol. 25, Parts 1, 2, and 3, Wiley-Interscience, New York, NY, 1972.
Saxton, J.E.: *Monoterpenoid Indole Alkaloids: Supplement to Part 4,* John Wiley & Sons, Inc., New York, NY, 1994.
Sundberg, R.J.: *The Chemistry of Indoles,* Academic Press, Inc., New York, NY, 1970.
Sundberg, R.J., O. Meth-Cohn, C.S. Rees, and Alan R. Katritzky: *Indoles,* Elsevier Science & Technology Books, New York, NY, 1996.

INDUSTRIAL BIOTECHNOLOGY.

In the broad sense, *industrial biotechnology* is the practical application of scientific principles of biology learned over many decades to the development and manufacture of useful products. In modern terms, this includes the use of knowledge gained from studies in cell biology, molecular biology, and gene-transfer techniques. See also **Molecular Biology**.

In a more restricted sense, industrial biotechnology is frequently referred to as *industrial microbiology*—because in a large number of bioprocesses, living substances in the form of yeasts, molds, bacteria, etc. are used as raw materials. For centuries, processes that depend upon living substances have existed, fermentation being a notable example. The scope of biology-related manufacturing, so to speak, was greatly expanded with the appearance of antibiotics, pioneered by Alexander Fleming in 1929 and soon followed by the bioprocessing of microorganisms to produce antibiotics in large quantities. The science of biology was impacted in the 1950s by a breakthrough in our knowledge of the DNA molecules, which led to gene recombination technology. The impact was felt early and continues today in industrial biological processes and products. As bioprocessing progressed from the traditional fermentation industries through the massive production of antibiotics and other pharmaceuticals to the current and potential application of gene-transfer techniques on an industrial scale, the nomenclature changed. A preferred term for the foregoing activities is now *industrial biotechnology.*

Starting Materials—Microorganisms

Four kinds of microorganisms make up the raw materials for most biochemical processes. They are:

A. *Eukaryotes*—cells or organisms whose DNA is organized into chromosomes with a protein coat and surrounded by a nuclear membrane. Eukaryotes contain organelles, such as mitrochondria. The latter function and furnish the cells with their main energy supply. Yeasts and molds, the eukaryotes in common use, are fungi.

   1. *Yeast (Ascomycetes)*, notably the *Saccharomyces* (called *sugar fungus* by Meyen, 1837). Among the most industrially important are *S. cerevisiae* (used in making alcoholic beverages and bread); *S. cerevisiae var. ellipsoideus, S. bayanus,* and *S. beticus* (used in making wine); *S. uvarium* (used in brewing); and *Kluyveromyces fragilis* (used in whey disposal). *K. fragilis* was formerly called *S. fragilis.*

      Yeasts play a major role in biochemical processes, but notably participate in (a) *fermentation,* where simple sugars and other chemicals are transformed into the desired intermediate or end-products of the process; and (b) *respiratory (oxidative) metabolism.* This respiratory activity of oxidative dissimilation is characteristic of many species of yeast. For example, during aerobic growth, sugar is oxidized to carbon dioxide and water, with the release of large amounts of energy. Other biochemical processes in which yeasts participate include amination, condensation, deamination, decarboxylation, esterification, hydrolysis, lipolysis, pectinolysis, and proteolysis, among others. See also **Yeasts and Molds.**

   2. *Molds* are filamentous and of numerous varieties. They grow as a branched system of threadlike hyphae rather than as single cells. In bioprocessing molds play a positive role, as in the instance of *Penicillium roqueforti* used as a culture in making certain cheeses; or *Penicillium chrysogenum* used in the production of penicillin antibiotics. Molds also have a significant negative side in biochemical processing because of the degradative roles they play in causing food spoilage and human disease, not to mention plant and crop damage. When food processing vessels and machines are not kept clean, so-called *dairy* or *machinery mold* may be found in the equipment. The presence of this mold may be indicative of serious microbial contamination.

      Molds are widely distributed throughout nature. Critical to the growth of molds is availability of sufficient moisture. Some molds are quite resistant to adverse conditions, including high temperature.

B. *Prokaryotes*—cells or organisms that have only one chromosome. They have no nuclear membrane or mitochondria.

   3. *Bacteria* (normally unicellular), of which many are used in bioprocessing (see also **Bacteria**), including: *Lactobacillus bulgaricus* (used in making yogurt); *Gluconobacter suboxidans* (vinegar); *Clostridium acetobutylicum* (acetone and butanol); *Corynebacterium glutamicum* (flavor enhancing nucleotides); *Methylophius methylotrophus* (single-cell proteins); *Propionibacterium* (vitamin B_{12}); *Bacillus* (enzymes—proteases); *Xanthomonas campestris* (polysaccharides—xanthan gum); *Mycobacterium* (pharmaceuticals—steroids); *Streptomyces* (anti biotics—amphotericin, streptomycin, tetracyclines, etc.); *Esc herichia coli* (by way of recombinant DNA technology to produce insulin, human growth hormone, somatostatin, interferon, etc.); and *Bacillus thuringiensis* (bioinsecticides).

   4. *Actinomycetes*—a group of branching unicellular organisms, which reproduce either by fission or by means of special spores or conidia. They usually form a mycelium which may be of a single kind, designated as substrate or vegetative,

or of two kinds, substrate and aerial. The actinomycetes are closely related to the filamentous bacteria and some authorities regard them as prototypes from which both fungi and bacteria were derived. These microorganisms have been prolific sources of several thousand antibiotics, although only a relatively few of these have been produced commercially in very large quantities.

Classes of Bioproducts Made from Microorganisms

The principal commercially important products made from the microorganisms previously mentioned include:

   1. *Microorganisms per se*—for use in a wide variety of bioprocesses.
   2. *Large molecules,* including enzymes.
   3. *Primary metabolic products* (compounds required for their growth).
   4. *Secondary metabolic products* (compounds that are not essential to their growth).

The primary and secondary metabolites that are important commercially normally are of relatively low molecular weight, usually 1500 daltons[1] or less. For comparison, an enzyme may range from about 10,000 to millions of daltons.

Commercial Applications of Microbial Cells

There are two major classes of the commercial use of microbial cells:

   1. *Protein sources,* of which the most important current product is called *single-cell protein,* used in animal feedstuffs. See also **Proteins.**
   2. *Bioactive ingredients* for use in bioprocessing. In bioprocessing, chemical reactions are usually referred to as *biological conversions* (even though they proceed at the molecular level) in processes where microorganisms are the major participants. The term *microbial transformation* is also used.

For large numbers of end-objectives, microorganisms as participants have a number of advantages over nonbiological reactants. For example, the latter involve substantial energy exchange (either requiring heating or cooling). Also, nonbiological processes usually are conducted in a solvent medium and often in the presence of inorganic catalysts. Both solvent and catalyst are possible sources of product pollution. Further, a large percentage of biological conversions do not yield undesirable byproducts which must be removed in separate purification operations. Once separated, profitable uses must be found for the byproducts, or they must be disposed in a costly, nonpolluting way.

In biological conversions, water is usually the solvent and temperatures and pressures are at reasonable levels, as dictated by the properties of most natural living substances. When a specific enzyme is used in a biological conversion, outstanding specificity maintains because a given enzyme usually will catalyze but one specific kind of reaction. As pointed out by Demain and Solomon (see reference), an enzyme can be caused to select one isomer, or molecular form of a compound, in a mixture of forms to produce a single isomer of the product. These characteristics account for the high yields that are typical of biological conversion, sometimes reaching nearly 100 percent.

The production of enzymes is the major target of numerous biological conversions. In years past, the principal sources of enzymes were extraction products from plant and animal sources. The application of DNA technology has made large inroads in the production of "synthetic" enzymes.

Products of commercial importance extend well beyond enzymes and include the production of large molecules, such as polysaccharides. Xanthan is an example. This is a gum widely used in food processing. See also **Xanthan Gum.**

Fermentation Industry. Primary metabolites of importance in the fermentation field include amino acids, purine nucleotides, vitamins, and organic acids. Specific products include citric acid, riboflavin (vitamin B_2), and cobalamin (vitamin B_{12}). Check alphabetical index pertaining to specific vitamins. Of the *secondary metabolites,* antibiotics are the most

---

[1] Dalton = atomic mass unit.

important. In the past, about three-quarters of all antibiotics have been obtained from the actinomycetes, and a very large percentage of these have stemmed from a single genus, *Streptomyces*. See also **Antibiotics**.

The process of fermentation has been known for at least 4000 years, notably in connection with the arts of making wine, leavened bread, brews (beer), and for decades, for example, in connection with naturally produced vinegar. In most fermentation, the microorganisms effect the desired biological transformations (serving in a way that is somewhat comparable to intermediate chemicals used in conventional organic synthesis and not present in the final product). Usually, at some point in the bioprocess, the earlier invaluable microorganisms must be killed because their presence in the final product would lead to numerous undesirable characteristics, including spoilage.

Examples of organic synthesis by way of fermentation are given in Figs. 1 and 2.

Bioprocessing Methodologies

Bioprocessing equipment must be designed to meet the environmental requirements of whatever microorganisms are used. As well understood for a century or more, enzymes created by the microorganisms catalyze the desired biological conversions in a highly efficient manner, as exemplified by the conversion of sugars into ethanol and carbon dioxide. A major advancement in fermentation occurred a number of years ago when brewers found that instead of relying on microorganisms to create the desired enzymes, they could add the enzymes manufactured separately

and available from commercial sources to the fermentation vessel directly. Many bioprocesses are not so simple as producing alcohol from sugar, but require cadres of enzymes to effect the biological transformation of the substrate by initiating a number of reactions (transformations) that involve numerous specific enzymes.

The environmental requirements of microorganisms are quite demanding. Process parameters must be carefully controlled to gain full efficiency, or indeed to permit the bioprocess to continue at all. This contrasts with a considerably greater flexibility permitted when dealing with nonliving reactants and catalysts. Such parameters include temperature control and pH (hydrogen ion concentration) within very narrow limits. Sufficient water of specified purity acts as the processing medium. Although enzymes can be preserved by drying, there is no catalytic activity in the absence of water.

An Additional requirement, not encountered in effecting nonbiological reactions, is the need to provide nutrition for the living microorganisms. They require a source of carbon that normally furnishes energy for metabolism. In some cases, this requirement is met by one of the starting raw materials, such as carbohydrates (sugar in the case of alcohol fermentations). Considerable investigation for certain bioprocesses has targeted other sources of carbon, including hydrocarbons. It has been learned that some industrially important microorganisms can exist on these nutritional carbon sources. Sometimes a period of adaptation is required and because of lack of efficiency, the use of such alternative carbon sources is frequently uneconomic. But with so many products now being produced by microorganisms, research continues and effective uses are expected to

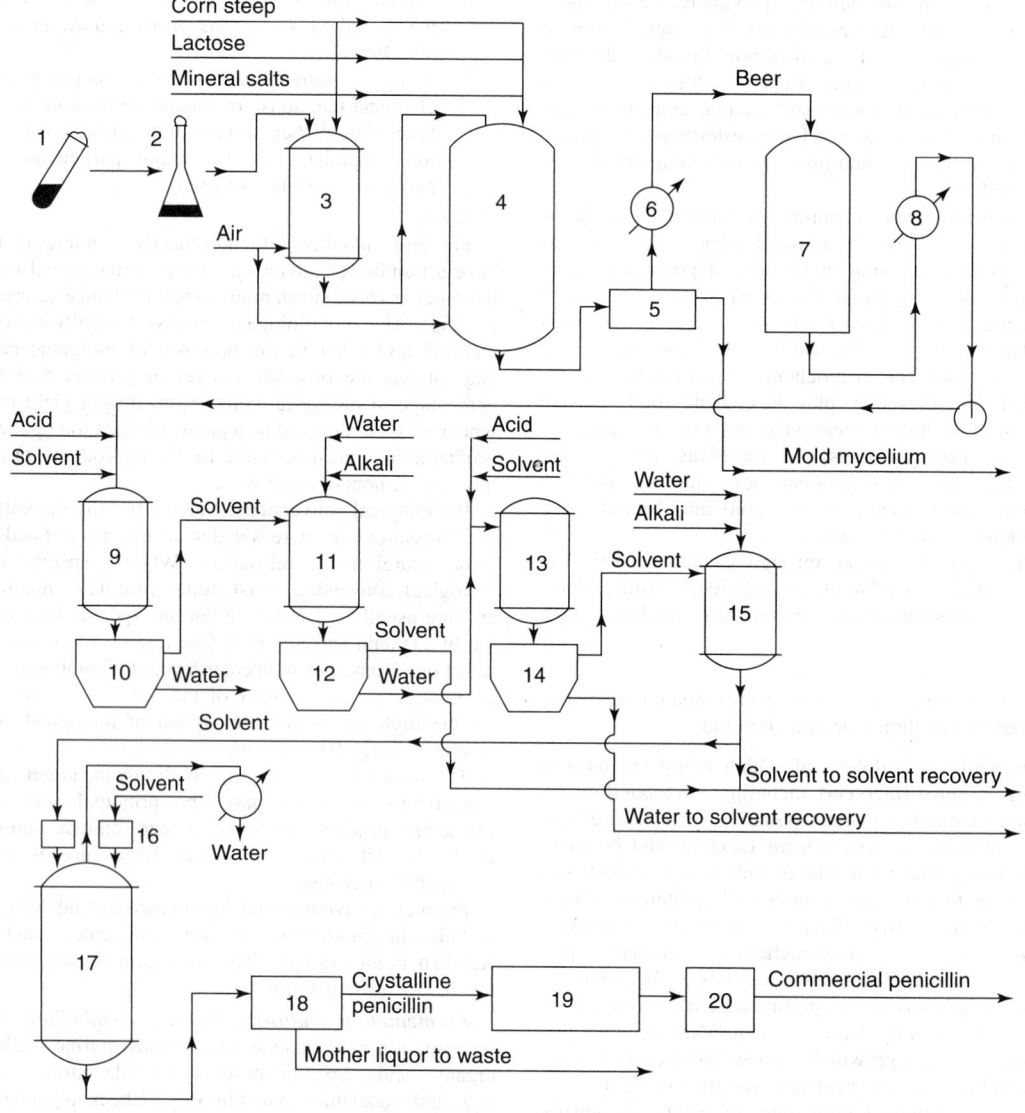

Fig. 1.   Schematic representation of materials flow in penicillin manufacturing: (1) agar slant culture, (2) bran spore culture, (3) seed tank, (4) fermentor, (5) filter, (6) brine cooler, (7) storage tank, (8) brine cooler, (9) mixing tank, (10–15) separation operations, (16) bacteriological filters, (17) crystallizer, (18) filter, (19) dryer, (20) finishing operations

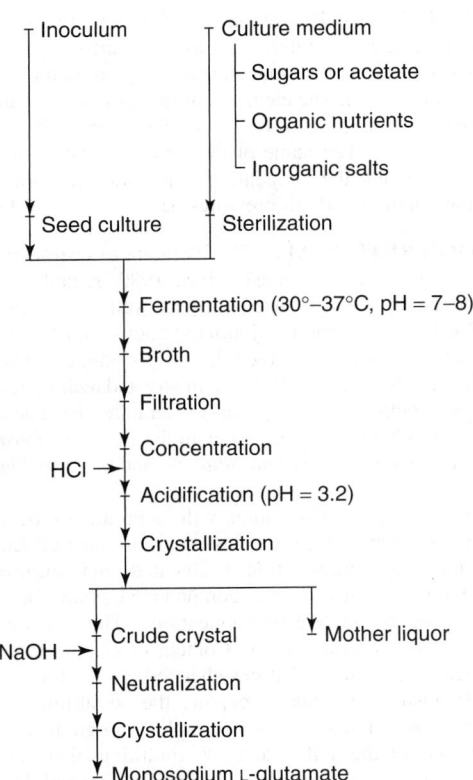

Fig. 2. Preparation of monosodium l-glutamate by fermentation. Although not involving genetic engineering, this process as pioneered by Japanese microbiologists was very advanced when introduced in 1956. Currently, almost all common amino acids can be produced by amino acid fermentation and in terms of very high tonnage production

be found. Hydrocarbon sources considered have included petroleum and natural fats (soybean oil, etc.). The important point is that the demonstrated versatility of enzyme mechanisms bodes well for much further exploitation of those characteristics in the future.

In addition to provision of carbon, other nutrients required by microorganisms embrace nitrogen, phosphorus, and oxygen, all elements of which are part of the structural and functional molecules of the cell. Smaller quantities of micronutrients are needed. The requirement for cobalt in the synthesis of cobalamin is one of these obvious requirements.

In connection with oxygen, there is an interesting situation because some microorganisms are *anaerobic* (requiring *absence* of oxygen), while others are *aerobic* (needing an ample supply of oxygen). Normally, oxygen is furnished by pumping large volumes of air through the mixture. An advancement in this procedure is the use of enriched air (over 21% oxygen).

A further important parameter in bioprocessing is that of providing adequate mixing to ensure that the several ingredients will be within immediate vicinity of each other. Depending upon the particular microorganism, they remain continually suspended in the watery medium, desirable from a processing standpoint, but some tend to collect in clusters; others tend to take the form of slimes.

To date, bioprocessing is primarily conducted on a batch basis. This provides flexibility in shifting the products to be made from time to time and, in particular, any failure to provide aseptic conditions may result in the condemnation of only one batch versus what could happen in the case of a continuous process. Notably, in the case of pharmaceutical biologicals, it is practical to keep track of the product by batch number from start to final use. Even with these kinds of problems, however, the many attractions of continuous processes are being thoroughly studied and applied in limited instances.

Genetic Engineering—State of the Art

As development of the concept of recombinant DNA progressed after its discovery in the early 1970s, great promise was given for the "engineering" of new plant species and varieties, new designer drugs, and new processes. In the 1970s, there were continual predictions of an "explosion" in such

new product development. As of the early 1990s, however, there have been fewer dramatically improved products developed than had been initially contemplated. Actually, only comparatively few products are now moving from the laboratory to the processing plant. Over 77 small-scale field trials of genetically engineered tomato, potato, alfalfa, cucumber, corn (maize), and cotton have been conducted in several growing regions. The first food processing aid produced by a genetically engineered microorganism (the enzyme rennet) was approved by the U.S. Food and Drug Administration in March 1990. Most scientists in the field, however, maintain their confidence that the applications of genetic engineering in the food system alone are seemingly unlimited. Because of hunger problems throughout so much of the world, genetically engineered agriculture may prove to be the ultimate solution.

The principal steps required to genetically engineer a "new plant" are illustrated very schematically in Fig. 3.

In 1991, an expert in food biotechnology observed the following. (1) The public, in general, is not enamored with food biotechnology. (2) Public concerns continue pertaining to the potential long-term unanticipated effects of modifications of food. (3) Although willing to take modest risks, the public becomes very conservative in terms of food modifications on infants, children, and the chronically ill. (4) The public challenges corporation-sponsored university research in this area of technology.

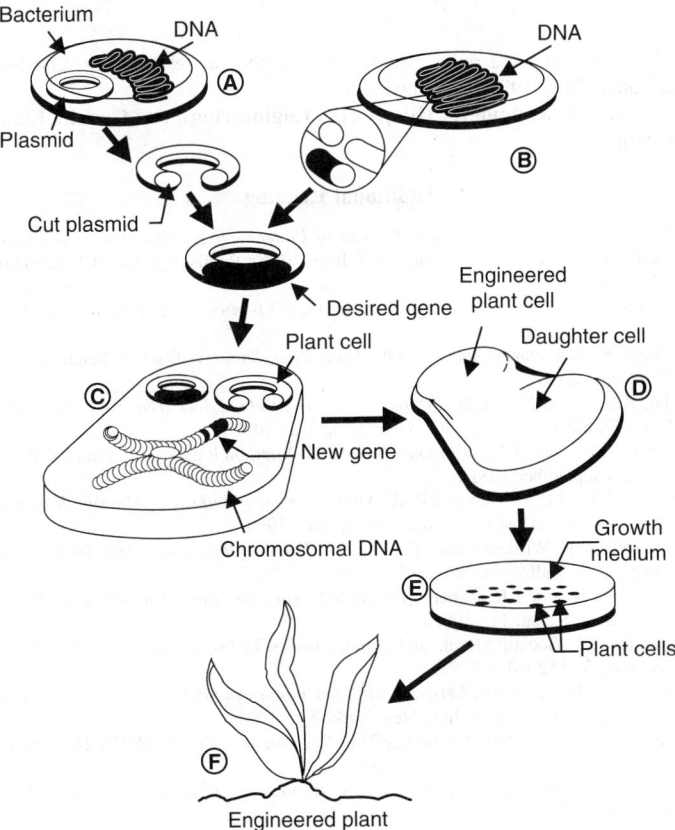

Fig. 3. Major steps in producing a genetically engineered plant: (**a**) Plasmid (a circular piece of DNA found outside the chromosome in bacteria) is removed from the bacterium and cut open by using restriction enzymes. Plasmids are the main tool for inserting new genetic information into the microorganisms of plants. Restriction enzymes are proteins that recognize specific gene sequences on a chromosome and cut DNA at these sites. (**b**) The gene of interest is cut out of the chromosomal DNA of another organism and "pasted" into the plasmid by using ligase enzymes. A ligase enzyme is one that splices segments of DNA together. (**c**) The plasmid is put back into the bacterium and mixed with plant cells. The bacterium duplicates the plasmid and transfers the new gene into the chromosomal DNA of the plant cell. (**d**) When the plant cell divides, each daughter cell receives the new gene, giving the whole plant a new trait or characteristic. (**e**) Plant cells are placed on special growth media to promote the formation of callus (unorganized tissue). After shoots grow from the callus, the plantlets are transferred to traditional media that stimulates roots to grow. (**f**) The plantlets are transferred to soil and grow to maturity. (*After Volpo and Monsanto*)

(5) The public has lost confidence in governmental regulatory actions pertaining to genetic engineering in the food field.

Attempts to provide sociological answers to scientific problems are beyond the purview of this encyclopedia. Provided such answers are forthcoming, biotechnology can be applied effectively to the solution of some of the following technical problems, particularly in terms of agriculture and food production:

- Develop temperature-tolerant plants that can survive in warmer or cooler climates. Frost damage causes more than $14 billion per year worldwide in crop losses.
- Engineer plants that can withstand drought conditions. For example, if salt-tolerant varieties can be developed, sea water could be used for irrigation.
- Improve ways for certain crop plants, such as corn (maize) and wheat to fix their own nitrogen from the atmosphere, thus reducing fertilization costs by $ billions per year.
- Engineer insect and pest resistance into plants, thus not only reducing the cost for chemical control applications, but also alleviating the environmental problems related to agricultural chemicals.
- Develop plants that have nutritional values superior to those obtainable from the existing natural varieties. One example, would be that of increasing various amino acids contained in the edible portions of the plant.

The foregoing are only part of a larger genetic research agenda. See Harlander (May 1991) reference listed.

See also **Biotechnology (Bioprocess Engineering)**; and **Genetic Engineering**.

Additional Reading

Acharya, R.: *The Emergence and Growth of Biotechnology: Experiences in Industrialised and Developing Countries,* Edward Elgar Publishing, Inc., Cheltenham, UK, 1999.

Bhamidimarri, R.: *Environmental Biotechnology,* Elsevier Science, New York, NY, 1998.

Bielecki, S., J. Tramper, and J. Polak: *Food Biotechnology,* Elsevier Science, New York, NY, 2000.

Cobb, A.B.: *Scientifically Engineered Foods: The Debate over What's on Your Plate,* Rosen Publishing Group, Inc., New York, NY, 2000.

Demain, A.L. and N.A. Solomon: "Industrial Microbiology," *Sci. Amer.,* **245**(3), 66–75 (September 1981).

Demain, A.L., J.E. Davies and R.M. Atlas: *Manual of Industrial Microbiology and Biotechnology,* ASM Press, Washington, DC, 1999.

Goldberg, I., R. Williams and E. Riemer: *Biotechnology and Food Ingredients,* Chapman & Hall, New York, NY, 1999.

Greenhalgh, R. and T.R. Roberts: *Pesticide Science and Biotechnology,* CRC Press, LLC., Boca Raton, FL, 1998.

Harlander, S.: "Social, Moral, and Ethical Issues in Food Biotechnology," *Food Technology,* 152 (May 1991).

Ives, C.L. and B. Bedford: *Agricultural Biotechnology in International Development,* Oxford University Press, Inc., New York, NY, 1998.

Kiely, T.: "Is Biotech Safe for the Big Time?" *Technology Review (MIT),* 24 (October 1991).

Klefenz, H.: *Industrial Pharmaceutical Biotechnology,* John Wiley & Sons, Inc., New York, NY, 2001.

Levin, M.A. and H.S. Strauss: *Risk Assessment in Genetic Engineering,* The McGraw-Hill Companies, New York, NY, 1991.

Morris, P.C. and J.H. Bryce: *Cereal Biotechnology,* CRC Press, LLC., Boca Raton, FL, 2000.

Persley, G.J.: *Agricultural Biotechnology,* CAB International, New York, NY, 2000.

Rhodes, P.M. and P.F. Stanbury: *Applied Microbial Physiology: A Practical Approach,* Vol. 183, Oxford University Press, Inc., New York, NY, 1997.

Sterckx, S.: *Biotechnology, Patents & Morality,* Ashgate Publishing Company, Brookfield, VT, 1997.

Uhlig, H.: *Industrial Enzymes and Their Applications,* John Wiley & Sons, Inc., New York, NY, 1998.

Welch, R.A.S., S.R. Davis, and A.I. Popay, et al.: *Milk Composition, Production and Biotechnology,* CAB International, New York, NY, 1997.

INERT. A term used to indicate chemical inactivity in an element or compound. Helium, neon, and argon are practically inert gaseous elements; carbon dioxide is a gaseous compound of low activity. Ingredients added to mixtures chiefly for bulk and weight purposes are said to be inert.

INERT GASES (The). The elements of group 18 of the periodic classification sometimes are referred to as the Inert Gases, or the Noble Gases. In order of increasing atomic number, they are helium, neon, argon, krypton, xenon, and radon. The elements of this group are characterized by their closed shells or subshells of electrons. They generally are considered as having zero valence. The name of this group derives from the lack of chemical activity which the elements display, forming compounds only under abnormal conditions (high pressures, strong electrical fields, etc.).

INFORMATION RETRIEVAL. The literature of chemistry and associated fields has increased enormously since 1980. Establishment of subspecialties and newly defined disciplines as well as increased research output have led to an explosion of journals, books, and on-line databases, all of which attempt to capture, record, and disseminate this plethora of knowledge. Tertiary reference tools in chemistry and technology (e.g., *Kirk-Othmer,* 4th ed.) help track the primary literature. Excellent references that discuss basic chemical information tools are *The Literature Matrix of Chemistry, Chemical Information Sources,* and *How to Find Chemical Information.*

Retrieval of chemical information will continue to be an issue of accessing what is available in the fastest, most cost-effective manner. Changes in the ways information is located and retrieved will be driven by technological advances in computer hardware, development of software, and progress in telecommunications. The resources available through Internet are increasing daily. Content of electronic databases has remained basically the same; what has changed are the tools to access the information. Publishers continue to explore the possibilities of electronic media. Ease of information access, for example, through the use of a natural language interface, or the ability to query multilingual databases using a single language, is another emerging issue. The Special Interest Group on Information Retrieval of the Association of Computing Machinery (SIGIR ACM) meets annually to address these and other information issues. Proceedings of their meetings provide an overview of advances in information technology and access. As technology becomes more complex, issues of ownership, copyright protection, reuse of retrieved information, and access costs will need to be examined and resolved.

Libraries and information centers are rapidly moving from purchase and ownership of print and on-line resources to rapid access to these resources; a change in philosophy from "just in case" to "just in time." Libraries are making hard decisions about what to purchase and what to exclude because of the magnitude and cost of a complete collection of available information. These factors have forced severe cutbacks in what is actually bought as well as increases in cooperative purchasing and loan agreements between libraries on the local, regional, national, and even international levels. Increased computer power and technology have made geographic boundaries and limitations obsolete. Documents, such as journal articles, can be obtained quickly from other libraries and commercial document delivery vendors. Paper copy remains the preferred format for delivery of such documents. Documents can be delivered by regularly scheduled mail, by overnight delivery, or by facsimile transmission.

Although electronic publishing of journals is in its infancy, more and more full text journals are becoming available on-line, allowing printing of a document from the user's computer. The limitation of ASCII format (text only, no graphics) is being addressed by vendors. The American Association for the Advancement of Science's publication *Online Journal of Current Clinical Trials,* which debuted in July 1992, supports text and nontext and publishes a paper within days of acceptance. The Research Libraries Group has produced ARIEL, a software package that allows image scanning of a document transmitted through Internet. ARIEL provides images and text of greater resolution than fax and uses standard personal computer (PC) hardware.

Location of and access to chemical and technical information other than journal articles is available through computerized information networks. Electronic bulletin board systems (BBS) provide a telecommunications tool to anyone who has a computer and a modem. Questions can be posted and read by thousands of bulletin board users worldwide, and files and software are easily transferred from virtually anywhere to one's computer.

Networks. The rise in popularity and use of Internet has dramatically changed the way information is disseminated. Internet is a worldwide link of thousands of separately administered computer networks of many sizes and types. Each of these networks is connected to as many as tens of thousands of computers; the total number of individual Internet users is in the millions.

The three basic Internet applications of remote login, electronic mail, and file transfer are building blocks of more sophisticated applications that offer increased functionality and ease of network use. Tools such as Gopher, Wide Area Information Servers (WAIS), and World Wide Web (WWW) go beyond the three basic Internet functions to make information on the network easier to locate and use. Detailed descriptions of these tools are available. This trend toward more powerful, user-friendly networked information resource access systems should continue as Internet grows and matures.

Copyright. Any discussion of emerging technologies in storage and retrieval of information leads to a question of copyright compliance. Copyright in the electronic and computer age, using the internal and external networks and document delivery mechanisms outlined above, is a complex and unresolved issue. One of the primary reasons for using document delivery services, other than for obtaining information that is not locally available, is copyright compliance. Document suppliers generally handle payment of any required copyright fees. In the electronic and computer age, establishing who owns a particular piece of information can be difficult, as can determining what can legally be done with information once it is obtained.

Budgeting. Changes in the storage and retrieval of chemical information require that libraries and information centers now consider not only what should be purchased but also what monies should be allocated for the purchase of information in nonprint formats such as CD-ROMs (compact disk read-only memory) and on-line databases. Coupled with this is budgeting for the cost of hardware and software to enable the rapid and cost-effective delivery of needed information. The geometric increase in sources, both printed and on-line, has increased the role of the information specialist as an expert in the delivery of chemical information. Retrieval from increasingly diverse and complex sources becomes the paramount issue for searchers of chemical literature in the 1990s.

On-Line Database Resources

The on-line information industry has grown dramatically since 1972 when Dialog Information Services, Inc. (Dialog) offered the first publicly available commercial databases. Databases covering virtually all important subject areas were developed, and significant publications became available via thousands of bibliographic, abstract, textual, directory, and numeric databases. For electronic databases to be made available publicly, development was required in three primary technologies: computers, communications, and databases themselves.

Database Producers. Producers of databases, also known as database publishers or information providers, determine the content of the databases, produce them, and typically lease or license them to private organizations or database vendors. Database producers may be categorized as government, not-for-profit, commercial/industrial, and mixed.

Primary database vendors that offer or are developing fixed price options include Mead Data Central, Dow Jones News/Retrieval and DataTimes, NewsNet, Dialog, and OCLC. The environment of the 1990s is one in which database vendors and producers are competing for survival, and end users are leading the way to what could represent significant growth in the on-line database industry. Vendors include BRS Online Products, Cambridge Crystallographic Data Centre, Chemical Information Systems, Data-Star, DIALOG Information Retrieval Service, ESA-IRS, MDL Information Systems, Inc, Mead Data Central, MEDLARS, ORBIT ONline Service, Questel, and STN International.

Chemical Information and Search Methods and Services

Chemical information is reported and recorded in many forms, and a wide variety of databases have evolved to collect the various types of information. Bibliographic, business, structure, numeric, spectra, and reaction databases currently are available.

Bibliographic–Technical. These include Agrochemical Handbook, Analytical Abstracts (AA), APILIT, APIPAT, CA File, CA Registry File, CA Search CAB ABSTRACTS, Ceramic Abstracts, Chemical Engineering and Biotechnology Abstract (CEBA), Chemical Journals of the American Chemical Society (CJACS), Chemical Journals of the Royal Society of Chemistry, Chemical Safety NewsBase (CSNB), Chinese Patent Abstracts in English Database, CLAIMS, COMPENDEX PLUS, CORROSION, Current Biotechnology Abstracts (CBA), Current Patents (Evaluation/Fast-Alert), Dissertation Abstracts Online, EMBASE, Energy Science & Technology, EPAT, European Directory of Agrochemical Products, GenBand, INPADOC, INSPEC, JAPIO, Rapra Abstracts, SciSearch, Thomas Register Online, U.S. Patents Fulltext, World Patent Index (WPI), World Surface Coatings, and World Textiles.

Business–Industrial. These include ABIINFORM, BIOBUSINESS BUSINESSWIRE, CENDATA, Chemical Industry Notes (CIN), Commerce Business Daily, Conference Papers Index, Dialog Journal Name Finder, Dialog Product Name Finder, DISCLOSURE DATABASE, Dun's Market Identifiers, ERIC, Federal Register, Food Science and Technology Abstracts, Harvard Business Review, Health Periodicals Database, Health Planning and Administration, ICC International Business Research, International Pharmaceutical Abstracts (IPA), INVESTEXT, MANAGEMENT CONTENTS, Marquis Who's Who (MWW), MATERIALS BUSINESS FILE, NTIS Bibliographic Database, Nursing and Allied Health, PHARMACEUTICAL NEWS INDEX (PNI), Pharmaceutical Business News, Pharmaprojects, Pollution Abstracts, PR Newswire, PTS News, PTS PROMPT, SEC ONline, Textile Technology Digest, Textline, THOMAS REGISTER ONLINE, TOXLINE, Trade and Industry, and World Textiles.

Structure. Structure searching involves matching a query compound against a machine-readable file of chemical structures. Structure searching determines if a compound is present in a file and retrieves it along with any associated information. Chemical structure files are compiled as novel chemicals, and compounds are registered and given unique identifiers, e.g., the CAS Registry Number, which is assigned sequentially to each new structure entering the system.

Substructure searching involves retrieval of all the compounds in a file containing some specified portion of a chemical structure, irrespective of the rest of the molecule in which the query substructure occurs.

Databases include the Beilstein File, the Gmelin File, the Registry File, the Description, Acquisition, Retrieval, and Correlation File, and the Structure and Nomenclature Search System.

Numeric. Researchers routinely use reported numeric measurements and data in their work. Numeric databases include the Beilstein Handbook of Organic Chemistry, the Gmelin Handbook of Inorganic and Organometallic Chemistry, property data networks [the Materials Property Data Network Inc. (MPD) and Chemical Property Data Network (CPDN)], and TDS NUMERICA.

Cambridge. The Cambridge Structural Database is an integrated system of programs for searching, retrieving, and analyzing data on more than 96,000 organic and organometallic structures, which were determined by x-ray and neutron diffraction. About 15,000 compounds a year are being added to the database.

Spectra. The ability to consult collections of standard spectra is crucial in the analysis of unknown compounds. A long history of data collection efforts has been aimed at these applications. Among the best known of the published handbooks are the Sadtler Spectral Data Sheets, which include ir, Raman, and nmr spectra. On-line sources include the Chemical Information System, SpecInfo, and The Canadian Scientific Numeric Database Service (CAN/SND).

Reactions. CASREACT File is a chemical reaction database containing over 118,500 records with reaction information derived from documents covered in the Organic Section of *Chemical Abstracts*. The file is available from STN.

Integrated Systems. Until recently, each of the numerous databases and sources of information available to chemists and technologists had to be searched individually, and selected results either printed for file storage or downloaded to an in-house or private computer system for easy future access.

Molecular Design Limited (MDL) has marketed an Integrated Scientific Information System (ISIS), which provides the capability to query multiple systems, including binary, text, proprietary, and relational databases across global networks, thereby providing transparent desktop access to multiple autonomous data sources.

Patent Information and Search Methods/Services

Patents are unique as primary source documents because of their stylized format, specialized language, presence of legally significant claims, descriptive drawings, and frequent disclosure of chemical compositions as generic (Markush) structures. A comprehensive review of electronic databases that contain patent information, including legal aspects, is available. Databases include bibliographic databases [World Patents Index, U.S. Patents, CLAIMS, APIPAT, EPAT, JAPIO, INPADOC], and PHARM.

Full-Text Databases. Two vendor-provided full-text patent databases are LEXPAT, produced by Mead Data Central, and PATFULL, produced by Dialog Information Services.

Chemical Substructure Databases. Several patent databases are searchable by chemical substructure. These are designed to give higher relevance of retrieval when searching chemical compounds than the bibliographic or full-text databases. They include MPHARM, WPIM (World Patents Index Markush), and MARPAT.

CD-ROM Databases. Since about 1989, CD-ROM format bibliographic and image patent databases have become available as current awareness, reference, or image storage and retrieval tools. The databases are designed for stand-alone PC or local network use, and in some cases they may be alternatives to retrieving patent information on-line. U.S. patent information resources available on CD-ROM include APS (Automated Patent Searching), CASSIS, FullText, OG/PLUS, Patent-Images, and PatentView.

The European Patent office ESPACE series of CD-ROM products are ESPACE-EP, ESPACE-FIRST, ESPACE-UK, ESPACE-WORLD, and ESPACE-ACCESS.

MARKUSH TOPFRAG. Derwent's TOPFRAG family of products is PC-based software that automates the selection of search codes and strategies.

Environmental and Safety Information and Search Methods and Services

There are public and private databases. Public databases are produced by the government and private enterprise and are commercially available through database vendors, such as STN, DIALOG, BRS, ORBIT, and NLM, and through various universities. Private databases are produced by government agencies, corporations, or other organizations for in-house use by their employees or others affiliated with them. The information in these databases may be made available on a need-to-know basis to individuals or corporations in the public or private sectors. Knowledge of the existence of private databases is usually obtained by personal contact within an organization or thorough disclosure in published literature. Private industrial databases are not usually accessible by the public, although information contained in them on hazardous materials must be reported to the EPA under provisions of TSCA (Toxic Substances Control Act) Section 8e.

Public Databases. The most comprehensive list of publicly available databases is the two-volume *Gale Directory of Databases*.

Private (EPA) Databases. The U.S. EPA maintains a list of approximately 600 current information systems, as well as some of the models and databases used within the organization. The list is published in *Information Systems Inventory* (ISI) which is updated yearly and maintained by the Information Management and Services Division of the Office of Information Resources Management.

On-Line Search Aids

MACCS-II. The Molecular Access System is a chemical information management system from Molecular Design Limited (MDL), San Leandro, California. It offers menu-driven graphical input for building, maintaining, and accessing chemical structures and any associated data, e.g., chemical and physical properties, biological activity, toxicity data, pricing, safety, and supplier information.

MACCS-II enables direct interface with other database management systems, such as the Relational Database Management System (RDBMS) and Oracle, so that databases that contain text and numeric data, for which special interfaces are normally needed, can be constructed.

Optical Disk-Based Information and Document Image Systems

Optical-based storage technology has joined paper, microfilm, and electronic/magnetic technologies as another medium for the storage, retrieval, and management of information. Optical media differ from magnetic media in that the information is encoded and read by means of laser optics. Information stored on optical disk may be either in a searchable text format (ASCII) or in a format containing only bit-mapped images, usually obtained as output from a scanner. Through the scanning and digitization process, pages that consist of printed text, graphics, photographs, drawings, handwriting, tables, etc., are converted to their binary representation and are stored as bit-mapped images on the optical media. The information stored on optical disk often has counterparts in other formats, such as printed publications or on-line databases or files. Advantages of document imaging systems for complementing, enhancing, or replacing traditional paper- or microfilm-based systems include increased storage capacity, ease of access via automated retrieval, simultaneous searching and viewing at multiple workstations, speed of access and delivery resulting in productivity gains, improved customer service, document security, document integrity via preservation and elimination of lost or misfiled documents, and networking and integration capabilities for these systems. Among the types of optical media that have been developed, the two most common for information storage and retrieval systems are both optical disk-based systems, namely CD-ROM and WORM (write once read many).

Several projects such as CORE (Chemistry On-Line Retrieval Experiment) at Cornell University, Project Mercury at Carnegie-Mellon University, Right-Pages at AT&T, and Red Sage at the University of California, San Francisco are in progress and illustrate the issues arising from implementation of on-line information systems that combine text and image.

Private Bibliographic and Text Databases

Personal computers have introduced new ways to handle private bibliographic and text files. The most important factors to consider to achieve satisfactory results in building a bibliographic or text database are the type of information to be stored and the needs of the user. Types of information include correspondence, research results and documentation, meeting notes, and bibliographic references. Needs of the user to be considered should include the potential number of users of the database, restrictions for the access and display of the information because of privacy or proprietary reasons, and the retrieval mechanisms (e.g., by keyword, authority list, controlled vocabulary, author, title, date, or other document or information attributes). In addition, criteria for selecting and encoding information for the database need to be established.

The type of hardware or computer system to be used and the potential size of the database should also be considered. Another factor frequently overlooked in private database creation is commitment to the support and maintenance of the database. Support involves training users, solving software and hardware problems, and upgrading the software when new features become available or are needed by the end user. Maintenance of the database include adding new information, deleting information no longer wanted, and correcting information in the database when errors are detected. Software packages available for building databases range from generic personal computer database management systems, which require customizing to software designed specifically for bibliographic files. Most of the software is for Macintosh or IBM compatible PCs. Some examples are PROCITE, NOTEBOOK II, ENDNOTE, LIBRARY MASTER, PERSONAL FILE SYSTEM, ASKSAM, BASISPLUS, and PERSONAL LIBRARIAN.

CYNTHIA S. BARCELON-YANG
EVELYN L. BROWNLEE
EMMETT D. CALHOUN
BRUNO A. CAPUTO
CHARLES C. CUMBO
JOSEPH P. DANISZEWSKI
DOUGLAS A. ECKEL
KENNETH H. GLASPEY
DARLYN C. GREEN-KOCHER
MARIANNE B. GRUBER
MARGARET M. ISSELMANN
THOMAS C. JOHNS
ALICIA P. KING
DAVID M. KRENTZ
FLUORENCE H. KVALNES
LURAY M. MINKIEWICZ
BEHROOZ NAZER
ANGELA K.G. PARSONS
CAROL R. PERROTTO
RITA D. RATLIFF
JEANETTE C. SIKES
AMIE H. WEBSTER
Du Pont Company

Additional Reading

Maizell, R.: *How to Find Chemical Information,* John Wiley & Sons, Inc., New York, NY, 1987.

Skolnik, H.: *The Literature Matrix of Chemistry,* John Wiley & Sons, Inc., New York, NY, 1982, p. vi.

Wiggins, G.: *Chemical Information Sources,* McGraw-Hill Book Co., Inc., New York, NY, 1991.

INFRARED RADIATION. The region of the electromagnetic spectrum between the wavelength limits 0.7 and 1,000 micrometers. The lower wavelength limit is set to coincide with the upper limit of the visible radiation region. Radiation of wavelength greater than 1,000 micrometers is generally considered of the microwave spectrum. Both limits are arbitrary. The infrared region is sometimes broken down into three subregions: (1) the *near-infrared region* (0.7–1.5 micrometers); (2) the *intermediate-infrared region* (1.5–20 micrometers); and (3) the *far-infrared region* (20–1,000 micrometers).

Infrared radiation is produced principally by the emission of solid and liquid materials as a result of thermal excitation and by the emission of molecules of gases. Thermal emission from solids is contained in a continuous spectrum, whose wavelength distribution is described by

$$_\lambda d\lambda = \frac{2\pi c^2 h \varepsilon_\lambda}{\lambda^5} \frac{1}{e^{ch/\lambda kT} - 1} d\lambda$$

where λ = spectral radiant emittance of the solid into a hemisphere in the wavelength range from λ to $(\lambda + d\lambda)$.

  c = velocity of light

  h = Planck's constant = $6.62 \times 10 - 27$ erg/second

  ε_λ = spectral emissivity

  k = Boltzmann's constant = $1.38 \times 10 - 16$ erg/K

  T = absolute temperature of the solid emitter, K

The spectral emissivity, ε_λ, is defined as the ratio of the emission at wavelength λ of the object to that of an ideal blackbody at the same temperature and wavelength. When ε_λ is unity, the foregoing equation becomes the Planck radiation equation for a black body.

Gaseous emission of infrared radiation differs in character from solid emission in that the former consists of discrete spectrum lines or bands, with significant discontinuities, while the latter shows a continuous distribution of energy throughout the spectrum. The predominant source of molecular radiation in the infrared is the result of vibration of the molecules in characteristic modes. Energy transitions between various states of molecular rotation also produce infrared radiation. Complex molecular gases radiate intricate spectra, which may be analyzed to give information of the nature of the molecules or of the composition of the gas.

Spectral Emittance

The spectral radiant excitance of a blackbody at various temperatures is shown in Fig. 1. It is apparent from the figure that blackbody radiation from emitters at temperatures below about 2000 K falls predominantly in the infrared region. An emitter that exhibits at all wavelengths is called a *gray-body* radiator. Most solid radiators show a general decrease in spectral emittance with increasing wavelength in the infrared; however, over limited spectral ranges, many materials are approximately gray-body

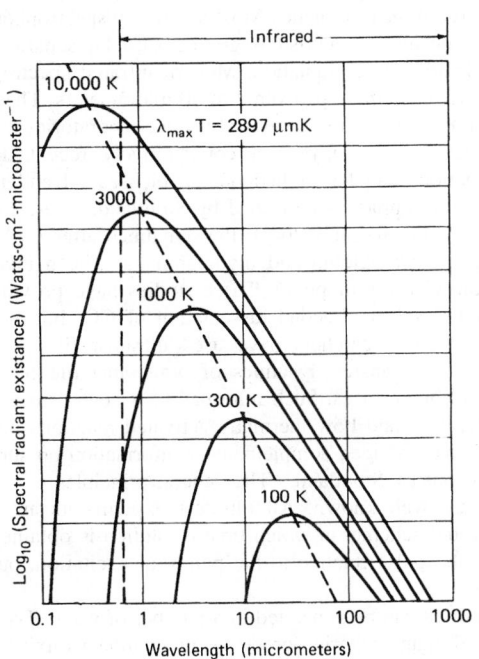

Fig. 1.   Spectral radiant excitance of a blackbody at various temperatures

radiators. Radiators that approach the characteristics of ideal blackbodies can be made in the form of uniformly heated cavities. A relatively small aperture, through which the cavity can be observed, serves as the source of blackbody radiation.

Propagation

Infrared radiation propagates through various media and, in general, is subject to absorption, which varies with the wavelength of the radiation. Molecular vibration and rotation in gases, which are related to the emission of radiation, are also responsible for resonance absorption of energy. The lesser gases in the atmosphere exhibit pronounced absorption throughout the infrared spectrum. However, nitrogen and oxygen do not absorb significantly in the infrared region. Water vapor, carbon dioxide, and ozone are responsible for strong absorption in the infrared. The absorption of radiation is so prevalent that those spectral bands in which relatively little absorption occurs are identified as atmospheric windows.

Solid and liquid materials show, as a rule, strong absorption in the infrared. There are, however, many solids that transmit well in broad regions of the infrared spectrum. Many materials, such as water and silica glasses, which show little absorption in the visible, are opaque to infrared radiation at wavelengths greater than a few micrometers. Many of the electrically insulating crystals, such as the alkali halides and the alkaline-earth halides, which transmit well in the visible, also are transparent to much of the near- and intermediate-infrared spectrum. Several of the semiconductor materials absorb strongly in the visible, but become transparent in the infrared beyond certain wavelengths characteristic of the semiconductor.

Detection of the presence, distribution and/or quantity of infrared radiation requires techniques which are, in part, unique to this spectral region. The frequency of the radiation is such that essentially optical methods may be used to collect, direct, and filter the radiation. Transmitting optical elements, including lenses and windows, must be made of suitable materials, which may or may not be transparent in the visible spectrum.

The detector for infrared represents the most unique component of the detection system. Photographic techniques can be used for part of the near-infrared region. Photoemissive devices, comparable to the visible- and ultraviolet-sensitive photocells, are available with sensitivity extending to about 1.3 micrometers. The intermediate-infrared region is most effectively detected by photoconductors. These elements, photosensitive semiconductors, are essentially photon detectors, which respond in proportion to the number of infrared photons in the spectral region of wavelength. This wavelength corresponds to the minimum photon energy necessary to overcome the forbidden gap of the semiconductor. All spectral regions from ultraviolet through visible, infrared, and microwaves, can be detected by an appropriately designed thermal element, which responds by being heated by the absorption of the incident radiation. In the infrared region, thermal detectors take the form of thermocouples, bolometers, and pneumatic devices. The thermal devices, in general, are not sensitive or as rapidly responding as photoconductors.

A very practical application of infrared radiation is found in radiant heating. Solid radiators, such as hot tungsten filaments, alloy wires, and silicon carbide rods are used widely as sources of infrared to provide surface heating by radiation. Commercially available infrared lamps are extensively used in specially designed ovens for drying painted and enameled surfaces.

Infrared Imagery

Infrared technology, sometimes called "night vision," has been used for several years in both military and commercial applications. The effectiveness of IR detectors was dramatically demonstrated during the military operation, Desert Storm, in 1991. These detectors, not previously demonstrated in actual war situations, were tremendous aids in achieving bombing target accuracy, not only from aircraft but from land vehicles as well. See series of images given in Figs. 2 and 3. Infrared imaging of the earth's surface from satellites also has become much more precise over the past decade. These topics are described in more detail in the articles on **Photography and Imagery**; and **Satellite (Scientific and Reconaissance)**.

Infrared Spectroscopy[1]

Scientists have long used infrared absorption as a means of probing the structure of molecules. Studying the manner in which specific wavelengths

---

[1] Information on infrared analytical instrumentation furnished by Rodney M. Durham, Instruments Division, Infrared Industries, Inc., Santa Barbara, California.

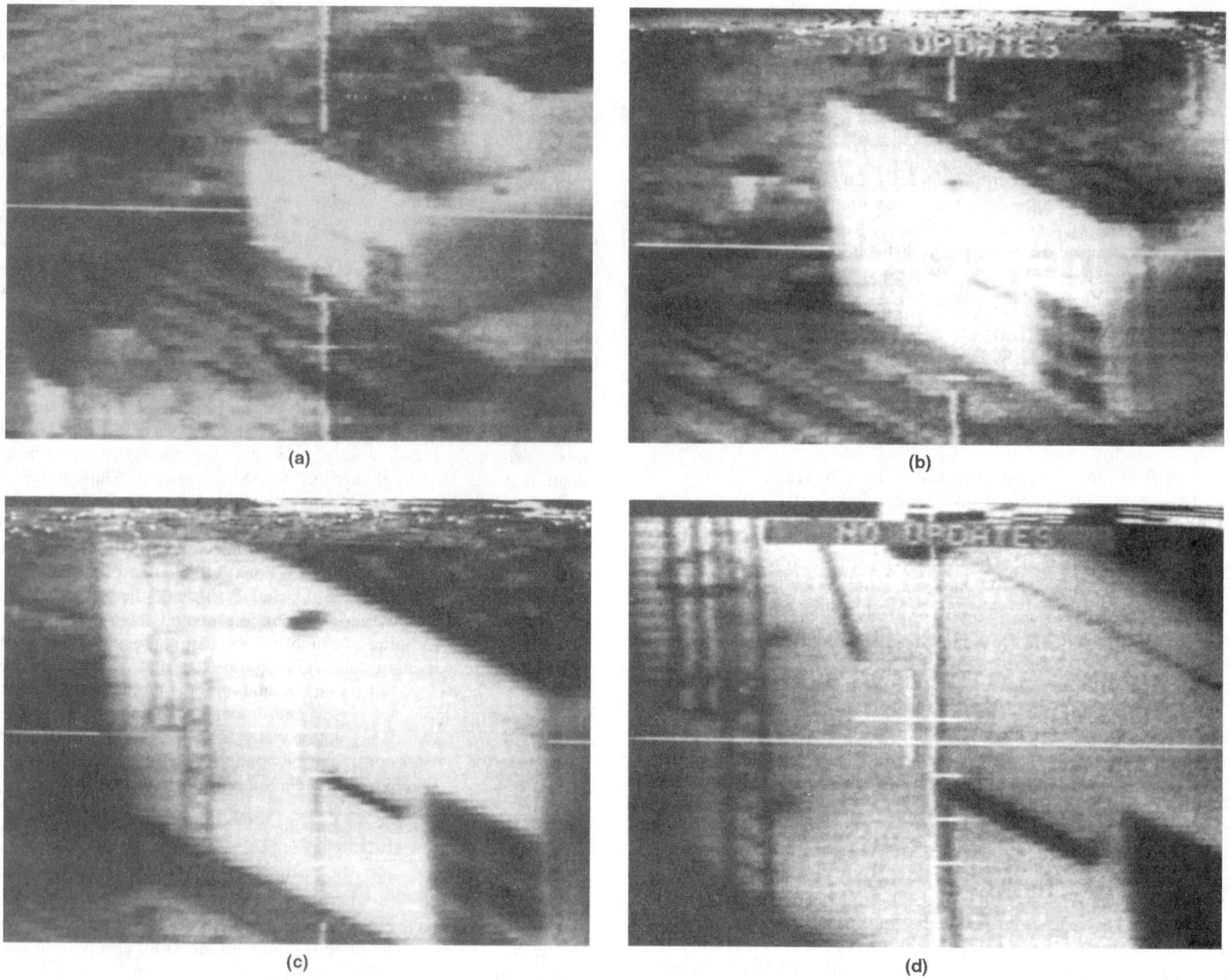

(a)

(b)

(c)

(d)

Fig. 2.   Second-by-second image of target taken by an attack aircraft. (*McDonnell Douglas photo*)

Fig. 3.   Infrared image of trawler in the North Atlantic

The *infrared spectrophotometer* is the principal instrument used by scientists for these measurements. Most laboratory spectrophotometers are of a dispersive design, i.e., a prism or grating is used to separate the spectral components in the source radiation. Modern infrared spectrophotometers have a wide wavelength range from 2 to 50 micrometers. They find use in research, quality control, and analytical service laboratories.

Ultrafast IR spectroscopy is a comparatively recent development. Chemical reactions can be studied on the picosecond and femtosecond time scale. For example, as described by Stoutland, Dyer, and Woodruff (Los Alamos National Laboratory), the dynamics after CO dissociation from CO-ligated hemoglobin and myoglobin have been investigated by monitoring the CO chromophore. These studies have provided not only evidence for the sub-picosecond dissociation of CO, but also structural information. Other researchers have used ultrafast IR spectroscopy to investigate energy transfer dynamics of organometallic molecules both in solution and on surfaces. Such information is central to understanding chemical activation and how thermally activated reactions occur. Recent experiments have provided complementary information on surfaces.

As pointed out by Stoutland, "The general principles of ultrafast laser experiments are well known. All ultrafast experiments are variants on the 'pump-probe' scheme, in which time resolution is obtained by spatial delay of a probe pulse relative to the pump, or excitation, pulse (1 ps = 3.0 mm)."

Short IR pulses can be generated in a number of ways. Typically, these are based on Raman scattering processes or nonlinear mixing schemes or the related optical parametric oscillator. Much more detail is given in the Stoutland reference listed.

of infrared energy excite vibration and rotation in molecules reveals information about the molecule that can be used to determine what and how many molecules are present.

Infrared Process Analyzer

This instrument has evolved from the laboratory spectrophotometer to satisfy the specific needs of industrial process control. While dispersive instruments continue to be used in some applications, the workhorse infrared analyzers in process control are predominantly nondispersive infrared (NDIR) analyzers. The NDIR analyzer can be used for either gas or liquid analysis. For simplicity, the following discussion addresses the NDIR gas analyzer, but it should be recognized that the same measurement principle applies to liquids. The use of infrared as a gas analysis technique is certainly aided by the fact that molecules, such as nitrogen (N_2) and oxygen (O_2), which consist of two like elements, do not absorb in the infrared spectrum. Since nitrogen and oxygen are the primary constituents of air, it is frequently possible to use air as a zero gas.

Many different analyzer configurations have been developed to address the diverse needs of the industrial process control industry. The basic constituents of an NDIR analyzer are: (1) a source of infrared radiation; (2) a means of restricting the wavelength range of the source radiation; (3) a means of detecting the infrared radiation; (4) a sample chamber to hold the gas or liquid to be measured; (5) a means of modulating the source radiation; and (6) electronics to process the signal generated by the source energy falling on the detector.

Microphone Detectors

Many IR process analyzers installed over the past several years have utilized *microphone detectors*. These detectors generally are the *Veingerov single-sided microphone* system or the *Luft balanced condenser microphone* system. These detectors are shown schematically in Figs. 4 and 5. See also Hill-Powell reference. The microphone detector uses an absorbing gas as its detecting medium. When radiation reaches the detector (that the sensitizing gas will absorb), the gas heats up and expands. This causes a diaphragm to distend. The diaphragm movement varies the condenser microphone capacity, which is part of an electric circuit that generates an electrical output signal. Both analyzers use dual sources, which are chopped to alternately allow energy to pass through a sample cell and a reference cell.

If the sample cell contains a nonabsorbing zero gas, such as nitrogen, the modulated beams reaching the detector through the two paths are of equal amplitude. In the case of the Veingerov single-sided detector, the chopper is configured so that, at any given time, the sum of the cross-sectional areas of the two beams as seen by the detector equals the total cross-sectional area of a single beam, so that, when no absorbing sample is present, a constant signal is produced, and the output is zero. When a sample is present, the sample and reference path signals become imbalanced and a signal at the chopper frequency is developed. The amplitude of this signal is a function of the concentration of the gas present in the sample cell.

The Luft detector (Fig. 5) operates similarly, but the detector has two chambers separated by a diaphragm. The signal generated by the presence of an absorbing gas in the sample cell is at twice the chopping frequency. This is an advantage over the single-sided microphone detection system since it is less susceptible to vibration caused by imbalance of the chopper motor. Having separate chambers does, however, allow for the possibility of a change occurring in one-half of the detector and not the other and thus resulting in zero drift. More recently, infrared process analyzers have been introduced which use a Luft-type detection system but replace the diaphragm with flow sensors. The flow of gas from one chamber to another is sensed by the flow sensor rather than by using a capacitance detection technique. This is claimed to eliminate one of the major modes of detector failure—failure of the thin diaphragm. For a given path length, process analyzers which use the microphone detector are more effective than those analyzers which use solid state detectors and optical filters at measuring low concentrations of gases which have a lot of structure in their absorption band. This structure results from the molecular rotation spectrum being superimposed on the vibration spectrum and is easily resolved in the simpler molecules such as carbon monoxide, methane, and ammonia.

Solid-State Detectors

The most recent generation of NDIR analyzers have evolved to satisfy the frequently harsh industrial environments encountered. These analyzers utilize solid-state sensors for the detection of infrared radiation. Most frequently used sensors are lead selenide (PbSe), thermopiles, or pyroelectric detectors. The gas analyzers generally are configured as single-path instruments, dual-beam with a reference path, or dual-channel with a reference filter.

Single-beam instruments find use where low cost is important, but where stability requirements are not stringent. Generally, this is true when the measurement period is short and frequent rezeroing is practical. Changes in source intensity due to power variation or changes in detector sensitivity due to temperature fluctuations are reflected directly in the output as zero drift. To avoid this, a reference path is commonly used. The dual-beam configuration is shown in Fig. 6. The source energy is modulated by a chopper blade. This allows the source to alternately pass through the reference and sample paths. The reference path is always free of absorbing gas so the detector is exposed to the source through a path unaffected by the presence of the sample. This signal is monitored by an automatic gain control circuit, which holds the reference signal level constant. If the source intensity or detector sensitivity change, the gain control will correct for it in both the reference and sample channels. Sync pickups monitor the chopper position and alert the electronics when the sample or reference path is irradiated. A narrow bandpass optical filter is located in front of the detector to limit the infrared energy and sensitize the analyzer to a particular gas absorption band. The signals generated by the two optical paths are synchronously demodulated. When an absorbing gas is introducing, the signal reaching the detector through the sample path is attenuated and the magnitude of the detected signal corresponds directly to the concentration of the sample gas present in the sample cell.

A reference optical filter can be used as an alternative to the reference path. This requires that a spectral window exist where is no interference from the sample. The 3.8 to 3.9 micrometer spectral region is frequently used in NDIR gas analyzers for this purpose. The spectral curves of Fig. 7

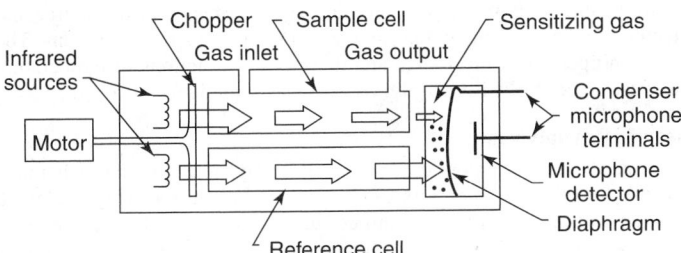

Fig. 4. Nondispersive infrared analyzer with a Veingerov-type detector

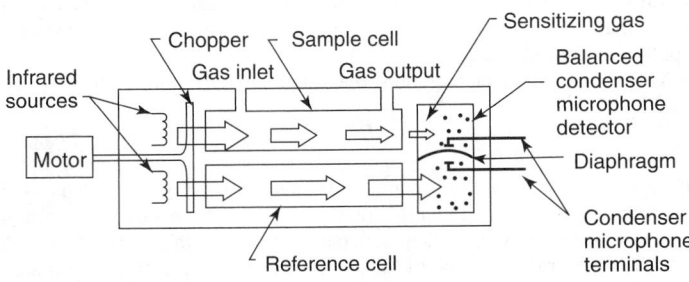

Fig. 5. Nondispersive infrared analyzer with a Luft-type detector

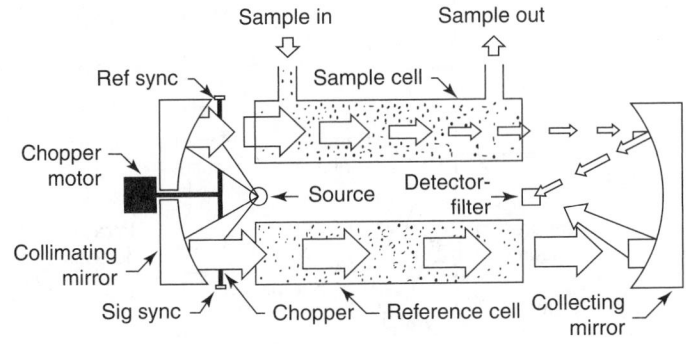

Fig. 6. Dual-beam infrared analyzer with a solid-state detector. (*Infrared Industries, Inc*)

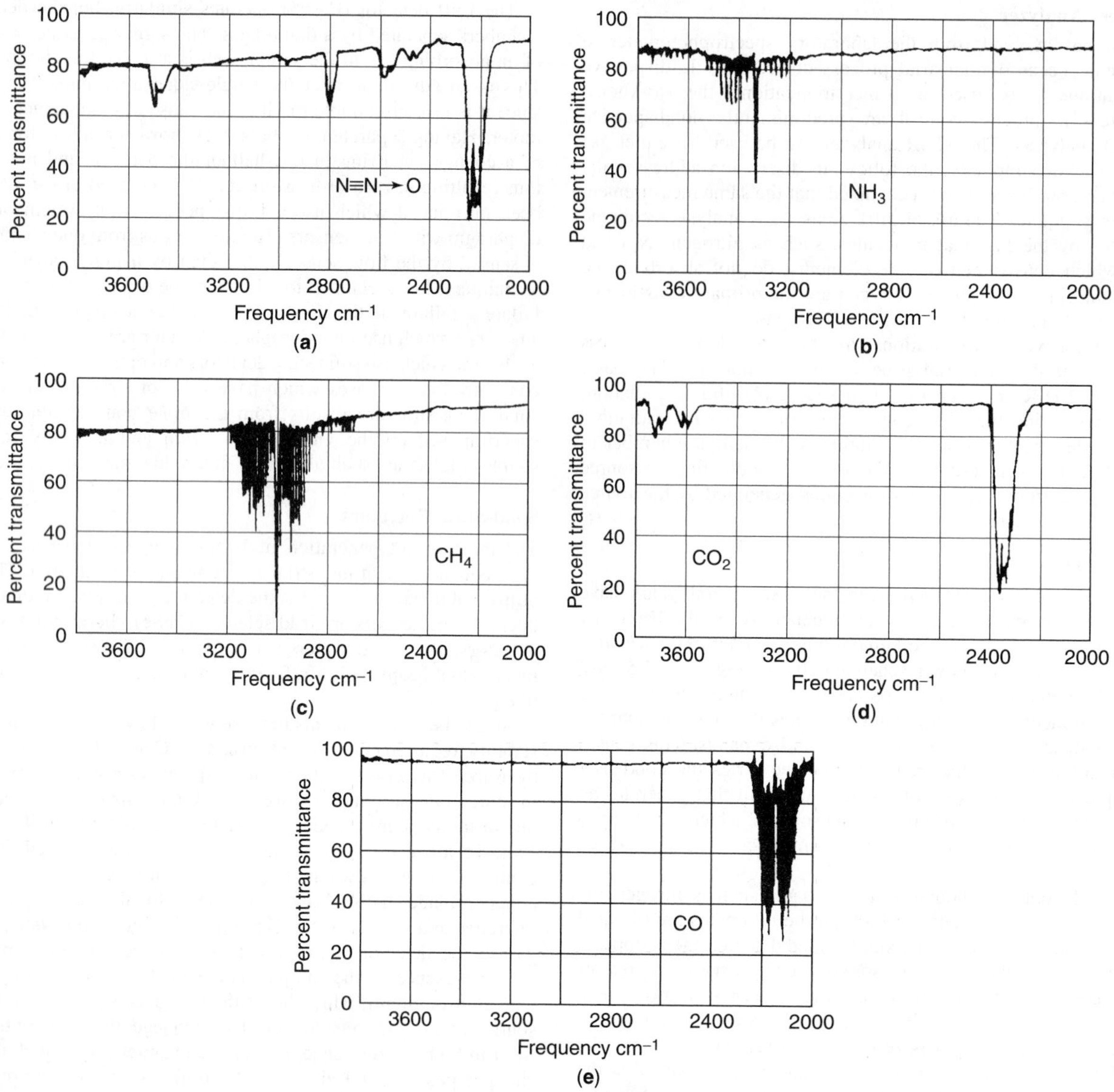

Fig. 7.   Infrared spectra: (**a**) nitrous oxide, (**b**) ammonia, (**c**) methane, (**d**) carbon dioxide, (**e**) carbon monoxide. (*Sadtler Research Laboratories*)

show that this spectral window will work well as a reference for measuring CO, CO_2, CH_4, or NH_3. It would be unsuited for nitrous oxide.

The dual-channel (reference filter) configuration is shown in Fig. 8. The reference filter and the sample filter, which define the spectral region of interest, are mounted on a spinning chopper wheel. As the chopper spins, it alternately positions the filters in the optical path. The signal is demodulated in a similar manner to the dual-beam approach. In order for the reference filter to be effective, it is important that the performance of the source and detector behave the same in the sample and reference spectral regions. The spectral properties of the source and detector are functions of temperature and must be controlled precisely, or zero and span drift will result.

Uses of IR Analyzers

Many gases can be monitored with IR analyzers. The petrochemical industry, for example, monitors CO_2 in the manufacture of ethylene oxide and ammonia. Acetylene is monitored during the production of acetylene and vinyl chloride. The metals industries use the analyzers to monitor CO_2 in steel converting and soaking pit operations. Carbon monoxide is monitored during heat treating, aluminum powder processing, and tin plant annealing. The food industry monitors CO_2 in greenhouses, storage facilities, and fermentation processes. The analyzers also find use in monitoring for explosive and toxic hazards, as well as for stack gases in pollution control systems.

Typical gases and concentrations measured by NDIR analyzers are given in Table 1.

Other Applications

Optical-electronic devices of many kinds have been designed to determine the direction of weakly radiating remote objects by means of detection of their infrared emission. Detailed maps of the earth's surface can be made from aircraft at night by observing the varying infrared emission of the ground. For security, personnel can be detected in total darkness by infrared radiation. Such devices require the detection of low-level

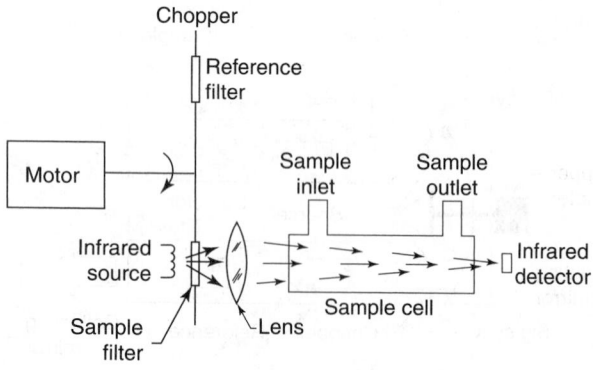

Fig. 8.   Dual-channel nondispersive infrared analyzer with a solid-state detector. (*Infrared Industries, Inc*)

TABLE 1. REPRESENTATIVE GASES AND CONCENTRATIONS MEASURED BY NDIR ANALYZERS

Gas	Range (Full scale concentration)	
	Minimum (ppm)	Maximum (percent)
Ammonia	1000	100
Butane	300	100
Carbon monoxide	500	100
Carbon dioxide	200	100
Ethylene	1000	100
Ethane	500	100
Ethylene oxide	1500	100
Hexane	400	5
Methane	400	100
Nitrous oxide	150	100
Propane	400	100
Sulfur dioxide	400	100
Water vapor	1500	10

Fig. 10.   Infrared radiation source with temperature controller. (*Infrared Industries, Inc*)

radiation in the intermediate-infrared region. Optical lenses or mirrors are used to collect the observed radiation and concentrate it onto the sensitive infrared detector. High-gain, low-noise amplifiers must be used to increase the weak signal from the detector. The wavelength of detection is such that angular resolution capability, as set by diffraction, is much greater with infrared devices than with radars. Infrared photography is described under **Photography and Imagery**.

Astronomy. The potential of using the infrared portion of the electromagnetic spectrum for investigating celestial bodies and interstellar space has been considered by some astronomers for a number of years. This concept was first proposed by William Herschel. It has only been within the last few decades, however, that serious experiments in infrared astronomy have been made.

Solar Energy

The sun's total radiation output is approximately equivalent to that of a blackbody at 10,350°R (5750 K). However, its maximum intensity occurs at a wavelength that corresponds to a temperature of 11,070°R (6150 K) as given by Wien's displacement law. A figure plotting solar irradiance versus spectral distribution of solar energy is given in Fig. 9. See also **Solar Energy**.

Infrared Laser Chemistry

The application of IF lasers to chemical reactions is described in entry on **Photochemistry and Photolysis**.

Infrared Radiation Sources

Blackbody radiation sources are accurate radiant energy standards of known flux and spectral distribution. They are used for calibrating other infrared sources, detectors, and optical systems. The radiating properties of a blackbody source are described by Planck's law. Energy distribution

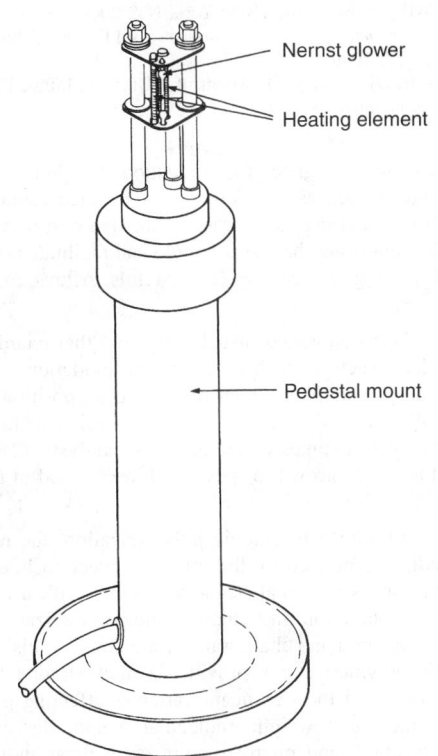

Fig. 11.   Nernst glower assembly. (*Infrared Industries, Inc*)

for blackbody sources at different temperatures is shown in previously mentioned Fig. 1. A blackbody radiation source with its temperature controller is shown in Fig. 10. The specifications of this unit include: temperature range, 100–1000°C; cavity diameter, 1 inch (25.4 mm); aperture diameter (in steps) from 0.0125 in (0.3175 mm) to 0.6 in (15.24 mm); emissivity, 0.99 ± .01; field of view, 20 degrees; internal thermocouple, platinum versus platinum-10% rhodium. Radiation sources are also available for use in the infrared region which have an emissivity somewhat less than one. The Nernst glower is a graybody source, which finds frequent use in spectrophotometers. The emissivity of a Nernst glower (Fig. 11) is a function of wave length, averaging approximately 0.6 from 2 to 15 micrometers. The glower element will not conduct electricity when cold, thus the unit must be heated to approximately 400°C before it will begin to conduct. The unit will operate at a temperature from 1500–1950 K with an expected life of several hundred hours.

Additional Reading

Caniou, J.: *Passive Infrared Detection: Theory and Applications*, Kluwer Academic Publishing, Norwell, MA, 1999.

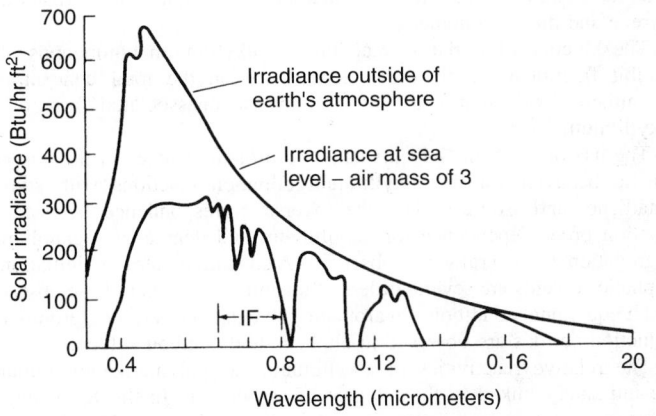

Fig. 9.   Spectral distribution of solar energy

Capper, P. and C.T. Elliott: *Infrared Detectors and Emitters: Materials and Devices,* Kluwer Academic Publishers, Norwell, MA, 2000.

Coleman, P.B.: *Practical Sampling Techniques for Infrared Analysis,* CRC Press, LLC., Boca Raton, FL, 1993.

Colthup, N.B., L.H. Daly and S.E. Wiberley: *Introduction to Infrared and Raman Spectroscopy,* 3rd Edition, Academic Press, Inc., San Diego, CA, 1999.

Drexage, M.G. and C.T. Moynihan: "Infrared Optical Fibers," *Sci. Amer.,* 110 (November 1988).

Durham, R.M.: "Infrared Process Analyzers," in *Process/Industrial Instruments and Controls Handbook* 5th Edition, (G.K. McMillan, Editor), McGraw-Hill Companies, Inc., New York, NY, 1999.

Eisenbud, M. and T.F. Gesell: *Environmental Radioactivity from Natural, Industrial, and Military Sources,* 4th Edition, Morgan Kaufmann Publishers, Orlando, FL, 1997.

Guelachvili, G. and K.N. Rao: *Handbook of Infrared Standards Two: With Spectral Coverage between 1.4 UM–4 UM and 6.2 UM–7.7 UM,* Academic Press, Inc., San Diego, CA, 1997.

Hill, D.W. and T. Powell: *Non-Dispersive Infrared Gas Analysis in Science, Medicine, and Industry,* Plenum Press, New York, NY, 1968.

Jacobs, P.A.: *Thermal Infrared Characterization of Ground Targets and Backgrounds,* SPIE International Society for Optical Engineering, Bellingham WA, 1996.

Jha, A.R.: *Infrared Technology: Applications to Electro-Optics, Photonic Devices and Sensors,* John Wiley & Sons, Inc., New York, NY, 2000.

Knoll, G.F.: *Radiation Detection and Measurement,* 3rd Edition, John Wiley & Sons, Inc., New York, NY, 1999.

Stoutland, P.O., R.B. Dyer and W.H. Woodruff: "Ultrafast Infrared Spectroscopy," *Science,* 1913 (September 25, 1992).

INGOT. A casting designed for reduction by hot working to a semifinished product such as a billet or to a finished product such as a bar, plate, or sheet. Steel ingots are cast in massive cast-iron ingot molds, which extract the heat faster than a sand mold and facilitate both the casting and handling of the ingots. See also **Iron Metals, Alloys, and Steels**.

INHIBITOR. (1) A compound (usually organic) that retards or stops an undesired chemical reaction, such as corrosion, oxidation or polymerization. Examples are acetanilide, which retards decomposition of hydrogen peroxide and salicylic acid, used to prevent prevulcanization of rubber. Such substances are sometimes called negative catalysts. (2) A biological antagonist used to retard growth of pests and insects and in medicine.

INITIATORS (Anionic). In anionic polymerization, the reactive propagating intermediate generated by the initiation reaction is an anion, i.e., a species which carries a formal negative charge, with a corresponding positively charged counterion. In living anionic polymerization, the kinetic steps of chain termination and chain transfer are absent. This unique aspect of many anionic polymerizations provides a methodology for preparing polymers with control of the significant variables affecting polymer properties including molecular weight, molecular weight distribution, block copolymer composition, and microstructure, as well as molecular architecture (linear, branched, and cyclic macromolecules). An important consideration for preparation of polymers with well-defined structures and low degrees of compositional heterogeneity is the choice of a suitable initiator.

In general, an appropriate initiator is a species which has approximately the same structure and reactivity as the propagating anionic species, i.e., the pK_a of the conjugate acid of the propagating anion should correspond closely to the pK_a of the conjugate acid of the initiating species. If the initiator is too reactive, side reactions between the initiator and monomer can occur; if the initiator is not reactive enough, then the initiation reaction may be slow or inefficient.

The general relationship between monomer structural type, pK_a and appropriate initiating species is shown in Table 1. Those monomers which form the least stable anions, i.e., which have the largest values of pK_a for the corresponding conjugate acids, are the least reactive monomers and require the use of the most reactive initiators as shown in Table 1.

Alkali Metals

The use of alkali metals for anionic polymerization of diene monomers is primarily of historical interest. The electron-transfer mechanism of the anionic polymerization of styrenes and 1,3-dienes initiated by alkali metals has been described in detail; the dimerization of radical anion intermediates is the important step.

TABLE 1. RELATIONSHIPS BETWEEN MONOMER REACTIVITY, CARBANION pK_A STABILITY, AND SUITABLE INITIATORS

Monomer type	pK_a[a] In DMSO[b]	In H_2O	Initiators[c]
ethylene	56		RLi
dienes and styrenes	44		NH_2^-, RLi, RMt
	43		aromatic radical anions,[d] cumyl K, Mt,
acrylonitrile	32		RMgX
alkyl methacrylates, alkyl acrylates	30–31	27–28	fluorenyl–, $RArC^{-2}$, ketyl radical anions[e]
vinyl ketones	26	19	
oxiranes	29–32	16–18	RO–
thiiranes	17	12–13	
nitroalkenes	17	10–14	
siloxanes		10–14	RO–, OH–
β-lactones	12	4–5	RCOO–
alkyl cyanoacrylates	12.8		HCO^{3-}, H_2O
vinylidene cyanide	11	11	

[a] pK_a of the conjugate acid of the anionic propagating intermediate.
[b] pK_a values in DMSO.
[c] Mt refers generally to alkali metals (Li, Na, K, Rb, Cs).
[d] For example, naphthalene radical anion [naphthalene structure]$^{\cdot-}$ with counterion (Li+, Na+, K+).
[e] $Ar_2CO^{\cdot-}$.

Aromatic Radical Anions

Many aromatic hydrocarbons react with alkali metals in polar aprotic solvents to form stable solutions of the corresponding radical anions. These solutions can be analyzed by uv-visible spectroscopy and stored for further use.

Sodium naphthalene and other aromatic radical anions react with monomers such as styrene by reversible electron transfer to form the corresponding monomer radical anions which rapidly dimerize.

Monomers which can be polymerized with aromatic radical anions include styrenes, dienes, epoxides, and cyclosiloxanes. Aromatic radical anions which are too stable do not efficiently initiate polymerization of less reactive monomers; thus the anthracene radical anion cannot initiate styrene polymerization.

Alkyllithium Compounds

Anionic polymerization of vinyl monomers can be effected with a variety of organometallic compounds; alkyllithium compounds are the most useful class. A variety of simple alkyllithium compounds are available commercially. Most simple alkyllithium compounds are soluble in hydrocarbon solvents such as hexane and cyclohexane and they can be prepared by reaction of the corresponding alkyl chlorides with lithium metal.

Simple alkyllithium compounds are aggregated in solution, in the solid state, and even in the gas phase. The important differences between the various alkyllithium compounds are their degrees of aggregation in solution and their relative reactivity as initiators for anionic polymerization of styrene and diene monomers.

The kinetics of initiation reactions of alkyllithium compounds often exhibit fractional kinetic order dependence on the total concentration of initiator, consistent with initiation by the unassociated form of the alkyllithium.

The use of aliphatic solvents causes profound changes in the observed kinetic behavior for the alkyllithium initiation reactions with styrene, butadiene, and isoprene, i.e., the inverse correspondence between the reaction order dependence for alkyllithium and degree of organolithium aggregation is generally not observed. Also, initial rates of initiation in aliphatic solvents are several orders of magnitude less than those observed, under equivalent conditions, in aromatic solvents. Furthermore, pronounced induction periods are observed in aliphatic hydrocarbon solvents.

The relative reactivities of alkyllithiums as polymerization initiators are intimately linked to their degree of association. In the following the average degree of association in hydrocarbon solution, where known, is indicated in brackets after the alkyllithium. For styrene polymerization, the relative reactivity of alkyllithium initiators is menthyllithium > sec-C_4H_9Li > i-C_3H_7Li > i-C_4H_9Li > n-C_4H_9Li > t-C_4H_9Li. For diene polymerization, menthyllithium > sec-C_4H_9Li > i-C_3H_7Li > t-C_4H_9Li > i-C_4H_9Li > n-C_4H_9Li.

Alkyllithium compounds are primarily used as initiators for polymerizations of styrenes and dienes.

Quantitative Analysis of Alkyllithium Initiator Solutions. The amount of carbon-bound lithium is calculated from the difference between the total amount of base determined by acid titration and the amount of base remaining after the solution reacts with either benzyl chloride, allyl chloride, or ethylene dibromide.

Copolymerization Initiators. The copolymerization of styrene and dienes in hydrocarbon solution with alkyllithium initiators produces a tapered block copolymer structure because of the large differences in monomer reactivity ratios for styrene ($r_s < 0.1$) and dienes ($r_d > 10$). In order to obtain random copolymers of styrene and dienes, it is necessary to either add small amounts of a Lewis base such as tetrahydrofuran or an alkali metal alkoxide (MtOR, where Mt = Na, K, Rb, or Cs).

Difunctional Initiators

These initiators are of considerable interest for the preparation of triblock copolymers, telecheclic polymers, and macrocyclic polymers.

Aromatic radical anions, such as lithium naphthalene or sodium naphthalene, are efficient difunctional initiators. However, the necessity of using polar solvents for their formation and use limits their utility for diene polymerization.

The methodology for preparation of hydrocarbon-soluble, dilithium initiators is generally based on the reaction of an aromatic divinyl precursor with two moles of butyllithium.

Although a plethora of divinyl aromatic compounds have been investigated as precursors for hydrocarbon-soluble dilithium initiators, the only system which has been demonstrated to produce a hydrocarbon-soluble dilithium initiator is based on 1,3-bis (1-phenylethenyl)benzene.

Functionalized Initiators

The use of alkyllithium initiators which contain functional groups provides a versatile method for the preparation of end functionalized polymers and macromonomers. For a living anionic polymerization, each functionalized initiator molecule produces one macromolecule with the functional group from the initiator residue at one chain end and the active carbanionic propagating species at the other chain end.

Other Initiators

Other initiators include cumyl potassium, 1,1-diphenylmethylcarbanions, fluorenyl carbanions, enolate initiators, and alkoxide-type initiators.

Health and Safety Factors

Hydrocarbon solutions of alkyllithium compounds are air and moisture sensitive and should be either handled in an inert atmosphere or by using syringes using recommended procedures for handling air-sensitive compounds. Alkyllithium reagents react with acidic compounds that contain reactive hydrogens such as water, alcohols, phenols, acids, and even primary and secondary amines. The reaction of butyllithium with water produces butane and lithium hydroxide, which can lead to spontaneous ignition in the presence of oxygen. Contact of alkyllithium solutions with air does not generally lead to spontaneous ignition; however, if large surface areas are formed, for example in a spill, spontaneous ignition can occur. Carbon dioxide fire extinguishers must not be used because carbon dioxide reacts exothermically with alkyllithium compounds. It is prudent to have an all-purpose fire extinguisher available when working with these organometallic compounds. Suitable fire-extinguishing chemicals include powdered limestone and powders containing sodium chloride and sodium bicarbonate.

RODERIC P. QUIRK
University of Akron
VICTOR M. MONROY
General Tire, Inc.

Additional Reading

Bywater, S.: in J.I. Kroschwitz, ed., *Encyclopedia of Polymer Science and Engineering*, Vol. 2, 2nd Edition, John Wiley & Sons, Inc., New York, NY, 1985.

Fontanille, M.: in G.C. Eastmond and co-eds., *Comprehensive Polymer Science*, Vol. 3, "Chain Polymerization I," Pergamon Press, Elmsford, NY, 1989.

Morton, M.: *Anionic Polymerization: Principles and Practice*, Academic Press, Inc., New York, NY, 1982.

Szwarc, M.: *Carbanions, Living Polymers and Electron Transfer Processes*, Wiley-Interscience, New York, NY, 1968.

INITIATORS (Cationic).

Cationic polymerization may be induced by a variety of physical (high energy radiation, direct or indirect uv radiation, electroinitiation) and chemical methods (protic acids, Friedel-Crafts acids, stable cation salts, cation donor in conjunction with a Friedel-Crafts acid). The most important initiating system is the cation donor (initiator)/Friedel-Crafts acid (coinitiator) system, which has found many applications. Butyl rubber, a copolymer of isobutylene and isoprene containing 0.5–2.5% isoprene to make vulcanization possible, is the most important commercial polymer made by cationic polymerization. Another important commercial application of cationic polymerization is the manufacture of polybutenes, low molecular weight copolymers of isobutylene, and a smaller amount of other butenes used in adhesives, sealants, lubricants, viscosity improvers, etc.

Unless one is working with superdried systems or in the presence of proton traps, adventitious water is always present as a proton source. Polymerization rates, monomer conversions, and to some extent polymer molecular weights are dependent on the amount of protic impurities; therefore, well-established drying methods should be followed to obtain reproducible results.

In place of a proton source, i.e., a Brønsted acid, a cation source such as an alkyl halide, ester, or ether can be used in conjunction with a Friedel-Crafts acid. Initiation with the ether-based initiating systems in most cases involves the halide derivative which arises upon fast halidation by the Friedel-Crafts acid, MX_n. The efficiency of the initiator/coinitiator system depends greatly on the monomer in question. As a general rule, the stability (reactivity) of the initiating cation should be close to that of the propagating chain end. Since initiation involves two subsequent events, i.e., ion generation and cationation, species on the two extremes are less active or may be completely inactive, because they form ionic species very slowly and/or in extremely low concentration, or would form ions in high concentration that are, however, too stable to cationate the monomer.

The activity of an initiating system is also affected by the nature of the Friedel-Crafts acid. The following Friedel-Crafts acidity scale can be established: $BF_3 < AlCl_3 < TiCl_4 < BCl_3 < SbF_5 < SbCl_5 < BBr_3$. The advantage of the $TiCl_4$ and the aluminum-based systems is their relative insensivity toward solvent polarity. The activity of the BCl_3- or BBr_3-based system is greatly solvent-dependent, i.e., sufficient activity only occurs in polar solvent.

Solvent polarity and temperature also influence the results. The dielectric constant and polarizability, however, are of little predictive value for the selection of solvents relative to polymerization rates and behavior. Evidently every system has to be examined independently. In cationic polymerization of vinyl monomers, chain transfer is the most significant chain-breaking process. The activation energy of chain transfer is higher than that of propagation; consequently, the molecular weight of the polymer increases with decreasing temperature.

Initiation by a carbocation source provides control of the head-group (controlled initiation) when used in conjunction with a Friedel-Crafts acid (e.g., $(C_2H_5)_3Al$, $(CH_3)_3Al$, $(C_2H_5)_2AlCl$, BCl_3 for isobutylene, or I_2 and zinc halides for vinyl ethers) where chain transfer to monomer is absent or negligible, or in the presence of a proton trap to abort chain transfer to monomer. That is, initiation from tertiary, allylic, and benzylic halides gives rise to macromolecules carrying tertiary, allylic, and benzylic head-groups. Initiation by halogens results in head-groups carrying the halogen. Controlled initiation, however, is achieved only when polymer formation from adventitious protic impurities is also absent or negligible.

A special case of controlled initiation is the inifer method. The word inifer (from *ini*tiator trans*fer* agents) describes compounds that function simultaneously as initiators and as chain transfer agents. The inifer technique provided the first carbocationic route toward the synthesis of telechelic (α, ω functional) polymers.

Although it was long believed that most Friedel-Crafts acids, particularly halides of boron, titanium, and tin, require an additional cation source to initiate polymerization, recent results show that in many systems Friedel-Crafts acids alone are able to initiate cationic polymerization.

The mechanism of initiation appears to be halometalation, as originally suggested by Sigwalt and Olah.

Many initiating systems used in the cationic polymerization of vinyl monomers can also be used to initiate ring-opening polymerization of cyclic monomers such as cyclic ethers, acetals, lactams, lactones, and siloxanes. Polymerization of cyclic monomers may involve different type of ionic as well as covalent growing species. Under certain conditions, termination processes may be absent. The polymerization of cyclic monomers, however, is almost always complicated by inter- and intramolecular chain transfer to polymer. The later results in cyclic oligomer formation. The extent of cyclic oligomer formation can be minimized in the polymerization of epoxides by the recently discovered activated monomer mechanism. Cyclic ether and acetal polymerizations are also important commercially. Polymerization of tetrahydrofuran is used to produce polyether diol, and polyoxymethylene, an excellent engineering plastic, is obtained by the ring-opening polymerization of trioxane with a small amount of cyclic ether or acetal comonomer to prevent depolymerization.

Recently a variety of initiating systems have been described that allow not only controlled initiation but also controlled propagation in the polymerization of vinyl monomers. In these living polymerization systems, chain braking (chain transfer and irreversible termination) is absent. The key to these living polymerizations is the high stability of the growing end, where the nucleophilic counteranion interacts strongly with the cationic active site. Living polymerizations have also been reported with initiating systems, forming nonnucleophilic counteranions in the presence of added Lewis bases (electron donors) and, in the presence of common ion salts, shifting the ionic dissociation equilibrium toward the nondissociated species. With these systems, rapid advances have been made toward the synthesis of well-defined materials with controlled architecture, molecular weight, molecular weight distributions and end-functionalities by cationic polymerization.

Since the discovery of living cationic systems, cationic polymerization has progressed to a new stage, where the synthesis of designed materials is now possible.

RUDOLF FAUST
University of Massachusetts, Lowell

Additional Reading

Kennedy, J.P. and E. Marechal: *Carbocationic Polymerization,* Wiley-Interscience, New York, NY, 1982.

Kennedy, J.P. and B. Ivan: *Designed Polymers by Carbocationic Macromolecular Engineering,* Hanser Publishers, Munich, Germany, 1991.

INITIATORS (Free-Radical).

Free-radical initiators are chemical substances that, under certain conditions, initiate chemical reactions by producing free radicals. Initiators contain one or more labile bonds that cleave homolytically when sufficient energy is supplied to the molecule. The energy must be greater than the bond dissociation energy (BDE) of the labile bond. Radicals are reactive chemical species possessing a free (unbonded or unpaired) electron. Radicals may also be positively or negatively charged species carrying a free electron (ion radicals). Initiator-derived radicals are very reactive chemical intermediates and generally have short lifetimes, i.e., half-life times less than 10^{-3} seconds.

The principal commercial initiators used to generate radicals are peroxides and azo compounds. Lesser amounts of carbon–carbon initiators and photoinitiators, and high energy ionizing radiation are also employed commercially to generate radicals.

There are three general processes for supplying the energy necessary to generate radicals from initiators: thermal processes, microwave or ultraviolet (uv) radiation processes, and electron transfer (redox) processes. Radicals can also be produced in high energy radiation processes. Once formed, radicals undergo two basic types of reactions: propagation reactions and termination reactions.

Radicals are employed widely in the polymer industry, where their chain-propagating behavior transforms vinyl monomers into polymers and copolymers. The mechanism of addition polymerization involves all three types of reactions discussed above, i.e., initiation, propagation by addition to carbon–carbon double bonds, and termination.

Two other important commercial uses of initiators are polymer cross-linking and polymer degradation.

Structure–Reactivity Relationships. Much has been written about the structure–reactivity of radicals. No single unifying concept has satisfactorily explained all radical reactions reported in the literature. A longstanding correlation of structure and reactivity involves comparisons of the energies required to homolytically break covalent bonds to hydrogen. It is assumed that this energy, the hydrogen bond dissociation energy (BDE), reflects the stability and the reactivity of the radical coproduced with the hydrogen atom. However, this assumption should really be limited to radical reactivity and selectivity in hydrogen atom abstraction reactions, and can be particularly misleading for reactions with polar transition states, in which radicals can behave either as nucleophiles or electrophiles. Nevertheless, the correlation of radical reactivity with BDE is quite useful. Table 1 shows some general BDE values for the formation of various carbon and oxygen radicals from various precursors. According to the theory, the higher the BDE, the higher the reactivity and the lower the stability of the radical formed by removal of a hydrogen atom.

The choice of an initiator for a given radical process depends on the reaction conditions and reactivity of the initiator. These two factors must be balanced so that the desired reaction is achieved.

Activation Parameters. Thermal processes are commonly used to break labile initiator bonds in order to form radicals. The amount of thermal energy necessary varies with the environment, but absolute temperature, T, is usually the dominant factor. The energy barrier, the minimum amount of energy that must be supplied, is called activation energy, E_a. A third important factor, known as the frequency factor, A, is a measure of bond motion freedom (translational, rotational, and vibrational) in the activated complex or transition state. E_a and A are known as the activation parameters and, along with T, are related to the decomposition rate, k_d, by the equations:

$$k_d = Ae^{(-E_a/RT)} \text{ or } \ln k_d = \ln(A) - E_a/RT$$

Half-Life. Once these activation parameters have been determined for a initiator, half-life times at a given temperature, i.e., the time required for 50% decomposition at a selected temperature, and half-life temperatures for a given period, i.e., the temperature required for 50% decomposition of an initiator over a given time, can be calculated. Half-life data are useful for comparing the activity of one initiator with another when the half-life data are determined in the same solvent and at the same concentration and, preferably, when the initiators are of the same class.

Commercial initiators are primarily organic and inorganic peroxides, aliphatic azo compounds, certain organic compounds with labile carbon–carbon bonds, and photoinitiators.

Organic Peroxides

Organic peroxides are compounds possessing one or more oxygen–oxygen bonds. They have the general structure ROOR′ or ROOH, and decompose thermally by the initial cleavage of the oxygen–oxygen bond to produce two radicals:

$$ROOR' \rightarrow RO \cdot + \cdot OR'$$

Following radical generation, the radicals produced (RO· and R′O·) can initiate the desired reaction. However, when the radicals are generated in commercial applications, they are surrounded by a solvent, monomer, or polymer "cage." When the cage is solvent, the radical must diffuse out of this cage to react with the desired substrate. When the cage is monomer, the radical can react with the cage wall or diffuse out of the cage. When the cage is polymer, reaction with the polymer can occur in the cage. Unfortunately, other reactions can occur within the cage and can adversely affect efficiency of radical generation and radical reactivity. If the solvent reacts with the initiator radical, then solvent radicals may participate in the desired reaction.

TABLE 1. BOND DISSOCIATION ENERGIES

Precursor	BDE, kJ/mol[a]
$(R)_3C-H$	381
$(R)_2CH-H$	406
RCH_2-H	418
CH_3-H	439
$RO-H$	439
$RCOO-H$	444
C_6H_5-H	469
$HO-H$	498

[a] To convert kJ/mol to kcal/mol, divide by 4.184.

TABLE 2. COMMERCIAL ORGANIC PEROXIDE CLASSES

Organic peroxide class	Structure[a]	10-h $t_{1/2}$[b,c], °C
diacyl peroxides	$R\text{—}\overset{\overset{O}{\|}}{C}\text{—OO—}\overset{\overset{O}{\|}}{C}\text{—}R$	21–75
dialkyl peroxydicarbonates	$RO\text{—}\overset{\overset{O}{\|}}{C}\text{—OO—}\overset{\overset{O}{\|}}{C}\text{—}OR$	49–51[d]
tert-alkyl peroxyesters	$R\text{—}\overset{\overset{O}{\|}}{C}\text{—OO—}t\text{-}R$	38–107
OO-tert-alkyl O-alkyl monoperoxycarbonates	$RO\text{—}\overset{\overset{O}{\|}}{C}\text{—OO—}t\text{-}R$	99–100
di(tert-alkylperoxy)ketals	$\overset{R'\quad OO\text{-}t\text{-}R}{\underset{R\quad OO\text{-}t\text{-}R}{\diagdown C \diagup}}$	92–110
di-tert-alkyl peroxides	$t\text{-}R\text{—OO—}t\text{-}R$	115–128
tert-alkyl hydroperoxides	$t\text{-}R\text{—OO—H}$	e
ketone peroxides	$HOO\text{—}\overset{\overset{R'}{\|}}{\underset{R}{C}}\text{—}(OO\text{—}\overset{\overset{R'}{\|}}{\underset{R}{C}})_x\text{—OOH}$	e
	+ other structures	

[a] $x = 0$ or 1.
[b] Temperature at which $t_{1/2} = 10$ h.
[c] In benzene, unless otherwise noted.
[d] In trichloroethylene (TCE).
[e] Not applicable.

Two secondary propagating reactions often accompany the initial peroxide decomposition: radical-induced decompositions and β-scission reactions. Both reactions affect the reactivity and efficiency of the initiation process.

Approximately 100 different organic peroxide initiators, in well over 300 formulations, are commercially produced throughout the world, primarily for the polymer and resin industries.

The eight classes of organic peroxides that are produced commercially for use as initiators are listed in Table 2. Included are the 10-h half-life temperature (i.e., the temperature at which 50% of the peroxide decomposes in 10 h) ranges for the members of each peroxide class.

Inorganic Peroxides

Inorganic peroxide–redox systems have been employed for initiating emulsion homo- and copolymerizations of vinyl monomers. These systems include hydrogen peroxide–ferrous sulfate, hydrogen peroxide–dodecyl mercaptan, potassium peroxydisulfate–sodium bisulfate, and potassium peroxydisulfate–dodecylmercaptan. Potassium peroxydisulfate, KSO (or the corresponding sodium or ammonium salt), is an inorganic peroxide that is used widely in emulsion polymerization (e.g., latexes, rubbers, etc), usually in combination with a reducing agent.

When handling and using peroxide initiators, care should be exercised because they are thermally sensitive and decompose (sometimes violently) when exposed to excessive temperatures, especially where they are in their pure or highly concentrated states. However, they are useful as initiators because of their thermal instability. What may be a safe temperature for one peroxide can be an unsafe temperature for another, since peroxide initiators encompass a wide activity range. Because some peroxides are shock- or friction-sensitive in the pure state, they are generally desensitized by formulating them into solutions, pastes, or powders with inert diluents. All manufacturers' literature should be carefully scrutinized and the peroxide safety literature should be reviewed before handling and using specific peroxide initiator compositions.

Azo Compounds

Generally, the commercially available azo initiators are of the symmetrical azonitrile type:

$$R\text{—}\overset{\overset{R'}{\|}}{\underset{CN}{C}}\text{—N}=\text{N—}\overset{\overset{R'}{\|}}{\underset{CN}{C}}\text{—}R$$

TABLE 3. COMMERCIAL AZO INITIATORS

Name	Structure
2,2′-azobis[4-methoxy-2,4-dimethyl]-pentanenitrile	$CH_3\text{—}\overset{\overset{OCH_3}{\|}}{\underset{CH_3}{C}}\text{—}CH_2\text{—}\overset{\overset{CH_3}{\|}}{\underset{C\equiv N}{C}}\text{—N}=\text{N—}\overset{\overset{CH_3}{\|}}{\underset{C\equiv N}{C}}\text{—}CH_2\text{—}\overset{\overset{OCH_3}{\|}}{\underset{CH_3}{C}}\text{—}CH_3$
2,2′-azobis[2,4-dimethyl]-pentanenitrile	$HC\text{—}CH_2\text{—}\overset{\overset{CH_3}{\|}}{\underset{C\equiv N}{C}}\text{—N}=\text{N—}\overset{\overset{CH_3}{\|}}{\underset{C\equiv N}{C}}\text{—}CH_2\text{—}CH$ (with CH_3 groups)
2,2′-azobis[isobutyronitrile]	$CH_3\text{—}\overset{\overset{CH_3}{\|}}{\underset{C\equiv N}{C}}\text{—N}=\text{N—}\overset{\overset{CH_3}{\|}}{\underset{C\equiv N}{C}}\text{—}CH_3$
2,2′-azobis[2 methylbutyronitrile]	$CH_3CH_2\text{—}\overset{\overset{CH_3}{\|}}{\underset{C\equiv N}{C}}\text{—N}=\text{N—}\overset{\overset{CH_3}{\|}}{\underset{C\equiv N}{C}}\text{—}CH_2CH_3$
1,1′-azobis[cyclohexanecarbonitrile]	(two cyclohexane rings each bearing $C\equiv N$, joined by $\text{N}=\text{N}$)
4,4′-azobis[4-cyanovaleric acid]	$\overset{\overset{COOH}{\|}}{CH_2CH_2}\text{—}\overset{\overset{CH_3}{\|}}{\underset{C\equiv N}{C}}\text{—N}=\text{N—}\overset{\overset{CH_3}{\|}}{\underset{C\equiv N}{C}}\text{—}CH_2CH_2\text{—}COOH$
dimethyl-2,2′-azobis-[2-methylpropionate]	$CH_3OOC\text{—}\overset{\overset{CH_3}{\|}}{\underset{CH_3}{C}}\text{—N}=\text{N—}\overset{\overset{CH_3}{\|}}{\underset{CH_3}{C}}\text{—}COOCH_3$
azobis[2-acetoxy-2-propane]	$CH_3\overset{\overset{O}{\|}}{C}O\text{—}\overset{\overset{CH_3}{\|}}{\underset{CH_3}{C}}\text{—N}=\text{N—}\overset{\overset{CH_3}{\|}}{\underset{CH_3}{C}}\text{—O}\overset{\overset{O}{\|}}{C}CH_3$
2,2′-azobis[2-amidinopropane]dihydrochloride	$H_2N\text{—}\overset{\overset{NH}{\|}}{C}\text{—}\overset{\overset{CH_3}{\|}}{\underset{CH_3}{C}}\text{—N}=\text{N—}\overset{\overset{CH_3}{\|}}{\underset{CH_3}{C}}\text{—}\overset{\overset{NH}{\|}}{C}\text{—}NH_2 \cdot 2\,HCl$

The symmetrical azonitriles are solids with limited solubilities in common solvents. Some commercial aliphatic azo compounds and their 10-h half-life temperatures are listed in Table 3.

Care should be exercised in handling and using azo initiators in their pure and highly concentrated states, because they are thermally sensitive and can decompose rapidly when overheated. Although azonitriles are generally less sensitive to contaminants, the same cautions that apply to peroxides also should be applied to handling and using azo initiators. The manufacturers' safety literature should be read carefully. The potential toxicity hazards of decomposition products must be considered when using azonitriles.

See also **Azo and Diazo Compounds**.

Carbon–Carbon Initiators

Carbon–carbon initiators are hexasubstituted ethanes that undergo carbon–carbon bond scission when heated to produce radicals. The thermal stabilities of the hexasubstituted ethanes decrease rapidly as the sizes of the alkyl groups increase. The 10-h half-life temperature range of this class of initiators is very broad, extending from about 100°C to well above 600°C. An extensive compilation of half-life data on carbon–carbon initiators has been published. The commercially available carbon–carbon initiators are tetrasubstituted 1,2-diphenylethanes which undergo homolyses to generate low energy, tert-aralkyl radical pairs. Three carbon–carbon initiators are currently available commercially, 2,3-dimethyl-2,3-diphenylbutane (**1**), 3,4-dimethyl-3,4-diphenylhexane (**2**), and 1,1,2,2-tetraphenyl-1,2-bis(trimethylsiloxy)-ethane (**3**).

(1)   (2)

(3)

Other Radical Generating Systems

There are many chemical methods for generating radicals reported in the literature that do not involve conventional initiators. Most of these radical-generating systems cannot broadly compete with the use of conventional initiators in industrial polymer applications owing to cost or efficiency considerations. However, some systems may be well-suited for initiating specific radical reactions or polymerizations, e.g., grafting of monomers to cellulose using ceric ion.

Initiation Through Radiation and Photoinitiators

High energy ionizing radiation sources (e.g., x-rays, γ-rays, α-particles, β-particles, fast neutrons, and accelator-generated electrons) can generate radical sites on organic substrates. If the substrate is a vinyl monomer, radical polymerization can occur. If the substrate consists of a polymer and a vinyl monomer, then polymer cross-linking, degradation, grafting of the monomer to the polymer, and homopolymerization of the monomer can all occur. Radical polymerizations of vinyl monomers with ionized plasma gases have been reviewed.

Initiation of radical reactions with uv radiation is widely used in industrial processes. In contrast to high energy radiation processes where the energy of the radiation alone is sufficient to initiate reactions, initiation by uv irradiation usually requires the presence of a photoinitiator, i.e., a chemical compound or compounds that generate initiating radicals when subjected to uv radiation. There are two types of photoinitiator systems: those that produce initiator radicals by intermolecular hydrogen abstraction [benzophenone, 4-phenylbenzophenone, xanthone, thioxanthone, 2-chlorothioxanthone, 4, 4'-bis(N, N'-dimethylamino)benzophenone (Michler's ketone), benzil, 9,10-phenanthraquinone, and 9,10-anthraquinone] and those that produce initiator radicals by photocleavage [α, α-dimethyl-α-hydroxyacetophenone, (1-hydroxycyclohexyl)-phenylmethanone, benzoin ethers (methyl, ethyl, isobutyl), α, α-dimethoxy-α-phenylacetophenone, α, α-diethoxyacetophenone, 1-phenyl-1,2-propanedione, 2-(O-benzoyl)-oxime, diphenyl(2,4,6-trimethylbenzoyl)phosphine oxide, and α-dimethyl-amino-α-ethyl-α-benzyl-3,5-dimethyl-4-morpholinoacetophenone].

Economic Aspects

The principal worldwide producers of organic peroxide initiators (and their trade names) include Elf Athochem (Luperco, Luperox, Lupersol, Lucidol (U.S.), Luchem, Alperox, Decanox, Peroximon, and Retilox), Akzo-Nobel (Trigonox, Perkadox, Cadox, Cadet, Laurox, Liladox, Kenodox, Lucidol (Europe), Butanox, and Cyclonox), Aztec (Aztec), Peroxid-Chemie (Interox), Witco (Esperox, Esperal, USP, Quickset, and Hi Point), Nippon Oil & Fats Company (Nyper, Perbutyl, Percumyl, Perhexa, Permek, and Peroyl), Norac (Superox), Hercules (DiCup, VulCup), and Sanken Kako (Sanperox). The principal worldwide producers of organic azo initiators are DuPont (Vazo), Elf Atochem (Ficel), and Wako. The worldwide market for organic azo initiators is small, being only about 10% of the market for organic peroxide initiators. Ciba Geigy is a significant supplier of photoinitiators (Darocur, Irgacure). The market for these initiators has been reviewed. Because most of the consumption of organic peroxides and azo initiators is in the developed countries, market growth in the 1990s is expected to be modest, i.e., 2–3% annually.

JOSÉ SANCHEZ
TERRY N. MYERS
Elf Atochem North America, Inc.

Additional Reading

Ando, W. ed.: *Organic Peroxides,* John Wiley & Sons, Inc., New York, NY, 1992.
Pappas, S.P. ed.: *UV Curing: Science and Technology,* Technology Marketing Corporation, Stamford, Conn., 1978, Chapt. 1.
Sheppard, C.S.: in J.I. Kroschwitz, ed., *Encyclopedia of Polymer Science and Engineering,* Vol. 2, Wiley-Interscience, New York, NY, 1985, pp. 143–157.
Swern, D. ed.: *Organic Peroxides,* Vols. I, II, and III, Wiley-Interscience, New York, NY, 1970, 1971, and 1972.

INORGANIC CHEMISTRY.

A major branch of chemistry that is generally considered to embrace all substances except hydrocarbons and their derivatives, or substances that are not compounds of carbon, with the exception of carbon oxides and carbon disulfide. The chemical compounds, which are based upon chains or rings of carbon atoms, which are termed organic compounds, are studied under the separate heading of organic chemistry. See also **Organic Chemistry**.

Inorganic chemistry covers a broad range of subjects, among which are atomic structure, crystallography, chemical bonding, coordination compounds, acid-base reactions, ceramics, and the various subdivisions of electrochemistry (electrolysis, battery science, corrosion, semiconduction, etc.). See also **Acids and Bases**; **Ceramics**; **Chemical Composition**; **Chemical Elements**; **Compound (Chemical)**; **Coordination Compounds**; **Corrosion**; and **Semiconductors**. It is important to state inorganic and organic chemistry often overlap, most importantly in the sub-discipline of organometallic chemistry. For example, chemical bonding applies to both disciplines; electrochemistry and acid-base reaction have their organic counterparts; catalysts and coordination compounds may be either organic or inorganic.

Inorganic chemistry is based upon physical chemistry and forms the basis for mineralogy and materials chemistry. It often overlaps with geochemistry, analytical chemistry, environmental chemistry and organometallic chemistry. See also **Mineralogy**; and **Physical Chemistry**.

The range of inorganic chemistry includes both molecular compounds, which exist as discrete molecules, and crystals, whose structures are described by infinite lattices of regularly-ordered atoms and which are studied by crystallography and solid-state chemistry. See also **Solid-State Chemistry**.

Regarding the importance of inorganic chemistry, R. T. Sanderson has written: "All chemistry is the science of atoms, involving an understanding of why they possess certain characteristic qualities and why these qualities dictate the behavior of atoms when they come together. All properties of material substances are the inevitable result of the kind of atoms and the manner in which they are attached and assembled. All chemical change involves a rearrangement of atoms. Inorganic chemistry [is] the only discipline within chemistry that ... examines specifically the differences among all the different kinds of atoms."

Major branches of inorganic chemistry include:

- Minerals, such as salt, asbestos, silicates, ...
- Metals and their alloys, like iron, copper, aluminum, brass, bronze, ...
- Compounds involving non-metallic elements, like silicon, phosphorus, chlorine, oxygen, for example water
- Metal complexes

Commercially important inorganic substances include silicon chips, transistors, LCD screens, fiber optical cables and many catalysts.

Organometallic chemistry combines aspects of organic chemistry with those of inorganic chemistry, and is formally defined as the study of compounds containing metal-carbon bonds, although many "organometallic compounds" contain no such bonds. See also **Organometallic Compounds**. Among the simplest organometallic compounds are the metal carbonyls, in which carbon monoxide binds to a metal through the carbon. Vitamin B12, whose active site is similar to that of hemoglobin, is a naturally-occurring, metabolically-important organometallic compound containing large organic components (corrin and protein) and a metal, cobalt, bonded to carbon.

See also **Alloys**; **Iron Metals, Alloys, and Steels**; **Metals (The)**; and **Periodic Table of the Elements**.

Additional Reading

Bard, A.J., E. Gileadi, and M. Urbakh: *Encyclopedia of Electrochemistry, Thermodynamics and Electrified Interfaces*, Vol. 1, John Wiley & Sons, Inc., New York, NY, 2002.
Cotton, F.A., G. Wilkinson and P. L. Gaus: *Basic Inorganic Chemistry*, 3rd Edition, John Wiley & Sons, Inc., New York, NY, 1995.

Cotton, F.A., G. Wilkinson, M. Bochmann, and C. A. Murillo: *Advanced Inorganic Chemistry*, 6th Edition, John Wiley & Sons, Inc., New York, NY, 1999.

Gaus, P.L.: *Student's Solution Manual to Accompany Basic Inorganic Chemistry*, 3rd Edition, John Wiley & Sons, Inc., New York, NY, 1995.

Miessler, G. L., and D. A. Tarr: *Inorganic Chemistry*, 3rd Edition, Prentice-Hall, Inc., Upper Saddle River, NJ, 2003.

INORGANIC HIGH-POLYMERS.

The most commercially successful inorganic polymers to date are the polysiloxanes, owing to their unique high temperature stability, low temperature flexibility, and a number of other advantageous properties such as low surface energy and room-temperature vulcanizability.

Because of increasing technological needs in the area of high performance materials there has been a growing interest in the synthesis and development of new inorganic polymers. The polyphosphazenes and the polysilanes have shown the most promise in this area during the last two decades. In addition to these two, other novel inorganic polymer systems have also recently been developed.

Polyphosphazenes

The polyphosphazenes, sometimes also referred to as polyphosphonitriles, are the most chemically versatile inorganic polymers known to date. Based on their method of synthesis, two different types of phosphazene polymers are now in existence and are undergoing parallel development. The first type, bearing substituents on phosphorus bonded mostly via phosphorus-oxygen and phosphorus–nitrogen linkages, was developed in the mid-1960s as soluble, hydrolytically stable polymers. Polymers of the second type, with substituents linked via direct phosphorus–carbon bonds were first reported in the early 1980s and since then they have also seen significant development.

Synthesis. The synthesis of poly(dichlorophosphazene), the parent polymer to over 300 macromolecules of types (**1**) and (**2**), is carried out via controlled, ring-opening polymerization of the corresponding cyclic trimer, $(N = PCl_2)_3$.

(**1**)   (**2**)   (**3**)

Properties. One of the characteristic properties of the polyphosphazene backbone is high chain flexibility which allows mobility of the chains even at quite low temperatures.

The thermal stability of polymers of types (**1**) and (**2**) is dependent on the nature of the substituents on phosphorous. Polymers with methoxy and ethoxy substituents undergo skeletal changes and degradation above about 100°C, but arloxy and fluoroalkoxy substituents provide higher thermal stability. Most of the P–N- and P–O-substituted polymers either depolymerize via ring-chain equilibration or undergo cross-linking reactions at temperatures much above 150–175°C.

Phosphazene polymers are inherently good electrical insulators unless side-group structures allow ionic conduction in the presence of salts. Polyphosphazenes also exhibit excellent visible and uv-radiation transparency when chromophoric substituents are absent.

Another valuable characteristic of many phosphazene polymers is their flame-retardant behavior and low smoke generation on combustion.

A remarkable feature of phosphazene polymers of types (**1**) and (**2**) is that appropriate substituents (which are readily attached) can be used as toggle switches to turn several properties, such as hydrolytic stability and electrical conductivity, on and off.

Applications. The P–O- and P–N-substituted polymers have so far shown the greatest commercial promise. The fluoroelastomers possess good rubber properties with the added advantages of being nonburning, hydrophobic, and solvent- and fuel-resistant. In addition to these, because of flexibility down to about −60°C, these polymers have been used in seals, gaskets, and hoses in army tanks, in aviation fuel lines and tanks, as well as in cold-climate oil pipeline applications. These polymers have also found application in various types of shock mounts for vibration dampening.

The aryloxyphosphazene polymers, on the other hand, have been used primarily in wire and cable coatings and jackets and as fire-resistant, low smoke, closed-cell foams and sound-barrier sheets.

Biomedical Applications. In the area of biomedical polymers and materials, two types of applications have been envisioned and explored. The first is the use of polyphosphazenes as bioinert materials for implantation in the body either as housing for medical devices or as structural materials for heart valves, artificial blood vessels, and catheters.

The second type of biomedical application utilizes the versatile chemistry of polyphosphazenes to generate bioactive polymers. Two approaches have been developed: one is to tie or physically entrap biologically active molecules using the phosphazene backbone as the carrier or encapsulant. The other is to attach bioactive molecules to a hydrolyzable (degradable) phosphazene backbone that releases the active species on breakdown of the backbone to harmless species that can be metabolized or directly excreted.

Two crucial aspects of the design of bioactive polyphosphazenes have been carefully developed. One involves the hydrophilicity or hydrophobicity of the polymer, and the other is the stability of the polymer or tactical substituent linkages that allow release of the active agent or ensure its potency to be retained in the bound form.

Polymers Bearing Metal Complexes. A large number of polymers with side groups containing metal complexes have been reported. The complexes are linked to the phosphazene backbone primarily through a ligand on a substituent, although linkages through the skeletal nitrogen or through direct metal-phosphorus bonding with the skeletal phosphorus atoms have also been utilized.

Solid Electrolyte Applications. Among other potentially useful polymers synthesized by the versatile macromolecular substitution process are polymers based on oligoether substituents or heterocyclic substituents that have been under intense investigation for solid electrolyte battery applications. The most promising of these is poly [bis(methoxyethoxyethoxy)phosphazene (MEEP).

Polymers with Alkyl and Aryl Substituents on P

Even though partially alkyl- and aryl-substituted polyphosphazenes are accessible via the ring-opening polymerization followed by the macromolecular substitution route, polymers in which all substituents are attached through direct phosphorus–carbon bonds are not yet accessible by this method.

Synthesis. The first fully alkyl/aryl-substituted polymers were reported in 1980 via a condensation–polymerization route. In addition to providing fully alkyl/aryl-substituted polyphosphazenes, the versatility of the process has allowed the preparation of various functionalized polymers and copolymers.

Properties. The condensation–polymerization reaction yields alkyl- and aryl-substituted polymers with average molecular weights in the range 40,000 to 250,000 (M_n ranges from 20,000 to 100,000). In general, the polymers are soluble in chlorinated solvents such as CH_2Cl_2 and $CHCl_3$. Polymers with phenyl substituents are also soluble in tetrahydrofuran.

The P–N backbone remains quite flexible with small, unbranched alkyl substituents on phosphorus.

Alkyl- and aryl-substituted polyphosphazenes exhibit onset of decomposition at between 350 and 400°C.

Applications. Polymers with small alkyl substituents are ideal candidates for elastomer formulation because of quite low temperature flexibility, hydrolytic and chemical stability, and high temperature stability. In light of the biocompatibility of polysiloxanes and P–O- and P–N-substituted polyphosphazenes, poly(alkyl/arylphosphazenes) are also likely to be biocompatible polymers. A third potential application is in the area of solid-state batteries.

Phosphazenes Containing Skeletal Carbon, Sulfur, and Metal Atoms

The first phosphazene polymers containing carbon, sulfur, and even metal atoms in the backbone have been reported. These were all prepared by the ring-opening polymerization of partially or fully chloro-substituted (or fluoro-substituted) trimers containing one hetero atom substituting for a ring-phosphorous atom in a cyclotriphosphazene-type ring.

An example of polyphosphazene incorporating metal atoms is (**4**), where M = M_0 or W.

(**4**)

Poly(alkyl/aryloxothiazenes)

The synthesis of a new class of inorganic polymers (**5**) with a backbone consisting of alternating sulfur(VI) and nitrogen atoms, and with variable aklyl or aryl substituents as well as a fixed oxygen substituent on sulfur, has recently been accomplished. These polymers are structurally analogous to poly(alkyl/arylphosphazenes).

(**5**)

Synthesis and Properties. The synthesis of (**5**) follows a straightforward route based on readily accessible starting materials and on some novel reactions in organo–inorganic sulfur chemistry, as well as on polycondensation chemistry analogous to that utilized in the preparation of poly(alkyl/arylphosphazenes). The polymers exhibit some interesting characteristics that appear to be related to their unusual repeat unit.

Polysilanes

The polysilanes (**6**) are a unique class of polymers that exhibit σ-conjugation along the backbone.

(**6**)

Synthesis of Polysilanes. The most commonly utilized method is based on the Wurtz-type alkali metal coupling of dichlorosilanes. Other synthesis methods include dehydrogenative coupling, ring-opening polymerization, polymerization of masked disilenes, electrochemical synthesis, and polymer modification.

Properties. Most unsymmetrically substituted dialkyl and alkyl/aryl homopolymers as well as copolymers are soluble in solvents such as tetrahydrofuran or toluene. The longer Si−Si bond length, compared with the C−C bond length, allows quite a bit of flexibility in the backbone such that glass-transition temperatures as low as $-76°C$ (for poly(n-hexylmethylsilane)), have been observed. On the other hand, as expected, aryl substitution brings about significant increases in T_g. Thus, polysilanes cover the range from rubbery elastomers to brittle solids. Polysilanes are chemically inert to air and water at ordinary temperature, but their reactivity increases in solvent.

Electronic Properties. What distinguishes polysilanes from virtually all other polymers is their backbone σ- conjugation. This lead to strong electronic absorption in the near-uv from a $\sigma - \sigma$ transition.

The polysilanes are normally electrical insulators, but on doping with AsF_5 or SbF_5 they exhibit electrical conductivity up to the levels of good semiconductors.

Polysilanes absorb electromagnetic energy and undergo chain scission. This is an extremely important property of these polymers in terms of applications. Photochemistry is exhibited both in solution and in the solid state.

Applications. Polysilanes are used in the manufacture of β-silicon carbide, microlithography, xerography, and photoinitiation.

Polygermanes

Soluble and well-characterized polygermane homopolymers, $(R_2Ge)_n$, and their copolymers with polysilanes have been prepared by the alkali metal coupling of diorgano-substituted dihalogermanes, via electrochemical methods, and by transition-metal catalyzed routes, as with the synthesis of polysilanes.

The polygermanes exhibit many of the same electronic properties as polysilanes, including near-uv photoabsorption, thermochromism, photobleaching, as well as nonlinear optical activity, and have seen a fair amount of theoretical and experimental investigation. However, despite similarities with polysilanes, polygermanes appear to be unlikely candidates for commercial exploitation.

AROOP K. ROY
Dow Corning Corporation

Additional Reading

Allcock, H.R.: in *Inorganic Polymers*, Prentice-Hall, Inc., Englewood Cliffs, NJ, 1992, pp. 95–118.

Mark, J.E., H.R. Allcock, and R. West: *Inorganic Polymers*, Prentice-Hall, Inc., Englewood Cliffs, NJ, 1992.

Wade, C.W.R. et al.: in C.E. Carraher, J.E. Sheats, and C.U. Pittman, eds., *Organometallic Polymers*, Academic Press, Inc., New York, NY, 1978, pp. 289–300.

INOSITOL. A constituent of body tissue. In purified form it is used as a nutrient and dietary supplement in some foods and feed-stuffs. The chemical name of inositol is hexahydroxycyclohexane, CAS: 87-89-8. $C_6H_6(OH)_6 \cdot 2H_2O$. There are nine isomeric forms of inositol. Myoinositol or meso-inositol (*cis*-1,2,3,5-*trans*-4,6-hexahydroxycyclohexane) is the isomer that possesses essential nutrient activity. The substance, often identified as a vitamin, is found in small amounts in many vegetables, citrus fruits, cereal grains, liver, kidney, heart and other meat. The commercial source is corn (maize) steep liquor. In addition to its use in nutrition, it finds use in medicine and as an intermediate for organic syntheses.

INSECTICIDE. A substance that kills or interferes in the life cycle of certain insects and is thus useful for reducing and controlling insect populations. The reduction or elimination of such populations is desirable for several reasons. (1) Preventing the spread of certain diseases by insects that serve as transmitters or carriers of infective organisms. Thus, insecticides are used widely as a public health measure. (2) Preventing or reducing the damage caused by insects eating and inhabiting food plants, trees, and other crops. Such damage, without control, sometimes approaches 100% and averages 25% or more in some regions of the world. (3) Preventing or reducing physical property damage, notably of wood, cloth, and other materials of an organic nature that attracts certain insects as a source of food and place of habitation. Termites and moths are examples. (4) Reducing or eliminating discomfort, annoyance, and sometimes injury that results from the immediate presence of insects on or near people and domestic and farm animals. Ticks and face flies are examples.

Classification of Insecticides

Insecticides may be classified and characterized in many ways. (1) By their *selectivity*, i.e., their ability to control or not to control different forms, varieties, and species of insect, they range from compounds which have a rather *narrow control spectrum* and thus enable the user to eradicate or reduce selected target insects, all the way to *wide spectrum* insecticides that will destroy practically all insects, including those of a beneficial nature that are of positive economic importance. (2) By the manner in which insecticides *interact* with the insect enemy or target. That is: (a) whether or not a chemical requires overwhelming contact with the insect, or where the chemical acts systemically within and throughout the insect once local contact is made; (b) how the chemical interferes with the life process of the insect; for example, is it a stomach poison, or does it interfere with the insect's nervous or respiratory system, etc.; (c) how the insecticide enters the body of the insect, i.e., via the alimentary or respiratory system, etc.; and (d) whether or not the primary function of the chemical is to kill the insect, or essentially to sterilize the insect sexually and thus reduce insect population in this manner. (3) By application—is the insecticide a solid, liquid, or gas? Can it be sprayed, dusted, or incorporated into bait? Can it be worked into the soil? It is miscible with oil, water, or both? Can it be applied by aircraft? (4) By the useful life or persistence of the chemical, i.e., a few days, weeks, or months. (5) In terms of safety to humans and livestock and pets. Insecticide dangers range from highly toxic, to moderate, light, and low toxicity. Quite important, is the insecticide harmful to bees, fishes, other wildlife, or to adjacent crops and orchards should a spray or dust of the material drift away in the wind from the immediate point of application? (6) By the chemical structure of the insecticide, i.e., chlorinated hydrocarbon, organic phosphate, etc., and how produced—extracted from natural bacterial and biological materials or synthesized from basic raw materials and chemical intermediates. (7) By cost, a very important and practical consideration for large food producers, health authorities, and other users.

Selectivity and Spectral Range of Insecticides. There is no practical universal insecticide because such a chemical, to successfully eradicate all forms of insect life, would be dangerous to other life forms. There are, however, multipurpose pesticides, such as hydrogen cyanide gas, which are used *most carefully and with the ultimate of safety provisions* to kill not only all insects within a given area or space, but all other animal life forms as well. Some insects are much easier to destroy than others. Thus, numerous compounds are available to control aphid, fly, leafhopper, and trip, whereas only a few chemicals are effective against fire ant and certain beetle and weevil species.

Action of Insecticides. The average user of insecticides is essentially interested in the results of insecticide application rather than the exact manner in which these chemicals act upon the life and life cycle of the insect. But, there are some important distinctions that govern both the timing and selection of a given insecticide. For example, some chemicals are much more effective in destroying larvae or nymphs than when applied to insects in the adult stage, or vice versa. Also, some perennial crop plants and orchards are undamaged by certain chemicals (dormant sprays, for example) during winter inactivity, whereas the same chemicals would cause severe damage if applied during springtime budding. Also, the food producer, by taking advantage of multifunction compounds, can reduce the number of control chemicals required and hence the number of applications.

Control chemical manufacturers frequently offer combined formulations. In addition to blending acaricides with insecticides, molluskicides, etc., it is not uncommon to blend in herbicidal compounds. Usually, however, as the blending becomes more complex, the selection becomes more difficult, because great care must be taken to ensure the following: (1) That the chemicals will mix well without destroying any of their intended effectiveness (many control chemicals are weakened or destroyed, for example, if mixed with strongly alkaline materials, such as lime, or with sulfur-bearing compounds). (2) That the mixture is truly customized to the crop at hand, and that the combination is not used on other crops without specifically checking. And (3) that there will be no unexpected environmental damage. A combination of chemicals may be entirely effective and safe when applied to a specific crop or area, whereas it may be inappropriate if used on other crops and other areas that may be adjacent to pastures and streams or lakes and thus be dangerously polluting. For crops of large economic value and commonly grown, manufacturers offer numerous crop-specific formulations.

Nomenclature of Insecticides. It is estimated that there are well over 100,000 pesticide formulations and perhaps half of these would be basically classified as insecticides. Although the basic chemicals used may number in the hundreds, variations arise from the many thousands of possible formulations, not only combinations of materials, but formats (sprays, wettable powders, emulsifiable concentrates, dusts, baits, etc.). Added to this are the scores of control chemical manufacturers worldwide. Each manufacturer markets products under trade names—names that are essentially coined for their marketing charisma and infrequently connoting the content or purpose of the product. Thus, there are scores of equivalent (or essentially equivalent) products, adding to the difficulty of selecting these chemicals. Unfortunately, the generic chemical names of the majority of insecticide chemicals are long and complex and essentially meaningless to persons who are not well versed in organic and biochemistry. There are also a number of frequently revised directories of control chemicals and considerable information is available from various government agencies and universities. See list of references at end of this entry. This situation of nomenclature is quite similar to that which applies to generic and trade name drugs and pharmaceuticals.

Toxicity of Insecticides. Even though insecticides and other pesticides vary greatly in toxicity, all may be considered hazardous if they are not handled properly and with precautions. The following elementary rules are always worthy of repetition. (1) Observe all directions, restrictions, and precautions on pesticide labels. It is dangerous, wasteful, and, in some regions and countries, illegal to do otherwise. (2) Store all pesticides behind locked doors in original containers with labels intact. (3) Use pesticides at correct dosage and intervals to avoid excessive residues and injury to plants and animals. (4) Apply pesticides carefully to avoid drifting of the compounds to nearby fields, lakes, and streams. (5) Bury surplus pesticides and destroy used containers so that contamination of water and other hazards will not result. (6) Certain pesticides must not be used during a specified period just prior to harvest because of the danger to workers and pickers who will be handling the product and because of the danger of inadequately washing away or otherwise removing residues prior to releasing the commodity to the consumer. (7) Do not mix two or more compounds without prior knowledge of their compatibility. (8) When in

doubt concerning the applicability of a given compound to a given situation, such as a specific crop, seek advice from local sources of expertise, such as extension service representatives, reliable suppliers, neighboring users who have faced similar problems, regional colleges and universities, and agricultural experiment stations. Above all, the final responsibility for safe usage rests with the person who ultimately applied the chemicals.

In the United States, when registering insecticides, pesticides, and other control chemicals, regulatory agencies have been using acute LD_{50} values to determine the toxicity category and the words or symbols that must be placed on labels and containers. For this purpose, the test animals usually are rats, mice, or rabbits, but other mammals are sometimes used. The LD_{50} value is the dosage of the chemical at which 50% of the test animals are killed. It is based on the body weight of the animal and is expressed in milligrams of the chemical per kilogram of animal (mg/kg). One mg/kg = 1 part per million (ppm). Thus, the lower the LD_{50} value, the higher the toxicity of the chemical. The usual way of administering chemicals to test animals is by mouth, application to the skin, and in some cases, by inhalation. Toxicity may be either acute or chronic. Acute refers to rather quick action from a single exposure, whereas chronic refers to the toxic effect of many exposures over a period of time.

Chemistry of Insecticides

In classifying compounds used as insecticides, from the standpoint of chemical characteristics and structure, the most fundamental division separates the inorganic from the organic chemicals. The latter are by far the most widely used. Acaricides, bactericides, fungicides, and nematicides, along with insecticides, are included in the following descriptions.

Inorganic Compounds. These control chemicals are comparatively simple compounds and include calcium and lead arsenates, elementary sulfur and inorganic sulfur compounds, such as calcium polysulfide (lime-sulfur) and sodium thiosulfate. Because of their effectiveness against certain fungus infections, these compounds are more frequently considered as fungicides than as insecticides. This is also true of a number of copper, zinc, and other metal inorganics, such as copper carbonate, copper oxychloride, copper sulfate, and copper-zinc sulfate. It should be pointed out that Paris Green (cupric acetoarsenite), one of the older and once widely used inorganic insecticides, essentially has been phased out in most regions of the world because of the effects of chronic arsenic poisoning of workers and users who come in contact with the substance. This was particularly true in the case of vineyard workers a number of years ago.

A number of other metals, such as iron and tin, enter into insecticide and pesticide compounds, but as part of an organic chemical structure, as exemplified by triphenyltin hydroxide. Such compounds are sometimes referred to as organometallics (or, specifically in the case of tin, as organotins). Mercury compounds are rapidly being phased out because of their long-term toxic residual effects as pollutants, particularly of fresh and saline waters. Regulations vary from one country to another.

A few other inorganic chemicals, more frequently identified as multi-purpose pesticides than as insecticides, do have very strong insecticidal properties. These compounds are used sparingly, often requiring special permits in some places, and with the greatest of safety cautions observed. Such compounds would include calcium cyanide, carbon bisulfide, carbon tetrachloride, hydrogen cyanide, paraformaldehyde, and phosphine.

Organic Compounds. A listing of categories of organic chemicals used reads like the table of contents of an organic chemistry text, with relatively few subfamilies of organic compounds not represented in one way or other.

Alcohols. The open straight or branched chain (alkyl, aliphatic) saturated alcohols, such as methanol, ethanol, up through tetradecanol, etc., are not powerful insecticides, although they are somewhat more effective than the related hydrocarbons (methane, ethane, etc.). As is evidenced in other areas of activity, these alcohols increase in insecticidal effectiveness roughly in proportion to their molecular weight (number of carbon atoms in compound). This is an effect, however, which levels off when from 9 to 12 carbons (molecular weight of 144 to 186) are present in the chain. For example, although not a direct measure of insecticidal activity, nonyl (9 carbons) and decyl (10 carbons) alcohols are most effective in reducing the sprouting of stored potatoes. Beyond this point, a greater number of carbon atoms does not increase the effectiveness.

Unsaturated and Cyclic Alcohols. Although these compounds show somewhat stronger insecticidal effectiveness, as compared with the saturated aliphatic alcohols, this added strength is still not sufficient to warrant their serious considerations as insecticides. Some of these compounds, however, do make effective herbicides.

Aldehydes. The aldehydes possess greater insecticidal effectiveness than the alcohols, formaldehyde, for example, serving as a stomach poison. The compound also exhibits strong bactericidal and fungicidal activity. Formaldehyde tends to polymerize into paraformaldehyde, the pesticidal properties of which are considerably less than those of formaldehyde. The same increasing effectiveness with increasing molecular weight exhibited by the alcohols also applies to the aldehydes.

Metaldehyde, the polymer of acetaldehyde, is a widely used molluskicide and is effective in the control of snails. Some of the unsaturated aldehydes are more potent in their pesticidal and herbicidal effectiveness. However, only a few compounds are of commercial significance, notably acrolein and related compounds, which are used as aquatic herbicides in connection with water reservoirs and systems. Some aldehydes also have growth-regulating properties. See also **Plant Growth Modification and Regulation**.

Amines. This class of organic compounds also exhibits the same relationship between molecular weight and insecticidal activity as previously described. In the case of aliphatic amines, studies of toxic potency against house fly larvae have shown that the most effective compound is di-*n*-octylamine, with the compounds of higher or lower molecular weight in the series proving less toxic. Effectiveness also improves as one proceeds from the aliphatic amines to the aromatic amines, noting, for example, the greater toxicity of aniline as compared with hexylamine. Although *o*-iodoaniline and 2,5-dichloroaniline demonstrate some toxicity for caterpillar and louse, respectively, generally and perhaps surprisingly, inclusion of halogen atoms within the aromatic amine nucleus does not promote greater toxicity. Some toxicity increase is shown, however, when nitro groups are introduced into the nucleus. Diphenylamine, a diarylamine, is effective against lice and at one time enjoyed wide use for troops during wartime.

The insecticidal usefulness of the amines is hindered by their tendency to severely injure plants (phytotoxic effects) to which they may be applied. Advantage of this property, however, is taken by using some amine compounds as herbicides. Examples include benefin, nitralin, and trifluralin.

Carbonic Acid Derivatives. Mixed esters of carbonic acid display toxic potency as acaricides and fungicides, including their inhibiting action against powdery mildew. When sulfur is introduced into the structure, as in the case of the mixed esters of thio- and dithiocarbonic acids, the acaricidal and fungicidal toxicity is further increased. Derivatives of thio- and dithiocarbonic acids used commercially include carbon bisulfide (CS_2) and 6-methylquinoxaline-2,3-dithiocyclocarbonate (*Morestan*), an effective acaricide, fungicide, and insecticide.

Carbamic Acid Derivatives. A rather large number of commercially available food crop control chemicals fall into this category of organic compounds. Carbamic acid (or aminoformic acid) is NH_2COOH, but is best known in the form of its salts and esters. Several insecticides are found among the aryl esters of *N*-methylcarbamic acid, whereas the alkyl esters of *N*-arylcarbamic acids possess strong herbicidal powers, particularly in connection with undesired monocotyledonous plants.

The biological and physiological actions involved in the toxicity of the carbamates to animal and plant life processes are quite complex and not fully understood. It has been established that the esters of *N*-alkylcarbamic acids with insecticidal properties inhibit cholinesterase. Of the *N*-methylcarbamic acid ester series, the most powerful insecticidal compound is 1-naphthyl-*N*-methylcarbamate, the basis for such commercially produced compounds as carbaryl, naphthyl carbamate, and Sevin. Other important carbamate-type acaricides, fungicides, insecticides, and nematicides include Aldicarb, Allyxycarb, Aminocarb, Bassa, Benomyl, Buffencarb, Carbendazim, Carbofuran, Ethiofencarb, Formetanate, Knockbal, Landrin, Mancozeb, Maneb, Meobal, Metacrate (Tsumacide), Metiram, Mexacarbate, Pirimcarb, Promecarb, and Propoxur (Baygon). Most of these names are proprietary.

Thio- and Dithiocarbamic Acid Derivatives. The derivatives of thio-carbamic acid are essentially herbicidal in nature. See also **Herbicides**. However, excellent nematocidal effectiveness is illustrated by the dithio-carbamic acid derivatives, notably by sodium N-methyldithiocarbamate (Vapam), which serves the multipurpose of not only eradicating nematodes, but many insects and weeds as well. It is frequently used as a soil sterilant.

In the case of the alkali metal salts of alkyldithiocarbamic acids, studies have shown that the fungicidal, nematocidal, and herbicidal effectiveness

decreases as the length of the alkyl radical is increased. In terms of nematicidal activity, the esters of alkyl- and dialkylcarbamic acids is greater than the salts. Toxicity is greatest in the methyl and ethyl esters and decreased as the number of carbons in the ester radical increases.

Commonly available (check regulations) bactericides, fungicides, insecticides, and nematicides that fall into this category include: Carbothion, Eptam, Ferban, Nabam, Propineb (zinc bearing), TEC, Thiram, Zineb, and Ziram. Most of the foregoing names are proprietary.

Aliphatic Carboxylic Acids. Studies indicate a very low pesticidal activity for the aliphatic monobasic acids (acetic, propionic, etc.) and the dibasic carboxylic acids (oxalic, fumaric, maleic, etc.). As is usually expected, the pesticidal activity of the acids increases when halogen atoms are introduced to displace hydrogens in the alkyl radicals. Examples of this effect include the monohaloacetic acids, which are sometimes used commercially. The salt *calcium propionate* is used in bread- and cheese making as a preservative for its mild antibactericidal and fungicidal effects. Generally, the fluorine-containing derivatives are more toxic than those compounds with chlorine atoms. As can be expected, the unsaturated compounds exhibit greater toxicity than their saturated counterparts.

Alicyclic Carboxylic Acids. With exception of copper-bearing compounds, such as copper naphthenate, which is strongly fungicidal, the free alicyclic acids are not important as control chemicals. This is not the case, however, for a number of their derivatives. The natural pyrethrins and their synthetic analogs are included in this category. Among them are Allethrin, Barthrin, Bioallethrin, Cinerin I, Cyclethrin, Dimethrin, Furethrin; as well as Neopyanimin, Pyresy, Pyrexcel, and Pyrocide. Most of the foregoing names are proprietary.

The alicyclical carboxylic acid derivatives also include a family of growth-regulating compounds, known as the *gibberellins*. See also **Gibberellic Acid and Gibberellin Plant Growth Hormones**; and **Plant Growth Modification and Regulation**.

The alicyclic carboxylic acid derivatives have been studied extensively and, to date, with the exception of the compounds already mentioned, relatively few have been found to possess commercially important potential as pesticides. An exception is dimethyl carbate (Dimelton), which finds use as a repellent for certain blood-sucking diptera. The compound frequently is mixed with other pesticides.

Aromatic Carboxylic Acids. As in the case of aliphatic and the alicyclic carboxylic acids, the free aromatic acids (benzoic, naphthenic, etc.), their halogen and nitro derivatives, as well as their alkali metal salts, possess a lower insecticidal effectiveness. However, for some species of mite, the benzyl ester of benzoic acid is very effective. As a general rule, the incorporation of chlorine or other halogen atoms in the benzoic acid and benzyl alcohol configurations accentuate the biological power. Introduction of chlorine into the para position of the benzyl radical appears to provide a maximum effect against both egg and adult mite. Further enhancement of acaricidal potency is obtained by the presence of amino, hydroxy, and nitro groups. For codling moth and body louse, the aliphatic esters of anisic, anthranoic, and salicylic acids have toxic effectiveness. Chlorobenzilate is an effective selective acaricide, useful against numerous species of mite, including the tracheal mite which is parasitic to honeybee.

Numerous other aromatic carboxylic acids possess some bactericidal, fungicidal, and herbicidal characteristics, as well as growth-regulating properties. Salicylanilide, the amide of salicylic acid, is effective against leaf mold and tomato brown spot.

Heterocyclic Compounds. Quite a large number of insecticides and other pesticides as well as herbicides are heterocyclic compounds. The variations pertaining to structure and composition are so many that this is strictly a generalized, umbrella like classification. For the biochemist, organic chemist, entomologist, or other professional concerned in the development and theoretical aspects of control compound chemistry, other more detailed classifications are required. A first step toward this was undertaken by Melnikov. See list of references. Among important fungicides and insecticides in this classification are copper quinolate and Phenazim fungicide.

Aliphatic Hydrocarbons. After extensive research into the biological activity of the aliphatic hydrocarbons, relatively few of the pure (nonderivative) compounds have been found worthy of commercial attention. Popular for use in orchard spraying are the petroleum derivative oil sprays, which possess a good combination of acaricidal and insecticidal activity with low phytotoxicity. These sprays are effective against San Jose scale and mite.

Ethylene. For a number of years, has been used to hasten the ripening of some fruits. Although difficult to apply, ethylene is also an excellent defoliant. The insecticidal and nematocidal effectiveness of the halogenated aliphatic hydrocarbons is in proportion with the general chemical activity of these compounds. A number of these compounds have been used in the form of fumigants, notably in treating stored commodities and storage areas. Some of these include methyl bromide and DD pesticide.

Aromatic Hydrocarbons. Compounds in this category, such as benzene, naphthalene, xylene, etc.) have undergone extensive investigation. Although many of their derivatives are important, the pure compounds find little if any pesticidal use. Some of them, however, and in particular the xylenes, are used as solvents for carrying other control chemicals. For a number of years, naphthalene did enjoy large usage as a control agent against moth species. This has largely been replaced by synthetic compounds.

Halogen derivatives of benzene vary considerably in acaricidal and insecticidal toxicity, depending upon type, location, and number of halogen atoms introduced. Bromine appears to impart maximum effectiveness, followed by chlorine and fluorine. Biological activity increases with halogen atom introduction up to a total of three such atoms. A larger number tends to decrease effectiveness. Dichlorobenzene is more powerful than hexachlorobenzene. Loading a compound with bromine exhibits a greater effect than chlorine in reducing effectiveness.

Although there are eight stereoisomers of benzene hexachloride, only one of these, 1,2,3,4,5,6-hexachlorocyclohexane, is important commercially.

DDT (1,1,1,-trichloro-2,2-*bis* (*p*-chlorophenyl)ethane), now banned in many countries, is a derivative of an asymmetrical diarylethane and an effective insecticide. Paradichlorobenzene is an effective multipurpose pesticide. The compound is useful against sugarbeet weevil and in the control of phylloxera.

Ketones. Because of their rather weak insecticidal effectiveness, the ketones are used mainly as solvents for control chemical formulations.

Mercaptans. The aliphatic mercaptans with four or fewer carbon atoms have rather powerful insecticidal properties and can be used as fumigants against certain insect species. This is not true of those compounds containing over four carbon atoms; and also not true of aromatic mercaptans.

The control chemical interest of the mercaptans essentially is in the derivatives that incorporate chlorine or bromine. Methyl mercaptan rivals hydrogen cyanide as a powerful and useful fumigant. This compound is also an important intermediate in the synthesis of Captan and Folpet fungicides.

Some of the closely related organic sulfides and thioacetals have found commercial pesticidal use. These formulations include Mikazin, Fluoroparacide, and Fluorosulfacide. Names are proprietary in most cases.

Nitro Compounds. Biological activity of organic nitro compounds compares favorably against their pure (nonderivative) hydrocarbon compounds. However, this activity is considerably enhanced in terms of those nitro compounds that contain one or more halogen atoms. Bromine imparts a greater toxic power than chlorine. Many of the halonitro compounds command a wide spectrum of functionality, ranging from acaricidal, bactericidal, fungicidal, herbicidal, insecticidal, and nematicidal attributes. Some of their action is ascribed to their strong oxidizing properties. Examples of effective nitro compounds include chloropicrin and several related compounds, such as dichloronitroethane and chloronitropropane, as well as Binapapacryl acaricide-fungicide, Dicloran fungicide, Dinobuton acaricide, Dinocap acaricide fungicide, and Dinoterb acetate pesticide. Most of these names are proprietary.

Mercury Compounds. The powerful biological activity of mercury in simple compounds, such as the inorganic mercuric chloride, particularly against molds and other bacterial and fungus infections, has been recognized for generations. Also, the toxicity not only to microorganisms, but to higher animal forms also has been known and of major concern for a long time. Because of emphasis on environmental factors and safety during the past few decades, mercury-containing chemicals have been undergoing a phaseout in many countries. In connection with mercury compounds, it is interesting to note that many of these compounds have a good chemotherapeutic rating (or index), i.e., the dosage required to control a plant disease organism is many, many times smaller than the dosage that would be harmful to the plant. Some mercury compounds stimulate plant growth and yield.

Tin Compounds. Several organotin compounds are quite biologically active. Some of the simple inorganic tin salts, such as stannous or stannic chloride, have little if any pesticidal value. The fungicidal effectiveness is achieved by substituting alkyl or aryl groups for the chlorine atoms. A peak of insecticidal activity is achieved with the trialkyl- and triaryltins. It is interesting to note that the tetraalkyl- and tetraaryltins are essentially ineffective. Research has indicated that tributyltin chloride and tributyltin fluoride are the most active of the possible combinations. Most popularly used (check regulations) commercially are triphenyltin hydroxide and triphenyltin acetate.

Copper Compounds. Principally used as fungicides, copper inorganic compounds are widely applied. Of the organic copper compounds, the most commonly used are copper linoleate, copper naphthenate, and copper quinolate.

Zinc Compounds. Zinc is associated in a number of organic pesticides, notably Propineb fungicide, Zineb fungicide, and Ziram fungicide. Names are proprietary.

Phenols. As compared with the aliphatic alcohols, the phenols are more active biologically, but even with this greater activity, most of these compounds are not of practical commercial importance. Introduction of halogen, nitro, thiocyano, and some other groups increase their activity, the nitro group appearing to have the greatest insecticidal power. Included among these compounds are some that date back 50 years or more—dinocap (Karathane) acaricide-fungicide and dinobuton (Dinoseb) acaricide. The phenols tend to severely burn plants, a property that led to their use, commencing in the late 1930s, as contact, selective-type herbicides and desiccants.

Phosphorus Organics. Possibly the most extensive of all categories of organic compounds used as control chemicals, the organophosphorus compounds, are derived from the inorganic acids of phosphorus. It is estimated that the very fundamental compounds in this category number well over one hundred and that the commercial formulations resulting may number into the several hundreds. For study it is sometimes convenient to classify these compounds as derivatives of (1) phosphorous acid, H_3PO_3; (2) phosphoric acid, H_3PO_4; (3) thiophosphoric acid, $PS(OH)_3$; (4) pyrophosphoric acid, $H_4P_2O_7$; and (5) phosphonic acids (phosphine = PH_3).

Phosphorous Acid Derivatives. The principal control chemical potential of these organic phosphite compounds lies with their abilities as herbicides. Acaricidal, fungicidal, insecticidal, and nematicidal powers are comparatively weak. But, as is nearly always the case, the toxic potential increases with many different kinds of derivatives that can be prepared. Commercial products based upon derivatives include: DDVP (Dichlorvos) acaricide-insecticides; *tris*-(2,4-dichlorophenoxyethyl) phosphite (Falone); Gestid (Mevinphos, Phosdrin) acaricide-insecticide; Naled (Dibrom) acaricide-insecticide; and Phosphamidon acricide-insecticide. Most of these names are proprietary.

Phosphoric Acid Derivatives. The biological activity of the phosphates is considerably greater than the phosphites and notably among the mixed esters of phosphoric acid where one of the ester radicals is acidic. The toxicity of the resulting derivative is roughly proportional to the dissociation constant of the parent alcohol, phenol, or acid, the toxicity decreasing as the dissociation constant decreases. Research on the phosphoric acid mixed esters shows that the methyl derivatives are the most toxic. The toxicity of these types of compounds is believed to be the result of (1) high alkylating potential as regards certain biological nitrogen and sulfur constituents; and (2) elevated rates of hydrolysis.

Numerous proprietary examples of the derivatives of phosphoric acid include: Bromophos insecticide; Chlorpyrifos (Dursban) insecticides; Demeton (Mercaptophos) acaricide-insecticide; Diazinon acaricide-insecticide; Fenitrothion insecticide; Fensulfothion insecticide-nematicide; Fenthion acaricide-insecticide; Kitazin fungicide; Methyl Parathion (Metafos) acaricide-insecticide; Oxydemeton-Methyl acaricide- insecticide; Parathion (Thiophos) acaricide-insecticide; Vamidothion acaricide-insecticide; and Zytron insecticide. Most of these names are proprietary. Some of the compounds are banned in some countries.

Thiophosphoric Acid Derivatives. A fortunate combination of characteristics occurs in the phosphoric acid derivatives upon substitution of a sulfur atom for one of the oxygens of the parent compound. Namely, the toxicity of the derivatives to higher forms of life is substantially diminished, while at the same time, the acaricidal and insecticidal powers, with few exceptions, remain strong. Derivatives of thiophosphoric acids may feature a thiono or a thiolo (most toxic) structure. Commercial preparations are usually mixed esters of thiophosphoric acid, as well as of dithio- and trithiophosphoric acids. The trithio compounds usually are markedly less effective than the dithio compounds.

Some commercial formulations based upon derivatives of the dithiophosphoric acids include: Azinphos-Methyl (Guthion) acaricide-insecticide; Carbophenothion (Trithion) pesticide; Dimethoate acaricide-insecticide; Disulfoton acaricide-insecticide; Malathion (Carophos) acaricide-insecticide; Mecarbam acaricide-insecticide; Menazon acaricide-insecticide; Phorate acaricide-insecticide; Phosalone acaricide-insecticide; and Phosmet (Imidan, Phthalodophos) acaricide-insecticide. Most of these names are proprietary. Some of these compounds are banned in some countries.

Pyrophosphoric Acid Derivatives. A pioneer among the phosphorus-containing organic control chemicals, tetraethyl pyrophosphate (Bladen, TEPP), was developed by Bayer AG in Germany in the early 1940s. The pyrophosphates are powerful contact-type acaricide-insecticides that have little or no tendency to function systemically. In addition to TEPP, some of the commercial formulations in this category include NPD, Pirophos, Schraden, and Sulfotepp. Most of these names are proprietary. Some of these compounds have been banned in some countries.

Phosphonic Acid Derivatives. Very important in this category of proprietary compounds are Trichlorfon (Chlorophos, Dipterex, Dylox) and Trichloronate.

Other Organic Bases.

Quinones. As compared with the alcohols and aldehydes, the biological activity of the quinones is greater, notably in their actions as fungicides. Although benzoquinones exhibit relatively low activity, the presence of halogen and hydrocarbon radicals in the ring structure, as is true of many other organic compounds, significantly increases the effectiveness. Some of the derivatives of the quinones used commercially include tetrachlorobenzoquinone (*Chloranil*), which is particularly effective for disinfecting seed, and 2,3-dichloronaphthoquinone-1,4 (Dichlone, Phygon).

Sulfonic Acid Derivatives. The sulfonic acids, including their salts, have found value as agents for treating wool fabrics against species of moth for a number of years. One of these products is Eulan, a number of variations of which have been produced. Mitin-FF, a derivative of urea, also has been used in this way. To date, the free sulfonic acids have not played an important role as insecticides for crops.

Impressive activity against mite larva and egg has been shown by some of the aromatic esters of arylsulfonic acids. Some of the commercial formulations in this category include CPCBS (Chlorofenson, Ester Sulfonate, Ovex, Ovotran) acaricide.

Thiocyanates and Isocyanates. The intense biological activity of hydrogen cyanide has been known for scores of years. Similarly, derivative compounds have been known and used for many years. The derivatives of thiocyanic acid are particularly powerful fungicides and pesticides. Research has indicated that the straight-chain thiocyanates are more effective than those with branched chains and, as can be expected, introduction of halogen atoms into the compounds increases their toxicity. Some of the commercial formulations that fall into this overall category include Thiophanate fungicide; and Thiophanate Methyl fungicide. Regulations over usage must be checked.

Urea and Thiourea Derivatives. The value of urea as a fertilizer is well known. See also **Fertilizer**. However, the most elementary derivatives of urea show strong phytotoxic effects. Thus urea derivatives are widely used as herbicides. A number of derivatives of thiourea also show strong bactericidal and fungicidal activity. In this category are found Dodine fungicide and Guazatine fungicide.

Bacterial and Botanical Compounds as Insecticides and Control Chemicals

One of the long-established and better known of the botanical compounds is pyrethrum or pyrethrin insecticide. Known since the early 1800s, the active ingredient of this formulation is obtained from pyrethrum plants found in Africa and South America. Pyrethrin is considered to be one of the safest insecticides, but it is costly. The first synthesis of allethrin, the analog of pyrethrin, was developed in the late 1940s and is now widely used.

A much more recent natural insecticide, developed in the early 1960s, is Bacillus Thuringiensis-Berliner and, as indicated by the name, the compound is developed from living spores of the *Bacillus thuringiensis*, a bacterial strain that causes disease among certain types of insects.

Other naturally derived commercial insecticides include Evisect, Hellebore, nicotine sulfate, and Sabadilla insecticide.

An interesting concept of using specific antibodies as a potential insecticide has been proposed by Nogge, Giannetti, and associates at the Institut für Angewandte Zoologie (Bonn, Germany). See reference listed. It has been learned that many insects are able to absorb orally administered antibodies. The researchers found that when tsetse flies are fed on human blood, the hemolymph of the flies contains human albumin. Then, if the flies ingest antibodies to human albumin, they perish within a short time. It is observed that the albumin fraction in the insect's hemolymph disappears and osmoregulation is severely disturbed. Thus, antibodies may be used as a biological insecticide.

Additional Reading

Ishaaya, I.: *Biochemical Sites of Insecticide Action and Resistance,* Springer-Verlag, Inc., New York, NY, 2000.

Jones, D.G.: *Piperonyl Butoxide: The Insecticide Synergist,* Academic Press, Inc., San Diego, CA, 1998.

McKenzie, J.A.: *Ecological and Evolutionary Aspects of Insecticide Resistance,* R.G. Landes Company, New York, NY, 1996.

Melnikov, N.N. (Gunter, F.A. and R.T. Huber): *Chemistry of Pesticides,* Springer-Verlag Inc., New York, NY, 1971.

Mullin, C.A. and J.G. Scott: *Molecular Mechanisms of Insecticide Resistance: Diversity Among Insects,* American Chemical Society, Washington, DC, 1992.

Narahashi, T. and J.E. Chambers: *Insecticide Action: From Molecule to Organism,* Plenum Publishing Corporation, New York, NY, 1990.

Nogge, G. and M. Giannetti: "Specific Antibodies: A Potential Insecticide," *Science,* **209,** 1028–1029 (1980).

Page, B.G.: *Insecticide, Herbicide, Fungicide Quick Guide, 2000,* Thomson Publications, Washington, DC, 2000.

Staff: *Insecticide and Acaricide Tests: 1993,* Entomological Society of America, Lanham, MD, 1993.

INSECTICIDE AND PESTICIDE TECHNOLOGY.

The technology of controlling pests, notably in the area of food production, is undergoing serious examination and reevaluation.

Chemicals are and have been the main weapon for controlling agricultural pests for well over a century and will continue to be important for the foreseeable future. But within the last 20 years and notably the last decade, the total chemical approach to pest control has been subject to questioning and alternative approaches have been sought.

It no longer can be taken for granted that progress in pesticide technology will be confined to the research and development of new and improved chemicals. Progress in pesticide chemistry will continue to be important, but for the long term, actions commenced just a few years ago toward development of a *total systems concept of pest management* may represent the technology of the future. In the long term, this new technology may provide more effective pest control, coupled with a progressively lessened dependence upon chemicals.

A Retrospective View. When an important technology is at a crossroads, it is in order to glance back in an effort to clarify the forces which are bringing about a major change in direction. As early as 1828, it is reported that Persian (Iranian) farmers used pyrethrum (obtained from *Chrysanthemum coccineum*) as an insect control on certain farm crops. Rotenone, another naturally-derived organic chemical, has been used for pest control for over a century. Bordeaux mixture (copper sulfate and hydrated lime) has been used as a fungicide since the early 1880s. Inorganic arsenic compounds were used in German vineyards at the turn of the century and not banned until 1942.

Prior to the period just preceding, but mainly following World War II, pesticide chemicals were either inorganics or naturally derived and extracted organic compounds. Although there was an early awareness of the poisonous nature of most of these compounds, there was not an immediate connection made between poisoning insects and other pests during the growing period of a crop and possible poisoning of persons who might consume the produce after harvest. During this period, most likely it was generally assumed that the chemical insecticide would be confined to the foliage surfaces of the plants, later to be washed off by rain and in preparing produce for market. And, of course, it is true that a number of these compounds are contact-type pesticides, that is, their area of influence is confined to the surfaces to which applied. There was little, if any, understanding or consideration of possible systemic actions of such toxic materials, that is, the absorption of the poisons by the plant. Poisons were transported throughout the plant by its vascular system and thus residuals remained in edible parts that, considering the analysis techniques then available, were extremely difficult to assay in minute quantities, even if suspected.

Numerous important findings of biochemistry, microbiology, and human and plant physiology were still unknown. The concepts of slow, prolonged poisoning processes were unknown and/or unappreciated. Based upon suspicions of arsenic-caused deaths, it was as recent as 1938 when a laboratory in Speyer, Germany studied 336 samples of wine bottled for sale that year and found to contain as much as 14.4 milligrams of arsenic per liter. Earlier vintages were found to contain as much as 24 milligrams per liter. It was later concluded that many vineyard workers, who regularly consumed "house wine" prepared from grape skins, succumbed to cancer of the liver after a latent period ranging from 10 to 35 years. Even in 1972, some 30 years after arsenic use in the vineyards of the Mosel and Kaiserstuhl regions had been banned, oldtime workers of the vineyards were expiring from arsenic-induced liver cancer.

Further, little was known and/or appreciated pertaining to the poisonous nature of the metabolites (compounds resulting from digestive processes) produced from certain pesticide chemicals, particularly those of an organic nature. Only during the last 40 to 50 years (a comparatively short period in terms of the total history of chemical pesticides) has the persistent nature of some chemicals been appreciated. This is also true of awareness of the large capacity of soils to retain chemical residuals for many years. When these two factors are coupled, excessive concentrations of chemicals can be built up over a period of years.

During the pre-World War II period, the world population was much smaller than today and the concentration of agricultural operations much less. Thus, the effect of certain pesticidal chemicals on birds, beneficial insects, livestock, fishes, and other forms of wildlife was less discernible. There certainly were various forms of warning, but these were noted only by comparatively few people. There was no popular awareness and concern as regards environmental problems.

A major alteration in insecticide and pesticide technology occurred as the result of what might be called a tremendous expansion in the field of organic chemicals. Pre-World War II efforts to synthesize dyes from coal tar chemicals were among the first efforts toward commercially expanding organic chemical synthesis. These efforts were shortly followed by the wartime needs for synthetic rubber, improved aircraft fuels, and an improvement of the earlier resins for the manufacture of plastic substitute materials for a host of applications. Knowledge and experience in organic chemistry multiplied several-hundredfold during the postwar period and spawned the petrochemical industry. Tens of thousands of new, previously unknown organic compounds were produced, for which in many instances uses had to be found.

Further expansion of commercial organic chemicals was greatly aided by the addition of natural gas as an almost ideal raw material. Although used for many years, a booming natural gas industry did not develop until shortly after World War II. The 1930s, 1940s, and 1950s became the age of miracle chemicals—with the introduction of new fibers, new plastics and resins, new coatings, and, importantly, new chemical pesticides. Mass production made it possible to produce many new chemical pesticides at a relatively low cost and their convenience in application was widely accepted by food producers. And the period was essentially without any major worries concerning possible deleterious side effects from their use. This period was also marked by a rapidly expanding worldwide population and greatly expanded food-producing operations. During this period, a great chemical pesticide industry and distribution system was created and, even more importantly, many food producers became highly dependent upon the use of pesticide chemicals. Many earlier farming practices were discarded in favor of wider application of chemicals and, at that time, apparently all for good reasons. Chemicals greatly reduced crop losses to insects and other pests, thus increasing effective yields from a given unit of land and labor. Chemicals still offer these advantages and, as of the late 1980s, this is still the general mode of food production operations in countries with advanced technology. The chemical trend, in fact, was again markedly accelerated a few decades ago by the introduction of herbicides, which, in turn, led to the concept of "no-till" farming.

The Apex of Conventional Chemical Pesticides. The wide acceptance of conventional chemical pesticides, not only by food producers for reasons previously given, but also by society in general, is exemplified by DDT (dichlorodiphenyltrichloroethane). This organic chemical was developed during the early 1930s by Paul Müller, a Swiss research scientist. In the early years of World War II, people were advised to use it without reservation on food and fodder plants, since it was said to be entirely harmless to warm-blooded animals. DDT was welcomed not only as an excellent control chemical for food production, but also as a public health measure against mosquitoes and other annoying and sometimes dangerous insects. Household preparations containing DDT and other related organic compounds were widely sold. Thus DDT in its early days exemplified the best of technology—a scientific breakthrough accompanied by completely positive economic and social benefits, as witnessed by the award of a Nobel Prize in 1948 to Müller.

Knowledge of the true nature of DDT was very slow in arriving. Not until the early 1950s did scientists at the U.S. Department of Agriculture find that although fodder treated with DDT caused no damage to the cows eating it, the health of their calves was severely impaired, sometimes with fatal results. The DDT was being passed along from cow to calf via the milk. These findings were confirmed in 1953 by experiments sponsored jointly by Swiss universities and pesticide manufacturers. The Swiss experimenters found that about one-tenth of the DDT sprayed from aircraft to control the May beetle settled on and was retained on the surface of pasture grass. Again, there was no apparent damage to the cows eating the grass, but their calves suffered the same effects as those previously described in the United States. It was also learned that the damage was principally to the nervous system of the animals.

In 1972, a group of German scientists discovered that a conversion product of DDT, a metabolite known as DDD, has a mutagenic effect. Some 30 or 40 years earlier, instrumental analytical techniques were not available to detect small residuals of highly complex organic chemicals and, further, the knowledge of human biology and biochemistry was but a fraction of that knowledge amassed during the period after World War II. In a way, possibly, the introduction of DDT marked the apex of conventional chemical pesticides, certainly not in terms of tonnages of chemicals produced, but in terms of their apparently trouble-free acceptance by society. Numerous chemical pesticides have been banned since the banning of DDT (in several countries) and the banning and tight governmental regulating trend continues at a rapid pace. In fact, some authorities believe that possibly the present period of questioning and suspicion of chemical pesticides may produce a net deficit for society. That is, society, on the one hand, might be fearful of the long-term effects of pesticidal chemicals, but, on the other hand, having learned to depend upon chemical pesticides for many years, may need all the assistance it can get from modern technology, including chemical pesticides, to feed an ever expanding world population.[1]

Possibly, the root of the fundamental biological problems resulting from widespread application of modern chemical pesticides extends back to the previously mentioned great expansion of organic chemicals that occurred during and just after World War II. With present knowledge of molecular biology and of proven carcinogens, although admittedly this knowledge is still extremely limited, it is almost certain that much more concern would have been expressed pertaining to the public release of many of the chemicals in use today. But once an entire business of chemical pesticide supply is established and once an industry (food production) takes on certain performance patterns, change becomes extremely difficult, particularly when satisfactory substitutes are not in view.

System Concept of Pest Management

The system approach to pest control involves not only a total look at the target pests and the plants to be protected, but an investigation of all elements which make up what might be termed a crop ecosystem. Particular

---

[1] A February 5, 1976 report of the study committee on pest control of the National Academy of Sciences expressed concern that "future agricultural productivity is threatened by a possible breakdown in chemical control of pests." Factors leading to this conclusion included: (1) the appearance of genetic resistance among "target" insect pests; (2) disruption of natural pest control mechanisms when beneficial insects as well as target pests are killed by a chemical compound toxic to a broad spectrum of insect life (the committee gave as an example the use of an organophosphate insecticide by California cotton growers for controlling the lygus bug that also kills certain predators which normally control the bollworm, a late-season pest); and (3) the effect of increasing constraints by laws and regulations on much needed chemical pesticides. Although not decrying the new laws and regulations, the study committee indicated that such regulations make more difficult and expensive the introduction of new pesticides to replace those already banned so that, if pesticide developers and suppliers are sufficiently discouraged, a serious gap in availability of effective control chemicals could result.

emphasis is placed upon the interactions of all elements; this emphasis, of course, involves a study of all feedback and any feedforward loops that may be present in the system. In essence, the systems approach represents applied ecology, which is basically a system-oriented science, but with the addition of numerous specialist viewpoints—physiology, biochemistry, engineering, entomology, botany, agronomy, economics, meteorology, and climatology, among others.

The long-range of objectives of the system approach are several: (1) to control food crop pests more effectively than is possible with chemical insecticides alone, even when these chemicals are used in excessive dosages; (2) to take full advantage of all natural factors that may act against the pests and in the favor of the plant; (3) to find new methods for combating high pest populations; (4) to improve crop yields as the result of optimizing favorable growth conditions; and (5) one hopes, to reduce the total cost of pest control by lowering the amount of control chemicals required for equivalent or better results.

It is obvious, of course, that the food producer cannot bring all the aforementioned skills together as they may pertain to a given crop ecosystem and make individual decisions as what to do next and when to do it in an effort to control pest populations. How is such very specialized information gathered in the first place? How is such information interpreted and translated into actions for the individual food producer?

The amount of data to be gathered just in the interest of applying the system approach to a limited number of major crops is staggering. A pioneering program developed by the Purdue University Agricultural Experiment Station (West Lafayette, Indiana) may, at some time in the future, point the way toward substituting information for hard chemicals in dealing with insect damage to crop plants. A program of this type requires the interest and intellectual cooperation of crop producers rather than simply depending upon chemical overkill of plant enemies.

In the early Indiana program, approximately one hundred alfalfa growers participated who essentially were dairy farmers who rely on alfalfa as the major source of protein for their herds. One of the initial tasks required was, to assess the level of alfalfa management practiced by program participants. This was accomplished by developing a 40-item questionnaire, which each cooperator completed and returned. A considerable variation in statewide insecticide and herbicide practices was found.

In developing the early data bank for the system, many actions were required: (1) collection of insect, alfalfa plant, and weed data from other cooperator's fields once each week for a season; (2) collection of weather data from appropriate agricultural weather stations on a daily basis, to be used as input into the alfalfa plant and weevil models to be developed later; (3) comparison of model output with actual grower field data and to make specific pest management recommendations based upon current insect, plant, and weather conditions; (4) input and storage of files of individualized alfalfa pest management advisories on a central computer; and (5) utilization of these data by cooperative extension agents by dissemination of advisories to the cooperating growers, using telephone, local radio station farm broadcasts, etc. Weather, obviously, plays a very important role in any pest management system. Thus, the Purdue program involved an excellent meteorological observing, forecasting, and communicating system.

Crop Ecosystem. What constitutes a crop ecosystem? This will vary considerably from one region within a state to another and probably can be determined practically only through making many observations. But it is possible, for example, that pest control information dispatched to producers in the southern half of one county may be applicable to producers of the same crop in the northern half of the adjoining county. In the long run, the system may take form of daily pest management advisories to food producers, possibly over local radio stations that for years have been featuring farm news of all kinds. Such an advisory might read along the following lines:

"Alfalfa weevil larval populations are on the increase and have reached sufficient levels in many cooperators' fields in southern Indiana to eventually result in economic losses. The alfalfa in these fields is still relatively short, averaging 4.6 inches and should (should not) be treated at this time. For growers in (such and such) areas, insecticide application should be delayed until more larvae have hatched so that a greater number can be controlled. Delay insecticide application for 7 to 10 days from this advisory. For growers in (such and such) areas, there are sufficient larval numbers and spraying should commence in accordance with previously given schedules."

Central to some pest management systems is a computerized simulator which, based upon the analysis of hundreds of past observations and experiments, can accept current weather information, for example, and read out the effects of the weather parameters. It thus provides directions for whatever pest control actions should or should not be taken at any given time. In essence, the simulator takes the place of numerous observers in the field and enables an information center to pass along directives in real time. A number of factors in addition to weather information, of course, can be input into the system. Needless to say, if such a network were established, all manner of other information pertaining to the crop ecosystem could be handled in addition to pest management data.

Such a system should be contrasted with the pest management means available today. Pest control information comes to the food producer from a number of sources: (1) pesticide chemical suppliers, who usually provide fundamental information on the application of their product, including precautionary information, safety measures to be taken, timing directions, etc.; (2) local extension personnel; and (3) special booklets, bulletins, etc., issued by state agriculture departments, giving specific directions as to types of acceptable pesticide chemicals, signs to watch for on the plant to diagnose pest conditions and stages, and general counsel pertaining to the timing of initial and repeat applications of chemicals—together, of course, with safety and precautionary information. But the prime limitation of this kind of information system is that it is not dynamic, it does not function in real time. Inputs of new information may range from one to several years and thus such bulletins do not always reflect the latest in chemical pesticide technology. And possibly the greatest weakness is the fact that this kind of information is based upon traditional pest control methods, as contrasted with the concepts of dynamic pest management.

Biological Pest Control Methods

The role of chemicals generated by insects in affecting the normal metabolism of plants has been known for many years. Less understood has been the role of chemicals generated by plants on the metabolic processes of insects. Research is beginning to demonstrate that some plants possess surprising chemical defense against attack by insects and other pests. Such defense chemicals operate in a variety of ways, and it has been mainly during the past 10 to 15 years that operation of these chemical defenses has been explained in a rudimentary way.

Pheromones. Some entomologists believe that insect pests may be controlled economically with a minimum of environmental disruption by exploiting the hormones and pheromones by which an insect regulates its growth, development, and behavior. Pheromones may be defined as chemicals which are secreted by one insect that affect the behavior of other individuals of the same species. Pheromones evoke several behavioral responses, but the sex-attractant pheromones are those most frequently mentioned by entomologists. It is believed that inasmuch as pheromones are natural substances the insects may be less likely to develop a resistance to them than to some synthetic organic insecticides. However, entomologists point out that insects are quite adaptable and that a change in their pheromones is a possibility.

Three approaches in the use of pheromones have appeared in the literature: (1) use of traps baited with sexual attractant material as a means for monitoring the infestation of areas with select insects; (2) similar use of traps, except on a massive scale, to attract males (female sex pheromone used as bait); and (3) "male confusion" technique, in which female sex pheromone is permeated in the air, frustrating the attempts of males to locate females.

The use of pheromones is particularly attractive because of the high selectivity of the method, enabling the destruction of pests by way of large reductions in future populations and doing this without interfering with the normal life and habits of beneficial insects.

Juvenile Hormones. These are organic chemicals that are present in insects during the greater part of the insect's development. It is only during metamorphosis (period when a larva changes into an adult) that these chemicals are absent. When juvenile hormones are applied to insects during metamorphosis, the adults produced are deformed and lack the capacity for further development and soon die. Because juvenile hormones are relatively simple compounds, synthetic analogs are not too difficult to prepare and thus can be used as effective insecticides. However, timing if very critical because effectiveness is limited to the relatively short period of metamorphosis. If applied before or after this period, the compounds are essentially ineffective. Wide use of these chemicals could become practical if tied into a computerized pest management system control center as previously described. See also **Juvenile Hormones**.

Antiallatotropins. As part of their defense mechanism, some plants contain chemicals with juvenile hormone activity. Plants also have been found that contain chemicals with antijuvenile hormone activity, known as antiallatotropins. Although still not fully understood, it is assumed that the biological control system of the plant distinguishes between the metamorphosis period of an insect (during which time the juvenile hormones would be used as weapons) and the other periods of the insect's life cycle (during which time juvenile hormones are required by the insect, hence use by the plant of the antijuvenile or antiallatotropin compounds). Two antiallatotropins have been isolated from a common bedding plant (*Ageratum houstoniatum*). Chemically, these are 7-methoxy-2,2-dimethylchromene and 6,7-dimethoxy-2,2-dimethylchromene.

Phytoalexins. First reported by K. Müller in Germany in 1940, phytoalexins are lipidlike chemicals that are synthesized by some plants. Research indicates that these compounds are toxic to fungi and bacteria, as well as some other pests. It has been found that the chemicals are produced as the result of an attack upon the plant and an analogy between these compounds and inteferon (an antiviral substance produced in humans in response to a viral infection) has been suggested. To date, about 100 phytoalexins have been isolated. One of their roles is believed to be prevention of germination of fungus spores.

Research by K. Uehara at the University of Osaka, Japan, in 1958 produced the concept of elicitors. Since then, at the University of Colorado (Boulder), the first phytoalexin elicitor was isolated. This was obtained from filtrates of cultures of *Phytophthora megasperma* var. *sojae*, a fungus that attacks soybean. This compound stimulates the accumulation of the phytoalexin glycerollin, previously characterized by research at the University of London.

Viruses. There are epidemics caused by viruses, which occur periodically in the insect populations and thus naturally help to check their spread. Entomologists would like to find ways of infecting pest insects before they can cause serious damage. Most insect viruses are specific for a few closely related hosts. The general structure of the viral particles includes DNA plus protein, all imbedded in a protein matrix. They are termed nuclear polyhedrosis virus (NPV). The viruses spread when larvae eat contaminated foliage.

Apparently viral infections in the past have helped to control the population of the Douglas-fir tussock moth. This insect undergoes population explosions at intervals of about 10 years. The outbreaks may last up to 3 years, during which time severe damage results. DDT helped to control the tussock moth population before it was banned. Attempts are now being made to combat a current outbreak of tussock moths with virus. Preliminary experiments in spraying virus on trees have been encouraging.

A virus called NPV is commercially produced and is used for the control of the cotton bollworm, a species closely related to the tobacco budworm. With present technology, the viruses can reproduce only in living cells. The insect viruses are usually grown in the appropriate hosts. Investigators are attempting to develop cell culture systems for propagating insect viruses that will eliminate the inconvenience of contamination of large numbers of insects or insect larvae for virus production.

The possible impact of virus insecticides on other forms of life, including humans, in the long term will obviously require some years of actual experience at a high level of usage. Regulatory agencies at this juncture are understandably extremely cautious in approving more than limited use. There appears to be evidence in support of the safety of the viruses and the negative aspects of this juncture appear to be in the category of theoretical possibilities in the absence of any substantive evidence. One scientist at the U.S. Department of Agriculture, with reference to the case of the cabbage looper caterpillar that succumbs to a virus attack, refers to the so-called coleslaw example. The insect body dissolves and sheds onto the leaf of cabbage large quantities of virus, which are not killed in the preparation of coleslaw. In mid-October, when mortality of the loopers is at the highest level, the average bowl of coleslaw will contain about 4 billion live particles of cabbage looper nuclear polyhedrosis virus. The scientist reasons that if the virus were harmful to people, this would have been long evident.

Microbial Agents. Considerable academic and governmental research has been conducted in this area for many years and is also an area to which industry has made significant contributions. The bacterial agents, *Bacillus thuringiensis* and *B. popilliae* are used commercially. Permission also has been granted for the experimental investigation of NPV (nucleopolyhedrosis virus) of *Heliothis zea* (corn earworm). Although it is not generally believed that microbial agents will replace chemical methods, it is felt that such agents will function importantly in integrated pest management programs of the future.

Insect Adaptability to Climate. Closely related to the foregoing discussion is the concept of genetic suppression of insect populations by their adaptations to climate. Many of the serious insect pests have very broad geographical distributions. For example, the codling moth (*Carpocapsa pomonella* (L.)) is distributed from Canada to Argentina and it occurs in Australia, South Africa, the Mediterranean, and throughout Europe north to the Scandanavian countries. Insects with such broad distributions adapt in various ways to climate.

A number of reports indicate that genetic differences occur between insect populations within a species with regard to: (1) ability to undergo a hibernal diapause; (2) response to diapause-inducing stimuli; (3) duration of diapause; (4) temperature limits of diapause termination; (5) temperature optima for diapause termination; (6) ability to develop cold hardiness; (7) thermal constants and temperature threshold for development; (8) choice of hibernal niches and other behavioral traits associated with surviving inhospitable seasons; and (9) ability for aestival (summer) diapause, its duration and response to conditions that induce or terminate it.

Such genetic differences must exist so that adaptations of insects to climate may be appropriate to their locality. Changes in climate from locality to locality require appropriate changes in adaptations. Insects may synchronize their life cycles with the seasons so that (1) frost-sensitive stages are passed in the frost-free season; (2) feeding stages occur when food is available; and (3) no actively developing stages occur in periods of intense heat or drought. Therefore, insects must be sensitive to stimuli that portend the change in seasons so that they may prepare for adverse periods.

If it is possible to genetically disrupt the seasonal regulations or other climatic adaptations of insects, the insects may not survive. For example, if an insect population at Fargo, North Dakota must respond to a photoperiod of 15 hours in order to enter diapause and be cold-hardy in time for dangerous frosts, and if the population is genetically modified, so that it does not diapause until the day has shortened to 13 hours, then the population may be destroyed by the winter. Further, it may be assumed that this population must remain in diapause until early May, when the danger of killing frost is past and when the host plant has again become available. If the diapause is genetically shortened so that the insects resume development in March, then they would be destroyed either by frost or lack of food.

Inappropriate adaptations to climate are lethal at certain times of the year; they are conditional lethal traits. A conditional lethal trait or combination of conditional lethal traits can be used to suppress or eradicate insect populations. The principle of suppressing insect populations by means of their adaptations to climate has been suggested by a number of investigators.

Related entries include **Herbicides**; and **Insecticide**.

Additional Reading

Alfassi, Z.B. and R.M. Joy: *Pesticides and Neurological Diseases,* CRC Press, LLC., Boca Raton, FL, 1994.

Baringa, M.: "Entomologists in the Medfly Maelstrom," *Science,* 1168 (March 9, 1990).

Best, G.: *Pesticides–Developments, Impacts, and Controls,* CRC Press, LLC., Boca Raton, FL, 1995.

Brooks, G.T. and T. Roberts: *Pesticide Chemistry and Bioscience: The Food Environment Challenge,* The Royal Society of Chemistry, London, UK, 1999.

Cheremisinoff, N.P. and J.A. King: *Toxic Properties of Pesticides,* Marcel Dekker, Inc., New York, NY, 1994.

Dodge, A.D. Editor: *Herbicides and Plant Metabolism,* Cambridge University Press, New York, NY, 1990.

Frehse, H.: *Pesticide Chemistry: Advances in International Research, Development, and Legislation,* John Wiley & Sons, Inc., New York, NY, 1991.

Gibbons, A.: "Moths Take the Field Against Biopesticide (Bacillus thuringiensis)," *Science,* 646 (November 1, 1991).

Greene, C.: "Environmental Concern Sparks Renewed Interest in Integrated Pest Management," *Food Review,* 8 (April–June 1991).

Hayes, W.J., Jr. and E.R. Laws, Jr.: *Handbook of Pesticide Toxicology,* Academic Press, Inc., San Diego, CA, 1991.

Ishaaya, I.: *Biochemical Sites of Insecticide Action and Resistance,* Springer-Verlag, Inc., New York, NY, 2001.

Jacobson, M.: *Glossary of Plant-Derived Insect Feeding Deterrents,* CRC Press, LLC., Boca Raton, FL, 1990.

Kearney, P.C., N.N. Ragsdale and J.R. Plimmer: *Eight International Congress of Pesticide Chemistry,* American Chemical Society, Washington, DC, 1994.

Kirchhoff, J. and Hans-Peter Their: *Manual of Pesticide Residue Analysis,* Vol. 2, John Wiley & Sons, Inc., New York, NY, 1992.

Laird, M., L.A. Lacey and E.W. Davidson: *Safety of Microbial Insecticides,* CRC Press, LLC., Boca Raton, FL, 1990.

Lynch, L.: "Consumers Choose Lower Pesticide Use Over Picture-Perfect Produce," *Food Review,* 9 (January–March 1991).

McKenzie, J.A.: *Ecological and Evolutionary Aspects of Insecticide Resistance,* Academic Press, Inc., San Diego, CA, 1996.

Milne, W.A.: *CRC Handbook of Pesticides,* CRC Press, LLC., Boca Raton, FL, 1994.

Nadasy, M. and GyForgy Matolcsy: *Pesticide Chemistry,* Elsevier Science, New York, NY, 1989.

Reganold, J.P., R.I. Papendick and J.F. Parr: "Sustainable Agriculture," *Sci. Amer.,* 112 (June 1990).

Schaub, J.R.: "Pesticides: How Safe and How Much?" *Food Rev.,* 2 (April–June 1991).

Sherma, J.: "Pesticides (Analysis of)," *Analytical Chemistry,* 118R (June 15, 1991).

Staff: *Handbook of Natural Pesticides, Vol. VI: Microbial Insecticides,* CRC Press, LLC., Boca Raton, FL, 1990.

Staff: "Organically Grown Foods," *Food Technology,* 26 (June 1990).

Strobel, G.A.: "Biological Control of Weeds," *Sci. Amer.,* 72 (July 1991).

Torgersen, T.R.: "Saving Forests the Natural Way," *Amer. Forests,* 31 (January/February 1990).

Van Rie, J., et al.: "Mechanism of Insect Resistance to the Microbial Insecticide Bacillus thuringiensis," *Science,* 72 (January 5, 1990).

Zilberman, D., et al.: "The Economics of Pesticide Use and Regulation," *Science,* 581 (August 2, 1991).

INSTRUMENTATION (Analytical).   See Analysis (Chemical).

INSULATION (Electric).   An electrical insulator, when placed between conductors at different potentials, will permit a negligible current (in phase with the applied voltage) to pass through it. In essence, electrical insulators are applied dielectrics. A perfect insulator (dielectric) will pass no current in the foregoing situation. The closest approach to a perfect dielectric is a perfect vacuum. The difference between low-resistance insulators and semiconductors is ill defined. Materials that can be considered as insulators have resistivities greater than 10^{20} ohms down to 10^6 ohms. Because of varying needs for insulators in the construction and use of electrical and electronic equipment and systems, the wide range of materials available, including a considerable span in cost, provides a convenient selection for the designer.

Relative Dielectric Constant

The capacitance between plane electrodes when in a vacuum, neglecting fringing, may be expressed by

$$C = k_0 A/t = 0.0884 \times 10^{-12} A/t \text{ farads}$$

where k_0 is dielectric constant of a vacuum; A is area, cm^2, t is spacing between plates, cm. If the vacuum is replaced by a dielectric material, the capacitance increases for the same applied voltage. With the dielectric between the plates, the capacitance is

$$C = kk_0 A/t$$

where k is the relative dielectric constant of the material. Capacitance relations change, of course, for other commonly occurring situations, such as coaxial conductors, concentric spheres, and parallel cylindrical conductors. The relative dielectric constant, then, is the key criterion in the functioning of an insulator. Values for several representative materials are given in Table 1. Numerous factors, including temperature, frequency, humidity, age, degree of cure (in case of plastics and polymers), and geometry, affect the relative dielectric constant. For this reason, the term dielectric permittivity is often used instead of relative dielectric constant.

Materials Used

With increasing frequency, the permittivity of dielectric decreases. A major factor in the selection of insulation is the ability of the insulation to resist the absorption of moisture. Moisture, of course, can greatly lower resistivity. For wire insulation, synthetic polymers and plastics essentially have replaced the use of natural rubber. Usually, prior to coating a wire with a plastic material, the wire must be treated to assure good contact and adhesion of the insulating material. Copper wire, for example, is treated with hydrogen fluoride, which creates a coating of copper fluoride; in the

TABLE 1. DIELECTRIC PERMITTIVITY OF REPRESENTATIVE MATERIALS*

Material	k	Material	k
Ceramics and Glasses		*Nonpolar Resins*	
Alumina	8.1–9.5	Polyethylene	2.3
Aluminum silicate	4.8	Polypropylene	2.2
Pyrex (Corning 7740)	5.1	Polystyrene	2.5–2.6
Fused silica	3.8	Polytetrafluoroethylene	2.0
Forsterite	6.2–6.3		
Steatite	5.5–7.0		
High-tension Porcelains		*Polar Resins*	
Beryl	4.5	Cellulose cotton (dry)	5.4
Magnesia	8.2	Cellophane (dry)	6.6
Mica (glass-bonded)	6.4–9.2	Cellulose triacetate	4.7
Titanates	50–10,000	Epoxies (unfilled)	3.0–4.5
Zirconia	8.0–10.5	Nylon	4.0–4.6
		[a]Phenolics (cellulose)	4–15
Crystals		[a]Phenolics (glass)	5–7
Aluminum oxide	10.0	[a]Phenolics (mica)	4.7–7.5
Calcium carbonate	9.2	Polyvinyl chloride	3.2–3.6
Boron nitride	4.2	Polyvinyl acetate	3.2
Barium titanate	4.100	Polyvinyl fluoride	8.5
Mica, synthetic	6.3	Methylmethacrylate	3.6
(fluorophlogopite)		Polycarbonate	2.9–3.0
Mica (muscovite)	7.0–7.3	[a]Silicone (glass)	3.1–4.5
Magnesium oxide	8.2	Polyethylene	3.25
Sodium chloride (dry)	5.5	terephthalate	

[a] Filled with material indicated in parentheses.
* Some insulations incorporating organic materials may be carcinogenic when subjected to high temperatures. Regulations on the use of these materials varies from one country to the next. Thus some of these products are not available throughout the world.

case of aluminum wire, aluminum fluoride. Thin films of fluoride possess high dielectric strength and resist heat well.

With the possible exception of common line insulators, electrical porcelain is especially formulated and may contain varying percentages of zirconia and beryllia. These ingredients increase both strength (mechanical) and resistance to high temperatures. Hard porcelains are especially formulated to resist thermal shock as well.

Liquid insulators are required for circuit breakers, transformers, and some cable applications. Natural hydrocarbon mineral oils are commonly used, as well as chlorinated aromatic liquids (desirable because of nonflammability). For high-temperature situations, silicone fluids may be used. Permittivities range between 2 and 7. Insulating liquids function both as electrical insulators and heat-transfer media. See also **Dielectric Theory.**

INSULATION (Thermal).   Thermal insulation is any substance or configuration of materials that resists the flow of heat. Thermal insulation does not stop heat flow, but retards it to rates that suit particular requirements. For example, in the case of buildings or residences in climates or during seasons when the ambient temperature of the atmosphere is uncomfortably hot or cold, thermal insulation will be used to retain heat within a structure during cold weather and to shield or insulate the structure from the penetration of external heat during hot weather. In the one case, by reducing the flow of heat from structure to atmosphere and near outer space during winter, less energy is required to maintain the desired temperature within the structure. In the other case, by reducing the flow of heat from atmosphere and sun to structure during summer, less energy is required to artificially cool the inside temperature. Many parallel instances occur in industry. By thermally insulating processing vessels, piping, etc., where it is desired to maintain warm or hot conditions, energy is not lost to ambient surroundings. The efficient maintenance of temperature is critical to many industrial situations because temperature affects the physical and chemical properties of materials, such as viscosity, and determines the rate at which chemical reactions occur, among numerous other temperature-sensitive properties. In cryo-processing, as in the liquefaction of gases, the freezing of foods, etc., the objective is the maintenance of low temperatures and thermal insulation in such cases restricts the flow of heat from ambient surroundings and thus reduces the amount of energy required to maintain desirable low temperatures.

Although the principles of heat flow have been understood and treated mathematically since the early 19th century (Fourier, LaPlace, Poisson, Peclet, Lord Kelvin, Riemann, and many others), it was not until nearly the beginning of the 20th century that major developments of commercial thermal insulating materials and systems were undertaken.

In addition to increasing thermal efficiency (conservation of energy), thermal insulation is frequently used to protect personnel from injury by burns and to shield adjacent structures from overheating, thus assisting in protection against fire. Thermal insulations are not usually suited to fire protection per se once a fire is in progress, but by restricting heat flow in the first place, insulation plays an important role in preventing some kinds of fires from starting. The behavior of thermal insulation once a fire has started is not necessarily positive in all situations and thus requires the attention of equipment and structure designers. A major concern in fires is thermal diffusivity rather than thermal resistance. Thermal insulation may not be suited for protection against high-velocity radiation.

Although not the primary function, the retardation of moisture migration into insulated spaces, where condensation may occur, is an important engineering consideration in the design of thermal insulation systems.

Insulation and Heat-Flow Principles. Heat flows from places of higher temperature to those of lower temperature by one or more of three modes: (1) Conductance through solids; (2) convection by induced motion of fluids carrying heat; and (3) radiation by heat waves emitted from a surface. The rate of heat flow in solids depends upon temperature difference $T_2 - T_1$ and the resistances encountered. The heat flow, under steady state, is expressed by:

$$Q_{\text{heat flow}} = \frac{T_2 - T_1}{R - \text{value}}$$

R-value is a measure of thermal resistance and varies from one insulating material to the next.[1] Q can be expressed Btu/square foot/hour or as Cal/square meter/hour, depending upon the units used in the equation. Helpful relationships between English and metric units are given in Table 1. See also **Heat Transfer**.

While convection may be a significant factor within processes, in general, convection affects thermal insulation primarily at the surface of the insulation or its jacket, where the air film is a resistance to heat flow from (or to) that surface. Wind reduces the air film resistance. To a lesser extent convection can occur within some low-density fibrous insulations, especially in walls, and to a greater extent within cavities and unfilled spaces within constructions. In walls, it is important that insulation that does not fill the space completely be installed so that remaining spaces are uniform, not skewed. Radiant heat flows through space, either vacuum or gaseous, from a higher-temperature surface toward a lower-temperature substance by the difference in absolute temperatures to the fourth power and the surface characteristic called emittance, e, as shown by the Stefan-Boltzman relation:

$$Q_{\text{rad}} = 0.174e\left[\left(\frac{T_2}{100}\right)^4 - \left(\frac{T_1}{100}\right)^4\right] \text{Btu/ft}^2 \text{ hr} \quad \text{(on Rankine scale)}$$

$$Q_{\text{rad}} = 5.670e\left[\left(\frac{T_2}{100}\right)^4 - \left(\frac{T_1}{100}\right)^4\right] \text{W/m}^2 \quad \text{(on Kelvin scale)}$$

A common use of low convection with high-reflectance/low-emittance surfaces is in the food and liquid containers (Dewar flask, Thermos™ bottle, etc.). The double glass-wall space is under high vacuum so convection is virtually eliminated, and the surfaces are coated with silver to reduce heat transfer by the low emittance on the outside of the inner wall and the high reflectance on the inside of the outer wall.

Low emittance can be observed if the hand is held close to a very hot silver teapot without feeling much heat despite the high temperature of the surface of the teapot.

In high-temperature process plants, men have been burned on low-emittance hot metal jackets on insulation because the low emittance did not give them a sense of heat. Yet the thermal resistance from the heat conservation standpoint was excellent.

In some materials, especially foams, the spaces may contain gases other than air and the performance in convective and radiative heat transfer in the spaces affects the overall performance of the material. Also, the emittance/reflectance performances of the walls of the spaces affect performance overall.

[1] Resistance, thermal, R-value—the mean temperature difference at equilibrium between two defined surfaces of material, or a construction, that induces unit heat flow rate through unit area. (From ASTM STD. C168-80A)

TABLE 1. CONVERSION FROM U.S. CUSTOMARY UNITS TO METRIC UNITS[1]

(Data are for thermochemical values unless noted. International table values differ slightly.)

W = watt; m = metre; J = joule; kg = kilogram;
C° = temperature difference Celsius

Multiply	By	To Obtain
Btu (mean) British thermal unit	$1.055\ 870 \times 10^3$	J
Btu ft/h ft² F°	1.729 577	W/m C°
(*k*-factor, thermal conductivity)		
Btu in/h ft² F°	$1.441\ 314 \times 10^{-1}$	W/m C°
(*k*-factor, thermal conductivity)		
Btu in/s ft² F°	$5.188\ 732 \times 10^2$	W/m C°
Btu/h	$2.928\ 751 \times 10^{-1}$	W
Btu/ft² h	3.152 481	W/m²
Btu/ft² min	$1.891\ 489 \times 10^2$	W/m²
Btu/ft² s	$1.134\ 893 \times 10^4$	W/m²
Btu/h ft² F°	5.674 466	W/m² C°
(*C*-factor, thermal conductance)		
(*U*-factor, overall thermal conductance)		
Btu/s ft² F°	$2.042\ 808 \times 10^4$	W/m² C°
Btu/lb	$2.324\ 444 \times 10^3$	J/kg
(Heat capacity)		
Btu/lb F°	$4.184\ 000 \times 10^3$	J/kg C°
(Specific heat capacity)		
Btu/ft³	$3.723\ 402 \times 10^4$	J/m³
Calorie (mean)	4.190 020	J
Calorie (kilogram)	$4.184\ 000 \times 10^3$	J
(Kilocalories)		
Calorie/cm²	$4.184\ 000 \times 10^4$	J/m²
Calorie/g	$4.184\ 000 \times 10^3$	J/kg
Calorie/g C°	$4.184\ 000 \times 10^3$	J/kg C°
Calorie/min	$6.973\ 333 \times 10^{-2}$	W
Calorie/s	4.184 000	W
Calorie/cm² min	$6.973\ 333 \times 10^2$	W/m²
Calorie/cm² s	$4.184\ 000 \times 10^4$	W/m²
Calorie/cm s C°	$4.184\ 000 \times 10^2$	W/m² C°
F° h ft²/Btu	$1.762\ 280 \times 10^{-1}$	C° m²/W
(*R*-value, thermal resistance)		
F° h ft²/Btu in.	6.928 113	C° m/W
(*ru*-value, thermal resistivity)		
Therm (100,000 Btu)	$1.055\ 056 \times 10^8$	J

[1] Most metric units shown are SI (the universally adopted designation for Le Systéme International d'Unités), except SI uses K (kelvin) for both absolute temperature and for temperature differences even though temperatures are determined on the Celsius scale. Since temperatures will usually be measured on the Celsius scale, the symbols used here are °C for temperature Celsius, as in the past, while C° is Celsius degrees difference. Similarly, °F is temperature Fahrenheit and F° is Fahrenheit degrees difference.

Technically, the performance of materials and systems depends upon all three modes of heat transfer to varying degrees in different materials, so it is the effective or apparent conductance that is to be evaluated. Although such terminology may be correct technically, and is appearing again in the literature, it was discussed many years ago and abandoned because it aroused too many questions by users of insulations with limited technical knowledge. It was felt that those with necessary technical competence would understand that a multimode heat transfer was involved, and the simple thermal conductance would satisfy users so long as the data were correct. R-values are even more readily understood by users, and technical analyses can be made by identifying by subscript that phase of the analysis being evaluated.

Thermal Insulation Systems and Materials

ASTM[2] Committee C-16 on Thermal Insulating Materials defines thermal insulation as a material or assembly of materials used primarily to resist heat flow. The reference to assembly of materials indicates that the concern is thermal insulating systems, because it is not until materials have been designed into systems that performance can be estimated. Thermal insulating systems include not only the basic materials, but also the auxiliary materials and the methods of application and protection in service.

[2] American Society for Testing and Materials, 1916 Race St., Philadelphia, Pa. 19103.

Time of exposure differentiates the needs for insulation performance when used in relatively continuous exposures, in cyclic increases and decreases of temperature, in processes with wide ranges of temperature in the various phases, especially in pipelines carrying fluids that must not fall below critical temperatures lest they solidify and necessitate dismantling and replacement of the lines.

A special short-time performance of thermal insulation is the ablative protection on the bottom of astronautical capsules returning from outer space when they are heated to sudden high temperatures by impact with the atmosphere. As principles stated below indicate, ablation is the process of resisting heat flow by using absorptance in changes of state from solid to liquid and to vapor of the ablative insulation, which is thereby lost, so that a one-time or at most a few times of exposure is practicable.

Classes of Insulating Materials. See Fig. 1. While glass has a high conductance, if it is fiberized and formed into wool-like masses, the high conductance of the fibers is counteracted by the still air that is held within the mass. Still air (no motion) has high thermal resistance, and at one time it was presumed that still air was the best insulator. A few other materials have been found with somewhat greater thermal resistance than still air, but they are so costly that they are suited only to very special applications. Other fiberized materials perform similarly, and rock, slag, and glass wools are collectively called mineral wools, but each has its own temperature limits.

All mass-type thermal insulations rely for their thermal resistance upon dispersion of the solid phase with air, or sometimes with gases. Plastics are reduced in density by foaming them into low-density material (0.5 to 2 lb/ft^3; 8 to 32 kg/m^3). Molten glass is foamed so that it performs as thermal insulation; its advantages are high compressive strength and virtually imperviousness to moisture, although thermal resistance is not as high as in some other materials.

Reacted materials, such as hydrous calcium silicate, are made with the solids dispersed to create density on the order of 12 lb/ft^3 (192 kg/m^3) and still provide high compressive strength. While the reflectance and emittance of the solid phase have an effect on heat transfer, it is the still air within the mass that gives it significant thermal resistance.

Still air is the important factor in dispersed solids, such as wood fibers, exfoliated mica, powdered diatomite, and expanded perlite.

The earliest reacted insulation was so-called 85% magnesium that had wide acceptance for many years, but since it was suited to less than 500°F (260°C), it has been replaced by other insulations with higher R-values or more desirable physical properties.

Although metals are good conductors of heat, they can also perform as good thermal insulators when their surface properties are used to advantage. While solid metal conducts heat readily, the emittance and reflectance of some metals are used to provide high overall R-values, especially when they are used in multiple sheets with spacers. Moreover, metals have relatively high temperature tolerances but have no absorption, so that all-metal thermal insulation may be suited to higher-temperature services than some other kinds of materials. While silver and gold are the best metals for high reflectance service, their use is limited to special applications where cost is not governing. For long-time exposure, gold surfaces would maintain high reflectance longest. However, aluminum sheets are used effectively when all-metal insulation is required. The first sheet of aluminum spaced about 12 mm (1/2 in.) from the hot surface reflects a large percentage of heat striking it. The unreflected heat passes through the metal rapidly by

conductance, but the low emittance of the reverse surface prevents much of the heat leaving to strike the next sheet of aluminum where, again, the reflectance prevents a large portion of the emitted heat to enter that sheet. The number of reflective sheets is designed for the R-value desired. If the service temperature is above the working temperature of aluminum, about 1000°F (540°C), the first one or two sheets can be made of polished stainless steel.

Since aluminum is also used for jackets, it should be noted that although the thermal performance may be acceptable, the low emittance of an aluminum surface on the outside may introduce a personnel hazard, mentioned above.

An opposite effect of emittance occurs when an attempt is made to use aluminum jackets on very cold (cryogenic) piping. In this case, heat reaching the surface from surroundings is reflected away so that the metal surface becomes so cold that it will condense and freeze moisture from the air. To overcome this surface condensation and freezing by use of thicker thermal insulation on the lines would require a very great increase in thickness over that needed if the jacket had been made of a heat-absorbing material that would keep the surface above the dew point.

In the high-temperature case, low emittance is desirable from an operating standpoint, whereas in the cryogenic case low emittance is undesirable.

Two general types of heat flow systems exist: one in which it is desired that heat flow as rapidly as practicable, and another in which heat flow must be resisted as much as practicable. The former is a high thermal conductance type of system, whereas the latter is a high thermal resistance, high R-value, type. Consequently, in heat conserving systems it is simpler to think in terms of thermal resistances, because resistances are additive whereas conductances are not.

However, sometimes both high and low and low R-value materials are needed simultaneously, as in traced lines. See Fig. 2. Although systems are designed with limitation on overall heat loss, high-conductance cements are used for process safety on pipelines carrying hot fluids that would solidify if flow was interrupted or fluids become so viscous that the pumps could not handle the material. In such cases, one or more small pipes called tracers carrying hot fluids are enclosed with the thermal insulation envelope, and high conductance cements are used to improve heat flow from tracers to the main pipe. The size of the pipe insulation must be large enough to enclose the main pipe and the tracers, and also the insulation on fittings.

Importance of Moisture Migration. Heat conservation has been treated from an energy standpoint as if it were an independent subject, whereas great costs and energy losses have been incurred by premature failures, because it was not realized that in many cases thermal insulations cannot be installed without inducing an effect on the migration of moisture. The problems are usually not great from the standpoint of solutions, but are great in getting people to realize that they have created problems that have been overlooked. In most constructions, heat and moisture performance must be considered jointly, because even high-temperature systems are shut down for alterations, maintenance, or repair. For example, in a house wall, adding thermal insulation makes the indoor wall warmer, as desired, but at the same time it makes the outdoor wall colder, and it is well known that when surfaces are colder than the dewpoint, condensation occurs. The old log cabin with its loose construction had no moisture problems, but it was often only the side of the body that faced the fireplace that felt warm while the rest did its best to accommodate the facts.

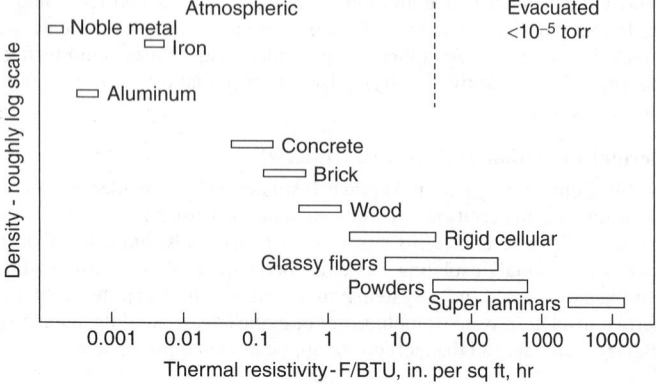

Fig. 1.   Thermal resistivity of materials in insulation systems

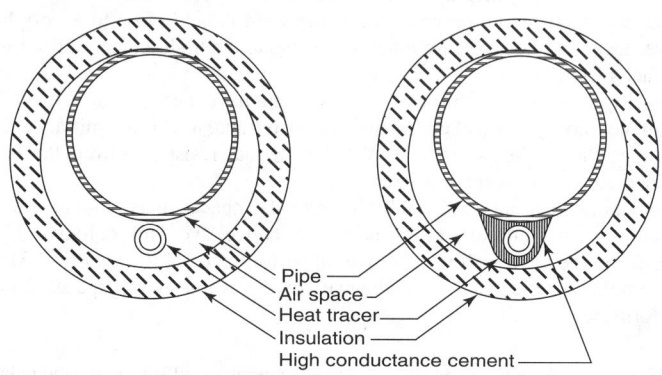

Fig. 2.   Pipe insulation over pipe and heat tracer line, with and without high-conductance cement

Economic Considerations. For many years, cost appraisal of insulation was based on a publication by L.B. McMillan in *ASME Proceedings*, December 1926, in which several factors enter into the analysis, as shown in Fig. 3. At present, the cost of money and the costs of materials and labor are so unstable, that such analyses are of little value except that they do indicate the factors that enter into actual costs to plant management.

To presume that thermal insulation is always necessary is false. For example, when cryogenic fluids are transported from a supply source to a vessel, even in bright sunshine, it is usually most economical not to insulate the line at all. The reason for this is twofold: (1) for the short time of transport the area of the pipe exposed to the sun is much smaller than the area that would be exposed if it were insulated; and (2) the heat that would be in the insulation when the transport starts would have to be removed by the cryogenic fluid. The combination of these two factors often makes the use of insulation undesirable for this type of cold fluid transport system. Moreover, the rapid formation of ice crystals from moisture in the air constitutes a thermal insulation.

While thermal insulations are selected first for their resistance to heat flow, their other properties need evaluation for each application. Hence there is no "best" insulation because a material well suited to one service may be poorly suited to another. Economics must be studied in detail because in some services the cost of a highly efficient material per unit of thickness may be overcome by additional thickness of a less efficient material, provided there is room for the greater thickness. All properties of materials must be considered for each exposure, even within the same system. Recognize that when different materials are used on different parts of insulation systems that are close together, the probability of using the wrong material on a particular surface is increased appreciably. Unless there is specific reason for wide use of multiple types of materials, it may be prudent to accept some compromises of properties.

In general, it is desirable to place thermal insulation on or near the outside of constructions, including basement walls, because this location reduces appreciably the temperature stresses in the structure induced by the changes in exposure.

Selection of Thermal Insulation

Test methods, specifications, and some recommended practices are in ASTM Book of Standards, Part 18. Producer's literature gives forms, properties and design data.

1. *Thermal Resistance—R-value*—Thermal data in general Tables, as Table 2, are for dry materials, and for one or two mean temperatures, but most materials are not linear with mean temperature, hence, for specific designs, the whole range of resistance should be obtained.

A factor often overlooked is that space to be occupied by the insulation must be made available; a 3-inch (76-millimeter) iron pipe with 3 inches (76 millimeters) of insulation covered with a protective jacket would have a diameter of about 10 inches (254 millimeters).

2. *Temperature Limit*—Usually the high temperature limit governs because shrinkage then becomes excessive. Generally, high-temperature materials are physically stable at low temperatures but another material may be preferable.

Note that shrinkage data are usually from tests in soaking heat, whereas the field condition is for heat on only one surface. Hence, such data should not be presumed to mean that the insulation will shrink that amount in service; a 1% shrinkage would imply a change of almost 3/8 inch (9.5 millimeters) in a standard 36-inch (914-millimeter) length and leave a wide crack. However, in service such a large shrinkage does not occur; shrinkage would be 1/16–1/8 inch (1.6–3.2 millimeters). Moreover, hot metals expand, and it is this expansion that must be considered in

compensating for openings between adjacent pieces of insulation, usually by use of double-layer insulation with staggered joints.

For cold-temperature service, moisture ingress is a problem that must be designed against and the material selected for its resistance to moisture and potential freezing.

3. *Corrosion*—While corrosion is usually not a concern on hot surfaces, recognize that all systems have shut down periods, with the probability that moisture will find ingress and condense. Many insulations are alkaline and have little adverse effect on iron and copper, but aluminum is affected adversely. A major concern is stress-corrosion of stainless steel induced by even trace amounts of soluble chlorides; an ASTM Test Method may be used.

4. *Density*—While the density of thermal insulation is low enough not be a problem in most cases, density (or mass-weight) must be considered in airplanes, balloons and other antiterrestrial constructions.

5. *Moisture-Wetting*—If moisture can enter an insulation, the high conductance of water (or ice) reduces the thermal resistance. However, all materials that admit moisture are not affected adversely to the same degree. Some fibrous insulations have a threshold moisture content below which the adverse effect on resistance is not significant; thresholds may be on the order of 8% by weight. The reason is that small droplets adhere to the contact points of one fiber against another so that the volume of relatively still air that provides the thermal resistance is not reduced significantly. Moreover, the temperature difference within the insulation drives the moisture to the cold side, and if it condenses there, the thickness of the wet insulation with decreased resistance is still a small portion of the total resistance.

In materials that readily absorb water, a decrease in thermal resistance will occur, but consideration must be given to the performance after the material is redried.

Caution is needed in interpreting absorption tests by immersion because the internal structure may resist displacement of air or gas so that a low absorption is indicated. However, if moisture is induced to flow through the specimen by vapor pressure differential and a condensing temperature is reached, high absorption may occur in service.

Materials with very high internal surface areas may absorb a measurable amount of moisture; even a monomolecular thickness of moisture on a large area becomes readily measurable. However, exposure to high relative humidity will not "saturate" the material, and as relative humidity changes so will the absorbed moisture. Some materials are tested wet to indicate wet to dry strength ratio.

6. *Handleability*—In the field, materials must have strength properties that enable them to be handled in application by the usual procedures of the industry without excessive breakage. ASTM tests for flexure, impact, and friability are aimed at indicating handling potential. Sometimes vibration tests are indicated if an unusual vibrating condition is to be encountered, although this test is receiving less attention than in the past because vibrations are usually designed against.

7. *Reflectances*—When an all-metal system is desired for temperatures to 1000°F (540°C) usually to avoid absorptions in case of leaks, multilayer sheets of aluminum are spaced on the order of 1/2 inch (13 millimeters). For temperatures to 1400°F (760°C) the first sheets are made of stainless steel.

8. *Fire Behavior*—While thermal insulations are not intended for fire protections (treated elsewhere) their behavior in fire is important, especially from the standpoint of contribution of combustible matter to a fire that has started at the site. Material behavior may be complex, e.g., an absorptive material that would hold a combustible fluid (say, kerosene) would not be a major contribution to fire intensity because the fluid would not flow to the surface to burn as rapidly as it would from a pool of the fluid. Materials that contain organic binders may not be a serious contribution in an open fire, but if they are totally enclosed they may contribute to persistence of fire by smoldering.

While some thermal insulation may not constitute a significant fire hazard, consideration must be given to jackets, coatings, or coverings, and to the methods of attachment, so that even in moderate fires the insulation will not readily fall off the construction they insulate.

Principal Types of Thermal Insulating Materials

Thermal insulations are made from natural or processed materials and combined to provide properties that meet the needs of specific installations. Obviously, all desired properties are not available in any one insulation. Hence, selection of thermal insulation for specific uses involves comparisons for each use, and some high thermal resistance (*R*-value) may

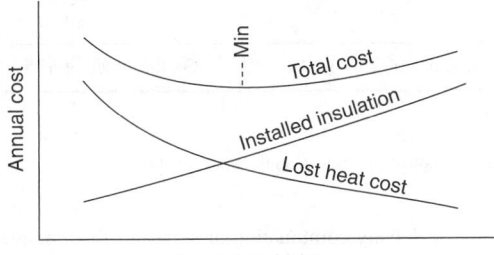

Fig. 3. Relation of incremental cost of Additional thickness of insulation to the resultant savings and total cost. (After *L.B. McMillan*)

TABLE 2. REPRESENTATIVE THERMAL INSULATIONS

Temperature Limit °F	°C	Materials and Usual Forms	Inorganic (I) or Organic (O)	Density lb/ft³	kg/m³	At Mean temp., °F	F° ft² h / Btu in	At Mean temp., °C	C° m / W
		HIGH-TEMPERATURE SERVICE							
2300	1265	Alumina-silica ceramic fiber, soft mass	I	3–12	48–192	300	3.2	160	22.2
						1000	1.2	540	8.3
2200	1200	Potassium titanate fiber, soft mass	I	15–18	240–288	300	3.0	160	20.8
1900	1040	Diatomaceous silica, bonded, semirigid, preformed block and pipe	I	23–25	368–400	300	1.5	160	10.4
						1000	1.3	540	9.0
1800	1000	Mineral fiber, rock, and slag, loose fill, preformed block and pipe	I	16–24	256–384	400	1.7	204	11.8
						1000	1.3	540	9.0
1600	875	Perlite, expanded, loose granules	I	4–10	64–160	0	3.0	−18	20.9
						1000	0.9	540	6.2
1200	650	Hydrous calcium silicate, may contain unexposed organic reinforcing fibers, preformed rigid block and pipe, compression over 100 psi (0.69 Mpa)	I/O	11–14	176–224	300	2.5	160	17.3
						700	1.7		11.8
1000	540	Glass fiber, no binder, loose mass	I	3–5	48–80	300	2.9	160	20.1
						800	1.4	430	9.7
500	275	Gilsonite, processed pure asphalt powder for underground fill, compacted, impervious	O	35–48	560–640	50	1.9	10	13.2
						3300	1.6	1820	11.1
800	425	Glass, cellular, preformed block, impervious, compression over 100 psi (0.69 Mpa)	I	10–18	160–288	300	1.8	160	12.5
						600	1.0	320	7.0
450	225	Glass fiber, organic binder, loose fill, blankets, batts, preformed block and pipe	I/O	0.5–3	8–24	75	3.8	25	26.4
						300	2.8	160	19.4
200	95	Cellulosic fibers of wood, cane, reused paper, as loose fill	O	0.3–5	4.8–80	75	3.8	25	26.4
		LOW-TEMPERATURE SERVICE							
−225	−140	Plastic Foams Polyurethane (under investigation)	O	1.8–2.2	28.8–35.2	−200	11.0	95	76.3
						100	6.0	40	41.6
−200	−130	Polystyrene	O	1.0–4.0	16–64	40	3.9	4.4	27.0
						75	2.6	25	18.0
−40	−40	Polyvinyl chloride	O	4–25	64–400	75	3.9	25	57.0
−40	−40	Rubber, cellular	O	3–20	48–320	25	4.3	−3.9	29.8
						75	3.3	25	22.9
−400	−245	Glass, cellular, preformed block	I	10–18	160–288	25	2.8	−3.9	19.4
						100	2.4	40	16.7
−450	−270	Mineral Fibers	I	0.5–10	8–160	25	3.7	−3.9	25.7
						100	3.3	40	22.9
−459	−273	Evacuated multilayer foil and fiber mats	I	Various		On the order of 25–100+		On the order of 175–700+	

[1] Most high-temperature insulations are usable also at low temperatures, but other materials may be preferable.
[2] Maximum temperatures apply to surface service, not soaking heat.
[3] Thermal resistivity data are approximate and vary widely with density (for specific designs and temperatures, consult current manufacturers' data).
[4] Celsius temperatures are rounded. °C is temperature; C° is temperature difference.

need to be sacrificed in favor of some other property, such as handleability, resistance to compression, thermal diffusivity, avoidance of stress-corrosion, toxicity, or in-fire performance. See Table 2. Based upon ASTM C 168–80a, the principal types of thermal insulation are:

Ablative. Heavy density combination of materials that change state from solid to liquid or vapor in high temperature so that heat absorbed through their change of state reduces substantially the heat transfer rate through the material. Suited to one-time or very few times use.

Calcium Silicate. Composed principally of hydrous calcium silicate, usually containing reinforcing fibers.

Cellular Elastomeric. Composed principally of natural or synthetic elastomers, or both, processed to form flexible, semirigid, or rigid foams which have a predominantly closed-cell structure.

Cellular Glass. Composed of glass processed to form a rigid foam usually having a predominantly closed-cell structure.

Cellular Polystyrene. Composed principally of polymerized styrene resin processed to form rigid foam having a predominantly closed-cell structure.

Cellular Polyurethane. Composed principally of the catalyzed reaction product of polyisocyanate and polyhydroxy compounds, processed usually with fluorocarbon gas to form a rigid foam having a predominantly closed-cell structure. Under investigation by regulators.

Cellulosic Fiber. Composed principally of cellulose fibers usually derived from paper, paperboard stock, or wood, with or without binders.

Diatomaceous Silica. Composed principally of diatomite (diatomaceous earth) with or without binders, usually containing reinforcing fibers.

Gilsonite. A pure form of asphalt processed into powdered form for use as the enclosing insulating mass around pipes or tanks underground.

Mineral Fiber. Composed principally of fibers manufactured from rock, slab, or glass, with or without binders.

Perlite. Composed of natural perlite ore expanded and processed to form particles of various sizes with a cellular structure.

Vermiculite. Composed of natural vermiculite ore expanded and processed to form particles of various sizes with an exfoliated structure.

Wood Fiber. Composed of wood fibers, with or without binders. This is a type of cellulosic fiber insulation.

Principal Forms of Thermal Insulation

Also based upon ASTM C 168–80a, the principal forms of thermal insulation are:

Blanket Insulation. A relatively flat and flexible insulation in coherent form furnished in units of substantial area. Some forms are called batts.

Blanket Insulation, Metal Mesh. Blanket insulation covered by flexible metal-mesh facings attached on one or both sides.

Block Insulation. Rigid insulation preformed into rectangular units.

Board Insulation. Semirigid insulation preformed into rectangular units having a degree of suppleness particularly related to their geometrical dimensions.

Cement, Finishing. A mixture of dry fibrous or powdery materials, or both, that when mixed with water develops a plastic consistency, and when dried in place forms a relatively hard, protective surface.

Cement, Insulating. A mixture of dry granular, flaky, fibrous, or powdery materials that when mixed with water develops a plastic consistency, and when dried in place forms a coherent covering that affords substantial resistance to heat transmission.

Fitting Covers. Manufactured or assembled segments of insulation to form covers for various pipe and vessel fittings such as elbows, tees, crosses, valves, etc. See ASTM Standard C450–76 (or later) for dimensions.

Loose-Fill Insulation. Insulation in granular, nodular, fibrous, powdery, or similar form designed to be installed by pouring, blowing, or hand placement.

Pipe Insulation. Insulation in a form suitable for application to cylindrical surfaces.

Reflective Insulation. Insulation depending for its performance upon reduction of radiant heat transfer across spaces by use of one or more surfaces of high reflectance and low emittance.

Roof Insulation. Rectangular boards or blocks of various thicknesses with properties for use beneath the roofing membrane protected from the weather, or with properties for use above the roofing membrane exposed to the weather.

Underground Systems. Systems that enclose insulated piping in small tunnels that include expansion arrangements and provide drainage. Since accidental general flooding may occur, the insulations should be capable of withstanding "boiling water" effects so that when the system has been dewatered the redried insulation will perform thermally essentially as it did prior to flooding.

Super Insulations. Several insulation systems have been developed that have very higher thermal resistances, such as multi-layer radiation

shields, specially selected and matted fibers with small interfiber distances, special powders, ceramic foams, honeycomb composites, often highly evacuated, but they are too costly for usual services with which the general public is familiar.

Insulating Concrete. This should be recognized as relative in performance to usual heavy density concrete, and does not provide thermal resistance in the range of materials understood to be thermal insulations.

E. C. SHUMAN
Consulting Engineer
State College, Pennsylvania

Additional Reading

ASHRAE: *Handbook of Fundamentals,* American Society of Heating, Refrigerating, and Air Conditioning Engineers, Inc., New York, NY (Published every 5 years). http://www.ashrae.org/

ASTM: *Annual Book of ASTM Standards: Thermal Insulation: Environmental Acoustics,* Vol. 6, American Society for Testing and Materials, Philadelphia, PA, 1997. http://www.astm.org/American Society for Testing and Materials.

ASTM: *Book of Standards,* Part 18, American Society for Testing and Materials, Philadelphia, Pennsylvania (Issued annually).

ASTM: *Thermal Insulation Performance,* STP718, American Society for Testing and Materials, Philadelphia, Pennsylvania (1981).

Bynum, R.T. Jr.: *Insulation Handbook,* The McGraw-Hill Companies, Inc., New York, NY, 2000.

Glaser, P.E., et al.: "Thermal Insulation Systems," National Aeronautics and Space Administration, NASA SP-5027, Washington, DC, 1967.

Staff: *How to Determine Economic Thickness of Insulation,* Thermal Insulation Manufacturers Association, Mt. Kisco, New York, NY, (Revised periodically).

Turner, W.C. and J.F. Malloy: *Thermal Insulation,* Krieger Publishing Company, Melbourne, FL, 1990.

Web Reference

North American Insulation Manufacturers Association. http://www.naima.org/

INSULIN. A polypeptide hormone having a molecular weight of 5733. It is formed in the isles of Langerhans located in the pancreas and was so named for this reason. Insulin is composed of 16 amino acids arranged in a coiled chain and cross-linked in several places by the disulfide bonds of cystine residues. The sequence of amino acids has been elucidated. The insulin molecule was synthesized in 1963. In 1977, rat insulin was produced in the bacterium *E. coli* by recombinant DNA techniques. A year later human insulin was generated after chemically synthesized genes were added to *E. coli.* This synthetic insulin is now in commercial production and has been approved by the FDA. Insulin regulates carbohydrate metabolism in the body by decreasing the blood glucose level. A systemic deficiency leads to diabetes.

See also **Carbohydrates.**

INTERCALATION COMPOUND. A compound composed of a crystalline lattice that acts as an electron donor, and "foreign" electron acceptor atoms interspersed of diffused between the planes of the lattice. An important group of intercalated compounds are composed of graphite, where bromine, for example, can act as electron acceptor. Graphite is particularly susceptible to this phenomenon because of its orderly stacked layers of crystals. Anhydrous metal nitrates such as copper and zinc nitrates also form intercalated compounds with graphite. A further example is trilithium nitride, whose structure consists of a series of layers of dilithium nitride, between which is a layer of lithium atoms. This markedly increases the conductivity, so that the material becomes an effective solid electrolyte in batteries. Other substances having this property are sodium β-alumina, titanium disulfide, and some metal dioxides. The phenomenon does not impair the crystalline structure and is reversible. Intercalated compounds are used for superconductors, synthetic lubricants, catalysts, and storage batteries. They are used in biochemical research; an acridine-based compound that can intercalate between stacked pairs of bases in a DNA helix is used in cancer research.

INTERFACE. The area of contact between two immiscible phases of a dispersion which may involve either the same or different states of matter. Five types are possible: (1) solid–solid (carbon black-rubber), (2) liquid–liquid (water-oil), (2) solid–gas (smoke-air), (4) solid–liquid (clay-water), (5) liquid–gas (water-air). At a fresh surface of either liquid or solid the molecular attraction exerts a net inward pull. Hence the characteristic property of a liquid is surface tension and that of a solid surface is adsorption. Both have the same cause, namely, the inward cohesive forces acting on the molecules at the surface. These phenomena provide to some degree the fundamental mechanism for many industrially important processes (catalysis, emulsification, mixing, alloying) and products (detergents, adhesives, lubricants, paints). Such properties as wettability of solid powders, spreading coefficients of liquids, and protective action of colloidal substances are intimately associated with interfacial behavior.

See also **Catalysis**; **Colloid Systems**; **Detergents**; **Surface**; **Surface Tension**; and **Wetting Agent**.

INTERMEDIATE (Chemical).

An intermediate generally is considered to be a material (usually a chemical compound) that occurs somewhere in a chemical manufacturing process between the introduction of the basic raw materials and the creation of the final end products. When two or more separate chemical reactions are involved, the intermediate may be the product of one of the *between reactions* and serve as a charge material for a subsequent reaction. For example, in the manufacture of aromatic polyester, several materials and reactions are required. The fundamental raw materials are nitric acid, xylene, methanol, and ethylene glycol. In one reaction, *p*-xylene and nitric acid yield terephthalic acid. The terephthalic acid then is esterified with methanol using sulfuric acid as a catalyst to yield dimethyl terephthalate. The dimethyl terephthalate then undergoes an ester interchange with ethylene glycol which yields *bis*-(β-hydroxyethyl)terephthalate, later condensed to polyethylene terephthalate. This low-molecular-weight polymer then is polymerized to a high-molecular-weight polyethylene terephthalate. In this operation, terephthalic acid and dimethyl terephthalate can be regarded as intermediates. In some instances, a producer will procure intermediate materials from the outside rather than produce them in-house, particularly in the cases of the pharmaceutical and dye industries. Thus, a number of intermediates are high-tonnage items of commerce. Some intermediates are of low-tonnage requirements and sometimes the economics is in favor of one producer who supplies a number of using firms. A representative list of intermediates would include: *o*-aminophenol-*p*-sulfonic acid; 2,6-dichloro-4-nitroaniline; 4-sulfophthalic acid; *o*-tolidine dihydrochloride; diphenylmethane; diphenylacetaldehyde; methyl cyclopentylphenylglycolate; and 2,3-dichloro-5,6-dicyano-benzoquinone. See also **Synthesis (Chemical)**.

INTERMEDIATE-MOISTURE FOODS.

Authorities have various defined intermediate-moisture foods (IMF) as having from 15–40% water at a water activity $a_w = 0.6 - 0.8$. Interest in intermediate-moisture foods for the human diet largely stemmed from the introduction and acceptance of "soft-moist" pet foods, which were first marketed in the 1970s. Research continues at a good pace and considerable progress has been made as, for example, in processed cheese foods with high moisture contents.

More important than moisture content, per se, is the objective of creating a preserved food substance that is stable and can be eaten directly. An idealized IMF system will have (1) a microbial stability at reduced water activity, (2) storage stability without special conditions, (3) reduction of weight and more compactness of product, and (4) can be consumed from the package without rehydration. Some authorities believe that IMF systems represent excellent potential in the development of the snack market.

Two factors have largely delayed expansion of IMF systems: (1) technology and (2) consumer acceptance. Some of the technical problems involved in formulating IMF systems include: (1) rates of lipid oxidation, (2) enzymatic deterioration, and (3) nonenzymatic deterioration. With some products, there are also problems associated with the desired texture.

INTERMETALLIC COMPOUND.

In certain alloy systems, distinct intermediate phases occur where the constituent atoms are in fixed integral ratios, e.g., CuZn (β-brass). Such a compound is held together by metallic bonding and may form a very complicated crystal structure. The constitution of such an alloy is often governed by the Hume-Rothery rules. In some cases, if the electron concentration is such as to just fill a band, the material may even be semiconducting (e.g., InAs). See also **Compound (Chemical)**.

INTERSTITIAL.

(1) Descriptive of a nonstoichiometric compound of a metal and a nometal whose structure conforms to a simple chemical formula, but exists over a limited range of chemical composition. Interstitial compounds are represented by borides, nitrides, and carbides of the transition metals. (2) Descriptive of an atom of an impurity that causes a defect or dislocation in a crystalline lattice, e.g., an atom of carbon or nitrogen in an iron crystal, or of arsenic in a semiconductor. (3) In a biological sense, the term describes cells located between or within layers of tissue.

IN VITRO.   An event or process occurring outside a living organism—in an unnatural environment, as in a test tube.

IN VIVO.   An event or process occurring naturally or spontaneously within a living organism.

IODINE.   [CAS: 7553–56–2]. Chemical element symbol I, at. no. 53, at. wt. 126.9045, periodic table group 17 (halogens), mp 113.5°C, bp 184.35°C, density 4.94 g/cm³ (20°C). Iodine has an orthorhombic crystal structure. Solid iodine is a violet-to-black color; vapor is a beautiful violet color. The element sublimes readily and is easily purified in this way. Iodine is insoluble in H_2O, soluble in alcohol, ether, CS_2, or carbon tetrachloride. The element was first identified by Courtois in 1812 when making a study of kelp. There is one stable isotope ^{127}I and fourteen radioactive isotopes ^{122}I through ^{126}I and ^{128}I through ^{136}I. The lengths of half-lives of the isotopes vary widely, the shortest ^{136}I with a half-life of 86 seconds; the longest ^{129}I with a half-life of 1.72×10^7 years. See also **Radioactivity**. In terms of abundance in the crust of the earth, the element ranks 53rd and is about as plentiful as tin, antimony, cesium, and barium. Considerable quantities of iodine have concentrated in the oceans. The average iodine content of a cubic mile of seawater is 230 tons (50 metric tons per cubic kilometer).

First ionization potential 10.44 eV; second, 19.4 eV. Oxidation potentials $I^- \rightarrow \frac{1}{2}I_2 + e^-$, -0.535 V; $I^- + H_2O \rightarrow HIO + H^+ + 2e^-$, -0.99 V; $I^- + 3H_2O \rightarrow IO_3^- + 6H^+ + 6e^-$, -1.085 V; $\frac{1}{2}I_2 + 3H_2O \rightarrow IO_3^- + 6H^+ + 5e^-$, -1.195 V; $\frac{1}{2}I_2 + H_2O \rightarrow HIO + H^+ + e^-$, -1.45 V; $IO_3^- + 3H_2O \rightarrow H_5IO_6 + H^+ + 2e^-$, ca. -1.7 V; $I^- + 6OH^- \rightarrow IO_3^- + 3H_2O + 6e^-$, -0.26 V; $I^- + 2OH^- \rightarrow IO^- + H_2O + 2e^-$, -0.49 V; $IO + 4OH^- \rightarrow IO_3^- + 2H_2O + 4e^-$, -0.56 V; $IO_3^- + 3OH^- \rightarrow H_3IO_6^{2-} + 2e^-$, ca. -0.70 V. Other important physical characteristics of iodine are given under **Chemical Elements**.

Sea plants, particularly kelp found in the waters around California and the Bay of Biscay, have been a source of iodine. Because of pollution, the kelp beds in California are no longer a major source. Iodine also is found in the petroleum oil well brine of California and, in small percentages, in sodium nitrate of Chile. The latter was once the primary source of the element. Brines now are the major source.

Uses

For many years, iodine tincture (3% to 7% dissolved in ethyl alcohol) has been an important antiseptic. The commercial tinctures also usually contain 5% potassium iodide to provide stability. This form produces a mild burning of the skin and stains both skin and fabrics. A milder preparation is available in which about 2% iodine is contained in an oil-water emulsion that also contains lecithin. The burning effect of the compound is greatly reduced and the mild stains produced usually wash off easily. There are a number of prepared medicines that contain iodine, although some of these have been removed from the market in recent years. At one time, iodoform CHI_3, a yellow, insoluble, crystalline powder with a very penetrating odor, was a very popular antiseptic and used widely in the preparation of gauzes and packings for infected cavities. Because of possible toxic effects with some individuals, the compound largely has been replaced by other less objectionable, less odorous materials.

The medical use of iodine compounds, particularly organo-iodine substances, has been researched with resultant limited applications. The dietary requirement for iodine was established many years ago as needed for the maintenance of cell growth in humans and animals. The largest concentration of iodine occurs in the thyroid gland where the hormone thyroxine $C_{15}H_{11}O_4I_4N$ is present. The waters and soils of some areas, as in the inland areas of the United States, do not contain the minimal trace quantities of iodine required by the normal diet. Thus for many years so-called iodized table salt with a content of about 0.01% potassium iodide has been available. This preventive practice has been accredited with forestalling goiter and associated glandular disturbances in untold thousands of instances. For health reasons, iodine supplements also are added to cattle feeds.

Iodine tablets provide an easy means for sterilizing drinking water in small portions, usually resulting in less odor and taste objections than chlorine compounds for the same purpose. Iodine chemicals are widely used in photography and printing reproduction processes. See also **Photography and Imagery**.

The production of vanadium metal essentially is the calcium reduction of vanadium pentoxide in the presence of iodine and is known as the McKechnie-Seybolt process. The reaction is carried out in a steel bomb at about 700°C. The end products are vanadium metal, lime, and calcium iodide. A similar iodide process also is used in the production of high-purity zirconium.

Chemistry and Compounds

Iodine exhibits in common with the other halogens a marked readiness to form singly charged negative ions, as would be expected from the fact that these atoms need only one electron to acquire an inert gas configuration. However, of the four common halogens, iodine has the lowest electron affinity (3.2 eV) due to more effective screening of the nucleus. The iodides range in character from ionic to covalent compounds, many of them, such as hydrogen iodide, having bonds of intermediate nature. Iodine is the most electropositive of the common halogens, functioning as the positive univalent "ion" I^+, as in the compound iodine perchlorate, which, however, is not a salt, as well as in forming trivalent complex radicals, such as IO^+. Iodine also forms essentially covalent linkages with negative elements, in which it has positive valences 1, 3, 5, or 7.

In binary combination with oxygen, however, only one simple compound has been isolated, iodine pentoxide, I_2O_5, a white compound. The yellow I_2O_4, prepared from sulfuric and iodic acids, is considered to be made up of 3+ and 5+ iodine, and the structure iodyl iodate, (IO^+) (IO_3^-), is assigned to it, in which the trivalent state is stabilized by the acid radical. Similarly, yellow tetraiodine enneaoxide, I_4O_9, is considered to have the structure (I^{3+}) $(IO_3^-)_3$, in which again the trivalent state is stabilized by the acid radicals.

Hydrogen iodide, HI, is the least stable of the four common hydrogen-halides, and correlatively, the best reducing agent, readily reducing vanadic acid, nitrous oxide to ammonium, nitrous acid to nitric oxide, and HNO_3 to nitrous acid. Because it is so readily oxidized, it cannot be prepared by action of H_2SO_4 on an iodide, but can be made by the action of weak acids, e.g., H_2S, upon iodine, or by hydrolysis of certain iodides. It can be prepared by direct combination of hydrogen and iodine vapor on a platinum catalyst. It is also liberated in many organic iodination reactions, such as the reaction of iodine and refluxing tetralin. The H—I bond is considered to be partly covalent. HI is a monoprotic acid, and is stronger than hydrogen chloride and hydrogen bromide.

The oxyacids of iodine are essentially covalent compounds, the known acids being hypoiodous acid, HIO, iodic acid, HIO_3, and the various periodic acids. Hypoiodous acid is formed, along with iodide ion, by dissolving iodine in dilute alkali, or by action of mercuric oxide upon iodine and water. On standing hypoiodous acid disproportionates to iodic acid and hydriodic acid. Hypoiodous acid is a powerful oxidant. It is also amphiprotic: $H^+ + IO^- \leftrightarrow HIO \leftrightarrow I^+ + OH^-$, $pK_A = 10.4$, $pK_B = 9.49$. The evidence for the formation of the I^+ ion is the existence of a number of compounds of composition $I(r)_nX$, where r is pyridine or some other nitrogen organic base, n is 1 or 2, and X is hydroxyl, nitrate, or chlorate ion. The conductivity of liquid iodine also indicates ionization into solvated I^+ and I^-.

Hydriodic acid (HI) is a colorless solution formed when hydrogen iodide gas is dissolved in water, commercially of strength 10% HI, frequently colored brown by iodine. There is a maximum constant boiling point 127°C (774 mm) at 57% HI (distillate) for mixtures of hydriodic acid and water. Hydriodic acid is used in the preparation of iodides, and as an important reagent in organic chemistry.

All metallic iodides except silver iodide, mercurous iodide, mercuric iodide, lead iodide, cuprous iodide, thallium iodide, and palladium iodide, are soluble. The iodides of antimony, bismuth, tin require a little free acid to keep them in solution.

Dilute hydriodic acid reacts with hydroxides, oxides, carbonates, sulfides, metals in a manner chemically analogous to dilute HCl; with solutions of some salts, e.g., silver nitrate, to yield the corresponding iodide, e.g., silver iodide, precipitate. Higher strengths of hydriodic acid react with oxygen of the air upon standing to yield free iodine, which imparts a brown color to the solution, thus indicating the reducing character of the acid.

Hydriodic acid is made by the reaction (1) of iodine and hydrosulfuric acid (or sulfurous acid), (2) of phosphorus plus iodine plus water, with subsequent distillation in all cases.

Iodic acid, HIO_3, is commonly prepared by oxidizing iodine with HNO_3. It is a strongly oxidizing acid, oxidizing iodide to iodine, sulfite to sulfate, and H_2S to sulfur. It reacts vigorously even with dry carbon, phosphorus, or organic matter.

Iodate ion in aqueous solution appears to be $I(OH)_6^-$, and the iodine atom in crystalline periodates always is coordinated to six oxygen atoms, three nearest neighbors on one side and three next nearest neighbors on the other side. Thus, potassium iodate, KIO_3, has distorted perovkite structure.

In the broad picture, the iodides, like the other halides, range in character from completely ionic structures to covalent ones. The transition is clearly exhibited by the iodides of the first four groups in the (extended) periodic table, potassium iodide, KI, being ionic and titanium(IV) tetraiodide, TiI_4, essentially covalent. On the right-hand side of the periodic table, even the group I elements form bonds with iodine that exhibit a considerable degree of covalence. In general, the ionic iodides are the most soluble of the halides of the given element, e.g., sodium iodide, NaI, is the most soluble sodium halide, while the covalent (or partly covalent) iodides are the least soluble, e.g., silver iodide, AgI. These effects are correlated with the size and polarizing power of the cation and the increasing size and polarizability of the halogen ion, as is increase of reducing character and stability of coordination complexes. A related fact is the readiness of formation of the complex ion, I_3^-, upon dissolving iodine in an aqueous solution of KI. The variation in character of iodine solutions in various organic solvents as well as water, is also attributed to complex formation.

Because iodine is the least electronegative of the four common halogens, it forms a relatively larger number of compounds with the other three. They include iodine trichloride, iodine chloride, iodine bromide, iodine fluoride, iodine trifluoride, iodine pentafluoride, and iodine heptafluoride. These are reactive compounds, especially the lower ones, which enter into reactions with many organic and some inorganic substances. With ICI and organic substances, the end product is usually an iodine or chlorine substitution product, the solvent being important in determining which is formed. With inorganic compounds, the addition product is often the final product, thus ICI and antimony (V) pentaiodide $SbCl_5$ give a product of composition $ISbCl_6$, which ionizes to I^+ and $SbCl_6^-$.

Several oxyfluorides exist: IOF_3 (iodyl hexafluoridate, $[IO_2][IF_6]$), iodyl fluoride, IO_2F, and periodyl fluoride, IO_3F.

Related to the complex ions, as well as the interhalogen molecules, are their association products, the polyhalide complexes. For iodine they include the alkali (and ammonium) triiodides, along with higher compounds such as that with cesium, Cs_2I_8, ammonium, NH_4I_5, tetraethylammonium iodide, $(C_2H_5)_4NI_7$, tetramethylammonium iodide, $(CH_3)_4NI_9$, and benzene, $KI_9 \cdot 3C_6H_6$. They also include compounds of iodine with alkali metals and one or two other halogens, such as NH_4IBr_2, $KICl_2$, $RbICl_2$, $HICl_4 \cdot 4H_2O$, $CsFIBr$, $RbClIBr$, $CsClIBr$ and $KIF6$. Most of these compounds hydrolyze readily, decompose on heating to give the metal halide of greatest lattice energy, and ionize to give ions such as $ClICl^-$, $BrIBr^-$, and $BrII^-$.

Structural studies show that the complex $HICl_4 \cdot 4H_2O$ contains a positive trivalent iodine atom, and the planar ion ICl_4^- is one of the most stable of the polyhalide anions, as might be expected from consideration of the large size of the iodine atom and the small size of the chlorine atoms.

The difluoroiodate ion. $IO_2F_2^-$, is a trigonal bipyramid with two apical fluorine atoms, two equatorial oxygen atoms and one equatorial nonbonding electron pair surrounding the central iodine atom.

Iodine forms many organic compounds, chiefly by replacement of hydrogen or by addition reactions at double bonds.

Periodic acid H_5IO_6 has been isolated as a colorless solid, mp about 130°C, and at 138°C begins to decompose, metaperiodic acid HIO_4 being formed and at higher temperatures iodine pentoxide plus oxygen plus water. $H_4I_2O_9$ and H_3IO_5 have been reported as fairly well established in identity; in solution the evidence points to the presence of HIO_4. Prepared by reaction of iodine and perchloric acid.

Paraperiodic acid H_5IO_6, is obtained from sodium paraperiodate, formed by action of chlorine upon a NaOH solution containing I_2. On vacuum drying, paraperiodic acid yields metaperiodic acid, HIO_4, and dimesoperiodic acid, $H_4I_2O_9$; which form heteropoly acids with a number of oxides and acids. The periodic acids and their salts are strong oxidizing agents both with inorganic and organic compounds. Although, in general, the iodates do not disproportionate to give periodates, barium iodate, $Ba(IO_3)_2$, on ignition gives barium paraperiodate, $Ba_5(IO_6)_2$, iodine and oxygen.

Sodium periodate $Na_2H_3IO_6$ is formed by reaction of sodium iodate plus sodium hydroxide plus chlorine (sodium chloride also formed), and the periodate separates as crystals from the medium. In solution, it is stated, periodate gradually forms ozone and iodate at the ordinary temperatures.

Metallic periodates are solids, slightly soluble in water. Periodates, when heated, evolve oxygen with simultaneous formation of iodate, which is decomposed at higher temperatures. Periodate in acid solution oxidizes hydrosulfuric acid or sulfurous acid to H_2SO_4, oxalic acid to CO_2, manganous to manganate, and with hydrogen peroxide yields oxygen and iodate.

The biological aspects of iodine are covered in **Iodine (In Biological Systems)**.

Additional Reading

Neilson, A.H.: *Organic Bromine and Iodine Compounds,* Springer-Verlag New York, Inc., New York, NY, 2003.

UNIPUB: *Control of Iodine in the Nuclear Industry,* Bernan Associates, Lantham, MD, 1973.

Varvoglis, A.: *The Organic Chemistry of Polycoordinated Iodine,* John Wiley & Sons, Inc., New York, NY, 1992.

Varvoglis, A.: *Hypervalent Iodine in Organic Synthesis,* Morgan Kaufmann Publishers, San Mateo, CA, 1996.

Wirth, T.: *Hypervalent Iodine Chemistry: Modern Developments in Organic Synthesis,* Springer-Verlag New York, Inc., New York, NY, 2003.

IODINE (In Biological Systems).   Iodine is not required by plants, but if iodine is present in the soil, it is taken up by most plants and moves on into the diets in forms that are effective in preventing goiter. In areas where the soils are high in iodine, ground water is also high in iodine, but the food supply is still the major source of iodine for people in these areas. Seafoods are good sources of dietary iodine.

Gout is no longer the single target of iodine deficiency concerns. The deficiency is implicated in mental retardation, deaf mutism, short stature, and an increased risk of death during childhood. Iodine deficiency is known to affect development of the central nervous system, particularly during the growing years. Intellectual impairment may range from a mild disorder to one of *cretinism* (significant and abnormal intellectual disturbance). Thus, there is a marked trend today among professionals to group all consequences of iodine deficiency under the category "iodine deficiency disorders."

Uneven Distribution of Iodine

Many of the iodine-deficient regions of the world have been identified. They are generally either mountainous or in the centers of continents, and distant from the oceans in the prevailing wind directions. Studies of the geochemistry of iodine indicate that this element is volatilized from oceans, carried overland by winds, and deposited on the soil by rain. The mountainous areas are low in iodine because little of that volatilized from the seas reaches sufficient altitude to be deposited in high altitudes. In some areas, the younger soils have less iodine than the older ones because of less time for the geochemical processes to build up the iodine level.

Although the amount of iodine in the soil is the primary factor determining iodine levels in food crops from various regions, the level of iodine in plants and the dietary requirements for iodine are modified to some extent by the plants themselves. There are important differences among plant species, and even among varieties of the same species, in their tendency to take up iodine from the soil. Certain plants, especially some of the *Brassica* genus, such as cabbage, contain compounds called *goitrogens*, which interfere with the effect of iodine on the thyroid gland. The amount of iodine required to protect animals, including humans, against goiter and other important iodine-deficiency disorders depends not only upon the kinds of plants in the diet, but also upon the iodine level characteristic of the soils in which the plants are cultivated. Iodine could be increased by adding iodine compounds to the soil, but that is a very inefficient way of insuring adequate dietary levels of the element. Much of the iodine added to the soil would be leached out and returned to the seas before it could be taken up by the crop plants. The use of iodized table salt is such an effective way of supplying this element that there is little need to include iodine in fertilizers.

A more recent method is that of using iodized oil, which has been found to be particularly effective in underdeveloped areas, such as several African countries.

Several comparative regional studies of iodine deficiency have been made. For example, one study shows that 1 of every 12 newborn infants in Freiberg, Germany, had an elevated serum thyrotropin concentration as contrasted with 1 of 1,428 infants born in Stockholm, Sweden, during an equivalent time frame. It has been found that iodine dietary supplementation reduced the incidence of congenital deafness in Switzerland by half. In countries that extend over broad and varied geographical regions, such as the United States, China, and Russia, the iodine content of soil can vary widely from one geographical community to the next. Dietary supplementation of iodine has increased the survival and birth weight of newborns in Zaire, prevented cretinism and decreased the rate of childhood mortality in Papua New Guinea, and advanced educability and economic productivity in China. Inasmuch as the facts leading to past successes have been well established, a meeting of the World Summit for Children in 1990 pledged to eliminate iodine deficiency by the year 2000. The Summit comprised responsible persons from the International Council for the Control of Iodine Deficiency Disorders, the United Nations Children's Fund, and the World Health Organization. Dunn (see reference listed) makes the interesting observation that, in 1923, David Marine wrote that "simple goiter is the easiest of all known diseases to prevent."

Dunn observes, "One intramuscular injection of iodinated vegetable oil containing 480 milligrams of iodine provides adequate amounts of iodine for up to three years. Oral administration of iodized oil is appealing for its simplicity and safety and millions of doses have been given worldwide. Still, the optimal dose and duration of effect have not been fully defined." Tonglet (see reference) and others have reported that as little as 48 milligrams of iodine administered orally as iodized oil is sufficient supplementation for 6 months. See also **Vegetable Oils (Edible)**.

One must be careful in dose determination because excessive doses of iodine can produce harmful effects.

Biochemistry of Iodine

In historical terms, 3,5-diidotyrosine (iodogorgoic acid) had been discovered in sponges and corals long before there was any hint concerning thyroid hormone structure even in mammalian forms. Since this time, iodotyrosines have been demonstrated in algae as well as in many animals possessing a horny skeleton. It has been suggested that iodide in the water is activated by peroxidases due to the presence of oxygen in the water and the resulting iodine is accepted by tyrosine in the protein molecule, similarly to the process in vertebrates. In animals possessing an exoskeleton, the presence of benzoquinones is part of the formation of scleroproteins by a quinone tanning process, and the iodination of tyrosine may be related to this in some way. In rotochordates, the endostyle, a structure secreting mucus, has been found capable of iodination of protein present in the mucus, which is then secreted into the alimentary canal.

The next phylogenetic development takes place in the vertebrates, in the most primitive members of which there is a structure located in the hypopharynx similar to a thyroid capable of collecting iodine and of forming iodinated protein, which is broken down by a protease, liberating iodinated amino acids. In some of these forms, small amounts of thyroxine are actually found in addition to the iodinated tyrosines. In other vertebrates, culminating in the amphibia and higher vertebrates, a thyroid gland is present in which the iodinated protein is held in a storage form known as colloid. In some, but not all, of these forms, hormonal material liberated by action of proteolytic enzymes is secreted into the bloodstream and plays an essential role in the development of the young animal as well as in the behavior and metabolic activity of the adult. Because of the process known as ontogeny, it is not surprising that the development of the thyroid gland in the human embryo commences with a structure located near the alimentary canal, which then separates and develops its peculiar follicular structure. Not until this type of development has occurred can a genuine function for the iodinated substances be demonstrated.

In these discrete glands, a series of specific chemical reactions can be demonstrated. See Fig. 1. Some of these reactions take place in the absence of any apparently specific synthesis of iodotyrosines, and it has not been fully explained how many of these steps require enzymes, even in mammalian forms, although these processes are usually considered as enzymatic. It is possible to iodinate tyrosine in soluble proteins in vitro by addition of elemental iodine and under these circumstances thyroxine and triiodothyroxine will also be formed, in company with small amounts of iodohistidine.

In amphibian vertebrates undergoing metamorphosis, thyroxine is known to be essential for this transition from an immature to a mature animal.

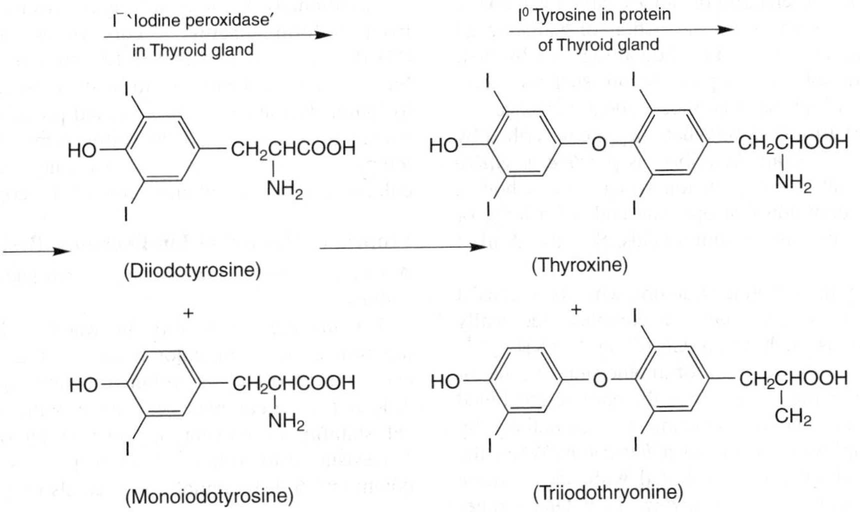

Fig. 1.   Series of reactions occurring in thyroid system

There is considerable evidence that after metamorphosis has occurred, the thyroid gland is no longer essential, although it may be involved in seasonal changes, such as molting.

There is no such clear cut differentiation as metamorphosis in the mammal, but development is an extremely complex process and has been shown to depend upon the presence of adequate amounts of thyroid hormones. Deficient development, especially of the central nervous system, is marked in children suffering from thyroid deficiency early in life, and this inadequacy cannot be overcome completely by medication commenced after the first few weeks. In the adult, thyroxine is important in the maintenance of energy turnover in most of the tissues of the body, such as the heart, skeletal muscle, liver, and kidney. Other physiological functions, most notably brain activity and reproduction, are also dependent upon thyroxine, although the metabolic rates of the tissues concerned in these functions do not seem to be altered.

A great deal of work has been done on determining the portions of the thyroxine molecule essential to biological activity. The fact that the hormone is an amino acid is almost certainly due to the widespread existence of the excellent iodine acceptor, tyrosine. Thus, the deaminated and decarboxylated metabolic product of thyroxine, tetraiodothyroacetic acid, has been shown to have appreciable biological activity, although quantitatively less than that of thyroxine itself. As for the halogens present on the diphenyl ether portion of the molecule, bromines and chlorines are also active, although diminishing considerably in that order from iodine.

Assuming there are some quite definite structural requirements for thyroid hormone activity, it is important to inquire into the specific actions of this material. It seems clear that thyroxine does not participate directly in any enzyme system, but rather affects the function of many systems, presumably by some far more general process. One of the earliest of such demonstrated actions was the uncoupling of oxidation from formation of high-energy phosphate compounds, such as adenosine triphosphate. However, such uncoupling is also produced by many other substances not showing thyroid hormone-like effects, and it has been shown that thyroxine is actually capable of accelerating coupled reactions under the proper conditions. From this evidence, it is suggested that the principal role of thyroxine may be the acceleration of enzyme processes ordinarily limiting the level of metabolic turnover. Mitochondria isolated from broken cell preparations by high-speed centrifugation have been shown to swell when placed in contact with thyroxine and similar substances. This may be evidence for a membrane function of the hormone, although it is not fully understood as to how this may alter cellular function in such specific manner as the hormone does *in vivo*.

An acceleration of protein turnover by thyroxine also has been shown, implying that the hormone may alter various processes by a specific effect on synthesis of certain key proteins involved in enzymatic reactions. Thus, not only does thyroxine increase the rate of formation of new protein material, but it also may be responsible for the transformation of non-enzymatically active protein into protein with enzymatic activity. The hormone has also been shown to be capable of acceleration of the synthesis of urea cycle enzymes and probably is essential for the production of a

sodium ion transporting mechanism, both of which are essential in the metamorphic transformation of larval forms into mature amphibia.

Steps in the synthesis of thyroid hormone include the following. (1) Active concentration of inorganic iodides within the thyroid epithelial cells. Concentration achieved is approximately 30 times that of plasma concentration. The so-called trapping of iodide is stimulated by thyroid-stimulating hormone (released by the pituitary gland). When present, this step is competitively opposed by thiocyanate and perchlorate ions. (2) Next, the inorganic iodide is oxidized to an organic form, in which peroxidase participates. The iodine becomes part of tyrosine residues in the thyroglobulin molecule. In this way, monoiodotyrosine (MIT) and diiodotyrosine (DIT) are formed. (3) The MIT and DIT are coupled by way of an ether linkage to form tetraiodothyronine (T4) and triiodothyronine (T3).

It has been observed that non-iodinated thyronine is not found in the thyroid gland.

The storage capacity and slow release mechanism of thyroid hormone appear to be unique among the endocrine glands. Usually a reserve of 100 days' needs (about 80,000 micrograms) are stored in the gland. For diseases related to malfunction of the thyroid gland. Sometimes iodides in drugs can cause a condition known as eosinophilia, in which there are reduced counts of eosinophil in the plasma. However, there are many other possible causes of this condition. Eosinophilic granulocytes are components of the blood in which the cytoplasm is filled with coarse acidophilic granules which may be spherical or rod-shaped, and the nucleus is bilobed and stains deeply.

Additional Reading

Braverman, L.E.: *Diseases of the Thyroid*, Vol. 2, Humana Press, Totowa, NJ, 1997.

Delange, F., J.T. Dunn and D. Glinoer: *Iodine Deficiency in Europe: A Continuing Concern*, Kluwer Academic Publishers, Norwell, MA, 1993.

Dunn, J.T. and F. van der Haar: *A Practical Guide to the Correction of Iodine Deficiency*, International Council for Control of Iodine Deficiency, Wageningen, the Netherlands, 1990.

Dunn, J.T.: "Iodine Deficiency—The Next Target for Elimination," *N. Eng. J. Med.*, 267 (January 23, 1992).

Hawkins, P.N., et al.: "Evaluation of Systemic Amyloidosis by Scintigraphy with 123I-Labeled Serum Amyloid P Component," *N. Eng. J. Med.*, 508 (August 23, 1990).

Lide, D.R.: *CRC Handbook of Chemistry and Physics*, 84th Edition, CRC Press, LLC., Boca Raton, FL, 2003.

Tonglet, R., et al.: "Efficacy of Low Oral Doses of Iodized Oil in the Control of Iodine Deficiency Disease," *N. Eng. J. Med.*, 236 (January 23, 1992).

Varvoglis, A.: *The Organic Chemistry of Polycoordinated Iodine*, John Wiley & Sons, Inc., New York, NY, 1992.

Varvoglis, A.: *Hypervalent Iodine in Organic Synthesis*, Morgan Kaufmann Publishers, Orlando, FL, 1996.

IODINE VALUE (Unsaturates).   See **Vegetable Oils (Edible)**

ION.   An atom or molecularly bound group of atoms which has gained or lost one or more electrons, and which has thus a negative or positive electric charge, and sometimes a free electron or other charged subatomic particle. Ions may be produced in gases by the action of radiation of sufficient energy; ionic solids are built up of ions bound together by their electrostatic forces, and when dissolved in a polar liquid, such as water, the salt dissociates into its ions, which have an independent existence.

Ions may be characterized in various ways. When they are described by the sign of their electric charge, they are described as *positive, negative* or *amphoteric* (or *zwitter*), the latter being an ion which carries both a positive and a negative charge, commonly at opposite ends of a long, or fairly long, chain, as in the case of ions of amino acids. See also **Amino Acids**.

Ions may also be described by their atomic structure, when they consist of more than one atom. Thus, a *complex ion* is a complex electrically charged radical or group of atoms such as $Ag(CN)_2^-$ or $Cu(NH_3)_2^{++}$, which may be formed by the addition to an ion of another ion or ions, or of an electrically neutral radical or molecule. When the particle combined with the ion is a large molecule, and the attachment is essentially by an adsorption process, the complex ion is called a *heteroion*. When the complex particle consists of a simpler ion combined with one or more molecules of water, it is known as an *aquoion* or *hydrated ion*. On the other hand, charged molecules, commonly produced by electrical discharges through gases, are often called molecular ions.

In meteorology, there are two special types of "ions" that enter into atmospheric processes: small "ions" and large "ions."

A *small ion* (also called a "light ion" or "fast ion") is the type that has the greatest mobility; hence, collectively, it is the principal agent of atmospheric conduction. The exact physical nature of the small ion has never been fully clarified, but much evidence indicates that each is a singly charged atmospheric molecule (or, rarely, an atom) about which a few other neutral molecules are held by the electrical attraction of the central ionized molecule. Estimates of the number of satellite molecules range as high as twelve. When freshly formed, by any of several atmospheric ionization processes, small ions are probably singly charged molecules; but after a number of collisions with neutral molecules, they acquire (actually, in a fraction of a second) their cluster of satellites.

A *large ion* (also called a "slow ion" or a "heavy ion") is an ion of relatively large mass and low mobility, produced by the attachment of a small ion to an Aitken nucleus.

ION-EXCHANGE RESINS. These materials are insoluble solid acids or bases that have the property of exchanging ions from solutions. During the ion-exchange reaction, the ion-exchange resins are converted into insoluble acids, bases, or salts. Cation-exchange resins contain fixed electronegative charges that interact with mobile counterions having the opposite, or positive, charge. Anion-exchange resins have fixed electropositive charges and exchange negatively charged anions. Ion-exchange resins are three-dimensional macromolecules or insoluble polyelectrolytes having fixed charges distributed uniformly throughout the structure.

Frequently, ion-exchange resins are used in fixed-bed processing equipment for softening and deionizing water. Equations for the removal of sodium chloride from water are as follows.

Exhaustion, or service step:

$$NaCl + RSO_3H \longrightarrow HCl + RSO_3Na \qquad (1)$$

Regeneration step:

$$2\ RSO_3Na + H_2SO \longrightarrow Na_2SO_4 + 2\ RSO_3H \qquad (2)$$

Equations (1) and (2) illustrate the reversible exchange of sodium ions for the hydrogen ion from the sulfonic cation-exchange resin. When the resin is depleted of hydrogen ions, it is regenerated with a dilute (5%) solution of H_2SO_4 [Eq. (2)]. The duration of the *service step* is usually a matter of hours; the *regeneration step* takes about 30 min.

The removal of NaCl from water (deionization) is completed by passage of the cation-exchanger effluent through a bed of anion-exchange resin.

Exhaustion, or service, step:

$$HCl + ROH \longrightarrow HOH + RCl \qquad (3)$$

Regeneration step:

$$RCl + NaOH \longrightarrow ROH + NaCl \qquad (4)$$

The effluent [Eq. (3)], which is free of NaCl, is deionized water. The resin is prepared for, the next service cycle by treatment with a 5% NaOH solution [Eq. (4)].

Traditionally, the ion-exchange resins used by industry are manufactured from uniform spheres of copolymers, such as styrene-divinylbenzene (DVB), having diameters 0.3–1.0 mm (20–30 mesh, U.S. standard screen). Such copolymer beads are formed by *pearl polymerization* and converted to ion-exchange resins by a second processing step. Sulfonic-type cation-exchange resins are made by sulfonation of the copolymer beads at elevated temperature. Strong-base anion-exchange resins are produced by means of chloromethylation and amination of the copolymer spheres.

Properties Desired of Ion-Exchange Resins

A well-performing ion-exchange resin should meet the following specifications:

1. Complete insolubility in water and solvents to prevent imparting tastes, odors, or color bodies to the solution being treated; 2. High exchange capacity per volumetric unit, with high regenerate efficiency; 3. Rapid and complete exchange with counter-ions; 4. Good chemical stability to prevent degradation by oxidizing and reducing agents; 5. Resistance to osmotic shock to prevent loss in use by physical breakdown; and 6. Low initial cost. See also Fig. 1.

Early Developments

The first ion-exchange resins were described by Adams and Holmes, a water-treatment expert and polymer chemist respectively, of the British Chemical Research Laboratory (1935). These ion-exchange resins were condensation products of phenol and formaldehyde. The granular-type cation-exchange resin contained sulfonic groups, and the anion exchanger contained aromatic amine groups. They are termed *strong-acid* and *weak-base* ion exchangers. A number of condensation-type ion-exchange resins were manufactured during 1935–1945. The first commercial deionization system was installed in 1939.

The next important step in ion-exchange resin technology was the synthesis of sulfonated styrene-DVB cation exchangers. Commercial quantities of strong-base styrene-DVB anion exchangers appeared in 1948. The first anion exchangers, the weak-base type, removed only strong mineral acids from water, such as HCl and H_2SO_4. The strong-base materials remove all acids, thus paving the way for production of water of

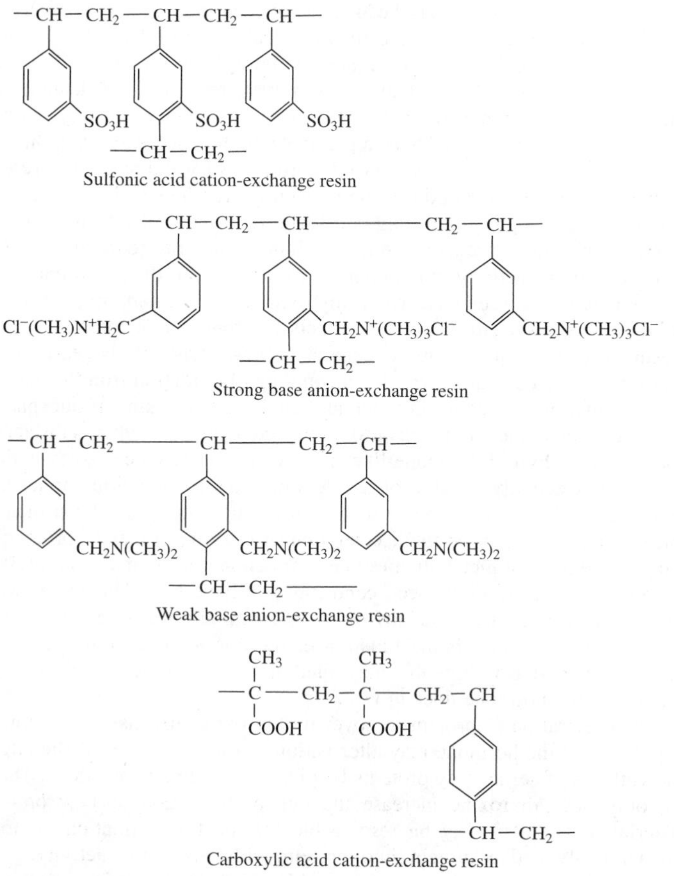

Sulfonic acid cation-exchange resin

Strong base anion-exchange resin

Weak base anion-exchange resin

Carboxylic acid cation-exchange resin

Fig. 1. Examples of ion-exchange resins

TABLE 1. CLASSIFICATION OF ION-EXCHANGE RESINS

Type	Active Group	Typical Configuration
Cation-Exchange Resins		
Strong acid	Sulfonic acid	benzene ring $—SO_3H$
Weak acid	Carboxylic acid	$\sim\!\sim\!CH_2CHCH_2\sim\!\sim$ with $COOH$
Weak acid	Phosphonic acid	benzene ring $—PO(OH)_2$
Anion-Exchange Resins		
Strong base	Quaternary ammonium	benzene ring $—CH_2N(CH_3)_3Cl$
Weak base	Secondary amine	benzene ring $—CH_2NHR$
Weak base	Tertiary amine (aliphatic matrix)	benzene ring $—CH_2NR_2$
Weak base	Tertiary amine (aliphatic matrix)	$—CHCH_2NCH_2—$ with OH and CH_2

equal or better quality than distilled water and at a much lower cost. The combination of the styrene strong-acid and strong-base exchange resins in a single tank (the mixed-bed deionizer), commercialized in 1949, produces water containing just a few parts per billion of dissolved salts and at a very low operating cost. The mixed-bed process produces ultrapure water from most freshwater supplies at a fraction of the cost of distillation. This is the basic method used for centralization high-pressure boilers (5,500 psig) in the power industry and for applications in the electronics, chemical, and pharmaceutical industries.

Research has continued apace over the years to develop new organic ion-exchange organic polymers—the details of most are proprietary. Both natural and synthetic zeolites also are used in ion-exchange processes, but their extreme importance as catalysts has tended to overshadow their applications for deionizing purposes. See also **Adsorption**; and **Zeolite Group**.

Classification of Ion-Exchange Resins

In Table 1, ion-exchange resins are classified by type, active exchange group, and configuration of the active group on the polymer. Some of the proprietary resins are not included.

Chemical Behavior of Ion-Exchange Resins

This is governed by the nature of the active exchange groups. The acid or basic strength of ion-exchange resins is determined by means of an acid-base titration. Strong-acid and strong-base ion-exchange resins have titration curves similar to H_2SO_4 and NaOH respectively. Weak-acid and weak-base ion-exchanger titration curves are very close to those of CH_3COOH and NH_4OH respectively.

The hydrogen-form strong acid and the hydroxyl-form strong-base anion exchangers convert a solution of a neutral salt into the corresponding acid

TABLE 2. CHEMICALS PURIFIED BY ION EXCHANGE

Chemical	Materials Removed	Ion-Exchange Method*
Formaldehyde	Formic acid	AE
Methanol	Ammonia	CE
Glycerin	Salts, acids, color	MB
Sorbitol	Salts, color	MB
Gelatin	Salts, color	MB
Sugars (sucrose, dextrose, lactose)	Salts, acids, color	MB
Citric acid	Acids, salts, color	CE
		AE
Uranium	Ionic impurities	AE
Chromic acid	Heavy-metal ionic impurities	CE
Copper	Ionic impurities	CE

*AE anion exchange; CE cation exchange; MB multibed ion exchange.

and base, while the ion-exchange resins are converted to the salt form.

$$RSO_3H + NaCl \longrightarrow RSO_3Na + HCl \qquad (5)$$

$$ROH + NaCl \longrightarrow RCl + NaOH \qquad (6)$$

Weak-acid and weak-base ion exchangers react with strong and weak bases and acids but do not split neutral salts.

$$RCOOH + NaOH \longrightarrow RCOONa + HOH \qquad (7)$$

$$RNH_3OH + HCl \longrightarrow RNH_3Cl + HOH \qquad (8)$$

Ion-exchange reactions are generally reversible and are analogous to reactions that occur in solution. When a cation-exchange resin with A as its counterion is in a solution containing B cations, the reaction is

$$R^-A^+ + B \longrightarrow R^-B^+ + A^+ \qquad (9)$$

where R is the cation-exchange resin.

After equilibrium is established according to the law of mass action, the reaction is

$$K_c = \frac{[\underline{B^+}][A^+]}{[\underline{A^+}][B^+]} \qquad (10)$$

A bar under the ion represents the ion in the resin phase; the absence of the bar indicates the ion in solution; the brackets indicate activities. Activity coefficients of ions in the resin phase cannot be precisely determined, and K_c is not constant with change in ionic concentration. The value K_c is considered a *selectivity coefficient* rather than an equilibrium constant. K_c is a useful measure of ion affinity. Values of K_c generally follow this order for strong-acid and strong-base ion exchangers: (1) divalent ions are preferred over monovalent ions, and (2) higher-molecular-weight ions are preferred over lower-molecular-weight ions of equal valence.

Applications of Ion-Exchange Resins

The use of ion-exchange resins fall into five categories: 1. Transformation of ionic constituents; 2. Removal of ionic impurities; 3. Concentration of ionic substances; 4. Fractionation of ionic substances; and 5. A variety of other applications.

Transformation of Ionic Constituent. Softening water with the sodium form of a cation-exchange resin is the prime example of transformation of ionic constituents. Calcium and magnesium ions (hardness) occur in all freshwater supplies, forming objectionable scale and precipitates in boilers, laundries, and home appliances. These ions react with soap to form Ca and Mg stearates and reduce the effectiveness of detergents. The softening process is accomplished by passing the hard water through a vessel containing the Na form of cation-exchange resin. When the Na ions on the resin are depleted by exchange with Ca and Mg ions, the resin is regenerated with a 10% solution of NaCl and rinsed, and the softening cycle is repeated.

Other examples of this type of reaction are the conversion of the antibiotic streptomycin sulfate to its corresponding chloride by means of anion exchange, the exchange of Na ions in milk for the K ion, and the conversion of Na_2CrO_4 to H_2CrO_4 by cation exchange. The latter process is used extensively in the plating industry to concentrate H_2CrO_4 from rinse waters, with subsequent reuse of a toxic chemical and reuse of the rinse water in what might be termed a *closed system.*

Removal of Ionic Impurities. The major use of various combinations of ion-exchange resins under this category is the deionization of water for many purposes. Municipal and industrial water supplies contain dissolved salts, such as Ca, Mg, $NaHCO_3$ chlorides, and sulfates, which must be removed before use. Deionized water must be used in supercritical boilers to prevent scale formation. Deionization is also used to remove dissolved silica from boiler feedwater and condensate. At operating pressures above 1,000 psig, silica is carried over with the steam and condenses on turbine blades in power plants, causing a marked reduction in efficiency. Condensate purification at flow rates of 50 gal/(ft^2)(min) of ion-exchange-resin bed area is used in most power plants having high-pressure boilers to reduce the Na, Fe, Cu, and silica concentrations to less than 50 ppb total contaminants.

A number of aqueous solutions of organic and inorganic chemicals are purified commercially. Dissolved salts, acids, bases, and color bodies are removed by ion-exchange resins.

Concentration of Ionic Constituent. Ion exchange is successfully applied to concentrate electrolytes from dilute solutions with subsequent elution by a more concentrated regenerate solution to obtain a more concentrated solution of the electrolyte. An example is the recovery of H_2CrO_4 from rinse waters in the metal-finishing industry. The rinse waters are passed through a two-bed strong-base deionizer, and the deionized water is recycled to the plating system. The H_2CrO_4 is recovered as Na_2CrO_4, which is converted to H_2CrO_4 by treatment with the hydrogen form of a

TABLE 3. SEPARATIONS MADE POSSIBLE BY ION EXCLUSION

Ionic	Nonionic	Resin
HCl	Acetic acid	RSO_3H
Salt	Ethanol	RSO_3Na
Salt	Glycerin	RSO_3Na
Salt	Sucrose	RSO_3Na

cation exchanger. Similar exchange processes are used to recover heavy and noble metals, such as Cu, Ni, Pt, and Au.

Fractionation of Electrolyte. Ionic species with opposite charges can be separated with either cation- or anion-exchange resins. The separation of ionic species of the same charge is possible if differences exist in acidic or basic strength, valence of the ion, or ionic radius. Examples of fractionation of electrolytes practiced on a commercial scale include: (1) removal of a strong acid from an organic acid, e.g., sulfuric acid from citric acid; (2) ion-exchange chromatography to produce pure rare earths from a mixture; and (3) concentration of copper and cobalt from dilute solutions and their fractionation by using a carboxylic-type cation exchanger, an example of concentration and fractionation done at the same time.

Miscellaneous Application. In the four categories just discussed the exchange of ions is common to all applications. It should be stressed that ion-exchange resins are reactive but insoluble acids, bases, and salts. These properties are used to advantage on an industrial scale for the adsorption of acidic and basic gases from gas streams. Gases which form an acid or a base with water can be removed by cation- or anion-exchange resins. Examples are SO_2, NH_3, CO_2, and H_2S. Ion-exchange resins have been used as catalysts for a number of years, some of the advantages being: (1) that catalyst-free products are obtained by means of simple filtration; (2) that catalyst can be reused for a number of cycles; (3) that continuous production is possible by passage of the solution through a bed of the material; and (4) that side reactions usually are kept to a minimum. Examples of ion-exchange resin catalysts are (1) sucrose inversion by means of the hydrogen form of a sulfonic-type cation exchanger, (2) ester hydrolysis with sulfonic-type cation exchangers, and (3) epoxidation of fats and oils with RSO_3H-type cation exchangers.

Ion Exclusion. Another process that uses ion exchange resins without exchanging ions is *ion exclusion*. This process uses an ion-exchange resin having high exchange capacity, which excludes free electrolytes from the inner phase. Low-molecular-weight soluble non-electrolytes distribute themselves equally between the resin and solution phases. If the solution of an electrolyte and a nonelectrolyte is passed through a column of a sulfonic-type cation exchanger whose exchangeable ions are the same as in the electrolyte, the nonelectrolyte will not be retarded and the electrolyte, or salt, will be *excluded*. The effluent from such a column can be collected as a relatively pure salt solution, followed by a solution of the mixture and then by a pure product cut. The middle cut usually is recycled. Ion exclusion has the advantage of separating nonionic materials from ionic species without the use of chemicals for regeneration. See Table 3.

The ion-exclusion process is particularly suited to sugar processing, e.g., sucrose recovery from molasses.

Developed in the early 1980s, tobermorites have selectivity properties intermediate between those of clay minerals and zeolites. They have been considered in catalysis and nuclear and hazardous waste disposal. Tobermorite, $Ca_5Si_6H_2O_{18} \cdot 4H_2O$, occurs naturally as a hydrous calcium silicate in calc-silicate rock. Tobermorites have layer structures similar to those of 2:1 clay minerals, but the structure varies with the chemical composition as well as with the nature of their synthesis. They have been synthesized from a number of starting materials.

Additional Reading

Alper, J.: "Archimedes, Plato Make Millions for Big Oil!: Zeolite Structure," *Science*, 1190 (June 8, 1990).

Bauman, W.C.: U.S. Patent 2,684,331 (1954).

Cavender, M.R., H.-L. Chiang and K. Myers: "Optimize Ion Exchange Resins Replacement," *Chem. Eng. Progress*, 56 (September 1992).

Kerr, G.T.: "Synthetic Zeolites," *Sci. Amer.*, 100 (July 1989).

Korkisch, J.: *Handbook of Ion Exchange Resins*, Vols. I–VI, CRC Press, LLC., Boca Raton, FL, 1989.

Korkisch, J.: *Handbook of Ion Exchange Resins: Their Application to Inorganic Anal,* CRC Press, LLC., Boca Raton, FL, 1999.

Korkisch, J.: *Concise Handbook of Ion Exchange Resins in Analytical Chemistry,* CRC Press, LLC., Boca Raton, FL, 2001.

Kunin, R.: *Ion Exchange Resins,* 2nd Edition, Krieger Publishing Company, Melbourne, FL, 1990.

Lawton, S.L. and W.J. Rohrbaugh: "The Framework Topology of ZSM-18, a Novel Zeolite Containing Rings of Three (Si, Al)-O Species," *Science,* 1319 (March 16, 1990).

Weitkamp, J. and H.G. Karge: *Zeolite Science, 1994: Recent Progress and Discussions,* Elsevier Science, New York, NY, 1995.

Weitkamp, J. and L. Puppe: *Catalysis and Zeolites: Fundamentals and Applications,* Springer-Verlag, Inc., New York, NY, 1999.

Zoccolante, G.V.: "Produce Ultrapure Process Water," *Chem. Eng. Progress,* 69 (December 1990).

ION EXCLUSION. The process in which a synthetic resin of the ion exchange type absorbs nonionized solutes such as glycerine or sugar while it does not absorb ionized solutes that are also present in a solution in contact with the resin. Thus, sodium chloride and glycerine can be separated by passage of their aqueous solution through a bed of particles of an ion exclusion resin.

IONIC CRYSTAL. A crystal that consists effectively of ions bound together by their electrostatic attraction. Examples of such crystals are the alkali halides, including potassium fluoride, potassium chloride, potassium bromide, potassium iodide, sodium fluoride, and the other combinations of sodium, cesium, rubidium or lithium ions with fluoride, chloride, bromide or iodide ions. Many other types of ionic crystals are known.

IONIC EQUILIBRIUM. In a system containing ions, at any particular temperature and pressure, the conditions at which the rate of dissociation of unionized molecules, or other particles to form ions, is equal to the rate of combination of the ions to form the unionized molecules, or other particles so that activities and concentrations remain constant as long as the conditions are unchanged.

IONIC MOBILITY. 1. The ratio of the average drift velocity of an ion in solution to the electric field. It is expressed by the relationship

$$\mu_+ \text{ or } \mu_- = \frac{\lambda_+ \text{ or } \lambda_-}{F}$$

in which μ_+ or μ_- is the mobility of the ion, λ_+ or λ_- is the ion conductance, i.e., the contribution of the particular ion to the equivalent conductance, and F is the Faraday constant.

2. For gaseous ions in an electric field, the quantity k defined by the relationship

$$k = vp/E$$

where v is the drift velocity, p, the gas pressure, and E, the electric field strength.

3. Conduction of electricity in ionic crystals is due to the motion of lattice defects, either of the Schottky or Frenkel type. The mobility is given by

$$\mu = (eD_0/kT)e^{-E/kT}$$

where D_0 is a numerical constant, and E is an activation energy, which depends on the energy required to make a defect and on the height of the energy barrier that must be surmounted in order that the defect may move.

ION IMPLANTATION. A process for introducing alloying elements into a host material by accelerating the ions to a high energy (at least tens of kilovolts) and allowing them to strike the surface of the host. The impinging atoms penetrate into the substrate material to a depth of 0.01 to 1 micrometer, depending on the atomic number and energy of the atom, and create a thin alloyed surface layer on the substrate. The process differs from others, such as electroplating, in that it does not produce a discrete coating, but rather it alters the chemical composition near the surface of the base material.

In recent years, the electronics industry has made increasing use of ion implantation as a method of doping semiconductors. Since the number of ions implanted is determined by the charge transferred to the substrate and their depth distribution by the incident energy, ion implantation has improved the controllability and reproducibility of certain semiconductor device processing operations. Also, ion implantation processes do not require the high temperatures needed to introduce impurities by diffusion. Thus the limitations arising from the changes produced in materials by high temperature are eased. Ion implantation also has been used in electronics to change the magnetic properties of substrates used for magnetic bubble devices.

Ion implantation also has promise in other fields involving surface technology; for example, new metallurgical phases with prior unknown properties can be formed. In some cases, such as heavy implantations of tantalum in copper of phosphorus in iron, amorphous or glassy phases can be formed. Or, if the implanted atoms are mobile, inclusions and precipitates can be formed as, for example, implanted argon and helium atoms are insoluble in metals and may form bubbles. The composition of a surface layer can be changed by differential sputtering caused by the implanted ions.

The damage and high concentrations of lattice defects, resulting from atomic displacements produced by the incident atoms, can change the chemical reactivity and mechanical hardness of a treated surface. Implantation can enhance the diffusion of impurities already deposited in a substrate, presumably through the motion of the high concentrations of lattice defects produced by the incident ions.

One of the most promising nonelectronic applications of ion implantation involves surface treatment to improve the hardness and wear resistance, as well as lowered susceptibility to corrosion, of metals. In some experiments, the benefits of ion implantation on wear may persist to a depth 103 times that of the implanted layer thickness. The implanted atoms are apparently transported into the metal as a tool wears. Thus, the technology is of large interest in connection with improving cutting tools and bearings. Some experiments have suggested that nitrogen implantation increases the fatigue life of carbon steel parts. The results are consistent with present understanding of the mechanisms of fatigue failure. It is well known that fatigue cracks start at the surface and that there is a close connection between surface hardness and fatigue life. Compressive stresses due to the presence of additional implanted ions may also play a role in the suppression of crack initiation.

The production of corrosion-resistant materials by alloying is well established, but the mechanisms are not fully understood. It is known, of course, that elements like chromium, nickel, titanium, and aluminum depend for their corrosion resistance upon a tenacious surface oxide layer (passive film). Alloying elements added for the purpose of passivation must be in solid solution. The potential of ion implantation is promising because restrictions deriving from equilibrium phase diagrams frequently do not apply (i.e., concentrations of elements beyond the limits of equilibrium solid solubility might be incorporated). This can lead to heretofore unknown alloyed surfaces which are very corrosion resistant.

Ion plating is another area of surface treatment. Ion plating is carried out in a gaseous electrical discharge in which the substrate to be plated is the cathode. The discharge is created by an applied potential of 500 to 5000 V. The primary component of the gaseous environment usually is an inert gas, most often argon. Atoms of the material to be plated are introduced into the gas by evaporation from a heated source. A fraction of the atoms injected by evaporation are ionized before striking the substrate. In ion plating, atoms arrive at the surface with energies of only a few hundred volts and penetrate no more than a few lattice constants into the substrate. Thus, ion implantation produces an alloyed surface layer whose composition varies continuously with depth because of the rather broad distribution of the ranges of the implanted ions, while ion plating produces a coating, the composition of which is independent of the nature of the substrate.

Semiconductor Applications

In semiconductor manufacture, the area of the workpiece into which ions are implanted is quite small. High homogeneity is sought in semiconductor applications, that is, the concentration of the implanted species should not vary by more than a few percent over the surface of a wafer. The implantation of ions into semiconductors is usually patterned, that is, some areas of the substrate are covered by a mask that stops the incident ions before they enter the substrate. A doped layer in which the implanted atoms are locally in an equilibrium phase is usually desired. Thus, implantation is usually followed by a high-temperature annealing treatment, which removes radiation damage through diffusion of lattice defects to defect sinks and the recrystallization of disturbed regions. Laser annealing has been used successfully. Laser annealing affects only a surface layer approximately equal to the depth of typical implantations, leaving the bulk of the piece unaltered.

See also **Semiconductors**.

IONIZATION. A process that results in the formation of ions. Such processes occur in water, liquid ammonia, and certain other solvents when polar compounds (such as acids, bases, or salts) are dissolved in them. Dissociation of the compounds occurs, with the formation of positively and negatively charged ions, the charges on the individual ions being due to the gain or loss of one or more electrons from the outermost orbits of one or more of their atoms. The ionization of gases is a process by which atoms in gases similarly gain or lose electrons, usually through the agency of an electrical discharge, or passage of radiation, through the gas.

Ionization by collision is an ionization process occurring by removal of an electron or electrons from an atom as the result of the energy gained in a collision with a particle (or quantum of radiation) possessing sufficient energy.

Specific ionization is the number of ion pairs formed per unit distance along the track of an ion passing through matter. This is sometimes called the total specific ionization to distinguish it from the primary specific ionization, which is the number of ion clusters produced per unit track length. The relative specific ionization is the specific ionization for a particle of a given medium relative either to that for (1) the same particle and energy in a standard medium, such as air at 15°C and 1 atmosphere, or (2) the same particle and medium at a specified energy, such as the energy for which the specific ionization is a maximum.

Total ionization is a term used to denote either the total specific ionization (defined above); or the total electric charge on the ions of one sign when the energetic particle that has produced these ions has lost all of its kinetic energy. For a given gas the total ionization is closely proportional to the initial energy and is nearly independent of the nature of the ionizing particle. It is frequently used as a measure of particle energy.

Minimum ionization is the smallest possible value of the specific ionization that a charged particle can produce in passing through a particular substance. When the specific ionization produced along the path of a charged particle is plotted as a function of the particle energy, minimum ionization appears as a broad dip, bound on one side by a rather sharp rise for decreasing particle energy, and on the other side by a gradual rise for increasing particle energy. For singly charged particles in ordinary air, the minimum ionization is about 50 ion pairs per centimeter of path. In general, it is proportional to the density of the medium and the square of the charge of the particle. It occurs for particles having velocities of 95% of the velocity of light, which corresponds to a kinetic energy of 1 MeV for an electron, 2 BeV for a proton and 8 BeV for an alpha-particle.

Ionization potential is the energy per unit charge, for a particular kind of atom, necessary to remove an electron from the atom to infinite distance. The ionization potential is usually expressed in volts, and is numerically equal to the work done in removing the electron from the atom, expressed in electron-volts. See also **Chemical Elements**.

IONIZED GASES. Various agencies, such as fast-moving electrons, alpha particles, various forms of radiation, and high temperature, are capable of dislodging electrons from atoms or molecules of a gas and thereby leaving them positively charged. Some of the dislodged electrons may attach themselves to other molecules and render them negatively charged. In some cases, two or more electrons may be removed from the same molecule, or a molecule with a double positive charge may unite with a singly charged negative molecule, forming a singly charged complex, etc. Such charged atoms, molecules or molecular groups are called ions, and their production from neutral molecules is called ionization. The complete separation of an electron from a molecule or an atom requires a definite amount of energy. This may be expressed in ergs, but is more commonly given in electron volts (1.59×10^{12} erg), its value being the ionization potential. A lesser amount of energy may excite the atom or molecule to emit radiation, but will not ionize it.

If an ionized gas is left to itself, the ions soon recombine and become neutral. But if it is subjected to an electric field, as in an ionization chamber, the ions pass to the electrodes, such a migration being an "ionization current." Such currents, commonly called electric discharges, are attended by diverse phenomena and vary widely in character from the silent glow discharge to the lightning stroke.

At ordinary pressures, discharges may be classified into four types: (1) If the voltage between two electrodes in open air is gradually increased, the electrodes become surrounded with a luminosity. This "glow" or "corona" gives way, at the negative electrode first, to (2), a "brush," composed of hair-like branches. (3) Finally, the disruptive spark passes. (4) Under

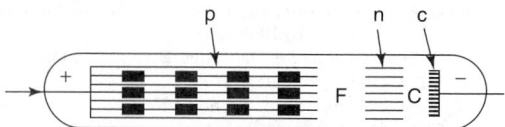

Fig. 1.   Elementary gas-discharge tube

other conditions, an arc may be formed. If, however, the electrodes are enclosed in a tube and the pressure reduced, a point is reached at which the tube becomes filled with a beautiful luminosity. Close examination shows this to have structure. Very close to and surrounding the cathode is a thin, luminous layer c, the cathode glow (Fig. 1); and outside this, the Crookes dark space C. Next, extending toward the anode, is the short negative glow n, then the Faraday dark space F. From this to the anode extends the long positive column p, with its regular, transverse striations. As the pressure is further reduced, the cathode dark space enlarges and the other features dwindle toward the electrodes until they finally disappear at about 0.001 mm pressure. From this point on, the cathode rays are the predominant feature.

Upon exploring the discharge in a Crookes tube with suitable probes, it is found that in certain regions the positive and negative ions are so nearly equal in number as to neutralize each other's effect. Such a region is called a "plasma." The plasma may be surrounded by a "sheath" of ionized gas in which ions of one sign greatly predominate, the effect being that of a space charge.

IONIZING PARTICLE. A particle that produces ion pairs in its passage through a substance. Ionizing particles may be divided into two groups: (1) *directly ionizing particles*—charged particles (electrons, protons, alpha particles, and so on) having sufficient kinetic energy to produce ionization by collision; and (2) *indirectly ionizing particles*—uncharged particles, such as neutrons and photons, which can liberate directly ionizing particles or can initiate a nuclear reaction.

Ionizing radiation is any radiation consisting of directly or indirectly ionizing particles, or a mixture of both. Ionizing radiation, unless controlled, poses a biological and environmental hazard.

ION MICROPROBE MASS ANALYZER. An instrument designed to provide an *in situ* mass analysis of microvolume of the surface of a solid sample. The analysis is accomplished by bombarding the surface with a high-energy beam of ions which causes the atoms at the surface to be sputtered away. A fraction of the sputtered particles is electrostatically charged and these sputtered ions are collected and analyzed according to their mass-to-charge ratio in a mass spectrometer.

Figure 1 represents a schematic diagram of the instrument. The ions used for sample bombardment are generated in a hollow cathode, dual plasmatron ion source capable of producing ions of a wide variety of gases including those of a highly electronegative character. The ions which can be either positively or negatively charged, are accelerated to energies ranging from 5.0 to 22.5 kilovolts and passed through the primary mass spectrometer. The spectrometer permits the analyst to select and purify, by mass separation, a specific chemical species from those produced in the ion source. The purified ion beam is focused to a small probe in an electrostatic lens column and allowed to impinge on the surface of the sample. The diameter of the ion probe may be varied continuously from about 2 to 500 micrometers. The sample and the point being analyzed can be viewed through an optical microscope while under bombardment.

Sputtered ions are collected and their masses analyzed in a double-focusing mass spectrometer in which the velocity dispersions of the magnetic and electric sectors are matched to permit the acceptance of a wide range of initial energies of the sputtered ions. No entrance slit is used and the bombarded area is stigmatically focused directly onto the resolving slit.

The ion beams are then detected with a high gain device that permits single ion counting. Sputtered ions from the sample eject secondary electrons at the conversion electrode and these are accelerated towards the scintillator of a photomultiplier tube where the light produced by their impact is detected. The resolved ion signals can be read as count rates from scalers which can accommodate rates in the megacycle range within significant dead time losses or as direct-currents on chart recorders.

Analytical Method

The analytical method applied with this instrument is based upon the observation that the yields of sputtered ions are greatly affected by the

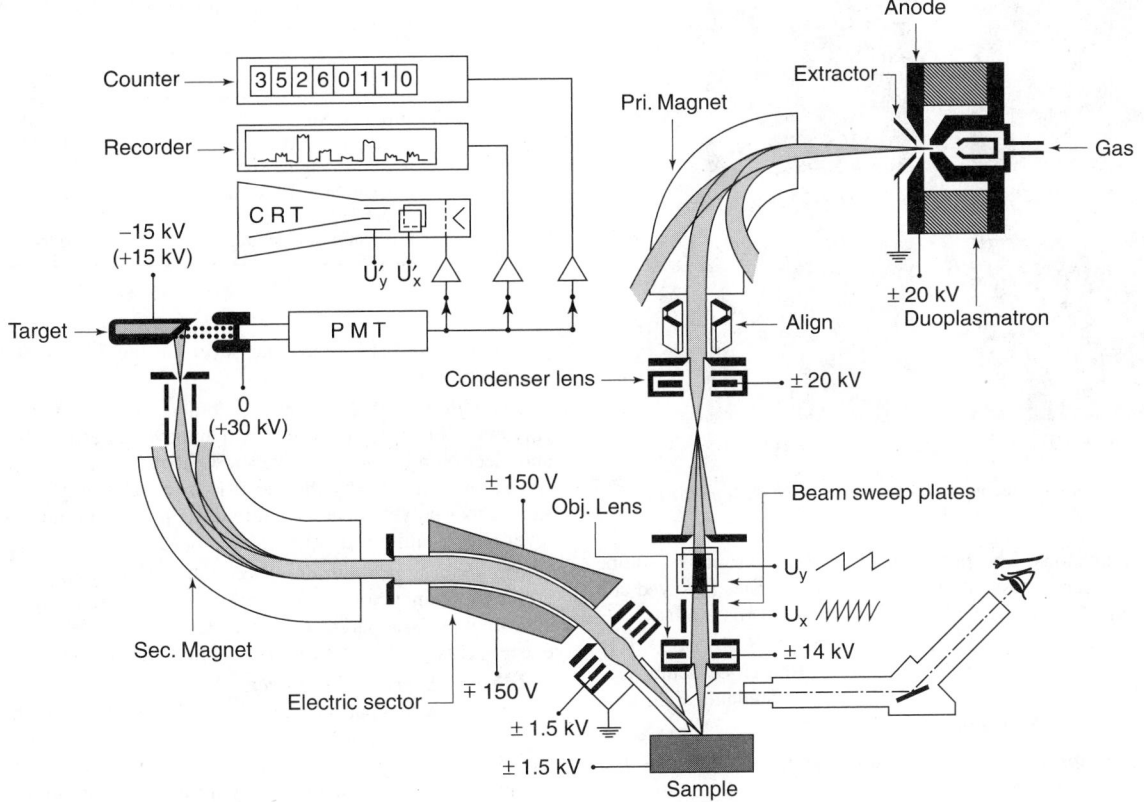

Fig. 1. Schematic representation of the ion microprobe mass analyzer. (*Bausch & Lomb/ARL*)

surface chemistry of the sample. When a metal such as aluminum is bombarded with ions of inert gas such as argon, the yield of positive aluminum ions falls exponentially with time. The ability of the sample to yield positive ions is progressively destroyed by the bombardment. On the basis of the similar behavior of many metals under bombardment by inert gases, it was postulated that the production of sputtered ions is a function of the electronic properties of the surface. The ability to extract positive ions from the sample diminishes as the strongly bonded compounds formed on the surface of the sample through the chemisorption of reactive gases are removed by the eroding action of the bombarding ion beam. It has been shown that the production of positive ions may be maintained at a higher level by controlling the surface chemistry through a proper selection of the species of bombarding ions. Instead of destroying the necessary chemical compounds with an inert gas, it is possible to reconstitute them by bombarding with a reactive gas. Enhanced stable yields of sputtered positive ions of many pure elements have been produced, by bombarding them with beams of carbon, nitrogen, oxygen, chlorine, and iodine ions.

Figure 2 illustrates the relative sputtered ion intensities of some pure elements subjected to bombardment by oxygen ions $^{16}O^-$. The relative intensity for each isotope has been corrected only for its natural abundance.

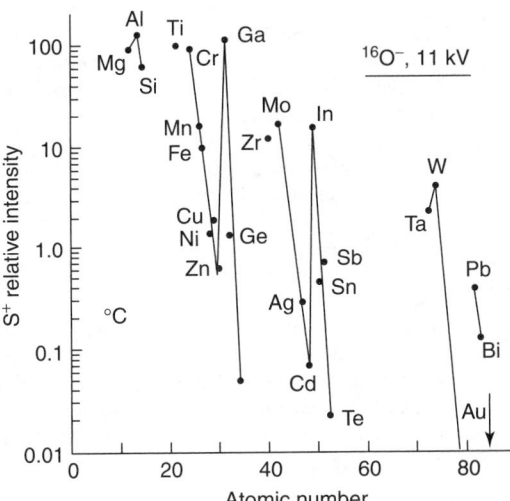

Fig. 2. Relative sputtered ion intensities of some pure elements subjected to bombardment by oxygen ions $^{16}O^-$

Application

Ion sputtering mass spectrometry has been applied to several problems in the analysis of solids with various types of instruments. These include studies of semiconductor devices as shown in Fig. 3, oxygen concentrations and concentration gradients and of processes of oxidation in a variety of metals, some catalytic and corrosion processes on metals, and the chemistry of trace elements in geologic specimens. The distribution of trace elements in lunar rocks has also been studied.

The ion microprobe has also been applied in a preliminary fashion to the rubidium-strontium dating technique. The correlation of the ion microprobe results with the independently determined isochron indicates that it may be possible to obtain useful results for samples on a micrometer scale from this dating technique.

The ion microprobe mass analyzer's unique features permit three dimensional microanalysis of all elements in the periodic table and in addition, the determination of their relative isotopic abundances in a given matrix. Both conductors and insulators may be analyzed. The instrument is applicable in many areas of the science of solid materials analysis. Most elements will have optimum yields in the spectrum of positive sputtered ions and will be detected in concentrations of parts per million in micrometer-sized sampling areas. Electronegative elements will be detected with similar sensitivities in the spectrum of negative sputtered ions but inert gases which are ionized with difficulty and have small electron affinities will be detected with considerably poorer sensitivities. In general, it is possible to measure isotope ratios without chemical separation of the constituent elements of the sample. A controlled sputtering process provides mono-layer resolution of depth profiles and also the ability to make precise in-depth analyses of thick and thin films. Detection efficiency of the instrument can yield quantitative accuracy in the parts per billion

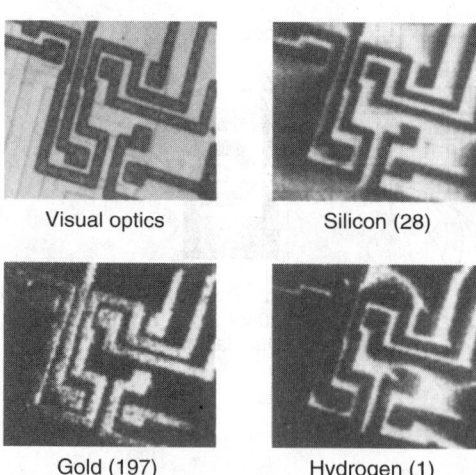

Visual optics   Silicon (28)

Gold (197)   Hydrogen (1)

Fig. 3.   Ion images of semiconductor device. (*Bausch & Lomb/ARL*)

range in many applications. The precision of an ion microprobe isotope ratio measurement depends basically upon the counting rates involved and its accuracy can approach its precision if auxiliary standards are used.

WINSTON G. SHEQUEN, P.E.
Bausch & Lomb/ARL
Sunland, California

Additional Reading

Breese, M.B.H., P.J. King and D.N. Jamieson: *Materials Analysis Using Nuclear Microprode,* John Wiley & Sons, Inc., New York, NY, 1995.
Goldstein, J.I. and H. Yakowitz: *Practical Scanning Electron Microscopy: Electron and Ion Microprobe Analysis,* Perseus Books, Boulder, CO, 1975.
Murr, L.E.: *Electron and Ion Microscopy and Microanalysis: Principles and Applications,* Marcel Dekker, Inc., New York, NY, 1991.

IONOMERS.

The generic term *ionomer* was introduced by DuPont in 1964 in conjunction with the commercialization of the new Surlyn resins to denote a thermoplastic polymer containing both covalent and ionic bonds, and having properties influenced to substantial effect by the ionic bonding. Since that time, the meaning has been expanded to include many compositions such as the glass ionomers used in dentistry which cannot be melt processed. In the interest of clarity and consistency, it is proposed that the term ionomer be reserved for polymers having melt viscosities suitable for conventional melt processing methods. Descriptions such as ion-containing or ion-linked are appropriate for highly viscous or true thermoset materials.

Despite the broad scope of the field and the unusual property combinations obtainable, commercial exploitation has been confined mainly to the original family based on ethylene copolymers. Within certain industries, such as flexible packaging, the word ionomer is understood to mean a copolymer of ethylene with methacrylic or acrylic acid, partly neutralized with sodium or zinc.

Ethylene-Based Ionomers

Physical Properties. The semicrystalline, ethylene-based ionomers of commerce are flexible, transparent polymers notable for high strength and elasticity in both solid and molten states. The ionic bonding is completely reversible and has a strong influence on properties, even at temperatures well above the melting point.

Mechanical Properties. Table 1 shows the general range of mechanical properties available in commercial Surlyn ionomers. The substitution of acrylic acid for methacrylic acid has only minor effects on properties.

The issue of mechanical property changes over time has been addressed and a structural model has been developed. A correlation was established between stiffness and the size of an endotherm (T_i), normally seen in dsc scans of ionomers at about 50°C. This endotherm increases in size with increasing neutralization. The T_i endotherm disappears completely when the dsc measurement is repeated immediately, but then gradually reappears during room-temperature storage.

In addition to time-related effects, the solid-state physical properties are also affected by adsorbed water, which functions as a plasticizer. Water pickup is affected by the nature of the cation, with sodium ionomers

TABLE 1. MECHANICAL PROPERTIES OF SURLYN IONOMERS

Property	Range
stiffness, MPa[a]	90–400
yield point, MPa[a]	8–20
tensile strength, MPa[a]	23–40
elongation at break, %	280–500
Shore D	54–70
brittleness temperature, °C	−100 to −140

[a] To convert MPa to psi, multiply by 145.

absorbing about 10 times the level of the zinc equivalent under the same conditions.

Crystallinity of Ionomers. Ionomers are much less hazy than the ethylene acid copolymers from which they are derived. Studies with optical and electron microscopes have shown that this is due to suppression of the spherulitic structure by the metal ions. Surprisingly, x-ray diffraction has shown that polyethylene crystallinity is present in the ionomers. A typical level of crystallinity is 30%.

Rheological Properties. The melt viscosity of an acid copolymer increases dramatically as the fraction of neutralization is increased.

Softening is apparent over a wide range, while the melt is strong and elastic. This gradual melting is beneficial in heat-sealing applications.

Infrared Spectra of Ionomers. Infrared absorption data, first published in 1964, show that partial neutralization of ethylene–methacrylic acid introduced new absorption bands at 1480–1670 cm^{-1} for the ionized carboxylate group while the 1698—cm^{-1} band of the free acid carboxyl diminishes in size. In addition to providing information on structural features, the numerous absorption bands are significant in applications technology, providing rapid warmup of film and sheet under infrared radiation.

Solubility of Ionomers. Ionic bonding with metal ions decreases solubility in organic solvents. At high neutralization levels with alkali metal ions, many ionomers spontaneously form colloidal suspensions in water when stirred vigorously at 100–150°C under pressure. These provide convenient methods for applying thin coatings of ionomers to paper and other substrates.

Electrical Properties. Due to the comparatively low content of polar groups, most commercial ionomers are very good insulating resins.

Permeability. Acid copolymers are less permeable to natural oils than conventional homopolymers, and this difference increases greatly when they are neutralized.

In the area of gas permeability, the low crystallinity of a typical ionomer (∼30%) results in relatively high permeability to oxygen. For packaging of fresh meat this is advantageous, but in other packaging areas, combination with a barrier layer may be required.

Manufacture and Processing. Most commercial processes involve copolymerization of ethylene with the acid comonomer followed by partial neutralization, using appropriate metal compounds.

Many methods for the conversion of acid copolymers to ionomers have been described by DuPont. The chemistry involved is simple when cations such as sodium or potassium are involved, but conditions must be controlled to obtain uniform products. Solutions of sodium hydroxide or methoxide can be fed to the acid copolymer melt, using a high shear device such as a two-roll mill to achieve uniformity. All volatile by-products are easily removed during the conversion, which is run at about 150°C.

Economic Aspects. Worldwide production is of the order of 110,000 t.

Health and Safety Factors

During processing at elevated temperatures, normal precautions are needed to prevent accidental burns. Surlyn ionomers have U.S. Food and Drug Administration clearance for food contact.

Uses

Flexible packaging is the largest commercial application area for ethylene ionomers. The unusual resilience and roughness of ionomers have resulted in sporting goods applications, including golf ball covers and bowling pin coatings. Ionomers are easily foamed due to high melt strength, and the foams are durable, leading to uses in construction, skilifts, and softball cores.

Noncommercial Ethylene-Based Ionomers

Noncommercial ethylene-based ionomers include amine-linked and complexed ionomers and ethylene–dicarboxylic acid copolymers.

Ionomers Not Based on Ethylene

Ionomers not based on ethylene include styrene-based ionomers, EPDM-derived ionomers, butadiene–methacrylic acid ionomers, telechelic ionomers, pentenamer ionomers, bitumen ionomers, and polyoxymethylene ionomers.

RICHARD W. REES
E. I. du Pont de Nemours & Co., Inc.

Additional Reading

MacKnight, W.J. and T.R. Earnest: *J. Macromol. Rev.* **16**, 41 (1981).

Rees, R.W.: in K.C. Frisch, ed., *Polyelectrolytes,* Technomic Publishing Co., Inc., Westport, Conn., 1976, pp. 177–197.

Rees, R.W.: in J.I. Kroschwitz, ed., *Encyclopedia of Polymer Science and Engineering,* 2nd Edition, Vol. 4, Wiley-Interscience, New York, NY, 1986, pp. 395–417.

Zutty, N.L., J.A. Faucher, and S. Bonotto: in N.M. Bikales, ed., *Encyclopedia of Polymer Science and Technology,* Vol. 6, Interscience Publishers, a Division of John Wiley & Sons, Inc., New York, NY, 1967, p. 420.

ION RETARDATION.   A process based on amphoteric (bifunctional) ion-exchange resins containing both anion and cation adsorption sites. These sites will associate with mobile anions and cations in solution and thus remove both kinds of ions from solutions. These ions may be eluted by rinsing with water. This process can make clean separations of ionic-nonionic mixtures. It has also been suggested for demineralization of salt solutions.

See also **Ion**.

IPATIEFF, VLADIMIR N. (1890–1952).   Born in Russia, Ipatieff was an army officer as well as a chemist. He was a member of the Academy of Science and carried out organic research at the Institute of Chemistry in Leningrad. He left the former U.S.S.R. under the Stalin regime and at the invitation of Gustav Egloff joined the Universal Oil Products Co. He and his close associate, Herman Pines, did basic development on catalytic alkylation and isomerization of hydrocarbons of the greatest importance for high-octane aviation gasoline.

IRIDESCENCE.   The exhibition of the colors of the rainbow, commonly by interference of light of the various wavelengths reflected from superficial layers in the surface of a substance.

IRIDIUM.   [CAS: 7439-88-5]. Chemical element symbol Ir, at. no. 77, at. wt. 192.22, periodic table group 9 (transition metals), mp 2,410°C, bp 4,130°C, density 22.42 g/cm^3 (solid at 17°C), 22.8 g/cm^3 (single crystal at 20°C). Elemental iridium has a face-centered cubic crystal structure. The two stable isotopes of iridium are ^{191}Ir and ^{193}Ir. The ten unstable isotopes are ^{187}Ir through ^{190}Ir, ^{192}Ir, and ^{194}Ir through ^{198}Ir. In terms of earthly abundance, iridium is one of the scarce elements. Also, in terms of cosmic abundance, the investigation by Harold C. Urey (1952), using a figure of 10,000 for silicon, estimated the figure for iridium at 0.0025. No notable presence of iridium in seawater has been found. The element was identified and named by Tennant (England) in 1804.

Electronic configuration

$$1s^2 2s^2 2p^6 3s^2 3p^6 3d^{10} 4s^2 4p^6 4d^{10} 4f^{14} 5s^2 5p^6 5d^9.$$

Metallic Ir is not attacked by any mineral acid unless it is very finely divided. It can be brought into solution by fusion with indium at 800–1000°C to give a soluble alloy. When fused with Na_2O_2 or an alkaline oxidizing flux, water-soluble irradiates(IV) are formed. The finely divided metal is oxidized by air or O_2 at red heat to the dioxide, which decomposes into its elements at higher temperature. The valences of Ir are 1–6, the 3 and 4 valences being most common.

Iridium black is only slightly soluble in aqua regia. When fused with alkalies and alkaline nitrates or Na_2O_2, the metal is converted to an acid-soluble form. The metal at red heat reacts to a small extent with O_2, S, and P. At elevated temperature, the metal is attacked by Cl_2 and F_2. When fused with NaCl and treated with Cl_2, the water-soluble sodium hexachloroiridate(IV), Na_2IrCl_6, is formed.

Iridium(III) hydroxide is a yellow-green or blue-black compound soluble in alkali and insoluble in water. It is made by adding KOH to a solution of potassium hexachloroiridate(III), K_3IrCl_6, in an inert atmosphere. When the trihydroxide is heated, a mixture of iridium(IV) oxide and the metal is formed. Iridium(III) oxide, Ir_2O_3, is made by fusing potassium hexachloroiridate(IV) with Na_2CO_3 and then leaching the mixture with water. At about 1100°C, both the oxide and the hydroxide decompose into the metal and O_2. When a solution of Ir is heated with $NaBrO_3$ at a pH of approximately 6, the dark-blue precipitate $Ir(OH)_4$ or $IrO_2 \cdot H_2O$ is formed. This water-insoluble compound, when heated to 350°C in N_2, loses its H_2O and is converted to the black oxide, IrO_2.

When iridium(III) chloride is heated in Cl_2 at 773–798°C, iridium(I) chloride is formed. The copper-red crystals are insoluble in acids and alkalies. The compound sublimes in Cl_2 at 790°C and decomposes into Cl_2 and metallic Ir. Iridium(II) chloride is stable from 763 to 773°C. The brown crystals are insoluble in H_2O, acids, and alkalies. Iridium(III) chloride is an insoluble green compound made by reacting the elements at about 600°C. The reaction is catalyzed by CO. Iridium(III) chlorides are prepared by reducing the corresponding iridium(IV) chlorides with oxalate or SO_2. Iridium(III) bromide is made by dissolving iridium(III) hydroxide in HBr. The blue solution yields olive-green crystals of $IrBr_3 \cdot 4H_2O$. When heated, the anhydride is formed. The triiodide is formed in analogous fashion as a trihydrate. Iridium(IV) chloride can be made in solution by the action of Cl_2 or aqua regia on ammonium hexachloroiridate(III). The relative insolubility of ammonium hexachloroiridate(IV), $(NH_4)_2IrCl_6$, makes it useful in the purification of Ir. The compound may be reduced in H_2 to the metal. The analogous sodium salt is very soluble, and the potassium salt is relatively insoluble.

Iridium(VI) fluoride is made from the element at 300–400°C. The bright yellow solid melts at 44°C and boils at 53°C. Potassium hexafluoroiridate(V) also has been prepared. Iridium fluoride can be made by heating the hexafluoride with the metal in a sealed tube at 150°C or heating it with glass above 200°C, at which temperature the glass reduces it to the tetrafluoride. This yellow solid melts at 106–107°C and boils above 300°C. Iridium(III) fluoride is formed by reducing the tetrafluoride with glass for 12–18 hrs. at 430–450°C.

Iridium(II) sulfide is formed by burning the metal in sulfur or by heating a higher sulfide at 700°C in N_2. This black solid is insoluble in H_2O, acids, and aqua regia. Iridium(III) sulfide, Ir_2S_3, is formed as a brown-black insoluble precipitate by passing H_2S through a hot acidic solution of an iridium(III) chloride. The precipitation is usually not quantitative. This amorphous black solid is not attacked by HNO_3 but is slowly dissolved by aqua regia or fuming HNO_3. The brown insoluble iridium(IV) sulfide, IrS_2, is partly formed by treating a tetravalent Ir solution with H_2S; when it is prepared in this way, some iridium(III) sulfide also is formed by reduction. Iridium(III) sulfate can be formed by dissolving iridium(III) hydroxide in H_2SO_4 in the absence of air. Trivalent iridium forms numerous cationic and anionic complexes in which it has a coordination number of 6. The amines are extremely stable and, once formed, difficult to destroy. Tetravalent Ir also forms complex ions, but to a lesser extent.

Iridium As a Key to Mass Extinctions

Airborne particles from the January 1983 eruption of Kilauea (Hawaii) volcano indicated exceptionally large concentrations of selenium, arsenic, indium, gold, and sulfur, as expected from a volcanic eruption. Unexpected were exceptionally high levels of iridium. Investigators found that the ratio of Ir to Al was about 17,000 times its value in the associated Hawaiian basalt. Inasmuch as Ir enrichments had not previously been detected in volcanic eruptions, Zoller et al. (1983) suggested that the Kilauea volcano may be part of an unusual volcanic system which may be fed by magma from the mantle. The researchers further suggested that the Ir enrichment may be linked with the high fluorine content of the volcanic gases, indicating that the Ir was released as volatile IrF_6.

In recent years, Zoller and associates (University of Maryland) have studied six active volcanoes (Augustine, Mount St. Helens, El Chichón, Arenal, Poas, and Colima) and have found no evidence of Ir enrichment. The new Kilauea evidence of volcanic action as an Ir source tends to conflict with that of other researchers who have generally attributed the Ir anomaly to an extraterrestrial source, such as resulting from a cataclysmic meteorite or asteroid impact, notably in connection with the Cretaceous-Tertiary (K-T) boundary layer.

Much of the theorizing pertaining to mass extinctions of flora and fauna during the past history of the earth has been based upon finding

Ir anomalies. Currently, scientists are attempting to establish the order of mass extinctions. Present opinions seem to suggest that the extinction of terrestrial fauna, including the dinosaurs, was an event that followed rather than occurring concurrently with the catastrophe met by many marine species. See also **Chemical Elements**; and **Platinum and Platinum Group**.

Additional Reading

Fox, L.S., et al.: "Gaussian Free-Energy Dependence of Electron-Transfer Rates in Iridium Complexes," *Science*, 1069 (March 2, 1990).

Greenwood, H.N. and A. Earnshaw: *Chemistry of the Elements,* 2nd Edition, Butterworth-Heinemann, Inc., Woburn, MA, 1997.

Krebs, R.E.: *The History and Use of Our Earth's Chemical Elements: A Reference Guide,* Greenwood Publishing Group, Inc., Westport, CT, 1998.

Lide, D.R.: *CRC Handbook of Chemistry and Physics,* 84th Edition, CRC Press, LLC., Boca Raton, FL, 2003.

Stwertka, A. and E. Stwertka: *A Guide to the Elements,* Oxford University Press, Inc., New York, NY, 1998.

Zoller, W.H., J.R. Parrington and J.M. Phelan Kotra: "Iridium Enrichment in Airborne Particles from Kilauea Volcano: January 1983," *Science*, **222**, 1118–1121 (1983).

IRON. [CAS: 7439-89-6]. Chemical element symbol Fe, at. no. 26, at. wt. 55.847, periodic table group 8 (transition metals), mp 1,535°C, bp approximately 2,750°C, density 7,874 g/cm3 for the pure solid (20°C); 7.92 for a single crystal of a-iron. Iron has a body-centered cubic crystal structure (a-iron).

Iron is a silver-white metal, capable of taking a high polish; ductile; malleable; can be welded when hot. Pure iron is attracted by a magnet, but does not retain the magnetism. Silicon steel is preferred for electromagnets because it retains magnetism even less than pure iron. See also **Magnetism**. The discovery of iron was prehistoric. There are nine isotopes of iron, 52Fe through 60Fe Isotopes 54, 56, 57, and 58 are fully stable, whereas four others have fairly short half-lives, ranging from 8.9 minutes (53) to 2.94 years (55). The half-life of 60Fe is approximately 3×10^5 years. Iron has valence numbers of 2+ (ferrous) and 3+ (ferric). The hardest of the ductile metals, iron is surpassed only by cobalt and nickel in tenacity. Iron is an extremely versatile construction and engineering material and serves both in relatively pure forms, such as malleable and wrought iron, and in many hundreds of iron-base alloys of major importance, including the numerous types of steel.

Electronic configuration is 1s 22s 22p 63s 23p 63d 64s 2. First ionization potential is 7.896 eV; second 16.5 eV. Oxidation potentials: $Fe \rightarrow Fe^{2+} + 2e^-$, 0.441 V; $Fe \rightarrow Fe^{3+} + 3e^-$, 0.036 V; $Fe^{2+} \rightarrow Fe^{3+} + e^-$, −0.771 V; $Fe + 2OH^- \rightarrow Fe(OH)_2 + 2e^-$, 0.877 V; $Fe(OH)_2 + OH^- \rightarrow Fe(OH)_3 + e^-$, 0.56 V; $Fe(OH)_4^- + 4OH^- \rightarrow FeO_4^{2-} + 4H_2O + 3e^-$; E = 0.55 V (80°C; 40% NaOH). Metallic radius, 1.2412 Å; ionic radius Fe 2+, 0.80 Å, Fe3+, 0.67 Å. Other important physical properties of iron are given under **Chemical Elements**. See also Table 1.

Three allotropic forms of iron are known: (1) *alpha iron*, which is present below 769°C; (2) *gamma iron*, which exists between 906° and 1,404°C, and (3) *delta iron*, which occurs between 1,404° and 1,536°C. On slow cooling, the reverse changes occur, but may be slowed or partly or entirely prevented in the presence of alloying elements.

In terms of abundance in the earth's crust, iron ranks fourth, being estimated as comprising about 5% of the weight of igneous rocks. Of course, only a very small portion of this large amount of the element in the earth's crust is obtainable as iron ore. In terms of cosmic abundance, the estimate of Harold C. Urey made in 1952 put iron as number 9 among the elements, having a figure of 7,250 related to a base for silicon of 10,000. Iron is ranked number 23 among the elements in terms of its presence in seawater, an estimated 47–48 tons per cubic mile (10.2–10.4 metric tons per cubic kilometer) of seawater. In this regard, it is approximately equal with aluminum, molybdenum, and zinc.

Iron Ores

Iron occurs abundantly in several materials, mainly in the form of oxides, carbonate, silicates, and sulfides. These are shown in Table 2. Most of the ores shown in the table are described under separate alphabetical listings in this volume. Briefly, the major iron-bearing materials are:

Magnetite, Fe_3O_4, corresponding to 72.4% Fe and 27.6% O_2, dark gray to black, sp gr 5.16–5.18, strongly magnetic, permitting magnetic

TABLE 1. SELECTED PHYSICAL PROPERTIES OF IRON

Electrical Properties	
Electrical conductivity, volume, % of annealed copper at 20°C	17.75
Electrical resistivity, microohm-cm,	
at 0°C	8.9
at 2s0°C	9.7
Temperature coefficient of electrical resistance (0–100°C)	0.65×10^{-4}
Electrode potential (standard hydrogen scale), at 25°C, volts	−0.44
Electrochemical equivalent, milligrams/second-absolute amperes	
Fe^{2+}—Fe	0.1929
Fe^{3+}—Fe	0.2893
Magnetic susceptibility, at 820°C	$1,000 \times 106$
Thermal Properties	
Emissivity, at 0.65 micrometers, %	40
Heat of combustion, cal/gram atom	88.355
Latent heat of fusion, cal/gram	65.5
Latent heat of vaporization, cal/gram	1.598
Specific heat, at 25°C, cal/(°C)(gram atom)	6.55
Thermal conductivity, at 0°C, (cal)(cm)/(second)(cm²)(°C)	0.18
Thermal expansion, linear coefficient,	
cm/(cm)(°C)	11.76×10^{-6}
in/(in)(°F)	6.53×10^{-6}
Average coefficient of linear expansion, at 77°F,	
cm/(cm)(°C)	12.3×10^{-6}
in/(in)(°F)	6.83×10^{-6}
Mechanical Properties	
Brinell hardness, at 25°C, 99.9% Fe	82–100
Percent elongation, at 25°C, 99.9% Fe	30–40
Yield strength, psi, at 25°C, 99.9% Fe	10,000–20,000
Tensile strength, psi, at 25°C, 99.9% Fe	30,000–40,000
Modulus of elasticity, psi	28.5×10^6
Modulus of rigidity, psi	11.64×10^6
Poisson's ratio	0.28
Other Properties	
Density, liquid, at 1564°C,	
g/cm³	7.00
lb/in³	0.253
Density, solid, at 20°C,	
g/cm³	7.874
lb/in³	0.284
Reflectivity (light from tungsten filament),	
2,500 Å	38
10,000 Å	65
Surface tension, at 1,550°C, dynes/cm	1,835–1,865
Thermal-neutron-absorption cross section, barns	2.53
Viscosity, cP,	
at 1,743°C	4.45
at 1,390°C	7.85

exploration methods. Some ores contain small amounts of titanium whereupon they are referred to as titaniferous magnetite.

Hematite, Fe_2O_3, corresponding to 69.94% Fe and 30.06% O_2, steel gray to dull red or bright red, earthy to compact or crystalline, sp gr 5.26, most important of iron ores, occurs widely in many types of rocks of varying origin.

Ilmenite, $FeTiO_3$, corresponding to 36.8% Fe, 31.6% Ti, and 31.6% O_2, iron-black, opaque, generally mined for titanium with iron as a byproduct, also called iron titanate.

Limonite, mineralogically composed of various mixtures of the minerals *goethite* and *lepidocrocite*, $HFeO_2$ and $FeO(OH)$, respectively. Goethite contains 62.9% Fe, 27% O_2, and 10.1% H_2O, sp gr 3.6–4.0, commonly yellow or brown to nearly black, compact to earthy and ocherous. Limonites are important sources of iron throughout the world.

Siderite, $FeCO_3$, corresponding to 48.2% Fe, 51.8% CO_2, sp gr 3.83–3.88, white to greenish-gray and brown, contains variable amounts of calcium, magnesium, and manganese, varies from dense, fine-grained and compact to crystalline, sometimes referred to as spathic iron ore, or black-band ore. The carbonate ores are calcined

TABLE 2. PRINCIPAL IRON-BEARING MINERALS

Class and mineralogical name	Chemical composition of pure mineral	Common designation
Oxides		
Magnetite	Fe_3O_4	Ferrous-ferric oxide
Hematite	Fe_2O_3	Ferric oxide
Ilmenite	$FeTiO_3$	Iron-titanium oxide
(Goethite)	$HFeO_2$	
Limonite		Hydrous iron oxides
(Lepidocrocite)	$FeO(OH)$	
Carbonate		
Siderite	$FeCO_3$	Iron carbonate
Silicates		
Chamosite		
Stilpnomelane	Various; often complex	Iron silicates
Greenalite		
Minnesotaite		
Grunerite		
Sulfides		
Pyrite (iron pyrites)	FeS_2	
Marcasite (white iron pyrites)	FeS_2	{Iron sulfides}
Pyrrhotite (magnetic iron pyrites)	FeS	

before they are charged into the blast furnace; frequently contain sufficient lime and magnesite to be self-fluxing.

Silicate Group Ores. There are comparatively few silicates with iron as the principal base. Often they have a rather complex chemical formula, with sp gr higher than 2.8, occurring in various shades of green or black tones. Important iron-silicate minerals are *chamosite, stilpnomelane, greenalite, minnesotaite,* and *grunerite.* Presently of minor importance as a source of iron.

Sulfide Group Ores. The principal materials in this group are *pyrite, pyrrhotite,* and *marcasite.* Pyrite, FeS_2, corresponding to 46.6% Fe, 53.4% S, sp gr 4.95–5.10, is pale brass yellow, and the most widespread of the iron sulfides. Pyrrhotite (magnetic pyrite), varies in composition from FeS to FeS +S, typically contains 59.4% Fe, 40.6% S, bronze yellow to copper red, frequently tarnished, and is often considered an indicator of nickel deposits because of common association with pentlandite. Marcasite (white iron pyrites), FeS_2, corresponding to 46.6% Fe, 53.4% S, is pale brass yellow, commonly associated with limestones, clays, and lignite deposits. It differs from pyrite only in its crystal structure and greater chemical instability. Iron sulfides sometimes are mined for their sulfur content; more commonly because of their association with other valuable metallic elements, such as copper, nickel, zinc, gold, and silver. Iron sometimes is recovered as a byproduct.

Geology and Genesis of Iron Deposits. Iron ores have a wide range of formation in geologic time as well as a wide geographic distribution. They are found in the oldest known rocks of the earth's crust, with an age in excess of 2.5 billion years, as well as in rock units formed in various subsequent ages. Iron ores are forming presently where iron oxides are being precipitated in marshy areas, and where magnetite placers are being formed on certain beaches. Thousands of iron deposits are known throughout the world. The deposits range in size from a few tons to many hundreds of millions of tons. Many of the world's largest deposits of iron ore are located in the oldest geologic series, the Pre-Cambrian.

Iron ores occur in igneous, metamorphic or sedimentary rocks, or as weathering products of various primary iron-bearing materials. For convenience of analysis, iron ores are grouped into (1) igneous ores, (2) contact ores, (3) hydrothermal ores, (4) sedimentary ores, with several subclassifications of the latter ores. Brief definitions follow:

Igneous Ores. These were formed by crystallization from liquid rock materials, either as layered-type deposits that possibly are the result of crystals of heavy iron-bearing minerals settling as they crystallize to form iron-rich concentrations, or as bodies which show intrusive relationship

with their wall rocks. These ore bodies may be tabular or irregular and are composed largely of magnetite with varying amounts of hematite.

Contact Ores. Iron-ore deposits formed at or near the contact between igneous rocks and sedimentary rocks, the latter usually limestones, are commonly composed of magnetite and hematite with associated carbonates and pyrite. The ore deposits are commonly in the sedimentary rocks as irregular or tabular replacement bodies.

Hydrothermal Ores. Iron-ore deposits formed by hot solutions which transported iron and replaced rocks of favorable chemical composition with iron minerals to form irregular ore bodies, commonly in limestones, are termed hydrothermal deposits. The iron often occurs as siderite, or sometimes as oxides.

Sedimentary Ores. There are six subclassifications:

Bedded Ores. These often are composed of oölites of hematite, siderite, iron silicate, or less commonly, limonite in a matrix of siderite, calcite, or silicate. They have a wide geographic distribution associated with other sedimentary rocks. They sometimes contain fossils and fine grains of sand. They often have a fairly high phosphorus content and may be self-fluxing.

Siderite Ores. These ores consist of beds of siderite or siderite nodules associated with shales. They are common in coal-associated beds and commonly contain associated sulfides, with a fairly high sulfur and phosphorus content.

Placer Ores. Iron oxides, when compact, are rather resistant to weathering and erosion, and under favorable conditions may form placer deposits, which, in relatively few instances, constitute iron ores. Generally they are of rather minor importance as sources of iron.

Bog Iron Ores. Bog ores occur in many swampy areas, particularly in glaciated areas in Europe, Asia, and North America. They occur commonly as dark-brown, cellular masses, or granular or fine particles of limonite. Once important when iron furnaces were local and small, they have ceased to be of commercial importance.

Metamorphic Ores. These ores include sedimentary iron-ore deposits, which have been metamorphosed, as well as ores associated with metamorphic rocks, in which the origin of the ore is obscured by recrystallization. Essentially all of the Pre-Cambrian sedimentary iron formations are of this type.

Residual Ores. These ores are commonly products of the surficial weathering of rocks, but may include ores formed by hydrothermal oxidation and leaching. Ores of this kind were formed extensively in Pre-Cambrian iron formations by leaching of silica, which commonly constituted in excess of 50% of the rock. Oxidation changes iron carbonate, silicate minerals, and magnetite to hematite or limonite.

Principal World Iron Deposits

Name and/or Type of Ore	Location
Kerchoölitic limonite	Crimea, Russia
Salzgitter limonite and hematite	Germany
Minette limonite and hematite	France, Germany, Luxembourg
Blackband ironstones	British Isles
Siegerland siderite	Germany
Clinton hematites	Alabama (U.S.A.)
Wabana oölitic hematites	Newfoundland
Minas Gerais hematite	Brazil
Krivoi Rog hematites	Ukraine, Russia
Bihar, Orissa, and Bastar hermatites	India
Labrador hematite	Quebec, Labrador (Canada)
Lake Superior taconites, jaspilites, hematites, and magnetites	Michigan, Wisconsin, Minnesota (U.S.A.) Ontario (Canada)
Cerro Bolivar and El Pao hematites	Venezuela
Kirunavaara magnetite	Sweden
Hematites	Australia

Iron Ore Processing. Two major developments have occurred during recent years because of the increasing needs for iron: (1) increased search for new supplies of high-grade iron ores, and (2) expansion of iron-ore pellet-plant production, particularly in those industrial nations where supplies of high-grade ores have diminished. More recently, iron ore has been shipped in slurry form in ocean-going tankers. Once the slurry, containing about 75% solids, settles, the excess water is pumped off, leaving a nonshifting cargo in the hold of about 92% solids. Upon arrival at the receiving port, the cargo is reslurried with high-pressure water jets. Considerable savings in dockside loading and unloading expense are thus effected.

Beneficiation. This is a term that describes all processes used to improve the chemical and physical characteristics of ore (not limited to iron ore) for later use. In the case of iron ore, beneficiation makes the ore better for handling by the blast furnace. The principal methods include crushing, screening, blending, grinding, concentrating, classifying, and agglomerating. Concentration operations include jigging, flotation, and magnetic separation. A blast furnace operates best with a permeable burden, which permits not only a high rate of gas flow, but also a uniform gas flow with minimum channeling of the gas. Agglomeration improves burden permeability and thus the gas-solid contact in the furnace. This reduces blast-furnace coke rates and increases the rate of reduction. Agglomeration also decreases the amount of fines blown out of the blast furnace, thus reducing the load on the gas-recovery system.

Practice has shown that the best agglomerate for a blast furnace will contain about 60% or more of iron, a very minimum of undesirable constituents, and a minimum of material larger than 1 inch. However, the agglomerate must have sufficient strength to withstand degradation while in stockpiles and during transportation and handling. The target is to have the material arrive at the blast furnace, after prior handling, with about 85–90% of the material over 1/4 inch. The agglomerate must be able to withstand the high temperature and the degradation forces within the furnace without slumping or decrepitating. Further, the agglomerate should be reasonably reducible so that a satisfactory reduction rate can be maintained in the furnace.

There are four major types of agglomerating processes: (1) sintering, (2) pelletizing, (3) briquetting, and (4) nodulizing. The first two processes have been most popular.

Sinter consists of small particles of iron-bearing materials, which are fused or fritted together at high temperature. The latter is achieved by burning carbon in the form of coke breeze in a sintering-machine feed mix. Optionally, fluxing material may be added to eliminate later additions to the blast furnace. A number of materials can be converted by sintering, such as flue dust, naturally fine ores, ore fines from screening operations, and

other iron-bearing material of small particle size. A continuous sintering process is shown in Fig. 1.

A traveling grate conveys a bed of ore fines or other finely divided iron-bearing material intimately mixed with approximately 5% of a finely divided fuel, such as coke breeze or anthracite. Near the head or feed end of the grate, the bed is ignited on the surface by gas burners and, as the mixture moves along the traveling grate, air is pulled down through the mixture to burn the fuel by downdraft combustion. As the grates move continuously over the windboxes toward the discharge end of the strand, the combustion front in the bed moves progressively downward. This creates sufficient heat and temperature (1,310–1,480°C) to sinter the fine ore particles together into porous, coherent lumps. Sinter plants with a suction area of up to 5,200 square feet (483 square meters) a strand width of 17 feet (5.2 meters) and a production capacity of about 20,000 tons per day have been built.

In the pelletizing process, the agglomeration of material is effected prior to heat treatment. A green, unbaked pellet or ball (glomerule) is formed and hardened by heating. The iron ores to be pelletized are ground finely to present an adequate surface area for the formation of green balls and mixed with water and a binding agent, such as bentonite clay. A small amount of fine solid fuel may be added to the pellet mix or coated on the pellets to furnish part of the heat required. Oxidation of a pelletized magnetite concentrate to hematite during the firing step may also furnish a substantial portion of the heat requirements. The optimum moisture content for pellets is a function of fineness and use of additives, but usually ranges from 9–12%. To improve pellet strength, soda ash, limestone, or dolomite may be added. Hardening of pellets is effected in several ways, including a traveling-grate, as shown in Fig. 2, a combined grate and rotary kiln, or a shaft furnace. As illustrated by Fig. 3, ore beneficiation operations usually require large amounts of water.

Chemistry of Iron

Reduction of iron ores in preparation of iron and steel making is described under **Iron Metals, Alloys, and Steels.**

With its $3d^6 4s^2$ electron configuration, iron forms Fe^{2+} and Fe^{3+} ions, the latter, involving the removal of one 3d electron. The ferrate ion, FeO_4^{2-}, containing hexavalent iron, is unstable in acidic solution, being a very strong oxidizing agent.

Three oxides of iron are known, FeO, Fe_3O_4, and Fe_2O_3 although pure FeO does not exist. The actual composition of iron(II) oxide may be

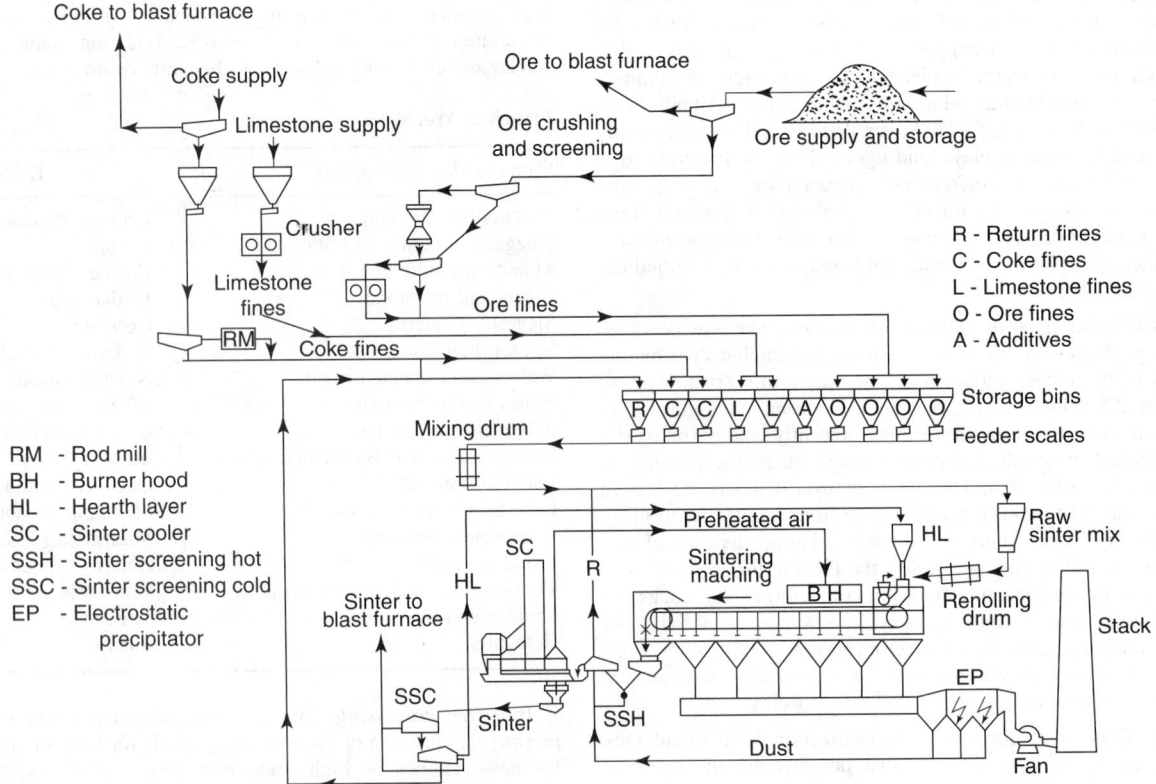

Fig. 1.   Schematic flow diagram of continuous iron-ore sintering process. (*Metallgessellschaft A.G*)

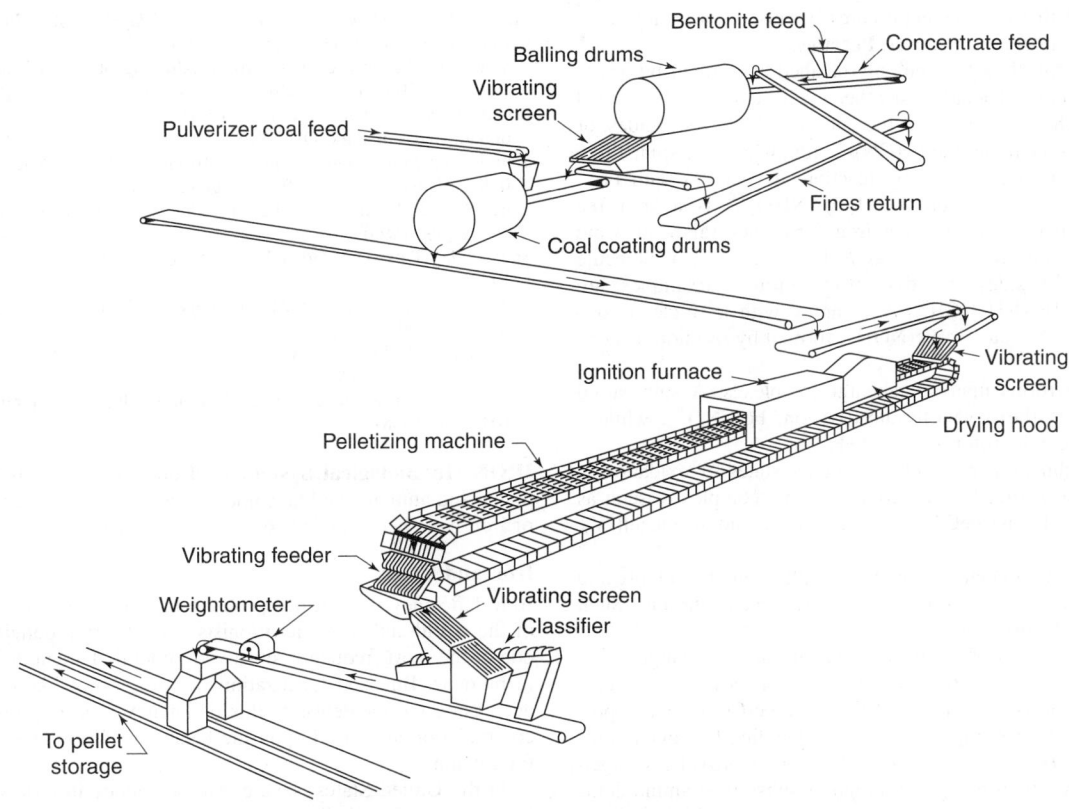

Fig. 2.   Pelletizing process in which a traveling grate is used

Fig. 3.   The beneficiation of taconite ore on the iron range requires large volumes of water in concentrating by magnetic separation. To eliminate massive waste-disposal problems, huge thickeners, such as the 300–foot (91.5–meter) diameter caisson unit shown here, are used. This system will handle over 70 million gallons (265 million liters) per day or 50,000 gallons (189,250 liters) per minute of liquid and 250 tons per day of suspended solids. Clarifying the waste tailing stream permits reclamation of water on a large scale for plant reuse

approximated by replacement of a small proportion of the Fe(II) atoms by two-thirds their number of Fe(III) atoms. If the operation is continued until three-quarters of the Fe(II) atoms have been replaced, then the composition Fe_3O_4 is reached, which may thus be described by the formula $Fe^{2+}(Fe^{3+}O^{2-}O^{2-})_2$. Continuation of the replacement until all the Fe(II) atoms have been replaced yields α-Fe_2O_3. The γ-allotrope of Fe_2O_3 and the compound Fe_3O_4 are both ferromagnetic.

Iron(II) sulfide, FeS, also may show considerable departure from stoichiometric proportions, exhibiting an electrical conductivity when in large crystals that resembles an alloy rather than a sulfide. Iron(III) sulfide, Fe_2S_3, cannot be prepared in solution in pure form, because of the oxidizing action of Fe^{3+} upon H_2S, and even upon S^{2-} ions in alkaline solution, and when Fe_2S_3 is prepared by reaction of dry H_2S and the hydrated Fe_2O_3, it breaks up into FeS and FeS_2. The latter is made up of Fe^{2+} and S_2^{2-} ions, and the mineral, pyrites, it has a cubical structure composed of these ions.

Iron forms dihalides with all four of the common halogens, and trihalides with all but iodine. FeF_2 and $FeCl_2$ are readily formed in anhydrous state by action of the hydrogen halide upon the heated metal, and the others can be made directly from the elements. The iron(III) halides, like iron(III) salts generally, are more readily hydrolyzed than the corresponding ferrous compounds, due to the smaller size and greater change of the Fe^{3+} than the Fe^{2+} ion.

Other elements with which iron forms binary compounds, especially at higher temperatures, are boron, carbon, nitrogen, silicon, and phosphorus. Like FeO, these compounds often depart slightly or even considerably from daltonide composition, frequently being interstitial compounds, and in higher elements of groups VB and VIB, merging into the interstitial compound-solid solution picture which iron exhibits with the transition metals.

The oxyacid salts of iron(III) are more numerous than those of iron(II). Among the former, the sulfates are of interest because of the readiness with which iron(III) sulfate replaces aluminum sulfate in the alums, which are hydrated double sulfates formed by certain trivalent and alkali metal (and other monovalent) sulfates. Iron(III) sulfate, $Fe_2(SO_4)_3$, is isomorphous with aluminum sulfate, $Al_2(SO_4)_3$, because the radius of the Fe^{3+} ion is so close to that of the Al^{3+} ion (0.57 Å). For that reason, the isomorphous relationship extends to other salts, i.e., the fluorides and some of the nitrates.

Like the neighboring elements of group 8, iron forms large numbers of complexes. This is due to the availability of two $3d$ orbitals in Fe^{2+} and Fe^{3+} to form hybrid orbitals with the $4s$ and $4p$ orbitals to yield spin-paired complexes (the so-called "covalent" or "inner" complexes).

Many ions contain iron combined with oxygen atoms or hydroxyl groups and are known, respectively, as ferrates or hydroxyferrates. Those of iron(II) include FeO_2^{2-}, $Fe(OH)_4^{2-}$, and $Fe(OH)_6^{4-}$, while those of iron(III) include FeO_2^-, $Fe_2O_4^{4-}$, and $Fe(OH)_8^{5-}$. Their compounds are also called ferrates, except that the name ferrite is used for such compounds of iron(III) as MFe_2O_4, in which M is a divalent metal, this last type of compound being used to form magnetic cores because of their low core losses when properly fabricated. See also **Ferrites**.

The complexes of iron(II) are commonly octahedral (formed by d^2sp^3 hybridization), and include chelate and other cyclic compounds as well as monocyclic ones. The most stable of the latter are the ferrocyanides, or hexacyanoferrates(II), containing the $Fe(CN)_6^{4-}$ ion, which is a spinpaired, diamagnetic complex. It is produced by reaction of cyanides with Fe^{2+} solutions. The Fe^{2+} diammine ion, $Fe(H_2O)_4(NH_3)_2^{2+}$ is a spin-free complex, and is paramagnetic. Divalent iron forms several pentacyano complexes, which contain an ion (such as NO_2^- or Cl^-) or a molecule (NH^3, CO, or H_2O) besides the five cyano groups. Examples are $[Fe(CN)_5NO_2]^{4-}$ and $[Fe(NH_3)_5Cl]^+$. A complex with a single nitroso group in addition to H_2O occurs in $Fe(NO)^{2+}$, formed by reaction of Fe^{2+} and NO.

The ferric ion also forms many octahedral complexes. A spin-paired type is the ferricyanide (hexacyanoferrate(III)) ion, $Fe(CN)_6^{3-}$, while a spin-free type is the hexafluoroferrate ion, FeF_6^{3-}.

The effect of coordination in stabilizing higher states of oxidation is seen in the occurrence of iron(IV) in certain cationic complexes, such as $[Fe(Cl)_2 \bullet 2C_6H_4(As(CH_3)_2)_2](FeCl_4)_2$ (cf. the corresponding nickel(IV), Ni(IV), complex).

Iron forms polydentate chelate compounds with a number of organic substances, including the oxalates, dipyridyl and orthophenanthroline. Such a structure is found in hemoglobin.

Iron forms a number of carbonyls in which its atomic charge number is zero, $Fe(CO)_5$ being produced by heating iron powder with carbon monoxide under pressure, and $Fe_2(CO)_9$ and $Fe(CO)_{12}$ being prepared from it. These iron carbonyls are reactive, yielding hydrogen compounds, e.g., $H_2Fe(CO)_4$ with alcoholic potassium hydroxide; halogen compounds, $Fe(CO)_4X_2$, with halogens; amino or substituted-amino compounds $Fe(CO_2)_n(Am)_{5-n}$, where Am is an amino group, with ammonia, pyridine or ethylenediamine. Iron carbonyls form compounds with R_3P, R_3As, R_3Sb, or diphosphines, etc., such as o-phenylenediarsine. They also form nitroso derivatives; such as $Fe(CO)_2(NO)_2$, and mercaptides, such as $Fe_2(C)_6(SC_2H_5)_2$. In strong acids (e.g., liquid hydrofluoric acid), $Fe(CO)_5$ gives $Fe(CO)_5H^+$ with the proton attached directly to the iron atom.

Iron also forms a series of similar nitrosyls, such as $Fe(NO)_4$, which is probably $[NO^+][Fe(NO)_3^-]$, $Fe_2(NO)_4I_3$, $Fe(NO)_2I$, $Fe(NO)I$, $Fe(NO)_3Cl$, $Fe_2(NO)_4(SA)_2$, where A=H, a metal, a sulfonate group, an alkyl or aryl group, $M^1[Fe(NO)_2S]$ (Roussin's red salt), $M^1(Fe_4(NO)_7S_3]$ (Roussin's black salt), obtained by treating the red salts with alkali, $Fe(NO)_2SR$, where $R=C_2H_5$ or C_6H_5, $M^1(Fe(NO) SSO_3]$ and others described under hexacyanoferrates.

By heating powdered iron with cyclopentadiene

$$\overline{CH{=}CH{-}CH{=}CH{-}CH_2}$$

the compound ferrocene, insoluble in water, soluble in organic solvents, consisting of two cyclopentadienyl radicals connected to an iron atom, C_5H_5-Fe-C_5H_5 is prepared. It is of great interest because the iron atom is sandwiched between the two parallel and symmetrical rings with delocalized bonding. Alkali metals produce salts of the type $M^1Fe(C_6H_5)_2]$, powerful reductants. Halogens produce water-soluble ferrocenium salts, such as $[Fe(C_5H_5)_2]Cl$. From both of these, sigma-bonded alkyl and aryl derivatives, such as $Fe(C_5H_5)_2CH_3$, can be obtained in which the alkyl group is attached directly to the iron atom.

See also **Iron Metals, Alloys, and Steels**; and **Iron (In Biological Systems)**.

The assistance of SUMAN C. DESAI, DAVY MCKEE IRON & STEEL Division, Ashmore House, Stockton-on-Tees, England, in preparation of parts of this entry, and of vital technical inputs received from Oak Ridge Associated Universities Institute for Energy Analysis, and the Electric Power Research Institute, are gratefully acknowledged.

Additional Reading

Greenwood, N.N., A. Earnshaw: *Chemistry of the Elements,* 2nd Edition, Butterworth-Heinemann, Inc., Woburn, MA, 1997.

Hill, H., et al.: *Metal Sites in Proteins and Models: Iron Centres,* Springer-Verlag New York, Inc., New York, NY, 1999.

Krebs, R.E.: *The History and Use of Our Earth's Chemical Elements: A Reference Guide,* Greenwood Publishing Group, Inc., Westport, CT, 1998.

Lide, D.R.: *CRC Handbook of Chemistry and Physics,* 84th Edition, CRC Pres, LLC., Boca Raton, FL, 2003.

Rawers, J.C.: "High-Pressure-Nitrogen Alloying of Steels," *Advanced Materials & Processes,* 50 (August 1990).

Rothman, M.: *High-Temperature Property Data: Ferrous Alloys,* A S M International, Materials Park, OH, 1988.

Schmidt, P.: *Iron Technology in East Africa: Symbolism, Science, and Archaeology,* Indiana University Press, Bloomington, IN, 1997.

Staff: "New Mill Technology Reshapes Industry," *Advanced Materials & Processes,* 26 (August 1990).

Staff: "Slag Puts on a Good Face," *Advanced Materials & Processes,* 6 (March 1990).

Staff: "Steelmaking in the 21st Century," *Advanced Materials & Processes,* 31 (August 1990).

Stwertka, A., E. Stwertka: *A Guide to the Elements,* Oxford University Press, Inc., New York, NY, 1998.

Van Noten, F., J. Raymaekers: "Early Iron Smelting in Central Africa," *Sci. Amer.,* 104 (June 1988).

IRON (In Biological Systems). Iron plays a number of vital roles in plants and animals. Unlike some of the minerals, research on the functions of iron have been studied for several decades.

Iron and Crops

Iron deficiency is a serious problem in crop production in certain regions of the world and some nutritionists consider iron deficiency anemia to be one of the most frequently observed mineral element deficiency conditions in humans. But iron fertilization of soils is not likely to be effective in decreasing the incidence of this deficiency. The reasons for this apparent contradiction are based upon the behavior of iron at several stages in the food chain.

In the United States, severe iron deficiency in crop plants occurs most frequently on the alkaline soils of the western states and on very sandy soils, although some plants, especially broad-leaved evergreens, are sometimes iron deficient on many other kinds of soils. Iron deficiency is rarely due to a total lack of iron in the soil. It is nearly always due to the low solubility of the iron that is present. For example, some soils that are red from iron compounds may contain too little available iron for normal plant growth. The relative susceptibilities of cultivated crops to iron deficiency are listed in Table 1.

To correct iron deficiency in plants, it is usually necessary to add a soluble form of iron to the soil or to spray the foliage. Since soluble iron

TABLE 1. RELATIVE SUSCEPTIBILITY OF CULTIVATED CROPS TO IRON DEFICIENCY

Crop	Highly Susceptible	Moderately Susceptible	Relatively Tolerant
Alfalfa		X	X
Barley		X	X
Berries	X		
Citrus	X		
Corn		X	X
Cotton		X	X
Field beans	X	X	
Flax	X	X	X
Forage sorghum	X		
Grain sorghum	X	X	
Grapes	X		
Grasses		X	
Groundnuts (Peanuts)	X		
Millet			X
Mint	X		
Oats		X	X
Potatoes			X
Rice		X	X
Soybeans	X	X	X
Sudangrass	X		
Sugar beets			X
Tree fruits	X	X	
Vegetables	X	X	X
Walnuts	X		
Wheat		X	X

Note: Crops listed in more than one category have a wide range of tolerance, depending upon variations in soils, crop varieties, and growing conditions.
Source: Martvedt/Wallace/Curley (1977).

added generally will revert to insoluble forms, these procedures for plants are only temporarily effective. Soil treatments that make alkaline soils more acid, such as incorporating large amounts of sulfur, may offer a more lasting correction. Incorporating large amounts of manure into soil makes the iron more soluble and may be effective in correcting iron deficiency, particularly in fine-textured alkaline soils.

Iron-deficient plants are generally stunted and chlorotic, i.e., normally green leaves are yellow or streaked with yellow. When the iron deficiency is treated by adding soluble iron to the soil, the plants turn green, grow larger and yield more, but sometimes the concentration of iron unit weight of plant material may be no higher than in the stunted, iron-deficient plants. Thus, in terms of forage plants, correction of iron deficiency in the plant does not necessarily improve the plant as a source of dietary iron. The iron-treated plants, however, may contain a higher concentration of carotene or provitamin A than the yellow, stunted, deficient plants. Thus, iron fertilization can be more useful in improving the vitamin A than the iron level in diets.

In livestock, iron deficiency is most common in young pigs raised in confinement on concrete floors. Injecting iron compounds and painting the sow's udder with iron compounds are measures used to prevent this deficiency. Grazing animals seldom suffer from iron deficiency unless they are heavily parasitized.

Synthetic chelates and some natural organic complexes tend to resist the adverse effects of soil reactions, climate, and management practices that change available iron into unavailable forms. Natural organic complexes, including some lignin sulfonates and polyflavonoids, or synthetic chelates of iron tend to remain in an available form during most of the growing season. Synthetic chelates are mobile, however, and may be lost from the root zone if subjected to excessive leaching due to overirrigation or high rainfall. Not all iron chelates behave the same in soils, since some are differentially fixed onto clay particles. Fixed chelates do not function in delivering iron to plants. Stability or resistance to decomposition of metal chelates is related to soil pH. Some chelates tend to release their iron more readily than others. For application to calcareous soils of pH 7.5 or higher, the principal iron chelate that will correct iron deficiency for most dicotyledonous plants is FeEDDHA or other similar compounds which have high metal chelate stability. Commercial sources of iron for agricultural application include:

Inorganic Sources: Ferrous sulfate; ferric sulfate; ferrous carbonate; ferrous ammonium sulfate; iron frits.

Chelates: FeEDTA (ethylenediaminetertraacetic acid); FeHEDTA (hydroxyethylethylenediaminetriacetic acid)

Organic Complexes: Lignin sulfonate; methoxy phenylpropane complex; polyflavonoid

During recent years, more emphasis has been placed on testing soils for iron deficiency. A commonly used test, developed at Colorado State University, employs a solution of the chelating agent DTPA (diethylenetriaminepentaacetate) and calcium chloride buffered at pH 7.3 for extracting the soil under test. For testing larger areas, a new method of detection and visual assessment of iron chlorosis or deficiency symptoms in growing crops has been developed. This involves aerial infrared photography which records distinctive color differences between chlorotic and normal green plants. Assessment of a spotty chlorotic field by proportion (percent) and three-dimensional projection makes possible economic evaluation of the iron deficiency problem existing in any given field on a large-scale basis. See also **Fertilizer**.

Iron in Human Physiology

The bioavailability of iron has been researched extensively. Investigators have found this to be quite variable because numerous factors influence absorption of iron, including the consumer's needs and the composition of the diet. Chemical factors that affect iron availability include the valence, solubility, and degree of chelation or complex formation of the iron. Researchers have shown that the ferrous valence is considerably more available than the ferric valence. Others have shown that prior to absorption in the gut, iron must be in solution. Further, it has been demonstrated that chelation may augment iron absorption by maintaining the iron in solution under conditions where it would otherwise be insoluble. Because of more ready availability, ferrous sulfate is used in bread and flour enrichment even though it has a greater reactivity with foods. Ascorbic acid has been shown to increase iron availability when in the diet.

Iron Absorption. The iron content of the normal adult human is dependent on the size of the individual and the hemoglobin concentration. The distribution of iron in a male weighing 155 pounds ($\sim$70 kilograms) has been estimated at just under 3.5 grams. About 64% of this iron is in hemoglobin as part of the peripheral blood; and 2.5% as hemoglobin in the bone marrow. Another 4% is present as myoglobin which also participates in oxygen transport and storage. Another 13% is present as ferritin and 16% as hemosiderin, which are storage forms. Extremely small amounts are found in cellular cytochrome and in the enzyme catalase.

It is generally agreed that iron is absorbed from the most part in its ferrous form directly into the bloodstream. Radioactive iron has been shown to be absorbed from any portion of the intestinal tract, but its uptake appears greatest in the duodenum. On the basis of experiments done on the absorption of iron from the intestinal tract of guinea pigs, it was earlier postulated that iron is taken into the mucosa cells and ferritin is formed by a combination of a protein, *apoferritin*, with iron. After the cell is saturated with ferritin, absorption no longer takes place until the iron of ferritin is transferred to plasma. For a number of years, this concept of a mucosal block was the accepted explanation for iron absorption. However, later research showed that there is no absolute block to iron absorption. It is found that the absorption of iron in patients and in experimental animals is greater than normal in iron deficiency and in cases where erythropoiesis is accelerated, even when the body iron reserves are elevated. Later evidence indicated that the ferritin concentration in the intestinal mucosa neither controls nor blocks absorption. An active transport mechanism requiring energy is concerned with iron transfer across the intestinal mucosa.

The factors involved in the absorption of iron in food are more complex than those involving inorganic iron. To obtain Fe^{59}-marked foods, radioactive iron has been injected into hens to obtain labeled eggs and meat; plants have been grown in media containing Fe^{59}, and Fe^{59}-enriched bread has been prepared. It has been shown that iron-deficient subjects absorb more food iron than normal subjects. Absorption from liver, hemoglobin, muscle, and "enriched" bread is greater than from eggs or plants. Most probably, the low absorption from egg yolk derives from the presence of a ferric iron-phosphate complex. In such research, large variations in results have been obtained.

In the presence of a large amount of ascorbic acid (vitamin C), the absorption of iron is appreciably enhanced, because of the reduction of Fe^{3+} to the Fe^{2+} form. In the presence of phosphates, carbonates, and phytates, insoluble iron compounds are formed, thus reducing absorption.

It has been estimated that normal subjects ingesting a mixed diet containing 12–15 milligrams of iron retain 5–10% (0.6–1.5 milligram), whereas iron-deficient patients retain 10–20% (1.2–3 milligrams) iron.

Iron Transportation. After iron enters the bloodstream, it is immediately bound by a specific plasma protein that is a β_1-globulin. This protein, *transferrin* (siderophilin), has a molecular weight of about 90,000 and binds two atoms of ferric iron. About 0.25 gram of transferrin in 100 milliliters of plasma is capable of binding about 300 micrograms Fe^{3+}, but normally it is only one-third saturated while the remaining two-thirds are unbound reserve. If a small amount of ionized iron is injected intravenously, it is bound by the transferrin, which may be completely saturated. If the binding limit is exceeded, ionized iron exhibits toxic effects. The transferrin concentration is increased in iron deficiency and during the latter half of pregnancy; it is decreased during infection and a variety of other disorders.

Electrophoretic studies show there are several genetically controlled variants of human transferrins. They all deliver iron in an equivalent manner for utilization and storage. Evidence indicates that iron may be transferred directly to the developing erythroblast. It has been demonstrated that transferrin-bound iron is utilized by reticulocytes for hemoglobin formation. The transfer of iron is not maximal until 25% of the transferrin is saturated.

Excretion. The total loss of iron from an adult is about 1 milligram daily and is distributed in sweat, feces, hair, and urine. Since approximately 1 milligram of iron is normally absorbed daily, the organism is in iron balance. The loss of red cells from the body in normal menstruation would account for 16–32 milligrams of iron, which would amount to an average daily loss of from 0.5–1.0 milligram during the 28-day menstrual cycle. Pregnancy would also represent a loss of iron from the body, but this is compensated by the absence of menstruation. During normal hemoglobin catabolism, about 20–25 milligrams of iron are released per day. The excretion of minute amounts of iron allows the body to conserve and reutilize the iron for the synthesis of hemoglobin. This tenacious conservation has been demonstrated repeatedly by radioactive techniques.

Enzymes. Heme serves as the prosthetic group for catalase, peroxidase, cytochrome oxidase, and the related cytochromes. Catalase and peroxidase iron are presumably present in the ferric form while the iron of the cytochromes may exist in the reduced or oxidized form. A number of flavoproteins, including succinic dehydrogenase, contain iron in the molecule. Iron appears to act as coenzyme for aconitase. A number of other enzymes require the presence of iron for their activities.

Storage Iron. Ferritin and hemosiderin represent practically all the iron present in the reticulo-endothelial cells of the liver, spleen, and bone marrow and in the parenchymal cells of the liver. Ferritin is an iron protein complex containing up to 23% iron. It is composed of a protein, which has a molecular weight of 450,000 and a colloidal ferric-hydroxide-phosphate complex. Preparations of hemosiderin granules contain up to 40% iron and are insoluble in water. It appears to be an iron-loaded organelle, such as mitochondrion. The granule contains a small amount of ferritin, but the remaining material is composed of heterogenous proteins.

Hemoglobin. The approximate formula (molecular weight >52,000) is $(C_{738}H_{1166}FeN_{208}S_2)_4$. Hemoglobin is the respiratory protein of the red blood cells. It transfers oxygen from the lungs to the tissues and carbon dioxide from the tissues to the lungs. Its affinity for carbon monoxide is over 200 times that for oxygen. Hemoglobin is a conjugated protein consisting of approximately 94% globin (protein portion) and 6% heme. Each molecule can combine with one molecule of oxygen to form oxyhemoglobin. The iron (in the heme portion) must be in the reduced (ferrous) state to enable the hemoglobin to combine with oxygen. Heme $(C_{34}H_{32}FeN_4O_4)$ is the nonprotein portion of hemoglobin and myoglobin, consisting of reduced (ferrous) iron bound to protoporphyrin. See also **Hemoglobin**.

Iron Deficiency Disorders

Iron deficiency may occur from several causes, including loss of blood through hemorrhage, an iron-poor diet, or an inability to metabolize iron in a normal fashion. There are a few more complex disorders, some of which have hereditary vectors.

Nutritional Anemias. These disorders may result from nutritional deficiencies or decreased bone marrow function, both of which cause defective blood formation. The least severe but most common of these anemias results from an inadequate amount of iron required for red cell formation. The result is *microcytic hypochromic anemia*. About 100 milligrams of iron per day are needed for hemoglobin manufacture. About 85% of this iron may be obtained from the iron released by breakdown of older red cells. However, some iron is always lost in the excretions and thus must be made up by the diet. Where there is chronic blood loss, as in cases of ulcers or hemorrhoids, or where the iron may not be properly absorbed from foods, the need for iron may be greater. Milk, cereals, and many refined foods, unless artificially supplemented, do not contain much iron. Better sources of iron include meat and leafy vegetables. Iron deficiency is not uncommon.

A common form of iron-deficiency anemia frequently seen in young women during the last century was sometimes called chlorosis, or "green sickness" because of the peculiar hue of the skin. With the discovery that iron salts can effect a cure, the disease almost completely disappeared. Idiopathic hypochromic anemia is another iron-deficiency anemia associated with a lack of proper stomach acidity. When hydrochloric acid in the stomach is lacking, iron cannot be liberated from foods and converted into a form that can be absorbed. Administration of iron in proper form also alleviates this condition.

Iron in Pregnancy. During pregnancy, the mother must furnish greater volumes of blood to support herself and the developing baby. Blood volume is increased and causes dilution unless sufficient iron is available. Vomiting in early pregnancy may increase the danger of an iron deficiency. Usually, babies are born with adequate supplies of iron in their tissues to last several months. However, infants born of a mother with an iron deficiency have low reserve stores of iron and will require a diet that is supplemented with the proper amounts of iron. Milk is a poor source of iron, and infants strictly on a diet of milk almost invariably develop hypochromic anemias. Anemic babies are much more subject to infections, which may in turn further increase the anemia. Thus, such children should be treated early.

Infants with Iron Deficiency. B. Lozoff (Case Western Reserve University) and associates reported that "Several consistent results have emerged from five studies of the behavior and development of infant with iron-deficiency anemia, a condition that affects at least 20 to 25 percent of the world's babies." All five studies used careful definitions of iron status and included comparison groups without anemia. All showed that infants with anemia scored lower on tests of mental development administered before treatment than infants without anemia did.

Iron Overload. An inborn error of metabolism leads to the absorption of excess iron from a normal diet. Hereditary *hemachromatosis* is found mainly within the white population. Black people who live in sub-Saharan Africa, however, show a high incidence of this disorder. At one time, this was attributed to the high content of iron in a home-brewed beer. Recent studies show that it is a combination of a high-iron diet and hereditary disposition.

Thalassemia Major. Transfusion-dependent thalassemia major patients have abnormal growth and sexual maturation at puberty, presumably as a result of pituitary iron overload. Still poorly understood, this disorder is reported to respond well to deferoxamine iron chelation therapy, particularly if administered before the age of maturity.

Mitochondrial Myopathy. A general deficiency of iron may be implicated in mitochondrial myopathy, which is a complex disorder that affects muscular activity. It has been suspected for a number of years that the disorder is caused by a defect of mitochondrial-protein transport. H.H.V. Scharpa and a team of researchers (Royal Free Hospital, London) postulate that a deficiency of an iron-sulfur protein in muscle dehydrogenase may be the specific cause.

Iron Chelation Therapy in Cerebral Malaria. It is estimated that over 1 million children die from severe forms of malaria annually, notably in sub-Saharan Africa. Cerebral malaria is one of the most severe complications of malaria infection (*Plasmodium falciparum*). It is known that iron is an essential nutrient for promoting the growth of the infectious agent. Victor Gordeuk and a team of investigators report that the iron-chelating agent deferoxamine enhances the clearance of the *P. falciparum* parasitemia. Iron chelation inhibits peroxidant damage to the central nervous system, as previously tested in animals. Iron also serves as a redox agent in the generation of free radicals that mediate ischemic and hemorrhagic tissue energy. A report issued in November 1992 concludes: "Iron chelation therapy may hasten the clearance of parasitemia and enhance recovery from deep coma in cerebral malaria."

Additional Reading

Bacon, B.R.: "Causes of Iron Overload," *N. Eng. J. Med.*, 126 (January 9, 1992).

Bronspiegel-Weintrob, N., et al.: "Effect of Age at the Start of Iron Chelation Therapy on Gonadal Function in Thalassemia Major," *N. Eng. J. Med.*, 713 (September 13, 1990).

Bullen, J.J. and E. Griffiths: *Iron and Infection: Molecular, Physiological and Clinical Aspects,* John Wiley & Sons, Inc., New York, NY, 1999.

Gordeuk, V., et al.: "Effect of Iron Chelation Therapy on Recovery from Deep Coma in Children with Cerebral Malaria," *N. Eng. J. Med.*, 1473 (November 19, 1992).

Gordeuk, V., et al.: "Iron Overload in Africa—Interaction between a Gene and Dietary Iron Content," *N. Eng. J. Med.*, 95 (January 9, 1992).

Loretta, J. and P.W. Atkins: *Chemistry: Molecules, Matter and Change,* W.H. Freeman and Company, New York, NY, 1999.

Lozoff, B., Jimenez, E. and A.W. Wolf: "Long-Term Developmental Outcome of Infants with Iron Deficiency," *N. Eng. J. Med.*, 687 (September 5, 1991).

Mielczarek, E.V. and S.B. McGrayne: *Iron, Nature's Universal Element: Why People Need Iron and Animals Make Magnets,* Rutgers University Press, Piscataway, NJ, 2000.

Mortvedt, J.J., A. Wallace and R.D. Curley: "Iron-The Elusive Micronutrient," *Fertilizer Solutions,* **21**, 1, 26–36 (1977).

Scharpa, A.H.V.: "Mitochondrial Myopathy with a Defect of Mitochondrial-Protein Transport," *N. Eng. J. Med.*, 37 (July 5, 1990).

Schwertmann, U.: *Iorn Oxides in the Laboratory,* 2nd Edition, John Wiley & Sons, Inc., New York, NY, 2000.

Staff: Institute of Medicine, *Iron Deficiency Anemia,* National Academy Press, Washington, DC, 1994.

Sullivan, J.L.: "Retinopathy of Prematurity," *N. Eng. J. Med.*, 648 (Aug. 27, 1992).

Sykes, A.G. and R. Cammack: *Advances in Inorganic Chemistry: Iron-Sulfur Proteins,* Academic Press, Inc., San Diego, CA, 1999.

Symons, M.C.: *Free Radicals and Iron: Chemistry, Biology, and Medicine,* Oxford University Press, Inc., New York, NY, 1998.

Wyler, D.J.: "Bark, Weeds, and Iron Chelators—Drugs for Malaria," *N. Eng. J. Med.*, 1519 (November 19, 1992).

IRON METALS, ALLOYS, AND STEELS. Chemically pure iron is used essentially in powder metallurgy and for chemical applications where the element serves as a catalyst, or as a base ingredient for ferrous and ferric chemicals. Iron principally is used as the dominant ingredient of cast irons and steels. Iron-base alloys notably are known for their physical strength and toughness and, when compared with most other metals for similar applications, reasonable cost. The following properties depend upon the nature and extent of the ingredients, such as carbon, present in the alloys and also upon the mechanical and heat treatments given to the formed metals: (1) impact strength or brittleness, (2) cohesive strength, (3) compressive strength, (4) creep, (5) fatigue, (6) ductility, (7) hardness, (8) malleability, (9) shear strength, (10) yield strength, (11) torsional strength, (12) electrical conductivity, (13) thermal conductivity, (14) thermal stability, (15) thermal expansion, (16) corrosion resistance, (17) magnetic properties, and (18) heat treatability.

The iron metals family of products may be classified into: (1) *the pure irons*, such as ingot iron and wrought iron, which have only traces of carbon (see Table 1) and other elements, and are very ductile; (2) *cast irons*, which are alloys of iron and carbon, with or without other elements, and normally containing from 2.4 to 4.5% carbon; (3) *steels*, which are alloys of iron and carbon, with or without other elements, in which the carbon content seldom exceeds 1.7%; (4) *alloy steels* whose properties mainly are attributed to the presence of one or more elements other than carbon. There are other groups and numerous subgroups in the total iron metals family.

Ironmaking

The starting ingredients for the numerous iron metals are obtained in three major ways: (1) by smelting run-of-mine or beneficiated iron ore in a blast furnace, low-shaft furnace, or electric smelter to yield a liquid, molten product; (2) by reducing run-of-mine or beneficiated iron ore via direct reduction processes to produce sponge iron; and (3) by melting ferrous scrap in a cupola, electric furnace, or fuel-fired furnace. Inasmuch as the majority of iron ores are in the form of oxides of iron, the iron is obtained by employing suitable reducing agents to reduce the oxides. Reducing agents most often used are carbon, carbon monoxide, hydrogen, and hydrocarbons, such as methane. With carbon monoxide, the reduction is exothermic; with the other reductants, it is endothermic.

Blast Furnace. For several decades, the blast furnace was the unchallenged producer of iron (pig iron) from iron ore. With the growing availability of steel scrap over the last several years, scrap as a source of iron for steelmaking became increasingly important, a fact which made electric furnace steel production, to be described later, very attractive. Blast furnaces are large, bulky structures that have been part of the integrated mill concept, a concept that has been threatened economically for several years. Blast furnaces are high-capacity producers and, consequently, essentially unsuited to the current philosophy of spreading steel production geographically (nearer the consumers). It is much easier to commence with scrap than to have to make pig iron. A number of the older style blast furnaces have been dismantled in recent years because of the growing dependability of scrap as a starting material. But, as pointed out later, all steel products do not represent potential usable scrap and, consequently, the iron ore reduction processes, including the blast furnace, will continue to be used in a number of locations.

A typical, traditional blast furnace is shown in Fig. 1. The furnace is a tall, refractory-lined vessel. Raw materials including iron ore (sinter or pellets), coke (the reducing and thermal agent), and limestone (for fluxing the gangue material) are charged into the top of the furnace. A blast of hot air is introduced at the bottom of the furnace to burn the coke and thus to heat, reduce, and melt the charge as it descends toward the bottom of the furnace. Liquid iron and slag collect in the furnace hearth. These materials are tapped at regular intervals. Although the furnace can be damped down for short periods, the process essentially is continuous. The waste gas contains about 28% carbon monoxide with a calorific value of about 90 Btu/cubic foot (800 kcal/cubic meter). After collection at the top of the furnace, dust is removed, and the gas is used as a fuel for heating the hot-blast stoves. Blast furnaces range from 100 to 10,000 tons/day in capacity; the hearth diameter may range from 9 to 46 feet (2.7 to 13.8 meters); the height from 50 to 150 feet (15 to 45 meters).

Air required for combustion is furnished by turboblowers and preheated in hot-blast stoves lined with refractory-brick checker-work. Commonly,

TABLE 1. TYPICAL ANALYSES OF PURE IRONS

Ingredient	Ingot iron	Electrolytic iron	Carbonyl iron	Hydrogen-purified iron
		PERCENT OF TOTAL		
Carbon	<0.020	0.006	0.0004	0.005
Manganese	<0.020	—	—	0.028
Phosphorus	0.005±	0.005	—	0.004
Sulfur	0.020±	0.004	—	0.003
Silicon	Trace	0.005	—	0.001
Copper	0.04±	—	—	—
Oxygen	Some	Some	<0.01	0.003
Nitrogen	0.004±	—	—	0.0001

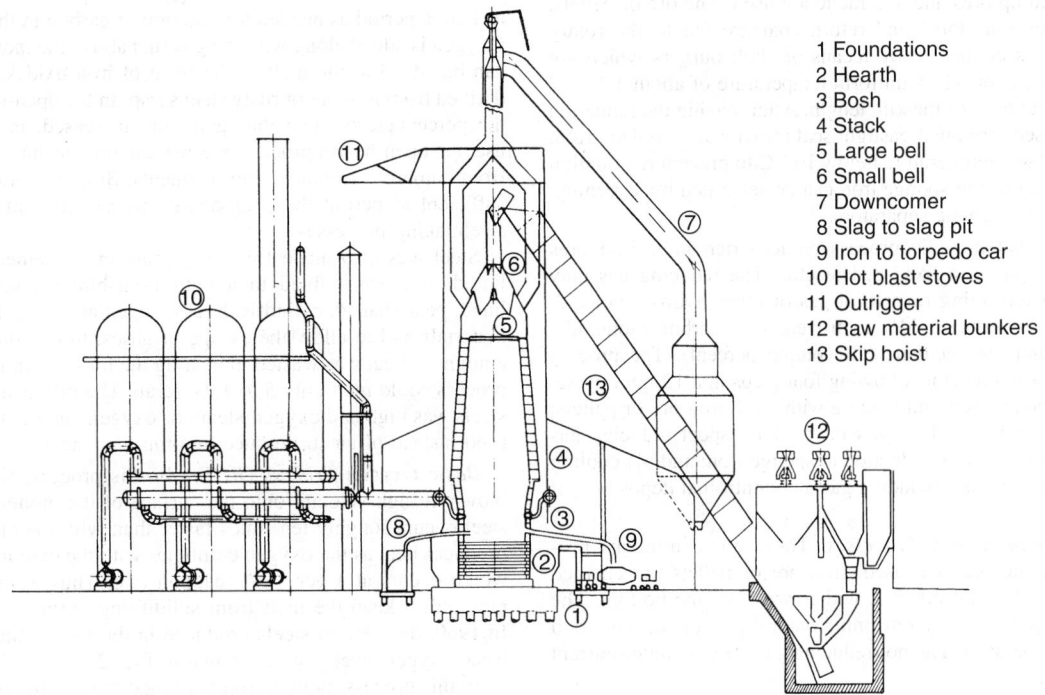

1 Foundations
2 Hearth
3 Bosh
4 Stack
5 Large bell
6 Small bell
7 Downcomer
8 Slag to slag pit
9 Iron to torpedo car
10 Hot blast stoves
11 Outrigger
12 Raw material bunkers
13 Skip hoist

Fig. 1.   Cross section of a traditional blast furnace plant

three stoves are used per furnace. The stoves are operated alternatively on a regenerative principle, one stove normally providing the hot blast while the other two stoves are heating up to temperature. The checkerwork provides the means for temporarily storing the rather large quantities of heat required. Fuel for the stoves includes the blast-furnace off-gas previously mentioned, augmented by coke-oven gas from a nearby coke-producing facility. The exact fuel-supply arrangement varies with local conditions. The hot blast is supplied by a gas main to a bustle pipe, which encircles the bosh and distributes it to water-cooled tuyeres located below the bosh for injection into the furnace. The metal is tapped into refractory-lined, open-top torpedo or kling-type ladles and thence transported to the pig-casting machine; or directly to the steel plant. A comparatively few blast furnaces are operated with charcoal instead of coke, but these are limited to a capacity of about 300 tons/day because of the low crushing strength of charcoal. Improvements in blast furnace operations over the last several years have resulted from better preparation of the charge, fuel injection through tuyeres, reducing-gas injection in bosh, and oxygen enrichment of blast.

Low-Shaft Furnace. These furnaces are circular or oval in cross section. The oval shape permits greater hearth area without increasing the required depth of penetration of the blast supplied through the tuyeres. Such furnaces are designed for finer raw materials and low-grade coke or lignite. Once considered ideal for small-scale iron production, only a limited number have been installed. The operating principles are essentially similar to those of a blast furnace.

Electric Smelter. Electric energy provides the heat in these designs. Low-grade coke can be used as the reducing agent. Electric smelters are generally limited to areas of low-cost electric power. The most commonly used furnaces of this type employ a submerged arc, using the Söderberg continuous self-baking electrodes. Developed by Tysland and Hole in Norway, the furnace is circular or rectangular in cross section with transformer ratings of up to 60,000 kVA. Production capacity can be increased and power and coke requirements lowered by preheating and prereducing the iron-ore charge.

Direct Reduction Process. Numerous schemes over the years have been attempted as an alternative to the blast furnace. These include rotary and stationary kilns and furnaces, reverberatory furnaces, retorts, fluid-bed reactors, pot furnaces, and jet smelting. Similarly a variety of reductants have been used, including lignite, coal, char, fuel oil, tar, and various gases. The direct reduction approach is essentially useful for producing a highly reduced product containing mostly metallic iron and little gangue material, thus providing a substitute for ferrous scrap for steelmaking operations.

SL/RN Process. In this process, using a rotary kiln and solid reductant, high-grade pellets or lump ores and anthracite are used. The ore or pellets, anthracite, dolomite or limestone, and return coal are fed to the rotary kiln. The temperature is controlled by means of shell burners which are furnished with air and gas or oil. A uniform temperature of about 1,100°C is maintained over about 60% of the kiln length. After leaving the reduction kiln, the charge is passed through a gastight seal into a water-cooled drum, whereupon it is cooled to a temperature below 100°C to prevent reoxidation of the sponge iron. Most of the sponge iron can be separated by screening, which is augmented by magnetic separators.

HyL Process. This batch-cyclic process reduces rich lump-iron ores by flowing a reducing gas in a fixed-bed reactor. The reducing gas may be prepared by steam reforming of natural gas or other hydrocarbons. A typical reducing gas may contain 74% hydrogen, 13% carbon monoxide, 8% carbon dioxide, and 5% methane (all volume percent). The process requires four reactors, each reactor following four steps in a 12-hour cycle: (1) removal of cold sponge iron and loading with fresh iron ore or pellets; (2) preheating and secondary reduction with partially spent reducing gas from another reactor; (3) primary reduction to sponge iron; and (4) cooling the sponge iron with fresh, cool reducing gas and controlled deposition of carbon where required.

Purofer, Midrex, and Armco Processes. These are continuous processes in which shaft furnaces are used. Iron ore or pellets are charged from the top and the reduced product is withdrawn from the bottom. The reducing gas may be generated by reforming natural gas, or methane rich gas from naphtha may be used. The hot reducing gas flows counter-current to the descending charge.

Fior and U.S. Steel Processes. Similar reducing gases are used in connection with a fluid-bed reactor. The iron ore may require further grinding and drying before introduction into the fluid bed. The reduced ore may be briquetted or used as fines.

Cupola. This type of furnace is widely used for making iron for casting in foundries and sometimes may be used to augment the iron required by steelmaking operations. A cupola may be operated for just a few hours/day, or up to 2 to 3 months continuously. A cupola is a vertical cylindrical shaft furnace that uses the countercurrent-flow principle to heat and melt the charge as it descends. Cupola capacity may range from 2 to 75 tons/hour. Unlike a blast furnace, a cupola is not a reducing unit, but is essentially used to melt ferrous scrap and cold pig iron or previously reduced sponge iron. The heat required is supplied by nearly complete combustion of coke. Air is injected through tuyeres near the hearth zone. Some cupolas are equipped for hot-blast supply through recuperators. The iron raw material is charged in alternate layers with coke. Limestone is added to flux the ash from the coke and form slag. The molten iron collects in the hearth and may be removed continuously or intermittently through a taphole.

Steelmaking

Raw materials for steelmaking include liquid iron, steel scrap, preproduced sponge iron, or mixtures of the foregoing ingredients. During the past quarter century, steelmaking has undergone numerous changes, essentially motivated by marked increases in energy costs, by the need for better production flexibility, and by the impact of worldwide competition, particularly by nations that could essentially commence serious production with modern technology rather than engaging in tremendously costly modernization of decades-old, large, and sprawling integrated mills. Essentially since the turn of this century, the traditional steelmaking countries such as the United States, United Kingdom, and Germany depended upon the integrated mill concept to produce vast tonnages of steel, largely a system that served them well during two world wars. The basic economics of the steel industry today are strikingly different, and altered economics, even more than the availability of technology, have brought about steel production changes. The competition of other metals and materials also has adversely affected the once-exclusive market for steels. The greater availability of steel scrap has also been a factor.

Historical Open-Hearth Process. For well over a century prior to the 1960s, the open-hearth process was the principal means for making steel. As of the late 1980s, open-hearth steels had dropped from well over 90% of total steel production to less than 10% of steel production. The attraction of the basic open-hearth system over many decades was versatility in handling a variety of raw materials required for most grades of steel. Raw material charges could be 100% scrap, 100% hot metal, or scrap and hot metal in all intermediate ratios.

In the open-hearth process, the iron is kept molten with gas burners for as long a period as needed for reaction of carbon in the metal with oxygen. Oxygen is added along with the gas fuel above the molten material. Oxygen can be added to the melt in the form of iron oxides, the latter sometimes derived from iron ore or rusty steel scrap. In the open-hearths still operating, the percentage of scrap has generally increased. In its many decades of use, the open-hearth process became a mature technology, having benefited from numerous technical improvements. But, these improvements were not sufficient to permit the open-hearth process to compete with other basic steelmaking processes.

Steel was also made for many years in Bessemer converters, wherein liquid iron was refined in a bottom-air-blown converter—a refractory-lined, pear-shaped, cylindrical vessel open at the top to permit charging of materials and to allow the escape of gases. In this process, a considerable amount of heat was wasted in heating the nitrogen in the air. The Bessemer process could melt only 5 to 10% scrap. The nitrogen content of Bessemer steels was high and oxygen-steam or oxygen-carbon dioxide mixtures were used instead of air to produce low-nitrogen steels.

Basic Oxygen Process (BOP). In this process, almost pure oxygen is blown at high velocity onto the surface of the molten iron. Conversion to steel occurs roughly ten times faster than with the open hearth. The BOP produces heat as the oxygen combines with the carbon in the molten metal, which occurs at a very fast reaction rate. Thus, supplemental fuel is not required to keep the melt from solidifying during its conversion to steel. In 1960, only 3% of steels produced in the United States were by BOP. A basic oxygen steel plant is shown in Fig. 2.

In this process, molten iron is refined to steel by top-blowing oxygen at high pressure onto the surface of the metal through a water-cooled lance contained in a tilting furnace. The oxidation of carbon, silicon, manganese,

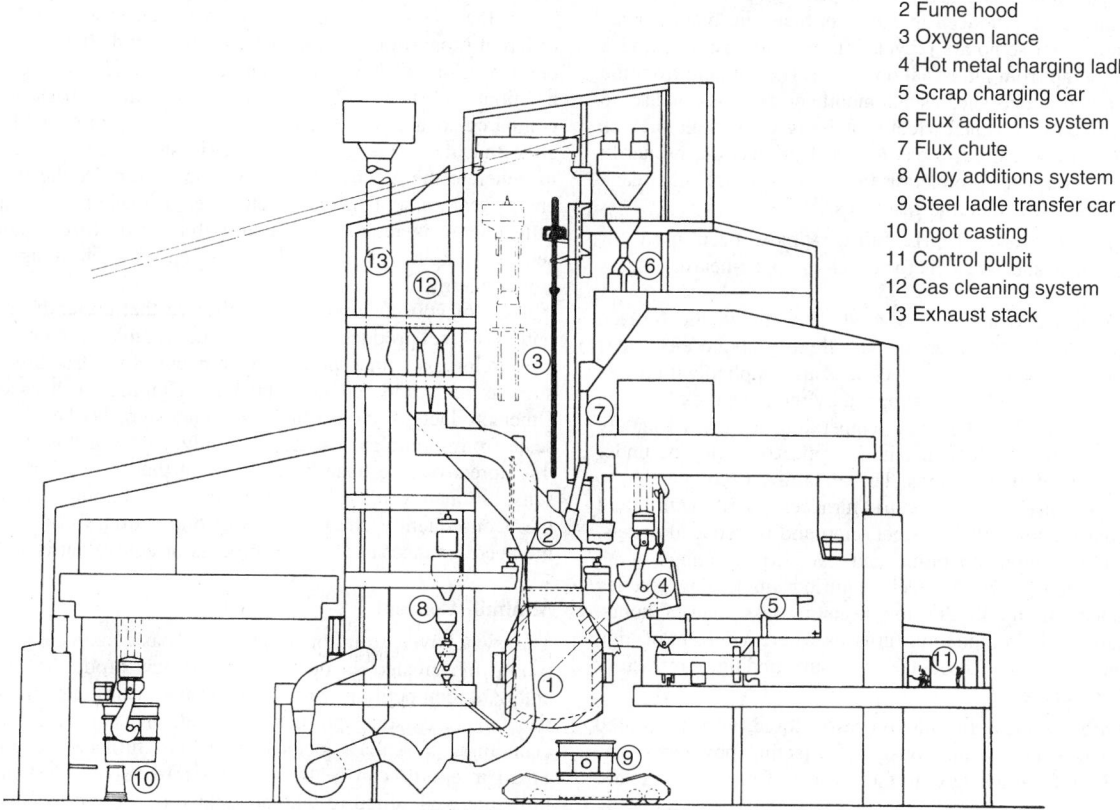

1 Basic oxygen furnace
2 Fume hood
3 Oxygen lance
4 Hot metal charging ladle
5 Scrap charging car
6 Flux additions system
7 Flux chute
8 Alloy additions system
9 Steel ladle transfer car
10 Ingot casting
11 Control pulpit
12 Cas cleaning system
13 Exhaust stack

Fig. 2.   Section through basic oxygen steelplant

and phosphorus provides sufficient heat for converting molten iron into steel. Because of the excess heat generated, up to about 30% scrap can be charged. A conventional basic oxygen process can refine iron containing up to 0.3% phosphorus into most grades of steel. Where the phosphorus content is higher, a modified process uses injection of powdered lime with the oxygen stream, or double slagging is required. A basic oxygen furnace with 300-ton capacity is shown in Fig. 3.

Numerous modifications of the BOP have appeared during recent years. For example, the OBM/Q-BOP, LWS, and SIP processes have been developed which use bottom blowing of oxygen and a shielding hydrocarbon through tuyeres in the bottom of the converter vessel. The endothermic dissociation of hydrocarbon by its cooling effect prevents excessive refractory wear in the tuyere area. The result is a substantial increase in the life of the bottom refractory plug. The OBM process was initially developed in Germany. The designation Q-BOP, introduced in the United States after further development work, is intended to emphasize the advantages of the new process compared with the Basic Oxygen Process (BOP). The letter, Q, stands for "quiet, quick, quality." Natural gas, propane, or liquefied petroleum gas are used as the gaseous hydrocarbon shield injected through an outer concentric gap around the central oxygen tuyere. A similar development, preferably using fuel oil as a hydrocarbon shield, is termed the LWS process. The OBM tuyere is successfully inserted in the bottom of an open hearth furnace for injecting oxygen into the metal bath for refining. This relatively new steelmaking technique is known as SIP (submerged injection process).

In the Kaldo process developed in Sweden, the refining of molten iron to steel is carried out in a titled pear-shaped basic-lined converter. Oxygen is blown at an oblique angle to the metal bath through a water-cooled lance. Much of the carbon monoxide produced by the carbon/oxygen reaction is burned inside the converter. The heat generated is absorbed by the rotating vessel and transferred to the bath, thus providing a high thermal efficiency and allowing up to 40% scrap in the charge. Both low- and high-phosphorus irons can be handled. Refractory consumption is high, tending to reduce the availability of the furnace.

In the Rotor process, a long cylindrical horizontal vessel, rotating at 1 to 5 rpm, is used. Oxygen is injected by two lances, one carrying high-purity oxygen into the bath, the other carrying low-purity oxygen for burning the

Fig. 3.   A basic oxygen furnace of 300-ton capacity. (*Davy McKee Iron & Steel Division, Stockton-on-Tees, England*)

carbon monoxide evolved by the refining reaction. While the configuration differs, the operating principle is essentially similar to the Kaldo process.

A drawback of the basic oxygen process is its limitation of about 30% scrap in its charge. This amount of scrap in the steel mix is often barely adequate to utilize the scrap produced at the manufacturing plant per se. As a result, other sources of scrap are recycled in electric furnaces. These other sources include scrap from industrial operations (i.e., arising from the manufacture of finished products such as automobile body parts), and scrap reclaimed from discarded or obsolete steel-containing equipment such as automobiles and rail cars. Thus, basic oxygen steel production, primarily from pig iron, and electric steel production from scrap complement each other in utilizing different raw material resources.

Electric Furnace Processes. Electric furnaces have been used for several decades to produce special steels for which the open-hearth process was not suitable.

Direct-arc electric furnaces are widely used. Essentially, the furnace is a tilting cylindrical bowl-shaped hearth with three graphite electrodes inserted vertically through the roof. The electrodes are supplied with three-phase current via a transformer. Heat is supplied by the arc struck between the charge and the electrodes. The arc temperature is approximately 3,400°C. The furnace is highly versatile, in that operation may be under oxidizing, reducing, or neutral conditions. The versatility is comparable to that of the open-hearth process. Some electric furnaces operate with liquid iron in the charge, but the majority use steel scrap and prereduced pellets.

With very high power input operation and transformer ratings of up to 100,000 kVA, common grades of steel, requiring single slag practice only, can be produced in up to 300-ton heats in less than 3 hours. Special steels also are made in induction furnaces where a current of high or medium frequency is passed through a coil surrounding a refractory crucible containing the charge.

Spray Steelmaking Process. In this process, liquid iron is poured through a tundish and refined continuously by injecting powdered lime and oxygen tangentially from a ring onto the surface of the metal stream.

FOS Process. In the fuel-oxygen-scrap process, a vessel similar in shape to an electric-arc furnace is used, but having a greater height-to-diameter ratio. There is a removable roof to permit rapid charging of scrap. Heat is supplied by an oxyfuel burner inserted through a central opening in the roof.

Additional processes include the Wocra and the Irsid processes which are based upon continuous melting and/or refining techniques. The cyclosteel and jet-smelting processes make liquid iron by flash smelting of iron ore.

TABLE 2. AVERAGE ENERGY REQUIREMENTS FOR PROCESSING STEPS IN THE PRODUCTION OF RAW STEEL

Operation	Energy requirement* (MBtu Ton)**
Ore beneficiation	1.7
Ore transport	0.5
Blast furnace	14.6
Steel production	
open hearth	4.1
basic oxygen process	1.3
electric process	5.3
Scrap processing and transport	0.6

* Use of electricity counted at 10,600 Btu/kWh.
** 1 Btu = 0.252 kilogram-calorie or 1055 joules.
Source: "Energy Expenditures Associated with the Production and Recycle of Metals," Bravard, Flora, and Portal, Oak Ridge National Laboratory (ORNL-NSF-EP-24).

TABLE 3. ENERGY REQUIRED PER TON OF RAW STEEL PRODUCED*

Steel process	Energy use, MBtu per Ton of raw steel**
Open hearth	14.9
Basic oxygen process	15.1
Electric process	6.3

* By taking into account the actual amounts of pig iron and raw steel used in each process, and the energy associated with those inputs, the total energy required to produce steel by the various methods can be computed.
** 1 Btu = 0.252 kilogram-calorie or 1055 joules.
Source: Electric Power Research Institute, Palo Alto, California, August 1986.

Comparison of Process Energy Requirements. Energy requirements for the various steps in raw steel production are shown in Table 2. When the energy requirement for each step is weighted according to the different proportions of raw materials used in the three basic processes just described, the total amount of energy per ton of raw steel is determined to be about 15 MBtu per ton for the open hearth and basic oxygen processes, compared to about 6.3 MBtu per ton for electric melting. See Table 3. Note that the value given for electric steel includes the energy required to generate the electricity. Part of the reason for the large difference in energy requirements is that electric steel is made from scrap and, therefore, little energy is required to reduce iron oxide to elemental iron. In this sense, stockpiles of iron and steel scrap represent a significant source of stored energy.

It is, of course, important to observe that currently about 30% of the output of steel products made is not recoverable as scrap. Examples include reinforcing bars incorporated within concrete structures, wire products, such as nails and fencing, and buried piping, such as oil well casings. Other products, such as "tin" cans, may someday be recovered on a large scale from municipal wastes. Presently, much of this steel is wasted. It is this unrecoverable quantity of material that in any long-term equilibrium sense ultimately must be derived from the mining and reduction of iron ore. Consequently, the reduction of ore to iron, as in a blast furnace, and the need for the basic oxygen process or equivalent is self-evident.

Minimill Concept

The switchover from open-hearth to basic oxygen processes for steel production created an opportunity for new producers of electric steel to enter the competition. Small "minimills" using electric furnaces are not tied to the logistical problems of coal, ore, and limestone supply or the economies of scale associated with blast furnace and coking operations. Small minimills can be built with a relatively modest investment. These mills are well suited to take advantage of the greater availability of local scrap and, because of their relatively small size, can be located virtually anywhere, thus avoiding the costs associated with transportation.

Casting Steel

Although steelmaking processes vary considerably, the liquid steel resulting is tapped from the furnace into a ladle. This is a refractory-lined cylindrical container with trunnion attachment for crane lifting to transport the steel. Generally, the bottom of the ladle is fitted with a stopper-rod nozzle or a sliding-gate nozzle for pouring. Lip-poured ladles are occasionally used. Additions of deoxidants, recarburizers, and alloying materials may be made to the ladle during tapping from the furnace so that final composition of the steel may be adjusted. Sometimes, vacuum or gas-stirring treatment is used prior to casting steel.

Treatment of liquid steel under vacuum makes it possible to reduce the amounts of hydrogen, nitrogen, and oxygen and some harmful non-metallic inclusions, thus improving the properties and qualities of the steel. Lengthy heat treatments of up to 2 and 3 weeks can be eliminated by the effective removal of hydrogen from certain forging steels. Vacuum degassing of steel prior to casting includes: (1) ladle degassing in a chamber; (2) stream degassing by pouring from ladle to ladle or ladle to ingot; (3) vacuum-lifter or circulation degassing; (4) mold degassing; (5) a combination of arc heating and degassing; (6) vacuum-furnace degassing; and (7) employing a consumable electrode under vacuum. To equalize temperature and improve steel quality, sometimes inert gases, such as argon or nitrogen, are bubbled through liquid steel.

Once tapped from the furnace, vacuum degassing, or stirring treatment, the liquid steel is teemed into molds as ingots, continuously cast or pressure-poured into semifinished shapes, or poured into molds as steel casting. In conventional casting-pit practice, steel is teemed from the ladle into iron ingot molds of square, rectangular, polygonal, or round cross section where it solidifies as blocks. The ingots can be top-poured directly into individual molds, or bottom-poured simultaneously into a cluster through a trumpet-and-runner arrangement. To counteract shrinkage during solidification of killed steels, hot tops are used on molds. Once teemed, the molds are stripped and the ingots charged into soaking pits for heating for subsequent processing; or allowed to go cold for placement in the stockyard.

Continuous Casting. This process, wherein liquid steel is poured directly into semifinished shapes, such as slabs, blooms, blanks, or billets,

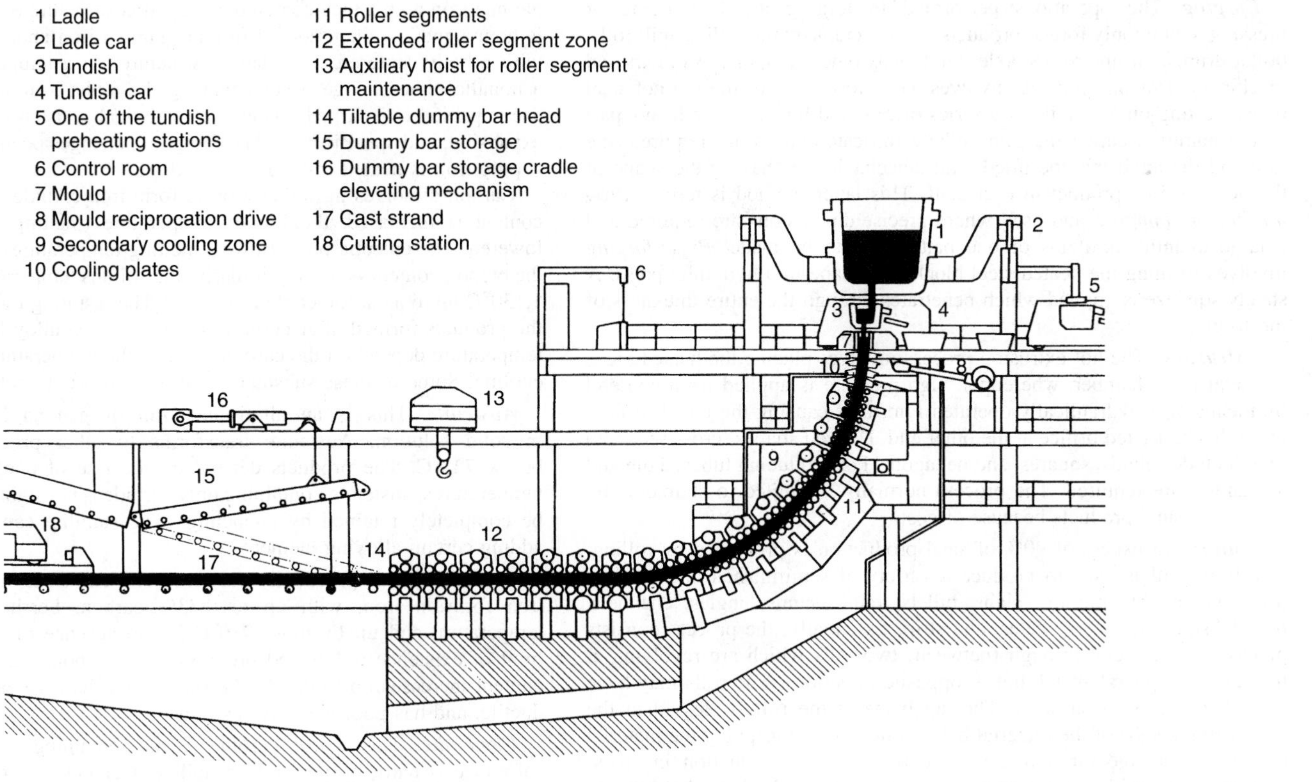

1 Ladle
2 Ladle car
3 Tundish
4 Tundish car
5 One of the tundish preheating stations
6 Control room
7 Mould
8 Mould reciprocation drive
9 Secondary cooling zone
10 Cooling plates
11 Roller segments
12 Extended roller segment zone
13 Auxiliary hoist for roller segment maintenance
14 Tiltable dummy bar head
15 Dummy bar storage
16 Dummy bar storage cradle elevating mechanism
17 Cast strand
18 Cutting station

Fig. 4. Continuous slab casting machine. (*Concast A.G. Zurich*)

is growing in use, mainly because it eliminates the need for heavy rolling-mill equipment.

In continuous casting, 90% or more of the molten steel ends up as finished product. This represents an enormous productivity gain compared with traditional practice. This gain is the result of converting steel finishing from a batch operation (with the ingot removed for reheating at several stages) to a continuous operation (with reheating applied as needed as the ingot moves along).

In principle, all steel could be delivered to finishing operations via continuous casting regardless of the process used to produce the steel. In practice, however, continuous casting operations are most easily and economically introduced in conjunction with new steelmaking facilities where the capacity and design of the steelmaking equipment can be matched with the capacity and layout of the finishing section and with the line of products planned for the operation. This is precisely the situation for the regional minimill specializing in the production of a small variety of high-volume simple shapes. In Japan, where more than 70% of the steelmaking capacity came on line after 1963, more than 80% of steel production was continuously cast by 1982. In that year, the United States by comparison continuously cast only about 25% of its steel production.

A continuous casting line is shown in Fig. 4. Steel is poured via a tundish into a water-cooled copper mold. As casting commences, the bottom of the mold is sealed with a dummy bar onto which the steel solidifies.

The solidified cast product is removed continuously by way of a direct spray-cooling withdrawal roll and cutoff system, maintaining a desired molten metal level in the copper mold. Once cut off, the cast product is discharged onto a cooling bank. Principal types of single or multistrand continuous-casting machines available include: (1) vertical mold with vertical cutoff; (2) vertical mold with blending rolls and horizontal cutoff; (3) curved mold with bending and horizontal cutoff; and (4) machines with direct strand-reduction units to reduce cross section by rolling before cutoff.

Pressure-Pouring Process. In the system shown in Fig. 5, the molten metal is forced up through a refractory tube into a mold by means of compressed air. To cast a number of molds in succession, two systems are used. The ladle may be placed in a stationary airtight pressure chamber and the molds moved over the chamber; or the molds may be stationary and the pressure chamber incorporating the ladle may be transported underneath the molds. The rate of pouring is determined by the rate of air-pressure

A
Ladle crane
Tube crane
Mobile pouring tank
Slab crane
Tube per-heater
Mold
A
Plan view
Ladle crane
Mold coating mach.
Mold
Mobile pouring tank containing ladle.

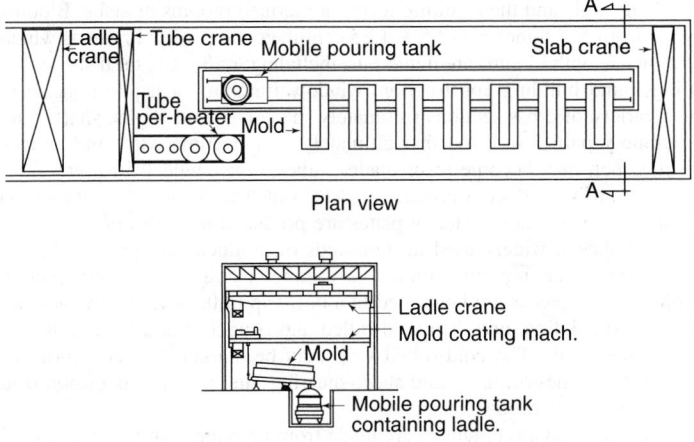

Fig. 5. Pressure pouring plant showing fixed mold and moving pouring tank layout. (*Davy McKee Iron & Steel Division, Stockton-on-Tees, England*)

increase; the height to which the liquid steel is raised is a function of the pressure applied.

Steel Castings. Where intricate shapes are involved or where mechanical working from standard shapes is not possible, the liquid steel can be poured into a mold of desired shape. Sand molds are commonly used.

Shaping Steel

With exception of a small tonnage of steel castings, most steel products are made from steel that is cast first as ingots or semifinished products, followed by mechanical working into desired sizes and shapes. This working processing reforms the cast structure, generally improving the physical properties of the steel.

There are three main ways for forming steel by hot working: (1) forging; (2) extrusion; and (3) rolling. For these processes, the steel must be heated until it is plastic. Soaking pits are used for heating ingots; reheating furnaces are used for heating semifinished products. Generally, the heating temperature is in the range of 1,150 to 1,350°C.

Forging. The operations performed in forging are hammering or pressing. Commonly forged products include crankshafts, rolling-mill rolls, boiler drums, turbine rotors, axles, and many other components of cars and machinery. *Hammer forging* involves the deformation of the red-hot steel block, resting on an anvil, by a series of repeated blows of the heavy part of the hammer, called the ram. Where intricate shapes are required, the ram and the anvil may be fitted with detachable dies having the shape of the desired final product in each half. This latter method is termed *drop forging* or *stamping* and used where precise dimensions are required and a large quantity of items of one pattern are to be made. *Press forging* involves forming the heated steel block into shape in a hydraulic press. A steady squeeze is applied which penetrates through the entire thickness of the forging.

Extrusion. The hot-extrusion process involves placing the heated piece of metal in a chamber, whereupon high pressure is applied from one end by means of a hydraulically operated ram, thus causing the metal to flow through a restricted orifice at the other end. Desired shapes (cross sections) may include rounds, squares, and hexagons. For producing tubes, a die and a mandrel are required. The process normally is limited to stainless and high-alloy steel products because of cost.

Rolling. In excess of 90% of steel production is processed by rolling. A rolling mill is used to produce a variety of semifinished and finished products, notably blooms, slabs, billets, rails, beams, angles, channels, rounds, squares, sheets, plates, and strip. Essentially, the process consists of passing the metal through (between) two rolls which are revolving at the same peripheral speed, but in opposite directions. The rolls may have smooth or grooved surfaces. The tap between the rolls is less than the height (thickness) of the material being rolled. While gripping the material during its passage through them, the rolls effect a reduction in cross-sectional area, with a corresponding increase in length. A final shape can be produced from a large block by making multiple passes between the rolls in a reversing action.

Rolling mills for processing large steel ingots are termed *blooming* or *slabbing* mills and the resulting forms are termed blooms or slabs. Blooms range from 5 × 5 inches (12.5 × 12.5 centimeters) upward. Products which commence with blooms are numerous, including: rails, structural shapes for bridges and building construction, window framing, steel partitions, bars of a variety of cross sections (ultimately made into nuts, bolts, shafts, and machinery parts, etc.), rod that ultimately may become wire, and narrow strip, which may become razor blades, tubes, pipes, wheel rims, etc.

Billet mills produce a product of 2 to 5 inches (5 to 12.5 centimeters) square in cross section. Heavy plates are produced in *plate mills*.

Steel sheets, widely used in thousands of products, are produced from slabs from a slabbing mill, continuous casting machine, or pressure-poured molds. These pieces are hot-rolled in a hot strip mill, descaled by pickling in an acid solution, and then cold-rolled and tempered in a cold-reduction and temper mill. The cold-rolled strip may be marketed in coil form, or carried to a side-trimming and sheet-shearing line, where it is customized for specific user needs.

Tin cans used as containers are made from tin plate, that is, a sheet steel thinly coated on each side by hot-dipping or the electrolytic process.

The hot strip is pickled and passed through cold and temper mills, after which it is passed through annealing and tinning lines. Further, zinc, terne, aluminum, or plastic coating can be applied for additional corrosion resistance.

Summary of Production Operations

The flow sheet of Fig. 6 indicates the principal processes and processing routes encountered in iron- and steelmaking, including major classes of products made. Traditionally, prior to the 1960s, these processes generally were effected in a huge central facility, referred to as an *integrated* iron and steelworks.

Metallurgy of Iron and Steel

The physical properties of ferrous products can be altered by cold working or heat treatment, both processes that affect the microstructure. The role of carbon in ferrous products is explained to some degree by the *iron-carbon diagram*. See Fig. 7. This diagram shows the relationship between carbon content and temperature and includes key information on microstructure and heat treatment.

When cooled, pure iron solidifies at about 1,536°C as delta iron, having a body-centered cubic lattice structure. This form changes allotropically to

gamma iron with a face-centered cubic lattice structure below 1,404°C, and is nonmagnetic. When cooled further, gamma iron changes to alpha iron, with a body-centered cubic lattice structure. At a temperature of 768°C, a nonallotropic change occurs, making the alpha iron strongly magnetic, accompanied by marked changes in electrical resistance, rate of thermal expansion, and specific heat. The foregoing changes occur in reverse order if pure iron is heated instead of cooled.

Carbon dissolves in molten iron to form iron carbide. When the carbon content is increased, the liquids temperature (melting point of iron) is lowered. The eutectic point (lowest melting temperature) is 1,130°C when the carbon content is 4.3%. Similarly, the solidus temperature is lowered to 1,130°C up to a carbon content of 1.7%. The resulting transformations and the products formed after cooling the iron-carbon alloy below the solidus temperature depend on the carbon content, the temperature, and the rate of cooling. Some of these substances are defined briefly below:

Austenite. This is an allotropic form of gamma iron with carbon in solid solution. Austenite transforms to other products on cooling below 723°C. The products depend on the rate of cooling. At ordinary temperatures, austenite containing only carbide is not stable and thus cannot be completely retained by quenching. The stability can be increased by adding certain alloying elements.

Ferrite. This is practically pure iron and can exist in magnetic alpha-iron form in iron, with up to 0.83% carbon. Ferrite exists at room temperature and up to about 910°C in the absence of carbon. Its upper limit of existence is lowered progressively to about 723°C as the carbon content increases up to 0.83%. Ferrite cannot dissolve carbon, is soft and ductile, and has poor abrasive resistance.

Cementite. This is iron carbide, Fe_3C, containing 6.67% carbon. The substance is hard, brittle, and crystalline. Cementite is precipitated when austenite cools.

Pearlite. This is a eutectoid comprised of a laminated structure of ferrite and cementite. Pearlite is formed by transformation of austenite upon cooling. The fineness or coarseness of the laminated structure is determined by the rate of cooling. The lamellar arrangement of ferrite and cementite produces a very tough structure. It is responsible for the mechanical properties of steels.

Graphite. This is the free or uncombined carbon usually found in cast irons. Because graphite occurs as flakes, cast irons are easily machinable even though they have a high resistance to abrasion.

Mystery of Damascus Steels Solved. The legendary steel used in Damascus swords, probably first used as early as 320 B.C., has puzzled historians and metallurgists for many centuries. Probably the first serious attempt to explain the superiority of Damascus steel was made by Anosoff in 1841. A two-volume monograph (*On the Bulat*) was published by Anosoff, who proclaimed, "Our warriors will soon be armed with bulat blades, our laborers will till the soil with bulat plowshares, our artisans will use tools fashioned of bulat, and bulat will supersede all steel employed for the manufacture of articles of special sharpness and endurance." This forecast was not realized. BrÉant and Faraday also had investigated the secret of making Damascus steel. The most recent and very serious study of the topic was made by Sherby and Wadsworth (Stanford University), who reported in 1985 their concept as just how the Damascus steel was processed. In outlining a typical manufacturing procedure, the investigators claim that a Damascus sword commenced with the casting of an ultrahigh-carbon steel, called *wootz*, in Indian foundries. (Damascus steels contained more carbon than most modern steels.) Iron ore and charcoal were mixed and heated to about 1200°C in a shallow stone hearth. The iron was reduced, that is, stripped of oxygen by virtue of reactions with carbon present in the charcoal. At this point, the metal had a spongy consistency. To remove impurities, the sponge iron was hammered, to produce bits of wrought iron with a low carbon content. To increase the carbon content, the investigators envision that pieces of the wrought iron were heated in a clay crucible along with charcoal. To prevent oxidation of the iron, the crucible was sealed. When there was indication of melting within the crucible, the latter was cooled slowly within the furnace. The product was wootz, which incidentally during the period was traded in the form of cakes. A Damascus blade was forged from an individual cake of wootz, which is estimated to have been heated to a temperature of 650–850°C. (Modern ultrahigh-carbon steels are ductile within that temperature range.) The finished blades were hardened, by reheating them and then quenching them in water, brine, or some other liquid. The investigators attribute their postulations to a study

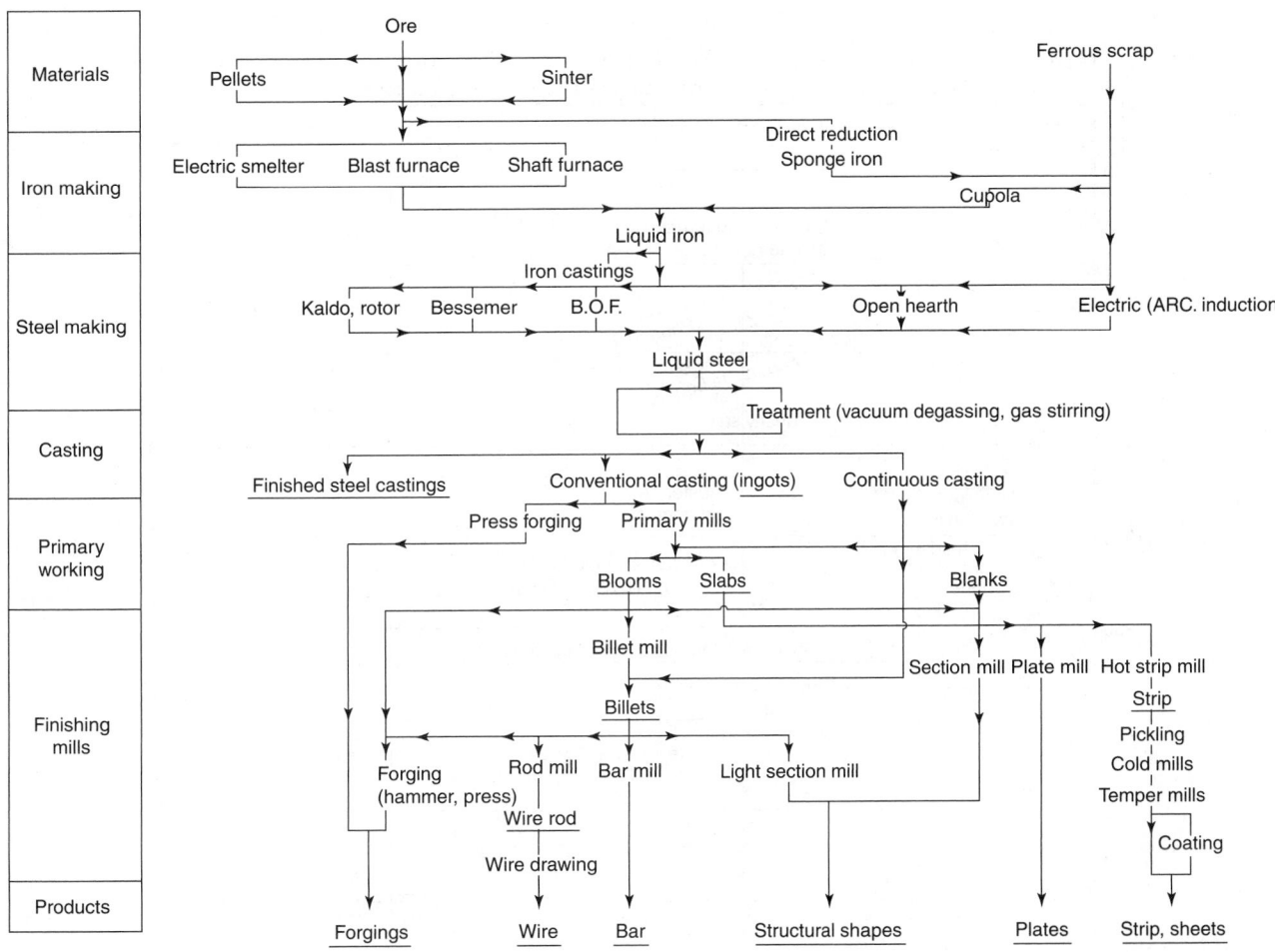

Fig. 6.   Principal processes and processing routes in iron- and steelmaking, indicating major classes of products made

of the iron phase diagram (previously shown in Fig. 7.). Quoting from their report. "When wrought iron and charcoal were heated to 1200°C in a crucible, the iron converted into face-centered *austenite*. Carbon from the charcoal could then dissolve in the iron, decreasing its melting temperature. Molten cast iron formed at the surface of the iron particles when the carbon content of the surface layer exceeded 2%. Slow cooling allowed the carbon to diffuse through the metal, producing a steel with an average carbon content between 1.5 and 2%. Slow cooling also allowed the austenite grains to grow to a coarse size. When the temperature fell below about 1000°C, carbon precipitated out of solution as cementite at the grain boundaries. The coarse *cementite* network was the source of the whitish damask markings. As the temperature fell below 727°C, face-centered austenite converted into alternating layers of cementite and carbon-poor, body-centered *ferrite*. The blades were hardened by being reheated above 727°C and then quenched, which converted austenite into martensite." It is assumed that medieval smiths estimated the metal's temperature from its color.

It is interesting to note that a description of the hardening procedure for Damascus steel (*bulat*) was located in the Balgala Temple in Asia Minor. "The bulat must be heated until it does not shine, just like the sun rising in the desert, after which it must be cooled down to the color of the king's purple then dropped into the body of a muscular slave.... the strength of the slave was transferred to the blade and is the one that gave the metal its strength."

Types of Steel

The carbon content of steels usually does not exceed 1.7%. In addition to carbon, *plain carbon steels* contain small amounts of silicon, manganese, phosphorus, and sulfur—derived from the raw materials and fuel used in the steelmaking process. Within limitations, silicon and manganese are beneficial and often are purposely added, mainly because they are deoxidants. Except in free-cutting steels, where sulfur is purposely added, sulfur and phosphorus are deleterious and their content is kept as low as

possible. Some of these terms are rooted in steelmaking technology prior to the growing adoption of continuous casting processes.

Killed Steel. Deoxidizing elements are used to remove oxygen by forming solid oxides. Thus the reaction to form carbon and oxygen gas is suppressed and the killed steel lies quiet in mold when poured, shrinking upon solidification, usually with formation of a conical cavity known as a pipe. The most commonly used deoxidizers are silicon and aluminum. Killed steels are used for forging, carburizing, heat treating, and other applications because of their superior uniformity and soundness.

Rimming Steel. This is produced by leaving sufficient oxygen to react with carbon to evolve bubbles of carbon monoxide, conditions best achieved in steels with low carbon and manganese contents through the controlled addition of deoxidants. The effervescing action of carbon monoxide evolution causes a pure outer skin of ferrite and a tougher inner core containing carbon, impurities, and inclusions on solidification of the poured block. The sandwich like macrostructure is retained during subsequent shaping operations. Thin sheets intended for deep drawing or deep pressing, as used in the manufacture of auto bodies and domestic appliances, are made from rimming steel because they provide a smooth surface with adequate strength. Rimming steels also are used for forgings requiring smooth surfaces.

When the top end of the ingot is sealed after pouring, the term *capped steel* is applied. Big-end-down bottle-top molds are used and, after a small addition of aluminum, are sealed by using a heavy metal cap. Steels of this type are used for sheet, strip, skelp, tin plate, wire, and bars.

Semikilled or Balanced Steel. This is produced by adjustment of the silicon and aluminum added to low-carbon steel before teeming. The character is intermediate between the killed and rimming steels. The steel is deoxidized less than the killed steel, leaving sufficient oxygen to react with carbon and form blowholes to compensate in all or in part for the shrinkage that accompanies solidification. Semikilled steels are used for less severe drawing and pressing than rimming steels and for structural shapes, plates, and merchant bar.

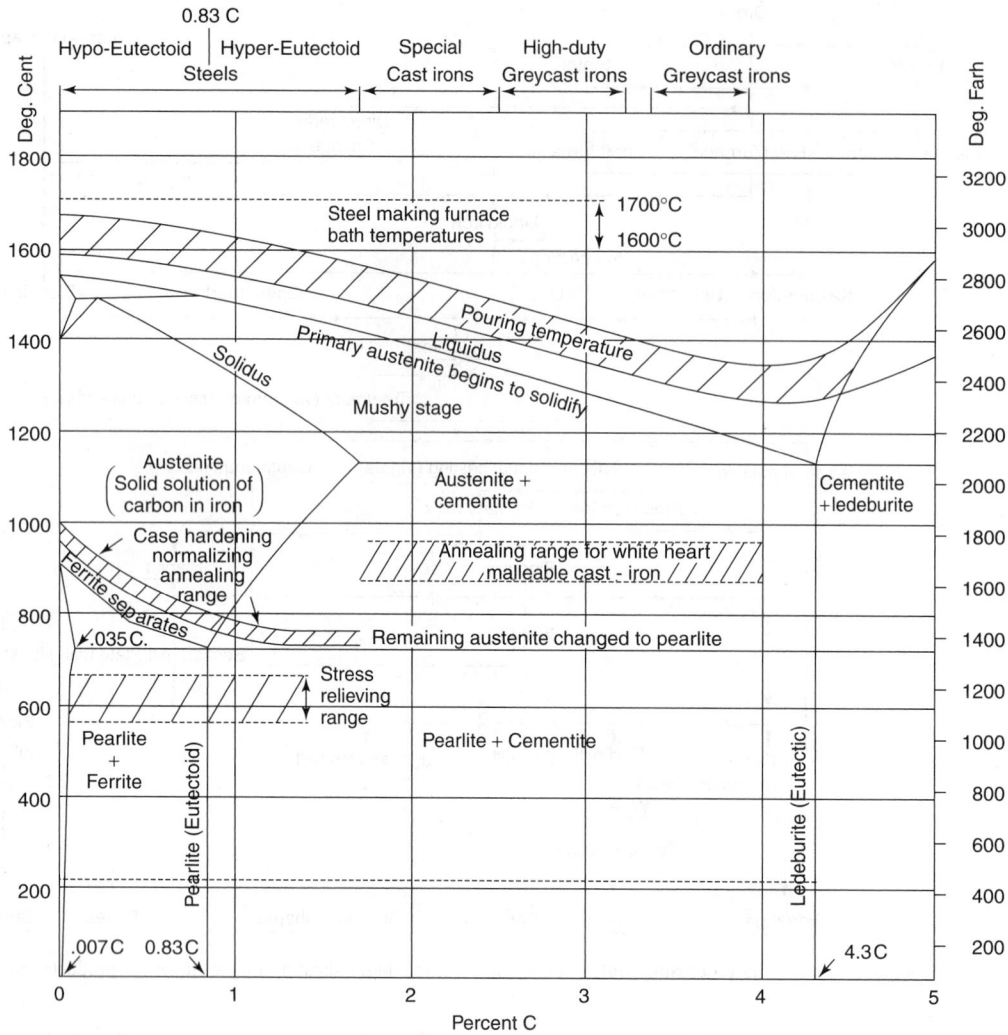

Fig. 7.   Iron-carbon equilibrium diagram

Effect of Ingredients on Steels

The iron-carbon diagram (Fig. 7) shows the effects of carbon in steel. Steels containing less than 0.83% carbon are known as *hypoeutectoid steels*. When cooled slowly, the microstructure of these steels consists of pearlite and ferrite. *Eutectoid steel* containing 0.83% carbon consists entirely of pearlite. Steels containing more than 0.83% carbon are known as *hypereutectoid steels*. When cooled slowly, their microstructure is comprised of pearlite and cementite. Each increase in the carbon content of the steel increases the hardness and tensile strength of the steel in the *as-rolled* or *normalized* condition up to 0.83% carbon. The effect is less pronounced above this figure. The maximum hardness attainable after quenching also increases with the carbon content up to about 0.60% carbon. The strength of quenched and tempered steels depends upon the tempering temperature. Ductility decreases as the carbon content increases, and weldability is impaired above certain levels.

Manganese. This element is generally added to bring the amount to between 0.5 and 1.0%. Normally some manganese is present since it occurs in so many iron ores. Manganese contributes to strength and hardness, but the effect is less than like additions of carbon. Manganese lowers difficulty, but again to an extent less than carbon. Surface quality improves with manganese in all carbon ranges, notably in resulfurized steels. Manganese also increases the rate of carbon penetration during carburizing.

A steel qualifies as an alloy steel when the manganese specified is within the limits of 1.65–2.10%. Manganese is of major importance in increasing hardenability—the depth of hardness penetration after quenching. Thirteen percent manganese steel is widely used as a wear-resistant steel.

Phosphorus. A high phosphorus content in some types of steel is undesirable because it decreases ductility and impact toughness. Because

of large loss of ductility, phosphorus is notably undesirable in the higher-carbon steels. In lower-carbon steels, phosphorus promotes machinability and, with copper, improves resistance to atmospheric corrosion.

Sulfur. This element is detrimental to surface quality, but beneficial to machinability, particularly in low-carbon and low-manganese steels. Sulfur decreases transverse ductility and impact resistance, but has only a small effect on longitudinal properties. As sulfur content increases, weldability decreases. Sulfur is added to the extent of 0.2–0.4% in free-cutting steels to improve machinability.

Silicon. Rimmed and capped steels contain no significant amounts of silicon. When specified within the limits of 0.60–5.00%, silicon qualifies a steel as an alloy steel. The resiliency of steel for spring applications is increased with silicon content. The element also raises the critical temperature for heat treatment. Silicon promotes the susceptibility of steel to decarburization. Because they have a low hysteresis loss and a high electrical resistance, very low carbon steels with 0.6–5.00% silicon are used as transformer steels. Silicon promotes the adherence of zinc coating on hot-dipped galvanized wire. Silicon is less effective than manganese in increasing strength and hardness.

Aluminum. The main use for this element is to deoxidize steels and to obtain a fine grain size. Aluminum also is used to obtain nonaging characteristics and to prevent the recurrence of stretcher strains in sheets and strip. When added in amounts of about 1%, aluminum promotes nitriding properties, that is, surface hardening by means of nitrogen-bearing gases at high temperatures.

Copper. This element is beneficial to atmospheric corrosion resistance if present in amounts in excess of 0.20%. Appreciable amounts of copper are detrimental to hot-working operations. Copper also adversely affects forge welding and is detrimental to surface quality. However, copper does not seriously affect arc or acetylene welding. Copper is not removed in the conventional steelmaking processes and hence, because of increasing accumulation in scrap, it is becoming increasingly difficult to control copper within low limits.

Nickel. Aside from manganese, nickel is the most common alloying element for steel. Nickel is used in amounts up to 5% to increase strength and improve shock resistance. The element counteracts the brittleness that develops in most pearlite steels at subnormal temperatures, lowers the critical temperature of steel, widens the temperature range for successful heat treatment, and promotes corrosion resistance. When nickel is used in quantities greater than 5%, the steels fall into the stainless and heat-resistant steel categories. These are described shortly.

Niobium. By addition of amounts of up to 1%, niobium stabilizes chromium and stainless steels. Additions of only about 0.02% increase the yield point of medium-carbon steels by about 50% without any loss of weldability.

Tungsten. When added to steels in amounts up to 20%, tungsten greatly improves the hardness of a steel, a hardness that is maintained at high temperature and very important to high-speed tool steels. Smaller amounts of tungsten are added to hot-working steels.

Zirconium. Small amounts of this element, when added to high-chromium steels, improve their machinability.

Cobalt. This element provides cutting efficiency to high-speed steels and also is a constituent of heat-resisting steels because it conveys a resistance to creep and scaling.

Chromium. This element increases hardness, improves hardenability, and promotes the formation of carbides and for these reasons is used in constructional steels. Chrome steels are relatively stable at elevated temperatures and have outstanding wear resistance. Chromium is an important constituent of stainless and heat-resistant steels to be described shortly.

Molybdenum. Steels with molybdenum are usually less susceptible to temper brittleness. Molybdenum has a major effect on increasing hardenability and a notable effect on increasing the high-temperature tensile and creep strengths of alloy steels, that is, the steels have less tendency toward deformation under stress at elevated temperatures.

Vanadium. A strong deoxidizing agent, vanadium promotes a fine austenitic grain size. Constructional steels contain about 0.03–0.25% vanadium. Larger quantities are used in tool steels. Vanadium additions of about 0.04–0.05% increase the hardenability of medium-carbon steels with a minimum effect on grain size. Further additions, however, decrease the hardenability with normal quenching temperatures. Where the austenizing temperatures are increased, however, the hardenability can be increased with higher vanadium contents.

Titanium. This element acts as a deoxidizer in pearlitic steels. The yield point of plain-carbon steels is increased with titanium in amounts of 0.02–0.05%. Titanium promotes weldability without the need for normalizing.

Boron. This element is added to increase hardenability, but is effective only when added to fully killed steels. Since only a few thousandths of 1% of boron usually remains in the steel, evaluation of boron steels is by increased hardenability rather than chemical content. The hardenability characteristics of elements already present in the steel are intensified by boron, making possible alloy ingredient conservation. Although effective with low-carbon steels, the effectiveness of the element decreases as the carbon content increases.

Industrial Classification of Steels

Numerous systems are used for classifying steels—some based on composition, others on physical properties, special properties, and so on. For convenience, there are two broad categories: (1) plain carbon steels; and (2) alloy steels. Plain carbon steels account for about 95% of all steel production. As described earlier, plain carbon steels are classed as hypoeutectoid or hypereutectoid steels, depending on whether the carbon content is above or below 0.83% (the eutectoid composition). *Low-carbon* steels have a carbon content below 0.20%. *Medium-carbon* steels have a carbon content in the range between 0.20 and 0.50%. *High-carbon* steels have a carbon content in excess of 0.50%.

Alloy Steels. An alloy steel as defined by The American Iron and Steel Institute is: "By common custom alloy steel is considered to be alloy steel when the maximum of the range given for the content of alloying elements exceeds one or more of the following limits: manganese, 1.65%; silicon, 0.60%; copper, 0.60%; or in which a definite range or a definite minimum quantity of any one of the following elements is specified or required within the limits of recognized field of constructional alloy steels: aluminum, boron, chromium up to 3.99%, cobalt, columbium (niobium), molybdenum, nickel, titanium, tungsten, vanadium, zirconium, or any other alloying element added to obtain a desired alloying effect."

High-strength, low-alloy steels have a twofold objective: (1) higher mechanical properties, and (2) greater resistance to atmospheric corrosion than achievable with structural-grade carbon- or copper-bearing steels. Often, these are proprietary steels with specific trade names. A representative steel in this class will have a tensile strength of about 70,000 psi (483 MPa) for a 1/2-inch (∼1.3-centimeter) thick section and have a yield point of about 50,000 psi (345 MPa).

Constructional alloy steels are a major part of the tonnage of alloy steels and are used mostly in the automotive and aircraft industries. These steels usually are quench-hardened and tempered with or without carburizing.

Stainless steels have a large degree of resistance to chemical attack. This property sometimes is referred to as *passivity*. This property results when iron is alloyed with at least 11% chromium. The corrosion resistance is further enhanced by higher chromium additions and by the addition of nickel. A steel with 12% chromium will stain, but will not exhibit progressive rusting in normal atmospheres. Under normal circumstances, a steel with 18% chromium will not stain, but may discolor, particularly in heavy industrial areas. When 8% nickel is added to an 18% chromium steel, the metal will be stain-resistant in all but the very worst of atmospheres. Even further enhancement of corrosion and heat resistance results with the addition of molybdenum.

Iron-chromium alloys and their general corrosion-resistant properties were known in England and France nearly 150 years ago, but the phenomenon of passivity was not formally recognized until 1910 (Borchers and Monnartz in Germany). This discovery led to rapid development of a series of commercial stainless steels. Stainless steels fall into three broad categories: (1) *martensitic types*—chromium-iron alloys with chromium in the lower range (12 to 17%) and with a wide range of carbon. A main characteristic is an ability to harden by heat treatment in a manner similar to carbon steels. Tensile strengths range from 70,000 to 105,000 psi (483 to 725 MPa) for annealed steels and 125,000 to 200,000 psi (863 to 1380 MPa) for hardened steels. They are particularly well suited for hot working and forging; the lower-carbon types can be cold-worked. (2) *ferritic types*—chromium-iron alloys with higher chromium in a range of 18 to 30% and with a lower carbon content. They have a microstructure that is predominantly ferritic. The steels are not hardenable by heat treatment. They are ferromagnetic. They have a relatively low coefficient of thermal expansion. These steels exhibit good resistance to oxidation and corrosion; they are frequently selected for high-temperature service, notably for applications involving intermittent heating and cooling, because of their ability to retain the oxide scale that has formed. (3) *austenitic types*—iron-chromium-nickel alloys, with a chromium content ranging from 8 to 30% and a nickel content ranging from 6 to 20%. They retain austenite at room temperature. They are characterized by high ductility of the austenite, work-hardening ability, good corrosion resistance, and superior high-temperature properties. Austenitic stainless steels are inherently tough; well adapted for fabrication by deep drawing. They are easily welded and soldered. Their tensile strength (annealed) approximates 90,000 psi (621 MPa) with a yield strength of about 35,000 psi (242 MPa).

Heat-Resisting Steels, as may be required for steam-generating boilers, pressure vessels, furnaces, distillation equipment, and internal-combustion engines must retain their specified physical properties at elevated temperatures. Where temperatures exceed 540°C, molybdenum is used along with chromium as an alloying ingredient. Only 2% chromium provides oxidation protection up to about 620°C. A chromium content of 10 to 14% is required for temperatures up to about 760°C. For higher temperatures, stainless steels are used. For service in the temperature range of about 815° to 1,095°C, steels containing 25% chromium and from 20% to 27% nickel are frequently used.

Electrical Steels. The properties required of a good electrical steel include high electrical resistance, high permeability, and low hysteresis loss. These properties are provided by the addition of 0.6–5.0% silicon to a relatively carbon-free steel. Such steels are used in power transformers, motor and generator rotors and stators, and communications equipment.

Cold Working and Heat Treating Steels

When steel is cold-drawn or cold-rolled, it is said to be *cold-worked*. This process significantly improves mechanical properties, such as increasing the tensile strength, yield strength, torsional strength, hardness, and wear resistance. By suitably combining chemical composition, cross section, method of steel production, and thermal treatment with cold-working, distinctive properties in steels can be achieved. Cold working can impart properties to some steels comparable to those of heat-treated bars. In low-carbon steels, cold-worked steel bars show greatly improved machinability. The ratio of yield strength to tensile strength influences machinability. A high yield-strength ratio, resulting from cold drawing, minimizes plastic flow during machining, thus permitting better utilization of machine tool energy.

Heat Treating Steels. Heat treatment enables the modification of mechanical properties of steels. Three fundamental operations are involved: (1) heating the steel above the critical range, to approach a uniform solid solution of austenite; (2) hardening by quenching in oil, water, or air, to induce the formation of martensite (the hardest micro-constituent of steel); and (3) tempering by reheating to a temperature below the critical range, to secure the desired combination of strength and ductility. Three types of steel generally do not respond to this form of heat treatment: (1) steels which contain very low amounts of carbon; (2) austenitic steels for which the critical ranges are below room temperature; and (3) ferritic stainless steels. Products which normally can be furnished in the quenched and tempered condition include carbon and alloy steel plates; carbon, alloy, and martensitic stainless steel bars; hot-rolled alloy steel sheets; alloy steel tubular products; and carbon steel wire.

Normalizing. This process consists of heating to an appropriate temperature above the critical range, followed by cooling to below that range in still air. This process promotes uniformity of structure. Products which can be normalized include: (1) carbon, alloy, and high-strength low alloy steel hot-rolled bars; (2) carbon, alloy, and high-strength low alloy steel hot-rolled plates; (3) carbon and alloy semifinished steel; (4) carbon alloy, and high-strength low alloy steel hot-rolled sheets; (5) carbon, alloy, and high-strength low alloy steel hot-rolled and cold-rolled strip; (6) carbon and alloy steel tubular products; and (7) carbon and alloy steel wire.

Annealing. For carbon steels, alloy steels, and martensitic and ferritic stainless steels, regular annealing consists in maintaining the steels at a temperature in or near the critical range, followed by cooling at a predetermined rate or cycle. In the case of austenitic stainless steels, these are generally annealed by holding at appropriate temperatures and rapidly cooling to minimize the precipitation of carbides. Annealing provides softness, improves machining, forming, or shearing, reduces stress, improves or restores ductility, and may modify other properties. Usually annealing is used on stainless and heat-resisting steels and may be performed on the same kinds of products as listed under normalizing. See also **Annealing**.

Box Annealing. A process of annealing steel in an appropriate metal container to shield the steel from objectionable oxidation. Sometimes, a reducing atmosphere is used.

Spheroidize Annealing. A process of prolonged heating at a suitable temperature, followed by slow or cyclic cooling to produce a globular condition of the carbide. The structure produced may be attractive for machining or cold-forming, cold-drawing operations, or it may be desirable for subsequent heat treatments.

Stress Relieving. In this process, internal stresses are reduced by heating to a temperature below the critical range and holding it for a sufficient time for equalization of the temperature throughout the piece.

Patenting. This is a continuous heating of individual strands to above the critical range, followed by relatively rapid cooling. The process applies to wire and wire rods and increases toughness for withstanding severe distortion or drawing without breakage.

Isothermal Annealing. This is a process of heating to the correct temperature above the critical for proper austenizing, followed by rapid cooling to a suitable temperature and holding sufficiently long for completion of the transformation.

Additional Reading

Babu, P.B., et al.: "Bar Steel: User Concerns," *Advanced Materials & Processes*, 35 (August 1990).

Baxter, D.F., Jr.: "Users Like Steel's New Look," *Advanced Materials & Processes*, 17 (August 1990).

Colling, D. and T. Vasilos: *Industrial Materials: Metals and Alloys, Vol. 1,* Prentice Hall, Inc., Upper Saddle River, NJ, 1994.

Davis, J.R.: *Metals Handbook,* 2nd Edition, ASM International, Materials Park, OH, 1998.

Decker, R.F.: "Maraging Steels," *Advanced Materials & Processes*, 45 (June 1988).

Dulski, T.R.: "Steel and Related Materials (Analysis of)," *Analytical Chemistry*, 65R (June 15, 1991).

Feinman, J. and D.R. Rae: *Direct Reduced Iron: Technology and Economics of Production & Use,* Iron & Steel Society, Warrendale, PA, 1999.

Fischer, J.J. and J.H. Weber: "Mechanical Alloying," *Advanced Materials & Processes*, 43 (October 1990).

Fromont, R.I.: "NODS Alloy Makes Better Heat-Exchanger Tube," *Advanced Materials & Processes*, 68 (October 1990).

German, R.M.: *Powder Metallurgy of Iron and Steel,* John Wiley & Sons, Inc., New York, NY, 1998.

Gordon, R.B.: *American Iron, 1607–1900,* Johns Hopkins University Press, Baltimore, MD, 1996.

Gupta, V.K.: "New Treatments Toughen Maraging Steels," *Advanced Materials & Processes*, 90 (September 1990).

Ho, C.: *Properties of Selected Ferrous Alloying Elements, Vol. 3,* Taylor and Francis, Inc., Philadelphia, PA, 1989.

Linstroth, R.L.: "Check for Atmospheric Corrosion When Using Stainless Steels," *Chem. Eng. Progress*, 49 (July 1991).

Molloy, W.J.: "Investment-Cast Superalloys a Good Investment," *Advanced Materials & Processes*, 23 (October 1990).

Polmear, I.: *Light Alloys: Metallurgy of the Light Metals,* Hodder Headline PLC, London, UK, 1995.

Pope, G.T.: "One Step Steel," *Sci. Amer.*, 79 (March 1990).

Staff: "Potpourri of New Steel Products," *Advanced Materials & Processes*, 19 (August 1990).

Staff: "Quick-Quenching Steels," *Pop. Mechanics*, 18 (October 1990).

Staff: "Steel Forecasts," *Advanced Materials & Processes*, 31 (January 1991).

Staff: *Annual Book of ASTM Standards 2000: Section 1, Iron and Steel Products: Ferrous Castings: Ferroalloys (Annual Book of ASTM Standards,"* Vol. 01.0, American Society for Testing & Materials, West Conshohocken, PA, 2000.

Wright, P.H.,: "Microalloyed Forging Steels," *Advanced Materials and Processes*, 29 (December 1988).

Web References

American Iron and Steel Institute (AISI): http://www.steel.org/
ASM International: http://www.asm-intl.org/
Iron & Steel Society: http://www.idis.com/aime/iss.htm
Steel Links: http://www.steel.org/hotlinks/

IRON MICA.   See **Biotite**.

IRON OXIDE.   See **Hematite**; **Limonite**; **Magnetite**.

ISOCYANTES, ORGANIC.   Isocyanates are derivates of isocyanic acid, $HN=C=O$, in which alkyl or aryl groups, as well as a host of other substrates, are directly linked to the NCO moiety via the nitrogen atom. Structurally, isocyanates (imides of carbonic acid) are isomeric to cyanates, $ROC\equiv N$ (nitriles of carbonic acid), and nitrile oxides, $RC\equiv N \rightarrow O$ (derivates of carboxylic acid).

Isocyanates are liquids or solids which are highly reactive. The basis for the high reactivity of the isocyanates is the low electron density of the central carbon.

Industrially, isocyantes have become large-volume raw materials for addition polymers, such as polyurethanes, polyureas, and polyisocyanurates. By varying the reactants (isocyanates, polyols, polyamines, and others) for polymer formation, a myriad of products have been developed, ranging from flexible and rigid insulation foams to the high modulus automotive exterior parts to high quality coatings and abrasion-resistant elastomers unmatched by any other polymeric material. The most significant mono-, di-, and oligomeric isocyanates, which constitute over 90% of global isocyanate production, are aromatic isocyanates [toluene 2,4-diisocyanate (TDI), toluene 2,6-diisocyanate (TDI), 4,4′-methylene diphenyl diisocyanate (MDI), 2,4′-methylene diphenyl diisocyanate, polymeric methylene diphenyl diisocyanate (PMDI), *p*-phenylene diisocyanate (PDI), naphthalene-1,5-diisocyanate (NDI)], aliphatic isocyanates [1,6-hexamethylene diisocyanate (HDI), isophorone diisocyanate (IPDI), 4,4′-dicyclohexylmethane diisocyanate ($H_{12}MDI$), 1,4-cyclohexane diisocyanate (CHDI), bis(isocyanatomethyl)cyclohexane (H_6XDI,DDI), tetramethylxylylene diisocyanate (TMXDI)] and monoisocyanates methyl isocyanate (MIC), *n*-butyl isocyanate (BIC), phenyl isocyanate (PIC), 3-chlorophenyl isocyanate, 3,4-dichlorophenyl isocyanate, *p*-toluenesulfonyl isocyanate.

Synthetic Methods

Preparation from Amines. The most common method of preparing isocyanates, even on a commercial scale, involves the reaction of phosgene and aromatic or aliphatic amine precursors.

Preparation from Nitrene Intermediates. A convenient, small-scale method for the conversion of carboxylic acid derivatives into isocyanates involves electron sextet rearrangements, such as the ones described by Hofmann and Curtius.

Nonphosgene Preparation. The term nonphosgene route is primarily used in conjunction with the conversion of amines (or the corresponding nitro precursor) to isocyanates via the use of carboxylation agents.

Chemical Properties

Addition Reactions. Isocyanates undergo addition reactions with a wide variety of substrates. Preferred addition occurs across the C=N bond of the NCO moiety.

Insertion Reactions. Isocyanates also may undergo insertion reactions with C—H bonds.

Cycloaddition Reactions. Isocyanates undergo cycloadditions across the carbon–nitrogen double bond with a variety of unsaturated substrates. Addition across the C=O bond is less common.

Oligomerization and Polymerization Reactions. One special feature of isocyanates is their propensity to dimerize and trimerize. Aromatic isocyanates, especially, are known to undergo these reactions in the absence of a catalyst. The dimerization product bears a strong dependency on both the reactivity and structure of the starting isocyanate.

Commercial Manufacturing Processes

Aromatic Isocyanates. A variety of methods are described in the literature for the synthesis of aromatic isocyanates. Only the phosgenation of amines or amine salts is used on a commercial scale. Much process refinement has occurred to minimize the formation of disubstituted ureas arising by the reaction of the generated isocyanate with the amine starting material.

Aliphatic Isocyanates. Conventional aliphatic isocyanates have historically been manufactured using the hydrogen chloride salt slurry approach. Exceptions to this are the longer chain aliphatics which, due to the increased solubility, have reaction rates conductive to the free amine process. An alternative approach, generally referred to as a two-phase phosgenation, has gained wide scale acceptance for the production of aliphatic isocyanates.

Low boiling isocyanates, such as methyl isocyanate, are difficult to prepare via conventional phosgenation due to the fact that the *N*-alkyl carbamoyl chlorides are volatile below their decomposition point. A convenient method for the synthesis of these low boiling materials consists of the reaction of *N*, *N'*-dimethylurea with toluene diisocyanate to yield an aliphatic–aromatic urea which is pyrolyzed to yield the desired isocyanate. Alternatively, an appropriate aliphatic–aromatic urea can be prepared by the reaction of diphenylcarbamoyl chloride with methylamine.

Specialty Isocyanates. Acyl isocyanates, extensively used in synthetic applications, cannot be directly synthesized from amides and phosgene. Reactions of acid halides with cyanates have been suggested. However, the dominant commercial process utilizes the reaction of carboxamides with oxalyl chloride. Cyclic intermediates have been observed in these reactions which generally give a high yield of the desired products.

Of the many other methods leading to isocyanates, only a few are practical enough in regard to availability of starting materials to be of general applicability. One of the more promising approaches utilizes olefinic substrates which add isocyanic acid in Markovnikov fashion to form alkyl isocyanates. One approach uses the slow addition of the olefin to an excess of solvent and isocyanic acid in the presence of a catalytic amount of inorganic acid. Another approach involves the formation of the dichloro intermediate. The dichloro compound reacts at low temperatures with an excess of isocyanic acid in the presence of a Lewis acid. Pyrolysis approaches can also be used to prepare substituted isocyanates which cannot be prepared using other methods.

Carbodiimide Formation. Carbodiimide formation has commercial significance in the manufacture of liquid MDI. Heating of MDI in the presence of catalytic amounts of phosphine oxides or alkyl phosphates leads to partial conversion of isocyanate into carbodiimide.

Health and Safety Factors

Isocyanates are classified as dangerous substances (EEC Guidelines). They are generally labeled toxic and should be handled with care. Exposure hazards increase substantially when handling vapors or mists. Isocyanate vapors or mists may be irritating to the nose, throat, and lungs. Sensitization may result from excessive exposure.

Repeated or prolonged skin contact may cause irritation, blistering, dermatitis, or skin sensitization. Contact with the eye has been reported to cause irritations in testing with rabbits. For these reasons, isocyanates must be handled in well-ventilated areas. Respirators should be worn whenever the possibility of vapor exposure exists. Chemical goggles should be worn when handling isocyanates. In the event of direct skin contact, use a safety shower immediately, removing all clothing while washing. In all cases, call a physician immediately.

The most overlooked hazard and contaminant is water. Water reacts with isocyanates at room temperature to yield both ureas and large quantities of carbon dioxide. The presence of water or moisture can produce a sufficient amount of CO_2 to overpressurize and rupture containers. For these reasons, the use of dry nitrogen atmospheres is recommended during handling.

Also, the presence of strong bases, even in trace amounts, can promote the formation of isocyanurates or carbodiimides.

Temperature control is important in the handling and storage of isocyanates. Storage at inappropriate temperatures can cause product discoloration, viscosity increases, and dimerization.

Most commercial isocyanates have a high flash point and are classified as Class IIIB combustible liquids. These materials, however, burn in the presence of an existing fire or heat source in the presence of oxygen. In the event of an isocyanate fire, use a carbon dioxide or dry chemical extinguisher. For fires covering large areas, use of a protein foam or water spray is recommended. Personnel engaged in fighting isocyanate fires must be protected against nitrogen dioxide vapors and isocyanate fumes. Firefighters should wear approved positive pressure, self-contained breathing apparatus, and fire-resistant clothing.

Economic Aspects and Applications

Since 1971, the overall demand for isocyanates has increased at a compounded rate of 12%. Although this level will not likely be sustained in the future due to the maturation of key application markets, it is probable that additional growth will occur through the year 2000. This trend will likely include a shift in emphasis from TDI to MDI and polymeric MDI-based materials. New growth opportunities in the construction industry, structural applications, and growth in the automotive industry exist. Third-world markets are also anticipated to provide growth opportunities.

Globally, BASF, Bayer (Miles in North America), Dow, and ICI historically have been the leading producers of aromatic isocyanates. In North America, Olin is a principal supplier of TDI and aliphatic isocyanates. Rhône-Poulenc and Hoechst are principal suppliers in Europe.

Aromatic Isocyanates. In North America, aromatic isocyanates are heavily used as monomers for addition and condensation polymers. The principal applications include both flexible and rigid polyurethane foam and noncellular applications, such as coatings, adhesives, elastomers, and fibers.

Aliphatic Isocyanates. Aliphatic diisocyantes have traditionally commanded a premium price because the aliphatic amine precursors are more expensive than aromatic diamines. They are most commonly used in applications which support the added cost or where the long-term performance of aromatic isocyanates is unacceptable. Monofunctional aliphatic isocyanates, such as methyl and *n*-butyl isocyanate, are used as intermediates in the production of carbamate-based and urea-based insecticides and fungicides.

A number of markets have been established for light-stable, aliphatic diisocyanates in the United States. The largest market is in high performance coatings. The largest coating market is in automotive refinishes. Other coating include uv-cured coating for vinyl tile and sheet flooring, electronic circuit boards, powder coatings, and paints. Hydrogenated MDI (H_{12}MDI), *m*-xylylene diisocyanate (XDI), and isophorone diisocyanate are currently used in many of these coating applications.

Aliphatic isocyanates have a small but growing market application in thermoplastic polyurethanes (TPU). Medical applications include wound dressings, catheters, implant devices, and blood bags. A security glass system using light stable TPU as an inner layer is under evaluation for shatterproof automotive windshield applications.

Developments in aliphatic isocyanates include the synthesis of polymeric aliphatic isocyanates and masked or blocked diisocyanates for applications in which volatility or reactivity are of concern.

Specialty Isocyanates. Specialty isocyanates are organic isocyanates having the isocyanate function attached to a carbonyl group or to elements other than carbon. *p*-Toluenesulfonyl isocyanate is used as a drying agent for organic solvents. Arenesulfonyl diisocyanates, such as *m*-phenylenedisulfonyl diisocyanate, are used as monomers for base-soluble polymers. Arenesulfonyl monoisocyanates are used as intermediates for pharmaceuticals and herbicides.

REINHARD H. RICHTER
RALPH D. PRIESTER, JR.
Dow Chemical

Additional Reading

Sayigh, A.A.A. H. Ulrich, and W.J. Farrissey: in J.K. Stilles and T.W. Campbell, eds., *Condensation Monomers,* John Wiley & Sons, Inc., New York, NY, 1972, pp. 369–476.

Richter, R. and H. Ulrich: in S. Patai, ed., *The Chemistry of Cyanates and their Thio Derivatives,* John Wiley & Sons, Inc., New York, NY, 1977, p. 619.

Rassmussen, J.K. and A. Hassner: *Chem. Rev.* **76**, 389 (1976).

Ulrich, H. in W.F. Gum, W. Riese, and H. Ulrich, eds.: *Reaction Polymers,* Hanser, New York, NY, 1992, p. 358.

ISODESMIC STRUCTURE. An ionic crystal structure in which there are no distinct groups formed within the structure, i.e., where no bond is stronger than all the others.

ISODIAPHERE. One of two or more nuclides having the same difference between the number of neutrons and protons in their nuclei. In alpha-particle decay, for example, the parent and daughter nuclides are isodiapheres.

ISODIMORPHISM. The condition, double isomorphism, in which both crystalline forms of a dimorphous substance which is isomorphous with a second dimorphous compound are isomorphous with both forms of the second compound. Example: arsenious oxide and antimonious oxide, which crystallize in rhombs and also in regular octahedra.

ISOELECTRIC POINT. See **Amino Acids**.

ISOLEUCINE. See **Amino Acids**.

ISOMER (Nuclear). One of two or more nuclides that are both isotopes (same atomic number) and isobars (same mass number) of each other, but which have some measurably different physical property, such as half life. Of any two isomeric states, one must be an excited metastable state of the other. Ultimately the nuclide in the excited state decays with a measurable lifetime to a lower energy state, usually its ground state. At present about 200 nuclear isomeric states with half-lives longer than 10^{-6} seconds are known. Metastable isomeric states are denoted by adding the letter *m* to the mass number where it appears in the nuclidic symbolism, as for example ^{80m}Br. In this particular case, ^{80m}Br and ^{80}Br are nuclear isomers. Ordinary excited states with lifetimes too short to be measured are generally *not* considered to be isomeric states, but this is only a matter of convention. On rare occasions, such as for ^{124}Sb, more than two isomeric states may exist for a single atomic number and a single mass number (isotopic isobar).

ISOMERISM. If two chemical compounds incorporate the same elements in exactly the same numbers, the compounds are referred to as *isomers* or *isomerides*. An excellent example of relatively simple isomers is the case of normal butane, $CH_3CH_2CH_2CH_3$, which is an open, straight chain of four carbon atoms, and of isobutane, $(CH_3)_2CHCH_3$, wherein one of the carbons lies in a short branch from the main chain of three carbon atoms. Obviously, as the number of carbon atoms in a compound increases, the possibility of branches and subbranches increases. Normally,

then, compounds with high carbon counts, at least theoretically, are capable of numerous isomers.

In *geometric isomerism*, the isomeric relationship can be explained in terms of two dimensions—as shown by the relationship of the two isomers, maleic acid and fumaric acid:

(Maleic acid)   (fumaric acid)

Where the identical atoms or groups are in juxtaposition, as in maleic acid, the compound is designated as the *cis* form. (*Cis* = "on this side" in Latin.) Where the identical atoms or groups are on the opposite sides, as in fumaric acid, the compound is designated as the *trans* form.

In *stereoisomerism*, three dimensions must be considered. In stereoisomerism (also termed optical isomerism), there is no plane of symmetry in the molecule, so that the two forms are mirror-images, and thus cannot be turned into a position of coincidence. Thus, compounds containing a carbon atom (or other tetravalent atom) to which four different atoms or radicals are bonded are optical isomers. They receive this name from the fact that one isomer rotates the plane of polarized light to the right (*dextro form*); the other rotates it to the left (*levo form*). Lactic acid is an example. See also **Lactic Acid**, and formulas below:

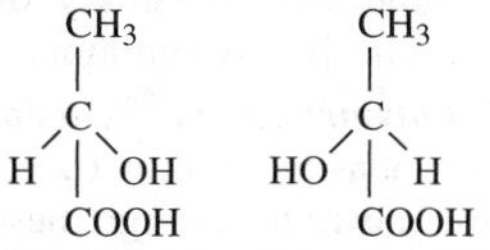

(*d*-lactic acid)   (*l*-lactic acid)

The carbon atom, to which are attached the four different groups to produce stereoisomerism, is known as *asymmetric*, and when written (not shown structurally), that carbon may be printed more prominently than the nonasymmetric carbon atoms.

The projection formulas of the four forms of tartaric acids are shown below. Note that the arrows indicate the direction of rotation of light by the asymmetric carbon atoms.

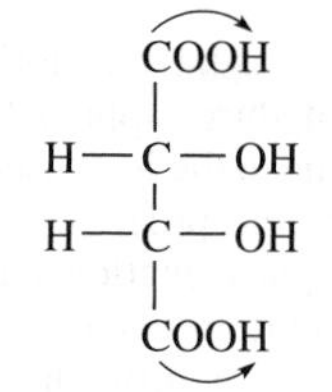

inactive or *meso*-tartaric acid (internally compensated; possesses a plane of symmetry; optically inactive).

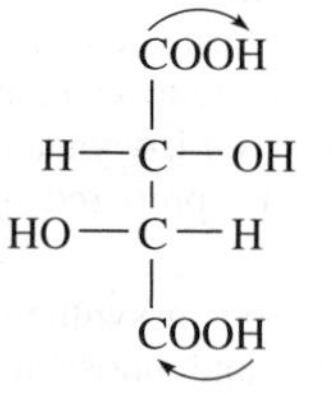

dextro-tartaric acid (arrangement of groups around each asymmetric carbon atom is cumulative; optically active; dextrorotatory).

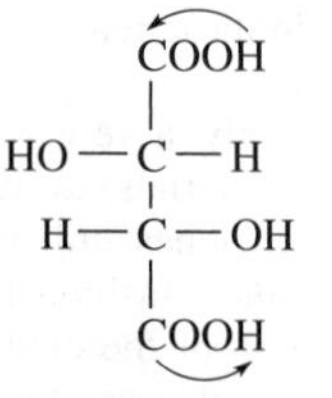

levo-tartaric acid (arrangement of groups around each asymmetric carbon atom is cumulative; optically active; levorotatory).

$\begin{cases} d\text{-tartaric acid} \\ l\text{-tartaric acid} \end{cases}$ *Dextrolevo*-tartaric acid, racemic tartaric acid (externally compensated; optically inactive; can beresolved into *d* and *l* components).

The two optically active tartaric acids when crystallized differ in the arrangement of the faces—one is the mirror image of the other. Pasteur

(1848), observing this difference, was able to separate the two optically active forms of ammonium sodium tartrate crystals made from racemic tartaric acid.

Tautomerism is a form of isomerism in which a substance exists in two forms which are in equilibrium and exhibit characteristic reactions; either one may predominate, depending upon the conditions. Thus acetoacetic ester may react as a ketone or an enol (a compound containing a carbon atom having both an alcoholic hydroxyl group and a double bond) depending upon the conditions:

$$H_3C - \overset{\displaystyle O}{\underset{\displaystyle O}{\overset{\|}{C}}} - H_2C - \overset{\displaystyle O}{\overset{\|}{C}} - O - H_2C - CH_3$$

Ketone form

$$H_3C - \underset{\displaystyle O-H}{\overset{\|}{C}} - CH - \overset{\displaystyle O}{\overset{\|}{C}} - O - H_2C - CH_3$$

Enol form

Acetoacetic ester

Chirality

In chemistry, *chiral* is a term used to describe asymmetric molecules that are mirror-images of each other, i.e., they are related to each other optically as right and left hands. Such molecules are also called enantiomers and are characterized by optical activity. An excellent summary of chirality in chemistry is given by Prelog, *Science*, **193**, 17–24 (1976).

ISOMERIZATION.

The rearrangement of the structural configuration of a molecule without changing its molecular weight. Although structural changes of this type occur in other processes, e.g., catalytic reforming and cracking, isomerization can be the principal reaction desired in some processes. In petroleum refining, isomerization processes are used to change the structural configuration of C_4 paraffins (alkanes), such as normal butane, into isobutane in order to supplement other sources to provide enough butane for alkylation with olefins (alkenes) in the production of motor fuel. C_5 and C_6 paraffins are isomerized to the more highly branched structures to improve their antiknock ratings. Isomerization is also applied to a lesser extent in C_8 aromatic hydrocarbons.

One isomerization process (UOP) is shown in the accompanying diagram. This unit is arranged to process a C_5/C_6 mixture with fractionating facilities to provide for the recycling of both *n*-pentane and *n*-hexane. A desulfurized C_5/C_6 blend first is fractionated to remove the native isopentane as a net product. The de-isopentanizer bottoms are desiccant-dried before being joined by *n*-hexane recycle abd brought to reaction temperature by heat exchange and suitable preheating. Before entering the reactor, the combined feed stream is joined by hydrogen recycle gas, which functions to suppress catalyst-deposit formation.

The fixed-bed reactor effluent is cooled and passed to a high-pressure separator. Gas from the separator, along with a small quantity of dried make-up hydrogen, is recycled to the reactor. The separator liquid is stabilized as a next step to remove any C_4 and lighter hydrocarbons that may be introduced with the make-up hydrogen, plus a very minor amount of light hydrocarbons formed by hydrocracking in the reactor. Hydrogen dissolved in the separator liquid is also removed by the stabilizer.

The next fractionator in series receives the stabilized liquid, from which it separates an equilibrium isopentane-*n*-pentane mixture that is routed back to the deisopentanizer for separating the isopentane as a net product. Thus, the *n*-pentane content of the feed is converted entirely to isopentane in the arrangement shown in Fig. 1.

As a final step in the fractionation sequence, the hexane fraction is separated into a dimethylbutanes concentrate as a net overhead product and an *n*-hexane-rich bottoms stream to be recycled for the further isomerization of the *n*-hexane and methylpentanes. With economically practical fractionation, the methylpentanes split between the overhead and bottoms of the deisohexanizer column. For the C_5 fraction, the boiling points of the two isomers are far enough apart to make a relatively clean split economically feasible. For the C_6 fraction, the greater number of

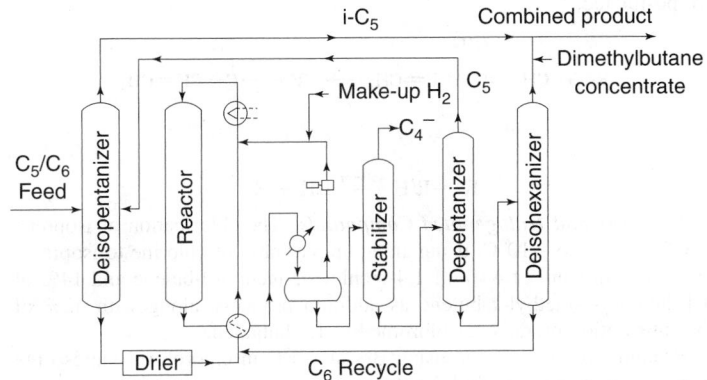

Fig. 1. C_5H_6 Isomerization unit. (*UOP Process Div.*)

isomers and the bunching of some of their boiling points preclude precision separation in columns having a reasonable number of plates.

Once-through processing of a typical C_5/C_6 (68–70 octane number) straight-run fraction results in a product having a Research Method octane number (clear) of about 83. By recycling the unconverted *n*-pentane, *n*-hexane, and most of the methylpentanes, a product having an octane number of about 93 would result. Obviously, any octane number between 83 and 93 could be produced, depending on the amount and quality of the equipment installed to separate the reactor effluent into net product and recycle streams.

ISOPRENE.

In the 1990s isoprene is used almost exclusively as a monomer for polymerization.

The isoprene unit exists extensively in nature. It is found in terpenes, camphors, diterpenes (e.g., abietic acid), vitamins A and K, chlorophyll, and other compounds isolated from animal and plant materials. The correct structural formula for isoprene was first proposed in 1884.

$$CH_2 {=} C{-}CH{=}CH_2$$
$$\underset{CH_3}{|}$$

Properties

Isoprene (2-methyl-1,3-butadiene) is a colorless, volatile liquid that is soluble in most hydrocarbons but is practically insoluble in water. Typical properties of isoprene are listed in Table 1.

Conformation. The exact conformation of the isoprene molecule is still in doubt. It is generally accepted that rotation is restricted around the central C—C single bond. Isoprene may be considered as an equilibrium of two conformations, namely a cisoid (*s-cis*) conformation in which both vinyl groups are located on the same side of the C—C bond, and a transoid (*s-trans*) one with the vinyl groups located on the opposite sides of the bond.

Reactions

Isoprene is highly reactive both as a diene and through its allylic hydrogens, and its reactions are similar to those of butadiene. Apart from polymerization, the most widely investigated isoprene reactions are the formation of six-membered rings by the Diels-Alder reaction.

Free-Radical Reactions. Free radicals attack isoprene, and two competing mechanisms, at the double bond or involving C—H bonds,

TABLE 1. PROPERTIES OF ISOPRENE

Property	Value
mol wt	68.11
density of liquid, gm/cm^3 at 25°C	0.6759
freezing point, °C	−145.95
bp at 101.3 kPa,[a] °C	34.067
n_D^{30}	1.41524
flash point, °C	−48

[a] To convert kPa to atm, divide by 101.3.

are postulated:

$$R\cdot + R'H \xrightarrow{\text{solvent}} RH + R'$$

Halogens and Halogenated Compounds. The chlorination of isoprene in CCl_4 at -5 to $-10°C$, using an equimolar ratio of chlorine to isoprene, gives us a mixture of 44% of 1,4-dichloro-2-methyl-2-butene and 14% of 3,4-dichloro-2-methyl-1-butene as addition products, along with 42% of the substitution product, 2-chloromethyl-1,3-butadiene.

Bromination of isoprene using Br_2 at $-5°C$ in chloroform yields only *trans*-1,4-dibromo-2-methyl-2-butene.

The reaction of dihalocarbenes with isoprene yields exclusively the 1,2- (or 3,4-) addition product, e.g., dichlorocarbene Cl_2C: and isoprene react to give 1,1-dichloro-2-methyl-2-vinylcyclopropane.

Isoprene reacts with α-chloroalkyl ethers in the presence of $ZnCl_2$ in diethyl ether from $0-10°C$.

Hydrocarbons. The reaction of isoprene with toluene, ethylbenzene, or isopropylbenzene is catalyzed by sodium or potassium. The products are chiefly monopentenylated in the side chain, and no information can be obtained on whether the addition is 1,4- or 1,2-, because under these conditions the double bond migrates.

Other Compounds. Primary and secondary amines add 1,4- to isoprene.

Polymerization. Isoprene polymerization can proceed by either 1,4- or 1,2- (vinyl) addition.

Of the many catalysts that polymerize isoprene, four have attained commercial importance. One is a coordination catalyst based on an aluminum alkyl and a vanadium salt which produces *trans*-1,4-polyisoprene. A second is a lithium alkyl which produces 90% *cis*-1,4-polyisoprene. Very high (99%) *cis*-1,4-polyisoprene is produced with coordination catalysts consisting of a combination of titanium tetrachloride, $TiCl_4$, plus a trialkylaluminum, R_3Al, or a combination of $TiCl_4$ with an alane (aluminum hydride derivative).

The polymerization of isoprene by alkali metal and organometallic compounds (other than organolithium) is a heterogeneous reaction both in bulk and hydrocarbon solvents.

Production

The largest capacity has been and remains in the CIS (former USSR) region. Several plants around the world have been shut down, and the trend appears to continue downward. On the other hand the use of isoprene in block copolymers has grown rapidly. This growth has tended to offset some of the decline of *cis*-1,4-polyisoprene.

The principal route for production of isoprene monomer outside of the CIS is recovery from ethylene by-product C_5 streams.

Synthesis. Because of the limited availability of by-product isoprene much effort has been devoted to synthesis of isoprene. Most routes tend to have marginal selectivity and require large amounts of energy. The choice of which route is preferable depends on availability and cost of raw materials and cost of energy. Several synthetic routes have been practiced commercially including propylene dimer, dehydrogenation of tertiary amylenes, isobutylene–formaldehyde, isopentane dehydrogenation and acetone–acetylene.

Health and Safety

Isoprene is not known to present serious toxicological hazards in handling; however, as is the case with many other chemicals, studies concerning the safety of isoprene are ongoing. In humans, a one minute inhalation of 0.16 mg isoprene per liter air is mildly irritating to the mucous membranes of the eyes, nose, and upper respiratory passages. It was proposed that the limit of isoprene concentration on industrial sites be set at 0.04 mg/L air; it was also recommended that the maximum concentration of isoprene in water be set at 0.005 mg/L.

Isoprene is classified by the ICC as a flammable liquid requiring a red label. Because of the potential hazards on its exposure to oxygen, isoprene should be stored in an inert atmosphere (nitrogen) in the presence of at least 50 ppm of *t*-butylcatechol.

Economic Aspects and Applications

Isoprene pricing tends to vary considerably due to a fairly thin commercial market. Because isoprene raw materials are primarily petroleum-based and synthesis or recovery is energy intensive, most pricing is indexed to petroleum and energy. For large-scale applications, monomer production is in tandem with application production. Generally isoprene availability is less than butadiene, and the price is higher. Isoprene is used where the unique properties of the products can command a premium over butadiene. Almost all isoprene produced is used for the preparation of polymers or copolymers. *cis*-Polyisoprene is the largest application, with SIS block polymers being a rapidly growing secondary application. Butyl rubber is a significant third application.

HUGH M. LYBARGER
The Goodyear Tire and Rubber Company

Additional Reading

Arpe, Hans-Jurgen: *Immobilized Biocatalysts to Isoprene*, Vol. 14, 5th Edition, John Wiley & Sons, Inc., New York, NY, 1996.
Adams, R. ed.: *Organic Reactions*, Vol. 4, John Wiley & Sons, Inc., New York, NY, 1948, pp. 60–173.
"Isoprene," Report No. 28, Stanford Research Institute, Menlo Park, CA, 1967.
Saltman, W. M. and E. Schoenberg: in Elliot, J. R. ed., *Macromolecular Syntheses*, Vol. 2, John Wiley & Sons, Inc., New York, NY, 1966, p. 50.

ISOPROPYL ALCOHOL. [CAS: 67-63-0]. Also called dimethyl-carbinol or secondary propyl alcohol, formula $(CH_3)_2$ CHOH, *isopropyl alcohol* is a colorless liquid at room temperature. Pleasant odor, bp 82.4°C, specific gravity 0.7863 (20/20°C), autoignition temperature, 400°C. The compound is soluble in water, ethyl alcohol, and ether.

Two basic methods of production are in commercial use: (1) absorption of propylene in sulfuric acid to form alkyl hydrogen sulfate, followed by the hydrolysis of the ester; and (2) by direct hydration with water, using a catalyst. An inherent disadvantage in the first process is the need to handle sulfuric acid. Further, the first process yields little more than 70% isopropanol as compared with the second process, in which liquid propylene is used as the charge stock. All direct-hydration processes can be represented by: $C_3H_6 + H_2O \rightarrow C_3H_7OH + heat$.

Isopropyl alcohol is a widely used chemical, finding use as a starting material for making acetone and its derivatives; in the manufacture of glycerol and isopropyl acetate; as a solvent for essential and other oils, alkaloids, gums, and resins; as a latent solvent for cellulose derivatives; as a deicing agent for liquid fuels; in pharmaceuticals, perfumes, and lacquers; in extraction processes; as a dehydrating agent; and as a preservative.

See also **Alcohol**; and **Organic Chemistry**.

ISOTHERMAL. A term used to denote the following: 1. Of constant temperature, with respect to either space or time. Isothermal processes are those conducted without temperature change.

2. A line or curve expressing a relationship between variables such as pressure and volume, for all values of which the temperature remains constant.

3. A line joining points at the same temperature.

ISOTONE. One of two or more nuclides having the same number of neutrons in their nuclei.

ISOTOPE. An isotope is one of two or more nuclides that have the same number of protons in their nuclei. Any two isotopes have the same atomic number, Z. However, their mass numbers, A, differ. Isotope is a term that stems from the Greek words, *isos* (same) and *topos* (place), to designate substances having different atomic weights and yet having chemical properties so much alike that in the early days of research it was not possible to perform a chemical separation of the isotopes of a given element.

Sometimes, the term *nuclide* is confused with isotope. A nuclide may be defined as a species of atoms, with a specified atomic number and mass number. Different nuclides having the same atomic number should be described as isotopes. This is evident from the accompanying table. Different nuclides having the same mass number are termed *isobars*.

The existence of isotopes first became evident in the early years of this century, from the investigation of natural radioactivity. Then it was found that the natural radio elements underwent successive nuclear

TABLE 1. RELATIONSHIP OF ATOMIC AND MASS NUMBERS IN DESIGNATING ISOTOPES

Element	Mass number A	Atomic number Z	Atomic number Z	Atomic number Z	Atomic number Z	
Hydrogen	1	1				
Hydrogen	2	1				
Hydrogen	3	1				
Helium	3		2			
Helium	4		2			
Helium	5		2			
Lithium	5			3		
Helium	6		2			
Lithium	6			3		
Lithium	7			3		
Beryllium	7				4	} All nuclides
Lithium	8			3		
Beryllium	8				4	
Beryllium	9				4	
Beryllium	10				4	

Isobars - - - - - - - - - - - - - - - Isotopes - - - - - - - - - - - -

Table indicates that hydrogen has three isotopes, each with same atomic number, but with differing mass numbers, and designated as ^{1}H, ^{2}H, and ^{3}H (thus using the mass number to designate a particular isotope. Similarly, the four lithium isotopes (all with atomic number 4) are designated by their mass number 4) are designated by their mass numbers, ^{5}Li, ^{6}Li, ^{7}Li, and ^{8}Li. Although redundant, because the element symbol implies the atomic number, symbols sometimes are sometimes are written to indicate both mass and atomic numbers, as $^7_3$Li.

disintegrations, that they could be arranged in radioactive series according to these changes, and that in these series there were several instances in which atoms of the same atomic number (or as then stated, atoms occupying the same place in the periodic table) differed widely in their radioactive behavior. For example, it was found that radium C, radium E, thorium C and actinium C were all identical in their chemical properties with bismuth (atomic number 83) but differed in their radioactive properties and origins.

In the long course of research that led to the conclusion that more than one stable isotope of an element may exist, an important milestone was the method of positive-ray analysis. As applied by J.J. Thomson, an electric discharge was passed through a vessel containing a gas at low pressure. The effect of the discharge was to produce ions in the gas, and these ions, because of their electric charge, could be formed into beams, deflected and otherwise directed by applied electric and magnetic fields. An experimental apparatus was designed so that the amount of this deflection would depend upon the masses of the particles of the gas. By using neon gas (atomic weight 20.183) in the tube, Thomson obtained photographs showing two beams of particles, one of them in a position calculated for particles having a mass of about 20, and the other for a mass of about 22. Although the conclusion was not immediately reached, it was later concluded (from the work of Aston) that neon (and other elements) consisted of atoms of more than one mass. Aston expressed this conclusion in the whole number rule, according to which, all atomic weights of individual atomic species are close to whole numbers, and the whole number plus decimal values calculated chemically for the atomic weights of elements are due to the presence of two or more isotopes, each of which has an atomic weight that is approximately a whole number. The fact that the chemically determined values for the elements of naturally occurring materials from different sources are the same is because the isotopic composition of naturally occurring materials (except those of radioactive origin) is essentially the same.

Aston's work was founded upon accurate measurements of the deflections of charged particles. These measurements were made in an instrument he devised, the mass spectrograph. Many later instruments were developed following Aston's work, or following the Dempster instrument, which was built before Aston's. The direction-focusing mass-spectrographs and the later velocity-focusing instruments and composite instruments facilitated the determination, not only of the masses (and hence mass numbers) of the isotopes of an element, but their quantities as well. As a result of the immense amount of research in this field, the isotopic composition of the stable elements has been closely determined. And this data can now be said to be subject to only slight revision, as more refined methods, and the possible discovery of stable isotopes present in very small quantities, are found.

It will be seen from Table 1 that naturally occurring elements differ widely in the number of isotopes they contain. Some (usually of odd atomic number) are composed entirely of atoms of one mass number. Others have many stable isotopes. For example, the element of atomic number 50, which is tin, has at least ten stable isotopes and many radioactive ones.

The great importance of isotopes is due to two facts. (1) Since the atomic number of an atom determines its chemical properties, all the isotopes of a given element exhibit essentially the same chemical behavior; that is, all atoms of the same atomic number undergo essentially the same reactions with atoms of other atomic numbers. Thus, all three of the isotopes of hydrogen (protium, deuterium and tritium) undergo essentially the same reactions with oxygen, carbon, and all the other elements. Therefore, if we add to the ordinary form of an element (which has a known isotope composition) a measured amount of an isotope of that element, we can follow the course of our sample through chemical reactions, especially those in which the same element enters at other points. This instance cited is representative of the many applications of isotopes which will be discussed in this article. (2) The other fact that accounts for the great present-day importance of the knowledge of isotopes is that, while the isotopes of an element exhibit similar chemical behavior, the nuclear characteristics of these isotopes often differ greatly. A most important example of this difference is that between ^{235}U and ^{238}U See also **Uranium**, where the separation of these and other isotopes is described.

The isotopes of all elements are tabulated under **Chemical Elements**. Also, there are descriptions of isotopes under the alphabetical entries for each element. The nature and importance of radioisotopes are described under **Radioactivity**.

Additional Reading

Clark, I. and P. Fritz: *Environmental Isotopes in Hydrogeology*, CRC Press, LLC, Boca Raton, FL, 1997.

Cook, P.: *Enzyme Mechanism from Isotope Effects*, CRC Press, LLC, Boca Raton, FL, 1991.

Fritz, P. and J. Fontes: *Handbook of Environmental Isotope Geochemistry*, Elsevier Science, New York, NY, 1989.

Jackson, M. and N. Lowe: *Advances in Isotope Methods for the Analysis of Trace Elements in Man*, CRC Press, LLC, Boca Raton, FL, 2000.

J

JACOBSEN REARRANGEMENT. Reaction of polymethylbenzenes with concentrated sulfuric acid to give rearranged polymethylbenzenesulfonic acids. Under identical conditions, halogenated polymethylbenzens undergo disproportionation.

JACQUEMART'S REAGENT. Analytical reagent used to test for ethyl alcohol. Consists of an aqueous solution of mercuric nitrate and nitric acid. See also **Ethyl Alcohol**.

JADE. Jade is a general term for a compact green mineral substance much prized for ornamental purposes. Jade is either a compact actinolite called nephrite, a variety of amphibole, or jadeite, a monoclinic pyroxene. It is easily worked, and many prehistoric implements have been found of this material in Mexico, Switzerland, France, Greece, and Egypt. The word jade is derived from the Spanish *pietra di hijada*, kidney stone, because it was supposed to be beneficial to diseases of the kidneys. Nephrite is derived from the Greek word for kidney, the allusion being the same as in the case of jade.

See also **Amphibole**; **Jadeite**; and **Pyroxene**.

JADEITE. The mineral jadeite, essentially sodium-aluminum silicate, $Na(Al, Fe^{2+})Si_2O_6$, is a monoclinic pyroxene usually appearing in crystalline masses, or may be granular, fibrous, or compact. It has a prismatic cleavage; splintery fracture; hardness, 6 in crystals, 6.5–7 massive variety; specific gravity, 3.24–3.43; luster, vitreous to pearly; color, various shades of green, bluish-green, greenish-white or almost white; translucent to opaque. The processes that have acted to form this mineral are little understood both because of the confusion that exists between jadeite and nephrite, and the fact that the localities are not well known. Jadeite is found in Myanmar and China and has been reported from Mexico. It has probably resulted from the metamorphism, at great depths, of rocks rich in soda and aluminum, such as nephelite syenites. Its association in Myanmar with serpentine suggests its origin in more basic igneous rocks. See also **Pyroxene**.

Jadeite is a tough and yet rather easily worked substance, and has long been used for ornamental purposes. Evidence has been found in Europe, Mexico, Egypt, and elsewhere that it was used in prehistoric times for both ornaments and implements. The word jadeite is the general term used for all green-colored tough compact stones that have been used as indicated above.

JAMESONITE. This mineral is commonly called brittle *feather-ore* from its brittle character and usual habit in acicular mats of crystals. It crystallizes in the monoclinic system and is of metallic luster, opaque with gray-black color and streak. It has a hardness of 2.5 and a specific gravity of 5.63–5.67. It is a sulfide of lead and iron antimony, $Pb_4FeSb_6S_{14}$.

Jamesonite occurs in low- to moderate-temperature hydrothermal veins with other lead sulfosalt minerals. Exceptional specimens are found in Bolivia, at Potosi and Oruro, and as choice felted masses on pyrite at Zacatecas in Mexico. Jamesonite is also found in Arkansas, Idaho, and Utah in the United States and in Ontario and British Columbia, Canada. It is named after the English mineralogist, Robert Jameson, from specimens obtained from Cornwall, England. It is a minor ore of lead.

JANOVSKY REACTION. Reaction of aldehydes and ketones containing α-methylene groups with m-dinitrobenzenes in the presence of a strong base, resulting in the formation of an intense purple coloration, used for the detection of carbonyl compounds. The color is due to the formation of Meisenheimer complex.

JAPP-KLINGEMANN REACTION. Formation of hydrazones by coupling of aryldiazonium salts with active methylene compounds in which at least one of the activating groups is acyl or carboxyl. This group usually cleaves during the process.

JAROSITE. The mineral jarosite is a basic hydrous sulfate of potassium and iron corresponding to the formula $KFe_3(SO_4)_2(OH)_6$. It is formed in the outcrops of ore deposits during oxidation of iron sulfides. It is a hexagonal mineral with basal cleavage; is brittle; hardness 2.5–3.5; specific gravity 2.9–3.26; luster vitreous to dull; color, dark yellow to yellowish-brown; pale yellow streak; translucent to opaque. Jarosite was originally reported from and named for Barranco Jaroso in the Sierra Almagrera, Spain. It has been found in Bohemia, France, the Island of Elba, Siberia, and Bolivia. In the United States it is found in Arizona, Colorado, Texas, New Mexico, Utah, Nevada, and South Dakota.

JASPER. A variety of chert, always quartz, that is associated with iron ores and thus contains iron-oxide impurities which give the rock a variety of colorations, often red, but ranging through yellow, green, grayish-blue, brown, and even black. The term also has been used with reference to any red chert or chalcedony. The material is dense, cryptocrystalline, and usually opaque, although it may be slightly translucent.

JATROPHONE. A diterpenoid growth inhibitor isolated from an alcohol extract of the plant *Jatropha gossypiifolia*. Its unique structure includes a 12-membered ring and is readily attacked by nucleophiles. *Jatrophone* is useful in the study of tumor growth inhibition and other biochemical research.

JELLY. A modified form of the word *gel* widely used in popular language but also used in chemical literature to refer to the mechanical strength of the gel structures occurring with pectins, gelatin, and various natural gums. "Jelly strength" is frequently specified in the food industry. Other uses of the word are found in "petroleum jelly" obtained as a distillation product of petroleum residues (petrolatum) and in the so-called *royal jelly*, a natural nutrient mixture of proteins and carbohydrates produced by bees as food for the queen bee. See also **Colloid Systems**; and **Petroleum**.

JENNER, EDWARD (1749–1823). An English physician, Jenner studied medicine in London and established his practice in the rural area of Gloucestershire. Here he discovered the technique of vaccination as a preventive of smallpox (1776). The idea of utilizing cowpox, a disease of cattle, as a protective medium was suggested by his observation that personnel working in dairies developed immunity to smallpox after contracting the much milder cowpox. Jenner's work not only led to almost complete elimination of smallpox in Europe, but also anticipated the development of immune reactions by Pasteur a century later. His success was no accident, but rather the result of detailed observations from which he drew correct conclusions. He was a scientist of the highest caliber and a noteworthy benefactor of mankind.

JET FUEL. A fuel for jet (turbine) engines, usually a petroleum distillate similar to kerosine. A number of types with somewhat different compositions and properties have been used. See also **Petroleum**.

JETSET. A fast-setting cement developed by the Portland Cement Association. Reported to harden in 20 minutes after pouring. Accelerating agent has not been disclosed.

Web Reference

Portland Cement Association: http://www.portcement.org/index.asp

JH. (methyl-*cis*-10,11-epoxy-7-ethyl-3,11-dimethyl-*trans,trans*-2,6-tridecadienoate). A synthetic hormone containing a 13-carbon chain; said to have possibilities as an insecticide. It acts by preventing insects from maturing. Its future depends on the possibility of large-scale production.

See also **Junvenile Hormones**.

JOLIOT-CURIE, FREDERICK (1900–1958). A French physicist who, along with his wife Irene Joliot-Curie, won the Nobel prize in chemistry in 1935. His important discoveries included artificial radioactivity. He did much work on atom structure, dematerialization of electrons, and inverse transformation. Work on hormone synthesis and thyroid substances containing radioactively labeled elements was significant. Sc.D. from the University of Paris was followed by a distinguished career filled with honors and appointments.

JOLIOT-CURIE, IRENE (1897–1956). A French nuclear scientist who won the Nobel prize for chemistry with her husband Frederick Joliet-Curie. Their joint work involved production of artificial radioactive elements by using α-rays to bombard boron. They discovered that hydrogen-containing material when exposed to what they considered γ-rays would emit protons. They were involved in many firsts: they gave the first chemical proof of artificial transmutation and of capture of alpha particles, and were the first to prepare positron emitter. Her career started with a Sc.D. at the University of Paris, and included scores of honors and awards.

JONES OXIDATION. The oxidation of primary and secondary alcohols to acids and ketones by the addition of the calculate amount of chromic anhydride in dilute sulfuric acid to a solution of the alcohol in acetone. This procedure does not attack triple bonds or shift double bonds into conjugation with the ketone formed in the oxidation.

JOSEPHSON, BRIAN DAVID (1940–). Josephson was born in Wales and was recognized even in his undergraduate studies at Cambridge University as having great natural insight on scientific ideas. In 1973, he won the Nobel Prize in Physics jointly with Leo Esaki and Ivar Giaever for his work in experimental superconductivity. He is remembered for the "Josephson effect" also explained as the tunneling phenomena, which is an important idea for modern physics. His discoveries helped advance technology, especially in the development of high-speed switching circuits used in instrumentation and computers.

Josephson was greatly influenced by transcendental meditation techniques of the 1960s and 1970s and believed that mystical experience should be incorporated into science. His latter research also included work in the theory of intelligence.

See also **Josephson Tunnel-Junction**; and **Thin Films**.

J.M.I.

JOSEPHSON TUNNEL-JUNCTION. As early as 1962, B. Josephson recognized the implications of the complex order parameter for the dynamics of the superconductor, in particular when one considers a system consisting of two bulk superconductors connected by a "weak link." The basic requirement for the weak link is that the amplitude of the order parameter at the link should be substantially smaller than in the bulk regions. In early experimentation, such a situation was realized in a variety of ways—two evaporated films separated by a thin (less than 20 angstroms) oxide layer; a light point contact between two bulk superconductors; a single hourglass-shaped evaporated film, with the constriction of dimensions small compared to the coherence length; or even a bare niobium wire with a pendant frozen blob of soft solder, where the weak links, indeterminate in number, are formed by solder bridges through pinholes in the surface oxide. Collectively, all such weak link junctions are referred to as *Josephson junctions*.

Both the dc and ac Josephson effects have found interesting and novel applications. The high sensitivity to magnetic field of the dc Josephson current in certain circuit configurations has been used to develop a family of devices called squids (super conducting *q*uantum *i*nterferometric *d*evices) which can be used to measure extremely small currents, voltages, and magnetic fields. The ac effect has been useful

in making precise measurements in connection with research toward improving the maintenance of the U.S. legal volt.

Much interest is presently exhibited in the use of Josephson superconducting devices of the tunnel-junction type (operated at near absolute zero) in connection with building superfast computers. In such devices, the electrical signals have only a millimeter or two to travel. Switching time is about 10^{-11} second. The power dissipation of Josephson devices permits high circuit density. It is estimated that the power dissipation of transistors is about 100 times that of Josephson devices. Different materials are being studied to improve the original lead alloy thin-film materials. These include lead-indium-gold alloys and niobium-tin alloys.

Additional Reading

Kircher, C. J. and J. Murakami: "Josephson Tunnel-Junction Electrode Materials," *Science*, **208**, 944–950, 1980.

Likharev, K. K.: *Dynamics of Josephson Junctions and Circuits,* Taylor & Francis, Inc., Philadelphia, PA, 1986.

JOULE, JAMES PRESCOTT (1818–1889). Joule was an English physicist. He was born into a family of brewers and as he grew up he became knowledgeable about steam engines used in the brewery. Also, the town he grew up in, Manchester, England had factories with machinery that depended on the steam engine and so it was almost natural that with his curiosity of science, Joule became interested in mechanical energies. By 1837, Joule published his first scientific paper on electric motors.

He is remembered for Joule's Law that describes the rate at which heat is produced by an electric current. Joule's work showed there were different kinds of energy, which can be changed into each other. He established the mechanical equivalence of heat. His work led to the law of conservation of energy. Also, he collaborated with William Thomson (Lord Kelvin) and verified experimentally the Joule-Thomson refrigeration effect.

See also **Joule Law**; **Joule-Thomson Effect**; and **Oxygen**.

JOULE LAW. The quantity of heat generated by a steady electric current is proportional to the resistance of the conductor in which the heat is generated, to the square of the current, and to the time of its duration: $H = KRI^2t$. If the resistance is in ohms, the current in amperes, the time in seconds, and the heat in calories, the constant K has the value 0.2390 calories/joule.

JOULE-THOMSON EFFECT. In passing a gas at high pressure through a porous plug or small aperture, a difference of temperature between the compressed and released gas usually occurs. This phenomenon is called the Joule-Thomson effect. The equation for this effect contains two partial derivatives and is

$$\left(\frac{\partial T}{\partial P} \right)_H = \frac{T \left(\frac{\partial V}{\partial T} \right) p - V}{C_P}$$

where the expression on the left is the rate of change of temperature with pressure at constant enthalpy (heat content) (since no heat is supplied to, or removed from the system). The expression on the right has in its numerator the difference between the product of the temperature and the rate of change of volume with temperature at constant pressure, from which the volume is subtracted; the denominator contains the molar specific heat at constant pressure. The term on the left of the equality sign is called the *Joule-Thomson Coefficient*. It varies with the temperature and pressure of the gas, passing from positive values through zero to negative values. The temperature at which it is zero is called the *Joule-Thomson Inversion Temperature* and varies with the particular gas. It is to be noted, however, that for hydrogen, and also for helium, the temperature is low, far below 0°C. For other gases, however, much higher values are found, the maximum value for oxygen being 1,058 K (785°C). At temperatures above their inversion temperature, gases are warmed, while at temperatures below it, they are cooled by the effect. For that reason, this type of expansion is often used in industrial processes for cooling gases. An interesting application is the application is the process for producing solid carbon dioxide by expanding carbon dioxide through an aperture.

For an ideal gas, $PV = RT$; thus, $(\partial V/\partial T)$ is equal to R/P, that is, to V/T, so that the numerator of the right-hand term in the equation above becomes 0; thus an ideal gas shows no Joule-Thomson effect.

JOURDAN-UIIMANN-GOLDBERG SYNTHESIS.   Synthesis of substituted diphenylamines. The reaction products can be used as intermediates in the synthesis of acridones.

JUNVENILE HORMONES.   One of several hormones, that retard the development of insects in the larval stage. So called because they prevent the insect from maturing by maintaining its juvenile characteristics. Obtained naturally from silk moths; various syntheses indicate possible use as insecticides, especially for fire ants. Composition of one type is $C_{18}H_{30}O_3$.

See also **JH**; and **Insecticide**.

K

KAOLINITE. The most common mineral of a group of hydrous aluminum silicates, which result from the breaking down of aluminum-rich silicate rocks, such as the feldspars and nepheline syenites, either through weathering or hydrothermal activity. Kaolinite, when pure, corresponds to the formula $Al_2Si_2O_5(OH)_4$, and occurs in white, clay-like masses. It has a perfect basal cleavage; is flexible but not elastic; hardness, 2–2.5; specific gravity, 2.6–2.63; luster, pearly to dull; color, white when pure, as described above, but may be yellow, red, blue, or brown; translucent to opaque.

Kaolinite is a mineral of widespread occurrence, well distributed throughout the world. The finest kaolinite locality in Europe is said to be in France, from whence the clay is obtained for porcelain ware. Cornwall and Devonshire in England supply large quantities of this mineral. In the United States, Pennsylvania, Virginia, Colorado, Georgia, and South Carolina contain deposits of kaolinite. The word kaolin or kaolinite is said to be a corruption of a Chinese word *kauling*, the name of a locality where this mineral is found. Kaolinite is very important commercially in the manufacture of china and pottery.

KARLE, JEROME (1918–). An American physical chemist who won the Nobel prize for chemistry along with Herbert A. Hauptman in 1985. He developed a series of mathematical equations that allow determination of phase information from X-ray crystallography intensity patterns. The advent of computers allowed the use of the equations to determine the conformation of thousands of chemicals. The work was done at the Naval Research Laboratory in Washington, DC., where Karle headed the Laboratory for the Structure of Matter.

KARRER, PAUL (1889–1971). A recipient of the Nobel prize for chemistry in 1937 with Walter N. Haworth. Although born in Moscow, he attended European universities and received his doctorate in Zurich. He Initiated work on flavins, carotenoids, and vitamins A and B, and accomplished work on structure and synthesis of vitamin B_2 as well as vitamins A and E.

KAURI-BUTANOL VALUE. A measure of the aromatic content and hence the solvent power of a hydrocarbon liquid. Kauri gum is readily soluble in butanol but insoluble in hydrocarbons. The kb value is the measure of the volume of solvent required to produce turbidity in a standard solution containing kauri gum dissolved in butanol. Naphtha fractions have a kb value of about 30, and toluene about 105.

KEKULE, AUGUST (1829–1896). Born in Darmstadt, Germany, Kekule laid the basis for the ensuing development of aromatic chemistry. His idea of a hexagonal structure for benzene in 1865 was a monumental contribution to theoretical organic chemistry. This had been preceded in 1858 by the remarkable notion that carbon was tetravalent and the carbon atoms could be joined to each other in molecules. The theory of the benzene ring has been called the "most brilliant piece of scientific prediction to be found in the whole field of organic chemistry, for besides promulgating the idea, he had predicted the number and types of isomers which might be expected in various substitutions on the ring" (L. B. Clapp).

KENDREW, JOHN C. (1917–1997). An English molecular biologist who won the Nobel prize for chemistry in 1962 with Max F. Perutz, for their studies of the structures of globular proteins. His work verified Pauling's earlier thesis concerning the α-helix structure of the polypeptide chain. After receiving his Ph.D. from Cambridge, he was science advisor to the allied air commander-in-chief during World War II. He was also editor of the journal *Molecular Biology*.

KERATIN. A class of natural fibrous proteins occurring in vertebrate animals and humans, they are characterized by their high content of several amino acids, especially cystine, arginine, and serine. They are generally harder than the fibrous collagen group of proteins. The softer keratins are components of the external layers of skin, wool, hair, and feathers, while the harder types predominate in such structures as nails, claws, and hoofs. The hardness is largely due to the extent of cross-linking by the disulfide bonds of cystine by the mechanism shown below:

$$\bullet \bullet \bullet R_1CH\text{–}CO\text{–}HN \bullet CH \bullet CO\text{–}NH\text{–}CHR_2 \bullet \bullet \bullet$$
$$|$$
$$CH_2$$
$$|$$
$$S$$
$$|$$
$$S$$
$$|$$
$$CH_2$$
$$|$$
$$\bullet \bullet \bullet R_3CH \bullet OC\text{–}HN \bullet CH \bullet CO\text{–}NH\text{–}CHR4 \bullet \bullet \bullet$$

Kertains are insoluble in organic solvents but do absorb and hold water. The molecules contain both acidic and basic groups and are thus amphoteric.

Uses for kertains include: tablet coatings which dissolve only in the intestines, foam extinguishers and protein hydrolyzates.

KERATINASE. A water-soluble, proteolytic enzyme having the ability to digest the keratin in wool and other forms of hair, converting a portion of it to a water-soluble form. It thus acts as a depilatory and is used in removing hair from pelts and hides, as well as from human skin. It is inactivated by heating to 100°C.

KERMA. Of ionizing particles, the kerma is $\Delta E_K / \Delta m$, where ΔE_K is the sum of the initial kinetic energies of all the charged particles liberated by indirectly ionizing particles in a volume element of the specified material, and Δm is the mass of the matter in that volume element. In these definitions, the symbol Δ precedes the letters E_K and m to denote that these letters represent quantities that can be deduced only from multiple measurements that involve extrapolation or averaging procedures. Since ΔE_K is the sum of the initial kinetic energies of the charged particles liberated by the indirectly ionizing particles, it includes not only the kinetic energy these charged particles expend in collisions, but also the energy they radiate in bremsstrahlung. The energy of any charged particles is also included when these are produced in secondary processes occurring within the volume element. Thus the energy of Auger electrons is part of ΔE_K. In actual measurements, Δm should be so small that its introduction does not appreciably disturb the radiation field. This is particularly necessary if the medium for which kerma is determined is different from the ambient medium; if the disturbance is appreciable, an appropriate correction must be applied.

KERNITE. This hydrated sodium borate mineral, $Na_2B_4O_7 \cdot 4H_2O$, occurs in a single known world locality at Kramer, Kern County, California. It is found here as veins and interbedded masses in clay, hundreds of feet thick, thought to be a product of heating and dehydration of a large buried body of borax by intrusive igneous rocks. Crystals are monoclinic and attain individual size to 3 by 8 feet. Cleavage is very perfect to both the macro and basal pinacoids, with hardness, 2.5–3, and specific gravity of 1.908; possesses vitreous luster grading to satiny on cleavage surfaces; colorless to white; transparent. Surface alters readily to tincalconite, a white powder. It is a major source of borax and boron compounds. See also **Borax**.

KEROGEN. The organic component of oil shale, it is a bitumen-like solid whose approximate composition is 75–80% carbon, 19% hydrogen, 2.5% nitrogen, 1% sulfur, and the balance oxygen. It is a mixture of aliphatic and aromatic compounds of humic and algal origin and comprises a substantial proportion of the shale; after fractionating and refining, the oil is reported to yield 18% gasoline, 30% kerosene, 27% gas oil, 15% light lube oil, and 10% heavy lube oil.

See also **Oil Shale**.

KEROSINE. See **Petroleum**.

KETENES. Members of a class of compounds which contain the functional group =C=C=O. Examples are ketene itself (CH_2CO), methylketene CH_3CH=C=O, dimethylketene $(CH_3)_2C$=C=O, diphenylketene $(C_6H_5)_2C$=C=O, and carbon suboxide O=C=C=C=O. Of these, ketene is by far the most important. It is a reactive, colorless gas of considerable industrial importance. Physiologically, it is extremely poisonous and care must be taken to avoid breathing it.

The availability of ketone by pyrolysis of acetone (or acetic acid) is the reason for the attention it has received, contrasted to other ketenes which are relatively unavailable. Ketene may be prepared also by pyrolysis of acetic anhydride or phenyl acetate or diketene. Other sources are quite unsatisfactory from a standpoint of yield. Small quantities may be made conveniently by heating acetone in a "ketene lamp." This is a glass apparatus containing a Nichrome filament, heated electrically to red heat. Larger amounts are made by passing acetone or acetic acid through a tube at 700°C. A very brief contact time is required, so that much of the acetone is undecomposed and has to be condensed and recycled. Also, it is imperative that the reaction tube be of inert material such as porcelain, glass, quartz, copper or stainless steel. A copper tube, if used, should be protected from oxidation by an iron sheath. Inert packing may be used (glass, vanadium pentoxide, porcelain), but just as good yields are obtained with empty tubes. No catalyst is known which accelerates this decomposition at significantly lower temperatures.

Methyl ethyl ketone is totally unsatisfactory as a source of methylketene by pyrolysis, but pyrolysis of propionic anhydride in a quartz tube at 400–600°C and low pressures does produce it in a stated yield of about 90%. Another synthetic approach is to prepare methylketene dimer by allowing a mixture of propionyl chloride and triethylamine to stand at 25° for 24 hours and then to pyrolyze the dimer.

The known disubstituted ketenes include dialkyl-ketenes, diarylketenes, and the ester analogs. Dimethylketene may be made from α-bromoisobutyryl bromide by reaction with zinc in boiling ether. Diphenylketene may be made similarly, but the usual way to prepare it is to oxidize benzil hydrazone with yellow mercuric oxide to benzoylphenyldiazomethane which, on heating in benzene solution, decomposes into the ketene.

The best way to make carbon suboxide is a pyrolytic method, starting with tartaric acid. The latter is converted into diacetyltartaric anhydride and then pyrolyzed at 625–650°C (either in an empty tube or in a ketene lamp) into acetic acid and carbon suboxide, the latter in 35–50% yields.

Some recent synthesis of ketenes involve interesting chemistry. A butadienylketene is obtainable at −100 to −150°C by photolysis (mercury lamp) of the appropriate cyclohexadienone:

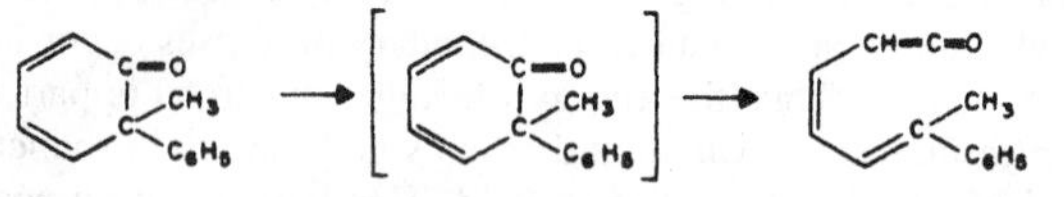

The reaction is reversed on warming, but the ketene may be captured by an amine to form an amide.

1-Ethoxy-1-alkyne, R–C≡C–OC_2H_5, pyrolyzes at 120° into ethylene and alkylketene, but the latter is consumed by the original alkynyl ether to form 2,4-dialkyl-3-ethoxy-2-cyclobutenone,

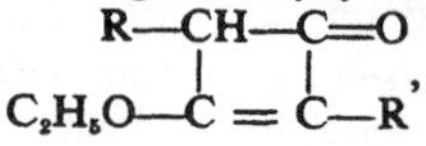

in high yield.

Ketene and diimide are formed during the alkaline decomposition of chloroacetic hydrazide. The diimide, however, spontaneously changes into hydrazine plus nitrogen, and the hydrazine consumes the ketene to form acetohydrazide, $CH_3CONHNH_2$. If an olefin is present in the reaction mixture it is reduced to a paraffin by the diimide, thus preventing hydrazine formation.

The formulas of acetic acid, acetic anhydride and ketene show that the anhydride differs from the acid by 0.5 mole of water, and that ketene differs by 1.0 mole. Ketenes, therefore, may be regarded as super acid anhydrides, and this viewpoint leads to a good appreciation of their reactions. Because of an original lack of understanding of this relationship, the monosubstituted ketenes were once classed as "aldoketenes," and the disubstituted ketenes as "ketoketenes." Such terms are quite misleading, however, and should be abandoned.

Ketene is absorbed in sodium hydroxide solution, yielding sodium acetate. Aniline adds to ketene to form acetanilide. Both of these reactions are quantitative and are used to assay the ketene in a gas stream.

Primary alcohols react readily with ketene to form acetic esters but tertiary alcohols require the catalytic help of sulfuric acid. Even with primary alcohols, as 1-butanol, it has been established that addition of ketene ceases at about the 75% conversion point unless a little sulfuric acid is present as catalyst. Phenol, which is inert toward ketene at ordinary temperature, may be converted into phenyl acetate by reaction at the boiling point of phenol or by reaction at room temperature if a trace of sulfuric acid is present.

An important industrial synthesis of acetic anhydride is via acetic acid and ketone. Mixed acetic anhydrides are made similarly by passing ketene into the acid in question: $RCOOH + CH_2CO \longrightarrow RCO$–$O$–$COCH_3$. This is the basis of a good method of synthesizing symmetrical anhydrides in view of their formation from the mixed anhydrides on heating:

$$2\ RCOOCOCH_3 \longrightarrow (RCO)_2O + (CH_3CO)_2O.$$

Comparable reactions of ketene are those with mercaptans to form thio esters (CH_3COSR), with amino acids (in water) to obtain N-acetyl derivatives ($CH_3CONHCHRCOOH$), with hydroxylamine to yield acetohydroxamic acid ($CH_3CONHOH$), dimethylchloroamine to form chloroacetic dimethylamide ($ClCH_2CON(CH_3)_2$), and Grignard reagents to form ketones ($RCOCH_3$).

Ketene adds to pyridine in a 4:1 ratio to form a yellow, crystalline compound,

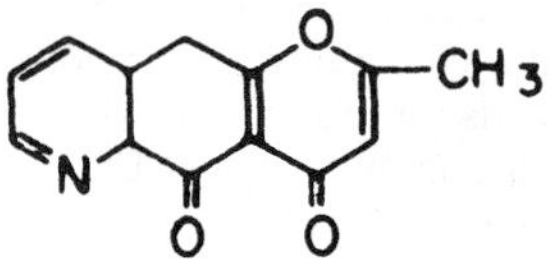

Quinoline, isoquinoline, and phenanthridine resemble pyridine in reacting comparably, and diketene may be substituted for ketene.

Aromatic aldehydes take up ketene in the presence of potassium acetate in the manner of a Perkin reaction. The product is a cinnamic acid:

$$ArCHO + CH_2CO \xrightarrow{AcOK} ArCH{=}CHCOOH.$$

Friedel-Crafts catalysts are effective in converting formaldehyde and ketene into β-propiolactone. This is a process of industrial importance. Similarly, furfural and ketene give rise to 3- (2-furyl)-propionolactone.

When aluminum chloride is used as catalyst, ketene reacts with benzene to form acetophenone: $C_6H_6 + CH_2CO \longrightarrow C_6H_5COCH_3$. Also, methyl chloromethyl ether under such conditions reacts to yield 3-methoxypropionyl chloride:

$$CH_3OCH_2Cl + CH_2CO \xrightarrow{AlCl_3} CH_3OCH_2CH_2COCl.$$

Ketones such as acetone, ethyl acetoacetate, ethyl levulinate or acetylacetone react smoothly with ketene if a trace of sulfuric acid is present. The products are enol acetates of the ketones, acetone giving rise to isopropenyl acetate,

$$CH_2{=}C(CH_3)-OCOCH_3,$$

and ethyl acetoacetate changing into ethyl 3-acetoxycrotonate. Cyclobutanone derivatives are made quite easily by addition of ketenes to styrene or to vinyl ethers or to enamines:

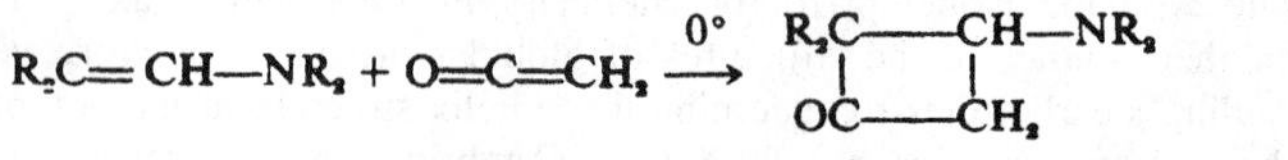

One of the most characteristic reactions of ketene or dialkyl-ketenes is that of polymerization into dimers. Diarylketenes do not display this tendency. The dimer from ketene, or "diketene," is a liquid that boils without decomposition at 43° (28 mm), but the compound tends to decompose (into dehydroacetic acid and resinous substances) on distillation (b.p. 127°C) at atmospheric pressure. The structure of diketene was in doubt for many years, but recent critical chemical and physical evidence indicates the structure as 3-buteno-β-lactone. Diketene is an acetoacetylating agent. Thus, with aniline it yields acetoacetanilide, and with methanol (catalyzed by sulfuric acid) it produces methyl acetoacetate. These reactions are useful industrially.

CHARLES D. HURD
Northwestern University
Evanston, Illinois

Additional Reading

Kroschwitz, J.I., and M. Howe-Grant: *Kirk-Othmer Encyclopedia of Chemical Technology,* John Wiley & Sons, Inc., New York, NY, 2001.

Lide, D.R.: *CRC Handbook of Chemistry and Physics,* 84th Edition, CRC Press, LLC., Boca Raton, FL, 2003.

Patai, S. *Chemistry of Ketenes, Allenes and Related Compounds,* John Wiley & Sons, Inc., New York, NY, 1980.

Tidwell, T.T.: *Ketenes,* John Wiley & Sons, Inc., New York, NY, 1994.

KETIMINE. A type or class of curing agent for epoxy resins that makes it possible to use very-high-solids content coatings in spray equipment. Reacts with epoxies very slowly and thus delays curing time, which prevents setting up of the resin during spraying operation. In presence of water or water vapor, ketimine breaks down to a polyamine and a ketone. Epoxy coatings cured with ketimine should not exceed a thickness of 10 mils.

KETONES. The homologous series of ketones (like aldehydes) has the formula $C_nH_{2n}O$. Structurally, the ketones consist of a carbonyl group (C:O) linkage between two radicals (R—CO—R'). R and R' may be the same as in acetone (dimethyl ketone) CH_3—CO—CH_3; or they may differ as in methylethyl ketone CH_3—CO—C_2H_5. The latter may be referred to as a *mixed ketone*. A ketone is isomeric with the aldehyde that contains the same number of carbon atoms. Thus, acetone C_3H_6O is isomeric with propaldehyde C_3H_6O. Where R and R' are alkyls, the ketone may be called an *alphyl ketone* and may be considered to be derived from the secondary alcohols. Ketones also may be formed from aromatic alcohols as in the case of benzophenone C_6H_5—CO—C_6H_5. The latter compound is a fully aromatic or *diaryl ketone*. Further, there are mixed *aryl-alphyl ketones* as in the case of acetophenone (phenylmethyl ketone) C_6H_5—CO—CH_3. The foregoing examples illustrate the use of trivial names for the ketones. In another system, the ketone may take its name from the alcohol from which it may be derived—thus, propione (from propyl alcohol); or the ketone may be named for the acid to which it may be oxidized—thus, acetone (acetic acid the oxidation product).

Essentially ketones exhibit the following properties: (1) all ketones up to C_{11} are neutral, mobile, volatile liquids. Ketones above C_{11} are solids under usual ambient conditions; (2) all ketones have a reasonably agreeable odor; (3) all ketones except those with a very high carbon count are soluble in H_2O, the solubility decreasing with a rise in formula weight; (4) most ketones are soluble in alcohol or ether, and (5) the specific gravity of ketones rises uniformly to about 0.83 with a rise in formula weight. The physical properties of some common ketones are listed in Table 1.

The presence of the double bond (carbonyl group C:O) markedly determines the chemical behavior of ketones. The hydrogen atom connected directly to the carbonyl group is not easily displaced. The chemical properties of the ketones may be summarized as follows: (1) they are readily reduced to form *secondary alcohols*, particularly in the presence of a catalyst. This property is used to advantage in the production of numerous organic compounds where ketones serve as a starting material or as an intermediate; (2) unlike the aldehydes, ketones are considerably more stable and do not combine readily with the alcohols, nor generally with NH_3 at ordinary temperatures and pressures; (3) they do not reduce alkaline solutions of metals and also, unlike aldehydes, ketones do not undergo polymerization; (4) they combine with hydroxylamine to yield *ketoximes*; (5) they react with hydrazine to form *hydrazones*; (6) they combine with semicarbazine to form *semicarbazones*; (7) with H_2SO_4 or HCl present, ketones can be induced to undergo a cyclic trimerization with the loss of H_2O; (8) when oxidized, ketones decompose to form two acids.

Each acid will contain fewer carbon atoms than the originating ketone; and (9) ketones react with HCN to form *cyanohydrins*.

When ketones are reduced to secondary alcohols, varying amounts of ditertiary alcohols (pinacols) are produced. In a reaction known as the pinacol-pinacoline rearrangement, effective in synthesis of difficult compounds, ketones when treated with magnesium amalgam, after hydrolysis, yield a 1, 2-glycol. Important to the purification of methyl and some cyclic ketones is their ability to form crystalline additive compounds with sodium bisulfite solutions. Ketones also react with phosphorus pentachloride and pentabromide to form dihalogen derivatives of the alkyls in which instances the oxygen atom of the carbonyl group is replaced by two hydrogen atoms.

Ketones are widely used as starting and intermediate ingredients in the production of numerous synthetics, such as resins, and they find wide application as solvents. Other important ketones produced on a tonnage basis include methylethyl ketone (MEK) and methylisobutyl ketone (MIBK). Commercially, MEK may be manufactured by the direct oxidation of butylene in which air is used, along with a catalyst solution comprising copper chloride and palladium chloride. The overall reaction is $C_4H_8 + \frac{1}{2}O_2 \longrightarrow CH_3COC_2H_5$. During the reaction, the palladium chloride is reduced to elemental palladium and HCl. Cupric chloride causes reoxidation. The resulting cuprous chloride is reoxidized to cupric chloride during the catalyst regeneration cycle. The process proceeds under moderate pressures at a temperature of about 100°C. The MEK product must be treated with sodium bisulfite and caustic soda, followed by distillation, to yield pure MEK.

In the production of MIBK, acetone may be the chargestock. In a first step, acetone is converted to diacetone alcohol (DAA) by condensation under pressure with an alkaline catalyst. The latter may be calcium or barium hydroxide, both of which are slightly soluble in H_2O. After cooling, in a second step, mesityl oxide (MSO) is produced by dehydrating the DAA. For this step, an acid catalyst is used and the step proceeds in the temperature range of 100 to 120°C. When the MSO is formed from the DAA, the latter partially decomposes into acetone. Thus, a distilling phase is required to separate and recover the acetone. In a third step, the MSO is hydrogenated and a mixture of MIBK and methylisobutyl carbinol (MIBC) is produced. These must be separated by a further fractionating phase. The hydrogenation reactions are: $(CH_3)_2C{=}CHCOCH_3 + H_2 \longrightarrow (CH_3)_2CHCH_2COCH_3$ (MIBK); and $(CH_3)_2CHCH_2COCH_3 + H_2 \longrightarrow (CH_3)_2CHCH_2CHOHCH_3$ (MIBC). The temperature of the process and the hydrogen mole ratio used determines the ratio of MIBK/MIBC produced and thus the manufacturer has effective control over the amounts of end-products (MIBK and MIBC) which can be manipulated in accordance with market requirements.

Related Compounds

A polyhydric ketone is refereed to as a *ketose*. Monosaccharoses are examples of open-chain polyhydroxyaldehydes or ketones. The aldehyde sugars are termed *aldoses*; the ketone sugars are called *ketoses*. Fructose is a ketose. Dihydroxyacetone HO · CH_2· CO · CH_2· OH is a very simple ketose. Compounds that contain both a carbonyl and a carboxylic group are termed *ketonic acids*. They display the reactive properties of both an acid and a ketone. Pyroracemic acid CH_3· CO · CO_2H, acetoacetic acid CH_3· CO · CH_2· CO_2H, and laevulic acid CH_3· CO · CH_2· CH_2· CO_2H are examples of monobasic ketonic acids. *Ketonic hydrolysis* is a term used to describe such actions as the hydrolysis of the ethyl ester of acetoacetic acid into acetone, CO_2, and ethyl alcohol. Compounds such as ketene CH_2:CO, methyl ketene CH_3CH:CO, dimethyl ketene $(CH_3)_2C$:CO, and diphenyl ketene $(C_6H_5)_2C$:CO are of the family known as *ketenes*.

Health and Safety Factors

Ketones are flammable substances that do not exhibit a known high degree of chronic toxicity. Low molecular weight (C_3–C_{12}) saturated aliphatic ketones, which represent the bulk of industrially important ketones, may be classified among the solvents of comparatively low toxicity hazard. The eight-hour threshold limit value is generally above 100 ppm, although the odor threshold is in the range 5–25 ppm. High vapor concentrations of these volatile ketones induce anesthesia, however, the vapors are so irritating to the eyes and mucous membranes of the respiratory system that the atmosphere generally becomes intolerable before toxic concentrations are achieved. Many ketones are also powerful drying and degreasing agents, and prolonged skin contact can cause dermatitis.

TABLE 1. PHYSICAL PROPERTIES OF KETONES

Systematic name (common name)	Mol wt	Fp, °C	Bp at 101.3 kPa,[a] °C	Refractive index, n^{20}_D	Sp gr 20/20, °C	Viscosity at 20°C, mPa·s(= cP)	Surface tension, mN/m (=dyn/cm) at 20°C	H_{vap} at 101.3 kPa,[a] kJ/mol[b]	Liquid specific heat capacity at (T) °C, J/(kg·K)[b]	Flash point, open cup, °C (closed)	Soly at 20°C, wt % In water	water in
METHYL ALKYL KETONES												
2-propanone (acetone)	58.08	−94.7	56.1	1.3590	0.7905	0.33	24.0	29.53	2224 (30)	−16 (−18)	complete	complete
2-butanone (methyl ethyl ketone)	72.10	−85.9	79.57	1.3780	0.8062	0.41	24.6	31.64	2203 (20)	−6 (−6)	26.8	11.8
2-pentanone (methyl propyl ketone)	86.13	−77.8	102.4	1.3902	0.8076	0.51 (28.3°C)	23.2	33.39		(7)	4.3	3.3
3-methyl-2-butanone (methyl isopropyl ketone)	86.13	−92	94.2	1.3882	0.8044	0.43 (25°C)	24.6 (25°C)	30.63		(6)	6.53	
4-methyl-2-pentanone (methyl isobutyl ketone)	100.16	−84.0	116.2	1.3957	0.8020	0.61	23.6	35.60	1920 (20)	23 (16)	1.6	1.9
2-hexanone (methyl n-butyl ketone)	100.16	−55.8	127.5	1.4007	0.8125	0.62	25.4	36.05	2228 (25)	(35)	1.75	3.7
3-methyl-2-pentanone (methyl sec-butyl ketone)	100.16	−83	117.4	1.4001	0.8142			35.12			2.26	
3,3-dimethyl-2-butanone (pinacolone)	100.16	−50	106.4	1.3986	0.8070			33.5			2.0	1.8
2-heptanone (methyl amyl ketone)	114.18	−35	151.5	1.4087	0.8166	0.77	26.1	39.25		47 (49)	0.43	1.45
5-methyl-2-hexanone (methyl isoamyl ketone)	114.18	−73.9	144.9	1.4069	0.8127	0.77	25.3 (25°C)			41 (35)	0.54	1.28
2-octanone (methyl hexyl ketone)	128.22	−20.5	173.3	1.4153	0.8197	0.95 (25°C)	26.6 (25.5)	40.88		(62)		
4-hydroxy-4-methyl-2-pentanone (diacetone alcohol)	116.16	−44.2	169.2	1.4226	0.9406	3.2	31	41.6	1883	61 (47)	complete	complete
DIALKYL KETONES												
3-pentanone (diethyl ketone)	86.3	−39.4	101.8	1.3923	0.8155	0.47	24.7	33.69	2215 (25)	(13)	3.4	2.6
2,4-dimethyl-3-pentanone (diisopropyl ketone)	114.19	−69	125.0	1.399								
2,6-dimethyl-4-heptanone (diisobutyl ketone)	142.24	−46	169.4	1.4172	0.8076	1.02	22.2	39.31		49 (49)	0.05	0.75
3-hexanone (ethyl propyl ketone)	100.16		123.2	1.4003	0.8174		25.04	35.66		35	1.57	
3-heptanone (butyl ethyl ketone)	114.19	−39	147.3	1.4088	0.8197	0.84	25.7	36.59		41 (46)	0.43	0.78
3-octanone (ethyl amyl ketone)	128.22	−46	167–168	1.4150	0.8220							
2,6,8-trimethyl-4-nonanone (isobutyl heptyl ketone)	184.32	−75	218.2	1.4257	0.8180	1.9		44.56		90 (88)	<0.01	0.2
UNSATURATED KETONES												
3-buten-2-one (methyl vinyl ketone)	70.09	−6	81.4	1.4130								
3-methyl-2-buten-2-one (methyl isopropenyl ketone)	84.12	−54	98	1.4236	0.855							
4-methyl-3-penten-2-one (mesityl oxide)	98.15	−53	129.5	1.4414	0.8521	0.6	28.4	43.1	2176 (20)	29 (31)	3.1	3.4
4-methyl-4-penten-2-one (isomesityl oxide)	98.15		121.5	1.4458	0.8548							
3,5,5-trimethyl-2-cyclohexen-1-one (isophorone)	138.21	−8.1	215.3	1.4775	0.9229	2.6	32	43.4	1799 (20)	104 (85)	4.3 (25°C)	1.2 (25°C)
3,5,5 trimethyl-3-cyclohexen-1-one (β-isophorone)	138.21		181–191		0.89						0.03	
DIKETONES												
2,3-butanedione (diacetyl)	86.09	−2.5	90.2	1.3938	0.9843			34.3				
2,3-pentanedione	100.12	−52	111				31	35.4	1983 (20)			
2,4-pentanedione (acetylacetone)	100.11	−23.5	140.4	1.4510	0.9753	0.58		36.55	1956.2			
2,5-hexanedione	114.15	−5.4	192.3	1.4256	0.9734	1.6			(15)			
CYCLIC KETONES												
cyclopentanone (adipic ketone)	84.12	−50.6	130.8	1.4359	0.9512	1.2	33.35	36.53			29	14
cyclohexanone (pimelic ketone)	98.15	−31.1	155.7	1.4510	0.9482	2.21	35.2	37.62	2039.8 (30.8°C)	46 (43)	2.5	8.0
cycloheptanone	112.17	−21	179	1.4611			26.4			72 (62.5)	0.3	1.4
3,3,5-trimethylcy-clohexanone	140.22	−10	188.8	1.4455	0.888	2.54						
AROMATIC KETONES												
acetophenone (methyl phenyl ketone)	120.15	19–20	201.7	1.5342	1.0296	0.93		45.69		93 (82)	0.55	1.65
benzophenone (diphenyl ketone)	182.22	48–49.5	305									
1-phenyl-2-propanone (phenylacetone)	134.18	−15		1.5158								
propiophenone (phenyl ethyl ketone)	134.17	18.2	218	1.5265	1.012		37.4	45.44		96 (85)	0.01 (25°C)	

[a] To convert kPa to mm Hg, multiply by 7.5.

[b] To convert J to cal, divide by 4,184.

The C_3-C_{12} ketones are all highly flammable liquids with flash points varying from $-18°C$ for acetone to $85°C$ for isobutyl heptyl ketone. Ketones float on water, and become only partially soluble in water with increasing molecular weight. Thus, ketones typically require copious quantities of water to extinguish pool fires. Saturated ketones are in general stable at ambient conditions, and do not undergo hazardous polymerization in normal environment. Most ketones are incompatible with strong oxidizing and reducing agents; some ketones are also incompatible with bases and/or acids.

Additional Reading

Bingham, E. et al.: *Patty's Toxicology: Ketones, Alcohols and Ester Compounds*, Vol. 6, John Wiley & Sons, Inc., New York, NY, 2000.
Gmehling, J. et al.: *Vapor-Liquid Equilibrium Data Collection: Ketones, Supplement 1*, Scholium International, Inc., Port Washington, NY, 1993.
Kroschwitz, J.I., and M. Howe-Grant: *Kirk-Othmer Encyclopedia of Chemical Technology*, John Wiley & Sons, Inc., New York, NY, 2001.
Lide, D.R.: *CRC Handbook of Chemistry and Physics*, 84th Edition, CRC Press, LLC., Boca Raton, FL, 2003.
McKetta, J.: *Encyclopedia of Chemical Processing and Design: Hydrogen Cyanide to Ketones Dimethyl (Acetone)*, Vol. 27, Marcel Dekker, Inc., New York, NY.

KETOSES. See **Carbohydrates**.

KEYES EQUATION. An equation of state for a gas, deduced from the concept of the nuclear atom. This equation is designed to correct the van der Waals equation for the effect upon the term b of the surrounding molecules. The equation is written as

$$P = \frac{RT}{V - Be^{-\alpha/V}} - \frac{A}{(V+l)^2}$$

in which P is pressure, T is absolute temperature, V is volume, R is the gas constant, e is the base of natural logarithms, $2.718\ldots$, and A, α, and B, and l are constants for each gas.

KEYES PROCESS. A distillation process involving the addition of benzene to a constant-boiling 95% alcohol-water solution to obtain absolute (100%) alcohol. On distillation, a ternary azeotropic mixture containing all three components leaves the top of the column while anhydrous alcohol leaves the bottom. The azeotrope (which separates into two layers) is redistilled separately for recovery and reuse of the benzene and alcohol.

KIMBERLITE. The name applied to a mica peridotite which occurs at Kimberley and other places in the Republic of South Africa, the source of rich deposits of diamonds. These valuable gem stones were originally found in the decomposed kimberlite which, being colored yellow by limonite, was termed "yellow ground." Deeper workings disclosed the less altered rock, kimberlite, which the miners call "blue ground."

KINETIC MEASUREMENTS. Kinetic measurements are studies of the rates at which chemical reactions occur. Generally, these studies involve preparing a chemical system using reagent concentrations different from the equilibrium values and then monitoring the concentration changes as the system approaches equilibrium, although other, less direct strategies are sometimes exploited. Chemical kinetic data are used in materials science, biochemistry and molecular biology, earth and atmospheric science, and many branches of engineering. Related concepts appear in nuclear physics, but presuppositions and methods are different there.

Kinetic information is acquired for two different purposes. First, data are needed for specific modeling applications that extend beyond chemical theory. These are essential in the design of practical industrial processes and are also used to interpret natural phenomena such as the observed depletion of stratospheric ozone. Compilations of measured rate constants are published in the United States by the National Institute of Standards and Technology (NIST). Second, kinetic measurements are undertaken to elucidate basic mechanisms of chemical change, simply to understand the physical world. The ultimate goal is control of reactions, but the immediate significance lies in the patterns of kinetic behavior and the interpretation in terms of microscopic models.

Explaining chemical change by postulating mechanisms in terms of macroscopic concentrations is expected to continue for the foreseeable future. For a fundamental understanding of very simple reactions, however, traditional kinetics is being challenged by theoretical and experimental methods that focus directly on the behavior of individual atoms.

Macroscopic Behavior and the Rate Law

Chemical Equations. Chemical changes are discussed with the aid of the equations used to treat equilibrium, i.e., the reaction of reactants A, B, C, and so on, to produce products P, Q, and so forth.

The essential information implied by the chemical equation is the stoichiometry at the macroscopic level, i.e., if α moles of A react, then b moles of B do also; p moles of P formed, etc.

A kinetic study typically prepares some set of initial concentrations not at equilibrium and describes the subsequent evolution of each. A basic assumption is that each component evolves according to some differential equation where t represents time.

$$d[A]/dt = f([A], [B], \ldots, [P], \ldots, \text{other conditions}) \quad (1)$$

In general, the differential equation could be very complicated, e.g., the concentrations may be functions of spatial coordinates as well as time. Experimental measurements are arranged to ensure that simplified equations apply.

The Well-Stirred Mixture. A key assumption of most kinetic measurements is that of a well-mixed solution of reactants. Then any component can be characterized by a single time-dependent concentration, applicable to the entire system.

In particularly simple cases, which occur frequently, one may assume a dependence on powers of reactants and ignore products

$$d[A]/dt = -h[A]^x[B]^y[C]^z \quad (2)$$

Experimental Verification of a Rate Law

It is possible to prepare a system having an initial concentration for each component, and then measure a finite, but small, change in the concentration of one component, $\Delta[A]$ for example, over a known interval of time, Δt. The experimental velocity $\Delta[A]/\Delta t$ and the concentrations can be substituted into a proposed rate law, like equation 2, along with postulated values for the exponents $x, y, \ldots$ to determine an observed constant k_{obs}. If this process is repeated for a reasonable range of concentrations, and the postulated rate law having the same exponents always yields the same k_{obs}, then it is asserted that the rate law has been verified for those concentration ranges and the rate constant determined. This approach is a reasonable strategy for an initial survey of a totally unknown system; but it is wasteful, in that it extracts very little data from each set of initial conditions. More often, the integrated form of the rate law is fit to multiple concentration measurements recorded at different times for each set of initial conditions.

Flooding and Pseudo-First-Order Conditions. Flooding is an experimental strategy that simplifies both measurement and analysis. For an example, consider a reaction that is independent of product concentrations and has three reagents. If a large excess of $[B_i]$ and $[C_i]$ are used, and the disappearance of a lesser amount of $[A]$ is measured, the rate law can be integrated with the assumption that all concentrations are constant excepts $[A]$. Consequently, simple expressions are derived for the time variation of $[A]$. Under flooding conditions and using equation 2, if x happens to be 1, the time-dependent concentration of A exhibits an exponential decrease from its initial value $[A_i]$ to its final equilibrium value, or endpoint, $[A\infty]$:

$$[A(t)] - [A_\infty] = ([A_i] - [A_\infty])\exp(-k_{obs}t) \quad (3)$$

The conditions chosen make the reaction appear to be first-order overall, although the reaction is really not first-order overall, unless y and z happen to be zero. The pseudo-first-order rate constant k_{obs} is related to the k in the originally postulated rate law by

$$k_{obs} = k[B_i]^y[C_i]^z \quad (4)$$

If x is not 1, equation 3 is replaced by a different, but still simple, integrated rate law.

The Initial Conditions. One of two very different strategies are used in kinetic measurements to produce the initial, nonequilibrium concentrations of reactants. Either the separate reagents are mixed or a system previously at equilibrium is perturbed.

Mixing known quantities of reagents to produce desired initial concentrations can be carried out with either a continuous flow or a stopped flow apparatus. Engineering details become important for fast reactions, especially those occurring in less than one millisecond.

Perturbation methods can be divided further into two categories. One uses a flash of light (flash photolysis) or other radiation (radiolysis) to create a homogeneous distribution of a desired reagent from some precursor molecule. This method is capable of very fast time resolution; recent advances in lasers allow the study of processes occurring in 10^{-14} s. The other perturbation method does not affect reagents but instead changes some intensive thermodynamic property, such as temperature, pressure, or an electric field. Concentrations are monitored as the system adjusts to the requirements of the new equilibrium.

Indirect and Novel Methods. A direct measurement is a record of changing concentrations as a function of time. In principle, the same information is available as a Fourier transform in the frequency domain. Since the latter part of the nineteenth century, it has been possible to measure the absorption of electromagnetic radiation as a function of frequency (or wavelength) and interpret lineshapes to yield kinetic information on the picosecond time scale. In gases, line widths increase at high pressure owing to collision broadening. Dissociation or ionization may also determine a linewidth. More subtle effects occur in liquids. More recently, chemical reactivity on slower time scales has been measured by lineshape analysis in Mossbauer spectroscopy and magnetic resonance.

Dramatic progress is being made in extending kinetic analyses to microscopic samples, such as single biological cells. One strategy infuses precursor molecules into the correct part of a cell and then uses flash photolysis to start a reaction that is monitored continuously by fluorescence microscopy. Within such small volumes, there may be only thousands of molecules of a given type and their number may change due to statistical fluctuations, even in a system at equilibrium. Such fluctuations obey the same laws of chemical kinetics as any other perturbation; and a study of fluctuations can determine kinetic parameters, although this is far from being a routine technique. Pioneering experiments are even being carried out for kinetic studies on single atoms, not in cells, but isolated in a magnetic or optical trap.

Rapid instrumental measurements have largely, but by no means completely, supplanted an earlier tradition that relied on a more chemical strategy of measuring concentrations of reactants or intermediates by intercepting these with a scavenger that reacted quickly to form a stable product that could be quantified later. Such trapping methods can be construed as being indirect in the sense that they rely on one chemical reaction being faster or slower than another.

Experimental Variation of Chemical Rates with Temperature and Pressure

The experimentally measured dependence of the rates of chemical reactions on thermodynamic conditions is accounted for by assigning temperature and pressure dependence to rate constants.

Microscopic Models in Kinetics

Mechanism is a technical term, referring to a relatively detailed, microscopic description of a chemical transformation, which, nevertheless, still falls far short of a complete dynamical description at the atomic level. A mechanism for a reaction is sufficient to predict the macroscopic rate law of the reaction. This deductive process is valid only in one direction, i.e., an unlimited number of mechanisms are consistent with any measured rate law. A successful kinetic study postulates a mechanism, derives the rate law, and demonstrates that the rate law is sufficient to explain experimental data over some range of conditions. New data may be discovered later that prove inconsistent with the assumed rate law and require that a new mechanism be postulated. Mechanisms state, in particular, what molecules actually react in an elementary step and what products these produce. An overall chemical equation may involve a variety of intermediates, and the mechanism specifies those intermediates.

DOUGLAS MAGDE
University of California at San Diego

Additional Reading

Claesson, S. ed.: *Fast Reactions and Primary Processes in Chemical Kinetics,* 5th Edition, Wiley-Interscience, New York, NY, 1967.

Kustin, K. ed.: *Fast Reactions, Methods in Enzymology,* Vol. 16, Academic Press, Inc., New York, NY, 1969.

Moore, J.W. and R.G. Pearson: *Kinetics and Mechanism,* 3rd Edition, Wiley-Interscience, New York, NY, 1981.

Strehlow, H.: *Rapid Reactions in Solution,* VCH, Weinheim, Germany, 1992.

KINETIC THEORY. A theory (proved by experiment) that explains the phenomena of heat and pressure as due to the kinetic motion and elastic collisions of atoms and molecules. The phenomena include gas and vapor pressure, evaporation, and diffusion of fluids.

Gases expand indefinitely when released, not because of repulsion between the molecules as formerly supposed (though the Joule-Thomson effect under certain conditions may involve this), but because the molecules are in rapid motion and do not stop unless they collide with something. Air is not "forced" out through a tire puncture; only those air molecules pass out which, in their aimless wanderings, happen to encounter the opening. Molecules also pass in from the outside; but since there are several times as many per unit volume inside as outside, many more pass out than in. This continues until, a statistical equilibrium being reached, the air inside is no more dense than that outside, and the tire is "flat." The rapidity with which this takes place emphasizes the speed of the molecular motion and the relative insignificance of the "internal friction" opposing it.

What appears to be a steady pressure is due to the incessant impacts of the gas molecules on any surface exposed to them. If n molecules of equal mass m are released in an enclosure of volume v, and if their speeds are $u_1, u_2, \ldots u_n$, it is easy to show that the average pressure set up by these impacts, neglecting the effects of collisions and gravity, is

$$p = \frac{m}{3v}(u_1^2 + u_2^2 + \cdots + \cdots u_n^2) \qquad (1)$$

This may be written

$$P = \frac{1}{3}\frac{nm}{v}\frac{\sum(u^2)}{n}$$

or, since nm/v is the gas density ρ, and $\rho(u^2)/n$ is the mean square molecular speed $\overline{u^2}$,

$$p = \frac{1}{3}\rho\overline{u^2} \qquad (2)$$

This relation gives the mean square speed as $\overline{u^2} = 3p/\rho$; which is the square of the effective speed, corresponding to average kinetic energy. From this it may be shown that the average speed is

$$\overline{u} = \sqrt{\frac{8p}{\pi\rho}} \qquad (3)$$

easily determined since p and ρ are measurable.

Again, Equation (1) may be written

$$pv = \frac{2}{3}\left(\frac{1}{2}mu_1^2 + \frac{1}{2}mu_2^2 + \cdots + \frac{1}{2}mu_n^2\right) = \frac{2}{3}E \qquad (4)$$

in which E is the total kinetic energy of linear motion of the molecules. From this it follows that the absolute temperature T of the gas bears a constant ratio to this total kinetic energy, and hence to the average translational kinetic energy of the molecules. See also **Ideal Gas Law**.

Further analysis shows that when gravity is considered, the pressure in an undisturbed pure gas of uniform temperature, at an elevation h, is given by

$$p = p_0 e^{-3gh/\overline{u^2}} \qquad (5)$$

in which p_0 is the pressure at the zero of elevation. This is a form of Laplace's "law of atmospheres," useful in barometric altitude determinations.

A quantity much used in kinetic theory is the "mean free path," which is the average distance traversed by a molecule between collisions. There are ways of calculating this and also the effective diameters of molecules, and these data lead to many conclusions as to frequency of collisions and rate of diffusion.

Irreversibility. The kinetic theory assumes that the velocity of a molecule may depend on the conditions in the region where it has just suffered a collision, but is otherwise random—in other words, independent of its previous history. This assumption permits one to use the methods of probability theory even though, in classical mechanics, the actual motions of the molecules are regarded as completely determined by their initial configurations. As long as one uses the theory only to calculate properties of a gas that can actually be measured during a relatively short time, the assumption of randomness leads to no serious errors. However, it introduces an element of irreversibility that is inconsistent with the reversibility of the laws of classical mechanics. A reversible process is one that can go equally well forwards or backwards, in contrast to an irreversible process. Lord Kelvin pointed out the importance of irreversible processes in 1852. The irreversible aspect of the kinetic theory is shown most clearly by Boltzmann's "H-theorem," which has led to a considerable controversy over the foundations of kinetic theory. Boltzmann showed in 1872 that a certain quantity, later called H, which depends on the velocity distribution, must always decrease with time, unless the velocity distribution is Maxwell's distribution, in which case H remains constant. In the latter case, which corresponds to the equilibrium state, H is proportional to the negative of the entropy. Thus, the H-theorem provides a molecular interpretation of the second law of thermodynamics or, in particular, the principle that the entropy of an isolated system must always increase or remain constant.

Irreversible processes are those in which entropy increases. The entropy itself can be regarded as a measure of the degree of randomness or disorder of the gas, although it must be recognized that disorder really means a lack of knowledge about the details of molecular configurations. The equilibrium state represents the maximum possible disorder; the H-theorem implies that a gas which is initially in a nonequilibrium (partly ordered) state will eventually reach equilibrium and then stay there forever if it is not disturbed.

If the long-term consequences of the H-theorem were applicable to all matter in the universe, one might expect that the universe would eventually "run down"—although the total energy might always remain the same, no useful work could be done with this energy, because all matter would be at the same temperature. This final state has been called the "heat death" of the universe.

The contradiction between the H-theorem and the laws of classical mechanics is shown by two well-known criticisms of the kinetic theory: (1) the reversibility paradox and (2) the recurrence paradox. The first paradox is based on the fact that Newton's laws of motion are unchanged if one reverses the time direction, so that it would seem to be impossible to deduce from these equations a theorem that predicts irreversible behavior. Kelvin discussed this paradox in 1874, and concluded that while any single sequence of molecular motions could be reversed, leading to an ordered state, the number of disordered states is so much greater than the number of ordered states that it is virtually impossible to stay in an ordered state for any period of time. Thus, irreversibility is a statistical, but not an absolute consequence of kinetic theory. Boltzmann gave a similar answer when the problem was pointed out to him by Loschmidt a few years later.

The second paradox is based on a theorem of Henri Poincaré—if a mechanical system is enclosed in a finite volume, then after a sufficiently long time it will return as closely as one likes to its initial state. Hence H must return to its original value; if it has decreased during some period of time, it must increase during some other period. The time between successive recurrences of the same state for the molecules in one cubic centimeter of air is much longer than the present age of the universe, so one does not have to be concerned about recurrences in any actual experiment. In his attempt to resolve the recurrence paradox, Boltzmann was finally led to a remarkable psycho-cosmological speculation; he suggested that the direction of time as perceived by an animate being is determined by the direction of irreversible processes in his environment and in his body. Thus, when the time comes for a recurrence, entropy will decrease, but subjective time will flow in the opposite direction. The concept of alternating time-directions in cosmic history was further explored by Reichenbach in 1956 and has been proposed again in recent theories of the expanding (and contracting) universe.

Updating of Kinetic Theory. Since the late 1940s, there has been a revival of interest in the classical kinetic theory of gases, based on the assumptions of Clausius, Maxwell, and Boltzmann, and ignoring quantum effects except insofar as these may determine the intermolecular force law. In part this interest grew out of applications involving high speed aerodynamics and plasma physics and, in part, from renewed attempts to construct reliable theories of liquids as well as dense gases. Methods for obtaining accurate solutions of the Boltzmann equation were developed by Grad, Pekeris, Ikenberry, and Truesdell, among others. These solutions were used to describe the behavior of gases in many circumstances more complex than those treated in the 1800s (including the interactions of charged particles and magnetic fields). Problems such as the propagation and dispersion of sound waves were treated by Uhlenbeck and his collaborators.

In 1946, three general formulations of kinetic theory were published (Born and Green; Kirkwood; and Bogoliubov). In each case, the goal was to derive a generalized Boltzmann equation in a form that would be valid when simultaneous interactions among more than two molecules have to be taken into account, and thence to obtain solutions of the equation from which transport properties of dense gases and liquids could be calculated. In each formulation, certain approximations had to be made in order to obtain practical results. Because of the difficulty in estimating the error involved in these approximations, and the great complexity of the equations involved, there was no clear evidence that the results for properties such as the viscosity coefficient would be significantly more accurate than those obtained by Enskog from his modified kinetic theory for dense gases, published in 1922. Eventually, in the early 1960s, attention was centered on the systematic derivation of series expansions for the transport coefficients in ascending powers of the density, together with attempts to calculate the first few terms in such series for special molecular models, such as elastic spheres.

In the meantime, an alternative and apparently more rigorous method for deriving theoretical expressions for transport coefficients, based on the fluctuation-dissipation theory introduced in 1928 by H. Nyquist in electrical engineering problems, was developed by Green, Mori, and Kubo. This method had the heuristic advantage of bringing out clearly the connection between transport theory and the description of fluctuations in equilibrium statistical mechanics. Later, it was proved that the Green-Mori-Kubo method gives results precisely equivalent to those that could be obtained from the Born-Green, Kirkwood, and Bogoliubov methods. Thus, just as in the case of quantum mechanics, several alternative approaches are equally valid in modern kinetic theory.

After intensive efforts to calculate terms in the density expansion of transport coefficients, it was finally discovered in 1965 that such a density expansion does not actually exist, for mathematical reasons associated with the persistence of weak correlations between colliding particles over very long times. The divergence of the expansion (and thus the inadequacy of the approximations on which most earlier theories had been based) was established by a number of investigators, including Dorfman and Cohen, Weinstock, and Goldman and Friedman. The result was an increased overall interest in kinetic theory.

Additional Reading

Brush, S.G., and N.S. Hall: *Kinetic Theory of Gases: An Anthology of Classic Papers with Historical Commentary,* World Scientific Publishing Company, Inc., Riveredge, NJ, 2003.

Liboff, R.L.: *Kinetic Theory,* 3rd Edition, Springer-Verlag New York, LLC., New York, NY, 2003.

Pauli, W., and C.P. Enz: *Thermodynamics and the Kinetic Theory of Gases,* Vol. 3, Dover Publications, Inc., Mineola, NY, 2000.

Perry, R.H., and D. Green: *Perry's Chemical Engineers' Handbook,* 7th Edition, The McGraw-Hill Companies, Inc., New York, NY, 1997.

KININS. Chemical substances that stimulate differentiation of plant cells that otherwise may have lost permanently the power of differentiation. The kinins contain adenine, which seems to give these substances their biological activity. Small bits of a carrot taken from a region containing no cambium can be stimulated to growth and differentiation to produce entire new carrot plants through the addition of kinins and a small amount of indoleacetic acid.

See also **Adenine**.

KISHNER CYCLOPROPANE SYNTHESIS. Formation of cyclopropane derivatives by decomposition of pyrazolines formed by reacting $\alpha\beta$-unsaturated ketones or aldehydes with hydrazine.

KISTIAKOWSKY, GEORGE B. (1900–1982). Born in Kiev, Russia, where he fought in the White Russian Army, he studied in Germany under Maxwell Bodenheim, where he obtained his doctorate in chemistry. In 1926, he came to the U.S. and became an American citizen in 1933. For 41 years, he was Professor of Chemistry at Harvard; in addition to many chemical awards, he was the recipient of the Priestley Medal as well as distinguished Medals of Honor from three Presidents of the U.S. President Eisenhower appointed him as his assistant for science and technology, and he was chairman of the Science Advisory Committee from 1957 to 1963. Among his many achievements in both chemistry and physics, he was a world-famous authority on explosives. A key member of the Manhattan

Project, he devised the detonating mechanism for the first experimental atomic bomb in New Mexico, at which time he was head of the Los Alamos Laboratory. Though he ranks high among those who developed the bomb, he perceived the awesome destructive potential of nuclear weapons and became an ardent opponent of their future use. Resigning from the Pentagon in 1967, he returned to teaching at Harvard. Among other distinguished organizations, he was a member of the Royal Society of London, the AAAS, and the ACS.

KJELDAHL TEST. An analytical method for determination of nitrogen in certain organic compounds. It involves addition of a small amount of anhydrous potassium sulfate to the test compound, followed by heating the mixture with concentrated sulfuric acid, often with a catalyst such as copper sulfate. As a result ammonia is formed. After alkalyzing the mixture with sodium hydroxide, the ammonia is separated by distillation, collected in standard acid, and the nitrogen determined by back-titration.

KLUG, AARON S. (1926–). A South African-born chemist who won the Nobel prize for chemistry in 1982 for his work with the electron microscope and research into the structure of nucleic and protein complexes. His use of crystallographic electron microscopy to analyze the structures of biologically important complex chemicals was noteworthy. He was cited in particular for his establishment of Fourier microscopy.

KOMAROWSKY REACTION. The reaction between certain alcohols and *p*-hydroxybenzaldehyde in dilute sulfuric acid solutions to give soluble colored complexes. 1,2-Propylene glycol gives a colored product while ethylene glycol does not. The reaction has also been employed to determine cyclohexanol in cyclohexanone.

KONOWALOFF RULE. The vapor over a liquid is relatively rich in the component whose addition to the liquid mixture results in an increase of the total vapor pressure.

KOPP'S LAW. The molecular heat of a solid compound is an additive function of the atomic heat capacities of its individual atoms. The molecular volume of a liquid is equal to the sum of the atomic volumes of its constituent atoms.

KRAFT PULP PROCESS. See **Pulp (Wood) Production and Processing**.

KREBS CITRIC ACID CYCLE. See **Carbohydrates**.

KROLL PROCESS. A widely used process for obtaining titanium metal. Titanium tetrachloride is reduced with magnesium metal at red heat and atmospheric pressure, in the presence of an inert gas blanket of helium or argon. Magnesium chloride and titanium metal are produced. The reaction is $TiCl_4 + 2Mg \longrightarrow Ti + 2MgCl_2$. Essentially the same process is also used for obtaining zirconium.

KROTO, SIR HAROLD W. (1939–). A British chemist who won the Nobel prize for chemistry along with Robert F. Curl, Jr, and Richard E. Smalley in 1996, the 100th anniversary of Alfred Nobel's death. The trio won for the discovery of the c_{60} compound called buckminsterfullerene. He received a Ph.D. from the University of Sheffield.

See also **Buckminsterfullerene (Buckyballs); Curl, Robert F., Jr. (1933–);** and **Smalley, Richard E. (1943–).**

KRYPTON. [CAS: 7439-90-9]. Chemical element, symbol Kr, at. no. 36, at. wt. 83.80, periodic table group 18 (inert or noble gases), mp $-156.6°C$, bp $-152.3°C$, density 3.4 g/cm^3 (solid at $-273°C$). Solid krypton has a face-centered cubic crystal structure. At standard conditions, krypton is a colorless, odorless gas and does not form stable compounds with any other element. Due to its low valence forces, krypton does not form diatomic molecules, except in discharge tubes. It does form compounds under highly favorable conditions, as excitation in discharge tubes, or pressure in the presence of a powerful dipole. Krypton forms a hydrate at 14.5 atmospheres pressure and 0°C. The element also forms addition compounds with a number of organic substances, such as $Kr \cdot 2C_6H_5OH$ with phenol, which has a dissociation pressure of 6 to 10 atmospheres at 0°C. Krypton also forms compounds, possibly clathrates, with certain

substances in nonstoichiometric proportions. Examples are its crystalline compounds with benzene and the compounds formed in aqueous solutions of hydroquinone, under 40 atmospheres pressure of krypton, which contains 15.8% krypton by weight. First ionization potential, 13.996 eV; second, 26.4 eV; third, 36.8 eV.

Krypton occurs in the atmosphere to the extent of approximately 0.000114% and thus is the second least abundant of the rare gases in ordinary air. In terms of abundance, krypton does not appear on lists of elements in the earth's crust because it does not exist in stable compounds. However, because of its limited solubility in H_2O, krypton is found in seawater to the extent of approximately 1.4 tons per cubic mile. Commercial krypton is derived from air by liquefaction and fractional distillation. With exception of very special applications, krypton usually is not prepared in pure form, but supplied along with other rare gases, such as argon and neon, for filling fluorescent and incandescent light. As a filter in lamps, the gas assists in reducing filament evaporation and enables higher operating temperatures for lamps. Very high-candlepower aircraft-approach lamps contain krypton. When contained in an electric-discharge tube, krypton, when pure, emits a characteristic pale-violet light; when impure, it emits a brilliant red color characteristic of so-called neon tubes. Krypton also has been used in lasers. There are five natural isotopes ^{78}Kr, ^{80}Kr, and ^{82}Kr through ^{84}Kr, and five radioactive isotopes ^{76}Kr, ^{77}Kr, ^{79}Kr, ^{81}Kr, and ^{85}Kr. The latter isotope is generated in atomic reactors, and is a beta emitter with a half-life of approximately 10.6 years. This isotope, in solid form combined with a hydroquinone, has been used for activating phosphors and in luminous paints. ^{81}Kr, has a particularly long half-life of 2.1×10^5 years. While investigating the properties of liquid air in 1898, Ramsay and Travers found krypton in the residue remaining after nearly all of the liquid air had boiled away. The element then was identified spectroscopically. The element emits a characteristic brilliant green and yellow line in its spectrum.

In 1989, physicists at the Los Alamos National Laboratory's High-Energy Physics facility developed a krypton-fluoride laser at a cost of about $1 million. The laser's energy beam is capable of striking a very small target. Key to the success of the small laser is the ability to compress modest energy into pulses of less than 1 picosecond duration. The laser is being used in research on new materials. With the laser, tiny targets can be heated to temperatures approaching those of the sun's core.

By international agreement in 1960, the *fundamental unit of length*, the *meter*, is defined in terms of the orange-red spectral line of ^{86}Kr. This corresponds to the transition $5p[O_{1/2}]_1 - 6d[O_{1/2}]_1$. One meter = 1, 650, 763.73 wavelengths (in vacuo) of the orange-red line of ^{86}Kr.

Meteorological Effects of Krypton-85 in the Atmosphere. As pointed out by Boeck (1976), projections indicate that ^{85}Kr, a radioactive, chemically inert gas, may be produced and released in such quantities that it will create atmospheric ions at rates comparable to the present ion production rate near the tropical ocean surface. Krypton-85 is a byproduct in nuclear fission reactors and explosions. The ^{85}Kr is sealed in fuel elements of reactors, but during reprocessing of the fuel or plutonium separation, ^{85}Kr is released in a controlled manner to the atmosphere. The radioactive half-life of ^{85}Kr is 10.76 years. With no apparent natural mechanism to remove it, the gas will accumulate in the atmosphere until a balance between the rate of release and decay is reached. Ultimately this could produce a unique form of atmospheric radiation as contrasted with such natural radioactivity as emanates from uranium, thorium, etc., which are limited to producing ions near ground level. In the atmosphere, of course, are several radioactive isotopes (half-lives are given in parentheses): ^{22}Na (2.6 years); ^{32}P (14.22 days); ^{7}Be (53.6 days); ^{33}P (24.4 days); ^{35}S (87.1 days), among others. Not much concern has been given to these isotopes because of their comparatively short half-lives. Boeck (1976) makes a case for the need for more fundamental research into the possible effects of ^{85}Kr in the atmosphere, including possible climatic alterations.

Krypton in Meteorites. Along with other rare gases, krypton has been found in meteorites, notably the Murchison carbonaceous chondrite. Some scientists have been predicting for years that meteorites may entrap materials that date back to the beginning of the universe. For many years, the field was dominated by the dogma of an isotopically and chemically uniform early solar system (Srinivassan and Anders, 1978). However, careful analyses of the Murchison meteorite have produced anomalies in the form of unexpected differences in the ratios of the various rare element isotopes present.

See also **Chemical Elements**.

Additional Reading

Alexander, E.C., Jr. and M. Ozima (editors): *Terrestrial Rare Gases,* Japan Scientific Societies Press, Tokyo, 1978.

Baker, H.: "Getting to the Heart of the Matter (Krypton Laser)," *Advanced Materials and Processes,* 8 (September 1989).

Birgeneau, R.J. and P.M. Horn: "Two-Dimensional Rare Gas Solids, *Science,* 232, 329–336 (1986).

Boeck, W.L.: "Meteorological Consequences of Atmospheric Krypton-85," *Science,* **193,** 195–198 (1976).

Lide, D.R.: *CRC Handbook of Chemistry and Physics,* 84th Edition, CRC Press, LLC., Boca Raton, FL, 2003.

Newton, D.E. and L.W. Baker: *Chemical Elements: From Carbon to Krypton,* Vol. 1–3, Gale Group, Inc., Farmington Hills, MI, 1999.

Srinivassan, B. and E. Anders: "Noble Gases in Murchison Meteorite," *Science,* **201,** 51–55 (1978).

Styra, B. and D. Butkus: *Geophysical Problems of Krypton-85 in the Atmosphere,* Hemisphere Publishing Corporation, New York, NY, 1991.

KUCHEROV REACTION.   The hydration of acetylenic hydrocarbons with dilute sulfuric acid in the presence of mercuric sulfate or boron trifluoride as catalyst.

KUHN, RICHARD (1900–1967).   A German chemist who won the Nobel prize for chemistry in 1938. He worked on cartinoids and synthetic vitamins and discovered the chemical formula for vitamin B_6. He also discovered a method for dissolving simplexes from plants, using invert soaps. He received his Ph.D. in Munich, and went on to teach in Switzerland.

KUHN-WINTERSTEIN REACTION.   Conversion of 1,2 glycols into *trans* olefins by reaction with diphosphotetraiodide (P_2I_4) or other halogenated reagents. This reaction is useful in the preparation of polyenes.

KYANITE.   The mineral kyanite is an aluminum silicate, corresponding to the formula Al_2SiO_5. It is triclinic, and has a good cleavage parallel to the macropinacoid. Its hardness varies considerably, depending on the crystallographic direction from 5 to 7.5; specific gravity, 3.56–3.67; luster, vitreous to pearly; color, commonly blue to white, but sometimes gray to green or nearly black; transparent to translucent; usually found in long-bladed crystals or columnar to fibrous structures. Kyanite is found in some metamorphic rocks as gneisses or mica schists. Of the many European localities for fine specimens might be mentioned the Ural Mountains of the former U.S.S.R.; the Czech Republic and Slovakia; Austria; Trentino, Italy; the St. Gotthard region of Switzerland; and France. In the United States, Chesterfield, Massachusetts; Litchfield, Connecticut; and Gaston County, North Carolina, have furnished fine specimens. Kyanite derives its name from the Greek word meaning *blue,* in reference to the delicate blue of the inner portions of the bladed crystals.

L

LACQUER. A protective or decorative coating that dries primarily by evaporation or solvent, rather than by oxidation or polymerization. Lacquers were originally comprised of high-viscosity nitrocellulose, a plasticizer (dibutyl phthalate or blown castor oil), and a solvent. Later, low-viscosity nitrocellulose became available; this was frequently modified with resins, such as ester gum or rosin. The solvents used are ethanol, toluene, xylene, and butyl acetate. Together with nitrocellulose, alkyd resins are used to improve durability. The nitrocellulose used for lacquers has a nitrogen content of 11–13.5% and is available in a wide range of viscosities, compatibilities, and solvencies. Chief uses of nitrocellulose-alkyd lacquers are for coatings for metal, paper products, textiles, plastics, furniture, and nail polish. Various types of modified cellulose are also used as lacquer bases, combined with resins, and plasticizers. Many noncellulosic materials such as vinyl and acrylic resins are also used, as are bitumens, with or without drying oils, resins, etc.

See also **Paints and Coatings**; and **Nitrocellulose**.

LACTIC ACID. [CAS: 50-21-5]. Alpha-hydroxy-propionic acid, $H-C_3H_5O_3$, formula weight 90.05, colorless or yellowish syrupy liquid, mp 18°C, bp 122°C, sp gr 1.248, miscible with water, alcohol, or ether in all proportions. It was first discovered in 1780 by the Swedish chemist Scheele. The substance exists in two forms: (1) *dextro* lactic acid, which rotates the plane of polarized light to the right; and (2) *levo* lactic acid which rotates the plane of polarized light to the left. A mixture of these two forms is ordinary lactic acid, which does not rotate the plane of polarized light. Ordinary lactic acid is termed dextrolevo lactic acid. Lactic acid is a product of corn refining.

Lactic acid was one of the first biological substances to be investigated from the standpoint of the existence of the two optically active forms.

Lactic Acidosis

Lactic acid is the cause of one of many possible disorders in human acid-base metabolism. Lactic acidosis represents an accumulation of lactic acid in the blood and tissues. This condition gradually depletes the natural buffers in the body and there is a consequent lowering of pH. As described in the entry on **Glycolysis**, lactic acid is the end product of that process. Lactic acid blood levels are determined by at least four factors. The rate of generation of lactic acid; the rate of transport from tissues to plasma and from plasma to the liver (point of utilization of lactic acid); the rate of utilization; and excretion of lactic acid by the kidneys. Normally, all of these functions are maintained in balance to give a normal blood lactate concentration of about 1 mEq/l.

On the generation side, three factors are involved. (1) The availability of oxygen is a major controlling determinant of lactic acid generation because, as adenosine triphosphate (ATP) generation from oxidative phosphorylation diminishes, the cells naturally respond with a greater rate of glycolysis. This increases tissue lactate levels and ultimately lactate blood levels. See also **Phosphorylation (Oxidative)**. (2) If, as may be caused by various factors, there is an increase in pH, the activity of phosphofructokinase will increase (this is the rate-limiting enzyme of glycolysis). With increases in pH, the enzyme is more active and more lactate is formed. (3) Factors that affect the biological oxidation-reduction potentials also influence the rate at which glucose is metabolized to lactate.

Fundamental predisposing conditions causing an increased generation of lactate include: decreased tissue perfusion associated with shock, which may occur in cardiac arrest; increased skeletal muscle activity (the rate of glycolysis increases with exercise; this also may be associated with convulsive states that may follow severe exercise—brought about by increased blood lactate concentrations); large tumors, since tumors (leukemias, lymphomas, etc.) may have an increased rate of glycolysis even in the presence of a sufficient supply of oxygen; and both cyanide and carbon monoxide poisoning, which can increase lactate levels because of insufficient oxygen supply.

On the utilization side, there are a number of influencing factors. The liver is the principal lactic acid utilization center. In liver failure, a surplus of lactate builds up, a condition which may be associated with reduced hepatic perfusion, hepatocyte failure, and hepatocytes replaced by tumor. Blood lactate concentrations are elevated in persons with diabetic ketoacidosis. The observed elevation of lactic acid levels in cases of alcoholism is not fully understood: the condition may increase generation or by decreasing utilization. The latter effect is now favored by many authorities, the theory being that ethanol completes for electrons in the liver, thus decreasing utilization of lactic acid in that organ.

Specifications, Quality Control, and Analytical Methods

Lactic acid is generally sold under four general product categories: synthetic: a highly purified product from a chemical synthesis process. It is water-white, has excellent heat stability, and can be used in both food and industrial applications; fermentation: a food-grade product from carbohydrate fermentation refined by ion exchange and activated carbon. The product contains residual carbohydrate or protein impurities, is not heat stable; heat-stable fermentation: a highly refined, heat-stable product from esterification of fermentation-derived lactic acid, followed by hydrolysis of the recovered ester to produce the acid; and technical: a crude product from either a synthetic or fermentation process, used in industrial applications where high purity is not required.

The fermentation-derived food-grade product is sold in 50, 80, and 88% concentrations; the other grades are available in 50 and 88% concentrations. The food-grade product meets the Food Chemicals Codex III and the pharmaceutical grade meets the FCC and the United States Pharmacopoeia XX specifications. Other lactic acid derivatives such as salts and esters are also available in well-established product specifications. Standard analytical methods such as titration and liquid chromatography can be used to determine lactic acid, and other gravimetric and specific tests are used to detect impurities for the product specifications. A standard titration method neutralizes the acid with sodium hydroxide and then back-titrates the acid. An older standard quantitative method for determination of lactic acid was based on oxidation by potassium permanganate to acetaldehyde, which is absorbed in sodium bisulfite and titrated iodometrically.

Lactic acid is generally recognized as safe (GRAS) for multipurpose food use. Lactate salts such as calcium and sodium lactates and esters such as ethyl lactate used in pharmaceutical preparations are also considered safe and nontoxic. The U.S. Food and Drug Administration lists lactic acid (all isomers) as GRAS and sets no limitations on its use in food other than current good manufacturing practice.

Uses

Currently, the principal use of lactic acid is in food and food-related applications, which in the United States accounts for approximately 85% of the demand. The rest (~15%) of the uses are for nonfood industrial applications. The expected advent of the production of low cost lactic acid in high volume can open new applications for lactic acid and its derivatives, because it is a versatile molecule that can be converted to a wide range of industrial chemicals or polymer feedstock's.

As a food acidulant, lactic acid has a mild acidic taste, in contrast to other food acids. It is nonvolatile, odorless, and is classified GRAS for general-purpose food additives by the FDA in the United States and by other regulatory agencies elsewhere. It is a good preservative and pickling agent for sauerkraut, olives, and pickled vegetables. It is used as acidulant, flavoring, pH buffering agent, or inhibitor of bacterial spoilage in a wide variety of processed foods such as candy, breads and bakery products, soft drinks, soups, sherbets, dairy products, beer, jams and jellies, mayonnaise, processed eggs, and many other processed foods, often in conjunction with other acidulants. An emerging new use for lactic acid or its salts is in the disinfection and packaging of carcasses, particularly those of poultry and fish, where the addition of aqueous solutions of lactic acid and its salts during processing increase shelf life and reduce the growth of anaerobic spoilage organisms such as *Clostridium botulinum*.

Polymers of lactic acids are biodegradable thermoplastics that can be made from a variety of renewable carbohydrate resources. A fairly wide range of properties is obtainable by copolymerization with other functional monomers such as glycolide, caprolactone, polyether polyols, etc. The polymers are transparent, which is important for packaging applications. They offer good shelf life because they degrade slowly by hydrolysis which can be controlled by adjusting the composition and molecular weight. The properties of lactic copolymers which approach that of large-volume petroleum-derived polymers such as poly-styrene, flexible poly(vinyl chloride) (PVC), vinylidene chloride, etc, have been summarized by Lipinsky and Sinclair. There are numerous patents and articles on lactic acid polymers and copolymers, their properties, potential uses, and processes, dating back to the early work by Carothers at Du Pont.

Additional Reading

Bozoglu, T., , Faruk, and B. Ray (editors): *Lactic Acid Bacteria: Current Advances in Genetics, Metabolism, and Application of Lactic Acid Bacteria,* Springer-Verlag New York, Inc., New York, NY, 1996.

Dave, P. and co-workers: *Polymer Prep.* **31**(1), 442–443, 1990.

Goldberg, I., and R. Williams: *Biotechnology and Food Ingredients,* Chapman & Hall, New York, NY, 1999.

Lipinsky, E.S. and R.G. Sinclair: *Chem. Eng. Progr.* **82**(8), 26–32, 1986.

Nakamura, T. and co-workers: "Adv. Biomater." *Biomater. Clin. Appl.* **7**, 759–764, 1987.

Salminen, S.: *Lactic Acid Bacteria: Microbiological and Functional Aspects,* 3rd Edition, Marcel Dekker, Inc., New York, NY, 2004.

Staff: *Code of Federal Regulations,* U.S. Food and Drug Administration, Washington, DC., April 1, 1992.

U.S. Pats. 4,045,418, 4,057,537 (1977), R.G. Sinclair (to Gulf Oil Corp.).

Wood, B., and P.J. Warner: *Genetics of Lactic Acid Bacteria,* Kluwer Academic Publishers, Norwell, MA, 2003.

Wood, B.: *The Lactic Acid Bacteria in Health and Disease,* Aspen Publishers, Inc., Gaithersburg, MD, 1999.

Wood, B. and W. Holzapfel: *The Genera of Lactic Acid Bacteria,* Blackie Academic & Professional, UK, 1999.

Zhu, J. and co-workers: *Proc. of C-MRS Intl. Symp.* **3**, 387–390. 1990.

LACTONE.   An inner ester of a carboxylic acid formed by intramolecular reaction of hydroxylated or halogenated carboxylic acids with elimination of water. They occur in nature as odor-bearing components of various plant products, also made synthetically.

LACTOSE.   [CAS: 63-42-3]. Also called milk sugar or saccharum lactis, formula $C_{12}H_{22}O_{11} \cdot H_2O$. White, hard, crystalline mass or white powder; sweet taste; odorless; stable in air. D 1.525, mp decomposes at 203.5C., soluble in water; insoluble in ether and chloroform; very slightly soluble in alcohol, formed from whey, by concentration and crystallization. Cows' milk contains about 5% lactose.

Uses for lactose include: Infant foods; bacteriology; backing and confectionery; margarine and butter manufacture; manufacture of penicillin, yeast, edible protein, and riboflavin; culture media; adsorbent in chromatography and pharmacy.

See also **Carbohydrates**; and **Milk and Milk Products**.

LACTULOSE.   See **Sweeteners**.

LADDER POLYMER.   An ordered molecular network of double-stranded chains connected by hydrogen or chemical bonds located at regular intervals along the chains. Many complex proteins, including DNA, are of this nature.

LAKES (Colors).   See **Colorants (Foods)**; **Dyes**.

LAMBDA PARTICLE.   A hyperon with a rest-mass energy of 1115.6 MeV, an isospin quantum number zero, an angular momentum spin quantum number $\frac{1}{2}$, and a strangeness quantum number 1. Symbol, λ.

LAMBERT.   See **Units and Standards**.

LAMBERT'S LAW.   See **Bouguer and Lambert Law**.

LAMINAR FLOW.   A condition of fluid flow in a closed conduit in which the fluid particles or "streams" tend to move parallel to the flow axis and not mix. This behavior is characteristic of low flow rates and high viscosity fluid flows. As the flow rate increases (or viscosity significantly decreases), the streams continue to flow parallel until a velocity is reached where the streams waver and suddenly break into a diffused pattern. This point is called the *critical velocity*. See also **Turbulent Flow**.

Laminar flow is characterized by a parabolic flow profile where the maximum velocity at or near the center of the conduit is approximately twice the average velocity in the profile. Laminar flow often is referred to as *viscous flow*, streamline flows, and *low-Reynolds number flow*. Special attention must be paid to the constancy of coefficient of most flowmeters in the region of laminar flow. See also **Reynolds Number**.

Laminar Sublayer

When a fluid is in turbulent flow past a rigid surface, fluctuations of velocity in the direction normal to the surface are inhibited, and very close to the surface they may be negligible. Then the Reynolds shear stress is small compared with the viscous stresses, and it has been common to describe the region as a laminar sublayer. In fact, turbulent fluctuations of velocity in planes parallel to the wall are considerable in comparison with the mean velocity.

See also **Fluid and Fluid Flow**.

LAMINATE.   A composite made of any one of several types of thermosetting plastic (phenolic, polyester, epoxy or silicone) bonded to paper, cloth, asbestos, wood, or glass fiber. High tensile and dielectric strength and low moisture absorption are characteristic of these products. Available as sheet, rod, or tubing in mechanical, electrical, and general-purpose grades (National Electrical Manufactures Association). Plywood is composed of a verneer with grain oriented at a 90-degree angle on successive layers and bonded with a thermosetting adhesive of the urea or phenol-formaldehyde type to give a high strength, dimensionally stable, weather-resistant construction material. It can be made nonflammable by treatment with salt solution. Polyvinyl butyral sheet is used in safety glass.

See also **Reinforced Plastics**.

LAMPBLACK.   A black or gray pigment made by burning low-grade heavy oils or similar carbonaceous materials with insufficient air and in a closed system such that the soot can be collected in settling chambers. Lampblack is markedly different from carbon black, strongly hydrophobic and nonflammable.

Uses for lampblack include: Black pigment for cements, ceramic ware, mortar, inks, linoleum, surface coating, crayons, polishes, carbon paper, soap, etc.; ingredient of insulating compositions, liquid-air explosives, matches, fertilizer, furnace lutes, lubricating compositions, carbon brushes; reagent in cementation of steel.

See also **Carbon Black**.

LANGMUIR, IRVING (1881–1957).   Langmuir was an American scientist whose fields of contribution include chemistry, physics, and technology. He graduated as a metallurgical engineer from the School of Mines at Columbia University in 1903. Postgraduate work in Physical Chemistry under Nernst in Göttingen earned him the degrees of M.A. and Ph.D. in 1906.

Returning to America, Dr. Langmuir became Instructor in Chemistry at Stevens Institute of Technology, Hoboken, New Jersey, where he taught until July 1909. In 1909, Langmuir began working for the General Electric Company in Schenectady, New York where he eventually became Associate Director.

His work on filaments in gases led directly to the invention of the gas filled incandescent lamp and to the discovery of atomic hydrogen. He later used the latter in the development of the atomic hydrogen welding process.

He was the first to observe the very stable adsorbed monatomic films on tungsten and platinum filaments, and was able, after experiments with oil films on water, to formulate a general theory of adsorbed films. He also studied the catalytic properties of such films.

Dr. Langmuir received twenty-three scientific medals and prizes, including the Nobel Prize in Chemistry in 1932 for his work on surface chemistry.

See also **Molecular and Supermolecular Electronics**.

J.M.I.

LANTHANIDE CONTRACTION.   The decreasing sequence of crystal radii of the tripositive rare-earth ions with increasing atomic number in the group of elements (57) lanthanum through (71) lutetium of the Lanthanide Series in the periodic table.

LANTHANIDE SERIES.   The chemical elements with atomic numbers 58 to 71 inclusive, commencing with cerium (58) and through lutetium (71) frequently are termed collectively, the Lanthanide Series. Lanthanum, the anchor element of the series, appears in group 3b of the periodic table. Some authorities consider lanthanum a part of the series. Members of the series, along with lanthanum and yttrium, are described under **Rare-Earth Elements and Metals**. See also **Actinide Series**.

LANTHANUM.   [CAS: 7439-91-0]. Chemical element, symbol La, at. no. 57, at. wt. 138.91, periodic table group 3, homolog of the Lanthanide Series of elements, mp $918°C$, bp $3464°C$, density 6.146 g/cm^3 ($20°C$). Elemental lanthanum has a double close-packed hexagonal crystal structure at $25°C$. The pure metallic lanthanum is silver-gray in color, but with a luster that remains only briefly upon exposure to air, rapidly oxidizing to a white powder. The oxide is hygroscopic and tends to spall, thus exposing fresh surfaces of the metal for oxidation. Thus, the metal must be handled in an inert atmosphere. Chips and powdered lanthanum are quite pyrophoric. Under required inert atmospheric conditions, the metal is easy to work with normal tools, paralleling tin in its workability. There are two natural isotopes ^{139}La and ^{138}La. The latter is mildly radioactive with a half-life of $10^{10}-10^{15}$ years. The element becomes a superconductor below 6 K. There are 19 known artificial isotopes, all radioactive. Of the light (or cerium-group) rare-earth metals, lanthanum is the second most plentiful and ranks 57th in abundance of elements in the earth's crust, exceeding gold, tantalum, platinum, mercury, bismuth, and several other commonly-used elements. The element was first identified by C.G. Mosander in 1839. Electronic configuration

$$1s^2 2s^2 2p^6 3s^2 3p^6 3d^{10} 4s^2 4p^6 4d^{10} 5s^2 5p^6 5d^1 6s^2.$$

Ionic radius La^{3+} 1.045 Å. Metallic radius 1.879 Å. First ionization potential 5.577 eV; second, 11.06 eV. Oxidation potentials La $\longrightarrow$ La^{3+} + 3e^-, 237 V; La + 3OH$^-$ $\longrightarrow$ La(OH)$_3$ + 3e^-, 2.76 V.

Other important physical properties of lanthanum are given under **Rare-Earth Elements and Metals**.

Much of the commercial lanthanum production uses bastnasite, a rare-earth fluorocarbonate found in Inner Mongolia and Southern California, as the source. See also **Bastnasite**. The element is separated from other rare-earth elements by a liquid solvent extraction or an ion-exchange process after acid leaching of bastnasite (or monazite) minerals. Pure lanthanum is obtained by (1) electrowinning from the oxide La$_2$O$_3$ in a molten fluoride electrolyte, (2) electrolysis of fused anhydrous LaCl$_3$, or (3) metallothermic reduction of LaF$_3$ by calcium in a reactor under an inert atmosphere.

Lanthanum metal dissolves readily in dilute mineral acids. The oxide is dissolved by concentrated mineral acids and acetic and formic acids. The metal is a component of mischmetal used for lighter "flints" and in the cores of carbon electrodes for high-intensity lighting. See also **Cerium**. Several of the best grades of optical glass require pure lanthanum oxide as an ingredient for lowering the dispersion of light and for improving the index of refraction. The oxide melts at 2,310°C. The oxide ranks eleventh among the most refractory metal oxides, but finds limited use because of its highly hygroscopic nature. The oxide also is used as a host matrix for fluorescent phosphors and in thermistors and capacitors and other elements of electronic circuitry. Lanthanum oxide combined with transition metal (e. g. Mn, Co, Cr) oxides are being used in solid oxide fuel cells (SOFC) as electrodes. By far, the largest use of lanthanum (mixed with other rare-earths) is for molecular-sieve catalysts for cracking crude petroleum.

As an alloying metal, lanthanum finds broad use. Although lacking mechanical strength, lanthanum has a high affinity for oxygen, sulfur, nitrogen, and hydrogen and thus makes an effective component for scavenging gases from molten metals. Cobalt-base alloys containing lanthanum have shown increased resistance to oxidation and hot corrosion. One of the intermetallic compounds of lanthanum LaCo$_5$ possesses excellent magnetic properties—well in excess of those of alnico and platinum cobalt permanent magnets. The intermetallic compound LaNI.C."eye"$_5$ shows exceptional properties for absorbing and desorbing large amounts of hydrogen at room temperature. It is the major component of rechargeable metal hydride batteries, which is a rapidly expanding major application.

See references listed at ends of entries on **Chemical Elements**; and **Rare-Earth Elements and Metals**.

K. A. Gschneidner, Jr.
B. Evans
Iowa State University
Ames, Iowa

LASER-ENRICHMENT PROCESS (Uranium).   See **Uranium**.

LASER-INDUCED FUSION.   See **Nuclear Reactor**.

LASER-INDUCED REACTION.   See **Photochemistry and Photolysis**.

LASERS.   An acronym for *light amplification by stimulated emission of radiation*. The device is identical in theory of operation to the maser except that it operates at frequencies in the optical region of the electromagnetic spectrum, rather than in the microwave. See also **Maser**. By common usage, these devices are all called lasers, although more precise terminology would aptly use such terms as ultraviolent maser, optical maser, infrared maser, etc. Although the original microwave maser offers an extremely stable frequency source, its main use has been as an amplifier with very low noise output. In contrast, one of the principal attributes of the laser is its ability to produce a single frequency at high intensity in the optical region. Not only may the output be a single monochromative wave, but the wave may be coherent, or in phase, over the whole face of the radiator. In this mode of operation, the laser is an oscillator whose output depends upon the selective amplification of one of the single-frequency modes of the resonant cavity containing the active laser medium.

The decade of the 1980s represented the period when laser technology received profound acceptance, to the point where lasers, like semiconductors and digital computers before them, are now considered important components (hardware) of larger systems. The development of laser technology continued apace during the early 1990s, with numerous new and practical applications occurring at an accelerated rate. lasers now range from microlasers that measure but a few millionths of a meter, for use in optical communications and information processing, to the building-sized x-ray research laser that can deliver 100,000 joules of energy in less than one-billionth of a second—that is, a 10^{14}-watt pulse. One other feat, accomplished in the mid 90's, is a veritable tour de force. A group at MIT managed to coax single excited barium atoms crossing a miniature resonator to serve as a laser. The crucial part is the extreme precision of the mirror arrangement that keeps the photons emitted by the atoms resonating. So far the main goal of this work has been pure physics: to test the theory of quantum electrodynamics (QED).

Historical Perspective

The fundamental theoretical concepts of the maser and laser date back many years, to the early workers in quantum mechanics who appreciated that an incident electromagnetic beam of an appropriate resonant frequency, passing through a medium, might stimulate molecules in an upper quantum energy state to return to a lower quantum energy state, and thus reinforce the primary beam by negative absorption. As early as 1940, Fabrikant (Russia) suggested that experiments be made to prove negative absorption. Fabrikant was the first scientist to introduce the term "collisions of the second kind" (later to prove of importance in laser patent litigation). By this he meant a collision in which some of the kinetic energy of motion of colliding particles is converted to internal energy (or a change in energy state of at least one of the colliding particles).

It is also interesting to note that, as early as 1950, researchers Lamb and Retherford (Columbia University), observed that, if an upper quantum energy state could be caused to be more highly populated (as compared with a lower quantum energy state), the result would be a net induced emission to an incident beam. They further indicated that such a population inversion would probably occur between the $2p$ and $2s$ levels in hydrogen. There shortly followed experiments by Purcell and Pound (Harvard University) who used magnetic techniques to invert the population of a pair of nuclear spin states in lithium fluoride and thus were the first researchers known to directly observe negative absorption of an applied pulse, a phenomenon which they called negative temperature.

Townes (Columbia University) is generally credited with first recognizing that stimulated emission could be utilized in the making of practical hardware. In 1951, Townes described an approach wherein an ammonia beam would be divided into two portions along the lines of experimentation carried out in Germany, wherein a quadrupolar focusing technique was used for separating a beam of molecules into two portions, one of which contained molecules predominantly in the upper of two energy states. Townes proposed to pass the high-energy portion through a cavity resonant at the frequency corresponding to the energy separation of the two states and so reported this proposal. Instead of a millimeter wave generator, a microwave oscillator was used. The latter was given the name maser. This led to the award of U.S. Patent No. 2,879,439, which covered the use of stimulated emission for the amplification and/or generation of oscillatory electromagnetic energy. This patent subsequently was licensed to laser manufacturers.

Following the development of the microwave maser, Schawlow (AT&T Bell Laboratories) and Townes in 1958 proposed that optical maser action could be obtained by placing an active medium in an optical cavity. The medium would be a gas or a solid excited electrically or by light in such a manner that any optical wave present would be amplified as it moved through the material. This work led to the ultimate awarding of the basic laser patent in March 1960 (U.S. Patent No. 2,929,922). During the 1960s, laser research was carried on at a rapid pace at AT&T Bell Laboratories, among others. Considering the work at AT&T Bell Laboratories, in 1960 a laser capable of emitting a continuous beam of coherent light (using helium-neon gas) was developed; in 1961, the continuous-wave solid-state laser (neodymium-doped calcium tungstate) was developed. As a refinement of the helium-neon laser, in 1962 the basic visible light helium-neon laser was developed, of which several hundred thousand are in use as of the early 1980s. In 1964, the carbon dioxide laser (highest continuous-wave power output system known to date) was developed. Other developments during 1964 included the neodymium-doped yttrium aluminum garnet laser, the continuously operating argon ion laser, the tunable optical parametric Oscillator, and the synchronous mode-locking technique, a basic means for generating short and ultrashort pulses. In 1967, the continuous wave helium-cadmium laser (utilizing the Penning ionization effect for high efficiency) was developed. These lasers find use in high-speed graphics and biological and medical applications. In 1969, the magnetically tunable spin-flip Raman infrared laser, used in high-resolution spectroscopy, as well as in pollution detection in both the atmosphere and the stratosphere, was developed. Laser developments continued and, in 1970, semiconductor heterostructure lasers capable of continuous operation at room temperature were introduced. The distributed feedback laser, a mirror-free laser structure compatible with integrated optics, was introduced, in 1971. These were followed by the tunable, continuous-wave color-center laser (1973), techniques for creating optical pulses of less than one-trillionth second duration (1974), and, in the late 1970s, long-life semiconductor lasers for lightwave communications.

Laser Principles

The basic concept of the laser may be described in general terms as follows. Optical maser action can be obtained by placing an active medium in an optical cavity. The medium may be a gas, solid, or liquid which can be excited electrically, by light, chemically, or thermally in such a manner that any optical wave present will be amplified as it moves through the material. One of the first cavities proposed, for example, is a Fabry-Perot resonator—two plane, parallel reflecting plates with a small transmission through which the radiation can escape. Upon excitation of the material, light is emitted with a band of frequencies determined by the particular material. In addition, the direction of emission is nominally random. In the presence of the cavity, some of the waves escape after several back-and-forth reflections from the parallel plates, "walking off" the edge of the reflectors, so to speak. Those waves that travel normal to the walls remain in the cavity and are amplified, provided that they reinforce each other after each round-trip reflection at the two surfaces. this reinforcement or resonance is only satisfied if the spacing of the plates is an integral multiple of one-half the wavelength in the medium. Thus, after a short time, only that frequency which satisfies the resonant condition and those waves traveling normal to the reflector will build up an appreciable intensity. The resultant

light which is partially transmitted through one of the reflectors will thus be a single frequency, or several discrete frequencies if there is more than one cavity resonance within the band of frequencies emitted by the laser material. In addition, the wave front will be in phase across the surface of the reflector since waves striking the surface at normal incidence are amplified most strongly. The resultant beam will then be diffraction limited, i.e., the beam will spread by an angle in radians given approximately by the ratio of the wavelength to the diameter of the beam. In actual practice, single-mode operation is obtained only under special conditions. Several frequency modes may be present because of the multiple resonances of the cavity and numerous "off-axis" modes may be found that correspond to resonant waves, which travel at small angles from the normal to reflectors. These waves "walk off" so slowly that they still are amplified appreciably. Refinements of the simple cavity may consist of concave reflectors that decrease the diffraction losses, or several parallel reflectors, which limit the oscillation to a frequency common to each pair in the set, among other approaches.

Basic Requirements of Lasing Media. The key to successful laser operation is the active medium that amplifies the wave. Qualitatively, a material that fluoresces or exhibits luminescence is an obvious candidate. In fluorescence, electrons are excited to an upper-energy state by short-wavelength light, such as ultra-violet, while luminescence is produced by passing an electron current through the medium, such as in a gaseous discharge. In either process, stimulated emission can occur only if more electrons are produced in the upper-energy state than in the lower or terminal state for the radiating transition. In this case, an incident photon will stimulate further transitions and amplification will result. If the final state were more heavily populated, then the photon would cause more upward or absorbing transitions and the net effect would be absorption.

Laser source requirements vary widely among specific applications. Practical systems require a large range of wavelengths, output powers, spatial and temporal beam characteristics, among other features. In many applications, the following factors apply: (1) an optimum wavelength exists; (2) a specific minimum power level is required; (3) capital and operating costs of the laser must be minimized; (4) size and weight constraints must be met; (5) the laser should be capable of operating for extended periods with little maintenance; (6) the laser output should have a specific temporal and spatial characteristics; and (7) operator safety must be considered in design and use of the overall laser system.

Wavelength characteristics of available laser types are given in Table 1.

Types of Lasers

From the above list of media requirements, it is obvious that a majority of materials are not candidates for making an effective laser. When the laser was first conceived, the possibility of finding numerous materials, as has turned out to be the case, initially was not envisioned by most researchers in the field.

Early Ruby Laser. It is recorded that the first optical laser was demonstrated by Maiman (Hughes Aircraft Research Laboratories) in 1960. Maiman used a ruby single crystal of aluminum oxide doped with chromium impurities. By applying semitransparent reflective coatings on the ends of a rod about 2 inches (5 centimeters) long, Maiman made the cavity and the crystal an integral unit. See Fig.1. Exposure to an intense exciting light from a xenon flashtube was found to invert the population between the red-emitting level and the ground or lowest-energy state of the electrons. The result was a burst of intense red light emanating in a beam through the end reflectors. This was a powerful laser.

TABLE 1. WAVELENGTH RANGE OF SOME LASERS

Type of laser	Wavelength (micrometers)
Solid state	1.06
Ion	0.514, 0.488
Carbon dioxide	10.6
Diode	0.65–1.8
Dye	Visible (tunable)
Helium-neon	0.63
Rare gas halide	0.35, 0.25, 0.19
Helium-cadmium	0.422

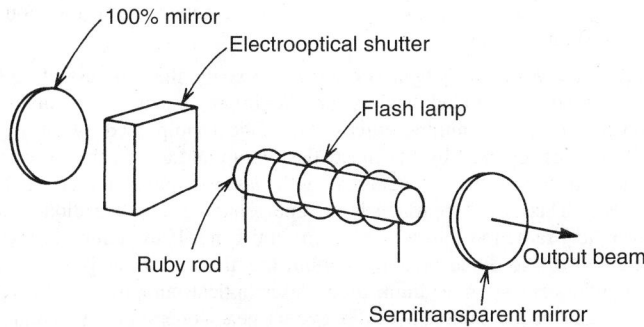

Fig. 1. Schematic diagram of early giant pulsed ruby laser

Because of off-axis modes and multiple resonances, the output is not a single frequency, single plane-wave mode, but generally consists of the order of 100 separate modes. The beam is still quite narrow, being of the order of 1 milliradian or 0.05 degree. As a comparison with conventional light sources, the energy radiated from 1 square centimeter of the brightest flash lamp is less than 10 kW and is distributed over the entire visible spectrum. In addition, the radiation is incoherent and is spread out uniformly in all angles from the source. Thus, the directivity and spectral purity of the laser source are many orders of magnitude superior to that of an incandescent source. The ruby laser suffers from low efficiency, about 1%, and except with elaborate cooling systems, only operates on a pulsed basis. The ruby laser was first reported in the literature (Maiman, 1960) and, in 1967, Maiman was awarded U.S. Patent No. 3,353,115 for the optically pumped ruby laser.

Other crystalline or glass systems with impurity ions have been developed, which yield wavelengths from the ultraviolet to approximately 3 micrometers wavelength in the infrared. Some, such as neodymium-doped yttrium aluminum garnet (YAG), operate in a continuous mode at the one-watt level, while peak powers have reached values as high as 10^{14} W, or greater, in pulses of the order of 10^{-12} second. These ultrahigh powers are obtained in neodymium-doped glass systems, using several stages of amplification and novel pulse-forming techniques.

Gas Lasers. In 1961, Javian, Bennet, and Harriott demonstrated laser action in a gaseous discharge of helium and neon. The parallel-plate reflector cavity was used, but with much greater spacing. Later, concave mirrors were used to decrease the loss of energy out the sides of the cavity. See Fig. 2. The gas laser operated continuously and delivered power up to about one watt. Pulsing the gas discharge yielded peak power as high as 100 watts. The first laser radiated at 1.15 micrometers in the infrared, while further development with different gases yielded output from the ultraviolet to 330 micrometers or 0.33 millimeters in the far infrared. In contrast to the ruby laser, the gaseous laser beam may be diffraction limited and the frequency is pure, i.e., oscillation may be limited to one mode. By careful design, the frequency may be stabilized to within a few thousand cycles per second, or approximately one part in 10^{13} or better. Although the original gas laser utilized electrical excitation of electronic transitions, later versions use vibrational transitions in molecules, such as carbon dioxide, and the excitation may be electrical, chemical, or thermal. The helium-neon laser is commonly used in supermarket bar-code scanners.

In the chemical laser, atomic species such as hydrogen and fluorine can be reacted to produce molecules in an excited vibrational state, which, in turn, yields amplification or oscillation. Electrically excited lasers, particularly those using carbon dioxide at 10 micrometers, can operate at atmospheric pressure, using spark discharges or pre-ionization voltages in the 100-kV range. The high pressure and the powerful electrical excitation result in peak powers in the 10 to 100 MW region. For continuous laser operation, the gas may be circulated rapidly to avoid excessive heating.

In a gas dynamic laser, an appropriate fuel is burned to produce carbon dioxide and nitrogen at high temperature and pressure. When released through a nozzle into the optical resonator region, the gas cools rapidly in terms of its kinetic or translational energy, but the population of the vibrational energy levels of the carbon dioxide molecules becomes inverted since the lower level of the laser transition relaxes much more rapidly. In addition, the vibrationally excited nitrogen molecules are in near resonance with the upper laser state of the carbon dioxide and transfer energy with high efficiency to maintain the inversion. Lasers of this type are capable of producing continuous power at a relatively high level.

Semiconductor Laser. In the semiconductor laser, a solid (semiconductor) material is used. The electron current flowing across a junction between *p*- and *n*-type material produces extra electrons in the conduction band. These radiate upon making a transition back to the valence-band or lower-energy states. If the junction current is large enough, there will be more electrons near the edge of the conduction band than there are at the edge of the valence band and a population inversion may occur. To utilize this effect, the semiconductor crystal is polished with two parallel faces perpendicular to the junction plane. The amplified waves may then propagate along the plane of the junction and are reflected back and forth at the surfaces. See Fig. 3.

A major advantage of the gallium arsenide (GaAs) laser is that it has the electron distribution of a semiconductor. The main difference between electrons in semiconductors and electrons in other laser media is that in semiconductors all of the electrons occupy and thus share the entire crystal volume. Although all semiconductors possess this property, not all of them can be used as lasers. See Fig. 4.

The gain in the material is high enough so that the reflection at the semiconductor-air interface is sufficient to produce oscillation without special reflective coatings. The first such device used gallium arsenide and radiated at 8400 angstroms, or just beyond the visible region in the infrared. The efficiency of this laser, first demonstrated in 1962, was high (about 40%) and the power source was low-voltage direct current.

The compactness and efficiency of the semiconductor laser make it particularly attractive for systems use. Other substances used have included indium arsenide, indium phosphide, indium antimonide, and alloys such as gallium-arsenide-phosphide. These lasers may be tuned over several percent of their normal frequency of operation by varying the current flow through the device. The tuning results from the variations in temperature with current, which, in turn, changes the index of refraction and the resultant resonant frequency of the cavity.

Free Electron Laser. Lasers of this type depart markedly from conventional lasers. Free electron lasers employ an electron beam and a magnetic field. A "free" electron may be defined as an electron that is not bound into atoms or molecules. Traditional lasers use bound electrons. Thus, the conventional laser is limited to producing light (radiation) that is consistent with those frequencies that are specific to the vibration of a

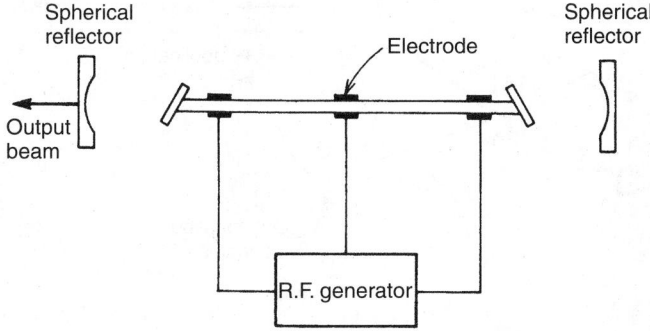

Fig. 2. Schematic diagram of early helium-neon laser with external spherical reflectors. Curvature of reflectors is exaggerated

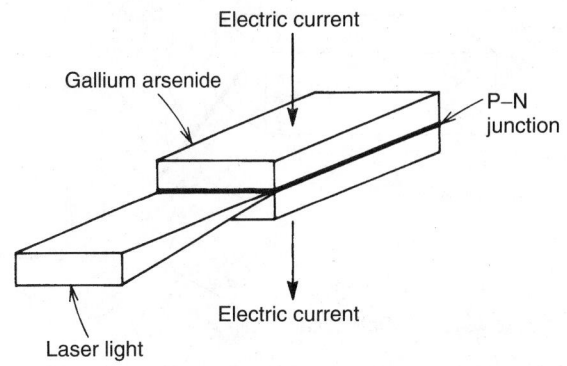

Fig. 3. Schematic diagram of early gallium arsenide laser

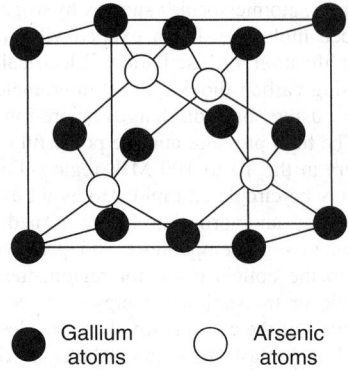

Gallium atoms   Arsenic atoms

Fig. 4. Schematic diagram of a gallium arsenide (GaAs) crystal, which is commonly referred to as a zinc-blende structure. The structure consists of a face-centered cubic lattice of gallium atoms with arsenic atoms positioned on the body diagonals. The arsenic atoms also lie on a face-centered cubic lattice displaced relative to the gallium lattice by one-fourth the body diagonal of the cube

given atom or molecule. Carbon dioxide and helium lasers, for example, are so inhibited.

Free electrons are caused to vibrate by passing them through an alternating magnetic field. By changing (tuning) the apparatus, a broader range of frequencies is obtainable. Coherent radiation can range from the far-infrared to the far-ultraviolet regions of the spectrum. See Fig. 5.

Currently, free electron lasers are large and costly, but are ideal for certain kinds of research. Developers believe that ultimately the free electron laser will find numerous applications beyond research. Kim and Sessler suggest that applications may include surgery, fixing polymers, pharmaceutical manufacture, and lithography. It is predicted that the free electron laser will continue to expand in research usage, including condensed matter studies, nonlinear plasma studies, nonlinear quantum electrodynamics, nonlinear optics, and nonlinear microwaves, as well as microscopy, DNA studies, and cell response research in biology. One study is in progress to determine the feasibility of adapting the free electron laser to perform precision radar measurements in space. Researchers will study the potential of the laser as a compact space radar transmitter for discriminating objects in space, as would be required, for example, in connection with the Strategic Defense Initiative (SDI) program. The program will take advantage of the electron laser's inherent tunability, high power and efficiency, and ability to operate in frequency bands of 100 GHz and higher. The program's ultimate goal is a space-based, multiband, adaptive laser capable of operating efficiently at randomly chosen, stable frequencies.

Researchers have found that electrostatic accelerators are well suited for the far-infrared spectrum. An early device of this type was built at the University of California at Santa Barbara for the main objective of free-electron research and studies in solid-state physics and biophysics. The accelerator has an operating range of 2–6 MV, corresponding to

wavelengths in the range from 100–800 micrometers. Pulse duration is from 3–30 microseconds.

Solid-State Laser Development. Over the years, the progress of solid-state lasers has depended heavily on the improvement of old and the establishment of new pump sources. The helical lamp used by the early ruby laser was replaced by the linear flash lamp and arc discharge lamps. The next step was that of using a diode laser to pump another solid-state laser. This latter approach is advantageous because the diode laser emits optical radiation into a narrow spectral band. If the emission of the wavelength of the diode laser lies within the absorption band of the ion-doped solid-state laser medium, diode laser optical pumping can be very efficient and accompanied by little excess heat generation. By contrast, flash lamp pumping is limited by the broad emission spectrum and by excess heat production.

As pointed out by R.L. Byer (Stanford University), "The diode laser is essentially a continuous wave device with low energy storage capability, whereas the solid-state laser can store energy in the long-lived metastable ion levels." Stored energy can be extracted by Q-switching (rapid switching) to provide peak power levels that are orders of magnitude greater than obtainable from the diode laser per se. Important, too, is the fact that a solid-state laser can collect output from several diode lasers and thus furnish greater average power than is obtainable from a single diode laser. Furthermore, the line width of the diode laser-pumped solid-state laser is many times less than that of the diode laser source. Finally, the solid-state laser source emits optical radiation in a diffraction-limited spiral beam that is easily focused into a fiber or small space.

As early as 1982, a diode laser-pumped miniature Nd:YAG laser with a linewidth of less than 10 kHz was demonstrated. The research in this area continued apace at Stanford University and by a number of commercial electronics firms, with emphasis placed on the development of three-level lasers, Q-switched and mode-locked operation, single-frequency operation (monolithic nonplanar ring oscillator), visible radiation by harmonic generation, and array-pumped solid-state lasers. See Fig. 6.

A new class of lasers, so called vibrionic solid state lasers, can emit—in contrast to other solid state lasers—a comparatively broad range of wavelengths. The lower level in these lasers is a band of energy levels that is caused by interaction between the electron motion and lattice vibrations. With the help of the usual tools (filters, etalons etc.) a narrow and tunable frequency bandwidth is selected for the laser output. The most popular materials are Ti-sapphire, i.e., titanium doped Al_2O_3 with output range from 600 to 1180 nm and Alexandrite, i.e., chromium doped $BeAl_2O_4$ lasing between about 700 and 825 nm.

Laser Microminiaturization—Target Optical Computer. For many years, scientists have accepted the fact that optical computers would not become a reality until optical components of micro size and exceptional performance equivalent to the already existing electronic switches and circuits could be developed. Thus, the optical computer became a major driving force toward the development of optical components. The problem was extraordinarily complex because size reductions of several orders of magnitude were mandatory.

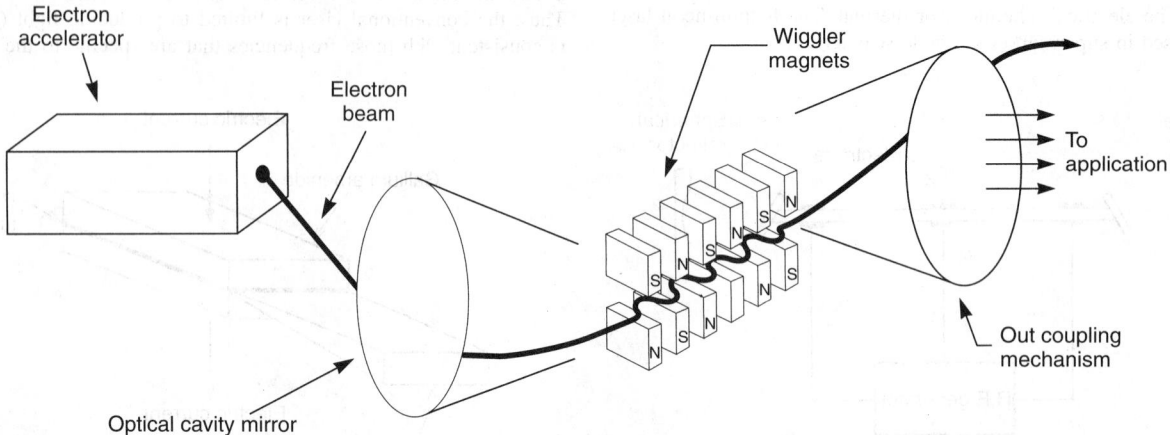

Fig. 5. Schematic diagram of a free-electron laser. Beam of accelerated electrons passes through a field of alternating magnetism (wiggler magnets). The coherent light is generated and contained in an optical cavity defined by mirrors. (*Kim and Sessler.*)

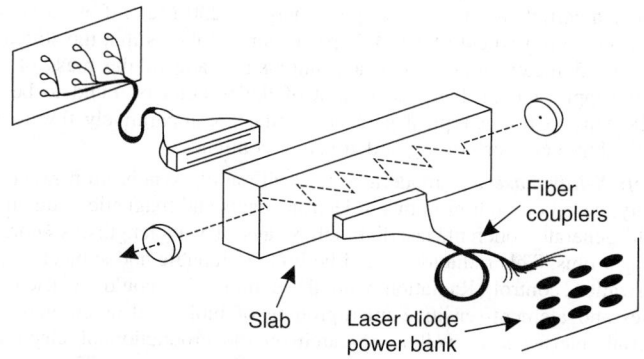

Fig. 6. Diagram of high-average power slab laser oscillator pumped by an array of diode lasers. Such an arrangement offers lower cost, ease of power scaling, and long-term reliability. (*After Byer*.)

As early as 1989, researchers at AT&T Bell Laboratories fabricated more than a million micron-size lasers (microlasers) on a single semiconductor chip, about 7 mm wide by 8 mm long. Individual devices ranged in size from 1 to 5 microns. Thus, these devices were two orders of magnitude smaller than conventional diode lasers. Researchers feel that it may be feasible to manufacture such devices that measure only between 1/2 and 1/4 micron. Traditional devices that measure a few microns wide by several hundred microns long have been well established for use in compact-disk players and fiber-optic communications. Thus, the *much, much smaller* devices coming out of research comprise a major step toward achieving the needs of optical computing. Much competitive research continues during the early 1990s because the market demand for microminiature lasers can be immense.

As pointed out by Jewell and associates (Bellcore), "The principles of operation underlying a diode laser are the same as those for any laser. Atoms in a part of the laser called the *amplifying medium*—typically a solid, liquid, or gas- are pumped, or energized, either electrically or with a source of electromagnetic radiation. When a light wave of a specific wavelength traveling through the amplifying medium encounters a pumped atom, it can induce the atom to release its energy in the form of a light wave at the same wavelength. The process is coherent, which is to say that the crests and troughs of the waves match up, and the intensity of the light increases. Mirrors on each end of the amplifying medium form a cavity, and they force the light to bounce back and forth many times through the medium, maximizing the increase in intensity.

The differences in construction of a microlaser from a conventional gas laser or conventional diode laser are shown in Fig. 7.

To date, manufacture of the very tiny microlasers depend upon critical production techniques. For example, molecular beam epitaxy allows the basic material of each laser to be built up from layers of semiconducting materials. A typical microlaser may comprise 500 or more individual layers.

It was reported in late 1991 that a semiconductor laser that emits a blue-green light had been developed (3 M Company). Manufacturers of optical disks and other consumer electronic devices have been seeking a laser that would produce light in this range of the spectrum. The detailed development procedures required are beyond the scope of this article. However, it has been reported that the layered device is comprised of a gold electrode, a *p*-type zinc selenide, a quantum well (cadmium zinc selenide), an *n*-type zinc selenide, a substrate (gallium arsenide), and an indium electrode. By the end of the decade, progress in semiconductor research has led to reliable diode lasers with output in the blue and green part of the spectrum. In particular, SiC and GaN based systems promise to yield successful applications (See further reading by Fasol and by Fasol, Nakamura and Davies).

Liquid Lasers. Lasers operating in liquid media utilize rare earth ions in such organic hosts as chelates. Laser action is obtained in liquids using a flash tube or another laser as the pump. Early versions used rare earths in an organic liquid, while organic dyes, introduced somewhat later, were found to be more efficient, but required a separate laser for the exciting radiation. The dye laser has the special attraction that one laser may be tuned over a significant fraction of the visible spectrum by using a reflection grating as one of the cavity mirrors. One of the major areas of applications for dye

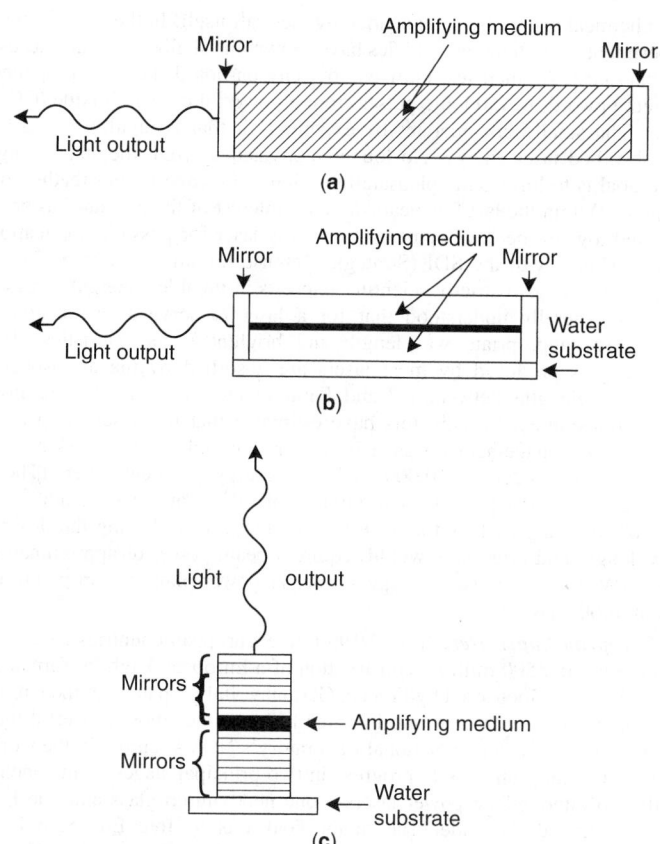

Fig. 7. Comparison of constructional configurations of (**a**) conventional helium-neon laser, (**b**) traditional diode laser, and (**c**) microlaser. Several orders of magnitude in size difference cannot be depicted here. The helium-neon laser ranges from 100 to 1,000 times longer than a traditional diode laser. The latter is some 100 times longer than a microlaser. (*After Jewell, Harbison, and Scherer*.)

lasers is the field of spectroscopy—for both basic research and applications such as combustion diagnostics or trace element analysis (See section on **Laser Spectroscopy**).

Another type of liquid laser utilizes a different principle and depends upon stimulated Raman scattering. Raman laser action was discovered by Woodbury in 1962, using a ruby laser and nitrobenzene. Here the laser excites the nitrobenzene, which in turn shows amplification at a frequency displaced from the ruby line by the vibrational frequency of the molecule. There is not true inverted population in this case. The incident photon is scattered by the molecule, which absorbs an amount of energy determined by its vibrational energy. The molecule is left in an excited state and the scattered photon is frequency-shifted by the energy loss. The process may be stimulated inasmuch as the rate at which the scattered photons are produced is proportional to the number of photons already present in the cavity at the scattering wavelength. As in the normal stimulated emission case, the frequency and phase of the output wave are identical with the wave that stimulates the scattering.

The Raman laser normally operates using the Stokes line, or the wavelength corresponding to the loss of one vibrational quantum. Other modes of operation utilize the second or third Stokes lines corresponding to double or triple vibrational absorptions. Similarly, higher-order effects in the medium may produce a series of anti-Stokes lines, which correspond to vibrational energy being added to the initial energy of the photons from the driving laser. The wavelength range of Raman lasers using different liquids is from the visible to the near-infrared.

X-Ray and Other Very High-Power Lasers. Almost since the inception of laser technology, the laser has been of interest to the military in connection with a variety of applications—weapons, radars, illuminators, rangers, etc. Among the highly sophisticated laser applications long envisioned by the military is the use of a laser to propel a spacecraft. In early concepts, a laser beam produced at ground level would vaporize an appropriate fuel, which could be water, and the supersonic jet caused by vaporization would be sufficient to place the vehicle into orbit without any chemical energy being expended by the craft itself. In the late nineties, experiments with tethered vehicles have shown initial albeit limited success in this quest. A small test craft without any on-board fuel was propelled 75 feet upward by 450 joule pulses from a powerful carbon dioxide (CO_2) laser. The small craft operates by expelling air that is rapidly heated by the absorbed laser energy. No longer motivated by SDI, the stated long-term goal is to loft small "picosatellites" into orbit (See further reading by Appell). Within the last few years, a major interest of the military has been directed toward the development of an x-ray laser for possible application in connection with the SDI (Strategic Defense Initiative) program. There are, of course, also strictly scientific interests in tunable coherent x-rays.

It is generally understood that for a laser to serve as a weapon it must have appropriate wavelength and brightness characteristics. The wavelengths produced by most lasers are absorbed by the atmosphere. Laser wavelengths between 0.3 and 1 micrometer are generally the most easily transmitted. Investigators have estimated that if a laser firing over a 3000 km engagement distance is to burn through a missile skin in 1 second, it must deliver 10,000 joules of energy per centimeter. (These requirements correspond to a brightness of 10^{21} watts per steradian, or unit of solid angle.) It is further estimated that a laser having the desired wavelength and brightness would require a beam power of approximately 100 MW. By comparison, a typical nuclear power plant has an output of about 1000 MW.

European Superlaser. In mid-1990, five European countries agreed to fund ($200 to $500 million) construction of a European High Performance Laser Facility. Sponsored by France, Germany, Italy, Spain, and the United Kingdom, the new laser would be three to four times more powerful than the Lawrence Livermore National Laboratory's NOVA, currently the world leader. The program was to progress in two principal stages, commencing with two intermediate power lasers—one neodymium glass and one KrF laser, built side by side. The major goal was to free European laser scientists from the dependence on high-powered machines in the United States and Japan. A superlaser of this type can be used to investigate some fundamental problems of physics. The intense pulse of the proposed laser would create conditions even hotter than the core of a burning star.

As of the late 1980s and early 1990s, four kinds of laser were under development. (1) *Chemical lasers* utilize chemical reactions between two gases to generate radiation. This technology is probably the most mature of the four kinds of lasers. It has been reported that the brightest of the chemical lasers is the MIRACL (mid-infrared advanced chemical laser), which in a demonstration at the White Sands Missile Test Range (New Mexico) destroyed a mock-up of a missile standing about a half-mile distant from the laser. The MIRACL was estimated to have a brightness of about 10^{17} watts per steradian, short of the SDI goal by a factor of about 10,000. (2) The *Excimer* (meaning excited dimer) consists of an unstable compound made up of two molecules. An electric discharge excites the molecules to form the dimer, and in breaking down, the dimer emits radiation. In some way, this radiation triggers a cascade of reactions that produce a laser beam. Radiation in the beam is estimated between 0.2 and 0.4 micrometer. A krypton-fluoride laser, tested at the Los Alamos National Laboratory, is estimated to operate at a wavelength of about 0.25 micrometer, delivering 10,000 joules of energy in a 380 nanosecond pulse. (One nanosecond = One billionth of a second.) It is reported that while the energy produced meets SDI goals, the pulse duration is off by a factor of about 3 million. (3) In the *free-electron laser*, a beam of electrons passes by a series of so-called wiggler magnets, which cause the electrons to vibrate and emit radiation. The wavelength can be tuned, theoretically, to any value between about 0.1 and 20 micrometers. The smaller the wavelength, the greater the energy. It is reported that a free-electron laser operated at wavelengths down to 10 micrometers in tests at Los Alamos. More recent research targets a 1-micrometer radiation of 100 microsecond pulses, containing 30 kW of power. As of the mid-1990s, both excimer and free-electron lasers were relatively poor converters of electric energy into beam energy, requiring massive power supplies—hence difficulties in locating the needed equipment in space. See prior description of free-electron laser in this article. (4) The *X-ray laser* also appears to be plagued with heavy and costly support equipment. Essentially, the x-ray laser consists of a nuclear explosive surrounded by an array (cylindrical configuration) of metal fibers. The emission of x-rays during the nuclear explosion stimulates the emission of a beam of x-rays from the fibers. This occurs within a microsecond prior to immolation of the device per se. For obvious reasons, further details remain sparse. It has been reported that to make an effective

weapon, a particle beam would require energy of 250 MeV. If one assumes an acceleration gradient of 10 MeV per meter, it follows that the structure must be 25 meters long. When accounting is made of the mass of the power supply and its fuel, the weight of the weapon is found to be 50 to 100 tons. (Current typical payloads weigh a comparatively few tons.) Much additional work of a guarded nature continues.

Soft X-Ray Laser. A modern 1- to 2-billion-eV synchrotron radiation facility, based on high-brightness-electron beams and magnetic undulators, would generate coherent, laserlike soft x-rays of wavelengths as short as 10 angstroms. This radiation would be broadly tunable and subject to full polarization control. Radiation with these properties could be used for phase- and element-sensitive microprobing of biological assemblies and material interfaces as well as research on the production of electronic microstructures with features smaller than 1000 angstroms. These short-wavelength capabilities, which extend to the K-absorption edges of C, N, and O, are neither available nor projected for laboratory XUV (soft x-ray and ultraviolet radiation) lasers. Higher-energy storage rings (5 to 6 million eV) would generate significantly less coherent radiation and would be further compromised by additional x-ray thermal loading of optical components. Synchrotron radiation is discussed further in the article on **Particles (Subatomic)**.

To extend scientific and technological opportunities, authorities suggest that a bright source of tunable, partially coherent, XUV radiation is needed. Coherence, in the limited sense used here, refers to the ability to form interference patterns when wave fronts are separated and recombined. The availability of a tunable source of coherent soft x-rays, combined with other developments in x-ray optical techniques, would make it possible to construct an x-ray microprobe of sufficient intensity to permit fundamentally new, phase-sensitive experimentation in a number of scientific and technological fields. Various imaging and scattering techniques would be enhanced by the greatly increased photon flux available to study small samples, as well as providing the capability of tuning the radiation to the wavelength of interest. For example, with soft X-rays well matched to the absorption edges of elements, such as carbon (284), nitrogen (400), and oxygen (532), as well as other elements of relatively low atomic numbers (Na, P, S, K, and Ca), it should be possible to study elemental distributions and motion within biological specimens without the need for dehydration, fixing, or staining. Three-dimensional imaging, made possible by combining partially coherent undulator radiation and x-ray microholographic techniques, would complement the information available from electron microscopes.

Extensive development and experimental designs of soft x-ray lasers have been underway at the Lawrence Livermore National Laboratory and the Princeton Plasma Physics Laboratory, among other research institutions. Researchers Suckewer and Skinner (Princeton University Plasma Physics Laboratory), in an early 1990 paper, observe, "Most of what is known about the internal structure of cells has been learned by the development and application of the techniques of electron microscopy. This knowledge rests on the premise that the intensive procedures necessary to prepare a specimen for electron microscopy do not significantly influence the structure, form, and high-resolution detail observed. Nonetheless, unanswered questions remain about the fidelity of the image of a cell that has been fixed, stained with heavy metals, and sectioned to the original living cell. X-ray microscopy offers a new way to look at unaltered cells in their natural state." X-ray laser microscopy can offer numerous advantages in this regard.

In summarizing the current status of soft x-ray laser research, the aforementioned researchers comment, "The general impact soft X-ray lasers will have in science and technology will depend on improvements in their performance and cost. It is necessary for their successful commercialization that these devices operate routinely at high gain-lengths ($GL > 4$), with the use of a low-cost driver laser, and this needs more system development and engineering. Most applications of visible-wavelength lasers are based on the fact that the brightness of these lasers is several orders of magnitude greater than that of conventional spontaneous emission sources, and this is achieved principally by the laser cavity mirrors. This technology is significantly more difficult in the x-ray region because of intrinsic limitations of x-ray absorption in materials and present limits in the soft x-ray laser pulse lengths. Nevertheless, a 'revolution' in x-ray optics is under way and the precedent of visible-wavelength lasers illustrates the potential benefits awaiting the creative inventor of applications of this technology to novel fields."

Laser Applications

During the early phases of laser development (late 1950s to early 1970s), there was a high tempo of research activity and confidence in the ultimate potential for practical applications. Some of the early suggestions for laser use included instrumental applications in metrology and spectroscopy, as well as working tools for industry, such as cutting, welding, and annealing. But during that period it was also observed by some researchers that the laser was an invention for nonexistent needs. The decade of the 1980s removed all such doubts, and as science entered the 1990s, the laser had become established as an essential component in numerous laboratory research programs and industrial and medical applications. Just a cursory inspection of the literature during the late 1980s and early 1990s is indicative of the wide scientific interest in the laser.

Atomic Cooling and Trapping. During the last few years, the ability to control the position and velocity of isolated atoms and microscopic particles has progressed markedly. By the end of 1998 molecules have also been laser cooled and several groups have obtained sufficiently low temperatures and high densities to achieve Bose-Einstein condensation of trapped atoms. In such a state of matter—predicted by Bose and Einstein—the quantum nature of the atoms causes them to lose their individual existence and to coalesce into one collective system. The tremendous success of these efforts has been recognized by the award of the 1997 Nobel Prize in physics to Steven Chu, Claude Cohen-Tanoudji and William Phillips for development of methods to cool and trap atoms with laser light. As pointed out by S. Chu (Stanford University) in a late 1991 paper, "Light can exert forces on an atom because photons carry momentum. The exchange of photon momentum with an atom can occur *incoherently*, as in the absorption and reemission of photons, or *coherently*, as in the redistribution (or lensing) of the incident field by the atom."

Coherent interaction is called the *dipole force*. The incoherent interaction that alters the momentum of an atom is called the *scattering force*.

Successful atom manipulation, however, often depends more upon cooling the atoms than upon exciting the aforementioned forces. Dramatic cooling of atoms to extremely low temperatures is accomplished by employing counterpropagating laser beams, arranged along x, y, and z axes—in essence, creating three-dimensional cooling. As pointed out by Chu, "Because the cooling force is viscous (linearly proportional to the velocities of the atom for low velocities), we named the laser beams that generate the drag force, 'optical molasses'."

In 1991, a research team (Ecole Normale Superieure, Paris) reported the cooling of a sample of cesium atoms to 2.5 μK. At about the same time, a research group (Joint Institute for Laboratory Astrophysics, Boulder, Colorado) reported the achievement of 5 μK. The aforementioned "optical molasses" technique was used in both cases.

Laser cooling, trapping, and related techniques are finding numerous research and practical applications. For example, practical laser-cooled atomic clocks are now possible, constituting a major improvement in accuracy over present atomic clocks. As mentioned by Chu, "A cesium time standard based on a sealed design for which the cooling, manipulation, and detection of the atoms are all done with diode lasers should exceed the stability of the best present-day time standards."

In an excellent paper, Chu (See reference) observes, "Perhaps the most exciting applications in the field of laser cooling and trapping will come out of the ability to study problems in polymer physics and biology on a single molecular basis. Normally one examines the behavior of a large number of molecules, and the fundamental chemistry of the molecules must be inferred from the average behavior of the entire ensemble. On the other hand, the processes that govern the behavior of a single molecule are important: for example, the nucleus of a cell has a single molecular copy of its genetic blueprint, and its chemistry depends in part on the chemistry of single molecules."

In a 1990 paper, Zewail (See reference) describes how atoms can collide, interact, and give birth to molecules in less than a trillionth of a second. As an example of how high-speed imagery has improved over the years, he compares photos of a galloping horse (10 meters per second) taken in 1887 with quantitative observations (made in 5 trillionths of a second) of hydrogen iodide colliding with carbon dioxide to form carbon monoxide, hydroxide, and iodine.

Lasers as Mini-Manipulators. As scientists continue to probe the very minute aspects of natural organs and substances (nanotechnology), small lasers have been found to possess "manipulative" abilities of a kind not envisioned in the early years of laser technology. Scientists (Massachusetts Institute of Technology) in 1990 reported of how lasers can be used effectively as manipulators at the microscopic level. In a study of "mechanoenzymes," which are responsible for the rotary motions of flagella, laser light was used to lift up, move, and position microscopic objects with the "pressure" of the laser light itself, a phenomenon that has been described by Amato as "akin to a blast of air levitating a plastic ball."

Laser mini-tweezers also have been used to clip off regions of chromosomes, moving organelles around inside cells, pushing molecules tiny distances within crystals, and, when used as tiny scalpels or scissors, to catch, trap, puncture, and splice subcellular structures. Recently, measurements have been made of the elastic properties of DNA. Also, it has been found that bacteria can be moved around in a water solution without apparent damage to the organism. Medical applications are described later.

Laser Spectroscopy. In the early years of laser technology, spectroscopy was one of its major uses, an application that has expanded markedly during the past few decades. The review by Gupta provides a good starting point for further reading since it includes an—at the time—up-to date resource letter featuring a large number of annotated references. The techniques of laser spectroscopy parallel those of microwave or radio-frequency spectroscopy, but because lasers are imbued with high spectral purity, they permit vastly improved resolution of fine detail. Early lasers were limited to molecular lines that were coincident with the laser wavelengths. Then lasers using fluorescent organic dyes appeared. These instruments had relatively wide emission bands, offering a tuning capability. Both continuous-wave and pulsing dye lasers have been widely used in most of the visible and near-visible ranges of the spectrum. During the interim, much progress has been made, particularly in providing tunability to lasers. For example, a methyl fluoride molecular gas laser is continuously tunable over broad portions of the far infrared, a region that previously had been difficult. A highly schematic diagram of the operating principle used in early laser-probe emission spectrography is given in Fig. 8.

A particularly interesting development is that of the so-called "atomic fountain." As noted by Chu in 1991, "The precision of a spectroscopic measurement depends on both the high Q (Q = quality factor of the resonance defined by $Q = V/\Delta V$) and the signal-to-noise ratio of the signal. Thus, it is important to create a high-flux source of cold atoms. Also, many applications would benefit from a continuous beam of atoms instead of the pulsed sources." An extreme limit of a slow beam is an "atomic fountain," which first was envisioned by Zacharias in the early 1950s. A group of Stanford University scientists has constructed an atomic fountain by first trapping atoms from a thermal beam in a magneto-optic trap and then pushing the atoms upward with a pulse of light from a continuous-wave laser. See Fig. 9.

One reason for the rapid advancement of chemical reaction dynamics research has been the availability of tunable laser sources that operate throughout the infrared, visible, ultraviolet, and vacuum ultraviolet regions of the spectrum. By using nonlinear optical techniques, the outputs from high-power, pulsed visible dye lasers can be summed and mixed to yield useful tunable ultraviolet and vacuum ultraviolet light, with wavelengths as short as 100 nm. Techniques have been developed to probe almost any kind of atomic or molecular state, quite often with sensitivities approaching number densities of 10^5 cm^{-3}, and, in special situations, with detection sensitivity for single atoms.

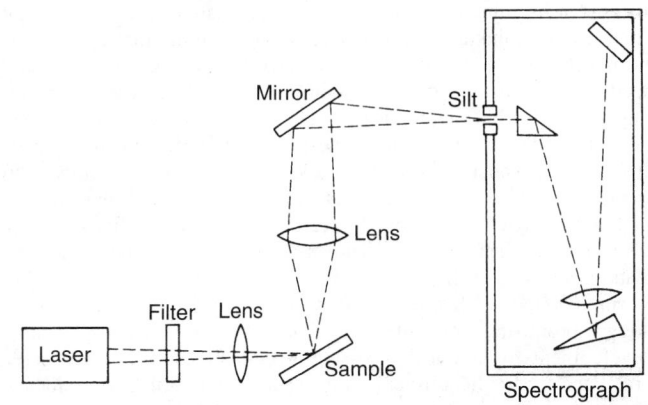

Fig. 8.   Operating principle of laser-probe emission spectrography

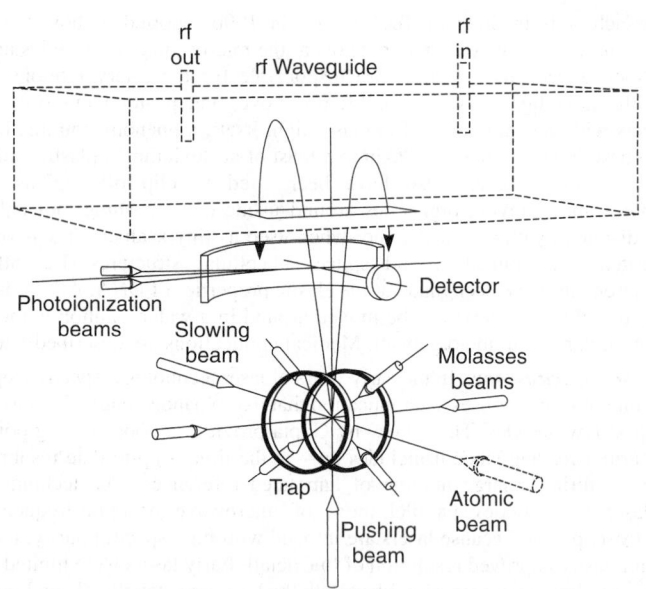

Fig. 9.   The atomic fountain makes it possible to determine precisely the energy states of atoms. Upon injection, the atoms in question are slowed down by a laser beam. Then, the atoms are captured and cooled by means of a magnetic field and several light beams. The cooled atoms follow a ballistic trajectory through a radio frequency (rf) waveguide and a resonant photoionization detection region. (*After Kasevich, Riis, Chu, and DeVoe.*)

In a detailed reference, Grant and Cooks (Purdue University) explain in considerable detail the combining of the latest advances in mass spectrometry with laser spectrometers. This technique is contributing in a major way to studies of chemical dynamics, cluster structures, and reactivity, and to the elucidation of the properties of highly excited molecules and ions.

Laser Remote Sensing of Atmospheric Properties

LIDAR (an acronym for light detecting and ranging) is analogous to radar. In lidar, the projection of a short laser pulse is followed by reception of a portion of the radiation reflected from a distant target or from atmospheric constituents, such as molecules, aerosols, clouds, or dust. As explained by Killinger and Menyuk (See reference), the incident laser radiation interacts with the aforementioned constituents to cause alteration in the intensity and wavelength in accordance with the strength of the optical interaction and the concentration of the interacting species in the atmosphere. Information on both composition and physical state of the atmosphere can be deduced from lidar data. The range of the interacting species can be determined from the temporal delay of the backscattered radiation. See Fig. 10.

Among specific uses of lidar have been: (1) measurement of movement and concentration of urban air pollution; (2) determinations of chemical emissions from and in the vicinity of industrial plants; (3) determination of atmospheric trace chemicals in the atmosphere; (4) measurement of the velocity and direction of winds near storms and airports, including windshear and gust fronts; and (5) determination of the global circulation of volcanic ash emitted into the atmosphere, relatively recent examples including Mount Pinatubo and Kilauea; among several other applications.

Shortly after the discovery of lasers (early 1960s), Fiocco and Smullin bounced a laser beam off the moon (1962). These researchers also investigated the turbid layers in the upper atmosphere. As early as 1963, Ligda used a ruby laser to obtain the first lidar measurements of cloud heights and tropospheric aerosols. In 1964, Scotland used a temperature-tuned ruby laser to detect water vapor in the atmosphere. Lidar, in recent years, has been greatly improved because of the availability of several kinds of laser sources and improvements made in optical instrumentation and data processing.

As summarized by Killinger and Kenyuk, the future of laser remote sensing is promising and will depend upon several factors, including: (1) development of practical, eye-safe laser sources that cover certain spectral gaps where lidar is currently weak; (2) a further simplification of lidar systems, including lowering size and cost of equipment needed; and (3) more experience to be gained from promising new applications.

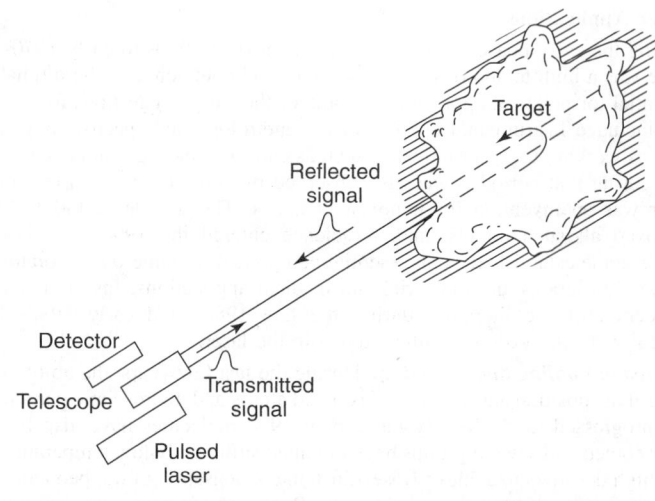

Fig. 10.   Basic components of lidar system used for remote sensing of the atmosphere. Backscattered information sometimes will contain spectral information useful for determining composition and physical characteristics of the cloud or of the intervening atmosphere

Among these new applications are: (a) detection of methane gas leaks in coal mines, using a diode laser lidar system; (b) detection of methane and natural gas leaks in industrial plants, using a laser coupled to a low-loss optical fiber network; (c) measurement of global wind fields through the use of Doppler lidar systems mounted in a satellite as a means for improving weather forecasting; and (d) the planned use of lidar on the NASA space-borne Earth Observing System for measurements of global temperature, water vapor, and pressure.

Classification of Lidar System. Lidar systems can be classified on the basis of particular optical interactions which they utilize. Classes of lidar include:

1. *Atmospheric backscatter lidar*, wherein the lidar system transmits one laser wavelength and detects changes in the backscatter due to the aerosols or dust in the atmosphere. This is the most common type of lidar and consists of a nontunable, high-power, pulsed laser. Atmospheric constituents having comparatively large optical scattering cross sections are relatively easy to detect. These systems are used in tracking turbid effluent and gas plumes from factories as well as for mapping rain, snow, ice crystals, and dense clouds in the atmosphere. This type of system was used for checking volcanic ash in the atmosphere.

2. *Differential-absorption lidar* (DIAL), a system which measures the concentration of a molecular species in the atmosphere. This is accomplished by transmitting two wavelengths, only one of which is absorbed. The difference in the intensity of the returns at the two wavelengths is measured. Backscatter in DIAL may come from a hard target or aerosols and dust. One wavelength will be absorbed by the target molecules; the other wavelength will not be absorbed. Many DIAL studies have been carried out in the infrared (IR) range, where almost all molecules of interest have extensive absorption bands. Molecules so far studied include SO_2, NH_3, O_3, CO, CO_2, HCl, NO, N_2H_4, N_2O, and SF_6.

3. *Fluorescence lidar* uses two wavelengths (as in DIAL) plus spectrometric techniques for separating the wavelength-shifted fluorescence signal from the strong Rayleigh backscatter in the atmosphere. The laser is tuned to an absorption line of the species to be measured. Reradiated fluorescence is detected by selective spectral filtering of the returned radiation. The fluorescence radiation may be at the same wavelength as the excitation wavelength, or it may have a longer wavelength because of the red-shift. The backscatter coefficient for fluorescence is greater in the ultraviolet (UV) than in the IR—this due to combined effects of absorption cross section, which is greater in the UV than in the IR. For some applications, fluorescence lidar is limited for remote sensing because of detector sensitivity coupled with solar background radiation. The latter tends to confine fluorescence measurements to nighttime studies and to wavelengths shorter than 1 micrometer, where photomultiplier detection can be used. Nevertheless, some investigators have been quite successful in using the method, particularly in the study of alkali metal (Na, K,

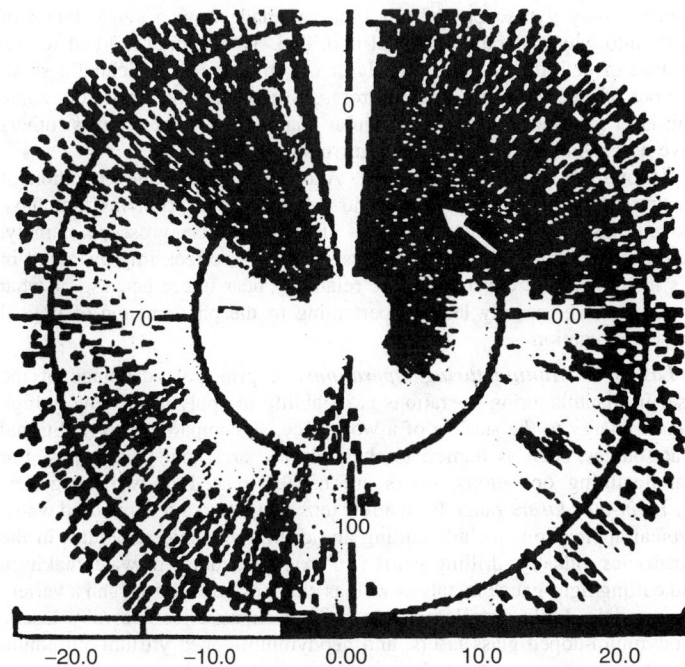

Fig. 11. Doppler lidar measurement of wind direction and velocity near an airport during a storm. White arrow points to presence of a strong, localized downburst-gustfront

Li, and Ca) profiles at altitudes of 80 to 100 km. The method also has been useful for studying the hydroxyl free radical (OH). This radical is of principal interest because of the catalytic role which it exerts in atmospheric chemistry. The OH radical, along with chlorine and nitrogen oxides, is involved in the ozone destruction cycle.

4. *Raman lidar*, a method that is limited by the small optical interaction strength for Raman scattering. High-energy pulsed lasers are employed in this method. The method is limited to the UV or visible regions to permit the use of sensitive photomultiplier tubes for detection. Raman lidar has been used effectively for species that either are at close range or in high concentration, such as N_2, O_2, and H_2O. As mentioned by Killinger and Menyuk, this method does have some attractive features, the most noteworthy of which is that the laser wavelength need not be tuned across an absorption line because the spectral information is given by the frequency shift of the emission (independent of laser wavelength).

5. *Doppler lidar*, a method that detects only a very narrow spectral range ($\sim 10^{-5}$ nm) that encompasses the Doppler-shifted backscatter lidar return. Doppler shifts in the return lidar signals have been used to measure wind velocities and to differentiate between molecular and aerosol returns in the atmosphere. Optical heterodyne techniques are used to detect the shifts which are very small—for example, a fractional change in frequency of about 10^{-8} for a velocity of 1 m/sec at a wavelength of 10 micrometers. Carbon dioxide lasers which provide high power and stable single-frequency operations, are commonly used in Doppler lidar systems. These are in the 10-micrometer range. The system has provided information on boundary layer flow near storm gust fronts and wind shears

near airports. They also have been used to measure aircraft vortices and clear air turbulence. See Fig. 11.

Doppler broadening effects also have been used to separate backscattered lidar signals into molecular and aerosol components. Characteristics of some lidar systems are summarized in Table 2.

Laser Techniques in High-Pressure Geophysics

Laser techniques used in conjunction with the diamond cell make it possible to study the high-pressure properties of material that heretofore had to be inferred from samples found on the surface. Spontaneous Raman scattering of crystalline and amorphous solids at high pressure demonstrates that dramatic changes in structure and bonding occur on compression. High-pressure Brillouin scattering is sensitive to the pressure variations of single-crystal elastic moduli and acoustic velocities. Laser heating techniques with the diamond anvil cell can be used to study phase transitions, including melting, under deep-earth conditions. Laser-induced ruby fluorescence has been essential for the development of techniques for generating maximum pressures now possible with the diamond anvil cell, and currently provides a calibrated *in situ* measure of pressure well above 100 gigapascals. Hemley, Bell, and Mao (Carnegie Institution of Washington) point out that applications of new spectroscopic techniques, such as double resonance, ultrafast kinetics, Fourier-transform, Raman, and nonlinear optical methods, are likely prospects in future work on geophysical problems with the diamond anvil cell. Recent high-pressure studies involving the use of picosecond spectroscopy and hyper-Raman scattering of perovskites may be representative of this trend. Time-resolved studies may permit the detailed investigation of the kinetics of high-pressure phase transitions and the rheology of minerals under *in situ* deep-earth conditions. The combination of spectroscopic and x-ray diffraction probes with laser-heating techniques may yield detailed structural information on earth materials at high temperatures and pressures, thus advancing an understanding of the connection between atomic-scale properties and global deep-earth processes.

Laser Metrology

Lasers are widely used for making precision measurements of geometric variables. In the early 1960s, laser pioneers demonstrated precise measurements with the device. Lasers introduced the concept of frequency metrology as contrasted with wavelength metrology. In an early experiment with a super-stabilized laser, scientists at the Massachusetts Institute of Technology during the early 1960s worked out a laser version of the famous Michelson-Morley experiment at Case Institute of Technology in Cleveland. The MIT scientists concluded that an advance in measurement sensitivity by a factor of 1000 over the Michelson-Morley data was potentially available through the use of frequency rather than length metrology. In 1962, the first laser measurement of the speed of light (c) was made, yielding a value of 299, 792, 462 ± 18 meters per second.

Since that time, several more sophisticated determinations have been made by leading metrology laboratories, including the National Institute of Standards and Technology (Boulder, Colorado), the National Physical Laboratory (United Kingdom), the Laboratoire de Physique des Lasers (Villetaneuse, France), and the Laboratory for Spectroscopy (Russia),

TABLE 2. SUMMARY OF LIDAR SYSTEM CHARACTERISTICS

Type of lidar	Type of laser used	Nominal accuracy	Range (km)	Atmospheric targets
Atmospheric backscatter	Ruby, Nd:YAG	1–10%	10–50	Dust, clouds, volcanic ash, smoke plumes
DIAL, Raman	Dye, CO_2, optical parametric amplifier, CO:MgF_2	1 ppb–100 ppm	1–5	H_2O, O_3, SO_2, NO, NO_2, N_2O_3, C_2H_4, CH_4, HCl, CO_2, CO, Hg, SF_6, NH_3
Fluorescence	Dye	10^2–10^7 atoms/cm	1–90	OH, Na, K, Li, Ca, Ca^+
DIAL, Raman	Dye, Nd:YAG	1 K, 5 mbar	1–30	Temperature, pressure
Doppler	CO_2	0.5 m/sec	15	Wind speed

After Leone (1987).

among others. Based on these measurements, c is now given as precisely 299,792,458 meters per second. Furthermore, the length unit has been abandoned as a fundamental unit and is now derived from the time unit and the above quoted value for the speed of light.

At the practical manufacturing level of metrology, laser guidance systems can be used. Mergler (Case Western Reserve University) introduced a machine in 1978 along these lines. In a conventional machining operation, a part is cut, then measured, then remachined until the required dimensions are obtained. Manual measuring methods are tedious, time consuming, and somewhat limited in accuracy. In Mergler's system, a small modulated gas laser beam follows the surface of the part being machined and, within a precision of 1/5000 inch (0.005 millimeter) measures the piece as it is being cut. In later systems, the gas laser was replaced by a solid-state laser which occupies less space.

Laser Doppler Flowmeter

As shown in Fig. 12 fluid flow can be determined by measuring the doppler shift in laser radiation scattered from particles in the moving fluid stream. No sensor is required in the moving stream. The laser radiation focal point can be moved across the flow tube to measure velocity profiles. Fluid linear flows from 0.01 to 5000 inches (0.03 centimeter to 127 meters) per second have been measured. Contaminants, such as smoke, may have to be added to gases to provide scattering centers for the laser beam.

Laser Gyroscope

As early as the beginning of the 20th century, some investigators suggested that light will exhibit gyroscopic behavior, that is, the time required by light to traverse a circular pathway depends on whether the pathway is stationary or rotating. Thus the time difference can be used as a measure of the amount of rotation. The practical application of this observation, however, had to await vast improvements in optical systems, including the discovery of the laser, advances in fiber-optics, and better reflective mirrors. Within recent years, this principle has been applied in two configurations—fiber gyroscopes and ring-laser gyroscopes. The latter is described briefly here. As of the late 1980s, several aircraft depend upon ring-laser gyroscopes instead of their mechanical counterparts. The ring-laser gyroscope is more sensitive, has virtually no moving parts, and is as accurate as the best mechanical instruments. The rotation-induced difference in length of light path traversed is called the Sagnac effect, after the researcher who first demonstrated the phenomenon in 1913.

As previously mentioned in this article, a laser usually incorporates a resonant cavity. C.V. Heer (Ohio State University) in 1958 proposed that a resonant cavity could be used to measure rotation rates. In such an instrument, light circulates many times around a given path, not just back and forth between two mirrors. The first gyroscopes of this kind were constructed on a large scale, consisting of four glass tubes, each a meter long and arranged in a square. Light was made to travel around the device by placing a mirror in each corner. Over the last several years, the device has been markedly reduced in size (fits in the palm of the hand).

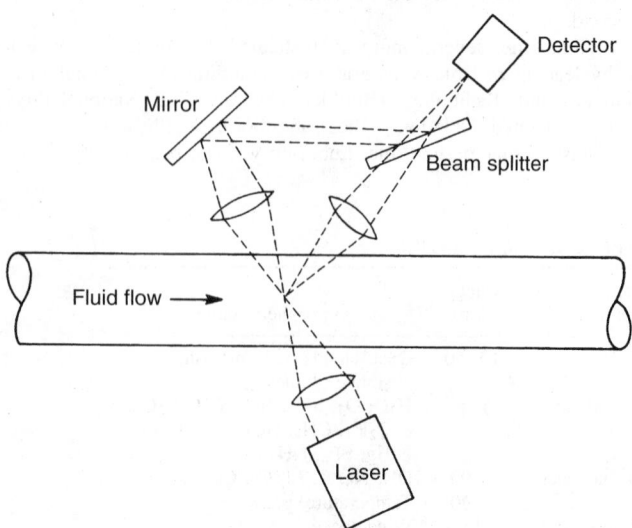

Fig. 12.   Operating principle of laser doppler flowmeter

Contemporary gyroscopes of this type are made from a single block of glass, into which a square channel is drilled. The channel is filled with a mixture of helium and neon. The laser is completed by attaching a small number of electrodes and four mirrors. As explained by Anderson, some ring-laser gyroscopes have a triangular channel and three mirrors; others have a hexagonal channel and six mirrors.

Beyond the scope of this article, Anderson explains the operation of the gyroscope in intimate detail and describes two problems that have proved most vexing to manufacturers of the ring-laser gyroscope, namely, frequency locking at low rotation rates and the bias effect. Improvements in this instrument are expected in the relatively near future because of what scientists have recently learned pertaining to the phenomenon of optical phase conjugation.

Lasers in Manufacturing Operations. A principal advantage of the laser in manufacturing operations is its ability to apply an extremely high flux of energy to the surface of a workpiece, as compared with traditional heat sources, such as flames, torches, electric arcs, and plasma jets. For manufacturing operations, lasers are usually placed in two categories. (1) *Light-duty lasers* range from a few tens of watts to a few hundred watts. Typical applications include cutting and drilling ceramic substrates in the electronics industry, drilling gems (for example, rubies in watchmaking), and cutting light-gauge metals as well as cloth, plastics, wood, and a variety of materials. Light-duty lasers that have been used include ruby lasers, neodymium-doped glass lasers, and neodymium-doped yttrium aluminum garnet lasers, among others. Depending upon the particular laser selected, the laser may operate in a pulsed or continuous mode. Argon and CO_2 lasers usually are operated in the continuous-wave mode. (2) *Heavy-duty lasers* range from a few kilowatts to a few tens of kilowatts. Typical applications include pipeline welding, automobile part welding, surface heat-treating of engine and other parts, with the applications expanding as experience is gained.

The high flux of electromagnetic energy applied to the surface of a workpiece by a laser is absorbed in an outer layer only about 10 nanometers thick. Thus the heat source is confined essentially to a thin film. Through careful design of equipment, the heat energy required is maintained in a comparatively small region, thus preventing or reducing thermal damage to the rest of a given part, and achieving a very high energy efficiency, estimated to range from 10 to 1000 times greater than can be achieved with conventional energy sources.

The electronics industry utilizes laser welders for joining dissimilar materials, fixing electrodes to batteries and connectors to a host of devices. A whole new area of laser technology, sometimes called laser microchemistry, has been exploited in the microstructure engineering of semiconductors. Lasers are used to initiate chemical reactions that result in deposition of material at a surface, for removing materials, and for alloying or diffusively mixing two or more solids on microscopic spatial scales. Lasers thus have played a major role in establishing new dimensions in microfabrication technology. It is possible to use a single laser to produce both gas-phase photolysis and surface heating. As described by Christensen (See reference), solar cells have been fabricated by using a UV laser to photodissociate trimethylboron, $B(CH_3)_3$, over a silicon surface in the manufacture of solar cells. The laser also heats the surface so that the boron atoms absorbed on the surface after the photolytic step rapidly diffuse into the bulk of the material. After irradiation, the silicon is heavily doped with boron near the surface, and the p-n junction thus formed functions as a photovoltaic cell. In some other applications, it has proved advantageous to use two lasers of different wavelengths to separately achieve photolysis and heating.

Perspective. The industrial applications for lasers developed comparatively slowly. As previously mentioned, lasers depend upon raising active molecules of the lasing medium to what might be called an *upper laser level* of energy, after which they relax to a *lower laser level*. Energy is given up during this process. Part of this energy is represented by photons of which the laser beam is composed. The other part is waste heat, which raises the temperature of the lasing medium. Thus, an excess of waste heat be removed so that the upper-level population can be maintained, a significant problem in the case of a continuously emitting high-power laser. Higher packets of energy in pulses can be attained, but waste heat must be removed by conduction between pulses. The end result is a pulsed high-power laser, but one that has a comparatively low average energy level simply because of the pauses in between.

In early laser designs, the quantity of waste heat generated was a limiting factor and consequently the average power output was low. Such lasers were excited by diffuse longitudinal electric discharges in long tubes with relatively large diameters. Heat generated at the center of the tube diffused to the side walls essentially by conduction, and the rate of heat transfer varied inversely with the tube radius and essentially directly with the length of the tube. Thus, the length of laser tubes increases as greater output power was sought.

Various component cooling schemes were proposed and used, but a major improvement was made when the concept of cooling a flowing laser medium was proposed, thereby taking advantage of the far more effective cooling by convection than by conduction. Gas lasers were considered the most apt for application of this concept and this led to the gas dynamic laser.[1]

With this concept, gas dynamic lasers increased in power outputs from less than 10 kilowatts by a factor of 13 to 14 within less than a decade (by the late 1960s). Success with the early gas lasers in this respect catalyzed a number of other refinements and improvements. However, the problem of maintaining a high-pressure glow discharge remained. Population inversion can be produced when electrons in an ionized gas are at a temperature relatively high as compared with the kinetic temperature of ions or molecules. This is a condition referred to as *glow discharge*. But an arc may form when the discharge is destabilized as the result of greatly increased gas pressure. Overheating of the molecules and ions destroys the population inversion. This problem was overcome by the concept of the ionizer/sustainer.

With the availability of high-power continuous electric discharge lasers capable of operating at up to 20 kilowatts output, a number of the previously predicted applications for lasers became practical. One of the first uses of a laser beam strictly for its power in cutting (exploding) a material was the fabric cutting system developed by Hughes Aircraft Company (circa 1966–1967) for which U.S. Patent No. 3,761,675 was awarded to W.J. Mason, D.W. Wilson, D.M. Considine, F.J. Viosca, and J.P. Wade on September 25, 1973. See Fig. 13. In this system cloth is carried in a single layer into a cutting area where a laser beam focused on the cloth is directed by computer commands to travel within the cutting area so as to cut many patterns in the cloth rapidly and accurately. The cut produced by the focused laser beam is sharp and narrow, leaving the fabric unfrayed. With synthetic materials, such as nylons and Dacrons®, the laser beam also serves to seal the cut edges by melting them during the cutting process. Unlike a mechanical blade, the laser beam does not dull; its cut remains uniform and is effective in cutting a wide range of materials, even those having metallized threads.

In the early 1980s, a helium-neon laser was used in a scanner system for inspecting textiles. The system uses laser output split into three beams, each of which scans the fabric independently in a pattern covering the entire surface. The system, moving at a rate of four meters per second, detects flaws through changes in reflected light and flags these areas for elimination or repair. In terms of economics, one laser system working one shift performs the same function as human inspectors at two plants working two shifts.

High-power lasers can perform many metalworking operations, including welding, cutting, surface hardening, and surface alloying. For small devices, the laser can perform much as a conventional electron beam, but without requiring the need for operation under a vacuum. High power densities can be achieved—up to 10^6 watts per square centimeter. It has been shown that a 16-kilowatt laser can make a 0.75-inch (1.9-centimeter) penetration weld in stainless steel at a rate of about 30 inches (76 centimeters) per minute. The laser beam can be directed by mirrors, thus making it effective for welding pipe from the inside. It also has been shown that a continuous-wave carbon dioxide device (15 kilowatts) can be used for welding half-inch (1.2-centimeter) thick steel plates at the rate of about 50

---

[1] Invented by Kantrowitz in the late 1960s. In essence, the device had two compartments separated by a nozzle. In the first compartment, gas was held at a temperature of about 1400 K and pressure of 17 atmospheres. This high-pressure compartment held about 10% of the active CO_2 molecules in the total system. Expansion of this gas through an orifice caused cooling. Because of the cooling, the lower-level population essentially vanished a few centimeters downstream from the nozzle. This occurred before the upper-level population had an opportunity to decline significantly. The population "inversion" resulting was adequate for effecting a laser beam of considerable power.

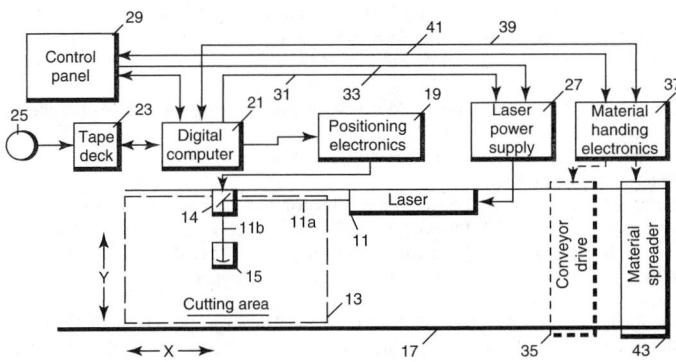

Fig. 13. A cloth cutting system wherein cloth is carried in a single layer into a cutting area, where a laser beam is focused on the cloth and is directed by computer commands to travel within the cutting area so as to cut a plurality of patterns through the cloth rapidly and accurately. Invented by W.J. Mason, D.W. Wilson, D.M. Considine, and J.P. Wade (*Hughes Aircraft Company*) in 1973, this was one of the very early and successful industrial applications of the laser. Diagram is part of U.S. Patent 3,761,675

inches (127 centimeters) per minute. If the laser is focused to a spot size of about 0.03 inches (0.08 centimeter) in diameter, power densities of some 2200 kilowatts per square centimeter are produced.

Laser Recording. For many years, it has been known that lasers can be used to encode information on materials that respond in an irreversible manner to exposure to high-intensity light. However, it is only comparatively recently that the concept has been reduced to commercial practice—with the almost sudden appearance of optical disk recording (compact disk) in the entertainment field. It is because of the coherence and relatively short wavelength of laser radiation that such large volumes of information can be written onto a very small space of the recording medium. The potential for microlasers in this field is discussed earlier in this article.

Additional Reading

Adams, C.S. and E. Riis: "Laser Cooling And Trapping Of Neutral Atoms," *Progress in Quantum Electronics*, 1–79 (1997).

Amato, J.: "Moving Tiny Things by Optical Tweezers," *Science News*, 148 (March 10, 1990).

Anderson, D.Z.: "Optical Gyroscopes," *Sci. Amer.*, 94–99 (April 1986).

Appell, D.: "High-Power Laser Beam Launches Fuel-Less Craft," *Laser Focus*, 90 (March 1998).

Attwood, D., K. Halbach, and K. Kwang-Je, Kim: "Tunable Coherent X-rays," *Science*, **228**, 1265–1272 (1985).

Byer, R.L.: "Diode-Laser-Pumped Solid-State Lasers," *Science*, 742 (February 12, 1988).

Cherfas, J.: "A European Superlaser?" *Science*, 1073 (June 1, 1990).

Christensen, C.P.: "New Laser Source Technology," *Science*, **224**, 117–123 (1984).

Chu, S.: "Laser Manipulation of Atoms and Particles," *Science*, 861 (August 23, 1991).

Chu, S.: "Laser Trapping of Neutral Particles," *Sci. Amer.*, 71 (February 1992).

Corcoran, E.: "Diminishing Dimensions," *Sci. Amer.*, 122 (November 1990).

Corcoran, E.: "True Blue (Laser)," *Sci. Amer.*, 171 (September 1991).

Corcoran, E.: "Tacky Lasers Are the Tiniest Yet," *Sci. Amer.*, 28 (January 1992).

Duley, W. and K. Shibata: *1996 International Congress on Applied Lasers and Electro-Optics Proceedings*, Laser Institute of America, Orlando, FL, 1997.

Duarte, F.J.: *Laser Optics*, Academic Press, Inc., San Diego, CA, 2003.

Fasol, G.: "Room-Temperature Blue Gallium Nitride Laser Diode," *Science*, 1751 (June 21, 1996).

Fasol, G., S. Nakamura, and I. Davies: *The Blue Laser Diode: GaN Based Light Emitters and Lasers*, Springer-Verlag New York, Inc., New York, NY, 1997.

Feld, M.S. and K. An: "The Single Atom Laser," *Scientific American*, 56–63 (July 1998).

Feng, S. and P.A. Lees: "Mesoscopic Conductors and Correlations in Laser Spackle Patterns," *Science*, 633 (February 8, 1991).

Freund, H.P. and R.K. Parker: "Free-Electron Lasers," *Sci. Amer.*, 84 (April 1989).

Grant, E.R. and R.G. Cooks: "Mass Spectrometry and Its Use in Tandem with Laser Spectroscopy," *Science*, 61 (October 5, 1990).

Hemley, R.J., P.M. Bell, and H.K. Mao: "Laser techniques in High-Pressure Geophysics," *Science*, **237**, 605–612 (1987).

Hannaford, P., H. Bachor, and A. Sidorov: *Laser Spectroscopy*, World Scientific Publishing Company, Inc., Riveredge, NJ, 2004.

Hirschfelder, J.O., R.E. Wyatt, and R.D. Coalson: *Lasers, Molecules, and Methods,* John Wiley & Sons, Inc., New York, NY, 1989.

Jewell, J.L., J.P. Harbison, and A. Scherer: "Microlasers," *Sci. Amer.*, 86 (November 1991).

Killinger, D.K. and N. Menyuk: "Laser Remote Sensing of the Atmosphere," *Science*, **235**, 37–45 (1987).

Kinoshita, J.: "Atomic Fountain: Laser Light Slows Atom Beam to a Trickle," *Sci. Amer.*, 26 (June 1990).

Kim, K.Kwang-Je, and A. Sessler: "Free-Electron Lasers: Present Status and Future Prospects," *Science*, 88 (October 5, 1990).

Lamb, W.E., Jr. and R.C. Retherford: *Phys. Rev.*, **79**, 549 (1950).

Langreth, R.N.: "Laser Cooling Made Simpler, Cheaper," *Science News*, 216 (October 6, 1990).

Maddox, John: "The Wonders Of The Microlaser," *Nature*, 101 (January 12, 1995).

Matthews, D.L. and M.D. Rosen: "Soft X-Ray Lasers," *Sci. Amer.*, 86 (December 1988).

Maiman, T.H.: *Br. Commun. Electron.*, **7**, 674 (1960).

Meyers, R. (Editor): *Encyclopedia of Lasers and Optical Technology,* Academic Press, Inc., San Diego, CA, 1990.

Misaelides, P.: *Application of Particle and Laser Beams in Materials Technology,* Kluwer Academic Publishers, New York, NY, 1995.

Morrison, D.C.: "An Unsung Legacy of the First Lunar Landing (Laser)," *Science*, 447 (October 27, 1989).

Murname, M.M. et al.: "Ultrafast X-ray Pulses from Laser-Produced Plasmas," *Science*, 531 (February 1, 1991).

Narayan, J.: "Surfaces, Interfaces, and Films: New Tools (Lasers) Aid Engineering," *Adv. Materials & Processes*, 51 (January 1988).

Numai, T.: *Fundamentals of Semiconductor Lasers,* Springer-Verlag New York, LLC., New York, NY, 2004.

Pepper, D.M., J. Feinberg, and N.V. Kukhtarev: "The Photorefractive Effect," *Sci. Amer.*, 62 (October 1990).

Phillips, W.D., P.L. Gould, and P.D. Lett: "Cooling, Stopping, and Trapping Atoms," *Science*, 877 (February 19, 1988).

Pool, R.: "Making Atoms Jump Through Hoops," *Science*, 1076 (June 1, 1990).

Pool, R.: "Laser Cooling Hits New Low," *Science*, 1077 (June 1, 1990).

Purcell, E.M. and V. Pound: *Phys. Rev.*, **81**, 279 (1951).

Ruthen, R.: "Surfing Photons," *Sci. Amer.*, 12D (August 1989).

Silfvast, W.T.: *Laser Fundamentals,* Cambridge University Press, New York, NY, 2004.

Suckewer, S. and C.H. Skinner: "Soft X-Ray Lasers and Their Applications," *Science*, 1553 (March 30, 1990).

Svelto, O. and D.C. Hanna: *Principles of Lasers,* 4th Edition, Perseus Publishing, Boulder, CO, 1998.

Staff: "Miniature Lasers Reach Mass Production," *Chem. Eng. Progress*, 15 (August 1991).

Taylor, N.: *Laser: The Inventor, the Nobel Laureate, and the Thirty-Year Patent War,* Simon & Schuster Trade, New York, NY, 2000.

Townes, C.H.: "Harnessing Light," *Science* **84**, 153–155 (November 1984).

Vander Been, M.R.: "Gallium Arsenide Sandwich Lasers," *Adv. Materials & Processes*, 39 (May 1988).

Waterbury, R.C.: "Catalysts Enable Sealed Carbon Dioxide Laser," *Instrumentation Technology*, 80 (April 1990).

Yamamoto, Y., M. Susumu, and W.H. Richardson: "Photon Number Squeezed States in Semiconductor Lasers," *Science*, 1219 (March 6, 1992).

Zewail, A.H.: "The Birth of Molecules," *Sci. Amer.*, 76 (December 1990).

LATENT HEAT. Heat gained by a substance or system without an accompanying rise in temperature during a change of state. As examples, the latent heat of fusion is the amount of heat necessary to convert a unit mass of a substance from the solid state to the liquid state at the same temperature, the pressure being that to allow coexistence of the two phases. A considerable part of the latent heat arises from the entropy increase consequent on the greater disorder of the liquid state. The latent heat of sublimation is the amount of heat necessary to convert a unit mass of a substance from the solid state to the gaseous state at the same temperature, the pressure being that to allow coexistence of the two phases.

LATERITE. The sub-aerial decay of rocks in tropical regions, having a distinctly moist or rainy climate, results in the development of a residual, reddish, and usually sticky soil frequently containing concretions. The principal products of laterization are the hydrated oxides of aluminum and iron either in the crystalline or amorphous form. If the concentration of iron oxide is sufficiently high the laterite may be valuable as an iron ore. If, on the other hand, the concentration of alumina is high the laterite may be valuable as an ore of that metal.

LATEX. Latex is a milky substance found in many plants. It is a complex emulsion in which such substances as proteins, alkaloids, starches, sugars, oils, tannins, resins, and gums are found. In most plants the latex is white; but in some it is yellow; in others, orange or scarlet.

The cells or vessels in which latex is found make up the laticiferous system. There are two very different ways in which this system may be formed. In many plants the laticiferous system is formed from cells laid down in the meristematic region of the stem or root. Rows of these cells are formed. The cell walls separating them are dissolved, so that continuous tubes, called latex vessels, are formed. This method of formation is found in the poppy family; in the rubber plant, *Hevea brasiliensis*; and in the *Cichorieae*, a section of the composite family distinguished by the presence of latex in its members. Dandelion, lettuce, hawkweed, and salsify are members of the *Cichorieae*. See also **Latex Technology**; and **Rubber (Natural)**.

LATEX TECHNOLOGY. Latex technology encompasses colloidal and polymer chemistry in the preparation, processing, and conversion of natural and synthetic latices into useful products.

Worldwide concern over the AIDS epidemic has sharply increased the demand for latex used in the preparation of rubber gloves and similar dipped goods. Estimates are for a 5–7% annual growth rate in this market.

Many synthetic latices exist. They contain butadiene and styrene copolymers (elastomeric), styrene–butadiene copolymers (resinous), butadiene and acrylonitrile, chloroprene copolymers, methacrylate and acrylate ester copolymers, vinyl acelate copolymers, vinyl and vinylidene chloride copolymers, ethylene copolymers, fluorinated copolymers, acrylamide copolymers, styrene–acrolein copolymers, and pyrrole and pyrrole copolymers. Many of these latices also have carboxylated versions.

Traditional applications for latices are adhesives, binders for fibers and particulate matter, protective and decorative coatings, dipped goods, foam, paper coatings, backings for carpet and upholstery, modifiers for bitumens and concrete, thread, and textile modifiers. More recent applications include biomedical applications as protein immobilizers, visual detectors in immunoassays, as release agents, in electronic applications as photoresists for circuit boards, in batteries, conductive paint, copy machines, and as key components in molecular electronic devices.

By 1935 emulsion polymerization became the method of choice in making synthetic rubber because of its many advantages: (*1*) the reaction mass viscosity remains low throughout polymerization, providing for improved heat transfer, agitation, and product handling; (*2*) the sensible heat of the water in the emulsion balances the heat of reaction generated by free-radical polymerization; and (*3*) the rate of reaction is rapid, while producing very high molecular weight.

Synthetic Latex Manufacture

Kinetics and Mechanisms. In 1945 the first recognized qualitative theory of emulsion polymerization was presented. This mechanism for classic emulsion preparation was quantified and the polymerization separated into three stages—stage I: particle nucleation; stage II: growth in polymer particles saturated with monomer; and stage III: growth in polymer particles with a decreasing monomer concentration.

Basic Components. The principal components in emulsion polymerization are deionized water, monomer, initiator, emulsifier, buffer, and chain-transfer agent. A typical formula consists of 20–60% monomer, 2–10 wt % emulsifier on monomer, 0.1–1.0 wt % initiator on monomer, 0.1–1.0 wt % chain-transfer agent on monomer, small amounts of various buffers and bacteria control agents, and the balance deionized water.

Process. Commercial processes manufacturing latex can be divided into batch, semibatch, and continuous methods.

Latex Properties

The observable properties of a latex, i.e., stability, rheology, film properties, interfacial reactivity, and substrate adhesion, are determined by the colloidal and polymeric properties of the latex particles. Important polymer properties include molecular weight distribution, monomer sequence distribution, glass-transition temperature, crystallinity, degrees of crosslinking, and free monomer. Methods for analyzing each of these properties exist, depending on the end use of the product. An overview of the various polymer colloid characterization methods is available.

Improving Properties Through Compounding. The potential value of most polymers can be realized only after proper compounding. Materials used to enhance polymer properties or reduce polymer cost include antioxidants, cross-linking reagents, accelerators, fillers, plasticizers, adhesion promoters, pigments, etc.

CHESTER H. GELBERT
MICHAEL C. GRADY
E. I. du Pont de Nemours & Co., Inc.

Additional Reading

Daniels, E., E.D. Sudol, and M.S. El-Aasser, eds.: *Polymer Latexes: Preparation, Characterization and Applications,* ACS Symposium Series, Vol. 492, American Chemical Society, Washington, DC, 1992.
Gelbert, C.H. and H.E. Berkheimer: *Paper C in Educational Symposium No. 18 on Latex Technology,* Rubber Division of the American Chemical Society, Montreal, Canada, 1987.
Richards, J.R., J.P. Congalidis, and R.G. Gilbert: *J. Applied Polym. Sci.* **37**, 2727–2756 (1989).

LATTICE COMPOUNDS. Chemical compounds formed between definite stoichiometric amounts of two molecular species that owe their stability to packing in the crystal lattice, and not to ordinary valence forces.

LATTICE ENERGY OF CRYSTAL. The decrease in energy accompanying the process of bringing the ions, when separated from each other by an infinite distance, to the positions they occupy in the stable lattice. It is made up of contributions from the electrostatic forces between the ions, from the repulsive forces associated with the overlap of electron shells, from the van der Waals forces, and from the zero-point energy.

LAURIC ACID. [CAS: 143-07-7]. Also called dodecanoic acid, formula $CH_3(CH_2)_{10}COOH$. A fatty acid that occurs in many vegetable oils and fats as the glyceride, especially in coconut oil and laurel oil. See also **Vegetable Oils (Edible).** Combustible. It takes the form of colorless needles at room temperature. Specific gravity 0.833; mp 44°C; bp 225°C (100 millimeters pressure). Insoluble in water; soluble in alcohol and ether. It is derived by the fractional distillation of coconut oil. Lauric acid is used in alkyd resins; wetting agents; soaps; detergents; cosmetics; insecticides; food additives.

LAVA. Molten material that has poured out on the surface of the earth and, due to relief of pressure, may have lost much of its original gas and water content during its relatively rapid consolidation. The term lava is used for both the liquid and the consolidated state of the igneous material. Lava may be erupted either by volcanoes or from fissures. Flowing lava is shown in Fig. 1. The most extensive lava flows are fissure eruptions, such as the Columbia Plateau basalts in Oregon or the plateau basalts of the Deccan, India, which are derived from basic magma. Had this magma, either basic or acid, cooled slowly beneath the surface of the earth under great pressure and with all its original gases, the resulting rock would have had a coarser texture and somewhat different mineral content.

Fig. 1.   Lava flow

LAVOISIER, ANTOINE, LAURENT (1743–1794). A French chemist generally regarded as the "father" of chemistry. His "Traite elementaire de chimie" (1789) listed 30 elements, clarified the nomenclature of acids, bases, and salts, and described the composition of numerous organic substances. He erroneously believed that oxygen is the characteristic element of acids. However, his fundamental work on combustion, as a result of which he identified and named nitrogen (azote), and on the separation of hydrogen from water by a unique reduction experiment carried out in a heated gun barrel, earned him a leading position among early chemists.

LAWRENCE, ERNEST O. (1901–1958). An American physicist who invented the cyclotron in 1929. Both the element lawrencium and the Lawrence Livermore Research Laboratory at the University of California were named after him.

See also **Cyclotron.**

LAWRENCIUM. [CAS: 22537-19-5]. Chemical element, symbol Lr, at. no. 103, at. wt. 257 (mass number of known isotope), radioactive metal of the Actinide series, also one of the Transuranium elements. ^{103}Lr was identified in 1961 by A. Ghiorso, T. Sikkeland, A. Larsh, and R. Latimer at the University of California at Berkeley.

This method used to produce and identify lawrencium was similar to that used in the later, direct-counting experiments performed in connection with the production of nobelium at Berkeley. About 3 micrograms of a mixture of californium isotopes were bombarded with boron ions accelerated in the heavy-ion linear accelerator. The atoms of lawrencium recoiled from the target into an atmosphere of helium, where they were electrostatically collected on a copper conveyor tape. This tape was then periodically pulled into place before radiation detectors to measure the emission rate and the energy of the alpha particles being emitted. By this means, it was possible to identify the lawrencium isotope ^{257}Lr, with a half-life of 8 seconds. At present, because of the short half-life and the lack of a suitable daughter isotope, available in the case of nobelium, it has not been possible to perform a chemical identification.

Another isotope, ^{256}Lr, half-life about 45 seconds, was reported by the Soviet Union in 1965. It was produced by impact of oxygen atoms (^{18}O) on americium (^{243}Am). It decayed by alpha-particle emission and electron capture to form ^{252}Fm. See also **Chemical Elements.**

Lawrencium has been found to behave quite differently from dipositive nobelium and, in fact, it is comparable to the tripositive elements that appear earlier in the Actinide series.

Additional Reading

Eskola, K., Eskola, P., Nurmia, M., and A. Ghiorso: "Studies of Lawrencium Isotopes with Mass Numbers 255 through 260," *Phys. Rev.,* **4**, 2, 632–642 (1971). (A classic reference.)
Fuger, J. and L.R. Morss: *Transuranium Elements: A Half Century,* American Chemical Society, Washington, DC, 1992.
Ghiorso, A., T. Sikkeland, Larsh, A.E., and R.M. Latimer: "New Element, Lawrencium, Atomic Number 103," *Phys. Rev., Lett.,* **6**, 9, 473–475 (1961). (A classic reference.)
Greenwood, N.N. and A. Earnshaw: *Chemistry of the Elements,* 2nd Edition, Butterworth-Heinemann, Inc., Woburn, MA, 1997.
Lide, D.R.: *CRC Handbook of Chemistry and Physics,* 84th Edition, CRC Press, LLC., Boca Raton, FL, 2003.
Seaborg, G.T. and W.D. Loveland: *The Elements beyond Uranium,* John Wiley & Sons, Inc., New York, NY, 1990.

LAWSONITE. Named for Andrew Cowper Lawson (1861–1952), a Scottish-American geologist. This calcium aluminum silicate mineral, $CaAl_2(Si_2O_7)$ $(OH)_2 \cdot H_2O$, is found as grains and veins within the metamorphic rocks, gneisses, and schists. It was found originally on the Tiburon Peninsula, San Francisco Bay, California, but also occurs in schistose rocks in France and New Caledonia. The mineral has a hardness of 7; specific gravity of 3.09. It is colorless, pale blue to bluish gray, translucent, with vitreous to greasy luster. The mineral crystallizes in the orthorhombic system.

LAZULITE. Lazulite is named from an Arabic word for *heaven* in allusion to its sky blue color. This mineral crystallizes within the monoclinic system, a basic phosphate of magnesium and aluminum, $MgAl_2(OH)_2$ $(PO_4)_2$. Ferrous iron can substitute for the magnesium and the isomorphous mineral scorzalite is the product. Usually occurs massive but acute pyramidal crystals are not uncommon. Color is azure-blue to bluish-green, usually translucent (rarely transparent), with vitreous luster. It has a hardness of 5.5–6, with specific gravity of 3–3.1.

Lazulite is a rare mineral found principally within high-grade metamorphic rocks. Notable world crystal occurrences are Salzburg, Austria; Syria; Hörnsjöberg, Sweden; Madagascar; Brazil; and Graves Mountain, Georgia. When transparent, the mineral can be cut into gem stones.

LAZURITE. The mineral lazurite or lapis lazuli has been used since ancient times for jewelry and other ornamental purposes. Ground to powder it forms the pigment ultramarine, now, however, largely superseded by artificial preparations. Lapis lazuli is a mixture of minerals, lazurite being the chief component. This mineral is isometric, and chemically a sodium, calcium, aluminum sulfo-chlorosilicate. A general formula is $(Na, Ca)_8(Al, Si)_{12}O_{24}(S, SO_4)$. Lapis lazuli has a hardness of 5–5.5; specific gravity, 2.4; color, various shades of blue; luster, vitreous to greasy; translucent to opaque. Localities are Afghanistan, Siberia, Chile, and California.

LEAD. [CAS: 7439-92-1]. Chemical element, symbol Pb. at. no. 82, at. wt. 207.2, periodic table group 14, mp 327.5°C. bp 1740°C, density 11.35 g/cm³. (20°C). Elemental lead has a face-centered cubic structure with an edge length of 4.950 Å.

Lead is a white to bluish-gray metal, soft, malleable, and slightly ductile; tarnishes in air, forming a film of oxide, forms oxide scum upon heating the molten metal in air; soluble in dilute HNO_3; HCP or H_2SO_4 attack lead only slightly, the extent depending markedly upon the concentration and the temperature; slowly dissolves in H_2O and consequently the use of lead constitutes a health hazard due to its toxic effect; attacked by solutions of organic acids or sodium hydroxide. Lead is one of the four most largely produced and utilized metals, and considerable scrap metal is recovered. Used (1) in construction and apparatus where workability is demanded, and definite resistance to corrosion is supplied by the metal, (2) as a constituent of various alloys, especially solder, type metal, pewter, and fusible alloys, (3) for storage battery plates, (4) for shot and bullets, and (5) as a protective coating for iron and steel.

Lead has four naturally occurring isotopes. In order of abundance, these are ^{208}Pb, ^{206}Pb, ^{207}Pb, and ^{204}Pb. There are ten unstable isotopes, 200–203, 205, and 209–214. See also **Radioactivity**. In terms of abundance, lead is scarcely represented in the earth's crust, the average composition of igneous rocks containing only 0.002% Pb by weight. In terms of cosmic abundance, an estimate made by Harold C. Urey in 1952, using silicon as a basis with the figure of 10,000, lead had an abundance figure of less than 0.02. In terms of presence in seawater, lead is 27th among the elements, with an estimated 14 tons per cubic mile (3 metric tons per cubic kilometer) of seawater. In this regard, it is comparable to tin, copper, arsenic, protactinium, and selenium.

The atomic weight varies because of natural variations in the isotopic composition of the element, caused by the various isotopes having different origins: ^{208}Pb is the end product of the thorium decay series, while ^{207}Pb and ^{206}Pb arise from uranium as end products of the actinium and radium series respectively. Lead-204 has no existing natural radioactive precursors. Electronic configuration $1s^2 2s$; $2p^6 3s^2 3p^6 3d^{10} 4s^2 4p^6 4d^{10} 4f^{14} 5s^2 5p^6 5d^{10} 6s^2 6p^2$. Ionic radius Ph^{2+} 1.18 Å. Pb^{44} 0.70 Å. Metallic radius 1.7502 Å. Covalent radius (sp^3) 1.44 Å. First ionization potential 7.415 eV; second, 14.97 eV. Oxidation potentials $Pb \longrightarrow Pb^{2+} + 2e^-$, 0.126 V; $Pb^{2+} + 2H_2O \longrightarrow PbO_2 + 4H^+ + 2e^-$, −1.456 V; $Pb + 2OH^- \longrightarrow PbO + H_2O + 2e^-$, 0.576 V; $Pb + 3OH^- \longrightarrow HPbO_2^- + H_2O + 2e^-$, 0.54 V. Other physical properties are given under **Chemical Elements**.

Lead is of interest as being the terminal product of radioactive decay. Thus, while ordinary lead has the atomic weight 207.19 (being composed of 1.37% ^{204}Pb, 26.26% ^{206}Pb, 20.8% ^{207}Pb and 51.55% ^{208}Pb), the isotopic composition, and hence the atomic weight, varies somewhat in lead from meteorites, from deep-seated rocks and from uranium ores (the last being somewhat less dense, as would be expected from the fact that ^{206}Pb is the end-product of the uranium series). These variations in isotopic composition of lead permit of calculations of the age of the earth (and the meteorites).

Lead Melting Point as a Standard

Melting, defined as the equilibrium transition between crystalline and liquid states, is of large concern in the development of the physical and materials sciences. To date, some of the purest crystals of silicon, diamond, and other technologically important materials have been produced from melts. Studies of melts also are of significance in understanding the interiors of terrestrial plants and, in fact, of Earth. In research at the University of California (Berkeley), studies of the effects of high pressure on the fusion temperature of lead have been underway. The advantages of studying lead are outlined by the investigators as: (1) the melting temperature of lead at ambient pressures is low and well determined, (2) lead is highly compressible and therefore should show the effects of pressure, (3) the behavior of lead under pressure is relatively simple, involving only one known polymorphic transition (from face-centered cube to hexagonal close-packed crystal structure), and (4) shock-wave experiments have been carried out previously to document the compression of both crystalline and molten lead at simultaneously high pressures and temperatures.

Occurrence and Processing

Galena, PbS, lead sulfide, is the source of over 95% of the lead currently produced. Bodies containing galena range from 3% to 30% lead. One of the most widely distributed sulfide minerals, galena frequently occurs along with sphalerite, ZnS. The lead-zinc ores processed usually contain recoverable quantities of copper, silver, antimony, and bismuth. Principal sources being worked are in Australia's Broken Hill area in New South Wales, the western United States, Canada, Mexico, Peru, former Yugoslav Republics, and the former Soviet Union. When groundwater reacts with galena, cerussite, $PbCO_3$, is formed; when galena is in contact with sulfate solutions generated by the oxidation of sulfide minerals, anglesite, $PbSO_4$, may be formed. See also **Anglesite**; **Cerussite**; and **Galena**.

In processing, the ore first is crushed, wet-ground, and classified to a point where it is at least 90% less than 200 mesh. Separation of the sulfide ore from the gangue is aided by flotation agents. The resulting concentrates contain from 45% to 60% lead, from zero to 15% zinc, and often a few ounces (∼50 grams) of gold and up to 50 ounces (1.4 kilograms) of silver per ton. Copper content may be as much as 3%, arsenic, 0.4%, and antimony, 2%. The sulfur content (10 to 30%) is reduced by roasting in a Dwight-Lloyd sintering machine. This sulfur reduction is necessary because PbS is not reduced by carbon or carbon monoxide at blast-furnace temperatures. Once formed, the sinter, together with limestone and coke, is fed into a blast furnace. Further oxidation and electrolytic methods may be used to refine the lead. Lead is commercially produced to standards of very high purity. The minimum lead content permitted by specifications for Pig Lead (7 classifications) is 99.73%. Fully refined lead averaging 99.99% lead is obtainable. Large quantities are used for production of chemicals. At one time, primary uses for lead chemicals were in the production of paint pigments and lead tetraethyl gasoline additive.

Lead Metals and Alloys

Lead is soft and ductile and is readily worked by common methods, predominantly by rolling and extruding. Lead is easily formed and readily joined by welding (burning), or by soldering and can be bonded to steel, or used as a liner for steel, wood, concrete, and other materials. Lead is widely used in this manner because of its excellent resistant to atmospheric and soil corrosion, and attack by sulfuric and phosphoric acids. Lead generally does not resist the action of the organic acids, nor the oxidizing mineral acids, such as HNO_3. Lead is attacked by alkalies.

Due to its low melting point, pure lead will very gradually flow or creep at room temperature. Thus, lead sheeting used as a roofing material on old buildings will usually be thicker at the lower edge than at the upper edge. Other examples of creep occur under low sustained stresses due to the oil pressure in lead-covered power conducting cable, for example, or due to the weight in the case of a deep tank lined with sheet lead. To counter the effects of creep, lead containing 0.06% copper (*chemical lead* or *acid lead*) is preferred.

The addition of antimony in amounts up to 12% greatly improves the casting properties and increases the hardness very materially. These properties make possible the casting of intricately shaped antimonial lead storage-battery grids which, including the weight of the lead oxide paste applied to them, constitute the largest single use for the metal.

Tin and lead in various proportions form a highly useful series of alloys generally known as the soft solders which are used for joining copper, iron, nickel, lead, zinc and even glass. The solders can be applied by means of a soldering tool, by wiping, by hot-dipping, or by special machines as in the tin-can industry. Numerous compositions are used, the most popular of which are listed in the accompanying table.

Further additions of bismuth, cadmium, and antimony to the tin-lead alloys result in the low melting or "fusible" alloys widely used as safety devices, the melting points of which can be varied to suit a wide range of requirements. The type metals of the printing industry are lead-tin-antimony alloys having the requisite hardness and good casting properties needed for high-fidelity reproduction.

Babbitt metals (white-metal bearing alloys) are generally classified as either tin-base or lead-base. The true tin-base Babbitts contain only tin, antimony and copper, and have been used for many years. The practice of adding up to 25% lead to the tin Babbitts to reduce their cost is to be avoided since the net result is an expensive series of alloys with inferior properties to the inexpensive lead-base Babbitts. The lead-base bearing alloys of the older type usually contain lead, antimony and tin, and while not considered the equal of the tin-base alloys for severe service have been widely employed due to their low cost. The lead-base alloy containing arsenic has found extensive use and has come to the fore of this group since it has successfully met many automotive and other severe service requirements. All of these alloys render their most efficient service when used in the form of a thin lining bonded to a bronze or steel shell. See Table 1.

Lead Eliminated from Free-cutting Alloys

Among the numerous efforts being made to eliminate lead from the environment, including the potable water plumbing systems, free-cutting copper alloys that contain no lead have been developed. As reported in late 1991, bismuth, as a replacement, has significant potential as a nontoxic alternative to lead to enhance the machinability of copper. When bismuth is used alone, however, the element embrittles copper because of its tendency to "set" grain boundaries. J.T. Plewes (see reference) ascribes this characteristic to the large difference in surface tension between copper and bismuth. It has been found that adding a third element in modest amounts removes this limitation of bismuth. Such elements include phosphorus, indium, and tin.

Chemistry of Lead

A number of oxides of lead are known, but not all are daltonide compounds. Thus, lead(I) oxide, Pb_2O, made by heating lead(II) oxalate, has been shown by x-ray analysis to be a mixture of the metal and lead(II) oxide, PbO. The latter is obtained by heating lead in air, which yields a yellow, rhombic material, which has a peculiar layer structure having each lead atom attached to four oxygen atoms all lying on the same side of it, forming a square pyramid with the lead at the apex. Each oxygen atom is surrounded tetrahedrally by four lead atoms. Another form of PbO, somewhat more stable and soluble in water, red in color, and tetragonal in structure, may be obtained along with the yellow form by alkaline dehydration of $Pb(OH)_2 \cdot PbO$ is amphiprotic, but only weakly acidic. Lead(IV) oxide, Pb_2O, is obtained by action of chlorine on alkaline solutions of lead(II) oxide or acetate. The reaction is $Pb(OH)_3{}^- + ClO^- \longrightarrow PbO_2 + Cl^- + OH^- + H_2O \cdot PbO_2$ can also be produced on a lead or platinum anode by electrolysis in acidic solution. Like the lower elements of main group 4, lead(IV) forms tetrahedral bonds exhibiting sp^3 hybridization. In its relatively more stable salts, however, the $6s^2$ electrons are unused, and Pb^{2+} ions are formed by loss of the $6p^2$ electrons. These facts explain the marked difference between the essentially covalent character of many of the tetravalent compounds and the essentially electrovalent character of the divalent compounds, as well as the peculiar structure of PbO and many other Pb(II) compounds.

The dioxide, Pb_2O, has rutile structure, and the compound is a strong oxidizing agent. It is also amphiprotic, giving unstable lead(IV) salts with acids, and orthoplumbates, $M_4^I PbO_4$, or metaplumbates, $M_2^I PbO_3$, upon fusion with alkalies. Lead dioxide dissolves in aqueous alkali with formation of the ion $Pb(OH)_6{}^{2-}$, the alkali salts of which are isomorphous with the corresponding stannates and platinates. Lead sesquioxide, Pb_2O_4, has been shown not to exist as a stable phase.

TABLE 1. REPRESENTATIVE LEAD AND TIN ALLOYS

Name	Pb	Sn	Sb	Cu	Bi	Ag	Cd	Typical application
				LEAD ALLOYS				
Chemical or acid lead	99.9			.06				Tank linings, coils, etc., power cable sheath.
Cable sheath	98.9		1.0					Telephone cable sheath.
Hard lead	96–92		4–8					Cast shapes, wrought sheet and pipe.
Battery grid metal	92–88	.25	8–12					Cast battery grids.
				SOLDERS				
Soft solder	50	50						General purposes, most popular solder.
Wiping solder	60	40						For wiping joints in cables, lead
	60	37.5	2.5					pipes, etc.
"Fine solder"	40	60						For making joints at low temperature.
Solder	95–97.5					5–2.5		High temperature solder.
				FUSIBLE ALLOYS				
Wood's metal	25	12.5			50		12.5	Melts in hot water at 154°F. Wets glass. Wide range of melting points possible with changes in composition for automatic sprinkler systems and other safety devices.
Matrix metal	28.5	14.5	9		48			For anchoring punches, etc., in jigs and fixtures. Expands on freezing.
Bending alloy	26.5	13.5			50		10	Filler for tubes, etc., during bending. Melts out in hot water.
				TYPE METALS				
Electrotype	93	3	4					
Linotype	84	4	12					
Stereotype	80.5	5.75	13.75					
Monotype	76	8	16					Single type.
				TIN BASE BABBITTS				
	89	7.5	3.5					General usage.
	83.3	8.3	8.3					Hard Babbitt.
				LEAD BASE BABBITTS				
	82.5	1.0	15	.5	1.0 As			General usage.
	80	5	15					General usage.
	75	10	15					General usage.

Notes: Figures given in percent. Wood's metal melts at ~68°C in water.

Lead orthoplumbate, Pb_2PbO_4, red lead, is similarly described as a salt, in this case an orthoplumbate of divalent lead, Pb_2PbO_4, because on treatment with nitric acid, two-thirds of the lead dissolves and one-third remains as PbO_2. It is prepared in the red form by atmospheric heating of PbO, and in a black form by reaction of PbO with pure oxygen. Red lead is formed of PbO_6 octahedra (with one common edge) linked by lead atoms covalently bonded to three oxygen atoms.

The lead dihalides are known for all four of the common halogens. They are not strictly ionic in the anhydrous state, but they dissolve in (hot) water to give Pb^{2+} ions, more or less hydrated. They are much less soluble in cold water. They also form complex compounds such as M_2PbCl_4, MPb_2Cl_5, M_4PbF_6, and $MPbF_3$, where M is an alkali metal. The compound formed, especially of the fluoroplumbates(II) depends somewhat on the alkali metal, some of which form/nondaltonide (berthollide) compounds. Of the lead tetrahalides, only PbF_4 and $PbCl_4$ are known, the fluoride being prepared by fluorination of PbF_2. The chloride, which easily loses chlorine, is made by careful acidification of a hexachloroplumbate(IV). $PbCl_4$ forms the complex compound ammonium hexachloroplumbate, $(NH_4)_2PbCl_6$, upon addition to its solution of solid ammonium chloride.

Lead(II) inorganic compounds and salts of organic acids are far more numerous than those of lead(IV), as is to be expected from the essentially covalent character of the latter. In addition to the oxides and halides already discussed, there are lead(II) compounds of essentially all of the common anions, including many basic compounds. Thus lead(II) chloride forms such basic compounds as $PbCl_2 \cdot Pb(OH)_2$, $PbCl_2 \cdot PbCl_2 \cdot 2PbO$, $PbCl_2 \cdot 3PbO$, and $PbCl_2 \cdot 7PbO$. In fact, a whole series of lead salts are derived from the hydroxide, some of which are double compounds, such as $PbX_2 \cdot 2Pb(OH)_2$ and some of which, of composition $Pb(OH)X$, have been shown to be dimeric of the general formula

$$\left[Pb \begin{pmatrix} HO \\ \\ HO \end{pmatrix} Pb \right] X_2$$

Other lead compounds include the following:

Acetates. Lead acetate, "sugar of lead." [CAS: 301-04-2]. $Pb(C_2H_3O_2)_2 \cdot 3H_2O$, white crystals, soluble, formed by reaction of lead oxide and acetic acid, and then crystallization. Used (1) to furnish a soluble lead salt, (2) as a mordant in dyeing and printing textiles, (3) as a paint and varnish drier, basic lead acetate, white crystals, soluble, formed by reaction of lead acetate solution and lead oxide, and then crystallization. Used as a coagulating, clarifying, and deacidifying agent for many organic solutions.

Arsenate. Lead arsenate. [CAS: 10031-13-7]. arsenate of lead $Pb_3(AsO_4)_2$, white precipitate, formed by reaction of soluble lead salt solution and sodium arsenate solution. Used as an insecticide. Banned or tightly controlled in some countries.

Azide. Lead azide. [CAS: 13424-46-9]. PbN_6, white precipitate, formed by reaction of soluble lead salt solution and sodium azide solution (white solid, formed by reaction of sodamide $NaNH_2$ upon heating in nitrous oxide N_2O gas). Used as a detonator.

Borate. Lead borate $Pb(BO_2)_2$, white crystals, insoluble, by reaction of lead oxide and boric acid solution. Used in preparing special types of glass.

Carbonates. Lead carbonate $PbCO_3$, white precipitate, formed by reaction of soluble lead salt solution and sodium carbonate solution in the cold; basic lead carbonate, formed by reaction of (1) soluble lead salt solution and hot sodium carbonate solution, and (2) lead sheets, carbon dioxide and acetic acid, and pigment, the quality depending largely upon the conditions of the reaction.

Chromates. Lead chromate, "chrome yellow" $PbCrO_4$, yellow precipitate, by reaction of soluble lead salt solution and sodium dichromate or chromate solution, melting point of lead chromate 844°C. Used as a pigment; basic lead chromate, red solid, insoluble, formed by heating lead chromate and sodium hydroxide solution.

Nitrates. Lead nitrate. [CAS: 10099-74-8]. $Pb(NO_3)_2$, white crystals, soluble, formed by reaction of lead oxide and nitric acid, and then crystallization, decomposes on heating leaving lead oxide residue. Used to furnish a soluble lead salt; basic lead nitrate, formed by reaction of lead nitrate solution and lead oxide.

Oxalate. Lead oxalate PbC_2O_4, white precipitate, formed by reaction of soluble lead salt solution and ammonium oxalate solution, yields lead suboxide on heating at 300°C out of contact with air.

Phosphate. Lead phosphate. [CAS: 7446-27-7]. $Pb_3(PO_4)_2$, white precipitate, by reaction of soluble lead salt solution and sodium phosphate solution.

Sulfates. Lead sulfate. [CAS: 7446-14-2]. $PbSO_4$, white precipitate, formed by reaction of soluble lead salt solution and sulfuric acid or sodium sulfate solution; basic lead sulfate, "sublimed white lead," white solid, formed (1) by reaction of lead sulfate and lead hydroxide in water (slow reaction), and (2) by roasting galenite in a current of air.

Sulfide. Lead sulfide, plumbous sulfide. [CAS: 1314-87-0]. PbS, brownish-black precipitate, formed by reaction of soluble lead salt solution and hydrogen sulfide or sodium or ammonium sulfide, soluble in dilute nitric acid.

In the great majority of organometallic compounds of lead, the metal is tetravalent and covalently bonded, although the organolead group includes many compounds with both organic radicals and halogen atoms attached to Pb which are not to be described merely as covalent compounds. More than five hundred organometallic compounds of lead have been reported, many of which are named as substituted plumbanes, although PbH_4 is not a starting point in their production. Tetraethyl lead, $Pb(C_2H_5)_4$, is made from a sodium-lead alloy and ethyl chloride.

Like carbon and silicon, and to a lesser extent, germanium and tin, lead forms binary compounds with metals, such as Na_4Pb_7 and Na_4Pb_9. These materials are essentially salt-like, and contain polyplumbide anions. They are of theoretical interest, because they are intermediate in character between stoichiometric compounds (daltonide compounds) and intermediate phases. The two compounds cited dissolve in liquid ammonia, electrolyze in such solutions to give the metals, and apparently form ions such as $[Pb_7]^{4-}$ and $[Pb_9]^{4-}$ which readily form amine complexes.

Lead in Biological Systems—Toxicity

Lead has been identified as a biological system deterrent for decades. It is only recently, however, that studies pertaining to low-dosage exposures of lead have been published despite the fact that probably millions of words have appeared in various publications on the overall topic of lead poisoning.

From a qualitative standpoint, exposure to lead results in a clinical picture of hypertensive encephalopathy, neuropathy, and hemolytic anemia characterized by coarse basophilic stippling in red blood cells. The mechanism of lead's action on human tissue is complex. For one thing, lead blocks heme synthesis. This leads to a build-up of red blood cell protoporphyrin. Lead interferes with cell metabolism by causing a deficiency of pyrimidine 5'-nucleotidase. Lead attacks erythrocyte membrane phospholipids with resultant loss of potassium and interference with the sodium-potassium balance. Diagnosing lead poisoning may involve a determination of the free erythrocyte protoporphyrin level as well as determination of blood and urine levels. Once confirmed, further exposure to lead must be stopped immediately. Chelating compounds, such as $CaNa_2EDTA$, may be administered intravenously over an 8-hour period for several days. This may be followed by treatment with oral penicillamine for several days.

Lead poisoning can lead to chronic renal failure. In its effect on kidney function, lead acts much like cadmium. Chronic exposure to or ingestion of practically any heavy metal, such as lead, is the most common path to polyneuropathy. Where effects of heavy metals on the peripheral nervous system are suspected, many physicians will require testing for metal in hair, fingernails, serum, and urine of the patient. Habitual sniffing of leaded gasolines can lead to lead poisoning. Scientists have compared the lead concentration in the diets of present Americans (0.2 part per million) with the diets of prehistoric peoples (estimated to be less than 0.002 part per million). Some investigators believe that the presence of "natural" lead contamination has been grossly overestimated and that what has appeared to be natural has been the result mainly of a gradual build-up of lead pollution in the air derived from anthropogenic sources. The principal sources of atmospheric lead contamination include (1) natural sources, such as wind-blow volcanic dust, sea spray, forest foliage, and volcanic sulfur compounds; and (2) anthropogenic sources, such as lead alkyls (present in fuels), iron smelting, lead smelting, zinc and copper smelting, and the burning of coal. Much remains by way of research into the sources of lead contamination, including the contributions of atmospheric pollution, of food containers, and of food processing equipment.

In 1990, H.I. Needleman and co-researchers (University of Pittsburgh, Boston University, and Harvard University) reported their findings on the long-term effects of exposure to low doses of lead in children. An abstract of the report is as follows:

> To determine whether the effects of low-level lead exposure persist, we reexamined 132 of 270 young adults who had initially been studied as primary school-children in 1975 through 1978. In the earlier study, neurobehavioral functioning was found to be inversely related to dentin lead levels. As compared with those we restudied, the other 138 subjects had somewhat higher lead levels on earlier analysis, as well as significantly lower IQ scores and poorer teachers' ratings of classroom behavior.
>
> When the 132 subjects were reexamined in 1988, impairment in neurobehavioral function was still found to be related to the lead content of teeth shed at the ages of six and seven. The young people with dentin lead levels >20 ppm had a markedly higher risk of dropping out of high school (adjusted odds ratio, 7.4; 95 percent confidence interval, 1.4 to 40.7) and of having a reading disability (odds ratio, 5.8; 95 percent confidence interval, 1.7 to 19.7) as compared with those with dentin lead levels <10 ppm. Higher lead levels in childhood were also significantly associated with lower class standing in high school, increased absenteeism, lower vocabulary and grammatical-reasoning scores, poorer hand—eye coordination, longer reaction times, and slower finger tapping. No significant associations were found with the results of 10 other tests of neurobehavioral functioning. Lead levels were inversely related to self-reports of minor delinquent activity.
>
> We conclude that exposure to lead in childhood is associated with deficits in central nervous system functioning that persist into young adulthood.

An interesting professional critique of the Needleman report is summarized by J. Palca (reference listed).

The lead elimination and clean-up problem has numerous parallels with the asbestos pollution problem. The main problem is not one of finding substitutes for these substances, because lead-free paints, for example, have been available for several years, just as substitute insulating materials for asbestos have been found. As pointed out in an excellent article by Pollack (reference listed), the problem (or dilemma) lies with the cleanup of old structures that have such materials installed; and to what limits must one go, within the limitations of financial resources, to remove such materials; and, once removed, how to dispose of them safely. While removing all lead-painted surfaces from schools, for example, ultimately can provide assurance that children will not be exposed to the long-term effects of lead, a great deal of new exposure to workmen and the immediate neighborhoods of such removal projects can occur. Modern technology has designed equipment to protect the safety of restoration or demolition crews, but there remains the enforcement of their using such equipment. One crux of the problem is that posed by the apparent effects of lead in very low dosages. Such pollutants of low concentration, as may be typified by the creation of dust and windborne aerosols, definitely exacerbates the problem. Obviously, the problem becomes one more hampered by socioeconomic measures than by technology.

Additional Reading

Bodwal, B.K. et al.: "Ultralight-Pressure Melting of Lead: A Multidisciplinary Study," *Science*, 462 (April 27, 1990).

Davis, J.R.: *Metals Handbook,* 2nd Edition, ASM International, Materials Park, OH, 1998.

Greenwood, N.N. and A. Earnshaw: *Chemistry of the Elements,* 2nd Edition, Butterworth-Heinemann, Inc., Woburn, MA, 1997.

Holden, C.: "Resurrected Lead," *Science*, 192 (October 11, 1991).

Holden, C.: "Toxic Waste Program Lacks Science Base," *Science*, 797 (November 8, 1991).

Krebs, R.E.: *The History and Use of Our Earth's Chemical Elements: A Reference Guide,* Greenwood Publishing Group, Inc., Westport, CT, 1998.

Lide, D.R.: *CRC Handbook of Chemistry and Physics,* 84th Edition, CRC Press, LLC., Boca Raton, FL, 2003.

Needleman, H.L. et al.: "The Long-Term Effects of Exposure to Low Doses of Lead in Childhood," *N. Eng. J. Med.,* 83 (January 11, 1990).

Palca, J.: "Get-the-Lead-Out Guru Challenged," *Science*, 842 (August 23, 1991).

Plewes, J.T. and D.N. Loiacono: "Free-Cutting Copper Alloys Contain No Lead," *Advanced Materials & Processes,* 23 (October 1991).

Pollack, S.: "Solving the Lead Dilemma," *Technology Review (MIT),* 22 (October 1989).

Raloff, J.: "Beverages Intoxicated by Lead in Crystal," *Science News*, 54 (January 26, 1991).

Stwertka, A. and E. Stwertka: *A Guide to the Elements,* Oxford University Press, Inc., New York, NY, 1998.

LEAVENING AGENTS. The generation of carbon dioxide for use as dough leavening is produced by reacting sodium carbonate (baking soda) with one of several leavening acids. In the case of an acidic phosphate salt (with two replaceable hydrogen atoms), the reaction is:

$$MH_2PO_4 + 2\ NaHCO_3 \longrightarrow MNa_2PO_4 + 2\ H_2O + 2\ CO_2$$

where M can be a hydrogen or an alkali metal ion. Claims for use of acidic phosphate salts, in addition to formation of carbon dioxide, are the buffering effects for providing an optimal pH for the baked product, as well as interactions with protein constituents of flour, with resulting optimal elastic and viscosity properties of the dough batter.

Other leavening acids used in modern bakeries include sodium aluminum sulfate, $Na_2SO_4 \cdot Al_2(SO_4)_3$; sodium aluminum phosphate hydrate (and anhydrous); potassium acid tartrate, $KHC_4H_4O_6$ (cream of tartar); and glucono-delta-Iactone. The baker is concerned with (1) *dough rate of reaction* (DRR), a measure of the rate at which the leavening acid reacts with the baking soda during both the mixing stage and the holding period after mixing (bench action); and (2) *neutralizing value* or neutralizing strength, i.e., the weight of leavening acid required to neutralize a given weight of sodium bicarbonate. This value is used to compute the amount of leavening acid required to yield the needed amounts of leavening gas as well as its effect upon the pH of the baked goods.

Properties of the principal leavening acids are given in Table 1, which shows the most appropriate baking applications for each.

Baking powders, as prepared for the home baker and for use in premixes, usually incorporate, along with sodium bicarbonate, one of the following leavening acids: (1) potassium hydrogen tartrate (2 parts for 1 part sodium bicarbonate); (2) tartaric acid (infrequent), 1 part; (3) calcium hydrogen phosphate (crystallized), 1.5 parts; or (4) sodium aluminum sulfate or ammonium aluminum sulfate, 1.8 parts. With 7 parts by weight of this finely powdered mixture, there is usually mixed about 3 parts by weight of starch to diminish the effects of moisture in storage. In some cases, dry powdered egg albumin is added to decrease the loss of carbon dioxide upon wetting the flour and baking powder mixture when used. For some purposes, ammonium carbonate can be used alone, since upon heating this material furnishes both ammonia and carbon dioxide gases to make

TABLE 1. PROPERTIES OF LEAVENING ACIDS

Chemical name and formula	Abbreviation	Relative speed at room temperature	Neutralizing value[1]
Sodium aluminum phosphate (anhydrous)	SALP	Medium	110
Sodium aluminum sulfate, $Na_2SO_4 \cdot Al_2(SO_4)_3$	SAS	Slow	100
Monocalcium phosphate (anhydrous), $CaH_4(PO_4)$	MCP	Slow	83
Monocalcium phosphate (monohydrate), $CaH_4(PO_4) \cdot H_2O$	MCP · H₂O	Quite fast	80
Sodium acid pyrophosphate, $Na_2H_2P_2O_7$	SAPP	Medium	72
Glucono-delta-lactone, $C_6H_{10}O_6$	GDL	Slow	55
Potassium acid tartrate, $KHC_4H_4O_6$	—	Medium to fast	50
Dicalcium phosphate dihydrate, $CaHPO_4 \cdot 2H_2O$	DCP	Very slow	33
Sodium aluminum phosphate hydrate	SALP · H₂O	Slow	100

[1] Values in this column indicate the parts of sodium bicarbonate that will be neutralized by 100 parts of the leavening acid under nominal conditions. Values vary with composition of dough.

the product light. These gases escape from the product during the baking process. In selecting a baking powder, one must keep in mind the speed with which the components react at room temperature: alum-containing baking powders act slowly; phosphate baking powders have a medium speed; and tartrate baking powders act quickly to produce carbon dioxide. Hence, when using the latter type, it is necessary to bake quickly after mixing to eliminate the loss of too much gas.

Within the last several years, advantage has been taken of mixing different leavening acids in premixes and household baking powders. Because the use of emulsifiers in most cake mixes reduces the need for early leavening action, it is common practice to use combinations of slow-acting leavening acids that retain much of their leavening reaction for the baking stage. In mixes, the leavening process must be regarded as a system because, in addition to gas generation, the leavening system controls the pH of the finished product and thus affects crumb and crust color, the intensity of flavor, as well as other properties. For various cakes, the optimum pH values are: white cakes, 6.9–7.2; yellow cakes, 7.2–7.5; chocolate or devil's food cakes, 7.1–8.0. Monocalcium phosphate (anhydrous) and sodium aluminum phosphate are frequently used together in white and yellow cake mixes; monocalcium phosphate and sodium acid pyrophosphate or dicalcium phosphate dihydrate are used in chocolate cake mixes. Generally, the combination will be comprised of 10–20% fast-acting leavening acid and 80–90% slow-acting leavening acid.

For pancake and waffle mixes, a common blend of leavening acids is 20–30% monocalcium phosphate monohydrate or monocalcium phosphate (anhydrous), combined with 70–80% sodium aluminum phosphate. A batter of this type can be prepared several hours in advance if retained under refrigeration. It has been observed that such a batter will sour before a serious loss of leavening power occurs.

Prepared biscuit mixes made of flour, shortening, and salt usually contain 30–50% monocalcium phosphate (anhydrous) and 50–70% sodium aluminum phosphate or sodium acid pyrophosphate. Self-rising flours and corn meals usually contain flour or corn meal, salt, soda, and leavening acid. Usually used in these products are combinations of sodium aluminum phosphate and monocalcium phosphate (anhydrous).

Refrigerated doughs available for preparation of biscuits, dinner rolls, and various sweet rolls, usually contain flour, water, shortening, nonfat milk solids (or dried whey solids), sugar (or corn sugar), salt, soda, and a leavening acid. Long-term refrigerated storage requires that only slow-acting leavening acids be used, frequently the sodium acid pyrophosphates. The latter have the disadvantage of possibly producing orthophosphates under certain conditions. The orthophosphates have a rather disagreeable, astringent flavor.

Unleavened Products

The principal unleavened bakery product is pie crust, which is low in moisture and high in fat content. The ingredients and method of preparation prevent the formation of a continuous gluten network through the dough mass. The porosity associated with leavened products is not desirable because the crust literally acts as a container and requires some strength.

Additional Reading

Amendola, J., and N. Rees: *Understanding Baking,* 3rd Edition, John Wiley & Sons, Inc., New York, NY, 2002.

Figoni, P.I.: *How Baking Works: Exploring the Fundamentals of Baking Science,* John Wiley & Sons, Inc., New York, NY, 2003.

LE CHÂTELIER'S PRINCIPLE.

Let us perturb a system that is initially in stable equilibrium to a neighboring nonequilibrium state. Since the initial equilibrium is supposed to be stable, the system will return to an equilibrium state.

Theorems governing the behavior of perturbed systems are often known as *theorems of constraint or theorems of moderation.* The best known thermodynamic theorem of moderation is that of Le Châtelier-Braun, which in the form stated by Le Châtelier is:

"Any system in chemical equilibrium undergoes, as a result of a variation in one of the factors governing the equilibrium, a *compensating* change in a direction such that, had this change occurred alone it would have produced a variation of the factor considered in the *opposite direction.*"

However, this principle suffers from a number of important exceptions. It is therefore preferable to study the "moderation" starting from the usual thermodynamic formalism without invoking a special principle.

LECITHIN.

Lecithin and other phospholipids are of universal occurrence in living organisms. They are constituents of biological membranes and are involved in permeability, oxidative phosphorylation, phagocytosis, and chemical and electrical excitation.

Lecithin is not only used in the strict scientific sense to describe pure phosphatidylcholine (Fig. 1), but also to describe crude phospholipid mixtures containing phosphatidylcholine (PC), phosphatidylethanolamine (PE), phosphatidylinositol (PI), other phospholipids, and a variety of other compounds such as fatty acids, triglycerides, sterols, carbohydrates, and glycolipids. Commercial lecithin is currently available in more than 40 different formulations varying from crude oily extracts from natural sources to purified and synthetic phospholipids. Many of these products are defined according to the stage of the purification process from which they are obtained and fall into three broad categories (Table 1) varying in their constituents both qualitatively and quantitatively.

Industrial lecithins from a variety of sources are utilized. The main sources include vegetable oils (e.g., soybean, cottonseed, corn, sunflower, rapeseed) and animal tissues (egg and bovine brain). However, egg lecithin and in particular soy lecithin are by far the most important in terms of quantities produced, so much so that the term soy lecithin and commercial lecithin are often used synonymously.

Physical Properties

Commercial crude lecithin is a brown to light yellow fatty substance with a liquid to plastic consistency. Its density is 0.97 g/mL (liquid) and 0.5 g/mL (granule). The color is dependent on its origin, process conditions, and whether it is unbleached, bleached, or filtered. Its consistency is determined chiefly by its oil, free fatty acid, and moisture content. Properly refined lecithin has practically no odor and has a bland taste. It is soluble in aliphatic and aromatic hydrocarbons, including the halogenated hydrocarbons; however, it is only partially soluble in aliphatic alcohols. Pure phosphatidylcholine is soluble in ethanol.

Commercial lecithin is soluble in mineral oils and fatty acids but is practically insoluble in cold vegetable and animal oils. It is insoluble

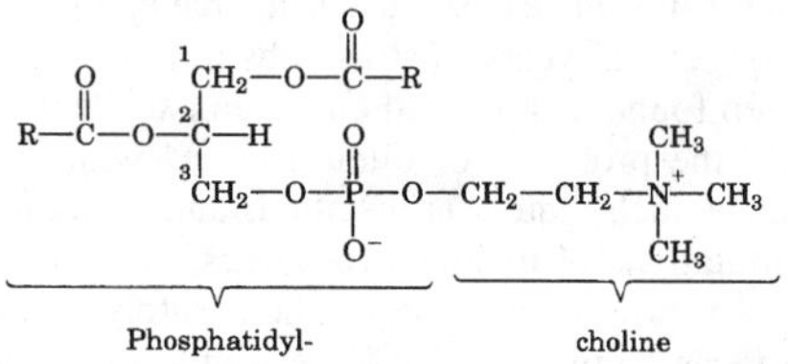

Fig. 1. Chemical structure of phosphatidylcholine (PC) (1) and other related phospholipids. $R-\overset{O}{\underset{||}{C}}-O-$ represents fatty acid residues. The choline fragment may be replaced by other moieties such as ethanolamine (2) to give phosphatidylethanolamine (PE), inositol (3) to give phosphatidylinositol (PI), serine (4), or glycerol (5). If H replaces choline, the compound is phosphatidic acid. The corresponding IUPAC-IUB names are (1), 1,2-diacyl-*sn*-glycero(3)phosphocholine; (2), 1,2-diacyl-*sn*-glycero(3)phosphoethanolamine; (3), 1,2-diacyl-**sn**-glycero(3)phosphoinositol; (4), 1,2-diacyl-**sn**-glycero(3)-phospho-L-serine; and (5), 1,2-diacyl-*sn*-glycero(3)phospho(3)-*sn*-glycerol

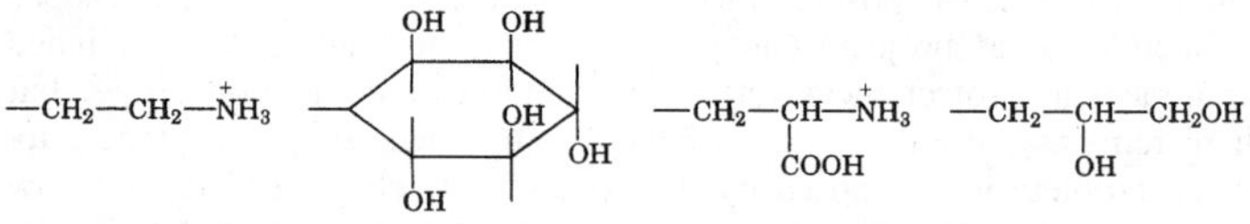

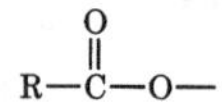

TABLE 1. CATEGORIES OF COMMERCIAL LECITHIN

Natural	Refined	Modified
Plastic	*Deoiled*	*Physically*
unbleached		custom-blended
bleached		natural and refined
double-bleached		
Fluid	*Fractionated*	*Chemically*
unbleached	alcohol-soluble	
bleached	alcohol-insoluble	
double-bleached		*Enzymatically*

but infinitely dispersible in water. Commercial lecithin is a wetting and emulsifying agent. Lecithin is one of the very few natural and edible surface-active agents of this type that is soluble or dispersible in oil.

Chemical Properties

In general, the presence of fatty acid groups in the phospholipid molecule permits reactions such as saponification, hydrolysis, hydrogenation, halogenation, sulfonation, phosphorylation, elaidinization, and ozonization.

Manufacture and Processing

Crude soy lecithin is obtained as a by-product during the degumming process of soy oil. Only a minor proportion of the total lecithin that is potentially available in the vegetable processing industry is produced.

Purification Processes

Separation of neutral and polar lipids, so-called deoiling, is the most important fractionation process in lecithin technology. A classic solvent for the deoiling is acetone.

Due to the possible environmental problems with acetone, new technologies are being developed for the production of deoiled lecithins like an ethanol-based extraction and fractionation or a process involving treatment of lipid mixtures with supercritical gases or supercritical gas mixtures.

Commercial Grades

There are six common grades of lecithin available including (*1*) clarified lecithins, (*2*) fluidized lecithins; (*3*) compounded lecithins; (*4*) hydroxylated lecithins; (*5*) deoiled lecithins and (*6*) fractionated lecithins. Fractions with different phosphatidylcholine content are commercially available. Besides these common commercial grades, more special products are available, e.g., enzymatically modified lecithin and phospholipids, semisynthetic phospholipids, and acetylated lecithins.

Economic Aspects

The total commercial lecithin potential if all vegetable oils were degummed worldwide would be 552,000 t. Although soybean, sunflower, and rape lecithins are available in the market, the principal commercial interest is only in soybean lecithin. The annual worldwide production is 130,000 t.

Health and Safety Factors

The phospholipids are biodegradable, but their presence in streams and water resources, especially in the form of soap stock, is undesirable. Fatty acid recovery from phospholipids is less than with neutral oils because of the lower fatty acid content. There are no known health hazards involved in the production of commercial lecithin from crude vegetable oils because the phospholipids are nonvolatile and are a nonirritating food material.

Uses

The worldwide uses of lecithin break down as follows: margarine, 25–30%; baking/chocolate and ice cream, 25–30%; technical products, 10–20%; cosmetics, 3–5%; and pharmaceuticals, 3%.

Cosmetics and Soaps. One to five percent lecithin moisturizes, emulsifies, stabilizes, conditions, and softens when used in products such as skin creams and lotions, shampoos and hair treatment, and liquid and bar soaps. Since the introduction of Capture in 1986, liposomes produced from phospholipids are commercially available worldwide.

Pharmaceuticals. Lecithin and especially purified phosphatidylcholine can act as excipients in pharmaceutical (drug) formulation to enhance and control the bioavailability of the active component. Moreover, phosphalidylcholine can be utilized as a diedelic source, as it involved in the cholesterol metabolism and the metabolism of fats in the liver; also, it can be utilized as a precursor of brain acetylcholine, as neurotransmitter.

ARMIN WENDEL
Rhône-Poulenc Rorer

Additional Reading

Hanin, I. and G.B. Ansell, eds.: *Lecithin: Technological, Biological and Therapeutic Aspects,* "Advances in Behavioral Biology," Vol. 33, Plenum Press, New York, NY, 1987.

Hanin, I. and G. Pepeu, eds.: *Phospholipids: Biochemical, Pharmaceutical, and Analytical Considerations,* Plenum Press, New York, NY, 1990.

Szuhaj, B.F. and G.R. List, eds.: *Lecithins,* American Oil Chemist's Society, Champaign, Ill., 1985.

Szuhaj, B.F. ed.: *Lecithins: Sources, Manufacture and Uses,* American Oil Chemist's Society, Champaign, Ill., 1989.

LEDUC'S RULE. States that the volume occupied by a gas mixture is equal to the sum of the volumes occupied separately by each constituent at the same temperature and pressure as the mixture.

LEE, YUAN T. (1936–). Awarded the Nobel prize in chemistry in 1986 jointly with John C. Polanyi and Dudley R. Herschbach for their contributions concerning the dynamics of chemical elementary processes. A former student of Herschbach, Lee refined molecular-beam and laser techniques, combining them with theory to perform definitive studies of reactions of individual complex molecules. Lee received his Doctorate from the University of California at Berkeley in 1965.

LEHN, JEAN-MARIE PIERRE (1939–). Awarded the Nobel prize for chemistry, together with Donald J. Cram, in 1987 for work in elucidating mechanisms of molecular recognition, which are fundamental to the enzymic catalysis, regulation, and transport. He also studied three-dimensional cyclic compounds that maintained a rigid structure, accepting substrates in a structurally preorganized cavity. Lehn named these compounds cryptands, while Cram called them cavitands. Awarded Doctorate by the University of Strasbourg, France in 1963.

LELOIR, LUIS F. (1906–1987). A French-born biochemist who won the Nobel prize for chemistry in 1970 for work in biosynthesis of carbohydrates. He discovered chemical compounds that affect the storage of chemical energy in humans and animals. He headed the Department of Biochemistry at the University of Buenos Aires for many years.

LEPIDOLITE. This member of the mica group of minerals is a silicate of potassium, lithium and aluminum, sometimes with sodium, fluorine, or rarely rubidium. A general formula is $K(Li, Al)_3(SiAl)_4O_{10}(F, OH)_2$. Crystals of lepidolite are monoclinic but often pseudo-hexagonal; cleavage, basal and perfect, being susceptible of splitting into thin laminae; hardness, 2.5–4; specific gravity, 2.8–3.3; luster, pearly; color, reddish to violet, grayish-blue, gray to white. A variety carrying rubidium is yellowish-gray; translucent. It usually is found as granular to scaly masses, in short stocky prisms or less often in easily cleavable sheets. Lepidolite is characteristic of pegmatite veins, frequently being associated with other lithium-bearing minerals such as tourmaline, spodumene, amblygonite, and others. It occurs in the Ural Mountains, the Czech Republic and Slovakia, the Island of Elba, and Madagascar, where it is often found in large sheets. In the United States, it is found in the pegmatites of New England, California, South Dakota, and New Mexico. The name lepidolite is derived from the Greek, meaning scale. See also **Lithium.**

LEPTONS. The electron, muon, and two kinds of neutrino are collectively called *leptons.* The leptons are considered to be point-like particles without structure and thus truly elementary. Leptons can interact with other particles through the weak interactions. Electrons and muons also can interact through electromagnetic and gravitational forces, but they appear to be without capability of interaction through the strong (nuclear) forces. The neutral members, the electron neutrino, the muon neutrino, and their antiparticles have extremely weak interaction with matter and do not participate in electromagnetic interactions. Leptons make excellent probes in particle physics experiments. The other major family of subatomic particles is referred to as *hadrons.* See also **Electron; Muon; Neutrino;** and **Particles (Subatomic).**

The name *lepton* from its derivation means "light," referring to the fact that the masses of the leptons are all lighter than that of the lightest meson. The properties of the electron are discussed in that entry; here it will merely be noted that the term *electron* is used to denote the negative electron (often called the "negatron" when ambiguity might arise). Its antiparticle is the positron (also called positive electron).

LEUCITE. The mineral leucite is a metasilicate of potassium and aluminum corresponding to the formula $KAlSi_2O_6$. It is isometric at a temperature of about 600°C (1112°F) and pseudoisometric at lower temperatures, at which leucite is tetragonal but retains an external isometric crystal form, usually trapezohedral. It has a conchoidal fracture; is brittle;

hardness, 5.5–6; specific gravity, 2.47–2.50; luster, vitreous; color, white or some shade of gray; translucent to opaque. It is commonly found in the more recent lavas of high alkali content. Leucite is seldom reported from plutonic rock types. It is a relatively rare mineral. It is found plentifully at Vesuvius and Monte Somma and elsewhere in Italy, and Germany in the Tertiary volcanic district of the Eifel. In the United States, leucite has been found in the Leucite Hills of Wyoming, the Highwood Mountains of Montana, and as pseudomorphs (pseudoleucites) representing a mixture of nepheline, orthoclase, analcime, and aegerine from New Jersey, Arkansas, and Montana. Its name is derived from the Greek word leukos, referring to its white color.

LEVENE-HUDSON PHENYLHYDRAZIDE RULE. The direction of rotation of the phenylhydrazides of the sugar acids indicates the configuration of the hydroxyl on the α-carbon atom. If the phenylhydrazide rotates to the right, the hydroxyl on the α-carbon is to the right, and vice-versa. The rule was shown to be valid for salts, amides, and corresponding acylated nitriles. In connection with the rule, Hudson mentioned that, "the sugar benzylphenylhydrazones rotate to the left when the asymmetric α-carbon atom of the configuration has its hydroxyl to the right, and vice-versa."

LEVOROTATORY COMPOUNDS. Having the property when in solution of rotating the plane of polarized light to the left or counterclockwise. Levorotatory compounds may have the prefix *l*-to distinguish them from their dextrorotatory or *d*-isomers, but the minus sign $(-)$ is preferred.

See also **Asymmetry (Chemical)**; **Isomerism**.

LEWIS ACID. Any molecule or ion (called and electrophile) that can combine with another molecule or ion by forming a covalent bond with two electrons from the second molecule or ion. An acid is thus an electron acceptor. Hydrogen ion (proton) is the simplest substance that will do this, but many compounds such as boron trifluoride, BF_3, and aluminum chloride, $AlCl_3$, exhibit the same behavior and are therefore properly called acids. Such substances show acid effects on indicator colors and when dissolved in the proper solvents.

LEWIS BASE. A substance that forms a covalent bond by donating a pair of electrons, neutralization resulting from reaction between the base and the acid with formation of a coordinate covalent bond. It is also called a nucleophile.

See also **Lewis Electron Theory**.

LEWIS ELECTRON THEORY. A theory involving acid and base formation, neutralization, and related phenomena on the basis of exchange of electrons between substances and the formation of coordinate bonds. It represented an important advance in chemical theory, largely replacing earlier concepts. Advanced in 1923 by Gilbert N. Lewis, it contributed much to the development of coordination chemistry in which the base is represented by the ligand and the acid by the metal ion.

LEWIS, GILBERT N. (1875–1946). An American chemist, native of Massachusetts, professor of chemistry at MIT from 1905 to 1912 after which he became dean of chemistry at the University of California at Berkeley. His most creative contribution was the electron-pair theory of acids and bases, which laid the groundwork for coordination chemistry. He was also a leading authority on thermodynamics.

LEWIS, WARREN P. (1882–1974). Born in Laurel, Maryland, graduated from MIT in 1905, and received is Ph.D. from the University of Breslau, Germany in 1908. He became professor of chemical engineering at MIT in 1910. He is often regarded as the "father" of chemical engineering in the U.S., as his outstanding books and other publications did much to establish the fundamental principles of this field.

LEYDEN TEMPERATURE SCALE. A low-temperature thermometer scale based on a boiling point of hydrogen equal to $-252.74°C$ and of Oxygen equal to $-182.95°C$.

LIBBY, WILLARD P. (1908–1980). An American chemist, famous for his role in the development of radiocarbon dating, a process which revolutionized archaeology.

After the start of World War II he worked on the Manhattan Project at Columbia University with Nobel laureate Harold Urey. Libby was responsible for the gaseous diffusion separation and enrichment of Uranium-235 which was used in the atomic bomb on Hiroshima.

In 1945 he became a professor at the University of Chicago. In 1954, he was appointed to the U.S. Atomic Energy Commission, and in 1959, he became Professor of Chemistry at University of California, Berkeley, a position he held until his retirement in 1976. He also started the first Environmental Engineering program at UCLA in 1972.

Libby won the Nobel prize for chemistry in 1960, for his method to use carbon-14 for age determination in archaeology, geology, geophysics, and other branches of science.

LIEBIG, JUSTUS Von (1803–1873). A German chemist who founded the *Annalen*, a world-famous chemical journal. He was a great teacher of chemistry, training such men as Hofmann, August Whilhelm, who did basic work on organic dyes. Liebig contributed original research in the fields of human physiology, plant life, soil chemistry, and was first to discover chloroform, chloral, and cyanogen compounds. He was the first to recommend the addition of nutrients to soils and thus may be considered the originator of the fertilizer industry.

LIFE, ORIGIN (Biogenesis). The succession of chemical events that led up to the appearance of living organisms on earth about 3.3 billion years ago. According to one theory, substantiated by experimental evidence, this occurred as follows. The inorganic compounds originally present were carbides, water, ammonia, and carbon dioxide. The carbides reacted with water to form methane, which in turn reacted with ammonia and water vapor as a result of an electric impulse to form amino acids, porphyrins, and nucleotides (or their precursors). All these compounds have been created artificially in the laboratory. It has further been shown that amino acids and nucleotides can be concentrated into proteins (and probably nucleic acids) by the action of zinc-bearing clays, which were present along the shores of the primeval oceans. Little or no free oxygen existed in the primordial atmosphere, which consisted chiefly of reducing gases. The complex chemical reactions that eventually resulted in the formation of DNA took place in an anaerobic aqueous environment, and the earliest living organisms developed in a nutrient solution in which free oxygen finally appeared as the result of photosynthesis by blue-green bacteria. Another theory advances the idea that essential life chemicals such as purines and amino acids were formed under primitive conditions from aqueous solutions of hydrogen cyanide. Both of these theories are based on research carried out by highly competent biochemists.

LIGAND. Any atom, radical, ion, or molecule in a complex (poly-atomic group) which is bound to the central atom. Thus, the ammonia molecules in $[Co(NH_3)_6]^{3+}$, and the chlorine atoms in $[PtCl_6]^{2-}$ are ligands. Ligands are also complexing agents, as for example, EDTA, ammonia, etc. See also **Chelates and Chelation**.

Ligand field theory incorporates elements from the valence bond theory of Pauling and the molecular orbital method of Hund, Mullikan, and others. As pointed out by Mortimer, the chemists of the late nineteenth century had difficulty in understanding how "molecular compounds" or "compounds of higher order" are bonded. The formation of a compound such as $CoCl_3 \cdot 6NH_3$, was baffling, particularly in this case since simple $CoCl_3$ does not exist. In 1893, Alfred Werner proposed a theory to account for compounds of this type. Werner wrote the formula of the cobalt compound as $[Co(NH_3)_6]Cl_3$. Werner assumed that the six ammonia molecules are symmetrically coordinated to the central cobalt atom by "subsidiary valencies" of cobalt, while the "principal valencies" of cobalt are satisfied by the chloride ions. Werner devoted over 20 years preparing and studying coordination compounds and perfecting and proving his theory. Although modern work has amplified his theory, it has required relatively little modification.

In ligand field theory, one is concerned with the origin and the consequences of splitting the inner orbitals of the central metal by the surrounding ligands. The most satisfactory correlations have been demonstrated with the first transition series, in which the $3d$-orbitals are split into different energy levels. To appreciate the effect of a ligand field, imagine that a symmetrical group of ligands is brought up to a charged ion from a distance. First, the electrostatic repulsions between the ligand electrons and those in the d-orbitals of the metal will raise the energy of all five d-orbitals equally. Then, as the ligands approach to within bonding distances, the repulsion interactions will take on a directional character that will vary with the particular d-orbitals under consideration. This arises because of the different shapes and orientations of the five d-orbitals in space along a Cartesian coordinate system. The splitting of the orbitals for a given central metal ion is dependent on the set of ligands.

Applications of ligand field theory to many transition metal complexes have played an important role in the interpretation of visible absorption spectra, magnetism, luminescence, and paramagnetic resonance spectra.

Additional Reading

Bohm, Hans-Joachim, G. Schneider, and R. Mannhold: *Protein-Ligand Interactions: From Molecular Recognition to Drug Design,* John Wiley & Sons, Inc., New York, NY, 2003.

Figgis, B.N.: *Ligand Field Theory and Its Applications,* John Wiley & Sons, Inc., New York, NY, 1999.

Harding, S.E., and B.Z. Chowdhry: *Protein-Ligand Interactions: A Practical Approach,* Oxford University Press, New York, NY, 2001.

Russo, N., M. Witko, and D.R. Salahub: *Metal-Ligand Interactions: Molecular-, Nano-, Micro-, and Macro-Systems in Complex Environments,* Kluwer Academic Publishers, Norwell, MA, 2003.

LIGHT-EMITTING DIODES.   See **Luminescence**.

LIGNIN.

Approximately 25% of the content of most woods is lignin. Lignin concentration in wood substance is greatest in the middle lamella (the zone around each individual fiber cell), decreasing in concentration through the cross section of the fiber, reaching a concentration of about 12% at the inner layer of the fiber adjacent to the fiber cavity, or lumen. Lignin and hemicellulose cement the fiber cells together, providing rigidity to the fibrous wood structure. In the destructive distillation of wood, the methanol produced is derived from the lignin. In the manufacture of paper pulp, it is necessary to remove the lignin, usually accomplished by treatment of the wood fibers with such agents as sulfur dioxide, calcium bisulfite, and sodium sulfate/sodium sulfide solutions. Sodium hydroxide is sometimes used. An important byproduct of the paper pulp industry is dimethyl sulfoxide, $(CH_3)_2SO$ which is produced from the lignin released during wood pulping by the Kraft process. Dimethyl sulfoxide has a number of industrial uses—as an intermediate in organic syntheses, as a solvent in spinning synthetic fibers, and in some pharmaceuticals.

The wall material of plant cells is one of their distinguishing characteristics. As a result, lignin, cellulose, and other wall constituents have been studied in many plant tissue cultures. Phenylpropanoids, for example, have been shown to be precursors of lignin formation in white pine, *Sequoia*, lilac, rose, carrot, and geranium tissue cultures. Moreover, the biosynthesis of lignin has been shown to be affected by kinetin, boron, and major elements, such as calcium.

Lignin is a major source of vanillin.

LIGNITE AND BROWN COAL.

Lignite and brown coal are common names for coals having properties intermediate between peat and bituminous coal as a result of limited coalification. In general, brown coal designates a geologically younger, i.e., less coalified, material than the firmer, fibrous lignite. In many English-speaking countries, the consolidated coals are termed *lignite*. In many English-speaking countries, the consolidated coals are termed lignite, and unconsolidated coals are termed brown coal. In Australia, and in Germany and a number of other European countries, the generic term brown coal is used for the whole class, including some coals that are included in the ASTM classification as subbituminous. Lignite signifies the firmer, fibrous, woody variety. Herein lignite is used as the comprehensive term.

Selection of coal for a particular use requires a knowledge of composition greater than that supplied from the ASTM classification. Progress is being made toward classifying all kinds of coal, including lignite, by correlating properties with composition and other qualities.

Lignite is less valuable than coals of higher rank, primarily because its much higher (30–70% as mined) water content and high chemically combined oxygen content result in a relatively low heating value (LHV). In the past, the expense of shipping limited the market largely to the vicinity of the mine. However, in the United States the low sulfur content of lignite has made long distance shipments economically feasible, in order to limit sulfur oxide emissions at electric power generation plants. The

increasing worldwide demand for energy together with desire for national self-sufficiency has increased the importance of low heating value coals.

Geology

Lignite was deposited relatively recently (ca $2.5-60 \times 10^6$ yr ago), mainly during the Tertiary era. U.S. deposits include those in the Dakotas, Alaska, Montana, and Wyoming. The Miocene period provided the brown coal deposits that are up to 300 m thick in the Latrobe Valley of Victory in Australia.

Composition, Properties, and Analysis

Macroscopic Appearance. Lignitic coals vary from brown to dull black when moist, although the color may appear considerably lighter when the coal is dried. Breakage is easiest for the unconsolidated coals. Strength and toughness increase as coalification increases.

Physicochemical Structure. Water-filled pores and capillaries of differing diameters permeate the organic gel material that makes up asmined lignite.

Properties. The apparent density of lignite is $0.8-1.35$ g/cm^3, which is lower than values given for higher ranking coals. Therefore, greater volume is required for storage, transportation, and lignite reactors than is needed for an equivalent weight of more mature coals. Lignite generally has lower elasticity and greater plasticity than more mature coals. The tar yield is usually higher for lignite than for more mature coals. Tar yields are important in determining selection for carbonization and for liquid fuel production by pyrolysis.

Oxidation. The high reactivity of lignites with oxygen requires special care during mining, transportation, and storage to avoid spontaneous combustion from heat generation.

Resources and Production

The importance of a coal deposit depends on the amount that is economically recoverable by conventional mining techniques. The world total recoverable reserves of lignitic coals were 3.28×10^{11} metric tons at the end of 1990, of which ca 47% was economically recoverable as of 1992. These estimates of reserves change as geological survey data improve and as the resources are developed.

The extent of lignite production is generally not proportional either to total resources or to known economic reserves. Lack of energy alternatives is a strong motive to developments in lignite production.

Main Deposits and Production Areas. The eastern European reserves of lignitic coals provide the primary solid fuel for the eastern part of Germany, the former Czechoslovakia, Hungary, the former Yugoslavia, and Bulgaria. The importance of lignite as an energy source is great enough in Germany to permit long-range planning that includes removal and relocation of towns or villages situated on deposits in order to permit more complete recovery of the lignite resource. Hard coal is more important in most of the western European countries, with the exception of Austria and Italy.

In the U.S., lignite deposits are located in the northern Great Plains and in the Gulf states. Subbituminous coal is found along the Rocky Mountains. The lignite deposits of North Dakota and Montana extend into Canada as far as Saskatchewan. Canadian deposits are also located in Alberta, Yukon, the Northwest Territories, Ontario, and Manitoba.

Production. The mining or winning of lignitic coal typically involves deposits near the surface. The open-cast, open-cut, or stripmining techniques employed involve mobile equipment built to provide a range of capacities to over 200,000 m^3/d. The rate of production can be increased rapidly, and the amount of labor per ton of coal mined is less than for underground mining. The quality of the coal, ratio of overburden thickness to seam thickness, stratigraphy, and distance to location of consumption are important in determining the cost to the consumer.

Concern about spontaneous ignition has led some operators to try to match the mining and consumption rates, so that there is little if any reserve, as in minemouth power generation stations. When the coal must be stockpiled, careful stacking minimizes oxygen reaction and overheating. To limit drying, spraying with cold water is useful.

For short distances from the mine, transportation is by truck or conveyer belt. Rail transportation is generally used for greater distances. Slurry pipelines are being considered as an alternative. Drying can be accomplished by evaporative, hydrothermal, or other thermal processes.

Health and Safety Factors

The principal hazards involve the tendency of the coal toward spontaneous combustion as the coal dries, especially at the exposed seam.

Economic Aspects

The price of lignite per mined ton or per heat unit is lower than that for higher rank coals. The market for all coals is primarily as boiler fuel for electric power production. Prices are generally established by contracts between utility and supplier before mining begins. Because of its sulfur content, lignite is becoming more important.

Uses

Most of the world's coal supply is used for combustion to generate steam for electric power production. This is especially true of lignitic coals. Other uses for lignite, such as briquetting, for domestic and industrial fuels; carbonization, to provide coke and liquid by-products; gasification, to provide gaseous fuels; chemical feedstocks, for making fertilizers and other liquid fuels; and direct liquefaction are being developed.

KARL S. VORRES
Argonne National Laboratory

Additional Reading

Durie, R.A. ed.: *The Science of Victorian Brown Coal: Structure, Properties and Consequences for Utilisation,* Butterworth Heinemann, Oxford, 1991. An excellent reference not only for Victorian Brown Coal, but for lignitic coals of the world.
Symposia on the Technology and Use of Lignite have been held in conjunction with the University of North Dakota Energy and Environmental Research Center and the preceding organizations.
The World Energy Council issues Conference reports on reserves, resources and production at six-year intervals. More limited reports are issued at two-year intervals.

LIME AND LIMESTONE.   The term *lime* includes a variety of chemicals manufactured from limestone or derived from chemical processes that utilize calcium compounds. According to the composition of the parent limestone, lime may be designated as *high calcium lime* or *dolomitic lime.* Both *quicklimes,* CaO and CaO · MgO, and *hydrated limes,* CaO · H$_2$O, Ca(OH)$_2$ · MgO, and CaO · MgO · 2H$_2$O, are conventionally called lime. Precise terminology requires complex wording, e.g., *dolomitic quicklime* to denote CaO · MgO. The various lime oxides and hydroxides are among the lowest-cost and most widely used sources of alkali for the chemical and metallurgical industries. About 80% of the lime used in the United States is used by the chemical and related industries, mostly as quicklime. About 10% is dead-burned dolomite, and less than 10% goes into construction uses, mostly as hydrate. Very little lime is imported into or exported by the United States.

Limestone

[CAS: 1317-65-3]. This is a rock containing chiefly calcium carbonate and variable quantities of magnesium carbonate. Limestone is classified along the lines of lime as previously mentioned. *High-calcium limestone* contains 5% or less of MgCO$_3$ and occurs in two mineral forms, calcite and aragonite. See also **Aragonite**; and **Calcite**. *Dolomitic limestone* usually contains over 35% MgCO$_3$, with the remainder CaCO$_3$. See also Dolomite. *Magnesian limestone* is predominantly CaCO$_3$, but contains from 5 to 35% MgCO$_3$. All limestones evolve carbon dioxide and bubble in dilute hydrochloric acid. Dolomite reacts only with dilute HCl, while calcite will decompose in cold dilute HCl.

Limestones vary greatly in color and texture, the latter ranging from dense and hard limestone, e.g., marble or travertine, which can be sawed and polished, to soft, friable forms, e.g., chalk and marl. Chalk is a very fine-grained white limestone, while marl is an impure deposition product that contains clay and sand. Texture, hardness, and porosity appear to be functions of the degree of cementation and consolidation during the formation of these materials. Color variations arise from the presence of impurities. Some impurities, such as sulfur and phosphorus, make limestone unattractive for metallurgical uses.

A high percentage of all limestone is quarried; the balance is mined underground. Although limestone occurs widely, good chemical- and metallurgical-grade limestone is less plentiful. Along the seacoasts, oyster or clam shells are dredged as a source of CaCO$_3$. Limestone is normally processed through a series of crushing, screening, and grinding operations.

Because of transportation costs, the proximity of limestone sources to points of use is highly desirable. The major uses of limestone are in construction (asphalt filler, road stone, riprap, and bituminous aggregate); in Portland cement; in agriculture; and in metallurgy.

Precipitated CaCO$_3$ is produced in a number of chemical processes. Sometimes it is economical to dry and calcine the byproduct to regenerate CaO or Ca(OH)$_2$. Some precipitated CaCO$_3$ is made to specific particle size and shape, whiteness, and purity for use as functional filler for paper coatings, paint, and polymers. These products command a premium price as compared with pulverized limestone fillers.

Manufacture of Lime

The basic processes are calcination and hydration. Commencing with high-calcium limestone, the reactions are:

$$CaCO_3 + heat \rightleftharpoons CaO + CO_2 \tag{1}$$

$$CaO + H_2O \rightleftharpoons Ca(OH)_2 + Heat \tag{2}$$

If dolomitic limestone is used, the reactions are:

$$CaCO_3 \cdot MgCO_3 + heat \rightleftharpoons CaO \cdot MgO + 2\ CO_2 \tag{3}$$

$$CaO \cdot MgO + H_2O_{(liq)} \rightleftharpoons Ca(OH)_2 \cdot MgO + heat \tag{4a}$$

or

$$CaO \cdot MgO + 2\ H_2O_{(gas)} + pres \rightleftharpoons Ca(OH)_2 \cdot Mg(OH)_2 + heat \tag{4b}$$

High-calcium limestone dissociates at 900°C (1650°F) in 100% carbon dioxide atmosphere at 1 atm pressure. Under similar conditions, dolomitic limestone dissociates over 727–900°C (1340–1650°F). The heat of reaction required to convert CaCO$_3$ to CaO is about 2.8 million Btu per ton (0.64 million kg-Cal/metric ton) of CaO. In practice, heat input may vary from 4 to 10 million Btu/ton (0.9 to 2.3 million kg-Cal/metric ton) of lime. Calcination of limestone particles proceeds by a receding-surface mechanism. To attain reasonable rates of heat transfer into the center of the rock or pebble-sized stone, operating temperatures in lime kilns are 980–1260°C (1800–2300°F). Reaction rate is increased and opportunity for recarbonation of the oxide is decreased by rapid removal of CO$_2$ from the kiln.

Except for very old mixed-feed vertical kilns, lime kilns operate with countercurrent flow of raw material and heat. Modern lime kilns utilize coolers to preheat air by recuperating heat from the hot quicklime. Lime kilns may be fired directly with coal, oil, or gas.

Two major types of lime calciners are the rotary (Fig. 1) and the vertical kiln. In North America, rotary kilns are widely used for lime calcination, whereas in Europe vertical kilns are most popular. Rotary kilns typically have higher output [up to 600 tons (545 metric tons)/day] and lower labor cost. Vertical kilns can be designed for higher fuel efficiency and lower capital investment. They handle down to about $\frac{3}{4}$-in. (19-cm) stone, but

Fig. 1.   Large rotary lime kiln designed to operate 24 hours per day year around. Rugged construction is required for handling abrasive limestone rock at very high temperature

normally the rock feed is at least 3×6 in. (7.5×15 cm). Rotary kilns can handle down to $\frac{1}{4}$-in. (0.6 cm) stone.

The long-established use of lime is as a structural material in masonry mortars, wall plasters, sand-lime brick, and for soil stabilization. Double-hydrated dolomitic lime or specially processed high-calcium lime mixed with gypsum plaster is troweled on interior walls or ceilings to provide a hard, white, finished surface. It is mixed with cement and sand to make exterior plaster or stucco. Mason's mortar used to lay up bricks or blocks usually contains lime. Lime provides plasticity, water retention, and easy troweling. Sand-lime bricks are more popular in Europe than in the United States. About 10% hydrated lime is mixed with graded sand and water, pressed into shape, and put into autoclaves for $4-8$ hours at $150-205°C$ ($300-400°F$). The reaction product, calcium silicate, results in a strong, white brick. Dead-burned dolomite, formed by calcining dolomite at about $1650°C$ ($300°F$) to convert MgO to periclase, is used as a *refractory*. Lime is used in the sulfate process for making paper. In the seawater process for producing magnesium metal, lime reacts with $MgCl_2$ to precipitate $Mg(OH)_2$. Calcium metal is made from lime by reducing CaO with coke. In water treatment, hydrated lime can be added to remove temporary hardness, or in the lime-soda process to remove permanent hardness. Dolomitic lime removes silica from boiler feedwater due to silica absorption by $Mg(OH)_2$. For acid neutralization of industrial wastes, lime is widely used.

Additional Reading

Boynton, R.: *Chemistry and Technology of Lime and Limestone,* 2nd Edition, John Wiley & Sons, Inc., New York, NY, 1980.

Gerhartz, W.: *Ullmann's Encyclopedia of Industrial Chemistry,* John Wiley & Sons, Inc., New York, NY, 1987.

Oates, J.: *Lime and Limestone: Chemistry and Technology, Production and Uses,* John Wiley & Sons, Inc., New York, NY, 1999.

Web Reference

National Lime Association: http://www.lime.org/

LIMONITE. The mineral limonite, hydrated oxide of iron, corresponds to the formula $Fe_2O_3 \cdot nH_2O$, but is often very impure due to the admixture of sand and clay. It is not found crystallized but grades from loose porous material to compact masses. Its hardness is variable but pure material is $5-5.5$; specific gravity $3.6-4$; usual luster, dull to earthy but may be silky to submetallic; color, various shades of yellowish-brown, sometimes nearly black; streak, yellowish-brown; opaque. Limonite is a secondary mineral from the alteration of various other iron-bearing ores or minerals; it is of widespread occurrence and used both as an ore of iron and as a pigment. Limonite has been formed in marshy and boggy areas and is frequently called bog iron ore. Limonite is an important ore of iron in Lorraine, Luxemburg, Bavaria and Sweden. It is found in Saxony, Austria and England. In the United States, limonite is found particularly in Connecticut, Massachusetts, Pennsylvania, New York, Virginia, Tennessee, Georgia and Alabama, but these deposits are of little economic importance at the present time.

LINOLEIC ACID. [CAS: 60-33-3]. Also called linoleic acid, formula $CH_3(CH_2)_4$ HC:CHCH$_2$CH:CH(CH$_2$)$_7$COOH. This is a polyunsaturated fatty acid (two double bonds) existing in both conjugated and unconjugated forms. It is a plant glyceride essential to the human diet. It is found in linseed oil, safflower oil, and tall oil. See also **Vegetable Oils (Edible)**. At room temperature linoleic acid is a colorless to straw-colored liquid. Specific gravity 0.905 (15/4°C); mp $-5°C$; bp $228°C$ (at 14 millimeters pressure). Insoluble in water; soluble in most organic solvents. Combustible. Sources are the oils previously mentioned. Linoleic acid is used in soaps; special driers for protective coatings; emulsifying agents; pharmaceuticals; livestock feeds; and margarine.

LINOLENIC ACID. [CAS: 463-40-1]. Also called 9,12,15-octadecatri-enoic acid, formula CH_3CH_2CH:CHCH$_2$CH:CHCH$_2$CH:CH(CH$_2$)$_7$COOH. This is a polyunsaturated fatty acid (three double bonds). It occurs as the glyceride in many seed fats. It is an essential fatty acid in the diet. See also **Vegetable Oils (Edible)**. At room temperature, linolenic acid is a colorless liquid; soluble in most organic solvents; insoluble in water. Specific gravity 0.916 (20/4°C); mp $-11°C$; bp $230°C$. Combustible. Linolenic acid finds use in various pharmaceuticals and drying oils.

LIPIDS. A heterogenous group of substances which occur ubiquitously in biological materials. They may be categorized as a group by their extractability in nonpolar organic solvents, such as chloroform, carbon tetrachloride, benzene, ether, carbon disulfide, and petroleum ether. Structural types within the group range from simple, straight-chain hydrocarbon molecules to complex ring structures with varying side chains. A useful classification of the lipids is: (1) fatty acids; (2) neutral fats; (3) phosphatides; (4) glycolipids; (5) aliphatic alcohols and waxes; (6) terpenes; and (7) steroids. See also **Steroids**.

Many lipids, especially the phospholipids, have a strong tendency to form complexes with each other and with various substances. Complex formation is due to the electrostatic attraction of polar groups and to the mutual solubility of the long hydrocarbon chains. Thus, the lipoproteins and proteolipids are complexes of proteins and a variety of lipids, such as cholesterol, phospholipids, glycerides, and glycolipids. The lipids are linked to the proteins by several types of forces. Electrostatic forces, Van der Waals forces, hydrogen bonding, and hydrophobic bonding hold these complexes together. Because of their attraction for water, the polar groups of the protein and phospholipid arrange themselves on the outside of the complex, while the hydrocarbon groups of the lipids are folded into the center. Thus, there is presented to the aqueous phase those groups which have an affinity for water. This arrangement accounts for the solubility of the complexes in water. The phospholipids, owing to their polar groups, act as water solubilizers for the nonpolar lipids. The arrangement may be different in the proteolipids of the brain and nerves, since they are not soluble in water. In these complexes, the lipids may completely envelop the protein.

Knowledge of lipid metabolism has increased at an accelerated rate during the past few decades. The detailed biochemical reactions whereby the fatty acids are synthesized and oxidized; how phospholipids, glycolipids, and cholesterol are synthesized; and how lipids are absorbed and transported have been elucidated. Fatty acids are synthesized from acetyl coenzyme A and malonyl coenzyme A thiol esters. The vitamin biotin plays a vital part in the fixing of carbon dioxide to form malonyl coenzyme A, an important intermediate in fatty acid synthesis. The hormone insulin also favors fatty acid synthesis. The oxidation of fatty acids occurs as their coenzyme A esters in the Krebs cycle of the mitochondria. Cholesterol is biosynthesized from acetyl coenzyme A. Cholesterol in humans is converted to bile acids, fecal sterols, and to steroid hormones. The synthesis of lecithin is mediated via phosphatidic acid and diglyceride precursors. Cytidine nucleotides play a role in the transfer of choline (as phosphorylcholine) to a diglyceride to form lecithin. Uridine nucleotides act to transfer sugar residues in the synthesis of glycolipids.

The transport of lipids in the blood plasma is effected by complex formation with proteins to yield lipoproteins. The liver is the major organ for the synthesis of the lipoproteins. Analysis of serum lipoprotein patterns is important in the understanding of vascular disease (atherosclerosis). The clearing of lipemic blood, such as may occur after a heavy fat meal, is brought about by enzyme known as lipoprotein lipase. This enzyme yields free fatty acids which combine immediately with the plasma albumin to form complexes known as NEFA (nonesterified fatty acids). NEFA act as important transport vehicles for transport of triglycerides and the levels of blood NEFA are very sensitive to hormonal control and neural control. Certain hormones, such as epinephrine, stimulate the membrane-bound adenyl cyclase which converts ATP (adenosine triphosphate) to cyclic AMP (adenosine monophosphate). The latter stimulates adipose tissue lipase and mobilizes depot fat.

The excess utilization of lipids and excess oxidation of fatty acids causes an increase in acetoacetic acid in the body. This condition is known as *ketosis* and can lead to *acidosis*. This situation is common in severe diabetes and can occur whenever carbohydrate utilization is severely decreased.

Research continues in an effort to gain a more thorough understanding of how lipoproteins are synthesized; how lipids are arranged and combined with proteins to form cell membranes; what specific role lipids play in transport across cell membranes; how hormones act to regulate lipid metabolism; the biochemical basis of such abnormal lipid metabolic states as Gaucher's disease, Niemann-Pick's disease, etc.; how lipids per se permeate cell membranes; and how many phenotypic lipoproteins occur in serum.

See also **Cholesterol**.

Additional Reading

Akoh, C.C.: *Food Lipids: Chemistry, Nutrition, and Biotechnology,* Marcel Dekker, Inc., New York, NY, 1998.

Bornscheuer, U.T.: *Enzymes in Lipid Modification,* John Wiley & Sons, Inc., New York, NY, 2000.

Chang, T.Y. and D.A. Freeman: *Intracellular Cholesterol Trafficking,* Kluwer Academic Publishers, Norwell, MA, 1999.

Feher, M.D. and W. Richmond: *Lipids and Lipid Disorders,* Mosby-Year Book, Inc., St. Louis, MO, 1997.

Fox, P.: *Advanced Dairy Chemistry: Lipids,* Vol. 2, Chapman and Hall, New York, NY, 1999.

Gatt, S., L. Douste-Blazy, and R. Salvayre: *Lipid Storage Disorders: Biological and Medical Aspects,* Perseus Books, Boulder, CO, 1988.

Gotto, A.M. and H.J. Pownall: *Manual of Lipid Disorders: Reducing the Risk for Coronary Heart Disease,* 2nd Edition, Lippincott Williams & Wilkins, Philadelphia, PA, 1999.

Gurr, J., J. Harwood, and K.N. Frayn: *Lipid Chemistry,* 5th Edition, Blackwell Science, Inc., Malden, MA, 2001.

Hainik, T. and V. Passechnik: *Bilayer Lipid Membranes: Structures and Mechanical Properties,* Kluwer Academic Publishers, Norwell, MA, 1995.

Katsaros, J. and T. Gutberlet: *Lipid Bilayers: Structure and Interactions,* Springer-Verlag, Inc., New York, NY, 2000.

Keane, W.F. and B.L. Kasiske: *Lipids and the Kidney,* S. Karger Publishers, Inc., Farmington, CT, 1997.

Vance, D. and J. Vance: *Biochemistry of Lipids Lipoproteins and Membranes,* 3rd Edition, Elsevier Science, New York, NY, 1997.

LIPMANN, FRITZ (1899–1986). A German-born biochemist who won the Nobel Prize in 1953 for the discovery of coenzyme A (CoA). He earned doctorates at the University of Berlin in both chemistry and medicine. He worked at Cornell and Harvard Universities. He founded the biochemistry department of Brandeis University and later joined the faculty of Rockfeller University.

LIPSCOMB, WILLIAM N. (1919–). An American chemist who won the Nobel prize for chemistry in 1976 for his studies on the structure and bonding mechanisms of boranes. Much of the research concerned structure and function of enzymes and natural products in organic and theoretical chemistry. He studied at the Universities of Kentucky, California, and Minnesota.

LIQUATION. 1. The separation of two or more compounds of a mixture by heating to a temperature at which one component melts, leaving the others as solids.

2. The separation of a more readily fusible substance from a less fusible one by controlled heating.

LIQUID. Matter in a fluid state but relatively incompressible. An ideal liquid offers no permanent resistance to a shear stress but is incompressible. It has then a constant volume and incompletely fills any container of less than this volume. A real liquid is appreciably compressible, and the liquid state of a substance might be defined as the denser, and less compressible, phase of the two-phase fluid system which can exist in equilibrium at temperatures below the critical temperature. X-ray diffraction experiments show that, near the melting-point, the molecules of a liquid show a considerable degree of short-range order and that, in small volumes, they are arranged much as in a solid crystal. This crystalline structure persists over volumes comparable with the intermolecular distances, but cannot be traced beyond. This local or short-range order means that the average molecule is at any moment surrounded by a number of molecules occupying nearly the same relative positions as they would in the solid state. The degree of short-range order is described by the radial distribution function.

This concept of a liquid as an imperfect crystal requires that the molecules in a liquid are packed sufficiently loosely for comparatively free movement, i.e., the energy required to move a molecule from a lattice site to a vacant space is not large compared with thermal energies. Under these conditions, shear flow of the liquid resembles closely the high temperature creep of crystalline solids. A number of theories of the liquid state have this concept as their starting point.

With a few exceptions, including helium, the accompanying universal phase diagram applies for all pure compounds. The triple point is the single point at which all three phases (crystal, liquid, and gas) are in equilibrium. The triple point pressure is normally below atmospheric. Those substances, such as carbon dioxide, where $P_t = 3,885$ millimeters, $T_t = -56.6°C$, sublime without melting at atmospheric pressure. From the triple point, the melting curve defines the equilibrium between crystal and liquid, usually rising with small but positive dT/dP, and presumably always with positive dT/dP at sufficiently high P values. The line is

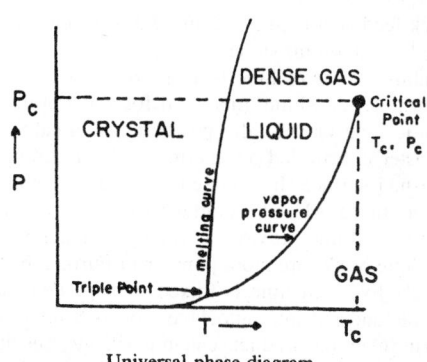

Universal phase diagram.

believed to extend infinitely without a critical point (it has been followed to $T \cong 16T_c$ for helium, and calculations indicate that hard spheres would show a gas-crystal phase change). The gas-liquid equilibrium line, the vapor pressure curve, has dT/dP always positive and greater than the melting curve. The vapor pressure curve always ends at a critical point, $P = P_c, T = T_c$, above which the liquid and gas phase are no longer distinguishable. Since the liquid can be continuously converted into the gas phase without discontinuous change of properties by any path in the $P - T$ diagram passing above the critical point, there is no definite boundary between liquid and gas. Two liquids of similar molecules are usually soluble in all proportions, but very low solubility is sufficiently common to permit the demonstration of as many as seven separate liquid phases in equilibrium at one temperature and pressure (mercury, gallium, phosphorus, perfluoro-kerosene, water, aniline, and heptane at 50°C, 1 atmosphere).

See also **Liquid State**.

Additional Reading

Fourkas, J.T.: *Liquid Dynamics: Experiment, Simulation, and Theory,* American Chemical Society, Washington, DC, 2002.

Hansen, J.P., and I.R. McDonald: *Theory of Simple Liquids,* 2nd Edition, Elsevier Science & Technology Books, New York, NY, 1990.

March, N.H.: *Chemical Physics of Liquids,* Taylor & Francis, Inc., Philadelphia, PA, 1990.

March, N.H., and M.P. Tosi: *Atomic Dynamics in Liquids,* Dover Publications, Inc., Mineola, NY, 1991.

March, N.H., and M.P. Tosi: *Introduction to Liquid State Physics,* World Scientific Publishing Company, Inc., Riveredge, NJ, 2002.

Stephan, K., and K. Lucas: *Viscosity of Dense Fluids,* Plenum Publishing Corp., New York, NY, 1979.

LIQUID CHROMATOGRAPHY. An analytical method based on separation of the components of a mixture in solution by selective adsorption. All systems include a moving solvent, a means of producing solvent motion (such as gravity or a pump), a means of sample introduction, a fractionating column, and a detector. Innovations in functional systems provide the analytical capability for operating in three separation modes: (1) liquid–liquid: partition in which separations depend on relative solubilities of sample components in two immiscible solvents (one of which is usually water); (2) liquid–solid: adsorption where the differences in polarities of sample components and their relative adsorption on an active surface determine the degree of separation: (3) molecular size separations which depend on the effective molecular size of sample components in solution.

Solvents, often referred to as carriers, include isooctane, methyl ethyl ketone, acetone/chloroform, tetrahydrofuran, hexane, and toluene.

Packing materials in columns of various lengths include silica gel, alumina, glass beads, polystyrene gel, and ion exchange resins.

High-performance liquid chromatography (HPLC) is the term applied to new and more effective instrumental techniques developed in recent years that have greatly increased the scope of this analytical method. It can now be applied to biological as well as chemical research. Among the separations possible are peptides (by reverse phase chromatography), proteins and enzymes (hydrophobic and size exclusion modes of chromatography), amino acids, and inorganic and organometallic compounds (neutral species, including, clusters, by adsorption and size exclusion, and ionic species including coordination compounds). A comparatively recent development is the use of supercritical fluids as solvents, e.g., carbon dioxide and sulfur hexafluoride.

See also **Gas Chromatography**; and **Thin-Layer Chromatography (TLC)**.

Additional Reading

Kromidas, S.: *Practical Problem Solving in HPLC,* John Wiley & Sons, Inc., New York, NY, 2000.

Patel, D.: *Liquid Chromatography: Essential Data,* John Wiley & Sons, Inc., New York, NY, 1999.

Sadek, P.C.: *HPLC Solvent Guide,* 2nd Edition, John Wiley & Sons, Inc., New York, NY, 2002.

Swadesh, J.K.: *HPLC: Practical and Industrial Applications,* 2nd Edition, CRC Press LLC., Boca Raton, FL, 2000.

LIQUID CRYSTALLINE MATERIALS. Liquid crystals represent a state of matter with physical properties normally associated with both solids and liquids. Liquid crystals are fluid in that the molecules are free to diffuse about, endowing the substance with the flow properties of a fluid. As the molecules diffuse, however, a small degree of long-range orientational and sometimes positional order is maintained, causing the substance to be anisotropic as is typical of solids. Therefore, liquid crystals are anisotropic fluids and thus a fourth phase of matter. There are many liquid crystal phases, each exhibiting different forms of orientational and positional order, but in most cases these phases are thermodynamically stable for temperature ranges between the solid and isotropic liquid phases. Liquid crystallinity is also referred to as mesomorphism.

Many thousands of organic substances, some rigid-rod polymers, and other macromolecules exhibit liquid crystallinity. The general common molecular feature is either an elongated or flattened, somewhat inflexible molecular framework, which is usually depicted as either a cigar- or disk-shaped entity. The orientational and positional order in a liquid-crystal phase is only partial, with the intermolecular forces striking a very delicate balance involving both attractive and repulsive interactions. As a result, liquid crystals are extraordinarily sensitive to external perturbations, e.g., temperature, pressure, electric and magnetic fields, shearing stress, or foreign vapors. For this reason, liquid crystals are used to design practical devices to either monitor ambient changes of various kinds or to transduce an environmental fluctuation into a useful electrical or optical output.

Besides being used in the scientific study of cooperative phenomena and complex fluid phases, liquid crystalline phenomena have received a good deal of attention due to the possibility of practical applications. Liquid crystals are widely used in electrooptic displays. Other applications include radiation and pressure sensors, optical switches and shutters, and thermography. The liquid crystalline structures formed by amphiphilic molecules form the basis for emulsions and are studied thoroughly by researchers in the food, drug, and oil industries. Polymers that form an anisotropic fluid phase are important in the fabrication of lightweight, ultrahigh strength, and temperature-resistant fibers, and are beginning to be used in electrooptic displays. Liquid crystals also appear to play an important role in the structure and biochemical function of living tissue, where the characteristic combination of order and flow mobility is particularly suited to life processes. Certain disease states, e.g., atherosclerosis, sickle cell anemia, or cancer, may be associated with physical changes in the liquid-crystalline order within biological structures.

Orientational and Positional Order in Fluids

Solids of mesogenic (liquid-crystal forming) molecules melt to form fluids in which some of the long-range molecular order is retained. At the simplest level, the elongation or flattening of the mesogenic molecules prevents the immediate dissolution of the parent, solid-state order. The loss of positional order of the centers of mass of the molecules in the parent solid may be either partial or complete upon melting, but some degree of orientational order is always retained. The fluid retains many solid-like properties, which are finally eliminated when the substance passes into the normal, isotropic liquid phase at a higher temperature (a second melting point). Solid-like features return if the substance is cooled from the isotropic state; this intermediate state is usually thermodynamically reversible, but in some cases it only forms upon cooling. Partial dissolution of solid-state order also may occur in certain substances by the use of solvents. In this case the molecules are either orientationally ordered in the solvent (some macromolecules), or form aggregates, in which the molecules exhibit long-range positional and/or orientational order. Liquid crystals that are established solely by the adjustment of temperature are referred to as thermotropics, whereas those that form through the addition of a solvent are called lyotropics.

Orientational Distribution Function and Order Parameter. In a liquid crystal a snapshot of the molecules at any one time reveals that they are not randomly oriented. There is a preferred direction for alignment of the long molecular axes. This preferred direction is called the director, and it can be used to define an orientational distribution function, $f(\theta)$, where $f(\theta)\sin\theta d\theta$ is proportional to the fraction of molecules with their long axes within the solid angle $\sin\theta d\theta$.

It is useful to describe the amount of orientational order with a single quantity. X-ray, uv, optical, ir, and magnetic resonance techniques are used to measure the order parameter in liquid crystals.

Positional Distribution Function and Order Parameter. In addition to orientational order, some liquid crystals possess positional order in that a snapshot at any time reveals that there are parallel planes which possess a higher density of molecular centers than the spaces between these planes. If the normal to these planes is defined as the z-axis, then a positional distribution function, $g(z)$, can be defined, where $g(z)dz$ is proportional to the fraction of molecular centers between z and $z + dz$. Since $g(z)$ is periodic, it can be represented as a Fourier series (a sum of a sinusoidal function with a periodicity equal to the distance between the planes and its harmonics). To represent the amount of positional order, the coefficient in front of the fundamental term is used as the order parameter. The more the molecules tend to form layers, the greater the coefficient in front of the fundamental sinusoidal term and the greater the order parameter for positional order.

Bond Orientational Order. In some cases, although the lattice of points of high density of molecular centers parallel to the planes are not correlated from layer to layer, the two principal directions of the lattice are the same for all layers. In these materials, the interactions between the planes do not prevent the planes from translating relative to each other, but do prevent them from rotating relative to each other.

Thermotropic Liquid Crystals

Thermotropic liquid crystals result from the melting of mesogenic solids due to an increase in temperature. Both pure substances and mixtures form thermotropic liquid crystals. In order for a mixture to be a thermotropic liquid crystal, the different components must be completely miscible. Examples include nematic liquid crystals (p-methoxybenzylidene-p′-n-butylaniline (MBBA); p-azoxyanisole (PAA); p-n-hexyl-p′-cyanobiphenyl; di-4-methoxyphenyl-*trans*-1,4-cyclohexane-dicarboxylate; and p-quinquephenyl), cholesteric liquid crystals ((−)-2-methylbutyl 4-(4′-methoxybenzylideneamino)cinnamate), and smectic liquid crystals (ethyl 4-(4′-phenylbenzylideneamino)benzoate; ethyl 4-(4′-ethoxybenzylideneamino)-cinnamate; p-n-octyloxybenzoic acid; 4-(4′-n-octadecyloxy-3′-nitrophenyl) benzoic acid; diethyl p-terphenyl-p, p″-carboxylate; 2-(p-pentylphenyl-5-(p-pentyloxyphenyl)-pyrimidine; and 4-ethyl-4′-butyloxybenzylodeneaniline). Much more is known about calamitic (rod-like) liquid crystals then discotic (disk-like) liquid crystals, since the latter were discovered only recently.

Nematic. In a nematic liquid crystal, the long axes of the molecules remain substantially parallel, but the positions of the centers of mass are randomly distributed. Therefore, there is orientational order and a nonzero orientational order parameter, but there is no positional order.

If the molecules of a liquid crystal are optically active (chiral), then the nematic phase is not formed. Instead of the director being locally constant as is the case for nematics, the director rotates in helical fashion throughout the sample. Within any plane perpendicular to the helical axis the order is nematic-like. In other words, as in a nematic there is only orientational order in chiral nematic liquid crystals, and no positional order.

Smectic. Smectic liquid crystals are distinguished from nematics by the presence of some positional order (a tendency to form layers) in addition to orientational order. The direction of preferred orientational order is perpendicular to the layers in a smectic A liquid crystal and at an angle with the layer normal in a smectic C liquid crystal.

In much the same way as a chiral compound forms the chiral nematic phase instead of the nematic phase, a compound with a chiral center forms a chiral smectic C phase rather than a smectic C phase. In a chiral smectic C liquid crystal, the angle the director is tilted away from the normal to the layers is constant, but the direction of the tilt rotates around the layer normal in going from one layer to the next.

Frustrated Phases. Chiral molecules normally form chiral phases, but in some cases this is done in an interesting way. For example, it is not unusual for a chiral molecule to form a smectic *A* phase, which is not chiral. If the molecule is highly chiral, however, twist is sometimes introduced into the smectic *A* phase by an array of grain boundaries which are perpendicular to the smectic *A* layers and parallel to the director. In one compound the normal to the layers is rotated by roughly 17° on either side of a grain boundary and the grain boundaries are separated by about 24 nm, giving this twist grain boundary (TGB) phase a pitch of a little more than 500 nm. In a sense the frustration of an achiral phase of chiral molecules has been relieved by the introduction of these twist grain boundaries.

Discotic Phases. Molecules which are disk-shaped rather than elongated also form thermotropic liquid-crystal phases. Usually these molecules have aromatic cores and six lateral substituents, although the predominance of six lateral substituents is solely historical; molecules with four lateral substituents also can form liquid-crystal phases. Although the flatness of these molecules creates a steric effect promoting alignment of the normal to the disks, the fact that disordered side chains are also necessary for the formation of these phases (as is often the case for liquid crystallinity in elongated molecules) should not be ignored. The most simple discotic phase is the nematic phase, in which the normal to the disks are preferentially aligned along a single direction (director). If the molecules are chiral or if a chiral dopant is added to a discotic liquid crystal, a chiral nematic discotic phase can form.

Metallomesogens. It is also possible to synthesize compounds based on metal atoms which possess liquid-crystal phases. The series based on dithiolene complexes (**1**), where M = Ni, Pd, or Pt, contains a number of compounds which show the liquid-crystal phases typical of rod-like molecules.

(**1**)

Disk-shaped molecules based on a metal atom possess discotic liquid-crystal phases.

Lyotropic Liquid Crystals. Some molecules in a solvent form phases with orientational and/or positional order. In these systems, the transition from one phase to another can occur due to a change of concentration, so they are given the name lyotropic liquid crystals. Of course temperature can also cause phase transitions in these systems, so this aspect of thermotropic liquid crystals is shared by lyotropics. The real distinctiveness of lyotropic liquid crystals is the fact that at least two very different species of molecules must be present for these structures to form.

Amphiphilic Molecules. In just about all cases of lyotropic liquid crystals, the important component of the system is a molecule with two very different parts, one that is hydrophobic and one that is hydrophilic. These molecules are called amphiphilic because when possible they migrate to the interface between a polar and nonpolar liquid.

Even more interesting phenomena occur when amphiphilic compounds are put into water–oil mixtures. If the oil concentration is low, the amphiphilic molecules form micelles and the oil collects inside the micelles. As the oil concentration is increased, the micelles continue to swell with oil until it is safe to say that the system is really composed of volumes of water and volumes of oil separated by a single amphiphilic layer. This type of system is called an emulsion, and thus amphiphiles can serve as emulsifiers.

When a highly polar liquid, a slightly polar liquid, and an amphiphile are mixed together at the right temperature and in the right concentrations, the micelles which form are not spherical. Within this vary narrow concentration range, the micelles are rod-shaped for one part of this range and disk-shaped for another part. In either case the micelles themselves orient their symmetry axes (the long axis for the rod-shaped micelles and the short axis for the disk-shaped micelles) just like a thermotropic liquid crystal.

Polymorphism

A liquid crystal compound in more cases than not takes on more than one type of mesomorphic structure as the conditions of temperature or solvent are changed. In thermotropic liquid crystals, transitions between various phases occur at definite temperatures and are usually accompanied by a latent heat.

An exception to the rule that lowering the temperature causes transitions to phases with increased order sometimes occurs for polar compounds which form the smectic A_d phase (a layered structure formed by molecular dimers). Decreasing the temperature causes a transition from nematic to smectic A_d, but a further lowering of the temperature produces a transition back to the nematic phase (called the reentrant nematic phase). Electric or magnetic fields also may induce mesomorphic phase transitions.

Synthesis

Just because a molecule is long, narrow, and meets the requirement of geometric anisotropy does not ensure that it will have a liquid crystal phase. The particular phase structure that occurs in a compound, i.e., smectic, nematic, or chiral nematic, not only depends on the molecular shape but is intimately connected with the strength and position of the polar or polarizable groups within the molecule, the overall polarizability of the molecule, and the presence of chiral centers.

Molecular interactions that lead to attraction include dipole–dipole interactions, dipole-induced dipole interactions, dispersion forces, and hydrogen bonding.

In order for dipole–dipole and dipole-induced dipole interactions to be effective, the molecule must contain polar groups and/or be highly polarizable. Ease of electronic distortion is favored by the presence of aromatic groups and double or triple bonds. These groups frequently are found in the molecular structure of liquid crystal compounds. The most common nematogenic and smectogenic molecules are of the type shown in Table 1. In general, if the X link is rigid, a liquid crystal phase is favored.

The importance of unsaturation is illustrated by the fact that 2,4-nonadienoic acid forms a liquid-crystal phase, whereas the *n*-aliphatic carboxylic acids do not. The two double bonds enhance the polarizability of the molecule and bring intermolecular attractions to a level that is suitable for mesophase formation. The overall linearity of the molecule must not be sacrificed in potential liquid-crystal candidates. Bulky, even if highly polarizable, functional groups or atoms that are attached anywhere but on the end of a rod-shaped molecule are usually less favorable for liquid-crystal formation.

In the case of carboxylic acids, hydrogen bonding can induce liquid-crystal phases by lengthening the molecular unit through dimerization:

On the other hand, hydrogen bonding may lead to nonlinear molecular associations that disrupt the parallelism. Hydrogen bonding associations may also be so strong that by the time the solid reaches its melting point the thermal energy is too intense to permit substantial order to remain within the fluid.

Although it is difficult to predict exactly which type of liquid-crystal phase will be formed by a molecule meeting the general requirements, rough trends can be recognized. The presence of functional groups that lead to strong lateral interactions, e.g., dipoles operating across the long

TABLE 1. SOME CENTRAL LINKAGES FOUND IN LIQUID CRYSTALLINE COMPOUNDS

$$R_1 - \bigcirc - X - \bigcirc - R_2$$

X	Series name
—CH=N—	Schiff bases
—N=N—	diazo compounds
—N=N— ↓ O	azoxy compounds
—CH=N— ↓ O	nitrones
—CH=CH—	stilbenes
—C≡C—	tolans
—OC— ‖ O	esters
—(nothing)	biphenyls

molecular axis, favor the layered smectic structure. When these structural elements are not present but the molecule is otherwise suitable for mesomorphism, i.e., is long and narrow, the nematic phase is likely. Longer terminal groups favor the smectic phase over the nematic phase. An asymmetric center on the molecule causes the chiral nematic and chiral smectic C phases in place of the nematic and smectic C phases.

Goals in liquid crystal synthesis include the design of room temperature thermotropics which are stable, colorless liquid crystalline over a wide range of temperature, and operate at low voltage and power levels.

A good deal of synthesis effort has been devoted to chiral liquid crystals, especially those with chiral smectic C phases. The chiral smectic C phase is ferroelectric, which gives it properties quite useful for applications. Perhaps the most important property of these phases is that a lateral dipole can produce a spontaneous polarization.

Polymer Liquid Crystals

Both polymer melts and polymer solutions sometimes form phases with orientational and positional order. Thermotropic polymer liquid crystals possess at least one liquid crystal phase between the glass-transition temperature and the transition temperature to the isotropic liquid. Lyotropic polymer liquid crystals possess at least one liquid crystal phase for certain ranges of concentration and temperature.

Polymer Melts. When a rigid, polarizable monomer forms wither a mainchain polymer with flexible segments in between or a side-chain polymer with flexible segments between the rigid segments and the flexible main chain, liquid-crystal phases are usually stable.

Examples of polymers which form anisotropic polymer melts include petroleum pitches, polyesters, polyethers, polyphosphazines, α-poly-p-xylylene, and polysiloxanes. Synthesis goals include the incorporation of a liquid crystal-like entity into the main chain of the polymer to increase the strength and thermal stability of the materials that are formed from the liquid-crystal precursor, the locking in of liquid crystalline properties of the fluid into the solid phase, and the production of extended chain polymers that are soluble in organic solvents rather than sulfuric acid.

Polymer Solutions. Perhaps the most extensively studied macromolecular liquid crystals are the synthetic polypeptides, such as poly(γ-benzyl L-glutamate) (PBLG). PBLG is a homopolymer of the L-enantiomorph of a single amino acid with the following repeat unit.

PBLG adopts the α-helical conformation in a number of solvents as a result of intramolecular hydrogen bonding and favorable stacking of the pendent side chains. Thus the polymer assumes an extended, relatively rigid geometry and may become ordered spontaneously at sufficiently high concentrations. The formation of this lyotropic liquid crystal phase occurs at a critical volume fraction of polymer ϕ^* which is inversely proportional to the length-to-diameter ratio of the macromolecule.

A variety of aromatic and extended-chain polyamides that spontaneously form a mesophase in concentrated solutions also have been synthesized.

The polyamides are soluble in high strength sulfuric acid or in mixtures of hexamethylphosphoramide, N,N-dimethylacetamide, and LiCl. The liquid-crystal phase is optically anisotropic and the texture is nematic. The nematic texture can be transformed to a chiral nematic texture by adding chiral species as a dopant or incorporating a chiral unit in the main chain as a copolymer.

Applications. The polyamides have important applications. The very high degree of polymer orientation that is achieved when liquid crystalline solutions are extruded imparts exceptionally high strengths and moduli to polyamide fibers and films. DuPont markets such polymers, e.g., Kevlar, and Monsanto has a similar product, e.g., X-500, which consists of polyamide and hydrazide-type polymers. Liquid-crystal polymers are also used in electrooptic displays.

Liquid Crystals in Biological Systems

Many biological systems exhibit the properties of liquid crystals. Considerable concentrations of liquid crystalline compounds have been found in many parts of the body, often as sterol or lipid derivatives. A liquid crystal phase has been implicated in at least two degenerative diseases, atherosclerosis and sickle cell anemia. Living tissue, such as muscle, tendon, ovary, adrenal cortex, and nerve, show the optical birefringence properties that are characteristic of liquid crystals. The liquid crystal state has been identified in many pathological tissues, particularly in areas of large lipid deposits. Massive deposits of liquid crystalline cholesterol derivatives have been found in the kidneys, liver, brain, spleen, marrow, and aorta walls. Certain living sperms possess a liquid crystalline state.

Cell Membrane. The fluid mosaic model of the cell membrane is one in which the phospholipids provide the basic order and integrity of the cell through amphiphilic interaction with the aqueous environment.

Microfilaments and Microtubules. There are two important classes of fibers found in the cytoplasm of many plant and animal cells that are characterized by nematic-like organization. These are the micro-filaments and microtubules which play a central role in the determination of cell shape, either as the dynamic element in the contractile mechanism or as the basic cytoskeleton.

Liquid Crystalline Structures. In certain cellular organelles, deoxyribonucleic acid (DNA) occurs in a concentrated form. Striking similarities between the optical properties derived from the underlying supramolecular organization of the concentrated DNA phases and those observed in chiral nematic textures have been described. Concentrated aqueous solutions of nucleic acids exhibit a chiral nematic texture *in vitro*.

PETER J. COLLINGS
Swarthmore College

Additional Reading

Bahadur, B.: *Liquid Crystals: Applications and Uses,* Vols. 1–3, World Scientific, Singapore, 1990–1992.

Collings, P.J.: *Liquid Crystals: Nature's Delicate Phase of Matter,* Princeton University Press, Princeton, NJ, 1990.

Collings, P.J., and M. Hird: *Introduction to Liquid Crystals: Chemistry and Physics,* Taylor & Francis, Inc., Philadelphia, PA, 1997.

de Gennes, P.G. and J. Prost: *The Physics of Liquid Crystals,* Clarendon Press, Oxford, UK, 1993.

Dierking, I.: *Textures of Liquid Crystals,* John Wiley & Sons, Inc., New York, NY, 2003.

Sonin, A.A.: *Freely Suspended Liquid Crystalline Films,* John Wiley & Sons, Inc., New York, NY, 1999.

White, A.M. and A.H. Windle: *Liquid Crystalline Polymers,* Cambridge University Press, Cambridge, UK, 1992.

LIQUID CRYSTAL POLYMERS. These materials (LCPs) exhibit a highly ordered structure in the melt, solution, and solid states. A tightly packed and highly ordered morphology particularly susceptible to orientation during processing is characteristic. Commercial applications for LCP resins include chemical pumps, tower packings, coil bobbins, connectors, sockets, etc. for electronic components, and various automotive parts. LCPs have an excellent combination of chemical and flame resistance, dimensional stability, and ease of processing. Their thermal stability makes them suitable for dual ovenable cookware and where thermal resistance for both conventional and microwave oven service is important.

Compared with other polymeric materials, LCPs have very high unidirectional properties. *Vectra*™ (Celanese Corp.) resins are primarily aromatic polyesters based on p-hydroxybenzoic acid and hydroxynaphthoic acid monomers. *Xydar*™ (Celanese Corp.) injection molding resins are polyesters based on terephthalic acid, p, p'-dihydroxybiphenyl and p-hydroxybenzoic acid. Differences in monomers are primarily responsible for the differences in specific properties and end uses. The fibrous nature of the polymers imparts good impact strengths.

LIQUID CRYSTALS. Liquids that have the structural character of cybotactic liquids (see also **Cybotaxis**), but which are considerably more viscous, with viscosities extending from that of a light glue to that of a glassy solid. They also exhibit much more definite evidences of structure than the cybotactic liquids.

Liquid crystals must be geometrically highly anisotropic—usually long and narrow—and revert to an isotropic liquid through thermal action (thermotropic mesomorphism) or by the influence of a solvent (lyotropic mesomorphism). Several thousand organic compounds are now known which meet these criteria, but significant molecular features found in thermotropic liquid crystals are among the following. The molecule will be elongated and rectilinear; if "flat segments," e.g. benzene rings, are present its liquid crystallinity will be enhanced. The molecule will be rigid along its long axis and double bonds will be common in this direction. The simultaneous existence is seen in the molecule of strong dipoles and easily polarizable groups. Of lesser importance are weak dipolar groups at the extremities of the molecule.

The present day classification of thermotropic liquid crystals is threefold. *Smectic liquid crystals*, such as *p*-ethyl azoxybenzoate, have their molecules arranged in definite strata, a variety of molecular arrangements being possible within each stratification. In *smectic Type A* crystals, the molecules may be considered to "stand on end" with their long axes perpendicular to the plane of the layer but with their centers irregularly spaced. When the molecular centers adopt hexagonal close packing, the crystals are considered *smectic Type B*, and when they adopt a titled form of Type A, they are classified as *smectic Type C*. See Fig. 1.

In *nematic liquid crystals*, the molecular structures possess a high degree of long-range orientation order, but no long-range translational order. The molecules are spontaneously oriented with their long axes approximately parallel, but without the stratification seen in smectic crystals. Nematic liquid crystals like *p*-azoxyanisole are generally optically uniaxial, positive, and strongly birefringent, and some are composed of hundreds of molecules (cytotactic groups), the molecular centers in each group arranged in layers. See Fig. 2.

Lyotropic liquid crystals possess at least two components. One of these is water and the other is amphible (a polar head group attached to one or more long hydrocarbon chains). In the lamellar form, water molecules are sandwiched between the polar heads of adjacent layers while the hydrocarbon tails lie in a nonpolar environment. Lyotropic liquid crystals have very complex structures, but occur abundantly in nature, particularly in living systems. See Fig. 3.

Polarized light is the most powerful tool for investigating liquid crystals, all of which exhibit characteristic optical properties. A smectic liquid crystal transmits light more slowly perpendicular to the layers than parallel to them. Such substances are said to be optically positive. Nematic liquid crystals are also optically positive, but their action is less definite than that of smectic liquid crystals. However, the application of a magnetic field to nematic liquid crystals lines up their molecules, changing their optical properties and even their viscosity.

Both smectic and nematic crystals split a beam of ordinary light into two polarized components whose transverse vibrations are at right angles

Fig. 3.   Cholesteric crystals

to each other. This is the well-known phenomenon of double refraction. Cholesteric liquid crystals exhibit the phenomenon of circular dichroism. That is, they break a beam of ordinary light into two components, one with the electric vector rotating clockwise and the other with it rotating counterclockwise. The first is usually transmitted, and the second is the one to be reflected. It is this property that gives cholesteric crystals their characteristic iridescent colors when illuminated by white light.

This ability to exhibit colors is one of the most useful attributes of liquid crystals. Many cholesteric substances behave as liquid crystals only in a certain temperature range. Above it they are colorless, but as they are cooled through it they assume a succession of colors, running down the spectrum from red to violet and finally becoming colorless. However, at this final stage they still retain their molecular orientation, but it is that of smectic liquid crystals rather than cholesteric. Some cholesteric liquid crystals do not exhibit all the colors mentioned and others, which are naturally colored, simply change to another color on heating or cooling. Since the exact temperatures at which these color changes occur are invariable, these substances can be used for measuring temperatures; in fact, combinations of them cover the range from −20 to +250°C.

Useful applications have been found for the varied effects of these crystal changes. One of the first came from the property of selectively reflecting visible light; because this is temperature-dependent, the property can be used as a temperature detector, and in gel form liquid crystals have been used for the early detection of those cancers which cause hot spots in the body. Applications of the smectic modifications arise from their ferroelectric properties; this phase can function as a fast-switching light-valve device with memory. This kind of application requires some

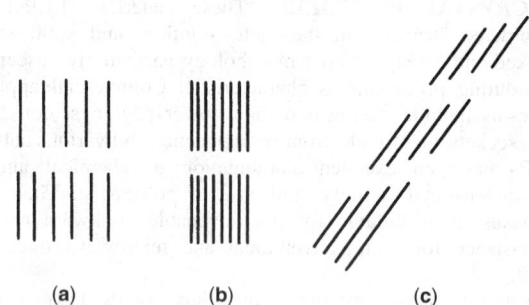

(a)   **(b)**   **(c)**

Fig. 1.   Smectic liquid crystals; types (a), (b), and (c)

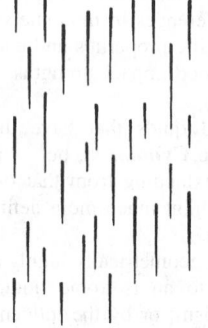

Fig. 2.   Nematic crystals

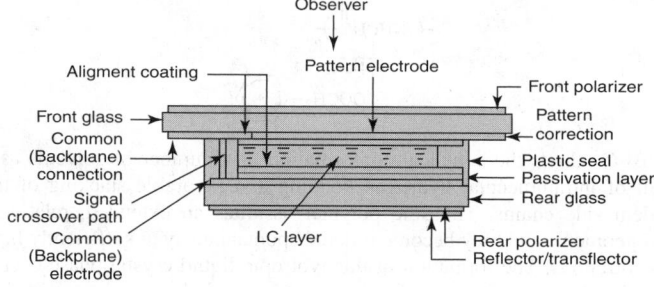

Fig. 4.   Liquid crystal display operation, unactivated

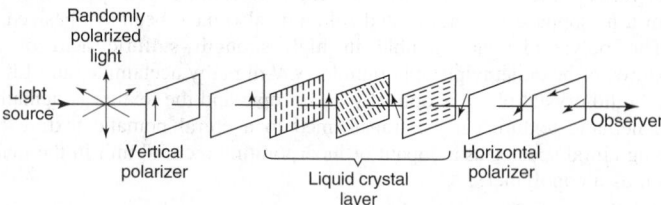

Fig. 5.   Liquid crystal display operation, activated

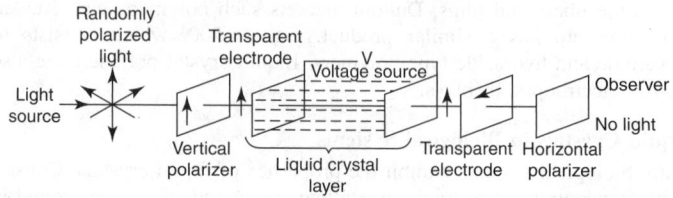

Fig. 6.   Basic liquid crystal display construction. (*Hamlin, Inc*)

control on the pitch of the polarized helix, which is obtained by blending together materials. Most twisted nematic field effect liquid crystal displays make use of a 90° twist between transparent conductive electrodes and crossed polarizers, as shown in Fig. 4. As randomly polarized light enters the device, only that portion which is vertically polarized may pass through the front polarizer. This, in turn, is rotated another 90° through the rear polarizer. If a reflective surface is placed behind the rear polarizer, the light will be passed back through the cell, its polarization again being rotated. By applying a voltage to the transparent electrodes, the molecules of the crystal will leave the nematic structure and align with the field, as shown in Fig. 5. Then the incoming light is no longer polarized—so extinction occurs and the area of extinction is defined very sharply by the shape of the electrode pattern, producing a dark area on a light reflective background. The reverse can also be achieved by using parallel rather than crossed polarizers. Fig. 6 shows a cross section of a typical display. A polymeric seal contains the liquid crystal material and holds the glass substrates together and the whole is laminated to the assembly.

See also **Liquid Crystalline Materials**.

Additional Reading

Anisimov, M.: *Critical Phenomena in Liquids and Liquid Crystals,* Gordon & Breach Publishing Group, Newark, NJ, 1991.

Cladis, P.E. and P. Palffy-Muhoray: *Dynamics and Defects in Liquid Crystals,* Gordon & Breach Publishing Group, Newark, NJ, 1998.

Collings, P.J.: *Liquid Crystals : Nature's Delicate Phase of Matter,* 2nd Edition, Princeton University Press, Princeton, NJ, 2001.

Collings, P.J., and M. Hird: *Introduction to Liquid Crystals: Chemistry and Physics,* Taylor & Francis, Inc., Philadelphia, PA, 1997.

Demus, D., G.W. Gray, J.W. Goodby et al.: *Handbook of Liquid Crystals: High Molecular Weight Liquid Crystals,* Vol. 3, John Wiley & Sons, Inc., New York, NY, 1998.

Demus, D., G.W. Gray, J.W. Goodby et al.: *Handbook of Liquid Crystals: Low Molecular Weight Liquid Crystals II,* Vol. 2, John Wiley & Sons, Inc., New York, NY, 1998.

Kumar, S.: *Liquid Crystals: Experimental Study of Physical Properties and Phase Transitions,* Cambridge University Press, New York, NY, 2000.

Langerwall, S.T.: *Ferroelectric and Antiferroelectric Liquid Crystals,* John Wiley & Sons, Inc., New York, NY, 1999.

Prigogine, I. et al.: *Advances in Chemical Physics Volume 113, Advances in Liquid Crystals,* John Wiley & Sons, Inc., New York, NY, 2000.

Singh, S., and D.A. Dunmur: *Liquid Crystals: Fundamentals,* World Scientific Publishing Company, Inc., Riveredge, NJ, 2003.

Sonin, A.A.: *Freely Suspended Liquid Crystalline Films,* John Wiley & Sons, Inc., New York, NY, 1999.

Yeh, P. and C. Gu: *Optics of Liquid Crystal Displays,* John Wiley & Sons, Inc., New York, NY, 1999.

LIQUID JUNCTION. To avoid the unknown liquid junction potential in measuring the potential of a half-cell against a reference electrode, the two half-cells are frequently connected via a salt bridge, usually a concentrated solution of potassium chloride. Since its anion and cation have almost the same velocity, a negligible diffusion potential is set up across the liquid junctions at the ends of the bridge.

LIQUID STATE. Because of the theoretical and practical importance to the era of electronics, which commenced nearly a half-century ago, the solid state of matter has become better known and understood than the physics of fluids (liquids and gases). Much practical engineering knowledge has been amassed pertaining to substances in the fluid state, but much research of a fundamental nature on fluids remains to be finished. Particularly, the transition of liquids to solids (and vice versa) at the theoretical level has not been fully explored and explained.

Prestigious scientists have commented on the mysteries that confront them. Russell J. Donnelly (University of Oregon) has observed, "Most flows of fluids, in nature and in technology, are turbulent. Since much of the energy expended by machines and devices that involve fluid flows is spent in overcoming the drag caused by turbulence, there is a strong practical motivation to understand the phenomenon. The study of turbulent flows, however, is one of the most formidably difficult subjects in physics and engineering. At present (1988), there is no substantial aspect of turbulent flow that can be understood fully from first principles."

Steve Granick (University of Illinois) has commented (1991), "Apart from structure, what are the dynamics of liquids in intimate contact with a solid boundary? This question has proven to be one of the most baffling aspects of liquids, in spite of long-standing interest."

Sir Samuel F. Edwards (Cavendish Laboratory, University of Cambridge) noted (1987), "Liquids are everywhere in our lives, in scientific studies and in our everyday existence. The study of their properties, in terms of the molecules of which they are made, has been the graveyard of many theories put forward by physicists and chemists. The modern student of liquids places his faith in the computer, and simulates molecular motion with notable success, but this still leaves a void where simple equations should exist, as are available for gases and solids. There is a powerful reason for the failure of analytical studies of liquids, i.e., the difficulty experienced in finding simple equations for simple liquids. We can explain the origin of the trouble and show that it does not apply to what at first might seem a much more complex system, that of polymer liquids where, instead of molecules like H_2O or C_6H_6, one has systems of molecules like $H_2(CH_2)_{10,000}$ or $H_2(CHC_6H_6)_{2,000}$ which behave like sticky jellies and yet have complex properties that can be predicted successfully."

Jacob N. Israelachvili and Patricia M. McGuiggan (University of California, Santa Barbara) observe, "The subtleties that can occur in the last few nanometers as two surfaces, particles, or solute molecules approach each other in a medium can be quite remarkable. Sometimes the forces are well described by 'continuum' or 'mean-field' theories, such as the DLVO (Derjaguin, Landau, Verwey, and Overbeck) theory, but more often they are not. Important fundamental questions remain concerning the origin of long-range attractive and repulsive hydration forces in water, the spontaneous nucleation of a bulk liquid or vapor phase between two surfaces close together, and the nature of entropic-fluctuation forces between two fluid-like interfaces. The elucidation of these interactions both at the fundamental level and when applied to specific systems (where a number of different interactions may be occurring simultaneously) present a challenge to experimentalists and theoreticians. On the purely experimental side, new techniques are constantly being introduced for extending the range and scope of surface force measurements. For example, one may anticipate that the atomic force microscope will soon provide the first direct measurements of the forces between molecules, as opposed to between surfaces."

General Properties of Liquids

A liquid is matter in a fluid state that is relatively incompressible. An *ideal liquid* offers no permanent resistance to a shear stress, but is incompressible. A liquid has a constant volume and incompletely fills any container of less than this volume. A *real liquid* is appreciably compressible, and the liquid state of a substance might be defined as the denser and less compressible phase of the two-phase fluid system that can exist in equilibrium at temperatures below the critical temperature. X-ray diffraction experiments show that, near the melting-point, the molecules of a liquid show a considerable degree of short-range order and that, in small volumes, they are arranged much as in a solid crystal. This crystalline structure persists over volumes comparable with the intermolecular distances, but cannot be traced beyond. This local or short-range order means that the average molecule is at any moment surrounded by a number of molecules occupying nearly the same relative positions as they would in the solid state. The degree of short-range order is described by the radial distribution function.

This concept of a liquid as an imperfect crystal requires that the molecules in a liquid are packed sufficiently loosely for comparatively free movement, i.e., the energy required to move a molecule from a lattice site to a vacant space is not large compared with thermal energies. Under these conditions, shear flow of the liquid resembles closely the high temperature creep of crystalline solids. A number of theories of the liquid state have this concept as their starting point.

With a few exceptions, including helium, the universal phase diagram shown in Fig. 1 applies for all pure compounds. The triple point is the single point at which all three phases (crystal, liquid, and gas) are in equilibrium. The triple point pressure is normally below atmospheric. Those substances, such as carbon dioxide, where $P_t = 3,885$ millimeters, $T_t = -56.6°C$, sublime without melting at atmospheric pressure. From the triple point, the melting curve defines the equilibrium between crystal and liquid, usually rising with small but positive dT/dP, and presumably always with positive dT/dP at sufficiently high P values. The line is believed to extend infinitely without a critical point (it has been followed to $T \cong 16T_c$ for helium, and calculations indicate that hard spheres would show a gas-crystal phase change). The gas-liquid equilibrium line, the vapor pressure curve, has dT/dP always positive and greater than the melting curve. The vapor pressure curve always ends at a critical point, $P = P_c$,

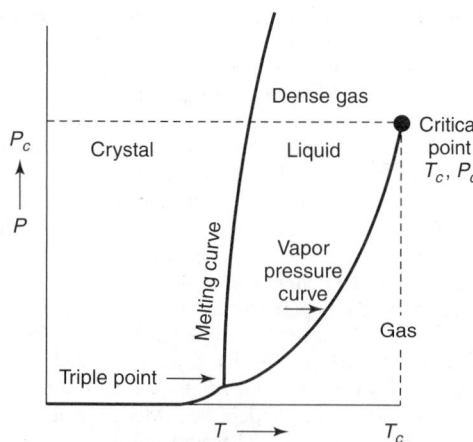

Fig. 1.   Universal phase diagram

$T = T_c$, above which the liquid and gas phase are no longer distinguishable. Since the liquid can be continuously converted into the gas phase without discontinuous change of properties by any path in the $P - T$ diagram passing above the critical point, there is no definite boundary between liquid and gas. Two liquids of similar molecules are usually soluble in all proportions, but very low solubility is sufficiently common to permit the demonstration of as many as seven separate liquid phases in equilibrium at one temperature and pressure (mercury, gallium, phosphorus, perfluoro-kerosene, water, aniline, and heptane at 50°C, 1 atmosphere).

Stability Limits[1]. With the exception of helium and certain apparent exceptions discussed below, Fig. 1 gives a universal phase diagram for all pure compounds. The triple point of one P and one T is the single point at which all three phases, crystal, liquid, and gas, are in equilibrium. The triple point pressure is normally below atmospheric. Those substances, e.g., CO_2, $P_t = 3885$ mm, $T_t = -56.6°C$, for which it lies above, sublime without melting at atmospheric pressure.

From the triple point, the melting curve defines the equilibrium between crystal and liquid, usually rising with small but positive dT/dP, and presumably always with positive dT/dP at sufficiently high P values. The line is believed to extend infinitely without a critical point (it has been followed to $T \cong 16T_c$ for He, and calculations indicate that hard spheres would show a gas-crystal phase change). The gas-liquid equilibrium line, the vapor pressure curve, has dT/dP always positive and greater than the melting curve. The vapor pressure curve always ends at a critical point. $P = P_c$, $T = T_c$ above which the liquid and gas phase are no longer distinguishable. Since the liquid can be continuously converted into the gas phase without discontinuous change of properties by any path in the $P - T$ diagram passing above the critical point, there is no definite boundary between liquid and gas.

The term *liquid* is commonly reserved for $T < T_c$, and "dense gas" is used for $T > T_c$. However, certain properties, such as the ability to dissolve solids, change rather abruptly at the critical density. In many respects, the dense gas resembles the low-temperature liquid of the same density more closely than it does the dilute gas.

The slope, dT/dP, of all phase equilibrium lines obeys the thermodynamic Clapeyron equation:

$$dT/dP = \Delta V/\Delta S = T \Delta V/\Delta H \qquad (1)$$

with ΔV, ΔS, and ΔH the differences, for the two phases, of volume, entropy, and heat content or enthalpy, respectively. The quantity ΔH is the heat absorbed in the phase change at constant P. Since always $S_{cr} < S_{liq} < S_{gas}$, and usually $V_{cr} < V_{liq} < V_{gas}$, one usually has $dT/dP > 0$; the relatively rare cases, including water, for which $V_{liq} < V_{cr}$ at low pressures leads to $dT/dP < 0$ for the melting curve near the triple point.

Figure 1 gives the $P - T$ boundaries of the stable liquid phase. Clean liquids can readily be superheated or supercooled, and, in vessels having walls to which the liquid adheres, they can be made to support negative pressures of several tens of atmospheres. Thus the properties

of the metastable liquid can be investigated outside the limits shown in the diagram.

Two apparent exceptions to the universality of the phase diagram of Fig. 1 deserve mention. First, many of the more complicated molecules decompose at temperatures below melting or boiling, and the diagram is unobservable. Secondly, some liquids, notably glycerine and SiO_2 and many multicomponent solutions, supercool so readily that crystallization is difficult to observe. In these cases, there is a continuous transition on cooling to a glass, which has the elastic properties of an isotropic solid. The structure of the glass is qualitatively that of the high-temperature liquid, lacking long-range order. Since glass and liquid are not sharply differentiated, the term *liquid* is sometimes used to include glasses, although common parlance reserves *liquid* for the state in which flow is relatively rapid.

Quantum Liquids. The one real exception to the phase diagram of Fig. 1 is that of helium, Fig. 2. Both isotopes, ^{4}He and ^{3}He, have no triple point, the liquid is stable to 0 K below about 20 atm for ^{4}He and below about 30 atm for ^{3}He. The liquids have zero entropy at 0 K in both cases. This is also the only case in which isotopic mixtures form two liquid phases at equilibrium, the isotopic solution separating below 1 K. The isotope ^{4}He has itself two phases, He I above the dotted λ-line of the diagram, and He II with remarkable properties of superfluidity, second sound, etc., below the λ-line. The phase transition along the λ-line is second order; that is, whereas S and V are continuous, heat capacity and compressibility change discontinuously across the λ-line.

Although no completely satisfactory single theory of liquid helium has yet been formulated, one can say that most of the remarkable properties are qualitatively understood and are due to the predominance of quantum effects, including the difference in the statistics of the even and odd isotopes. Thus helium is the one example in nature of a quantum liquid, all other liquids showing only minor deviations from classical behavior.

Structure. Considerable confusion in the description of liquid structure exists, due primarily to difficulties of precise formulation of verbal concepts. The geometric arrangement of any small number (say 10 to 12) of close-lying molecules resembles the arrangement in the crystal, but the order rapidly disappears as larger groups are considered. Long-range order is lacking. The fact that numerical theories based on a lattice or cell structure have some success is evidence only that most properties depend on the configuration of near neighbors alone. Insofar as the arrangement of nearest neighbors is describable in terms of that of the crystal, the structure of the normal liquid is probably characterized best by a somewhat closer spacing than the crystal of the same molecules. The reduced density arises from a considerable number of vacancies in the lattice; the coordination number, or number of nearest neighbors, is lower than in the crystal. The exception is water, in which the low coordination number, 4, of the crystal, is increased by interstitial molecules in the liquid, leading to a higher density of the liquid.

Structural descriptions of this nature usually lack the possibility of precise formulation. It is, however, possible to define for any disordered array of molecules in three-dimensional space an arrangement of contiguous cells, each containing one and only one molecule, the faces of the cells being the loci of the midpoints of neighboring molecules. The statistics of the fraction of cells with n faces and of the distances of the

[1]The following several paragraphs by Joseph E. Mayer are part of a large article that appears in "The Encyclopedia of Physics" (Robert M. Besancon, Editor), Van Nostrand Reinhold, New York, 1984.

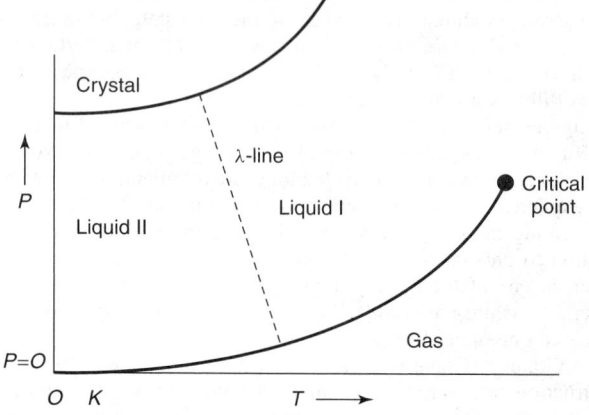

Fig. 2.   Quantum liquid exception to phase diagram of Fig. 1

faces from the molecules would give the fraction of molecules having a given number of nearest neighbors and the distance distribution of these in a precisely defined manner. Neither present experimental information nor present theories lend themselves to analysis in such terms.

The only clearly defined manner of describing liquid structure in use at present involves the concept of a set of probability density functions, ρ_n, for ascending numbers, n, of molecules. The function ρ_n depends on the vector coordinates $\mathbf{r}_1, \mathbf{r}_2, \ldots \mathbf{r}_n$ of n molecules, and

$$\rho_n \mathbf{r}_1, \mathbf{r}_2, \ldots, \mathbf{r}_n, d\mathbf{r}_1, \ldots, d\mathbf{r}_n$$

is defined as being the probability that in the liquid of definite P and T, there will be, at any instant of time, one molecule at each position, $\mathbf{r}_i$, within the volume element, $d\mathbf{r}_j$. For a fluid, unlike a perfect single crystal, $\rho_i(\mathbf{r})$ is a constant independent of $\mathbf{r}$ and equal to the number density: the number, ρ, of molecules per unit volume. The first significant member of the set is then the pair density function, $\rho_2(\mathbf{r}_1, \mathbf{r}_2)$, which depends only on the distance, $\mathbf{r} = |\mathbf{r}_1 - \mathbf{r}_2|$, between the two molecules. At large distances $\rho_2(\mathbf{r} \to \infty) = \rho^2$. This function can be found experimentally from the x-ray scattering intensities of the liquid (it is the three-dimensional Fourier transform of the scattering intensity at angle θ vs $(4\pi/\lambda)/\sin; (\theta/2))$. A typical plot is shown in Fig. 3. The area under the ill-defined first peak integrated over $4\pi \mathbf{r}^2 d\mathbf{r}$ is the average number of nearest neighbors, and is of order 10 to 11 for normal liquids.

The quantity of dimensions of energy,

$$W_n(\mathbf{r}_1, \ldots, \mathbf{r}_n) = -kT \ln[\rho^{-n} \rho_n(\mathbf{r}_1, \ldots, \mathbf{r}_n)]$$

can be shown to be the potential of average force of n molecules located at the positions $\mathbf{r}_1, \ldots, \mathbf{r}_n$. That is, if there are n molecules at these positions, there will be some average force, f_{xi}, along the x-coordinate of molecule i. This average is the sum of the direct force due to the other $n - 1$ plus the average of a fluctuating force due to the others, whose average position is affected by that of the n specified ones. This average force is

$$f_{xi} = -(\partial/\partial x_i) W_n(\mathbf{r}_i, \ldots, \mathbf{r}_n)$$

One frequently assumes that W_n is a sum of pair forces only,

$$W_n(\mathbf{r}_i, \ldots, \mathbf{r}_n) = \sum_{n \geq i >} \sum_{j \geq l}' W_2(\mathbf{r}_{ij})$$

although this assumption is known to be only approximate. With this assumption, the pair average force potential, $W_2(\mathbf{r}_{ij})$, can be computed as the solution of an integral equation, and the solutions agree quite well with the experimental curves.

The knowledge of the complete set of functions ρ_n plus that of the intermolecular forces would permit the computation of all equilibrium properties of the liquid, and indeed if the intermolecular forces are the sum of pair forces, only a knowledge of ρ_2 at all P, T values is necessary. An adequate, although numerically difficult, theory of the transport properties also exists, using the equilibrium functions, ρ_n. At present, only qualitative success is obtained in the completely *a priori* use of the equations.

Associated Liquids. The description given above is adequate only for liquids composed of spherically symmetric molecules or molecules that are nearly so. These constitute the so-called normal liquids, which obey reasonably well the law of corresponding states, for which the entropy of vaporization at the boiling point has the Trouton's rule value of approximately 21 cal/deg. For molecules containing large dipole moments, or those forming mutual hydrogen bonds, the concept of the probability

density functions must be extended to include angles or other internal degrees of freedom in the coordinates. Such inclusion is conceptually easy, but incredibly complicates the already difficult numerical evaluation of any equations. However, certain qualitative statements may be made.

Liquids composed of molecules with large dipole moments are frequently referred to as associated. Although in some instances relatively stable dimer or definite polymer units of relatively fixed orientation may exist, in many cases, notably water, it is extremely doubtful if an exact knowledge of the structure would reveal any distinguishable entities of associated molecules other than that of the whole liquid. In such cases, one would, however, expect that certain mutual angular orientations between neighboring molecules will be highly preferred, whereas in the dilute gas this will not be the case. The effect of this restriction on the internal coordinates will be to decrease the entropy of the liquid markedly compared to the gas. This effect is qualitatively the same as in association, and the properties of these liquids, particularly the high entropy of vaporization, will simulate those of a liquid composed of definite associated complexes.

Traditional Views of Forces Between Surfaces in Liquids

For many years, four kinds of forces have been recognized to operate between surfaces or particles in liquids:

1. *Van Der Waals forces*—Normally, these are monotonically attractive and occur between all molecules. See also **Van Der Waals Forces**.
2. *Repulsive electrostatic (double-layer) forces*—These forces are apparent when ionizable surfaces have a net electric charge, the common case in water.
3. *Structural, hydration, or solvation forces*—These forces may be attractive, repulsive, or oscillatory and depend upon the structuring or ordering of liquid molecules. (Solvation may be defined as the adsorption of a microlayer of film of water or other solvent on individual dispersed particles of a solution or dispersion.)
4. *Repulsive entropic (steric or fluctuation) forces*—As defined by Israelachvili and McGuiggan, these are "forces which arise from the thermal motions of protruding surface groups (such as polymers or lipid head groups) or from the thermal fluctuations of flexible fluid-like interfaces (or surfactant or lipid bilayers)."

Although, in a vacuum, only the Van Der Waals forces are important; in liquids, all forces may operate simultaneously. In liquids, it is extremely difficult to separate the effects of each of the aforementioned forces.

In the 1950s, the DLVO theory was based largely upon forces (1) and (2) defined above. The DLVO theory became the basis for studying the properties of colloidal and biocolloidal systems.

Electrorheological Fluids

Complications continue in the theoretical exploration of liquid behavior, but, in attempting to learn about the complexities, leads toward a more fundamental understanding of liquids may emerge. One of these complexities is a class of fluids referred to as *electrorheological*.

In a normal setting, these liquids are liquid in the conventional sense, but, when they are subjected to a strong electric field, they become solids. A common example is a mixture of cornstarch and vegetable oil. The viscous, sticky starting mixture of these two components can be converted into a hardened solid material with the application of an appropriate electric field. In recent years, through random researching, investigators have found numerous combinations of materials that qualify as electrorheological fluids, but to date no satisfactory explanation of the effect from a theoretical standpoint has been developed. The effect of the ER effect can be observed readily by microscopic examination. In a normal liquid or in an electrorheological mixture not in an electric field, examination shows particles moving in random fashion throughout a container, as may be expected. But, when a field is applied, long strands of particles appear to be "solidly" linked together, thus providing rigidity to the mix. No retentivity is involved, however, because liquid normalcy is returned immediately upon cessation of the electric field. This action intrigues a number of designers of equipment, as in the electronics and valve fields, where such materials may be used in future circuits, valves, and any number of other "on-off" devices.

Artificial Magnetic Fluids. The concept of magnetizable liquids dates back over several decades. The first breakthrough occurred in the mid-1960s, with the production of stable colloids of subdomain solid

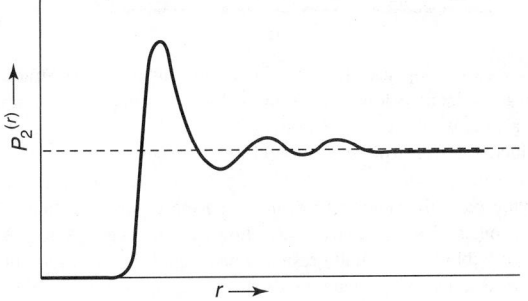

Fig. 3. Fourier transform of scattering intensity

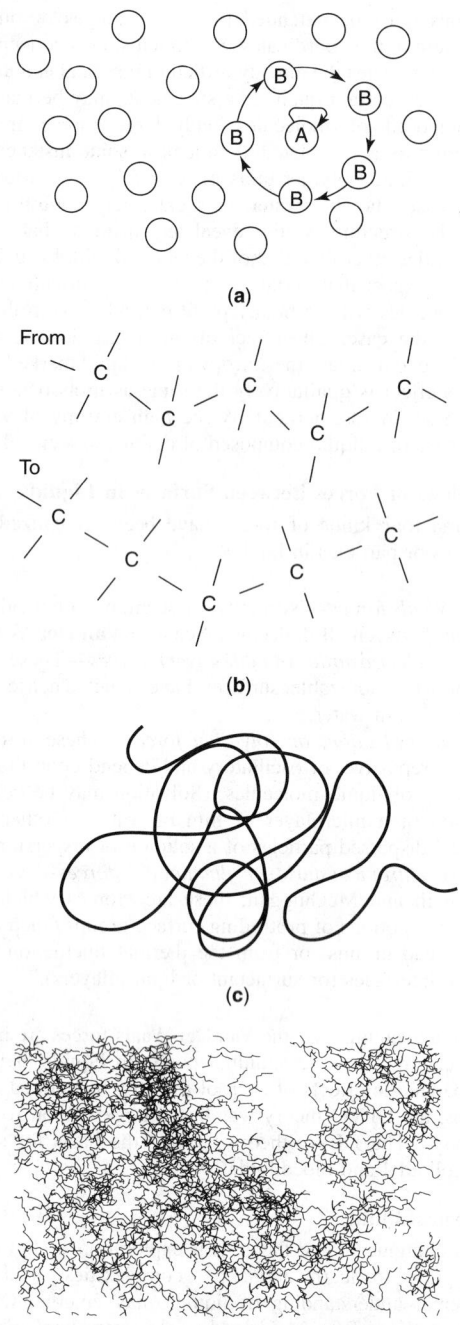

(a)

(b)

(c)

(d)

Fig. 4. Sketches by Edwards for schematically illustrating the behavior of polymer liquids. (**a**) Central molecule A moves around a (temporary) average position, occasionally escaping the barrier of the molecules B when some fluctuation of their positions permits it to do so. But, in addition to this single molecule motion, the molecules B can move around cooperatively, one of many cooperative motions that the observer may "invent." The numerous possibilities immediately derail the formulation of a simple theory. As explained by Edwards, "If the motion only involved one molecule at a time, quite a reasonable theory can be put together, but it will always be inadequate to describe the whole motion, and possibly, this always will be the case. The very apparatus of mathematical equations cannot handle this level of complexity, even though the human mind has no difficulty in seeing where the trouble is. The best current way to deal with this problem is to put all the molecules onto the computer and study their molecular dynamics." Continual changes are possible, and the long-chain molecule can be regarded as a flexible string in continual motion. A single coil may be drawn, as shown in (**c**). A computer model could be generated to show many polymers projected onto a plane, as indicated in (**d**). Edwards observes, "So the liquid looks like a seething pit of wriggling motion, a kind of living spaghetti, but with less smooth shapes than real spaghetti. How can one describe such a system? How do the molecules move, and how does the material flow?" (*Sketches after Sir Samuel F. Edwards*)

ferromagnets, which variously were called magnetic fluids, magnetizable fluids, or simply ferrofluids. These may be prepared by size reduction or precipitation. It has been a rather remarkable feat to grind bulk material down to a size of 100 micrometers. Grinding is done in the presence of a dispersing agent and a solvent. In chemical precipitation, iron(II) and iron(III) ions in aqueous solution are coprecipitated in the molar ratio of about 2:1 using ammonium hydroxide. To maintain the magnetic product in a small colloidal size range, a peptization step is included in which the particles are transferred to a heated organic phase containing the dispersing agent. The behavior of these artificial fluids offers techniques for achieving efficient heat and mass transfer, drag reduction, wetting, fluidization, sealing, damping, and other process and product potential uses.

Polymer Liquids

Sir Samuel F. Edwards of the Cavendish Laboratory, previously mentioned, observes, "Polymer liquids are liquid because the temperature is high enough to change the configuration of the molecules easily." See Fig. 4.

Scientists and engineers experienced in the production and application of polymeric liquids have learned from thousands of examples of how polymeric liquids behave, and can even classify some of them into behavioral families. But the complexity of these substances to date has eluded the achievement of precise designs and predictive behavior.

Mixing of Liquids

In the chemical process industries and in food manufacturing, the mixing of liquids is an important and frequently used operation. Over the years, the mixing operation has been poorly understood and essentially remains so. Progress in equipment design has been achieved mainly through the development of extensive empirical information rather than upon the creation of precise mathematical and theoretical relationships. J.M. Ottino (University of Massachusetts) observes, "There are many fundamental questions regarding mixing in slow three-dimensional flows, and unfortunately some of the intuition we have obtained from our study of two-dimensional flows does not necessarily carry over to flows in three

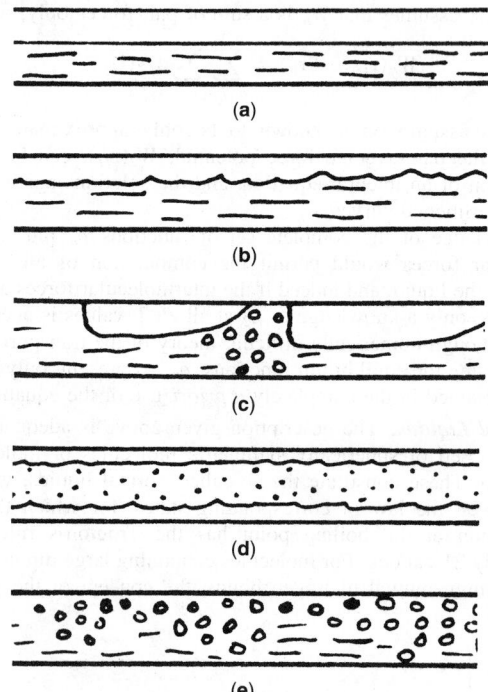

(a)

(b)

(c)

(d)

(e)

Fig. 5. Types of two-phase flow in a horizontal pipeline: (**a**) Stratified smooth flow where gas velocity is low. Liquid flows along bottom portion of pipelines with essentially a smooth surface. (**b**) Stratified flow with a wavy surface, the waviness caused by increased gas flow velocity. (**c**) Liquid bridges the pipeline cross section, thus causing slugs or plugs of liquid, which move at a velocity approximately that of the flowing gas. (**d**) Annular flow, in which the liquid essentially flows as an annular film on the pipe wall while gas flows as in a central core of the pipe. (**e**) Dispersed bubble flow usually results when liquid flow rates are high and gas rates are low. Because of comparative density differences, most bubbles are found above the pipe center line. Conditions vary somewhat when the pipeline is in a vertical orientation. (*After Cindric, Gandhi, and Williams*)

dimensions." The Ottino reference listed describes studies of chaotic and nonchaotic flows in laboratory setups.

Multiphase Fluid Flow

It is quite common in industrial and cross-country liquid transport problems to encounter mixtures of liquids, vapors, and gases—that is, the presence of two phases of matter. As pointed out by D.I. Koch (Cornell University), "Research programs in this area at Cornell involve studies that fall outside of the traditional realm of chemical engineering, including blood flow in capillaries, the transport of contaminants in groundwater, the dynamics of geothermal reservoirs, enhanced oil recovery, the processing of fibrous composites, melt-spinning processes, and the growth of silicon crystals."

As previously mentioned in this article, liquid behavior presents an immense variety of puzzling problems that are difficult to comprehend and hence difficult to forecast precisely. As just one example, in the study of large drops of liquid at high flow rates, the inertia of the drops and the surrounding fluids play an important role. When such drops are propelled toward one another, they may coalesce into a single drop or may rebound like a pair of elastic balls, a phenomenon that is partially (but not fully) dependent upon the comparative velocities of the two drops.

In traditional industrial two-phase flow situations, as occur in process piping, engineers classify flow patterns, as indicated in Fig. 5. This type of classification and the development of empirical data from past experimentation and practice assist much in simplifying the problems from a practical, if not from a theoretical, standpoint. Multiphase flow behavior considerations are essential in calculating pipe diameters, pumping capacities, and energy consumption. Multiphase flows also exhibit different characteristics when being pumped uphill or flowing downhill. See also **Fluid and Fluid Flow**.

Additional Reading

Amato, I.: "Liquids That Tiptoe on the Edge of Solidity," *Science News*, 342 (December 1, 1990).

Cindric, D.T., S.L. Gandhi, and R.A. Williams: "Designing Piping Systems for Two-Phase Flow," *Chem. Eng. Progress*, 51 (March 1987).

Coker, A.K.: "Understand Two-Phase Flow in Process Piping," *Chem. Eng. Progress*, 60 (November 1990).

Donnelly, R.J.: "Superfluid Turbulence," *Sci. Amer.*, 100 (November 1988).

Edwards, S.F.: "Polymer Liquids," *Review (University of Wales)*, 58 (March 1987).

Egelstaff, P.A.: *Introduction to the Liquid State*, Oxford University Press, Inc., New York, NY, 1994.

Granick, S.: "Motions and Relaxations of Confined Liquids," *Science*, 1374 (September 20, 1991).

Grimmett, G., B. Eckmann, S.S. Chern, and H. Hironaka: *Percolation*, 2nd Edition, Vol. 321, Springer-Verlag, Inc., New York, NY, 1999.

Heyes, D.M.: *The Liquid State: Applications of Molecular Simulations*, John Wiley & Sons, Inc., New York, NY, 1998.

Israelachvili, J.N. and P.M. McGuiggan: "Forces Between Surfaces in Liquids," *Science*, 795 (August 12, 1988).

Koch, D.L.: "Fluid Dynamics in Multiphase Systems," *Chem. Eng. Progress*, 74 (November 1989).

Langer, J.S.: "Dendrites, Viscous Fingers, and the Theory of Pattern Formation," *Science*, 1150 (March 3, 1989).

Lounasmaa, O.V. and G. Pickett: "The ^{3}He Superfluids," *Sci. Amer.*, 104 (June 1990).

Luessen, L.H., L.G. Christophorou, and E.E. Kunhardt: *The Liquid State and Its Electrical Properties*, Perseus Books, Boulder, CO, 1988.

March, N.H., M.P. Tosi and R.A. Street: *Amorphous Solids and the Liquid State*, Kluwer Academic Publishers, Norwell, MA, 1985.

McComb, W.D.: *The Physics of Fluid Turbulence*, Oxford University Press, Inc., New York, NY, 1992.

Monastersky, R.: "Stretching Liquid to Its Physical Limit," *Science News*, 87 (August 11, 1990).

Ottino, J.M.: "The Mixing of Fluids," *Sci. Amer.*, 56 (January 1989).

Ottino, J.M. et al.: "Morphological Structures Produced by Mixing in Chaotic Flows," *Nature*, 419 (June 2, 1988).

Pool, R.: "The Fluids with a Case of Split Personality," *Science*, 1180 (March 9, 1990).

Schmidt, W.F.: *Liquid State Electronics of Insulating Liquids*, CRC Press, LLC., Boca Raton, FL, 1997.

Snedden, R.: *States of Matter: Solids, Liquids and Gases*, Heinemann Library, Oxford, UK, 2001.

Stixrude, L. and M.S.T. Bukowinski: "A Novel Topological Compression Mechanism in a Covalent Liquid," *Science*, 541 (October 26, 1990).

Tabor, D.: *Gases, Liquids, and Solids: And Other States of Matter*, 3rd Edition, Cambridge University Press, New York, NY, 1991.

Thompson, P.A. and M.O. Robbins: "Origins of Stick-Slip Motion in Boundary Lubrication," *Science*, 792 (November 9, 1990).

LITHIUM. [CAS: 7439-93-2]. Chemical element, symbol Li, at. no. 3, at. wt. 6.941, periodic table group 1, mp 180.54°C, bp 1342°C (at 760 torr), density 0.534 g/cm^3 (20°C). Lithium is lightest in weight of all the chemical elements that are solid at standard conditions. Elemental lithium in the solid phase has a body-centered cubic crystal structure. In comparison with other members of the alkali metal series, lithium has the smallest ionic radius, the highest ionization potential, the highest electronegativity, and the greatest heat capacity. Generally, lithium is the least reactive of the alkali metals. Lithium is a silver-white metal, harder than sodium, but softer than lead. It is tough and may be drawn into wire or rolled into sheets. The element tarnishes rapidly in air and often is preserved under naphtha. The reaction with H_2O is vigorous, producing LiOH (lithium hydroxide) and hydrogen. There are two naturally occurring isotopes, ^{6}Li and ^{7}Li. They are not radioactive. Two radioactive isotopes have been identified, ^{5}Li and ^{8}Li, both with very short half-lives, measured in fractions of seconds. Among elements occurring naturally in the earth's crust, lithium ranks 28th with an estimated average content of about 10–20 ppm. In terms of content in seawater, lithium ranks 17th with an estimated content of approximately 950 tons of lithium per cubic mile of seawater. The element was first identified by Johann August Arfvedson in 1817 in the laboratory of Berzelius. The name of the element is accredited to Berzelius.

First ionization potential 5.39 eV. Oxidation potential Li $\longrightarrow$ Li$^+$ + e$^-$, 3.02 V. Other physical properties of lithium are given under **Chemical Elements**.

The main sources of lithium are pegmatites and brines. The most important pegmatite mineral is *spodumene*, $LiAlSi_2O_6$, which contains a theoretical content of 8.03% Li_2O. *Petalite*, $LiAlSi_4O_{10}$, contains between 4 and 4.5% Li_2O. *Lepidolite*, a complex mica, contains between 3 and 4% Li_2O. See also entries on **Lepidolite**; **Petalite**; and **Spodumene**. Brines contain normally a few hundred to a few thousand parts per million (ppm) of lithium. The only commercial source of spodumene in North America is located in North Carolina. Abundant resources of lithium pegmatites occur in Canada, the African continent, and unconfirmed sources in Russia and China. Significant quantities of lithium (as carbonate) are produced from the brines of Clayton Valley, Nevada. A recently discovered, lithium-rich brine deposit has been located in the Atacama desert of Chile. More detail on lithium resources is given in the next entry.

There are three major processes for extracting lithium from pegmatite ores. (1) An acid process, wherein the spodumene concentrate, after calcining at about 1095°C, is reacted with sulfuric acid, followed by water leaching of the resulting lithium sulfate, Li_2SO_4. The sulfate is then converted to the carbonate with soda ash. (2) An alkaline process, wherein the ore is reacted with lime or limestone at high temperatures followed by water leaching of the resulting lithium hydroxide. (3) A base exchange method, whereby the ore is reacted with an alkaline chloride or sulfate at a high temperature in an aqueous phase to yield a soluble lithium salt. The sulfuric acid leaching method is the only commercial process for extraction of lithium from spodumene in practice today.

Lithium metal was first prepared by Sir Humphry Davy in 1818 by electrolyzing lithium oxide. At about that same time, Brande also isolated the metal. In 1855, R. Bunsen and A. Matthiessen prepared gram amounts by electrolyzing fused lithium chloride. Modest commercial quantities were first made in Germany during World War I when the metal was considered as a potential alloying material. Limited production did not commence in the United States until the early 1930s. Present commercial methods were pioneered by Guntz in 1893 and involve electrolyzing a low-melting mixture of LiCl and KCl. Graphite anodes and mild steel cathodes are used. Lithium is formed at the cathode and rises to the surface, from which it is skimmed periodically. Pure lithium chloride is added to the bath as required. Chlorine gas is liberated at the anodes. The process yields a lithium metal of about 99.8% purity. The metal normally is cast into ingots of different sizes, but is also available as extruded rod, ribbon, or wire. The metal also is available as "sand"—fine dispersions in the 10–30 μm range.

Lithium in Metallurgy and Alloys

In metallurgy, lithium metal is used as a deoxidizer, desulfurizer, and degasifier in the production of a number of molten metals, notably copper and copper alloys. Lithium also is an ingredient of an increasing number of alloys, particularly with aluminum. Early alloys included aluminum alloy

X2020 (1% Li), which is a structural alloy with improved high-temperature strength. In another early Li alloy, about 14% Li is alloyed with magnesium in the LA 141 alloy, designed for very light-weight structural applications, notably in aerospace applications.

In late 1989, a new proprietary (Martin Marietta Corp.) family of weldable, high-strength Al-Li alloys was introduced. With a 690-MPa (100×10^3 psi) yield strength, the material is claimed to be twice as strong as the previous leading Al-Li alloys. This alloy was developed specifically for space launch systems. The alloy is claimed to maintain a high strength under thermal conditions ranging from cryogenic to elevated temperatures. A primary use is for fuel and oxidizer tanks, where its weldability is a marked advantage. Sheet, plate, extrusion, and ingot products of the new alloy also are available.

The addition of lithium to aluminum castings has been found to be particularly advantageous. Lithium produces a lower density and higher stiffness over conventional aluminum alloys used in aerospace applications. Lithium has one of the highest solubilities of any aluminum alloying element. About 4.2% Li can be dissolved in Al at the 602°C (1116°F) eutectic temperature. The hardness of Al-Li alloys improves with aging temperature. Al_3Li precipitates are formed, producing higher hardness. Yield strength also increases with higher aging temperature and higher Li content.

Lithium alloyed with silver has been used for fluxless brazing.

Lithium Batteries. For many years, lithium has been considered for use in batteries, particularly with the growing emphasis on the electric car. See also **Batteries.**

Chemistry and Compounds

Lithium has the highest ionization potential (i.e., Li $\longrightarrow$ Li$^+$ in the vapor) of the alkali metals. However, the measured value of its oxidation potential against a normal aqueous solution of its ion is 3.02 V, which does not differ from those of the other main group I metals by as much as the difference in ionization potentials. That difference, attributed to the high heat of hydration of Li$^+$, explains why lithium is a vigorous reductant in aqueous systems, but reacts slowly with H_2O, and not at all with dry oxygen except above 100°C.

The single $2s$ electron in the outer shell of lithium is easily removed to form the positive ion, and stability of the remaining $1s^2$ electron pair requires too high a potential (75.62 eV) for any further ionization (by chemical means) so that lithium is exclusively monovalent in its compounds.

Because of the reactivity of lithium with water to form its hydroxide, LiOH, and hydrogen, its properties when dissolved in other solvents have been studied extensively. It does not decompose liquid NH_3, but does form a blue solution, which decomposes to yield its amide, $LiNH_2$, and hydrogen, when catalyzed by metallic salts. With the elements of main groups 2 to 7, lithium in liquid NH_3 reacts to form binary compounds, which may vary from simple halides, as with the halogens, to intermetallic phases, as with cadmium and mercury. Lithium amide in liquid NH_3 is regarded in the same class as a hydroxide in aqueous solution.

Many other lithium compounds not obtainable in aqueous solution can be produced from the solution of lithium in liquid ammonia. Thus the acetylide is obtained by action of acetylene.

$$C_2H_2 + LiNH_2 \longrightarrow LiC_2H + NH_3$$

$$2LiC_2H \longrightarrow Li_2C_2 + C_2H_2$$

The amide, as stated above, is produced by catalyzed decomposition of the liquid NH_3 solution, and the nitride, Li_3N, by heating the amide or by direct combination of the elements.

Lithium salts exhibit general high solubility and a high degree of dissociation in other nonaqueous solvents than liquid ammonia, such as liquid sulfur dioxide and acetic acid.

Like the other alkali metals lithium forms compounds with virtually all of the anions, inorganic as well as organic. The lithium salts are in many instances different in their solubility properties from the corresponding salts of the other alkali metals. Thus lithium fluoride, phosphate, and carbonate are the least soluble alkali metal fluoride, phosphate, and carbonate, the solubilities for the other alkali metals increasing with increasing ionic radius. Lithium chlorate and dichromate are, on the other hand, the most soluble alkali chlorate and dichromate, the solubilities for the other alkali metals decreasing with increasing ionic radius. These differences are partly

explained, as was that in the oxidation potential, by the considerable hydration of the lithium ion, which also explains the fact that lithium salts generally crystallize as hydrates. Lithium salts, probably because of the small size of the lithium ion, do not form mixed crystals with the other alkali salts, but they do form double salts, notably the two series of lithium-sodium and lithium-potassium sulfates.

Lithium forms several organic compounds. Most of them are lithium salts or lithium acid salts of organic acids or other oxygen-connected lithium compounds. The number of lithium-carbon bonded compounds that have been reported is very small including, in addition to the carbide, methyllithium, CH_3Li, ethyllithium, C_2H_5Li, *n*-propyllithium, C_3H_7Li, *n*-butyllithium, C_4H_9Li, benzyllithium, $C_6H_5 \cdot CH_2Li$, and methylenedilithium $LiCH_2Li$.

The alkyllithium compounds are usually colorless, soluble in organic solvents, and capable of distillation or sublimation. They are nonelectrolytes and are widely used in synthetic organic chemistry, since, like other lithium compounds, they resemble in their properties the corresponding magnesium compounds.

Lithium carbonate. [CAS: 554-13-2]. Li_2CO_3, mp 72.6°C, slightly soluble in H_2O. Used in glass, enamel, and ceramic formulations, in the electrowinning of aluminum, and in the manufacture of other lithium compounds. The compound also has been used in the treatment of manic-depressive psychoses.

Lithium hydride. [CAS: 7580-67-8]. LiH, mp 686.4°C, reacts vigorously with H_2O. With NH_3, it forms the amide. The compound is used to produce $LiAlH_4$ and other double hydrides. Lithium hydride is an excellent light-weight source of hydrogen. One pound yields 45 cubic feet of hydrogen (one kilogram yields 2.8 cubic meters of hydrogen) at standard conditions. The compound also can serve as a light-weight shield for thermal neutrons.

Lithium hydroxide monohydrate. [CAS: 1310-65-2]. LiOH · H_2O loses water at 101°C. LiOH melts at 450°C. The compound is soluble in water. The compound is used in the formulation of lithium soaps used in multipurpose greases; also in the manufacture of various lithium salts; and as an additive to the electrolyte of alkaline storage batteries. LiOH also is an efficient, light-weight absorbent for carbon dioxide.

Lithium bromide. [CAS: 7550-35-8]. LiBr, mp 550°C, soluble in H_2O or alcohols. The compound is very hygroscopic and forms four hydrates. Major use has been in absorption-refrigeration air-conditioning systems in which H_2O is the refrigerant—strong LiBr is used to absorb H_2O vapor.

Lithium chloride. [CAS: 7447-41-8]. LiCl, mp 608°C, soluble in H_2O or alcohols. Very hygroscopic and forms four hydrates like the bromide. The compound is a component of brazing fluxes for aluminum and magnesium. It is used in dehumidification systems, as an additive to the electrolyte of dry cells for low-temperature applications; and it is used in low-freezing fire-extinguishing systems; as an ingredient of fused-salt baths to lower fusing temperature; and, as a coating, in humidity-sensing instruments.

Lithium fluoride. [CAS: 7789-24-4]. LiF, mp 848°C, soluble in H_2O (slight). Used in enamel and glass formulations; as a component of welding and brazing fluxes; in the electrowinning of aluminum; and as an ingredient of molten salts.

Lithium in Biological Systems

Although much remains to be learned, there is considerable evidence that lithium can play an active role (positive and negative) in biological systems. Possibly most widely known is the use of lithium salts, notably lithium carbonate, in the therapy for mania (a condition where the patient is mentally and physically hyperactive, associated with an elevated mood and disorganized behavior). Frequently associated with mania is the broad swinging of the patient's mood (*bipolar disorder*). Studies have shown that persons with mania have a defect in the transmissions of impulses between and along nerve cells in the brain, which depends upon the regulated movement of ions across the membranes of those cells. Lithium antagonizes synaptic transmission of catecholamines in the brain by inhibiting norepinephrine and dopamine release. This is the result of weakly increasing their re-uptake by the presynaptic neuron and by decreasing storage. Lithium also interferes with the ability of several hormones to stimulate adenylate cyclase, a property that is believed to decrease the action of catecholamines at the postsynaptic receptor sites.

Additional Reading

Birch, N.J., V.S. Gallicchio, and R.W. Becker: *Lithium: 50 Years of Psychopharmacology: New Perspectives in Biomedical and Clinical Research,* Weidner Publishing Group, Riverton, NJ, 1999.

Carr, S. et al.: "Increase in Glomerular Filtration Rate in Patients with Insulin-Dependent Diabetes and Elevated Erythrocyte Sodium-Lithium Countertransport," *N. Eng. J. Med.*, 500 (February 22, 1990).

Davis, J.R.: *Metals Handbook,* 2nd Edition, ASM International, Materials Park, OH, 1998.

Greenwood, N.N. and A. Earnshaw: *Chemistry of the Elements,* 2nd Edition, Butterworth-Heinemann, Inc., Woburn, MA, 1997.

Julien, C. and Z. Stoinov: *Materials for Lithium-Ion Batteries,* Kluwer Academic Publishers, Norwell, MA, 2000.

Krebs, R.E.: *The History and Use of Our Earth's Chemical Elements: A Reference Guide,* Greenwood Publishing Group, Inc., Westport, CT, 1998.

Kubel, E.J., Jr.: "New Al-Li Alloy," *Advanced Materials 7 Processes,* 10 (October 1989).

Lewis, R.J. and N.I. Sax: *Sax'x Dangerous Properties of Industrial Materials,* 10th Edition, John Wiley & Sons, Inc., New York, NY, 1999.

Lide, D.R.: *CRC Handbook of Chemistry and Physics,* 84th Edition, CRC Press, LLC., Boca Raton, FL, 2003.

Sapse, Anne-Marie and P.von Rague Schleyer: *Lithium Chemistry: A Theoretical and Experimental Overview,* John Wiley & Sons, Inc., New York, NY, 1994.

Schou, M.: *Lithium Treatment of Manic Depressive Illness: A Practical Guide,* S. Karger Publishers, Inc., Farmington, CT, 1993.

Swartz, C.M.: "Serum Lithium During Treatment of Bipolar Disorder," *N. Eng. J. Med.*, 1159 (April 19, 1990).

Taketani, H.: "Properties of Al-Li Alloy 2091-T3 Sheet," *Advanced Materials and Processes,*" 113 (April 1990).

Wakihara, M. and O. Yamamoto: *Lithium Ion Batteries: Fundamentals and Performance,* John Wiley & Sons, Inc., New York, NY, 1998.

LITTLE, ARTHUR D. (1863–1935). Born in Boston, Little was a pioneer in the field of industrial research and chemical consulting. Originally an authority on paper technology, he established a consulting industrial chemical laboratory in 1886, which has since become a large institution of worldwide reputation, located in Cambridge, MA. It has served as a prototype of many industrially oriented consulting firms that have become a significant factor in the growth of research in the last half century. It has made significant contributions in such fields as flavors, food chemistry and acceptability, paper chemistry, and rubber chemistry, as well as in corporate management.

LOW-TEMPERATURE TECHNOLOGY.   See **Cryogenics**.

LUBRICANT. A material used to diminish friction between the moving surfaces of machine parts; also to decrease friction between a cutting tool and the material being cut. A wide variety of materials is used for manufacturing lubricants. Animal lubricants are obtained from the fat of common animals and can be classified as hard fats (stearin) and soft fats (lard) or naturally occurring combinations. Vegetable lubricants include rape seed oil, cottonseed oil, soybean oil, castor oil, and linseed oil. They range in properties from solid to liquid. Petroleum and mineral oil lubricants, because of their greater stability, are usually preferred for machine applications. Lubricants range from light oils to very heavy solid greases. Graphite, a solid, is also used as a lubricant.

Because of increased requirements for lubricants, including higher temperature and pressure applications, greater durability, and tolerance to wide changes in ambient temperature conditions, numerous synthetic lubricants have been developed. These include synthetic hydrocarbons, carboxylic acid esters, silicones, polyethers (polyalkylene glycols), phosphate esters, silicate esters, highly fluorinated compounds, and polyaromatics (polyphenyls and polyphenyl ethers). In selecting a lubricant, the following characteristics are considered: (1) lubricity and antiwear properties; (2) fluid range; (3) viscosity index; (4) additive response of base oil; (5) oxidation stability; (6) thermal stability; (7) hydrolytic stability; (8) fire resistance; (9) compatibility with petroleum products; (10) compatibility with paints, plastics, and elastomers; and (11) cost.

Over a number of years, the early polyol esters, the formulas of some of which are shown below, appeared to be adequate for coping with the increasing rigorous properties required for increasingly difficult lubrication problems. They continue to be used, but some professionals in the field have developed a number of proprietary formulations that are claimed to be superior to the polyolesters.

Polyaromatics (polyphenyls)

Diester

Polyether (polyalkylene gloycols)

$Si-(O-R)_4$
Silicate ester

$[-CH_2-CH_2-CH_2-]_2$
Synthetic hydrocarbon

$C-(CH_2-O-C-R)_4$
Neopentyl polyol ester

More attention has been given to tribology, the scientific discipline of friction, wear, and lubrication. The underlying principles of tribology have been investigated intensively by a number of research groups, including the U.S. Naval Research Laboratory and physicists at the Georgia Institute of Technology, Atlanta, Georgia. Researchers are attempting to develop a better theoretical basis for understanding the processes that occur when two solid bodies move past each other at close to overlapping distances. Most of this new knowledge has stemmed from working models as well as from computer models. Researchers at the Georgia Institute of Technology have used a large Cray computer to predict what occurs when the tip of a thin nickel needle, for example, is pressed repeatedly onto a flat gold surface. The investigators initially employed quantum mechanics for answering such questions as adhesion, cohesion, and the making and breaking of chemical bonds. Researchers at the Naval Research Laboratory have used an atomic force microscope as a tool to check their experiments. The present goal is to model more complex systems. As pointed out by one researcher, "To make progress in the molecular engineering of lubricants, you need to know the molecular details of the process."

Additional Reading

Klaman, D.: *Lubricants and Related Products: Synthesis, Properties, Applications, International Standards,* John Wiley & Sons, Inc., New York, NY, 1984.

Mang, T., and W. Dresel: *Lubricants and Lubrication,* John Wiley & Sons, Inc., New York, NY, 2000.

Pirro, D.M., A.A. Wessol, and J.G. Wills: *Lubrication Fundamentals,* 2nd Edition, Marcel Dekker, Inc., New York, NY, 2001.

Rudnick, L.R.: *Lubricant Additives: Chemistry and Applications,* Marcel Dekker, Inc., New York, NY, 2003.

Rudnick, L.R. and R.L. Shubkin: *Synthetic Lubricants and High-Performance Functional Fluids,* 2nd Edition, Marcel Dekker, Inc., New York, NY, 1999.

Sequiera, A.: *Lubricant Base Oil and Wax Processing,* Marcel Dekker, Inc., New York, NY, 1994.

Staff: Society of Automotive Engineers: *Heavy Duty Diesel Engine Lubricants,* Society of Automotive Engineers, Warrendale, PA, 1996.

LUBRICATING AGENTS. There are two fundamental classes of lubricating agents—natural and synthetic.

Lubricating oils are fluids whose function is the reduction of friction and wear between solid surfaces (generally metals), in relative motion. This function is accomplished in either of two ways: (1) by formation of adsorbed films on the two opposed surfaces, which can be more easily sheared than the solid substrate, or (2) by interposition of a fluid film between the two opposed surfaces. In the former case the shear strength of the film, and in the latter, the viscosity of the fluid determine the magnitude of the work which must be done to maintain the opposed surfaces in relative motion. In most cases, bearings are designed to operate under fluid film conditions and thus the viscosity of an oil is its most important property in classifying it for lubricating purposes. The range of viscosities required is approximately 1 centistoke at 100°F (38°C) for high-speed, lightly loaded spindle bearings to approximately 4000 centistokes at 100°F (38°C) for some gear units. The great majority of oils used fall within a much narrower viscosity range of about 20–200 centistokes.

Aside from the primary function of friction and wear control, lubricating oils are often called on to serve other purposes, such as corrosion prevention, electrical insulation, power transmission and cooling. This last is particularly important in metal cutting and grinding.

As petroleum consists of a highly complex mixture of hydrocarbons and other organic compounds, the raw material must be carefully processed to separate the fractions which are desired for lubricating purposes. This is generally accomplished by a combination of fractional distillation, solvent extraction and adsorption treatments to obtain various cuts which are blended if necessary, to produce the desired viscosity. A typical sequence of operations would be:

(1) Preliminary distillation of the crude petroleum to strip off volatile matter, i.e., dissolved gases and light hydrocarbons through the fuel oil range.

(2) Secondary vacuum distillation to separate the various lubricating oil fractions, which in some cases, may include the *residuum*. Since hydrocarbons tend to undergo thermal decomposition above 600°F (316°C), the distillation is generally carried out under vacuum to enable the higher-boiling fractions to be distilled.

(3) Solvent extraction to remove waxes, unsaturates, aromatics, asphalt, and nonhydrocarbon material.

(4) Contact with an adsorbent, e.g., activated clay, for final removal of polar impurities.

(5) Incorporation of special-purpose additives and blending to obtain the desired viscosity.

The lubricating oil fractions thus obtained consist of some of the largest and most complex hydrocarbon molecules to be found in crude petroleum. Their molecular weight ranges from about 250–1000 or more, based on structures containing 20–70 carbon atoms. They may be classified broadly as (1) straight-chain paraffins, (2) branched chain paraffins, (3) naphthenes (one or more saturated 5 or 6 membered rings with paraffin side chains), (4) aromatics (benzene ring structures with paraffinic side chains), and (5) mixed aromatic-naphthene-paraffin.

Of the above classes, the most desirable are the branched chain paraffins and the naphthenes. The refining processes are directed to increasing the concentration of these and removal of the others, as well as of organic compounds containing sulfur, oxygen, and nitrogen which are found in varying proportions in all crude petroleum.

Most hydrocarbons in the lubricating oil range are subject to low-temperature oxidation (180°F; 82°C and above) by atmospheric oxygen, particularly under the influence of catalysts such as metals (copper is particularly active), moisture, and occasionally nonhydrocarbon impurities. The paraffinic types show the greatest resistance to oxidation, followed by the naphthenes and the aromatics. An empirical rule states that the rate of oxidation approximately doubles for each 20°F (11°C) rise in temperature. Thus, in practice where long life of the oil is desired, bearing temperatures are held to around 150°F (66°C) as a maximum. In an internal combustion engine where temperatures at the sliding surfaces of the order of 400–500°F (204–260°C) are encountered, shorter life will result, e.g., 100 hrs. in an engine, compared with several years in a water-wheel generator.

Besides viscosity, the viscosity-temperature variation of lubricating oils is a most important property. It has been found that this behavior can be accurately expressed by the relation: $\log_{10}\log_{10}(v + 0.8) = b\log_{10}(t°F + 460) + c$ where v is the viscosity in centistokes, t the temperature in °F, and b and c are constants. In most cases an oil with the smallest viscosity-temperature variation is considered the more desirable. As a generalization, it may be said that the oils with the highest paraffin-to-naphthene ratios have the smallest change, while low paraffin-to-naphthene ratios correspond to a higher viscosity-temperature change. However, it is possible to dissolve materials in oil (e.g., olefinic polymers) which will effect a decrease in this variation.

While most bearings depend on the viscosity of the oil for the formation of a lubricating film (0.0001″ to 0.002″; 0.0025 to 0.051 mm thick), there are many cases where this is not feasible and successful lubrication depends on the formation of very thin films (10 to 10^3 Å thick). This was generally thought to occur through preferential adsorption by the solid surface of polar compounds (e.g., naphthenic acids, sulfur compounds) present in small amounts in the oil.

Another factor possibly contributing to the lubricating ability of very thin films is the variation of viscosity with pressure. The viscosity-pressure relationship is approximately $\log_{10}v = kP + c$ where P is the pressure and k and c are constants. Thus, at very high unit pressures of the order of 10^5 psi (690 MPa), viscosity increases of the order of 1000–10,000 times are encountered. This can account for a significant increase in load-carrying capacity. More recent work on rolling contact bearings has shown that these extremely high viscosities may be partially responsible for the observed formation of true hydrodynamic lubricating films in the 10^{-6} to 10^{-5} inch (0.00254–0.000254 mm) range.

Though the major portion of industrial lubricants is derived from petroleum, a small but significant portion is being obtained from other sources.

(1) Natural liquid fatty esters, such as lard oil, palm oil, sperm oil, etc. These are good lubricants but have poor chemical stability.

(2) Synthetic hydrocarbons, prepared by polymerization of olefinic hydrocarbons; these have good stability when saturated and good viscosity-temperature coefficients.

(3) Polyalkylene glycol oils, made by reaction of alcohols with polymerized ethylene and propylene glycols. They are either water-soluble or -insoluble. Fair stability (improved with additives), good viscosity-temperature coefficient, and good lubricating qualities.

(4) Synthetic esters, (a) primarily esters of dibasic acids such as adipic and sebacic, though in Europe some monobasic acid esters have been prepared and used; (b) organic esters of phosphoric and silicic acid, which have some advantage of being more fire-resistant than the other organic compounds but which are subject to hydrolysis on exposure to water.

(5) Silicone oils, which are linear and cyclic siloxane polymers of the formula $(-SiR_2O)_n$. They generally possess good thermal and oxidative stability, and good viscosity-temperature coefficients. When $-R$ is a hydrocarbon substituent their lubricating ability is poor. This can be somewhat improved by the introduction of aromatic halogen in $-R$.

(6) Halogenated hydrocarbons (chlorinated or fluorinated). These have good lubricating properties but very poor viscosity-temperature relations. The fluorinated materials are extremely stable.

(7) Perfluorinated polyalkylene glycols which have better viscosity-temperature behavior and temperature behavior and temperature resistance.

(8) Polyphenyl ethers are very stable organic fluids which can be used in the 500–700°F (260–371°C) temperature range although, like the silicones, they do not have good boundary lubrication properties. These reportedly, however, can be improved with suitable additives.

Generally speaking, these synthetic materials are at present used primarily as specialty lubricants where the need for a particular property outweighs high cost, e.g., diesters as military aircraft engine lubricants. As synthetic processes are improved more extensive application may be expected.

Synthetic Lubricants

Chemistry has been the guiding science in creating the various classes of chemical lubricants. Each class is named in terms of the chemical structure involved and each has at least one property that is unobtainable with naturally occurring materials. In varying degrees and combinations, synthetic lubricants provide excellent lubricity over extremely wide temperature ranges, high thermal and oxidative stability, fire-resistance, and outstanding resistance to nuclear radiation. Other noteworthy advantages of synthetic lubricants are the ease with which physical or performance properties can be altered by chemical modification or the incorporation of additives. The facility with which modifications can be achieved makes it possible for the chemists to tailor individual members of the family to meet exactly the requirements of a particular application. Compromises in properties are likewise made to obtain a satisfactory lubricant at lowest cost.

Polyglycols. $RO(CH_2CHR'O)_xR''$, where R's can be hydrogen and/or organic groups. Terminal groups determine the type of polyglycol such as diol, monoether, diether, ether-ester, etc. Various alkylene oxides including mixtures are the basic raw materials. The major attributes of the polyglycols are viscosity-temperature characteristics, volatility of products of decomposition, relatively low cost, and the wide variety of properties (water-soluble to organic-soluble) obtainable by structural variations. Major uses are as lubricants in automotive hydraulic brake fluids and water-based hydraulic fluids, as compressor lubricants, textile lubricants and rubber lubricants, and in greases where the polyglycol is the carrier for solid lubricants.

Phosphate Esters. $R'OP(O)(OR'')(OR'')$ where at least one R represents an organic group while the remaining represent organic groups or hydrogen. These products are prepared from phosphorus oxychloride or phosphoryl chlorides plus phenols, alcohols or their sodium salts. The tertiary phosphate esters are often classified as triaryl, trialkyl, and alkyl aryl phosphates. The primary and secondary phosphates are used extensively as lubricant additives in various chemical forms. Phosphate esters are best known for their fire-resistant characteristics, and as a result have found extensive application as industrial and aircraft lubricants and hydraulic fluids.

Dibasic Acid Esters. These can include both simple and complex materials. The simple dibasic acid esters, ROOCR'COOR'' are made by reacting a dibasic acid, such as sebacic acid, with a primary branched alcohol, such as ethyl hexanol. Complex esters are prepared by reacting a dibasic acid with a polyglycol, such as polyethylene glycol, and capping the chain with a branched primary alcohol or a monobasic acid. The outstanding characteristics of the dibasic acid esters are favorable viscosity-temperature characteristics, excellent lubricating ability, and high stability. Because of this combination of properties, these products are now used as lubricants in almost all aircraft turbine engines.

Chlorofluorocarbons. The polymerization or telomerization of chlorotrifluoroethylene is the route to these relatively low-molecular-weight synthetic lubricants. Manufacture involves elaborate polymerization techniques and stabilization methods to eliminate all of the hydrogen and terminal chlorine introduced into the polymer chain by peroxide fragments or the chain transfer agent. The major characteristics of the chlorofluorocarbons are chemical inertness and thermal stability. Industrial and aerospace applications involving exposure to corrosive or oxidizing atmospheres are the largest uses for these lubricants.

Silicones.

$$\begin{matrix} & R & & R & & R \\ & | & & | & & | \\ & SiO & -\left(\!\! SiO \!\!\right)- & SiR \\ & | & & | & & | \\ & R & & R & & R \end{matrix}$$

where the R's may be the same or different organic groups. The properties are varied by the use of different types of organic substituents; the most popular are methyl, phenyl, and chlorophenyl groups. Recent advances involve fluorine-containing substituents. Manufacture entails the preparation of organochlorosilane intermediates, hydrolysis and condensation of these intermediates, and polymer finishing. In addition to good stability and low volatility, silicones have the best viscosity-temperature characteristics of any lubricant. Although they perform well under many conditions of lubrication, silicone lubricants are generally unsatisfactory for situations involving sliding contact of steel-on-steel. The many and varied uses include lubricating electric motors, precision equipment, plastic and rubber surfaces, and as greases for antifriction bearings.

Silicate Esters. ROSi(OR')(OR'')(OR'''), where the R's may be similar or dissimilar groups. The best-known types are the tetraalkyl, tetraaryl, and mixed alkylaryl orthosilicates. The classic means of preparation is through the reaction of phenol or alcohol with silicon tetrachloride. A closely related group of products, the hexaalkoxy- and hexa-aryloxydisiloxanes, is also generally included in the silicate esters classification. These products, the so-called "dimer silicates," are conveniently made by the reaction of an alcohol or phenol with hexachlorodisiloxane. Notable characteristics of the silicate esters are low volatility, low-temperature fluidity, and thermal stability. The hydrolytic stability varies from poor to good, depending upon chemical structure. The products are used as high-temperature heat-transfer fluids, wide-temperature range hydraulic fluids, electronic coolants, and automatic weapon lubricants.

Neopentyl Polyol Esters. These polyesters are prepared by the esterification of 5-carbon polyfunctional alcohols with mono-functional acids. Because the beta carbon of the starting alcohol does not contain hydrogen, these esters are superior in thermal stability to the diesters. Most of the other characteristics are similar to those of the diesters. As a result of their superior stability, the neopentyl esters are finding increasing use as the lubricant for aircraft turbine engines.

Polyphenyl Ethers. Both alkyl-substituted and unsubstituted polyphenyl ethers are included in this class of synthetic lubricants. General preparation involves the Ullman ether synthesis. The unsubstituted polyphenyl ethers have outstanding thermal, oxidative and radiation resistance, however, poor low-temperature characteristics are a major drawback. Alkyl substitution improves low-temperature viscosity, but detracts from stability. Most lubricant uses are developmental in nature and involve aircraft and aerospace applications.

Lubrication

Lubrication is the process of separating two rubbing solid surfaces by means of a layer which is effectively "softer" than either surface. Depending upon circumstances, the "soft" layer (the lubricant) may be

a gas, a liquid, a solid, or a combination of various phases. The many ways of accomplishing lubrication can be conveniently grouped into two categories: hydrodynamic and solid lubrication.

Hydrodynamic Lubrication. The separation of moving surfaces by a "fluid" (gas, liquid, or gel) is accomplished according to the laws of hydrodynamics. One of the surfaces "swims" on the lubricant, i.e., it is lifted through the simultaneous possession of velocity relative to the lubricating fluid and of an acute angle of attack. This so-called glider bearing is the basis of design of nearly all fluid-lubricated bearings, including the journal bearing, which is just a circular glider bearing.

The basic relation which expresses the load-carrying ability of a glider bearing can be given in the form:

$$W/\eta U = a(r/h) \tag{1}$$

where W is the load on the glider, U the velocity of the glider relative to the stator, h is the minimum distance between the rubbing surfaces, r is a dimension characterizing the angle of attack, a is a numerical coefficient (somewhere between 1 and 10) and η is the viscosity of the lubricating fluid under the temperature and pressure conditions of the application. The energy expenditure required for the service performed by the lubricant film is usually expressed in terms of the friction coefficient, f, defined as the ratio of the frictional force required to move the glider (or the journal) to the load carried by it. For the case under discussion:

$$f = K\sqrt{\eta U/W} \tag{2}$$

where K is a numerical coefficient which depends upon the geometry of the system and varies between about 2 and 6.

While Equations (1) and (2) provide only a qualitative guide— the detailed calculations for specific bearings are quite complicated— they clearly indicate that the only variable at the disposal of the chemist, the viscosity, will be chosen such as to give the maximum load-carrying ability of the bearing consistent with a reasonable amount of frictional energy lost in the lubricant.

The low friction losses (f is usually of the order 10^{-3}) and the naturally wear-free operation generally make the maintenance of hydrodynamic lubrication the primary aim of bearing design. This goal is attained perfectly only with journal and with glider (pad) bearings and to some extent in roller bearings. It can be achieved by a special mechanism (thermal expansion of the flowing fluid) in parallel thrust bearings. Its attainment is uncertain (and not readily subject to calculation) in ball bearings and in gear lubrication.

Since the viscosity of readily available fluid varies over a 10^{10}-fold range between gases and the thickest liquids, and the temperature and pressure coefficients of viscosity vary similarly over an about 10^4-fold range, the choice of a suitable lubricant is usually determined by the ancillary conditions of temperature, volatility, chemical stability, etc. The use of sulfuric acid as lubricant for oxygen compressors is a typical illustration of a choice dictated by chemical conditions. The evaporation and deterioration of liquid lubricants in high energy particle fluxes and/or in hard vacua ($<10^{-9}$ torr) has led to increased use and rapid development of gas lubricated bearings. These call generally for extremely high and therefore very costly standards of workmanship. Hence the use of gas lubrication is at present largely restricted to precision instruments and to military and space applications.

The most important source of lubricants is petroleum. There is hardly a chemical species (esters, ethers, sulfides, metal-or-ganic compounds, etc.) which has not contributed to the array of synthetic lubricants now available. Some chemicals, such as the perfluorocarbon compounds and the siloxane polymers were originally synthesized for just this service.

Elastohydrodynamic Lubrication. Well designed and equally well built machine elements with very smooth bearing surfaces permit the imposition of very high bearing loads. The resulting elastic deformation of the bearing may then change the geometry of the load-bearing surfaces substantially. Hence the elastic properties of the load-bearing materials enter the bearing calculations. The change of lubricant properties with pressure, temperature, and with the duration of exposure to the pressure and shear regime also enter the bearing calculations. The required viscoelastic properties of lubricants at high-frequency deformations are only beginning to be determined.

Hydrostatic Lubrication. Slow-moving or even static "gliders" can be separated from the bearing surface by a fluid film if the hydrostatic pressure required to carry the load is provided by an external source, as for instance by a pump. The very large journals of turbo generators, or heavy thrust bearing pads are generally lifted by these means before the onset of rotation in order to avoid scoring damage to the valuable bearings. Exceedingly low friction coefficients are obtained in these hydrostatic bearings, the most spectacular being that of the Mt. Palomar 200-inch (508 cm) telescope mirror pad bearings, where $f = 0.000004$, such that the 500-ton (450 metric ton) structure is easily moved by a $\frac{1}{2}$ HP clock motor.

Solid Lubrication. While fluid film separation of rubbing surfaces is the most desirable objective of lubrication, it is often unattainable, especially when bearings are too small or unsuitable for liquid lubricants. Even bearings built for full fluid lubrication during most of their operating periods experience solid-to-solid contact when starting and stopping.

Solid surfaces in rubbing contact are characterized by friction coefficients varying between 0.04 (Teflon on steel) and >100 (pure metals *in vacuo*). Solid lubrication, in contrast to fluid lubrication, is generally accompanied by a certain amount of wear of the rubbing parts. Optical inspection of the surfaces after rubbing reveals macroscopic (i.e., bulk) damage of the metal both when unlubricated and when lubricated. Quantitative differences between the two cases are easily measured radiographically when a radioactive glider has been used. In this manner it is found that effective solid lubrication can reduce the amount of wear—compared to the dry case— by a factor as high as 10^5.

Typical solid lubricants are the soft metals lead, indium, and tin, the layer lattice crystals graphite and molybdenum disulfide, many soft organic solids, such as metallic soaps, and waxes as well as the crystalline polymers Teflon (polytetrafluorethylene), polythene (polyethylene), and nylon. The integral bonding of these solids to the surface of the hard solid to be lubricated is essential for good performance. The bonding is accomplished either by alloying (copper-lead bearing metals), by flash coating, by introduction of the lubricating solid into the interstices of the sintered metal bearing (Teflon emulsion into sintered bronze), by chemical coating (phosphate coatings), or by anchorage to a phosphate or bonded plastic coating.

Special cases of solid lubrication are boundary and EP (extreme pressure) lubrication. In both cases the solid lubricant is formed by chemical reaction of special compounds, usually applied as oil solutions, with the metallic rubbing surfaces. Typical boundary lubricants are the fatty acids which react with the metal surface to form metallic soaps which then carry the load. Strongly adsorbed but nonreacting substances of linear structure, such as long chain fatty alcohols, can also act as boundary lubricants but only under very mild conditions.

Under the very severe conditions, sometimes encountered in automobile transmissions and especially in hypoid gear differentials as well as in machining operations, only those substances act as lubricants which contain chemically active chlorine, sulfur, or phosphorus to form the corresponding iron chloride, sulfide or phosphide by instantaneous attack on the surface hot spots resulting from the collisions of surface asperities. The chemical stability of these so-called E.P. agents is designed to permit activity at the temperature near the rubbing surface, say 200°C and above, but not be corrosive under normal conditions.

Mixed Film Lubrication. Mixed film lubrication is almost invariably the true state of affairs when boundary and EP lubrication are encountered, i.e., an appreciable fraction of the load is carried by the fluid film in the "valleys" of the surface while the asperities in contact are permitted to carry the balance of the load without seizure through the beneficent intervention of the boundary or EP lubricant. The very important breakin process of rubbing surfaces consists in the controlled reduction of the number and the size of the surface asperities so that fluid lubrication will prevail for most of the time.

"Real" Lubrication. In "real" lubrication of machinery, such as automotive engines, turbines, etc., the lubricating oil has to perform many functions besides lubrication. The most important of these is the cooling of the bearings. It also has to keep internal combustion engines clean by dispersing the partial combustion products of the fuel and its own degradation products, it has to carry chemicals to counteract wear, and— in common with many other lubrication applications— it must prevent corrosion of the equipment and be inhibited against its own deterioration in service. A relatively large amount of synthetic organic chemicals is therefore carried by many oils to perform the additional functions which one must expect from a modern lubricant.

Ideally one should always have full fluid separation of rubbing surfaces. But in inaccessible locations, or reactive environments, or under conditions of very slow motion or of intermittent operation, recourse must be had to solid lubrication.

HENRY E. MAHNCKE
King of Prussia, Pennsylvania. (Lubricating Oils)

REIGH C. GUNDERSON
The Dow Chemical Co.
Midland, Michigan. (Synthetic Lubricants)

A. A. BONDI
Shell Development Co.
Houston, Texas. (Lubrication)

LUMINESCENCE. A characteristic nonthermal emission of electromagnetic radiation by a material upon some form of excitation. Some luminescent materials are called phosphors. E. Wiedemann defined the term in 1888 as "all those phenomena of light not solely conditioned by the rise in temperature."

Whereas the output from blackbody radiators consists of broad-band emissions which follow the Stefan-Boltzmann temperature relationships, luminescence emission from phosphors consists of relatively narrow bands, which do not follow the blackbody laws. Thus, light emission due solely to the temperature of a source is referred to as *incandescence*, while *luminescence*, unlike incandescence, is a function of the specific material involved. Although *fluorescence* and *phosphorescence* are sometimes used synonymously with luminescence, a more rigid definition of *fluorescence* would be luminescence having a persistence (afterglow) shorter than about 10^{-8} second, with *phosphorescence* being longer than 10^{-8} second.

The luminescence process itself involves (1) absorption of energy; (2) excitation; and (3) emission of energy, usually in the form of radiation in the visible portion of the spectrum. The *type* of luminescence is usually defined by the excitation means, i.e., *cathodo*luminescence where excitation is by cathode rays, as in a television kinescope. The most commonly encountered types of luminescence are listed in Table 1.

The luminescent material may be considered as a transformer of energy, i.e., from ultraviolet photons to photons of lower energy; from cathode rays to photons; from electric fields to photons, etc. An inorganic luminescent material, or phosphor, usually consists of a crystalline host material to which is added a trace of an impurity (activator and coactivator).

Chemiluminescence

Numerous chemical reactions produce heat, but relatively few release their energy as light. This latter phenomenon is termed *chemiluminescence*. As pointed out by researchers at the University of Wales College of Medicine, Cardiff, "Absorption of energy by an atom or molecule raises an electron to a higher energy level. This is known as an 'excited state' and is inherently unstable. When the electron drops back to its ground state the energy must either be transferred to another atom or molecule, be released as heat or be emitted as light. The decay of the electron to ground state is very fast, occurring within $1-10$ nanoseconds ($10^{-9}-10^{-8}$ second)."

In chemiluminescence, the chemical reaction raises an electron to a higher level, which then decays back to the ground state, releasing a photon of light, the energy of which is predictable by Einstein's equation. When the energy drop is large, the light is blue; if small, the light is red. Inasmuch as the electronic excitation-decay process is extremely fast, the intensity of light in chemiluminescence is determined by the kinetics of the chemical reaction.

The distinction of chemiluminescence from fluorescence results from two factors:

1. When an atom or molecule fluoresces, it remains chemically unchanged and can be immediately be reexcited once light emission has occurred.

TABLE 1. TYPES OF LUMINESCENCE

Luminescence Type	Excitation Source	Example
Photoluminescence	Photons	ZnS Ag
Cathodoluminescence	Cathode Rays	$Zn_2SiO_4 \cdot Mn$
Electroluminescence	Electric Fields	$Zn (S \cdot Se) \cdot Cu$
Chemiluminescence	Chemical Reactions	Oxidation of Luminol
Bioluminescence	Biochemical Reactions	Luciferin
Triboluminescence	Mechanical Disruption	$ZnS \cdot Mn$

2. In chemiluminescence, each molecule only reacts once to form an excited state, while the excited product (actual emitter) has a different chemical structure from the initial substrate.

See also **Bioluminescence**.

Light-Emitting Diode (LED)

Recombination of injection electroluminescence was first observed in 1923 by Lossew, who found that when point electrodes were placed on certain silicon carbide crystals and current passed through them, light was often emitted. Explanation of this emission has been possible only with the development of semiconductor theory. If minority charge carriers are injected into a semiconductor, i.e., electrons are injected into *p*-type material or "positive holes" into *n*-type material, they recombine spontaneously with the majority carriers existing in the material. If some of these recombinations result in the emission of radiation, electroluminescence results. Minority-carrier injection may occur not only at point contacts, but also at broad area rectifying junctions; in this case, the junction must be biased in the forward or "easy flow" direction, and the electric field in the junction is lower when the voltage is applied than in its absence. This type of emission has been observed in several materials, including SiC, diamond, Si, Ge, CdS, ZnS, ZnSe, ZnO, and some of the so-called III–V compounds, such as AlN, GaSb, GaAs, GaP, InP, and InSb. The emission of many of these materials lies in the infrared region of the spectrum. For radiation in the visible region (instead of the infrared), the energy difference between the holes and electrons (band gap of the semiconductor) must be more than 1.8 eV. Numerous materials satisfy this requirement, notably those used for cathode-ray tube phosphors, but the materials present difficulties in fabricating *p-n* junctions and thus are not candidates for light-emitting diodes.

The list of materials for LEDs includes GaP, GaAsP, GaAlAs, GaN, and SiC. The two materials of choice to date have been GaP and GaAsP. Early commercial LEDs were made from $GaAs_{0.6}P_{0.4}$ deposited epitaxially as a thin layer on a GaAs crystal substrate. With these, *p-n* junctions were made, using diffusion techniques similar to those used in making silicon diodes. The band gap is 1.92 eV. There is an emission band of red light with a peak at about 650 nm, resulting from direct recombination of electrons and holes.

GaAsP has a high index of refraction, and consequently only light emitted toward the surface (4%) is usable—the remainder is reflected back. A diode can be encapsulated in epoxy material to take on the shape and form of a lens. These diodes are particularly effective where a number are fabricated in close proximity on a single-crystal chip.

Diodes that emit light in shorter wavelengths (green, yellow, etc.) can be made by increasing the phosphorus content, but only up to about 40% because of rapid decrease in efficiency. Efficiency can be increased by incorporating nitrogen atoms into crystals. The N atoms act as isoelectronic centers, trapping electron-hole pairs in an excited state. Three types of nitrogen-doped diodes have gained some importance: $GaAs_{0.65}P_{0.35}$ (orange light); $GaAs_{0.85}P_{0.15}$ (yellow light); and GaP (green light). Zinc and oxygen doping are also used. Diodes operate more efficiently if driven with periodic pulses of high current rather than with constant current. The short response time of junction diodes to current pulses (a fraction of a microsecond) and their rectifying property (they block current flow in the reverse, nonemitting direction) combine to make the diodes a good choice for *X-Y* addressing arrangements.

Additional Reading

Aitken, M.J.: *An Introduction to Optical Dating: The Dating of Quaternary Sediments by the Use of Photon-Stimulated Luminescence,* Oxford University Press, Inc., New York, NY, 1998.

Burgess, C. and D. Jones: *Spectrophotometry, Luminescence and Colour: Science and Compliance: Papers Presented at the 2nd Joint Meeting of the UV Spectrometry Group of the U.K. and the Council for Optical Radiation Measurements of the U.S.A., Rindge, NH, U.S.A., 20–23,* Elsevier Science, New York, NY, 1995.

Mueller, G.: *Electroluminescence,* Vol. 64, Elsevier Science & Technology Books, New York, NY, 1999.

Ropp, R.: *Luminescence and the Solid State,* Elsevier Science, New York, NY, 1991.

Schubert, E.F.: *Light-Emitting Diodes,* Cambridge University Press, New York, NY, 2003.

Schulman, S.G.: *Molecular Luminescence Spectroscopy: Methods and Applications,* Vol. 3, John Wiley & Sons, Inc., New York, NY, 1993.

Stanley, P.E. and L.J. Kricka: *Bioluminescence and Chemiluminescence: Fundamental of Applied Aspects,* John Wiley & Sons, Inc., New York, NY, 1996.

Vij, D.R. and N. Singh: *Luminescence and Related Properties of II-VI Semiconductors,* Nova Science Publishers, Inc., Huntington, NY, 1997.

Vij, D.R.: *Luminescence of Solids,* Kluwer Academic Publishers, Norwell, MA, 1998.

Ziegler, M.M. and T.O. Baldwin: *Bioluminescence and Chemiluminescence, Part C,* Vol. 305, Academic Press, Inc., San Diego, CA, 2000.

LUTETIUM. [CAS: 7439–94–3]. Chemical element symbol Lu, at. no. 71, at. wt. 174.97, fourteenth and last element in the Lanthanide Series in the periodic table, mp. 1,663°C, bp 3.402°C, density 9.841 g/cm³ (20°C). Elemental lutetium has a close-packed hexagonal crystal structure at 25°C. The pure metallic lutetium is silver-gray in color and retains its luster at room temperature indefinitely. Although not as extensively studied as most of the other lanthanides, most of the basic properties of lutetium are known, and it behaves as a normal trivalent metal with no magnetic transitions because its $4f$ levels are completely filled. There are two natural isotopes ^{175}Lu and ^{176}Lu. The latter isotope is radioactive with a half-life of 2.2×10^{10} years. Fourteen artificial isotopes are known. The element was first identified by G. Urban in 1907 and independently by C.A. von Welsbach in 1908. Although not investigated fully, lutetium is classified with a low acute-toxicity rating. Lutetium is the least abundant of the Lanthanide elements, estimated as present on the average of 0.5 ppm in the earth's crust. Potentially, however, it is more plentiful than mercury, cadmium, or any of the precious metals. Electronic configuration

$$1s^2 2s^2 2p^6 3s^2 2p^6 3d^{10} 4s^2 4p^6 4d^{10} 4f^{14} 5s^2 5p^6 5d^1 6s^2$$

Ionic radius Lu^{3+} 0.848 Å. Metallic radius 1.735 Å. Other important physical properties of lutetium are given under **Rare-Earth Elements and Metals**.

The source of lutetium to date has been the processing of the other heavy rare-earth metals. Because of very limited availability, little research was conducted on lutetium until the mid-1960s. Most of these studies now are concentrating on prospective uses in phosphors, semiconductor, and other electronic circuitry components. A lutetium dithalocyanine complex has received much consideration recently for application in large, thin screens for television projection.

See references listed at ends of entries on **Chemical Elements**; and **Rare-Earth Elements and Metals**.

K. A. GSCHNEIDNER, JR.
B. EVANS
Iowa State University
Ames, Iowa

LUTIDINES. See **Coal Tar and Derivatives**; **Pyridine and Derivatives**.

LYOPHILIC. Characterizing a material that readily goes into colloidal suspension in a liquid; if into water, it is called *hydrophilic*. The colloid is stabilized by the formation of an adsorbed layer of molecules of the dispersing medium about the suspended particles. Systems of this type are said to be lyocratic. Examples include glue, gelatin, and milk-fat particles.

LYOPHOBIC. Characterizing a material that exists in the colloidal state but with a tendency to repel liquids: if the liquid is water, the material is called *hydrophobic*. Such colloids are generally stabilized by the adsorption of ions and coagulate when the charge is neutralized. Examples include colloidal gold, and colloidal arsenic sulfide.

LYSINE. See **Amino Acids**.

M

MACERATE. (1) To soften or break up a fibrous substance by long soaking in water at or near room temperature, often accompanied by mechanical action, as in the preparation of paper stock in the beater. (2) In the plastics industry, to comminute a fabric so that it can be used as a filler in a plastics composition. (3) The term is also used in pharmacy to describe a method of preparing medicinal compositions.

MACROMOLECULAR SCIENCE. Macromolecules or polymers are now such a common feature of modern life that it is sometimes forgotten that the rapid development of such materials in industry has taken place in the last 60 years. In fact, the idea that such molecules exist was proposed by Staudingser, Mark, and coworkers in the 1920s and it was some time before these concepts were accepted by the scientific community.

Polymers are long-chain molecules synthesized by the linking together of a large number of identical or similar small units termed *monomers*. The large size of macromolecules leads to their many useful properties. Macromolecular science is the study of these properties in relation to their chemical and physical structure. Naturally occurring or biological macromolecules, such as proteins, carbohydrates, nucleic acid, and natural rubber, are studied by the same methods as synthetic polymers, resulting in a number of industrial and biomedical applications.

In response to the needs of industry, a number of centers for graduate study in macromolecular science have developed. The first such center was at Brooklyn Polytechnic Institute (now Polytechnic University) and large centers now exist at Case Western Reserve University in Cleveland Ohio (Fig. 1), University of Massachusetts (Amherst) and University of Akron (Ohio). The first accredited engineering undergraduate degree program in polymer science was developed in the Department of Macromolecular Science at Case Western Reserve University.

In the past, research emphasis has centered on the synthesis of new macromolecules. This is still a major emphasis, especially as applied to the production of new polymers or composite systems having unique electrical and/or mechanical properties and to the synthesis and fabrication of unique transport and barrier membranes. Such research areas relate directly to the current emphasis on high tech, high price products rather than on bulk commodity polymers.

At least equally important in aiding in the solution of current problems of society has been the development, primarily over the past ten years, of the fields of polymer physics and engineering. The general thrust of this development has been to gain an understanding of the physical structure and morphology of macromolecules and to develop methods to modify these parameters to produce useful properties in both polymer and composite systems. See accompanying illustration. See also **Macromolecule**; **Molecular and Supermolecular Electronics**; and **Colloid Systems**.

JOHN BLACKWELL
Case Western Reserve University
Cleveland, Ohio

MACROMOLECULE. A molecule, usually organic comprised of an aggregation of hundreds or thousands of atoms. Such giant molecules are in general of two types: (1) Individual entities (compounds) that cannot be subdivided without losing their chemical identity. Typical of these are proteins, many of which have a molecular weights running into the millions. (2) Combinations of repeating chemical units (monomers) linked together into chain or network structures called polymers. Each monomer has the same chemical constitution as the polymer, e.g., isoprene (C_5H_8) and polyisoprene (C_5H_8)$_x$. Synthetic elastomers (plastics) are typical of this kind of macromolecule. Cellulose is the most common example found in nature. Most macromolecules are in the colloidal size range.

See also **Colloid Systems**; and **Polymerization**.

Additional Reading

McPherson, A.: *Introduction to Macromolecular Crystallography,* John Wiley & Sons, Inc., New York, NY, 2002.
Munk, P. and T.M. Aminabhavi: *Introduction to Macromolecular Science,* 2nd Edition, John Wiley & Sons, Inc., New York, NY, 2002.
Pittman, C.U., M. Zeldin, and J.E. Sheats: *Macromolecules Containing Metal and Metal-like Elements: Biomedical Applications,* Vol. 3, John Wiley & Sons, Inc., Hoboken, NJ, 2004.
Sun, S.F.: *Physical Chemistry of Macromolecules: Basic Principles and Issues,* 2nd Edition, John Wiley & Sons, Inc., Hoboken, NJ, 2004.
Wohrle, D. and A.D. Pomogailo: *Metal Complexes and Metals in Macromolecules: Synthesis, Structure and Properties,* John Wiley & Sons, Inc., New York, NY, 2003.

MAGNESITE. The mineral magnesite is carbonate of magnesium, $MgCO_3$. It is a hexagonal mineral, but usually found massive. It has a rhombohedral cleavage; conchoidal fracture; brittle; hardness, 3.5–4.5; specific gravity, 3.75–4.25; luster, vitreous to dull; color, white, gray, yellow, or brown; transparent to opaque. Most magnesite is believed to have been derived from the action of carbonated waters upon rocks rich in magnesium. Magnesium-bearing waters, on the other hand, may have in some cases acted upon calcite or dolomite. Magnesite deposits are known in Greece, Austria, Norway, India, Australia, and the Republic of South Africa. In the United States, magnesite is found in California and Nevada, some of which deposits seem to be of original sedimentary character. Magnesite is in demand for the manufacture of refractories and various compounds of magnesium.

MAGNESIUM. [CAS: 7439-95-4]. Chemical element, symbol Mg, at. no. 12, at. wt. 24.305, periodic table group 2, mp 649°C, bp 1,090°C, critical temperature (calculated) 1,867°C, density 1.74 g/cm³ (20°C), 1.64 g/cm³ (solid at 650°C), 1.57 g/cm³ (liquid at 650°C). Elemental magnesium has a close-packed hexagonal crystal structure, as do the common alloys of magnesium except those that contain lithium in excess of 11%.

Fig. 1. Automated Langmuir-Blodgett deposition station used for production of ultrathin films and located in a clean room at the Polymer Microdevice Laboratory, Case Western Reserve University, Cleveland, Ohio

Magnesium is a silver-white metal, malleable and ductile when heated; unattacked by dry oxygen, by H_2O or alkalis at room temperature; when heated to about 800°C reacts in air or steam and emits a brilliant white light of high actinic power; reactive with acids including carbonic at room temperature; reactive upon heating with nitrogen, phosphorus, arsenic, sulfur, in some cases with such vigor as to constitute a hazard.

Magnesium occurs extensively in the earth's crust, ranking 8th among the chemical elements in terrestrial abundance. An average composition of igneous rocks contains 2.09% magnesium. Of the elements present in seawater, magnesium ranks 5th with an estimated 6,125,000 tons of magnesium per cubic mile (1,323,000 metric tons per cubic kilometer) of seawater, its content exceeded only by hydrogen, oxygen, sodium, and chlorine. Magnesium is a constituent of over 150 minerals and also is found in bitterns and subterranean brines and salt beds. Only a few magnesium minerals are important commercially, notably dolomite, $CaO \cdot MgO \cdot 2CO_2$, as a source of magnesium. See also **Dolomite**. More than half of metallic magnesium produced is extracted from seawater. There are three naturally occurring isotopes, ^{24}Mg through ^{26}Mg; and three radioactive isotopes have been identified, ^{23}Mg, ^{27}Mg, and ^{28}Mg, all with comparatively short half-lives measured in seconds, minutes, or hours. The first known magnesium compound to be isolated was Epsom salt, $MgSO_4$, which Nehemiah Grew obtained in 1695 by evaporating the mineral waters at Epsom, England. In 1754, Joseph Black demonstrated that magnesia and lime were two different substances, but the exact identify of magnesia was not reported until 1808 by Sir Humphrey Davy who demonstrated that magnesia was an oxide of a heretofore unknown element. He first termed the element magnium. Metallic magnesium was first isolated by A. Bussy in 1828 when he fused magnesium chloride with potassium. Michael Faraday produced the first magnesium metal electrolytically in 1883. First ionization potential 7.64 eV; second, 14.97 eV. Oxidation potential $Mg \longrightarrow Mg^{2+} + 2e^-$, 2.375 V; $Mg + 2OH^- \longrightarrow Mg(OH)_2 + 2e^-$, 2.67 V. Other important physical properties of magnesium are given under **Chemical Elements**.

Production

There are two principal magnesium production processes: (1) electrolytic, and (2) metallothermic reduction. Electrolytic processes account for 80% of commercial production. In this process, seawater is pumped into large settling tanks where it is treated with lime. Roasted oyster shells sometimes are used if a convenient source is nearby. The lime precipitates the magnesium as the insoluble hydroxide. The hydroxide is filtered and then converted into a slurry with fresh H_2O. Subsequent treatment with HCl converts the $Mg(OH)_2$ into $MgCl_2$. The latter compound is dried and then electrolyzed in the fused state to produce molten magnesium and chlorine gas. The latter is recycled. The magnesium is cast into ingots. In the thermal or ferrosilicon process, used in some European countries, a mixture of magnesium oxide and powdered ferrosilicon (an iron-silicon alloy) is fed into a retort and heated under vacuum to about 1,200°C. The magnesium is freed in the form of vapor and condenses into crystals at the cool end of the retort. The crystals then are remelted and cast into pigs.

Uses of Magnesium

Magnesium finds principal uses as a primary metal to which other metals are added in various alloying amounts to enhance the properties of magnesium. Magnesium is the lightest of all structural metals and consequently the metal has enjoyed much attention over the years in connection with the transportation industry, notably for applications in the aircraft, aerospace, and automotive industries. Vehicle designers constantly are aware of the additional power requirements for simply moving "dead weight" that wastes fuel and contributes to air pollution.

In addition to its use as a structural metal, magnesium is an important metallurgical chemical in the form of a deoxidizer and desulfurizer and as the constituent of numerous industrial and laboratory chemical compounds.

Magnesium Alloys. Even prior to the use of magnesium as a structural metal in the aerospace field, in 1921, Louis Chevrolet put a set of magnesium-alloy pistons in the Ford racing car that won the Indy 500 for him that year. The magnesium pistons gave racing and sports cars faster acceleration and deceleration. This application of magnesium was not intended so much as a dead-weight savings feature for the car, but rather more in terms of inertia (obviously also relative to weight). Although the magnesium pistons provided better acceleration/deceleration because of smaller inertia, the early designers encountered what is known as piston

slap, which results when the piston material has a considerably higher coefficient of thermal expansion than the cylinder material does.

The use of magnesium castings for auto wheels was introduced a few years later and also serves the principal purpose of reducing inertia. With wheels, it is not just faster acceleration/deceleration that can be achieved, but also minimizing the amount of unsprung weight for a smoother and easier-to-control ride and minimizing the problem of gyroscopic action of the rapidly spinning wheels. Designers of racing cars switched from wire-spoke wheels to magnesium-alloy wheels in the early 1950s. The use of magnesium has increased not only in racing cars, but some passenger cars, both for the purpose of reducing inertia and weight. Today, magnesium is used for transmission and differential housings and a variety of other racing car parts. Serious attention continues to be given to major engine components, such as the cylinder block, head, and oil sump, all of which are candidates for reducing dead weight and increasing fuel economy.

The most extensive use of magnesium castings in automobiles commenced in 1936, with the introduction of the Volkswagen Beetle. Each Beetle used from 40 to 50 pounds (18 to 23 kg) of primary magnesium ingot plus scrap metal.

Magnesium has a density only $\frac{2}{3}$ that of aluminum, $\frac{1}{4}$ that of zinc, and about $\frac{1}{5}$ that of irons and steels. In addition to the obvious aerospace and automotive applications, other applications include hand trucks, containers, materials-handling equipment, portable electric and pneumatic tools (such as chain saws), hand tools, luggage, sporting goods, dockboards, and tooling jigs and fixtures. It has been found that lighter-weight equipment significantly reduces accidents and lost time due to injuries. On an arbitrary scale, where the power required to machine magnesium alloys is 1.0, the Figures for other metals are: aluminum alloys, 1.8; brass, 2.3; cast iron, 3.5; mild steel, 6.3; and nickel alloys, 10.0.

Some magnesium alloys are listed in Tables 1 and 2.

Magnesium in Other Metal Alloys. Magnesium is an important alloying ingredient in the production of other base metal alloys. When added during metallurgical processing, magnesium in small amounts has a marked effect on final properties of the metals:

Aluminum—Magnesium increases resistance to corrosion, facilitates heat treatment, and increases most mechanical properties. If magnesium-containing aluminum is remelted, the magnesium may be lost and should be replaced by adding pure magnesium to the casting ladle or pot.

Copper—Magnesium improves tensile strength and allows age hardening. Magnesium is used mainly as a deoxidizer, notably in copper-nickel-zinc alloys and in leaded brasses and bronzes. The magnesium is added during melting.

Lead—Magnesium increases hardness, strength, and resistance to creep. Magnesium also is used as a debismuthizer in refining primary lead.

Nickel—Magnesium, in combination with carbon, forms an age-hardenable alloy. The main use of magnesium is to deoxidize and desulfurize the melts, including pure nickel, nickel-chrome, and nickel-copper alloys.

Tin—Magnesium increases hardness and tensile strength. The effect of magnesium on tin can be dramatic. However, too much magnesium will reduce corrosion resistance and ductility.

Zinc—Magnesium improves dimensional stability and reduces the intergranular corrosion of zinc die castings. Magnesium refines the grain and increases hardness and creep strength of zinc sheet. Magnesium also is used in zinc-base bearing metals and in zinc alloy metalworking dies.

Magnesium alloy extrusions have become very popular for numerous items in recent years. Extrusion is particularly attractive as a parts making method—when extruded parts and sheet can be easily joined to form an assembly, where the desired shapes are too costly to machine from castings, and where pieces cut from extrusions can replace individually cast or forged parts. Final products with outstanding performance qualities coupled with light weight include concrete hand finishing tools, tennis racquets, portable shelters for the military, snowshoes, and improved luggage, among others.

The use of magnesium composites has become popular for rotary engine parts. Rotary engines remain attractive for business aircraft, boats, industrial equipment and compressors, and well over a million rotary-engine-powered cars have been built. In a research program (NASA

TABLE 1. REPRESENTATIVE MAGNESIUM ALLOYS

Alloy designation	Elements added	Tensile strength 1,000 psi	Brinell hardness	Melting point°C	Forms available	Features
AZ31B	3% Al 1% Zn	29	49	627	Sheet, plate, extrusions, forgings.	Moderate strength, good formability, general-purpose alloy. Dent resistant, weldable.
AZ91B	9% Al 0.6% Zn	33	67	596	Die casting alloy.	Good strength and castability. Popular for portable tools, business machines, vehicles.
AZ91C	8.7% Al 0.7% Zn	40	53	596	General-purpose sand and permanent-mold casting alloy.	Good castability, pressure tightness, and weldability. Moderate strength.
HK31A	3% Th 0.7% Zn 0.7% Zr	38	57	649	Sheet and plate for aerospace uses. (200–370°C). Sand and permanent-mold castings.	Good short-time, elevated temperature characteristics. Weldable without stress relief. Low microporosity in cast form.
HM21A	0.6% Mn 2% Th	35	56	650	Sheet, plate, forgings for aerospace uses. (200–425°C)	Very stable at elevated temperatures. Good creep strength and formability. Weldable without stress relief.
HM31A	1.2% Mn (min) 3% Th	44	63	605	Extrusions for aerospace uses. (200–425°C)	Excellent elevated temperature properties. Weldable without stress relief.
QE22A	2% Pr 0.7% Zr 2.5% Ag	40	78	549	Castings for aerospace uses. (up to 260°C)	Superior tensile strength plus excellent creep and fatigue strength.
ZK60A	5.7% Zn 0.5% Zr	47	—	635	Highly stressed parts of aerospace and military uses. Used as a forging alloy.	High strength, good toughness, good spot-weldability. Limited arc-weldability.

Note: 1 psi(pounds/square inch) = 0.0069 megapascal (MPa). *Designation of Magnesium Alloys (an ASTM system now accepted by the SAE)*. A four-part system is used:

1. Letters indicate the two principal alloying elements: A, Aluminum; E, Rare-Earth; H, Thorium; K, Zirconium; M, Manganese; Q, Silver, S, Silicon; T, Tin; Z, Zinc. Thus HK signifies a thorium-zirconium magnesium alloy.
2. The approximate amounts (percent, wt) of the two principal alloying materials follow to the immediate right of the alloying element letters. Thus HK31 indicates approximately 3% thorium, and 1% zirconium.
3. The next two letter symbols to the right are used to distinguish two different alloys of the same chemical composition. Any letter may be used except I and O.
4. A fourth part of the designation (not indicated in this table) is separated by a dash from the foregoing parts and is used to indicate temper and other characteristics, such as F (as fabricated), 0 (annealed), H10 and H11 (slightly strain hardened), H23, H24, and H26 (strain hardened and partially annealed), T4 (solution heat treated), T5 (artificially aged only), T6 (solution heat treated and artificially aged), and T8 (solution heat treated, cold worked, and artificially aged). Thus, the complete designation may appear as: AZ91C-T6 for an aluminum-zinc-magnesium alloy containing 9% Al, 1% zinc, C indicating that this is the third alloy standardized with the same percentages of Al and Zn and T6 indicating that the alloy is solution treated and artificially aged.

TABLE 2. MAGNESIUM CASTING ALLOYS FOR AUTOMOTIVE APPLICATIONS

AM60B[2]   Die-casting alloy for uses needing toughness and ductility.

5.5–6.5% Al	0.25% Mn (min)	0.002% Ni (max)
0.010% Cu max	0.22% Zn (max)	0.10% Si (max)
	0.005% Fe (max)	

AZ91D[2]   Provides an optimum combination of properties with die castability.

8.3–9.7% Al	0.15% Mn (min)	0.02% Ni (max)
0.030% Cu (max)	0.35–1.0% Zn	0.10% Si (max)
	0.005% Fe (max)	

AZ91E[2]   A sand and permanent-mold casting alloy with properties and castability similar to AZ91B.

8.1–9.3% Al	0.17–0.5% Mn	0.0010% Ni (max)
0.015% Cu (max)	0.40–1.0% Zn	0.20% Si (max)
	0.005% Fe (max)	

ZE41A   A sand and permanent mold casting alloy for applications to 175°C (350°F). Low microporosity and good pressure tightness.

0.0% Al	0.15% Mn (max)	0.40–1.0% Zr
0.010% Cu (max)	0.75–1.75% Re	0.01% Ni (max)
	3.5–5.0% Zn	

ZC63   A proprietary sand and permanent-mold casting with properties Similar to ZE41A, but less expensive.

0.0% Al	0.25–0.75% Mn
2.4–3.0% Cu	5.5–6.5% Zn

Lewis Research Center) rotary engine parts are made from graphite-fiber-reinforced magnesium. An AZ91C magnesium alloy is reinforced by 30% (vol) graphite fibers.

Progress has been in the early 1990s toward the development of metal-matrix composites (MMCs) that blend liquid magnesium alloys with ceramic particles, such as silicon carbide (SiC) and alumina (Al_2O_3). The method is similar to methods that have been developed for aluminum composites in that blending is accomplished by way of a high-shear process. Major differences of the new process result from increased general reactivity of magnesium and the difference in surface chemistry between the Al-SiC and Mg-SiC systems.

The particulate-reinforced MMCs are lightweight and demonstrate a significant increase in modulus and tensile strength at both ambient and elevated temperatures of the unreinforced material. The process was announced in late 1992 by Magnesium Elektron Ltd., Manchester, U.K.

Chemistry and Compounds

The behavior of magnesium is intermediate between that of beryllium and the higher alkaline earths. While it reacts readily with halogens, oxygen, and sulfur to form halides, oxide, and sulfide, it reacts with cold water only when the formation of protective oxide is prevented by amalgamation. All its compounds are divalent. Its oxide does not react with water to form the hydroxide, and it does not normally form a peroxide. Its major difference from the higher elements of the group is its much greater number of complexes. Anhydrous magnesium halides, especially, combine easily with many oxygen-functional organic compounds to form addition compounds. These reactions usually suggest covalent or dative bonding (both electrons from the oxygen) of the magnesium. Magnesium salts often form amines and amine complexes, though these are less stable than beryllium complexes. Magnesium also forms some basic salts, and many more of its salts are hydrated than are those of the higher alkaline earths. The metal reacts with alkyl and aryl halides to form the Grignard reagents, through which many organic reactions are conducted. The Grignard reagents themselves form complexes with ethers, tertiary amines, tertiary phosphines and many other type compounds. See also **Grignard Reactions**.

Important compounds of magnesium include the following:

Magnesium Acetate. [CAS: 142-72-3]. (1) $Mg(C_2H_3O_2)_2$ or (2) $Mg(C_2H_3O_2)_2 \cdot 4H_2O$, colorless, crystalline aggregate or nonoclinic crystals; acetic acid odor. (1) Mp 323°C, d 1.42 (2) mp 80°C, d 1.45, soluble in water and dilute alcohol, formed by reaction of magnesium carbonate and acetic acid. Uses: The largest use of magnesium acetate is in the production of rayon fiber, a dye fixative in textile printing, as a deodorant, a disinfectant, and antiseptic in medicine, and as a reagent chemical.

Magnesium Acetylacetonate. $Mg(C_5H_7O_2)_2$, crystalline powder, slightly soluble in water, resistant to hydrolysis, a chelating nonionizing compound.

Magnesium amide, $Mg(NH_2)_2$, whitish to gray crystals, d1.40, decomposes when heated, formed by reaction of magnesium and ammonia under elevated pressure. Use: Catalyst for polymerization.

Magnesium Arsenate. (arsenic acid, magnesium salt), [CAS: 10103-50-1]. $Mg_3(AsO_4)_2 \cdot xH_2O$, white powder, when pure it is insoluble in water.

Magnesium Ammonium Arsenate. $MgNH_4AsO_4$, white precipitate, solubility 0.0013 molar, formed by reaction of soluble magnesium salt solution and sodium arsenate in the presence of excess ammonium hydroxide, and upon igniting yields magnesium pyroarsenate, $Mg_2As_2O_7$, white solid.

Magnesium Benzoate. [CAS: 553-70-8]. $Mg(C_7H_5O_2)_2 \cdot 3H_2O$, white crystalline powder, loses $3H_2O$ at 110°C, mp approximately 200°C, soluble in water and alcohol.

Magnesium Borate. $3MgO \cdot B_2O_3$ (orthoborate) or $Mg(BO_2)_2 \cdot 8H_2O$ (metaborate) transparent, colorless crystals or white powder, soluble in alcohol, acetic acid and inorganic acids, slightly soluble in water, formed by heating magnesium oxide and boric anhydride. Uses: Magnesium borate is used as a preservative, antiseptic in medicine, and fungicide. Magnesium boride, Mg_3B_2, brown solid, by reaction of boron oxide and magnesium powder ignited.

Magnesium Bromide. [CAS: 7789-48-2]. $MgBr_2 \cdot 6H_2O$, white solid, soluble, formed by reaction of magnesium carbonate and hydrobromic acid. Magnesium bromide is found in seawater, some mineral springs, natural brines, inland seas and lakes such as the Dead Sea and the Great Salt Lake, and salt deposits such as the Stassfurt deposits. The solubility of magnesium bromide is 101 g/100 mL of water at 29°C; soluble in alcohols and forms addition compounds with numerous organic substances such as alcohols. Uses: Magnesium bromide is used in medicine as a sedative in treatment of nervous disorders, in electrolyte paste for magnesium dry cells, and as a reagent in organic synthesis reactions. Magnesium carbonate, $MgCO_3$, white solid, K_{sp} 4.0×10^{-5}, formed by reaction of soluble magnesium salt solution and sodium carbonate or bicarbonate solution. Present in carbonate minerals and rocks, magnesite (more or less pure magnesium carbonate), dolomite (magnesium-calcium carbonate mixtures), dolomitic limestone. When ignited yields magnesium oxide and CO_2; when treated with acids yields the corresponding magnesium salt and CO_2, but with carbonic acid yields soluble magnesium bicarbonate. Magnesium bicarbonate, $Mg(HCO_3)_2$, colorless solution, by reaction of magnesium carbonate and carbonic acid, yields, upon boiling, magnesium carbonate and CO_2; magnesium ammonium carbonate, $(MgCO_3 \cdot NH_4)_2CO_3 \cdot 4H_2O$, white precipitate (soluble in ammonium chloride solution) by reaction of soluble magnesium salt solution and excess ammonium carbonate. Uses: Magnesium salts, heat insulation and refractory, rubber reinforcing agent, inks, glass, pharmaceuticals, dentrifice and cosmetics, free-running table salts, antacid, making magnesium citrate, filtering medium.

Magnesium Chlorate. [CAS: 10326-21-3]. $Mg(ClO_3)_2 \cdot 6H_2O$, white powder; very hygroscopic. D 1.8, mp 35°C (decomposes at 120°C), soluble in water; slightly soluble in alcohol. Uses: Defoliant, desiccant.

Magnesium Chloride. [CAS 7786-30-3]. (1) $MgCl_2$ (2) $MgCl_2 \cdot 6H_2O$, white solid, soluble, formed by reaction of magnesium carbonate (or hydroxide, oxide, or metal) and HCl, loses hydrogen chloride when heated, yielding magnesium oxychloride; anhydrous magnesium chloride, $MgCl_2$, white solid, soluble, formed: (1) by heating hydrated magnesium chloride crystals in a current of dry hydrogen chloride, and (2) by heating magnesium ammonium chloride, mp 712°C. Magnesium ammonium chloride, $MgCl_2 \cdot NH_4Cl \cdot 6H_2O$, white solid, soluble, when heated yields anhydrous magnesium chloride; magnesium potassium chloride, $MgCl_2 \cdot KCl \cdot 6H_2O$, white solid, soluble, when heated fuses to anhydrous magnesium potassium chloride; magnesium oxychloride, white solid, insoluble, formed (1) by heating hydrated magnesium chloride crystals, and (2) by mixing magnesium chloride solution and magnesium oxide. Uses: Source

of magnesium metal, disinfectants, fire extinguishers, fireproofing wood, magnesium oxychloride cement, refrigerating brines, ceramics, cooling drilling tools, textiles (size, dressing and filling of cotton and woolen fabrics, thread lubricant, carbonization of wool), paper manufacture, road dust-laying compounds, floor-sweeping compounds, flocculating agent, catalyst.

Magnesium Chromate. [CAS: 13423-61-5]. $MgCrO_4 \cdot 5H_2O$, small readily soluble, yellow crystals, formed by reaction of magnesium carbonate and chromic acid solution. Use: Since it does not produce a fusible alkaline residue when thermally decomposed, it is used as a corrosion inhibitor in the water coolant of gas turbine engines. Insoluble basic magnesium chromates also are available. Their potential applications are in the treatment of light metal surfaces.

Magnesium Citrate. $Mg_3(C_6H_5O_7)_2 \cdot 4H_2O$, white solid, soluble, formed by reaction of magnesium carbonate and citric acid.

Magnesium Fluoride. [CAS: 7783-40-6]. MgF_2, white precipitate, K_{sp} 6.5×10^{-9}, formed by adding sodium fluoride or hydrofluoric acid to a solution of magnesium salt. Uses: Ceramics, glass, single crystals for polarizing prisms, lenses and windows. Magnesium formate, $Mg(CHO_2)_2 \cdot 2H_2O$, colorless crystals, soluble in water; insoluble in alcohol and ether; combustible. Used in Analytical chemistry.

Magnesium Gluconate. [CAS: 3632-91-5]. $Mg(C_6H_{11}O_7)_2 \cdot 2H_2O$, white powder or fine needles, odorless, almost tasteless, soluble in water, combustible, formed by reaction of magnesia or magnesium carbonate dissolved in gluconic acid. Uses: Medicine, and vitamin tablets.

Magnesium Hydroxide. [CAS: 1309-42-8]. $Mg(OH)_2$, occurs naturally as the mineral brucite, white precipitate, K_{sp} 9.0×10^{-12}, formed by reaction of soluble magnesium salt solution and NaOH solution. Use: The principal use of magnesium hydroxide is in the pulp (qv) and paper (qv) industries. The main captive use is in the production of magnesium oxide, chloride, and sulfate. Other uses include, ceramics, sugar refining, pharmaceuticals (antacid, laxative), plastics, flame retardants/smoke supressants, residual fuel oil additive, sulfate pulp, uranium processing, dentrifices, in foods as frying agent, color retention agent, frozen desserts, and the expanding environmental markets for wastewater treatment and So_x removal from waste gases.

Magnesium Hypophosphite. [CAS: 10377-57-8]. $Mg(H_2PO_2)_2 \cdot 6H_2O$, white solid, soluble, formed by reaction of magnesium carbonate and hypophosphorous acid.

Magnesium Iodide. $MgI_2 \cdot 8H_2O$, white solid, soluble, formed (1) by reaction of magnesium carbonate and hydriodic acid, (2) anhydrous, by heating magnesium metal and iodine.

Magnesium Lactate. [CAS: 18917-93-6]. $Mg(C_3H_5O_3)_2 \cdot 3H_2O$, white solid, soluble, formed by reaction of magnesium carbonate and lactic acid.

Magnesium Iodide. MgI_2 can exist as two deliquescent and heat-sensitive compounds: the octahydrate, $MgI_2 \cdot 8H_2O$, and the hexahydrate, $MgI_2 \cdot 6H_2O$, soluble in alcohols and many other organic solvents, and forms numerous addition compounds with alcohols, esters, aldehydes, esters, and amines. Uses: Used in the deoxygenation of oxiranes into olefins and iodine. Anhydrous MgI_2 is used in a process for producing organometallic and organobimetallic compositions, which are important in the preparation of pharmaceutical and special chemicals.

Magnesium Methoxide. (magnesium methylate), $(CH_3O)_2 Mg$, colorless, crystalline solid, decomposes on warming, formed by reaction of magnesium and methanol. Uses: Dielectric coatings, a cross-linking agent to form stable gels, and catalyst.

Magnesium Molybdate. $MgMoO_4$, crystalline powder, soluble in water. Use: Electronic and optical applications.

Magnesium Nitrate. [CAS: 10377-60-3]. $Mg(NO_3)_2 \cdot 2H_2O$, white crystals, D1.45, mp 95–100°C, decomposes at 330°C, soluble in water and alcohol, deliquescent. Uses: A soluble form of magnesium nitrate is used as a fertilizer in states such as Florida, where drainage through the porous, sandy soil depletes the magnesium. Used as a prilling aid in the manufacture of ammonium nitrate and in Pyrotechnics. Another use is as an alternative to sulfuric acid in the purification of nitric acid.

Magnesium Nitride. [CAS: 12057-71-5]. Mg_3N_2, yellow solid, with moist air or water yields ammonia and magnesium hydroxide, formed by heating magnesium to a high temperature in nitrogen or NH_3 (hydrogen gas evolved).

Magnesium Oleate. [CAS: 1555-53-9]. $Mg(C_{18}H_{33}O_2)_2$, yellowish mass, soluble in linseed oil, hydrocarbons, alcohol, and ether, insoluble in water, combustible, formed by reaction of soluble magnesium salt solution and sodium oleate. Uses: Varnish driers, in dry-cleaning solvents (to prevent spontaneous ignition), emulsifying agent, and lubricant for plasticizers.

Magnesium Oxalate. $MgC_2O_4 \cdot 2H_2O$, white solid, insoluble, K_{sp} 8.6×10^{-5}, formed by reaction of soluble magnesium salt solution and ammonium oxalate solution.

Magnesium Oxide. [CAS: 1309-48-4]. MgO, white solid, reacts slowly with H_2O to form magnesium hydroxide, has cubic structure, absorbs CO_2 from the air to form magnesium carbonate, is readily soluble in acids, insoluble in alkalies; formed (1) by heating magnesium carbonate to high temperature (CO_2 gas evolved), (2) by heating magnesium hydroxide, nitrate, sulfate, or oxalate, (3) by burning magnesium metal in air or oxygen. Uses: Refactories, especially for steel furnace linings, polycrystalline ceramic for aircraft windshields, electrical insulation, pharmaceuticals and cosmetics, inorganic rubber accelerator, oxychloride and oxysulfate cements, paper manufacture, fertilizers, removal of sulfur dioxide from stack gases, adsorption and catalysis, semiconductors, and food and feed additive.

Magnesium Peroxide. [CAS: 14452-57-4]. MgO_2, white solid, insoluble in water, soluble in dilute acids with formation of hydrogen peroxide, formed by reaction of soluble magnesium salt solution and sodium or barium peroxide. Uses: Bleaching and oxidizing agent, in medicine for treating hyperacidity in the gastric intestinal tract and in the treatment of metabolic diseases such as diabetes and ketonuria.

Magnesium Ammonium Phosphate. $MgNH_4PO_4$, white precipitate, K_{sp} 2.5×10^{-12}, by reaction of soluble salt solution and sodium phosphate in the presence of excess ammonium hydroxide, upon igniting yields magnesium pyrophosphate, $Mg_2P_2O_7$, white solid.

Magnesium Phosphate, Dibasic. (Dimagnesium orthophosphate; dimagnesium phosphate; magnesium phosphate, secondary; magnesium hydrogen phosphate), [CAS: 7782-75-4]. $MgHPO_4 \cdot 3H_2O$, white, crystalline powder, D 2.13, loses water at 205°C, decomposes at 550–650°C, decomposes to pyrophosphate on heating, soluble in dilute acids, slightly soluble in water. Uses: Stabilizer for plastics, food additive, and medicine (laxative).

Magnesium Phosphate, Monobasic. (Magnesium biphosphate; acid magnesium phosphate; magnesium tetrahydrogen phosphate), $MgH_4(PO_4)_2$ $\cdot 2H_2O$, white, hygroscopic, crystalline powder, decomposes to metaphosphate on heating, soluble in water and acids, insoluble in alcohol. Uses: Fireproofing wood, and as a stabilizer for plastics.

Magnesium Phosphate, Tribasic. (Magnesium phosphate, neutral; trimagnesium phosphate), $Mg_3(PO_4)_2 \cdot 8H_2O$ or $4H_2O$, soft, bulky, white powder, odorless, tasteless, loses all water at 400°C, soluble in acids, insoluble in water, formed by reaction of magnesium oxide and phosphoric acid at high temperatures. Uses: Dentifrice polishing agent, pharmaceutical antacid, adsorbent, stabilizer for plastics, food additive and dietary supplement. Magnesium salicylate, $Mg(C_7H_5O_3)_2 \cdot 4H_2O$, white solid, soluble in water and alcohol, formed by action of salicylic acid on magnesium hydroxide. Use: Medicine (antiinfective).

Magnesium Silicide. [CAS: 22831-39-6]. Mg_2Si, bluish crystals, mp 1085°C, d 1.9, decomposes on heating above 500°C, also by water and hydrochloric acid, formed by heating magnesium powder with silicon in ratio of 20:6. Uses: Semiconductor technology, and electrical equipment.

Magnesium Stannate. [CAS: 12032-29-0]. $MgSnO_3 \cdot 3H_2O$, white crystalline powder, decomposes at 340°C, soluble in water. Use: Additive in ceramic capacitors.

Magnesium Stannide. Mg_2Sn, blue-white crystals, Mp 775°C, soluble in water and dilute hydrochloric acid, has electrical and magnetic properties. Use: Semiconductor technology, magnetochemistry, thermoelectric research.

Magnesium Stearate. [CAS: 557-04-0]. $Mg(C_{18}H_{35}O_2)_2$ or with one H_2O, soft, white, light powder, tasteless, odorless, insoluble in water and alcohol. Uses: Dusting powder, lubricant in making tablets, drier in paints and varnishes, flatting agent, in medicines, stabilizer and lubricant for plastics, emulsifying agent in cosmetics, and dietary supplement.

Magnesium Sulfate. [CAS: 7587-88-9]. (1) $MgSO_4$, (epsom salts) $MgSO_4 \cdot 7H_2O$, colorless crystals, very soluble in water, soluble in glycerol, sparingly soluble in alcohol, (1) formed by action of sulfuric acid on magnesium oxide, hydroxide, or carbonate, (2) mined in a high degree of purity. Uses: Primarily in the chemical and pharmaceutical industries, fireproofing, textiles (warp-sizing and loading cotton goods, weighting silk, dyeing and calico printing), mineral waters, catalyst carrier, ceramics, fertilizers, paper (sizing), cosmetic lotions, and as a dietary supplement.

Magnesium Sulfide. [CAS 1032-36-9]. MgS, red-brown, crystalline solid, decomposes above 2000°C, decomposes in water. Use: Source of hydrogen sulfide, laboratory reagent.

Magnesium Sulfite. $MgSO_3 \cdot 6H_2O$, white, crystalline powder, D 1.725, mp loses $6H_2O$ at 200°C, bp (decomposes), slightly soluble in water, insoluble in alcohol, formed by action of sulfurous acid on magnesium hydroxide. Use: Manufacture of paper pulp (as bisulfite).

Magnesium Tungstate. [CAS: 13573-11-0]. (magnesium wolframate) $MgWO_4$, white crystals, D 5.66, soluble in acids, insoluble in water and alcohol, formed by interaction of solutions of magnesium sulfate and ammonium tungstate. Use: Fluorescent screens for X-rays, luminescent paint.

Magnesium Zirconium Silicate. $MgZrSiO_5$, or $MgO \cdot ZrO_2 \cdot SiO_2$, white solid, mp 1760°C, d 80 lb/ft³, insoluble in water and alkalies, slightly soluble in acids. Use: Electrical resistor, ceramics, glaze opacifier.

Additional Reading

Avedesian, M.M. and H. Baker: *Magnesium and Magnesium Alloys,* ASM International, Materials Park, OH, 1999.

Davis, J.R.: *Metals Handbook,* 2nd Edition, ASM International, Materials Park, OH, 1998.

Greenwood, N.N. and A. Earnshaw: *Chemistry of the Elements,* 2nd Edition, Butterworth-Heiemann, Inc., Woburn, MA, 1997.

Kainer, K.U.: *Magnesium Alloys and Their Applications,* John Wiley & Sons, Inc., New York, NY, 2000.

Kainer, K.U.: *Magnesium Alloys and Technologies,* John Wiley & Sons, Inc., New York, NY, 2003.

Kainer, K.U.: *Magnesium: Proceedings of the 6th International Conference Magnesium Alloys and their Applications,* John Wiley & Sons, Inc., New York, NY, 2004.

Kaplan, H.I., J. Hryn, and B. Clow: "Magnesium Technology 2000: Proceedings of the Symposium Sponsored by the Light Metals Division of the Minerals, Metals and Materials Society (TMS) and the International Magnesium," Warrendale, PA, 2000.

Krebs, R.E.: *The History and Use of Our Earth's Chemical Elements,* Greenwood Publishing Group, Inc., Westport, CT, 1998.

Lide, D.R.: *CRC Handbook of Chemistry and Physics,* 84th Edition, CRC Press, LLC., Boca Raton, FL, 2003.

MAGNESIUM (In Biological Systems). Magnesium is an integral part of the molecule of chlorophyll, the green pigment in plants that absorbs solar energy. See also **Chlorophylls.** Magnesium deficiency is a fairly common cause of poor crop yields, especially among crops produced on sandy soils. Magnesium is a prosthetic ion in enzymes that hydrolyze and transfer phosphate groups. Hence it is essential for energy-requiring biological functions, such as membrane transport, generation and transmission of nerve impulses, contraction of muscles, and oxidative phosphorylation. See also **Phosphorylation (Oxidative).** Magnesium is essential for the maintenance of ribosomal structure and thus protein synthesis. Magnesium may be related to the incidence of ischemic heart disease among Western populations.

The accumulation of magnesium from the soil by plants is strongly affected by the species of plant. The leguminous plants, such as clovers, beans, and peas, usually contain more magnesium than grasses, tomatoes, corn (maize), and other nonleguminous plants, regardless of the level of available magnesium in the soil where they grow.

A very high level of available potassium in the soil interferes with the uptake of magnesium by plants, and magnesium deficiency in plants is often found in soils that are very high in available potassium. High levels of available potassium may occur naturally, especially in soils of subhumid and semiarid regions; or they may be caused by heavy applications of certain commercial fertilizers or animal manure. On sandy and loamy soils, applications of magnesium fertilizers are often effective in increasing crop yields and the concentration of magnesium in the crop, but on fine-textured, clay-containing soils, especially those with substantial reserves of potassium, the application of a magnesium fertilizer may not cause higher magnesium concentration in crops. Since magnesium is not a highly toxic element in either plants or animals, precautions against its overuse are rarely necessary. When animals are fed diets primarily of grains, a proper balance among magnesium, calcium, and phosphorus should be maintained to minimize danger from urinary calculi.

The biological functions of magnesium, such as its essential role as a nutrier, its activation of enzyme systems, and its pharmacological properties, have been widely investigated. Nevertheless, some aspects of its critical physiological role remain obscure.

Distribution in System

Magnesium, primarily an intracellular ion, is distributed among all tissues. It constitutes about 0.05% of the animal body and, of this, 60% occurs in the skeleton and only 1% in extracellular fluids.

Reported serum magnesium values for most species range from 1.0 to 3.5 meg/liter, with a mean value of about 2. Between 65% and 80% of the plasma magnesium is ultrafilterable, and most of this exists as the free ion. The nonfilterable portion is reversibly bound to plasma protein. Cerebrospinal fluid contains slightly more than plasma. Interstitial fluid is similar to plasma ultrafiltrate.

The magnesium content of soft tissues varies from 0.06 to 0.13% of dry weight and remains remarkably constant regardless of the magnesium status of the animal. Normally, the intracellular concentration is more than 20 times that of the interstitial fluid, and the highest concentration occurs in the cell nucleus. Maintenance of such a large concentration gradient across the cell membrane suggests an active transport mechanism.

In late 1990's, R.R. Preston (University of Wisconsin–Madison) reported that recent reappraisals of the role of ionized magnesium in cell function suggests that many cells maintain intracellular free Mg^{2+} at low concentrations and that external agents can influence cell functions via changes in intracellular Mg^{2+} concentration. There is considerable evidence to suggest that intracellular free magnesium ions may be a key physiological regulator of cell activity.

The relatively large proportion of magnesium found in the skeleton, which amounts to about 0.6% of dry, fat-free bone, serves in part as a body reserve. It occurs largely as Mg^{2+} and $MgOH^+$ ions held by electrostatic attraction to the apatite crystal surface. During deficiency in young animals, 30% or more of bone magnesium can be mobilized for metabolic functions. Calcium ions appear to replace the magnesium that occupied the original adsorption sites.

Metabolism

The rate of absorption from the intestine exerts an important role in magnesium metabolism. Whereas *in vitro* studies show that magnesium absorption is positively correlated with the concentration of magnesium, it does not appear to be a purely passive process. Magnesium absorbed in excess of body needs is excreted primarily by way of the kidney. Urinary excretion is controlled primarily by a filtration-reabsorption mechanism so that magnesium appears in the urine only when glomerular filtration exceeds tubular reabsorption. Acute renal failure is accompanied by hypermagnesemia. In some species, considerable endogenous magnesium is lost by way of the feces, the amount depending upon the magnesium status of the animal and upon other dietary factors, such as the digestibility of the diet. The endogenous fecal magnesium in calves has been estimated at 3.5 milligrams/kilogram of body weight.

In contrast with the metabolism of calcium, no one endocrine gland exerts a primary regulatory function on magnesium. Thyroparathyroidectomy in dogs causes only a temporary lowering of plasma magnesium. Adrenalectomy causes a rise, whereas hyper-aldosteronism produces a fall in the plasma level. Administrative of deoxycorticosterone or aldosterone to sheep lowers the magnesium concentration in plasma. Magnesium-deficient animals exhibit a higher metabolic rate than normal, and the toxic effect of excess thyroxine is partially overcome by increasing the dietary level of magnesium.

Function

Although magnesium activates isolated enzymes, in most cases an absolute requirement is difficult to establish because the enzymes are partially active without added magnesium. The stimulating effect is not always specific for magnesium. In some cases, manganese or calcium will also activate the system.

Magnesium is particularly concerned with enzyme-catalyzed reactions involving the cleavage of phosphate esters and the transfer of phosphate groups. Magnesium ions activate phosphatases and the phosphorylation reactions involving adenosine triphosphate (ATP). Among the latter group may be mentioned glucokinase, phosphoglucokinase, phosphofructokinase, myokinase, creatine transphosphorylase, arginine transphosphorylase, and flavokinase. It has been suggested that an ATP-Mg complex is the active substrate inasmuch as ATP forms a 1:1 complex with magnesium and maximal activation occurs when the ATP: Magnesium ration is 1. Alkaline phosphatases, pyrophosphatases, and ATPase are activated by magnesium, as are enolase, certain peptidases, and pyruvic oxidase. Since magnesium is tied to ATP utilization, it follows that magnesium plays a role in important metabolic processes, including the synthesis of protein, fat, and nucleic acids, and in the trapping and utilization of energy derived from catabolism of carbohydrate and fat.

There is little change in magnesium concentration of soft tissues from deficient animals even at the point of expiration. This does not preclude the possibility that a small component of the cell, such as the nucleus or a cell particulate, is deprived of its critical level, but the dramatic drop in extracellular magnesium suggests that a function outside the cell is of greatest significance. It appears that tetany and convulsions in deficient animals result from a derangement of neuromuscular transmission. Magnesium ion possesses strong pharmacological properties, depressing both the central and peripheral nervous systems. These effects are counteracted by calcium. In the presence of normal calcium levels, a reduction of extracellular magnesium is believed to increase the release of acetylcholine and to decrease the rate of its hydrolysis. Such effects would increase the irritability of the neuromuscular system.

Magnesium generally has not been considered a major factor in bone formation and strength, but recent studies suggest closer attention be given to dietary levels of magnesium in this regard. Because of the close interrelationship with calcium, it is not surprising to see research findings of magnesium interfering with calcium entry into cells of the islets of the pancreas in studies of diabetes. The recognized presence of magnesium as part of numerous enzyme systems has led to observations of the reduction in carbohydrate metabolism associated with a deficiency and to beneficial effects in reducing blood cholesterol and lipids associated with other dietary agents when supplemental magnesium is added to the diet. The relationship to calcium also shows up in a study showing that adding magnesium to the rations of laying hens causes an increase in shell thickness, with a consequent reduction in the number of broken eggs.

Magnesium-Induced Diarrhea

It is well established that an excessive intake of magnesium causes diarrhea. This source of diarrhea is difficult to differentiate from other causes of diarrhea. Consequently, the diagnosis may be long and costly unless the physician questions a patient on possible excessive magnesium intake, which may result from large dosages of antacids or off-the-shelf food supplements.

Pathology of Magnesium Deficiency

Although there are numerous clinical symptoms, two cardinal aspects of pathology have been observed in all species of higher animals. These are hyperirritability and soft tissue calcification. While there are species differences as to the dominating syndrome, this is determined in part by the severity of the deficiency. Metastatic calcification is more likely to occur in a chronic deficiency in which the animal does not succumb at an early age. Hyperirritability, terminating in convulsions and death, has been observed in rat, rabbit, pig, calf, chick, and duck. Magnesium deficiency in humans is characterized by muscle tremors and twitching, often accompanied by delirium and occasionally by convulsions. The guinea pig, calf, dog, and cotton rat are prone to metastatic calcification and develop grossly visible deposits in and around joints, along the muscles of the rib cage, and also in the heart, great vessels, and other critical organs. Most soft tissues show an elevated ash content and marked histopathology. Some researchers have hypothesized that long-term intakes of marginal dietary levels of magnesium may be related to the incidence of ischemic heart disease.

The first clinical symptom of magnesium deficiency is a hypomagnesemia which occurs in cattle and less frequently in sheep and is described by such names as grass tetany, grass staggers, lactination tetany, and wheat pasture poisoning. It is observed most frequently when animals are first grazed on lush grass or wheat pastures. The disease is characterized by irritability, tetany, and convulsions, and all animals have subnormal plasma magnesium. Symptoms can be relieved by administration of magnesium salts and cam be prevented by providing extra magnesium in the diet.

Hypomagnesemia, often associated with hypocalcemia, is frequently encountered in heavy users of alcohol. Alcoholism sometimes is ascribed to impaired intestinal calcium absorption.

Nutritional Requirements and Dietary Supplementation

As is true of many mineral nutrients, the requirement for magnesium is affected by other dietary constituents, by the age and species of the animal, and by the criterion of adequacy applied. An allowance for magnesium

TABLE 1. DISTRIBUTION OF MAGNESIUM IN VARIOUS FOOD GROUPS

Group (types of samples tested)	Magnesium concentration (milligrams/ 100 grams (wet))	Magnesium-calorie ratio (Micrograms/kilocalorie)
Milk products (cheeses, ice cream, milk, puddings)	6.8–25.7	18–198
Meat and meat alternates (chicken, dried beef, eggs, fish, sausage)	9.8–37.6	20–353
Vegetables (cabbage, carrot, onion, turnip)	6.7–20.6	196–1000
Breads and cereals (buns, cereals, cornbread, crackers, croutons, English muffins, pasta, taco shells)	10.6–126.0	27–325
Baked desserts (cakes, cookies, doughnuts, pastries, sweet rolls)	4.6–53.2	18–307
Candies	21.8–89.9	63–225

has been included in the Recommended Dietary Allowance since 1968. Calcium and magnesium have an important effect upon magnesium availability. Either of these ions in excess increases the requirements for magnesium, and their effects are additive. Since calcium is known to compete with magnesium pharmacologically, it is reasonable to believe that it also competes with magnesium for absorption sites in the intestine. It is believed that phosphate decreases magnesium absorption by formation of insoluble magnesium phosphates, and excess of calcium aggravates the effect of creating a more alkaline intestinal medium. Excess magnesium can be considered toxic, but this effect is largely due to the induction of a calcium deficiency. Magnesium deficiency in humans generally has not been fully documented except in cases of predisposing and complicating disease states.

Distribution of magnesium in food classes is summarized in Table 1.

Additional Reading

Sigel, H. and A. Sigel: *Metal Ions in Biological Systems: Compendium on Magnesium and Its Role in Biology, Nutrition, and Physiology,* Marcel Dekker, Inc., New York, NY, 1990.

Theophanides, T.M. and J. Anastassopoulou: *Magnesium, Current Status and New Developments: Current Status and New Developments: Theoretical, Biological, and Medical Aspects,* Kluwer Academic Publishers, Norwell, MA, 1997.

Tsang, R.C.: *Calcium and Magnesium Metabolism in Early Life,* CRC Press, LLC., Boca Raton, FL, 1995.

Vedral, J.L.: *Dietary Reference Intakes: For Calcium, Phosphorus, Magnesium, Vitamin D, and Fluoride,* National Academy Press, Washington, DC, 1997.

Web References

NIH Clinical Center National Institutes of Health: http://www.cc.nih.gov/ccc/supplements/magn.html

International Magnesium Association: http://www.intlmag.org/
Magnesium Update Information: http://www.krispin.com/magnes.html
The Magnesium Home Page: http://www.members.tripod.com/Mg/

MAGNETIC MATERIALS.

BULK

All materials that are magnetized by, i.e., exhibit a response in magnetic field are magnetic materials and are classified according to the nature of the response, e.g., as ferromagnetic or ferrimagnetic, the latter typified by the ferrites. Most commercially important magnetic materials are ferromagnets and ferrimagnets (see also **Ferrites**).

Soft Magnetic Materials

Soft magnetic materials are characterized by high permeability and low coercivity. There are six principal groups of commercially important soft magnetic materials: iron and low carbon steels, iron–silicon alloys, iron–aluminum and iron–aluminum–silicon alloys, nickel–iron alloys, iron–cobalt alloys, and ferrites. In addition, iron–boron-based amorphous soft magnetic alloys are commercially available. Table 1 summarizes the properties of some of these materials. Table 2 summarizes properties of some ferrites. Properties of amorphous soft magnetic alloys are listed in Table 3.

Uses of Soft Magnetic Materials. Because of low coercivity and high magnetic permeability, iron and low carbon steels tend to be used in static applications. Low carbon steels and the lower grade Fe–Si alloys are used in small motors and generators. The higher grade Fe–Si alloys have traditionally been used in power distribution transformers and large rotating machinery, but certain economical amorphous iron–metalloid alloys, because of their lower resistivity, are increasingly being used in the manufacture of distribution transformers by General Electric, Westinghouse, and Osaka. Ni–Fe alloys, used widely in high quality relays, electronic transformers, converters, and inverters in the electronics industry, have much higher permeability and much lower resistivity than Fe–Si alloys. Soft ferrites (oxides) are suitable for high frequency applications. The Co–Fe alloys are used because of their higher saturation polarization (flux density) electrical resistivity and Curie temperature compared to the iron–nickel alloys, but have the disadvantage of poorer workability and higher cost. Thus, they are used in special applications.

Soft magnetic ferrites are oxides and they are electrical insulators. Because of their exceptionally higher resistivities, ferrites are particularly suitable for high frequency applications, of about 100,000 cycles (10 kHz).

Hard Magnetic Materials

Hard or permanent magnetic materials are characterized by high coercivity and high energy product. The important commercial hard magnetic materials are hard ferrites such as ferroxdure (Table 4), rare-earth (R)-cobalt alloys, and the ternary alloys based on $Nd_2Fe_{14}B$. The last exhibits the highest coercivities and energy products. The use of Alnico and the

TABLE 1. MAGNETIC PROPERTIES OF FULLY ANNEALED IRON AND IRON ALLOYS

Iron and alloys	B_s, T[b]	d, g/cm³	Resistivity, $\mu\Omega \cdot cm$	H_c (B_m = 1 T),[b] A/cm[a]	Permeability, A/cm[a] $H = 0.8$	Permeability, A/cm[a] $H = 8$	Core loss (1.5 T,[b] 60 Hz), W/kg[c] 0.35 mm	Core loss (1.5 T,[b] 60 Hz), W/kg[c] 0.46 mm	Core loss (1.5 T,[b] 60 Hz), W/kg[c] 0.64 mm
magnetic ingot iron cast	2.15	7.85	10.7	0.68	3,500	1,500			
0.2-cm sheet	2.15	7.85	10.7	0.88	1,800	1,575			13.20
electromagnet iron, 0.2-cm sheet	2.15	7.85	12.0	0.81	2,750	1,575			
hydrogen-annealed iron	2.15	7.85	10.1	0.04	14,000	1,580			
low carbon steel, decarburized	2.14	7.85	12.5	0.70	2,000	1,530	8.10	9.2	11.44
cold-rolled M36 Si–Fe	2.04	7.75	41.0	0.36	7,400	1,485		3.85	4.73
M22 Si–Fe	1.98	7.65	49.0	0.31	8,100	1,450		3.63	4.29
M6 (110)[001] 3.2% Si–Fe	2.03	7.65	48.0	0.06	16,000	1,820	1.45		

[a] To convert A/cm to Oe, divide by 0.7958.

[b] To convert T to G, multiply by 10^4.

[c] At thickness shown.

TABLE 2. CHARACTERISTICS OF FERRITES

Property	MnZn ferrites									NiZn ferrites		
code	H5A	H5B	H5C2	H5E	H6F	H6H3	H6K	H7C1	H7C2	K5	K6A	K8
practical frequency, MHz	<0.2	<0.1	<0.1	<0.01	0.2–2.0	0.01–0.8	0.01–0.3	<0.3	<0.2	<8	1–50	<200
initial permeability, μ_0	3,300	5,000	10,000	18,000	800	1,300	2,200	2,500	3,900	290	25	16
relative loss factor, $\tan\delta/\mu_i \times 10^6$, at (kHz)	<2.5 (10)	<6.5 (10)	<7.0 (10)		<17 (1,000)	<1.2 (100)	<3.5 (100)			<28 (1,000)	<150 (10,000)	<250 (100,000)
temperature coefficient of $\mu_i \times 10^6$ from -30 to $20°C$, $(\mu_2-\mu_1)/\mu_1^2(T_2-T_1)$	−0.5 to 2.0	−0.5 to 2.0	−0.5 to 1.5	−0.5 to 2.0		0.3 to 2.0	0.4 to 1.2			−4.0 to 2.0		
Curie temperature, °C	>130	>130	>120	>115	>200	>200	>130	>230	>200	>280	>450	>500
saturation flux density, T[a]	0.41	0.42	0.40	0.44	0.40	0.47	0.39	0.51	0.48	0.33	0.30	0.27
disaccommodation factor, $D \times 10^6$ (from 1–10 min), $(\mu_1-\mu_2)/\mu_1^2 \log(t_2/t_1)$ where t = time	<3	<3	<1	<1	<12	<5	<2			<30	<20	
resistivity, $\Omega \cdot m$	1	1	0.15	0.05	4	25	8	10	2	20×10^5	2.5×10^5	1.0×10^5
applications			transformers			inductors			power supplies		inductors	

[a] To convert T to G, multiply by 10^4.

TABLE 3. PROPERTIES OF AMORPHOUS MAGNETIC ALLOYS

Alloy	Composition	Saturation induction, B_s, T[a]	Coercive force H_c, A/m	Magnetostriction, $\lambda_s \times 10^{-6}$
	IRON-BASED			
Metglas 2605SC	$Fe_{81}B_{13.5}Si_{3.5}C_2$	1.61	3.2	30
Metglas 2605S-2	$Fe_{78}B_{13}Si_9$	1.56	2.4	27
Metglas 2605CO	$Fe_{67}Co_{18}B_{14}Si_1$	1.80	4.0	35
Metglas 2605S-3	$Fe_{79}B_{16}Si_5$	1.58	8.0	27
	IRON–NICKEL-BASED			
Metglas 2826MB	$Fe_{40}Ni_{38}Mo_4B_{18}$	0.88	1.2	12
	COBALT-BASED			
Metglas 2705M	$Co_{67}Ni_3Fe_4Mo_2B_{12}Si_{12}$	0.72	0.4	0.5

[a] To convert T to G, multiply by 10^4.

TABLE 4. MAGNETIC PROPERTIES OF COMMERCIAL PERMANENT MAGNET MATERIALS

Material	T_C, °C	$(BH)_{max}$, kJ/m³[a]	B_r, T[b]	H_c, kA/m[c]
Ferroxdure ($SrFe_{12}O_{19}$)	450	36	0.42	250
Alnico 9	850	72	1.05	120
$SmCo_5$	724	144	0.87	600
$Sm(Co_{0.68}Cu_{0.10}Fe_{0.21}Zr_{0.01})_{7.4}$	800	240	1.10	510
$Nd_2Fe_{14}B$	312	290	1.23	880

[a] To convert kJ/m³ to G·Oe, multiply by 12.57×10^4.
[b] To convert T to G, multiply by 1×10^4.
[c] To convert kA/m to Oe, divide by 7.958×10^{-2}.

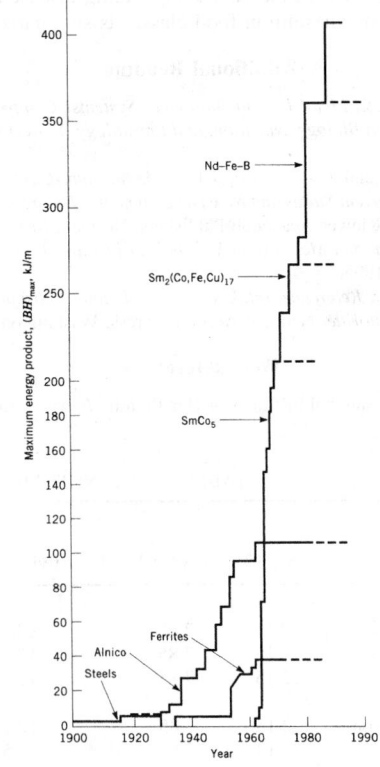

Fig. 1. Progress in energy product for hard magnetic materials. To convert J to cal, divide by 4.184

binary R–Co alloys has continually decreased because of the high cost of cobalt. These are being replaced by the ternary NdFeB materials, including $Nd_2Fe_{14}B$. The progress in energy products for hard magnetic materials is shown in Figure 1.

Uses. Hard ferrites are used widely in electromechanical devices, e.g., generators, relays, motors, and magnetos; electronic applications, e.g., loudspeakers, traveling-wave tubes, and telephone ringers and receivers; antitheft tags, holding devices such as door closers, seals, and latches;

and are perennial favorites in various toy designs. Loudspeakers are the largest use of permanent magnets (ca 50%). Strontium ferrites exhibit higher coercivities and are increasingly being produced.

The commercial development of magnets based on $Nd_2Fe_{14}B$ boride proceeded rapidly, and they are now being used in many diverse applications, for example, servo devices for machine tools, for over 30 dc motors for fully equipped automobiles (windshield wipers, cooling fans, window and antenna lift motors, etc), magnetic resonance imaging (mri), computer disk drives, and medical device applications. Their largest use is in positioning motors for computer hard disk drives. New designs of electrical machines are now taking place.

Economic Aspects

The manufacturers of permanent magnets include Hitachi which produces Alnico Grades 5–9, in cast form in various shapes, of Grades 2 and 5 in sintered form as well as the rare-earth–cobalt (Hicorex); IG Technologies Inc., which produces Alnico, cast grades (Hyflux) 5, 8, 9 in various shapes, sintered, grades 2, 5, 8, cunife magnets, ceramic magnets (Indox) grades 1 and 5, and the rare-earth–cobalt Incor; Crucible Magnetics, which produces Alnico, cast grades 5, 7, and 8, ceramic magnets (Ferrimag), and the rare-earth–cobalt Crucore; Arnold Engineering, which produces Alnico, cast grades 5 and 8, sintered grades 2, 5, 8; GM, Delco Remy Division which produces Nd–Fe–B, limited to several shapes only (Magnaquench), isotropic and anisotropic, many grades, under the name of Permag, in the form of disks, rectangles, and squares, supplied by the Magnetic Materials Division of the Dexter Corp.; Sumitomo, which produces Nd–Fe–B and sintered magnets; 3M Co., which produces magnetic oxides, ferrites rubber bonded to form flexible permanent magnet tape with or without adhesive (Plastiform); and AlliedSignal, which produces the amphorous magnetic alloys known as Metglas.

THIN FILMS AND PARTICLES

The largest use of magnetic films and particles, in the form of tapes and disks for recording and retention of audio, visual, and digital information, is in memory and storage technologies. Price per bit of information, including the cost of the peripheral electronics, and performance, as denoted by access time, generally are used to characterize the various memory technologies. Power, modular capacity, reliability, nonvolatility, etc, are also factors describing the efficacy of memories.

Magnetic Properties and Structure

The static or low frequency magnetic properties pertinent to thin-film materials generally are utilized to characterize magnetic materials. As a first approximation, these properties serve to suggest utility for device applications. Saturation magnetization M_s and Curie temperature T_C are intrinsic (structure insensitive) properties and are equal to the bulk values when thick films are made properly. For very thin highly paramagnetic films, such as those of platinum, sandwiched between ferromagnetic, e.g., Co, or antiferromagnetic, e.g., Cr, thin films, in the form of multilayered structures being developed for recording heads, a magnetization can be induced in the normally paramagnetic material. The surface area-to-volume ratio of the individual layers is so large that the atomic moments at the interfaces play an important role.

Fabrication

Fabrication methods include thermal evaporation, sputtering, magnetron sputtering, pulsed laser evaporation, molecular beam epitaxy, chemical vapor deposition, electrolytic and electroless deposition, and growth from solution.

Materials

Magnetic storage materials for storage of audio and video information as well as of digital data are in the form of tape and disks. There are two states of remanent magnetization for recording: longitudinal, in which the magnetization is in the plane of the recording medium; and perpendicular, in which the magnetization is normal to the plane. For particulate media in which the acicular submicronic particles are single domain and embedded in plastic, the magnetization is confined in the direction of the long dimension.

Multilayer materials exhibiting high magnetization and permeability are undergoing considerable research and development for advanced recording heads. The discovery of giant magnetoresistance in multilayered nano-thick magnetic materials is expected to become important for advanced read heads.

Particulate Materials. There are three principal classes of particulate magnetic materials: γ-ferric oxide, γ-Fe_2O_3, and its modifications; chromium dioxide, CrO_2; and iron. A comparison of the remanent magnetization, B_r, and coercivity, H_c, for several material systems is shown in Table 5.

Recording Heads. Materials that are suitable for read/write recording heads for tapes and disks are characterized by high saturation flux density, low remanent induction to avoid erasure of information when the writing current ceases, and low hysteresis and low eddy-current loss, particularly for high data rates or high frequency operation. In addition, because of the small air gap between the head and recording medium, the head material should be abrasion resistant. Dust particles and the magnetic attraction between head and tape can lead to abrasion.

For general-purpose audio recording, laminated Ni–Fe alloys exhibit the required high saturation and low remanence and eddy-current losses; moreover, abrasion is low. Head wear is improved by use of precipitation hardened material. The spinel structure oxides, manganese–zinc and nickel–zinc ferrites, exhibit good abrasion resistance and high frequency characteristics and in some cases are the preferred material despite their relatively low saturation.

For high quality audio and video recording where the recording medium is CrO_2 or Co impregnated γ-Fe_2O_3, sputtered Sendust alloy films (9.6 wt % Si, 5.4 wt % Al, balance Fe; see also Table 6) and ferrites are used as head materials.

Thin-Film Magnetic Metallic Media. Advanced magnetic recording media are in the form of thin films. The metallic media are typically sputtered films having carbon overcoats for protection. Cobalt-based alloys have been developed for use as longitudinal, i.e., c-axis of the crystalline Co-alloy parallel to the plane of the substrate, magnetic recording media (see **Cobalt** and **Cobalt alloys**). Magnetic disks are presently fabricated alloys on NiP coated aluminum alloy disk substrates.

Magnetooptic Materials. The application of magnetooptic effects to optical memory systems, such as for laser beam writing and magnetooptic read, has been the subject of much research.

Memory systems based on laser writing and reading through the interaction of electromagnetic radiation, either through reflection utilizing the Kerr effect or by transmission utilizing the Faraday effect, have begun to appear in the marketplace.

The magnetic storage media being employed are ternary amorphous alloys (Table 6) composed of the rare-earth elements gadolinium, Gd, and terbium, Tb, with Fe and Co for use in the near infrared. These materials are compatible with GaAs-based lasers. These alloys are ferrimagnetic.

The rare-earth (R) garnets, $R_3Fe_5O_{12}$, which are ferrimagnetic, are being investigated for magnetooptic recording.

Amorphous single-domain CoTaZr cores having Al_2O_3 interlayers where the CoTaZr thickness is from 0.23–0.9 μm, depending on the number of layers, and Al_2O_3 is 0.01-μm thick, were evaluated for use as thin-film heads. This material combination is attractive for low noise heads operating at frequencies up to 40 MHz.

Magnetic Superlattices. The discovery in the late 1980s of giant magnetoresistance (GMR) in antiferromagnetically coupled Fe/Cr superlattices stimulated great interest (see Table 7). Properties of metallic superlattices consisting of thin alternate single-crystal layers of different magnetic materials as well as alternate layers of magnetic and nonmagnetic materials were examined.

TABLE 5. MAGNETIC PROPERTIES OF COMMON MAGNETIC RECORDING MEDIA

Material	B_r, T[a]	H_c, kA/m[b]
γ-Fe_2O_3	0.11	26
Fe_2O_3–Fe_3O_4	0.15	37
Co–γ-Fe_2O_3	0.15	52
CrO_2	0.15	45
$BaFe_{12}O_{19}$	0.12	64
Fe	0.30	120

[a] To convert T to G, multiply by 10^4.

[b] To convert kA/m to Oe, divide by 7.958×10^{-2}.

TABLE 6. MAGNETIC METALS INVESTIGATED FOR THIN-FILM DEVICES

Material	Composition, wt %	Application[a]
iron		MR
iron–nickel alloys	Permalloys	MR
		MD, T, RH
cobalt–nickel	82 Co, 18 Ni	
cobalt–phosphorus	98 Co, 2 P	R
cobalt–nickel–phosphorus	75 Co, 23 Ni, 2 P	R
iron–nickel–chromium	76 Fe, 12 Ni, 12 Cr	
	74 Fe, 8 Ni, 18 Cr	
Vicalloy II	13 V, 35 Fe, 52 Co	R
Cunife I	60 Cu, 20 Ni, 20 Fe	R
Cunife II	50 Cu, 20 Ni, 27.5 Fe, 2.5 Co	R
Cunico I	50 Cu, 21 Ni, 29 Co	R
Cunico II	35 Cu, 24 Ni, 41 Co	R
manganese bismuth (1:1)		TH
manganese aluminum germanide		TH
manganese gallium germanide		TH
Sendust alloy	85 Fe, 9.6 Si, 5.4 Al	TS, 12-μm films, RH, MRM
RCo(Fe) amorphous alloys[b]	variable	MRM
Co–Fe–Cr–P–C–B amorphous alloys	variable	SMB, H_c < 8 A/m[c]
Co–Cr	18–22 Cr	PR, LR
Co–Cr–Ta/Cr, Co–Pt–Cr/Cr		LR
Pt/Co or Pd/Co multilayers	ultrathin alternating layers	MRM
Pt/Fe epitaxial multilayers	3 nm Pt/2.3 nm Fe	MRM, PR
CoTa–Zr amorphous multilayers on Al_2O_3 separators	variable	HFRH
FeTaN multilayers on Al_2O_3	0.5 μm alloy/0.1 μm Al_2O_3	RH/HDTV
Co/Ni multilayers	variable	MRM, PR
Co/Au multilayers	variable	PR
$Co_{1-x}Pt_x$ multilayers	$x = 0.45 - 0.9$	PA, MRM

[a] MR = magnetic recording; MD = magnetoresistive detectors; T = transducers; RH = recording heads; R = recording; TH = thermomagnetic or Curie-point writing; TS = tetrode sputtering; MRM = magnetooptic recording media or magnetooptical recording; SMB = soft magnetic behavior; PR = perpendicular recording; LR = longitudinal recording; HFRH = high frequency recording heads; RH/HDTV = recording heads for high definition television; and PA = perpendicular anisotropy.
[b] Where R = Gd or Tb.
[c] 8 A/m = 0.10 Oe.

Magnetoresistive recording heads offer much more sensitivity than inductive heads and there is strong evidence as of this writing that such heads will be used exclusively by the year 2000. The trend in the development of head materials is toward thin-film media. Although Permalloy films ($Ni_{18}Fe_{19}$) are used for magnetoresistive sensors, the change in resistance is only about 2.5%. Higher magnetoresistive materials are needed.

Magnetic Fluids. Magnetic fluids are stable colloidal suspensions of ferromagnetic particles, such as Fe_3O_4, and of subdomain size (ca 10 nm) in aqueous or organic bases. The fluid behaves as a homogeneous Newtonian liquid and reacts to a magnetic field. These materials are used in bearings, rotary-shaft seals, and feedthroughs.

JACK WERNICK
Murray Hill, New Jersey

Additional Reading

Buschow, K.H.: *Handbook of Magnetic Materials,* Vol. 15, Elsevier Science, New York, NY, 2003.
Buschow, K.H. and F.R. De Boer: *Physics of Magnetism and Magnetic Materials,* Kluwer Academic Publishers, Norwell, MA, 2002.
Buschow, K.H. and E.P. Wohlfarth: *Handbook of Magnetic Materials: A Handbook on the Properties of Magnetically Ordered Substances,* Vol. 5, Elsevier Science, New York, NY, 1990.

TABLE 7. MAGNETIC SUPERLATTICES EXHIBITING GIANT MAGNETORESISTANCE

Material system	Multilayer information	Result[a,b]
Fe/Cr	var $((001)\ Fe/Cr(001))_n$, 0.9–9.0-nm single-crystal layer thickness on single-crystal GaAs	resistivity lowered by factor of 2 at 4.2 K, switching fields on order of 1 T req'd
Co/Cu	var	strong antiparallel coupling through nonferromagnetic Cu; high (>800 kA/m) fields req'd to change antiferromagnetic spin structure into ferromagnetic
Co–Fe/Cu	1.0 nm Co_9Fe/1.0 nm Cu ion-beam sputtering on MgO(110) substrates	Co–Fe/Cu grew having inplane uniaxial anisotropy, easy axis parallel to cube direction in the MgO(110) plane; saturation field (240 kA/m) at RT for GMR = 45%
NiFe/Cu	$Ni_{81}Fe_{19}$/Cu/Co	GMR enhanced by presence of thin Cu layer; magneto-resistance of >17% for field changes of ±8 kA/m at RT
NiFe/Cu/Co	$[Ni_{80}Fe_{20}$/Cu/Co/Cu] r-f diode sputtering on Si(100) single-crystal wafers at RT	resistance changes ≤70% within a few amperes/m
NiFe/Cu/NiFe/FeMn		resistance changes of 3–4% in fields of 40–800 A/m
Co/Ag	Co(15 nm)/Ag(6.0 nm) electron-beam evaporation on top of a 5.0-nm Cr buffer layer on Si(111) substrates	interface roughness important in understanding connection between GMR and antiferromagnetic coupling

[a] To convert T to G, multiply by 10^4.
[b] To convert kA/m to Oe, divide by 7.958×10^{-2}.

Chen, C.W.: *Magnetism and Metallurgy of Soft Magnetic Materials,* North-Holland, New York, NY, 1977.
Freund, B. and S. Suresh: *Thin Film Materials: Stress, Defect Formation and Surface Evolution,* Cambridge University Press, New York, NY, 2003.
Harper, J.M.E.: "Ion Beam Techniques in Thin Film Deposition," *Solid State Technol.,* 129 (Apr. 1987).
Herbst, J.F.: "Permanent Magnets," *Am. Sci.,* 251 (May–June 1993).
Heirich, B. and J.A.C. Bland, eds.: *Ultrathin Magnetic Structures,* Springer, Berlin, 1994.
Kryder, M.H.: "Data Storage in 2000: Trends in Data Storage Technologies," *IEEE Trans. Magn.* **25**(6), 4358 (1989).
Luborsky, F.E. ed.: *Amorphous Metallic Alloys,* Butterworths, London, 1983.
Nalwa, H.S.: *Handbook of Thin Films,* Five-Volume Set, Elsevier Science, New York, NY, 2001.
O'Handley, R.C.: *Modern Magnetic Materials: Principles and Applications,* John Wiley & Sons, Inc., New York, NY, 1999.
Ohring, M.: *Materials Science of Thin Films,* 2nd Edition, Elsevier Science, New York, NY, 2001.
Sharrock, M.P.: "Particulate Magnetic Recording Media: A Review," *IEEE Trans. Magn.* **25**(6), 4374 (1989).
Spaldin, N.: *Magnetic Materials: Fundamentals and Device Applications,* Cambridge University Press, New York, NY, 2003.
Venables, J.A.: *Introduction to Surface and Thin Film Processes,* Cambridge University Press, New York, NY, 2000.
Wolfarth, E.P. ed.: *Ferromagnetic Materials—A Handbook on the Properties of Magnetically Ordered Substances,* Vols. 1 and 2, Elsevier, New York, NY, 1980; E.P. Wolfarth and K.H.-J. Buschow, eds., Vols. 3 and 4, Elsevier, 1988.

MAGNETIC SEPARATION. Use of a magnetic field to remove unwanted magnetic particulates from solid or liquid mixtures of nonmagnetic materials, e.g., sands, and mineral processing. Low-gradient fields are suitable for separation of strongly magnetic materials, whereas high-gradient fields can separate particles of materials that are weakly magnetic, such as coliform bacteria from municipal wastes and sulfur from coal.

Removal of magnetic impurities from industrial wastewater is called magnetic filtration, e.g., reconditioning of boiler water and regeneration of condensate in power plants. See also **Electromagnetic Separation**; **Magnetic Materials**; and **Mass Spectrometry**.

MAGNETISM.   A magnet is a body possessing the property of attracting magnetic substances. The so-called permanent magnet should be used where a constant magnetic field is to be produced, inasmuch as a well-made permanent magnet loses its magnetism very slowly, and then only up to a certain value, after which it is said to be aged. Therefore, it essentially maintains a constant degree of magnetism unless subjected to strong demagnetizing effects. Permanent magnets for precision apparatus are artificially aged during manufacture.

A bar that has been magnetized is found to have poles, which are centers where magnetic attraction is strongest. If the magnet is free to turn, the pole which points northerly is aptly termed the north pole; the other, the south pole. Like poles repel; unlike poles attract. Thus, because of the magnetic properties of the earth, a magnet can serve as a compass. The poles, of course, have no physical reality, but provide a convenient concept for describing certain magnetic phenomena.

Magnetic Field

The region surrounding a magnet (or an electric current) is endowed with specific properties. The most familiar is the torque experienced by a small magnet when placed in such a region. For any point of the field, there is only one direction in which the small magnet will reach stable equilibrium. The direction in which the north pole of the magnet points when in equilibrium is termed the direction of the field. By moving a small magnet in a magnetic field, it is found that the field direction will follow curved lines of force. If the field is due to the current flowing in a conductor, the lines form completely closed curves enclosing the conductor. If the field is due to a magnet, they appear to enter the iron at the south pole and emerge at the north pole, inferring that they complete themselves through the iron as they do through a coreless, current-carrying helix. A magnetic field may be defined as a vector function field described by the magnetic induction. The term magnetic field is used interchangeably to refer to magnetic induction and magnetizing force.

The difference in the magnetic potential at two points in a magnetic field is measured by the work necessary to move a unit magnetic pole against the field from one point to the other. This difference is sometimes called a magnetomotive force, in analogy to electromotive force.

The lines of magnetic intensity around a current-carrying wire are circular, having the plane of the circle perpendicular to the axis of the wire. The direction of the lines of force is determined by the "right-hand rule." When the wire is grasped by the right hand, the fingers encircling the wire, and the thumb pointing along the wire in the direction of the current, the fingers encircle the wire in the direction of the lines of force. This rule is used to determine in which direction the north pole would lie in a helix or solenoid of wire, for, instead of having a straight wire encircled by magnetic lines, the wire itself is bent into a circular form by being wound in a solenoid or helix. The lines of force will then produce an axial magnetic field, so that one end of the solenoid is equivalent to a north pole, the other to a south pole.

In the vicinity of magnets, electric currents, or time-varying electric fields, it is found that a small current-carrying wire loop, if free to turn, will come to an equilibrium orientation. This behavior is attributed to an auxiliary field vector **B** called the magnetic induction or the magnetic flux density. The magnetic moment of a current loop or a magnetized body is a measure of the magnetizing force **H**, produced by the current or magnetized body. The magnetic moment of a plane current loop is a vector (**m**), normal to the plane of the loop and directed so that the current has a clockwise rotation around **m**. The magnitude of **m** is the product of the current and loop area. The magnetic moment of a magnetized body is the vector summation of the magnetic moments of the internal current loops and spins of the body. The magnetizing force produced at a displacement **r** from a small source of moment **m**

$$\mathbf{H} = -\nabla \frac{\mathbf{m} \cdot \mathbf{r}}{r^3}$$

where ∇ is del. the vector differential operator.

The ratio of the magnetic moment of a magnetized body to its volume, expressed by the equation $\mathbf{B} = \mu_0(\mathbf{H} + 4\pi\mathbf{M})$, where μ_0 is the permeability of free space, **B** is magnetic induction, **H** is the external magnetizing force, and **M** is the magnetization. (The symbol I is sometimes used for intensity of magnetization.) The application of an increasing magnetizing force to a ferromagnetic substance yields a resulting intrinsic induction that asymptotically approaches a constant value known as the saturation magnetization. Spontaneous magnetism is the magnetic saturation of the domains of a ferromagnetic material, even in the absence of an applied magnetizing force. Each domain is magnetized to saturation at all temperatures below the Curie point, although the material as a whole may be unmagnetized because of the differing orientations of the various domains. See also **Ferromagnetism**.

The intensity of magnetization I induced at any point in a body is proportional to the strength of the applied field H:

$$I = \kappa H \,(\text{or } \kappa = I/H)$$

where κ is a constant of proportionality depending on the material of the body. It is called the magnetic susceptibility per unit volume, and may be defined qualitatively as the extent to which a material is susceptible to induced magnetism. For an isotropic body, the susceptibility is the same in all directions. However, for anisotropic crystals, the susceptibilities along the three principle magnetic axes are different, and measurements on their powder samples give the average of the three values.

Magnetic susceptibility is obviously related to magnetic permeability, and the following relationships may be derived:

$$B = \mu_H = 4\pi I + H$$

Therefore,

$$\mu = 4\pi I/H + 1 = 4\pi\kappa + 1$$

or

$$\kappa = (\mu - 1)/4\pi$$

Magnetic Flux.   The magnetic flux through any closed figure, such as a circle, a rectangle, or a loop of wire, is the product of the area of the figure by the average component of magnetic induction normal to that area. Thus, if a rectangle 5 cm × 8 cm is placed in a region where there is a uniform magnetic induction of 2,500 gauss (other units defined shortly), and at an angle of 30° with the lines of induction, the magnetic flux through it is 2,500 gauss ° 40 cm² × sin 30° = 50,000 gauss-cm² or "maxwells." The magnitude of this quantity is often conventionally represented by imagining the lines of induction to be so spaced that the number of them through a given area is equal to the number of gauss-cm² or maxwells of flux through that area. The flux in the above example would be commonly expressed as 50,000 "lines." When a coil has several (n) turns and each turn has approximately the same flux (ϕ) through it. This product, which is called the "linkage," is expressed in "maxwell-turns" or "line-turns."

The magnetic flux or the linkage through a loop or a coil may be measured by putting into the circuit a ballistic (undamped) galvanometer and then suddenly removing the flux (or the coil). If the resistance of the whole circuit and the constant of the galvanometer are known, the flux may be calculated from the "throw" of the galvanometer.

The *gauss* (G) is the electromagnetic CGS unit of magnetic flux density. The *tesla* (T) is the **SI** unit. The *maxwell* (**Mx**) is the electromagnetic CGS unit of magnetic flux. The weber (Wb) is the SI unit.

Magnetic Circuit.   The flux from a bar magnet or from a straight electromagnet issues from one end of the magnet or coil, bends around, and reenters at the other end. As mentioned before, this can be exhibited by exploring the region with a small magnet or compass needle. If there is an iron frame or ring extending from one pole of the magnet or coil around to the other, and in the case of the coil, running clear through it, the magnetic flux is not only concentrated largely in the iron, but is much greater in total amount than if the induction is entirely in the air. Even a short gap in the iron reduces the flux considerably.

The analogy of such a magnetic path to an electric circuit is easily seen. The magnetic flux corresponds to a current. The magnet or coil corresponds to a battery, and provides magnetomotive force just as a battery supplies electromotive force. The amount of flux produced by a given magnetomotive force depends upon the dimensions and material of the "magnetic circuit," e.g., the length and cross section of the iron ring followed by the flux and the permeability of the iron; just as the dimensions and material of the electric conductor determine its resistance. The attribute of the magnetic circuit (corresponding to resistance) is called its reluctance.

It must be remembered that this analogy is purely mathematical, not physical. In magnetism, there is no flow of charge, as in electricity.

For a single magnetic circuit consisting of two parts, magnetic material having a permeability μ_1, length l_1, and A_1, and an air gap of permeability μ_0, length l_2 and area A_2, Ampere's circuital law gives $\int H \cos\theta dl = H_1 l_1 + H_2 l_2 = NI$, N is the number of turns and I the current through the coil producing the magnetic flux Φ_m. Now as

$$H_1 l_1 = B_1 l_1 / \mu_1 = \Phi_m l_1 / A_1 \mu_1$$

and

$$H_2 l_2 = B_2 l_2 / \mu_0 = \Phi_m l_2 / A_2 \mu_0$$

So

$$\Phi_m [l_1 / A_1 \mu_1 + l_2 / A_2 \mu_0] = NI$$

or

$$\Phi_m = \frac{NI}{[l_1 A_1 \mu_1 + l_2 / A_2 \mu_0]} = \frac{F}{R}$$

where F is the magnetomotive force in ampere-turns and R, the reluctance of the magnetic circuit. This relation is a magnetic analogy of Ohm's law known as Bosanquet's law.

The *gilbert* (Gb) is the electromagnetic CGS unit of magnetomotive force. The *ampere (or ampere turn)* (A) is the SI unit. The *oersted* (Oe) is the electromagnetic CGS unit of magnetic field strength. The *ampere per meter* (A/m) is the SI unit.

Magnetic Materials. The electrical industry is dependent upon four basic types of materials: (1) good conductors of electricity; (2) high-resistivity conductors capable of withstanding high temperatures; (3) insulators; and (4) magnetic materials. The development of better magnetic materials has contributed to marked improvements in motors, generators, transformers, and instruments of all kinds. See also **Magnetic Materials**.

The ferromagnetic elements are iron, nickel, and cobalt. Of these, iron is the only one that has important magnetic applications in pure, or commercially pure form. All three elements, and many others which themselves are nonmagnetic, are used in special alloys in which certain magnetic characteristics are developed to a high degree. It is even possible, by alloying, to develop magnetic material from certain nonmagnetic elements.

There are two distinct groups of ferromagnetic materials, those that are easily demagnetized, and those that are not. These are often designated as soft and hard magnetic materials. In many respects, iron is an excellent soft, magnetic material. It attains the highest saturation value of magnetic induction found in any material, with the exception of certain cobalt alloys. However, its maximum permeability (maximum ratio of magnetic induction to magnetizing force) is surpassed by other materials. When used as a core in a field induced by ac, the so-called "iron losses" are relatively high. Iron losses consist of energy losses related to the area of the hysteresis loop, and of eddy-current losses caused by induced current circuits within the metal. High values of residual induction (B_r on the typical hysteresis curve) and coercive force (H_c) are indications of high hysteresis loss. See also **Hysteresis**. A low value of electrical resistivity results in high eddy current losses. However, other factors such as the thickness of the sheets in laminated cores, and to a lesser degree grain size also affect eddy-current losses. By special heat treatments and by the use of specially purified iron, much higher values of permeability and lower hysteresis losses can be obtained.

With a good grade of silicon electrical steel, the hysteresis loss can be cut in half and the eddy-current loss reduced even more because of the high resistivity compared with iron. This results in much more efficient operation of alternating-current equipment.

Certain high-nickel alloys of the Permalloy type have very high permeabilities at low and moderate inductions and give low hysteresis losses. For many special applications in instruments and communications equipment, these higher cost alloys are economically justified. Power transformers, motors and generators, etc., are designed for silicon electrical steels.

In the use of any of these materials for ac applications, it is desirable to reduce the thickness of individual laminations to reduce eddy-current losses. For special applications at higher than usual power frequencies, extremely thin strips are used, approaching $\frac{1}{1000}$ of an inch (0.025 millimeter) in thickness. The most common thickness at 60-cycle power frequencies is about $\frac{14}{1000}$ inch (0.35 millimeter). In the communications field, where the use of very high frequency current greatly aggravates the problem of eddy-current losses, alloys of the Permalloy type are used. They are produced in powdered form, given a very thin insulating film, and compacted under high pressure into a suitable core shape.

In the production of magnetically soft materials, it is important that elements such as carbon and sulfur, which disturb the continuity of the base-metal crystal structure, be reduced to low residual amounts, and that the material be used in a strain-free condition. High annealing temperatures resulting in coarse grain structures are used. These conditions are, for the most part, reversed when magnetically hard materials are sought. These materials must be capable of reaching high inductions, but upon removal of the direct magnetizing force the magnetic flux remaining in the circuit should be high, therefore a hysteresis loop of large area is desired. A good overall measure of the quality of a permanent magnet is the "maximum external energy." This is the maximum value which can be obtained for the product of the coordinates of a point on the curve between B_r and H_c on the hysteresis curve.

Progress in Development of Magnetic Alloys. During the last 100 years of experience in producing and using permanent magnets, alloy magnets as of the early 1980s are the most powerful by a factor of 30. Until the late 1930s, magnets were produced from steel. Then considerable research was devoted to producing and testing various magnetic alloys. Alnico magnets became available in the mid-1940s. representative present-day Alnico magnets contain a number of other metals. For example, Alnico 5 contains 8% aluminum, 14% nickel, 24% cobalt, and 3% copper, the remainder being iron. Alnico 9 contains 7% aluminum, 15% nickel, 35% cobalt, 4% copper, 5% titanium, the remainder being iron. Work on the development or magnetic alloys containing various rare-earth elements commenced in the late 1950s. A number of magnetic alloys containing rare-earth elements and cobalt are available today. As pointed out by Chin (1986), the magnetism of the R-Co compounds is due to the interatomic exchange between the spins of the two sublattices plus the spin-orbit coupling within the rare-earth atoms. The spins of the lighter rare-earth elements, such as cerium, praseodymium, neodymium, and samarium, align parallel with the cobalt atoms, resulting in high saturation magnetization values at room temperature. Other rare-earth elements align antiparallel with the cobalt atoms and thus the values are lower. In the mid-1970s, the alloy $Sm_2(Co, Fe, Cu)_{17}$ was introduced. See also **Rare-Earth Elements and Metals**.

The importance of improved magnetic alloys is sometimes underestimated. Magnets are used widely in telecommunications and in scores of electrical and electronic instruments and appliances, and it has been estimated that the demand for permanent magnets has increased by a factor of three since 1970. Chromium-cobalt-iron alloys have increased markedly in application. It is interesting to note that these magnets have essentially about the same coercive force and maximum energy product as Alnico 5, but with only half the cobalt content.

See also **Magnetostriction**.

Molecular Magnets. Ordinary iron and alloy magnets are made of atomic constituents that are difficult to modify. By contrast, molecular magnets may be customized much more easily, and they account for the past decade of research into formulating superior molecular magnets. In mid-1991, A.I. Epstein (Ohio State University) and J.S. Miller (DuPont Central Research and Development Facility) appear to have synthesized an organic-based magnet that retains its magnetism at room temperature and up to and including an elevated temperature of 350°K. At the latter temperature, the polymeric material commences to break down chemically. M. Hoffman (Northwestern University) observes that, if this is a molecular magnet, it represents an enormous leap. No researcher in the field has come even close to 350°K.

This particular magnet, although of exceeding interest in the laboratory, is not considered for practical applications because the material deteriorates rapidly unless it is contained within an inert atmosphere. According to a paper released by J.M. Manriquez, et al., the magnet is the result of a reaction of bis (benzene) vanadium with tetrocyanoethylene (TCNE). The material is an amorphous black solid that exhibits field-dependent magnetism and hysteresis at room temperature.

Quantum Effects and Tiny Magnets. Since early computer designs of the 1950s, there has been tremendous pressure exerted to microminiaturize magnetic information systems. This marked trend continues as designers strive for increased information density. But is there a limit? Some scientists now are showing serious concern that, as magnetic devices become smaller and smaller, the point may be reached where quantum effects could cause upheavals in the magnetic storage system. It is interesting to note that, in the computers of the 1950s, an estimated 100 billion atoms were needed to store one bit of information. In the early 1990s, this figure had dropped to an estimated 1 billion atoms. Experts then predicted that, by the year 2000, the storage of one bit of information would require only 100,000 or fewer atoms. Researchers at the IBM Thomas J. Watson Research Center and other researchers in the computer field have pioneered techniques for studying magnets composed of 100,000 atoms at temperatures approaching absolute zero. Some of these researchers, who have instrumentation that can measure a magnetic field with a million times the sensitivity of traditional instruments, have observed what they believe may be evidence of tunneling and other effects of quantum mechanics in experimental magnetic systems. One group claims that, when a tiny magnet is cooled to absolute zero, the north and south poles of this magnets can be reversed effortlessly, causing havoc to enter the magnetic storage system. Another group clings to the concept of classical mechanics and the idea that "spins" cannot occur without receiving energy—thus a so-called "energy barrier." By contrast, consistent with quantum mechanics, there is a "chance" that spins will be capable of breaking through the energy barrier. As pointed out by R. Ruthen, "This phenomenon is analogous to the ability of an electron to tunnel through an energy barrier."

Thus, a profound difference of opinion remains and will require continued research to find the truth.

An excellent description of the early investigations of electricity and magnetism is given by L. Pearce Williams (reference listed).

Additional Reading

Aharoni, A.: *Introduction to the Theory of Ferromagnetism,* Oxford University Press, Inc., New York, NY, 2000.

Amato, I.: "Some Molecular Magnets Like It Hot," *Science,* 1379 (June 7, 1991).

Beeteson, J.S.: *Visualising Magnetic Fields,* Academic Press, Inc., San Diego, CA, 2000.

Campbell, W.H.: *Earth Magnetism,* Academic Press, Inc., San Diego, CA, 2000.

Chin, G.Y.: *Magnetic Materials,* in Encyclopedia of Materials Science and Engineering, MIT Press, Cambridge, MA, 1986.

Comstock, R.L.: *Magnetic Recording,* John Wiley & Sons, Inc., New York, NY, 1999.

Craik, D.J.: *Electricity, Relativity and Magnetism,* John Wiley & Sons, Inc., New York, NY, 1999.

Edmonds, D.: *Electricity and Magnetism in Biological Systems,* Oxford University Press, Inc., New York, NY, 2001.

Hamilton, D.P.: "A Reprieve for MIT's Magnet Lab," *Science,* 850 (August 23, 1991).

Kubler, J.: *Theory of Itinerant Electron Magnetism,* Oxford University Press, Inc., New York, NY, 2000.

Majils, N.: *The Quantum Theory of Magnetism,* World Scientific Publishing Company, Inc., Riveredge, NJ, 2000.

Manriquez, J.M. et al.: "A Room-Temperature Molecular/Organic-Based Magnet," *Science,* 1415 (June 7, 1991).

Miller, J.S., A.J. Epstein, and W.M. Reiff: "Molecular/Organic Ferromagnets," *Science,* 40 (April 1, 1988).

Miller, J.S., M. Drillo: *Advances in Magnetism: From Molecules to Materials,* John Wiley & Sons, Inc., New York, NY, 2000.

Prinz, G.A.: "Hybrid Ferromagnetic-Semiconductor Structures," *Science,* 1092 (November 23, 1990).

Ruthen, R.: "Quantum Magnets," *Sci. Amer.,* 28 (July 1991).

Williams, L.P.: "AndrÉ-Marie AmpÈre," *Sci. Amer.,* 90 (January 1989).

Winters, A.J., et al.: "Large-Scale Superconducting Separator for Kaolin Processing," *Chem. Eng. Progress,* 36 (January 1990).

MAGNETITE. The mineral magnetite, ferroferric oxide, Fe_3O_4, is isometric, commonly occurring in octahedrons, dodecahedrons, and massive, granular, and laminated forms. It is brittle with an uneven fracture; cleavage is not distinct, but with pressure an octahedral parting may develop; hardness, 5.5–6.5; specific gravity, 5.18; luster, metallic to dull; color, iron black, streak, black. It is opaque and strongly magnetic; when possessing polarity it is known as lodestone. Important large ore bodies are products of magmatic segregations, with titanium a prominent constituent of such deposits. Magnetite is a common mineral in the igneous rocks, especially those of the ferromagnesian varieties, and is found in many metamorphic types. It is associated with corundum in emery.

In northern Sweden are located what may be the largest magnetic deposits in the world, believed to have been formed by segregation in the magma. Magnetite is also found in Norway, in the Urals, Italy,

Switzerland, Australia, and Brazil. In the United States, the Precambrian rocks of the Adirondacks contain large beds of magnetite, as well as extensive deposits of titaniferous magnetite, and the mineral is found also in New Jersey, Arkansas, and Utah. In Canada it is found in Quebec and Ontario. The Iodestone or natural magnet is found in Siberia, the Harz Mountains, Germany; the Island of Elba; and at Magnet Cove Arkansas. The name magnetite is said to be derived from the district of Magnesia, near Macedonia. There is, however, a fable that it was named for a shepherd, Magnes, whose iron-bound staff and shoes with iron nails struck to the ground in which magnetite was present.

This mineral is an important ore of iron, 72% being metallic iron. Magnetite sometimes is referred to as magnetic iron ore.

See also **Ocean Resources (Mineral).**

ELMER B. ROWLEY
Union College
Schenectady, New York

MAGNETOCHEMISTRY. A subdivision of chemistry concerned with the effect of magnetic fields on chemical compounds; analysis and measurement of these effects, (e.g., magnetic moment and magnetic susceptibility) are important tools in crystallographic research and determination of molecular structures. Substances that are repelled by a magnetic field are diamagnetic (water, benzene); those that are attracted are paramagnetic (oxygen, transition element compounds). Diamagnetic materials have only induced magnetic moment; paramagnetic materials have permanent magnetic moment. Magnetochemistry has been useful in detection of free radicals, elucidation of molecular configurations of highly complex compounds, and in its application to catalytic and chemisorption phenomena. See also **Nuclear Magnetic Resonance (NMR) and Magnetic Resonance Imaging (MRI)**, and **Magnetism.**

MAGNETOHYDRODYNAMICS (MHD). The behavior of high-temperature ionized gases passed through a magnetic field. A power-generating method using MHD involves an open cycle in which hot combustible gases from coal, seeded with cerium or potassium to increase electrical conductivity, constitute the working fluid. These are sent through a nozzle surrounded by a magnet; the electricity induced by movement of the ionized gas through the magnetic field is passed to electrodes and the gas sent to a steam generator. Efficiency is rated at 50–60% compared with 40% for conventional fossil fuel plants and 33% for plants using nuclear fuels. Two-phase liquid-metal systems are being studied as auxiliary units for a number of energy converters. MHD is an important field of expansion of research activity on new sources of energy; its high efficiency and low pollution factor indicate that it may have a significant future in electric power supply.

MAGNETOSTRICTION. When a polycrystalline nickel sample is placed in a magnetic field, it contracts along the field direction by about 30 parts per million and elongates in the transverse direction by about half that amount. There is also a small volume change. Such changes in dimension of magnetic materials with variation of magnetic field strength or direction are termed *magnetostriction.* See also **Magnetic Materials.** They are measured by strain gages, optical dilatometers, capacitance variation, and x-ray analysis. Below the Curie Temperature, magnetostriction in weak fields is caused by domain rotation, becoming appreciable at fields near the knee of the $B - H$ curve. See also **Hysteresis.** The saturation magnetostriction of a single crystal depends upon the direction of the (sublattice) magnetization and the direction of measurement with respect to the crystal axes. Magnetostriction coefficients vary greatly, depending upon the material, temperature, and the magnetization state. The source of magnetostriction is the dependence of magnetic energy on strain. Because the elastic energy is quadratic in strain while the magnetoelastic energy is linear in strain, the minimum free energy occurs at nonzero strain.

Magnetostriction has been put to practical use in the magnetostrictive resonator. This is essentially an iron rod maintained in longitudinal elastic vibration by a high-frequency current in a helix wound upon it, and used, through the joint operation of the Joule and Villari effects, to control the frequency of the current, somewhat after the manner of the familiar piezoelectric (crystal) resonator. It is also used in band-pass electrical wave filters.

MALACHITE. The mineral malachite is a basic carbonate of copper corresponding to the formula $Cu_2(CO_3)(OH)_2$. It is monoclinic, crystals tending to be acicular, but usually found massive. It is a brittle mineral; hardness, 3.5–4; specific gravity, 4.05; luster, vitreous to silky or dull; color, green; streak, green; translucent to opaque. Malachite is an alternation product found associated with other copper-bearing minerals. It is a rather common mineral and is found quite widely distributed. Large quantities have been found in the Ural Mountains; it is also found in Germany, France, England, Zaire, Rhodesia, and Australia. In the United States, beautiful radiated masses of fibrous crystals have been found in Berks County, Pennsylvania, as well as in Tennessee at Ducktown, and in Arizona, Nevada, and Utah. Malachite, besides being an ore of copper, has been used for various ornamental purposes. The word malachite is derived from the Greek, meaning a *mallow*, because of its green color.

MALAPRADE REACTION.
Compounds containing two hydroxyl groups or a hydroxyl and an amino group attached to adjacent carbon atoms undergo cleavage of the carbon-to-carbon bond when treated with periodic acid.

MALEIC ANHYDRIDE, MALEIC ACID, AND FUMARIC ACID.
Maleic anhydride [CAS: 108-31-6]. (1), maleic acid [CAS: 110-16-7]. (2), and fumaric acid [CAS: 110-17-8]. (3) are multifunctional chemical intermediates that find applications in nearly every field of industrial chemistry. Each molecule contains two acid carbonyl groups and a double bond in the α, β position. Maleic anhydride and maleic acid are important raw materials used in the manufacture of phthalic-type alkyd and polyester resins, surface coatings, lubricant additives, plasticizers, copolymers, and agricultural chemicals (see **Alkyd Resins**; **Polymers**, and **Lubricant**). Both chemicals derive their common names from naturally occurring malic acid.

(1)   (2)

(3)

Fumaric acid occurs naturally in many plants and is named after *Fumaria officinalis*, a climbing annual plant, from which it was first isolated. It is used as a food acidulant and as a raw material in the manufacture of unsaturated polyester resins, quick-setting inks, furniture lacquers, paper sizing chemicals, and aspartic acid.

Physical Properties

Physical constants for maleic anhydride, maleic acid, and fumaric acid are given in Table 1. From single crystal x-ray diffraction data, maleic anhydride is a nearly planar molecule with the ring oxygen atom lying 0.003 nm out of the molecular plane.

Maleic and fumaric acids have physical properties that differ due to the cis and trans configurations about the double bond.

Chemical Properties

Extensive descriptions of the chemistry of maleic anhydride and its derivatives are available in the literature. The broad industrial applications for

this chemistry derive from the reactivity of the double bond in conjugation with the two carbonyl oxygens. Reactions include acid chloride formation, acylation, alkylation, amidation, concerted nonpolar reactions, decomposition and decarboxylation, electrophilic addition, esterification, free-radical reactions, grignard-type reactions, halogenation, hydration and dehydration, hydroformylation, isomerization, ligation to metal atoms, nucleophilic addition, oxidation, polymerization, reduction, and sulfonation.

Manufacture

Process Technology Evolution. Growth in the worldwide maleic anhydride industry is exclusively in the butane-to-maleic anhydride route, often at the expense of benzene-based production. Table 2 shows 1995 and estimated 1996, 1997, and 2000 worldwide maleic production capacity broken down in categories of benzene, butane and phthalic anhydride coproduct. As can be seen from this table butane routes are expected to grow at the expense of benzene-based processes.

Recovery and Purification. All processes for the recovery and fining of maleic anhydride must deal with the efficient separation maleic anhydride from the large amount of water produced in the reaction process. Recovery systems can be separated into two general categories: aqueous- and nonaqueous-based absorption systems. Solvent-based systems have a higher recovery of maleic anhydride and are more energy efficient than water-based systems.

Fumaric Acid. Fumaric acid for commerce is derived from maleic acid through catalytic isomerization. Purified maleic anhydride is the main source of maleic acid. High purity fumaric acid is produced through crystallization of the aqueous mixture, washing, and drying. Decolorizing and crystallization techniques are used to treat impure maleic solutions.

Shipment

Molten maleic anhydride is shipped in tank rail cars, tank trucks and isotanks (for overseas shipments). Solid form maleic anhydride is produced from molten maleic anhydride as briquettes or pastille weighing 0.5 to 20 g. Fumaric acid is shipped in solid form, the particles size varying based upon the specification.

Economic Aspects

The switch of feedstock from benzene to butane was completed in the U.S. in 1985, being driven by the lower unit cost and lower usage of butane in addition to the environmental pressures on the use of benzene. Worldwide, the switch to butane was continuing with 58% of the total world maleic anhydride capacity based on butane feedstock in 1992. This capacity percentage for butane increased from only 6% in 1978. In 1995, 31% of the total world maleic anhydride capacity was based on benzene feedstock and 2% was derived from other sources, primarily phthalic anhydride by-product streams.

Another characteristic of the maleic anhydride supply picture has been the emergence of newer capacity, predominantly in Asia, which has resulted in a significant oversupply to that area of the world. The historically largest and second largest maleic anhydride markets are North America and Western Europe, respectively. Although the names of the producers in both of these areas of the world have changed rather dramatically between the late 1970s and the early 1990s, the total capacity in these world areas has only declined a small amount. Beginning in the late 1980s, small commercial plants based on butane fluidized-bed technologies were built in Asia. As of the end of 1992, fixed-bed butane-based technology accounted for 74% of the butane-based maleic anhydride facilities worldwide with butane-based fluidized-bed technology accounting for the remainder. The butane-based transport-bed process announced by DuPont is scheduled to start up in Spain in 1996.

TABLE 1. PHYSICAL PROPERTIES OF MALEIC ANHYDRIDE, MALEIC ACID, AND FUMARIC ACID

Property	Maleic anhydride	Maleic acid	Fumaric acid
formula	$C_4H_2O_3$	$C_4H_4O_4$	$C_4H_4O_4$
formula weight	98.06	116.07	116.07
mp, °C	52.85	138–139	287
bp, °C	202	ca 138 dec	290
sp gr, at 20/20°C, solid	1.48	1.590	1.635
molar volume		81	79

TABLE 2. PROJECTED WORLD MALEIC ANHYDRIDE CAPACITY BY FEEDSTOCK

Feedstock	1995		1996		1997		2000	
	10^3 t/yr	%	10^3 t/yr	%	10^3 t/yr	%	10^3 t/yr	%
butane	662.2	67.3	712.2	68.6	756.2	71.2	912.7	75.8
benzene	301.0	30.6	301.0	29.0	282.0	26.5	261.5	21.7
phthalic anhydride coproduct	20.5	2.1	24.5	2.4	24.5	2.3	29.5	2.5
Total	*983.7*	*100.0*	*1037.7*	*100.0*	*1062.7*	*100.0*	*1203.7*	*100.0*

Health and Safety Factors (Toxicology)

Maleic Anhydride. The ACGIH threshold limit value in air for maleic anhydride is 0.25 ppm and the OSHA permissible exposure level (PEL) is also 0.25 ppm. Maleic anhydride is a corrosive irritant to eyes, skin, and mucous membranes. Pulmonary edema (collection of fluid in the lungs) can result from airborne exposure. Maleic anhydride is combustible when exposed to heat or flame and can react vigorously on contact with oxidizers.

Maleic Acid. Maleic acid is produced by the hydration of maleic anhydride. The hazards of its use are analogous to those of maleic anhydride. It is a skin and severe eye irritant. It is combustible when exposed to heat or flame and can react vigorously with oxidizing agents.

Uses

Maleic anhydride itself has few, if any, consumer uses but its derivatives are of significant commercial interest. The majority of the maleic anhydride produced is used in unsaturated polyester resin (see **Polyesters, Unsaturated**). Unsaturated polyester resin is then used in both glass-reinforced applications and unreinforced applications.

Fumaric acid and malic acid are produced from maleic anhydride. The primary use for fumaric acid is in the manufacture of paper sizing products. Fumaric acid is also used to acidify food as is malic acid. Malic acid is a particularly desirable acidulant in certain beverage selections, specifically those sweetened with the artificial sweetener aspartame.

Lubrication oil additives represent another important market segment for maleic anhydride derivatives. Maleic anhydride is used in a multitude of applications in which a vinyl copolymer is produced by the copolymerization of maleic anhydride with other molecules having a vinyl functionality. The use of maleic anhydride in the manufacture of agricultural chemicals has declined in the U.S. since the early 1980s.

There are numerous further applications for which maleic anhydride serves as a raw material. These applications prove the versatility of this molecule. The popular artificial sweetener aspartame is a dipeptide with one aminoacid (L-aspartic acid) which is produced from maleic anhydride as the starting material. An important future use for maleic anhydride is believed to be the production of products in the 1,4-butanediol-γ-butyrolactone–tetrahydrofuran family. This technology can be used to produce the product mix of the three molecules as needed by the producer.

TIMOTHY R. FELTHOUSE
JOSEPH C. BURNETT
SCOTT F. MITCHELL
MICHAEL J. MUMMEY
Huntsman Corporation

Additional Reading

Cooley, S.D. and J.D. Powers: *Maleic Acid and Anhydride,* in "Encyclopedia of Chemical Processing and Design," Vol. 29, Marcel Dekker, Inc., New York, NY, 1988, pp. 35–55.

Culbertson, B.M.: *Maleic and Fumaric Polymers,* in J.I. Kroschwitz, ed., "Encyclopedia of Polymer Science and Engineering," 2nd Edition, Vol. 9, Wiley-Interscience, New York, NY, 1987, pp. 225–294.

Lohbeck, K. and co-workers: *Maleic and Fumaric Acids,* in "Ullmann's Encyclopedia of Industrial Chemistry," Vol. A16, 5th Edition, VCH, Weinheim, Germany, 1990, pp. 53–62.

Trivedi B.C. and B.M. Culbertson: *Maleic Anhydride,* Plenum Press, New York, NY, 1982; ~500 pp. on polymers and their applications.

MALEIC HYDRAZIDE GROWTH INHIBITOR.

This compound (1,2-dihydro-3,6-pyridazinedione) is used to inhibit the growth of certain food commodities when in storage, including onion and potato. Maleic hydrazide is also used to promote dormancy in citrus trees as well as increasing protection from frost.

MALONIC ACID AND DERIVATIVES

Malonic Acid

[CAS: 141-82-2]. Malonic acid, $HOOC-CH_2-COOH$, was discovered and isolated in 1858 as a product of malic acid oxidation. The physical properties of malonic acid are listed in Table 1.

Reactions. Malonic acid is a useful tool for synthesizing α-unsaturated carboxylic acids because of its ability to undergo decarboxylation and condensation with aldehydes or ketones at the methylene group. Cinnamic acids are formed from the reaction of malonic acid and benzaldehyde derivatives. If aliphatic aldehydes are used acrylic acids result. Similarly this facile decarboxylation combined with the condensation with an activated double bond yields α-substituted acetic acid derivatives. Reactions of the carboxylic acid groups include monoesterification, diesterification, or conversion with thiols.

Preparation. The industrial production of malonic acid is much less important than that of the malonates. Malonic acid is usually produced by acid saponification of malonates. Further methods which have been recently investigated are the ozonolysis of cyclopentadiene, the air oxidation of 1,3-propanediol, or the use of microorganisms for converting nitriles into acids.

Economic Aspects. Malonic acid is produced by Juzen and Tateyama in Japan as well as Lonza Ltd. in Switzerland and Riedel-De Haen Ltd. in Germany.

Health and Safety Factors (Toxicology). Due to its acidity malonic acid is classified as a mild irritant (skin irritation, rabbits).

Meldrum's Acid. Meldrum's acid is commercially used for the production of monoesters of malonic acid and beta-keto acids.

Malonates

Physical Properties. Industrially, the most important esters are dimethyl malonate. [CAS: 108-59-8]. $CH_3OCO-CH_2-COOCH_3$, and diethyl malonate. [CAS: 105-53-3]. $C_2H_5-OCO-CH_2-COOC_2H_5$, whose physical properties are summarized in Table 2. Both are sparingly soluble in water (1 g/50 mL for the diethyl ester) and miscible in all proportions with ether and alcohol.

Reactions. The chemical properties of malonates are highlighted by the acidity of the methylene group ($pK_a \sim 13$) to such an extent that a proton can be easily detached by a strong base, usually alkoxides.

Manufacture. The predominant manufacturing processes are the hydrogen cyanide process and carbon monoxide process.

Economic Aspects. Dimethyl and diethyl malonates are produced via the carbon monoxide process at Hüls (Germany), Juzen (Japan), and Korean Fertilizers (S. Korea); they are produced via the hydrogen cyanide process at Lonza (Switzerland) and Tateyama (Japan). Total capacity is estimated to be about 12,000 t/yr. Furthermore, producers are also reported in the People's Republic of China and in Romania.

Health and Safety Factors. Dimethyl malonate and diethyl malonate do not present any specific danger of health hazard if handled with the usual precautions. Nevertheless, inhalation and skin contact should be avoided.

Diisopropyl Malonate. This dialkyl malonate has gained industrial importance for the synthesis of the fungicide isoprothiolane through condensation with carbon disulfide and ethylene dichloride. Diisopropyl malonate is produced by Mitsubishi Chemical (Japan) using the carbon monoxide process.

TABLE 1. PROPERTIES OF MALONIC ACID[a]

Property	Value
mol wt	104.06
melting point, °C	135 dec
solubility	
in water at 20°C	139 g/100 mL
in pyridine at 15°C	15 g/100 g

[a] Also called propanedioic acid or methanedicarboxylic acid.

TABLE 2. PHYSICAL PROPERTIES OF DIMETHYL AND DIETHYL MALONATE

Property	Dimethyl malonate[a]	Diethyl malonate[b]
mol wt	132.12	160.17
mp, °C	−62	−50
bp, °C[c]	181.4	199
d_4^{20}, g/mL	1.1544	1.0551
refractive index at 20°C	1.4140	1.4143

[a] Also called propanedioic acid dimethyl ester.

[b] Also called propanedioic acid diethyl ester.

[c] At 101 kPa = 1 atm.

TABLE 3. PHYSICAL PROPERTIES OF CYANOACETIC ACID, METHYL CYANOACETATE, AND ETHYL CYANOACETATE

Property	Cyanoacetic acid	Methyl cyanoacetate	Ethyl cyanoacetate
mol wt	85.06	99.09	113.12
mp, °C	66	−22	−22
bp °C/kPa[a]	108°/1.5 kPa	203°/101 kPa	206°/101 kPa

[a] To convert kPa to mm Hg, multiply by 7.5.

Cyanoacetic Acid and Cyanoacetates

Physical Properties. The physical properties of cyanoacetic acid $N{\equiv}C-CH_2COOH$ are summarized in Table 3. The industrially most important esters are methyl cyanoacetate and ethyl cyanoacetate. Both esters are miscible with alcohol and ether and immiscible with water.

Reactions. The chemical properties of cyanoacetates are quite similar to those of the malonates.

Manufacture. Cyanoacetic acid and cyanoacetates are industrially produced by the same route as the malonates using a modified hydrogen cyanide process, starting from a sodium chloroacetate solution via a sodium cyanoacetate solution.

Economic Aspects. In order to avoid the extraction and evaporation steps, most of the cyanoacetic acid derivatives are made directly from solution; therefore, only a small portion of the acid produced is traded. Cyanoacetic acid is produced by Boehringer-Ingelheim and Knoll (Germany), Juzen (Japan), as well as Hüls (U.S.).

Methyl cyanoacetate and ethyl cyanoacetate are produced by Lonza (Switzerland) and Huls (U.S.), as well as Juzen and Tateyama (Japan). The total production capacity is estimated to be in the range of 10,000 metric tons per year.

Health and Safety Factors. Handling of cyanoacetic acid and cyanoacetates do not present any specific danger or health hazard if handled with the usual precautions.

Uses. In many cases cyanoacetic acid, cyanoacetates, or cyanocetamide can be used alternatively. The traded cyanoacetic acid is mainly intended for the synthesis of the cough remedy dextromethorphan and of the fungicide cymoxanil See also **Fungicides**. Otherwise cyanoacetic acid is directly converted as a solution with 1,3-dimethylurea into 2-cyano-N,N'-dimethylcarbamoyl acetamide which is further upgraded into the diuretics theophylline and caffeine.

The largest application of methyl and ethyl cyanoacetate is the production of the cyanoacrylate adhesives widely used within the car and electronic industries.

Malononitrile

Physical Properties. The physical properties of malononitrile $N{\equiv}C-CH_2C{\equiv}N$ are listed in Table 4.

Reactions. As in the case of malonates and cyanoacetates, the chemical properties of malononitrile are determined by two reactive centers, namely the methylene group and the two cyano functions. A peculiar reaction of malononitrile is the base-catalyzed dimerization leading to 2-amino-1,1,3-tricyanopropene.

Manufacture. Malononitrile can be produced batchwise by elimination of water from cyanoacetamide with phosphorous pentachloride. Most of it is now produced continuously starting from cyanogen chloride and acetonitrile in a high temperature gas phase reaction.

Removal of maleic and fumaric acids from the crude malononitrile by fractional distillation is impractical because the boiling points differ

only slightly. The impurities are therefore converted into high boiling compounds in a conventional reactor by means of a Diels-Alder reaction with a 1,3-diene.

Economic Aspects. Malononitrile is produced by Lonza Ltd. (Switzerland) using the cyanogen chloride process.

Health and Safety Factors. Malononitrile is usually available as a solidified melt in plastic-lined drums. Remelting has to be done carefully because spontaneous decomposition can occur at elevated temperatures, particularly above 100°C, in the presence of impurities such as alkalies, ammonium, and zinc salts. Occupational exposure to malononitrile mainly occurs by inhalation of vapors and absorption through the skin. Malononitrile has a recommended workplace exposure limit of 8 mg/m^3.

Uses. Malononitrile is extensively used in the life sciences industry. The most important products are vitamin B_1 (thiamine) and bensulfuronmethyl, a sulfonyl urea herbicide. Most other product uses fall under the N-containing heterocycles.

PETER POLLAK
GÉRARD ROMEDER
Lonza Ltd.

Additional Reading

Fatiadi, A.J.: *Synthesis* **3**, 165 (1978).
Freeman, F.: *Synthesis* **12**, 925 (1981).
Gawronski, J., and K. Gawronska: *Tartaric and Malic Acids in Synthesis: A Source Book of Building Blocks, Ligands, Auxiliaries, and Resolving Agents,* John Wiley & Sons, Inc., New York, NY, 1998.
Ger. Offen. 2,329,251 (Dec. 13, 1973), H. Marketz (to Lonza Ltd.).
Ger. Offen. 2,741,383 (Feb. 2, 1979), E. Catalucci (to Lonza AG).
Solomons, T.W.G., M.M. Shenkman, and C.B. Fryhle: *Organic Chemistry,* 8th Edition, John Wiley & Sons, Inc., New York, NY, 2003.
Yurkanis Bruice, P.Y.: *Organic Chemistry,* 4th Edition, Prentice Hall, Inc., Upper Saddle River, NJ, 2003.

MALT. Unless otherwise specified, malt usually connotes barley malt. Malt, however, can be prepared from other cereal grains. Between 75 and 80% of the malt produced in the United States goes into the manufacture of beer and associated beverages. Nearly 15% of the production goes to the manufacture of distilled alcohol products; and the remainder (about 5–6%) goes into the preparation of malt syrups, breakfast foods, malted milk concentrates, and coffee substitutes.

Barley malt is barley that has been germinated by moisture under controlled conditions and for a specified time, after which the germinated plants are carefully dried under controlled conditions. The drying or kilning operation and other operations in the total malting process are customized to the final product in which the malt will be used. The principal stages of the malting process are diagrammed in Fig. 1.

Upon receipt, the barley must be inspected carefully to make certain that it fully meets the minimum acceptable specifications established by the malt manufacturer. The qualities desired in grain for malting are described in the entry on Barley.

Prior to processing, the barley is stored for up to 6 weeks to permit any further ripening (after-ripening) to take place. This enhances later germination. The cleaning operation that follows removes impurities, unwanted foreign seeds, and damaged and broken kernels. Because the size of the kernels affects handling during the germinating operation, the barley is graded into 2 or 3 size ranges, each of which is malted separately.

The role served by malt in the later production of beer and distilled spirits is that of furnishing enzymes, which convert starches and other

TABLE 4. PHYSICAL PROPERTIES OF MALONONITRILE[a]

Property	Value
mol wt	66.06
mp, °C	31
bp at 76 kPa[b], °C	218–219
d_4^{35}, g/mL	1.0494
solubility in water, g/mL	0.133

[a] Also called propanedinitrile and dicyanomethane.

[b] To convert kPa to mm Hg, multiply by 7.5.

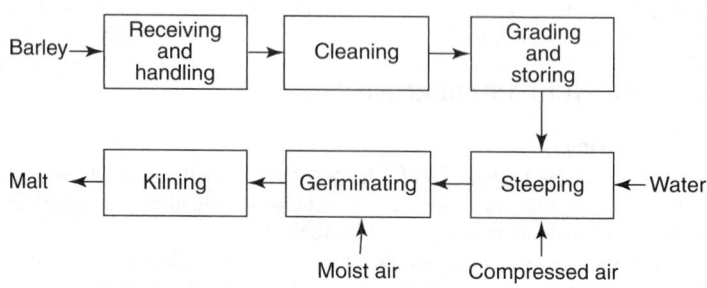

Fig. 1. Simplified flowsheet of operations involved in preparing malt

ingredients during the brewer's and distiller's mashing operations. J. de Clerck states: "Quantitatively the most significant chemical constituent of the grain is starch which constitutes some two-thirds of the dry weight of barley. Apart from a comparatively small contribution from other sugars in barley it is the starch which eventually furnishes some 85–90% of malt extracts, of which some 70% is fermented in brewing and sometimes nearly all in distilling. Starch, therefore, occupies a key position in relation to the brewing and distilling industries and its chemistry is, apart from its intrinsic biochemical interest as the final product of photosynthesis, of outstanding importance." The starch granules of barleys are first seen in the cells of the endosperm a few days after development of the seed begins as small spheres later developing into the bean-shaped and lenticular forms characteristic of mature starch. Luers observed that barley starch undergoes enzymic degradation during malting and that malt starch is different from barley starch. This was recognized as early as 1902. The loss in starch that occurs during malting has been estimated by various researchers. Luhder (1908) first established that the starch content of barley declined considerably during malting. For example, Moravian barley before malting contains 65.57% starch and after malting, 54.89%. Of the two starch constituents amylopectin and amylose, the former is more susceptible to enzymolysis during malting. Several theories have been proposed to explain this difference. The principal factors that influence the amount of starch in barley are environmental during growth and also relate to the variety planted.

Brewers' and Distillers' Malts

Barleys of large grain size are desired for preparing brewers' malt. The larger sizes usually have a greater percentage of starch and relatively little protein. The grains, however, do contain adequate amylolytic enzymes for solubilizing native starch and also some of the other ingredients of the mashing operation at the brewery.

In contrast, the primary objective of distillers' malt is to furnish enzymes in larger quantities for later use in converting grains and starch present in other substances. Barley best suited for this purpose is of a relatively high nitrogen content, a factor that relates to the ability to produce amylases. In the United States, certain barley varieties, such as *Kindred, Manchuria, Montcalm*, and *Odessa*, are specifically selected for making distillers' malt.

Steeping

Once size-graded, the barley is transferred to large steeping tanks equipped with conical or funnel-shaped bottoms. Cool, clean water is added to partially fill the tank. Air agitation assists in cleaning the barley. Any lightweight kernels present automatically rise to the surface of the water and are skimmed off. During a period of from 45 to 60 hours, the barley soaks up water (steeps). At the end of the steeping period, the moisture content of the barley will range between 45 and 48%, representing sufficient moisture to commence germination. Distillers' malt usually requires a somewhat longer period of steeping in order to bring the moisture content up to a minimum of 50%. The process that takes place during steeping is, not surprisingly, essentially the same as the changes that occur when the seed is planted in moist soil.

Germinating

During the next process step, germination, enzymes are produced or liberated in a structure situated between the germ and endosperm of the barley kernel. During germination, the cell wall is made more permeable by the action of the enzyme cytase. Mellowness and friability of the finished malt is determined by this enzymatic action. In barley malt, the amylase enzymes are the most important. These enzymes convert starch into maltose sugars and dextrins in later brewing processes. During germination, only a small part of the starch present should be converted to maltose or other sugars, recalling the prior mention of starch losses. The principal conversion of starch occurs later in mashing operations at the brewery or distillery. Other enzymes that transform proteins also are produced or activated during malting.

Over the years, three principal malting procedures have been developed. The oldest of these is the germinating floor or floor malting, still practiced in some European countries. The steeped barley, once removed from the steeping tanks, is spread in heaps about 1 foot (0.3 meter) in height. The first spreading occupies perhaps only 40% of the available floor area. The barley commences to dry out and sprout, producing small hairlike fibers (rootlets). It is necessary to aerate the early-sprouting barley (known as

green malt), which is accomplished by turning the heap over and over with forks. The barley is spread over a larger floor area. These actions, plus the use of forced aeration, cool the green malt. This procedure is repeated at regular intervals over a period of 5 to 7 days. The process is confined to the cooler seasons of the year because germinating temperature should be maintained within a span between 60° and 70°F (15.6° and 21.1°C). Excessive temperature accelerates germination and generates waste, increasing starch losses and excessively large rootlets. Technically, the optimum cut-off time for germination is when the plumule or acrospire reaches the length of the kernel. An extension of the floor malting process to avoid any seasonal limitations on the process is pneumatic floor or compartment malting. Instead of placing the barley on the floor in heaps, it is placed in box-shaped compartments, which hold the grain during the entire germinating period. Purified air is circulated through these compartments. The air is both temperature and humidity controlled. The green malt is turned by mechanical screws within the compartments. Such compartments will contain up to 5000 bushels (108 metric tons) of green malt.

In *drum malting*, widely used in modern malting plants, there are two concentric hollow metal cylinders. Grain is placed in the spaces between the cylinders, which are perforated for introduction and circulation of humidity-and-temperature-controlled air. To permit maximum movement of the grain, the equipment is only partially filled. The revolving action of the cylinders keeps the grain tumbling, constantly exposing new surfaces of the grain to the air stream. Capacity of these drums is up to 650 bushels (about 13.5 metric tons).

Drying or Kilning

These are not high-temperature or rotary kilns as one may visualize in connection with a cement or ore plant, but they are usually 2-story buildings. Drying is commenced on the upper story by spreading the green malt in layers of 2 to 3 feet (0.6 to 0.9 meter) in depth. The primary purpose of the kilning operation is to halt further germination, although the conditions of kilning also affect the final end-use properties of the malt. Hot air is drawn through the malt at a relatively low temperature for the first 24 hours to accomplish a partial drying. The malt is then transferred to the lower story, where the air temperature is increased. In the case of brewers' malt, the final kilning temperature is in the range of 160° to 180°F (71° to 82°C), and the malt is retained in the kiln until the moisture has reached a content of about 4%. The entire process requires from 48 to 72 hours. In the case of distillers' malt, the final kilning temperature is lower—in the range of 120° to 140°F (49° to 60°C) and the final moisture content is from 5 to 7%. The lower temperature preserves higher enzymatic activities. The higher temperature for brewers' malt introduces a more intense malt flavor and aroma, usually desired for the brewing process. In the case of porter, bock beer, and stout, the kilning temperature is higher and a caramel malt is produced. This imparts a dark color and distinctive flavor during the brewing of these malted beverages.

After kilning, the malt is cleaned to remove rootlets and any broken kernels that may remain in the batch.

Malting Process Innovations

In recent years, several steps to modernize and improve the operations just described have been made. Sophisticated control and conveying systems have assisted in more exacting control of processing conditions and reducing hand labor costs. There is a growing trend toward continuous operations. Some consideration has been given to the use of gibberellic acid as part of the steepwater or in the form of a spray during the germinating process. The objective of these steps would be that of increasing yield. It also has been found that potassium bromate in the steepwater will depress respiration and rootlet formation.

A nutritional profile is given in Table 1.

Malt from Other Grains

Sometimes wheat malt is used as a flour supplement for bakery products. The malt provides a source of a-amylase, which degrades starch to sugars, the fermentation of which causes rising of the dough. Of course, barley malt is also used in some bakery product flours. Wheat malt imparts a characteristic flavor to beer. Most consumers find this undesirable. However, *Weissbier*, made with wheat malt, is quite popular in certain regions of Germany. In producing wheat malt, kilning temperatures are more moderate.

TABLE 1. NUTRITIONAL PROFILE OF MALT (100-GRAM SAMPLES)

	Dry malt	Dried malt extract
Water	18.7 g	11.5 g
Food energy	374	374 cal
Protein	1.3 g	1.7 g
Fat	1.8 g	trace g
Carbohydrate	78.8 g	91.1 g
Calcium	—	50 mg
Phosphorus	—	299 mg
Iron	4.0	9.0 mg
Sodium	—	83 mg
Potassium	—	234 mg
Vitamin A	—	—
Thiamine	0.50	0.36 mg
Riboflavin	0.32	0.47 mg
Niacin	9.4	10.1 mg
Vitamin C (ascorbic acid)	—	—

Source: U.S. Dept. of Agriculture, Washington, DC.

Additional Reading

Hough, J.S.: *The Biotechnology of Malting and Brewing,* Cambridge University Press, New York, NY, 1992.

Hough, J.S., D.E. Briggs, R. Stevens, and T.W. Young: *Malting and Brewing Science,* Vol. 2, Chapman & Hall, New York, NY, 1999.

Hough, J.S. and T.W. Young: *Malting and Brewing Science,* Vol. 1, Kluwer Academic Publishers, Norwell, MA, 1999.

Staff, Briggs Corporation: *Malts and Malting,* Aspen Publishers, Inc., Gaithersburg, MD, 2000.

MALTHA. A black, viscous, natural bitumen consisting of a complex mixture of hydrocarbons. Its viscosity and rheological properties lie between those of crude oil and semisolid asphalt. It is the chief component of Athabaska oil sands.

MALTITOL. See **Sweeteners**.

MALTOSE. See **Carbohydrates**.

MANGANESE. [CAS: 7439-96-5]. Chemical element, symbol Mn, at. no. 25, at. wt. 54.9380, periodic table group 7, mp $1,244 \pm 3°C$, bp $1,962°C$, density 7.3 g/cm^3 (solid), 7.21 (single crystal) $(20°C)$. Manganese has a cubic (complex) crystal structure.

Manganese is a little-known element other than to a small circle of technical specialists who are predominantly metallurgists and chemists. Yet it is the fourth most used metal in terms of tonnage, being ranked behind iron, aluminum and copper, with in the order of 20 million tons of ore being mined annually (2000).

Manganese has numerous applications which impact on our daily lives, whether it be as consumers of objects made of steel, of portable batteries, or of beverage cans based on aluminum. In each case, manganese plays a vital role in improving the properties of the alloys and compounds involved in each specific application.

In the mid-17th century, the German chemist Glauber obtained permanganate, the first usable manganese salt. Nearly a century later, manganese oxide became the basis for the manufacture of chlorine. Yet manganese was only recognized as an element in 1771, by the Swedish chemist Carl Wilhelm Scheele. It was isolated in 1774 by one of his collaborators, J.G. Gahn. At the beginning of the 19th century, both British and French scientists began considering the use of manganese in steelmaking, with patents granted in the U.K. in 1799 and 1808. In 1816, a German researcher observed that manganese increased the hardness of iron, without reducing its malleability or toughness.

In 1826 Prieger in Germany produced a ferromanganese containing 80% manganese in a crucible. J.M. Heath produced metallic manganese in England in about 1840. The following year, Pourcel began industrial-scale production of "spiegeleisen", a pig-iron containing a high percentage of manganese, and in 1875 he started the commercial production of ferromanganese with a 65% manganese content. The major breakthrough in the use of manganese occurred in 1860. At that time, Sir Henry Bessemer was trying to develop the steelmaking process which was to bear his name. But he experienced difficulty with an excess of residual oxygen and sulphur

in the steel. The problems were overcome thanks to the beneficial effect of manganese, disclosed in a patent granted to Robert Mushet in 1856. Mushet suggested adding "spiegeleisen" after the blow to introduce both manganese and carbon and remove oxygen. This procedure made the Bessemer process possible, and thus paved the way for the modern steel industry. Ten years later, in 1866, Sir William Siemens patented the use of ferro-manganese in steelmaking so as to control the levels of phosphorus and sulphur.

Subsequently, and in contrast to all the early work involving manganese and steelmaking, Leclanché in 1868 developed the dry cell battery. This uses manganese dioxide as a depolariser in a simple yet effective dry cell and the battery market today is the second largest consumer of manganese. The history of manganese in the 20th century has been a stream of new processes and metallurgical/chemical applications developed with a significant impact on markets as diverse as beverage cans, agricultural pesticides and fungicides and electronic circuitry used in consumer products. Details of these applications are analysed later.

Manganese is a silver-white metal, not notably hard (becomes hard on alloying with carbon), brittle, capable of taking a brilliant polish but readily oxidized upon heating, reacts with water upon boiling, soluble in dilute acids. In terms of abundance, manganese is present in igneous rocks to an average extent of 0.10% (weight). In terms of cosmic abundance, in the estimate by Harold C. Urey (1952), using a base figure of 10,000 for silicon, the figure for manganese is 75. Manganese is estimated as 34th among the elements in its content in seawater, an estimated 9.5 tons per cubic mile of seawater. There are eight isotopes of manganese, ^{50}Mn through ^{57}Mn, all radioactive with exception of ^{55}Mn. Half-lives range from a fraction of a second for ^{50}Mn to approximately 140 years for ^{53}Mn. Electronic configuration $1s^2 2s^2 2p^6 3s^2 3p^6 3d^5 4s^2$. Ionic radius Mn^{2+} 0.83 Å. Metallic radius 1.365 Å. First ionization potential 7.32 eV; second, 15.7 eV. Oxidation potentials $Mn \longrightarrow Mn^{++} + 2e^-$, 1.18 V; $Mn^{2+} + 2H_2O \longrightarrow MnO_2 + 4H^+ + 2e^-$, -1.28 V; $Mn^{2+} \longrightarrow Mn^{3+} + e^-$, 1.51 V; $Mn^{2+} + 4H_2O \longrightarrow MnO_4^- + 8H^+ + 5e^-$, -1.52 V; $MnO_2 + 2H_2O \longrightarrow MnO_4^- + 4H^+ + 3e^-$, -0.168 V; $Mn(OH)_2 + OH^- \longrightarrow Mn(OH)_3 + e^-$, 0.40 V; $MnO_4^= \longrightarrow MnO_4^- + e^-$, -0.54 V; $MnO_2 + 4OH^- \longrightarrow MnO_4^- + 2H_2O + 3e^-$, -0.58 V. Other important physical properties are given under **Chemical Elements**.

Occurrence. The most common manganese ore is pyrolusite, MnO_2. Other commercial ores include braunite, Mn_2O_3; hausmannite, Mn_3O_4; and rhodochrosite, $MnCO_3$. Although not of industrial value, manganese also exists in nature as the silicate, sulfate, sulfite, and tungstate. See also **Pyrolusite**; and **Rhodochrosite**.

Manganese Nodules. These are rocks composed largely of ferromanganese oxides formed by precipitation at the bottom of lakes and the oceans. They range in size from micrometers to meters. Their morphology is highly variable. They contain up to 55% manganese, 35% iron, and 2% nickel, cobalt, and copper. Manganese nodules were first discovered in the open ocean by Thompson, Murray, and Renard during the *Challenger* expedition (1873–1876). Buchanan reported the occurrence of nodules in the Firth of Clyde, a shallow-water area, and by the end of the century at least five additional occurrences of manganese nodules in shallow marine environments had been discovered. Early workers chemically analyzed about a score of manganese nodules and hypothesized about their mechanism of growth. Two principal concepts emerged: (1) they grow by the slow precipitation of manganese from seawater; and (2) they are formed by the rapid precipitation of manganese released in submarine volcanism.

Until the 1950s, little additional work was done except for some early measurements of manganese nodule growth rates. During recent years, however, there has been a strong revival of interest in manganese nodules, stimulated both by the expansion of oceanographic facilities and the realization of the economic importance of the nodules as ores. It has been found that in large areas of the ocean floor, manganese nodules may be absent. In other areas, they may cover nearly 100% of the area. In all of the Pacific Ocean, nodules have been estimated to cover approximately 10% of the ocean floor. The estimated coverage in the Indian and Atlantic Oceans is less. The local variability in manganese nodule concentration is large. Two ocean bottom photographs only a few meters apart may show very different nodule concentrations. In some locations, the weight concentration of nodules ranges up to 5 g/cm^2.

Manganese nodules are composed of cryptocrystalline minerals. They are known to consist of three major manganese phases: (1) δMnO_2 (birnessite); (2) 10-Å manganite; and (3) 7-Å manganite. The first is the most highly oxidized form, and has a chemical composition of about

$Mn_{1.9}$. Barnes (1967) examined the depth dependence of the mineralogy in nodules taken from the Pacific. His data indicate that above 3,500 m in depth, the only important manganese phase is δ MnO_2, but, below the 3,500 m depth, both 10-Å manganite and 7-Å manganite coexist with the δ MnO_2. The observed phase changes may be pressure induced.

During recent years, the growth rates of manganese nodules have been determined by various methods. Results all indicate that the nodules measured grow at a rate of a few millimeters per million years. This does not exclude the possibility that nodules in certain areas evolve more rapidly, but it appears that most deep-sea nodules grow slowly. There is some belief that the nodules are primarily the result of bacterial fixation of manganese. Other investigators believe that the nodules are formed by inorganic precipitation of metals supersaturated in sea water. There is some experimental and theoretically tenable evidence to support both concepts.

Research gathered during the DeepSea Drilling Project (DSDP) and the International Decade of Ocean Exploration (IDOE) programs is described in the entry **Ocean Resources (Mineral)**.

Hypotheses continue to evolve concerning the manner in which manganese nodules develop. In the early 1980s, in an effort to sweep away some of the mysteries concerning the Mn nodules, a consortium of American researchers participated in the MANOP (Manganese Nodule Project) program. The program involved the creation of mathematical models, not a simple task because it has been estimated that nodule attrition is about one atomic layer of the Mn-O structure per year. As reported by J. Dymond and colleagues (Oregon State University), two different processes operate within sediments, the particular process depending on whether any oxygen remains below the sediment surface. The amount of oxygen present, of course, is a function of the biological productivity of the overlying surface waters. Where winds and currents mix the sea in the needed manner, microscopic plants and animals do well and part of their inorganic skeletons and probably about one percent of their organic tissues will sink to the bottom—along with clay washed from the land. Regardless of how deep the sea floor is, bacteria and animals dwelling on and in the sediment will oxidize organic matter. However, a small percentage of their organic matter does reach the bottom. If the falling organic matter is light, not all of the sediment oxygen will be consumed. It is proposed that toxic chemical alterations (oxic diagenesis) of the sediment can then supply metals to nodules. It is further reasoned that oxic sediments must be altered chemically to produce nodules because under such oxidizing conditions, Mn and Fe are tied up as insoluble oxides, which cannot move and thus cannot be incorporated in nodules. Several possible diagenetic reactions, such as those involving volcanic ash and skeletal opal, have been suggested. As reported by Kerr (1984), the Oregon State University team concluded that the nodule composition most typical of growth under oxic diagenesis is that of the nodule bottom most rich in trace metals from a siliceous sediment in the tropical North Pacific. In terms of the overall picture, some scientists have observed that manganese nodules should not be there, but are! Obviously Mn nodule research will require many more years in supplying answers. See also **Ocean Resources (Mineral)**.

Todorokites may be defined as calcium-bearing manganese oxides, which are found in terrestrial Mn ore deposits, in weathering products of Mn-bearing rocks, and in some Mn nodules. In some cases, todorokites are the principal constituents of Mn nodules. Knowledge collected concerning todorokites has contributed and will continue to contribute to a better understanding of Mn nodule formation in ocean waters. See also **Todorokites**.

Processing

Manganese metal can be obtained from oxide ores by reduction with carbon, aluminum, magnesium, or sodium in an electric furnace. The main form in which manganese is used is *ferromanganese*. This material contains approximately 80% manganese and 20% iron. Ferromanganese is generally made in a blast furnace or an electric-arc furnace. Usually a mixture of ores is used, proportioned to yield the final desired specifications of the alloy. To reduce slag volume, low-silica ores are preferred. It is also desirable to maintain a low phosphorus content in the alloy. The charge to the electric furnace process for making ferromanganese is the manganese ore, coke, and limestone. The loss of metal to the slag is determined by the silica present. Usually about 75% of the metal is recovered. Where high-purity manganese is produced, the ore is first roasted to MnO, then leached with H_2SO_4 to form the sulfate. The solution is then neutralized to precipitate iron and aluminum. Other impurities are removed as the sulfides. Electrolysis of the resulting solution yields a 99.94% pure manganese metal.

The high-purity (electrolytic) manganese is used as a deoxidizing agent and sometimes as a constituent of nonferrous metals where it improves strength, ductility, and hot-rolling properties. Because of their very high temperature thermal coefficient of expansion, manganese-base alloys with 72% manganese (balance is copper and nickel) are used in bimetals for switching applications. Manganese (60–80% Mn) and copper alloys find application because of their vibration-damping properties.

The standard ferromanganese (7% carbon; 74–78% Mn) is used both to produce a manganese alloy steel and as a deoxidant. As early as 1856, Robert Mushed used *spiegeleisen* (10–23% Mn; 4–5% C) in alloys. Where the carbon content of steel is critical, low-carbon ferromanganese is added. Silicomanganese is used as a blocking agent to stop the reaction of carbon and oxygen in steel. Developed in 1888, Hadfield steel contains about 13% manganese. It finds use where a very hard material is needed and it has the interesting property of increasing its hardness when subjected to repeated impacts. In the 200 series of stainless steels, manganese is replacing nickel in order to achieve more economical austenitic materials.

Manganese Inorganics. A number of chemical processes have been developed to upgrade Mn ore which produce an intermediate Mn compound. These intermediates usually are free of most siliceous matter. Although these processes were designed to convert the compound to an oxide for use in metallurgical applications, the purity of the compounds often renders them suitable for commercial use.

The ammonium carbonate process (developed by Manganese Chemicals Corp.) is the first such upgrading process that has reached commercial application. The high-grade manganese carbonate produced is sold to the chemical industry. The process involves reducing the ore to MnO by roasting with gases rich in CO as the initial step. The calcine is then ground and leached in an aqueous solution containing 18 moles of NH_3 and 3 moles of CO_2. The resulting product is decomposed to yield $MnCO_3$ and NH_3.

The manganese nitrate process is the second upgrading process that has reached commercial application. The high-purity manganese oxides produced are sold to the chemical and ferrite industries. The process involves the reaction between NO_2 and manganese ore to form manganese nitrate solution. The resulting aqueous solution is then thermally decomposed to produce MnO_2 and nitrogen oxides. The nitrogen oxides are recycled to the leaching step, while the MnO_2 is recovered and processed by reduction to Mn_2O_3, Mn_3O_4, or MnO. Processes of lesser importance include the chloride and sulfur oxide processes and bacterial leaching.

Chloride Process. In 1985, investigators at the Argonne National Laboratory reported on the success of a two-step process that extracts cobalt and manganese from low- and medium-grade ores that are mined mainly for other metals. Inasmuch as cobalt and manganese are strategically critical minerals for the United States and several other industrial nations, a viable process for secondary sources of Co and Mn is attractive. A molten salt is used to dissolve more than 90% of the Co and Mn found in common nickel- and copper-bearing ores. The salt mixture contains the chlorides of sodium, potassium, and magnesium (mp 750°C; 1382°F). Approximately one part (wt) of the ore requires four parts of the chloride mixture. The latter is recyclable. The desired metals are subsequently separated electrolytically.

Chemistry of Manganese

Manganese has a $3d^5 4s^2$ electron configuration, and compounds in all oxidation states from 0 to 7+ are known, although those of 1+ and 5+ are uncommon. The reducing power of the manganese atom (Mn $\longrightarrow$ Mn^{2+}, 1.18 V) is less than that of magnesium, although the first and second ionization potentials are closely similar, due to the higher heat of sublimation of manganese. However, manganese is oxidized by the halogens, H+ or even H_2O to the dipositive state.

Like so many other metals manganese forms compounds with nitrogen, carbon and even oxygen that exhibit unusual valences, or are even of nonstoichiometric character. With nitrogen manganese combines with unusual valence of 5+ to form Mn_3N_5; with carbon it forms Mn_3C, while with free oxygen it forms first MnO, then Mn_3O_4, and finally Mn_2O_3. An exception to this rule is the MnO_2 produced by thermodecomposition of concentrated manganese nitrate solutions where the oxygen-to-manganese ratio is 1.99+.

Manganese (0) compounds are exemplified by the carbonyls discussed below.

Manganese(I) is found chiefly in the few complex ions, such as the hexacyanomanganate(I) ion $[Mn(CN)_6]^{5-}$, produced by vigorous reduction (e.g., by aluminum in alkaline solution) of the corresponding manganese(II) ion $[Mn(CN)_6]^{4-}$, or in isocyanide complexes (formed by reduction of the diiodide with alkyl isocyanides) $[Mn(RNC)_6]^+$ where R is an alkyl radical.

Manganese(II) (manganous) compounds are obtained, as stated above, by action of water, halogens (except fluorine) or acids upon the metal, or by reduction of more highly oxidized compounds in acid solution. Many salts of Mn^{2+} are known, including all four of the common halides, the nitrate, the sulfate, the sulfite, various phosphates, the arsenate, and many salts of organic acids, e.g., the acetate, butyrate, citrate, lactate, oleate, and tartrate. The manganese(II) compounds are in general relatively resistant to oxidation, due to the stability of the half-filled $3d$ subshell. However, the oxide, MnO, and the hydroxide, $Mn(OH)_2$, are rather easily oxidized by air.

This stability of the Mn(II) state is also reflected in the relatively strong oxidation potential of Mn^{3+} (manganic) ion (the value for Mn^{3+}/Mn^{2+} being -1.51 V), and the readiness with which Mn(III) compounds disproportionate. Manganese(III) fluoride, produced by the action of fluorine on lower compounds, reacts with H_2O to produce the difluoride, hydrogen fluoride, and MnO_2. In general, however, the manganic compounds such as dimanganese trisulfate and manganese triacetate, decompose in H_2O to divalent manganese ions and Mn_2O_3, forming the MnO_2 only if the pH is definitely below 7. The phosphate, $MnPO_4$, is easily formed by action of nitric acid on manganese(II) phosphate in concentrated phosphoric acid. The fluoro salt K_3MnF_6 is formed by reduction of potassium permanganate, $KMnO_4$, in 40% hydrofluoric acid by an excess of diethyl ether or manganese(II) salt. Manganese(III) also forms a variety of complexes with chelating agents, e.g., oxalate, glycine, acetylacetone, and the like. Like other tripositive transition metal ions it forms alums. The cyanide $K_3Mn(CN)_6$ is stable. All Mn(III) compounds undergo hydrolysis except the complexes.

In addition to the dioxide and the manganites, formed by fusion of MnO_2 with alkali, manganese(IV) forms a number of complexes, such as K_2MnF_6 by reduction of potassium permanganate in 40% hydrofluoric acid with a limited amount of diethyl ether or manganese(II) salts; and Cs_2MnCl_6, by action of cold concentrated HCl containing cesium chloride on MnO_2. Complex iodates are known, e.g., $M_2^1[Mn(IO_3)_6]$, as are cyanides, formed by the action of potassium cyanide on potassium permanganate and said to be $K_4Mn(CN)_8$ (cf. $K_4Mo(CN)_8$ and $K_4Re(CN)_8$).

Manganese rarely occurs with an oxidation number of 5+. In addition to the nitride, there is another compound of interest, in that it can be formed in solution. It is an oxyanion of pentavalent manganese, MnO_4^{3-}, which occurs in the compound, $Na_3MnO_4 \cdot 7H_2O$, formed by reduction of the manganate in strongly alkaline formate or sulfite solutions or by heating MnO_2 in alkali hydroxide at very high temperature. Upon neutralization, it disproportionates to the manganate (and MnO_2).

The manganates, containing MnO_4^{2-}, and produced by alkaline oxidation of MnO_2, are the principal known compounds of hexavalent manganese. They are unstable in neutral or acidic solution, undergoing disproportionation to permanganate (MnO_4^-) and MnO_2. In basic solution, the reaction is reversible. The equilibrium is displaced toward the MnO_4^- by the action of strong oxidants.

The permanganates are strong oxidizing agents, and are usually reduced down to Mn^{2+} under acidic conditions, but to MnO_2, manganate, MnO_4^{2-}, or even hypomanganate, MnO_4^{3-}, under progressively more alkaline conditions. Permanganic acid, $HMnO_4$, and its anhydride, Mn_2O_7, can be obtained at lower temperatures, but they are unstable, decomposing above $0°C$. Permanganyl fluoride, MnO_3F, formed by the action of liquid hydrogen fluoride on potassium permanganate, decomposes at about $0°C$. In strongly acidic media, such as 100% H_2SO_4, manganese (VII) appears to exist as permanganyl ion, MnO_3^+. The sigma bond hybridization in MnO_4^-, MnO_4^{2-} and MnO_4^{3-} is best represented as d^3s.

The only compound of manganese with carbon monoxide alone is the decacarbonyl dimanganese, $(CO)_5MnMn(CO)_5$, but several hydrogen-containing carbonyls, such as $HMn(CO)_5$, halogen-containing carbonyls, such as $Mn(CO)_5Br$, alkyl carbonyls, such as $C_2H_5Mn(CO)_5$ and oxygen-function organometallic compounds, such as

$$[CH_3C(=O)O]_3Mn$$

are known. With the exception of the dicyclopentadienyl compounds, $C_5H_5MnC_5H_5$ manganese does not combine with unsubstituted hydrocarbons or their radicals.

It is interesting to note that some bacteria found near manganese ore plants have the ability to dissolve manganese oxides in solutions of pH 5–6 by the slow addition of H_2SO_4. The only requirement other than the organisms is a nutrient solution. The extraction of manganese as a sulfate is on the order of 71.7–99.9%, depending on the ore. The action of the bacteria is not fully understood.

See also **Manganese (In Biological Systems)**.

Health and Safety Factors

Health and Environment. Manganese in trace amounts is an essential element for both plants and animals and is among the trace elements least toxic to mammals, including humans. Exposure to abnormally high concentrations of manganese, particularly in the form of dust and fumes, is, however, known to have resulted in adverse effects to humans. Two kinds of disease owing to manganese are known in humans: manganic pneumonia and manganism.

Airborn manganese concentrations in the U.S. range from 0.02 to 0.57 $\mu g/m^3$ in urban areas and 0.0017–0.047 $\mu g/m^3$ in non-urban areas. The ACGIH (American Conference of Governmental Industrial Hygienists: http://www.acgih.org/) recommends a TLV (threshold limit values) of 5 mg/m^3.

Plant Safety. Of the many ferroalloy products produced in electric furnaces, ferromanganese has the greatest potential for furnace eruptions or the more serious furnace explosions.

Most of the serious eruptions of manganese furnaces can be traced to a set of conditions that cause bridging or hang-up of the charge materials so that the normal downward movement through the furnace is disrupted or retarded. Safe operation of ferromanganese furnaces requires careful control of raw material particle size, oxygen content of the ore blend, and charge stoichiometry.

The major portion of this entry was furnished by J. Y. WELSH and D. F. DE CRAENE, Chemals Corporation Baltimore, Maryland

Additional Reading

Davis, J.R.: *Metals Handbook,* 2nd Edition, ASM International, Materials Park, OH, 1998.

Greenwood, N.N. and A. Earnshaw: *Chemistry of the Elements,* 2nd Edition, Butterworth-Heinemann, Inc., Woburn, MA, 1997.

Kerr, R.A.: "Manganese Nodules Grow by Rain from Above," *Science,* **223**, 576–577 (1984).

Klimas-Tavantzis, D.: *Manganese in Health and Disease,* CRC Press, LLC., Boca Raton, FL, 1994.

Krebs, R.E.: *The History and Use of Our Earth's Chemical Elements,* Greenwood Publishing Group, Inc., Westport, CT, 1998.

Lewis, R. and N. Sax: *Sax's Dangerous Properties of Industrial Materials,* 10th Edition, John Wiley & Sons, Inc., New York, NY, 2000.

Lide, D.R.: *CRC Handbook of Chemistry and Physics* 84th Edition, CRC Press, Boca Raton, FL, 2003.

Varentsov, I.: *Manganese Ores of Supergene Zone: Geochemistry of Formation,* Kluwer Academic Publishers, Norwell, MA, 1996.

MANGANESE (In Biological Systems). Manganese is required by both plants and animals. Although its deficiency is normally a problem in small areas of fields, it has caused economic losses in the production of cereal small grains on some alkaline soils. In acid soils, manganese is more soluble and plants may be damaged by excessive uptake of the element. Reduced crop yields due to manganese toxicity on acid soils are probably responsible for greater economic losses in a number of regions of the world than are reduced crop yields as the result of manganese deficiency. Measurement of the total manganese concentration in any soil is of little value for predicting possible manganese deficiency or toxicity. The amounts of soluble manganese are more directly related to the level of manganese in plants, but soluble manganese in the soil may fluctuate over short periods because of flooding or drying of the soil or the addition of fresh organic matter. The concentration of manganese in food and in feed plants varies widely; it is more dependent upon the acidity or alkalinity of the soil than on the amount of manganese used in fertilizers.

Although established as an essential trace element, manganese is less well understood than many of the other trace elements. The evidence for its essentiality rests extensively on the consequences of limiting or curtailing the supply of the element of various organisms. Manganese deficiency has induced in most organisms studied a diminished life expectancy. The element is associated with reproductive processes.

Additional manifestations depend upon the kind of organism under observation, its age, the degree and duration of manganese deficiency, as well as the coexistence of still another deficiency. For example, in plants, a striking manifestation is chlorosis in which the leaves become pale or yellow, while the veins of the leaves remain green. Manganese plays a significant role in photosynthesis by plants, but it also participates in the regulation of several other enzymic processes.

In poultry, manganese deficiency causes a different clinical picture when it affects the egg than when it affects the hatched bird. In the case of the egg, the embryo become swollen and deformed, and their skeletons become defective and fragile ("chondrodystrophy"). Adult birds develop perosis (slipped tendon) which is an enlargement and malfunction of the tibial metatarsal joint, followed by slipping of the Achilles tendon from the condyles. The bone deformities seen in poultry also can be induced in mammals.

Manganese deficiency also results in the birth of "crooked calves" that are born with enlarged joints, stiffness, and twisted legs.

If the deficiency is induced prior to birth, there is a high intrauterine mortality and whatever young are born alive tend to suffer from an inability to coordinate their muscles (ataxia). These young also have convulsions, delayed growth, and defective bone formation. Adequate manganese intake after birth will correct many of these anomalies, but not the ataxia. If the deficiency is imposed on adult female mammals, ataxia develops infrequently. Instead there appear anemia, defective bone formation, infertility, a tendency to miscarry, and a tendency to absorb the embryo which die within the uterus. The sickly offspring are jeopardized after birth by a disinterest on the part of the manganese-deficient mothers. These avoid nursing their young even when they produce adequate milk. In males, in addition to poor growth, bone deformities and anemia, impotence and infertility develop. Adult animals also develop defects in metabolizing body fat, which are reflected in abnormal amounts and abnormal distribution of body fat. This liptropic effect of manganese extends also to the metabolism of cholesterol and, in this particular role, it can be antagonized by vanadium. The bone deformities are ascribed primarily to poor synthesis of the mucopolysaccharides that make up the matrix of the bones. The infertility is a consequence of death of the testicle's germ cells.

On the other hand, manganese in excessive amounts can cause manganese toxicity. In the past, this disease has mainly affected miners who work either in manganese mines or in ore crushing mills. The manganese ore enters the body by inhalation of the dust. Among the many miners exposed throughout the world, some develop brain symptoms. Involvement of the brain manifests itself first in mental aberrations. Later, neurological changes occur in the form of trembling, rigidity, salivation, mask-like face, and a general appearance of a person afflicted with Parkinson's disease. Chronic manganese poisoning has occurred in epidemics. The condition is incurable, but not necessarily life-limiting.

It is believed that high manganese diets in cattle will decrease fiber digestion. Manganese interferes with iron metabolism by antagonizing the enzyme system that oxidizes or reduces iron at the absorption site, thus affecting iron availability. It is also suggested that manganese may cause a condition in ruminant cattle (hypomagnesia). Most of the foregoing observations can be explained on the basis of manganese activating various cellular enzymes. Much importance has been given to the particular enzyme systems responsible for oxidative phosphorylation. These systems determine the generation and utilization of energy from foodstuffs by the cells. Additionally, manganese appears to activate many other enzymes (arginase, enolase, peroxidases). It also appears to participate in the structure of the nucleic acids responsible for the manufacture of enzymes and other proteins. Manganese probably plays a number of unique roles. No other metal replacement for the element in biological functions has been identified to date.

Manganese occurs in the liver of the animal body. Even though the amount of manganese present in mammalian tissues is very small, its concentration seems to be accurately controlled by elaborate mechanisms. These mechanisms function primarily by promoting the excretion of excesses of the element from the body rather than by regulating the amounts of manganese the body absorbs. The mechanisms are located in the liver and on the mucosa of the gut. In cases of manganese toxicity, it is assumed that these mechanisms become saturated.

One vital feature of manganese, which is not widely appreciated, is its role as an essential element in maintaining human health. Recommended daily dietary intake levels have been established by US regulatory authorities in an effort to ensure the maintenance of good health.

The exact role of manganese is not fully understood, but complex cellular reactions involving metallo-enzymes have been identified. Humans have well-developed homeostatic control mechanisms whereby manganese levels are regulated to keep them in the desired range. Medical research into conditions arising from an excess or deficit of body manganese is being carried out in a number of institutions.

Cereals and pulses (peas and beans) are the major sources of manganese in human diets, and diets containing these foods can be expected to provide adequate manganese. Dietary supplements for manganese (feeds and foodstuffs) include: manganese chloride, manganese gluconate, manganese glycophosphate, manganese hypophosphate, and manganese sulfate. The usual intake of this mineral is 2 to 5 mg/day, and absorption is 5 to 10%.

Additional Reading

Adriano, D.C.C.: *Biogeochemistry of Trace Metals,* Lewis Publishers, Boca Raton, FL, 1992.

O'Dell, B.L., R.A. Sunde: *Handbook of Nutritionally Essential Minerals,* Vol. 2, Marcel Dekker, Inc., New York, NY, 1997.

Klimas-Tavantzis, D.: *Manganese in Health and Disease,* CRC Press, LLC., Boca Raton, FL, 1994.

Staff: *Handbook of Inorganic and Organometallic Chemistry,* Gmelin Institute Series, Springer-Verlag Inc., New York, NY, 1997.

Staff: *Dietary Reference Intakes: Vitamin A, Vitamin K, Arsenic, Boron, Choromium, Copper, Iodine, Iron, Manganese, Molybdenum, Nickel, Silicon, Vanadiu,* National Academy Press, Washington, DC, 2001.

Underwood, E.J. and W. Mertz: *Trace Elements in Human and Animal Nutrition,* 5th Edition, Academic Press, Inc., San Diego, CA, 1990.

Web Reference

The Linus Pauling Institute: Micronutrient Information Center: http://lpi.oregonstate.edu/infocenter/minerals/manganese/index.html

MANGANITE. The mineral manganite is a hydrous oxide of manganese corresponding to the formula MnO(OH). It occurs in prismatic monoclinic crystals, sometimes in massive columnar forms, granular, concretionary, and stalactitic. It is a brittle mineral, with perfect prismatic cleavage; hardness, 4; specific gravity, 4.33; luster, submetallic; color, steel gray to iron black; streak, red-brown to almost black; opaque. Manganite is of secondary origin and it may itself alter to pyrolusite. It is usually associated with other manganese minerals. It is found in the Harz Mountains, Germany; Sweden; Cornwall and Cumberland, England; and in the United States in Michigan. It is an ore of manganese. See also **Pyrolusite.**

MANNANS. See **Sweeteners.**

MANNICH REACTION. Reaction of active methylene compounds with formaldehyde and ammonia of primary or secondary amines to give β-aminocarbonyl compounds.

MANNITOL. See **Sweeteners.**

MANY-BODY FORCE. An interaction between two particles that becomes modified when a third particle is present, e.g., the forces between polarizable molecules. A large part of the experimental data of physics is concerned with natural objects that may be looked upon as being made up from smaller bodies. Thus, results the so-called many-body problem; there are several approaches to its solution. This highly theoretical topic is beyond the scope of this volume, but reference is suggested to the "Many-Body Problem," in *The Encyclopedia of Physics* (R.M. Besancon, editor), 3rd Edition, Van Nostrand Reinhold, New York, 1985.

MARATHON-HOWARD PROCESS. A treatment of waste sulfite liquor from sulfite pulp manufacture to recover chemicals and reduce steam pollution. The waste sulfite is treated with line and precipitates. (1) calcium sulfite for use in preparing fresh cooking acid for the sulfite pulp process, and (2) a basic calcium salt of lignin sulfonic acid (lignin sulfonates) that can be pressed and used as a fuel of used as raw material for vanillin, lignin plastics, and other chemicals. The remaining liquor with its BOD reduced 80% is the effluent.

MARBLE. A metamorphic form of calcium carbonate, usually containing admixtures of iron and other minerals that impart variegated color patterns. Marble chips are often used as source of carbon dioxide in laboratory experiments.

MARCASITE.

The mineral marcasite, sometimes called white iron pyrites, is, like ordinary pyrites, disulfide of iron corresponding to the same formula, FeS_2. Marcasite, however, crystallizes in the orthorhombic system often yielding serrate, spear-shaped twins, hence the name "cock's comb pyrites." It is a brittle mineral; hardness, 6–6.5; specific gravity, 4.92; luster, metallic; color, light bronze-yellow; streak, greenish-black; opaque. Marcasite alters very easily and may disintegrate with the formation of sulfuric acid and iron sulfate. Fossils replaced by marcasite are therefore often destroyed after being placed in collections. Marcasite is found in numerous places in Europe, notably in Czechoslovakia, France, and England; in Mexico; and in the United States in the lead districts of Illinois, Wisconsin, and Missouri. The name marcasite is believed to be of Arabic origin and formerly was applied to common pyrite.

MARCUS, RUDOLPH A. (1923–).

Professor Marcus from the California Institute of Technology, Pasadena, California, won the Nobel prize for chemistry in 1992 for his contributions to the theory of electron transfer reactions in chemical systems.

The processes Marcus has studied, the transfer of electrons between molecules in solution, underlie a number of exceptionally important chemical phenomena, and the practical consequences of his theory extend over all areas of chemistry. The Marcus theory describes, and makes predictions concerning, such widely differing phenomena as the fixation of light energy by green plants, photochemical production of fuel, chemiluminescence ("cold light"), the conductivity of electrically conducting polymers, corrosion, the methodology of electrochemical synthesis and analysis, and more.

From 1956 to 1965 Marcus published a series of papers on electron transfer reactions. His work led to the solution of the problem of greatly varying reaction rates.

MARK-HOUWINK EQUATION.

Defines the relationship between the intrinsic viscosity and molecular weight for homogeneous linear polymers.

MARKOWNIKOFF RULE.

When a halogen acid adds to an asymmetrical ethylenic compound, the halogen usually appears on the carbon atom carrying the smaller number of hydrogen atoms; this order of addition is frequently reversed with hydrogen bromide if peroxides are present (peroxide effect).

MARTENSITE.

The chief constituent of hardened carbon tool steels. It is a solution of carbon or Fe_3C in β-iron, or and exceedingly fine-grained α-iron with carbon or Fe_3C in atomic or molecular dispersion. Carbon content up to 1%; easily obtained by quenching small bodies of hypereutectoid steel in cold water; more difficult to obtain in low-carbon steels.

MARTIN, ARCHER J. P. (1910–2002).

An English chemist and engineer who won the Nobel prize for chemistry in 1952 along with Richard L.M. Synge. His work involved partition chromatography in analysis, which led to development of new antibiotics and amino acids. He received his doctorate from Cambridge University.

MASER.

An acronym for microwave amplification by stimulated emission of radiation. The device is identical in theory of operation to the laser except that it operates at frequencies in the microwave region of the electromagnetic spectrum, rather than in the light range. See also **Lasers**.

Consider a stream of atoms in equilibrium for a given energy transition, that is, some are emitting radiation at the frequency of the transition and others are absorbing it. This process can be represented by the equation

$$h\nu_{21} = E_2 - E_1$$

where E_2 represents the energy of an atom in an excited state, E_1 represents its energy in a lower state, such as the ground state, h is Planck's constant, ν_{21} represents the frequency of a photon which would excite the atom from the state E_1 to state E_2, or conversely, the frequency of a photon that would be emitted in the reverse transition. Now consider a large number of atoms in thermal equilibrium in a closed box. The radiation inside the box will be black body radiation; in other words, the number of photons of given energy will be uniquely determined by the temperature. In addition,

the temperature will fix the proportion of atoms in the excited state, which will be given by the Boltzmann distribution

$$\frac{N(E_2)}{N(E_1)} = \exp\left(\frac{-(E_2 - E_1)}{kT}\right)$$

where $N(E_2)$ and $N(E_2)$ are the populations of the states E_1, E_2, respectively. These two assertions, governing the number of excited atoms and the number of photons, respectively, are both fundamental thermodynamic principles and should fit together into a consistent picture. To bring about the energy balance necessary for equilibrium, an extra term is introduced, corresponding to a second process of radiation emission. The second term is quite different from the first in that it represents a process whose rate (or probability) is proportional to the intensity of radiation of frequency ν falling on the atom. The radiation field stimulates the excited atom to emit, and we can contrast this process of stimulated emission with the more familiar process of spontaneous (random) emission. The latter is more familiar because it is overwhelmingly dominant so long as $h\nu/kT$ is large, and this is so in the optical and infrared regions of the spectrum, where most of our accumulated experience lies. On the other hand, if $h\nu/kT$ is small (as it is in the microwave region), stimulated emission is the more important; however, stimulated emission in the visible region may be significant under certain conditions.

In terms of photons, stimulated emission appears as the interaction of the photon with an excited atom, which leads to the emission by the atom of a second photon, identical with the primary. In other words the photon has multiplied.

To accomplish amplification it is only necessary to produce more photons by stimulated emission than are lost by absorption. The details of the balance at equilibrium show that the probability of an atom in the ground state absorbing a photon, and of one in the excited state emitting one by stimulation, are equal. Hence, if amplification is to be possible, the number of atoms in the excited state must be greater than the number in the ground state. A look at the Boltzmann equation shows that this cannot be attained at any (positive) value of temperature, which means that it cannot be attained in thermal equilibrium. The problem then is essentially one of disturbing equilibrium so as to bring about the population inversion required.

Although having a number of limitations, the ammonia maser can be used to describe the principles involved. What is involved is picking out the excited atoms from the unexcited ones, and segregating them. This implausible procedure is in certain cases possible, the most notable involving the ammonia molecule, NH_3. The structure of this molecule is pyramidal, with the nitrogen atom at the vertex and the three hydrogen atoms forming the base. It is possible for the molecule to execute vibrations in which the nitrogen atom vibrates back and forth through the plane of the hydrogen atoms. The energy difference between this excited state and the ground state (no vibration) corresponds to a wave-length in the microwave region, i.e., about 3 centimeters. Because of its lack of symmetry, the molecule in the ground state will have a dipole moment, but the average dipole moment of the molecule executing the vibrations is described as zero. If ammonia molecules are formed into a beam by allowing the gas to stream out of a collimating tube into a vacuum, and this beam is passed through a nonuniform electric field, the separation can be effected. The ground state molecules will be deflected by the nonuniform electric field, and lost to the beam; the excited molecules, however, by virtue of their zero dipole moment, experience no deflecting force, and are introduced into a resonant cavity resonating at the appropriate frequency. The cavity will then contain a preponderance of excited molecules, provided the lifetime of the excited state is long enough (as it will be at microwave frequencies). See Fig. 1.

If now an input signal of the resonant frequency is fed into the cavity, it can bring about spontaneous emission, and amplification will occur. The shortcoming of this arrangement is that the ammonia resonance cannot be tuned and it represents an impracticably narrow bandwidth. On the other hand, like any other amplifier, it can be made to oscillate, so that it can be used as a very stable frequency standard (ammonia clock).

The three-level maser provides a more versatile arrangement for achieving inversion. Three energy levels are associated with the same atomic or molecular system. See Fig. 2. A strong microwave signal of frequency ν_{31} corresponding to $E_3 - E_1$ raises some of the atoms to E_3. A limit is reached when the populations of E_3 and E_1 are the same, since then the radiation absorbed is just balanced by stimulated emission, a state of affairs known as "saturation." Not all the atoms in state E_3 will return

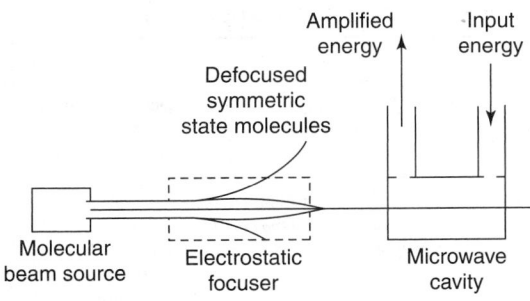

Fig. 1. Schematic representation of an ammonia beam maser. (*After Gordon, Zeiger, Townes*)

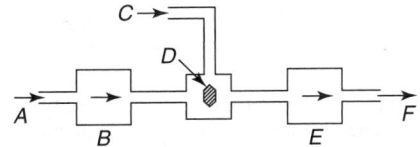

Fig. 2. Three-level maser amplifier, where A is signal input; B is the first isolator; C is the power input for pumping; D is the crystal; E is the second isolator, usually operated at low temperature as a noise reflector; and F is the signal output

directly to E_1—some will return via E_2. By the choice of states of suitable lifetimes ($\tau_2 > \tau_3$) it is possible to arrange that the number of atoms in E_2 exceeds that in E_1 because the "pump" frequency maintains a "head" of atoms in E_3. Thus E_2 and E_1 are inverted and amplification is possible.

Maser amplifiers are used where the requirement for a very low noise amplifier outweighs the technological problems of cooling to low temperatures. They have been used in passive and active radiostronomical work, in satellite communications, and as preamplifiers for microwave spectrometry. The ammonia and the atomic hydrogen masers have been studied as frequency standards and have been used in accurate tests of special relativity.

Additional Reading

Arecchi, F. et al.: *Instabilities and Chaos in Quantum Optics II,* Perseus Publishing, Cambridge, MA, 1988.
Bertolotti, M.: *Masers and Lasers: An Historical Approach,* Institute of Physics (IoP), London, UK, 1988.
Elitzur, M.: *Astronomical Masers,* Kluwer Academic Publishers, Norwell, MI, 1992.

MASS.   The physical measure of the principal inertial property of a body, i.e., its resistance to change of motion. At speeds small compared with the speed of light, the mass of a body is independent of its speed. Under these circumstances, the masses m_1 and m_2 of two bodies may be compared by allowing the two bodies to interact. Then

$$m_1/m_2 = |a_2|/|a_1|$$

where $|a_1|$ and $|a_2|$ are the magnitudes of the respective accelerations of the two bodies as a result of the interaction. This permits the measurement of the mass of any particle with respect to a standard particle (for example, the standard kilogram). At higher speeds, the mass of a body depends on its speed relative to the observer according to the relation:

$$m = m_0/\sqrt{1 - v^2/c^2}$$

where m_0 is the mass of the body as found by an observer at rest with respect to the body, v is the speed of the body relative to the observer who finds its mass to be m, and c is the speed of light in empty space as prescribed by the theory of relativity.

When relativistic mechanics is appropriate, e.g., when speeds comparable to the speed of light are involved, mass may be converted into energy and vice versa, hence the energy of the system must be converted into mass through the Einstein equation

$$E = mc^2$$

where c is the speed of light in empty space, before the conservation law may be applied.

MASS (Center of).   The center of mass is that point in a collection of mass-particles which moves as if the total mass of the collection were concentrated there and the resultant of all the external forces were acting there. The position vector of such a point is given by

$$\mathbf{r} = \frac{\sum_{i=1}^{n} m_i \mathbf{r}_i}{\sum_{i=1}^{n} m_i}$$

where m_i is the mass of ith discrete particle, $\mathbf{r}_i$ is the position vector of ith discrete particle.

MASS DEFECT.   The difference Δ between the atomic number A and the atomic mass M of a nuclide, $\Delta = A - M$. The negative of the mass defect, $-\Delta$, is known as the mass excess.

MASS NUMBER.   The total number of nucleons in the nucleus of an atomic species is its mass number, which then is numerically equal to the sum of the atomic number and the neutron number of the species. See also **Chemical Elements**.

MASS SPECTROMETRY.   A general quantitative and qualitative analyzer for most components in all types of samples—gas, liquid, or solid, but with some volatility limitations. A complete analysis is obtained from nanogram samples in a few seconds or minutes. The range is from parts per billion to 100% purity. Accuracy is $\pm 1\%$; the specificity is good. Conventional methodology is first described after which recent trends are presented.

Operating Principle.   With reference to Fig. 1, the sample to be studied is introduced into an evacuated area (ion source), where it is ionized, accelerated by an electrostatic field, and separated according to mass. The various masses are collected and measured.

Production of Ions.   Several methods are used: (1) by bombardment with electrons from a heated filament; (2) by application of a strong electrostatic field (field ionization, field desorption); (3) by reaction with an ionized reagent gas (chemical ionization); (4) by direct emission of ions from a solid sample that is deposited on a heated filament (surface ionization); (5) by vaporization from a crucible and subsequent electron bombardment (e.g., Knudsen cell for high-temperature studies of solids; and (6) by radio-frequency spark bombardment of sample for parts-per-billion (ppb) elemental analysis of solids as encountered in metallurgical, semiconductor, ceramics, and geological studies. Ions also are produced by photoionization and laser ionization.

Fragmentation.   Ionization usually is accompanied by partial fragmentation of the molecule. The fragmentation pattern is constant for a specific molecule and operating conditions. Fragmentation complicates computation, but permits distinguishing between isomers and gives molecular-structure information.

Ion Separation.   After acceleration, the ions are focused and separated according to mass. The most common separating means is a magnetic field,

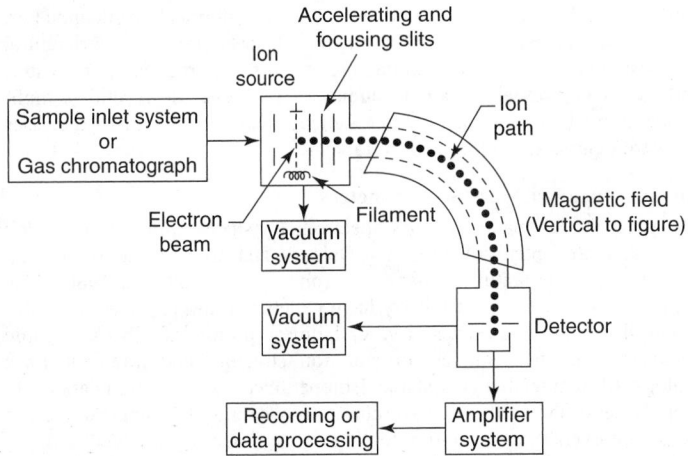

Fig. 1. Basic elements of conventional mass spectrometer system

which causes the ions to follow curved paths of radii proportional to their masses. Many different geometrics are used. The masses may be scanned by varying the accelerating voltage or magnetic field. Other separation means include: (1) combinations of electrostatic fields (double focusing cycloidal focusing); (2) crossed alternating electrical fields (quadrupole mass filter); and (3) use of a filled-free drift tube combined with pulsed ion source and gated detection (time-of-flight). Still additional means are omegatron, radio frequency, and cyclotron resonance.

Detection. Commonly used detection means are: (1) *electrical*—ion beams are successively scanned across a collector where they pick up electrons. The resulting current is amplified with an electron multiplier and/or electrometer and recorded or computer-processed. (2) *photographic plate*—ions strike a photographic plate, activating the emulsion and thus giving a line for each mass, after development. Line intensity is proportional to ion abundance. All masses are recorded simultaneously.

Sample Introduction. A variety of inlet systems is available for gases, liquids, and solids: (1) *heated batch inlet*. The sample is expanded into a volume at about 50 micrometers pressure and bled into the ion source through a molecular or viscous leak. This method will handle gases, liquids, and solids with vapor pressures above approximately 1 torr at 350°C. (2) *Direct introduction system*. The sample is inserted directly into the ion source through a vacuum lock on a heatable probe. Liquids and solids with vapor pressures above approximately 10^{-7} torr at 350°C can be handled. (3) *Direct insertion by venting*, or through a vacuum lock (spark or surface ionization). This method is well suited for involatile samples. (4) *Gas chromatograph interface*. A continuous inlet which permits mass spectrometric analysis of the separated components as they emerge from the gas chromatograph.

The combination of gas chromatograph and mass spectrometer provides a separating and identifying capability not achievable by other means, particularly for very small samples.

Data Processing. The high data output rate of many mass spectrometer systems requires data processing to fully utilize the capability of the instrument. The present trend is toward systems using small dedicated computers with digital tape, core, or disk memory, and printer, plotter, and cathode-ray tube output. Typical outputs available are: (1) mass and abundance printouts; (2) mass versus abundance plots; (3) elemental composition printouts (high-resolution mass spectrometer); (4) "total ionization chromatogram" plot (summed ion plot, similar to gas chromatogram); (5) mass chromatogram (similar to total ionization chromatogram, but for a selected mass or masses only); and (6) quantitative analysis printout.

Outputs are available from raw data, data with background or other spectra subtracted, and normalized data. A number of other options of data selection, manipulation, and output are usually available from the instrument manufacturers.

Uses and Applications. Common uses for mass spectrometers include: (1) compound identification; (2) elemental formula determination; (3) molecular-structure determination; (4) quantitative mixture analysis; (5) ppb solids elemental analysis; (6) isotope ratio determination; (7) leak detection (helium tracer); (8) residual-gas analysis in vacuum systems; and (9) age dating.

Mass spectrometers are widely used for studies and determinations in: organic chemistry; petroleum and biological laboratories; nuclear investigations; geochemistry and cosmochemistry; metallurgical, semiconductor, and ceramics investigations; pollution control; space programs; agricultural and pesticide research and manufacture; flavors and fragrances chemistry; and as a basic research tool in studies of ion-molecule reactions, high-temperature chemistry kinetics, free radicals, and thin films. Broad classes of instruments are summarized in Table 1.

Advancements in Mass Spectrometers

The great strides made by mass spectrometry since its inception several years ago are aptly put forth by Delgass and Cooks (see reference), who describe the status of mass spectrometry as of the late 1980s. The applications of mass spectrometry have penetrated into physical chemistry (bond dissociation energies, ion enthalpies, proton affinities); organic chemistry (structure studies, organic ion structure and fragmentation); biology (drug metabolism, stable isotope tracer work, modifications in biopolymers); the earth sciences (chronology/dating of geological and life extinction events); and environmental science (trace organic analysis).

Mass spectrometry is undergoing a rapid development that shows little indication of abating either in the areas of instrument refinement or of

TABLE 1. TYPES AND CHARACTERISTICS OF MASS SPECTROMETERS

General class	Type	Major uses
Leak detector	Magnetic analyzer (helium only)	Detects small leaks, using helium gas tracer.
Residual-gas	Magnetic or quadrupole analyzer	Analysis of gases in vacuum systems
Low-resolution	Magnetic, cycloidal, or quadrupole analyzer	Analysis of gases or light liquids
Medium-resolution	Magnetic analyzer	Identification, molecular-structure studies, mixture analysis, isotope ratios.
High-resolution	Magnetic and electric (mass and energy) focusing	Molecular-structure determinations, ppb solids analysis, isotope ratios of solids, high temperature chemistry

extended applications. Mass spectrometry is expected to play a major role in the revitalized science of materials and surface phenomena.

Tandem Mass Spectrometry. Coupling mass spectrometers in series has many advantages for the analysis of specific organic compounds in complex mixtures. Sensitivity to picograms of targeted compounds can be achieved with high specificity and almost instantaneous response. As reported by McLafferty (see reference), the targeted compound is selectively ionized, and its characteristic ions are separated from most others of the mixture in the *first* mass spectrometer. The selected primary ions then are decomposed by collision and, from the resulting products, the *final* (or second) mass analyzer selects secondary ions characteristic of the targeted compound. Tandem mass spectrometry (MS-MS) can achieve specificities and sensitivities equivalent to those of methods such as radioimmunoassay and gas chromatography/mass spectrometry, while performing analyses in much shorter times. Just a few of the materials and systems successfully studied by tandem mass spectrometry include: polynuclear aromatics, DNA pyrolysis, steroid mixtures, ion structures, stereoisomers, pyrolysis of bacteria, alkaloids in plants, penicillins, polychlorodibenzodioxins, petrochemicals, drug metabolites, enkephalins, peptide mixtures, diesel exhaust, odors in the air, concealed drugs, parathion in lettuce, and ion plasmas.

Tandem mass spectrometry has been particularly effective in molecular structure determinations. To increase the number and absolute abundance of peaks in the secondary mass spectrum, it is necessary to add energy to the separated primary ions. Collisionally activated dissociation (CAD) is frequently used.

Fourier Transform Mass Spectrometry (FTMS). This technique enables chemists to use mass spectrometry in expanding and new ways. The technique enables the measurement of high molecular masses and, by application of ion manipulation (MS-MS), molecules can be degraded into more manageable pieces for which accurate mass measurement is feasible. As observed by Gross and Rempel, the unique applications of FTMS result from its ability to store ions. Ions in the trap can be reacted in very specific ways (chemical ionization) or activated by using lasers to give photodissociation spectra. Instead of operation at 1 torr of pressure as in the case of conventional mass spectrometers, FTMS chemical ionization must be conducted at 10^{-6} torr of reagent gas and 10^{-8} torr of sample.

Coupling Lasers with Mass Spectrometers. The analysis of inorganic atomic species can be facilitated by using lasers with mass spectrometers. A tunable dye laser, by itself or in combination with a pump laser, ionizes atoms by resonant excitation processes. The ions are then analyzed in the mass spectrometer. This combination of techniques has much potential for overcoming traditional limits of sensitivity and selectivity and will lead to increasing applications in analytical chemistry. As observed by Fassett, et al. (see reference), a large potential exists for the application of *multiphoton resonance ionization mass spectrometry*, ranging from basic spectroscopic studies of atoms and molecules to the detection of solar neutrinos and quarks. Discovery of the optogalvanic effect (Green, et al.) and earlier work (Hurst, et al.) led to the development of *laser-enhanced ionization* (LEI). It has been determined that elements suitable for resonance ionization by the one-photon-resonant, two-photon ionization scheme (wavelengths between 260 and 355 nm) include: Na, Ca, Ba, Cr, Mn, Sc, Cr, Mn, Ru, Rh, Pd, Pt, Au, Ga, Ge, Sn, Bi, Eu, Gd, Tb, Ho, Tm,

be discarded. See also **Filtration**. (2) In *crossflow*, the influent unprocessed stream is separated into two effluent streams, known as the *permeate* and *concentrate*, respectively. The permeate is that fraction which has passed through a semipermeable membrane; the concentrate is that stream which has been enriched with the solutes or suspended solids, i.e., those materials which have not passed through the membrane. This design permits the membrane medium to operate continuously in a self-cleaning mode, with solutes and solids swept away by the concentrate stream which is running parallel to the membrane (hence the term "crossflow"). As with conventional filtration, sometimes the trapped material (filter cake) is the principal desired end product; in other cases, the desired product is clarified effluent. However, in contrast with conventional filtering, membrane, methodologies not only separate solids (or gases) from liquids (or gases), but proper selection of the membrane will allow separation of solids (particles) by size range. Thus, the permeate (what passes through the membrane) will contain much smaller particle sizes (in the molecular range) than will the concentrate. Again, depending upon the objective of the process, the permeate or the concentrate will be the principal product of interest. In some instances, both products may be of vital interest. See Fig. 1.

Membranes may be *isotropic*, i.e., their pore structure and material are the same throughout the membrane; or they may be *anisotropic*, i.e., they have a dense skin layer on top which defines the degree of separation effected, with a spongy support layer underneath. The dimensions of the pores range widely as detailed later in this article. Membranes are made by a number of processes, including solvent casting and mechanical stretching to form pores in an otherwise impervious film. Irradiation, followed by acid etching, has been used to create pores. The cellulose acetate membrane (Fig. 2) is made by casting thin sheets of polymer dissolved in a water-miscible solvent on a flat plate, usually glass. Shortly after casting, the cast solution is immersed in water. The water diffuses into the solution and causes the polymer to coagulate at a rate that is a complex function of the polymer and solvent properties. These membranes are porous throughout, possessing a thin, relatively dense skin near one surface.

The range of small particles, molecules, and ions dealt with by membrane separation technology, particularly as encountered in the biochemical field are as follows.

Microfiltration

This process effects separations in the 0.02–2.0 micron range and historically has been run in the perpendicular flow mode, requiring disposal of the membrane medium as a result of binding by the retained material.

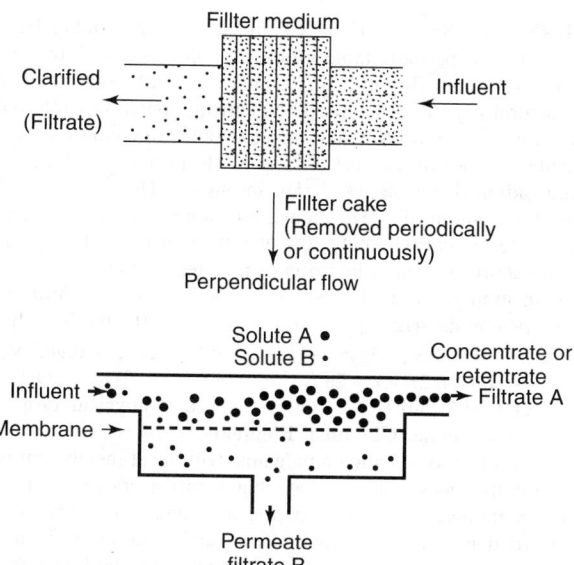

Fig. 1. Perpendicular flow contrasted with crossflow. Perpendicular flow is shown here in the familiar terms of conventional filtering, but the principle is the same if a membrane is used. There is one influent and a single effluent. In crossflow, there are two effluents, a concentrate (or retentate) and a permeate, usually both liquids of different concentrations in terms of particle size. However, membranes are also used for separating acid gases and hydrocarbons

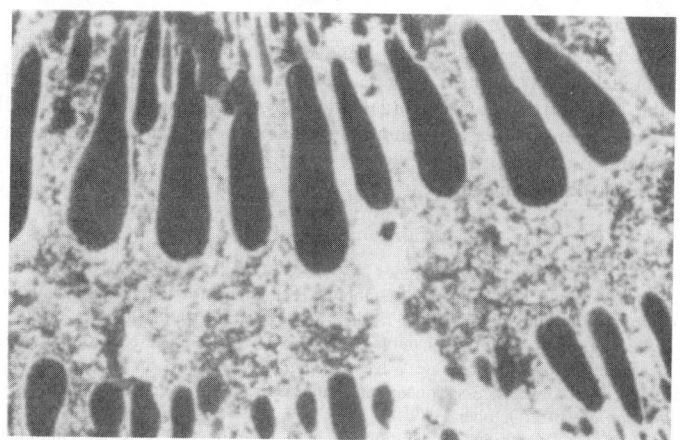

Fig. 2. Facsimile of photomicrograph of cellulose acetate membrane, prepared by solvent casting process, clearly showing pore structure

Crossflow technology is increasing, as it proves practical. Microfiltration membranes are of an isotropic and homogeneous morphology, i.e., the pore structure is consistent throughout. There is some movement, however, toward the use of "skinned" anisotropic membranes. Microfiltration membranes are available in a wide variety of polymers, including some that are quite chemically inert. They also are available as tubular, hollow fiber, or capillary fiber elements.

Ultrafiltration

This process effects separations in the 0.002 to 0.2 micron range—more specifically described as the 500–300,000 molecular-weight cutoff range, requiring pore sizes of from 15 to 1000 angstroms. For practical reasons, ultrafiltration almost always requires a crossflow configuration. Because of the size and the gelatinous nature of many of the solutions and particles handled by this process and that the membrane retains, an ultrafiltration membrane would have a very short life if used in the perpendicular flow mode. Nearly all membranes used in ultrafiltration are anisotropic, as previously defined. Membranes usually are of a homogeneous material, in that they consist of the same polymer or copolymer throughout their structure. Membranes must be made of tough, relatively inert materials, such as polysulfone polymer, or cellulose acetate polymer, the latter used more extensively in the earlier days of ultrafiltration.

Reverse Osmosis

Sometimes called *hyperfiltration*, this is the most technically complex class of membrane technology. Reverse osmosis effects separations both at the micromolecular and ionic size ranges. Pore sizes range from 5 to 15 angstroms. These membranes can effect separation of solutions down to a molecular weight of 150 and sometimes lower. Reverse osmosis membranes are anisotropic. Cellulose acetate has long been the favorite, especially for many industrial and medical applications. However, homogeneous polyamide-type membranes are finding a large share of the market and are increasingly favored for seawater desalting purposes.

Applications of Membrane Separations Technology

The food, biochemical, and petrochemical industries are the largest users. In the food processing industry, there are three main categories of use—processing, waste treatment, and pure water make-up. Processing applications include milk concentration and fractionation, numerous fermentation operations, and the production of colorants, among many others. Waste handling operations include corn processing byproducts, soybean protein reclamation, and handling meat processing oils and fish processing proteins and oils.

Membrane separations technology is also widely used in the pharmaceutical industry in connection with recovering antibiotic products. The list of references given at the end of this article is rather long. There the reader will find a wealth of information on applications, including the growing recognition by the petroleum and petrochemical industries of the viability of membrane technology as a replacement or companion for the more traditional separation processes.

Membranes Configurations for Use in Separation Equipment

For insertion into separation equipment, frequently with large throughputs, membranes are available in several configurations: (1) the tubular form ($\frac{1}{2}$ in; 12.5 mm) is common; (2) hollow-fiber; (3) plate-and-frame; and (4) spiral wound. The Paulson reference covers membrane formats in considerable detail.

See also **Desalination**.

Additional Reading

Abelson, P.H.: "Synthetic Membranes," *Science*, 1421 (June 23, 1989).

Beaudry, E.G. and K.A. Lampi: "Membrane Technology for Direct-Osmosis Concentration of Fruit Juices," *Food Technology*, 121 (June 1990).

Bedzyk, M.J. et al.: "Diffuse-Double Layer at a Membrane-Aqueous Interface Measured with X-ray Standing Waves," *Science*, 52 (April 6, 1990).

Carroll, L.E.: "New Process Concentrates Juices, Preserving 'Fresh Notes,'" *Food Technology*, 148 (October 1989).

Dziezak, J.D.: "Membrane Separation Technology Offers Processors Unlimited Potential," *Food Technology*, 107 (September 1990).

Friedman, R.: "Seawater to Drink," *Technology Review (MIT)*, 14 (August/September 1989).

Hsieh, H.P.: *Inorganic Membranes for Separation and Reaction*, Elsevier Science, New York, NY, 1996.

Kosenoglu, S.S., J.T. Lawhon, and E.W. Lusas: "Use of Membranes in Citrus Juice Processing," *Food Technology*, 90 (December 1990).

Kosenoglu, S.S., J.T. Lawhon, and E.W. Lucas: "Vegetable Juices Produced with Membrane Technology," *Food Technology*, 124 (January 1991).

Matsurra, T.: *Synthetic Membranes and Membrane Separation Processes*, CRC Press, LLC., Boca Raton, FL, 1994.

Noble, R.D., C.A. Koval, and J.J. Pellegrino: "Facilitated Transport Membrane Systems," *Chem. Eng. Progress*, 58 (March 1989).

Noble, R.D. and S.A. Stern: *Membrane Separations Technology: Principles and Applications*, Elsevier Science, New York, NY, 1995.

Paulson, D.J., R.L. Wilson, and D.D. Spatz: "Crossflow Membrane Technology and Its Applications," *Food Technology*, 77–87 (December 1984).

Rousseau, R.W.: *Handbook of Separation Process Technology*, John Wiley & Sons, New York, NY, 1987.

Singh, R.: "Surface Properties in Membrane Filtration," *Chem. Eng. Progress*, 59 (June 1989).

Spillman, R.W.: "Economics of Gas Separation Membranes," *Chem. Eng. Progress*, 41 (January 1989).

MENDELEVIUM. [CAS 7440-11-1]. Chemical element, symbol Md, at. no. 101, at. wt. 256 (mass number of known isotope), radioactive metal of the Actinide series, also one of the Transuranium elements. The element was produced synthetically and first identified by A. Ghiorso, B.G. Harvey, G.R. Choppin, S.G. Thompson, and G.T. Seaborg at the University of California at Berkeley in 1955. ^{256}Md was produced by the bombardment of ^{253}Es on gold foil with 48-MeV alpha particles in the 60-inch cyclotron at Berkeley. By ion exchange treatment of the dissolved gold foil, only one or two atoms of ^{256}Md were obtained, which decayed (half-life 1.3 hours) by K-electron capture to ^{256}Fm, which underwent its characteristic spontaneous fission.

Probable electronic configuration $1s^2 2s^2 2p^6 3s^2 3p^6 3d^{10} 4s^2 4p^6 4d^{10} 4f^{14} 5s^2 5p^6 5d^{10} 5f^{13} 6s^2 6p^6 7s^2$. Ionic radius Md^{3+} 0.96 Å.

Another isotope, ^{255}Md, is also formed during the bombardment of ^{253}Es by alpha particles. It also decays by electron capture, and has a half-life of 30 minutes.

Regarding the first identification, scientists considered it notable in that only in the order of 1 to 3 atoms per experiment were produced, thus making Md the first element to be discovered on an atom-at-a-time basis. The techniques developed in the search for Md served as a prototype for the discovery of subsequent elements in the Transuranium series.

See also **Chemical Elements**.

Additional Reading

Fuger, J. and L.R. Morss: *Transuranium Elements: A Half Century*, American Chemical Society, Washington, DC, 1992.

Ghiorso, A., B.G. Harvey, G.R. Choppin, S.G. Thompson, and G.T. Seaborg: "New Element Mendelevium, Atomic Number 101," *Phys. Rev.*, **98**, 5, 1518–1519 (1955). (A classic reference.)

Hulet, E.K. et al.: "Mendelevium: Divalency and Other Chemical Properties," *Science*, **158**, 486–488 (1967).

Lide, D.R.: *CRC Handbook of Chemistry and Physics* 84th Edition, CRC Press, LLC., Boca Raton, FL, 2003.

Seaborg, G.T.: *Transuranium Elements*, Dowden, Hutchinson & Ross, Stroudsburg, Pennsylvania, 1978. (A classic reference.)

Seaborg, G.T. and W.D. Loveland: *The Elements beyond Uranium*, John Wiley & Sons, Inc., New York, NY, 1990.

MENDELEYEV, DIMITRI (1834–1907). Born in Siberia, Mendeleyev made a fundamental contribution to chemistry in 1869 by establishing the principle of periodicity of the elements. His first periodic table recognized the regular variation in the chemical and physical properties of the elements and classified the 63 elements known at the time into groups placing the elements in ascending order of atomic weight and grouping them by similarity of properties. So accurate was Mendeleyev's thinking that he predicted the existence and atomic weights of several elements that were not actually discovered until years later. The original table has been modified and corrected several times, notably by Moseley, but it has accommodated the discovery of isotopes, rare gases, etc. Its importance in the development of chemical theory can hardly be overestimated. See also **Becquerel, Antoine Henri (1852–1908)**; **Mosley, Henry (1887–1915)**; and **Periodic Table of the Elements**.

MENISCUS. The curved surface of a liquid, particularly noticeable in vessels of tubes of small diameter and due to the surface tension of the liquid. If the liquid wets the containing vessel, the meniscus is concave; otherwise it is convex. The meniscus of mercury in glass is convex.

MERCAPTANS. Hydrogen sulfide yields two classes of organic compounds: (1) hydrosulfides, and (2) sulfides. The hydrosulfides are termed mercaptans, a name derived from *mercurium captans*, because of their ability to react with mercuric oxide to form crystalline compounds. Mercaptans also are termed thioalcohols and sulfur-alcohols. A more general term *thiols* also is used. This term not only embraces mercaptans, but also covers thioethers, sulfhydrates, and thiophenols.

Ethyl mercaptan C_2H_5SH, (legal label name for ethanethiol) one of the better known mercaptans, is an odorous liquid, mp $-121°C$, bp $36–37°C$, sp gr 0.839. The compound is very slightly soluble in H_2O; soluble in alcohol and ether. It is prepared by distilling ethyl potassium sulfate with potassium hydrogen sulfide. Additional mercaptans can be prepared in a similar manner with the corresponding proper ingredients. All mercaptans have unpleasant garlic-type odors; when oxidized with HNO_3 they yield sulfonic acids.

Some formulations for styrene-butadiene rubber (GR-S) contain dodecyl mercaptan which plays the role of a chain-transfer agent used to control the molecular weight of the final synthetic product.

MERCURY. [CAS: 7439-97-6]. Chemical element, symbol Hg, at. no. 80, at. wt. 200.59, periodic table group 12, mp $-38.84°C$, bp $356.58°C$, density 13.546 g/cm^3 (liquid), 14.193 g/cm^3 (solid). Solid mercury has a rhombohedral crystal structure. The element, sometimes referred to as quicksilver, is a silver-white liquid metal at standard conditions. There are seven stable isotopes of mercury, ^{196}Hg, ^{198}Hg through ^{202}Hg, and ^{204}Hg, and seven radioactive isotopes, ^{192}Hg through ^{195}Hg, ^{197}Hg, ^{203}Hg, and ^{205}Hg. With exception of ^{194}Hg (half-life of approximately 130 days) and ^{203}Hg (half-life of approximately 46 days), the half-lives of the radioactive isotopes are short, measured in terms of minutes or hours.

First ionization potential 10.434 eV; second, 18.65 eV; third, 34.3 eV. Oxidation potentials Hg $\longrightarrow \frac{1}{2}Hg_2^{2+}$ + e$^-$, -0.7986 V; Hg $\longrightarrow$ Hg^{2+} + 2e$^-$, -0.852 V; Hg$_2^{2+}$ $\longrightarrow$ 2Hg^{2+} + 2e$^-$, -0.905 V. Hg + 2OH$^-$ $\longrightarrow$ HgO + H$_2$O + 2e$^-$, -0.098 V; 2Hg + 2OH$^-$ $\longrightarrow$ Hg$_2$O + H$_2$O + 2e$^-$, -0.123 V. Other important physical properties of mercury are given under **Chemical Elements**.

Mercury forms alloys, called amalgams, with most metals, but not with iron or platinum; does not wet glass but forms a convex surface when in a glass container; is slightly volatile at ordinary temperatures and a health hazard due to its poisonous effect; slowly tarnishes in moist air; upon heating in air or oxygen, somewhat below its boiling temperature of 357°C, forms mercuric oxide slowly, as in the classical experiment by Lavoisier on the composition of air; may be purified by distillation and condensation (health hazard); unattacked by dilute HCl or H_2SO_4, but dissolved by dilute or concentrated HNO_3 with the formation of mercurous and mercuric nitrates, respectively, and by hot concentrated H_2SO_4 with the formation of both mercurous and mercuric sulfates; unattacked by alkalis. Discovery ancient.

When cooled to sufficiently low temperatures, mercury becomes *superconducting*, virtually conducting electricity with no resistance. See also **Superconductivity**.

Mercury was mined as early as 500 B.C. and currently ranks tenth in worldwide production of nonferrous metals. The unusual combination of physical properties possessed by the element give it an importance exceeding its production rating. The chief source of mercury is cinnabar, HgS, the red sulfide that contains 86.2% mercury. See also **Cinnabar**. Although the mineral occurs widely throughout the world, relatively few deposits are of commercial importance, notably those in China, Italy, Mexico, the Philippines, Peru, Spain, the former Soviet Union, the former Yugoslavia, and the United States. World reserves of mercury currently are estimated at 4 million flasks, of which sources in the United States account for about 300,000 flasks. A flask contains 76 pounds (34.5 kilograms) of the liquid metal.

The ore is concentrated to about 25–50% mercury by flotation. Beneficiation of mercury ores is not commonly practiced. The concentrate is roasted: $HgS + O_2 \longrightarrow Hg + SO_2$. The process is essentially one of distillation because the freed mercury quickly volatilizes, after which it is condensed with a resulting purity of 95% (furnace plants) to 98% (retort plants). The mercury is further refined, through filtering, oxidation, acid leaching of the impurities, or by distillation, to yield prime or virgin mercury with an average purity of 99.9%. This purity is satisfactory for all but the most exacting requirements. Some special mercury chemicals may require that the virgin mercury be triple distilled. Significant quantities of secondary mercury are recovered from waste products, such as dental amalgams, sludges, used batteries, used instruments, and other mercury-bearing materials. Secondary recovery accounts for close to 20% of the total domestic production of mercury in the United States.

Uses of Mercury

Because of the poisonous nature of Hg, wherever possible the industrial uses of the element have been reduced and research to find suitable substitutes for Hg for many applications has accelerated during recent years. Traditionally, mercury's largest usage has been in connection with electrical apparatus (switching, etc.), the electrolytic preparation of chlorine and caustic soda (mercury cells), in antifouling and mildew-proofing agents and paints, in industrial and clinical thermometers, in pharmaceutical preparations, in agricultural herbicides and pesticides, in dental preparations, in the preparation of amalgams, and in use of Hg as a catalyst for certain chemical reactions.

The consumption of mercury by the chlor-alkali industry may range from 15% to as high as 35% of the total in any given year, depending upon new construction. Large amounts of mercury are required for the start-up of mercury cell operations, whereas replacement requirements are quite low. Several areas of mercury use are declining gradually, particularly in the pharmaceutical field where sulfa drugs, iodine, and various antiseptics and disinfectants have made inroads on mercury chemicals. Mercury compounds, used for many years in the treatment of syphilis, for example, largely have been displaced by antibiotics and other treatments. Because of the fundamentally toxic nature of mercury and its compounds, the agricultural uses of mercury formulations are being deemphasized, with constant research for substitute materials. In the dental field, a number of metal powders, porcelain, and plastic materials have displaced mercury amalgams in many dental applications. In the explosives field, several compounds, such as lead azide, diazodinitrophenol, and other organic initiators, are serving the same function as mercury fulminate. The use of mercury as a heat-transfer medium in boilers was essentially abandoned a number of years ago. On the other hand, mercury-base catalysts are increasing in application and, to date, suitable substitutes for mercury in the antifouling and mildew-proofing formulations area have not been found. The essential properties of mercury that will be difficult to replace are its high specific gravity, fluidity at room temperatures, and excellent electrical conductivity.

In addition to some diminishment of mercury usage for various products because of increasing awareness of its toxicity potential, conservation-minded technologists also have pointed to the relatively limited world resources of the metal. Considerable ingenuity has been used to replace mercury. For example, diaphragm cells can be used in caustic-chlorine production, organic biocides are replacing mercury-containing compounds, gold recovery is accomplished by the cyanide process rather than by the amalgamation process, and plastic paints and copper oxide paints can be used in place of mercuric biocidal paints. The use of the diaphragm cell for chlorine-alkali production possibly is the most dramatic substitution in terms of mercury conservation. Diaphragm cells require no mercury, whereas the traditional mercury cells once accounted for as much as 35% of the mercury use in some years.

Chemistry and Compounds

The apparent anomaly between mercury and the lighter elements of transition group 2, in that mercury regularly forms both univalent and divalent compounds, while zinc and cadmium do so very rarely, is partly understood from the observation that mercury(I) salts ionize even in the gaseous state to Hg_2^{2+}, rather than Hg^+. Evidence for this double ion is provided by its Raman spectral line, by the lineal $Cl-Hg-Hg-Cl$ units in crystals of mercury(I) chloride, and by the emf of mercury(I) nitrate concentration cells. The anomaly is further removed by the observation that cadmium also forms a (much less stable) diatomic ion Cd_2^{2+}, e.g., in $Cd_2(AlCl_4)_2$.

Oxides. Heating of mercury in air yields the (divalent) oxide HgO, which at higher temperatures decomposes into its elements. Mercury(II) oxide is also precipitated from solutions of mercury(II) salts by alkaline solutions. Alkalies precipitate a yellow form, while alkali carbonates give a red one. The yellow is apparently a finely divided form of the red, since they are crystallographically identical, but differ slightly in certain chemical and physical properties, including solubility. Mercury(II) oxide exhibits solubility in solutions of alkali salts, which is attributed to formation of complex ions such as $[Hg(OH)_2NO_3]^-$ and $[Hg(OH)_2SO_4]^{2-}$.

Halogen Compounds. All eight compounds of univalent and divalent mercury with the single halogens are known, as well as several compounds of mercury(II) with two halogens, such as HgBrI and HgClI. The mercury(I) halides are insoluble in water, with the exception of the fluoride which, like mercury(II) fluoride, is hydrolyzed by water. Like the zinc and cadmium halides, mercury(II) halides behave anomalously in aqueous solution, and for similar reasons, i.e., the presence of complex ions and un-ionized molecules. In the case of mercury(II) halides, with their more covalent character than zinc or cadmium halides, the ionization is somewhat less, and the concentration of Hg^{2+} relatively low. Thus, in aqueous solution, mercury(II) chloride, $HgCl_2$, is present largely as un-ionized molecules, but also ionizes $HgCl^+$ and Cl^-, and only secondarily and to a slight extent to give Hg^{2+}. In the presence of added Cl^-, an $HgCl_2$ solution is a complex system involving equilibria between $HgCl_2$, $HgCl^+$, Cl^-, Hg^{2+}, and the complex ions $HgCl^{3-}$ and $HgCl_4^{2-}$. Similarly, the hydrolysis of $HgCl_2$, though slight, involves several equilibria whose relative importance varies with the concentration of the solution. In more concentrated solutions the hydrolysis of $HgCl_2$ to $HgOHCl$, Cl^- and H^+ is prominent, while in more dilute solutions the most important equilibria involve the ionization $HgCl_2$ to $HgCl^+$ and Cl^-, and the hydrolysis of $HgCl_2$ to $[HgOHCl_2]_2^{2-}$ and H^+, and that of $HgCl^+$ to $HgOHCl$ or $[HgOHCl]_2$ and H^+. Finally, oxyhalides, such as $HgBr_2 \cdot 3HgO$, $HgCl_2 \cdot 2HgO$, $HgCl_2 \cdot 3HgO$ and $HgCl_2 \cdot 4HgO$ are also obtainable, usually by action of alkali hydroxides upon mercury halides. Mercury(II) iodide, like the oxide, is polymorphic. It has three forms, yellow, red, and white, the second being the most stable up to 127°C, where it undergoes a definite transition to the yellow. The colorless HgI_4^{2-} is very stable, especially to alkalies, and is used in Nessler's reagent.

Salts. Mercury forms many salts, both of mercury(I) and mercury(II). In general, action of oxidizing acids upon the metal yields the latter, while the former requires either a limited amount of the oxidant or indirect methods. Mercury(I) salts are made by treating a solution of a soluble mercury(II) salt with metallic mercury. Thus, heating mercury with H_2SO_4 or HNO_3 yields mercury(II) sulfate or nitrate, $HgSO_4$ or $Hg(NO_3)_2$, respectively, crystallizing as hydrates, while mercuric(I) nitrate results from the use of cold acid in limited amount, and mercuric(I) sulfate is produced by the last method as well as from mercury(I) nitrate and sulfuric acid. The other salts of both univalent and divalent mercury include the acetates, antimonates, arsenates, bromates, carbonates, chlorates, chromates, fluorosilicates, iodates, oxalates, perchlorates, periodates, phosphates, tartrates, thiocyanates, tungstates, uranates, and vanadates. Also known in both valences are the arsenides (from arsine and the mercury solutions), azides (from hydrozoic acid and the mercury solutions), nitrides and phosphides. Only mercury(II) selenide and telluride exist. Hg_2S_2 has been reported to be obtained as a black powder, but is believed to be a mixture of Hg and HgS. The latter exists in two forms, the black form that is usually precipitated by hydrogen sulfide, and the red, cinnabar, precipitated by H_2S from a solution of mercury(II) acetate and ammonium thiocyanate. The black changes to the red in liquid H_2S, and the red to the black on heating to $386°C$. Cinnabar is the thermodynamically stable form at room temperature.

Compounds with Nitrogen and Sulfur. The reactions of mercury(I) compounds and NH_3 are complex, and published results vary. Recent (x-ray) studies show that this reaction, modified by the presence of ammonium chloride, NH_4Cl, yields three aminobasic compounds containing divalent mercury: $Hg_2NCl \cdot H_2O$, $HgNH_2Cl$, and $Hg(NH_3)_2Cl_2$. The first of these is the chloride of Millon's base, $Hg_2NOH \cdot 2H_2O$, which is produced by warming HgO with aqueous ammonia.

Mercury is the least active of the elements of its group as an electron acceptor from oxygen; however, with sulfur it is more active, the mercury halides forming dialkyl sulfide addition products, $R_2S \cdot 2HgX_2$, and HgS dissolves in alkali sulfides forming $[HgS_2]^{2-}$ or $Hg(SH)_4^{2-}$. Like zinc and cadmium, mercury forms a series of ammines, which with the mercury halides are principally the diammines, $[Hg(NH_3)_2]X_2$, where X is a covalently bonded halogen atom, and with more ionic mercury compounds, e.g., the nitrate and sulfate, especially in the presence of high concentrations of ammonium salts, the tetrammines, e.g., $[Hg(NH_3)_4](NO_3)_2$. These complexes also form with amines and diamines, e.g., ethylenediamine, which contributes three molecules per Hg^{2+} ion. Besides forming univalent and divalent cyanides, mercury forms complex cyanide ions, $[Hg(CN)_3]^-$ and $[Hg(CN)_4]^{2-}$, as well as additional compounds with the mercury(II) halides of the structure HgCNX. Mercury forms insoluble thiocyanates and complex ions, e.g., $Hg(SCN)_3^-$, $Hg(SCN)_4^{2-}$. Mercury cyanates and fulminates are insoluble. Mercury(II) mercaptides, $Hg(SR)_2$, decompose on heating to give HgS and R_2S. Mercury(II) hydrogen sulfite, $Hg(HSO_3)_2$, is actually mercuridisulfonic acid, $Hg(SO_3H)_2$, with Hg−S bonds. Like the halides, compounds such as $Hg[C(NO_2)_3]_2$. $Hg[C(CN)_3]_2$, $Hg(NO_2)_2$ (dinitromercury), $Hg(CF_3)_2$ are noteworthy for their lack of ionic dissociation.

Organometallic Compounds. Mercury also forms a large number of organometallic compounds of the type HgR_2, where R may be not only an alkyl radical, but an alkoxy radical, an acyl radical, a halogenated alkyl radical, an alkylthio radical, an aryl radical or a perfluoroalkyl radical. In addition, mercury also forms numerous organometallic compounds of structure RHgX, where X is a halogen atom, and R one of the foregoing organic radicals.

The organic compounds containing mercury that are used as disinfectants, germicides, and antiseptics are known as mercurials. Among these are Merthiolate, Mercurochrome, and Metaphen. Merthiolate is the sodium salt of ethylmercurithiosalicylic acid, $C_2H_5HgS \cdot C_6H_4 \cdot COONa$. It contains 49.5% of mercury. It is a crystalline, cream-colored powder which is very soluble in water, for about 1 gram dissolves in 1 milliliter of water. It is much less soluble in alcohol, 1 gram in 8 milliliters. It is insoluble in organic solvents like benzene and ether. It is used as an antiseptic for tissues in concentrations of the order of 1:1,000 to 1:30,000. It is commonly used as an antiseptic in biologics.

Mercurochrome, also known by the name of *Merbromin* and by many other trade names, is the disodium salt of 2,7-dibromo-4-hydroxymercurifluorescein, $C_{20}H_8Br_2HgNa_2O_6$. It forms green, iridescent scales or granules, which are freely soluble in water, yielding a bright red solution with dilute solutions having a yellow-green fluorescence. It is generally used in a 2% aqueous solution as a mild antiseptic. It is nearly insoluble in alcohol and is insoluble in organic solvents like acetone and ether. Most other common mercurials and iodine solution are considered to be better antiseptics. See also **Organometallic Compounds.**

Health and Safety Factors

One of the early examples of mercury poisoning may have involved the so-called "dark year" (1693) in the life of Sir Isaac Newton, when it was reported that Newton broke away from friends and associates, accusing them of plotting against him, when he slept very little, and when he reported conversations that actually did not occur. Initial explanations of Newton's erratic behavior included his failure to obtain certain appointive government positions, overwork, and the traumatic fire that destroyed a number of his valuable manuscripts. This evidence was not convincing, however, to investigators P.E. Spargo and C.A. Pounds. Through contacts with Newton's descendants, this team obtained four hairs taken from the head of the master and subjected them to laboratory tests. In the very interesting account by Broad (1981), the investigators concluded that Newton's madness was "due principally to poisoning by the metals which he used so frequently and with such cavalier disregard for his own safety." Locks of hair were located at Trinity College, Cambridge. One hair was found to contain 197 ppm Hg (even quite high in terms of modern-day Hg poison cases). Broad concludes his article, "Perhaps poisoning by mercury was not only the cause of Newton's brief lunacy, but was also the pivotal event that nudged the superstitious genius away from his researches in the laboratory to the seemingly less dangerous ways of the world." There are several doubters of the Spargo-Pounds hypothesis and the controversy is likely to persist.

Exposure. The exposure of humans and animals to mercury from the general environment occurs mainly by inhalation and ingestion of terrestrial and aquatic food chain items. Fish generally rank the highest (10–300 ng/g) in food chain concentrations of mercury.

In occupational settings mercury exposure results predominantly from direct dermal contact or inhalation of mercury vapors. Chloralkali plants are one of the principal sources for occupational exposure to mercury. Mining and refining of mercury contribute to exposure, and the processing of cinnabar can result in high exposures to the skin and lungs, producing poisoning in a relatively short time.

Toxicity. The toxic effects of mercury and mercury compounds are well known, and several detailed discussions on mercury toxicity are available. Toxicity to the central nervous system is more prominent after exposure to mercury vapor than to divalent mercury.

Short-term exposure to mercury vapor may produce symptoms within several hours. These symptoms include weakness, chills, metallic taste, nausea, vomiting, diarrhea, labored breathing, cough, and a feeling of tightness in the chest. Pulmonary toxicity may progress to an interstitial inflammation of the lung with severe compromise of respiratory function. Recovery, although usually complete, may be complicated by residual interstitial growth of excess fibrous tissue.

Chronic exposure to mercury vapor produces an insidious form of toxicity that is manifested by neurological effects and is referred to as the asthenic vegetative syndrome.

The biochemical basis for the toxicity of mercury and mercury compounds results from its ability to form covalent bonds with sulfur. Even in low concentrations divalent mercury is capable of inactivating enzymes containing sulfhydryl (−SH) groups, causing interference with cellular metabolism and function.

The affinity of mercury for sulfhydryl groups provides the basis for treatment of mercury poisoning using chelating agents (qv) such as dimercaprol (for high level exposures or symptomatic patients), or penicillamine (for low level exposure or asymptomatic patients).

Additional Reading

Agocs, M.M. et al.: "Mercury Exposure from Interior Latex Paint," *N. Eng. J. Med.*, 1096 (October 18, 1990).
Broad, W.J.: "Sir Isaac Newton: Mad as a Hatter," *Science*, **213**, 1341–1344 (1981).
Cai, Y. and O.C. Braids: *Biogeochemistry of Environmentally Important Trace Elements,* American Chemical Society (ACS), Washington, DC, 2002.
Clarkson, T.W.: "Mercury—An Element of Mystery," *N. Eng. J. Med.*, 1137 (October 18, 1990).
Fackelmann, K.A.: "Painting a Perilous Picture of Mercury," *Science News*, 244 (October 20, 1990).
Greenwood, N.N. and A. Earnshaw: *Chemistry of the Elements Revised and Updated,* 2nd Edition, Butterworth-Heinemann, Inc., Woburn, MA, 1997.
Krebs, R.E.: *The History and Use of Our Earth's Chemical Elements: A Reference Guide,* Greenwood Publishing Group, Inc., Westport, CT, 1998.
Lide, D.R.: *CRC Handbook of Chemistry and Physics* 84th Edition, CRC Press, LLC., Boca Raton, FL, 2003.
Raloff, J.: "Mercurial Risks from Acid's Reign," *Science News*, 154 (March 8, 1991).

Stone, R.: "Mercury's Metabolic Fingerprint," *Science*, 29 (April 3, 1992).

Stwertka, A. and E. Stwertka: *A Guide to the Elements*, Oxford University Press, Inc, New York, NY, 1998.

Web Reference

U. S. Environmental Protection Agency (EPA): http://www.epa.gov/mercury/information2.htm

MERESBURG REACTION.   See **Fertilizer**.

MERRIFIELD, R. BRUCE (1921–).   An American chemist who won the Nobel prize for chemistry in 1984. Merrifield was cited for work on the use of solid matrix as an aid to chemical synthesis of complex peptides and proteins. His synthesis techniques have been used in the development of solid matrix-bound inorganic and organic agents. Awarded doctorate from U.C.L.A.

MESITYL OXIDE (MSO).   See **Ketones**.

MESODESMIC STRUCTURE.   A type of ionic crystal in which one of the cation-anion bonds is equal in strength to all the bonds from the cation to the other anions. The silicates are important members of this class.

MESONS.   The *mesons* are subatomic particles of the hadron family. See also **Hadrons**. Fermionic hadrons are called *baryons*; the others are called *mesons*. The meson family consists of eight members which fall into a triplet of *pions*, a singlet *eta*, a doublet of *kaons*, and a doublet of *antikaons*. (They are all pseudoscalar (spin zero and odd parity) and exhibit strong interactions.) The charged particles are coupled to the photon, but even the neutral members can participate in electromagnetic interaction by virtue of the large probability for virtual dissociation into charged particles. They participate in a variety of weak interactions including the nuclear beta decay interaction.

It is found that the kaons, the hyperons (baryons other than the neutron and proton) and their antiparticles, collectively known as *strange particles*, can decay by weak interactions not involving leptons or photons, with a lifetime which is large compared to the natural periods appropriate to strong interactions. On the other hand, these particles are produced copiously in high-energy nuclear collisions. These two circumstances can be understood in terms of the existence of another additive quantum number (*hypercharge*) which is conserved in strong and electromagnetic interactions, but violated in weak interactions.

The meson-baryon system exhibits further regularities as far as strong interactions are concerned. The neutron and the proton have very nearly the same mass and similar nuclear interactions although their electromagnetic properties are quite different. The three *pions* have different electric charges, but again they have approximately equal masses and similar nuclear interactions. This kind of multiplet structure is evident for other strongly interacting particles; the *kaons* form a doublet, the *sigma hyperons* form a triplet, the *xi hyperons* form a doublet, and the *lambda hyperon* remains a singlet. See also **Particles (Subatomic)**.

The pion (*pi meson*) was first recognized in 1947 in photographic films made by C.F. Powell, P.S. Occhialini, and their collaborators of cloud chamber tracks made by cosmic rays high in the Andes. The masses of these pions were greater than those of the previously discovered muons, corresponding more closely to those predicted by K. Yukawa in his theory of the nuclear structure of the atom. The positive or negative pion has a mass 273 times that of the electron, and a charge equal in magnitude to that of the electron. Both positively charged and negatively charged pions are found in cosmic rays. Neutral pions may also be present in cosmic rays, but are produced in much greater abundance by high-energy particle accelerators and are therefore more easily detected in these laboratories. The first artificially produced pions were made in 1948 by the impact of 380 MeV alpha particles, from the Berkeley synchrocyclotron, on a target of carbon or certain metals. These were charged pions. Others were produced later by beams of protons and deuterons. The first evidence of neutral pions were the gamma-rays produced in 1950 by the impact of 175 MeV protons upon similar targets (carbon, beryllium, etc.). These gamma-rays had a minimum energy of about 140 MeV, which would be expected from the decay of a (neutral) pion into two gamma-rays. The same method is used today to produce beams of charged pions. The life of the neutral pion is so much shorter (about 10^{-16} seconds against 2.6×10^{-8} seconds) that beams cannot be produced. Unless it is captured by an atom, or reacts with another particle, a charged pion decays into a muon of the same sign and a neutrino or antineutrino. Like the other mesons the pion is a boson. It has zero spin.

Evidence of the *kaon* was found in 1944 by L. Laprince-Ringuet and M. LhÉritier in a cloud chamber photograph of a cosmic ray event. It was found again in 1947 by Rochester and Butler as a V-shaped track in a cloud chamber, the particle forming the other side of the V being probably a pion. For that reason it was first called a charged V-particle, which has been superseded by kaon See also **Quarks**.

Additional Reading

Ejiri, H. and H. Toki: *Nucleon-Hadron Many-Body Systems: From Hadron-Meson to Quark-Lepton Nuclear Physics*, Oxford University Press, New York, NY, 1999.

METABOLISM.   The chemical transformations occurring in an organism, from the time a nutrient substance enters it until it has been utilized and the waste products eliminated. In animals and humans, digestion and absorption are primary steps, followed by complicated series of degradations, syntheses, hydrolyses, and oxidations, in which agents such as enzymes, bile acids, and hydrochloric acid take part. These transformations are often localized with respect to organs, tissues, and types of cells involved.

Basal metabolism is the rate of total heat production of an individual who is awake but in complete mental and physical repose, at comfortable temperature and without having had food for at least 12 hours. See also **Basal Metabolism**. Under these conditions, oxidation of stored nutrients provides the sole source of energy expended and heat is measurable by calorimetry. See also **Digestion**.

META COMPOUNDS.   See **Organic Chemistry**.

METAL ANODES.   In electrolytic processes, the anode is the positive terminal through which electrons pass from the electrolyte. Most materials used in metal anode fabrication are characteristically expensive; use has, however, been justified by enhanced performance and reduced operating cost. An additional consideration that has had increasing influence on selection of the appropriate anode is concern for the environment.

Industrial metal anodes can generally be classified in one of two groups. The first group, chlorine-generating anodes, find application primarily in the manufacture of chlorine and caustic, sodium chlorate, and sodium hypochlorite. The second group, consisting of the oxygen-evolving anodes, do not generate a saleable product directly, but rather facilitate the desired cathodic reaction. Commercial uses include high speed electrogalvanizing of steel (qv), electrowinning of base metals, plating operations, cathodic protection, electrophoretic painting, copper foil treatment, and, more recently, the primary production of copper foil itself. See also **Electrochemistry**; and **Electroplating**.

Coating Structure and Morphology

The crystal structure of metal anode coatings has been investigated using x-ray diffraction studies, x-ray fluorescence analysis, and microprobe studies in conjunction with scanning electron micrographs. However, the role of the titanium metal substrate, or that of the oxides formed when the substrate is anodized or heated in air, is not completely clear for coatings that contain titanium in their solution. Most coatings, even those not containing titanium in the formulation, exhibit traces of titanium dioxide, present either in rutile or anatase form.

Ruthenium–Titanium Oxides.   The x-ray diffraction studies of ruthenium–titanium oxide coatings show that the coating components are present as the metal dioxides, each in the rutile form as well as in solid solution with each other.

It appears that the titanium metal substrate on which the coating is deposited plays an important role in the structure and morphology of the coating. The surface layer of rutile titanium dioxide normally found on oxidized titanium metal apparently acts as a seed to initiate growth of the rutile form of the oxide, rather than the anatase form. Interfacial layers of titanium suboxides, known to be electrically conductive, also act to effect a gradual transition from pure metal to pure rutile oxides. Scanning electron micrographs of ruthenium–titanium oxide coatings show a characteristic microcracked surface.

Iridium Oxide. Iridium dioxide coatings, typically used in combination with valve metal oxides, are quite similar in structure to those of ruthenium dioxide coatings.

Platinum–Iridium. There are two distinct forms of 70/30-wt % platinum–iridium coatings. The first, prepared as prescribed in British patents, consists of platinum and iridium metal. The surface morphology of a platinum–iridium metal coating is cracked, but not in the regular networked pattern typical of the DSA oxide materials. The second form consists of Pt metal but the iridium is present as iridium dioxide. Iridium metal may or may not be present, depending on the baking temperature.

Spinel Cobalt Oxides. The cobalt mixed oxide, Co_3O_4, containing Co(II) and Co(III) ions, has the spinel structure. In these coatings containing zirconium, a separate, partially crystalline phase of zirconium dioxide occurs. The coating has large microscopic pores, providing a high surface area which appears to be related to the presence of zirconium. Spinel oxides applied without the presence of zirconium are dense and closely packed coatings.

Operating Performance of Coatings

The key yardstick of performance for a coated metal anode is the period of time, measured in ampere-hours, that the coating operates before it reaches an unacceptable voltage, as measured by the single electrode potential (SEP). Factors influencing the escalation include accumulated average lifetime, operating current density, electrolyte conditions, exposure to oxygen in the anolyte compartment, and cell design.

Chlorine Anodes. In chlorine manufacture, anode operating life is limited by coating wear. The wear rate varies according to cell type. The longest anode coating life has been achieved in diaphragm chlorine cells.

Because mercury chlorine cells operate at higher current densities and because the mercury cell anode can be adjusted during operation to minimize the anode-to-cathode gap, anode coating life in these cells is much shorter. Because of limited commercial experience with anode coatings in membrane cells, commercial lifetimes have yet to be defined. Expected lifetime is 7–12 years.

Metal anode coatings commercially used for manufacture of sodium chlorate include not only the ruthenium oxide coatings, but also platinum–iridium coatings. Whereas ruthenium oxide coatings might be preferred for a longer performance life and higher resistance to process upsets, platinum–iridium coatings generally operate at higher efficiencies during the first months of operation.

Oxygen-Evolving Anodes. In the case of oxygen-evolving anodes employing iridium dioxide-based coatings, the most significant commercial experience has been in the electrogalvanizing industry where coated titanium anodes have supplanted lead and zinc anodes, and thus lead contamination of the product and waste streams.

In contrast to the coating wear limitation on anode life experienced in chlorine cells, passivation of the substrate beneath the coating is typically the limiting factor for oxygen-evolving anodes. As a result, technology has been introduced to either maintain or modify the titanium surface to increase coating adhesion and significantly improve lifetimes.

Structure Design

Each electrolytic application demands a unique approach to anode structure design and fabrication. Factors such as current distribution, gas release, ability to maintain structural tolerances, electrical resistance, and the practicality of recoating must be taken into account. The most commercially accepted design for diaphragm chlorine cells is that of the expandable anode (Fig. 1).

In mercury chlorine cells, it has been found that cells operate at lower voltages when fitted with anode structures comprised of triangular rods or of vertical blades.

The dimensionally stable characteristic of the metal anode made the development of the membrane chlorine cell possible. These cells are typically arranged in an electrolyzer assembly which does not allow for anode-to-cathode gap adjustment after assembly. Also, very close tolerances are required. The latitude that titanium affords the cell designer has made a wide variety of monopolar and bipolar membrane cell designs possible.

When used in radial cells such as for the production of copper foil and in some electrogalvanizing operations, the anode must be curved to meet the shape dictated by the cathodic drum. The anode must also achieve

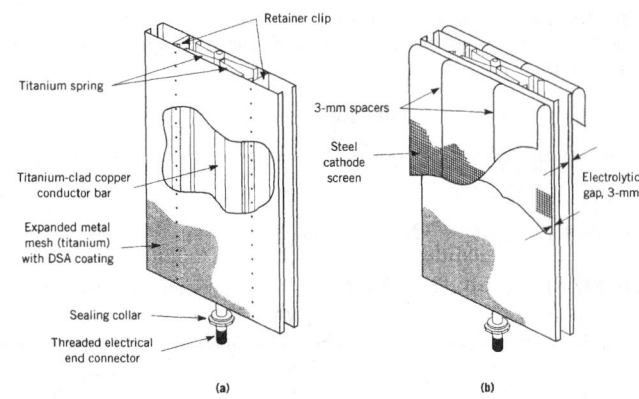

Fig. 1. Expandable anode for diaphragm chlorine cells: (**a**) clamped and (**b**) expanded

exact tolerances to assure that a constant anode-to-cathode gap is created and maintained.

Manufacturing Technology

Manufacturing techniques for metal-coated anodes have been developed to a high level of sophistication. By the mid-1990s commercial plants were producing or recoating in excess of 50,000 anodes annually. This scale of operation requires continuous coating processes (qv) the use of electrostatic spray application, robotics, and strict process control (qv). The high cost of the coating itself demands high utilization efficiency. Quality requirements include consistent coating distribution, strong adherence, and a good surface appearance.

Environmental and Safety Factors. The primary environmental concern for the coating plant is actually the residual material on the anode structures being returned for recoating. Therefore the anode user must enact effective cleaning procedures prior to shipment. Overall, the DSA (Electrode Corp.) has made chlorine manufacture cleaner, more consistent, simpler, and therefore safer.

Economic Aspects. For the most part, coated metal anodes are made available by long-term lease rather than through outright sale. There has been a single-source supply protected by patents in most geographic areas. As the cost-saving and environmental value of this technology continues to increase, upward expectation in respect to prices is created. However, the appearance of new anode producers, together with expiration of some early patents, has created a competitive situation in some applications. This competition can be expected in turn to create a counter-balancing pressure to lower prices.

Thomas A. Liederbach
Electrode Corporation

Additional Reading

Hardee, K.L., L.K. Mitchell, and R.C. Carlson: in *Proceedings of the AESF Sixth Continuous Steel Strip Coating Symposium*, May 1990.

O'Leary, K.J. and T.J. Navin: *Morphology of Dimensionally Stable Anodes,* paper presented at the Chlorine Bicentennial Symposium, San Francisco, Calif., May 1974.

U.S. Pat. 3,632,498 (Jan. 4, 1972), H.B. Beer (to Chemnor Aktiengesellschaft).

U.S. Pat. 3,711,385 (Jan. 16, 1973), H.B. Beer (to Chemnor Corp.).

METALLIC COATINGS. Metallic coatings provide an inexpensive way to modify or control the properties of a base material. Coatings may be functional or decorative, permanent or temporary, sacrificial, or noble. The metallic coating may be continuous, or it may be patterned into discontinuous functional or decorative areas. The base material may be metallic, nonmetallic inorganic, or organic such as plastic, paper, or fiber. The common criterion for the purposes herein is that the metallic coating is functionally bonded to the base material. Galvanized zinc items, gold clad jewelry, electroplated materials, and semiconductor chips are among the most common types of materials having metallic coatings. See also **Electroplating**; and **Semiconductors**.

Metallic coatings are most often selected for protective function. Decorative ability is a common secondary function. Most coatings applied by hot dipping, such as galvanizing or aluminizing, use a film of a more chemically reactive metal over a less reactive material such as iron alloy. These are sacrificial coatings because the zinc or aluminum slowly dissolves instead of the underlying steel.

Decorative coatings must be sufficiently corrosion resistant to maintain an attractive appearance during the anticipated life of the composite product. Composites of multiple metal layers or mixtures of metallic and nonmetallic coatings are becoming increasingly common as material properties are selectively tailored. Another important function of metallic coatings is to provide wear resistance. The most technically demanding types of coatings are those grown in precisely defined multiple layers.

Plating and galvanizing are the most common methods of applying metallic coatings, but many other processes, such as hot dipping, cementation, thermal spraying, sputtering, and chemical vapor disposition, have been developed. All metallizing techniques are potential sources of pollution, a subject of increasing ecological, economic, legislative, and public health interest.

There are a plethora of commercially useful methods for applying a metallic coating. Many more techniques have been demonstrated in the laboratory. Each method has different critical parameters, i.e., maximum and minimum coating thickness producible, substrate temperature, bulk or imagewise deposition, coating adhesion value, type and cost of coating equipment, labor requirement, scrap rate, rework capability, and safety and waste disposal aspects. For convenience, metallic coating methods may be divided into classes defined by the general way in which the metal is applied, e.g., liquid-phase, gas-phase, and vacuum-phase metallizing or by direct physical or thermal bonding.

The two largest-volume processes, in terms of amount of metal used and amount of surface area coated, are the liquid-phase metallizing processes hot dip coating and electroplating.

Many of the newer coating methods give metallized coatings that are quite distinct from simple single or multiple coatings of metals or alloys. Many of these newer coatings, which cannot be made except by one or a small family of processes, are defined as much by process method as physical and chemical structure. Examples of these coatings include ion-plated surfaces, laser heat-treated surfaces, modifications of existing surfaces to predetermined depths by selective additions of atoms or heat; and Teflon–electroless nickel composite coatings.

Liquid-Phase Metallizing Techniques

Hot Dip Galvanizing. The largest single type of metallic coating process in terms of amount of metal used is hot dip galvanizing. Half of all galvanized zinc is used on steel coils, one-third is employed for coating of parts after fabrication, and the remainder for wires and tubes. The other uses, in approximate order, are electroplating, zinc-filled paints, zinc spray processes, zinc foils having conductive adhesive, and mechanical zinc plating.

In hot galvanizing, zinc is applied to iron and steel parts by immersing the parts into a bath of molten zinc. Whereas in principle almost any metal could be coated with molten zinc, this coating serves no worthwhile purpose on most metals. The combination of zinc and ferrous materials are almost uniquely suited to each other. Aluminum and cadmium are the only other similar combinations. Zinc provides iron parts with better corrosion protection by developing a coating of zinc and zinc compounds on the base metal surface.

Hot Dipping Using Other Metals. Processes include coating with aluminum, coating with 55% aluminum–zinc alloy, hot dip tin coating of steel and cast iron, hot dip terne coating, and solder coatings.

Electroplating. Many developments in electroplating have been the direct result of increased coating functionality and economy, and others of the increasing need for environmental and legislative compliance. The field of electroplating has expanded rapidly.

The most common electroplated metal is zinc, followed by nickel, copper, and chromium. Many other metals and metaloids can be electroplated, including manganese, iron, cobalt, gallium, germanium, arsenic, selenium, ruthenium, rhodium, palladium, silver, cadmium, indium, tin, lead, bismuth, mercury, antimony, gold, iridium, and platinum. Types of electroplating include fused salt plating, plating from nonaqueous solvents, and brush plating.

Chromium metal seems to be biologically harmless owing to inertness. Trivalent chromium Cr(III), one of the essential mineral nutrients, has not been shown to be harmful. Hexavalent chromium, Cr(VI), found in chromic acid and used in electroplating baths, is very toxic as well as a suspected carcinogen. Two basic types of chromium are plated: hard chromium up to 7100 μm or more, and decorative chromium at 0.25–1.0 μm. No replacement coating has been found for thick hard chromium deposits for wear resistance and parts salvage, although electroless nickel can partially substitute for chromium plating baths. Many decorative applications are being converted to Cr(III). Products from the newest Cr(III) baths are essentially equivalent to those from the older decorative Cr(VI) baths, although some cosmetic color differences exist which have prevented complete conversion to the Cr(III) baths.

Cyanide solutions formerly were ubiquitous in electroplating shops. The best cleaners contained sodium cyanide. The removal of cyanides from all plating baths has been a general goal since the latter part of the 1980s. As of this writing, cyanide-containing cleaners are rare and are only used for special purposes. Several noncyanide copper strike baths have been introduced.

The greatest tonnage decrease in cyanide plating has occurred for zinc plating. As of this writing, less than one-third of all zinc is plated from cyanide baths. The remainder is plated from alkaline noncyanide, zinc potassium chloride, and zinc ammonium chloride baths.

Chlorinated and fluorinated cleaners have been widely used as degreasers during surface preparation prior to plating. Recognition of the role of many of these solvents as either global warming gases or ozone depletion agents has led to prohibitions against their continued use. See also **Pollution (Air)**. Many new and improved cleaning systems are being developed, such as alcohol-based cleaners, emulsion cleaners, and cleaners based on natural bioproducts such as limonene. During this process a massive reformulation of many aqueous cleaners has also occurred.

Cadmium usage, illegal in most of Europe, is being discouraged elsewhere. The U.S. military has cadmium specifications for electronic, fastener, and marine equipment, which requires only cadmium. Tin is being substituted for tin–lead as a metallic etch resist during printed circuit board production.

Electroless Plating. The metallizing process known as electroless plating is mainly used for deposition of copper on plastics and for nickel–phosphorus alloy on plastics and metals. Formaldehyde is the most common copper reducing agent, giving pure coatings. Sodium hypophosphite is the agent mainly used for electroless nickel. An unusual nickel–phosphorus alloy or solid solution is formed. Unlike electroplating, electroless plating can be used on almost any substrate, metallic or nonmetallic. Often electroless plating is used as the first coating to make glass ceramic, or plastic conductive, followed by conventional electrolytic plating.

Immersion Plating. A simplified aqueous metal deposition process which does not use electric current, immersion plating works only when a metal of higher electromotive force, such as copper, is deposited on a metal of lower electromotive force, such as iron or aluminum. The coatings are typically very thin and porous. One important application is immersion plating of copper sulfate bath onto steel wire, to use as a drawing agent. Very thin immersion gold deposits are used for inexpensive jewelry.

Miscellaneous Techniques. Lasers have been used for both electroless and electrolytic plating; selective dissolution has been used from ancient times to give the appearance of a thin plated coating of precious metal; and mercury layers plated onto the surface of analytical electrodes serve as liquid metal coatings.

Gas-Phase Metallizing Techniques

Metal or Thermal Spray Coatings. These specific processes include plasma arc spray, flame spray, laser spray, and electric arc spray, depending on the energy input source. Rods, wires, and powders are used as coating material sources. Thermal spray metal and ceramic coatings have diverse properties suitable for numerous applications, including corrosion resistance, e.g., zinc and aluminum, especially against oxidation or salt water corrosion; high temperature oxidation, e.g., nickel, cobalt, and chromium alloys; electrical conductivity, e.g., radio frequency interface (RFI) shielding by zinc or tin on nonconductors; electrical resistance, e.g., insulating layers in induction heating coils and high temperature strain gauges; wear resistance, e.g., chromium–nickel–boron alloys and carbide-containing coatings; catalytic surfaces; nuclear moderators; and dimensional buildup for salvaging worn metal parts.

A great advantage of zinc arc spray is that it can be applied to almost any plastic. Zinc arc spray, also suitable for prototypes and small lots of materials, is less suited for very small parts and parts having blind holes or complex interior surfaces, or where warpage is a problem.

Zinc arc spraying or flame spray equipment is hardly more hazardous than a welding torch, and only safety goggles and gloves are required. Safety aspects emphasize reduction of noise and vapor inhalation.

Thermal spray processes can be used to give coatings of chromium carbide or nickel chromium for erosion resistance, copper nickel indium for fretting resistance, tungsten carbide cobalt for wear and abrasion resistance, and even aluminum silicon polyester mixtures for abradability.

Carburizing and Nitriding. Several commonly used metallurgical surface treatments are applied by gas-phase reactions in a reducing atmosphere for carburizing, and in a nitrogen atmosphere for nitriding. These treatments are used to increase the surface hardness of ferrous alloys by diffusion of carbon and nitrogen at high temperature.

Pack Diffusion. Pack diffusion or cementation processes are similar to pack carburizing, and are used to coat iron, nickel, cobalt, and copper with chromium, boron, zinc (Sheradizing), aluminum, silicon, titanium, molybdenum, and other metals.

Chromizing and Related Diffusion Processes. Chromizing is similar to aluminizing. A thin corrosion and wear resistant coating is applied to low cost steels such as mild steel, or to a nickel-based alloy. In the related boronizing process, a thin boron alloy is produced for extreme hardness, wear, and corrosion resistance. Siliconizing is yet another process used especially for coating of the refractory metals Ti, Nb, Ta, Cr, Mo, and W.

Metals. Aircraft and space vehicles, turbine generators, and other such applications require high strength at high temperatures along with excellent oxidation resistance. Superalloys, i.e., complex nickel and cobalt-based alloys, and refractory metals, e.g., niobium, tungsten, molybdenum, tantalum, and their alloys, are used for applications at temperatures above 1000°C. In many case the coatings must be resistant both to oxidation and to hot corrosion by sulfidation from sulfur-bearing gases.

Two types of coatings have been used for superalloys: diffusion coatings, in which a layer of nickel, cobalt, platinum, or palladium aluminide, i.e., NiAl, CoAl, PtAl, or PdAl, is formed on the surface by diffusion; and overlay coatings, in which a complex coating material such as nickel–cobalt–chromium–aluminum–yttrium, NiCoCrAlY, is applied to the surface. Pack cementation is the most widely used process for applying diffusion coatings to superalloys.

Vacuum-Phase Metallizing Techniques

Vacuum-phase metallizing techniques all depend on the use of a vacuum as part of the metallizing process. The tonnage of metal deposited by these techniques is insignificant compared to hot dip galvanizing or plating, and the total surface area metallized is much less than that done by plating. However, the economic added value of vacuum metallizing probably exceeds that of either plating or galvanizing because of the extremely rapid growth and deep market penetration of semiconductor devices. Projections call for at least a 10% compounded annual growth rate into the twenty-first century. The state of the art in vacuum metallizing is in semiconductors processing.

Thermal Evaporation. Thermal evaporation is done in a high vacuum to minimize chemical side reactions of the evaporated active metal. Thermal evaporation is inexpensive and efficient. It is used for low cost items such as aluminized plastic sheet and decorative Christmas tinsel, second surface coating of transparent plastic auto parts, and metallization of glass and ceramics. Aluminum is the predominant metal used. Color effects such as gold or brass are achieved by applying a dyed translucent organic protective over the aluminum.

Sputtering or Glow Discharge. Sputtering can be done using both conductive and nonconductive items. A low pressure atmosphere of argon is used in the deposition vessel. This method gives excellent coating adhesion and more consistent coatings than simple vacuum deposition. It can also be applied to a wider range of materials, including complex metal alloys and oxides. The equipment and operating costs are higher than vacuum deposition, and deposition rates are lower. Reactive sputtering is a variation in which two or more deposition sources are used. This process can be used to give compounds such as silicon carbide on the surface.

Chemical Vapor Deposition. This process is distinct from simple thermal evaporation because chemical vapor deposition (CVD) depends on a chemical reaction at the surface of the part and in that way is analogous to electroless plating. This process uses a gas of one or more chemical species, which react at a heated substrate to form an appropriate film. The film can be metallic, nonmetallic, single element, or compound. Among the coatings available are copper and aluminum conductors, diamond and tantalum oxide dielectrics, lead zirconium titanate piezoelectrics, and bismuth strontium calcium cuprate superconductors. Some of the largest applications are for coating TiC and TiN on cutting tools.

Ion Implantation. Also known as ion plating, ion implantation is a high vacuum process for modifying the surface properties of any material. This is a surface modification technique rather than a surface coating technique. Semiconductors are commonly modified using this technique, because ion implantation can be done using total or imagewise scanning of the beam over the surface. Nitrogen, chromium, and phosphorus can be used to harden metals and increase the corrosion resistance. Metastable alloys can be produced that are difficult to make by other processes.

Laser Hardening and Modification. Lasers are used to surface harden ductile steels and improve the toughness to a depth of 0.35 mm or more. Lasers can also be used to bond solid or powder coatings to a surface. Typical coatings are nickel or titanium carbide on iron, and nickel, cobalt, manganese, and titanium carbide, TiC, on aluminum. Use of lasers with other specialized coating methods is common.

Metallizing by Direct Physical or Thermal Bonding

Direct bonding techniques are among the oldest types of metallizing, and the most versatile. Many methods depend on heat or pressure and an adhesive layer to glue the coating to the substrate. Methods for metallizing on a metallic surface often depend on removal or displacement of a preexisting surface oxide layer. Many metals form intermetallic alloys or self-diffuse into one another even at room temperature, but the surfaces in contact must be clean and oxide-free. Processes include lamination, mechanical plating, slurry coatings, and roll bonding or strip roll welding.

Environmental Concerns

Each type of metallic coating process has some sort of hazard, whether it is thermal energy, the reactivity of molten salt or metal baths, particulates in the air from spray processes, poisonous gases from pack cementation and diffusion, or electrical hazards associated with arc spray or ion implantation. Vacuum or inert gas operations can produce flammable dusts or powders when opened for cleaning. Most of the hazards are confined to the operator and immediate environs in the operating plant. OSHA is the primary regulator of these hazards in the United States, although many local and state agencies, especially fire departments, also regulate coatings plants. Adequate training, documentation, and protective equipment are the minimum requirements. The principal regulatory burden falls on wastes and discharges which leave the plant.

The U.S. EPA regards metallizers such as platers, surface finishers, and printed circuit board producers as among the most important source polluters for metals. Much production of electroplated items has, however, shifted from the United States to less environmentally stringent countries.

The surviving U.S. plants have embraced all types of waste treatment processes. (See also **Wastes and Pollution**). The most desired pollution prevention processes are those which reduce the total amount of waste discharged. Treatment and disposal are less strongly emphasized options. Zero wastewater discharge facilities and water recycling processes are becoming more common.

Discharge limits vary between localities and among plants. Table 1 shows federal EPA maximum discharge limits for a number of metals for a new metal-finishing installation.

Newer federal limits on metals contents of sewage sludge combined with laws on fuller treatment of the sewage sludge and its allowed disposal methods have affected limits. The metallic content of sludges from municipal waste treatment facilities is becoming of great concern.

Air pollution is recognized as a significant problem for coating facilities.

Many types of waste treatment and waste minimization processes are in common use in the metallization industry. Air scrubbers are commonly required, even for general acid fume removal from plating shops. Additionally, many newer technologies have been adapted for use in metallizing operations. These include air stripping, antimisting agents, biological destruction, carbon absorption, countercurrent rinsing, crystallization,

TABLE 1. EPA PRETREATMENT STANDARDS FOR AQUEOUS DISCHARGE[a,b]

Material, mg/L	Existing sources, PSES[c]		New source, PSNS[c]	
	1 Day	30 Days	1 Day	30 Days
cadmium	0.69	0.26	0.11	0.07
chromium, total	2.77	1.71	2.77	1.71
copper	3.38	2.07	3.38	2.07
lead	0.69	0.43	0.69	0.43
nickel	3.98	2.38	3.98	2.38
silver	0.43	0.24	0.43	0.24
zinc cyanide	2.61	1.48	2.61	1.48
Total	*1.2*	*0.65*	*1.2*	*0.65*
Treatable	*0.86*	*0.32*	*0.86*	*0.32*
Total toxic organics	*2.13*		*2.13*	

[a] Captive manufacturers performing metal finishing, including electroplating, discharging to POTWs.

[b] pH equals 6–10.

[c] Maximum value. PSES = pretreatment standards for existing sources; PSNS = pretreatment standards for new sources.

distillation, Donnan dialysis, electrodialysis, electrowinning, evaporation, filtration, fluocculation, hydrolysis, incineration, ion exchange, metallic replacement, neutralization, oxidation, pH adjustment, photolysis, precipitation, process modification, reduction, reverse osmosis, salt splitting, sedimentation, solidification, and spray rinsing.

GERALD A. KRULIK
Applied Electroless Concepts, Inc.

NENAD V. MANDICH
HBM Engineering Company

Additional Reading

Durney, L.J. ed.: *Electroplating Engineering Handbook,* 4th Edition, Van Nostrand Reinhold Co., Inc., New York, NY, 1984.

Metal Finishing Guidebook and Directory, Vol. 92, No. 1A, Elsevier Publishing, New York, 1944; collected vols., Publications, Inc., Hackensack, NJ, updated yearly.

Plating and Surface Finishing, Semiconductor International, and *Metal Finishing* provide some of the best information on innovative metallizing methods.

Safranek, W.H.: *The Properties of Electrodeposited Metals and Alloys,* 2nd Edition, American Electroplaters and Surface Finishers Society, Orlando, FL, 1986.

METALLOBIOMOLECULES. Natural products, the biologically active forms of which contain one or more metallic elements. Metallobiomolecules may be transport and storage proteins, such as cytochromes (Fe), ferritin (Fe), transferrin (Fe), cruloplasmin (Cu), myoglobin (Fe), or hemoglobin (Fe); or they may be enzymes, such as carboxypeptidases (Zn), aminopeptidases (Mg, Mn), phosphatases (Mg, Zn, Cu), hydroxylases (Fe, Cu, Mo), or isomerases and synthetases, such as coenzymes (Co). Metallobiomolecules also may be nonproteins, such as siderophores (Fe) or chlorophyll (Mg). Ibers and Holm provide an excellent review of metallobiomolecules in *Science*, **209**, 223–235 (1980).

METALLOCENES. A class of neutral transition metal compounds containing two cyclopentadienyl (C_5H_5) ligands π-bonded to a central metal atom in a "sandwich" structure, as exemplified by ferrocene (see below).

The original metallocene, ferrocene, was first reported simultaneously and independently by two groups of workers in 1951. Miller, Tebboth, and Tremaine at British Oxygen Ltd. passed cyclopentadiene over iron powder and an ammonia catalyst; Kealy and Pauson at Duquesne University treated ferric chloride with cyclopentadienyl magnesium bromide. Both groups obtained the same orange, air-stable, hydrocarbon-soluble crystals (m.p. 173°) and proposed a sigma bonded structure for this first organometallic derivative of iron. Within a year groups at Harvard University including Woodward, Wilkinson, Rosenblum and Whiting revealed that $C_{10}H_{10}Fe$ exhibits unusual thermal stability, has no dipole moment, and undergoes typical aromatic substitution reactions; whereupon, they proposed the novel sandwich structure and advanced the name *ferrocene* to reflect both the iron content of the material and its aromaticity. X-ray crystallographic studies subsequently supported the sandwich structure, showing the preferred staggered, antiprismatic conformation in the solid state. In solution or the vapor phase the barrier to rotation of the rings has been shown to be very small.

One proposed bonding scheme, using a molecular orbital model, calls for utilizing the π orbitals of the $C_5H_5^-$ group and the $d_{xz}, d_{yz}, s, p_x, p_y,$ and p_z iron orbitals to obtain a total of six bonding orbitals with the twelve π-electrons from the rings to fill them.

After the discovery of ferrocene many additional metallocenes and their derivatives were prepared, including titanocene (green), vanadocene (purple), chromocene (scarlet), cobaltocene (purple-black) and nickelocene (green). Dicyclopentadienyl manganese (amber) is ionic, and dicyclopentadienyl mercury, tin and lead are sigma bonded; their derivatives are not properly called metallocenes. In the second and third transition series, only ruthenocene and osmocene are known as simple, neutral metallocenes; the other metallocenes are known as cations (e.g. Cp_2Ti^+) or with additional ligands such as halide or hydride attached to the metal atom, as in $Cp_2TiCl_2, Cp_2ReH, Cp_2TaH_3,$ etc. Though several preparative approaches are known, the most general reaction leading to metallocenes is the treatment of a metal halide with sodium cyclopentadienide in tetrahydrofuran. Ideally a +2 salt of the metal is used, although excess of the cyclopentadienide will often reduce a higher valent metal halide to a lower oxidation state during the reaction.

Chemically, ferrocene undergoes many typical aromatic substitution reactions such as Friedel-Crafts acylation or alkylation, sulfonation, mercuriation, lithiation, the Villsmaier reaction and the Mannich reaction (with dimethylamine, formaldehyde and acetic acid). Its reactivity is very great ("superaromatic") and is comparable in rate to that of phenol. Mono- and disubstitution on one or both rings can be realized, though some measure of control to predominately mono- or disubstitution can be exercised by adjusting conditions.

However, the central metal atom in ferrocene imposes some limitations on the chemistry of the system or provides in some cases additional reaction pathways. For example, the central iron atom is readily and reversibly oxidized from the Fe(II) state to Fe(III) in the form of the water-soluble red-blue dichroic ferrocinium ion. This oxidation occurs with halogens or with nitric acid so that direct aromatic halogenation and nitration cannot be realized; however, halo- and nitroferrocene have been prepared by indirect methods. Aminoferrocene cannot be diazotized, presumably due to oxidative destruction of the system. Coupling with diazonium salts is anomalous; ferrocene reduces diazonium salts to phenyl radicals in aqueous or nonaqueous solution to yield Gomberg (phenylation) products. Condensation with aldehydes in acid solution gives rearranged products due to a role that the iron atom can play. The central iron atom is readily protonated in strong acid media.

In the solid state ruthenocene and osmocene prefer the eclipsed, pentagonal prismatic structure and in solution exhibit a chemistry similar to that of ferrocene. The other metallocenes, on the other hand, are quite different chemically, none of them showing the typical aromatic substitution reactions of ferrocene.

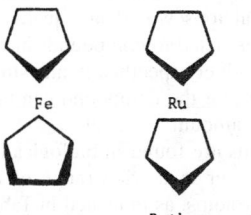

Ferrocene   Ruthenocene

Cobaltocene is rapidly oxidized in air or in solution (it liberates H_2 from water, slowly) to yield the very stable yellow +1 cation, which has been reported to be stable to aqua regia. Neutral cobaltocene reacts with alkyl and acyl halides to give adducts $(C_5H_5)(C_5H_5R)Co^+X^-$ in which the substituted ring is π-bonded to the cobalt as a diene.

Nickelocene, on the other hand, is more slowly oxidized and undergoes addition of suitable activated olefins to one ring converting it to a bicyclic ligand. Some ligand replacement reactions are also known for nickelocene.

Titanocene derivatives undergo substitution reactions at the metal atom:

$$Cp_2TiCl_2 + C_6H_5Li \longrightarrow Cp_2Ti(C_6H_5)_2 + LiCl_2$$

$$Cp_2TiCl_2 + 2OCH_3^- \longrightarrow Cp_2Ti(OCH_3)_2 + 2Cl^-$$

The utility of these materials is limited and disappointing. Ferrocene has been used to promote the burning of fuel oils and as a catalyst in rocket fuels. Its extreme thermal stability (to over 470°) spurred an interest in its derivatives as high-temperature fluids. Some polymers have been made, but they have found no substantial application. Titanocene dichloride has been shown to react with molecular nitrogen in systems containing added Grignard reagents or similar other materials.

It is generally agreed that the real importance of metallocenes lies in the effect of their discovery twenty years ago along with Ziegler's catalyst on the course and direction of organometallic chemistry. There has been a veritable explosion of interest and effort in this field during the past two decades that is unabated today. Many new materials and new insights into bonding of transition metals have evolved. Great strides in understanding of homogeneous catalysis and the design of new catalysts for hydrogenation, coupling, carbonylation, polymerization, etc., have been made. Even a closer understanding of the action of vitamin B_{12} has resulted from the new brand of organometallic chemistry, and we are moving closer to a practical method for fixation of nitrogen in solution.

See also **Osmium**.

WILLIAM F. LITTLE
University of North Carolina
Chapel Hill, North Carolina

Additional Reading

Rosenblum, M.: *Chemistry of the Iron Group Metallocenes: Ferrocene, Ruthenocene, Osmocene,* Vol. 1, Krieger Publishing Company, Melbourne, FL, 1965.
Togni, A. and T. Hayashi: *Ferrocenes: Homogeneous Catalysis, Organic Synthesis, Materials Science,* John Wiley & Sons, Inc., New York, NY, 1995.

METALLOGRAPHY. Study of the structure and properties of metals and alloys, principally by microscopic and x-ray diffraction methods. The term is also used in a broader sense to include the processing of metals by mechanical and heat treatments and the fabrication and testing of finished products. In this usage it is synonymous with physical metallurgy. See also **Metallurgy**.

METALLOID. A chemical element that may exhibit physical and chemical properties both of a metal and a nonmetal sometimes is referred to as a metalloid. Antimony, arsenic, and tellurium are examples. Less frequently, metalloid refers to elements, such as carbon, silicon, phosphorus, and sulfur, which are added in small amounts in the manufacture of iron and steel.

METALLOPROTEINS. Proteins, especially in solution, readily participate in a greater variety of chemical reactions than any other class of compounds of biological interest. This reactivity is a function primarily of the many polar side chains containing −OH, −COOH, −NH₂, −SH, and other groups, all of which can, to varying extents, interact with metal ions. Proteins can bind metals, some of them very tightly. However, relatively specific and nonspecific binding should be differentiated.

A negatively charged protein molecule exerting a nonspecific electrostatic attraction on metal ions would not qualify as metalloprotein. The term metalloprotein is restricted to compounds in which under natural conditions a metal ion is relatively specifically and strongly bound to a protein molecule in such a way that the compound can be isolated and shown to contain a stoichiometric amount of metal.

A variety of metal ions are found in biologically important metalloproteins. Metalloproteins occur in a wide range of biological systems. The function of the metalloproteins, as indicated in Table 1, varies widely from one compound to another.

The chemical properties of the metal in these compounds may be greatly affected by bonding to a protein ligand. The bound metal can play one of many roles. Thus, in an enzyme, the metal ion may permit the formation of a ternary complex between protein, metal and substrate or coenzyme. An instance of this role is provided by the enzyme enolase, which is unable to catalyze the equilibrium between 2-phosphoglycerate and 2-phosphopyruvate in the absence of Mg ions. In other enzymes, the metal may actually participate in electron transport by cyclic oxidation and reduction. Such is probably the case with the Cu in polyphenol oxidase. The metal may serve primarily for the maintenance of a specific spatial folding of the polypeptide chains in the protein molecule.

The strength of metal-protein bonds in metalloproteins may vary from relatively loose association to very tight binding. When the metal ion is able to dissociate with some ease from the protein, it is usually a single ligand responsible for the metal binding. Such ligand groups are mainly

TABLE 1. REPRESENTATIVE METALLOPROTEINS AND THEIR FUNCTIONS

Name	Metal	Source	Function
Hemocuprein	Cu	Erythrocytes	Unknown
Ceruloplasmin	Cu	Serum	Oxidase (probable)
Hepatocuprein	Cu	Liver	Unknown
Polyphenol oxidase	Cu	Mushroom	Enzyme
Hemocyanin	Cu	Mollusks	Respiratory pigment
Tyrosinase	Cu	Mushroom	Enzyme
Metallothionein	Cd + Zn	Kidney	Na Reabsorption (probable)
Xanthine oxidase	Mo	Liver	Enzyme
Carbonic anhydrase	Zn	Erythrocytes	Enzyme
Alcohol dehydrogenase	Zn	Yeast	Enzyme
Ferritin	Fe	Spleen	Fe storage
Transferrin	Fe	Plasma	Fe transport
Conalbumin	Fe	Eggs	Fe storage
Ferredoxin	Fe	Bacteria	Electron transport
DPNH-cytochrome *c* reductase	Fe	Heart muscle	Electron transport
Hemovanadin	V	Tunicates	Respiratory pigment

found in the amino acid side chains of the protein molecule (e.g., −NH₂ or −OH groups). The interaction of metal and ligand may exhibit strong pH dependence because of competition between metal and hydrogen ions. Of the single ligand groups, by far the strongest is the −SH group in the amino acid cysteine. Even stronger metal bonding to protein may be observed when a divalent or trivalent metal forms chelate complexes with the protein. Chelation is often indicated not only because of the strength of the bond, but also because of the specificity of the reacting site on the protein molecule for one particular metal. Such a specificity may reflect the coordination requirements of the various metals. The preferred electron donor in the formation of protein-metal coordination compounds is N, such as that of the imidazole nucleus of histidine, but S and O may also participate in this process. If the protein contains carboxyl or phosphoryl groups, strong ionic bonds between metal and protein may be formed. A completely different type of protein-metal interaction is illustrated by the Fe-containing protein ferritin. Basically, this compound consists of a coat of protein (apoferritin) surrounding a micelle of hydrated iron hydroxide. The metal can be readily and reversibly removed from the apoprotein.

See also **Chelates and Chelation**.

METALLOTHIONEINS. These are low-molecular weight, cysteine-rich proteins that bind metal ions. As reported by Furey, et al. (*Science,* **231**, 704, 1986), metallothioneins and their genes have several potential kinds of physiological activity, including: (1) the genes are induced by metal ions and glucocorticoid hormones; (2) transcription is modulated during embryonic development; (3) the genes may be involved in control of cell differentiation and proliferation; (4) the proteins may function to activate Zn requiring apo-enzymes and regulate cellular metabolism; and (5) the proteins may act as free-radical scavengers. It appears that metallothioneins are synthesized in response to ultraviolet radiation. Cadmium, in addition to zinc, is found in metallothioneins.

METALLURGY. The science and technology of metals and alloys. Process metallurgy is concerned with the extraction of metals from their ores and with the refining of metals; physical metallurgy, with the physical and mechanical properties of metals as affected by composition, processing, and environmental conditions; and mechanical metallurgy, with the response of metals to applied forces. (*From Glossary of Metallurgical Terms and Engineering Tables, American Society for Metals, with permission.*)

Early sources define metallurgy as the process of extracting metal from ores. For many metals, the primary source materials as of the 1990s are still crude metalliferous ores. For some metals however, recycled materials contribute significantly to total metal production. For example, in the United States the recycling (qv) rate of all-aluminum used beverage cans is over 50%. For an energy-intensive metal such as aluminum, this represents a substantial energy saving. Recycled aluminum requires only 5% of the energy needed to make aluminum from bauxite ore.

Metallurgy includes not only the treatment of crude ore and scrap, but also the processing of intermediates, i.e., concentrates, and wastes such as slags, tailings, etc, for contained metal values.

Definitions

The field of metallurgy has a unique and frequently very specialize vocabulary. Understanding this language helps to clarify certain concepts and processing steps. Definitions of key terms follow.

To **concentrate** is to take an action to intensify in strength or purity by the removal of valueless or unneeded constituents.

Electrometallurgy covers the various electrical processes for the working metals.

Flotation is the method of mineral separation in which a froth created in water by a variety of reagents floats some finely crushed minerals.

Gangue consists of the undesired minerals associated with ore, mostly nonmetallic.

Hydrometallurgy refers to the treatment of ores, concentrates, and other metal-bearing materials by wet processes.

Leaching is the extracting of a soluble metallic compound from an ore by selectively dissolving it in a suitable solvent.

A **mineral** is an inorganic substance occurring in nature and having a definite chemical composition or a characteristic range of chemical composition, and distinctive physical properties or molecular structure.

Mineral dressing is the physical or chemical concentration of raw ore into a product form which a metal can be recovered at a profit.

Ore is a mineral or aggregate of minerals from which a valuable constituent, especially a metal, can be profitably extracted.

Pyrometallurgy is metallurgy involved in winning and refining metals where heat is used, as in roasting and smelting.

Roasting is the heating of solids, frequently to promote a reaction with a gaseous constituent in the furnace atmosphere.

Smelting is any metallurgical operation in which metal is separated by fusion from those impurities with which it may be chemically combined or physically mixed. Scores of entries in this encyclopedia deal directly or indirectly with various aspects of metallurgy. Check the alphabetical index. In addition to each of the individual metals (aluminum, cadmium, iron, vanadium, etc.) described in separate articles in this encyclopedia, also check the following key words and phrases: alloys; amalgam; annealing; brazing; brittle fracture; calorizing (and other metal treating processes); corrosion; creep; metallography; phase diagram; powder metallurgy; temper; welding; wire drawing; wrought iron; etc.

METALS (The).

In terms of classification, several of the chemical elements are referred to as metals, principally because of the metallic qualities they exhibit. The new group designating system is used here. There are several subclassifications:

Group 11: In order of increasing atomic number, these are copper, gold, and silver. Sometimes, these metals also are referred to as noble metals, principally because they sometimes occur in nature in elemental form. Gold and silver also are frequently referred to as "coinage" metals. The elements of this group are characterized by the presence of one electron in an outer shell. Although copper and gold also have other valences, all of the elements in this group have a 1+ valence in common.

Group 12: In order of increasing atomic number, these are zinc, cadmium, and mercury. The elements of this group are characterized by the presence of two electrons in an outer shell. Although mercury also has a valence of 1+, all of the elements in this group have a 2+ valence in common.

Group 4: In order of increasing atomic number, these are titanium, zirconium, and hafnium. The elements of this group are characterized by the presence of two electrons in an outer shell. Although titanium and zirconium also have other valences, all of the elements in this group have a 4+ valence in common.

Group 5: In order of increasing atomic number, these are vanadium, niobium (sometimes called columbium), and tantalum. Vanadium and tantalum have two electrons in an outer shell; niobium has one electron in its outer shell. Although niobium and vanadium also have other valences, all of the elements in this group have a 5+ valence in common.

Group 6: In order of increasing atomic number, these are chromium, molybdenum, and tungsten. Chromium and molybdenum have one electron in their outer shells; tungsten has two electrons in its outer shell. Although chromium and molybdenum also have other valences, all of the elements in this group have a 6+ valence in common.

Group 7: In order of increasing atomic number, these are manganese, technetium, and rhenium. Manganese and rhenium have two electrons in their outer shells; technetium has one electron in its outer shell. Although

manganese and rhenium also have other valences, all of the elements in this group have a 7+ valence in common.

Groups 8, 9, 10: In order of increasing atomic number, these are iron, cobalt, nickel, ruthenium, rhodium, palladium, osmium, iridium, and platinum. Ruthenium, rhodium, and platinum have one electron in their outer shells; iron, osmium, cobalt, and nickel have two electrons in their outer shells; iridium has 17 outer electrons and palladium 18 outer electrons. Although all of these elements fall into one group, they appear in the classification in three subgroupings (hence the sometimes-used term *triads*): (1) iron, cobalt, and nickel each have valences of 2+ and 3+; (2) ruthenium, rhodium, and palladium each have valences of 4+, in addition to other valences; (3) osmium, iridium, and palladium each have valences of 4+, in addition to other valences.

In terms of the periodic classification, all elements here designated as the metals fall between highly alkaline elements (alkali metals and alkaline earths) at the left end of the table and the acidic elements, ending with the halogens at the right end of the table. Thus, the term "transition elements" sometimes is used to describe these in-between elements. Actually, the term transition can be applied to the differences between any series of elements within the overall classification, or between individual elements within a group—because of the gradual alteration in chemical behavior that takes place between groups and between elements.

METASTABLE NUCLEI.

Nuclei in excited nuclear states that have measurable lifetimes (exceeding 10^{-10}–10^{-9} second).

METASTABLE STATE.

Three common uses of this term denote: (1) A peculiar state of pseudo-equilibrium, in which the system has acquired energy beyond that for its most stable state, yet has not been rendered unstable. Thus, by using great care, water at 760 millimeters pressure may be heated several degrees above its normal boiling point, say to 105°C, yet not boil. In this condition it has received heat energy beyond that normally required for liquid-vapor equilibrium, energy which it might be expected to release by spontaneously exploding into steam; and only a slight disturbance will precipitate that change, but the disturbance must come from some external source.

(2) The term has been used in atomic physics for various excited states, but its most general usage today is for an excited state from which all possible quantum transitions to lower states are forbidden transitions by the appropriate selection rules.

(3) In nuclear physics, the term is used to denote the states in which metastable nuclei are found.

METHACRYLIC ACID AND DERIVATIVES.

Methacrylic acid. [CAS: 79-41-4]. (MAA) was first prepared in 1865 by the hydrolysis of ethyl methacrylate, which was in turn obtained by dehydrating ethyl α-hydroxyisobutyrate. The polymerizability of methacrylic acid was first noted in 1880 when a white powder was obtained in a distillation of methacrylic acid. The acetone cyanohydrin process for the synthesis of methyl methacrylate via the formation of methacrylamide sulfate, which was patented in 1934, still forms the basis for the bulk of the methyl methacrylate (MMA) currently produced.

Physical Properties

Selected physical properties of various methacrylates are given in Table 1.

Reactions

Methacrylic acid and its ester derivatives are α,β-unsaturated carbonyl compounds and exhibit the reactivity typical of this class of compounds, i.e., Michael and Michael-type conjugate addition reactions and a variety of cycloaddition and related reactions. Although less reactive than the corresponding acrylates as the result of the electron-donating effect and the steric hindrance of the α-methyl group, methacrylates readily undergo a wide variety of reactions and are valuable intermediates in many synthetic procedures.

Polymerization

The vast majority of commercial applications of methacrylic acid and its esters stem from their facile free-radical polymerizability. Solution, suspension, emulsion, and bulk polymerizations have been used to advantage. Although of much less commercial importance, anionic

TABLE 1. SELECTED PROPERTIES OF METHACRYLATES

Compound	Mol wt	Mp, °C	Viscosity, mPa(= cP)	Flash point, °C	Autoignition temperature, °C
methyl[a]	100.11	−48	0.53	9[b]	435
ethyl[a]	114.14	−17	0.92	16[c]	393
butyl[a]	142.19	−50	0.92	49[d]	294
lauryl[e]	262	−22		110	277
2-dimethyl-aminoethyl[f]	157.2	−30	1.1	75	
2-hydroxyethyl[g]	130.14	−12		66	
2-hydroxypropyl[g]	144.17	−89	7.1	98	
glycidyl	142.1	<−60		5	76

[a] Heat of polymerization = 57.5 kJ/mol (13.7 kcal/mol).
[b] Lower explosion limit (LEL) = 2.1%; upper explosion limit (UEL) = 12.5%z.
[c] LEL = 1.8%; UEL to saturation.
[d] LEL = 2.0%; UEL = 8%.
[e] Made from a mixture of higher alcohols, predominantly C-12.
[f] pK_a = 8.4.
[g] Heat of polymerization = ~50 kJ/mol (12 kcal/mol).

polymerizations of methacrylates have also been extensively studied. Strictly anhydrous reaction conditions at low temperatures are required to yield high molecular weight polymers in anionic polymerization. Side reactions of the propagating anion at the ester carbonyl are difficult to avoid and lead to polymer branching and inactivation. Polymerization of methacrylates is also possible via what is known as group-transfer polymerization.

Higher Alkyl and Functional Methacrylates

Most large-scale industrial methacrylate processes are designed to produce methyl methacrylate or methacrylic acid. In some instances, simple alkyl alcohols, e.g., ethanol, butanol, and isobutyl alcohol, may be substituted for methanol to yield the higher alkyl methacrylates. In practice, these higher alkyl methacrylates are usually prepared from methacrylic acid by direct esterification or transesterification of methyl methacrylate with the desired alcohol.

Hydroxy functional methacrylates are accessible by the reaction of methacrylic acid and ethylene oxide or propylene oxide in the presence of chromium, iron, or ion-exchange catalysts.

Manufacture and Processing

The basic feedstock for the manufacture of methyl methacrylate and methacrylic acid is, ultimately, natural gas or crude oil. It is convenient to categorize the various manufacturing routes in terms of the specific hydrocarbon raw material used. Propylene (C-3) routes require the addition of one carbon atom. Ethylene (C-2)-based routes require the addition of two carbon atoms to create the four-carbon methacrylate backbone. The commercial viability of a process is determined by the aggregate of raw material cost and utilization (process yield), operating costs (energy), waste disposal costs, environmental impact, and plant capital investment.

Research is currently directed toward development of novel technologies that may present economic advantages with respect to the conventional acetone cyanohydrin (ACH) route. Mitsubishi Gas Chemical Co. has developed and patented a modified acetone cyanohydrin-based route that does not use sulfuric acid and therefore presents the opportunity for reduced waste costs. A novel C-3 route based on the palladium-catalyzed carbonylation of methylacetylene has been developed by Shell Oil Co. There have been significant improvements in catalysts and resulting yields for key transformations in many routes since the 1980s.

MMA from Acetone Cyanohydrin (ACH). [CAS:75-86-5]. The process for conversion of acetone cyanohydrin–H_2SO_4 to methyl methacrylate through methacrylamide sulfate has been practiced commercially since 1937 and is based on technology patented by ICI in 1934. Acetone cyanohydrin is prepared via base-catalyzed reaction of acetone and hydrogen cyanide. Acetone and hydrogen cyanide are obtained as by-products from the commercial production of phenol and acrylonitrile, respectively. Hydrogen cyanide is also manufactured directly by catalytic ammoxidation of methane. Sulfuric acid is used in excess and serves as both reactant and solvent in the reaction with acetone

cyanohydrin to form methacrylamide sulfate through an α-sulfatoamide intermediate.

Inhibitors are introduced at specific points in the process to prevent polymerization. Sulfuric acid serves as catalyst in a combined hydrolysis-esterification of methacrylamide sulfate to a mixture of methyl methacrylate and methacrylic acid. Conversion of methacrylamide sulfate to methyl methacrylate can be carried out using a variety of procedures for the recovery of crude methyl methacrylate and for separation of methanol and methacrylic acid for recycling. A schematic of the overall process is given in Figure 1. The overall yield based on acetone cyanohydrin is approximately 90%. Most of the world supply of MMA is still produced by this process.

Ethylene-Based (C-2) Routes. MMA and MAA can be produced from ethylene as a feedstock via propanol, propionic acid, or methyl propionate as intermediates. Propanal may be prepared by hydroformylation of ethylene over cobalt or rhodium catalysts. The propanal then reacts in the liquid phase with formaldehyde in the presence of a secondary amine and, optionally, a carboxylic acid. The reaction presumably proceeds via a Mannich base intermediate which is cracked to yield methacrolein. Alternatively, a gas-phase, crossed aldol reaction with formaldehyde catalyzed by molecular sieves (qv) may be used to form methacrolein. The methacrolein is then oxidized to methacrylic acid.

Isobutylene-Based (C-4) Routes. Isobutylene or *tert*-butyl alcohol can be converted to methacrylic acid in a two-stage, gas-phase oxidation process via methacrolein as an intermediate. The alcohol and isobutylene may be used interchangeably in the processes since *tert*-butyl alcohol readily dehydrates to yield isobutylene under the reaction conditions in the initial oxidation. Variations of this process have been commercialized.

Methacrylonitrile Process. MAA and MMA may also be prepared via the ammoxidation of isobutylene to give methacrylonitrile as the key intermediate. A mixture of isobutylene, ammonia, and air are passed over a complex mixed metal oxide catalyst at elevated temperatures to give a 70–80% yield of methacrylonitrile. Relatively modest yields are obtained in the ammoxidation reaction and the generation of a considerable acid waste stream combine to make this process economically less desirable than the ACH or C-4 oxidation to methacrolein processes.

Uses

Methacrylic acid and methacrylic esters are used in a wide variety of polymers with a broad spectrum of applications. Poly(methacrylic acid) or its neutralized salts are used as additives for detergent builders and rheology modifiers. Methacrylic esters, the most important of which is methyl methacrylate, yield hard, tough polymers in contrast to the softer

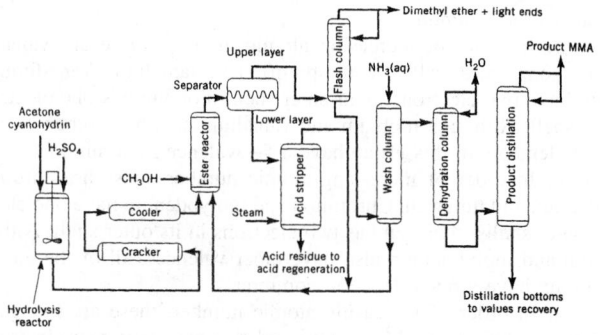

Fig. 1.   MMA from acetone cyanohydrin via methacrylamide sulfate

acrylates. Copolymerization of methacrylic esters with acrylates allows the preparation of hard but flexible polymers for use in paints, polishes, and many other coatings. Methacrylate polymers are prized for their clarity, colorability, color compatibility, weatherability, and ultraviolet light stability, which allows them to be used for both indoor and outdoor applications. The principal end uses of methacrylates are in acrylic sheet and molding resins which find commercial application in signs, displays, glazing compounds, lighting fixtures, building panels, automotive components, plumbing fixtures, and appliances. They are also used as impact modifiers in poly(vinyl chloride) (PVC) siding, film, sheet, and plastic bottle manufacture.

Storage and Handling

Polymerizations of methacrylic acid and derivatives are very energetic (MAA, 66.1 kJ/mol; MMA, 57.5 kJ/mol = 13.7 kcal/mol). The potential for the rapid evolution of heat and generation of pressure presents an explosion hazard if the materials are stored in closed or poorly vented containers. To prolong usable shelf-life, commercially available methacrylic monomers are inhibited with the methyl ether of hydroquinone (MEHQ). Other commonly used inhibitors are alkylphenols and hydroquinone. Once inhibited, methacrylic acid or its esters may be handled as flammable materials. To avoid photoinitiation of polymerization, all methacrylates should be stored with minimal exposure to light.

Most unwanted polymerization events of methacrylic monomers occur because of overheating, leading to inhibitor depletion or oxygen depletion, and in turn to inhibitor inactivation. Care should be taken to avoid stagnant areas in transfer lines or pump heads where polymerization may begin and which may then act to seed polymerization of the bulk material.

The relatively high freezing point of methacrylic acid (15°C) is a problem because the inhibitor tends to partition into the liquid phase upon freezing. Thawing of the material tends to create localized pools of uninhibited methacrylic acid which are extremely susceptible to polymerization. Care should be taken to limit thawing temperatures to less than 40°C and to ensure good mixing of the thawed material. For most polymer applications the removal of the inhibitors from the monomer is unnecessary.

Health and Safety Factors

Methacrylates are slightly to severely irritating to skin and eyes and are considered potential skin sensitizers. Several lifetime exposure studies by a variety of routes (oral capsule, drinking water, inhalation) in rodents have shown it to be noncarcinogenic. Methyl methacrylate also is nonteratogenic (did not produce birth defects) after inhalation exposures. Headaches, vomiting, and drowsiness are symptomatic of overexposure to vapors. In the workplace, the pungent odor and irritant nature of the methacrylate monomers serves as a warning property and tends to keep exposures low.

Appropriate protective clothing and equipment should be worn to minimize exposure to methacrylate liquids and vapors. The working area should be adequately ventilated to limit vapors.

ANDREW W. GROSS
JOHN C. DOBSON
Rohm and Haas Company

Additional Reading

Salkind, M., E.H. Riddle, and R.W. Keefer, *Ind. Eng. Chem.* **51**, 1232 (1959).

Yocum, R.H. and E.B. Nyquist, eds.: *Functional Monomers,* Marcel Dekker Inc., New York, NY, 1973.

METHACRYLIC POLYMERS. The nature of the R group in methacrylic acid ester monomers having the generic formula $CH_2=C(CH_3)COOR$ generally determines the properties of the corresponding polymers. Methacrylates differ from acrylates in that the α-hydrogen of the acrylate is replaced by a methyl group. See also **Acrylic Ester Polymers.** This methyl group imparts stability, hardness, and stiffness to methacrylic polymers. The methacrylate monomers are extremely versatile building blocks. They are moderate-to-high boiling liquids that readily polymerize or copolymerize with a variety of other monomers. All of the methacrylates copolymerize with each other and with the acrylate monomers to form polymers having a wide range of hardness; thus polymers that are designed to fit specific application requirements can be tailored readily.

The uniqueness of methyl methacrylate as a plastic component accounts for its industrial use, and it far exceeds the combined volume of all of the other methacrylates. In addition to plastics, the various methacrylate polymers also find application in sizable markets as diverse as lubricating oil additives, surface coatings (qv), impregnates, adhesives (qv), binders, sealers, and floor polishes.

Physical Properties

The nature of the alkyl group from the esterifying alcohol, the molecular weight, and the tacticity determine the physical and chemical properties of methacrylate ester polymers. The physical properties of amorphous methacrylic polymers evidence a principal change in the glass-transition region. Chemical reactivity, mechanical and dielectric relaxation, viscous flow, load bearing capacity, hardness (qv), tack, heat capacity, refractive index, thermal expansivity, creep, and diffusion differ markedly below and above the transition region.

The properties of methacrylic polymers are also affected by molecular weight. Typically, mechanical properties increase as the molecular weight increases. However, beyonds some critical value, about 100,000, the increase is slight and levels off asymptotically.

Mechanical Properties. Methacrylates are harder polymers of higher tensile strength and lower elongation than their acrylate counterparts because substitution of the methyl group for the α-hydrogen on the main chain restricts the freedom of rotation and motion of the polymer backbone.

At room temperature, the first member of the linear aliphatic methacrylate series, poly(methyl methacrylate), is a hard, fairly rigid material which can be sawed, carved, or worked on a lathe. When heated above its T_g, on poly(methyl methacrylate) is a tough, pliable, extensible material that is easily bent or formed into complex shapes, and can be molded or extruded.

Optical Properties. Poly(methyl methacrylate) transmits light in the range of 360–1000 nm almost perfectly (92% compared to the theoretical 92.3%). The wavelength of visible light falls approximately between 400 and 700 nm. At a thickness of 2.54 cm or less, poly(methyl methacrylate) absorbs virtually no visible light. Beyond 2800 nm, essentially all infrared radiation is absorbed. Commercial grades of poly(methyl methacrylate) often contain uv radiation absorbers that block light in the 290–350 nm range. The absorber thus screens the user from sunburn and protects the polymer against long-term degradation from light. Poly(methyl methacrylate)'s transparency to x-rays and radiation has been found to be about the same as that of human flesh or water. Sheets of poly(methyl methacrylate) are opaque to alpha particles, and for thicknesses above 6.35 mm (0.250 in.), the polymer is essentially opaque to beta radiation; poly(methyl methacrylate) is used as a transparent neutron stopper. Most formulations of colorless sheet have high transmittance to standard broadcast and television waves as well as to most radar bands.

Many items, such as magnifiers, reducers, camera lenses, prisms, and especially complex reflex lenses widely used in automotive tail-lights, are made from poly(methyl methacrylate).

Electrical Properties. The surface resistivity of poly(methyl methacrylate) is higher than that of most plastic materials. Weathering and moisture affect poly(methyl methacrylate) only to a minor degree. High resistance and nontracking characteristics have resulted in its use in high voltage applications, and its excellent weather resistance has promoted the use of poly(methyl methacrylates) for outdoor electrical applications.

Chemical Properties

Methacrylate polymers have a greater resistance to both acidic and alkaline hydrolysis than do acrylate polymers; both are far more stable than poly(vinyl acetate) and vinyl acetate copolymers. There is a marked difference in the chemical reactivity among the noncrystallizable and crystallizable forms of poly(methyl methacrylate) relative to alkaline and acidic hydrolysis. Conventional (i.e., free-radical), bulkpolymerized, and syndiotactic polymers hydrolyze relatively slowly compared with the isotactic type. Polymer configuration is unchanged by hydrolysis.

The chemical resistance of poly(methyl methacrylate) may be summarized as follows. PMMA is not affected by most inorganic solutions, mineral oils, animal oils, low concentrations of alcohols; paraffins, olefins, amines, alkyl monohalides; and aliphatic hydrocarbons and higher esters, i.e., >10 carbon atoms. However, PMMA is attacked by lower esters, e.g., ethyl acetate, isopropyl acetate; aromatic hydrocarbons, e.g., benzene, toluene, xylene; phenols, e.g., cresol, carbolic acid; aryl halides, e.g., chlorobenzene, bromobenzene; aliphatic acids, e.g., butyric acid, acetic acid; alkyl polyhalides, e.g., ethylene dichloride, methylene chloride; high

TABLE 1. RELATIVE OUTDOOR STABILITY OF POLY(METHYL METHA-
CRYLATE)

Material	Light transmittance, %	After exposure,[a] %	Haze, %	After exposure,[a] %
poly(methyl methacrylate)	92	92	1	2
polycarbonate	85	82	3	19
cellulose acetate butyrate	89	68	3	70

[a] Three-yr outdoor.

concentrations of alcohols, e.g., methanol, ethanol; 2-propanol; and high concentrations of alkalies and oxidizing agents.

The chemical resistance and excellent light stability of poly(methyl methacrylate) compared to two other transparent plastics is illustrated in Table 1.

Manufacture and Processing

Free-radical polymerization processes are used to produce virtually all commercial methacrylic polymers. Usually free-radical initiators (qv) such as azo compounds or peroxides are used to initiate the polymerizations. Photochemical and radiation-initiated polymerizations are also well known. At a constant temperature, the initial rate of the bulk or solution radical polymerization of methacrylic monomers is first-order with respect to monomer concentration, and one-half order with respect to the initiator concentration. Methacrylate polymerizations are markedly inhibited by oxygen; therefore considerable care is taken to exclude air during the polymerization stages of manufacturing.

A substantial fraction of commercially prepared methacrylic polymers are copolymers. Monomeric acrylic or methacrylic esters are often copolymerized with one another and possibly several other monomers. Copolymerization greatly increases the range of available polymer properties. The all-acrylic polymers tend to be soft and tacky; the all-methacrylic polymers tend to be hard and brittle. By judicious adjustment of the amount of each type of monomer, polymers can be prepared at essentially any desired hardness or flexibility. Small amounts of specially functionalized monomers are often copolymerized with methacrylic monomers to modify or improve the properties of the polymer directly or by providing sites for further reactions.

Bulk Polymerization. This is the method of choice for the manufacture of poly(methyl methacrylate) sheets, rods, and tubes, and molding and extrusion compounds. Three bulk polymerization processes are commercially important for the production of methacrylate polymers: batch cell casting, continuous casting, and continuous bulk polymerization. Approximately half the worldwide production of bulk polymerized methacrylates is in the form of molding and extrusion compounds, a quarter is in the form of cell cast sheets, and a quarter is in the form of continuous cast sheets.

Solution Polymerization. The solution polymerization of methacrylic monomers to form solution polymers or copolymers is an important commercial process for the preparation of polymers for use as coatings, adhesives, impregnates, and laminates. Typically the polymerization is done batchwise by adding monomer to an organic solvent in the presence of a soluble peroxide or azo initiator.

Emulsion Polymerization. The principal markets for aqueous dispersion polymers made by emulsion polymerization of methacrylic esters are the paint, paper, textile, floor polish, and leather industries where they are used principally as coatings or binders. Copolymers of methyl methacrylate with either ethyl acrylate or butyl acrylate are most common.

Suspension Polymerization. This method yields polymethacrylates in the form of tiny beads, which are primarily used as molding powders and ion-exchange resins. Most suspension polymers prepared as molding powders are poly(methyl methacrylate); copolymers containing up to 20% acrylate for reduced brittleness and improved processibility are also common. Suspension polymers of poly(methyl methacrylate) copolymerized with an amino or acid functional monomer, and with a di- or trivinyl monomer for cross-linking, are useful as ion-exchange resins.

Graft Polymerization. Graft copolymers are prepared by attaching one polymer as a branch to the chain of another polymer of different composition. This is usually accomplished by generating radical sites on the first polymer onto which monomer of the second polymer is grafted. The grafting may be accomplished in bulk, solution, or dispersion systems. The presence of distinct, but chemically bonded segments of two polymers often confers interesting and useful properties. Commercially, the most important methacrylate graft copolymers are the MABS and MBS polymers.

The MABS copolymers are prepared by dissolving or dispersing polybutadiene rubber in a methyl methacrylate—acrylonitrile—styrene monomer mixture. MBS polymers are prepared by grafting methyl methacrylate and styrene onto a styrene—butadiene rubber in an emulsion process. The product is a two-phase polymer useful as an impact modifier for rigid poly(vinyl chloride).

Ionic Polymerization. The anionic polymerization of methacrylic monomers to stereoregular or block copolymers is well known. These polymerizations are conducted in organic solvents, primarily using organometallic compounds as initiators. This technology is of minor commercial significance, but is of interest for the preparation of polymers of narrow molecular weight distribution and controlled molecular architecture. Methacrylate monomers do not generally polymerize by a cationic mechanism.

Health and Safety Factors

In general, methacrylate polymers are considered nontoxic. Various methacrylate polymers are used in food packaging (qv) and handling, in dentures and dental fillings (see **Dental Materials**), and as medicine dispensers and contact lenses. However, care must be exercised because additives or residual monomers present in various types of polymers can display toxicity.

During manufacture, considerable care is exercised to reduce the potential for violent polymerizations, and to reduce exposure to flammable and potentially toxic monomers and solvents.

Dust explosions ignited by static discharge are a recognized hazard encountered in the handling of poly(methyl methacrylate) powders or in the fabrication of poly(methyl methacrylate) plastic sheet. Methacrylic solution polymers are treated as flammable mixtures; latex polymers are nonflammable.

Uses

The principal U.S. market for methacrylate resins is for glazing and skylights. Other significant markets include consumer products, transportation signs and lighting fixtures, plumbing (spas, tubs, showers, sinks, etc), and panels and siding. The resins are also used in medicine, optics, and as oil additives.

RONALD W. NOVAK
PATRICIA M. LESKO
Rohm and Haas Company

Additional Reading

Brandrup, J. and E.H. Immergut: *Polymer Handbook*, 3rd Edition, Wiley-Interscience, New York, NY, 1989.

Harrington, M.: in I.I. Rubin, ed., *Handbook of Plastic Materials and Technology*, John Wiley & Sons, Inc., New York, NY, 1990.

Lesko P.M. and P.R. Sperry: in P.A. Lovell and M.S. El-Aasser, eds., *Emulsion Polymerization and Emulsion Polymers*, John Wiley & Sons, Ltd., Chichester, UK, 1997.

METHANATION.   See **Coal**; **Substitute Natural Gas (SNG)**.

METHANE.   [CAS: 74-82-8]. CH_4, formula weight 16.04, colorless, odorless (when pure) gas, mp $-182.6°C$, bp $-161.4°C$, sp gr 0.415 (at $-164°C$).

Sometimes referred to as *marsh gas* or *fire damp*, methane is practically insoluble in H_2O, and moderately soluble in alcohol or ether. The gas burns when ignited in air with a pale, faintly luminous flame, forming an explosive mixture with air between gas concentrations of 5% and 13%. Methane is the principal constituent of natural gas, averaging 75% by weight. Natural gas from the Pennsylvania fields is almost 99% methane, but some gas from Kentucky fields contains as little as 23% methane. See also **Natural Gas**. Pipeline gas from several fields typically will contain about 78% methane, 13% ethane, 6% propane, 1.7% butane, and 0.6% pentane. The remaining fraction consists of gases higher in the alkyl series. While generally not referred to as such, methane can be classified as a major fuel. The heating value of pure methane is 995Btu/ft^3 (8856 Calories per cubic meter).

Methane, as the major constituent of natural gas, is an extremely important raw material for numerous synthetic products. For most processes, it is not required to isolate and purify the methane, but the natural gas as received may be used. The high percentage of CH_4 in various feedstocks makes possible the formation of synthesis gas: $CH_4 + H_2O \longrightarrow CO + 3H_2$. The percentages of CO and H_2 in synthesis gas vary depending on the end product to be made. Synthesis gas is used widely in the manufacture of NH_3, oxo-chemicals, and methyl alcohol. See also **Synthesis Gas**.

In addition to the preparation of synthesis gas, which is used so widely in various organic syntheses, methane is reacted with NH_3 in the presence of a platinum catalyst at a temperature of about 1,250°C to form hydrogen cyanide: $CH_4 + NH_3 \longrightarrow HCN + 3H_2$. Methane also is used in the production of olefins on a large scale. In a controlled-oxidation process, methane is used as a raw material in the production of acetylene.

Most artificial gases, such as producer gas, coal gas, water gas, manufactured gas, and town gas contain a high content of methane. In addition to its use as a basic chemical and fuel, methane is of notable interest because of its role as the anchor compound of the *alkanes* (paraffin or aliphatic hydrocarbons). All of these compounds may be considered derivatives of methane.

Some methane is manufactured by the distillation of coal. Coal is a combustible rock formed from the remains of decayed vegetation. It is the only rock containing significant amounts of carbon. The elemental composition of coal varies between 60% and 95% carbon. Coal also contains hydrogen and oxygen, with small concentrations of nitrogen, chlorine, sulfur, and several metals. Coals are classified by the amount of volatile material they contain, that is, by how much of the mass is vaporized when the coal is heated to about 900°C in the absence of air. Coal that contains more than 15% volatile material is called bituminous coal. Substances released from bituminous coal when it is distilled, in addition to methane, include water, carbon dioxide, ammonia, benzene, toluene, naphthalene, and anthracene. In addition, the distillation also yields oils, tars, and sulfur-containing products. The non-volatile component of coal, which remains after distillation, is coke. Coke is almost pure carbon and is an excellent fuel. However, it may contain metals, such as arsenic and lead, which can be serious pollutants if the combustion products are released into the atmosphere.

Carbon monoxide and hydrogen react to form CH_4 in the presence of a nickel catalyst. Methane also is formed by reaction of magnesium methyl iodide in anhydrous ether (Grignard's reagent) with substances containing the hydroxyl group. See also **Grignard Reactions**. Methyl iodide (bromide, chloride) is preferably made by reaction of methyl alcohol and phosphorus iodide (bromide, chloride).

Additional Reading

Clever, H. et al.: *Methane*, Elsevier Science, New York, NY, 1987.
Lee, S.: *Methane and Its Derivatives*, Marcel Dekker, Inc., New York, NY, 1996.
Mastalerz, M. et al.: *Coalbed Methane: Scientific, Environmental, and Economic Evaluation*, Kluwer Academic Publishers, Norwell, MI, 1999.

METHANOGENS. Cells that resemble bacteria in a superficial way, but that have unique genetic and metabolic characteristics. Methanogens are anaerobic, methane-producing microorganisms that occur in a wide variety of places—the gastrointestinal tract of animals, including humans, in the sediments of natural waters, in sewage treatment plant vessels and piping, and in natural hot springs. As proposed by Woese (University of Illinois) and Fox (University of Houston), the methanogens probably make up a third line of descent of cells in addition to the prokaryotes (bacteria and blue-green algae cells which do not have a well-defined nucleus) and eukaryotes (more complex cells with a nucleus). These researchers also have suggested that there may be still other kinds of cells that do not meet the criteria set down for prokaryotes and eukaryotes.

Methanogens are distinguished from bacteria on at least three counts: (1) The cell walls do not contain muramic acid, the characteristic constituent of the peptidoglycans that form bacterial cell walls. (2) Their metabolism differs markedly from bacteria. A number of coenzymes apparently unique to methanogens have been identified. Some of these enzymes are involved in methyl transfer reactions, including the formation of methane. One of the coenzymes is possibly the smallest coenzyme yet

to be discovered. The methanogens also differ in the manner in which carbon dioxide is fixed into cellular carbon. However, the pathway has not been clearly identified. (3) The RNA sequences of methanogens differ from those of other organisms. These observations have indicated to Woese and Fox that although the methanogens share a common ancestor with prokaryotes and eukaryotes, an independent line of descent branched off at possibly about the same time the other cell types diverged.

Although not fully understood, the methanogens place new challenges to the evolutionary biologists for further explanation in terms of the development of early life on earth and may be very valuable toward understanding life on extraterrestrial bodies as these may be explored over future years.

Barker (University of California at Berkeley) and Huntgate (University of California at Davis) as early as the mid-1950s noted that methanogens differ radically from bacteria.

Methanogens take part in the terminal stages of organic matter degradation and survive on carbon dioxide and hydrogen yielded by anaerobic bacteria and converting them to methane.

METHANOL. See **Methyl Alcohol**.

METHIONINE. See **Amino Acids**.

METHYL ALCOHOL. [CAS: 67-56-1]. CH_3OH, formula weight 32.04, colorless, mobile liquid with mild characteristic odor, mp −97.6°C, bp 64.6°C, sp gr 0.792. Also known as *methanol*, the compound is miscible in all proportions with H_2O, ethyl alcohol, or ether. When ignited, methyl alcohol burns in air with a pale blue, transparent flame, producing H_2O and CO_2. The vapor forms an explosive mixture with air. The upper explosive limit (% by volume in air) is 36.5 and the lower limit is 6.0.

Methyl alcohol possesses distinct narcotic properties. It is also a slight irritant to the mucous membranes. The principal toxic effect is exerted on the nervous system, particularly the optic nerves and possibly the retinae. The effect upon the eyes has been attributed to optic neuritis, which subsides, but is followed by atrophy of the optic nerve. Once absorbed, methyl alcohol is only very slowly eliminated. Coma resulting from massive exposures may last as long as 2 to 4 days. In the body, the products formed by its oxidation are formaldehyde and formic acid, both of which are toxic.

Chemical Properties

Methyl alcohol is a versatile material, reacting (1) with sodium metal, forming sodium methylate, sodium methoxide CH_3ONa plus hydrogen gas, (2) with phosphorus chloride, bromide, iodide, forming methyl chloride, bromide, iodide, respectively, (3) with H_2SO_4 concentrated, forming dimethyl ether $(CH_3)_2O$, (4) with organic acids, warmed in the presence of H_2SO_4, forming esters, e.g., methyl acetate CH_3COOCH_3, methyl salicylate $C_6H_4(OH) \cdot COOCH_3$, possessing characteristic odors, (5) with magnesium methyl iodide in anhydrous ether (Grignard's solution), forming methane as in the case of primary alcohols, (6) with calcium chloride, forming a solid addition compound $4CH_3OH \cdot CaCl_2$, which is decomposed by H_2O, (7) with oxygen, in the presence of heated smooth copper or silver forming formaldehyde. The density of pure methyl alcohol is 0.792 at 20°C compared with H_2O at 4°C (the corresponding figure for ethyl alcohol is 0.789), and the percentage of methyl alcohol present in a methyl alcohol-water solution may be determined from the density of the sample.

A common test for methyl alcohol is by its oxidation in air with a hot copper wire to form formaldehyde.

At one time, most methyl alcohol was obtained by the destruction distillation of hardwoods (hence the name *wood alcohol*) at about 350°C, along with a yield of acetic acid and small percentages of acetone in the water condensate. Interest in returning to wood as a source has revived because of fossil fuel shortages.

Production of Methyl Alcohol[1]

Synthetic methanol is one of the major raw materials of the organic chemical industry. Methanol has economic stability and a steady growth rate owing to the low costs of production and diversity of applications. Nearly all the methanol producers also make formaldehyde, which is

---

[1] Remainder of this entry prepared by J.R. Masson, Process Engineering Consultant, Davy McKee (Oil & Chemicals) Ltd., London, England.

the main end use (more than 50%) of methanol. The other main end uses are dimethyl terephthalate, methacrylates, methylamines (for resins, herbicides, and fungicides), methyl halides (for silicones, tetramethyl lead, butyl rubbers, paint removers, photographic films, aerosol propellants, (diminishing use), and degreasing compounds), acetic acid, and solvents.

An important process for production of synthetic protein uses methanol as feedstock. The use of methanol as a fuel, either as pure methanol, as a mixture (approximately 15%) with gasoline, or as a feedstock for synthetic gasoline is envisaged for possible large-scale application; as well as use in gas turbines for electricity generation. See also **Wastes as Energy Sources**.

There are three principal commercial grades of methanol (as defined in U.S. Federal Specification O-M-232f: June 5, 1975): *Grade A*, synthetic, 99.85% by weight (solvent use); *Grade AA*, synthetic, 99.85% by weight (hydrogen and carbon dioxide generation use); and *Grade C*, wood alcohol (denaturing use).

The most recent advances in methanol synthesis are the low- and intermediate-pressure processes of the type shown in Fig. 1. The synthesis step of this process[2] relies upon a copper-based catalyst, which gives good yields of methanol at pressures of 50 and 100 atmospheres. These pressures are substantially below those of the 250–350 atmospheres required by earlier processes. The high catalyst activity allows the synthesis reaction to take place at a relatively low temperature of 250–270°C. As a result, methanation is avoided, and byproduct formation is lower, giving increased process efficiency.

The development of this low-pressure technology has caused a major reassessment of the economics of methanol production. The energy required to compress the synthesis gas from its production pressure to the synthesis unit is reduced by a factor between 2 and 3. The lower synthesis pressure allows the exclusive use of centrifugal compressors in plants with capacities as low as 15 million gallons (0.57 million hectoliters) per year. Small producers find attractive the savings in investment, operating, and maintenance costs made possible by low-pressure operation. Plants range in capacity from 15 million gallons (0.57 million hectoliters) to 250 million gallons (9.46 million hectoliters) per year.

Synthesis gas is prepared by the steam reforming or partial oxidation of a liquid or gaseous hydrocarbon feedstock, or by direct combination of carbon dioxide with purified hydrogen-rich gases. Economic considerations usually favor the steam-reforming route for a naphtha or natural gas

[2] Developed by Imperial Chemical Industries, Ltd.

feedstock. In this instance, desulfurized feedstock is preheated, mixed with superheated steam, and reacted over a conventional catalyst (normally nickel-based) in multitubular reformer. The reformer usually is operated at between 15 and 30 atmospheres and at a tube outlet temperature of 840–900°C. The reforming conditions are chosen to give the most economic overall production costs. Methane slip (amount of unconverted methane) usually is greater than for conventional high-pressure synthesis processes, since the cost of compressing the additional methane is less significant with the low-pressure process. With a naphtha feedstock, an almost exact stoichiometric ratio of carbon oxides to hydrogen in the synthesis gas is achieved, but when natural gas is the feedstock, there is an inherent deficiency of carbon. Established practice for many years has been to add carbon dioxide from an external source in preparing a stoichiometric synthesis gas. Development of the low-pressure process has shown that this addition of carbon dioxide is not required and that, depending upon the cost of carbon dioxide production, the production of methanol from natural-gas feedstock alone is economic.

After heat recovery and cooling, the synthesis gas is compressed to the required synthesis pressure and passed into the synthesis loop at the suction of a circulator. The circulator, which boosts the pressure of the circulating gases to make up the total loop pressure drop, also is a centrifugal machine. Feed-gas preheating is carried out by heat exchange with the hot gases leaving the converter. Heat recovery is incorporated into the loop to recover the heat of reaction of methanol synthesis.

Synthesis takes place in a hot-wall converter over the low-pressure methanol-synthesis catalyst at 250–270°C. Temperature control of the converter is effected by injecting cold gas at appropriate levels in the catalyst bed, using specially developed distributors that provide excellent gas mixing while allowing free passage of the catalyst for easy charging and discharging. After leaving the converter and passing through the feed-gas preheater the converted gases are cooled, and crude methanol is condensed and separated from the uncondensed gases, which are recycled with makeup synthesis gas to the converter. A continuous gas purge is taken from the synthesis loop in order to remove an accumulation of inert gases. This purge is recycled to the synthesis-gas preparation section as reformer fuel. The crude methanol is reduced in pressure before passing forward to the methanol-purification section, where methanol of the required purity is produced by conventional distillation methods.

Economics in fuel gas consumption are achieved by use of recovered heat in reboiling in the distillation columns. In addition, distillation schemes

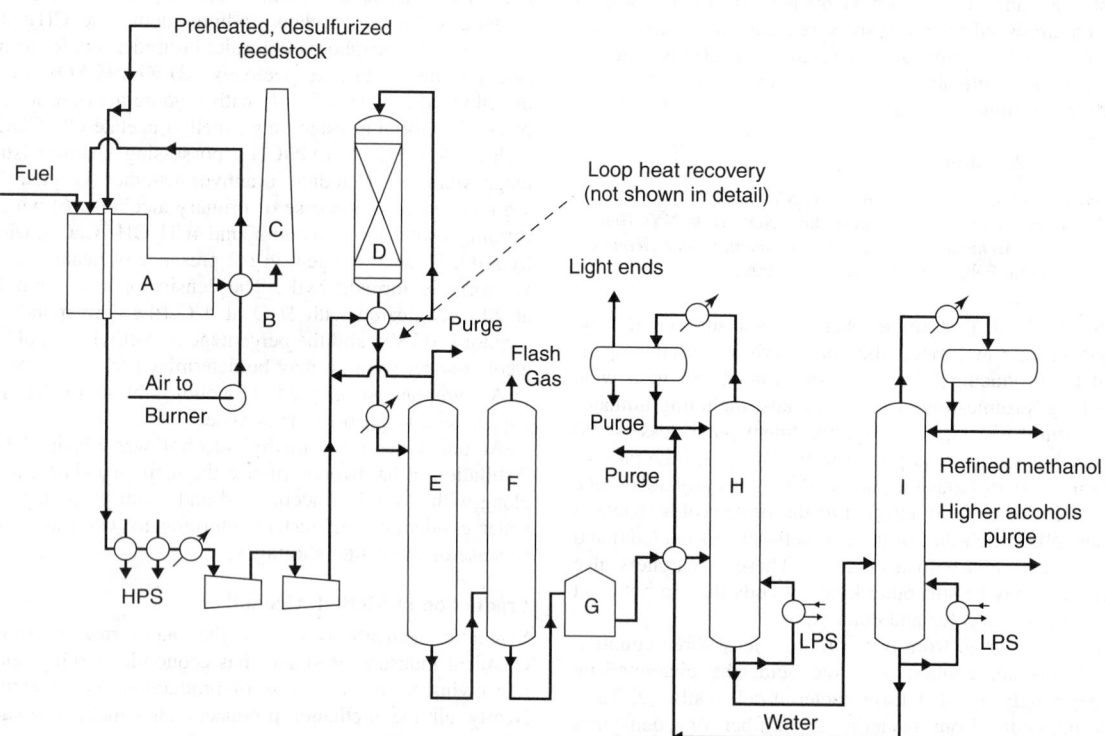

Fig. 1.   Low-pressure methanol production. (A) Burner and superheater; (B) air preheater; (C) stack; (D) methanol converter; (E) separator; (F) flash vessel; (G) crude storage; (H) topping column; (I) refining column. HPS = high-pressure steam; LPS = low-pressure steam. (*Imperial Chemical Industries, Ltd*)

involving three or four columns have also been developed with reduced reboil heat requirements.

Health and Safety

Methanol is not classified as carcinogenic, but can be acutely toxic if ingested: 100–250 mL may be fatal or result in blindness. The principal physiological effect is acidosis resulting from oxidation of methanol to formic acid. See also **Acidosis**. Methanol is a general irritant to the skin and mucous membranes. Methanol vapor can cause eye and respiratory mucous membranes. Methanol vapor can cause eye and respiratory tract irritation, nausea, headaches, and dizziness.

Methanol does not pose an undue toxicity hazard if handled in well-ventilated areas, and is rated as a slight health hazard by the National Fire Protection Association (NFPA): http://www.nfpa.org/.

Storage and Handling. Methanol is stable under normal storage conditions. Methanol is not subject to hazardous polymerization reactions, but can react violently with strong oxidizing agents. The greatest hazard involved in handling methanol is the danger of fire or explosion. The NFPA classifies methanol as a serious fire hazard.

Additional Reading

Chang, C.: *Hydrocarbons from Methanol,* Marcel Dekker, Inc., New York, NY, 1983.
Cheng, W. and H. Kung (Editors): *Methanol Production and Use,* Marcel Dekker, Inc., New York, NY, 1994.
Murrell, J. and H. Dalton: *Methane and Methanol Utilizers,* Kluwer Academic Publishers, Norwell, MI, 1992.

Web Reference

Methanol Institute: http://www.methanol.org/

METHYL CHLORIDE.   See **Chlorinated Organics**.

METHYLENE CHLORIDE.   See **Chlorinated Organics**.

METHYLISOBUTYL CARBINOL (MIBC).   See **Ketones**.

METHYLISOBUTYL KETONE (MIBK).   See **Ketones**.

METHYLPRIDINE.   See **Pyridine and Derivatives**.

MEVALONIC ACID.   See **Steroids**.

MEYER REACTION.   Preparation of alkylstannonic acids by reacting alkali stannite with an alkyl iodide. When applied to alkali arsenites or plumbites, the reaction yields alkylarsonic and alkylplumbonic acids, respectively.

MEYER-SCHUSTER REARRANGEMENT.   Acid catalyzed rearrangement of secondary and tertiary α-acetylenic alcohols to α,β-unsaturated carbonyl compounds: aldehydes when the acetylenic group is terminal, ketones when it is internal.

MICA.   Mica is a generic term that applies to a wide range of hydrous aluminum silicate minerals characterized by sheet or plant-like structure, and possessing to varying degrees, depending on composition and weathering, flexibility, elasticity, hardness (qv), and the ability to be split into thin (1 μm) sheets. All micas form flat six-sided monochromic crystals, and possess cleavage parallel to the basal plane.

Mica exists in nature in a wide variety of compositions. Muscovite and phlogopite are the only natural micas of commercial importance. Vermiculite, although not considered a true mica by most mineralogists, is a micaceous mineral formed from the weathering of phlogopite or biotite and is also of commercial importance. Fluorophlogopite, $K_2Mg_6(Al_2Si_6O_{20})F_4$, is a synthetic mica made from pure chemical oxides.

Mica has been classified into three groups: *(1)* the mica group proper, *(2)* the clintonite or brittle micas group, and *(3)* the chlorite group. Supplementary to these are the vermiculites, which are hydrated compounds that result from the alteration of any one of the micas, but usually biotite. All minerals in these groups belong to the monoclinic crystal system, and all show plane angles of 60 and 120° on the basal section. The crystals usually form in hexagonal or rhombohedral-shaped

scales, prisms, or plates. The basic structural unit of mica is a layer composed of two silicon tetrahedral sheets with a central octahedral sheet.

Muscovite is dioctahedral, having a theoretical composition of 11.8% K_2O, 45.2% SiO_2, 38.5% Al_2O_3, and 4.5% H_2O. Muscovite mica formed as a primary mineral in pegmatites and granodiorite differs in physical properties compared to muscovite mica formed by secondary alteration (mica schist). The main differences are in flexibility and ability to be delaminated. Primary muscovite is not as brittle and delaminates much easier than muscovite formed as a secondary mineral. Mineralogical properties of the principal natural micas are shown in Table 1.

Mining

Flake Mica. Flake mica is mined from weathered and hard rock pegmatites, granodiorite, and schist and gneiss by conventional openpit methods. In soft, residual material, dozers, shovels, scrapers, and front-end loaders are used to mine the ore. Often kaolin, quartz, and feldspar are recovered along with the mica. See also **Clays**; and **Silicon**.

Hard rock mining of these ore bodies requires drilling and blasting with ammonium nitrate and dynamite. After blasting, the ore is reduced in size with a drop ball and then loaded on trucks for transportation to the processing plant. Mica, quartz, and feldspar concentrates are separated, recovered, and sold from the hard rock ore.

Sheet Mica. Pockets of mica crystals are found in pegmatite stills and dikes or grandonite ore bodies. Sheet mica is mined by both underground and open-pit mining procedures. Underground mining is accomplished by driving a shaft, formed with tungsten carbide-tipped air drills, hoists, and explosives. After blasting, the mica is placed in boxes or bags for transporting to the trimming shed where it is graded, split, and cut to various specified sizes for sale.

Sheet mica is no longer mined in the U.S. Most sheet mica is mined in India.

Beneficiation Processes

Flake or Scrap Mica. In the early to mid-1900s, flake or scrap mica was mainly processed by a jigging procedure which consists of hydraulically washing a pile of bulldozed ore across a series of roll crushers and Trommel screens gaped at different size openings.

The grade of mica produced by jigging is very poor, usually about 75% concentrate, and recovery of available mica low (50%). Specifications on mica have become more stringent, therefore a more efficient processing method has been devised that provides higher quality mica, as well as more efficient recovery.

Because of improved mica processing operations, low cost earthen waste impoundment ponds have been built to store solid waste and thereby provide for a relatively cheap means of meeting new federal and state environmental laws. There are several methods of preparing ore for beneficiation after it arrives at the plant site (Fig. 1).

Flake mica is also produced as a by-product from processing feldspar ore (hard granodiorite) from mica schist which normally contains from 30–60% recrystallized muscovite mica along with quartz and iron minerals. The quartz is usually not suitable for glass sand or high purity material, however.

Sheet Mica. The preparation of sheet mica for feedstock for various punching and machining operations involves cobbing mica blocks or books

TABLE 1. MINERALOGICAL PROPERTIES OF MICAS[a]

Properties	Muscovite	Biotite	Lepidolite	Phlogopite
sp gr	2.76–3.0	2.7–3.1	2.8–3.3	2.8–2.9
luster	vitreous–pearly	splendent, sometimes submetallic	pearly	pearly
crystal	rhombic or hexagonal	pseudo-rhombohedral	hexagonal	hexagonal
colors	gray, brown, pale green, violet, yellow, dark olive-green, ruby	green, black, yellow	rose red, violet-gray, lilac, yellowish, grayish white, white	yellow-white, gray–green, pearly, brown, black

[a] Optical signs of micas are negative, crystal system is monoclinic, streak is colorless.

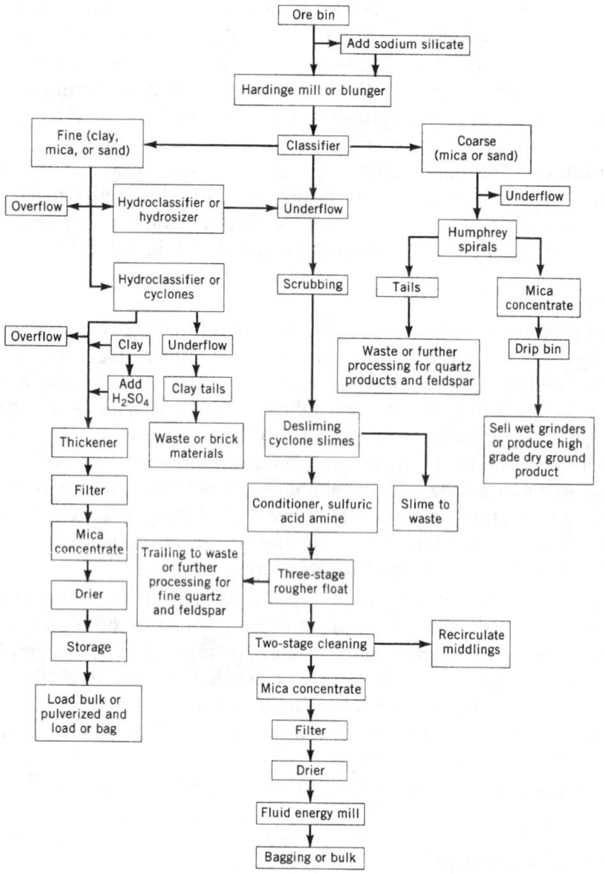

MICA

Ore bin → Add sodium silicate → Hardinge mill or blunger → Classifier

Fine (clay, mica, or sand) — Coarse (mica or sand)

Overflow — Hydroclassifier or hydrosizer — Underflow — Underflow — Humphrey spirals

Hydroclassifier or cyclones — Scrubbing — Tails — Mica concentrate

Overflow — Clay — Underflow — Waste or further processing for quartz products and feldspar — Drip bin

Add H₂SO₄ — Clay tails

Thickener — Waste or brick materials — Desliming cyclone slimes — Sell wet grinders or produce high grade dry ground product

Filter

Mica concentrate — Conditioner, sulfuric acid amine — Slime to waste

Drier — Trailing to waste or further processing for fine quartz and feldspar — Three-stage rougher float

Storage — Two-stage cleaning — Recirculate middlings

Load bulk or pulverized and load or bag — Mica concentrate — Filter — Drier — Fluid energy mill — Bagging or bulk

Fig. 1. Flow sheet for the acid circuit processing and recovery of mica from weathered granodiorite ore. An alkaline—cationic circuit may be used by inserting a second conditioner containing lignin sulfonate, adjusting the pH to 8.0, and adding NaOH and DRL (distilled tall oil) fatty acid to the first conditioner

to remove dirt, rock, and defective mica, trimming and splitting into sizes and thicknesses suitable for punching and milling to desired shapes, and grading the finished mica sheets according to size and quality. The waste mica resulting from cobbing and trimming (scrap mica) is often mixed with flake mica for processing by dry or wet ground procedures.

The grade determines whether the mica can be used in high technology electronic instruments, e.g., computer-aided tomography (CAT) scan, or in low technology devices, e.g., a toaster. Many types of insulators, as well as the base for electronic circuits, are formed from the high quality sheets of mica by a punch pressing operation.

Procedures for Production

The general pieces of equipment used in grinding flake mica or mica concentrate into saleable mica products are hammer mills of various types, fluid energy mills, Chaser or Muller mills for wet grinding, and Raymond or Williams high side roller mills. Another method is being developed, called a Duncan mill (J. M. Huber, Inc.), that is similar in many respects to an attrition mill. All of these mills are used in conjunction with sieves, and all but some types of hammer mills-incorporate air classifiers as a part of the circuit.

Ground Mica. This constitutes by far the largest commercial use for mica. It is largely produced from the beneficiation of weathered and unweathered pegmatites, granodiorite, and metamorphic schists, although some higher grades are produced from trimmings of sheet mica or Type A (low quality) mica blocks.

Wet ground mica products account for approximately 15% of the total mica market. Exact sizing of mica products coupled with surface treatment procedures have led to a greater use for wet ground mica in plastic compositions, particularly automobile bodies. These quality products demand a high dollar value.

Dry ground mica concentrate is processed into usable products by several different grinding methods. Relatively coarse particle sizes (1.651–0.147 mm (10–100 mesh)) are used in oil-well drilling muds, some types of welding rod coatings, asphalt (qv), roofing shingles, and some other types of fillers (qv). These products are ground on a hammer mill in closed circuit with a sieve. Roofing micas produced from mica schist are often ground in a Raymond or Williams high side roll mill in closed circuit with an air classifier and a sieve. The finer particle-size micas ≤0.147–0.044 mm (−100 to −325 mesh), used mainly in textured paints and joint cement compounds, are ground on several types of fluid energy mills, but generally a mill of the Majacs type. The finest dry ground mica product is ground with superheated steam (Micronized, KMG Minerals).

Testing of Mica

There are several conventional tests required by consumers of ground mica. They include screening and the determination of bulk density, true specific gravity, chemical analysis, moisture, free silica, refraction index, oil absorption, brightness, grit content, and aspect ratio.

By-Products of Mica

The main by-products of mica processing plants are kaolin, quartz, and feldspar. Some plants produce all of these products for sale.

Mica Market

Sheet Mica. Good quality sheet mica is widely used for many industrial applications, particularly in the electrical and electronic industries, because of its high dielectric strength, uniform dielectric constant, low power loss (high power factor), high electrical resistivity, and low temperature coefficient. Mica also resists temperatures of 600–900°C, and can be easily machined into strong parts of different sizes and shapes.

Built-Up Mica. When the primary property needed for a particular application is insulation, built-up mica made by binding layered mica splittings together serves as a substitute for the more expensive sheet mica. The principal uses for built-up mica are segment plate, molding plate, flexible plate, heater plate, and tape.

Wet Ground Mica. Wet ground mica is used because of its unique properties, i.e., luster, slip and sheen, and high aspect ratio. It is used in wallpaper and coated paper, nacreous pigments, as a coating for rubber, in outdoor house paint, and in aluminum paints. Mica is used in all types of sealers for porous surfaces, such as wallboard masonry, and concrete blocks, to reduce penetration and improve holdout and as a filler in plastics to improve its electrical and thermal resistance and its insulating qualities. See also **Sealants**.

Dry Ground Mica. Dry ground mica produced by hammer milling and screening is used in oil-well drilling, coatings for roofing shingles, roofing felt, and for some types of welding rod flux.

The largest use for fine, dry ground mica is in the manufacture of wallboard joint cements. Ground mica that is essentially ≤0.147 mm-100 mesh and ~70% passing a 0.044 mm (325 mesh) Tyler sieve is used in the joint compound mixture as a filler and extender. These compounds are used to fill joints between panels of gypsum plaster board. Mica contributes to making a nonabsorbing smooth surface that reduces shrinkage and eliminates cracks. It is also used in the finished coating on ceilings and to prepare thermal insulation and acoustical qualities of ceiling tile and prefabricated concrete. See also **Calcium**.

Fine particle-size dry ground mica is also used as an extender and filler in certain texture and traffic paints. Mica particles are stronger than iron and not brittle like other inerts. It is an antifriction, antifouling, antisettling, anticorrosive, antitarnish, and antisiege agent. It is a superior reinforcing pigment that acts as a sealer over porous surfaces and reduces penetration and flushing; moreover, it improves the moisture resistance of protective coatings and adhesion to all types of surfaces.

Micronized mica is a trade name (KMG Minerals, formerly English Mica Co.) for a very fine particle-size dry ground product, usually ground with superheated steam in a special fluid energy mill and used as a replacement for wet ground mica in certain types of paints. Micronized mica, preferably calcined, is also used in cosmetic applications, i.e., nail varnishes, lipsticks, eyeshadows, and barrier cream, because is has the advantages of high ultraviolet light stability, excellent lubricity, skin adhesion, and compressibility. Some of these micas are coated with oxides like titanium and iron.

Environmental and Health Regulations

Mica mining is subjected to local, state, and federal laws. The Mining, Safety and Health Administration (MSHA) regularly monitors mica mining operations for safety violations.

Both state air and water environmental departments together with the U.S. EPA regulate and oversee air and water quality associated with mica mining operations. Most states have land management departments that regulate dam safety, erosion, sedimentation, and reclamation. The mica mines must control erosion and sedimentation and restore the mined out areas. This is accomplished either by backfilling or contouring and seeding operations, or in cases where this is impractical or undesirable, lakes for water-related recreation may be built. The Corps of Engineers have jurisdiction over laws governing wetlands.

Health regulations are supervised by county and state health departments. There are no known health problems caused by the mica crystal, however, most industrial mica products contain some free silica particles that can cause silicosis and some states require employees who work in mica plants to receive an annual x-ray.

JAMES T. TANNER
North Carolina State University

Additional Reading

Davis, L.L.: *Minerals Yearbook*, U.S. Bureau of Mines, Washington, DC, 1991–1993, p. 4, 5, 7–9.

Grim, R.W.: *Clay Mineralogy*, McGraw-Hill Book Co., Inc., New York, NY, 1968, 596 pp.

Preston, J.B.: "Mica," *Pigment Handbook*, John Wiley & Sons, Inc., New York, NY, 1971, 30 pp.

Rajgarhia, M.L.: *Ground Mica*, Mica Manufacturing Co., Private Ltd., Calcutta, India, 1987, p. 30; *British Standards*, British Standards Institute, London.

MICELLE.   An electrically charged colloidal particle, usually organic in nature, composed of aggregates of large molecules, such as found in surfactants and soaps. The term is also applied to the casein complex in milk.

See also **Colloid Systems**.

MICHEL, HARTMUT (1948–).   Awarded Nobel prize for chemistry in 1988, along with Johann Deisenhofer and Michel Huber, for work that revealed the three-dimensional structure of closely-linked proteins that are essential to photosynthesis. Doctorate awarded in 1977 by the University of Wurtzburg, Germany.

MICROCHEMISTRY.   A branch of analytical chemistry that involves procedures that require handling of very small quantities of materials. Specifically, it refers to carrying out various chemical operations (weighing, purification, quantitative and qualitative analysis) on samples ranging from 0.1 to 10 mg; this often involves use of a microscope, and still more often chromatography. See also **Microscopy (Chemical)**.

MICROCRYSTALLINE.   A form in which a number of high-polymeric substances have been prepared. They include cellulose, chrysotile asbestos, amylose (starch), collagen, nylon, and certain mineral waxes. On the microscopic level, these substances are composed of colloidal microcrystals connected by molecular chains. The process involves breaking up the network of microcrystals (by acid hydrolysis in the case of cellulose) and separating them by mechanical agitation. The size range of the microcrystals is from 2.5 to 500 nanometers (millimicrons). The products form extremely stable gels that have a number of commercial use possibilities. Petroleum-derived waxes of high molecular weight have been available in microcrystalline form for many years. Chlorophyll has a naturally microcrystalline structure.

MICROEMULSIONS.   There is no official or universally accepted definition of what constitutes a "microemulsion."

However, the concept of microemulsions holds a central role within the field of surfactant technology. Perhaps the most fundamental fact captured by the term is that, contrary to a popular saying, oil and water can mix. See also **Surfactants**.

Definition of a Microemulsion

The term *micro*emulsion implies a system which (like an emulsion) contains droplets of oil or water, but in which the droplets are too small to scatter light see also **Colloid Systems**.

A microemulsion is a true, thermodynamically stable, liquid solution that contains water, oil, and at least one amphiphile. Typically the oil is a

mineral oil or hydrocarbon, but it may be almost any nonpolar compound. The oil and the amphiphile may be single, pure components; or (as in most commercial formulations), the oil, amphiphile, or both may contain an indefinitely large number of compounds.

Microemulsions and Phase Diagrams

The existence or nonexistence of a microemulsion depends not only on the presence of certain classes of compounds (i.e., components), but also on the concentrations of these components. The number of phases present and their compositions, when presented in graphical form, constitute a phase diagram. Moreover, as specified by Gibbs' phase rule, amphiphile–oil–water–phase diagrams form characteristic patterns that change in qualitatively similar ways when the temperature, pressure, concentrations, or molecular structures of the components are changed. Thus, phase diagrams offer not only another way to define microemulsions, but also a rigorous way to clarify differences in terminology and usage.

Figure 1 illustrates the phase diagram of an amphiphile–oil–water system such as $C_4H_9OC_2H_4OH$ ("4E1")–decane–water or $C_6H_{12}(OC_2H_4)_2$ OH ("C6E2")–tetradecane–water. For a real surfactant, such as C12E4, the diagram would be more complicated, because of the occurrence of liquid crystalline phases. Samples whose compositions fall within the tie-triangle of Figure 1 form three liquid phases, of compositions, T, M, and B (corners of the tietriangle). Each pair of adjacent corners of the tietriangle is connected by a binodal curve (as well as by a side of the tietriangle). Compositions between a side of the tietriangle and the adjacent binodal curve form two conjugate phases in equilibrium with each other; each such pair of phases is connected by a tieline. For two of the binodals the compositions of the conjugate phases can become closer and closer and their connecting tielines shorter and shorter, until the phase compositions and the end points of the tielines become identical at a plait point. Any composition outside of the tietriangle and the three binodal curves forms only a single liquid phase. On the amphiphile–water side of the phase diagram these single-phases contain only amphiphile and water (no oil); on the amphiphile–oil side of the phase diagram these single-phases contain only amphiphile and oil (no water).

By the most general definition of a microemulsion, every phase described by Figure 1 would be a microemulsion. When two (or even three) phases are simultaneously present, little (except confusion) is gained by giving the different phases the same name. Accordingly, along the binodal curves (which describe the compositions of conjugate phases) only compositions between the two plait points are termed microemulsions. Other conjugate phases are called oleic phases or aqueous phases, respectively, depending on whether their main component is oil or water.

In Figure 1, the pairs (or triad) of phases that form in the various multiphase regions of the diagram are illustrated by the corresponding

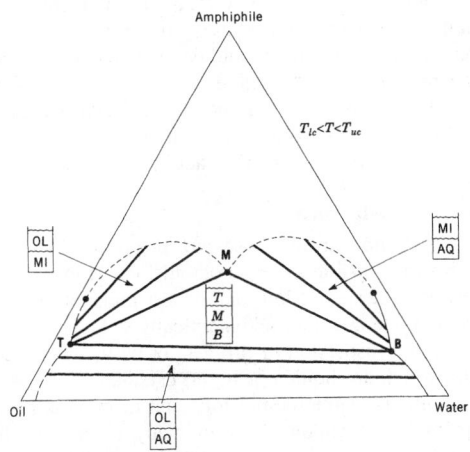

Fig. 1. Phase diagram of an amphiphile-oil-water system that forms a middle-phase microemulsion, definition of microemulsion, and illustration of the pairs (and triad) of phases formed in the various multiphase regions of the diagram. Boundaries:——, aqueous (AQ); ————, oleic (OL); ————, limiting microemulsion (MI)

test-tube samples. Except in rare cases, the densities of oleic phases are less than the densities of conjugate microemulsions and the densities of microemulsions are less than the densities of conjugate aqueous phases. Thus, for samples whose compositions lie within the oleic phase-microemulsion binodal, the upper phase (i.e., layer) is an oleic phase and the lower layer is a microemulsion. For compositions within the aqueous phase-microemulsion binodal, the upper layer is a microemulsion and the lower layer is an aqueous phase. When a sample forms two layers, but the amphiphile concentration is too low for formation of a middle phase, neither layer is a microemulsion. Instead the upper layer is an oleic phase ("oil") and the lower layer is an aqueous phase ("water").

In three-phase systems the top phase, T, is an oleic phase, the middle phase, M, is a microemulsion, and the bottom phase, B, is an aqueous phase. Microemulsions that occur in equilibrium with one or two other phases are sometimes called "limiting microemulsions," because they occur at the limits of the single-phase region.

Temperature and Salinity Scans

The locations of the tietriangle and binodal curves in the phase diagram depend on the molecular structures of the amphiphile and oil, on the concentration of cosurfactant and/or electrolyte if either of these components is added, and on the temperature (and, especially for compressible oils, on the pressure).

Often the identities (aqueous, oleic, or microemulsion) of the layers can be deduced reliably by systematic changes of composition or temperature. Thus, without knowing the actual compositions for some amphiphile and oil of points T, M, and B in Figure 1, an experimentalist might prepare a series of samples of constant amphiphile concentration and different oil–water ratios, then find that these samples formed the series (a) 1 phase, (b) 2 phases, (c) 3 phases, (d) 2 phases, (e) 1 phase as the oil–water ratio increased. As illustrated by Figure 1, it is likely that this sequence of samples constituted (a) a "water-continuous" microemulsion (of normal micelles with solubilized oil), (b) an upper-phase microemulsion in equilibrium with an excess aqueous phase, (c) a middle-phase microemulsion with conjugate top and bottom phases, (d) a lower-phase microemulsion in equilibrium with excess oleic phase, and (e) an oil-continuous microemulsion (perhaps containing inverted micelles with water cores).

Physical Properties and Applications

The current or potential industrial applications of microemulsions include metal working, catalysis, advanced ceramics processing, production of nanostructured materials, dyeing, agrochemicals, cosmetics, foods, pharmaceuticals, and biotechnology. Environmental and human-safety aspects of surfactants are receiving considerable attention. See also **Nanotechnology (Molecular)**.

Microemulsions became well known from about 1975 to 1980 because of their use in "micellar-polymer" enhanced oil recovery (EOR). This technology exploits the ultralow interfacial tensions that exist among top, microemulsion, and bottom phases to remove large amounts of petroleum from porous rocks, that would be unrecoverable by conventional technologies. Since about 1990, interest in the use of this property of microemulsions has focused on the recovery of chlorinated compounds and other industrial solvents from shallow aquifers. The latter application is sometimes called surfactant-enhanced aquifer remediation (SEAR).

Microemulsions and Macroemulsions

Operationally, it is not always easy to determine whether a given sample is a microemulsion or macroemulsion. However, the formal differences between microemulsions and macroemulsions are well defined. A microemulsion is a single, thermodynamically stable, equilibrium phase; a macroemulsion is a dispersion of droplets or particles that contains two or more phases, which are liquids or liquid crystals.

From the definitions of microemulsions and macroemulsions and from Figure 1, it immediately follows that in many macroemulsions one of the two or three phases is a microemulsion. Until recently, it was thought that all nonmultiple emulsions were either oil-in-water (O/W) or water-in-oil (W/O). However, the phase diagram of Figure 1 makes clear that there are six nonmultiple, two-phase morphologies, of which four contain a microemulsion phase. These six two-phase morphologies are oleic-in-aqueous (OL/AQ, or O/W) and aqueous-in-oleic (AQ/OL,

or W/O), but also, oleic-in-microemulsion (OL/MI), microemulsion-in-oleic (MI/OL), aqueous-in-microemulsion (AQ/MI), and microemulsion-in-aqueous (MI/AQ).

DUANE H. SMITH
Technical Solutions and West Virginia University

Additional Reading

Degiorgio, V. and M. Corti: *Physics of Amphiphiles: Micelles, Vesicles, and Microemulsions*, Elsevier Science Publishing Co., New York, NY, 1985.

Friberg, S.E. and P. Bothorel: *Microemulsions: Structure and Dynamics*, CRC Press, Boca Raton, FL., 1987.

Ganguli, D. and M. Ganguli: *Inorganic Particle Synthesis Via Macro- and Microemulsions: A Micrometer to Nanometer Landscape*, Plenum Publishing Corporation, New York, NY, 2003.

Gelbart, W.M., D. Roux, and A. Ben-Shaul: *Micelles, Membranes, Microemulsions and Monolayers*, Springer-Verlag New York, Inc., New York, NY, 1995.

Mittal, K.L. and P. Kumar: *Handbook of Microemulsion Science and Technology*, Marcel Dekker, Inc., New York, NY, 1999.

Robb, D.: *Microemulsions*, Plenum Press, New York, NY, 1982.

Shah, D.O.: *Macro and Microemulsions: Theory and Applications*, American Chemical Society, Washington, DC, 1985.

Shah, D.O.: *Micelles, Microemulsions, and Monolayers: Science and Technology*, Marcel Dekker, Inc., New York, NY, 1998.

Solans, C. and H. Kunieda: *Industrial Applications of Microemulsions*, Marcel Dekker, Inc., New York, NY, 1996.

MICROENCAPSULATION. Microencapsulation is the coating of small solid particles, liquid droplets, or gas bubbles with a thin film of coating or shell material. Here, the term microcapsule is used to describe particles with diameters between 1 and 1000 µm. Particles smaller than 1 µm are called nanoparticles; particles greater than 1000 µm can be called microgranules or macrocapsules.

Many terms have been used to describe the contents of a microcapsule: active agent, actives, core material, fill, internal phase (IP), nucleus, and payload. Many terms have also been used to describe the material from which the capsule is formed: carrier, coating, membrane, shell, or wall. In this article the material being encapsulated is called the core material; the material from which the capsule is formed is called the shell material.

Table 1 lists representative examples of capsule shell materials used to produce commercial microcapsules along with preferred applications.

Microcapsules can have a wide range of geometries and structures. Figure 1 illustrates three possible capsule structures. Parameters used to characterize microcapsules include particle size, size distribution, geometry, actives content, storage stability, and core material release rate.

Encapsulation Process

Classification of the many different encapsulation processes is useful. Previous schemes employing the categories chemical or physical are unsatisfactory because many so-called chemical processes involve exclusively physical phenomena, whereas so-called physical processes can utilize chemical phenomena. An alternative approach is to classify all encapsulation processes as either Type A or Type B processes. Type A processes are defined as those in which capsule formation occurs entirely in a liquid-filled stirred tank or tubular reactor. Emulsion and dispersion stability play a key role in determining the success of such processes. Type B processes are processes in which capsule formation occurs because a coating is sprayed or deposited in some manner onto the surface of a liquid or solid core material dispersed in a gas phase or vacuum. This category also includes processes in which liquid droplets containing core material are sprayed into a gas phase and subsequently solidified to produce microcapsules. Emulsion and dispersion stabilization can play a key role in the success of Type B processes also.

Many Type A and Type B processes are similar. For example, solvent evaporation is a key step in most spray dry encapsulation protocols (Type B) and protocols involving solvent evaporation from an emulsion (Type A). The difference in these protocols is that evaporation in the former case occurs directly from a liquid to a gas phase, whereas in the latter case evaporation involves transfer of a volatile liquid from a dispersed phase to a continuous liquid phase from which it is subsequently evaporated. Another example is encapsulation by gelation. In Type A gelation processes, the droplets that are gelled and become microcapsules are formed by dispersion in a liquid phase and are gelled in this phase. In Type B gelation processes, droplets formed by atomization or extrusion into a gas phase are subsequently gelled either in the gas phase or a liquid gelling bath.

TABLE 1. SHELL MATERIALS USED TO PRODUCE COMMERCIALLY SIGNIFICANT MICROCAPSULES

Shell material	Regulatory status	Chemical class	Encapsulation process	Applications
gum arabic	edible	polysaccharide	spray drying	food flavors
gelatin	edible	protein	spray drying	vitamins
gelatin-gum arabic[a]	nonedible[b]	protein-polysac-charide complex	complex coacervation	carbonless paper
ethylcellulose	edible	cellulose ether	Wurster process or polymer—polymer incompatibility	oral pharma-ceuticals
polyurea or polyamide	nonedible	cross-linked polymer	interfacial polymeri-zation	agrochemicals and carbonless paper
aminoplasts	nonedible	cross-linked polymer	*in situ* polymeri-zation	carbonless paper, fragrances, and adhesives
maltodextrins	edible	low molecular weight carbohydrate	spray drying and desolvation	food flavors
hydrogenated vegetable oils	edible	glycerides	fluidized bed	assorted food ingredients

[a] Treated with glutaraldehyde.

[b] For intended application, i.e., carbonless paper.

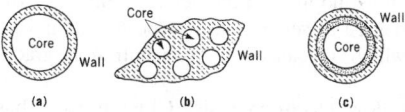

Fig. 1. Schematic diagrams of several possible capsule structures: (**a**) continuous core/shell microcapsule in which a single continuous shell surrounds a continuous region of core material; (**b**) multinuclear microcapsule in which a number of small domains of core material are distributed uniformly throughout a matrix of shell material; and (**c**) continuous core capsule with two different shells.

Most Type A processes might be classified as chemical processes, whereas most Type B processes are classified as mechanical processes. Representative examples of both types of processes follow. Type B processes tend to be promoted by organizations that sell and service equipment for producing microcapsules. Most Type A processes are not promoted by equipment manufacturers, but are developed and used by organizations that produce microcapsules.

Type A processes	Type B processes
complex coacervation	spray drying
polymer–polymer incompatibility	fluidized bed
interfacial polymerization at liquid–liquid and solid–liquid interfaces	interfacial polymerization at solid–gas or liquid–gas interfaces
in situ polymerization	centrifugal extrusion
solvent evaporation or in-liquid drying	extrusion or spraying into a desolvation bath
submerged nozzle extrusion	rotational suspension separation (spinning disk)

Applications

Microcapsules are used in a number of pharmaceutical, graphic arts, food, agrochemical, cosmetic, and adhesive products. Other specialty products also exist, thus the concept of microencapsulation has been accepted by a wide range of industries. In order to illustrate how microcapsules are used commercially, it is appropriate to describe a number of commercial microcapsule-based products and the role that microcapsules play in these products.

Carbonless copy paper is by far the largest single commercial application of microcapsules. This product consumes thousands of tons of capsules annually. Figure 2, a schematic diagram of a three-part business form, illustrates the concept of carbonless copy paper.

Success of all carbonless paper products depends on the microcapsules, leuco dyes, and reactive coating. A number of leuco dyes are available.

The concept of microencapsulation has intrigued the pharmaceutical industry for many years, because it offers the possibility of providing a number of important new oral and parenteral dosage forms. Microcapsules in oral dosage forms could conceptually taste-mask bitter pharmaceuticals, provide extended release *in vivo*, provide enteric release, improve the stability of incompatible drug mixtures, provide resistance to oxidation, reduce volatility, and distribute a drug in many small carrier particles so that effects of the drug on the sensitive walls of the stomach are minimized. Microencapsulated parenteral formulations could provide prolonged delivery of drugs with short half-lives *in vivo* and perhaps even achieve targeted drug delivery. For these reasons, microencapsulation has received much attention by pharmaceutical scientists. Several microcapsule-based oral pharmaceutical formulations which offer some of these features are available.

The use of microcapsules for a variety of biomedical and biological applications has been promoted for many years. Several biomedical microcapsule applications are in clinical use or have approached clinical use. One application is the use of air-filled human albumin microcapsules as ultrasound contrast agents. Another biomedical application of microcapsules is the encapsulation of live mammalian cells for transplantation into humans. The purpose of encapsulation is to protect the transplanted cells or organisms from rejection by the host.

A number of food ingredients or additives have been encapsulated and are available commercially. Solid ingredients encapsulated are typically water-soluble and are encapsulated with a hydrophobic or hydrophilic coating material usually applied by the Wurster process. Both types of coating materials are well-accepted food-grade products (see **Food Additives**).

The microencapsulation of pesticides (qv) and herbicides (qv) has been an active area of development that has produced several commercial products. The function of the microcapsules is to prolong activity while reducing mammalian toxicity, volatilization losses, phytotoxicity, environmental degradation, and movement in the soil. Ideally, encapsulation would also reduce the amount of agrochemical needed.

Advertising inserts that utilize encapsulated perfumes and flavors contain a coating of scent-filled capsules which break and release scent when the insert is torn open are widely used as a marketing tool, primarily for new perfumes. Children's crayons loaded with encapsulated scents are appearing on the market. The capsules break during the drawing process thereby releasing a scent characteristic of the drawn object.

Microcapsules are used in several film coatings other than carbonless paper. Encapsulated liquid crystal formulations coated on polyester film are used to produce a variety of display products including thermometers. Polyester film coated with capsules loaded with leuco dyes analogous to those used in carbonless copy paper is used as a means of measuring line and force pressures. Encapsulated deodorants that release their core contents as a function of moisture developed because of sweating represent

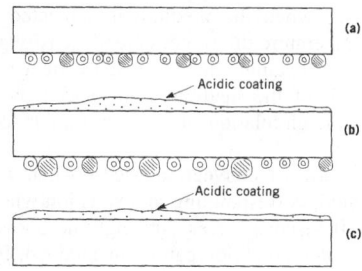

Fig. 2. Cross section of a three-part business form prepared from carbonless copy paper where ⊚ are microcapsules and ⊘ are starch: (**a**), CB sheet; (**b**), CFB sheet; and (**c**), CF sheet.

another commercial application. Microcapsules are incorporated in several cosmetic creams, powders, and cleansing products.

A majority of the fasteners used in automobiles in the U.S. are coated with microcapsules loaded with an adhesive. Other uses include encapsulated ammonium polyphosphate incorporated in plastics that acts as a fire-retardant and microencapsulated oil-field chemicals for use by the oil industry.

CURT THIES
Washington University

Additional Readings

Bakan, J.A.: in L. Lachman, H.A. Lieberman, and J.L. Kanig, eds., *The Theory and Practice of Industrial Pharmacy*, 3rd Edition, Marcel Dekker, New York, 1986.

Cohen, S. and H. Bernstein: *Microparticulate Systems for the Delivery of Proteins and Vaccines*, Marcel Dekker, Inc., New York, NY, 1996.

Deasy, P.B.: *Microencapsulation and Related Drug Processes*, Marcel Dekker, Inc., New York, 1984.

Kondo, A.: *Microcapsule Processing and Technology*, Marcel Dekker, Inc., New York, 1979.

Thies, C.: *How-to-Make Microcapsules: Lecture and Lab Manual*, Thies Technology, St. Louis, Mo., 1994.

Whateley, T. L.: *Microencapsulation of Drugs*, Taylor & Francis, Inc., Philadelphia, PA, 1992.

MICROGRAVITY AND MATERIALS PROCESSING.

The development of space transportation systems during the past several years has drawn attention to materials processing in a reduced gravity environment. Actually, exploratory work in this area has been proceeding since the early *Apollo* missions to the moon. For example, materials processing experiments were carried out during the *Skylab* program and on the *Apollo-Soyuz* test program fight. More recently, tests have been conducted on some of the *Space Shuttle* projects and the International Space Station (ISS).

Effects of Reduced Gravity. The reduced gravity of space offers a unique environment for materials processing. It is untrue, of course, to claim that a state of zero gravity is achieved as, for example, on space flights as currently experienced. Rather, the gravity is greatly reduced—to about 10^{-6} g. Hence, the term *micro* rather than *zero* gravity is more appropriate to use here. It is interesting to note that spacecraft in low earth orbit may experience changes in the gravitational field, referred to as *g-jitter*, which are caused by such factors as maneuvering the craft while in orbit, atmospheric drag, and even movement of the crew within the spacecraft. The *g-jitter* phenomenon can cause "spikes" in the gravitational field, ranging from 10^{-2} g to 10^{-1} g. These may have an adverse effect upon on-board experiments and processes.

Manned or unmanned orbital spaceflight is not the only possibility for reduced-gravity studies of materials processing. Drop tubes and towers, aircraft, and sounding rockets also offer opportunities for varying levels and time periods of reduced gravity. Time spent in low gravity and payload size are the limiting factors for earth-based facilities. Drop tube experiments provide 2.5 to 4.5 seconds at 10^{-8} g to 10^{-9} g for free falling droplets, whereas a drop tower allows an entire experimental package, weighing 100 kg, to be tested. By flying in parabolic flight paths, aircraft ranging in size from a KC-135 to a single-seat F-104 can provide 10^{-1} g to 10^{-2} g for 15 to 60 seconds for payloads ranging from 10 to 35 kg. Sounding rockets provide up to 300 seconds at 10^{-5} g. When compared with the duration time provided by an orbiting space platform, the earthbound methods are quite brief.

One effect of the reduction of gravitational forces experienced in space is the virtual elimination of *buoyancy-driven convection*. Convection within a fluid medium arises when the medium is subjected to a nonvertical thermal gradient. Temperature differences create density gradients as gases or liquids expand upon heating. In the presence of a gravitational field, the less dense volume of the medium is displaced by a denser, cooler volume, resulting in the circulation of gases or liquids, commonly known as convective flow.

A more unstable form of convection occurs when a denser fluid lies above a less dense fluid, corresponding to a situation where the medium is heated from below. If viscous forces outweigh the buoyant forces within the medium, this unstable condition can be maintained. If the buoyancy is greater than the viscosity, however, the volume element rises too quickly, resulting in spontaneous flow, which takes the form of cells or vortex rolls.

Concentration gradients caused by chemical reaction within a fluid medium may also cause convection. In this case, density gradients occur when a particular chemical component is consumed or produced by reaction.

Convective flow is mathematically characterized by the Grashof number *Gr*, which represents the ratio of buoyant to viscous forces. The Grashof number is given by the expression

$$Gr = \frac{gl^3 \beta \Delta T}{\mu^2}$$

where *g* is the gravitational acceleration; β the coefficient of expansion; ΔT the temperature gradient; and μ the kinematic viscosity. In an earth-based experiment, convection can be reduced somewhat within practical limits by altering the geometry of a given system or minimizing temperature gradients. Conducting a similar experiment in low earth orbit can reduce the value of *Gr* by six orders of magnitude.

Because convection and diffusion occur simultaneously in an earth-based process, corresponding studies in a microgravity environment can help identify the effects of these two phenomena on a given process such as crystal growth.

The reduced gravity of space can also be used to process materials in a container-free environment. This feature of microgravity processing is particularly advantageous when a material of high purity is desired, or for achieving a high degree of supercooling in a sample. Containerless processing is especially useful for obtaining glasses and alloys.

The presence of a gravitational field causes substances of differing densities to separate out. In microgravity, however, this gravity-induced separation is eliminated, thereby producing a more uniform mixture. This effect is useful in the processing of alloys and organic polymers.

Representative Microgravity Experiments

Biological Materials. The degree of purity of biological materials severely limits their usefulness. Electrophoresis is a commonly used method of separation and purification of substances such as cells, enzymes, and proteins. This technique relies upon the fact that surface charge distribution, and thus mobility in an electric field, vary from one material to another. The degree of separation, product yield, and purity are limited by convection which is caused by concentration gradients within the process medium.

A continuous flow electrophoresis (CFE) process has been used to effect separation of biological materials in microgravity. The absence of convection permits continuous processing of relatively large volumes of material, higher yields, finer separation, and higher product purity than are possible on earth. Erythropoietin, which is produced by the kidneys and controls red blood cell production in the body, has been produced by CFE on board the *Shuttle*. The first CFE experiment performed in 1982 in *Spacelab* yielded 463 times more material than comparable earth-based processes. Separation rates were boosted in later flights to yield 700 times more material having a fourfold increase in purity over products obtained on earth.

Polymers. Research efforts in the area of organic polymer growth in space seek to take advantage of the absence of phase separation due to density differences. On earth, density differences in nonhomogeneous mixtures of organic liquids produce buoyancy-driven convection and cause immiscible liquids to separate. These phenomena affect the growth of organic polymers, causing flaws in the final product. In the absence of phase separation and convection, more uniform mixtures can be produced. Secondary effects such as surface tension can also be utilized to obtain more perfect polymers and organic compounds.

One type of experiment performed in space was the diffusive mixing of organic solutions (DMOS) study conducted by the 3M Corporation. The DMOS experiments mixed different types of organic solutions to yield crystalline material. The purpose of the study was to determine the effect of microgravity upon the ordering of organic molecules upon crystallization. The crystals grown in the experiment were not only significantly larger than similar crystals grown on earth, but possessed much better optical and electrical properties as well.

The dominance of surface tension, due to the lack of convection in microgravity, has been used to produce perfectly round spheres of polystyrene-latex. The spheres are grown in space by the coalescence of an emulsion. Under conditions where surface tension controls the process, droplets do not readily break up, thereby allowing large spheres to coalesce. As a result, large, perfect spheres, having a diameter of up to 30 mm, can be produced in space. In comparison, a maximum diameter of 5 mm can be produced on earth. These spheres are offered commercially for reference and calibration applications. As such, they are the first commercial products to be made in space.

Physical Metallurgy. One obstacle to the processing of alloys on earth is that components of a given mixture are often immiscible. As a result, density differences cause the components to separate as the bulk melt cools. By eliminating this gravity-driven separation, the manufacture of alloys can benefit from a microgravity environment.

Several alloy systems have been studied in space. In general, these experiments have yielded promising results, showing that finer, more homogeneous mixtures of components can be obtained in microgravity. In space, reduced convection in the melt apparently reduces microsegregation and heat transfer. This allows materials possessing highly directional physical properties, such as magnetic coercivity and microstructure, to be produced.

Containerless processing is also of interest in physical metallurgy, as it provides opportunities to study thermophysical properties of high temperature metals and alloys, avoid sample contamination due to contact with container walls, and observe the solidification of materials that have been rapidly cooled from the melt.

Containerless processing is accomplished on earth under the influence of gravity by using electrostatic, acoustic, or electromagnetic energy to levitate a sample. Sample size is limited, however, by power requirements for levitation. The application of the forces necessary to levitate a substance also induces a certain amount of mixing and heating. Gravity-driven convection is also present in this situation and can cause unwanted mixing of liquid samples.

To date, containerless processing in near zero gravity has been limited to drop tube experiments, which provide a few seconds of low gravity for small drops of material. Alloys studied in this fashion have been undercooled as much as 500°C, which corresponds to a cooling rate of greater than 10^6 K/sec. Samples obtained in these experiments have exhibited metastable or peritectic phases which are extremely difficult to obtain under normal conditions.

Some levitation of samples will still be necessary to carry out containerless processing in space, although the magnitude of the forces necessary to do so will be small relative to those required in earth-based work. Levitation would be necessary only to avoid contact between sample and container, so larger samples could be used in space-based experiments.

Other metallurgy experiments scheduled for space will examine the role of macrosegregation in the processing of metals, the feasibility of using directional solidification in the processing of different classes of alloys, and the manufacture of alloys which cannot be produced on earth due to density induced separation.

Glasses. The manufacture of glasses also benefits from containerless processing in space. As with alloys, the purity of glasses can be affected by contact and subsequent reaction with container walls. In the case of glass processing, however, contact between sample and reactor wall also causes crystallization, and hence loss of the amorphous glassy state. In addition, less viscous glasses require high cooling rates in order to prevent crystallization. By avoiding contact-induced nucleation during cooling, containerless processing may be used to obtain larger, high purity samples of such glasses.

As mentioned previously, some levitation is required in microgravity to maintain sample positioning. Several designs, some of which are capable of processing temperatures up to 1600°C, have been developed for use in space experiments.

Glass processing experiments that have already been flown on the *Shuttle* have been concerned with melt homogenization, bubble behavior in molten glass spheres, preparation of glass microballoons, and comparing properties of space-produced glasses with those manufactured on earth. Results to date indicate that glasses having different microstructures than those of glasses processed on earth can be obtained. Galliacalcia, sodium-borate, and lead-silica glasses have been selected for the above experiments. The list of glasses to be studied will be expanded to include materials that are particularly difficult to produce on earth. Among these materials are heavy cation (Zr, Hf, Th) glasses that tend to react with containers, and silica glasses that must be processed at high temperatures.

Crystal Growth. Single crystals of both organic and inorganic substances can be grown from either the vapor or liquid phase, using several different experimental techniques. Gravity-driven convection affects the motion of these fluid media, greatly affecting the mixing and transport

of individual chemical components. Experiments aimed at examining the effects of eliminating buoyancy-driven convection upon different crystal growth techniques have been performed on *Skylab*, the *Apollo-Soyuz* mission, and several *Space Shuttle* flights.

Growth of crystals from the vapor relies upon the presence of a temperature gradient. Concentration gradients of vapor species and subsequent migration from a source region to a seed crystal, substrate, or deposition region are caused by this temperature difference. In the case of physical vapor transport (PVT), solid source material vaporizes at one temperature. The gaseous vaporization products migrate, usually through an atmosphere of inert gas, to another temperature where solid material condenses. In space, the PVT method has been used to grow highly ordered organic thin films onto silicon wafer substrate and large single crystals of germanium selenide.

Another PVT experiment examined the growth of HgI_2 onto a seed crystal. This substance has potential for use as a radiation detector. Due to its high density, however, the HgI_2 crystal structure readily deforms during earth-based processing. Large crystals of HgI_2 have been grown by the PVT method aboard *Spacelab 3*. Growth times in space were considerably less than those normally required on earth. Performance of the space-grown crystals as radiation sensors is matched by only the very best crystals obtained on earth.

Unlike PVT, chemical vapor transport (CVT) utilizes a highly reactive gaseous substance—such as a halogen or metal halide—to transport source material to a region of the reaction container where single crystals condense from the vapor. In earth-based CVT studies, under conditions where buoyancy-driven convection drives the overall transport process, crystal size is generally small. The morphology of crystals grown under these conditions is often poor; surfaces are marked by large numbers of defects and irregular growth steps. In contrast, space-grown crystals grown by chemical vapor transport are much larger, have smoother growth steps, and fewer defects. Chemical homogeneity within these crystals is also considerably better than in similar crystals grown on earth.

The presence of convection also affects crystal growth from the melt. Single crystals of Te-doped InSb were grown from the melt on *Skylab*. The crystals obtained in space were free of striations caused by convection-driven growth rate fluctuations that are normally seen on earth. Future space experiments will examine the growth of electronic materials such as GaAs from a solution subjected to an electric current.

ROBERT P. SANTANDREA, Ph.D.
Los Alamos National Laboratory
Los Alamos, New Mexico

MICRONUTRIENTS (Soil).   See **Fertilizer**.

MICROSCOPY (Chemical).   Use of a microscope primarily for study of physical structure and identification of materials. This is especially useful in forensic chemistry and police laboratories. Many types of microscopes are used in industry; most important are the optical, ultra-, polarizing, stereoscopic, electron, and X-ray microscopes. Organic dyes of various types are used to stain samples for precise identification.

MICROWAVE SPECTROSCOPY.   A type of adsorption spectroscopy used in instrumental chemical analysis that involves use of that portion of the electromagnetic spectrum having wavelengths in the range between the far infrared and the radiofrequencies, i.e., between 1 mm and 30 cm. Substances to be analyzed are usually in the gaseous state. Klystron tubes are used as microwave source.

MIDGLEY, THOMAS, JR. (1889–1944).   An American chemist and inventor. One of the most creative and brilliant chemists of his era. Midgley's early work was in the field of rubber chemistry and technology, especially in the development of synthetic and substitute rubbers that were being introduced in the 1930s. He worked with Kettering at General Motors and then became vice president of Ethyl Corporation, as well as of the Ohio State University Research Foundation. His innovative genius was responsible for the development of organic lead compounds for antiknock gasoline and later for the discovery of fluorocarbon refrigerants for which he did the basic research. He was recipient of many of chemistry's highest honors including the Nichols medal, the Perkin medal, and the Priestly medal.

MIESCHER DEGRADATION. Adaptation of the Barbier-Wieland carboxylic acid degradation to permit simultaneous elimination of three carbon atoms, as in degradation of the bile acid side chain to the methyl ketone stage. Conversion of the methyl ester of the bile acid to the tertiary alcohol, followed by dehydration, bromination, dehydrohalogenation, and oxidation of the diene yields the required degraded ketone.

MIGNONAC REACTION. Formation of amines by catalytic hydrogenation of aldehydes and ketones in liquid ammonia and absolute ethanol in the presence of a nickel catalyst.

MILK AND MILK PRODUCTS. Intricate chemical and microbiological problems arise in the production, processing, and distribution of milk products. Milk is a complex mixture of fat (4%), protein (3.5%), carbohydrate (4.8%) and mineral components (0.7%) and is an excellent bacterial growth medium; hence the need for care and cleanliness in handling.

Milk Fat

Milk fat or butterfat is a mixture of triglycerides of various fatty acids. Milk fat also contains a small percentage of cholesterol (0.37%), a substance characteristic of fats and oils of animal origin in contrast to those of plant origin which contain plant sterols. The phospholipids, lecithin, caphalin, and sphingomyclin are present in milk within the range of 0.03–0.04%. These are fatlike substances containing phosphorus and nitrogen. They have emulsifying properties and are associated with the fat-globule surfaces: hence their tendency to concentrate in butter and buttermilk.

Milk fat is distinguished from all other fats in that it is the only one containing butyric acid (C_4) as a component of glycerides. This acid occurs in the fat in a concentration of approximately 10 mole % and the total of C_6, C_8, and C_{10} fatty acids accounts for another 10 mole % of the component fatty acids. These acids ($C_4–C_{10}$) are often called the volatile fatty acids of milk fat. In the free form, they have a pungent characteristic flavor which is important in many types of cheese.

The deterioration of milk fat is an important cause of off-flavor development in dairy products and its control requires technical understanding of the processes involved. Three major types of deterioration associated with milk fat are recognized:

(1) *Rancidity*, due to free volatile fatty acids liberated from the glycerides by enzymic (lipase) hydrolysis. Lipases are normal components of raw milk, and are inactivated by the heat of pasteurization.

(2) *Tallowiness or oxidation*, due to autoxidation of unsaturated fatty acids with the production of flavorful unsaturated aldehydes. These reactions are accelerated by oxygen, high storage temperatures, and copper catalysts. Oxidation is usually the primary cause for spoilage of dried whole milk, cream, butter, and butteroil.

(3) *Heat-generated flavors*, due to the formation of lactones and methyl ketones from hydroxy and keto acid precursors, which occur in trace quantities in milk fat. These flavors are considered to be desirable in fried and baked goods and are partly responsible for the unique condiment properties of butter in food preparation. However, they are undesirable in dried whole milk and evaporated milk where the objective is to make a bland product as much like fresh milk as possible.

Proteins

The proteins of milk fall into two groups: casein, precipitable by both acid and proteolytic enzymes such as rennin; and whey proteins, which are acid soluble but heat-denaturable. There is about 3% casein in milk, the removal of which leaves a whey of approximately 1.0% nitrogenous matter. Of this 0.6% is heat-coagulable protein and 0.4% is nonheat-coagulable. The latter fraction is composed of protein-like fragments of a proteose or peptone nature plus other nitrogenous substances. Among these are small percentages of urea, creatin, creatinine, uric acid, and various forms of amino nitrogen.

Casein exists in milk as a calcium caseinate-calcium phosphate complex; the ratio of these components is approximately 95.2 to 4.8. The dispersed casein particles appear to be spherical in shape and of various sizes. The size distribution of the casein micelles is not constant, but varies with aging, heating, concentration, and other processing treatments. Processing alters the water-binding of casein and this in turn affects the apparent viscosity of products that contain casein. Changes in hydration have not been measured quantitatively although the casein particles of raw milk

appear to consist of one volume of water-free protein and three volumes of solvate liquid.

The whey or serum proteins have been partially resolved into three relatively homogeneous, crystallizable proteins: (1) β-lactoglobulin (50% of total serum protein), (2) an albumin resembling the albumin of bovine blood (5% of the serum protein), and (3) α-lactalbumin (12% of the serum protein).

On heat denaturation, the serum proteins show decreased solubility at pH 4.7 and in concentrated salt solutions. There is some variability in response toward heat treatment, but complete denaturation will occur during heating within the range of 60 to 80°C for periods of time up to two hours. There is practically no denaturation during normal pasteurization. Heat increases the activity of the sulfhydryl groups and the sulfhydryl titer can be employed as a measure of denaturation. The −SH groups are readily oxidized in liquid systems and consequently they appear to act as antioxidants to protect milk fat in dairy products. The fat of fluid milk, heated to produce a high −SH titer, shows increased resistance to oxidation and this carries through to the dried product which exhibits superior storage stability, if it is made from high heat milk. The high sterilization temperature to which evaporated milk is subjected and the low oxygen content in the can protect this product from development of oxidized and tallowy flavors during storage. See **Proteins**.

Milk is widely used as an ingredient for bread and other baked goods to which it adds substantial nutritional value. Milk is heated when used in bread to avoid softening of the dough and reduction of loaf volume. Why heat improves the baking properties of milk is not clear, but good baking properties have long been associated with low whey-protein-nitrogen values.

Lactose

The sugar of milk, lactose ($C_{12}H_{22}O_{11}$), occurs in the milk of all mammals. It is mildly sweet with a final solubility in water of 10.6% at 0°C, 17.8% at 25°, 29.8% at 49°, 58.2% at 89°. Lactose, on hydrolysis by acid or the enzyme lactase, yields a mixture of approximately equal parts of glucose and galactose, together with a small but variable quantity of oligosaccharides. The products of lactose hydrolysis are much more soluble than the original disaccharide. Lactose is a reducing sugar which is converted to lactobionic acid on mild oxidation. Two forms which differ in solubility and optical rotation are known. Alpha-lactose hydrate crystallizes at ordinary temperatures with one molecule of water, but this is lost with the formation of the anhydrous form during heating to a temperature between 149 and 200.3°F. Anhydrous β lactose, more soluble than alpha, crystallizes from supersaturated lactose solutions above 200.3°F. Solid beta-hydrate has never been prepared. The crystalline alpha-hydrate is stable in dry air at room temperatures, but both anhydrous forms readily absorb moisture and change to alpha-hydrate at ordinary temperatures. Alpha-lactose crystallizes out in some dairy products and because of the hardness of its crystals and their slow and limited solubility, "sandy" products may result.

The crystallization of lactose in frozen concentrated milk has been associated with a denaturation of casein which ultimately appears as a gel structure in the thawed product. Gelation in frozen milk can be retarded by enzymic hydrolysis of part of the lactose before freezing or by addition of a polyphosphate salt.

Lactose, when fermented by lactic bacteria, is the source of the lactic acid formed in sour milk and whey. Lactose is helpful in establishing a slightly acid reaction in the intestine, which assists in calcium assimilation.

Mineral Components

When milk is heated to a temperature high enough to volatilize the water and oxidize the organic constituents, the residue of inorganic oxides that remains is called the milk ash; its major components are: K_2O, CaO, Na_2O, MgO, Fe_2O_3, P_2O_5, Cl, and SO_3. The calcium and phosphorus of the ash are of special interest because of their nutritional importance and because calcium phosphate is part of the casein micelle, influencing its physiochemical behavior toward coagulation with rennin, acid, and heat. Minor inorganic constituents are present in milk in trace amounts, i.e., iron, copper, zinc, aluminum, manganese, iodine, and cobalt.

Miscellaneous Components

The hydrogen-ion concentration of milk increases slightly with age, after milking, as natural carbon dioxide escapes. Most samples of cow's milk vary within the range of pH 6.5–6.7. Titratable acidity of fresh milk which may vary from 0.13 to 0.16%, expressed as lactic acid is an arbitrary measurement influenced by the protein and salt-buffer systems present in the particular sample. Citrates, phosphates, and carbonates are the principal buffers in milk.

Milk contains some important vitamins. The vitamin D content may vary from 30 I.U. per quart in summer to 6 in winter, depending upon the feed and the sunlight which reach the cow. Both pasteurized and evaporated milk are often fortified by the addition, on a fluid basis, of 400 I.U. of vitamin D per quart. Vitamins A, D, and E (alpha-tocopherol) are fat-soluble and stable at the heat treatments used in processing milk and milk products. The remaining vitamins are water-soluble and of varying stability. Vitamins B_1 (thiamine) and C (ascorbic acid) are partially destroyed by heat, while B_6 and B_{12}, are relatively heat-stable. Vitamin B_2 (riboflavin) is heat-stable but it is quickly destroyed by light. In spite of the varying sensitivity of the water-soluble vitamins toward heat, pasteurized milk is a good source of all the milk vitamins except C.

Two types of enzymes in milk are important: those useful as an index of heat treatment and those responsible for bad flavors. Phosphatase is destroyed by the heat treatments used to pasteurize milk; hence its inactivation is an indication of adequate pasteurization. Lipase catalyzes the hydrolysis of milk fat which produces rancid flavors. It must be inactivated by pasteurization or more severe heat treatment to safeguard the product against off-flavor development. Other enzymes reported to have been found in milk include catalase, peroxidase, protease, diastase, amylase, oleinase, reductase, aldehydrase, and lactase.

BYRON H. WEBB
U.S. Department of Agriculture
Washington, District of Columbia

Additional Reading

Considine, D.M., Editor: *Foods and Food Production Encyclopedia,* Van Nostrand Reinhold, New York, 1982.

Early, R.: *Technology of Dairy Products,* 2nd Edition, Blackie Academic & Professional, London, UK, 1999.

Ernstrom, C.A. and N.P. Wong: *Fundamentals of Dairy Chemistry,* AVI, Westport, Connecticut, 1974.

Fox, P.F. and P. McSweeney: *Dairy Chemistry and Biochemistry,* Chapman & Hall, New York, NY, 1999.

Harper, W.J. and C.W. Hall: *Dairy Technology and Engineering,* AVI, Westport, Connecticut, 1976.

Lewis, M.J. and N.J. Heppell: *Continuous Thermal Processing of Foods: Pasteurization and Uht Sterilization,* Aspen Publishers, Inc., Gaithersburg, MD, 2000.

Marth, E.H., Editor: *Standard Methods for Examination of Dairy Products,* American Public Health Association, Washington, DC, (Revised periodically).

Mazza, G.: *Functional Foods: Biochemical and Processing Aspects,* CRC Press LLC., Boca Raton, FL, 1998.

Robinson, R.K.: *Dairy Microbiology Handbook,* 3rd Edition, John Wiley & Sons, Inc., New York, NY, 2002.

Selitzer, R., Editor: "The Dairy Industry in America," *Dairy and Ice Cream Field* (publishers), New York, 1977.

Spreer, E. and A. Mixa: *Milk and Dairy Product Technology,* Marcel Dekker, Inc., New York, NY, 1998.

Varnam, A.H. and J.P. Sutherland: *Milk and Milk Products: Technology, Chemistry, and Microbiology,* Vol. 1, Aspen Publishers, Inc., Gaithersburg, MD, 2001.

Wong, N.P.: *Fundamentals of Dairy Chemistry,* 3rd Edition, Chapman & Hall, New York, NY, 1999.

MILLER INDICES.   In mineral crystallography the identity of a crystal face consists of a series of whole numbers which are the products of their parameters relating to that face by their inversion, and where required the clearing of fractional values. A parameter is the relative intercept of a crystallographic axis on a given crystal face.

Assuming parameter values on a given crystal face to be $1a$, $1b$, $\frac{1}{2}c$ would on inversion yield 1, 1, $\frac{2}{1}$ parameters $1a$, $1b$, $2c$ would on inversion yield 1, 1, $\frac{1}{2}$ and parameters of $3a$, $2b$, $6c$ would on inversion yield $\frac{1}{3}$, $\frac{1}{2}$, $\frac{1}{6}$. Clearing the fractions in each instance would yield Miller indices of (112), (221) and (231) respectively.

The three Miller indices for a crystal face in all systems except the hexagonal, which requires four indices, are always given in the same order as their crystallographic axes, a, b, c, respectively; a^1, a^2, a^3, in the isometric system; a^1, a^2, c, in the tetragonal system; and a^1, a^2, a^3, c, in the hexagonal.

If the parameter intercepts for a given face are unknown, general indices (hkl) may be used if that face intercepts all three axes; four in the hexagonal, with general indices $(hkil)$. If a crystal face cuts two axes and parallels the third, general indices would be identified as $(h0l)$, $(0kl)$, or $(hk0)$ as applicable to that face; in the hexagonal system as $(h0hl)$, etc.

See also **Crystal**; and **Mineralogy**.

MILLERITE.   The mineral millerite is nickel sulfide, NiS, whose slender hexagonal interwoven crystals so suggestive of hairs has led to the application of the name "capillary pyrites." It occurs also as radiated masses and coatings. It is brittle; hardness, 3–3.5; specific gravity, 5.48–5.52; luster, metallic; color, brass-yellow, often with an iridescent tarnish. Millerite is found in association with other nickel-bearing minerals and other sulfides. European localities are Bohemia, Westphalia, Wales, etc.; and in the United States at Antwerp, New York; with pyrrhotite in Lancaster County, Pennsylvania; at St. Louis, Missouri; Keokuk, Iowa; and Milwaukee, Wisconsin. In Canada millerite occurs in Oxford, Quebec, and in the famous Sudbury District, Ontario. It is used as an ore of nickel. Millerite was named for the English mineralogist, W.H. Miller.

MIMETITE.   The mineral mimetite is a chloro-arsenate of lead corresponding to the formula $Pb_5(AsO_4)_3Cl$. It is monoclinic (pseudohexagonal); brittle; hardness, 3.5; specific gravity, 7.0–7.25; luster, resinous; color, usually yellow to brown but may be colorless or white; translucent. Mimetite is a rather rare secondary mineral occurring in altered lead deposits. Found in Bohemia; Saxony; Cornwall and Cumberland, England; South West Africa; Mexico; and in the United States, in Pennsylvania and Utah. The name mimetite is derived from the Greek word meaning imitator, because of the similarity of mimetite and pyromorphite. See also **Pyromorphite**.

MINERAL NUTRIENTS.   Minerals that are essential to life are the source of metals and other inorganic elements involved in the most fundamental processes. For example, oxygen, required by the cells of animals, is utilized with the aid of metal complexes. In humans both iron-containing hemoglobin and zinc-containing carbonic anhydrase play pivotal roles in binding oxygen and delivering it to the cells. Moreover, enzymes developed to protect cells from high levels of oxygen also contain metals. One such class of protective enzymes is known as the superoxide dismutases (SODs). These contain metals such as manganese, copper, zinc, and iron. Mutations in the copper- and zinc-containing superoxide dismutase gene have been linked to amyotrophic lateral sclerosis.

As for other biological substances, states of dynamic equilibrium exist for the various mineral nutrients as well as mechanisms whereby a system can adjust to varying amounts of these minerals in the diet. In forms usually found in foods, and under circumstances of normal human metabolism, most nutrient minerals are not toxic when ingested orally. Amounts considerably greater than the recommended dietary allowances (RDAs) can generally be eaten without concern for safety (Table 1).

Some elements found in body tissues have no apparent physiological role, but have not been shown to be toxic. Examples are rubidium, strontium, titanium, niobium, germanium, and lanthanum. Other elements are toxic when found in greater than trace amounts, and sometimes in trace amounts. These latter elements include arsenic, mercury, lead, cadmium, silver, zirconium, beryllium, and thallium. Numerous other elements are used in medicine in non-nutrient roles. These include lithium, bismuth, antimony, bromine, platinum, and gold. The interactions of mineral nutrients with carbohydrates, fats, and proteins, minerals with vitamins (qv), and mineral nutrients with toxic elements are areas of active investigation.

The amount of each element required in daily dietary intake varies with the individual bioavailability of the mineral nutrient.

The Principal Elements

Calcium.   Calcium, the most abundant mineral element in mammals, comprises 1.5–2.0 wt % of the adult human body, over 99 wt % of which is present in bones and teeth. About 48% of serum calcium is ionic, ca 46% is bound to blood proteins, the rest is present as diffusible complexes, e.g., of citrate. The calcium ion level must be maintained within definite limits.

Bones act as a reservoir of certain ions, in particular Ca^{2+} and PO_4^{3-}, which readily exchange between bones and blood. Bone structure comprises a strong organic matrix combined with an inorganic phase which is principally hydroxyapatite, $3\ Ca_3(PO_4)_2 \cdot Ca(OH)_2$. Bones contain two forms of hydroxyapatite. The less soluble crystalline form contributes to the rigidity of the structure. The crystals are quite stable, but because of the small size present a very large surface area available for rapid

TABLE 1. ESSENTIAL MINERAL NUTRIENTS

Element[a]	Body content, mg/kg body wt	Daily requirement, mg
PRINCIPAL ELEMENTS		
calcium	14,000–20,000	800–1,200[b,c]
phosphorus	11,000–12,000	800–1,200[b,c]
sulfur	1,600–2,500	[d]
potassium	2,000–3,500	2,000
sodium	1,500–1,600	500
chlorine	1,200–1,500	750
magnesium	270–500	280[b,c,e]; 350[b,f]
TRACE AND ULTRATRACE ELEMENTS		
iron	60–66	10[b,f]; 15[b,e]
fluorine	37	1.5–4.0[b]
zinc	33–50	12[b,c,e]; 15[b,f]
silicon	15–16	5–20
copper	1.0–2.5	1.5–3.0
boron	0.69	0.5–1.0
selenium	0.2–0.3	0.055[b,c,e]; 0.07[b,f]
iodine	0.2–0.4	0.15[b,c]
manganese	0.2–4.0	2.0–5.0[b]
molybdenum	0.1–0.5	0.075–0.25[b]
chromium	0.06–0.2	0.05–0.2[b]
cobalt	0.02	0.003[g]
tin	0.2	
vanadium	0.14	<0.01
nickel	0.07–0.14	<0.10

[a] Generally not ingested in elemental form.
[b] Values are for adults.
[c] Increased amounts are required during pregnancy and lactation.
[d] Adequate intake with adequate intake of protein.
[e] Value for females.
[f] Value for males.
[g] As vitamin B_{12}.

exchange of ions and molecules with other tissues. There is also a more soluble intercrystalline fraction. Bone salts also contain small amounts of magnesium, sodium, carbonate, citrate, chloride, and fluoride. Osteoporosis is reported to result when bone resorption is relatively faster than bone formation. The calcium ion, necessary for blood-clot formation, stimulates release of bloodclotting factors from platelets. See also **Blood**; and **Anticoagulants**.

In normal adults, the blood Ca_{2+} level is established by an equilibrium between blood Ca^{2+} and the more soluble intercrystalline calcium salts of the bone. Additionally, a subtle and intricate feedback mechanism responsive to the Ca^{2+} concentration of the blood that involves the less soluble crystalline hydroxyapatite comes into play. The thyroid and parathyroid glands, the liver, kidney, and intestine also participate in Ca^{2+} control.

In addition to hypocalcemia, tremors, osteoporosis, and muscle spasms (tetary), calcium deficiency can lead to rickets, osteomalacia, and possibly heart disease. These, as well as Paget's disease, can also result from faulty utilization of calcium. Calcium excess can lead to excess secretion of calcitonin, possible calcification of soft tissues, and kidney stones when combined with magnesium deficiency.

Phosphorus. Eighty-five percent of the phosphorus, the second most abundant element in the human body, is located in bones and teeth. Whereas there is constant exchange of calcium and phosphorus between bones and blood, there is very little turnover in teeth. The Ca:P ratio in bones is constant at about 2:1. Every tissue and cell contains phosphorus, generally as a salt or ester of mono-, di-, or tribasic phosphoric acid, as phospholipids, or as phosphorylated sugars. Phosphorus is involved in a large number and wide variety of metabolic functions. Examples are carbohydrate metabolism, adenosine triphosphate (ATP) from fatty acid metabolism, and oxidative phosphorylation.

The formation of phosphate esters is the essential initial process in carbohydrate metabolism. See also **Carbohydrates**. The glycolytic, i.e., anaerobic or Embden-Meyerhof pathway comprises a series of nine such esters. The phosphogluconate pathway, starting with glucose, comprises a succession of 12 phosphate esters. Cyclic adenosine monophosphate (cAMP), produced from ATP, is involved in a large number of cellular reactions including glycogenolysis, lipolysis, active transport of amino

acids, and synthesis of protein. Inorganic phosphate ions are involved in controlling the pH of blood. The principal anion of intercellular fluid is HPO_4^{2-}.

Phospholipids, components of every cell membrane, are active determinants of membrane permeability. They are sources of energy, components of certain enzyme systems, and involved in lipid transport in plasma. Because of their polar nature, phospholipids can act as emulsifying agents. The structure of most phospholipids resembles that of triglycerides except that one fatty acid radical has been replaced by a radical derived from phosphoric acid and a nitrogen base, e.g., choline or serine.

Phosphorus is an essential component of nucleic acids, polymers consisting of chains of nucleosides, a sugar plus a nitrogenous base, and joined by phosphate groups. In ribonucleic acid (RNA), the sugar is D-ribose; in deoxyribonucleic acids (DNA), the sugar is 2-deoxy-D-ribose.

Phosphorus nutrient deficiency can lead to rickets, osteomalacia, and osteoporosis, whereas an excess can produce hypocalcemia. Faulty utilization of phosphorus results in rickets, osteomalacia, osteoporosis, and Paget's disease, and renal or vitamin D-resistant rickets.

Sulfur. Sulfur is present in every cell in the body, primarily in proteins containing the amino acids methionine, cystine, and cysteine. Inorganic sulfates and sulfides occur in small amounts relative to total body sulfur, but the compounds that contain them are important to metabolism. Sulfur intake is thought to be adequate if protein intake is adequate and sulfur deficiency has not been reported.

Although sulfur is in the same group of the Periodic Table, Group 16(VIA), as oxygen, sulfur functions much more like phosphorus, Group 15(VA), in biological systems. In fat metabolism, sulfur plays a key role analogous to that of phosphorus in carbohydrate metabolism. Fatty acid synthesis and degradation begin and end with the same compound, acetyl-S coenzyme A (acetyl–SCoA).

Detoxification systems in the human body often involve reactions that utilize sulfur-containing compounds. For example, reactions in which sulfate esters of potentially toxic compounds are formed, rendering these less toxic or nontoxic, are common as are acetylation reactions involving acetyl–SCoA. Another important compound is S-adenosylmethionine (SAM), the active form of methionine. SAM acts as a methylating agent, e.g., in detoxification reactions such as the methylation of pyridine derivatives, and in the formation of choline (qv), creatine, carnitine, and epinephrine. Sulfur nutrient deficiency results in retarded growth, and faulty utilization in homocystinuria.

Sodium and Potassium. Whereas sodium ion is the most abundant cation in the extracellular fluid, potassium ion is the most abundant in the intracellular fluid. Small amounts of K^+ are required in the extracellular fluid to maintain normal muscle activity. Some sodium ion is also present in intracellular fluid.

Sodium ion acts in concert with other electrolytes, in particular K^+, to regulate the osmotic pressure and to maintain the appropriate water and pH balance of the body. Homeostatic control of these functions is accomplished by the lungs and kidneys interacting by way of the blood. Sodium is essential for glucose absorption and transport of other substances across cell membranes. It is also involved, as is K^+, in transmitting nerve impulses and in muscle relaxation. Potassium ion acts as a catalyst in the intracellular fluid, in energy metabolism, and is required for carbohydrate and protein metabolism.

Maintenance of the appropriate concentrations of K^+ and Na^+ in the intra- and extracellular fluids involves active transport, i.e., a process requiring energy. Sodium ion in the extracellular fluid (0.136–0.145 M Na^+) diffuses passively and continuously into the intracellular fluid (<0.01 M Na^+) and must be removed. This sodium ion is pumped from the intracellular to the extracellular fluid, while K^+ is pumped from the extracellular (ca 0.004 M K^+) to the intracellular fluid (ca 0.14 M K^+). The energy for these processes is provided by hydrolysis of adenosine triphosphate (ATP) and requires the enzyme Na^+–K^+ ATPase, a membrane-bound enzyme which is widely distributed in the body. In some cells, e.g., brain and kidney, 60–70 wt % of the ATP is used to maintain the required Na^+–K^+ distribution.

Sodium and potassium ions are actively absorbed from the intestine. As a consequence of the electrical potential caused by transport of these ions, an equivalent quantity of Cl^- is absorbed. The resulting osmotic effect causes absorption of water.

Selective excretion and reabsorption of Na^+ and K^+ are accomplished by means of the kidney tubular cell membranes. The volume of extracellular fluid is directly related to the Na^+ concentration which is closely controlled by the kidneys. Homeostatic control of Na^+ concentration depends on the hormone aldosterone. The kidney secretes a proteolytic enzyme, rennin, which is essential in the first of a series of reactions leading to aldosterone. In response to a decrease in plasma volume and Na^+ concentration, the secretion of rennin stimulates the production of aldosterone resulting in increased sodium retention and increased volume of extracellular fluid.

Salt-free or low salt diets often are prescribed for hypertensive patients. However, sodium chloride increases the blood pressure in some individuals but not in others. Conversely, restriction of dietary NaCl lowers the blood pressure of some hypertensives, but not of others. Genetic factors and other nutrients, e.g., Ca^{2+} and K^+, may be involved. The optimal intakes of Na^+ and K^+ remain to be established.

Potassium and/or sodium deficiency can lead to muscle weakness and sodium deficiency to nausea. Hyperkalemia resulting in cardiac arrest is possible from 18 g/d of potassium combined with inadequate kidney function. Faulty utilization of K^+ and/or Na^+ can lead to Addison's or Cushing's disease.

Chlorine. The chlorides are essential in the homeostatic processes maintaining fluid volume, osmotic pressure, and acid–base equilibria. Most chloride is present in body fluids; a little is in bone salts. Chloride is the principal anion accompanying Na^+ in the extracellular fluid. Less than 15 wt % of the Cl^- is associated with K^+ in the intracellular fluid. Chloride passively and freely diffuses between intra- and extracellular fluids through the cell membrane. If chloride diffuses freely, but most Cl^- remains in the extracellular fluid, it follows that there is some restriction on the diffusion of phosphate.

Some of the blood Cl^- is used for formation in the gastric glands of hydrochloric acid, HCl, required for digestion. Hydrochloric acid is secreted into the stomach where it acts with gastric enzymes in the digestive processes. The chloride is then reabsorbed with other nutrients into the blood stream. Chloride is actively transported in gastric and intestinal mucosa. In the kidney, chloride is passively reabsorbed in the thin ascending loop of Henle and actively reabsorbed in the thick segment of the ascending loop, i.e., the distal tubule. In the chloride shift, Cl^- plays an important role in the transport of carbon dioxide.

Numerous neurotransmitter receptors, e.g., glutamate, γ-aminobutyric acid (GABA), and benzodiazepine (called the valium receptor), have been identified as chloride channel proteins. The genetic defect in cystic fibrosis involves defective-functioning chloride channel proteins with excessive Cl^- loss. Deficient Cl^- during development adversely affects language skills in humans, as well as impaired growth in infants and metabolic alkalosis.

Fruit and vegetable juices high in potassium have been recommended to correct hypokalemic alkalosis in patients on diuretic therapy. Apparently the efficacy of this treatment is questionable. A possible reason for ineffectiveness is the low Cl^- content of most of these juices. Because Cl^- is high only in juices in which Na^+ is high, these have to be excluded.

Magnesium. In the adult human, 50–70% of the magnesium is in the bones associated with calcium and phosphorus. The rest is widely distributed in the soft tissues and body fluids. Most of the nonbone Mg^{2+}, like K^+, is located in the intracellular fluid where it is the most abundant divalent cation. Magnesium ion is efficiently retained by the kidney when the plasma concentration of Mg^{2-} falls; in this respect it resembles Na^+. The functions of Na^+, K^+, Mg^{2+}, and Ca^{2+} are interrelated so that a deficiency of Mg^{2+} affects the metabolism of the other three ions.

Magnesium is essential in numerous metabolic processes. It is the activator of many enzymes, e.g., adenyl cyclase, alkaline phosphatases, and the phosphokinases, pyrophosphatases, and thiokinases. Because the phosphokinases are required for the hydrolysis and transfer of phosphate groups, magnesium is essential in glycolysis and in oxidative phosphorylation. The thiokinases are required for the initiation of fatty acid degradation. Magnesium is also required in systems in which thiamine pyrophosphate is a coenzyme.

As an activator of the phosphokinases, magnesium is essential in energy-requiring biological processes, such as activation of amino acids, acetate, and succinate; synthesis of proteins, fats, coenzymes, and nucleic acids; generation and transmission of nerve impulses; and muscle contraction.

Regulation of serum Mg^{2+} appears to result from a balance among intestinal absorption, renal reabsorption, and excretion. The controlling factor is probably the renal threshold.

A severe magnesium deficiency in humans is seldom encountered except as a secondary effect resulting from numerous disease states, e.g., chronic alcoholism with malnutrition, acute or chronic renal disease, long-term Mg^{2+}-free parenteral feeding, protein–calorie malnutrition, and hyperthyroidism. In these situations, it is difficult to attribute specific clinical manifestations to magnesium deficiency. The specific role of magnesium in cardiovascular disease, e.g., arrythmia, spasms, or ischemia, remains a subject of conflicting research findings.

Neuromuscular irritability, convulsions, muscle tremors, mental changes such as confusion, disorientation, and hallucinations, heart disease, and kidney stones have all been attributed to magnesium deficiency. Excess Mg^{2+} can lead to intoxication exemplified by drowsiness, stupor, and eventually coma.

Trace Elements and Ultratrace Elements

Iron. The total body content of iron, i.e., 3–5 g, is recycled more efficiently than other metals. There is no mechanism for excretion of iron and what little iron is lost daily, i.e., ca 1 mg in the male and 1.5 mg in the menstruating female, is lost mainly through exfoliated mucosal, skin, or hair cells, and menstrual blood.

A large percentage of the iron in the human body is in hemoglobin: 85 wt % in the adult female, 60 wt % in the adult male. The remainder is present in other iron-containing compounds involved in basic metabolic functions, or in iron transport or storage compounds.

Absorption of iron from food to maintain homeostasis, tightly controlled, increases in instances of increased demands, such as during pregnancy and lactation, and iron-deficiency states which are the result of blood loss or iron-deficiency anemia resulting from inadequate iron intake. Iron absorption is greatly reduced in the normal individual when iron stores are adequate or excessive. Absorption is enhanced by acid conditions and reducing agents. Heme iron from animal sources is absorbed more readily than nonheme iron from cereals and vegetables.

A system of internal iron exchange exists which is dominated by the iron required for hemoglobin synthesis. For formation of red blood cells, iron stores can furnish 10–40 mg/d of iron, as compared to 1–3 mg from dietary sources. Only ca 10 wt % of ingested iron actually is absorbed. Transferrin is essential for movement of iron and without it, as in genetic absence of transferrin, iron overload occurs in tissues. This hereditary atransferrinemia is coupled with iron-deficiency anemia. The iron overload in hereditary or acquired hemochromatosis results in fully saturated transferrin and is treated by phlebotomy.

Iron deficiency is a significant worldwide nutritional problem and cause of anemia which can also lead to a decreased resistance to infection. Insufficient dietary iron intake; iron losses, e.g., bleeding and parasite infestation; and malabsorption of iron are the principal causes. The groups at greatest risk for developing iron-deficiency anemia are menstruating females, pregnant or nursing females, and young children. Children can experience impaired psychomotor development and intellectual performance.

Iron toxicity resulting from excess absorbable iron ingestion is rare except in Africa where fermented beverages made in large iron pots have levels of iron approaching 80 mg/L in a brew where the pH is very low. This results in Bantu siderosis which can result in hemochromatosis, i.e., damage to various organs from excessive storage of iron. This condition can cause numerous disease states, e.g., hepatic fibrosis and diabetes in 80% of the cases of idiopathic hemochromatosis patients. Iron overload is frequently a complication of repeated blood transfusions in anemias, e.g., thalassemia. The lethal dose of ferrous sulfate for a two-year old is 2 g; for an adult the lethal dose is from 200–300 g.

Fluorine. Fluoride is present in the bones and teeth in very small quantities. Human ingestion is from 0.7–3.4 mg/d from food and water.

Fluoridation of public water supplies, a common practice throughout much of the United States, may be an effective means of significantly reducing the incidence of dental caries. See also **Fluorine**. Concern regarding the narrow range of safety between effective and toxic fluoride concentrations has been expressed and poisoning from excessive fluoride, fluorosis, added to public water has been reported. Assertions that fluoridation of water supplies increases the incidence of cancer have not been substantiated.

Excess fluoride ingestion damages developing teeth, causing mottling, chalky-white coloration, and pitting.

Zinc. The 2–3 g of zinc in the human body are widely distributed in every tissue and tissue fluid. About 90 wt % is in muscle and bone; unusually high concentrations are in the choroid of the eye and in the prostate gland. Almost all of the zinc in the blood is associated with carbonic anhydrase in the erythrocytes. Zinc is concentrated in nucleic acids, and found in the nuclear, mitochondrial, and supernatant fractions of all cells.

Zinc is essential for the function of many enzymes, either in the active site, i.e., as a nondialyzable component, of numerous metalloenzymes or as a dialyzable activator in various other enzyme systems.

Zinc-hormone interactions include hormonal influence on absorption, distribution, transport, and excretion of zinc and zinc influence on synthesis, secretion, receptor binding, and function of numerous hormones. Zinc enhances pituitary activity by increasing circulating levels of growth hormone, thyroid-stimulating hormone, luteinizing hormone, follicle-stimulating hormone, and adrenocorticotropin. The role of zinc in insulin action is recognized but not well understood. Zinc is required for maintenance of normal plasma concentrations of vitamin A and for normal mobilization of vitamin A from the liver.

Zinc was confirmed as essential for humans in 1956 and deficiency symptoms were reported in 1961. The size of the human fetus is correlated with zinc concentration in the amniotic fluid and habitual low zinc intake in the pregnant female is thought to be related to several congenital anomalies in humans. Low zinc intakes result in hypogonadism, dwarfism, mental retardation, low serum and red blood cell zinc in humans and animals, and retarded growth and teratogenic effects on the nervous system in rats.

In children suffering from marginal zinc deficiency, impaired taste acuity, poor appetite, and suboptimal growth can be reversed upon zinc supplementation. Accelerated wound healing occurs in humans upon zinc supplementation, suggesting that marginal zinc deficiency in humans may be more widespread than has been thought. Zinc supplementation has also been effective in alleviating symptoms of active rheumatoid arthritis in clinical trials. Acrodermatitis enteropathica, a hereditary disease that involves aberrant zinc metabolism, responds to oral zinc supplementation. Excessive zinc intake may interfere with copper metabolism.

Silicon. Silicon comes mainly from ingestion of silicates, primarily from vegetables. It is found in the serum as silicic acid, $Si(OH)_4$, and normal blood serum levels are ca 1 mg/100 mL regardless of intake because of efficient kidney excretion of excess. Silicon is necessary for calcification, growth, and as cross-linking material in mucopolysaccharide formation. Silicon is especially helpful in situations where the diet is low in calcium or high in aluminum, or thyroid function is inadequate. The human requirement may be 5–20 mg/d. Silicon deficiency may lead to altered metabolism of connective tissue and bone and/or aluminum accumulation in the brain.

Copper. All human tissues contain copper. The highest amounts are found in the liver, brain, heart, and kidney. In blood, plasma and erythrocytes contain almost equal amounts of copper, i.e., ca 110 and 115 mg/100 mL, respectively.

In plasma, ca 90 wt % of copper is in the metalloprotein ceruloplasmin, also known as a_2-globulin, mol wt 151,000, which contains 8 atoms of copper per molecule. Ceruloplasmin has been identified as a ferroxidase(I) which catalyses the oxidation of aromatic amines and of Fe^{2+} to Fe^{3+}. The ferric ion is then incorporated into transferrin which is necessary for the transport of iron to tissues involved in the synthesis of iron-containing compounds, e.g., hemoglobin. Lowered levels of ceruloplasmin interfere with hemoglobin synthesis.

Copper deficiency is characterized by poorly formed collagen which leads to bone fragility and spontaneous bone fractures in animals, and also results in cardiac hypertrophy. Abnormal electrocardiographs have been noted when low copper diets were fed to humans. Anemia, neutropenia, and bone disease have been reported in children having protein calorie malnutrition (PCM) and accompanying hypocupremia. At least two genetic diseases involving copper are known: Wilson's disease, an autosomal recessive disease, usually detected in adulthood, and Menke's kinky-hair syndrome.

Analytical data indicate that many diets contain less than the RDA for copper. Excessive copper has been reported to be fatal for oral dose levels of copper sulfate of 200 mg/kg body weight for a child and 50 mg/kg for adults.

Boron. The essentiality of boron, first accepted for higher plants in 1923, then for animals, was recognized in 1981 for human metabolism.

Boron is reported to help maintain function or stability of cell membranes and is thought to be involved with hormone reception and transmembrane signaling. Enhanced need for boron may develop with nutritional or metabolic stress involving other nutrients, e.g., magnesium deprivation or physiological changes in calcium metabolism. Boron depletion impairs cognitive function. Organs known to contain the highest levels of boron are bone, spleen, and thyroid. An excess of boron can, however, cause seizures in infants, riboflavinuria, and gastrointestinal upset.

Selenium. Selenium, thought to be widely distributed throughout body tissues, is present mostly as selenocysteine in selenoproteins or as selenomethionine. Animal experiments suggest that greater concentrations are in the kidney, liver, and pancreas and lesser amounts are in the lungs, heart, spleen, skin, brain, and carcass.

The most clearly documented role for selenium is as a necessary component of glutathione peroxidase. Selenium is also involved in the functions of additional enzymes, e.g., type 1 iodothyronine deiodinase, leukocyte acid phosphatase, and glucuronidases. A role for selenium in electron transfer has been suggested as has involvement in nonheme iron proteins. Selenium and vitamin E appear to be necessary for proper functioning of lysosomal membranes. A role for selenium in metabolism of thyroid hormone has been confirmed.

Alkali disease and blind staggers of grazing livestock in the western U.S. were reported as the result of selenium poisoning. There are unusually high concentrations of selenium in certain plants, because of selenium-accumulating properties, or in ordinary plants growing on highly seleniferous soils. No treatment for this type of poisoning is known, thus excess selenium in the animal diet must be avoided. Prolonged ingestion of up to 600 mcg/d of selenium did not produce toxic effects in humans. Toxic effects in humans have been reported, however, from chronic ingestion of food in China supplying 5 mg/d of Se, and from supplements in the U.S. of 27–2387 mg/d of Se. Effects include hair loss, changes in the nails, gastrointestinal upset, and peripheral neuropathy.

Pure selenium deficiency, without concurrent vitamin E deficiency, is not generally seen except in animals on experimental diets. In China, selenium deficiency in humans has been associated with Keshan disease, a cardiomyopathy seen in children and in women of child-bearing ages, and Kashin-Beck disease, an endemic osteoarthritis in adolescents. Selenium may have anticarcinogenic effects possibly because of the antioxidant properties of selenium compounds.

Iodine. Of the 10–20 mg of iodine in the adult body, 70–80 wt % is in the thyroid gland. See also **Iodine (In Biological Systems)**. The essentiality of iodine, present in all tissues, depends solely on utilization by the thyroid gland to produce thyroxine and related compounds. Well-known consequences of faulty thyroid function are hypothyroidism, hyperthyroidism, and goiter. Dietary iodine is obtained from eating seafoods and kelp and from using iodized salt.

The functions of the thyroid hormones and thus of iodine are control of energy transductions. These hormones increase oxygen consumption and basal metabolic rate by accelerating reactions in nearly all cells of the body. A part of this effect is attributed to increase in activity of many enzymes. Additionally, protein synthesis is affected by the thyroid hormones.

In many parts of the world, simple goiter is endemic and usually results from dietary iodine deficiency or goiterogens in foods which bind iodine. Iodine deficiency disorders (IDD) include cretinism, myxedema, hypothyroidism, and goiter. In technologically advanced countries, the problem of iodine deficiency has been minimized by the use of iodized salt. Faulty utilization of iodine can lead to Grave's disease. A sodium iodide excess can also produce goiter and an excess of 500 mg/kg body weight can be fatal.

Manganese. The adult human body contains ca 10–20 mg of manganese widely distributed throughout the body. The largest Mn^{2+} concentration is in the mitochondria of the soft tissues, especially in the liver, pancreas, and kidneys. Manganese concentration in bone varies widely with dietary intake.

Manganese is essential for normal body structure, reproduction, normal functioning of the central nervous system, and activation of numerous enzymes. An excess of manganese can lead to neural damage and possible impaired insulin production.

In animals, manganese deficiency results in wide-ranging disorders, e.g., impaired growth, abnormal skeletal structure, disturbances of reproduction, and defective lipid and carbohydrate metabolism. Although overt manganese deficiency has not been induced in humans, some forms of epilepsy in humans and animals and a decrease in glucose tolerance in animals have been linked to low levels of manganese in the tissue.

Molybdenum. Molybdenum is a component of the metalloenzymes xanthine oxidase, aldehyde oxidase, and sulfite oxidase in mammals. Two other molybdenum metalloenzymes present in nitrifying bacteria have been characterized: nitrogenase and nitrate reductase. The molybdenum in the oxidases, is involved in redox reactions. The heme iron in sulfite oxidase also is involved in electron transfer. Foods rich in molybdenum include legumes, dark green vegetables, liver, whole-grain cereals, and milk. Xanthine oxidase, mol wt ca 275,000, present in milk, liver, and intestinal mucosa, is required in the catabolism of nucleotides. Xanthine oxidase is also involved in iron metabolism.

A copper–molybdenum antagonism involving sulfate occurs in animals, i.e., large amounts of molybdenum and sulfate can depress copper absorption. Cattle grazing on pasturage of high Mo content succumb to teart or peat scours, characterized by diarrhea and general wasting. Control involves increasing copper intake. The Cu–Mo antagonism has been observed in humans. Significant increases in urinary copper excretion have been observed with increasing Mo intake.

Molybdenum deficiency in humans results in deranged metabolism of sulfur and purines and symptoms of mental disturbances. Toxic levels produce elevated uric acid in blood, gout, anemia, and growth depression. Faulty utilization results in sulfite oxidase deficiency, a lethal inborn error.

Chromium. Chromium(III) potentiates the action of insulin and may be considered a cofactor for insulin. Chromium is thought to form a complex with insulin and insulin receptors.

Studies of elderly people and mildly diabetic patients showed significant improvement in the glucose tolerance test (GTT) when chromium supplementation of 150–200 μg/d was given. In other tests, these positive results were not obtained. It is possible that not all subjects are capable of utilizing inorganic chromium to the same extent. Some may require a preformed GTF (glucose tolerance factor). Chromium chloride supplementation has been effective in normalizing impaired glucose tolerance in malnourished children and in patients receiving total parenteral nutrition for a long time. The most available form of chromium is GTF obtained from brewer's yeast. Chromium deficiency may also lead to atherosclerosis and peripheral neuropathy.

Cobalt. Cobalt is nutritionally available only as vitamin B_{12}. Although Co^{2+} can function as a replacement *in vitro* for other divalent cations, in particular Zn^{2+}, no *in vivo* function for inorganic cobalt is known for humans. In ruminant animals, B_{12} is synthesized by bacteria in the rumen.

In pernicious anemia, the bone marrow fails to produce mature erythrocytes as a result of defective cell division, a consequence of impaired DNA synthesis which requires vitamin B_{12}. If the disease goes untreated, extensive neurological damage, e.g., irreversible degeneration of the spinal cord by demyelinization, may occur because of faulty fatty acid metabolism.

Vitamin B_{12} deficiency commonly is caused by inadequate absorption resulting from a lack or insufficient intrinsic factor (IF). Intrinsic factor is a glycoprotein, mol wt ca 50,000, which binds vitamin B_{12} in a 1:1 molar ratio. The B_{12}–IF complex, formed in the stomach, is absorbed in the ileum. Absorption in this part of the intestine occurs because of the specific characteristics of the cells of the microvilli (brush border) of the ileum. The IF remains in the intestine attached to the epithelial cells. Transport of B_{12} into the blood stream requires Ca^{2+}. In the blood, B_{12} is bound to transcobalamin II (transport protein). Whatever bound B_{12} is not utilized immediately is stored in the liver. With increasing quantities of dietary B_{12}, the fraction that is absorbed decreases. Generally, vitamin B_{12} is excreted in the urine, but with large intake, some is excreted in the bile. A nutritional excess of cobalt can lead to polycythemia.

Tin. The widespread use of canned foods results in a daily intake of tin that is ca 1–17 mg for an adult male. At this level it has not been shown to be toxic. Some grains also contain tin. Too much tin can adversely affect zinc balance and iron metabolism. Essentiality has not been confirmed for humans. It has been shown for the rat. An enhanced growth rate results from tin supplementation of low tin diets. Animals on deficient diets exhibit poor growth and decreased feed efficiency.

Vanadium. Vanadium is essential in rats and chicks. Estimated human intake is less than 4 mg/d. In animals, deficiency results in impaired growth, reproduction, and lipid metabolism, and altered thyroid peroxidase activities. Vanadium may play a role in the regulation of (NaK)–ATPase, phosphoryl transferases, adenylate cyclase, and protein kinases.

Nickel. There is considerable evidence for the essentiality of nickel in animals. Various pathological manifestations of nickel deficiencies have been observed in chicks, cows, goats, pigs, rats, and sheep. Average intake is reported to be about 60–260 μg/d, and a dietary requirement for humans of less than 100 μg/d has been suggested. *In vitro* studies have shown nickel to be an activator of several enzymes.

Arsenic. Arsenic is under consideration for inclusion as an essential element. No clear role has been established, but arsenic, long thought to be a poison, may be involved in methylation of macromolecules and as an effector of methionine metabolism.

Health and Safety Factors

Under unusual circumstances, toxicity may arise from ingestion of excess amounts of minerals. This is uncommon except in the cases of fluorine, molybdenum, selenium, copper, iron, vanadium, and arsenic. Toxicosis may also result from exposure to industrial compounds containing various chemical forms of some of the minerals. Aspects of toxicity of essential elements have been published.

Efficient homeostatic controls of mammalians generally prevent serious toxicity from ingestion of the mineral nutrients. Toxicity may occur under conditions far removed from those of nutritional significance or for individuals suffering from some pathological conditions. Because of very low concentrations in foods, the trace elements are not toxic under normal nutritional conditions. Exceptions are selenium and iron.

See also **Calcium (In Biological Systems); Cobalt (In Biological Systems); Copper (In Biological Systems); Magnesium (In Biological Systems); Molybdenum (In Biological Systems); Nickel (In Biological Systems); Phosphorus (In Biological Systems); Potassium and Sodium (In Biological Systems); and Zinc (In Biological Systems).**

MARY G. ENIG
Enig Associates, Inc.

Additional Reading

Considine, D.M., and G.D. Considine: *Foods and Food Production Encyclopedia,* Van Nostrand Reinhold Company Inc., New York, NY, 1982.

Ensminger, A.H., M.E. Ensminger, J. Konlande, and J.R.K. Robson: *Food and Nutrition Encyclopedia,* 2nd Edition, CRC Press, Boca Raton, FL, 1994.

Linder, M.C., ed.: *Nutritional Biochemistry and Metabolism with Clinical Applications,* 2nd Edition, Appleton & Lange, Norwalk, CT, 1991.

Mertz, W., ed.: *Trace Elements in Human and Animal Nutrition,* Vols. 1 and 2, Academic Press, San Diego, CA, 1987.

Seelig, M.S., and A. Rosanoff: *The Magnesium Factor,* Avery Publishing Group, Inc., Garden City Park, NY, 2003.

Shils, M.E., J.A. Olson, and M. Shike, eds.: *Modern Nutrition in Health and Disease,* 8th Edition, Vols. 1 and 2, Lea & Febigher, Philadelphia, PA, 1993.

MINERALOGY. The science of mineralogy is concerned with the formation, occurrence, properties, composition, and classification of minerals. Various definitions of a mineral have been proposed. Possibly, the most acceptable may be, "a naturally occurring inorganic substance, usually crystalline, possessing a relatively definite chemical composition and physical characteristics." It should be pointed out that some naturally formed organic substances, particularly of an economic resource nature, are sometimes classified as minerals.

Although in its broadest application, mineralogy is as ancient as human civilization, mineralogy is a modern science, the mineralogist taking full advantage of all modern tools and instruments for exploration, analysis, testing, and study of minerals. Several major scientific advances in the materials field have stemmed from the study of minerals as will be pointed out shortly.

Presumably in the early ages man used minerals as weapons. Through the passage of time and attainment of knowledge regarding certain mineral characteristics man learned, notably initially by accident but later by design, that the content of minerals provided essential materials for his expanding needs. Very early, the natural form and beauty of certain minerals became objects for personal adornment. Later it was found that both the form and the innate beauty of minerals were enhanced by cutting and polishing them. Although the science of mineralogy touches the life of every person, a fundamental understanding of minerals is not common.

Modern mineralogy is the product of research and discovery by many persons. Robert Hooke (1665) foretold the atomic theory by constructing models of alum crystals out of leaden musket balls. Nicolaus Steno (1669) discovered the constancy of interfacial angles between corresponding faces of quartz crystals from many localities. This was later formalized by Rome de I'Isle (1764) under his *law of constancy of crystal interfacial angles*. In 1784, René Just Haüy proposed the theory of *integral molecules* by stacking calcite rhombs to show that structural units could produce exact external facial planes of various forms of calcite crystals. Haüy is known as the "father" of geometrical crystallography. A great advance in mineral studies was made in 1828 when Nicol invented the nicol prism for investigating the behavior of polarized light in crystallized minerals. In 1912, Max von Laue, a student of Roentgen, theorized that the wavelengths of x-rays and atomic spacing of crystals may be of the same magnitude. Laue found that the diffracted rays, when passed through a crystal, substantiated his theory. This discovery opened up an entirely new field of mineral research, i.e., *crystal chemistry*.

Origin of Minerals

Minerals are products of formation and deposition from Earth's natural open solution systems, as opposed to substances formed as products of closely controlled laboratory systems, and thus may vary in many instances from an exact chemical content and formula. Minerals crystallizing from such open solution systems utilize effectively the ions required to form the mineral, but may incorporate within that structure other nonessential ions foreign to it. The incorporation of those foreign ions may, under favorable conditions of chemical affinity, replace certain specific elements named in the given formula. Or there may be defects in the space lattice of the mineral, e.g., vacancies in the crystalline structure comparable to a classroom with X number of seats orderly arranged where, under ideal conditions, each seat would be occupied by a student. Owing to some extraneous circumstance one or more students may be absent, yet the total seating capacity and orderly arrangement would remain constant. Deviations of the first type may be a product of included impurities in the structure; of the second type, by occupancy of a foreign compatible element in the vacant space(s) of the host crystal structure. See "**Isomorphism and Diadochy**" later in this article.

In other minerals, the formula may vary within restricted limits, such as $(Zn, Fe)S$ for sphalerite, where ferrous iron substitutes for zinc within the sphalerite structure. Two factors prevail in such chemical substitution within a given mineral structure: (1) reasonably comparable ionic radii (approximately, but not strictly limited to ($\pm 15\%$); and (2) maintenance of electrical neutrality of the compound. In the event the substituting ions have a different valence or charge, electrical neutrality may be obtained by an accompanying substitution elsewhere within the crystal structure. Basically, a mineral is a homogeneous inorganic solid with an ordered atomic arrangement, which places it in the category of a crystalline material and possessing a definite, though not a fixed chemical formula.

Natural systems strive toward a state of equilibrium when all component units attain their lowest energy level. The respective energy level of the elements within any mineral is dependent upon the physical environment, principally the temperature, pressure, and chemical substances present, at the time and place of its formation. Any later change in its environment may cause a change in the mineral's composition and form. Whatever the primary or intermediate environmental conditions may have been, the mineral as observed, represents its present equilibrium energy state or crystal structure.

Matter exists in three states—gaseous, liquid, and solid.* In the gaseous state, the elements move freely about within their environment, their only contact being haphazard collisions. Elements in the liquid state are in closer contact with each other, but still retain freedom of mobility. The solid state is characterized by the chemical elements combining under atomic bonds of various types and strengths into a structured system. For any element within the structure, there are equivalent elements in a definite three-dimensional crystalline pattern. Under restricted circumstances, two or more minerals may form in contact with one another in such a manner as to preclude their complete development as single crystals. Those minerals which develop fully and which exhibit well-formed polyhedral (many sides) facial planes are referred to as being *euhedral*. Those minerals that exhibit no facial planes are referred to as *anhedral*. Imperfectly formed crystals are referred to as *subhedral*. Crystalline minerals, where the crystallinity can be determined only by aid of a microscope, are said to be *microcrystalline*.

---

* Traditional concepts. Details on more advanced concepts on the states of matter are given elsewhere in this volume.

Where the materials are so finely divided as to be discernible only by x-ray analysis they are referred to as *cryptocrystalline*. The rate of crystallization plays an important part in the resultant crystalline character of minerals.

Crystalline solids are bonded together by electrical forces, which originate in the constituent atoms. The position of the respective unit particles within the crystal structure are determined by geometric factors, along with considerations of electrical neutrality and lattice energy. One type of bond is produced by the lending or borrowing of electrons in an attempt to complete the outer electron shell of the atom. A stable compound is formed when the outer shell is complete and the mutual attraction is termed an *ionic bond*. For example, sodium with one electron in its outer electron shell lends that electron to a neutral chlorine atom with seven electrons in its outer electron shell. The result is a stable electron shell for each ion (charged atom) which then join by electrostatic attraction in a crystal structure to form sodium chloride, $NaCl$. Minerals so bonded are of moderate hardness, readily soluble, and poor conductors of electricity and heat.

The bond produced by the sharing of electrons by which a stable configuration is achieved is termed a *covalent bond*. For example, chlorine with seven electrons in its outer shell needs one additional electron to achieve stability. If its nearest neighbor is another chlorine atom, the two atoms combine in such a way that the one electron serves in the outer shell of each atom, thus achieving a stable configuration. Diamond is another example of this type. Carbon with four electrons in its outer shell mutually shares the four electrons with four adjoining carbon atoms, thus achieving a stable configuration by completing the shell with eight electrons. This type of electron-sharing or covalent bonding is the strongest of all chemical bonds. Minerals so bonded are generally insoluble, possess a high degree of stability, and are nonconducting.

Metallic bonds consist of a structure of positive ions through which free electrons can drift. The structure of metals may be envisioned as a mesh of electrons that surround the atomic nuclei and bind them together. This electron mobility produces minerals of generally a low hardness and of high electrical and thermal conductivity.

The *Van der Waals* bond is a very weak attraction between atoms, usually between essentially neutral atoms or groups of atoms. These neutral and essentially uncharged structure units are held together within the crystal lattice by virtue of small residual charges on their surfaces. This type of bonding is rare in minerals. Minerals so bonded often display ready cleavage and low hardness. For example, the foliated scales of graphite are linked weakly together by Van der Waals bonds.

Every mineral is a product of the redistribution or recombination of its component chemical elements to form a stable substance. The process is known as *crystallization*. The process may involve *precipitation* of chemical elements from aqueous solutions at the earth's surface; or from siliceous melts (magmas) from the earth's interior. In either situation, the process is dependent upon the degree of concentration of the constituent chemical elements present and the temperature/pressure conditions. Precipitation from vapor also is possible. An example is the hot vapor, rich in sulfur dioxide, which is emitted from vents associated with volcanoes. Upon becoming exposed to the cooler atmosphere, crystal sulfur is deposited around those vents. Snow crystals are another example of precipitation from vapor.

Crystal Structure

Associated chemical units become systematically arranged in the crystal structure, which is constructed from a single motif that develops repetitively. The resulting three-dimensional array is called the *space lattice* of the crystal. The lattice or framework is defined by three directions and by the distance along those directions where the motif repeats itself. Because the units within the structure adhere to a strict arrangement, the external facial planes of a crystal represent the limiting surfaces of that growth and are an external expression of its internal atomic order. Crystals are formed, therefore, where constituent atoms or ions are free to combine in constant chemical proportions and are an expression of the environmental conditions that promote their formation.

Periodic repetitions of a space lattice cell in three dimensions from the original cell will completely partition space without overlapping or omissions. It is possible to develop a limited number of such three-dimensional patterns. Bravais, in 1848, demonstrated geometrically that there were but fourteen types of space lattice cells possible, and that these fourteen types could be subdivided into six groups called *systems*. Each system may be distinguished by symmetry features, which can be related to four symmetry elements:

1. *Symmetry with respect to a point*. If through a central point in a geometric figure, lines are drawn from a point on one side of the figure to a similar point equidistant on the other side, the figure is symmetrical to a point.

2. *Symmetry with respect to a plane*. A geometrical figure is symmetrical with respect to a plane when for each edge, solid angle, or face on one side, there is a corresponding edge, solid angle, or face on the other side of that plane. One side is, in fact, a mirror image of the opposite side. That plane is called a *plane of symmetry*.

3. *Symmetry with respect to a line*. If during a complete revolution of 360° about a given axis, a geometrical figure repeats itself in appearance two or more times, it is said to be symmetrical with respect to a line, or to an *axis of symmetry*. Possible axes of symmetry are twofold, threefold, fourfold, and sixfold. A rotation of onefold or 360° is equivalent to no rotation at all.

4. When a crystal is rotated about an axis and inverted about the central point and at that point repeats itself, it is said to have an *axis of rotary inversion*. It is a twofold axis of rotary inversion if the geometrical figure is rotated 180° and then inverted. Additionally, there are threefold, fourfold, and sixfold axes of rotary inversion possible.

Within each of the six crystal systems, there are specific *crystal classes*. Each class displays distinctive symmetry elements. There are 32 possible classes distributed among the six crystal systems. One of the crystal classes within each system possesses all of the symmetry elements that are characteristics of its space lattice cell. These are called the *holohedral* class of that system. Other classes within each system possess somewhat fewer symmetry elements and are called *merohedral* classes.

It is significant to note that in most geometrical situations regarding crystal study, the distinguishing characteristic is symmetry, not geometry. This is especially true in the case of a mineral such as pyrite (FeS_2) which assumes a cubic shape, but its symmetry, controlled by the three sets of opposing striations on its crystal faces, identify it as belonging to a much lower order symmetry than that of a true cube.

The six crystal systems are identified by hypothetical lines of reference known as *crystallographic axes*, and their angular relationship to each other. Crystal orientation and axial order are given as: (1) front to back; (2) right side to left side; and (3) top side to bottom side. The front, right, and top side axial ends represent the positive ends of those respective axes. The back, left, and bottom side ends are designated as the negative ends. The relative intensity of crystal growth along those axes gives to each crystal face a distinctive identifying character. This character is evidenced by the physical similarity of equivalent facial planes on the crystal and their relationship to the crystallographic axes of that form. The six systems, their crystallographic axes, and interaxial angular relationship are defined and illustrated in the Fig. 1.

Crystallographic identification of facial planes on a crystal become possible through assignment of numerical values to a face which represents its relationship to the crystal axes. *Parameters*, or relative intercepts, are obtained by plotting coordinates of crystal faces with respect to their crystallographic axes. The actual distances of axial intercepts of the crystal face are determined and expressed as a unit of measurement. The product of these values is known as *Miller indices*.

A Miller indices face on the front face of a cube would be (100), signifying that face intersects the a^1 axis at 1 unit length from the center of the crystal, and is parallel to axes a^2 and a^3, or intersects those axes at infinity. In this system, zero (0) is the numerical substitute for infinity. A (111) Miller indices face identifies that facial plane as intersecting each of the three crystallographic axes of that form at 1 unit length from the crystal center.

The Miller Indices identify the orientation of a face in relationship to its axes of reference regardless of its size and position on the crystal. Haüy first proposed this basic law of crystallography—"crystal faces make simple

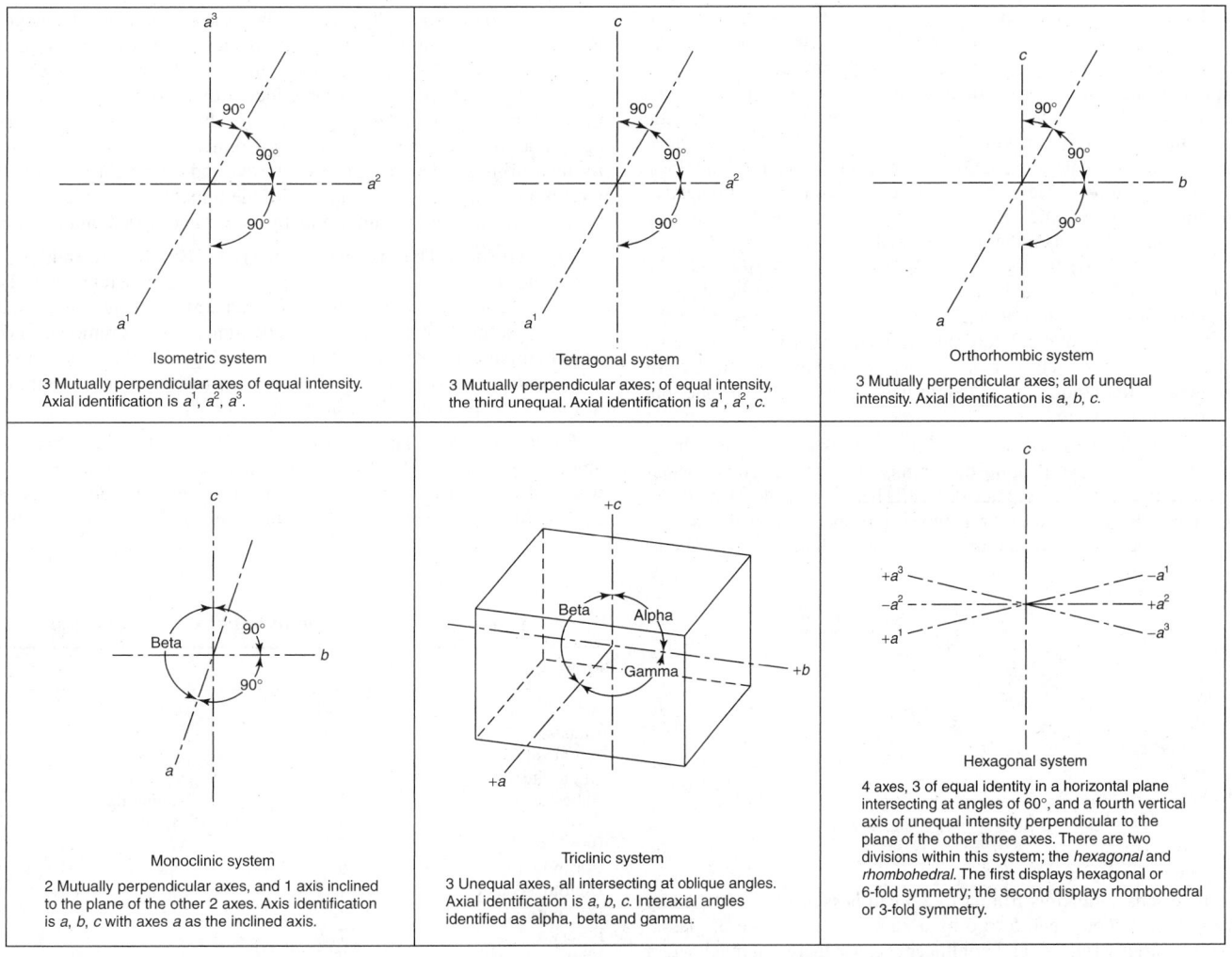

Isometric system
3 Mutually perpendicular axes of equal intensity. Axial identification is a^1, a^2, a^3.

Tetragonal system
3 Mutually perpendicular axes; of equal intensity, the third unequal. Axial identification is a^1, a^2, c.

Orthorhombic system
3 Mutually perpendicular axes; all of unequal intensity. Axial identification is a, b, c.

Monoclinic system
2 Mutually perpendicular axes, and 1 axis inclined to the plane of the other 2 axes. Axis identification is a, b, c with axes a as the inclined axis.

Triclinic system
3 Unequal axes, all intersecting at oblique angles. Axial identification is a, b, c. Interaxial angles identified as alpha, beta and gamma.

Hexagonal system
4 axes, 3 of equal identity in a horizontal plane intersecting at angles of 60°, and a fourth vertical axis of unequal intensity perpendicular to the plane of the other three axes. There are two divisions within this system; the *hexagonal* and *rhombohedral*. The first displays hexagonal or 6-fold symmetry; the second displays rhombohedral or 3-fold symmetry.

Fig. 1.   Crystallographic axes and interaxial angular relationship of six crystal systems

rational intercepts on suitable crystal axes." Inasmuch as the intercepts are simple, it follows that the Miller indices should likewise be simple whole numbers.

See also **Crystal**; and **Miller Indices**.

Crystal Forms

Form is used here to designate the general outward appearance of a crystal, specifically to a group of crystal faces that bear identical relationships to the crystal's symmetry elements. It is essentially a geometric form with equivalent facial planes in their relationship to the crystal lattice symmetry elements. In this regard, an octahedron in the isometric system would be identified as a "closed form," inasmuch as its eight equivalent faces totally enclose the crystal space; its form identification would be {111}, enclosed in braces. Even though the crystal faces may be equivalent they may vary widely in size and distance from the crystal center, owing to irregular or distorted development during their formation. Braces shown around the Miller indices, e.g., {111} signify form identification, as opposed to Miller indices shown in parentheses, e.g. (111), which identify a specific crystal face only. In crystals possessing different lengths of their crystallographic axes general form $\{hkl\}$ would be used. The hexagonal system with four axes requires a four-unit form identification.

In the isometric system, crystals are recognized by their geometrical forms, e.g., cube, octahedron, trapezohedron, pyriteohedron, and so on. In all other systems, crystal faces are given specific names, which refer to their relationship with their respective crystallographic axes. The most common facial forms in these systems are *pedion, pinacoid, dome, prism*, and *pyramid*. Crystal forms bear specific relationships to symmetry axes or planes, and not to the general shape of the crystal.

Physical Mineralogy

This aspect of mineralogy is concerned with several observable physical characteristics, such as color, hardness, lustre, fracture, cleavage, magnetic properties, radioactivity, fluorescence, and specific gravity.

Color. The color of a mineral is a product of selective absorption of certain wave lengths of visible white light by atoms within the mineral structure. A mineral color may be indicative of a species, but more often is of descriptive value only. There are two broad classification categories of color in minerals—*allochromatic* and *idiochromatic*. Allochromatic minerals are those which occur with variable colors, such as quartz, SiO_2; corundum, Al_2O_3; and calcite, $CaCO_3$. Color in such minerals may be a product of included foreign elements, rate of crystallization, or a defective lattice structure. Idiochromatic minerals are those in which the color is a characteristic constant, such as the green of malachite, $Cu_2CO_3(OH)_2$; the blue of azurite, $Cu_3(CO_3)_2(OH)_2$; the black of magnetite, Fe_3O_4; and the red of cinnabar, HgS. The actual color of most minerals can be obtained by rubbing the mineral on an unglazed porcelain plate and observing the powder (streak) of that mineral left on the plate. This test is essential in certain instances inasmuch as many minerals display a physical color foreign to their actual color, e.g., hematite may appear black or blue, but its streak color is always red.

Hardness. The hardness of a mineral is its resistance to scratching. Testing for hardness is based upon the premise that all minerals possess such resistance to a lesser or greater degree. The Mohs scale of hardness has been universally adopted to test this physical property. In the list below, the ascending numeric order of the mineral named will scratch all of those of lower order.

1	Talc
2	Gypsum
3	Calcite
4	Fluorite
5	Apatite
6	Orthoclase
7	Quartz
8	Topaz
9	Corundum
10	Diamond

In a relative sense, minerals from 2 to 2.5 hardness can be scratched by a fingernail; 4 by a penny, and 5 to 6 by a knife blade or piece of glass.

This scale is strictly relative and nonlinear because there is a much wider differential, for example, between corundum and diamond than between topaz and corundum. Hardness also varies according to crystallographic direction in certain minerals. This test should always be made on a freshly broken area because certain minerals are subject to surface alteration. A hardness test on an altered area can be misleading.

The hardness (Mohs scale) of several minerals is given in Table 1.

Lustre. The lustre of a mineral is a product of both light reflection and refraction. Reflection is the governing factor in translucent to opaque minerals; refraction in transparent minerals. There are two broad categories in describing lustre—*metallic* and *nonmetallic*. Minerals of nonmetallic lustre are further subdivided into categories, such as vitreous (glass); adamantine (diamond); resinous (sphalerite); silky (asbestos); waxy (chalcedony and opal); greasy (some quartz and diamonds); and pearly (talc or mica).

Fracture. The product of irregular breaking of a mineral is termed fracture. Categories include conchoidal or conch-like (quartz); uneven (serpentine); and hackly (copper). Fracture in a mineral is unrelated to its crystal structure.

Cleavage. The cleavage of a mineral is the product of regular breaking of the mineral along specific surfaces related to the mineral's internal structure. The cleavage plane is always parallel to a possible crystal face and, therefore, is a reflection of the atomic structure of the mineral. Three planes of cleavage are present in the cube and rhombohedron. Four cleavage planes produce an octahedron, six planes a dodecahedron, both of which fall within the isometric system.

Magnetic Properties. A few minerals possess the property of being attracted by a magnet and they are known as *ferromagnetic* minerals. The more common ferromagnetic minerals are magnetite, Fe_3O_4, and pyrrhotite, $Fe_{2-x}S$. Those minerals which are natural magnets are known as *lodestones*. Most minerals are affected to some extent in a magnetic field. The minerals that are repelled are known as *diamagnetic*; those that are weakly attracted are known as *paramagnetic*.

Radioactivity. Basically, this involves the spontaneous disintegration of uranium and thorium minerals. Both of these mineral species disintegrate at a steady rate that is completely unaffected by external chemical or mechanical conditions. The end-product of this disintegration is lead within the mineral structure. Geophysicists are able to determine the geologic age of a specific uranium or thorium mineral and its host environment by measuring the amount of lead present and computing the known time required to produce this amount. With the impetus of atomic energy, many previously unknown uranium minerals were uncovered and identified.

Fluorescence. This is the property of certain minerals to absorb ultraviolet radiation and convert that energy into visible light. The rays energize certain elements within the mineral, causing the excitation of electrons in the orbital shell. If the activating source is removed and visible light continues for a period of time following, the phenomenon is known as *phosphorescence*. The visible light continues until the electrons return to their normal orbital electron shell.

Specific Gravity. The density or specific gravity of a substance is the amount by which that substance decreases in apparent weight when first weighed in air, followed by weighing in water. The density is obtained by dividing the weight of the substance in air by the loss of its weight in water.

TABLE 1. HARDNESS OF REPRESENTATIVE MINERALS

Mohs scale

Agate	6–7	Galena	2.5
Alabaster	1.7	Garnet	6.5–7
Andalusite	6.5–7.5	Graphite	0.5–1
Aragonite	3.5	Kaolinite	2.0–2.5
Asbestos	5.0	Magnetite	6
Barite	3–3.5	Marble	3–4
Beryl	7.8	Mica	2.8
Corundum	9	Opal	5.5–6.5
Diatomaceous earth	1–1.5	Pumice	6
Dolomite	3.5–4	Pyrite	6–6.5
Emery	7–9	Serpentine	3–4
Flint	7	Tourmaline	7–7.5

Isomorphism and Diadochy

Isomorphism is the substitution of an atom, ion, or radical for another within a mineral structure. The degree of substitution is controlled largely by two factors: (1) temperature, and (2) ionic size. Only ions of similar size will readily substitute for one another.

Under certain conditions, isomorphism may be either partial or complete. The spinel, garnet, and amphibole mineral groups represent a complete isomorphous series. Partial isomorphism is exemplified by chemical components displaying extensive ionic substitution at high temperature levels, as evidenced by the nearly complete random mixing of potassium and sodium in microcline $KAlSi_3O_8$ and albite $NaAlSi_3O_8$ feldspars. Reducing temperature causes segregation of their component content according to their respective radii, which then group together, forming two separate mineral phases. This phenomenon is known as *exsolution* and is shown by the mixed feldspar known as perthite, common in igneous and high-grade metamorphic rocks. Other isomorphous examples include calcite and siderite; magnesite and siderite.

Diadochy is a term applied to the substitution of foreign atoms/ions within the same space lattice of a crystal atomic structure. Diadochic substitution refers only to specific crystal structures, as substitution elements may be diadochic in one structure and not in another. For example, excess Al^{3+} (0.57kX) may substitute diadochically for Fe^{3+} (0.64kX) in the epidote structure, thereby producing clinozoisite; when within the dolomite structure, Fe^{2+} (0.83kX) substitutes for Mg^{2+} (0.78kX), ankerite is the product; marmatite (ferroan sphalerite) results when Fe^{2+} (0.83kX) replaces Zn^{2+} (0.83kX) within the sphalerite structure. Diadochy involves the actual substitution/replacement of a given compatible element for another within a crystal structure, which, as in isomorphism, may be either partial or complete. In like manner, high temperature contributes materially to such substitution.

Polymorphism

Polymorphism is the capability of a substance to exist in more than one crystal form. The basic controlling factors appear to be temperature-pressure conditions at the time of formation, which controls the type of atomic packing within the structure. Examples of polymorphism are given in Table 2.

Pseudomorphism

Pseudomorphism is the change of the original chemical composition of a substance into some other equally definite compound by the action of natural agencies. Pseudomorphism exists when the external crystalline form of a mineral is inconsistent with its internal chemical composition and atomic structural arrangement. It is always a secondary process. The altered substance is known as a *pseudomorph*.

Types of pseudomorphic alteration include:

Substitution. This is a process whereby silica replaces wood fiber to form silicified (petrified) wood. Quartz replacing fluorite is another example. In the latter instance, the original fluorine in the fluorite is removed by silica-rich solutions that first remove the fluorine and then substitute silica in its place.

Incrustation. This is a process whereby one mineral forms a crust over another mineral, e.g., prehnite over anhydrite crystals. Later solutions may remove the anhydrite, but the space occupied by the anhydrite remains as a cast surrounded by the prehnite.

Alteration results from the addition of new material or partial removal of original material from a mineral, e.g., anglesite after galena; or gypsum after anhydrite.

Paramorphism results when the internal crystal structure of a mineral is changed to a polymorphous form, yet retaining the external crystal form of the original mineral, e.g., aragonite after calcite.

TABLE 2. REPRESENTATIVE POLYMORPHIC FORMS

Chemical composition	Mineral	Crystal system	Specific gravity
Carbon	Diamond	Isometric	3.5
Carbon	Graphite	Hexagonal	2.2
Al_2SiO_5	Sillimanite	Orthorhombic	3.2
Al_2SiO_5	Kyanite	Triclinic	3.7
FeS_2	Pyrite	Isometric	5.0
FeS_2	Marcasite	Orthorhombic	4.9

Twinning

Twinning in crystals results from the intergrowth of two or more individuals in such a way as to yield parallelism in the case of certain parts of the different individuals and, at the same time, other parts of the different individuals are in reverse positions in respect to each other. For example, an octahedral crystal of magnetite is twinned when one-half of the crystal is rotated 180° parallel to an octahedral facial plane. This type of twinning is known as *spinel twinning*, owing to its common occurrence in the spinel group of minerals.

Mineral aggregates are often grouped together to form compound crystal structures. Such groupings may be a product of irregular and accidental growth which do not conform to basic twinning laws. Two or more intergrown crystals should not arbitrarily be labeled as twins unless their twin relationship can be established.

Rock Types, Associated Minerals and Their Uses

Minerals are the basic building blocks of all rock types found in the earth's crust. Rocks are classified into three broad categories: *igneous, sedimentary*, and *metamorphic*.

Igneous rocks have their origin deep within the earth's interior from molten magmas of siliceous (silica-rich) melts. The rocks resulting from deep-seated solidification are known as *plutonic*; while those that have been extruded on the surface as lava flows are termed *extrusives*. Plutonic rocks are products of slow cooling and possess well-formed crystalline structure, e.g., granite. Extrusive rocks are products of fast cooling and occur as glassy or microcrystalline masses, e.g., obsidian and basalt.

Minerals occurring in igneous rocks possess crystalline character, but their rate of precipitation from the parent melt prevented their development as euhedral crystals and thus they occur as granular (anhedral) aggregates within these rocks. The more prominent component minerals include quartz, feldspar, and feldspathoid family members, micas, pyroxenes, amphiboles, and olivine. Zircon, magnetite, ilmenite, hematite, apatite, pyrite, and garnet are commonly associated in these rock types.

Pegmatites represent a residual phase of igneous depositions, characterized by extremely coarse crystalline material, that results from the presence of associated volatiles, e.g., water vapor, carbon dioxide, sulfur dioxide, and others, which decrease the viscosity and facilitate crystallization. Quartz, feldspar, and mica are the more common minerals found in this environment, but such bodies are also hosts for many rare minerals and several types of gem stones, e.g., beryl, tourmaline, and topaz.

Sedimentary rocks are products of deposition from either the mechanical or chemical breakdown of all preexisting rock types. Precipitation from aqueous bodies rich in soluble salts produces economically valuable beds of halite and gypsum. Weathering of iron-bearing rocks has produced extensive deposits of hematite. Limestone, chalk, and diatomite are products of biochemical precipitation. Minerals commonly associated with sedimentary rocks include calcite, galena, sphalerite, pyrite, marcasite, fluorite, barite, celestite, and quartz.

Metamorphic rocks are those that have undergone a reconstitution or redistribution of the chemical elements contained in the original formations to new mineral species. The process of change involves attaining a state of equilibrium of the constituent elements with the newly imposed environment. Chlorite, biotite, garnet, staurolite, andalusite, kyanite, and sillimanite, with ubiquitous quartz, are a few of the more common minerals associated with metamorphic rocks.

Minerals of Primary Industrial and Economic Importance

Economically valuable mineral occurrences are the products of particular types of worldwide geological formations that are, or have been, hosts to such minerals. In each occurrence, the minerals found represent the products of geological processes, which include:

1. The character and concentration of the mineral components within the source magmas from which the minerals were formed.

2. Secondary deposition from either percolating solutions or gaseous emanations from intrusive formations, causing chemical reactions within the intruded formations.

3. Precipitation from chemically supersaturated solutions.

4. Alluvial deposits resulting from the erosion weathering of the original host rocks.

Minerals that are a product of the first type of formation include the basic native metals, gold, silver, and copper.

Gold is primarily a product of deposition from ascending hydrothermal solutions associated genetically with siliceous-rich igneous rocks. Pyrite and other sulfide minerals are common associates within which the gold is often physically admixed. Surficial weathering of such deposits removes the sulfides, leaving free gold as a residual deposit. Erosion of these deposits results in alluvial deposits of placer gold, both as flakes and nuggets. Other characteristics and the uses for gold are described under **Gold**.

Silver occurs both as native ore and in combination with various silver sulfide minerals. Native silver is predominantly a product of primary deposition from hydrothermal solutions. Minor occurrences are products of oxidation of silver sulfide minerals with which the native ore is secondarily associated. See also **Silver**.

Native copper commonly occurs in the oxidized zones of copper deposits in association with cuprite, malachite, and azurite. The native copper deposits on the Michigan Keeweenaw Peninsula represent an exceptional occurrence. The copper occurs there as veins within igneous trap rocks interbedded with conglomerates. See also **Copper**.

Similar and valuable minerals include cinnabar (mercury); antimony (for type metal and battery plates); galena (lead and silver); argentite, pyrargyrite, and proustite (silver); sphalerite (zinc); chalcocite, chalcopyrite, bornite, malachite, azurite, and cuprite (copper); nickeline and pentlandite (nickel); bauxite (aluminum); magnetite, hematite, and goethite (iron).

Diamonds were found originally as loose crystals in geologically ancient alluvial stream beds. Later, their host formations were found to be a basic igneous rock (kimberlite) in the Republic of South Africa. Diamonds are the products of extremely high-temperature, high-pressure environment and are composed of pure carbon. See also **Diamond**.

Trap rocks (basalts) are products of volcanic action, either as extensive lava flows, or as intrusive dikes in preexisting rocks. Secondary mineralization within such rocks from circulating waters produces interesting suites of zeolitic minerals, such as analcime, heulandite, natrolite, stilbite, mesolite, and others.

These minerals possess the ability to exchange ions contained in the mineral structure for those in solutions. This facility promotes the use of zeolitic minerals (or their synthetic counterparts) as water softeners. Water rich in calcium (hard water), when passed in solution through a tank containing zeolites, loses the calcium ions by absorption in the zeolite structure, with substitution of calcium ions by sodium ions. A reverse process may be initiated with the sodium ions replacing calcium ions in the structure, thereby reconstituting the original zeolite composition. See also **Zeolite Group**.

Halite, gypsum, and anhydrite are products of precipitation from large bodies of supersaturated salt water. The salt and gypsum of commerce are derived from such deposits. See also **Gypsum**; and **Sodium Chloride**.

Landlocked inland seas and lakes become enriched with various soluble elements from waters draining into those basins. Sylvite and carnallite are valuable for their potassium content. They represent the final evaporation products of landlocked bodies of supersaturated sea water. Two famous localities are Stassfurt, Germany and near Carlsbad, New Mexico. Minerals formed from the evaporation of boron-rich waters include borax, kernite, colemanite, and ulexite. See also separate alphabetical entries for these minerals. The only known locality for kernite is the Mohave Desert in California, where a deposit of great extent exists—with potential reserves of millions of tons beneath the desert floor.

Pegmatites are valuable mineral sources. These formations represent the residual phase of igneous crystallization from magmas rich in siliceous content. As crystallization of their component elements proceeds, these magmas become increasingly enriched with volatile substances (mineralizers), such as water vapor, carbon dioxide, chlorine, fluorine, phosphorus, and others. The volatiles reduce the viscosity of the residual magmas and facilitate crystallization, as previously mentioned. When the residual liquids are injected into cooler rocks, they crystallize from their peripheral borders inward. The great mobility of the constituents enhances the growth of large mineral crystals—a characteristic feature of pegmatite bodies. Beryl and spodumene crystals from pegmatites attain sizes in terms of feet and tons. Feldspar, quartz, and mica crystals of comparable character are not uncommon.

There are two genetic pegmatite types—*simple* and *complex*. Simple pegmatites are recognized by their coarse texture and normal granite components, e.g., quartz, feldspar, and mica. Pegmatites produce the feldspar of commerce and mica for industrial and commercial uses. See also **Feldspar**; and **Mica**. Complex pegmatites are characterized by the presence of rare elements, in addition to the normal feldspar, quartz, and mica. Such bodies are also hosts for many semiprecious gem stones, such as amethyst, rose quartz, topaz, tourmaline, beryl, and chrysoberyl. Many rare-earth minerals obtained from complex pegmatite minerals include columbite/tantalite (columbium or niobium; tantalum), lepidolite, triphylite, spodumene, amblygonite (lithium), zircon (zirconium), and monazite (thorium oxide).

A most unusual pegmatite occurs near Ivigtut, Greenland. This consists of a cryolite with subordinate siderite, chalcopyrite, galena, and sphalerite. Cryolite is a fluoride of sodium and aluminum. For many years, cryolite was mined from this single occurrence for use as a flux in the electrolytic recovery of aluminum from bauxite, the major ore source of aluminum. Synthetic sodium aluminum fluoride essentially has replaced the need for natural cryolite. See also **Aluminum**; **Bauxite**; and **Cryolite**.

Quartz and tourmaline crystals once were commercially important for their piezoelectric properties as radio oscillation wafers and other electronic and instrumental uses. Synthesized quartz crystals have largely replaced the need for natural quartz for such applications.

Nuclear fission reactors are supplied with materials from uranium-bearing minerals of primary origin, e.g., uraninite/pitchblende, and other uranium-bearing minerals of secondary origin, e.g., carnotite, tyuyamunite, torbernite, and autunite.

Minerals of economic importance within sedimentary formations include, but are not limited to fluorite, barite, phosphorite, and oolitic hematite. Fluorite is utilized as a flux in steelmaking and when of high quality as lenses and prisms in the optical industry. Barite is an essential mineral used in gas- and oil-well drilling. Phosphorite, a product of chemical precipitation from seawater, when treated with sulfuric acid, produces superphosphate fertilizer. Oolitic hematite deposits of extensive size are important sources of iron ore.

Garnet is a common mineral component in metamorphic rocks. A major occurrence of this type is at the summit of Gore Mountain, North River, in Warren County, New York. The garnet is a composite of almandine and pyrope with a hardness exceeding that of most world garnets. Gore Mountain garnet retains sharp cutting edges even when crushed to sub-micron size, making it an outstanding abrasive. It is used extensively as an abrasive (garnet paper) and as a glass-polishing agent in the optical industry.

Titaniferous iron ores, represented by the mineral ilmenite, occur within crystalline metamorphic environments. These ores are the major source of titanium. See also **Titanium**.

Major sources of industrial (non-fuel) minerals are shown in Table 3.

Ocean Sources of Minerals

Mineral requirements for future world needs has focused increased attention to potential ocean resources. Major attention has been directed to petroleum resources. Associated with the petroleum are salt domes or bedded salt deposits, often with anhydrite and sulfur. Their potential is dependent upon development of economically feasible recovery methods. Not enough is known at this time about the origin of sulfur to satisfactorily predict precise occurrences. The Frasch process is presently being utilized in certain offshore deposits of this type. The economics of sulfur also may be affected by availability of large quantities of the element from Claus recovery units used in connection with the desulfurization of flue gases.

Valuable deposits of detrital sands and lime muds occur on the continental shelves of many world areas. These can be recovered by dredging operations. Diamonds are presently being recovered by means of vacuum suction tubes from detrital subsea sands adjacent to the Orange River section off the south-western African coast. Dependent upon the nearshore geology, it is known that iron, copper, and coal deposits extend into the subsea areas. In several world areas (Scotland and Japan for coal; Finland and Canada for iron ore; the English coast for tin and copper) deposits have been mined from underground entrances from the adjacent land areas. Sphene and zircon, plus other heavy minerals, have been noted in Texas offshore sediments.

Phosphorite, a major source of phosphorus, is known to occur both as nodular masses and crusts on rocks in subsea areas. Although enormous amounts of phosphorite are accessible in relatively shallow water, marine phosphorites have not been economically competitive with terrestrial supplies.

Metallic sulfides of copper, zinc, and iron have been found in central oceanic rocks and muds under conditions that indicate their deposition

TABLE 3. MAJOR SOURCES OF INDUSTRIAL (NONFUEL) MINERALS

ALUMINUM
Major source is bauxite (gibbsite and boehmite). Deposits are found worldwide, except in Antarctica. Major producers: Australia: Caribbean countries; Venezuela; Brazil; Indonesia.

CHROMIUM
Nearly all chromium ores are found in the Republic of South Africa and Zimbabwe.

COBALT
Zaire, Zambia, and Russia.

COPPER
Chile, United States, Russia, Canada, Zambia, and Zaire.

GOLD
Republic of South Africa: Russia, Brazil, and the United States (not extensive).

IRON
Russia, the United States, Brazil, Australia, and China.

LEAD
United States, Russia, Australia, and Canada

MANGANESE
United States, Japan, and Western Europe. (Major reserves are in South Africa and Russia)

NIOBIUM (Columbium)
Brazil and Canada

NICKEL
Russia, Canada, Australia, and Indonesia.

PLATINUM-GROUP METALS
Russia and Republic of South Africa.

SILVER
United States. (Silver is mined in more than 53 countries.)

TANTALUM
Thailand, Australia, and Brazil.

TITANIUM
Russia, Japan, United States

ZINC
Canada, Russia, Australia, Peru, and United States.

Source: U.S. Bureau of mines.

from hydrothermal solutions. Such solutions, rich in carbon dioxide, leached metallic elements from both basic rock masses and sedimentary formations with which they came in contact. When such solutions ascended, with concomitant cooling, the minerals were precipitated in the overlying sediments.

Manganese and iron oxides occur as nodular masses in many subsea world areas. They presently are of more interest for their copper, nickel, and cobalt content than for the manganese. Most extensive occurrences are at great ocean depths, as much as 3,500 to 4,500 meters. Fullest exploitation of these deposits and the metallic sulfides will require not only additional technical knowledge for their initial recovery, but also for their refinement to a marketable form. Again, the economics depends upon future demands for the metals as the continental deposits become depleted. Beyond these considerations, the persistent problem of ownership of oceanic resources must be solved as a condition of large-scale recovery. See also **Manganese**.

Lunar Rocks

Geological specimens collected on the *Apollo* lunar missions are indicative of an anhydrous igneous origin. There are three major rock types: (1) a potassium-rich basalt; (2) anorthosite; and (3) an iron, titanium-rich basalt. The first two types are prevalent in the highland areas; the latter in the maria terrain. They occur as crystalline vescicular masses, breccias, and regolithic mantle dust. The absence of an atmosphere and weathering processes on the moon has left the rocks and their component minerals unchanged through eons of time since their formation. Secondary mineralization, therefore, is generally absent and the rocks exhibit a rather limited mineralogy.

Lunar rocks differ in their chemical content rather than type of rock from their terrestrial counterparts. They consistently contain more titanium

and chromium, less sodium, and most are richer in iron content. Lunar plagioclase, the major mineral component of anorthosite, is almost always the calcium-rich anorthite, $CaAl_2Si_2O_8$, indicating extreme magmatic differentiation. Lunar basalts are olivine-rich and have been found to be from 3 to 10 times richer in ilmenite (titaniferous iron) as compared with terrestrial basalts.

Clinopyroxene materials, which are common in terrestrial basalts, are well represented in lunar rocks. They include diopside, hedenbergite, johannsenite, aegerine-augite, spodumene, jadeite, augite, pigeonite, omphacite, and fassaite. The more prominent lunar mineral species noted include ilmenite, with rutile intergrowths in certain subfloor maria basalts, cristobalite/tridymite, and pyroxferroite, a mineral closely related in both structure and composition to terrestrial pyroxmangite. Accessory minor minerals include troilite, chromite, ulvospinel, apatite-whitlockite, potash feldspar, quartz, hafnium-rich baddeleyite, and perovskite. Two newly classified species have been recorded—armalcolite, an iron-magnesium titanate, named after *Apollo* astronauts *Arm* strong, *Al* drin, and *Col* lins; and zirkelite, an oxide of calcium, iron, thorium, uranium, with titanium, niobium, and zirconium. Euhedral iron crystals in a pyroxene-rich vug of recrystallized breccia were recovered on the *Apollo 15* mission.

Tiny translucent-to-opaque glassy spherules are prominent in the lunar regolith, within which ilmenite (as thin plates) with minor olivine are present.

Lunar mineralogy is generally analogous to that of terrestrial basalts, the major difference being the lack of oxygen during crystallization which has resulted in the presence of free iron and the exotic minerals, such as troilite, pyroxferroite, armalcolite, and zirkelite. The bulk mineralogy, however, is quite similar to terrestrial rocks with pyroxene, plagioclase, ilmenite, and olivine as the dominant minerals.

Phase equilibrium research of mineral solids has revealed vital information regarding their molecular structure. Application of knowledge gained from this research has extended into the fields of metallurgy, glass and ceramics, and a more adequate interpretation of mineral geology.

Mariner 9 fly-by of Mars revealed a surface terrain of massive blocks of tumbled character cut by ridges and graben-type troughs. Huge volcanic peaks dominate a pock-marked landscape. Extensive channels characteristic of concentrated erosive powers of torrential floods were also evident, as were braided stream systems emanating from what were resolved to be plateau-type elevations.

Viking spacecraft equipped with a hoe-type scoop and spectrometer analyzed the surface soil to be of a character suggestive of an igneous mafic rock origin, rich in magnesium and iron. The *Viking Landers'* spacecraft analysis of the surface soil chemical composition by x-ray fluorescence revealed a low SiO_2 concentration (~45%) with iron as Fe_2O_3 near 20%. Further analysis revealed that the regolith mantle soil consisted essentially of iron-rich clay mineral with iron hydroxide, and minerals of sulfate and carbonate content with approximately 1% water by weight. Magnets attached to *Viking's* scoop attracted magnetic material aggregates. It is quite probable that the magnetic material represents a component part of that regolith and possibly the soil is enriched by both magnetite (Fe_3O_4, color black) and maghemite (γ-Fe_3O_4, color yellowish-brown). The yellowish-brown surface color may be the product of thin coating of hydrated iron oxides, with nontronite/montmorillionite as the host soil.

Much remains to be resolved before final definitive answers can be given in this area of planetary investigation and evaluation.

Classification of Minerals

Minerals are classified in groups, according to their chemical composition, based upon the dominant anion or anionic group. The system works well in various ways. Generally, the dominant anion or anionic group brackets minerals of corresponding characteristics which tend to occur in quite similar environments. The dominant chemical subdivisions are:

Elements. Minerals composed of uncombined chemical elements, e.g., Au, gold; Ag, silver; Cu, copper; although minor impurities may be present within the structure.

Sulfides. Minerals composed of compounds of metals with sulfur.

Sulfosalts. Minerals composed of compounds of semimetals with sulfur.

Halides. Minerals composed of compounds of metals with fluorine, chlorine, bromine, and iodine.

Oxides and Hydroxides. Minerals composed of compounds of the metallic elements with oxygen.

Carbonates. Minerals composed of compounds of a metal with the carbonate radical CO_3.

Borates. Minerals composed of compounds of a metal with the borate radical, BO_3.

Nitrates. Minerals composed of compounds of a metal with the nitrate radical, NO_3.

Sulfates. Minerals containing the sulfate radical, SO_4.

Chromates. Minerals containing the chromate radical, CrO_4.

Molybdates. Minerals containing the molybdate radical, MoO_4.

Tungstates. Minerals containing the tungstate radical, WO_4.

Phosphates. Minerals containing the phosphate radical, PO_4.

Arsenates. Minerals containing the arsenate radical, AsO_4.

Vanadates. Minerals containing the vanadate radical, VO_4.

Silicates. This mineral classification encompasses the largest group of mineral species and includes most of the important rock-forming minerals, such as the feldspars, feldspathoids, pyroxenes, amphiboles,

---

TABLE 4. MINERALS DESCRIBED IN THIS ENCYCLOPEDIA

ARSENATES	OXIDES AND	SILICATES (Inosilicates)	SILICATES (Sorosilicates)	Calaverite
Annabergite	HYDROXIDES	Acmite-Aegerine	Allanite	Chalcopyrite
Erythrite	Alabandite	Actionolite	Clinozoisite	Chalcopyrite
Mimetite	Alexandrite	Aegerine	Epidote	Cinnabar
Scorodite	Anatase	Amphibole	Hemimorphite	Cobaltite
	Bauxite	Anthrophyllite	Lawsonite	Covellite
BORATES	Brookite	Augite	Prehnite	Galena
Boracite	Brucite	Babingtonite	Vesuvianite	Gersdorffite
Borax	Cassiterite	Bustamite	Zoisite	Greenockite
Colemanite	Cat's-Eye	Crocidolite		Hessite
Inyoite	Chromite	Cummingtonite	SILICATES (Tectosilicates)	Krennerite
Kernite	Chrysoberyl	Diallage	Agate	Mercasite
Ulexite	Columbite	Diopside	Amethyst	Millierite
	Corundum	Enstatite	Analcime	Molybdenite
CARBONATES	Cuprite	Glaucophane	Bloodstone	Nickeline
Aragonite	Diaspore	Hornblende	Cairngorm Stone	Orpiment
Azurite	Emery	Hypersthene	Cancrinite	Pentlandite
Barytocalcite	Fergusonite	Jade	Carnelian	Pyrite
Bastnasite	Franklinite	Jadeite	Chabazite	Pyrrhotite
Calcite	Gahnite-Zinc Spinel	Pyroxene	Chalcedony	Realgar
Cerussite	Geikielite	Rhodonite	Citrine	Skutterudite
Chalk	Goethite	Riebeckite	Danburite	Sperrylite
Dolomite	Hematite	Serandite	Desmine	Sphalerite (Biende)
Magnesite	Ilmenite	Spodumene	Feldspar	Stannite (Mineral)
Malachite	Limonite	Tremolite	Flint	Stibnite
Phosgenite	Magnetite	Uralite	Harmotome	Sylvanite
Rhodochrosite	Manganite	Wollastonite	Heulandite	Tetradymite
Siderite	Perovskite		Hyalite	Wurtzite
Smithsonite	Pitchblende	SILICATES (Nesosilicates)	Jasper	
Strontianite	Psilomelane	Andalusite	Lazurite	SULFOSALTS
Travertine	Pyrolusite	Chrondrodite	Leucite	Boulangerite
Witherite	Pyrophanite	Datolite	Natrolite	Bournonite
	Rutile	Dumortierite	Nepheline	Enargite
CHROMATES	Spinel	Fayalite	Opal	Geocronite
Crocoite	Tantalite	Forsterite	Perthite	Jamesonite
ELEMENTS	Tenorite	Garnet	Petalite	Polybasite
Amalgam	Thorianite	Kyanite	Phillipsite	Poustite
Antimony	Uraninite	Olivine	Pollucite	Pyrargyrite
Bismuth	Wad	Phenacite	Quartz	Stephanite
Copper	Zincite	Silimanite	Scolecite	Tetrahedrite
Diamond		Sphene	Sodalite	
Electrum	PHOSPHATES	Staurolite	Stilbite	TUNGSTATES
Gold	Amblygonite	Thorite	Tridymite	Scheelite
Graphite	Apatite	Topaz	Wernerite	Wolframite
Mercury	Autunite	Willemite	Zeolite Group	
Platinum	Lazulite	Zircon		VANADATES
Plumbago	Monazite		SULFATES	Carnotite
Quicksilver	Pyromorphite	SILICATES (Phyllosilicates)	Alabaster	Tyuyamunite
Silver	Torbernite		Alunite	Vanadinite
Sulfur	Triphylite	Apophyllite	Anglesite	
	Turquois	Asbestos	Anhydrite	OTHER MINERALOGICAL
HALIDES	Vivianite	Biotite	Antlerite	TERMS
Atacamite	Wavelite	Chlorite	Barite	Abrasion pH
Carnallite		Chloritoid	Brochantite	Carbonado
Chlorargyrite	SILICATES (Cyclosilicates)	Chrysotile	Celestite	Diamond
Cryolite	Axinite	Garnierite	Chalcanthite	Diatomite
Fluorite	Beryl	Glauconite	Epsomite	Clay
Halite	Chrysocolla	Kaolinite	Glauberite	Fuller's Earth
Sylvite	Cordierite	Lepidolite	Gypsum	Gangue
	Dioptase	Mica	Jarosite	Gem Stones
MOLYBDATES	Dravite	Muscovite	Polyhalite	Kimberlite
Wulfenite	Elbaite	Phlogopite		Peridotite
	Emerald	Pyrophyllite	SULFIDES	Tripolite
NITRATES	Euclase	Sepiolite	Argentite	Vitrophyre
Niter	Iolite	Serpentine	Arsenopyrite	
Soda-Niter	Liddicoatite	Talc	Bismuthinite	
	Tourmaline		Bornite	

micas, olivine, and quartz. Silicon is the basic chemical element, as the name implies. The small silicon cation combines with four oxygens to form an SiO_4 tetrahedral structure. The SiO_4 formula leaves a net negative charge, which requires additional combinations with other tetrahedra or anions to effect a neutral balance. The type and degree of such tetrahedral combinations control the final structural character and act as a convenient classification of the silicate mineral family.

Subclassification of silicates are:

Nesosilicates, with each tetrahedron existing within the structure as isolated SiO_4 units.

Sorosilicates involve the pairing of SiO_4 tetrahedra. The shared-oxygen anion represents the link between these tetrahedra.

Cyclosilicates involve two oxygens from each SiO_4 tetrahedron combining with oxygen in adjacent tetrahedral units to form *ring* structures.

Inosilicates are the product of oxygen sharing between adjacent tetrahedra to form *single* or *double* chains. In the single-chain structure, two oxygens from each tetrahedra combine with adjacent tetrahedra. In the double chain, half of the tetrahedra share three oxygens, while the other half share only two.

Phyllosilicates involve the sharing of three oxygens in each tetrahedron with adjacent tetrahedrons to form *sheet* structures. Minerals in this classification are usually flaky in character and relatively soft.

Tectosilicates involve the sharing of all four oxygens in each tetrahedral unit with adjacent tetrahedrons to form a three-dimensional *framework* of SiO_4 units linked together. The product is a strongly bonded structure with a silicon-oxygen ratio of 1:2. The greater portion of the earth's crust is composed of minerals found within this classification.

Most minerals exhibit a variation in their chemical composition, with the exception of the elements (see preceding list). The substitution of one ion for another is common, since minerals crystallize in solutions of complex composition.

A full listing of all minerals described separately in this encyclopedia is given in Table 4.

ELMER B. ROWLEY
F.M.S.A., Union College
Schenectady, New York

Additional Reading

Arem, J.E.: *Color Encyclopedia of Gemstones,* 3rd Edition, Chapman & Hall, New York, NY, 1994.

Boyd, F.R. and J.J. Gurney: "Diamonds and the African Lithosphere," *Science,* **232,** 472–477 (1986).

Boyle, R.W.: *Gold: History and Genesis of Deposits,* Chapman & Hall, New York, NY, 1990.

Brierley, C.L.: "Microbiological Mining," *Sci. Amer.,* 44–53 (August 1982).

Brown, W.L., Editor: *Feldspars and Feldspathoids,* Reidel, Boston, 1984.

Campbell, A.N. et al.: "Recognition of a Hidden Mineral Deposit by an Artificial Intelligence Program," *Science,* **217,** 927–929 (1982).

Carmichael, R.S. and S. Robert: *Practical Handbook of Physical Properties of Rocks and Minerals,* CRC Press, LLC., Boca Raton, FL, 1990.

Cornelis Klein, C. and C.S. Hurlbut: *Manual of Mineralogy,* 21st Edition, John Wiley & Sons, Inc., New York, NY, 1998.

Crowson, P.: *Minerals Handbook: 1996–1997,* Groves Dictionaries, Inc., New York, NY, 1996.

Derry, D.R.: *A Concise World Atlas of Geology and Mineral Deposits,* John Wiley & Sons, Inc., New York, NY, 1981.

Dietrich, R.V., B.J. Skinner, and R. Vincent: Cambridge University Press *Gems, Granites, and Gravels: Knowing and Using Rocks and Minerals,* Cambridge University Press, New York, NY, 1990.

Glusker, J.P.: *Structural Crystallography in Chemistry and Biology,* John Wiley & Sons, Inc., New York, NY, 1982.

Goeller, H.E. and A. Zucker: "Infinite Resources: The Ultimate Strategy," *Science,* **223,** 456–462 (1984).

Golden, D.C., C.C. Chen, and J.B. Dixon: "Synthesis of Todorokite," *Science,* **231,** 717–719 (1986).

Hein, J.R.: *Siliceous Sedimentary Rock-Hosted Ores and Petroleum,* John Wiley & Sons, Inc., New York, NY, 1987.

Holland, H.D. and M. Schidlowski: *Mineral Deposits and the Evolution of the Biosphere,* Springer-Verlag, Inc., New York, NY, 1982.

Kahle, A.B. and A.F.H. Goetz: "Mineralogic Information from a New Airborne Thermal Infrared Multispectral Scanner," *Science,* **222,** 24–27 (1983).

Kelly, E.G. and D.J. Spottiswood: *Introduction to Mineral Processing,* John Wiley & Sons, Inc., New York, NY, 1982.

Lide, D.R.: *CRC Handbook of Chemistry and Physics* 84th Edition, CRC Press, LLC., Boca Raton, FL, 2003.

Meyer, C.: "Ore Metals Through Geologic History," *Science,* **227,** 1421–1428 (1985).

Nancollas, G.H.: *Biological Mineralization and Demineralization,* Springer-Verlag, Inc., New York, NY, 1982.

Nesse, W.D.D.: *Introduction to Mineralogy,* Oxford University Press, Inc., New York, NY, 1999.

Newton, R.C., A. Navrotsky, and B.J. Wood: *Thermodynamics of Minerals and Melts,* Springer-Verlag, Inc., New York, NY, 1981.

O'Reilly, W.: *Rock and Mineral Magnetism,* Chapman and Hall, New York, NY, 1984.

Ozima, M. and S. Zashu: "Primitive Helium in Diamonds," *Science,* **219,** 1067–1068 (1983).

Park, C.F.: *The Geology of Ore Deposits,* 4th Edition, W.H. Freeman Company, New York, NY, 1998.

Parker, S.P.: *McGraw-Hill Dictionary of Geology and Mineralogy,* The McGraw-Hill Companies, Inc., New York, NY, 1997.

Robinson, E.S.: *Basic Physical Geology,* 3rd Edition, John Wiley & Sons, Inc., New York, NY, 1991.

Rona, P.A.: "Mineral Deposits from Sea-Floor Hot Springs," *Sci. Amer.,* 84–92 (January 1986).

Sawkins, F.J.: *Metal Deposits in Relation to Plate Tectonics,* 2nd Edition, Springer-Verlag, Inc., New York, NY, 1989.

Sohn, H.Y. et al.: *Processing of Energy and Metallic Minerals,* American Inst. of Chemical Engineers, New York, NY, 1982.

Swanson, E.A., D.F. Strong, and J.G. Thurlow: *The Buchans Orebodies,* Geological Association of Canada, Toronto, Ontario, 1981.

Touloukian, Y.S. et al.: *Physical Properties of Rocks and Minerals,* Vol. 2, The McGraw-Hill Companies, Inc., New York, NY, 1989.

Wills, B.A.: *Mineral Processing Technology: An Introduction to the Practical Aspects of Ore Treatment and Mineral Recovery,* Butterworth-Heinemann, Inc., Woburn, MA, 1997.

Web Reference

Mineralogical Society of America: http://www.minsocam.org/

MIRROR NUCLIDES. Pairs of nuclides, having their numbers of protons and neutrons so related that each member of the pair would be transformed into the other by exchanging all neutrons for protons and vice versa.

MISCIBILITY. The ability of two or more substances to mix, and to form a single, homogeneous phase.

MISSLE PROPELLANTS. See **Rocket Propellants**.

MITCHELL, PETER D. (1920–1992). A British biochemist who was the recipient of the Nobel prize for chemistry in 1978 for his contribution to the understanding of biological energy transfer through the formulation of the chemiosmotic theory.studies of cellular energy transfer. A graduate of Cambridge and recipient of many awards, he was the Director of Research, Glynn Research Institute.

MITSUNOBU REACTION. Intermolecular dehydration reaction occurring between alcohols and acidic components on treatment with diethyl azodicarboxylate and triphenyl phosphine under mild neutral conditions. The reaction exhibits stereospecificity and regional and functional selectivity.

MIXING AND BLENDING. These operations are important to chemical research and processing. In exploratory work in the laboratory the effects of mixing may be very great, and it is essential that the desired type and amount of mixing can be reproduced or can be varied by known amounts. The primary purpose of mixing is to distribute components as uniformly as possible; temperature distribution is frequently a major purpose. These may be followed by a chemical reaction or a transfer of matter between phases, and by a transfer of heat for temperature control. The mixer produces mechanical effects only. Molecules of themselves will diffuse, but mixing impellers produce flow which results in forced convection and mixing. Hence, reactants can be brought to an interface as rapidly as desired by controlling the fluid motion. Most fluid mixing is done by rotating impellers.

Both large scale (mass flow) motion and small scale (turbulent) motion are ordinarily required to bring about rapid mixing. The discharge stream from an impeller initiates the large scale flow pattern. Turbulence is generated mostly by the velocity discontinuities adjacent to the stream of fluid flowing from the impeller, and also by boundary and form separation effects. Turbulence spreads throughout the mass flow and is carried to all parts of the container. Some mixing operations require relatively large mass flows for best results, whereas others require relatively large amounts of turbulents. There is usually an optimum ratio of flow to turbulence for a desired mixing operation, whether it is a simple blending of immiscible liquids or a mass transfer followed by chemical reaction.

In the research laboratory it is important to recognize the effect of mixing on reaction rate or on other performance criteria. Energy must be supplied to produce fluid motion, thus, to compare mixing with different equipment or with different sizes of the same type impeller, it is essential that the comparisons be made on the basis of equal power input.

For the same power, the ratio of flow to turbulence from mixing impellers can be varied by changing the size and speed of the impeller. Figure 1 illustrates the differences in mass flow and turbulence which can be achieved for the same power input for dimensionally similar impellers. A large-diameter low-speed impeller produces a large ratio of flow to turbulence, whereas a small-diameter high-speed impeller will give a small ratio. Curve A, Figure 2, illustrates a reaction best accomplished by large flow and small turbulence. This curve, which is typical of blending operations, shows that the rate of blending increases to a maximum with a large impeller as impeller diameter is increased (and impeller speed is decreased) with power input constant.

Curve B of Figure 2 is typical of gas-liquid contacting operations. Here the rate pf mass transfer between phases increases to a maximum at small impeller diameter and then decreases as impeller diameter is increased. The significance is that more turbulence is available with the small impeller and that turbulence is more important than flow in this operation.

In all bench-scale and pilot plant work where mixing is important, the effect of the impeller diameter-turbine diameter ratio should be determined so that the type of flow motion best suited to the operation can be found. If an optimum ratio is found, it becomes the basis for large scale design.

Mixing Vessels, Flow Patterns, and Impellers. Flow motion is dependent upon the shape and fitting of the container, the shape and position of the rotating impeller, and the physical properties of the fluid. The best mixing is usually one which produces lateral and vertical flow currents, and these currents must penetrate to all portions of the fluid;

swirling motion should be avoided. Cylindrical vessels provide the best environment for mixing.

The most useful impellers are the simple flat paddle, the marine-type propeller, and the turbine. If any of these are on a vertical shaft rotating on the center line of a cylindrical vessel, the fluid motion will be one of rotation. A vortex forms around which the liquid swirls. A minimum of turbulence and of vertical and lateral flow motion will result. very little power can be applied.

Rotary motion (and surface vortex) can always be stopped by inserting projections in the body of the fluid; when these are at the side if the tank they are called baffles, and this is the method most commonly used to obtain good mixing in large industrial equipment. The propeller with baffles will produce an axil flow pattern, Fig. 3, and the paddle and turbine will produce radial flow, Fig 4.

Motionless Mixers. Mixing of molten polymers has been a problem in the plastics industry for many years mainly because of the high viscosity

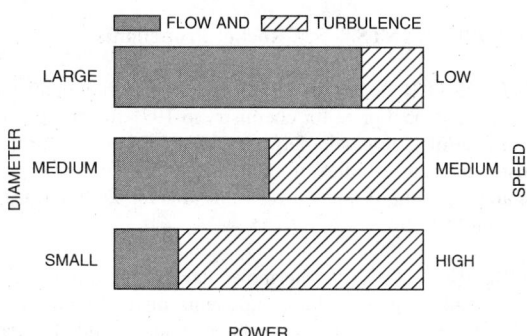

Fig. 1.   Constant power, effect of impeller size, and speed on flow and turbulence

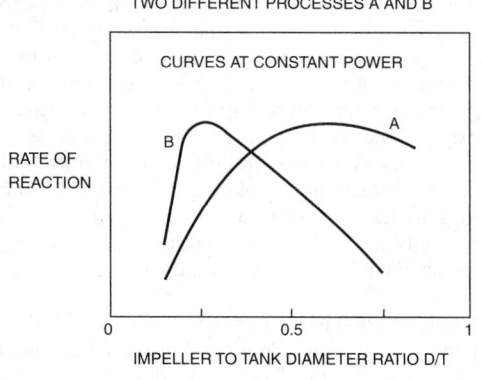

Fig. 2.   Effect of impeller size on reaction rate at equal power output

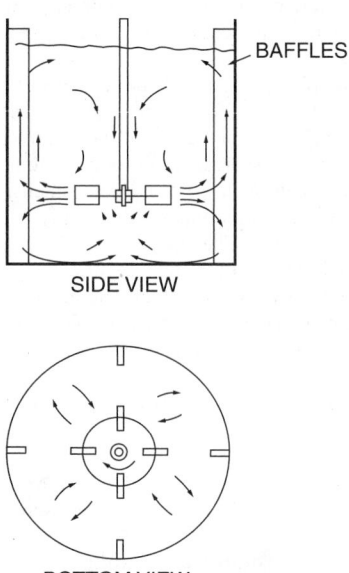

Fig. 3.   Typical flow patterns from axial flow impeller in baffled tank

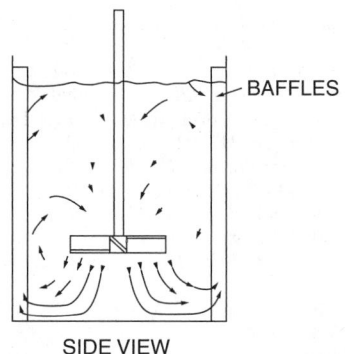

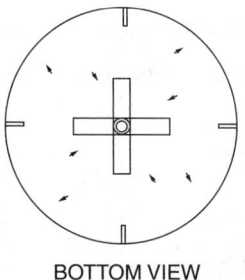

Fig. 4.   Radial flow pattern for flat blade turbine positioned on center in baffled tank

of the melt. Blending of color concentrations, fillers, stabilizers, and other additives have sometimes been difficult to achieve efficiently with traditional mixing approaches, such as the extruder screw and other rotating devices. These problems have been partially solved through the application of so-called motionless mixers. A stationaly baffle installed in a pipe can utilize the energy of the flowing fluid to produce mixing. Turbulent conditions are required for this simplistic approach. In recent years, more sophisticated motionless mixers have been developed. One design consists of a number of short right- and left-hand helices. The opposite hand helices are welded together so that their leading edges are 90 degrees to the trailing edge of the preceeding element. The mixing unit is housed in a tube or pipe for in-line installation. Materials entering the static mixer experiences flow division at the leading edge of each element. As the flow divides, it follows the semicircular channel of each element and repeatedly divides at succeeding junctions of additional elements, resulting in flow division and radial mixing. Other approaches involving complex geometric stationary elements have been developed. Some of these units also can function for heating and cooling as well and mixing. [See "Motionless Mixers in Plastic Processing," Schott, Weinstein, and LaBombard, *Chem. eng. Progress*, **71**(11) 52–58 (1975)].

Automatic Blending Systems. For a number of industrial applications, in-line blending of liquids and solids has replaced former batch-type operations. In these systems, all components flow together simultaneously to a central collection point where they combine to form the finished product. A modern in-line blending system is shown in Fig. 5. The blend controller will normally utilize microprocessor technology with a cathode-ray tube (CRT) display. Each fluid component is pumped from a storage tank, through a strainer, and then through a flowmeter, with the meter and valve carefully selected for prevailing process conditions (viscosity, temperature, pressure, flow rates). The signal from the flowmeter is fed to the blend controller which compares the actual flow rate to the desired flow rate. For minor components, such as dyes or additives, it is sometimes

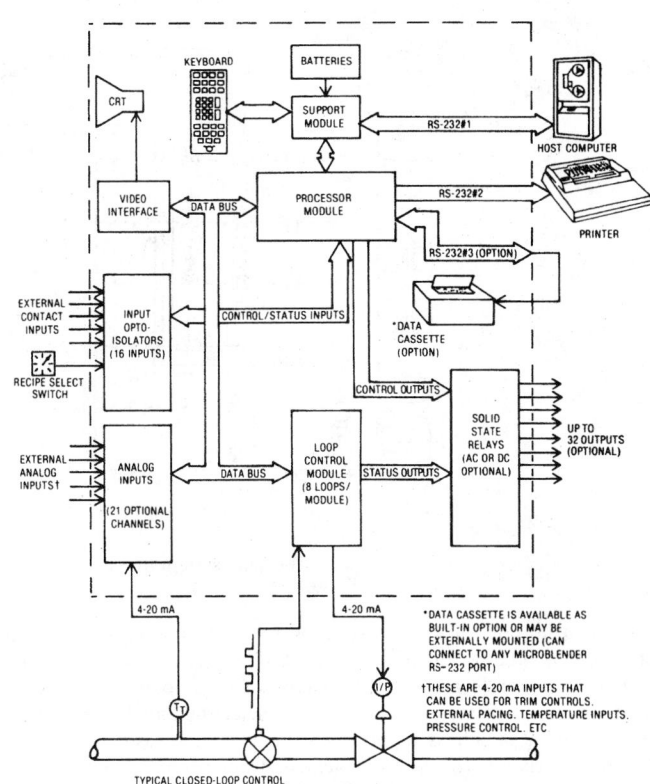

Fig. 6.   Blend controller block diagram

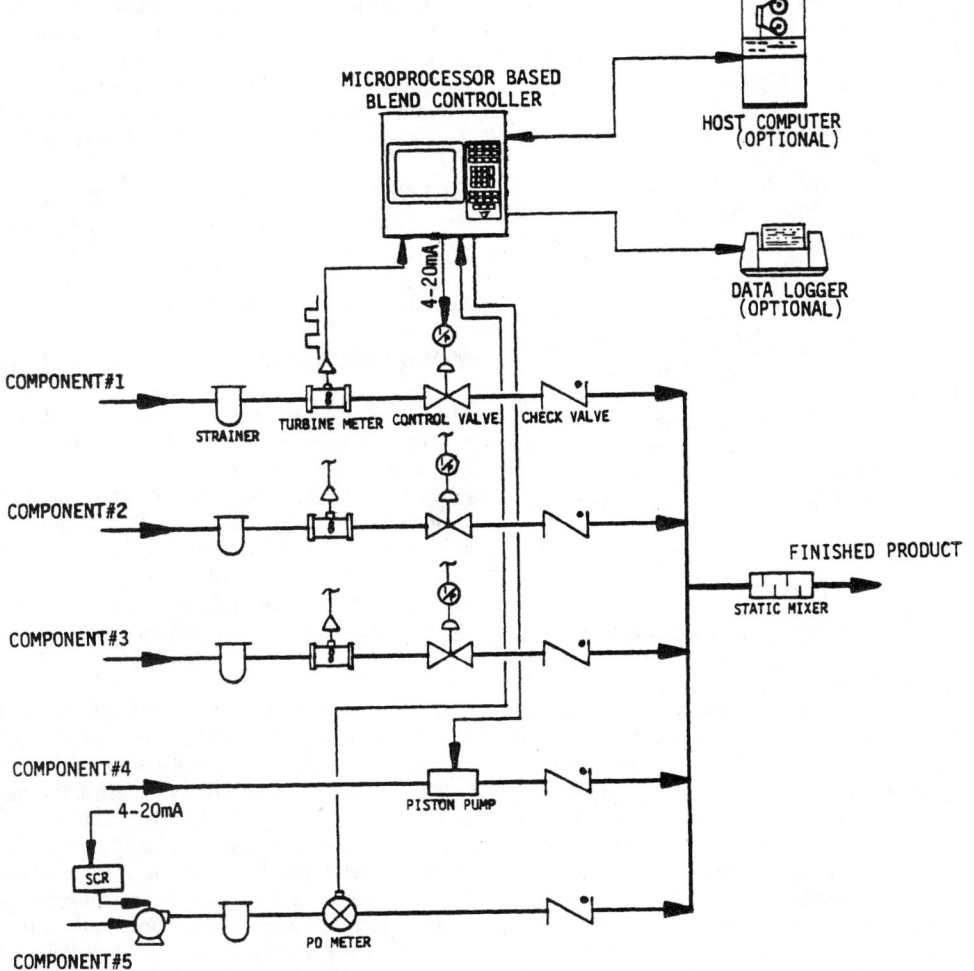

Fig. 5.   Typical blender configutaion. (*Waugh Controls Corporation.*)

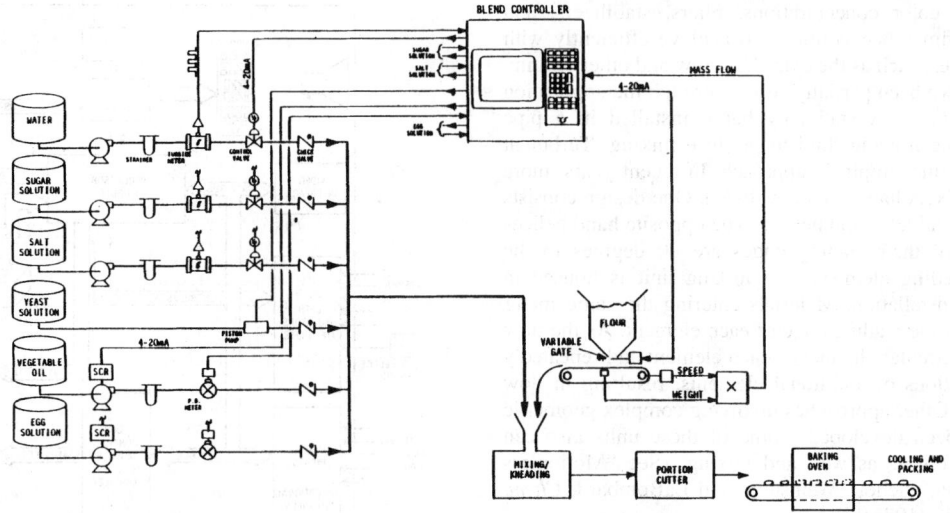

Fig. 7.   Blending system for preparing bread and partry dough. (*Waugh Controls Corporation.*)

most practical to control the flow rate by means of proportioning pumps which inject a precise amount of the fluid when a pulse signal from the blend controller is received. This type of open-loop control is cost efficient, but some means for assuring flow (detecting any dry line) should be considered inasmuch as there is no direct fluid measurement device used. Other variations of measurement and control involve the use of variable-speed pump motor controllers (SCRs) for flow control; adding a flowmeter in series with an injection pump. Weigh-belt feeders with variable-feed/speed control and tachometer/load-cell outputs are frequently used for blending powders and aggregates.

A blend controller block diagram is shown in Fig. 6. A system for preparing bread and pastry dough is shown in Fig. 7. Applications for continuous blending systems are frequently found in the petroleum, petrochemical, food and beverage, building materials, pharmaceutical, automotive, and chemical industries, among others.

MODACRYLIC FIBERS.   See **Fibers: Acrylic; Fibers**

MODERATOR.   A substance used to slow down neutrons by means of collisions. Moderators play an important role in the design and operation of nuclear reactors. Moderators *thermalize* neutrons to an energy of about 0.025 eV.

MOHS SCALE.   An empirical scale of the hardness of mineral or mineral-like materials originally consisting of 10 values, ranging from talc, with a rating of 1, to diamond, with a rating of 10. The rating is based on the ability to each material to scratch the one directly below it in the series. The number of materials has been expanded from 10 to 15 with the addition of several synthetically produced substances (e.g., silicon carbide) between the original 9 to 10 positions. The scale is named after the German mineralogist, Friedrich Mohs (1773–1839). See also **Hardness**; and **Mineralogy**.

MOIETY.   An indefinite portion of a sample.

MOISSAN, HENRI (1852–1907).   A Native of Paris, Moissan was a professor at the School of Pharmacy from 1886 to 1900 and at the Sorbonne from 1900 to 1907. At the former institution, he first isolated and liquefied fluorine in 1886 by the electrolysis of potassium acid fluoride and anhydrous hydrogen fluoride. His work with fluorine undoubtedly shortened his life as it did that of many other early experimenters in the field of fluorine chemistry. He won great fame by his development of the electric furnace and pioneered its use in the production of calcium carbide, making acetylene production and use commercially feasible in the preparation of pure metals, such as magnesium, chromium, uranium, tungsten, etc., and in the production of many new compounds, e.g., silicides, carbides, and regractories. In 1906, he was awarded the Nobel prize in chemistry.

MOISTURE RETAINING SUBSTANCES.   See **Humecants and Moisture-Retaining Agents**.

MOLAL CONCENTRATION.   A one molal solution contains one mole of a particular substance (the solute) in 1,000 grams of solvent. Thus, a 0.5 molal solution of potassium chloride in water contains $0.5 \times$ (gram-molecular weight of KCl = 74.555), or 37.278 grams of the salt in 1,000 grams of H_2O. See also **Molar Concentration**; and **Normal Concentration**.

MOLAR CONCENTRATION.   A one molar solution contains one mole of a particular substance (the solute) in 1,000 milliliters of solution. Thus, a 0.5 molar solution of potassium chloride in water will be prepared by placing $0.5 \times$ (gram-molecular weight of KCl = 74.555), or 37.278 grams of the salt in a vessel and then adding H_2O, while thoroughly mixing to assure complete solution of the salt, until a total volume of 1,000 milliliters of solution is obtained. Molar is abbreviated M. Thus, the solution in the foregoing example would be 0.5 M KCl. Molar solutions sometimes are referred to as *formal* solutions, not to be confused with normal solutions. See also **Molal Concentration**; and **Normal Concentration**.

MOLAR HEAT.   The product of the gram-molecular weight of a compound and its specific heat. The result is the heat capacity per gram-molecular weight

MOLD.   See **Yeasts and Molds**.

MOLD INHIBITORS.   See **Antimicrobial Agents (Foods)**.

MOLE (Stoichiometry).   Sometimes spelled *mol*, a mole is a quantity of a substance, expressed in specified mass units, that is equal to the molecular weight of the substance. For example, a *gram-mole* or *gram-molecular mass* of hydrogen H_2 will have a mass of $2 \times$ (atomic weight of hydrogen), or 2.016 grams. A gram-mole of carbon dioxide CO_2 will have a mass of $1 \times$ (atomic weight of carbon) plus $2 \times$ (atomic weight of oxygen), or 12.011 plus $31.998 = 44.009$ grams. A pound-mole of ammonia gas NH_3 will have a mass of $1 \times$ (atomic weight of nitrogen) plus $3 \times$ (atomic weight of hydrogen), or 17.031 pounds. See also **Avogadro Constant**.

MOLECULAR AND SUPERMOLECULAR ELECTRONICS.   Professor Gareth Roberts FRS, Director of Research, Thorn EMI plc, and Professor of Engineering Science at the University of Oxford.

The microelectronics and optoelectronics industries will continue to grow vigorously well into the 21st Century. Until now, they have relied largely on inorganic materials such as silicon and lithium niobate in single crystal form. However, as the perceived limitations inherent in these materials begin to restrict the realization of more complex system designs, more attention is being focussed on the *organic* solid state. The richness of the variety of organic molecular materials offers enormous potential compared with the relative paucity of structures achievable with inorganic compounds, even when due allowance is made for the recent exciting developments in inorganic quantum well semiconductors (Kelly and Weisbach 1986).

The ability to enlist the assistance of synthetic organic chemists to produce organic materials with tailored properties has, of course, already been used to advantage in several applications. The best known is that of liquid crystals and their use in displays and digital thermometers. New phenomena and types of molecule are still being discovered and seem likely to lead to successful large area displays for high definition television and to high density information stores. Other examples are piezoelectric polymers as very sensitive hydrophones for submarine detection, photoconducting polymers for electrocopying, and photochromic molecules for reversible high density optical storage and signal processing. Biosensors and chemical sensors for converting specific biochemical or chemical solute or gas interactions into electrical signals for use in industrial or medical diagnostics can also be mentioned. All are examples of 'Molecular Electronics,' that is, they are fields in which organic molecular materials perform an *active* function in the processing of information and its transmission and storage. This definition does not embrace their use in possible roles such as insulation, adhesion or encapsulation. Thus, molecular electronics is interpreted broadly and is not limited to phenomena concerning the movement of electrons only. Electromagnetic radiation, polarization phenomena, and various forms of electromechanical and electrochemical energy transfer are also included in the definition. A common feature of all the examples cited and of the area in general is that progress is achieved *via* 'molecular engineering' that is, using the ability to manipulate the architecture of a material to optimize a specific physical parameter.

An alternative definition exists for molecular electronics; this is formulated in terms of switching on a molecular scale and is aimed more at the long term problem of fabricating molecular electronic devices suitable for assembly into a computer (Carter 1986). It is interesting to note that only a modest diminution in the size of electronic circuit components is required before the scale of individual molecules is reached; in fact, many existing circuit elements could already be accommodated within the area occupied by a leukaemia virus. An illustration of the rapid evolution of silicon based microelectronics may be gained by studying Figure 1. If this systematic reduction in feature size suggested by the good log-linear graph is sustained, then the extrapolated line indicates device geometries with nanometer dimensions in approximately thirty years' time! The requirements of reliability and testing of complex structures suggest a system approach rather than the traditional one, which uses the properties of individual circuit elements. it appears likely that sequential designs, because of their vulnerability, will be abandoned in favor of supermolecular arrays acting as concurrent processor networks. For this reason, and to differentiate it from 'molecular electronics,' signal transport and control in nanometer scale assemblies is referred to as 'supermolecular electronics'.

Animals can solve with little apparent effort the tasks for which advanced supermolecular information processors are required; accordingly, some enthusiasts have speculated that the thirty-year time scale could be foreshortened using biological molecules. However, such thoughts are misplaced, for nature constructs organic materials for purposes other than those required for logic and memory functions. What can be learnt from studying biological systems are the scientific principles of organization and assembly; eventually, it may be possible to apply these concepts to construct synthetic supermolecular arrays. These will require nonlinear interactions between neighbors so that configurations of 'on' and 'off' elements propagate with time. However, it will be a difficult task to control the correlations between the parallel processing elements whether they be based on electronic, vibrational or magnetic effects. A great deal of fundamental research will be required to identify molecular systems with the necessary degree of cooperativity and to develop the routine analytical techniques designed for investigations in three dimensions with molecular scales of resolution.

Three-dimensional integration is extremely difficult using silicon technology while chemically nonspecific methods such as molecular beam epitaxy are relatively crude. A self-assembly technique is a far more attractive alternative if regular, three-dimensional, ordered structures are required. This involves the construction of unique assemblies whose architectures depend on the shapes and charge distributions of the units from which they are built, as distinct from the methods used to assemble them. There is a considerable degree of self assembly associated with organic monomolecular films deposited using the Langmuir trough technique (Roberts 1985). About half a century after the report of their discovery, intense interest is now being displayed in these so called Langmuir-Blodgett films. Their precise thickness, coupled with the degree of control over their molecular architecture has now firmly established a role for such layers in thin film technology. It seems likely that an understanding of their physical properties and utilization in molecular based devices will assist generally in the transition from molecular to supermolecular electronics. The likely pattern for this evolutionary process is given in Figure 2. The plan is speculative and envisages three main stages before the advent of applied, complex supermolecular systems. In the near term it predicts that current research and development will result in organic materials ousting inorganic materials in existing applications, e.g. optoelectronics. Hybrid technologies, comprising a novel device partly based on conventional solid state materials and partly on organic compounds, could possibly follow in the mid-term, say 5 or 8 years. Thereafter, at some stage dictated by the emergence of reliable, stable supermolecular assemblies, a true watershed will occur. When this occurs, materials and process technologies of conventional solid state devices will be superceded by radically new types of devices. This era will be equivalent to that witnessed about forty years ago when inorganic semiconductor materials were developed. Just as then, novel effects should be discovered and these in turn will lead to the fabrication of novel devices that can be integrated in novel systems.

Molecular and supermolecular electronics are broad and interesting subjects. Moreover, they require a multidisciplinary approach where collaboration between biologists, chemists, computer scientists, electronic engineers and physicists is of paramount importance. To illustrate these features, this article concentrates on Langmuir-Blodgett films. Equal emphasis is placed on their importance in basic science and on their potential applications, especially in the area of electronics.

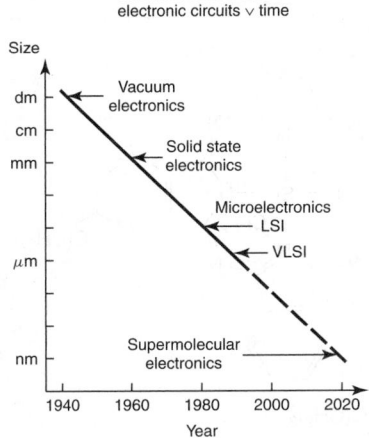

Minimum feature size in electronic circuits ∨ time

Fig. 1.   Linear feature size of commercial electronic circuits versus time; the bottom arrow speculatively points to an era where switches on a molecular scale will have application in computer systems

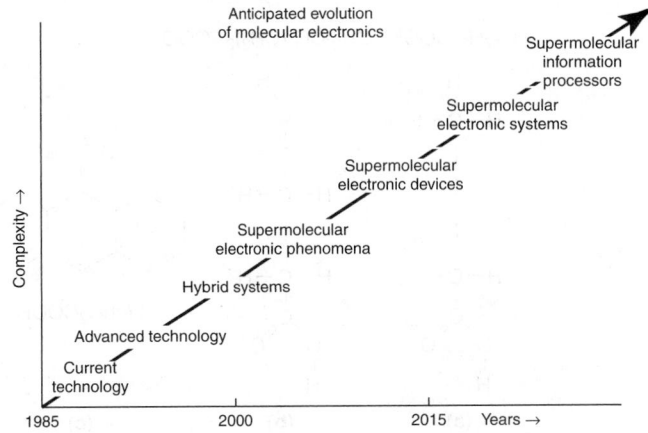

Fig. 2.   The anticipated evolution of molecular electronics to supermolecular electronics

Historical Review of Langmuir-Blodgett Films

According to Tabor (1980) the earliest written record of observations of the spread of oil on water is in cuneiform on clay tablets, dating from Hammurabi's period (18th century B.C.) in Babylonia. The earliest technical application of organic monolayer films is believed to be the Japanese printing art called subminagashi, involving a suspension of submicron carbon particles and protein molecules spread on the surface of water. The distinctive patterns so formed can be transferred by lowering a sheet of paper onto the water surface. There are also many references dating from the classical times of Plutarch. Aristotle and Pliny describing the ability of oil spread on water to dampen surface waves and ripples. It was this property which attracted Benjamin Franklin, the versatile American statesman, to the subject. During his frequent visits to Europe in the 18th century A.D. to negotiate the sovereignty of his country with the French and the British he carried out his famous 'teaspoonful of oil' experiment. Often quoted and picturesque, Franklin's (1774) account to the Royal Society included the following phrases: "At length, being at Clapham where there is, on the common, a large pond, which I observed to be one day very rough with wind, I fetched out a cruet of oil, and dropped a little of it on the water. I saw it spread itself with surprising swiftness upon the surface ... I then went to the windward side, where (the waves) began to form and there the oil, though not more than a teaspoonful, produced an instant calm over a space of several yards square, which spread amazingly, and extended itself gradually till it reached the lee side, making all that quarter of the pond, perhaps half an acre, as smooth as a looking glass. After this, I contrived to take with me, whenever I went into the country, a little oil in the upper hollow joint of my bamboo cane, with which I might repeat the experiment as opportunity should offer; and I found it constantly to succeed."

Franklin must have been too preoccupied with political affairs to place his observations on a quantitative basis. Had he done so he might well have calculated that a volume of one teaspoonful (approximately 2 ml) spread over an area of nearly half an acre (200 m²) leads to a surface coating approximately 1 nm thick. However, it is Lord Rayleigh (1890) who is given the distinction of first suspecting that the maximum extension of an oil film on water represents a layer one molecule in thickness. For a direct measurement on molecular sizes he was indebted to Pockels (1891) whose simple apparatus later became the model for what is now called a Langmuir trough; using very simple equipment he calculated the precise thickness of a monomolecular layer of castor oil on water to be 1 nm. This significant observation was not fully followed up until the pioneering work by Langmuir on the adsorption of gases or solutes by solids. In order to test the general applicability of his hypothesis about the involvement of short range forces, he turned his attention to liquids and he essentially repeated (Langmuir 1917; 1920) the earlier measurements of Pockels and Rayleigh, and extended them to include the transfer of molecules from a water surface to a solid support. The first detailed description of sequential monolayer transfer was given by Blodgett (1935),

his collaborator at the General Electric Company laboratories. These built up monolayer assemblies are now called Langmuir-Blodgett (LB) films, while the floating monolayer is referred to as a Langmuir film. The extensive list of publications resulting from the pioneering experiments carried out by these two investigators during the period 1934 to 1952 has been compiled by Gaines (1983).

The Chemistry and Preparation of Langmuir Films

Most of the experiments reported by Langmuir and Blodgett were on a well-defined series of fatty acids and their salts. Figure 3(a) shows stearic acid, a molecule in which sixteen CH_2 groups form a long hydrophobic chain; the other end of the molecule terminates in a hydrophilic carboxylic acid group. When dissolved in a suitable solvent and spread on the surface of water, molecules may be compressed with the aid of a barrier. Figure 4 shows a plot of the surface pressure (differential surface tension) versus area occupied per molecule for stearic acid. The monolayer undergoes a number of phase transformations during compression; the well-defined sequence can be viewed as the two-dimensional analogue of the classical transitions observed with pressure-volume isotherms. However, it should be emphasized that some materials, while forming acceptable quality LB films, do not display the well-defined break points shown in Fig. 4.

Generally speaking, the approach to the synthesis of suitable molecules for examination with a Langmuir trough has been an ad hoc one and has relied on the modification of known materials. For example, the alkyl group of fatty acids may be replaced by chains containing one or more double bonds. The ω-tricosenoic acid (Barraud 1983) molecule shown in

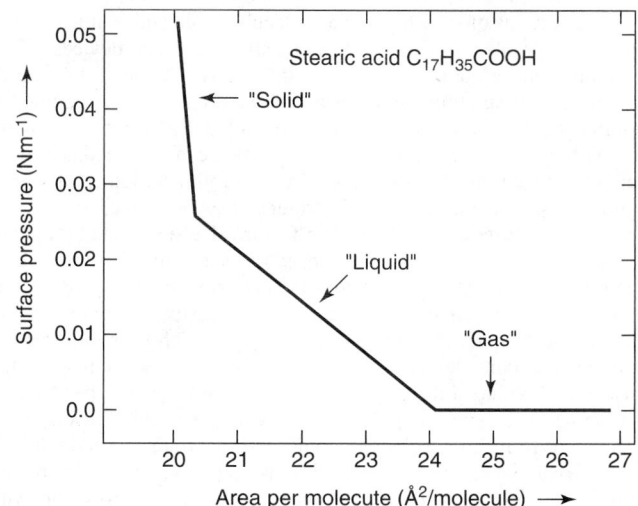

Fig. 4.   Surface pressure versus area characteristic for stearic acid

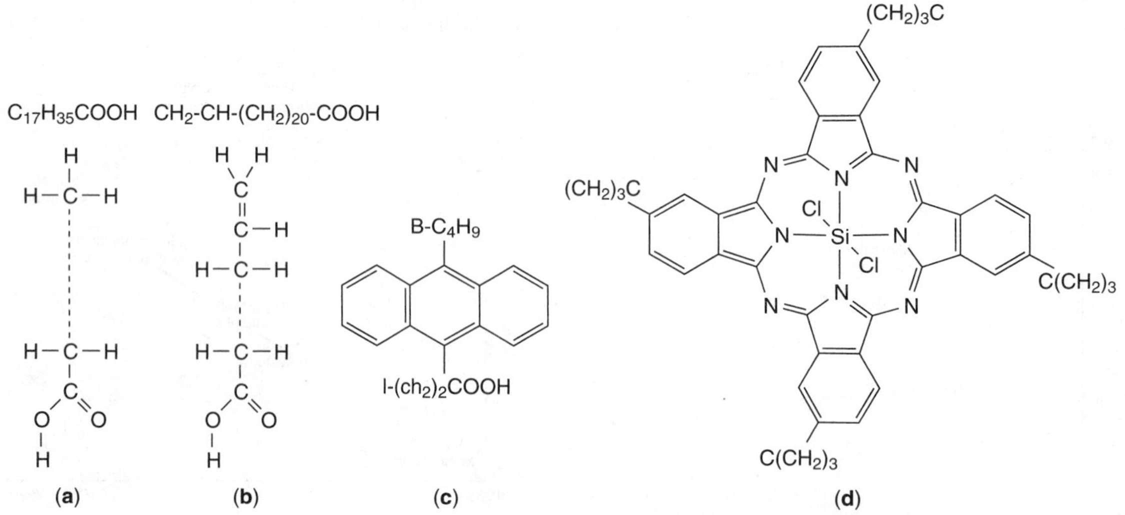

Fig. 3.   A selection of molecules used to form LB films: (**a**) fatty acid; (**b**) ω-tricosenoic acid; (**c**) 9-butyl-10-anthrylpropionic acid; (**d**) tetra-4-tert-butyl-10-phthalocyaninato silicon dichloride

Figure 3(b), which is similar to stearic acid but contains a terminal double bond, displays all the essential film-forming qualities including solubility in convenient organic solvents, stability at the surface of water, shear resistance, stability against collapse, and suitable orientation features. It is relatively straightforward to attach long aliphatic chains to a molecule and spread a monolayer. However, this may well dilute the desirable properties of the basic molecule; moreover, for stability reasons, the presence of long side groups will severely restrict their practical applicability. It has therefore been recognised that the scope of the Langmuir trough technique would be considerably enhanced if interesting materials containing only short, stable, side groups could be formed into LB films. A good example is provided by the anthracene derivative (Vincett et al., 1979) shown in Figure 3(c); multilayers of excellent quality can be obtained, even though the alkyl group contains only four aliphatic carbons and the hydrophobic group is attached to the ring structure *via* only two methylene groups. Extremely robust monolayer assemblies can be constructed using dye molecules such as the porphyrins and phthalocyanines. In general their quality is relatively imperfect compared with those of the classic film forming materials but their significant advantages lie in their thermal and mechanical stabilities. An example of a substituted phthalocyanine molecule (Hue et al., 1986) that can be deposited in monolayer form is shown in Figure 3(d).

The molecules shown in Figure 3 represent only a few of the materials that have been studied in LB film form. Nonetheless, a great deal more needs to be done to tap the vast wealth of opportunities available with organic systems. There will inevitably be short-term opportunistic attempts aimed at discovering molecules for specific devices. However, there is a more pressing need for a systematic approach that will yield rules governing structure-property correlations, so as to enable scientists confidently to predict the molecular architecture of monolayer assemblies. See also **Macromolecular Science**.

Langmuir-Blodgett Film Deposition

An LB film is formed by transferring a floating monolayer onto a solid substrate. The quality of the Langmuir film and the surface pressure at which 'dipping' occurs is established using the type of isotherm shown in Figure 4. The subphase is normally ultra-pure water, because it is readily available and it has an exceptionally high value of surface tension. The composition of the subphase, including its purity, pH, and ionic strength, can have a profound influence on factors such as the solubility of the monolayer and segregation effects resulting in molecular aggregates or domains.

Using conventional LB film technology, the substrate is raised and lowered vertically through a compact floating monolayer; the surface pressure at which this occurs is normally just above the 'knee' in the steeply rising sector of the isotherm indicating low compressibility in the monolayer. At this stage, if conditions are carefully controlled and appropriate molecules are used, one monolayer is transferred during each excursion through the subphase surface. The most common deposition mode (Y-type) is illustrated in Figure 5(a), where the molecules can be seen to stack in a head to head and tail to tail configuration. The floating molecules on a water surface are shown at the top left in this diagram. With a hydrophobic substrate (for example, a group III-V compound semiconductor), no pick-up occurs during the first immersion and the first monolayer is therefore deposited during the first withdrawal as shown in Figure 5(b). The surface is now hydrophobic and deposition does now occur during the next immersion into the water. Thus, one monolayer coverage is obtained on each traversal of the liquid surface. With a hydrophobic surface such as freshly etched silicon, pick up also occurs during the first insertion. Sometimes, the common deposition mode illustrated in Figure 5 is not followed and one of the other two possible configurations, X and Z-type, is observed, where transfer occurs only during immersion or withdrawal, respectively. The surface quality and chemical composition of the substrate is bound to control the nature of the deposited layer. When adhesion is poor, some researchers have resorted to the less satisfactory method of placing the substrate flat on the liquid surface, a technique first used by Langmuir and Schaefer.

Many modifications of the very early film balances have been described by Gaines (1983). However, the upsurge in interest in LB films has led to greater attention being placed on trough design and control systems to meet the stringent requirements of scientists and engineers. Therefore, modern instruments are relatively sophisticated and, for device-related

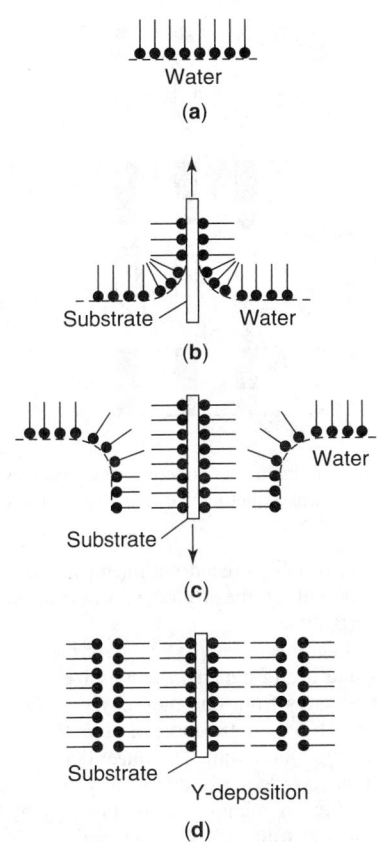

Fig. 5. Langmuir-Blodgett film deposition (Y-type) on a hydrophilic substrate: (**a**) monolayer on the surface of water; (**b**) first layer on withdrawal; (**c**) second layer (second insertion); (**d**) substrate with three layers (after second removal)

work, need to be situated on anti-vibration Tables in clean environments. Although it is possible to automate most features, the primary benefit at the present time lies in efficient data collection and the ease with which data can be manipulated. For example, phase transitions are more apparent when the differential of the pressure-area isotherm is plotted. No difficulties are envisaged in scaling up the Langmuir trough or in the design of continuous fabrication arrangements. When an important practical application is discovered there will be a need to produce a specially designed trough capable, for example, of coating a moving belt or multiple wafers of silicon.

A recent development in trough design is worthy of special attention as it could have important commercial significance. It has arisen because of the need to produce non-centrosymmetric structures that display interesting non-linear physical effects. The conventional Y-type films are symmetrical in character and experience has shown that X and Y type layers, although non-symmetrical, are usually imperfect. Therefore, an alternative approach to producing noncentrosymmetric structures is to use alternate layers of two different materials where the contributions of adjacent molecules do not cancel. See Fig. 6. The additions of a fixed beam and a revolving center section to an automated constant perimeter barrier Langmuir trough enables the formation of an alternating Y-type structure of two different molecules spread in the two distinct areas of the subphase. The structural qualities of the LB films prepared in this way can be of high quality (Holcroft et al., 1985); another advantage of the rotating substrate arrangement, which is conducive to fast dipping, is that the meniscus, unlike that in the vertical dipping method, is always in the same direction.

Many different experimental techniques indicate that carefully prepared films of appropriate molecules do indeed possess a high degree of structural order. The reader is referred to the proceedings of the two international conferences on LB films for literature references describing the vast range of characterization experiments that have been employed (Roberts and Pitt 1983; Gaines 1985). These include ellipsometry, electron spin resonance, infrared dichroism, photoacoustic spectroscopy, secondary ion mass spectroscopy, surface potential, polarized X-ray and electron diffraction, neutron reflection and diffraction. Most of the electrical data for LB films are suspect in that they have been obtained for films deposited

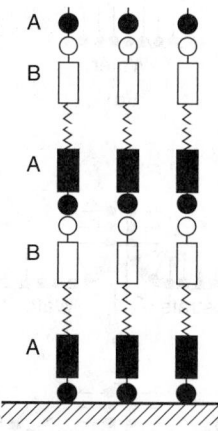

Fig. 6.   Alternating organic multilayer structure which enables a Y-type LB film to be produced of non-centrosymmetric character (molecular lego!)

onto metals that are invariably coated with semi-insulating native oxides. A comprehensive account of these studies was presented in a review by Vincett and Roberts (1980).

The four separate diagrams in Figure 7 all describe results for fatty acids or their derivatives and are designed to emphasize the reproducibility of various physical parameters from one monolayer to the next. Figure 7(a) shows the capacitance (C) as a function of film thickness for cadmium arachidate deposited onto aluminium. The linear dependence of C^{-1} versus the number of monolayers demonstrates clearly the repeatability of the dielectric thickness of each monolayer. In Figure 7(b) it is a band in the infrared reflection spectrum of the same material that has been used to demonstrate the uniformity of successive monolayers. The reason for the scatter around the origin is not understood but it probably reflects in this case that the structure of the first few monolayers is affected by the metal underlay. Figure 7(c) is based on experiments using barium stearate as the absorber for L shell Auger electrons. By labelling the molecules in these overlays with ^{14}C and examining their autoradiographs it is possible to confirm the uniformity of the deposition process by

plotting the count of ^{14}C rays versus the number of monolayers. The final diagram in the set, Figure 7(d), illustrates another powerful tool for investigating organic coatings on metals. In this case different thicknesses of cadmium dimethyl arachidate have been used to attenuate the substrate X-ray photoemission signal.

During the next few years many techniques-oriented scientists will be attracted to work on LB films because they provide interesting novel structures whose molecular architecture can be systematically controlled. The quality of the floating monolayer is also important and needs to be characterized as does the interface between the first deposited monolayer and the substrate. Fluorescence microscopy and Brillouin and Fourier transform infrared spectroscopies are currently being used to address these problems.

Applications of Langmuir-Blodgett Films

Following the pioneering work of their famous employees, the General Electric Company introduced several simple applications of LB films including step-thickness gauges, anti-reflection coatings and soft X-ray gratings. Since that time, stimulated no doubt by the availability of well engineered troughs and a wider range of suitable materials, researchers have suggested other applications for monolayer and multilayer films. A selection of areas where LB films may find practical use is given in Table 1. Further details are given in the review by Roberts (1985). However, it should be remembered that one of the principal virtues of LB films is their usefulness in fundamental research. Therefore, before discussing more applied areas we shall mention a few areas of science that can benefit from investigations of model systems based on monomolecular assemblies.

Model System in Basic Research

(a) Energy transfer in complex monolayers. The Langmuir trough technique provides a method of constructing simple artificial systems of co-operating molecules on a substrate. The pioneer in this field has been Kuhn (1983); the elegance of his and his colleagues' work is evident in their reviews of the subject. These describe the use of LB films to investigate intermolecular interactions and various photophysical and photochemical processes. Their supermolecular structures have mainly involved long chain fatty acids as matrices for appropriate synthetic dyes and have been ingeniously designed to clarify the different interactions that can occur between various molecules via photon, electron and proton transfer. An example of their research, designed to investigate the Förster type of energy transfer from a sensitising molecule, S, to acceptor molecule, A, is given in Figures 8 and 9. If S is a compound that absorbs in the ultraviolet part of the spectrum and fluoresces in the blue, while A absorbs in the blue and fluoresces in the yellow, then interesting effects are observed when the system is irradiated with ultraviolet light. If there is a sufficient distance between S and A, as in Figure 8(a), the fluorescence of S appears since A does not absorb UV radiation. However, below a certain threshold distance, as in Figure 8(b), the excitation energy of S is transferred to A and the yellow fluorescence of A is expected. Similar experiments based on fluorescence quenching indicate that the rate constant of the electron transfer decreases exponentially with increasing barrier thickness separating a donor chromophore and an electron acceptor. In the example shown in Figure 9, N, N'-dioctadecylthiacyanine has been used in conjunction with a viologen acceptor layer to observe the steady fluorescence intensities of the cyanine dye monolayer in the absence (I_0) and in the presence of the acceptor layer (I). The quantity, $[I_0/I] - 1$ is proportional to the rate constant of the electron transfer; its linear dependence with d, the distance between the chromophores, is evidence of electron tunnelling. In a similar series of experiments it has been possible to investigate the energy transfer mechanism responsible for spectral sensitization.

(b) Biological membranes. The physical structure and chemical nature of classical LB films gives them a close resemblance to naturally occurring biological membranes. For example, because the two ends of a lipid molecule have incompatible solubilities, they spontaneously organize themselves in the form of a bilayer, or essentially a two layer LB film. Scientists have suggested that they might provide a suitable model of the lipid membrane for probing the cooperative interactions between its constituents. However, caution must be exercised in assessing the biological relevance of this type of work and associated studies aimed at incorporating ionophores into phospholipid layers. Much of this research is targeted at novel integrated solid state devices incorporating biological molecules such as enzymes.

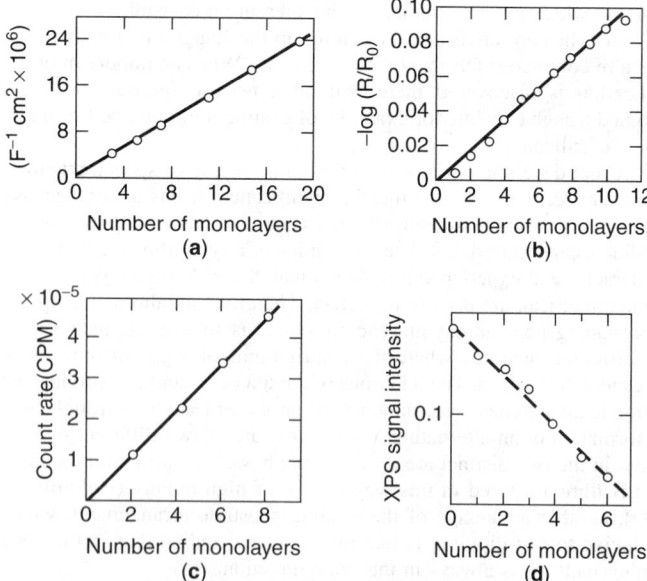

Fig. 7.   These diagrams are designed to emphasize the reproducibility of various physical parameters in monolayer assemblies of different thicknesses: (**a**) reciprocal capacitance per unit area versus number of monolayers of cadmium arachidate on an aluminum substrate (Roberts et al., 1978); (**b**) absorption intensity versus number of monolayers for the symmetric carboxylate stretching mode of cadmium arachidate at 1432 cm^{-1} (Allara and Swalen, 1982); (**c**) count rate of ^{14}C rays versus number of layers of barium stearate labelled with ^{14}C (Mori et al., 1980); (**d**) X-ray photoelectron signal intensity versus number of layers of cadmium dimethyl arachidate on silver (Brundle et al., 1979)

TABLE 1. PROMISING APPLIED RESEARCH AREAS

Topic	Molecular electronics applications
Model Systems in Fundamental Research	Spectroscopy of Complex Monolayers: spectral sensitization, fluorescence quenching, energy transfer between excited states. Model membranes to mimic photosynthetic systems. Modification of solid surface properties. Examination of lipids, proteins and membrane phenomena; organic semiconductors.
Applied Chemistry	Surface chemistry and behavior of surfactants: catalysis; filtration/reverse osmosis membranes; adhesion; surface lubrication, e.g. magnetic tape; encapsulation.
Electron Beam Microlithography	Good sensitivity and contrast, acceptable plasma etching resistance; less scattering of electrons and therefore better resolution; negative and positive resists possible.
Integrated Optics and Storage Optics	Film thickness plus refractive index of film and hence guided wave velocity can be controlled with great precision; acceptable attenuation loss. Possible uses in conventional optics and optical data storage: photochromic and ablative systems. Optical sensors e.g. based on coated fibres.
Nonlinear Physics	Control of molecular architecture to produce asymmetric structures with high non-linear coefficients, e.g. in electro-optics, pyroelectric detectors, or acoustoelectric devices.
Dilute Radioactive Sources	Radioactive nuclide incorporated in conventional LB film; used to measure the ranges of low energy electrons.
Electronic Displays	Large area capability of LB films is an advantage; the monolayers can either be the active electroluminescent layer or used to enhance efficiency of an inorganic diode; passive application to align liquid crystal displays. Deposition of liquid-crystal type molecules also possible.
Photovoltaic Cells	Used as a tunnelling layer in an MIS solar cell or as an active layer in p-n junction diode, perhaps involving an inorganic/organic junction.
Two Dimensional Magnetic Arrays	Magnetic atoms e.g. Mn, periodically spaced in LB film; possible applications include magnetic control of superconducting junctions and bubble and magneto-optical devices.
Field Effect Devices	Accumulation, depletion and inversion regions possible with a variety of semiconductors; can therefore form the basis of several devices e.g. CCD, bistable switch, gas detector or pyro/piezo FET, is suitable LB films are used.
Biological Membranes	Attractive supporting membranes for commercial exploitation of biological material, e.g. immobilization of membrane bound enzymes in solid state sensors; ISFET type structures.
Supermolecular Structures	Speculative work aimed at superconductors, organic metals, 3D memory storage, molecular switches.

In some cases, LB films are useful to facilitate physical studies of biological molecules e.g. to measure the ionic permeability of reconstituted membranes. Supermolecular structures have also been designed to mimic the primary process in photosynthesis and for achieving an efficient photoinduced charge separation by appropriate modelling of potential

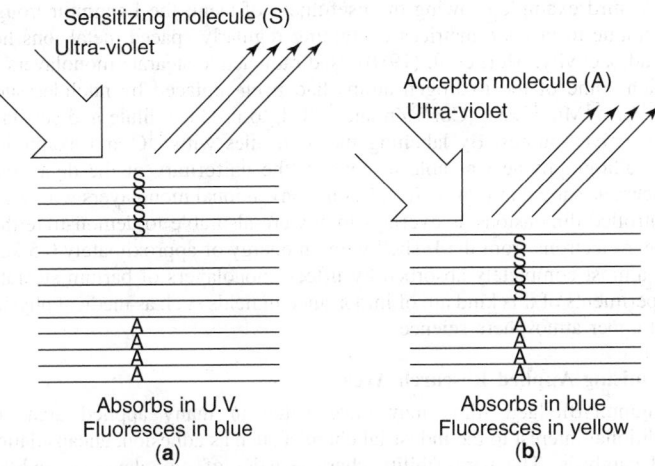

Fig. 8. Schematic diagram showing basis of experiments designed to investigate energy transfer from a sensitizing molecule (S) to an acceptor molecule (A). The number of monolayers separating the two species governs the spectral response of the fluorescence spectrum. In (b) the separation distance is sufficiently small for the excitation energy of S to be transferred to A. In (a) the acceptor molecules does not absorb the ultraviolet radiation

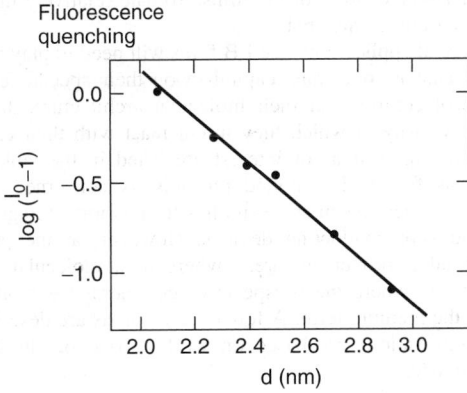

Fig. 9. The fluorescence intensity I_0 of a donor dye is reduced to a value I in the presence of an acceptor dye. The logarithm of $[(I_0/I)-I]$ is shown as a function of d, the spacing between the donor and acceptor planes. (Mobius, 1981)

profiles. Chlorophyll has been studied in this context. Some of the results may have relevance to solar photochemical conversion devices.

(c) Metal-ion incorporation. The addition of divalent ions into the liquid subphase in a Langmuir trough can increase both the shear resistance and cohesion of the monolayer. For this reason it is more common to find reports of studies on fatty acid salts than their acids. By adjusting the pH of the subphase it is possible to assemble multilayers containing metal ions separated by the width of an integral number of monolayers (for Y-type deposition, two monolayers). The ability to do this has been capitalized upon in several fundamental investigations, three of which will be mentioned here. The first of these relates to two-dimensional magnetic monolayers involving iron or manganese ions. Using electron spin resonance Pomerantz (1980) has demonstrated that at temperatures near 2K, the resonance field and line-shapes were affected, thus signifying the rapid development of a large internal magnetic field. His results have been interpreted in terms of a predominantly anti-ferromagnetic state, but with a weak ferromagnetic component. Further experiments are required to clarify magnetic ordering in two-dimensional space.

In the area of surface science, X-ray photoemission is now used extensively to study organic materials on surfaces. In such experiments it is important to establish electron mean free path lengths as a function of kinetic energy. An example, of this type of investigation in LB films is illustrated in Figure 7(d); generally it is found that the mean free paths for ordered multilayers are significantly longer than those for conventionally produced polymers (Clark et al., 1981).

A third example showing the usefulness of using the Langmuir trough technique to provide matrices containing regularly spaced metal ions lies in radioactivity. Mori et al. (1980) used radioactive stearate monolayers in which some of the hydrogen atoms had been replaced by nuclides such as ^{51}Cr, ^{54}Mn, ^{55}Fe, ^{57}Co, ^{65}Zn and ^{109}Cd, to produce dilute and standard radioactive sources. By labelling the molecules with ^{14}C and examining autoradiographs he was able to confirm the uniformity of the deposition process as shown in Figure 7(c). Using conventional monolayers with well-controlled dimensions as overlays they were also able to demonstrate that Auger electrons from the L shell with an energy of approximately 0.5 keV are almost completely absorbed by fifteen monolayers of barium stearate. Experiments of this kind are of importance in fields such as medical physics and upper atmosphere science.

Promising Applied Research Areas

Langmuir-Blodgett films may have value in many applied areas of traditional interest to the industrial chemist such as adhesion, encapsulation and catalysis. The permeability characteristics of monolayer assemblies may also find application as synthetic membranes for ultra fine filtration, gas separation and reverse osmosis. For example, Albrecht et al. (1985) have proved the efficiency of polymeric diacetylene monolayers on semi-permeable supports in reducing the flow of CH_4. One interesting possibility lies in using LB monolayers as lubricants in magnetic tape technology. Unpublished reports have indicated that frictional coefficients can be reduced markedly when the tape is coated with a few monolayers. In applications such as those listed above, difficulties may well be encountered with the mechanical stability of the films. To date relatively little research has been carried out in this area.

For commercial applications, the LB films will need to play an essential, integral role. That is, one must capitalize on their special features such as the degree of control over their molecular architecture, their thinness and the selective way in which they might react with their environment. Some of the potential areas of interest are listed in the table. The long term interest as far as the applied physicist is concerned, lies in the possible uses of supermolecular assemblies for memory storage, molecular switching, and superconducting devices. However, at the present time it is in potential improvement areas where monomolecular films show most promise and where the prospects of commercial exploitation seem reasonable in the medium term. A few of these areas are described below; most of the illustrations are based on work carried out in the author's research laboratory:

(a) Nonlinear physics. There is evidence that many organic molecules possess very high non-linear coefficients and therefore, if LB films with the required architecture can be formed, these could form the basis of novel devices. In order to avoid the symmetry inherent with conventional Y-type deposition, X and Z type films have been studied; some have displayed a permanent polarization with a strong component in a direction perpendicular to the substrate. However, as has already been mentioned, films produced in this way with their dipoles supposedly aligned in a common direction, are invariably of poorer quality than Y-type layers. a possible method of improving the structure is to use electric or magnetic fields to help align the molecules but efforts to orient films on the subphase and substrate have met with only limited success. The problem can be overcome by using organic superlattices based on alternating layers of two different materials (See Fig. 6). A good example is given in Figure 10 which shows a superlattice comprising acid and amine molecules whose dipole moments are in opposite senses but when deposited in Y-type LB film form are aligned in the same direction. Two areas of particular interest that would capitalize on this feature of organic superlattices are pyroelectricity and optoelectronics. Each of these will now be considered in turn.

Pyroelectric devices respond to a rate of change of temperature rather than to changes of temperature as in other, types of thermal detector. This gives them inherent advantages but their full potential has yet to be realized. For applications where both high speed and sensitivity are required, conventional materials have been unsuccessful. The desirable pyroelectric properties of inorganic materials appear to vanish for thicknesses less than 10 μm, and pyroelectric organic single crystals, such as triglycine sulfate, which possess a high pyroelectric coefficient, cannot be produced in thin film form. Thus the future development of relatively cheap thermal imaging systems with reasonable performance and an optimum thickness of approximately 0.5 μm, requires a materials breakthrough. A normal

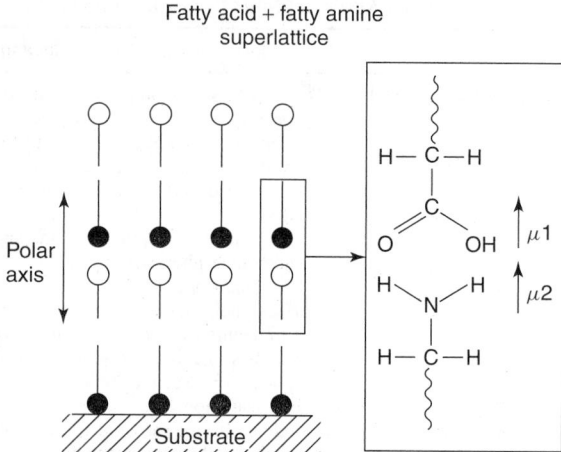

Fatty acid + fatty amine superlattice

Fig. 10. The left-hand diagram shows an organic superlattice with a unique polar axis. The two types of molecule involved could be a fatty acid and a fatty amine. The insert is designed to show that these two materials have dipole moments in opposite senses with respect to the hydrophobic chain. Thus, the Y-type film has a resultant dipole moment

detector consists of a capacitor whose dielectric is an oriented pyroelectric material; using this type of device Christie et al. (1986a, b) have observed encouraging results using the alternate layer structure shown in Figure 10. The pyroelectric coefficient, p, has been determined using both dynamic and static detection techniques. For the simple fatty acid/fatty amine superlattice $p \cong 1$ C cm^{-2} K^{-1} but more recent results involving a system where proton transfer occurs from the acid to the amine have yielded higher values, comparable with those observed for triglycine sulfate. The exploitation of this work will depend not only on the properties of the LB films but also on our ability to deposit the layers onto reticulated structures with low thermal mass.

Although many of the potential optical applications of LB films are in transmission optics employing the linear response properties of molecules, it is in the area of non-linear optics where the most exciting applications are perceived. Highly efficient nonlinear optic materials permit functions such as those illustrated in Figure 11 to be performed in a totally optical manner without the need for electron-photon conversion processes. Second harmonic generation and parametric amplification can be obtained using inorganic single crystal materials such as lithium niobate, but recently, organic crystals such as 3-methyl-4-nitroaniline have been shown to possess exceptionally large second order electrooptic coefficients (Zyss 1982). However, a thin film geometry is preferred; then it will be possible to integrate nonlinear interactions, linear filtering and transmission functions into one precision monolithic structure. Therefore, researchers are currently substituting such molecules with appropriate side groups to enable LB film deposition to occur. To date, second harmonic generation has been reported in multilayers of nitrooctadecylazobenzene and mercocyanine and hemicyanine dyes. Nonlinear coefficients comparable to those of inorganic materials have been achieved but there are many other considerations involved (e.g. phase matching, suitable spectral response and refractive index, good optical damage threshold, low scattering coefficients and mechanical stability) before practical objectives can be accomplished. The most interesting observation to date is that of second harmonic generation in LB organic superlattices of the two molecules displayed in Figure 12. Neal et al. (1986) have shown that the nonlinear response of a hemicyanine/nitrostilbene layer is greater than that expected from the simple addition of contributions arising from the individual (separated) monolayers. The coefficient for second harmonic generation of the alternate layer structure is approximately five times the average value of the same parameter measured for hemicyanine and nitrostilbene. This superadditive effect is best explained in terms of improved film structure with adjacent molecules influencing each other to orient more vertically with respect to the substrate.

(b) Enhanced device processing. In integrated circuit technology the quest for faster speeds and larger memories has led to a gradual refinement of microlithographic methods for producing smaller and more closely spaced circuit elements. Sub-micron resolution is now

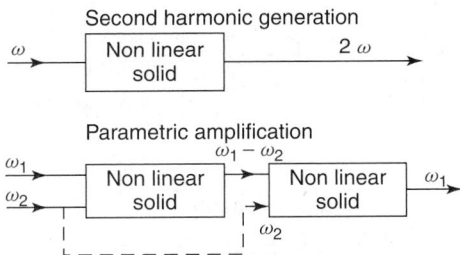

Fig. 11. Schematic diagram illustrating the ability of nonlinear materials to double the frequency of incident laser radiation and also function as a parametric amplifier when two beams of different frequency are involved. Most of the non-centrosymmetric solids currently in use are inorganic, but they may be replaced by organic single crystal thin films

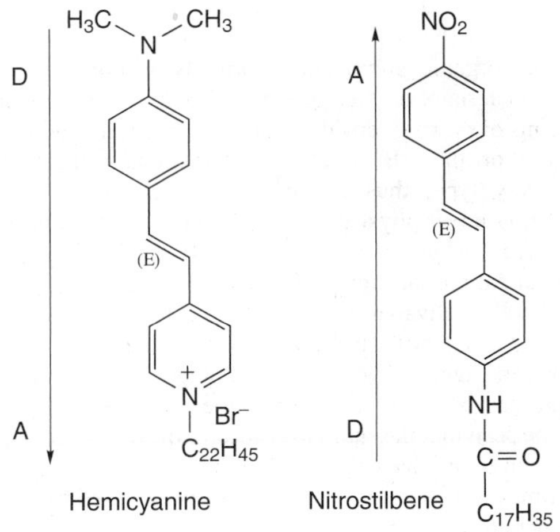

Fig. 12. These two molecules, one a hemicyanine and the other a nitrostilbene dye, can be used to form an organic superlattice displaying a high coefficient for second harmonic generation

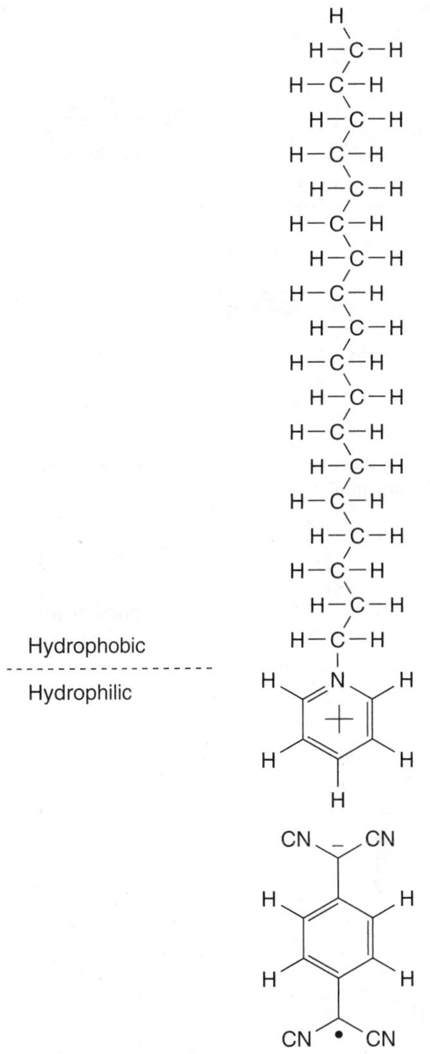

Fig. 13. A substituent, pyridinium tetracyanoquinodimethane (TCNQ) molecular system

required and this has necessitated a move away from conventional photolithography to techniques involving electron or ion beams and x-rays. The main disadvantage of electron beam systems lies with their scattering characteristics; this enforces the requirement that the resist materials be pinhole free and less than 1 μm thick. Conventional spin-coated polymers display unacceptably large pin-hole densities and variations of thickness; however, LB films have already demonstrated their capability in this regard. There are good examples of both positive and negative resists but the best material reported to date is the ω-tricosenoic acid molecule shown in Figure 3(b). In purified form this material has adequate sensitivity, may be deposited at a rate of 0.5 cm s^{-1} to produce uniform coatings in the range 30–90 nm, and has been shown capable of a line resolution of 60 nm. The main property which requires some improvements is the etch resistance in plasma processing (Barraud et al. 1983).

Interconnects become increasingly important as the size of electronic circuits reduces. It has been suggested that protein layers might be useful in this regard. The method involves first depositing a synthetic protein and patterning it using a conventional resist. The NH$_2$ groups in the exposed protein layer can then be used to adsorb silver ions from a silver nitrate solution; this physical development stage converts the ions to metallic silver. An alternative approach is to fabricate supermolecular assemblies that conduct. The most successful attempt to date uses the molecule shown in Figure 13. Barraud et al. (1985) have succeeded in producing close packed multilayers of this TCNQ derivative, where the polar planes are separated by insulating lamellar regions. The resistivity of this N-stearylpyridinium$^+$ 1TCNQ$^-$ film is approximately 1 Ω cm. Of course, there are many other attractions in producing conducting LB films, including their uses.

There are several niches where LB films could have a useful role in semiconductor technology. Probably the most important is the ability of an oriented monolayer to change the effective barrier height at a semiconductor surface. Researchers at the University of Durham first

demonstrated the effect on cadmium telluride and showed how it could be used to improve the efficiency of a photovoltaic diode (Dharmadasa et al., 1980). Similar effects have now been confirmed on a variety of materials including ZnSe, InP, GaP and GaAs and related III-V semiconductor alloys. The control afforded by the Langmuir-Blodgett technique permits the degree of band bending to be adjusted to suit the particular application, e.g. the increase in efficiency of an electroluminescent diode. In most applications it has been necessary to use robust monolayers of phthalocyanine which can withstand the large current densities involved. Figures 14 and 15 show data for gallium phosphide (Batey et al., 1983; Petty et al., 1985); for convenience the error bars have been removed. The LB thickness required to optimize the electroluminescence efficiency is found to be approximately 21 nm; this value is determined by the ability of minority carriers to cross the semi-insulating phthalocyanine film. Similar results have recently been achieved using zinc selenide layers grown using MOCVD. Blue electroluminescence is observed provided the organic film is present.

There are many methods of producing a surface layer on a semiconductor. However, experience has shown that when an energetic process such as evaporation, sputtering, or growth from a plasma, is used to deposit a thin film onto a semiconductor, a surface damaged layer is produced which invariably dominates the electrical characteristics of the junction so formed. However, the Langmuir trough technique, being a low temperature deposition process, provides a means of circumventing this particular difficulty. On the other hand, it does mean that how the substrate is prepared before dipping is of considerable importance in determining the quality of the interface produced. That is, the nascent 'oxide' layer formed during the etching procedure remains relatively undisturbed and this can play a vital

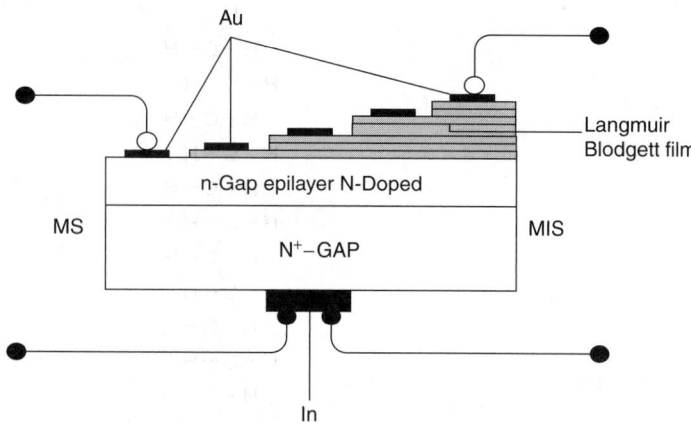

Fig. 14. Schematic diagram (not to scale) of a Gold-LB film—Gallium Phosphide device structure

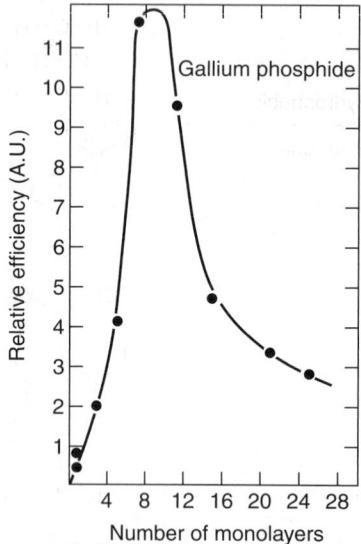

Fig. 15. The electroluminescent efficiency versus number of monolayers or substituted phthalocyanine for the device shown in Fig. 14

role even after it has been coated with an LB film. For this reason it is important first to carry out a systematic study of the surface chemistry of the semiconductor substrates.

(c) Sensors. The good insulating properties of LB films suggest their possible use in field effect devices, not so much to compete with existing semiconductor technology but more to capitalize on the advantages of being able to incorporate an organic layer within a semiconductor structure. Figure 16 shows schematic diagrams of both a field effect transistor (FET) and the 'heart' of this device, which is the metal-insulating-semiconductor (MIS) diode. Conventionally these devices are made using inorganic materials; silicon holds a pre-eminent position mainly because of the insulating qualities of its native oxide. The first transistor incorporating LB monolayers as the insulator was reported several years ago (Roberts et al., 1978); using the type of three terminal device shown in Figure 16, on indium phosphide and cadmium stearate, they showed that the channel conductivity between the source and the drain could be modulated by the action of a gate electrode.

Subsequently, other semiconductors have been used and results have confirmed the ease with which a range of single crystal surfaces can be accumulated, depleted or inverted with an applied voltage. It is recognised that in all cases, the LB film is deposited on top of a nascent 'oxide' layer and that the insulation is provided essentially by a double dielectric structure. In the case of silicon, this can be used to advantage, in that a closely packed LB film layer can seal its surface from the atmosphere and thereby greatly retard the development of interface states. The organic film is also efficient in increasing the dielectric strength of a leaky silicon oxide film.

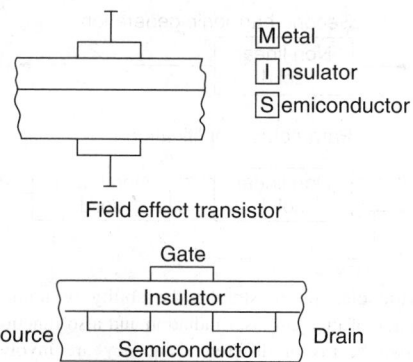

Fig. 16. Schematic diagrams (not to scale) showing, in the upper diagram a metal-insulator-semiconductor structure which forms an integral part of the field effect transistor shown in the lower diagram

The fact that organic compounds normally respond more positively than inorganic materials to external stimuli such as pressure or radiation, provides a means of making sensitive transducers. Moreover, by controlling the architecture of the LB film, the interactions can be designed to be of a lock-key type, thus enhancing the selectivity of the device. Following the non-linear physics work described earlier, it is now possible to envisage pyro- or piezo- FETs based on insulating LB films with an inbuilt polarization, or field effect devices incorporating biological membranes. Another advantage of using ultra-thin organic films is their fast response and recovery times because so little material is present. In the example shown in Figure 17, an eight monolayer LB film of a substituted phthalocyanine has been exposed to volume parts per million of nitrous oxide. It may be seen that the saturation current of the device scales linearly with gas concentration. Even though operation was at room temperature, the recovery times were shorter than those for evaporated film devices used at higher temperatures.

It is not necessary to confine the discussion of microelectronic LB film based sensors to MIS or FET structures. For example, the switching voltage of a bistable switch or the characteristics of a gate-controlled diode could be very sensitive to an ambient. Moreover, optical and acoustic devices frequently show interesting threshold or resonance effects which could form the basis of a sensor. One area receiving particularly strong attention is that of surface plasmon resonance (SPR). The principle of this optical detection method is illustrated in Figure 18. A surface plasmon is a surface charge density wave at a metal surface. If the metal is sandwiched between

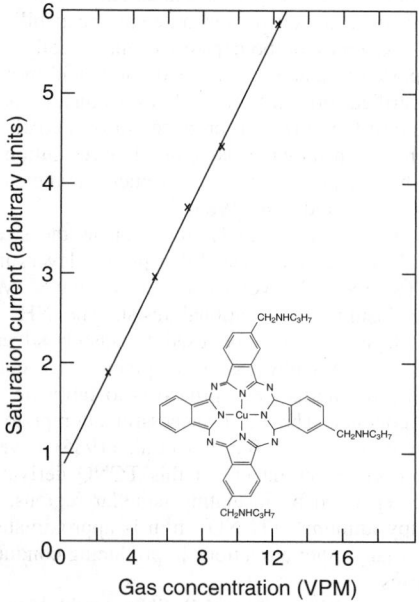

Fig. 17. Saturation current versus NO_2 gas concentration for a device incorporating eight monolayers of the asymmetrically substituted phthalocyanine molecule shown in the insert (Baker et al., 1983)

two materials of different dielectric constant, then resonance can occur; this is observed as a very sharp minimum of the light reflectance when the angle of incidence is varied. The resonance angle is ultrasensitive to variations in the refractive index of the medium adjacent to the metal film. For example, the small change in an organic material due to gas absorption can easily be monitored even for concentrations in the part per billion range. In a practical situation, one normally selects an angle of incidence approximately half way down the reflectance minimum curve when no special gas is present; the change in intensity of the reflected light is then monitored at a constant angle.

There is widespread interest in the potential of LB films as biosensors as many believe that the incorporation of biological molecules such as enzymes will lead to novel devices. Some are exploring the deposition of biologically active molecules onto the gate electrodes or oxides of field effect transistors but optical sensors, probably based on fiber optics, are the most favored technique. In all cases the aim is to couple the specificity of interaction of chemicals or biochemicals with proteins or enzymes e.g., the change in their molecular conformation, with the sensitivity and signal transduction properties of the device. It is recognized that stability and lifetime may be problem areas and, for this reason, cross linked polymers are being explored as the hosts for the active species.

Some of the most convenient types of sensor are based on acoustoelectric devices; these can either be conventional bulk piezoelectric oscillators normally made of quartz, or surface acoustic wave (SAW) devices. The resulting change in the quiescent resonant frequency of a quartz oscillator coated with LB films of different thickness provides a very simple and elegant way of monitoring the reproducibility of monolayer deposition. Figure 19 illustrates how the quartz oscillator functions as a microgravimetric sensor; from the change in frequency it is possible to determine the density of the thin films. There is a further change when the organic films are exposed to minute concentrations of gas. The results presented in Figure 20 are for ω-tricosenoic acid in the presence of acetic acid, a species used for the detection of heroin. Greater effects can be obtained if the organic film is specially sensitised. Another way of increasing the sensitivity is to use acoustic surface waves. In such devices, input and output interdigitated electrodes are formed on a piezoelectric substrate usually made of quartz or lithium niobate. These perform the conversion between electric and acoustic energy. The single

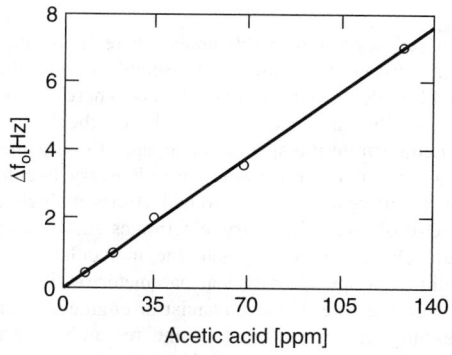

Fig. 20. Response characteristics of ω-tricosenoic acid coated quartz crystal oscillator at 22°C exposed to acetic acid. A greater change in the resonance frequency can be obtained if the film is specially sensitized to detect acetic acid

crystal surface region between the transducers serves as a propagation path for acoustic waves and thus forms a delay line which has application in electronic signal processing. Figure 20 shows a dual line configuration specially designed for sensing purposes. Basically, the device comprises two identical SAW (surface acoustic wave) oscillators positioned alongside each other. The hatched regions are earth shields to minimize reflections and cross talk between the two oscillators. The selective coating is placed in the propagation path of one of the oscillators thus affecting the delay time; the relative shift in frequency between the two oscillators is measured using a mixer circuit to obtain the difference frequency and then passing the resulting signal through a low pass filter. The change in frequency (Δf) between the two channels is then directly attributable to the sensor layer and other extraneous effects such as those due to temperature changes, are eliminated or very much reduced. Other geometries, for example, a surface acoustic wave resonator, are also possible. The device shown in Figure 21 has been constructed by Roberts et al. (1987) using quartz and interdigitated electrodes 24 μm wide separated by gaps of 25 μm; the operating frequency of the device is 98.4 MHz. Results have been obtained for both insulating (ω-tricosenoic acid) and conducting (pyridinium TCNQ) LB films as the sensor layer. In the case of the ω TA, only a mass loading effect is observed but with the TCNQ films, electric field effects are also apparent; these are associated with interactions between the surface acoustic wave and mobile charge carriers in the LB film. The combination of a device capable of measuring mass changes as low as femtograms per square centimeter, and the ability to detect minute changes in the electrical characteristics of a monolayer, augers well for sensors based on surface acoustic waves and monolayers.

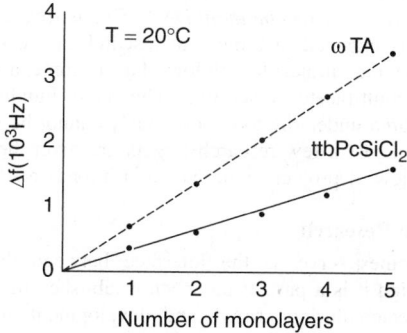

Fig. 18. Schematic diagrams illustrating the basis of the surface plasmon resonance technique. (Left) a beam of radiation striking the back surface of a glass prism coated with a metal film (usually evaporated silver). The reflected intensity and angle of reflection are extremely sensitive to variations in the dielectric on the metal surface. (Right) The shift in the R/θ plot when the organic film is exposed to a gas

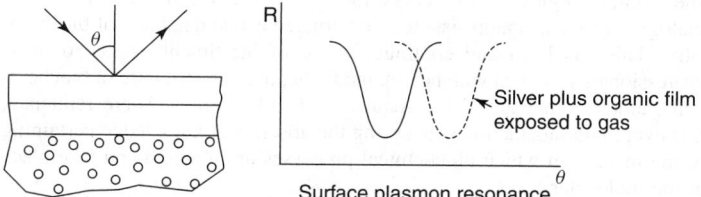

Fig. 19. The change in resonance frequency of 18 MHz piezoelectric quartz crystals coated with LB films of different thickness. The molecular structures of the organic films are shown in Figs. 3(b) and 3(d)

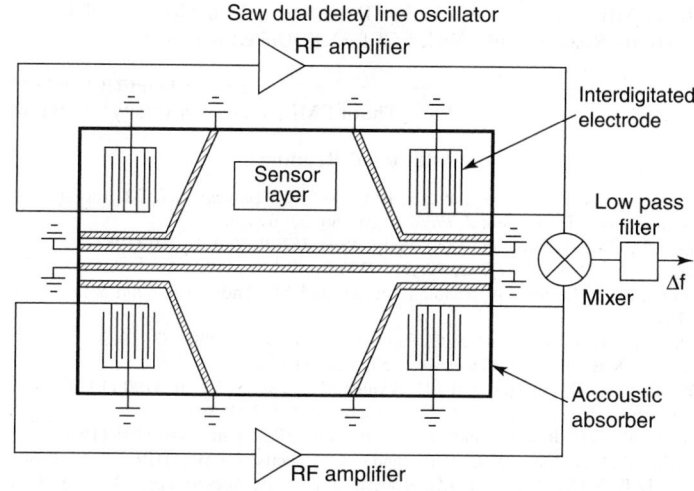

Fig. 21. A surface acoustic wave dual-delay line oscillator. The sensitive layer is placed in the propagation path of one of the two SAW devices. The difference in frequency (Δf) between the two channels provides a direct result of the mass loading and electric field effects associated with the sensor layer

Conclusions

We have discussed several research areas where there appears to be a tangible benefit in using monomolecular assemblies rather than organic or inorganic thin films deposited by other means. There are many ways of producing organic films and the onus will be on the Langmuir-Blodgett enthusiasts to demonstrate the special advantages to be gained using their technique for a particular application. There is every likelihood that this will occur but it will require the combined efforts of high caliber teams with knowledge of physics, chemistry, electronics and biology. The ability of the synthetic chemist to manipulate the molecular architecture of a material to optimize a specific physical parameter or figure of merit will be vital. However, the role of the physicist or engineer in identifying the targets and guiding the main thrust of the research program will also be essential. Industry is already well able to organize interdisciplinary activities. The relatively inexpensive equipment requirements associated with the Langmuir trough technique coupled with the elegance of the fundamental science and the interesting applied prospects for LB films provides an excellent opportunity for the academic community to similarly break down the traditional barriers between disciplines. At the present time it would appear that non-linear optics, electron beam microlithography, magnetic tape lubrication, sensors, and optical storage, are the areas most likely in the short term to benefit from LB film technology.

The above comments referring to the need for multidisciplinary activity apply equally well to other fields covered by the definition of Molecular and Supermolecular Electronics. The goal of a supermolecular information processor based on organic assemblies is a very ambitious one. Nevertheless, pursuing this target should serve to enumerate technologies and identify areas of basic scientific research that would otherwise remain dormant and unexplored. Far simpler but novel devices based on organic molecular materials should appear as a result of this research effort. It seems likely that Langmuir-Blodgett films will play a key role in helping scientists identify basic physical phenomena in supermolecular assemblies and at the same time enable engineers to become more familiar with devices incorporating organic films.

Author's Acknowledgments

The author is indebted to his many colleagues and former colleagues at the Universities of Durham and Oxford for their important contributions to the research described in this article. Particular thanks are due to Dr. M.C. Petty and Mr. B. Holcroft.

Editor's Acknowledgments

This article, originally published in the *University of Wales Science and Technology* Review, is reprinted here with approval of J.H. Purnell, Honorary Editor. This prestigious journal, issued quarterly, contains dissertations on various topics of science and technology, written by professionals who have refreshingly new viewpoints. The *Review*, which commenced publication in March 1987, is a welcome contribution to the scientific community throughout the English-speaking world. Publication Office: MBN2 Cardiff, Marketing Department, 17th Floor, Pearl House, Greyfriars Road, Cardiff, Wales CF1 3XX, United Kingdom.

GARETH ROBERTS
FRS, Thorn EMI plc and University of Oxford

Additional Reading

Albrecht O., A. Laschewsky, and H. Ringsdorf: *J. Membrane Sci.*, **22**, 186 (1985).
Allara D. and J.D. Swalen: *J. Phys. Chem.* **86**, 2700 (1982).
Baker S., G.G. Roberts, and M.C. Petty: *Proc. IEE Pt. 1*, **130**, 260 (1983).
Barraud, A.: *Thin Solid Films*, **99**, 317 (1983).
Barraud A., P. Lesieur, A. Ruaudel-Teixier, and M. Vandevyver: *Thin Solid Films*, **134**, 195 (1985).
Batey J., G.G. Roberts, and M.C. Petty: *Thin Solid Films*, **99**, 283 (1983).
Blodgett, K.B.: *J. Amer. Chem. Soc.*, **57**, 1007 (1935).
Brundle C.R., H. Hopster, and J.D. Swalen: *J. Chem. Phys.*, **70**, 5190 (1979).
Carter, F.L.: *Superlattices and Microstructures*, **2**, 113 (1986).
Christie P., G.G. Roberts, and M.C. Petty: *Appl. Phys. Letts.*, **48**, 1101 (1986a).
Christie P., C.A. Jones, M.C. Petty, and G.G. Roberts: *J. Phys.*, **D19**, L167 (1986b).
Clark D.T., Y.C.T. Fok, and G.G. Roberts: *J. Electron. Spectroscopy*, **22**, 17 (1981).
Dharmadasa I.M., G.G. Roberts, and M.C. Petty: *Electronics Letts.*, **16**, 201 (1980).
Franklin, B.: *Phil. Trans. Roy. Soc.*, **64**, 445 (1974).
Gaines, G.L.: *Thin Solid Films*, **99**, ix (1983).
Gaines, G.L.: *Insoluble Monolayers at Liquid-Gas Interfaces*, Interscience, New York, NY, 1966.
Gaines, G.L.: *Thin Solid Films*, **132, 133, 134** (1985).
Holcroft B., M.C. Petty, G.G. Roberts, and G.J. Russell: *Thin Solid Films*, **134**, 83 (1985).
Hua Y.L., G.G. Roberts, M.M. Ahmed, M.C. Petty, M. Hanack, and M. Rein: *Phil. Mag.*, **B53**, 105 (1986).
Kelly M.J. and Weisbach, C.: *The Physics and Fabrication of Microstructures and Microdevices*, Springer-Verlag, Berlin (1986).
Kuhn, H.: *Thin Solid Films*, **99**, 1 (1983).
Langmuir, I.: *J. Amer. Chem. Soc.*, **39**, 1848 (1917).
Langmuir, I.: *Trans. Faraday Soc.*, **15**, 62 (1920).
Mobius, D.: *Accts. Chem. Res.*, **14**, 63 (1981).
Mori C., H. Noguchi, M. Mizuno, and T. Watanabe: *Jap. J. Appl. Phys.*, **19**, 725 (1980).
Neal D., M.C. Petty, G.G. Roberts, M.M. Ahmed, W.J. Feast, I.R. Girling, N.A. Cade, P.V. Kolinsky, and I.R. Peterson: *Proc. Int. Symp. on the Applications of Ferroelectric Materials*, Philadelphia (1986).
Petty M.C., J. Batey, and G.G. Roberts: *IEE Proc. Pt. 1*, **132**, 133 (1985).
Pockels, A.: *Nature*, **43**, 437 (1891).
Pomerantz, M.: *Phase Transitions in Surface Films*, Plenum Press, New York, NY, 1980, p. 317.
Rayleigh, Lord: *Proc. Roy. Soc.*, **47**, 364 (1890).
Roberts, G.G., K.P. Pande, and W.A. Barlow: *Proc IEE Pt. 1*, **2**, 169 (1978).
Roberts, G.G. and C.W. Pitt: *Thin Solid Films*, **99** (1983).
Roberts, G.G.: *Advances in Physics*, **34**, 475 (1985).
Roberts, G.G., B. Holcroft, J. Ross, and A. Barraud: *British Polymer Journal*, to be published (1987).
Tabor, D.: *J. Colloid and Interface Science*, **75**, 240 (1980).
Vincett P.S., W.A. Barlow, F.T. Boyle, J.A. Finney, and G.G. Roberts: *Thin Solid Films*, **60**, 265 (1979).
Vincent, P.S. and G.G. Roberts: *Thin Solid Films*, **68**, 135 (1980).
Zyss, J.: *J. Non. Cryst. Solids*, **47**, 211,–(1982).

MOLECULAR BEAM. A unidirectional stream of neutral molecules passing through a vacuum, generally with thermal velocity. Such a beam may be produced by emergence from a pinhole in a chamber containing low pressure gas or vapor, and it may be defined by a system of slits. By passing the beam through known electric or magnetic fields, quantities such as nuclear magnetic moments can be determined.

MOLECULAR BIOLOGY. A self-defining term—the study of biological substances and phenomena at the molecular level.

For many years, biologists, aided by numerous laboratory instruments and methodologies (X-ray crystallography, electron microscopy, chromatography, electrophoresis, etc.) constructed a vast databank of biochemistry. This has been and continues to be of inestimable value to those professionals who deal with health, medicine, the industrial use of biochemicals, among other areas that require such information. There remained, however, a dissatisfaction concerning the absence of knowledge pertaining to the manner in which biochemical processes are directed and take place at the molecular level.

The first break occurred early in the 1950s when very illuminating findings were made concerning the complex biomolecule, DNA. The research of DNA, in essence, constituted the beginnings of *molecular biology*. This pioneering work altered the thrust of biochemical research, as previously typified by traditional biochemistry, to a concentration and emphasis on studying biological events at the molecular level.

A second break occurred in the early 1970s when it was discovered that a strand of the DNA molecule can be cut by restriction enzymes and that the sticky ends can be reassembled by a new technology known as *genetic recombination* or *recombinant DNA*. This is frequently the starting point for projects engaged in biological research at the molecular level. It is no surprise that molecular biology has become a highly diverse, wide-spectrum, multifaceted discipline. The many hundreds of research projects in this area underway today obviously cannot be delineated here. The highlights of just a few research targets are given here to provide a representative view of projects underway and inroads made to date.

Cell Membrane Research

Cellular membranes serve as the interface between the cell and the organism of which it is a part. It has been established that the movement of cells, as directed during growth and development, must involve, in some way, the plasma membranes. It also has been observed that plasma membranes play some role in cancerous growth, where cell multiplication and migration proceed at an uncontrolled rate. Membranes are composed of lipids that interact with each other in a watery medium to form a closed and flexible compartment. This formation process occurs at a relatively rapid rate. Bretscher reports that an area of membrane equivalent to the area of the entire surface of the cell requires less than an hour to form. A good start has been made by a number of investigators toward understanding the detailed molecular structure of membranes. In retrospect, it is interesting to note that recent findings pertaining to the basic frame of membranes was qualitatively proposed as early as 1925 by Gorter and Grendel (University of Leiden). This framework is composed of a double layer of lipid molecules. One end of the molecule is hydrophilic, that is, soluble in water; the other end is hydrophobic, a hydrocarbon that is oily and insoluble in water. Most commonly, membrane lipids are phospholipids. See also **Lipids**.

Cytoplasm Research

The existence of fibers in cytoplasm was first observed microscopically about a century ago. It has been learned during the last few years as the result of biochemical and immunological studies, that a distinct set of proteins characterizes each filament system. Proteins in the cytoplasm make up a highly structured, yet changeable matrix. The matrix determines the cell shape, its division and motion, as well as the transport of vesicles and organelles. Research in molecular genetics is targeted to reveal the function of cytoskeletal proteins. Weber and Osborn report that current knowledge of the cell matrix is contributing to diagnosis and research in human pathology. For example, typing of intermediate filaments by way of immunofluorescence microscopy will distinguish the major tumor groups. Keratins are found in carcinomas. What is missing is detailed knowledge at the molecular level of why and how the protein molecules in the cytoplasm function.

Immune System Research

This is one of the most active research areas as of the late 1980s. It has been observed that proteins responsible for recognizing foreign invaders are the most diverse of all known proteins. Tonegawa observes that these proteins are encoded by hundreds of scattered gene fragments and these can be combined in millions, probably billions of ways. This gives some idea of how complex research in this area really is. It is one of the most difficult of tasks for molecular biologists, but is proceeding apace.

Cellular Intercommunications Research

It is well known that the majority of higher organisms utilize hormones or systems of neurons for intercellular communication. Snyder observes that chemical messengers mediate long-range communication via hormones; short-range communication via nerve cells. The two systems differ in directness, but some messenger molecules are common to both. Hormonal molecules are usually peptides and steroids. Communication between cells or groups of cells is mandatory for the survival of all multicellular organisms. Snyder also observed that as investigators come to know the properties and functions of highly specialized messenger molecules, it will be possible to develop better therapeutic agents for a number of diverse conditions, including hormonal abnormalities, heart disease, and mental illness.

Cellular Intracommunications Research

The communication within a cell is being investigated at the molecular level. The mechanism by which a cell receives external signals by way of receptors has been reasonably well understood for a number of years. But what comprises the cell's internal mechanism for reacting to received signals by way of the receptors? Berridge refers to receptors as "molecular antennas." Obviously, the barrier to the flow of information from receptor to mechanisms within the cell is the plasma membrane, previously mentioned. How are external signals acted upon internally within the cell to cause it to secrete, to contract, to metabolize, or to grow, among several other biological reactions? What are the "second messengers" within the cell? How do they function? A second messenger, cyclic adenosine monophosphate (cyclic AMP) has been identified. Other second messengers include Ca^+ ions, inositol triphosphate (IP_3), and diacylglycerol (DG). Berridge also reported that some evidence has been gained to the effect that the aforementioned substances are cannibalized from the plasma membrane itself. Obviously, the cell must include genes required for the synthesis of the proteins used in the internal pathways. Aberration of these gene functions could lead to abnormalities of cellular growth and hence to uncontrolled growth and structural transformations

typical of cancer. Cells related in some way to tumor growth revealed thus far are called *oncogenes*. Berridge and other researchers are attempting to identify additional second messengers that may be present and how they function in internal communication.

Biological Development and Growth Research

For many years, the processes which take place within an organism to cause it to grow (the early embryonic phase), leading to maturity (adulthood) and later to initiate the aging phase have been among the least understood of biological phenomena. As aptly observed by Gehring: How is the basic architecture of an embryo laid down? How does the linear information of DNA generate a three-dimensional organism? One of the most recent and intriguing discoveries is that of the so-called *homeobox*, which is found as a short stretch of DNA. When the gene containing the homeobox is translated into a protein, the homeobox yields a string of amino acids that are believed to bind to the DNA double helix. By binding to the DNA of particular genes, the protein may be able to turn them on or off. The homeobox was first noted in studies of *Drosophila*; subsequent studies have revealed a homeobox within a range of organisms extending from worms to humans. It is well established, of course, that animals develop in numerous ways. Researchers now question whether or not the molecular mechanisms underlying development may be more universal than previously suspected. Again, it appears that the answers to still another of the biological secrets may lie in better understanding the process at the molecular level.

Adaptive and Evolutionary Processes

Wilson observes that in the past most biologists working in this area have concentrated their attention at the level of the whole organism. Mutation frequently has been suggested as the cause of adaptive changes. Prior to studies at the molecular level, however, it was *not* known that mutations may accumulate at steady rates over time in the genes of all lineages. Wilson, a proponent of the molecular evolution concept, reports on two assumptions: (1) Heritable differences among organisms result from differences in their DNAs, and (2) molecular evolutionists must not only measure differences in DNA, but also explain the origin of the differences and their relation to organismal differences. Wilson also points out that two critical elements of molecular evolution are (1) *point mutations*, specifically those occurring in the genes coding, and (2) *regulator mutations*. A point mutation is defined as a single replacement of a DNA base; a regulatory mutation is any change in a gene or within the vicinity of a gene that determines whether the gene is active or inactive. Research on point mutations led to the concept of a *molecular clock* and in the discovery of still another kind of genetic change, which Wilson calls a *neutral mutation*. This mutation is neither advantageous nor disadvantageous for an organism. Wilson also refers to the "pressure" to evolve, noting that this stems from geologic forces as well as from the brains of mammals and birds. It is obvious that the study of molecular evolution occupies a special position in contemporary biology.

Additional Reading

Ausubel, F. et al.: *Short Protocols in Molecular Biology: A Compendium of Methods from Current Protocols in Molecular Biology,* John Wiley & Sons, Inc., New York, NY, 1999.

Berridge, M.J.: "The Molecular Basis of Communication within the Cell," *Sci. Amer.,* 142–148 (October 1985).

Lackie, J. and J. Dow: *The Dictionary of Cell and Molecular Biology,* 3rd Edition, Academic Press, Inc., San Diego, CA, 1999.

Stahl, W. and A. Hershey: *We Can Sleep Later: Alfred D. Hershey and the Origins of Molecular Biology,* Cold Spring Harbor Laboratory Press, Cold Spring Harbor, NY, 2000.

Synder, S.H.: "The Molecular Basis of Communication Between Cells," *Sci. Amer.,* 132–137 (October 1985).

Tonegawa, S.: "The Molecules of the Immune System," *Sci. Amer.,* 122–128 (October 1985).

Weber, K. and M. Osborn: "The Molecules of the Cell Matrix," *Sci. Amer.,* 110–118 (October 1985).

Wilson, A.C.: "The Molecular Basis of Evolution," *Sci. Amer.,* 164–170 (October 1985).

MOLECULAR DISTILLATION. A special form of distillation conducted at pressures of 1–7 micrometers in the laboratory and 3–30 micrometers in industrial applications. Compared with conventional laboratory vacuum distillations carried out between 1 and 10 millimeters of mercury pressure, this is a very high vacuum. One micrometer equals 0.001 millimeter of mercury pressure. The other feature of the molecular still is that the condenser is located within a distance less than the mean free path of the evaporating molecules from the evaporator portion of the apparatus. Thus, although a molecule may return to the distilland many hundred times before reaching the exit of a conventional vacuum still, 50% of the molecules in a properly functioning molecular still will reach the exit on their first try. Thus, efficiency is remarkably high.

Because of the absence of convection due to ebullition and because high viscosities and high molecular weights may impede diffusion within the distilland, the surface of the distilland in a molecular still may not always represent the total liquid. Therefore, efficient molecular distillation requires the mechanical renewal of the surface film. This is achieved by vigorous agitation, as in the *stirred-pot still*; by employing a *falling film*; or by using centrifugal force, as in the *centrifugal* molecular still. Commercial installations of falling-film stills achieve throughputs of many tens of liters per hour, whereas the centrifugal still is capable of several hundred liters per hour. Centrifugal stills usually are arranged in groups of from three to seven. This permits fractionation by multiple redistillation. Among the uses of molecular distillation are: the separation of mono and diglycerides for bread and paraffin wax for milk cartons; the distillation of plasticizers, fatty acid dimers, and synthetics; the distillation of vitamin A esters and intermediates; and the stripping of α, β, γ, and δ-tocopherols and sitosterols from vegetable oils.

See also **Distillation**.

MOLECULAR MODELING.

Molecular modeling refers boldly to any study of molecules utilizing physical or theoretical models to explain an observed or predicted behavior. Whereas molecular modeling as a practice has its roots in the development of quantum theory at the turn of the twentieth century, it was the exponential growth in computing power between the mid-1970s and the mid-1990s that catalyzed the development and application of molecular modeling methods during that period. The spectrum of software systems available covers all aspects of modeling.

Molecular modeling can be defined as the application of computational techniques, grounded in theory, to predict or explain observable biological or physical chemical properties. Wherever molecular modeling is practiced using a computer, the technique then becomes computer-assisted (aided) molecular modeling, or CAMM. CAMM, is often used synonymously with CAMD, or computer-assisted molecular (materials) design/discovery. CADD refers to computer-assisted drug design/discovery. A computational technique as used herein is a mathematical model derived from principles of chemistry, physics, or statistics which facilitates molecular modeling. An entire branch of chemistry, i.e., computational chemistry, is devoted to developing, benchmarking, and applying computational techniques in order that researchers may be able to better understand and predict properties. Some of the properties which may be calculated either exactly or approximately by computational methods are the following:

boiling points	dipole moments
melting points	quadrupole moments
crystallization energy	octupole moments
heat capacity	infrared spectra/intensities
heat of formation	nmr spectra/chemical shifts
heat of fusion	optical rotary dispersion
heat of sublimation	raman spectra
heat of vaporization entropy	ultraviolet spectra
molar refractivity	ionization potentials
molar volume	electron affinities
partition coefficients	protonation energies and pK_as ionic strength
radius of gyration elasticy	conformational energies
tensile strength	Boltzmann distributions

Molecular properties can be classified according to their end-point observables, such as chemical (reactivity, solubility, acid–base), physical (a function of physical state: gas, liquid, solid; thermodynamic), or biological (ligand or enzyme; agonist or antagonist). These properties reflect macroscopic, or bulk, properties, which exist only for the bulk material, e.g., heat of crystallization, or microscopic properties, which exist for an ensemble of the molecule. As use of CAMM methods

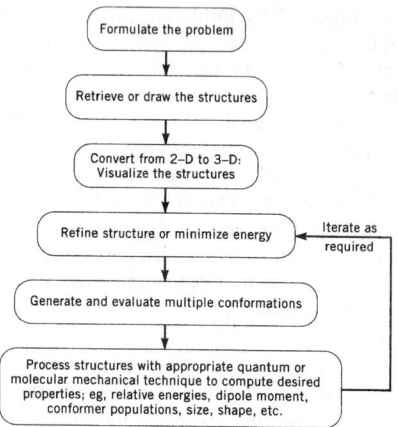

Fig. 1.   Flow chart for a typical small molecule modeling project

expands to address a broader horizon of applications beyond those in organic, medicinal, and biological chemistry, calculations on metals, semiconductors, and magnetic systems have become more common.

The specific process involved in a given molecular modeling study depends significantly on the nature of the primary objective of the task. However, for many small-molecule and even macromolecular studies, a number of authors have diagrammed the individual steps in the process, and the flowchart in Figure 1 summarizes their efforts. An initial step that is critical to any CAMM project is the generation or retrieval of the pertinent structures themselves. Structures may range from simple organic molecules or monomers of more complex polymers, to full proteins, enzymes, metal surfaces, or zeolites. The modeling process can be influenced by the initial structure and its geometry. Thus, the selection and development of the starting molecular geometry needs to be given particular attention. An important component of the quality control process, as well as in gaining an understanding of the molecules themselves, is the visual examination of structures involved in a modeling study.

Computer Graphics In Molecular Modeling

The goal of molecular modeling is to define clearly the relationship between chemical constitution, i.e., the molecular formula or a topographic representation thereof, its geometric constitution or 3-D topology (the disposition of its atoms in Cartesian space), and its observed (or predicted) properties. The representation and facile manipulation of 3-D arrays of atoms comprise the domain of molecular or computer graphics. The growth of graphics tools has paralleled the evolution of computing hardware. Computing software and hardware systems have had a profound impact on the ability of modelers to compose a modeling study and address all aspects of the work, ranging from generating 2-D drawings of structures to statistical quality control of computed properties via visualization of multidimensional data.

One facet of chemistry which has benefited greatly from the use of computer graphics to enhance its own development is x-ray crystallography; computers are also used in x-ray structure refinement.

Among the available software systems which together help to exemplify the promise of computer graphics, Advanced Visualization Systems (AVS), stands out as one of the more extensible and practical of them to use. The premise behind development of the system was to provide modelers with a toolkit of modules having sophisticated intrinsics that would enable even casual programmers to link together multiple simple functionalities into a complex construct with which to accomplish exactly the types of visualization and manipulations that their work required.

Computational Methods for Molecular Modeling

The ultimate goal of quantum mechanical calculations as applied in molecular modeling is the *a priori* computation of properties of molecules with the highest possible accuracy (rivaling experiment), but utilizing the fewest approximations in the description of the wave-function. *Ab initio*, or from first principles, calculations represent the current state of the art in this domain. *Ab initio* calculations utilize experimental data on atomic systems to facilitate the adjustment of parameters such as the exponents of the Gaussian functions used to describe orbitals within the formalism.

The performance of *ab initio* techniques distinguishes them significantly from their predecessors, semiempirical methods. Their consistent reproduction of data from structural, thermodynamic, and reaction sources to a range falling within the error limits of the experimental values provides scientists with an important tool with which to address various modeling problems. Whereas the absolute value of the relative performance of *ab initio* techniques varies for each structural or energetic feature examined, it is not unreasonable to suggest that if the quantity can be computed by both semiempirical and *ab initio* methods, the *ab initio* value will be closer to experiment or an ideal value than any other method.

Molecular Mechanics and Molecular Dynamics. In the realm of quantum mechanics, researchers deal explicitly with electrons, with their interactions, and with their attraction to nuclei, albeit in varying degrees depending on the rigor of the method chosen to solve the particular problem involved. These techniques were somewhat limited in their application prior to the age of large-scale computers and super workstations. Computers made possible the development of molecular mechanics, or empirical force field methods and, ultimately, the application of these methods to a full spectrum of studies in structure and energetics.

Molecular Mechanics. Molecular mechanics (MM), or empirical force field methods (EFF), are so called because they are a model based on equations from Newtonian mechanics. This model assumes that atoms are hard spheres attached by networks of springs, with discrete force constants. The force constants in the equations are adjusted empirically to reproduce experimental observations. The net result is a model which relates the "mechanical" forces within a structure to its properties. Force fields are made up of sets of equations each of which represents an element of the decomposition of the total energy of a system (not a quantum mechanical energy, but a classical mechanical one). The sum of the components is called the force field energy, or steric energy, which also routinely includes the electrostatic energy components. Typically, the steric energy is expressed as

$$E_{\text{Total}} = E_{\text{steric}} + E_{\text{electrostatic}} = E_{\text{bonds}} + E_{\text{angles}} + E_{\text{vdW}}$$
$$+ E_{\text{torsion}} + E_{\text{charge/dipole}}$$

The overall form of each of these equations is fairly simple, i.e., energy = a constant times a displacement. In most cases the focus is on differences in energy, because these are the quantities which help discriminate reactivity among similar structures. The computational requirement for molecular mechanics calculations grows as n^2, where n is the number of atoms, not the number of electrons or basis functions. These calculations will be much faster than an equivalent quantum mechanical study. The size of the systems which can be studied can also substantially eclipse those studied by quantum mechanics.

In a force field calculation, a molecule in three dimensions is constructed using either Cartesian coordinates x, y, and z, or via an internal coordinate matrix consisting of bond distances, bond angles, and dihedral angles to specify the atoms' unique positions. Then the initial structure is evaluated to determine the extent to which each degree of freedom (bonds, angles, etc) deviates from the ideal (the zero-energy value) for the particular element and its hybridization. An energy minimization process follows wherein the energy associated with the distortions from ideal is minimized as the individual atomic positions or degrees of freedom are adjusted. Iteratively, this converges on a "minimum energy" or an "optimized" structure. This structure represents the best attempt of the minimization algorithm to render the smallest deviations in position of each of the atoms such that either the derivatives of the change in energy associated with the deviations are the smallest, or they satisfy either energetic convergence or coordinate change criteria from iteration to iteration. This process is analogous to the geometry optimization process within a quantum mechanical program, except that there the objective is to converge on a structure which yields the smallest energy derivatives and lowest total energy from solution of the SCF equations. Most simple molecular mechanics force fields include terms (Fig. 2) for bond stretching, bond angle distortion (bending), dihedral angles, van der Waals nonbonded interactions, Coulombic interactions (dipoles or charges).

Force Fields, Molecular Dynamics, and Vibrational Spectroscopy. The link between molecular mechanics and molecular dynamics comes about through the force field itself. In molecular mechanics, the main interest is in computing the energy of molecules in the gas phase at room temperature in a single, discrete configuration and conformation; time is not a variable

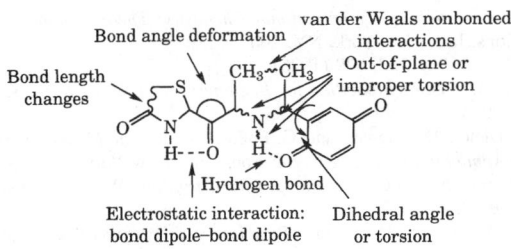

Fig. 2. Structural representation of the energetic components of a typical molecular mechanics force field

in the equations. From molecular dynamics, the objectives are properties which represent a time-averaged ensemble of states, including, for example, conformationally excited states, rotational states, interconversion rates, or inversion barriers for amines or amides. From this ensemble of states, characterizing the existence of the excited states and their contribution to the total energy of the system (from a Boltzmann distribution) is next. From an understanding of the vibrational spectroscopic roots of molecular mechanical force fields used in dynamics simulations, molecular dynamics may be seen as an extension bridge between theory and experiment, linking "static" molecular mechanical representations of properties with "dynamical" experimental properties. Applications of force field techniques to problems in environmental chemistry, materials science, and molecular biology demand that the methods go substantially beyond those for which reliable experimental data is available. Only since the time computers have made larger-scale *ab initio* quantum mechanical calculations practical for appropriate model systems have these techniques been reliable for such a broad spectrum of important applications.

Molecular Dynamics and Monte Carlo Simulations. At the heart of the method of molecular dynamics is a simulation model consisting of potential energy functions, or force fields. Molecular dynamics calculations represent a deterministic method, i.e., one based on the assumption that atoms move according to laws of Newtonian mechanics. Molecular dynamics simulations can be performed for short time-periods, e.g., 50–100 picoseconds, to examine localized very high frequency motions, such as bond length distortions, or, over much longer periods of time, e.g., 500–2000 ps, in order to derive equilibrium properties. Those that can be evaluated by performing molecular simulations are conformational states and energetics, kinetic properties: rates of reaction and interconversion, reaction pathways, solubilities, diffusion rates, binding and complexation data, folding processes, transition temperatures, free energies for point mutations, and free energies of binding. The molecular simulations can also serve as an adjunct to x-ray and nmr for structure refinements.

As noted, force fields are a set of equations relating the total energy of the system to its individual interaction components. After selection of a force field simulation program which is appropriate to a given problem, the general procedure includes (*1*) initial structure equilibration, and (*2*) structure refinement.

Monte Carlo (MC) techniques for molecular simulations have been used to a great extent in studying the chemical physics of polymers. The majority of molecular modeling studies today do not involve the use of MC methods; however, the sampling capability provided by MC methods has gained some popularity among computational chemists.

Combined Quantum and Molecular Mechanical Simulations. In this technique, a molecular dynamics simulation includes the treatment of some part of the system with a quantum mechanical technique. This approach, QM/MM, is similar to programs that use quantum mechanical methods to treat the π-systems of the structures in question separately from the sigma framework. The results are combined at the end to render a structure which is optimized and energy-refined to satisfy both self-consistent field (SCF) and force field energy convergence.

SALVATORE PROFETA, JR.
Monsanto

Additional Reading

Bolcer, J.D. and R.B. Hermann: in K.B. Lipkowitz and D.B. Boyd, eds., *Reviews in Computational Chemistry*, Vol. 5, VCH Publishers, New York, NY, 1994, pp. 1–63.

Cramer, C.J.: *Essentials of Computational Chemistry: Theories and Models,* John Wiley & Sons, Inc., New York, NY, 2002.

Heisenberg, W.: *Z. Phys.* **33**, 879 (1925).

Hinchliffe, A.: *Molecular Modelling for Beginners,* John Wiley & Sons, Inc., New York, NY, 2003.

Holtje, Hans-Dieter, D. Rognan, and G. Folkers: *Molecular Modeling: Basic Principles and Applications,* John Wiley & Sons, Inc., New York, NY, 2003.

Jensen, F.: *Introduction to Computational Chemistry,* John Wiley & Sons, Inc., New York, NY, 1998.

Leach, A.R.: *Molecular Modelling: Principles and Applications,* 2nd Edition, Prentice Hall Professional Technical Reference, Upper Saddle River, NJ, 2001.

Schrödinger, E.: *Ann. Phys.* **79**, 361, 489; **80**, 437; **81**, 109 (1926).

Schlick, T.: *Molecular Modeling and Simulation,* Springer-Verlag New York, Inc., New York, NY, 2002.

MOLECULAR ORBITALS.   See **Orbitals**.

MOLECULAR RECOGNITION.   Molecular recognition implies complementary lock-and-key type fit between molecules. The lock is the molecular receptor and the key is the substrate that is recognized and selected to give a defined receptor-substrate complex, a coordination compound or a supermolecule. Hence molecular recognition is one of the three main pillars, fixation, coordination, and recognition, that lay the foundation of what is known as supramolecular chemistry.

Supramolecular chemistry, the chemistry beyond the molecule, is a highly interdisciplinary field of science covering the chemical, physical, and biological features of chemical species of greater complexity than molecules themselves that are held together and organized by means of intermolecular (nonbinding) interactions. The chemistry of molecular recognition is also the core of host–guest chemistry, which is a subdiscipline or a particular aspect of supramolecular chemistry mostly involving inclusion and complex formation. See also **Inclusion Compounds**.

Principles of Receptor Design

Information Storage and Read Out. Molecular recognition is defined by the energy and the information involved in the binding and selection of substrates by a given receptor molecule that may also involve a specific function. Mere binding is not recognition, although it is often taken as such. Instead, one may say that recognition is binding with a purpose, like receptors are ligands with a purpose. It implies a pattern recognition process through a structurally well-defined set of intermolecular interactions. Molecular recognition, thus, deals with the molecular storage and supramolecular readout of molecular information.

Information may be stored in the architecture of the receptor, in its binding sites, and in the ligand layer surrounding the bound substrate such as specified in Table 1. It is read out at the rate of formation and dissociation of the receptor-substrate complex. The success of this approach to molecular recognition lies in establishing a precise complementarily between the associating partners, i.e., optimal information content of a receptor with respect to a given substrate.

TABLE 1. STRUCTURAL PARAMETERS FOR STORAGE OF INFORMATION IN A CHEMICAL RECEPTOR

Receptor	Parameter
architecture	size
	shape
	connectivity
	cyclic order
	conformation
	chirality
	dynamics
binding sites	electronic properties (charge, polarity, polarisability, van der Waals attraction and repulsion)
	size
	shape
	number
	arrangement
	reactivity (protonizable, deprotonizable, reducible, oxidizable)
surrounding ligand layer	thickness
	overall polarity (lipophilic, hydrophilic)
	specific polarity (exo/endo-lipo/polarophilic)

Complementarity. To a first approximation, complementarity should take two forms. Firstly, the shape and size of the receptor cavity must complement the form of the substrate. Secondly, there must be a chemical complementarity between the binding groups lining the interior of the cavity and the external chemical features of the substrate.

The weak intermolecular forces that are principally involved in stabilizing receptor-substrate interactions and involved in molecular recognition processes are summarized in Table 2. Examples are shown in Figure 1.

Reorganization and Preorganization. On principle there are two different modes of receptor behavior. One of them is the so-called lock-and-key image involving complementary fit concept between rigid substrate and rigid receptor or rigid guest and rigid host relating to conformational flexibility of the molecular constituents forming the receptor–substrate (host–guest) complex. Receptors of this type are expected to present very efficient recognition between complementary partners, i.e., both high stability and high selectivity of the receptor–substrate complex.

However, in most biological systems there is a degree of flexibility in the receptor. The approach of the substrate leads to conformational changes and an organization of the binding site around it. With this induced fit mechanism of binding, a higher entropy price is paid, but there are several advantages. A flexible receptor will permit a more wraparound interception or even complete encapsulation with the substrate involving many more potential binding interactions. This may lead to high selectivity of binding involving the amplification of molecular recognition interactions.

The balance between rigidity and flexibility is of particular importance for the binding and the dynamic properties of a receptor. It is, thus, a decisive structural design parameter of the receptor depending on the use. For instance, processes of exchange regulation, cooperativity and allostery connected with molecular recognition require a built in flexibility so that the receptor may adapt and respond to changes, unlike rigid receptors.

Topology. This parameter may have reference to either the receptor as an individual molecular structure or to the receptor–substrate complex on a higher level of organization that is directly related to the mode and efficiency of molecular recognition.

A concave receptor is a favorable case. Under these circumstances the receptor cavity is lined with binding sites directed towards the bound species (see Fig. 1). This corresponds to Cram's definition of a receptor (host) molecule providing binding sites that are convergent, as contrasted with the bound substrate (guest) featuring divergent complementary sites, i.e., the substrate is more or less completely surrounded by the receptor, forming an inclusion complex. This widely used principle of convergence defines a convergent or endo-supramolecular chemistry (host–guest chemistry) with endo-receptors (endo-hosts) effecting endo-recognition (Fig. 2**a**).

The opposite procedure consists in making use of an external receptor surface rather than an internal cavity as substrate receiving site. This

TABLE 2. TYPES OF INTERACTIONS IN MOLECULAR RECOGNITION

hydrogen bonding between basic and acidic center
elastrostatic attraction between anionic and cationic centers
metal-ligand interaction
dipole-dipole interaction
π-stacking and charge-transfer interaction between aromatic residues in the receptor and delocalized regions of the substrate
van der Waals attraction between hydrophobic regions on the two components
covalent bonds, that can be reversibly formed and broken (e.g., disulfides, borate esters).

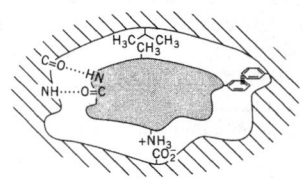

Fig. 1.   Schematic representation of a receptor–substrate (host–guest) complex involving cavity inclusion of the substrate and the formation of different types of weak supramolecular interactions between receptor (hatched) and substrate (dotted)

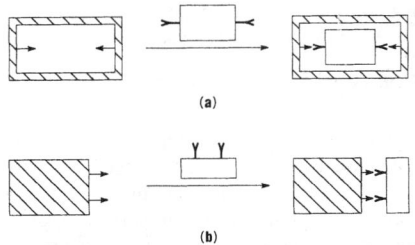

Fig. 2. Diagram of (**a**) endo- and (**b**) exo-receptor recognition of substrates (dotted rectangles)

amounts to the passage from a convergent endo-supramolecular chemistry to a divergent or exo-supramolecular chemistry, and from endo- to exo-receptors (Fig. 2**b**). Here receptor–substrate binding occurs by surface-to-surface interaction which may be termed affixation as contrasted with inclusion. Exo-recognition with strong and selective binding, in particular, requires a large enough contact area and a sufficient number of complementary interactions along the interface. Such a mode of molecular recognition also finds biological analogies, for instance at the antibody–antigen interface of immunological importance. Metallo-exoreceptor aggregation, molecular recognition at organic and inorganic monolayers, films and solid surfaces bearing recognition groups, as well as the design of supramolecular solid architectures and materials, are other important instances of the exo-recognition principle.

Simple Modes of Molecular Recognition

Substrates involved in molecular recognition may feature a particular shape, size, state of charge, chemical affinity or optical specification. In general most of these parameters share. Nevertheless there may be dominating

features of a certain substrate molecule to be used by a complementary receptor in the recognition process.

Size and Shape Dominated Substrate Recognition. Perhaps the simplest recognition process is that of a spherical substrate, in its most elementary form a ball-shaped metal ion of defined diameter. Three main classes of receptors provide the spherical recognition property. They are (*1*) macrocyclic polyethers, the well-known crown ethers and their derivatives; (*2*) the macropolycyclic cryptands; and (*3*) the acyclic analogues of crown compounds and cryptands, usually designated as podands. Prototypical compounds for each substance class are given by compounds (**1**)–(**3**) (Fig. 3). They all possess a spherical or quasispherical negatively polarized cavity prepared for the accommodation of alkali- and alkaline-earth metal ions that have complementary size, giving rise to a feature known as spherical recognition. Coronates, cryptates or podates are the names of the respective inclusion complexes. See also **Inclusion Compounds**.

Although acyclic podands do not provide a permanent cavity, they may create one by encircling a spherical cation with the length of the receptor molecular thread being the controlling parameter. Nevertheless, from what has already been said, low preorganization and topology of the podands handicap the substrate recognition which is increasingly higher in the circular crown and spheroidal cryptand case, but is most pronounced for the spherand type of receptor (e.g., **4** in Fig. 3).

Natural macrocycles displaying antibiotic properties are also very efficient in the recognition of alkali metal ions. For instance, valinomycin (**5** in Fig. 3) gives a strong and selective complex in which a K$^+$ ion is included in the macrocyclic cavity in octahedral environment of six carbonyl oxygens (Fig. 4).

Recognition of a tetrahedral substrate geometry requires the construction of a receptor molecule with a tetrahedral recognition site. This may be realized by positioning four suitable binding sites at the corners of a tetrahedron and incorporating them into a bridged molecular framework.

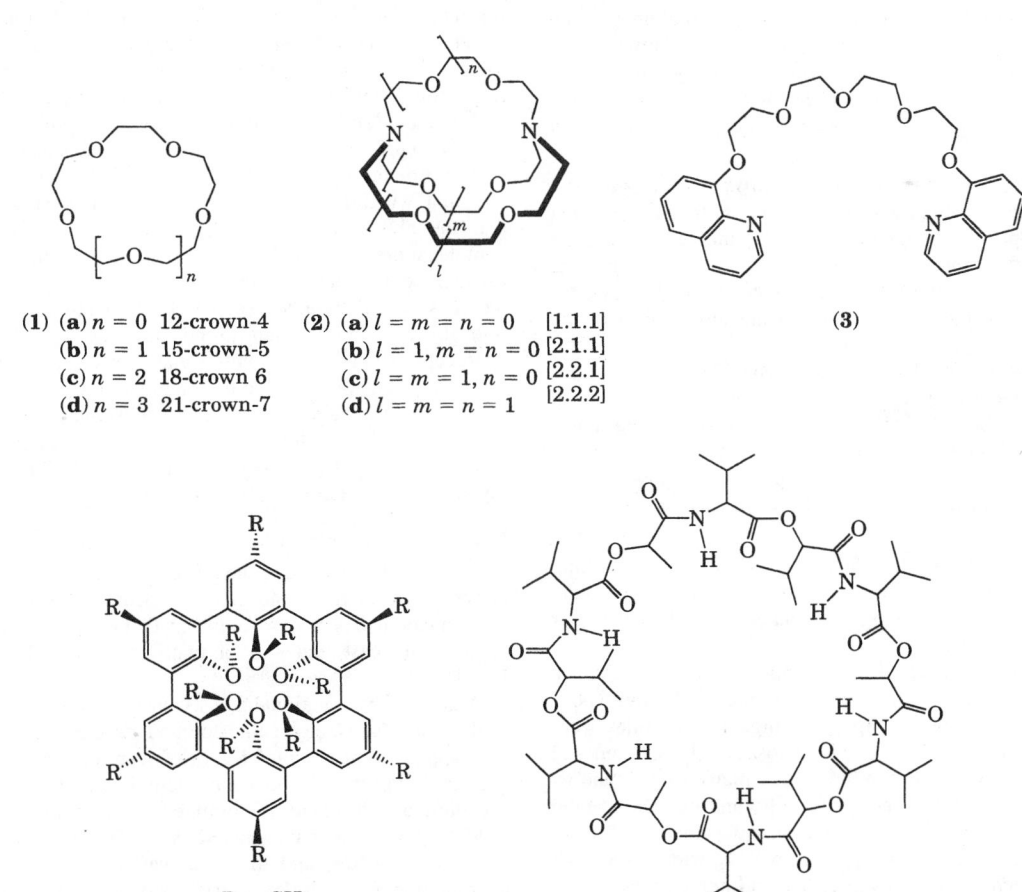

Fig. 3. Crown type and analogous receptor molecules of different varieties; (**1**) crown ethers; (**2**) cryptands; (**3**) a podand; (**4**) a spherand; and (**5**) the natural depsipeptide valinomycin

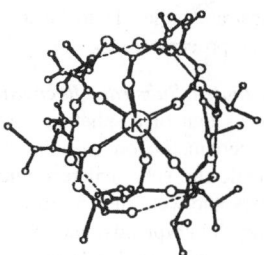

Fig. 4.   Spherical recognition of K^+ complex of valinomycin (**5**)

Charge Attraction Dominated Recognition. Thus far, for recognition sizes and shapes of the substrate have been the focus. Nevertheless, charge attraction between the substrate and the receptor has also played a part, because cations such as metal ions or ammonium ions were complexed by negatively polarized cavities. But metal ions involved hard alkali and alkaline-earth cations rather than weaker transition metal ions. Replacing the oxygen sites of crown compounds and cryptands with nitrogen and sulfur atoms yields receptors that show marked preference for transition metal ions and may allow highly selective recognition of toxic heavy metal ions such as cadmium, lead or mercury according to the hard and soft acid and base (HSAB) principle. Others containing internally directed functionalized units form very strong and selective complexes with Fe^{3+} or actinide and lanthanide ions, while a similar receptor with hard endocarboxylic acid groups is efficient for hard Ca^{2+} and Mg^{2+} ions, showing again responsibility of a charge density effect in the receptor–substrate recognition. Thus, recognition of hard alkali and alkaline earth metal ions is determined by coulombic attraction, whereas the weak transition metal ions are mainly controlled by geometrical parameters of orbital overlap.

Hydrogen Bond Dominated Recognition. Recognition of bioactive compounds is largely determined by the use of hydrogen bonding between polar sites. Here substrate recognition results from the formation of specific hydrogen bonding pattern between complementary subunits, in a way reminiscent of base pairing in nucleic acids. Hence hydrogen bonding has also been determined an important parameter for the design of artificial receptors.

π-Stacking and Charge-Transfer Dominated Substrate Recognition. Nature's strategy for the recognition of substrates featuring a flat aromatic frame-work (planar recognition) affords another recognition element, namely $\pi - \pi$ stacking interactions between aromatic rings, i.e., aromatic groups of receptor and substrate that meet a parallel face-to-face orientation, apart from hydrogen bonding being also typical of the nucleotide recognition.

In a way, $\pi - \pi$ stacking and charge transfer type of recognition have something in common. For example, a certain macrobicyclic intercaland and related receptors have been found to recognize flat shaped substrates through $\pi - \pi$ stacking and bind them to form a molecular cryptate, in particular if electron donating substrate species are involved to allow charge-transfer interaction, such as planar molecular anions or nucleic acids.

Lipophilic Interaction Dominated Substrate Recognition. Making recognition through lipophilic interaction possible requires receptors presenting large and more or less rigidly connected architectures of macrocyclic or cage-like nature.

The naturally occurring cyclodextrins having endo-lipophilic cone-shape are perhaps the most important and also the first receptor molecules whose selective inclusion properties towards lipophilic organic molecules were recognized. They comprise a family of cyclic oligosaccharides, composed of 6, 7, and 8 glucose units in its most familiar representatives (α, β, and γ-cyclodextrin, respectively) providing endo-lipophilic and exo-hydrophilic cone-shaped molecular cylinders of increasing size. Cyclodextrins form size and shape selective inclusion compounds with a wide variety of substrates including benzene derivatives, paraffins and noble gases.

Calixarenes (from the Latin *calix*) may be understood as artificial receptor analogues of the natural cyclodextrins. In its prototypical form they feature a macrocyclic metacyclophane framework bearing protonizable hydroxy groups made from condensation of *p*-substituted phenols with formaldehyde. Dependent on the ring size, benzene derivatives are the

substrates most commonly included into the calix cavity, but other interesting substrates such as C_{60} have also been accommodated.

Multiple and Multisite, Coreceptor- and Coupled-System Substrate Recognition

Once recognition units for specific groups and individual features of a substrate have been identified, one may consider combining several of them within the same receptor. Thus far, though not carefully directed, the previous receptors in many cases already possess this property of nonindividual interaction modes. More carefully directed, this leads to multiple and multisite recognition depending on the design of binding subunits which may cooperate for the simultaneous complexation of several substrates or of a multiply bound polyfunctional species to yield polynuclear complexes (homo- or heteronuclear) and mononuclear polyhapto-type complexes, respectively.

Chiral Recognition

Enantiomers are perhaps the substrate type most difficult to distinguish. They are stereochemical species that have exactly the same structure except for their mirror image (chirality) relationship See also **Chiral Separations.** Chiral (enantiomer) recognition in complexation is one of the most important means by which receptor sites of biological systems such as in genes or enzymes act and regulate. From the principle point of view, recognition of a substrate enantiomer from racemic mixture (50:50% mixture of enantiomers) requires an enantiomeric optically resolved receptor structure in order to make possible two diastereomeric receptor–substrate complexes allowing differentiation.

Following this line, a great variety of optically resolved (optically active) crown compounds were prepared for the resolution of racemic cationic substrates.

Artificial Receptors for Particular Substrate Recognition

Some particular substrates are biorelevant species or play central roles as drugs. Barbiturates are such an important family of drugs and are the target for molecular recognition. According to their structure, the barbiturate moiety essentially fuses two imide groups within a six-membered ring. Thus, two diaminopyridine units correctly positioned in a macrocyclic ring should bind to all six of the accessible hydrogen bonding sites in barbiturates. A crystal structure of a respective receptor–substrate complex has been performed that comes up to the expectations.

The structural and synthetic relationships shared between barbiturates and urea, which is another substrate of high physiological interest, suggest that the above receptor strategy could be modified for the selective complexation of urea. The designed modification for urea recognition involves replacement of the H-bond donating pyridine-6-amido groups in the previous barbiturate receptor by two H-bond accepting groups that differ by 120° in alignment to the substrate.

Receptors that are monomolecular species possessing a monomolecular cavity, pocket, cleft, groove or combination of it including the recognition sites to yield a molecular receptor–substrate complex can be assembled and preserved in solution. By way of contrast, molecular recognition demonstrated in the following comes from multimolecular assembly and organization of a nonsolution phase such as polymer materials and crystals.

Molecular Recognition in Polymers and Solids

If a polymer is prepared in the presence of molecules, the "print molecules" of which are extracted after polymerization, the remaining polymer may contain cavities, prints, or footprints that can recognize the print molecule. Actually, the cast relates to the matrix molecule like lock-and-key fit. (see Fig. 1).

A great many functionalized styrenes, including carboxylic acids, amino acids, Schiff bases, or specific compounds, e.g., L-DOPA, have successfully been applied as print templates. Moreover, it has also been shown that silica gel can be imprinted with similar templates, and that the resulting gel has specific recognition sites determined by the print molecule.

Microporous inorganic materials dominated historically by the zeolites and alumosilicates, and the great variety of more recent nonoxide and coordination framework materials should also be mentioned here. This type of molecular recognition is usually known as molecular sieving.

Molecular Recognition at Interfaces and Surface Monolayers

There are three advantages to study molecular recognition on surfaces and interfaces (monolayers, films, membranes or solids): (*1*) rigid receptor sites can be designed; (2) the synthetic chemistry may be simplified; (3) the surface can be attached to transducers, which makes analysis easier and may transform the molecular recognition interface to a chemical sensor. This kind of molecular recognition involves outside directed interaction sites, i.e., exo-receptor function (see Fig 2**b**).

Molecular recognition of crystal interfaces makes possible the control of crystal growth processes in that suitably designed auxiliary molecules act as promoters or inhibitors of crystal nucleation inducing, for instance, the resolution of enantiomers or the crystallization of desired polymorphs and crystal habits.

Following another direction, it has previously been shown that alkanethiols spontaneously adsorb to Au from dilute solutions of ethanol and other nonaqueous solvents, and that the resulting self-assembling monolayers (SAMs) assume a close-packed over-layer structure on Au and other textured Au surfaces, being quite robust in aqueous solutions and vapor-phase ambients. This mode of self-assembly chemistry has been used to synthesize monolayer assemblies that function as molecular recognition interfaces based on the presence of recognizer end groups.

Self-Recognition

This mode of molecular recognition, on principle, is defined as the recognition of like from unlike or self from unself molecules, embodied in the spontaneous selection and preferential assembly of like components in a mixture.

So far this article has been concerned with interactions among chemically different species, which is true for most of the chemical recognition processes, including supramolecular and biomolecular processes. With crystals it is usually the other way around. Although some crystals, co-crystals, crystalline complexes, and crystalline inclusion compounds are built from more than one kind of molecule and are exceptions, most crystals are built from identical (or enantiomeric) copies of the same molecule. Thus, a usual one-component crystal is a macro-supramolecular assembly where one should more properly speak of molecular self-recognition. This is not a fact contradictory to the basic principles of molecular recognition, since the case might occur in which the two complementary structures happen to be identical in dealing with a self-complementary relationship. Even when all the molecules are identical (or enantiomeric), an acceptor part of one molecule can interact with a donor part of a second, and the acceptor part of the second can interact in exactly the same manner with the donor part of a third, and so on, giving rise to periodicity of the crystal and to the limited number of space groups used in molecular crystals. For instance, it is very uncommon for molecules in a crystal structure to be related by rotation axis or mirror planes, because identical parts of molecules avoid one another, except for molecular sites having a so-called self-complementary donor-acceptor group. Self-complementary groups form finite, one-dimensional tape, two-dimensional layer, or three-dimensional motifs of organic molecules mostly obtained from hydrogen bonding.

In solution, highly ordered structures created via self-recognition and self-assembly of a programmed H-bonding molecular component are also possible. With respect to inorganic self-recognition and self-assembly this would involve preferential binding of like metal ions by like ligands in a mixture of ligands and ions.

A particular point of interest included in these helical complexes concerns the chirality. The helicates obtained from the achiral strands are a racemic mixture of left- and right-handed double helices. This special mode of recognition where homochiral supramolecular entities, as a consequence of homochiral self-recognition, result from racemic components is known as optical self-resolution. It appears in certain cases from racemic solutions or melts (spontaneous resolution) and is often cited as one of the possible sources of optical resolution in the biological world.

EDWIN WEBER
Technische Universität Bergakademie
Freiberg Institut für Organische Chemie

Additional Reading

Behr, J.-P. ed.: *The Lock and Key Principle, Perspectives in Supramolecular Chemistry,* Vol. 1, Wiley, Chichester, UK, 1994.
Hulme, E.C.: *Receptor Biochemistry,* Oxford University Press, New York, NY, 1990.
Page, M.I.: *The Chemistry of Enzyme Action,* Elsevier, Amsterdam, 1984.
Roberts, S.M. ed.: *Molecular Recognition–Chemical and Biochemical Problems,* The Royal Society of Chemistry, Cambridge, 1989.
Stanford, C., and R. Horton: *Receptors: Structure and Function,* 2nd Edition, Oxford University Press, New York, NY, 2002.

MOLECULAR SIEVES. In this article, the term *molecular sieve* is restricted to inorganic materials that possess uniform pores with diameters in either the micro- (<2 nm) or meso- (2–20 nm) size range. The most technologically important molecular sieves are zeolites, i.e., crystalline silicate or aluminosilicate framework structures with channels of diameters <1.2 nm. Several of these topologies, with boron, gallium, or iron replacing aluminum, or germanium replacing silicon, have also been prepared. The chemical composition of microporous framework structures has been expanded considerably with the substitution of phosphorus for silicon, and new families of aluminophosphate and silicoaluminophosphate structures have been synthesized in the laboratory. Some of these frameworks have zeolite analogues, whereas others are unique. The addition of elements such as Mg, Ti, Mn, Co, Fe, or Zn into these structures has made it possible to generate metalloaluminophosphates, metallosilicoaluminophosphates, etc. Microporous sulfide-based framework structures are also possible.

Considerable synthesis effort has been devoted to developing frameworks with pore diameters within the mesoporous range; the largest synthesized are the phosphate-based AlPO-8 (14-membered ring), VPI-5 (18-MR), and cloverite (20-MR), which have pore diameters within the 0.8–1.3 nm range. A new family of mesoporous molecular sieves designated M41S has been discovered. Although not framework structures like zeolites, silicate, and aluminosilicate, M41S materials possess very uniform mesopores.

The technological applications of molecular sieves are as varied as their chemical makeup. Heterogeneous catalysis and adsorption processes make extensive use of molecular sieves. The utility of the latter materials lies in their microstructures, which allow access to large internal surfaces, and cavities that enhance catalytic activity and adsorptive capacity.

Zeolites

Molecular-sieve zeolites of the most important aluminosilicate variety can be represented by the chemical formula $M_{2/n}O \cdot Al_2O_3 \cdot ySiO_2 \cdot wH_2O$, where y is 2 or greater, M is the charge balancing cation, such as sodium, potassium, magnesium, and calcium, n is the cation valence, and w represents the moles of water contained in the zeolitic voids. The zeolite framework is made up of SiO_4 tetrahedra linked together by sharing of oxygen ions. Substitution of Al for Si generates a charge imbalance, necessitating the inclusion of a cation. The structures contain channels or interconnected voids that are occupied by the cations and water molecules. The water may be removed reversibly, generally by the application of heat, which leaves intact the crystalline host structure permeated with micropores that may account for >50% of the microcrystal's volume. In some zeolites, dehydration may produce some perturbation of the structure, such as cation movement, and some degree of framework distortion.

Zeolite minerals are formed over much of the earth's surface, including the sea bottom. Most zeolites occurring in cavities of basaltic and volcanic rocks are exceedingly rare. However, several zeolite minerals generated by the natural alteration of volcanic ash in alkaline environments over long periods of time occur in recoverable deposits. Although such minerals are rarely useful for catalytic application, mainly because of iron impurities (exception: Chabazite of Bowie, Arizona), more abundant zeolites, e.g., clinoptilolite, have found use as soil conditioners, additives to animal feed, as animal litter, in aquaculture to remove ammonia (phillipsite), and for ion exchange to remove heavy metals from industrial and mining effluents (chabazite).

Synthetic and mineral zeolites of primary importance are listed in Table 1.

Structure

Of the approximately 120 known framework aluminosilicates, ~50 occur naturally, the rest being synthetic. There are 56 structural types of zeolites known. Understanding the complexities of zeolite structures is made easier by recognizing three important structural keys: the basic arrangement of the individual structural units in space, which defines the framework topology; the location of the charge-balancing cations; and the channel-filling material, such as water or an organic template, which is incorporated as the zeolite is formed. After the channel-filling material is removed, the void space can be used for the adsorption of gases, liquids, salts, elements, metal complexes, etc. In turn, this void-filling property makes zeolites commercially useful in ion exchange, catalysis, etc.

TABLE 1. ZEOLITE COMPOSITIONS

Zeolite	Typical formula
Natural	
chabazite	$Ca_2[(AlO_2)_4(SiO_2)_8] \cdot 13H_2O$
mordenite	$Na_8[(AlO_2)_8(SiO_2)_{40}] \cdot 24H_2O$
erionite	$(Ca, Mg, Na_2, K_2)_{4.5}[(AlO_2)_9(SiO_2)_{27}] \cdot 27H_2O$
faujasite	$(Ca, Mg, Na_2, K_2)_{29.5}[(AlO_2)_{59}(SiO_2)_{133}] \cdot 235H_2O$
clinoptilolite	$Na_6[(AlO_2)_6(SiO_2)_{30}] \cdot 24H_2O$
phillipsite	$(0.5Ca, Na, K)_3[(AlO_2)_3(SiO_2)_5] \cdot 6H_2O$
Synthetic	
zeolite A	$Na_{12}[(AlO_2)_{12}(SiO_2)_{12}] \cdot 27H_2O$
zeolite X	$Na_{86}[(AlO_2)_{86}(SiO_2)_{106}] \cdot 264H_2O$
zeolite Y	$Na_{56}[(AlO_2)_{56}(SiO_2)_{136}] \cdot 250H_2O$
zeolite L	$K_9[(AlO_2)_9(SiO_2)_{27}] \cdot 22H_2O$
zeolite omega	$Na_{6.8}TMA_{1.6}[(AlO_2)_8(SiO_2)_{28}] \cdot 21H_2O^a$
ZSM-5	$(Na, TPA)_3[(AlO_2)_3(SiO_2)_{93}] \cdot 16H_2O^b$

[a] TMA = tetramethylammonium.

[b] TPA = tetrapropylammonium.

There are two types of structures: one provides an internal pore system comprising interconnected cage-like voids; the second provides a system of uniform channels which, in some instances, are one-dimensional and in others intersect with similar channels to produce two- or three-dimensional channel systems. The preferred type has two- or three-dimensional channel systems to provide rapid intercrystalline diffusion in adsorption and catalytic applications.

In most zeolite structures, the primary structural units, tetrahedra, are assembled into secondary building units, which may be simple polyhedra such as cubes, hexagonal prisms, or truncated octahedra. The final framework structure consists of assemblages of the secondary units.

Zeolite Minerals. Crystal structures of zeolite minerals are illustrated by the zeolite chabazite. The structure of chabazite is hexagonal and the framework consists of double six-membered rings of $(Si,Al)O_4$ tetrahedra arranged in parallel layers in an AABBCC sequence. These tetrahedra are cross-linked by four-membered rings, as shown in Figure 1.

Synthetic Zeolites. Many new crystalline zeolites have been synthesized and several fulfill important functions in the chemical and petroleum industries and in consumer products such as detergents. The structural formula of a zeolite is based on the crystal unit cell, the smallest unit of structure, represented by $M_{x/n}[(AlO_2)_x(SiO_2)_y] \cdot wH_2O$, where n is the valence of cation M, w is the number of water molecules per unit cell, x and y are, respectively, the number of AlO_4 and SiO_4 tetrahedra per unit cell, and y/x usually has values of 1–5. Examples of important synthetic zeolites are shown in Table 1.

The secondary structure unit in zeolites A, X, and Y is the truncated octahedron. These polyhedral units are linked in three-dimensional space through the four- or six-membered rings. The former linkage produces the zeolite A structure, and the latter the topology of zeolites X and Y and of the mineral faujasite.

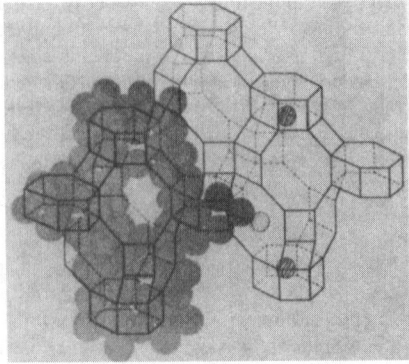

Fig. 1.   Structure of the mineral zeolite chabazite is depicted by packing model, left, and skeletal model, right. The silicon and aluminum atoms lie at the corners of the framework depicted by solid lines. In this figure the solid lines do not depict chemical bonds. Oxygen atoms lie near the midpoint of the lines connecting framework corners. Cation sites are shown in three different locations referred to as sites I, II, and III. Courtesy of *Scientific American*

The structure of the high silica zeolite ZSM-5 (y/x = 10 to>5000) contains a high concentration of five-membered rings and has two intersecting channels. It has become a very important catalyst for petrochemical reactions.

Structure Modification. Several types of structural defects or variants can occur which figure in adsorption and catalysis: *(1)* surface defects due to termination of the crystal surface and hydrolysis of surface cations; *(2)* structural defects due to imperfect stacking of the secondary units, which may result in blocked channels; *(3)* ionic species, e.g., OH^-, AlO_2^-, Na^+, SiO_4^-, may be left stranded in the structure during synthesis; *(4)* the cation form, acting as the salt of a weak acid, hydrolyzes in aqueous suspension to produce free hydroxide and cations in solution; and *(5)* hydroxyl groups in place of metal cations may be introduced by ammonium ion exchange, followed by thermal deammoniation.

Properties

Adsorption. Although several types of microporous solids are useful as adsorbents for the separation of vapor or liquid mixtures, the distribution of pore diameters does not enable separations based on the molecular-sieve effect. The most important molecular-sieve effects are shown by crystalline zeolites. The sieve effect may be total or partial.

Activated diffusion of the adsorbate is of interest in many cases. As the size of the diffusing molecule approaches that of the zeolite channels, the interaction energy becomes increasingly important. If the aperture is small relative to the molecular size, then the repulsive interaction is dominant and the diffusing species needs a specific activation energy to pass through the aperture. Similar shape-selective effects are shown in both catalysis and ion-exchange, two important applications of these materials.

In order to utilize the absorption properties of the synthetic zeolite crystals in processes, the commercial materials are prepared as pelleted aggregates combining a high percentage of the crystalline zeolite with an inert binder. The formation of these aggregates introduces macropores in the pellet which may result in some capillary condensation at high adsorbate concentrations. In commercial materials, the macropores contribute diffusion paths. However, the main part of the adsorption capacity is contained in the voids within the crystals.

Zeolites are high capacity, selective adsorbents capable of separating molecules based on the size and shape of the structure. They adsorb molecules, in particular those with a permanent dipole moment which show other interaction effects, with a selectivity that is not found in other solid adsorbents. Separation may be based on the molecular-sieve effect or may involve the preferential or selective adsorption of one molecular species over another. These separations are governed by several factors. The basic framework structure, or topology, of the zeolite determines the pore size and the void volume. The exchange cations, in terms of their specific location in the structure, their population density, their charge and size, affects the molecular-sieve behavior and adsorption selectivity of the zeolite. By changing the cation types and number, the selectivity of the zeolite in a given separation can be tailored or modified, within certain limits.

The high silica version of ZSM-5, also known as silicalite, is a hydrophobic adsorbent capable of adsorbing, e.g., ethanol from an aqueous solution.

Catalytic Properties. In zeolites, catalysis takes place preferentially within the intracrystalline voids. Catalytic reactions are affected by aperture size and type of channel system, through which reactants and products must diffuse. Modification techniques include ion-exchange, variation of Si/Al ratio, hydrothermal dealumination or stabilization, which produces Lewis acidity, introduction of acidic groups such as bridging Si(OH)Al, which impart Brønsted acidity, and introducing dispersed metal phases such as noble metals. In addition, the zeolite framework structure determines shape-selective effects. Several types have been demonstrated, including reactant selectivity, product selectivity, and restricted transition-state selectivity. Nonshape-selective surface activity is observed on very small crystals, and it may be desirable to poison these sites selectively, with bulky heterocyclic compounds unable to penetrate the channel apertures, or by surface silation.

Some current and possible future zeolite catalyst applications are as follows: alkylation, cracking, hydrocracking, dewaxing, isomerization, hydrogenation and dehydrogenation, hydrodealkylation, methanation, shape-selective reforming, dehydration, methanol to gasoline, methanol to olefins, organic catalysis, inorganic reactions, H_2S oxidation, NH_3 reduction of NO, $H_2O \longrightarrow 1/2\ O_2 + H_2$, and CO oxidation.

Ion Exchange. The exchange behavior of nonframework cations in zeolites, e.g., selectivity, and degree of exchange, depends on the nature of the cation, e.g., the size and charge of the hydrated cation, on the temperature, the concentration, and, to some degree, on the anion species. Cation exchange may produce considerable change in various other properties, such as thermal stability, adsorption behavior, and catalytic activity.

Framework Modification

The zeolite framework can be stabilized by hydrothermal treatment, which removes aluminum from the framework and forms aluminum cations. During this steaming process, the tetrahedral vacancies left behind in the framework are gradually refilled with silicon, which appears to migrate as a form of silicic acid from other parts of the framework and contributes to stabilization by repairing the damaged framework. Simultaneously, such other parts of the framework disappear under formation of mesopores. Cationic aluminum can be extracted with an acid, and a subsequent steaming causes further dealumination of the framework and migration of silicon into the vacancies. Carefully controlled conditions can produce high silica forms of zeolites, e.g., zeolite Y. Since the Si—O bond is shorter than the Al—O bond, hydrothermal dealumination causes the unit cell parameter to decrease. Mesopores can be avoided by replacing the aluminum directly with external silicon, e.g., by treatment with silicon tetrachloride. In a reversal of the reaction with $SiCl_4$, aluminum can be introduced into the framework by reaction of the hydrogen or ammonium form with gaseous $AlCl_3$.

Manufacture

Zeolites are formed under hydrothermal conditions, defined here in a broad sense to include zeolite crystallization from aqueous systems containing various types of reactants. Most synthetic zeolites are produced under nonequilibrium conditions, and must be considered as metastable phases in a thermodynamic sense.

Many important types of zeolites have no natural mineral counterpart. Conversely, synthetic counterparts of many zeolite minerals are not yet known. The conditions generally used in synthesis are reactive starting materials such as freshly co-precipitated gels, or amorphous solids; relatively high pH introduced in the form of an alkali metal hydroxide or other strong base, including tetraalkylammonium hydroxides; low temperature hydrothermal conditions with concurrent low autogenous pressure at saturated water vapor pressure; and a high degree of supersaturation of the gel components, leading to nucleation of a large number of crystals.

A gel is defined as a hydrous metal aluminosilicate prepared from either aqueous solutions, reactive solids, colloidal sols, or reactive aluminosilicates such as the residue structure of metakaolin and glasses.

The gels are crystallized in a closed hydrothermal system at temperature varying from room temperature to about 200°C. The time required for crystallization varies from a few hours to several days. When prepared, the aluminosilicate gels differ in appearance, from stiff and translucent to opaque gelatinous precipitates and heterogenous mixtures of an amorphous solid dispersed in an aqueous solution. The alkali metals form soluble hydroxides, aluminates, and silicates. These materials are well suited for the preparation of homogeneous mixtures.

Gel preparation and crystallization is represented systematically using the $Na_2O-Al_2O_3-SiO_2-H_2O$ system as an example.

$$NaAl(OH)_4(aq) + Na_2SiO_3(aq) \xrightarrow{25°C} [Na_a(AlO_2)_b$$

$$(SiO_2)_c \cdot NaOH \cdot H_2O]gel \xrightarrow{25-175°C} Na_x[(AlO_2)_x(SiO_2)_y] \cdot mH_2O$$

Typical gels are prepared from aqueous solutions of reactants such as sodium aluminate, NaOH, and sodium silicate; other reactants include alumina trihydrate ($Al_2O_3 \cdot 3H_2O$), colloidal silica, and silicic acid. Some synthetic zeolites prepared from sodium aluminosilicate gels are given in Table 2. The temperature strongly influences the crystallization time of even the most reactive gels.

Synthesis mechanisms of the typical low silica zeolites, such as A, X, and Y, are apparently different from the high silica zeolites such as ZSM-5. In the low silica zeolites, nuclei are formed consisting of alkali metal-ion complexes of the aluminosilicate species. Structural units consisting of four-membered rings, six-membered rings, and cages coordinated with

TABLE 2. SOME SYNTHETIC ZEOLITES PREPARED FROM SODIUM ALUMINOSILICATE GELS

Zeolite type	Typical composition, mol/mol Al_2O_3			Reactants	Reactant temp,°C	Zeolite product composition, mol/mol Al_2O_3		
	Na_2O	SiO_2	H_2O			Na_2O	SiO_2	H_2O
A	2	2	35	$NaAlO_2$ NaOH sodium silicate	20–175	1	2	4.5
X	3.6	3	144	$NaAlO_2$ NaOH sodium silicate	20–120	1	2.0–3.0	6
Y	8	20	320	$NaAlO_2$ colloidal SiO_2 NaOH	20–175	1	3.0–6.0	9
mordenite, Zeolon	6.3	27	61	$NaAlO_2$ diatomite sodium silicate	100	1	9–10	6.7
omega	5.60[a]	20	280	colloidal SiO_2 $Al(OH)_3$ $TMAOH$[b] NaOH	100	0.71 0.36 TMA	7.3	6.3
ZSM-5	10[c]	7.7	453	$NaAlO_2$ SiO_2 $TPAOH$[e]	150	0.89	31.1[d]	2.0

[a] Also 1.4 TMA_2O.
[b] TMA = tetramethylammonium.
[c] Also 8.6 TPA_2O.
[d] After calcination at 1000°C.
[e] TPA = tetrapropylammonium.

cations are thought to be involved in the nucleation and crystallization. In the high silica zeolites, the mechanism appears to be a templating type where an alkylammonium cation complexes with silica by hydrogen bonding. These complexes cause the structures to replicate by hydrogen bonding of the organic cation with framework oxygen atoms.

Processes. Manufacturing processes for commercial molecular sieve products may be classified into three groups, as shown in Table 3.

Health and Safety Factors and Toxicology

Zeolites have applications in food, drugs, cosmetic products, and detergents. Thus, extensive toxicological and environmental studies have been carried out. Feeding of 5.0 g/kg of body weight (powder form of type 4A, 5A, 13X, and Y) for seven days produced no ill effect in rats. There is no contraindication to the use of zeolite A (Sasil) in detergents. No negative effect on biological wastewater treatment was found, and zeolite A showed no evidence of acute toxicity to four species of freshwater fish.

Uses

In most cases, the water content of the commercial product is below 1.5–2.5 wt %; certain products, however, are sold as fully hydrated

TABLE 3. PROCESSES FOR MOLECULAR SIEVE ZEOLITES

Process	Reactants	Products
hydrogel	reactive oxides soluble silicates soluble aluminates caustic	high purity powders gel preform zeolite in gel matrix
clay conversion	raw kaolin *meta*-kaolin calcined kaolin acid-treated clay soluble silicate caustic sodium chloride	low to high purity powder binderless, high purity preform zeolite in clay-derived matrix
other	natural SiO_2 amorphous minerals volcanic glass caustic	low to high purity powder zeolite on ceramic support binderless preforms

crystalline powders. Molecular sieve products are used for adsorption (e.g., purification of water, carbon dioxide, and sulfur compounds); bulk separation of normal and isoparaffins, xylene, olefin, and oxygen from air; catalysis e.g., catalytic cracking, hydrocracking for fuels production, dewaxing of distillate fuel and lube basestocks, paraffin isomerization, catalysis of aromatic reactions (in selective toluene disproportionation, xylene isomerization, and ethylbenzene synthesis), synthesis of *p*-ethyltoluene, and the methanol-to-gasoline process; and ion exchange (e.g., cesium and strontium radioisotopes, ammonium ion removal, and detergent builders).

New Trends

Aluminosilicate of faujasite topology, but higher Si/Al ratio (≤ 5), has been synthesized with the use of a crown ether, 15-crown-5, as the directing agent. The same Si/Al ratio was obtained when 18-crown-6 was applied, but the topology, although related to faujasite, had a different stacking order of the sodalite cages, so that the structure has hexagonal instead of cubic symmetry. This product, sometimes called hexagonal faujasite, has the designation EMT.

New directions in the preparation of framework structures of different chemical composition and of large-pore molecular sieves include the development of phosphate-containing molecular sieves and mesoporous molecular sieves.

GÜNTER H. KÜHL
CHARLES T. KRESGE
Mobile Research and Development Corporation

Additional Reading

Auerbach, S.M., K.A. Carrado, and P.K. Dutta: *Handbook of Zeolite Science and Technology,* Marcel Dekker, Inc., New York, NY, 2003.

Barrer, R.M.: *Zeolites and Clay Minerals as Sorbents and Molecular Sieves,* Academic Press, London, 1978.

Breck, D.W.: *Zeolite Molecular Sieves, Structure, Chemistry, and Use,* John Wiley & Sons, Inc., New York, NY, 1974.

Guisnet, M., and Jean-Pierre Gilson: *Zeolites for Cleaner Technologies,* Imperial College Press, London, UK, 2002.

Meier, W.M., D.H. Olson, and Ch. Baerlocher: *Atlas of Zeolite Structure Types,* 4th Revised Edition, Elsevier, New York, NY, 1996.

Occelli, M.L., and H. Kessler: *Synthesis of Porous Materials: Zeolites, Clays and Nanostuctures,* Vol. 69, Marcel Dekker, Inc., New York, NY, 1997.

Rabo, J.A. and G.J. Gajda: *Catal. Rev.-Sci. Eng.* **31**, 385 (1989–1990).

MOLECULAR STRUCTURE (Organic Compounds). See Organic Chemistry.

MOLECULAR WEIGHT.

The sum of the atomic weights of the atoms in a molecule. That of methane (CH_4) is 16.043, the atomic weights being carbon = 12.011, hydrogen = 1.008. The chemical formula used in such a calculation must be the true molecular formula of the substance designated. For example, the molecular formula of oxygen is O_2 and its molecular weight is 31.998 (atomic weight of oxygen = 15.999). For ozone the molecular formula is O_3 and the molecular weight is 47.997. The true molecular weight of a gas or vapor is found by calculating the weight of 22.4 L at 0C and 760 mm Hg. The molecular weight of many complex organic molecules runs as high as a million or more (proteins and high polymers). See also **Avogadro Law**; and **Atomic Mass (Atomic Weight)**.

MOLECULE.

In the traditional sense, a molecule is the smallest particle of a chemical substance capable of independent existence with retention of all its chemical properties. Molecules comprise one or more atoms which need not be of the same kind. Only the rare, or noble gases form single-atom or monatomic molecules. All other elements form bi-, tri-, quadri-, etc. atomic molecules, e.g., hydrogen, H_2; ozone, O_3; phosphorus, P_4; and sulfur, S_8; or hydrogen chloride, HCl; sodium sulfide, Na_2S, aluminum chloride, $AlCl_3$, carbon tetrachloride, CCl_4, and so on.

Structurally, a more specific definition would be that a molecule is a local assembly of atomic nuclei and electrons in a state of dynamic stability. The cohesive forces are electrostatic, but, in addition, relatively small electromagnetic interactions may occur between the spin and orbital motions of the electrons, especially in the neighborhood of heavy nuclei. The internuclear separations are of the order of $1-2 \times 10^{-10}$ meter, and the energies required to dissociate a stable molecule into smaller fragments

fall into the 1–5 eV range. The simplest diatomic species is the hydrogen molecule, H_2^+, with two nuclei and one electron. At the other extreme, the protein ribonuclease contains 1876 nuclei and 7396 electrons per molecule.

Another form of molecule is known, however, and this is formed by atomic nuclei alone. Although these have not yet been found in nature, they may well play a role in stellar evolution. Under special conditions in high-energy interactions, they can be momentarily held together by effective bonds. Whether these bonds are the result of exchange or sharing of valence protons and neutrons is as yet moot. In these *nuclear molecules* a somewhat unstable balance is attained between long-range electrostatic repulsion of positively charged nuclei and the much stronger short-range nuclear force which determines the motions of protons and neutrons. Nuclear molecules are significant entities because they live much longer ($\sim 10^{-21}$ second) than the time usually taken for nuclei to collide ($\sim 10^{-23}$ second).

The molecular, or kinetic, theory of matter makes four assumptions: (1) that the molecules of which matter is composed are constantly in motion; (2) that their energy is increased by the addition of heat; (3) that they undergo elastic collision with each other and with the walls of a containing vessel; and (4) that they exert forces upon each other. As first developed by Heisenberg, Schrödinger, and Dirac, reduction of these theoretical assumptions to mathematical bases is somewhat inadequate and can relate only to interaction between hypothetical electron clouds. Except for the simplest of systems, the Schrödinger equation cannot be solved exactly. On the other hand, a more manageable understanding of atom interactions in molecules is afforded by the Valence Shell Electron Pair Repulsion Theory, independently enunciated by Nyholm and Gillespie. This theory proposes that both bonding and nonbonding pairs of outer atomic shell electrons in a molecule repel each other and establish themselves as far apart as possible.

Historically, molecules were regarded as being formed by the association of individual atoms. This led to the concept of *valency*, i.e., the number of individual chemical bonds or linkages with which a particular atom can attach itself to other atoms. When the electronic theory of the atom was developed, these bonds were interpreted in terms of the behavior of the valence, or outer shell, electrons of the combining atoms. Each atom with a partly-filled valence shell attempts to acquire a completed octet of outer electrons, either by electron transfer, as in (a) shown below, to give an electrovalent bond, resulting from coulombic attraction between the oppositely charged ions; or, as in (b) and (c) to give a covalent bond. The concept of (a) was proposed by Kossel in 1916; that of (b) and (c) by Lewis, also in 1916.

$$Na^+ \; [:\!\overset{\cdots}{\underset{\cdots}{Cl}}\!:]^- \qquad :\!\overset{\cdots}{\underset{\cdots}{Cl}}\!:\!\overset{\cdots}{\underset{\cdots}{Cl}}\!: \qquad R:\!\overset{\overset{\textstyle R}{\cdots}}{\underset{\underset{\textstyle R}{\cdots}}{N}}\!^+\!:\!\overset{\cdots}{\underset{\cdots}{O}}\!:^- \qquad R = CH_3$$

(a)   (b)   (c)

In (b), each chlorine atom donates one electron to form a *homopolar bond*, which is written Cl−Cl where the bar denotes on this theory one single bond, or shared electron pair. In (c), the nitrogen-oxygen bond is formed by two electrons donated by only the nitrogen atom, giving a *semipolar*, or *coordinate-covalent bond*, which is written $R_3N \longrightarrow O$, and which is electrically polarized. Double or triple bonds result from the sharing of 4 to 6 electrons between adjacent atoms. More information on these bonding theories is given in the entries on **Chemical Composition**; **Chemical Elements**; and **Compound (Chemical)**.

However, difficulties arise in describing the structures of many molecules in this manner. For example, in benzene, C_6H_6, a typical aromatic compound, the carbon nuclei form a plane regular hexagon, but the electrons can only be conventionally written as forming alternate single and double bonds between them. Furthermore, an electron cannot be identified as coming specifically from any one of these bonds upon ionization. Such difficulties disappear in the quantum-mechanical theory of a polyatomic molecule, whose electronic wave function can be constructed from nonlocalized electron orbitals extending over all of the nuclei. The concept of valency is not basic to this theory, but is simply a convenient approximation by which the electron density distribution is partitioned in different regions in the molecule.

Molecular compounds consist of two or more stable species held together by weak forces. In *clathrates*, a gaseous substance, such as SO_2, HCl, CO_2, or a rare gas is held in the crystal lattice of a solid, such as beta-quinol, by Van der Waals-London dispersion forces. The gas hydrates, e.g., $Cl_2 \cdot 6H_2O$, contain halogen molecules similarly trapped in ice-like structures. The hydrogen bond, with energy ~0.25 eV, is responsible not only for the high degree of molecular association in liquids, such as water, but also for such molecules as the formic acid dimer, which contains two hydrogen bonds indicated by dashed lines.

Molecular complexes vary greatly in their stability; in donor-acceptor complexes, electronic charge is transferred from the donor (e.g., NH_3) to the acceptor (e.g., BF_3), as in a semipolar bond. The $BF_3 \cdot NH_3$ complex has a binding energy with respect to dissociation into NH_3 and BF_3 of 1.8 eV. The bond here is relatively strong; the electron transfer can occur between the components in their electronic ground states. On the other hand, in weaker complexes, such as $C_6H_6 \cdot I_2$, with binding energy of about 0.06 eV, there is only a fractional transfer of charge from benzene to iodine. The actual ionic charge-transfer state lies at much higher energy than the ground state of the complex.

Complete pairing of all electrons present in a molecule and absence of any bonding orbitals was long taken to be a stable, unreactive state exemplified by the inert, or rare gases. In 1962, however, Bartlett unequivocally synthesized $XePtF_6$, and this was rapidly followed by the synthesis of other rare gas compounds whose existence was not predicted by classical valency theories. Compounds such as XeF_2, XeF_4, XeF_6, and $XeOF_4$ are quite stable, the average Xe-F bond energy in the square planar XeF_4 being 1.4 eV.

A molecule is characterized by (1) a stoichiometric formula; (2) the spatial distribution of the nuclei in their mean equilibrium or "rest" positions; and (3) the dynamical state.

The ratio $a : b : c : \ldots$ in a formula $A_a B_b C_c$, where $a, b, c, \ldots$ are the numbers of atoms of elements $A, B, C, \ldots$ that it contains is found by chemical analysis for these elements. The absolute values of $a, b, c, \ldots$ are then fixed by determination of the molecular weight. This principle is further described under the entry on **Compound (Chemical)**.

The spatial distribution of the nuclei in their mean equilibrium positions, at an elementary level, is described in geometrical language. For example, in carbon tetrachloride, CCl_4, the four chlorine nuclei are disposed at the corners of a regular tetrahedron, and the carbon nucleus is at the center. In the $[CoCl_4]^{2-}$ ion, the arrangement of the chlorine nuclei about the central metal nucleus is also tetrahedral, whereas in $[PdCl_2]^{2-}$, it is planar. For example, the pyramidal ammonia molecule NH_3 has a threefold rotation axis C_3 through the nitrogen nucleus and three reflection planes σ_v intersecting at the axis, and belongs to the $C_{3v}(3m)$ point group. Tetrahedral molecules CX_4 belong to the $T_d(\overline{4}3m)$ point group. Linear diatomic and polyatomic molecules belong to either of the continuous point groups $D_{\infty h}$ or $C_{\infty v}$ according to whether a center of symmetry is present or not.

The symmetry classification does not define the geometry of a molecule completely. The values of certain bond lengths or angles must also be described. In carbon tetrachloride, it is sufficient to give the C—Cl distance $(1.77 \times 10^{-10}$ meters), since classification under the T_d point group implies that all four of these bonds have equal length and the angle between them is $109°28''$. In ammonia, both the N—H distance $(1.015 \times 10^{-10}$ meters) and the angle HNH $(107°)$ must be specified. In general, the lower the molecular symmetry, the greater is the number of such independent parameters required to characterize the geometry. Information about the symmetry and internal dimensions of a molecule is obtained experimentally by spectroscopy, electron diffraction, neutron diffraction, and x-ray diffraction.

The dynamic state is defined by the values of certain observables associated with orbital and spin motions of the electrons and with vibration and rotation of the nuclei, and also by symmetry properties of the corresponding stationary-state wave functions. Except when heavy nuclei are present, the total electron spin angular momentum of a molecule is separately conserved with magnitude Sh, and molecular states are classified as singlet, doublet, triplet ... according to the value of the multiplicity $(2S + 1)$. This is shown by a prefix superscript to the term symbol, as in atoms.

The Born-Oppenheimer approximation permits the molecular Hamiltonian H to be separated into a component H_e that depends only on the coordinates of the electrons relative to the nuclei, plus a component depending upon the nuclear coordinates. This in turn can be written as a sum $H_v + H_r$ of terms for vibrational and rotational motion of the nuclei,

translation being ignored. The eigenfunctions Ψ of H may correspondingly be factorized as the product $\Psi_e \Psi_v \Psi_r$ of eigenfunctions of these three operators, and the eigenvalues of E decomposed as the sum $E_e + E_v + E_r$. In general, $E_e > E_v > E_r$.

Molecular Spectra

The spectra of substances in the molecular state, like atomic spectra, are made up of lines, although more complex. The transitions in a molecule which release the most energy (largest quanta) are due to electron changes, as in atoms, and the results of these changes are observed as lines in the ultraviolet region. But there are other ways in which a molecule can release or absorb energy. Thus, the component atoms oscillate with reference to each other within the molecule, and this motion apparently is "quantizied," i.e., changes abruptly from one state to another of different energy. But these "vibrational" energy changes are much less than the electronic, so that the resulting quanta and spectrum lines are of much lower frequency, and appear in the extreme red or near-infrared. Again, the molecule rotates, and the quantization of its rotational energy results in the emission of quanta of still lower frequency, appearing as lines in the far-infrared.

Molecular and laser spectroscopic approaches are making possible a deeper resolution of the dynamics of atomic and molecular motions and the potential energy surfaces governing energy transference within a molecule or group of molecules as chemical bonds are made and broken.

Integration of evolutionary biology with molecular biology, epitomized by the later work of Szent-Györgyi, is, in effect, translating morphogenesis into molecular language. Along metabolic pathways, intermediate molecules usually appear in such an order that thermodynamic stabilities increase progressively from starting materials to final products. Such stabilities are also associated with the influence of solvent water on chemical systems. The qualitative effects of this have been known for many years, but the quantitative data have so far been scarce. The solvent molecules tend to be reorganized in the neighborhood of the interacting groups, but not all molecules are attracted enough to overcome the self-cohesive properties of water.

Free energies of two or more moderately polar groups are often approximately additive and departures from this suggest special interactions between parts of the solute molecules and the surrounding solvent.

Polymerization

This term is used to designate a reaction in which a complex molecule of high molecular weight (or macromolecule) is formed from a number of simpler molecules. Thus the monomer formaldehyde, CHOH, can form the trimer trioxane $(CHOH)_3$, or the long-chain polymer paraformaldehyde, $HO(CHOH)_nH$, where $n = 8-100$. But the combining molecules may be of the same or different sorts. An *additional polymerization* is one in which like or unlike molecules combine without the elimination of any atoms or molecules. A *simple polymerization* involves only one species of molecule. *Copolymerization* is an addition polymerization in which two or more distinct molecular species are involved, each one of which is capable of polymerizing by itself. The high polymer formed contains each molecular constituent or an essential portion, as a distinct unit in the structure of the polymer. *Heteropolymerization* is an addition polymerization in which two or more molecular species are involved, one of which species will not polymerize by itself. It does, however, form distinct units in the high polymer.

A condensation polymerization is one in which the molecules undergoing polymerization react with the elimination of simple molecules like water, ammonia, and the like.

Polymerizations are also characterized by the state in which they are carried out such as: *gaseous polymerization* or those carried out in the vapor or gaseous phase; *mass polymerizations* in the liquid state; *solution polymerizations* carried out by first dissolving the material to be polymerized in an adequate solvent; *emulsion polymerizations* carried out in which one of the components is in an emulsion as in the case of rubber polymerizations; and *bulk polymerizations* in which the polymerization takes place without the use of a solvent or other medium. The wide variety of methods by which the process of polymerization can be used in the manufacture of plastics is exemplified by the production of polystyrene from styrene (vinylbenzene, $CH_2 \cdot CH \cdot C_6H_5$). One factor that conditions all these processes is the highly exothermic (high heat production) nature of this polymerization. This fact, together with the low heat conductivity of polystyrene, determines certain characteristics of an industrial process.

This is particularly the case because the extent of the polymerization of styrene, like that of many other plastics, depends on the temperature. When the process is conducted at higher temperatures, the resulting polymer has a low molecular weight and low physical strength properties (i.e., it is weak and brittle). Extremely high molecular weight polymers, although mechanically tough, are more difficult to fabricate. Since the higher polymerization temperatures also result in faster polymerization rates, economic considerations dictate a compromise between faster production and better physical properties. Practical experience indicates a temperature range from 60 to 150°C.

The polymerization of styrene illustrates the application of several of the methods defined above.

In *batch mass polymerization* the reaction vessel is loaded with styrene monomer and heated to a temperature sufficient to initiate polymerization within a reasonable time. As polymerization proceeds, the temperature within the vessel rises, thus increasing the rate.

In *continuous mass polymerization* the reaction vessel contains monomer and polymer. As more polymer is formed it is drawn off the bottom of the reaction vessel while monomer is added to the top of the vessel. The temperature of the reaction is controlled by cooling coils within the vessel.

In *solution polymerization* the styrene monomer is diluted with a solvent. The solvent acts as a diluent, which decreases the rate of polymerization and also serves as a heat transfer medium for removing the excess heat developed by the reaction.

In *suspension polymerization* water is used as a diluent and as a heat transfer aid. Suspending agents such as starch and methylcellulose are used to keep the styrene monomer particles in suspension. The more efficient heat transfer of this process also allows for a narrower molecular weight distribution.

In *emulsion polymerization* the styrene monomer is emulsified with water by the addition of certain emulsifying agents. This results in very small particles and rapid polymerization rates. The heat of polymerization is dissipated by the water ingredient.

See also **Fibers**; and **Macromolecular Science**.

Additional Reading

Barondes, S.: *Molecules and Mental Illness,* Scientific American Library, New York, NY, 1999.

Eisberg, R. and R. Resnik: *Quantum Physics of Atoms, Molecules, Solids, Nuclei, and Particles,* 2nd Edition, John Wiley & Sons, Inc., New York, NY, 1990.

Parr, R. and Y. Weitao: *Density-Functional Theory of Atoms and Molecules,* Oxford University Press, New York, NY, 1994.

Suchocki, J.: *Conceptual Chemistry: Understanding Our World of Atoms and Molecules,* 2nd Edition, Pearson Education, Boston, MA, 2003.

MOLE FRACTION. As applied to a system, the *mole fraction* (sometimes spelled *mol fraction*) of a given substance in the system is found by dividing the number of moles of that substance by the total number of moles in the system. For a mixture of ideal gases, the mole fraction is equal numerically to the volume fraction. The *volume fraction* of a component in a mixture is found by dividing the volume of that component at the total pressure and at the temperature of the mixture by the volume of the mixture at the same pressure and temperature.

In a binary solution consisting of components X, and Y, the mole fractions, F_X and F_Y, respectively, are:

$$\text{Mole fraction of } X = F_X \frac{f_X}{f_X + f_Y}$$

$$\text{Mole fraction of } Y = F_Y \frac{f_Y}{f_X + f_Y}$$

where f = number of moles of specific component present.

It is apparent, of course, that the mole fraction of X plus the mole fraction of Y must equal unity, or, if expressed as a percentage (mole percent), must equal 100. In the instance of three or more components, the denominators of the prior expressions must reflect the additional moles present.

In considering a solution containing 50 grams of methyl alcohol (m.w., 32) in 1,000 grams of H_2O (m.w., 18), the mole percent of each component will be:

Moles of $CH_3OH = 50/32 = 1.562$

Moles of $H_2O = 1,000/18 = \underline{55.556}$

Total Moles   57.118

Mole percent $CH_3OH = 1.562/57.118 \times 100 = 2.735\%$

Mole percent $H_2O = 55.556/57.118 \times 100 = 97.265\%$

The same solution expressed in terms of weight percentage, is:

Weight percent $CH_3OH = 50/1,050 = 4.762\%$

Weight percent $H_2O = 1,000/1,050 = 95.238\%$

In nuclear chemistry, the term *mole fraction* may be used to indicate the number of atoms of a given isotope in an isotopic mixture, as a fraction of the total number of atoms of that element in the mixture.

MOLE VOLUME. A mole of gas will occupy a definite volume under definite conditions regardless of the nature of the gas. This definite volume is called the *mole volume*. Under a pressure of 760 torr and at a temperature of 0°C, a gram-mole of gas will occupy 22.41 liters. This situation also applies to a mixture of gases. A pound-mole of gas will occupy 359 cubic feet at a pressure of 760 torr and at a temperature of 32°F (0°C).

Because the volume of one mole of gas at any specific pressure and temperature contains the same number of molecules even though there may be several different gases in the mixture, the percent by volume of any given gas is equal to the percent pressure exerted by that gas and is also equal to the mole percent of that gas. Mole percent equals volume percent equals pressure percent.

MOLINA, MARIO (1943–). A Mexican who won the Nobel prize for chemistry along with Paul Crutzen and Frank Sherwood Rowland in 1995 for their work in atmospheric chemistry, particularly concerning the formation and decomposition of ozone. See also **Rowland, Frank Sherwood (1927–)**.

MOLYBDENITE. The mineral molybdenite is sulfide of molybdenum, MoS_2. Its hexagonal crystals are usually tabular to short prismatic, but if in massive form it may be foliated or granular. Has a perfect basal cleavage; is sectile; hardness, 1–1.5; specific gravity, 4.52–5.06; luster, metallic; color, very slightly bluish, lead gray; streak, greenish-gray; opaque. Molybdenite is one of the few minerals soft enough to give a distinctly greasy feel. Molybdenite is found as a contact mineral with cassiterite and wolframite, in granite pegmatites and sometimes in granites, syenites, or gneisses. It is found associated with tin ore in Saxony and Bohemia; in Norway; England; Australia; and in the United States in Colorado, Washington County and Oxford County, Maine, in New Hampshire, Connecticut, Pennsylvania, and Washington. Its name, derived from the Greek meaning lead, was formerly applied to minerals containing lead, to graphite and to molybdenite as well. Later the term was restricted to the latter mineral. It is an ore of molybdenum.

MOLYBDENUM. [CAS: 7439-98-7]. Chemical element symbol Mo, at. no. 42, at. wt. 95.94, periodic table group 6b, mp 2610°C, bp 5560°C, density, 9.01 g/cm^3 (solid, 20°C), 10.2 g/cm^3 single crystal. Molybdenum has a body-centered cubic crystal structure. Molybdenum is silvery-white, tough, malleable, softer than glass, not oxidized by air at ordinary temperatures, but above 600°C burns to form white molybdenum oxide. The metal is dissolved by dilute HNO_3 and aqua regia, but is made passive by concentrated HNO_3. Like chromium, molybdenum exhibits the phenomenon of passivity to a marked degree and even though it shows a strong reducing action when offering a fresh surface or in potentiometric determinations, it may be quite resistant to chemical action. The metal is attacked by fused alkalis. Seven isotopes occur naturally, ^{92}Mo, ^{94}Mo through ^{98}Mo, and ^{100}Mo; and five radioactive isotopes have been identified, ^{90}Mo, ^{91}Mo, ^{93}Mo, ^{99}Mo, and ^{101}Mo. With exception of ^{93}Mo, which has a half-life of approximately 10^4 years, the other radioactive isotopes have short half-lives, measured in terms of minutes and hours. The element does not appear on the list of the first 36 most frequently occurring elements in the earth's crust and thus may be considered a scarce element. Molybdenum is listed as the 42nd element in terms of estimated occurrence in the universe. The element ranks 25th among the elements occurring in seawater, there being an estimated 50 tons of molybdenum per cubic mile (10.8 metric tons per cubic kilometer) of seawater. Molybdenum was one of the first of the uncommon elements to be identified; reported by Carl Wilhelm Scheele in 1778 when he was investigating lead ores.

First ionization potential 7.18 eV. Oxidation potentials Mo $\longrightarrow$ $Mo^{3+} + 3e^-$, ca. 0.2 V; Mo(IV) $\longrightarrow$ Mo(V) $+ e^-$, (0.1 M KCl, pH 3) -0.01 V.

Other physical properties of molybdenum are given under **Chemical Elements**. See also summary of properties of refractory metals under **Niobium**.

The main sources of molybdenum are MoS_2 (molybdenite), $Ca(MoW)O_4$ (powellite), and $PbMoO_4$ (wulfenite), of which the first is by far the most important. See also **Molybdenite**; **Wulfenite**. A representative molybdenite ore contains about 0.2% molybdenum. A concentrate containing about 90% MoS_2 is prepared through crushing, grinding and flotation operations. Most of the concentrate so produced is roasted in air to form technical molybdic oxide (about 90% MoO_3), the product used to add molybdenum in most steelmaking processes. The prevalent industrial method for producing ferromolybdenum, the material used to add molybdenum to some cast irons and steels, is a thermit process wherein aluminum and silicon reduce a charge of iron oxide and molybdic oxide. Ferromolybdenum contains about 60% molybdenum.

Pure molybdic oxide (99.95% MoO_3) is prepared by sublimation of technical oxide or by calcining ammonium molybdate. Metallic molybdenum powder is prepared commercially by hydrogen reduction of either pure molybdic oxide or ammonium molybdate in a two-step process. The first step, carried out at about 500°C yields MoO_2. The second step, reduction of MoO_2 to Mo, is carried out at about 1100°C.

Uses

Molybdenum is added to a number of alloy steels because its presence increases hardenability, toughness, cold-formability, and weldability. Even though the steels so treated may contain less than 1% Mo, these uses account for about 95 million pounds ($\sim$43 million kilograms) of the metal per year. One of the more recent special steels contains manganese, molybdenum, and niobium for use in pipelines under arctic conditions. Stainless steels account for over 40 million pounds ($\sim$13.6 million kilograms) of Mo per year. In these steels, molybdenum increases corrosion resistance, elevated-temperature strength, and weldability. However, ferritic stainless steels that contain 18–26% chromium and 1–2% molybdenum are displacing the traditional 18–8 type stainless steels for many applications. For example, a steel containing 18% chromium and 2% molybdenum is finding wide use in solar energy panels.

The use of molybdenum in tool steels accounts for about 18 million pounds ($\sim$8.2 million kilograms) of Mo per year. In these steels, molybdenum provides better hot strength and improved resistance to softening and thermal cycling effects. A number of the earlier tungsten-grade tool steels have been replaced by molybdenum steels because of improved performance, lower metal density, and greater price stability. About 15 million pounds ($\sim$6.8 million kilograms) of Mo per year are consumed by the foundry industry where the metal improves the strength and abrasion resistance of cast iron. About 6 million pounds ($\sim$2.7 million kilograms) of Mo per year go into superalloys for high-temperature environments as required, for example, by jet engines. One of the later-developed superalloys is a nickel-base alloy that contains about 18% Mo and has the advantages of a very high melting point, low density, and low coefficient of thermal expansion. Uses of molybdenum compounds are described below.

Chemistry and Compounds

In keeping with its $4d^5 5s^1$ electron configuration, molybdenum forms many compounds in which its oxidation state is 6+, to an even greater extent than chromium. Also, like chromium, it forms compounds in which it is divalent and those in which it is trivalent; unlike chromium, it forms a number of pentavalent compounds, and a few more tetravalent compounds, especially complexes.

Among the divalent compounds of molybdenum are the dibromide, $MoBr_2$, and the dichloride, $MoCl_2$. There is also a complex ion of divalent molybdenum $Mo_6Cl_8^{4+}$, that is of particular interest because it does not yield Mo^{2+} ions. The chloride salt of the ion, Mo_6CL_{12} has been shown, by precipitation with Ag^+, to have two-thirds of its chlorine content present in a complex, so its structure is established as $[Mo_6Cl_8]Cl_4$. It is obtained by higher temperature dismutation of $MoCl_3$, while the corresponding dibromide can be produced by direct reaction of the elements.

The relatively small number of trivalent molybdenum compounds generally exhibit the marked reducing action of the Cr^{3+} compounds,

though not quite so strongly. They include the trichloride $MoCl_3$ and the tribromide $MoBr_3$, as well as the sesquioxide Mo_2O_3.

In addition to Mo_2O_3, the other oxides include MoO_2, Mo_2O_5, MoO_3, the tetravalent and pentavalent oxides being obtainable from the hexoxide by hydrogen reduction, the MoO_3 being formed by direct combination of the elements.

In addition to the dioxide MoO_2 and the disulfide, MoS_2, example of tetravalent molybdenum compounds include the tetrachloride, $MoCl_4$, and tetrabromide, $MoBr_4$, both of which are hydrolyzed in hot H_2O. Other tetravalent compounds include a few of the complexes.

The complexes also exist among the pentavalent molybdenum compounds, which include a number of simple compounds as well. In addition to the pentoxide, Mo_2O_5 and pentasulfide, Mo_2S_5, pentahalides and oxyhalides, such as $MoCl_5$ and $MoOCl_3$ are also known. Direct reaction with fluorine, yields MoF_6; with chlorine, $MoCl_5$, and with bromine, $MoBr_3$.

As stated above, hexavalent compounds of molybdenum constitute the most numerous group. The hexavalent molybdenum oxyhalides include $MoOF_4$, MoO_2F_2, $MoOCl_4$, MoO_2Cl_2, and MoO_2Br_2. $Mo_2O_3Cl_5$ may contain tetravalent and hexavalent molybdenum. Hexavalent molybdenum also exists in the sulfide, MoS_3, and in various oxyacid salts, and molybdates.

Molybdenum trioxide dissolves in alkaline solutions to yield, in more or less hydrated form, the molybdate ion, MoO_4^{2-}. However, the ionic species that exist in solutions of low pH are more complex than hydration of MoO_4^{2-} would indicate, and this is especially true of the compounds obtained from such solutions. Thus from strongly ammoniacal solution, $(NH_4)_2MoO_4$ is obtained, but from nearly neutral solutions containing NH_4^+ and MoO_4^{2-}, the complex $(NH_4)_6Mo_7O_{24} \cdot 4H_2O$ crystallizes. The process of complex anion formation is considered to occur upon neutralization, or particularly upon acidification, of a solution, by the addition of a proton to an MoO_4^{2-} ion, forming $HMoO_4^-$ ions, which condense with other oxyanions to form complexes, which can be crystallized as salts. Such salts are considered to be derived from "poly" acids. When such acids have one kind only of metal atom (e.g., Mo) in their anions, like the complex ammonium salt cited above, they are called *isopoly acids*; with more than one kind they are called *heteropolyacids*. The latter group comprises an entire field of molybdenum chemistry (as well as that of tungsten and related elements); the molybdophosphates are important in analysis and other applications. Other examples are the heteropolyacid salts formed by molybdates with oxyanions of boron, silicon, germanium, tin, arsenic, titanium, zirconium, and hafnium. In such compounds, one important group is the "12-group," containing 12 atoms of molybdenum, but many other proportions are known. Other interesting compounds are the "molybdenum blues," complex oxides of colloidal nature obtained by reduction of molybdates.

Molybdenum forms many other complexes. Of particular interest are the octacyano complexes, containing eight cyanide ions, CN^-, coordinated to a single tetravalent of pentavalent molybdenum ion, $Mo(CN)_8^{2-}$ and $Mo(CN)MO_8^{3-}$, the latter being exceptionally stable, and both form octacyanomolybdic acids $H_3[Mo(CN)_8] \cdot 3H_2O$ and $H_4[Mo(CN)_8] \cdot 6H_2O$. Molybdenum(III) forms salts of

$$H_3MO(CN)_6$$

A fluoro complex of Mo(VI) has the structure, $[MoF_8]^{2-}$.

In liquid NH_3 solution, potassium amide reacts with MoO_3 to form the salt K_3MoO_3N, completely hydrolyzed by H_2O, in which the molybdenum atom is the center of a monomeric tetrahedral anion, being surrounded by three oxygen and one nitrogen atoms.

Like chromium, molybdenum forms a number of cyclopentadienyl compounds, many of which are carbonyls, e.g., $C_5H_5Mo(CO)_2NO$, $C_5H_5Mo(CO)_3X$ (where X may be Cl, Br, I, H, CH_3, C_2H_5 or C_3H_7). Molybdenum also forms a simple carbonyl, $Mo(CO)_6$.

Molybdates. Ammonium polymolybdate, also referred to as ammonium dimolybdate or molybdic acid (85%), is the highest-purity molybdenum compound commercially available (up to 99.97% purity). The compound is a source of very high-purity MoO_3 which is used as a catalyst in hydrogen-treating and hydrocracking processes. The compound also is used in electroplating baths and as an important laboratory reagent for determinations of phosphates, arsenates, and lead. Sodium molybdate Na_2MoO_4 or $Na_2MoO_4 \cdot 2H_2O$ is made by dissolving MoO_3 in excess NaOH. The compound is widely used in the manufacture of molybdate-chromate orange pigments and several phosphomolybdic acid-organic pigments. The compound also is used as a condensation catalyst for phthalocyanine pigments, as a synthetic cutting fluid, and as a corrosion inhibitor in circulating cooling water systems and in glycol-based anti-freeze formulations. Several of the soluble molybdates react with organic intermediates to form dyes that are used for furs and hair and have the advantage of being very colorfast. Zinc molybdate, because of its nontoxicity and excellent corrosion-inhibiting properties, is an excellent white pigment. Lithium molybdate is an additive for porcelain enamel coatings. Iron, cobalt, and nickel molybdates are used as catalysts in hydrogenation, desulfurization, denitrification, and hydrocracking processes. Lead molybdate is used in connection with applying vitreous designs to glass bottles.

Sulfides. In addition to serving as the primary natural source of molybdenum, purified molybdenum disulfide MoS_2 is an excellent lubricant when in the form of a dry film, or as an additive to oil or grease. The compound also is used as a filler in nylons, and as an effective catalyst for hydrogenation-dehydrogenation reactions. Molybdenum also combines with sulfur as the sesquisulfide Mo_2S_3 and the trisulfide MoS_3, uses for which are under study.

Halides. Molybdenum pentachloride $MoCl_5$ is used as a catalyst for several polymerization reactions involving olefins, vinyl monomers, trioxane, ethylene, vinylcyclohexane, cyclopentene, and butadiene. Vapor-phase coatings of molybdenum on metallic or ceramic substrates are prepared, using $MoCl_5$ as the starting compound.

Organomolybdenum Compounds. Soluble molybdates, molybdenum hexacarbonyl $Mo(CO)_6$, and several molybdenum halides form complex compounds with many organic oxygen, nitrogen, and sulfur compounds. Some of the oxygen-coordinated compounds include alkoxides, acetonates, oxalates, carboxylates, phenoxides, and organic chlorides. Nitrogen-coordinated compounds include several organic molybdates and chlorides. Sulfur-coordinated compounds include dialkyldithiophosphates, cysteine complexes, α-diketone complexes, and dialkyldithiocarbamates. Examples of industrial use of organomolybdenum compounds include pyrogalol-molybdate complexes in dyes, molybdenum oxalate in photochemicals, molybdenum dithiocarbamate as a lubricant additive, and molybdenum acetylacetonate as a catalyst for the polymerization of ethylene and the formation of polyurethane foam.

A discussion of the biological aspects of molybdenum is given in the next entry.

ROBERT Q. BARR
Director Technical Information, Climax
Molybdenum Company
Greenwich, Connecticut

Additional Reading

Barry, H.F., and P.C.H. Mitchell, eds.: "Chemistry and Uses of Molybdenum," *Proceedings of Symposium*, sponsored by Climax Molybdenum Co., and University of Michigan, Ann Arbor, Michigan (August 1979).

Braithwaite, E.R.: "The Chemical Uses of Molybdenum and Its Compounds," *Chem. and Ind.*, **12**, 405–412 (1978).

Chianelli, R.R., et al.: "Molybdenum Disulfide in the Poorly Crystalline 'Rag' Structure," *Science*, **203**, 1105–1107 (1979).

Climax: Various product data sheets revised periodically, Climax Molybdenum Company, Greenwich, Connecticut 06830.

Lander, H.N.: "Technological Progress with Molybdenum," *Molybdenum Mosaic*, **3**, 1, 2–11 (1978).

Sutulov, A.: *International Molybdenum Encyclopedia*, Vol. III, Intermet Publications, Santiago, Chile (1980).

MOLYBDENUM (In Biological Systems). Molybdenum is required in very low amounts by both plants and animals. Nutrient imbalances involving molybdenum and copper have caused serious problems in cattle and sheep production.

Molybdenum deficiencies are found in plants grown on certain acid soils, and sometimes the deficiency can be corrected by adding either small quantities of manganese compounds or larger quantities of limestone to the soil. The limestone makes the soil more alkaline and increases the availability of the native molybdenum in the soil. In certain parts of the world (including the eastern United States), small amounts of molybdenum fertilizer are used regularly for producing some vegetables, notably cauliflower. In Australia, large areas have been changed from near-desert conditions to productive agriculture through the application of molybdenized superphosphate.

In alkaline soils, molybdenum is more available to plants. Forage crops growing on some alkaline soils (as in the western United States) may take up high concentrations of molybdenum. The element is not toxic to the plants. They grow normally and may produce excellent yields. But cattle and sheep that eat these forages may suffer from molybdenum toxicity. It is now well established that what appears to be molybdenum toxicity is actually a copper deficiency that is induced by the molybdenum. Thus, the symptoms of molybdenum toxicity are the same as those of copper deficiency and include fading of the hair and diarrhea. The condition may be prevented by supplementing the animal diet with extra copper, or by injecting copper compounds into the animal body, usually by an experienced veterinarian. Cattle are more susceptible to molybdenum-induced copper deficiency than other types of livestock. Horses and pigs are rather tolerant of high levels of dietary molybdenum.

High levels of molybdenum are generally considered to be 20 parts per million (ppm) or more in dry forage. Some symptoms of interference with copper metabolism in cattle may be evident when the forage contains as little as 5 ppm molybdenum if the forage is also low in copper. The effects of high-molybdenum forage in interfering with copper metabolism in animals are generally more severe if the animal diet is also high in sulfates.

In terms of humans, some research in New Zealand and the United Kingdom indicates that diets containing moderately high levels of molybdenum help to prevent dental decay. The high-molybdenum soils in the United States are seldom used for production of food crops and thus the effects of molybdenum toxicity from food substances are not well known.

Restriction of the molybdenum intake by young rats in a synthetic purified casein diet results in a decreased level of tissue, particularly small intestinal, xanthine oxidase. The enzyme levels are restored to normal by the inclusion of sodium molybdate and other molybdate compounds. Sodium tungstate is a competitive inhibitor of molybdate, and dietary intakes of tungstate greatly reduce the molybdenum and xanthine oxidase concentrations in tissues.

Legumes, cereal grains, and some green leafy vegetables are good sources of molybdenum, whereas fruits, berries, and most root or stem vegetables are poor sources. Vertebrate tissues are generally low in molybdenum, with concentrations in liver and kidney being higher than in other organs and cells. Excess molybdenum intake by cattle causes the disease known as "teart," characterized by severe diarrhea and degradation of general health.

For many years, it has been established that molybdenum is a catalyst for biological nitrogen fixation, that is, reducing molecular nitrogen to ammonia by a nitrogenase which exists in soil- and water-dwelling microorganisms. This was considered an exclusive process. However, in 1985, some British investigators (University of Sussex) confirmed that a second nitrogen-fixation scheme, one not involving molybdenum, also exists. Their experiments were conducted with *Azotobacter vinelandii*. As reported by Marx, the genes that code for the enzymes of the alternative system are being sought. Recent work has shown that the alternative system is activated in soils where there is a molybdenum deficiency. The detailed role of Mo in the principal pathway still is not well understood. Whether or not further research may lead to better growth of some plants in molybdenum-deficient soils remains unanswered.

Additional Reading

Allaway, W.H.: *The Effect of Soils and Fertilizers on Human and Animal Nutrition*, Cornell University Agricultural Experiment Station, Agriculture Information Bulletin 378, U.S. Department of Agriculture, Washington, DC, 1975.

Braithwaite, E.R. and J. Haber: *Molybdenum: An Outline of Its Chemistry and Uses*, Elsevier Science, New York, NY, 1994.

Considine, D.M. and G.D. Considine: *Foods and Food Production Encyclopedia*, 633–634, 674, 1307–1308, Appendix 2, Table 1, Van Nostrand Reinhold, New York, NY, 1982.

Kirchgessner, M. (editor): *Trace Element Metabolism in Man and Animals*, Institut für Ernahrungsphysiologie, Technische UniversitÄt München, Freising-Weihenstephan, Germany, 1978.

Lewis, R.J. and N.I. Sax: *Sax's Dangerous Properties of Industrial Materials*, 10th Edition, John Wiley & Sons, Inc., New York, NY, 1999.

Marx, J.L.: "Fixing Nitrogen without Molybdenum," *Science*, **229**, 956–957 (1985).

Mertz, W.: "The Essential Trace Elements," *Science*, **213**, 1332–1338 (1981).

Underwood, E.J.: *Trace Elements in Human and Animal Nutrition*, 5th Edition, Academic Press, Inc., San Diego, CA, 1990.

MONATOMIC GASES. See **Chemical Elements**.

MONAZITE. The mineral monazite is essentially a phosphate of the rare-earth metal cerium. (Ce, La, Nd, Th)PO$_4$; but other rare-earth metals are usually present. So constant is the presence of thorium that monazite is the chief source of thorium dioxide. It is monoclinic, but found ordinarily as translucent yellow to brown grains with a resinous luster, often as sand. Its hardness is 5.0–5.5; specific gravity, 4.6–5.4. Monazite is found in granites, pegmatites and similar rocks, but rarely in any concentration. The commercial deposits are residual sands. The Ilmen Mountains in the U.S.S.R., Norway, India, Madagascar, the Republic of South Africa, and Brazil are well known for their monazite deposits. In the United States monazite is known in Connecticut, New York, Virginia, North Carolina and Idaho. Monazite derives its name from the Greek word meaning solitary, in reference to the relative rarity of this mineral.

MOND PROCESS. See **Nickel**.

MONEL. See **Nickel**.

MONOBASIC. Descriptive of acids having one displaceable hydrogen atom per molecule. Acids having two, three, or more displaceable hydrogen atoms are called dibasic, tribasic, and polybasic, respectively.

MONOMER. A single molecule, or a substance consisting of single molecules. The term monomer is used in differentiation of dimer, trimer, etc., terms designating polymerized or associated molecules, or substances composed of them, in which each free particle is composed of two, three, etc., molecules.

MONOMOLECULAR LAYER. The early work of Rayleigh, Langmuir, Hardy and others has shown that it is possible to deposit on solid or liquid surfaces films which are one molecule thick. Any such layer is called a monomolecular layer, *unilayer* or *monolayer*. See also **Molecular and Supermolecular Electronics**.

MONOSODIUM GLUTAMATE (MSG). See **Flavor Enhancers and Potentiators**.

MOORE, STANFORD (1913–1982). An American biochemist who won the Nobel prize for chemistry in 1972, with Christian B. Anfinsen and William H. Stein, for enzyme studies. He was involved with the analysis of the action of the complex enzyme deoxyribonuclease. His Ph.D. was granted from the University of Wisconsin.

MORDANT. A substance capable of binding a dye to a textile fiber. The mordant forms an insoluble lake in the fiber, the color depending on the metal of the mordant. The most important mordants are trivalent chromium complexes, metallic hydroxides, tannic acid, etc. Mordants are used with acid dyes, basic dyes, direct dyes, and sulfur dyes. Premetallized dyes contain chromium in the dye molecule. A mordant dye is a dye requiring use of a mordant to be effective. See also **Dyes**.

MORPHINE. [CAS: 57-27-2]. About 10% of the weight of opium is morphine which was the first of the vegetable alkaloids to be isolated in 1805 by Sertürner. Since the source of the natural alkaloids is opium, all narcotics whose actions resemble those of morphine are sometimes referred to as opiates. Semisynthetic agents are usually made by altering the morphine molecule, and include such agents as diacetylmorphine (heroin), ethylmorphine (*Dionin*), dihydromorphinone (*Dilaudid*), and methyldihydromorphinone (metopon). Synthetic narcotics include agents with a wide variety of chemical structures. Some of the important synthetic agents are meperidine (piperidine type), levorphanol (morphinian type), methadone (aliphatic type), phenaxocine (benzmorphan type), and their derivatives. The structures of the various narcotics are given in Fig. 1.

Since morphine is responsible for the major actions of opium and the actions of all narcotics are qualitatively similar, morphine can be used as a model for discussing narcotic agents. The most prominent effects of morphine in the human body are on the central nervous system and the gastroenteric tract. The principal central action of morphine is the

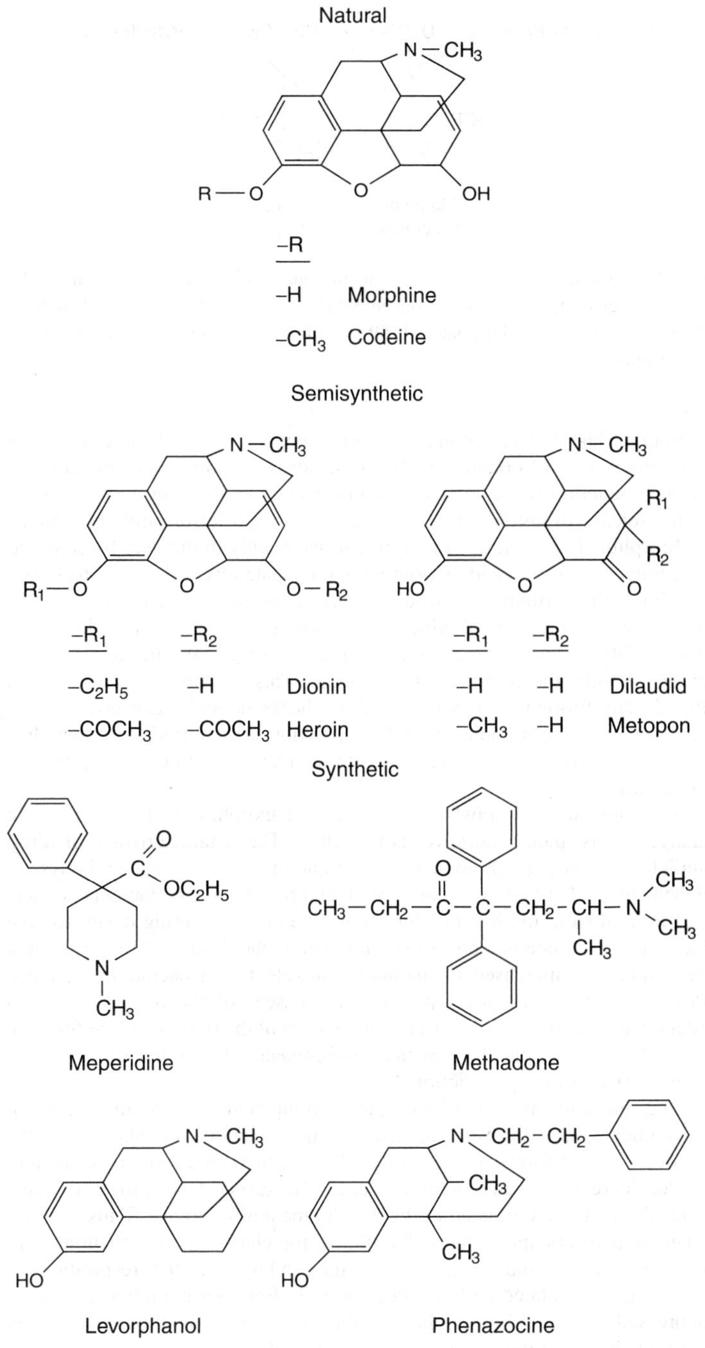

Fig. 1. Chemical structures of various narcotics

relief of pain, and this occurs in at least three ways: (1) morphine reduces central perception of pain probably at the thalamic level; (2) it alters the reactions to pain probably at the level of the cerebral cortex; and (3) it elevates the pain threshold by inducing sedation or sleep. In the medulla, morphine depresses the respiratory, cough and vasomotor centers and indirectly stimulates the vomiting center. The nuclei of the occulomotor (III) and vagus (X) nerves are stimulated by sufficient doses of morphine causing myosis (constriction of the pupils), bradycardia (slowing of the heart rate), and increased gastroenteric tone. The overall effect of morphine on the gastroenteric tract is spasmogenic and constipative. Morphine causes the constipative action by several means, including increased segmental movement of the large bowels, spastic tonus of the sphincters, decreased defecation reflex, and increased reabsorption of water in the large intestines to cause drying of feces.

The metabolic effects of morphine are not marked and are clinically unimportant. The metabolic rate may be decreased slightly due to the lowered activity and tone of the skeletal muscles resulting from the central depression. A rise in blood sugar may be observed after the injection of

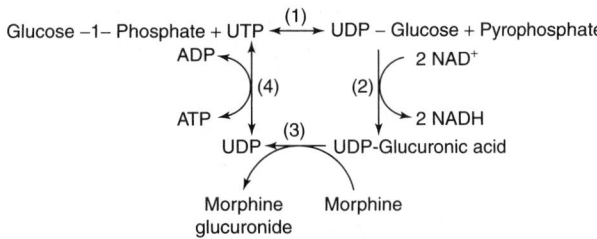

Fig. 2. Formation of morphine glucuronide. NAD^+ = nicotinamide adenine dinucleotide; NADH = reduced NAD^+; ATP = adenosine triphosphate; ADP = adenosine diphosphate; UTP = uridine triphosphate; UDP = uridine diphosphate

morphine. The hyperglycemia is due to glycogenolysis in the liver resulting from the release of epinephrine from the adrenal medulla. The lowering of urine production noted after the administration of the drug is due mainly to the release of antidiuretic hormone from the posterior pituitary gland.

Morphine is detoxified or biotransformed mainly in the liver by conjugation with glucuronic acid. Morphine is conjugated by a series of reactions involving the formation of uridine diphosphoglucose (UDP-glucose), the oxidation of carbon-6 of glucose to form uridine diphosphoglucuronic acid (UDP-glucuronic acid) and the transfer of glucuronic acid to morphine to form the morphine glucuronide. This reaction is diagramed in Fig. 2. The following enzymes catalyze the sequential reactions; reaction (1), UDP-glucose pyrophosphorylase; reaction (2), UDP-glucose dehydrogenase; reaction (3), glucuronyl transferase; reaction (4) nucleoside diphosphokinase.

The most serious drawback in the use of morphine and other narcotic analgesics is their addictive potentiality. The characteristics of drug addiction include psychological need or habituation, tolerance and physical dependence. Habituation consists of an emotional and psychic dependence, and in addiction, the habituation becomes an overpowering desire to take the drug. Tolerance is a phenomenon whereby the dosage of the drug must be continually increased to maintain equivalent pharmacological effects. Physical dependence develops when the tissues of the body become so adapted to the effects of the drug that the cells of the tissues cannot function normally without the drug in the environment. This is the most vicious characteristic of drug addiction.

The mechanisms underlying the development of tolerance are not fully understood. Biochemically, it may be attractive to explain tolerance by decreased absorption, altered distribution, increased biotransformation, and/or increased excretion of the drug. However, these processes have been shown to be unrelated to the development of tolerance. Thus, cellular adaptation offers the greatest likelihood for clarifying the phenomenon. Evidence for cellular adaptation is the finding that the respiration of chemically stimulated cortical slices of brain from normal rats is markedly depressed by morphine, whereas the respiration of those from rats chronically dosed with morphine is unaffected.

Heroin is diacetylmorphine (diamorphine hydrochloride) and is prepared by the action of acetic anhydride on morphine, possessing four times the analgesic affect of morphine, but having a considerably less depressant effect. Addiction is common, the drug being taken in the form of snuff, or by injection.

Nalorphine, the allyl ($-CH_2-CH=CH_2$) derivative of morphine (N-allylnormorphine) is remarkable in that it is antagonizing to almost all the effects of narcotics. The antagonizing action is specific for the narcotic analgesics. For instance, nalorphine will antagonize the respiratory depression due to morphine or other narcotics, but not that caused by other depressants, such as hypnotics or anesthetics. This property of nalorphine makes it a particularly useful antidote in cases of acute morphine poisoning. The agent can also precipitate acute withdrawal symptoms if administered to persons addicted to narcotics. The agent has become a useful biochemical tool for studying the mechanism of action of narcotics and tolerance. Since the chemical structures between morphine and nalorphine are so similar, it has been suggested that nalorphine acts by competing with morphine for the receptor site. The antagonistic effect of nalorphine cannot be explained by a simple competitive inhibition if equal affinity for the receptor site with the agonist and antagonist is assumed, because small doses of nalorphine antagonize the effects of much higher doses of the narcotic. Nalorphine also antagonizes the effects of synthetic narcotics of varying chemical structures, such as methadone and meperidine.

Morphine, $C_{17}H_{19}NO_3 \cdot H_2O$, is a white powder melting at 253°C and is derived from opium which is the dried juice obtained from unripe capsules of the poppy plant (*Papaver somniferum*), variously cultivated in the Near East and Far East. The opium poppy is an annual. When the petals drop from the white flowers, the capsules are cut. The juice exudes and hardens, forming a brownish mass which is crude opium. It contains a total of about 20 narcotics, including morphine. See also **Alkaloids**; and **Analgesics, Antipyretics, and Antiinflammatory Agents**.

Additional Reading

Courtwright, D.: *Dark Paradise: Opiate Addiction in America before 1940,* Harvard University Press, Cambridge, MA, 1982.

Gianino, J. et al.: *Intrathecal Drug Therapy for Spasticity and Pain: Practical Patient Management,* Springer-Verlag New York, Inc., New York, NY, 1995.

Web Reference

U.S. Food and Drug Administration Center for Drug Evaluation and Research. http://www.fda.gov/cder

MOSLEY, HENRY (1887–1915). A British chemist who studied under Ernest Rutherford and brilliantly developed the application of X-ray spectra to the study of atomic structure; his discoveries resulted in a more accurate positioning of elements in the periodic table by closer determination of atomic numbers. Tragically for the development of science, Mosely was killed in action at Gallipoli in 1915.

MOSLEY'S LAW. The square root of the frequency of a given line of an element in the X-ray spectrum is directly proportional to the atomic number of the element.

MOSSBAUER EFFECT. The phenomenon of recoilless resonance fluorescence of gamma rays from nuclei bound in solids. It was first discovered in 1958 by R.L. Mössbauer. The extreme sharpness of the recoilless gamma transitions and the relative ease and accuracy in observing small energy differences make the effect an important tool in nuclear physics, solid-state physics, and chemistry.

If a gamma ray is emitted by an atomic nucleus, the system to which the emitting atom belongs must recoil, in order to conserve momentum, in a direction opposite to that in which the gamma rays is emitted. Similarly, if an atomic nucleus absorbs a gamma ray, the system must continue to move, following absorption, in such a way that momentum is conserved. If the recoiling system is a single atom, such as in a gas and shown schematically in Fig. 1(a), the emitting atom carries away enough energy from the transition for the observed energy $E_0 - R$ of the emitted gamma ray to be measurably less than the energy E_0 of the nuclear transition that caused the gamma ray to be emitted, also indicated in (a) of the diagram. Furthermore, a gamma ray that is absorbed by a single atom must transfer a measurable kinetic energy to that atom, as well as the energy of the nuclear transition. On the other hand, if the emitting or absorbing nucleus belongs to an atom that is bound into a crystalline structure, such that the structure as a whole can recoil, and as indicated schematically in Fig. 1(b), the kinetic energy that must be given to the crystalline system to conserve momentum is greatly reduced, compared to the energy that must be given to a single atom, because of the much larger mass of the system. The recoil energy is then so small that the gamma ray carries away essentially the full energy E_0 of the transition in the case of emission, transferring such a small fraction of its energy to the absorbing system that emission and absorption appears to be recoil-free.

This process is observed, of course, in the analogous case of the resonance radiation in atomic transitions, in which case, the photons have energies in the range of light, commonly visible light. However, the protons of gamma radiation are so much more energetic that their energy loss by recoil of the nucleus emitting them is great enough, in the case of free atoms, for the resonance effect not to occur.

Mössbauer discovered, however, that in the case of atoms which are not free, but bound in a solid, the effect can often be observed. It is easily demonstrated when the normal, free-atom recoil energy is comparable to the energy of the quantizied lattice vibrations. Under these conditions, zero-phonon processes are possible in which the entire energy of the nuclear

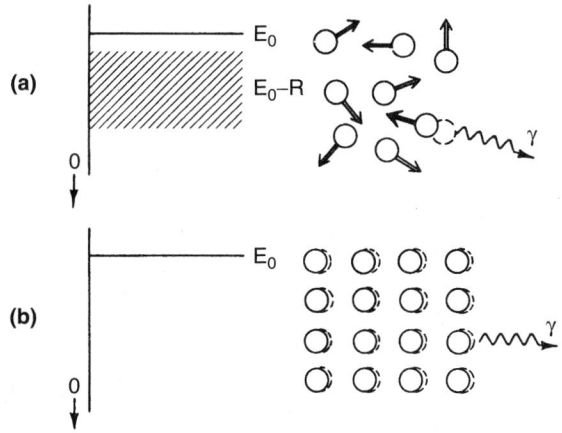

Fig. 1. (a) Emission of a gamma-ray by a single atom moving randomly in a gas transfers appreciable energy to the emitting atom in the form of recoil kinetic energy, and reduces the energy of the gamma ray from the transition energy E_0 to some lower energy E_0-R. (b) Emission of a gamma ray by an atom bound into a crystalline structure may sometimes cause recoil of the whole crystal, in which case, the loss of energy in the form if kinetic energy of the crystal is negligibly small and the gamma ray appear to be emitted with an energy E_0

transition goes into the gamma ray and the recoil momentum is taken by the solid as a whole. The resulting gamma rays then have the proper energy to be resonantly absorbed or scattered in an analogous zero-phonon process.

The Mössbauer effect is useful in determining nuclear level widths and Doppler effects. Another application is based upon the measurement of nuclear hyperfine structure, a measurement which is possible when the line-width of the gamma ray is smaller than the hyperfine interaction (that due to the coupling of the nuclear moments with external fields). In this application, the Mössbauer effect is almost unique because one obtains the splitting of both the ground and the first excited nuclear states. This effect, in turn, makes possible the determination of nuclear moments of the excited states, which can be important tests of nuclear models. Another important feature is the so-called isomer or chemical shift (terms used interchangeably) which measures the simple electrostatic interaction of the nucleus with its own s-electron and has given information about the difference in the nuclear radii of the ground and excited states.

MOTOR OCTANE NUMBER.   See **Petroleum**.

MUCILAGES.   See **Gums and Mucilages**.

MULLIKEN, ROBERT S. (1896–1986).   An American chemist, physicist, and educator who won the Nobel prize for chemistry in 1966 for his fundamental work concerning chemical bonds and the electronic structure of molecules by the molecular orbital method. Mulliken received is B.Sc. Degree in 1917 at the Massachusetts Institute of Technology, Cambridge, MA. and a Ph.D. degree at the University of Chicago, IL., in 1921.

MULLIS, KARY BANKS (1944–).   An American who won the Nobel prize for Chemistry in 1993 for his invention of the polymerase chain reaction (PCR) method.

MUON.   The *muon* (μ^-) is an elementary particle of the lepton family. Properties include: Spin, $\frac{1}{2}$; mass (MeV), 105.66; lifetime, 2.20×10^{-6} second. The antiparticle is the positive muon (μ^+). The muon neutrino (ν) has spin, $\frac{1}{2}$; 0 mass; and is stable. The muon family appears to be simply a duplicate of the electron family except for a change in the unit of mass. See also **Particles (Subatomic)**.

The positive muon was discovered in cloud chamber photographs made by C.D. Anderson and S.H. Neddermyer on Pike's Peak in 1935, and the negative muon almost simultaneously in cloud chamber photographs made by J.C. Street and E.C. Stevenson. These particles have long been called mu-mesons, but since they are fermions (spin $\frac{1}{2}$ while all other mesons are bosons, the name *muon* is preferred, as is their classification with the leptons because of their small rest mass, which is about 206 m_e, where m_e is the mass of the electron. Another reason is their inability to interact with other particles through the nuclear forces.

Their charges are equal in magnitude to that of the electron. They are produced by the decay of pions (*pi mesons*) and (to a limited extent) by the decay of kaons and hyperons. Positive-negative muon pairs also can be generated by the action on matter of gamma-rays of energy greater than the rest masses of the particles, i.e., exceeding 211 MeV. Their lives are short, about 2.2×10^{-6} seconds in the free state, and the negative muon usually decays into an electron, a neutrino, and an antineutrino, while the positive muon usually gives a positron, as well as a neutrino and antineutrino. As explained in the entry on neutrino, there are two types of neutrinos and antineutrinos (ν_e or $\overline{\nu_e}$) like that produced in the decay of radionuclides, and a muon-associated neutrino or antineutrino (ν_μ or $\overline{\nu}_\mu$) so that these reactions would be written

$$\mu^- \to e^- + \nu_\mu + \overline{\nu}_e$$

$$\mu^+ \to e^+ = \overline{\nu}_e + \nu_e$$

Muons can easily penetrate many meters of iron and can sometimes cause problems in particle physics research. For example, the upsilon experiment at Fermilab in 1977, conducted by L.M. Lederman and others, required building a simple magnetic system that would remeasure each muon's energy after it emerged from the main detector. See also **Upsilon Particle**; and **Particles (Subatomic)**.

MUONIUM.   The atom consisting of a positive muon and an electron. Thus, muonium may be regarded as a light isotope of hydrogen in which the positive muon replaces the proton. When a beam of positive muons is stopped in a gas (argon under such pressures as 50 atmospheres has been used in much of this research), muonium is formed directly in its ground state by the capture of an electron by a positive muon. The reaction is important because of its bearing upon the nature of the muon-electron interaction and the muon itself. The study has included measurement of the hyperfine structure interval in the ground state of muonium, and measurement of muon polarization as a function of time and impurity concentration. By adding such gases as oxygen (O_2) and nitric oxide (NO) as impurities to the argon, data on spin exchange of electron and muon is obtained, while with impurities such as nitrogen dioxide (NO_2) and ethylene (C_2H_2), evidence of such reactions as $NO_2 + M \longrightarrow NO + OM$ and $C_2H_4 + M \longrightarrow C_2H_4M$ is obtained.

MUTATION.   See **RECOMBINANT DNA**.

MYOSIN.   See **Contractility and Contractile Proteins**.

MYRISTIC ACID.   [CAS: 544-63-8]. Also called tetradecanoic acid, formula $CH_3(CH_2)_{12}COOH$. At room temperature, it is an oily, white crystalline solid. Soluble in alcohol and ether; insoluble in water. Specific gravity 0.8739 (80°C); mp 54.4°C; bp 326.2°C. Combustible. The acid is derived by the fractional distillation of coconut oil. Myristic acid is used in soaps; cosmetics; in the synthesis of esters for flavorings and perfumes; and as a component of food-grade additives. Myristic acid is a constituent of several vegetable oils. See also **Vegetable Oils (Edible)**.

N

NACREOUS PIGMENT. A pigment, containing guanine crystals obtained from fish scales or skin, that produces a pearly luster. May be applied as surface coatings, as in simulated pearls, or incorporated into plastics. The pigment particle is generally a very thin platelet of high index of refraction. The crystals are readily oriented into parallel layers because of their shape. Being transparent, each crystal reflects only part of the incident light reaching it and transmits the remainder to the crystal below. The nacreous effect is obtained from the simultaneous reflection of light from the many parallel microscopic layers.

NANOTECHNOLOGY (Molecular). Molecular nanotechnology is the production of functional materials and structures in the 0.1 to 100 nm range (the nanoscale) by any of a variety of physical and chemical methods. These methods include nanolithography, direct atomic and molecular manipulation with nanoscale probes, biotechnological selection and production of useful nanomaterials, and chemical synthesis and self-assembly of functional molecules and molecular aggregates. Nanoscale structures and devices may be constructed synthetically from their atomic or molecular constituents (the synthetic or "bottom-up" approach, also referred to as nanochemistry), or by fabrication techniques that use methods to form small structures from larger ones (the reductive or "top-down" approach). A natural analogy to nanotechnology can be found in the biosphere, wherein small and large molecules interact to form complex structures necessary for all the functions of living organisms.

Synthetic vs Reductive Technologies

Many of the devices that have thus far been envisioned as products of nanotechnology (e.g., nanoscale environmental sensors, information processors, and actuators) cannot be produced by the large-scale microfabrication techniques currently in use. The further development of nanotechnology hinges on the understanding and manipulation of physical laws and processes at the nanometer level, such as electronic, interatomic, and intermolecular interactions that can be manipulated to allow efficient assembly of nanostructures.

Biological systems employ a variety of synthetic strategies and processes that make efficient use of these interactions to form highly ordered molecular and supermolecular units that perform a wide variety of functions. The distribution and functional success of biological nanostructures in living organisms are an evolutionarily directed balance between the internal stability of the nanostructures (by atomic and molecular interactions) and their responses to the external environment. Thus, biological systems form important existence theorems and design criteria for the production of functional, nanoscale materials through nanotechnology employing the synthetic approach to nanostructures.

An important nature-mimicking methodology involves the use of covalent synthesis followed by molecular self-assembly of the synthesized molecules. These molecules are generally small mono- or oligomers that interact with each other and with other kinds of mono- or oligomers to form thermodynamically stable, nanoscale structures. See also **Molecular Recognition**.

Existence Theorem: Nanobiology

Analogies: Structure and Function. Biology is replete with complex, functional nanoscale structures formed by directed synthesis and subsequent self-assembly of the component molecules. Initially, the small molecules are produced by covalent synthesis, with the larger, functional structures resulting from many weak, noncovalent interactions that energetically overcome interactions with the solvent and the entropic advantages of disintegration of the ordered aggregates. The final structure represents a thermodynamic minimum, and incorrect subunits are rejected in the dynamic, equilibrium assembly. In this process, complementarity in shape and polarity provides the foundation for the association (binding) between components (e.g., phospholipids, polypeptide chains, proteins, nucleic acids). Shape-dependent association based on the nonspecific van der Waal's and hydrophobic interactions are made stronger and more specific by hydrogen bonds and electrostatic interactions, as well as, in some cases, covalent bonds (e.g., disulfides). Often, positive cooperativity is displayed, i.e., conformations of individual subunits change upon binding in such a manner that their affinity for other components of the final structure increases. Moreover, the amount of information required to execute the assembly of a particular structure (e.g., a protein) is minimized by use of only a few types of molecules and a limited number of binding interactions.

For example, a polypeptide is synthesized as a linear polymer derived from the 20 natural amino acids by translation of a nucleotide sequence present in a messenger RNA (mRNA). The mature protein exists as a well-defined three-dimensional structure. The information necessary to specify the final (tertiary) structure of the protein is present in the molecule itself, in the form of the specific sequence of amino acids that form the protein. This information is used in the form of myriad noncovalent interactions (such as those in Table 1) that first form relatively simple local structural motifs (helix and sheet structures associated through networks of hydrogen bonds); these motifs then tend to aggregate in ways that associate hydrophobic regions with one another and out of reach of water, and to place hydrophilic regions close to each other and to water. Thus, proteins self-assemble by two types of processes after synthesis: formation of relatively simple local structures from an unfolded polypeptide chain; and more complex structure-specific association (i.e., associations involving some form of molecular recognition) of these local structures.

In addition to self-assembly of protein structures, in living systems the complex maneuvers needed to achieve properly folded tertiary structures are facilitated by the function of a pre-existing protein machinery, of which, the molecular chaperones are an illustrative example. Chaperones are proteins that bind to and stabilize an otherwise unstable conformer of another protein, and by controlled binding and release, facilitate its correct fate *in vivo*. Molecular chaperones may be said to be the natural

TABLE 1. TYPES OF BONDS AND INTERACTIONS THAT ARE POTENTIALLY USEFUL IN THE ENGINEERING OF FUNCTIONAL NANOSCALE MATERIALS

Bond type	Examples	
	Natural	Synthetic
covalent bonds	peptide (amide), disulfide (RSSR), and phosphodiester (DNA) bonds	palytoxin, oligomers of peptides, nucleotides, and thiophenes
electrostatic interactions	salt bridges in proteins, zinc fingers	metal-coordinated crowns
hydrogen bonds	nucleotide base pairs, amide hydrogen bonds in proteins	peptide cylinders, DNA objects, replicating systems
hydrophobic interactions	hydrophobic cores of proteins, lipid–lipid and lipid–protein interactions	pockets within cyclodextrins and cyclophanes
van der Waals interactions	membranes, vesicles	Langmuir and self-assembled monolayers, lipid bilayers
aromatic π-stacking and charge transfer	nucleic acids	porphyrins

counterparts of assemblers and transporters envisaged as products of nanotechnology.

Analogies: Molecular Devices. Among the many examples in biology of complex nanoscale structures and devices, one of the most ubiquitous and versatile classes is that of the membrane proteins. Membrane proteins, i.e., proteins that are associated with cellular and organellular membranes, serve myriad functions including recognition, adhesion, chemical triggering, ion and molecular transport, light harvesting, chemical and mechanical sensing, and actuation. Humans have attempted to mimic the function of some of these structures for a wide variety of uses, including chemical sensing, selective chemical transport (pumping), and tissue engineering.

An interesting class of proteins which may prove to be an illustrative archetype for future nanotechnological devices, if not integral components of such, are a class of molecular devices collectively known as motor proteins. Motor proteins are enzymatic protein complexes whose catalytic function results in a distinct mechanical function; a variety of examples are known, including those that perform functions analogous to levers, rotary motors, pumps and springs. The structures and mechanisms of action of a variety of motor proteins are being investigated.

Reductive Approaches

Conventional methods of microfabrication of integrated circuits and devices constitute the reductive (top-down) approach to the construction of micrometer and submicrometer-scale structures. The smallest features in commercial integrated circuits have measured 0.35 μm across, and technologies for further reduction in size have succeeded in forming feature sizes as small as 0.18 μm. Further miniaturization, however, will require major technological breakthroughs in the processes underlying microfabrication, especially photolithography, the heart of microfabrication. The technological barriers faced by the lithographers may not be completely insurmountable.

Limitations. The number of transistors present on a chip has doubled approximately every 18 months since the integrated circuit was first developed. The main reason for this continuing decrease in the minimum feature sizes of transistors (and consequent increase in density of transistors on the chip) has been the development of photolithography. The most important limitation for further size reduction remains the development of new photo- and other lithographic techniques.

Current Advances. New light sources are currently being developed, such as the krypton—fluoride ultraviolet laser (wavelength 0.248 μm for features as small as 0.25 μm) and excimer lasers (wavelength 0.193 μm for features below 0.2 μm). But these technologies still need to overcome several obstacles before they can be implemented by the semiconductor industry. For example, for wavelengths below 0.2 μm, the current photoresists absorb so much light that throughput suffers. Also, the fused silica glass lenses used to demagnify the image absorb light and heat up, resulting in a degradation of the image. Problems associated with depth of focus become more acute, requiring further innovation in planarization technology. New photoresist materials, based on the deposition and patterning of monomolecular layers (e.g., self-assembled monolayers and Langmuir-Blodgett films), are also being developed to address the issues of further reduction in the limits of photolithography.

Further reduction in feature size to achieve nanoscale structures by photolithography will necessitate the use of ever smaller wavelengths of light.

Focused particle-based lithography, such as ion-, neutral atom-, and electron-beam lithography, are capable of achieving very high resolutions. However, the method of writing each circuit feature separately is serial and inherently slow so that the technology cannot be used for simultaneous fabrication of many chips. To speed up the process of electron-beam lithography, methods are being explored to scan a broad electron beam across the entire chip by projecting the beam through an appropriate mask (in mimicry of photolithography).

The most recent approach to reductive nanofabrication that can indeed construct nanoscale structures and devices uses microscopic tools (local probes) that can build the structures atom by atom, or molecule by molecule. Optical methods using laser cooling (optical molasses) are also being developed to manipulate nanoscale structures.

Atomic and Molecular Manipulation

Scanning Probe Microscopy. The scanning tunneling microscope (STM) can image and manipulate matter on the atomic scale. In general,

when a small conducting probe (the tip, consisting of one or a few atoms, or a metal) is placed close (less than 10 nm) to the surface of a conducting substrate, an electronic current results under a suitable bias due to the overlapping of the electronic wave functions of the probe and the surface. Because this tunneling current is exponentially dependent on the separation of the probe from the surface, imaging resolutions of fractions of an angstrom can be obtained. These images reflect both the topography and the electronic structure of the surface. The STM can also be used to modify surfaces locally. As a result, individual atoms and molecules can be manipulated with atomic-scale precision.

In parallel processes, e.g., field-assisted diffusion and sliding, the bond between the surface and the adatom is never completely broken. Field-assisted diffusion of an adatom on the surface occurs due to the presence of the intense, inhomogeneous electric field between the probe tip and the surface, which gives rise to a potential gradient.

The second class of atomic manipulations, the perpendicular processes, involves transfer of an adsorbate atom or molecule from the STM tip to the surface or vice versa. The tip is moved toward the surface until the adsorption potential wells on the tip and the surface coalesce, with the result that the adsorbate, which was previously bound either to the tip or the surface, may now be considered to be bound to both. For successful transfer, one of the adsorbate bonds (either with the tip or with the surface, depending on the desired direction of transfer) must be broken. The fate of the adsorbate depends on the nature of its interaction with the tip and the surface, and the materials of the tip and surface. Directional adatom transfer is possible with the application of suitable junction biases. See also **Scanning Tunneling Microscope**.

Optical Manipulation. Laser beams provide another means of capturing individual atoms or molecules. When an atom is irradiated from both sides by laser light at a frequency slightly lower than the frequency at which the atom absorbs photons, then the atom loses some of its momentum. In particular, the laser beam propagating in a direction opposite that of the motion of the atom increases in frequency due to the Doppler effect, resulting in light absorption with subsequent isotropic emission (scattering). The light propagating in the same direction as the atom is not absorbed, so that the atom is pushed in a direction opposite its motion and slows down. By surrounding the atom with three sets of counterpropagating laser beams orthogonal to each other the atom can be cooled (i.e., slowed) in all three dimensions. Because the light field acts as a viscous drag force, the combination of laser beams is known as optical molasses.

Synthetic Approaches

The synthetic (bottom-up) approach offers a level of control over the selection and placement of atoms and molecules that is ultimately much higher than that offered by other methods of large-scale microfabrication (e.g., fabrication of integrated circuits). Such synthesis can employ a variety of chemical methods utilizing some or all of the forces and interactions listed in Table 1 to produce nanoscale molecules. Most chemical synthetic reactions that produce large molecules (e.g., collections of atoms that can act as nanodevices) generate polydisperse materials. These materials are mixtures of oligo- or polymeric chains of varying molecular weights. For nanotechnological applications, it is important to have synthetic strategies that yield compounds of uniform length, size, and shape; inhomogeneities can be detrimental to the designated function of the molecule. Thus, the specific objectives of synthesis are to discover and develop rapid and efficient methods for the precise control of composition, molecular weight, stereochemistry, aggregation, and placement of functional molecules. Four strategies are currently in use (either separately or in combination with one another) for the fabrication of large molecules (substances with molecular weights of a few hundred to a few million): biotechnological synthesis, sequential covalent synthesis, covalent polymerization, and molecular self-assembly.

In nature, complex functional molecules are produced by using the chemical synthetic approach. The molecules are first formed by covalent synthesis. In particular, the polypeptides are formed by the directional joining of amino acids through amide bonds. The primary structure of the amino acids (i.e., sequence) during synthesis is specified by the mRNA (a transcript of the original gene encoded in the DNA). Once a polypeptide is produced, it then undergoes many conformational changes that reduce its size to a compact, native form that is the functional protein. These conformational changes are termed folding.

RNA is also capable of folding into specific shapes for ligand recognition and catalysis. Although there are only four naturally occurring bases available for the formation of tertiary structures through noncovalent interactions (while proteins have twenty different amino acids to choose from), myriad RNA shapes can be produced. Metal ions have been shown to confer extraordinary stability to RNA and RNA fragments.

Thus, protein and RNA folding studies form a fundamental paradigm for the design and synthesis of new, functional macromolecules with both final structure and function built into the primary structure of the macromolecule. Such synthetic strategies could be applied to the rational design of functional nanostructures, including drugs, sensing elements, photonic and electronic components, catalysts, and even mechanical devices.

Biotechnological Synthesis. Biotechnological synthesis of new nanomaterials (e.g., proteins) and biotechnological modification of living systems (e.g., conferring specific therapeutic properties, new genetic traits) exploit the many ways in which this protein manufacturing machinery can be modified. In particular, the DNA of an organism can be altered in a specific manner, resulting in a modified mRNA, and consequently, a modified or new protein.

Recombinant DNA technology provides a powerful tool for analysis, synthesis, and alteration of genes and proteins. It is based on the ability to rapidly synthesize polynucleotides with any sequence using nucleic acid enzymology (e.g., using DNA polymerases, restriction enzymes or endonucleases, DNA ligases). The unique base-pairing attributes of the constituents of the DNA and the ability to express the modified or synthetic DNA in microorganisms and eukaryotic cells result in a powerful tool for the production of synthetic molecules with specific biological functions.

Covalent Synthesis. The first strategy for chemical synthesis employs elaborate and sophisticated methods for assembling atoms into molecules based on the general strategy of sequential formation of covalent bonds. The atoms can also be assembled into subunits that are then reacted to form more complex, designed molecules (convergent synthesis).

Covalent synthesis of complex molecules involves the reactive assembly of many atoms into subunits with aid of reagents and established as well as innovative reaction pathways. These subunits are then subjected to various reactions that will assemble the target molecule. Very complex molecules can be synthesized in this manner.

Molecular Self-Assembly. Reductive techniques, such as those used in the microelectronics industry, can produce structural features smaller than about 200 nm. The use of proximal probes and other nanomanipulative techniques can be considered to be a hybrid of the reductive lithographic techniques and the synthetic strategies of assembling functional nanostructures atom by atom, or molecule by molecule. The organization of nanostructures and devices by the self-assembly of the component atoms and molecules, a ubiquitous phenomenon in biological systems, forms the noncovalent synthetic approach to nanotechnology.

In this approach, well-defined subunits (small molecules similar to, e.g., nucleotides) are first formed through covalent synthesis. Second, these subunits aggregate with themselves or with other subunits through covalent or noncovalent (or both) interactions to form large, stable, structurally defined assemblies. For the final supramolecular structure to be stable and to have a well-defined shape, the noncovalent connections must be collectively stable. Therefore, molecules must be stabilized by many noncovalent interactions.

Conclusions and Outlook

Much of the current progress in nanotechnology is confined to understanding and harnessing pre-existing, exquisitely evolved nanomachinery, i.e., natural living systems. Table 2 summarizes the various reductive and synthetic strategies that are employed.

Support for this work was provided by ONR Multidisciplinary University Research Initiative Grant N00014-95-1-1315, by ONR Grant N00014-95-1-0901, and by NSF Grant HRD-9450475.

RAJESH VAIDYA
GABRIEL LÖPEZ
University of New Mexico
JOSÉ A. LOPEZ
Baylor College of Medicine

Additional Reading

Ball, P.: *Designing the Molecular World: Chemistry at the Frontier,* Princeton University Press, Princeton, NJ, 1994.

TABLE 2. SOME APPROACHES IN REDUCTIVE (TOP DOWN) AND SYNTHETIC (BOTTOM UP) NANOFABRICATION

Approach	Examples
reductive (also known as top-down)	conventional microfabrication photolithography x-ray lithography e-beam lithography imprint lithography local probe lithography
synthetic (also known as bottom-up)	nanochemical synthesis biotechnological expression molecular templating molecular self-assembly
hybrid (reductive/synthetic)	local probe-assisted synthesis microcontact printing optical-laser manipulation

Jones, F. J. and C. P. Poole, Jr.: *Introduction to Nanotechnology,* John Wiley & Sons, Inc., New York, NY, 2003.

Mulhall, D.: *Our Molecular Future: How Nanotechnology, Robotics, Genetics, and Artificial Intelligence Will Transform Our World,* Amherst, NY, 2002.

Ratner, M. and D. Ratner: *Nanotechnology: A Gentle Introduction to the Next Big Idea,* Prentice Hall Professional Technical Reference, Upper Saddle River, NJ, 2002.

Regis, E.: *Nano: The Emerging Science of Nanotechnology,* DIANE Publishing Company, Colllingdale, PA, 1998.

Rieth, M.: *Nano-Engineering in Science and Technology: An Introduction to the World of Nano-Design,* World Scientific Publishing Company, Inc., Riveredge, NJ, 2003.

Stroscio, J. A. and D. M. Eigler: *Science,* **254,** 1319–1326 (1991).

Tolles, W. M.: in G. M. Chow and K. E. Gonsalves, eds., *210th National Meeting of the ACS,* ACS, Chicago, Ill., 1996, pp. 1–15.

Whitesides, G. M. and co-workers: *Acc. Chem. Res.* **28,** 37–44 (1995).

Wilson, M., K. Kannangara, and M. Wilson: *Nanotechnology: Basic Science and Emerging Technologies,* CRC Press LLC., Boca Raton, FL, 2002.

NAPHTHALENE. [CAS:91-20-3]. Naphthalene $C_{10}H_8$, is a white solid with a strong smell; is also called mothballs, moth flakes, white tar, and tar camphor. Naphthalene is a natural component of fossil fuels such as petroleum and coal; it is also formed when natural products such as wood or tobacco are burned.

The accepted configuration of naphthalene, i.e., two fused benzene rings sharing two common carbon atoms in the ortho position, was established in 1869 and was based on its oxidation product, phthalic acid. Based on its fused-ring configuration, naphthalene is the first member in a class of aromatic compounds with condensed nuclei. Naphthalene is a resonance hybrid:

In chemical reactions, naphthalene usually acts as though the bonds were fixed in the positions, as shown in the first structure above at the left.

Some selected chemical and physical properties of naphthalene are given in Table 1. Naphthalene is very slightly soluble in water but is appreciably soluble in many organic solvents, e.g., 1,2,3,4-tetrahydronaphthalene, phenols, ethers, carbon disulfide, chloroform, benzene, coal-tar naphtha, carbon tetrachloride, acetone, and decahydronaphthalene.

TABLE 1. PROPERTIES OF NAPHTHALENE

Property	Value
molecular wt	120.1732
mp, °C	80.290
normal bp at 101.3 kPa[a], °C	217.993
flash point (closed cup), °C	79
ignition temperature, °C	526
heat of vaporization, kJ/mol[b]	43.5
heat of fusion at triple point, kJ/mol[b]	18.979
heat of combustion, at 15.5°C and 101.3 kPa[a], kJ/mol[b]	−5158.41
density at 25°C, g/mL	1.175

[a] To convert kPa to atm, divide by 101.3.
[b] To convert J to cal, divide by 4.184.

The ir, uv, mass, nmr and ^{13}C-nmr spectral data for naphthalene and other related hydrocarbons have been reported in the literature. Additionally, information regarding the properties of naphthalene has been published.

Reactions

Substitution. Substitution products are formed by the substitution of one or more hydrogen atoms with other functional groups. Substituted naphthalenes of commercial importance have been obtained by sulfonation and alkali fusion, alkylation, nitration and reduction, and chlorination.

Sulfonation. Sulfonation of naphthalene with sulfuric acid produces mono-, di-, tri-, and tetranaphthalenesulfonic acids. Naphthalenesulfonic acids are important starting materials in the manufacture of organic dyes. They are also intermediates used in reactions.

Nitration. Naphthalene is easily nitrated with mixed acids, e.g., nitric and sulfuric, at moderate temperatures to give mostly 1-nitronaphthalene and small quantities, 3–5%, of 2-nitronaphthalene.

Halogenation. Under mild catalytic conditions, halogen substitution occurs, and all of the hydrogen atoms of the naphthalene molecule can be replaced. The only commercially significant halogenated naphthalene products are the mixed chlorinated naphthalenes. Uses for the chlorinated naphthalenes include solvents, gauge and instrument fluids, capacitor impregnants, components in electric insulating compounds and electroplating stop-off compounds.

Alkylation. Naphthalene can be easily alkylated. Isopropylnaphthalenes produced by alkylation of naphthalene with propylene have gained commercial importance as chemical intermediates, e.g., 2-isopropylnaphthalene, and as multipurpose solvents, e.g., mixed isopropylnaphthalenes.

Chloromethylation. The reactive intermediate, 1-chloromethylnaphthalene, has been produced by the reaction of naphthalene in glacial acetic acid and phosphoric acid with formaldehyde and hydrochloric acid.

Addition. The most important addition products of naphthalene are the hydrogenated compounds used in solvents, paints and other products. Of less commercial significance are those made by the addition of chlorine.

Oxidation. The vapor-phase reaction of naphthalene over a catalyst based on vanadium pentoxide is the commercial route used throughout the world to form phthalic anhydride. In the United States, the one phthalic anhydride plant currently operating on naphthalene feedstock utilizes a fixed catalyst bed, the preferred route worldwide.

Manufacture

Two sources of naphthalene exist in the U.S.; coal tar and petroleum. Coal tar was the traditional source until the late 1950s, when it was in short supply. In 1960, the first petroleum-naphthalene plant accounted for over 40% of total naphthalene production. The availability of large quantities of *o*-xylene at competitive prices during the 1970s affected the position of naphthalene as the prime raw material for phthalic anhydride. Production for 1992 was less than 50% of the levels in the early 1980s. The last dehydroalkylation plant for petroleum naphthalene was shut down late in 1991. Coal tar has stabilized at around 85×10^3 t/yr, and petroleum-naphthalene production is around $6-8 \times 10^3$ t/yr. The reduction of petroleum production has opened the door for imported naphthalene, mainly from Canada.

Coal-Tar Process. The largest quantities of naphthalene are obtained from the coal tar that is separated from the coke-oven gases. The coal tar first is processed through a tar-distillation step where ca the first 20 wt% of distillate, i.e., chemical oil, is removed. The chemical oil contains practically all the naphthalene present in the tar. It is processed to remove the tar acids by contacting with dilute sodium hydroxide and, in a few cases, is next treated to remove tar bases by washing with sulfuric acid. Principal U.S. producers obtain their crude naphthalene product by fractional distillation of the tar acid-free chemical oil.

Economic Aspects

Total nameplace capacity for all U.S. naphthalene producers in 1993 was 124×10^3 t, with 114×10^3 t produced from coal tar and 10×10^3 from petroleum.

The economics of naphthalene recovery from coal tar can vary significantly, depending on the particular processing operation used. A significant factor is the cost of the coal tar. As the price of fuel oil increases, the value of tar also increases.

The high price of the petroleum product results from its higher quality. The price of the crude coal-tar naphthalene is primarily associated with that of *o*-xylene, its chief competitor as phthalic anhydride feedstock.

The preferred route to higher purity naphthalene, either coal-tar or petroleum, is crystallization. This process has demonstrated significant energy cost savings and yield improvements. There are several commercial processes available: Sulzer-MWB, Brodie type, Betz, and Recochem.

Health and Safety Factors

Handling. Naphthalene is generally transported in molten form in tank trucks or tank cars that are equipped with steam coils. Storage tanks containing molten naphthalene have a combustible mixture in the vapor space and care must be taken to eliminate all sources of ignition. Naphthalene dust also can form explosive mixtures with air, which necessitates care in the design and operation of solid handling mixtures. Perhaps the greatest hazard to the worker is the potential for operating or maintenance personnel to be accidentally splashed with hot molten naphthalene while taking samples or disassembling process lines (ASTM D3438). Molten naphthalene tank vents must be adequately heated and insulated to prevent the accumulation of sublimed and solidified naphthalene.

Toxicology. The acute oral and dermal toxicity of naphthalene is low, with LD$_{50}$ values for rats from 1780–2500 mg/kg orally and greater than 2000 mg/kg dermally. The inhalation of naphthalene vapors may cause headache, nausea, confusion, and profuse perspiration, and if exposure is severe, vomiting, optic neuritis, and hematuria may occur. Chronic exposure studies conducted by the NTP in mice for two years showed that naphthalene caused irritation to the nasal passages, but no overt toxicity was noted. Rare cases of such corneal epithelium damage in humans have been reported. Naphthalene can be irritating to the skin, and hypersensitivity does occur. In other chemical carcinogen tests, little cancer risk was indicated. No incidents of chronic effects have been reported as a result of industrial exposure to naphthalene. Threshold limit value of 10 ppm (50 mg/m^3) has been set by the ACGIH.

Alkylnaphthalenes

Methyl- and dimethylnaphthalenes are contained in coke-oven tar and in certain petroleum fractions in significant amounts. In the U.S., separation of individual isomers is seldom attempted; instead, a methylnaphthalene-rich fraction is produced for commercial purposes. Such mixtures are used for solvents for pesticides, sulfur, and various fluids. They also can be used as low freezing, stable heat-transfer fluids. Mixtures that are rich in monomethylnaphthalene content have been used as dye carriers (qv) for color intensification in the dyeing of synthetic fibers, e.g., polyester. They also are used as the feedstock to make naphthalene in dealkylation processes. Phthalic anhydride also can be made from methylnaphthalene mixtures by an oxidation process that is similar to that used for naphthalene. A mixed monomethylnaphthalene-rich material can be produced by distillation and can be used as feedstock for further processing. Applications include use in solvents, drugs, polyesters, surfactants, and detergent products.

Acenaphthene. [CAS: 83-32-9]. Acenaphthene is a hydrocarbon, C$_{12}$H$_{10}$, present in high temperature coal tar. Acenaphthene may be halogenated, sulfonated, and nitrated in a manner similar to naphthalene. Oxidation first yields acenaphthenequinone, followed by 1,8-naphthalenedicarboxylic acid anhydride and pesticides.

ROBERT T. MASON
Koppers Industries, Inc.

Additional Reading

Donaldson, N.: *The Chemistry and Technology of Naphthalene Compounds,* Edward Arnold Publishers, London, 1958, pp. 455–473.

Naphthalene, American Petroleum Institute Monograph Series, Publication 707, API, Washington, DC, Oct. 1978.

Sachanen, A.N.: *Conversion of Petroleum,* 2nd Edition, Reinhold Publishing Corp., New York, NY, 1948, pp. 550–565.

Sandmeyer, E.E.: in G.D. Clayton and F.E. Clayton, eds., *Patty's Industrial Hygiene and Toxicology,* 3rd Revised Edition, Vol. II, Wiley-Interscience, New York, NY, 1981, Chapt. 46.

Web References

Naphthalene: http://www.jtbaker.com/msds/englishhtml/n0090.htm
U.S. Environmental Protection Agency: http://www.epa.gov/iris/subst/0436.htm

NAPHTHALENE DERIVATIVES

Naphthalenesulfonic Acids

Naphthalenesulfonic acids are important chemical precursors for dye intermediates, wetting agents and dispersants, naphthols, agricultural formulations, leather tanning agents, photographic materials, and air-entrainment agents for concrete. The production of many intermediates used for making azo, azoic, and triphenylmethane dyes involves naphthalene sulfonation and one or more unit operations, e.g., caustic fusion, nitration, reduction, or amination.

Generally, the sulfonation of naphthalene leads to a mixture of products. Naphthalene sulfonation at less than ca 100°C is kinetically controlled and produces predominantly 1-naphthalenesulfonic acid (**1**). Sulfonation of naphthalene at above ca 150°C provides thermodynamic control of the reaction and 2-naphthalenesulfonic acid as the main product. Reaction conditions for the sulfonation of naphthalene to yield desired products are given in Figure 1; alternative paths are possible. A list of naphthalenesulfonic acids and some of their properties are given in Table 1.

Nitronaphthalenes and Nitronaphthalenesulfonic Acids

The nitro group does not undergo migration of the naphthalene ring during the usual nitration procedures. Therefore, mono- and polynitration of naphthalene is similar to low temperature sulfonation. The nitronaphthalenes and some of their physical properties are listed in Table 2. Many of these compounds are not accessible by direct nitration of naphthalene but are made by indirect methods, e.g., nitrite displacement of diazonium halide groups in the presence of a copper catalysts, decarboxylation of nitronaphthalenecarboxylic acids, or deamination of nitronaphthalene amines. They are used in the manufacture of chemicals, dye intermediates, and colorants for plastics.

A by-product of some of these naphthalene derivatives, 2-naphthylamine, is carcinogenic. Respirators, protective clothing, proper engineering, controls and medical monitoring programs for workers involved with them should be used. The National Institute of Occupational Safety and Health (NIOSH) has published recommendations for working with this product.

Naphthaleneamines and Naphthalenediamines

Selected physical properties of naphthaleneamines and naphthalenediamines are listed in Table 3. They are used in rodenticides, rubber antioxidants, dye intermediates, insecticides, herbicides, pharmaceuticals, chemical manufacture, and colorants.

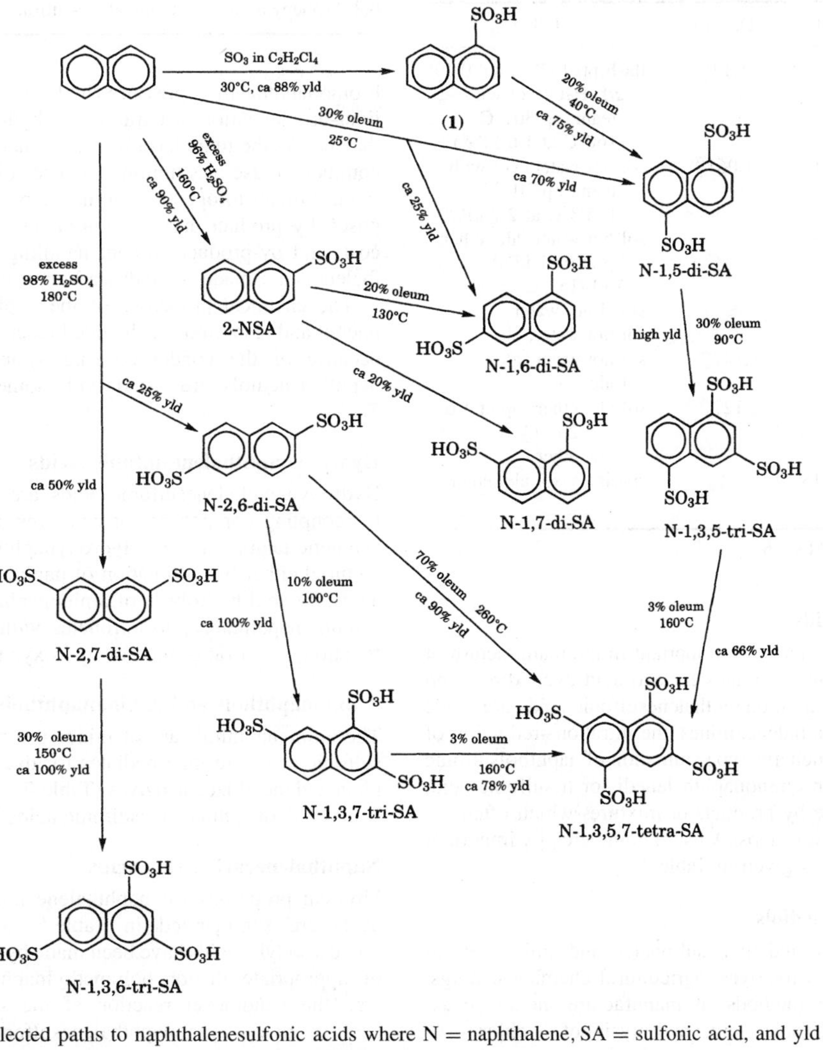

TABLE 1. MELTING POINTS OF NAPHTHALENESULFONIC ACIDS

Compound	Mp, °C	Mp of corresponding sulfonyl chloride, °C
1-naphthalenesulfonic acid	139–140	68
1-naphthalenesulfonic acid dihydrate	90	
2-naphthalenesulfonic acid	139–140	76
2-naphthalenesulfonic acid hydrate	124–125	
2-naphthalenesulfonic acid trihydrate	83	
1,2-naphthalenedisulfonic acid		160
1,3-naphthalenedisulfonic acid		137.5
1,4-naphthalenedisulfonic acid	240–245 dec	162
1,5-naphthalenedisulfonic acid	125 dec	183
1,6-naphthalenedisulfonic acid		129
1,7-naphthalenedisulfonic acid		123
2,6-naphthalenedisulfonic acid	199 dec	228–229
2,7-naphthalenedisulfonic acid		159.5
1,3,5-naphthalenetrisulfonic acid		146
1,3,6-naphthalenetrisulfonic acid		194–197
1,3,7-naphthalenetrisulfonic acid		165–166
1,4,5-naphthalenetrisulfonic acid		156–157
1,3,5,7-naphthalenetetrasulfonic acid		261–262

Fig. 1.   Selected paths to naphthalenesulfonic acids where N = naphthalene, SA = sulfonic acid, and yld = yield

TABLE 2. MELTING POINT OF NITRONAPHTHALENES

Compound	Mp, °C
1-nitronaphthalene	52[a]; 57.8[b]
2-nitronaphthalene[c]	78.7[d]
1,2-dinitronaphthalene[c]	161–162
1,3-dinitronaphthalene[c]	148
1,4-dinitronaphthalene[c]	134
1,5-dinitronaphthalene	219
1,6-dinitronaphthalene[c]	166.5[e]
1,7-dinitronaphthalene[c]	156
1,8-dinitronaphthalene	172
2,3-dinitronaphthalene[c]	174.5–175
2,6-dinitronaphthalene[c]	279
2,7-dinitronaphthalene[c]	234
1,2,3-trinitronaphthalene[c]	190
1,2,4-trinitronaphthalene[c]	258
1,3,5-trinitronaphthalene	122
1,3,6-trinitronaphthalene[c]	186
1,3,8-trinitronaphthalene	218
1,4,5-trinitronaphthalene	149
1,3,5,7-tetranitronaphthalene[c]	260
1,3,5,8-tetranitronaphthalene	194–195
1,3,6,8-tetranitronaphthalene	203
1,4,5,8-tetranitronaphthalene	340–345 dec

[a] Metastable form.
[b] Bp 304°C (169°C at 1.6 kPa (12 mm Hg)).
[c] Made by indirect methods, not by the direct nitration of naphthalene or naphthalene-nitration products.
[d] Bp 312.5°C at 97.8 kPa (733 mm Hg) and 165°C at 2.0 kPa (15 mm Hg).
[e] Bp 370°C (235°C at 1.3 kPa (9.75 mm Hg)).

TABLE 3. PHYSICAL PROPERTIES OF NAPHTHALENEAMINES AND NAPHTHALENEDIAMINES

Compound	Mp, °C	Density	Other
1-naphthaleneamine	50	1.13_4^{14}	flash pt, 157°C; sol 0.496 g/L H$_2$O; vol with steam; bp 301°C (160°C at 1.6 kPa[a])
2-naphthaleneamine	111–113	1.061_4^{96}	sol hot water; vol with steam; bp 306°C (175.8°C at 2.7 kPa[a])
1,2-naphthalenediamine	96–98		sol hot water, alc, ether; bp at 0.01 kPa[a] 150–151°C
1,4-naphthalenediamine	120		sl sol hot water
1,5-naphthalenediamine	189.5		sol hot water, alc
1,6-naphthalenediamine	78	$1.147_4^{99.4}$	sol hot water, alc
1,7-naphthalenediamine	117.5		sol alc
1,8-naphthalenediamine	66.5	$1.127_4^{99.4}$	sol alc, ether; bp at 1.6 kPa[a] 205°C
2,3-naphthalenediamine	191		sol alc, ether
2,6-naphthalenediamine	216–218		sparingly sol alc, ether
2,7-naphthalenediamine	159		

[a] To convert kPa to mm Hg, multiply by 7.5.

Aminonaphthalenesulfonic Acids

Many aminonaphthalenesulfonic acids are important in the manufacture of azo dyes or are used to make intermediates for azo acid dyes, direct and fiber-reactive dyes. Usually, the aminonaphthalenesulfonic acids are made by either the sulfonation of naphthaleneamines, the nitration–reduction of naphthalenesulfonic acids, the Bucherer-type amination of naphtholsulfonic acids, or the desulfonation of an aminonaphthalenedi- or trisulfonic acid. Most of these processes produce by-products or mixtures which often are separated in subsequent purification steps. A list of commercially important aminonaphthalenesulfonic acids is given in Table 4.

Naphthalenols and Naphthalenediols

Naphthalenols, naphthalenediols, and their sulfonated and amino derivatives are important intermediates for dyes, agricultural chemicals, drugs, perfumes, and surfactants. The methods of manufacture include caustic fusion of naphthalene-1-sulfonic acid, hydrolysis of 1-chloro- or

TABLE 4. MANUFACTURE, PRODUCTION, AND APPLICATION DATA FOR SELECTED AMINONAPHTHALENESULFONIC ACIDS

Acid	Trivial name
1-amino-2-naphthalenesulfonic	
4-amino-2-naphthalenesulfonic	
4-amino-1-naphthalenesulfonic	Piria's acid; naphthionic acid
5-amino-1-naphthalenesulfonic	Laurent's acid
5-amino-2-naphthalenesulfonic	1,6-Cleve's acid
8-amino-2-naphthalenesulfonic	1,7-Cleve's acid
5- and 8-amino-2-naphthalenesulfonic	Cleve's acid (mixed)
8-amino-1-naphthalenesulfonic	Peri acid
8-phenylamino-1-naphthalenesulfonic	Phenyl Peri acid
2-amino-1-naphthalenesulfonic	Tobias acid
6-amino-1-naphthalenesulfonic	Dahl's acid
6-amino-2-naphthalenesulfonic	Broenner's acid
7-amino-2-naphthalenesulfonic	F-acid
7-amino-1-naphthalenesulfonic	Badische acid
1-amino-2,7-naphthalenedisulfonic	Kalle's acid
4-amino-2,7-naphthalenesulfonic	1,3,6-Freund's acid
4-amino-2,6-naphthalenedisulfonic	1,3,7-Freund's acid
8-amino-1,6-naphthalenedisulfonic	amino-ε-acid
4-amino-1,7-naphthalenedisulfonic	Dahl's acid II
4-amino-1,6-naphthalenedisulfonic	Dahl's acid III
8-amino-1,5-naphthalenedisulfonic	
5-amino-1,3-naphthalenedisulfonic	
3-amino-2,7-naphthalenedisulfonic	amino-R-acid
3-amino-1,5-naphthalenedisulfonic	Cassella acid
6-amino-1,3-naphthalenedisulfonic	amino J-acid
7-amino-1,3-naphthalenedisulfonic	amino G-acid
4-amino-1,3,5-naphthalenetrisulfonic (as the sultam)	
8-amino-1,3,6-naphthalenetrisulfonic	Koch's acid
8-amino-1,3,5-naphthalenetrisulfonic	B-acid
6-amino-1,3,5-naphthalenetrisulfonic	
7-amino-1,3,6-naphthalenetrisulfonic	2R amino acid
6,8-di(phenylamino)-1-naphthalene-sulfonic	diphenyl-ε-acid

bromonaphthalene, pressure hydrolysis of 1-naphthaleneamine, oxidation–aromatization of tetralin, and hydroperoxidation of 2-isopropylnaphthalene. As the toxic hazard of the 1-naphthaleneamine was recognized, its commercial use was minimized. The sulfonation–caustic fusion process is more difficult to operate than in the past because of increasing difficulties posed by product purity requirements, high investment and replacement cost, and by-product effluent handling problems. In the U.S., the naphthalenols are made by hydrocarbon oxidation routes.

The chemical properties of the naphthalenols are similar to those of phenol and resorcinol, with added reactivity and complexity of substitution because of the condensed ring system. Some of the naphthols and naphthalenediols are listed with some of their physical properties in Table 5.

Hydroxynaphthalenesulfonic Acids

Hydroxynaphthalenesulfonic acids are important as intermediates either for coupling components of azo dyes or azo components, as well as for synthetic tanning agents. Hydroxynaphthalenesulfonic acids can be manufactured either by sulfonation of naphthols or hydroxynaphthalenesulfonic acids, by acid hydrolysis of aminonaphthalenesulfonic acids, by fusion of sodium naphthalenepolysulfonates with sodium hydroxide, or by desulfonation or rearrangement of hydroxynaphthalenesulfonic acids (Table 6).

Aminonaphthols and Aminonaphtholsulfonic Acids

The aminonaphthols are of minor use but the aminohydroxynaphthalenesulfonic acids are intermediates for dyes, e.g., fiber-reactive azo dyes and plain and metallized azo dyes (Table 7). A number of N-acyl-, N-alkyl-, and N-arylaminonaphthaleneosulfonic acids are used as couplers for azo dyes.

Naphthalenecarboxylic Acids

Physical properties for naphthalene mono-, di-, tri-, and tetracarboxylic acids are summarized in Table 8. Most of the naphthalene di- or polycarboxylic acids have been made by simple routes such as the oxidation of appropriate di- or polymethylnaphthalenes, or by complex routes, e.g., the Sandmeyer reaction of the selected aminonaphthalenesulfonic acid, to give a cyanonaphthalenesulfonic acid followed by fusion of the

TABLE 5. PROPERTIES OF NAPHTHALENOLS AND NAPHTHALENEDIOLS

Compound	Mp, °C	Density	Other
1-naphthalenol	95.8–96.0	1.224_4^4	sublimes; sol 0.03 g/100 mL H_2O at 25°C; readily sol alc, ether, benzene; bp 280°C (158°C at 2.6 kPa^a)
		1.099_4^{99}	
2-naphthalenol	122	1.078_4^{130}	sublimes; sol 0.075 g/100 mL H_2O at 25°C; readily sol alc, ether, benzene; flash pt 161°C; bp 295°C (161.8°C at 2.6 kPa^a)
		1.22_4^{25}	
1,2-naphthalenediol	103–104		
1,3-naphthalenediol	124		
1,4-naphthalenediol	195		heat of combustion 4.77 MJ^b
1,5-naphthalenediol	258		sublimes; sparingly sol water; readily sol ether, acetone
1,6-naphthalenediol	137–138		
1,7-naphthalenediol	181		
1,8-naphthalenediol	144		
2,3-naphthalenediol	159		
2,6-naphthalenediol	222		
2,7-naphthalenediol	194	sol boiling water	

[a] To convert kPa to mm Hg, multiply by 7.5.
[b] To convert MJ to kcal, divide by 4.184×10^{-3}.

latter with an alkali cyanide, with simultaneous or subsequent hydrolysis of the nitrile groups. These acids are used in the manufacture of dyes, photographic materials, pharmaceuticals, rodenticides, chemicals, polymers, and synthetic materials.

TABLE 7. SELECTED AMINONAPHTHALENOLS AND AMINOHYDROXYNAPHTHALENESULFONIC ACIDS

Compound	Trivial name	Manufacturing method[abc]	Intermediate for
5-amino-1-naphthalenol	Purpurol	a	azo dyes, e.g., CI Acid Blue 70; sulfur dyes
7-amino-2-naphthalenol	Cyanol	a	azo, dyes, e.g., CI Mordant Brown 65
3-hydroxy-4-amino-1-naphthalenesulfonic acid	1,2,4-acid; Boeniger acid	b	azo dyes, e.g., CI Acid Red 186, Mordant Red 7; chrome complex dyes
5-amino-6-hydroxy-2-naphthalenesulfonic acid	Amino-Schaeffer acid	c	photographic developer; rarely used for dyes
4-hydroxy-8-amino-2-naphthalenesulfonic acid	M-acid	a	azo dyes, e.g., CI Direct Green 42
4-hydroxy-7-amino-2-naphthalenesulfonic acid	J-acid	a	azo dyes, e.g., CI Direct Blue 71, Direct Red 16; direct dyes using N-phenyl J-acid and J-acid imide
4-hydroxy-6-amino-2-naphthalenesulfonic acid	γ-acid	a	azo dyes, e.g., CI Direct Black 22
4-amino-5-hydroxy-2,7-naphthalenedisulfonic acid	H-acid	a	azo dyes, e.g., CI Direct Black 19, Direct Blue 15
4-amino-5-hydroxy-1,3-naphthalenedisulfonic acid	Chicago acid; SS-acid; 2S-acid	a	azo dyes, e.g., CI Acid Blue 42

(continued overleaf)

TABLE 6. MANUFACTURE, PRODUCTION, AND APPLICATION DATA FOR SELECTED HYDROXYNAPHTHALENESULFONIC ACIDS

Compound	Trivial name	Manufacturing method[abcdef]	Intermediate for
4-hydroxy-2-naphthalenedisulfonic acid	Armstrong & Wynne's acid; 1,3-oxy-acid	a,b	azo dyes, e.g., CI Direct Blue 127
4-hydroxy-1-naphthalenesulfonic acid	Nevile-Winther acid; 1,4-oxy-acid	c,d	azo dyes, e.g., CI Acid Red 14; tanning agents
5-hydroxy-1-naphthalenesulfonic acid	L-acid	d,e	azo dyes and pigments, e.g., CI Pigment Red 54, toner; 1,5-naphthal enediol
8-hydroxy-1-naphthalenesulfonic acid		f	metallized o, o'-dihydroxyazo dyes, e.g., CI Acid Blue 58
2-hydroxy-1-naphthalenesulfonic acid	oxy-Tobias acid	c	Tobias acid; J-acid
6-hydroxy-2-naphthalenesulfonic acid	Schaeffer's acid	c	azo dyes, e.g., CI Acid Orange 12; synthetic tanning agents
7-hydroxy-2-naphthalenesulfonic acid	F-acid	e	azo dyes, e.g., CI Direct Blue 128
7-hydroxy-1-naphthalenesulfonic acid	Crocein acid; Baeyer's acid	c	azo dyes, e.g., CI Acid Red 70
4,5-dihydroxy-1-naphthalenesulfonic acid	dioxy S-acid	e	azo dyes, e.g., CI Direct Blue 26
6,7-dihydroxy-2-naphthalenesulfonic acid	dioxy R-acid	e	2,3-dihydroxy-naphthalene
5-hydroxy-2,7-naphthalene-disulfonic acid	RG-acid; violet acid	e	azo dyes, e.g., CI Acid Red 99
8-hydroxy-1,6-naphthalene-disulfonic acid	ε-acid; Andresen's acid	f	azo dyes, e.g., CI Direct Blue 98
4-hydroxy-1,6-naphthalene-disulfonic acid	Dahl's acid; D-acid	a,d	nitro coloring matter, e.g., CI Acid Yellow 1
4-hydroxy-1,5-naphthalene-disulfonic acid	Schoellkopf's acid; CS-acid; δ-acid	f	azo dyes, e.g., CI Acid Blue 169
3-hydroxy-2,7-naphthalene-disulfonic acid	R-acid	c	azo dyes, e.g., CI Acid Red 115, Acid Red 26
7-hydroxy-1,3-naphthalene-disulfonic acid	G-acid	c	azo dyes, e.g., CI Acid Red 73; triphenylmethane dyes
4,5-dihydroxy-2,7-naphthalenedisulfonic acid	chromotropic acid	a,e	azo dyes, e.g., CI Acid Violet 3
8-hydroxy-1,3,6-naphthalene-trisulfonic acid	oxy-Koch's acid	a	azo dyes, e.g., CI Direct Blue 27; chromotropic acid
7-hydroxy-1,3,6-naphthalene-trisulfonic acid		c	azo dyes, e.g., CI Acid Red 41

[a] By hydrolysis of corresponding aminonaphthalenesulfonic acid.
[b] By desulfonation of 8-hydroxy-1,6-naphthalenedisulfonic acid.
[c] By sulfonation of appropriate (1- or 2-) naphthalenol.
[d] By Bucherer reaction (with sulfite) of appropriate aminonaphthalenesulfonic acid.
[e] By alkali fusion or alkaline hydrolysis under pressure of appropriate naphthalenedisulfonic or naphthalenetrisulfonic acid or hydroxynaphthalenedisulfonic acid.
[f] By alkaline hydrolysis of sulfone formed on boiling aqueous solution of diazonium salt of 8-amino-1-naphthalenesulfonic acid or appropriate derivatives.

TABLE 7. (*continued*)

Compound	Trivial name	Manufacturing method[abc]	Intermediate for
4-amino-5-hydroxy-1,7-naphthalene-disulfonic acid	K-acid	[a]	azo dyes, e.g., Sulfon Acid Blue G, CI 13400
3-amino-5-hydroxy-2,7-naphthalene-disulfonic acid	RR-acid; 2R-acid	[a]	azo dyes, e.g., CI Direct Brown 31

[a] By alkali fusion or hydrolysis of appropriate aminonaphthalenesulfonic acid.
[b] By nitrosation of 2-naphthalenol and reaction of nitroso compound with sodium bisulfite.
[c] By nitrosation/reduction of 6-hydroxy-2-naphthalenesulfonic acid.

1- and 2-Naphthalenecarboxylic Acids. Naphthalenecarboxylic acids are useful intermediates for dyes and photographic materials. These acids are also used in the preparation of antitumor agents and also in the preparation of cholecystokinin-agonist tetrapeptide. The acids are prepared readily by the oxidation of 1- or 2-alkylnaphthalenes with dilute nitric acid, chromic acid, or permanganate. The oxygen or air oxidation of alkylnaphthalenes in an alkanoic acid solvent in the presence of Ce-, Co-, or Mn-containing catalyst and a Br-containing catalyst gives good

TABLE 8. SELECTED PROPERTIES OF NAPHTHALENECARBOXYLIC ACIDS

Compound	Mp, °C	Other
1-naphthalenecarboxylic acid	162	sol ethanol; sparingly sol water; $K_a = 2.04 \times 10^{-4}$ at 25°C; bp at 6.7 kPa[a] 231°C
2-naphthalenecarboxylic acid	184–185	sol ethanol, ether, chloroform; $K_a = 6.78 \times 10^{-5}$ at 25°C; bp >300°C
1,2-napthalenedicarboxylic acid	175 dec	sol ethanol, ether, acetic acid; mp anhydride 168–169°C
1,3-naphthalenedicarboxylic acid	267–268	
1,4-naphthalenedicarboxylic acid	309	sol ethanol; insol boiling water
1,5-naphthalenedicarboxylic acid	315–320 dec	insol common solvents
1,6-naphthalenedicarboxylic acid	310	sol hot ethanol, acetic acid
1,7-naphthalenedicarboxylic acid	308	sol common organic solvents
1,8-naphthalenedicarboxylic acid	converts to anhydride (mp 274°C)	sol warm ethanol; bp anhydride at 440 Pa[b] 215°C
2,3-naphthalenedicarboxylic acid	239–241 dec	sol hot ethanol; mp anhydride 246°C
2,6-naphthalenedicarboxylic acid	310–313 dec	sol aq alc
2,7-naphthalenedicarboxylic acid	>300	sol ethanol
1,2,5-naphthalenetricarboxylic acid	270–272	sol methanol
1,3,8-naphthalenetricarboxylic acid		mp 1,8-anhydride 289–290°C
1,4,5-naphthalenetricarboxylic acid	forms anhydride (mp undefined)	mp 4,5-anhydride 274°C
1,2,4,5-naphthalenetetra-carboxylic acid	263	mp dianhydride 263°C
1,4,5,8-naphthalenetetra-carboxylic acid	forms anhydride	sol acetone; dianhydride sublimes >300°C

[a] To convert kPa to mm Hg, multiply by 7.5.
[b] To convert Pa to mm Hg, divide by 133.3.

TABLE 9. SELECTED PROPERTIES OF HYDROXYNAPHTHALENECARBOXYLIC ACIDS

Carboxylic acid	Mp, °C	Other
2-hydroxy-1-naphthalene-	157–159	sparing sol H_2O; sol alcohol, benzene
3-hydroxy-1-naphthalene-	248–249	
4-hydroxy-1-naphthalene-	188–188	
5-hydroxy-1-naphthalene-	236	
6-hydroxy-1-naphthalene-	213	
7-hydroxy-1-naphthalene-	256–257	
8-hydroxy-1-naphthalene-	1691	acetone, mp 108°C
1-hydroxy-2-naphthalene-	2000.55 wt %	sol in boiling water, alcohol, ether, benzene
3-hydroxy-2-naphthalene-	222–2230.1 wt %	sol in water at 25°C, ether, benzene chloroform
4-hydroxy-2-naphthalene	225–226	
5-hydroxy-2-naphthalene	215–216	
6-hydroxy-2-naphthalene	245–248	
7-hydroxy-2-naphthalene	274–275	
8-hydroxy-2-naphthalene	229	

results. The direct carboxylation catalyst naphthalene with CO and oxygen in the presence of Pd-carboxylate catalysts has been patented. The photo carboxylation of naphthalene in the presence of carbon dioxide and an electron donor has been described. About 67% naphthoic acids were obtained by this method, upon visible light irradiation with phenazine as a sensitizer. Over 90% of the naphthoic acids was 1-naphthoic acid.

Hydroxynaphthalenecarboxylic and Aminonaphthalenecarboxylic Acids

Some properties of selected hydroxynaphthalenecarboxylic acids are presented in Table 9. These acids are used in the manufacture of dyes (most importantly, naphthol AS dye stuffs), color film, and polyester.

MANNAN TALUKDER
CURTIS R. KATES
Advanced Aromatics, Inc.

Additional Reading

Arpe, H.-J.: *Naphthalene and Hydronaphthalenes to Nuclear Technology,* Vol. 17, 5th Edition, John Wiley & Sons, Inc., New York, NY, 1996.
Cerfontain, H.: *Mechanistic Aspects in Aromatic Sulfonation and Desulfonation,* Wiley-Interscience, New York, NY, 1968.
Donaldson, N.: *The Chemistry and Technology of Naphthalene Compounds,* E. Arnold Ltd., London, 1958.
Radt, F.: in F. Radt, ed; *Elsevier's Encyclopedia of Organic Chemistry,* Vol. 12B, Elsevier Publishing Co., New York, NY, 1948, pp. 132–161.
Venkataraman, K. ed.: *The Chemistry of Synthetic Dyes,* 8 vols., Academic Press, Inc., New York, NY, 1952–1978.

NAPHTHENIC ACIDS. The term *naphthenic acid*, as commonly used in the petroleum industry, refers collectively to all of the carboxylic acids present in crude oil. Naphthenic acids are classified as monobasic carboxylic acids of the general formula RCOOH, where R represents the naphthene moiety consisting of cyclopentine and cyclohexane derivatives. Naphthenic acids are composed predominantly of alkyl-substituted cycloaliphatic carboxylic acids, with smaller amounts of acyclic aliphatic (paraffinic or fatty) acids. Aromatic, olefinic, hydroxy, and dibasic acids are considered to be minor components. Commercial naphthenic acids also contain varying amounts of unsaponifiable hydrocarbons, phenolic compounds, sulfur compounds, and water. The complex mixture of acids is derived from straight-run distillates of petroleum, mostly from kerosene and diesel fractions. See also **Petroleum**.

Chemical Structure

Naphthenic acids are based on saturated single or multicyclic condensed ring structures. The low molecular weight naphthenic acids contain alkylated cyclopentane carboxylic acids, with smaller amounts of cyclohexane derivatives occurring. The carboxyl group is usually attached to a side chain rather than directly attached to the cycloalkane. The simplest naphthenic acid is cyclopentane acetic acid (**1**, $n = 1$).

$$\text{H}_2\text{C} \overset{\displaystyle\text{CH}_2}{\underset{\displaystyle\text{H}_2\text{C} \longrightarrow \text{CH}_2}{\diagup}} \text{CH}-(\text{CH}_2)_n-\text{COOH}$$

(1)

Naphthenic acids are represented by a general formula $C_nH_{2n-z}O_2$, where n indicates the carbon number and z specifies a homologous series. The z is equal to 0 for saturated, acyclic acids and increases to 2 in monocyclic naphthenic acids, to 4 in bicyclic naphthenic acids, to 6 in tricyclic acids, and to 8 in tetracyclic acids.

Physical and Chemical Properties

Naphthenic acids are viscous liquids, with phenolic and sulfur impurities present that are largely responsible for their characteristic odor. Their colors range from pale yellow to dark amber. Naphthenic acids have wide boiling point ranges at high temperatures (250–350°C). They are completely soluble in organic solvents and oils but are insoluble (50 mg/L) in water. Commercial naphthenic acids are available in various grades and are marketed by acid number, impurity level, and color. Chemically, naphthenic acids behave like typical carboxylic acids with similar acid strength as the higher fatty acids.

Naphthenic acid corrosion has been a problem in petroleum-refining operations since the early 1990s. Refineries processing highly naphthenic crudes must use steel alloys; 316 stainless steel is the material of choice. Conversely, naphthenic acid derivatives find use as corrosion inhibitors in oil-well and petroleum refinery applications.

Occurrence

Not all crudes contain sufficient quantities of usable acids to make recovery an economic process. Heavy crudes from geologically young formations have the highest acid content, and paraffinic crudes usually have low acid content. Typical concentrations of acids are shown in Table 1.

Manufacture

The commercial production of naphthenic acid from petroleum is based on the formation of sodium naphthenate. Naphthenic acids are recovered by caustic extraction of petroleum distillates rather than from crude petroleum. Crude naphthenic acid is obtained by acidulating the sodium naphthenate, and can be further refined to remove impurities.

Interest in synthetic naphthenic acid has grown as the supply of product has fluctuated. Oxidation of naphthene-based hydrocarbons, free-radical addition of carboxylic acids to olefins, and addition of unsaturated fatty acids to cycloparaffins have been studied but not commercialized.

Production

Nameplate capacities of naphthenic acid producers in North America are 9000 metric tons of crude and refined acid at Merichem (Tuscaloosa, Ala.), and 3600 t of crude acid at Hewchem (Gulfport, Miss.). However, actual production capacity may vary widely as a result of the mix of feedstocks being processed. Naphthenic acid products are shipped in tank cars, tank trucks, and drums under DOT 9137/UN 3082 identifications numbers.

Economic Aspects

Naphthenic acid availability exceeds demand, although some minor market disruptions occurred in North America during the early 1990s. Long-term

TABLE 1. ACID CONTENT OF VARIOUS CRUDES

Crude oil source	Petroleum acids, wt %
Pennsylvania	0.03
West Texas	0.4
Gulf Coast	0.6
California	1.5
Russia, Balakhany light	1.0
Russia, Balakhany heavy	1.6
Romania, waxy	0.2
Romania, asphaltic	1.6
Venezuela, Lagunillas	1.2

yearly feedstock is expected to meet market growth projections with recent discoveries of high naphthenic-content oil.

Health and Safety Factors

Naphthenic acids are only slightly toxic to mammals but are toxic to fish, bacteria, and wood-destroying insects. The lethal oral dose for humans is approximately 1 L. Naphthenic acid is not listed as a carcinogen.

Commercial Uses

More than two-thirds of the naphthenic acids produced is used to make metal salts, with the largest volume being used for copper naphthenate consumed in the wood preservative industry. Oil field uses are primarily imidazolines for surfactant and corrosion inhibition. See also **Petroleum**. Besides the lubrication market for metals salts, the miscellaneous market is comprised of free acids used in concrete additives, motor oil lubricants, and asphalt-paving applications. See also **Lubricant**; and **Lubricating Agents**.

Naphthenic acid is ideal for synthesizing metal carboxylates that require a ligand with some oxidative stability, solubility in hydrocarbons and oils, and insolubility in water.

Another market application for naphthenic acid is the tire industry, where cobalt naphthenate is used as an adhesion promoter. Naphthenic acid esters have been repeatedly cited as surfactants, lubricants, and replacements for phthalates as plasticizers for PVC resins.

Naphthenyl alcohols are formed by reduction of the acids or their simple esters. They are valuable as surfactants, solvents, and components of lubricants. The acid halides are of value mainly as chemical intermediates.

JAMES A. BRIENT
PETER J. WESSNER
Merichem Company
MARY NOON DOYLE
Shepard Chemical Company

Additional Reading

Lochte, H. L. and E. R. Littmann: *The Petroleum Acids and Bases,* Chemical Publishing Co., Inc., New York, NY, 1955.
Lower, E. S.: *Specialty Chem.* **7**, 76 (1987); **7**, 282 (1987); **8**, 174 (1988); **9**, 135 (1989); **9**, 267 (1989).
Maass, W., E. Buchspiess-Paulentz, and F. Stinsky: *Naphthensäuren und Naphthenate,* Verlag für Chemische Industrie Ziolkowsky, H. Augsburg, Germany, 1961.
Narmetova, G., B. Khamidov, N. Ryabova, and E. Aripov: *Purification, Identification, and Use of Naphthenic Acid,* Fan, Tashkent, former USSR, 1983.

NARCOTICS. See **Alkaloids**; **Analgesics, Antipyretics, and Antiinflammatory Agents**; **Hallucinogens**; **Morphine**.

NASCENT. Descriptive of the abnormally active condition of an element, for example, the atomic oxygen released from hydrogen peroxide and sulfur atoms evolved from thiuramsulfide accelerators. The term is now obsolete.

NATROLITE. The mineral natrolite, one of the zeolites, is a sodium aluminum silicate corresponding to the formula $Na_2Al_2Si_3O_{10} \cdot 2H_2O$. It is orthorhombic, crystallizing in slender prisms of nearly square cross-section which are terminated by relatively flat pyramids. There are also fibrous to compact varieties. Natrolite is a brittle mineral; hardness, 5–5.5; specific gravity, 2.2; luster, vitreous; color, red, yellow, white, or colorless; transparent to opaque. Natrolite is found with other zeolites in fissures and cavities in basaltic and related rocks. Czechoslovakia, France, Italy, Norway, Scotland, Ireland, Iceland, Greenland, and South Africa contain well-known localities for natrolite. In the United States it is found in the Triassic traps of New Jersey; also from Oregon, Washington, Montana, Colorado, and as exceptional crystals from San Benito County, California. Superb crystals occur at Mt. St. Hilaire, Quebec, Canada, and from an asbestos mine in Quebec, crystals up to 3 feet (0.9 meters) long and 4 inches (10 centimeters) in diameter have been found. The name natrolite refers to its soda content.

NATTA, GIULIO (1903–1979). An Italian chemist along with Karl Ziegler won the Nobel prize for chemistry in 1963 for his fundamental work on catalytic polymerization. In 1954 he developed isotactic polypropylene at his laboratory at the Polytechnic Institute of Milan, which led to wide application of various stereospecific polymers with organometallic catalysts such as triethylaluminum. He was for many years consultant for the Montecatini chemical firm. The researchers of Natta, together with those of Karl Ziegler, made possible the chemical manipulation of monomers to form specifically ordered 3-dimensional polymers having predetermined properties, to which the term *tailor-made* is often applied.

NATURAL GAS. A major source of energy for industrial, commercial, and domestic needs, natural gas is consumed by numerous countries worldwide. Because natural gas is comparatively easy to transport over long distances, usage is not confined to regions that produce it. In addition to energy, natural gas also is a critically important source of industrial chemicals, including numerous hydrocarbon-based organics that find ultimate usage in plastics, films, fibers, solvents, and coatings.

In terms of interest as a fossil fuel for generating electrical power and other energy-conversion processes, natural gas is gaining favor. Compared with coal, natural gas frequently is termed the "clean-burning" fuel. As compared with coal as an energy raw material, the "add-on" costs for treating combustion effluents to satisfy environmental requirements are less for natural gas than for coal, and, consequently, the lower cost benefits of coal are eroding. Further, improvements in natural gas production technology and a brighter outlook for natural gas reserves are contributing to the expansion of natural gas consumption. In addition, large advances have been made in the combustion efficiency of natural gas. For example, the efficiency of some domestic heating appliances has increased to about 95% as compared with 60% or 70% efficiency a relatively few years ago. The cogeneration of heat and electricity in industrial utilities is tending to favor natural gas as the raw fuel. In terms of local natural gas distribution, utilities are finding the use of polyethylene pipe an important cost-saving factor.

In a summary of the Gas Research Institute (GRI)[1], the following observation is made: "The United States, the gas industry, and the gas consumer have entered a dynamic new decade filled with change, challenge, and opportunity. These include the reemergence of the environmental movement, the continuing deregulation of the U.S. natural gas market, the expansion of global trade in the former Soviet Bloc, the emergence of more competitive international and national energy markets, and the rapid expansion of technology options. Three strategic needs are likely to dominate the 1990s for the U.S. gas industry and the gas consumer: (1) ensuring gas deliverability while controlling costs to the consumer; (2) responding to increased concern for the environment; and (3) satisfying a demand for higher quality of energy service."

Composition of Natural Gas

The composition of natural gas varies with the source, but essentially is made up of methane, ethane, propane, and other paraffinic hydrocarbons, along with small amounts of hydrogen sulfide, carbon dioxide, nitrogen, and, in some deposits, helium. Natural gas is found underground at various depths and pressures, as well as in solution with crude-oil deposits. Principal gas deposits are found in the United States, Canada, the former Soviet Bloc, and the Middle East. The analysis of a gas sample taken from the Panhandle natural gas field in Texas is given in Table 1. Because numerous parts of the earth do not have natural gas at all, or where supply is less than demand, much natural gas is transported, notably by pipeline in the gaseous or liquid phase and across the seas in specially-designed LNG (liquefied natural gas) carriers.

Origin and Geology of Natural Gas

The most commonly accepted theory concerning the formation of natural gas is the organic theory. Methane is a product of decaying vegetable matter and in areas of stagnant water is found as *marsh gas* or *swamp gas*. It is theorized that over millions of years, the remains of plants and animals were washed down into lakes, the accumulations covered with layers of mud and stone. The latter became stone while the organic matter decayed through the action of heat and pressure and perhaps from effects of bacteria and radioactivity, forming various hydrocarbons. The hydrocarbons were held in tiny spaces between the particles of sand and porous rock and formed natural gas and petroleum. Often the natural gas so formed made its way through the rock to the surface and escaped. In some areas, however, the layers of sand and porous rock were covered by impermeable rock

[1] "1993–1997 Research & Development Plan," Gas Research Institute, Chicago, Illinois.

TABLE 1. ANALYSIS OF NATURAL GAS FROM NATURAL GAS FIELD IN TEXAS PANHANDLE

Component	Mole percent
Methane	76.2
Ethane	6.4
Propane	3.8
Normal butane	1.3
Isobutane	0.8
Normal pentane	0.3
Isopentane	0.3
Cyclopentane	0.1
Hexane plus other hydrocarbons	0.35
Nitrogen	9.8
Oxygen	Trace
Argon	Trace
Hydrogen	0.0
Hydrogen sulfide	0.0
Carbon dioxide	0.2
Helium	0.45

Note: Heating value of various natural gases averages between 975 and 1180 Btu/cubic foot (8678–10,502 Calories/cubic meter) at 60°F (15.6°C) and 30 inches (76.2 centimeters) mercury pressure.

to form huge reservoirs of natural gas at various levels of pressure. See also **Petroleum**.

The organic theory as usually presented is rather general and vague in many respects. As observed by Ourisson (UniversitÉ Louis Pasteur, Strasbourg) and colleagues, natural gas, as well as coal and petroleum, are fossil fuels, but fossils of what? Fossil fuels form only if the organic matter is buried before it can become completely oxidized to carbon dioxide by microorganisms. According to the microbial origin concept, as the carbon compounds sink deeper into the Earth under accumulating sediments, they are subjected to high temperatures and undergo chemical reaction, during which oxygen and most other elements are eliminated. This yields a mixture composed in the case of gas and petroleum almost entirely of hydrocarbons (carbon in the case of coal). Since the beginnings of photosynthesis on Earth, it is estimated that 10 quadrillion (1016) tons of carbonaceous material has been stored in sediments. Most of this material is stored in very dilute form and only under exceptional geologic conditions, is it concentrated to become a viable fuel source. In a twenty-year study, which might be called molecular paleontology, Ourisson and coworkers have been studying the detailed genesis of fossil fuels. Thus far, chemical analysis of the most varied organic sediments reveals a surprising commonality—all appear to derive much of their organic matter from once unknown microbial lipids. This topic is presented in more detail in entry on **Petroleum**.

A few scientists, notably Thomas Gold (Cornell University), have proposed that, in contrast with the organic sediments theory, the prime source of natural gas is primordial, abiotic methane rising from deep within the Earth's mantle. This is sometimes referred to as the "deep-earth gas" hypothesis. In this view, methane flows up around the edges of the shield and is responsible for the oil and gas fields in the North Sea and the southern Baltic. Admittedly, this reservoir of natural gas is presently out of the reach of any foreseeable drilling technology. In some areas, it is suggested that the granite crust may have been fractured and subsequently became porous to the extent that methane may have risen into the crust and have become trapped at accessible depths. Other scientists point out that no evidence supports the concept that a large amount of methane was incorporated in the Earth when it was formed. Available geochemical evidence suggests that the early atmosphere, produced by outgassing of the planetary interior, could not have been rich in hydrogen. Further, if the Earth did at one time contain primordial methane or other hydrocarbons, most of that volatile material would have long since escaped by way of volcanism and diffusion. It is also suggested that the analysis of volcanic basalts shows that the rock in the upper mantle is highly oxidizing, in which case any methane present would have been converted to carbon dioxide.

Some experimental drilling programs underway in Sweden, including a well some 5000 meters in granite bedrock, may shed further light on Gold's hypothesis.

In searching for new fields during the 1960s and 1970s, drillers seeking gas and oil in traditional suspect source reservoirs, would find gas and/or oil in only about 9 of 100 wells drilled. Usually, only 2 or 3 of these wells produced sufficient gas and/or oil to be of commercial value. Whereas the average depth of gas well drilled during this period ranged between 5000 and 6000 feet (1524–1829 meters), some drillers are now aiming at the 30,000-foot (9144-meter) depth. For many years, the deepest gas well in Texas was 28,600 feet (8717 meters). In the early days of offshore drilling, operations were conducted in water only 20–25 feet (6–7.5 meters) deep. There are now many platforms in waters that are deeper by a factor of 20–30 times.

In terms of the conventional or traditional sources, natural gas is found in areas close to exposed or buried mountain ranges. Major deposits of natural gas are found in inclined strata where the rock formations dip away from the crest of a buried hill or the ridge of a buried mountain. Some of the common types of formations in which natural gas is found are shown in Fig. 1. Although natural gas and crude oil are frequently found together, the largest natural gas reserves (about 70% of the estimated reserves) are in deposits neither in contact with, nor dissolved in, oil.

Forecasting Natural Gas Reserves

Since the 1920s, natural gas reserves have been based upon the amount of gas that most likely will be found with oil. Although numerous experts in the field have claimed over the years that this methodology overlooks large amounts of "unassociated" gas, the professionals have been slow to revise their procedures. The concept that much natural gas occurs quite apart from oil is now taking hold. The U.S. Geological Survey, which has proclaimed a gas resource base at about 400 trillion cubic feet, undertook a major study to be completed in the late 1990s. Survey techniques have undergone several dramatic improvements, including the use of computer techniques that will project three-dimensional survey information, as contrasted with past reliance on two-dimensional mapping. A number of experts now believe that the reserves are well over 1,200 trillion cubic feet (34 trillion cubic meters). One independent gas producer has estimated the figure at about 1,500 trillion cubic feet (42.5 trillion cubic meters).

Traditional estimates of oil-associated natural gas reserves historically have rated the former Soviet Bloc as holding nearly 40% of the reserves, Iran about 14%, and the United States nearly 6%, with additional fairly high reserves in Qatar, Algeria, and Saudi Arabia. At one time, North America was attributed to have a 60-year supply, but with revisions in estimating procedures, coupled with increased efficiency in natural gas production and gas combustion efficiency, the future is now believed to be in terms of at least a few centuries. Interest continues, however, in upgrading low-Btu natural gases and developing so-called *substitute natural gas*.

The principal gas fields within the continental United States are indicated in Fig. 2.

Ultimate Recovery of Natural Gas Reserves

It has been traditional for many years to categorize ultimate recovery of gas reserves by reservoir lithology which involves the compilation of such data by three types of reservoirs: (1) *Sandstone reservoir*—consisting of sedimentary rock composed predominantly of quartz grains or other noncarbonate mineral or rock detritus. Included in this reservoir type are unconsolidated sand, sandstone, siltstone, graywacke, arkose and granite wash, conglomerate, and breccia. (2) *Carbonate reservoir*—composed of sedimentary rock made up predominately of calcite (limestone) and/or dolomite. (3) *Other reservoirs*—including igneous and metamorphic rocks and some sedimentary rocks, such as fractured shale.

Estimated ultimate recovery is also reported by type of entrapment, of which there are two major types: (1) *Structural trap*—an entrapment in which migration of hydrocarbons in the reservoir rock has terminated

Fig. 2.   Location of principal natural gas fields in the lower 48 of the United States. Alaska, not shown, ranks third in terms of holding estimated reserves. (*Batelle Memorial Institute*)

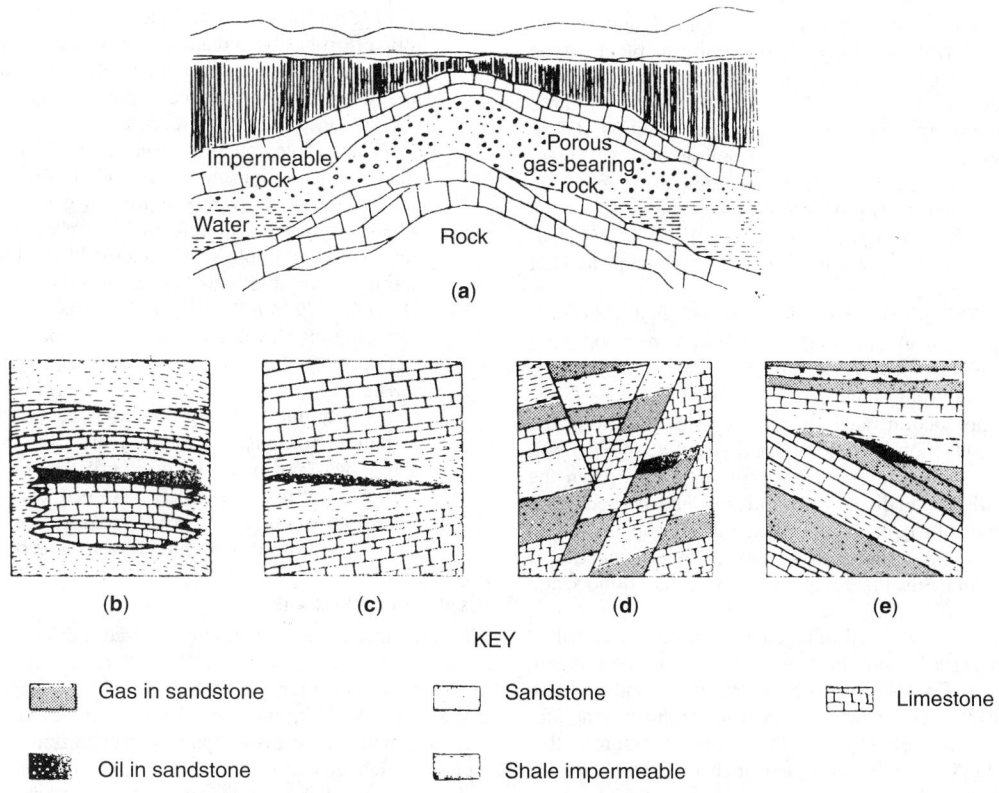

KEY

Gas in sandstone	Sandstone
Oil in sandstone	Shale impermeable
	Limestone

Fig. 1.   Types of natural gas reservoirs and entrapments: (**a**) anticlinal trap; (**b**) coral reef trap; (**c**) stratigraphic trap; (**d**) fault trap; and (**e**) unconformity

primarily because of closure induced by structural deformation, such as folding or faulting. Within this category should also be included entrapments attributed to hydrodynamic forces. (2) *Stratigraphic trap*—an entrapment in which migration of hydrocarbons has terminated because of the pinchout of reservoir rock due either to truncation or to nondeposition or to a facies change in the form of diminished permeability of reservoir rock. Also included in this category are entrapments in which a pinchout of facies change provides part of the barrier to migration of hydrocarbons, with structural elements providing the remaining closure for the entrapment. In these cases, it is recognized that the dominant cause of the accumulation is the lenticularity of truncation of the reservoir rock.

Some estimates indicate that about 65.8% of ultimately recoverable natural gas in the United States will be found in structural traps; the remaining 34.2% in stratigraphic traps.

The estimated ultimate recovery is also reported by the geologic age of the reservoir. It is recognized that problems may arise where the geologic age of a reservoir cannot be determined specifically, such as Permo-Pennsylvanian and Cambro-Ordovician, or where production from reservoirs of different geologic age is combined.

Natural gas liquids occur in either the gaseous phase or in solution with crude oil in the reservoir. They are recovered at the surface as liquids by separation from produced natural gas, by such processes as condensation and absorption in field separators, gasoline plants, and other surface facilities. In this processing, valuable by-products are recovered, such as light oils, natural gasoline, and other petroleum gases such as ethane, propane, and butane. Natural gasoline is blended with gasoline from petroleum refineries to improve starting properties, especially desirable in cold weather. Ethane is a major petrochemical raw material. Propane and butane are made available as LPG (liquefied petroleum gas). Processing of natural gas also removes unwanted material, such as nitrogen, sulfur compounds, carbon dioxide, and water vapor. Some gas fields produce helium, which is extracted cryogenically.

Unconventional Sources of Natural Gas

These include: (1) tight sandstones, (2) Devonian shales, (3) geopressured zones, (4) deep basins, (5) gas associated with coal seams, and (6) gas in the form of methane hydrates.

1. *Tight Sandstones.* In the United States, tight sandstones of the western basins range from the northern tier states to the Mexican border. Some tight gas sands also occur in the eastern United States. To date, resource development has occurred only in the limited areas characterized by thick, fairly uniform, blanket-type formations which, when hydraulically fractured, provide sufficient gas production rates to merit commercial exploitation. In these areas, as pointed out by Sharer (Gas Research Institute, GRI), state-of-the-art technologies can be used because only a limited knowledge of the formation characteristics is required for economic production. However, a majority of the resource base is associated with lower permeability and more complex blanket and lenticular sand formations, for which current technology is not adequate. GRI is concentrating research in these areas.

2. *Devonian Shales.* The large eastern Devonian gas shales resource base underlies approximately 174,000 square miles (453,000 km^2) of the eastern U.S. Estimates of recoverable gas range from 2 to 15% of the gas in place. Natural gas has been produced from these shales for decades. Well production rates are relatively low, but after the first few years of production it does not usually decline rapidly with time. A major constraint to present-day exploitation has been the extraordinary inability to predict with confidence the gas production rates that may be obtained in wells drilled outside the traditional production areas. Presently, the GRI is studying the systematics of historically successful fields, including the Appalachian, Illinois, and Michigan Basins.

3. *Geopressured Zones.* A test well in a geopressure zone was drilled some years ago in Tigre Lagoon in the coastal marshes of southern Louisiana. Known as Edna Delcambre #1, this well produced at a rate of up to 10,000 barrels of water per day from a sandstone aquifer some 12,600 feet (3840 meters) below the surface. Pressure at that depth is nearly 11,000 pounds per square inch (748 atmospheres) and the temperature is 116°C. Quite an elaborate manifold system is required to collect the gas. The water is disposed by forcing it by

its own pressure into another well bore, which penetrates to a depth of 2500 feet (762 meters). Scientists associated with this project had expected about 20 cubic feet/barrel (42 gallon); about 0.6 cubic meter/barrel (159 liters). In actuality, reports indicate that the yield of gas was about 2.5 times that amount.

As explained by specialists in geopressure technology, at great depths (in terms of present technology), the solubility of natural gas in water may be as much as 1000 cubic feet/barrel (28.3 cubic meters/159 liters) at depths of 30,000 feet (9144 meters), whereas that solubility will be reduced by a factor of ten at a depth of 20,000 feet (6096 meters). Under the right combination of geologic and hydrologic conditions, this gas-laden water will move toward the surface, during which process some of the gas will be released from the water in the form of very small bubbles. Ultimately, this gas collects beneath a geologic trap, where conventional free-gas reservoirs are formed. Some authorities now believe that the very deep aquifers are much more extensive than the free-gas reservoirs. It is this gas-saturated water that some scientists believe will be a great source of future natural gas.

The GRI has been investigating the coproduction of gas and water for a number of years. Natural gas from watered-out reservoirs, geopressured aquifers, and high-water-saturated gas-bearing reservoir strata are prime targets. This natural gas is trapped by water such that special production techniques must be used to move the water and remobilize the gas. Although some gas is also dissolved in the water, it is of less significance than the free gas trapped as dispersed bubbles or in pockets or stringers of various sizes in the reservoir rock matrix.

4. *Deep Basins.* These are found at depths between 15,000 and 30,000 feet (4572–9144 meters) and are estimated to contain significant quantities of gas, but generally await the development of advanced production technology and economic incentive.

5. *Gas Associated with Coal Seams.* Methane, the principal constituent of natural gas, is generated during the geologic process of coal formation. A significant portion of this gas is trapped by impermeable strata, and it is present within the fractures and micropores of the coal. (The presence of methane is an ever-present hazard in coal mining.) Major variations in resource estimates are due to uncertainties in the gas content and size of the deeper, not minable coal deposits in the western states that form the major portion of the resource base. Seeking such gas may involve depths as great as 6000 feet (1829 meters) underground. Except for reasons of safety, little effort has been made to recover any of this resource due to high recovery costs, potential uncertainties in production, and deficiencies in state-of-the-art equipment, particularly for the deeper coals. The GRI is concentrating its research efforts on unminable coal because of its large potential as a resource base. While the gas resource associated with mining amounts to about 10% of the energy value of the coal, producers rarely apply new gas recovery technology except where safety is a requirement. Targets of the GRI program are deep coal seams, multiple seams interbedded with shales and sandstones, and deep multiple beds that are too thin to mine.

6. *Methane Hydrates.* Within a certain range of pressures and temperatures, methane and water form hydrates. Described as icelike substances, these hydrates are believed to occur in very substantial quantities, particularly beneath permafrost and in deep-ocean bottoms. Although slush has occurred in gas pipelines under certain conditions for many years, the existence of hydrates in nature was not made known until the mid-1960s. Geologists and hydrologists had previously assumed that gas of this type would have dissipated during earlier geologic ages. This is another area of natural gas resource research awaiting economic incentives.

Exploratory Methods

The principal exploratory methods used are: (1) *Airborne magnetometers*, which seek out anomalies in the magnetic field. Experienced geologists relate these irregularities to the probability of gas reservoirs below the surface. (2) *Satellite imagery*, from which surface structures and patterns can be related to previous pattern recognition studies made of surfaces below which gas reservoirs exist. (3) *Gravitometers* are used to detect subtle variations in gravitational pull inasmuch as this is less for a gas reservoir than for continuous dense rock formations. (4) *Seismic methods*, which constitute the most widely-used of exploration methods. (5) *Data logging methods*, wherein an instrument is lowered into the borehole and which telemeters back to the surface readings of sonic absorption in an effort to determine the nature and thickness of rock formations. Data loggers operate on the basis of several physical phenomena. (6) *Fossil inspection*. The careful examination of microfossils can assist in fixing the age of rocks that are being penetrated. The condition of the fossils also can be related to probable temperatures to which they have been exposed over geologic periods and these, in turn, can be advantageous in locating possible gas deposits. Usually a combination of two or more exploratory techniques is used.

Liquefied Natural Gas (LNG)

The liquefaction of natural gas for storage and transportation and regasification for final distribution dates back several decades. A few major accidents in the handling of LNG thwarted the progress of the field for a while, but in the early 1970s, LNG was again considered in a major way because of energy-short nations. One of the more serious LNG accidents occurred in Cleveland, Ohio on October 20, 1944, when a storage tank developed a leak with spillage and subsequent fires in the surrounding neighborhood in which 135 persons lost their lives. While liquefaction offers marked storage space savings and convenience, the predominant advantage occurs in connection with both pipeline and ship transportation. Energy-short nations, such as Japan, and some of the European nations, have turned in recent years to the concept of shipping LNG by ship. For example, a large LNG plant at Lumut, Brunei, Borneo went onstream in mid-1974 essentially to furnish LNG to Japan.

Oil- and gas-rich nations, which at one time flared to the atmosphere much of the natural gas that accompanied the production of crude oil, have turned toward conservation—either through reinjection of much of the natural gas underground or through constructing LNG production facilities for shipment of the product overseas. Concurrent with such planning was reevaluation by a number of nations of their own valuable resources and a growing reluctance toward exporting inordinate quantities of gas and oil strictly for money. As of the early 1990s, the shipment of LNG overseas competes with other ways and means for alleviating energy shortages, including coal conversion and gasification, nuclear energy, solar energy, etc.

Three types of liquefaction processes may be used for production of LNG. The standard cascade process, which uses three refrigerants—methane, ethylene, and propane—all circulating in closed cycles, is shown in Fig. 3. There is a separate compressor for each of these refrigerants. The methane and propane are available from the feed gas (natural gas). The ethylene must be furnished separately. Ethane may be used in place of ethylene at a subatmospheric suction pressure. The cascade process has the highest rank in terms of thermal efficiency. As a possible improvement over the cascade process, the mixed refrigerant process was developed in the early 1960s. A single-pressure mixed refrigerant cascade (MRC) system is shown in Fig. 4. In one plant using this process, a hydrocarbon-plus-nitrogen mixture of relatively wide boiling range (N_2 through C_5) is used as the refrigerant. All of these components can be recovered from natural gas in separate apparatus. In still another system, shown in Fig. 5 a propane and mixture-refrigerant cycle is used. In this process, the cooling load is divided horizontally at about $-34.4°C$ into an upper portion absorbed by propane and a lower portion absorbed by the mixed refrigerant. In essence, the system is a dual refrigerant cascade in which the lower boiling fluid is a mixture refrigerant. The cascade combination with propane makes it possible to reduce the boiling range of the mixture refrigerant substantially, which improves the thermodynamic efficiency over that of the straight MRC process.

Cryogenic Upgrading of Low-Btu Natural Gases

Worldwide, there are substantial reserves of natural gas in which the reservoir formation hydrocarbons are contaminated with nonburning components. The presence of components, such as helium, nitrogen, or carbon dioxide, reduces the heating value of the gas mixture. This can result in the gas being unsuitable for existing transmission and distribution systems. Such contaminated mixtures are termed low-Btu gases if their heating values fall below the minimum standards, regulations, or contract heating value requirements.

Cryogenic processing can be used to upgrade some of these low-Btu gases so as to produce an acceptable high-Btu product. Cryogenic upgrading is a physical process in which subambient temperatures are

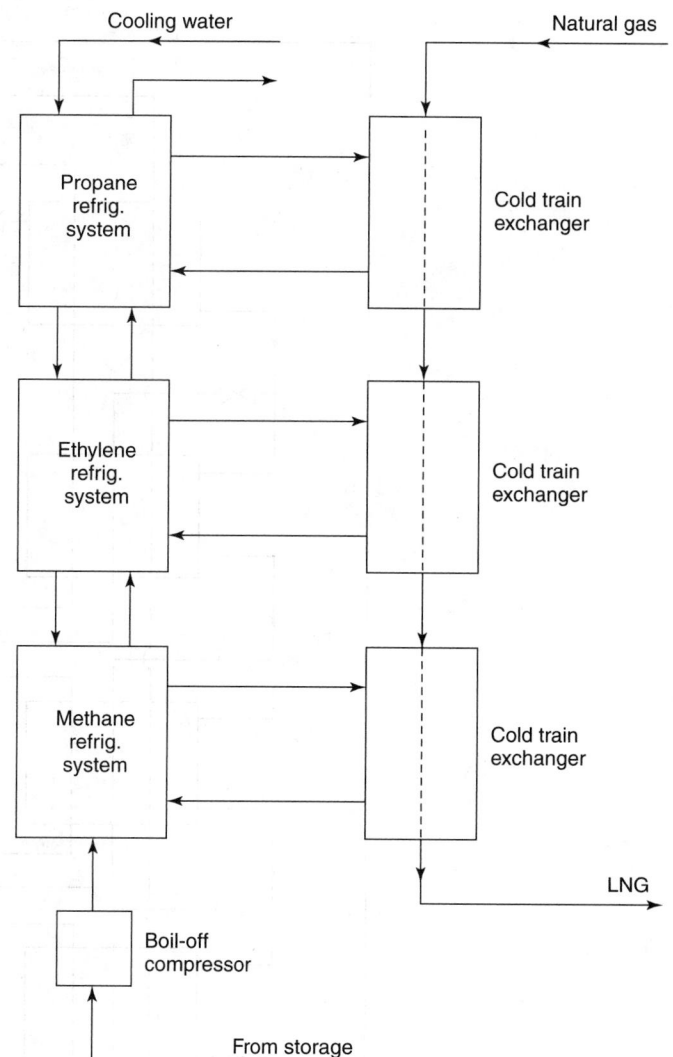

Fig. 3. Conventional or standard cascade system for producing liquefied natural gas (LNG)

employed to bring about a separation between the hydrocarbons and nonhydrocarbons in the mixture. The reduction of temperature occurring during cryogenic processing produces a two phase (gas-liquid) mixture. The relative volatilities between the components in the mixture result in selective mass transfer between the two phases. One phase becomes enriched with hydrocarbons and then has a heating value higher than the original gas. The second phase becomes denuded of hydrocarbons and has a heating value below that of the original gas mixture. Frequently, the mass transfer operation requires several theoretical stages in order to achieve the desired product heating value and high hydrocarbon recoveries. While cryogenic upgrading can be applied to gas mixtures containing carbon dioxide or hydrogen sulfide, it has so far only been applied commercially to those hydrocarbon mixtures contaminated with nitrogen and helium.

One of the main considerations in the design and operation of cryogenic upgrading plants is to identify and remove any component from the gas that could adversely affect the operation of the cold sections of the plant. Such components are carbon dioxide, water vapor, and heavy hydrocarbons that have high solidification temperatures and low solubilities. In general, if these components are allowed to remain in the gas to the cryogenic unit, they will form solids during the cooling process that will be deposited on the heat exchanger surfaces. This will lead to fall-off in performance and, possibly, to blockages and plant shutdown.

Particular attention should be paid to identifying any high freezing point components in the low-Btu gas. There exists a range of absorption and adsorption processes to pretreat the low-Btu gas to remove these undesirable components.

A simplified flowsheet of cryogenic upgrading plant is given in Fig. 6. The plant, consisting of two identical trains, is capable of processing 260

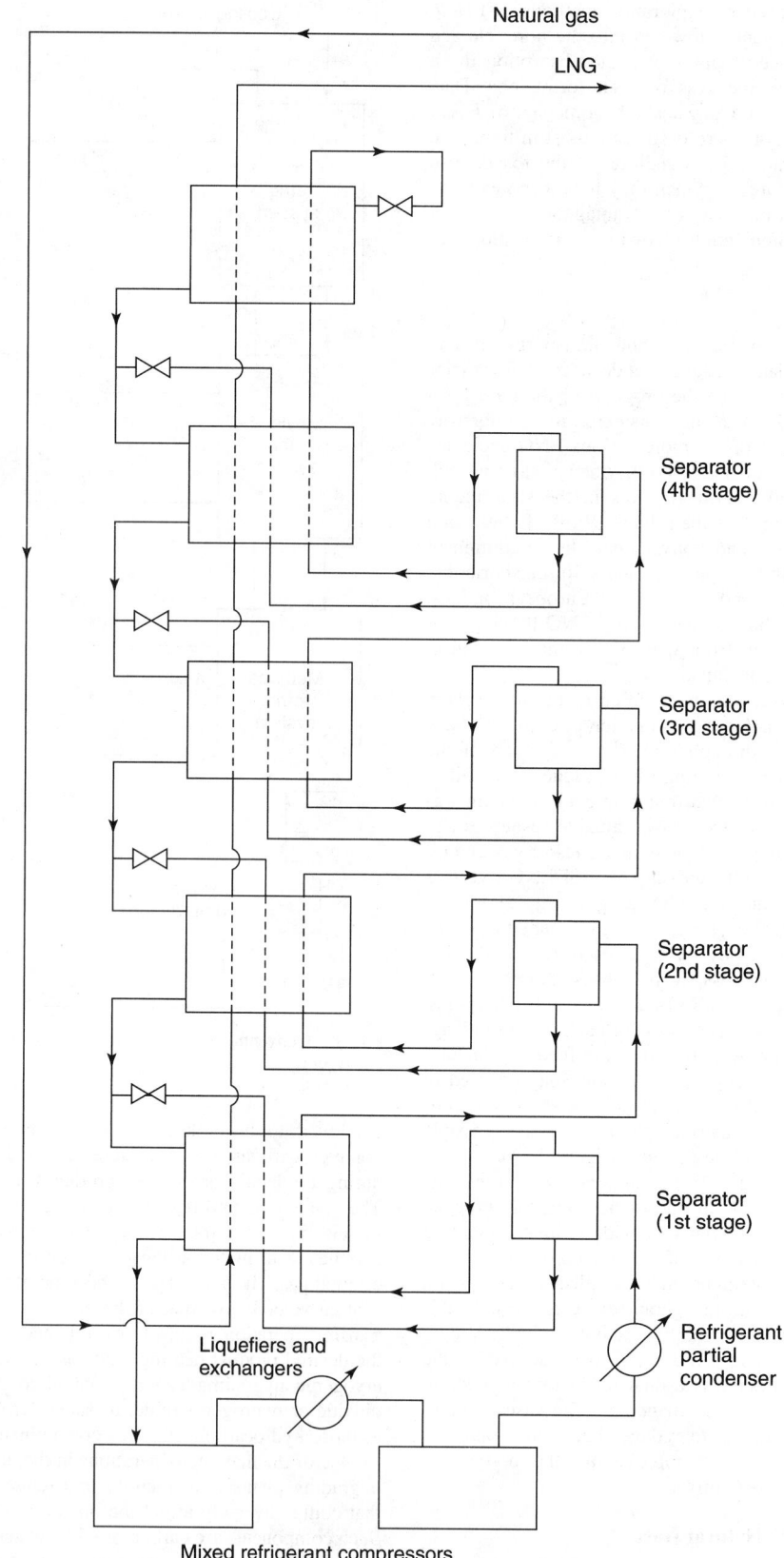

Natural gas

LNG

Separator (4th stage)

Separator (3rd stage)

Separator (2nd stage)

Separator (1st stage)

Refrigerant partial condenser

Liquefiers and exchangers

Mixed refrigerant compressors

Fig. 4. Single-pressure, mixed refrigerant cascade system for producing LNG

million standard cubic feet (7.3 million cubic meters) per day of low-Btu gas (580 Btu/standard cubic foot) (5162 Calories/cubic meter) and upgrading the gas into 143 million standard cubic feet (4 million cubic meters) of high-Btu gas (980 Btu/standard cubic foot; 8722 Calories/cubic meter). The plant stream parameters are indicated in Table 2.

The low-Btu gas is available at 800 psig (54 atmospheres) and is mainly a nitrogen-methane mixture. In addition to a small quantity of helium, the gas also contains small quantities of carbon dioxide, water vapor, and heavy hydrocarbons. The carbon dioxide is removed by washing with monoethanolamine (MEA), the water is taken out on molecular sieve, and the heavy hydrocarbons by adsorption on activated carbon. The gas is then cooled in aluminum plate-fin exchangers against the returning high-Btu product gas and vent gas. The gas is then expanded to 380 psig (26 atmospheres) and a vapor-liquid mixture passes into the H.P. (high

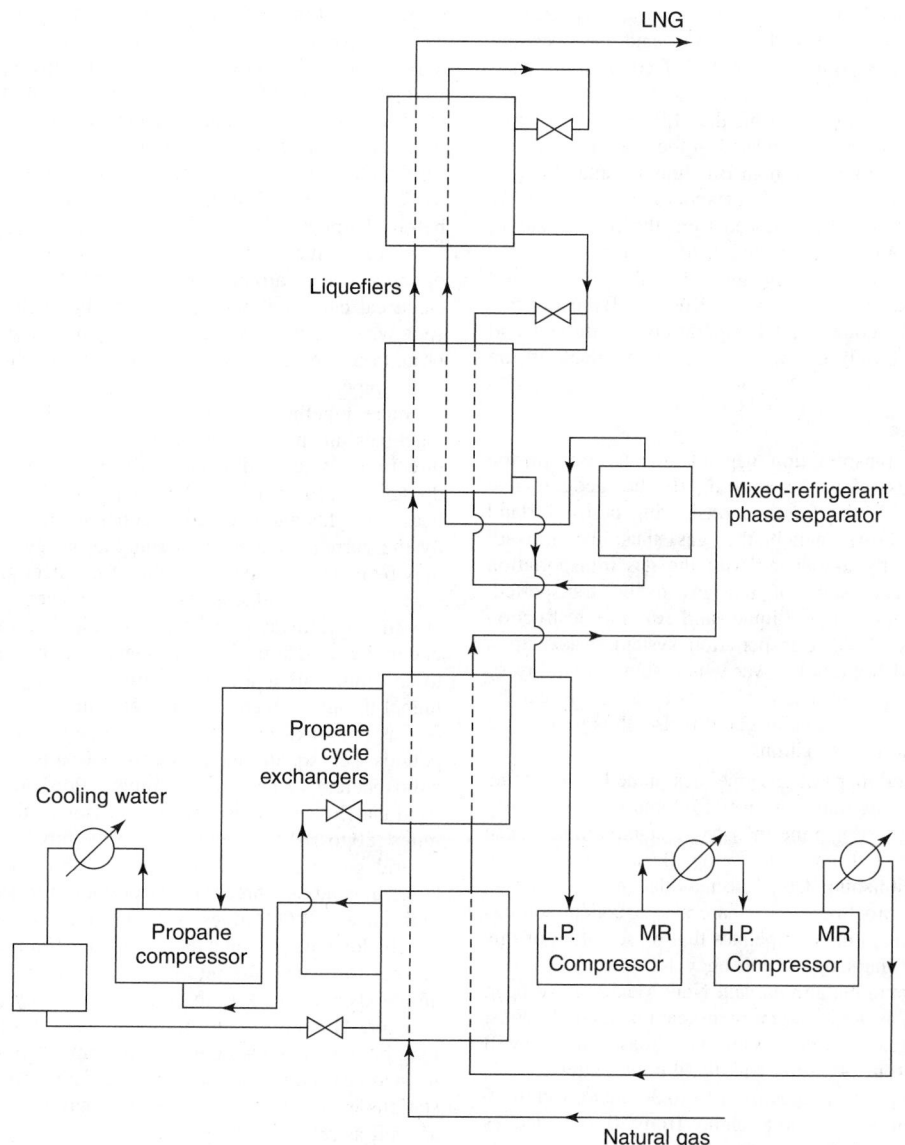

Fig. 5. Propane-mixed-refrigerant liquefaction system for producing LNG. L.P. = low pressure; H.P. = high pressure; MR = mixed refrigerant

pressure) fractionator. The purpose of this fractionator is to bring about an initial separation of the nitrogen-methane and to produce a liquid reflux for the L.P. (low pressure) fractionator.

TABLE 2. CRYOGENIC UPGRADING OF NATURAL GAS—STREAM PARAMETERS COMPOSITION—MOL.%

	Low-Btu gas	High-Btu gas	Vent gas	Helium
Helium	0.40	—	0.09	100.00
Nitrogen	42.75	4.00	98.95	—
Methane	56.02	95.09	0.96	—
Ethane +	0.53	0.91	—	—
CO_2	0.30	—	—	—
Flow (million standard cubic feet/day)	246	143	100	0.43
Flow (million cubic meters/day)	7	4	2.8	0.012
Heating value (Btu/standard cubic foot)	580	980	—	—
Heating value (Calories/cubic meter)	5162	8722	—	—

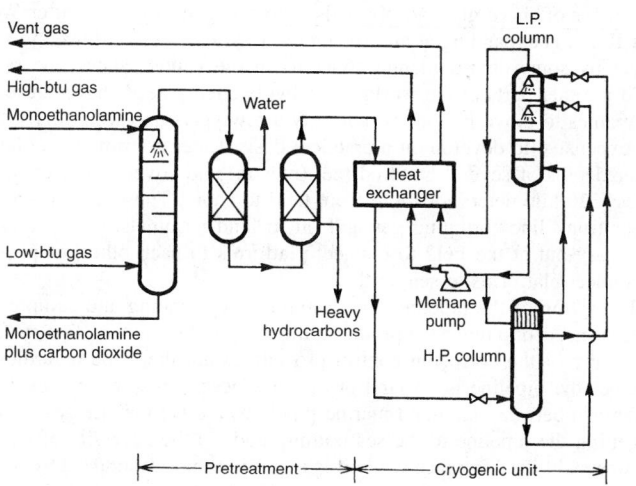

Fig. 6. Plant for nitrogen removal from natural gas using cryogenic upgrading. (*Petrocarbon Developments, Ltd*)

A nitrogen-enriched vapor flows up the H.P. fractionator, while methane is returned to the sump of this column by a nitrogen reflux stream produced in the tubes of the overhead condenser. The refrigeration required to produce this nitrogen reflux is provided by evaporating some of the liquid methane, from the L.P. column, in the shell of the overhead condenser. Two liquid streams are taken from the H.P. fractionator, and these become the feed and reflux for the L.P. fractionator. The L.P. feed is an enriched methane stream taken from the base of the H.P. fractionator. The L.P. reflux is a high purity nitrogen liquid taken off the H.P. fractionator just below the condenser. The upgrading is completed in the L.P. fractionator. The feed stream is stripped to produce a high-Btu liquid containing 4% nitrogen and having a heating value of 980 Btu/standard cubic foot (8722 Calories per cubic meter). The liquid is pumped from the column sump, evaporated, and superheated against the incoming low-Btu gas. The gas from the top of the L.P. column is mainly nitrogen and is also heated to ambient temperature against the incoming low-Btu gas. By using this arrangement of two distillation columns, the separation of nitrogen and methane can be achieved using only the pressure energy available in the low-Btu gas.

Transportation of Natural Gas

The mode selected for gas transportation depends mainly on: (a) the distance over which the gas must be moved; (b) the geographical and geological characteristics of the terrain (considering both overland and overseas [underseas]) across which the gas must be moved; (c) environmental factors directly associated with the gas transportation mode; (d) the physical characteristics of the gas to be transported, notably, the phase—whether gaseous or liquid; and (e) the construction and projected operating costs of the transportation system, based upon trading off the advantages and limitations over which some flexibility of selection may be present. Aside from economic factors, a system can be engineered to transport either the gaseous or liquid phase, thus giving rise to considerable flexibility in certain situations.

Overland Pipelines. Detailed maps of gas pipelines in the United States and other parts of the world can be found in several references, particularly among the periodicals serving the pipeline industry. Notable among these references is the international petroleum encyclopedia and atlas issued periodically by Petroleum Publishing Co., Tulsa, Oklahoma. Numerous trade associations serving the pipeline industry are also excellent sources on pipeline statistics. There are so many pipelines that presentation of this type of information is beyond the scope of this encyclopedia.

Historically, Texas, Louisiana, Oklahoma, and New Mexico have been large producers of natural gas, as well as some significant fields in the West Virginia-Ohio-Pennsylvania area. New developments in Alaska are and will continue to influence the gas transportation and distribution pattern.

Much of the installed gas pipeline ranges from 14 to 30 inches (36 to 76 centimeters) in diameter, the most common ranging from 20 to 35 inches (51 to 89 centimeters) but there is a strong trend toward larger-diameter lines, from 42 inches (107 centimeters) upward. Line pipe is made from high-strength plates, 3/38 inch to 1 inch (1 to 2.5 centimeters) in thickness. Sections of pipe are usually 40 feet (12 meters) long, minimum, ranging up to 60 or 80 feet (18 or 24 meters). Lengths of pipe arrive at the scene most often by truck and are strung out by special pipe carriers along the right-of-way so that the construction crews will find them near the place where they are to be installed. Helicopter delivery of pipe is sometimes used where it is impossible for trucks to do the job. The total weight of steel going into a long-distance pipeline is impressive. For example, a pipe with a wall thickness of 1/2 inch (13 millimeters) and a diameter of 30 inches (76 centimeters) will weigh more than 400 tons (360 metric tons) per mile.

In building very long pipelines, the pipeline company usually employs several construction contractors. The total length of line is divided into a number of sections with separate equipment and crews. Usually, each crew works on not more than 100 miles (161 kilometers). By partitioning the construction task, the entire operation can be speeded up, particularly important in areas where freezing temperatures or rain and mud may interfere with the work.

The numerous machines needed to dig the trench, weld the sections of pipe, apply protective coating to prevent corrosion, lower the pipe into the trench, and cover the trench with earth are known collectively as a *main line spread*. The trench is usually 3 or more feet (1 meter or more) in depth, sufficiently deep to prevent damage by plowing and earth-moving equipment. Depending on the size of the pipe, the trench will range from 2 to 4 feet (0.6 to 1.2 meters) or more in width.

Teams of welders join the pipe sections into a continuous tube. The most modern welding techniques involve automatic welding machines. X-ray equipment is used to inspect welds. When several sections of pipe have been welded together, the continuous tube is lowered gently into the trench by *sideboom tractors*. These machines have cranes or derricks slanted over to one side so that they can pick up the pipe and lower it several feet away from the tractor itself. Pipe purchased from steel mills may come with a coating and wrapping already applied. The thick coating may be of coal tar or asphaltic material, which is then covered with heavy paper or fiberglass. This protective coat-and-wrap is needed to prevent rusting. If bare pipe is used, there are special machines that coat and wrap right on the job just before the pipe is lowered into the trench.

A special piece of equipment, known as the *holiday detector*, is a hoop of metal placed around the pipe after it is coated and wrapped. A small electrical current flows through the hoop. If there is a "holiday," i.e., a spot where there is no coating, the detector alerts the operator. This is brought to the attention of a special crew that coats and wraps bare spots in the pipe.

Since pipelines do not follow an absolutely straight line, bending machines are used to curve the pipe in the vertical, horizontal, or both directions. When a pipeline must cross a river, the contractor will dig or dredge a deep trench in the river bed. The pipe is then surrounded by heavy weights and encased in concrete so that it will not be carried away by the current. If there is a suitable bridge across the river, the pipeline may be hung from the underside of the steel girders of the bridge. In some cases, a special bridge is constructed to carry the pipe across the stream. In crossing a highway or railroad, the pipeline must be put through a tunnel under the structure. A giant auger will be used to bore under the road to accommodate a section of somewhat larger-diameter pipe, forming the tunnel through which the main pipeline passes.

Gas pressures in long-distance pipelines may range from 500 to 5,000 pounds per square inch (34 to 340 atmospheres) with 1,000 psi (68 atmospheres) being quite common. Pressure is boosted to make up for frictional losses by use of compressor stations located every 50 to 100 miles (80 to 161 kilometers) along the pipeline. In terms of lineal velocity, natural gas may travel at a rate of about 15 miles (24 kilometers) per hour; thus, about three days are required to move a molecule of gas over a distance of 1,000 miles (1609 kilometers).

All along the pipeline, there are valves and regulators that may be opened or shut to control the internal pressure, or to cut off the flow entirely if an unexpected break in the line is caused by a flood, earthquake, or other disaster. The valves and regulators can be operated by microwave radio long before any crew could reach them. Stations for reducing the pressure, located near points of consumption, frequently are called *city gates*. These stations measure the amount of gas leaving the main pipeline at this point as well as reducing the pressure.

Marine Pipelines for Gas

With some alterations, the techniques that apply to construction and laying of marine pipelines for gas also apply to fluids, such as oil. Marine pipelines can be underwater in a river, marsh, or ocean, but the predominant industry effort in recent years is the construction of pipelines in the open ocean at increasingly deeper levels. The trend toward deepwater pipelining and construction in harsher environments naturally follows the expansion of the search for offshore gas and oil. This search began in earnest after World War II and is expanding at an ever-increasing rate; even if slowed to some extent by some environmental concerns in the United States, the rate is rapid in other parts of the world. Worldwide energy needs have caused oil companies to move into areas that only a few years ago would have been too expensive to develop on a practical basis. Lines are now being laid in water depths of several hundred feet (meters) and cover distances of 200 miles (320 kilometers) or more from field to shore. These longer lines are major trunk lines bringing gas and oil to land terminals. Other lines are necessary out at the field to connect platforms to each other; or possibly to connect platforms to sea berths.

The sizing of the pipeline, the design of the pumping and compression systems needed to move the products, the design of the automation systems, and many of the corrosion control procedures are the same regardless of whether the pipeline is on land or at sea. The two major areas of design difference between land and marine pipelines are (1) the stresses incurred in getting the pipeline to the sea bottom, and (2) the necessity of keeping the line stable and in place while it is exposed to forces induced by current and wave.

The stability problem is theoretically simple but is complicated somewhat by the uncertainty of precise values for some of the coefficients used in the calculations. Basically, it is a matter of providing enough weight in the pipe and pipe coating system to provide a net downward force when balanced against the buoyance and the lift force caused by the seawater moving by the pipe. This net downward force, in conjunction with the coefficient or friction for the particular pipe-soil combination under examination, can then mobilize a horizontal resisting force. This should be somewhat larger than the drag force exerted on the pipe by the water motion in order to give the desired safety factor.

Different safety factors or horizontal water velocities may be utilized depending on the operating conditions that will be encountered during the life of the pipeline. For example, many pipelines will be buried beneath the sea bottom at some time interval, ranging from a few weeks to a year or two, after their construction. The exposure of this line to maximum horizontal water velocities caused by storm current and waves is obviously much less than that of a line that will remain on the surface of the sea bottom. It is also obviously necessary to consider whether the line will contain gas, oil, or other substance at the time the design loads may occur.

It is important to carefully consider the foregoing points in the design of the weight coating since the ability of a contractor to safely construct the line relates very closely to the negative buoyancy of the pipe and coating.

The most common method of marine pipeline construction utilizes a floating vessel on which the pipe is assembled in a horizontal position. As additional joints or sections of pipe are added to the already-completed segment, the barge is moved forward, actually moving out from under the completed pipeline. This is sometimes called the "stovepipe" method, named after the manner the pipe sections are added, one after another. This pipeline extends off the stern of the vessel and spans down to the sea bottom. It is supported part of the way down by a construction aid called a "pontoon" or "stinger." This is basically a slender structure pinned to the vessel on one end and with built-in buoyancy that can be controlled so that it floats at the proper angle to the water surface to provide support to the pipeline.

In shallow water, the pipeline is then allowed to span from the end of the pontoon to the sea bottom as a simple beam. As water depths increase, it becomes necessary to add tension to the pipe on the barge. This, of course, changes the analytical problem from one of a simple beam to one of a beam under tension. This analysis must take into account the weight of the pipe, the wall thickness, the type of steel in the pipe, the tension on the pipe, the support of the pontoon, the geometrical configuration of the tension on the pipe, the geometrical configuration of the pipe-pontoon-barge system, and the pipe end condition at the sea bottom.

There are three basic configurations of pipelay vessels in common usage: (1) the barge-type hull; (2) the ship-shape hull; and (3) the semi-submersible vessel. The barge-type hull (Fig. 7) is the most common because of its economy and simplicity, its ability to provide the space and stability for heavy lifts and deck cargo, including pipe, and its shallow draft, permitting work close inshore. The primary disadvantage is its relative sensitivity to sea conditions. In particular, roll and heave motions will shut down pipelay operations in 6-foot to 14-foot (1.8- to 4.2-meter) waves, depending on wave direction and period.

Overseas Shipping of LNG. A key feature of most LNG carriers in operation is the insulation system, which maintains the cargo at $-162°C$. In one type of ship, the cargo is carried in five tanks constructed of a thin welded membrane of special steel. Each tank is separated from the inner hull by insulating material. The small fraction of the cargo that boils off

because of heat leakage is used as boiler fuel for the propulsion of the ship. On a loaded voyage, this may provide about 90% of the fuel needed. The ships are ballasted for return voyage with seawater carried in separate wing tanks. Some LNG is left in the cargo tanks to ensure a nonexplosive gaseous atmosphere and to keep the tanks cool for the next voyage. Again, boil-off gas provides part of the propulsion fuel. One configuration of an LNG ship-loading system is shown in Fig. 8.

Safety in handling LNG in ships and at loading and unloading terminals of large scale has been a matter of constant concern. The observation has been made that the LNG gas carried would bury a football field under 125 feet (38 meters) of liquefied gas, or, after conversion to the gaseous phase, 600 football fields to the same depth. One factor that has not been routinely considered in the past is a phenomenon called a *flameless vapor explosion.* It is well known that, if water, for example, could be heated without nucleation occurring on the sides of the vessel, the water temperature could be raised well above the boiling point of $100°C$. If this could be done, and with the continued application of heat, the liquid would suddenly explode in its transition from the liquid to the vapor phase. Although not probable, it is possible that conditions favoring flameless vapor explosion could occur if liquefied natural gas were permitted to escape over a water surface. One scientist has observed that an explosion of this kind is possible when a liquid is 4 to 6% (no more, no less) above its normal temperature of vaporization. It is further observed that an explosion of this nature would not occur when LNG first spreads across a volume of water, but with time and the warming of the LNG such a hazard could occur. While an explosion of this type is not comparable to that from a chemical reaction, the explosion could greatly disperse the LNG over a greater area, thus spreading the zone of risk.

Underground Storage

The largest additional supply of natural gas for peak demands comes from underground storage reservoirs located, for example, close to the northern cities, as compared with the producing wells which may be located in the southwestern area of the country. Some of the storage pools are operated by pipeline companies, but most of the gas in underground storage is owned by the local gas companies that serve metropolitan areas.

The underground reservoirs are filled with gas from the pipelines during the summer months, when all of the fuel that the lines can deliver is not consumed. This method allows the producing wells and the pipelines to operate at fairly steady rates at all times of the year. Also, it is established that a gas field will produce more gas over a longer period if the gas is withdrawn at a steady rate.

Four states—Michigan, Pennsylvania, Illinois, and Ohio—have half of the total underground gas storage pools in the United States, with a total

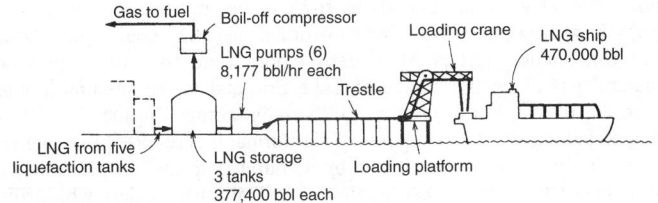

Fig. 8. LNG ship-loading system. Trestle is 2.6 miles (4.3 km) long

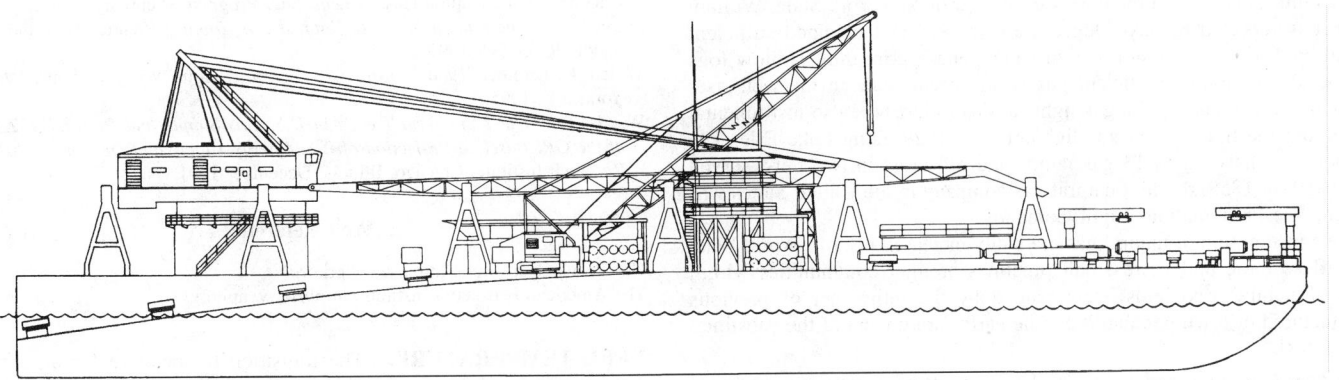

Fig. 7. Pipelay vessel with barge-type hull

capacity of over 5 trillion cubic feet (142 billion cubic meters). In a typical year, about one-fourth of this volume will be used during cold waves to furnish the additional gas needed to supply homes and apartments.

The most common type of underground reservoir now storing gas is a previously producing gas or oil field. The supplies remaining in these pools are too small, and at too low a pressure, to justify continued production. But, the reservoir rock can hold gas pumped down through the same wells that once took gas out of the ground.

About 90% of the storage pools being used once produced gas or oil. In Pennsylvania there are over 60 such pools close to the large industries and centers of population. There are over 30 such pools in West Virginia, Michigan, Ohio, Kansas, Indiana, New York, and Kentucky, as well as smaller numbers in 13 other states. The gas to be stored is pumped into the old wells by compressors similar to those used to move gas in pipelines. The gas is stored under about the same pressure as originally existed in the field. In developing a gas storage reservoir, a company obtains a lease from the landowners in much the same manner that gas producers do.

The gas industry has been developing underground storage reservoirs for more than 60 years. The first known experiment in storing gas underground was conducted in 1915 in Welland County, Ontario, Canada by the National Fuel Gas Company. The success of this effort prompted the Iroquois Gas Corporation, a subsidiary of National Fuel, to develop, in 1916, the Zoar field south of Buffalo, New York. It was the first storage operation in the United States and is the oldest continuously used reservoir.

During the past 60 years, over 80 companies have invested several billions of dollars in underground storage facilities.

Another kind of underground storage reservoir is called an aquifer. An aquifer is an underground rock structure holding large quantities of water. The underground rock is porous and permeable. The pore spaces are filled with water, and impermeable rock covers the porous rock. Wells are drilled into such formations, and gas is forced into the pores under pressure. As the gas pressure increases, the gas pushes the water farther down into the porous rock, making room for the gas.

There are over 40 aquifers in the United States, located in Illinois, Indiana, Iowa, Kentucky, Minnesota, Missouri, Utah, and Washington. Three unusual reservoirs have been developed: an abandoned coal mine in Colorado; and salt domes in Michigan and Mississippi.

History of Natural Gas as an Energy Resource

It is reported that, perhaps 2,000 years ago, the Chinese piped natural gas from shallow wells through bamboo poles, for burning under large pans to evaporate seawater for salt. The first commercial use of natural gas in the western world was for lighting the streets of Genoa, Italy, circa 1802. The first evidence of natural gas deposits in the United States is found in reports of "burning springs" in various parts of New York, Pennsylvania, Ohio, and West Virginia. As early as 1626, French missionaries visiting the Indians in northwestern New York recorded that they could ignite gases rising from shallow waters. Many early reports were given of the presence of natural gas along the shores of Lake Erie and in the streams flowing into it. There also are references to "burning springs" in the Ohio River valley and along the Pacific shores of California. It is reported that General George Washington was fascinated by a "burning spring" in the Kanawha Valley, near Charleston, West Virginia, in 1775. Early settlers who drilled wells for water often reported the presence of traces of natural gas. The generally accepted birthplace of the natural gas industry in the United States is Fredonia, New York. Fredonia is located on Canadaway Creek, which empties into Lake Erie in the northwest corner of New York State. William A. Hart is reported to have dug a well in 1825 and obtained sufficient natural gas to light two stores, two shops, and a grist mill. Hollow logs were used for piping. Sufficient gas would accumulate in the well riser during the day to supply the gas lights at dusk. Hart was also instrumental in building the first natural gas lighthouse in 1829 along Lake Erie. The lighthouse, consisting of 13 gas lamps and reflectors in two tiers, served until 1859. In 1858, the first natural gas company in the United States was formed, the Fredonia Gas Light Company.

The consumption of natural gas gradually increased prior to World Wars I and II as more and more small pipelines brought communities within reach of natural gas fields accompanied by the retirement of previous manufactured or town gas facilities (the early forerunners of the substitute natural gas).

Natural Gas–powered Vehicles. The concept of natural gas–powered vehicles has become a *limited* reality in terms of the millions of gasoline-

and diesel-fluid-power vehicles. In 1993, Mack Trucks (Allentown, Pennsylvania) and the Gas Research Institute have teamed to research and develop a natural-gas version of the Mack E7™ heavy-duty engine. A prototype of the design was scheduled for testing on a refuse vehicle in the Boston area. If successful, the engine also could be applied to a variety of heavy-duty vehicles, including long-haul tractor/trailers, construction equipment, and road maintenance trucks. The development was propelled by the needs of truck fleet owners who may be required by legislation to operate alternatively fueled vehicles. The engine will be required to meet applicable U.S. Environmental Protection Agency and California Air Resources Board emissions standards while maintaining the performance and reliability of its diesel-fueled counterpart. A 6-cylinder, 12-liter engine will be developed. The vehicle's onboard gas storage will hold the energy equivalent of about 45 gallons (170 liters) of diesel fuel.

Substitute Natural Gas. The oil crisis of the 1970s spawned a number of attempts to create synthetic natural gas. Some of these processes reached pilot and demonstration plant stages and beyond in their development. For example, substitute natural gas (SNG) from sewage wastes has enjoyed impressive success. The anaerobic digestion of a solid waste and water or sewage sludge slurry will produce a methane-rich gas.

Additional Reading

Abelson, P.H.: "The Gas Research Institute," *Science*, 1715 (December 11, 1992).

Bethke, C.M. et al.: "Supercomputer Analysis of Sedimentary Basins," *Science*, 261 (January 15, 1988).

Burnett, W.M. and S.D. Ban: "Changing Prospects for Natural Gas in the United States," *Science*, 305 (April 21, 1989).

Castaneda, C.J.: *A History of the Natural Gas Industry,* Macmillan Library Reference, New York, NY, 1999.

Caton, J. (Editor): *Alternative Fuels and Natural Gas, Volume 3,* American Society of Mechanical Engineers, New York, NY, 1995.

Considine, D.M.: *Energy Technology Handbook,* The McGraw-Hill, Companies, Inc., New York, NY, 1977.

Fischetti, M.: "There's Gas in Them Thar Hills!" *Technology Review (MIT)*, 17 (January 1993).

Fulkerson, W., R.R. Judkins, and M.K. Sanghvi: "Energy from Fossil Fuels," *Sci. Amer.*, 136 (September 1990).

Holtberg, P.: *1993 Policy Implications of the GRI Baseline Projection of U.S. Energy Supply and Demand to 2010,* Gas Research Institute (Washington Operations), Washington, DC, 1993.

Jensen, B.A.: "Improve Control of Cryogenic Gas Plants," *Hydrocarbon Processing*, 109 (May 1991).

Lyons, W.C.: *Standard Handbook of Petroleum and Natural Gas Engineering,* Vol. 1, Butterworth-Heinemann, Inc., Woburn, MA, 2001.

Lyons, W.C.: *Standard Handbook of Petroleum and Natural Gas Engineering,* Vol. 2, Butterworth-Heinemann, Inc., Woburn, MA, 2001.

McCabe, K.A., S.J. Rassenti, and V.L. Smith: "Natural Gas Pipeline Networks," *Science*, 534 (October 25, 1991).

Melvin, A.: *Natural Gas: Basic Science and Technology,* Adam Hilger (London), Taylor & Francis (Philadelphia), 1988.

Ourisson, G., P. Albrecht, and M. Rohmer: "The Hopanoids: Paleochemistry and Biochemistry," *Pure and Applied Chemistry*, 51(4), 709–729 (April 1979).

Ourisson, G., P. Albrecht, and M. Rohmer: "Predictive Microbial Biochemistry: From Molecular Fossils in Procaryotic Membranes," *Trends in Biochemical Sciences*, 7, 236–238 (1982).

Ourisson, G., P. Albrecht, and M. Rohmer: "The Microbial Origin of Fossil Fuels," *Sci. Amer.*, 44–51 (August 1984).

Sharer, J.C. and P. O'Shea: "Gas Research Institute's Research Program on Unconventional Natural Gas," *Chem. Eng. Progress*, (February 1986).

Sweetser, R.: *The Fundamentals of Natural Gas Cooling,* Prentice-Hall, Inc., Upper Saddle River, NJ, 1997.

Willett, R. (Editor): *1996 Natural Gas Yearbook,* John Wiley & Sons, Inc., New York, NY, 1995.

Woods, T.J.: *The Long-Term Trends in U.S. Gas Supply and Prices: 1992 Edition of the GRI Baseline Projection of U.S. Energy Supply and Demand to 2010,* Gas Research Institute, Chicago, Illinois, December 1991.

Web References

Gas Research Institute: http://www.gri.org/
The American Petroleum Institute: http://www.api.org/

NÉEL TEMPERATURE. The transition temperature for an antiferromagnetic material. Maximal values of magnetic susceptibility, specific heat, and thermal expansion coefficient occur at the NÉel temperature.

NEF REACTION. Formation of aldehydes and ketones from primary and secondary nitroparaffins, respectively, by treatment of their salts with sulfuric acid.

NEF SYNTHESIS. Addition of sodium acetylides to aldehydes and ketones to yield acetylenic carbinols; occasionally and erroneously referred to as the Nef reaction.

NEGATOL. A condensation product of *m*-cresolsulfonic acid with formaldehyde. A polymerized dihydroxydimethyldiphenylmethanedisulfonic acid. It is dispersible in water, forming very acidic colloidal solutions. The pH of a 5% dispersion is approximately 1.0.

NEGATRON. A term sometimes applied to the normally occurring negatively charged electron when it must be distinguished from a positron. In many parts of the world the name *negaton* is used instead of negatron. The word negatron is used in this encyclopedia wherever distinction is made between positively and negatively charged electrons.

NEMATIC LIQUID CRYSTALS. See **Liquid Crystals**.

NENCKI REACTION. The ring acylation of phenols with acids in the presence of zinc chloride, or the modification of the Friedel-Crafts alkylation-acylation procedure by substitution of ferric chloride for aluminum chloride.

NENITZESCU INDOLE SYNTHESIS. Hydrogenative acylation of cycloolefins with acid chlorides in the presence of aluminum chloride; with five- and six-membered rings, no change in ring size occurs, but with seven-membered rings, rearrangement takes place with formation of a cyclohexane derivative.

NEODYMIUM. [CAS: 7440-00-8]. Chemical element symbol Nd, at. no. 60, at. wt. 144.24, third in the Lanthanide Series in the periodic table, mp 1,016°C, bp 3,068°C, density 7.004 g/cm^3 (20°C). Elemental neodymium has a close-packed hexagonal crystal structure at 25°C. The pure metallic neodymium is silver-gray in color, the luster becoming dull upon exposure to moist air at room temperatures. When pure, the metal is soft and malleable and may be worked with ordinary equipment. Because the metal is pyrophoric, it must be stored in an inert atmosphere or vacuum. There are seven natural isotopes, ^{142}Nd through ^{146}Nd, ^{148}Nd, and ^{150}Nd. ^{144}Nd is mildly radioactive with a half-life of 10^{10}–10^{15} years. Seven artificial isotopes have been produced. Of the light (or cerium-group) rare-earth metals, neodymium is the third most plentiful and ranks 60th in abundance of elements in the earth's crust, exceeding tantalum, mercury, bismuth, and the precious metals, excepting silver. The element was first identified by C.A. von Welsbach in 1885. Electronic configuration $1s^2 2s^2 2p^6 3s^2 3p^6 3d^{10} 4s^2 4p^6 4d^{10} 4f^3 5s^2 5p^6 5d^1 6s^2$. Ionic radius Nd^{3+} 0.995 Å. Metallic radius 1.821 Å. First ionization potential 5.49 eV; second 10.72 eV.

Other important physical properties of neodymium are given under **Rare-Earth Elements and Metals**.

Primary sources of the element are bastnasite and monazite, which contain from 15 to 25% neodymium. Plant capacity involving liquid-liquid or solid-liquid organic ion-exchange processes for recovering the element is in excess of 200,000 pounds (90,720 kilograms) Nd$_2$O$_3$ annually. Metallic neodymium is obtained by electrolysis of fused anhydrous NdCl$_3$ or by the electrolytic reduction of the oxide in molten NdF$_3$.

Use of elemental neodymium as a colorant for glass was one of the early applications. The color ranges from pure violet to purple and finds use in sunglasses, protective glasses for industry, art objects of glass, tableware, and decorative fiber optics. Use of neodymium in amounts of 3–5% by weight imparts dichroic properties to glass. Neodymium-doped single-crystal yttrium-aluminum oxide garnets (Nd:YAG) have been used in lasers. Research has shown the Nd ion to exhibit laser characteristics in a wide range of compounds and glasses. A formulation of 75% neodymium and 25% praseodymium, frequently called didymium, is used as a metallurgical additive. Within the last several years, it has been found that the use of Nd$_2$O$_3$ in barium titanate capacitors increases the dielectric strength of these electronic components over a wider temperature range. Neodymium also has been used as an ingredient of phosphate-type phosphors. Investigations continue into further electronic and optical uses of the element and its compounds.

EDITOR'S NOTE: Extensive research during the early 1980s led to the development of a new and powerful magnet material with the probable composition, R$_2$Fe$_{14}$B (where R = a light rare earth). The rare earth predominantly used thus far is neodymium. The recent neodymium-iron-boron material exhibits extremely powerful magnetic qualities as compared with traditional magnet materials. More detail is given under **Rare-Earth Elements and Metals**. Also see **Magnetism**.

Scientists (California Institute of Technology) reported that the isotopic composition of Drake Passage (Antarctica) seawater had been determined. The Antarctic Circumpolar Current, which controls interocean mixing, flows through the Drake Passage. The ratio, ^{143}Nd/^{144}Nd, was found to be uniform with depth at two experiment stations—with an intermediate value between those of the Atlantic and Pacific Oceans. Further, Piepgras and Wasserburg determined that the Antarctic Circumpolar Current is made up of approximately 70% Atlantic water. It was further reported that cold bottom water from a site in the south-central Pacific has the Nd isotopic signature of the water in Drake Passage. The investigators used a box model to emulate the exchange of water between the Southern Ocean and ocean basis to the north with the isotopic results. An upper limit of about 33 million cubic meters/second was calculated for the rate of exchange between the Pacific and the Southern Ocean. Further determinations of samarium and neodymium were made and found to increase approximately linearly with depth. In essence, the findings suggest that Nd may be a valuable tracer in oceanography and possibly useful in paleo-oceanographic studies. See also Ocean; and Polar Research.

K. A. GSCHNEIDNER, JR.
B. EVANS
Rare-Earth Information Center
Institute for Physical Research and Technology
Iowa State University
Ames, Iowa

Additional Reading

Anderson, D.L.: "Composition of the Earth," *Science*, 367 (January 20, 1989).

Cherfas, J.: "Proton Microbeam Probes the Elements," *Science*, 11500 (September 28, 1990).

DePaolo, D.: *Neodymium Isotope Geochemistry,* Springer-Verlag New York, Inc., New York, NY, 1988.

Greenwood, N.N. and A. Earnshaw: *Chemistry of the Elements,* 2nd Edition, Butterworth-Heinemann, Inc., Woburn, MA, 1997.

Letokhov, V.S.: "Detecting Individual Atoms and Molecules with Lasers," *Sci. Amer.*, 54 (September 1988).

Lewis, R.J., Sr.: *Hawley's Condensed Chemical Dictionary,* 13th Edition, John Wiley & Sons, Inc., New York, NY, 1999.

Lide, D. (Editor): *CRC Handbook of Chemistry and 84th Edition - A Ready- Book of Chemical Reference and Physical Data,* CRC Press, LLC., Boca Raton, FL, 2003.

Lugmair, G.W. et al.: "Samarium-146 in the Early Solar System: Evidence from Neodymium in the Allende Meteorite," *Science*, **222**, 1015–1017 (1983).

Piepgras, D.J. and G.J. Wasserburg: "Isotopic Composition of Neodymium in Waters from the Drake Passage," *Science*, **217**, 207–214 (1982).

Robinson, A.L.: "Powerful New Magnet Material Found," *Science*, **223**, 920–922 (1984).

Staff: *ASM Handbook—Properties and Selection: Nonferrous Alloys and Pure Metals,* ASM International, Materials Park, OH, 1990.

White, R.M.: "Opportunities in Magnetic Materials," *Science*, **229**, 11–15 (1985).

NEON. [CAS: 7440-01-9]. Chemical element, symbol Ne, at. no. 10, at. wt. 20.183, periodic table group 18, mp −248.68°C, bp −246.01°C, density 1.204 g/cm^3 (liquid). Specific gravity compared with air is 0.674. Solid neon has a face-centered cubic crystal structure. At standard conditions, neon is a colorless, odorless gas and does not form stable compounds with any other element. Due to its low valence forces, neon does not form diatomic molecules, except in discharge tubes. It does form compounds under highly favorable conditions, as excitation in discharge tubes, or pressure in the presence of a powerful dipole. However, the compound-forming capabilities of neon, under any circumstances, appear to be far less than those of argon or krypton. No known hydrates have been identified, even at pressures up to 260 atmospheres. First ionization potential, 21.599 eV.

Neon occurs in the atmosphere to the extent of approximately 0.00182%. In terms of abundance, neon does not appear on lists of elements in the earth's crust because it does not exist in stable compounds. However, because of its limited solubility in H$_2$O, neon is found in seawater to the extent of approximately 1.5 tons per cubic mile (324 kilograms per cubic kilometer). Commercial neon is derived from air by liquefaction and fractional distillation. For most applications, the gas need not be in a highly pure form, but may be supplied along with small quantities of the other rare gases, such as argon and krypton. The gas finds principal applications in various electronic devices and lamps, but the most familiar application is the neon tubes used mainly in signs. The use of neon signs for identification and advertising signs reached the Iron Curtain countries at a date much later than in the Western countries. Neon emits the familiar orange light. Neon also has been used in certain lasers.

In the 1983 Luberoff reference, the author observes that neon signs, once considered vulgar symbols of a consumer society, are fast becoming icons of a bygone era. However, in recent years a group of preservationists, people who formerly decried the impact of neon advertising, now often defend it. Luberoff points out how a blue-lettered sign (5878 neon-filled glass tubes) became an integral part of the Boston skyline. A study of Boston's signs and lights by the Boston Redevelopment Authority showed that this sign (Citgo) was the only commercial sign that the public thought should remain.

There are three natural isotopes, ^{20}Ne through ^{22}Ne, and four radioactive isotopes, ^{18}Ne, ^{19}Ne, ^{23}Ne, and ^{24}Ne, all with half-lives of less than 5 minutes. Ramsay and Travers first found the element when investigating the properties of liquid air in 1898. The element is easily identified spectroscopically. Neon emits characteristic red and green lines in its spectrum.

Neon in Meteorites

As pointed out by Lewis and Anders, the noble gases are unique among the elements found in meteorites. They are highly volatile and unreactive and they did not condense in even the most primitive meteorites and thus are present at only a minute fraction of their proportion in the sun, ranging from about 10^{-5} for xenon to 10^{-9} for neon and helium. However, very small quantities of these gases are tightly bound in the meteorite and are freed when the host mineral begins to melt or decompose at high temperatures.

Scientists have found three types of neon in meteorites: (1) Primordial or planetary neon (called neon A); (2) solar neon (neon B), which consists of solar-wind neon ions implanted in meteorites that happen to have been at the surface of their parent body; and (3) cosmogenic neon (neon S), formed when cosmic rays passing through the meteorite spall, or shatter, atomic nuclei in their path. Each type has different proportions of the three isotopes of neon. Although the procedure is too detailed for inclusion here, Lewis and Anders explain how, through the use of stepped heating of meteorite materials, the types of neon can be measured. Their ratios to each other provide clues as to what type of star may have been the source of a given meteorite.

Additional Reading

Anderson, D.L.: "Composition of the Earth," *Science*, 367 (January 20, 1989).

Cherfas, J.: "Proton Microbeam Probes the Elements," *Science*, 11500 (September 28, 1990).

Greenwood, N.N. and A. Earnshaw: *Chemistry of the Elements,* 2nd Edition, Butterworth-Heinemann, Inc., Woburn, MA, 1997.

Letokhov, V.S.: "Detecting Individual Atoms and Molecules with Lasers," *Sci. Amer.*, 54 (September 1988).

Lewis, R.S. and E. Anders: "Interstellar Matter in Meteorites," *Sci. American*, **249**(2), 66–77 (1983).

Lewis, R.J., Sr.: *Hawley's Condensed Chemical Dictionary,* 13th Edition, John Wiley & Sons, Inc., New York, NY, 1999.

Lide, D.R.: *CRC Handbook of Chemistry and Physics* 84th Edition," CRC Press, LLC., Boca Raton, FL, 2003.

Luberoff, D.: *"But Is It Art? (Neon Signs), Technology Review (MIT)*, **86**(5), 76–77 (July 1983).

NEPHELINE.

NEPHELINE. Nepheline, of hexagonal crystallization, is a sodium-potassium aluminum silicate (Na, K)(AlSiO$_4$). It is found in silica-poor geological environments, where there had been insufficient silica to form feldspar. Nepheline rocks are characterized by the absence of quartz within them. They constitute a mineral family group known as the *feldspathoids*. Crystals are extremely rare; usually occurs massive to compact. Luster, is greasy in the massive varieties; vitreous in crystals. Color grades from yellowish to colorless in crystals; gray, green and reddish in massive material. It ranges from transparent to translucent. Hardness is of 5.5–6, specific gravity of 2.55–2.65.

Immense masses of nepheline-rich rocks occur on the Kola Peninsula, the former U.S.S.R., in Norway and in the Republic of South Africa; also in the Bancroft, Ontario, Canada region. Smaller deposits are found in Maine and Arkansas in the United States. Fine crystals are found in lavas on Mt. Vesuvius, Italy.

Nepheline is used extensively in the manufacture of glass.

ELMER B. ROWLEY
Union College
Schenectady, New York

NEPHELOMETRY. Sir John Tyndall noted that particles that are invisible when directly in the path of a strong light become discernible when viewed from the side. Now known as the Tyndall effect, the phenomenon derives from reflection of part of the incident light by the particles. The reflected light is directly proportional to the number of particles in suspension. An instrument for measuring the intensity of reflected light so produced in a nephelometer and may be used for the quantitative determination of small amounts of diverse materials that have the ability to reflect light when in liquid suspension. Examples include the measurement of traces of silver wherein the chloride ion is added to a solution of material containing silver to produce insoluble silver chloride in suspension form. Small amounts of calcium in titanium alloys may be determined by measuring suspensions of the stearate formed in a suitable medium. Nephelometry also finds application in the measurement of bacterial growth rates; for the analysis of cholesterol, glycogen, and enzymes; for controlling the clarity of beverages, water, and wastewater; for solution control in tanning operations; and for any measurement situation where an unknown composition may be transformed into, or related to, a form of suspension.

Nephelometric methods are similar to fluorometric methods in that both involve measurement of scattered light. However, the scattering is inelastic in nephelometry and elastic in fluorometry. Thus, the scattered light measured in fluorometry is of a longer wavelength than the incident light, and both incident and scattered light are of the same wavelength in a nephelometric determination. In fact, the two functions sometimes are combined into one instrument, which may be termed a nefluoro-photometer. When the instrument operates as a nephelometer, it utilizes two Tyndall windows, located opposite each other in a cylindrical sample cell and with their common axis perpendicular to the path of the entering light. The concentration of suspended particles is determined by summing the photocurrents of the two cells. When used as a fluorometer, the instrument measures light emitted by a sample that is excited by incident radiation in the appropriate spectral band. Further, the same instrument can be set up for use as a photometer to measure light transmitted by the sample. Three light sources may be used—an incandescent source for colorimetric or nephelometric applications; a mercury-arc source for fluorometry; and a sodium-arc source with principal emission at 320 and 590 nanometers, when a sharp peak at either of these wavelengths is required, as in the instance of vitamin A determinations.

See also **Analysis (Chemical)**; **Fluorometers**; and **Turbidimetry**.

NEPTUNIUM. [CAS: 7439-99-8]. Chemical element, symbol Np, at. no. 93, at. wt. 237.0482 (predominant isotope), radioactive metal of the Actinide series, also one of the Transuranium elements. Neptunium was the first of the Transuranium elements to be discovered and was first produced by McMillan and Abelson (1940) at the University of California at Berkeley. This was accomplished by bombarding uranium with neutrons. Neptunium is produced as a by-product from nuclear reactors. ^{237}Np is the most stable isotope, with a half-life of 2.20×10^6 years. The only other very long-lived isotope is that of mass number 236, with a half-life of 5×10^3 years.

^{237}Np is parent of the neptunium $(2n + 1)$ alpha decay series. Other isotopes include those of mass numbers 229–235 and 238–241; metastable forms of ^{236}Np, ^{240}Np and two of ^{237}Np are known. Electronic configuration $1s^2 2s^2 2p^6 3s^2 3p^6 3d^{10} 4s^2 4p^6 4d^{10} 4f^{14} 5s^2 5p^5 5d^{10} 5f^5 6s^2 6p^6 \text{-}6d^1 7s^2$. Ionic radii Np^{4+} 0.88 Å; Np^{3+} 1.02 Å (Zachariasen). Oxidation potential Np $\longrightarrow$ Np^{3+} + 3e$^-$, 1.85 V; Np^{3+} $\longrightarrow$ Np^{4+} + e$^-$, −0.155 V; Np^{4+} + 2H$_2$O $\longrightarrow$ NpO$_{2+}$ + 4H$^+$ + e$^-$, −0.739 V; NpO$_2{}^+$ $\longrightarrow$ NpO$_2{}^{2+}$ + e$^-$, −1.137 V. See also **Chemical Elements**.

Neptunium has the oxidation states (VI), (V), (IV), and (III) with a general shift in stability toward the lower oxidation states as compared to uranium. The compounds which are formed are very similar to the corresponding compounds of uranium.

The ionic species corresponding to the oxidation states vary with the acidity of the solution; in acid solution of moderate strength the species are Np^{3+}, Np^{4+}, NpO_2^+, and NpO_2^{2+} as in the case of uranium and plutonium. The potential scheme in 1-M HCl is as follows:

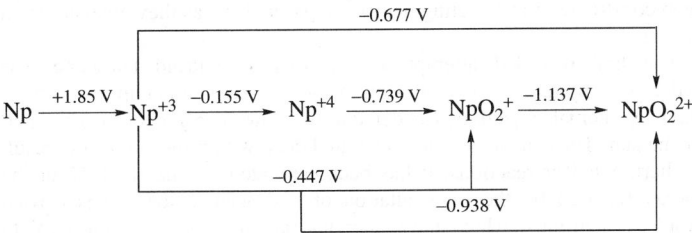

It will be seen that the metal is highly electropositive, in common with the other actinide elements. The $Np^3 \longrightarrow Np^{4+}$ couple is reversible and this oxidation can be accomplished by the oxygen of the air. The (IV) state is stable, not oxidized by air, and only slowly oxidized to NpO_2^+ by nitric acid. The $Np^{4+} \longrightarrow NpO_2^+$ couple is not readily reversible, whereas the $NpO_2^+ \longrightarrow NpO_2^{2+}$ couple is reversible; this is reasonable on the basis that the former involves making or breaking the neptunium-oxygen bonds, whereas the latter does not. The oxidation of NpO_2^+ to NpO_2^{2+} requires moderately strong oxidizing agents. Neptunium differs from uranium and plutonium in that its potential relations are such as to render NpO_2^+ moderately stable with respect to disproportionating, even in solutions containing moderate concentrations of hydrogen ion.

The potentials are altered extensively by change in the hydrogen ion concentration and by the presence of any of a number of anions capable of forming complex ions.

Neptunium ions in aqueous solution possess characteristic colors: pale purple for Np^{3+}, pale yellow-green for Np^{4+} green-blue for Np^{5+}, while NpO_2^{2+} varies from colorless to pink or yellow-green depending on the acid present.

The precipitation reactions of Np^{3+} are similar to those of the tripositive rare earths, those of Np^{4+}, to the other tetrapositive actinides and to Ce^{4+}, and those of NpO_2^{2+} to the corresponding ions of uranium and plutonium. All of the simple salts of NpO^{2+} appear to be soluble.

The neptunium oxide system exhibits complexity similar to that found in the uranium oxide system. Thus, the important oxide is NpO_2 and there exists a range of compositions, depending upon conditions, up to Np_3O_8.

As a metal, Np has a relatively low melting point ($\sim 640°C$), is very dense (20.45 g/cm^3), and is ductile. The alpha form reacts with hydrogen, carbon, oxygen, sulfur, the halogens, and phosphorus to yield a number of binary compounds.

The important halides of neptunium are the trifluoride, NpF_3, purple or black and hexagonal, the hexafluoride, NpF_6, brown and orthorhombic, the trichloride, $NpCl_3$, white and hexagonal, the tetrachloride, $NpCl_4$, red-brown and tetragonal, and the tribromide, $NpBr_3$, α-form green and hexagonal, β-form green and orthorhombic.

In research at the Institute of Radiochemistry, Karlsruhe, West Germany, during the early 1970s, investigators prepared alloys of neptunium with iridium, palladium, platinum, and rhodium. These alloys were prepared by hydrogen reduction of the neptunium oxide in the presence of finely divided noble metals. The reaction is called a *coupled reaction* because the reduction of the metal oxide can be done only in the presence of noble metals. The hydrogen must be extremely pure, with an oxygen content of less than 10^{-25} torr.

Industrial utilization of neptunium has been very limited. The isotope ^{237}Np has been used as a component in neutron detection instruments. Neptunium is present in significant quantities in spent nuclear reactor fuel and poses a threat to the environment. A group of scientists at the U.S. Geological Survey (Denver, Colorado) has studied the chemical speciation of neptunium (and americium) in ground waters associated with rock types that have been proposed as possible hosts for nuclear waste repositories. See Cleveland reference.

Additional Reading

Cleveland, J.M., K.L. Nash, and T.F. Rees: "Neptunium and Americium Speciation in Selected Basalt, Granite, Shale, and Tuff Ground Waters," *Science*, **221**, 271–273 (1983).

Fuger, J. and L.R. Morss: *Transuranium Elements: A Half Century*, American Chemical Society, Washington, DC, 1992.

Keller, C. and B. Erdmann: Preparation and Properties of Transuranium Element–Noble Metal Alloy Phases, Proc. 1972 Moscow Symp. Chem. Transuranium Elements, 1976.

Krot, N.N. and A.D. Gel'man: "Preparation of Neptunium and Plutonium in the Heptavalent State," *Dokl. Chem.* **177**, 1–3, 987–989 (1967).

Lewis, R.J., Sr.: *Hawley's Condensed Chemical Dictionary,* 13th Edition, John Wiley & Sons, Inc., New York, NY, 1999.

Lide, D.R.: *CRC Handbook of Chemistry and Physics* 84th Edition," CRC Press, LLC., Boca Raton, FL, 2003.

Magnusson, L.B. and T.J. LaChapelle: "The First Isolation of Element 93 in Pure Compounds and a Determination of the Half-life of 93Np237," *Amer. Chem. Soc. J.*, **70**, 3534–3538 (1948).

Marks, T.J.: "Actinide Organometallic Chemistry," *Science*, **217**, 989–997 (1982).

McMillan, E. and P.H. Abelson: "Radioactive Element 93," *Phys. Rev.*, **57**, 1185–1186 (1940).

Seaborg, G.T.: "The Chemical and Radioactive Properties of the Heavy Elements," *Chem. Engng. News*, **23**, 2190–2193 (1945).

Seaborg, G.T. and W.D. Loveland: *The Elements beyond Uranium*, John Wiley & Sons, Inc., New York, NY, 1990.

Thayer, J.S. and F.E. Brinckman: *Environmental Chemistry of the Heavy Elements: Hydrido and Organo Compounds,* John Wiley & Sons, Inc., New York, NY, 1995.

NERNST EFFECT. If heat is flowing through a strip of metal and the strip is placed in a magnetic field perpendicular to its plane, a difference of electric potential develops between the opposite edges. This phenomenon, discovered by Nernst in 1886, is analogous to the Hall effect, but with a longitudinal flow of heat replacing the longitudinal electric current. See also **Hall Effect and Quantized Hall Effect**.

NERNST HEAT THEOREM. For a homogeneous system, the rate of change of the free energy with temperature, as well as the rate of change of heat content with temperature, approaches zero as the temperature approaches absolute zero.

NERNST, HERMANN WALTHER (1864–1941). Nernst was a German Chemist and Physicist. In 1894 he received invitations to the Physics Chairs in Munich and in Berlin, as well as to the Physical Chemistry Chair in Göttingen. He accepted this latter invitation. At Göttingen Nernst founded the Institute for Physical Chemistry and Electrochemistry and became its Director. In 1905 he was appointed Professor of Chemistry, later of Physics, in the University of Berlin, becoming Director of the newly-founded "Physikalisch-Chemisches Institut" in 1924. He remained in this position until his retirement in 1933.

He made major contributions to electrochemistry, thermodynamics, and photochemistry. Nernst's early studies in electrochemistry were inspired by Arrhenius' dissociation theory which first recognized the importance of ions in solution. His heat theorem, known as the Third Law of Thermodynamics, was developed in 1906. In 1918 his studies of photochemistry led him to his atom chain reaction theory. In later years, he occupied himself with astrophysical theories, a field in which the heat theorem had important applications.

He is remembered best for the Nernst Effect, named after him. For his work in thermochemistry Nernst won the Nobel Prize in Chemistry in 1920.

See also **Nernst Effect**; **Nernst Heat Theorem**; **Nernst-Thompson Rule**; and **Thermodynamics**.

J.M.I.

NERNST-THOMPSON RULE. A solvent of high dielectric constant favors dissociation by reducing the electrostatic attraction between positive and negative ions, and conversely a solvent of low dielectric constant has small dissociating influence on an electrolyte.

NERVE BLOCK. See **Anesthetics**.

NEUROREGULATORS. Neuroregulators represent a diverse group of compounds that include both neurotransmitters and neuromodulators. Receptors for neuroregulators on both the cell surface and within the cell represent the molecular targets of the majority of drugs in clinical use. In order to classify an endogenous agent as a neurotransmitter, it must possess a number of general characteristics, which are described. Neurotransmitter receptors are grouped according to sequence homology and structures are given. In many cases human receptors have been cloned.

NEUTRALIZATION. A chemical reaction in which water is formed by mutual interaction of the ions that characterize acids and bases when both are present in an aqueous solution, i.e., $H^+ + OH^- \longrightarrow H_2O$, the remaining product being a salt. R. T. Sanderson states: "An aqueous solution containing an excess of hydronium ions is called acidic. It readily releases protons to electron-donating substances. An aqueous solution containing an excess of hydroxyl ions is called basic. It readily accepts protons from substances that can release then, and is in general an excellent donor. No aqueous solution can contain an excess of both hydronium and hydroxyl ions, because when these ions collide, a proton is immediately transferred from the hydronium to the hydroxyl ion, and both become water molecules."

Neutralization occurs with both (1) $Ca(OH)_2 + H_2SO_4 \longrightarrow CaSO_4 + 2H_2O$; (2) $HCOOH + NaHCO_3 \longrightarrow HCOONa + CO_2 + H_2O$. It should be noted that neutralization can occur without formation of water, as in the reaction $CaO + CO_2 \longrightarrow CaCO_3$. Neutralization does not mean the attaining of pH 7.0; rather it means the equivalence point for and acid-base reaction. When a strong acid reacts with a weak base, the pH will be less than 7.0, and when a strong base reacts with a weak acid, the pH will be greater than 7.0. See also **Acids and Bases**.

NEUTRINO. A neutral particle of very small (presumed zero) rest mass and of spin quantum number $\frac{1}{2}$. This particle was initially postulated to account for the continuous energy distribution of beta particles and to conserve angular momentum in the beta-decay process. Experimental evidence indicates that, for the linear momentum to be conserved in the beta process, there must be a contribution from a departing neutrino. Presumably, a neutrino (or antineutrino) is emitted in every beta transition. The energy of a neutrino emitted in a beta disintegration is assumed equal to the difference between the energy of the particular beta particle and the energy corresponding to the upper limit of the continuous spectrum for that beta transition. The neutrino has also been postulated as one of the particles in pion (π) decay and as two of the particles in muon (μ) decay. These processes, however, lead to two types of neutrinos: an electron-associated neutrino ν_e and a muon-associated neutrino ν_μ. For example, $\pi^+ \rightarrow \mu^+ \rightarrow \overline{\nu_\mu} + e^+ + \nu_e$ or $\pi^+ \rightarrow e^+ + \nu_e$, whereas neutron decay obeys only the process $n \rightarrow p^+ + e^- + \overline{\nu_e}$, where the bar over ν indicates an anti-particle. The difference between neutrinos was established at Brookhaven in 1962 when it was shown that a beam of neutrinos from the process $\pi^+ \rightarrow \mu^+ + \nu_\mu$ gave rise to the process $\nu_\mu + n \rightarrow p + \mu^-$ but not to $\nu_\mu + n \rightarrow p + e^-$. There is also a neutrino associated with the tau particle. Because of its properties, the neutrino has negligible interactions with matter and has proved difficult to detect. It was first positively identified experimentally in 1956 by Reines and Cowan, Jr. See also **Particles (Subatomic)**.

The term antineutrino usually denotes an antiparticle whose emission is postulated to accompany radioactive decay by negatron emission, such as, for example, in neutron decay into a proton p^+, negatron e^- and antineutrino $\overline{\nu_e}$, expressed by the equation $n \rightarrow p^+ + e^- + \overline{\nu_e}$. Capture of a neutrino by the neutron, $\nu_e + n \rightarrow p^+ + e^-$ would be an equally good description of the process. Positron emission is accompanied by a neutrino, as in the decay $^{64}Cu \longrightarrow ^{64}Ni + e^+ + \nu_e$. Orbital electron capture also involves a neutrino, as for example, $e^- + ^{64}Cu \longrightarrow ^{64}Ni + \nu_e$. Since there is no possibility of charge differentiation between the antineutrino and the neutrino, differentiation between these two particles can be made only on the basis of such properties as the sign of the ratio of magnetic moment to angular momentum.

In the past the terms neutrino and antineutrino were sometimes used in reverse sense to that stated above, i.e., the neutrino is said to accompany negatron emission and the antineutrino, positron emission. The preferred usage has been accepted in order to provide conservation of leptons in the conservation laws.

Neutrino Astronomy. For several years, there has been a marked trend in astronomy to expand the use of the electromagnetic spectrum beyond the visual range in investigating the universe. It now appears that, in addition to infrared, ultraviolet, gamma ray, x-ray, and radio astronomy, among others, increasing attention will be given to neutrino astronomy, i.e., the detection and measurement of neutrinos emanating from such celestial bodies as supernovas and x-ray double stars. Traditionally, neutrinos have been created in large accelerators and used for investigating the characteristics of other elementary particles. It is now reasoned that neutrinos entering the earth's atmosphere from celestial distances may furnish new, heretofore unavailable information.

Search for Neutrinos

Certain mysteries continue to surround the neutrino, emphasizing how far physicists still may be from a full comprehension of the complete array of subatomic particles. Numerous detectors in different locations have been constructed for detecting neutrinos, particularly as they emanate from the sun.

The first recorded attempt to construct a neutrino telescope was undertaken by Davis (Brookhaven National Laboratory) in the 1960s. Davis' principal objective was detection of low-energy neutrinos emitted by the sun. These neutrinos are generated deep within the sun as the result of thermonuclear reactions. It has been estimated that nearly 10% of the energy released by the transmutation of hydrogen in the sun is carried away by neutrinos, which have energies ranging from $\frac{1}{2}$ million eV to about 14 million eV. The solar flux of neutrinos is tremendous. It is estimated that 10^{14} solar neutrinos pass through the human body every second. Davis' detector consisted of a large tank containing 610 tons of tetrachloroethylene, C_2Cl_4. Of the chlorine atoms present, 25% are of the isotope chlorine 37. When this atom captures a neutrino, it is transformed to an atom of argon 37.

An early detector was established in the shaft of the Homestake Mine (South Dakota) in 1968. The data gathered did not correspond with the scientists' expectation. The neutron flux measured was less than one-third that predicted. This discrepancy challenged researchers to question perhaps the "established" concepts of particle physics and indeed the manner in which the sun functions.

Some years later, a Japanese-built detector (*Kamiokande II*), designed to detect the more energetic solar-emitted neutrinos (~5 mil electron volts), came up about 50% short of the expected counts, thus reconfirming a shortage of solar neutrons.

Neither of the aforementioned detectors was designed to sense comparatively low-energy (proton-proton) neutrinos, which result from the fusion of two protons.

Subsequently, two additional detectors were built with the objective of seeking lower-energy neutrinos. One of the detectors was located in the Caucasus of the former Soviet Bloc and was named the Soviet-American Gallium Experiment (SAGE). The detector used a gallium metal detector believed to be sensitive to neutrinos, with energy as low as 0.23 mil electron volts. Another similar detector (*Gallox*) was constructed in Italy. At both sites, difficulties with calibration were expected and did occur. Some tenuous observations later were made to suggest a correlation of neutrinos detected with the occurrence of sun spots and with solar acoustic oscillations.

Based upon the foregoing experiences, some researchers observed that the same reluctance to interact with matter is responsible for the neutrino's long range and ability to resist detection. Thus, it was reasoned that an apparatus for detecting neutrinos should be massive and shielded from the interference of other particles and radiation. As a solution to these problems, some researchers proposed a deep underwater muon and neutrino detector (acronym DUMAND).

Fiber optic data cable will stretch from the shore for 30 km to a connector box some 4800 meters below the ocean surface. Strings of nine separate cables will rise vertically about 280 meters above the ocean floor. Each cable, held up by a float, will contain 24 detectors. The apparatus, referred to as a neutrino telescope, will have to pick up at least ten muon events per year from any given 1° patch of sky for the DUMAND scientists to be confident that they have a significant neutrino source, not just a few background pulses from non-neutrino cosmic rays.

Construction of DUMAND, with Japanese and German collaborators, got underway off the west coast of Kona (Hawaii) in 1990 and was expected to be operational within a few years. V. Stenger (University of Hawaii) observes, "The idea of using the ocean floor as a detector actually goes back to the 1960s and people started taking it seriously back in the mid 1970s."

A number of scientists are hopeful that the neutrino telescope will ultimately yield much additional information on neutron stars, supernovas, quasars, etc., which are presumed to be large emitters of neutrinos. Some astronomers estimate that SS 433 puts out 1000 times more energy per second than the brightest stellar object known in the galaxy. Why should such a powerful object have been discovered so recently? The fact is that SS 433 is a comparatively weak source of photons. The accreting matter that gives SS 433 its great power also serves to screen its bright central region from view. Much remains to be learned concerning SS 433 and perhaps neutrino astronomy may supply some answers at a future date.

Laboratory Experimentation on Neutrinos

During the 1960s, L.M. Lederman, M. Schwartz, and J. Steinberger conducted the well-known two-neutrino experiment, which established a relationship between particles, muon and muon neutrinos, electron and electron neutrino. This later evolved into the standard model of particle physics. The Nobel prize in physics was shared by these researchers in 1988.

Massive Neutrino Proposed. J. Simpson (University of Guelph, Ontario) in the late 1980s presented evidence of the existence of a *heavy neutrino* having a mass of 17,000 electron volts (keV, the units of energy that are interchangeable with mass). A renowned neutrino physicist for several years prior, in 1985 Simpson conducted a series of experiments in his laboratory. The objective of his initial experiment was that of measuring the energy of electrons emitted from heavy hydrogen (tritium) in the radioactive process of beta decay. In this process, the energy of the emitted electrons should appear as a spectrum (smooth curve) from zero to a maximum endpoint. However, Simpson noted "occasional" aberrations in the plotted data. This so-called "kink" corresponded with an energy of 17 KeV of the normal plot. This indicated the probable presence of an unknown massive force. In addition to repeated experiments with tritium, Simpson also used Sulfur-35. Simpson then suggested that the aberration, when it occurred, could be attributed to a heavy 17 KeV neutrino.

Simpson's carefully prepared data, when published, attracted wide attention and drew a wide spectrum of professional reactions. Meanwhile, experiments by other laboratories have confirmed Simpson's results. Researchers at the Lawrence Berkeley Laboratory, using Carbon-14, have reported evidence for a 17.2 KeV neutrino in 1.4% of their experiments. Researchers at the Ruder Boskovic Institute in Zagreb (formerly Czechoslovakia), using Iron-55 and Germanium-71, also have found the evidence appearing in 1.5% of their experiments. Thus, confidence in Simpson's claim appeared to be gaining momentum in the early 1990s.

As pointed out by M. Turner (University of Chicago), a massive neutrino would "violate every theoretical prejudice we have in particle physics, astrophysics, and cosmology." J. Bahcall (Institute of Advanced Study at Princeton) observes, "It's a surprise. If it's true, then it's pointing us in a different direction than previous physics suggested."

See also **Particles (Subatomic)**.

Additional Reading

Bahcall, J.N.: *Neutrino Astrophysics,* Cambridge University Press, New York, NY, 1990.
Breuker, H. et al.: "Tracking and Imaging Elementary Particles," *Sci. Amer.,* 58 (August 1991).
Brown, L.M., M. Dresden, and L. Hoddeson: *Pions to Quarks: Particle Physics in the 1950s,* Cambridge University Press, New York, NY, 1989.
Caldwell, D.O.: *Current Aspects of Neutrino Physics,* Springer-Verlag New York, Inc., New York, NY, 2001.
Cence, R.J. et al.: *Neutrino 81,* University of Hawaii, Honolulu, HI, 1981.
Dar, A.: "Astrophysics and Cosmology Closing in on Neutrino Masses," *Science,* 1529 (December 14, 1990).
Dehmelt, H.: "Experiments on the Structure of an Individual Elementary Particle," *Science,* 539 (February 2, 1990).
Florini, E.: *Neutrino Physics and Astrophysics,* Plenum, New York, NY, 1982.
Gutbrod, H. and H. Stöcker: "The Nuclear Equation of State," *Sci. Amer.,* 58 (November 1991).
Horgan, J.: "Three Americans Honored for 1960's Neutrino Experiment," *Sci. Amer.,* 31 (December 1988).
Kim, C.W. and A. Pevsn: *Neutrinos in Physics and Astrophysics,* Gordon & Breach Publishing Group, Newark, NJ, 1993.
Lederman, L.M.: "Observations in Particle Physics from Two Neutrinos to the Standard Model," *Science,* 664 (May 12, 1989).
Mohapatra, R.N. and P.B. Pal: *Massive Neutrinos in Physics and Astrophysics,* World Scientific Publishing Company, Inc., River Edge, NJ, 1997.
Powell, C.S.: "Looking for Nothing: The Taciturn Neutrino Keeps Physicists Guessing," *Sci. Amer.,* 22 (April 1991).
Schwartz, M.: "The First High-Energy Neutrino Experiment," *Science,* 1445 (March 17, 1989).
Selvin, P.: "Is There a Massive Neutrino?" *Science,* 1426 (March 22, 1991).
Stenger, V.J. and J.G. Learned: *High Energy Neutrino Astrophysics: Proceedings of the Workshop,* World Scientific Publishing Company, Inc., Riveredge, NJ, 1992.

Traweek, S.: *Beamtimes and Lifetimes. The World of High Energy Physics,* Harvard University Press, Cambridge, MA, 1992.
Waldrop, M.M.: "A Nobel Prize for the Two-Neutrino Experiment," *Science,* 669 (November 4, 1988).
Waldrop, M.M.: "Solar Neutrino Deficit Confirmed?" *Science,* 1607 (June 29, 1990).
Waldrop, M.M.: "Astrophysics in the Abyss," *Science,* 208 (October 12, 1990).
Winter, K. and G. Altarelli: *Neutrino Mass,* Springer-Verlag New York, Inc., New York, NY, 2003.

Web Reference

The Sudbury Neutrino Observatory: http://www.sno.phy.queensu.ca/

NEUTRON. The discovery of the neutron by Chadwick in 1932 represented a great step forward in the investigation of nuclei of atoms. Chadwick found that a radiation emitted when α-rays from polonium reacted with beryllium could project protons from a thin sheet of paraffin wax. Although the radiation itself produced no observable ionization when passing through a gas, the protons released from the paraffin were detected in an ionization chamber. Inability to produce ionization was interpreted as a lack of electric charge. From measurements of the ionization from the protons, Chadwick deduced that the so-called beryllium radiation must consist of neutral particles with a mass very nearly equal to that of the proton. He announced the discovery of the neutron, a previously unknown particle. It has been confirmed that the neutron has no charge and a mass of 1.088665 atomic mass units. Thus, it is heavier than the proton by 0.00139 mass unit. The introduction of the neutron into nuclear structure produced a sharp change in previously held concepts. Lacking knowledge of the neutron, masses of atomic nuclei had been attributed solely to protons. The number of protons required on this basis for most nuclei greatly exceeded the known charge number. In an attempt to solve this dilemma, a number of electrons were assigned to each nucleus to adjust the charge number to the proper value. This compromise created an even greater problem, that of accommodating so many electrons in the small space occupied by a nucleus. Bringing the neutron into the picture meant that a nucleus contains only enough protons to equal the charge number, with the rest of the mass contributed by neutrons. No additional electrons were required.

Decay. The neutron in the free state undergoes radioactive decay. Elaborate experiments by Robson were required to identify the products of the decay and to measure the half-life of the neutron. He showed that the neutron emits a β-particle and becomes a proton. The half-life was found to be 12.8 minutes. In stable nuclei, neutrons are stable. In radioactive nuclei, decaying by β-emission, the neutrons decay with a half-life characteristic of the nuclei of which they are a part. See also **Radioactivity**.

Detection. Because it is a neutral particle the neutron is detected by means of a secondary charged particle which it releases in passing through matter or by means of the radioactivity which the neutron can induce in stable elements. Protons may be projected by collisions with neutrons in hydrogenous material and the ionization from the protons can be measured in an ionization chamber, as in the original experiment with neutrons. Secondary charged particles may be the direct result of nuclear disintegration produced by neutrons, as in the case of the reaction $^{10}B + {}^{1}n \rightarrow {}^{7}Li + \alpha$. Commonly, the radioactivity induced in a stable element by neutron capture serves to detect neutrons, and this technique is known as the *activated foil method*. Also, fission may be utilized for detection of neutrons by placing fissionable material inside an ionization chamber and observing the ionization generated by the fission fragments.

Energies. The kinetic energy of neutrons has an important bearing on their behavior when interacting with nuclei. These kinetic energies may range from near zero to as much as 50 MeV. It is therefore natural to classify neutrons in terms of energy according to their properties in each range of energy. For example, energies from zero to about 1000 eV are usually called *slow neutrons.* Because they are more readily captured by nuclei than faster neutrons, slow neutrons are responsible for a large number of nuclear transformations. When slow neutrons have velocities in equilibrium with the velocities of thermal agitation of the molecules of the medium in which they are situated, they are called *thermal neutrons.* The distribution of these velocities approaches the Maxwell distribution

$$dn(v) = Av^2 e^{-(Mv^2/2kT)} dv$$

where v is the neutron velocity, M its mass, k is Boltzmann's constant, and T is the absolute temperature. In the slow neutron range of energies, various atomic nuclei show strong absorption (capture) of neutrons at fairly well-defined energies. Neutrons having energies corresponding to those of the absorption bands are called *resonance neutrons*. Frequently, neutrons with energies greater than 1000 eV and less than 0.5 MeV are termed *intermediate neutrons*. In more general terms, all neutrons with energies greater than 0.5 MeV are called *fast neutrons*. The practical upper limit of neutron energy is set by the devices thus far developed for accelerating charged particles to extremely high energies.

Magnetic Moment and Spin. Alvarez and Bloch succeeded in measuring the moment of the magnetic dipole associated with the known spin of $\frac{1}{2}$ possessed by the neutron. More refined measurements by Cohen, Corngold, and Ramsey of the magnetic moment μn yielded a value of

$$\mu n = 1.913148 \text{ nuclear magnetons}$$

Interactions with Nuclei. Neutrons may be scattered or captured by heavy nuclei. Scattering may be elastic, resulting only in the change of direction of the neutrons, or inelastic, in which the neutron loses part of its energy to the scattering nucleus. Collisions with light nuclei, in the absence of capture, result in communicating considerable fractions of the neutron energy to the target nucleus. A neutron colliding head-on with a proton will give practically all its kinetic energy to the proton. As the mass of the target nucleus increases, the transfer of energy decreases, in accordance with the laws of conservation of energy and momentum. The loss of energy by mechanical impact is utilized in slowing down fast neutrons, a process known as *moderation*. Slow neutrons are most useful, for example, in the production of radioelements from stable elements by neutron capture. A good moderator should have low mass and a small capture cross section. The rate r of capture of neutrons from a neutron flux F (neutrons cm^{-2} sec^{-1}) incident on a layer of matter having N nuclei per square centimeter is given by

$$r = F\sigma N$$

where σ is the complete probability of capture. Replacing r by dN/dt and writing the flux as nv, where n is the number and v is the velocity of the neutrons, we have

$$dN/dt = nv\sigma N$$

which integrated gives

$$N = N_0 e^{-nv\grave{e}t}$$

where N is the number of unchanged nuclei in the target area at time t and N_0 is the number at time $t = 0$. The cross section σ is so named because it has the dimensions of an area. The unit for the cross section is the *barn*, equal to 10^{24} cm^2. When, as is often the case, σ is proportional to $1/v$, the advantage of slow neutrons in capture interactions becomes apparent. When the value of s departs sharply from that predicted by the $1/v$ law, it usually increases over a narrow range of energies, and we have what is called a *resonance*. Slow neutron cross sections are customarily quoted for thermal neutrons at $20°C$, corresponding to a value of v of 2200 m/sec. Under these conditions, representative thermal neutron capture cross sections are: boron, 759 barns; cobalt, 38 barns; cadmium, 2450 barns; gadolinium, 46,000 barns; gold, 99.8 barns; helium, 0; lead, 0.170 barn; and oxygen <0.0002 barn.

Additional interactions of neutrons with nuclei include the release of charged particles by neutron-induced nuclear disintegration. Commonly known reactions are $n - p$, $n - d$, and $n - \alpha$. In these cases, the incident neutrons may contribute part of their kinetic energy to the target nucleus to effect the disintegration. Hence, more than mere neutron capture is involved. Then, there is usually a lower threshold for the neutron energy below which the reaction fails to occur. Another important reaction involving neutrons is fission, which may occur under different conditions for either slow or fast neutrons with appropriate fissionable material.

Sources of Neutrons. Any nuclear reaction in which neutrons are released might serve as a source of neutrons. In the initial experiments with neutrons, an $\alpha-n$ reaction was used. Because of the charge on the α-particle, it must have a high kinetic energy to penetrate a nucleus. Thus, polonium α-particles could release neutrons from beryllium. Such a natural source produces relatively few neutrons. The yield of neutrons from charged particle reactions can be increased manyfold by the use of particle accelerators. Here large numbers of charged particles of high energy can be used in the bombardment of the target to release numerous neutrons. Frequently deuterons or protons are used for the bombardment. A far more prolific source is the nuclear reactor. Fission of uranium is usually the source of the neutrons in this case. A nuclear reactor as usually constructed generates neutrons of different energies in various parts of its structure. Neutrons of suitable energy for a given experiment may be brought outside the reactor through channels into appropriate sections of the reactor. See also **Nuclear Power Technology**.

Traditionally, the neutron is regarded as a particle which is a component of nuclei and which exists only briefly in the free state. For many purposes, this view is sufficient. However, it became obvious some years ago from various experiments, for example, in very high energy accelerators, that the neutron must have a complex structure. This view was reinforced by the nature of the decay of the neutron. A ς particle is ejected from the neutron on decay, but it is quite certain that the electron did not exist within the neutron prior to the decay. Rearrangements of an internal structure of the neutron must provide the energy for the formation and ejection of the ς particle. An early theory would have the neutron consist of a proton and a π^- meson bound together so that they oscillate between a completely bound state and a more loosely bound state. This concept might explain the feeble interaction that has also been observed between electrons and neutrons at very short range.

Fermi Age Model. This is a model for the study of the slowing down of neutrons by elastic collisions. It is assumed that the slowing down takes place by a very large number of very small energy changes. Phenomena due to the finite size of the individual losses are ignored. In this model, the word age is somewhat of a misnomer, since its units are those of area rather than time. The name arises because the variable τ, the Fermi age, appears in the Fermi age equation in the same way that time appears in the standard heat-diffusion equation. The equation for the Fermi age in a unit volume of nuclear reactor is $\tau = D\phi/q$, in which D is the diffusion coefficient for fast neutrons, $\phi = nv$ is the fast neutron fluence, and q is the number of neutrons thermalized per second per cubic centimeter. For this purpose fast neutrons are inclusively all neutrons with energies between those acquired at fission and that energy at which they are thermalized.

Ultracold Neutrons. As pointed out by King (Massachusetts Institute of Technology), there is probably as much to be learned between the lowest energy yet reached and zero energy as there is between the highest energy attained and infinite energy. Ultracold neutrons may provide an avenue to very-low-energy research. It should be recalled that a neutron with an energy of 10^{-7} electron volt is at the low end of the energy scale. A neutron has the energy that would be imparted to an electron by a potential difference of one ten-millionth of a volt (0.1 microvolt). As pointed out by Golub et al. in 1979, this is the amount of energy of a particle in a gas whose temperature is one millidegree K. Unlike high-energy particles, ultracold neutrons move at a rate measured in a few meters per second. Golub et al. have proposed that inasmuch as ultracold neutrons cannot penetrate a solid surface, they can be confined in a metal bottle and by storing over long periods, it may be possible to measure the fundamental properties of the neutron.

Outside the nucleus, the neutron is an unstable particle. Free neutrons are rare in nature. By the process of beta decay, the neutron breaks down into a proton, an electron, and a neutrino (a massive particle). For probing atomic and molecular structures, thermal neutrons have traditionally been used. As explained by Golub et al., ultracold neutrons can be employed in a similar way, but their low energy and long wavelength adapt them to the examination of materials on a somewhat larger scale. At the Technical University (Munich), Steyeri and associates have developed a neutron spectrometer. In conventional neutron spectrometers, particles are analyzed by a magnet that bends their trajectories. In an ultracold-neutron spectrometer, the earth's gravitational field is used. In the device, the neutrons enter the spectrometer in a horizontal movement and are accelerated as they fall a fixed distance to a specimen. Those neutrons that rebound from a target are collected by an exit slit of the instrument. An exchange of energy with the specimen is reflected by the maximum height to which the neutrons rebound.

During the past decade, since their detection, much research has been directed toward methods of extracting, storing, and manipulating ultra-cold neutrons. It now appears that the next period will be one of investigating the neutron per se and possibly of using this new knowledge for the study of other systems of particles.

See **Particles (Subatomic)** for more recent views. See also **Proton**.

Glossary of Neutron Terms

Neutrons are designated according to their energies, including the following:

Thermal neutrons, or neutrons in thermal equilibrium with the substance in which they exist; most commonly, neutrons of kinetic energy about 0.025 eV, which is about $\frac{2}{3}$ of the mean kinetic energy of a molecule at 15°C.

Epithermal neutrons, or neutrons having energies just above those of thermal neutrons; the epithermal neutrons energy range is between a few hundredths eV and about 100 eV.

Slow neutrons (a less definite classification), which may mean either neutrons having energies up to about 100 eV, or thermal neutrons.

Intermediate neutrons, which are neutrons having energies in a range that extends roughly from 100 to 100,000 eV. This range is above that of epithermal neutrons and below that of fast neutrons.

Fast neutrons, which are neutrons with energies exceeding 10^5 eV, although sometimes a lower limit is given.

Resonance neutrons may be either of the following: (1) for a specified nuclide or element, neutrons that have energies in the region where the cross section of the nuclide or element is particularly large because of the occurrence of a resonance. For example, cadmium resonance neutrons have energies between 0.05 and about 0.3 eV. (2) Neutrons having kinetic energies in the region of values for which prominent resonances are encountered in many nuclides; loosely, epithermal neutrons.

Prompt neutrons are those neutrons released coincident with a fission process.

Delayed neutrons are those neutrons released subsequently in a fission process, or, more generally, neutrons emitted by excited nuclei formed in any radioactive process (beta disintegration, in all cases so far known). The neutron emission itself is prompt, so that the observed half life is that of the preceding beta emitter. The situation is similar to that involving gamma-ray emission, which is a competing process. Delayed neutron emission is possible only if the excitation energy of the product nucleus exceeds the neutron binding energy for that nucleus. The chemistry of the delayed neutron emitter is that of the beta activity; thus ^{87}Br, ^{137}I, and ^{17}N are delayed neutron precursors, although the neutron emission actually takes place from excited nuclei of the products ^{87}Kr, ^{137}Xe, and ^{17}O.

Neutron cycle is the average life history of a neutron in a nuclear reactor. The gain in the number of neutrons in a reactor during any individual neutron cycle is given by $n(k-1)$, where n is the number of neutrons in the reactor of the beginning of the cycle and k is the multiplication factor.

Neutron excess is the difference between the number of neutrons and the number of protons in an atomic nucleus. This is found by subtracting the atomic number of that nuclide from the neutron number; or by subtracting twice the atomic number from the mass number.

Neutron flux density is the number of neutrons that enter a sphere of unit cross-sectional area per unit of time. This quantity is sometimes defined in terms of a unidirectional beam of neutrons incident perpendicularly upon a unit area, but this definition is less general. It is also sometimes called neutron current density.

Neutron number is the number of neutrons in a nucleus. Its symbol is N. The neutron number for a given nuclide is equal to the difference between the mass number and the atomic number for that nuclide.

Additional Reading

Breuker, H. et al.: "Tracking and Imaging Elementary Particles," *Sci. Amer.*, 58 (August 1991).

Brown, A.: *The Neutron and the Bomb: A Biography of Sir James Chadwick*, Oxford University Press, Inc., 1997.

Gutbrod, H. and H. Stöcker: "The Nuclear Equation of State," *Sci. Amer.*, 58 (November 1991).

Hamilton, J.H.: *Fission and Properties of Neutron-Rich Nuclei*, World Scientific Publishing Company, Inc., River Edge, NJ, 1998.

Harding, A.K.: "Physics in Strong Magnetic Fields Near Neutron Stars," *Science*, 1033 (March 1, 1991).

Ruthen, R.: "Out of Its Field: How the Neutron Responds to A Field That Exerts No Force," *Sci. Amer.*, 26 (October 1989).

Schopper, H.: *Low Energy Neutrons and Their Interaction with Nuclei and Matter: Low Energy Neutron Physics*, Springer-Verlag, Inc., New York, NY, 2000.

Shepherd, G.M.: *Foundations of the Neutron Doctrine*, Oxford University Press, Inc., New York, NY, 1998.

Soloway, A.H., R.F. Barth, and D.E. Carpenter: *Advances in Neutron Capture Therapy*, Kluwer Academic Publishers, Norwell, MA, 1993.

Traweak, S.: *Beamtimes and Lifetimes: The World of High Energy Physics*, Harvard University Press, Cambridge, MA, 1992.

Web Reference

Neutron Scattering Web: http://www.neutron.anl.gov/

NEUTRON ACTIVATION ANALYSIS. This is a method of elemental analysis based upon the quantitative detection of radioactive species produced in samples via nuclear reactions resulting from neutron bombardment of samples. The neutron-induced reactions are of two main types: (1) those induced by very slow (thermal) neutrons, having energies of about 0.025 eV; and (2) those induced by fast neutrons having energies in the range of MeV. The method is used in two different forms. The purely instrumental form is fast and nondestructive and is based upon the quantitative detection of induced gamma-ray emitters by means of multichannel gamma-ray spectrometry. The amount of the element present is usually computed from the photopeak (total absorption peak) height or area of its gamma ray, or one of its principal gamma rays, compared with that of the standard. Where interferences from other induced activities are serious and cannot be removed by decay, spectrum subtraction, or computer solution, one must turn to the radiochemical separation method. Here the activated sample is put into solution and equilibrated chemically with measured amounts (typically 10 milligrams) of added carrier of each of the elements of interest, before chemical separations are carried out. The element to be detected needs then to be recovered in chemically and radiochemically pure form, but it need not be quantitatively recovered, since the carrier recovery is measured and the counting data are then normalized to 100% recovery. This form is slower, but it applies to pure beta emitters, as well as to gamma emitters, and it does eliminate interfering activities.

NIACIN. [CAS: 59-67-6]. Sometimes referred to as nicotinic acid or nicotinamide and earlier called the P-P factor, antipellagra factor, antiblacktongue factor, and vitamin B_4, niacin is available in several forms (niacin, niacinamide, niacinamide ascorbate, etc.) for use as a nutrient and dietary supplement. Niacin is frequently identified with the B complex vitamin grouping. Early in the research on niacin, a nutritional niacin deficiency was identified as the cause of pellagra in humans, blacktongue in dogs, and certain forms of dermatosis in humans. Niacin deficiency is also associated with perosis in chickens as well as poor feathering of the birds.

Varying in degree in relationship to the length and severity of diet deficiency of niacin, pellagra is clinically manifested by skin, nervous system, and mental conditions. The disease occurs most frequently among the economically deprived and particularly in areas where the diet may be high in maize (corn) intake. The disease was first described by Gaspar Casal in 1735 and was common in many areas, including Europe, Egypt, Central America, and the southern portion of the United States for many years. The largest outbreak occurred in the United States during the period 1905–1915 and resulted in a high mortality. The medical awareness and understanding of vitamins and dietary deficiencies, coupled with the availability of dietary supplements in staple foods, have resulted in a great lessening in the occurrence of pellagra. Where the disease is found, niacin is a specific for the treatment of acute pellagra. Those afflicted are accustomed to a diet low in protein and made up largely of carbohydrates. Predisposing causes are found idiosyncrasies, chronic alcoholism, and diseases that interfere with the assimilation of a proper diet.

Huber first synthesized nicotinic acid in 1867. In 1914, Funk isolated nicotinic acid from rice polishings. Goldberger, in 1915, demonstrated that pellagra is a nutritional deficiency. In 1917, Chittenden and Underhill demonstrated that canine blacktongue is similar to pellagra. In 1935, Warburg and Christian showed that niacinamide is essential in hydrogen transport as diphosphopyridine nucleotide (DPN). In the following year, Euler et al. isolated DPN and determined its structure. In 1937, Elvhehjem et al. cured blacktongue by administration of niacinamide derived from liver. In the same year, Fouts et al. cured pellagra with niacinamide. In 1947, Handley and Bond established conversion of tryptophan to niacin by animal tissues.

In the physiological system, niacin and related substances maintain nicotinamide adenine dinucleotide (NAD) and nicotinamide adenine dinucleotide phosphate (NADP). Niacin also acts as a hydrogen and electron transfer agent in carbohydrate metabolism; and furnishes coenzymes for dehydrogenase systems. A niacin coenzyme participates in lipid catabolism, oxidative deamination, and photo synthesis.

Nicotinic acid can be converted to nicotinamide in the animal body and, in this form, is found as a component of two oxidation-reduction coenzymes, NAD and NADP, as previously mentioned. Structurally, these are:

Nicotinic acid   Nicotinamide

Nicotinamide adenine dinucleotide (NAD) R* = H

Nicotinamide adenine dinucleotide phosphate (NADP)

The nicotinamide portion of the coenzyme transfers hydrogens by alternating between an oxidized quaternary nitrogen and a reduced tertiary nitrogen as shown by:

NAD   NADH + H+
(oxidized)   (reduced)

Enzymes that contain NAD or NADP are usually called dehydrogenases. They participate in many biochemical reactions of lipid, carbohydrate, and protein metabolism. An example of an NAD-requiring system is lactic dehydrogenase which catalyzes the conversion of lactic acid to pyruvic

acid. Numerous NAD-dependent enzyme systems are known.

Lactic acid   Pyruvic acid

Distribution and Sources. In plants, niacin production sites occur in leaves, germinating seeds, and shoots. In humans, niacin is not available from intestinal bacteria, but some conversion is made from tryptophan which occurs in tissues.

High niacin content (10–100 milligrams/100 grams) Chicken (white meat), groundnut (peanut), halibut, heart (calf), kidney (beef, pork), liver (beef, calf, chicken, pork, sheep), meat extracts, rabbit (white meat), swordfish, tuna, turkey (white meat), yeast

Medium niacin content (1–10 milligrams/100 grams) Almond (dry), asparagus, avocado, barley, bean (kidney, lima, snap, wax), beef, broccoli, cashew, cheeses (camembert, roquefort, Swiss), chestnut, chicken (dark meat), clam, date (dry), duck, fig (dry), fishes (except those listed under "High"), kale, lamb, lentil (dry), maize (corn), molasses, mushroom, oats, oyster, parsley, pea, potato, prune (dry), rice (brown), rye, soybean (dry), shrimp, walnut, wheat, wheat germ

Low niacin content (0.1–1.0 milligram/100 grams) Apple, apricot, banana, beet, beet greens, berries (black-, blue-, cran-, rasp-, straw-), Brussels sprouts, cabbage, carrot, cauliflower, celery, cherry, chicory, coconut, cucumber, currant, dandelion greens, eggs, eggplant, endive, fig, grape, kohlrabi, lettuce, lemon, melons, milk, onion, parsnip, peach, pear, pecan, pepper, pineapple, plum, pumpkin, radish, raisin (dry), rhubarb, spinach, sweet potato, tangerine, tomato, turnip, watercress

Precursors in the biosynthesis of niacin: In animals and bacteria, tryptophan; and in plants, glycerol and succinic acid. Intermediates in the synthesis include kynurenine, hydroxyanthranilic acid, and quinolinic acid. In animals, the niacin storage sites are liver, heart, and muscle. Niacin supplements are prepared commercially by: (1) Hydrolysis of 3-cyanopyridine; or (2) oxidation of nicotine, quinoline, or collidine.

Bioavailability of Niacin. Factors which cause a decrease in niacin availability include: (1) Cooking losses; (2) bound form in corn (maize), greens, and seeds is only partially available; (3) presence of oral antibiotics; (4) diseases which may cause decreased absorption; (5) decrease in tryptophan conversion as in a vitamin B_6 deficiency. Factors that increase availability include: (1) alkali treatment of cereals; (2) storage in liver and possibly in muscle and kidney tissue; and (3) increased intestinal synthesis.

Antagonists of niacin include pyridine-3-sulfonic acid (in bacteria); 3-acetylpyridine, 6-aminonicotinamide, and 5-thiazole carboxamide. Synergists include vitamins B_1, B_2, B_6, B_{12}, and D, pantothenic acid, folic acid, and somatotrophin (growth hormone).

In humans, overdosage of niacin causes a limited toxicity (1 to 4 grams/kilogram) with individual variations in sensitivity.

Additional Reading

FAO United Nations: *Requirements of Vitamin A, Thiamine, Riboflavin and Niacin: Report of a Joint FAO-Who Expert Group*, FAO United Nations, Geneva, Switzerland, 1967. *http://www.fda.gov/ (United States Food and Drug Association).*

Institute of Medicine, Food Nutrition Board Staff: *Dietary Reference Intakes: Thiamin, Riboflavin, Niacin, Vitamin B6, Folate, Vitamin B12, Pantothenic Acid, Biotin, and Choline,* National Academy Press, Washington, DC, 1998.

Web Reference

United States Food and Drug Association. http://www.fda.gov/

NICKEL. [CAS: 7440-02-0]. Chemical element, symbol Ni, at. no. 28, at. wt. 58.69, periodic table group 10, mp 1453°C, bp 2732°C, density 8.9 g/cm³ (solid, 20°C), 9.04 g/cm³ (single crystal). Elemental nickel has a face-centered cubic crystal structure. Nickel is a silver-white metal, harder than iron, capable of taking a brilliant polish, malleable and ductile, magnetic below approximately 360°C. When compact, nickel is not oxidized on exposure to air at ordinary temperatures. The metal is soluble in HNO_3 (dilute), but becomes passive in concentrated HNO_3. The metal does not react with alkalis. Finely divided nickel dissolves 17× its own volume of hydrogen at standard conditions. There are five naturally occurring stable isotopes, ^{58}Ni, ^{60}Ni through ^{62}Ni, and ^{64}Ni. Six radioactive isotopes have been identified, ^{56}Ni, ^{57}Ni, ^{59}Ni, ^{63}Ni, ^{65}Ni, and ^{66}Ni. ^{59}Ni has a half-life of 8×10^4 years, and ^{63}Ni has a half-life of 80 years. The half-lives of the remaining radioactive isotopes are relatively short, expressed in hours and days. The element ranks 21st among the elements in terms of abundance in the earth's crust, the estimated average content of igneous rocks being about 0.02%. In terms of cosmic abundance, nickel ranks 28th among the elements. Nickel ranks 40th in terms of concentration in seawater, the estimated content being about 2.5 tons of nickel per cubic mile (540 kilograms per cubic kilometer) of seawater. Awareness of nickel probably dates back to antiquity, but the element was not firmly identified until 1751 when Axel Fredric Cronstedt isolated the metal from the sulfide ore NiAsS.

First ionization potential 7.33 eV, second, 18.13 eV. Oxidation potentials Ni $\longrightarrow$ Ni^{2+} + 2e$^-$, 0.230 V; Ni^{2+} + 2H$_2$O $\longrightarrow$ NiO$_2$ + 4H$^+$ + 2e$^-$, −1.75 V; Ni + 2OH$^-$ $\longrightarrow$ Ni(OH)$_2$ + 2e$^-$, 0.66 V; Ni(OH)$_2$ + 2OH$^-$ $\longrightarrow$ NiO$_2$ + 2H$_2$O + 2e$^-$, −0.49 V.

Other physical properties of nickel are given under **Chemical Elements**. In the early 1800s, the principal sources of nickel were in Germany and Scandinavia. Very large deposits of lateritic (oxide or silicate) nickel ore were discovered in New Caledonia in 1865. The sulfide ore deposits were discovered in Sudbury, Ontario in 1883 and, since 1905, have been the major source of the element. The most common ore is pentlandite, (FeNi)$_9$S$_8$, which contains about 34% nickel. Pentlandite usually occurs with pyrrhotite, an iron-sulfide ore, and chalcopyrite, CuFeS$_2$. See also **Chalcopyrite**; **Pentlandite**; and **Pyrrhotite**. The greatest known reserves of nickel are in Canada and Russia, although significant reserves also occur in Australia, Finland, the Republic of South Africa, and Zimbabwe.

Principal producers and/or exporters of nickel include, in diminishing order, Canada, Russia, the United Kingdom, Norway, and Indonesia. Main consumers are the United States, Japan, the United Kingdom, Norway, Germany, Canada, and France.

After beneficiation of the raw ore to form a sulfide concentrate, the latter is roasted to achieve partial oxidation of iron and partial removal of sulfur. The roasted material then is smelted with a flux to eliminate the rock content. At this point, part of the iron goes into the slag. The remaining material is a copper-bearing nickel-iron matte, made up mainly of the sulfides of these metals. The matte is then treated in a Bessemer converter to achieve further removal of iron and sulfur. After controlled cooling, which assists separation, the Bessemer product is finely ground and subjected to magnetic separation and differential flotation. The separated product is an impure nickel sulfide. The sulfide then is sintered to nickel oxide. This product may be marketed for some applications, but the majority of the oxide is cast into anodes for refining into nickel metal by one of two major processes.

In (1) the electrolytic process, a nickel of 99.9% purity is produced, along with slimes which may contain gold, silver, platinum, palladium, rhodium, iridium, ruthenium, and cobalt, which are subject to further refining and recovery. In (2) the Mond process, the nickel oxide is combined with carbon monoxide to form nickel carbonyl gas, Ni(CO)$_4$. The impurities, including cobalt, are left as a solid residue. Upon further heating of the gas to about 180°C, the nickel carbonyl is decomposed, the freed nickel condensing on nickel shot and the carbon monoxide recycled. The Mond process also makes a nickel of 99.9% purity.

Uses. The three main commercial forms of primary nickel are: (1) electrolytic sheets, (2) pellets resulting from the decomposition of nickel carbonyl, and (3) ferronickel. Traditionally, pellets are favored in Europe, whereas electrolytic nickel is favored in North America. Additional forms of commercial nickel are powder, ingots, shot, and briquettes. Ferronickel, containing 24−48% nickel with the remainder iron, is used mainly in the production of stainless steel. More than half of the nickel produced is used in stainless steels and high-nickel alloys. Additional uses include nickel plating, iron and steel castings, coinage, and copper and brass products.

The main consumer of nickel is austenitic stainless steel, which contains from 3.5 to 22% nickel and 16 to 26% chromium. In these steels, nickel stabilizes the austenite and enhances the ductility of the steel. Nickel, along with chromium, contributes to corrosion resistance. Up to amounts of about 9%, nickel adds strength, hardness, and toughness to many alloy steels.

Alloys in the 9% nickel range remain stable at low temperatures and are capable of handling liquefied gases. The lower-nickel steels (0.5 to 0.7%) are ductile, strong, and tough, and find use for many automobile parts, in power machinery, and construction equipment. There are hundreds of nickel-containing alloys, running the gamut from hardenable silver alloy (0.02% Ni) up to malleable nickel (99% Ni).

Wrought Nickel and High-Nickel Alloys. Some of the major nickel alloys, along with wrought nickel, are described in Table 1.

Commercially pure wrought nickel in the form of sheets, wire, and tubing has many uses because of its corrosion resistance. These uses include utensils, food-processing equipment, marine hardware, coinage, and chemical equipment. Electroplated nickel also is used as a protective coating on steel. *Nimonic* alloys, not shown on the table, are based on an 80% Ni, 20% Cr composition. They are high-strength, heat-resistance metals that are age-hardened to increase strength at elevated

TABLE 1. WROUGHT NICKEL AND REPRESENTATIVE NICKEL ALLOYS

Melting range °C	Poisson's ratio
Wrought nickel	1,435−0.31
99% Ni, 0.25% Cu, 0.15% C	1,445
Duranickel 301	1,400−0.31
93.9% Ni, 0.05% Cu, 0.15% C, 0.15% Fe, 0.5% Ti, 4.5% Al	1,440
Monel 400	1,300−0.32
66.0% Ni, 31.5% Cu, 0.12% C, 1.35% Fe	1,350
Hastelloy B	1,320−−
63.5% Ni, 0.05% C, 5.0% Fe, 2.5% Co, 1.0% Cr, 28.0% Mo, 0.3% V	1,460
Hastelloy F	1,290−0.305
45.5% Ni, 0.05% C, 20.5% Fe, 2.5% Co, 22.0% Cr, 6.5% Mo, 1% W, 2% (Nb + Ta)	1,295
Inconel 600	1,370−0.29
72% Ni, 0.5% Cu, 0.15% C, 8.0% Fe, 15.5% Cr	1,425
Incoloy 800	1,355−0.30
32.5% Ni, 0.75% Cu, 0.10% C, 45.6% Fe, 21.0% Cr	1,390
Illium G	1,255−0.29
56.0% Ni, 6.5% Cu, 22.5% Cr, 6.5% Mo q	1,340

Note: Recently introduced new or improved nickel alloys include:

Inconel alloy 625 —Low-cycle fatigue resistance has been increased from 70−80,000 to 110−120,000 psi at 10 cycles. This has been achieved through grain size control and improved product cleanliness. Major applications are bellows and expansion joints.

Inconel alloy 725 —An age-hardenable alloy for deep sour gas well service, combining height strength with the attributes of Inconel alloy 625, such as pitting resistance and stress corrosion cracking resistance to salt, hydrogen sulfide, and sulfur at temperatures up to about 230°C (450°F) and to sulfide stress corrosion cracking.

Inconel alloy 622 —Modified composition and special thermal mechanical processing give this alloy superior thermal stability and resistance to intergranular attack and localized corrosion. The alloy is particularly suited to acidified halide environments, especially those containing oxidizing acids.

Inconel alloy 925 —An age-hardenable nickel-iron-chromium alloy providing high strength up to 540°C (1000°F). Developed for use in gas production applications, such as tubular products, tool joints, and equipment for surface and downhole hardware in gas industry.

Inco alloy 25−6MO —Used for its corrosion resistance in many environments, this is an austenitic nickel-iron-chromium alloy with a substantial (6%) addition of molybdenum. Especially useful for resisting pitting and crevice corrosion in media containing chlorides or other halides. Applications include equipment for handling sulfuric and phosphoric acids, offshore platforms and other marine equipment, and for bleaching circuits in pulp and paper plants.

temperatures—with a useful range of 700–825°C. *Monel* metal (several types) is a high-strength corrosion-resistant alloy available in many wrought and cast forms for use in processing equipment, marine construction, and household appliances. *K Monel* can be heat treated by precipitation hardening to about 2× the strength of annealed *Monel*. *Hastelloy*-type alloys are well known for their excellent resistance to HCl, H_2SO_4, and other acids. The *Incoloy*-type alloys (35% Ni approximately) are heat-resistant alloys used mainly as castings for furnace parts. The lower-nickel/higher-chromium alloys generally are classified as stainless steels. See also **Iron Metals, Alloys, and Steels**.

Although not of high-tonnage production, several nickel metals serve important uses, such as:

Permalloy, 78.5% Ni, 21.5% Fe; *Hipernik,* 50% Ni, 50% Fe; and *Perminvar,* 45% Ni, 30% Fe, 25% Co—are representative of a group of high-nickel magnetic alloys.

Constantan, 45% Ni, 55% Cu, has high electrical resistivity and a very low temperature coefficient of resistivity. It is extensively used with copper as a thermocouple element.

Nichrome, 80% Ni, 20% Cr (several types with variations of these percentages and additions of other elements, such as silicon in small amounts), is used as resistance wire for heating elements.

Calorite, 65% Ni, 8% Mn, 12% Cr, 15% Fe, also is used in electric heating elements.

Alumel, 94% Ni, 2.5% Mn, 0.5% Fe plus small amounts of other elements, is used in thermocouples

Chromel, 35–60% Ni, 16–19% Cr, generally with the balance Fe, also is used as resistance wire and for thermocouples.

Invar, 36% Ni, 64% Fe, has a very low temperature coefficient of expansion and is used for measuring tapes, instruments, and bimetallic thermostats.

Elinvar, 34% Ni, 57% Fe, 4% Cr, 2% W, has a very low temperature coefficient of elasticity which makes it useful for springs in watches and precision instruments.

There are hundreds of special nickel-bearing alloys of proprietary formulations and tradenames.

Alloy with Memory. In seeking a way to reduce the brittleness of titanium, U.S. Navy researchers serendipitously discovered a nickel-titanium alloy having an amazing memory. Previously cooled clamps made of the alloy (*nitinol*) are flexible and can be placed easily in position. When warmed to a given temperature, the alloy hardware then exerts tremendous pressure. Use of conventional clamps for holding bundles of wires or cables in a ship or aircraft structure requires special tools. For this and other applications in industry and medicine, nitinol has been in demand. The alloy, however, is not easy to produce because only minor variations in composition can affect the "snap back" temperature by several degrees of temperature.

Nickel Powders. The use of nickel in powdered form has increased markedly during the last few years. As shown by Fig. 1, nickel powders are available in several types and are used in a variety of products.

Nickel in Nanometer Materials. Coating a metal with an ultrathin layer of another metal creates properties not found separately in either of the materials. Considerable recent research has been directed toward improving the mechanical properties of bimetallic laminates, sometimes called *composition modulated films,* which have interlayers only a few nanometers thick. Attractive properties also have been found for similar systems, called *nanometer materials.* Nickel has been used in combination with copper, ruthenium, and other metals for producing these new materials.

Production of High-Performance Nickel Alloys. In the production of high-performance alloys, the critical first step of alloying requires sophisticated equipment, stringent controls, and expertise. Several production methods are used.

Air melting in electric-arc or induction furnaces is used for many alloys, sometimes for final alloying, with further refining by argon-oxygen decarburization. Melting in air can result in impurities in some alloys, a problem eliminated by vacuum induction melting, used to produce ingots for direct rolling or for remelting. Remelting is accomplished by two methods, both with precise, computerized control. *Electroslag* remelting uses electrical resistance heating to remelt an ingot (electrode) under molten slag containing fluxes that remove impurities. *Vacuum arc* melting refines

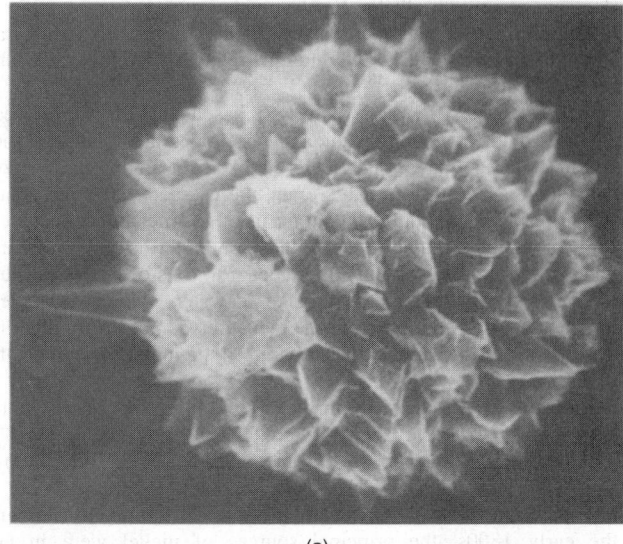

(a)

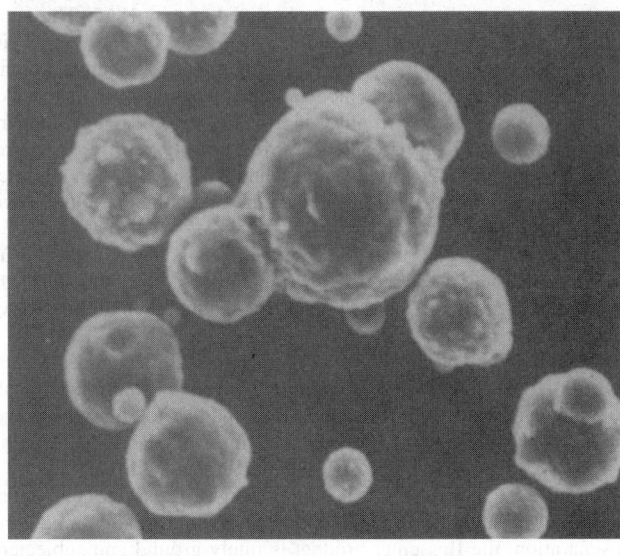

(b)

Fig. 1. Types of available nickel powders. (**a**) With a surface area of 0.4 m^2/g. this is a spiked nickel powder of single particles 3–7 microns in diameter, with a bulk density of 2 g/cc. The powder is used in both metal and chemical systems for powder metal parts, getters, magnets, electronic strip, flake, and organo-nickel compounds and nickel salts and soaps. (**b**) High-density nickel powder consisting of 8–12 micron semismooth particles, offering mixability with both metallic and nonmetallic powder systems. Applications include welding rods, nickel aluminide, nickel-columbium additives, abradable seals, powder metal parts, carbide binders, and conductive plastics. (**c**) Spherically shaped, high- purity nickel powder, with a surface area of 0.15 m^2/g and a Fisher particle size of 8–9 microns. Applications include friction materials, plasma spraying, metal injection molding, welding electrodes, magnets, cemented carbides, and powder metal nickel steels. (*Source: INCO Specialty Powder Products, Saddle Brook, New Jersey*)

the structure of cast electrodes in a contaminant-free chamber. Remelting yields alloys of the highest level of refinement.

Nickel Chemistry and Compounds

With its $3d^84s^2$ electron configuration, nickel forms Ni^{2+} ions. Having a nearly complete $3d$ subshell, nickel does not yield a $3d$ electron as readily as iron and cobalt, and trivalent and tetravalent forms are known only in the hydrated oxides, Ni_2O_3 and NiO_2, and a few complexes.

Nickel(II) oxide, NiO, produced by heating the carbonate, is thermally stable. Higher oxides of nickel, including Ni_2O_3 and NiO_2, are known only as hydrates, being prepared by vigorous oxidation of NiO in alkaline solution.

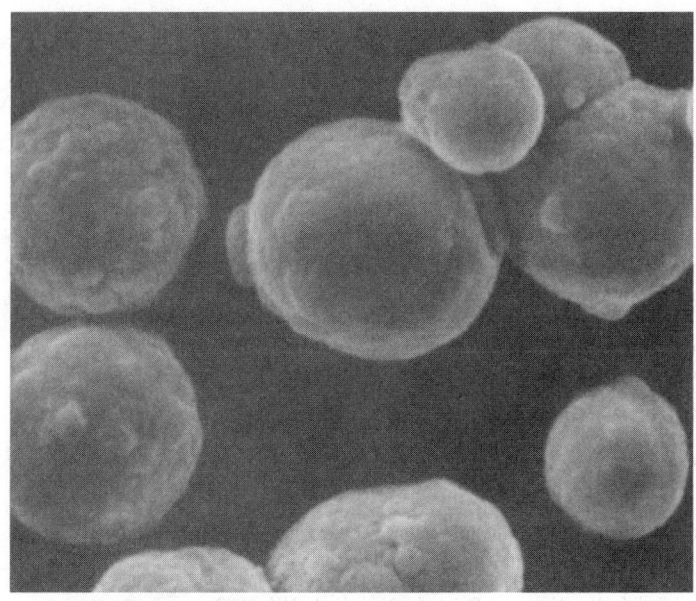

(c)

Fig. 1.   (*continued*)

Nickel(II) sulfide, precipitated from Ni^{2+} solutions by ammonium sulfide, may show quite a little departure from stoichiometric composition. Like iron(II) and cobalt(II) FeS and CoS, it has in crystal form an electrical conductivity and other properties similar to a metal or alloy. There is no conclusive evidence that Ni_2S_3 can be prepared, but NiS_2 is known and believed to be, like FeS_2, a compound of Ni^{2+} and the S_2^{2-} ion.

All four dihalides of nickel with the common halogens are known: NiF_2, formed by reaction of hydrofluoric acid or nickel(II) chloride or by thermal decomposition of $[Ni(NH_3)_6][BF_4]_2$, is greenish yellow, while the other three dihalides, formed directly from the elements, are green for the chloride, yellow for the bromide, and black for the iodide. In general, anhydrous Ni^{2+} salts are yellow and the ion $Ni(H_2O)_6^{2+}$ in aqueous solution is green.

Other elements with which nickel forms binary compounds, especially at higher temperature, are boron, carbon, nitrogen, silicon, and phosphorus. Like NiO, these compounds may depart slightly or even considerably from daltonide composition, frequently being interstitial compounds, and with higher elements of transition groups 5 and 6, merging into the interstitial compound-solid solution picture which nickel exhibits with the other transition metals.

Divalent nickel forms two main types of complexes. The first consists of complexes of the spin-free ("ionic" or outer orbital) octahedral type (see also **Ligand** for their discussion) in which the ligands are principally H_2O, NH_3, and various amines such as ethylenediamine and its derivates, e.g., $Ni(H_2O)_6^{2+}$, $Ni(NH_3)_6^{2+}$, $Ni(en)_6^{2+}$. These complexes usually have colors toward the high-frequency side of the spectrum, i.e., violet, blue, and green. The other class consists of tetracovalent square complexes with ligands such as CN^-, the dioximes and their derivatives, and other chelates, which usually have colors on the low frequency side of the spectrum, i.e., red, orange, and yellow. The structure of the nickel-dimethylglyoxime complex is

This compound is of interest not only in analysis, but because by limited oxidation with the halogens it yields a unipositive ion containing trivalent

nickel and also because the hydrogen bonds formed to the oxygen atoms are among the shortest known. Similarly, the tetracyanide complex of nickel, $Ni(CN)_4^{2-}$, may be reduced by sodium amalgam to give an ion of composition $Ni(CN)_4^{3-}$, or $(NC)_3Ni-Ni(CN)_3^{4-}$ containing Ni(I). This latter ion forms a potassium salt of nickel(I) of the formula $K_4Ni_2(CN)_6$ which is reduced in liquid NH_3 by metallic potassium to give the compound $K_4Ni(CN)_4$ in which the nickel has an effective valence of zero. Of course, this zero valence also exists in the carbonyls of nickel (and other elements) which, however, are covalent. $Ni(CO)_4$ is prepared by reaction of carbon monoxide with freshly reduced nickel, which occurs at ordinary temperatures and pressures. As with the carbonyls of other metals, the CO groups may be directly or indirectly, partially or completely, replaced by other groups. Derivatives of trivalent phosphorus form many such compounds of general formula $Ni(CO)\times_{4-x}(PR_3)_x$, where R may be one or more of such groups as F, Cl, Br, I, alkyl, aryl, alkoxy, aryloxy, etc.

See also **Nickel (In Biological Systems)**.

Health and Safety Factors

Eye and Skin Contact. Some nickel salts and aqueous solutions of these salts, e.g., the sulfate and chloride, may cause a primary irritant reaction of the eye and skin. The most common effect of dermal exposure to nickel is allergic contact dermatitis.

Protective equipment and clothing such as face shields and gloves should be worn and safety showers should be available wherever there is a possibility of being splashed or otherwise contacted by nickel containing solutions. If dermatitis should occur, the possibility that it is nickel related should be brought to the attention of a physician.

Inhalation. Nickel carbonyl is an extremely toxic gas. The permissible exposure limit (PEL) in the United States is 1 part per billion (ppb) in air. Nickel carbonyl may form wherever carbon monoxide and finely divided nickel are brought together. Nickel carbonyl should be used in totally enclosed systems or under good local exhaust.

The potential chronic toxicity is of concern. Based on epidemiological and experimental results, the International Agency for Research on Cancer (IARC) http://www.iarc.fr/, has concluded that all nickel compounds are Category 1, i.e., known human carcinogens, and has classified metallic nickel as a Category 2B carcinogen, i.e., possibly carcinogenic to humans.

It is good practice to keep concentrations of airborne nickel in any chemical form as low as possible and certainly below the relevant standard. Local exhaust ventilation is the preferred method, particularly for powders, but personal respirator protection may be employed.

Additional Reading

Carter, G.F. and D.E. Paul: *Materials Science and Engineering,* ASM International, Materials Park, OH, 1991.

Greenwood, N.N. and A. Earnshaw: *Chemistry of the Elements,* 2nd Edition, Butterworth-Heinemann, Inc., Woburn, MA, 1997.

Hanson, A.: "The Metal That Remembers," *Technology Review(MIT),* 26 (May–June 1991).

Houston, J. and P. Feibelman: "Ultrathin Metal Coatings Yield Unique Properties," *Advanced Materials & Processes,* 31 (March 1991).

Lancaster, J.R., Jr.: *The Bioinorganic Chemistry of Nickel,* John Wiley & Sons, Inc., New York, NY, 1988.

Lide, D.R.: *CRC Handbook of Chemistry and Physics 84th Edition,* CRC Press, LLC., Boca Raton, FL, 2003.

Sax, N.R. and R.J. Lewis, Sr.: *Sax'x Dangerous Properties of Industrial Materials,* 10th Edition, John Wiley & Sons, Inc., New York, NY, 1999.

Staff: "A Quick Reference Guide to Nickel and High-Nickel Alloys," *Advanced Materials and Processes,* 54 (October 1991).

Staff: *ASM Handbook—Properties and Selection: Nonferrous Alloys and Pure Metals,* ASM International, Materials Park, OH, 1991.

Staff: "Metals Forecast," *Advanced Materials & Processes,* 17 (January 1991); 24 (January 1992); 18 (January 1993).

Web References

Nickel Institute: http://www.nidi.org/

Nickel Producers Environmental Research Association (NiPERA): http://www. nipera.org/

NICKEL (In Biological Systems).   Despite its many pharmacological and in vitro actions, convincing evidence showing that nickel is an essential element for some animal species did not appear until the early 1960s. There has been considerable further and more convincing research during the 1970s and early 1980s.

Like most trace elements, nickel can activate various enzymes in vitro, but no enzyme has been shown to require nickel, specifically, to be activated. However, urease has been shown to be a nickel metalloenzyme and has been found to contain 6 to 8 atoms of nickel per mole of enzyme (Fishbein et al., 1976). RNA (ribonucleic acid) preparations from diverse sources consistently contain nickel in concentrations many times higher than those found in native materials from which the RNA is isolated (Wacker-Vallee, 1959; Sunderman, 1965). Nickel may serve to stabilize the ordered structure of RNA. Nickel may have a role in maintaining ribosomal structure (Tal, 1968, 1969). These studies and other information have led to the suggestion that nickel may play a role in nucleic acid and/or protein metabolism.

Nickel also may act to stimulate or inhibit the release of various hormones (Nielsen, 1971, 1972; Dormer et al., 1973; Clay, 1975; Horak-Sunderman, 1975). Nickel has been found to inhibit insulin release from the pancreas (Dormer et al., 1973; Clay, 1975), and stimulates glucagon secretion (Horak-Sunderman, 1975).

Nickel as an essential element in ruminant nutrition has not been proved conclusive as of the early 1980s. However, with nonruminants, some evidence indicates that certain species fed low-nickel diets have a greater infant mortality rate and a general degradation of the reproductive process (Nielsen, 1975; Anke et al., 1973).

Zinc and nickel appear to behave similarly at certain sites in the biological system. Both elements are capable of activating certain enzymes; for example, arginase is an enzyme which can be activated by either element (Parisi-Vallee, 1969). Stimulation of enzyme activity is at a site at which trace element substitutions or interactions may occur. However, some sacrifice of activity usually results when normally occurring metal is replaced by a trace metal. Nucleic acids as well as the ribosomes are likely sites of interaction between nickel and zinc. Both metals are consistently found in high concentrations firmly bound to RNA. It has been suggested that they function in maintaining the structure of RNA, thus preventing conformational changes. Nickel appears to be as effective as zinc at equal concentrations in this respect. Nickel and zinc are also found in ribosomal ash and studies have indicated that both can contribute to ribosomal conformation. The white blood cell is another possible site at which nickel and zinc may interact. Leukocytes are high in zinc and total leukocyte counts as well as differential white cell counts change drastically during a zinc deficiency. The interrelationship between nickel and zinc has been studied in vitro primarily in swine and rats. Their relationship has been studied largely from a substitution standpoint. Nickel appears to substitute for zinc to a certain extent in both species.

Similarly, the relationship between nickel and copper has been under study. One of the major functions of copper is in hemoglobin formation. Hemoglobin and hematocrit values decline rapidly during a copper deficiency. Copper is currently believed to exert its effect on hemoglobin metabolism through ceruloplasmin. Early work also indicated that nickel might be involved in hematopoiesis. Investigators in 1974 found a decreased concentration of copper in the lung and spleen of rats receiving 5 parts per million of nickel in drinking water. High levels of dietary nickel in rats and mice have been reported to decrease the activity of cytochrome oxidase, a copper-containing enzyme.

As pointed out by Eskew, Welch, and Cary in 1983, in contrast with the situation in animals, for which four new essential trace elements were identified in recent years, no new generally essential micronutrient for higher plants has been found since the mid-1950s. When it was found that urease is a nickel-metalloenzyme, this suggested that Ni may play a role in higher plants. Nickel has evidenced a stimulation of growth when urea is the sole source of nitrogen, but has slight or no effect when other nitrogen enrichment sources are used. The aforementioned investigators claim that Ni is essential for nitrogen metabolism in soybeans (*Glycine max* (L.) Merr.), either when nitrogen is furnished as NO_3^- and NH_4^+ or when plants depend upon nitrogen fixation. In experiments, soybean plants deprived of Ni accumulated toxic concentrations of urea (2.5%) in necrotic lesions on their leaflet tips. This occurred regardless of whether the plants were furnished with inorganic N or were dependent on N fixation. Nickel deprivation resulted in delayed nodulation and in a reduction of early growth. The addition of Ni (1 microgram/liter) to the nutrient media prevented urea accumulation, necrosis, and growth reductions. Extrapolating these findings, it is suggested that Ni may be essential for other higher plants.

Toxicity

Nickel contact dermatitis can occur among wearers of nickel-containing jewelry, more common among females than males. This is particularly true of nickel sulfate present in some jewelry. Localization of sites unexpectedly involves the ear lobes, neck, fingers, and wrists. Nickel is a major offender in connection with AECD (allergic eczematous contact dermatitis).

As mentioned earlier, nickel carbonyl is a volatile intermediate in the Mond process for nickel refining. This compound also is used for vapor plating of nickel in the semiconductor industry, and as a catalyst in the chemical and petrochemical industries. The toxicity of the compound has been known for many years. Exposure of laboratory animals to the compound has induced a number of ocular anomalies, including anophthalmia and microphthalmia, and has been shown to be a carcinogenic for rats.

Additional Reading

Anke, M. et al.: "Low Rations for Growthe and Reproduction in Pigs," in *Trace Element Metabolism in Animals,* (W.G. Hockstra et al., editors), University Park Press, Baltimore, MD, 1073.

Clay, J.J.: "Nickel Chloride-induced Metabolic Changes in the Rat and Guinea Pig," *Toxicol. Appl. Pharmacol.,* **31**, 55 (1975).

Considine, D.M. and G.D. Considine, Eds.: "Foods and Food Production Encyclopedia," in *Van Nostrand Reinhold,* New York, NY, 1982.

Dormer, R.L. et al.: "The Effect on Nickel on Secretory Systems," *Biochem.J.,* **140**, 135 (1973).

Eskew, D.L., R.M. Welch, and E.E. Cary: "Nickel: An Essential Micronument for Legumes and Possibly All Higher Plants," *Science,* **222**, 621–623 (1983).

Fishbein, W.N. et al.: "The First Natural Nickel Metalloenzyme: Urease," *Fed. Proc.,* **35**, 1680 (1976).

Hausinger, R.P.: *Biochemistry of Nickel,* Kluwer Academic Publishers, Norwell, MA, 1993.

Horak, E. and F.W. Sunderman, Jr.: "Effects on Ni(II) upon Plasma Glucagon and Glucose in Rats," *Toxicol. Appl. Pharmacol.,* **33**, 388 (1975).

Neilsen, F.H.: "Studies on the Essentiality of Nickel," in *Newer Trace Elements in Nutrition,* (W. Mertz and W.E. Cornatzer, editors), Marcel Dekker, New York, NY, 1971.

Parisi, A.F. and B.L. Vallee: "Zinc Metalloenzymes: Characteristics and Significance in Biology and Medicine," *Amer. J. Clin. Nutr.,* **22**, 1222 (1969).

Spears, J.W and E.E. Hatfield: "Role of Nickel in Animal Nutrition," *Feedstuffs,"* 24–28 (June 13, 1977).

Sunderman, F.W., Jr.: "Measurements of Nickel in Biological Materials by Atomic Absorption Spectrometry," *Amer. J. Clin. Path.,* **44**, 182 (1965).

Sunderman, F.W., Jr. et al.: "Eye Malformations in Rats:Induction by Prenatal Exposure To Nickel Carbonyl," *Science,* **203**, 550–552 (1979).

Tal, M.: "On the Role of Zn^{2+} and Ni^{2+} in Ribosome Structure, *Biochem. Biophys. Acta,* **169**, 564 (1968).

Wacker, E.E.C. and B.L. Vallee: "Nucleic Acids and Metals. I. Chromium, Manganese, Nickel, Iron and Other Metals in Ribonucleic Acid from Diverse Biological Sources," *J. Biol. Chem.,* **234**, 3257 (1959).

Web Reference

Biochemical Periodic Table—Nickel: http://umbbd.ahc.umn.edu/periodic/elements/ni.html

NICKELINE. A nickel arsenide mineral, NiAs, crystallizes in the hexagonal system but is usually found massive. Color, light copper; hardness, 5.0–5.5; specific gravity, 7.784; luster, metallic; opaque. Found in several European localities and in the Province of Ontario, Canada; in the United States at Franklin, New Jersey, and Silver Cliff, Colorado. It is an ore of nickel.

NICOTINE. See **Alkaloids**.

NINHYDRIN REACTION. See **Amino Acids**.

NIOBIC ACID. Any hydrated form of Nb_2O_5. It forms as a white, insoluble precipitate when a potassium hydrogen sulfate fusion of a niobium compound is leached with hot water or when niobium fluoride solutions are treated with ammonium hydroxide. Soluble in concentrated sulfuric acid, concentrated hydrochloric acid, hydrogen fluoride, and bases. Important in analytical determination of niobium. See also **Niobium**.

NIOBIUM. [CAS: 7440-03-1]. Chemical element, symbol Nb, at. no. 41, at. wt. 92.906, periodic table group 5, mp 2,458–2,468°C, bp 4,742°C, density 8.6 g/cm³ (20°C). Elemental niobium has a body-centered cubic crystal structure. The metal has a slightly bluish tinge, is ductile and malleable, and when polished resembles platinum. The metal burns upon being heated in air. There is one natural isotope ^{93}Nb. Seven radioactive isotopes have been identified ^{90}Nb through ^{92}Nb and ^{94}Nb through ^{97}Nb, with a wide range of half-lives. ^{94}Nb has the longest half-life (2×10^4 years). The element was first identified by C. Hatchett in 1801 and was originally called columbium which name persisted for many years. The name still appears widely in the literature, particularly in connection with alloys bearing the element, such as columbium steels.

First ionization potential 6.77 eV; second 13.895 eV; third 24.2 eV. Oxidation potential Nb $\longrightarrow$ Nb^{3+} + 3e$^-$, ca. 1.1 V; 2Nb + 5H$_2$O $\longrightarrow$ Nb$_2$O$_5$ + 10H$^+$ + 10e$^-$, 0.62 V.

Other important physical properties of niobium are given in the Table 1 and under **Chemical Elements**.

Niobium occurs, usually with tantalum, in columbite Fe(NbO$_3$)$_2$, (80% Nb$_2$O$_5$), pyrochlore (50% Nb$_2$O$_5$), samarskite (50% Nb$_2$O$_5$), chiefly found in western Australia, and South Dakota. Recovered along with tantalum by fusion with potassium bisulfate, and obtained in the residue after subsequent extraction with H$_2$O. Niobium and tantalum are separated by fractional crystallization of the potassium fluorides, niobium concentrating in the mother liquid and tantalum in the crystals.

The principal uses for the element are in alloys. Niobium also has gained prominence in research as a superconducting material. At the temperatures of liquid helium, niobium becomes a superconductor and, in the form of a fine wire, has been incorporated in a superconducting cell. The element has both size and cost advantages over electronic materials. The alloy Nb$_3$Sn becomes superconducting at a somewhat higher temperature. Niobium-titanium and niobium-zirconium alloys also have potential as superconductors.

Alloys. Niobium is used in steel, notably stainless steels, to stabilize the carbon present (as carbide) and for preparing niobium carbide, used for dies and cutting tools. Ferroniobium is a strong carbide-forming material and, when added to 18-8 stainless steel, stabilizes areas that may be heat-affected during welding and thus cause subsequent intergranular corrosion. Niobium steels are used for rotors in gas turbines where temperatures up to 700°C must be withstood. Niobium-base alloys find application in fast reactors. Superalloys for very demanding use, as in military applications contain niobium with cobalt and zirconium. When alloyed with titanium, molybdenum, and tungsten, the elevated-temperature hardness of niobium in enhanced, whereas when alloyed with vanadium and zirconium, the strength of niobium up to temperatures of 500°C is increased. Metallurgically, niobium is attractive because of its density, good workability, retention of tensile strength at high temperatures, and its high melting point. In the temperature range 920–1,200°C, niobium has been found superior to most other metals on a strength-to-weight basis for aerospace applications. In multicomponent alloys, zirconium and hafnium when added with niobium add effectively to strength, even more so than molybdenum or tungsten, but there is some sacrifice in ductility.

In metallurgy, niobium is classified as a refractory metal, along with tungsten, tantalum, and molybdenum. A comparison of the four metals is given in the accompanying table.

Niobium in Tool Steels. In the matrix method of tool-steel development, the composition of the heat-treated matrix determines the steel's initial composition. Carbide volume-fraction requirements then are calculated, based upon historical data, and the carbon content is adjusted accordingly. This approach has been used to design new steels in which niobium is substituted for all or part of the vanadium present as carbides in the heat-treated material. Niobium provides dispersion hardening and grain refinement, and forms carbides that are as hard as vanadium, tungsten, and molybdenum carbides.

Chemistry and Compounds

Elemental niobium is insoluble in HCl or HNO$_3$, but soluble in hydrofluoric acid or a mixture of hydrofluoric and HNO$_3$.

As might be expected from its $4d^4 5s^1$ electron configuration, niobium forms pentavalent compounds. However, the stability of its compounds of lower valence is greater than that of the corresponding tantalum compounds, in keeping with the group 5 position of niobium and tantalum. Nevertheless the similarity of the properties of the compounds of the two metals is so great that special methods are required for their separation, such as solvent extraction of the pentachlorides or chromatographic removal of adsorbed TaF$_5$ with an ethylmethyl ketone-water system. In addition, divalent and tetravalent compounds are known, and an interstitial, nonstoichiometric hydride.

Niobium forms a divalent oxide, NbO, insoluble in water, but readily soluble in acids or NH$_4$OH. It also gives by direct combination of the metal on heating with oxygen, the pentoxide, Nb$_2$O$_5$, which can be reduced by hydrogen at high temperature to NbO$_2$, and on heating with magnesium to Nb$_2$O$_3$.

Niobium(III) halides are known, notably the chloride, NiCl$_3$, which is of particular interest because its solution has been shown to contain Nb^{3+} ions (in equilibrium with NbCl$_6$$^{3-}$ complex ions).

Tetravalent niobium is believed to occur in the form of NbOCl$_4$$^{2-}$ ions in a solution obtained, with color change, by reduction of HCl solution of NbCl$_5$, and by inference in similarly reduced solutions of the other pentahalides. Tetravalent niobium also is found in the dioxide (see above) and the carbide, NbC.

Four pentahalides of niobium, NbF$_5$, NbCl$_5$, NbBr$_5$, and NbI$_5$ have been prepared by heating the pentoxide with carbon in a current of the halogen. They are hydrolyzed in H$_2$O, and even in concentrated aqueous solution of the respective halogen acids; the Nb^{5+} ion is apparently not present, but rather complex ions such as [NbOCl$_4$]$^-$ or [NbOCl$_5$]$^{2-}$. The products of partial hydrolysis of the pentahalides are oxyhalides, such as NbOF$_3$, NbOCl$_3$, and NbOBr$_3$. They are designated in the older literature as columbyl or columboxy compounds. The more stable oxyhalogen compounds of niobium are complexes, such as NbOF$_3 \cdot$ 3NaF, NbOF$_3 \cdot$ ZnO $\cdot$ 6H$_2$O, and NbOF$_3 \cdot$ 2KF $\cdot$ H$_2$O.

TABLE 1. REPRESENTATIVE PROPERTIES OF REFRACTORY METALS

Property	Tungsten	Tantalum	Molybdenum	Niobium
Density, g/cm^3	19.3	16.6	10.2	8.7
Melting point, °C	3,390–3,420	2,996	2,617	2,458–2,468
Boiling point, °C	5,660	5,325–5,525	4,612	4,742
Linear coefficient of expansion per °C	4.3×10^{-6}	6.5×10^{-6}	4.9×10^{-6}	7.2×10^{-6}
Thermal conductivity, 20°C (cal/cm^2/cm/°C/s)	0.40	0.13	0.35	0.13
Specific heat, 20 °C (cal/g/°C)	0.032	0.036	0.061	0.065
Working temperature, °C	1,700	ambient	1,600	ambient
Electrical conductivity, % IACS	31	13	30	12
Nuclear cross section (thermal neutrons, Barns/atom)	19.2	21.3	2.4	1.1
Tensile strength, 1000 psi				
20°C	100–500	100–150	120–200	75–150
500°C	175–200	35–45	35–65	35
1000°C	50–75	15–20	20–30	13–17
Young's Modulus of Elasticity, psi				
20°C	59×10^6	27×10^6	46×10^6	14×10^6
500°C	55×10^6	25×10^6	41×10^6	7×10^6
1000°C	50×106	22×106	39×10^6	—
Poisson's Ratio	0.284	0.35	0.32	0.38
Corrosion resistance, 100°C				
Dilute HNO$_3$		N	R	N
Dilute H$_2$SO$_4$		N	S	VS
Concentrated H$_2$SO$_4$		N	S	R
Dilute HCl	See Tungsten	N	S	—
Concentrated HCl		N	SL	SL
Concentrated Hydrofluoric acid		R	SL	R
Phosphoric acid, 85%		N	SL	VS
Concentrated NaOH		R	N	R

N = no appreciable corrosion.
VS = < 0.0005 inch (0.013 millimeter) per year.
SL = 0.0005–0.005 inch (0.013–0.13 millimeter) per year.
S = 0.005–.01 inch (0.13–0.25 millimeter) per year.
R => 0.01 inch (0.25 millimeter) per year.

Further complexes of Nb(V) are formed with oxygen-function compounds, such as *o*-dihydroxybenzene and acetylacetone.

The so-called niobic acid is the hydrated pentoxide, $Nb_2O_5 \cdot xH_2O$, insoluble in H_2O.

The metaniobates of the alkali metals, $MNbO_3$, the orthoniobates M_3NbO_4 and the pyroniobates, $M_4Nb_2O_7$, where M is an alkali metal, can be prepared by various alkali carbonate or hydroxide fusion processes.

Niobium forms a nitride, NbN, and a carbide, NbC.

Niobium forms a diamino compound, $(NH_2)_2NbCl_3$, and an ammine complex, $NbCl_5 \cdot 9NH_3$. It forms two cyclopentadienyl compounds, $(C_5H_5)_2NbBr_3$ and $(C_5H_5)Nb(OH)Br_2$. Its other organometallic compounds are essentially oxygen-functional ones, such as $Nb(OCH_3)_5$, $Nb(OC_2H_5)_5$, $Nb(O)(OC_5H_{11})_3$, and $Nb(OC_5H_{11})_5$. These compounds are named as substituted niobanes (thus, the last is pentabutoxy niobane) or as alkyl niobate esters.

Health and Safety Factors

Toxicity data on niobium and its compounds are sparse. The most common materials, e.g., niobium concentrates, ferroniobium, niobium metal and niobium alloys, appear to be relatively inert biologically. Limited animal experiments show high toxicity for some salts, which are related to disturbance of enzyme action. Niobium hydride has moderate fibrogenic and general toxic action. Recommended maximum allowable concentrations are 6 mg/m³. Recommended maximum permissible concentration of Nb in reservoir water is 0.01 mg/L. The threshold for affecting clarity and biological oxygen demand (BOD) is 0.1 mg/L.

Unstable niobium isotopes that are produced in nuclear reactors or similar fission reactions have typical radiation hazards. See also **Radioactivity**.

Fire fighting procedures for niobium and niobium hydride powder, suggest letting the fire burn itself out. Small fires can be controlled by smothering with dry table salt or using Type D dry powder fire-extinguishing material. Under no circumstances should water be used, as a violent explosion may result.

Additional Reading

Carter, G.F. and D.E. Paul: *Materials Science and Engineering,* ASM International, Materials Park, OH, 1991.

Gupta, C.K. and A.K. Suri: *Extractive Metallurgy of Niobium,* CRC Press, LLC., Boca Raton, FL, 1994.

Lide, D.R.: *CRC Handbook of Chemistry and Physics* 84th Edition," CRC Press, LLC., Boca Raton, FL, 2003.

Staff: *ASM Handbook—Properties and Selection: Nonferrous Alloys and Pure Metals,* ASM International, Materials Park, OH, 1990.

Staff: "Tool-Steel Developers Take Note of Niobium," *Advanced Materials 7 Processes,* 15 (June 1991).

Titran, R.H.: "Niobium and Its Alloys," *Advanced Materials & Processes,* 34 (November 1992).

NITER. This potassium nitrate mineral KNO_3 of orthorhombic crystallization usually occurs as thin crusts, or as silky acicular crystals. It has a hardness of 2, and specific gravity of 2.09–2.14, is of white color, translucent with vitreous luster. It occurs as a surface efflorescence, or in soils rich in organic material in arid regions. World occurrences include Spain, Italy, Egypt, Arabia, India, Russia, the western United States, the Republic of South Africa, and Bolivia, South America. Large quantities were recovered from limestone caves in Tennessee, Kentucky, Alabama and Ohio during the Civil War for use in the manufacture of gunpowder. It is used as a source of nitrogen compounds, for explosives and fertilizers.

NITRATION. Nitration is defined in this article as the reaction between a nitration agent and an organic compound that results in one or more nitro ($-NO_2$) groups becoming chemically bonded to an atom in this compound. Nitric acid is used as the nitrating agent to represent *C*-, *O*-, and *N*-nitrations. *O*-Nitrations result in esters. *N*-nitrations are often used as a first step for production of nitramines.

For example, a nitro group is substituted for a hydrogen atom, and water is a by-product. Nitro groups may, however, be substituted for other atoms or groups of atoms. Nitro compounds can also be produced by addition reactions, e.g., the reaction of nitric acid or nitrogen dioxide with unsaturated compounds such as olefins or acetylenes.

Nitrations are highly exothermic, i.e., ca 126 kJ/mol (30 kcal/mol) However, the heat of reaction varies with the hydrocarbon that is nitrated. The mechanism of a nitration depends on the reactants and the operating conditions. The reactions usually are either ionic or free-radical. Ionic nitrations are commonly used for aromatics; many heterocyclics; hydroxyl compounds, e.g., simple alcohols, glycols, glycerol, and cellulose; and amines. Nitration of paraffins, cycloparaffins, and olefins frequently involves a free-radical reaction.

Ionic Nitration Reactions

Acid mixtures containing nitric acid and a strong acid, e.g., sulfuric acid, perchloric acid, selenic acid, hydrofluoric acid, boron trifluoride, or an ion-exchange resin containing sulfonic acid groups, can be used as the nitrating feedstock for ionic nitrations. These strong acids are catalysts that result in the formation of nitronium ions, NO_2^+. Sulfuric acid is almost always used industrially since it is both effective and relatively inexpensive.

Mechanism. The NO_2^+ mechanism has been accepted since about 1950 for the nitration of most aromatic hydrocarbons, glycerol, glycols, and numerous other hydrocarbons in which mixed acids or highly concentrated nitric acid are used. The mechanism has been discussed in detail and critically analyzed. NO_2^+ attacks an aromatic compound (ArH) as follows:

$$ArH + NO_2^+ \longrightarrow \left[Ar \begin{array}{c} H \\ \\ NO_2 \end{array} \right]^+ \longrightarrow ArNO_2 + H^+ \tag{1}$$

Nitrosonium ions, NO^+, are, however, the ions employed to start the nitration sequence for easily nitratable aromatic compounds such as phenol.

Kinetics of Aromatic Nitrations. The kinetics of aromatic nitrations are functions of temperature, which affects the kinetic rate constant, and of the compositions of both the acid and hydrocarbon phase. In addition, a larger interfacial area between the two phases increases the rates of nitration since the main reactions occur at or near the interface. Larger interfacial areas are obtained by increased agitation and by the proper choice of the volumetric % acid in the liquid–liquid dispersion. The viscosities and densities of the two phases and the interfacial tension between the phases are important physical properties affecting the interfacial area.

Increased agitation of a given acid–hydrocarbon dispersion results in an increase in interfacial areas owing to a decrease in the average diameter of the dispersed droplets. As the droplets decrease in size, the ease of separation of the two phases, following completion of nitration, also decreases.

Industrial Applications. Significant process changes have occurred in many nitration plants. Continuous-flow units are now widely used in the 1990s, replacing batch nitrations. A well-designed continuous-flow plant often offers all of the following advantages, per unit weight of product, as compared to batch units; increased safety, decreased energy requirements, reduced amounts of undesired by-products, fewer environmental problems, reduced labor requirements, and lower operating expenses.

Many nitrated products are explosives, including DNT, TNT, and nitroglycerine (NG). To minimize the potential for run away reactions and explosions, the compositions of the feed acids and reaction conditions are currently better controlled than formerly. In some processes, 99% or more of the feed HNO_3 reacts. Dispersions (or mixtures) of such a waste acid and the nitration product are relatively safe to handle. Also, centrifugal separators are used in many modern processes to rapidly separate the hydrocarbon and used acid phases. Rapid separation greatly reduces the amounts of nitrated materials in the plant at any given time and reduces undesired reactions of the nitrated products.

Considerable effort has been made to minimize energy requirements in the nitrations plants too.

A significant concern in all nitration plants using mixed acid centers on the disposal method or use for the waste acids. They are sometimes employed for production of superphosphate fertilizers. Processes have also been developed to reconcentrate and recycle the acid.

Nitrations Using N_2O_5. Considerable worldwide interest has occurred in the late 1980s and the early and mid-1990s for nitrations using N_2O_5. Production of nitramines (or *N*-nitrations) is particularly promising, since these compounds are more stable in the presence of $N_2O_5–HNO_3$ solutions as compared to mixed acids containing H_2SO_4. Good results have been obtained for the production of the high explosives, cyclotetramethylenetetranitramine or HMX and DADN. Another high explosive, polynitrofluorene, has been produced via *C*-nitrations, for the first time. The overall exothermicities of the reactions are less when N_2O_5 is used, as compared to mixed acids.

Solutions of CH_2Cl_2 and N_2O_5 have only mild nitrating power. Yet some nitrations are rapid: the N_2O_5 reacts on an almost stoichiometric basis, and only minimal residual nitric acid is present upon completion of the nitration. Some nitrations having unique characteristics can be accomplished.

Free-Radical Nitrations of Paraffins

Both vapor-phase and liquid-phase processes are employed to nitrate paraffins, using either HNO_3 or NO_2. The nitrations occur by means of free-radical steps, and sufficiently high temperatures are required to produce free radicals to initiate the reactions steps.

Chemistry. Free-radical nitrations consist of rather complicated nitration and oxidation reactions. When nitric acid is used in vapor-phase nitrations, the main initiating reaction (eq. 2) produces either $\cdot NO_2$ or $\cdot ONO$. Temperatures of > ca350°C are required to obtain a significant amount of initiation, and equation 2 is the rate-controlling step for the overall reaction. Reactions 3 and 4 are chain-propagating steps.

$$HNO_3 \longrightarrow \cdot OH + NO_2 \tag{2}$$

$$RH + \cdot OH \longrightarrow R \cdot + H_2O \tag{3}$$

$$R \cdot + HNO_3 \longrightarrow RNO_2 + \cdot OH \tag{4}$$

When nitrogen dioxide is used, the main reaction steps are as in equations 5 and 6.

$$RH + NO_2 \longrightarrow R \cdot + HNO_2 \longrightarrow R \cdot + \cdot OH + NO \tag{5}$$

$$R \cdot + \cdot NO_2 \longrightarrow RNO_2 \tag{6}$$

An important side reaction in all free-radical nitrations is production, of unstable alkyl nitrites (eq. 7). They decompose to form nitric oxide and alkoxy radicals (eq. 8) which form oxygenated compounds and lower molecular weight alkyl radicals which can form lower molecular weight nitroparaffins by reactions 4 or 6. The oxygenated hydrocarbons often react further to produce carbon oxides and water.

$$R \cdot + \cdot ONO \longrightarrow RONO \tag{7}$$

$$RONO \longrightarrow NO + RO \cdot \tag{8}$$

Processes for Paraffin Nitrations. Propane is thought to be the only paraffin that is commercially nitrated by vapor-phase processes. Temperature control is a primary factor in designing the reactor, and several approaches have been investigated. A spray nitrator in which liquid nitric acid is sprayed into hot propane is used industrially. Relatively small-diameter tubular reactors, fluidized-bed reactors, and molten salt reactors have all been successfully used in laboratory units.

Health and Safety Factors

The danger of an explosion of a nitrated product generally increases as the degree of nitration increases. Nitroaromatics and some polynitrated paraffins are highly toxic when inhaled or when contacted with the skin. All nitrated compounds tend to be highly flammable.

LYLE F. ALBRIGHT
Purdue University

Additional Reading

Albright, L. F., R. V. Carr, and R. J. Schmitt: *Nitration: Recent Laboratory and Industrial Developments,* American Chemical Society (ACS), Washington, DC, 1996.

Fischer, J. F.: in H. Feuer and A. T. Nielsen, eds., *Nitro Compounds: Recent Advances in Synthesis and Chemistry,* VCH Publishers, New York, NY, 1990, Chapt. 3.

Gilbert, E.: in S. M. Kaye, ed., *Encyclopedia of Explosives and Related Items,* Vol. 9, U.S. Army Armament Research and Development Command, Dover, NJ, 1980, T235–286.

Hill, M. E. and co-workers: in *ACS Symposium Series No. 22,* American Chemical Society, Washington, DC, 1976, Chapt. 17, pp. 253–271.

Urbanski, T.: *Chemistry and Technology of Explosives,* Vols. 1–3, The Macmillan Co., New York, NY, 1964, 1965; Vol. 4, Permagon Press, Elmsford, NY, 1983.

NITRIC ACID. [CAS: 7697-37-2]. This important industrial chemical has been known for at least 1000 years. The acid was known to alchemists as *aqua fortis* (strong water) or *aqua valens* (powerful water). Nitric acid was of particular interest to the early experimenters because of its ability to dissolve a number of metals, including copper and silver. Early chemists were also fascinated by the fact that addition of sal ammoniac (ammonium chloride) gave *aqua regia* (royal water) which dissolves gold as well as silver.

Nitric acid is a colorless liquid, sp. gr. 1.503 (25°C), freezing point −41.6°C, and boiling point 86°C. The 100% acid is not entirely stable and must be prepared from its azeotrope (constant-boiling mixture) by distillation with concentrated sulfuric acid. Reagent grade HNO_3 is a water solution containing about 68% HNO_3 (weight). This strength corresponds to the constant-boiling mixture of the acid with water, which is 68.4% HNO_3 (weight) at atmospheric pressure and boils at 121.9°C. Nitric acid is completely miscible with water. It forms two solid hydrates, $HNO_3 \cdot H_2O$ and $HNO_3 \cdot 2H_2O$, with corresponding melting points of approximately −38 and −18.5°C. Nitric acid is a strong acid and a powerful oxidizer. In dilute solutions, it is almost completely ionized to H^+ and NO_3^- ions and behaves like a strong acid.

With organic compounds, HNO_3 may act as a nitrating agent, as an oxidizing agent, or simply as an acid. The classic example of nitration is its reaction with benzene or toluene in the presence of concentrated H_2SO_4 to form nitrobenzene or nitrotoluene (TNT). An example of oxidation properties is in the oxidation of cyclohexanol by HNO_3 to produce adipic acid, an intermediate of nylon. Behaving like an acid, it forms nitroglycerin by esterification of glycerol in the presence of concentrated sulfuric acid.

An interesting property of HNO_3 is its ability to passivate some metals, such as iron and aluminum. This property is of significant industrial importance, since modern processes for producing the acid depend on it. Modern suitability formulated stainless steel alloys are usefully resistant to nitric acid through a wide range of conditions. The acid's passivity or the metal's resistance to attack is attributed to the formation of a protective oxide layer on the surface of the metal.

Nitric acid is a high tonnage industrial chemical. Much of the production is used in the manufacture of agricultural fertilizers, largely in the form of ammonium nitrate, NH_4NO_3. See also **Fertilizer.** About 15% of the nitric acid produced is used in explosives (nitrates and nitro compounds), and about 10% is consumed by the chemical industry. As the red fuming acid or as nitrogen tetroxide, HNO_3 is used extensively as the oxidizer in propellants for space rockets and missiles.

Production of Nitric Acid

Three commercial methods have been developed for nitric acid production: (1) the reaction between sulfuric acid and sodium nitrate, (2) the thermal combination of oxygen and nitrogen in air, and (3) the catalytic oxidation of ammonia and absorption of the gaseous products in waters. There are numerous variations of these fundamentals processes. The principal process used today is based on the catalytic oxidation of ammonia and absorption of the gaseous products in water. This process was developed by Ostwald (Germany) and based on earlier work of Kuhlmann (France). In the Ostwald process, HNO_3 is produced in a 3-stage operation: (1) Ammonia is oxidized to nitric oxide, (2) the nitric oxide is further oxidized to nitrogen dioxide, and (3) the gases are absorbed in water to yield HNO_3 according to

$$4NH_3 + 5O_2 \longrightarrow 4NO + 6H_2O$$

$$2NO + O_2 \longrightarrow 2NO_2$$

$$3NO_2 + H_2O \longrightarrow 2HNO_3 + NO$$

The nitric oxide formed in the last equation returns to the gas phase, is reoxidized to nitrogen dioxide, and reabsorbed. These reactions are highly exothermic. In actuality, numerous complex reactions occur in addition to the main reactions just outlined.

In a manufacturing plant, air is preheated, mixed with superheated ammonia vapor, and reacted catalytically over a gauze composed of 90% platinum and 10% rhodium at a temperature of 800–960°C and operating pressures between 1 and 8.2 atmospheres. The reaction produces nitrogen dioxide, NO_2 and nitric oxide, NO. The latter is oxidized to NO_2 in the reaction train. The NO_2 actually exists in equilibrium with its dimer, N_2O_4. This equilibrium mixture, sometimes referred to as nitrogen peroxide, is absorbed in water in a cooled absorber tower to form HNO_3 at a strength of 55–60% HNO_3.

Health and Safety Factors

Nitric acid and the oxides of nitrogen found in its fumes are highly toxic and capable of causing severe injury and death. It is corrosive, and can destroy human tissue. Nitric acid is regulated by OSHA, which lists it as a Process Safety Hazardous Chemical and Air Contaminant. Under SARSH, the EPA lists it as an Extremely Hazardous Substance and Toxic Chemical.

First-aid practices for the treatment of exposure to nitric acid should be obtained from a current version of the Material Safety Data Sheet or other appropriate safety literature.

Web Reference

Chemical and Other Safety Information: http://physchem.ox.ac.uk/MSDS/

NITRIDES.

At elevated temperatures and pressures, nitrogen combines with most elements to form nitrogen compounds. In the presence of metals and semimetals, it forms nitrides where nitrogen has a nominal valence of -3. Atomic nitrogen, which reacts much more readily with the elements than does molecular nitrogen, forms nitrides with elements that do not react with molecular nitrogen even at very high pressures. The binary compounds of nitrogen may be classified, according to their chemical and physical properties, into four groups: saltlike, metallic, nonmetallic or diamondlike, and volatile nitrides.

Properties

Saltlike Nitrides. The nitrides of the electropositive metals of Group 1 (IA), 2 (IIA), AND 3 (IIIB) form saltlike nitrides having predominantly heteropolar (ionic) bonding and are regarded as derivatives of ammonia. The composition of these nitrides is determined by the valency of the metal. The thermodynamic stability of the saltlike nitrides increases with increasing group number. The saltlike nitrides are generally electrical insulators or ionic conductors. The nitrides of the Group 3 (IIIB) metals are metallic conductors or at least semiconductors, and thus, represent a transition to the metallic nitrides. The saltlike nitrides are characterized by sensitivity to hydrolysis.

Metallic Nitrides. The nitrides of the transition metals of Groups 6 and 7 (IVB–VIIB) are generally termed metallic nitrides because of metallic conductivity, luster, and general metallic behavior. These compounds, characterized by a wide range of homogeneity, high hardness, high melting points, and good corrosion resistance, are grouped with the carbides (qv), borides and silicides as refractory hard metals. Metallic nitrides can be alloyed with other nitrides and carbides of the transition metals to give solid solutions.

Although there are several hundred binary nitrides, only a relative few ternary bimetallic nitrides are known. A group of ternaries of the composition $M_x M'_y N_z$, where M is an alkali, alkaline-earth, or a rare-earth metal and M is a transition or post-transition metal, have been synthesized.

Metallic nitrides are wetted and dissolved by many liquid metals and can be precipitated from metal baths.

Nonmetallic (Diamondlike) Nitrides. The nitrides of some elements of Groups 13 (IIIA) and 14 (IVA) e.g., BN, Si_3N_4, AlN, GaN, and InN, are characterized by predominantly covalent bonding. These are stable chemically, have high degrees of hardness (e.g., cubic BN) and high melting points, and are nonconductive or semiconductive. The structural elements of diamondlike nitrides are tetrahedral, M_4N, which are structurally related to diamond.

Volatile Nitrides. The nitrogen compounds of the nonmetallic elements are generally not very stable. Exceptions are $(SN)_x$, which is polymeric, chemically stable, and has semimetallic properties; and $(PNCl_2)_x$, which has attracted some scientific interest as inorganic rubber. None of the volatile nitrides has obtained any substantial industrial application except ammonia (hydrogen nitride) and nitrogen oxide (oxygen nitride).

Preparation

Nitriding Metals or Metal Hydrides. Metals or metal hydrides may be nitrided using nitrogen or ammonia. Pure metal powders or pure metal hydride powders yield nitride products that are nearly as pure as the precursors.

Metal Oxides. A process based on the reaction of metal oxides rather than more expensive metal powders and nitrogen or ammonia in the presence of carbon is economical and has possibilities for large-scale production. However, the products, which contain oxygen and carbon, are not very pure.

Metal Compounds. Many nitrides, e.g., BN, AlN, TiN, ZrN, HfN, CrN, Re_2N, Fe_2N, Fe_4N, and Cu_3N, may be prepared by the reaction of the corresponding metal halide and ammonia. Nitrides may also be obtained by the reaction of ammonia and oxygen-containing compounds, ammonium-oxo complexes, or oxides and ferrous metal oxides. These nitrides, however, are not very pure and may contain residual oxygen and halogen.

Precipitation from the Gas Phase. The van Arkel gas decomposition process gives especially pure nitrides and nitride films, which under certain conditions may precipitate as single crystals. The nitrides include TiN, ZrN, HfN, VN, NbN, BN, and AlN.

Other Methods of Preparation. The nitrides, Si_3N_4, Ge_3N_4, Zn_3N_2, Cd_3N_2, and Ni_3N, may also be produced by thermal decomposition of the corresponding metal amide of imide. Rb_3N and Cs_3N are obtained by azide decomposition. AlN and Si_3N_4 can be produced by the carbothermal reduction of intercalation compounds, magadiite- and montmorillonite-polyacrylonitrile. Nitrides low in nitrogen can be synthesized from nitrides having a higher nitrogen content by decomposition in a vacuum or by reduction with hydrogen.

The formation of nitrides from gaseous halides, ammonia, and nitrogen (atomic and molecular) in a plasma processing torch is possible by means of a specific type of plasma processing called cathodic arc plasma deposition (CAPD). Plasma nitriding offers several advantages: It is nonpolluting and energy efficiency, provides flexible deposition conditions without sacrifice of quality, minimizes distortion, and is easily applicable to compound film deposition. Ion implantation directly inserts nitrogen into metal surfaces.

Nitride-Containing Layers. The hardening, i.e., increase in nitrogen content, achieved by nitriding special alloy steels is technologically significant in the heat treatment of high quality parts, such as gears. Hardness properties are imparted by the resulting coatings of needle-shaped precipitates of the nitrides and carbonitrides of iron, aluminum, chromium, molybdenum, etc. The hardness of these coatings exceeds that of the precipitation-hardened parts by ca 30%.

In nitriding or carbonitriding of condensed materials, molten cyanides are used at ca 570°C. This method produces fairly thick coatings of nitrides or carbonitrides after ca 1 hr without the risk of distortion during surface hardening.

Wear-resistant layers can be deposited on the surface of nearly every kind of material (e.g., steel, cast iron, and cemented carbides) by a chemical vapor deposition (CVD) process. See also **Thin Films**.

Manufacture and Processing

Nitride Coatings. Carbide tips coated with titanium nitride or titanium carbonitride are usually manufactured by a CVD process using $TiCl_4$, H_4, and N_2 in a hot-wall reactor.

Silicon Nitride. Silicon nitride is manufactured either as a powder as a precursor for the production of hot-pressed parts or as self-bonded, reaction-sintered, silicon nitride parts.

Health and Safety Factors

As a chemical group, toxicity of nitrides generally stems from the possible reactions with water to form toxic fumes (especially ammonia) rather than from the nitride. There are, of course, exceptions: Fine powder or dust of the nitrides of the transition metals can be pyrophoric; nitrides of the actinide metals are carcinogenic.

The diamondlike nitrides, especially as dust, can irritate the lungs or cause scratching of the eyes owing to mechanical means. Nitrides of the 11(IB) and 12(IIB) metals and especially the volatile nitrides have to be handled with extreme care because of their instability and high degree of toxicity.

Uses

Nitrides are used for their high strength and hardness, in nuclear applications, solid electrolytes, refractories, abrasives, coatings and lubrication, catalysis, and electronic and optoelectronic applications.

ERIC J. MARKEL
M. E. LEAPHART II
University of South Carolina

Additional Reading

Albrecht, M., and J. Neugebauer: *Nitride Semiconductors: Handbook on Materials and Devices,* John Wiley & Sons, Inc., New York, NY, 2003.

Freer, R. ed.,: NATO ASI Series: *The Physics and Chemistry of Carbides, Nitrides, and Borides,* Vol. 185, Kluwer Academic Publishers, Boston, MA, 1990.

Gil, B.: *Group III Nitride Semiconductor Compounds: Physics and Applications,* Oxford University Press, New York, NY, 1998.

Goldschmidt, H.: *Interstitial Alloys,* Butterworths, London, 1967.

Moustakas, T. D., and J. I. Pankove: *Gallium-Nitride (GaN) II,* Vol. 5, Elsevier Science & Technology Books, New York, NY, 1998.

Rabenau, A.: *Solid State Ionics* **6**, 277 (1982).

Samsonov, G. V.: *Nitridij,* Naukova Dumka, Kiev, USSR, 1969.

Toth, L. E.: *Transition Metal Carbides and Nitrides,* Academic Press, Inc., New York, NY, 1971.

NITRIDING. Surface hardening of alloy steels by heating the metals to a temperature of $490-650°C$ in an atmosphere of partially dissociated NH_3 (ammonia). As in cyaniding, hardening results from the formation of nitrides of iron and of certain alloying elements that may be present in the steel. Much longer heating time is required than in carburizing practice, and while the depth of penetration is generally less, the maximum hardness at the surface is higher, $900-1,100$ D.P.H. (Vickers Brinell) compared to $800-900$ D.P.H. for an average carburized case. Nitriding also differs from carburizing in that the parts are fully heat-treated to develop the required core properties before the nitriding treatment. Because of the comparatively low temperature of the process, distortion and dimensional changes are at a minimum. Nitrided steels have good corrosion-resistance when used for valves, pump parts, shafting, and bearing surfaces operating in steam, crude oil, gasolines, and gaseous products of combustion. The fatigue strength is also improved by nitriding.

Other typical applications are piston pins, crankshafts, cylinder liners, timing gears, gauges, and ball and roller bearing parts.

NITRILE RUBBER. See **Elastomers**.

NITRILES. Nitriles, or organic cyanides, are organic compounds which contain the cyano (i.e., $-CN$) group. Nitriles are often considered derivatives of carboxylic acids and are named according to the carboxylic acid which is produced upon hydrolysis of the nitrile. For example, cyanomethane (methyl cyanide) is named acetonitrile, because hydrolysis of its cyano group yields acetic acid. Nitriles which contain additional functional groups are typically named as cyano-substituted compounds, (e.g., cyanoacetic acid). Nitriles which contain a hydroxy ($-OH$) group on the carbon atom that is bonded to the cyano moiety are known as cyanohydrins (qv). Aliphatic nitriles are named as derivatives of the longest carbon chain and the carbon of the nitrile is included.

General Preparations and Chemical Properties

While nitriles may be prepared by several methods, the reaction of alkyl halides with sodium cyanide to produce nitriles (eq. 1) is a general reaction with wide applicability:

$$RX + NaCN \longrightarrow RCN + NaX \tag{1}$$

where $X = Cl$, Br, or I. If dimethyl sulfoxide is used as solvent, high yields of nitriles can be obtained with both primary and secondary alkyl chlorides. See also **Sulfoxides**.

Ammoxidation, a vapor-phase reaction of hydrocarbon with ammonia and oxygen (air) (eq. 2), can be used to produce hydrogen cyanide (HCN), acrylonitrile, acetonitrile (as a by-product of acrylonitrile manufacture), methacrylonitrile, benzonitrile, and toluinitriles from methane, propylene, butylene, toluene, and xylenes, respectively. See also **Acrylonitrile**; and **Methacrylic Acid and Derivatives**.

$$RCH_3 + NH_3 + O_2 \xrightarrow{\text{catalyst}} RCN + H_2O \tag{2}$$

Addition of HCN to unsaturated compounds is often the easiest and most economical method of making organonitriles. However, the addition of HCN to unactivated olefins and the regioselective addition to dienes is best accomplished with a transition metal catalyst.

Chemistry and Uses of Nitriles

As a class of compounds, nitriles have broad commercial utility that includes their use as solvents, feedstocks, pharmaceuticals, catalysts, and pesticides. The versatile reactivity of organonitriles arises both from the reactivity of the $C\equiv N$ bond, and from the ability of the cyano substituent to activate adjacent bonds, especially $C-H$ bonds. Nitriles can be used to prepare amines, amides, amidines, carboxylic acids and esters, aldehydes, ketones, large-ring cyclic ketones, imines,

heterocycles, orthoesters, and other compounds. Some of the more common transformations involve hydrolysis or alcoholysis to produce amides, acids and esters, and hydrogenation to produce amines, which are intermediates for the production of polyurethanes and polyamides.

Acrylonitrile is an important monomer both for plastics and synthetic fibers. Acetonitrile, a by-product of acrylonitrile manufacture, is commercially important for solvent extraction, reaction media, and as an intermediate in the preparation of pharmaceuticals (qv) and other organic chemicals. See also **Extraction (Liquid–Liquid)**. Propionitrile, a by-product of the electrodimerization of acrylonitrile to adiponitrile, is used as a chemical intermediate. Hydrogenation of organonitriles to amines provides important intermediates both for polyurethanes (by way of isocyanates) and polyamides (nylons); adiponitrile is used almost exclusively by the manufacturers in the production of 1,6-diaminohexane (hexamethylenediamine), an intermediate for nylon 6,6. Other nitriles that are produced in thousands of metric tons per year include acetone cyanohydrin, 2-amino-2-methylpropionitrile, and fatty acid nitriles. Acetone cyanohydrin is an intermediate for the preparation of methyl methacrylate and acrylic resins, (e.g., lucite and plexiglas) and for 5,5-dimethylhydantoin, which is used to make commercial water treatment chemicals. 2-Amino-2-methylpropionitrile is an intermediate for the preparation of azobis(isobutyronitrile), which is a widely used polymerization initiator, (e.g., Vazo 64) and in the production of some agrichemicals. Other aminonitriles are unisolated intermediates in the production of chelants such as ethylenediaminetetraacetate (EDTA) and nitrilotriacetate (NTA). The fatty acid nitriles are intermediates in the production of a large variety of commercial amines and amides.

General Health and Safety Factors

As a class of compounds, the two main toxicity concerns for nitriles are acute lethality and osteolathyrsm. Nitriles vary broadly in their ability to cause acute lethality and subtle differences in structure can greatly affect toxic potency. The biochemical basis of their acute toxicity is related to their metabolism in the body.

The propensity of nitriles to release cyanide subsequent to metabolism is the basis of their acute toxicity. Cyanohydrins are acutely toxic because they are unstable and release cyanide quickly. Persons handling nitriles should take precautions to prevent inhalation of fumes or skin contact.

Acetonitrile

[CAS: 75-05-8]. Acetonitrile (methyl cyanide), CH_3CN, is a colorless liquid with a sweet, ethereal odor. It is completely miscible with water and its high dielectric strength and dipole moment make it an excellent solvent for both inorganic and organic compounds including polymers. Many gases also are highly soluble in acetonitrile. It forms low boiling azeotropes with many organics and high boiling azeotropes with BF_3, $SiCl_4$, and $(CH_3)_4Pb$.

Although acetonitrile is one of the more stable nitriles, it undergoes typical nitrile reactions and is used to produce many types of nitrogen-containing compounds.

Most, if not all, of the acetonitrile produced commercially in the United States recently was isolated as a by-product from the manufacture of acrylonitrile by propylene ammoxidation. The acetonitrile is recovered as the water azeotrope, dried, and purified by distillation.

Uses. Because of its good solvency and relatively low boiling point, acetonitrile is used widely as a recoverable reaction medium, particularly for the preparation of pharmaceuticals. Its largest use is for the separation of butadiene from C_4 hydrocarbons by extractive distillation.

Acetonitrile also is used as a catalyst and as an ingredient in transition-metal complex catalysts. There are many uses for it in the photographic industry and for the extraction and refining of copper. It also is used as a reagent for the preparation of a wide variety of compounds.

Adiponitrile

Adiponitrile (hexanedinitrile, dicyanobutane, ADN), $NC(CH_2)_4CN$, is manufactured mainly for use as an intermediate for hexamethylenediamine (1,6-diaminohexane), which is a principal ingredient for nylon-6,6. BASF has announced the development of a process to make caprolactam from adiponitrile. Caprolactam is used to produce nylon-6.

Pure adiponitrile is a colorless liquid and has no distinctive odor. It is soluble in methanol, ethanol, chloroalkanes, and aromatics but has low solubility in carbon disulfide, ethyl ether, and aliphatic hydrocarbons. At $20°C$, the solubility of adiponitrile in water is ca 8 wt %; the solubility increases to 35 wt % at $100°C$.

Adiponitrile undergoes the typical nitrile reactions, e.g., hydrolysis to adipamide and adipic acid and alcoholysis to substituted amides and esters.

Adiponitrile is made commercially by several different processes utilizing different feedstocks. The reaction of adipic acid with ammonia in either liquid or vapor phase produces adipamide as an intermediate, which is subsequently dehydrated to adiponitrile. The most widely used catalysts are based on phosphorus-containing compounds. Vapor-phase processes involve the use of fixed catalyst beds; whereas, in liquid–gas processes, the catalyst is added to the feed. DuPont currently practices a butadiene-to-adiponitrile route based on direct addition of HCN to butadiene.

Uses. The principal use of adiponitrile is for hydrogenation to hexamethylene diamine leading to nylon-6,6. Adipoquanamine, prepared by the reaction of adiponitrile with dicyandiamide (cyanoguanidine), has typical liquid nitrile properties that suggest its use as an extractant for aromatic hydrocarbons.

α-Aminonitriles

α-Aminonitriles are compounds containing both cyano and amine substituents attached to the same carbon atom. They are versatile synthetic intermediates that are used to make amino acids, agrichemicals, chelants, radical initiators, and water-treatment chemicals. In some cases, aminonitriles produced as intermediates are not isolated, but immediately further reacted, for example by hydrolysis, as is the case in producing ethylenediaminetetraacetate (EDTA) or nitrilotriacetate (NTA). Isolated and commercially available aminonitriles include 2-amino-2-methylpropanenitrile (aminoisobutyronitrile, AN-64), 2-amino-2-methylbutanenitrile (AN-67), 2-amino-2,4-dimethylpentanenitrile (AN-52), and 1-aminocyclohexane carbonitrile (AN-88). The designations in parentheses arise from their identity as intermediates in the production of azo radical initiators.

In 1990, DuPont began practicing a one-step process in which a ketone is treated simultaneously with both HCN and ammonia at 40–60°C. This process (Fig. 1) is both faster and more selective than previous two-step processes.

Physical Properties. α-Aminonitriles are stable at modest temperatures (<70°C) in the absence of water; in the presence of water, they can degrade to their original constituents, i.e., ketone (aldehyde), ammonia and hydrogen cyanide if insufficient ammonia is present. The aminonitriles based on ketones are clear colorless liquids, but sometimes appear yellow to brown depending on the synthetic procedure and the amount of decomposition. They are soluble in polar organic solvents and in aromatic solvents.

Uses. α-Aminonitriles may be hydrolyzed to amino acids, such as is done in producing ethylenediaminetetracetate (EDTA) or nitrilotriacetate (NTA). In these cases, formaldehyde is utilized in place of a ketone in the synthesis. The principal use of the ketone-based aminonitriles is in the production of azobisnitrile radical initiators.

Azobisnitriles

Azobisnitriles are efficient sources of free radicals for vinyl polymerizations and chain reactions, e.g., chlorinations. See also **Initiators (Free-Radical)**; **Initiators (Anionic)**; and **Initiators (Cationic)**. These compounds decompose in a variety of solvents at nearly first-order rates to give free radicals with no evidence of induced chain decomposition. They can be used in bulk, solution, and suspension polymerizations; and because no oxygenated residues are produced, they are suitable for use in pigmented or dyed systems that may be susceptible to oxidative degradation.

The structures of several members of this class of compounds are shown below. They are crystalline solids that are produced by hypochlorite oxidation of α-aminonitriles.

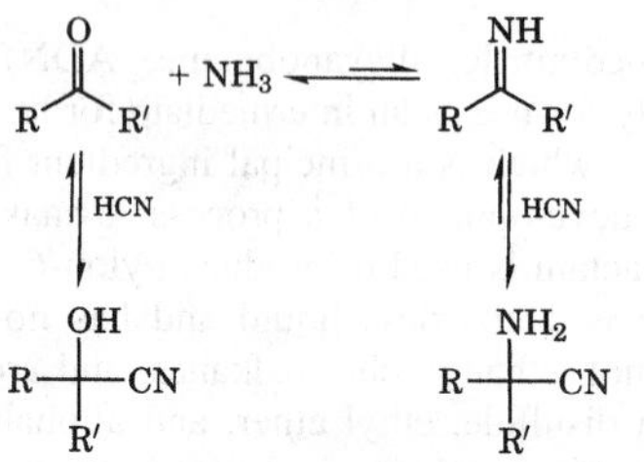

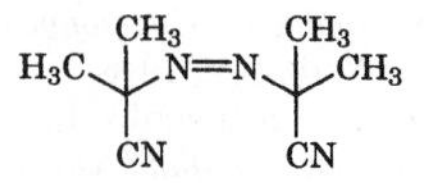

2,2′-Azobis(isobutyronitrile)

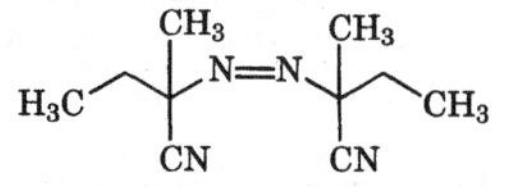

2,2′-Azobis(2-methylbutanenitrile)

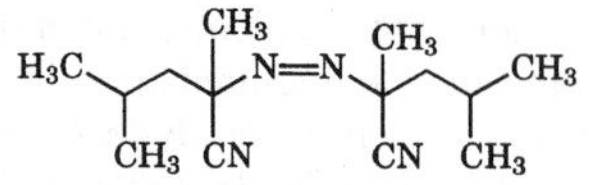

2,2′-Azobis(2,4-dimethylpentanenitrile)

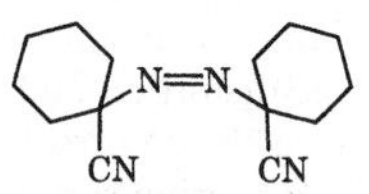

1,1′-Azobis(cyanocyclohexane)

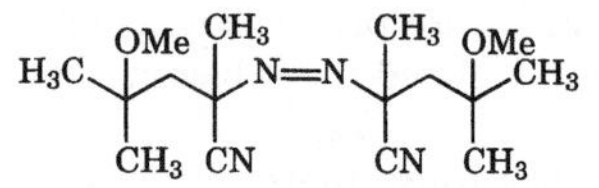

2,2′-Azobis(4-methoxy-2,4-dimethylpentanenitrile)

These compounds are essentially insoluble in water, sparingly soluble in aliphatic hydrocarbons, and soluble in functional compounds and aromatic hydrocarbons.

In solution, the azobisnitriles decompose on heating to form two free radicals with the liberation of nitrogen (eq. 1):

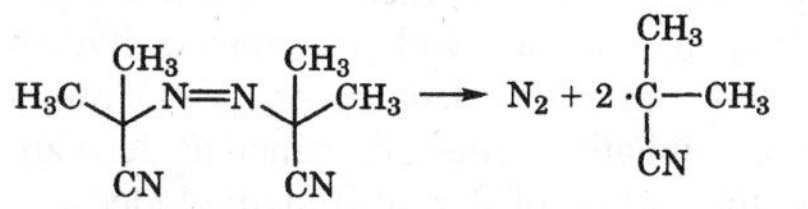

(1)

Uses. The azobisnitriles have been used for bulk, solution, emulsion, and suspension polymerization of all of the common vinyl monomers, including ethylene, styrene vinyl chloride, vinyl acetate, acylonitrile, and methyl methacrylate. The polymerizations of unsaturated polyesters and copolymerizations of vinyl compounds also have been initiated by these compounds.

Benzonitrile

[CAS: 100-47-0]. Benzonitrile, C_6H_5CN, is a colorless liquid with a characteristic almondlike odor. It is miscible with acetone, benzene, chloroform, ethyl acetate, ethylene chloride, and other common organic solvents but is immiscible with water at ambient temperatures and soluble to ca 1 wt% at 100°C. It distills at atmospheric pressure without decomposition, but slowly discolors in the presence of light.

Like acetonitrile, benzonitrile is a powerful solvent for many inorganic and organic materials including some polymers. It can be converted to a large number and variety of derivatives by simple syntheses; e.g., by hydrolysis, it can be converted to either benzoic acid or benzamide. The most important reaction is with dicyandiamide to produce 2,4-diamino-6-phenyl-1,3,5-triazine (benzoguanamine):

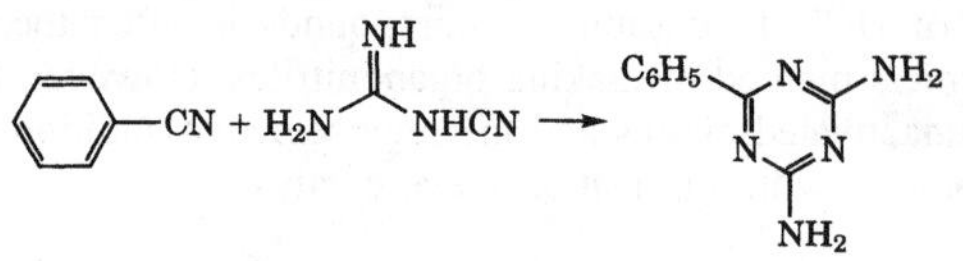

Benzonitrile can be produced in high yield by the vapor-phase catalytic ammoxidation of toluene:

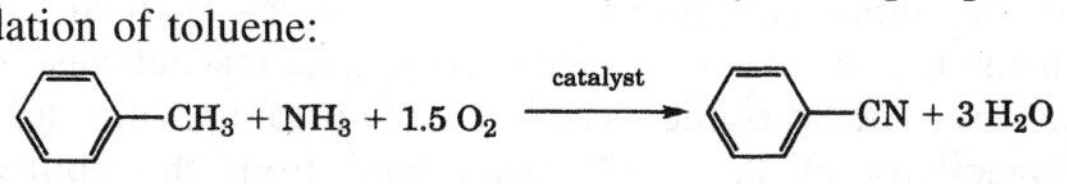

A more recent process involves the reaction of benzoic acid (or substituted benzoic acid) with urea at 220–240°C in the presence of a metallic catalyst.

Fig. 1.   Reactive pathway for α-aminonitriles synthesis

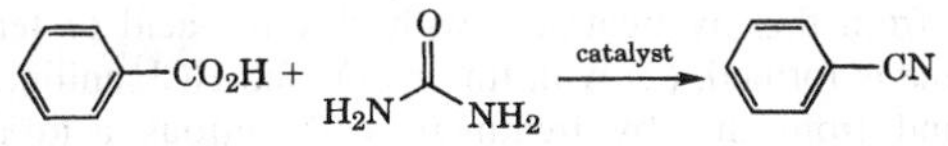

Uses. The most important commercial use for benzonitrile is the synthesis of benzoguanamine, which is a derivative of melamine and is used in protective coatings and molding resins. See also **Amino Resins**; and **Cyanamides**.

Cyanoacetic Acid and Esters

Cyanoacetic acid, $CNCH_2COOH$, is a strong organic acid with a dissociation constant at $25°C$ of 3.36×10^3. It is prepared by the reaction of chloroacetic acid with sodium cyanide. It is hygroscopic and highly soluble in alcohols and diethyl ether but insoluble in both aromatic and aliphatic hydrocarbons. It undergoes typical nitrile and acid reactions but the presence of the nitrile and the carboxylic acid on the same carbon cause the hydrogens on C-2 to be readily replaced. The resulting malonic acid derivative decarboxylates to a substituted acrylonitrile:

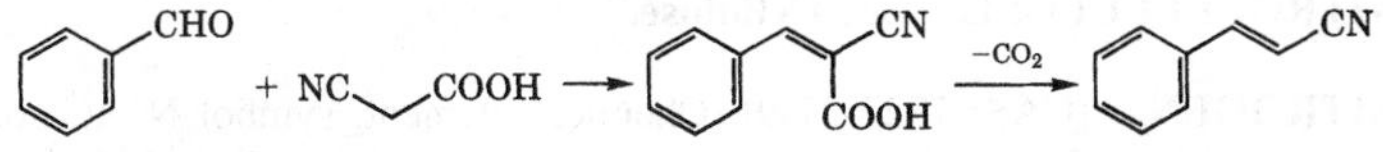

The methyl and ethyl esters of cyanoacetic acid are slightly soluble in water but are completely miscible in most common organic solvents including aromatic hydrocarbons. The esters, like the parent acid, are highly reactive, particularly in reactions involving the central carbon atom. They are prepared by esterification of cyanoacetic acid and are used principally as chemical intermediates.

Uses. Although cyanoacetic acid can be used in applications requiring strong organic acids, its principal use is in the preparation of malonic esters and other reagents used in the manufacture of pharmaceuticals. See also **Alkaloids**; and **Vitamin**.

Isophthalonitrile

Isophthalonitrile (1,3-dicyanobenzene, IPN), is a white solid which melts at $161°C$ and sublimes at $265°C$. It is slightly soluble in water but readily dissolves in dimethylformamide, *N*-methylpyrrolidinone and hot aromatic solvents. IPN undergoes the reactions expected of an aromatic nitrile. It is prepared by vapor-phase ammoxidation of *meta*-xylene. Its principal use is as an intermediate to amines. As a reagent, IPN can be used to convert aromatic acids to nitriles in near quantitative yields.

2-Methylglutaronitrile

Methylglutaronitrile (2,3-dicyanobutane) MGN, is a by-product of Du-Pont's adiponitrile process.

Uses. Methylglutaronitrile is readily hydrogenated to give 2-methyl-1,5-pentanediamine (DYTEK A, MPMD), used as a comonomer in polyamide fibers and resins, as a curing agent for epoxy coatings, and as its isocyanate in specialty urethanes. A co-product of the DYTEK A process is 3-methylpiperidine, which can be used to produce vulcanization accelerators for rubber curing.

Pentenenitriles

Pentenenitriles are produced as intermediates and by-products in DuPont's adiponitrile process. 3-Pentenenitrile is the principal product isolated from the isomerization of 2-methyl-3-butenenitrile.

Uses. 3-Pentenenitrile (3PN) is used entirely by the manufacturers to make adiponitrile. *cis*-2-Pentenenitrile (2PN) can be cyclized catalytically at high temperature to produce pyridine, a solvent and agricultural chemical intermediate. 2PN is also used in the manufacture of pentachloropyridine, an intermediate in the insecticide Dursban, and 1,3-pentadiamine, which is used as a curing agent for epoxy coatings and as a chain modifier in polyurethanes.

Fatty Acid Nitriles

Fatty acid nitriles are produced as intermediates for a large variety of amines and amides. See also **Carboxylic Acids**. Fatty acid nitriles are produced from the corresponding acids by a catalytic reaction with ammonia in the liquid phase. They have little use other than as intermediates.

Disclaimer

This article has been reviewed by the Office of Pollution Prevention and Toxics, U.S. Environmental Protection Agency, and approved for publication. Approval does not signify that the contents necessarily reflect the views and policies of the Agency, nor does mention of commercial products or synthesis constitute endorsement or recommendation for use.

RONALD J. MCKINNEY
E. I. du Pont de Nemours & Co., Inc.

STEPHEN C. DEVITO
U.S. Environmental Protection Agency

Additional Reading

Barton, D. H. R. and W. D. Ollis: *Comprehensive Organic Chemistry,* Vol. 2, Pregamon Press, Oxford, UK, 1979, pp. 528–562.
Hoffman, R. V.: *Organic Chemistry: An Intermediate Text,* 2nd Edition, John Wiley & Sons, Inc., Hoboken, NJ, 2004.
Lide, D. R.: *CRC Handbook of Chemistry and Physics,* 84th Edition, CRC Press, Boca Raton, FL, 2003.
Moss, R. A., M. S. Platz, and M. Jones: *Reactive Intermediate Chemistry,* John Wiley & Sons, Inc., Hoboken, NJ, 2004.
Moury, D. T. *Chem. Rev.* **42**, 192 (1948).
Solomons, T. W. Graham, and C. B. Fryhle: *Organic Chemistry,* 8th Edition, John Wiley & Sons, Inc., New York, NY, 2003.
Torssell, K.: *Nitrile Oxides, Nitrones and Nitronates in Organic Synthesis: Novel Strategies in Synthesis,* Vol. 20, John Wiley & Sons, Inc., New York, NY, 1988.
U.S. Pat. 2,481,826 (Sept. 13, 1949), Cosby J. N. (to Allied Chemical).
U.S. Pat. 2,915,455 (Nov. 10, 1959), Smiley R. A. (to DuPont).

NITRO- AND NITROSO-COMPOUNDS. Nitro-compounds contain the nitro-group ($-NO_2$) attached directly to a carbon atom; nitroso-compounds contain the nitroso-group ($-NO$) similarly attached. A very important member of this group is nitrobenzene, which upon reduction yields a variety of products, important in the synthesis of drugs and dyes. See Table 1.

TABLE 1. REPRESENTATIVE NITRO- AND NITROSO COMPOUNDS

Compound	Formula	Melting point,°C	Boiling point,°C
REPRESENTATIVE NITRO COMPOUNDS			
Nitrobenzene	$C_6H_5 \cdot NO_2$	6	211
1,3-Dinitrobenzene	$C_6H_4(NO_2)_2$ (1,3)	90	302
2-Nitrotoluene	$CH_3C_6H_4(NO_2)$ (2)	−11	222
2,4-Dinitrotoluene	$CH_3C_6H_3(NO_2)_2$ (2,4)	70	300
Trinitrotoluene (TNT)	$CH_3C_6H_2(NO_2)_3$ (2,4,6)	81	240 expl.
3-Nitrophenol	$HOC_6H_4 \cdot NO_2$ (3)	96	194 (70 torr)
2,4,6-Trinitrophenol (picric acid)	$HOC_6H_2(NO_2)_3$ (2,4,6)	122 expl.	>300
4-Nitrobenzaldehyde	$C_6H_4(COH)(NO_2)$ (1,4)	58	164 (23 torr)
4-Nitrobenzoic acid	$C_6H_4(COOH)(NO_2)$ (1,4)	240	subl.
4-Nitrobenzyl alcohol	$C_6H_4(CH_2OH)(NO_2)$ (1,4)	93	185 (12 torr)
2-Nitronaphthalene	$C_{10}H_7(NO_2)$ (2)	79	165 (15 torr)
1-Nitroanthraquinone	$C_6H_4(CO)_2C_6H_3(NO_2)$ (1)	230	subl.
2-Nitropropane	$(CH_3)_2CHNO_2$	−93	120
Nitroethyl alcohol	$CH_2OHCH_2NO_2$	< −80	194
Nitrobromoform (bromopicrin)	NO_2CBr_3	10	expl.
Nitrochloroform (chloropicrin)	NO_2CCl_3	−64	112
Nitrofurane	$C_4H_3O \cdot NO_2$	28	
Nitrourea		155 dec.	
Nitroguanidine		246	
1,3-Nitroaniline	$C_6H_4(NO_2)(NH_2)$ (1,3)	114	>285

(continued overleaf)

TABLE 1. (*continued*)

Compound	Formula	Melting point, °C	Boiling point, °C
REPRESENTATIVE NITROSO COMPOUNDS			
Nitrosobenzene	C_6H_4NO	68	58 (18 torr)
4-Nitrosophenol (4-quinoneoxime)	$C_6H_4(OH)(NO)$ (1,4)	125	144 dec.
4-Nitrosonaphthol-1 (4-naphthaquinoneoxime)	$C_{10}H_6(OH)(NO)$ (1,4) or $C_{10}H_6(O)(NOH)$ (1,4)	193	
2-Nitrosonaphthol-1	$C_{10}H_6(OH)(NO)$ (1,2)	163 dec.	
N-Nitrosomethylaniline	$C_6H_5N\diagdown^{CH_3}_{NO}$	13	128 (20 torr)
4-Nitrosophenylaniline	$C_6H_5NH \cdot C_6H_4NO$	145	
1-Nitrosonaphthylamine-2	$C_{10}H_6(NH_2)(NO)$ (2,1)	151	
Diphenylnitrosamine	$(C_6H_5)_2N \cdot NO$	66	

dec., decomposes; expl., explodes; sub., sublimes

Alkylnitro-Compounds:

Primary	Secondary	Tertiary
$CH_3CH_2 \cdot NO_2$	$(CH_3)_2CH \cdot NO_2$	$(CH_3)_3C \cdot NO_2$
Nitroethane	Nitrodimethylmethane (2 - nitropropane)	Nitrotrimethylmethane

Isomeric Nitrites:

$CH_3CH_2 \cdot ONO$	$(CH_3)_2CH \cdot ONO$	$(CH_3)_3 \cdot ONO$
Ethylnitrite	Isopropylnitrite	1,1-dimethylethyl nitrite

Alkylnitroso-Compounds:
$(CH_3)_3C \cdot NO$
Nitrosotrimethylmethane

Nitrates

		:
$CH_3CH_2 \cdot ONO_2$	$(CH_3)_2CH \cdot ONO_2$	$(CH_3)_3C \cdot ONO_2$
Ethylnitrate	Isopropylnitrate	1,1-dimethylethylnitrate

Nitrosamine:
$(C_2H_5)_2N$:NO
Diethylnitrosamine

Benzenoid Nitro- and Nitroso-Compounds:

Mononitro-compound	Dinitro-compound	Trinitro-compound
Nitrobenzene	1,3-Dinitrobenzene	1,3,5-Trinitrobenzene

Nitroso-compounds

Nitrosobenzene	Diphenylnitrosamine

Under the proper conditions of concentration of HNO_3 and of temperature, benzene forms mainly nitrobenzene, nitrobenzene forms mainly 1,3-dinitrobenzene, and 1,3-dinitrobenzene, mainly 1,3,5-trinitrobenzene.

When nitrobenzene is treated (1) with zinc and calcium chloride or ammonium chloride solution, beta-phenylhydroxylamine, C_6H_5NHOH, is formed, and from this by treatment with chromic acid or ferric chloride nitrosobenzene is formed, (2) with tin or iron and HCl, aniline, $C_6H_5NH_2$, is formed and from this by treatment with nitrous acid followed by treatment with stannous chloride plus HCl phenylhydrazine, $C_6H_5NH \cdot NH_2$, is formed.

Mono- or poly-substituted nitro-compounds are changed in whole or in part to the corresponding amino-compounds by proper choice of reducing agent and temperature, e.g., in acid medium 1,3-dinitrobenzene yields 1,3-phenylenediamine, $C_6H_4(NH_2)_2(1, 3)$, and with ammonium sulfide yields 3-nitroaniline $(1)H_2NC_6H_4NO_2(3)$. When diphenylnitrosamine is reduced, 1,1-diphenylhydrazine, $(C_6H_5)_2N \cdot NH_2$, is formed.

See also **Nitration**.

Additional Reading

Nishimura, S.: *Handbook of Heterogeneous Catalytic Hydrogenation for Organic Synthesis,* John Wiley & Sons, Inc., New York, NY, 2001.
Patai, S.: *Chemistry of Amino, Nitroso and Nitro Compounds,* Vol. 1, John Wiley & Sons, Inc., New York, NY, 1997.

NITROCELLULOSE. See **Cellulose**.

NITROGEN. [CAS: 7727-37-9]. Chemical element, symbol N, at. no. 7, at. wt. 14.0067, periodic table group 15, mp $-209.86°C$, bp $-195.8°C$, critical temperature $-147.1°C$, critical pressure 33.5 atmospheres, density 1.14 g/cm^3 (solid), 1.25057 g/L (0°C, 760 torr), 0.9675 (air = 1.0000). Solid nitrogen has a hexagonal crystal structure. Nitrogen at standard conditions is a colorless, odorless, tasteless gas. The gas is slightly soluble in H_2O (2.35 parts nitrogen in 100 parts H_2O at 0°C), the solubility decreasing with increasing temperature (1.55 parts nitrogen in 100 parts H_2O at 20°C). Nitrogen is slightly soluble in alcohol and is essentially insoluble in most other known liquids. There are two naturally occurring isotopes, ^{14}N and ^{15}N, with ^{14}N by far the most abundant (99.635%). Four radioactive isotopes have been identified, ^{12}N, ^{13}N, ^{16}N, and ^{17}N, all with extremely short half-lives measured in seconds or minutes. In terms of abundance in igneous rocks in the Earth's crust, nitrogen does not appear among the first 37 most abundant elements. In terms of abundance in seawater, nitrogen ranks 16th, with an estimated 2,300 tons of nitrogen per cubic mile of seawater. In terms of cosmic abundance, nitrogen ranks 7th. For comparison, assigning a value of 10,000 to silicon, the figure for nitrogen is 160,000 and that for hydrogen, estimated the most abundant, a figure of 3.5×10^8. Of dry air in the earth's atmosphere, disregarding pollutants, 78.09% is nitrogen by volume and 75.54% by weight. In the atmosphere, the nitrogen is mixed with oxygen, argon, the rare gases, CO_2, and H_2O vapor. Nitrogen was first identified as an element by Daniel Rutherford in 1772. Lavoisier further confirmed Rutherford's findings in 1776. Like oxygen, nitrogen is essential to practically all forms of life, making some of the compounds of this element extremely important as foods and fertilizers. Nitrogen serves the important function of diluent in the earth's atmosphere, controlling natural burning and respiration rates that otherwise would proceed much faster with higher concentrations of oxygen. Nitrogen is an important ingredient of numerous inorganic and organic compounds, including alkaloids, amides, amines, cyanides, cyanogens, diazo compounds, hydrazines, imides, nitrates, nitrides, nitrites, nitriles, oximes, purines, pyridines, and ureas. In terms of high-tonnage production, the nitrogen compound NH_3 (ammonia) ranks first with worldwide production exceeding 50 million tons annually.

First ionization potential 14.84 eV; second, 29.47 eV; third, 47.17 eV; fourth, 73.5 eV; fifth, 97.4 eV. Oxidation potentials $H_2N_2O_2 + 2H_2O \longrightarrow 2HNO_2 + 4H^+ + 4e^-$, -0.80 V; $N_2O_4 + 2H_2O \longrightarrow 2NO_3^- + 4H^+ + 2e^-$, -0.81 V; $HNO_2 + H_2O \longrightarrow NO_3^- + 3H^+ + 2e^-$, -0.94 V; $NO + 2H_2O \longrightarrow NO_3^- + 4H^+ + 3e^-$, -0.96 V; $NO + H_2O \longrightarrow HNO_2 + H^+ + e^-$, -0.99 V; $2NO + 2H_2O \longrightarrow N_2O_4 + 4H^+ + 4e^-$, -1.03 V; $2HNO_2 \longrightarrow N_2O_4 + 2H^+ + 2e^-$, -1.07 V. $N_2O + 3H_2O \longrightarrow 2HNO_2 + 4H^+ + 4e^-$, -1.29 V; $N_2O + H_2O \longrightarrow 2NO + 2H^+ + 2e^-$, -1.59 V; $N_2 + H_2O \longrightarrow N_2O + 2H^+ + 2e^-$, -1.77 V; $N_2O_4 + 4OH^- \longrightarrow 2NO_3^- + 2H_2O + 2e^-$, 0.85 V; $NO + 2OH^- \longrightarrow NO_2^- + H_2O + e^-$, 0.46 V; $N_2O_2^{2-} + 4OH^- \longrightarrow 2NO_2^- + 2H_2O + 4e^-$, 0.18 V; $NO_2^- + 2OH^- \longrightarrow NO_3^- + H_2O + 2e^-$, -0.01 V; $N_2O_2^{2-} \longrightarrow 2NO + 2e^-$ -0.10 V; $N_2O + 6OH^- \longrightarrow 2NO_2^- + 3H_2O + 4e^-$ -0.15 V; $N_2O + 2OH^- \longrightarrow 2NO + H_2O + 2e^-$ -0.76 V. $2NO_2^- \longrightarrow N_2O_4 + 2e^-$ -0.88 V.

Other physical properties of nitrogen are given under **Chemical Elements**

Industrial Nitrogen

Like many of the elements, the compounds of nitrogen by far exceed the use of elemental nitrogen (discounting its important role as diluent in the atmosphere). Industrially, nitrogen gas is produced as a by-product in the liquefaction of air to produce pure oxygen. For some applications, nitrogen provides an excellent inert atmosphere for electric furnace operations and for the gaseous insulation of transformers. An inert atmosphere is required where air must be excluded. Nitrogen is one of the three main gases used for such atmospheres, the other two being carbon monoxide and hydrogen. In providing an inert atmosphere, nitrogen reduces the velocities of reactions, lowers the partial pressure and reduces the flammability of any active gases that may be present. Since commercial nitrogen usually contains traces of oxygen, H_2O vapor, and CO_2, sufficient to cause some oxidation at high temperatures, methane may be added to make the gas fully inert.

Nitrogen gas also is required for nitriding certain alloy steels, but pure gas is not required. The nitrogen is provided by dissociating ammonia at the process temperatures ranging from 475–650°C. Metals treated in this manner are hardened by the formation of nitrides on their surface (casehardening). In cyaniding, iron-base alloys simultaneously absorb carbon and nitrogen by heating the metals in a cyanide salt. Again, the nitrogen is not required in initial gaseous form. See also **Nitriding**. Several powder metallurgy techniques also utilize dissociated NH_3 atmospheres.

Environmental Aspects of Nitrogen

The oxides of nitrogen are among the most critical of air pollutants—both in their effects and in their abatement. These aspects of nitrogen are discussed under **Pollution (Air)**

Chemistry and Compounds

Most of the high-tonnage nitrogen-bearing compounds are described elsewhere in this volume. See also **Ammonia**; **Ammonium Chloride**; **Ammonium Hydroxide**; **Ammonium Nitrate**; **Ammonium Phosphates**; **Ammonium Sulfate**; and **Fertilizer**.

In the laboratory, nitrogen, mixed with argon, neon, krypton, and xenon, is obtained from the air by passing it over heated copper to remove the oxygen, or pure by fractional distillation of liquid air whereby the nitrogen distills off before the oxygen. Pure nitrogen may also be obtained by heating such compounds as ammonium nitrite and ammonium dichromate, and collecting the gas. Mixed with carbon monoxide in producer gas, nitrogen may be utilized without separation by first making methyl alcohol from carbon monoxide and hydrogen and then using hydrogen and nitrogen for ammonia. When nitrogen at low pressure is subjected to the silent electric discharge, activated nitrogen is produced. Activated nitrogen displays a golden yellow afterglow upon cessation of the current, increased by cooling and decreased by heating. This form of nitrogen is very active with phosphorus, with alkali metals (forming azides), with the vapor of zinc, mercury, cadmium, arsenic (forming nitrides), with many metallic chlorides (forming a green fluorescence), and with hydrocarbons (forming hydrocyanic acid and cyanides). The transformation of nitrogen to activated nitrogen is partial, and its return to ordinary nitrogen takes place rapidly, in about one minute.

The metal amides and imides are important in the nitrogen system. The amides of the active metals are produced by (1) reaction of the metal with NH_3, (2) reaction of the metal hydride with NH_3, (3) reaction of the metal nitride with ammonia, (4) reaction with another amide, as $KNH_2 + NaI \longrightarrow NaNH_2 + KI$ (in liquid NH_3). This last method is generally useful for the preparation of the heavy metal amides and imides from halides and binary halogenoids of the heavy metals. Cadmium amide, $Cd(NH_2)_2$ and lead imide, PbNH, for example, are readily prepared in this way. In some cases neither the amide nor the imide is stable, and the reaction proceeds to the nitride.

$$3\ HgBr_2 + 6\ KNH_2 NH_3 liq.Hg_3N_2 + 6\ KBr + 4\ NH_3$$

The metal amides and imides are very reactive with oxygen, and are often unstable or even explosive. Some nitrides (e.g., of silver, gold, and mercury) are explosive, but others are stable. The latter may be obtained, (1) by reaction with the metal with nitrogen or ammonia at higher temperatures, e.g., aluminum nitride and magnesium nitride, AlN and Mg_3N_2, (2) by deamination of the metal amide or azide on heating, e.g., Ba_3N_2. The great thermal stability of certain nitrides, e.g., those of boron, silicon and phosphorus, BN, Si_3N_4 and P_3N_5, is attributed

to polymerization. Many of the transition metal nitrides are interstitial compounds and are hard and metal-like in their properties.

In the nitrogen system, hydrazine is analogous to hydrogen peroxide in the oxygen system, its structure being

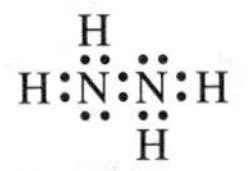

$$H\ddot{:}\ddot{N}\ddot{:}\ddot{N}\ddot{:}H$$

It is readily oxidized, even undergoing auto-oxidation under many conditions, and it is a powerful reducing agent. Like hydrogen peroxide it readily disproportionates (e.g., with a platinum catalyst), giving nitrogen and NH_3. Its reactivity (and other properties) makes it, and its derivative, unsymmetrical dimethylhydrazine, important rocket fuels. It forms addition compounds with many substances, including a monohydrate with H_2O. Hydrazine ($pK_{B1} = 6.04$, $pK_{B2} = 14.88$) forms hydrazinium(1+) compounds, containing the $N_2H_5^+$ ion, analogous to ammonium, and hydrazinium(2+) compounds containing the $N_2H_6^{2+}$ ion.

Hydroxylamine is related in its structure both to hydrazine (see formula above) and to hydrogen peroxide.

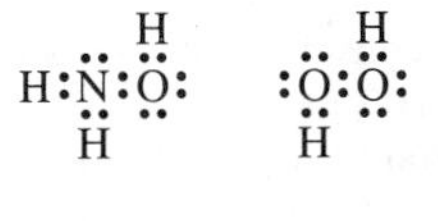

The chemical properties of hydroxylamine also suggest a compound intermediate between hydrazine and hydrogen peroxide. Its bond lengths are, N−, 1.46 Å, N−H, 1.01 Å, O−H, 0.96 Å, and its angles are H−O−N, 103°, H−N−O, 105°, and H−N−H, 107°. It is a base ($pK_B = 9.02$), forming salts containing the hydroxylammonium ion $HONH_3^+$.

Hydrazoic acid, HN_3, $pK_A = 4.72$, and most of its covalent compounds (including its heavy metal salts) are explosive. It is formed (1) in 90% yield by reaction of sodium amide with nitrous oxide, (2) by reaction of hydrazinium ion with nitrous acid, (3) by oxidation of hydrazinium salts, (4) by reaction of hydrazinium hydrate with nitrogen trichloride (in benzene solution). Hydrazoic acid forms metal azides with the corresponding hydroxides and carbonates. It reacts with HCl to give ammonium chloride and nitrogen, with H_2SO_4 to form hydrazinium acid sulfate, with benzene to form aniline, and it enters into a number of oxidation-reduction reactions.

The azides, except those of mercury(I), Hg(I), thallium(I), Tl(I), copper, Cu, silver, Ag, and lead, Pb, are readily prepared from hydrazoic acid and the oxide or carbonate of the metal, or by metathesis of the metal sulfate with barium azide. They are all thermally unstable, giving nitrogen and free metal or occasionally nitride. The azide ion appears to resonate between four structures:

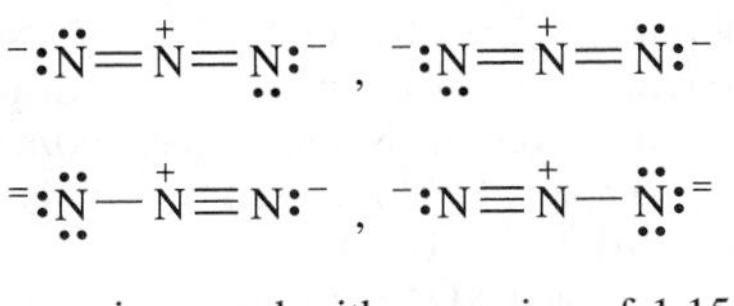

These structures are in accord with a spacing of 1.15 Å and electronic charges of −0.83, 0.66, and 0.83 on the three nitrogen atoms.

N(I) Compounds. Hydration of nitrogen(I) oxide, N_2O to hyponitrous acid, $H_2N_2O_2$, is not possible. However, the latter decomposes (in three steps) to yield the former, which is thus its anhydride. Spectroscopic studies indicate a linear structure for N_2O, resonating between

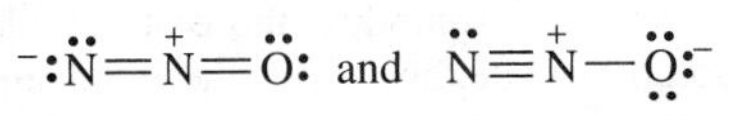

However, heat capacity measurements give a higher entropy at low temperatures than spectroscopic studies do, which is explained by a partial randomness of the structure at low temperatures.

Hyponitrous acid ($pK_{A1} = 7.05$, $pK_{A2} = 11.0$) and its salts are obtained by: (1) reduction of sodium nitrite with (a) sodium amalgam, (b) by electrolysis, (c) by stannous or ferrous salts; (2) reduction of alkyl nitrates; (3) reduction of hydroxylamine by noble metal oxides; and (4) by reduction of sodium hydroxylamine monosulfonate in alkaline solution.

Explosive salts such as NaNO can be prepared by the reaction of NO and liquid ammonia solutions of alkali metals. The unstable free acid, HNO, is thought to be an intermediate in many redox reactions of nitrogen compounds.

Nitramide, NO_2NH_2, a weak acid ($pK_A = 6.59$), is relatively more stable than its isomer hyponitrous acid.

N(II) Compounds. Nitrogen(II) oxide is formed in many reductions of nitrous acid, but is best prepared pure by reduction with ferrous ions, Fe^{2+}, or iodide ions, I^- It undergoes many types of addition reactions, but its very slight tendency to dimerize and its low reactivity under ordinary conditions suggest that its odd electron lies in an antibonding orbital of very low energy; and the molecular orbital formulation is

$$NO[KK(z\sigma)^2(y\sigma^*)^2(x\sigma)^2(w\pi)^4(v\pi^*)]$$

The nitrosyl compounds can be readily classified on the basis of three modes of reaction of the NO molecule in accordance with the above formulation.

1. It can lose (or partly lose) the odd electron to form an ion of the formula

$$:N\equiv O:^+$$

This formula gives rise to ONF, ONCl and ONBr by direct reaction of NO and the halogen. These are covalent compounds. Such salts as $NOBF_4$, $NOPF_6$, $NOAuF_4$, $NOSO_3F$, and $NOHSO_4$, on the other hand, are ionic. These may be considered the salts of nitrous acid acting as a base, $ONOH \square NO^+ + OH^-$, $pK_B = 18.2$.

2. It can gain an electron to form a negative ion of the formula

$$N\equiv\ddot{O}:^-$$

Thus dry NO reacts with sodium in liquid ammonia to form sodium nitrosyl, NaNO (empirical formula).

3. It can share a pair of electrons to form a coordinate link, as it does in coordination compounds. In most of these, it appears to coordinate as the positive ion, by transfer of an electron to an acceptor metal, which is thereby reduced by 1 unit in oxidation state. This causes, in some cases, the need for placing a negative charge on the metal. To avoid this, Pauling assumed the presence of four bonding electrons, involving structures of the type

$$M=\overset{+}{N}=\ddot{O}:$$

Nitrogen(III) Compounds. Nitrogen(II) oxide, NO, readily enters into equilibrium with NO_2 to form N_2O_3, nitrogen sesquioxide. The latter is unstable even at room temperature and consists of an equilibrium mixture of the three compounds. Its structure appears to be $O=N-NO_2$. If an equimolar mixture of NO and NO_2 is cooled and condensed, a blue liquid, bp $3.5°C$, largely N_2O_3, is obtained. The latter readily combines with H_2O to form nitrous acid, HNO_2 ($pK_A = 3.29$). Nitrous acid is unstable, forming the equilibrium mixture, $3HNO_2 \square NO_3^- + 2NO + H_3O^+$, which in concentrated solution or on warming is largely displaced to the right ($K = 39.6$ at $30°C$). Moreover, the NO undergoes further reactions, so that the actual system is complex. One of these reactions is: $NO + OH \longrightarrow NO^+ + OH^- \square NO \cdot OH \rightleftharpoons HNO_2$.

The existence of NO^+ and NO^- helps to explain the kinetics of nitrous acid as an oxidizing agent. It oxidizes I^-, Sn^{2+}, Fe^{2+}, Ti^{3+}, $S_2O_3^{2-}$, SO_2, and H_2S. It reacts with NH_3, urea, sulfonates and some other nitrogen compounds to produce nitrogen. With aromatic amines in the cold, it gives diazo compounds, while with secondary amines it gives nitroso compounds. Nitrous acid also functions as a reducing agent, as in the reactions with permanganate and hydrogen peroxide, in which nitrate ion is formed.

The nitrites vary widely in solubility, those of the alkalies and alkaline earths being very soluble, while those of the heavy metals are only slightly so. Moreover, the latter are relatively unstable, some decomposing at room temperature. The nitrites, like nitrous acid, function either as oxidizing or reducing agents. X-ray and spectroscopic studies give a triangular structure for the nitrite ion, with the N—O bond length 1.13 Å and the O—N—O angle 120–130°. Values of 1.23 Å and 116° have also been reported. Complex ions containing the NO_2 group may be either nitrito complexes (e.g., $Co(NH_3)ONO^{2+}$) or nitro complexes (e.g., $Co(NH_3)NO_2^{2+}$). The former of these two examples readily isomerizes to the latter.

Nitrosyl fluoride, NOF, and nitrosyl chloride, NOCl, are quite stable, but the bromide decomposes at room temperature. They are prepared by direct union of NO and the halogen, among other methods. Three trihalides, NF_3, NCl_3, and NI_3, are known. The first is a colorless stable gas; NCl_3 is a yellow liquid and NI_3, a brown solid; both are explosive. The contrast in

stability is attributed to the large amount of ionic resonance energy of the N—F bond, which gives NF_3 a negative heat of formation.

Nitrogen(IV) Compounds. Nitrogen dioxide, NO_2, readily associates to form the tetroxide, N_2O_4, so that at ordinary temperatures and pressures both forms are present in equilibrium. Since nitrogen dioxide has an unpaired electron, it is paramagnetic and colored (red). N_2O_4 is diamagnetic and colorless. As with NO, the odd electron is in an antibonding orbital but of higher energy so that NO_2 is more reactive and more readily undergoes dimerization. The N—O bond length is 1.20 Å and the angle is 132° (electron diffraction). The structure of N_2O_4 is, on the basis of spectral and entropy considerations,

$$\ddot{\underset{\cdot\cdot}{O}}:^- \quad \overset{+}{N}-\overset{+}{N} \quad :\ddot{O}:^-$$

This formula is at variance with Pauling's stability argument, but is supported by Ingold's evidence (Nature, **159**, 743, 1947). Longuet–Higgins has proposed the structures

$$\bar{O}-N\overset{O}{\underset{O}{\diamond}}\overset{+}{N}=O \quad \text{and} \quad O=\overset{+}{N}\overset{O}{\underset{O}{\diamond}}N-\bar{O}$$

Nitrogen dioxide molecules react with NO to form N_2O_3, in an equilibrium mixture. The equilibrium mixture of NO_2 and N_2O_4 also reacts with water in a series of reactions

$$2NO_2 + H_2O \rightleftharpoons H^+ + NO_3^- + HNO_2$$
$$3HNO_2 \rightleftharpoons H^+ + NO_3^- + NO + H_2O$$

In warm solution, at high acidity, the second reaction is very rapid. In basic solutions the simple disproportionation $N_2O_4 + 2OH^- \longrightarrow NO_2^- + H_2O$ takes place.

Nitrogen(V) Compounds. Nitrogen(V) oxide, N_2O_5, the anhydride of nitric acid, is a white solid subliming at $32.4°C$ and 760 mm. It hydrates readily to HNO_3, is a strong oxidizing agent, and decomposes at $20°C$ slowly into NO_2 and O_2. Its structure in the gas state consists of the molecules

$$:\ddot{O}: \quad :\ddot{O}: $$
$$\overset{+}{N}-\ddot{O}-\overset{+}{N}$$
$$:\ddot{O}: \quad :\ddot{O}:^-$$

However, x-ray, Raman, and infrared spectra show the crystalline solid to consist of NO_2^+ and NO_3^- ions.

Pure nitric acid, HNO_3, is a colorless liquid boiling with decomposition at $86°C$ and 760 torr. Upon continued heating it decomposes into NO_2, O_2 and H_2O. It is a fairly strong acid ($K_A = 22$), showing dissociation in concentrated solutions, and the presence of nitryl cation, NO_2^+ (nitronium ion). Solutions of HNO_3 in H_2SO_4 owe many of their properties to ions such as NO_2^+ and NO^+, as well, of course, as to HSO_4^- and oxonium ions.

The properties of HNO_3 are in accordance with resonance between the three electronic structures:

$$H:\overset{..}{\underset{..}{O}}=\overset{+}{N}\overset{\ddot{O}:^-}{\underset{\ddot{O}:^-}{\diagup}} \quad H-\overset{..}{\underset{..}{O}}-\overset{+}{N}\overset{\ddot{O}:}{\underset{\ddot{O}:^-}{\diagup}} \quad H-\overset{..}{\underset{..}{O}}-\overset{+}{N}\overset{\ddot{O}:^-}{\underset{\ddot{O}:}{\diagup}}$$

in which the last formula contributes a relatively small proportion to the overall structure. The two N—O bond lengths are 1.22 Å, and N—O—H bond lengths 1.41 Å and 0.96 Å. The N—O—H angle is 90° and the O—N—O angle 130°.

The reactions of nitric acid are of three types: (a) acid-base reactions which are typical of a strong acid; (b) oxidation reactions, such as those with metals and organic materials, the latter often involving carbonization; (c) substitution reactions such as the replacement of —H by —NO_2 in aromatic hydrocarbons, to form nitro compounds, or of hydroxyl hydrogen by —NO_2 to produce esters of HNO_3.

These esters of nitric acid form one of the two groups of nitrates, the covalent group, which are also exemplified by nitryl hypofluorite and hypochlorite ($FONO_2$ and $ClONO_2$), often called fluorine and chlorine nitrate. Most nitrates, however, are ionic, e.g., salts of HNO_3. All metal nitrate are soluble in H_2O. Anhydrous metal nitrates, such as $Cu(NO_3)_2$. $Ti(NO_3)_4$, $VO(NO_3)_3$, $CrO_2(NO_3)_2$, $Si(NO_3)_4$, can be made by the action of liquid N_2O_4 on the metal (e.g., Cu) or of $ClONO_2$ on the corresponding chloride (e.g., the other examples given above).

The nitrate ion is considered to resonate between three equivalent structures of the form:

$$\begin{array}{c} :\!\ddot{O}\!: \\ \| \\ N^+ \\ :\!\ddot{O}\!: \quad \ddot{O}\!:^- \end{array}$$

Two nitryl halides, NO_2F and NO_2Cl, are known, as well as nitryl salts, such as NO_2AsF_6, NO_2SbF_6, $(NO_2)_2SiF_6$, NO_2ClO_4, etc. Nitrogen also forms higher oxides, such as NO_3, and possibly NO_4, under action of the electric discharge.

Nitrate Losses from Disturbed Forest Ecosystems

Nutrient losses occur following a forest harvest or other disturbance, whether natural or anthropogenic. Studies have shown a variety of patterns of such losses. Vitousek et al. (1979) report on a systematic examination of nitrogen cycling in disturbed forest ecosystems and show that at least 8 processes, operating in 3 stages in the nitrogen cycle, can delay or prevent solution losses of nitrate from disturbed forests. The study involved 19 forest sites in the United States, including Pack Forest, Findley Lake, and Cascade Head in the northwest; Tesuque Watersheds in the southwest; Lake Monroe in southern Indiana; Coweeta in southwestern North Carolina; and Harvard Forest, Mount Mossilauke, and Cape Cod in the northeastern United States.

The 3 stages and 8 operative processes identified are:

Stage 1. Processes preventing or delaying ammonium accumulation.
(a) Nitrogen immobilization
(b) Ammonium fixation
(c) Ammonia volatilization
(d) Plant nitrogen uptake

Stage 2. Processes preventing or delaying nitrate accumulation.
(e) Lag in nitrification
(f) Denitrification to: $-N_2$, N_2O, or NO_x, $-NH_4$

Stage 3. Processes preventing or delaying nitrate mobility.
(g) Lack of water
(h) Nitrate sorption
(i) Denitrification at depth

The researchers stress that the net effect of all of these processes, except uptake by regrowing vegetation, is insufficient to prevent or delay losses from relatively fertile sites and thus such sites have the potential for very high nitrate losses following disturbance.

Nitrogen Fixation

A positive balance of usable nitrogen on earth depends upon nitrogen fixation which is the process by which atmospheric nitrogen, N_2, is converted either by biological or chemical means to a form of nitrogen, such as ammonia, NH_3, that can be used by plants and other biological agents. Insofar as the total amount of N_2 fixed, the biological processes for converting from N_2 to NH_3 are the most significant. In biological nitrogen fixation, microorganisms, either free-living or in symbiosis with plants (mainly in root nodules), reduce N_2 to NH_3 at atmospheric pressure and within the temperature range of 20–37°C. This natural process is to be contrasted with industrial chemical conversion processes, which may require up to 300 atmospheres of pressure and a reaction temperature range of 200–300°C.

Biological Nitrogen Fixation. The occurrence and importance to soil fertility of biological nitrogen fixation have been known since the early 1800s. The first major finding did not occur until 1960, however, when it was shown that cell-free extracts of the anaerobic bacterium *Clostridium pasteurianum* could be made to fix nitrogen if molecular oxygen, O_2, were

rigorously excluded—and also if pyruvic acid, a source of energy and electrons, was supplied. This finding demonstrated that studies no longer were restricted to whole cells, as previously indicated, but that it should be possible to isolate and chemically identify the components of the nitrogen-fixing system.

The first demonstrable product of cell-free N_2 fixation is NH_3, as had been strongly suggested by previous whole-cell studies. Since the reduction of N_2 to $2NH_3$ requires six electrons and since most electron transfer systems known in biochemical pathways involve either a one-or a two-electron transfer, it could be expected that either six one-electron or three two-electron transfer steps would be involved in nitrogen fixation. This would also suggest the existence of nitrogen compounds of valence states (reduction states) intermediate between N_2 and NH_3. However, no such intermediates have been found even in systems using cell-free extracts.

Because of failure to detect intermediates, attention was focused on the mechanism in extracts of *Clostridium pasteurianum* through which electrons were transferred from pyruvic acid to the nitrogen-fixing system. These investigations led to the discovery and isolation of the new electron carrier ferredoxin (Fd) which functioned by accepting electrons released during pyruvate oxidation by enzymes present in the clostridial extracts. The electrons from reduced Fd were transferred to a variety of different acceptors as directed by the cell. For example, some of the electrons from reduced Fd were transferred to hydrogenase, an enzyme which combined the electrons with protons (H^+) to produce molecular hydrogen, H_2, a major by-product of this anaerobe. Other electrons from reduced Fd were transferred via a flavoprotein carrier to nicotinamide adenine dinucleotide phosphate ($NADP^+$) to yield NADPH, a reduced electron carrier shown to be important in the metabolism of all biological agents. It was also found that electrons from Fd were required for nitrogen fixation when pyruvate was present as supporting substrate.

A major finding was that H_2, through hydrogenase, would act as an electron source for reducing ferredoxin. Thus, in these extracts, H_2 could be used to reduce $NADP^+$ to NADPH and NO_{2-} to NH_3, and Fd was necessary as an intermediary electron carrier. Since Fd is required for pyruvate-supported N_2 fixation, it may be expected that H_2 would support nitrogen fixation, since reduced Fd is readily produced from H_2 in these extracts. Molecular H_2 alone, however, did no support N_2 fixation. This suggested either that a component other than reduced Fd was required, or that H_2, although capable of reducing Fd, was inhibitory to N_2 fixation as prior whole-cell studies had indicated. If an additional component were required, it appeared that it was produced from pyruvic acid, since pyruvic acid supported active N_2 fixation.

Several unsuccessful attempts were made to obtain N_2 fixation in extracts to which H_2, N_2, and one of the other products of pyruvate metabolism, ATP, were added. Active N_2 fixation did occur, however, when another product of pyruvate metabolism, acetyl phosphate, was added in addition to H_2 and N_2. When compounds such as ADP were removed from cell extracts by dialysis, no N_2 fixation occurred unless ADP was added together with phosphate, H_2, and N_2. Acetyl phosphate then was acting as a source of ATP. The reason ATP did not work directly was that a continuous supply of ATP was required, and a high concentration of ATP, if added directly to a cell-free extract, was highly inhibitory to N_2 fixation. In whole cells that are fixing N_2, a continuous supply of ATP is made available during sugar metabolism.

Genetic Manipulation. High on the list of many researcher's agendas for projects using the practical application of recombinant DNA research has been the possible development of a living organism that will produce ammonia—in an effort to lessen dependence upon costly and highly energy-consuming synthetic ammonia fertilizers. However, at symposia held on this topic, these achievements are considered by most researchers as quite long-range. There are fundamental problems difficult to overcome, including: (1) the possibility that increasing biological nitrogen fixation, for which the plant furnishes the energy, can cause a net decrease in crop yields by depriving the plant of nitrogen for the production of certain critical growth elements; and (2) the very rapid-acting inactivation by oxygen of nitrogen-fixation mechanisms. Cloning techniques may be a path toward introducing nitrogen-fixation genes into certain bacteria. One objective is that of developing new forms of bacteria that will enter into symbiotic relationships with crop plants, such as corn (maize) and wheat, that do not possess their own nitrogen fixation symbionts.

In addition to recombinant DNA and molecular cloning techniques, some scientists have combined their research with more conventional genetic techniques. An *E. coli* plasmid capable of carrying nitrogen-fixation genes of *K. pneumoniae* has been developed. Some researchers also believe that nitrogen-fixation genes may be introduced directly into plant cells to result in a plant that requires no nitrogen fertilizer.

In research activities such as these, much knowledge has been gained concerning the energy needs for biological nitrogen fixation. More energy is used than originally contemplated; for example, 20 moles of adenosine triphosphate (ATP) are required to fix one mole of nitrogen. This contributes largely to the first problem mentioned earlier, namely, the great amount of energy required for the plant to fix its own nitrogen, possibly leading to yield reduction.

The well-known nitrogen fixation by rhizobia depends on photosynthesis by the plant. Although the method is essentially impractical, photosynthesis can be increased by blanketing the plant with an atmosphere enriched in carbon dioxide. When this is done in the laboratory, increased legume yields are reported. Some investigators postulate that this effect is the result of a reduction photorespiration, a rather wasteful process in which carbon dioxide gained through photosynthesis is diverted into a series of less productive pathways in the plant. Investigators have also found that 30% of the energy used by the nitrogenase of most rhizobial species goes to producing hydrogen rather than ammonia. Research has also shown that the organisms that perform the nitrogen fixation function in plants are indeed quite diverse in themselves. Thus, new combinations of plants and organisms may increase efficiency in some cases.

As pointed out by Evans-Barber (1977), nitrogen is fixed by a variety of microorganisms in addition to those associated with legumes. Some of these include bacteria located in soils, in decaying wood, and on the surfaces of plant roots. They also include free-living blue-green algae, with fungi, ferns, mosses, liverworts, and higher plants (Hardy-Havelka, 1975). Reviews of numerous nitrogen-fixing organisms are given by Silvester (1976), Dalton (1974), Bond (1974), and Stewart (1974).

Role of Molybdenum in Nitrogen Fixation. Traditionally, molybdenum has been considered a key to the reduction of molecular nitrogen to ammonia by soil- and water-dwelling microorganisms. The metal is believed to be a part of the catalytically active site of nitrogenase, which is the enzyme that accomplishes the reduction. As early as 1980, researchers (North Carolina State University) suggested that the bacterium *Azobacter vinelandii* may have an alternative system for nitrogen fixation, a mechanism that may not require molybdenum. The proposal was regarded with some skepticism until the findings by researchers (Agriculture and Food Research Council Unit of Nitrogen Fixation, University of Sussex, England) in 1985 that confirmed the fact that *A. vinelandii* does have a second fixation system. Mutants of *A. vinelandii* were studied. Genes coding for the nitrogenase proteins were specifically deleted or inactivated. It was found that deletion of all three nitrogenase structural genes did not interfere with the fixation process. However, the process was effective only when molybdenum was not present. It was also found that the "wild" type of bacterium must have Mo in order to reduce nitrogen. Thus, it appears that the alternative mechanism is activated only when the system is subjected to molybdenum starvation. It has been suggested that the alternative system represents an adaptation to molybdenum-poor soils. The extension of these findings to other nitrogen-fixing microorganisms remains to be accomplished. Some further details pertaining to the Sussex investigation are given by Marx (1985).

Madigan (1979) and associates found that photosynthetic purple bacteria can grow with dinitrogen gas as the only source of nitrogen under anaerobic conditions, with light as the energy source. They also found that *Rhodopseudomonas capsula* can fix nitrogen in darkness with alternative energy conversion systems.

See also **Fertilizer**.

Health and Safety Factors

Gaseous nitrogen is nontoxic and nonflammable, but does not support life. Nitrogen should be stored and used only in well-ventilated areas. Special care must be taken entering an enclosed area, which may be enriched in nitrogen.

Liquid nitrogen and its vapor are extremely cold and can rapidly freeze human tissue. Liquid nitrogen spills should be flushed with water to accelerate evaporation. When exposed to liquid nitrogen, carbon steel, rubber, and plastic become embrittled and may fracture under stress.

Copper, brass, bronze, Monel, aluminum, and 300 series austenitic stainless steels remain ductile and are acceptable for cryogenic service. Liquid nitrogen in poorly insulated containers can concentrate and condense atmospheric oxygen on the exterior surfaces, which may cause a serious fire hazard. Storage vessels or handling equipment should be provided with multiple pressure relief devices to prevent the buildup of high pressure. A pressure relief valve for primary protection and a frangible disk for secondary protection are commonly provided for on commercial liquid storage vessels.

Additional Reading

Bingham, E., B. Cohrssen, and C. Powell: *Patty's Toxicology, Hydrocarbons—Organic Nitrogen Compounds, Vol. 4,* 5th Edition, John Wiley & Sons, Inc., New York, NY, 2000.

Bond, G.: in *The Biology of Nitrogen Fixation,* (A. Quispel, editor), North-Holland, Amsterdam, 1974.

Cheung, H. and J.H. Royal: "Efficiently Produce Ultra-High-Purity Nitrogen On-Site," *Chem. Eng. Progress,* 64 (October 1991).

Clark, J.S.: *Nitrogen, Oxygen and Sulfur Ylide Chemistry (The Practical Approach in Chemistry Series 2002),* Oxford University Press, New York, NY, 2002.

Dalton, H.: *Crit. Rev. Microbiol,* **3**, 183 (1974).

Evans, H.J. and L.E. Barber: "Biological Nitrogen Fixation for Food and Fiber Production," *Science,* **197**, 332–339 (1977).

Graham, P.H. and S.C. Harris: *Biological Nitrogen Fixation,* Unipub, New York, NY, 1984.

Greenwood, N.N. and A. Earnshaw: *Chemistry of the Elements,* 2nd Edition, Butterworth-Heinemann, Inc., Woburn, MA, 1997.

Golterman, H.L.: *Chemistry of Phosphate and Nitrogen Compounds in Sediments,* Kluwer Academic Publishers, Norwell, MA, 2004.

Hardy, R.W.F. and M.D. Havelka: *Science,* **188**, 633 (1973).

Knowles, R. and T.H. Blackburn: *Nitrogen Isotope Techniques,* Academic Press, Inc., San Diego, CA, 1997.

Legocki, A., H. Bothe, and A. Puhler: *Biological Fixation of Nitrogen for Ecology and Sustainable Agriculture,* Springer-Verlag, Inc., New York, NY, 1997.

Leigh, G.J.: *Nitrogen Fixation at the Millennium,* Elsevier Science, New York, NY, 2002.

Lide, D.R.: *CRC Handbook of Chemistry and Physics 84th Edition,* CRC Press, LLC., Boca Raton, FL, 2003.

Madigan, M.T., J.D. Wall, and H. Gest: "Dinitrogen Fixation by Photosynthetic Microorganisms," *Science,* **204**, 1429–1430 (1979).

Marx, J.L.: "Fixing Nitrogen without Molybdenum," *Science,* **229**, 956–957 (1985).

Metz, C.B.: *Biology of Fertilization,* Academic Press, Inc., San Diego, CA, 1985.

Meyers, R.A.: *Handbook of Chemicals Production,* The McGraw-Hill Companies, Inc., New York, NY, 1986.

Postgate, J.R.: *Nitrogen Fixation,* Cambridge University Press, New York, NY, 1998.

Silvester, W.B.: in Proceedings of the 1st International Symposium on Nitrogen Fixation, (W.E. Newton and C.J. Nyman, editors), Washington State University Press, Pullman, WA, 1976.

Staff: "Nitrogen Plant Opens in South Carolina," *Chem. Eng. Progress,* 10 (February 1990).

Staff: "Can Catalytic Combustion in Jet Engines Zap NOx?" *Chem. Eng. Progress,* 19 (March 1992).

Staff: "Advanced Catalyst Zaps Nitrogen," *Chem. Eng. Progress,* 21 (July 1992).

Stevenson, F.J. and M.A. Cole: *Cycles of Soils: Carbon, Nitrogen, Phosphorus, Sulfur, Micronutrients,* 2nd Edition, John Wiley & Sons, Inc., New York, NY, 1999.

Stewart, W.D.P.: in *The Biology of Nitrogen Fixation,* (A. Quespel, editor), North-Holland, Amsterdam, 1974.

Vitousek, P.M. et al.: "Nitrae Losses from Disturbed Ecosystems," *Science,* **204**, 469–474 (1979).

Web References

Nitrogen: http://www.praxair.com/nitrogen

Nitrogen Fixation: http://academic.reed.edu/biology/Nitrogen/

The Nitrogen Cycle: http://www.physicalgeography.net/fundamentals/9s.html

NITROGEN FIXATION. See **Nitrogen**.

NITROGEN GROUP (The). The elements of group 15 of the periodic classification sometimes are referred to as the Nitrogen Group. In order of increasing atomic number, they are nitrogen, phosphorus, arsenic, antimony, and bismuth. The elements of this group are characterized by the presence of five electrons in an outer shell. The similarities of chemical behavior among the elements of this group are less striking than hold for some of the other groups, e.g., the close parallels of the alkali metals or alkaline earths. Although all of the elements of this group have valences in addition to 5+, all do have the 5+ valence in common. Unlike the alkali metals or alkaline earths, for example, the elements of the nitrogen group are not so similar chemically that they comprise a separate group in classical qualitative chemical analysis separations. Three of the five, however, antimony, arsenic, and bismuth are members of the second group in terms of qualitative chemical analysis.

NITROGLYCERIN.   See **Explosives**.

NOBEL, ALFRED B. (1833–1896).   Nobel was a Swedish industrialist and European munitions maker. He was born in Stockholm in 1833. Most of his early education was from tutors and what he learned as he traveled during his teenage years through much of North America and Europe. He learned to speak several languages fluently and studied mechanical engineering.

After his travels, he joined his father's business and worked developing mines, torpedoes, explosives, and other war materials for the Russian czar. The Nobel factory was financially successful until the end of the Crimean War when it fell into bankruptcy. Alfred then went into business with his brother manufacturing drilling tools for oilfields.

In the 1860's Alfred revolutionized the explosives industry by developing the Nobel detonator, a new fuse for nitroglycerin followed by a safe way to handle nitroglycerin by using an organic packing material to reduce its volatility and producing dynamite. In the late 1880's, Nobel produced ballistite, a smokeless powder.

Nobel's inventions brought him a monetary fortune as he controlled most of the explosive manufacturing factories throughout the world. He was, however, disillusioned that his inventions had mostly applications for war. Nobel personally desired world peace. Throughout his life, he had acted in a humanitarian manner. In his will, Nobel directed that the great majority of his estate be used for the purpose of giving yearly prizes to persons whose personal efforts made outstanding contributions to the advancement of physics, chemistry, medicine and physiology, literature, and world peace. The first Nobel Prizes were awarded in 1901 to Wilhelm K. Roentgen (discovery of X-rays in 1895), J.H. van't Hoff (chemical thermodynamics and osmotic pressure), and E.A. von Behring (diphtheria antitoxin). See also **Nobel Prizes**.

J.M.I.

Web Reference

Nobel Foundation: http://www.nobel.se/nobel/nobel-foundation/index.html

NOBELIUM.   [CAS: 10028-14-5]. Synthetic radioactive chemical element, symbol No, at. no. 102, at. wt. 254 (mass number of ^{254}No), radioactive metal of the Actinide series, also one of the Transuranium elements. Nobelium has valences of 2^+ and 3^+. In 1957, a group of American, English, and Swedish scientists bombarded a target of several curium isotopes (largely ^{244}Cm) with a beam of ^{13}C ions from the cyclotron at the Nobel Institute for Physics. They obtained a few alpha particles of 8.5 MeV energy and half-life of 10 minutes. This was considered to indicate the presence of element 102 with a probable mass number of 251 or 253. At that time, the element was named nobelium with assignment of the symbol, No. Further experiments at the University of California, however, failed to confirm this discovery. In April 1958, Ghiorso, Sikkeland, Walton, and Seaborg, working with the heavy ion linear accelerator (HILAC) at Berkeley, showed the isotope 102^{254} to be a product of the bombardment of ^{246}Cm with ^{12}C ions. Confirming experiments at Berkeley in 1966 showed the existence of ^{254}No with a 55-second half-life; ^{252}No with a 2.3 second half-life; and ^{257}No with a 23-second half-life. Four other isotopes are now recognized, including ^{255}No with a half-life of 3 minutes.

In 1973, scientists at Oak Ridge National Laboratory and Lawrence Berkeley Laboratory, produced a relatively long-lived isotope of nobelium through the bombardment of ^{248}Cm with ^{18}O ions. A total half-title of 58 ± 5 minutes was computed from the combined data of both laboratories. See also **Chemical Elements**

Additional Reading

Note: The following classic references as listed in prior editions of this encyclopedia are preserved here.

Ditmer, P.F. et al.: "Identification of the Atomic Number of Nobelium by an X-ray Technique," *Phys. Rev. Lett.*, **26**, 17, 1037–1040 (1971).

Fields, P.R. et al.: "Production of the New Element 102," *Phys. Rev.*, **107**, 5, 1460–1462 (1957).

Flerov, G.N. et al.: "Experiments to Produce Element 102," *Sov. Phys. Dokl.*, **3**, 3, 546–548 (1958).

Ghiorso, A., T. Sikkeland, J.R. Walton, and G.T. Seaborg: "Attempts to Confirm the Existence of the 10-minute Isotope of 102," *Phys. Rev. Lett.*, **1**, 1, 17–18 (1958).

Ghiorso, A., T. Sikkeland, J.R. Walton, and G.T. Seaborg: "Element No. 102," *Phys. Rev. Lett.*, **1**, 18–20 (1958).

Hammond, C.R.: "The Elements," in *Handbook of Chemistry and Physics,* 67th Edition, CRC Press, Boca Raton, Florida, 1986–1987.

Maly, J., T. Sikkeland, R. Silva, and A. Ghiorso: "Nobelium: Tracer Chemistry of the Divalent and Trivalent Ions," *Science*, **160**, 1114–1115 (1968).

Marks, T.J.: "Actinide Organometallic Chemistry," *Science*, **217**, 989–997 (1982).

Mikheev, V.L. et al.: "Synthesis of Isotopes of Element 102 with Mass Numbers 254, 253, and 252," *Sov. At. Energy*, **22**, 93–100 (1967).

Seaborg, G.T. (editor): *Transuranium Elements,* Dowden, Hutchinson & Ross, Stroudsburg, PA, 1978.

Silva, R.J. et al.: "The New Nuclide Nobelium-259," *Nucl. Phys.*, **A216**, 97–108 (1973).

NOBEL PRIZES.   In 1895, the will of Alfred Bernhard Nobel, a successful Swedish industrialist and European munitions maker, directed that the great majority of his estate be invested for the purpose of yielding annual prize money to be awarded to persons who, as the result of their personal efforts, made outstanding contributions to the advancement of science, literature, and peace. Initially, the awards were confined to five domains—Physics, Chemistry, Physiology or Medicine, Literature (of an idealistic tendency), and Peace (to promote the fraternity of nations and the abolition or diminution of standing armies and the formation and increase of peace congresses). In later years, the governors of the fund added Mathematics as a qualifying discipline and, in 1968, a Nobel Prize for the Economic Sciences was established. Although a prize for each of the foregoing categories is awarded each year, it is not mandatory that an award be made for each category every year, a factor determined by the governors of the fund. The will became effective when Nobel died on December 10, 1886, but the first prizes were not awarded until 1901 because of the need for legal clarification of Nobel's wishes as demanded by family members who also were mentioned in the will.

In studying the will, attorneys recognized the possible requirement for splitting a given award among two or even three individuals, but with a maximum of three persons per award. Nobel also mentioned certain criteria for selection of award recipients in each field. In connection with the prize for Physics, Nobel mentioned "discovery" or "invention," whereas the words "discovery" or "improvement" were stipulated concerning the prize for Chemistry. In terms of the prize for Physiology and Medicine, the key word was "discovery." These were guiding factors for application by the governors of the fund, especially during the early years.

If adjusted for inflation over the years, the original fund of 27,716,243 Swedish kroners ($7,427,953) was quite significant, especially when allowing for capital gains through investments realized over subsequent years. Money available for the annual prizes is essentially determined by the annual capital generated by the fund, but with the stipulation that at least 10% of that gain be reinvested each year. It is interesting to note that the honor associated with the prize is never split—that is, a Nobel Laureate is so designated even though a prize may be shared by two or three persons. The original will specified that the Royal Academy of Science (Sweden) select winners in Physics and Chemistry, that the Karolinska Institute of Medicine (Stockholm) make the selections in Physiology and Medicine, and that the Swedish Academy (of Letters) select winners in Literature. A committee of the Norwegian Parliament was specified to select winners of the Peace prize, noting "no consideration whatever be paid to the nationality of the candidates."

Over the years, with the addition of new prize categories, the governors of the Nobel Foundation necessarily have amended certain procedures. The Foundation is governed by a five-person board of control made up of one appointment by the King of Sweden and one each as appointed by the aforementioned organizations (now referred to as Nobel Institutes). Members of the selection board are appointed for a period of $4\frac{1}{2}$ years. The science awards are presented on December 10 (anniversary of Nobel's death) of each year at the Stockholm Concert Hall, with personal felicitation of the King of Sweden. The Peace prize is presented in a formal ceremony in Oslo.

The first Nobel prizes were awarded in 1901 to Wilhelm K. Roentgen (discovery of X-rays in 1895), J.H. van't Hoff (chemical thermodynamics and osmotic pressure), and E.A. von Behring (diphtheria antitoxin). Listings of scores of Nobel prizes awarded over a century of progress can be found in a number of references, such as "The Information Please Almanac," 45th Edition, 701–709, Houghton Mifflin Company, Boston, Massachusetts. Specific 1992 winners are described in "U.S. Researchers Gather a Bumper Crop of Laurels," Science, 542 (October 23, 1992). An excellent review of the first half-century of the Nobel Foundation is given in an article by George W. Gray, "The Nobel Prizes," *Sci Amer.*, 1 (September 1949).

Although a chronological listing of Nobel prizes in the sciences provides a good source of tracing the progress of science over the years, not all major discoveries and inventions have been so honored. There are several outstanding scientists who have not been included. Traditionally, Nobel prizes are given for achievements that date back a few to several years rather than for discoveries of the immediate past.

Web Reference

The Nobel Foundation: http://www.nobel.se/nobel/nobel-foundation/index.html

NOMENCLATURE. Chemical nomenclature embraces several subcategories; names for chemical elements and compounds; names for classes of compounds and substances, such as mixtures and composites; names for particles, processes and transformations, properties, effects, units of measurements, techniques, instruments and apparatus, and even for theories and concepts. Only the first three are considered to be the heart of chemical nomenclature. The largest part of the subject is the nomenclature of organic compounds, simply because there are so many of them, and of such diverse nature. Concern with chemical nomenclature has grown on a broad international scale as the importance of consistent, uniform nomenclature is increasingly recognized. Various committees, both national and international, are working toward a consistent, systematic nomenclature. Among the areas in which nomenclature plays a key role are patent law, trade and customs regulations, identification of controlled substances, pharmaceutical and health information, and studies of the environment and pollution.

In the United States, the Committee on Nomenclature of the American Chemical Society is the clearinghouse for nomenclature recommendations and adoptions, aided by various divisional nomenclature committees of the Society. Close liaison is maintained with the various nomenclature bodies of the International Union of Pure and Applied Chemistry. Progress is being made not only in improved nomenclature, but also in the extension of nomenclature recommendations to newly developing areas of chemistry.

Important as names are, they cannot serve all purposes. There are other, complementary means of identifying chemical compounds, e.g., structural formulas, notation systems, and registry numbers. None of these are nomenclature, however.

Although symbols are not a part of nomenclature, the two are closely related, and the former have played an extremely important role in chemistry. Because of the difficulty of establishing priority of discovery for most of the elements of atomic number above 100, and because of the need to refer to hypothetical elements with higher atomic numbers, IUPAC has developed interim systematic symbols and names for such elements.

Inorganic Nomenclature

Perhaps no subject in chemistry has undergone less change over the twentieth century than inorganic nomenclature. This longevity attests to the fundamental soundness of the original proposals of Guyton de Morveau that established it, but it also suggests why inconsistencies and confusions have remained as well, which have continued to disconcert chemists. The development of inorganic nomenclature has again accelerated, however, and the inconsistencies are being eliminated.

The System of Guyton de Morveau, Lavoisier, and Co-Workers. The first attempt toward a convenient nomenclature belongs to Guyton de Morveau. His pioneer work led to publication in 1787 of *Methode de Nomenclature Chimique*, written in collaboration with Lavoisier, Berthollet, and Fourcroy, which proved to be a landmark in the development of chemistry.

The fundamental principle of the new nomenclature was that the name of a compound should exhibit the elements involved and their relative proportions, if known. The combinations of oxygen with other elements played a dominant role. Thus, the product of the union of a simple nonmetallic substance with oxygen was called an acid, whereas that of the union of a metal with oxygen was called an oxide. The union of an acid and an oxide produced salt. The acids or oxides were given names in which the generic part was the word "acid" or "oxide" and the specific part

was an adjective derived from the name of the other element. The same principle supplied names for sulfides and phosphides.

The names adopted for salts consisted of a generic part derived from the acid and a specific part from the metallic base: The names for salts of acids containing an element in different degrees of oxidation were given different terminations.

Berzelius divided the elements into metalloids (nonmetals) and metals according to their electrochemical character, and the compounds of oxygen with positive elements (metals) into suboxides, oxides and peroxides. His division of the acids according to degree of oxidation has been little altered. He introduced the terms anhydride and amphoteric and designated the chlorides in a manner similar to that used for the oxides.

Established Practice in the English Language. The nearly literal translation of the French terms in English, Russian and other languages resulted in the system whose use has become standard practice in English-speaking as well as other countries. The system has been molded by the fact that elemental composition and valence (or oxidation number) are the principal variables for most inorganic compounds other than the most complex, whereas connectivity and the possibility for isomers have been of little concern.

Modified Forms in Common Use. There are numerous situations in which the foregoing system does not meet all requirements. In the formation of binary compounds, several elements exhibit more than two states of oxidation. One method, recommended by the IUPAC, of handling these situations is the use of prefixes derived from Greek to indicate stoichiometric composition, e.g., titanium dichloride, $TiCl_2$; and dinitrogen oxide (nitrous oxide) N_2O. Other accepted methods of indicating proportions of constituents are the Stock system (oxidation number) and the Ewens-Bassett (charge number) system.

Some elements form acids with more than four oxidation states, requiring other combinations of prefixes and suffixes: $H_4P_2O_6$, intermediate between H_3PO_3 and H_3PO_4, is known as hypophosphoric acid. Here again, the oxidation-number and charge-number systems offer advantages. Ortho-, meta-, and pyro- prefixes or numerical prefixes to denote stages of hydroxylation of acids also find use. In many instances, special names have been created to deal with unusual situations.

Systems of Compounds. The nomenclature system of Guyton de Morveau and co-workers was designed specifically for oxygen compounds. As early as 1826, it became evident that the halogens could play much the same role in many other compounds as oxygen does in the familiar oxygen salts. By 1840, Hare was writing of chloro acids and chloro bases, and recognized classes of salts: oxy-, sulfo- (now called thio-), seleni-, telluri-, chloro-, fluoro-, cyano-, etc. Remsen was a proponent of this system of nomenclature, but it received its fullest treatment from Franklin in connection with his concept of systems of compounds.

The analogies are shown by the following reactions:

$$K_2O + B_2O_3 \longrightarrow K_2O \cdot B_2O_3 \text{ or } KBO_2$$

$$K_2S + B_2S_3 \longrightarrow K_2S \cdot B_2S_3 \text{ or } 2\ KBS_2$$

$$KF + BF_3 \longrightarrow KF \cdot BF_3 \text{ or } KBF_4$$

$$K_3N + BN \longrightarrow K_3N \cdot BN \text{ or } K_3BN_2$$

The products resulting from such reactions should, therefore, have analogous names. If KBO_2 is a borate, then KBS_2 is a thioborate, and KBF_4 is a fluoroborate. Similarly, the replacement of an oxygen atom by a sulfur atom or two fluorine atoms is understandable. However, the relationship of K_3BN_2 is less obvious, until one considers the dehydration and deammoniation schemes:

$$\overset{-H_2O}{B(OH)_3 \longrightarrow OBOH}$$

$$\overset{-NH_3}{B(NH_2)_3 \longrightarrow HNBNH_2}$$

His scheme of nomenclature for nitrogen compounds, and the names thio-, chloro-, etc, did become widespread, especially for sulfur and halogen compounds.

Although the foregoing pattern of nomenclature is useful, it does lead to some difficulties. Many quaternary compounds contain oxygen and another electronegative element. In the series M_2CO_3, M_2CO_2S, M_2COS_2, and

M_2CS_3, the names are carbonates, (mono)thiocarbonates, dithiocarbonates, and trithiocarbonates, respectively. However, in practice both the prefixes mon- and tri- are often omitted, and its is uncertain whether the omission signifies the mono- or the completely substituted compound. The situation is somewhat more complicated when oxygen and fluorine are present in the same compound, because one is bivalent and the other univalent, and the coordination number toward fluorine is different from that toward oxygen: H_3PO_4, H_2PO_3F, HPO_2F_2, and HPF_6. Furthermore, investigators have not always been consistent in choosing the same reference state for the names of the oxygen salts and the halogen salts.

Coordination Compounds. The approach of Werner to the problem of naming ternary and higher order compounds is based on an entirely different point of view. By considering all such substances as complex or coordination compounds, he succeeded in making a wide variety of them according to a single general pattern. To designate the oxidation state of the element serving as the center of coordination, Werner chose the characteristic endings suggested by Brauner, but these have been totally superseded by the oxidation-number and charge-number systems.

The Stock Oxidation-Number System. Stock sought to correct many nomenclature difficulties by introducing Roman numerals in parentheses to indicate the state(s) of oxidation.

The oxidation-number system is easily extended to include other coordination compounds. Even substances represented by the formulas $Na_4Ni(CN)_4$ and $K_4Pd(CN)_4$ create no nomenclature problem; they become sodium tetracyanonickelate(0) and potassium tetracyanopalladate(0), respectively.

The Charge-Number (Ewens-Bassett) System. The oxidation state of an atom as expressed by the oxidation number is a formal concept for partitioning the electric charge between atoms in a molecule or chemical structure. For many chemical structures, this formal procedure may lead to representations of charge distribution that are inconsistent with experiment. Therefore, Ewens and Bassett proposed to express only the total charge on an ion without representing valence and its associated arbitrariness of assigning electronic distribution within a given structure, e.g., titanium(2+) chloride for $TiCl_2$; titanium(3+) for $TiCl_3$; potassium tetrachloroplatinate(2−) for K_2PtCl_4; and sodium tetracyanonickelate(4−) for $Na_4Ni(CN)_4$.

International Agreement. The first report of the Commission for the Reform of the Nomenclature of Inorganic Chemistry was written in 1926 by Délépine. Subsequent rules (1940, 1959) were expanded and improved in 1990 to provide the basis for naming inorganic compounds. They retain most of the well established names for binary and pseudobinary compounds and for the oxoacids of the nonmetals and derivatives.

The IUPAC Commission on Nomenclature of Inorganic Chemistry continues its work, which is effectively open-ended. Guidance in the use of the IUPAC rules as well as explanations of their formulation are available. A second volume on nomenclature of inorganic chemistry is in preparation; it will be devoted to specialized areas. Some of the contents have had preliminary publication in the journal, *Pure and Applied Chemistry*, e.g., "Names and Symbols of Transfermium Elements" in 1944.

Organic Nomenclature

Modern organic nomenclature is such that it can be better understood by first tracing how it developed. Organic substances played a minor role in the *Methode de Nomenclature Chimique*. Eighteen organic acids were given their present names (succinic, malic, etc), and several other substances were mentioned, such as alcohol, ether (including esters as well as true ethers), starch, gluten, and camphor. Gaseous hydrocarbons, the only ones included, were lumped together as carbonated hydrogen gas. Thus a few common names were incorporated into the new method, but no systematic organic names were possible because of lack of knowledge. Little else could be done, for the basis for determining elemental composition, as in empirical formulas, did not yet exist.

The practice of assigning ad hoc names to organic compounds was neither avoidable, nor burdensome when only a small number of compounds were recognized. Such ad hoc names are termed "trivial" or "traditional," to indicate that they contain no encoded structural information. They are useful for common compounds, and many of them are retained to this day, but they are not helpful in understanding chemical relationships. As they proliferated, the number and variety of them became unmanageable. The development of systematic nomenclature was driven

by this circumstance, and was made possible by advances in understanding and determining the structure of molecules.

Systematic nomenclature is in essence a scheme for encoding structural information in a name. For organic chemistry, it probably began in 1832, when Justus Liebig's journal, *Annalen der Chemie*, was born and when Liebig and Woehler published their memorable article on the radical of benzoic acid. This radical (C_6H_5CO in the modern formula) they termed "benzoyl," thus coining -yl (from the Greek *hyle*, meaning stuff or material), one of the most useful suffixes in chemistry. By radical or compound radical, they meant a group of atoms that remains unaltered in chemical transformations. The word group is used for almost any portion of a molecule considered as a unit for convenience in naming or otherwise. The name ethyl soon followed. These two names, one of an acid group (radical) and the other of a hydrocarbon group (radical), may be regarded as the progenitors of the host of group names used in the 1990s. From them it was an easy step to the combinations of benzoyl chloride, ethyl iodide, ethyl oxide, etc, many of which still survive. These binary names are analogous to the binary inorganic names introduced in 1787.

It was many years before organic nomenclature shook off the influence of electrochemical theory and its binary names. Gradually, as facts accumulated, it became clear that this theory must give way to a unitary conception of the molecule. At the same time, the phenomenon of substitution, or replacement of one atom or group of atoms by another, was recognized to be of central importance. Some binary names are still used, either as a true expression, as for salts, or for convenience, as in ethyl sulfide or acetyl chloride, but for the most part the principle of substitution is used without regard to whether such replacement can actually be effected experimentally. Usually the atom replaced is hydrogen, and the replacement may be indicated by either a prefix or a suffix. Thus, in naming CH_3Cl chloromethane rather than methyl chloride, the replacement of one atom of hydrogen in methane, CH_4, by chlorine is indicated. A third group of names is formed by combining a class name with a specific word, as in ethyl alcohol or benzophenone oxime. Whatever the method or combination of methods used, there must be a name for a parent compound to form a basis for it.

By 1866, it was possible for Hofmann to arrange hydrocarbons in series by their empirical formulas, i.e., methane, CH_4, and methene, CH_2; ethane, C_2H_6, ethene, C_2H_4, and ethine, C_2H_2; propane, C_3H_8, propene, C_3H_6, and propine, C_3H_4; and quartane, C_4H_{10}, quartene, C_4H_8, and quartine, C_4H_6. These are known as the Hofmann-Gerhardt names.

Hofmann's scheme has been modified by replacing quartane with butane and continuing the homologous series with the Greek forms pentane, hexane, etc, which are still used. For the C_nH_{2n} series, i.e., the olefins, the names methylene, ethylene, propylene, etc, came into use instead of Hofmann's terms, but the names propene, butene, pentene, etc were revived in the Geneva system and are the preferred terms. For C_2H_2, ethine has never replaced the older term acetylene, but propine, butine, etc, reappeared in the Geneva system. The ending -yne is used, as in propyne, etc, to avoid confusion with the ending -ine of organic bases such as aniline.

The Hofmann-Gerhardt names did not distinguish between isomers. Different methods of distinguishing isomers arose: $CH_3CH_2CH_2CH_3$ became normal butane (abbreviated to *n*-butane), and $CH_3CH(CH_3)_2$ isobutane or trimethylmethane. Of olefins, $CH_2=CHCH_2CH_3$ became α-butylene or ethylethylene; $CH_3CH=CHCH_3$, β-butylene or symmetrical dimethylethylene; and $CH_2=C(CH_3)_2$, isobutylene or unsymmetrical dimethylethylene. It thus becomes evident that as the number of carbon atoms and therefore the number of isomers increases, the coining of such names meets with insuperable difficulties. The situation with regard to hydrocarbons had its parallel in the nomenclature of alcohols and other types of compounds.

The Geneva System. The Geneva Conference was strongly influenced by the need for names that would be suitable for systematic indexing of organic compounds. The groundwork was laid by a French subcommission. One of the chief principles of the system was the selection of the longest straight chain of carbon atoms in the molecule as a parent structure. Thus, the names butane, pentane, etc, would refer to the normal (unbranched) isomers only. The parent hydrocarbon could then be modified by attaching to its name one or more prefixes or suffixes to specify chemically characteristic features commonly termed functional groups. A representative selection is given in Lance 1 when two or more different positions of attachment of a prefixed or suffix exist, a position designator, called a locant, is necessary. These are arabic numerals, set off by hyphens,

TABLE 1. PREFIXES AND SUFFIXES FOR SOME PRINCIPAL FUNCTIONAL GROUPS[a]

Formula	Class name	Prefix	Suffix	Radicofunctional form
$-COOH$	carboxylic acids	carboxy	-carboxylic acid[b] or -oic acid	
$-SO_3H$	sulfonic acids	sulfo	-sulfonic acid	
$-COOR$	esters	alkoxy carbonyl	alkyl -oate or carboxylate[b]	
$-COX$	acid halides	halocarbonyl	-oyl halide or carbonyl halide[b]	
$-CONH_2$	amides	carbamoyl cyano	-amide or carboxamide[b]	
$-CN$	nitriles		-nitrile or -carbonitrile[b]	alkyl[c] cyanide
$-CH=O$	aldehydes	formyl	-al or -carbaldehyde[b]	
$-C=O$	ketones[d]	oxo	-one	dialkyl[c] ketone
$-OH$	alcohols	hydroxy	-ol	alkyl[c] alcohol
$-SH$	thiols (or mercaptans[e])	sulfanyl (mercapto[e])	-thiol	alkyl mercaptan
$-NH_2$	amines	amino	-amine	
$=NH$	imines	imino	-imine	
$-OR$	ethers	alkoxy or aryloxy		(di)alkyl ether
$-Cl$	chlorides	chloro		alkyl[c] chloride
$-NO_2$	nitro compounds	nitro		
$-SO_2$	sulfones	alkyl[c] sulfonyl		(di)alkyl[c] sulfone

[a] In order of precedence.

[b] The shorter form implies no additional carbon atoms, and is used when the group is part of a chain. The long form implies one more carbon atom than the parent structure, and is used when the group is attached to a ring, or for other reasons is not conveniently named as part of a carbon skeleton.

[c] Or aryl.

[d] Both bonds from the carbonyl group must be to a carbon atom.

[e] Has been widely used, but is no longer officially recommended.

starting with 1 at an end of the chain. Accordingly, $CH_3CH=CHCH_3$ became 2-butene, and $CH_3CH_2CHOHCH_2CH_3$ became 3-pentanol. The position of the locants at the beginning or at the end was considered equally acceptable. In the 1990s however, the official IUPAC recommendation is to place the locant immediately before the feature that it locates, as in but-2-ene and pentan-3-ol for the foregoing examples.

The International Union and the Definitive Report. The next important step was the *Definitive Report of the Commission on the Reform of the Nomenclature of Organic Chemistry* in 1930 at a meeting in Liège. This report used the Geneva rules as a basis for modification, and many of the 68 Liège rules deal with topics not touched in the original Geneva report.

An important modification of the Geneva system is that the fundamental chain used as a basis in an aliphatic compound is not necessarily the longest chain in the molecule, but must be the longest chain of those containing the maximum number of occurrences of the principal functional group (Rule 18). This shifts the importance for naming from side chains such as methyl and ethyl to functional groups such as $-COOH$ and $-OH$.

The concept of the principal function raises the question of how priority is determined when two or more different functional groups are present. No arbitrary rule can be entirely satisfactory, but an order has been codified in IUPAC recommendations, and an essentially similar order is used by Chemical Abstracts Service. In general, a higher state of oxidation takes precedence over a lower one (Table 1).

The use of prefixes and suffixes for distinguishing the various radicals, groups, and functions has caused some problems, because some groups, e.g., HS—, have borne more than one name, and some names, e.g., anisyl, have had more than one meaning. The *Definitive Report* included only a limited number of prefixes and suffixes. Chemical Abstracts Service publishes its own lists the most recent version can be found in the comprehensive *Guide to the Use of IUPAC Nomenclature of Organic Compounds*. An important departure from earlier recommendations is that the systematic names of acyl groups derived from carboxylic acids must end in -oyl, common traditional or trivial names, such as acetyl and oxalyl, excepted. The purpose of this rule is to distinguish unambiguously between hydrocarbyl groups and acyl groups. Thus anisyl can only mean methoxyphenyl, whereas anisoyl refers only to methoxybenzoyl.

The ending -yl (or -oyl) is standard for univalent groups (with certain traditional exceptions, such as succinoyl). It may be combined with a sign for unsaturation, as in propenyl, $CH_3CH=CH$ thynyl, $CH\equiv C-$. The ending -ylene is one device for denoting a bivalent group in which the two free valences are on different atoms, but, with the exception of methylene, $-CH_2-$, and ethylene, $-CH_2CH_2-$, the ending -diyl, with locants as appropriate, is preferred, as in propane-1,3-diyl, $-CH_2CH_2CH_2-$. When the two free valences of a bivalent group form a double bond, the ending is -ylidene, as in ethylidene, $CH_3CH=$. For a trivalent group forming a triple bond, the ending is -ylidyne, as in ethylidyne, $CH_3C\equiv$.

For indicating the number of groups of the same kind, the prefixes di-, tri-, tetra-, etc are used when the expressions are simple, and bis-, tris-, tetrakis-, etc when they are complex; for example, "dichloro," but "bis(dimethylamino)." The prefix bi- is used to denote the joining of two groups of the same kind together, as in biphenyl, $C_6H_5-C_6H_5$, or the doubling of a compound with loss of two hydrogen atoms, as in biarsine, H_2AsAsH_2.

A historical account of the development of organic nomenclature from the time preceding the Geneva Conference to fairly recent times is available.

The IUPAC Commission on Nomenclature of Organic Chemistry has continuing responsibility for revising and expanding the rules that appeared in the *Definitive Report*.

A considerable number of trivial or semitrivial (traditional) names have been retained by IUPAC for compelling practical reasons; the approved ones are available. A very brief selection is shown in Table 2. Other contributions to organic nomenclature are available.

Biochemical Nomenclature

The IUPAC Commission of Nomenclature of Biological Chemistry was established in 1921, along with the organic and inorganic commissions. It worked actively and closely with the organic commission. Early subjects of

TABLE 2. COMPOUNDS USED AS PARENT STRUCTURES WITH TRIVIAL NAMES

Chemical formula	Name	Systematic equivalents
C_6H_6	benzene	
$C_{10}H_6$	naphthalene	
C_5H_5N	pyridine	
C_4H_4S	thiophene	
C_4H_4O	furan	$-COOH$
C_6H_5OH	phenol	benzenol
$C_6H_5NH_2$	aniline	benzenamine
$H_2N=NH_2$	hydrazine	diazane
NH_3	ammonia or amine	azane
$HO=OH$	hydrogen peroxide	dioxidane
CH_3COCH_3	acetone	propan-2-one
H_2NCONH_2	urea	carbamide

concern were carbohydrates, proteins, enzymes, and fats. More recently, this Commission shared its work with a corresponding Commission of the International Union of Biochemistry, which is not the International Union of Biochemistry and Molecular Biology (IUBMB); this led to the establishment of the Joint Commission on Biochemical Nomenclature (JCBN) in 1964.

The Joint IUPAC/IUBMB Commission has published many recommendations dealing with the nomenclature of natural products. The IUBMB Commission on Nomenclature has also issued a number of recommendations dealing with areas of a more biochemical nature and for naming enzymes.

The presence of many chiral centers in compounds of biochemical significance or natural-product interest has led to the use of stereoparents. These are parent structures having trivial names that imply (without explicitly expressing) a particular steric configuration. Common examples are the names of simple sugars, exemplified by glucose.

Although it is not strictly within the subject of biochemical nomenclature, it is appropriate to mention the existence of standardized generic names for pharmaceutical drugs. Such names are essentially coined, or trivial, names, but often include syllables from the systematic organic names, and endings that reflect a structural class, e.g., -cillin (from penicillin), or an important area of medical application, e.g., -vir (antiviral). Glossaries of these generic names are published periodically and a glossary of United States Approved Names (USAN) is published annually.

Macromolecular Nomenclature

In 1967, the Polymer Nomenclature Committee of the American Chemical Society published proposals for naming linear polymers on the basis of their chemical structure, which were then introduced into *Chemical Abstracts (CA) Indexes* and published in their final form in 1968.

A Macromolecular Division of IUPAC was created in 1967, and it created a permanent Commission on Macromolecular Nomenclature, parallel to the other nomenclature commissions. The Commission over the years has issued recommendations on basic definitions, stereochemical definitions and notations, structure-based nomenclature for regular single-strand organic polymers and regular single-strand and quasi-single-strand inorganic and coordination polymers, source-based nomenclature for polymers and abbreviations for polymers. All of these are collected in a compendium referred to as the IUPAC Purple Book.

Recommendations on additional aspects of macromolecular nomenclature such as that of regular double-strand (ladder and spiro) and irregular single-strand organic polymers continue to be published in *Pure and Applied Chemistry*. Recommendations on naming nonlinear polymers and polymer assemblies (networks, blends, complexes, etc) are expected to be issued in the near future.

Examples of the two macromolecular nomenclature systems are as follows. For source-based names for homopolymers and copolymers: polyacrylonitrile, poly(methyl methacrylate), poly(acrylamide-*co*-vinylpyrrolidinone), polybutadiene-*block*-polystyrene, and poly(propyl methacrylate)-*graft*-poly(1-vinylnaphthalene). Structure-based examples are as follows: poly(oxy-1,4-phenylene) (**1**), poly(oxyethyl-eneoxytereph-thaloy) (**2**) and poly[imino(1-oxo-1,6-hexanediyl)] (**3**).

(1)   (2)

(3)

Nomenclature in Other Areas of Chemistry

A number of glossaries of terms and symbols used in the several branches of chemistry have been published. They include physical chemistry, physical-organic chemistry, and chemical terminology (other than nomenclature) treated in its entirety. IUPAC has also issued recommendations in the fields of analytical chemistry, colloid and surface chemistry, ion exchange, and spectroscopy, among others.

PETER A. S. SMITH
University of Michigan

Additional Reading

American Chemical Society, *Macromolecules* **1**, 193 (1968).

Block, B.P., W.H. Powell, and W.C. Fernelius: *Inorganic Chemical Nomenclature: Principles and Practices,* American Chemical Society, Washington, DC, 1990.

International Union of Biochemistry and Molecular Biology, *Biochemical Nomenclature and Related Documents,* 2nd Edition, Portland Press, London, UK, 1992.

Richer, J.-C., R. Panico, and W.H. Powell: *Guide to the Use of IUPAC Nomenclature of Organic Compounds,* Blackwell, Oxford and London, UK, 1994.

Web References

International Union of Biochemistry and Molecular Biology (IUBMB): http://www.chem.qmul.ac.uk/iubmb/

International Union of Pure and Applied Chemistry (IUPAC): http://www.chem.qmul.ac.uk/iupac/

A Nomenclature of Junctions and Branchpoints in **Nucleic Acids**: http://www.chem.qmul.ac.uk/iubmb/misc/bran.html

Abbreviations and Symbols for **Nucleic Acids, Polynucleotides and their Constituents**: http://www.chem.qmul.ac.uk/iupac/misc/naabb.html

Abbreviations and Symbols for the Description of the Conformation of **Polypeptide Chains**: http://www.chem.qmul.ac.uk/iupac/misc/ppep1.html

Abbreviations and Symbols for the Description of Conformations of **Polynucleotide Chains**: http://www.chem.qmul.ac.uk/iupac/misc/pnuc1.html

Basic Terminology of **Stereochemistry**: http://www.chem.qmul.ac.uk/iupac/stereo/

Enzyme Nomenclature: http://www.chem.qmul.ac.uk/iubmb/enzyme/

Glossary of Terms Used in **Bioinorganic Chemistry**: http://www.chem.qmul.ac.uk/iupac/bioinorg/

Glossary of Terms Used in **Physical Organic Chemistry**: http://www.chem.qmul.ac.uk/iupac/gtpoc/

Glossary of Terms Used in **Medicinal Chemistry**: http://www.chem.qmul.ac.uk/iupac/medchem/

Nomenclature and Symbolism for **Amino Acids and Peptides**: http://www.chem.qmul.ac.uk/iupac/AminoAcid/

Nomenclature of **Carbohydrates**: http://www.chem.qmul.ac.uk/iupac/2carb/

Nomenclature of **Carotenoids**: http://www.chem.qmul.ac.uk/iupac/carot/

Nomenclature and Symbols for **Folic Acid and Related Compounds**: http://www.chem.qmul.ac.uk/iupac/misc/folic.html

Nomenclature of **Glycolipids**: http://www.chem.qmul.ac.uk/iupac/misc/glylp.html

Nomenclature of **Glycoproteins, Glycopeptides** and **Peptidoglycans**: http://www.chem.qmul.ac.uk/iupac/misc/glycp.html

Nomenclature of **Lignans** and **Neolignans**: http://www.chem.qmul.ac.uk/iupac/lignan/

Nomenclature of **Lipids**: http://www.chem.qmul.ac.uk/iupac/lipid/

Nomenclature for **Multienzymes**: http://www.chem.qmul.ac.uk/iubmb/misc/menz.html

Nomenclature of **Multiple Forms of Enzymes**: http://www.chem.qmul.ac.uk/iubmb/misc/isoen.html

Nomenclature for Incompletely Specified Bases in **Nucleic Acid Sequences**: http://www.chem.qmul.ac.uk/iubmb/misc/naseq.html

Nomenclature of **Peptide Hormones**: http://www.chem.qmul.ac.uk/iubmb/misc/phorm.html

Nomenclature of **Phosphorus-Containing Compounds** of Biochemical Importance: http://www.chem.qmul.ac.uk/iupac/misc/phospho.html

Numbering of atoms in **Myo-inositol**: http://www.chem.qmul.ac.uk/iupac/cyclitol/myo.html

Recommendations for Nomenclature and Tables in Biochemical Thermodynamics: http://www.chem.qmul.ac.uk/iubmb/thermod/

Nomenclature of **Cyclitols**: http://www.chem.qmul.ac.uk/iupac/cyclitol/

Nomenclature of **Electron-Transfer Proteins**: http://www.chem.qmul.ac.uk/iubmb/etp/

Nomenclature for **Vitamins B-6 and Related Compounds**: http://www.chem.qmul.ac.uk/iupac/misc/B6.html

Nomenclature of **Quinones with Isoprenoid Side-Chains**: http://www.chem.qmul.ac.uk/iupac/misc/quinone.html

Nomenclature of **Retinoids**: http://www.chem.qmul.ac.uk/iupac/misc/ret.html

Nomenclature of Steroids: http://www.chem.qmul.ac.uk/iupac/steroid/

Nomenclature of **Tetrapyrroles**: http://www.chem.qmul.ac.uk/iupac/tetrapyrrole/

Nomenclature of **Tocopherols and Related Compounds**: http://www.chem.qmul.ac.uk/iupac/misc/toc.html

Prenol Nomenclature: http://www.chem.qmul.ac.uk/iupac/misc/prenol.html

Prokaryotic and Eukaryotic translation factors: http://www.chem.qmul.ac.uk/iubmb/misc/trans.html

Revised Section F: **Natural Products and Related Compounds**: http://www.chem.qmul.ac.uk/iupac/sectionF/

Section H: **Isotopically Modified Compounds**: http://www.chem.qmul.ac.uk/iupac/sectionH/

Symbols for Specifying the Conformation of **Polysaccharide Chains**: http://www.chem.qmul.ac.uk/iupac/misc/psac.html

Symbolism and Terminology in **Enzyme Kinetics**: http://www.chem.qmul.ac.uk/iubmb/kinetics/

The Nomenclature of **Corrinoids**: http://www.chem.qmul.ac.uk/iupac/misc/B12.html

NOMENCLATURE (Fertilizer). See **Fertilizer**.

NOMENCLATURE (Organic Chemistry). See **Organic Chemistry**.

NONDESTRUCTIVE TESTING (NDT).

The examination of materials and objects for the purpose of detecting defects without in any way harming the test object. NDT contrasts vividly with destructive testing methods, which chemically consume or physically damage the test object, rendering it unfit for use. Whereas destructive testing must be confined to statistical sampling procedures, NDT enables 100% on-line inspection if desired. The trend in recent years has been in this direction, with emphasis on automating and increasing the speed of NDT operations. Another significant trend has been that of *testing work in progress*, as contrasted with earlier procedures which concentrated on testing raw materials and final products. In this way, very helpful information for step-by-step quality control can be provided. There remain, of course, numerous examples of where statistical destructive testing is needed—for example, in determining the ultimate compressive and tensile strengths of materials and parts or checking the corrosion resistance of materials. In recent years, it has proven possible to combine the results of NDT with computerized simulation in some instances to predict failure of test objects under certain conditions. For obvious economic reasons, NDT is preferred by manufacturers over destructive testing this accounts for high acceptance and many advancements which have occurred in NDT methods. For research and development applications, nondestructive methods are sometimes referred to as NDE (nondestructive evaluation).

Traditionally, NDT has been associated with metals and materials of construction for finding potentially unsafe conditions, such as cracks, voids, holes, inclusions, and other inconsistencies, as may be found in metal sheets, plates, bars, tubes, castings, forgings, and weldments. Such defects may arise from faulty manufacturing, or from later use, as the result of corrosion, abrasion, vibration, mishandling, and inattention to required maintenance procedures. In recent years, the applications for NDT have broadened to include all manner of materials—films, coatings, polymers, composites, and ceramics as encountered in a wide variety of industries, including numerous uses in the electronics manufacturing industry. Also, NDT is widely used for on-site inspection of large and heavy equipment, which cannot be detached for testing, after installation, but where periodic checks are required. Examples include the inspection of weldments in pipelines, aircraft engines and structural components, military equipment, bridge structures, etc.

During the last half of the 1980s and well into the 1990s, NTD enjoyed the benefits of measurement and computer technologies that contributed immensely to the speed, accuracy, and reliability of NTD, even though instrument costs have risen markedly as a result. However, the ability to make more measurements within shorter periods of time probably has not increased the unit costs proportionately. A number of measurement techniques entirely new to the NDT field have been added to increase the variety of choices.

Radiographic Methods

Radiographic (X- and gamma-ray technology) method using film was one of the earliest NTD schemes used. Although early systems are undergoing modernization, this comparatively simple method still enjoys acceptance for certain applications. This basic technique has taken on a number of new formats.

Film Images. Images made by the traditional film technique are shown in Fig. 1. To reduce costs and meet environmental restrictions, a dry-silver system was introduced in 1991. The system produces a silver-based image without the use of wet chemistry, using photothermographic technology. The image is developed on exposed film by thermal energy rather than by the traditional method of immersing film in a liquid developer and fixer. Three elements required for dry processing are a specially coated film, fluorescent exposing screen, and a thermal processor. The film has a translucent polyester base similar to that of conventional film. Its ultrafine grain produces detailed images of archival quality.

Traditional radiographic methods use two-dimensional film to record the attenuation of X-rays passing through a three-dimensional object. The

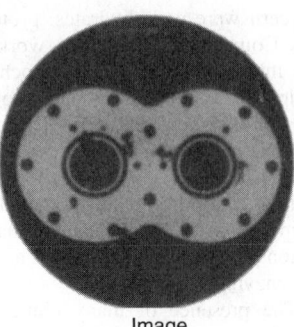

| Basic method | Image |

Fig. 1. X-rays or gamma rays are used to create a shadow image of light and dark that reveals any flaws or inclusions in a test part

result is a shadowgraph in which all object features are superposed. To improve the totality of information obtained, backscattering methods were introduced several years ago.

Principal applications for radiography include the inspection of castings, electrical assemblies, weldments, small, thin, and complex wrought products, some nonmetallics, solid propellant rocket motors, cans or containers, composites, and nuclear reactor fuel rods, among many others.

Chronology of Radiographic Methods. In 1985, researchers at John Hopkins University described a flash X-ray system that uses increased-power X-ray sources to generate very intense short pulses. High-gain X-ray intensifier detectors are used. Exposure times as short as 30 ns are possible and thus microstructural changes due to explosions, heat pulses, and shock waves can be detected. An indirect and direct method are used. In the indirect method, the X-ray diffraction image is converted into a visible light image by a fluorescent screen. The researchers have found that for the indirect method, a multiple-stage image-intensifier system coupled to an external fluorescent screen is the most sensitive and almost instantaneous system. Multiple stages of amplification allow individual X-ray photons to be detected. In more advanced systems, there is inclusion of a microchannel plate where electrons strike the output phosphor and are converted into a strong, visible image.

In the direct method, an X-ray-sensitive vidicon TV camera directly converts the X-ray image into an electronic charge pattern on a photoconductive target, which is read out by a scanning electron beam and displayed visually on a TV monitor.

In addition to testing uses per se, flash-X-ray techniques have been used to study the orientation of single crystals, to study lattice rotation accompanying plastic deformation, to measure the grain boundary migration during recrystallization annealing, and to determine the physical state of exploding materials.

In another technique known as X-ray transmission asymmetric crystal topography, changes in defect structure during polymerization of single crystals have been studied.

Digital Radiography. In this technique, the traditional film is replaced by a linear array of detectors and the X-ray beam is collimated into a fan beam. The object is moved perpendicularly to the detector array, and the attenuated radiation is sampled digitally by the detectors. Data are processed by stored information in the computer's memory to yield a two-dimensional image of the part being inspected.

X-Ray Computed Tomography (CT). In computed tomography, penetrating radiation from many angles is used to reconstruct cross-sectional images of an object. The advantages of CT are exemplified by the inspection of aircraft/aerospace castings for internal defects. Advantages of CT include increased reliability, elimination of unnecessary rejects, and wider use of castings instead of forgings and parts machined from wrought stock. CT has been found to have greater sensitivity (dependent on part size and geometry) than conventional film. CT can spatially define flaw distribution. Aerospace test engineers claim that castings can be measured with an accuracy of better than 0.05 mm (0.002 in), but is adversely affected by the amount of image noise and the edge-detection method used. Computed tomography systems are costly. The general principles of CT are described in article on **X-Ray Scan and Other Medical Imagery**.

Ultrasonic Methods

Typically, ultrasonic images are produced by mechanically scanning an ultrasonic transducer in a raster pattern over an area of a structure and then displaying the reflected or transmitted energy in a suitable format. Usually, the scan is performed in a tank of water or with some form of squirter nozzle. The liquid medium serves to transmit the ultrasonic energy from the transducer into the test material. Conventionally, the data are displayed as C-scans (a plan view image where a color scale is used to display signal amplitude or depth information) or as B-scans (image of a cross section at one particular location of interest, typically with a color indicating signal amplitude).

As indicated by Fig. 2, there are several testing modes: (1) pulse-echo mode, (2) through-transmission mode, (3) reflector-plate mode, and (4) angle-beam mode.

Sonic (< 0.1 MHz) and ultrasonic (0.1 to 25 MHz) radiation have been used for many years in NDT. In a simple testing scheme, sonic or ultrasonic vibrations are generated and sent by way of a pulse beam through the part to be tested. The beam travels unimpeded through large parts, may be angled for testing sheet stock, and can impact materials immersed in a liquid. Any flaw reflects vibrations back to the instrument, which indicates the location and size of the discontinuity on a CRT (cathode-ray tube). Access is required to only one side of the material being tested. Although energy can be lost from the ultrasonic beam due to geometrical effects, these can be controlled to increase the sensitivity of attenuation measurements. Hence, microstructural alterations, such as microcracks, foreign particles, precipitates, grain boundaries, interphase boundaries, and dislocation defects, can be detected. Research has shown that attenuation measurements have detected microstructural change during fatigue testing, therefore giving early warning to fatigue-induced failure, as well as measuring oxygen content in titanium welds.

Acoustic methods are applicable to numerous kinds of materials. The method can be used, for example, to reveal fiber/matrix bond strength in polymer-matrix composites. In one method (Wan-li Wu, National Institute of Standards and Technology), a continuous wave argon-ion laser is used to heat a very small area of the composite. The resulting thermal expansion between fiber and resin produces a measurable change due to debinding. Conventional methods of evaluating bond strength are time consuming and tedious. Instead of measuring the thermal stress, the laser power level at which debonding occurs is used as the index of debonding stress. Although sonic scanning techniques can be used to detect voids and cracks at interfaces in polymer-matrix composites, they do not measure the strength of interfacial bonds.

As pointed out by D. Sturges (General Electric Aircraft), "Modern ceramic materials offer many attractive physical and mechanical properties for use in a rapidly growing variety of industrial applications. The critical nature of many applications, however, imposes technical challenges in manufacturing and inspection. One nondestructive evaluation (NDE) technique of major relevance to inspecting ceramics is ultrasonic microscopy (also termed acoustic microscopy), that is, the use of tightly-focused, high-frequency sound beams to form images of the point-to-point reaction of a material to periodic stress waves. This technique offers high sensitivity for the detection of small defects, and often is a complementary technique to X-ray inspection."

Computer-assisted ultrasonic microscopy (CAUM) has been of particular significance in the testing of new materials developed for more fuel-efficient engines, wherein one objective is that of maximizing the high thermal efficiency of gas turbine engines by way of incorporating high-temperature ceramic components and exhaust-heat recovery. The object of NDE is that of assuring that ceramic components are free of both surface and internal flows that limit component life. Surface flaws can be generated during production by machining and normal handling.

Penetrant Method. This method does not depend upon radiation interactions with the test object and is essentially noninstrumental. A special penetrant substance is applied freely on the test object and allowed to work into tight cracks. See Fig. 3. The penetrant is removed from all surface areas and the piece is sprayed with a developer. The developer dries to an even white coating, while the penetrant bleeds up from any flaws through the developer, forming bright-red or fluorescent indications on the white surface. The size of the defect is indicated by the richness of color, speed of bleed-out, and dimensions observed.

Because of environmental concerns, a new generation of biodegradable penetrants having sensitivity levels ranging from 1 to 4 has been developed. The new penetrants are water washable and, in most instances, can be directly discharged into sewers. They are free of petroleum-based solutions.

Magnetic-Particle Method. This method makes use of iron powder to reveal the leakage magnetic field created at a flaw or break when any part is magnetized. The familiar horseshoe magnet best illustrates this principle. (1) If a horseshoe magnet is bent into a circle, the field between the ends attracts and holds magnetic iron powder. (2) If a magnet is made completely closed, the field will be contained entirely within the ring and no iron powder will be attracted. (3) However, if the round magnet is cracked, poles are created at the break, and iron powder is instantly attracted to the cracked area to pinpoint the defect. See Fig. 4.

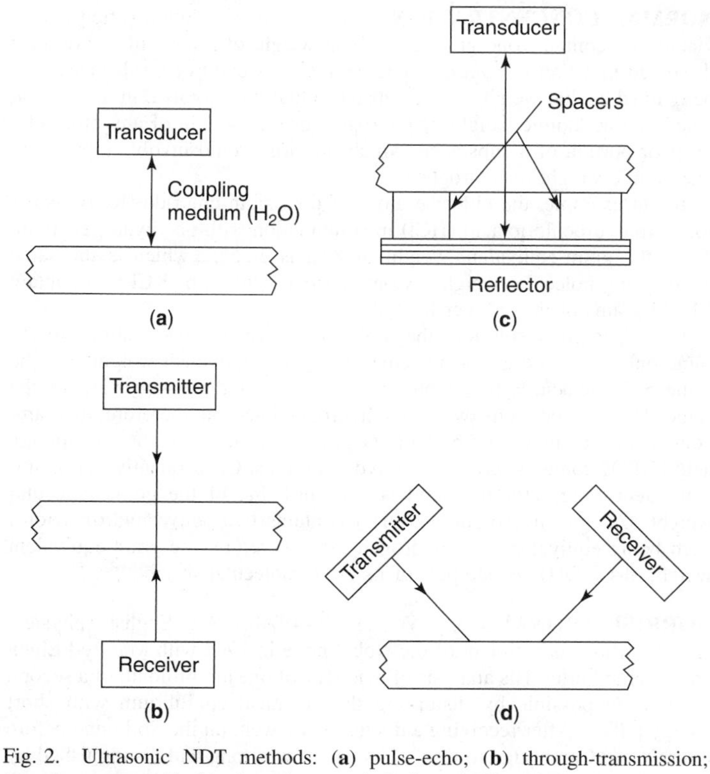

Fig. 2.   Ultrasonic NDT methods: (**a**) pulse-echo; (**b**) through-transmission; (**c**) reflector-plate (double-through transmission); and (**d**) angle-beam

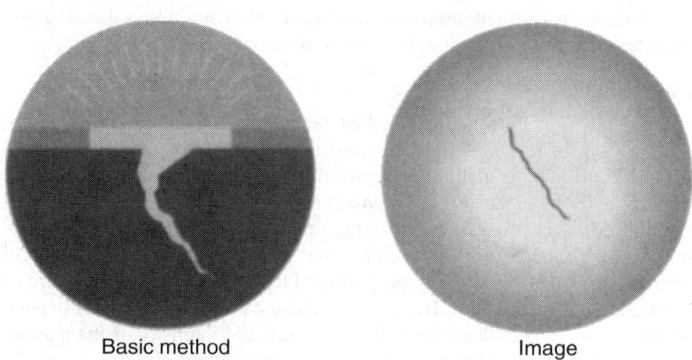

Basic method   Image

Fig. 3.   Penetrant method for detecting flaws

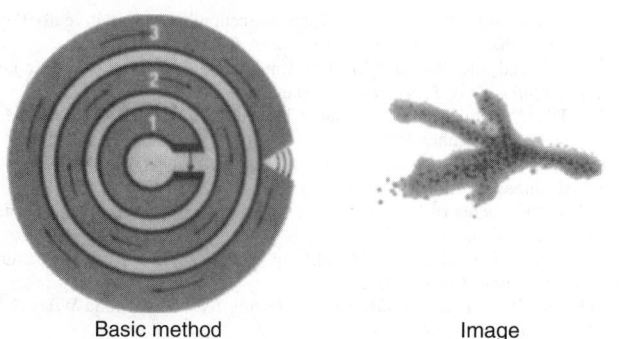

Basic method   Image

Fig. 4.   Magnetic particle method for detecting flaws

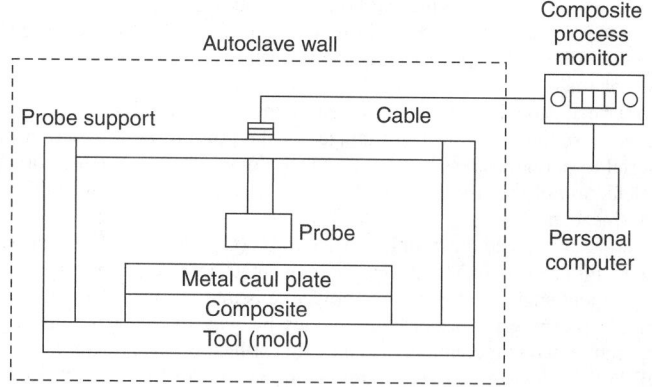

Fig. 5. Use of eddy-current testing for monitoring a composite cure. Typically, eddy-current testing uses an electronic instrument having a small probe on the end of a flexible electrical cable. The probe is placed either against or close to the target. The target must be electrically conductive to allow the generation of eddy currents. The time-variant nature of the probe's magnetic field causes electric currents to flow in the target material. Higher field frequencies or a more electrically conductive material increase the depth of the eddy-current penetration into the material. The concentration of eddy currents near the surface of the material is referred to as the "skin effect." Eddy currents generate their own magnetic field that opposes the probe's magnetic field. Detection circuits in the instrument sense the impedance changes in terms of phase/amplitude changes in probe-coil voltage. (*Suggested by Bar-Cohen and Nguyen*)

Eddy-Current Methods. This is one of the earliest NDT methods and is still used. Basically, this method reveals any differences in electrical impedance between parts to be tested and a reference sample. Parts to be examined are passed through a coil or explored with a probe, and a trace appears on a CRT. Since magnetic and electrical characteristics are closely related to metallurgical quantities, a trace position or pattern or a meter reading clearly shows variations in metal hardness and composition, as well as defects. Both ferrous and nonferrous parts can be tested, and various coils, probes, and detector tips are available.

Aside from more sophisticated electronics, a major contribution to improve eddy-current instrumentation has come from the development of the eddy current resonance digitizing (ECRD) method. With this method, eddy-current instrumentation can separate nonferrous alloys based upon characteristics other than simply their conductivity. See Fig. 5.

NDT Outlook

Improvements in current, established technologies and the introduction of new ways to test materials, nondestructively are expected to continue apac. One promising method is *positron annihilation*. The positron is the antiparticle of the electron; thus a positron/electron pair is unstable and will annihilate. In this process, two gamma rays at approximately 180° to one another are emitted from the center of the mass of the pair. A very slight departure from 180° is directly proportional to the transverse component of the momentum of the pair. The momenta of the electrons involved in such collisions can be calculated from the geometry and intensity of the gamma rays. The dynamics of the electron/positron system underlie the use of the technique for the study of defects in materials.

Additional Reading

Adams, T.E. and A.C. Wey: "Nondestructive Sectioning: Alternative to Physical Sectioning," *Advanced Materials & Processes*, 54 (February 1992).

Akuezue, H.C. and S.K. Verma: "Positron Annihilation: NDE at the Atomic Level," *Advanced Materials & Processes*, 26 (March 1992).

Altshuler, T.L.: "Atomic-Scale Materials Characterization," *Advanced Materials & Processes*, 18 (September 1991).

Bar-Dohen, K.H. Nguyen, and R. Botsco: "Eddy Currents Monitor Composites Cure," *Advanced Materials & Processes*, 41 (April 1991).

Bindell, J.B.: "Elements of Scanning Electron Microscopy," *Advanced Materials & Processes*, 20 (March 1993).

Blitz, J.: *Electrical & Magnetic Methods of Nondestructive Testing,* Institute of Physics Publishing, London, UK, 1991.

Bray, D.E. and D. McBride: *Nondestructive Testing Techniques,* John Wiley & Sons, Inc., New York, NY, 1992.

Carter, G.F. and D.E. Paul: *Materials Science and Engineering,* ASM International, Materials Park, OH, 1991.

Cartz, L.: *Nondestructive Testing: Radiography, Ultrasonics, Liquid Penetrant, Magnetic Particle, Eddy Current,* ASM International, Materials Park, OH, 1995.

Cormia, R.D.: "Problem-Solving Surface Analysis Techniques," *Advanced Materials & Processes*, 16 (December 1992).

Dulski, T.R.: "Residual-Element Analysis: Measuring the Minuscule," *Advanced Materials & Processes*, 20 (February 1992).

Engl, H.W. and W. Rundell: *Inverse Problems in Medical Imaging and Nondestructive Testing,* Springer-Verlag, Inc., New York, NY, 1997.

Evans, N.J.: "Impedance Spectroscopy Reveals Materials Characteristics," *Advanced Materials & Processes*, 41 (November 1991).

Hauk, V. and H. Behnken: *Structural and Residual Stress Analysis by Nondestructive Methods: Evaluation, Application, Assessment,* Elsevier Science, New York, NY, 1997.

Hellier, C.J.: *Handbook of Nondestructive Evaluation,* McGraw-Hill Professional Book Group, New York, NY, 2000.

Malhotra, V. and N. Carino: *CRC Handbook on Nondestructive Testing of Concrete,* CRC Press, LLC, Boca Raton, FL, 1990.

McGonnagle, W.: *International Advances in Nondestructive Testing, Vol. 16,* Gordon & Breach Publishing Group, Newark, NJ, 1991.

Michaels, T.E. and B.D. Davidson: "Ultrasonic Inspection Detects Hidden Damage in Composites," *Advanced Materials & Processes*, 34 (March 1993).

Prask, H.J.: "Neutron Probes Tackle Industrial Problems," *Advanced Materials & Processes*, 26 (September 1991).

Staff: "Testing for Materials Selection," *Advanced Materials & Processes*, 5 (June 1990).

Staff: "Computed Tomography Details Casting Defects," *Advanced Materials & Processes*, 54 (November 1990).

Staff: "Nondestructive Examination," *Advanced Materials & Processes*, 63 (January 1992).

Staff: *Nondestructive Testing,* American Society for Testing & Materials, West Conshohocken, PA, 1999.

Sturges, D.: "Sounding Out Ceramic Quality," *Advanced Materials & Processes*, 35 (April 1991).

Webb, S.C.: "PCs Help Optimize Materials Testing," *Advanced Materials & Processes*, 21 (November 1991).

Wu, Wen-li: "Acoustic Emissions Reveal Fiber/Matrix Bond Strength," *Advanced Materials and Processes*, 39 (August 1991).

Xavier Maldague, P.V.: *Theory and Practice of Infrared Technology for Nondestructive Testing,* John Wiley & Sons, Inc., New York, NY, 2001.

Web References

Nondestructive Testing Information Analysis Center: http://www.ntiac.com/
The American Society for Nondestructive Testing (NDT): http://www.asnt.org/
The online Journal of Nondestructive Testing: http://www.ndt.net/

NONNUTRITIVE SWEETENERS.   See **Sweeteners**.

NORMAL CONCENTRATION.   A one normal solution (often abbreviated 1N) contains one gram-equivalent weight of a particular substance dissolved in 1 liter of *solution*. The equivalent weight of a substance may be defined as that weight of the substance that will involve, in a chemical reaction, one atomic weight of hydrogen, or that weight of any other element or portion of a substance, which, in turn, would involve in reaction one atomic weight of hydrogen.

As an example, the chlorine atom of potassium chloride (KCl) also is found in hydrochloric acid (HCl) in combination with one hydrogen atom. Thus, the gram-equivalent weight of KCl is 74.555, which is the same as its gram-molecular weight. A one normal solution of KCl will contain 74.555 grams of the salt per liter of solution.

For a particular solution, the molar and normal concentration are the same only when the gram-molecular and gram-equivalent weights are the same. Sulfuric acid H_2SO_4 represents a case where these values are not the same. This acid contains two active hydrogen ions and, therefore, its gram-equivalent weight is one-half of its gram-molecular weight. Phosphoric acid H_3PO_4 contains three active hydrogen ions. Consequently, the gram-equivalent weight for this acid is one-third that of the gram-molecular weight. Calcium hydroxide $Ca(OH)_2$ contains two active hydroxyl ions, each being equivalent to a hydrogen ion. Therefore, the gram-equivalent weight of $Ca(OH)_2$ is one-half of its gram-molecular weight.

NORRISH, RONALD G. W. (1897–1978).   An English physical chemist who was recipient of the Nobel prize in 1967 with Manfred Eigen and George Porter. His analysis of reactions of one ten-billionth of a second were made possible by disturbing the chemical equilibrium with short energy pulses. After receiving a doctorate, he went on the Sorbonne before returning to Cambridge to teach. His career was long and distinguished by many awards.

NORTHRUP, JOHN H. (1891–1987). An American chemist who won a Nobel prize in chemistry in 1946 along with James B. Sumner and Wendell M. Stanley. His work was primarily concerned with isolation and crystallization of enzymes. Many first included the production of the enzyme trypsin in the laboratory and isolation of the first bacterial virus. He was also responsible for producing diphtheria antitoxin in crystalline form. His education was at eastern schools including Harvard, Yale, and Princeton.

NOXIOUS GAS. Any natural or by-product gas or vapor that has specific toxic effects on humans or animals (military poison gases are not included in this group). Examples of noxious gases are ammonia, carbon monoxide, nitrogen oxides, hydrogen sulfide, sulfur dioxide, ozone, fluorine, and vapors evolved by benzene, carbon tetrachloride, and a number of chlorinated hydrocarbons. Gases that act as simple asphyxiants are not classified as noxious. See also **Pollution (Air)**.

NUCLEAR CHEMISTRY. The division of chemistry dealing with changes in or transformations of the atomic nucleus. It includes spontaneous and induced radioactivity, the fission or splitting of nuclei, and their fusion or union; also the properties and behavior of the reaction products and their separation and analysis. The reactions involving nuclei are usually accompanied by large energy changes, far greater than those of chemical reactions; they are carried out in nuclear reactor for electric power production and manufacture of radioactive isotopes for medical use, also (in research work) in cyclotrons. See also **Nuclear Fusion**; **Nuclear Fusion**; **Radiochemistry**; and **Nucleus**

NUCLEAR FISSION. A type of nuclear reaction in which the compound nucleus splits into two nearly equal parts, rather than ejecting one or a few small nuclear particles, as in most nuclear reactions. Our knowledge of nuclear fission dates back to the mid-1930s when Fermi and his coworkers showed that the number of distinctly different radioactive nuclides that could be induced by neutron bombardment of uranium far exceeded the number expected, unless some previously unknown pattern of isomerism could be found. Furthermore, the radiochemical properties of many of these radio-elements different quite markedly from expectations. For example, both Hahn and Strassman in Germany and Curie and Savitch in France found that certain unknown activities, thought to be radioactive radium, always followed the chemically separated barium fraction rather than the radium fraction. Hahn and Strassman found several other similar examples and were able to show that uranium, when bombarded by neutrons, undergoes what then appeared to be a very unusual nuclear reaction in that the products are radio-elements with about half the atomic number of uranium. These findings were interpreted by Meitner and Frisch as the division of an excited nucleus into nuclei of medium mass, a process that was given the name *nuclear fission*.

The first such process to be extensively studied was fission induced in ^{235}U by thermal neutrons (neutrons with energies of about 0.03 eV). This reaction, symbolically represented by the equation

$$^{235}\text{U} + n \longrightarrow {}^{236}\text{U} \longrightarrow \text{fission},$$

produces an unstable system which achieves stability by splitting into two large fragments, not by ejecting one or a few small particles.

An individual fission does not produce a unique pair of fragments, but in a large number of such processes, the mass distribution of the fragments can be predicted with reasonable certainty, leading to predictable fission yields. A fission yield, usually expressed as a percentage, describes that fraction of nuclear fission processes that give rise to a specified nuclide or group of isobars. The yields of single nuclides are known as independent yields and those of a set of isobars as mass yields or chain yields. Since two fragments are produced by each fission, the total of all fission yields for a given fission process is 200%. The fission yield curve is different for each mode of induced fission, the most commonly known one being that for thermal neutron induced fission of ^{235}U, shown in Fig. 1. The chemical characteristics of the two fragments vary within limits, so that many elements are formed. Analysis of the fission products shows that most of them are in two mass groups, a "light" group consisting of elements having mass numbers between 85 and 104, and a "heavy" group consisting of elements having mass numbers between 130 and 149. Fragment mass numbers that have been detected range from around A = 70 to around A = 160. The determination of independent yields is made more difficult

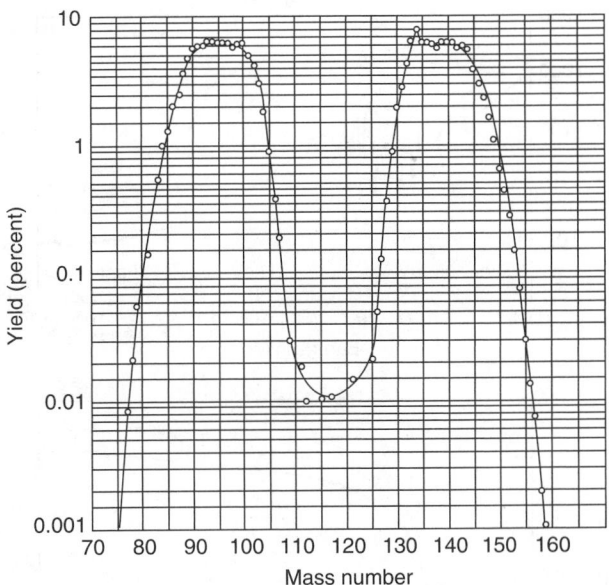

Fig. 1.   Mass fission yield curve for ^{235}U + n (thermal)

by the fact that many of the products are highly radioactive and undergo extensive secondary changes, sometimes in extremely short times, a very small fraction of a second.

A most significant aspect of nuclear fission is its great release of energy. The source of this energy is the loss of mass between the initial and final products of the reaction. The total mass of all atoms and nuclear particles produced in a single fission process in less than the original mass of the ^{235}U atom and the neutron that combined with the ^{235}U to induce fission. During fission of ^{235}U, the total energy released because of loss of mass is about 200 MeV. In practical units, the fissioning of 1 gram of ^{235}U yields 24,000 kilowatt-hours of energy.

Another important feature of fission is the presence of neutrons among the reaction products, slightly more than two for each fission of a ^{235}U atom. These neutrons are not an immediate consequence of fission, but are boiled off the original fission products, their release being possible because of the very large amount of available energy. If all these neutrons were captured by other ^{235}U nuclides, the number of available neutrons would multiply by factors of two for every generation of fission processes, a very rapid increase. However, some neutrons escape from the region containing the ^{235}U and others are absorbed in nonfission capture processes. The minimum conditions for a self-sustaining chain reaction is that at least one neutron from each nucleus undergoing fission must cause fission of another nucleus, a multiplication factor of one or greater. Maintenance of a chain reaction is essential to the proper functioning of both nuclear weapons and nuclear reactors.

The probability that fission can occur (generally called the cross section for fission) varies widely among different nuclides. Only a few nuclides, such as ^{235}U, have a high probability of undergoing fission when they capture a neutron. In other nuclides, the probability of fission is generally much smaller. As an example, the cross section as a function of incident neutron energy is shown in Fig. 2 for fission of ^{235}U and of ^{238}U. Although fission can be induced in ^{238}U, such a process is possible only if the incident neutron has an energy greater than 1 MeV, whereas neutrons of any energy can induce fission in ^{235}U. The characteristic double hump yield curve of Fig. 1 (asymmetric fission) is common only for low neutron excitation energy and targets consisting of highly fissile elements. For either higher excitation energies or less fissile elements, such as actinium or radium, symmetric fission becomes much more important, creating a triple humped fission-yield curve, shown in Fig. 3. Slightly fissile elements, such as lead and bismuth, or very high excitation energies further emphasize the symmetric mode of fission, also illustrated in Fig. 3. Nuclear fission may be induced by particles other than neutrons, such as alpha particles and photons. In some nuclides, it also occurs spontaneously, although the probability of such occurrence is so low that it has almost no effect on the radioactive decay characteristic of the nuclide.

Nuclear fission has generally been explained theoretically in terms of the liquid-drop model of the nucleus. In this model, the incident neutron

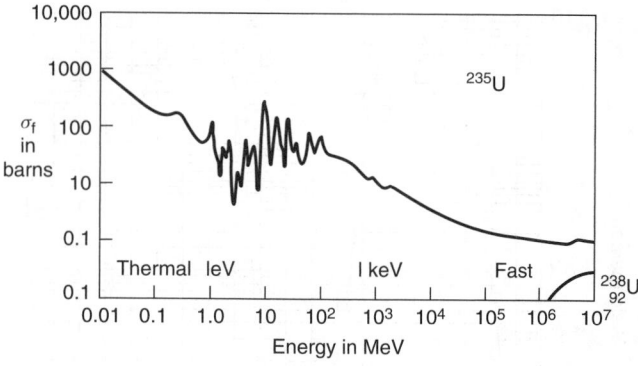

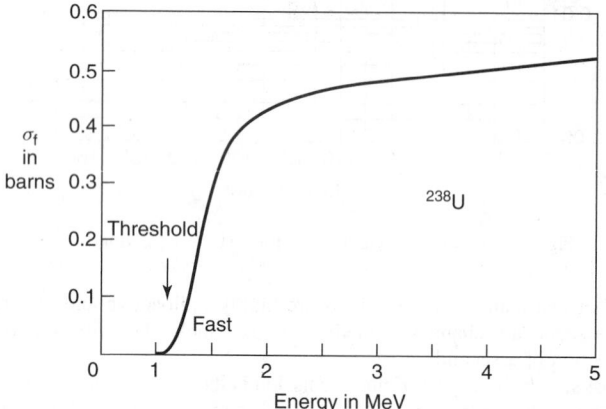

Fig. 2.   Fission cross section as function of energy for ^{235}U and ^{238}U

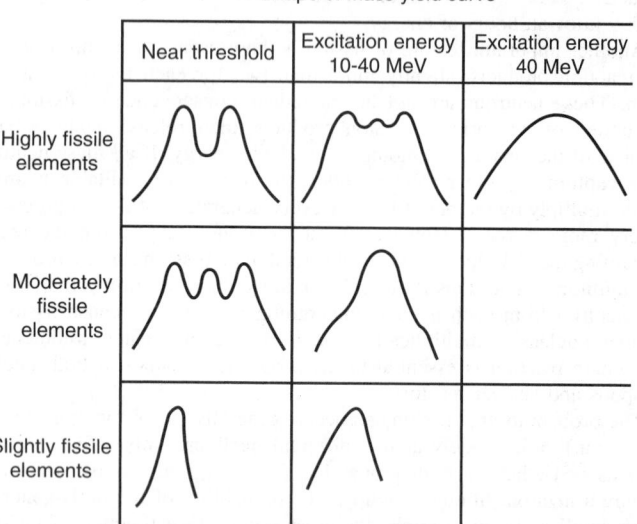

Fig. 3.   Mass fission yield curves as function of excitation energy and degree of fission probability

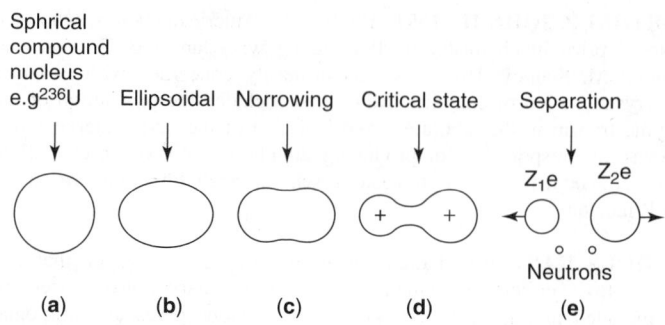

Fig. 4.   Fission mechanism according to liquid-drop model of the nucleus

the effects of the closed shells of the nucleus. See also **Nuclear Power Technology**; and **Nuclear Structure**.

C. SHARP COOK
The University of Texas at El Paso

NUCLEAR FORCES.   Strong, short-range, attractive forces that interact between the individual nucleons of an atomic nucleus. Unfortunately, despite several decades of research, a clear and unambiguous description cannot be given for the forces that hold individual protons and neutrons together in an atomic nucleus. Unlike the electrostatic force that holds electrons in an atom, no equation can be written that completely describes the nature of the force that holds an atomic nucleus together, or the nature of its associated potential energy. A description of the detailed structure of a nucleus cannot, therefore, be derived directly from calculations based on knowledge of nuclear forces. Instead, detailed knowledge of the structure of atomic nuclei has been derived from nuclear models. These models have been constructed by using results from other fields of physical science which display the same or similar characteristics as those observed in nuclear reactions and in radioactive decay. From such analogies, construction of a partial description of nuclear structure and of the nature of nuclear forces has been possible.

Because of the unknown characteristic of nuclear forces, many different suppositions have been made, using available experimental evidence, regarding the nature of the potential energy V of a nuclear particle as a function of its position in the field of a nucleus, or of another nuclear particle. To a first approximation, the nuclear potential is assumed to be spherically symmetric, such as V is a function only of the distance r from the center of the field, thus being the same in all directions, and is representable by a curve as in Fig. 1 curves (a) to (f).

A *potential well* is the name given to a region in which a minimum in the potential is formed; it results from attractive forces. A *potential barrier* is the name given to a region in which there is a maximum in the potential; it results from repulsive forces, either alone or in combination with attractive forces. Some central potentials commonly used as approximations to nuclear potentials are illustrated in the curves. Curve (a) shows a square well potential, which has a constant negative value $-V_0$ for $r \leq r_0$ and zero value for $r \geq r_0$. When this curve represents the potential between two nucleons, r_0 is called the *range of nuclear forces*; when it represents the potential of a nucleus, as this nucleus interacts with an individual nucleon, r_0 is called the *nuclear radius*. Curve (b) shows a square well potential for $r \leq r_0$ with a Coulomb potential resulting from repulsive electrostatic forces, for $r > r_0$. The resulting barrier is called a *Coulomb barrier*, and the maximum energy b is called the *barrier height*. Such a potential approximates that of a positively charged particle in the field of a nucleus, and is often used in the theory of alpha particle disintegration and nuclear reactions. Curve (c) shows an exponential well, $V = V_0 e^{-r/b}$; curve (d) shows a Gaussian well, $V = V_0 e^{-r/b2}$. Curve (e) shows a Yukawa potential, $V = -(V_0/r)e^{-r/b}$ used in the meson theory of nuclear forces for the interaction between two nucleons; and (f) shows a wine-bottle potential, characterized by a low central elevation. If a high central elevation is present, the resulting barrier is called a *central barrier*, or *repulsive core*.

Although the predominant part of the nuclear potential is the part described above that is derived from the central force produced by the average effects of all other nucleons in the system on the individual nucleon under observation, evidence exists that nonsymmetric tensor and spin-orbit coupling terms must be included in the description of the nuclear potential.

combines with the target nucleus to form a compound nucleus at a high excitation energy. A small part of this excitation energy can be attributed to the kinetic energy of the incident neutron, but most of it usually comes from the binding energy of the incident neutron. This added energy initiates oscillations in the drop, which then sometimes assumes an elongated shape, similar to B in Fig. 4. If oscillations become sufficiently violent that a form similar to D is reached, fissioning (form E) becomes inevitable, since the positive charge at the two ends of the dumbbell-shaped nucleus then produces an electrostatic repulsive force greater than the attractive nuclear force holding the neck of the dumbbell together. The reason for asymmetric fission is not clearly understood. The liquid drop model predicts symmetric fission. Most people believe that asymmetric fission results because of

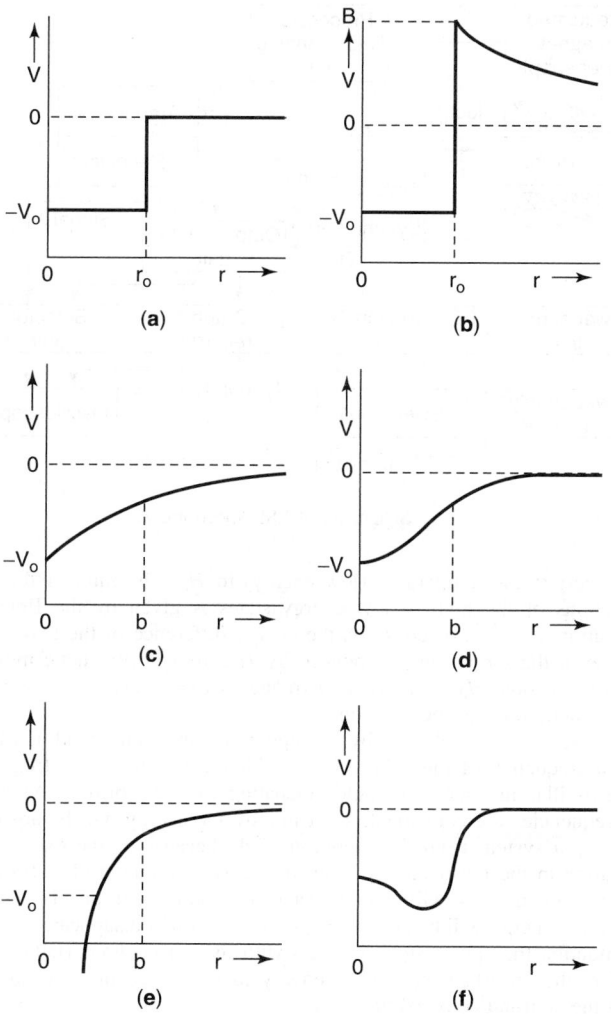

Fig. 1. Potential energy of a nuclear particle versus distance from the center of the field

These are derived from a tensor force resulting from a coupling between individual pairs of nucleons and from the coupling between spin and orbital angular moments of the individual nucleus, as described by the shell model of the nucleus.

A considerable amount of evidence indicates that nuclear forces are charge-independent, i.e., the neutron-neutron, neutron-proton, and proton-proton forces are identical. The meson theory of nuclear forces, originated by Yukawa, postulates the atomic nucleus being held together by an exchange force in which particles, now called mesons, are exchanged between individual nucleons within the nucleus.

C. SHARP COOK
The University of Texas at El Paso

NUCLEAR FUSION. The character of the atomic nucleus is such that the individual nuclear particles are most tightly bound in elements of intermediate atomic number. When energy is sought, attention is focused on the more loosely assembled elements, releasing energy by splitting (*fissioning*) the heavy isotopes, or by joining (*fusing*) the lighter ones. There is less energy release per fusion reaction than there is per fission reaction, but the reactants are more plentiful and, in many respects, easier to handle. A particular fusion reaction is of interest if the power produced can be sufficiently large to offset the power consumed in generating and maintaining the reacting medium, and if the relevant rates can be large enough so that economically interesting regimes are accessible to modern technology. There are over thirty such reactions possible. The most appealing of the fusion reactions as possible routes to fusion energy are (1) those which involve the heavy hydrogen isotopes, deuterium, $_1^2$H or D; and (2) those which involve tritium $_1^3$H or T. These tend to have the largest fusion reaction probability (cross section) at the lowest energies.

Deuterium is abundant, naturally occurring and in wide use now as D_2O in heavy-water-moderated reactors. Tritium is a radioactive isotope with a 12.3-year half-life and does not occur in nature. Tritium emits an electron and decays to stable helium-3.

The deuterium (D-D) reaction chain may be represented by:

$$D + D \longrightarrow {}^3He + n + 3.2 \text{ MeV}$$

$$D + D \longrightarrow T + p + 4.0 \text{ MeV}$$

$$D + T \longrightarrow {}^4He + n + 17.6 \text{ MeV}$$

$$D + {}^3He \longrightarrow {}^4He + p + 18.3 \text{ MeV}$$

. . . .

$$6D \longrightarrow 2{}^4He + 2p + 2n + 43.1 \text{ MeV}$$

The first two equations represent the fact that the D-D reaction can follow either of two paths, producing tritium and one proton; or helium-3 and one neutron, with equal probability. The products of the first two reactions form the fuel for the third and fourth reactions and are burned with additional deuterium. The net reaction consists of the conversion of six deuterium nuclei into two helium nuclei, two hydrogen nuclei, and two neutrons along with a net energy release of 43.1 MeV. The reaction products—helium, hydrogen, and neutrons—are harmless as contrasted with the myriad fission products obtained in a fission reactor. The neutrons produced may be absorbed in sodium to produce an additional 0.25 MeV per cycle. Therefore, the D-D reaction produces at least 7 MeV per deuterium atom (deuteron) and, with absorption in sodium, more than 10 MeV per fuel atom.

The peak reaction rate coefficient of the D-D reaction is considerably less than that of the deuterium-tritium (D-T) reaction occurring within the (D-D) cycle. Thus, attention tends to focus on the latter. Because tritium does not occur naturally, the reaction must be supplemented by one using lithium to reproduce the tritium fuel:

$$D + T \longrightarrow {}^4He + n + 17.6 \text{ MeV}$$

$$n + {}^6Li \longrightarrow {}^4He + T + 4.8 \text{ MeV}$$

. . . .

$$D + {}^6Li \longrightarrow 2{}^4He + 22.4 \text{ MeV}$$

This reaction is tritium-regenerating and produces only helium as a reaction product.

The D-T reactor is technologically more complex than the D-D reactor because of the need to facilitate the second reaction (which takes place outside the plasma) and because very energetic neutrons must be slowed down to allow the reaction with lithium to take place. However, the conditions needed to achieve net power output are less demanding than for the D-D fuel reactor. The D-T reaction will probably be exploited first, but its ultimate, very long term use may be limited by the availability of lithium. See also **Lithium (For Thermonuclear Fusion Reactors)**.

Fusion reactions can take place only when the nuclei of the fuel atoms are brought into close enough conjunction. The nuclei are positively charged and so repel each other. This repulsion is equivalent to an energy barrier which can be penetrated with reasonable efficiency only if the reacting nuclei have kinetic energy comparable to the barrier height. The level of kinetic energy required depends upon the particular reaction and the desired reaction rate, but in general, plasmas of interest have average energy per particle in excess of 5 keV. A collection of particles with average energy 5 keV has an effective temperature of at least 10^8 degrees Kelvin. At these temperatures, the gas is completely dissociated into its constituent positively charged nuclei and free electrons. The density ranges between 10^{13} to 10^{14} cm^{-3}. The electrical charge density is such that the behavior of the collection of particles is completely dominated by electrostatic and electromagnetic phenomena. Such a charge-dominated collection of ionized matter is known as *plasma*. This plasma at such extremely high temperatures cannot be confined by walls made of materials, known or imagined. But confinement, even for a nanosecond or less, is required if fusion reactions are to occur. Ways to confine the plasma have been researched for 20–30 years by scientists in a number of countries.

After nearly three decades of effort, fusion ignition, that is, the efficient burnup of deuterium and tritium has been accomplished only in one way, namely, the thermonuclear or hydrogen bomb. In this instance, obviously tremendously greater amounts of energy are released than are required to trigger the fusion reactions. In the case of the hydrogen bomb, an atomic bomb was used to generate the extremely high temperature and the degree of confinement required for hydrogen nuclei to fuse. The methods researched to date for confining the plasma include containment within magnetic fields; inertial confinement methods in which the fuel is pelletized in a special way and fusion reactions are initiated either by laser beams or beams of particles; and by heating the plasma with high-power microwave radiation.

NUCLEAR MAGNETIC MOMENT. An electrically charged particle of finite size that possesses angular momentum, acts like a small magnet and thus possesses a magnetic moment. For atomic nuclei, the magnitude of this moment is within a range of values between zero and a few nuclear magnetons.

NUCLEAR MAGNETIC RESONANCE (NMR) AND MAGNETIC RESONANCE IMAGING (MRI). Since the discovery of electron spin resonance (ESR) by Zavoiski in 1945 and the codiscovery of nuclear magnetic resonance (NMR) in 1946 by Purcell, Pound, and Torrey (Harvard University) and Bloch, Hansen, and Packard (Stanford University), the field of magnetic resonance has reached a high degree of sophistication and versatility.

Soon after its discovery, the principle was applied by physicists to a variety of research programs. Chemists also became interested in NMR because they recognized its potential for elucidating structure and dynamics in pure and applied chemistry. NMR has gained exceptional acceptance over the intervening years as a tool for yielding details of molecular structure, such interests coinciding with other developments in molecular biology. The use of NMR in medical research dates back about 40 years, commencing with proton NMR measurements on cells and organs of both humans and laboratory animals. During the past 20 years, the applications of NMR to clinical medicine have expanded rapidly. NMR permits the analysis of protein structures, for which the former X-ray methods were inadequate. By the late 1980s, the structures of some 50 important proteins had been determined. NMR allows the investigation of proteins and peptides and other macromolecules in an aqueous medium. Consequently, the effects of pH (hydrogen ion concentration) can be observed and the binding of water at the interior and surface of proteins can be determined.

The molecular weights of proteins studied thus far range from about 5000 to 15,000. The use of three-dimensional Fourier transform NMR may ultimately permit the study of macromolecules up to a molecular weight of 40,000.

In analytical chemistry in the early 1990s, NMR was routinely used to study: (1) polymers, polymer networks, and copolymers, and (2) asphalt, bitumens, tars, and pitches, among numerous examples that could be cited.

Commencing with clinical medicine (circa 1981), which led to the development of NMR and magnetic resonance imaging (MRI) for medical research and diagnostics, the medical applications have captured the "headlines," particularly since the early 1990s. It is interesting to note that in 1981 there were only two MRI devices in the United States. By 1987, the installations had reached about 600 in the United States and about 100 MRI facilities in Europe. As of 1994, these Figures expanded into several hundred installations, representing a tremendous investment by the health care field.

Fundamentals of the Technology

Discovery of the principles of ESR and NMR led to the development of several subtechnologies.

Electron Spin Resonance (ESR). Fundamentally, ESR spectrometers consist of three main parts: (1) a magnet with corresponding power supply to provide a steady dc magnetic field at about 3300 G; (2) a microwave bridge capable of producing an oscillating electromagnetic field at a frequency of about 0.5 GHz (*X* band), which is coupled via waveguide to a high-*Q* microwave reflection cavity; and (3) the associated signal detection including dc field modulation, amplification, and display systems. See Fig. 1.

Observation of ESR from a particular sample is contingent upon the presence of a macroscopic spin magnetic moment $\overline{\mu}$; i.e., the sample under investigation must contain some minimum number of unpaired electron spins. Upon insertion into the cavity, the sample is subjected to the dc magnetic field H_0, and the unpaired electrons align themselves both parallel

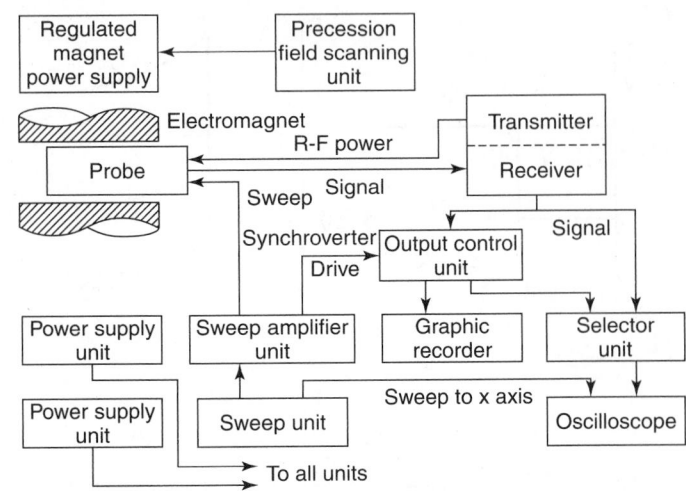

Fig. 1.   Schematic of ESR spectrometer

(high energy) and antiparallel (low energy) to H_0. The ratio of the spin populations of the high to low energy states is given by the Botzman distribution, $e^{-h\nu/kT}$, where $h\nu$ is the energy difference of the two states. Because of the torque on $\overline{\mu}$ produced by H_0, the spin magnetic moment will precess about H_0, at the Larmor frequency $\omega = \gamma_e H_0$, where γ_e is the gyromagnetic ratio of the electron.

Analysis is accomplished by sweeping the magnetic field until the Larmor frequency of the spin system is identical to the fixed frequency of the oscillating microwave field emanating from the bridge. When the two frequencies are coincident, a net microwave energy will be absorbed by the spin system from the oscillating field because of the excess spin population in the lower energy state. If energy continues to be absorbed, the spin populations will equalize and saturation will occur—no net microwave energy will be absorbed and the signal will disappear. To avoid this situation the spin system interacts with its surroundings (lattice) and transfers the absorbed microwave energy to the lattice in some interval called the spin-lattice relaxation time.

This phenomenon, known as relaxation, acts to restore the spin system to its original Botzman distribution of populations. At equilibrium, microwave energy is being absorbed by the spin system and then transferred to the lattice. This process is monitored via microwave rectification of the reflected cavity signal by a diode detector, preamplification, and lock-in phase detection at the dc field modulation frequency. The signal which appears on the recorder may be a single line from an unpaired electron or a group of lines. The latter are caused by neighboring nuclei with nonzero nuclear spin (hyperfine structure) and by surrounding electric field gradients (quadrupole interactions).

In this manner the unpaired electron spins may be used as a probe for analysis of their immediate microscopic surroundings. Typical instrument sensitivity is such that approximately 10^{-11} mole of a paramagnetic species can be detected, but generally concentrations of 10^{-4} mole give optimum ESR spectra.

Nuclear Magnetic Resonance (NMR). This technique is essentially based on the same principle as ESR, but NMR is capable of detecting nuclei (MHz) instead of electrons (GHz). (Lack of a standardized nomenclature has resulted in numerous modifiers in connection with magnetic resonance instrumentation—electron, proton, nuclear, etc., plus application-related terms, such as silicon-29, oxygen-17, ^{13}C, ^{31}P NMR, etc.)

In nuclear magnetic resonance spectroscopy, a nucleus possessing a magnetic moment when placed in a homogeneous magnetic field will precess about the field axis at a rate which is dependent upon the strength of the field. See Fig. 2. If these nuclei are then brought into contact with an oscillating electromagnetic field of the same frequency as the Larmor precessional frequency of the nuclei, energy will be absorbed by the nuclei spin system from the oscillating ratio frequency field. As stated above, the net energy absorption is proportional to the Botzman distribution of spin populations and the effective nuclear relaxation times.

Different nuclei will have different precessional frequencies and therefore at a particular field will absorb energy at certain characteristic radio frequencies. Also, nuclei of the same nuclear species (such as hydrogen) will absorb energy at slightly different frequencies, dependent

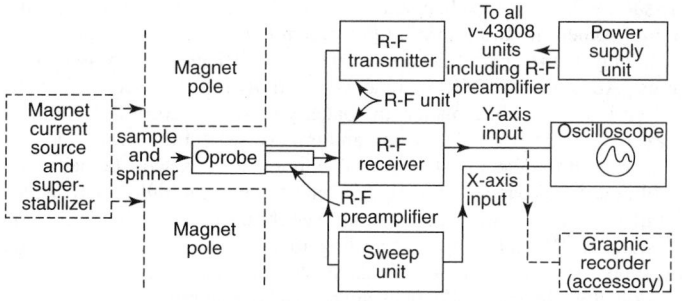

Fig. 2. Schematic of NMR spectrometer

upon their molecular environment. The latter observation makes possible an entire subfield of NMR spectroscopy, termed high-resolution NMR, which is an appropriate method for determining chemical structure and identifying and measuring similar nuclei in two or more different compounds (mixtures).

The major components of a magnetic resonance spectrometer are: (1) a magnet capable of producing a very strong homogeneous field which may be continuously varied over a very small range; (2) a low-power radio frequency oscillator which supplies rf power to a small transmitter coil surrounding the sample; (3) a small receiver coil which also surrounds the sample (but is orthogonal with respect to the transmitter), and feels (4) a sensitive radio receiver (tuned to the same frequency as the transmitter) capable of amplifying any signal which might be induced in the receiver; and (5) a recorder of oscilloscope which can display the resulting spectra.

Various decoupling techniques are used to simplify complex NMR spectra. For example, in a simple two-spin system (homonuclear) giving rise to four NMR lines it is possible to saturate a nucleus at a particular rf frequency and collapse the remaining doublet to a single NMR line. This occurs because the second nucleus sees an averaged interaction from the first rather than two distinct interactions from spins in the high and low energy states. Noise decoupling is predominantly used to decouple different nuclei—for example, hydrogen nuclei C^{13} spectra (heteronuclear). Here the entire hydrogen spectrum is saturated over a range of frequencies (noise) leaving behind the much simplified C^{13} spectrum.

In internuclear double resonance one nuclear line at a given frequency is observed while a second frequency is swept through the remainder of the NMR spectrum. All nuclei which are in some way coupled to the line being observed will enhance or de-enhance the latter as the second frequency passes through the resonance condition. Also, the Fourier transform technique has nicely complimented NMR. Any complex waveform can be converted to a spectrum of frequencies by Fourier transformation.

In NMR, the waveform is a superposition of a set of nuclear precession frequencies with amplitudes decaying due to relaxation and field inhomogeneity. The transformation may be carried out by analog means (spectrum analyzer) or on a small dedicated laboratory computer. The latter appears to be the most convenient solution. Since the free induction decay signal decays with time, whereas the instrumental noise remains constant, the noise content is higher in the tail of the transient signal, and it is possible to improve the overall signal-to-noise ratio by weighting the transient signal with an exponentially decaying function of time. The shorter the time constant of this exponential, the greater the improvement in sensitivity, but this increases the linewidths of the transformed spectrum. The reversal of this procedure can enhance resolution at the cost of sensitivity.

High-Resolution Nuclear Magnetic Resonance of Solids. Important developments in solid-sample NMR techniques of the early 1980s have made NMR of significant interest as a tool for characterizing solid samples—as it has been in the past for the study of liquids. As observed by Maciel, the development of line-narrowing techniques, such as magic-angle spinning (MAS) and high-power decoupling, has led to powerful high-resolution NMR for studying solids. In favorable cases (for example, where high abundances of protons are present) cross polarization (CP) provides a means of circumventing the time hurdle caused by inefficient spin-lattice relaxation in many solids. Combining the CP and MAS approaches for carbon-13 with proton decoupling has become a popular and routine experiment for organic solids. For many nuclides with spin quantum number $1 > \frac{1}{2}$, the central nuclear magnetic resonance transition can be used in high-resolution experiments that involve rapid sample spinning. A continuing stream of other advances in NMR technology bodes well for the

characterization of solids by a wide range of nuclides. The complexities of this topic unfortunately are beyond the scope of this encyclopedia.

It is interesting to note that several of the concepts for improving NMR technology, as listed by Levy and Craik, in 1988, already have been partially or fully achieved: (1) two-dimensional Fourier transform (FT NMR); (2) high-resolution NMR in solids; (3) new types of pulse sequences; (4) chemically induced dynamic nuclear polarization; (5) multiple quantum NMR; and (6) NMR imaging (MRI).

Two-Dimensional NMR. Bax and Lerner report on how two-dimensional Fourier transform pulse NMR (2-D FT NMR) has extended the range of applications of NMR spectroscopy into the area of large, complex molecules, such as DNA and proteins. Great spectral simplification can be obtained by spreading the conventional one-dimensional NMR spectrum in two independent frequency dimensions, thus removing spectral overlap, facilitating spectral assignment, and providing additional information. Conformational information related to interproton distances is available from resonance intensities in certain types of two-dimensional experiments. Two-dimensional NMR spectroscopy also has been applied to the study of ^{13}C and ^{15}N to provide connectivity information and greatly improving the sensitivity of these determinations. A traditional NMR spectrum of a sugar and a 2D spectrum are contrasted in Fig. 3.

Correlative spectroscopy (COSY) is an approach that involves correlating groups believed to be coupled to each other and to prove that this coupling does exist. *Spin-echo correlation spectroscopy* (SECSY) is a variation of the COSY technique. Sometimes called *J-resolved spectroscopy*, it is a technique that allows a separation of the chemical shift of a nucleus from the coupling to other nuclei. This simplifies the spectrum and permits one to assign each resonance to a specific nucleus. Sometimes referred to as *nuclear Overhauser effect* (NOESY), this is a technique that makes it possible to determine distances between nonadjacent residues in a peptide chain. Another application, *multiple quantum transitions*, takes advantage of the fact that molecules in a sample are forced to absorb or emit several quanta of energy at one time. For example, the technique can be used to determine which carbon atoms in a molecule are connected to other specific atoms in the molecule.

Magnetic Resonance Imaging (MRI). In 1973, Lauterbur added a new aspect to NMR basics, namely that of image formation based upon NMR

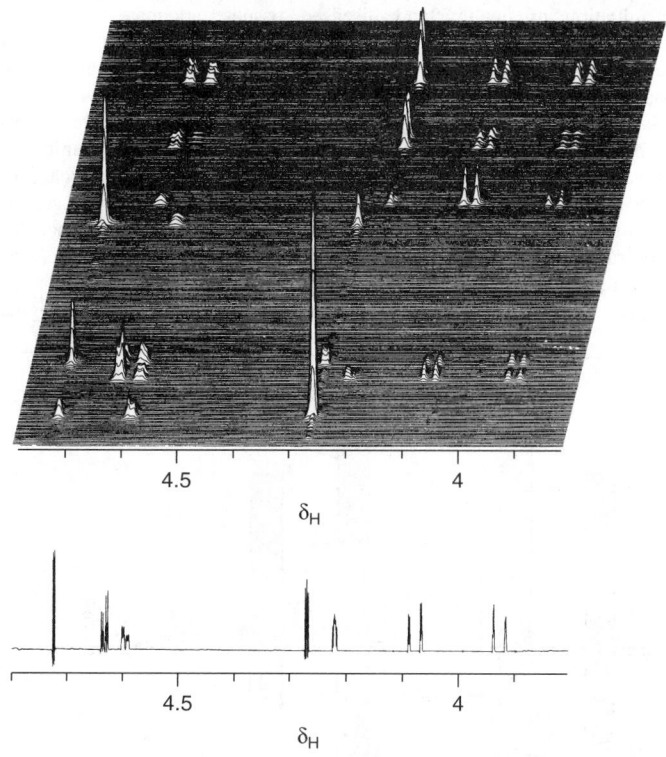

Fig. 3. (*Top*) A two-dimensional nuclear magnetic resonance spectrum of a sugar; (*Bottom*) conventional NMR spectrum of same sugar. The two-dimensional spectrum also can be plotted as a contour map with intensities denoted by color. (*JEOL Inc*)

principles. This led to NMR imaging, now commonly referred to as magnetic resonance imaging (MRI).

The source of the imaging photons is not a Roentgen tube, nor an unstable isotope, nor an electron storage ring, but rather it is a radio transmitter. The commonly referred *radio photon* radiation allows photons to pass into the tissues under study, where they are absorbed by nuclear particles rather than by electrons. The nucleons are momentarily excited by the process and then return to their resting state by emitting photons of the same or nearly the same energy that they had absorbed. Because the nuclei of the different chemical elements absorb radio photons of different frequencies, it is possible to use the method to detect the presence of a single element in a sample. When protons are irradiated by radio photons of a frequency that precisely matches their own precessional frequency, resonance occurs. The resonant frequency (Larmor frequency) is determined by the natural rotational velocity (gyromagnetic ratio) characteristic of each species of atomic nucleus and by the strength of the applied static magnetic field, as expressed by: Larmor frequency =Magnetic field ×Gyromagnetic ratio. See Fig. 4.

Magnetic resonance imaging is well suited for the imaging of soft tissues. It has been found particularly effective in (1) studying skeletal musculature, notably in the male and female pelvic regions, (2) delineating tumor development and assisting in planning surgery or therapy, and (3) precisely locating tumors, particularly in the brain where, with time, the evolution of hematomas can be studied. For clinical work, MRI installations may be used in two staff shifts per day, holidays, and weekends.

The magnetic fields used are usually 1.5 to 2 teslas (15,000 to 20,000 gauss). Barriers to using higher magnetic fields are magnet cost, additional expense for building the site, and notation by some persons (volunteers) to discomfort when over 4 teslas are used, which also causes the patient to move.

In 1986, Fossel and McDonough (Beth Israel Hospital, Boston) proposed that proton NMR spectroscopy of human plasma possibly could be a "potentially valuable approach to the detection of cancer and the monitoring of therapy." Subsequent investigations through 1990 failed to develop a convincing correlation.

Traditional MRI of the human body relies mainly on the detection of the most abundant type of nuclei, the hydrogens in water, and, to some extent, fat.

As pointed out by C. Moonen (National Institutes of Health), "For discrimination of healthy and diseased tissues, adequate contrast is essential. Such contrast depends not only on differences in water concentration, but also on the NMR relaxations, which, in turn, are related to local mobilities and interactions."

Magnetic resonance imaging, in addition to providing detailed information about the macroscopic structure and anatomy, also permits the noninvasive spatial evaluation of various biophysical and biochemical processes in living systems. These include the motion of water in processes, such as vascular flow, capillary flow, diffusion, and exchange. Further, the concentrations of various metabolites can be determined for the assessment of regional regulation of metabolism. In the scholarly Moonen paper, examples are given of flow imaging, diffusion imaging, and the imaging

of tissue perfusion, of exchange, and of metabolites. These aspects of MRI imaging sometimes are referred to as *functional MRI*.

Imaging of the central nervous system by MRI technology is of great interest. Advantages of MRI include its sensitivity to soft tissue, the contrast between gray and white matter, the paucity of signals from the skull, and the availability of coronal, sagittal, and transverse sections. The procedure has been used to detect posterior fossa tumors, such as acoustic neuromas, and pituitary and parasellar tumors, orbital tumors, multiple sclerosis, and a number of other lesions in the craniovertebral junction, and of the cord and spine. MRI technology, according to a number of authorities, has large potential for use in *noninvasive* cardiological studies, particularly of blood flow imaging. Contrast agents are not required because nuclei in rapidly flowing blood move out of the volume of interest during the interval required for application of the rf pulse and of gradient magnetic fields. The parameters that influence MRI signals at various flow rates are being studied.

In addition to hydrogen atoms, MRI can create, for example, ^{31}P images which are excellent indicators of energy metabolism. ^{23}Na images reflect extracellular and intracellular fluid fluxes.

Echo-Planar Imaging. A major problem with MRI in the past has been the long data acquisition times (up to several minutes). Consequently, MR images are subject to so-called *motional artifacts*, caused by physiological motions (heartbeat, blood flow, bowel peristalsis, breathing) as well as by voluntary movements in severely ill and uncooperative patients, including children. Echo-planar imaging (EPI) permits faster scan times, thus effectively reducing imaging to a fraction of a second as compared with minutes. Although a technical description of how EPI is implemented is too complex for coverage here, echo-planar imaging uses only one nuclear spin excitation per image.

EPI has broadened the use of MRI to include the evaluation of cardiac function in real time, mapping of organ blood pool and perfusion, functional imaging of the central nervous system, depiction of blood and cerebrospinal fluid flow dynamics, and motion picture imaging of the mobile fetus in utero. EPI also has the practical advantages of increasing patient throughput at a lower cost per MRI examination. With these advantages, it is expected that EPI will become an established tool for early diagnosis of some common and potentially treatable diseases, such as ischemic heart disease and stroke.

Comparison with Ultrasonography. The health care community is aware of the high costs of MRI and continues to seek other techniques that may be effective at lower cost. One such comparison was made in connection with prostate cancer. The approach to treatment varies and depends on the extent of cancer at the time of diagnosis.

In a specific comparative study over a period of 15 months, 230 patients were evaluated with identical imaging techniques. It was found that MRI correctly staged 77% of cases of advanced disease and 57% of cases of localized disease. The corresponding Figures for ultrasonography were 66% and 46%. MRI identified only 60% of all malignant tumors measuring more than 5 mm on pathological analysis, while ultrasonography identified only 59%. The study concluded, "The MRI and ultrasonography equipment is not highly accurate in staging early prostate cancer, mainly because neither technique has the ability to identify microscopic spread of disease."

Superconducting Quantum Interference Device (SQUID). These devices are sensitive detectors of magnetic fields. Low-temperature superconductors have been used in the past to sense weak magnetic signals from the brain for medical diagnosis. Such devices, however, required cooling with liquid helium and thus have resulted in costly, unwieldy apparatus. Research is now underway toward using higher-temperature superconductors, such as $YBa_2Cu_3O_7$. Although the higher-temperature semiconductor still requires refrigeration, this can be accomplished with liquid nitrogen, thus resulting in a less costly, simpler, and more portable detecting device.

Additional Reading

Abelson, P.H.: "New Horizons in Medicine," *Science*, 1109 (November 25, 1988).

Ancreasen, N.C.: "Brain Imaging: Applications in Psychiatry," *Science*, 1381 (March 18, 1988).

Bain, L.: "MRI—Safety Issues Stimulate Concern," *Science*, 1245 (May 31, 1991).

Bax, A. and L. Lerner: "Two-Dimensional Nuclear Magnetic Resonance Spectroscopy," *Science*, **232**, 960–967 (1986).

Bushong, S.C.: *Magnetic Resonance Imaging: Physical and Biological Principles,* 3rd Edition, Elsevier Science, New York, NY, 2003.

Edelman, R. et al.: *Clinical Magnetic Resonance Imaging,* 2nd Edition, W.B. Saunders, Philadelphia, PA, 1996.

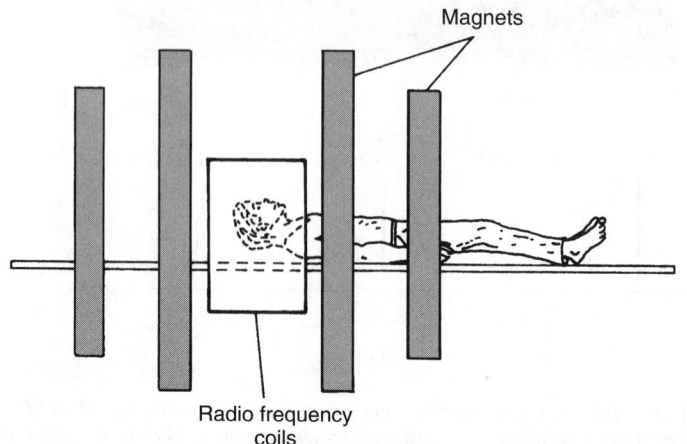

Fig. 4.   General arrangement of MRI equipment in a medical setting

Grant, D.M., R.K. Harris, E. Becker: *Encyclopedia of Nuclear Magnetic Resonance, Advances in NMR,* Vol. 9, John Wiley & Sons, Inc., New York, NY, 2002.

Hari, R. and O.V. Lounasmaa: "Recording and Interpreting Cerebral Magnetic Fields," *Science,* 432 (April 28, 1989).

Kirkwood, J.R.: *Essentials of Neuroimaging, 2nd Edition,* Churchill-Livingstone, Inc., New York, NY, 1995.

Krestel, E. (Editor): *Imaging Systems for Medical Diagnosis: Fundamentals and Technical Solutions—X-Ray Diagnostics—Computed Tomography—Nuclear Medical Diagnostics—Magnetic Resonance Imaging—Ultrasound Technology,* John Wiley & Sons, Inc., New York, NY, 1990.

Levy, G.C. and D.J. Craik: "Developments in Nuclear Magnetic Resonance Spectroscopy," *Science,* **214,** 291–299 (1981).

Magin, R. et al. (Editors): *Biological Effects and Safety Aspects of Nuclear Magnetic Resonance Imaging and Spectroscopy, Vol. 649,* New York Academy of Sciences, New York, NY, 1992.

Mitchell, T.N., and B. Costisella: *NMR: From Spectra to Structures,* Springer-Verlag New York, Inc., New York, NY, 2004.

Mitchell, D.G., and M. Cohen: *MRI Principles,* 2nd Edition, Elsevier Science, New York, NY, 2003.

Moonen, C.T. et al.: "Functional Magnetic Resonance Imaging in Medicine and Physiology," *Science,* 53 (October 6, 1990).

Nelson, J.H.H.: *Nuclear Magnetic Resonance Spectroscopy,* Prentice Hall, Inc., Upper Saddle River, NJ, 2001.

Pool, R.: "Putting SQUIDs to Work," *Science,* 862 (August 24, 1990).

Randal, J.: "NMR: The Best Thing Since X-Rays," *Technology Review (MIT),* 59 (January 1988).

Rifkin, M.D. et al.: "Comparison of Magnetic Resonance Imaging and Ultrasonography in Staging Early Prostate Cancer," *New Eng. J. Med.,* 623 (September 6, 1990).

Shriner, R.L., T. Morrill, and D. Curtin: *The Systematic Identification of Organic Compounds,* 8th Edition, John Wiley & Sons, Inc., New York, NY, 2003.

Shulman, R.: "NMR—Another Cancer-Test Disappointment," *New Eng. J. Med.,* 1002 (April 5, 1990).

Sochurek, H. and P. Miller: "Medicine's New Vision," *National Geographic,* 2 (January 1987).

Stehling, M.K., R. Turner, and P. Mansfield: "Echo-Planar Imaging in a Fraction of a Second," *Science,* 43 (October 4, 1991).

Sutton, D. and J.W.R. Young: *A Short Textbook of Clinical Imaging,* Springer-Verlag, Inc., New York, NY, 1991.

Tamraz, J. and Y. Comair: *Atlas of Regional Anatomy of the Brain Using MRI: With Functional Correlations,* Springer-Verlag New York, Inc., New York, NY, 2000.

Tycko, R.: *Nuclear Magnetic Resonance Probes of Molecular Dynamics,* Kluwer Academic Publishers, Norwell, MA, 2003.

Wilson, M. and F. Ruzicka (Editor): *Modern Imaging of the Liver: Applications of Computerized Tomography Ultrasound, Nuclear Medicine and Magnetic Resonance Imaging,* Marcel Dekker, Inc., New York, NY, 1989.

NUCLEAR POTENTIAL.   The potential energy V of a nuclear particle as a function of its position in the field of a nucleus or of another nuclear particle. A central potential is one that is spherically symmetric, so that V is a function only of the distance r of the particle from the center of force. A noncentral potential, on the other hand, is one that is not spherically symmetrical, or one that depends upon the relative directions of the angular momenta associated with the particle and the center of force, as well as upon the distance r. A negative potential corresponds to an attractive force, while a positive potential corresponds to a repulsive force.

Although the expression can certainly be applied to the problem of nuclear forces, the usual meaning of a nuclear potential refers to the interaction of a nucleon (neutron or proton) with a complex nucleus. Although the potential energy of a single nucleon inside a nucleus is clearly a rapidly varying function of position and time (since it represents the interaction with a large number of closely packed, fast-moving particles), one may nevertheless speak of the average potential energy, and one may regard this as a smoothly varying function. For a neutron, the nuclear potential is essentially negative inside the nucleus, rising rapidly to zero outside the nuclear radius R. For a proton, the long-range electrostatic repulsion must, of course, be added. Owing to the Pauli exclusion principle, and to the exchange nature of nuclear forces, however, such a potential cannot in general be regarded simply as a function of position, $V = V(r)$; it depends in addition upon the momentum of the particle, which in quantum mechanics does not commute with the position. Hence, the potential must be regarded as a nondiagonal matrix operator $V = \langle r|V|r' \rangle$ configuration space, or a similar operator in momentum space.

Although the concept of a nuclear potential in this latter sense cannot be defined in a precise way, it has nevertheless been useful, both qualitatively and quantitatively, in the investigations of nuclear structure and nuclear reactions. It has been of particular usefulness in the optical model of nuclear reactions.

NUCLEAR POWER TECHNOLOGY.   After a terse review of the physics and chemistry of nuclear fission and the nature of these reactions, some of the design features and operating parameters of the four families of reactors currently installed are described. These units, some of which are approaching their retirement are, of course, the starting basis for the design of next-generation reactors, which appeared prior to the end of the 20th century. Design improvements that constitute these forthcoming systems are described in terms of their status as of the mid-1990s. The topic of radioactive waste handling is summarized. Based upon available information, nuclear power technology in the United States as well as in Canada, France, Japan, the U.K., and other leading industrialized nations is covered.

The first nuclear power plant was installed at Shippingport, Pennsylvania in 1957 and, after serving as a test facility for several years, was dismantled because of "old age." The plant had a capacity of 60,000 kilowatts. Indeed, the plant was extremely small by comparison with hundreds of units installed today in the United States and other major countries of the world.

NATURE OF NUCLEAR FISSION REACTIONS

The energy of a nuclear fission reaction can be computed from the change in mass between reactants and products according to Einstein's law:

$$\Delta E = \Delta mc^2$$

where E is the energy in ergs, m is mass in grams, and c is the velocity of light in centimeters per second. For example, the mass difference in this equation is $\Delta m = 0.2058$ amu (atomic mass units). Therefore, $\Delta E = 931$ MeV/amu $\times 0.2058$ amu $= 191.6$ MeV. The average amount of energy released in the various fission reactions is about 200 MeV (million electron volts). This energy is distributed in the fission process as:

	MeV
Kinetic energy of fission fragments	165
Radioactive-decay energy	23
Kinetic energy of neutrons	5
Prompt gamma-ray energy	7

The energy of a chemical reaction, approximately 3–4 eV, is dramatically lower than that of a nuclear reaction. Hence, the fission of ^{235}U yields 2.5 million times as much energy as the combustion of the same weight of carbon.

The importance of fission in energy (power) production lies in two facts: (1) an exceedingly large amount of energy is released in the fission reaction; and (2) the production of excess neutrons permits a chain reaction. These two circumstances make it possible to design nuclear reactors in which self-sustaining reactions occur with the continuous release of energy. Although described later, it may be pointed out here that nuclear fission is not the only energy-releasing nuclear reaction. The fusion of light nuclides, like hydrogen, into heavier elements is also an energy-producing process.

The heat generated in nuclear power plants is transferred to a working fluid and from this point on the nuclear power plant and the conventional fossil-fueled power plant are essentially similar.

Fission Reaction.   In nuclear fission, the nucleus of a heavy atom is split into two or more fragments. The reaction is initiated by the absorption of a neutron. A typical reaction is

$$^{235}_{92}\text{U} + ^{1}_{0}n \longrightarrow ^{137}_{56}\text{Ba} + ^{97}_{36}\text{Kr} + 2^{1}_{0}n + \Delta E$$

In this reaction, a ^{235}U atom absorbs one neutron, becomes unstable, and subsequently fissions into two fission fragments plus two neutrons. This is just one of the many ways in which ^{235}U might fission. The number of neutrons produced in a fission reaction is usually 2 or 3. The excess neutrons produced by the fission reaction provide the means of self-sustaining the chain reaction. Nuclides including ^{233}U, ^{235}U, and ^{239}Pu, which are fissionable by neutrons of all energies, are termed fissile nuclides.

Nuclear Fuels.   There are two broad categories of nuclear fuels: (1) the fissile nuclides previously mentioned; and (2) the fissionable nuclides,

^{232}Th (thorium) and ^{238}U. Thermal reactors use fissile nuclides as fuel, while fast reactors are designed to burn fissionable materials. In fast reactors, only a small portion of the ^{232}Th and ^{238}U are fissioned directly. A larger portion of these materials is converted into ^{235}U and ^{239}Pu, respectively, through neutron absorption. Thus, this type of reactor not only consumes fuel, but also produces (breeds) new fuel material. Hence, the term *breeder reactor* is used for reactors designed to take advantage of this phenomenon. Breeding is possible in thermal reactors also, but to a lesser extent. The fuel material in a fast reactor must contain a significant amount (about 10%) of one of the fissile materials. The remainder of the fuel must have a high mass number in order to avoid slowing down the neutrons. The natural reserves of fissionable materials are more than 100 times greater than the reserves of fissile materials. Consequently, from the viewpoint of utilization of available energy resources, fast reactors are of great importance. Breeder reactors are described later.

Moderators. The most important slow-down mechanism is elastic scattering on elements of low mass number. Materials like light and heavy water, beryllium oxide, and graphite are used to slow down, or *thermalize* the neutrons to an energy of about 0.025 eV. As neutrons collide with the nuclei of these atoms, their kinetic energy and speed are gradually reduced until thermal equilibrium is achieved with the reactor structure. The fewer such collisions before deceleration is complete, the less chance of ^{238}U atoms absorbing neutrons.

Critical Mass. Thermal neutrons, which move like atoms in a low-pressure gas, diffuse throughout the reactor. They must be absorbed by a nucleus of the reactor structure, in which case they merely make that nucleus radioactive. Or, they may strike a fissionable atom of ^{235}U, causing fission and, in turn, releasing more neutrons to maintain the reaction. Should the number of neutrons absorbed by the moderator and ^{238}U be greater than about 1.5 excess neutrons emitted from each fission, the chain reaction will not be maintained. Therefore, the reactor core must be designed so that the mass of fuel will be just sufficient to ensure one neutron from each fission causing fission in another atom. A mass and configuration of fissionable material in which this occurs is termed the *critical mass*—or a reactor in which this condition is achieved is said to have "gone critical."

To measure a chain reaction, a multiplication factor k is used to indicate the ratio of neutrons in one generation to those in the preceding generation. Thus, in a constant chain reaction where the total number of neutrons neither increases nor decreases, the heat output is constant and $k = 1$. Should k rise above unity, the rate of fission, and hence the rate of heat productivity, steadily rises. This is so even if k is held constant at its new value. Here lies one *major difference between nuclear reactors and conventional steam generators*. In the latter, heat output is proportional to firing rate. If the firing rate is increased, the steam output is increased; but it remains constant at its new level. In a nuclear reactor, an increase in k results in continuously rising heat output. Only by returning the rate of neutron production to its original ratio can heat output be maintained at its new level.

Reactivity Control. Absorption of excess neutrons, above those needed to maintain a constant reactivity level, provides close control over the degree of reactivity. This is accomplished by inserting materials having a high neutron-capture rate into the core. Control rods of special alloy metals are moved into and out of the cores as required. To start the reactor from shutdown (black start), control rods are partially withdrawn until k becomes greater than one. Neutron flux and heat output grow until the desired level is reached. At this point, control-rod movement is quickly reversed to keep k at unity. The reactor is shut down by inserting the rods to their full extent. In this position, the rods absorb more than 1.5 excess neutrons per fission and the chain reaction quickly stops. Heat production continues for a time, but is usually dissipated by an auxiliary cooling system.

It is interesting to contrast a nuclear power reactor and a nuclear bomb. The designer of a nuclear fission bomb seeks to release as much fission energy as possible within the shortest possible time (milliseconds). Thus, a bomb is designed to favor *prompt neutrons*. By contrast, the normal operating mode of a nuclear power reactor is one in which prompt neutrons alone cannot sustain a chain reaction, but prompt neutrons together with delayed neutrons can. Only the delayed neutrons are controllable. A power reactor is designed to release fission energy *slowly and smoothly* and in just the right amounts to convert water into steam. Whereas the "fuel" in a bomb is used up essentially in an instant, in a power reactor the energy release is spread over months and years. It has been agreed by physicists for

many years that it is physically impossible for a power reactor to explode in the manner of an atomic bomb.

TYPES AND MAJOR CHARACTERISTICS OF NUCLEAR POWER REACTORS

In order, the following types of nuclear fission reactors are described in this section: (1) light water reactors, (a) pressurized water reactors, (b) boiling-water reactors; (2) high-temperature gas-cooled reactors; (3) heavy water reactors; and (4) fast breeder reactors. Military reactors are not described.

Contemporary Light-Water Reactors (LWRs)[1]

These reactors are of two principal designs: (1) *pressurized water reactors* (PWR), and (2) *boiling-water reactors* (BWR). In a PWR, heat generated in the nuclear core is removed by water (reactor coolant) circulating at high pressure through the primary circuit. The water in the primary circuit both cools and moderates the reactor. Heat is transferred from the primary to the secondary system in a heat exchanger, or boiler, thereby generating steam in the secondary system. The BWR differs from the PWR primarily in that boiling takes place in the reactor itself. Comparable steam temperatures are possible at pressures of about 1000 pounds per square inch (6.9 mPa) as contrasted with 2000 psi (13.8 mPa) for pressurized reactors.

Contemporary Boiling Water Reactor (BWR)

Aside from its heat source, the boiling water reactor (BWR) generation cycle is substantially similar to that found in fossil-fueled power plants. One of the first BWRs was the Vallecitos BWR, a 1000 psi (6.9 mPa) reactor which powered a 5 MW electric generator and provided power to the Pacific Gas & Electric Company grid through 1963. Power output capabilities have increased many times during the intervening years as shown by tabular summaries given later in this entry.

The direct-cycle boiling water reactor nuclear system (Fig. 1) is a steam generating system consisting of a nuclear core and an internal structure assembled within a pressure vessel, auxiliary systems to accommodate the operational and safeguard requirements of the nuclear reactor, and necessary controls and instrumentation. Water is circulated through the reactor core producing saturated steam which is separated from the recirculation water, dried in the top of the vessel, and directed to the steam turbine-generator. The turbine employs a conventional regenerative cycle with condenser de-aeration and condensate demineralization. The direct-cycle system is used because of its inherently simple design, contributing to reliability and availability.

The steam from a BWR is, of course, radioactive. The radioactivity is primarily ^{16}N, a very short-lived nitrogen isotope (7 seconds half-life) so that the radioactivity of the steam system exists only during power generation. Extensive generating experience has demonstrated that shutdown maintenance on a BWR turbine, condensate, and feedwater components can be performed essentially as a fossil-fuel plant.

The reactor core, the source of nuclear heat, consists of fuel assemblies and control rods contained within the reactor vessel and cooled by the recirculating water system. A 1,220-MWe BWR/6 core consists of 732 fuel assemblies and 177 control rods, forming a core array 16 feet (4.8 meters) in diameter and 14 feet (4.2 meters) high. The power level is maintained or adjusted by positioning control rods up and down within the core. The BWR core power level is further adjustable by changing the recirculation flow rate without changing control rod position, a feature that contributes to excellent load-following capability.

The BWR is the only light water reactor system that employs bottom-entry control rods. From the very first BWRs, bottom-entry control rods have been used because reactivity and moderator density is highest in the lower part of the core. They provide optimum power shaping characteristics for the type of core where moderator density is varied as a function of power level. Bottom-entry and bottom-mounted control rod drives also allow refueling without removal of rods and drives, and allow drive testing with an open vessel prior to initial fuel loading, or at each refueling operation. The hydraulic system, using reactor system pressure, provides rod insertion forces that are greater than gravity or mechanical systems.

The BWR requires substantially lower primary coolant flow through the core than pressurized water reactors. The core flow of a BWR is the sum of the feedwater flow and the recirculation flow, which is typical

---

[1] In nuclear power technology, ordinary water, in contrast with *heavy water*, is termed *light water*.

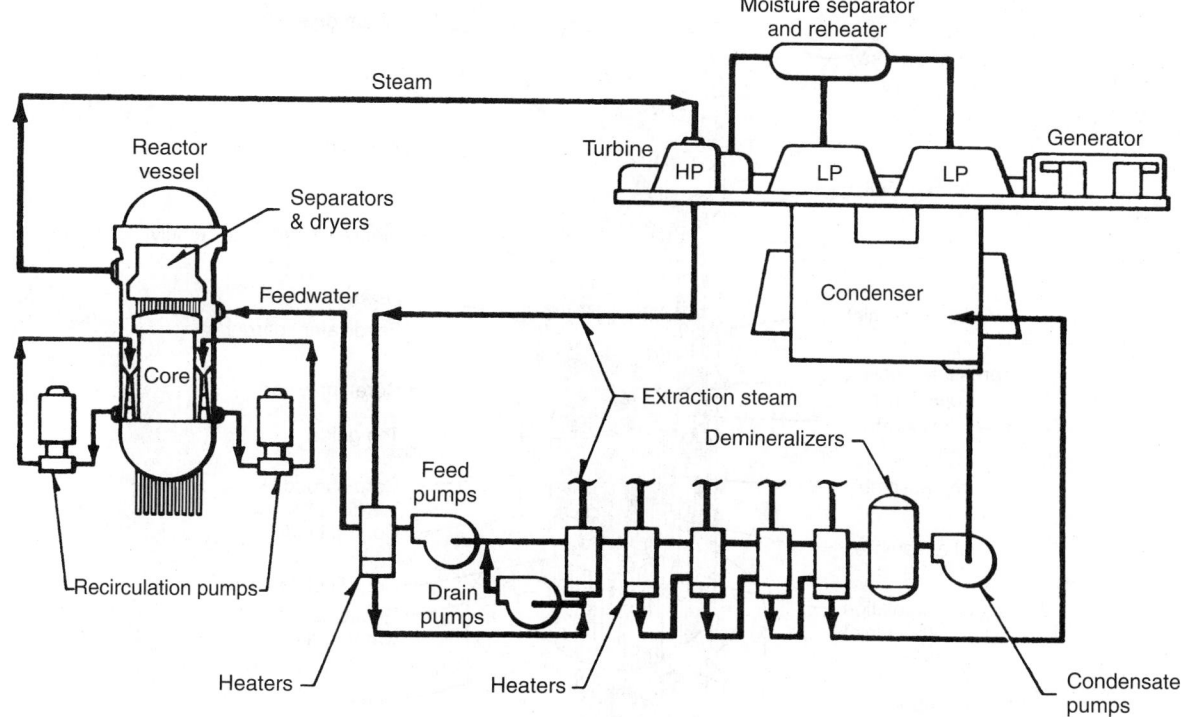

Fig. 1.   Contemporary direct-cycle reactor system used in boiling water reactor. (*General Electric*)

of any boiler. Unique to the BWR is the application of jet pumps inside the reactor vessel. See Fig. 2. The jet pumps deliver their driving force from the external recirculation pumps and generate about two-thirds of the recirculation flow within the reactor vessel. The jet pumps also contribute to the inherent safety of the BWR design under loss-of-coolant emergency conditions because they continue to provide internal circulation with one or both external recirculation loops out of service. The BWR can deliver about one-third power through this natural jet pump circulation mode, a vital capability in effecting a "black start" (a fully fresh start-up of a reactor) of the plant without external power.

The BWR operates at constant pressure and maintains constant steam pressure similar to most fossil-fueled boilers. The BWR primary system operates at pressure about one-half that of a pressurized water reactor primary system, while producing steam of equal pressure and quality.

The integration of the turbine pressure regulator and control system with the reactor water recirculation flow control system permits automated

changes in steam flow to accommodate varying load demands on the turbine. Power changes of up to 25% can be accomplished automatically by recirculation flow control alone, at rate of 15% per minute increasing and 60% per minute decreasing. This provides a load-following capability that can track rapid changes in power demand.

Nuclear Boiler Assembly. This assembly consists of the equipment and instrumentation necessary to produce, contain, and control the steam required by the turbine-generator. The principal components of the nuclear boiler are: (1) reactor vessel and internals—reactor pressure vessel, jet pumps for reactor water circulation, steam separators and dryers, and core support structure; (2) reactor water recirculation system—pumps, valves, and piping used in providing and controlling core flow; (3) main steam lines—main steam safety and relief valves, piping, and pipe supports from reactor pressure vessel up to and including the isolation valves outside of the primary containment barrier; (4) control rod drive system—control rods, control rod drive mechanisms and hydraulic system for insertion and withdrawal of the control rods; and (5) nuclear fuel and in-core instrumentation.

Reactor Assembly. This assembly (Fig. 3) consists of the reactor vessel, its internal components of the core, shroud, top guide assembly, core plate assembly, steam separator and dryer assemblies and jet pumps. Also included in the reactor assembly are the control rods, control rod drive housings and the control rod drives.

Each fuel assembly that makes up the core rests on an orificed fuel support mounted on top of the control rod guide tubes. Each guide tube, with its fuel support piece, bears the weight of four assemblies and is supported by a control rod drive penetration nozzle in the bottom head of the reactor vessel. The core plate provides lateral guidance at the top of each control rod guide tube. The top guide provides lateral support for the top of each fuel assembly.

Control rods occupy alternate spaces between fuel assemblies and may be withdrawn into the guide tubes below the core during plant operation. The rods are coupled to control rod drives mounted within housings welded to the bottom head of the reactor vessel. The bottom-entry drives do not interfere with refueling operations. A flanged joint is provided at the bottom of each housing for ease of removal and maintenance of the rod drive assembly.

Except for the Zircaloy in the reactor core, the reactor internals are stainless steel or other common corrosion-resistant alloys. The reactor vessel is a pressure vessel with a single full-diameter removable head. The base material of the vessel is low alloy steel, which is clad on the

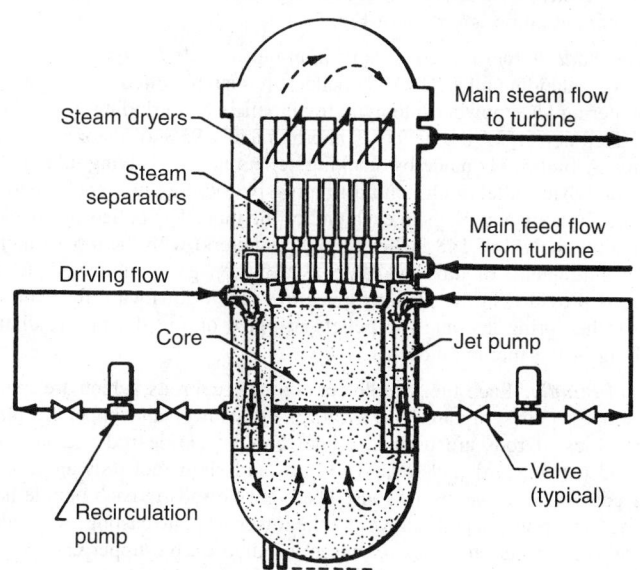

Fig. 2.   Steam and recirculation water flow paths of contemporary boiling water reactor. (*General Electric*)

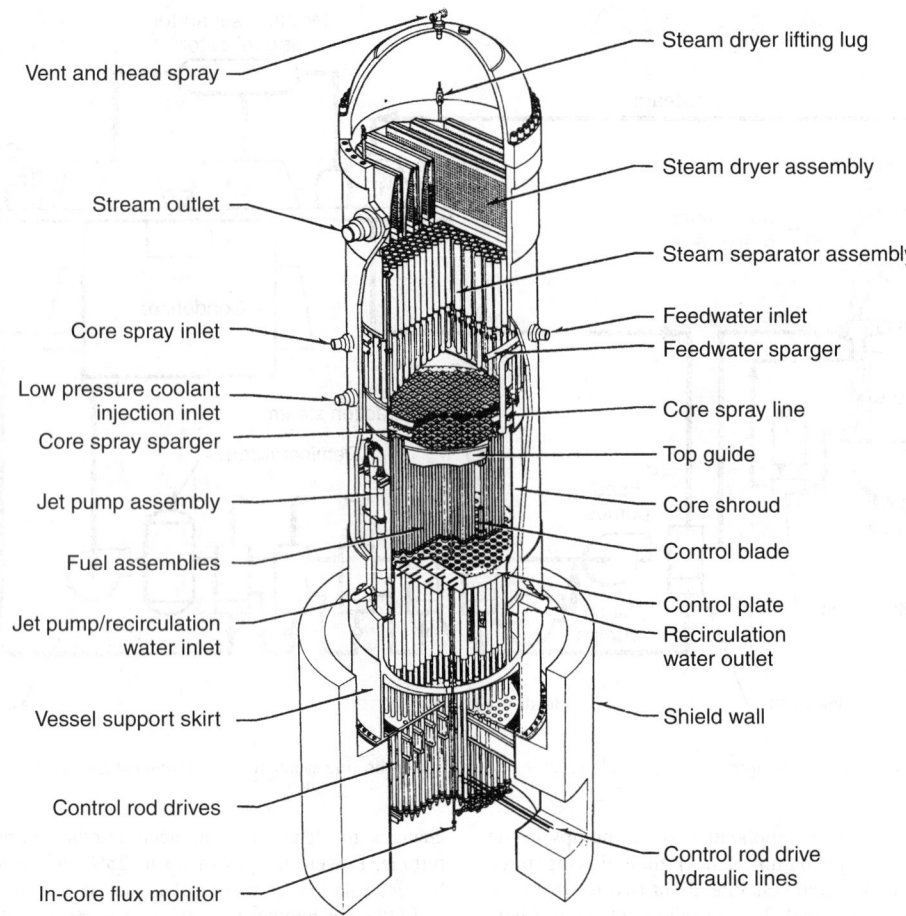

Fig. 3.   Reactor assembly of contemporary boiling water reactor. (*General Electric*)

interior, except for nozzles with stainless steel weld overlay to provide the necessary resistance to corrosion.

The shroud is a cylindrical, stainless steel structure that surrounds the core and provides a barrier to separate the upward flow through the core from the downward flow in the annulus. Two ring spargers, one for low pressure core sprays and the other for high pressure core spray are mounted inside the core shroud in the space between the top of the core and steam separator base. The core spray ring spargers are provided with spray nozzles for the injection of cooling water under emergency conditions. A nozzle for the emergency injection of neutron absorber (sodium pentaborate) solution is mounted below the core in the region of the recirculation inlet plenum.

The steam separator assembly consists of a domed base, on top of which is welded an array of standpipes with a 3-stage separator located at the top of each standpipe. The steam separator assembly rests on the top flange of the core shroud and forms the cover of the core discharge plenum region. In each separator, the steam-water mixture rising through the standpipe impinges on vanes which give the mixture a spin to establish a vortex wherein the centrifugal forces separate the water from the steam in each of three stages. Steam leaves the separator at the top and passes into the wet steam plenum below the dryer. The separated water exits from the lower end of each stage of the separator and enters the pool that surrounds the standpipes to join the downcomer annulus flow.

The steam dryer assembly is mounted in the reactor vessel above the separator assembly and forms the top and sides of the wet steam plenum. Vertical guides on the inside of the vessel provide alignment for the dryer assembly during installation. The dryer assembly is supported by pads extending inward from the vessel wall and is held down in position during operation by the vessel head. These vanes are attached to a top- and bottom-supporting member forming a rigid, integral unit. Moisture is removed and carried by a system of troughs and drains to the pool surrounding the separators and then into the recirculation downcomer annulus between the core shroud and reactor vessel wall.

Control Rod Drive System. Positive core reactivity control is maintained by the use of movable control rods interspersed throughout the core.

These control rods thus control the overall reactor power level and provide the principal means of quickly and safely shutting down the reactor. The rods are vertically moved by hydraulically actuated, locking piston type drive mechanisms. The drive mechanisms perform both a positioning and latching function, and a scram function with the latter overriding any other signal (scram signifies prompt shutdown).

Core Configuration. The reactor core of the BWR is arranged as an upright cylinder containing a large number of fuel assemblies and located within the reactor vessel. The coolant flows upward through the core. The plan of a typical core arrangement of a large BWR is shown in Fig. 4. The lattice configuration is shown in Fig. 5.

Fuel Rod. A fuel rod consists of uranium dioxide (UO_2) pellets and a Zircaloy-2 cladding tube. The UO_2 pellets are manufactured by compacting and sintering UO_2 powder into cylindrical pellets and grinding to size. The immersion density of the pellets is approximately 95% of theoretical UO_2 density. A fuel rod is made by stacking pellets into a cladding tube that is evacuated, back-filled with helium to atmospheric pressure, and sealed by welding Zircaloy end plugs in each end of the tube. The pellets are stacked to an active height of 148 inches (376 centimeters) with the top 12 inches (30.5 centimeters) of tube available as a fission gas plenum. A plenum spring is provided in the plenum space to exert a downward force on the pellets, the spring keeping the pellets in place during the pre-irradiation handling of the fuel bundle.

Fuel Bundle. Each fuel bundle contains 63 fuel rods, which are spaced and supported in a square (8×8) array by a lower and upper tie plate. Three types of rods are used in a fuel bundle: (1) tie rods; (2) a water rod; and (3) standard fuel rods. The third and sixth fuel rods along each outer edge of a bundle are tie rods. The eight tie rods in each bundle have threaded-end plugs, which screw into the lower tie plate casting. A stainless steel hexagonal nut and locking tab is installed on the upper end plug to hold the assembly together. The water rod not only serves as a spacer support rod, but also provides a source of moderator material near the center of the fuel bundle. This flattens the neutron flux across the bundle,

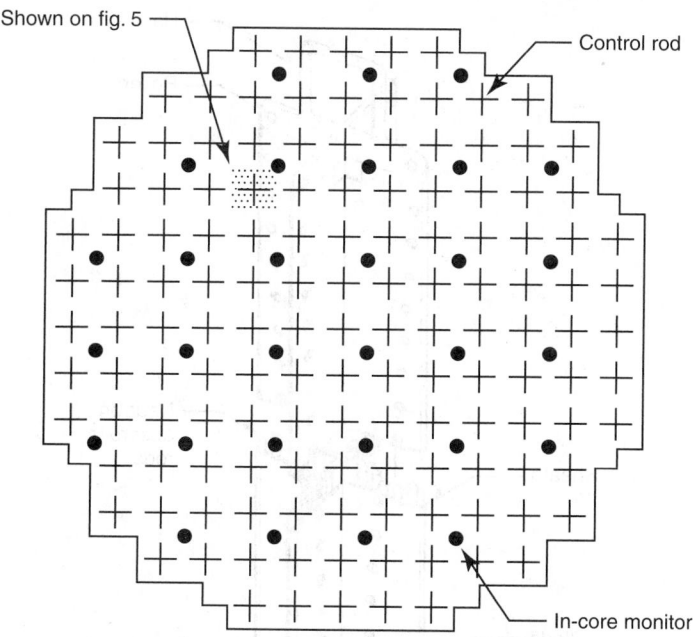

Fig. 4. Typical core arrangement in contemporary boiling water reactor. (*General Electric*)

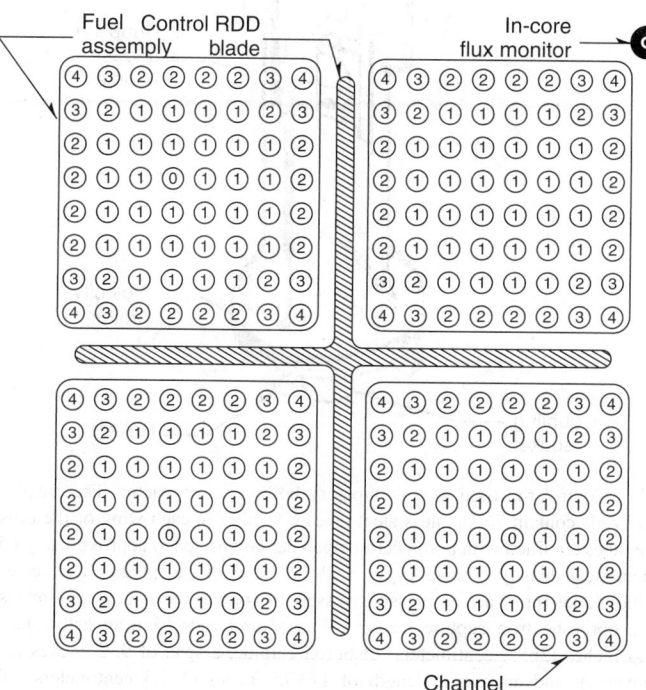

Fig. 5. Core lattice arrangement in contemporary boiling water reactor. (*General Electric*)

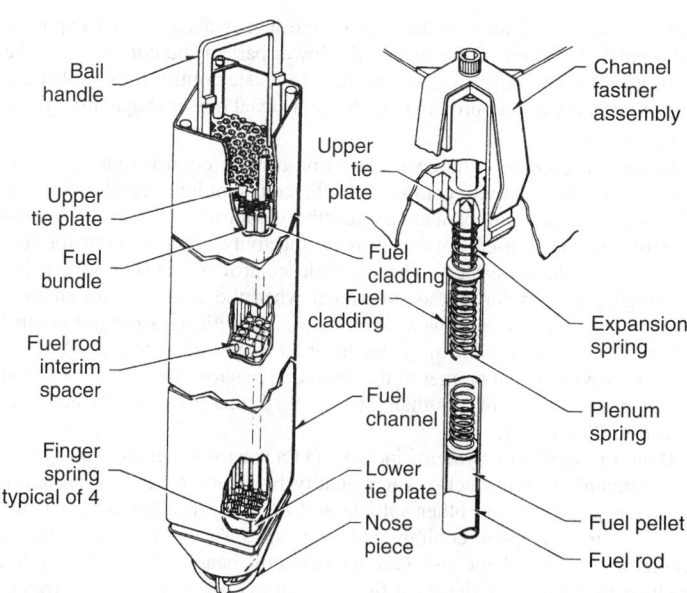

Fig. 6. Fuel assembly of contemporary boiling water reactor. (*General Electric*)

The channel is a square-shaped tube fabricated from Zircaloy 4. The outer dimensions are 5.518 inches (14 centimeters) by 5.518 inches (14 centimeters) by 166.9 inches (424 centimeters) long. The reusable channel makes a sliding seal fit on the lower tie plate surface. It is attached to the upper tie plate by the channel fastener assembly, consisting of a spring and a guide, and a cap screw secured by a lock washer. The fuel channels direct the core coolant flow through each fuel bundle and also serve to guide the control rods.

Neutron Sources. Several antimony-beryllium start-up sources are located within the core. They are positioned vertically in the reactor by "fit up" in a slot (or pin) in the upper grid and a hole in the lower core support plate. The active portion of each source consists of a beryllium sleeve enclosing two antimony-gamma sources. The resulting neutron emission strength is sufficient to provide indication on the source range neutron detectors for all reactivity conditions equivalent to the condition of all rods inserted prior to initial operation. The active source material is entirely enclosed in a stainless steel cladding with an outside diameter of approximately 0.7 inch (1.8 centimeter). The source is cooled by natural circulation of the core leakage flow in the annulus between the beryllium sleeve and the antimony-gamma sources.

Core Design Margins. The reactor core is designed to operate at rated power with sufficient design margin to accommodate changes in reactor operations and reactor transients without damage to the core. In order to accomplish this objective, the core is designed, under the most limiting operating conditions and at 100% rated power, to meet the following bases. (1) The maximum linear heat generation rate, in any part of the core, is always less than 13.4 kW/foot (43.97 kW/meter). (2) The minimum ratio between critical heat flux and fuel operating heat flux, in any part of the core, is always greater than 1.9.

Power Distribution. The design power distribution is divided for convenience into several components: (1) relative assembly power; (2) local; and (3) axial. The relative assembly power peaking factor is the maximum fuel assembly average power divided by the reactor core average assembly power. The local power peaking factor is the maximum fuel rod average heat flux in an assembly divided by the assembly average fuel rod heat influx. The axial power peaking factor is the maximum heat flux of a fuel rod divided by the average heat flux in that rod. Peaking factors vary throughout an operating cycle, even at steady-state full-power operation, since they are affected by withdrawal of control rods to compensate for fuel burnup. The design peaking factors represent the values of the most limiting power distribution that will exist in the core throughout its life.

Because of the presence of steam voids in the upper part of the core, there is a natural characteristic for a BWR to have the axial power peak in the lower part of the core. During the early part of an operating cycle, bottom-entry control rods permit a partial reduction of this axial peaking by locating a larger fraction of the control rods in the lower part of the

and leads to lower local peaking factors and better utilization of uranium in the interior rods of the fuel assembly.

The initial core will contain fuel assemblies having a common average enrichment ranging from approximately 1.6% (weight) of ^{235}U to 2.2%, depending upon initial cycle requirements. Each assembly will contain different enrichment rods. Selected rods in each assembly will, in addition, be blended with gadolinium burnable poison. The reload fuel will also contain four different enrichment rods with an average enrichment in the range of 2.4 to 2.8%. Different ^{235}U enrichments are used in fuel assemblies to reduce the local power peaking. Low enrichment rods are used in the corner rods and in the rods nearer the water gaps; higher enrichment uranium is used in the central part of the fuel bundle.

Fuel Channel. A fuel channel encloses the fuel bundle. The combination of a fuel bundle and a fuel channel is called a fuel assembly. See Fig. 6.

core. At the end of an operating cycle, the higher accumulated exposure and greater depletion of the fuel in the lower part of the core reduces the axial peaking. The operating procedure is to locate control rods so that the reactor operates with approximately the same axial power shape throughout an operating cycle.

Reactivity Control. The movable boron-carbide control rods are sufficient to provide reactivity control from the cold shutdown condition to the full-load condition. Supplementary reactivity control in the form of solid burnable poison is used only to provide reactivity compensation for fuel burnup or depletion effects. The movable control rod system is capable of bringing the reactor to the subcritical when the reactor is an ambient temperature (cold), zero power, zero xenon, and with the strongest control rod fully withdrawn from the core. In order to provide greater assurance that this condition can be met in the operating reactor, the core is designed to obtain a reactivity of less than 0.99, or a 1% margin on the "stuck rod" condition. See Fig. 7.

Reactor Auxiliary Systems include: (1) a reactor water cleanup system for maintaining high reactor water quality by removing fission products, corrosion products, and other soluble and insoluble impurities; (2) a fuel and containment pool cooling and cleanup system—a system which accommodates the beta and gamma radiation heating from the fission products that remain in the spent fuel, as well as drywell heat transferred to the upper containment pool; (3) a closed cooling water system for reactor service consisting of a separate, force circulation loop; (4) emergency equipment cooling system; (5) standby liquid control system; (6) reactor core isolation cooling system; (7) emergency core cooling system; (8) high-pressure core spray system; and (9) residual heat removal system.

Contemporary Pressurized Water Reactor (PWR)

In a typical pressurized water reactor (PWR), heat generated in the nuclear core is removed by water (reactor coolant) circulating at high pressure through the primary circuit. The water in the primary circuit cools and moderates the reactor. The heat is transferred from the primary to the secondary system in a heat exchanger, or boiler, thereby generating steam in the secondary system. The steam produced in the steam generator, a tube-and-shell heat exchanger, is at a lower pressure and temperature than the primary coolant. Therefore, the secondary portion of the cycle is similar to that of the moderate-pressure fossil-fueled plant. In contrast, in boiling-water or direct-cycle systems, steam is generated in the core and is delivered directly to the steam turbine.

The similarities of basic pressurized water reactor design from one manufacturer to the next are more striking than the differences. Therefore, the description of one particular configuration (Combustion Engineering, Inc.) can suffice to convey the general operating principles. The major components of a PWR are: (1) the reactor vessel which contains the oxide fuel core, core intervals, control element assemblies, and in-core instruments; (2) the electrically-heated pressurizer; (3) the electric-motor-driven primary coolant pumps; and (4) the U-tube type steam generators. See Fig. 8. The primary coolant system layout can be fitted into a variety of containment types and concepts. A prestressed cylindrical containment is common. Figure 9 shows the arrangement in a spherical containment. This type of building lends itself to separation of safeguards equipment, steam lines, and emergency power supplies.

Steam Generators. The basic geometry is shown in Fig. 10. With the nuclear steam supply system operating at 3,817 MW, two steam generators produce a total of 17.18×10^6 pounds (7.89×10^6 kilograms) of steam per hour at 1,070 psia (72.8 atmospheres). The steam generators are constructed, using carbon steel pressure-containing members and Inconel-600 tubes. The tube-sheet is clad by weld deposit for maximum strength; tongue and groove construction of the divider plate places no stress on the tube-sheet cladding. Fusion welding of the end of each tube to the tube-sheet primary cladding provides an effective seal for leakage control, and "expanding" (explosively expanding) the tubes in the full length of the tube-sheet eliminates corrosion-prone crevices. An economizer section on the units improves heat transfer by preheating the incoming feedwater, using the low (primary side) temperature heat transfer area of the U-tubes. Multiple feed nozzles allow the economizer flow distribution to be optimized for each power level.

Reactor Coolant Pumps. As indicated by Fig. 8, four reactor coolant pumps are used, two for each steam generator. The pumps are vertical, single-bottom-suction, horizontal-discharge, motor-driven centrifugal units. The pump impeller is keyed and locked to its shaft. A complex system of

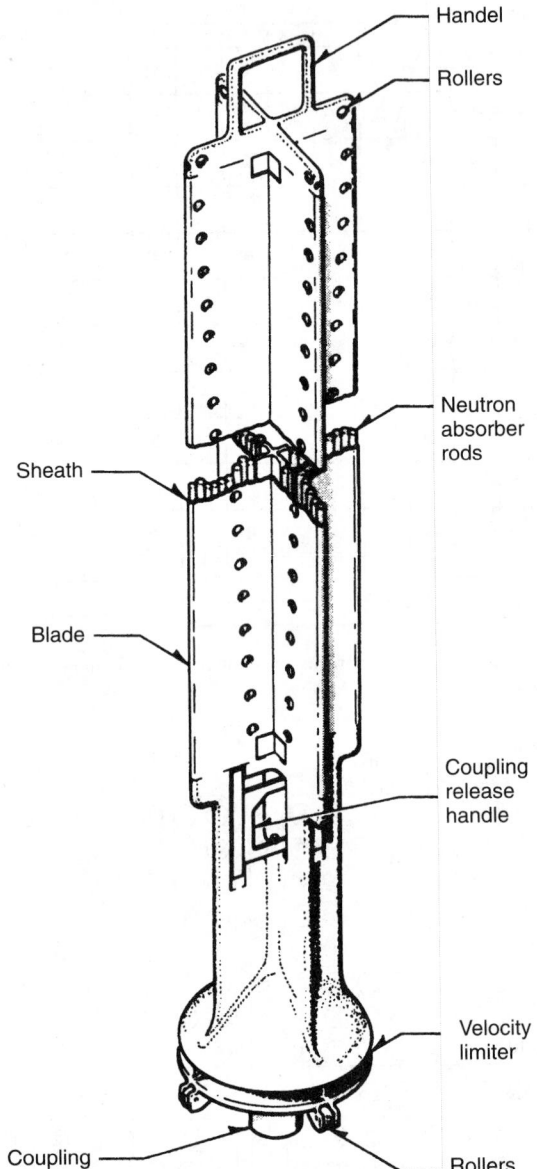

Fig. 7. Control rod used in contemporary boiling water reactor. The cruciform control rods contain 76 stainless steel tubes (19 tubes in each wing of the cross). These tubes are filled with boron carbide powder compacted to approximately 65% of theoretical density. The tubes are seal-welded with end plugs on either end. The individual tubes act as pressure vessels to contain the helium gas released by the boron-neutron capture reaction. The control rods have an active length of 144 inches (365.8 centimeters) of boron carbide, a span of 9.75 inches (24.8 centimeters), and an overall length of 173.75 inches (441.3 centimeters). The control rods can be positioned at 6-inch (15-centimeter) steps and have a nominal withdrawal and insertion speed of 3 inches (7.5 centimeters) per second. Control rods are cooled by the core leakage (bypass) flow. In addition to satisfying initial control effectiveness requirements, it is expected that the control rods will have an average lifetime of approximately 15 full-power years. (*General Electric*)

seals is used to prevent any leakage. The motors are designed to start and accelerate to speed under full load with a drop to 80% of normal rated voltage at the motor terminals. Each motor is provided with an anti-reverse rotation device. Each reactor coolant pump is provided with four vertical support columns, four horizontal support columns, and one vertical snubber. The structural columns provide support for the pumps during normal operation, earthquake conditions, and any hypothetical loss-of-coolant accident in either the pump suction or discharge line.

Pressurizer. The pressurizer is a cylindrical pressure vessel, vertically mounted and bottom supported. Energy to the water is supplied by replaceable direct-immersion electric heaters, which are inserted from the bottom head of the pressurizer. Nozzles are provided for spray, surge,

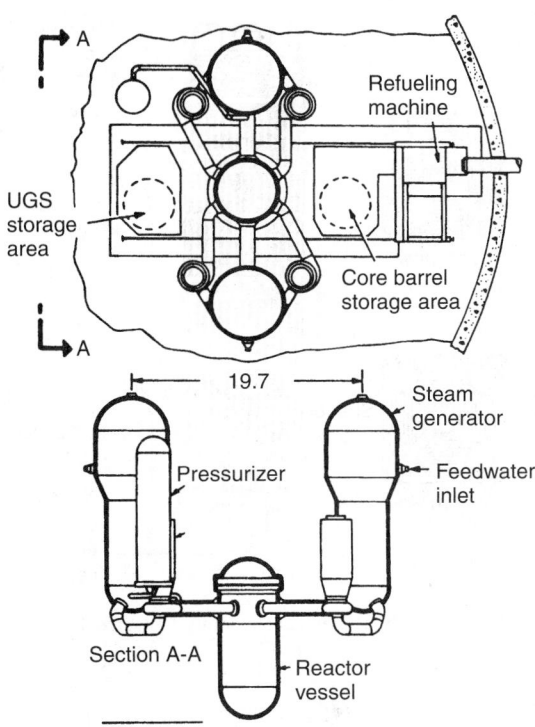

Fig. 8. Nuclear steam supply system for contemporary pressurized water reactor. (*Combustion Engineering*)

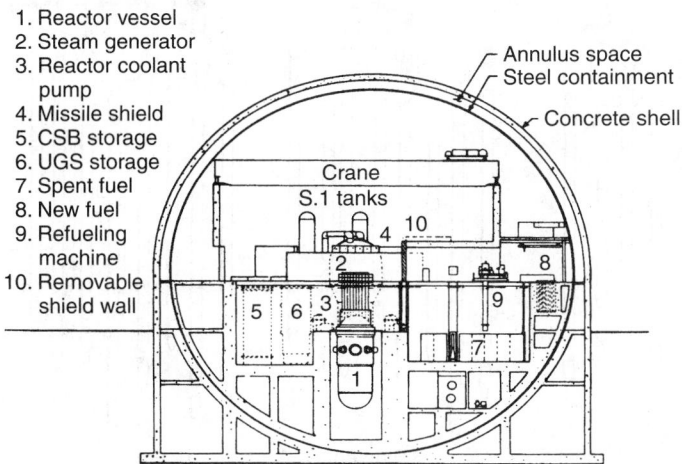

1. Reactor vessel
2. Steam generator
3. Reactor coolant pump
4. Missile shield
5. CSB storage
6. UGS storage
7. Spent fuel
8. New fuel
9. Refueling machine
10. Removable shield wall

Fig. 9. Spherical containment for contemporary pressurized water reactor. (*Combustion Engineering*)

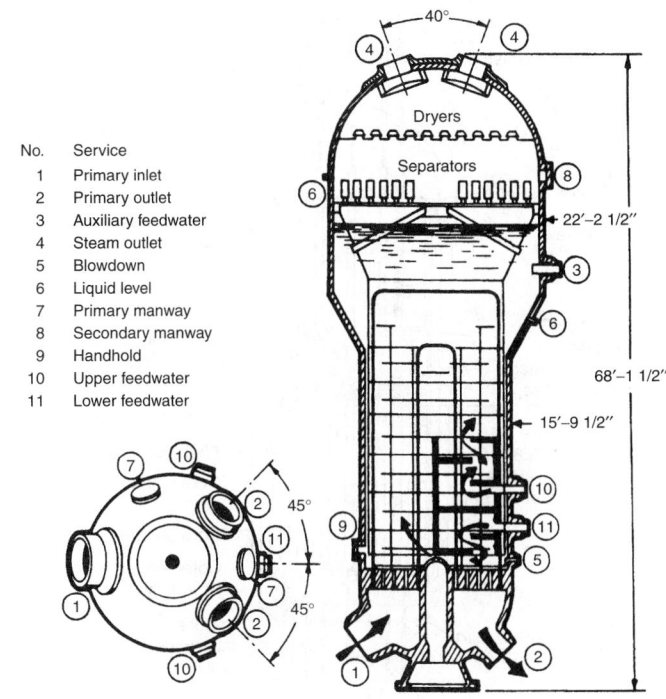

No.	Service
1	Primary inlet
2	Primary outlet
3	Auxiliary feedwater
4	Steam outlet
5	Blowdown
6	Liquid level
7	Primary manway
8	Secondary manway
9	Handhold
10	Upper feedwater
11	Lower feedwater

Fig. 10. Steam generator for contemporary pressurized water reactor. (*Combustion Engineering*)

relief, and instrumentation connections. The pressurizer maintains reactor coolant system operating pressure and, in conjunction with the chemical and volume control system, compensates for changes in reactor coolant volume during load changes, heat-up, and cool-down. During full-power operation, the pressurizer is about $\frac{1}{3}$ full of saturated steam.

Reactor Vessel. This vessel is designed to contain the fuel bundles, the control element assembles, and the internal structures necessary for support of the core. The reactor is a stainless clad, thick-walled, carbon steel pressure vessel comprised of a cylinder with two hemispherical heads. The lower head is integrally welded to the vessel shell and contains in-core instrumentation nozzles. The upper closure head, containing the control element drive mechanism nozzles, is attached to the vessel by means of a bolted flange, thus permitting the head to be removed to provide access to the reactor internals. The head flange is drilled to match the vessel flange stud bolt locations.

The vessel flange is a forged ring with a machined ledge on the inside surface to support the core support barrel. The flange is drilled and tapped to receive the closure studs and is machined to provide a mating surface for the reactor vessel closure seals. Sealing is accomplished by using two

silver-plated, NiCrFe-alloy, self-energizing O-rings. The space between the two rings is monitored to detect any inner-ring coolant leakage. The inlet and outlet nozzles are located radially on a common plane below the vessel flange. Extra thickness in the vessel course provides the reinforcement required for the nozzles. Snubbers built into the lower portion of the vessel shell limit the amplitude of any displacement of the core support barrel. Core stops are also built into the reactor vessel to limit the downward displacement of the core support barrel.

Cladding for the reactor vessel is a continuous integral surface of corrosion-resistant material, having $\frac{7}{32}$-inch (0.56 centimeter) nominal thickness, and a $\frac{1}{8}$-inch minimum thickness. The reactor vessel is supported by four vertical columns located under the vessel inlet nozzles. These columns are designed to flex in the direction of horizontal thermal expansion and thus allow unrestrained heat-up and cool-down. The columns also act as a hold-down device for the vessel. The supports are designed to accept normal loads and seismic and pipe rupture accident loads.

The reactor arrangement is shown in Fig. 11. The barrel-calandria guide structure is a rugged (3-inches thick, barrel section) unit that can withstand and protect all control element fingers from the combined effects of seismic and blowdown loads that may result from a loss-of-cooling accident. The calandria structure fits over the control element guide tubes of the fuel assemblies, aligning all fuel assemblies, and laterally restraining the top ends of the fuel assemblies. With the upper guide structure in place, a continuous guide tube for each control finger is formed, extending from the top of the tube-sheet to the bottom of the fuel assembly. Because of this feature, which isolates every control finger from the coolant crossflow, flexibility is obtained in the number of control fingers that can be attached to one control assembly, i.e., one control element assembly can serve more than one fuel assembly.

Severe emergency core cooling system criteria require that the builders of water reactors increase the linear feet of fuel in the reactor core for the same power in order to reduce LOCA (loss-of-cooling accident) fuel temperatures. In the unit described here, an assembly with a 16 × 16 fuel rod array of smaller diameter rods is used in the same assembly envelope that was occupied by a 14 × 14 assembly in earlier designs. This results in a maximum linear heat rate decrease in the assembly of about 25%.

As shown in Fig. 12, the active core is made up of 241 fuel assemblies, all of which are mechanically identical. As indicated by Fig. 13, each fuel assembly contains 236 Zircaloy-clad, UO_2 fuel rods retained in a structure consisting of Zircaloy spacer grids welded at about 15-inch (38.1-centimeter) intervals to five Zircaloy control element assembly guide tubes which, in turn, are mechanically fastened at each end to stainless

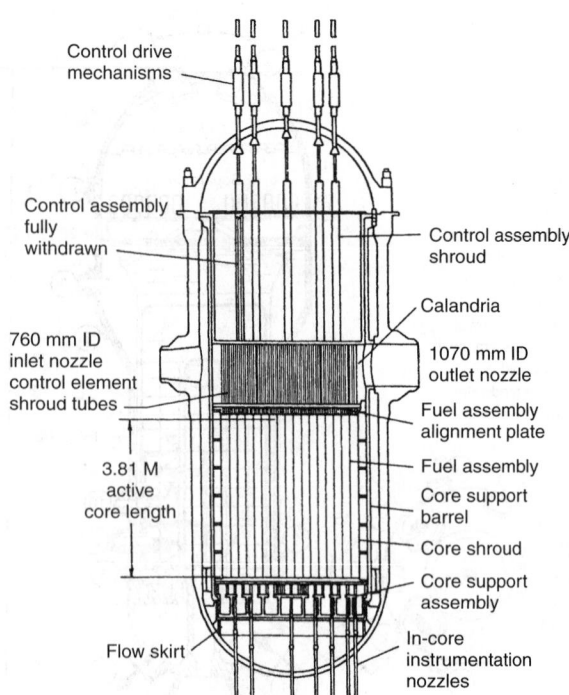

Fig. 11.   Reactor arrangement in contemporary pressurized water reactor

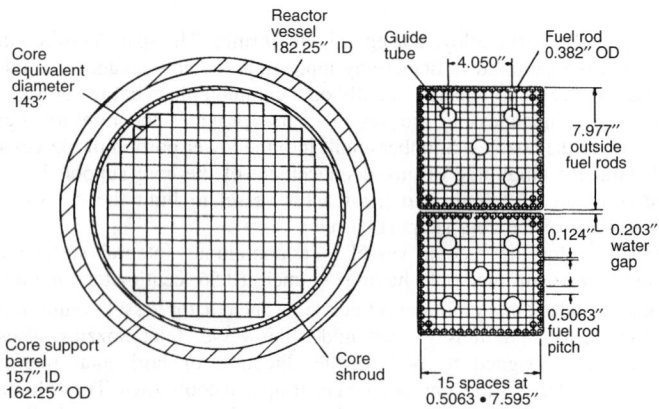

Fig. 12.   Reactor core cross section of contemporary pressurized water reactor with 241 fuel assemblies. (*Combustion Engineering*)

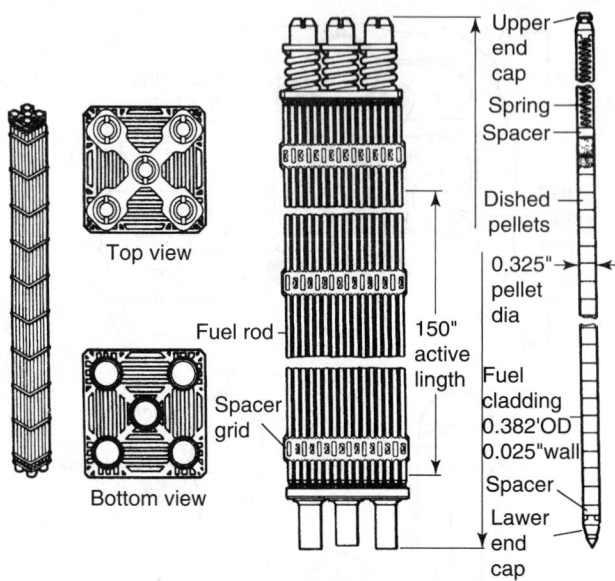

Fig. 13.   Fuel assembly used in contemporary pressurized water reactor

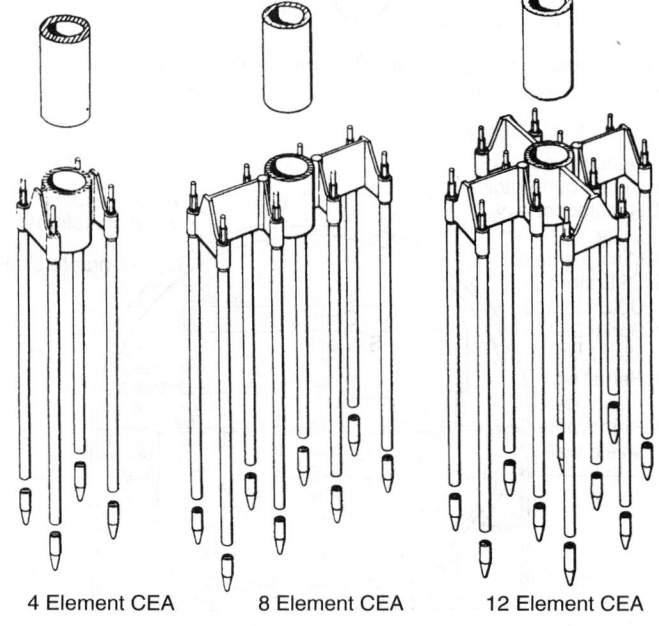

Fig. 14.   Four 8- and 12-element control element assemblies. (*Combustion Engineering*)

steel end fittings. The overall length of the fuel assembly is about 177 inches (450 centimeters) and the cross section is about 8 inches (20.3 centimeters) by 8 inches (20.3 centimeters). Each fuel assembly weighs about 1,450 pounds (657.7 kilograms). With reference to Fig. 13, fuel rods, consisting of uranium dioxide (UO_2) pellets of low enrichment canned in thin-walled Zircaloy-4 tubing, are designed to achieve average burnups of about 33,000 MWD/MTU (thermal megawatt days/metric tons of uranium) and peak burnups of about 50,000 MWD/MTU. The design factors limiting burnup of the fuel are the effects on the clad of volumetric changes of the fuel pellet and fission gas release.

As indicated in Fig. 13, the fuel rod consists essentially of 0.325-inch (0.82-centimeter) diameter, 0.390-inch long UO_2 pellets canned in a 0.382-inch (0.97-centimeter) outside diameter Zircaloy-4 tube. The high density fuel pellets are dished at both ends to allow for axial differential thermal expansion and fuel volumetric growth with burnup.

The control element assemblies consist of an assembly of 4, 8, or 12 fingers approximately 0.8-inch (2-centimeter) outside diameter and arranged as shown in Fig. 14. The use of cruciform control rods, as in boiling water and early pressurized water reactors, necessitates large water gaps between the fuel assemblies to ensure that the control rods will scram (prompt shutdown) satisfactorily. These gaps cause peaking of the power in fuel rods adjacent to the water channel compared to fuel rods some distance from the channel.

A five-hole assembly design was evolved from consideration of the lower peaking effect of smaller (removal of one fuel rod) water holes versus the mechanical advantages of larger (four fuel rods removed) water holes. The larger water holes allowed the use of rugged, 0.9-inch (2.3-centimeter) outside diameter by 0.035-inch (0.89-centimeter) thick Zircaloy guide tubes for the fuel assembly structure. These water holes are distributed relatively uniformly in the reactor core when placed in the 16×16 fuel rod lattice. The particular arrangement of water holes was selected in consideration of the water gap between fuel assemblies; the effect of the central water hole in the fuel assembly is balanced by the water gap between fuel assemblies. The mechanical simplicity and ruggedness outweigh the advantage of obtaining a small decrease in local peaking by using very small fingers. The slightly higher peaking associated with the design can be compensated, to a large extent, by varying the enrichment of the fuel in the rods adjacent to the control channel and/or by using water displacers in local hot spots.

The control element assembly, shown in Fig. 15, consists of 0.8-inch (2-centimeter) outside diameter Inconel tubes containing boron carbide pellets as the neutron absorbing material. A gas plenum is provided in order to limit the maximum stress due to generation of internal gas pressure.

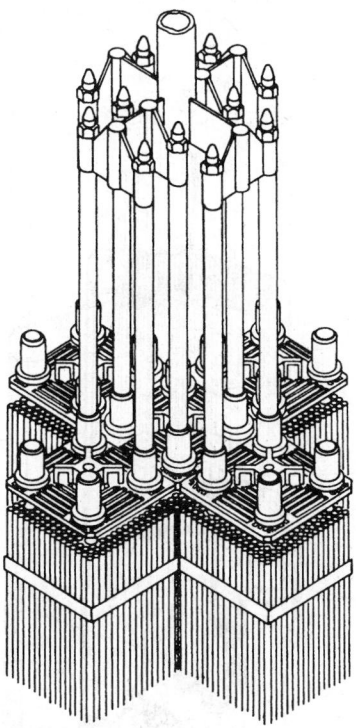

Fig. 15. Control element assembly and fuel for contemporary pressurized water reactor

The individual control fingers are attached mechanically and locked to the various spider assemblies. This allows for simplifications in manufacture, shipping, and assembly of the control element assembly. Because all fingers are removable and replaceable, servicing and disposal problems are decreased. It is intended that the spider assembly and its extension shaft be reused whenever possible.

Design of the upper guide structure permits flexibility in the number of control fingers that can be attached to one control assembly. The standard pattern of control assemblies is shown in Fig. 16. In the standard design, power changes at close to full power, shaping of the radial power distribution, and control of the axial power distribution are best handled by the low worth 4-finger control element assembly entering a single fuel

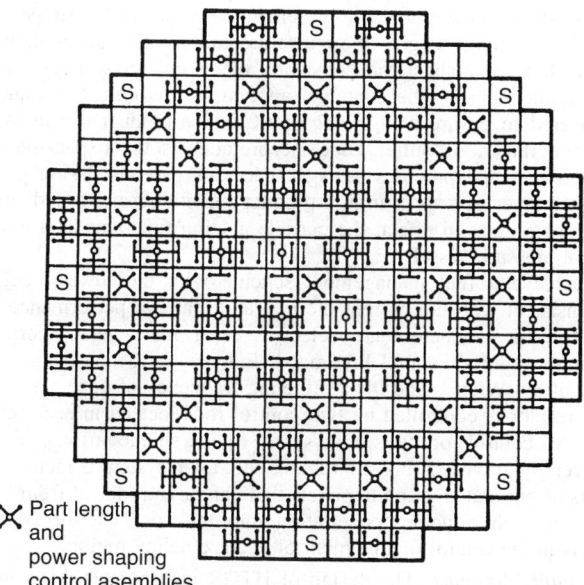

✕ Part length and power shaping control asemblies

Fig. 16. Standard pattern of control assemblies in contemporary pressurized water reactor core. The pattern provides more-than-sufficient control for self-generated plutonium recycle. For complete open-market plutonium recycle, 4-element control assemblies are added in positions marked S

assembly. Shutdown reactivity control in the peripheral region of the core is handled by the 8-finger control element assembly and in the central region of the core it is handled by the 12-finger control element assembly. The need for the two types of shutdown control element assemblies is to obtain "stuck rod worths" in the high reactivity fuel on the periphery of the core which are about equal to the control element assembly control worth in the lower reactivity central zone of the core.

Instrumentation. The large size of present water reactors and the nuclear effects which can occur, such as xenon redistribution, stuck rods, and reactivity anomalies require that emphasis be placed on instrumentation and control systems if high plant availability is to be maintained, while providing the necessary protection due to abnormal occurrences. Because there are many reactivity effects that can produce changes in the reactor power distribution, more reliable operation can be obtained by on-line monitoring of the reactor. This is best achieved with in-core instrumentation. The system described here has provision for up to 61 in-core instrument (ICI) assemblies, which enter from the bottom of the reactor vessel. Radial distribution of the ICI is such that every type of fuel assembly, rodded and unrodded, is instrumented, assuming symmetrical core power distribution. Five sets of four symmetrically located ICI assemblies are included to monitor core power tilts. Also, every instrumented fuel assembly is either immediately adjacent to or diagonally adjacent to an instrumented fuel assembly to obtain good radial coverage of the core. Each of the ICI assemblies contains five self-powered fixed detectors distributed axially along the length of the core, a thermocouple at the end of the assembly to monitor outlet temperature, and a dry-well instrument tube which can accommodate a movable detector. This allows for high measurement accuracy of the on-line fixed detector.

The continuous monitoring and processing of the data from over 300 fixed detectors by the core monitoring computer provides the operator with information on core power distribution, maximum linear heat rate in the fuel, departure from nucleate boiling ratio, and fuel exposure. These data can then be used to obtain improved maneuvering of core power using the relative low worth 4-finger control element assemblies for power changes and power distribution control.

High-Temperature Gas-cooled Reactor (HTGR)

Although there have been comparatively few gas-cooled reactors installed for generating commercial nuclear electric power, the concept has a number of operating advantages over light-water reactors and could play an important role in the reactor designs for the next century.

The high-temperature gas-cooled reactor (HTGR) is a thermal reactor that produces desired steam conditions. Helium is used as the coolant. Graphite, with its superior high-temperature properties, is used as the moderator and structural material. The fuel is a mixture of enriched uranium and thorium in the form of carbide particles clad with ceramic coatings.

The high-temperature conditions and high thermal efficiency (approximately 39%) of the (HTGR) result in high performance. The amount of cooling water required to carry away the waste heat is significantly less than in a light-water reactor (LWR). The use of thorium in the fuel cycle decreases fuel cost, improves the conservation of fuel, and adds the large deposits of thorium to available fuel reserves. The HTGR has significant environmental advantages, including: (1) lower thermal discharge because of its high efficiency; (2) low release of radioactive waste because of the high-integrity fuel and the inert coolant; and (3) low consumption of raw materials because of high efficiency and use of thorium in the fuel cycle.

High operating temperatures at moderate pressures are achieved through the use of helium as the coolant. Helium is attractive as a coolant because it: (1) is chemically inert; (2) absorbs essentially no neutrons; and (3) makes no contributions to the reactivity of the system. Carbon dioxide also has been used as a coolant.

Graphite is used as the moderator and core structural material because of (1) excellent mechanical strength at high temperatures; (2) very low neutron-capture cross section; (3) good thermal conductivity; and (4) high specific heat. Graphite has a long history of use in thermal reactors. Because of low neutron-capture cross section, no neutrons are lost within the core through absorption in metallic fuel cladding or structural supports. Graphite also is well suited to high-temperature operations, increasing in strength with temperature up to a point (2,482°C) well beyond the operating range of the HTGR.

The use of the thorium-uranium fuel cycle in the HTGR provides improved core performance over the plutonium/uranium low-enrichment cycle used in LWRs. The principal reason for this is that fissile ^{233}U produced from neutrons captured in thorium during reactor operation is neutronically a better fuel than ^{239}Pu, produced from ^{238}U in the low-enrichment cycle. The excellent neutronic characteristics of the graphite-moderated thorium/uranium cycle leads directly to high conversion ratios and low fuel inventories. Reduced ^{235}U inventories and make-up requirements spell reduced sensitivity to uranium prices.

Early Development of the HTGR

Work on the gas-cooled reactor has been underway essentially since the dawning of the nuclear power industry. The earliest developments were in Britain and France, at which time carbon dioxide gas was used as the coolant. In 1965, Britain opted for an advanced gas-cooled reactor (AGR). In 1969, France swung away from the HTGR (because of high construction costs) and targeted to the employment of more LWRs as well as commencing a concerted effort to develop a fast breeder type reactor. West Germany has been active in the development and testing of HTGRs since the early 1960s, but only recently (late 1980s) have the Germans indicated serious efforts toward commercialization of the HTGR. In the United States, a HTGR was installed at Peach Bottom, Pennsylvania, commencing commercial operation in 1974. As of the late 1980s, only one other HTGR was installed in the United States, the Fort St. Vrain plant near Denver, Colorado.

A simplified flow diagram of this station, which generates 842 MW (thermal) to achieve a net output of 330 MW (electrical), is given in Fig. 17. The helium coolant, at a pressure of about 700 psi (47.6 atmospheres), flows downward through the reactor core, where it is heated to 777°C. The coolant flow can be trimmed by the use of orifice valves located at the top of the core that are integral with the control rod drive mechanisms. From the reactor core, the coolant flows through the steam generators. After passing through the steam generators, the helium is returned to the core at a temperature of about 404°C by four steam-turbine-driven helium circulators. Two identical loops are used, each including a six-module steam generator and two helium circulators. Each loop contributes half of the total output of the nuclear steam supply system, which produces steam at 2,400 psig (163.3 atmospheres) and 538°C with single reheat to 538°C. The helium circulators are driven by the exhaust steam from the high-pressure turbine. This steam is then reheated and returned to the intermediate-pressure turbine. The circulators are also equipped with a Pelton water wheel drive so that they may be driven using the boiler feed pumps for emergency conditions.

The general reactor arrangement is shown in Fig. 18. The prestressed concrete reactor vessel (PCRV) is 31 feet in internal diameter with a 75-foot (23 meters) internal height. The upper and lower heads are nominally 15 feet (4.5 meters) thick, and the walls have a nominal thickness of 9 feet (2.7 meters). Thus, the PCRV provides the dual function of containing the coolant at operating pressure and also providing radiological shielding.

Reactor Core. The HTGR fuel element is a graphite block, hexagonal in cross section and having a grid of longitudinal fuel holes and coolant channels. The fuel element blocks are stacked in columns of eight blocks each and grouped into fuel regions consisting of a central column surrounded by six columns. Each region rests on a large core support block

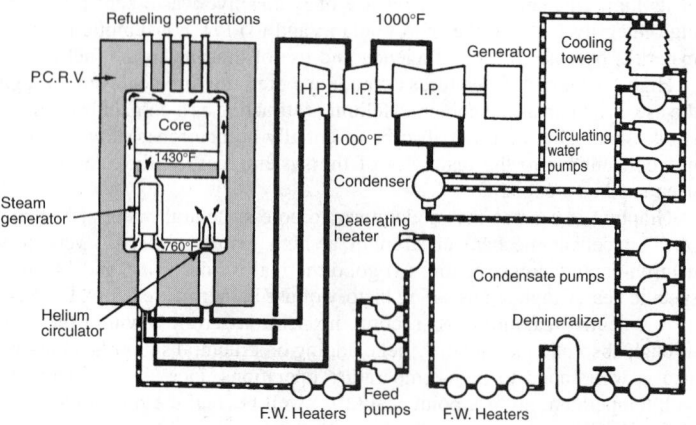

Fig. 17.   Simplified flow diagram of Fort St. Vrain Nuclear Generating Station. (*GA Technologies*)

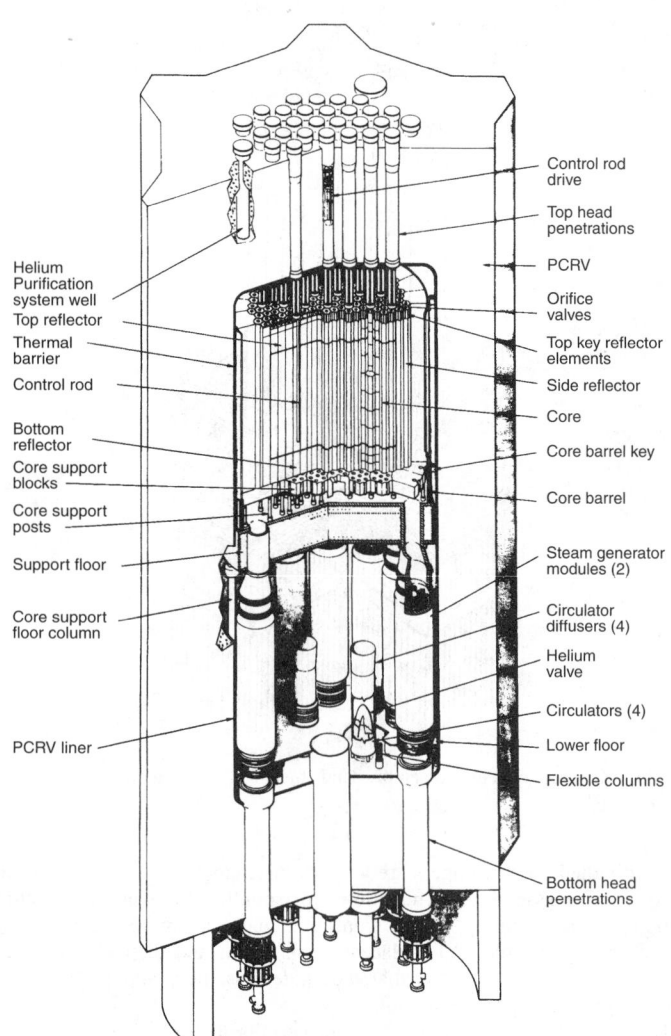

Fig. 18.   General reactor arrangement of Fort St. Vrain high-temperature gas-cooled reactor. (*GA Technologies*)

which, in turn, rests on graphite posts standing on the liner of the central cavity. Hexagonal graphite reflector elements are located above, below, and around the active core. These elements are surrounded by permanent side-reflector blocks to give the entire assembly a circular configuration. The fuel holes contain a rod consisting of ceramic-coated fuel particles in a graphite matrix. The coatings, applied by pyrolitic techniques, are multilayered to ensure a high degree of fission-product confinement. A porous interlayer, or buffer zone, accommodates the expansion of the irradiated fuel and provides storage space for gaseous fission products. The outer layer acts as a fission-product retention barrier and provides structural strength. In effect, the particle coating functions as a miniature spherical pressure vessel.

To achieve a fuel management scheme with the lowest fuel cycle cost consistent with the current thermal and material performance limits, the following parameters are selected: (1) a fuel cycle incorporating uranium/thorium; (2) a fuel lifetime of four years; (3) an average power density of 8.4 W/cm^3; and (4) a refueling frequency of once a year.

The reactor is controlled by two control rods located in each refueling region. All control rod pairs have scram (quick shutdown) capability and are driven by gravity. A backup reserve shutdown system is included. This consists of boronated graphite pellets that can be introduced from hoppers located in each refueling penetration into the core via the cylindrical channels in the central fuel element of each refueling region.

Safeguard Systems. The design of HTGR incorporates many inherent safety features and a number of engineered safeguards. The inherent safety characteristics include negative power and temperature coefficients, assured by the thorium content of the fuel. In addition, the high heat capacity of the large mass of graphite ensures that any core temperature transient resulting from reactivity insertions or interruptions in cooling will be slow and readily controllable. This important safety feature eliminates the need for an emergency core cooling system. Only a residual heat removal system is required for the long-term decay heat, and control of the HTGR is inherently easier than in reactors in which the coolant functions as the moderator. The uranium/thorium fuel contained in the ceramic-coated particles is not susceptible to sudden release of the stored-up fission products as a result of melting. Since the entire primary coolant system is contained within the PCRV, external piping, which might be subject to sudden rupture, is eliminated. Structural strength and integrity of the PCRV is enhanced by the redundant reinforcing steel and prestressed wire tendons. At the maximum credible pressure, the prestressing elements are not stressed above levels experienced during their initial tensioning. As a result, sudden loss of coolant due to prestress failure is not credible.

Second-Generation HTGRs

In addition to upgrading the HTGR at the Fort St. Vrain nuclear power station, efforts have been underway for several years to make both larger and smaller gas-cooled high-temperature reactors. Smaller, modular units could provide the flexibility needed by the public utilities as they plan their expansions for projected increases in electricity requirements. Inherent safety, already a feature of the HTGR, would be enhanced because of the smaller size and low power density of modular units. For example, it is estimated that the power density of a modular gas-cooled reactor would be only 3 kW/liter, as compared with 6 kW/liter for a large reactor and 100 kW/liter for a conventional pressurized-water reactor (PWR) as previously described.

In the new designs, if coolant were lost, the nuclear chain reaction would be terminated by the reactor's negative temperature coefficient after a modest temperature rise. Core diameter of the modular units would be limited so that decay heat could be conducted and radiated to the environment without overheating the fuel to the point where fission products might escape. Thus, inherent safety would be realized without operator or mechanical device intervention.

Large Commercial HTGRs[2]

Following construction of the Fort St. Vrain facility, the HTGR was marketed commercially in direct competition with large pressurized water reactors (PWRs) and boiling water reactors (BWRs). Between 1971 and 1975, ten such reactors were ordered by U.S. utilities. The designs of the commercial HTGR were similar to Fort St. Vrain in that they used the graphite based core structure, helium coolant, prestressed concrete reactor vessel (PCRV) and superheated steam cycle. However, the designs differed in that power outputs were significantly larger and the reactor system was rearranged to accommodate the larger-size components.

The large HTGRs had power ratings of 2000 and 3000 MWt which corresponded to net electrical outputs of 770 and 1160 MWe, respectively. An example of the rearranged reactor system is shown in Fig. 19. A multi-cavity PCRV was used to enclose the reactor system instead of the single-cavity PCRV used in Fort St. Vrain. This was a major advancement in PCRV technology and necessitated the development of a circumferential wire-wrap prestressing system instead of circumferential tendons, although the longitudinal tendons were retained. The sizes of the multi-cavity PCRV were approximately 100 feet (30 m) high by 120 feet (36 m) in diameter for the 3000 MWt plant and 100 feet (30 m) high by 105 feet (32 m) in diameter for the 2000 MWt plant.

The graphite reactor core was located in the central cavity of the PCRV. The steam generators and steam-driven main helium circulators were located in vertical cavities arranged around the periphery of the core. The 2000 MWt system had four steam generator-circulator side cavities while the 3000 MWt unit had six such cavities. The hot primary coolant helium (1366°F; 741°C) exiting from the bottom of the core collected in the lower core plenum from which it was distributed to the steam generators through the lower cross-ducts. The circulators, located above the steam generators, returned the cool helium (710°F; 377°C) through the upper cross-ducts to the upper core plenum. The helium in the upper plenum then flowed down through the core where it was heated.

Another feature of the large HTGRs was the core auxiliary cooling system which provided an independent means of core afterheat removal in the event that the main coolant loops (i.e., steam generators and

[2] This portion of article on HTGRs contributed by R.A. Dean, Sr. Vice President, GA Technologies Inc., San Diego, California.

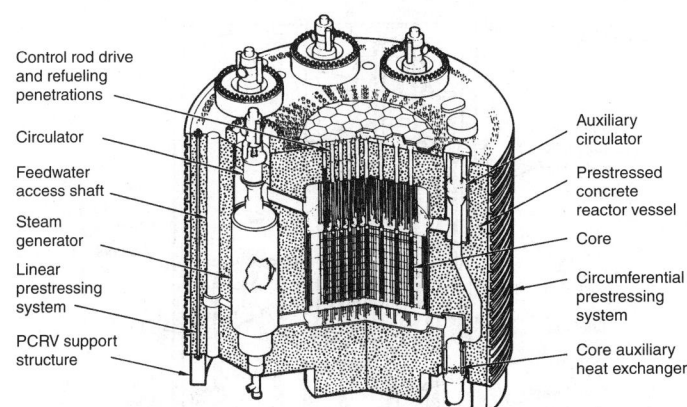

Fig. 19. Integrated HTGR nuclear steam system (1170 to 3360 MW thermal). (*GA Technologies*)

steam-driven circulators) were shut down. The core auxiliary cooling system consisted of two redundant cooling loops, each capable of removing 100% of the afterheat, for the 2000 MWt plant and three redundant cooling loops, each capable of removing 50% of the decay heat, for the 3000 MWt plant. Each loop contained a motor-driven circulator and water-cooled heat exchanger and circulated flow through the core just at the main loops. Shutoff valves were located in both auxiliary and main loops to assure that helium would not bypass the core through the one system while the other system was in operation.

All ten large commercial HTGRs were ordered during the early 1970s as an indirect consequence of the energy crisis brought about by the oil embargo; all were cancelled by 1976. The combination of the recession plus new emphasis on conservation brought about a rapid reduction in electric energy demand which, in turn, resulted in cancellation of over 100 nuclear power plant orders, including the large HTGRs.

Small Modular HTGRs

In the early 1980s, the major influence on new designs was the renewed emphasis on safety brought about by the accident at the Three Mile Island nuclear plant. The experience from licensing and operation of nuclear plants during the 1970s indicated a need to reduce design complexity and develop passive approaches to reactor safety rather than rely on complex emergency safety systems. HTGR designers in the United States and Europe determined that a substantial reduction in plant size could enable the HTGR to be entirely inherently safe by virtue of the high temperature structural integrity of the graphite core and ceramic coated fuel. This means that a small HTGR would not require any active safety equipment or any action by the operator in order to prevent release of radioactivity for any accident condition.

A major concern with reducing plant size was the economic impact from reversing the economy of scale. However, it was learned that economy-of-scale effects could be offset by several beneficial factors that apply to smaller nuclear plants. This includes the shift of major portions of the work from the site to the factory; the learning effects appreciated by replication of a larger number of smaller units in a factory environment and the elimination or simplification of many components/systems no longer required for smaller plants.

These considerations led to the reconfiguration of the HTGR plant into a system of one or more downsized 350 MWt modular reactors. The physical arrangement of a single reactor module, designed for installation in a below-grade silo, is shown in Fig. 20. The primary components are contained within two vertically oriented metal pressure vessels connected by coaxial cross-duct. Thus, the field-erected PCRV, which was used on previous large HTGRs, was eliminated in favor of shop-fabricated metal pressure vessels. The use of metallic pressure vessels also facilitates installation in underground silos, which enhances the safety of the plant.

The reactor vessel, which is approximately 72 feet (22 m) high by 22.5 feet (6.9 m) in diameter, contains the graphite core, reflector, and shutdown heat removal system (non-safety). The other vessel contains a single helical coiled steam generator and motor-driven circulator with magnetic bearing instead of water-lubricated bearings as in previous concepts. The size of both vessels is within allowable limits for barge, rail, and overland transportation.

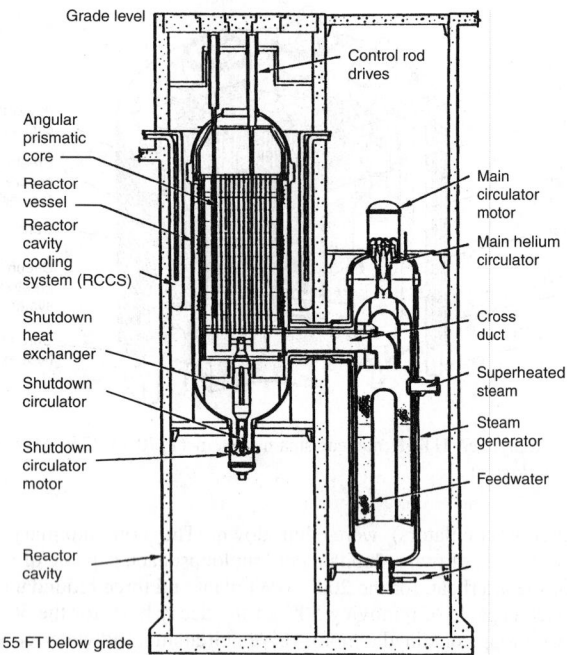

Fig. 20. Elevation of 350 MW modular HTGR. (*GA Technologies*)

During normal operation, the main circulator transports hot helium at 1266°F (686°C) from the bottom of the core to the steam generator which, in turn, produces superheated steam at 1005°F (541°C) and 2500 psia. The cold helium at 496°F (258°C) is returned to the top of the reactor core. During normal shutdown and refueling, the non-safety auxiliary shutdown heat removal system removes core afterheat if the main heat transport system is not operational.

A principal feature of the modular HTGR is its capacity for safely rejecting core afterheat in a completely passive manner (i.e., without the need for any active core cooling systems) such that any release of fission products from the fuel is prevented during severe accident conditions. This feature is a result of both the reactor system configuration and the high temperature capability of the fuel. In the event of a loss of forced circulation cooling of the core via either the main circulator/steam generator or auxiliary shutdown circulator/heat exchanger, core afterheat will continue to be safely removed by direct conduction through the core and reflector to the reactor vessel wall. The heat is then dissipated from the reactor vessel surface by radiation and natural convection to cooling panels surrounding the interior surface of the reactor cavity. See Fig. 20, previously mentioned. These panels are part of the Reactor Cavity Cooling System (RCCS) which consists of natural convection air ducts that ultimately transport the core afterheat directly to the atmosphere.

In order to achieve this passive core cooling capability in a reactor with a power level and power density that are economically attractive, the annular core arrangement shown in Fig. 21 was adopted. The active core consists of an annular region of hexagonal graphite fuel blocks containing standard HTGR fuel. Unfueled graphite reflector blocks make up the region inside the active core annulus and the region surrounding the outside of the annulus. This arrangement results in a higher radial heat conductance for fuel at the innermost radius than for a solid cylindrical active core. Thus, the annular core arrangement permits operation at a higher power for a given volume than a solid active core.

The modular HTGR uses the same form of fuel as the large HTGR and Fort St. Vrain installation except for the important difference that the ^{235}U enrichment was reduced to about 20% from the previous value of 93%. The fuel is in the form of coated fuel and fertile particles which are bonded into graphite rods and inserted into the hexagonal graphite fuel blocks. See Fig. 22. The fuel and fertile particles consist of uranium oxycarbide and thorium oxide kernels (about 350 micrometers in diameter), respectively, first coated with a porous graphite buffer, followed by three successive layers of pyrolytic carbon, silicon carbide, and pyrolytic carbon. The outer diameter of the coated particles is about 800 micrometers for the uranium particles and slightly larger for the thorium particles. The coatings essentially form a high-temperature refractory-based pressure vessel around each fuel/fertile kernel for the purpose of retaining fission products. Extensive operation and test data on these particles confirm that essentially no failure of the refractory coating occurs if the fuel is maintained below 3272°F (1800°C). As previously mentioned, the reactor design parameters were selected such that passive core afterheat removal will prevent this temperature from being reached during any credible accident condition. Thus, the modular HTGR is inherently, passively safe.

The reference plant arrangement features four 350 MWt HTGR modules supplying steam to two turbine generators that produce a net electrical output of 558 MWe at a new plant efficiency of 39.9%. Each reactor module is housed in an independent, vertical, cylindrical, concrete confinement, which is fully embedded in the earth. The four reactor modules share common systems for fuel handling, helium processing, and other essential services. A common control room is used to operate all four reactors and the turbine plant. Operation of the entire complex is completely automated. Human operator actions are not required for control during power production or to assure safe shutdown during hazardous conditions.

Potential for HTGRs

The HTGR's use of ceramic-coated fuel and graphite moderator enables operation of the reactor core at much higher temperatures than are required for electric power production via a steam Rankine cycle. Core outlet helium temperatures in excess of 1800°F (982°C) are achievable without impacting the integrity of the HTGR fuel or core structures. This very high temperature capability opens up the possibility for more efficient methods of power production or direct use of high-temperature thermal energy for process heat applications. See also **Cogeneration (Electricity and Thermal Energy)**.

An attractive electric power producing concept for the 21st century is the HTGR gas-turbine (HTGR-GT) which has the potential for thermal efficiencies over 50% by taking advantage of the high HTGR core outlet temperature. Although several variations in system configuration are possible, the most straightforward HTGR-GT concept is the direct Brayton cycle, illustrated in Fig. 23. This cycle is closed and the helium

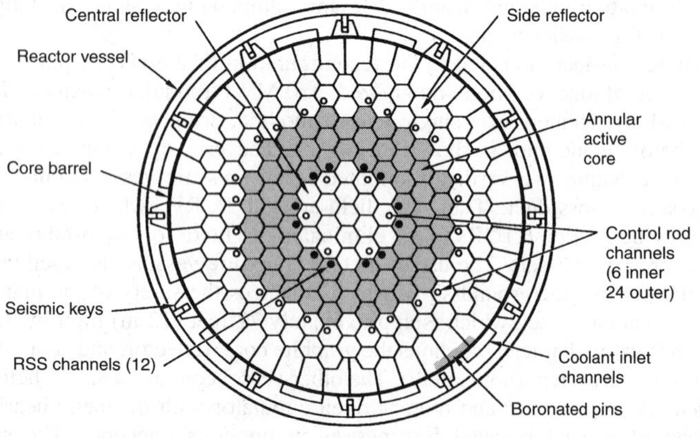

Fig. 21. Reactor core cross section of 350 MW modular HTGR. (*GA Techno logies*)

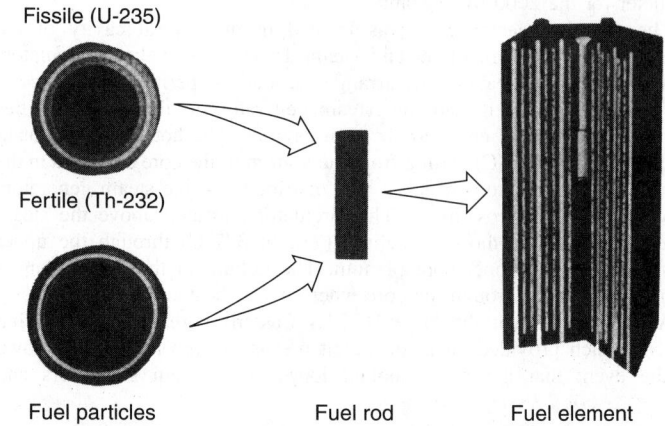

Fig. 22. Fuel components of HTGR. (*GA Technologies*)

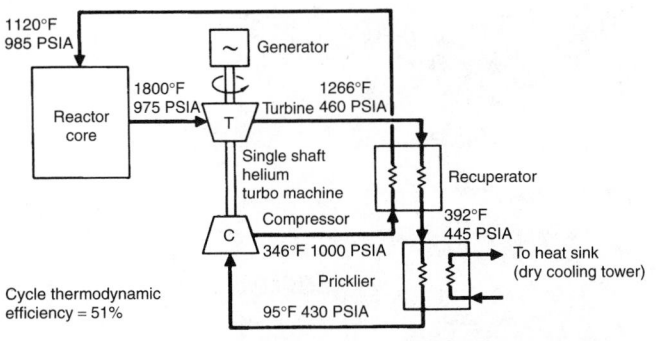

Fig. 23. HTGR gas turbine system with exceptional cycle thermodynamic efficiency. (*GA Technologies*)

primary coolant is also the working fluid for the power conversion system. The entire heat source and power conversion system of an HTGR-GT, which is capable of a net electric output of 170 MWe, can be enclosed within the two pressure vessels and cross-duct arrangement similar to the modular steam cycle HTGR previously shown in Fig. 20. The recuperator and precooler would occupy the same space as the steam generator and the turbo-compressor would replace the circulator.

Perhaps the most significant potential use of the HTGR's high temperature capability is the production of synthetic fuels from coal. The HTGR is an important option for supplanting the current consumption of oil and natural gas with synthetic natural gas (SNG). The HTGR can supply the necessary energy for this endothermic process and, therefore, increase the recoverable energy in the SNG product by at least 60% over traditional coal combustion processes.

The most effective method of SNG production with an HTGR is the steam-carbon reforming process in which superheated steam reacts with pulverized coal to form methane-rich SNG. A system for accomplishing this process is shown in Fig. 24. In this system, an intermediate heat exchanger (IHX) has been used to isolate the nuclear heat source from the process steam, thus allowing the use of conventional equipment for

the SNG production portion of the plant. The IHX and reactor can be configured in the same arrangement as previously shown in Fig. 23, except that the IHX would occupy the space allocated to the steam generator.

Both gas turbine and process heat versions of the HTGR are based on the demonstrated high-temperature capability of the fuel and core structure. However, some development in the metallic components, such as the turbine, hot ducts and intermediate heat exchanger is necessary. Present commercial alloys would have limited lifetime under service conditions at 1650°F (899°C) and above. However, currently envisioned advancements in ceramics and carbon-carbon composites indicate that high-temperature nonmetallic substitutes for metallic alloys will soon be available. These materials advances are the key to making future application of the HTGR a reality.

Heavy Water Reactor (HWR)

During the atomic energy developments in the World War II years and for a period thereafter, the United States, the United Kingdom, and Canada cooperated closely and many of the nuclear scientists of these countries appreciated the merits of heavy water as a moderator. Each of these countries pursued some development of HWRs for commercial power generation, but at different paces and dedication. Only Canada took to the HWR for commercial power generation. See Figs. 25 and 26.

One of the first high-priority nuclear applications of the United States was for naval propulsion. Because of a very tight minimal physical size criterion, LWRs offered advantages over the HWR. The United Kingdom placed emphasis on the production of plutonium for weapons programs. Gas graphite reactors were a reasonable early choice. When commercial nuclear power was recognized as a needed source of energy, because of the accumulated operating experience it was reasonable to adapt the reactors which had already been developed in the United States and the United Kingdom for military purposes. Long-term savings at that time was not a major criterion.

In the postwar years, hydroelectric power amply met a large portion of Canada's power needs and its abundance made nuclear power quite noncompetitive. Canadian utility operators were used to capital-intensive plants combined with low operating costs. In analyzing the prospects for nuclear power in Canada, utility planners and engineers placed a

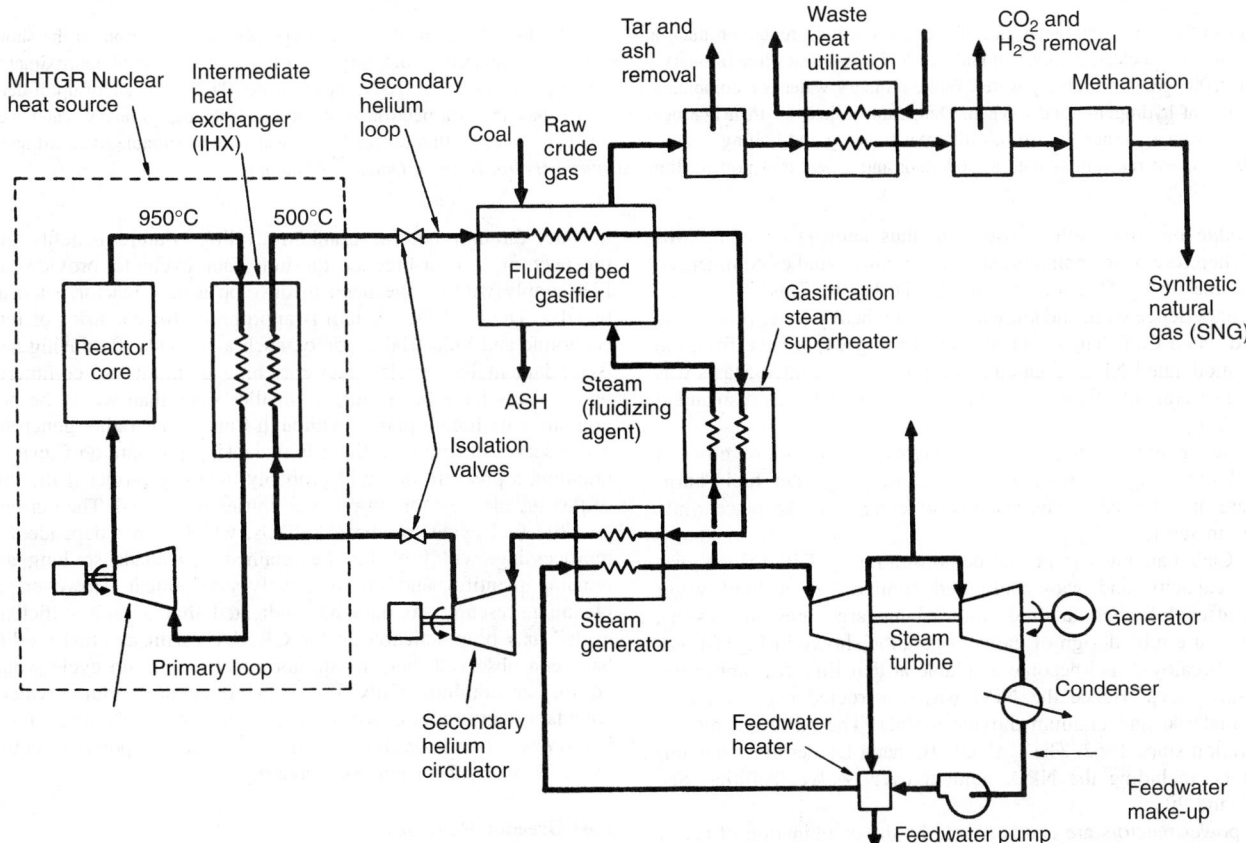

Fig. 24. Advanced process heat HTGR with intermediate loop for producing synthetic natural gas (SNG) by steam gasification of coal. (*GA Technologies*)

Fig. 25.   Series of towers comprising part of the heavy water production plant at Ontario Hydro's Bruce nuclear power complex near Tiverton on the shores of Lake Huron. Heavy water is a clear, colorless liquid that looks and tastes like ordinary water. It occurs naturally in ordinary water in the proportion of approximately one part heavy water to 7000 parts of ordinary water. While ordinary water is a combination of hydrogen and oxygen (H_2O), heavy water (D_2O) is made of up of deuterium—a form, or isotope, of hydrogen—and oxygen. Deuterium is heavier than hydrogen in that it has an extra neutron in its atomic nucleus, so heavy water weighs about 10% more than ordinary water. It also has different freezing and boiling points. It is the extra neutron that makes heavy water more suitable than ordinary water for use in CANDU nuclear reactors as both a moderator and a heat transport medium. (*Ontario Hydro, Toronto, Ontario, Canada*)

significant value on low fueling costs, and thus neutron economy was paramount. Therefore, when commercial nuclear power studies commenced in Canada in the mid-1950s, the choice was the HWR. This choice was bolstered by experience with, and knowledge about, heavy water production plants gained when Canadian scientists were trading experience from the heavy water-moderated NRX research reactor when the United States was developing the Savannah River production facility, which was dismantled in the early 1990s.

Principal advantages of heavy-water reactors are: (1) more efficient absorption of the energy released in the reactor, (2) greater fuel "burn-up" and, therefore, fuel economy, and (3) refueling can take place while the reactor is in service.

The first Canadian nuclear power demonstration (NDP) reactor was of 20-MWe capacity and was configured similarly to a light water reactor. Because of limited facilities for making large pressure vessels, a modular pressure-tube design of the configuration shown in Fig. 24 was investigated. Zircaloy-2 had become available at that time for fabrication of the pressure tubes. Hence the NPD was constructed using Zircaloy as cladding material and uranium dioxide as fuel. The NPD reactor has been in operation since 1962. The CANDU (Canada Deuterium Uranium) power reactors, including the NPD, number over twelve facilities. See Figs. 27, 28, and 29.

CANDU power reactors are characterized by the combination of heavy water as moderator and pressure tubes to contain the fuel and coolant. Their excellent neutron economy provides the simplicity and low costs

of once-through natural uranium cycling. Future benefits include the prospect of a near-breeder thorium fuel cycle to provide security of fuel supply without the need to develop a new reactor, such as the fast breeder. The CANDU system is appropriate for countries of intermediate economic and industrial capacity, such as Canada. Producing heavy water is fundamentally simpler than enriching uranium and commercial heavy water plants have been built in smaller sizes than would be possible for uranium enrichment plants. Although Canada has rather generous supplies and reserves of uranium, there is increasing pressure on Canada to export uranium, a pressure that will probably intensify further if the introduction of fast breeder reactors in other countries is delayed. The current simplest possible fuel cycle for the CANDUs, which is not dependent upon fuel reprocessing, will probably be retained in Canada so long as uranium remains plentiful and comparatively economical. However, for future planning, research to date has indicated that a "self-sufficient thorium cycle" may be practicable in the CANDUs with minimal modification. It has been observed that, at equilibrium, the thorium cycle would require no further uranium. Only small quantities of thorium, which is more abundant than uranium, would be required. Also of interest for the future is *electronuclear breeding*, i.e., the use of electric power to convert fertile to fissile material for neutron economy.

Fast Breeder Reactors

The fast breeder reactor derives its name from its ability to breed, that is, to create more fissionable material than it consumes. This ability stems from

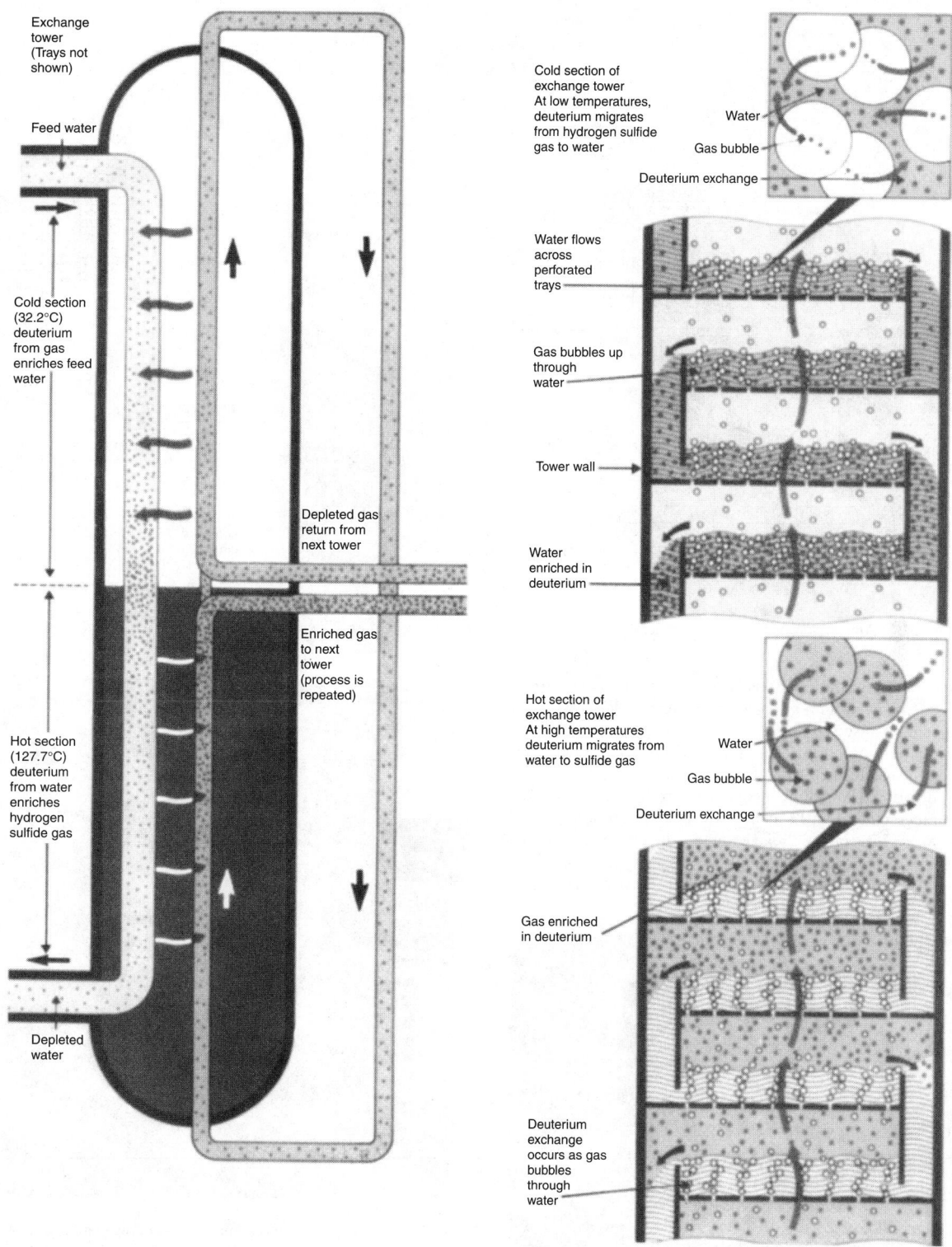

Fig. 26. The production of heavy water is based upon the behavior of deuterium in a mixture of water and hydrogen sulfide. When liquid H_2O and gaseous H_2S are thoroughly mixed, the deuterium atoms exchange freely between the gas and the liquid. At high temperatures, the deuterium atoms tend to migrate toward the gas, while they concentrate in the liquid at lower temperatures. In the first and second stages of production, the towers of a heavy water plant are operated with the top section cold and the lower section hot. Hydrogen sulfide gas is circulated from bottom to top and water is circulated from top to bottom through the tower. In the cold section, the deuterium atoms move toward the water and are carried downward, while in the hot section, they move toward the gas and are carried upward. The result is that both gas and liquid are enriched in deuterium at the middle of the tower. A series of perforated trays are used to promote mixing between the gas and water in the towers. A portion of the H_2S gas, enriched in deuterium, is removed from the tower at the juncture of the hot and cold sections and is fed to a similar tower for the second stage of enrichment

the fact that neutrons travel faster than they do in a thermal reactor. The breeding process depends, in part, upon the neutrons maintaining a high speed, or high energy. If their speed or energy is allowed to degrade as occurs in thermal reactors, the number of neutrons produced per absorption in uranium or plutonium decreases. Furthermore, at lower velocities, neutrons tend to be captured in various structural materials of the reactor, and this further reduces the breeding potential. It is important, therefore, in fast reactors to keep the velocity of the neutrons high. Water, which is used as a coolant in some thermal reactors, tends to slow the neutrons down and thus prevent efficient breeding. Therefore, it is necessary to use

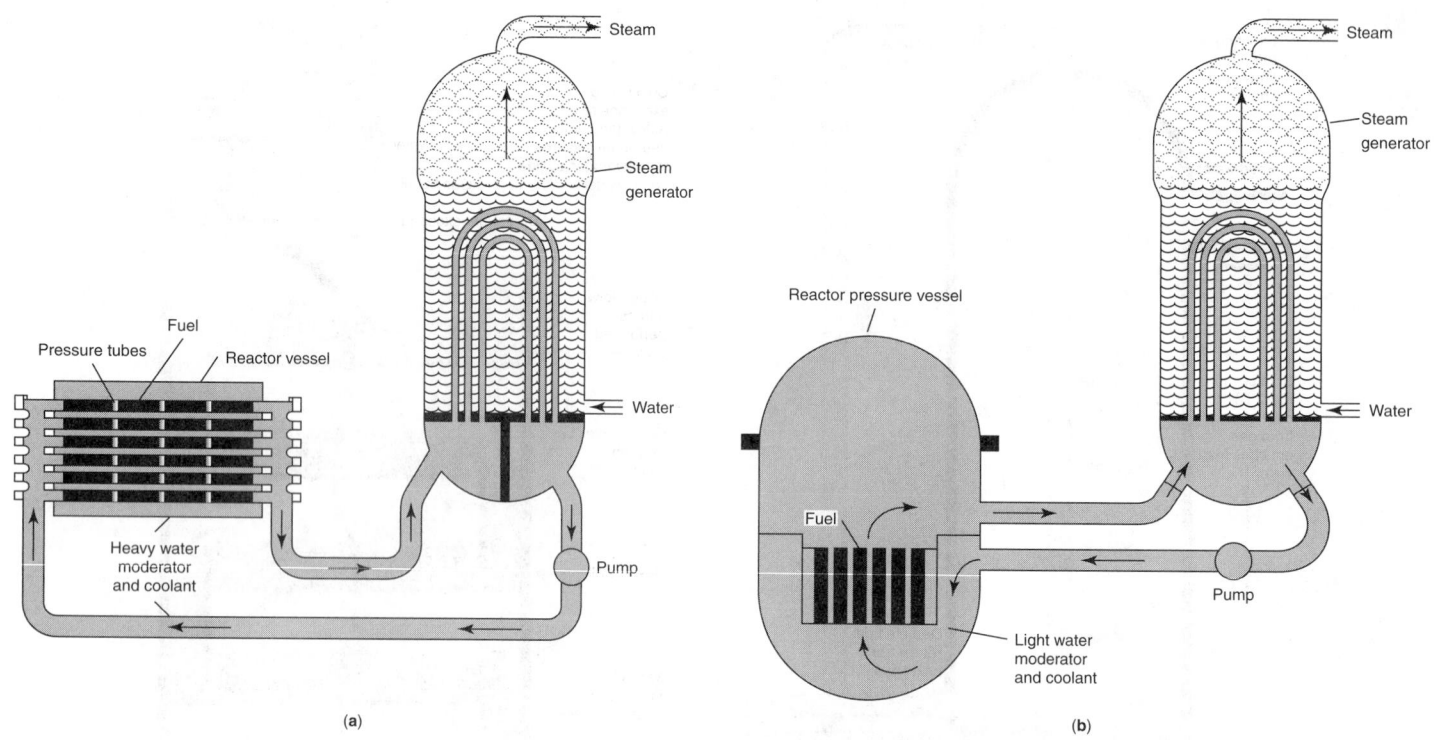

Fig. 27. Comparison of heavy water reactor (**a**) with light water reactor (**b**). (*After Robertson, Atomic Energy of Canada Limited*)

a coolant that does not slow the neutrons or capture them as they travel through the coolant. Liquid sodium and gaseous helium under pressure are the two principal coolants used to date.

The first stage of the process enriches the gas from 0.015% deuterium to 0.07%. A second stage further enriches it to about 0.35%. Again, the enriched gas is fed forward to a third stage. The product from this third stage, now in the range of 10 to 30% heavy water, is sent to a distillation unit for finishing to 99.75% purity "reactor-grade" heavy water. Because the production of heavy water uses a toxic gas, H_2S, safety is a top priority at heavy water plants. H_2S is a colorless gas, slightly heavier than air. To

Fig. 29. Darlington, Ontario, nuclear generating station, one of the largest energy projects undertaken in North America. The plant is shown in late stages of construction in 1986. It is located on the shores of Lake Ontario, 5 km southwest of Bowmanville in the Town of Newcastle. Currently nuclear power plants provide one-third of Ontario's electricity needs (the other two-thirds comes from hydro installations). Selected because of its proximity to the residential, industrial, and commercial energy markets of Ontario, the 1200-acre (485-hectare) site has good transportation access, an abundant supply of cooling water from Lake Ontario, relative isolation, and excellent bedrock for station foundations. When the four-unit station is completed it will provide 3,524,000 KW of electricity, enough to serve a city of 3 million people. Electricity from the four units (each 881,000 KW net) will be fed into the Ontario Hydro electricity grid system through 500,000 V transmission lines already crossing the site. The Darlington generating station was approved by the provincial government in July 1977. Site preparation began in 1978 and by 1981, the first concrete was poured for the station's foundations. Construction activity peaked in 1986–1987 with approximately 6800 workers on site. (*Ontario Hydro, Toronto, Ontario, Canada*)

Fig. 28. Pickering (Ontario) "A" generating station's initial performance was outstanding and the station was hailed as a major Canadian technological achievement at the time of its commissioning in February 1971. It reached full power in 3 months, well ahead of schedule. The final Unit 4 went on line in May 1973. At full output (2,160,000 kW), Pickering "A" generates enough power to supply more than 1.5 million homes. In 1974, construction was begun on a twin station, Pickering "B," also shown in this view. The "B" station has the same capacity as its forerunner and all four units became operational in 1986. (*Ontario Hydro, Toronto, Ontario, Canada*)

safely expel H_2S from the system, it is directed to a flare tower where it is burned off. Initially, each of Ontario Hydro's reactors requires about 800 megagrams, or one year's production from a heavy water plant. After that, less than 1% of the heavy water is lost and has to be replenished each year. (*Ontario Hydro, Toronto, Ontario, Canada.*)

Fuel Cycle Considerations. Approximately 99.3% of uranium as it is found in nature is the isotope ^{238}U and 0.7% is ^{235}U. Uranium-235 is a fissile isotope, that is, if it is struck by a neutron it will split; this fission yields on the average approximately two neutrons and 200 MeV of energy. This amount of energy corresponds to approximately 78 million Btu for every gram of uranium which fissions (3.5×10^{10} Btu/pound) (1.95 Calories/kilogram). Most reactors today are largely dependent upon ^{235}U for their energy. However, some of the neutrons released in fission of ^{235}U also are absorbed in nonfissionable ^{238}U. As the ^{238}U absorbs a neutron, it is transformed into fissionable ^{239}Pu (plutonium). Thus, while the reactor is sustaining the fission process and thereby creating energy, it is also generating fresh fuel, which can later be used to create more energy. Unfortunately, this is an inefficient process in present thermal reactors where the neutron velocity is established by the temperature, or thermal energy, so only limited amounts of additional energy are made available by transformation of ^{238}U into ^{239}Pu.

The fast breeder reactor makes possible the recovery of most of the available energy in uranium. This occurs because during fission in the fast breeder nearly three neutrons are released for every neutron absorbed as compared with only approximately two neutrons in a thermal reactor. On the average, between one and two neutrons are necessary for sustaining the fission process, and the extra neutron in a fast reactor can be absorbed in nonfissionable ^{238}U and thereby transformed into fissionable ^{239}Pu. Reactors which have a breeding ratio greater than one create more fuel than they need for their own purposes, and the extra plutonium can be used to fuel new breeder reactors. By this means, 80% or more of the available energy in uranium can be recovered and used in reactors.

In a typical fast breeder, most of the fuel is ^{238}U (90 to 93%). The remainder of the fuel is in the form of fissile isotopes, which sustain the fission process. The majority of these fissile isotopes are in the form of ^{239}Pu and ^{241}Pu, although a small portion of ^{235}U can also be present. Normally, the fissile isotopes are located in a central "core" region that is surrounded by the fertile isotopes in the "blanket" region. This is illustrated in Fig. 30.

When the fuel is initially loaded into the reactor, the core region will typically contain from 10 to 15% fissile isotopes with the remainder being ^{238}U. Essentially all of the blanket will be ^{238}U. As energy is extracted from the fissile isotopes, they become depleted (the initial plutonium is gradually used up). However, in a breeder reactor, new plutonium will be formed in the core and blanket regions faster than it is consumed. Additionally, undesirable fission products are formed which must ultimately be removed. This process is schematically illustrated in Fig. 31. The "before" chart

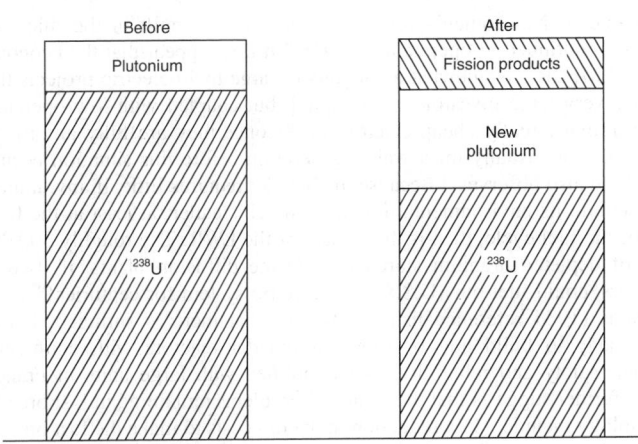

Fig. 31. Basic operation of the breeder reactor. The illustration does not include geometrical disposition of fuel in the core and blanket system. (*General Electric*)

represents the new fuel condition and the "after" chart corresponds to the situation when the fuel is removed for reprocessing. Typically, the fuel removed for reprocessing will contain from 1 to 3% new plutonium. It is in this manner that the fast breeder can recover from 80 to 90% of the available energy in uranium resources. Most present reactors require some enrichment of the ^{235}U isotope used to fuel them. This enrichment process requires a plant, which, in turn, uses large amounts of electrical energy. Because the fast breeder converts the fertile isotope ^{238}U into the fissile isotope ^{239}Pu, no enrichment plant is necessary. The fast breeder serves as its own enrichment plant. The need for electricity for supplemental uses in the fuel cycle process is thus reduced.

Fast Breeder Reactors in Perspective

Of the several fundamental ways to use nuclear fission reactions to generate electric power, the fast breeder reactor (notably the liquid-metal-cooled fast breeder reactor, LMFBR) probably has the most checkered history. The fast breeder reactor received its early impetus when there was serious concern over what appeared to be a limited supply of uranium and consequent increasing prices of uranium fuels. With the fast breeder concept offering up to a 100-fold increase in the utilization efficiency of uranium, it appeared to be the logical replacement (second generation) for light-water reactors. It would present a technical solution in time for the expected tight supplies of uranium. The United States funded LMFBR research quite generously until a serious reduction in 1982. Then, it was determined that the shortage of uranium fuels no longer posed a serious threat in the short-term, thus establishing the general consensus in the U.S. that the breeder, if needed at all, could be delayed until well into the next century. The interest of the French and Japanese in breeders also stemmed from early concerns with uranium shortages and prices, but was much more serious because these countries are extremely uranium poor and, further, these countries have fewer energy options. For example, the cost of coal ranges from 1.6 to 2.5 times the cost of coal in the United States. Continued progress in fast breeder development, particularly in France, also has been accelerated by a very heavy past investment in the technology coupled with a desire to be fully self-contained as regards the generation of electric power.

Particularly, in the United States, because of its several energy fuel options and its current "rethinking of nuclear power," the principal nuclear power research targets no longer include uranium supplies. Rather, the targets are the lowering of capital costs for existing light-water reactor technology, shortening the plant construction and licensing lead times, achieving higher plant availability (essentially eliminating long power outages), and increasing plant safety. From this "rethinking" process, the U.S. Congress suspended funding for the Clinch River Breeder Reactor.

It is interesting to note that, as of the early 1990s, most authorities were not quite willing to "forget" the fast breeder altogether. The differences in views essentially reside in timing. The fast breeder, even though costly to build today, does offer several temptations. For example, experience gained from the Experimental Breeder Reactor (Idaho Falls, ID) which commenced operation in 1964 is reported to operate better now than when it was first built, showing no evidence of corrosion. It is suggested that if such experience were extrapolated the useful life of an LMFBR could be between 100 and 150 years! Weinberg (1986 reference listed) recalls,

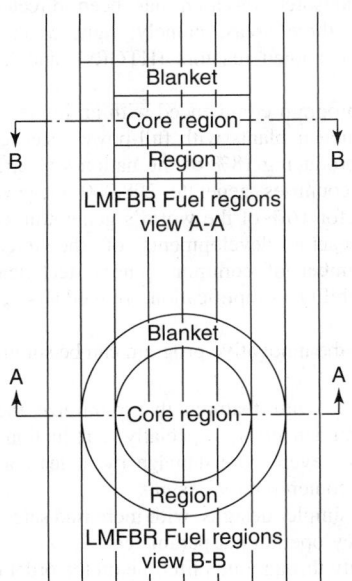

Fig. 30. Liquid-metal fast breeder reactor core and blanket arrangement

"that one of Newcomen's original steam engines, built in the mid-18th century, continued to operate until 1918." It does appear that the economy of the LMFBR is somewhat analogous to large hydroelectric projects that require very large investments of capital, but coupled with low operating costs, provide really cheap electric power once they are fully amortized. However, one usually must wait a generation before this situation occurs. Weinberg also observes, "Because the breeder requires little, if any, mining of uranium, its environmental impact is much smaller, at least at the front end of the fuel cycle, than is the impact of the LWR. The roughly 300,000 tons of depleted uranium stored outside the diffusion plants, if used in breeders, could fuel our (U.S.) entire electric system for centuries!"

Davis (1984 reference listed) observes that there is sufficient know-how today to build and operate fast breeder reactors with confidence of their safety and reliability, as exemplified with large units in France. Davis further suggests that the principal problems remaining in fast breeder technology include: (1) a reduction in overall costs to make the fast breeder competitive with coal and the light-water reactors—an estimated reduction factor of 1.5 to 2; (2) assurance of adequate safeguard on the plutonium fuel cycle; (3) more demonstrations of commercial-scale operations, such as the engineering scale-up of important system components—pumps and steam generators; (4) implementing a large, overall system which must include parallel facilities for fuel processing and refabrication; and (5) setting in place the reprocessing of light-water reactor fuel to provide the "start up" plutonium for the fast breeders.

Some technical successes with the fast breeder have occurred in France: (1) demonstrations of a positive breeding gain of 0.15 ± 0.04 in a complete breeder fuel cycle, and (2) demonstration at several laboratories of uranium and plutonium oxide fuel elements that can sustain more than 10% burnup of the original mixture of ^{238}U/^{239}Pu before the fuel has to be reconstituted, representing a tenfold improvement (Weinberg, 1986). It has also been observed that, compared with a number of so-called alternative fuels, the new Super-Phoenix (France) plant can produce electricity at costs that are markedly less than electrical energy from solar-powered photovoltaics, for which funding remains significant. Another proposed inexhaustible power source, nuclear fusion, still remains in an early stage of development. See **Fusion Energy**.

LMFBR Design Principles. There are many design differences among the reactor designs, including: (1) primary coolant system arrangement; (2) refueling mechanism design and arrangement; (3) steam generator type and arrangement; (4) core support method; (5) structural material choices; and (6) safety features. Perhaps the most noticeable difference is that of the primary system arrangement. This difference is schematically illustrated in Figs. 32 and 33. The system of Fig. 32 corresponds to a "loop" or "piped" arrangement where the reactor, pumps, and intermediate heat exchangers are located separate from each other and piping carries the sodium from one point to the other. The "pool" or "tank" arrangement of Fig. 33 includes the reactor, intermediate heat exchangers and pumps in one large pool of sodium which is contained in a separate tank. Each concept has advantages and disadvantages. The pool concept is somewhat easier to design for certain hypothetical accident situations. The loop concept is easier to construct and to maintain.

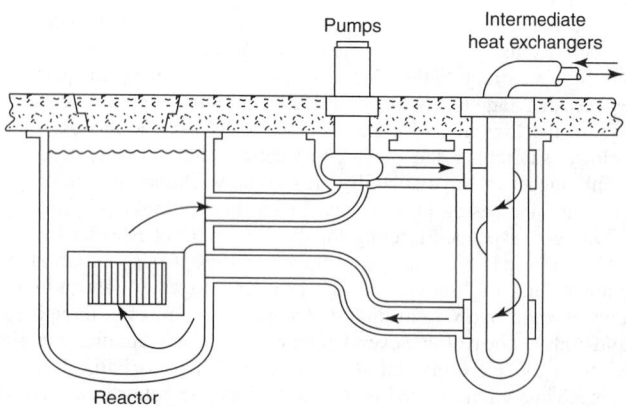

Fig. 32. Loop arrangement in the liquid-metal fast breeder reactor. (*General Electric*)

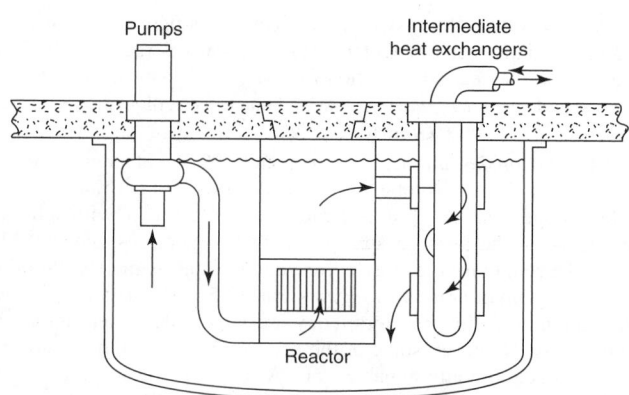

Fig. 33. Pool arrangement in the liquid-metal fast breeder reactor. (*General Electric*)

The flow circuit for an LMFBR where two sodium circuits are included is shown schematically in Fig. 34. The reactor is cooled by the primary sodium, which becomes radioactive as it picks up heat in passing through the core or fueled region. In this particular arrangement, the sodium is heated to 560°C and flows through pipes (schematically shown as a single line in the figure) to the intermediate heat exchangers. In the heat exchangers, the primary sodium transfers heat to the nonradioactive sodium. After being cooled to 393°C in the heat exchangers, the primary sodium is pumped back into the reactor where it again repeats the circuit. The nonradioactive secondary sodium is circulated from the intermediate heat exchangers through steam generators where the heat from the sodium is transferred to water, which becomes superheated steam for use in the turbine. The cooled secondary sodium is pumped back through the intermediate heat exchangers where the process is repeated. Steam from the steam generators is used to turn the rotor of the turbine generator to generate electricity. In the arrangement shown, 1,200 MW of electricity are generated at a net overall efficiency of 39%. This relatively high efficiency is possible because of the excellent thermal characteristics of sodium.

Nuclear Power Innovations for 1995 and Beyond

Increasing concerns over the impact of fossil fuel-burning electric power generating plants on the environment, the accelerating demands for electric power, and a growing dependence on foreign oil supplies brought about a resurgence of interest in nuclear power reactor research during the mid-1980s. Advanced programs were established to create new plant designs that would reflect past experience to achieve an extremely high degree of operating safety, competitive construction, and operating costs, including the streamlining of the licensing process and consumer confidence in nuclear power technology. Thus, during the past decade, much private and government-sponsored research has been directed toward nuclear reactor research in three areas, namely, light water reactors (LWRs), high-temperature gas-cooled reactors (HTGRs), and liquid metal-cooled reactors (LMRs).

The innovative program commenced with an impressive nuclear power base—some 107 nuclear plants with full-power licenses operating in the United States and producing 18% of the nation's electricity—worldwide, 414 plants in 26 countries generate 298,000 megawatts of electricity (MWe), accounting for 16% of the world's generating capacity.

A survey of reactor developments of the three aforementioned types reveals a number of common generic technical features. These include passive stability, simplification, ruggedness, ease of operation, and modularity.

Overall goals for the innovative program can be summarized as follows:

1. Assured safety with features that minimize the negative consequences of human error, especially a reduction in the chance of occurrence of severe core damage by at least a factor of 10 less than former, contemporary designs.
2. Significantly simpler designs, with increased safety and performance margins in key operational parameters.
3. High reliability throughout a lifetime on the order of 60 years and an increase in plant availability to 85% or greater than the contemporary average of less than 70%.

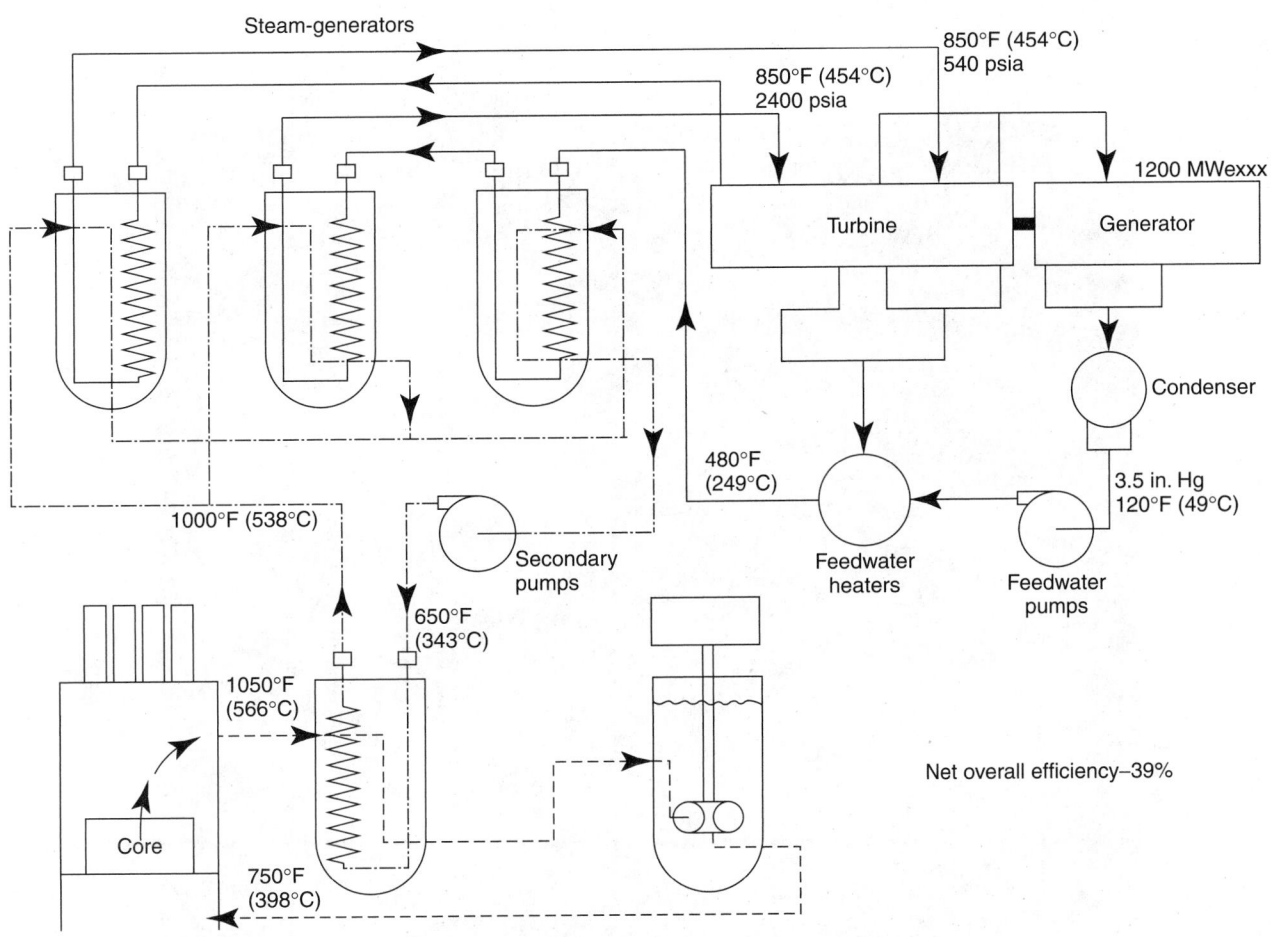

Fig. 34.   Liquid-metal fast breeder reactor flow circuit. (*General Electric*)

4. Reduction in capital, operating, maintenance, and fuel costs to meet the economic competition with coal-burning generators. A reduction in construction time to the range of 3 to 5 years as compared with more than 10 years, which has been the experience with some of the later contemporary reactors.

5. A modular design that is standardized at a highquality level and thus predictably licensable.

Passive Stability. Passive design characteristics ensure core stability by eliminating the potential for a runaway chain reaction. In the innovative program, this has been a hallmark from the outset of the program. Passive characteristics are internal governors—that is, physical laws ensure that the reaction rate decreases instantaneously as the temperature of the coolant or fuel or the power of the reactor increases, without the need for external control devices.

Ruggedness. In some past designs, long-term reliability has been impaired by attempts to achieve the highest in efficiency and economic performance. In response to this negative experience, the margin in certain key performance parameters is being increased in order to lessen the burden on the equipment. By reducing power densities and coolant temperatures, higher reliability will be achieved over a longer lifetime. Past field experience has identified more effective methods of coolant chemistry control and materials selection, factors that will contribute to the long-lived reliability of the components of future systems. Greater emphasis is being given to the selection of proven, high-quality materials and components and on improved methods and quality control over assembly and construction.

Ease of Operation. Thorough investigation of the Three Mile Island incident some years ago showed that a lack of attention had been given to the human factor. The innovation program is addressing this problem in several ways. The computer and telecommunication revolution has made it more practical to use improved technology and human engineering methodologies to revamp the control room and the reactor instrumentation system. These improvements will make the plant easier to operate and provide the operator with a greatly increased amount and quality of

information on plant conditions. Graphic displays, diagnostic aids, and expert systems are being developed for such advanced control rooms.

The other design goals complement the new technology to make the operator's task even easier. The passive safety features substantially extend the response time required of the operators in an emergency condition. The margin being built into the systems provides broader normal operating regimes and longer response times for operator action. Greater emphasis is being placed on simplification of operating procedures.

Modularity. Economic competitiveness requires that the construction time be shortened dramatically. Modular construction techniques are a key contributor to achieving this goal and are a proven approach to cost control in major construction projects. Modularization provides for a larger percentage of factory construction, rather than field construction. New innovative concepts will rely heavily on modularization and will be centered around lower unit power outputs, factory assembly, and transportation of modules to the plant site. The overall plant size target is 600 MWe net electrical output.

The AP600 Advanced Plant

As of 1994, the AP600 (Westinghouse) plant was in the most advanced stage of the innovative program. The first unit may be put on line just prior to the expiration of the last century. This design satisfies all of the previously mentioned goals of the innovative program. A sectional view of the AP600 is shown in Fig. 35. The site plan is shown in Fig. 36.

AP600 Passive Safety System Details. These features reduce operator responsibilities and add an extra margin of safety over contemporary PWR designs. See Fig. 37 on p. 1121. Large volumes of water stored in the containment eliminate the need for operator action to assure make-up water, either for small leaks that may occur during normal operation or for a major loss of coolant accident (LOCA). A passive plant is a system that assures public safety even if the operators fail to act.

The passive residual heat removal heat exchangers remove core decay heat if steam generator heat removal is not available. Passive residual heat removal heat exchangers in the in-containment refueling water storage

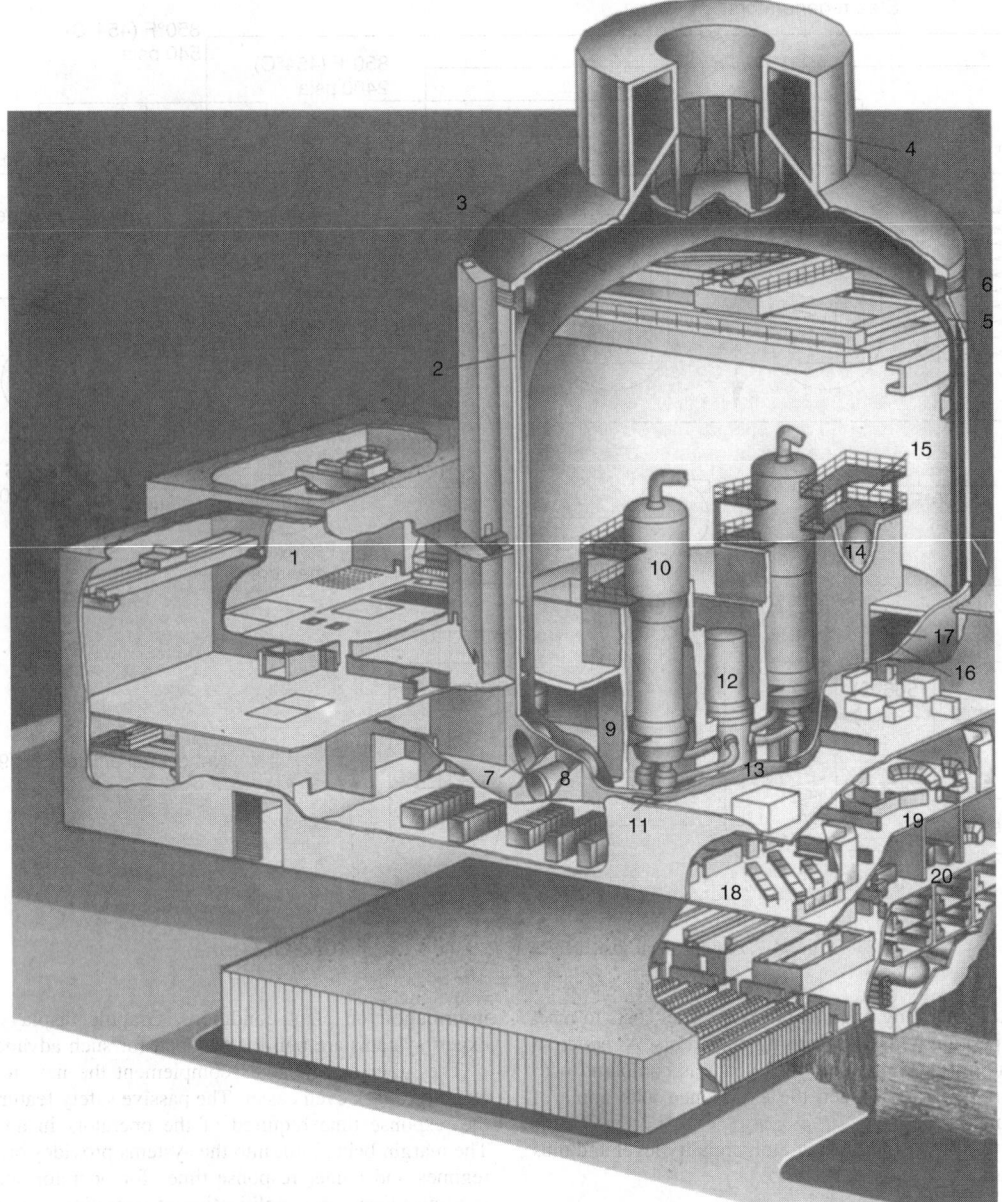

Fig. 35. Sectional view of the 600 MWe pressurized water reactor (PWR) expected to be operational by 1995. Considered the PWR of the future, the plant is designed for a minimum useful life span of 60 years and features numerous economic and safety features, including passive systems for ultimate protection. (*Joint project of Westinghouse, the Electric Power Research Institute, and the U.S. Department of Energy*)

Legend:

1. Fuel Handling Area	11. Reactor Coolant Pumps (4)	21. High Pressure Feedwater Heaters
2. Concrete Shield Building	12. Integrated Head Package	22. Feedwater Pumps
3. Steel Containment	13. Reactor Vessel	23. Deaerator
4. Passive Containment Cooling Water Tank	14. Pressurizer	24. Low Pressure Feedwater Heaters
5. Passive Containment Cooling Air Baffles	15. Depressurization Valve Module Location	25. Turbine/Generator
6. Passive Containment Cooling Air Inlets	16. Passive Residual Heat Removal Heat Exchangers	26. Spargers (2)
7. Equipment Hatches (2)	17. Refueling Water Storage Tank	27. Accumulators (2)
8. Personnel Hatches (2)	18. Technical Support Center	28. Main Steam Line
9. Core Make-up Tanks (2)	19. Main Control Room	29. Feedwater Line
10. Steam Generators (2)	20. Integrated Protection Cabinets	30. Passive Containment Cooling Air Flow

tanks are connected to the reactor coolant system (RCS) piping, forming a full-pressure, closed, natural circulation cooling loop.

Two core make-up tanks provide borated make-up water whenever the normal make-up system is unavailable. The tanks are located above the reactor coolant system loop piping and kept at system pressure by steam lines from the pressurizer. These tanks function at any system pressure, using only gravity as a motive force. If the reactor protection system detects a need for make-up water, core make-up tanks discharge and

isolation valves open automatically, allowing the tanks to drain into the reactor vessel.

Two accumulators provide the high make-up flows initially required by a large LOCA. These tanks contain 1700 cubic feet (about 48 cubic meters) of borated water pressurized with 300 cubic feet (about 8.5 cubic meters) of nitrogen at 700 psi (4.8×10^6 Pa). The accumulators are isolated from the RCS by check valves. Each accumulator is paired with a core make-up tank, the pair sharing an injection line to the reactor vessel downcomer.

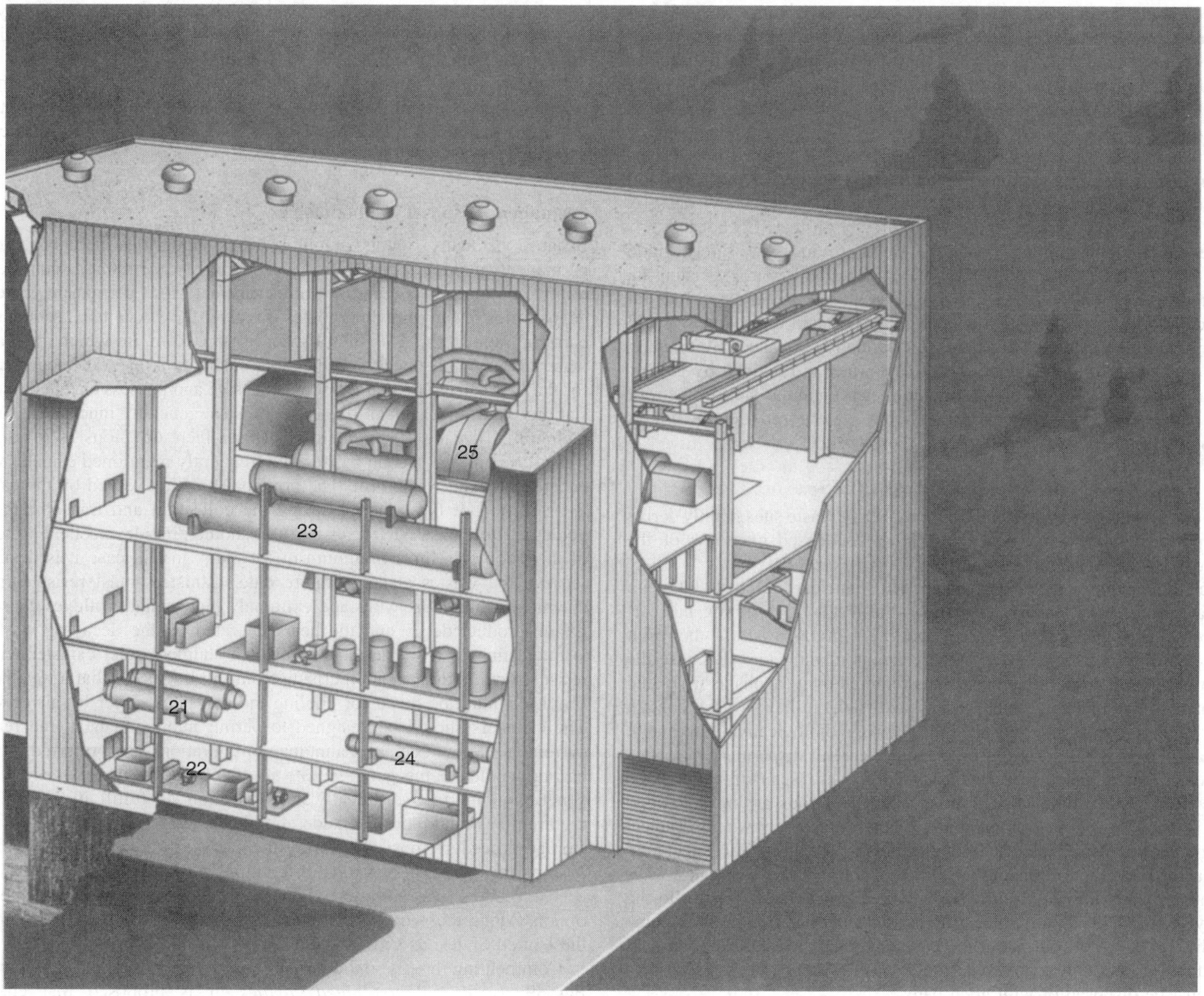

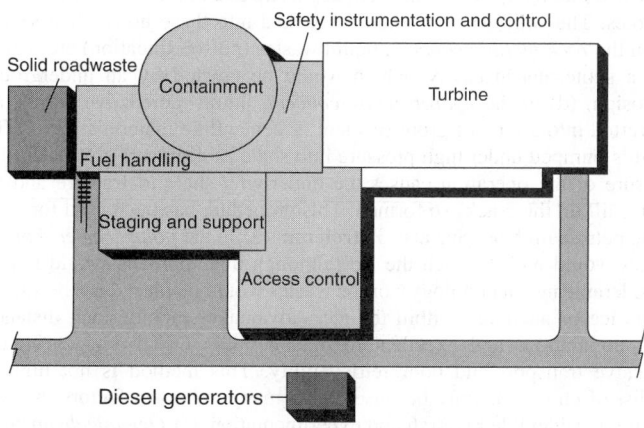

Fig. 35.   (*continued*)

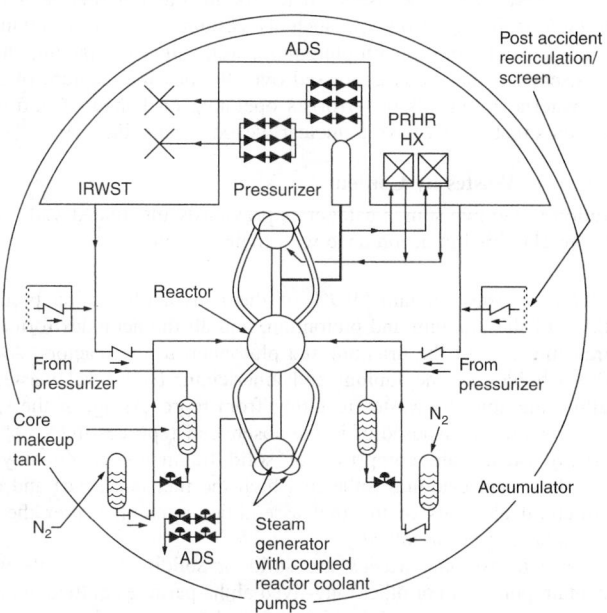

Fig. 36.   Site plan for the advanced pressurized water reactor (PWR). (*Westinghouse Electric Corporation, Energy Systems*)

This ensures that at least one accumulator/core make-up tank pair would be available following an injection line LOCA.

The in-containment refueling water storage tank provides 500,000 gallons (about 1900 cubic meters) of water with a gravity head above the core. This water inventory is sufficient to flood the containment above the level of the reactor core and provide decay heat removal by natural circulation.

Fig. 37.   Schematic representation of the in-containment passive safety injection system (PSIS). IRWST = in-containment refueling water storage tank. PRHR-HX = passive residual heat removal heat exchanger. ADS = automatic depressurization system (four stages). (*Westinghouse*)

The automatic depressurization system depressurizes the RCS if core make-up tank level is low. Depressurization allows gravity injection from the in-containment refueling water storage tank, which is at atmospheric pressure. To ensure that the automatic depressurization system works when needed, while minimizing the consequences of spurious valve operation, the system provides phased depressurization with two redundant sets of valves connected to the pressurizer. The discharge is sparged into the in-containment refueling water storage tank. The automatic depressurization system valves are arranged in three stages to reduce peak flow rates. A fourth depressurization stage is provided directly on the RCS hot leg.

The passive containment cooling system provides the safety grade ultimate heat sink that prevents the containment shell from exceeding its design pressure of 45 psig [3.1×10^5 Pa (g)]. The system uses natural air circulation between the steel containment shell and the concrete shield building. During postulated accidents, air cooling is enhanced by draining water onto the steel containment shell. The water is provided by gravity from a 350,000-gallon (about 1300 cubic meters) annular tank in the roof of the shield building. This tank has sufficient water to provide three days of cooling.

RADIOACTIVE WASTES

Decisions pertaining to the location of radioactive waste sites mainly derive from political and sociological sources. A condensed overview of the technical aspects is given here.

The radioactive wastes associated with nuclear reactors fall into two categories: (1) *commercial wastes*—the result of operating nuclear-powered electric generating facilities; and (2) *military wastes*—the result of reactor operations associated with weapons manufacture. Because the fuel in plutonium production reactors, as required by weapons, is irradiated less than the fuel in commercial power reactors, the military wastes contain fewer fission products and thus are not as active radiologically or thermally. They are nevertheless hazardous and require careful disposal.

Nuclear power plants use fuel rods with a life span of about three years. Each year, roughly one-third of spent fuel rods are removed and stored in cooling basins, either at the reactor site or elsewhere. Typical modern nuclear power plants discharge about 30 tons of the spent fuel per reactor per year. Comparatively little of the radioactive wastes, as is currently reliably known worldwide, has been processed for return to the fuel cycle. Actually, fuel reprocessing causes a net increase in the volume of radioactive wastes, but, as in the case of military wastes, they are less hazardous in the long term. Nevertheless, the wastes from reprocessing also must be disposed of with great care.

Spent fuel from a reactor contains unused uranium as well as plutonium-239 which has been created by bombardment of neutrons during the fission process. Mixed with these useful materials are other highly radioactive and hazardous fission products, such as cesium-137 and strontium-90. Since reprocessed fuels contain plutonium, well suited for making nuclear weapons, concern has been expressed over the possible capture of some of this material by agents or terrorists operating on behalf of unfriendly governments that do not have a nuclear weapons capability.

Categories of Wastes by Content

In addition to the two source categories previously mentioned, radioactive wastes are classified in accordance with their content:

High-level wastes contain 99.9% of the nonvolatile fission products, 0.5% of the uranium and plutonium, and all the actinides formed by transmutation of the uranium and plutonium in the reactors. Among the actinides are neptunium and americium. High-level wastes are either the aqueous wastes resulting from reprocessing; or the spent-fuel rods to be disposed of in the absence of reprocessing.

Cladding wastes are comprised of solid fragments of Zircaloy and stainless steel cladding (tube in which the fuel is placed) and other structural elements of the fuel assemblies remaining after the final cores have been dissolved.

Low-level transuranic wastes are solid or solidified materials which contain plutonium or other long-lived alpha-particle emitters in known or suspected concentrations higher than 10 nanoCuries per gram and external radiation levels after packaging sufficiently low to allow direct handling.

Intermediate level transuranic wastes are solids or solidified materials that contain long-lived alpha-particle emitters at concentrations greater than 10 nanoCuries per gram and which have, after packaging, typical surface dose rates between 10 and 1000 mrems/hour due to fission product contamination.

Nontransuranic low-level wastes are diverse materials which are contaminated with low levels of beta- and gamma-emitting isotopes, but which contain less than 10 nanoCuries of long-lived alpha activity per gram.

Permanent Disposal Methodologies

Because of many doubts among various authorities pertaining to the permanent geologic depositories, as previously mentioned, considerable effort continues as regards semipermanent and permanent types of depositories. Many methodologies involve the so-called "sequence of barriers" approach. The first barrier is the form in which radioactive materials are embedded—vitrification, calcination, etc. The requirements for the first barrier are that it not be corrosive and possess excellent thermal stability and mechanical integrity. Wastes generate much heat during their initial decade of confinement. This affects decisions as regards the wasteform and the second barrier, the frequently mentioned canister which encapsulates the wasteform. The principal function served by the canister is protection of the material during the collection and transportation (to geologic site) phases. The canisters also should provide excellent protection of their contents for a minimum of 50 years, just in case it is desired to retrieve the wastes at some future date. Canisters must resist corrosive chemicals, they must withstand extremely high radiation fluxes caused by fission-product decay and the heat generated by the decaying wastes. It is interesting to note that an unprotected stainless steel canister will not resist structural deterioration arising from salt brines for that long a period. Provision must be made for cooling the canisters, either by air or water. The canisters should be designed to permit maximum heat transfer and, currently, the cylinder and annulus configurations are preferred. A third barrier would be the geologic site itself, obviously impervious to water penetration and in a seismically stable location. To fulfill all the foregoing requirements (and more), consideration is being given to phasing the waste storage procedure, possibly storing the canisters for water cooling during the first few years, after which air cooling would suffice.

The predominant preference appears to be the underground depository option. Along these lines, it is interesting to review the rock formations in the United States, as shown in Fig. 38.

Commenting briefly on other proposed methodologies, as shown in Fig. 39: (a) In *solution-mined cavities*, it is proposed that chemical solutions would be used to mine cavities in appropriate media, such as rock salt; (b) in the *drilled-hole matrix*. A series of large-diameter holes would be drilled into the geologic media to depths up to 2 kilometers to form a grid of holes. The solid wastes would be packed into these holes, then sealed. (c) In the *rock-melting concept*, liquid wastes (no solidification) are poured into a subterranean cavity, which would be created by an underground explosion. (d) In the *hydrofracture concept*, liquid radioactive wastes are converted into a type of grout (cement or cementlike materials used). This grout is pumped under high pressure into shale as deep as 1 kilometer. The pressure of the operation causes the underlying shale to fracture and the wastes fill up the cracks so formed. This procedure has been used for years in the petroleum field. See also **Petroleum**. (e) In the *polar ice concept*, the wastes would melt through the ice (although this approach would require considerable new technology); or the wastes would be placed on the surface of the ice or anchored within the ice. Advantages include long distances from populations and excellent thermal cooling. Disadvantages include extensive transport and poor retrievability. This method is not high on the list of choices mainly because of too many unknown factors that will require considerable research and experimentation. (f) *Oceanic disposition*, in addition to the polar ice cap concept, are subduction zones and other deep sea trenches and rapid sedimentation areas. A "Seabed Disposal Program Annual Report" states, "Placing high-level wastes on the seafloor, i.e., in the water column, effectively puts the waste contained directly into the biosphere. Since it is difficult to conceive of a practical man-made wasteform/container system that would survive without releasing radionuclides for hundreds of thousands of years in a marine environment, one must assume the radioactive material would eventually enter the ecosystem." The sub-seabed sediments in the central North Pacific are loosely packed, fine-grained, deepsea "red clays" and have not been fully dismissed from continuing investigation.

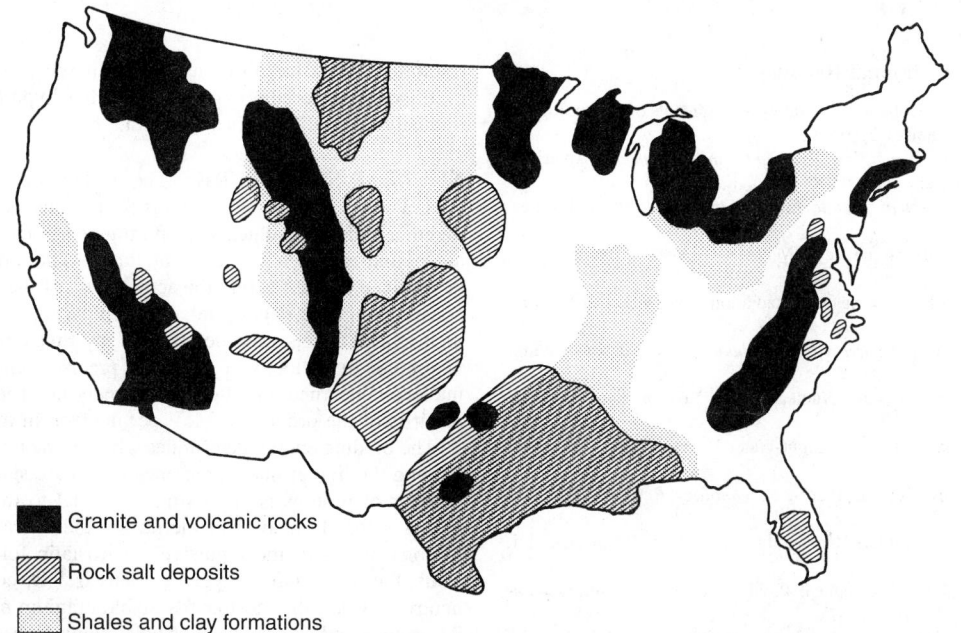

Granite and volcanic rocks

Rock salt deposits

Shales and clay formations

Fig. 38. Some authorities prefer rock salt deposits for the permanent disposal of nuclear wastes on the assumption that the heat generated by radioactive decay would fuse salt and wastes into an impermeable mass. Other experts question the integrity of salt formations. Increasing attention has turned to hard media, such as the granitic and basaltic rocks, and to shale and clay formation, with the hope that the extensive occurrence of such formations would minimize the need to transport wastes over long distances

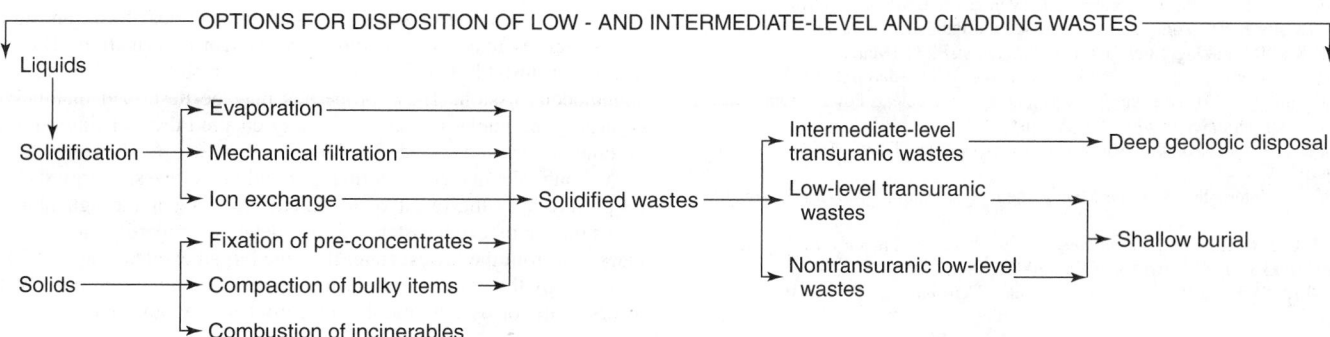

OPTIONS FOR DISPOSITION OF LOW - AND INTERMEDIATE-LEVEL AND CLADDING WASTES

Liquids

Solidification
- Evaporation
- Mechanical filtration
- Ion exchange

Solids
- Fixation of pre-concentrates
- Compaction of bulky items
- Combustion of incinerables

Solidified wastes
- Intermediate-level transuranic wastes → Deep geologic disposal
- Low-level transuranic wastes
- Nontransuranic low-level wastes
→ Shallow burial

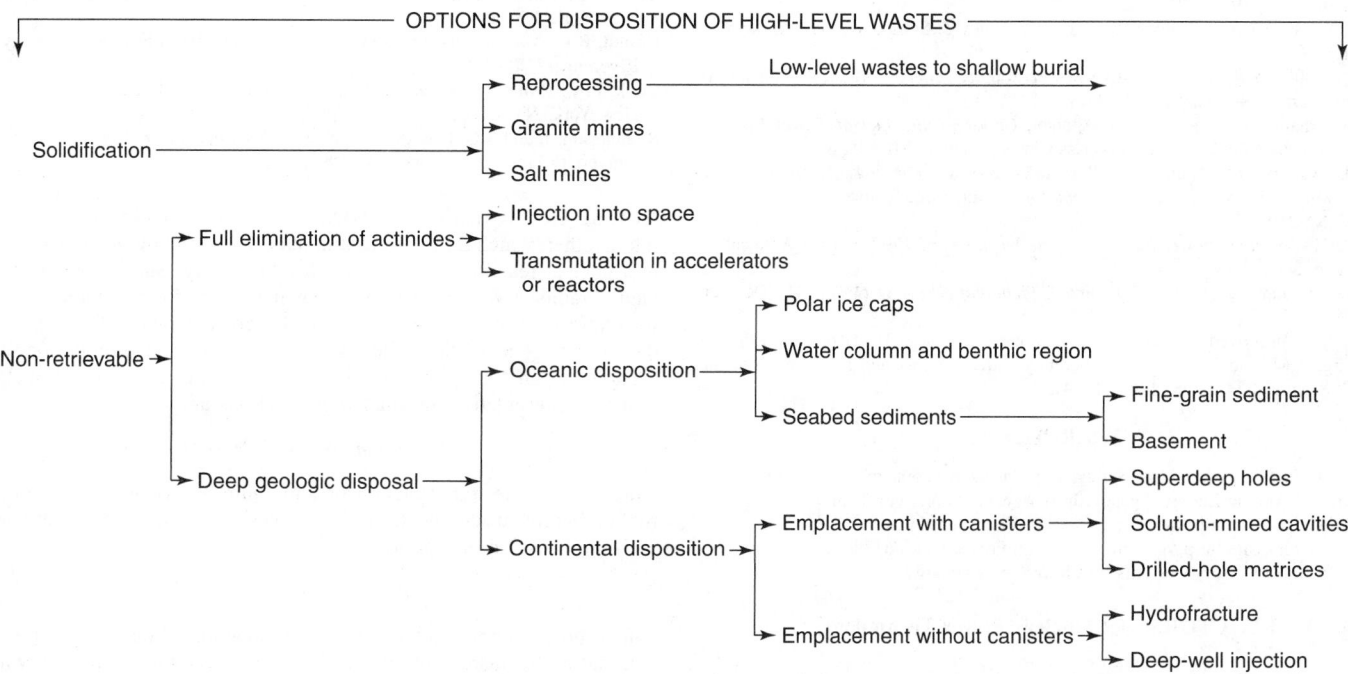

OPTIONS FOR DISPOSITION OF HIGH-LEVEL WASTES

Solidification
- Reprocessing → Low-level wastes to shallow burial
- Granite mines
- Salt mines

Non-retrievable
- Full elimination of actinides
  - Injection into space
  - Transmutation in accelerators or reactors
- Deep geologic disposal
  - Oceanic disposition
    - Polar ice caps
    - Water column and benthic region
    - Seabed sediments
      - Fine-grain sediment
      - Basement
  - Continental disposition
    - Emplacement with canisters
      - Superdeep holes
      - Solution-mined cavities
      - Drilled-hole matrices
    - Emplacement without canisters
      - Hydrofracture
      - Deep-well injection

Fig. 39. Panorama of options for consideration in the disposition of radioactive wastes. (*Top portion of chart*: Options for disposition of low- and intermediate-level and cladding wastes. *Bottom portion of chart*: Options for disposition of high-level wastes)

Additional Reading

Apostolakis, G.: "The Concept of Probability in Safety Assessments of Technological Systems," *Science*, 1359 (December 7, 1990).

Bodansky, D.: *Nuclear Energy: Principles, Practices, and Prospects,* American Institute of Physics, College Park, MD, 1996.

Bothwell, R.: *Nucleus. The History of Atomic Energy Limited,* Univ. of Toronto, Toronto, Canada, 1988.

Bruschi, H. and T. Andersen: "Turning the Key," *Nuclear Engineering International,* 1 (November 1991).

Cobb, C.E., Jr. and K. Kasmauski: "Living with Radiation," *National Geographic,* 403 (April 1989).

Cottrell, A.: "Safe Energy," *Review (University of Wales),* 38 (RefSeasonsAutumn 1987).

Davis, W.K.: "Problems and Prospects for Nuclear Power," *Chem. Eng. Progress,* **80**(6), 11–16 (June 1984).

Golay, M.W. and N.E. Todreas: "Advanced Light-Water Reactors," *Sci. Amer.,* 82 (April 1990).

Golay, M.W.: "Longer Life for Nuclear Plants," *Technology Review (MIT.),* 25 (May/June 1990).

Goldschmidt, B.: *Atomic Rivals,* Rutgers University Press, New Brunswick, NJ, 1990.

Green, S.J.: "Solving Chemical and Mechanical Problems of PWR Steam Generators," *Chem. Eng. Progress,* 31 (July 1987).

Hansen, K. et al.: "Making Nuclear Power Work: Lessons from Around the World," *Technology Rev. (MIT),* 30 (February 1989).

Hodgson, P.E.: *Nuclear Power, Energy and the Environment,* World Scientific Publishing Company, Inc., River Edge, NJ, 2000.

Jerome, F.: "Yo-Yo Journalism and Nuclear Power," *Technology Review (MIT),* 73 (April 1989).

Kairi, S.P.: "Outage Risk Management," *EPRI J.,* 34 (April/May 1992).

Kurdsunoaeglu, B., A. Perlmutter, and S.L. Mintz: *The Challenges to Nuclear Power in the Twenty-First Century,* Kluwer Academic Publishers, Norwell, MA, 2000.

Lester, R.K.: "Rethinking Nuclear Power," *Sci. Amer.,* 31 (March 1986).

Marshall, E.: "Counting on New Nukes," *Science,* 1024 (March 2, 1990).

Miller, P. and R.H. Ressmeyer: "A Comeback for Nuclear Power? Our Electric Future," *National Geographic,* 60 (August 1991).

Roberts, L.: "British Radiation Study Throws Experts into Tizzy," *Science,* 24 (April 6, 1990).

Shoup, R.L.: "International Waste Management Symposium," *Nuclear Safety,* **18**(4) (1977).

Shultis, J.K. and R.E. Faw: *Fundamentals of Nuclear Science and Engineering,* Marcel Dekker, Inc., New York, NY, 2002.

Skerret, P.J.: "Will the Public Say Yes to Nukes?" *Technology Review MIT,* 8 (April 1991).

Slovic, P., J.H. Flynn, and M. Layman: "Perceived Risk, Trust, and the Politics of Nuclear Waste," *Science,* 1603 (December 13, 1991).

Spinard, B.I.: "U.S. Nuclear Power in the Next Twenty Years," *Science,* 707 (December 12, 1988).

Stacey, W.M.: *Nuclear Reactor Physics,* John Wiley & Sons, Inc., New York, NY, 2001.

Staff: "AP600—A Cost Competitive Power Source," *Energy Digest,* 4 (Pittsburgh, Pennsylvania) (RefSeasonsFall 1991).

Staff: International Atomic Energy Agency, *Choosing the Nuclear Power Option: Factors to Be Considered,* Bernan Associates, Lanham, MD, 1998.

Staff: University Press Cambridge, *Resistance to New Technology: Nuclear Power, Information Technology and Biotechnology,* Cambridge University Press, New York, NY, 1997.

Staff: IAEA, *Nuclear Power Reactors in the World, April 1997,* Bernan Associates, Lanham, MD, 1997.

Suzuki, T.: "Japan's Nuclear Dilemma," *Technology Review (MIT),* 41 (October 1991).

Taylor, J.J.: "Improved and Safer Nuclear Power," *Science,* 318 (April 21, 1989).

Weinberg, A.M. and I. Spiewak: "Inherently Safe Reactors and a Second Nuclear Era," *Science,* **224**, 1398–1402 (1984).

Web References

History of Nuclear Power Plant Safety: http://users.owt.com/smsrpm/nksafe/
International Atomic Energy Agency: http://www.iaea.org/worldatom/
NRC Short History: http://www.nrc.gov/SECY/smj/shorthis.htm
Nuclear History Site: http://geocities.com/RainForest/Andes/6180/
U.S. Nuclear Regulatory Commission: http://www.nrc.gov/

NUCLEAR REACTOR. See **Nuclear Power Technology**.

NUCLEAR SPIN. The intrinsic angular momentum of the atomic nucleus due to rotation about its own axis. It is usually designated I and has the magnitude, $\sqrt{I(I+1)}h/2\pi \approx I(h/2\pi)$, where I is the nuclear spin quantum number which has different (integral or half-integral) values (including zero) for different nuclei. In spectroscopy, the nuclear spin is of importance for the explanation of the hyperfine structure, and of the intensity alternation in band spectra.

NUCLEAR STRUCTURE. The nucleus of an atom of atomic number Z and mass number A contains Z protons and $A - Z$ neutrons, bound together under the influence of shortrange nuclear forces much as molecules are bound together in a drop of liquid. The strength of binding may be determined by subtracting the actual mass of the atom from the mass of its constituent particles considered as free particles. The binding energy E_B is then related to this mass defect ΔM by Einstein's relation, $E_B = \Delta M c^2$, where c is the velocity of light. The precise value of E_B depends upon the nucleus concerned, and upon how many neutrons and protons it contains but it is of the order of 8 MeV per nucleon in most nuclei.

The binding energy determines whether the nucleus is stable or unstable. Among the lighter nuclei, the ones which are stable are those in which the number of protons is approximately equal to the number of neutrons, so that $A \approx 2Z$. In heavier stable nuclei, there is an excess of neutrons over protons owing to the repulsive electrostatic forces between the protons. Thus, the most stable oxygen nucleus is ^{16}O, containing 8 protons and 8 neutrons, while the most nearly stable uranium nucleus is ^{238}U, containing 92 protons and 146 neutrons. Nuclei containing a disproportionate number of neutrons tend to be unstable and decay radioactively by emission of electrons whereby neutrons are converted into protons; those containing an excess of protons similarly tend to decay by emission of positrons or by capture of orbital electrons.

It has been shown that the nucleus is approximately spherical in shape and of volume proportional approximately to its mass. It is, however, capable of executing oscillations about the spherical form, and in certain circumstances may even acquire a permanent deformation. The heaviest nuclei are unstable under deformation, as a result of which they undergo spontaneous fission. These properties may be described qualitatively by regarding the nucleus as an electrically charged drop of liquid possessing volume energy and surface tension.

Although the nucleus is normally found in its lowest energy state, it may be produced as the result of a nuclear reaction, or through radioactivity in a number of excited states whose detailed properties may differ quite markedly from the lowest state. If formed in an excited state, it will decay, normally by the emission of electromagnetic radiation (gamma rays) to the lowest state, or by the emission of particles to another nucleus.

Additional Reading

Berrios, M. *Nuclear Structure and Function,* Elsevier Science & Technology Books, New York, NY, 1997.

Casten, R.F.: *Nuclear Structure Physics,* World Scientific Publishing Company, Inc., Riveredge, NJ, 2001.

Cook, P.R.: *Principles of Nuclear Structure and Function,* John Wiley & Sons, Inc., New York, NY, 2001.

Ragnarsson, I. and S.G. Nilsson: *Shapes and Shells in Nuclear Structure,* Cambridge University Press, New York, NY, 1994.

NUCLEAR TRANSMUTATION. The transformation of one nuclide into another, which differs from it in nuclear charge, mass, or stability, i.e., in a nuclear reaction or a process of radioactivity. Such changes occur in natural radioactive processes, but the general need for a systematic notation for expressing them came only with the great number of transmutations discovered after particle accelerators provided high-energy ion beams capable of penetrating the Coulomb barrier of all stable atomic nuclei.

Two representative transmutation equations are

$$^{27}Al + n \longrightarrow \, ^{27}Mg + {}^1H$$

which shows the transmutation of aluminum atoms of mass number 27, by bombardment with neutrons, to magnesium atoms of mass number 27, with the emission of a proton, and

$$^9Be + \gamma \longrightarrow \, ^8Be + n$$

which shows the transmutation of beryllium atoms of mass number 9, under gamma-ray bombardment, to beryllium atoms of mass number 8, with the emission of a neutron.

The two reactions above may also be expressed in condensed from as

$$^{27}Al(n,\, p)^{27}Mg$$

and

$$^9\text{Be}(\gamma, n)^8\text{Be}.$$

NUCLEIC ACIDS. Nucleic acids are polymeric materials formed from nucleotides and essential to all organisms. Deoxyribonucleic acid (DNA), most often a double-helical biopolymer, encodes the genetic information contained in each cell. Ribonucleic acid (RNA) constitutes a more diverse class of biopolymers that are able to adopt both helical and other more complex tertiary structures. The structural diversity of RNAs enables these molecules to carry out a variety of intracellular functions, including transmitting the genetic message to the site of protein synthesis. See also **Proteins**. Both DNA and RNA interact with a host of other molecules, e.g., proteins, drugs, as well as other RNAs and DNAs. The specificity of these interactions is thought to be related to local sequence-dependent structural variation.

Development of techniques to synthesize oligonucleotides, i.e., short, well-defined sequences of DNA or RNA, has provided the opportunity to study nucleic acid structure in detail. In addition, oligonucleotides have proved invaluable in analytical procedures used in genetic engineering (qv), protein engineering (qv), affinity chromatography, and forensics, as well as in medicine. See also **Chromatography**; and **Forensic Chemistry**. The unique ability of nucleic acids to bind to self-complementary sequences has been exploited in the design of oligonucleotide probes and in antisense drug strategies.

DNA Structure

The structure of DNA is characterized by its primary sequence, secondary helical structure, and higher order structure or topology. The primary sequence of DNA refers to the atomic connectivities required to construct the polynucleotide chain. The helical conformation of these polynucleotide chains constitutes the secondary structure of DNA. Sequence-dependent structural diversity and flexibility are important DNA characteristics and play a crucial role in biological processes. The organization of helical DNA in topologically distinct three-dimensional conformations represents the higher order structure. Higher order structural features, in particular the supercoiling of DNA, are thought to have a profound influence on the dynamic processes and biology of nucleic acids within living cells.

The DNA double helix was first identified by Watson and Crick in 1953. Not only was the Watson-Crick model consistent with the known physical and chemical properties of DNA, but it also suggested how genetic information could be organized and replicated, thus providing a foundation for modern molecular biology.

The primary structure of DNA is based on repeating nucleotide units, where each nucleotide is made up of the sugar, i.e., 2'-deoxyribose, a phosphate, and a heterocyclic base, N. The most common DNA bases are the purines, adenine (A) and guanine (G), and the pyrimidines, thymine (T) and cytosine (C) (see Fig. 1). The base, N, is bound at the 1'-position of the ribose unit through a heterocyclic nitrogen.

The nucleotides are linked together via the phosphate groups, which connect the 5'-hydroxyl group of one nucleotide and the 3'-hydroxyl group of the next to form a polynucleotide chain (Fig. 1a). DNA is not a rigid or static molecule; rather, it can adopt a variety of helical motifs.

In A- or B-form DNA, two self-complementary polynucleotide strands associate with one another to form a right-handed double helix. The two polynucleotide chains are antiparallel.

In addition to A- and B-form DNA, several other helical conformations have been identified. Among these, the most well-studied is Z-DNA, a left-handed helix first characterized by x-ray crystallographic analysis of the oligonucleotides d(CGCGCG) and d(CGCG). Other alternating purine—pyrimidine sequences, in particular alternating CG sequences, have been shown to adopt the Z-conformation at high ionic strength.

RNA Structure

RNA has a variety of functions within a cell; for each function, a specific type of RNA is required. Messenger RNA (mRNA) serves as intermediaries for carrying genetic messages from the DNA to the ribosomes where protein synthesis takes place. Ribosomal RNA (rRNA) serves both structural and functional roles in the ribosome; it is diverse, both in terms of its size and structure. Transfer RNAs (tRNAs) are small molecules that have a central role in protein synthesis. Other RNA molecules, called ribozymes, function as enzymes to catalyze chemical transformations. Although ribozymes most often catalyze cleavage of the

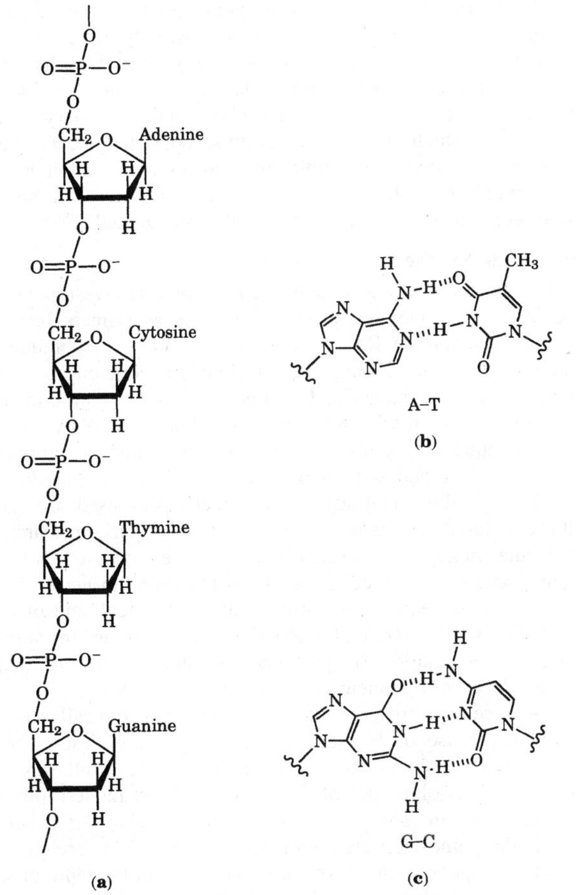

Fig. 1. Elements of DNA structure: (**a**) a deoxypolynucleotide chain, which reads d(ACTG) from $3' \rightarrow 5'$ or d(GTCA) from $3' \rightarrow 5'$; and (**b**) and (**c**) the Watson-Crick purine—pyrimidine base pairs, A–T and G–C, respectively, where—ξ represents attachment to the deoxyribose

RNA phosphodiester backbone, they have also been shown to participate in cleavage of DNA, replication of RNA, and reactions with phosphate monoesters. Other RNAs are associated with enzymes to form riboprotein complexes involved in many biological processes. The multifunctional character of RNA, particularly the involvement of RNA in enzymatic processes, has led to the hypothesis that life on earth evolved from RNA, and that RNA had both the genetic and catalytic functions commonly associated with DNA and proteins, respectively.

The primary structure of RNA is similar to that of DNA, but with a few notable exceptions. First, in RNA, instead of thymine, the pyrimidine base uracil (U) occurs, forming a complementary base pair with adenine in regions of double-stranded RNA. Also, a wide variety of ribonucleotides having modified or minor bases are found in naturally occurring RNA, one of the most common of which is pseudouridine. In human tRNAs, as many as 25% of the bases are nonstandard. Over 80 modified bases have been characterized in naturally occurring tRNA; although the role of base modification is not clear, it may be important for biological recognition.

The other important feature of the primary structure of RNA is the presence of the 2'-hydroxyl group in ribose. Although this hydroxyl group is never involved in phosphodiester linkages, it does impose restrictions on the helical conformations accessible to double-stranded RNA.

RNAs are single-stranded molecules that fold, allowing different regions of the ribonucleotide to form distinct secondary structural elements. When self-complementary regions of the RNA strand are aligned, duplex regions, which may have Watson-Crick base pairs, are formed. In contrast to DNA, double-stranded regions in RNA are much more likely to have unusual base-pairing between noncomplementary bases and to incorporate non-Watson-Crick base-pairing. Owing to the steric requirements of the 2'-hydroxyl group on the ribose sugar, these duplex regions are constrained to an A-form helix, i.e., a 3'-endo sugar conformation. Although double-stranded RNA has the general features of an A-form helix, actual duplex characteristics, such as rise per base pair, groove dimensions, and base pair displacement from the helical axis, may vary.

The functional diversity of RNA is directly related to its structural diversity. In contrast to DNA, RNA molecules are synthesized as single-stranded polynucleotides that fold to give complex tertiary structures. These structures, which incorporate hairpins, loops, bulges, and junctions between single-stranded and double-stranded regions, exhibit long-range interactions within the folded tertiary structure. Long-range intramolecular interactions serve to stabilize the three well-characterized RNA structures.

Oligonucleotide Synthesis

Synthetic oligonucleotides are widely used in scientific investigations. Most synthetic oligonucleotides are produced for use as primers in the polymerase chain reaction (PCR), a widely used analytical technique having commercial applications in diagnostic medicine, genetic engineering (qv), and forensics. See also **Forensic Chemistry**. A large volume of oligonucleotides are also synthesized for use as primers in DNA-sequencing. Demands for sequencing primers have increased rapidly to support large-scale DNA mapping and sequencing efforts such as the human genome project. Although the quantities of oligonucleotides used are relatively small, these materials are essential in many areas of basic research.

In molecular biology, synthetic oligonucleotides are used as linkers in gene-cloning and to introduce site-directed mutations in genes. Synthetic oligonucleotides are required for structural, biochemical, and biophysical studies of DNA and RNA. Oligonucleotides also are important for examining the association of proteins and small molecules, e.g., for intercalating drugs with nucleic acids on a molecular level.

The first procedures for oligonucleotide synthesis, typically carried out in solution, made use of *H*-phosphonate and phosphotriester chemistry. These approaches are useful in some large-scale syntheses and in syntheses of various oligonucleotide analogues. Most modern procedures, however, are based on solid-phase phosphoramidite chemistry. Automated oligonucleotide synthesizers are commercially available, as are the required reagents and phosphoramidites. Together these permit the rapid production of custom oligonucleotides and oligonucleotide analogues.

Modified Oligonucleotides. Much of the interest in modified oligonucleotides is related to use as antisense agents. Antisense agents are typically short (15–30 base pairs in length) oligonucleotides having sequences that are complementary to coding or regulatory regions within mRNA, although some antisense oligonucleotides have also been designed to target DNA. The antisense sequence recognizes and binds to a complementary sequence via the formation of a double-stranded duplex that have normal Watson-Crick base-pairing. Antisense oligonucleotides can inhibit gene expression at the translational level. The potential to design oligonucleotides having the ability to recognize and inhibit specific genes makes the antisense approach promising in the development of new therapeutic agents. In addition, antisense oligonucleotides can be used in research to elucidate gene function by providing a mechanism for regulating a gene artificially.

Although all natural antisense oligonucleotides are short RNA sequences, most of the synthetic antisense oligonucleotides are deoxyoligonucleotides. In the design of an effective antisense oligonucleotide, several factors must be considered. First, the oligonucleotide must be specific, binding with high affinity to a single sequence within the target RNA. A second consideration is stability within the cellular environment. Thus all unmodified oligonucleotides are degraded too rapidly to be used effectively as therapeutic agents. A significant research effort has been directed toward discovering chemical modifications that can increase the nuclease resistance of the oligonucleotide backbone.

An effective therapeutic agent must also have the ability to reach its target sequence *in vivo*. In order to enhance membrane transport, antisense oligonucleotides are frequently modified by covalent attachment of carrier molecules or lipophilic groups.

Antisense oligonucleotides are usually designed to inhibit gene expression by interfering with the translation of mRNA. One mechanism for this type of inhibition involves binding the oligonucleotide to the translation-initiation sequences of the mRNA, which prevents ribosome association and protein synthesis. Another potential mechanism involves hybrid formation at some other sequence within the mRNA, thus impeding translocation of the ribosome along the mRNA strand by steric blocking. These two mechanisms are based on blocking a sequence of RNA or DNA by double-stranded duplex formation using a specific antisense oligonucleotide.

A less specific mechanism based on the action of Rnase H, an enzyme catalyzing single-strand cleavage of RNA, may be predominate for unmodified, thioate and dithioate oligonucleotides. In the Rnase H mechanism, the duplex formed by the antisense oligonucleotide and the target RNA is a substrate for Rnase H. The enzyme cleaves the RNA at the complexes site rendering the RNA vulnerable to further degradation and inactivation by cellular exonucleases. The oligonucleotide, which is probably not a substrate for Rnase H, can target multiple copies of complementary RNA. Where applicable, antisense oligonucleotide action mediated by the Rnase H mechanism has been shown to be a potent inhibitor of gene expression.

Modified oligonucleotides can also be designed for binding to double-stranded DNA by forming a triple helix.

Oligonucleotides can also inhibit gene expression at the transcriptional level by binding to a single-stranded or open sequence of DNA. In this mechanism, the antigene oligonucleotide is designed to be complementary to a regulatory sequence preceding a gene. For example, an oligonucleotide complementary to the *lac* operator sequence (repressor protein binding site) has been found to inhibit specifically β-galactosidase synthesis in *E. coli*. Normally, expression of the lactose-metabolizing enzyme, β-galactosidase, is blocked at the transcriptional level by the repressor binding to the operator sequence. In the presence of a lactase metabolite, the repressor is converted to a nonbinding form and dissociates from the operator, which results in the transcription of the gene and β-galactosidase production. However, in the presence of the antisense oligonucleotide, the synthesis of β-galactosidase is inhibited. The antisense oligonucleotide can then act as a repressor by binding to an open or single-stranded region within the operator.

Although development of modified oligonucleotides as antisense and antigene agents is a principal focus of research in the 1990s, there are many other interesting applications that have greater immediate commercial significance. Included among these are applications using nucleic acid probes, which are oligonucleotides that have been modified by the attachment of a detectable chemical group. Probes can be designed to recognize RNA or DNA sequences characteristic of specific eukaryotic genes, viruses, or bacteria. Several analytical and diagnostic procedures have been developed based on the hybridization of the probe with its target sequence and the subsequent detection of the hybrid by the group attached to the oligonucleotide. Probes, particularly useful in automated sequencing protocols, may contain fluorescent groups, phosphors, radioactive tracers (qv), etc. In addition, probes can be designed to help elucidate the structure of biological molecules. DNA-binding molecules, including intercalators, alkylating agents, and photosensitive molecules, have also been linked to oligonucleotides as a way of directing a drug to a specific DNA sequence. These modifications often enhance binding as well.

Oligonucleotide Bioconjugates

Although many molecules, including proteins and small intercalating and groove-binding ligands, bind to RNA and DNA, only nucleic acids are able to bind with the high specificity required to recognize a single sequence within the 3×10^9 base-pairs of the human genome. The unique specificity of oligonucleotides can be exploited to direct a multitude of other chemical agents to a sequence of interest by attaching these agents to oligonucleotides through molecular linkers. Oligonucleotides labeled with fluorescent or other detectable groups provide nucleic acid probes that can be used to screen a large pool of DNA for a specific sequence. Cationic or lipophilic groups can be attached to improve the binding and bioavailability of antisense oligonucleotides. Bioconjugates have also been widely used in research because they enable scientists to learn more about the structure and function of nucleic acids and ligands that bind to them.

Several strategies have been devised to attach various chemical groups to oligonucleotides. Groups can be attached to the 3′- or 5′-terminus of the oligonucleotide, along the backbone through the phosphate or the 2′-hydroxy group of ribose, or to modified purines or pyrimidines.

See also **Adenosine**; **Adenosine Phosphates**; **Amino Acids**; **Human Genome Project (The)**; and **Proteins**.

JILL REHMANN
Fordham University

Additional Reading

Caruthers, M. H. and co-workers: *Methods in Enzymology*, Vol. 154, Academic Press, Inc., New York, NY, 1987, pp. 287–313; Vol. 211, 1992, pp. 3–20.

Gesteland, R. F. and J. F. Atkins, eds.: *The RNA World*, Cold Spring Harbor Laboratory Press, Cold Spring Harbor, New York, NY, 1993.

Lodish, H., P. Matsudaira, and A. Berk: *Molecular Cell Biology,* 5th Edition, W. H. Freeman and Company, New York, NY, 2003.

Moldave, K.: *Progress in Nucleic Acid Research and Molecular Biology: Subject Index Volume (40–72),* Elsevier Science & Technology Books, New York, NY, 2003.

Neidle, S.: *Nucleic Acid Structure and Recognition,* Oxford University Press, New York, NY, 2002.

Ronot, X. and Y. Usson: *Imaging of Nucleic Acids and Quantification in Phototonic Microscopy,* CRC Press LLC., Boca Raton, FL, 2001.

Rosemeyer, H. and M. Volkan Kisakurek: *Perspectives in Nucleoside and Nucleic Acid Chemistry,* John Wiley & Sons, Inc., New York, NY, 2000.

Saenger, W.: *Principles of Nucleic Acid Structure,* Springer-Verlag, New York, NY, 1984.

NUCLEONICS. The applications of nuclear science in physics, chemistry, biology, and other sciences, including military science, and in industry, and the techniques associated with these applications.

NUCLEONS. Two nuclear particles, the proton and the neutron, and their antiparticles are known as *nucleons.* The rest mass of the proton is 1.0076 amu; that of the neutron, 1.0089 amu. The antiproton bears the same relation to the proton that the electron does to the positron, i.e., its charge is equal and opposite and its mass is the same, the charge being equal in magnitude to the electronic charge. Protons and antiprotons also annihilate each other when they collide, the reaction of a single pair producing positive and negative pions or kaons. If the proton and antiproton do not collide, but experience a "near miss," then an exchange of charge can occur, resulting in the formation of a neutron-anti-neutron pair. The four nucleon particles are fermions and have a spin angular momentum quantum number of $\frac{1}{2}$. See also **Neutron**; **Particles (Subatomic)**; and **Proton**.

NUCLEOPHILE. An ion or molecule that donates a pair of electrons to an atomic nucleus to form a covalent bond. The nucleus that accepts the electrons is called an *electrophile.* This occurs, for example, in the formation of acids and bases according to the Lewis concept, as well as in covalent bonding in organic compounds.

NUCLEOPHILIC REACTION. A reaction in which a nucleophilic reagent attacks an electrophilic compound. The reagent is taken to be the inorganic substance (in the case of reactions of inorganic and organic substances) or the simpler of two reacting organic compounds. The electron pair for the bond formed is furnished by the nucleophilic reagent.

NUCLEOPROTEINS AND NUCLEIC ACIDS. Nucleic acids are compounds in which phosphoric acid is combined with carbohydrates and with bases derived from purine and pyrimidine. Nucleoproteins are conjugated proteins consisting of a protein moiety and a nucleic acid. Originally, nucleoproteins were thought to occur only in the nuclei of cells, but it was later established that they are far more widely distributed, being found in cells of all types, animal and plant. They are found in the chromosomes, in the genes, in viruses, and bacteriophages.

The protein portion of the nucleoproteins is basic in nature and being complex in structure may form several types of linkage, depending upon the type of nucleic acid. In gastric digestion or hydrolysis with weak acid, nucleoproteins yield protein and nuclein. The latter in pancreatic digestion or hydrolysis with weak alkali yields additional protein and nucleic acid. See also **Nucleic Acids**.

NUCLEOSIDE. A compound of importance in physiological and medical research, obtained during partial decomposition (hydrolysis) of nucleic acids and containing a purine or pyrimidine base linked to either *d*-ribose, forming ribosides, or *d*-deoxyribose, forming deoxyribosides. They are nucleotides minus the phosphorus group. See also **Adenosine**; and **Nucleic Acids**.

NUCLEUS. 1. The nucleus of an atom is the positively charged core, with which is associated practically the entire mass of the atom, but only a minute part of its volume.

2. The nucleus of a molecule is a group of atoms connected by valence bonds so that the atoms and their bonds form a ring or closed structure, which persists as a unit through a series of chemical changes.

3. A group of cell bodies in the central nervous system of vertebrates. Examples of such groups are the red nucleus in the midbrain, through which impulses are routed for the control of subconscious muscular movements,

and Deiter's nucleus, lying at the junction of the medulla with the hindbrain. Through this center impulses pass for muscular action involved in the maintenance of equilibrium.

4. A somewhat spherical or oblong body in most living cells. This nucleus contains the chromosomes, which, in turn, bear the genes of heredity. The nucleus also contains a nucleolus, or sometimes two or more nucleoli and a basic ground substance, the nucleoplasm. A nuclear membrane surrounds it on the outside, but this membrane is very porous, allowing materials to pass through rather freely.

5. Condensation nuclei take part in phase changes, as in the formation of clouds; in seeding concentrated solutions to bring about crystallization, precipitation, etc.

NUCLIDES. See **Chemical Elements**.

NUTRIENTS (Soil). See **Fertilizer**.

NYLON. [CAS: 63428-83-1]. $(C_6H_{11}NO)_n$. Generic name for a family of polyamide polymers characterized by the presence of the amide group $-CONH$. By far the most important are nylon 66 (75% of U.S. consumption) and nylon 6 (25% of U.S. consumption). Except for slight difference in melting points, the properties of the two forms are almost identical, though their chemical derivations are quite different. Other types are nylons 4, 9, 11, and 12.

The first nylon developed (type 6/6) was discovered in 1938 by W. H. Carothers. Since that time, nylons have filled an important role for industry and the consumer in various formulations, shapes, and forms, e.g., oriented fibers, which are subsequently processed into fabrics, fishing line, and other monofilament uses; injection-molded nylons, used as bearings, gears, and other parts subjected to wear and impact; extruded nylon tubing and hose, used in large quantity because of its chemical inertness, high strength, and flexibility; oriented nylon strip used as strapping for packaging, displacing traditional steel strapping; and heavy cast-nylon parts, frequently used in the textile, paper-making, and bottle-handling fields.

Most nylons exhibit a combination of high melting point, high strength, impact resistance, wear resistance, chemical inertness, and a low coefficient of friction.

Types of Nylon. Type 6/6 and type 6 nylons are widely used, dominating the field of textile fibers. Nylon 6/10, a lower-strength material produced in less volume, is used for industrial applications requiring improved moisture stability and high dielectric strength. It also has a lower melting point, lower specific gravity, and higher cost than types 6/6 and 6. Nylons 11 and 12 appeared considerably later than the other formulations. Generally, these nylons have a lower order of moisture absorption and are thus preferred where consistent properties are required in the presence of moisture. They are also more chemically inert, flexible, and in certain cases, are transparent.

Formulations. Types 6/6 and 6/10 are formed by the condensation of diamines with dibasic organic acids into linear chains containing amide groups. Types 6, 11, and 12 are self-condensed amino acids.

Type 6/6:

$$NH_2(CH_2)_6 NH_2 + HOOC(CH_2)_4COOH \longrightarrow$$

Hexamethylenediamine   Adipic acid

$$[NH(CH_2)_6NHCO(CH_2)_2CO]_n + H_2O$$

Polyhexamethyleneadipamide

Type 6/10:

$$NH_2(CH_2)_6 NH_2 + HOOC(CH_2)_8COOH \longrightarrow$$

Hexamethylenediamine   Sebacic acid

$$[NH(CH_2)_6NHCO(CH_2)_8CO]_n + H_2O$$

Polyhexamethylenesebacamide

Type 6:

$$NH(CH_2)_5CO \longrightarrow [NH(CH_2)_5CO]_n$$

ε−Caprolactam   Polycaprolactam

Type 11:

$$NH_2(CH_2)_{10}COOH \longrightarrow [NH(CH_2)_{10}CO]_n + H_2O$$

Aminoundecanoic acid   Polyaminoundecanamide

Type 12:

$$NH(CH_2)_{11}CO \longrightarrow [NH(CH_2)_{11}CO]_n$$

Laurolactam   Polydodecanolactam

In addition to the basic nylons, a variety of copolymers can be manufactured, some of which are commercially available. Nylons and nylon copolymers can be blended to form alloys with specific customized properties.

Nylons can be modified by the addition of certain plasticizers, fillers, reinforcements, and stabilizers. Ordinarily, nylons used for injection molding, such as type 6/6, have relatively low molecular weights (on the order of 15,000 to 20,000). High molecular weights are available to provide higher melt viscosity for nylon resins which are to be extruded into tubing or shapes. The molecular weight of nylon generally is determined by the ASTM relative-viscosity test.

Nylon resins usually are supplied in the form of cylindrical or rectangular diced pellets. Most commercial nylon molding. resins are nontoxic. If a large amount of residual monomer is retained in the resin, as can occur with certain unextracted formulas, the material should not be in prolonged contact with food because of the possibility of monomer leaching.

Nylons require modification or stabilization to improve their resistance to certain environmental effects. Unstabilized nylon is degraded by ultraviolet light. The most widely used stabilizer has been approximately 2% well-dispersed carbon black, which has proved effective in the absorption of ultraviolet light. The nylons are considered adequate for outdoor applications if they are not exposed to direct sunlight.

See also **Caprolactam** and **Fibers**.

O

OCCUPATIONAL SAFETY AND HEALTH ADMINISTRATION (OSHA).

A federal agency (established on December 29, 1970) responsible for establishing and enforcing standards for exposure of workers to harmful materials in industrial atmospheres, and other matters affecting the health and well-being of industrial personnel.

Web Reference

U.S. Department of Labor: Occupational Safety & Health Administration: http://www.osha.gov/

OCEAN RESOURCES (Mineral).

Since antiquity, the oceans and seas have been a major source of salt (sodium chloride) and continue to be so. Today, solar sea salt is produced in about 60 countries. The People's Republic of China, Australia, Mexico, India, Brazil, the Bahamas, Spain, and France are among the leading producers. At present, about 38% of the sodium chloride produced is evaporated from seawater. The value is estimated at over $400 million per year. Solar salt is very important to many countries, such as Japan, where there are few or no salt deposits. For several decades, seawater has been a significant source for bromine (production commenced by DuPont in 1931); iodine (from kelp, once very important, but no longer an economic source); magnesium (production started by Dow in 1941); potassium; sulfur; and several other elements and their compounds. Today, over 13% of the requirements for bromine come from seawater, as do over 70% of magnesium metal and 33% of magnesium compounds required by industry. Sulfur, associated with the cap rock of salt domes, has been produced from two salt domes just off the coast of Louisiana for many years. For a few decades, the continental shelves under the oceans have been producing large volumes of natural gas and petroleum. And also, for a few decades, the oceans have provided fresh water for many regions through various desalination processes. There are over 500 desalination plants in operation or under construction, with plants in arid and semiarid locations, such as the Middle East, but also in some highly urbanized regions and cities where fresh water is in short supply, such as in Italy and the Netherlands.

Many of the ore deposits found on the continents are the result of ancient oceans. Tin is found in offshore deposits, such as in Indonesia, in Cornwall (Saint Ives Bay), and Phuket Island off the west coast of the Malay Peninsula.

Within the past 20–30 years, much interest has been shown in manganese nodules on the seafloor in various locations. More recently, the discovery of hot brines in the Red Sea, "black smokers" on the East Pacific Rise and suspected in many other locations, and ophiolites has excited the scientific community and attracted industrialists because these phenomena are associated with metals, such as cadmium, copper, nickel, and zinc. These findings have largely resulted from the funding provided for geological and oceanographic research as part of the Deep Sea Drilling Project (DSDP) and the International Decade of Ocean Exploration (IDOE), projects which have been in place since the late 1960s.

More details pertaining to most of these ocean raw materials will be found in a number of specific entries in this encyclopedia. See also **Bromine**; **Chemical Elements**; **Desalination**; **Magnesium**; **Manganese**; **Natural Gas**; **Petroleum**; **Sodium Chloride**; and **Sulfur**.

The resource potential of the oceans awaits further technological development. It is interesting to note that the famous German chemist, Fritz Haber, spent more than eight years after World War I in attempts to recover gold from seawater in order to pay the German war debt. The results were disappointing, but large quantities of gold are in very large quantities of seawater. Currently, there is considerable interest in attempts to recover uranium from seawater, particularly by nations with no assured supply. Should fusion power come to fruition, after a few years, the ocean may be looked to as a source of lithium. Beach sands also have received considerable attention in recent years as sources of metals and other materials. Marine beaches may contain gold, silver, platinum, and diamonds in addition to magnetite, cassiterite, chromite, columbite, ilmenite, rutile, scheelite, zircon, monazite, and wolframite. Heavy-mineral beach sands are usually commercially worked for the titanium content of the rutile and the ilmenite. The same sands may also be processed to recover thorium from monazite and zircon for use in foundry sands. Currently, marine beaches are mined for heavy mineral production in Australia, Brazil, India, Madagascar, Mozambique, Sierra Leone, South Africa, and Sri Lanka, among other countries.

Diamonds are found in the seafloor sediments on the coast of the Kalahari Desert in southwest Africa.* The origin of the diamonds is obscure, but it is generally believed that basaltic and kimberlite pipes exist on the ocean floor as on the nearby land. There is a relative abundance of gemstones in the marine deposits and a few large stones have been recovered. Dredging began in 1961, using suction dredges capable of operating in waters to depths of 50 meters. Because of rough seas on this exposed coast, a number of barges were lost and the operation was concluded. However, in recent years a subsidiary of DeBeers is using a dredge protected by a seawall, thus permitting mining offshore about 120 meters at depths of 90 meters.

Calcium carbonate often precipitates from tropical or subtropical waters when the water becomes supersaturated due to enrichment of the carbonate content by intense biological photosynthesis and by solar heating of carbon dioxide-rich cooler waters. The aragonite precipitates as single needles in the shallow waters at a rate of about one millimeter of wet sediment per year. Continuing deposition leads to cementation and the formation of successive concentric sheaths known as oöids. The most extensively studied oölithic aragonite deposit is that distributed over the 250,000-square-kilometer (96,525-square-mile) Great Bahama Bank on the continental shelf near islands of the Bahamas. Most of the areas are less than 5 meters deep and are composed of quite pure calcium carbonate containing higher levels of strontium and uranium than are found in limestones of biological origin. Similar deposits occur in the Gulf of Batabanó (Cuba) and in the Mediterranean Sea off Egypt and Tunisia, as well as on the Trucial coast of the Persian Gulf.

Iron is a common constituent of marine sediments. Magnetite is found in beach sands and iron is common in glauconitic marine silicates. Iron oxides and sulfides occur where anaerobic conditions and elevated temperatures are found, as in the hot, salty brines found near rifts. Iron is a major constituent of the ferromanganese nodules.

Magnetite-rich iron sands have been dredged from the ocean floor just off Kagoshima Bay (Japan) in water averaging from 15 to 40 meters in depth. Iron sand concentrates were produced in Japan as recently as 1976, although a major marine iron sand operation in Kyushu ceased operation in 1966.

Marine sand and gravel for fill and for aggregate have been produced on all coasts of the United States, particularly from San Francisco and San Pedro Bays in California and from Long Island Sound. Marine sand and gravel are found in significant quantities in the United Kingdom.

Phosphorites (marine apatites) are dense, light-brown-to-black concretions, ranging in size from sands to nodules and irregular masses. Phosphorites have been found off Argentina, Chile, Japan, Mexico, Peru, South Africa, and Spain, and several islands in the Indian Ocean. Some also have been found off the west coast of North America and on the eastern North American continental shelf. These deposits occur where water upwelling transports phosphorus and where the rate of sedimentation is slow. The

* Acknowledgement of assistance obtained from W.F. McIlhenny, The Dow Chemical Company, Freeport, Texas in preparation of several of the following paragraphs is hereby made.

nodules are usually found as a monolayer on the surface. The mineralogy of the marine phosphorites is similar to western U.S. land deposits, which were almost certainly marine in origin. Phosphorites are quite constant in composition, containing 45–47% calcium oxide and 29–30% phosphorus trioxide. Seawater is generally saturated with tricalcium phosphate so that, under the oxidative conditions normally present, the phosphates precipitate in colloidal form and accrete to existing surfaces, rather than forming a phosphorite suspension. Although most of the phosphorite is believed to have formed during the Miocene epoch, it is believed that precipitation is currently taking place. The largest known seafloor phosphorite deposit is off the coast of California from Point Reyes to the Gulf of California along the inner edge of the continental shelf. Additional deposits have been found on the edges of the Blake Plateau east of Florida. A recovery project was commenced in 1962–1963, but failed to materialize.

Glauconite or green sands (a hydrated silicate with potassium, iron, and aluminum as cations) is widely distributed on the ocean floor in both ancient and more recent marine sediments. Glauconite is often found with phosphorite and occurs on the tops of banks, submerged hillcrests, and on slopes in water from 50 to 2000 meters in depth. Glauconites are known off the coasts of Africa, Australia, China, Japan, Portugal, South America, the United Kingdom (Scotland), the United States (California and the Atlantic shelf), and New Zealand. A 130-square-kilometer (50-square-mile) deposit has been identified on the Santa Monica shelf off California.

Submarine Hydrothermal Deposits

Discovery of the East Pacific Rise hot springs has created extensive interest and plans are underway to commence a four-year, multi-institutional project to explore the East Pacific Rise for additional areas of hot spring activity and ore deposition. The major objective of the program will be to examine the nature of hydrothermal processes along the mid-ocean ridge system from the slow-spreading to the very fast-spreading segments, such as at 10–30° South. The project will involve the use of surface ships, deeply towed instrument packages, new high-precision multibeam echo sounding for making highly accurate topographic maps of the seafloor, and ultimately manned submersibles, such as *Alvin*.

The knowledge of submarine hydrothermal deposits was advanced by a large measure in 1979 when the hot springs on the East Pacific Rise at 21° north were discovered. Unlike the warm springs discovered on the Galápagos Spreading Center a few years ago, the springs on the East Pacific Rise are hot, with water venting at temperatures as high as 350°C and at velocities of several meters per second. These formations are precipitating large quantities of sulfide ore and minerals rich in copper, zinc, and iron. The precipitates form chimneys around the individual vents that spout black or white smoke composed of precipitated crystals of sulfides and other minerals. The discovery is the most exciting and significant in this field since the discovery of the Red Sea hot brines and metal deposits (Mottl, 1980).

By *hydrothermal* is meant hot water. When deposits are formed by chemical precipitation from hot solutions, they are termed hydrothermal. Hydrothermal deposits on land represent a very important class of economically retrievable ore deposits and provide a significant percentage of various metals, such as copper, zinc, lead, silver, gold, tin, molybdenum, among others. Mottl (1980) suggests that five factors are involved in forming hydrothermal ore deposits: (1) a source of the ore metals; (2) a source of water that dissolves and later precipitates the metals, concentrating them during the total cycle; (3) a source of heat; (4) a pathway between the site where metals are dissolved and precipitated and the site where they are finally deposited which is permeable and permits solution flow; and (5) the ultimate collection or deposition site. For preservation, it is also important that ores be deposited in places where they will not be eroded away, as by weathering. Because so many factors are involved, there is a wide variety of hydrothermal ore deposits.

In terms of submarine hydrothermal ore deposits, the source of heat is the thermal energy associated with the formation of new oceanic lithosphere along the mid-ocean ridge system, where the seafloor is spreading apart and basaltic magma wells up. Because of tensional forces present, the newly formed crust becomes fractured, allowing seawater to percolate down through the fractures. During this percolation, the seawater is heated by contact with hot rock and commences to react with the rock, leaching metals that may be present. Because of the lesser density of the seawater (due to temperature), it rises and ascends to the seafloor and exists at submarine hot springs. Because there are several factors involved, there is,

as on land, a wide variety of submarine hydrothermal ore deposits. Much remains to be understood and to confirm some of the early postulates, as given above, pertaining to the actual formation of submarine deposits.

For example, the concentrations of ore metals in most natural waters are quite low, particularly so in "normal" seawater. Measuring these low concentrations has been a problem of marine chemistry for many years. It is interesting to note that when artificial seawater, made up from pure reagent chemicals, is exposed to metallic elements, the ultimate solutions produced will contain from 100 to 1000 times the concentrations of these metals as compared with natural seawater.

The first submarine hot springs discovered along a mid-ocean ridge, those at the Galápagos Spreading Center, were emitting water at only 20°C, but the chemistry of this water indicated that it had reacted with basalt at 350–400°C. Then came the discovery of the 350° springs on the East Pacific Rise. Currently, the chemistry of this water is being studied at the Massachusetts Institute of Technology. To date, no submarine hydrothermal deposit has been sufficiently studied that all components contributing to its formation are known. Nevertheless, data at hand as of the early 1980s suggest some intriguing relationships among known deposits along mid-ocean ridges and point out the importance of special situations in producing and preserving large deposits.

Offshore Oil and Gas Resources

Although oil and gas exploration and production activities which occur offshore involve an extension of continents (the continental margins), they are nevertheless considered more in the general terms of oceanological rather than continental resources (the latter generally considered land or above-sea-level resources). Various aspects of offshore oil and gas production are described briefly in the entries on **Natural Gas**; and **Petroleum**.

As we go into the next century, because of an apparent glut of petroleum on world markets and because of much greater optimism pertaining to the ultimate natural gas reserves in the world, the emphasis on the exploration for petroleum and natural gas in underwater locations has markedly diminished.

Manganese Nodules

Deep-sea nodules, comprised mainly of manganese and iron oxides, have been found in abundance over large areas of the deep ocean floor that have been examined to date. In some locations, the nodules have been found to contain generous proportions of nickel, copper, cobalt, molybdenum, and vanadium, as well as manganese and iron. From the standpoint of potential commercial exploitation, a deposit is not considered promising unless the nickel-copper content is about 1.8% (weight) or greater. On this basis, one of the most promising areas is the Clarion and Clipperton fracture zone, an immense area some 4400 kilometers (2730 miles) long and 900 kilometers (560 miles) wide at its widest point. This zone is located southeast of Hawaii and southwest of Baja California. See Fig. 1.

An investigation of the origin and distribution of manganese nodules and the processes by which they selectively concentrate copper, nickel, and other metals was one of the first major projects under the sea beds Assessment project of the IDOE program. At a workshop attended by over a hundred scientists from various countries, the most likely locations

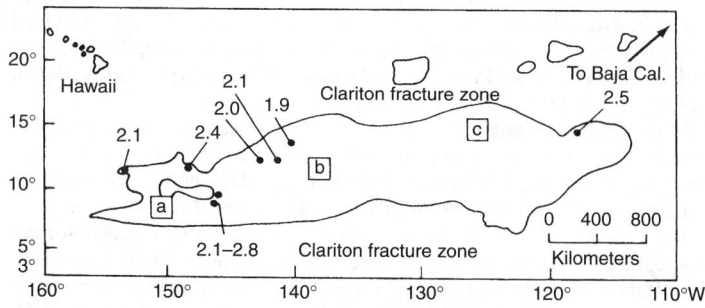

Fig. 1. Regions where manganese nodules containing more than 1.8% nickel-copper occur in the northeastern equatorial Pacific Ocean. Numbers indicate average percent of nickel-copper in one-degree squares. Areas a, b, and c indicate locations of activity carried out as part of Deep Ocean Mining Environmental Studies Program. (*After McKelvey, U.S. Geological Survey*)

for the exploration and study of manganese nodules were selected. The north central Pacific was identified as the zone where the nodules have the highest metal content. A team of American scientific investigators proposed that a comprehensive field and laboratory program be initiated to relate the high metal content to the local geological conditions. Data gathering was concerned along a transect that both academic and industrial scientists agreed could serve as a potential mining site. In addition to dredge sampling and piston cores, bathymetric measurements, sidescan sonar, and high-resolution television pictures were obtained. See Fig. 2. The results provided a broad-scale picture of the conditions under which nodules form, but the mechanisms for concentrating specific metals are still not well defined.

Although the nodules vary widely in their composition over the world oceans, metals are concentrated in three distinct types. One type comprises the nickel-copper-rich nodules of the Clarion-Clipperton variety, which is mainly formed in the equatorial regions. Another type, high in cobalt (1% or more) and low in nickel and copper, appears to be most commonly formed on sea mounts. The third type is high in manganese (35% or more), but low in other metals; it is known mainly on the eastern side of the Pacific Basin. As of the early 1980s, the most economically attractive were the cobalt-rich nodules.

The nodules form in a layered structure around a nucleus, which may be almost any material on the ocean floor. Most deep-sea nodules tend to be spherical or oblate in form. Nodules may occur up to 25 centimeters in diameter, but they average about 5 centimeters. The deposits may occur as slabs or agglomerates, or as incrustations on rocks or as pavement in some areas. The nodules are disorderly crystalline materials with layers of MnO_2 (mixed Mn^{2+}-Mn^{4+} oxides) alternating with $Mn(OH)_2$ and $Fe(OH)_3$. Excess iron appears as a mixture of goethite ($Fe_2O_3 \cdot H_2O$) and lepidocrocite.

The nodules are formed by the oxidation and precipitation of iron and manganese. The oxidation of Mn^{2+} is catalyzed by a reaction surface to a tetravalent state that absorbs additional Fe^{2+} or Mn^{2+} which, in turn, becomes oxidized. A surface is required and the initial deposition may be of iron oxide, possibly from volcanic or geothermal sources. Proper conditions of pH, redox potential, and metal ion concentration are found in deep ocean waters. The rate of accumulation appears to be very slow. The growth also may be discontinuous, and is estimated at a faster rater rate near the continental margins.

Precious Coral

Only a few species of coral have a combination of beauty, hardness, and luster, such the black coral species of Hawaii (*Antipathes dichotoma* and *Antipathes grandis*). These are highly valued by the jewelry trade. There are also a few red, pink, gold, and bamboo varieties that are in demand. Black coral also occurs in the Gulf of California and in the Pacific Ocean off Baja California, plus a few scattered locations in the Pacific Ocean east of Australia and north of New Zealand. Traditional sources of red and pink corals have been the Mediterranean Sea, various locations in the western

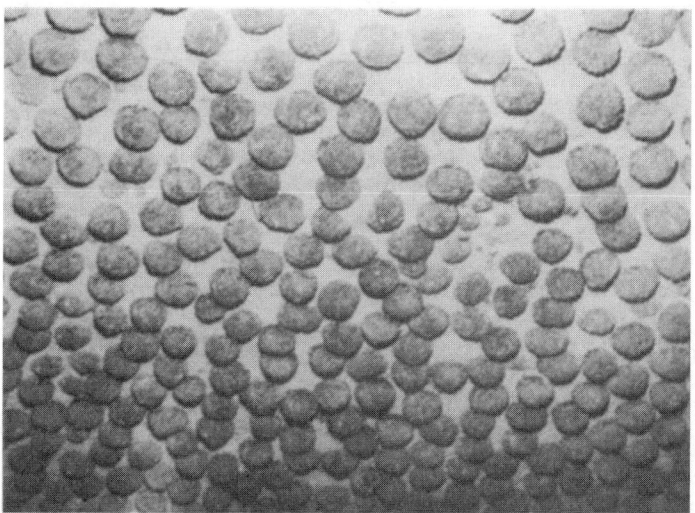

Fig. 2. Type of manganese nodules found in north central Pacific Ocean zone. (*Woods Hole Oceanographic Institution*)

Pacific Ocean, ranging from the Philippines, the Ryuku and Bonin islands and south of Japan. There is also a string of precious red and pink coral beds northwest of Hawaii and off the Cape Verde islands in the Atlantic Ocean off west Africa. A submersible vessel, the *Star II*, operated by Maui Divers of Hawaii, Ltd., is used to harvest pink coral (*Corallium secundum*) from the Makapuu bed. State regulations permit the collection of only 4400 pounds (1996 kilograms) within a 2-year period.

Additional Reading

Amsbaugh, J.K. and J.L. Van der Voort: "The Ocean Mining Industry: A Benefit for Every Risk?" *Oceanus*, 22–27 (Fall 1982).

Andreae, M.O. and H. Raemdonck: "Dimethyl Sulfide in the Surface Ocean and the Marine Atmosphere: A Global View," *Science*, **221**, 744–747 (1983).

Burroughs, T.: "Ocean Mining," *Technology Review (MIT)*, **87**(3), 54–60 (1984).

Clark, J.P.: "The Rebuttal: The Nodules are Not Essential," *Oceanus*, 18–21 (Fall 1982).

Cooke, R.: "Metals in the Sea," *Technology Review (MIT)*, **87**(3), 61–65 (1984).

Cronan, D.S.: *Underwater Minerals,* Academic Press, Inc., San Diego, CA, 1980.

Curtis, C.: "The Environmental Aspects of Deep Ocean Mining," *Oceanus*, 31–36 (Fall 1982).

Dordrecht, Y.: *Transfer of Technology for Deep Sea-Bed Mining: The 1982 Law of the Sea Convention and Beyond,* Kluwer Academic Publishers, Norwell, MI, 1995.

Fitzgerald, W.F., Gill, G.A., and J.P. Kim: "An Equatorial Pacific Ocean Source of Atmospheric Mercury," *Science*, **224**, 597–599 (1984).

Heath, G.R.: "Manganese Nodules: Unanswered Questions," *Oceanus*, 37–41 (Fall 1982).

Johnson, K.S.: "In situ Measurements of Chemical Distributions in A Deep-Sea Hydrothermal Vent Field," *Science*, **231**, 1139–1141 (1986).

Knecht, R.W.: "Introduction: Deep Ocean Mining," *Oceanus*, 3–11 (Fall 1982).

Koski, R.A., et al.: "Metal Sulfide Deposits on the Juan de Fuca Ridge," *Oceanus*, 42–46 (Fall 1982).

MacLeish, W.H.: *The Struggle for Georges Bank,* Atlantic Monthly Press, New York, NY, 1985.

Manheim, F.T.: "Marine Cobalt Resources," *Science*, **232**, 600–608 (1986).

Moore, J.G. and G.W. Moore: "Deposit from a Giant Wave on the Island of Lanai, Hawaii," *Science*, **226**, 1312–1315 (1984).

Mortlock, R.A. and P.N. Froelich: "Hydrothermal Germanium Over the Southern East Pacific Rise," *Science*, **231**, 43–45 (1986).

Mottl, M., Holland, H.D., and R.F. Corr: "Chemical Exchange During Hydrothermal Alteration of Basalt by Seawater," *Geochim. Cosmochim. Acta*, **43**, 869–884 (1980).

Mottl, M.J.: "Submarine Hydrothermal Ore Deposits," *Oceanus*, **23**, 2, 18–27 (1980).

Pendley, W.P.: "The Argument: The U.S. Will Need Seabed Minerals," *Oceanus*, 12–17 (Fall 1982).

Post, A.: *Deepsea Mining and the Law of the Sea,* Kluwer Academic Publishers, Norwell, MI, 1983.

Riggs, S.R.: "Paleoceanographic Model of Neogene Phosphorite Deposition, U.S. Atlantic Continental Margin," *Science*, **223**, 123–131 (1984).

Rona, P.A.: "Mineral Deposits from Sea-Floor Hot Springs," *Sci. Amer.*, 84–92 (January 1986).

Siegel, M.C. and S. Turner: "Crystalline Todorokite Associated with Biogenic Debris in Manganese Nodules," *Science*, **219**, 172–174 (1983).

Vogt, P. and B. Tucholke: *The Western North Atlantic Region,* Geological Society of America, Inc., Boulder, CO, 1986.

Web References

Scripps Institution of Oceanography: http://www.sio.ucsd.edu/

Scripps Research Oceanography: http://www.sio.ucsd.edu/research/oceanography. html

The National Oceanic & Atmospheric Administration (NOAA): http://www.noaa. gov/

Woods Hole Oceanographic Institution (WHOI): http://www.whoi.edu/science/ science.html

Woods Hole Department of Marine Chemistry and Geochemistry: http://www.whoi. edu/science/MCG/dept/

Woods Hole Department of Geology & Geophysics: http://www.whoi.edu/science/ GG/dept/

OCEAN THERMAL ENERGY CONVERSION (OTEC). Utilization of ocean temperature differentials between solar-heated surface water and cold deep water as a source of electric power. In tropical areas such differences amount to 35–40°F. A pilot installation now operating near Hawaii utilizes a closed ammonia cycle as a working fluid, highly efficient titanium heat exchangers, and a polyethylene pipe 2000 feet long and 22 inches inside diameter to handle the huge volume of cold water required. Alternate uses for such a system, such as electrolysis of water, ammonia production, and desalination, are envisaged. There has been active interest in the possibilities of this energy source in France from the time of d'Arsonval (1881) that continues, especially in Japan and Hawaii. Ongoing research indicates that OTEC may be harder to commercialize than once projected.

Web References

Hawaii OTEC: http://www.hawaii.gov/dbedt/ert/otec_hi.html

NREL: Ocean Thermal Energy Conversion: http://www.nrel.gov/otec/what.html

The Australian Renewable Energy Site: http://acre.murdoch.edu.au/ago/ocean/ocean.html

OCEAN WATER. An electrolyte solution containing minor amounts of nonelectrolytes and composed predominantly of dissolved chemical species of fourteen elements O, H, Cl, Na, Mg, S, Ca, K, Br, C, Sr, B, Si, and F (Table 1). The minor elements, those that occur in concentrations of less than 1 ppm by weight, although unimportant quantitatively in determining the physical properties of sea water, are reactive and are important in organic and biochemical reactions in the oceans.

Dissolved Species

The form in which chemical analyses of sea water are given records the history of our thought concerning the nature of salt solutions. Early analytical data were reported in terms of individual salts NaCl, CaSO$_4$, and so forth. After development of the concept of complete dissociation of strong electrolytes, chemical analyses of sea water were given in terms of individual ions Na$^+$, Ca^{++}, Cl$^-$, and so forth, or in terms of *known* undissociated and partly dissociated species, e.g., HCO$_3^-$. In recent years there has been an attempt to determine the thermodynamically stable dissolved species in sea water and to evaluate the relative distribution of these species at specified conditions. Table 1 lists the principal dissolved species in sea water deduced from a model of sea water that assumes the dissolved constituents are in homogeneous equilibrium, and (or) in equilibrium, or nearly so, with solid phases.

Both associated and nonassociated electrolytes exist in sea water, the latter (typified by the alkali metal ions Li$^+$, Na$^+$, K$^+$, Rb$^+$, and Cs$^+$) predominantly as solvated free cations. The major anions, Cl$^-$ and Br$^-$, exist as free anions, whereas as much as 20% of the F in sea water may be associated as the ion-pair MgF$^+$, and IO$_3^-$ may be a more important species of I than I$^-$. Based on dissociation constants and individual ion activity coefficients the distribution of the major cations in sea water as sulfate, bicarbonate, or carbonate ion-pairs has been evaluated at specified conditions by Garrels and Thompson (1962).

About 10% each of Mg and Ca is tied up as the sulfate ion-pair. It is likely that the other alkaline earth metals, Sr, Ba, and Ra, also exist in sea water partly as undissociated sulfates; about 60% and 21%, respectively, of the total SO$_4^=$ and HCO$_3^-$ are complexed with cations, and two-thirds of the CO$_3^=$ is present as the ion-pair MgCO$_3^0$.

The activities of Mg^{++} and Ca^{++} obtained from the model of sea water proposed by Garrels and Thompson have recently been confirmed by use of specific Ca^{++} and Mg^{++} ion electrodes, and for Mg^{++} by solubility techniques and ultrasonic absorption studies of synthetic and natural sea water. The importance of ion activities to the chemistry of sea water is amply demonstrated by consideration of CaCO$_3$ (calcite) in sea water. The total molality of Ca^{++} in surface sea water is about 10^{-2} and that of CO$_3^=$ is 3.7×10^{-4}; therefore the ion product is 3.7×10^{-6}. This value is nearly 600 times greater than the equilibrium ion activity product of CaCO$_3$ of 4.6×10^{-9} at 25°C and one atmosphere total pressure. However, the activities of the free ions Ca^{++} and CO$_3^=$ in surface sea water are about 2.3×10^{-3} and 7.4×10^{-6}, respectively; thus the ion activity product is 17×10^{-9} which is only 3.7 times greater than the equilibrium ion activity product of calcite. Thus, by considering activities of sea water constituents rather than concentrations, we are better able to evaluate chemical equilibria in sea water; an obvious restatement of simple chemical theory but an often neglected concept in sea water chemistry.

Constancy and Equilibrium

The concept of constancy of the chemical composition of sea water, i.e., that the ratios of the major dissolved constituents of sea water do not vary geographically or vertically in the oceans except in regions of runoff from the land or in semienclosed basins, was first proposed indirectly in 1819 by Marcet and expanded later by Forchammer and Dittmar. The concept

TABLE 1. ABUNDANCES OF THE ELEMENTS AND PRINCIPAL DISSOLVED CHEMICAL SPECIES OF SEAWATER, RESIDENCE TIMES OF THE ELEMENTS

Element	Abundance (mg/l)	Principal species	Residence time (years)
O	857,000	H$_2$O; O$_2$(g); SO$_4^{2-}$ and other anions	
H	108,000	H$_2$O	
Cl	19,000	Cl$^-$	
Na	10,500	Na$^+$	2.6×10^8
Mg	1,350	Mg^{2+}; MgSO$_4$	4.5×10^7
S	885	SO$_4^{2-}$	
Ca	400	Ca^{2+}; CaSO$_4$	8.0×10^6
K	380	K$^+$	1.1×10^7
Br	65	Br$^-$	
C	28	HCO$_3^-$; H$_2$CO$_3$; CO$_3^{2-}$; organic compounds	
Sr	8	Sr^{2+}; SrSO$_4$	1.9×10^7
B	4.6	B(OH)$_3$; B(OH)$_2$O$^-$	
Si	3	Si(OH)$_4$; Si(OH)$_3$O$^-$	8.0×10^3
F	1.3	F$^-$; MgF$^+$	
A	0.6	A(g)	
N	0.5	NO$_3^-$; NO$_2^-$; NH$_4^+$; N$_2$(g); organic compounds	
Li	0.17	Li$^+$	2.0×10^7
Rb	0.12	Rb$^+$	2.7×10^5
P	0.07	HPO$_4^{2-}$; H$_2$PO$_4^-$; PO$_4^{3-}$; H$_3$PO$_4$	
I	0.06	IO$_3^-$; I$^-$	
Ba	0.03	Ba^{2+}; BaSO$_4$	8.4×10^4
In	<0.02		
Al	0.01	Al(OH)$_4^-$	1.0×10^2
Fe	0.01	Fe(OH)$_3$(s)	1.4×10^2
Zn	0.01	Zn^{2+}; ZnSO$_4$	1.8×10^5
Mo	0.01	MoO$_4^{2-}$	5.0×10^5
Se	0.004	SeO$_4^{2-}$	
Cu	0.003	Cu^{2+}; CuSO$_4$	5.0×10^4
Sn	0.003	(OH)?	5.0×10^5
U	0.003	UO$_2$(CO$_3$)$_3^{4-}$	5.0×10^5
As	0.003	HAsO$_4^{2-}$; H$_2$AsO$_4^-$; H$_3$AsO$_4$; H$_3$AsO$_3$	
Ni	0.002	Ni^{2+}; NiSO$_4$	1.8×10^4
Mn	0.002	Mn^{2+}; MnSO$_4$	1.4×10^3
V	0.002	VO$_2$(OH)$_3^{2-}$	1.0×10^4
Ti	0.001	Ti(OH)$_4$?	1.6×10^2
Sb	0.0005	Sb(OH)$_6^-$?	3.5×10^5
Co	0.0005	Co^{2+}; CoSO$_4$	1.8×10^4
Cs	0.0005	Cs$^+$	4.0×10^4
Ce	0.0004	Ce^{3+}	6.1×10^3
Kr	0.0003	Kr(g)	
Y	0.0003	(OH)?	7.5×10^3
Ag	0.0003	AgCl$_2^-$; AgCl$_3^{2-}$	2.1×10^6
La	0.0003	La^{3+}; La(OH)$^{2+}$?	1.1×10^4
Cd	0.00011	Cd^{2+}; CdSO$_4$	5.0×10^5
Ne	0.0001	Ne(g)	
Xe	0.0001	Xe(g)	
W	0.0001	WO$_4^{2-}$	1.0×10^3
Ge	0.00007	Ge(OH)$_4$; Ge(OH)$_3$ O$^-$	7.0×10^3
Cr	0.00005	(OH)?	3.5×10^2
Th	0.00005	(OH)?	3.5×10^2
Sc	0.00004	(OH)?	5.6×10^3
Ga	0.00003	(OH)?	1.4×10^3
Hg	0.00003	HgCl$_3^-$; HgCl$_4^{2-}$	4.2×10^4
Pb	0.00003	Pb^{2+}; PbSO$_4$	2.0×10^3
Bi	0.00002		4.5×10^5
Nb	0.00001		3.0×10^2
Tl	<0.00001	Tl$^+$	
He	0.000005	He(g)	
Au	0.000004	AuCl$_2^-$	5.6×10^5
Be	0.0000006	(OH)?	1.5×10^2
Pa	2.0×10^{-9}		
Ra	1.0×10^{-10}	Ra^{2+}; RaSO$_4$	
Rn	0.6×10^{-15}	Rn(g)	

Adapted from Goldberg, E. D., "Minor elements in sea water," in "Chemical Oceanography," v. 1, pp. 164–165, J. P. Riley and G. Skirrow, Eds., Academic Press, New York, 1965.

was established on a purely empirical basis whereas in actual fact there is a theoretical basis for the concept.

Barth (1952) proposed the concept of residence (passage) time of an element in the oceanic environment and formalized this concept by the equation

$$\lambda = \frac{A}{dA/dt},$$

where λ is the residence time of the element, A is the total amount of the element in the oceans, and dA/dt is the amount of the element introduced or removed per unit time. Sea water is assumed to be a steady-state solution in which the number of moles of each element in any volume of sea water does not change; the net flow into the volume exactly balances the processes that remove the element from it. Complete mixing of the element in the ocean is assumed to take place in a time interval that is short compared to its residence time. Table 1 shows the residence times of the elements, and Table 2 compares the residence times of some elements on the basis of river input and removal by sedimentation. For the major elements the results are strikingly similar and suggest that at least as a first approximation sea water is a steady-state solution with a composition fixed by reaction rates involving the removal of elements from the ocean approximately equalling rates of element inflow into the ocean. Thus, as a first approximation the steady-state oceanic model implies a fixed and constant sea water composition and provides a theoretical basis for the concept of the constancy of the chemical composition of sea water. However, it is possible that at any time, t_0, for example, the present, the ratios of the major dissolved constituents in the open ocean may be nearly invariant simply because the amounts of new materials introduced by streams and other agents to the ocean are small compared to the amounts in the ocean, and these new materials are mixed into the oceanic system relatively rapidly. But over time periods of 1000 to 2000 years or more the major ionic ratios can only remain constant if the ocean is a steady-state solution whose composition is controlled by mechanism(s) other than simple mixing.

Further insight into the constancy concept can be gained by exploring possible mechanisms governing the steady-state composition of sea water. The steady-state solution could be simply a result of the rates of major element inflow into the oceans being equal to rates of outflow by biologic removal, flux through the atmosphere, absorption on sediment particles, and removal in the interstitial waters of marine sediments. For example, Ca^{++} carried to the oceans by streams is certainly removed, in part, in sea aerosol generated at the atmosphere-ocean interface and transported into the atmosphere, later to fall as rain or dry fallout on the continents. However, recent theoretical and experimental work suggests that sea water may be modeled as a steady-state solution in equilibrium with the solids

TABLE 2. THE RESIDENCE TIMES OF ELEMENTS IN SEA WATER CALCULATED BY RIVER INPUT AND SEDIMENTATION

Element	Amount in ocean (in units of 10^{20} g)	Residence time in millions of years	
		River input	Sedimentation
Na	147.8	210	260
Mg	17.8	22	45
Ca	5.6	1	8
K	5.3	10	11
Sr	0.11	10	19
Si	0.052	0.035	0.01
Li	0.0023	12	19
Rb	0.00165	6.1	0.27
Ba	0.00041	0.05	0.084
Al	0.00014	0.0031	0.0001
Mo	0.00014	2.15	0.5
Cu	0.000041	0.043	0.05
Ni	0.000027	0.015	0.018
Ag	0.0000041	0.25	2.1
Pb	0.00000041	0.00056	0.002

After Goldberg, E. D., "Minor elements in sea water," in "Chemical Oceanography," v. 1, p. 173, J. P. Riley and G. Skirrow, Eds., Academic Press, New York, 1965.

that are in contact with it. Sillén has modeled the oceanic system as a near-equilibrium of many solid phases and sea water. Experimental work has shown that aluminosilicate minerals typical of those in the suspended load of streams and in marine sediments react rapidly with sea water containing an excess or deficiency of dissolved silica. Reactions involving these aluminosilicates may control on a long-term basis the activities of H_4SiO_4 and other constituents in sea water. Thus it has begun to emerge that the composition of the oceans represents an approximation of dynamic equilibrium between the water and the solids that are carried into it in suspension or are precipitated from it by the continuous evaporation and renewal by streams. Therefore, if sea water is a solution in equilibrium with solid phases, or even closely approaches such a system, then the *ion activity ratios* of the major dissolved species would be fixed and the chemical composition of the ocean would be "constant." Consequently, the activity of Ca^{++} in the ocean is not simply a result of removal processes involving sea aerosol, adsorption and so forth but is controlled by solid-solution equilibria. A model leading to nearly invariant ion activity ratios geographically and vertically at any time, t_0, in the oceans based on mixing rates alone may be sufficient to explain the constancy of sea water composition but is somewhat misleading and uninformative when

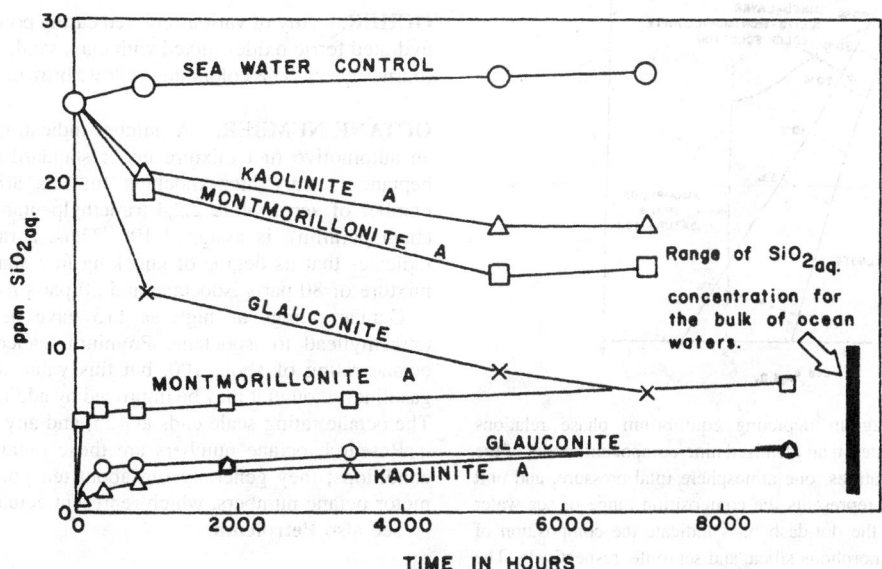

Fig. 1. Concentration of dissolved silica as a function of time for suspensions of silicate minerals in sea water. Curves are for 1-g (<62 μ) mineral samples in silica-deficient (SiO_2 in water was initially 0.03 ppm) and silica-enriched (SiO_2 was initially 25 ppm) sea water at room temperature. Notice that the minerals react rapidly and that the dissolved silica concentration for individual minerals becomes nearly constant at values within or close to the range of silica concentration in the oceans (from Mackenzie, F. T., Garrels, R. M., Bricker, O. P., and Bickley, F., "Silica in sea water: Control by silica minerals" *Science*, **155**, 1404 (1967)).

considered in light of the recent advances in treating the oceans as an equilibrium system.

Some limitations of the equilibrium model of sea water do exist. Sillén has pointed out that based on equilibrium calculations all the nitrogen in the ocean-atmosphere system should be present as NO_3^- in sea water; however, most of the nitrogen is present as N_2 gas in the atmosphere. Also, the concentrations of the major alkaline earth elements, Mg, Ca, and Sr, in sea water may vary slightly with depth or geographic location.

Buffering and Buffer Intensity of Sea Water

The view has long been held that hydrogen-ion buffering in the oceans is due to the $CO_2-HCO_3^--CO_3^=$ equilibrium. Within recent years this view has been challenged, and the importance of aluminosilicate equilibria in maintaining the pH of sea water emphasized. The buffer intensity of a system is of thermodynamic nature and is defined as

$$\beta_{c_j}^{c_i} = \frac{dC_i}{dpH},$$

where $\beta_{c_j}^{c_i}$ is the pH buffer intensity for incremental addition of C_i to a closed system of constant C_j at equilibrium. Homogeneous buffer intensities are defined for systems without solid phases, e.g., the addition of a strong acid to a carbonate solution, whereas heterogeneous buffer intensities are defined for systems with solid phases, e.g., the addition of a strong acid to a solution in equilibrium with calcite, $CaCO_3$, or with kaolinite and muscovite. The homogeneous buffer intensities for the range of sea water and interstitial marine water pH values (7.0 to 8.3) are about 10- to 100-fold less than the heterogeneous intensities involving equilibrium between calcite and sea water or kaolinite, muscovite, and sea water. Both of these heterogeneous equilibria represent large capacities for resistance to sea water pH changes. Unfortunately, the kinetic aspects of these buffer systems have not been investigated quantitatively. However, it is apparent that aluminosilicate equilibria have buffer intensities equal to and perhaps greater than (the buffer intensities of most aluminosilicate equilibria in natural waters have only been qualitatively evaluated) the $CO_2-CaCO_{3(s)}$ equilibria in sea water. Small additions of acid or base to the oceans could be buffered by the homogeneous equilibrium

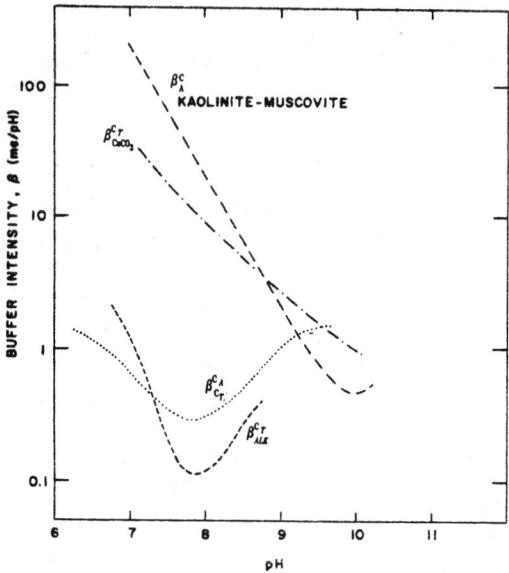

Fig. 3. Buffer intensity as a function of pH for some homogeneous and heterogeneous chemical systems. The buffer intensities are defined for $\beta^{C_A}_{Kaolinite-muscovite}$, addition of a strong acid (or base) to sea water in equilibrium with kaolinite and muscovite; $\beta^{C_T}_{CaCO}3$, addition of CO_2 in a sea water system of zero noncarbonate alkalinity in equilibrium with $CaCO_3$; $\beta^{C_A}_{C_T}$, addition of a strong acid (or base) to a sea water solution of constant total dissolved carbonate; and $\beta^{C_T}_{Alk}$, addition of total CO_2 to a sea water solution of constant alkalinity (data from Morgan, J. J., preprint, "Applications and limitations of chemical thermodynamics in natural water systems").

$CO_2-HCO_3^--CO_3^=$. However, large incremental additions of acid or base or additions over a duration of time would involve the heterogeneous carbonate and aluminosilicate equilibria; the relative importance of each would depend on the buffer intensities of the various equilibria and the relative rates of aluminosilicate and carbonate reactions.

For geologically short-term processes on the order of a few thousands of years, it is likely that the carbon dioxide-carbonate system regulates oceanic pH. The long-term pH is controlled by an interplay of various near-equilibria involving carbonates and silicates.

FRED T. MACKENZIE
Northwestern University
Evanston, Illinois

OCHER. Any of various colored earthy powders consisting essentially of hydrated ferric oxides mixed with clay, sand, etc. Some grades are calcined (burnt ocher). The colors are yellow, brown, or red.

OCTANE NUMBER. A number indicating the antiknock properties of an automotive fuel mixture under standard test conditions. Pure normal heptane (a very high-knocking fuel) is arbitrarily assigned an octane number of zero, while 2,2,4-trimethylpentane, or isooctane (a branched-chain paraffin), is assigned 100. Thus, a rating of 80 for a given fuel indicates that its degree of knocking in a standard test engine is that of a mixture of 80 parts isooctane and 20 parts n-heptane.

Octane ratings as high as 115 have been obtained by addition of tetraethyllead to isooctane. Premium leaded gasolines have a research octane rating of about 100, but this value drops to 85–90 for unleaded gasolines, though it may be improved by addition of methyl-*tert*-butyl ether. The octane rating scale ends at 125, and any higher figure is meaningless.

Research octane numbers are those obtained under test or laboratory conditions; they generally run about ten points higher than the so-called motor octane numbers, which represent actual road operating conditions.

See also **Petroleum**.

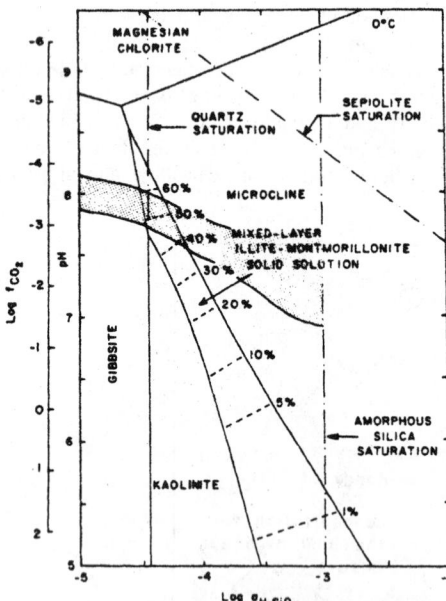

Fig. 2. Logarithmic activity diagram depicting equilibrium phase relations among aluminosilicates and sea water in an idealized nine-component model of the ocean system at the noted temperatures, one atmosphere total pressure, and unit activity of H_2O. The shaded area represents the composition range of sea water at the specified temperature, and the dot-dash lines indicate the composition of sea water saturated with quartz, amorphous silica, and sepiolite, respectively. The scale to the left of the diagram refers to calcite saturation for different fugacities of CO_2. The dashed contours designate the composition (in % illite) of a mixed-layer illitemontmorillonite solid solution phase in equilibrium with sea water (from Helgeson, H. C. and Mackenzie, F. T., 1970, Silicate-sea water equilibria in the ocean system: Deep Sea Res.).

Web References

ASTM International:
http://www.astm.org/DATABASE.CART/REDLINE_PAGES/D2700.htm
Exxon Company, U.S.A.: http://www.prod.exxon.com/exxon_productdata/lube_encyclopedia/octane_number.html

ODOR. An important property of many substances, manifested by a physiological sensation caused by contact of their molecules with the olfactory nervous system. Odor and flavor are closely related, and both are profoundly affected by submicrogram amounts of volatile compounds. Attempts to correlate odor with chemical structure have produced no definitive results. Objective measurement techniques involving chromatography are under development. Even potent odors must be present in a concentration of 1.7×10^7 molecules/cc to be detected. It has been authentically stated that the nose is 100 times as sensitive in detection of threshold odor values as the best analytical apparatus.

Many compounds have a characteristic odor that is an effective means of identification. Toxic and noxious gases have distinctive odors often utilized for warning purposes. An important exception is carbon monoxide, which is almost odorless. The penetrating, banana-like odor of amyl acetate has been used in mine rescue work. Among the most powerful unpleasant odors are those of organic sulfur compounds, especially ethyl mercaptan (skunk). Organic substances having a pleasant odor are broadly designated as aromatic, regardless of chemical nature. The cyclic aromatic (benzene) series of essential oils have a pleasant odor and are the basis of perfumes and fragrances. Odor research, including evaluation by test panels, is conducted at the Olfactronics and Odor Sciences Center at Illinois Institute of Technology, Chicago, IL.

See also **Odor Modification**.

Web Reference

Illinois Institute of Technology: http://www.iit.edu/

ODORANT. A substance having a distinctive, sometimes unpleasant odor that is deliberately added to essentially odorless materials to provide warning of their presence. For example, mercaptan derivatives may be added to natural gas for this purpose. In a broad sense, perfumes are odorants that are added to cosmetics, toilet goods, etc, largely for consumer appeal.

See also **Odor**; and **Odor Modification**.

ODOR MASKING. Addition of a substance with strong odor to obtain a less-offensive effect without changing the composition of the original odorous substance.

See also **Odor Modification**.

ODOR MODIFICATION

Olfaction

Olfaction begins when an odorant stimulates the olfactory receptor cells, triggering the opening or closing of the ion channels, which in turn convert this stimulus to an electrical response to the olfactory bulb and ultimately to other parts of the brain. The olfactory neurons send messages to the olfactory bulbs, structures about the size and shape of peach pits, located on the underside of the large overhanging frontal lobes of the cerebrum. The fact that there is a gene responsible for odor receptor proteins was discovered in 1991.

Perception of odor is therefore a physical mechanism by which information is processed in the brain. Day by day new odors are appearing, all of which are immediately accepted and sorted within the seemingly unlimited categories of the olfactory brain. The brain not only recognizes this information, but evaluates it, sorts it, and associates it with experiences, events, likes, and dislikes.

Some substances are odorous, others are not. Humans can smell at a distance; if one smells the roses in a garden, it is not ordinarily considered that part of the rose is in contact with the nose. Substances of different chemical constitution may have similar odors.

The sense of smell is rapidly fatigued. Fatigue for one odor does not affect the perception of other dissimilar odors, but will interfere with the perception of similar odors. Two or more odorous substances may cancel each other out; this compensation means that two odorous substances smelled together may be inodorous.

Odor travels downwind. Many animals have a keener sense of olfaction than humans. Insects have such extraordinary keenness of smell that it may be a different modality of the chemical sense from that known to humans.

Odors

Odors have been classified according to Carolus Linnaeus, the eighteenth century Swedish botanist who proposed seven odoriferous qualities:

aromatic, fragrant, musky, garlicky, goaty, repulsive, and nauseous. Later in the twentieth century, ethereal (fruity) and empyreumatic (burnt organic matter), together with subdivisions of Linnaeus' classification, were added. In the 1990s, researchers concentrate less on categorizing odors, and more on how people detect and interpret them. Although the average person can name only a handful of common odors, this limitation results from memory retrieval failure, rather than a failure to detect the differences.

Odors are measured by their intensity. The threshold value of one odor to another, however, can vary greatly. Detection threshold is the minimum physical intensity necessary for detection by a subject where the person is not required to identify the stimulus, but just detect the existence of the stimulus. Accordingly, threshold determinations are used to evaluate the effectiveness of different treatments and to establish the level of odor control necessary to make a product acceptable. Concentration can also produce different odors for the same material.

Evaluation Methodologies

Industry has standardized procedures for the quantitative sensory assessment of the perceived olfactory intensity of indoor malodors and their relationship to the deodorant efficacy of air freshener products. Synthetic malodors are used for these evaluation purposes. These malodors should be hedonically associated to the "real" malodor, and must be readily available and of consistent odor quality. These malodors should be tested in various concentrations and be representative of intensities experienced under normal domestic conditions. Panelists are trained to evaluate malodor intensity and the degree of modification.

Modification

Masking. Masking can be defined as the reduction of olfactory perception of a defined odor stimulus by means of presentation of another odorous substance without the physical removal or chemical alteration of the defined stimulus from the environment. Masking is therefore hyperadditive; it raises the total odor level, possibly creating an overpowering sensation, and may be defined as a reodorant, rather than a deodorant. Its end result can be explained by the simple equation of $1 + 1 => 2$.

Odor masking does little or nothing to control malodors; it merely covers them up. Many materials used in masking odors are aldehydes, which are very chemically reactive and usually comprise the top note of a fragrance. Odor masking is used in many areas of household, industrial, and institutional use via products that mask such malodors as pet smells, smoke, cooking, and numerous other odors. The forms by which masking is executed vary, and can be solid, liquid, and aerosol.

Counteraction. Counteraction, sometimes referred to as neutralization, occurs when two odorous substances are mixed in a given ratio and the resulting odor of the mixture is less intense than that of the separate components. The acceptable term to describe this occurrence is compensation. Materials that can accomplish this are basically organic odors which are highly polarized, have a strong affinity for each other, and may also have a low vapor pressure. Some of these molecules have the ability to compensate physiologically for certain malodor materials; others to react chemically with them. Counteraction occurs when the compensating substrate is able to form a coordinate bond with osmophoric sites unique to malodor molecules, such as amino- and thio- moieties. The result is overall reduction in odor; the malodor is transformed into an acceptable state, often with some residual freshening odor. This result lowers the total odor perception and can be exemplified by $1 + 1 =< 2$.

Commercial Aspects

Translating odor modifiers into consumer products results in forms, such as solids, liquids, and aerosols, for a market defined as products "for the nose." This includes products that cover up or eliminate odors, perfume the home, or cleanse the air. The categories of this market can be broken out as traditional air fresheners, cat litter products, aroma care, air purification, and disinfectant in both consumer and industrial applications.

Behavior Modification by Odor

Although odorous materials no doubt impact each other, much discussion centers around the ability of odorous materials to influence human behavior. In articles ranging from scientific journals to trade magazines, there is discussion on the potential of fragrances, i.e., essential oils, to affect people's moods, their ability to focus and maintain attention, to relax and sleep, and even their sexual capability.

The words aromatherapy, aromachology, and aromakinetics are coinages of the 1990s. Aromatherapy, once based on a tradition of folklore and herbal medicine, is being investigated scientifically.

A technique known as contingent negative variation (CNV) measures brain-wave reaction to olfaction. These types of studies have shown the effect of materials such as lavender and nutmeg in reducing stress or anxiety, and the ability of oils such as peppermint to stimulate brainwave activity. CNV research was incorporated into the development of the fragrance for a consumer personal care product launched in the late 1980s.

The interrelationship between fragrance and psychology has been the subject of systematic investigation only recently. Consumer moods can be calculated in both positive and negative directions, and changes can be measured subjectively after exposure to fragrance.

Odors play a much greater role in human behavior than previously thought. The sense of smell provides a direct link with the function of the brain; therefore, the further study of olfaction can only advance the learning of causes and effects of stimuli to the brain.

The future in research will certainly lead to a better understanding of how odors are recognized, sorted, and classified. Studies promise, among other things, to determine whether perceptually similar, but structurally different, odors share the same class of receptor proteins, whether responses to odors can be modified, and possibly why olfactory neurons regenerate but other neurons do not.

YVETTE BERRY
Reckitt & Colman Inc.

Additional Reading

Lawless, H.T.: *Chem. Senses Flavor*, **14**(3), 349–360 (1989).
Lord, S.: *Vogue*, 171 (Dec. 1991).
McCord, C.P. and W.N. Witheridge: *Odors Physiology and Control,* McGraw-Hill Book Co., Inc., New York, NY, 1949.
Zwaardemaker, H.C.: *Arch. Anat. Physical*, 423–432 (1900).

OIL BLACK. A carbon black made from oil, usually and aromatic-type petroleum oil. See also **Petroleum**.

OIL BLUE. Violet-blue copper sulfide pigment use in varnishes.

OIL CAKE. The residue obtained after the expression of vegetable oils from oil-bearing seeds, used as cattle feed and fertilizer.

OIL GAS. A gas made by the reaction of steam at high temperature on gas oil or similar fractions of petroleum, or by high-temperature cracking of gas oil. One typical analysis is heating value 554 Btu/ft^3, illuminants 4.2%, carbon monoxide 10.4%, hydrogen 47.6%, methane 27.0%, carbon dioxide 4.6%, oxygen 0.4%, nitrogen 5.8%, autoign temp 637°F (336°C).

OILINESS. That property of a lubricant that causes a difference in coefficient of friction when all the known factors except the lubricant itself are the same. This concept is also expressed by the term *lubricity*.

OIL (Petroleum). See **Petroleum**.

OIL SANDS. See **Tar Sands**.

OILS, ESSENTIAL. The volatile etherial fraction obtained from a plant or plant part by a physical separation method is called an essential oil. The physical method involves either distillation (including water, steam, water and steam, or dry) or expression (pressing). For the most part, essential oils represent the odorous part of the plant material, and therefore these oils have traditionally been associated with the fragrance and flavor industry (See **Perfume**). Since essential oils frequently occur as a very small percentage by weight of the original plant material, the processing of large quantities is often required to obtain usable amounts of oil. As a result, expression of an essential oil is only employed in those cases where both the form of the natural plant material, such as a citrus peel, and the quantity of oil present make the process feasible.

It has frequently been observed that the aroma of an essential oil is substantially different from that of the plant before processing. Because this phenomenon is largely the result of the treatment of the plant material with heat or hot water, various other methods have evolved over the years in an attempt to obtain a concentrate of the volatiles which more truly represents the aroma of the original. With the exception of the method of expression, almost all of these involve treatment of the plant material with one or more organic solvents (or mixtures thereof) followed by concentration of the extracted solute. Solvent extraction frequently yields, in addition to the volatile oil, various quantities of semi- or nonvolatile organic material such as waxes, fats, fixed oils, high molecular weight acids, pigments, and even alkaloidal material. However, because solvent extraction often results in a product with superior and more representative odor properties to that of a distilled oil, many natural products critically important to the flavor and fragrance industry are available as various extracts in addition to an essential oil.

Some of the commonly used botanical extracts include the following.

Absolute. This is concentrated extract obtained by treatment of a concrete or other hydrocarbon-type extract of a plant or plant part with ethanol.

Absolute Oil. This is the steam distillable portion of an absolute.

Aroma Distillate. Used by the flavor industry, aroma distillates are the product of continuous extraction of the plant material with alcohol at temperatures between ambient and 50°C followed by steam distillation, and, lastly, concentration of the combined hydro-alcoholic mixture.

Concrete. Hydrocarbon extracts of plant tissue, concretes are usually solid to semisolid waxy masses often containing higher fatty acids such as lauric, myristic, palmitic, and stearic as well as many of the nonvolatiles present in absolutes.

Infusion. Infusion botanical extracts are tinctures that have been concentrated by either total or partial removal of the alcohol by distillation.

Oleoresin. Natural oleoresins are exudates from plants, whereas prepared oleoresins are solvent extracts of botanicals, which contain oil (both volatile and, sometimes, fixed), and the resinous matter of the plant. Natural oleoresins are usually clear, viscous, and light-colored liquids, whereas prepared oleoresins are heterogeneous masses of dark color.

Pommade. These are botanical extracts prepared by the enfleurage method wherein flower petals are placed on a layer of fat which extracts the essential oil.

Resin and Resinoid. Natural resins are plant exudates formed by the oxidation of terpenes. Many are acids or acid anhydrides. Prepared resins are made from oleoresins from which the essential oil has been removed. A resinoid is prepared by hydrocarbon extraction of a natural resin.

Tincture. This is prepared by aqueous alcoholic extraction of the raw plant material. Since the extract is not further concentrated, the plant extract is not exposed to heat.

Essential oils are isolated from various plant parts, such as leaves (patchouli), fruit (mandarin), bark (cinnamon), root (ginger), grass (citronella), wood (amyris), heartwood (cedar), gum (myrrh oil), balsam (tolu balsam oil), berries (pimento), seeds (dill), flowers (rose), twigs and leaves (thuja oil), and buds (cloves).

Exceptions to the simple definition of an essential oil are, for example, garlic oil, onion oil, mustard oil, or sweet birch oils, each of which requires enzymatic release of the volatile components before steam distillation. In addition, the physical process of expression, applied mostly to citrus fruits such as orange, lemon, and lime, yields oils that contain from 2–15% nonvolatile material.

Economic Aspects

Essential oils are used as flavoring and fragrance agents in every possible application. Combinations have raised greatly the total sales volume; e.g., mint and cinnamon are used in toothpaste, mouthwash, or lozenges. Combinations can be found in every fragranced product, such as fine fragrances, soaps, detergents, room fresheners, paper, printing ink, paint, candles, condiments, floor polishes, etc. Convenience foods and frozen foods are flavored best by essential oils or oleoresins. Although citronella oil was used as such as an insect repellant, synthetic repellants have, for the most part, taken their place. Flavor essential oils are encountered in baked goods, snack foods, soft drinks, liqueurs, tobacco, sauces, gravies, salad dressings, and other food products.

Composition

The volatile components of essential oils, for the most part, are made up of relatively low molecular weight ($\leq\sim300-350$) organic molecules of carbon, hydrogen, and oxygen, and occasionally nitrogen and sulfur. By far the largest class of natural volatiles of plants is the terpenes, which consist of head-to-tail condensation products of unsaturated five-carbon isoprene units.

Other commonly occurring chemical groups in essential oils include aromatics such as β-phenethyl alcohol, eugenol, vanillin, benzaldehyde, cinnamaldehyde, etc; heterocyclics such as indole, pyrazines, thiazoles, etc; hydrocarbons (linear, branched, saturated, or unsaturated); oxygenated compounds such as alcohols, acids, aldehydes, ketones, ethers; and macrocyclic compounds such as the macrocyclic musks, which can be both saturated and unsaturated.

An essential oil may contain >200 components, and often the trace substances ($\leq$ppm) are essential to the odor and flavor of the oil. The absence or decreased presence of even one component may be cause for odor or flavor rejection of the oil. The same species of plant grown in different parts of the world usually contains the same chemical components, but the relative percentages may be different. Climatic and topographical conditions affect plant chemistry and can alter the essential oil content both qualitatively and quantitatively.

Commercial Essential Oils

Commercial essential oils include rose, jasmin, orange flower (neroli) oil, lavender and lavandin, geranium oil, citronella oil, bergamot oil, lime oil, orange oil, grapefruit oil, sandalwood oil (East Indian), patchouli oil, vetiver oil, galbanum oil, myrrh, oakmoss, tonquin musk, ambergris, tobacco, osmanthus, olibanium, amyris oil, anise oil, anise oil (star), sweet basil oil, bay oil, bitter orange oil, black pepper, bois de rose oil, cannaga oil, caraway oil, cardamom oil, cassia oil, cedarleaf oil, cedarwood oil, Roman chamomile oil, cinnamon bark oil, citronella oil, clove bud oil, coriander oil, cornmint oil, eucalyptus oil, ginger oil, juniper oil, labdanum oil, lemon oil, nutmeg oil, oregano oil (Spanish), orris, palmarosa oil, peppermint oil, petitgrain bigarade oil, pimento berry oil, pine oil, rosemary oil, sage oil (Dalmatian), sage (clary) oil, spearmint oil (native), tagetes oil, thyme oil, turpentine oil, wintergreen oil, and ylang ylang.

Safety and Regulatory Aspects

Essential oils possess a variety of biological properties which may result in varying responses by humans on exposure. An important factor in these effects is the dose to which one is exposed. Thus, essential oils may have both beneficial and toxic effects, depending on their dose. The potential for biological effects from essential oils is not surprising; many botanical species are known to contain substances that possess biological properties, and their identification has contributed significantly to knowledge of biochemistry and physiology as well as the development of therapeutic agents, e.g., quinine and digitalis.

The toxicities of many essential oils have been reported in monographs. Most essential oils used by the flavor and fragrance industries are relatively nontoxic or slightly toxic on acute oral or dermal exposure, and are considered safe when used at levels present in consumer products. In general, the levels of fragrances and flavors in consumer products, and thus the levels of any essential oil ingredients, are relatively low. For example, a fragrance oil may typically be used in a soap at 0.5%. The oil may contain 5% of orange oil distilled. The final concentration of orange oil distilled in the soap therefore is 0.025%.

Because essential oils are used predominantly by the flavor and fragrance industries, these commercial oils must undergo the same scientific scrutiny as all other flavor and fragrance substances and must be in compliance with all applicable health, safety, and environmental regulations. Guidelines and regulations on the use of essential oils in fragrances differ from those applying to essential oils used in flavors.

Many essential oils have been designated by the FDA or by the expert panel of FEMA as Generally Recognized As Safe (GRAS) for their intended use in foods and flavors. The use and safety of these GRAS substances are continuously being reviewed and the list of GRAS substances updated. New essential oils intended to be used as a flavor ingredient must undergo extensive safety evaluations and scrutiny by one or more of these groups of experts before they may be used in flavors.

Many countries have adopted chemical substance inventories in order to monitor use and evaluate exposure potential and consequences. In the case of essential oils used in many fragrance applications, these oils must be on many of these lists. New essential oils used in fragrances are subject to premanufacturing or premarketing notification (PMN). PMN

requirements vary by country and predicted volume of production. They require assessment of environmental and human health-related properties, and reporting results to designated governmental authorities.

Essential oils are also influenced by legislation that regulates specific products that may contain these oils, e.g., the U.S. Food, Drug, and Cosmetic Act and the European Community Cosmetic Directive. Essential oils would not be anticipated to be of environmental concern, considering that they originate from botanical sources. Thus, natural processes exist to degrade essential oils and recycle their components effectively in the environment.

Braja D. Mookherjee
Richard A. Wilson
International Flavors & Fragrances, Inc.

Additional Reading

Arctander, S.: *Perfume and Flavor Materials of Natural Origin,* 1960.
Bauer, K., D. Garbe, and H. Surburg: *Common Fragrance and Flavor Materials: Preparation, Properties and Uses,* 4th Edition, John Wiley & Sons, Inc., New York, NY, 2001.
Kubeczka, K.H. and V. Formaecek: *Essential Oils Analysis by Capillary Gas Chromatography and Carbon-13 NMR Spectroscopy.* 2nd Edition, John Wiley & Sons, Inc., New York, NY, 2002.
Mookherjee, B.D. and C.J. Mussinan, eds.: *Essential Oils,* Allured Publishing Corp., Wheaton, IL., 1981.
Mookherjee, B.D. and R.A. Wilson: in *On Essential Oils,* Synthite Industrial Chemicals Private Ltd., Synthite Valley, Kolenchery, India, 1986, pp. 281–329.
Srinivas, S.R.: *Atlas of Essential Oils,* the Bronx, NY, 1986.

OIL SHALE.

This term refers to a carbonaceous rock that can produce oil when heated to pyrolysis temperatures of 427 to 538°C (800–1,000°F). The oil cannot be extracted with organic solvents at room and moderate temperatures with known technology. The oil precursor in the rock is a high-molecular-weight organic polymer, *kerogen*. An elemental analysis of kerogen derived from the upper zones of Colorado and Utah oil shales is: Carbon, 80.5% (weight); hydrogen, 10.3%; nitrogen, 2.4%; sulfur, 1.0%; and oxygen 5.8%. Usually, the host rock is comprised of dolomite, calcite, quartz, and clays.

Status of Oil Shale Technology

As with most other alternative sources of energy and notably hydrocarbons, the so-called oil glut of the 1980s essentially brought to a halt the further development of oil shale technology. With the fading of federal support in connection with the demise of the Synthetic Fuels Corporation (SFC) in the mid-1980s and the abandonment by most petroleum companies of their earlier interest in alternative energy sources, the shale oil program is essentially a "dead issue" as of the late 1980s. It is estimated that few researchers are now engaged in advancing the technology. One of the last major investments made in oil shale retorting was sponsored by Union Oil Company in connection with its plant located near Parachute Creek, Colorado.

Resources

Within a radius of about 200 kilometers (124 miles) of a center point where the borders of Colorado, Utah, and Wyoming intersect, an area comprised of four great basins—the Green River, Piceance, Uinta, and Washakie Basins as shown on the map in Fig. 1—some authorities estimate there is the equivalent of 2 trillion barrels of oil to be found—more than 50 times the total reserves for petroleum in the United States as presently known. These shales can be expected to yield from 25 to 30 gallons of oil per ton of shale. Of the explored basins, it is estimated that about 83% of these shales are in Colorado; 8.8% in Utah; and 8.2% in Wyoming. These deposits are generally referred to as occurring in the Green River formation. These oil shale resources occur in beds at least 30 meters (98 feet) in thickness. A comparatively small percentage may be mined by surface techniques. For the deeper veins, essentially underground mining technology or in-situ recovery technology will be required. Geologists consider that these deposits date from the Eocene era.

There are also Eastern oil shales, which represent petroleum locked in older shales dating back to the Devonian and Mississippian eras. It is estimated that these shales contain some 2 trillion barrels of oil and underlie an area of over a million square kilometers (400,000 square miles) in Michigan, western Pennsylvania, eastern Ohio, southwestern Indiana,

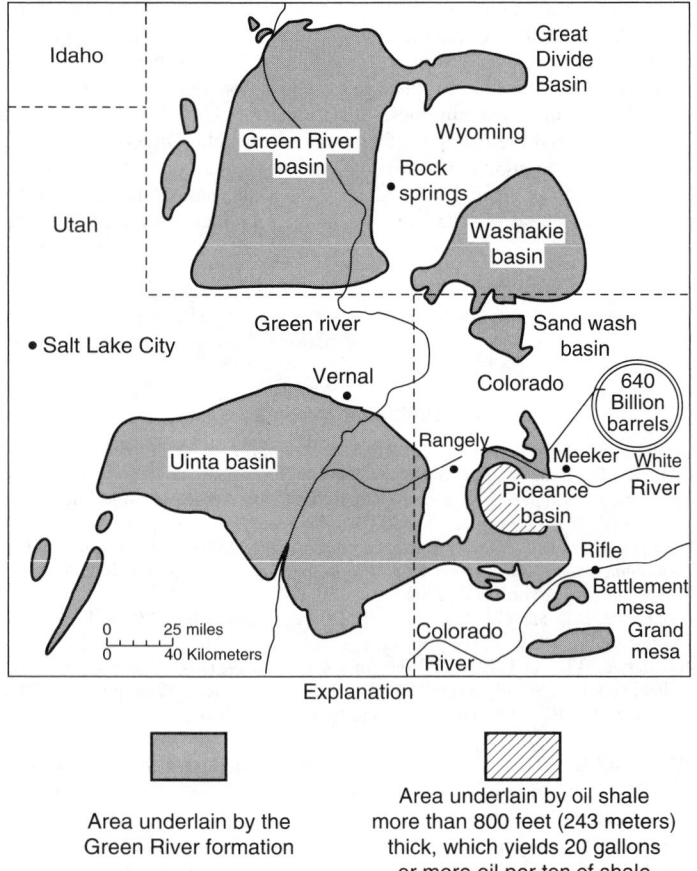

Explanation

Area underlain by the Green River formation	Area underlain by oil shale more than 800 feet (243 meters) thick, which yields 20 gallons or more oil per ton of shale

Fig. 1.   Distribution of oil shale in the Green River formation. (*U.S. Geological Survey*)

eastern and southern Illinois, most of Kentucky and Tennessee, much of Oklahoma, and northern Alabama. They are sometimes broken down into what are known as the Eastern and Midwestern oil shale deposits. In Michigan, the shale is approximately 60 meters (197 feet) thick and is in a basin at depths ranging from about 0.8 kilometer (0.5 mile) to outcroppings in three of the northern counties. In general, the Eastern oil shales become thicker and deeper toward the east.

Processing Oil Shale

Work on the production of petroleum-like materials from oil shale in the Green River formation dates back many years. One of the first major efforts was that undertaken by the U.S. Bureau of Mines in 1944. This program involved two 40-ton capacity retorts that operated between 1947 and 1951, with a production of some 920 runs and a total consumption of 37,500 tons of raw shale. This was a batch process and much was learned from this experience.

In a status report (1983), one of the largest (12,800 tons/day) and the last of numerous oil shale retorting processes (Union Oil Company) is described in considerable detail by Duir, Griswold, and Christolini in the February 1983 issue of *Chemical Engineering Progress* (pp. 45–50). This paper includes several diagrams and provides an excellent starting point for the reader who may be interested in this topic.

Additional Reading

Berkowitz, N.: *Fossil Hydrocarbons: Chemistry and Technology,* Elsevier Science, New York, NY, 1997.

Lee, S.: *Oil Shale Technology,* CRC Press LLC., Boca Raton, FL, 1990.

Selly, R.C.: *Elements of Petroleum Geology,* 2nd Edition, Elsevier Science, New York, NY, 1997.

OIL (Vegetable).   See Vegetable Oils (Edible).

OLAH, GEORGE A. (1927–).   Born in Hungary, now an American citizen, he won the Nobel prize for chemistry in 1994 for his work with carbocations. These are positively charged hydrocarbons with lifetimes

on the order of microseconds. Olah developed methods of studying carbocations with different physical techniques, changing the direction of this field. He received a Ph.D. from the Technical University of Budapest in 1949.

OLEATE.   Salt made up of a metal or alkaloid with oleic acid. It is used for external medications and in soaps and paints.

See also **Oleic Acid.**

OLEFIN FIBERS.   Olefin fibers, also called polyolefin fibers, are defined as manufactured fibers in which the fiber-forming substance is a synthetic polymer of at least 85 wt % ethylene, propylene, or other olefin units. Several olefin polymers are capable of forming fibers, but only polypropylene (PP) and, to a much lesser extent, polyethylene (PE) are of practical importance. Olefin polymers are hydrophobic and resistant to most solvents. These properties impart resistance to staining, but cause the polymers to be essentially undyeable in an unmodified form.

Advances in olefin polymerization provide a wide range of polymer properties to the fiber producer. Inroads into new markets are being made through improvements in stabilization, and new and improved methods of extrusion and production, including multicomponent extrusion and spunbonded and meltblown nonwovens.

Properties

Physical Properties. Table 1 shows that olefin fibers differ from other synthetic fibers in two important respects: (1) olefin fibers have very low moisture absorption and thus excellent stain resistance and almost equal wet and dry properties, and (2) the low density of olefin fibers allows a much lighter weight product at a specified size or coverage. Thus one kilogram of polypropylene fiber can produce a fabric, carpet, etc, with much more fiber per unit area than a kilogram of most other fibers.

Tensile Strength. Tensile properties of all polymers are a function of molecular weight, morphology, and testing conditions. Lower temperature and higher strain rate result in higher breaking stresses at longer elongations, consistent with the general viscoelastic behavior of polymeric materials. Similar effects are observed on other fiber tensile properties, such as tenacity or stress at break, energy to rupture, and extension at break. Under the same spinning, processing, and testing conditions, higher molecular weight results in higher tensile strength.

Creep, Stress Relaxation, Elastic Recovery. Olefin fibers exhibit creep, or time-dependent deformation under load, and undergo stress relaxation, or the spontaneous relief of internal stress. High molecular weight and high orientation reduce creep.

Elastic recovery or resilience is the recovery of length upon release of stress after extension or compression. A fiber, fabric, or carpet must possess this property in order to spring back to its original shape after being crushed or wrinkled, Polyolefin fibers have poorer resilience than nylon; this is thought to be partially related to the creep properties of the polyolefins.

Chemical Properties. The hydrocarbon nature of olefin fibers, lacking any polarity, imparts high hydrophobicity and consequently resistance to soiling or staining by polar materials, a property important in carpet and upholstery applications. Unlike the condensation polymer fibers, such as polyester and nylon, olefin fibers are resistant to acids and bases. At room temperature, polyolefins are resistant to most organic solvents, except for some swelling in chlorinated hydrocarbon solvents. At higher temperatures, polyolefins dissolve in aromatic or chlorinated aromatic solvents, and show some solubility in high boiling hydrocarbon solvents. At high temperatures, polyolefins are degraded by strong oxidizing acids.

Thermal and Oxidative Stability. In general, polyolefins undergo thermal transitions at much lower temperatures than condensation polymers; thus, the thermal and oxidative stability of polyolefin fibers are comparatively poor. Preferred stabilizers are highly substituted phenols such as Cyanox 1790 and Irganox 1010, or phosphites such as Ultranox 626 and Irgafos 168.

Ultraviolet Degradation. Polyolefins are subject to light-induced degradation; polyethylene is more resistant than polypropylene. Because polyolefins readily form hydroperoxides, the more effective light stabilizers are radical scavengers. Hindered amine light stabilizers (HALS) are favored, especially high molecular weight and polymeric amines that have lower mobility and less tendency to migrate to the surface of the fiber. This migration is commonly called bloom.

TABLE 1. PHYSICAL PROPERTIES OF COMMERCIAL FIBERS

Polymer	Standard tenacity, GPa[a]	Breaking elongation, %	Modulus, GPa[a]	Density, kg/m³	Moisture regain[b]
olefin	0.16–0.44	20–200	0.24–3.22	910	0.01
polyester	0.37–0.73	13–40	2.1–3.7	1,380	0.4
carbon	3.1	1	227	1,730	
nylon	0.23–0.60	25–65	0.5–2.4	1,130	4–5
rayon	0.25–0.42	8–30	0.8–5.3	1,500	11–13
acetate	0.14–0.16	25–45	0.41–0.64	1,320	6
acrylic	0.22–0.27	35–55	0.51–1.02	1,160	1.5
glass	4.6	5.3–5.7	89	2,490	
aramid	2.8	2.5–4.0	113	1,440	4.5–7
fluorocarbon	0.18–0.74	5–140	0.18–1.48	2,100	
polybenzimidazole	0.33–0.38	25–30	1.14–1.52	1,430	15

[a] To convert GPa to psi, multiply by 145,000.
[b] At 21°C and 65% rh.

Flammability. Most polyolefins can be made fire retardant using a stabilizer, usually a bromine-containing organic compound, and a synergist such as antimony oxide. However, the required loadings are usually too high for fibers to be spun. Fire-retardant polypropylene fibers exhibit reduced light and thermal resistance.

Dyeing Properties. Because of their nonionic chemical nature, olefin fibers are difficult to dye. A broad variety of polymeric dyesites have been blended with polypropylene; nitrogen-containing copolymers are the most favored. In apparel applications where dyeing is important, dyeable blends are expensive and create problems in spinning fine denier fibers. Hence, olefin fibers are usually colored by pigment blending during manufacture, called solution dying in the trade.

Manufacture and Processing

Olefin fibers are manufactured commercially by melt spinning, similar to the methods employed for polyester and polyamide fibers.

Slit-Film Fiber. A substantial volume of olefin fiber is produced by slit-film or film-to-fiber technology. For producing filaments with high linear density, above 0.7 tex (6.6 den), the production economics are more favorable than monofilament spinning. The fibers are used primarily for carpet backing and rope cordage applications.

Bicomponent Fibers. Polypropylene fibers have made substantial inroads into nonwoven markets because they are easily thermal bonded. Further enhancement in thermal bonding is obtained using bicomponent fibers. In these fibers, two incompatible polymers, such as polypropylene and polyethylene, polyester and polyethylene, or polyester and polypropylene, are spun together to give a fiber with a side-by-side or core–sheath arrangement of the two materials. The lower melting polymer can melt and form adhesive bonds to other fibers; the higher melting component causes the fiber to retain some of its textile characteristics. Bicomponent fibers have also provided a route to self-texturing (self-crimping) fibers.

Meltblown, Spunbond, and Spurted Fibers. A variety of directly formed nonwovens exhibiting excellent filtration characteristics are made by meltblown processes, producing very fine, submicrometer filaments. A stream of high velocity hot air is directed on the molten polymer filaments as they are extruded from a spinnerette. This air attenuates, entangles, and transports the fiber to a collection device. Because the fiber cannot be separated and wound for subsequent processing, a nonwoven web is directly formed.

In the spunbond process, the fiber is spun similarly to conventional melt spinning, but the fibers are attenuated by air drag applied at a distance from the spinnerette. This allows a reasonably high level of filament orientation to be developed. The fibers are directly deposited onto a moving conveyor belt as a web of continuous randomly oriented filaments.

Pulp-like olefin fibers are produced by a high pressure spurting process developed by Hercules Inc. and Solvay, Inc. Polypropylene or polyethylene is dissolved in volatile solvents at high temperature and pressure. After the solution is released, the solvent is volatilized, and the polymer expands into a highly fluffed, pulp-like product. Additives are included to modify the surface characteristics of the pulp. Uses include felted fabrics, substitution in whole or in part for wood pulp in papermaking, and replacement of asbestos in reinforcing applications.

High Strength Fibers. The properties of commercial olefin fibers are far inferior to those theoretically attainable. A number of methods, including superdrawing, high pressure extrusion, spinning of liquid crystalline polymers or solutions, gel spinning, and hot drawing produce high strengths, but these methods are tedious and uneconomical for olefin fibers. A high modulus commercial polyethylene fiber with properties approaching those of aramid and graphite fibers is prepared by gel spinning.

Hard-Elastic Fibers. Hard-elastic fibers are prepared by annealing a moderately oriented spun yarn at high temperature under tension. They are prepared from a variety of olefin polymers, acetal copolymers, and polypivalolactone.

Economic Aspects

In the United States, olefin fiber consumption has risen steadily since its introduction in 1961. Olefin fiber is the only synthetic fiber showing market growth in recent years.

Applications

Olefin fibers are used for a variety of purposes from home furnishings to industrial applications. These include carpets, upholstery, drapery, rope, geotextiles, and both disposable and nondisposable nonwovens. Fiber mechanical properties, relative chemical inertness, low moisture absorption, and low density contribute to desirable product properties.

C. J. WUST, JR.
L. M. LANDOLL
Hercules Incorporated

Additional Reading

Ahmed, M. *Polypropylene Fibers: Science and Technology,* Elsevier Science Publishing Co., Inc., New York, NY, 1982, pp. 344–346.
Landoll, L.M.: "Olefin Fibers," in J.I. Kroschwitz, ed., *Encyclopedia of Polymer Science and Engineering,* 2nd Edition, Vol. 10, John Wiley & Sons, Inc., New York, NY, 1987, pp. 373–395.

OLEFIN POLYMERS

POLYETHYLENE

Polyethylene (PE) is a generic name for a large family of semicrystalline polymers used mostly as commodity plastics. PE resins are linear polymers with ethylene molecules as the main building block; they are produced either in radical polymerization reactions at high pressures or in catalytic polymerization reactions. Most PE molecules contain branches in their chains. In very general terms, PE structure can be represented by the following formula:

$$(CH_2-CH_2)_x-branch_1-(CH_2-CH_2)_y-branch_2-(CH_2-CH_2)_z$$

$$-branch_3 \cdots$$

where the $-CH_2-CH_2-$ units come from ethylene, and x, y, and z values can vary from 4 or 5 to over 100. This allows the industry to produce a large variety of PE resins with different molecular weights and branching characteristics.

The total number of monomer units (which approximately equals $x + y + z + \ldots$ in the above formula) in PE chains is called the degree of polymerization. It can vary from small (about 10–20 in PE waxes) to very large (over 100,000 for PE of ultrahigh molecular weight (UHMW)).

TABLE 1. COMMERCIAL CLASSIFICATION OF POLYETHYLENES

Designation	Acronym	Density, d, g/cm^3
high density polyethylene	HDPE	≥ 0.941
ultrahigh molecular weight polyethylene[a]	UHMWPE	0.935–0.930
medium density polyethylene	MDPE	0.926–0.940
linear low density polyethylene	LLDPE	0.915–0.925
low density polyethylene[b]	LDPE	0.910–0.940
very low density polyethylene	VLDPE	0.915–0.880

[a] Linear polymer with molecular weight of over 3×10^6.
[b] Produced in high pressure processes.

In high pressure ethylene polymerization processes, the branches are formed spontaneously according to peculiarities of radical polymerization reactions. These branches are either linear or branched alkyl groups. Their lengths vary widely, sometimes even within a single polymer molecule, and can be both short (from methyl to isooctyl group, collectively known as short-chain branching) and long, up to several thousands of carbon atoms (long-chain branching). In catalytic polymerization processes, the branches are introduced deliberately by copolymerizing ethylene with α-olefins. The structure of these branches is determined by the type of olefin used in the copolymerization reaction.

Some PE molecules, on the other hand, contain no branches at all. From a chemical standpoint, such resins can be regarded as polymethylene, $H(CH_2)_n H$.

The distributions of branches among different polymer molecules in PE resins can be quite different, depending on the method of PE production. Some PE resins have uniform branching distributions, which means that any given polymer molecule contains the same relative fraction of branches as all others. PE prepared in high pressure processes and most ethylene copolymers produced with metallocene catalysts belong to this group. In contrast, PE resins produced with heterogeneous titanium- and chromium-based catalysts are mixtures of polymer molecules with very different branching degrees.

Classification of PE resins

The classification of PE resins has developed in conjunction with the discovery of new catalysts for ethylene polymerization as well as new polymerization processes and applications. The classification (given in Table 1) is based on two parameters: the resin density and its melt index. This classification provides a simple means for a basic differentiation of PE resins, even though it cannot easily describe some important distinctions between the structures and properties of various resin brands.

Synthesis Technologies

A variety of technological processes are used for polyethylene manufacture. They include polymerization in supercritical ethylene at a high ethylene pressure and temperature above the PE melting point (110–140°C), polymerization in solution at 120–150°C or in slurry, and polymerization in the gas phase.

Control of PE Properties

The tailoring of PE properties in commercial processes is achieved mostly by controlling the density, molecular weight, and molecular weight distribution, or by cross-linking. Successful control of all reaction parameters enables the manufacture of a large family of PE products with considerable differences in physical properties, such as the softening temperatures, stiffness, hardness, clarity, impact, and tear strength.

The Market

PE resins command a wide range of applications, both as commodity resins and as specialty polymers. Their uses include numerous film grades of LDPE, HDPE, and LLDPE for bags and packaging; coatings for paper, metal, wire, and glass; household and industrial containers such as bottles for various fluids, i.e., water, food products, detergents, liquid fuels, etc; toys; and different types of pipe and tubing. Because of its versatility, PE has become the largest commercially manufactured polymer in the world.

YURY V. KISSIN
Mobil Chemical Company

LOW DENSITY POLYETHYLENE

The first high molecular weight crystalline polyolefin was produced in 1933 through the high pressure process. Initial production was targeted for use in specialized applications such as insulation for high voltage cable. Since that time, the number of applications of various grades of homopolymers and copolymers have expanded to cover many areas once dominated by paper (qv), glass (qv), steel (qv), and other polymers.

The molecular weight of low density polyethylene (LDPE) ranges from waxy products at about 500 mol wt to very tough products at about 60,000 mol wt. One unique feature of LDPE, as opposed to high density polyethylene (HDPE) or linear low density polyethylene (LLDPE), is the presence of both long- and short-chain branching along the polymer chain. Another important feature of LDPE is its ability to incorporate a wide range of comonomers that can be polar in nature along the polymer chain. Disadvantages of LDPE include the high capital investment for commercial plant construction, engineering problems related to high pressure operation, and high energy costs in production.

LDPE, also known as high pressure polyethylene, is produced at pressures ranging from 82–276 MPa (800–2725 atm). Operating at 132–332°C, it may be produced by either a tubular or a stirred autoclave reactor. Reaction is sustained by continuously injecting free-radical initiators, such as peroxides, oxygen, or a combination of both, to the reactor feed.

Traditionally, LDPE has been defined as homopolymer products having a density between 0.915–0.940 g/cm^3 (products having a density above 0.940 g/cm^3 are considered HDPE). However, with the commercialization of LLDPE via the fluidized-bed or solution processes, this distinction is no longer valid.

Properties

The mechanical properties of LDPE fall somewhere between rigid polymers such as polystyrene and limp or soft polymers such as polyvinyls. LDPE exhibits good toughness and pliability over a moderately wide temperature range. It is a viscoelastic material that displays non-Newtonian flow behavior, and the polymer is ductile at temperatures well below 0°C. Table 2 lists typical properties.

Structure. The physical properties of LDPE depend on the molecular weight, the molecular weight distribution, as well as the frequency and distribution of long- and short-chain branching.

Molecular Weight Distribution. MWD offers a general picture of the range of long, medium, and short molecular chains in the polymer; the broad molecular weight distribution in LDPE, however, is attributed only to the presence of long branches on the polymer molecule. LDPE may have molecules that range in length from a few thousand carbons to a million or more carbons.

With increasing molecular weight, certain properties increase: melt viscosity, abrasion resistance, tensile strength, resistance to creep, flexural stiffness, resistance to brittleness at low temperature, shrinkage, warpage, and film impact strength. On the other hand, increasing mol wt results in reduced film transparence, freedom from haze, and gloss; drawdown rate; neck-in and beading; and adhesion.

Melt Index or Melt Viscosity. Melt index describes the flow behavior of a polymer at a specific temperature under specific pressure. If the melt index is low, its melt viscosity or melt flow resistance is high; the latter is a term that denotes the resistance of molten polymer to flow when making film, pipe, or containers. ASTM D1238 is the designated method for this test.

Film Clarity. Slight haziness is a characteristic of all polyethylene resins. It may be caused either by a surface roughness which diffuses light passing through the film (surface roughness is a function of extrusion conditions and the fundamental structure of the polymer), or it may be caused by the partly crystalline structure of the polymer which has a larger index of refraction than the surrounding amorphous material.

Stress Crack Resistance. Failure caused by environmental stress cracking may be attributed to stored stresses acquired in the molding or extruding operation. These dormant stresses may release themselves by cracking under the combined influence of an adverse environment and polyaxial stretching during use. Polyethylene of narrow molecular weight distribution tends to crack less under environmental stress.

Yield Strength. Yield strength, tensile strength, and elongation are all functions of the basic molecular properties. In other words, a higher

TABLE 2. PROPERTIES OF LOW DENSITY HIGH PRESSURE POLYETHYLENE

Property	ASTM method	LDPE[a]
tensile yield stress, kPa[b]	D638	80–180
yield elongation, %	D638	10–40
tensile ultimate stress, kPa[b]	D638	100–170
ultimate elongation, %	D638	100–700
secant modulus of elasticity at 1% strain, kPa[b]	D638	900–5,000
hardness, Rockwell	D785	D41–D60
dart drop 38-μm thickness, at g/25 μm	D1709	50–300
low temperature brittleness at F_{50}[c], °C	D746	<−76°C
dielectric constant at 60 Hz	D150	2.25–2.35
density, g/cm^{3}[d]	D1505	0.912–0.940
refractive index, n_D^{25}[d]	D542	1.51
thermal expansion, 10^{-5} cm/cm per °C[d]	D696	10–22
narrow angle scatter	D1746	4–80
haze, %	1003	40–50
gloss, %	2457	0–80
water absorption 24 h, %[d]	D570	<0.02

[a] LDPE homopolymers in the 0.2–150 melt index or 100,000–20,000 mPa · s(= cP) viscosity range. Specialty polymers such as greases and waxes or highly cross-linked polymers are not included.
[b] To convert kPa to psi, divide by 6.895.
[c] F_{50} = number of hours at which 50% fail.
[d] Taken at 25°C.

density LDPE homopolymer has higher yield strength, slightly lower tensile strength, and lower elongation; in contrast, a higher average molecular weight polymer has higher tensile strength and slightly higher yield strength.

Low Temperature Brittleness. Brittleness temperature is the temperature at which polyethylene becomes sufficiently brittle to break when subjected to a sudden blow. Because some polyethylene end products are used under particularly cold climates, they must be made of a polymer that has good impact resistance at low temperatures; namely, polymers with high viscosity, lower density, and narrow molecular weight distribution.

Film Appearance. It is important to minimize film imperfections or defects when LDPE is blown into film. The common defects are arrow-heads, pinpoint gels, gels or "fish eyes", and oxidized gels or colored specks.

Film imperfections are one of the more serious quality problems for both the producer and the film converter. Film imperfections can arise from many sources. Most commonly, they are the result of contamination in the reaction system, in post-reactor handling, during shipping and unloading, or in the end users equipment.

Chemical Resistance. LDPE is highly resistant to penetration by most chemically neutral or reactive substances. This is a property of prime importance for all kinds of packaging applications. Because of its high impermeability, polyethylene containers can store and transport many kinds of chemicals without leak hazards. Likewise, easily spoiling foods such as vegetables or meats can be shelved and sold in polyethylene bags without the danger of water infiltration from the outside or irreplaceable moisture being lost from the inside; exchange of gases through the film can also be kept to a minimum. In addition, LDPE is resistant to penetration from most polar liquids, water, aqueous acids, and alkalies, as well as most metal plating solutions. However, it can be easily penetrated by nonpolar liquids such as hydrocarbons, and animal and vegetable oils.

Electrical Properties. LDPE's electrical properties make it extremely well suited for wire and cable insulation for electrical power supplies at high transmission, lower domestic voltage, and high frequency, very high frequency, or ultrahigh frequency applications in electronics. As a dielectric, i.e., electric insulator, LDPE for all practical purposes does not transmit electrical current.

Manufacture

LDPE is produced in either a stirred autoclave or a tubular reactor; total domestic production, divided between the two systems at 45% for tubular and 55% for autoclave, is estimated to be 3.4 million metric tons per year. Neither process has gained a clear advantage over the other, although all new or added capacity production in the 1990s has been through the autoclave.

Recycle and Polymer Collection. Due to the incomplete conversion of monomer to polymer, it is necessary to incorporate a system for the recovery and recycling of the unreacted monomer. Both tubular and autoclave reactors have similar recycle systems.

Additives. Compounds are often added to the polymer at the extruder or melt homogenizer. Common additives are antioxidants, thermal stabilizers, slip agents, antiblock agents, and uv stabilizers.

Blending and Purging. Polymer pellets are air-conveyed to holding silos where they are purged with heated, filtered air and then blended to ensure uniformity. Purging is necessary for removing the entrapped monomer and comonomer.

Environmental Considerations. Good progress has been made in reducing emissions during production. Ethylene is by far the largest fugitive gas; purge from the recycle system is commonly recycled to the ethylene cracking unit for recovery. Owing to improved operation techniques and control equipment, runaway reactions or "decomps" that are vented to the atmosphere have also decreased substantially. In addition, many plants utilize particulate and noise reduction devices on the emergency venting systems.

Many units have waste heat recovery systems that generate low pressure steam from reaction heat. Such steam is often employed to drive adsorption refrigeration units to cool the reactor feed stream and to increase polymer conversion per pass, an energy-saving process that reduces the demand for electrical power.

Pellet degassing, however, still presents a problem. The volume of purge air required to degas the pellets makes recovery of ethylene from the purge stream using conventional methods uneconomical.

Applications

LDPE is used in a wide range of applications, the largest segment of which is taken up by end uses requiring processing into thin film.

Blown Film. Blown film has a number of advantages over flat film. For instance, in blown film extrusion, molecular orientation is achieved in both the machine and transverse direction, the relative degree depending primarily on the drawdown ratio (the die opening to film gauge) and the blow-up ratio. The result is a film with more uniform strength in both directions. Moreover, with proper extrusion conditions, the physical properties of the film are equal in both the machine and transverse direction; such even distribution provides maximum toughness. Another advantage of blown film or tubing is the absence of a lengthwise seam.

In blown film extrusion, the molten polymer enters a ring-shaped die either through the bottom or from the side. It is then forced around a mandrel inside the die, shaped into a sleeve, and extruded through the die opening in the form of a comparatively thick-walled tube.

Flat Film Extrusion. In flat film extrusion, the melt is extruded through a long slot in a "T" or coat hanger-type die, past the die lands. In this setup, the polymer melt is forced into the slot die at its center; it reaches the slot opening by way of a manifold and over the lands. The principal advantages of film casting are substantial improvements in the film's transparency, freedom from haze, improved gloss, and other optical properties.

Extrusion Coating. In extrusion coating, a thin film of molten polymer is pressed onto or into the substrate. Coating thickness may range from 6.5 μm or less to more than 100 μm. In polymer lamination, a related operation, two or more substrates, such as paper or aluminum foil, are combined by using the polymer film as adhesive and moisture barrier. In order to coat a substrate, the polymer must be extruded through a narrow slit in the extrusion coating die by an extruder screw.

Molding Applications. Molding is accomplished by three different methods: blow molding, injection molding, and rotational molding, although the use of LDPE in these applications has been declining since the introduction of LLDPE on the market.

Adhesives and Sealants. Dominated by copolymers, adhesives and sealants remain somewhat of a specialty market. These polymers usually contain high copolymer content and low viscosity, and often require blending with other compounds prior to final application. Their uses are numerous, e.g., as seals for bottled drinks, as tie layers between incompatible polymers, or as automotive adhesives.

Health and Safety Factors

LDPE is nontoxic, and is commonly used in food packaging where Food and Drug Administration requirements must be met. It is also used in packaging pharmaceuticals and other medical applications such as iv bags and tubes.

Waste management of LDPE can be approached in several different ways, i.e., via recycling, energy recovery, biodegradability, increased packaging efficiency, or uv degradability. Ultraviolet degradable grades for can carriers are available to processors; these products are mandated in several states.

Environmental Degradability. It has become increasingly clear that, for certain end uses at least, plastics with a limited lifetime are preferable because they help to solve problems of litter and waste disposal. These types of plastics are subject to biodegradation, which occurs when microorganisms such as fungi or bacteria secrete enzymes and chemically break down the polymer structure into small fractions that can be digested by other microorganisms.

Biodegradability has been approached in several areas. One attempt is to add fillers such as starches into the product.

Another approach is to replace petrochemical-based polymers with polymers made from carbohydrates. Unfortunately, approaches of this type have yet to produce economically competitive polymers.

LLOYD W. PEBSWORTH
Polyethylene Technology

HIGH DENSITY POLYETHYLENE

High density polyethylene (HDPE) is defined by ASTM D1248-84 as a product of ethylene polymerization with a density of 0.940 g/cm³ or higher. This range includes both homopolymers of ethylene and its copolymers with small amounts of α-olefins. The first commercial processes for HDPE manufacture were developed in the early 1950s and utilized a variety of transition-metal polymerization catalysts based on molybdenum, chromium, and titanium. Commercial production of HDPE was started in 1956 in the United States by Phillips Petroleum Co. and in Europe by Hoechst. HDPE is one of the largest volume commodity plastics produced in the world. The term HDPE embraces a large variety of products differing predominantly in molecular weight, molecular weight distribution (MWD), and crystallinity.

Molecular Weight. The range of molecular weights of commercially produced HDPE is wide, from several hundreds for polyethylene (PE) waxes to several millions for ultrahigh molecular weight PE resins (UHMWPE). A parameter that is widely accepted, easily measured, and which provides information on molecular weight, is the rheological parameter called the melt index. Different HDPE resins have melt indexes ranging from over 500 (low molecular weight polymers) to less than 0.001.

Molecular Weight Distribution. The width of the molecular weight distribution (MWD) of PE resins is usually represented by the ratio of the weight-average and the number-average molecular weights, $\overline{M}_w/\overline{M}_n$, or by MFR value, which is the ratio of two melt indexes measured at two melt pressures that differ by a factor of 10. The range of MFR values for commercial HDPE resins is wide, from around 25 for injection molding resins with a narrow MWD, to over 150 for some HDPE film resins with a broad MWD.

Crystallinity and Density. Crystallinity and density of HDPE resins depend primarily on the extent of short-chain branching in polymer chains and, to a lesser degree, on molecular weight. The density range for HDPE resins is between 0.960 and 0.941 g/cm³. UHMWPE is a completely nonbranched ethylene homopolymer, but, due to its very high molecular weight, it crystallizes poorly and has a density of 0.93 g/cm³.

Molecular Structure and Chemical Properties

HDPE is a linear polymer with the chemical composition of polymethylene, $(CH_2)_n$. Depending on application, HDPE molecules either have no branches at all, as in certain injection molding and blow molding grades, or contain a small number of branches which are introduced by copolymerizing ethylene with α-olefins, e.g., ethyl branches in the case of 1-butene and *n*-butyl branches in the case of 1-hexene. The number of branches in HDPE resins is low, at most 5 to 10 branches per 1000 carbon atoms in the chain.

HDPE is a saturated linear hydrocarbon and exhibits very low chemical reactivity. The most reactive parts of HDPE molecules are the double bonds at chain ends and tertiary CH bonds at branching points in polymer chains. Because its reactivity to most chemicals is reduced by high crystallinity and

low permeability, HDPE does not react with organic acids, most inorganic acids, or with alkaline solutions.

At room temperature, HDPE is not soluble in any known solvent, but at a temperature above 80–100°C, most HDPE resins dissolve in some aromatic, aliphatic, and halogenated hydrocarbons.

HDPE is relatively stable under heat. Chemical reactions at high temperature in the absence of oxygen become noticeable only above 290–300°C. Thermocracking of HDPE is a free-radical C–C bond scission reaction.

At elevated temperatures, oxygen attacks HDPE molecules in a series of radical reactions. These reactions reduce the molecular weight of HDPE and introduce oxygen-containing groups, such as hydroxyl and carboxyl groups, into polymer chains. Other oxidation products are low molecular weight compounds such as water, aldehydes, ketones, and alcohols. Oxidative degradation in HDPE is initiated by impurities, which are mainly catalyst residues containing transition metals, e.g., titanium and chromium. The protection from thermooxidative degradation is provided by antioxidants such as naphthylamines or phenylenediamines, hindered phenols, quinones, and alkyl phosphites, which are used in 0.1–1.0 wt % concentration.

Many commercial processes involving surface dyeing and printing (e.g., on film and containers) employ thermooxidation as a pretreatment step. Dyes adhere poorly to HDPE surfaces but their adhesion can be improved by thermooxidation of the surface layer by treatment with an open flame or in a strong electric field.

Even though degradation of HDPE initiated by oxygen and light resembles thermooxidative degradation, it proceeds at a much lower temperature. Photooxidative degradation of HDPE causes aging, development of surface cracks, brittleness, change in color, and drastic deterioration of mechanical and dielectric properties. The reaction can be slowed down or prevented by utilizing light stabilizers that protect the resin and absorb uv radiation. Photooxidative degradation is the principal process responsible for gradual disintegration of discarded PE litter. The chemical industry manufactures a large number of antioxidants as well as uv stabilizers; their mixtures with other additives are used to facilitate resin processing.

Crystalline Structure and Physical Properties

HDPE is a semicrystalline plastic, whose crystallinity varies from 40 to 80%, depending on the degree of branching and molecular weight. Polymer chains in crystalline HDPE have a flat zigzag configuration. The principal crystalline form of HDPE is orthorhombic, with a density of 1.00 g/cm³ and the cell parameters $a = 0.740$, $b = 0.493$, and $c = 0.2534$ nm. The polymer chains are aligned in the *c*-axis direction. HDPE crystallizes from the melt under typical conditions as densely packed morphological structures known as spherulites. Spherulites are small spherical objects (usually from 1 to 10 μm) visible only under high magnification. They are composed of even smaller structural subunits: rod-like fibrils that spread in all directions from the spherulite centers, filling the spherulite volume. These fibrils, in turn, are made up of the smallest morphological structures distinguishable, small planar crystallites called lamellae. Crystalline lamellae offer the spherulites rigidity and account for their high softening temperature, whereas the amorphous regions between lamellae provide flexibility and high impact strength to HDPE articles.

The extrapolated equilibrium melting point of orthorhombic HDPE crystals is 146–147°C. Actual measurements of slowly crystallized samples give the highest melting point, T_m, at 133–138°C.

Owing to the high crystallinity of HDPE, most articles made from HDPE resins are opaque. Thin HDPE film, in contrast, is translucent, but its transparency is significantly lower than that of LDPE or LLDPE film. The ultraviolet transmission limit of HDPE is around 230 nm.

Various properties of HDPE are listed in Table 3.

Catalysts for HDPE Production

HDPE resins are produced in industry with several classes of catalysts, i.e., catalysts based on chromium oxides (Phillips), catalysts utilizing organochromium compounds, catalysts based on titanium or vanadium compounds (Ziegler), and metallocene catalysts.

Polymerization Processes and Processing

All technologies employed for catalytic polymerization processes in general are widely used for the manufacture of HDPE. The three most often

TABLE 3. PHYSICAL, THERMAL, ELECTRICAL, AND MECHANICAL PROPERTIES OF HDPE

Property	Highly linear	Low degree of branching[a]
PHYSICAL		
density, g/cm^3	0.962–0.968	0.950–0.960
refractive index, n_D at 25°C	1.54	1.53
THERMAL		
melting point, °C	128–135	125–132
brittleness temperature, °C	−140 to −70	−140 to −70
heat resistance temperature, °C	~122	~120
specific heat capacity, kJ/(kg · K)[b]	1.67–1.88	1.88–2.09
thermal conductivity, W/(m · K)	0.46–0.52	0.42–0.44
temperature coefficient of linear expansion	$(1–1.5) \times 10^{-4}$	$(1–1.5) \times 10^{-4}$
of volume expansion	$(2–3) \times 10^{-4}$	$(2–3) \times 10^{-4}$
heat of combustion, kJ/g[b]	46	46
ELECTRICAL		
dielectric constant at 1 MHz	2.3–2.4	2.2–2.4
dielectric loss angle, 1 kHz–1 MHz	$(2–4) \times 10^{-4}$	$(2–4) \cdot 10^{-4}$
volume resistivity, $\Omega \cdot$ m	10^{17}–10^{18}	10^{17}–10^{18}
surface resistivity, Ω	10^{15}	10^{15}
dielectric strength, kV/mm	45–55	45–55
MECHANICAL		
yield point, MPa[c]	28–40	25–35
tensile modulus, MPa[c]	900–1,200	800–900
tensile strength, MPa[c]	25–45	20–40
notch impact strength, kJ/m^2[d]	~120	~150
flexural strength, MPa[c]	25–40	20–40
shear strength, MPa[c]	20–38	20–36
elongation, %		
at yield point	5–8	10–12
at break point	50–900	50–1,200
hardness		
Brinell, MPa[c]	60–70	50–60
Rockwell	R55, D60–D70	

[a] 2–3 CH$_3$ per 1000 carbons.
[b] To convert J to cal, divide by 4.184.
[c] To convert MPa to psi, multiply by 145.
[d] To convert kJ/m^2 to ft · lbf/in.^{2}, divide by 2.10.

used technologies are slurry polymerization, gas-phase polymerization and solution polymerization. Catalysts are usually fine-tuned for a particular process.

Most high density polyethylene processing technologies require the melting of HDPE. Typical HDPE melt viscosities are between 1,000 and 100,000 Pa · s(10,000 − 10^6 P); the melt viscosity of HDPE strongly depends on temperature and on the resin molecular weight. Some resins can have a viscosity 250 times greater than that of others. The effect of temperature on the HDPE melt viscosity is described by an exponential dependence similar to the Arrhenius equation with an activation energy of 25–29 kJ/mol (6–7 kcal/mol).

Because of its low melting point and high chemical stability, HDPE is easily processed by most conventional techniques (injection molding, blow molding, rotational molding, and extrusion). Blown HDPE film is manufactured on high stalk film lines; specialized techniques have also been described.

Because of high molecular weight and extremely high melt viscosity, UHMWPE cannot be readily processed by any technique involving melt extrusion or thermoforming. Instead, these resins are processed either by compression molding into sheet, block, and precision parts; by ram extrusion into board, rods, pipe, and profiles; or by forging into parts of complex configuration.

Recycling of HDPE. Polyolefins, including HDPE are the second most widely recycled thermoplastic materials after PET. A significant fraction of articles made from HDPE (mostly bottles, containers, and film) are collected from consumers, sorted, cleaned, and reprocessed. Processing of post-consumer HDPE includes the same operations as those used for virgin resins: blow molding, injection molding, and extrusion.

Specifications, Standards, and Quality Control

According to ASTM D1248, HDPE materials are divided into various classifications based on properties. Two of the most easily measured characteristics are density and melt index; the former determines the type of HDPE, the latter its category.

Health and Safety Factors

HDPE by itself is a safe plastic material on account of its chemical inertness and lack of toxicity. Film and containers made from HDPE are used on a large scale in food and drug packaging and HDPE has been used in prosthetic devices including hip and knee joint replacements. All these applications underscore polymer safety. If articles made of HDPE contain fillers, processing aids, and colorants, their toxic effects must be estimated separately.

HDPE can present health hazards when it burns. Heavy smoke, fumes, or potentially toxic decomposition products can result from incomplete combustion. Large-scale fire testing has shown that the products formed from HDPE present no greater hazard than those from cellulosic materials, wood, felt, or rubber.

A significant part of HDPE is collected from consumers for recycling; uncollected HDPE can be disposed of by landfill or incineration. In landfill, HDPE is completely inert, degrades very slowly, does not produce gas, and does not leach any pollutants into groundwater. When incinerated in commercial or municipal facilities, HDPE produces a large amount of heat (the same as heating fuel) and therefore should constitute less than 10% of the total trash.

Uses

Blow-molded products represent the biggest use of HDPE resins, around 40%. Packaging applications account for by far the greatest share of this market. These include such products as bottles (especially for milk, juice, and soap), housewares, toys, pails, drums, and tanks.

Injection-molding products are the second largest application with approximately 20% of the HDPE market. These products includes housewares, toys, food containers, pails, crates, and cases.

Film is the third most important application for HDPE resins accounting for nearly 15% of the total HDPE market. Its share of the market is increasing rapidly as it gradually replaces paper and glass and competes with LLDPE film for many uses.

HDPE pipes are used for transporting water, sewer wastes, and gas; they are also widely used in the chemical industry. Other significant applications include wire and cable coatings, foam, insulation for coaxial and communication cables, as well as those areas where high resistance to oil and chemicals is desirable.

Low molecular weight HDPE with molecular weight of several thousand (waxes) is widely used for paper coatings, spray coatings emulsions, printing inks, crayons, and wax polishes. Waxes are also used as additives to butyl rubber and various higher molecular weight PE grades for improving melt flow characteristics, hardness, and resistance to abrasion and grease.

UHMWPE possesses a unique combination of mechanical and technological properties and enjoys a variety of special applications based on low friction (solid lubricant), wear resistance (protection of metal surfaces), excellent chemical stability, as well as radiation and neutron resistance. UHMWPE is used in chemical processing, food and beverage industries, foundries, the lumber industry; the electrical industry, as medical implants; and in mining and mineral processing sewage treatment, and transportation.

YURY V. KISSIN
Mobil Chemical Company

LINEAR LOW DENSITY POLYETHYLENE

The chemical industry manufacturers a large variety of semicrystalline ethylene copolymers containing small amounts of α-olefins. These copolymers are produced in catalytic polymerization reactions and have densities lower than those of ethylene homopolymers known as high density polyethylene (HDPE). Ethylene copolymers produced in catalytic polymerization reactions are usually described as linear ethylene polymers, to distinguish them from ethylene polymers containing long branches which are produced in radical polymerization reactions at high pressures.

TABLE 4. BASIC CLASSIFICATION OF COPOLYMERS

Resin	Designation	α-Olefin, mol %	Crystallinity, %	Density, g/cm³
PE of medium density	MDPE	1–2	55–45	0.940–0.926
linear PE of low density	LLDPE	2.5–3.5	45–30	0.925–0.915
PE of very low density	VLDPE	>4	<25	<0.915

Densities and crystallinities of ethylene–α-olefin copolymers mostly depend on their composition. The classification in Table 4 is commonly used (ASTM D1248-48).

The large number of commodity and specialty resins collectively known as LLPDE are in fact made up of various resins, each different from the other in the type and content of α-olefin in the copolymer, compositional and branching uniformity, crystallinity and density, and molecular weight and molecular weight distribution (MWD).

Four olefins are used in industry to manufacture ethylene copolymers: 1-butene, 1-hexene, 4-methyl-1-pentene, and 1-octene. Copolymers containing 1-butene account for approximately 40% of all LLDPE resins manufactured worldwide, 1-hexene copolymers for 35%, 1-octene copolymers for about 20%, and 4-methyl-1-pentene copolymers for the rest. The type of α-olefin exerts a significant influence on the copolymer properties.

Two classes of LLDPE resins are on the market. One has a predominantly uniform compositional distribution (uniform branching distribution); that is, all copolymer molecules in these resins have approximately the same composition. Most commercially produced LLDPE resins, in contrast, have pronounced nonuniform branching distributions; there are significant differences in copolymer compositions among different macromolecules in a given resin.

Crystallinity and density of LLDPE, which are closely related, depend mostly on the amount of α-olefin in the copolymer. Both density and crystallinity of ethylene copolymers are also influenced by their compositional uniformity. As a rule, lower α-olefin content is needed in a uniformly branched ethylene copolymer to decrease its crystallinity and density to a given level.

The range of molecular weights of commercial LLDPE resins is relatively narrow, usually from 50,000 to 200,000. One accepted parameter that relates to the resin molecular weight is the melt index, a rheological parameter which, broadly defined, is inversely proportional to molecular weight. A typical melt index range for LLDPE resins is from 0.1 to 5.0, but can reach over 30 for some applications.

Molecular Structure and Properties

LLDPE resins are copolymers of ethylene and α-olefins with low α-olefin contents. Molecular chains of LLDPE contain units derived both from ethylene, $-CH_2-CH_2-$, and from the α-olefin, $-CH_2-CHR-$, where R is C_2H_5 for ethylene–1-butene copolymers, n-C_4H_9 for ethylene–1-hexene copolymers, $-CH_2-CH(CH_3)_2$ for ethylene–4-methyl-1-pentene copolymers, and n-C_6H_{13} for ethylene–1-octene copolymers. In a typical copolymer molecule containing 2–4 mol % of α-olefin, the majority of olefin units stand alone in the chain but most ethylene units form long sequences. As a rule, LLDPE resins do not contain long-chain branches. However, some copolymers produced with metallocene catalysts in solution processes can contain about 0.002 long-chain branches per 100 ethylene units.

LLDPE resins produced with different catalysts vary greatly in their compositional uniformity. The fastest way to evaluate the branching distribution of an LLDPE resin is to measure its melting point. Melting points of copolymers with uniform compositional distributions show a noticeable dependence on their composition and may vary widely: from ~120°C for copolymers containing 1.5–2 mol % of α-olefin to ~110°C for copolymers containing 3.5 mol % of α-olefin. A copolymer with a nonuniform compositional distribution, in contrast, is a mixture that contains copolymer molecules with a broad range of compositions, from almost linear macromolecules (usually of higher molecular weights) to short macromolecules with quite high α-olefin contents. Melting of such mixtures is dominated by their low branched fractions which are highly crystalline. They are not too sensitive to copolymer composition and usually fall in the temperature range of 125–128°C.

LLDPE is a saturated branched hydrocarbon. The most reactive parts of LLDPE molecules are the tertiary CH bonds in branches and the double bonds at chain ends. LLDPE is nonreactive with both inorganic and organic acids and is stable in alkaline and salt solutions. At room temperature, LLDPE resins are not soluble in any known solvent; at temperatures above 80–100°C, however, the resins can be dissolved in various aromatic, aliphatic, and halogenated hydrocarbons.

LLDPE is relatively stable to heat. Thermal degradation starts at temperatures above 250°C and results in a gradual decrease of molecular weight and the formation of double bonds in polymer chains.

Oxidation of LLDPE starts at temperatures above 150°C. To protect molten resins from oxygen attack, antioxidants must be used in concentrations of 0.1–0.5 wt %.

Photooxidative degradation of LLDPE at ambient temperature under sunlight is also a radical oxidation reaction. It causes change in color and drastic deterioration of mechanical and dielectric properties of LLDPE articles. Photooxidation can be prevented by using light stabilizers.

Physical Properties. LLDPE is a semicrystalline plastic whose chains contain long blocks of ethylene units that crystallize in the same fashion as paraffin waxes or HDPE. The degree of LLDPE crystallinity depends primarily on the α-olefin content in the copolymer and is usually below 40–45%. The principal crystalline form of LLDPE is orthorhombic (the same as in HDPE); the cell parameters of nonbranched PE are $a = 0.740$ nm, $b = 0.493$ nm, and c (the direction of polymer chains) = 0.2534 nm.

LLDPE rapidly crystallizes from the melt with the formation of spherulites, small spherical objects 1–5 μm in diameter visible only in a microscope. The elementary structural blocks in spherulites are lamellae, small flat crystallites formed by folded linear segments in LLDPE chains, which are interconnected by polymer chains that pass from one lamella to another (tie molecules). Crystalline lamellae within spherulites give LLDPE articles necessary rigidity, whereas the large amorphous regions between lamellae, constituting over 60% of the spherulite volume, provide flexibility.

The size of the lamellae for a copolymer of a given composition depends on the degree of branching uniformity. If an LLDPE resin is compositionally uniform, all its macromolecules crystallize poorly due to branching, forming very thin lamellae. Such materials have low rigidity (low modulus) and high flexibility. On the other hand, if an LLDPE resin is compositionally nonuniform, its least-branched components are able to form thicker lamellae; consequently, more branched fractions of the resin remain amorphous and fill the voids between the lamellae. Articles made from such resins are more rigid.

Optical properties of LLDPE resins also depend on the degree of branching uniformity. Resins with a uniform branching distribution make highly transparent film with haze as low as 3–4%. In contrast, film manufactured from compositionally nonuniform copolymers is much more opaque, with haze of over 10–15%; this is due to the presence of large crystalline lamellae consisting of nearly nonbranched PE chains.

Because it is a saturated aliphatic hydrocarbon, LLDPE does not conduct electricity, and so is widely used for wire and cable insulation. LLDPE is poorly permeable to water and inorganic gases and only slightly more so to organic compounds, whether liquid or gas.

Mechanical Properties. Mechanical characteristics of ethylene copolymers are functions of their structural characteristics, such as content and type of α-olefin, branching uniformity, molecular weight and width of molecular weight distribution (MWD), and orientation (see Table 5 for properties of films made from three grades of LLDPE).

Catalysts and LLDPE Polymerization Processes

LLDPE resins are produced in industry with three classes of catalysts titanium-based catalysts (Ziegler), metallocene-based catalysts (Kaminsky and Dow), and chromium oxide-based catalysts (Phillips).

Ziegler catalysts account for by far the greatest share of LLDPE resins manufactured. They consist of two components: the first contains as its active ingredient a derivative of a transition metal (usually titanium); the second is an organoaluminum compound such as triethylaluminum. The molar [Al]:[Ti] ratio in these catalyst systems is usually in the 50–500 range. Most commercially important catalysts are heterogeneous; a large

TABLE 5. PROPERTIES OF COMMERCIAL LLDPE FILM OF RESINS WITH NONUNIFORM COMPOSITIONAL DISTRIBUTION[a]

Property	Copolymer of ethylene and:				
	1-butene		1-hexene		1-octene
density g/cm^3	0.918	0.918	0.918	0.918	0.919
melt index, g/10 min	2.0	1.0	1.0	0.5	1.0
dart impact strength, g[b]	110	150	250	300	350
puncture energy J/mm[c]	60	70	85	94	61
tensile strength,[d] MPa[e]					
MD	33	38	38	43	43
TD	25	31	32	43	34
elongation at break,[e] %					
MD	690	620	570		550
TD	740	760	790		660
modulus,[d] MPa[e]					
MD	210	230			
TD	250	260			

[a] Film thickness 37 μm.

[b] Average weight of the dart sufficient to break the film in 50% of tests (ASTM D4272-90, Method A).

[c] To convert J/mm to ft·lbf/in., multiply by 18.73.

[d] MD = machine direction; TD = transverse direction.

[e] To convert MPa to psi, multiply by 145.

number of them are supported. Both inorganic and organic supports are used, the most important are MgCl$_2$ and silica.

Typical heterogeneous Ziegler catalysts operate at temperatures of 70–100°C and pressures of 0.1–2 MPa (15–300 psi). The polymerization reactions are carried out in an inert liquid medium (e.g., hexane, isobutane) or in the gas phase. Molecular weights of LLDPE resins are controlled by using hydrogen as a chain-transfer agent. Reactivities of α-olefins in copolymerization with ethylene depend on two factors: the size of the alkyl groups attached to their double bonds and the type of catalyst.

Three types of metallocene catalysts are presently used in industry: Kaminsky, ionic, and Dow constrained-geometry catalysts.

Chromium oxide-based catalysts, which were originally developed for the manufacture of HDPE resins, have been modified for ethylene–α-olefin copolymerization reactions. These catalysts use a mixed silica–titania support containing from 2 to 20 wt % of Ti.

Polymerization

The technologies suitable for LLDPE manufacture include gas-phase fluidized-bed polymerization, polymerization in solution, polymerization in a polymer melt under high ethylene pressure, and slurry polymerization. Most catalysts are fine-tuned for each particular process.

Processing

All LLDPE processing technologies involve resin melting; viscosities of typical LLDPE melts are between 5000 and 70,000 Pa·s (50,000–700,000 P). The main factor that affects melt viscosity is the resin molecular weight; the other factor is temperature. Its effect is described by the Arrhenius equation with an activation energy of 29–32 kJ/mol (7–7.5 kcal/mol).

LLDPE melts in the 140–250°C range (a typical processing range) are non-Newtonian liquids; their effective viscosity is significantly reduced when the melt flow speed is increased. This phenomenon is called shear-thinning; it plays an important role in resin processing. The resins with an expressed shear-thinning capability have decreased viscosities and hence a greatly reduced energy demand at high speed processing.

LLDPE is easily processed by most conventional techniques due to its low melting point and high chemical stability.

Most LLDPE produced worldwide is made into thin film, either melt blown or cast from the melt. Blown film is produced by extrusion of LLDPE melt through a circular die with a large diameter, up to 100–120 cm, and a narrow gap, usually less than 1 mm. The tube of molten polymer expands from internal air pressure and forms a bubble of a larger diameter. Typically, the blow-up ratio, the ratio of the final film tube diameter to the circular-die diameter, is between 1.5:1 and 4:1. LLDPE cast film is manufactured by depositing polymer melt on a rotating heated drum with a highly polished surface. Compositional uniform LLDPE resins produced with metallocene catalysts can be easily processed into film by using standard equipment with minor modifications.

Injection molding is used for the manufacture of LLDPE articles of complex shapes. The duration of the molding cycle depends on the melt viscosity and the rate of polymer crystallization. Because LLDPE crystallizes rapidly, molding cycles are short, typically from 10 to 30 seconds. Molds usually accommodate up to 50 or more articles formed in a single shot. Bottles and simple containers are manufactured in large quantities by the blow molding technique. LLDPE resins with high molecular weights and high melt viscosities are used in this method. Large containers and some toys are manufactured from LLDPE powder with a specialized technique called rotational molding. Extrusion applications include resin pelletization after LLDPE synthesis, and manufacture of thick film, sheet, pipe, tubing, and insulated wire.

Health and Safety Factors

LLDPE by itself does not present any health-related hazard on account of its chemical inertness and low toxicity. Consequently, film, containers, and container lids made from LLDPE are used on a large scale in food and drug packaging. Some LLDPE grades produced with unsupported metallocene catalysts have an especially high purity due to high catalyst productivity and a low contamination level of resins with catalyst residue. FDA approved the use of film manufactured from these resins for food contact and for various medical applications. However, if LLDPE articles contain fillers, processing aids, or colorants, their health factors must then be judged separately.

LLDPE can present a certain health hazard when it burns, since smoke, fumes, and toxic decomposition products are sometimes formed in the process.

LLDPE can be disposed of by landfill or incineration. In landfill, the material is completely inert, degrades very slowly, does not produce gas, and does not leach any pollutants into ground water. When incinerated in commercial or municipal facilities, LLDPE produces a large amount of heat (the same as heating fuel) and should constitute less than 10% of the total trash.

Uses

By far the largest application for LLDPE resins (over 60% in the United States) is film. Because LLDPE film has high tensile strength and puncture resistance, it is able to compete with HDPE film for many uses. Bags manufactured from thin LLDPE film have excellent tensile strength, puncture resistance, and seal strength at thin gauges; they can be used either for packaging or as garment bags, laundry and dry-cleaning bags, and ice bags. Several LLDPE films are currently competing for a special film market: elastic stretch film for packaging. Both blown and die-case film can be used for this application. The resins based on ethylene–1-hexene and ethylene–1-octene copolymers are particularly suited for these purposes. A significant volume of LLDPE film is used to manufacture large-size packaging material for food (e.g., grocery sacks) and textiles, in addition to such applications as industrial sheeting and agricultural mulch film. LLDPE bags are thinner than LDPE bags; their thickness can be further reduced to around 25 μm, which makes them price-competitive with paper.

High oxygen-barrier properties of metallocene-derived resins make their film especially attractive for packaging poultry, frozen foods, and vegetables. Such properties, moreover, make these resins and their blends with HDPE suitable for producing film for blood bags, surgical disposable bags, and medical gowns.

An important property recommending the use of LLDPE in many packaging applications is their sealability. Compositionally uniform resins are especially attractive for such use because their melting and softening points are 15–20°C lower than those of commodity LLDPE resins.

Injection molding is the second largest market for LLDPE, accounting for over 10% of its total consumption. Over half of the LLDPE consumed in injection-molding applications is used for housewares.

LLDPE resins formulated for the blow molding applications have superior environmental stress-cracking resistance and low gas permeability. These features opened new bottle markets where such properties are important. A large variety of molded articles with a complex configuration are manufactured from LLDPE resins, including toys, large square-edged containers, as well as tanks for agriculture and water treatment.

The same qualities that make LLDPE attractive for blow molding applications also play a crucial role in its being adapted for pipe manufacture, an area that accounts for about 1% of the LLDPE market. LLDPE pipes provide flexibility, high burst strength, and high environmental stress-cracking resistance, and a high heat-distortion temperature. LLDPE tubing is used for drip piping, swimming pool tubing, household hoses, and in such specialty markets as medical tubing applications.

LLDPE is also widely used for wire and cable coating in electrical and telephone industry, which amounts to 2.5% of LLDPE production.

YURY V. KISSIN
Mobil Chemical Company

POLYPROPYLENE

Propylene polymerization processes have undergone a number of revolutionary changes since the first processes for the production of crystalline polypropylene (PP) were commercialized in 1957 by Montecatini in Italy and Hercules in the United States. These first processes were based on Natta's discovery in 1954 that a Ziegler catalyst could be used to produce highly isotactic polypropylene. The stereoregular, crystalline polymers produced by this technology had sufficiently attractive economic and property performance that they became significant commercial thermoplastics in a remarkably short period.

Properties

Structure and Crystallinity. The stereochemistry of propylene polymers was first studied by Natta, who defined three possible structures of polypropylene by the location of the pendent methyl groups relative to the polymer backbone. Isotactic polypropylene consists of molecules in which all methyl groups have the same stereochemistry as a result of all insertions of propylene monomer being identical. Syndiotactic polypropylene is produced by regular alternating stereochemistry of monomer insertion, resulting in alternating locations of the pendent methyl groups. Atactic polypropylene, which is noncrystalline, is the result of nonstereospecific monomer insertion and random location of the pendent methyl groups (Fig. 1).

Crystallinity of polypropylene is usually determined by x-ray diffraction. Isotactic polymer consists of helical molecules, with three monomer units per chain unit, resulting in a spacing between units of identical conformation of 0.65 nm.

Syndiotactic polypropylene also forms helical molecules; however, each chain unit consists of four monomer units having a spacing of 0.74 nm. The unit cell is orthorhombic and contains 48 monomer units having a crystallographic density of 0.91 g/cm^3.

Molecular Weight. The molecular weight of polypropylene is typically determined by viscosity measurements. The melt viscosity, or melt flow rate, measured under standard conditions, can also be correlated to molecular weight. Because of the distribution of molecular weights in polypropylene, and the relationship of this distribution to important polymer properties, considerable effort has been exerted to measure the molecular weight distribution of these polymers.

Thermodynamic Properties. The thermodynamic melting point for pure crystalline isotactic polypropylene obtained by the extrapolation of melting data for isothermally crystallized polymer is 185°C. Under normal thermal analysis conditions, commercial homopolymers have melting points in the range of 160–165°C. The heat of fusion of isotactic polypropylene has been reported as 88 J/g (21 cal/g). The value of 165 ± 18 J/g has been reported for a 100% crystalline sample.

The value of the glass-transition temperature, T_g, is dependent on the stereoregularity of the polymer, its molecular weight, and the measurement techniques used. Transition temperatures from −13 to 0°C are reported for isotactic polypropylene, and −18 to 5°C for atactic.

Syndiotactic polypropylene has an ultimate melting point of 174°C, and extrapolated heat of fusion of 105 J/g (25.1 cal/g); both lower than those of isotactic polymer. The heat of fusion of the polymer produced using a metallocene catalyst is reported as 79 J/g (19 cal/g).

Physical Properties. Properties of various homopolymer grades are given in Table 6.

Polypropylene polymers are typically modified with ethylene to obtain desirable properties for specific applications. Specifically, ethylene-propylene rubbers are introduced as a discrete phase in heterophasic copolymers to improve toughness and low temperature impact resistance.

Catalysts Used in Manufacture

TiCl$_3$-Based Catalysts. Isotactic polypropylene was first synthesized by Natta in 1954, using a catalyst system consisting of TiCl$_4$ and Al(C$_2$H$_5$)$_3$. This system, based on Ziegler's catalyst for polyethylene, produced a large fraction of polymer with poor structural uniformity and properties. These catalysts, activated with Al(C$_2$H$_5$)$_2$Cl or Al(C$_2$H$_5$)$_3$, dramatically increased the percentage of isotactic polymer. Although all four crystal forms of TiCl$_3$, α, β, γ, and δ, are active as catalysts, the best results were obtained using δ-TiCl$_3$ activated by Al(C$_2$H$_5$)$_2$Cl. These catalysts enabled rapid commercialization of the production of isotactic polypropylene.

TiCl$_3$ catalysts produced by the reduction of TiCl$_4$ with Al(C$_2$H$_5$)$_2$Cl, and subsequently treated first with an electron donor (diisoamyl ether), then with TiCl$_4$, are highly stereospecific and four to five times more active than δ-TiCl$_3$. These catalysts were a significant advance over the earlier TiCl$_3$ systems, because removal of atactic polymer was no longer required. They are often referred to as second-generation catalysts. The life of many older slurry process facilities has been extended by using these catalysts to produce "clean" polymers with very low catalyst residues.

MgCl$_2$-Supported Catalysts. Magnesium chloride, in active form as a support for TiCl$_4$, has been found to significantly increase catalyst activity, enabling the design of processes in which removal of catalyst residues from the polymer is not required. The use of electron donors as stereoregularity control agents reduces formation of the undesirable atactic polymer, giving stereoregularity sufficient for commercial production of polypropylene.

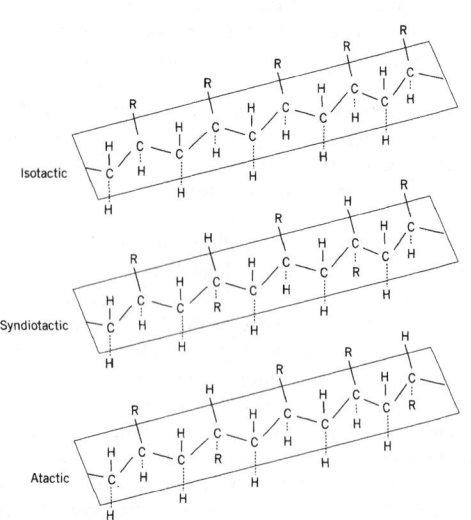

Fig. 1.   Polypropylene stereoisomers.

TABLE 6. PROPERTIES OF HOMOPOLYMERS

Properties	ASTM test	Extrusion, sheet	General-purpose injection molding		Injection molding thin complex parts	
melt flow, g/10 min	D1238L	0.8	4	12	20	35
density, g/cm^3	D792A-2	0.903	0.903	0.903	0.903	0.902
tensile strength,[a] MPa[b]	D638	35	35	35	32	33
elongation,[a] %	D638	13	12	11	11	12
flexural modulus 1% secant, MPa[c]	D790B	1700	1700	1600	1500	1450
Rockwell hardness, R scale	D785A	95	99	100	100	98
deflection temperature at 455 kPa,[b] °C	D648	95	97	92	91	90
notched Izod impact at 23°C, J/m[d]	D256A	130	40	35	21	32

[a] At yield.
[b] To convert kPa to psi, multiply by 0.145.
[c] To convert MPa to psi, multiply by 145.
[d] To convert J/m to ft · lb/in., divide by 53.38.

Active catalyst systems have also been produced using silica as a support for MgCl$_2$, and TiCl$_4$. A number of magnesium-containing compounds such as alkylmagnesium, magnesium hydrocarbyl carbonates, magnesium alkanoates, magnesium alkoxides and aryloxides, and alkyl magnesium chloride, can be used as the source of activated MgCl$_2$ in addition to the alcoholates of MgCl$_2$ or anhydrous MgCl$_2$. In all cases, these systems contain active centers consisting of TiCl$_4$ supported on activated MgCl$_2$.

Metallocene Catalysts. The use of the dicyclopentadienyl titanium dichloride—diethyl aluminum chloride system to catalyze the polymerization of ethylene homogeneously was first reported in 1957. Dramatic improvements in the performance of these metallocene catalysts were obtained by using methylalumoxane as cocatalyst and dicyclopentadienyl zirconium dichloride as catalyst. Modification of the zirconocene by the addition of substituents to the cyclopentadiene rings provided the capacity of achieving high molecular weight polyethylene at economical process temperatures. Modifications of the organic substituents have resulted in the development of metallocene catalysts capable of producing isotactic polypropylene with similar melting points and molecular weights to commercial Ziegler-Natta propylene.

Manufacturing Processes

Early Processes. The first commercial processes for the production of polypropylene were batch polymerization processes using TiCl$_3$ catalysts activated by Al(C$_2$H$_5$)$_2$Cl in a hydrocarbon medium. As the demand for polypropylene increased, these batch polymerization processes were rapidly replaced by continuous ones.

Polymerization in liquid monomer was pioneered by Rexall Drug and Chemical and Phillips Petroleum (United States). Gas-phase polymerization of propylene was pioneered by BASF, who developed the Novolen process which uses stirred-bed reactors. Eastman Chemical has utilized a unique, high temperature solution process for propylene polymerization. In the 1970s, Solvay introduced an advanced TiCl$_3$ catalyst with high activity and stereoregularity.

Montedison and Mitsui Petrochemical introduced MgCl$_2$-supported high yield catalysts in 1975. These third-generation catalyst systems reduced the level of corrosive catalyst residues to the extent that neutralization or removal from the polymer was not required. Stereospecificity, however, was insufficient to eliminate the requirement for removal of the atactic polymer fraction. These catalysts are used in the Montedison high yield slurry process.

Current Processes. The development of superactive third-generation supported catalysts enabled the introduction of simplified processes, without sections for catalyst deactivation or removal of atactic polymer. By eliminating the waste streams associated with the neutralization of catalyst residues and purification of the recycled diluent and alcohol, these processes minimize any potential environmental impact. Investment costs are reduced by approximately one-third over slurry process plants. Energy consumption is minimized by elimination of the distillation of recycled diluent and alcohol. The total plant cost for the production of polymer is less than 130% of the monomer price, when a modern process is used, compared to 175% for a slurry process.

Processing

Polypropylene is produced in a variety of molecular weights, molecular weight distributions, and crystallinities; consequently, it can be used in most polymer processing technologies. The physical and mechanical properties of polypropylene in the end use product are a function of both the molecular structure and the processing conditions. The final crystalline morphology is a function of the melt temperatures, polymer orientation, and cooling temperatures and rates. The most commonly used processes are injection molding, blow molding, extrusion, and thermoforming (see **Plastics**).

Melt Spinning. This process is used to produce a broad range of polypropylene fibers ranging from fine, dtex (one denier) staple coarse continuous filaments. Homopolymers are almost exclusive used to produce fibers, although copolymer blends are used in some special applications. Processing conditions and polymer melt flow vary with the desired fiber type.

Slit and Split Films. Thick industrial-grade yarns are often produced by slitting films, providing a less expensive alternative to melt spun fiber. Cast film is slit in the machine direction by parallel rotary knives. The resulting tape can then be cold drawn in an oven in a manner similar to melt spun fibers to produce the final fiber.

Melt Blowing. The melt blowing process uses very high melt flow polymers, sometimes in excess of 400 dg/min, and extrusion and temperatures above 300°C to produce very fine fibers (<5 μm dia).

Spun Bonded Fabrics. Spun bonded fabrics are produced by depositing extruded, spun filaments onto a collecting belt in a uniform randomized manner. The fibers are separated during the web laying process by air jets and the collecting belt is usually perforated to prevent the air stream from deflecting and carrying the fibers in an uncontrolled manner. Spun bonded fabrics are used in a variety of applications requiring nonwoven fabrics (qv), competing with thermally bonded staple fiber.

Polypropylene (PP) films were first produced by extrusion casting. Polymer is extruded through a slit or tubular die and quenched by cooling on chill rolls or in a water bath. Cast films can be sealed over a wide range of temperatures and do not shrink in a steam autoclave. Polymers with melt flow rates below 5 dg/min are usually used to maintain the stability of the extrudate. Higher clarity films are produced using random copolymers.

Biaxially oriented polypropylene (BOPP) films have higher stiffness than cast films and consequently can be used in much thinner gauges. Homopolymers are used almost exclusively to provide maximum stiffness and water-vapor barrier. Oriented films are produced by the tenter frame and bubble processes.

Polypropylene is subject to attack by oxygen, radiation, and excessive heat causing a loss of molecular weight and physical properties. Stabilizers are added to the polymer to minimize these effects. Small quantities of hindered phenolic antioxidants (qv) are added in the polymerization plant, usually in the drying section, to protect the polymer against degradation during short-term storage. The bulk of the stabilizer is added during pelletization or fabrication to protect the polymer during processing or in the final application.

Uses

Polypropylene is extensively used in injection molding because of the wide range of physical properties and melt flow rates available. The principal markets served include transportation (primarily automotive), appliances, consumer products, rigid packaging, and medical products. Polypropylene use has increased in the automotive industry because of the wide availability of high melt flow rate impact copolymers for use in large thin parts, such as interior trim.

Polypropylene is frequently utilized in the design of container closures, lunchboxes, and similar articles. Polypropylene closures are used on a wide variety of containers, including child-proof caps and screw caps on plastic beverage bottles. Medical devices injection molded from polypropylene include syringes, pans, trays, and a variety of utensils. A large variety of toys, cups, dishes, and other household articles are molded from polypropylene. The development of high melt flow rate copolymer grades for thin-wall injection molding has increased the use of polypropylene in food containers.

Polypropylene blow-molded bottles are used to package a wide variety of products including foods, cleansers, shampoo, pharmaceuticals, and mouthwash.

Improved equipment and polymers have increased the capability to extrude and thermoform polypropylene. Drinking straws are commonly extruded from polypropylene, however most larger diameter tubes, such as pipes and conduits, are predominantly extruded from other thermoplastics. Extruded sheet is thermoformed into food containers and trays; polypropylene is used when microwavability is desired.

Polypropylene fibers are extensively used in carpeting.

Disposable polypropylene nonwoven fabrics are widely used as the cover-stock for disposable baby diapers, baby wipes, adult incontinence, and feminine hygiene products. Use of polypropylene nonwovens in disposable medical apparel, such as surgical gowns, has increased as a means of reducing the spread of infection.

Oriented polypropylene films are widely used in the packaging of snack foods, candy, and other products. The most common method of packaging using these films is the form and fill process in which the package is formed filled with product, and sealed in a continuous process.

Cast films provide a high clarity, heat sealable film and are primarily used as an overwrap for boxes and other packaging. These films have a lower density than cellophane and provide a longer product shelf life.

RICHARD B. LIEBERMAN
Montell Polyolefins

POLYMERS OF HIGHER OLEFINS

Crystalline polymers of α-olefins, i.e., those with carbon numbers of four or higher, are stereoregular, i.e., isotactic or syndiotactic polymers, such as polypropylene (PP). These polymers are produced with a number of different catalysts. Heterogeneous Ziegler-Natta catalysts and some soluble bridged metallocene catalysts produce isotactic polymers, other types of bridged metallocene catalysts produce crystalline syndiotactic polymers whereas nonbridged metallocene catalysts yield amorphous atactic polymers.

The synthesis of isotactic polymers of higher α-olefins was discovered in 1955, syndiotactic polymers of higher α-olefins were first prepared in 1990. The first commercial production of isotactic poly(1-butene) (PB) and poly(4-methyl-1-pentene) (PMP) started in 1965.

Higher α-olefins can also be polymerized with cationic initiators to liquid oligomeric materials with isomerized structures. These liquids are manufactured commercially and used as lubricating oils.

Cycloolefins are polymerized by means of two different mechanisms. In the first, catalysts based on tungsten and molybdenum compounds induce ring-opening polymerization (metathesis) of monocycloolefins with the formation of linear elastomers containing regularly spaced double bonds in polymer chains. If cyclodienes are used in this reaction, these catalysts then produce cross-linked resins. In the second mechanism, metallocene-based catalysts polymerize cycloolefins without ring opening into linear, stereoregular, highly crystalline polymers. Polymers of several cycloolefins, polydicyclopentadiene, polyoctenamers, and norbornene elastomers are all produced commercially.

Monomers

The monomers of the greatest interest are those produced by oligomerization of ethylene and propylene. Some olefins are also available as by-products from refining of petroleum products or as the products of hydrocarbon thermal cracking.

Commercial production of 1-butene, as well as the manufacture of other linear α-olefins with even carbon atom numbers, is based on the ethylene oligomerization reaction.

4-Methyl-1-Pentene is produced commercially by dimerization of propylene in the presence of potassium-based catalysts at 150–160°C and ~10 MPa.

Linear α-olefins, such as 1-hexene and 1-octene, are produced by catalytic oligomerization of ethylene with triethylaluminum or with nickel-based catalysts. Olefins with branched alkyl groups are usually produced by catalytic dehydration of corresponding alcohols.

1-Butene is a colorless, flammable, noncorrosive gas. Because 1-butene has a very low flash point, it poses a strong fire and explosion hazard.

4-Methyl-1-pentene is a light, colorless, flammable liquid. It is an irritant and, in high concentrations, a narcotic. Like 1-butene, this chemical compound has a low flash point and represents a significant fire hazard when exposed to heat, flame, or oxidizing agents.

Higher α-olefins are exceedingly reactive in radical and ionic reactions. These olefins participate in numerous reactions, such as oxidations, hydrogenation, double-bond isomerization, complex formation with transition-metal derivatives, polymerization, and copolymerization with other olefins in the presence of Ziegler-Natta, metallocene, and cationic catalysts. All olefins readily form peroxides by exposure to air.

Polymer Properties

Chemical properties of most polyolefins resemble those of polypropylene. The resins resist most inorganic or organic acids and bases below 90°C as well as most salt solutions, solvents, soaps, and detergents. Properties of polyolefins rapidly deteriorate in contact with strong oxidizing agents. All polyolefins undergo peroxidation, halogenation, and halosulfonation reactions.

Commercially produced crystalline polyolefins, PB, and PMP exhibit high stability to inorganic substances, and excellent resistance in nonoxidative inorganic environments. PMP easily withstands prolonged boiling and autoclave treatment required for medical and pharmaceutical applications. Prolonged exposure of polyolefin specimens under stress to some hydrocarbon solvents and aqueous detergent solutions can cause cracks and eventual failure, a phenomenon usually referred to as environmental stress cracking.

Polymers of α-olefins are susceptible to thermal and thermooxidative degradation. Reactivity in degradation reactions is especially significant in the case of polyolefins with branched alkyl side groups.

Thermooxidative degradation of PMP is noticeable even at 140–150°C, and its photooxidative degradation, especially at wavelengths below 400 nm, also proceeds at a relatively high rate, thus limiting some of its outdoor applications.

Both thermooxidation and photooxidation of polyolefins can be prevented by using the same antioxidants as those employed for the stabilization of polypropylene, i.e., alkylated phenols, polyphenols, thioesters, and organic phosphites in the amount of 0.2–0.5%.

Polybutene can be cross-linked by irradiation at ambient temperature with γ-rays or high energy electrons in the absence of air. PMP is relatively stable to β- and γ-radiation employed in the sterilization of medical supplies.

Highly crystalline isotactic polyolefins are not soluble in organic solvents at room temperature. However, most amorphous polyolefins and oligomers of α-olefins are easily soluble in saturated and aromatic hydrocarbons at ambient temperature. This difference in solubility can be used to separate amorphous atactic components of polyolefins from crystalline isotactic material in crude polyolefins mixtures.

Above 100°C, most polyolefins dissolve in various aliphatic and aromatic hydrocarbons and their halogenated derivatives.

Crystalline PMP is relatively highly permeable to various organic and inorganic gases. Permeabilities to oxygen, nitrogen, and light hydrocarbons are 20–30 times higher than those of HDPE.

Physical Properties. Table 7 lists physical properties of stereoregular polymers of several higher α-olefins.

Thermal Properties. Melting points of stereoregular PO resins depend on the size and shape of side groups in the polymer chains (Table 7). In the case of isotactic polymers of linear α-olefins, the melting points of the crystalline phase rapidly decrease with increasing side-chain length, isotactic poly(1-hexene) is amorphous, and isotactic polyolefins with longer linear side groups derive their crystallinity from the side chains rather than from the polymer backbone. Polymers of α-olefins with branched alkyl groups generally exhibit much higher melting points (Table 7); the melting points of isotactic poly(3-methyl-1-butene) and poly(vinylcyclohexane) are the highest, over 350°C.

Mechanical Properties. The side-group type in the polymer chains determines mechanical properties of stereoregular polyolefins. Resins with long linear side groups have low crystallinities and exhibit mechanical behavior typical for elastomers. On the other hand, PB and most polymers with branched side groups are highly crystalline and exhibit mechanical properties similar or superior to those of isotactic PP. Excellent mechanical and optical properties contribute to the industrial importance of PB and PMP. Although poly(3-methyl-1-butene) and poly(vinylcyclohexane) also exhibit good mechanical characteristics, their high melting points, poor oxidative stability, and brittleness still preclude them from finding industrial application.

All polyolefins have low dielectric constants and can be used as insulators; in particular, PMP has the lowest dielectric constant among all synthetic resins. As a result, PMP has excellent dielectric properties and a low dielectric loss factor, surpassing those of other polyolefin resins and polytetrafluoroethylene (Teflon). These properties remain nearly constant over a wide temperature range. The dielectric characteristics of poly(vinylcyclohexane) are especially attractive: its dielectric loss remains constant between –180 and 160°C, which makes it a prospective high frequency dielectric material of high thermal stability.

Although most polyolefins are highly opaque, isotactic PMP possesses the outstanding feature: it has low haze (1.2–1.5%) and high optical transparency (~90–92%) comparable to that of polystyrene (88–92%) and acrylics (90–92%). Light transmittance of PMP in the near uv region is also excellent, higher than that of glass and inferior only to quartz. Optical clarity accounts for many applications of PMP.

Polymers of monocyclic olefins (cyclopentene, cyclooctene) produced by ring-opening metathesis are linear elastomers. Their properties are somewhat similar to those of poly(cis-1,4-butadiene). Polymers of dicyclopentadiene produced with the same catalysts are heavily cross-linked resins displaying high toughness and tensile strength as well as excellent impact strength at low temperatures.

Polymerization of α-Olefins

Ziegler-Natta Catalysts. All isotactic polymers of higher α-olefins are produced with the same type of heterogeneous, titanium-based Ziegler-Natta catalyst systems as that used for the manufacture of isotactic PP.

TABLE 7. PROPERTIES OF POLYOLEFINS

Polymer[a]	Melting point, °C	Crystal type	Helix type	Crystalline density, g/cm³
POLYMERS OF α-OLEFINS				
iso-polybutene				
form I	138–142	hexagonal	3_1	0.951
form II	120–130	tetragonal	11_3	0.902
form III	101–109	orthorhombic		0.905
syndio-polybutene				
form I	~50		4_1	
form II	~50		10_3	
iso-poly(1-pentene)	105–115	monoclinic	3_1	0.92
	75–80	pseudo orthorhombic	4_1	0.90
iso-poly(3-methyl-1-butene)	350	monoclinic	4_1	0.93
iso-poly(1-hexene)	<20	monoclinic	7_2	0.83
iso-poly(3-methyl-1-pentene)	200		7_2	
iso-poly(4-methyl-1-pentene)	235–240	tetragonal	7_2	0.813
syndio-poly(4-methyl-1-pentene)	197		24_7	
iso-poly(1-heptene)	18		3_1	
iso-poly(4-methyl-1-hexene)	188–200	tetragonal	7_2	0.845
syndio-poly(4-methyl-1-hexene)	147			
iso-poly(5-methyl-1-hexene)	110–130	monoclinic	3_1	0.84
iso-poly(1-octene)	~20			
iso-poly(5-methyl-1-heptene)	130	tetragonal	3_1	
iso-poly(vinylcyclo-hexane)	376–385	tetragonal	4_1	0.95
syndio-poly(vinylcy-clohexane)		amorphous		
iso-poly(1-decene)	22–27	side chain		
POLYMERS OF CYCLOOLEFINS				
diiso-polycyclobutene	485			
diiso-polycyclopentene	395			
polynorbornene	>600			

[a] *iso* designates isotactic; *syndio* designates syndiotactic.

The catalyst systems have two components, a solid catalyst containing a titanium compound and a co-catalyst containing an organoaluminum compound. Both 1-butene and 4-methyl-1-pentene are three to four times less reactive in the polymerization reactions with Ziegler-Natta catalysts than propylene. Modern, highly active catalysts are supported on $MgCl_2$. Some contain aromatic esters such as ethyl benzoate. These catalysts are employed with co-catalyst mixtures containing $Al(C_2H_5)_3$ or $Al(i\text{-}C_4H_9)_3$ and aromatic esters, ethyl benzoate or ethyl anisate. Another highly active type of the supported catalysts uses aromatic diesters (phthalates) and mixtures of $Al(C_2H_5)_3$ and phenylalkoxysilanes as co-catalysts.

Polymerization reactions with Ziegler-Natta catalysts are carried out at 40–80°C in pure monomers or in monomer mixtures with aliphatic solvents. Molecular weights of polymers are controlled by the addition of hydrogen, an effective chain-transfer agent. Some Ziegler-Natta catalysts polymerize linear α-olefins, such as 1-hexane or 1-decene, into linear polymers with ultrahigh molecular weights which are used as drag-reducing agents for hydrocarbon flow.

All higher α-olefins, in the presence of Ziegler-Natta catalysts, can easily copolymerize both with other α-olefins and with ethylene. In these reactions, higher α-olefins are all less reactive than ethylene and propylene.

Isotactic PB and PMP are produced commercially in slurry processes in liquid monomers or monomer mixtures (optionally diluted with light inert hydrocarbons) at 50–70°C.

Metallocene Catalysts. Higher α-olefins can be polymerized with catalyst systems containing metallocene complexes. The first catalysts of this type (Kaminsky catalysts) include metallocene complexes of zirconium such as biscyclopentadienylzirconium dichloride, activated by methylaluminoxane. These catalysts polymerize α-olefins with the formation of amorphous atactic polymers. Polymers with high molecular weights are produced at decreased temperatures and have rubber-like properties.

Zirconocene complexes containing two indenyl or tetrahydroindenyl groups bridged with short links such as $-CH_2-CH_2-$ or $-Si(CH_3)-$ produce isotactic polymers of higher α-olefins. To synthesize syndiotactic PO, bridged zirconocene complexes with rings of two different types are required, one example of which is isopropyl(cyclopentadienyl)(1-fluorenyl)zirconocene. These complexes are used for the synthesis of syndiotactic PB, PMP, and poly (4-methyl-1-hexene).

Cationic Polymerization Reactions. α-Olefins with linear and branched alkyl groups can be readily polymerized with cationic initiators. Olefins containing linear alkyl groups (1-pentene, 1-hexene, 1-octene, 1-decene, etc) and their mixtures are oligomerized by using BF_3, mixtures of BF_3 and alcohols, as well as $AlCl_3$ or $AlBr_3$-HBr systems at low temperatures with the formation of low molecular weight oils of an irregular structure. These oligomers are used as base stocks for synthetic lubricating oils.

Polymerization of Cycloolefins

Depending on the type of catalyst used, polymerization of cycloolefins proceeds through either ring opening or by opening of the double bond with the preservation of the ring.

Ring-Opening Polymerization. Ring-opening polymerization of cycloolefins in the presence of tungsten- or molybdenum-based catalysts proceeds by a metathesis mechanism.

Dicyclopentadiene is also polymerized with tungsten-based catalysts. Because the polymerization reaction produces heavily cross-linked resins, the polymers are manufactured in a reaction injection molding (RIM) process, in which all catalyst components and resin modifiers are slurried in two batches of the monomer. The first batch contains the catalyst (a mixture of WCl_6 and $WOCl_4$), additives, and fillers; the second batch contains the co-catalyst (a combination of an alkylaluminum compound and a Lewis base such as ether), antioxidants, and elastomeric fillers. Mixing two liquids in a mold results in a rapid polymerization reaction.

Metallocene Catalysts. Polymerization of cycloolefins with combinations of metallocenes and methylaluminoxane produces polymers with a completely different structure. The reactions proceeds via the double-bond opening in cycloolefins. If the metallocene complexes contain bridged and substituted cyclopentadienyl rings, such as ethylene(bisindenyl)zirconium dichloride, the polymers are stereoregular and have the *cis*-diisotactic structure.

Processing

Both PB and PMP melts exhibit strong non-Newtonian behavior: their apparent melt viscosity decreases with an increase in shear stress. Melt viscosities of both resins depend on temperature. Equipment used for PP processing is usually suitable for PB and PMP processing as well; however, adjustments in the processing conditions must be made to account for the differences in melt temperatures and rheology.

Extrusion. The main applications of this method include the production of film, sheet, pipe, and tubing. PB is usually extruded by using the same equipment (single- or twin-screw extruders) as that used for PP and HDPE, at melt and die temperatures of 170–190°C. PMP is processed on extruders with a high length-to-diameter ratio at temperatures of 240–300°C.

Injection molding of PB is carried out under conditions similar to those for PP at 145–190°C. Injection molding of PMP is carried out at melt temperatures of 260–330°C, mold temperatures of 30–80°C, injection pressures of around 30 MPa (300 atm), and at relatively low injection rates.

Film. The blown film process is most commonly used in the production of PB film from resins with melt indexes from 0.3 to 10 g/10 min at a melt temperature of 200–215°C using conventional equipment. Mechanical properties of blown PB film depend on the degree of orientation and other processing parameters. PB film can be sealed at 160–220°C. Another technique for the PB film production consists of film casting from the melt on polished chilled rolls and co-extrusion or lamination with other films.

Health and Safety Factors

Polymers and higher α-olefins are not toxic; their main potential health hazards are associated with residual monomer, antioxidants, and catalyst residues. In particular, PB and PMP are inert materials and usually present no health hazard. PMP is employed extensively for a number of medical and food packaging applications.

Uses

Polybutene. The largest share of commercially produced crystalline PB is used for manufacturing pipe and tubing. The advantages of PB in pipe applications include high flexibility, toughness, and high resistance to creep, environmental stress-cracking, wet abrasion, and various chemicals. Pipes manufactured from PB retain their properties at temperatures up to 85°C. PB pipe is used in residential and commercial hot- and cold-water plumbing (including chlorine-containing hot water), water wells, water manifolds, and fire sprinklers. Black-pigmented pipe grades are suitable for outdoor use.

Blown film manufactured from PB has a high tensile strength and exhibits good resistance to tear, impact, and puncture.

Poly(4-methyl-1-pentene). Most PMP applications capitalize on the resin's high optical transparency, excellent dielectric characteristics, high thermal stability, and good chemical resistance. The manufacture of medical equipment comprises about 40% of PMP production, including such articles as hypodermic syringes, needle hubs, blood collection and transfusion equipment, pacemaker parts, blood analysis cells, and respiration equipment. It is also used in chemical and biomedical laboratory equipment, e.g., cells for spectroscopic and optical analysis, laboratory ware, and animal cages, and in a variety of injection-molded articles, such as caps for enclosures, ink cartridges for printers, light covers, tableware, and sight-glasses. PMP is suitable also for microwave oven cookware and service, and is used in food packaging (qv). In many applications, PMP replaces stainless steel trays. PMP is also utilized for wire and cable coating, as well as for film and paper coatings with good release properties.

Synthetic Lubricating Oils. Liquid oligomers of higher linear α-olefins such as 1-decene are produced with cationic initiators. They are the most versatile of all synthetic lubricants. They exhibit not only good lubricating properties over a wide temperature range (including excellent low temperature properties), but also high frictional and oxidative stability as well as low volatility, and are miscible with all mineral oils and most synthetic lubricants. They have found wide application as synthetic base oils in the formulation of various lubricants, including lubricating oils for cars, transformer oils, transmission and crankcase fluids, hydraulic fluids, and compressor oils. See **Lubrication and Lubricants**.

Polymers of Cycloolefins. Polyoctenamer elastomers are processed by extrusion, injection molding, and calendering into hoses, rubber coatings, and tire components. They are mostly used as components in rubber-, PVC-, and PS-based compositions.

Cured liquid-molding resins based on polydicyclopentadiene are uses in the manufacture of automotive parts for trucks, snowmobiles, wheel loaders, recreational vehicles, and also in other areas that require toughness and good all-weather impact resistance.

Yury V. Kissin
Mobil Chemical Company

OLEFINS, HIGHER

Higher olefins are versatile chemical intermediates for a number of important industrial and consumer products, providing a better standard of living with low environmental impact (qv) in many commercial uses. These uses can be characterized by carbon number and by chemical structure.

The even-numbered carbon alpha olefins (α-olefins) from C_4 through C_{30} are especially useful. For example, the C_4, C_6, and C_8 olefins impart tear resistance and other desirable properties to linear low and high density polyethylene; the C_6, C_8, and C_{10} compounds offer special properties to plasticizers used in flexible poly (vinyl chloride). Linear C_{10} olefins and others provide premium value synthetic lubricants; linear C_{12}, C_{14}, and C_{16} olefins are used in household detergents and sanitizers. In addition, many carbon numbers from C_4 to C_{30+} are also utilized in specialty applications such as sizing agents to produce longer-lasting paper.

The C_6-C_{11} branched, odd and even, linear and internal olefins are used to produce improved flexible poly (vinyl chloride) plastics.

Physical Properties

For a listing of selected physical properties of linear alpha olefins, see Table 8.

Chemical Properties

The general reactivity of higher α-olefins is similar to that observed for the lower olefins. However, heavier α-olefins have low solubility in polar solvents such as water; consequently, in reaction systems requiring the addition of polar reagents, apparent reactivity and degree of conversion may be adversely affected. Reactions of α-olefins typically involve the carbon-carbon double bond and can be grouped into two classes: (1) electrophilic or free-radical additions; and (2) substitution reactions.

Commercial Olefin Reactions. Some of the more common transformations involving α-olefins in industrial processes include the oxo reaction (hydroformylation), oligomerization and polymerization, alkylation reactions, hydrobromination, sulfation and sulfonation, and oxidation.

Commercial α-Olefin Manufacture

Most linear α-olefins are produced from ethylene.

Ethylene oligomerization can be accomplished in the following commercial processes: (1) stoichiometric chain growth on aluminum alkyls followed by displacement (Albemarle); (2) catalytic chain growth on aluminum alkyls (Chevron-Gulf); (3) catalytic chain growth using a nickel ligand catalyst (Shell); and (4) catalytic chain growth using a modified zirconium catalyst (Idemitsu). In the Albemarle (formerly Ethyl) process, stoichiometric quantities of aluminum alkyls are used with subsequent displacement of α-olefins from the aluminum, followed by separation of the α-olefins from the aluminum alkyls. In the Chevron-Gulf process, catalytic amounts of aluminum alkyl are used. The operating temperatures

TABLE 8. PROPERTIES OF C_4 TO C_{20} LINEAR 1-OLEFINS

Compound	Mol wt	Density, g/mL 20°C	Viscosity, mm²/s(= cSt) 20°C	Viscosity, mm²/s(= cSt) 100°C	Free energy of formation, kJ/mol[a]
1-butene	56.11	0.6012			72.09
1-pentene	70.13	0.6402	0.202		78.67
1-hexene	84.16	0.67317	0.39		87.61
1-heptene	98.19	0.69698	0.50		96.02
1-octene	112.2	0.71492	0.656	0.363	104.4
1-nonene	126.2	0.72922	0.851	0.427	112.8
1-decene	140.3	0.74081	1.09	0.502	121.3
1-undecene	154.3	0.75032	1.38	0.587	129.6
1-dodecene	168.3	0.75836	1.72	0.678	138.0
1-tridecene	182.3	0.7653	2.14	0.782	146.4
1-tetradecene	196.4	0.7713	2.61	0.894	154.8
1-pentadecene	210.4	0.7765	3.19	1.019	163.3
1-hexadecene	224.4	0.78112	3.83	1.152	171.7
1-heptadecene	238.4	0.7852	4.60	1.30	180.1
1-octadecene	252.5	0.7888	5.47	1.46	188.5
1-nonadecene	266.5	0.7920 f[b]		1.63	196.9
1-eicosene	280.5	0.7950 f[b]		1.82	205.3

[a] To convert kJ/mol to kcal/mol, divide by 4.184.
[b] f = frozen.

are higher than those in the stoichiometric process, thus favoring displacement reactions after a finite amount of chain growth. In the Shell process, a three-phase system is employed, which gives a high linearity at higher carbon numbers. A nickel ligand catalyst dissolved in a solvent forms one liquid phase, the produced olefins form a second liquid phase, and the ethylene forms a third. Once formed, the olefins usually do not engage in further reactions because most of them are not in contact with the catalyst. Shell practices isomerization and disproportionation to produce a narrow range of internal linear olefins for feed to their oxo-alcohol unit. In the Idemitsu process, a zirconium oligomerization catalyst is modified by adding an aluminum alkyl and a Lewis base or an alcohol in a solvent. Variations in the catalyst mix thus offer a variety of carbon-number distributions, some of which resemble those in the catalytic processes, others approaching those in the stoichiometric process. Although operating at lower pressures than the other ethylene-based oligomerizations, the Idemitsu process still produces high quality linear α-olefins.

Vista has offered for license a stoichiometric process, which has not yet been commercialized, although the related primary alcohol process has been described.

One-Step Ziegler Process. Gulf Research and Development Corp. developed the one-step Ziegler process. This process is now owned by Chevron, which has two plants at Cedar Bayou, Texas. Plants based on licensing the technology are operated by Mitsubishi in Japan and at Neratovice in the Czech Republic. Chevron further improved the process around 1990 by reducing paraffin impurities.

Uses

The principal outlets for higher olefins are in the polymer, surfactant, and detergent industries. See also **Alcohols, Higher Aliphatic-Synthetic Processes**. Generally, higher olefins are seldom incorporated directly into a product as an ingredient; rather, they are processed through at least one chemical reaction step before appearing in a finished product.

Polymers. The manufacture of alcohols from higher olefins via the oxo process for use in plasticizers is a significant outlet for both linear α-olefins and branched olefins such as heptenes, nonenes, and dodecenes. These olefins are converted into alcohols containing one more carbon number than the original olefin. The alcohols then react with dibasic anhydrides or acids to form PVC plasticizers. The plasticizers produced from the linear olefins have superior volatility and cold-weather flexibility characteristics, making them an ideal product to use in flexible PVC for automobile interiors.

Detergents. The detergent industry consumes a large quantity of α-olefins through a variety of processes. Higher olefins used to produce detergent actives typically contain 10–16 carbon atoms because they have the desired hydrophobic and hydrophilic properties.

Lubricants. Lubricants represent a significant and growing outlet for higher olefins. Both basestocks and lube additives are produced from higher olefins by a variety of processes. See **Lubrication and Lubricants**.

Other Uses. A small but growing outlet for C_{16} and higher linear olefins is the production of alkenylsuccinic anhydride (ASA) for the paper industry. ASA is an effective alkaline sizing agent and competes with alkylketene dimer (AKD) in this application.

Additional uses for higher olefins include the production of epoxides for subsequent conversion into surface-active agents, alkylation of benzene to produce drag-flow reducers, alkylation of phenol to produce antioxidants, oligomerization to produce synthetic waxes (qv), and the production of linear mercaptans for use in agricultural chemicals and polymer stabilizers. Aluminum alkyls can be produced from α-olefins either by direct hydroalumination or by transalkylation. In addition, a number of heavy olefin streams and olefin or paraffin streams have been sulfated or sulfonated and used in the leather (qv) industry.

Health and Safety Factors

Toxicological Information. The toxicity of the higher olefins is considered to be virtually the same as that of the homologous paraffin compounds. Based on this analogy, the suggested maximum allowable concentration in air is 500 ppm. Animal toxicity studies for hexene, octene, decene, and dodecene have shown little or no toxic effect except under severe inhalation conditions.

Handling. The main hazard associated with these olefins, especially the lighter homologues, is their low flash point. Although no special precautions are necessary with regard to fire extinguishing, these olefin

products should be stored and shipped under an inert atmosphere to maintain product purity.

G. R. LAPPIN
L. H. NEMEC
J. D. SAUER
J. D. WAGNER
Albemarle Corp.

Additional Readings

Billmeyer, F.: *Text Book of Polymer Science,* 3rd Edition, Wiley-Interscience, New York, NY, 1984.

Boenig, H.V.: *Polyolefins: Structure and Properties,* Elsevier, Amsterdam, the Netherlands, 1966.

Dragutan, V., A.T. Balaban and M. Dimonie: *Olefin Metathesis and Ring-Opening Polymerization of Cycloolefins,* John Wiley & Sons, Ltd., Chichester, UK, 1985.

Dyachkovsky F.S. and A.D. Pomogailo: *J. Polym. Sci. Polym. Symp.* **68**, 97 (1980).

Godrej, N.B. and co-workers: *Proceedings from Alpha Olefins from Oleochemical Raw Materials,* The Third World Detergent Conference, Montreux, Switzerland, Sept. 1993.

Kissin, Y.V.: in N.P. Cheremisinoff, ed., *Encyclopedia of Engineering Materials,* Part A, Vol. I, Marcel Dekker, Inc., New York, NY, 1988, p. 103.

Krentsel, B.A., Y.V. Kissin, V.I. Kleiner, and L.L. Stotskaya: *Polymers and Copolymers of Higher α-Olefins,* Hanser Publishers, Munich, Germany, 1997.

Krentsel, B.A., Y.V. Kissin, V.I. Kleiner, and L.L. Stotskaya: *Polymers and Copolymers of Higher α-Olefins,* Hanser Publishers, Germany, Munich, 1997, Chapt. 8.

Lappin, G.R. and J.D. Sauer, eds.: *Alpha Olefins Applications Handbook,* Marcel Dekker, Inc., New York, 1989.

Maier, C. and T. Calafut: *Polypropylene, The Definitive User's Guide and Databook,* Plastics Design Library, New York, NY, 1998.

Moore, E.P. Jr.: *Polypropylene Handbook,* Hanser Publishers, Munich, Germany, 1996.

Nowlin, T.E.: *Prog. Polym. Sci.* **11**, 29 (1985).

Oosterwijk, H. and H. Van Der Bend: *Akzo Chemie America Bulletin,* Initiations Seminar, 1980, New York, NY, pp. 87–35.

Petrothene Polyolefins: A Processing Guide, 5th Edition, Quantum Chemical Co., Cincinnati, OH, 1986.

Raff, R.A.V. and K.W. Doak: *Crystalline Olefin Polymers,* John Wiley & Sons, Inc., New York, NY, 1965, pp. 307, 495, 682.

Read, C.S., R. Wilhalm, and Y. Yoshida: *SRI Chemical Economics Handbook: Linear Alpha Olefins,* SRI International, Menlo Park, CA, Oct. 1993.

Rubin, I.D.: *Poly (1-Butene): Its Preparation and Properties,* Gordon & Breach, New York, 1968.

Shubkin, R.L.,: *Synthetic Lubricants and High Performance Functional Fluids,* Marcel Dekker, Inc., New York, 1993.

Technical data, UOP Inc. Division of AlliedSignal, Des Plaines, Ill., 1966 to 1993.

Van der Ven, S.: *Polypropylene and Other Polyolefins: Polymerization and Characterization,* Elsevier Science Publishers, B. V., Amsterdam, the Netherlands, 1990.

OLEIC ACID. [CAS: 112-80-1]. $CH_3(CH_2)_7CH:CH(CH_2)_7 \cdot COOH$, formula weight 282.45, colorless liquid, mp 14°C, bp 286°C, sp gr 0.854. Sometimes referred to as red, oil, elaine oil, or octadecenoic acid, this compound is insoluble in H_2O, but miscible with alcohol or ether in all proportions. Oleic acid solidifies into colorless needle crystals.

Oleic acid differs from stearic acid chemically by possessing 33 instead of 35 hydrogen atoms in the radical $C_{17}H_{33} \cdot COOH$ (oleic acid), $C_{17}H_{35}COOH$ (stearic acid). It is possible to convert oleic acid and oleate esters into stearic acid and stearate esters by treatment with hydrogen gas in the presence of finely divided nickel as a catalyzer at 250°C under pressure as in the hydrogenation of oils and fats. Either by careful oxidation, or by addition of ozone and splitting, oleic acid yields products of 9 carbon atoms, thus leading to the conclusion that the double bond is in the center of the carbon chain. Oleic acid adds bromine or iodine in definite amounts to confirm the conclusion that one double bond is contained. Nitric acid converts oleic acid into elaidic acid, $C_{17}H_{33}COOH$, mp 51°C (oleic and elaidic acids are related, cis- and trans-, as maleic and fumaric acids).

Oleic acid may be obtained from glycerol trioleate, present in many liquid vegetable and animal nondrying oils, such as olive, cottonseed, lard, by hydrolysis. The crude oleic acid after separation of the water solution of glycerol is cooled to fractionally crystallize the stearic and palmitic acids, which are then separated by filtration, and fractional distillation under diminished pressure. Oleic acid reacts with lead oxide to form lead oleate, which is soluble in ether, whereas lead stearate or palmitate is insoluble. From lead oleate oleic acid may be obtained by treatment with H_2S (lead sulfide, insoluble solid, formed). With sodium oleate, a soap is formed. Most soaps are mixtures of sodium stearate, palmitate, and oleate.

Representative esters of oleic acid are: methyl oleate, $C_{17}H_{33}COOCH_3$, bp 190°C at 10 millimeters pressure; ethyl oleate, $C_{17}H_{33}COOC_2H_5$, bp 205°C at 10 millimeters pressure; glyceryl trioleate (triolein), $C_3H_5(COOC_{17}H_{33})_3$, bp 240°C at 18 millimeters pressure.

Oleic acid is used in the preparation of metallic oleates, such as aluminum oleate for thickening lubricating oils, for water-proofing materials, and for varnish dryers. The glyceryl ester of oleic acid is one of the constituents of many vegetable and animal oils and fats.

See also **Vegetable Oils (Edible)**.

OLEORESIN.   Any of a number of mixtures of essential oils and resins characteristic of the tree or plant from which they are derived. Most types are semi-solid and tacky at room temperature, becoming soft and sticky at high temperatures. They have various distinctive odors. See also **Balsam**; **Oils, Essential**; and **Rosin**.

OLEUM.   The Latin word for oil, applied to fuming sulfuric acid. (Sulfuric acid was originally called oil of vitriol.)

OLIVINE.   The mineral olivine is a silicate of magnesium and ferrous iron corresponding to the formula $(Mg,Fe)_2(SiO)_4$. Olivine is the group name for the isomorphous series forsterite, Mg_2SiO_4, and fayalite, Fe_2SiO_4. The ratio of magnesium to iron varies considerably but the more common olivines are richer in Mg than in Fe. Olivine crystallizes in the orthorhombic system, usually in flattened prismatic forms, also granular and massive. It has a conchoidal fracture and is rather brittle; hardness, 6.5–7; specific gravity, 3.22–4.39; luster, vitreous; color, olive to gray-green; may be yellowish-brown from the oxidation of the iron. It is transparent to translucent. Olivine occurs both in igneous rocks as a primary mineral as well as in certain rocks of metamorphic origin. It has also been discovered in meteorites.

Olivine crystallizes from magmas that are rich in magnesia and low in silica and which form such rocks as gabbros, norites, peridotites and basalts. The metamorphism of impure dolomites or other sediments in which the magnesia content is high and silica low seems to produce olivine.

Transparent olivines of good color are sometimes used as a gem, often called *peridot*, the French word for olivine; it is also called chrysolite from the Greek meaning gold, and stone. Olivine occurs in the lavas of Vesuvius and Monte Somma and in the Eifel district of Germany. Gem material comes from St. John's Island in the Red Sea, Upper Burma, and from Minas Geraes, Brazil. In the United States olivine localities are Orange County, Vermont; Webster and Jackson counties, North Carolina. Arizona and New Mexico have also furnished some gem material.

See also **Peridotite**.

ONSAGER, LARS (1903–1976).   A Norwegian chemist who won the Nobel prize for chemistry in 1968. He studied and wrote on the theory of electrolytic conduction and theory of dielectrics. He also worked with superfluids and crystal statistics and reciprocal relations in irreversible processes. After receiving his doctorate in Norway, he came to the U.S. and became a citizen. See also **Dielectric Theory**.

OPACITY.   The optical density of material, usually a pigment; the opposite of transparency. A colorant or paint of high opacity is said to have a good hiding power or covering power, by which is meant its ability to conceal another tint or shade over which it is applied. See also **Paints and Coatings**.

OPAL.   The mineral opal, long classified as an amorphous mineral gel, has been found by X-ray analysis to consist of a microcrystalline aggregate of crystallites of cristobalite. On this basis, opal may be considered as a variety of cristobalite bearing the same relationship to that mineral as chalcedony does to quartz. Opal is hydrous silica, $SiO_2 \cdot nH_2O$, with variable water content. It never occurs in crystal form; usually as irregular veins or masses, or as pseudomorphous replacements after wood or fossilized material such as bones and shells. Opaline silica occurs in many forms: geyserite from geyser deposits, siliceous sinter (fiorite) form siliceous waters of hot springs, and diatomite (diatomaceous earth) from siliceous shells of diatoms and comparable microscopic species. It has a conchoidal fracture; hardness 5.5–6.5; specific gravity 2.1–2.3; luster, vitreous or greasy to dull; color very variable, colorless, white, milky-blue, gray, red, yellow, green, brown, and black. Often a beautiful play of colors may be observed in the gem varieties. The color play in opals is attributed to three different mechanisms: finely divided pigmentation of foreign material; light interference by open-spaced grid of cristobalite crystallization; and reflected light. It may well be that two or all three causes may contribute to the color effect in any given opal specimen. Before a more complete understanding of opal color is established these phases seem to be of prime significance.

Besides the gem varieties, which show the delicate play of colors, there are other kinds of common opal, such as: the milk opal, a milky bluish to greenish kind; resin opal, which is honey-yellow with a resinous luster; wood opal, resulting from the replacement of the organic matter of wood by opal, and hyalite, a colorless glass-clear opal sometimes called Muller's Glass. Opal is deposited at relatively low temperatures and may occur in the fissures of almost any type of rock. Hungary, Australia, Honduras, Mexico and in the United States Nevada and Idaho, have been sources of gem opals. Hyalite comes from Czechoslovakia, Mexico, Japan and British Columbia. Other common varieties of opal are widespread in their occurrence. The word opal is derived from the Latin *opallus*.

OPTICAL BRIGHTENER.   Also referred to as optical bleach; colorless dye; fluorescent brightener. A colorless fluorescent, organic compound that absorbs UV light and emits it as visible blue light. The blue light masks the undesirable yellow of textiles, paper, detergents, and plastics. Some examples are derivatives of 4,4′-diaminostilbene-2,2′-disulfonis acid, coumarin derivatives such as 4-methyl-7-diethylaminocoumarin.

OPTICAL CRYSTAL.   A comparatively large crystal, either natural or synthetic, used for infrared and ultraviolet optics, piezoelectric effects, and shortwave radiation detection. Examples are sodium chloride, potassium iodide, silver chloride, calcium fluoride, and (for scintillation counters) such organic materials as anthracene, naphthalene, stilbene, and terphenyl.

OPTICAL EMISSION SPECTROCHEMICAL ANALYSIS.   In this analytical technique, an optical device is used to analyze radiation from electrically excited sample atoms. The analyzing device provides monochromatic images whose intensities are measured and related to the concentration of the elements within the sample that produces the specific radiation measured. The technique is precise and rapid, and adaptable to solid, powder, or liquid samples.

More than seventy elements may be detected by standard procedures. Atomic gases, such as O, N, H, He, Ar, Ne, Kr, Xe, and Rn and the halogens are excluded. Nonmetallic substances, such as C, S, and Se, require vacuum path spectrometers for optimum detection and measurement. Analytical ranges may extend from fractional parts per million to about 40% concentration. Computer-controlled photoelectric optical emission spectrometers will output printed percent concentrations for 30 to 50 elements per sample in just a few minutes. This form of analytical instrumentation is used widely in production and quality control, as well as for research studies.

A schematic diagram of an optical emission spectrometer is given in Fig. 1. Various means are used to introduce the sample, whether solid or liquid, into an excitation stand where energy is imparted to it by some form of excitation source. The atoms composing the sample are excited and therefore emit their characteristic radiations, which are then separated by a grating in the spectrometer into line spectra. The light of selected element lines is isolated by slits and focused on phototubes. The sensitivity is adjusted by attenuating the high voltage from the high voltage supply. The intensity of a spectrum line can be correlated with the concentration of the element producing it. It is therefore necessary to measure intensities with very high precision.

Sample atoms may be excited by absorbing specific energies from an electric discharge. These atoms, raised to higher-than-usual energy levels, are unstable, and revert to their stable states by emitting the absorbed radiation according to the relation:

$$E_2 - E_1 = hv = hc/l \qquad (1)$$

where E = energy, eV

h = Planck's constant $(6.624 \times 10^{-27}$ erg-second)

v = frequency, Hz

$\lambda = c/v$ = wavelength (in Å = 10^{-8} centimeter)

c = velocity of light $(3 \times 10^{10}$ centimeters/second)

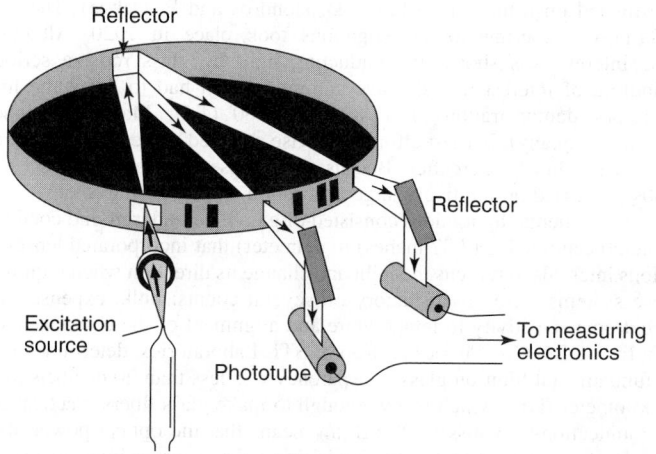

Fig. 1.   Operating principle of optical emission spectrometer

Atomic transitions may be expressed in terms of wavelength, and qualitative analysis may be performed by wavelength determination and identification.

The commonly used dispersive device is the diffraction grating, which produces spectra by light interference according to the relation:

$$N\lambda = d(\sin\alpha \pm \sin B) \qquad (2)$$

where N = an integer
λ = wavelength
d = grating constant (width of single groove)
α = angle of incident light
B = angle of diffracted light.

For constant a and the same sin B, integer values of N produce spectra of $\frac{1}{2}\lambda$, $\frac{1}{3}\lambda$, etc., called *spectral orders*.

A grating ruled on a spherical surface combines the properties of the diffraction grating with the focusing ability of the optical surface. Such a device, with radius of curvature R, focuses spectra as images of the entrance (primary) slit on the circumference of a circle of diameter R, when the entrance slit is also located on the circumference of the circle.z

The usual measure of how well a grating separates individual wavelengths is given by the reciprocal linear dispersion, in angstroms per millimeter, as follows:

$$Å/mm = \frac{d\cos B}{Nf} \qquad (3)$$

Thus, dispersion is governed by the fineness of the grating ruling d and the focal length f of the focusing element.

The concentration C of an irradiating element is related to the intensity I of the emitted spectral line, according to the relationship:

$$I = kC^n \qquad (4)$$

where k and n depend on the excitation conditions employed. Accuracy and precision are improved by use of an internal standard reference line of another element of constant concentration. The relationship becomes:

$$\frac{I_x}{I_r} = k_1 C_x^{n_1} \qquad (5)$$

where I_x and I_r are the intensities of spectral lines emitted by elements x and r; C_x is the concentration of element x; and k_1 and n_1 are constants depending on the line pair and on the excitation conditions. The relative intensities of lines having different excitation energies depend on the temperature of the spark discharge column.

The source unit must vaporize and excite a portion of the sample, which is generally used as one of the electrodes between which the electric discharge takes place. No single excitation source is ideally suited for all applications of emission spectrochemistry. Trace impurities in metals, alloying constituents in high concentrations, biological substances, ceramics, slags, oils, nonconductors, refractories—all may require different excitation techniques and sample preparation procedures. Table 1 summarizes the important characteristics of the commonly used spectrochemical source units.

Photographic radiation detection may be used, but film emulsion response is not linear. Film calibrations are required to relate measured densities with the intensities producing these densities before *intensity* versus *concentration* working curves can be formulated. Although quite general in application at one time, the photographic technique is slower than photoelectric radiation detection wherein each beam whose intensity is to be measured is directed onto a photomultiplier detector through a suitably sized exit (secondary) slit. The output of the detectors is transmitted to the measuring console, where it is translated into the readout format of the system.

Many dramatic changes in the development of readout electronics have occurred over the last 15 to 20 years. Modern systems use integrated circuit and digital computing devices. The engineer is no longer required to design complex circuitry to perform the basic tasks of control. Software now becomes the tool by which timing, sequencing, and logic control are accomplished.

Generally, spectrometer systems fall into two major functional categories—system control and data handling. The digital controller with its controlling and computing capabilities is ideal for handling both tasks with a minimum of effort required by the circuit design engineer.

Particularly during the last decades, several new analytic techniques have been developed that, when appropriate, have a tendency to displace former traditional methodologies.

Additional Reading

Crouch, S.R.: *Spectrochemical Analysis,* Pearson Custom Publishing, Boston, MA, 1988.
Thorne, A.P., S. Johansson, and U. Litzen: *Spectrophysics: Principles and Applications,* Springer-Verlag New York, Inc., New York, NY, 2001.

OPTICAL FIBER SYSTEMS. Optical fibers are hair-thin structures (usually cylindrical in shape) capable of transmitting light signals with extremely low signal loss and at very high digital pulse rates. Fibers are available in a variety of sizes and material compositions and with a wide range of optical performance. Although commercially available fibers are solid structures, they function as "light pipes" that guide rays of light and are therefore sometimes called *lightguides*. When used to connect a light

TABLE 1. SPECTROCHEMICAL EXCITATION SOURCE UNITS

Type	Voltage	Current, A		Characteristics
Dc arc	220	3–30		Most sensitive, least reproducible, quantitative analysis; trace element quantitative analysis.
Ac arc	2500	5		Good sensitivity, more reproducible, best use for self-electrode metal analysis.
	5000	2–5		
High-voltage spark interrupted auxiliary gap	15–40,000	3–20 RF		Least sensitive, most precise, ±1% or better. Excites higher energy lines. Parameter selection allows variations between arc-like and spark-like in spectral excitation.
Multisource	1000	Peak discharge currents from 5 to in excess of 500 A at time constants ranging from 8 to less than 1 ms.		Sensitive and precise. Parameter selection allows wide variety of controlled unidirectional and oscillatory charges variable from arc-like to spark-like in spectral excitation.

source to a light receiver (photodetector) to form a communication system, the fiber carries *photons* instead of the *electrons* used in traditional metal-conductor communication links. Although a laser light source containing a narrow range of optical wavelengths is preferred for carrying signals the farthest and fastest, light-emitting diodes (LEDs) are also used, especially for short distance communication links within buildings. Thus, the three key elements of a lightwave communication system are the light source, the optical fiber, and the photodetector.

Fiber-Optic Systems in Perspective

As early as 1841, D. Colladon demonstrated light guiding by a jet of water, and in the following year J. Babinet showed the phenomenon in a bent glass rod. However, these experiments did not receive wide publicity until John Tyndall duplicated and popularized the same effect in 1854. In 1880, shortly after the telephone was invented, Alexander Graham Bell proposed telecommunications using lightwaves. See Fig. 1. Patented as the "Photophone," the concept depended on the free propagation of light through the atmosphere. Of course, a century ago Bell had no powerful steady light source such as the laser, and even on very clear days atmospheric disturbances severely limited the practical distance over which undisturbed light could travel.

These experiments were followed by the development of light pipes to illuminate homes (W. Wheeler), bent glass rods to illuminate body cavities for dentistry and surgery (Roth and Reuss), and surgical lamps (D. Smith).

As early as 1910, the possibility of guiding electromagnetic waves by internal reflection within long cylinders of dielectric material was investigated on a theoretical basis (D. Hondros and P. Debye). The first quantitative experimental investigations took place in 1920. Although some interest was shown in conducting light by glass rods, a serious rekindling of interest in lightwave communication had to await the first laboratory demonstration of a laser in 1960 (T.H. Maiman, Hughes Aircraft Company). Earnest efforts to devise shielded waveguide structures were made shortly thereafter. Because the glasses available in the early 1960s possessed prohibitively large absorption and scattering losses, some early experimental lightguides consisted of gas-filled underground conduits (some 20 centimeters (7.87 inches) in diameter) that incorporated lenses at various intervals to refocus the light and change its direction when required. These systems were unsatisfactory on several counts: bulk, expense, and the extreme sensitivity to temperature and alignment of the components.

In the mid-1960s, Charles K. Kao of STL Laboratories, determined that the fundamental limit on glass transparency was less than 20 decibels (dB) per kilometer (km), which is low enough to make glass fibers practical for communications. (A loss of 20 dB/km means that the optical power after 1 km is 1% of the amount at the beginning.) This spurred investigators to explore methods for making purer glass. In 1970, fibers with attenuation less than 20 dB/km were made and demonstrated in the laboratory (Maurer, Keck and Schultz, Corning), and this was reduced to 4 dB/km by 1972.

Combining these fibers with emerging semiconductor lasers with lifetimes reaching 1000 hours (Bell Labs, 1973) and photodetectors, enabled lightwave communications to become practical in the mid-1970s. The first nonexperimental fiber-optic link became operational in Dorset, England (1975) to service a police station. In early 1976, Bell Labs started tests at 45 million bits per second on multimode fibers installed in Norcross, Georgia. Fibers continued to improve as their attenuation decreased to 0.47 dB/km (M. Horiguchi).

In 1980, video signals were carried by optical fibers $2\frac{1}{2}$ miles (4 kilometers) for the Winter Olympic Games in Lake Placid, New York. The first long-haul intercity installations (AT&T, Washington-New York; New York-Boston) were made in 1983. After that, the capacity of fiber optic transmission systems increased exponentially. Despite this progress, the fundamental limits predicted by the physics of photonics materials, devices, and systems have not yet been approached (Kogelnik). The challenge of future research and development continues to be a fuller exploitation of the ultimate capacity of optical fibers.

Fig. 2 shows a schematic history of the development of optical communication systems.

Seldom does a new technology have as many practical advantages as found in optical fiber systems and become available in only a decade of concentrated research and development. As solid state physics and semiconductors created the electronics industry, fiber-optic systems have revolutionized the telecommunications industry.

Fiber optic systems are more economical than their alternatives—copper wire, radio relay, and satellite. The regeneration of signals sent on copper cables is necessitated at several mile intervals, whereas the distance on optical fibers can be over a thousand miles by using optical amplifiers approximately every 50 miles.

The first fiber optic systems used light in the 850 to 875 nanometer (nm) wavelength region (the so-called "first window"), and were followed with systems operating near 1300 nm (the "second window") where fiber loss is smaller, and then near 1550 nm (the "third window") where fiber loss is even smaller. In the year 2000, fibers were carrying digital signals at 10 billion bits per second on one or more (as many as 40) wavelengths in the 1550 nm window.

Because optical fibers are nonconducting, fiber optic systems provide excellent electrical isolation and immunity from electrical interference. Signal losses are much lower in fibers (as low as 0.20 dB/km) compared to other guided transmission media, such as twisted copper pairs, coaxial cable, and metallic waveguides. In addition, the bandwidth or information carrying capacity of fibers is far greater. When one or more optical fibers are packaged into cables, the cables are smaller and more flexible than their metallic counterparts.

In 1984, British Telecom laid the first submarine fiber optic cable to carry regular traffic, to the Isle of Wight. In 1988, service began on the first transatlantic fiber-optic cable (TAT-8) from Tuckerton, New Jersey to France and England. Similar undersea communication links between Japan and Guam and other Pacific locations began in 1989. With the development of optical fiber technology proceeding at such a rapid rate, decisions as to whether or not to delay undersea cables for further advancements sometimes are difficult to make.

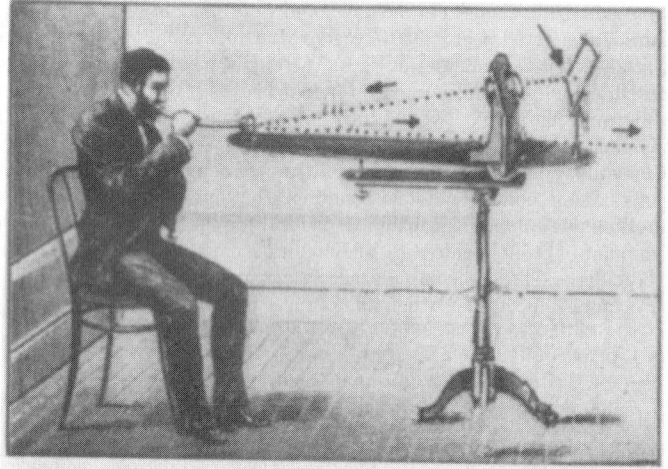

Fig. 1.   Old woodcuts showing photophone patented by Alexander Graham Bell in 1880, representing the first attempt to utilize light for the transmission of sound. (*Top*) Sunlight was reflected and focused by a lens onto a mechanism that was vibrated by sound waves (speech), thus modulating the intensity of the exiting light beam. (*Bottom*) At the receiving end of the system, variations of intensity of the light changed the resistance of a selenium photocell, thus controlling an electric current input to the receiving telephone

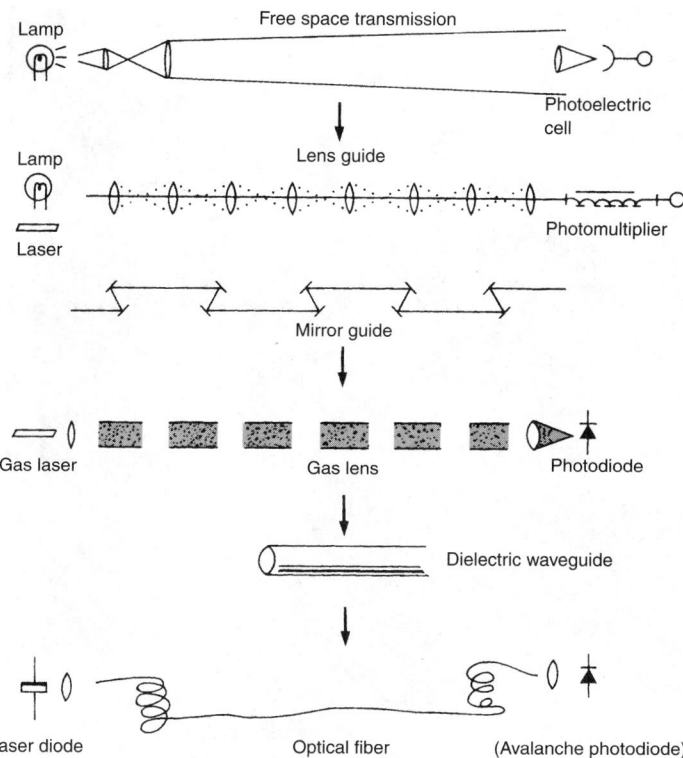

Fig. 2.   Schematic representation of the development of optical communication systems. (*Adapted from Suematsu and Iga*)

Earlier "fantastic" claims that 24,000 simultaneous telephone calls on a single pair of fibers at rates of up to 1.7 billion bits per second could be made, that full-length motion pictures such as *Gone with the Wind* could be fed to a home memory unit in one second, or that major symphonies, such as Beethoven's Fifth, could be transmitted in less than 1/50 of a second have already been surpassed. For example, here is listing of some "hero" experiments illustrating an optical fiber's capability as of the year 2000.

- Using 160 wavelengths of light with each wavelength carrying 40 billion bits per second of information, NEC researchers transmitted 6.4 trillion bits of information per second over a special fiber 186 kilometers long.
- Using a single wavelength of light on an experimental TrueWave® fiber, Bell Labs scientists transmitted 320 billion bits of information per second over a distance of 200 kilometers.
- Bell Labs demonstrated 3.28 trillion bits per second over a 300 km of experimental TrueWave fiber. This experiment used 82 wavelengths of light with each operating at 40 billion bits per second.

Types of Optical Fibers

Optical telecommunications fibers fall into two main categories: *multimode fiber*, and *single-mode* (also called *monomode*) *fiber*. Multimode fibers receive their name because they can propagate many (hundreds) of light modes, which can be thought of as paths taken by the light as it travels in the fiber. Single-mode fibers, on the other hand, propagate only one mode. Multimode and single-mode fibers can be further divided into subset categories. For example, multimode fibers can be either *step-index* or *graded-index*, and single-mode fibers can be *dispersion-unshifted, dispersion-shifted,* or *nonzero-dispersion shifted*. Each of these, in turn, can be further differentiated. For example, two types of graded-index multimode fibers have either 50 micron or 62.5 micron core diameters. Two types of dispersion-unshifted fiber have either depressed clad or matched clad refractive index profiles. Or they may have a typical high loss at the "water peak" wavelength near 1385 nm or low loss at this wavelength, such as with Lucent's AllWave™ fiber.

Single-mode fibers have many advantages over multimode fibers. See Fig. 3. Compared to the 50 or 62.5 micron core diameter of typical graded-index multimode fibers, some single-mode fibers have a core diameter near

8 micrometers. Single-mode fibers also have a core-cladding refractive index difference of a few tenths of a percent compared to 1 or 2% for multimode fibers. Their smaller core diameter and refractive index difference allow single-mode fibers to propagate light in only one clearly defined path (or mode). Because this reduces multipath effects that spread the arrival times of a short input pulse, single-mode fibers have high bandwidth, meaning that they can carry signals at higher bit rates than multimode fibers.

Various types of optical fibers are used for specific applications. For example, multimode fibers are used primarily in enterprise systems: buildings, offices, campuses. Special single-mode transmission fibers exist for submarine applications, and for metropolitan and long-haul terrestrial applications. And in addition to these transmission fibers, there are various "specialty" fibers for performing dispersion compensation (dispersion compensating fiber), optical amplification (erbium-doped fiber), and other special functions.

Characteristics Affecting Optical Fiber Performance

Phase Index of Refraction. Phase index of refraction is the ratio of the phase velocity of light in a vacuum to its velocity in another medium, such as glass. The value of this parameter depends on wavelength, and the composition, temperature and pressure of the medium. The higher the refractive index of a material, the lower the phase velocity of light in the material, and the more a light ray is bent as it enters the material from air.

Numerical Aperture (NA). Numerical aperture describes an angle just outside the a fiber's end face that determines the largest angle that a light ray can have to the fiber axis and still be captured and propagate within the fiber. The formula from Snell's law governing the numerical aperture number of a fiber is

$$NA = \sqrt{n_1^2 - n_2^2}$$

where n_1 and n_2 are the phase refractive indices of the fiber's core and cladding, respectively.

Most optical fibers have numerical apertures between 0.15 and 0.4, and these correspond to light acceptance half-angles of about 8 and 23 degrees. Typically, fibers having high *NAs* exhibit greater loss and lower bandwidth.

Light Loss or Attenuation through a Fiber. The amount of light at the output of a fiber is smaller than at the input. This attenuation is expressed in decibels per kilometer (dB/km), which is a relative power unit according to the formula

$$\alpha(dB) = -10 \log \frac{P_o}{P_i}$$

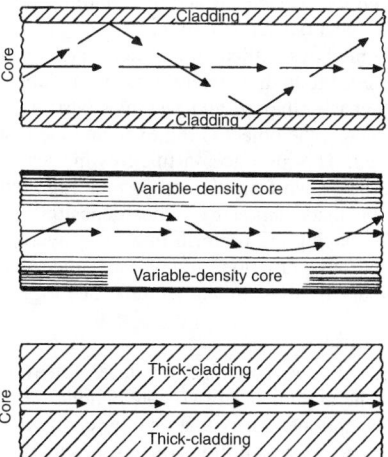

Fig. 3.   Fundamental classes of optical fibers: (**a**) Step-index fiber made up of glasses of two different densities, the core and the cladding. Lightwaves travel in a zig-zag fashion down the core, bouncing off one side and then the other side of the core-cladding interface; (**b**) Graded-index fiber where the glass in the core varies in density—hence the light travels in a smooth, curving path, causing less distortion of transmitted information; and (**c**) single- or monomode fiber where the core is very small relative to the wavelength of the transmitted light causing the light to move down the fiber in a straight line—hence resulting in very low distortion. In modern designs, the distortion has been reduced to almost the theoretical low limit

where P_o/P_i is the ratio of optical power at the output to the optical power launched into the fiber at the input. This ratio is smaller than 1, and the logarithm of a number smaller than 1 is negative. Consequently, the minus sign in the equation makes the attenuation value a positive number. A comparison of the ratio of output power to input power in percent to the quantity in dB is as follows:

80% transmission = a loss of : 1 dB

50% transmission = a loss of 3 dB

10% transmission = a loss of : 10 dB

1% transmission = a loss of : 20 dB

Bandwidth. Bandwidth is a measure of information-carrying capacity. An optical fiber's bandwidth can be expressed either in the time domain as pulse dispersion in nanoseconds per kilometer (ns/km), or in the frequency domain as frequency passband in megaHertz-kilometers (MHz-km). Light pulses spread or broaden as they pass through a fiber depending on the material used and its design. A fiber's bandwidth limits the rate at which optical pulses can be transmitted and decoded without error at the terminal end of the optical fiber. In general, optical fibers with small core diameter and low numerical aperture have higher bandwidth and lower loss.

Glass Processing

Depending on a fiber's application, a number of glass compositions may be used. However, for low-loss applications, the options become increasingly limited. Multicomponent glasses containing a number of oxides are not suited to making very low-loss fibers. Multicomponent glasses are prepared by essentially standard optical melting procedures, but with special attention given to details for increasing transmission and controlling defects from later fiber drawing steps. In contrast, low-loss fibers are usually made from pure fused silica doped with minor constituents. These require special manufacturing techniques.

Several methods have been used for manufacturing low-loss optical fibers, including the rod-in-tube method, the double-crucible method, and the more recent and widely used chemical vapor deposition (CVD) methods. CVD methods can be divided into two main categories: inside processes and outside processes. With the inside processes, such as modified chemical vapor deposition (MCVD) and plasma-activated chemical vapor deposition (PCVD), the chemical reactions that form the glass occur inside a glass starting tube (sometimes called a "substrate" tube). This starting tube serves as a containment vessel during deposition and eventually becomes part of the fiber. With outside processes, such as outside vapor deposition (OVD) and vapor axial deposition (VAD), vapor deposition occurs on the outside surface of a starting rod, and the reaction that forms the glass occurs near a burner.

With all CVD methods, silica and other glass-forming oxides and dopants are deposited at high temperatures on an object. In the inside processes, the object is collapsed and becomes part of the fiber, whereas in the outside processes, the object is removed and the resultant *soot blank* is dried and sintered. This produces a thick cylindrical *preform*. A long, thin fiber can then be drawn from this preform at high temperature, or the preform can first be made larger by either depositing more glass on its outer surface or by placing the preform in a glass overclad tube.

There are numerous variations on these processes and these are considered proprietary by most manufacturers. See Fig. 4.

Improvements in Lasers

Just a few years ago, the so-called C^3 laser (cleaved-coupled-cavity) appeared, wherein the alignment of two conventional semiconductor lasers yields a beam of exceptional purity that enables communication systems to send signals at rates as great as billions of bits, or binary digits per second. Just as recently as the late 1980s, commercial lightwave systems were limited to somewhat less than 2 million bits per second, but nevertheless a rate that permits the transmission of 24,000 simultaneous telephone calls on a single pair of fibers.

The cleaved-coupled-cavity laser developed by W.T. Tsang and colleagues (AT&T Bell Laboratories) was designed especially for use in fiber optics telecommunication systems and probably as much as the optical fibers themselves will contribute to the gross claims, as mentioned earlier in the article, made for faster, higher-capacity optical links.

Traditionally, semiconductor lasers have been used in optical communication systems. They are about the size of a grain of salt and produce

Fig. 4. Experimental optical fiber designs are tested by creating glass preforms, heating them in a special furnace located above the research fiber drawing tower (as shown) and drawing the fiber from the preform onto a collection drum shown in the foreground. The fiber is then used in tests conducted with light generators (lasers and LEDs) and photodetectors. The single-mode fiber is rated for high-capacity transmission, while lower-capacity multimode fibers are well matched to various economical sources and detectors. (*AT&T Bell Laboratories*)

pulses of light from pulses of electric current. Their electrical requirements are minimal (generally a few mA at 1 or 2 V). These lasers can generate infrared (IR) light where optical fibers are most nearly transparent. Unlike a gas laser, for example, the semiconductor laser is mechanically stable and reliable.

For comparison, a simple semiconductor laser is illustrated and briefly described in Figs. 5 and 6. A buried-heterostructure laser is shown in Fig. 7. The alignment of two lasers to form the C^3 laser is shown in Fig. 8. The turnability of a C^3 laser is illustrated in Fig. 9.

As observed in Tsang, the C^3 laser configuration offers at least three advantages in optical communication systems: (1) The exceedingly monochromatic output of the laser eliminates the problem of chromatic dispersion in optical fibers. This facilitates the transmission of digital information in a single-mode fiber at a wavelength of 1.55 micrometers, that is, the wavelength at which a silica fiber is most nearly transparent to electromagnetic radiation. In a test, using the C^3 laser, scientists at AT&T Bell Laboratories transmitted digital information in an optical fiber more than 120 kilometers long at a rate of one gigabit (10^9 bits) per second without reamplification along the path. The frequency of error was less than two bits in 10^{10}. (2) By coupling several C^3 lasers, each tuned to a different wavelength, to a single optical fiber, wavelength-division multiplexing can make it possible to carry several independent messages on the fiber. (3) As proposed by Tsang, at rates on the order of a billion

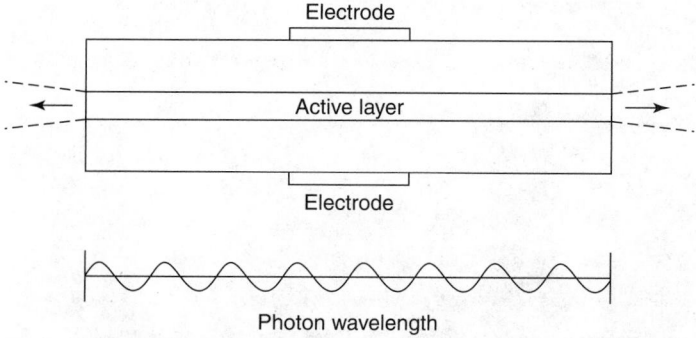

Fig. 5.   A simple semiconductor laser. It is a *p−n* junction in a semiconductor crystal the end faces of which are flat and perfectly parallel. Thus, the faces form a pair of semireflecting mirrors that bounce photons back and forth through the *active layer* of the crystal. Current injection causes photons to arise by spontaneous emission. Those photons traversing the semiconductor cause an avalanche of *stimulated* emission. Reflections at the mirrors are self-reinforcing provided that the wavelength of the photon fits evenly into the *length* of the laser. See also Fig. 6

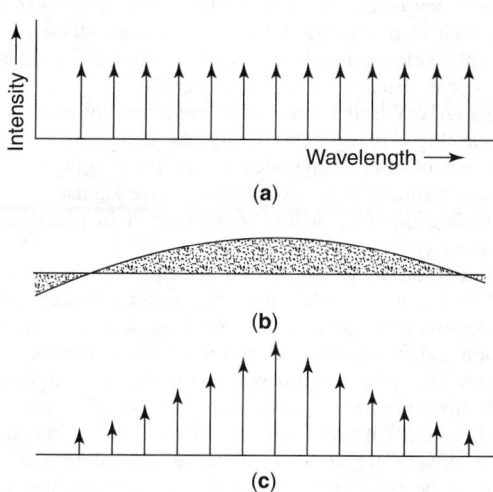

Fig. 6.   Simple semiconductor laser energy diagram. This schematic diagram illustrates why the output beam of the laser jumps randomly among several wavelengths. Fundamentally, the laser resonates at the *infinite number* of wavelengths that fit evenly into the length of the laser as indicated in (**a**). On the other hand, the *p−n* junction produces photons in only a narrow range of wavelengths—the *gain profile*, as indicated in (**b**). Therefore, the beam emitted by the laser includes only the resonant wavelengths positioned within the profile, as shown in (**c**). (*After Tsang*)

switchings per second, it is possible to shunt the output wavelength of a C³ laser among as many as 15 modes spaced about 2 nm apart. Thus, the single-wavelength transmission of data, with high-power and low-power pulses representing the binary digits 1 and 0, respectively, yields to multiple-wavelength transmission.

In the future, it is expected that the C³ laser may become part of optical logic circuitry. This will be feasible because of the laser's tunability and from the electrical isolation of the two half-lasers. See Fig. 10.

One problem of the past results from the fact that much energy is wasted at the beginning of an optical link. H.M. Presby (Bell Laboratories) asserts in connection with the loss of about half the laser light, "It seemed like an awful waste, like throwing away half a tank of gas." Through redesigning the tips of optical fibers into carefully sculpted microlenses, a surprising amount of light was captured. Optical problems could be solved by designing an aspheric, hyperbolic shape. In a process that is akin to shaping a piece of metal in a lathe, the researchers rotate silicon fibers in and around a carbon dioxide laser beam.

Researchers also have been investigating microlasers. These lasers range from 1 to 5 microns in diameter and are "carved" from a multilayered

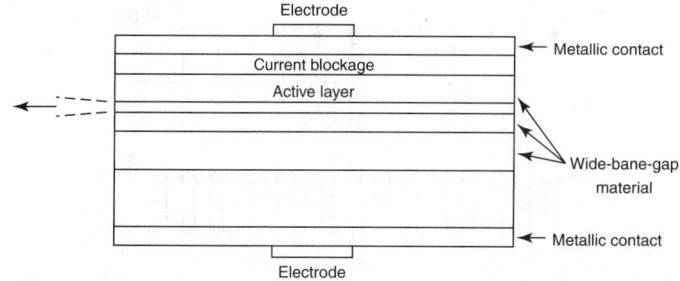

Fig. 7.   In an effort to improve the performance of the basic semiconductor laser, researchers developed the buried-heterostructure laser. In this configuration, the *p−n* junction is reduced to a "tube" that runs the length of the semiconductor crystal. This tube is surrounded by layers of semiconductor whose wide band gap raises the electrical barrier confining charge carriers within the tube. The wide-band-gap material also confines the photons produced at the junction. The laser beam spreads because of diffraction occurring where the beam emerges from the face of the device

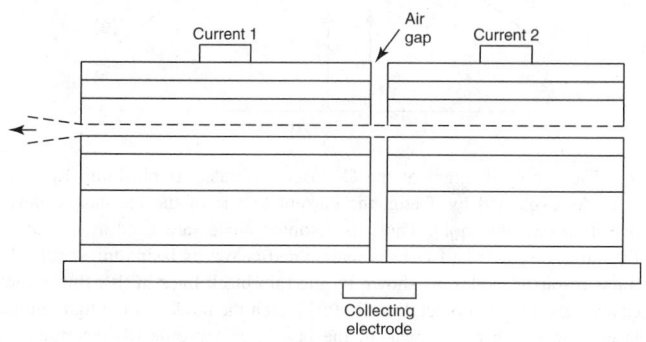

Fig. 8.   The C³ laser consisting of two aligned lasers. The half-lasers have different lengths and thus their resonant wavelengths are differently spaced. Only a few of them match. The mismatches are suppressed. Among the matches, only one is near the peak gain. Thus, the C³ laser beam is made up almost exclusively of that wavelength. Tests have shown that the probability of the beam "jumping" to another wavelength is less than 1 in 10 billion beam samplings

semiconductor substrate. It is estimated that a million such lasers occupy but a square centimeter on a chip. See Fig. 11.

Light-Emitting Diode for Telecommunications

In telecommunications, a light-emitting diode (LED) operates similarly to a laser in that it sends pulses of light signals representing speech or data through an optical fiber. Lasers are powerful and are preferred for sending light signals over great distances, such as between cities. Lasers require a measurable amount of power and require temperature regulation. An LED system developed by GTE Laboratories (LOC-LED System) operates on less than one-tenth the power required for a typical laser transmitter and requires no temperature control or means to adjust light signals, thus greatly simplifying the circuitry. The LOC-LED system generates sufficient optical power to fulfill most needs of fiber optic systems over a range of about 6 to 8 miles (9.7–12.9 km), which is a typical span for a local service loop. The system has been tested at temperatures from −4 to +185°F (−20 to +85°C). No prior commercially available LED device generates as much power while using so little energy as the LOC-LED.

Erbium-doped Fibers for Amplification

The use of repeaters to amplify signals dates back to the earliest days of telegraphy and telecommunications. Although optical fibers offered numerous improvements over copper wire communications, the need to use repeaters was not overcome. Since the serious introduction of optical fibers in the mid-1980s, long-range research has been directed toward maintaining signals over long distances and minimizing the requirement for repeater stations. As aptly pointed out by E. Corcoran, "Repeaters have become as endemic—and as constraining—on the telecommunications freeway as tollbooths on a turnpike." Lightwaves, of course, are capable of many frequencies, but electronic repeaters can handle only one frequency at a

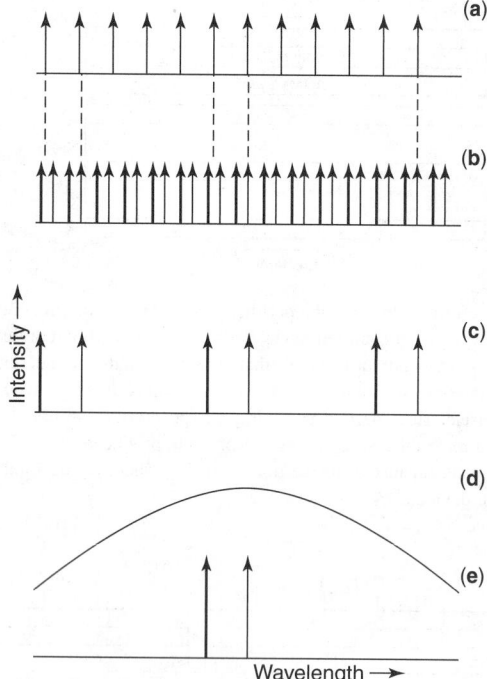

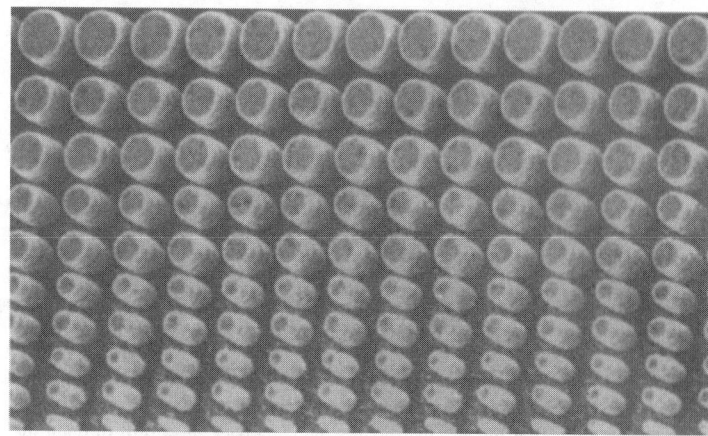

Fig. 11.   An array of microscopic lasers sculpted from a multi-layered semiconductor substrate. The lasers range from 1 to 5 microns in diameter, and a million of them occupy a square centimeter on a chip. (*AT&T Bell Laboratories*)

Fig. 9.   The energy diagram of the C^3 laser illustrates its tunability by current injection. As explained by Tsang, the current to one of the half-lasers elevates it above its *lasing* threshold. Thus, its resonant modes are fixed as indicated in (**a**). The other (second) half-laser is maintained below its lasing threshold. Thus, it still has resonant modes, as shown by the thin black lines in (**b**). But, some of the resonant modes [thin black lines in (**c**)] match the modes of the light-emitting half-laser. The match at the peak of the laser's gain profile (**d**) determines the wavelength of the C^3 laser's output (**e**). Now the current applied to the second half-laser is changed. The change alters its resonant modes [heavy black lines in (**b**)]. Thus, a new set of matching resonances is established [heavy black lines in (**c**)] and with it a new output wavelength [heavy black line in (**e**)]. (*After Tsang*)

population inversion) than remain at lower levels. As a transmission signal carrying information enters the erbium fiber, some of the excited erbium ions give up their energy to the information signal—thereby amplifying them. The amplification is purely optical. The information signal does not first need to be converted to an electrical signal.

Researchers (AT&T Bell Laboratories) have demonstrated that erbium-doped devices (Fig. 12) can "boost signals traveling at any bit rate, transmission networks can be upgraded by simply changing the transmitters (and receivers). Virtually any video data or voice signal can be dumpled 'like marbles' into one end of the 'transparent light pipes' and roll out intact at the other end."

Earlier research on the erbium-doped principle had been conducted by University of Southampton, U.K., and Japanese communications engineers. Currently, scientists envision that the new devices will be applied first to cable television and transoceanic telecommunications network. As pointed out by E. Corcoran, "Fiber amplifiers could enable a video broadcaster to lay one cable from the transmission center to a neighborhood, then split up the signal, amplify it and route the fibers directly into the homes." The Japanese already are at work on an all-optical network under the Pacific Ocean to be operable by 1996. For local communications loops, L.J. Andrews (GTE) observes, "Cost is the critical issue." John Mellis (BTD, Ipswich, U.K.) says, "The challenges are all engineering now... Because optical amplifiers are being seriously considered in transoceanic submarine systems, they will certainly happen, if not in this decade then the next." Another professional in the field observes, "There's absolutely no doubt that this is the thing to work on. It's changed how we think about fiber-optic systems."

There are three different locations in which optical amplifiers can be located in a network. See Fig. 13. Used immediately after a laser, booster amplifiers deliver output powers higher than 100 mW, and can therefore increase the output power of a laser by more than an order of magnitude. Preamplifiers increase the strength of an optical signal before it enters a conventional receiver.

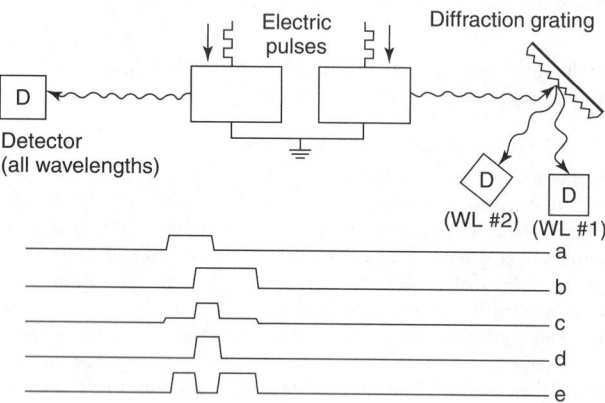

Fig. 10.   Use of the C^3 laser in an optical logic circuit. The laser's tunability and the electrical isolation of the two half-lasers make this application possible. Visualize the application of independent trains of electric pulses applied to the half-lasers, as indicated in (**a**) and (**b**). Simultaneous pulses cause the emission of light at wavelength #1. A pulse to one of the half-lasers causes the emission of light at wavelength #2. Detection of light at both wavelengths is equivalent to the logic operation OR (**c**). Detection of wavelength #1 is equivalent to the logic operation AND (**d**). Detection of wavelength #2 is equivalent to the logic operation EXCLUSIVE OR (**e**). (*After Tsang*)

time and thus decrease data throughput through a network. Elias Snitzer (Rutgers University) found that erbium-doped glass fibers could be used to provide optical amplification and thereby increase the number of light frequencies that can be amplified. When pumped with a light source, more erbium ions in these fibers are forced to a higher energy level (known as

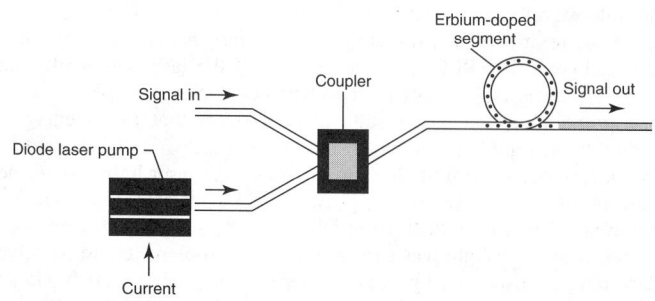

Fig. 12.   Erbium-doped light amplifier. Light from a diode laser pump excites erbium ions in segment of an optical fiber. These ions then emit light and thus boost a passing optical signal. (*After AT&T Bell Laboratories*)

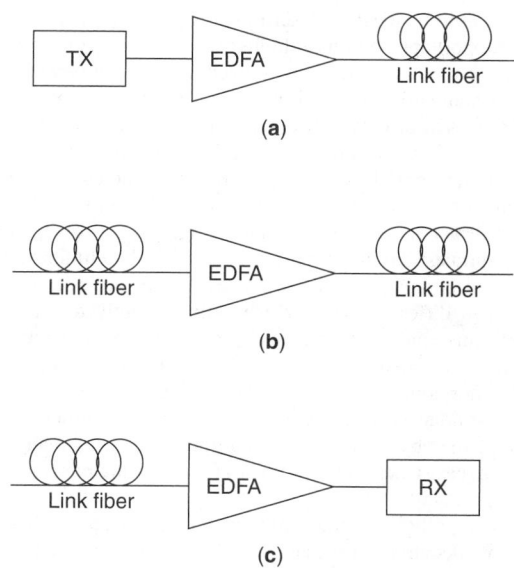

Fig. 13. Three types of optical amplifier as distinguished by their location in a network: (**a**) Postamplifier (booster), (**b**) in-line amplifier, and (**c**) optical preamplifier. (*From J. Augé*)

Solitons. A *soliton* may be defined as a wave that retains its shape indefinitely. Bright-pulse solitons are pulses of light that travel over long distances without dispersing or broadening. Each of the wavelengths that comprise an *ordinary* light pulse tend to travel at a slightly different speed. D.R. Grischkowsky (IBM Thomas J. Watson Research Center) observes that a pulse can be prevented from dispersing by taking advantage of an optical property of glass fibers—that is, proportional to the received light intensity. This output current is AC coupled to the load to deliver the RF signal.

Optical Fibers in Aircraft

Traditionally, communications and control in modern aircraft have been accomplished by as much as 240 km (in a typical wide-body jet) of electric wires, adding significantly to the weight of the craft. Even with insulation and shielding, such wire connections constantly are subject to electromagnetic interference (EMI), not to mention increased vulnerability to lightning during storm conditions and other sources of electromagnetic radiation of particular note to military aircraft. The potential of an "electronic blizzard" over a war zone long has been recognized as a hazard by military planners. Numerous crashes of the Sikorsky Aircraft Black Hawk helicopter have been attributed to EMI. R.J. Baumbick (National Aeronautics and Space Administration, Cleveland, Ohio) notes, "Wires are becoming the dominant antennae in aircraft." By replacing copper lines with optical fibers, the weight of cabling required can be reduced by an estimated 50%. See Fig. 14.

For the use of light systems in aircraft, researchers recognize the need for developing improved light sources for transmitters. Lasers do not serve well because of the high temperatures encountered in supersonic flight and of intense heat from aircraft engines. Powerful light-emitting diodes are required. A difficult problem is that of using optical components in place of traditional electric control of hydraulic actuators (control of flaps,

rudders, and other flight-control surfaces). Sensors that heat a small amount of hydraulic fluid and thus build up pressure that can be amplified by the actuator's hydraulic system are being developed and tested.

Fiber Optics in Biomedicine

Sensors incorporating glass or plastic optical fibers have demonstrated several advantages over electrosensors for biomedical applications. These sensors involve no electrical connections and hence are safe from that standpoint; the leads are quite small and flexible: they can be incorporated in catheters for multiple sensing; where required, they can be implanted for relatively long periods. The fibers are considerably less than 1 millimeter in diameter. Where designed for simplicity, they often can be considered disposable.

As reported by Peterson (National Institutes of Health) and Vurek (Sorenson Research Corp.), there are three principal types of *in vivo* fiber-optic sensing configurations—*photometric* (or bare-ended fiber), *physical parameter sensors* in which a transducer at the end of a fiber alters the light signal in accordance with the values of the parameter measured, and *chemical probes*. In the latter, a suitable reversible reagent fixed at the end of a fiber provides spectrophotometric or fluorometric analysis. The earliest use of fiber optic sensors was in connection with reflectometry, spectrophotometry, and fluorometry where no transducer was used. In oximetry, the hemoglobin content of blood is measured spectrophotometrically. Where dyes are injected into the blood, fiber optic sensors can be used for measuring blood flow, cardiac output, and perfusion. When a microtransducer is attached at the end of a fiber optic conductor, temperature and pressure measurements can be made. A notable application is the use of a temperature sensor in connection with the hyperthermal treatment of cancer. Accuracy of $\pm 0.1°C$ can be obtained. Sometimes such devices are used in multiple locations. For example, a layer of liquid crystals at the end of optical fibers will produce changes in light scattering due to temperature change. Fiber optic sensors have been designed for monitoring intracranial and intracardiac pressure. Chemical sensors have been developed to measure pH, PO_2, PCO_2, and glucose, among other chemical variables. The details of these biomedical applications for fiber optic sensors are well developed in the Peterson-Vurek paper (reference listed).

Fiber Optic Communications in Botany

It has been known for centuries that the position of a plant relative to its source of light will affect the manner in which the plant grows and develops, including its shape. Thus, there arose expressions such as "a plant seeks or reaches for the light." Horticulturists and farmers carefully lay out their greenhouses and fields so that maximum advantage can be taken of available sunlight or, in some cases, artificial light. Only in recent years, however, has it been suspected that plants may possess means for further utilizing the radiation which they receive. Researchers Mandoli (Stanford University) and Briggs (Carnegie Institution) have observed that, in addition to depending upon light as a source of energy (photosynthesis), light is also used as a means of communication. For example, the tissues of plant seedlings can guide light through distances measured in centimeters (a distance of 4.5 centimeters has been demonstrated in the laboratory). The exact method of light transmission within plants has not been fully established, but researchers suspect that natural fibers are involved—so called "light pipes."

In the initiation and control of several physiological processes in plants, lightwave communications may serve in a way comparable to the nervous

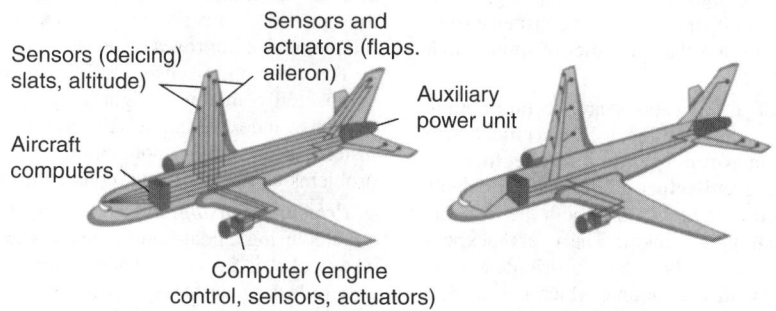

Fig. 14. Contrast of (left) hard-wire aircraft communications and controls and (right) use of fiber optics. (*After United Technologies Research Center*)

system of animals. Some of the factors influenced by light receptors include the time for a seed to germinate, the angle a shoot should take to counter gravitational forces, the rate with which a leaf should develop, and the time when a plant should bloom. Laboratory findings show that the amount of energy needed for light signaling is several orders of magnitude less than the light used for photosynthesis.

Light-sensitive detectors are pigment cells, of which the molecules making up a substance known as *phytochrome* are among the most important. These photoreceptors are sensitive to various parts of the light spectrum and thus play distinctive roles in managing different parts of a plant's physiology. In addition to the spectral distribution of light as received by the photosensitive cells, other important factors include the amount of light received, the direction from which the light is received, and the duration of the light signal.

This relatively recent area of botany and biology admittedly is in an early stage of development.

Fiber Optics in the Private Network

Private Networks, new and existing, must be well planned and carefully structured. Their network cabling solution may require a balance of both copper and fiber to cost effectively meet today's needs and support the high-bandwidth applications of the future, such as multimedia and full motion digital video-conferencing. Lucent Technologies, SYSTIMAX® SCS product families, supports these new and existing networks.

Rapidly evolving applications and technologies will drastically increase the speed and volume of traffic on LAN/WAN networks. Ensuring that your structured cabling solution is designed to accommodate the higher transmission rates associated with these evolving bandwidth intensive applications will be critical. Some examples include:

- Multimedia workstation
- Networked Scientific Modeling
- Imaging, Radiography, Computer Aided Design/Computer Aided Manufacturing (CAD/CAM)
- Asynchronous Transfer Mode (ATM)
- High-Definition Television (HDTV)
- Array processor workstations
- Mass memory database transfer
- Videophone
- Photonic (lightwave) switches and processors.

Local Area Networks (LANs). A LAN is a data communications system that enables users to access common data processing (PCs, minicomputers, and mainframe computers) and peripheral equipment (printers and fax machines). LANs, are created by using workstations with adapter cards and connecting them to file servers (where the operating system/software resides) and printers. Gateways are used to connect LANs to other LANs or operating systems like large mainframes where there is a need to share departmental or corporate computing systems. A LAN can be as simple as a few workstations working off a file server or as complex as putting hundreds of workstations on a network that runs between floors of a building or between a number of buildings in a campus environment. LANs, which were originally designed so that users could share and access a few expensive printers or controllers, have expanded into essential telecommunications networks. Today, LANs are used for file and printer sharing, electronic mail, shared databases, point-of-sale, and order entry systems.

LANs deployed on different floors or buildings are typically connected with multimode fiber. However, newer high-speed LAN topologies like full motion video do utilize single-mode fiber in some long-distance route applications where the excellent transmission characteristics of single-mode fiber are required.

LAN Topologies. SYSTIMAX SCS offers cabling architecture options for Fiber-to-the-Desktop installations: the traditional **Hierarchical Star architecture** and the new **Single Point Administration architecture**.

The traditional **Hierarchical Star architecture** is designed for maximum flexibility. Cross-connect facilities are provided in both the telecommunications closets and the main equipment room. The riser backbone cables can be sized with low counts which allow only distributed active equipment, or for greatest flexibility, with high counts which permit both distributed and centralized active equipment. The horizontal cross-connect facility helps ensure the greatest life span for the system by allowing the

active equipment to be located closer to the work areas. The short horizontal runs can support applications at higher speeds than the longer combined horizontal/riser runs used with centralized active equipment. Also, this architecture is standards-compliant with both riser design approaches.

Single Point Administration architecture is designed for simplicity and cost-effectiveness for centralized equipment. This approach provides direct connections from all work areas (offices) to the cross-connect in the main equipment room, forming a single point of administration optimized for centralized active equipment. The single point of administration provides the simplest circuit management possible by eliminating the need to cross-connect circuits in multiple places. It also provides the ability to connect users in different areas of a building directly to the same LAN segment, reducing traffic on bottleneck-prone bridges and routers. Single point administration also provides three cost benefits. First, it eliminates the need for horizontal cross-connects, saving passive hardware costs. Second, it consolidates active equipment, reducing the number of idle ports in the system, thereby saving active equipment costs. Finally, this type of architecture eases network administration and maintenance, reducing technical support staff effort.

The LAN topology is the physical layout of a LAN—how the controllers, workstations (primarily PCs), and other equipment are connected by the cable. The three basic LAN topologies are: Bus; Star; and Ring.

The bus topology has all the workstations on the network attached (via the information outlet) to a single cable that carries the signal in both directions through the network. The bus network, can be expanded by adding several segments together with bridges, routers, and repeaters. The Institute of Electrical and Electronics Engineers Inc. (IEEE) Standard 802.3 is an example of a bus topology. In a star topology, all of the nodes or workstations are connected with unshielded twisted pair (UTP) or fiber optic cable to a centrally located common controller or concentrator. The central control point permits centralized network administration, management, and troubleshooting. StarLAN is an example of a star topology, as is IEEE 802.3. In a ring topology, the network forms a ring. A token carries data through the network. Workstations are connected as in a star topology to a central administration point, such as a multistation access unit (MAU). Token Ring (IEEE 802.5) is an example of a ring topology.

The LANs, which were introduced back in the 1980s, were based on copper cable, with very few fiber applications being used. Most of the more recently introduced LANs (FDDI, 10BASE-F, and DFON) primarily use fiber optics but offer interfaces (bridges and routers) with the older copper-based networks. The more recent trend is in the use of "smart hubs" or concentrators, which allow both bus and ring (Ethernet and Token Ring) topologies to be mixed in the same electronics within the communications closet. In addition, the backbone network, which links the hubs/concentrators together, could be on a higher speed LAN topology such as FDDI running at 100 Mbps.

The advent of new lower cost optoelectronics for LAN applications has spurred a growing interest in using fiber all the way to the workstation. These total fiber networks offer easy migration to even higher speed applications like ATM.

Fiber Optics in Industrial Instrumentation and Networks

The combination of light-transmitting optical cable and miniature silicon sensors has resulted in the development of a new measurement technology for various industrial processes. Three variables may be measured with this technique—temperature, pressure, and refractive index. The new systems are immune to electromagnetic and radio-frequency interference, they provide more accuracy in electrically noisy environments, and their miniature size improves response and causes minimal process disturbances.

The fiber optic sensors utilize an extrinsic Fabry-Perot interferometer to spectrally modulate light in proportion to pressure, temperature, or refractive index variations. Because they are based on spectral modulation instead of amplitude modulation, they are not affected by such common problems as fiber bending, connector losses, and aging.

Pressure Sensing. As shown by Fig. 15, a cavity resonator is constructed using a float-bottom pocket-etched into a refractory glass substrate. The pocket is 2–3 wavelengths deep and covered with a diaphragm. The two reflecting surfaces (bottom of pocket and the diaphragm) form an interferometer. The cavity is evacuated, thus permitting the diaphragm to deflect. This deflection is a function of absolute pressure. In effect, the path length changes. A parallel can be made with the action of a soap bubble. As the bubble changes size and hence film thickness, interference effects occur. These reinforce reflected light at certain wavelengths, enhancing transmission of light through the film at other wavelengths. Color (frequency) changes occur as a result.

The cavity resonator (sensor) is 0.015 in (0.4 mm) in all three dimensions. The resonator is joined to a multimode fiber by a glass capillary 0.032 in (0.8 mm) in diameter. The light source's spectrum is centered at 850 nm. As the effective cavity depth changes as a function of the measured variable, the cavity's reflectance spectrum shifts. This spectral shift modulates (skews) the incoming light spectrum. A dichroic filter and photodiodes are then used to discern differences in the returning light's spectrum strength within two wavebands.

Temperature Sensing. As shown in Fig. 16, a layer of silicon (whose refractive index changes with temperature) is placed in the optical path in place of the evacuated cavity, as previously described. The second reflector (glass) is rigid. The effective path length thus changes with temperature.

Refractive Index. This property is the ratio of the velocity of light in a vacuum to the velocity of light in a transparent material. The property is used extensively in the processing industries for measuring component concentrations or for tracking changes in molecular makeup, such as materials that are reacting. The hydrogenation of food oils is an example of such a process.

The extent of hydrogenation (the degree of saturation of a double bond in the ester chain of an edible oil) may be indicated by conventional laboratory titrations of the iodine number. This is a time-consuming, grab-sampling method. Continuous measurement and control via refractive index measurement is far more efficient. As shown by Fig. 17, the fluid of interest is drawn by capillary action into a duct through a glass substrate, and the effective path length varies in proportion to the refractive index of the fluid. Excellent correlation of refractive index with the iodine number of hydrogenated oil is shown in Fig. 18.

Note: The foregoing information on fiber optic sensors was furnished by Gene Yazbak, Foxboro, MA.

Optical Fiber Terminology

An abridged glossary of terms used to describe optical fiber products and processes would include:

Absorption—A physical mechanism in fibers that attenuates light by converting it into heat-thereby raising the fiber's temperature. In practice the temperature increase is slight and difficult to measure. Absorption arises from tails of the ultraviolet and infrared absorption

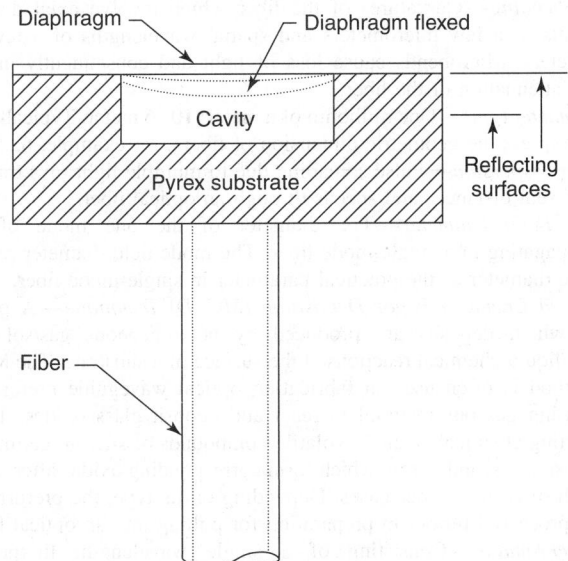

Fig. 15.   Fiber-optic pressure sensor. As light passes into a cavity resonator formed by a glass substrate and a flexible diaphragm, it is reflected from both surfaces, forming an interference pattern that changes as the diaphragm flexes with pressure changes. (*Yazbak, Foxboro, Massachusetts*)

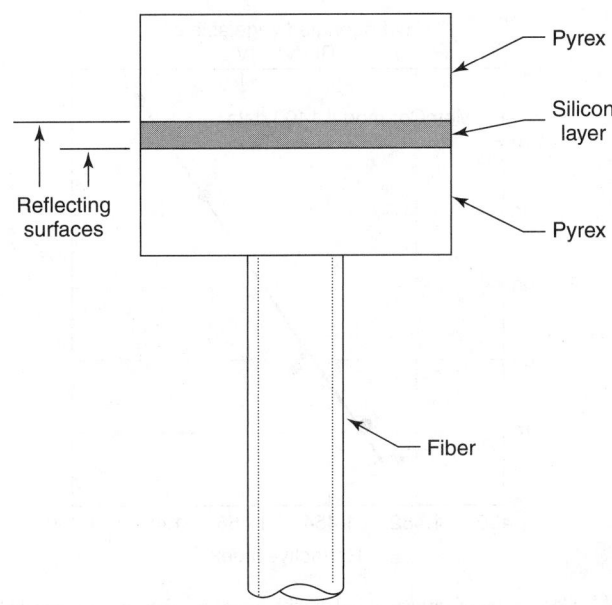

Fig. 16.   Fiber-optic temperature sensor. A thin layer of silicon placed in the optical path exhibits a large change in refractive index with temperature, changing the effective path length. (*Yazbak, Foxboro, Massachusetts*)

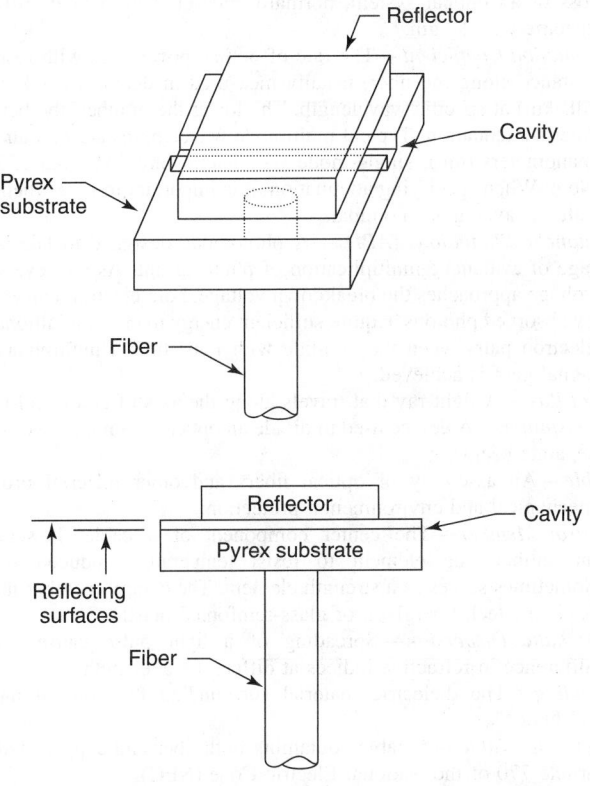

Fig. 17.   Fiber-optic refractive index sensor. Fluid of interest is drawn by capillary action into a duct through a glass substrate. The effective path length varies in proportion with the refractive index. (*Yazbak, Foxboro, Massachusetts*)

bands, from impurities such as the OH−ion, and from defects in the glass structure.

Adapter Loss—The power loss suffered when coupling light from one optical device to another.

Aramid Yarn—Strength elements that provide tensile strength and provide support and additional protection of the fiber bundles. Kevlar is a particular brand of aramid yarn.

Armor—Additional protective element beneath outer jacket to provide protection against severe outdoor environments. Usually made of plastic-coated steel, it may be corrugated for flexibility.

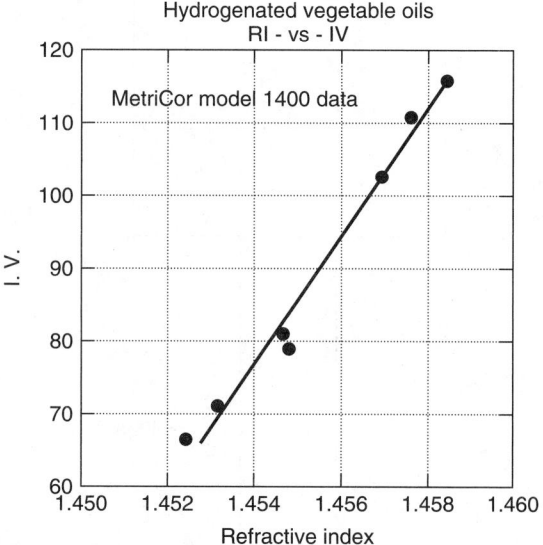

Fig. 18. Correlation of refractive index with iodine number of hydrogenated oil. (*Yazbak, Foxboro, Massachusetts*)

Attenuation—The decrease in magnitude of power, of a signal in transmission between points. A term used for expressing the total loss of an optical system, normally measured in decibels (dB) at a specific wavelength.

Attenuation Coefficient—The rate of optical power loss with respect to distance along the fiber, usually measured in decibels per kilometer (dB/km) at specific wavelength. The lower the number, the better the fiber's attenuation. Typical multimode wavelengths are 850 and 1300 manometers (nm); single-mode wavelengths are 1310 and 1550 nm. Note: When specifying attenuation, it is important to note whether the value is average or nominal.

Avalanche Photodiode (APD)—A photodiode designed to take advantage of avalanche multiplication of photocurrent. As the reverse-bias voltage approaches the breakdown voltage, hole-electron pairs created by absorbed photons acquire sufficient energy to create additional hole electron pairs when they collide with ions; thus a multiplication or signal gain is achieved.

Axial Ray—A light ray that travels along the axis of an optical fiber.

Beamsplitter—A device used to divide an optical beam into two or more separate beams.

Cable—An assembly of optical fibers and other material providing mechanical and environmental protection.

Central Member—The center component of a cable. It serves as an antibuckling element to resist temperature induced stresses. Sometimes serves as a strength element. The central member material is either steel, fiberglass, or glass-reinforced plastic.

Chromatic Dispersion—Spreading of a light pulse caused by the difference in refractive indices at different wavelengths.

Cladding—The dielectric material surrounding the core of an optical fiber.

Composite Cable—A cable containing both fiber and copper media per article 770 of the National Electric Code (NEC).

Core—The central region of an optical fiber through which light is transmitted.

Dielectric—Nonmetallic and, therefore, nonconductive. Glass fibers are considered dielectric. A dielectric cable contains no metallic components.

Ferrule—A mechanical fixture, generally a rigid tube, used to protect and align a fiber in a connector. Generally associated with fiber optic connectors.

Fiber—Any filament or fiber, made of dielectric materials, that guides light.

Fiber Bend Radius—Radius a fiber can bend before the risk of breakage or increase in attenuation.

Fiber Optics—The branch of optical technology concerned with the transmission of radiant power through fibers made of transparent materials such as glass, fused silica, or plastic.

Fresnel Reflection—The reflection of a portion of the light incident between two homogeneous media having different refractive indices. Fresnel reflection occurs at the air/glass interfaces at entrance and exit ends of an optical fiber.

Fresnel Reflection Losses—Reflection losses that are incurred at the input and output of optical fibers due to the differences in refraction index between the core glass and immersion medium.

Index Matching Fluid—A fluid with an index of refraction close to that of glass that reduces reflections caused by refractive-index differences.

Index Matching Material—A material, often a liquid or cement whose refractive index is nearly equal to the core index. Used to reduce Fresnel reflections from a fiber end face.

Intermediate Cross-Connect (IC)—A secondary crossconnect in the backbone cabling used to mechanically terminate and administer backbone cabling between the main cross-connect and horizontal cross-connect.

Irradiance—Power density at a surface through which radiation passes at the radiating surface of a light source or at the cross section of an optical waveguide. The normal unit is Watts per centimeters squared, or W/cmu2d.

Laser Diode (LD)—Light Amplification by Stimulated Emission of Radiation. An electro-optic device that produces coherent light with a narrow range of wavelengths, typically centered around 780 nm, 1320 nm, or 1550 nm. Lasers with wavelengths centered around 780 nm are commonly referred to as CD Lasers.

Lasing Threshold—The lowest excitation level at which a laser's output is dominated by stimulated emission rather than spontaneous emission.

Light—In the laser and optical communication fields, the portion of the electromagnetic spectrum that can be handled by the basic optical techniques used for the visible spectrum extending from the near ultraviolet region of approximately 0.3 micron, through the visible region and into the mid infrared region of about 30 microns.

Lightwaves—Electromagnetic waves in the region of optical frequencies. The term "light" was originally restricted to radiation visible to the human eye, with wavelengths between 400 and 700 manometers (nm). However, it has become customary to refer to radiation in the spectral regions adjacent to visible light (in the near infrared from 700 to about 2000 nm) as "light" to emphasize the physical and technical characteristics they have in common with visible light.

MDPE—Abbreviation used to denote medium density polyethylene. A type of plastic material used to make cable jacketing.

Material Dispersion—The dispersion associated with a non-monochromatic light source due to the wavelength dependence of the refractive index of a material or of the light velocity in this material.

Microbending—Curvatures of the fiber which involve axial displacements of a few micrometers and spatial wavelengths of a few millimeters. Microbends cause loss of light and consequently increase the attenuation of the fiber.

Micrometer (gm)—One millionth of a meter; 10−6 meter. Typically used to express the geometric dimension of fibers, for example, 62.5 μm.

Modal Dispersion—Pulse spreading due to multiple light rays traveling different distances and speeds through an optical fiber.

Mode Field Diameter—The diameter of the one mode of light propagating in a single-mode fiber. The mode field diameter replaces core diameter as the practical parameter in single-mode fiber.

Modified Chemical Vapor Deposition (MCVD) Technique—A process in which deposits are produced by heterogeneous gas/solid and gas/liquid chemical reactions at the surface of a substrate. The MCVD method is often used in fabricating optical waveguide preforms by causing gaseous material to react and deposit glass oxides. Typical starting chemicals include volatile compounds of silicon, germanium, phosphorus, and boron, which form corresponding oxides after heating with oxygen or other gases. Depending on its type, the preform may be processed further in preparation for pulling into an optical fiber.

Monochromatic—Consisting of a single wavelength. In practice, radiation is never perfectly monochromatic but, at best, displays a narrow band of wavelengths.

Multimode Fiber—An optical waveguide in which light travels in multiple modes. Typical core/cladding size (measured in micrometers) is 62.5/125.

Nanometer (nm)—A unit of measurement equal to one billionth of a meter; 10−9 meters. Typically used to express the savelength of light, for example, 1300 nm.

Near Field Radiation Pattern—Distribution of the irradiance over an emitting surface; in other words, over the cross section of an optical waveguide.

Optical Waveguide—Dielectric waveguide with a core consisting of optically transparent material of low attenuation (usually silica glass) and with cladding consisting of optically transparent material of lower refractive index than that of the core. It is used for the transmission of signals with lightwaves and is frequently referred to as fiber. In addition, there are planar dielectric waveguide structures in some optical components, such as laser diodes, which are also referred to as optical waveguides.

Optoelectronic—Pertaining to a device that responds to optical power, emits or modifies optical radiation, or utilizes optical radiation for its internal operation. Any device that functions as an electrical-to-optical or optical-to-electrical transducer.

PE—Abbreviation used to denote polyethylene. A type of plastic material used for outside plant cable jackets.

PVC—Abbreviation used to denote polyvinyl chloride. A type of plastic material used for cable jacketing. Typically used in flame-retardant cables.

PVDF—Abbreviation used to denote polyvinyl difluoride. A type of material used for cable jacketing. Often used in plenum-rated cables.

Preform—A glass structure from which an optical fiber waveguide may be drawn.

Prefusing—Fusing with a low current to clean the fiber end. Precedes fusion splicing.

Single-Mode Fiber—Optical fiber with a small core diameter (typically 9 μm) in which only a single-mode, the fundamental mode, is capable of propagation. This type of fiber is particularly suitable for wideband transmission over large distances, since its bandwidth is limited only by chromatic dispersion.

Spontaneous Emission—This occurs when there are too many electrons in the conduction band of a semiconductor. These electrons drop spontaneously into vacant locations in the valence band, a photon being emitted for each electron. The emitted light is incoherent.

Step Index Fiber—A fiber having a uniform refractive index within the core and a sharp decrease in refractive index at the core/cladding interface.

Stimulated Emission—This occurs when photons in a semiconductor stimulate available excess charge carriers to the emission of photons. The emitted light is identical in wavelength and phase with the incident coherent light.

Lucent Technologies, Optical Fiber Solutions
Norcross, Georgia

Additional Reading

Adrian, P.: "Technical Advances in Fiber-Optic Sensors: Theory and Applications," *Sensors* 23 (September 1991).

Agrawal, G.: *Fiber-Optic Communication Systems,* John Wiley & Sons, Inc., New York, NY, 1997.

Amato, I.: "The Natural Roots of Fiber Optics," *Science News* 414 (December 23–30, 1989).

AugÉ, J., et al.: "Progress in Optical Amplification," *Microwave J.* 62 (June 1993).

Baumbick, R.J. and J. Alexander: "Fiber Optics Sense Process Variables," *Control Eng.* **27**, 3, 75–77 (1980).

Bobb, L.C. and P.M. Shankar: "Tapered Optical Fiber Components and Sensors," *Microwave J.* 219 (May 1992).

Corcoran, E.: "Light Talk: U.S. and Japanese Compete to Put Optical Fibers in the Home," *Sci. Amer.* 74 (October 1989).

Corcoran, E.: "Light Traffic: Optical Amplifiers Promise to Unclog Lightwave Communication," *Sci. Amer.* 106 (March 1991).

Corcoran, E.: "Avoiding the Potholes on Optical Highways," *Sci. Amer.* 143 (April 1992).

Desurvire, E.: "Lightwave Communications: The Fifth Generation," *Sci. Amer.* 114 (January 1992).

Dutton, H.: *Understanding Optical Communications,* Prentice-Hall, Inc., Upper Saddle River, NJ, 1999.

Furse, C. and R. Haupt: "Down to the Wire," *IEEE Spectrum* (February 2001).

Gabel, D.: "Fiber Optics on the Rise," *Electronic Buyers' News* 36 (January 28, 1991).

Grimes, G.: "Microwave Fiber-Optic Delay Lines: Coming of Age in 1992," *Microwave J.* 61 (August 1992).

Hamilton, K.J.: "Fiber Optic Sensors Grow Into Networks," *InTech* 20 (February 1991).

Hecht, J.: *City of Light,* Oxford University Press, New York, NY, 1999.

Henkel, S.: "Single Optical Fiber Does It All for Smart Transmitters," *Sensors* 8 (January 1992).

Holden, C.: "Plugging Into the Pacific Ocean," *Science* 599 (August 11, 1989).

Horgan, J.: "Dark Solutions: Physicists Generate Durable Pulses of Darkness," *Sci. Amer.* 24 (May 1988).

Ito, T., K. Fukuchi, K. Sekiya, D. Ogasahara, R. Ohhira and T. Ono: *6.4 Tb/s (160 × 40 Gb/s) WDM Transmission Experiment with 0.8 bits/Hz Spectral Efficiency,* European Conference on Optical Communication," post-deadline paper, September 2000.

Jones, W.B. Jr.: *Introduction to Optical Fiber Communication Systems,* Oxford University Press, Inc., New York, NY, 1995.

Kazovsky, L., et al.: *Optical Fiber Communication Systems,* Artech House, Inc., Norwood, MA, 1996.

Kogelnik, H.: "High-Speed Lightwave Transmission in Optical Fibers," *Science* **228**, 1043–1048 (1985).

Ledwith, A.: "Glasses for Fibre Optic Communications," *Review (University of Wales)* 15 (RefSeasonsSpring 1988).

Mandoli, D.F. and W.R. Briggs: "Fiber Optics in Plants," *Sci. Amer.* 90–98 (August 1984).

McHugh, P.: "Fiber Optics Extend the Reach of Photoelectric Sensors," *Instruments & Control Systems* 57 (August 1989).

Nicholson, P.J.: "An Introduction to Fiber Optics," *Microwave J.* 26 (June 1991).

Nicholson, P.J.: "An Overview of the Synchronous Optical Network," *Microwave J.* 24 (December 1991).

Nielsen, T.N., et al.: *3.28 Tb/s (82 × 40 Gb/s) Transmission Over 3 × 100 km of Nonzero-dispersion Fiber Using Dual C- and L-band Hybrid Raman/Erbium-doped Inline Amplifiers,* Optical Fiber Communication Conference, post-deadline paper 29, March 2000.

Papannareddy, R.: *Introduction to Lightwave Communication Systems,* Artech House, Inc., Norwood, MA, 1997.

Peterson, J.I. and G.G. Vurek: "Fiber-Optic Sensors for Biomedical Applications," *Science* **224** 123–127 (1984).

Pratsinis, S.E. and S.V.R. Mastrangelo: "Material Synthesis In Aerosol Reactors (Optical Fiber Manufacture)," *Chem. Eng. Progress* 65 (May 1989).

Raybon, G., et al.: *320 Gbit/s Single-channel Pseudo-linear Transmission over 200 km of Non-zero Dispersion Fiber,* Optical Fiber Communication Conference, post-deadline paper 29, March 2000.

Refi, J., *Fiber Optic Cable—a LightGuide,* abc TeleTraining, Inc., Geneva, IL, 1991.

Stix, G.: "Light Flight: Optical Fibers May Be the Nerves of New Aircraft," *Sci. Amer.* 120 (May 1991).

Suematsu, Y. and K.I. Iga: *Introduction to Optical Fiber Communications,* John Wiley & Sons, Inc., New York, NY, 1982.

Tsang, W.T., N.A. Olsson and R.A. Logan: "High-Speed Direct Single-Frequency Modulation with Large Tuning Rate and Frequency Excursion in Cleaved-Coupled-Cavity Semiconductor Lasers," *Applied Physics Letters* **42**(8), 650–652 (April 15, 1983).

Woracek, D.: "Fiber Optic Sensors Endure Microwaves," *InTech* 24 (February 1991).

Yazbak, G.: "Fiberoptic Sensors Solve Measurement Problems," *Food Technology* 76 (July 1991).

Web References

CoreTek Inc: http://www.coretekinc.com/
General Cable: http://www.generalcable.com/
Lucent Technologies: http://www.lucent.com/ofs/
Nanoptics, Inc: http://www.nanoptics.com/

OPTICAL GLASS. Glass to be useful for lenses, prisms and other optical parts through which light passes, as distinguished as for mirrors, must be completely homogeneous. This includes freedom from bubbles, striae, seeds, strains, etc. In order to reduce aberrations, the optical designer needs many different kinds of glass. A few typical types are described in Table 1. The v-number is the reciprocal of the dispersive power of the glass.

TABLE 1. VARIOUS GLASSES

	Type	n_D	v-Number
Borosilicate	Crown	1.5170	64.5
Barium	Crown	1.5411	59.5
Spectacle	Crown	1.5230	58.4
Light	Flint	1.5880	53.4
Ordinary	Flint	1.6170	38.5
Dense	Flint	1.6660	32.4
Extra dense	Flint	1.7200	29.3

OPTICAL ISOMER. Either of two kinds of optically active three-dimensional isomers (stereoisomers). One kind is represented by mirror-image presence of one or more asymmetric carbon atoms in the compound (glyceraldehyde, lactic acid, sugars, tartaric acid, amino acids). The other kind is exemplified by diastereoisomers, which are not mirror images. These occur in compounds having two or more asymmetric carbon atoms; thus, such compounds have 2_n optical isomers, where n is the number of asymmetric carbon atoms.

See also **Optical Rotation**.

OPTICAL MICROSCOPE. A magnifying lens system that utilizes light in the visible wavelength range of the electromagnetic spectrum (5000 Å). A convex glass lens bends or focuses light waves because of the difference in density between glass and air. Invented in 1590 by the Janssen brothers and later improved by van Leeuwenhoek, the compound microscope has three lenses: a condenser lens, which concentrates the incident light; an objective lens, which gives an enlarged reverse image of the specimen; and a projector lens, which further enlarges the image and return it to normal position. Its maximum resolving power is 0.5 micron, compared with 100 microns for the human eye. The compound microscope is particularly useful in studying bacterial and other microorganisms in their natural state without interfering with their behavior. It has been of untold benefit to biologists and bacteriologists and also has innumerable uses in chemical and metallurgical research, as well as in forensic chemistry.

OPTICAL ROTATION. The change of direction of the plane of polarized light to either the right of the left as it passes through a molecule containing one or more asymmetric carbon atoms, e.g., sugars. The direction of rotation, it to the right, is indicated by either a plus sign (+) or a d-; if to the left, by a minus sign (−) or an l-. Molecules having a right landed configuration (D) usually are dextrorotatory, D(+), though they may be levorotatory, D(−); those having a left-handed configuration (L) are usually levorotatory, L(−), but may be dextrorotatory d(+). Compounds having this property are said to be optically active and are isomeric. The amount of rotation varies with the compound but is the same for any two isomers, though in opposite directions.

See also **Optical Isomer**.

ORBITALS. This article embraces both atomic and molecular orbitals.

Atomic. From spectroscopic studies, it is known that when an electron is bound to a positively charged nucleus only certain fixed energy levels are accessible to the electron. Before 1926, the old quantum theory considered that the motion of the electrons could be described by classical Newtonian mechanics in which the electrons move in well defined circular or elliptical orbits around the nucleus. However, the theory encountered numerous difficulties and in many instances there arose serious discrepancies between its predictions and experimental fact.

A new quantum theory called wave mechanics (as formulated by Schrödinger) or quantum mechanics (as formulated by Heisenberg, Born and Dirac) was developed in 1926. This was immediately successful in accounting for a wide variety of experimental observations, and there is little doubt that, in principle, the theory is capable of describing any physical system. A strange feature of the new mechanics, however, is that nowhere does the path or velocity of the electron enter the description. In fact it is often impossible to visualize any classical motion that could be consistent with the quantum mechanical picture of the atom.

In this theory the electron is viewed as a three-dimensional standing wave. The pattern of the wave is described by a wave function ϕ (analogous to the amplitude of a water wave). This one-electron wave function is called an atomic orbital. Since the wave function can be positive or negative (and real or complex) it does not describe an observable property of the electron. However, the square of the wave function (ϕ^2 or ϕ times its complex conjugate) is always positive and real, and can be identified with the probability of finding the electron at any point. This was first suggested by Born, but has now received ample experimental support. Hence, when the wave function is calculated for any electron we can determine the regions in space where the electron is most likely to be found, though we cannot say what type of motion results in that particular probability pattern.

Atomic orbitals are usually labeled by a set of designating numbers called quantum numbers. The one that determines the energy of the resulting state (for hydrogen) is given the symbol n and called the "principal quantum number." It assumes the values 1, 2, 3, 4, 5, ... to infinity, with increasing electron energies. The second quantum number is given the symbol l. It can be identified with the angular momentum of the electron due to its orbital motion, and assumes values of 0, 1, 2, 3, ... to $(n-l)$. For historical reasons the orbitals with these values are referred to as s, p, d, f, ... orbitals respectively. Hence a $3d$ orbital is one for which $n = 3$ and $l = 2$. The third quantum number is usually given the symbol m and is difficult to define in the absence of an external field. However, under all conditions it can assume $2l + 1$ values.

From the rules given in the previous paragraph it can easily be shown that for $n = 1$, 2 and 3 there are a total of 14 allowed orbitals. For an isolated hydrogen atom all orbitals must be spherically symmetrical and have the shapes shown in series (a) of Fig. 1. These are plots of the probability function (ϕ^2) for the orbitals and hence the darkest areas represent regions where the electron is most likely to be found. Though there are actually three $2p$ orbitals, three $3p$ orbitals and five $3d$ orbitals, in the absence of an external field the orbitals within a given set are identical in shape and energy, and are said to be degenerate. If a direction is defined by the presence of a magnetic field or the approach of another atom, the degeneracy is removed. The shapes of the allowed orbitals under these conditions are shown in series (b) of Fig. 1. The quantum number m is well defined for these orbitals and its value is given beside each orbital. For the discussion of bonding in polyatomic molecules another set of orbitals is useful. These are shown in series (c) of Fig. 1. The quantum number m is not well-defined for these orbitals, and they are usually labeled according to the axis along which they lie.

In addition to the orbitals shown in Fig. 1 there are "hybrid" orbitals that are not stationary states for the electron in an isolated atom. They can be obtained by taking a linear combination of the standard orbitals in Fig. 1. Since the electron distribution is "off center" they are useful only for atoms that are perturbed by an electric field (Stark-effect) or by the approach of other atoms as occurs in chemical-bond formation.

In addition to the three quantum numbers discussed above, experimental evidence requires an additional quantum number m_s, which by analogy to classical mechanics is attributed to an intrinsic (i.e., position-independent) property of the electron called "spin." Unlike the other quantum numbers, however, it can assume only two values ($\pm\frac{1}{2}$). As we shall see, this fact determines the orbital population of the many-electron atom.

It is an unfortunate consequence of the mathematical complexity of the quantum mechanical equations that the hydrogenic atom (i.e., one-electron atom) is the only system for which an exact probability distribution (ϕ^2) can be obtained. Approximate methods must be used to calculate the wave functions for the many electron atoms. We begin with the natural assumption that the wave function for such a more complex atom (ψ) can be obtained by taking a product of the appropriate one electron functions (ϕ). However, even if we ignore for the moment the coulombic repulsion between electrons, there are two fundamental postulates of quantum mechanics that complicate the picture. One is that electrons are indistinguishable. Hence, when their positions are interchanged, the probability function (ψ^2) must remain unchanged. This means that the wave function (ψ) must either remain the same or only change signs when electron positions are interchanged. A second postulate (called the Pauli principle) is that the total wave function must change signs when electron positions are interchanged. From these requirements it can be seen that when two electrons are placed in the same orbital (i.e., assigned the same orbital wave function ϕ) the total wave function will change signs (as required by the Pauli principle) only if the electrons have different spin quantum numbers ($+\frac{1}{2}$ and $-\frac{1}{2}$). It is hence a general rule that hydrogenic orbitals can only be occupied by a maximum of two electrons and these must have opposite spin.

If we begin with the most tightly bound orbital and add two electrons to each orbital the so called "ground state configuration" of the atom is obtained. For example the electronic configuration of the silicon atom is written $1s^2, 2s^2, 2p^2; 3s^2, 3p^2$. The orbital shapes and energies are, however, considerably altered by electron-electron repulsion, especially between electrons whose orbitals overlap appreciably. This has several marked effects. For a given value of n, all the orbitals no longer have the same energy. The binding energy now decreases with increasing values of the quantum number l. Secondly, because of the coulombic repulsion between electrons, they will tend to occupy separate orbitals whenever feasible (for example the $3p$ orbitals configuration in the silicon atom is actually $3p^1, 3p^1$). Furthermore, when electrons are forced together into the region of one hydrogenic orbital it is quite likely that electron-electron

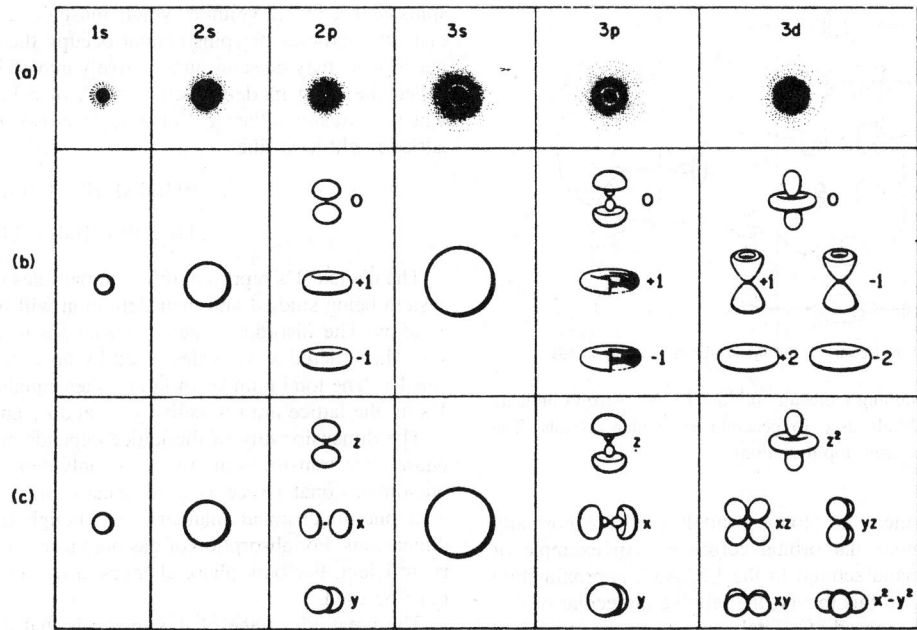

Fig. 1.   Hydrogen atomic orbitals: (a) for isolated atoms; (b) with one direction defined; (c) with three directions defined.

repulsion (always greater than 25 kcal) leads, in effect, to slightly different orbitals for each electron.

For chemical purposes we are most interested in the shapes of orbitals in which the valence (outermost) electrons reside. By assuming that the inner electrons act only to screen some of the positive charge on the nucleus the valence electrons can be shown to assume the shape of the appropriate hydrogenic orbital, with the insignificant difference that the inner nodes in the orbitals shown in Fig. 1 are drawn in closer to the nucleus because the shielding is poorer in this region. It is worth noting that for the many-electron atom, the indistinguishability of electrons has the effect of making only the total probability function physically meaningful. For this reason and because of the repulsion between electrons, the individual one-electron wave functions are so correlated that, though the "independent orbital concept" remains a very useful approximation, it is not fundamental to the problem.

Molecular. By analog with atomic orbitals, the wave function (ψ) for one electron in a molecule is called a molecular orbital, and the probability of finding the electron at any point is similarly given by the value of ψ^2 at that point. Just as in the case of atomic orbitals, an exact solution of the equations is possible only for the one-electron molecule (H_2^+) and it is only for this species that accurate molecular orbitals can be obtained.

The shapes of 10 of the most tightly bound orbitals of H_2^+ are shown in Fig. 2. Three quantum numbers can be used to label these orbitals. The one that is always well defined is given the symbol λ and can be identified with the component of the orbital angular momentum along the internuclear axis. It can take on values 0, 1, 2, 3 ... for which the orbitals are called $\sigma, \pi, \delta, \ldots$ respectively. The other two quantum numbers are defined differently depending on the nuclear separation. When the internuclear distance is short, as the case of H_2^+, it is convenient to consider the atomic orbital that would result from a given molecular orbital if the two nuclei were made to coalesce (in our imaginations). In the case of homonuclear diatomic molecules, for example, the molecular orbital designated $3d\ \sigma$ has $\lambda = 0$ and correlates to a $3d$ atomic orbital when the nuclei coalesce (i.e., the atomic quantum numbers n and l become well-defined at short internuclear distance). On the other hand, when the internuclear distance is large, a more useful and significant label is one which identifies the atomic orbitals with which a given molecular orbital correlates when the distance between the two nuclei approaches infinity. For this "separate atom" designation an additional symbol must be used to distinguish between the two molecular states that can arise from a given pair of atomic states. Chemists find it most useful to use the superscript * to indicate the higher energy (antibonding) orbital and the absence of * to indicate the lower energy (bonding) orbital. When the difference in symmetry between the two states is important the symbols g (gerade-symmetric) and u (ungerade-symmetric) are used. Thus, the orbital $\sigma^*(2p_x)$

is an antibonding orbital with $\lambda = 1$ that correlates with two $2p_x$ orbitals on the separated atoms. Similarly, a $\sigma_g 2(sp)$ orbital is a bonding orbital with $\lambda = 0$ that correlates with two atomic orbitals on the separated atoms. It is worth noting that the "antibonding" orbitals possess a nodal surface (i.e., a region of zero electron-probability density) between the two nuclei.

Exact solutions such as those given above have not yet been obtained for the usual many-electron molecules encountered by chemists. The approximate method which retains the idea of orbitals for individual electrons is called "molecular-orbital theory" (M. O. theory). Its approach to the problem is similar to that used to describe atomic orbitals in the many-electron atom. Electrons are assumed to occupy the lowest energy orbitals with a maximum population of two electrons per orbital (to satisfy the Pauli exclusion principle). Furthermore, just as in the case of atoms, electron-electron repulsion is considered to cause degenerate (of equal energy) orbitals to be singly occupied before pairing occurs.

It has not proved mathematically feasible to calculate the electron-electron repulsion that causes this change in orbital-energies for many-electron molecules. It is even difficult to rationalize the qualitative changes in sequence on the basis of the shapes of the H_2^+ orbitals. Greater success has been achieved by an approximate method which begins with orbitals characteristic of the isolated atoms present in the molecule, and assumes that molecular orbital wave functions can be obtained by taking linear combinations of atomic orbital wave functions (abbreviated L.C.A.O.). For

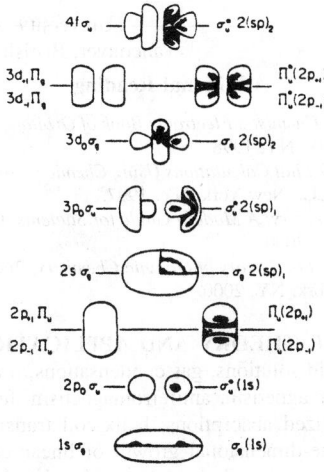

Fig. 2.   Molecular orbitals of H_2^+. The "united atom" designation is given on the left-hand side and the "separate atom" designation on the right-hand side.

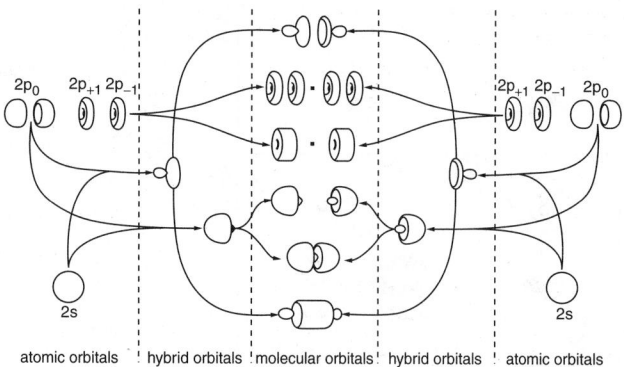

Fig. 3.   L.C.A.O. method of obtaining molecular orbitals for N_2. Arrows indicate the combinations of atomic orbitals used to generate molecular orbitals. The stabilities of the orbitals increase from top to bottom.

homonuclear diatomic molecules, the atomic orbitals used are normally those with which a given molecular orbital correlates. An example of how molecular orbitals are manufactured in the L.C.A.O. approximation is shown in Fig. 3. According to this theory, the relative molecular-orbital energies will depend on (a) the spread of orbital energies on the individual atoms (since this determines the extent of hybridization) and (b) on the internuclear distance which is largely determined by the relative number of bonding and antibonding orbitals that are filled.

The L.C.A.O. approximation has also proved useful for the description of heteronuclear diatomic molecules, although valence-bond theory has been somewhat more successful in its quantitative calculations of bond energies. For these molecules the selection of appropriate atomic orbitals to be used in linear combination is governed by considerations of symmetry, energetics and overlap. These considerations also apply to the formation of molecular orbitals in polyatomic molecules by the L.C.A.O. approximation. In the case of polyatomic molecules it is always possible to develop either (a) molecular orbitals that extend between only two atoms (localized M.O.s) or (b) molecular orbitals that extend over the entire molecule (delocalized M.O.s). In the case of saturated molecules, the two descriptions are practically equivalent. However, for molecules with conjugated double bonds and especially for aromatic compounds, delocalized molecular orbitals which extend over several atoms must be used. When the L.C.A.O. method is used to develop molecular orbitals in complexes between metal ions (possessing d orbitals) and ligands such as CO, OH^-, NH_3, CN^- etc., the procedure is called Ligand Field Theory.

A thorough test of the L.C.A.O. molecular orbitals has been possible only for H_2^+ where both the wave functions and orbital energies are known accurately. Such a comparison shows that though the shapes of the wave functions are reasonably represented by the approximation, their energies can be appreciably in error, especially for excited state. Nevertheless, the L.C.A.O. approximation has proved the most fruitful method of obtaining molecular orbitals that has been developed to date.

E. A. OGRYZLO
University of British Columbia
Vancouver, British Columbia, Canada

Additional Reading

Clark, T., and R. Koch: *Chemist's Electronic Book of Orbitals,* Springer-Verlag New York, LLC., New York, NY, 1998.

Dias, J. R.: *Molecular Orbital Calculations Using Chemical Graph Theory,* Springer-Verlag New York, LLC., New York, NY, 1997.

Gil, V.: *Orbitals in Chemistry: A Modern Guide for Students,* Cambridge University Press, New York, NY, 2000.

Rauk, A: *Orbital Interaction Theory of Organic Chemistry,* 2nd Edition, John Wiley & Sons, Inc., New York, NY, 2000.

ORDER-DISORDER THEORY AND APPLICATIONS.   Phase transitions in binary liquid solutions, gas condensations, order-disorder transitions in alloys, ferromagnetism, antiferromagnetism, ferroelectricity, antiferroelectricity, localized absorptions, helix-coil transitions in biological polymers and the one-dimensional growth of linear colloidal aggregates are all examples of transitions between an ordered and a disordered state.

The two quantities which apparently must be used to explain or describe these phenomena are the presence of a potential between the particles or

spins and a small volume which must be assigned to each particle so that two particles or spins cannot occupy the same space. All the above phenomena may be semi-quantitatively treated by a single statistical model called the Ising model which consists of a lattice in space, each site of which possesses either a 0 or a 1. Thus two rows of a two dimensional lattice might look like

$$\ldots 011000011011110011 \ldots$$

$$\ldots 110110010100110111 \ldots$$

The 0's and 1's represent different particles or spins depending upon the system being studied and their definition will be given below for different systems. The disordered state corresponds to a random array of 0's and 1's. The ordered state is described by an ordered arrangement of the 0's and 1's. The total number of lattice sites equals N. The number of 0's and 1's on the lattice are respectively n_0 and n_1 and $n_1 + n_0 = N$.

The dimensionality of the lattice depends on the physical nature of the phase. For transitions in linear biopolymers and associative colloids, a one-dimensional lattice is used because these materials become ordered in a one-dimensional manner even though they actually exist in three dimensions. For absorption of gas onto a surface, a two-dimensional lattice is sufficient. For bulk phase changes, however, a three-dimensional lattice must be used.

The great advantage of this model is that it gives essentially the same explanation for a large number of seemingly completely diverse physical and chemical phenomena, often with quantitative success. This permits a deeper insight into the statistical thermodynamic behavior of different types of matter.

The calculation of the thermodynamic properties begins by selecting values for w_{00}, w_{11}, w_{10}, which are the potential energies between like and unlike particles or spins occupying nearest neighbor sites. Since two particles cannot occupy one lattice point, the energy of repulsion of two particles on the same site is considered infinite. The energy of a given configuration is then given by

$$E_{Conf} = n_{11}\, w_{11} + n_{10}\, w_{10} + n_{00}\, w_{00} \qquad (1)$$

where n_{11}, n_{00} and n_{10} are the number of nearest neighbor pairs. The probability of a given configuration is given by Boltzmann's theorem as

$$P(E_{Conf}) = \frac{e^{(-E_{Conf}/kT)}}{\sum\limits_{Conf} e^{(-E_{Conf}/kT)}} \qquad (2)$$

where the summation occurs over all possible configurations, keeping the number of 0's and 1's constant. The restriction of a constant number of 0's and 1's may be removed by letting the lattice interact with its surroundings. The probability that the lattice possesses a given number of 1's and a configurational energy, E_{Conf}, is given by

$$P(E_{Conf}, n_1) = \frac{e^{-Xn_1/kT}\, e^{-E_{Conf}/kT}}{\sum\limits_{n_1=0}^{N} e^{-Xn_1/kT} \sum\limits_{Conf} e^{-E_{Conf}/kT}} \qquad (3)$$

where X is a suitable thermodynamic quantity such as the chemical potential or magnetic field, etc.

For a lattice gas, the 0 and 1 stand for an empty and filled site respectively. Consequently, w_{11} is the attractive potential energy between two gas molecules when they occupy adjacent sites on the lattice, and $w_{00} = w_{10} = 0$.

| HELIX | → | RANDOM COIL | EXAMPLES
POLYLBENZY - L - GLUTAMATE
COLLAGEN
POLYLGLUTAMIC ACID
DNA
AMYLOSE - IODINE HELIX |

| HELICAL AGGREGATE | → | MONOMER | BENZOPURPURIN - 4B
PSUEDOCYANINE CHLORIDE
SODIUM DEOXYCHOLATE
GUANOSINE MONOPHOSPHATE
FLAVONE - IODINE COMPLEX |

Diagrammatic illustration of the helix-coil transitions in biopolymers (*top*). Helix-monomer transition in associative biocolloids (*bottom*)

For binary solutions including linear associative colloids, the 0 and 1 represent solvent and solute respectively; w_{11} and w_{00} are the potential energies between like molecules and w_{10} is the potential energy between unlike molecules.

For a ferromagnet the 0 and 1 represent spins of $-\frac{1}{2}$ and $+\frac{1}{2}$ respectively. It is generally assumed that the potential energy between like spins is always the same, i.e., $w_{11} = w_{00}$. To show how the quantity, X of Eq. (3), is obtained, consider a ferromagnet interaction with an external magnetic field. The total energy is

$$E = (n_0 - n_1)H \cdot d + E_{\text{Conf}}$$

where H is the magnetic field strength, and d is the magnetic dipole moment of a spin. Thus for a ferromagnet $X = 2H \cdot d$ since $N - 2n_1 = n_0 - n_1$.

The above described order-disorder transitions are all three-dimensional phase transitions and occur with essentially infinite sharpness unless the condensed phase exists in a colloidally dispersed state. Recently, it has been shown that certain polymers and associative colloids, particularly those of biological interest, have one-dimensional order-disorder transitions which may be explained in exactly the same terms that describe the three-dimensional phase changes discussed above. However, because the condensed state is colloidally dispersed and the ordering occurs only in one dimension, it may be shown that such transitions cannot be infinitely sharp.

The accompanying figure illustrates the transition from an ordered helical to a disordered state which occurs in a large number of colloidal systems. For the helix-coil transition in polymers, the 1 and 0 represent, respectively, a hydrogen bonded turn of the helix and an unhydrogen bonded section of randomly fluctuating polymer. In an associative colloid, the 0 or 1 represent respectively a cell containing a solvent or colloid monomer. In either case, there exists an energy of attraction between 1's which leads to the formation of a one dimensional ordered helix. The fact that the transition tends to be rather sharp comes about because the 01 and 10 configurations which always occur at the beginning and end of a helical sequence are energetically unfavorable. This means that configurations with

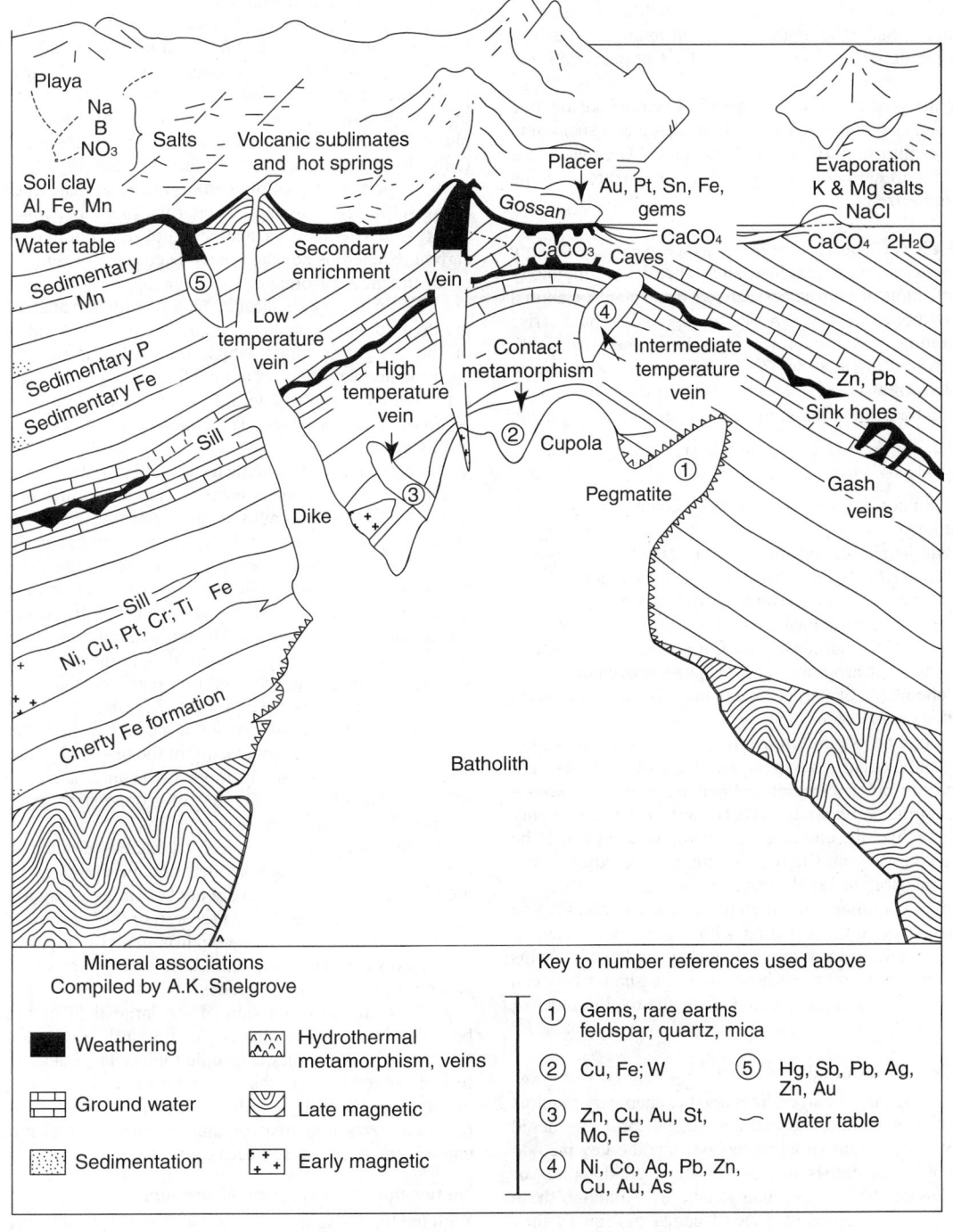

Fig. 1.   Diagrammatic illustration of the origin of ore deposits. (*Field*)

a lot of ends containing many short segments are suppressed since the configurational energy is large when n_{10} is large and this leads to a small Boltzmann factor for this configuration.

The fact that so many diverse phenomena can be correlated and explained by such a simple model makes it possible to develop tables of equivalent thermodynamic properties. This was first done by Yang and Lee who showed the equivalence between the properties of the lattice gas and Ising ferromagnet. Hill extended their analogies to the case of binary liquid mixtures. After the development of the helix-coil transition theory by Zimm and Bragg and others, Peticolas gave a corresponding table of equivalent thermodynamic properties between polymers and associative colloids. Thus the force-length curve for a helical polymer is the one dimensional analogue of the three dimensional pressure-volume curve for gas condensation and is equivalent to the chemical potential-mole-fraction curve for the associative linear colloid.

W. L. PETICOLAS
University of Oregon
Eugene, Oregon

ORES. Mineral aggregates in which the valuable metalliferous minerals are sufficiently abundant to make the aggregates worth mining. Types and origins of ore deposits are illustrated in Fig. 1. See also **Copper**; and **Iron**.

ORGANELLE. A portion of a cell having specific functions, distinctive chemical constituents, and characteristic morphology; it is a unit subsystem of a cell. Examples are mitochondria and chromosomes. Organelles are often closely associated with enzymes. The lysosome (an enzyme-bearing organelle) has been synthesized.

ORGANIC CHEMISTRY. The term *organic*, which means *pertaining to plant or animal organisms*, was introduced into chemical terminology as a convenient classification of substances derived from plant or animal sources. Early it was believed that organic compounds could arise only through the operation of a vital force inherent in the living cell. However, Wöhler discovered in 1828 that the organic compound urea (see also **Wöhler, Friedrich (1800–1882)**), identified in urine by Rouelle in 1773, could be produced by heating the inorganic salt ammonium cyanate:

$$NH_4{}^+OCN^- \longrightarrow H_2NCONH_2$$
Ammonium cyanate   Urea

Subsequently, the association of organic compounds only with living organisms was discontinued.

The term *organic* has persisted, but the modern definition of *organic chemistry* has changed to mean the *chemistry of carbon compounds*. Sometimes a few carbon compounds are excluded from this category, such as carbon dioxide, CO_2; metal carbonates, e.g., Na_2CO_3; carbonyls, e.g., $Ni(CO)_4$; cyanides, e.g., KCN; carbides, e.g., CaC_2; and a few others, but this exclusion is somewhat arbitrary. The designation *organic* is still pertinent because the chemistry of carbon compounds is more important to everyday life than that of any other element.

The uniqueness of carbon stems from its ability to form strong carbon-carbon bonds that remain strong when the carbon atoms are simultaneously bonded to other elements. Whereas both the carbon-hydrogen and carbon-fluorine compounds CH_3CH_3 and CF_3CF_3 are highly stable and relatively unreactive, the corresponding compounds in which the carbon atoms are replaced by boron, silicon, phosphorus, and others either are thermodynamically unstable or highly reactive.

Theoretically, an infinite number of different carbon compounds can exist. Carbon atoms alone or in combination with other atoms, such as oxygen, nitrogen, etc., can join to form linear, branched, and cyclic chains of nearly any length. One, two, or three bonds may be shared between two carbon atoms. In most stable organic compounds, the total number of bonds to each carbon atom is four.

Classification of Organic Compounds

A subject with the wide scope that characterizes organic chemistry requires a logical approach to its organization so that knowledge can be gathered and applied manageably. Molecular structure has become the key method for classifying this subject. Scientists use two-dimensional diagrams or three-dimensional models to depict molecular structures. Although these analogies are sometimes crude representations of actual molecules, they are useful for communicating information about the molecules. Structural diagrams characteristic of some basic types of organic molecules are shown in Table 1. A more detailed discussion on naming organic compounds is given later.

The compounds shown in Table 1 contain only carbon and hydrogen and are called hydrocarbons. Most organic compounds that contain other kinds of atoms in addition to carbon and hydrogen are considered as formally derived from hydrocarbons in which a hydrogen atom has been replaced by another atom or collection of atoms. However, these derivatives usually are not formed directly from the hydrocarbons. The other atoms usually are referred to as functional groups. A group of compounds having the same functional group is referred to as a family. Table 2 shows a number of such families. The R and R' in the formulas of Table 2 represent any hydrocarbon in which a hydrogen atom has been removed from the position to which the functional group is attached.

Molecules with common features also are grouped into more specialized families. The compounds in one such family, carbohydrates, contain only carbon, hydrogen, and oxygen. The hydrogen and oxygen atoms are in the same ratio as in water (2H:1O), hence the suffix *-hydrate*. Other specialized families include terpenes, alkaloids, steroids, lipids, proteins, enzymes, vitamins, and organometallic compounds.

Although these classifications are important for education and documentation in organic chemistry, research usually is rather specialized.

It is often oriented in a practical way, not to the structure of compounds or to the kinds of atoms they contain, but to the manner in which the compounds are used. A partial list of such uses includes plastics, pharmaceuticals, insecticides, fungicides, herbicides, paints, petroleum, fuels, dyes, photography, and adhesives. See Fig. 1.

An important area of organic chemistry is that which deals with life and living substances. Organized under the title biochemistry, this is a subfield of organic chemistry, since most of the compounds involved contain carbon. Numerous advances have been made recently in biochemistry, and of all the areas of organic chemistry it probably will produce the greatest progress in the next decade. Noteworthy advances can be expected in the areas of biochemistry relating to medicine and human health. Some of the categories along which the study of biochemistry is organized are proteins, peptides, amino acids, nucleoproteins, enzymes, nucleotides, carbohydrates, lipids, steroids, carotenoids, porphyrins, nucleic acids, vitamins, and hormones. These topics are described in detail elsewhere in this volume.

Theoretical organic chemistry is another field that has progressed rapidly in recent years. Chemists have derived molecular orbital symmetry rules that allow understanding and predicting the stereochemistry and relative rates of organic reactions in electronic ground and excited states. In a ground state molecule, all electrons are in their lowest energy levels, whereas, in an excited state molecule, at least one electron is in a higher energy level. For example, the Woodward-Hoffman orbital symmetry rules for concerted reactions predict that ground state (thermal) cycloaddition reactions involving $4n + 2$ (where n is an integer) π-electrons, and excited state (photochemical) cycloaddition reactions that involve $4n$ π-electrons, may occur via a concerted process (Scheme 1).

A concerted process or reaction occurs without the involvement of an intermediate. The stereochemistry of the reactants is retained in the product, and the reaction is usually more facile than a comparable nonconcerted reaction. The other combinations of ground state, excited state, $4n + 2$, $4n$ reactions cannot be concerted reactions.

Another area that has received increased attention is environmental organic chemistry. Reactions that organic compounds undergo when they are released to the environment are becoming as significant as the reactions by which the compounds are prepared or the reactions that take place in the use of the compounds. Some environmentally important types of reactions are hydrolysis, oxidation, sunlight-initiated photochemical decomposition, and biodegradation by microbes.

Only a limited discussion of the large field of organic chemistry can be given in the space allotted for this article. Consult alphabetical index for further information on specific topics. The reader is also directed to the references at the end of this article for examples of more detailed treatments of organic chemistry. Refs. Schmerling; Richey; and Morrison et al. represent texts that treat organic chemistry at elementary, intermediate, and advanced levels, respectively.

Nomenclature of Organic Compounds

With the foregoing review in mind, the reader will appreciate the difficulties associated with naming organic molecules. Originally, chemical names

TABLE 1. EXAMPLES OF STRUCTURAL DIAGRAMS OF HYDROCARBONS

Type of compound	Formula	Name
Linear alkane	or $CH_3CH_2CH_2CH_2CH_2CH_3$ or $CH_3(CH_2)_4CH_3$ or $n\text{-}C_6H_{14}$ or	*n*-Hexane
Branched alkane	or	2-Methylpentane
Monocyclic alkane	or	Cyclohexane
Bicyclic alkane		Bicyclo[2.2.1] heptane or Norbornane
Polycyclic alkane		Pentacyclo [5.3.0.0.^{2,5} − 0^{3,9}0^{4,8}]decane or 1,3-Bishomocubane
Alkene	or	*trans-* 2-Pentene
Alkyne	$CH_3C \equiv CCH_2CH_3$ or	2-Pentyne
Aromatic	or	Benzene
Polymer	$X(CH_2)_n$ Y ($n \geq 1000$, X and Y vary according to how polymer was prepared)	Polymethylene or Polyethylene

TABLE 2. EXAMPLES OF SOME MAJOR FAMILIES OF ORGANIC COMPOUNDS

Family	General structure	Family	General structure
Alcohol, phenol	ROH	Isocyanate	RN=C=O
Ether	RoR′	Thiol	RSH
Aldehyde		Sulfide	RsR′
Ketone		Sulfoxide	
Carboxylic acid		Sulfone	
Ester		Sulfonic acid	
Amine	RNH$_2$	Chloride	RCl
Amide		Bromide	RBr
Nitrile	$RC \equiv N$	Iodide	RI
Isonitrile	$\overset{+}{RN} \equiv \bar{C}$	Organolithium	Rli
Nitro compound		Heterocycle	
Nitroso compound	RN=O		(Y is an atom othan than carbon such as N, O, Si, P, S, etc.: Y is bonded to R at two or more positions.) Examples:
Imine			
Azo compound	RN=NR′		, pyridine;
Diazo compound	$RR'C=\overset{+}{N}=\bar{N}$		, thiophene.
Diazonium salt	$RN \equiv N^+X^-$ (anion)		

As the number of known organic molecules increased, a systematic approach to nomenclature was required. To minimize confusion in communicating chemical information, a name should be consistent with other systems in use and should clearly define the structure of a molecule. Specialists in organic chemistry have developed nomenclatures that are logical for their disciplines, thus devising systems for naming alcohols, antibiotics, carboxylic acids, etc.

Systematic nomenclature on a worldwide scale began in 1892 when a committee of the International Chemical Congress established a set of standards known as the Geneva Rules for naming organic compounds. The International Union of Pure and Applied Chemistry (IUPAC) http://www.iupac.org/dhtml_home.html was formed in 1919 and further developed this nomenclature system. In 1886 in the United States, the American Chemical Society (ACS) established a Committee on

were indicative of the sources of compounds. For example, *catechol* was the name given to a compound isolated from the natural product *gum catechu*. Chemists have coined other nonsystematic names such as *cubane* or *basketene* to pictorially describe molecules. These nonsystematic names are called common or trivial names. The names *phenol, acetic acid*, and *styrene* are also nonsystematic but are widely understood in chemistry.

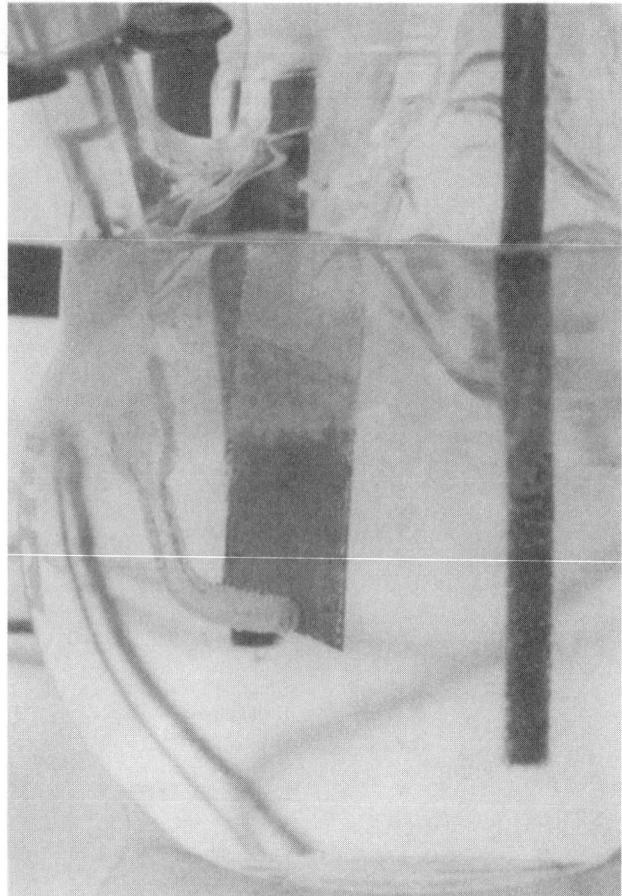

Fig. 1.   Apparatus for testing corrosion of metal coated with a plastic. (*The Dow Chemical Company*)

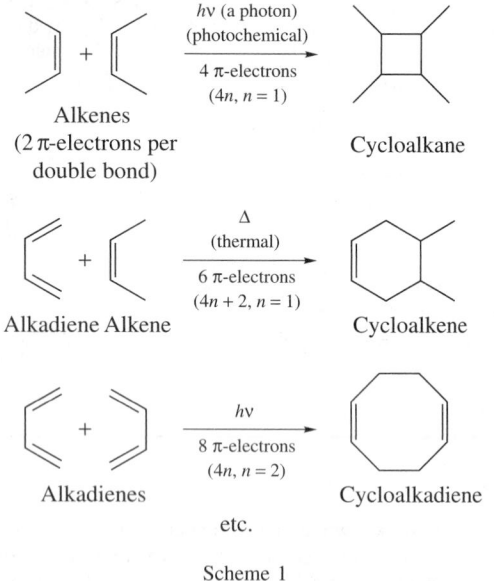

Scheme 1

nomenclature systems are complex and sometimes inconvenient, some chemists have retained the older methods of naming molecules. These older names are used in the parts of this section that do not deal specifically with nomenclature.

The following paragraphs introduce some basic areas of organic chemical nomenclature. Further details can be found in *Chemical Substances Index Names* and *Nomenclature of Organe Chemistry* listed and end of this entry.

In addition to trivial names, some of the other categories of organic compound names are:

Generic name: one that indicates a class of compounds; e.g., *alkanes, esters*.

Parent name: a base from which other names are derived; e.g., *ethanol*, from *ethane; butanoic acid*, from *butane*.

Systematic name: a name composed of syllables defining the structure of a compound; e.g., *chlorobenzene, 2-methylhexane*.

Substitutive name: one describing replacement of hydrogen by a group or element; e.g., *1-methylnaphthalene, 2-chloropropane*.

Replacement name: a name describing compounds that have carbon replaced by a hetero atom; also called "a" nomenclature; e.g., *2-azaphenanthrene*.

Subtractive name: one that indicates removal of specified atoms; e.g., in the aliphatic series names ending in *-ene* or *-yne*, such as *ethene* or *ethyne*, and names involving *anhydro-, dehydro-, deoxy-, nor-*, etc.

Additive name: one that signifies addition between molecules and/or atoms without replacement of atoms; e.g., *styrene oxide*.

Conjunctive name: a combination of two names, one of which represents a cyclic structure and the other an acyclic chain, with one hydrogen atom removed from each; e.g., *benzenemethanol*.

Fusion name: a combination that results from linking with an "o" two names of cyclic systems fused by two or more common atoms; e.g., *benzofuran*.

Multiplicative name: nomenclature describing the symmetrical repetition of radicals about a central unit; e.g., 2, 2'-*oxybis* (*ethanol*).

Hantzsch-Widman name: a name devised by Hantzsch and Widman for describing heterocyclic systems, in which the prefix denotes a hetero atom(s) and the suffix denotes the ring size and degree of saturation; e.g., *oxirene, aziridine*.

Von Baeyer name: a name that describes alicyclic bridged systems; e.g., *bicyclo* [2.2.1]*heptane*.

The procedure for naming a compound involves some or all of the following steps, depending on the structure of the molecule under study: (1) the type of nomenclature to be used (conjunctive, multiplicative, etc.) is chosen; (2) the parent structure is named; (3) the prefixes, suffixes, and names of functional and substituent groups that were not included in (2) are attached; (4) the numbering is completed.

Hydrocarbons

Acyclic Hydrocarbons. A knowledge of the structural features of hydrocarbon skeletons is basic to the understanding of organic chemical nomenclature. The generic name of saturated acyclic hydrocarbons, branched or unbranched, is *alkane*. The term *saturated* is applied to hydrocarbons containing no double or triple bonds.

The simplest saturated acyclic hydrocarbon is called methane. Names of the higher, straight-chain (*normal*) homologs of this series contain the termination "-ane," as shown in Table 3. The structures of the first four members of the series in Table 3 are: CH_4, CH_3-CH_3, $CH_3-CH_2-CH_3$, and $CH_3-CH_2-CH_3$. The structures of subsequent members of this series are formed by inserting additional CH_2 units.

Univalent groups derived from the preceding acyclic hydrocarbons by removal of one hydrogen atom from a terminal carbon atom are named by replacing the ending "-ane" with "-yl."

EXAMPLES: ethyl CH_3-CH_2-
             butyl $CH_3-CH_2-CH_2-CH_2-$

A saturated branched acyclic hydrocarbon is named by numbering the longest chain from one end to the other, and the positions of the side chains are indicated by the lowest possible numbers. The numbers precede the group, and are separated from them by a hyphen.

Nomenclature. The ACS and IUPAC have developed parallel rules for naming organic compounds.

Alternative rules within the latter system may allow assigning more than one unambiguous name to a compound. The controlled alphabetic listing in the *Chemical Substances Index* of Chemical Abstracts Service (CAS), a part of the ACS, requires that each compound have a unique name. This convention ensures that all information for a single compound such as $H_2n-CH_2-CH_2-OH$, which can be unambiguously named 2-aminoethanol, 2-aminoethyl alcohol, 2-hydroxyethylamine, etc., will appear in the index under one name: ethanol, 2-amino-. Because universal

TABLE 3. EXAMPLES OF SATURATED ACYCLIC HYDROCARBONS

Molecular formula	Name	Molecular formula	Name
CH_4	methane	$C_{18}H_{38}$	octadecane
C_2H_6	ethane	$C_{19}H_{40}$	nonadecane
C_3H_8	propane	$C_{20}H_{42}$	icosane
C_4H_{10}	butane	$C_{21}H_{44}$	henicosane
C_5H_{12}	pentane	$C_{22}H_{46}$	docosane
C_6H_{14}	hexane	$C_{23}H_{48}$	tricosane
C_7H_{16}	heptane	$C_{30}H_{62}$	triacontane
C_8H_{18}	octane	$C_{31}H_{64}$	hentriacontane
C_9H_{20}	nonane	$C_{32}H_{66}$	dotriacontane
$C_{10}H_{22}$	decane	$C_{40}H_{82}$	tetracontane
$C_{11}H_{24}$	undecane	$C_{50}H_{102}$	pentacontane
$C_{12}H_{26}$	dodecane	$C_{100}H_{202}$	hectane
$C_{13}H_{28}$	tridecane	$C_{101}H_{204}$	henhectane
$C_{14}H_{30}$	Tetradecane	$C_{102}H_{206}$	dohectane
$C_{15}H_{32}$	pentadecane	$C_{110}H_{222}$	decahectane
$C_{16}H_{34}$	hexadecane	$C_{120}H_{242}$	icosahectane
$C_{17}H_{36}$	heptadecane	$C_{132}H_{266}$	dotriacontahectane
		$C_{200}H_{402}$	dictane

EXAMPLES:

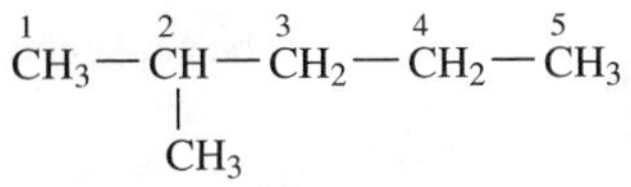

$$\overset{1}{CH_3}-\overset{2}{CH}-\overset{3}{CH_2}-\overset{4}{CH_2}-\overset{5}{CH_3}$$
$$\underset{CH_3}{|}$$

2-methylpentane (not 4-methylpentane)

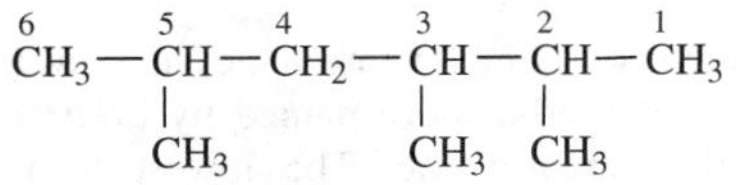

$$\overset{6}{CH_3}-\overset{5}{CH}-\overset{4}{CH_2}-\overset{3}{CH}-\overset{2}{CH}-\overset{1}{CH_3}$$
$$\quad\;\;|\qquad\quad\;\;|\quad\;\;|$$
$$\quad\;CH_3\qquad CH_3\;\;CH_3$$

2,3,5-trimethylhexane (not 2,4,5-trimethylhexane)

If two groups are attached to the same carbon atom, the number is repeated.

EXAMPLES:

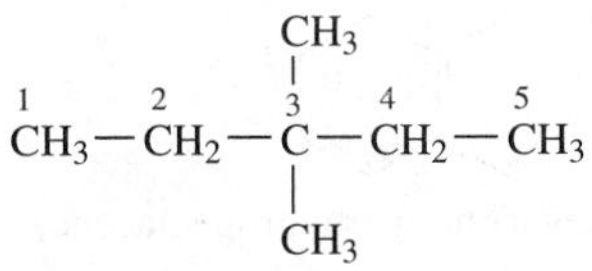

$$\qquad\qquad\overset{CH_3}{\underset{|}{}}$$
$$\overset{1}{CH_3}-\overset{2}{CH_2}-\overset{3}{C}-\overset{4}{CH_2}-\overset{5}{CH_3}$$
$$\qquad\qquad\underset{CH_3}{|}$$

3,3-dimethylpentane (not 3-dimethylpentane)

For some purposes, such as alphabetical listing of basic skeleton names, inverted word order is used.

EXAMPLES: pentane, 2-methyl-
benzene, chloro-

The two propyl groups are distinguished by calling them *normal*-propyl or *n*-propyl, $CH_3-CH_2-CH_2-$, and isopropyl or i-propyl, $[(CH_3)_2-CH-]$, in common usage. The latter is called 1-methylethyl in systematic nomenclature. The butyl groups are named as follows:

Structure	Systematic name	Trivial name
$CH_3-CH_2-CH_2-CH_2-$	butyl	*n*-butyl
$CH_3-CH-CH_2-$ $\quad\quad\;\|$ $\quad\quad CH_3$	2-methylpropyl	*i*-butyl
CH_3-CH_2-CH- $\quad\quad\quad\;\;\|$ $\quad\quad\quad\;CH_3$	1-methylpropyl	*s*-butyl (*s* = secondary)
$\quad\quad CH_3$ $\quad\quad\;\|$ CH_3-C- $\quad\quad\;\|$ $\quad\quad CH_3$	1,1-dimethylethyl	*t*-butyl (*t* = tertiary)

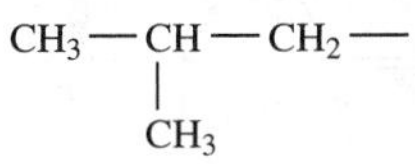

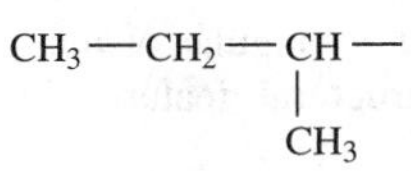

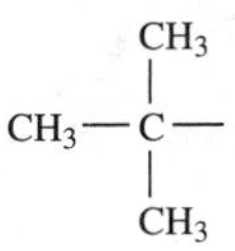

Hydrocarbons that contain one or more double bonds are called "unsaturated," and are named by replacing the ending "-ane" of the corresponding saturated hydrocarbon with the ending "-ene," "-adiene," or "-atriene," etc. The generic names of unsaturated hydrocarbons are *alkene, alkadiene, alkatriene*, etc. The double bonds receive the lowest possible numbers.

In the following examples, the names listed in the *Chemical Substances Index* of the CAS system are given first. Other names are given in parentheses.

EXAMPLES: 2-butene
(2-butylene)

$$\overset{1}{CH_3}-\overset{2}{CH}=\overset{3}{CH}-\overset{4}{CH_3}$$

1,4-hexadiene

$$\overset{1}{CH_2}=\overset{2}{CH}-\overset{3}{CH_2}-\overset{4}{CH}=\overset{5}{CH}-\overset{6}{CH_3}$$

ethene (ethylene) $\quad CH_2=CH_2$

2-methyl1-1,3-
butadiene
(isoprene)

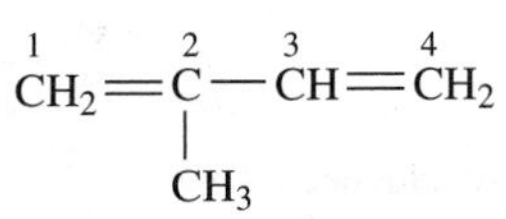

$$\overset{1}{CH_2}=\overset{2}{C}-\overset{3}{CH}=\overset{4}{CH_2}$$
$$\qquad\;\underset{CH_3}{|}$$

Hydrocarbons containing one or more triple bonds are named by replacing the ending "-ane" of the corresponding saturated hydrocarbon with the ending "-yne," "-adiyne," "-atriyne" etc. The triple bonds receive the lowest possible numbers. Double bonds take precedence over triple bonds when there is a choice in numbering.

EXAMPLES: 1-butyne

$$\overset{1}{CH}\equiv\overset{2}{C}-\overset{3}{CH_2}-\overset{4}{CH_3}$$

1-hexane-3,5-diyne

$$\overset{1}{CH_2}=\overset{2}{CH}-\overset{3}{C}\equiv\overset{4}{C}-\overset{5}{C}\equiv\overset{6}{CH}$$

Unsaturated branched acyclic hydrocarbons are numbered in the same manner as alkanes. The longest chain is chosen as the parent. If the alkene or alkyne contains two or more chains of equal length, the chain containing the maximum number of double bonds is chosen as the parent.

Univalent or multivalent groups derived from alkenes and alkynes are named as follows:

EXAMPLES: ethenyl (vinyl)

$$\overset{2}{CH_2}=\overset{1}{CH}-$$

ethynyl

$$\overset{2}{CH}\equiv\overset{1}{C}-$$

2-propynyl

$$\overset{3}{CH}\equiv\overset{2}{C}-\overset{1}{CH_2}-$$

methylidyne $\quad CH\equiv$
ethylidene $\quad CH_3-CH=$
ethylidyne $\quad CH_3-C\equiv$

Alicyclic Hydrocarbons. Saturated monocyclic hydrocarbons, or "cycloalkanes," are named by attaching the prefix "cyclo" to the name of the acyclic unbranched alkane.

EXAMPLES: cyclopropane

cyclohexane

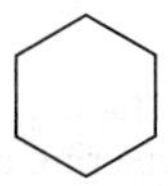

Univalent groups derived from unsubstituted cycloalkanes are named *cyclopropyl, cyclohexyl*, etc., in a manner analogous to that used for naming acyclic alkanes. The carbon atom with the free valence is numbered as 1.

EXAMPLES: cyclopropyl

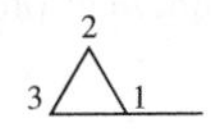

cyclohexyl

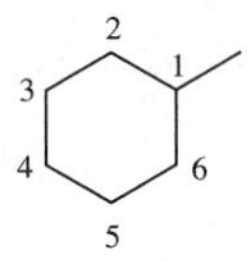

Unsaturated monocyclic hydrocarbons are named by substituting "-ene," "-adiene," "-atriene," "-yne," "-adiyne," etc., in the name of the corresponding cycloalkane. Double and triple bonds are given numbers as low as possible.

EXAMPLES: cyclohexene

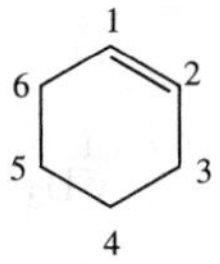

1,3-cyclohexadiene

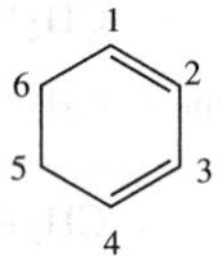

1-cyclodecen-4-yne

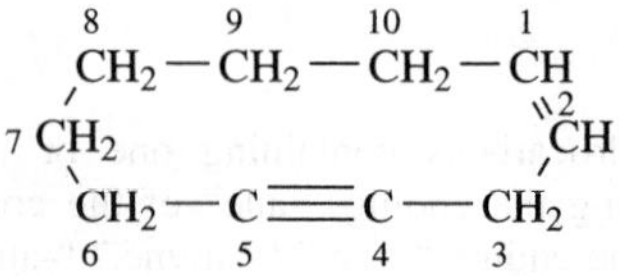

$$CH_2-CH_2-CH_2-CH$$

Aromatic Hydrocarbons. Aromatic hydrocarbons generally are considered those which have the characteristic chemical properties of benzene. Many such compounds are known more commonly by their trivial names than by their systematic names.

EXAMPLES: benzene

methylbenzene (toluene)

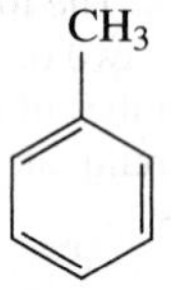

1,2-dimethylbenzene (*o*-xylene)

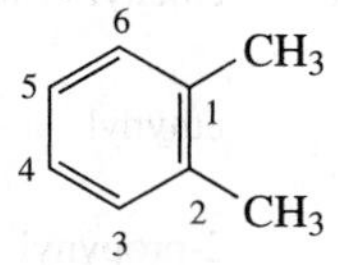

ethenylbenzene (styrene)

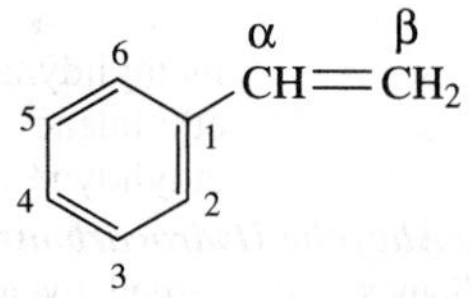

(1-methylethyl)benzene (cumene)

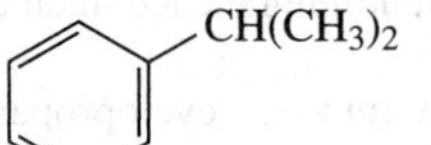

The terms *ortho, meta,* and *para* (abbreviated *o, m,* and *p*) refer to the location of substituents on the benzene ring and are equivalent to 1,2-, 1,3-, and 1,4-substitution in systematic nomenclature, respectively. The lowest numbers possible are given to substituents.

Fused aromatic systems are named by prefixing the largest parent trivial names with combining forms such as benz(o)- and naphth(o)-. Hydrocarbons that contain five or more fused benzene rings in a linear arrangement are named from a numerical prefix followed by "-acene."

EXAMPLES: naphthalene

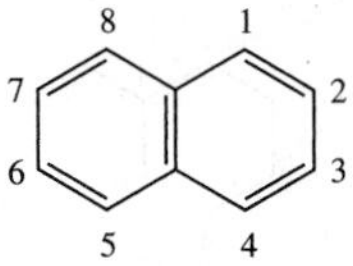

anthracene

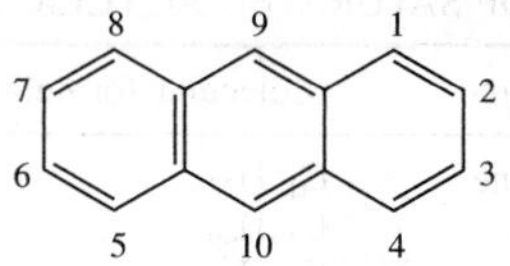

hexacene

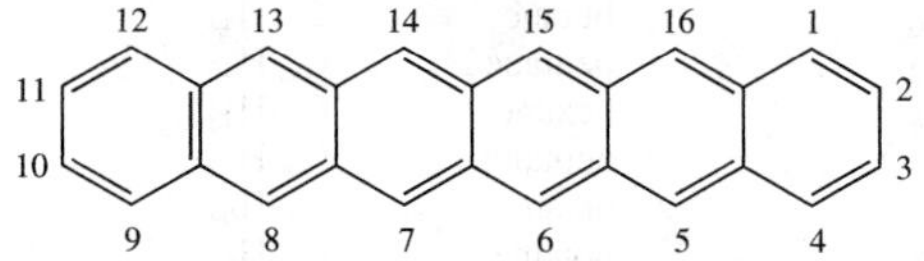

Fusion prefixes are used to designate to which side of the parent hydrocarbon a substituent ring is attached.

EXAMPLES:

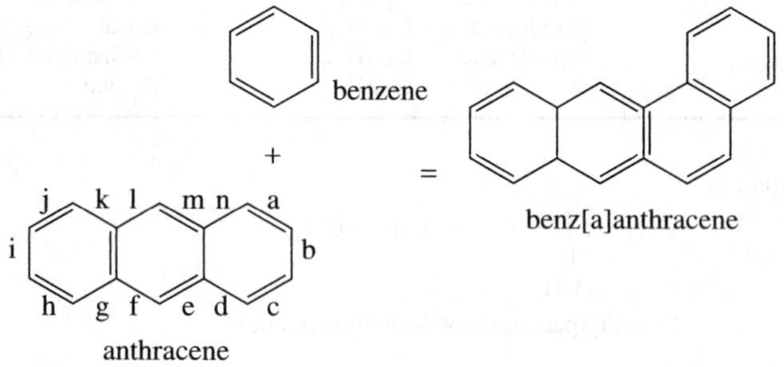

benzene

+

anthracene

=

benz[a]anthracene

Hydrogenation products of complex aromatic ring systems that are not treated as alicyclic hydrocarbons are named by prefixing "dihydro," "tetrahydro," etc., to the parent name. The lowest locants are used. "Perhydro" is used in trivial nomenclature to indicate a fully hydrogenated compound.

EXAMPLES: 1,2-dihydronaphthacene

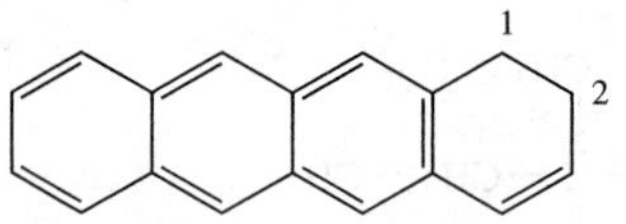

docosahydropentacene (perhydropentacene)

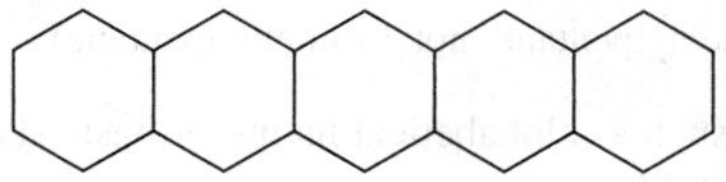

Multiple unsubstituted assemblies of benzene rings are named by using the appropriate prefix with the radical name "phenyl."

EXAMPLES: 1,1'-biphenyl

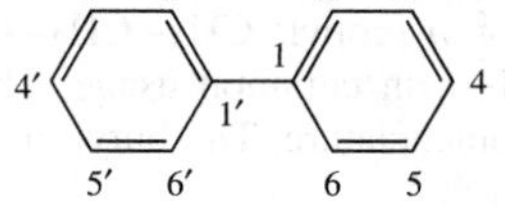

1,1':4',1''-terphenyl
(*p*-terphenyl)

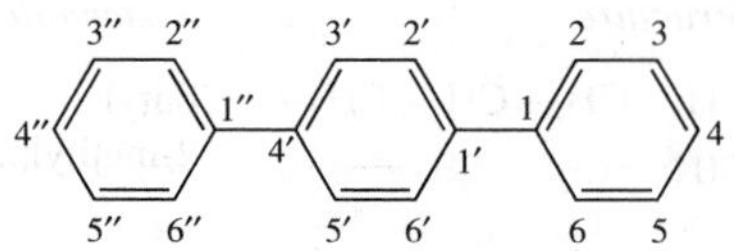

Indicated hydrogen in aromatic systems is assigned to angular or nonangular positions when needed to accommodate structural features in systematic nomenclature. The lowest locants are used.

EXAMPLE: 1-*H*-indene (not 3*H*-)

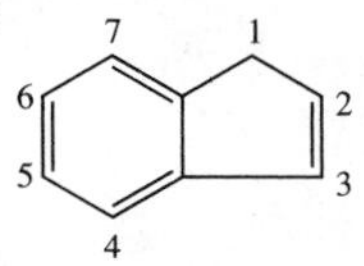

Bridged Hydrocarbon Ring Systems. These compounds are named by prefixing the parent ring with "bicyclo," "tricyclo," etc., or in some complex cases by using prefixes that denote the nature of the bridges. The three numbers in brackets denote the number of carbon atoms in each of the three bridges in descending order.

EXAMPLES:

bicyclo[3.1.0]hexane

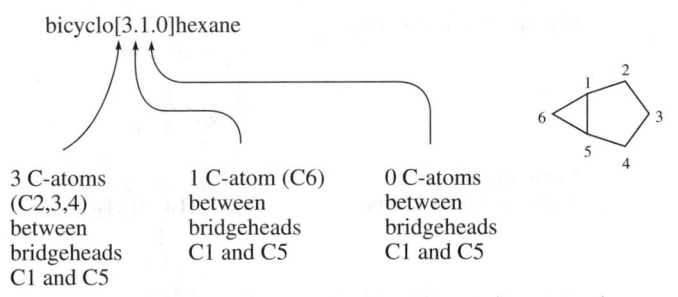

3 C-atoms (C2,3,4) between bridgeheads C1 and C5

1 C-atom (C6) between bridgeheads C1 and C5

0 C-atoms between bridgeheads C1 and C5

1,2,3,4-tetrahydro-1,4-methanonaphthalene

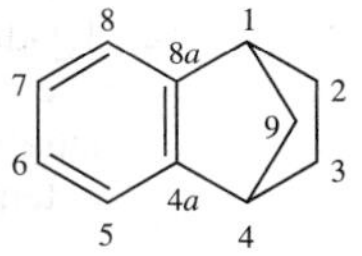

Spiro Hydrocarbon Ring Systems. Spiro systems contain pairs of rings or ring systems that have only one atom (a "spiro atom") in common. The name of the simplest monospiro system is formed by prefixing the acyclic hydrocarbon name with "spiro" and numerals separated by periods. The numerals are given in ascending order to define the number of atoms in each ring linked to the spiro atom. Numbering begins at the atom next to the spiro atom in the smallest ring for monospiro systems.

EXAMPLES: spiro[3,4]octane

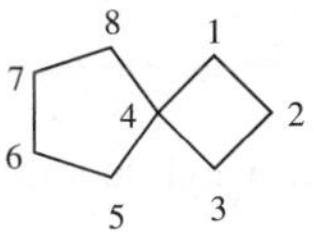

dispiro[5.1.7.2]heptadecane

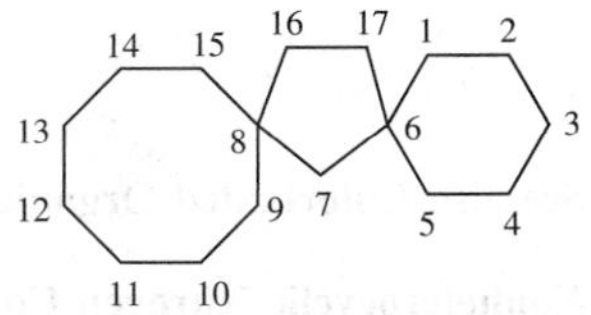

Carboxylic Acids and Their Anhydrides

Acids are named according to the Geneva ("-oic") or "-carboxylic" system. They are regarded as derived from parent hydrocarbons having the same number of carbon atoms so that CH_3 is replaced by COOH. The carbon atom of the carboxyl group is assigned number one in aliphatic monocarboxylic acids. In an alternative numbering system, the Greek letters *alpha, beta*, etc., are assigned to the second, third, etc., carbon atoms, respectively, leading away from the −COOH group. When chain branching is present, the longest chain containing the carboxylic acid group at one end is chosen for naming the molecule. Unsaturated aliphatic acids are named so the longest chain includes the maximum number of unsaturated linkages. Double bonds are given preference over triple bonds. Trivial names are retained for some common molecules.

EXAMPLES:

formic acid   HCOOH

acetic acid $\overset{2}{C}H_3 \overset{1}{C} OOH$

hexanoic acid (caproic acid) $CH_3(CH_2)_4COOH$
octadecanoic acid (stearic acid) $CH_3(CH_2)_{16}COOH$
2-propenoic acid (acrylic acid) $\overset{3}{C}H_2 = \overset{2}{C}H\overset{1}{C}OOH$

2-methylbutanoic acid $\overset{4}{C}H_3\overset{3}{C}H_2\overset{2}{C}H\overset{1}{C}OOH$
                                            |
                                          CH_3

cyclobutanecarboxylic acid   ☐—COOH

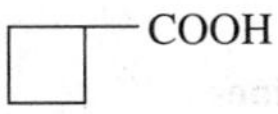

benzoic acid

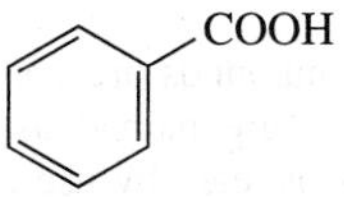

ethanedioic acid (oxalic acid)   HOOC−COOH
butanedioic acid (succinic acid)   HOOC−CH_2CH_2−COOH
1,2-benzenedicarboxylic acid (phthalic acid)

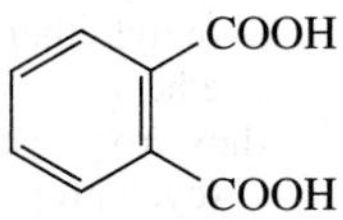

Conjunctive nomenclature may be used for naming cyclic acids. It is applied to any ring system attached by a single bond to one or more acyclic hydrocarbon chains, each of which bears only one principal functional group.

EXAMPLE: cyclohexaneacetic acid CH_2COOH

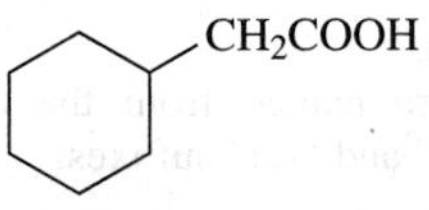

Acid anhydride names are formed from systematic, Geneva, trivial, conjunctive, or other type of acid names.

EXAMPLES: propanoic acid anhydride CH_3CH_2CO
                        (propionic anhydride)
                                                              CH_3CH_2CO

                    benzoic acid anhydride
                        (benzoic anhydride)

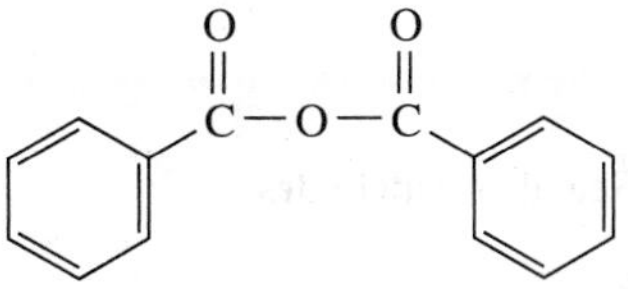

See also **Carboxylic Acids**.

Alcohols

Monohydric alcohols are named by adding "-ol" to a molecular skeleton name. Carbon chains, unsaturation, etc., are numbered in a manner analogous to that used for carboxylic acids. (See preceding section.)

EXAMPLES: methanol (methyl alcohol) CH_3OH
                    cyclohexanol (cyclohexyl alcohol)

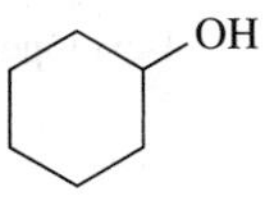

                    2-propen-1-ol (allyl alcohol) $\overset{3}{C}H_2 = \overset{2}{C}H\overset{1}{C}H_2OH$

See also **Alcohols**.

The simplest aromatic hydroxy compound is called phenol:

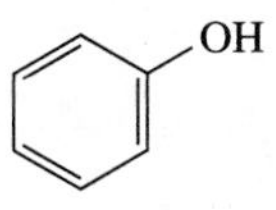

Esters

Simple esters are named on the basis of their alcohol and acid functions. Carbon chains, unsaturation, etc., are numbered in a manner analogous to that used for carboxylic acids.

EXAMPLES: ethyl acetate $CH_3COOC_2H_5$
                        (inverted name: acetic acid, ethylester)
                    propyl 2-butenoate
                                            $\overset{4}{C}H_3\overset{3}{C}H = \overset{2}{C}H\overset{1}{C}OOCH_2CH_2CH_3$

                    methyl benzoate $COOCH_3$

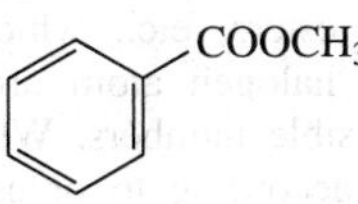

See also **Esters, Organic**.

Ethers

Alkoxy compounds are commonly called ethers. In current CAS nomenclature they are named as derivatives of functional parent compounds, hydrocarbons, etc., by use of "oxy" radicals.

EXAMPLES: 1,1'-oxybis(ethane) $C_2H_5OC_2H_5$
(ethyl ether or diethyl ether)
(hexyloxy)cyclopropane
(cyclopropyl hexyl ether)

methoxybenzene (anisole)

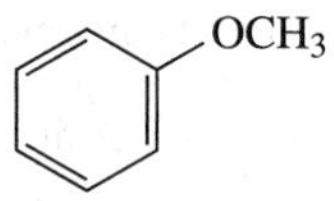

See also **Ethers**.

Aldehydes are named from the corresponding acids by use of "-carboxaldehyde" and "-al" suffixes.

EXAMPLES: acetaldehyde CH_3CHO
2-butenal

$$\overset{4}{C}H_3\overset{3}{C}H=\overset{2}{C}H\overset{1}{C}HO$$

3-methylbenzaldehyde

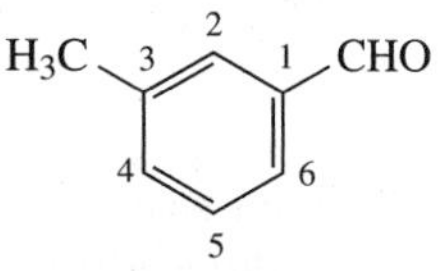

See also **Aldehydes**.

Ketones

Ketones are named by use of the characteristic suffix "-one."

EXAMPLES: 2-propane (acetone)

$$\overset{1}{C}H_3\overset{2}{C}O\overset{3}{C}H_3$$

1-hexen-1-one (butylketene)

$$CH_3(CH_2)_3\overset{2}{C}H=\overset{1}{C}O$$

4-cyclopentyl-2-butanone

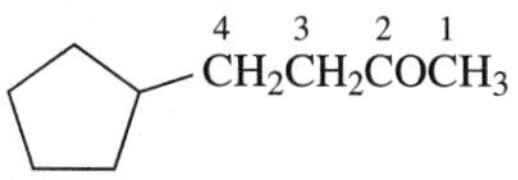

$$\overset{4}{C}H_2\overset{3}{C}H_2\overset{2}{C}O\overset{1}{C}H_3$$

diphenylmenthanone (benzophenone)

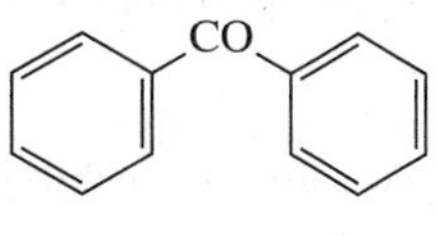

See also **Ketones**.

Peroxides

Simple peroxides are named as follows:

EXAMPLES: ethyl methyl peroxide $C_2H_5OOCH_3$
benzoyl peroxide

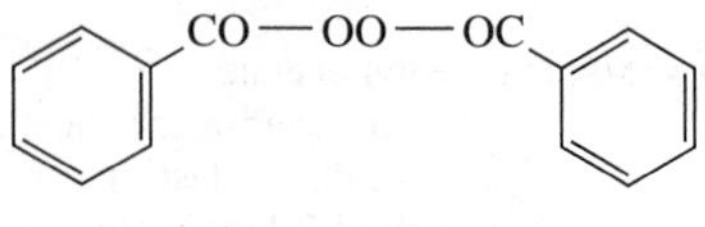

Halogenated Compounds

Hydrocarbons, esters, etc., which have one or more hydrogen atoms replaced by a halogen atom are named so that the substituents have the lowest possible numbers. When multiple functions are present, they are numbered according to precedences established by IUPAC or CAS nomenclature rules. Both trivial and systematic nomenclatures are used.

EXAMPLES: chloromethane (methyl chloride) CH_3Cl
2-iodobutane (sec- butyl iodide)

$$\overset{1}{C}H_3\overset{2}{C}H\overset{3}{C}H_2\overset{4}{C}H_3$$
$$|$$
$$I$$

1-bromo-2-fluorobenzene

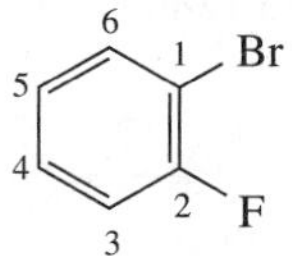

3-chloropentanoic acid (β-chlorovaleric acid)

$$\overset{5}{C}H_3\overset{4}{C}H_2\overset{3}{C}H\overset{2}{C}H_2\overset{1}{C}OOH$$
$$|$$
$$Cl$$

chloroethane (vinyl chloride) $CH_2=CHCl$
1,1-dichloroethene (vinylidene chloride) $CH_2=CCl_2$
tetrabromomethane (carbon tetrabromide) CBr_4

Acid halides are named as follows:

EXAMPLES: acetyl chloride CH_3COCl
6-heptenoyl chloride

$$\overset{7}{C}H_2=\overset{6}{C}H(CH_2)_4\overset{1}{C}OCl$$

cyclohexanecarbonyl bromide

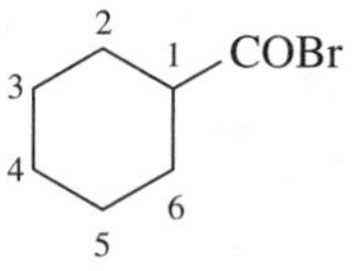

benzoyl fluoride

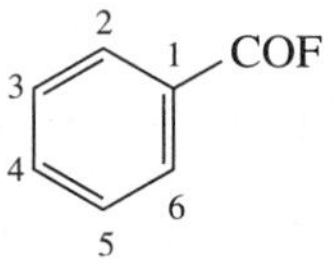

See also **Chlorinated Organics**.

Nonheterocyclic Nitrogen Compounds

Amines are named by adding the suffix "-amine" either to the name of the hydrocarbon or to the hydrocarbon radical. A second system names all amines as derivatives of primary amines.

EXAMPLES: methanamine (methylamine) CH_3NH_2
N, N-dipropyl-1-propanamine (tripropylamine) $(CH_3CH_2CH_2)_3N$
benzenamine (aniline)

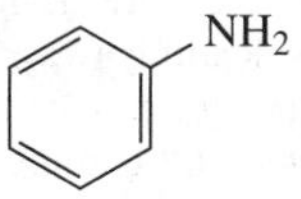

N-ethylcyclohexanamine (N-ethylcyclohexylamine)

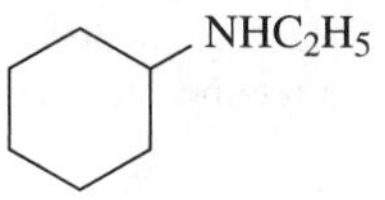

Imines are named from the hydrocarbon by the addition of the suffix "-imine."

EXAMPLES: ethanimine

$$\overset{2}{C}H_3\overset{1}{C}H=NH$$

2,4-cyclopentadien-1-imine

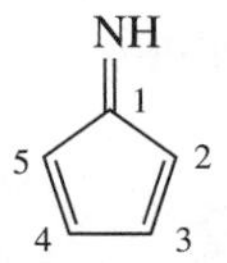

See also **Amines**.

Names of amides are based on the corresponding acids. Thus, "-oic acid" becomes "-amide," and "-carboxylic acid" becomes "-carboxamide."

EXAMPLES: acetamide

$$\overset{2}{C}H_3\overset{1}{C}ONH_2$$

benzamide

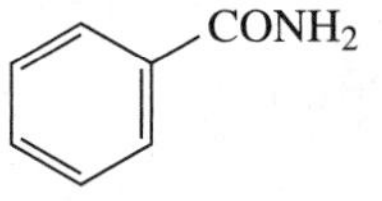

cyclohexanecarboxamide

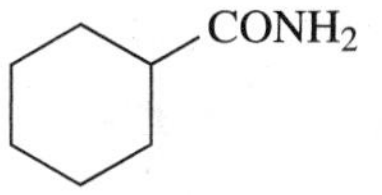

N-methylpentadecanamide

$$\overset{15}{C}H_3(CH_2)_{13}\overset{1}{C}ON\overset{N}{H}CH_3$$

N,N-dimethyl-2,4-pentadienamide

$$\overset{5}{C}H_2=\overset{4}{C}H\overset{3}{C}H=\overset{2}{C}H\overset{1}{C}ON\overset{N}{(CH_3)_2}$$

See also **Amides**.

Nitro and nitroso compounds are named so that the substituents have the lowest possible numbers.

EXAMPLES: 2-nitrobutane

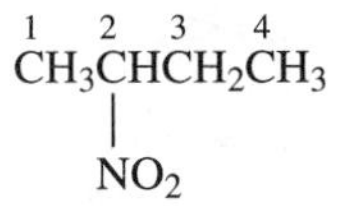

$$\overset{1}{C}H_3\overset{2}{C}H\overset{3}{C}H_2\overset{4}{C}H_3$$
$$|$$
$$NO_2$$

4-nitrosobenzoic acid

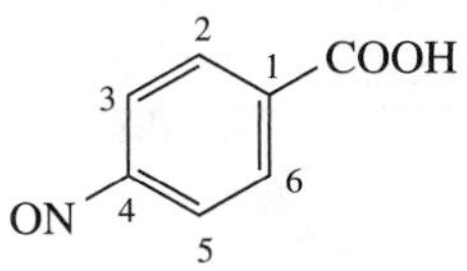

See also **Nitro- and Nitroso-Compounds**

Nitrile names are formed from common names of carboxylic acids.

EXAMPLES: acetonitrile CH_3CN

benzoinitrile (phenyl cyanide)

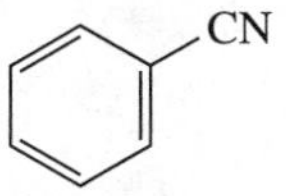

3-butenenitrile

$$\overset{4}{C}H_2=\overset{3}{C}H\overset{2}{C}H_2\overset{1}{C}N$$

ethenetetracarbonitrile (tetracyanoethylene)

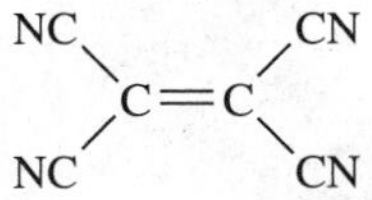

In the presence of more senior functional groups, the nitrile function is expressed by the prefix "cyano."

EXAMPLES: 4-cyanobenzamide

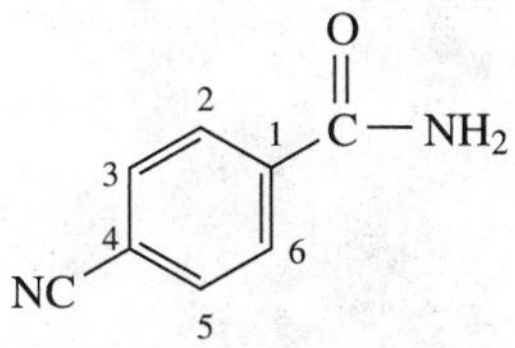

3-cyanobutanoic acid

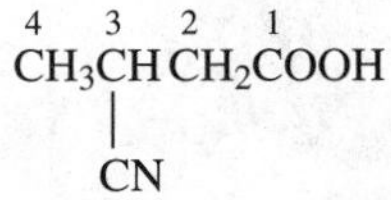

$$\overset{4}{C}H_3\overset{3}{C}H\overset{2}{C}H_2\overset{1}{C}OOH$$
$$|$$
$$CN$$

Nonheterocyclic Sulfur Compounds

Sulfur compounds are named similarly to oxygen compounds.

EXAMPLES:
(methylthio)benzene (methyl phenyl sulfide)

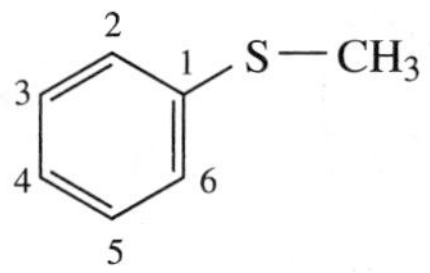

1,1'-sulfinylbis(benzene) (diphenyl sulfoxide)

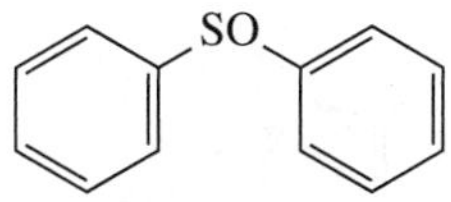

2-(propylsulfonyl)naphthalene (2-naphthyl propyl sulfone)

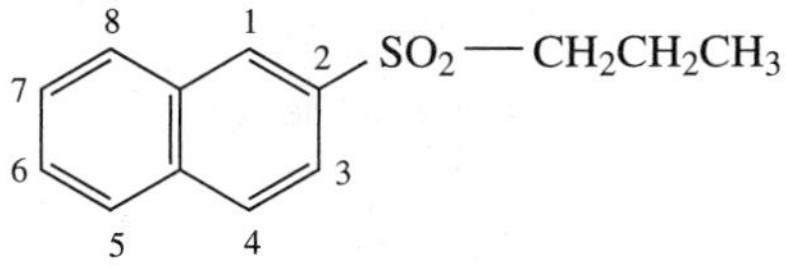

2-butanethiol (2-mercaptobutane)

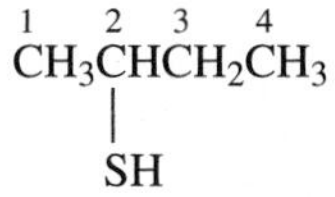

$$\overset{1}{C}H_3\overset{2}{C}H\overset{3}{C}H_2\overset{4}{C}H_3$$
$$|$$
$$SH$$

2-propanethione (thioacetone)

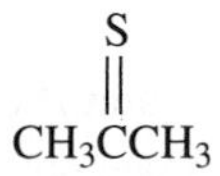

$$S$$
$$\|$$
$$CH_3CCH_3$$

N-methylbenzenesulfonamide

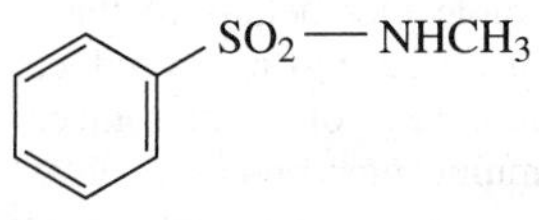

benzenesulfonic acid

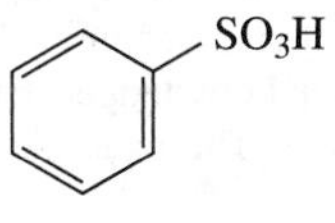

See also **Sulfonic Acids**.

Heterocyclic Compounds

Some of the common heterocyclic nitrogen, oxygen, and sulfur compounds and their numbering systems are shown:

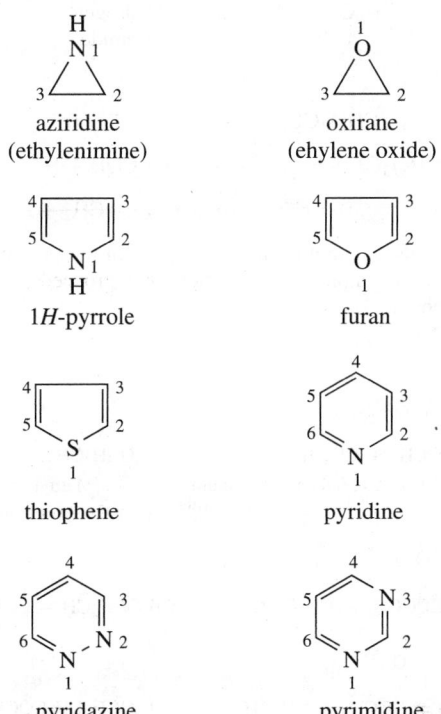

aziridine (ethylenimine)

oxirane (ehylene oxide)

1H-pyrrole

furan

thiophene

pyridine

pyridazine

pyrimidine

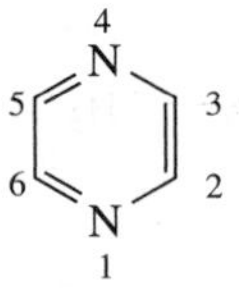

pyrazine

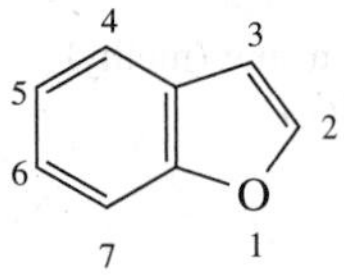

benzofuran

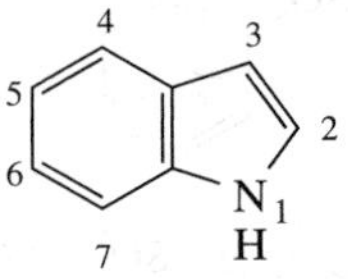

1*H*-indole

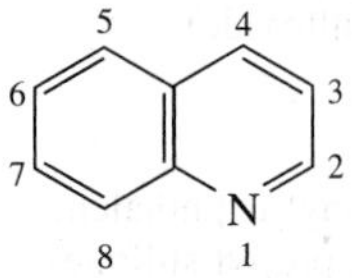

quinoline

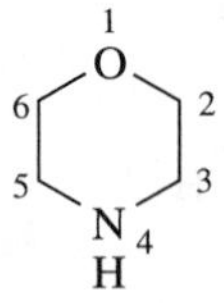

morpholine

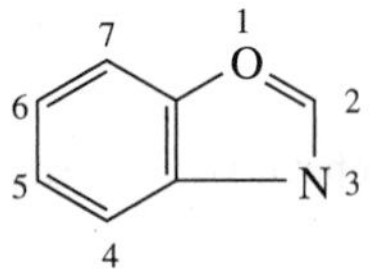

benzoxazole

Organic Reaction Mechanisms and Processes

Many reactions occur in which one organic compound is converted into another. The molecular details of the intermediate steps by which compounds are converted into new products are called reaction mechanisms. The four broad classes of reaction mechanisms are: cationic, anionic, free radical, and multicenter processes in which neither charged species nor odd electron species is involved. Examples of each type will be given, but many variations can exist within each type. Also, varying degrees of sophistication exist in our knowledge of the exact reaction pathways that organic compounds follow. The examples discussed show only the major steps involved.

Cationic and anionic mechanisms involve species that have either positive or negative charges, respectively (heterolytic reactions). An example of a reaction that proceeds via a cationic mechanism is the hydrolysis of *t*-butyl bromide (Scheme 2).

SCHEME 2

$$(CH_3)_3CBr + H_2O \longrightarrow (CH_3)_3COH$$
t-Butyl bromide   Water   *t*-Butyl alcohol

$$+ \ CH_3)_2 \ C = CH_2 + \text{Hydrogen}$$
Isobutylene   bromide
HBr

Mechanism:

$$(CH_3)_3CBr \longrightarrow (CH_3)_3C^+ + Br^-$$

$$(CH_3)_3C^+ + H_2O \longrightarrow (CH_3)_3COH + H^+$$

$$(CH_3)_3C^+ \longrightarrow (CH_3)_2C{=}CH_2 + H^+$$

The addition of methanol to methyl acrylate in the presence of sodium methoxide is an example of a reaction that proceeds by an anionic mechanism (Scheme 3).

SCHEME 3

$$CH_2{=}CHCOCH_3 + CH_3OH \xrightarrow[\text{Sodium methoxide}]{CH_3O-Na^+} CH_3OCH_2CH_2COCH_3$$
Methyl acrylate   Methanol   Methyl
β-methoxypropionate

$$CH_2{=}CHCOCH_3 + CH_3O^- \longrightarrow CH_3OCH_2CH{=}COCH_3$$

$$CH_3OCH_2CH{=}COCH_3 + CH_3OH \longrightarrow CH_3OCH_2CH_2COCH_3 + CH_3O^-$$

Mechanism:

Free radical (homolytic) reactions involve species with an unpaired electron. The ultraviolet-light-initiated reaction of methane with chlorine is an example (Scheme 4).

SCHEME 4

$$CH_4 + Cl_2 \xrightarrow{h\nu} CH_3Cl + HCl$$
Methane   Chlorine   Methyl chloride   Hydrogen Chloride

$$Cl_2 \xrightarrow{h\nu} 2Cl\cdot$$

$$Cl\cdot + CH_4 \longrightarrow CH_3\cdot + HCl$$

$$CH_3\cdot + Cl_2 \longrightarrow CH_3Cl + Cl\cdot$$

A number of reactions do not seem to belong to any of the above mechanistic types. Such processes are referred to as multicenter reactions. The Diels-Alder cycloaddition reaction of 1,3-butadiene with maleic anhydride is an example (Scheme 5). No charged or odd electron intermediates seemingly are involved in this reaction.

SCHEME 5

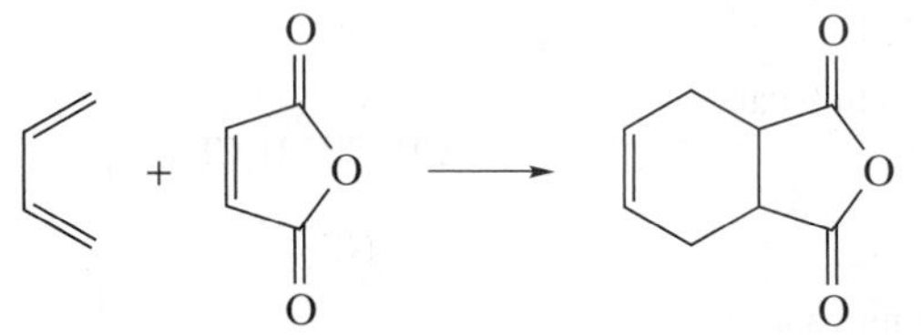

1,3-Butadiene   Maleic
anhydride

Cyclohex-4-ene-1,2-dicarboxylic
anhydride

Fig. 2.   Excited-state photochemical reaction being performed with a laser light source. (*The Dow Chemical Company*)

The reactions shown in Schemes 2–5, with the exception of the photodissociation of Cl_2 to Cl atoms (Scheme 4), occur in molecules that are in electronic ground states. Reactions also can occur in molecules existing in excited electronic states. Commonly, these excited states are produced by irradiating the reactants with ultraviolet or visible light, hence the term *photochemistry*. When a molecule is in an excited state, its reactions are often different from those it normally exhibits in its ground state. An organic compound often can exist in more than one excited state, as shown in Scheme 6.

SCHEME 6

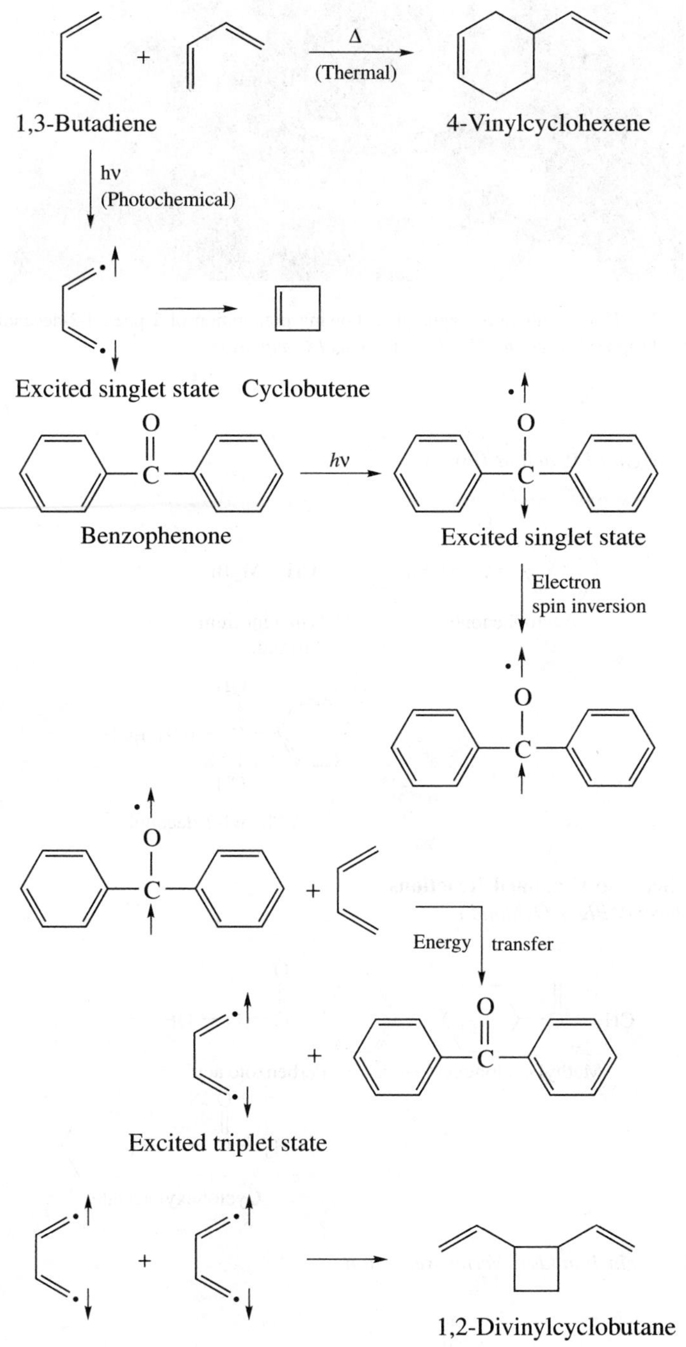

Δ
(Thermal)
1,3-Butadiene
4-Vinylcyclohexene
hν
(Photochemical)
Excited singlet state
Cyclobutene
O
C
hν
Benzophenone
Excited singlet state
Electron
spin inversion
Energy
transfer
Excited triplet state
1,2-Divinylcyclobutane

Excited state molecules usually differ from their ground state counterparts by having dissimilar electronic and geometric configurations, and shorter lifetimes, e.g., 10^{-2} to 10^{-10} second. See Fig. 2.

A few additional examples of organic reactions are shown in Scheme 7. Many novel and complex compounds can be prepared by these and similar reactions.

SCHEME 7. Examples of Organic Reactions.

Free Radical Addition

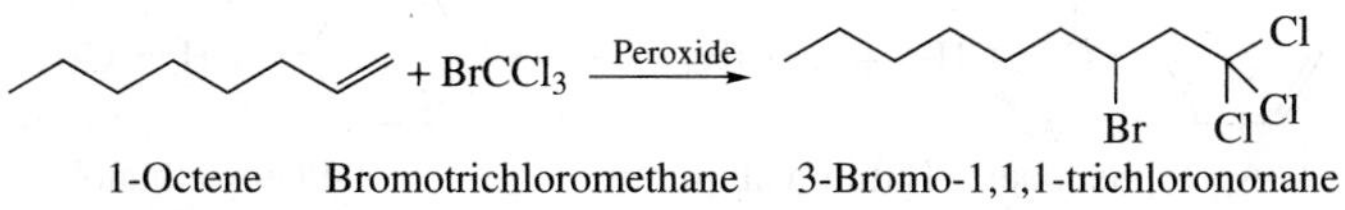

+ BrCCl3
Peroxide
Br
Cl
1-Octene
Bromotrichloromethane
3-Bromo-1,1,1-trichlorononane

Hydrogenation

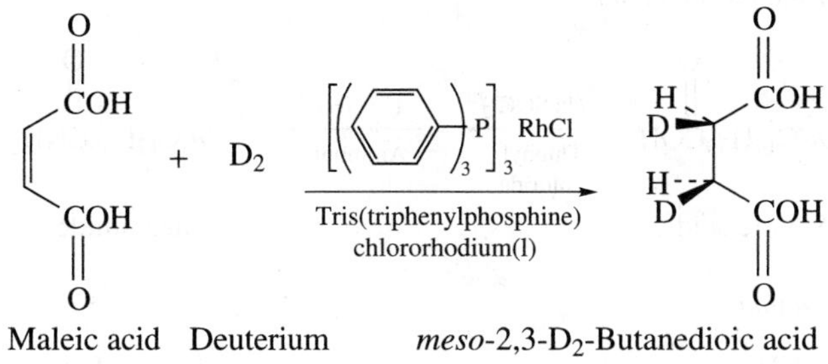

COH
+ D2
P RhCl
Tris(triphenylphosphine)
chlororhodium(I)
Maleic acid
Deuterium
meso-2,3-D2-Butanedioic acid

Carbene Addition

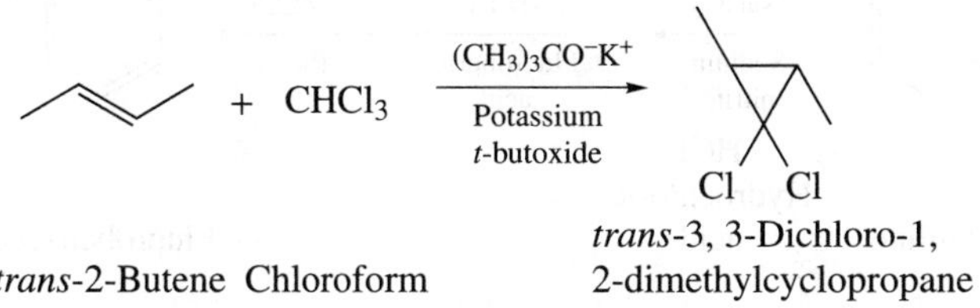

+ CHCl3
(CH3)3CO⁻K⁺
Potassium
t-butoxide
Cl Cl
trans-2-Butene
Chloroform
trans-3, 3-Dichloro-1,
2-dimethylcyclopropane

Oxidative Cleavage

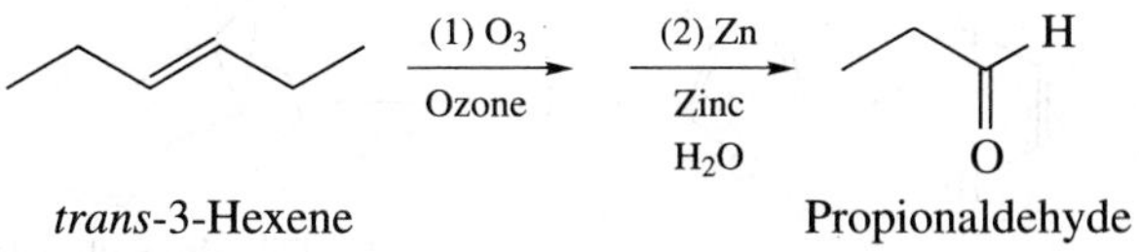

(1) O3
Ozone
(2) Zn
Zinc
H2O
H
O
trans-3-Hexene
Propionaldehyde

Hydroboration-Oxidation

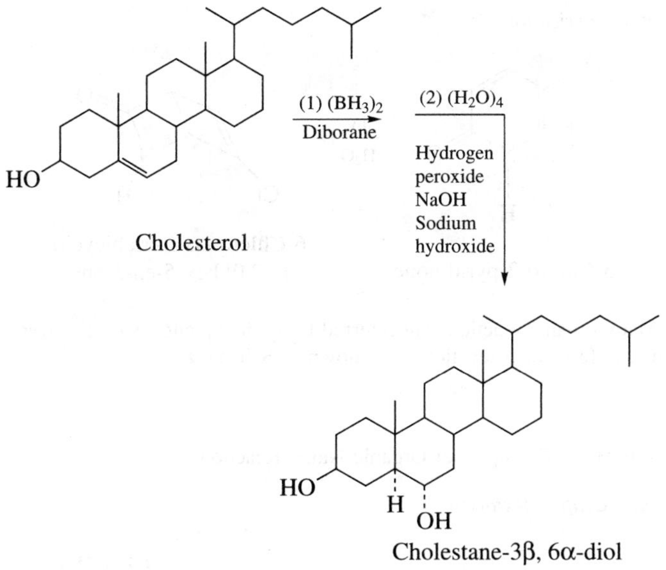

HO
(1) (BH3)2
Diborane
(2) (H2O)4
Hydrogen
peroxide
NaOH
Sodium
hydroxide
Cholesterol
H
OH
Cholestane-3β, 6α-diol

Oxidation

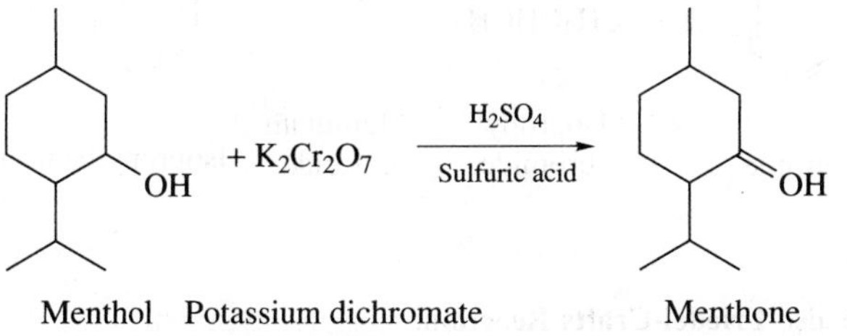

OH
+ K2Cr2O7
H2SO4
Sulfuric acid
Menthol
Potassium dichromate
Menthone

Reduction

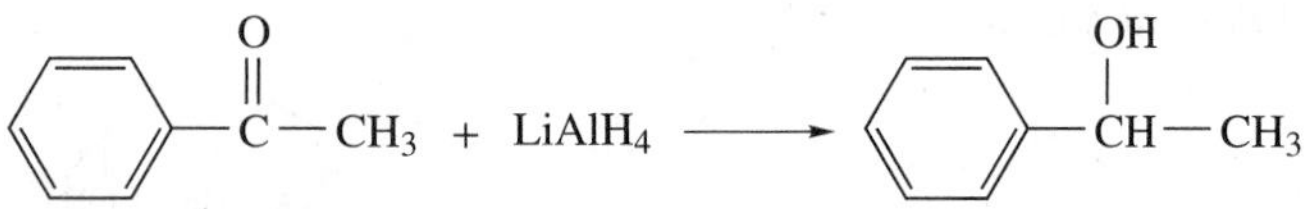

Acetophenone + LiAlH$_4$ → 1-Phenylethanol

Acetophenone   Lithium aluminum hydride   1-Phenylethanol

Chlorination-Amination

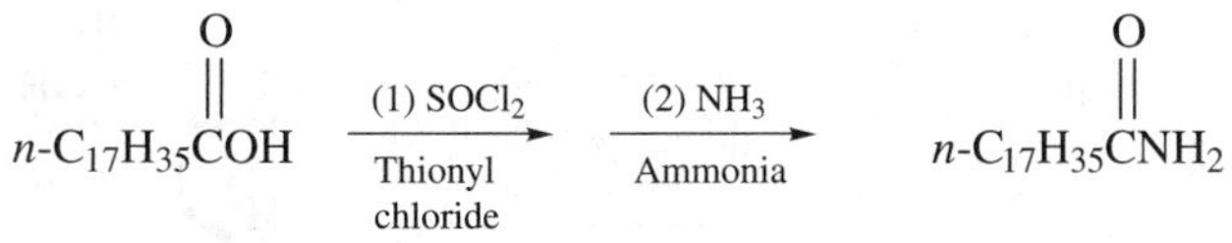

n-C$_{17}$H$_{35}$COH $\xrightarrow[\text{Thionyl chloride}]{(1)\ SOCl_2}$ $\xrightarrow[\text{Ammonia}]{(2)\ NH_3}$ n-C$_{17}$H$_{35}$CNH$_2$

Stearic acid   Stearamide

Substitution

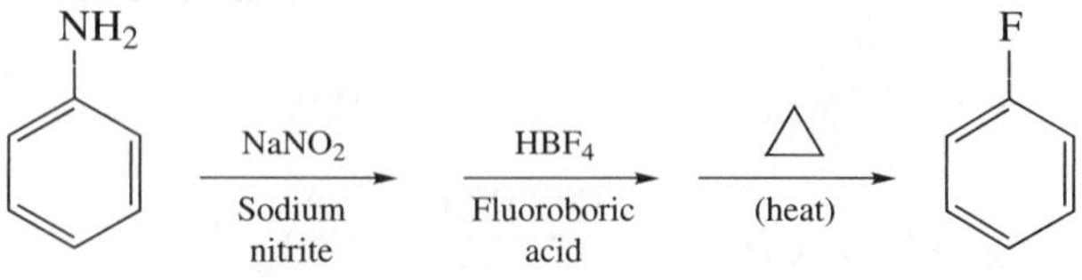

Aniline $\xrightarrow[\text{Sodium nitrite}]{NaNO_2}$ $\xrightarrow[\text{Fluoroboric acid}]{HBF_4}$ $\xrightarrow[\text{(heat)}]{\triangle}$ Fluorobenzene

HCl
Hydrochloric acid

Aniline   Fluorobenzene

Benzyne Cycloaddition

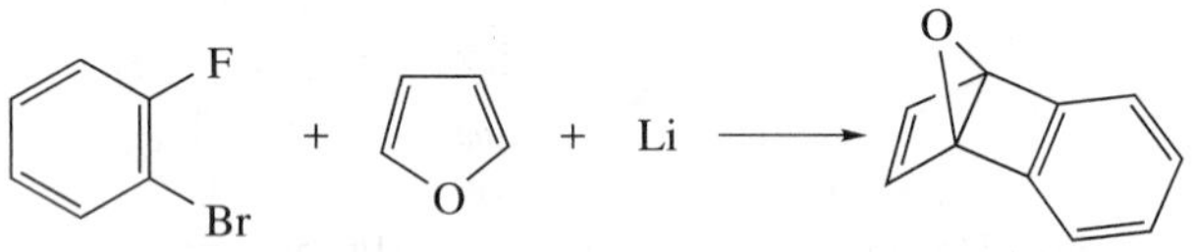

o-Bromofluorobenzene + Furan + Li → 1,4-Dihydronaphthalene-1,4-endoxide

o-Bromofluorobenzene   Furan   Lithium   1,4-Dihydronaphthalene-1, 4-endoxide

Photoisomerization

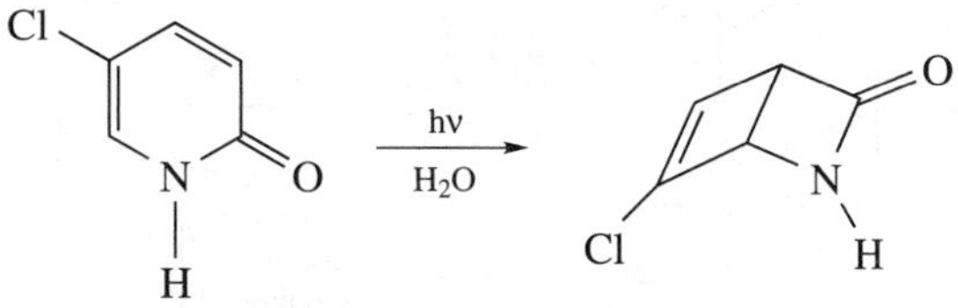

5-Chloro-2-pyridinone $\xrightarrow[H_2O]{h\nu}$ 6-Chloro-*cis*-2-azabicyclo [2.2.0] hex-5-en-3-one

5-Chloro-2-pyridinone   6-Chloro-*cis*-2-azabicyclo [2.2.0] hex-5-en-3-one

Many organic reactions are referred to by the inventor's or discoverer's name. A few name reactions are shown in Scheme 8.

SCHEME 8. Examples of Organic Name Reactions.

Friedel-Crafts Alkylation

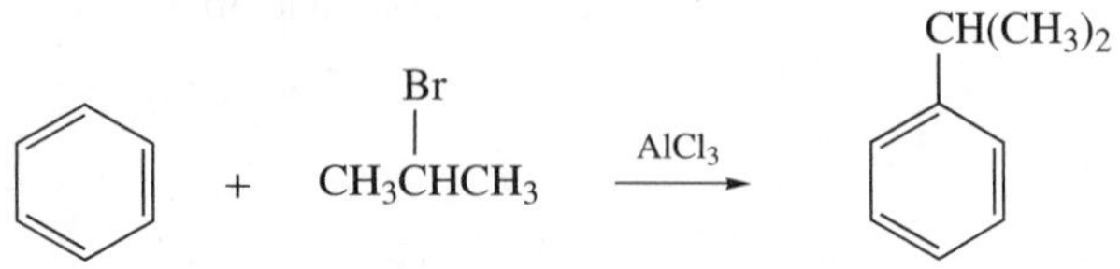

Benzene + CH$_3$CHCH$_3$ (Br) $\xrightarrow{AlCl_3}$ Isopropylbenzene CH(CH$_3$)$_2$

Benzene   Isopropyl bromide   Aluminum chloride   Isopropylbenzene

See also **Friedel-Crafts Reaction**.

Fig. 3. Typical laboratory apparatus showing preparation of 2-phenyl-2-decanol by a Grignard reaction. (*The Dow Chemical Company*)

Grignard Reaction (Fig. 3).

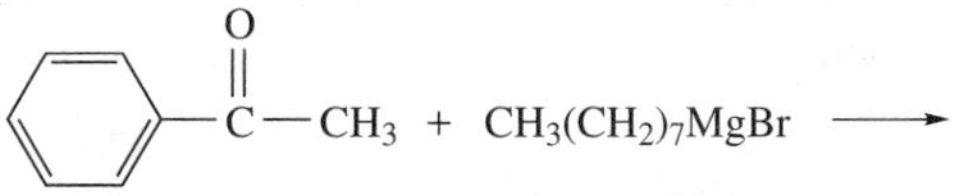

Acetophenone + CH$_3$(CH$_2$)$_7$MgBr →

Acetophenone n-Octylmagnesium bromide

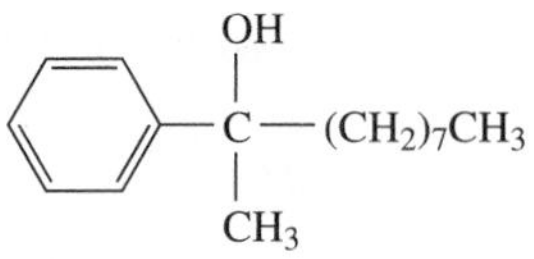

2-Phenyl-2-decanol

See also **Grignard Reactions**.

Baeyer-Villiger Oxidation

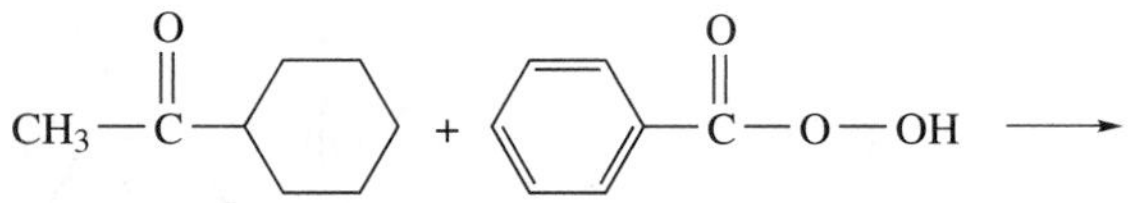

CH$_3$-C-cyclohexyl + Perbenzoic acid →

Methyl cyclohexyl ketone   Perbenzoic acid

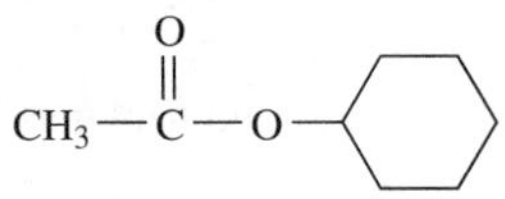

CH$_3$-C-O-cyclohexyl

Cyclohexyl acetate

Meerwein-Ponndorf-Verley Reduction

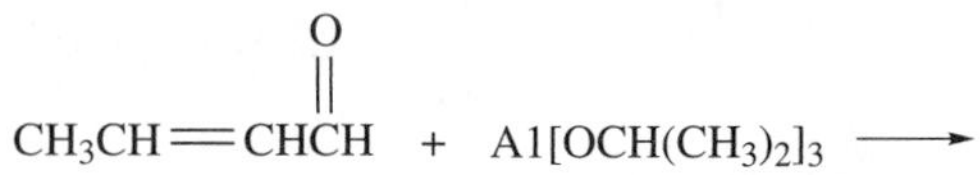

CH$_3$CH=CHCH=O + Al[OCH(CH$_3$)$_2$]$_3$ →

Crotonaldehyde   Aluminum isopropoxide

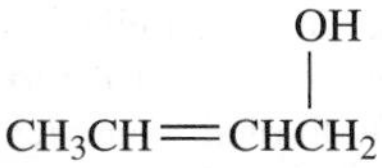

CH$_3$CH=CHCH$_2$ (OH)

But-2-en-1-ol

Skraup Quinoline Synthesis

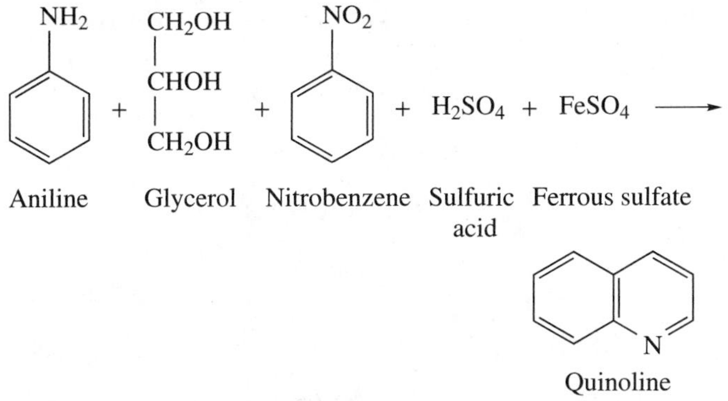

$$NH_2 \quad CH_2OH \quad\quad NO_2$$
$$\text{Aniline} + \overset{|}{\underset{|}{\text{CHOH}}}\text{CH}_2\text{OH} + \text{Nitrobenzene} + H_2SO_4 + FeSO_4 \longrightarrow$$

Aniline   Glycerol   Nitrobenzene   Sulfuric   Ferrous sulfate
                                        acid

Quinoline

A very active area of organic chemistry is the synthesis of complex natural products. In these syntheses, numerous reactions, of which those in Schemes 7 and 8 are examples, are often employed serially to convert a starting compound into a final product that occurs in nature.

These syntheses are useful because they serve to verify structure that has been assigned, provide an alternate source of the compound if a larger supply is needed, or provide a route to derivatives or analogs of the natural material. The derivatives may possess enhanced properties, such as biological activity, that are not found in the natural product.

Some important industrial organic processes are shown in Scheme 9. Although most of these reactions involve mixtures of isomers or homologs as reactants and products, for simplicity only the major components are shown.

SCHEME 9. Examples of Industrial Organic Processes.

Alkylation

$$(CH_3)_3CH + CH_2=CHCH_3 \xrightarrow[\text{catalyst}]{HF} \overset{CH_3 \quad CH_3}{\underset{}{CH_3CH=CHCH_2CH_3}}$$

Isobutane   Propylene   2,3-Dimethylpentane

See also **Alkylation**.

Isomerization

$$CH_3(CH_2)_2CH_3 \xrightarrow[\text{catalyst}]{AlCl_3} \overset{CH_3}{\underset{}{CH_3CHCH_3}}$$

n-Butane   *i*-Butane

Cracking

$$CH_3(CH_2)_{14}CH_3 \xrightarrow[\text{catalyst}]{\text{Silica alumina}} CH_3(CH_2)_4CH=CH_2$$

n-Hexadecane   1-Heptene

$$+ CH_3CH_2CH=CH_2 + \overset{CH_3}{\underset{}{CH_3CH_2CHCH_3}}$$

1-Butene   Isopentane

See also **Cracking Process**.

Oxidation

$$CH_2=CH_2 + O_2 \xrightarrow[\text{catalyst}]{Ag} \overset{O}{\underset{}{H_2C-CH_2}}$$

Ethylene   Oxygen   Ethylene oxide

Chlorination

$$\text{Benzene} + Cl_2 \xrightarrow[\text{catalyst}]{FeCl_3} \text{Chlorobenzene}$$

Benzene   Chlorine   Chlorobenzene

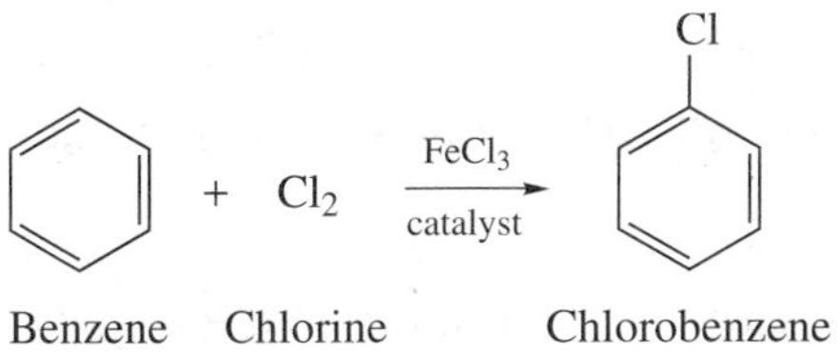

See Fig. 4; also separate entry on **Chlorinated Organics**.

Hypochlorination

$$CH_3CH=CH_2 + HOCl \longrightarrow \overset{OH}{\underset{}{CH_3CHCH_2Cl}}$$

Propylene   Hypochlorous   1-Chloro-
             acid   2-hydroxypropane

Dehydrochlorination

$$CH_3CF_2Cl \xrightarrow{\text{Pyrolysis}} CH_2=F_2 + HCl$$
1-chloro-1,1-difluoroethane   Vinylideneflouride   Hydrogen chloride

Bromination

$$CH_2=CH_2 + Br_2 \longrightarrow CH_2BrCH_2Br$$
Ethylene   Bromine   Ethylene dibromide

Fig. 4.   Typical industrial reactor for the preparation of chlorobenzene by the chlorination of benzene. (*The Dow Chemical Company*)

Hydrogenation

$$CH_2OC(CH_2)_7CH=CHCH_2CH=CH(CH_2)_4CH_3$$
$$O$$

$$CHOC(CH_2)_7CH=CHCH_2CH=CH(CH_2)_4CH_3$$
$$O$$

$$CH_2OC(CH_2)_7CH=CHCH_2CH=CH(CH_2)_4CH_3$$

Trilinolein

Fermentation

Sucrose → $\xrightarrow[\text{(a microorganism)}]{Aspergillus\ niger}$ → Citric acid

$$HOC-CH_2CH_2-COH$$
$$OH$$

Coupling

p-Sulfobenzenediazonium chloride + Sodium 2-naphtholate →

$$NaO_3S-\bigcirc-N=N-\bigcirc\bigcirc-OH$$

Orange II (a dye)

Sulfonation

Naphthalene + H_2SO_4 → α-Naphthalenesulfonic acid
(Naphthalene)   (Sulfuric acid)

Hydrolysis

$$CH_2=CHCH_2Cl + NaOH \longrightarrow CH_2=CHCH_2OH$$
Allyl chloride   Sodium hydroxide   Allylalcohol

Ammonolysis

p-Chloronitrobenzene + NH_3 → p-Nitroaniline
    Ammonia

See also **Amination.**

Hydration

$$CH_2=CH_2 + H_2SO_4 + H_2O \longrightarrow CH_3CH_2OH$$
Ethylene   Sulfuric acid   Water   Ethanol

$$+ H_2 \xrightarrow[\text{catalyst}]{Ni}$$
Hydrogen

$$CH_2OC(CH_2)_{16}CH_3$$
$$OH$$
$$CHOC(CH_2)_{16}CH_3$$
$$OH$$
$$CH_2OC(CH_2)_{16}CH_3$$

Tristearin

See Fig. 5 and also the separate entry on **Fermentation.**

Hydrogenolysis

$$CH_3(CH_2)_{10}COCH_3 + H_2 \xrightarrow[\text{catalyst}]{\text{Copper-chromic oxide}} CH_3(CH_2)_{11}OH$$
Methyl laurate   Hydrogen   Lauryl alcohol

Fig. 5. Laboratory fermentation process equipment. (*The Dow Chemical Company*)

Dehydrogenation

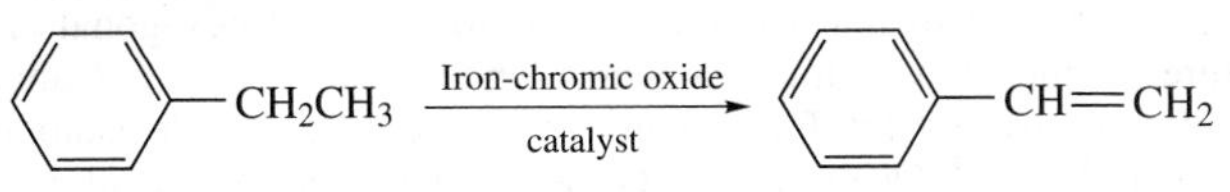

Ethylbenzene → Styrene

Nitration

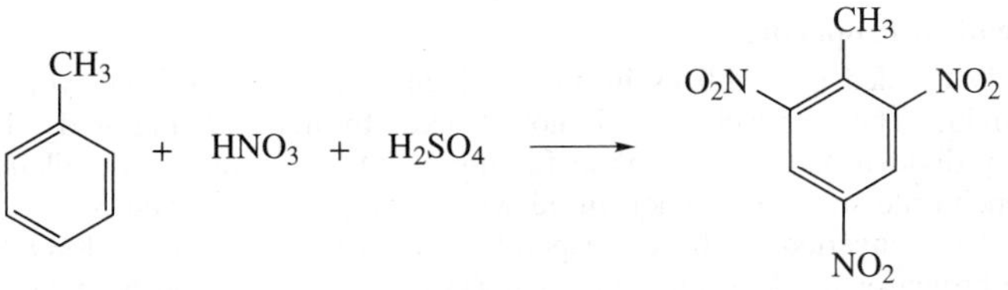

Toluene + Nitric acid + Sulfuric acid → 2,4,6-trinitrotoluene (TNT)

See also **Nitration**.

Hydroformylation

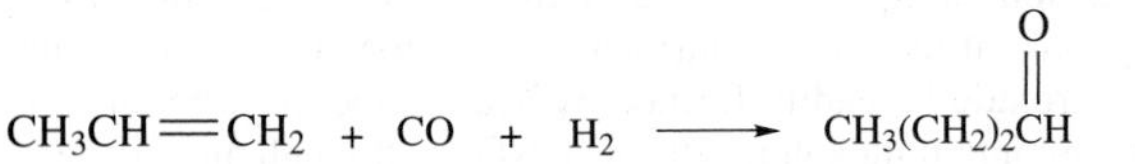

$$CH_3CH = CH_2 + CO + H_2 \longrightarrow CH_3(CH_2)_2CH$$

Propylene + Carbon monoxide + Hydrogen → *n*-Butyraldehyde

Esterification

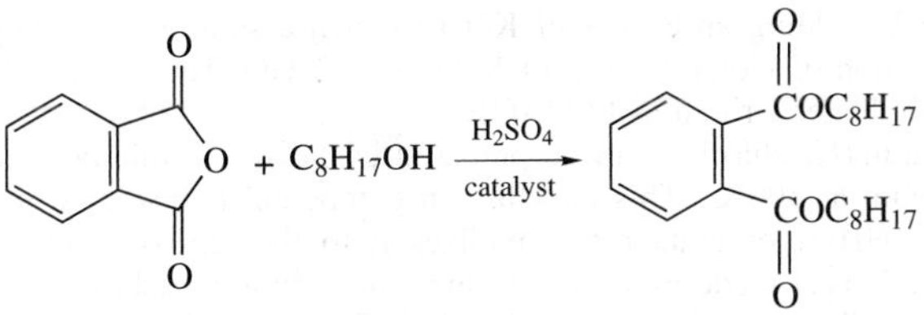

Phthalic anhydride + Octanol → Dioctyl phthalate

Vinyl Polymerization

$$CH_2 = CHCl \xrightarrow[\text{initiator}]{K_2S_2O_8} (CH_2CHCl)_{\sim 1,000-10,000}$$

Vinyl chloride → Polyvinyl chloride

Condensation Polymerization

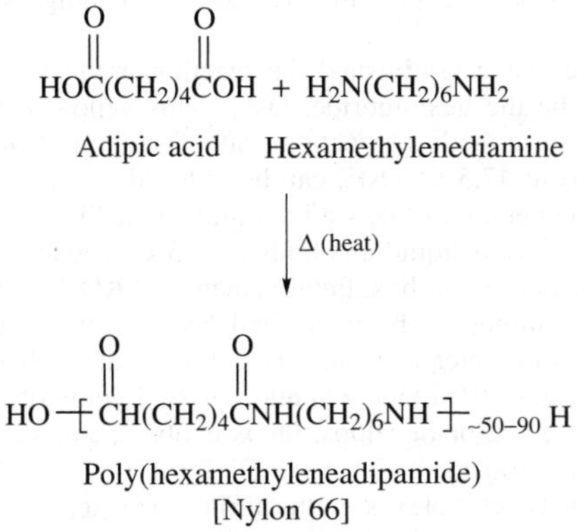

$$HOC(CH_2)_4COH + H_2N(CH_2)_6NH_2$$

Adipic acid + Hexamethylenediamine

↓ Δ (heat)

$$HO \left[CH(CH_2)_4CNH(CH_2)_6NH \right]_{\sim 50-90} H$$

Poly(hexamethyleneadipamide)
[Nylon 66]

WENDELL L. DILLING
MARCIA L. DILLING
Dow Chemical Company
Midland, Michigan

Additional Reading

Bickford, M., J.I. Kroschwitz, M. Howe-Grant: *Kirk-Othmer Encyclopedia of Chemical Technology, Concise,* 4th Edition, John Wiley & Sons, Inc., New York, NY, 2003.

Chemical Substance Index Names: Section IV from the Chemical Abstracts Index Guide, American Chemical Society, Columbus, Ohio, 1985—and references cited therein.

Green, M.M., and H.A. Wittcoff: *Organic Chemistry Principles and Industrial Practice,* John Wiley & Sons, Inc., New York, NY, 2003.

Gokel, G.W.: *Dean's Handbook of Organic Chemistry,* 2nd Edition, The McGraw-Hill Companies, Inc., New York, NY, 2003.

Graham Solomons, T.W.: *Fundamentals of Organic Chemistry,* John Wiley and Sons, Inc., New York, NY, 1999.

Graham, Solomons T.W., and C.B. Fryhle: *Organic Chemistry, Study Guide,* 8th Edition, John Wiley & Sons, Inc., New York, NY, 2003.

Hellwinkel, D.: *Systematic Nomenclature in Organic Chemistry: A Directory to Comprehension and Application on Its Basic Principles,* Springer-Verlag, Inc., New York, NY, 2001.

Hoffman, R.V.: *Organic Chemistry: An Intermediate Text,* 2nd Edition, John Wiley & Sons, Inc., Hoboken, NJ, 2004.

McMurry, J.: *Fundamentals of Organic Chemistry,* Thomson Learning Publications, Fresno, CA, 1997.

McMurry, J. and M. Castellion: *Fundamentals of Organic and Biological Chemistry, 2nd Edition,* Prentice-Hall, Inc., Upper Saddle River, NJ, 1998.

Morrison, R.T. and R.N. Boyd: *Organic Chemistry,* 6th Edition, Prentice-Hall, Inc., Upper Saddle River, NJ, 1992.

Nomenclature of Organic Chemistry; Sections A, B, C, D, E, F and H combined; Pergamon Press, Oxford, 1979. (IUPAC nomenclature.)

Richey, H.G., Jr.: *Fundamentals of Organic Chemistry,* Pentrice-Hall, Inc., Upper Saddle River, NJ, 1983.

Schmerling, L.: *Organic and Petroleum Chemistry for Nonchemists,* Penn-Well Publishing, Tulsa, OK, 1981.

Smith, M.B. and J. March: *March's Advanced Organic Chemistry: Reactions, Mechanisms, and Structure,* 5th Edition, John Wiley & Sons, Inc., New York, NY, 2001.

Wade, L.G. Jr.: *Organic Chemistry,* 4th Edition, Prentice-Hall, Inc., Upper Saddle River, NJ, 1998.

Weissermel, K., and Hans-Jurgen Arpe: *Industrial Organic Chemistry,* 4th Edition, John Wiley & Sons, Inc., New York, 2003.

ORGANIC SYNTHESIS. See **Synthesis (Chemical)**.

ORGANOBORANE. A compound composed of an unsaturated organic group and a borane obtained by the hydroboration reaction. Such compounds are useful catalytic reagents in organic syntheses of some complexity, e.g., *cis-* or *trans-*olefins, optically pure alcohols, alkanes, and ketones. Prostaglandins and insect pheromones have been synthesized by this means. A particularly versatile example is triphenylboron $B(C_6H_5)_3$.

See also **Borane**; **Carborane**; and **Hydroboration**.

ORGANOCLAY (Organopolysilicate). A clay such as kaoliln or montmorillonite, to which organic structures have been chemically bonded; since the surfaces of the clay particles, which have a latticelike arrangement, are negatively charges, they are capable of binding organic radicals. When this type of structure is in turn reacted with a monomer such as styrene, a complex known as a polyorganosilicate graft polymer results.

ORGANOLEPTIC. A term widely used to describe consumer testing procedures for food products, perfumes, wines, and the like in which samples of various products, flavors, etc. are submitted to groups or panels. Such tests are a valuable aid in determining the acceptance of the products and thus may be viewed as a marketing technique. They also serve psychological purposes and are an important means of evaluating the subjective aspects of taste, odor, color, and related factors. The physical and chemical characteristics of foods are stimuli for the eye, ear, skin, nose, and mouth, whose receptors initiate impulses that travel to the brain, where perception occurs.

ORGANOMETALLIC COMPOUNDS. An organic compound composed of a metal attached directly to carbon (RM); such compounds have been prepared of practically all the metals, as well as with such non-metals as silicon and phosphorus. Metallic salts (soaps) of organic acids are excluded. Examples are diethylzinc (the first known organometallic). Grignard compounds such as methy magnesium iodide (CH_3MgI), and metallic alkyls such as butyllithium (C_4H_9Li), tetraethyllead, triethyl aluminum, tetrabutyl titanate, sodium methylate, copper phthalocyanine, and metallocenes. Some are highly toxic or flammable; others are coordination compounds.

Reactive and moderately reactive organometallic compounds will react with all functional groups; two major types of reaction in which they are involved are oxidation and cleavage by acids. Probably the most important organometallic reactions are those involving addition to an unsaturated linkage. Many of them are powerful catalysts and form useful coordination complexes.

See also **Catalysis**; **Coordination Compounds**; and **Metallocenes**.

ORGANOPHOSPHORUS COMPOUND. Any organic compound containing phosphorus as a constituent. These fall into several groups, chief of which are the following: (1) phospholipids, or phosphatides, which are widely distributed in nature in the form of lecithin, certain proteins, and nucleic acids; (2) plasticizers, insecticides, resin modifiers, and flame-retardants; (3) pyrophosphates, e.g., tetraethyl pyrophosphate, which are the basis for a broad group of cholinesterase inhibitors used as insecticides; (4) phosphoric esters of glycerol, glycol, sorbitol, etc., which are components of fertilizers. While many of these compounds play an important part in animal metabolism, those in group (3) are toxic and should be handled with extreme care. See also **Flame-Retarding Agents**.

ORGANOSILICON. An organic compound in which silicon is bonded to carbon (organosilane). Such compounds were first made by Friedel and Crafts in 1863. Silicon was found to have a remarkable chemical similarity to carbon, which it can replace in organic compounds. The silicon-carbon bond is about as strong as the carbon-carbon bond, and compounds containing them are similar in properties to all-carbon compounds. Organosilicon oxides (organosiloxanes or silicones) were discovered by F. S. Kipping in England in 1900; he found that Grignard reagents would react with silicon tetrachloride to form silicon-carbon-bonded polymers or both ring and chain types. See also **Grignard Reactions**. These were named silicones because of the similarity of their empirical formula (R_2SiO) to that of ketones (R_2CO).

An organosilicon compound (tetramesityldisilene) containing a silicon to silicon double bond has been synthesized. It is a crystalline solid, mp 176C, and has reactive properties similar to olefins. Compounds of the type are silylenes.

ORGANOSOL. Colloidal dispersion of any insoluble material in an organic liquid; specifically the finely divided or colloidal dispersion of a synthetic resin in plasticizer in which dispersion the volatile content exceeds 5% of the total.

See also **Plastisol**.

ORGANOTIN COMPOUNDS. A family of alkyl tin compounds widely used as stabilizers for plastics, especially rigid vinyl polymers used as piping, construction aids, and cellular structures. Some have catalytic properties. They include butyl tim trichloride, dibutyltin oxide, etc., and various methyltin compounds. They are both liquids and solids.

ORNITHINE. See **Amino Acids**.

ORNITHINE CYCLE. See **Urea**.

ORPIMENT. This mineral, like realgar, its frequent associate, is an arsenic sulfide. Orpiment, however, is the trisulfide corresponding to the formula As_2S_3. It is monoclinic, usually in foliated, granular, or powdery aggregates; hardness, 1.502; specific gravity, 3.49; with a resinous to somewhat pearly luster; color, various shades of lemon yellow; translucent to nearly opaque. Orpiment is found in association with realgar, although a somewhat rarer mineral. It is believed to be formed from the alteration of other arsenic-bearing minerals. It occurs in Czechoslovakia, Romania, Macedonia, Japan, and in the United States in Utah, Nevada, and Wyoming. The name orpiment is derived from a corruption of the Latin *auripigmentum*, meaning golden paint, because of its color as well as the belief that it contained gold.

ORTHO COMPOUNDS. See **Organic Chemistry**.

ORTHO-STATE. 1. In diatomic molecules, such as hydrogen molecules, the ortho-state exists when the spin vectors of the two atomic nuclei are in the same direction (i.e., parallel), whereas the para-state is the one in which the nuclei are spinning in opposite directions.

2. In helium the ortho-state is characterized by a particular mode of coupling of the electron spins. See also **Helium**.

OSMIUM. [CAS: 7440-04-2]. Chemical element, symbol Os, at. no. 76, at. wt. 190.2, periodic table group 8, mp 3,015° to 3,075°C,

bp 4,927° to 5,127°C, density 22.6g/cm³ (solid), 22.8g/cm³ (single crystal) (20°C). Elemental osmium has a close-packed hexagonal crystal structure. Compact osmium is a bluish-white metal and is not attacked by acids. Discovered by Tennant in 1804. The seven stable isotopes of osmium are ^{184}Os, ^{186}Os through ^{190}Os, and ^{192}Os. Electronic configuration $1s^2 2s^2 2p^6 3s^2 3p^6 3d^{10} 4s^2 4p^6 4d^{10} 4f^{14} 5s^2 5p^6 5d^6 6s^2$. Ionic radius $Os_4{}^+$ 0.65 Å. Metallic radius 1.3377 Å. Oxidation potential $Os + 4H_2O \longrightarrow OsO_4 + 8H^+ + 8e^-$, -0.85 V.

Chemical Properties

Finely divided Os oxidizes in air, producing the poisonous and volatile tetroxide. The compact metal is not attacked by nonoxidizing acids. The finely divided metal dissolves in fuming HNO_3, aqua regia, and alkaline hypochlorite solutions. When fused with Na_2O_2 or KNO_3 and KOH, the metal is converted to the corresponding water-soluble osmate, K_2OsO_4. The brown or black insoluble osmium(IV) oxide, OsO_2, can be made by heating Os with a limited amount of O_2 or with osmium(VIII) oxide. This compound forms a brown to black-blue dihydrate that can be prepared by reducing a solution of the tetroxide or by hydrolyzing a solution of sodium hexachloroosmate, Na_2OsCl_6.

Osmium(VIII) oxide, the most important compound, is formed in one of the reactions unique to the platinum metals. Its ease of formation and volatility make it useful in a purification step for the refining or analysis of Os. The tetroxide is readily formed by heating the metal in air or distilling an osmium-containing solution from HNO_3. Although an aqueous solution of osmium(VIII) oxide is neutral to litmus, it is a weak acid with first dissociation constant of about 8×10^{-13}. Osmium(VIII) oxide is soluble in water, alcohol, and ether. The compound is widely used as a stain for tissues. When an alkaline solution of osmium(VIII) oxide is reduced with alcohol or KNO_2, an osmate(VI) is formed. Potassium osmate(VI) is formed by adding an excess of KOH to such a solution, resulting in the precipitation of violet crystals of $K_2OsO_4 \times 2H_2O$. The osmate(VI) ion is probably better written as $OsO_2(OH)_2$.

Osmium(II) chloride can be prepared by heating osmium(III) chloride in vacuum at 500°C. This dark-brown compound is insoluble in HCl or H_2SO_4. NHO_3 or aqua regia oxidizes it to the tetroxide. Osmium(III) chloride is best made by decomposing ammonium hexachloroosmate(IV), $(NH_4)_2OsCl_6$, in a Cl_2 stream at 350°C. The brown hygroscopic powder sublimes above 350°C, and at about 560°C it disproportionates into the tetrachloride and dichloride. Osmium(IV) chloride is formed from the elements at 650–700°C. The black compound slowly dissolves in water, eventually forming the dioxide. The free acid, H_2OsCl_6, is stable in solution and can be made by refluxing osmium(VIII) oxide with HCl and alcohol. The ammonium salt can be precipitated by adding NH_4Cl to such a solution. This salt is reduced to the metal when heated in H_2. The potassium salt is well known. Both are brownish-red solids yielding orange solutions in water.

Recent studies have established the reaction product of Os metal and F_2 at 300°C to be the hexafluoride, OsF_6. This yellow volatile solid had previously been described as an octafluoride. Osmium(VI) fluoride melts at 33.4°C and boils at 47.5°C. OsF_6 can be reduced to a pentafluoride and a tetrafluoride. The pentafluoride is a blue-gray crystalline solid that melts at 70°C to a green viscous liquid and boils at 226°C. The tetrafluoride distills at about 290°C. Potassium hexafluoroosmate(V), $KOsF_6$, can be made by reacting KBr, osmium(IV) bromide, and bromine trifluoride. The white powder dissolves in water to form a colorless solution that hydrolyzes to yield some osmium(VIII) oxide. On addition of 1 equiv of KOH to a fresh solution, an orange color develops, O_2 is evolved, and yellow crystals of potassium hexafluoroosmate(IV), K_2OsF_6, form.

Os forms many complexes with nitrite, oxalate, carbon monoxide, amines, and thio ureas. The latter are important analytically. Osmium forms the interesting aromatic "sandwich" compound, osmocene. A *metallocene* is described under **Ruthenium**. See also **Chemical Elements**; and **Platinum and Platinum Group**.

Proton nuclear magnetic resonance (NMR) is a widely used tool for researching biomolecules. Although much too detailed to report here, Zai-Wei Li and Henry Taube (Stanford University) reported in 1992 that they have had success in analyzing for certain molecules by using a dihydrogen osmium complex on a versatile 1H NMR recognition probe.

Osmium Isotopic Ratios in Paleogeology

As pointed out in 1983 by J.M. Luck and K.K. Turekian (Yale University), one of the most creative concepts regarding the cause of the many paleontologic extinctions at the Cretaceous-Tertiary boundary is the one put forward by Alvarez et al. in 1980 involving the impact of a large asteroid or comet with the earth at that geologic time period. The Alvarez hypothesis stemmed from the finding of an exceptional chemical signature (high iridium concentration) at the Cretaceous-Tertiary boundary at Gubbio, Italy, a signature that was later found in other marine as well as continental sections which bracket the Cretaceous-Tertiary boundary. Because of the radioactive decay of rhenium-187 (4.6×10^{10} years), the osmium-187/osmium-186 ratio changes in planetary systems as a function of time and the rhenium-187/osmium-186 ratio. For a value of the $^{187}Re/^{186}Os$ ratio of about 3.2, typical of meteorites and the earth's mantle, the present $^{187}Os/^{186}Os$ ratio is about one. The earth's continental crust has an estimated $^{187}Re/^{186}Os$ ratio of about 400. Thus for a mean age of the continent of 2×10^9 years, a present $^{187}Os/^{186}Os$ ratio of about 10 is expected. Marine manganese nodules show values (6 to 8.4), which are compatible with this expectation if an allowance for a 25% mantle osmium supply to the oceans is made. The Cretaceous-Tertiary boundary iridium-rich layer in the marine section at Stevns Klint, Denmark, yields a $^{187}Os/^{186}Os$ ratio of 1.65 and the one in a continental section in the Raton Basin, Colorado, is 1.29. The investigators conclude that the simplest explanation is that these represent osmium imprints of predominantly meteoritic origin.

As reported in 1992 by M.F. Horan, J.W. Morgan and J.N. Grossman (U.S. Geological Survey), and R.J. Walker (University of Maryland), "Rhenium and osmium concentrations and the osmium isotopic compositions of iron meteorites were determined by negative thermal ionization mass spectrometry. Data for the IIA iron meteorites define an isochron with an uncertainty of approximately ±31 million years for meteorites ~4500 million years old. Although an absolute rhenium-osmium closure age for this iron group cannot be as precisely constrained because of uncertainty in the decay constant of ^{187}Re, an age of 4460 million years ago is the minimum permitted by combined uncertainties. These age constraints imply that the parent body of the IIAB magmatic irons melted and subsequently cooled within 100 million years after the formation of the oldest portions of chondrites. Other iron meteorites plot above the IIA isochron, indicating that the planetary bodies represented by these iron groups may have cooled significantly later than the parent body of the IIA irons."

Additional Reading

Anderson, D.L.: "Composition of the Earth," *Science*, 367 (January 20, 1989).

Cherfas, J.: "Proton Microbeam Probes the Elements," *Science*, 1150 (September 28, 1990).

Davis, J.R.: *Metals Handbook,* 2nd Edition, ASM International, Materials Park, OH, 1998.

Greenwood, N.N. and A. Earnshaw: *Chemistry of the Elements,* 2nd Edition, Butterworth-Heinemann, Inc., Woburn, MA, 1997.

Horan, M.F., et al.: "Rhenium-Osmium Isotope Constraints on the Age of Iron Meteorites," *Science*, 1118 (February 18, 1992).

Krebs, R.E.: *The History and Use of Our Earth's Chemical Elements: A Reference Guide,* Greenwood Publishing Group, Inc., Westport, CT, 1998.

Lewis, R.J., Sr.: *Hawley's Condensed Chemical Dictionary,* 13th Edition, John Wiley & Sons, Inc., New York, NY, 1997.

Lide, D.R.: *CRC Handbook of Chemistry and Physics*, 84th Edition, CRC Press, LLC., Boca Raton, FL, 2003.

Luck, J.M. and K.K. Turekian: "Osmium-187/Osmium-186 in Manganese Nodules and the Cretaceous-Tertiary Boundary," *Science*, **222**, 613–615 (1983).

Sinfelt, J.H.: "Bimetallic Catalysts," *Sci. American*, **253**(3), 90–98 (1985).

Staff: *ASM Handbook—Properties and Selection: Nonferrous Alloys and Pure Metals,* ASM International, Materials Park, OH, 1990.

Zai-Wei, Li and H. Taube: "Use of a Dihydrogen Osmium Complex as a Versatile 1 H NMR Recognition Probe," *Science*, 210 (April 10, 1992).

OSMOSIS.   Passage of a pure liquid (usually water) into a solution (e.g., of sugar and water) through a membrane that is permeable to the pure water but not to the sugar in the solution. This passage can also occur when the two phases consist of solutions of different concentration. The membrane is called semipermeable when the molecules of the solvent, but not those of the solute, can penetrate it. This pushing of water through a membrane into a solution results from the greater tendency of water molecules to escape from water than from a solution. The term *osmosis* us usually restricted to movement through a solid or liquid barrier that prevents the phases from mixing rapidly. In test apparatuses parchment or collodion membranes are used; in plants and animals the cell wall acts as a diffusion barrier. The pressure exerted by osmosis is substantial and accounts for the elevation of sap from root systems to the tops of trees. Osmosis is considered an essential characteristic of growth.

Reverse osmosis is used as a method of desalting seawater, recovering wastewater from paper mill operations, pollution control, industrial water treatment, chemical separations, and food processing. This method involves application of pressure to the surface of a saline solution, thus forcing pure water to pass from the solution through a membrane that is too dense to permit passage of sodium and chlorine ions. Hollow fibers of cellulose acetate or nylon are used as membranes, since their large surface area offers more efficient separation.

See also **Desalination**; **Diffusion**; and **Membrane Separations Technology**.

OSMOTIC COEFFICIENT.   A factor introduced into equations for nonideal solutions to correct for their departure from ideal behavior, as in the equation:

$$\mu = \mu_x^0 + gRT \ln x_1$$

in which μ is the chemical potential, μ_x^0 is a constant, representing a standard value of the chemical potential, R is the gas constant, T the absolute temperature, x_1 is the mole fraction of solvent, and g is the osmotic coefficient.

OSMOTIC PRESSURE.   Pressure that develops when a pure solvent is separated from a solution by a semipermeable membrane which allows only the solvent molecules to pass through it. The osmotic pressure of the solution is then the excess pressure which must be applied to the solution so as to prevent the passage into it of the solvent through the semipermeable membrane.

Because of the similarity in the relations for osmotic pressure in dilute solutions and the equation for an ideal gas, van't Hoff proposed his bombardment theory in which osmotic pressure is considered in terms of collisions of solute molecules on the semipermeable membrane. This theory has a number of objections and has now been discarded. Other theories have also been put forward involving solvent bombardment on the semipermeable membrane, and vapor pressure effects. For example, osmotic pressure has been considered as the negative pressure which must be applied to the solvent to reduce its vapor pressure to that of the solution. It is, however, more profitable to interpret osmotic pressures using thermodynamic relations, such as the entropy of dilution.

A number of methods have been developed for measurement of osmotic pressure.

In the Berkeley and Hartley method, a porous tube with a semipermeable membrane such as copper ferrocyanide deposited near the outer wall, and a capillary tube attached to one end contains the pure solvent. The solution surrounds the tube and is enclosed in a metal vessel to which a pressure may be applied which is just sufficient to prevent the flow of solvent into the solution. Berkeley and Hartley also developed a dynamic method for measuring osmotic pressure.

Simple osmometers have also been developed by Adair particularly for aqueous colloidal solutions. A thimble-type collodion membrane is attached to a capillary tube and contains the solution. When equilibrium is established the difference in level inside and outside the capillary is measured. Capillary corrections are made. For organic solvents a dynamic type osmometer may be used. A membrane of large surface area is clamped between two half cells and attached to each half cell is a fine capillary observation tube. With such an apparatus, equilibrium is rapidly established between solution and solvent contained in the half cells. The volume of the half cell may be small (about 20 cubic centimeters). The level of the solvent is usually arranged to be a little below the equilibrium position, and the height of the solvent in the capillary as a function of time is measured. This procedure is repeated with the level of the solvent just above the equilibrium position. A plot is then made of the half sum of these readings.

Since the osmotic pressure is related to the concentration of dissolved solute particles, it is related to the lowering of the freezing point and elevation of the boiling point.

The relation between osmotic pressure and lowering of the freezing point and the elevation of the boiling point may be expressed by the relation:

$$\Pi = \frac{LT}{vT_0^2} \Delta T$$

where L is the molar heat of fusion or of vaporization of the solvent, T the temperature at which the osmotic pressure is measured, T_0 the freezing point or the boiling point of the solvent, $\bar{v}$, the partial molar volume of the solvent, and Π, the osmotic pressure.

Moreover, the relation to lowering of the vapor pressure is

$$\Pi = \frac{RT}{\bar{v}} \ln \frac{p_0}{p}$$

or

$$\Pi = -\frac{RT}{\bar{v}} \ln x_0 = -\frac{RT}{v} \ln(1 - x)$$

For dilute solutions,

$$\Pi = cRT$$

where Π is the osmotic pressure, $\bar{v}$ *the partial molar volume of the solvent in the solution*, p_0 the vapor pressure of pure solvent, p the partial vapor pressure of the solvent in equilibrium with the solution, x_0 the mole fraction of the solvent, x the mole fraction of the solute, c the concentration of the solution in moles per liter, R, the gas constant, and T, the absolute temperature.

OSTWALD, WIHELM (1853–1932).

A German chemist who won the Nobel prize for chemistry in 1909. He was considered to be a founder of modern physical chemistry. His work involved research in catalysis, the rates of chemical reactions, equilibrium, and conductivity of organic acids. He was an admirer of Mach and did not readily accept the atomic theory. He received honorary doctorates from several universities in Germany, Great Britain and the USA, and was made an honorary member of learned societies in Germany, Sweden, Norway, the Netherlands, Russia, Great Britain and the USA.

OUTGASSING.

The removal of gas from a metal by heating at a temperature somewhat below melting, while maintaining a vacuum in the space around the metal.

OXALIC ACID.

[CAS: 144-62-7]. Oxalic acid, HOOC–COOH, or ethanedioic acid, mol wt 90.04, is the simplest dicarboxylic acid. It is soluble in water, and acts as a strong acid. This acid does not exist in anhydrous form in nature and is available commercially as a solid dihydrate, $C_2H_2O_4 \cdot 2H_2O$, mol wt 126.07. The commercial product is packed in polyethylene-lined paper bags or flexible containers. Anhydrous oxalic acid can be efficiently prepared from the dihydrate by azeotropic distillation in a low boiling solvent that can form a water azeotrope, such as benzene and toluene.

Oxalic acid was synthesized for the first time in 1776 by Scheele through the oxidation of sugar with nitric acid. Then, Wöhler synthesized it by the hydrolysis of cyanogen in 1824.

The potassium or calcium salt form of oxalic acid is distributed widely in the plant kingdom. Oxalic acid is found in spinach, rhubarb, etc. Oxalic acid is a product of metabolism of fungi or bacteria and also occurs in human and animal urine; the calcium salt is a principal constituent of kidney stones.

Oxalic acid is used in various industrial areas, such as textile manufacture and processing, metal surface treatments leather tanning, cobalt production, and separation and recovery of rare-earth elements. Substantial quantities of oxalic acid are also consumed in the production of agrochemicals, pharmaceuticals, and other chemical derivatives.

Physical Properties

The physical and thermochemical constants of anhydrous oxalic acid and oxalic acid dihydrate are summarized in Table 1.

Reactions

The reactions of oxalic acid, including the formation of normal and acid salts and esters, are typical of the dicarboxylic acids class. Oxalic acid, however, does not form an anhydride.

On rapid heating, oxalic acid decomposes to formic acid, carbon monoxide, carbon dioxide, and water (qv). In aqueous solution, it is decomposed by uv, x-ray, or γ-radiation with the liberation of carbon dioxide. Photodecomposition also occurs in the presence of uranyl salts.

Oxalic acid is a mild reducing agent, and is oxidized by potassium permanganate in acid solution to give carbon dioxide and water. Oxalic

TABLE 1. PHYSICAL AND THERMOCHEMICAL PROPERTIES OF OXALIC ACID AND ITS DIHYDRATE

Property	Value
OXALIC ACID, ANHYDROUS, $C_2H_2O_4$	
melting point, °C	
α	189.5
β	182
density d_4^{17}, g/mL	
α	1.900
β	1.895
refractive index, β, n_4^{20}	1.540
vapor pressure (solid, 57–107°C), kPa[a]	$\log_{10} P = -(4726.95/T) + 11.3478$
specific heat (solid, −200 to 50°C), J/g	$C_p{}^b = 1.084 + 0.0318t$
heat of combustion, ΔE_c (at 25°C), kJ/mol[c]	−245.61
standard heat of formation, ΔH_f (at 25°C), kJ/mol[c]	−826.78
standard free energy of formation, ΔG_f (at 25°C), kJ/mol[c]	−697.91
heat of solution (in water), kJ/mol[c]	−9.58
heat of sublimation, kJ/mol[c]	90.58
heat of decomposition, kJ/mol[c]	826.78
specific entropy, S (at 25°C), J/(mol·K)[c]	120.08
logarithm of equilibrium constant, $\log_{10} K_f$	122.28
thermal conductivity (at 0°C), W/(m·K)[d]	0.9
ionization constant	
K_1	6.5×10^{-2}
K_2	6.0×10^{-5}
coefficient of expansion (at 25°C), nL/(g·K)	178.4
OXALIC ACID DIHYDRATE, $C_2H_2O_4 2H_2O$	
mp, °C	101.5
density d_4^{20}, g/mL	1.653
refractive index, n_4^{20}	1.475
standard heat of formation, ΔH_f (at 18°C), kJ/mol[c]	−1422
heat of solution (in water), kJ/mol[c]	−35.5
pH (0.1 M soln)	1.3

[a] To convert $\log_{10} P_{kPa}$ to $\log_{10} P_{mm\ Hg}$, add 0.875097 to the constant, $T = K$.
[b] To convert C_p, J/g, to C_p cal/g, divide both terms of the equation by 4.184.
[c] To convert J to cal, divide by 4.184.
[d] To convert W/(m·K) to (Btu·in.)/(h·ft²·°F), divide by 0.1441.

acid is catalytically reduced by hydrogen in the presence of ruthenium catalyst to ethylene glycol, and electronically reduced to glyoxylic acid.

Oxalic acid reacts with various metals to form metal salts, which are quite important as the derivatives of oxalic acid. It also reacts easily with alcohols to give esters.

Manufacture

Many industrial processes have been employed for the manufacture of oxalic acid since it was first synthesized. The following processes are in use worldwide: oxidation of carbohydrates, the ethylene glycol process, the propylene process, the dialkyl oxalate process, and the sodium formate process. Sodium formate process is no longer economical in the leading industrial countries, except for China.

Nitric acid oxidation is used where carbohydrates, ethylene glycol, and propylene are the starting materials. The dialkyl oxalate process is the newest, where dialkyl oxalate is synthesized from carbon monoxide and alcohol, then hydrolyzed to oxalic acid. This process has been developed by UBE Industries in Japan.

Many attempts have been made to synthesize oxalic acid by electrochemical reduction of carbon dioxide in either aqueous or nonaqueous electrolytes.

Health and Safety Factors

Oxalic acid is caustic and corrosive to humans. The severity of symptoms associated with oxalic acid poisoning is related to the concentration and quantity ingested. Oxalic acid removes calcium in the blood, forming calcium oxalate, and severe damage to the kidney may occur because of the insoluble calcium oxalate.

Uses

Because rare-earth oxalates have low solubility in acidic solutions, oxalic acid is used for the separation and recovery of rare-earth elements. The oxalic acid process for anodizing aluminum was developed in Japan. In addition to oxalic acid, inorganic oxalate salts are also used in coloring anodic coatings. Oxalic acid is a constituent of cleaners that are used for automotive radiators, boilers, and steel plates before phosphating. As a chelating agent, oxalic acid forms water-soluble complexes on metal surfaces during cleaning and rinsing.

In pulp bleaching, oxalic acid serves as a bleaching agent, but is often used together with other bleaching agents because of its relatively high cost. Oxalic acid is also used for the bleaching of cork, wood (particularly veneered wood), straw, cane, and natural waxes.

Oxalic acid has various uses in fabric cleaning, application of dyestuff, and modifying properties of cellulose fabrics. Oxalic acid is used as a pH modifier in leather tanning by tannin and basic chromium sulfate. It also functions as a bleaching agent for leather. It is used for marble polishing especially in Italy. It not only removes iron veins by forming water-soluble iron oxalate, but also serves as a polishing auxiliary. Starch powder is heated together with oxalic acid and hydrolyzed to produce millet jelly. Oxalic acid functions as a hydrolysis catalyst, and is removed from the product as calcium oxalate. This application is carried out in Japan. Oxalic acid is also used for the production of cobalt, as a raw material of various agrochemicals, and pharmaceuticals, for the manufacture of electronic materials, for the extraction of tungsten from ore, for the production of metal catalysts, as a polymerization initiator, and for the manufacture of zirconium and beryllium oxide.

Derivatives

Oxalic acid forms neutral and acid salts, as well as complex salts.

Ammonium Oxalate. Anhydrous ammonium oxalate is obtained when the monohydrate is dehydrated at 65°C. It is used for textiles, leather tanning, and precipitation of rare-earth elements.

Ammonium Iron(III) Oxalate. This mixed salt is produced as an emerald-green crystalline trihydrate. The compound is not stable to light. It was once used extensively in the manufacture of blueprinting papers.

Potassium Hydrogen Oxalate. Potassium acid oxalate, exists as a monohydrate. It is of historical interest because it is the salt of sorrel found in vegetation and the first oxalate isolated.

Potassium Oxalate. The monohydrate is produced as a colorless crystalline material or a white powder. The anhydrous salt is obtained when the monohydrate is dehydrated at 160°C. The monohydrate is preferred as a reagent in analytical chemistry and in miscellaneous uses principally because of its high solubility as compared with other simple neutral oxalates; the saturated solution, at 0°C, contains about 20 wt %, and at 20°C, about 25 wt % $K_2C_2O_4$.

Sodium Oxalate. This salt is obtained in such high purity and is so stable that it is used as a titrimetric standard.

Calcium Oxalate. The monohydrate is of importance principally as an intermediate in oxalic acid manufacture and in analytical chemistry; it is the form in which calcium is frequently quantitatively isolated.

Nickel Oxalate. This salt is produced as a greenish white crystalline dihydrate. Nickel oxalate is used for the production of nickel catalysts and magnetic materials.

Yttrium Oxalate. This compound exists as a trihydrate, nonahydrate, or heptadecahydrate. The compound is used for the production of a red fluorescent material for color television.

Dialkyl Oxalates. Oxalic acid gives various esters. Dialkyl esters ROOC—COOR, are industrially useful, but monoalkyl esters, ROOC—COOH, are not. The dialkyl esters are characterized by good solvent properties and serve as starting materials in the synthesis of many organic compounds, such as pharmaceuticals, agrochemicals, and fine chemicals (qv). Among the diesters, dimethyl, diethyl, and di-*n*-butyloxalates are industrially important.

Oxamide. This diamide is sparingly soluble in water and insoluble in various organic solvents. It melts at about 350°C, with accompanying decomposition. Because of the low solubility in water, the compound is granulated and used as a slow-release nitrogen fertilizer. Conventional nitrogen fertilizers such as ammonium sulfate, urea, ammonium nitrate, and ammonium phosphate, are soluble in water, and thus are easily lost as run-off when it rains. On the contrary, oxamide stays in the soil longer. Therefore, it is gradually decomposed by microorganisms in the soil and utilized by plants for longer periods.

Oxalyl Chloride. This diacid chloride is produced by the reaction of anhydrous oxalic acid and phosphorus pentachloride. The compound vigorously reacts with water, alcohols, and amines, and is employed for the synthesis of agrochemicals, pharmaceuticals, and fine chemicals.

Reduction Products. Glyoxylic acid is produced as aqueous solution by the electrolytic reduction of oxalic acid. It is used for the manufacture of vanillin.

Glycolic acid can be obtained by the electrolytic reduction of oxalic acid or the catalytic reduction of oxalic acid with hydrogen in the presence of a ruthenium catalyst. Because of its acidity it is used as a cleaning agent for metal surface treatments and for boiler cleaning. It also serves as an ingredient in cosmetics.

Hiroyuki Sawada
Toru Murakami
UBE Industries, Ltd.

Additional Reading

Jpn. Pat. 61-26977-B (1986), S. Tahara and co-workers (to UBE Industries).
Sarver, L.A. and P.H.M.P. Briton: *J. Am. Chem. Soc.* **49**, 943 (1927).
U.S.Pat. 2,057,119 (1936), G.S. Simpson (to General Chemical (Allied Chemical)).
Werneck S. and R. Pinner: *The Surface Treatment and Finishing of Aluminum and Its Alloys*, Robert Draper, Ltd., Teddington, UK, 1972.

OXIDATION AND OXIDIZING AGENTS.

Many years ago, the term *oxidation* signified a reaction in which oxygen combines chemically with another substance. The term now has a much broader meaning and includes any reactions in which electrons are transferred. Oxidation and reduction always occur simultaneously (redox reactions), and the substance which gains electrons is termed the *oxidizing agent*. For example, cupric ion is the oxidizing agent in the reaction: Fe (metal) + Cu^{++} $\longrightarrow$ Fe^{++} + Cu (metal). Here, two electrons (negative charges) are transferred from the Fe atom to the Cu atom. Thus the Fe becomes positively charged (is oxidized) by loss of two electrons, while the Cu receives the two electrons and becomes neutral (is reduced). Electrons may also be displaced within the molecule without being completely transferred away from it. Such partial loss of electrons likewise constitutes oxidation in its broader sense and leads to the application of the term to a large number of processes which at first inspection might not be considered to be oxidations. Reactions of a hydrocarbon with a halogen, for example: $CH_4 + Cl_2 \longrightarrow CH_3Cl + HCl$, involves partial oxidation of the methane. Also, when a halogen addition to a double bond is made, this is regarded as an oxidation.

Dehydrogenation is also a form of oxidation, when two hydrogen atoms, each having one electron, are removed from a hydrogen-containing organic compound by a catalytic reaction with air or oxygen, as in oxidation of alcohols to aldehydes. See also **Dehydrogenation**.

Oxidizing agents are widely used throughout the chemical and petrochemical industries. It is also interesting to note that, while a primary thrust in food processing is to prevent oxidation (associated with rancidity and spoilage in foods), some powerful oxidizing agents are required by some processes to perform the function of bleaching. See also **Bleaching Agents**.

As one example, after a crude fat or oil is refined to remove its impurities, it must be further treated by bleaching to remove coloring materials that are typically present. Sulfuric and metaphosphoric acids and hydrogen peroxide have been used for this purpose. Calcium hypochlorite is used in bleaching sugar syrup prior to crystallization. Not all effective bleaching agents can be used because of their toxicity and inability to remove them completely. The aim is to remove all traces of the bleaching agent during subsequent processing so that they do not occur in the final product. Calcium peroxide, acetone peroxide, and benzoyl peroxides are other bleaching agents sometimes used in the food industry.

The use of oxidizers, such as potassium bromate, potassium iodate, and calcium peroxide as dough modifiers in the baking industry dates back many years. However, their mechanism has not been fully explained. Among authorities, there are at least two major viewpoints. It has been proposed that oxidizers inhibit proteolytic enzymes present in flour. It has also been proposed that the number of —S—S—bonds between protein chains is increased, forming a tenacious network of molecules. This action leads to a tougher, drier, and more extensible dough. The need for oxidizers is less with well aged flours and in connection with supplemented flours that have been effectively *brominated* at the mills. When used properly, oxidizers contribute to improve appearance, brighter crumb, and better texture of breads. Oxidizers do not appear to interfere with the generation of gas by yeast or leavening chemicals. However, oxidizers used to excess can destroy the desirable properties of the dough and end-products.

Additional Reading

Afanas'ev, I.B.: *Oxidation and Antioxidants in Organic Chemistry and Biology,* Marcel Dekker, Inc., New York, NY, 2004.

Gitler, C., and A. Danon: *Cellular Implications of Redox Signaling,* Imperial College Press World Scientific, London, UK, 2003.

Grabke, H.J.: *Oxidation of Intermetallics,* John Wiley & Sons, Inc., New York, NY, 1998.

Hodnett, B.K.: *Heterogeneous Catalytic Oxidation: Fundamental and Technological Aspects of the Selective and Total Oxidation of Organic Compounds,* John Wiley & Sons, Inc., New York, NY, 2000.

St. Angelo, A.J.: *Lipid Oxidation in Food,* American Chemical Society (ACS), Washington, DC, 1992.

OXIDATION NUMBER. In its original and restrictive sense, the number of electrons which must be added to a cation to neutralize the charge. The concept has been extended to anions by assignments of negative oxidation numbers. Moreover, it has been further extended, first to all atoms or radicals joined by electrovalent bonds, and then to covalent compounds in which the shared electrons are distributed equally. For the broadest use of the concept, the expression "oxidation state" is often used.

OXIDATION POTENTIAL. The potential drop involved in the oxidation (i.e., ionization) of a neutral atom to a cation, of an anion to a neutral atom, or of an ion to a more highly charged state (e.g., ferrous to ferric).

OXIDATION-REDUCTION. See **Phosphorylation (Oxidative)**; **Phosphorylation (Photosynthetic)**.

OXIDATION-REDUCTION EQUATIONS. See **Chemical Equation**.

OXIDATION-REDUCTION INDICATOR. A substance that has a color in the oxidized form different from that of the reduced form and can be reversibly oxidized and reduced. Thus, if diphenylamine is present in a ferrous sulfate solution to which potassium dichromate is being added, a violet color appears with the first drop of excess dichromate.

See also **Indicator (Chemical)**.

OXIDATIVE COUPLING. A polymerization technique for certain types of linear high polymers. Oxidation of 2,6-dimethylphenol with an amine complex of copper salt as catalyst forms a polyether, with splitting off of water. The product is soluble in aromatic and chlorinated hydrocarbons; insoluble in alcohols, ketones, and aliphatics. It is thermoplastic and unaffected by acids, bases, and detergents. It has a very broad useful temperature range (from -170 to $+190°C$). It is also dimensionally stable and has good electrical resistance. Oxidative coupling of diacetylenes and dithiols also yields promising polymers.

OXIDE. A mineral in which metallic atoms are bonded to oxygen atoms.

OXIDIZING MATERIAL. Any compound that spontaneously evolves oxygen either at room temperature or under slight heating. The term includes such chemicals as peroxides, chlorates, perchlorates, nitrates, and permanganates. These can react vigorously at ambient temperatures when stored near of in contact with reducing materials such as cellulosic and other organic compounds. Storage areas should be well ventilated and kept as cool as possible.

OXIMES. One of a number of compounds that result from the interaction of aldehydes, ketones, and other carbonyl-containing substances with hydroxylamine, e.g., acetone yields acetoxime

$$\begin{array}{ccc} CH_3 & & CH_3 \\ | & & | \\ C{=}O + H_2NOH \longrightarrow & C{=}NOH + H_2O \\ | & & | \\ CH_3 & & CH_3 \end{array}$$

OXIME TEST (Rheinboldt). An ethereal solution of an aldoxime gives a blue color with aqua regia.

OXONIUM COMPOUNDS. Coordination compounds, commonly of certain oxygen-containing organic substances, with mineral acids, of the general type $[R_2O]HCl$. These compounds bear a strong resemblance to the oxonium (hydronium) ion, which is a proton in combination with a water molecule, and is the form in which protons commonly exist in aqueous solutions.

OXO PROCESS. The oxo process, also known as hydroformylation, is the reaction of carbon monoxide and hydrogen with an olefinic substrate to form isometric aldehydes as shown in equation 1. The ratio of isomeric aldehydes depends on the olefin, the catalyst, and the reaction conditions.

$$RCH{=}CH_2 + CO + H_2 \xrightarrow{\text{catalyst}} RCH_2CH_2CHO + R(CH_3)CHCHO \quad (1)$$

If a double-bond shift occurs, the number of aldehyde isomers is increased.

Synthesis gas, a mixture of CO and H_2, also known as syngas, is produced for the oxo process by partial oxidation (eq. 2) or steam reforming (eq. 3) of a carbonaceous feedstock, typically methane or naphtha. The ratio of CO to H_2 may be adjusted by cofeeding carbon dioxide, CO_2, as illustrated in equation 4, the water gas shift reaction.

$$2CH_4 + O_2 \longrightarrow 2CO + 4H_2 \quad (2)$$

$$CH_4 + H_2O \longrightarrow CO + 3H_2 \quad (3)$$

$$CO_2 + H_2 \Longleftrightarrow CO + H_2O \quad (4)$$

$$2CH_4 + CO_2 + O_2 \longrightarrow 3CO + 3H_2 + H_2O \quad (5)$$

The overall process for producing a 1:1 CO to H_2 ratio by partial methane oxidation and the water gas shift reaction is represented by equation 5.

The oxo reaction proceeds most frequently in the presence of a Group 8–10 (VIII) metal catalyst in the liquid phase, most particularly with members of Group 9, the Co-Rh-Ir triad. The earliest catalyst, hydrocobalt tetracarbonyl, $HCo(CO)_4$ was an outgrowth of Fischer-Tropsch investigations carried out prior to World War II on the effect of olefins on hydrocarbon synthesis. The hydroformylation reaction, as practiced in the early days using cobalt catalysis, presented formidable requirements of high pressure, containment of the hydrogen, containment of carbon monoxide, and handling of the toxic and unstable metal carbonyls.

The search for catalyst systems which could effect the oxo reaction under milder conditions and produce higher yields of the desired aldehyde resulted in processes utilizing rhodium. Oxo capacity built since the mid-1970s, both in the United States and elsewhere, has largely employed tertiary phosphine-modified rhodium catalysts.

Propylene is the predominant oxo process olefin feedstock. Ethylene as well as a wide variety of terminal, internal, and mixed olefin streams, are also hydroformylated commercially. Branched-chain olefins include octenes, nonenes, and dodecenes from fractionation of oligomers of $C_3–C_4$ olefins as well as octenes from dimerization and codimerization of isobutylene and 1- and 2-butenes.

Linear terminal olefins are the most reactive in conventional cobalt hydroformylation.

Oxo aldehyde products range from C_3 to C_{15}, ie, detergent range, and are employed principally as intermediates to alcohols, acids, polyols, and esters formed by the appropriate reduction, oxidation, or condensation chemistry.

The classic challenges in oxo technology are simultaneously to achieve high reaction rate, high selectivity to the desired aldehyde, and to utilize a highly stable catalyst.

Catalysts

Unmodified Cobalt. Typical sources of the soluble cobalt catalyst include cobalt alkanoates, cobalt soaps, and cobalt hydroxide (see **Cobalt**). These are converted *in situ* into the active catalyst, $HCo(CO)_4$, which is in equilibrium with dicobalt octacarbonyl.

Although largely supplanted by low pressure ligand-modified rhodium-catalyzed processes, the unmodified cobalt oxo process is still employed in some instances for propylene to give a low, e.g., $\sim$3.3–3.5 : 1 isomer ratio product mix, and for low reactivity mixed and/or branched-olefin feedstocks, e.g., propylene trimers from the polygas reaction, to produce isodecanol plasticizer alcohol.

Ligand-Modified Cobalt. The ligand-modified cobalt process, commercialized in the early 1960s by Shell, may employ a trialkylphosphine-substituted cobalt carbonyl catalyst, $HCo(CO)_3P(n\text{-}C_4H_9)_3$, to give a significantly improved selectivity to straight-chain product. There has been large industrial usage of the Shell process since the 1960s, particularly for the preparation of detergent range alcohols. 2-Ethyl-1-hexanol can be produced in a single step from propylene by conducting the hydroformylation in the presence of caustic.

Ligand-Modified Rhodium. The triphenylphosphine-modified rhodium oxo process, termed the LP Oxo process, is the industry standard for the hydroformylation of ethylene and propylene as of this writing (ca 1995). It employs a triphenylphosphine (TPP) modified rhodium catalyst. The process operates at low (0.7–3 MPa (100–450 psi)) pressures and low (80–120°C) temperatures. Suitable sources of rhodium are the alkanoate, 2,4-pentanedionate, or nitrate. A low (60–80 kPa (8.7–11.6 psi)) CO partial pressure and high (10–12%) TPP concentration are critical to obtaining a high (e.g., 10:1) normal-to-branched aldehyde ratio.

The first commercial LP Oxo process flow scheme (Fig. 1) used syngas and propylene feed.

Rhodium Modified with Ionic Phosphine Ligands. In 1984, a rhodium catalyst process employing a water-soluble ligand, triphenylphosphine-*m*-trisulfonic acid trisodium salt (TPPTS) was commercialized. Product recovery is achieved by decantation from the aqueous phase containing rhodium and ligand. An isomer ratio of 20:1 is obtained with the TPPTS-modified rhodium catalyst, but the catalyst activity is significantly lower, so higher temperatures, higher rhodium concentrations, and higher propylene pressures are employed.

Other Rhodium Processes. Unmodified rhodium catalysts, e.g., $Rh_4(CO)_{12}$ have high hydroformylation activity but low selectivity to normal aldehydes.

Functional Olefin Hydroformylation. There are two commercially practiced oxo processes employing functionalized olefin feedstocks: allyl alcohol hydroformylation, and the production of 1,4-butanediol by successive hydroformylation of allyl alcohol aqueous extraction of the intermediate 2-hydroxytetrahydrofuran, and subsequent hydrogenation.

Hydroformylation Using Other Metals. Ruthenium, as a hydroformylation catalyst, has an activity significantly lower than that of rhodium and even cobalt.

Platinum catalysts that utilize both phosphine and tin(II) halide ligands give good rates and selectivities, in contrast to platinum alone, which has extremely low or nonexistent hydroformylation activity.

A further improvement in platinum catalysis is claimed from use of tin(II) halide and phosphine ligands which are rigid bidentates, e.g., 1,2-bis(diphenylphosphinomethyl)cyclobutane.

Future Trends. In addition to the commercialization of newer extraction/decantation product/catalyst separations technology, there have been advances in the development of high reactivity oxo catalysts for the conversion of low reactivity feedstocks such as internal and α-alkyl substituted α-olefins. These catalysts contain (as ligands) ortho-*t*-butyl or similarly substituted arylphosphites, which combine high reactivity, vastly improved hydrolytic stability, and resistance to degradation by product aldehyde, which were deficiencies of earlier, unsubstituted phosphites.

Uses

n-Propanol and *n*-propyl acetate account for about 70% of the U.S. propionaldehyde derivative market. These compounds are used principally in flexographic and gravure inks which require volatile solvents to prevent smearing and ink accumulation on the printing presses. Some propanol is also converted into *n*-propylamines which are important pesticide intermediates. *n*-Propanol is also employed as a precursor for glycol ethers.

The highest volume oxo chemical in the United States, *n*-butyraldehyde, is converted mainly into *n*-butanol, employed chiefly to produce butyl acrylate and methacrylate. In contrast, the principal *n*-butyraldehyde derivative in Europe and Japan is 2-ethylhexanol, the precursor to the poly(vinyl chloride) (PVC) plasticizer, DOP.

1,4-Butanediol (BDO) goes primarily into tetrahydrofuran (THF) for production of polytetramethylene ether glycol (PTMEG), used in the manufacture of polyurethane fibers.

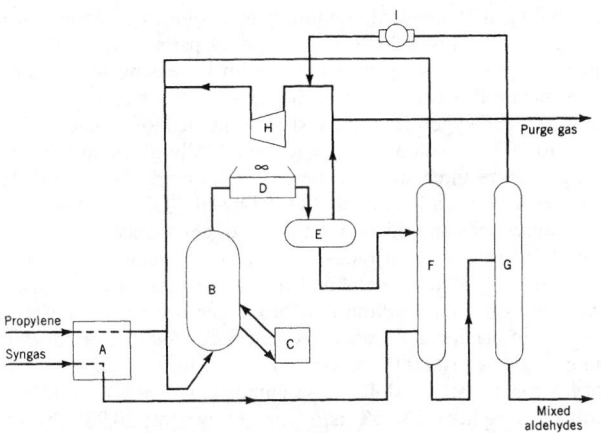

Fig. 1. LP Oxo gas recycle flow scheme: A, feedstock pretreatment; B, reactor; C, catalyst preparation and treatment systems; D, condenser; E, separator; F, stripper; G, stabilizer; H, cycle compressor; and I, stabilizer overhead gas compressor.

The principal C_5 valeraldehyde derivatives, *n*-amyl and 2-methylbutyl alcohols, are used predominantly to make zinc diamyldithiophosphate lube oil additives.

C_7–C_9 oxo-derived acids are the principal derivatives of the C_7–C_9 oxo aldehydes, and in analogy to C_5 oxo aldehyde market applications, are used chiefly to make neopolyol esters which are employed almost entirely in aeromotive applications.

Several alcohols in the C_6–C_{13} range are produced by oxo reactions and are used in both plasticizer and detergent applications. Linear C_{12}–C_{15} alcohols are employed primarily in detergent applications.

Safety, Health, and Environmental Concerns

Oxo plants employ mixtures of highly toxic, flammable gases under pressure at high temperatures and require strict adherence to established operating safety codes and emergency reporting procedures to local, state, and federal authorities. In the United States, carbon monoxide is classified as both an acute, fire, and sudden release hazard.

The carbon monoxide component of the oxo reactant gases presents the most immediate human health hazard.

ERNST BILLIG
DAVID R. BRYANT
Union Carbide Corporation

Additional Reading

Brown, C.K. and G. Wilkinson: *J. Chem. Soc. (A)*, 1392 (1970).
Chemical Economics Handbook, Oxo Chemicals Report, SRI International, Menlo Park, Calif., Jan. 1991 and preliminary 1994 draft.
Frohning, C.D. and C.W. Kohlpaintner: in *Applied Homogeneous Catalysis with Organometallic Compounds*, Vol. 1, B. Cornils and W.A. Herrmann, eds., VCH Publishers, Weinheim, Germany, 1996, pps. 29–90.
Pruett R.L.: *Adv. Organometal. Chem.* **17**, 1 (1979).

OXY. The radical —O— in organic compounds, performing in a manner similar to the oxo radical (O=) and the epoxy radical (—O—).

OXYACID. An acid that contains oxygen, such as chloric acid ($HClO_3$).

OXYAZO COMPOUNDS. Compounds of the type $RN=NC_6H_4OH$, containing both the azo group —N=N—, and a hydroxyl group —OH, both attached to carbon atoms in the same ring. These compounds are commonly produced by the action of diazo compounds upon phenols in alkaline solution. They constitute a class of dyes. See **Dyes**.

OXYCHLORINATION. See **Chlorinated Organics**.

OXYGEN. [CAS: 7782-44-7]. Chemical element, symbol O, at. no. 8, at. wt. 15.9994, periodic table group 16, mp −218.4°C, bp 182.96°C, critical temperature 118.8°C, critical pressure 49.7 atmospheres, density 1.568 g/cm³ (solid), 1.429 g/L (0°C). Solid oxygen has a cubic crystal structure. Oxygen at standard conditions is a colorless, odorless, tasteless gas. Oxygen is slightly soluble in H_2O (4.89 parts oxygen in 100 parts H_2O at $0°C$), the solubility decreasing with increasing temperature (2.6 parts oxygen in 100 parts H_2O at $30°C$; 1.7 parts oxygen in 100 parts H_2O at $100°C$). Oxygen is slightly soluble in alcohol. Molten silver dissolves up to $10\times$ its volume of oxygen, but easily gives up the gas upon cooling. There are three stable isotopes, ^{16}O through ^{18}O. Three radioactive isotopes have been identified, ^{14}O, ^{15}O, and ^{19}O, with short half-lives measured in seconds and minutes. In terms of abundance in igneous rocks in the earth's crust, oxygen ranks first, with an average composition by weight of 46.6%. In terms of abundance in seawater, oxygen also ranks first, with an estimated 4 billion tons of oxygen per cubic mile of seawater. In terms of cosmic abundance, oxygen ranks eighth. For comparisons, assigning a value of 10,000 to silicon, the figure for oxygen is 220,000 and that for hydrogen, estimated the most abundant, is 3.5×10^8. Of dry air in the earth's atmosphere, 23.15% is oxygen by weight; 20.98% by volume. In the atmosphere, the oxygen is mixed with nitrogen, argon, the rare gases, CO_2, and H_2O vapor. Oxygen first was identified by Priestly in 1774 when he was experimenting with mercuric oxide. In the same year, Scheele also identified the element. Oxygen is required for burning and combustion, although the conditions of combustion vary widely. For example, phosphorus burns in air at the low temperature of $34°C$ when ignited. The temperature if ignition for ether in air is $340°C$, for ethyl alcohol in air, $560°C$, kerosene in air, about $300°C$, and hydrogen in air, about $600°C$. The oxidation process may occur with the rapidity and violence of an explosion, or may be as slow as the rusting of iron. Nearly all known species of living things require oxygen in some form, either free or chemically bound. First ionization potential 13.614 eV; second, 34.93 eV; third, 54.87 eV. Oxidation potentials $H_2O_2 \longrightarrow O_2 + 2H^+ + 2e^-$, -0.68 V; $3H_2O \longrightarrow \frac{1}{2}O_2 + 2H_3O^+(10^{-7}M) + 2e^-$, -0.815 V; $3H_2O \longrightarrow \frac{1}{2}O_2 + 2H^+ + 2e^-$, 1.229 V; $4H_2O \longrightarrow H_2O_2 + 2H_3O^+ + 2e^-$, -1.77 V; $3H_2O \longrightarrow O(g) + 2H^+ + 2e^-$, -2.42 V; $HO_2^- + OH^- \longrightarrow O_2 + 2H_2O + 2e^-$, 0.075 V; $4OH^- \longrightarrow O_2 + 2H_2O + 4e^-$, -0.401 V; $3OH^- \longrightarrow HO_2^- + H_2O + 2e^-$, -0.87 V; $OH^- \longrightarrow OH + e^-$, -1.4 V. Other physical properties of oxygen are given under **Chemical Elements**.

Allotropic Forms

The three known allotropic forms of oxygen are (1) the ordinary oxygen in the air, with two atoms per molecule, O_2, (2) ozone, O_3, with three atoms per molecule, and (3) the rare, very unstable, nonmagnetic, pale-blue O_4. The latter breaks down readily into two molecules of O_2.

When oxygen is subjected to the silent electric discharge, activated atomic oxygen is produced. Atomic oxygen displays an afterglow upon cessation of the current, and the oxygen is notably active with hydrogen bromide, forming bromine; with H_2S, forming sulfur, SO_2, sulfur trioxide, and H_2SO_4; with CS_2, forming carbon monoxide, CO_2, and SO_2, and, strangely, reduced molybdenum trioxide to a white oxide not reducible with hydrogen. The concentration of atomic oxygen obtainable by the silent electric discharge through oxygen is estimated at 20%.

The normal electron distribution of the electrons of the oxygen atom is $1s^2 2s^2 2p_x^2 2p_y^1 2p_z^1$, with 2 unpaired electrons in the $2p$ orbitals. The covalent or partly covalent compounds of oxygen would be expected to have 90° bonding angles. But in many cases they have values significantly greater (ca. 104° for R_2O and 105° for H_2O). This suggests the promotion of a $2s$ electron to a $2p$ orbital (i.e., $2p_y$ orbital), still leaving two unpaired electrons (a $2s$ and a $2p_z$ electron), and permitting partial sp^3 hybridization (which is incomplete because sufficient energy is not available) but producing bond angles between 90° and 109° for the sp^3 tetrahedral structure, with covalent-polar bonds.

The oxygen molecule is paramagnetic with a moment in accord with two unpaired electrons. In molecular orbital terms, the configuration is written

$$O_2[KK(z\sigma)^2(y\sigma^*)^2(x\sigma)^2(w\pi^*)^4(v\pi^*)^2]$$

in which KK designates the complete $1s$ shells of the two atoms, which are nonbonding, the term $(z\sigma)^2$ denotes the bonding effect of one pair of $2s$ electrons, one from each of the O atoms, $(y\sigma^*)^2$ denotes the antibonding effect of the second pair, the $(x\sigma)^2$ term represents the s-bond formed by one pair of p-electrons, $(w\pi)^4$ represents the 2 π-bonds formed by the other two pairs of p-electrons, while the $(v\pi^*)^2$ term denotes the last pair of p-electrons, which go into the next π subshell (two orbitals) with unpaired spins, and are antibonding.

Ozone

[CAS: 10028-15-6]. O_3, obtained by electrical discharge through oxygen or high-current electrolysis of sulfuric acid, is considered on the basis of electron diffraction studies to have an $O–O–O$ bond angle of $127 \pm 3°$ and $O–O$ bond length of 1.26 ± 0.02 Å. Its structure is considered to resonate among several forms, chiefly

Ozone is an unstable blue gas, of characteristic odor, liquefiable at $-12°C$, formed when ordinary oxygen is subjected to electrostatic discharge, density 1.5 times that of oxygen gas, mp $-251.4°C$, bp $-111.5°C$. It is explosive by percussion or under variations of pressure. Ozone reacts (1) with potassium iodide, to liberate iodine, (2) with colored organic materials, e.g., litmus, indigo, to destroy the color, (3) with mercury, to form a thin skin of mercurous oxide causing the mercury to cling to the containing vessel, (4) with silver film, to form silver peroxide, Ag_2O_2, black, produced most readily at about $250°C$, (5) with tetramethyldiaminodiphenylmethane $(CH_3)_2N \cdot C_6H_4 \cdot CH_2 \cdot C_6H_4 \cdot N(CH_3)_2$, in alcohol solution with a trace of acetic acid to form violet color (hydrogen peroxide, colorless; chlorine or bromine, blue; nitrogen tetroxide, yellow). In contrast to hydrogen peroxide, ozone does not react with dichromate, permanganate, or titanic salt solutions. Ozone reacts with olefin compounds to form ozonide addition compounds. Ozonides are readily split at the olefinin-ozone position upon warming alone, or upon warming their solutions in glacial acetic acid, with the formation of aldehyde and acid compounds which can be readily identified, thus serving to locate the olefin position in oleic acid, $C_{17}H_{33} \cdot COOH$, as midway in the chain $(CH_3(CH_2)_7CH:CH(CH_2)_7COOH$. Ozone is used (1) as a bleaching agent, e.g., for fatty oils, (2) as a disinfectant for air and H_2O, (3) as an oxidizing agent. See also **Aerosols**; and **Ozone**.

The protective effects of an ozone layer in the stratosphere of the earth have been known for many years. Ozone prohibits full penetration of ultraviolet radiation from the sun to the surface of the earth. Much research has been conducted and is still underway to determine the extent to which certain chemical pollutants may be destroying the ozone layer gradually and, among other factors, causing marked warming of the earth.

Role of Oxygen in Water

The solvent properties of H_2O are due in great part to the dipole moment of its molecules (1.8 debye units) and its high dielectric constant (ca. 78). Its hydrogen atoms form hydrogen bonds with electronegative atoms such as fluorine, nitrogen, or oxygen. In fact, the H_2O molecules associate in H_2O by this mechanism. Also, the oxygen atoms of H_2O because of their residual negative charges are electrically attracted by cations, so that the H_2O molecules arrange themselves around cations, facilitating solution and ionization. In the same way, H_2O molecules surround anions by attraction of the positive ends of the dipoles. By these two processes, as well as the dissociation of water into oxonium and hydroxide ions, it forms hydrates with many compounds. Moreover, H_2O readily reacts with large numbers of compounds because of these properties. Thus the hydrolysis of covalent halides that have at least one lone pair of electrons is initiated by the donation of a proton by the H_2O, followed by splitting off of hydrogen chloride.

Oxides

Oxygen forms oxides with all the elements except some inert gases. Oxides are said to be normal when they contain no oxygen atoms that are bonded to each other, as in the peroxides. The normal oxides may be divided into three groups, basic, acidic, and neutral. The basic oxides, which react with or dissolve in H_2O to produce alkaline solutions, are formed by the alkali and alkaline earth elements (except beryllium) by the lighter Lanthanides and actinium, by silver(I), thallium(I) and lead(II). The oxides of the nonmetals and of the transition metals in their higher oxidation states are in general acidic. The oxides lying in the positions between the two groups exhibit both basic and acidic properties (amphiprotic or amphoteric) such as those of aluminum, tin(II) and iron(III), Al_2O_3, SnO, and Fe_2O_3.

The known facts about the structure of hydrogen peroxide, H_2O_2, are that the O—H distances are 0.97Å, the O—O distances 1.47Å, the HOO angles 94°, and the dihedral angle between the planes of the two O—H radicals 97°. The O—O bond is essentially a single one. In the liquid, H_2O_2 is somewhat more self-ionized than water. In water $pK_A = 11.75$, $pK_B = 17$. Its reactions may be oxidizing or reducing. Thus, it oxidizes Fe(II) to Fe(III), Ti(III) to Ti(IV) and SO_3^{2-} to SO_4^{2-}; but it reduces MnO_4^- (acid solution) to Mn^{2+}. Peroxides are known for the alkali and alkaline earth metals, as well as zinc, cadmium, mercury, thorium, uranium, plutonium, etc. However, not all compounds of formula MO_2 (where M is a metal atom) are peroxides; some are merely dioxides, as MnO_2, PbO_2, etc., others are superoxides, such as NaO_2, KO_2, RbO_2, CsO_2, CaO_4, SrO_4, and BaO_4. These last compounds contain the group O_2^-, as evident from their paramagnetism and crystal structure. Perhydroxyl, the free acid corresponding to the superoxides, is unstable ($H_2O_2 \longrightarrow HO_2 + H^+ + e^-$, $E° = 1.5$ V; $HO_2 \longrightarrow O_2 + H^+ + e^-$, $E° = +0.13$ V). It is a moderately strong acid, $pK_A = 2.2$.

The peroxyacids containing —O—OH groups, are formed with all the transition elements in groups 4, 5, 6 of the periodic table, with main group elements 4 and 5 as well as elements of atomic numbers from boron to sulfur, inclusive. Representative peroxyacids are peroxymonosulfuric acid,

$$H:\overset{\overset{\displaystyle ..}{O}:}{\underset{\underset{\displaystyle ..}{O}:}{O:S:O:O:H}}$$

and peroxychromic acid,

$$H:\overset{\overset{\displaystyle ..}{O}:}{\underset{\underset{\displaystyle ..}{O}:}{O:O:Cr:O:O:H}}$$

The only peroxydiacids are formed by sulfur, phosphorus, carbon and boron, of which the most important is peroxy disulfuric acid

$$H:\overset{\overset{\displaystyle ..}{O}:\quad\quad\overset{\displaystyle ..}{O}:}{\underset{\underset{\displaystyle ..}{O}:\quad\quad\underset{\displaystyle ..}{O}:}{O:S:O:O:S:O:H}}$$

although peroxy bridge compounds are also formed by certain transition element complexes, e.g., $[Co(NH_3)_5OOCo(NH_3)_5]^{4+}$ and $[Co(NH_3)_5OOCo(NH_3)_5]^{5+}$.

Industrial Oxygen

As with hydrogen, the electrolysis of water offers one approach to the production of pure oxygen. However, the economics are as unfavorable for oxygen production in this manner as for hydrogen. See also **Hydrogen**. For industrial oxygen production, air is the raw material. Using air, processes are of two major types: (1) liquid-oxygen processes wherein the oxygen is fractionally distilled from liquid air, and (2) gaseous-oxygen processes. See also **Cryogenics**.

Because of the relatively high energy costs of compressing and refrigerating involved in oxygen production, many processes have been developed and tested over the years, a high percentage of these later abandoned. An idea of the alternatives which face the process designer can be gathered from scanning the methods available specifically in the area of producing refrigeration for these processes: (1) Joule-Thomson effect only; (2) Joule-Thomson effect plus auxiliary refrigeration with an ordinary liquid-vapor cycle at moderate- or high-temperature levels, i.e., relative to liquid-air temperature; (3) Joule-Thomson effect plus approximately reversible expansion of the air or products in an expander; (4) refrigeration essentially due only to approximately reversible expansions of auxiliary fluid or fluids operating in liquid-vapor cycles, i.e., the cascade process; and (5) processes using an auxiliary nitrogen-liquefaction cycle.

Designers also face the choice of capacity of an oxygen plant. Costs per unit weight of oxygen made are lowered as the capacity of the plant goes up. For example, a plant with a capacity of 2000 tons (1800 metric tons) per day will produce oxygen at approximately 50% of the cost per unit weight as a plant with a 200-ton (180-metric ton) capacity per day.

The demand for industrial oxygen has created the need for several new plants during the past 20 years. Capacities for most recent plants range from about 1100 tons (990 metric tons) per day to 2500 tons (2250 metric tons) per day.

An oxygen pipeline system was established in western Europe in the late-1970s that is 592 miles (956 kilometers) long. The Eastern Network of this system serves 30 consumers in France, Luxembourg, and West Germany; the Northern Network serves some 40 additional users in France, Belgium, and the Netherlands.

Uses

In addition to the requirements by the chemical industry for oxygen as a reactant, either directly from the air or in purer, more concentrated form as from a separation plant, significant quantities of purified oxygen are used for welding and cutting metals. Oxygen of a purity of 99.5% is required for oxyacetylene and oxyhydrogen torches. When combined in proper proportions, acetylene and oxygen yield a flame with a temperature of about 3,480°C. Oxyhydrogen flames are somewhat lower in temperature, but they are particularly useful for welding light-gage aluminum and magnesium allows and for underwater cutting. In welding applications, a reduction in purity of oxygen used from 99.5% to 99.0% will cut welding efficiency by over 10%. During the past several years, basic oxygen steelmaking has increased requirements for pure oxygen. In this process, nearly pure oxygen is introduced by means of a lance into molten iron and scrap. The oxygen combines with carbon and other unwanted elements and refines raw steel in much less time than the older open-hearth furnaces. The basic oxygen process exceeded the open-hearth process in terms of output to the United States in 1970 for the first time. On the total scale of consumption, relatively limited amounts of oxygen go into medical and life-support applications, as required for emergency situations in aircraft at high altitudes.

Role of Oxygen in Corrosion

Oxygen and oxidizing agents exert both a positive and negative influence on corrosion of metals. On the one hand, an oxidizing agent may form a protective oxide film on the surface of certain metals, aluminum being an excellent example, which essentially arrests corrosion by many external agents. On the other hand, the presence of oxidants may increase the rate of corrosion by supporting cathode reactions. As an example, Monel metal fully resists attack by oxygen-free 5% H_2SO_4 at room temperature. The corrosion rate rises, however, in almost direct proportion to oxygen content. A 20% oxygen content will cause a corrosion rate of about 150 mdd (milligrams of metal corroded per square decimeter per day). A concentration of 40% will increase the rate to about 250 mdd; a concentration of 80% to about 450 mdd. The oxygen need not be present in all of the acid contained in the metal vessel, but simply present in that concentration at the interface of metal, acid, and surrounding atmosphere. The effect of oxidizing salts on corrosion can be dramatic. Several factors, in addition to oxygen, affect corrosion, including the presence of other metals (electromotive-force displacements of one metal by another), temperature, acidity, and velocity. These factors are discussed further under **Corrosion**.

Oxygen Toxicity of Plants

As early as 1801, Huber and Senebier observed that grains develop more satisfactorily in an atmosphere containing a mixture of 3 parts nitrogen and 1 part oxygen than in an atmosphere containing 3 parts of oxygen and 1 part nitrogen. Considerably later, in 1878, Bert noted that the earlier observations also apply to the development of many plant species and are not peculiar to grains. Bert further suggested that excessive oxygen may slow down various reactions involving fermentation. It was not until much later, in the mid-1940s, that scientists (Dickens, Haugaard, and Stadie) further confirmed that enzymes are inactivated by oxygen excesses. They particularly stressed this fact in connection with enzymes that contain a sulfhydryl group in the active site. Machaelis (1946), Barron (1946), and Gilbert (1963) later pointed out that molecular oxygen alone acts in a rather sluggish manner in this regard and that, therefore, a special process or phenomenon must be involved. Molecular oxygen can be reduced only by accepting one electron at a time.

A number of scientists in the late 1960s through the mid-1970s pointed out that many sources in biological systems produce oxygen *free radicals*. For example, some oxidative enzymes which contain flavin as a prosthetic group proceed by a radical mechanism. When illuminated, chloroplasts produce superoxide ions and singlet oxygen. Because of its singlet configuration, the latter is not hindered in its interactions with biological materials. As pointed out by Griffiths and Hawkins, singlet oxygen can be formed from the ground state when energy, usually in the form of light, is supplied n the presence of a photosensitizer. The compounds that are photosensitized include many dyes and pigments, such as chlorophyll, flavins, and hematoporphyrins. The interaction between the sensitizer and oxygen results in the transfer of electrons, with the formation of superoxide ion. McCord and Fridovich (1969) discovered the enzyme superoxide dismutase. Their later findings show that aerobic organisms contain it, giving further credence to the proposal that all oxygen-metabolizing organisms from superoxide free radicals as a result of a univalent reduction of oxygen. As pointed out by Kon (1978), those free radicals that are toxic to the organism, by themselves or through interaction with other active forms of oxygen, are dismutated by the action of this enzyme.

There is a close relationship between oxygen toxicity and radiation on enzymes, DNA, and fats. Gerschman et al. (1954) showed that the same substances that afford protection against oxygen poisoning also increase resistance to radiation. Their results were further strengthened by experiments that demonstrated that additive nature of the two effects. Work on the effects that free radicals have on some of the polysaccharides used in food processing was commenced by Kon and Schwimmer and reported in 1977.

Environmental Aspects of Oxygen

Gaseous oxides, notably those of carbon, nitrogen, and sulfur which result from the combustion of fossil fuels and numerous industrial processes, comprise a large portion of the air pollution problem. These compounds are discussed under the specific elements and, in particular, are described under **Pollution (Air)**. In connection with the pollution of water in streams, lakes, ponds, rivers, etc., the content of dissolved oxygen in water is of prime concern. Dissolved oxygen must be available to support fish and other desirable living species in natural waters, and sufficient additional oxygen must be available in the water to effect biological degradation of both natural and manufactured materials which reach the water. The overuse of streams for disposal purposes in many instances has almost fully depleted the dissolved oxygen available for life support and hence has given rise to the term "dead" lakes or streams. Two terms are widely used: (1) BOD (biological oxygen demand) which is the requirement for dissolved oxygen in water to degrade or decompose organic matter within a measured time period at a given temperature, and (2) COD (chemical oxygen demand) which is the requirement for dissolved oxygen in water to combine with chemicals, essentially of an inorganic nature, which are introduced into a stream as the result of disposal operations. These aspects of oxygen are discussed under **Water Pollution**.

Earth's Oxygen Supply

The manner in which the earth's present oxygen system and reserves were formed has been the subject of much postulation for many years. Many of the details remain unclear and unconfirmed. In a theory proposed by Berkner-Marshall (1964, 1965), as the earth's atmosphere evolved, there was a slow buildup of the concentration of oxygen—proceeding from a trace to the present content of 23.15% (weight). This theory also proposes that the oxygen content of the atmosphere fluctuated from time to time in a major and relatively rapid manner. There is speculation that these major alterations may have accounted for the extinctions of animal and life forms that took place at the ends of the Paleozoic and Mesozoic eras. For example, there was a great reduction in life in the latter part of the Permian period (Paleozoic era) when many kinds of strange reptiles and trilobites disappeared and seem to have left no descendents. Plant life declined greatly too during the late Paleozoic. From thousands of species in the Pennsylvanian period, there remained only a few hundred during the late Permian. Numerous explanations, particularly of a climatic nature, have been offered for these periods of reduction in life.

As pointed out by Van Valen (1971), photosynthesis does not produce a net change in oxidation. Except in bacterial photosynthesis, oxygen production is accompanied by a stoichiometrically equal quantity of reduced carbon. Thus, almost all of the oxygen is eventually used to oxidize reduced carbon. Predominantly, this oxidation occurs as the result of respiration in animals and plants. Further oxidation occurs as the result of forest fires. As observed by Borchert (1951), the only net gain in

oxygen equals the amount of reduced carbon buried, as in the form of peat, black mud, and similar sediments. It has been estimated that most individual molecules of carbon remain reduced only for relatively short periods (months or years) because animals and plants have geologically very short lives. Plants respire and so oxidize some reduced carbon almost immediately. Other net sources of oxygen include nitrogen fixation and the photolysis of water in the upper atmosphere. Some investigators have considered these sources quantitatively unimportant, although Brinkmann (1969) suggests that this process would produce, over the earth's history (4.5×10^9 years), about seven times the present mass of oxygen in the atmosphere.

Numerous ways have been proposed to explain a net loss of molecular oxygen. Oxidation of volcanic gases, ferrous iron, sulfur, sulfide, and manganese, and the accretion of hydrogen from the solar wind are among these. Such processes are sometimes referred to as *oxygen sinks*. Estimates by Holland (1962) indicate that the net gain and net loss over geologic time are essentially in balance.

Van Valen has posed the question, "What can happen if photosynthesis is suddenly and drastically reduced?" Under such conditions, at a new steady state, production of oxygen and its consumption in the oxidation of carbon would be equal. But, before the new steady state occurs, would animals and decomposers use up much of the previously stored carbon in plants, thus creating a new loss of oxygen? Several investigators have observed that even if all the carbon in all organisms now alive were oxidized, this would decrease the atmospheric concentration of oxygen by less than 0.1% of its present value. And, further, still less than 1% of the present oxygen concentration would be used if all the reduced carbon available in soils and the like were reduced.

Much more detailed explanation of the stability of atmospheric oxygen is contained in the excellent review by Van Valen (1971).

As pointed out by Broecker (1970), the earth's oxygen supply is frequently included in lists of concerns over alterations in the environment, particularly as brought about by anthropogenic activities. Several investigators have made a number of observations which tend to invalidate any claims that oxygen is in danger of serious depletion. Broecker observes that each square meter of the earth's surface is covered by 60,000 moles of oxygen gas. Further, plants living in the ocean and on land produce about 8 moles of oxygen per square meter of surface each year. It is also observed that animals and bacteria destroy nearly all of the products of this photosynthetic activity—thus they use an amount of oxygen nearly equal to that generated by plants. Using the rate at which organic carbon enters the sediments of the ocean as a measure of the amount of photosynthetic product preserved each year, Broecker estimates this to be about 3×10^{-3} mole of carbon per square meter per year. This corresponds to approximately 1 part in 15 million of the oxygen present in the atmosphere. It is estimated, however, that this small amount of oxygen is probably being destroyed by a number of processes, including oxidation of reduced carbon, iron, and sulfur (weathering mechanisms). Broecker points out that the oxygen content of the atmosphere is thus well buffered, particularly in terms of relatively short time spans (100 to 1000 years).

Over a period of time, people have recovered about 10^{16} moles of fossil carbon and the fuels containing this carbon have been oxidized as sources of energy. Byproduct carbon dioxide from this combustion represent about 18% of the carbon dioxide content of the atmosphere. Two moles of atmospheric oxygen are used to liberate each mole of carbon dioxide from fossil fuel sources. Broecker points out that this process uses up only 7 out of every 10,000 available oxygen molecules. It is estimated that if these fuels are burned at an accelerating rate (5% per year), by the end of this century, only about 0.2% of available oxygen (20 molecules in every 10,000) will be used. It is estimated that if all known fossil fuels were ultimately burned, only 3% of available oxygen would be consumed. In terms of urban oxygen needs, particularly for automotive combustion needs, it is estimated that carbon monoxide levels in the atmosphere (in terms of physiological damage) would reach intolerable levels before the oxygen content of the atmosphere would have decreased by 2%.

The case of anthropogenic alterations of photosynthetic rates and its possible effects on oxygen supply has been covered previously by the observation that stoppage of all photosynthetic activity would require less than 1% of the present oxygen concentration.

Sverdrup et al. (1942) estimated that the oxygen content of deep sea water averages about 2.5 cubic centimeters at standard temperature and pressure per liter (0.1 mole per cubic meter). Thus, there are about 250 moles of oxygen gas in the deep sea for each square meter of earth surface. The oxygen content of the deep-sea waters is renewed about every 1000 years. The magnitude of this oxygen reservoir is tremendous. Broeker emphasizes this by observing that if the entire terrestrial photosynthetic product were dumped each year into the deep sea, the supply of deep-sea oxygen would last 50 years. But, if the waste products of 1 billion people were limited to 100 kilograms of dry organic waste per year, this would consume 0.01 mole of oxygen per square meter of earth surface and the deep-sea oxygen supply would last some 25,000 years.

In the summary of this report, Van Valen (1971) states, "There are three processes weakly concentration-dependent that keep changes in concentration of atmospheric pressure from being a random walk—inhibition of net photosynthesis by oxygen, the passage of hydrogen through the oxidizing part of the atmosphere before it escapes from the earth, and burial of reduced carbon in anaerobic water. A stronger regulator seems desirable but remains to be found. The cause of the initial rise in oxygen concentration presents a serious and unresolved quantitative problem."

And, in summary of his report, Broeker (1970) states, in part, "It can be stated with some confidence that the molecular oxygen supply in the atmosphere and in the broad expanse of open ocean are not threatened by human activities in the foreseeable future. Molecular oxygen is one resource that is virtually unlimited."

Additional Reading

Baukal, C.: *Oxygen Enhanced Combustion,* CRC Press, LLC, Boca Raton, FL, 1998.

Berkner, L.V. and L.C. Marshall: in *The Origin and Evolution of Atmospheres and Oceans,* (P.J. Brancazio and A.G.W. Cameron, editors), pages 102–126, Wiley, New York, NY, 1964.

Berkner, L.V. and L.C. Marshall: *J. Atmos Sci.,* **22**, 225 (1965).

Borchert, H.: *Geochim. Cosmochim. Acta,* **2**, 62 (1951).

Brinkmann, R.T.: *J. Geophys. Res.,* **74**, 5355 (1969).

Brocker, W.S.: "Man's Oxygen Reserves," *Science,* **168**, 1537–1538 (1970).

Dickens, R.: "The Toxic Effects of Oxygen on Brain Metabolism and on Tissue Enzymes," *Biochem. J.,* **40**, 145, 170 (1946).

Gerschman, R., et al.: "Oxygen Poisoning and X-irradiation: A Mechanism in Common," *Science,* **119**, 623 (1954).

Gilbert, D.L.: "The Role of Pro-Oxidants and Anti-Oxidants in Oxygen Toxicity," *Radiation Res. Suppl.,* **3**, 44 (1963).

Griffiths, J. and C. Hawkins: "Mechanistic Aspects of the Photochemistry of Dyes and Their Immediates," *J. Soc. Dyers Colorists,* **89**, 173 (1973).

Holland, H.D.: in *Petrologic Studies: A Volume in Honor of A.F. Buddington,* (A.E.J. Engel, H.L. James, and B.F. Leonard, editors), pages 447–477, Geological Society of America, Washington, DC, 1962.

Kon, S. and S. Schwimmer: "Depolymerization of Polysaccharides by Active Oxygen Species Derived from Xanthine Oxidase Systems," *Food Biochem.,* **1**, 141 (1977).

Kon, S.: "Effects of Oxygen Fee Radicals on Plant Polysaccharides," *Food Technol.,* **32**, 5, 84–94 (1978).

Kruk, I.: *Environmental Toxicology and Chemistry of Oxygen Species: The Handbook Of Environmental Chemistry,* Springer-Verlag, New York, Inc., New York, NY, 1997.

Lide, D.R.: *CRC Handbook of Chemistry and Physics 84th Edition,* CRC Press, LLC., Boca Raton, FL, 2003.

McCord, J.M. and I. Fridovich: "Superoxide Dismutase: An Enzymatic Function for Erythrocuprein," *J. Biol. Chem.,* **244**, 6046 (1969).

Michaelis, L.: *Fundamentals of Oxidation and Reduction,* in Currents in Biochemical Research (D.E. Green, editor), pages 207, Wiley, New York, NY, 1946.

Sawyer, D.: *Oxygen Chemistry,* Oxford University Press, New York, NY, 1999.

Sundquist, E.T. and W.S. Broecker, Eds.: *The Carbon Cycle and Atmospheric CO2,* American Geophysical Union, Washington, DC, 1985.

Sverdrup, H.U., M.W. Johnson, and R.H. Fleming: *The Oceans, Their Physics, Chemistry and General Biology,* Prentice-Hall, Englewood Cliffs, New Jersey, 1942.

Van Valen, L.: "The History and Stability of Atmospheric Oxygen," *Science,* **171**, 439–443 (1971).

OXYGEN BALANCE. Oxygen content relative to the total oxygen required for oxidation of all carbon, hydrogen, and other easily oxidizable elements to carbon dioxide, water, etc.

OXYGEN CELL. An electrolytic cell whose emf is due to a difference in oxygen concentration at one electrode compared with that at another electrode of the same material.

OXYGEN CONSUMED (OC; COD; DOC). A measure of the quantity of oxidizable components present in water. Since the carbon and hydrogen, but not the nitrogen, in organic matter are oxidized by chemical oxidants, the oxygen consumed is a measure only of the chemically oxidizable components and is dependent on the oxidant, structure of the organic compound, and manipulative procedure. Since this value does not differentiate stable from unstable organic matter, it does not necessarily correlate with the biochemical oxygen demand value. It is also known as chemical oxygen demand (COD) and dichromate oxygen consumed (DOC).

See also **Biochemical Oxygen Demand (Bod)**; and **Disolved Oxygen (DO)**.

OXYGEN DEBT. A term used to refer to the buildup of a need for oxygen through anaerobic respiration of muscle cells in a higher vertebrate animal during violent exercise. When the energy demands are too great to be satisfied by the aerobic respiration, the cells turn to anaerobic respiration. Lactic acid is an end product of such respiration; this acid tends to accumulate in the muscles and some of it diffuses out into the blood and accumulates in the liver. When the activity ceases, deep breathing continues and the extra oxygen is used to reconvert the lactic acid back to pyruvic acid and to carry the pyruvic acid on through the tricarbocyclic acid cycle. It may also be reconverted back to glucose and glycogen. We say that the muscles have built up an oxygen debt during the very active exercise and this is repaid in the continued deep breathing during rest following the exercise.

OXYGEN GROUP (The). The elements of group 16 of the periodic classification sometimes are referred to as the Oxygen Group. In order of increasing atomic number, they are oxygen, sulfur, selenium, tellurium, and polonium. The elements of this group are characterized by the presence of six electrons in an outer shell. The similarities of chemical behavior among the elements of this group are less striking than hold for some of the other groups, e.g., the close parallels of the alkali metals or alkaline earths. With exception of oxygen, all elements of the group have a valence of 4+, in addition to other valences. All of the elements with the exception of polonium also have a valence of 2−. Unlike the alkali metals or alkaline earths, for example, the elements of the oxygen group are not so similar chemically that they comprise a separate group in classical qualitative chemical analysis separations. Tellurium and selenium do appear together among the rarer metals of the second group in terms of qualitative chemical analysis.

OXYGEN SINK. A reservoir consisting of a chemical element or compound that combines readily with oxygen and thus removes it from the atmosphere. During the early part of Precambrian time, sulfur iron, and other elements and compounds served as important oxygen sinks, preventing oxygen from accumulating in the atmosphere.

OXYL PROCESS. A method for directly producing higher alcohols by catalytically reducing carbon monoxide with hydrogen.

OXYTOCIC HORMONE. See **Hormones**.

OXYTOCIN. A polypeptide hormone which is secreted by the posterior lobe of the pituitary gland of mammals and other vertebrates. Oxytocin exerts a stimulating effect upon the muscles of the breast (milk-ejection) and those of the uterus of mammals. It is sometimes used medically to stimulate labor in cases of difficult childbirth and to time the onset of labor.

OZOCERITE. Sometimes spelled ozokerite, this is a natural, brown to jet black mineral (paraffin) wax comprised mainly of hydrocarbons. The melting point is variable. The material is soluble in chloroform. When heated with sulfuric acid (20–30%) from 120–200°C, ozocerite yields ceresine. Sometimes called earth wax, fossil wax, mineral wax, and native paraffin.

OZONE. [CAS: 10028-15-6]. Ozone, O_3, is an allotropic form of oxygen first recognized as a unique substance in 1840. Its pungent odor is detectable at ~0.01 ppm. It is thermally unstable and explosive in the gas, liquid, and solid phases. In addition to being an excellent disinfectant, ozone is a powerful oxidant not only thermodynamically, but also kinetically, and has many useful synthetic applications in research and industry. Its strong oxidizing and disinfecting properties and its innocuous by-product, oxygen, make it ideal for the treatment of water. Indeed, the most important application of ozone is in the treatment of drinking water, which began in Europe. The treatment of swimming pool water was also developed in Europe. Another important ozone application is for odor control in industrial processes and municipal wastewater-treatment plants. Ozone also is used on a large scale for the treatment of municipal secondary effluents. Industrial high quality water supplies are also treated with ozone. In addition, ozone has applications in the treatment of cooling-tower water and in pulp bleaching. Advanced oxidation processes employing ozone in combination with uv, H_2O_2, and/or solid catalysts such as TiO_2 greatly improve the reactivity of ozone toward organic contaminants.

Ozone, which occurs in the stratosphere (15–50 km) in concentrations of 1–10 ppm, is formed by the action of solar radiation on molecular oxygen. It absorbs biologically damaging ultraviolet radiation (200–300 nm), prevents the radiation from reaching the surface of the earth, and contributes to thermal equilibrium on earth.

Properties

At ordinary temperatures, pure ozone is a pale blue gas ($d = 2.1415$ g/L at 0°C and 101.3 kPa (1 atm)) that can be condensed to an indigo blue liquid, which freezes to a deep blue-violet solid. The solubility of gaseous ozone at atmospheric pressure and 0°C is 1.1 g/L H_2O. Gaseous ozone can be adsorbed by porous solid substrates such as silica gel and is often used in this form in organic synthesis.

Ozone is endothermic, thus it can burn or detonate by itself and represents the simplest combustible and explosive system. The concentration threshold for spark-initiated explosion of liquid ozone in oxygen at −183°C is 18.6 mol % O_3; the concentration limit for shock wave-initiated detonation of gaseous ozone-oxygen at 25°C is 9.2 mol % O_3. Gaseous ozone exhibits three principal absorptions in the infrared at 710, 1043, and 2105 cm^{-1}.

Ozone is a triangular molecule; its bond angle (116.8°) was established by microwave spectroscopy. The bond length of the ozone molecule (0.1278 nm) is intermediate to that of a single and double oxygen bond, corresponding to a bond order of 1.7. Ozone is diamagnetic with C_{2v} symmetry and has a low dipole moment of 1.77×10^{-30} C·m (0.53 D). Based on Pauling resonance concepts, the structure of ozone is a hybrid, principally of form (1), with a small contribution from (2).

Thermal Decomposition

Gas Phase. The decomposition of gaseous ozone is sensitive not only to homogeneous catalysis by light, trace organic matter, nitrogen oxides, mercury vapor, and peroxides, but also to heterogeneous catalysis by metals and metal oxides.

The calculated half-life of 1 mol % (1.5 wt %) of pure gaseous ozone diluted with oxygen at 25, 100, and 250°C is 19.3 yr, 5.2 h, and 0.1 s, respectively. Although pure ozone-oxygen mixtures are stable at ordinary temperatures in the absence of catalysts and light, ozone produced on an industrial scale by silent discharge is less stable due to the presence of impurities; however, ozone produced from oxygen is more stable than that from air. At 20°C, 1 mol % ozone produced from air is ~30% decomposed in 12 h.

Aqueous Phase. In pure water, the decomposition of ozone at 20°C involves a complex radical chain mechanism, initiated by OH$^-$ and propagated by O_2^- radical ions and HO radicals.

Hydrogen peroxide greatly accelerates the decomposition of ozone in alkaline solutions because of formation of HO_2^-, which reacts rapidly with ozone to form the radical ion O_2^-.

Photochemical Decomposition

Gas Phase. Gaseous ozone is decomposed to oxygen atoms and molecules by absorbing radiation in the visible and uv spectrum: $O_3 + h\nu \longrightarrow O_2 + O$.

Aqueous Phase. In contrast to photolysis of ozone in moist air, photolysis in the aqueous phase can produce hydrogen peroxide initially because the hydroxyl radicals do not escape the solvent cage in which they are formed. Hydrogen peroxide is photolyzed slowly to hydroxyl radicals, which decompose ozone.

Chemistry of Ozone

The inorganic chemistry of ozone is extensive, encompassing virtually every element except most noble metals, fluorine, and the inert gases.

Ozone reacts rapidly with various free radicals and radical ions such as O, O_2^-, H, HO, N, NO, Cl, and Br. Some of these radicals (HO, NO, Cl, and Br) can initiate the catalytic decomposition of ozone.

The strong electrophilicity of ozone is manifested in its reaction with a wide variety of organic and organometallic functional groups, e.g., olefins, acetylenes, aromatics (carbocyclic and heterocyclic), activated C—H bonds (acetals, alcohols, aldehydes, ethers, and glycosides), unactivated C—H bonds (alkanes, cycloalkanes, and alkyl aromatics), deactivated C—H bonds (carboxylic acids and ketones), C=N and N=N bonds, Si—H and Si—C bonds, organometallic bonds (e.g., Grignard reagents), and nucleophiles (e.g., ammonia, amines, amino acids, arsines, disulfides, hydroxylamines, nitriles, phosphites, selenides, sulfides, and thioethers). Ozone also acts as a nucleophile, e.g., in its reaction with carbocations.

Atmospheric Ozone

Stratosphere. Ozone is formed rapidly in the stratosphere (15–50 km) by the action of short-wave ultraviolet solar radiation (<240 nm) on molecular oxygen, $O_2 + h\nu \longrightarrow 2O$. At wavelengths above 175 nm, only ground-state (^3P) atoms are formed; whereas at wavelengths below 175 nm, one ground-state and one excited (^1D) atom are formed. Ground-state atoms also can be formed by the pre-dissociation of electronically excited O_2. The oxygen atoms can react with molecular oxygen to yield ozone: $O + O_2 + M \longrightarrow O_3 + M$. Ozone can be destroyed photochemically: $O_3 + h\nu \longrightarrow O + O_2$; at 226 nm, however, this reaction also can produce vibrationally excited O_2 capable of forming ozone. In addition, ozone can be destroyed by reaction with oxygen atoms, as well as with excited O_2 molecules and other free radicals. Since the early 1960s, it has been recognized that radicals such as NO, OH, Cl, and Br affect the abundance and distribution of ozone in the stratosphere. Earlier studies simulating stratospheric chemistry concluded that ozone formation is significantly less than its destruction, hence the ozone deficit problem. However, studies indicate that this may not be the case.

Most ozone is formed near the equator, where solar radiation is greatest, and transported toward the poles by normal circulation patterns in the stratosphere. Consequently, the concentration is minimum at the equator and maximum for most of the year at the north pole and about 60°S latitude. The equilibrium ozone concentration also varies with altitude; the maximum occurs at about 25 km at the equator and 15–20 km at or near the poles. It also varies seasonally, daily, as well as interannually. Absorption of solar radiation (200–300 nm) by ozone and heat liberated in ozone formation and destruction together create a warm layer in the upper atmosphere at 40–50 km, which helps to maintain thermal equilibrium on earth.

Troposphere. Ozone and nitric oxide are transported from the stratosphere to the troposphere, the region of the atmosphere below 15 km. Though only about 10% of the atmospheric ozone is present in the troposphere, this small fraction plays a fundamental role in atmospheric chemistry because it leads to the formation of hydroxyl radicals. Hydroxyl radicals initiate the oxidation and prevent the buildup of many organic and inorganic pollutants in the atmosphere. The radical-dominated chemistry of the troposphere is complex, involving intertwining cycles of gas-phase, condensed-phase, and multiple-phase reactions. In unpolluted atmosphere, HO radicals react with naturally occurring CO and methane, resulting in a net increase in ozone concentration.

Although the naturally occurring concentration of ozone at the earth's surface is very low, this distribution has been altered by the emission of anthropogenic pollutants which increase the production of ozone via the above mechanism. Photochemical smog, an aerosol irritant gas mixture, occurs in urban industrialized areas where heavy motor vehicle traffic is common, especially those areas where temperature inversions are common. It forms at low altitudes by photolytic reactions involving nonmethane hydrocarbons, NO, and CO, resulting in low but potentially harmful concentrations of ozone and other irritating substances, such as aldehydes, ketones, acids, H_2O_2, organic peroxides, and peroxyacetyl nitrate.

Although the background concentration of ozone in surface air is ~0.01–0.03 ppm, during severe smog days in the Los Angeles area, for example, it has often reached 0.5 ppm, and a maximum of 1 ppm in 1957. In the early morning hours, NO is removed slowly by the oxygen atom chain, which is initiated by the photolysis of NO_2 and subsequently by the photolysis of ozone. Later in the day when the light intensity is higher, the hydroxyl chain causes the NO conversion to accelerate.

Ozone can react rapidly with NO to produce NO_2, which re-enters the ozone formation cycle: $O_3 + NO \longrightarrow O_2 + NO_2$. This is the main ozone-depleting reaction in the absence of sunlight. Ozone also reacts with NO_2 (to form NO_3, which in turn reacts with NO_2 to form N_2O_5), C_2H_4, as well as HO and HO_2 radicals. Nitric acid formed by the reaction $HO + NO_2 \longrightarrow HNO_3$ is removed from the atmosphere by rain-out.

Ozone Generation

Ozone can be generated by a variety of methods, the most common of which involves the dissociation of molecular oxygen electrically (silent discharge) or photochemically (uv). The short-lived oxygen atoms (lifetime $\sim 10^{-5}$ s) react rapidly with oxygen molecules to form ozone. The widely employed technique of electric discharge produces much higher concentrations than the ultraviolet technique and is more practical and efficient for production of large quantities. A less common method of ozone formation is electrochemical generation.

Silent Electric Discharge. Commercial production and utilization of ozone by silent electric discharge consists of five basic unit operations: gas preparation, electrical power supply, ozone generation, contacting (i.e., ozone dissolution in water), and destruction of ozone in contactor off-gases.

Ultraviolet Light. The mechanism of the practical photochemical production of ozone is similar to that in the stratosphere; that is, oxygen atoms, formed by the photodissociation of oxygen by short-wavelength uv radiation (≤ 240 nm), react with oxygen molecules to form ozone. In practice, ozone concentrations obtained by commercial uv devices are low. This is because the low intensity, low pressure mercury lamps employed produce not only the 185-nm radiation responsible for ozone formation, but also the 254-nm radiation that destroys ozone, resulting in a quantum yield of ~ 0.5 compared to the theoretical yield of 2.0. The low concentrations of ozone available from uv generators preclude their use for water treatment because the transfer efficiencies of ozone from air into water is low and large volumes of carrier gas must be handled.

Uses

Ozone is used in the treatment of drinking water and in industries where high purity water is required (e.g., breweries, pharmaceuticals, and electronics). Ozone is also used in industrial wastewater pollution control, wastewater disinfection, and odor control; in the treatment of process water, such as cooling tower water; in the treatment of swimming pools and spas; in pulp bleaching; and in organic synthesis, as a selective oxidant.

Among other uses, ozone therapy, employing O_3-O_2, is increasingly being employed and studied in dentistry, veterinary and sports medicine, and proctology. Ozone is used as an aquatic oxidant and disinfectant in zoos, large aquariums, as well as fish and shrimp hatcheries. Ozone also is used for food preservation, in cold storage rooms, brewery cellars, hotel and hospital air ducts, and air conditioning systems. Ozone has also been used in textile bleaching and in the bleaching of esters, oils, fats, waxes (qv), starch, flour, ivory, etc. Oxidation of Ag^+ by ozone is employed commercially to produce high purity AgO. The use of ozone as a chemical agent decontaminant has been patented.

Health and Safety

As a constituent of the atmosphere, ozone forms a protective screen by absorbing radiation of wavelengths between 200 and 300 nm, which can damage DNA and be harmful to life. Consequently, a decrease in the stratospheric ozone concentration results in an increase in the uv radiation reaching the earth's surfaces, thus adversely affecting the climate as well as plant and animal life. For example, the incidence of skin cancer is related to the amount of exposure to uv radiation. Ozone can be toxic to plants, animals, and fish.

The toxicity of ozone to humans is largely related to its powerful oxidizing properties. The odor threshold of ozone varies among individuals but most people can detect 0.01 ppm in air, which is well below the limit for general comfort. The symptoms experienced on exposure to 0.1–1 ppm ozone are headache, throat dryness, irritation of the respiratory passages, and burning of the eyes caused by the formation of aldehydes and peroxyacyl nitrates. Exposure to 1–100 ppm ozone can cause asthma-like symptoms such as tiredness and lack of appetite. Short-term exposure to higher concentrations can cause throat irritations, hemorrhaging, and pulmonary edema.

Ozonation of drinking water produces various by-products such as aldehydes, ketones, carboxylic acids, organic peroxides, epoxides, nitrosamines, *N*-oxy compounds, quinones, hydroxylated aromatic compounds, brominated organics, and bromate ion. Although some of these compounds are potentially toxic or carcinogenic, most bioassay-screening studies have shown that ozonated water induces substantially less mutagenicity than chlorinated water.

J. A. WOJTOWICZ
Consultant

Additional Reading

Bailey, P.S.: *Ozonation in Organic Chemistry*, Vols. 1 and 2, Academic Press, Inc., New York, 1978–1982.

Beltran, F.J.: *Ozone Reaction Kinetics for Water and Wastewater Systems*, CRC Press LLC., Boca Raton, FL, 2003.

Dessler, A.: *Chemistry and Physics of Stratospheric Ozone*, Elsevier Science, New York, NY, 2000.

Hoigné J. and H. Bader: *Water Res.* **17**, 185 (1983).

Kogelschatz, U. B. Eliasson, and M. Hirth: *Ozone Sci. Eng.* **10**, 367 (1988).

Lide, D.R.: *CRC Handbook of Chemistry and Physics, 84th Edition*, CRC Press LLC., Boca Raton, FL, 2003.

Parker, L. and W.A. Morrissey: *Stratospheric Ozone Depletion*, Nova Science Publishers, Inc., Huntington, NY, 2003.

Seinfeld, J.H. and S.N. Pandis: *Atmospheric Chemistry and Physics: From Air Pollution to Climate Change*, John Wiley & Sons, Inc., New York, NY, 1997.

Wennberg, P.O. and co-workers: *Science*, **266**, 398 (1994).

OZONOLYSIS. (1) Oxidation of an organic material, i.e., tall oil, oleic acid, safflower oil, cyclic olefins, carbon treatment, peracetic acid production by means of ozone. (2) The use of ozone as a tool in analytical chemistry to locate double bonds in organic compounds and a similar use in synthetic organic chemistry for preparing new compounds. Under proper conditions, ozone attaches itself at the double bond of an unsaturated compound to form an ozonide. Since many ozonides are explosive, it is customary to decompose them in solution and deal with the final product.

See also **Oxygen**.

P

PAAR TURBIDIMETER. A visual-extinction device for measurement of solution turbidity. The length of the column of liquid suspension is adjusted until the light filament can no longer be seen.

See also **Nephelometry.**

PAINT AND FINISH REMOVERS. The term finish denotes the final process of manufacturing. Finishing operations include such processes as clear coating (varnishes and lacquers), painting, plating, anodizing, phosphatizing, galvanizing, and blueing, all of which take place at the terminal point of manufacturing. Finishing is defined as the process of coating or treating a surface for the purpose of protecting and/or decorating the product. The useful life of most usable objects is greater than the finish. This results in a periodic need to remove and replace the finish.

The physical properties of finish removers vary considerably due to the diverse uses and requirements of the removers. Finish removers can be grouped by the principal ingredient of the formula, method of application, method of removal, chemical base, viscosity, or hazardous classification. Except for method of application, a paint remover formulation usually has one aspect of each group, by which it can be used for one or more applications.

Finish removers are applied by brushing, spraying, troweling, flowing, or soaking. Removal is by water rinse, wipe and let dry, or solvent rinse. Removers may be neutral, basic, or acidic. The viscosity can vary from water thin, to a thick spray-on, to a paste trowel-on remover. The hazard classification, such as flammable or corrosive, is assigned by the U.S. Department of Transportation (DOT) for the hazardous materials contained in the remover.

Organic Finish Removers

Methylene Chloride Finish Removers. Methylene chloride formulas are the most common organic chemical removers. The low molar volume of methylene chloride allows it to rapidly penetrate the finish by entering the microvoids of the finish. When the solvent reaches the substrate, the remover releases the adhesive bond between the finish and the substrate and causes the finish to swell. The result is a blistering effect and an efficient rapid lifting action. Larger molecule solvents generally cannot cause this lifting action and must dissolve the finish. When methylene chloride is used in amounts of 78% or more, even with flammable cosolvents, the mixture is nonflammable. A typical methylene chloride base remover includes cosolvents, activators, evaporation retarders, corrosion inhibitors, thickeners, and wetting agents.

Typical cosolvents include methanol, ethanol, isopropyl alcohol, or toluene. The selection of cosolvents depends on the requirement of the formula and their interaction with other ingredients. Methanol is a common cosolvent in methylene chloride formulas since it has good solvency and is needed to swell cellulose-type thickening agents. A typical methylene chloride formula used to strip wood is as follows: methylene chloride 81.1%; 1toluene 2.1%; paraffin wax (ASTM 50–53°C mp), 1.6%; methycellulose,1.2%; methanol, 7.8%; and mineral spirits 6.2%.

The rate of stripping or the stripability on catalyzed urethane and epoxy resin finishes can be increased by adding formic acid, acetic acid, and phenol. Sodium hydroxide, potassium hydroxide, and trisodium phosphate may be added to the formula to increase the stripability on enamel and latex paints. Other activators include oleic acid, trichloroacetic acid, ammonia, triethanolamine, and monoethanolamine. Methylene chloride-type removers are unique in their ability to accept cosolvents and activators that allow the solution to be neutral, alkaline, or acidic. This ability greatly expands the number of coatings that can be removed with methylene chloride removers.

Paraffin wax vapor barriers are used in water rinse removers that can disperse the wax without coating the substrate. In soak tank applications, water is sometimes floated on top of an all-solvent, neutral pH, nonwater rinse remover to prevent evaporation. Flotation devices that cover the exposed surface area may be used with other formulas.

Health and Safety. Remover formulas that are nonflammable may be used in any area that provides adequate ventilation. Most manufacturers recommend a use environment of 50–100 parts per million (ppm) time weighted average (TWA). The environment can be monitored with passive detection badges or by active air sampling and charcoal absorption tube analysis. The vapor of methylene chloride produces hydrogen chloride and phosgene gas when burned. Methylene chloride-type removers should not be used in the presence of an open flame or other heat sources such as kerosene heaters

Persons exposed to methylene chloride removers should wear protective clothing and eye protection.

Environmental Impact. Methylene chloride is nonphotochemically reactive and is not listed as an ozone (qv) depleter. Methylene chloride removers can easily be recovered from paint chips and other residue sedimentation, thus allowing recovery of remover and its continued use. This greatly increases the useful life of the remover and, when mixed with fresh remover, eliminates the need for disposing of the used remover. This process requires no special recovery equipment. The high volatility of methylene chloride allows the waste residue from the stripping process to be easily dried. The resulting waste is normally considered hazardous because of the amounts of heavy metals from old finishes.

Petroleum and Oxygenate Finish Removers. Many older finishes can be removed with single solvents or blends of petroleum solvents and oxygenates. Varnish can be removed with mineral spirits, shellac can be stripped with alcohols, and lacquers can be removed with blends of acetates and alcohols (lacquer thinners). The removal mechanism is one of dissolving the coating, then washing the surface or wiping away the finish. This method is often used to reamalgamate or liquefy old finishes on antique items of furniture.

In petroleum and oxygenate finish removers, the major ingredient is normally acetone, methyl ethyl ketone, or toluene. Cosolvents include methanol, n-butanol, sec-butyl alcohol, or xylene. Sodium hydroxide or amines are used to activate the remover. Paraffin wax is used as an evaporation retarder though its effectiveness is limited because it is highly soluble in the petroleum solvents. Cellulose thickeners are sometimes added to liquid formulas to assist in pulling the paraffin wax from the liquid to form a vapor barrier or to make a thick formula. Corrosion inhibitors are added to stabilize the formula for packaging (qv).

Wetting agents are used to make a water rinse remover. Water rinse removers are normally used for removing paint, where the surfactants help remove paint and remover from the substrate. Solvent rinse removers or wipe and dry formulas may be used for stripping clear finishes. A typical petroleum and oxygenate formula is as follows: toluene 21%; acetone19%; alkyl acetate 31%; methyl ethyl ketone19%, and butyl alcohol10%.

This is a liquid scrape-off remover for brush or soak applications. Clean up is with a solvent that is compatible with finish to be used, or wipe and dry.

Health and Safety. Petroleum and oxygenate formulas are either flammable or combustible. Flammables must be used in facilities that meet requirements for hazardous locations.

Adequate ventilation that meets the exposure level for the major ingredient must be attained. Extreme caution must be taken to prevent the possibility of fire when using flammable removers.

Environmental Impact. Most petroleum and oxygenate removers are photochemically reactive and classed as volatile organic compounds (VOCs). Disposal of this type of remover is difficult because the dissolved finish cannot be separated from the spent remover and the whole mixture must be disposed as a liquid hazardous waste. Distillation to recover the solvents is dangerous because the nitrocellulose from lacquer finish may cause autoignition in the still. Several states restrict the use of these products.

Other Organic Removers. Concerns over the reported toxicity and carcinogenicity of methylene chloride have stimulated research for alternative solvents in remover formulas. N-Methylpyrrolidinone and dibasic esters (dimethyl glutarate or dimethyl adipate) have been used in removers. They remove single-component finishes but work much more slowly than methylene chloride, petroleum, and oxygenate group removers. They have little success on epoxy and catalyzed finishes.

Health and Safety. Both N-methylpyrrolidinone and dibasic esters have very low vapor pressure which limits worker exposure to vapors. Manufacturers recommend that the same safety precautions be taken as with other organic solvents. Hazardous location requirements must be considered if the formula is flammable. Ventilation that reduces vapors to manufacturer's recommended exposure levels should be used. Protective clothing must be worn during use.

Environmental Impact. The volume of waste remover from these products is remarkably increased when compared to methylene chloride, petroleum, and oxygenate removers, since both N-methylpyrrolidinone and dibasic esters have low vapor pressures. Recovery of the remover after use is difficult because the finish is resolubilized by the remover.

Inorganic Finish Removers

Liquid Alkaline Removers. This group consists of alkaline materials that are dissolved in water then heated to an appropriate temperature to remove finishes. In a typical application, a hot water bath large enough to submerge an item is used. Various alkaline materials may be used to provide the desired alkalinity. Of these, sodium compounds are preferred, such as sodium hydroxide, sodium carbonate, sodium silicates, mono-, di-, and trisodium phosphates, tetrasodium pyrophosphate, and sodium tripolyphosphate. Compounds of other metals, such as potassium or lithium, may be used.

This aqueous alkaline remover is used for stripping the finish from wood or ferrous metals at a mix ratio of 30–600 g/L (0.25–5 lbs/gal).

Paste-Type Alkaline Removers. Sodium hydroxide, potassium hydroxide, or other caustic compounds are blended to make these types of removers. Polymer-type thickeners are added to increase the viscosity that allows the remover to be applied with a brush, trowel, or spray. Some of these products use a paper or fabric covering to allow the remover finish mixture to be peeled away. The most common application for this group of removers is the removal of architectural finishes from the interior and exterior of buildings. The long dwell time allows for many layers of finish to be removed with one thick application of remover.

Sodium hydroxide and salt can be heated to a fused state in baths to allow the removal of finishes from ferrous metals. The most common use of this method is the removal of heavy concentrations of paint on conveyer parts and hangers used in production spray systems.

Health and Safety. Protective clothing that is compatible with the remover formula must be worn. Caustic soda baths should be ventilated to remove vapors from the work area. Most caustic removers are corrosive and cause severe burns with minimal contact to the skin.

The liquid from spent caustic soda baths must be disposed of or treated as a hazardous waste. The finish residue may contain heavy metals as well as caustic thus requiring treatment as a hazardous waste.

Manufacturing and Processing

Finish removers are manufactured in open or closed kettles. Closed kettles are preferred because they prevent solvent loss and exposure to personnel. To reduce air emissions from the solvents, condensers are employed on vent stacks.

Standard 0.25 or 0.50 lb (227 g) tin coated cans are used for packaging liquid with neutral and mildly alkaline base formulas; polypropylene is used for acid–base removers. Steel and polypropylene drums are used for industrial removers. Viscous removers are packaged in removable top containers. Dry caustic removers are packaged in bag-lined boxes or fiber drums.

The DOT has established standards for the packaging and labeling of hazardous materials offered for shipment by public transportation. The Consumer Product Safety Commission (CPSC) has set standards for retail labeling and packaging. OSHA and the U.S. EPA have labeling requirements.

See also **Paints and Coatings**.

David L. White
Kwick Kleen Industrial Solvents, Inc.
Jay A. Bardole
Vincennes University

Additional Reading

Staff: *Industrial Users of Paint and Finish Removers,* Paint Remover Manufacturer's Association, 1992.

Staff: *Solvents Used in Paint Removers,* Paint Remover Manufacturer's Association, Sept. 1991.

Staff: *Chemical Protective Clothing for Furniture Stripping,* Department of Health and Human Services, Washington, D.C., Mar. 11, 1991.

Staff: *Methylene Chloride Consumption By Paint and Coating Removal Groups,* Paint Remover Manufacturer's Association, 1992.

PAINTS AND COATINGS.

Traditionally, paints and other coatings have been considered in terms of protecting and decorating buildings, houses, furniture, automobiles, toys, boats, machines, and the like. These products represent a large industry estimated at about $15 billion per year worldwide. Nevertheless, these kinds of coatings account for only 10–15% of the total coatings industry.

One researcher[1] has classified coatings in two technically defined categories.

Type I—Products manufactured and sold with several coating layers. Examples are the aforementioned applications, but also include color photographic film, graphic arts films for printing, coated papers for printing, coated containers for food packaging, magnetic storage media for computers and audio/visual equipment, optical disks for digital data storage, adhesive tapes, wallpaper, and metallized films.

Type II—Coatings that become an integral part or a key intermediary for a device or piece of equipment. Sometimes, the term *core technology* is used for such applications. There is a wide variety of coating products in this classification. They would include photoresists for circuit board manufacture, thick- and thin-film coatings on integrated circuits, adhesives for bonding metals, phosphor coatings on electronic display screens, stain repellents on fibers, dyes on fibers, ceramic glazes, encapsulated time-release drugs, and thin-film photovoltaic cells.

The foregoing categories of coatings immediately dramatize the versatility and specialization of a huge industry that has developed essentially during the past three decades. With the continuing interest in materials composites, continued rapid growth is expected. Coating technology is a complex undertaking fraught with numerous design and manufacturing problems. These are particularly difficult in scaling up laboratory procedures to high production rates on the factory floor. Consequently, a number of successful processes remain proprietary. The technology has demanded much of chemists and chemical and mechanical engineers, with particular emphasis on an understanding of fluid dynamics.

Although the two foregoing classifications may serve a very useful purpose at the scientific level, for the purposes of this article, the first part is devoted to traditional surface coating products that serve protective and decorative purposes. The second part, in less depth, addresses coatings that are used for special purposes as represented, (e.g., by uses in electronic devices).

PROTECTIVE AND DECORATIVE COATINGS[2]

Paint, coating, and finish are terms used to describe a wide variety of materials designed to adhere to a substrate and act as a thin, plastic-like layer. Paints are available for decorative, protective, and other purposes. They can be decorative by covering defects (being opaque), by changing color, or by providing a desired gloss or sheen. Protective uses include shielding metals from corrosion, protecting plastic from degradation caused

[1] E.D. Cohen, E.I. Du Pont De Nemours & Co., Parlin, New Jersey.

[2] Some of this information was provided by The Sherwin-Williams Company, Cleveland, Ohio.

by ultraviolet light, acting as a moisture barrier and providing mar and scratch resistance for wood or plastic surfaces. A paint can also be used for its special spectral properties (e.g., light absorbing for heating swimming pools; radar absorbing for military vehicles; etc.), or unusual physical properties (e.g., strippable coatings for the interior of paint spray booths, insulating coatings for electrical parts, etc.).

Paints are generally liquids before they are applied to a surface. When applied, they should completely replace the substrate/air interface with a substrate/paint interface (called wetting). The forces of attraction created by this wetting process are responsible for the paint's adhesion. The paint then dries and/or cures to form a hard film. Drying is the physical action of solvent leaving the film, while curing refers to a chemical reaction which connects polymer chains of the paint together.

Composition

Thousands of raw materials are used to manufacture coatings, but they can generally be classified into four categories: (1) binders or resins, (2) pigments, (3) solvents, and (4) additives. Binders and resins are generally organic compounds, usually polymeric or oligomeric in nature, which provide a continuous matrix in the final film and have a major influence on the toughness, flexibility, gloss, chemical resistance and cure/dry properties of the coating.

Pigments. These are finely divided powders (particles between 0.1 and 50 micrometers in diameter) which are dispersed throughout the binder. In addition to reinforcing the final film, much as they do in composite plastics, they influence a coating's resistance to abrasion and corrosion, and they also are the major factor in the gloss, color, and opacity of a coating.

Solvents or Thinners. These substances have a major effect on a paint's application viscosity and also affect a coating's cure dry properties, and often its toxicity. Most solvents evaporate and leave the film during the drying process, although certain "reactive dilutents" are designed to react with the resin and become part of the binder system. Water is a popular solvent because of its low cost, low toxicity, and nonpolluting nature. Almost half of the coatings currently produced use water as a major solvent. Disadvantages of using water include the effect of humidity on the drying characteristics of the paint and the difficulty of making a water-resistant film from materials suspended in water. Major organic solvents include mineral spirits, ketones, acetates, alcohols, and xylene.

Additives. Among the more important classes of additives used in coatings are: (1) *surfactants*, which are used to suspend pigment and binder particles; (2) *thickeners* to obtain proper rheology (especially in latex paints); (3) *plasticizers*, which lower the glass transition temperature of the binder and increase the flexibility of the coating; (4) *antifoam agents* to prevent bubbles in aqueous paints; (5) *antiskin agents*, which prevent the formation of a dry layer on top of the paint while it is still in the can; (6) *preservatives*, such as biocides and mildewcides to protect the binder from microscopic organisms both before and after application; (7) *ultraviolet light absorbers* to protect the binder and/or substrate from degradation due to sunlight; and (8) a variety of surface *conditioners* and *lubricants*, which help the film adhere to the substrate or protect the film by giving it a lubricated surface. Additives will often interact and coating formulators must be careful to watch for synergistic and antagonistic effects.

Ratio of Components. The ratio of the four aforementioned components (binders, pigments, solvents, and additives) greatly influences the properties of a coating. The volume fraction of the solid film (which is pigment) is referred to as the *pigment volume concentration*. The concentration where there is barely sufficient binder to fill in the voids between the pigment particles is referred to as the *critical pigment volume concentration* (or *CPVC*). When the composition of a coating is changed from below its CPVC to above it, the properties of the coating begin to dramatically degrade (except for opacity, which increases). The performance changes are primarily due to the pockets of air which form in the final film because of the shortage of binder. With the exception of flat architectural (house) paints, fillers, and certain primers, almost all coatings have a pigment volume concentration less than the CPVC. The CPVC is a function of a pigment combination's oil adsorption and particle size distribution.

The ratio of solvent to nonvolatiles is important since the volatile organic compound (VOC) composition of coatings is increasingly a target of government regulations. Usually the VOC of a coating is expressed in units of *mass per volume* and is calculated by multiplying the density of

the coating by that fraction of the coating that is volatile. For coatings using water (or certain chloroalkanes) as a solvent, the effect of these "exempt" solvents is subtracted out before the calculation is made. Current VOC limits in the United States range from 250 grams per liter for some California architectural alkyds to 450 grams per liter for some furniture finishes.

Binder Classifications

Paints are often classified by the type of binder they include. The most common classifications (with percent of total coatings used in 1985) are: Latexes (31%); waterborne (10%); non-aqueous dispersions (2%); solvent-borne (55%); and one hundred percent solids coatings (2%). A small volume of paint is made with silane binders.

Latexes. These are dispersions of high-molecular-weight polymer particles in an aqueous medium. Since the polymer is in a suspended form, the viscosity of the mixture is almost exclusively a function of the viscosity of the continuous phase (i.e., water) and is not affected by the molecular weight of the polymer. This permits the use of higher-molecular-weight material than can be used with a solution-type approach. The film forms (i.e., the paint dries) when the water evaporates and the spherical latex particles are forced together, overcoming the steric and ionic forces which had been stabilizing them. Once the particles touch, the surface tension of the latex causes the individual particles to coalesce (i.e., the polymers in adjoining particles entangle), aided by the capillary forces created by the evaporating water. Latex particles are generally 0.1 to 0.5 micrometer in diameter, although particle sizes as much as an order of magnitude on either side of these values is used for specialized purposes.

The two most commonly used latex systems are acrylic systems (40% of usage), which perform very well, but are relatively expensive, and the vinyl-acrylic copolymer systems (57% of usage and growing), which do not perform as well, but are less expensive.

Latexes are usually considered separate from other waterborne binders because the method employed for synthesizing and suspending them is very different. For latexes, emulsion polymerization is used and no organic solvent is required to obtain or stabilize the emulsion. In contrast, most waterborne binders are synthesized in organic solvent solutions and then "let down" with water. Often, some quantity of organic solvent must remain or the resulting aqueous mixture will not be stable. Most waterborne resins need to be cured. Reactions commonly used include the oxidative polymerization of unsaturated aliphatic chains, the reaction with aminoplast resins, and the reaction of epoxies to form ether or ester bonds. Water-borne resin compositions include acrylics, polyesters (including alkyds), urethanes, phenolics, and epoxies, among others.

Non-Aqueous Dispersions (NADs). These are the solvent-borne analogues of latexes. They were used as automotive finishes in the 1970s, but their use is now declining. The use of NADs as an auxiliary binder, however, is increasing as it has been found that the addition of a small amount of NAD can improve a coating's drying characteristics and rheology.

Solvent-Borne Coatings. These cover a wide variety of resins including alkyds. Alkyds are polyesters made from soya, linseed, or other oils and are a major factor in architectural, automotive, and industrial maintenance usages, among other applications. Acrylics are known for their exterior durability and are the major binder in automotive coatings. Epoxies are used mostly in automotive, industrial maintenance, metal container, and coil coatings. Polyurethanes are isocyanate-based binders and are used where excellent properties are required, such as in the magnetic media, magnet wire, industrial maintenance, and deck coatings fields. Polyesters, other than alkyds, are used in coil, metal furniture, metal container, appliance, and automotive coatings. Amino crosslinkers primarily are modified melamines and are important in metal container, automotive, coil, and metal furniture applications, among others.

In addition to drying, most solvent-borne coatings undergo some type of cure. Chemical reactions used for the cure of solvent-borne coatings include the oxidative crosslinking of unsaturated carbon bonds (alkyds and polyesters), the reaction of melamine derivatives to form ureas (polyesters, alkyds, acrylics, and amino resins), ether and ester formation by epoxies (epoxies, polyesters and acrylics), and urethan formation by isocyanates. Curing is especially important in low VOC coatings because the viscosity of a resin is a function of the molecular weight of the material. The demand for lower VOC (i.e., higher solids) coatings has resulted in a move toward lower-molecular weight and more reactive solvent-borne coatings. About one-quarter of the solvent-borne coatings currently used are considered "high solids" (i.e., they have a VOC of 350 grams/liter or less).

The thrust for higher solids also has been a factor in the growth of *100% solids* coating technologies. These technologies include polymerization initiated by ultraviolet radiation or an electron beam; the use of powdered coatings which coalesce when sufficient heat is applied; and vapor cure technology, where a reactive resin is crosslinked by exposure to a reactive vapor. While the capital investment required by these technologies is high, their use is expected to continue to grow because of their efficient use of material and the superior properties that can be obtained. Another means of using a *solventless system is hot melt coatings*. These are applied at high temperatures without solvent and dry by cooling them to room temperature.

Silanes. Silanes and silane derivatives dominate the small market for inorganic binders. These materials are used both in combination with organic binders and by themselves. As co-binders, they increase the chemical and moisture resistance of a film. When used by themselves, they form brittle, very chemical-resistant films. Silane coatings are usually more expensive than their organic counterparts and must be kept dry before application.

Pigment Classifications

Pigments used in the paint industry are commonly classified by function: (1) *Hiding* (or *prime*) pigments scatter light and are used to obtain opacity; (2) *extender* (or "inert") pigments are used to reinforce the binder, increase the pigment volume concentration, lower gloss, and lower the cost of a paint; (3) *colored* pigments which are used to tint a paint can be either inorganic or organic; (4) *metallic* pigments are used for corrosion prevention and appearance reasons; and (5) *protective* and other *functional pigments* can be used to add special features to a coating.

Hiding Pigment. The refractive index of hiding pigments must be sufficiently different from the refractive index of the binder (usually about 1.5) if light is to be effectively scattered. Two crystal structures of titanium dioxide (commonly referred to as titanium) are the most widely used hiding pigments in the paint industry with the rutile version being more popular than anatase because the refractive index of rutile (2.76) is higher than that of anatase (2.55). Rutile is also more thermally stable and photostable, although the chalking property of the anatase pigment has been used to advantage in making "self-cleaning" paints. Like most inorganic pigments, titanium dioxide is a naturally occurring compound that must be mined, crushed and processed before it is a suitable raw material for paint. The optimum diameter for light scattering is about 0.2 micrometer.

Other naturally occurring hiding pigments include zinc oxide (refractive index = 2.01) and zinc sulfide (refractive index = 2.37). At one time, lead carbonate (refractive index = 2.0) was a leading hiding pigment. In addition to hiding, zinc oxide is a fungistat (i.e., it inhibits the growth of fungi).

Pockets of air (refractive index = 1.0) encapsulated in plastic are also used as light-scattering pigments. The size of these synthetic pigments range from 0.6 to 20 micrometers, depending upon the number of 0.5-micrometer bubbles per particle.

Extender Pigment. The major classes of extender pigments are as follows:

Calcium carbonate (also called *whiting*). *Calcium carbonate* is inexpensive, has a low binder demand, and is not colored, but is acid sensitive.

Clay (or *kaolin*) covers a wide variety of materials which are inert and inexpensive, but it is more yellow than calcium carbonate.

Talcs are very inexpensive and easily suspended pigments.

Silicas (mostly silicon dioxides) are low cost, inert, and very hard.

Other extender pigments include barium sulfate, feldspar, diatomite, and mica.

Color Pigments. Chrome yellow is the leading inorganic color pigment. It is used primarily in traffic-marking paint for roads and highways. The yellow, red, and brown versions of iron oxide also are important and are used in a variety of industrial and architectural coatings.

There are hundreds of organic color pigments used in the paint industry, the vast majority of which are synthetic. Since many of these are vulnerable to ultraviolet light or chemical degradation, great care must be taken in choosing pigments that are suitable for a given paint and its intended usage.

The most commonly used black pigments are carbon blacks. In addition to being efficient light absorbers, some varieties of these small-particle-size materials impart electrical conductivity and thixotropy to paints. The leading inorganic black pigment is black iron oxide. This material is used primarily because it is easier to disperse than the carbon blacks. See also **Carbon Black**.

Metallic Pigments. The leading metallic pigment is zinc dust, which is used mostly in zinc-rich primers, where it acts as a *passivating agent*. Aluminum flake is used for the silvery metallic appearance that it imparts.

Anticorrosive Pigments. Several pigments are used primarily because of their anticorrosive properties. Chromates (zinc, strontium, and lead, if permitted) are the most effective anticorrosive pigments, but the chronic toxicity danger associated with them is a matter of serious concern. Barium metaborate, red lead, and borosilicates have been the principal nonchromate materials used. Other pigments used for special purposes include iron oxide for magnetic media and copper oxide in marine coatings to prevent barnacle and algae growth.

Tributyltin acetate CAS: 56-36-0, $(C_4H_9)_3SnOOCCH_3$, gained widespread usage in marine paints. The loss of ship performance and efficiency resulting from the growth of barnacles, seaweeds, tubeworms, and other organisms on boat bottoms has been known since the time of the Phoenicians, when copper strips were fastened to hulls to prevent fouling. Various navies and ocean shippers worldwide have found tributyltin acetate (TBT) effective, particularly in tropical waters. Unfortunately, the ingredient in paint has been found to be toxic. One authority in ocean chemistry has observed that TBT is the most toxic compound man has introduced into the marine environment. In recent years, numerous studies and evaluations of TBT-based paints have been made. Some countries have regulations against its use. Details on the adverse growth of oysters, for example, have been reported. Some of these observations are covered in the Champ reference listed.

Coating Manufacturing Process

There are usually two steps in the paint-making process: (1) dispersing the pigment (called *grinding*) and (2) mixing in the raw materials not used in grinding (called *letting down*). Except for NADs, solvent-borne coatings are made by grinding the pigment into a binder/solvent solution. The polymer serves a dual purpose in the step: (1) It thickens the mixture, increasing the dispersing efficiency of energy put into the system, and (2) it adsorbs onto the surface of the pigment particles, stabilizing them in suspension. Once stabilized, the suspension is let down by stirring in the remaining raw materials. For most waterborne paints (including all latexes), the grinding step consists of dispersing the pigment in water. The binder is not included since it is not stable enough to withstand the grinding process. Instead, surfactants are used to help wet and stabilize the pigment. Paints prepared in this manner add the binder in the letdown.

Mills. Several types of machines (called *mills*) are used for grinding pigment. *Media mills* have a chamber where the pigment, binder and a solid media are all ground together. The grinding action of the media particles on one another provides the shear needed to breakdown the pigment agglomerates. Some media mills have chambers that hold an entire batch at one time, while others have smaller chambers through which a batch is passed in a continuous flow. Batch mills (those falling into the former category) include those using pebbles, ceramic beads, or steel shot as media. Continuous mills (where the batch flows through the chamber) usually use sand or ceramic beads as the media.

Roller mills have large, closely placed rollers capable of grinding very thick pigment suspensions. Adjoining rollers are turned in opposite directions and at high speeds. The point where adjoining roller surface separate is subjected to sufficient shear to pull the pigment agglomerate apart.

High-speed dispersers (HSDs) are a third general type of mill. HSDs use blades attached to rotating shafts to disperse the pigment, much as an egg beater is used to disperse flour. HSDs are currently the most widely used grinding equipment. HSDs or similar stirring devices are generally used for the letdown regardless of the type of mill used in the grinding step. See also **Ball, Pebble, And Rod Mills**.

Application Methods

Coatings can be applied in a variety of ways depending upon the nature of the substrate and the viscosity of the coating. Brush, roller and pressure pads are popular methods of application for architectural and industrial maintenance coatings. Advantages of these methods include lack of capital investment and the ability to apply coatings on site. Disadvantages include their labor intensity and their limitation to use *ambient cure* coatings. Air, and especially airless, spray equipment can apply coatings much faster than a brush or roller, but this method requires more equipment and is not as adaptable. Spraying is used for architectural, industrial maintenance, wood furniture, automotive refinish and other coatings.

Electrostatic Spraying. In this method, a paint with a negative charge is applied to a substrate with a positive charge. This is the method of choice for automotive, metal furniture, appliance, machinery, and metal container coatings. The equipment for electrostatic spray costs more than regular spray equipment, but the transfer efficiency can be much higher. A drawback is that electrostatic spraying can be used only if the object to be painted can hold an electrical charge and does not have deep crevices.

Electrodeposition. Another way of using electricity to paint is electrodeposition (ED). In this method, the object to be coated is dipped into a vat of charged aqueous coating. An opposite electrical charge is then applied to the object and the paint is attracted to the surface of the charged object. Having been painted, the object is removed from the vat, rinsed, and baked. Electrodeposition requires a very large capital investment, does not allow for color changes, works only for conductive (metal) objects, and is only suitable for coatings that are baked. Ambient cure paints do not have the long-term stability needed for ED. Nevertheless it is the greatly preferred application method for automotive primers and many other metal products because of its high transfer efficiency—desirable for both environmental and economic reasons.

Dip Tanks. This method, which does not use electricity, is especially suitable for small objects. The main disadvantage to dipping is the difficulty of getting an even coat without *drips*.

Roller Coaters and Sheet Coaters. These are often used in coil coating where very large, flat surface areas need to be painted quickly. In these methods, the objects to be painted are passed between a doctor blade and applicator rolls. These methods are suitable only for coatings that are to be baked.

Surface Preparation of the Substrate. This is extremely important for all methods of paint and coatings application. The failure of a paint system is often due *not* to the paint itself, but because of a failure in surface preparation. For example, an anticorrosive paint applied to a rusty surface will not be effective if the rust falls off taking the new paint with it. For wood and plastic surfaces, old paint or a weathered surface layer may have to be removed. For older metal objects, the removal of corrosion is often required. Sandblasting is one method to remove both the old paint and any corrosion. For new metal objects, a phosphate or chromate layer is often chemically bonded to the metal to provide a surface to which a coating can easily adhere.

Paints for Specific Functions

Paints are often separated into the types of jobs they perform. *Primers* are meant to be applied to bare substrates and then covered with a topcoat. As such, they must have good adhesion and good recoatability, but color and light stability are usually unimportant. When used over metal, primers are usually expected to provide corrosion protection. *Sealers* are similar to primers except they are used over porous substrates. Sealers eliminate the leaching of material from the substrate and also prevent paint components from migrating to the substrate. *Surfacers* are highly pigmented paints that are applied in a thick layer to mask surface irregularities and to allow good adhesion by a topcoat. *Fillers* are a type of surfacer that is used to fill holes. *Stains* are low-solids coatings applied to wood to accentuate its grain. Most interior stains require a topcoat.

Some finishes do not require a primer. Varnishes are clear, tough, and usually have a glossy finish. Exterior varnishes give wood some protection from sunlight. *Shellacs* are a type of varnish which offers the advantages of a quick dry and easy sanding, but which is sensitive to water and has a limited shelf life.

Topcoats are usually applied over an undercoat (i.e., primer, stain, etc.). A topcoat must protect its undercoat from environmental damage and should provide the appearance characteristics desired for the particular application. Topcoats are available in a variety of colors, textures, and glosses. One specialized topcoat is *lacquer*, which is a solution of resin in organic solvents. Lacquers dry, but do not cure. Because of their high VOCs, the use of lacquers is declining. *Enamel* is a term used to describe a glossy, opaque topcoat.

A unique two-coat, topcoat system is used in the automobile industry. A basecoat is applied to a primed surface to provide opacity, color, and a metallic appearance, while the clearcoat provides gloss and a mirrorlike finish (referred to as *distinctness of image*).

Various regulatory agencies in the United States and other countries have set limits on the use of low-volatile organic coatings. These are widely used in the chemical, petrochemical, and metallurgical industries to resist corrosion. Presently, high-temperature paints and coatings are largely silicone in nature. These are considered the best-performing coatings for smokestacks, boilers, mufflers, furnaces, incinerators, combustion chambers, and jet engines. High solids polyorganosiloxane polymers (silicone resins) are used to protect steel piping and other equipment where high temperatures accelerate deterioration of ferrous substrates. They also are used on storage tanks where appearance is an issue, as in the case of oil refinery tanks.

In addition to their high-temperature properties, high solids polyorganosiloxane polymers are resistant to numerous chemicals. For example, electric power generation frequently creates sulfuric acid as a byproduct. The process often involves extremely high temperatures as well. Silicone-based coatings currently appear to be the only materials capable of withstanding such conditions.

Special-purpose Coatings

These coatings include the use of materials (generic coatings) for purposes other than their contribution to protecting and decorating a structure or product. There is a multitude of such uses, and only a few can be described here because of space limitations.

Electronic Microstructure Fabrication. A microstructure may be defined as a pattern formed on or imbedded in the surface of some substrate material. Microstructure implies that the transverse dimensions of the patterns are in the microscopic and submicroscopic range, factors that determine the scale of electronic circuit integration. With the progress of electronic component miniaturization, such ranges have advanced from small-scale integration (SSI) to medium-scale integration (MSI) to large-scale integration (LSI) and to very-large-scale integration (VLSI). As of the early 1990s, such integration has progressed to millions and billions of transistors (e.g., on one integrated circuit). See also **Microelectronics.**

Patterns on microstructures may be formed in layers or insulators deposited on a surface or may consist of chemical or physical modification of shallow regions of the substrate. Traditionally, the most important use of microstructure fabrication has been the creation of large numbers of transistors, diodes, resistors, and capacitors fabricated with the interconnections that enable them to perform useful electronic functions on a single piece or "chip," usually silicon. Recent trends in miniaturization include two additional classifications (i.e., *application-specific* integrated circuits (ASICs) and *very-high-speed* integrated circuits (VHSICs).

Since the beginning of the concept of integrated circuits several years ago, the "yardstick" of dimension has been the micron (micrometer). A micrometer equals 1/1,000,000 of a meter, or 1/25,000 of an inch, or 1000 nanometers, or 10,000 angstroms. The wavelength of light, by comparison, extends from approximately 4000 to 7000 angstroms. Approximately 150 half-micron-wide lines would fit within the width of a human hair. The "resists" used in the fabrication of microminiaturized components fall into the "coatings" category in the parlance of modern coatings technology.

A *simplified* example will illustrate the process of microstructure fabrication. With reference to Fig. 1, an *n*-type region has been created by diffusion of a donor impurity into a surface of *p*-type silicon, forming a *p* – *n* junction diode. There is a metal contact to the *n*-region, and the contact line is insulated from the *p*-type surface by a layer of silicon dioxide. The diameter of the diode is on the order of 10 micrometers.

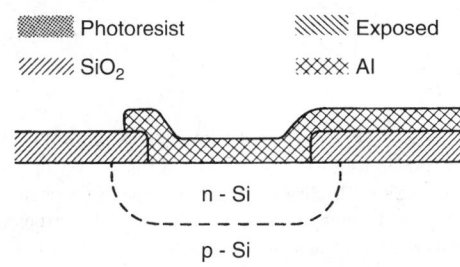

Fig. 1. A diode fabricated on the surface of a wafer of silicon. An *n*-type region has been created by diffusing a donor impurity through an opening in a layer of SiO₂ on the silicon. Electrical contact is made to the *n* region by a deposited aluminum conductor. The SiO₂ insulates the silicon from the aluminum

The fabrication begins with the application of a layer of photoresist to the oxidized surface of a silicon wafer. The photoresist is then exposed to light in the region where the diode is to be formed. Photoresist is a polymeric mixture that is deposited as a thin layer, perhaps 1 μm thick, upon an SiO_2 film on a silicon wafer. Irradiation with light in the near UV region of the spectrum modifies the chemical properties of the photoresist, and, in "positive" photoresist, makes it more soluble in certain developers. Thus, one step frequently employed in microstructure fabrication is the projection of the image of a mask onto the photoresist layer. It becomes possible to remove the exposed region of the photoresist by dissolving it with a suitable developer. The SiO_2 layer can then be removed from the areas that were exposed to light by hydrogen fluoride etches. The photoresist is resistant to HF etches and the SiO_2 in the unexposed areas is not affected by the etch. After etching, the remaining photoresist can be removed by a solvent, leaving a silicon substrate covered with SiO_2 only in the unexposed areas. The SiO_2 film acts as a barrier to the contact of impurities in a gaseous phase with the silicon. Thus, when the silicon wafer covered by the patterned SiO_2 film is exposed, to, for example, a gas containing phosphorus at high temperatures, the phosphorus, being very soluble in silicon, diffuses into the exposed areas rapidly. An idealized description of this sequence of process steps is shown in Fig. 2. The effect of this doping is very important in electronics, since phosphorus is a donor impurity and a region of n-type or electron conductivity is produced where it is present.

Many physical phenomena, however, obstruct the formation of the ideal structure depicted in Fig. 2. The technical literature is well supplied with papers devoted to each of the steps illustrated in Fig. 2. None is as straightforward as appears at first sight. It is instructive to discuss them further, since much of the essence of microstructure fabrication is revealed by examining them in detail.

Figure 2(**a**) suggests that the thickness of the photoresist and SiO_2 layers are independent of position. While the layer thickness is not an extremely critical process parameter, its control cannot be entirely neglected, as the time needed for the subsequent developing or etching steps depend on it. The wafers used in modern silicon technology have diameters of three or more inches, and maintaining uniformity of layers and process parameters across a wafer is not a trivial task. Also, very high standards of cleanliness must be maintained, as any particulate contamination will affect the resist adversely.

Figure 2(**b**) shows a well-defined boundary between the exposed and unexposed areas of the photoresist. In fact, the dimensions of the structures produced in modern microelectronics are comparable to the wavelength of the exposing light, so that diffraction prevents such sharp contrast from being achieved. Furthermore, high-resolution projection exposure schemes

require that the light be monochromatic to avoid the problems of chromatic aberration in the lenses. The photoresist must be reasonably transparent to insure that its full thickness is exposed to the light. The silicon surface is, however, reflective, so that the interference of the incident and reflected light produces standing waves in the photoresist and nonuniform exposure of the photoresist in the vertical direction. Complicated effects of this kind are clearly important to microstructure fabrication. It must also be apparent that, as the amount of exposure received is a continuous function of position, the time of development required to remove a given region of photoresist will also be a continuous function, and that the profile of the developed photoresist will depend on the time of development. In particular, the size of the opening in the photoresist, to which Fig. 2 is oriented, will depend on the time.

Resists can also be exposed with focused electron beams, as used in electron microscopes, instead of light. The electrons, however, pass through the resist layer into the substrate, where they are scattered, and some eventually return from the substrate to the resist, exposing it at a distance from the intended opening. Great care is needed to allow for the backscattering phenomenon in calculating exposures for nearby openings. Development of the photoresist proceeds somewhat as shown in Fig. 3, with simultaneous lateral and vertical removal of material. The tapered edge of the resist film may be a disadvantage, because the exact point at which the film is thick enough to protect the underlying SiO_2 layer during the succeeding etching step, and thus the size of the hole that will be produced in the SiO_2, is not clearly defined. Prolonging the development beyond the point shown in Fig. 3(**b**) allows continued lateral development and increase in the size of the opening. Also, however, achieving perfect adhesion of the photoresist film to the SiO_2 is difficult, and the developer may invade the interface between the two layers, producing the undesirable result shown in Fig. 3(**d**).

The developed photoresist, Fig. 2(**c**), is then used as a mask to etch the SiO_2 layer. Again, perfection is hard to achieve. Etching for too short a time will leave a certain amount of photoresist in the hole. Etching for too long a time can cause undercutting, as shown in Fig. 4(**c**). After removal of the photoresist, the wafer is exposed to a diffusant, affording additional opportunities for deviations from idealized behavior. Time and temperature of diffusion are important and can produce results resembling Figures 5(**a**) or 5(**b**). Diffusion can proceed rapidly along interfaces in certain cases, leading to junction profiles of the kind shown in Fig. 5(**c**).

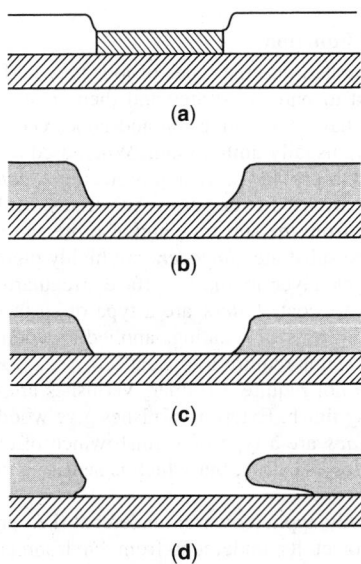

Fig. 3. Development of the photoresist: (**a**) Although exposure increases the solubility of photoresist in the developer, there is only a finite ratio of dissolution of the exposed photoresist to that of the unexposed region. In addition, the exposure received is not a perfect step function at the boundary of the exposed areas. Thus, dissolution proceeds laterally as well as vertically. (**b**) The opening in the photoresist has penetrated to the surface of the SiO_2. (**c**) With continued development the opening continues to enlarge. (**d**) Poor adhesion of the photoresist to the SiO_2 has allowed the development to penetrate the interface. (See Fig. 1 for legend)

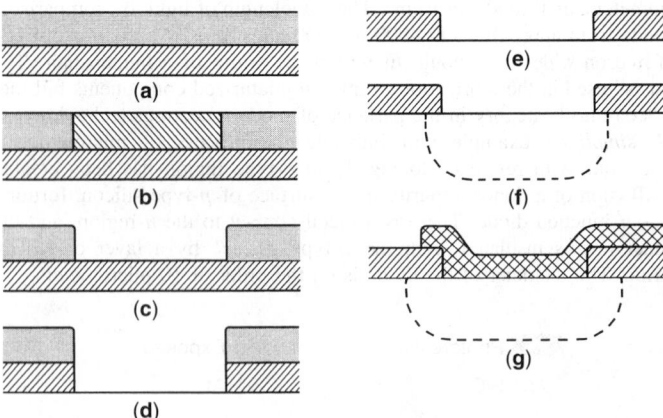

Fig. 2. Process steps used to produce the structure shown in Fig. 1: (**a**) A film of SiO_2 has been formed by oxidizing the silicon and a layer of photoresist has been deposited on the SiO_2. (**b**) Shading shows a region of the photoresist that has been exposed to light and thereby made more soluble. (**c**) The exposed photoresist has been removed. (**d**) An etchant that reacts with the SiO_2, but not with the photoresist, has been removed. (**e**) Another solvent has been used to remove the unexposed photoresist. (**f**) Donor atoms have diffused into the silicon through the opening in the SiO_2 to produce an *n*-type region. (**g**) Additional masking steps, not shown, have permitted aluminum to be evaporated onto the diode in a pattern that forms a contact to the n region of the diode. (See Fig. 1 for legend)

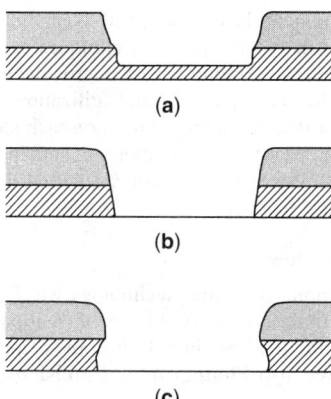

Fig. 4.   The opening in the photoresist is used to mask the etching of a hole in the SiO₂: (**a**) An early stage of development. (**b**) The opening in the SiO₂ has reached the silicon. (**c**) Prolonged etching can lead to removal of SiO₂ underneath the photoresist masks. (See Fig. 1 for legend)

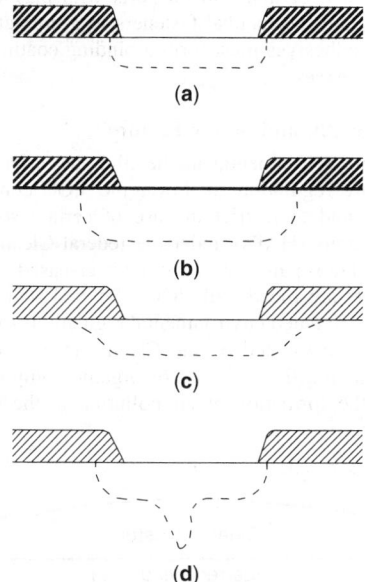

Fig. 5.   A donor impurity is diffused into the silicon from a gaseous phase. (**a**) A shallow n region has been created. (**b**) Continued diffusion, longer times, or higher temperatures increase the extent of the n region. (**c**) Surface diffusion has caused spreading of the n region along the SiO₂-silicon interface. (**d**) A crystal defect, such as a dislocation, has provided a path for anomalously high diffusion and led to penetration of the junction to unanticipated distance from the surface. (See Fig. 1 for legend)

Preferential diffusion along crystalline defects can give rise to a profile resembling that shown in Fig. 5(**d**).

Next, a metal connection is to be made to the diffusion-doped region. Aluminum is frequently used for this purpose, as it has high electrical conductivity and does not enter the silicon and alter its properties. The aluminum is also evaporated through a mask that defines the shape and location of the conductor. Examples of the region of contact between the aluminum and the doped semiconductor are shown in Fig. 6. It is seen that the current will be forced to flow through a narrow constriction if the profiles are as shown in Fig. 6(**b**). The high current densities may cause electromigration of the aluminum atoms, leading to the open circuit shown in Fig. 6(**c**).

Also, silicon is somewhat soluble in aluminum. One can thus encounter the situation shown in Fig. 6(**d**), where enough silicon has been dissolved to allow the metal to completely penetrate the doped region, shorting the junction.

None of the problems illustrated in Figs. 2, 3, 4, 5, and 6 is insurmountable. A great many ingenious ways to avoid the difficulties described are

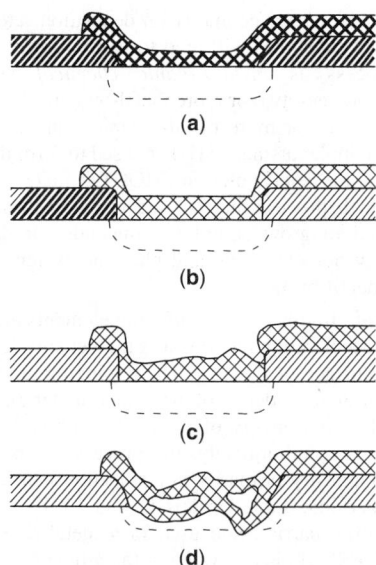

Fig. 6.   (**a**) Masking steps, not shown, have permitted the deposition of aluminum in selected areas to form a contact to the n region. (**b**) A sharp vertical profile in the SiO₂ opening may cause a reduction of the cross section of the aluminum conductor where it passes from the SiO₂ insulator to the silicon surface. (**c**) The high current density in the constriction shown in (**b**) has led to electromigration of aluminum atoms and opening of the conductor. (**d**) Solution of silicon in the aluminum has resulted in deformation of the metal-semiconductor interface and penetration of the aluminum through the n region. (See Fig. 1 for legend)

known. Sometimes these are guided by physical or chemical knowledge; frequently they are empirical fixes.

The microstructure engineer must also be aware of constraints that have little to do with chemistry and materials science, but tend to be more closely related to mechanical technology. One of the most difficult of these is registration or alignment. A substrate is usually passed through several process steps in order to form a desired structure. For example, fabricating a transistor may involve diffusing a base dopant through a window in SiO₂, subsequently diffusing an emitter dopant through a somewhat smaller opening in an SiO₂ layer, and finally, using masking layers to form contacts to these transistor elements. It is necessary to ensure that the emitter region is created in the correct position within the previously diffused base region. The mask that is used to define the emitter region lithographically must be precisely located with respect to the geometrical structure already established on the substrate by previous processing steps. This positioning is known as alignment. It may require that separate structures that can be easily located but have no electronic function be provided on the substrate.

Alignment all over a large substrate is made more difficult by dimensional changes that may take place during processing at high temperatures. Materials soften at high temperatures and can deform under the force of gravity or the stresses that accompany temperature gradients and contacts between different materials.

Further, economic factors also constrain the utilization of microstructure fabrication technology. These are the factors that control the cost of production, such as throughput, the rate at which substrates can be processed by the fabrication tools, capital investment required, and demands on operator time and skill. Electron beam exposure, for example, provides high resolution but uses expensive equipment that works slowly. Naturally, all of the elements of cost must be weighed against the value of the product produced.

Chemical Vapor Deposition.   Deposition of tungsten, molybdenum, and their silicides by chemical vapor deposition (CVD) is of relatively recent interest in the microelectronics industry. These materials are useful for gates and interconnects in metal oxide semiconductors (MOS) devices. Aluminum, the widely used interconnect material, has a comparatively low melting point (600°C) and a markedly different coefficient of thermal expansion (compared to silicon), so that over a period of years researchers have been seeking an alternative for aluminum.

In using CVD for microelectronics applications, the deposit thickness must be as uniform as possible. This can best be achieved by conducting the deposition in a surface-controlled, not a diffusion-controlled, regime.

In this way, effects of reactor geometry on deposition rates are minimized and irregular-shaped substrates will tend to be uniformly coated.

An allied process is *metal-organic chemical vapor deposition* (MOCVD). In this process, two or more metal-organic chemicals (example: trimethylgallium) or one or more metal-organic sources and one or more hydride sources (example: arsine, AsH_3) are used to form the corresponding intermetallic crystalline solid solution. MOCVD materials technology is a vapor-phase growth process that is used to study the basic physics of novel materials and to grow complex semiconductor device structures, particularly for new optoelectronic and photonic systems. The process is reported in some detail by Dupuis.

Ion Implantation. In this process, alloying elements are introduced into a host material by accelerating the ions to a high energy level and allowing them to strike the surface of the host. The impinging atoms penetrate into the substrate material to a depth of one micrometer or less, depending upon atomic number and energy of the atom. Although it has a number of other applications metallurgically, the process has been of interest in the semiconductor industry primarily in connection with doping substrates with elements in Periodic Table Groups 13 and 15. Use of the process in fabricating a Schottky barrier gate used in a metal-semiconductor field-effect transistor (MESFET) is described in the article on **Semiconductors**. **Multidisciplinary Characteristics of Microstructure Fabrication.** Upon observing the practice of microstructure fabrication, one cannot fail to notice a resemblance to certain aspects of modern metallurgy and chemical engineering. For example, the precipitates produced by metallurgical processing have a dimensional scale similar to that of electronic microstructures. Inhomogeneities on a scale of 0.01 ϕm to 10 ϕm control the desirable properties of a structure. This preoccupation with the properties of solids on a microscopic scale produces a common interest in techniques and in interactions with basic science. Thus, both microstructure fabrication and physical metallurgy: (a) rely on phenomena that take place in the solid state; (b) depend on analytical tools that are capable of chemical analyses with the highest possible spatial resolution; (c) involve the motion of atoms through solids, controlled by diffusion, solution, nucleation, and precipitation; (d) involve interface phenomena at the contact between different solids; and (e) are sensitive to crystal defects.

It must further be noted that both metallurgy and microstructure fabrication are practical disciplines, they are oriented toward the economic production of structures that have a useful role in commerce and industry. In this respect both are engineering rather than scientific disciplines. On the other hand, their deep probing of phenomena on an atomic scale and under unusual conditions produce new discoveries and lead to new concepts that enhance basic science.

This is not to say that microstructure fabrication is a branch of metallurgy. The detailed motivation of the two disciplines is rather different, metallurgy concentrating on the mechanical properties of solids, while microstructure fabrication controls the electronic properties of structures made from magnetic and optical materials, semiconductors, metals, and insulators. The basic difference, of course, is that microstructure fabrication involves control of the fabrication process in detail at the dimensional level of the structure, while metallurgical processing exercises control at a much grosser level. The application of lithography, with the attendant use of exposure tools, clean rooms, resists, masks, and etchants is the province of microstructure fabrication. Crystalline defects are usually undesirable in microstructures; the metallurgist can frequently use them to advantage. Metallurgy also encompasses its extractive aspects. There is no doubt that microstructure fabrication is a distinct activity.

The technique of microstructure fabrication has grown up as an art in response to a continuous economic and functional motivation to push to the smaller and smaller. Chemistry, physics, and empiricism are combined into the creation of novel physical structures that have enormous economic impact, that, indeed, form the basis of whole new industries. Progress has been made by adaptation and invention to meet the needs of the moment. By and large, the art has grown rapidly, adapting and improving old methods to new situations and inventing as needed and possible. It is common experience that obtaining reproducible results requires very careful control of all aspects of the fabrication processes, such as temperatures, pressures and time. High standards of cleanliness and reagent purity must be enforced. Seemingly identical apparatuses and starting materials often yield different results. Recipes must be carefully followed, with little understanding of which aspects of a process are critical or what contaminants are important, and why.

The rapid development has outstripped basic understanding of the fundamental mechanisms underlying the techniques. It must be recognized, however, that the optimal exploitation of microstructural technology, the maximization of performance, yields, and utilization of available silicon area will depend on a detailed interpretation of each step in fabrication as a chemical or physical process. Furthermore, unanticipated phenomena are encountered and inject surprises into basic science. A new interdisciplinary field of applied science has emerged.

Photographic Color Film

One of the early examples of coating technology was for building up layers in color film. These films exemplify the use of multiple layers of coatings that are "built in" to the final product. In use, the camera lens also will be coated. See Fig. 7. See also **Photographic Emulsions.**

Adhesives

Commonly, adhesives are applied by coating products and parts that must be held together. Some adhesives require application at the time of usage; others are pre-applied in semi-dry or dry form, but with a "sticky" surface. Adhesives range from natural products, such as glue to plastics and polymers. The epoxies are exemplary of modern adhesives. With the continuing replacement of many metal parts with plastics, adhesives are displacing welding and mechanical fasteners to a dramatic degree. In the final essence, most adhesives must form a binding coating on one form or other. See also **Adhesives.**

Environmental, Health, and Safety Factors

The most significant environmental and health issues affecting the paint and coatings industry are regulations to lower the VOC content for virtually all types of paints and to restrict the use of certain solvents known as hazardous air pollutants (HAPs) under the federal Clean Air Act. Except for the water in a latex paint or in other water-based coatings, solvents used in house paints are mostly all VOCs. Several states, along with the U.S. EPA, have implemented environmental regulations to restrict the VOC content of paints, as mandated by the Clean Air Act. These regulations are aimed at minimizing the emission of organic compounds from paints that contribute to the formation of air pollution in the form of smog or ground-level ozone.

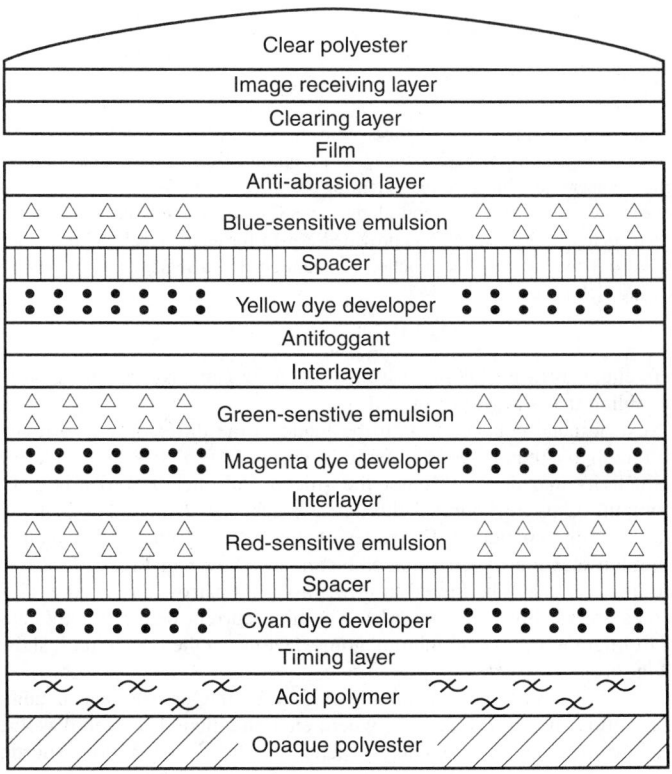

Fig. 7.   Schematic sectional view of layers composing the "thickness" of a color photographic film. (*After Cohen*)

Health and safety issues affect both the professional painter and paint manufacturer who as part of the occupation can be exposed to high concentrations of organic solvent for extended periods of time. Environmental issues focus on the contribution of organic solvents to air pollution and other issues such as hazardous waste disposal. Paint companies are reducing their use of organic solvents in the manufacture and development of architectural coatings by offering more water-borne and higher solids alternatives to conventional solvent-borne paints. Some paint manufacturers are offering solvent-free latex paint alternatives to water-borne paints that contain organic solvents. A typical solvent-borne paint may contain 30–60% of organic solvent. By contrast, a water-based paint may only contain 5–10% of organic solvent. This is a significant reduction in solvent content, as water-borne paints are becoming more and more predominant in the architectural coatings market, and are part of the focus for reducing the use of organic solvents in paints.

Restriction on the use of certain types of solvents, listed as HAPs under the Clean Air Act, are forcing paint manufacturers not only to lower the limits on the amount of organic solvents in a paint, but also to eliminate certain types of solvents. Thus paint manufacturers are challenged to comply simultaneously with both VOC and HAP regulations. These Clean Air Act mandates are expected to affect most types of paints and paint manufacturers beginning in 1996.

Another issue affecting the architectural paint industry is the remediation of homes, buildings, and structures that contain lead-based paint. Lead poisoning in children has been linked to ingestion of paint dust or paint chips that contain lead pigments and this has resulted in U.S. government regulations to reduce the lead content in paint to no more than 0.06%.

Additional Reading

Anderson, D.G.: "Coatings (Analysis of)," *Analytical Chemistry*, 87R (June 15, 1991).

Champ, N.A. and F.L. Lowenstein: "TBT: The Dilemma of High-Technology Antifouling Paints," *Oceanus*, 69 (Fall 1987).

Cohen, E.D.: "Coatings: Going Below the Surface," *Chem. Engineering Progress*, 19 (September 1990).

Cohen, E.D. and E.J. Lightfoot: "A Primer on Forming Coatings," *Chem. Engineering Progress*, 30 (September 1990).

Coyle, D.J., C.W. Macosko, and L.E. Scriven: "Fluid Dynamics of Reverse Roll Coating," *Amer. Inst. of Chem. Eng's J.*, 16 (February 1990).

Dupuis, R.D.: "Metalorganic Chemical Vapor Deposition of III—V Semiconductors," *Science*, **226**, 623–629 (1984).

Elliott, D.: *Microlithography: Process Technology for IC Fabrication*, The McGraw-Hill Companies, Inc., New York, NY, 1986.

Fettis, G.: *Automotive Paints and Coatings*, John Wiley & Sons, Inc., New York, NY, 1995.

Finzel, W.A.: "Use Low-VOC (Volatile Organic Compounds) Coatings," *Chem. Engineering Progress*, 50 (November 1991).

Freitag, W. and D. Stoye: *Paints, Coatings and Solvents*, 2nd Edition, John Wiley & Sons, Inc., New York, NY, 1999.

Glass, J.E.: *Technology for Waterborne Coatings*, Vol. 663, American Chemical Society, Washington, DC, 1997.

Herman, H.: "Plasma-Sprayed Coatings," *Sci. Amer.*, 112 (September 1988).

Lambourne, R.: *Paint and Surface Coatings: Theory and Practice*, Prentice-Hall, Inc., Upper Saddle River, NJ, 1993.

Marrs, J.M.: "Ultraviolet Light for Photochemical Deposition," *Chem. Eng. Progress*, 31–34 (January 1986).

Satas, D. and A. Tracton: *Coatings Technology Handbook*, 2nd Edition, Marcel Dekker, Inc., New York, NY, 2000.

Schweizer, P.M.: "Visualization of Coating Flows," *J. Fluid Mechanics*, 193,285 (1988).

Scriven, L.E. and W.J. Suszynski: "Take a Closer Look at Coating Problems," *Chem. Engineering Progress*, 24 (September 1990).

Shumay, W.C., Jr.: "High-Performance Adhesives on the Line," *Adv. Material & Processes*, 82 (August 1987).

Staff: *Paints Related Coating and Aromatics*, American Society for Testing & Materials, West Conshohocken, PA, 1998.

Sturge, J.M., V. Walworth, and A. Shepp: *Imaging Processes and Material: Neblette's*, 8th Edition, John Wiley & Sons, Inc., New York, NY, 1997.

Tobiason, T.L.: "Choosing a Finish for Your Electronic Package," *Instruments & Control Systems*, 85 (May 1990).

Various: "6th International Coating Process Science and Technology Symposium Papers," *Amer. Inst. of Chem. Engineers*, New York, NY, 1992.

Weldon, D.G.: *The Failure Analysis of Paints and Coatings*, John Wiley & Sons, Inc., New York, NY, 2001.

PALLADIUM. [CAS: 7440-05-3]. Chemical element symbol Pd, at. no. 46, at. wt. 106.4, periodic table group 10, mp 1554°C, bp 2970°C, density 12.02 g/cm^3 (solid), 12.25 g/cm^3 (single crystal) (20°C). Elemental palladium has a face-centered cubic crystal structure. The six stable isotopes of palladium are ^{102}Pd, ^{104}Pd, through ^{106}Pd, ^{108}Pd, and ^{110}Pd. The seven unstable isotopes are ^{100}Pd, ^{101}Pd, ^{103}Pd, ^{107}Pd, ^{109}Pd, ^{111}Pd, and ^{112}Pd. In terms of earthly abundance, palladium is one of the scarce elements. Also, in terms of cosmic abundance, the investigation by Harold C. Urey (1952), using a figure of 10,000 for silicon, estimated the figure for palladium at 0.0091. No notable presence of palladium in seawater has been found. Discovered by Wollaston in 1803.

Electronic configuration $1s^2 2s^2 2p^6 3s^2 3p^6 3d^{10} 4s^2 4p^6 4d^{10}$. Ionic radius Pd^{2+} 0.50 Å (Wyckoff). Metallic radius 1.3755. First ionization potential 8.33 eV; second 19.8 eV. Oxidation potentials Pd $\rightarrow$ Pd^{2+}+2e$^-$ 4MHClO$_4$, -0.83 V, Pd + 2OH$^-$ $\rightarrow$ Pd(OH)$_2$ + 2e$^-$, -0.1 V. Further physical properties are given under **Platinum and Platinum Group**.

Pd has some similarities with both Ni and Ag and many with Pt. Pd dissolves more readily in acids than any other member of the platinum group of metals. In aqua regia, the metal dissolves quickly. Even the compact metal dissolves slowly in HCl. In finely divided form, it is quite soluble in all acids. When heated in air at red heat, the monoxide, PdO, is formed. Pd is similarly converted to the dihalides under the same conditions when it is exposed to F$_2$ or CL$_2$. The metal is not affected by H$_2$S.

The black compound, palladium(II) oxide is formed by fusing palladium(II) chloride with NaNO$_3$ to 600°C and then leaching out the salts with water. This strong oxidizing agent is easily reduced to the metal by H$_2$. The compound is insoluble in water and acids, including aqua regia. The hydroxide, Pd(OH)$_2$, is made by the hydrolysis of palladium(II) nitrate. The compound is soluble in acids, and water is evolved on heating, but even at 500–600°C some water still remains. At this temperature, the compound starts to lose O$_2$.

Palladium(III) oxide, P$_2$O$_3$, is made as a hydrate by careful oxidation of a solution of palladium(II) nitrate either by anodic oxidation or ozone treatment at -8°C. This unstable brown powder reverts to the monoxide in about 4 days. When heated, the compound loses water and may explode as it changes to the monoxide.

Palladium(II) chloride is formed by direct combination of the elements at 500°C. It is the only stable chloride over 500–1500°C. The red crystals are partly soluble in water and completely soluble in HCl. The fraction insoluble in water is probably a polymer. Palladium(II) chloride also is the product obtained by evaporation of a solution of Pd in HCl. Palladium(II) bromide can also be made from the elements.

When KI is added to a solution of palladium(II) chloride, an insoluble diiodide is precipitated. The dark red-black crystals are soluble in excess iodide with formation of the tetraiodide complex ion. Palladium(II) iodide evolves iodine at 100°C, the decomposition to the elements being complete at 330–360°C. The black compound, palladium(III) fluoride, is made by direct combination of the elements. On reduction, the brown difluoride is formed.

Divalent Pd forms many planar complexes with a coordination number of 4. The tetrachlorides are quite soluble. When a solution of palladium(II) chloride is oxidized with chlorite or chlorate ion, Pd(IV) is formed, which has a coordination number of 8. The addition of NH$_4$Cl to such a solution precipitates ammonium hexachloropalladate(IV) as a red compound. It is somewhat less stable than the platinum analog.

The soluble yellow-brown palladium(II) nitrate is formed by dissolving finely divided Pd in warm HNO$_3$ and then crystallizing the compound from this solution. The analogous sulfate is similarly formed from H$_2$SO$_4$. It crystallizes as a red-brown dihydrate. Both these compounds easily hydrolyze.

Palladium(II) sulfide is precipitated as a brown powder by adding H$_2$S to a solution of palladium(II) ion. When this sulfide is heated with sulfur at 400°C, the insoluble disulfide is formed. The excess sulfur can be extracted with CS$_2$ to yield the gray-black crystalline palladium(IV) sulfide. This compound is not soluble in single acids but is soluble in aqua regia.

Some Pd complexes are important analytically or in the refining of Pd. The yellow dimethylglyoxime compound is quantitatively precipitated from a HCl solution of palladium(II) chloride by the addition of an alcoholic solution of dimethylglyoxime. Palladium(II) has a great affinity for nitrogen-containing ligands. The di- and tetramine find use in refining.

Palladium, as with other members of the platinum group, exhibits catalytic activity for various reactions. One of its best known uses is in conjunction with other platinum metals in the catalytic converters of present-day automobiles.

As reported by Chung-Chiun Liu et al. (*Science*, **207**, 188–189, 1980), a palladium-palladium oxide miniature pH electrode has been developed for pH measurement. The miniature wire-form electrode exhibits a super-Nernstian behavior and gives a mean pH response of 71.4 mV per [pH] (standard deviation, 5.3 mV). The electrode may find applications in biological, medical, and clinical studies.

Uses

Alloys for electrical relays and switching systems in telecommunication equipment, catalyst for reforming cracked petroleum fractions and hydrogenation, metallizing ceramics, "white" gold in jewelry, resistance wires, hydrogen valves (in hydrogen separation equipment), aircraft spark plugs, and protective coatings.

See also **Chemical Elements**.

LINTON LIBBY
Chief Chemist, Simmons Refining Company
Chicago, Illinois

Additional Reading

Carter, G.F. and D.E. Paul: *Materials Science and Engineering,* ASM International, Materials Park, OH, 1991.

Davis, J.R.: *Metals Handbook,* 2nd Edition, ASM International, Materials Park, OH, 1998.

Greenwood, N.N. and A. Earnshaw: *Chemistry of the Elements,* 2nd Edition, Butterworth-Heinemann, Woburn, MA, 1997.

Krebs, R.E.: *The History and Use of Our Earth's Chemical Elements: A Reference Guide,* Greenwood Publishing Group, Inc., Westport, CT, 1998.

Lide, D.R.: *CRC Handbook of Chemistry and Physics 84th Edition,* CRC Press, Boca Raton, FL, 2003.

Parker, P.: *McGraw-Hill Encyclopedia of Chemistry,* 2nd Edition, The McGraw-Hill Companies, Inc., New York, NY, 1993.

Sinfelt, J.H.: "Bimetallic Catalysts," *Sci. Amer.*, 90–98 (September 1985).

Staff: *ASM Handbook—Properties and Selection: Nonferrous Alloys and Pure Metals,* ASM International, Materials Park, OH, 1990.

Stwertka, A. and E. Stwertka: *A Guide to the Elements,* Oxford University Press, Inc., New York, NY, 1998.

PALMITIC ACID. [CAS: 57-10-3]. $CH_3(CH_2)_{14}COOH$, formula weight 256.42, white crystalline powder, mp 64°C, bp -271.5°C, sp gr 0.849. The acid is insoluble in H_2O, moderately soluble in alcohol, soluble in ether. About 60% of the content of palm oil is palmitic acid.

Palmitic acid is present as cetyl ester in spermaceti from which, by hydrolysis, the acid may be obtained; it is present in bee's wax as the melissic ester; and in most vegetable and animal oils and fats, in greater or lesser amounts, as glyceryl tripalmitate or as mixed esters, along with stearic and oleic acids. Palmitic acid is separated from stearic and oleic acids by fractional vacuum distillation and by fractional crystallization. With NaOH, palmitic acid forms sodium palmitate, a soap. Most soaps are mixtures of sodium stearate, palmitate, and oleate.

Representative esters of palmitic acid are: methyl palmitate, $C_{15}H_{31}$ $COOCH_3$, mp 30°C, bp 195°C at 15 mm pressure; ethyl palmitate, $C_{15}H_{31}COOC_2H_5$, mp 24°C, bp 185°C at 10 mm pressure; cetyl palmitate, $C_{15}H_{31}COOC_{16}H_{33}$, mp 54°C; glyceryltripalmitate (tripalmitin), $C_3H_5(COOC_{15}H_{31})_3$, mp 65°C, bp 310°C, approximately.

As the glyceryl ester, palmitic acid is one of the constituents of many vegetable and animal oils and fats.

Palmitic acid finds use in the production of cosmetics, food emulsifiers, pharmaceuticals, plastics, and soaps. One commercial formulation contains 95% palmitic acid, 4% stearic acid, and 1% myristic acid; another preparation contains 50% palmitic acid and 50% stearic acid.

See also **Vegetable Oils (Edible)**.

PALMITOLEIC ACID. [CAS: 373-49-9]. Also called *cis*-9-hexadecanoic acid, formula $CH_3 (CH_2)_5CH:CH(CH_2)_7COOH$. This is an unsaturated fatty acid found in nearly every fat, especially in marine oils (15–20%). At room temperature, it is a colorless liquid. Insoluble in water; soluble in alcohol and ether; mp 1.0°C; bp 140–141°C (5 millimeters pressure). Combustible. Palmitoleic acid is used in organic synthesis, and as a standard in chromatographic analysis. See also **Vegetable Oils (Edible)**.

PALM OIL. See **Vegetable Oils (Edible)**.

PANETH TECHNIQUE. Method demonstrating the existence of free radicals (e.g., methyl) or atoms, which is based on the removal of metallic "mirror" by a stream of gas containing the radicals. The reaction products can be collected and assayed.

PANTOTHENIC ACID. Pantothenic acid (PA), also known as vitamin B_5, is essential to all forms of life. Its name is derived from the Greek word *pantos* that means "everywhere", which is appropriate for this widely, distributed vitamin. Pantothenic acid is a constituent of coenzyme A, which participates in numerous enzyme reactions. CoA was discovered as an essential cofactor for the acetylation of sulfanilamide in the liver and of choline in the brain. CoA is particularly important in the initial reaction of the TCA cycle (citric acid cycle) of carbohydrate metabolism and energy production. These factors are described in greater detail in the entry on **Coenzymes**. Pantothenic acid is unique among the vitamin group, in that it was one of the first to be isolated, using as a basis a microbiological assay method. Even more unique is the fact that its structure was largely determined, using a highly quantitative biological yeast test, long before it was isolated or obtained in concentrated form. R.J. Williams and coworkers described it as an acid with an ionization constant lower than that of an alpha-hydroxy acid, but about right for a hydroxy acid in which the hydroxyl group was farther removed from the carboxyl group.

In 1901, Wildiers described Bios, an essential for yeast growth. In 1933, Williams isolated crystalline Bios from yeast and named it pantothenic acid. In 1938, Williams isolated pantothenic acid from liver; and, in 1939, Jukes determined liver antidermatitis factor (chick) to be identical with yeast factor. Also, in 1939, Woolley et al. demonstrated beta-alanine as a vital part of pantothenic acid:

$$H-O-\overset{\overset{\displaystyle H}{|}}{\underset{\underset{\displaystyle H}{|}}{C}}-\overset{\overset{\displaystyle CH_3}{|}}{\underset{\underset{\displaystyle CH_3}{|}}{C}}-\overset{\overset{\displaystyle OH}{|}}{\underset{\underset{\displaystyle H}{|}}{C}}-\overset{\overset{\displaystyle O}{\|}}{C}-\overset{\overset{\displaystyle H}{|}}{\underset{\underset{\displaystyle H}{|}}{N}}-\overset{\overset{\displaystyle H}{|}}{\underset{\underset{\displaystyle H}{|}}{C}}-\overset{\overset{\displaystyle H}{|}}{\underset{\underset{\displaystyle H}{|}}{C}}-COOH$$

(Pantoic acid)   (Beta-alanine)

d (+) Pantothenic acid ($C_6H_{17}O_5N$)

In 1940, Harris, Folkers, et al. reported structure determination and synthesis and crystallization of pantothenic acid. In 1950, Lipmann et al. discovered coenzyme A; and, in 1951, Lynen characterized the coenzyme A structure.

Pantothenic acid participates as part of coenzyme A in carbohydrate metabolism (2-carbon transfer-acetate, or pyruvate), lipid metabolism (biosynthesis and catabolism of fatty acids, sterols, +phospholipids), protein metabolism (acetylations of amines and amino acids), porphyrin metabolism, acetylcholine production, isoprene production.

Distribution and Sources. Particularly high in pantothenic acid content are yeasts, animal glands and organs. Fruits have a low content.

High pantothenic acid content (2.0–10.0 milligrams/100 grams): Beef (brain, heart, kidney, liver), chicken (liver), cod ovary, groundnut (peanut), herring, lamb (kidney), pea (dry), pork (kidney, liver), royal jelly, sheep (liver), wheat bran and germ, yeast.

Medium pantothenic acid content (0.5–2.0 milligrams/100 grams): Avocado, bean (lima), beef, broccoli, carrot, cauliflower, cheese, chicken, clam, kale, lamb, lentil (dry), mackerel, mushroom, oats, pea, pork (bacon, ham), rice, salmon, soybean, spinach, walnut, wheat.

Low pantothenic acid content (0.1–0.5 milligram/100 grams): Almond, apple, banana, bean (kidney), cabbage, grape, grapefruit, honey, lemon, lettuce, lobster, milk, molasses, onion, orange, oyster, peach, pear, pepper (white and sweet), pineapple, plum, potato, shrimp, tomato, turnip, veal, watercress.

Pantothenic acid is produced commercially by synthesis involving the condensation of *d*-pantolactone with salt of *β*-alanine. Some of the dietary supplement forms include calcium pantothenate, dexpanthenol, and panthenol.

Precursors in the biosynthesis of pantothenic acid include *α*-ketoiso valeric acid (pantoic acid), uracil (*β*-alanine), and aspartic acid. Intermediates in the synthesis include ketopantoic acid, pantoic acid, and *β*-alanine.

Some of the unusual features of pantothenic acid noted by investigators include: (1) it promotes amino acid uptake; (2) it is potentiated by zinc in preventing graying of hair in rats; (3) it promotes resistance to stress of cold immersion; (4) there is a deficiency of pantothenic acid in tumors; (5) it is required for chick hatchability; (6) it is useful in treating vertigo, postoperative shock, and poisoning with isoniazid and curare; (7) it is useful in accelerating wound healing; and (8) it is useful in treating Addison's disease, liver cirrhosis, and diabetes.

See also **Vitamin**.

Additional Reading

Williams, R.J.: "Pantothenic Acid," in *The Encyclopedia of Biochemistry*: (R.J. Williams and E.M. Lansford, Jr., editors), Van Nostrand Reinhold, New York, NY, 1967.

PAPERMAKING AND FINISHING. One of the most important factors in the progress of civilization has been paper, a thin flat tissue composed of closely matted fibers obtained almost entirely from plant sources. In modern life paper finds a variety of uses, for writing, for containers, wrappers, wall covering, and—perhaps most important—in all forms of printing: newspapers, magazines, books.

The art of making paper seems to have been discovered first by the Chinese, who were making paper as early as the beginning of the Christian era. From China the process was carried to Arabia and thence to Europe. Paper was not an important article at first and, since it is not a very durable substance under ordinary conditions, could not compete with parchment or vellum as a medium for the written word. In the fifteenth century writing became more general and the demand for a cheaper material increased. Paper became an important product. At this time paper was made largely from vegetable fibers reclaimed from cloth (especially linen), as had been done since the invention of paper in China. This paper was made entirely by hand, as is done even today in the manufacture of certain expensive types of paper. In making handmade paper, a pulp is formed by soaking the vegetable fibers in water in a vat. From this vat the pulp is dipped out in a mold, the bottom of which is a fine screen. By a deft motion of this mold, the soft pulp is spread over the screen in a thin layer of matted fibers. The water in the pulp drains off, leaving a rather firm mass, which is turned out on a piece of felt. More pieces of half-dried pulp spread on felt are added. The whole pile is then pressed to squeeze out more of the water, press the fibers closer together and form a firm sheet. These are then removed from between the felts, pressed again, and dried. During the final treatment surface sizing is added to render a surface more suitable to receive ink. Sheets of hand-made paper are naturally of limited size and expensive.

To meet the great demand for paper, machine methods were developed. This increased demand for paper also led to the utilization of material that could be obtained in quantities much greater than rags. Out of this developed the vast pulp industry, which today converts vegetable material, mostly soft woods such as spruce and fir, as well as poplar, into a white felt-like mass of fibrous substance, known as pulp. See also **Pulp (Wood) Production and Processing**.

Wet End

A modern paper machine begins with a flow spreader or distributor, conveying a dilute fiber suspension (0.1–1% fibers) to a headbox which delivers a jet of the suspension or slurry through a slice (sluice) across the full width of the machine, almost 400 inches (~10.2 meters) in some large machines. In the headbox, the fibers are dispersed, and the flow is rectified as well as possible so that the jet is delivered onto a moving endless fine-mesh wire screen with uniform composition, flow rate, and velocity. The pressure in the headbox and its slice opening are adjusted so that the jet velocity matches the speed of the wire screen, which may be up to 4000 feet (~1220 meters) per minute for newsprint. The proper stock flow per unit width corresponds to the desired *basis weight* of the paper. (Basis weight is weight per unit area and varies with grades and sizes of papers.)

The dispersion of fibers in the headbox is brought about by subjecting the slurry or suspension to shear stresses, usually with turbulence. Various designs have been developed to accomplish this.

As shown in Fig. 1, the most common type of paper machine is the Fourdrinier, in which the moving wire screen is in the form of an endless conveyor belt stretched between two large rolls. The roll situated under the headbox slice is called the *breast roll*. The roll located generally at the end of the straight wire run is the *couch roll*. Drainage of the slurry through the wire screen is induced by several types of driving forces. In the early, slow-speed machines, the principal force was gravity. Later, the hydrodynamic action of table rolls, which support the wire and rotate with it, began to play an important part in drainage as speed increased. More recently, foils came into use, i.e., rigid stationary, hydrodynamically shaped elements which support the wire and exert a pumping action through the wire screen. Other means are perforated or slotted boxes with vacuum over which the wire runs. When only water is drained, they are called *wet boxes*. When applied toward the dry end of the wire screen, they also draw air through the wet paper mat and are called *suction boxes*. Other equipment configurations have been developed to meet these objectives. On all modern Fourdriniers (Fig. 2), a forming board located close to the breast roll is used to scrape off the water, drained initially by gravity, from the bottom of the wire.

An important development in paper forming is the use of a top wire dewatering unit placed above the wire of the Fourdrinier. The top wire units use various dewatering elements such as rolls, foils, and vacuum

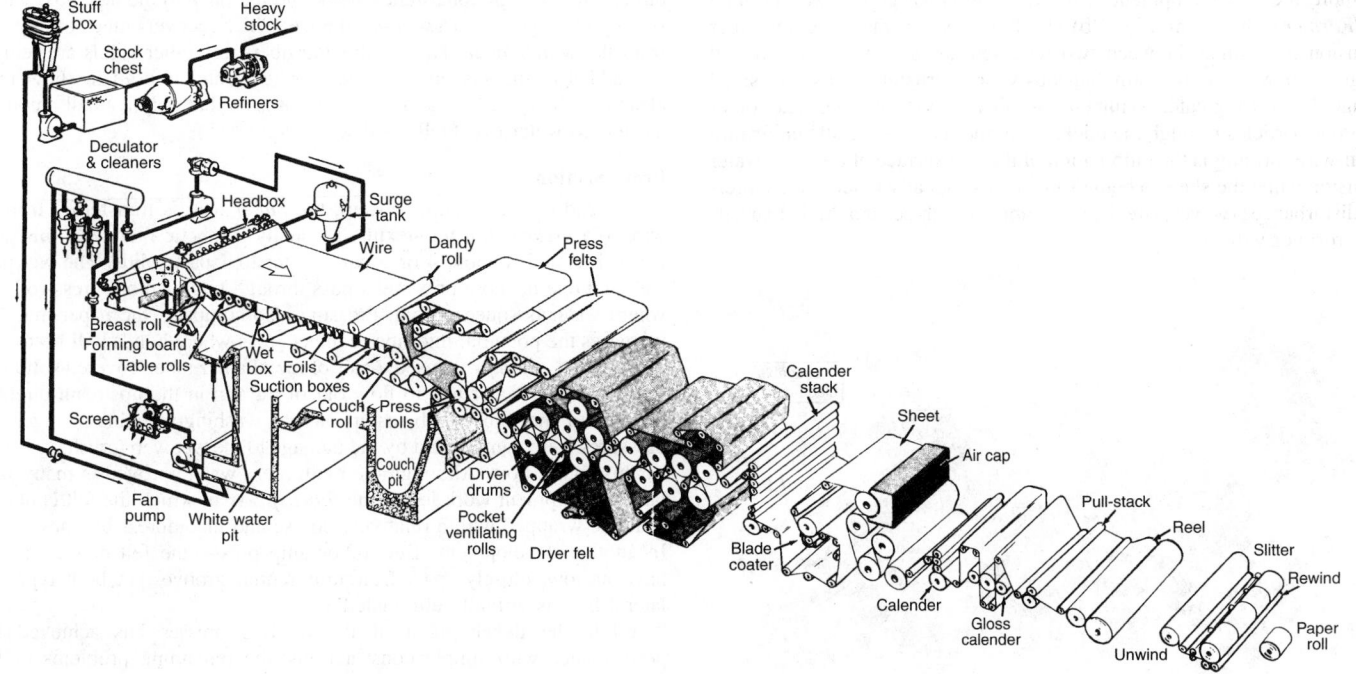

Fig. 1. Fourdrinier machine for producing printing-grade paper. (*Beloit Corporation*)

Fig. 2.   Fourdrinier machine located at Blandin Paper Company, Grand Rapids, Michigan. (*Beloit Corporation*)

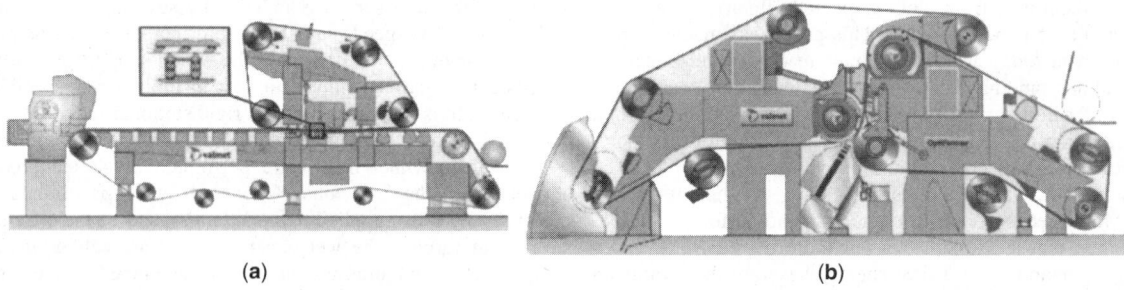

(a)   (b)

Fig. 3.   Paper formers: (**a**) top-wire former; (**b**) Roll and Blade gap former. (*Valmet Corporation*)

boxes. The top wire units add drainage capacity to an existing Fourdrinier plus improved symmetry through the thickness of the sheet similar to the twin wire formers. An example of this type of former is shown in Fig. 3(a) (*Valmet, SymFormer MB*). See also Fig. 4.

A more recent development is the *roll and blade gap former* (*Valmet SpeedFormer*) shown in Fig. 3(b). In this type of machine, the fiber suspension is confined between two wire screens, and water is removed through both wires either simultaneously or alternately. This two-sided drainage leads to greater symmetry of distribution of fines and other nonfibrous particles through the thickness of the sheet. A significant feature of twin-wire forming is the elimination of the free surface of the fiber-water suspension while the sheet is being formed. This greatly reduces the larger-scale disturbances (waves, streaks, and jumps) which occur at higher speeds on Fourdrinier wires.

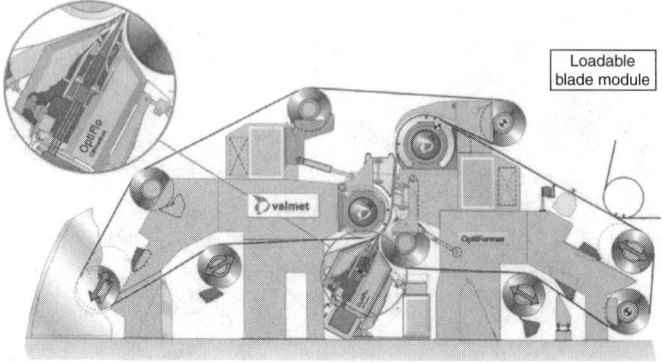

Loadable blade module

Fig. 4.   Roll and blade gap. (*Valmet Corporation*)

Not all the fiber and other solid materials are retained by the forming wire. For this reason and because so much water is used in the papermaking process, the *white water* removed in the sheet-forming process is recirculated in the overall system. A large part of it is added directly to the high-consistency stock and fed back to the headbox, while a small portion goes into a *save-all* device, which recovers much of the solids from the white water. These extracted fibers and other solids are returned and added to the suspension. The clarified water is used in showers for cleaning wires and felts and other purposes so that only a small amount of the reused water eventually is discharged.

Press Section

At the end of the forming system, the *paper web* is transferred from the wire to a *press felt*, a fine-textured, usually synthetic fabric. At this point, the web contains about 4 or 5 parts water to 1 part solids. The wet paper web and one or more press felts pass through two or more press-roll nips, where water is squeezed out. Pressing also compacts the paper mat. This increases the potential interfiber contact areas where bonds will be formed.

The early *plain press* used a pair of metal and rubber-covered solid rolls. The expressed water had to flow out of the nip in the upstream direction, parallel to the paper web, as in an old washing-machine wringer. Nip pressures were then limited by the damage to the wet web (crushing) caused by this lateral flow. Although the plain press was improved in many ways, later development work led to the *fabric press* in which the felt contacting rolls are wrapped with a relatively coarse and incompressible mesh fabric. In another development (Beloit Ventanip press), the felt contacting rolls have narrow, closely spaced circumferential grooves. In both types, the lateral flow is virtually eliminated.

While the development of the modern presses has achieved high performance with simple constructions, the remaining problems of flow resistance and web rewetting leave room for improvement. It is generally recognized that mechanical removal of water is much less costly than

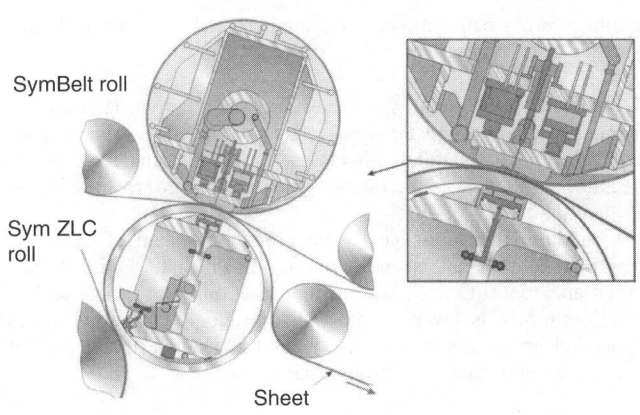

Fig. 5.   The basic belt nip press. (*Valmet Corporation*)

drying. This has led to the development of presses with very long nips (*Beloit Extended Nip Press*) as shown in Fig. 5. These presses use a shoe typically of 10 in. (25 cm) width to replace one roll in the roll press configuration and use oil lubrication between the stationary shoe and a moving impervious belt. An additional 20–40% of the remaining water in the sheet can be mechanically removed by these presses with their longer residence time under high pressure.

As the machine speed has continuously has increased the dynamic forces to the web have increased and it has forced to support the web with felts to avoid web breaks. Before 30's typically the web had open draw between press and wire section which was soon replaced with pick-up suction roll arrangement. In this arrangement web is picked up with a suction roll from wire and supported by press felt to the first nip. Most recent fast paper machines are using the same principle of suction roll and supporting fabric not only through press but also through dryer section.

Dryers

After water removal by pressing has been done to the extent practical with present technology, the paper web leaves the press section with 1–2 parts of water to 1 part fibers. Most of this remaining water, down to 5–10%, must be removed by evaporative drying. In the most common method, the paper web is passed over a series of staggered cast-iron drums internally heated by condensing steam at pressures ranging up to approximately 10.2 atmospheres. The paper web is held in contact with the rotating drums by means of dryer felts or fabrics under tension. The diameter of the dryer drums is typically 5–6 feet (1.5–1.8 meters). There may be as many as 100 of them in heavyweight paperboard machines. These dryer drums are shown in the panoramic view. See Fig. 1.

Ventilating devices that blow air of controlled temperature and humidity through the dryer felts into the spaces between adjacent dryers are used. Here the air is confined by the sheet and felt runs. These pocket ventilating systems, together with greater control of the flow patterns within the dryer hood (which usually encloses the entire dryer section) have led to significant improvements in cross-machine uniformity of paper drying. This results in paper and board of improved suitability for modern high-speed converting and printing operations.

Other types of dryers, including radiant heating, dielectric and microwave heating, and high-velocity, hot air impingement, have been developed. These devices are generally applied to drying coated paper where sheet contact to a solid surface may be detrimental during drying. Wider application has been limited because of low thermal efficiencies and high capital costs.

Size Press and Coaters

Many printing grades of paper and paperboard are coated with an aqueous suspension of pigments (such as clay) in adhesives (such as starch) to provide a smoother surface, control the penetration of inks, and improve the pick resistance, appearance, brightness, and opacity. These and other materials are also applied, such as *functional coatings*, to provide such features as water resistance, pressure sensitivity for carbonless copying, and a wide variety of other properties. The appropriate materials may be added to the papermaking furnish during some stage of stock preparation (called *internal sizing*). Application of sizing or coating to one or both surfaces

of the formed and dried sheet, rather than as internal sizing, simplifies the sheet-forming process and provides better control of surface properties.

The principal methods of surface coating may be classified as roll, blade, and air-knife coating, according to the method used to apply and control the final coating-layer thickness and smoothness. One version of coater (Beloit Billblade) simultaneously coats and smooths both surfaces of the paper web by running it down through the nip between a blade and a roll while maintaining two puddles, one between the web and the roll and the other between the flexible blade and the web, thus eliminating the necessity for two coating stations.

A more recent development is the short dwell coater (*Beloit Short Dwell*). The coater consists of a captive pond just before the blade that limits the contact time between sheet and coating material as shown in Fig. 6. The back flow assists in removing the boundary layer of air coming in with the sheet. The shorter contact time results in less coating penetration. Superior coating quality and improved runnability, due to fewer web breaks, have been achieved.

After sizing or coating, the solvent, usually water, must be removed from the coating by evaporative drying. With some coating formulations and paper grades, drying can be done on ordinary steam-heated drums without damage to the coated surface, particularly if the surface of the first drum is smooth (sometimes chrome-plated). However, it is often desirable to do the initial drying with air impingement or radiant heating. Surface coating can be done on the machine as a step in the paper-machine operation, as shown in Fig. 1.

Calenders and Winders

Nearly all paper grades are calendered after they have been dried to the desired final moisture content. Ordinary calendering involves passing the paper web through one or more nips between metal rolls with high linear pressures. The calendering process flattens out the paper structure by virtue of the high pressure and "irons" the sheet. Calendering causes bulk reduction, which often is not desired, and surface smoothing, which is desired. The results strongly depend upon moisture content, calender-roll temperature, roll pressure, and speed.

In *supercalendering*, an off-machine operation, the calender rolls consist of alternating chilled-steel and paper-filled rolls, i.e., paper disks clamped on a steel shaft. These roll fillers have to be replaced periodically. Very high pressures are used. The increased pressure and shear forces associated with deformation of the relatively soft paper roll and the very high roll pressures impart a smoother, glossier surface to the web than ordinary calendering with all-metal rolls. This type of calendering is frequently used on coated sheets to provide a glossy coated surface. Recently polymer covered rolls have been replacing paper-filled rolls and due to better durability have given an opportunity to locate this calender as part of papermachine. Fig. 7.

There are other process configurations for the various coating effects and specifications desired.

Other Types of Machines

Although the Fourdrinier machine is used for making almost all grades of paper and board, other designs are sometimes more advantageous. The *cylinder machine*, invented at about the same time as the Fourdrinier, consists of a rotating cylindrical mold covered with a wire screen and partially submerged in a vat. The stock flows into the vat, and a mat is formed on the cylinder under a hydraulic head difference between the stock level in the vat and the white-water level inside the cylinder. The

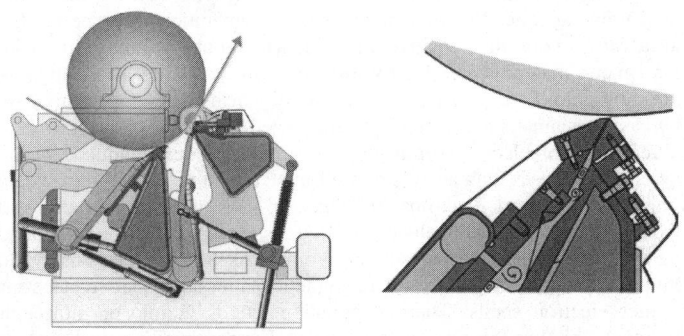

Fig. 6.   Jet coater head. (*Valmet Corporation*)

Fig. 7. Multinip calender as part of paper machine. (*Valmet Corporation*)

wet mat is picked up by a felt running through the nip between a couch roll and the cylinder. The cylinder machine is used for making multiply board, employing several vats in series. Because of slow speed and other limitations, the cylinder machine is becoming obsolete. In recent years, several new types of machines have emerged.

The most recent multiply machines use headboxes with simultaneous delivery of two different stocks from the headbox slice (*Beloit Strataflo Headbox*), followed by top wire dewatering units (*Beloit Bel Bond*). A secondary headbox with additional top-wire dewatering follows the first unit on the forming wire. Other versions of multiply formers use mini-fourdiniers on top of the primary forming wire in various configurations.

ROBERT A. DAENE
(original preparer and formerly of the Beloit Corporation);
revised and updated by Jipi Jaakkola,
Paper Machine Product Manager,
Valmet Corporation,
Charlotte, North Carolina

PARABENS. See **Antimicrobial Agents (Foods)**.

PARA COMPOUNDS. See **Organic Chemistry**.

PARAFFIN. (1) Also called alkane. A class of aliphatic hydrocarbons characterized by a straight or branched carbon chain; generic formula C_nH_{2n+2}. Their physical form varies with increasing molecular weight from gases (methane) to waxy solids. They occur principally in Pennsylvania and mid-continent petroleum. (2) Paraffin Wax.

See also **Organic Chemistry**.

PARAFFINS (Chlorinated). See **Chlorinated Organics**.

PARAMAGNETISM. A physical characteristic of some matter, advantage of which is taken in certain instrumental systems and scientific apparatus. The paramagnetic qualities of oxygen, due to two unpaired electrons per molecule, is used as the basis for some oxygen analyzers. See also **Analysis (Chemical)**. Of the other common gases, only nitric oxide and nitrogen dioxide exhibit paramagnetism.

Three basic magnetic forces exist: (1) paramagnetic, (2) ferromagnetic, and (3) diamagnetic. Ferromagnetic materials are much more permeable than a vacuum and thus positive, aligning with an applied magnetic field. Diamagnetic materials are slightly less permeable, thus negative, aligning across the field. Para- and ferromagnetism diminish with temperature rise, whereas diamagnetism essentially is unaffected.

Precession of electron orbits in atoms and molecules induced by an applied field causes diamagnetism in all matter even when paramagnetism dominates. Unpaired electrons in atoms or molecules cause para- and ferromagnetism. Normally, electrons of opposite spin pair off, netting zero magnetism per pair. Unbalanced spin moment of unpaired electrons yield both para- and ferromagnetism. Paramagnetic matter has unpaired electrons in outer electron shells. Thermal agitation retards atomic or molecular alignment, and thus the net moment is weak. Ferromagnetic matter, such as iron, cobalt, and nickel, has unpaired electrons in the next-to-outer shell.

Interatomic forces cause molecular alignment and permanent magnetism results. See also **Magnetism**.

PARA-STATE. 1. In diatomic molecules, such as hydrogen molecules, the para-state exists when the spin vectors of the two atomic nuclei are in opposite (i.e., antiparallel) directions, forming a singlet state, $S = 0$, whereas the ortho-state is the one in which the nuclei are spinning in the same direction.

2. In helium, the para-state is one group or system of terms in the spectrum of helium that is due to atoms in which the spin of the two electrons are opposing each other. Another group of spectral terms, the orthohelium terms, is given by those helium atoms whose two electrons have parallel spins. Because of the Pauli Exclusion Principle, a helium atom in its ground state must be in a para-state.

PARKES PROCESS. A standard process for the separation of silver from lead. From 1 to 2% molten zinc is added to the lead-silver mixture, heated to above the melting point of zinc. A scum containing most of the silver and zinc forms on the surface; this is separated and the silver recovered. The separation of silver is not complete, and the process is repeated several times.

PARTIAL PRESSURE. The pressure exerted by each component in a mixture of gases. In a mixture of perfect gases

$$p_i = \frac{n_i RT}{V}$$

The partial pressure of i is then the same as if component i occupies the same volume at the same temperature in the absence of the other gases. This is the Dalton law, which is treated more fully under that heading.

PARTICLE. Any discrete unit of material structure; the particulate basis of matter is a fundamental concept of science. The size ranges of particles may be summarized as follows: (1) Subatomic: protons, neutrons, electrons, deuterons, etc. These are collectively called fundamental particles. (2) Molecular: includes atoms and molecules with size ranging from a few angstroms to half a micron. (3) Colloidal: includes macromolecules, micelles, and ultrafine particles such as carbon black, resolved via electron microscope, with size ranges from I millimicron up to lower limit of the optical microscope (1 micron). (4) Microscopic: units that can be resolved by an optical microscope (includes bacteria). (5) Macroscopic: all particles that can be resolved by the naked eye.

See also **Carbon Black**; and **Particles (Subatomic)**.

PARTICLE ACCELERATOR. A device in which the speed of charged subatomic particles (protons, electrons) and heavier particles (deuterons, alpha particles) can be greatly increased by application of electric fields of varying intensity, often in conjunction with magnetic fields. It is possible to accelerate electrons and protons to speeds approaching the speed of light if sufficiently high voltage is used. Straight-line (linear) accelerators are used for protons, and doughnut-shaped betatrons for electrons; other types are the Van de Graaf electrostatic generator, the synchrotron, and the cyclotron. Before the development of nuclear reactors, the cyclotron was used to accelerate deuterons for use in bombarding stable nuclei to produce neutrons for inducing artificial radioactivity, fission and formation of synthetic (transuranic) elements.

See also **Cyclotron**; and **Particles (Subatomic)**.

PARTICLE SIZE. This term refers chiefly to the solid particles of which industrial materials are composed (carbon black, zinc oxide, clays, pigments, and the like). The smaller the particle, the greater will be the total exposed surface area of a given mass. Activity is a direct function of surface area; i.e., the finer a substance is, the more efficiently it will react, both chemically and physically. A colloidal pigment is a more effective colorant than a coarse one because of the greater surface area of its particles. A pound of channel carbon black has a surface area of 18 acres, which largely accounts for its powerful reinforcing effect in rubber. Thus, ultrafine grinding of powders is of utmost importance in such products as paints, cement, plastics, rubber, dyes, pharmaceuticals, printing inks, and numerous others.

See also **Carbon Black**; **Colloid Chemistry**; **Particle**; **Sedimentation**; and **Surface Chemistry**.

PARTICLES (Subatomic)

PARTICLES (Subatomic). For many years, the atom was traditionally described as having a central positively charged nucleus possessing considerable mass, but of minute dimension—this nucleus surrounded by a number of electrons in orbits at a relatively great distance from the nucleus. The number of electrons and their orbital arrangement determined the chemical properties of the atom, with the atoms of each chemical element possessing their own unique configuration. Recognition of the electron, the first elementary (presumably indivisible) particle, by J.J. Thomson and his associates in the 1890s ushered in an era of interest in *subatomic particles*. Ultimately, this led to the discipline of *high-energy physics*.

Organization of Matter—A Chronology[1]

A better understanding of the building blocks of nature has been a goal for many centuries, extending back to the period in Greek history of Anaxagoras of Ionia (500–428 B.C.), who held that "there was an infinite number of different kinds of elementary atoms, and that these, in themselves motionless and originally existing in a state of chaos, were put in motion by an eternal, immaterial, spiritual, elementary being, from which motion the world was produced." The concept of atoms appeared from time to time in medieval works, although the concepts expressed now seem vague. However, they seem to have been based on the idea that there could be a limit to the divisibility of matter and, consequently, the idea of a final indivisible particle out of which large pieces of matter could be built.

Atoms and Molecules. In the early 1800s, it became clear that chemical reactions could be most simply explained if each chemical element was thought of as composed of very small, identical entities characteristic of the chemical element. Thus there arose a rather well-defined idea of a chemical element composed of identical atoms, as distinguished from a compound composed of groups of different atoms combined into molecules. During the later part of the 1800s, the kinetic theory of gases made use of the idea of atoms and molecules in explaining the behavior of gases. During this period, few scientists still doubted the actual material existence and reality of atoms.

Electrons. It is perhaps rather curious that the idea of atoms became really well established only after it became clear that the atoms were not in any true sense indivisible, but that instead they probably had a complex structure that should be investigated. Since these investigations required equipment and methods that had been developed by physicists rather than chemists, the physicists took the lead and the work became known over a long period as *atomic physics*, or the physics of atomic structure. As mentioned previously, this era was inaugurated by Thomson, who first isolated and established the existence of electrons. He showed that electrons have only about 1/2000 the mass of the lightest known atom, hydrogen. He also showed that these particles, as indicated by their name, carry negative electrical charges. It was later shown by Millikan that all the electronic charges are the same. Thus, the identification of electrons as small electrically charged pieces of matter, and as constituents of all matter, became firmly established.

Electrical Neutrality. Since it was clear that normal matter is electrically neutral, it had to be assumed that each atom contained a positive electrical charge, as well as negative electrons. J.J. Thomson developed the picture of a somewhat spherical, jelly-like mass of positive electricity in which electrons are located at various positions, bound by "quasi-elastic" forces.

A principal means of investigating the structure of atoms was the examination of light emitted by the material in the gaseous state. This light was found to consist of a number of discrete wavelengths, or colors. Each of these wavelengths was associated, in the early days of the last century, with a mode of vibration of the electrons in the positive jelly. In particular, Lorentz (University of Leiden) was able to show that such electrons, when placed in a magnetic field, would have their modes of vibration changed in a way that explained the findings of Zeeman, who had made early observations of the wavelengths of the light emitted by a radiating gas in a magnetic field.

During 1910–1911, Sir Ernest Rutherford suggested an experiment, carried out by Geiger and Marsden, in which alpha particles from a radioactive source were scattered from thin foils. The angles at which the alpha particles were scattered were found to be such as could best be described by the close approach of a heavy positively charged particle, the alpha particle, to another heavier and more highly positively charged particle, representing the scattering atom.

Nuclear Atom. From the results of the experiments, Rutherford concluded that the mass in the positive charge of an atom, instead of being distributed throughout the volume of a sphere of the order of 10^{-8} centimeter in radius, was concentrated in a very small volume of the order of 10^{-12} centimeter in radius. He thus developed the idea of a nuclear atom. The atom was pictured as a small solar system with the very heavy and highly charged nucleus occupying the position of the sun, and with electrons moving around it, as planets in their respective orbits.

Although this picture of nuclear atoms served to describe the alpha-particle scattering experiments, it still left many questions unsolved. One of these questions referred to the apparent stability of the atoms. An electron moving around the nucleus would tend to emit radiation, to lose its energy, and thereby to spiral into the nucleus. Why did it not do so? Why did the atoms all seem to be quite stable, and all to be of approximately the same size, even though some contain 90 or more electrons, while hydrogen contains only one?

Electron Motion Around the Nucleus. The first approach to a treatment of these problems was made by Niels Bohr in 1913 when he formulated and applied rules for "quantization" of electron motion around the nucleus. Bohr postulated states of motion of the electron, satisfying these quantum rules, as peculiarly stable. In fact, one of them would be really permanently stable and would represent the ground state of the atom. The others would be only approximately stable. Occasionally an atom would leave one such state for another and, in the process, would radiate light of a frequency proportional to the difference in energy between the two states. By this means, Bohr was able to account for the spectrum of atomic hydrogen in a spectacular way. Bohr's paper in 1913 may well be said to have set the course of atomic physics on its latest path.

Correlation with Chemical Properties. Out of the experimental work on the scattering of alpha particles and the theoretical work of Bohr, there grew a fairly definite picture of an atom that could be correlated with its chemical properties. The chemical properties were determined in the first place by the nuclear charge. The nucleus contained most of the atomic mass and carried an electric charge equal to an integral number of positive charges, each of the same magnitude as an electronic charge. This positive nucleus then accumulated around itself a number of electrons just sufficient to neutralize its positive charge and form a neutral atom.

Atomic Number. The number of positive charges or the number of negative electrons around the nucleus was designated as the atomic number of the atom. These showed a close parallelism with the arrangement of atoms in the periodic system. Through the formulation of a number of rules based upon Bohr's picture of quantized orbits, the periodic system of the elements could be understood. Hydrogen was given one electron, and helium two. The two electrons in helium constituted a "closed shell" which exhibited almost perfect spherical symmetry and chemical inactivity.

Thus, during the years after 1913, the feeling grew that the chemical properties of atoms could be pretty well understood. The idea that there were undiscovered elements, as indicated by gaps in the periodic system, was reinforced. These elements and more have since been discovered.

Quantum Mechanics. It was not until 1925 that Bohr's ideas were developed into a mathematical form complete enough and precise enough to permit their general application, under the name *quantum mechanics*. This development associated with the names of Dirac, Heisenberg, and Schrödinger, provided the basic laws which permit, in principle, the complete and quantitative description of an atom consisting of a heavy positively charged nucleus, and surrounded by enough electrons to make the whole system electrically neutral. See also **Quantum Mechanics**.

Electron Spin. One of the properties of electrons that became evident during the study of optical spectra of atoms was that of *electron spin*. The suggestion was made by Uhlenbeck and Goudsmit in 1925 that one of the features of such spectra could be understood if each electron had associated with it a quantity called spin, which is similar in many ways to angular momentum. Each electron also has a certain magnetic moment which affects the energy in the presence of a magnetic field.[2] This property also has been incorporated into the wave concepts of quantum mechanics.

---

[1] Some of the concepts mentioned in this brief historical review have long since been abandoned or altered.

[2] Even with acceptance of the spin concept, in terms of high-energy experiments, most scientists believed that spin effects would be observed only in low-energy atomic

In order to learn more about the role of spin in colliding protons, in 1973, using the Zero Gradient Synchrotron (ZGS) at the Argonne National Laboratory, Krisch and colleagues (University of Michigan) scattered beams of polarized protons from targets in which the protons were also polarized, i.e., the spinning was all in the same direction. During the series of experiments, it was found that when the beam and the target were polarized in the same direction, violent proton-proton collisions occurred with much greater frequency than where the beam and target were spinning in a like direction. Under the latter circumstance, it appeared that the particles would pass each other, but would not interact.

Some years later (in the late 1970s), this research group further investigated the spin-collision phenomenon of protons, but used a different accelerator, i.e., the Alternating Gradient Synchrotron (AGS) located at the Brookhaven National Laboratory. The latter apparatus made it possible to study the particles at much higher energy levels: (1) an energy level up to 18.5 GeV (beam and target both polarized; and (2) up to 28 GeV (with only the target polarized). Several unexpected findings were yielded by the experiments:

- Effects of spin appear to oscillate with an increase of collision energy of the protons.
- Spin directions of the particles continue to make a difference even at high energy levels. (Normally, one would reason that at high energy levels, the difference in spin directions and the effects of spinning would become smaller simply because the spin of the proton is believed to be constant and this would tend to be overwhelmed by the higher collision energy. This was not the case and provoked suspicion that much less is known about the proton than formerly believed.)

Out of clues gained from experiments with the AGS, one central question, as posed by researcher Krish in his 1987 paper (reference listed), is posed: What does the observed difference between scattering to the left and to the right mean? Krisch observed that perhaps (as some theorists suggest) both the violence and the energy (28 GeV) of the experiments were much too low for a fundamental theory, such as quantum chromodynamics (QCD), to apply. With higher energies, the scattering difference between left and right may soon be measured at the 70- to 800-GeV proton synchrotrons at Seupukhov, CERN, and Fermilab, and even higher levels of the proposed Superconducting Supercollider (SSC).

See also **Quantum Mechanics**; and **Quarks**.

Neutrons and Protons. By 1932, it had been established that atomic nuclei are made of comparatively small numbers of neutrons and protons. Even prior to the use of particle accelerators and the birth of high-energy physics, other experiments continued to "hint" at the need of additional subatomic particles to satisfy any theory that would unify scientists' understanding of the atom's infrastructure.

A quantum theory of nuclei was made possible by the discovery of the proton and the neutron. The nuclear interaction responsible for holding the nucleus together (against disruptive electrostatic repulsion of the protons) was found to be of an entirely new kind, much stronger than the electric interaction at short distances, but decreasing very much more rapidly with distance. The various complex nuclei differ in the number of protons and neutrons they contain.

By that time, the theory of the interactions between electrons and photons had developed to the point where the electrostatic repulsion or attraction between electrically charged particles could be understood in terms of the exchange of photons between them. In the lowest nontrivial approximation, it gave the Coulomb law for small velocities. The basic interaction was the emission and absorption of "virtual" photons by charged particles.

Pions and Other Particles. A similar mechanism could be invoked to explain the short-range nuclear interaction—i.e., it is due to the exchange of particles, which have nonzero masses which are a fraction of nuclear mass. These theoretical considerations predicted the existence of a set of three particles called pions, which were ultimately discovered.

Another kind of particle and another kind of interaction were discovered from a detailed study of beta radioactivity in which electrons with a continuous spectrum of energies are emitted by an unstable nucleus. The corresponding interactions could be viewed as being due to the virtual transmutation of a neutron into a proton, an electron, and a new neutral particle of vanishing mass called the neutrino. The theory provided such a successful systematization of beta decay rate data for several nuclei that the existence of the neutrino was well established more than 20 years before its experimental discovery. The beta decay interaction was very weak even compared to the electron-photon interaction.

Meanwhile, the electron was found to have a positively charged counterpart called the positron; the electron and positron could annihilate each other, with the emission of light quanta. The theory of the electron did in fact predict the existence of such a particle. It was later found that the existence of such "opposite" particles (antiparticles) was a much more general phenomenon than once surmised.

With intensification of particle physics research, many more particles were discovered and a classification of these particles into five families was proposed—the photon family, electron family, muon family, meson family, and baryon family. Most of these particles are unstable and decay within a time which is often very small by normal standards, but which is many orders of magnitude larger than the time required for any of these particles to traverse a typical nuclear dimension. There is a wide variety of reactions between them, but they could be understood in terms of three basic interactions—the *strong* (or nuclear), *electromagnetic*, and *weak* interactions.

The Nuclear Force. The nuclear forces and the interactions between pions and nucleons are strong; the electron-electron and electron-photon interactions are electromagnetic; the beta decay interactions are weak.

As mentioned previously, by 1932 it was known that nuclei are made of comparatively small numbers of neutrons and protons. A new force was discovered (in addition to the electromagnetic and gravitational forces) that held the positive protons and electrically uncharged neutrons together in the nucleus. This nuclear force was very strong, but of limited range. Its "quantum," the particle analogous to the photon in the electromagnetic field, was of nonzero rest mass. This particle, later called the π-meson or pion, was predicted by Yukawa in 1936 and discovered by Lattes, Occhialini, and Powell in 1947.[3] For a short time, it appeared that physicists had achieved a clear, simple, and correct theory of the fundamental constitution of matter. However, shortly thereafter, two new and unpredicted particles were reported. The first of these was another

---

[3] Even though the meson was first predicted by the Japanese scientist Yukawa in 1935, the development of high-energy physics in Japan proceeded slowly from an experimental standpoint. It was not until 1975 that Japan established the National Laboratory for High Energy Physics (KEK). Rather than following traditional research approaches by way of building a proton synchrotron, for example, scientists at the University of Tokyo, in collaboration with physicists in the United States and Canada and later with CERN, decided to construct a meson facility with a powerful superconducting muon channel. In his report on the evolution of meson science in Japan, Yamazaki (1986 reference cited) lists at least four advantages of opting for a meson facility: (1) It made possible the measurement of μ-e decay time spectra in a much wider time range (0 to 20 μsec) than previously possible without background, enabling muon-spin relaxation functions (mainly long-time behavior) to be determined precisely; (2) extreme external conditions pulsewise (pulsed RF, laser, high magnetic fields) could be applied; (3) because the time of muon arrival is uniquely defined, any time-dependent transient phenomena could be investigated; and (4) rare events could be selected from continuous backgrounds. Beyond the editorial scope here, Yamazaki describes in considerable detail the numerous accomplishments of KEK during the 1980s, including an improved understanding of nuclear structure from the viewpoint of quark structure. As mentioned, since nucleons and mesons are composed of quarks and since nucleons are densely packed in a nucleus, whether the nucleons in nuclei keep their free identities (mass, size, magnetic moment) is a rewarding problem to investigate. Recently, a new type of hypernuclear spectroscopy has emerged from KEK. Yamazaki defines meson science as an interdisciplinary since it uses "second generation" particles (muons and K mesons) for the creation and detection of exotic states in matter. To study this interesting frontier, scientists strongly sense the need for experimental facilities that will provide meson beams a hundred times as strong as those available today. Plans are underway along these lines. See also **Mesons**.

---

collisions. In the 1950s, C.L. Oxley (University of Rochester) noted large spin effects in high-energy collisions (several hundred million eV). Experimentation in this area, however, was rather limited until the late 1950s, when researchers (University of California Berkeley) proposed constructing polarized proton targets. In a technique involving low temperatures and strong magnetic fields, it was possible to cause electrons to spin in the same direction and, using another technique (microwave radiation) to cause neighboring protons also to spin in one direction. Interesting experiments followed at some laboratories, but by and large many high-energy physicists in the field considered spin as relatively unimportant and it would be even less important as collision energy levels were increased. This assumption has been disproved. Spin direction does seem to be important to collisions even at high energy levels.

meson, somewhat like the pion but more massive. The second was a hyperon, i.e., a strongly interacting particle heavier than the neutron.

With the continuing discovery of more particles, investigators began to suspect that these particles were not in themselves fundamental or elementary, but that they had an internal structure. This paralleled the experiences of the 1800s when the large number of different types of atoms discovered suggested that atoms must have structure. Properties of particles also suggested an internal structure. For example, the neutron's total electric charge is indistinguishable from zero down to very fine limits, yet the neutron has a sizeable magnetic moment.

The discoverers of neptunium (1940), plutonium (1940), americium (1944), berkelium (1949), californium (1950), einsteinium (1952), fermium (1953), mendelevium (1955), and lawrencium (1961) gained much knowledge in the area of high-energy physics.

Research directed toward creating a nuclear bomb also contributed to an improved understanding of high-energy physics.

Exotic atomic nuclei may be described as structures that do not occur in nature, but are produced in collisions. These nuclei have abundances of neurons and protons that are quite different from the natural nuclei. In 1949, M.G. Mayer (Argonne National Laboratory) and J.H.D. Jensen (University of Heidelberg) introduced a spherical-shell model of the nucleus. The model, however, did not meet the requirements and restrains imposed by quantum mechanics and the Pauli exclusion principle. Hamilton (Vanderbilt University) and Maruhn (University of Frankfurt) reported on additional research of exotic atomic nuclei in a paper published in mid-1986 (see reference listed). In addition to the aforementioned spherical model, there are several other fundamental shapes, including other geometric shapes with three mutually perpendicular axes—prolate spheroid (football shape), oblate spheroid (discus shape), and triaxial nucleus (all axes unequal).

In 1964, M. Gell-Mann and G. Zweig (California Institute of Technology) independently pointed out that all the known hadrons (i.e., particles that interact via the strong nuclear force) could be constructed out of simple combinations of three particles (and their antiparticles). These hypothetical particles had to have slightly peculiar properties (the most peculiar being a fractional electric charge). Gell-Mann called these hypothetical particles *quarks* (referring to a sentence in James Joyce's work *Finnegan's Wake*, "Three quarks for Muster Mark"). The theory proposed postulated that three quarks bind together to form a baryon, while a quark and an antiquark bind together to form a meson. With supposition that the binding is such that the internal motion of the quarks is nonrelativistic (which requires the quarks be massive and sit in a broad potential well), then many quite detailed properties of the hadrons could be explained.

The purpose of the quark model was that of explaining the diversity of the hadrons; not to deal with the internal structure of any particle. But awareness of the model created a natural tendency among investigators to associate newly observed particles (among the poorly understood debris from particle experiments) with the hypothetical quarks. A number of properties of *partons* (a name given by Feynman, California Institute of Technology) were measured, including intrinsic spin angular momentum, and these were found to be consistent with the predictions of the quark model. Such observations, of course, added credence to the model.

In the 1960s, the quest for a grand unification theory—a theory that would explain all elementary particles and all forces acting between them—grew in intensity among most investigators who had the good fortune of discovering so many new particles, accompanied by the realization that the ultimate structure of matter was more complex than envisioned in the earlier years. The instrumental means for research (accelerators with higher and higher energies) were getting ahead of the theoretical aspects of the topic. Many particles resulting from collisions were found in the debris of experiments—their presence without plausible explanations. Many questions were posed—for instance, why four kinds of force, each with its own characteristic strength, the strengths differing by nearly 40 orders of magnitude (electromagnetism with its infinite range, the weak force for all practical purposes extending out only 10^{-15} centimeter)? For a while, prospects of a unified theory were dim, but a number of theories were proposed and given sufficient serious attention to warrant planning of experimental tests. As pointed out by Glashow (Nobel Prize, Physics, 1979, shared with Salam and Weinberg), in his Nobel Lecture, "In 1956, when I began doing theoretical physics, the study of elementary particles was like a patchwork quilt. Electrodynamics, weak interactions, and strong interactions were clearly separate disciplines, taught and separately studied. There was no coherent theory that described them

all. Developments such as the observation of parity violation, the successes of quantum electrodynamics, the discovery of hadron resonances, and the appearance of strangeness were well-defined parts of the picture, but they could not be easily fitted together."

In the early years of investigation, the weak force and the electromagnetic force were regarded as indistinguishable. They were of the same strength and possessed the same infinite range, and they were transmitted by four bosons, all of which were massless. The forces manifested a symmetry, that is, they could be interchanged freely. It was believed that no matter which of these forces was applied, the net effect was the same. These views were later to be altered in the light of the process called *spontaneous symmetry breaking*.[4]

The principle of charge conjugation symmetry states that if each particle in a given system is replaced by its corresponding antiparticle, then it would not be possible to tell the difference. For example, if in a hydrogen atom the proton is replaced by an antiproton and the electron is replaced by a positron, then this antimatter atom will behave exactly like an ordinary atom—if observed by "persons also made of antimatter." In an antimatter universe, the laws of nature could not be distinguished from the laws of an ordinary matter universe.

However, it turns out that there are certain types of reactions where this rule does not hold, and these are just the types of reactions where conservation of parity breaks down. For example, consider a piece of radioactive material emitting electrons by beta decay. The radioactive nuclei are lined up in a magnetic field which is produced by electrons traveling clockwise in a coil of wire, as seen by an observer looking down on the coil. Because of the asymmetry of the radio-active nuclei, most of the emitted electrons travel in the downward direction. If the same experiment were done with similar nuclei composed of antiparticles and the magnetic field were produced by positron current rather than an electron current, then the emitted positrons would be found to travel in the upward, rather than in the downward, direction. Interchanging each particle with its antimatter particle has produced a change in the experiment.

However, the symmetry of the situation can be restored if we interchange the words "right" and "left" in the description of the experiment at the same time that we exchange each particle with its antiparticle. In the above experiment, this is equivalent to replacing the word "clockwise" with "counterclockwise." When this is done, the positrons are emitted in the downward direction, just as the electrons in the original experiment. The laws of nature are thus found to be invariant to the simultaneous application of charge conjugation and mirror inversion.

Time reversal invariance describes the fact that in reactions between elementary particles, it does not make any difference if the direction of the time coordinate is reversed. Since all reactions are invariant to simultaneous application of mirror inversion, charge conjugation, and time reversal, the combination of all three is called *CPT* symmetry and is considered to be a very fundamental symmetry of nature.

A relatively recent type of space-time symmetry has been introduced to explain the results of certain high-energy scattering experiments. This is *scale symmetry* and it pertains to the rescaling or "dilation" of the space-time coordinates of a system without changing the physics of the system. Other symmetries, such as chirality, are more of an abstract nature, but aid the theorist in an effort to bring order into the vast array of possible elementary particle reactions.

A feature of quantum field theory is that the quanta of the fields are initially massless. Spontaneous symmetry breaking offers a mechanism by which weak field quanta, for example, can acquire masses. Unification of weak and electro-magnetic forces may be viewed thus in the following manner at short distances (high energies), the masses of the weak field quanta become unimportant and thus original symmetry is restored. Symmetry in this context refers to the properties of the equations of motion of particles in the field theories. Spontaneous symmetry breaking occurs when solutions of the equations do not display full symmetry. Some physicists have likened this to a ball moving on a roulette wheel, whose

---

[4] Prior to 1956, it was believed that all reactions in nature obeyed the law of conservation of parity, so that there was no fundamental distinction between left and right in nature. However, Yang and Lee pointed out that in reactions involving the weak interaction between particles, parity was not conserved, and that experiments could be devised that would absolutely distinguish between right and left. This was the first example of a situation where a spatial symmetry was found to be broken by one of the fundamental interactions.

equations of motion are symmetrical about the axis of rotation even though it always stops in an asymmetric position.

The first direct evidence that the proton has not only size, but structure was provided by an experiment at the Stanford Linear Accelerator Center (SLAC) in 1970. Previously it had been established that the proton is not a point-like particle, but has a finite size—a diameter of about 10^{-13} centimeter. Although it is only about 1/100,000 the size of an atom, it is still measurable. This is unlike certain other particles, notably the electron, for which no extension has been noted, so that the electron can be regarded as a mathematical point. In the experiment, electrons were raised to an energy of some 20 billion electron volts and struck protons and neutrons in the atoms of a stationary target. The angular distributions of the scattered electrons and of other particles created in the collisions were carefully monitored. Most of the electrons, as expected, passed through the target with little change in direction. An unexpected excess of widely scattered particles was produced, however—much greater than if the proton were diffuse and homogeneous. The excess of the widely scattered particles was attributed to a mass embedded within the proton, estimated at no more than $\frac{1}{50}$ the diameter of the proton. In later experiments, a target was illuminated by means of muons (like electrons but with a mass 200 times greater); and by a beam of neutrinos (which lack both mass and electric charge). The results of the original and later experiments were consistent and the deep scattering of particles was attributed to collisions between the incident leptons and some "hard" constituent of the proton.

Theories and postulations continue to be developed concerning the symmetry of nature. For example, in a 1986 paper, H.E. Haber (University of California, Santa Cruz) and G.L. Kane (University of Michigan) observe that *supersymmetry* could represent the next step in the quest for a few simple laws that explain the nature of matter. Physicists are seeking evidence to test the theory. As described, in supersymmetry, for every ordinary particle that exists there is a so-called "superpartner" having similar properties, with exception of the quantity referred to as *spin*, previously discussed here. In a 1985 paper, C. Quigg (Fermi National Accelerator Laboratory) further describes current theories and hypotheses pertaining to elementary particles and forces and observes that a coherent view of the fundamental constituents of matter and the forces governing them is beginning to emerge. A present goal of physicists is to merge disparate theories into a single comprehensive description of natural events. It is agreed among most high-energy specialists that to reach that goal, greater and greater energy must be brought to the experiments. The Superconducting Supercollider (SSC) would be one of these. The complex concept of CP invariance is described in a 1988 paper by R. Adair (Brookhaven National Laboratory). In this postulate, the claim is made that without CP invariance there would be no matter in the universe. Adair observes that if the approximate symmetry between matter and antimatter that has been observed were perfect, the universe would be elegantly simple but virtually empty of matter and of creatures made up of that matter who could contemplate that elegance. It is proposed that the existence of the universe as currently known comes from a flaw in a symmetry exhibited by a universal mirror (the CP mirror), i.e., a symmetry that requires that the outcomes of some events in nature should remain the same on changing matter to antimatter and viewing the result in a mirror.

Before discovery of the hadron particle (designated *psi* or *J*), and after much experimental and theoretical effort, physicists had about concluded that three massive, fractionally charged entities (quarks) were the primary building blocks of the universe. However, discovery of the psi particles in 1974 indicated a fourth quark was required. Previously, in the three-quark model, all mesons were made up of one quark and one antiquark; baryons, of three quarks; and all anti-baryons, of three antiquarks. Prior to 1974, all of the known hadrons could be accommodated within this basic scheme. Three of the possible meson combinations of quark-antiquark could have the same quantum numbers as the photon, and hence could be produced abundantly in e^+e^- annihilation. These three predicted states had all been found.

As pointed out by Richter (1977), the first proposal of a theory based on four quarks rather than three was published in 1964 by Amati and others. The motivation at that time was more esthetic than practical, and these models gradually expired for want of an experimental fact that called for more than a three-quark explanation. In 1970, Glashow explained in a paper that a fourth quark was required to explain the nonoccurrence of certain weak decays. The fourth or *c* quark was assumed to have a charge of $+\frac{2}{3}$, like the *u* quark, and also to carry +1 unit of a previously unknown

quantum number, called *charm* by Glashow, which was conserved in both the strong and electromagnetic interactions, but not in the weak interactions. Discovery of the psi particles demonstrated a more compelling need for the fourth quark. Richter observes that the four-quark model of hadrons seemed to account, in at least a qualitative fashion, for all of the main experimental information that had been gathered about the psions, and by the early part of 1976, the consensus for charm had become quite strong.

In 1977, the *upsilon particle* was found as the result of energetic collisions between protons and copper nuclei. The upsilon particle has a mass three times greater than any other subatomic entity yet detected (early 1980s). Researchers on this experiment from Columbia University, the State University of New York (Stony Brook), and the Fermi National Accelerator Laboratory reported that, with a mass at its lower energy state equivalent to 9.0 GeV and masses in excited states equivalent to 10 and 10.4 GeV, the upsilon particle has been interpreted as consisting of a massive new quark (the fifth) bound to its antiquark. Confirming experiments were also conducted at the Deutsches Elektronen Synchrotron (DESY) located near Hamburg, Germany. The quantum attribute of the fifth quark was named "bottom." With a fifth quark reported, many physicists felt that finding a sixth quark ("top") was highly probable.

In 1979, the Nobel Prize (Physics) was awarded to Glashow and Weinberg (both of Harvard) and Salam, a Pakistani physicist, in recognition of the significance of the theory which unites the weak force with the electromagnetic force. But most scientists recognize this finding as only a milestone in a series of predictions that include the existence of new particles so massive that they cannot be expected to appear at the energies thus far available to physicists.

The chemists of the nineteenth century once thought that all material substances were comprised of only 36 elements.[5] Over the years, these expanded to over 100 elements. Fifty years ago, it was proclaimed that the elements were made of electrons, protons, and neutrons. Then, commencing in the 1940s, many other particles were found, as described here. Then for a while it seemed that elementary matter could be reduced to three particles—the quarks. But quarks have multiplied in number, with a sixth quark now seriously proposed. Will there be too many quarks? Perhaps hypothetical particles will be proposed from which the quarks are comprised. Possibly the ultimate answer will lie with the "mathematical groups that order the particles rather than in truly elementary objects."

In the late 1980s and thereafter, the *superstring* theory was drawing the attention of most theoretical physicists. Out of this theory may emerge the long sought concept that will account for all four fundamental forces. Quantum theories have been formulated for three of the four known forces of nature—strong, weak, and electromagnetic interactions. This comprises one of two basic foundations. The other, of course, is Einstein's general theory of relativity, which relates the force of gravity to the structure of space and time. A quantum theory of gravity is missing. Some scientists currently feel that there is much potential for the string theory to bring this unification about. String theory was first proposed by Y. Nambu (University of Chicago) in 1970. Traditional models of elementary particles are based on quantum field theory, which involves dimensionless points and quantum numbers, but with no specified internal structure. In studying one of the alternatives to the foregoing, a dual resonance model, Nambu observed that it was equivalent mathematically to the interaction of bits of string. The dual resonance model had been constructed to show how one hadron should scatter off another. The theory envisions the strings about 10^{-35} meter long (10^{20} times smaller than the diameter of a proton). As observed by M. Green (Queen Mary College, University of London) in a 1986 paper (reference listed), string has extension; it can vibrate like a violin string. The harmonic, or normal, modes of vibrations are determined by the tension of the string. In quantum mechanics, waves and particles are dual aspects of the same phenomenon. Thus, each vibrational mode of a string corresponds to a particle. It is envisioned that two strings interact when they touch their tips together and are fused into one—or one string may split into two parts. Strings can absorb energy in a collision and may ripple and rotate (at the speed of light). The original string theory initially was limited to providing a satisfactory "explanation" for describing bosons (pi meson and rho meson), i.e., particles that have integral numbers of spin angular momentum units. Attempts to describe particles with half-integral spin, such as fermions (proton, neutron), failed. However, in 1976,

---

[5] Line of thought suggested by L.M. Lederman (Columbia University) in a paper on "The Upsilon Particle," *Sci. Amer.*, **239**, 4, 80 (1978).

a suggestion was made by Scherk, a French physicist, to the effect that a string model could represent a fermion, but with the proviso that a fermion was matched by a corresponding boson. Incidentally, this is the type of correspondence required by the principle of supersymmetry. One of several difficulties that lie ahead for the superstring theory is that of matching its mathematical concepts (26 dimensions—25 space and 1 time) with the real world of 4 dimensions. As one physicist has observed (E. Witten, Princeton University), it is a complete mystery what string theory is at a fundamental level.

State of the Science (Early 1990s)

In terms of comprehending the ultimate nature of matter, most scientists would agree that tremendous progress has been made over the past century and a half, but that a grand unifying theory of particles and their structures and interaction may continue to elude researchers for many years.

Number of Particle Families. How many families of matter may exist? Three, four, or more? An acceptable number among researchers today is three. Three family entities make up matter—the stars, the planets, molecules, and the atoms in the paper upon which this is printed. These fundamental particles are the "up" quark, the "down" quark, and the electron. Some other researchers are not quite so confident. One is reminded of the quotation from Jonathan Swift:

> So, naturalists, observe, a flea
> Hath smaller fleas that on him prey:
> And these have smaller still to bite'em
> And so proceed ad infinitum.

Within the limitations of contemporary knowledge, there are three families of fundamental particles, the approximate properties of which are given in Table 1. In 1991, G. Feldman (Harvard University) and J. Steinberger (CERN and 1988 Nobelist for discovery of the muon nutrino) observed, "Many questions remain unanswered. Why are there just three families of particles? What law determines the masses of their members, decreeing that they shall span 10 powers of 107? These problems lie at the center of particle physics today. They have been brought one step closer to solution by the numbering of the families of matter."

In their reference listed, Feldman and Steinberger describe how experiments at CERN and SLAC, using electron-positron collisions, showed that there are only three families of fundamental particles in the universe.

By contrast, as D. Cline (University of California) observed in 1988, "Several theorists think a new quark should exist in the vicinity of 246 GeV. One of the notable features of the standard model is its prediction that at high enough energies the various forces begin to unify. In particular, the electromagnetic force and the weak and strong nuclear forces should become a single 'grand unified' force. The forces should be unified at the incredible energy of 10^{15} GeV, considerably beyond what can ever be attained by an accelerator on earth. The extrapolation of measured values of fundamental parameters from low energies to the grand-unified energy scale would require the existence of a new massive quark for consistency."

Nuclear Equation of State. In a late 1991 paper, H. Guthrod (Institute of Heavy-Ion Research, Darmstadt) and H. Stöcker (University of Frankfurt) are developing a nuclear equation of state in order to clarify the

"new" states of matter and conditions that may occur inside a supernova and the organization of the universe. The authors observe, "Nuclear matter in its normal phase resembles a liquid. Increasing the temperature or density 'boils' nuclei into the hadron gas phase. Under extreme density but low temperature, nucleons could become 'frozen,' forming condensates, further heating or compression may produce the plasma phase, which would consist of free quarks and gluons. The gas and plasma phases may exist simultaneously over a wide region. Particles that have strange quarks, such as multistrange, metastable objects ('memos') and strangelets, may also form."

This approach parallels our consideration of the equation of state that applies to "ordinary" matter, such as gases, liquids, and solids that exist in macrostructured materials. See also **Equation of State**.

Particle Accelerators

Subatomic particles, such as electrons, positrons, and protons, can be accelerated to high velocities and energies, usually expressed in terms of center-of-mass energy, by machines that impart energy to the particles in small stages or nudges, ultimately achieving in this way very high-energy beams, measured in terms of billions and even trillions of electron volts. Thus, in terms of their scale, particles can be made to perform as powerful missiles for bombarding other particles in a target substance or for colliding with each other as they assume intersecting orbits. Because the particles are empowered with high energy, their smashing encounters are conducive to breaking the particles into their constituents. Instruments or machines used to arrange these particle encounters are known as *particle accelerators* and are very large, their dimensions frequently measured in terms of a few miles or kilometers. Thus, in a sense, it is ironic that the largest tools of science are required to seek knowledge concerning the smallest particles of matter that make up the universe and the earth. Theoretically, if at some future date sufficient energy can be imparted to two particles, their head-on collision will yield a complete fireball of disintegration, such that the absolute, indivisible particles of matter will be lain bare. A further division of matter then would be theoretically impossible. This is a goal of many particle physicists and perhaps achievable within a relatively few decades. Or, on the other hand, would then new theories be formulated to show that the true absolutes have indeed not yet been found?

Electromagnetic forces are used to accelerate particles, requiring that the particles must have an electric charge. Protons $(+1)$ and electrons (-1) are commonly used as the media for particle-physics experiments, although not exclusively. The particles must be accelerated within a vacuum because otherwise they would collide with the molecules of air. The tube-shaped enclosures for the speeding particles are maintained under a vacuum of about 10^{-9} torr. The particles are set in motion by an electric field, which in the simplest configuration is an applied high voltage across a pair of electrodes. The positive electrode attracts electrons; the negative electrode attracts protons. It may be pointed out that an acceleration of this kind occurs in the ordinary television receiver. This basic principle, which applied to the early accelerators, remains the basis of the current operating and planned accelerators of the future. A simple electrode arrangement cannot sustain a potential of over a few million volts because of breakdown of insulation and arcing, and thus is confined to the simplest kind of accelerator where high energies are not required. In a practical sense, particles must be provided with energy in a large number of stages or "nudges." Thus, as in a *linear accelerator* (linac), the stages are strung out along a straight path, each stage requiring a radio-frequency oscillator which sets up an alternating electric field which is connected to each set of electrodes. Many radio-frequency cavities are formed along the line, and to provide the correct parity so that the particles will be accelerated rather than retarded the oscillators for the successive cavities must be synchronized. Effectively, an electromagnetic wave that travels continuously through the evacuated chamber is set up. It has been suggested by one physicist that the particle "rides the electrical wave as a surfer rides a water wave." In a less costly arrangement, known as the *synchrotron*, the particles are made to follow a circular or closed curve rather than a linear course. Groups of particles may circle a ring of this kind several million times while they are increasing their energy and velocity. Only one or a few radio-frequency cavities are required because energy is picked up each time the particle completes a revolution.

In addition to linear accelerators and synchrotrons, the accelerated particles of which ultimately are directed to strike a fixed target, there are *colliding-particle machines* in which particles are made to collide head-on.

TABLE 1. THREE FAMILIES OF FUNDAMENTAL PARTICLES

	Charge	Mass in billions of electron volts (GeV)		
		Electron Family	Muon Family	Tau Family
Quarks	2/3	*Up* ~0.01 GeV	*Charm* ~1.5 GeV	*Top* Est. 89 GeV*
	−1/3	*Down* ~0.01 GeV	*Strange* ~0.15 GeV	*Bottom* ~5.5 GeV
Leptons	0	*Electron neutrino** < 2 × 10⁻⁸ GeV	*Muon Neutrino** < 2 × 10⁻⁴ GeV	*Bottom* <5.5 GeV
	−1	Electron 5.11 × 10⁻⁴ GeV	*Muon* 0.106 GeV	*Tau* 1.78 GeV

→ = Relative Increase in Mass →

(*Mass unknown)

Data Source: G.J. Feldman and J. Steinberger (see reference listed).

These machines are similar to synchrotrons except one bunch, or cluster of particles, travels in one direction while another bunch travels in the opposite direction. Where the colliding particles have the same rest mass and, after acceleration, have the same energy, the center-of-mass energy is the sum of the two beam energies. Thus, two beams with energies of 50 GeV (billion electron volts), for example, can provide the colliding force of one beam of 100 GeV against a fixed target. As will be shown later, head-on collision apparatus requires the use of storage rings, and systems are arranged in various configurations. Electron-positron storage rings are particularly efficient for creating new elementary particles from high-energy collisions.

The energy acquired by the particles in an accelerator is expressed in electron volts (eV), the amount of energy gained by any particle bearing a charge equal to that of an electron when it falls through a potential difference of 1 volt. Thus, 10^3 eV = 1 keV (kiloelectron volt); 10^6 eV = 1 MeV (megaelectron volt or one million eV); 10^9 eV = 1 GeV (gigaelectron volt or one billion eV); and 10^{12} eV = 1 TeV (teraelectron volt or one trillion eV). The large accelerators have been in the GeV range. Plans call for machines in the TeV range. Inasmuch as the mass and energy can be freely interconverted, the mass of a particle is usually expressed in terms of its energy equivalent in eV. The mass of the proton thus is 938 MeV.

High-energy machines require a supply not only of the type of particle desired, but also particles that have been preliminarily accelerated. Electrons are comparatively easy to generate as inputs to accelerators, as by the Cockcroft-Walton generator. Protons are obtained by ionizing hydrogen atoms.

As early as 1928, E.O. Lawrence constructed one of the first particle accelerators out of laboratory glassware. This was only a few inches in diameter. Principles remain essentially the same even today, but the size of accelerators has increased tremendously. They no longer are parts of laboratories, but rather the detection and laboratory aspects of modern accelerators are appended around the accelerator. Modern accelerators cost many millions of dollars and a major installation requires numerous scientists and scores of support people, aided by a number of digital computers. In the future, costs may become so high that several countries, working together, will have to support particle-physics research facilities, as already exemplified by the European Organization for Nuclear Research (abbreviation CERN for the former name, *Conseil EuropÉen pour la Recherche NuclÉaire*) with facilities located in Geneva, Switzerland, and adjacent land in France.

Particle Generators. Direct voltage for charging particles may be obtained in two basic ways: (1) by means of a cascade process, such as the cascade rectifiers or voltage-multiplying circuits; and (2) by charging up a terminal through actual transportation of the charge.

The Cockcroft-Walton generator is of the first type and consists of several stages of a voltage-doubling circuit together with an ion source and a suitably designed discharge tube. Although it is possible to use electrons, these accelerators are usually used as positive-ion sources and can provide dc currents up to about 10 mA and energies up to about 1.5 MeV without special pressure tanks. With this type of accelerator, Cockcroft and Walton were able, in 1932, to induce the first nuclear reaction using artificially accelerated protons. The reaction was: $^1H + ^7Li \longrightarrow ^8Be \longrightarrow 2\,^4He$. See Fig. 1.

The second method is employed in the *electrostatic*, or *Van de Graaff*, generator where a row of corona points sprays charge onto a moving belt that carries the charge to the field-free region inside a spherical metal terminal. Currents of about 1 mA for electrons, and up to about 500 μA for positive ions, are obtained in the range 1 to 5 MeV with a precision of about 0.1%. The whole apparatus is enclosed in a pressure tank and operated at about 10 atmospheres. In this form, the maximum energy is close to 10 MeV and is limited by breakdown between the terminal and its surroundings. However, double the energy can be attained by accelerating negative ions to the positively charged terminal; then, through electron-stripping, positive ions are created which can be accelerated again as they pass from the terminal to ground. Such tandem Van de Graaff generators, in two- and three-stage variations, can provide particles in the 10 to 30 MeV range, with high precision. See Fig. 2.

An electric field may also be produced by a time-varying magnetic field. The changing magnetic flux in the central core of a pulsed cylindrical electromagnet induces a transverse electric field that accelerates the particles. These travel in a doughnut-shaped vacuum chamber located

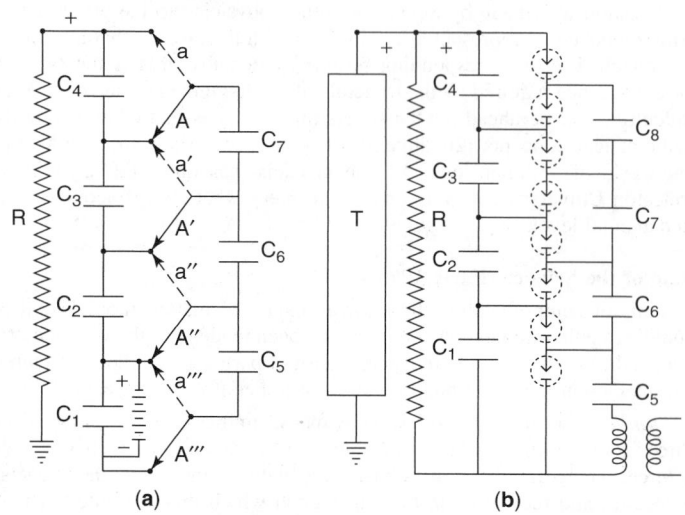

(a)   **(b)**

Fig. 1.   Cockcroft and Walton scheme in which rectified A.C. voltage is used to charge a series of capacitors to a high D.C. potential, which is then applied to the acceleration of charged nuclear particles. The principle of operation can best be described by the apparatus using mechanical switches shown in A. Condensers of equal capacity are represented by C_1, C_2, C_3, etc. If the switch blades are down, the battery B is connected across condensers C_1 and C_5. On moving the blades up (dashed lines), condenser C_5 transfers half its charge to C_2. When the blades are moved down again, C_2 transfers half its charge to C_6 and C_5 is recharged to full capacitance. A continuous up and down motion of the blades results in a transfer of charge up the condenser bank until every condenser is charged to the voltage of the battery. The total voltage applied to the discharge tube, symbolized here by the resistance R, is the sum of the voltage across the condensers C_1, C_2, C_3 and C_4 in series. In the actual apparatus (part B), an alternating voltage was applied through a transformer and the switching action was accomplished by the use of rectifier tubes

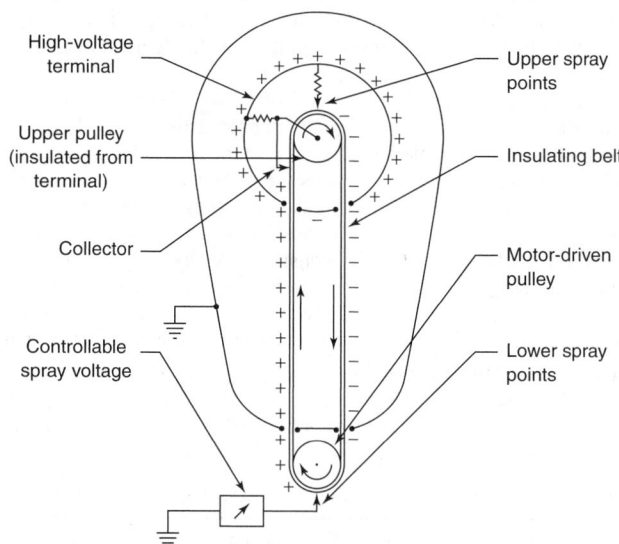

Fig. 2.   Diagram of Van de Graaff electrostatic belt generator

between the poles of the magnet surrounding the core. The magnetic field between these poles keeps the particles traveling in a circle, but it must be carefully designed to keep the particle orbits within the vacuum chamber during each pulse. Although betatrons can accelerate positively charged particles, they have been used for electrons. The electrons can be extracted, but they usually bombard an inner target to produce beams of X-rays, which can be as intense as 1400 roentgens/minute at a distance of 1 meter from the target. Pulsing rates vary from 30 to 60 times per second.

Other types of accelerators use various forms of rf electric fields, at relatively low voltage, which are applied many times in a given direction to the particles and are prevented from influencing them when the rf field is reversed.

Linear Accelerators. The linear accelerator (*linac*) has the advantage that the accelerated beam is easily extracted for experimental use. In principle, it is capable of producing well-focused beams of higher intensity than are available from circular machines of the synchrotron type. It does, however, require very high power levels at frequencies where conversion equipment is relatively expensive. For a given final energy, a linear accelerator will usually be materially more expensive than a synchrotron.

The rf fields used for acceleration are set up in a long cylindrical cavity whose axis is to be the axis of the accelerated beam. Hence, for acceleration the field pattern must have a major electric field component parallel to the axis. This requirement is satisfied by the TM_{01} waveguide mode in which a paraxial electric field has its maximum strength at the axis and falls to zero at the cavity wall. Azimuthal magnetic fields lie in planes normal to the axis, have small values near the axis and increase to maximum values at the cavity walls. Usually the field pattern is maintained by coupling to these magnetic fields by loops or apertures excited by external power sources. Corresponding to the high rf magnetic field at the wall, paraxial currents flow in the walls and are responsible for a major function of the power loss in the system. When high electric fields are required on the axis to accelerate to high energy in reasonable distances, the wall currents are correspondingly high.

Both standing wave and traveling wave patterns can be used in linear accelerators. If traveling waves are used, the phase velocity of the waves must be made equal to the velocity of the particles accelerated; as the particle velocity increases, the phase velocity must also increase. But, phase velocities in simple waveguides always are greater than the velocity of light, and loading must be introduced to reduce the phase velocity to the desired value. This can be accomplished by the introduction at intervals of washer-shaped irises.

The operating principles of a linear accelerator are shown in Fig. 3.

Synchrotrons. A particle is made to follow a circular (or other closed curve) orbit by arranging a number of magnets in a ring. The principle is illustrated by Fig. 4. Two kinds of magnets are required. Dipole magnets (two poles) generate a uniform magnetic field. Spaced around the ring, these magnets bend the particle trajectory. To keep a concentrated beam of particles, quadrupole magnets (two north poles and two south poles), which have no effect on deflecting the particles, are used to focus them. Acting like lenses, these magnets form the particles into a narrow beam. Depending upon the size and general configuration of a synchrotron, radio-frequency cavities may be variously interspersed among the magnets where the actual acceleration occurs. Special magnets are used for injecting particles into the ring and for extracting the accelerated beam of particles.

The first substantial synchrotron was built as early as 1952. Known as the *Cosmotron*, it was installed at Brookhaven National Laboratory on Long Island, New York. The device achieved energies up to 3 GeV. Two years later, the *Bevatron*, with energies up to 6.2 GeV, was installed at the University of California at Berkeley. A shortcoming of these earlier designs was the magnet system, which provided inadequate focusing of the beam. A system of strong focusing was introduced to later-generation synchrotrons. As pointed out by R.R. Wilson (1980). "The shape of the

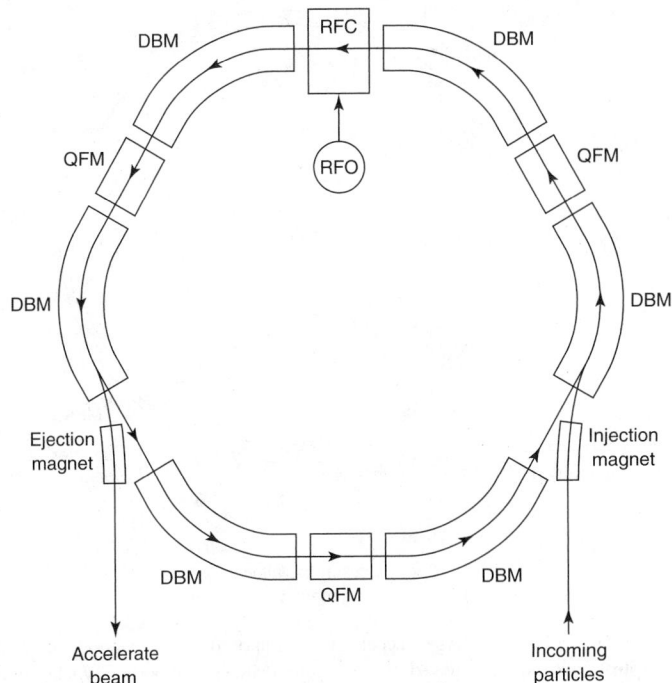

Fig. 4. Principle of synchrotron. One radio-frequency cavity (there may be several) provides a small "push" each time a particle passes through it. Unlike the linear accelerator, the synchrotron requires only one or few rf cavities. Dipole bending magnets keep the particles on their proper course. Focusing magnets keep the particles in a narrow beam, thus preventing undesired scattering. Particles enter the machine through an injection magnet and leave through an ejection magnet. DBM = dipole bending magnet; QFM = quadrupole focusing magnet; RFC = radio-frequency cavity; RFO = quadrupole focusing magnet; RFC = radio-frequency cavity; RFO = radio-frequency oscillator

magnetic field can be described mathematically as being partly uniform (the dipole component) and partly a gradient in a direction transverse to the orbit of the beam (the quadrupole component). The quadrupole component was made stronger and was alternated in sign, so the oscillations of the particles around the desired orbit were more frequent, but of smaller amplitude. As a result of this alternating gradient, the aperture of the magnets and the bore of the vacuum chamber could be smaller. It is the invention of the synchrotron and of strong focusing that has made the very large accelerators of today economically feasible." Synchrotrons with this design are known as *alternating gradient synchrotrons* (AGS).

The operation of a synchrotron is cyclic, i.e., a bunch of particles will be injected, with the bending magnets precisely adjusted to cause the particles to closely follow the curvature of the evacuated chamber. But, as the particles increase in energy, the field strength in the bending magnets must also be increased. Upon achieving their desired or maximum possible energy level, the particles are extracted—possibly going directly to bombard a target, or to supply an even more powerful synchrotron for further acceleration. When experiments with the first group or bunch of particles have been completed, the magnetic field is reduced to its original level, after which the next bunch or group of particles is added. The term *synchrotron* stems from the fact that the particles automatically synchronize their motion with the rising magnetic field and the rising frequency of the accelerating voltage. It is interesting to note that some accelerators (linacs or synchrotrons), which in their day may have been regarded as most powerful, may later be used as preliminary particle accelerators, as feeders to larger machines of later designs. Thus, instead of dismantling the older machines, in some cases the cost of a more powerful machine can be reduced. An example is shown in Fig. 5. Another example at CERN in Geneva, Switzerland will be cited later. Once the particles in a synchrotron have reached full energy, they are nudged out of their orbit by a special ejection magnet.

The linacs and synchrotrons described thus far are for use with *fixed targets*. The advantages of colliding-particle machines will be described shortly. What, then, are the reasons for continuing to bombard fixed targets? The fixed-target configuration provides the creation of a variety of

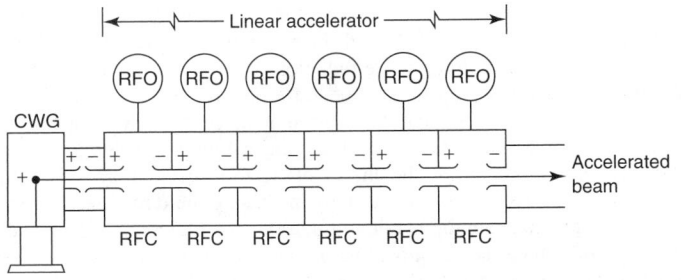

Fig. 3. Principle of linear accelerator (linac). Partially accelerated electrons from a source, such as a Cockcroft-Walton generator, are further accelerated by stages as the electrons pass through radio-frequency cavities, powered by rf oscillators. Each particle receives a small "push" as it passes from one cavity to the next until the final desired accelerated beam is produced. The machine must be carefully synchronized. CSG = Cockcroft-Walton generator; RFO = radio-frequency oscillator; RFC = radio-frequency cavity

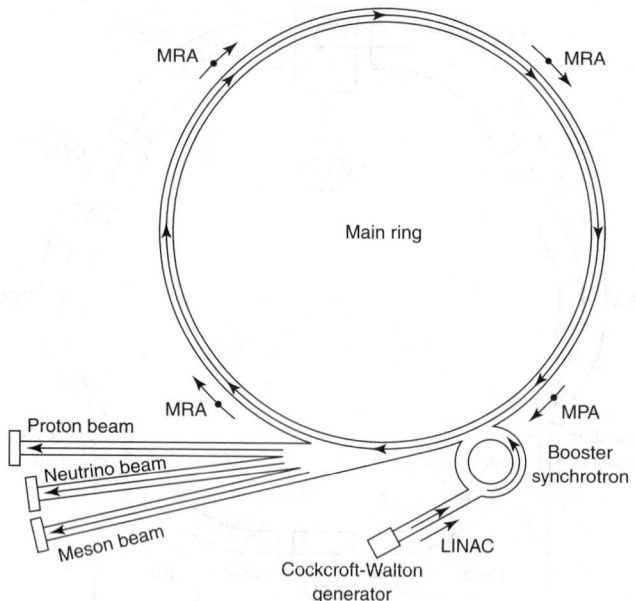

Fig. 5. An early fixed-target accelerator comprised of a large main-ring synchrotron with four stages of acceleration (MRA); a booster synchrotron; a linear accelerator (linac), and a Cockcroft-Walton generator. Protons are accelerated to 0.75 MeV in the Cockcroft-Walton generator; to 300 MeV in the linac; to 8 GeV in the booster synchrotron; and to 400–500 GeV in the main-ring synchrotron. Experiments are not limited to accelerated protons, but also can be conducted with beams of secondary particles (mesons and neutrinos) which are knocked out of the target by impacting protons

secondary beams (neutrinos, muons, pions and other mesons, antiprotons, and massive particles called hyperons). Fixed-target machines also are effective in furnishing particles to larger accelerators and storage rings. When particles are beamed at a fixed target, an interaction is assured. In contrast, since in a colliding-particle configuration the great majority of particles will simply pass each other without colliding, the number of events of interest may only occur at the rate of comparatively few per day.

Colliding-Particle Machines. As mentioned earlier, the effective beam energy essentially can be doubled when particles can be made to

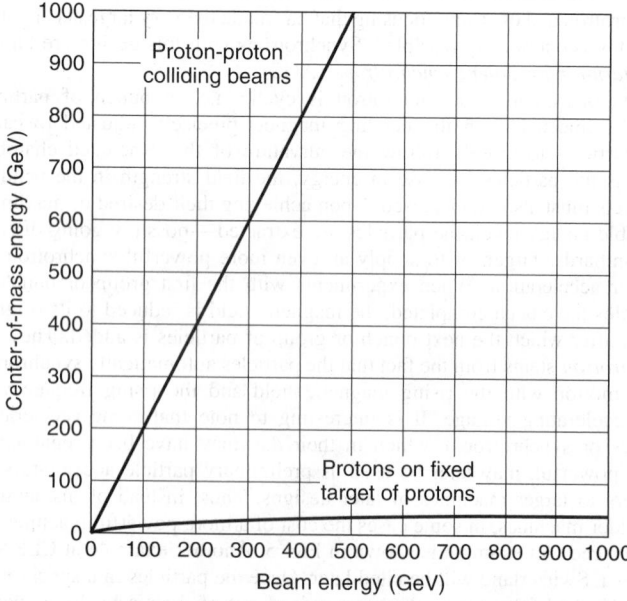

Fig. 6. Comparison of effective collision energy. Colliding-beam devices double the beam energy effectiveness. In fixed-target accelerators, the center-of-mass energy is proportional to the square root of the beam energy (at low energy) and at higher-energy levels, this rises even more slowly because of relativistic effects. (*After R.R. Wilson*)

collide head-on with each other. This is shown dramatically by the graph of R.R. Wilson (1980) and given in Fig. 6. Wilson had made an effective comparison of the resolving power of an accelerator with that of a microscope. In a microscope, the ultimate limit to resolution is the wavelength of radiation that illuminates the specimen. Thus, a visual lightwave microscope is limited to distinguishing objects of 10^{-5} centimeter or larger. Since a particle can be described (quantum mechanics) as a wave, the wavelength is inversely proportional to the momentum of the particle. Thus, one objective in accelerator design is improvement of resolution by reduction of particle wavelength. The early large accelerators had an effective resolution of about 10^{-16} centimeter (1/1000 the diameter of the proton).

When a particle and its antiparticle, such as an electron and a positron, or a proton and an antiproton, are used in head-on collision experiments, acceleration of the particles can be accomplished in one ring. This is because electrons and positrons, for example, behave in the same way in terms of their response to magnetic and electric fields. Thus, both particles can be injected into the same ring, one to follow an orbit in a clockwise direction; the other in a counterclockwise direction. Upon injection of a cluster of each type of particle, collisions occur at two points diametrically opposed. This arrangement provides maximum utilization of the equipment.

Other advantages of using a particle and its antiparticle, particularly the electron and the positron, is that when collisions occur, a less confusing splash of debris occurs.

When advantage of the single-ring, head-on collision approach cannot be taken, then *storage rings* are required. These rings are designed much like synchrotrons. Their primary purpose usually is not to accelerate the particles, but rather to maintain or "store" their energy while they continue to circulate in orbits in the ring. Just sufficient energy (usually furnished by a single radio-frequency cavity) is added to overcome losses, mainly due to synchrotron radiation. The storage rings are physically arranged tangentially to the main synchrotron ring so that precision transfer of particles from the storage rings to the main ring can be affected. Another concept was introduced in the early 1970s at the European CERN facility, where two interlaced rings that store counter-rotating proton beams are used. These rings cross over at eight points around their circumference. There are seven interaction zones that will accommodate detectors.

Scientists at the National Laboratories of the National Committee for Nuclear Energy (C.N.E.N.) at Frascati, Italy pioneered the first electron-positron ring in 1959. Each beam of the machine was 0.25 GeV and yielded a center-of-mass energy of 0.5 GeV. This ring was later moved to the Orsay laboratory outside Paris, France. The number of such rings grew quite rapidly.

Synchrotron Radiation

The electromagnetic radiation emitted as a result of continual acceleration toward the axis of rotation of charged particles moving in a magnetic field is generally known as synchrotron radiation because it was first observed during the operation of electron synchrotrons. The rate of emission of synchrotron radiation varies with the fourth power of the particle energy, inversely with the radius of curvature, and inversely with the fourth power of the rest mass. Because of the rest mass relationship, no significant quantity of this radiation is formed from proton trajectories in magnetic fields, but an easily measurable loss of energy to synchrotron radiation occurs when electrons with high kinetic energies spiral through a magnetic field. Many microwave radiations observed in radio-astronomy measurements are believed to have been formed as synchrotron radiation.

At the beam energies that have been attained to date, synchrotron radiation is not a major factor in designing a proton accelerator, but it is the principal limitation on the energy of electron accelerators. Lessening the curvature of a synchrotron and thus increasing the circumference will reduce synchrotron radiation. For this reason, the radiation is practically negligible in a linear accelerator (linac). Synchrotron radiation energy loss in a storage ring must be made up by providing additional energy from one or more radio-frequency cavities. However, synchrotron radiation is not entirely wasted energy. The effect of the radiation is to damp out small excursions of the electrons away from their main trajectory, making a beam of electrons easier to control than a beam of protons.

Aside from its effects in high-energy experimentation, synchrotron radiation is of interest and value primarily as a source of tunable coherent x-rays. As summarized in a 1985 paper by Atwood, Halbach, and Kim

(Lawrence Berkeley Laboratory), a modern 1- to 2-billion-eV synchrotron radiation facility (based on high-brightness electron beams and magnetic undulators) would generate coherent, laser-like, soft x-rays of wavelengths as short as 10 angstroms. The radiation would also be broadly tunable and subject to full polarization control. Radiation with these properties could be used for phase- and element-sensitive microprobing of biological assemblies and material interfaces as well as research on the production of electronic microstructures with features smaller than 1000 angstroms. These short wavelength capabilities, which extend to the K-absorption edges of carbon, nitrogen, and oxygen, are neither available nor projected for laboratory XUV lasers. Higher-energy storage rings (5 to 6 billion eV) would generate significantly less coherent radiation and would be further compromised by additional x-ray thermal loading of optical components.

Much interest and limited application of synchrotron radiation by medical professionals appeared in the late 1980s. Synchrotron radiation is intense and its brilliance extends continuously over a broad bandwidth from infrared to hard x-rays. For example at the Stanford Synchrotron Radiation Laboratory, the output in the x-ray region is more than 100,000 times that of the most powerful x-ray tube. The natural collimation of synchrotron radiation can be monochromatized by different crystals and by adjusting the angle at which the crystal intercepts the radiation, the energy of the monochromatic beam can be selected. The tunability and intensity of the monochromatic radiation are currently making transvenous angiographic applications possible. See also **X-Ray Scan and Other Medical Imagery**.

Third-Generation Synchrotrons. The first run of experiments at the new, $500-million European Synchrotron Radiation Laboratory (ESRF), located in the DauphinÉ Alps, was scheduled to commence in the early 1990s. The facility will provide the brightest continuous x-rays in the world, but have been eclipsed by a new Japanese facility in the mid-1990s. The ESRF will generate x-rays two orders of magnitude more brilliant than any prior facility. Stronger x-ray sources are of an advantage to biologists, physicists, chemists, and other researchers because of their shortened exposure time and increased resolution. As pointed out by one American physicist who plans to use the facility, "We'll be able to get a picture in minutes, seconds, or fraction of a second and there will be much less specimen damage when the photons enter and leave a specimen in a millisecond."

A large variety of experiments was scheduled, representing numerous scientific disciplines. Another similar but slightly larger facility was scheduled for the Riso National Laboratory in Roskilde, Denmark, sometime in 1996.

The most powerful synchrotron now is at Harima Science Garden City, approximately 60 mi (96 km) west of Osaka, Japan. It will be known as the SPring-8 (Super Photon Ring-8 GeV). It is envisioned that it may be able to fulfill the x-ray crystallographer's dream—that is, x-ray holography. Hiromichi Kamitsubo, project head, states, "X-ray holography would enable the direct visualization, not just see complex x-ray diffraction patterns of the structure of materials—as if we were seeing them with a microscope." Three-dimensional holograms created in a photosensitive material require high-powered and coherent x-rays. The SPring-8 will have an energy rating of 8 GeV, with a main ring circumference of 4708 ft (1435 m) and 80 beam lines. Estimated cost is $1 billion.

Superconducting Super Collider (SSC)

As mentioned earlier, the ultimate fate of the SSC will be determined by economic and political factors. Generally, the scientific community in the United States and worldwide, in fact, looks to the SSC for important information on the structure of matter. Specific goals include:

(1) *The origin of mass.* As explained by Quigg and Schwitters, the current model of the weak interactions, previously mentioned, suggests that a similar situation may be realized in the universe. The field involved is not a magnetic field, but rather what is called a Higgs field. A major objective of the SSC will be to clarify the exact nature of the Higgs field and its interactions with other matter. It is currently assumed that the Higgs field pervades the universe because the total density is minimized in its presence. To change the magnitude of the field significantly (or to manifest one of its particles), sufficient energy density must be supplied to overcome an assumed natural tendency of the field to return to its normal universal background value. The energy levels of the SSC is expected to be sufficient for this.

(2) *Families of particles.* Quigg and Schwitters observe that one theoretical approach to understanding particle families postulates new symmetries under which different families are interchanged. A facet of this concept involves combining family symmetry and gage symmetries of the strong, weak, and electromagnetic interactions into one all-encompassing symmetry, but hidden by the presence of suitable background Higgs fields. Implementing such symmetries could lead to a new class of "family interactions" mediated by heavy bosons that may be created at the levels made possible by the SSC.

(3) *Chirality* (a peculiar asymmetry in the weak interactions of the observed quarks and leptons) can be investigated.

(4) *Supersymmetry* will be investigated, and may lead to unified field theories that include gravity.

(5) *Compositeness* is a characteristic of some particles initially investigated at CERN. Better understanding of the cosmos through SSC experiments is also envisioned. The SSC will make it possible to simulate the conditions that prevailed about 10^{-15} second after the primordial Big Bang explosion when the temperature of the universe was estimated about 10^{17} K.

All of the foregoing factors are explained in considerable detail in the Quigg and Schwitters reference listed.

The new accelerator complex will be based on the accelerator principles and technology that were developed for construction of the Fermilab Tevatron, coupled with extensive experience with superconducting magnets gained over the past two decades. The energy of the SSC will be equal to 20 times that of the Tevatron collider, with a total collision energy of 40 TeV. Racetrack in shape with a circumference of 87 km (52 mi), the SSC will utilize 10,000 helium-cooled superconducting magnets. The SSC will utilize an injector system consisting of a linear accelerator followed by two circular accelerators. The diameter of the main ring used in the final acceleration phase will be about 30 km (18.7 mi), depending upon details yet to be developed pertaining to the magnets that will be used to guide the protons. The superconducting magnet system will require many hundreds of miles of cryogenic plumbing, including several hundred thousand vacuum joints to assure the establishment and maintenance of superconductivity conditions. The linear dimensions of the SSC will be roughly 15 times those of the Tevatron and about 4 times those of the electron-positron collider being constructed by CERN. As suggested by Quigg and Schwitters (Fermi National Accelerator Laboratory) in a 1986 paper, technological fallout from construction of the SSC would include: (1) large-scale industrialization of superconducting wire fabrication and cryogenic refrigerator manufacture, making such technologies available to future power distribution and transportation systems; (2) improved tunneling techniques for future application to public works and transportation projects; (3) large-volume storage of helium, a potentially critical and nonrenewable resource; and (4) computer control and mechanical alignment systems extending over very large areas.

The SSC may furnish answers to critical scientific questions.

Additional Reading

Adair, R.K.: "A Flaw in a Universal Mirror," *Sci. Amer.*, 50 (February 1988).

Alfassi, Z.B. and M. Peisach: *Elemental Analysis by Particle Accelerators,* CRC Press, LLC., Boca Raton, FL, 1991.

Amato, I.: "New Superconductors: A Slow Dawn," *Science*, 306 (January 15, 1993).

Ando, M., C. Uyama, M. Ibaraki, and M. Osaka: *Medical Applications of Synchrotron Radiation,* Springer-Verlag Inc., New York, NY, 1998.

Atutov, S.N.: *Trapped Particles and Fundamental Physics,* Kluwer Academic Publishers, Norwell, MA, 2002.

Bertsch, G.F., L. Frankfurt, and M. Strikman: "Where Are the Nuclear Pions?" *Science*, 773 (February 5, 1993).

Bertschinger, E.: *Uniting Cosmology and Particle Physics,* Freeman, Salt Lake City, UT, 1992.

Branco, G.C., Q. Shafi, and J.I. Silva-Marcos: *Recent Developments in Particle Physics and Cosmology,* Kluwer Academic Publishers, Norwell, MA, 2001.

Breuker, H., et al.: "Tracking and Imaging Elementary Particles," *Sci. Amer.*, 58 (August 1991).

Brown, F.R. and N.H. Christ: "Parallel Supercomputers for Lattice Gauge Theory," *Science*, 1393 (March 18, 1988).

Brown, L.M., M. Dresden, and L. Hoddeson: *From Pions to Quarks: Particle Physics in the 1950s,* Cambridge University Press, New York, NY, 1989.

Chanowitz, M.S.: "The Z Boson," *Science*, 36 (July 6, 1990).

Chupp, E.L.: "Transient Particle Acceleration Associated with Solar Flares," *Science*, 229 (October 12, 1990).

Cline, D.B.: "Beyond Truth and Beauty: A Fourth Family of Particles," *Sci. Amer.*, 50 (August 1988).

Cline, D.B., C. Rubbia, and S. van der Meer: *The Search for Intermediate Vector Bosons.* (March 1982). A classic reference in The Laureates' Anthology, 133, Scientific American, Inc., New York, NY, 1990.

Conte, M. and W.M. MacKay: *An Introduction to the Physics of Particle Accelerators,* World Scientific Publishing Company, Inc., River Edge, NJ, 1991.

Dawson, J.M.: "Plasma Particle Accelerators," *Sci. Amer.,* 54 (March 1989).

Dehmelt, H.: "Experiments on the Structure of an Individual Elementary Particle," *Science,* 539 (February 2, 1990).

Donoghue, J.F., E. Golowich, and B.R. Holstein: *Dynamics of the Standard Model,* Cambridge University Press, New York, NY, 1994.

Dunning, F.B. and R.G. Hulet: *Atomic, Molecular, and Optical Physics: Charged Particles,* Vol. 29, Academic Press, Inc., San Diego, CA, 1995.

Ericson, T. and W. Weise: *Pions and Nuclei,* Oxford University Press, Inc., New York, NY, 1988.

Ezhela, V.V., B. Armstrong, and J.D. Jackson: *Particle Physics: One Hundred Years of Discoveries: An Annotated Chronological Bibliography ANNOTATED,* Springer-Verlag Inc., New York, NY, 1996.

Feldman, G.J. and J. Steinberger: "The Number of Families of Matter," *Sci. Amer.,* 70 (February 1991).

Flam, F.: "CERN's New Detectors Take Shape," *Science,* 180 (April 10, 1992).

Flam, F.: "Neural Nets: A New Way to Catch Elusive Particles?" *Science,* 1282 (May 29, 1992).

Gottfried, K. and V.F. Weisskopf: *Concepts of Particle Physics,* Oxford University Press, Inc., New York, NY, 1997.

Graham, D.: "Testing Physicists' GUTS (Grand Unified Theory)," *Technology Review (MIT),* 10 (May–June 1988).

Gutbrod, H. and H. Stocker: "The Nuclear Equation of State," *Sci. Amer.,* 58 (November 1991).

Helliwell, J.R.: *Macromolecular Crystallography with Synchrotron Radiation,* Cambridge University Press, New York, NY, 1992.

Hermann, A., et al.: *History of CERN,* North-Holland AQ: 2 Different Publishers? Elsevier Science, New York, NY, 1990.

Hoddeson, L., M. Riordan, M. Dresden, and L.M. Brown: *Rise of the Standard Model: Particles Physics in the 1960s and 1970s,* Cambridge University Press, New York, NY, 1997.

Leader, E.: *Spin in Particle Physics,* Cambridge University Press, New York, NY, 2001.

Lederman, L.M.: "Observations in Particle Physics from Two Neutrinos to the Standard Model," *Science,* 664 (May 12, 1980).

Lederman, L.M.: "The Tevatron," *Sci. Amer.,* 48 (March 1991).

Martin, B.R. and G. Shaw: *Particle Physics,* 2nd Edition, John Wiley & Sons, Inc., New York, NY, 1997.

Month, M. and M. Dienes: *The Physics of Particle Accelerators,* Vol. 2, American Institute of Physics, College Park, MD, 1997.

Myers, S. and E. Picasso: "The LEP Collider," *Sci. Amer.,* 54 (July 1990).

Olive, K.A.: "The Quark-Hadron Transition in Cosmology and Astrophysics," *Science,* 1194 (March 8, 1991).

Peterson, I.: "Quantum Interference," *Science News,* 363 (December 2, 1989).

Peterson, I.: "Protons and Antiprotons Held in the Balance," *Science News,* 38 (July 21, 1990).

Peterson, R.J. and D.D. Strottman: *Pion-Nucleus Physics,* American Institute of Physics, College Park, MD, 1997.

Peterson, I.: "Beyond the Z," *Science News,* 204 (September 29, 1990).

Polchinski, J.G.: *String Theory: Superstring Theory and Beyond,* Vol. 2, Cambridge University Press, New York, NY, 1998.

Pool, R.: "The Hunting of the Quark—Computer Style," *Science,* 46 (April 3, 1992).

Quigg, C.: *Gauge Theories of the Strong, Weak and Electromagnetic Interactions,* Perseus Publishing, Boulder, CO, 1997.

Rees, J.R.: "The Stanford Linear Collider," *Sci. Amer.,* 58 (October 1989).

Rice, T.M.: "Can Europe Keep up the Pace in Condensed Matter Physics?" *Science,* 482 (April 24, 1992).

Riordan, M.: "The Discovery of Quarks," *Science,* 1287 (May 29, 1992).

Rothman, T.: "Ambidextrous Universe: New Particles Blur Distinction Between Fermions and Bosons," *Sci. Amer.,* 26 (May 1989).

Rubbia, V.: "The European Strategy in Particle Physics," *Science,* 484 (April 24, 1992).

Ruthen, R.: "Quark Quest," *Sci. Amer.,* 32 (March 1993).

Ruthen, R.: "Attractive and Demure," *Sci. Amer.,* 30 (May 1993).

Sarkar, S.: *Big Bang Laboratory for Particle Physics,* Cambridge University Press, New York, NY, 2002.

Schmidt, V.: *Electron Spectrometry of Atoms Using Synchrotron Radiation,* Cambridge University Press, New York, NY, 1997.

Schramm, D.N. and G. Steigman: "Particle Accelerators Test Cosmological Theory," *Sci. Amer.,* 66 (June 1988).

Selvin, P.: "How Do Particles Put on Weight?" *Science,* 173 (January 8, 1993).

Shifman, M., M.A. Shifman, and B.L. Ioffe: *At the Frontier of Particle Physics: Handbook of QCD,* World Scientific Publishing Company, Inc., River Edge, NJ, 2001.

Staff: *Particle Physics Phenomenology,* World Scientific Publishing Company, Inc., River Edge, NJ, 1997.

Sundaresan, M.K.: *Handbook of Particle Physics,* CRC Press, LLC., Boca Raton, FL, 2001.

Taubes, G.: "Are Neutrino Mass Hunters Pursuing a Chimera?" *Science,* 731 (May 8, 1992).

Waldrop, M.M.: "SLAC Feels the Thrill of the Chase," *Science,* 771 (May 10, 1989).

Weinberg, S.: *Discovery of Subatomic Particles,* 2nd Edition, Cambridge University Press, New York, NY, 2003.

Wilcek, F.: "Anyons," *Sci. Amer.,* 58 (May 1991).

Willeke, K.: *Physics of Particle Accelerators: An Introduction,* Oxford University Press, Inc., New York, NY, 2000.

Wilson, E.J.N.: *An Introduction to Particle Accelerators,* Oxford University Press, Inc., New York, NY, 2001.

Yam, P.: "Spin Cycle: Rotating Nucleii Share A Few Moments of Inertia," *Sci. Amer.,* 26 (October 1991).

Yan, Y.T., J.P. Naples, and M.J. Syphers: *Accelerator Physics at the Superconducting Super Collider,* Springer-Verlag Inc., New York, NY, 1995.

Zotter, B.W. and S. Kheifets: *Impedances and Wakes in High Energy Particle Accelerators,* World Scientific Publishing Company, Inc., River Edge, NJ, 1998.

Pre−1988 References

Adair, R.: *The Great Design: Particles, Fields, and Creation,* Oxford University Press, Inc., New York, NY, 1987.

Atwood, D., K. Halbach, and Kwange-Je Kim: "Tunable Coherent X-rays," *Science,* **228**, 1265−1272 (1985).

Brambilla, N. and G. Prosperi: *Quark Confinement and the Hadron Spectrum: Proceedings of the 5th International Conference,* World Scientific Publishing Company, Inc., Riveredge, NJ, 2003.

Barnett, R.M., H.E. Haber, and G.L. Kane: "Supersymmetry—Lost or Found?" *Nuclear Physics,* **B267**(3, 4) 625−678 (April 21, 1986).

Bengtsson, T., et al.: "Nuclear Shapes and Shape Transitions," *Physica Scripta,* **29**(5) 402−430 (May 1984).

Black, J.K., et al.: "Measurement of the CP-Nonconservation Parameter e 1/e," *Physical Review Letters,* **54**(15) 1628−1630 (April 15, 1985).

Broglia, R.A., C.H. Casso: *Frontiers in Nuclear Dynamics,* Plenum, New York, NY, 1985.

Court, G.R., et al.: "Energy Dependence of Spin Effects," *Physical Review Letters,* **57**(5), 507−510 (August 4, 1986).

Crosbie, E.A., et al.: "Energy Dependence of Spin-Spin Effects in p-p Elastic Scattering at 90°," *Physical Review,* **23**(3) 600−603 (February 1, 1981).

de Rujula, A.: "Superstrings and Supersymmetry," *Nature,* **320**(6064), 678 (April 24, 1986).

Eichten, E., et al.: "Supercollider Physics," *Reviews of Modern Physics,* **56**(4), 579−707 (October 1984).

Ellis, J.: "Hope Grows for Supersymmetry," *Nature,* **313**(6004), 626−627 (February 21, 1985).

Glashow, S.L.: "Toward a Unified Theory: Threads in a Tapestry," in *Nobel Lectures,* Elsevier Science, Amsterdam and New York, NY, 1981.

Green, M.B.: "Unification of Forces and Particles in Superstring Theories," *Nature,* **314**(6010), 409−414 (April 4, 1985).

Green, M.B.: "Superstrings," *Sci. Amer.,* 48−60 (September 1986).

Haber, H.E. and G.L. Kane: "The Search for Supersymmetry: Probing Physics Beyond the Standard Model," *Physics Reports,* **117**(2, 3), 75−263 (January 1985).

Haber, H.E. and G.L. Kane: "Is Nature Supersymmetric?" *Sci. Amer.,* 52−60 (June 1986).

Hamilton, J.H., P.G. Hansen, and E.F. Zganjar, *Reports on Progress in Physics,* **48**(5) 631−708 (May 1985).

Hamilton, J.H.: "Magic Numbers, Reinforcing Shell Gaps and Competing Shapes in Nucleii," *Progress in Particle and Nuclear Physics,* **15**, 107−134 (1985).

Krisch, A.D.: "Collisions between Spinning Protons," *Sci. Amer.,* 42−50 (August 1987).

Letessier, J. and J. Rafelski: *Hadrons and Quark Gluon Plasma,* Cambridge University Press, New York, NY, 2002.

Lipkin, H.J.: "Colour Theory in a Spin," *Nature,* **324**(6092), 14−16 (November 6, 1986).

Martin, J.A., W. Greiner: "Potential Energy Surface Model of Collective States," in *High-Angular Momentum Property of Nuclei* (N.R. Johnson, Ed.) Harwood Academic Publishers, New York, NJ, 1983.

Mulvey, J.H.: *The Nature of Matter,* Oxford University Press Inc., New York, NY, 1981.

Nadis, N.: "Anti-Proton Fishing," *Technology Review (MIT),* 15 (July 1987).

News: "Antiprotons Captured at CERN," *Science,* **233**, 1383−1384 (1986).

News: "Bright Synchrotron Sources Evolve," *Science,* **235**, 841−842 (1987).

News: "CERN Panel Backs New Accelerator," *Science,* **235**, 1567 (1987).

News: "Soviets Plan Huge Linear Collider," *Science,* **238**, 16−17 (1987).

Quigg, C.: "Elementary Particles and Forces," *Sci. Amer.,* 84−95 (April 1985).

Quigg, C. and R.F. Schwitters: "Elementary Particle Physics and the Superconducting Super Collider," *Science,* **231**, 1522−1527 (1986).

Richter, B.: "From the Psi to Charm: The Experiments of 1975 and 1976," *Science,* **196**, 1286−1297 (1977).

Sachs, R.G.: *The Physics of Time Reversal,* University of Chicago Press, Chicago, IL, 1987.

Scherk, J.: "An Introduction to the Theory of Dual Models and Strings," *Reviews of Modern Physics,* **47**(1), 123–164 (January 1975).

Schwartz, J.H., E. Witten: *Superstring Theory,* Cambridge University Press, New York, NY, 1987.

Schwarzschild, B.M.: "Polarized Scattering Data Challenge Quantum Chromodynamics," *Physics Today,* **38**(8), 17–20 (August 1985).

Smith, T.P.: *Hidden Worlds: Hunting for Quarks in Ordinary Matter,* Princeton University Press, Princeton, NJ, 2003.

Sutton, C.: *The Particle Connection,* Simon and Schuster, New York, NY, 1984,

van der Meer, S.: "Stochastic Cooling and the Accumulation of Antiprotons," *Science,* **230**, 900–906 (1985).

Waldrop, M.M.: "String as a Theory of Everything," *Science,* **229**, 1251–1253 (1985).

Weinberg, S.: *The Discovery of Subatomic Particles,* W.H. Freeman, New York, NY, 1983.

Wilson, R.R.: "The Next Generation of Particle Accelerators," *Sci. Amer.,* 42–57 (January 1980).

Yamazaki, T.: "Evolution of Meson Science in Japan," *Science,* **233**, 334–338 (1986).

Zichichi, A.: *From Quarks and Gluons to Quantum Gravity: Proceedings of the International School of Subnuclear Physics,* World Scientific Publishing Company, Inc., Riveredge, NJ, 2003.

Zweig, G.: "Quark Catalysis of Exothermal Nuclear Reactions," *Science,* **201**, 973–979 (1978).

PARTICULATE MATTER.   Solid or liquid matter that is dispersed in a gas, or insoluble solid matter dispersed in a liquid, that gives a heterogeneous mixture.

PASCHEN-BACK EFFECT.   In a strong magnetic field, the anomalous Zeeman effect changes into a pattern similar to the normal effect, and this is known as the Paschen-Back effect. Each energy level with a given value of L, the electronic orbital angular momentum splits into $(2L + 1)$ components characterized by the magnetic quantum numbers $M_L = L, L - 1, \ldots, -L$, and each level with a given value of M_L splits into $(2S + 1)$ components with quantum numbers $M_s = S, S - 1, \ldots, -S$, where S is the resultant electron spin. The selection rules are $\Delta M_L = 0, \pm 1, \Delta M_s = 0$. Lines with $\Delta M_L = 0$ are plane polarized with electric vector parallel to the direction of the applied magnetic field; those with $\Delta M_L = \pm 1$ are plane polarized with components perpendicular to the field. See also **Atomic Spectra**; and **Hyperfine Structure**.

PASSIVITY.   When iron is immersed in concentrated nitric acid, there is no visible reaction (Keir, 1790), although dilute nitric acid results in a marked reaction with iron. Upon removal of the iron from the concentrated nitric acid and immersion in copper sulfate solution, the iron is not plated by copper, although this occurs with ordinary iron. Iron in such a condition is described as passive iron, and the phenomenon is known as passivity. See also **Iron Metals, Alloys, and Steels**

PASTEURIZATION.   Heat treatment of milk, fruit juices, canned meats, egg products, etc. for the purpose of killing or inactivating disease-causing organisms. For milk, the minimum exposure is 62°C for 30 min or 72°C for 15 sec, the latter being called flash pasteurization. Although this treatment kills all pathogenic bacteria and also inactivates enzymes that cause deterioration of the milk, the shelf life is limited. To prolong storage life, temperatures of 80–88°C for 20–40 sec must be used. Complete sterilization requires ultrahigh pasteurization at from 94°C for 3 sec to 150°C for 1 sec. In-can heating at 116°C for 12 min and 130°C for 3 min is also employed for maximum stability and long storage life. Some meat products are pasteurized by α-radiation.
See also **Milk and Milk Products**.

PASTEUR, LOUIS (1822–1895).   Pasteur was a French chemist and microbiologist who made important contributions to biology, medicine, chemistry, and industry.

As a small child, Pasteur showed traits of becoming a scientist. He was fascinated by the local chemist, that made medicine for sick customers. He patiently and carefully observed the things the chemist did and then went home and made drawings of the herbs and roots the man used. Even before finishing high school, Pasteur's study of chemical crystals won attention of the scientific world. Most of what Pasteur is famous for is his work concerning the effects of microbes. He found that living

organisms, microbes, cause fermentation. His discovery was important both for theoretical science and for industry. Pasteur's studies showed microbes could be killed by heat. His discovery made winemaking a more scientific process. Pasteur applied the same idea to milk. The process of keeping milk free from bacteria is named pasteurization after him.

During the 1800s the theory of spontaneous generation was raging in the scientific circles. Pasteur's work proved that food and other organic matter does not spontaneously generate microbes and settled the controversy.

Pasteur also discovered a vaccine to prevent rabies and another vaccine to prevent anthrax. Pasteur's greatest achievement was the founding of the science of microbiology.

See also **Fermentation**.

J.M.I.

PATENTABILITY.   The qualifications for obtaining a patent on an invention of chemical process. These are (1) the invention must not nave been published in any country or in public use in the U.S., in either case for more than 1 year before the date of filing the application; (2) it must not have been known in the U.S. before date of invention by the applicant; (3) it must not be obvious to an expert in the art; (4) it must be useful for a purpose not immoral and not injurious to the public welfare; (5) it must fall within the five statutory classes on which patents may be granted, i.e., (a) composition of matter, (b) process of manufacture or treatment, (c) machine, (d) design (ornamental appearance), or (e) a plant produced asexually. Special regulations relate to atomic energy developments and subjects directly affecting national security (Robert Calvert). Note: In 1980, the Supreme Court in al landmark decision upheld the patentability of synthetic bacteria created by recombinant DNA techniques.

PATERNO-BUCHI REACTION.   Formation of oxetanes by photo-chemical cycloaddition of carbonyl compounds to olefins.

PATHFINDER ELEMENT.   An element present in small proportions less than 1% generally metallic in nature, associated with ore deposits at the time of formation. Mapping of the concentration variation of the selected element serves to locate the main ore deposit. Examples are zinc as the pathfinder for lead, copper, and silver ores, and molybdenum associated with copper deposits.

PATHWAY.   A sequence of reactions, usually of a biochemical nature, in which more-complex substances are converted to simple end products, as in the degradation of the components of foods to carbon dioxide and water. Its course is determined largely by preferential factors involving coenzymes and other catalysts. An example is the TCA cycle, which is the common pathway in the degradation of foodstuffs and cell constituents to carbon dioxide and water.

PATINA.   Variously used to refer to an ornamental and/or corrosion-resisting film on the surface of copper, copper alloys, including bronzes, and also sometimes iron and other metals. Such a film is formed by exposure to the air or by a suitable chemical treatment.
See also **Copper**.

PATTINSON PROCESS.   Process for the removal of silver from lead. The silver-lead mixture is melted in one of a series of pots and allowed to cool slowly. The lead, that is free from silver or poorer in silver separates out as crystals, which are removed, leaving the silver-rich lead in the molten state. From a number of such operations in series, a lead rich in silver is obtained, collected, and the silver recovered.
See also **Parkes Process**.

PAULI EXCLUSION PRINCIPLE.   The statement that any wave function involving several identical particles must be antisymmetric (must change sign) when the coordinates, including the spin coordinates, of any identical pair are interchanged. If the particles in a system can be considered as occupying definite quantum states, it follows from the principle that no more than one particle of a given kind can occupy a particular state; hence the name, exclusion principle. The principle applies to fermions, but not to bosons. Since electrons, protons, and neutrons are fermions, the Pauli exclusion principle must be used in the assignment of particles to quantum states in theories of atomic and nuclear structure.

PAULING, LINUS CARL (1901–1994). Dr. Linus Pauling was a famous American chemist. He was born in Portland, Oregon and his father was a pharmacist. When he was four, his family moved to Condon, Oregon. On the south edge of town, there was a creek and Pauling sent much time exploring the creek bed and collecting minerals. Eventually Pauling would establish structures for these minerals at the California Institute of Technology.

Pauling enrolled in Oregon Agricultural College (Oregon State University) and was majoring in chemical engineering, but when his mother became ill, he quit school and began working. Because he had show outstanding promise, he was offered a position to be an instructor of quantitative analysis at the college. He taught the class and graduated with a degree in chemical engineering. He attended Cal Tech graduate school. His Ph.D. dissertation was on the crystal structure of different minerals. Pauling received a Guggenheim fellowship to study at the University of Munich where he began applied the concept of quantum mechanics to chemical bonding. Then he took an assistant professorship in chemistry at Cal. Tech and published his paper "The Nature of the Chemical Bond." In 1931, he received the Langmuir Prize of the American Chemical Society for his noteworthy work. In 1933, he was made a member of the National Academy of Sciences. He was 32 years old at that time and he was the youngest appointment ever made.

His serious interest molecular biology began about 1935. He was intrigued by the question of how protein molecules were constructed. As a professor at the California Institute of Technology, he was known for giving "baby toy lectures" because he made models of molecules out of string, rod- and-ball structures, and plastic bubbles in different colors, shapes, and sizes. One day, working with paper, he sketched atoms and chemical bonds and folded them in different ways and discovered the basic structure of the protein molecule.

Pauling, with the help of C.D. Coryell, analyzed the effects of the oxygenation of hemoglobin molecules by measuring their magnetic susceptibility. In 1936, along with A.E. Mirsky, Pauling developed a theory of native, denatured, and coagulated proteins. And in 1950, he and R.B. Corey described the structure of several molecules including muscles, fingernails, hair, and other tissues.

During World War II, Pauling chose not to work on the Manhattan Project for the development of the atomic bomb. He became worried about the radiation the atomic bomb produced and helped organize the Pasadena Federation of Atomic Scientists, a group of scientists working for safe control of nuclear power. He also joined the Emergency Committee of Atomic Scientists, known as the Einstein Committee since it was chaired by Albert Einstein. The aim of the committee was to educate people about atomic weapons. In 1947, he was awarded the presidential Medal of Merit by President Truman for his work on crystal structure, nature of chemical bond, and his efforts to bring about world peace.

In 1954, Pauling received the Nobel Prize in chemistry for his research into the nature of the chemical bond and its application to the elucidation of the structure of complex substances. On October 10, 1962 he was awarded the Nobel Peace Prize for his efforts towards nuclear test ban treaty.

Pauling, in recent years, researched the chemistry of the brain and mental illness, the cause of sickle-cell anemia, and the effects of large doses of Vitamin C on the common cold and on cancer. On August 19, 1994, Pauling, himself, died of cancer at the age of 93.

See also **Electronegativity**.

J.M.I.

PAULI, WOLFGANG ERNST (1900–1958). Pauli was an Austrian theoretical physicist. After WWII, he became an American citizen. When just 20 years of age he wrote "The Theory of Relativity." Later he wrote articles on "Quantum Theory" and "Principles of Wave Mechanics." He is most remembered for formulating the "Pauli exclusion principle". This principle says that two electrons in an atom can never exist in the same state. This is important concept for modern physics. Pauli was awarded the Nobel Prize in physics in 1945 for this discovery.

See also **Pauli Exclusion Principle**; and **Quantum Mechanics**.

J.M.I.

PCB. See **Biphenyl and Terphenyls**.

PEARSON'S SOLUTION. A dilute sodium arsenate solution containing 0.1% anhydrous sodium arsenate.

PEAT. Semicarbonized residue of plants formed in water-saturated environments (bogs and marshes). It occurs in surface layers 3–10 ft thick and has a water content of 85%. Before peat can be used for chemical or fuel purposes it must be field-dried to a water content of 30–40%. Since the dried product is susceptible to autoignition, storage conditions must minimize this risk. Peat is easily converted to hydrocarbons and is an excellent source of natural gas; when dry it can be used directly as a fuel. The U.S. has peat sources second only to those of the former U.S.S.R., located in Alaska, the north-central states, and Maine, where processing on a large scale is planned. Their total energy content is said to be equivalent to 240 billion barrels of petroleum. The peat can be gasified for production of methanol after mechanical dewatering. Experimental conversion studies have been under way for some time. Substantial quantities of oil, ammonia, and sulfur can be obtained as by-products. See also **Coal**.

PEBBLE MILL. A jacketed steel cylinder rotating on a horizontal axis and containing flint or porcelain pebbles as the grinding medium. Its operation is similar to that of a ball mill. It is used for grinding and mixing of dry chemicals, pigments, food products, and the like. Pebble mills are usually lined with alumina, buhrstone, or similar material to protect the walls from wear.

PECHMANN PYRAZOLE SYNTHESIS. Formation of pyrazoles from acetylenes and diazomethane. The analogous addition of diazoacetic esters to the triple bond yields pyrazolecarboxylic acid derivatives.

PECTINS. A *pectic substance* is a group designation for those complex carbohydrate derivatives that occur in or are prepared from plants and contain a large proportion of anhydrogalacturonic acid units, which are thought to exist in a chainlike combination. The carboxyl groups of polyglacturonic acids may be partially esterified by methyl groups and partly or completely neutralized by one or more bases. The general term *pectin* (or *pectins*) designates those water-soluble pectinic acids of varying methyl ester content and degree of neutralization which are capable of forming gels with sugar and acid under suitable conditions. The term *protopectin* is applied to the water-insoluble parent pectic substances which occur in plants and which upon restricted hydrolysis yield pectin or pectinic acids. *Pectic acids* is a term that is applied to pectic substances mostly composed of colloidal polyglacturonic acids and essentially free from methyl ester groups. The salts of pectic acids are either normal or acid *pectates*. The term *pectinic acids* is used for colloidal polygalacturonic acids containing more than a negligible proportion of methyl ester groups. Pectinic acids, under suitable conditions, are capable of forming gels with sugar and acid, or, if suitably low in methosyl content, with certain metallic ions. The salts of pectinic acids are either normal or acid *pectinates*.

Pectins occur commonly in plants, particularly in succulent tissues, and are characterized by the polygalacturonic acids that are fundamental to their structure. The pectins are important emulsifying, gelling, stabilizing, and thickening agents used in the preparation of numerous food products. About 75% of the pectins produced are used in making fruit jams, jellies, marmalades, and similar products. Additional uses include the preparation of mayonnaise, salad dressings, malted milk beverages, frozen dessert mixes, and frozen fruits and berries (to prevent leakage upon thawing), among others. The addition of a dilute pectin solution to milk coagulates the casein. In many food products, the use of pectins as stabilizers is preferred, since they blend better into the flavor complex than do many gums, starches, or a number of carbohydrate derivatives. Pectin jellies do not melt at temperatures below 49°C, a distinct advantage over gelatin gels that require refrigeration. Pectins also have a number of nonfood uses, including pharmaceuticals and cosmetics.

The location of various pectin substances in plant tissues is well established. Pectins make up most of the middle lamella in unripe fruit and are to be found in the cell walls and in small proportion in all plant tissues. The genesis and fate of pectins in plant tissues have not been fully determined.

Citrus peel, apple pomace from juice manufacture, and beet pulp left over from the manufacture of sucrose are common commercial sources of pectins. After some preliminary purification of the raw material, the extraction is usually performed with hot dilute acid (pH = 1.0–3.5 in a temperature range of 70–90°C). The pectin is then precipitated from the extract with ethanol or isopropanol, or with metal salts (copper or aluminum). The metal ions have to be subsequently removed by washing with water or acid ethanol. Specific formulas for denatured ethanol for use in pectin manufacture are used. The precipitates are purified, dried, and pulverized to form the yellowish-white powder of commerce.

Pectin substances in solution behave as typical colloids. See also **Colloid Systems**. Dry, purified pectins are light in color and soluble in hot water to the extent of 2–3%. The pH of pectin solutions is usually 2–3.5.

The proportion of sugar which pectin will form into a firm jelly determines the *jelly grade* of the pectin. In a jelly, jam, or marmalade, the proportions of total solids of sugars, the pH, and the proportion and nature of the pectin used will determine the extent of jellification obtained. The use of pectin in fruit jams and related products is approved by most food regulators because the addition is believed to compensate for an incidental natural deficiency.

In pectic acids, all carboxyl groups are free, or at least not present as the methyl ester. Under suitable conditions, pectins will form jellies with sugar and acid, whereas the low-ester pectins will form *gels* with traces of polyvalent ions. The general structure of pectin is:

Additional Reading

Fishman, M.L. and J.J. Jen: *Chemistry and Function of Pectins,* American Chemical Society, Washington, DC, 1986.

Lutz, P.L., H. Schols, R. Visser, and F. Voragen: *Advances in Pectin and Pectinase Research,* Kluwer Academic Publishers, Norwell, MA, 2003.

Quilici-Timmecke, J.: *New Nutrients against Cancer: Modified Citrus Pectin, Soybeans, Lycopene and Other '90's Cancer Fighters,* Keats Publishing, Inc. Chicago, IL, 1998.

Seymour, G.B. and P. Knox: *Pectins and Their Manipulation,* CRC Press LLC., Boca Raton, FL, 2002.

Walter, R.H.: *The Chemistry and Technology of Pectin,* Academic Press, Inc., San Diego, CA, 1997.

Wood, W.A. and S.T. Kellogg: *Biomass: Lignin, Pectin, and Chitin,* Vol. 161, Academic Press, Inc., San Diego, CA, 1988.

PEDERSEN, CHARLES JOHN (1904–1989). An American born in Korea, Pedersen received a M.S. from M.I.T. in 1927. In 1987 he was awarded the Nobel Prize for Chemistry for his work in elucidating mechanisms of molecular recognition, which are fundamental to the enzymic catalysis, regulation and transport. He reported that alkali metal ions could be bound by crown ethers into a more rigid, layered structure, in which the alkali metal ion was bound into the center of the ring. This field of study is called host-guest chemistry.

PEGMATITE. The term *pegmatite,* derived from the Greek word meaning "joined together," was first applied by Haüy in 1822 to a peculiar interpenetrating growth of quartz and feldspar sometimes called graphic granite from its resemblance to written characters, particularly those of the Hebrew language. Pegmatite is also used to designate those coarse-grained dikes and sheets, chiefly of granite or syenite, that are apophyses of stocks or batholiths, or of the residual magma, during their congelation. The individual minerals may often reach great size. Granite pegmatites are chiefly composed of alkali feldspar and quartz with some muscovite or biotite but may carry such minerals as tourmaline, topaz, beryl, fluorite, apatite, garnet, lepidolite, etc. See also **Mineralogy**.

PELLET. A small unit of a light, bulky material compressed into any of several shapes and sizes, usually either spherical of rectangular. The operation is performed on a pellet mill, which consists essentially of a pair of steel rollers around which rotates a circular perforated metal die. Material is fed into the chambers above and below the inner face of the die. As the die turns, in contact with the rollers, the latter also turns, thus compressing the material and forcing it through the holes in the die at the point of tangency, where the extruded segment is sheared off by knives. Pelletizing is advantageous for fluffy particulates that are difficult to handle in loose form, e.g., carbon black, clays, plastic molding powders, etc. Binding materials called excipients are often used.

PELLIZZARI REACTION. Formation of substituted 1,2,4-triazoles by the condensation of amides and acylhydrazines. When the acyl groups of the amide and acylhydrazine are different, interchange of acyl groups may occur, with formation of a mixture of triazoles.

PELOUZE SYNTHESIS. Formation of nitriles from alkali cyanides by alkylation with alkyl sulfates of alkyl phosphates.

PENETRANT. Any agent used to increase the speed and ease with which a bath or liquid permeates a material being processed by effectively reducing the interfacial tension between the solid and liquid. Penetrants are widely used in the textile, tanning, and paper industries for improving dyeing, finishing, etc., operations. Sulfonated oils, soluble pone oils, and soaps are popular among the older penetrants, and the salts of sulfated higher alcohols are typical of the synthetic organics developed for this purpose.

See also **Wetting Agent**.

PENICILLIN. See **Antibiotics**.

PENNING DISCHARGE. A direct-current discharge where electrons are forced to oscillate between two opposed cathodes and are restrained from going to the surrounding anode by the presence of a magnetic field. It is sometimes referred to as a pig discharge since the device was originally used as an ionization gage (Penning ionization gage). It is used as a plasma-beam source by permitting the plasma to stream out along the magnetic field through a hole in one of the cathodes.

PENNING EFFECT. An increase in the effective ionization rate of a gas due to the presence of a small number of foreign metastable atoms. For instance, a neon atom has a metastable level at 16.6 volts and if there are a few neon atoms in a gas of argon which has an ionization potential of 15.7 volts, a collision between the neon metastable atom with an argon atom may lead to ionization of the argon. Thus, the energy which is stored in the metastable atom can be used to increase the ionization rate. Other gases where this effect is used are helium, with a metastable level at 19.8 volts, and mercury, with an ionization level at 10.4 volts.

PENTAERYTHRITOL TETRANITRATE (PETN). See **Explosives**.

PENTLANDITE. The mineral sulfide of iron and nickel corresponding to the formula $(Fe, Ni)_9S_8$. It is isometric, appears in granular masses; hardness, 3.5–4; specific gravity, 5.0; color, bronze-yellow; opaque. Occurs with pyrrhotite, millerite, and nickeline. The best known deposit of pentlandite is at Sudbury, Ontario, Canada, where it is associated with a nickel-bearing pyrrhotite

PENTOSAN. A complex carbohydrate (hemicellulose) present with the cellulose in many woody plant tissues, particularly cereal straws and brans, characterized by hydrolysis to give five-carbon-atom sugars (pentoses). Thus the pentosan xylan yields the sugar xylose $(HOH_2C·CHOH·CHOH·CHOH·CHO)$ that is dehydrated with sulfuric acid to yield furfural $(C_5H_4O_2)$.

See also **Carbohydrates**.

PENTOSE PHOSPHATE CYCLE. See **Carbohydrates**.

PEPTIZATION. Stabilization of hydrophobic colloidal solutions by addition of electrolytes which provide the necessary electric double layer of ionic charges around each particle. Such electrolytes are known as peptizing agents. The ions of the electrolyte are strongly adsorbed on the particle surfaces. Stable solutions of nonionizing substances acquire a charge in contact with water by preferential adsorption of the hydroxyl ions, which may be considered peptizing agents. The term is also loosely applied to the softening or liquefaction of one substance by trace quantities of another, analogous to the digestion of a protein by an enzyme (pepsin).

PEPTONE. A secondary protein derivative that is water-soluble, not coagulated by heat, and not precipitated on saturation of its solutions with ammonium sulfate.

PERCHLORIC ACID AND PERCHLORATES. When in a +7 valence state and combined with oxygen, chlorine forms a family of compounds known as the perchlorates. The perchlorate anion, ClO_4^-, as progenitor is derived from perchloric acid. [CAS: 7601-90-3]. $HClO_4$ is one of the strongest of the mineral acids. The perchlorates are more stable than the other chlorine oxyanions, i.e., chlorates, ClO_3^-; chlorites, ClO_2^-; or hypochlorites, OCl^-. See also **Hypochlorites**. Essentially, all of the commercial perchlorate compounds are prepared either directly or indirectly by electrochemical oxidation of chlorine compounds. The perchlorates of practically all the electropositive metals are known, except for a few cations having low charges.

The most outstanding property of the perchlorates is their oxidizing ability. On heating, these compounds decompose into chlorine, chlorides, and oxygen gas. Aqueous perchlorate solutions exhibit little or no oxidizing power when dilute or cold. However, hot concentrated perchloric acid is a powerful oxidizer and whenever it contacts oxidizable matter extreme caution is required. The acidified concentrated solutions of perchlorate salts must also be handled with caution. Ammonium perchlorate (AP), [CAS: 7790-98-9], is one of the most important perchlorates owing to its high (54.5%) O_2 content and the absence of residue on decomposition. These properties, along with a long shelf life, make it a useful rocket propellant. See also **Rocket Propellants**.

Actual perchlorate production is difficult to determine in any given year, because AP is classified as a strategic material. Future production is expected to depend mostly on space programs.

Properties

Chlorine Heptoxide. The anhydride of perchloric acid is chlorine heptoxide. [CAS: 10294-48-1]. Cl_2O_7 is also known as dichlorine heptoxide. It is obtained as a colorless oily liquid by dehydration of perchloric acid using a strong dehydrating agent such as phosphorus pentoxide, P_2O_5.

$$2 \, HClO_4 + P_2O_5 \longrightarrow Cl_2O_7 + 2 \, HPO_3 \qquad (1)$$

The Cl_2O_7 decomposes spontaneously on standing for a few days. The acid dehydration reaction requires a day for completion at $-10°C$ and explosions can occur. Upon ozonation of chlorine or gaseous ClO_2 at $30°C$, Cl_2O_7 is formed.

Chlorine heptoxide is more stable than either chlorine monoxide or chlorine dioxide; however, the Cl_2O_7 detonates when heated or subjected to shock. It melts at $-91.5°C$, boils at $80°C$, has a molecular weight of 182.914. It is soluble in benzene, slowly attacking the solvent with water to form perchloric acid; it also reacts with iodine to form iodine pentoxide and explodes on contact with a flame or by percussion. Reaction with olefins yields the impact-sensitive alkyl perchlorates.

Perchloric Acid. Pure anhydrous perchloric acid, $HClO_4$, is quite unstable. In aqueous solution, however, $HClO_4$ is a familiar and useful reagent. The acid is the strongest simple acid and the perchlorate ion the least polarizable negative ion known. Perchloric acid is commonly obtained as an aqueous solution, although the pure anhydrous compound can be prepared by vacuum distillation as a colorless liquid, which freezes at $-112°C$ and boils at $16°C$ at 2.4 kPa (18 mm Hg) without decomposition. The pure acid cannot be distilled at ordinary pressures and explodes at $90°C$ after standing at room temperature for $10–30$ days. The aqueous solution can be concentrated by boiling at 101 kPa (1 atm) at $203°C$, at which point an azeotropic solution is attained which contains 72.4% $HClO_4$.

A number of hydrates of perchloric acid, $HClO_4 \cdot nH_2O$, where $n = 1$, 2, 2.5, 3, and 3.5, are known. These are commonly referred to as the hydronium or oxonium perchlorates, $H_3O^+ClO_4^-$, because of the analogy between the x-ray patterns of these species and ammonium perchlorate.

The combination of oxidizing effect, acidic strength, and high solubility of salts makes perchloric acid a valuable analytical reagent. It is often employed in studies where the absence of complex ions must be ensured.

Ammonium Perchlorate. Ammonium perchlorate is a colorless, crystalline compound having a density of 1.95 g/mL and a molecular weight of 117.5. It is prepared by a double displacement reaction between sodium perchlorate and ammonium chloride, and is crystallized from water as the anhydrous salt. The perchlorates, especially those of the light metals and ammonium ion, are favored as solid oxidizers for rocket propellants.

A newer approach developed for producing commercial quantities of high purity AP involves the electrolytic conversion of chloric acid, [CAS: 7790-93-4] to perchloric acid, which is neutralized by using ammonia gas:

Alkali Metal Perchlorates. The anhydrous salts of the Group 1 (IA) or alkali metal perchlorates are isomorphous with one another

as well as with ammonium perchlorate. Crystal structures have been determined by optical and x-ray methods. With the exception of lithium perchlorate, the compounds all exhibit dimorphism when undergoing transitions from rhombic to cubic forms at characteristic temperatures. Potassium perchlorate, [CAS: 7778-74-7]. $KClO_4$ is discovered, is used in pyrotechnics and has the highest percentage of oxygen (60.1%).

The alkali metal perchlorates are either white or colorless, and have increasing solubility in water in the order of $Na > Li > NH_4 > K > Rb > Cs$. The high solubility of sodium perchlorate, $NaClO_4$, makes this material useful.

Group 11 (IB) Perchlorates. Copper and silver perchlorates have been studied quite extensively. Copper(I) perchlorate, [CAS: 17031-33-3], $CuClO_4$, and copper(II) perchlorate, [CAS: 13770-18-8], $Cu(ClO_4)_2$, form a number of complexes with ammonia, pyridine, and organic derivatives of these compounds. The copper perchlorate is an effective burn-rate accelerator for solid propellants.

Gold forms organic perchlorate, [CAS: 42774-61-8], complexes as well as complexes with silver, e.g., $(C_6H_5)_3AgAu(C_6F_5)_2ClO_4$.

Alkaline-Earth Perchlorates. Anhydrous alkaline-earth metal perchlorates can be prepared by heating ammonium perchlorate in the presence of the corresponding oxides or carbonates. The alkaline-earth perchlorates are unusually soluble in organic solvents.

Group 12 (IIB) Perchlorates. The zinc perchlorate, [CAS: 13637-61-1], cadmium perchlorate, [CAS: 13760-37-7], mercury(I) perchlorate, CAS: 13932-02-0, and mercury(II) perchlorate, [CAS: 7616-83-3], all exist.

Group 13 (IIIA) Perchlorates. Perchlorate compounds of boron and aluminum are known. Boron perchlorates occur as double salts with alkali metal perchlorates. Aluminum perchlorate, [CAS: 14452-95-3], $Al(ClO_4)_3$, forms a series of hydrates.

Group 3 (IIIB) and Inner Transition-Metal Perchlorates. The rare-earth metal perchlorates of yttrium and lanthanum have been reported, as have tetravalent cerium perchlorate, [CAS: 14338-93-3], $Ce(ClO_4)_4$, and uranium perchlorate.

Group 14 (IVA) Perchlorates. Perchlorates containing organic carbon have been reported, as have diazonium perchlorates, oxonium perchlorates, and the perchlorate esters. Extreme caution must be used in working with organic perchlorates; many decompose violently when heated, contacted with other reagents, or subjected to mechanical shock.

Group 4 (IVB) Perchlorates. Titanium tetraperchlorate, [CAS: 13498-15-2], is known.

Group 15 (VA) Perchlorates. Nitrogen perchlorates have been used as oxidizers in rocket propellants. Hydrazine perchlorate, [CAS: 13762-80-6], $NH_2NH_3ClO_4$, and hydrazine diperchlorate, $ClO_4NH_3NH_3ClO_4$, have been investigated as oxidizers for propellant systems.

Other Group 15 perchlorates include nitronium perchlorate, NO_2ClO_4, also called nitryl or nitroxyl perchlorate; nitrosyl perchlorate, [CAS: 15605-28-4], $NOClO_4$; and phosphonium perchlorate, $P(OH)_4ClO_4$.

Group 5 (VB) Perchlorates. Vanadyl perchlorate, [CAS: 67632-69-3], $VO(ClO_4)_3$, has been prepared.

Group 16 (VIA) Perchlorates. A perchlorate compound perchloryl sulfate, [CAS: 43059-05-8], $SO_4(ClO_4)_2$ was produced by the low temperature electrolysis of a $12-NH_2SO_4$ and $3-NHClO_4$ solution. This compound is a strong oxidizer; reaction with toluene, acetone, benzene, or alcohol at room temperature produces an exothermic and explosive reaction. The $SO_4(ClO_4)_2$ is soluble in Freon and CCl_4 without reaction.

Group 6 (VIB) Perchlorates. Both divalent and trivalent chromium perchlorate compounds, [CAS: 13931-95-8]; [CAS: 13527-21-9], have been reported. Chromyl perchlorate has been suggested for a gas-generating system operating at $-45°C$.

Group 17 (VIIA) Perchlorates. Fluorine perchlorate, [CAS: 37366-48-6], $FClO_4$, is formed by action of elemental fluorine and $60–70\%$ aqueous perchloric acid solution. The compound is normally a gas. It melts at $-167.5°C$ and boils at $-15.9°C$. It is extremely reactive and explosive in all states.

The perchloryl fluoride, [CAS: 7616-94-6], $FClO_3$, the acyl fluoride of perchloric acid, is a stable compound. Normally a gas having a melting point of $-147.7°C$ and a boiling point of $-46.7°C$, it can be prepared by electrolysis of a saturated solution of sodium perchlorate in anhydrous hydrofluoric acid. Some of its uses are as an effective fluorinating agent, as an oxidant in rocket fuels, and as a gaseous dielectric for transformers.

Other Transition Element Perchlorates. Both divalent and trivalent manganese perchlorate compounds, [CAS: 13770-16-6]; [CAS: 13498-03-8] are known. Perchlorates of Fe, Co, Ni, Rh, and Pd have been produced as colored crystals.

Manufacture

Perchloric Acid. Several techniques have been employed in the manufacture of perchloric acid, including thermal decomposition of chloric acid, anodic oxidation of chloric acid, irradiation of chlorine dioxide solutions, electrolysis of hydrochloric acid, oxidation of hypochlorites by ozone, ion exchange, and electrodialysis of perchlorate salts.

Perchlorates. Historically, perchlorates have been produced by a three-step process: (1) electrochemical production of sodium chlorate; (2) electrochemical oxidation of sodium chlorate to sodium perchlorate; and (3) metathesis of sodium perchlorate to other metal perchlorates. The advent of commercially produced pure perchloric acid directly from hypochlorous acid means that several metal perchlorates can be prepared by the reaction of perchloric acid and a corresponding metal oxide, hydroxide, or carbonate.

Shipping and Handling

Perchloric acid and perchlorates are classified as strong oxidizers and emit toxic fumes when decomposed; contact with combustible, flammable, or reducing materials must be avoided. Perchloric acid and perchlorates must be shipped in accordance with the U.S. Department of Transportation hazardous material regulations. Handling these compounds requires the procedures and safety precautions specified by the product supplier. Perchlorates contain a self-sustaining source of oxygen, thus fires involving perchlorates must be extinguished with water. A class of more hazardous compounds is formed by mixing inorganic perchlorates with finely divided metals, sulfur, or organic compounds and must be handled with the same precautions as explosives.

Economic Aspects

Anhydrous perchloric acid is not sold commercially. Aqueous solutions of perchloric acid are sold at low concentrations for analytical standard applications and at concentrations up to 70%. The price for 70% perchloric acid varies and starts at $2.70/kg, depending on the quantity and level of impurities. The U.S. domestic capacity of ammonium perchlorate is roughly estimated at 31,250 t/yr. The actual production varies, based on the requirements for solid propellants. Environmental effects of the decomposition products, which result from using solid rocket motors based on ammonium perchlorate-containing propellants, are expected to keep increasing public pressure until consumption is reduced and alternatives are developed. Approximately 450 t/yr of NH_4ClO_4-equivalent cell liquor is sold to produce magnesium and lithium perchlorate for use in the production of batteries.

Uses

Perchloric acid is used in analytical chemistry for the determination of trace metal constituents in oxidizable substances as well as in the production of high purity metal perchlorates; it has also been introduced as a stable reaction media in the thermocatalytic production of chlorine dioxide. Perchlorates are primarily used in ammonium perchlorate as an oxidizer in the formulations of propellant for solid rocket motors. Perchlorates are used in the production of explosives, pyrotechnics, and in solid, slurried, and gelled blasting formulations. Both magnesium and lithium perchlorates are used in dry batteries. Other perchlorates have found application in oxygen-generation systems, adhesive bonding of steel plates, and the recovery of potassium from brines such as $KClO_4$.

SUDHIR K. MENDIRATTA
RONALD L. DOTSON
ROBERT T. BROOKER
Olin Corporation

Additional Reading

Keller, C.L.: *Hazardous Materials Regulations,* Tariff No. BOE, U.S. Department of Transportation, Washington, DC., Apr. 23, 1992.

Lide, D.R.: *CRC Handbook of Chemistry and Physics,* 84th Edition, CRC Press LLC., Boca Raton, FL, 2003.
Schumacher, J.C.: *Perchlorates, Their Manufacture & Uses,* ACS Monograph 146, Reinhold Publishing Corp., New York, NY, 1960.
Smith, G.F.: *Perchloric Acid,* 2nd Edition, G. F. Smith Chemical Co., Columbus, OH, 1951.
Woodard, K.E. Jr., D.W. Cawlfield, and S.K. Mendiratta, *Chloric Acid: A New Electrochemical Product,* 183rd Electrochemical Society Meeting, Honolulu, Hawaii, May 1993.

PERFECT GAS. A perfect gas may be defined by the following two laws: The Joule law: the energy per mole, U, depends only on the temperature; the Boyle law: at constant temperature, the volume V occupied by a given number of moles of gas varies in inverse proportion to the pressure.

By combination of these two laws we obtain the equation of state for perfect gas:

$$pV = nRT \qquad (1)$$

where R is the gas constant, T, the absolute temperature. (It is also called the *perfect gas law.*)

The perfect gas is an abstraction to which any real gas approximates according to the nature of the gas and the conditions. For a given temperature and composition, the perfect gas condition is approached when the density tends to zero. From a molecular point of view, the perfect gas laws correspond to the behavior of a system of molecules whose interactions may be neglected in expressing the thermodynamic equilibrium properties. However, even at a low density, the transport properties depend essentially on the interactions.

The thermodynamic properties of a perfect gas are, of course, especially simple. For example, the difference between the molar heat capacities at constant pressure and constant volume is equal to the gas constant R,

$$C_p = C_v = R \qquad (2)$$

The value of R is 0.08205 liter-atm. degree^{-1} mole^{-1}, which in cgs units is equal to 8.314×10^7 g cm^2 sec^{-2} degree^{-1} mole^{-1}. This relationship, Formula (2), applies only approximately to real gases.

However, the way in which either C_p or C_v depends on the temperature can be calculated only from statistical mechanics.

PERFUME. A blend of pleasantly odorous substances (usually liquids) obtained from the essential oils of flowers, leaves, fruit, roots, or wood of a wide variety of plants, either by steam distillation or solvent extraction. Flower oils (rose, jasmine) are extracted with a nonpolar solvent to give a waxy mixture called a *concrete*; the wax is then removed by a second solvent (an alcohol), which is then in turn removed to form an absolute. It is necessary that all the solvent be eliminated to obtain the finest perfumes. The center of this industry has long been in Grasse, France. Perfume materials are also derived from animal sources (musk, ambergris) and from resinous extracts (terpenes and balsams); they are also made synthetically. Cologne and toilet water are weak alcoholic solutions (5% or Less) of perfumes. Fine perfumes may contain as many as 30 ingredients, and their blending is an art rather than a science. The largest volume use of perfumes is in soaps, lotions, shaving creams, and cosmetics.

See also **Odorant**.

PERIDOTITE. The term peridotite is derived from peridor, the French word for olivine.

It is a coarse-grained igneous rock related to gabbro, which consists of olivine and pyroxene in varying proportions. Certain peridotites contain spinel, chromite, or mica as accessories.

Rocks consisting essentially of olivine alone are known as dunites, the name coming from the occurrence of this rock in the Dun mountains of New Zealand. In the United States, this mineral is found in North Carolina, South Carolina, and Georgia, where corundum is associated with the dunite in commercial quantities. The olivine of peridotites alters readily to the mineral serpentine, often to such an extent that the rock itself is called a serpentine. As mentioned above, the peridotites may contain chromite or other valuable minerals, often to such an extent that they may be commercially exploited, for nickel, platinum, and precious garnet.

Kimberlite from which diamonds are secured is commonly called a mica peridotite but is more closely related to the lamprophyres. See also **Kimberlite**.

PERIODIC LAW. Originally stated in recognition of an empirical periodic variation of physical and chemical properties of the elements with atomic *weight*, this law is now understood to be based fundamentally on atomic *number* and atomic *structure*. A modern statement is; the electronic configurations of the atoms of the elements vary periodically with their atomic number. Consequently, all properties of the elements that depend on their atomic structure (electronic configuration) tend also to change with increasing atomic number in a periodic manner.

PERIODIC TABLE OF THE ELEMENTS. When the chemical elements are arranged in a matrix on the basis of increasing atomic numbers, a pattern of periodicity among the physical and chemical characteristics emerges. See Fig. 1. By no means is the resulting matrix perfect, but the resemblance of characteristics among groups of elements arranged in this manner is indeed both striking and illuminating. Attempts to classify the elements date back to the early work of de Chancourtois (1862) and Newlands (1863), but the discovery of the relationship between atomic-number groupings and characteristics was made by Dmitri Mendeleev in 1869. One year later, Lothar Meyer independently showed the periodicity of the elements in terms of atomic volumes. Meyer defined the latter characteristic as the atomic weight divided by the specific gravity of the element in the solid state.

Although there have been numerous refinements to Mendeleev's early tabulation, fortified by the discovery and isolation of several elements then unknown the fundamental principles of the matrix are the same. The conventional table is shown in the upper right in Fig. 2. The information also can be presented in polar fashion as shown in Fig. 2. It is interesting to note that as one proceeds clockwise around the circle the atomic numbers appear consecutively and that 18 sectors of the circle become the bases for families or groups of elements.

Notation for designating the grouping of the elements was changed and officially accepted in the mid-1980s. The new notations are used in Fig. 2. They are summarized in Table 1.

Thus, the members of the alkali metals (Group 1), alkaline earths (Group 2), halogens (Group 17), and so on, all bear resemblance, one element to the other, within any given group. There are two significant breakpoints in any representation of periodicity, namely, commencing with atomic number 57 (lanthanum) and atomic number 89 (actinium). Attempts to place the elements which follow—in the one case, atomic numbers 59 through 71, and in the other case, atomic numbers 90 through 103,

TABLE 1. REVISED CHEMICAL ELEMENT GROUP NOTATION VERSUS PRIOR SCHEMES

Revised notation	Elements in revised grouping	Prior IUPAC form	Prior CAS version
1	H, Li, Na, K, Rb, Cs, Fr	IA	IA
2	Be, Mg, Ca, Sr, Ba, Ra	IIA	IIA
3	Sc, Y, La, Ac	IIIA	IIIB
4	Ti, Zr, Hf, 104	IVA	IVB
5	V, Nb, Ta, 105	VA	VB
6	Cr, Mo, W, 106	VIA	VIB
7	Mn, Tc, Re, 107	VIIA	VIIB
8	Fe, Ru, Os	VIIIA	VIII
9	Co, Rh, Ir	VIIIA	VIII
10	Ni, Pd, Pt	VIIIA	VIII
11	Cu, Ag, Au	IB	IB
12	Zn, Cd, Hg	IIB	IIB
13	B, Al, Ga, In, Tl	IIIB	IIIA
14	C, Si, Ge, Sn, Pb	IVB	IVA
15	N, P, As, Sb, Bi	VB	VA
16	O, S, Se, Te, Po	VIB	VIA
17	F, Cl, Br, I, At	VIIB	VIIA
18	He, Ne, Ar, Kr, Xe, Rn	VIIIA	VIIIA

Lanthanides and Actinides not included in group numbering.

in the underlying geometric matrix (whether tabular or circular) do not succeed. These separate groups are known as the *Lanthanides* (rare earths) and the *Actinides*, respectively. Upon completion of the Lanthanide series (with lutetium, atomic number 71), the orderly geometry resumes with hafnium, atomic number 72 (Group 4) and continues through actinium, atomic number 89. The probable positions of elements 104 and 105 are indicated in Groups 4 and 5, respectively.

An amazing result of Mendeleev's pioneering classification was the prediction of elements yet to be discovered. Mendeleev found that he could maintain geometric logic of his table only if he allowed for some blank spaces in his table. He further reasoned that elements later would be discovered that would occupy these vacant positions and, thus, Mendeleev predicted the existence of gallium, scandium, and germanium. In fact, Mendeleev gave a preliminary name to scandium, calling it eka-boron and predicted the probable properties of the element. The element later

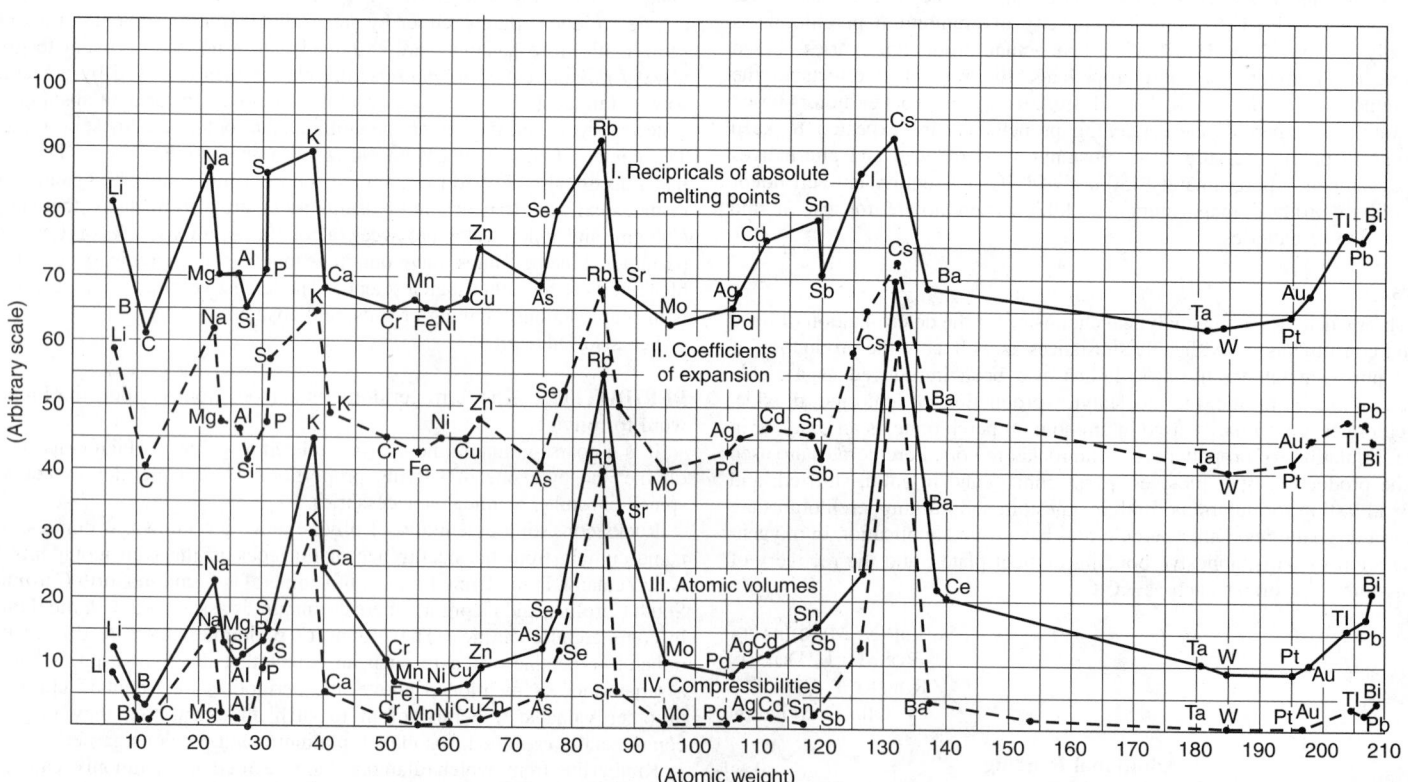

Fig. 1. Pattern obtained when various parameters are plotted against increasing atomic weight of chemical elements

Conventional representation of periodic table

1	2	3	4	5	6	7	8	9	10	11	12	13	14	15	16	17	18
1 H	2											13	14	15	16	17	2 He
3 Li	4 Be											5 B	6 C	7 N	8 O	9 F	10 Ne
11 Na	12 Mg	3	4	5	6	7	8	9	10	11	12	13 Al	14 Si	15 P	16 S	17 Cl	18 Ar
19 K	20 Ca	21 Sc	22 Ti	23 V	24 Cr	25 Mn	26 Fe	27 Co	28 Ni	29 Cu	30 Zn	31 Ga	32 Ge	33 As	34 Se	35 Br	36 Kr
37 Rb	38 Sr	39 Y	40 Zr	41 Nb	42 Mo	43 Te	44 Ru	45 Rh	46 Pd	47 Ag	48 Cd	49 In	50 Sn	51 Sb	52 Te	53 I	54 Xe
55 Cs	56 Ba	57 La	72 Hf	73 Ta	74 W	75 Re	76 Os	77 Ir	78 Pt	79 Au	80 Hg	81 Ti	82 Pb	83 Po	84	85 At	86 Rn
87 Fr	88 Ra	89 Ac															

Lanthanides

58 Ce	59 Pr	60 Nd	61 Pm	62 Sm	63 Eu	64 Gd	65 Tb	66 Dy	67 Ho	68 Er	69 Tm	70 Yb	71 Lu

Actinides

90 Th	91 Pa	92 U	93 Np	94 Pu	95 Am	96 Cm	97 Bk	98 Cf	99 Es	100 Fm	101 My	102 No	103 Lw

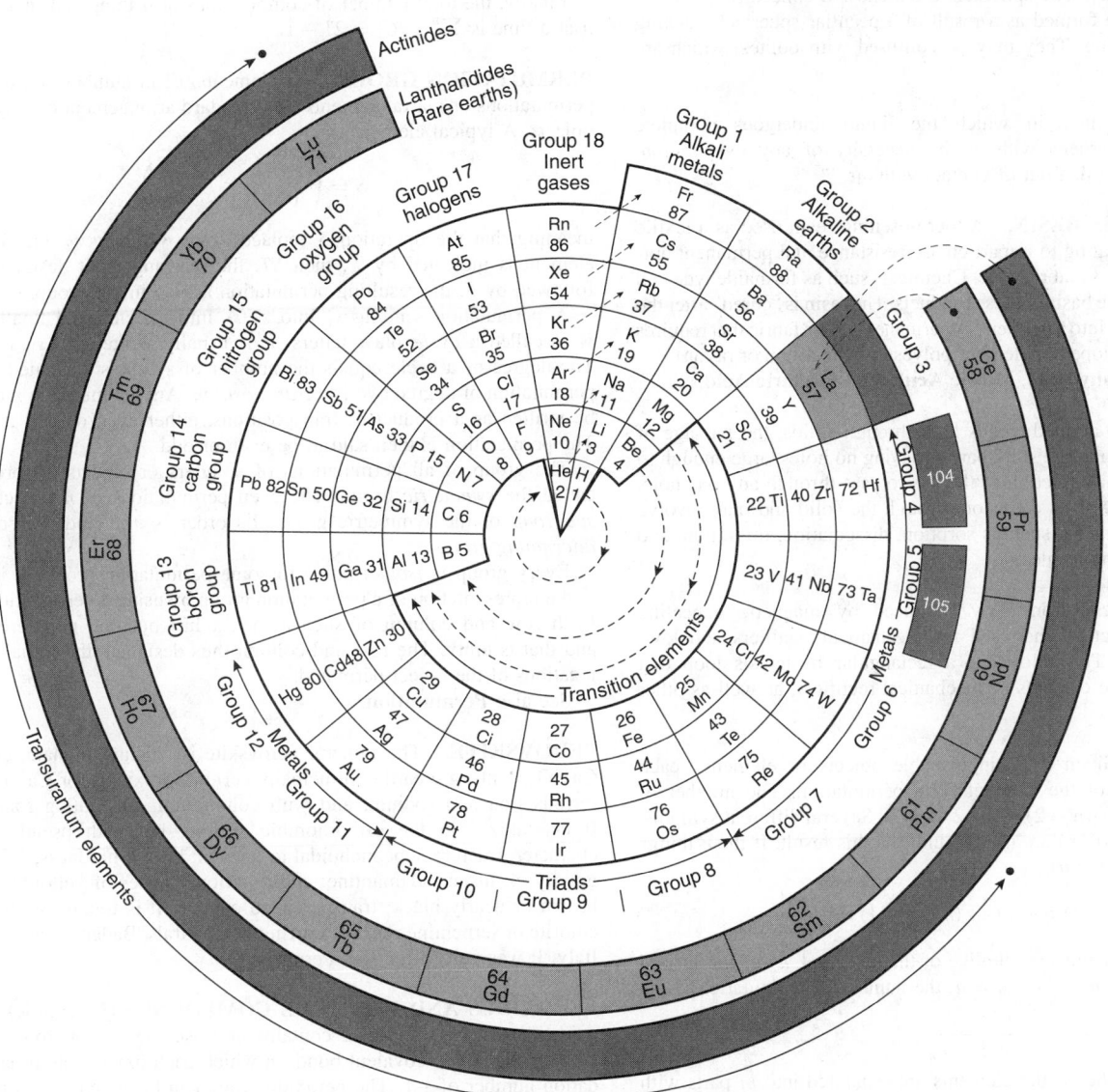

Fig. 2.   Polar representation of periodic relationships of the elements. (*Source: Omnibix, U.S.A.*) At upper right is shown the conventional representation, see the front matter for the current representation

was isolated by Lars Fredrik Nilson in 1879. Mendeleev lived to see his prediction confirmed.

In retrospect, with a much fuller understanding of the underlying electronic and particle structure of the elements, most aspects of the periodicity of the elements come as no surprise, but the fact remains that Mendeleev, Meyer, and others made these striking observations without benefit of over 100 years of additional knowledge. The periodicity of the elements is demonstrated by the accompanying graph which plots atomic weights along the abscissa versus an arbitrary ordinate for various observed physical characteristics. See also **Chemical Elements**.

PERITECTIC TEMPERATURE. Temperature in a peritectic system at which there is equilibrium between the solid and the remaining melt, the composition of which conforms to the peritectic point, at which the temperature line meets the liquids curve.

PERKIN, SIR WILLIAM HENRY (1838–1907). An English chemist who was the first to make a synthetic dyestuff (1856). He studied under Hofman at the Royal College of London. Perkin's first dye was called mauveine, but he proceeded to synthesize alizarin and coumarin, the first synthetic perfume. In 1907 he was awarded the first Perkin Medal, which has ever since been awarded by the American Division of the Society of Chemical Industry for distinguished work in chemistry. Not withstanding the fact that Perkin patented and manufactured mauve dye in England, the center of the synthetic dye industry shifted to Germany, where it remained until 1914.

See also **Hofmann, August Whilhelm (1818–1892)**.

PERLITE (or Pearlstone). An unusual form of siliceous lava composed of small spherules of about the size of bird shot or peas. It is grayish in color with a soft pearly luster. The spherules often show a concentric structure and are believed to be formed as a result of a peculiar spherical cracking developed while cooling. They may be confused with oölites, which are classified as concretions.

PERMAFIL. A mixture in which the liquid undergoes complete polymerization and hardens without the necessity of any evaporation. Anaerobic permafils harden out of contact with air.

PERMANENT-PRESS RESIN. A thermosetting resin used as a textile impregnant or fiber coating to impart crease resistance and permanent hot-creasing to suitings, dress fabrics, etc. Chemicals such as formaldehyde and maleic anhydride are the basis of these products. The resin is "cured" after the fabric has been tailored into a garment. A permanent-press fabric that requires no resin has been developed (a blend of polyester with cotton or rayon).

See also **Maleic Anhydride, Maleic Acid, And Fumaric Acid**.

PERMEATION. As applied to gas flow through solids, the passage of gas into, through, and out of a solid barrier having no holes large enough to permit more than a small fraction of the gas to pass through any one hole. The process always involves diffusion through the solid and may involve various surface phenomena, such as sorption, dissociation, migration, and desorption of the gas molecules.

PERMENORM. Nickel-iron alloy produced by magnetic annealing and drastic cold reduction and used for mechanical rectifiers and low-frequency amplifiers. This alloy has a rectangular hysteresis loop that eliminates arcing at the contacts of mechanical rectifiers, as well as other desirable properties.

PERMUTATION. Given n distinguishable objects or elements, each different arrangement of the elements is a permutation. The number of permutations is $n(n-1)(n-2)\cdots 3 \cdot 2 \cdot 1 = n!$ Several different symbols, such as $_nP_n$, $P_{n,n}$ or $P(n,n)$ are used to indicate this result. If the n things are taken r at a time ($r < n$),

$$_nP_r = n(n-1)(n-2)\cdots(n-r+1) = \frac{n!}{(n-r)!}$$

When n_1 of the elements are all alike of the first kind, n_2 of the second kind, etc., so that $n_1 + n_2 + \cdots n_m = n$, the number of permutations is

$$\frac{n!}{n_1!n_2!\cdots n_m!}$$

This result also applies if the elements are separated into m parts with n_i elements in the i th part. If the number of elements in each of the m parts is not specified, but each part must contain at least one element, the number of permutations is

$$\frac{n!(n-1)!}{(n-m)!(m-1)!}$$

This number is increased to

$$\frac{(m+n-1)!}{(m-1)!}$$

if empty parts are permitted.

A *combination* is an arrangement of objects or elements, where the order of arrangement is not distinguished. Thus, given the three letters a, b, c, the possible permutations are (abc), (acb), etc., six in number, but there is only one combination. With symbols as before,

$$_nC_r = \frac{n(n-1)(n-2)\cdots(n-r+1)}{r!} = \frac{n!}{r!(n-r)!}$$

This number is identical with the $\binom{n}{r}$, the coefficient in the binomial series. Moreover, $_nC_r = {}_nP_{n-r}$; $_nP_r = r!{}_nC_r$.

The number of combinations of n different elements into m specified parts, with empty parts allowed, is m^n; of n identical elements into m different parts with empty parts is

$$\frac{(n+m-1)!}{n!(m-1)!}$$

but when at least one element is in each part, the number is

$$\frac{(n-1)!}{(m-1)!(n-m)!}$$

Finally, the total number of combinations of n things taken 1, 2, 3, …, n at a time is $\sum_{i=1}^{n} {}_nC_i = 2^n - 1$.

PERMUTATION GROUP. Its elements, $n!$ in number, are the various permutations or rearrangements of a standard arrangement of n symbols of objects. A typical element is

$$S = \begin{pmatrix} s_1 & s_2 & \cdots & s_n \\ 1 & 2 & \cdots & n \end{pmatrix}$$

meaning that the operation S replaces 1 by s_1, 2 by s_2, etc. If another element is indicated by T, then ST, the rearrangement designated by T followed by S, the resulting permutation is also in the group.

A permutation sending s_1 into s_2, s_2 into s_3, etc., and finally s_n into s_1 is called a *cycle* on n letters. It is usually written as $(s_1, s_2, \ldots, s_n)$. The degree of a cycle equals the number of symbols permuted. A cyclic permutation of degree two is a *transposition*. Any permutation may always be written as a product of transpositions, either even or odd in number. The permutation is then said to be even or odd.

The group of all permutations of n letters or objects, of order $n!$, is called the *symmetric group*. The even permutations of n objects form a *subgroup* of the symmetric group. Its order is $n!/2$ and it is called the *alternating group*.

Every group is *isomorphous* to some permutation group. It is easy to find a representation of a permutation group by using a permutation matrix. Each row and column of such a matrix has but one non-zero element and that is unity. The row and column thus designate the initial and final locations of the object permuted.

See also **Permutation**.

PEROVSKITE. The mineral perovskite is calcium titanate, essentially $CaTiO_3$, with rare earths, principally cerium, proxying for Ca, as do both ferrous iron and sodium, and with columbium substituting for titanium. It crystallizes in the orthorhombic system, but with pseudo-isometric character; fracture subconchoidal to uneven; brittle; hardness, 5.5; specific gravity, 4; luster, adamantine; color, various shades of yellow to reddish-brown or nearly black; transparent to opaque. It is found associated with chlorite or serpentine rocks occurring in the Urals, Baden, Switzerland, and Italy. It was named for Von Perovski.

PEROXIDES AND PEROXIDE COMPOUNDS (Inorganic). A peroxide or peroxo compound contains at least one pair of oxygen atoms, bound by a single covalent bond, in which each oxygen atom has an oxidation number of -1. The peroxide group can be attached to a metal, M, through one (**1**) or two (**2**) oxygen atoms, or it can bridge two metals (**3**):

$$\text{M—O—O—} \qquad \begin{array}{c} \text{O} \\ \| \\ \text{M} \\ \| \\ \text{O} \end{array} \qquad \text{M—O—O—M}'$$

(**1**) \qquad\qquad (**2**) \qquad\qquad (**3**)

Peroxides should be distinguished from several other types of compounds having similar names. The higher oxides of lead, manganese, and other elements, although sometimes called peroxides, are not peroxides as defined herein because these contain no oxygen–oxygen bond. Similarly, compounds such as the perchlorates and permanganates are not peroxides. It is preferable for true peroxides to be designated by the prefixes *peroxo* or *peroxy*. In the IUPAC nomenclature, *peroxo* is used for inorganic compounds, *peroxy* for organic compounds.

All the simple peroxides form hydrogen peroxide on contact with water. Many inorganic peroxides tend, as does H_2O_2, to decompose evolving oxygen.

Group 1 (1A) Peroxides

Peroxides of all the alkali metals having the formula M_2O_2 are known. There are several general methods of preparation: reaction of the metal and oxygen, reaction of the metal monoxide and oxygen, thermal decomposition of the superoxide, and reaction of alkaline solutions of the metal and hydrogen peroxide.

Alkali metal peroxides are stable under ambient conditions in the absence of water. They dissolve vigorously in water, forming hydrogen peroxide and the metal hydroxide. They are strong oxidizing agents and can react violently with organic substances. Only lithium peroxide and sodium peroxide have been commercialized.

Lithium Peroxide. Lithium peroxide, Li_2O_2, is used in space technology because it absorbs carbon dioxide and liberates oxygen. This peroxide, also used for hardening certain plastics, is a white or pale yellow solid, stable at ambient temperature, and not hygroscopic.

Lithium peroxide is a strong oxidizer and can promote combustion when in contact with combustible materials. It is a powerful irritant to skin, eyes, and mucous membranes. Five grams of many lithium compounds can be fatal.

Sodium Peroxide. Sodium peroxide, Na_2O_2, is a pale yellow solid, stable at ambient temperature, and hygroscopic. Its melting point is 460°C.

The commercial product is a powder containing a minimum of 96% Na_2O_2 and approximately 20% active oxygen. It is made commercially by oxidizing the molten metal with either oxygen or air enriched in oxygen. As of the mid-1990s, sodium peroxide has only a few special applications, including chemical analysis and the extraction of platinum from its ores by the Leidie process.

Although neither inflammable nor self-igniting, sodium peroxide is highly inflammable when mixed with oxidizable substances. Such mixtures burn violently, even in the absence of air.

Sodium peroxide is a powerful irritant to skin, eyes, and mucous membranes; protective clothing should be worn when handling it.

Group 2 (IIA) Peroxides

All the elements of Group 2 form peroxides, with the exceptions of beryllium and radium. There are two general methods of preparation: reaction of the metal or monoxide with oxygen, and reaction of the hydroxide with aqueous hydrogen peroxide. These peroxides are more stable in the presence of water than the Group 1 peroxides, primarily because of insolubility in water. Calcium peroxide is used on a large scale; magnesium, strontium, barium, and zinc peroxides have small-scale uses; whereas cadmium and mercury peroxides have no commercial uses at all. A general account of these peroxides is available.

The materials are generally made by triturating the oxides, or hydroxides, with aqueous hydrogen peroxide and drying the solid products. The commercial products are typically mixtures of the peroxides with varying amounts of hydroxides, oxides, carbonates, hydrates, and peroxohydrates.

Magnesium Peroxide. Magnesium peroxide and MgO_2, used in medicine as a stomach antacid and as an antiseptic, has not been prepared in the pure state. The product is a white powder containing about 25% MgO_2 and 7% active oxygen. This material is sparingly soluble in water but reacts with water slowly, forming hydrogen peroxide and liberating oxygen gas.

There are minor uses for magnesium peroxide in household products, veterinary medicine, and metallurgy. Magnesium peroxide is a strong oxidizer and can cause fire when in contact with combustible materials. It is a powerful irritant to skin, eyes, and mucous membranes.

Calcium Peroxide. Commercial material contains either 60 or 75% CaO_2, the remainder is a poorly defined mixture of calcium oxide, hydroxide, and carbonate.

An important application of calcium peroxide is for curing the polysulfide sealants used in double-glazing window units. Calcium peroxide is also used at several gold mines in Australia to increase the recovery of gold and reduce the consumption of cyanide. Solid calcium peroxide can also be used in the heap-leaching of lean gold ores. A proprietary form of calcium peroxide for this purpose is sold by FMC (United States) under the trademark PermeOx.

PermeOx is also used to improve the bioremediation of soils contaminated with creosote or kerosene, to deodorize sewage sludges and wastewater, and to dechlorinate wastewater and effluents. See also **Odor Modification**.

Calcium peroxide has several horticultural and agricultural applications, particularly in Japan. Usually used in the form of granules, it acts by providing extra oxygen for germinating plants and other organisms.

Calcium peroxide has been used for many years as a dough conditioner in the United States, but not in Europe, where this use is not permitted. Another industrial application of calcium peroxide is as an oxidizing agent in the production of certain titanium–aluminum alloys.

Calcium peroxide is among the safest of the inorganic peroxides, presenting no significant hazard with regard to skin contact or absorption, inhalation, and ingestion; but it may be irritating to the skin under humid conditions. Airborne dust is irritating to the eyes, nose, throat, and lungs, but poses no significant long-term inhalation hazard.

Strontium Peroxide. Commercial strontium peroxide contains about 85% SrO_2 and 10% active oxygen. The only substantial application for this compound is in pyrotechnics. Strontium peroxide produces a red color in flames.

Strontium peroxide is a strong oxidizer and can cause fire when in contact with combustible materials. It is a powerful irritant to skin, eyes, and mucous membranes.

Barium Peroxide. The commercial product is a dull yellow powder containing about 90% BaO_2 and about 8.5% active oxygen. The principal use is in pyrotechnics, but there are also small uses in the curing of polysulfide rubbers and in the production of certain titanium–aluminum alloys.

Barium peroxide is a strong oxidizer and can cause fire when in contact with combustible materials. It is a powerful irritant to skin, eyes, and mucous membranes. Consequently, it is also toxic via the subcutaneous route; protective clothing should be worn during handling. The LD_{50} value (mouse, oral) is 50 mg/kg.

Group 12 (IIB) Peroxides

Zinc Peroxide. The commercial product is a pale yellow powder containing about 55% ZnO_2 and 9% active oxygen. It is stable in dry air but loses its oxygen in moist air and on heating. It is insoluble in water but dissolves in dilute acid, liberating hydrogen peroxide.

Zinc peroxide is used as an accelerator in rubber-compounding, as a curing agent for synthetic elastomers, and as a deodorant for wounds and skin diseases. Zinc peroxide is a powerful irritant to skin, eyes, and mucous membranes. The systemic toxicity is similar to that of zinc oxide, for which the LD_{50} (rat, oral) is 7950 mg/kg.

Zinc peroxide is a strong oxidizer and can cause fire when in contact with combustible materials.

Group 13 (IIIB) Peroxides

Boron Compounds

Nomenclature. The naming of sodium perborate, one of the most important commercial boron compounds, has long been confused. The tetrahydrate has more recently come to be called the hexahydrate. The crystallographically derived names are used to avoid confusion. The commercial or common names are also given.

Sodium Peroxoborate Hexahydrate. The compound sodium peroxoborate hexahydrate (sodium perborate tetrahydrate), $Na_2[B_2(O_2)_2(OH)_4] \cdot 6H_2O$, was formerly written as $NaBO_3 \cdot 4H_2O$. This material has been an important commercial bleaching agent for many years. The commercial product is a white, crystalline powder having an active oxygen content of at least 10%. It melts at about 60°C; however, if water vapor is free to escape during heating, the crystals do not melt but are converted to the anhydrous peroxoborate.

Sodium peroxoborate hexahydrate is an important ingredient of many household detergents, working best at temperatures above 60°C. It is also used in dishwasher detergents, denture cleaners, as well as foot and bath salts. Organic chemists have been using sodium peroxoborates as oxidants since the 1980s.

The toxicity of sodium peroxoborate hexahydrate in solution is equivalent to those of sodium borate and hydrogen peroxide. The LD_{50} (mouse, oral) is 1060 mg/kg. Local use of high concentrations in the mouth can cause chemical burns and other problems.

The product is considered nonhazardous for international transport purposes. However, it is an oxidizing agent sensitive to decomposition by water, direct sources of heat, catalysts, etc.

Sodium Peroxoborate Tetrahydrate. The compound sodium peroxoborate tetrahydrate (sodium perborate trihydrate), $Na_2B_2(O_2)_2[(OH)_4] \cdot 4H_2O$, was formerly written as $NaBO_3 \cdot 3H_2O$.

Sodium peroxoborate tetrahydrate is the most stable of the three peroxoborate hydrates under ambient conditions. It has, however, never been commercialized because it is slow to dissolve in water.

Sodium Peroxoborate. Sodium peroxoborate (sodium perborate monohydrate), $Na_2[B_2(O_2)_2(OH)_4]$, formerly written as $NaBO_3 \cdot H_2O$, is known only as a microcrystalline powder, made by dehydrating the hexahydrate.

The commercial product has an active oxygen content of at least 15%. This product has replaced the hexahydrate in some household detergents and other domestic products because it dissolves faster and has a greater content of active oxygen per unit volume of granular product.

The toxicity of sodium peroxoborate is similar to that of the hexahydrate. Sodium peroxoborate is a severe eye irritant, but not a skin irritant. Absorption through large areas of abraded or damaged skin can give systemic boron poisoning. The maximum eight-hour time-weighted average exposure is 5 mg/m³.

The product is considered nonhazardous for international transport purposes. However, it is an oxidizing agent sensitive to decomposition by water, direct sources of heat, catalysts, etc. Decomposition in the presence of organic material is rapid and highly exothermic.

Anhydrous Sodium Perborate. Anhydrous sodium perborate, $NaBO_3$, is an ill-defined, powdery material. It should perhaps be regarded more as an amorphous assemblage of radicals than as a defined compound.

Anhydrous sodium perborate effervesces in water. It is used mainly as an ingredient in denture-cleaning formulations. No toxicological data have been reported on this product, except that in humans, swallowing large amounts can cause nausea, vomiting, and diarrhea. Anhydrous sodium perborate is irritating to eyes, skin, and mucous membranes. It is also mutagenic to *E. coli.*

Group 14 (IVB) Peroxides

Peroxocarbonates. Peroxocarbonates contain the C—O—O— group and should be distinguished from the carbonate peroxohydrates.

There are international transport regulations controlling the transport of sodium percarbonate, which assigned it to Class 5.1, oxidizing substances, however, no such compound has ever been commercialized, and sodium carbonate peroxohydrate is treated as nonhazardous. The origin of this item is not known.

Peroxosilicates. No solid peroxosilicates are known.

Peroxotin Compounds. Older literature records some tin peroxides or peroxohydrates, but these claims have not been substantiated. In contrast, organometallic peroxotin compounds are well established.

Group 15 (VB) Peroxides

Peroxonitrous Acid and Its Salts. Peroxonitrous acid HOONO, is an isomer of nitric acid, HNO_3, to which it rapidly converts. The half-life of peroxonitrous acid at 0°C is 10 s; at 27°C, 0.23 s. It has been known since 1904 that the yellow solution made by mixing nitrous acid and hydrogen peroxide at low temperature contains a stronger oxidant than either ingredient alone, but the chemistry involved was not put on a sound basis until 1994. Additional preparatory methods are also available.

Peroxonitrous acid can decompose by two pathways: isomerization to nitric acid, and dissociation into the hydroxyl radical and nitrogen dioxide.

Peroxonitrite is believed to be present in the crystals of nitric trihydrate that form in the stratosphere and in Martian soil. See also **Extraterrestrial Materials.** Peroxonitrous acid may be present in mammalian blood and other biochemical systems.

Peroxophosphoric Acids and Their Salts. In its usual impure form (H_3PO_4 is the main contaminant), peroxomonophosphoric acid is a viscous, colorless liquid. It is not produced or used commercially and the salts that have been prepared are unstable and impure.

Pure peroxodiphosphoric acid, $H_4P_2O_8$, has not been obtained, but its properties in aqueous solution are understood.

Tetrapotassium peroxodiphosphate, $K_4P_2O_8$, is a colorless, crystalline solid, soluble in water to 42.2 wt % at 0°C and 51.2 wt % at 40°C.

Tetrapotassium peroxodiphosphate is being investigated as an ingredient in toothpaste as an anticalculus agent and bactericide. However, the peroxodiphosphates are not useful commercial products at present.

Arsenic Peroxides. Arsenic peroxides have not been isolated; however, elemental arsenic, and a great variety of arsenic compounds, have been found to be effective catalysts in the epoxidation of olefins by aqueous hydrogen peroxide. Transient peroxoarsenic compounds are believed to be involved in these systems.

Group 16 (VIB) Peroxides

Peroxosulfuric Acids and Their Salts. Two kinds of peroxosulfuric acid are known: peroxomonosulfuric and peroxodisulfuric acids. Neither is available commercially in the pure state.

Peroxomonosulfuric acid, H_2SO_5, when pure, forms colorless crystals that melt with decomposition at 45°C. Peroxomonosulfuric acid is a strong oxidizing agent. It hydrolyzes rapidly at pH <2 to hydrogen peroxide and sulfuric acid. It is usually made and used in the form of Caro's acid.

Caro's Acid. Caro's acid is the equilibrium mixture that results from mixing hydrogen peroxide and sulfuric acid. These liquids mix instantly, generating a considerable amount of heat.

Because the product is decomposed by heat, it is essential either to remove the heat of reaction quickly or to use the product quickly. The first option is known as the isothermal process; the second option, perfected and commercialized in the early 1990s, is known as the adiabatic process.

Caro's acid is finding increasing application in hydrometallurgy, pulp bleaching, effluent treatment, and electronics.

Peroxomonosulfates. When oleum is mixed with hydrogen peroxide and the mixture is partially neutralized by potassium hydroxide, a triple salt crystallizes out. The commercial product is a white, finely crystalline powder containing a minimum of 4.7% active oxygen. It is used because it is stable and safe, despite being a powerful oxidant. Its main use is in denture cleaners. It is also used in dishwashing detergents and toilet bowl cleaners, in the metal-fabricating industry as a mild etchant and pickling agent, in the electroplating industry for detoxifying cyanide solutions, and in the textile industry for rendering wool shrink-resistant and nonfelting.

In general, peroxomonosulfates have fewer uses in organic chemistry than peroxodisulfates. However, the triple salt is used for oxidizing ketones to dioxiranes, which in turn are useful oxidants in organic chemistry.

Potassium hydrogen monoperoxosulfate monohydrate $KHSO_5 \cdot H_2O$, related to the triple salt, is not made commercially. This compound is reported as toxic and irritating to eyes, skin, and mucous membranes. Although undoubtedly correct, this description probably better relates to the triple salt.

Peroxodisulfuric Acid. Also called persulfuric acid, and Marshall's acid, peroxodisulfuric acid, $H_4S_2O_8$, when pure, forms colorless crystals that melt with decomposition at 65°C. Peroxodisulfuric acid is a strong acid but not stable. It is seldom isolated but is synthesized and used in solution.

Peroxodisulfates. The salts of peroxodisulfuric acid are commonly called persulfates, three of which are made on a commercial scale: ammonium peroxodisulfate, $(NH_4)_2S_2O_8$; potassium peroxodisulfate, $K_2S_2O_8$; and sodium peroxodisulfate, $Na_2S_2O_8$. The peroxodisulfates are all colorless, crystalline solids, stable under dry conditions at ambient temperature but unstable above 60°C. All the peroxodisulfates are made commercially by electrolytic processes.

The peroxodisulfate ion in aqueous solution is one of the strongest oxidizing agents known. The principal use of the peroxodisulfate salts is as initiators for olefin polymerization in aqueous systems, particularly for the manufacture of polyacrylonitrile and its copolymers. See also **Acrylonitrile Polymers.** Etching of printed circuit boards and removal of photoresists are also important applications. Bleaching of textiles and natural fibers and finishing of furs are both long-established applications. Other established applications include curing grouts for soil stabilization initiating polymerization of graphite filament coatings, cleaning metal surfaces prior to plating or adhesive bonding, and regenerating active carbon.

An expanding development is the use of peroxodisulfates as oxidants in organic chemistry.

The three peroxodisulfates are all toxic and irritating to skin, eyes, and mucous membranes. The LD_{LO} value for sodium peroxodisulfate using iv administration in rabbits is 178 mg/kg.

Other Metal Peroxides

Transition-Metal Peroxides. Transition-metal peroxides, as isolated species, have no place in chemical technology because they are too dangerously explosive.

Transition metals can be divided into two groups according to the characteristics of their peroxides. The first group comprises those metals that, in their highest oxidation states, have no d electrons, e.g., Ti^{4+} and W^{6+}. These metals form peroxides from hydrogen peroxide. The peroxo species act as electrophiles.

The other group of transition metals comprises those metals that retain d electrons in their normal valence states, e.g., Co^{3+} and Pt^{2+}. These metals form peroxides from dioxygen or from hydrogen peroxide.

Actinide Peroxides. Many peroxo compounds of thorium, protactinium, uranium, neptunium, plutonium, and americium are known. Uranium peroxide has found several applications in the nuclear energy industry.

Peroxohydrates

Peroxohydrates are crystalline adducts containing molecular hydrogen peroxide. Peroxohydrates are usually made by simple crystallization from solutions of salts or other compounds in aqueous hydrogen peroxide. They are fairly stable under ambient conditions, but traces of transition metals catalyze the liberation of oxygen from the hydrogen peroxide.

Sodium Carbonate Peroxohydrate. Known commercially as sodium percarbonate, sodium carbonate peroxohydrate does not contain the C–O–O–C group and is not a peroxocarbonate. The stoichiometry is $2Na_2CO_3 \cdot 3H_2O_2$. The material is made commercially by three processes: batch crystallization, continuous crystallization, and fluid-bed reaction.

The commercial product is a white powder containing a minimum of 13% of active oxygen and up to 15% of anhydrous sodium carbonate. The solubility in water at 20°C is about 150 g/L.

The principal use of sodium carbonate peroxohydrate is as a bleaching agent in domestic and laundry detergents. It is used also for industrial textile-bleaching, tripe-bleaching, and in denture cleaners. It can also be used as a convenient oxidant in organic chemistry.

The LD_{50} (rat, oral) of sodium carbonate peroxohydrate is 1034 mg/kg. The occupational exposure limit is 10 mg/m^3 per 40-hour week. The compound is a skin and eye irritant; inhalation of dust can cause irritation to the mucous membranes and the respiratory system. Decomposition in the presence of organic material can be rapid and highly exothermic.

Other Peroxohydrates. Other peroxohydrates include those of potassium, rubidium, and cesium carbonates, $M_2CO_3 \cdot 3H_2O$; ammonium carbonate peroxohydrate, $(NH_4)_2CO_3 \cdot H_2O_2$; and urea peroxohydrate, $CO(NH_2)_2 \cdot H_2O_2$.

Urea peroxohydrate is an irritant to skin, eyes, and mucous membranes. The U.S. Food and Drug Administration approves it as an over-the-counter drug.

Peroxopolyoxometallates

Polyoxometallates, derived from both isopoly acids and heteropoly acids, are important homogeneous oxidation catalysts. The metals involved are vanadium, niobium, tantalum, molybdenum, and tungsten. The reactions involved are the oxidation of a wide range of organic compounds by hydrogen peroxide or organic hydroperoxide.

Superoxides

The superoxides are ionic solids containing the superoxide, O_2^-. Superoxides of all of the alkali metals have been prepared. Alkaline-earth metals, cadmium, and zinc all form superoxides, but these have been observed only in mixtures with the corresponding peroxides. The tendency to form superoxides in the alkali metal series increases with increasing size of the metal ion.

Metal superoxides are yellow-to-orange solids. Strong oxidizing agents, they react vigorously with most organic materials and reducing agents, and oxidize many metals to their highest oxidation states.

Sodium superoxide, NaO_2, is a yellow solid. No applications are known. Potassium superoxide, KO_2, is a canary yellow solid that melts at 450–500°C when pure. Potassium superoxide, a strong oxidizing agent, is similar to the Group 1 metal peroxides. Potassium superoxide is produced commercially by spraying molten potassium into an air stream, which may be enriched with oxygen.

Mine Safety Appliances Co. (MSA) manufactures potassium superoxide in the United States for use in self-contained breathing equipment. There are several published uses for potassium superoxide in organic chemistry, e.g., for oxidizing aromatic compounds and for initiating anionic polymerization.

On contact with skin and mucous membranes, potassium superoxide is converted to potassium hydroxide, which is corrosive and irritating. The reaction with moisture is exothermic and may induce further decomposition with the production of oxygen.

Other superoxides include rubidium superoxide, RbO_2; cesium superoxide, CsO_2; calcium superoxide, $Ca(O_2)_2$; strontium superoxide, $Sr(O_2)_2$; and barium superoxide, $Ba(O_2)_2$. These superoxides are not produced commercially.

Ozonides

The ozonides are characterized by the presence of the ozonide ion, O_3^-. They are generally produced by the reaction of the inorganic oxide and ozone. Sodium ozonide, NaO_3; potassium ozonide, KO_3; rubidium ozonide, RbO_3; and cesium ozonide, CsO_3, have all been reported. Ammonium ozonide, NH_4O_3, and tetramethylammonium ozonide, $(CH_3)_4 \cdot NO_3$, have been prepared at low temperatures. Whereas the inorganic ozonides are of potential importance as solid-oxygen carriers in breathing apparatus, they are not produced commercially.

Economic Aspects

All of the large-tonnage peroxo compounds, e.g., sodium peroxoborate hexahydrate, sodium peroxoborate, and sodium carbonate peroxohydrate, are made by hydrogen peroxide producers using captive hydrogen peroxide. The world demand for active oxygen provided by these products is fairly stable, rising with the gross national product.

ALAN E. COMYNS
Solvay Interox

Additional Reading

Bertsch-Franck, B. and co-workers: in R. Thompson, ed., *Industrial Inorganic Chemicals: Production and Use,* Royal Society of Chemistry, Cambridge, U.K., 1995, pp. 188–198.

Gerhartz, W. ed.: *Ullman's Encyclopedia of Industrial Chemistry,* 5th ed., Vol. A19, VCH, Weinheim, Germany, 1991.

Lide, D.R.: *CRC Handbook of Chemistry and Physics,* 84th Edition, CRC Press LLC., Boca Raton, FL, 2003.

McKillop, A. and W.R. Sanderson: *Tetrahedron* **51**(22), 6145 (1995).

Morgan, C.A.: *Mellor's Comprehensive Treatise on Inorganic and Theoretical Chemistry,* Suppl., Vol. 5. Longman, London, U.K., 1980.

Waldemar, A.: *Peroxide Chemistry,* John Wiley & Sons, Inc., New York, NY, 2003.

PEROXIDES AND PEROXIDE COMPOUNDS (Organic). Organic peroxides are compounds possessing one or more oxygen-oxygen bonds. They are derivatives of hydrogen peroxide, HOOH, in which one or both hydrogens are replaced by a group containing carbon (R, R'), i.e., ROOH or ROOR'. The ultimate source of the oxygen-oxygen linkage in organic peroxides is oxygen; either from direct air oxidation or from reactions of organic compounds with peroxidic materials derived from oxygen, e.g., hydrogen peroxide, alkali metal peroxides, ozone, or other organic peroxides. Organic peroxides are intermediates or products in air oxidation of many synthetic and natural organic compounds. They are involved in many biological processes including development of rancidity in fats, loss of activity of vitamin products, and firefly bioluminescence. Some biological products contain a peroxide group. Organic peroxides are also involved in gum formation in lubricating oils, prepolymerization of some vinyl monomers, and degradation of olefin polymers.

Almost all organic peroxides are thermally and photolytically sensitive owing to the facile cleavage of the weak oxygen–oxygen bond. This cleavage is a unimolecular (first-order) reaction. The thermal decomposition rates are affected by the structure of the organic peroxide and the decomposition conditions.

Thermal decomposition of peroxides initially forms oxygen-centered free radicals from the oxygen–oxygen bond homolysis. These radicals are reactive intermediates generally having very short lifetimes, i.e., half-life times less than 10^{-3} s. Because they form useful free radicals, they are used commercially as initiators for free-radical reactions.

Approximately 100 different organic peroxides in well over 300 formulations are commercially produced throughout the world as free-radical initiators for polymerizing vinyl monomers, grafting of monomers onto polymers, curing agents for unsaturated resins, rubber, and elastomers, crosslinking of thermoplastics (e.g., polyethylene), modification/degradation of polypropylene, halogenations, anti-Markovnikov additions to terminal olefins (e.g., formation of primary mercaptans), and telomerizations. Some are used as bleaching agents (i.e., for grain flours and fabrics), olefin epoxidizing agents, and active species in a variety of other applications, e.g., the use of BPO as the active antibacterial component in acne medications.

Organic peroxides can be classified according to peroxide structure. There are seven principal classes: hydroperoxides; dialkyl peroxides; α-oxygen substituted alkyl hydroperoxides and dialkyl peroxides; primary and secondary ozonides; peroxyacids; diacyl peroxides (acyl and organosulfonyl peroxides); and alkyl peroxyesters (peroxycarboxylates, peroxysulfonates, and peroxyphosphates).

Hydroperoxides

There are two main subclasses of hydroperoxides: organic (alkyl) hydroperoxides, i.e., ROOH, and organomineral hydroperoxides, i.e., $R_mQ(OOH)_n$, where Q is silicon, germanium, tin, or antimony. The alkyl group in ROOH can be primary, secondary, or tertiary. Except for ethylbenzene hydroperoxide, only *tert*-alkyl hydroperoxides are commercially important.

Physical Properties. Some physical properties of alkyl hydroperoxides (in order of increasing carbon content) are listed in Table 1.

Alkyl hydroperoxides can be liquids or solids. Those having low molecular weight are soluble in water and are explosive in the pure state. Alkyl hydroperoxides are stronger acids than the corresponding alcohols and have acidities similar to those of phenols. *tert*-Alkyl hydroperoxides can be purified through their alkali metal salts.

Hydroperoxides exist as hydrogen-bonded dimers in nonpolar solvents and readily form hydrogen-bonded associations with ethers, alcohols, amines, ketones, sulfoxides, and carboxylic acids. Other physical properties of hydroperoxides have been reported in the literature.

Chemical Properties. Hydroperoxides can react with or without cleavage of the oxygen–oxygen bond. Reactions resulting in scission of the oxygen–oxygen bond involve heterolytic, homolytic, or metal-promoted oxidation–reduction reactions.

Alkyl hydroperoxides are reduced readily to the corresponding alcohols; many such reductions are quantitative and useful for analytical methods. Alkyl hydroperoxides have been used as oxidizing or hydroxylating reagents in organic syntheses.

Bases, such as potassium or sodium hydroxide, piperidine, and pyridine, react with primary and secondary hydroperoxides to form aldehydes or ketones. *tert*-Alkyl hydroperoxides form stable alkali metal salts with caustic; however, when equimolar amounts of the hydroperoxide and its sodium salt are present in aqueous solution, rapid decomposition to *tert*-alcohol and oxygen occurs.

Acids react with alkyl hydroperoxides in two different ways, depending on the hydroperoxide structure and the acid strength.

$$R_3COOH \xrightarrow[H]{+} R_3C\text{-}OOH \longrightarrow H_2O_2 + R_3C \xrightarrow{+} \textit{tert}\text{-alcohol or olefin}$$

$$R_3COOH \xrightarrow{H^+} R_3CO\text{-}OH \longrightarrow H_2O$$

$$+ R_2 \overset{+}{C} \quad OR \longrightarrow \text{alcohol or phenol and a carbonyl compound}$$

Hydroperoxides are photo- and thermally sensitive and undergo initial oxygen–oxygen bond homolysis, and they are readily attacked by free radicals undergoing induced decompositions.

Hydroperoxides are decomposed readily by multivalent metal ions, i.e., Cu, Co, Fe, V, Mn, Sn, Pb, etc., by an oxidation-reduction or electron-transfer process. Depending on the metal and its valence state, metallic cations either donate or accept electrons when reacting with hydroperoxides. Either one or two electrons may be transferred depending on the metal. With most transition metals, e.g., Cu, Co, and Mn, both valence states react with hydroperoxides via one electron transfer. Thus, a small amount of transition-metal ion can decompose a large amount of hydroperoxide and, consequently, inadvertent contamination

of hydroperoxides with traces of transition-metal impurities should be avoided.

The reactions of *tert*-alkyl hydroperoxides with ferrous ion generate alkoxy radicals. These free-radical initiator systems are used industrially for the emulsion polymerization and copolymerization of vinyl monomers, e.g., butadiene–styrene. Alkyl hydroperoxides are among the most thermally stable organic peroxides. However, hydroperoxides are sensitive to chain decomposition reactions initiated by radicals and/or transition-metal ions. Such decompositions, if not controlled, can be autoaccelerating and sometimes can lead to violent decompositions when neat hydroperoxides or concentrated solutions of hydroperoxides are involved.

Organomineral hydroperoxides undergo thermal and photolytic homolyses:

$$R_3QOOH \xrightarrow[\Delta \text{ or}]{h\nu} R_3QO + OH$$

Synthesis. Hydroperoxides have been prepared from several types of peroxygen compounds including hydrogen peroxide or sodium peroxide, ozone, oxygen, and other organic peroxides. Hydrogen peroxide (H_2O_2) and its anions are powerful nucleophiles and react with reagents RX to form ROOH and HX, where X can be sulfate, acid sulfate, alkane- and arenesulfonate, chloride, bromide, hydroxyl, alkoxide, perchlorate, etc. RX can also be an alkyl orthoformate or *tert*-alkyl carboxylate.

Electron-rich olefins react with hydrogen peroxide under acidic conditions to form hydroperoxides, presumably by means of a carbonium ion intermediate, e.g., *tert*-butyl hydroperoxide from isobutylene.

Organomineral hydroperoxides have been prepared from hydrogen peroxide and organomineral halides, hydroxides, oxides, peroxides, and amines. If HX is an acid, ammonia is used to prevent acidic decomposition.

$$R_mQX_n + nH_2O_2 \longrightarrow R_mQ(OOH)_n + nHX$$

Many hydroperoxides have been prepared by autoxidation of suitable substrates with molecular oxygen. These reactions can be free-radical chain or nonchain processes, depending on whether triplet or singlet oxygen is involved.

Many organic peroxides of metals have been hydrolyzed to alkyl hydroperoxides. Saponification of *tert*-alkyl peroxyesters yields alkyl hydroperoxides and carboxylic acids or their alkali metal salts.

Dialkyl Peroxides

Dialkyl peroxides have the structural formula R–OO–R′, where R and R′ are the same or different primary, secondary, or tertiary alkyl, cycloalkyl, and aralkyl hydrocarbon or hetero-substituted hydrocarbon radicals. Organomineral peroxides have the formulas $R_mQ(OOR)_n$ and R_mQOOQR_m, where at least one of the peroxygens is bonded directly to the organo-substituted metal or metalloid, Q. Dialkyl peroxides include cyclic and bicyclic peroxides where the R and R′ groups are linked, e.g., endoperoxides and derivatives of 1,2-dioxane. Also included are polymeric peroxides, which usually are called poly(alkylene peroxides) or alkylene–oxygen copolymers, and poly(organomineral peroxides), where Q = As or Sb.

Physical Properties. The structures and the boiling and melting points of several dialkyl peroxides are listed in Table 2; a comprehensive list is given in the literature.

Metalloid peroxides behave as covalent organic compounds and most are insensitive to friction and impact but can decompose violently if heated rapidly. Most solid metalloid peroxides have well-defined melting points and the more stable liquid members can be distilled (Table 3).

Chemical Properties. Acyclic di-*tert*-alkyl peroxides efficiently generate alkoxy free radicals by thermal or photolytic homolysis. Primary and secondary dialkyl peroxides undergo thermal decompositions more rapidly than expected owing to radical-induced decompositions. Such radical-induced peroxide decompositions result in inefficient generation of free radicals.

The low molecular weight primary dialkyl peroxides are shock-sensitive and explosive, with sensitivity decreasing with increasing molecular weight. Decomposition products from primary and secondary dialkyl

TABLE 1. PROPERTIES OF SOME ALKYL HYDROPEROXIDES

Hydroperoxide	Structure	Bp, °C (kPa)[a]	Mp, °C	n_D^{20}
methyl	CH_3-OOH	45.5–46.5 (24.53)		1.3654[b]
ethyl	C_2H_5-OOH	43–44 (6.67)		
isopropyl	$(CH_3)_2CH-OOH$	38–38.5 (2.67)		
n-butyl	$n\text{-}C_4H_9-OOH$	40–42 (1.07)	1.4057	
sec-butyl	$sec\text{-}C_4H_9-OOH$	41–42 (1.47)	1.4050	
tert-butyl	$t\text{-}C_4H_9-OOH$	33–42 (2.27)	4.0–4.5	1.3983[c]
2-methoxy-2-propyl	$CH_3O-C(CH_3)_2$ with OOH	61–63 (2.40)		
tert-amyl	$t\text{-}C_5H_{11}-OOH$	34–35 (0.93)	1.4120[c]	
1,1-di-methylpropynyl	$HC{\equiv}CC(CH_3)_2$ with OOH	42 (2.27)		
3-hydroxy-1,1-dimethylbutyl	$HO-CHCH_2-C(CH_3)_2$; CH_3 and OOH			1.4418[d]
cyclohexyl	cyclohexyl–OOH	57 (0.16)	−20	1.4622
n-heptyl	$n\text{-}C_7H_{15}-OOH$	42–43 (0.008)	35.5	1.4269
3-ethyl-3-pentyl	$(C_2H_5)_3C-OOH$	71–73 (2.27)	2–3	1.4379
1-methylcyclohexyl	cyclohexyl with CH_3, OOH	38 (0.004)		1.4652
1-methoxycyclohexyl	cyclohexyl with OCH_3, OOH	54.5–55 (0.027)		
ethylbenzene	$C_6H_5CH-OOH$; CH_3	48.2 (0.027)		1.5265
1,1,3,3-tetramethylbutyl	$(CH_3)_3CCH_2\,C(CH_3)_2$ with OOH	44–45 (0.12)		
2,5-dimethyl-2,5-dihydroperoxyhexane	$(CH_3)_2C(CH_2)_2C(CH_3)_2$ with two OOH		105	
2,5-dimethyl-2,5-dihydroperoxy-3-hexyne	$(CH_3)_2CC{\equiv}CC(CH_3)_2$ with two OOH		107–109	
α-cumyl	$C_6H_5C(CH_3)_2$ with OOH	60 (0.027)		1.5242
1,2,3,4-tetrahydronaphthalene	tetrahydronaphthalene with OOH	120–125 (0.027)	56	
p-menthane	cyclohexane ring with CH_3[e] (position *) and $(CH_3)_2C-OOH$ (position *)			1.4558[f]
pinane	bicyclic pinane with CH_3, H_3C, H_3C, OOH	57 (0.013)		
p-diisopropylbenzene mono-hydroperoxide	$(CH_3)_2CH-$ C6H4 $-C(CH_3)_2$ with OOH		33–34	1.5134
p-diisopropylbenzene dihydroperoxide	$(CH_3)_2C-$ C6H4 $-C(CH_3)_2$ with two OOH		140–141.5	
stearyl	$n\text{-}C_{18}H_{37}-OOH$		49–50	

[a] To convert kPa to mm Hg, multiply by 7.5.
[b] At 21°C.
[c] At 25°C.
[d] 94% assay material. Courtesy of Elf Atochem North America, Inc.
[e] OOH group may alternatively be at positions marked by asterisk.
[f] At 25°C for 54% p-menthane hydroperoxide in p-menthane.

TABLE 2. PROPERTIES OF SOME DIALKYL PEROXIDES

Dialkyl peroxide	Structure	Bp, °C (kPa)[a]	Mp, °C
dimethyl peroxide	$CH_3-OO-CH_3$	13.5 (98.66)	
perfluoro dimethyl peroxide	$CF_3-OO-CF_3$	−37 (101.32)	
diethyl peroxide	$C_2H_5-OO-C_2H_5$	62–63 (101.32)	
1,2-dioxane	(ring structure, O–O)	61.5 (14.67)	
tert-butyl methyl peroxide	$t\text{-}C_4H_9-OO-CH_3$	23 (2.53)	
tert-butyl 2-hydroxyethyl peroxide	$t\text{-}C_4H_9-OO-CH_2CH_2OH$	37–38 (0.27)	
diisopropyl peroxide	$i\text{-}C_3H_7-OO-i\text{-}C_3H_7$		
3,3,5,5-tetramethyl-1,2-dioxolane	(ring structure, H_3C, CH_3, O–O, CH_3, CH_3)	55–58 (29.73), 46 (3.33)	14
di-tert-butyl peroxide	$t\text{-}C_4H_9-OO-t\text{-}C_4H_9$	109 (101.32)	−18
perfluoro-di-tert-butyl peroxide	$(CF_3)_3C-OO-C(CF_3)_3$	99 (101.32)	
3,3,6,6-tetramethyl-1,2-dioxane	(ring structure, H_3C, CH_3, O–O, CH_3, CH_3)	44–45 (1.5)	−26
di-tert-amyl peroxide	$t\text{-}C_5H_{11}-OO-t\text{-}C_5H_{11}$	44 (1.33)	
tert-butyl tert-cumyl peroxide	$t\text{-}C_4H_9-OO-C(CH_3)_2C_6H_5$	40 (0.027)	13
9,10-dihydro-9,10-epidi-oxyanthracene	(anthracene endoperoxide ring structure)		120[b]
2,5-dimethyl-2,5-di(tert-butyl-peroxy) hexane	$[t\text{-}C_4H_9-OO-C(CH_3)_2CH_2-]_2$	42 (0.008)	8
2,5-dimethyl-2,5-di(tert-butylperoxy)-3-hexyne	$t\text{-}C_4H_9OOC-\underset{\underset{CH_3}{\mid}}{\overset{\overset{CH_3}{\mid}}{C}}\equiv C-\underset{\underset{CH_3}{\mid}}{\overset{\overset{CH_3}{\mid}}{C}}-COO-t\text{-}C_4H_9$	65–67 (0.27)	
dicumyl peroxide	$C_6H_5(CH_3)_2C-OO-C(CH_3)_2C_6H_5$		40–41
1,4-di(2-tert-butylperoxyisopropyl)benzene	$1,4\text{-}[t\text{-}C_4H_9-OO-C(CH_3)_2-]_2C_6H_4$		79

[a] To convert kPa to mm Hg, multiply by 7.5.
[b] Explodes at 120°C.

TABLE 3. PROPERTIES OF SOME ORGANOMINERAL PEROXIDES

Organomineral peroxide	Structure	Bp, °C (kPa)[a]	Mp, °C
diethoxyaluminum tert-cumyl peroxide	$(C_2H_5O)_2Al-OO-C(CH_3)_2\text{-}C_6H_5$		113 dec
tri (tert-butylperoxy)borane	$(t\text{-}C_4H_9-OO)_3B$	60–70 (0.0013)	18
tert-butyl triethylgermanium peroxide	$(C_2H_5)_3Ge-OO-t\text{-}C_4H_9$	78 (1.87–2.0)	
dioxybis[triethylgermane]	$(C_2H_5)_3Ge-OO-Ge(C_2H_5)_3$	56–57 (0.0067)	
(tert-butyl-dioxy)triethylplumbane	$(C_2H_5)_3Pb-OO-t\text{-}C_4H_9$		34–36
tetra(tert-butylperoxy)silane	$(t\text{-}C_4H_9OO)_4Si$	78 (0.067)	35–40
dioxybis[trimethylsilane]	$(CH_3)_3Si-OO-Si(CH_3)_3$	36–38 (4.0)	
tert-butylperoxytrimethylsilane	$(CH_3)_3Si-OO-t\text{-}C_4H_9$	78 (28.66)	
dioxybis[triethylstannane]	$(C_2H_5)_3Sn-OO-Sn(C_2H_5)_3$		60[b]
tert-butylperoxytrimethylstannane	$(CH_3)_3Sn-OO-t\text{-}C_4H_9$	56 (1.60)	

[a] To convert kPa to mm Hg, multiply by 7.5.
[b] Explodes at 60°C.

peroxides include aldehydes, ketones, alcohols, hydrogen, hydrocarbons, carbon monoxide, and carbon dioxide.

Because di-tert-alkyl peroxides are less susceptible to radical-induced decompositions, they are safer and more efficient radical generators than primary or secondary dialkyl peroxides. They are the preferred dialkyl peroxides for generating free radicals for commercial applications.

The susceptibility of dialkyl peroxides to acids and bases depends on peroxide structure and the type and strength of the acid or base. In acidic environments, unsymmetrical acyclic alkyl aralkyl peroxides undergo carbon–oxygen fission, forming acyclic alkyl hydroperoxides and aralkyl carbonium ions. The latter react with nucleophiles, X⁻.

Substitution reactions on dialkyl peroxides without concurrent peroxide cleavage are known e.g., the nitration of dicumyl peroxide and the chlorination of di-tert-butyl peroxide.

The polymeric peroxides, $(-OOCH_2CXH-)_n$, where X = H, C_6H_5, $CH=CH_2$, etc., are viscous liquids or amorphous solids having as many as 10 repeating units. These compounds usually explode when heated. The products obtained from the thermal or photodecomposition show that cleavage of both oxygen–oxygen and carbon–carbon bonds occurs.

The type and amounts of products formed depend on the decomposition conditions and the structure of the peroxide.

Unsaturated aliphatic endoperoxides form bis(epoxides) and/or epoxy aldehydes upon thermolysis. The endoperoxides of polynuclear aromatic compounds are crystalline solids that extrude singlet oxygen when heated, thus forming the parent aromatic hydrocarbon. Endoperoxides undergo carbon–oxygen cleavage in acids and oxygen–oxygen bond cleavage in bases, and they are more easily reduced than dialkyl peroxides.

1,2-Dioxetanes have very low activation enthalpies (ca 109 kJ/mol), therefore, they are unstable at low temperatures and generally cleave thermally or photochemically at the oxygen–oxygen and carbon–carbon bonds. Upon fragmentation, chemiluminescence occurs and two carbonyl compounds are produced in the absence of trapping agents. 1,2-Dioxetanes are reduced to diols, epoxides, or allylic alcohols; the dioxetane structure and the reducing system determine which product forms or predominates.

Dioxiranes are three-membered cyclic ring peroxides that are expected to be very unstable owing to ring strain. They are effective oxygenating agents for epoxidations of olefins, allenes, polycyclic aromatic hydrocarbons, enols, and α, β-unsaturated ketones; for insertions of oxygen into X–H

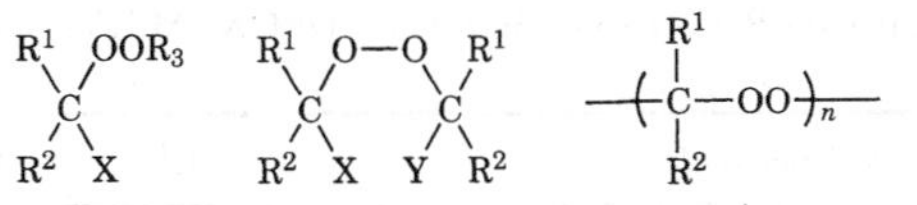

$$\begin{array}{ccc}
\underset{\underset{(1)}{R^2\ \ X}}{\overset{R^1\ \ OOR_3}{\underset{|}{\overset{|}{C}}}} &
\underset{\underset{(2)}{R^2\ \ X\ \ Y\ \ R^2}}{\overset{R^1\ \ O-O\ \ R^1}{\overset{|\ \ \ \ \ \ |}{C\ \ \ \ \ \ C}}} &
\underset{\underset{(3)}{R^2}}{\left(\!\!\!\overset{R^1}{\underset{|}{\overset{|}{C}}}\!-\!OO\!\!\right)_{\!n}}
\end{array}$$

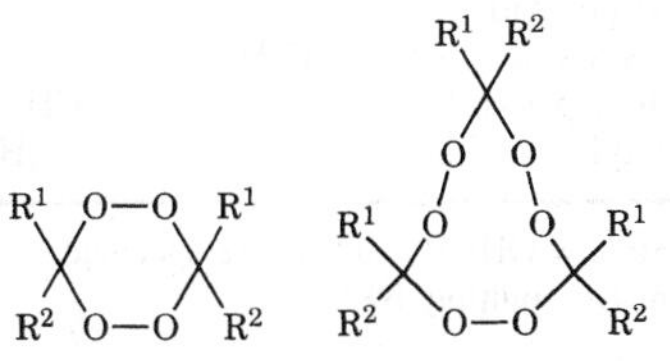

$$\begin{array}{cc}
\underset{\underset{(4)}{R^2\ \ O-O\ \ R^2}}{\overset{R^1\ \ O-O\ \ R^1}{C\ \ \ \ \ \ C}} &
\overset{R^1\ \ R^2}{\underset{\underset{(5)}{R^2\ \ O-O\ \ R^2}}{\overset{\overset{O}{\underset{O\ \ \ \ O}{}}}{C}}}
\end{array}$$

Fig. 1. Varieties of α-oxygen-substituted hydroperoxides and dialkyl peroxides. R^1, R^2, $R^3 = H$ or alkyl; X, Y = OH, OOH, OR^4, OSiR$_3$, or OOR^5; R^4, R^5 = alkyl; and R^3 and R^5 may also be acyl, C(=O)R^6

bonds of alkanes, primary and secondary alcohols, aldehydes, and silanes; and for oxidations of sulfides (to sulfoxides and sulfones), imines (to nitrones), and primary amines (to nitro compounds). In these reactions, the dioxirane transfers oxygen to the substrate and generates the ketone from which the dioxirane was derived.

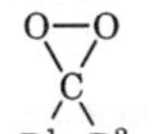

$$\underset{R^1\ \ R^2}{\overset{O-O}{\underset{|}{\overset{\diagdown\diagup}{C}}}}$$

Most organomineral peroxides are hydrolytically unstable and readily hydrolyze to alkyl hydroperoxides or hydrogen peroxide:

$$R_mQ-OO-QR_m \xrightarrow{H_2O} 2R_mQ-OH + H_2O_2$$

Consequently, most organomineral peroxides must be prepared and stored under anhydrous conditions.

Basic hydrolysis of secondary alkyl-substituted silicon and germanium peroxides results in oxygen–oxygen bond cleavage.

The reduction of alkyl-substituted silicon and tin peroxides with sodium sulfite and triphenylphosphine has been reported. Alkyl-substituted aluminum, boron, cadmium, germanium, silicon, and tin peroxides undergo oxygen-to-metal rearrangements, as in the following equations:

$$R_3Si-OO-SiR_3 \longrightarrow R_2Si(OR)OSiR_3$$

$$R_2B-OO-R \longrightarrow RB(OR)_2$$

Organomineral peroxides also undergo thermal and photo-induced homolysis, yielding free radicals that are effective for initiating polymerization of vinyl monomers.

Synthesis. Dialkyl peroxides are prepared by the reaction of various substrates with hydrogen peroxide, hydroperoxides, or oxygen. They also have been obtained from reactions with other organic peroxides.

α-Oxygen-Substituted Hydroperoxides and Dialkyl Peroxides

Dialkyl peroxides and hydroperoxides which have either a hydroxy, hydroperoxy, alkoxy, or alkylperoxy group on the carbon adjacent to the parent peroxide group are considered separately from the parent compounds due to their unique reactions and properties, but mainly because of their unique syntheses. Their primary preparation from aldehydes and ketones via reaction with hydrogen peroxide, alkyl hydroperoxides and peroxyacids is unique and makes it almost impossible to discuss them without referring to the parent carbonyl compound(s).

The α-oxygen-substituted hydroperoxides and dialkyl peroxides comprise a great variety as shown in Figure 1. When discussing peroxides derived from ketones and hydrogen peroxide, (1) is often referred to as a ketone peroxide monomer and (2) as a ketone peroxide dimer.

Syntheses, Physical and Chemical Properties

An example of the complex equilibrium that exists for mixtures of carbonyl compounds and hydrogen peroxide is that from aldehydes and hydrogen peroxide. Hydroxyalkyl hydroperoxides (1, X = OH, R^3 = H) and di(hydroxyalkyl) peroxides (2, X = Y = OH) are formed; cyclic

diperoxides (4) are formed in some cases, e.g., from benzaldehyde with concentrated sulfuric acid. Hydroxyalkyl hydroperoxides are the principal products when equimolar amounts of aldehyde and hydrogen peroxide are used at low temperatures. Di(hydroxyalkyl) peroxides are obtained by using excess aldehyde or higher temperatures. These reactions occur without catalysts but occur at much faster rates in the presence of acids. The peroxides (1) and (2) from most straight-chain aldehydes, i.e., C$_1$–C$_{11}$, have been characterized, and a few of these and some from other aldehydes are listed in Table 4.

Starting with ketones and hydrogen peroxide in the presence of a catalytic amount of acid, mixtures of up to eight components have been identified, i.e., (1, X = OH, R^3 = H), (1, X = OOH, R^3 = H), (2, X = Y = OH), (2, X = Y = OOH), (2, Y = OH, Y = OOH), (3), (4), and (5). The ketone structure and reaction conditions, i.e., acid strength, reactant molar ratios, temperature, and time, determine which compounds form and predominate. Mixtures of several peroxide structures usually are present. Individual peroxides have been isolated from several ketones under different conditions (Table 5). The pure peroxides should be handled with extreme caution since most, especially those derived from the low molecular weight ketones, are shock- and friction-sensitive and can explode violently. Methyl ethyl ketone peroxide (MEKP) mixtures are produced commercially only as solutions containing <40 wt% MEKPs in solvents, commonly dialkyl phthalates.

Hydroxyalkyl Hydroperoxides. These compounds, represented by (1, X = OH, R^3 = H), may be isolated as discreet compounds only with certain structural restrictions, e.g., that one or both of R^1 and R^2 are hydrogen, i.e., they are derived from aldehydes, or that R^1 or R^2 contain electron-withdrawing substituents, i.e., they are derived from ketones bearing α-halogen substituents. Other hydroxyalkyl hydroperoxides may exist in equilibrium mixtures of ketone and hydrogen peroxide.

Alkoxyalkyl Hydroperoxides. These compounds (1, X = OR^4, R^3 = H) have been prepared by the ozonization of certain unsaturated compounds in alcohol solvents. Alkoxyalkyl hydroperoxides are more commonly called ether hydroperoxides. They form readily by the autoxidation of most ethers containing α-hydrogens, e.g., dioxane, tetrahydrofuran, diethyl ether, diisopropyl ether, di-n-butyl ether, and diisoamyl ether. From certain ethers, e.g., diethyl ether, the initially formed ether hydroperoxide can yield alcohol on standing, or with acid treatment form dangerously shock-sensitive and explosive polymeric peroxides.

Hydroxyalkyl Alkyl Peroxides and Hydroxyalkyl Peroxyesters. Hydroxyalkyl alkyl peroxides (1, X = OH, R^3 = alkyl) are reasonably stable and usually can be distilled under a vacuum; the boiling points and structures of representative compounds are listed in Table 6.

Alkoxyalkyl Alkyl Peroxides. *tert*-Butyl tetrahydropyran-2-yl peroxide (1, where R^3 = *tert*-butyl, H = OR^4, R^1 = H, R^2 and R^4 = 1,4-butanediyl, has been isolated. This is one of many examples of alkoxyalkyl alkyl peroxides which may be prepared by reaction of hydroperoxides with vinyl ethers.

1,2,4-Trioxacycloalkanes. 1,2,4-Trioxanes (1), X = OR^4; R^3 and R^4 = alkylene) are generally prepared by the interaction of aldehydes with zwitterionic intermediates made from reaction of singlet oxygen with olefins. They can also be prepared by catalyzed reaction of ketones or

TABLE 4. MELTING POINTS OF SOME PEROXY COMPOUNDS FROM ALDEHYDES AND HYDROGEN PEROXIDE

Peroxy compound	R^{1a}	Mp, °C
hydroxymethyl hydroperoxide	H	oil
1-hydroxyethyl hydroperoxide	CH$_3$	oil
2,2,2-trichloro-1-hydroxyethyl hydroperoxide	CCl$_3$	122
1-hydroxypentyl hydroperoxide	n-C$_4$H$_9$	oil
1-hydroxyoctyl hydroperoxide	n-C$_7$H$_{15}$	46
1-hydroxynonyl hydroperoxide	n-C$_8$H$_{17}$	50–54
di(hydroxymethyl) peroxide	H	63–64
di(1-hydroxyethyl) peroxide	CH$_3$	
di(2,2,2-trichloro-1-hydroxyethyl) peroxide	CCl$_3$	
di(1-hydroxypentyl) peroxide	n-C$_4$H$_9$	
di(1-hydroxyoctyl) peroxide	n-C$_7$H$_{15}$	72
di(1-hydroxynonyl) peroxide	n-C$_8$H$_{17}$	74

^{a} See Fig. 1; R^2 = R^3 = H and X = Y = OH.

TABLE 5. PEROXY COMPOUNDS FROM KETONES AND HYDROGEN PEROXIDE

Peroxy compound	Structure	Mp, °C
2-chloro-1-hydroperoxycyclohexanol[a]		76
1,1-dihydroperoxycyclododecane[b]		140
3,5-dihydroxy-3,5-dimethyl-1,2-dioxolane[c]		90–91
di(1-hydroxycyclohexyl) peroxide[d]	**(6)** X = Y = OH	69–71
1-hydroxycyclohexyl 1-hydroperoxycyclohexyl peroxide	**(6)** X = OH Y = OOH	76–77
di(1-hydroperoxycyclohexyl) peroxide	**(6)** X = Y = OOH	82–83
di(2-hydroperoxy-2-butyl) peroxide	**(2)** R^1 = CH$_3$ R^2 = C$_2$H$_5$ X = Y = OOH	39–42
3,3,6,6-tetramethyl-1,2,4,5-tetroxane	**(4)** R^1 = R^2 = CH$_3$	131–133
3,6-diethyl-3,6-dimethyl-1,2,4,5-tetroxane	**(4)** R^1 = CH$_3$ R^2 = C$_2$H$_5$	[e]
7,8,15,16-tetraoxadispiro-[5.2.5.2]-hexadecane[f]		127–128
3,3,6,6,9,9-hexamethyl-1,2,4,5,7,8-hexoxonane	**(5)** R^1 = R^2 = CH$_3$	96–97
3,6,9-triethyl-3,6,9-trimethyl-1,2,4,5,7,8-hexoxononane	**(5)** R^1 = CH$_3$ R^2 = C$_2$H$_5$	30–32
7,8,15,16,23,24-hexaoxatrispiro-[5.2.5.2.5.2]tetracosane[g]		93

[a] Type (1) R^1 and R^2 are the ring; R^3 = H; X = OH.
[b] Type (1) R^1 and R^2 are the ring; R^3 = H; X = OOH.
[c] Type (2) R^1 = CH$_3$; R^2 is –CH$_2$–; X = Y = OH.
[d] Structure (6) is type (2) wherein R^1 and R^2 are the ring and X and Y are specified.
[e] C is compound has an mp of 12–14°C; trans compound has mp = 23–25°C.
[f] Type (4) R^1 and R^2 are the ring.
[g] Type (5) R^1 and R^2 are the ring.

TABLE 6. BOILING POINTS OF SOME HYDROXYALKYL ALKYL PEROXIDES[a]

Hydroxyalkyl alkyl peroxide	R^3	R^1	Bp, °C (kPa)[b]
hydroxymethyl methyl peroxide	CH$_3$	H	45 (2.27)
tert-butyl hydroxymethyl peroxide	t-C$_4$H$_9$	H	52–53 (1.07)
1-hydroxyethyl methyl peroxide	CH$_3$	CH$_3$	25–27 (2.27)
1-hydroxyethyl ethyl peroxide	C$_2$H$_5$	CH$_3$	48–50 (8.67)
tert-butyl 1-hydroxyethyl peroxide	t-C$_4$H$_9$	CH$_3$	30–31.5 (0.13)
tert-butyl 1-hydroxybutyl peroxide	t-C$_4$H$_9$	n-C$_3$H$_7$	34–37 (0.13)

[a] Structure (1); R^2 = H; X = OH; R^1 and R^3 are specified.
[b] To convert kPa to mm Hg, multiply by 7.5.

aldehydes with 1,2-dioxetanes or endoperoxides, and they can be prepared directly from certain hydroperoxides.

Geminal Dihydroperoxides. These dihydroperoxides as described previously (**1**, X = OOH, R^3 = H) can be made from many different carbonyl compounds. These peroxides can also be synthesized by perhydrolysis of ketals. Low molecular weight dihydroperoxides are soluble in water and are explosive when pure. They have been reduced to the corresponding ketones with hydriodic acid or zinc and acetic acid. Hydrolysis also gives the corresponding ketones. In the presence of catalytic amounts of acids or on prolonged storage, solutions of dihydroperoxides form equilibrium amounts of hydrogen peroxide and di(hydroperoxyalkyl) peroxides and ultimately equilibrium amounts of cyclic triperoxides.

Diperoxyketals and Diperoxyacetals. Aromatic aldehydes react with alkyl hydroperoxides in the presence of strong acid catalysts such as sulfuric acid to form diperoxyacetals (**1**, X = OOR^5; R^1 = H, R^2 = Ar, R^3 = R^5 = alkyl). Diperoxyketals (**1**, X = OOR^5; R^1, R^2, R^3, R^5 = alkyl) are generally prepared by acid-catalyzed reaction of a ketone with two equivalents of an alkyl hydroperoxide.

Diperoxyketals are solids or colorless liquids and are soluble in common organic solvents and insoluble in water. The physical properties and structures of some diperoxyketals are listed in Table 7. In the pure state, the low molecular weight compounds can decompose violently when heated, and addition of concentrated sulfuric acid can result in flaming decompositions. There are many commercial diperoxyketals, and they are usually diluted with solvents for improved safety.

Tertiary diperoxyketals (**1**, X = OOR^5, R^1, R^2 = alkyl, R^3, R^5 = tertiary alkyl) are excellent free-radical initiators. Such diperoxyketals are stable, especially those with R^3 = R^5 = *tert* − butyl. Less thermally stable diperoxyketals are those derived from cyclic ketones and those with bulkier *tert*-alkyl groups, e.g., *tert*-amyl, *tert*-octyl, *tert*-cumyl. Commercial members of this group all have R^3 = R^5, and thermally decompose to free radicals by cleavage of only one oxygen–oxygen bond initially, usually followed by β-scission of the resulting alkoxy radicals. For acyclic diperoxyketals, β-scission produces an alkyl radical and a peroxyester. Owing to similarity of thermal stability, the peroxyester decomposes almost simultaneously.

Diperoxyketals, and many other organic peroxides, are acid-sensitive, therefore removal of all traces of the acid catalyst must be accomplished

TABLE 7. BOILING POINTS OF SOME DIPEROXYKETALS

$$R^3OO-\overset{\displaystyle R^1}{\underset{\displaystyle R^2}{C}}-OOR^5$$

Diperoxyketal	R^3, R^5	R^1	R^2	Bp, °C (kPa)[a]
2,2-di(t-butylperoxy)propane	t-C$_4$H$_9$	CH$_3$	CH$_3$	69–70 (2.0)
2,2-di(t-amylperoxy)propane	t-C$_5$H$_{11}$	CH$_3$	CH$_3$	68 (0.17)
2,2-di(t-butylperoxy)butane	t-C$_4$H$_9$	CH$_3$	C$_2$H$_5$	50 (0.27)
1,1-di(t-butylperoxy)cyclohexane	t-C$_4$H$_9$	(CH$_2$)$_5$		52–54 (0.02)
2,2-bis[4,4-di(t-butylperoxy)cyclohexyl]propane				117–120[b]

[a] To convert kPa to mm Hg, multiply by 7.5.
[b] Mp, °C.

before attempting distillations or kinetic decomposition studies. The low molecular weight diperoxyketals can decompose with explosive force and commercial formulations are available only as mineral spirits or phthalate ester solutions.

Di(hydroxyalkyl) Peroxides. The lowest molecular weight member of this group (**2**, X = Y = OH), di(hydroxymethyl) peroxide ($R^1 = R^2 =$ H) is a dangerously explosive solid. With increasing molecular weight, di(hydroxyalkyl) peroxides become liquids and eventually solids of decreasing explosive nature and water solubility. In solution, these dialkyl peroxides exist in equilibrium with other α-oxygen-substituted peroxides, carbonyl compounds, and hydrogen peroxide.

Formaldehyde reacts with di(hydroxymethyl) peroxide and phosphorus pentoxide to form di(hydroxymethoxymethyl) peroxide (**2**), where X = Y = OCH_2OH, $R^1 = R^2 =$ H.

Reaction of 1,3- and 1,4-diketones ($n = 1$ or 2) with hydrogen peroxide yields cyclic di(hydroxyalkyl) (X = OH) or di(hydroperoxyalkyl) (X = OOH) peroxides (**7**).

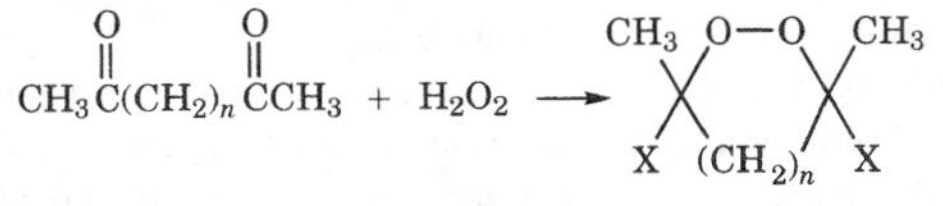

(**7**)

The di(hydroxyalkyl) peroxide (**2**) from cyclohexanone is a solid which is produced commercially. The di(hydroxyalkyl) peroxide (**2**) from 2,4-pentanedione (**7**, $n = 1$; X = OH) is a water-soluble solid which is also produced commercially (see Table 5). Both these peroxides are used for curing cobalt-promoted unsaturated polyester resins.

Hydroxyalkyl Hydroperoxyalkyl Peroxides. There is evidence that hydroxyalkyl hydroperoxyalkyl peroxides (**2**, X = OH, Y = OOH) exist in equilibrium with their corresponding carbonyl compounds and other α-oxygen-substituted peroxides. Thermal decomposition of hydroxyalkyl hydroperoxyalkyl peroxides produces mixtures of starting carbonyl compounds, mono and dicarboxylic acids, cyclic diperoxides, carbon dioxide, and water.

Di(hydroperoxyalkyl) Peroxides. Low molecular weight di(hydroperoxyalkyl) peroxides (**2**, X = Y = OOH) are dangerously prone to explosive decomposition when they are pure. Some have been characterized by acylation to the corresponding diperoxyesters.

Cyclic Peroxides. Cyclic diperoxides (**4**) and triperoxides (**5**) are solids and the low molecular weight compounds are shock-sensitive and explosive. The melting points of some characteristic compounds of this type are given in Table 5.

Polymeric α-Oxygen-Substituted Peroxides. Polymeric peroxides (**3**) are formed from the following reactions: ketone and aldehydes with hydrogen peroxide, ozonization of unsaturated compounds, and dehydration of α-hydroxyalkyl hydroperoxides; consequently, a variety of polymeric peroxides of this type exist. Polymeric peroxides are generally viscous liquids or amorphous solids, are difficult to characterize, and are prone to explosive decomposition.

Miscellaneous α-Substituted Peroxides. 3-Aryl-3-(*tert*-alkylperoxy)-phthalides (**8**) are prepared from the corresponding 3-chlorophthalides and *tert*-alkyl hydroperoxide. 2-Methyl-2-(*tert*-alkylperoxy)-1,3-benzodioxan-4-ones (**9**) are obtained from *o*-acetylsalicyloyl chloride and *tert*-alkyl hydroperoxides. Trisubstituted 2-(*tert*-alkylperoxy)-1,3-dioxolan-4-ones (**10**) are synthesized from sterically favored α-acyloxy acid chlorides and *tert*-alkyl hydroperoxides.

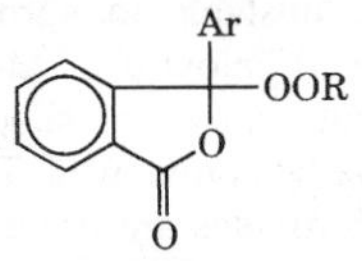

(**8**) 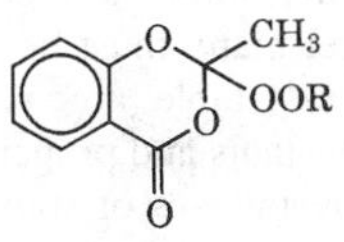   (**9**) 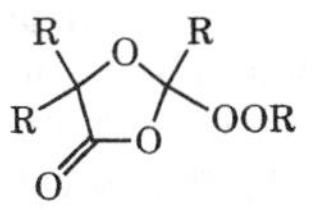   (**10**)

Ozonides and Ozonization

Unsaturated compounds undergo ozonization to initially produce highly unstable primary ozonides (**11**), i.e., 1,2,3-trioxolanes, also known as molozonides, which rapidly split into carbonyl compounds (aldehydes and ketones) and 1,3-zwitterion (**12**) intermediates. The carbonyl compound-zwitterion pair then recombines to produce a thermally stable secondary ozonide (**13**), also known as a 1,2,4-trioxolane.

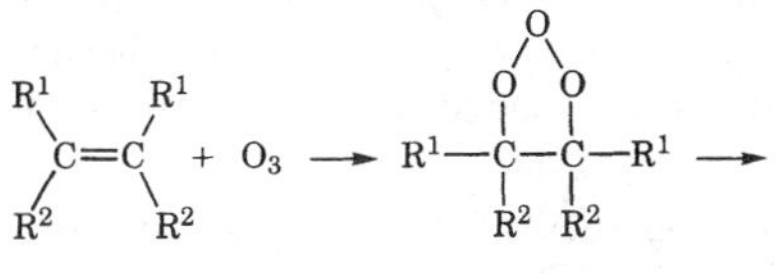

(**11**)

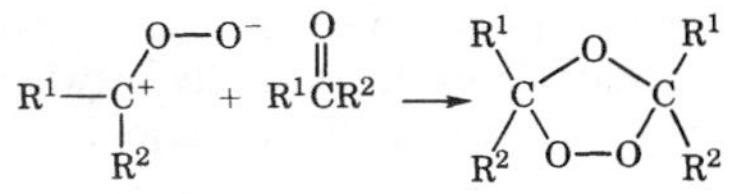

(**12**)   (**13**)

Most ozonolysis reaction products are postulated to form by the reaction of the 1,3-zwitterion with the extruded carbonyl compound in a 1,3-dipolar cycloaddition reaction to produce stable 1,2,4-trioxanes (ozonides) (**13**) as shown; with itself (dimerization) to form cyclic diperoxides (**4**); or with protic solvents, such as alcohols, carboxylic acids, etc., to form α-substituted alkyl hydroperoxides. The latter can form other peroxidic products, depending on reactants, reaction conditions, and solvent.

In the presence of alcohols, the ozonization products are alkoxyalkyl hydroperoxides (**1**, X = OR^4, R = H_3):

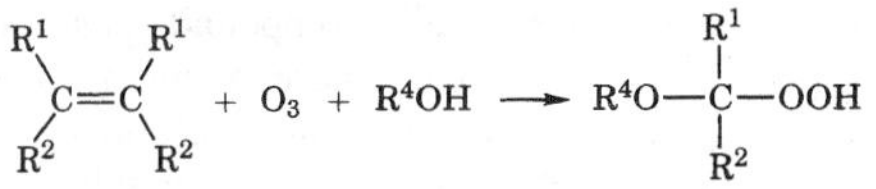

By-products include ozonides (**13**). Other peroxidic products including polymeric peroxides and polymeric ozonides can form, depending on reaction conditions, solvent, and olefin used. A variety of cyclic diperoxides (**4**) have been obtained by ozonolysis of olefins. Boiling point data for several 1,2,4-trioxanes are listed in Table 8.

Cyclic 1,2,4-trioxanes (**14** and **15**) have been obtained from the photosensitized oxidation of furans. These compounds are 2,3,7-trioxabicyclo [2.2.1] hept-5-ene (**14**) and 2,3,7-trioxabicyclo [2.2.1] heptane (**15**).

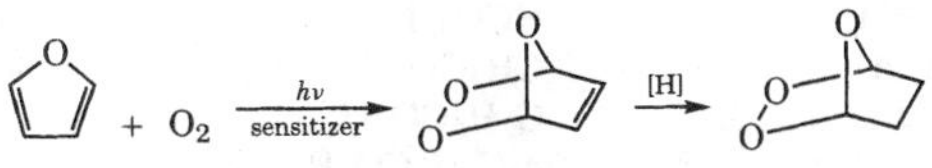

(**14**)   (**15**)

Peroxyacids

There are two broad classes of organic peroxyacids: peroxycarboxylic acids, $R[C(O)OOH]_n$, where R is an alkyl, aralkyl, cycloalkyl, aryl,

TABLE 8. BOILING POINTS OF SOME 1,2,4-TRIOXOLANES[a]

1,2,4-Trioxolanes	Structure	Bp, °C (kPa)[b]
1,2,4-trioxolane	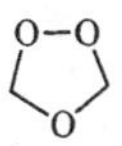	18 (2.13)
3,5-dimethyl-1,2,4-trioxolane	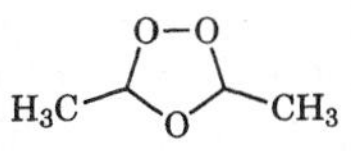	15 (2.67)
3,3-dimethyl-1,2,4-trioxolane	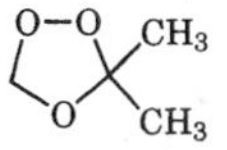	42–42.5 (18.67)
1,5-dimethyl-6,7,8-trioxabicyclo [3.2.1] octane	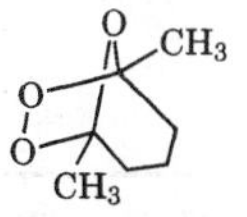	58.8 (2.0)
3,5-diphenyl-1,2,4-trioxolane	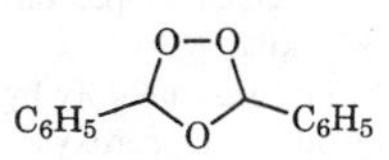	

[a] Secondary ozonides.

[b] To convert kPa to mm Hg, multiply by 7.5.

or heterocyclic group and $n = 1$ or 2, and organoperoxysulfonic acids, RSO_2-OOH.

Three peroxyacids are produced commercially for the merchant market: peroxyacetic acid as a 40 wt % solution in acetic acid, *m*-chloroperoxybenzoic acid, and magnesium monoperoxyphthalate hexahydrate. Other peroxyacids are produced for captive use, e.g., peroxyformic acid generated *in situ*, as an epoxidizing agent.

Physical Properties. Physical properties of peroxyacids have been extensively reviewed in the literature. The melting points of some peroxycarboxylic acids are listed in order of increasing number of carbon atoms in Table 9. Aliphatic peroxyacids are characterized by sharp unpleasant odors, the intensity of which decreases with increasing chain length. They also are irritating to the skin and mucous membranes.

Chemical Properties. Organic peroxyacids are not noted for their stability and many lose active oxygen during storage at room temperature. Those that are water soluble hydrolyze slowly to the parent acid and hydrogen peroxide; however, peroxyformic acid hydrolyzes more rapidly. The longer-chain aliphatic members decompose rapidly in methanol. Stabilizers are commonly used for peroxycarboxylic acid solutions, e.g., dipicolinic acid, phytic acid, and pyro- and metaphosphates. Stability of peroxycarboxylic acids increases with increasing molecular weight. The stabilities of peroxybenzoic acids are enhanced when ring substituents are present.

Peroxycarboxylic acids and precursors to peroxycarboxylic acids are used as bleaches for removal of stains and soils from textiles. Precursors to peroxycarboxylic acids are nonperoxidic compounds possessing a reactive acyl group and a good leaving group, L. These precursors react under basic conditions with hydrogen peroxide, inorganic perborates, and inorganic percarbonates to generate peroxycarboxylic acids or salts:

$$R-\overset{O}{\overset{\|}{C}}-L + H_2O_2 \xrightarrow[\text{conditions}]{\text{basic}} R-\overset{O}{\overset{\|}{C}}-OO^- \text{ or } R-\overset{O}{\overset{\|}{C}}-OOH + L-H$$

TABLE 9. PROPERTIES OF SOME ORGANIC PEROXYACIDS

Peroxyacid	Structure	Mp, °C
peroxyformic acid	HCO_3H	$-18^{a,b}$
peroxyacetic acid	CH_3CO_3H	0^d
peroxypropionic acid	$C_2H_5CO_3H$	-13^d
peroxybutyric acid	$n\text{-}C_3H_7CO_3H$	-10^e
monoperoxysuccinic acid	$HO_2C(CH_2)_2CO_3H$	107, dec
peroxyhexanoic acid	$n\text{-}C_5H_{11}CO_3H$	15^f
peroxybenzoic acid	$C_6H_5CO_3H$	41–42
m-chloroperoxybenzoic acid	$m\text{-}Cl\text{-}C_6H_4CO_3H$	88
diperoxyhexanedioic acid	$HO_3C(CH_2)_4CO_3H$	116–117 dec
peroxyoctanoic acid	$n\text{-}C_7H_{15}CO_3H$	31
4-methylperoxybenzoic acid	$4\text{-}CH_3C_6H_4CO_3H$	95–96
monoperoxyphthalic acid	$2\text{-}HO_2CC_6H_4CO_3H$	110^b
peroxynonanoic acid	$n\text{-}C_8H_{17}CO_3H$	35
peroxycinnamic acid	$CHCHnf > H56$	67–68 dec
diperoxynonanedioic acid	$HO_3C(CH_2)_7CO_3H$	90
peroxydecanoic acid	$n\text{-}C_9H_{19}CO_3H$	41
diperoxydecanedioic acid	$HO_3C(CH_2)_8CO_3H$	98
peroxydodecanoic acid	$n\text{-}C_{11}H_{23}CO_3H$	50
peroxytetradecanoic acid	$n\text{-}C_{13}H_{27}CO_3H$	56
peroxyhexadecanoic acid	$n\text{-}C_{15}H_{31}CO_3H$	61
peroxyoctadecanoic acid	$n\text{-}C_{17}H_{35}CO_3H$	65
magnesium monoperoxyphthalate hexahydrate		93

a 90 wt % melts at the given temperature.
b Bp = 50°C at 13.33 kPa^c.
c To convert kPa to mm Hg, multiply by 7.5.
d Bp = 25°C at 1.6 kPa^c for peroxyacetic acid, and 2.67 kPa^c for peroxypropionic acid.
e Bp = 26–29°C at 1.6 kPa^c.
f Bp = 41–43°C at 0.067 kPa^c.

Thermal decompositions of peroxycarboxylic acids and their salts can proceed by free-radical and nonradical paths. Often the decomposition products and the rate are affected by the nature of the solvent. Peroxycarboxylic acids undergo photodecomposition and radical-induced decomposition. They also are decomposed by a variety of metals, metal ions, and complexes.

Peroxycarboxylic acids are among the most powerful organic peroxide oxidizing agents. The main industrial uses of these acids are in the manufacture of epoxides, synthetic glycerol, and epoxy resins. They also have been used as disinfectants, fungicides, and bleaching agents and for shrink-proofing wool.

Synthesis. Many different methods for the preparation of peroxyacids have been described. The most widely used method is the direct, acid-catalyzed equilibrium reaction of 30–98 wt % hydrogen peroxide with carboxylic acids: The equilibrium also can be shifted to the right by removing water azeotropically and/or under a vacuum. Chelating agents may be added during processing to reduce metal-catalyzed decompositions. Sulfuric acid, methanesulfonic acid, and sulfonic acid ion-exchange resins are the most commonly used acid catalysts.

Other methods for preparing peroxycarboxylic acids include: (*1*) autoxidation of aldehydes, (*2*) reaction of acid chlorides, anhydrides, or boric-carboxylic anhydrides with hydrogen or sodium peroxide, and (*3*) basic hydrolysis or perhydrolysis of diacyl peroxides.

Organoperoxysulfonic acids and their salts have been prepared by the reaction of arenesulfonyl chlorides with calcium, silver, or sodium peroxide; treatment of metal salts of organosulfonic acids with hydrogen peroxide; hydrolysis of di(organosulfonyl) peroxides, $R^2S(O)-OO-S(O)R^2$, with hydrogen peroxide; and sulfoxidation of saturated, nonaromatic hydrocarbons, e.g., cyclohexane.

Other Peroxyacids. Benzeneperoxyseleninic acid has been prepared *in situ* from benzeneseleninic acid and hydrogen peroxide and is used to epoxidize terpenic olefins and Baeyer-Villiger oxidation of cyclic ketones.

$$C_6H_5\overset{O}{\overset{\|}{Se}}-OOH$$

Acyl Peroxides

The acyl peroxide class is characterized by the following structures:

Acyl peroxides of structure (**16**) are known as diacyl peroxides. In this structure R^1 and R^2 are the same or different and can be alkyl, aryl, heterocyclic, imino, amino, or fluoro. Acyl peroxides of structures (**17**), (**18**), (**19**), and (**20**) are known as dialkyl peroxydicarbonates, *OO*-acyl *O*-alkyl monoperoxycarbonates, acyl organosulfonyl peroxides, and di(organosulfonyl) peroxides, respectively. R^1 and R^2 in these structures are the same or different and generally are alkyl and aryl. Many diacyl peroxides (**16**) and dialkyl peroxydicarbonates (**17**) are produced commercially and used in large volumes.

Physical Properties. Almost all liquid diacyl peroxides (**16**) and concentrated solutions of the solid compounds are unstable to normal ambient temperature storage; many must be stored well below 0°C. Most of the solid compounds are stable at ca 20°C but many are shock-sensitive. Other physical constants and properties have been reviewed. The melting points and refractive indexes of some acyl peroxides are listed in Tables 10–12.

Chemical Properties. Diacyl peroxides (**16**) decompose when heated or photolyzed (<300 mm). Although photolytic decompositions generally produce free radicals, thermal decompositions can produce nonradical and radical intermediates, depending on diacyl peroxide structure. Symmetrical aliphatic diacyl peroxides of certain structures, i.e., diacyl peroxides (**16**, $R^1 = R^2$ = alkyl) without α-branches or with a mono-α-methyl substituent,

TABLE 10. PROPERTIES OF SOME DIACYL PEROXIDES

$$R^1\!-\!\overset{\overset{\textstyle O}{\|}}{C}\!-\!OO\!-\!\overset{\overset{\textstyle O}{\|}}{C}\!-\!R^2$$

Diacyl peroxide	R group	Mp,°C[a]
SYMMETRICAL, $R^1 = R^2$		
diacetyl peroxide	CH_3	30
di(chloroacetyl) peroxide	$ClCH_2$	85
dipropionyl peroxide	C_5H_2	oil
diisobutyryl peroxide	$i\text{-}C_7H_3$	80–80.5
di(3-carboxypropionyl) peroxide	$H_2OC(CH)_2$	132–133
dipentanoyl peroxide	$n\text{-}C_9H_4$	oil
di(4-carboxybutyryl) peroxide	$H_2OC(CH)_3$	104
di(2-furanylcarbonyl) peroxide	[2-furanyl ring]	86–87
di(2-thienylcarbonyl) peroxide	[2-thienyl ring]	92–93
dinicotinoyl peroxide	$3\text{-cyclo-}C_4H_5N$	88–89
diheptanoyl peroxide	$n\text{-}C_{13}H_6$	[b]
di(cyclohexylcarbonyl) peroxide	$cyclo\text{-}C_{11}H_6$	oil
dibenzoyl peroxide	C_5H_6	106–107
di(4-chlorobenzoyl) peroxide	$4\text{-}ClC_4H_6$	137–138
di(4-nitrobenzoyl) peroxide	$4\text{-}NO_4C_6H_2$	157–158
di(2-methylbenzoyl) peroxide	$2\text{-}CH_4C_6H_3$	54
di(2-carboxybenzoyl) peroxide	$2\text{-}HOC_4H_6H_2$	156
dioctanoyl peroxide	$n\text{-}CH$	29
di(phenylacetyl) peroxide	$CH_2C_5H_6$	41
dinonanoyl peroxide	$n\text{-}C_7H_8$	13.0–13.5
di(3,5,5-trimethylhexanoyl) peroxide		[c]
dicinnamoyl peroxide	$CHCH=C_5H_6$	133–134
di(benzocyclobutene-4-carbonyl) peroxide	[benzocyclobutene structure]	130–132
didecanoyl peroxide	$n\text{-}C_{19}H_9$	44–45
di(2-naphthaleny carbonyl) peroxide	$2\text{-}C_7H_{10}$	138–140
didodecanoyl peroxide	$n\text{-}C_{23}H_{11}$	54.7–55
dihexadecanoyl peroxide	$n\text{-}C_{31}H_{15}$	71.4–71.9
dioctadecanoyl peroxide	$n\text{-}C_{35}H_{17}$	76.5–76.9
UNSYMMETRICAL, $R^1 = CH_3$; R^2 AS GIVEN		
acetyl propionyl peroxide	C_5H_2	[d]
acetyl cyclohexyl carbonyl peroxide	$cyclo\text{-}C_{11}H_6$	42
acetyl benzoyl peroxide	C_5H_6	37–39
adipoyl bis(acetyl peroxide)	$CH_3\overset{\overset{\textstyle O}{\|}}{C}OO\overset{\overset{\textstyle O}{\|}}{C}(CH_2)_4\!-\!$	61–62

[a] Most of these peroxides decompose on melting, some violently.
[b] $n_D^{25} = 1.4340$.
[c] $n_D^{20} = 1.4382$.
[d] $n_D^{20} = 1.4069$.

TABLE 11. PROPERTIES OF SOME DIALKYL PEROXYDICARBONATES

$$R^1O\!-\!\overset{\overset{\textstyle O}{\|}}{C}\!-\!OO\!-\!\overset{\overset{\textstyle O}{\|}}{C}\!-\!OR^1$$

Peroxydicarbonate	R^1	n_D^{20}	Mp, °C[a]
diethyl	C_5H_2-	1.4065	
di-n-propyl	$n\text{-}C_7H_3-$	1.4091	
diisopropyl	$i\text{-}C_7H_3-$	1.4034	8–10
dibutyl	$n\text{-}C_9H_4-$	1.4129	
di-sec-butyl	$sec\text{-}C_9H_4-$	1.4112	
dicyclohexyl	[cyclohexyl]		46
dibenzyl	$CH_2C_5C_6-$		101–102
di(2-ethylhexyl)	$CH_3(CH_2)_3\overset{\overset{\textstyle C_2H_5}{\|}}{C}HCH_2-$	1.4366	
di(2-phenoxyethyl)	$CH_2OCH_5CH_6-$		97–100
di(cis-3,3,5-trimethylcyclohexyl)	[3,3,5-trimethylcyclohexyl structure]		78–79
di(4-tert-butylcyclohexyl)	[4-tert-butylcyclohexyl, $(CH_3)_3C$]		91–92
di(isobornyl)	$isobornyl-$		92–93
didodecyl	$n\text{-}C_{25}H_{12}-$		28–30
ditetradecyl	$n\text{-}C_{29}H_{14}-$		40–42
dihexadecyl	$n\text{-}C_{33}H_{16}-$		50–53

[a] All listed peroxides are unstable in liquid state >20°C; some decompose violently.

TABLE 12. MELTING POINTS OF SOME ORGANOSULFONYL PEROXIDES

Organosulfonyl peroxide	Structure	Mp, °C[a]
di(methanesulfonyl) peroxide	$[CH_3-S(O)_2-O-]_2-$	77
acetyl tert-butanesulfonyl peroxide	$t\text{-}C_4H_9-S(O)_2-OO-\overset{\overset{\textstyle O}{\|}}{C}CH_3$	35–37
acetyl cyclohexanesulfonyl peroxide	[cyclohexyl]$-S(O)_2-OO-\overset{\overset{\textstyle O}{\|}}{C}CH_3$	35–36
acetyl sec-heptanesulfonyl peroxide	$sec\text{-}C_7H_{15}S(O)_2-OO-\overset{\overset{\textstyle O}{\|}}{C}CH_3$	liquid
acetyl (1-methycyclohexane)-sulfonyl peroxide	[1-methylcyclohexyl] $\overset{CH_3}{}S(O)_2-OO-\overset{\overset{\textstyle O}{\|}}{C}CH_3$	liquid
di(benzenesulfonyl) peroxide	$[C_5H_6-S(O)_2-O-]_2-$	53–54
di(p-toluenesulfonyl) peroxide	$[p\text{-}C_4H_6-CH_3-S(O)_2-O-]_2-$	50

[a] Most listed peroxides decompose at mp; some decompose violently.

and diaroyl peroxides (**16**, $R^1 = R^2$ = aryl) thermally decompose almost exclusively by homolysis.

Diaroyl peroxides and diacyl peroxides without α-branches are significantly more thermally stable than those with mono- or di-α-substituents. The primary use of most commercial diacyl peroxides (**16**, $R^1 = R^2$ = alkyl or aryl) is initiation of free-radical reactions.

Diacyl peroxides (**16**, $R^1 = R^2$ = alkyl or aryl) also undergo radical induced decomposition either by direct radical displacement on the oxygen–oxygen bond, or by radical addition to, or abstraction from, the hydrocarbyl group adjacent to the peroxide.

Diacyl peroxide decompositions also are catalyzed by the metal ions of copper, iron, cobalt, and manganese:

$$R^1\!-\!\overset{\overset{\textstyle O}{\|}}{C}\!-\!OO\!-\!\overset{\overset{\textstyle O}{\|}}{C}\!-\!R^2 + Cu^+ \longrightarrow R^1\!-\!\overset{\overset{\textstyle O}{\|}}{C}O\cdot + R^2\!-\!\overset{\overset{\textstyle O}{\|}}{C}O^- + Cu^{2+}$$

This radical-generating reaction has been used in synthetic applications, e.g., aroyloxylation of olefins and aromatics, oxidation of alcohols to aldehydes, etc.

Hydrolysis and perhydrolysis of diacyl peroxides yields peroxycarboxylic acids. Carbanions react by displacement on oxygen.

Amines also react with diacyl peroxides by nucleophilic displacement on the oxygen–oxygen bond forming an ion pair intermediate.

$$R^1\!-\!\overset{\overset{\textstyle O}{\|}}{C}\!-\!OO\!-\!\overset{\overset{\textstyle O}{\|}}{C}\!-\!R^1 + :N\!\!<\; \longrightarrow \left[\,R^1\!-\!\overset{\overset{\textstyle O}{\|}}{C}\!-\!O^- \quad {}^+N\!-\!O\overset{\overset{\textstyle O}{\|}}{C}\!-\!R^1\right]$$

Dialkyl peroxydicarbonates (**17**) undergo thermolysis to form two alkoxycarbonyloxy radicals that subsequently undergo β-scission to form CO_2 and alkoxy radicals:

$$R^1O\!-\!\overset{\overset{\textstyle O}{\|}}{C}\!-\!OO\!-\!\overset{\overset{\textstyle O}{\|}}{C}\!-\!OR^1 \longrightarrow 2\,R^1O\!-\!\overset{\overset{\textstyle O}{\|}}{C}\!-\!O\cdot \longrightarrow 2\,R^1O\cdot + 2\,CO_2$$

These low temperature peroxides are susceptible to radical-induced decompositions. This susceptibility largely accounts for the hazards associated with their production and storage. In contrast to diacyl peroxides (**16**); the true first-order decomposition rates for dialkyl peroxydicarbonates (**17**) are not affected by the nature of the R group. In free-radical scavenging solvents, e.g., trichloroethylene, the decomposition rates of di-*n*-propyl, diisopropyl, di-*sec*-butyl, dicyclohexyl, di-2-ethylhexyl, and deheaxadecyl peroxydicarbonates are all essentially the same.

Dialkyl peroxydicarbonates are used primarily as free-radical initiators for vinyl monomer polymerizations. Dialkyl peroxydicarbonate decompositions are accelerated by certain metals, concentrated sulfuric acid, and amines. Violent decompositions can occur with neat or highly concentrated peroxides.

Acyl organosulfonyl peroxides (**19**) such as acetyl cyclohexanesulfonyl peroxide are efficient radical initiators for vinyl chloride polymerization.

Di(arenesulfonyl) peroxides (**20**, $R^2 = R^1 = $ aryl) react with aromatic solvents to form aryl arenesulfonates. These peroxides also form 1:1 adducts with styrene and form hydrobenzoin diarenesulfonates with stilbenes. Di(benzenesulfonyl) peroxide decomposes in water to phenol and sulfuric acid.

Synthesis. Symmetrical diacyl peroxides (**16**, $R^2 = R^1 = $ alkyl or aryl) are prepared by the reaction of an acyl chloride or anhydride with sodium peroxide or hydrogen peroxide and a base. Unsymmetrical diacyl peroxides (**16**, $R^2 \neq R^1 = $ alkyl or aryl) are prepared by the reaction of acid chlorides or anhydrides with peroxycarboxylic acids in the presence of a base. Polymeric diacyl peroxides can be prepared from the reaction of dibasic acid chlorides, e.g., succinoyl, fumaryl, sebacoyl, and terephthaloyl chlorides, with sodium or hydrogen peroxide. Cyclic diacyl peroxides can be generated from suitable dibasic acid chlorides and sodium or hydrogen peroxide, especially in dilute solutions. Symmetrical or unsymmetrical diacyl peroxides (**16**, R^1, $R^2 = $ alkyl or aryl) can be synthesized directly from carboxylic acids and hydrogen peroxide or from peroxycarboxylic acids with dicyclohexylcarbodiimide or *N,N*-dicarbonyldiimidazole as condensing agents.

Diacyl peroxides (**16**, $R^2 = R^1 = $ alkyl or aryl) have been obtained from the oxidation of carboxylic acid potassium salts by Kolbe electrolysis or by elemental fluorine.

Dialkyl peroxydicarbonates (**17**) are produced by reaction of alkyl chloroformates with sodium peroxide. *OO*-Acyl *O*-alkyl monoperoxycarbonates (**18**) are obtained from the reaction of alkyl chloroformates with peroxycarboxylic acids in the presence of a base. Symmetrical di(organosulfonyl) peroxides (**20**, $R = R21$) have been prepared by the reaction of organosulfonyl chlorides with sodium peroxide or hydrogen peroxide in the presence of a base. Acyl organosulfonyl peroxides (**19**) are prepared from the organosulfonyl chlorides and a metal salt of a peroxycarboxylic acid. Acetyl cyclohexanesulfonyl peroxide has been produced commercially by the sulfoxidation of cyclohexane, C_6H_{12}, in the presence of acetic anhydride.

Potassium salts of the peroxides (**21–23**) are prepared from the reaction of Caro's acid, H_2SO_5, with acyl chlorides, chloroformates, or organosulfonyl chlorides in the presence of potassium hydroxide.

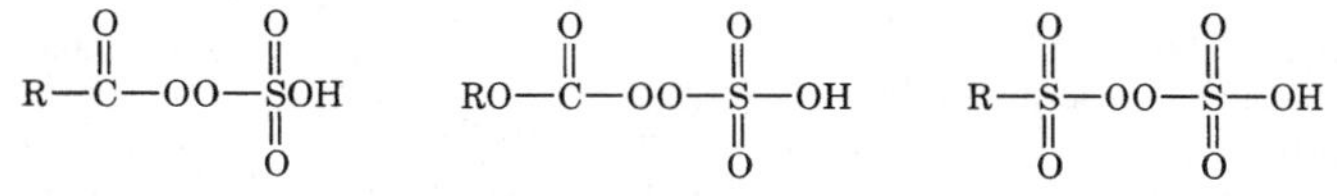

(**21**)   (**22**)   (**23**)

Alkyl Peroxyesters

Peroxyesters include the alkyl esters of peroxycarboxylic acids; monoperoxydicarboxylic acids; diperoxycarboxylic acids; monoperoxy- (**24**) and diperoxycarbonic (**25**) acids; monoperoxy- (**26**) and diperoxyoxalic (**27**) acids; peroxycarbamic acids (**28**); peroxysulfonic acids (**29**); and peroxyphosphoric acids.

Synthesis. Peroxyesters are prepared by the reaction of alkyl hydroperoxides R′OOH, with acylating agents, e.g., acid chlorides, anhydrides, ketenes, organosulfonyl chlorides, phosgene, alkyl chloroformates, oxalyl chloride, alkyl chlorooxalates, isocyanates, carbamoyl chlorides, carboxylic acids, and esters, under appropriate reaction conditions, according to Figure 2. Reactions with acylating agents that generate hydrogen chloride are carried out in the presence of a base, e.g., pyridine or sodium hydroxide, or by using the sodium or potassium salt of the hydroperoxide.

Physical Properties. Properties of some *tert*-alkyl peroxyesters are listed in Table 13 and the properties of some *tert*-alkyl areneperoxysulfonates are given in Table 14.

Chemical Properties. Alkyl peroxyesters are hydrolyzed more readily than the analogous nonperoxidic esters and yield the original acids and hydroperoxides from which they were prepared rather than alcohols and

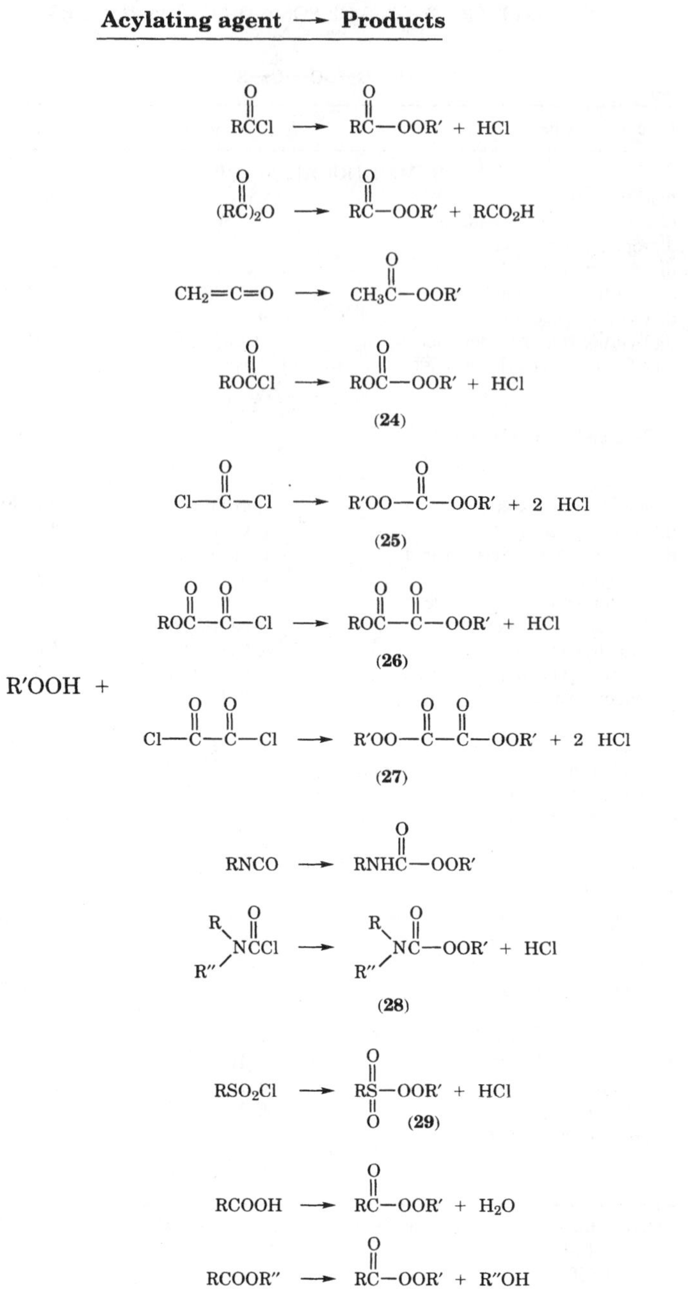

Acylating agent ⟶ Products

R′OOH +

(**24**)

(**25**)

(**26**)

(**27**)

(**28**)

(**29**)

Fig. 2.   Synthetic routes to alkyl peroxyesters. The acylating agent reacts with R′OOH in each case

peroxyacids. The *tert*-alkyl peroxyesters undergo homolysis, thermally and photochemically, to generate free radicals.

Primary and secondary alkyl peroxyesters thermally decompose by a nonradical process, giving almost quantitative yields of carboxylic acids and carbonyl compounds. *tert*-Alkyl peroxyesters are much less sensitive to radical-induced decompositions than diacyl peroxides. Induced decomposition is only significant in peroxyesters containing nonhindered α-hydrogens or α, β-unsaturation.

Peroxyesters decompose by an electron-transfer process catalyzed by transition metals. This reaction has been used synthetically to bond an acyloxy group to appropriate coreactive substrates.

Criegee rearrangement competes with homolysis in *tert*-alkyl peroxyesters, $RC(O)-OOCR^1R^2R^3$, in which R is strongly electron-withdrawing and the *tert*-alkyl group, i.e., $CR^1R^2R^3$, contains a group with high migratory aptitude and ability to stabilize adjacent carbonium ions. The rearrangement converts the peroxyester to a nonperoxidic ester.

The main industrial use of *tert*-alkyl peroxyesters is in the initiation of free-radical chain reactions, primarily for vinyl monomer polymerizations.

TABLE 13. PROPERTIES OF SOME *TERT*-ALKYL PEROXYESTERS

Name	R	Mp, °C	Bp, °C (kPa)[a]
	R' = *tert*-butyl(CH$_3$)$_3$C—		
tert-butyl peroxyacetate	CH$_3$—		22 (0.133)
tert-butyl *N,N*-dimethyl peroxycarbamate	(CH)$_2$N$_3$—		43–45 (0.013–0.027)
O,O-tert-butyl *O*-isopropyl monoperoxycarbonate	i-C$_7$H$_3$O—		52–55 (0.133)
O,O-tert-butyl *O*-hydrogen monoperoxymaleate	HOCCH=CH$_2$—	114–116	
tert-butyl peroxypivalate	i-C$_9$H$_4$—	oil	
di-*tert*-butyl diperoxycarbonate	i-C$_9$H$_4$OO—		54–55 (0.067)
di-*tert*-butyl diperoxyoxalate	i-C$_9$H$_4$OO-C(O)—	50.5–51.5	
tert-butyl peroxybenzoate	C$_5$H$_6$—	8	75–77 (0.267)
tert-butyl 2-ethyl-peroxy-haxonoate	CH(CH$_3$)C$_5$H$_2$(C$_2$H$_3$)—	< −30	
tert-butyl 2-carboxy-peroxy-benzoate	2-HOC$_4$C$_6$H$_2$—	104–104.5	
tert-butyl peroxydecanoate	n-C$_{19}$H$_9$—	−6.5	
di-*tert*-butyl diperoxyadipate	$t\text{-C}_4\text{H}_9\text{—OO—}\overset{\displaystyle O}{\overset{\displaystyle \|}{\text{C}}}\text{(CH}_2)_4\text{—}$	42–43	
di-*tert*-butyl diperoxyphthalate	$2\text{-}(t\text{-C}_4\text{H}_9\text{—OO—}\overset{\displaystyle O}{\overset{\displaystyle \|}{\text{C}}})\text{C}_6\text{H}_4\text{—}$	57.0–57.5	
tert-butyl peroxystearate	n-C$_{35}$H$_{17}$—	38.9–39.3	
	R' = *tert*-cumyl(C$_6$H$_5$C(CH$_3$)$_2$—		
tert-cumyl peroxyacetate	CH$_3$—		67–68 (0.0067)
tert-cumyl peroxypivalate	i-C$_9$H$_4$—	−18	
tert-cumyl peroxybenzoate	C$_5$H$_6$—	45	
	R' = *other*		
tert-amyl[b] peroxyacetate	CH$_3$—		65–66 (2.0)
3-hydroxy-1,1-dimethyl-butyl peroxyneodecanoate[c]	i-C$_{19}$H$_9$—	oil	
2,5-dimethyl-2,5-di(benzoyl-peroxy)-hexane[d]	C$_5$H$_6$—	118	

[a] To convert kPa to mm Hg, multiply by 7.5.

[b] R = *tert*-amyl (t-C$_5$H$_{11}$).

[c] R' = 3-hydroxy-1, 1-dimethylbutyl(CH$_3$CHOHCH$_2$C(CH$_3$)$_2$—).

[d]
$$R' = \text{—}\overset{\displaystyle \text{CH}_3}{\underset{\displaystyle \text{CH}_3}{\text{C}}}\text{—CH}_2\text{CH}_2\text{—}\overset{\displaystyle \text{CH}_3}{\underset{\displaystyle \text{CH}_3}{\text{C}}}\text{—in the diester } R\text{—}\overset{\displaystyle O}{\overset{\displaystyle \|}{\text{C}}}\text{—OO—}\overset{\displaystyle \text{CH}_3}{\underset{\displaystyle \text{CH}_3}{\text{C}}}\text{—(CH}_2)_2\text{—}\overset{\displaystyle \text{CH}_3}{\underset{\displaystyle \text{CH}_3}{\text{C}}}\text{—OO—}\overset{\displaystyle O}{\overset{\displaystyle \|}{\text{C}}}\text{—R.}$$

TABLE 14. *TERT*-BUTYL ARENEPEROXYSULFONATES

Benzeneperoxy sulfonate	Structure	Mp, °C
tert-butyl	$\text{C}_6\text{H}_5\overset{\displaystyle O}{\underset{\displaystyle O}{\overset{\displaystyle \|}{\underset{\displaystyle \|}{\text{S}}}}}\text{–OO–}t\text{–C}_4\text{H}_9$	[a]
tert-butyl *p*-chloro-	$p\text{–Cl–C}_6\text{H}_4\overset{\displaystyle O}{\underset{\displaystyle O}{\overset{\displaystyle \|}{\underset{\displaystyle \|}{\text{S}}}}}\text{–OO–}t\text{–C}_4\text{H}_9$	30–35[b]
tert-butyl *p*-methyl-	$p\text{–CH}_3\text{–C}_6\text{H}_4\overset{\displaystyle O}{\underset{\displaystyle O}{\overset{\displaystyle \|}{\underset{\displaystyle \|}{\text{S}}}}}\text{–OO–}t\text{–C}_4\text{H}_9$	36.5–37
tert-butyl *p*-methoxy-	$p\text{–CH}_3\text{O–C}_6\text{H}_4\overset{\displaystyle O}{\underset{\displaystyle O}{\overset{\displaystyle \|}{\underset{\displaystyle \|}{\text{S}}}}}\text{–OO–}t\text{–C}_4\text{H}_9$	47[b]

[a] $n_\text{D}^{25} = 1.4629$

[b] Explodes at mp.

Manufacture and Processing

Owing to the inherent hazards of organic peroxides, they are almost never distilled or confined during manufacture. Generally, open reactors are employed that can be easily vented and deluged with water if an unanticipated exotherm occurs. The preferred materials of reactor construction are 316 stainless steel, plastic, and glass. Significant cooling capacity is required to handle reaction exotherms and to maintain temperature. Because over 100 different organic peroxides are produced commercially, organic peroxide producers manufacture many organic peroxides in the same equipment. Batch processing is generally employed when relatively small production volumes are required, whereas semicontinuous and continuous processing are employed when larger production volumes are required and when safety is a primary issue. Continuous processes are significantly safer to operate than batch processes as smaller amounts of organic peroxides are continuously in process. Besides safer continuous processing, another trend has been the use of reactants of higher purity. These process improvements have resulted in reduced environmental impact as unplanned process decompositions have been decreased and waste streams have been reduced.

Economic Aspects

Prices of commercial organic peroxides range from ca $2.50 to >$35/kg, depending on peroxide type, production volume, assay, nature of formulation, cost of raw materials, and degree of special processing and handling requirements.

Health and Safety Factors

Toxicology. In general, organic peroxides are characterized by a low order of acute toxicity. Most organic peroxides have some oxidizing properties and are irritants. Most of the available toxicity data on commercial organic peroxides are summarized in the literature. There is limited evidence to suggest that organic peroxides are carcinogenic.

Decomposition Hazards. The main causes of unintended decompositions of organic peroxides are heat energy from heating sources and mechanical shock, e.g., impact or friction. In addition, certain contaminants, i.e., metal salts, amines, acids, and bases, initiate or accelerate organic peroxide decompositions at temperatures at which the peroxide is normally stable. These reactions also liberate heat, thus further accelerating the decomposition. Commercial products often contain diluents that desensitize neat peroxides to these hazards. Commercial organic peroxide decompositions are low order deflagrations rather than detonations.

The organic peroxides and peroxide compositions produced commercially are those that can be manufactured, shipped, stored, and used safely. Organic peroxides can be thermally and mechanically desensitized by wetting or by dilution with suitable solvents, inert solid fillers, or insoluble liquids (suspension of solid peroxides in liquid plasticizers or water, and emulsions of liquid peroxides in water).

Recommendations for safe handling and storage of commercial organic peroxides are available from organic peroxide manufactures.

In 1984 the United Nations (UN) Committee on the Transportation of Dangerous Goods, made up of experts from Prins Maurits Laboratory (TNO), Bundesanstalt für Materialprüfung (BAM), and the organic peroxide producers, developed a test procedure for the classification of organic peroxide compositions for transport purposes. The test procedure was accepted by most of the industrial countries of the world. The Department of Transportation (DOT) mandated that the United States peroxide industry would comply with the UN classification system by October 1993. Material Safety Data Sheets (MSDS) and the organic peroxides producers' recommendations should be followed carefully for handling and storage of organic peroxide compositions.

Uses

There are more than 100 commercially available organic peroxides in well over 300 formulations, e.g., neat liquids and solids, and pastes, powders, solutions, dispersions, and emulsions, that have utility in many commercial applications.

Excluding the peroxyacids, which are used primarily as epoxidizing and bleaching agents, approximately 90% of the commercial organic peroxides are consumed by the polymer industry.

They are used in the polymer industry as thermal sources of free radicals. They are used primarily to initiate the polymerization and copolymerization of vinyl and diene monomers.

Organic peroxides also are used as flame-retardant synergists for polystyrene, for preparing block and graft copolymers, for reactive processing, for reducing the molecular weight of polypropylene (i.e., controlled rheology or vis-breaking), for curing adhesives, for drying alkyd resin films, and for initiating cationic polymerization with cyclic ethers and maleic anhydride.

BPO is the preferred bleaching agent for flour and has been used to bleach gums, waxes, fats, and oils. It is the active ingredient in many acne medications. Diacyl peroxides have been used as burnout agents for acetate yarn, drying agents for Chinawood oils, and as free-radical sources in many organic syntheses. Di-*tert*-butyl peroxide is used as an ignition accelerator for diesel fuels and has been used in many organic syntheses either as a source of *tert*-butoxy (photo) or methyl (thermal) radicals.

JOSE SANCHEZ
TERRY N. MYERS
Elf Atochem North America, Inc.

Additional Reading

Ando, W. ed.: *Organic Peroxides,* John Wiley & Sons, Inc., New York, NY, 1992.
Lide, D.R.: *CRC Handbook of Chemistry and Physics,* 84th Edition, CRC Press LLC., Boca Raton, FL, 2003.
Patai, S. ed.: *The Chemistry of Peroxides,* John Wiley & Sons, Inc., New York, NY, 1983; S. Patai, ed., *Supplement E2: The Chemistry of Hydroxyl, Ether and Peroxide Groups,* John Wiley & Sons, New York, NY, 1993.
Swern, D. ed.: *Organic Peroxides,* Vols. I–III, Wiley-Interscience, New York, NY, 1970–1972.
Waldemar, A.: *Peroxide Chemistry,* John Wiley & Sons, Inc., New York, NY, 2003.

PERRHENIC ACID. $HReO_4$. Exists only in solution, commercially available as aqueous syrup. Strong very stable, monobasic acid; extremely soluble in water and organic solvents.

PERRIN RULE. Ions of charge opposite to that of a diaphragm have by far the greatest effect on endosmosis. The higher their valence (or opposite sign) the greater the reduction of electroosmotic flow.

PERTHITE. An alkali feldspar comprising parallel or subparallel intergrowths. The potassium-rich phase, usually microcline, seems to be the host from which the sodium-rich phase, usually albite inclusions, exsolved. The exsolved areas typically form blebs, films, lamellae, small strings, or irregular veinlets and usually are visible to the naked eye.

PERTURBATION. 1. A small contribution to a physical quantity, such that the problem into which the quantity enters can be solved exactly or in a far simpler manner than otherwise if the perturbation is neglected.

The form in which a perturbation is most frequently used in both classical and quantum mechanics is a small additional energy, called the *perturbation energy.*

2. Any departure introduced into an assumed steady state of a system. The magnitude of the departure is often assumed to be small so that product terms in the dependent variables may be neglected; the term *perturbation* is therefore sometimes used as synonymous with small perturbation. The perturbation may be concentrated at a point or in a finite volume of space; or it may be a wave (sine or cosine function); or, in the case of a rotating system, it may be symmetric about the axis of rotation.

3. In molecular spectra, perturbations cause the displacement of a band from its regular position in the band system (vibrational perturbation) or the displacement (and/or weakening) of corresponding lines in the different branches of a band (rotational perturbation). A perturbation observed in the spectrum is indicative of the presence of a perturbation (shift) of one of the energy levels involved due to interaction with another level of the same, or nearly the same, energy.

4. A much more frequent use is made of perturbation methods in quantum mechanics. The mathematical complexity of many quantum mechanical problems is such that one cannot hope to obtain exact solutions. However, good predictions can sometimes be obtained by means of perturbation theory, if one can assume that the actual system differs only slightly from a simpler system for which the problem can be solved, and the neglected difference can be dealt with as a perturbation of this simpler unperturbed system. The effect of a weak electromagnetic field on an atom, for instance, can be dealt with as a perturbation, and the transition probabilities between the energy states of the unperturbed atom can be calculated by means of perturbation theory. A weak interaction between two particles can be dealt with as a perturbation in the collision process of the two particles. The perturbation methods can be time-independent or time-dependent, according to whether the unperturbed states are described by time-independent wave functions or time-dependent ones. If the strength of a weak interaction between two systems is proportional to a constant parameter, the wave functions and energy values of the wave equation can be expanded in powers of this constant. The zero-order approximation is given by the unperturbed wave functions and energy values which are independent of this parameter. These determine the first-order approximations together with that part of the wave equation that is linear in the coupling parameter. By successive approximations one obtains expressions for second, third, and higher order perturbations in the wave function and the energy. If an unperturbed energy state is degenerate, that is, if two or more states have the same unperturbed energy, the effect of the perturbation has to be taken into account first between these degenerate states.

See also **Quantum Chemistry**; and **Quantum Mechanics**.

PERUTZ, MAX F. (1914–2002). An Austrian molecular biologist and recipient of the Nobel prize for chemistry in 1962 along with John C. Kendrew. His work was concerned with crystalline protein structure, particularly the molecular structure of hemoglobin and myoglobin. His education was in England and Austria.

PETALITE. The mineral petalite, lithium aluminum silicate ($LiAl Si_4O_{10}$), is monoclinic, although crystals are rare, this mineral usually occurring in cleavable, foliated masses, whence the name petalite, from the Greek meaning a *leaf.* Its hardness is 6–6.5; specific gravity 2.39–2.46. It is brittle with subconchoidal fracture; perfect basal cleavage; luster, vitreous, colorless to white or gray but may be greenish or reddish; transparent to translucent. Petalite occurs in granite pegmatites with sodium-rich feldspar, quartz, and lepidolite; has been found in Sweden; the former U.S.S.R.; on the Island of Elba; and in the United States at Bolton, Massachusetts, and Peru, Maine. It is interesting to note that lithium was first discovered in this mineral. See also **Lithium**.

PETROCHEMICALS. Chemicals derived from petroleum and, more specifically, substances or materials manufactured from a component of crude oil or natural gas. See Fig. 1. In this sense, ammonia and synthetic rubber made from natural gas components are petrochemicals. Many of these chemicals are described in separate articles in this encyclopedia. Check alphabetical index.

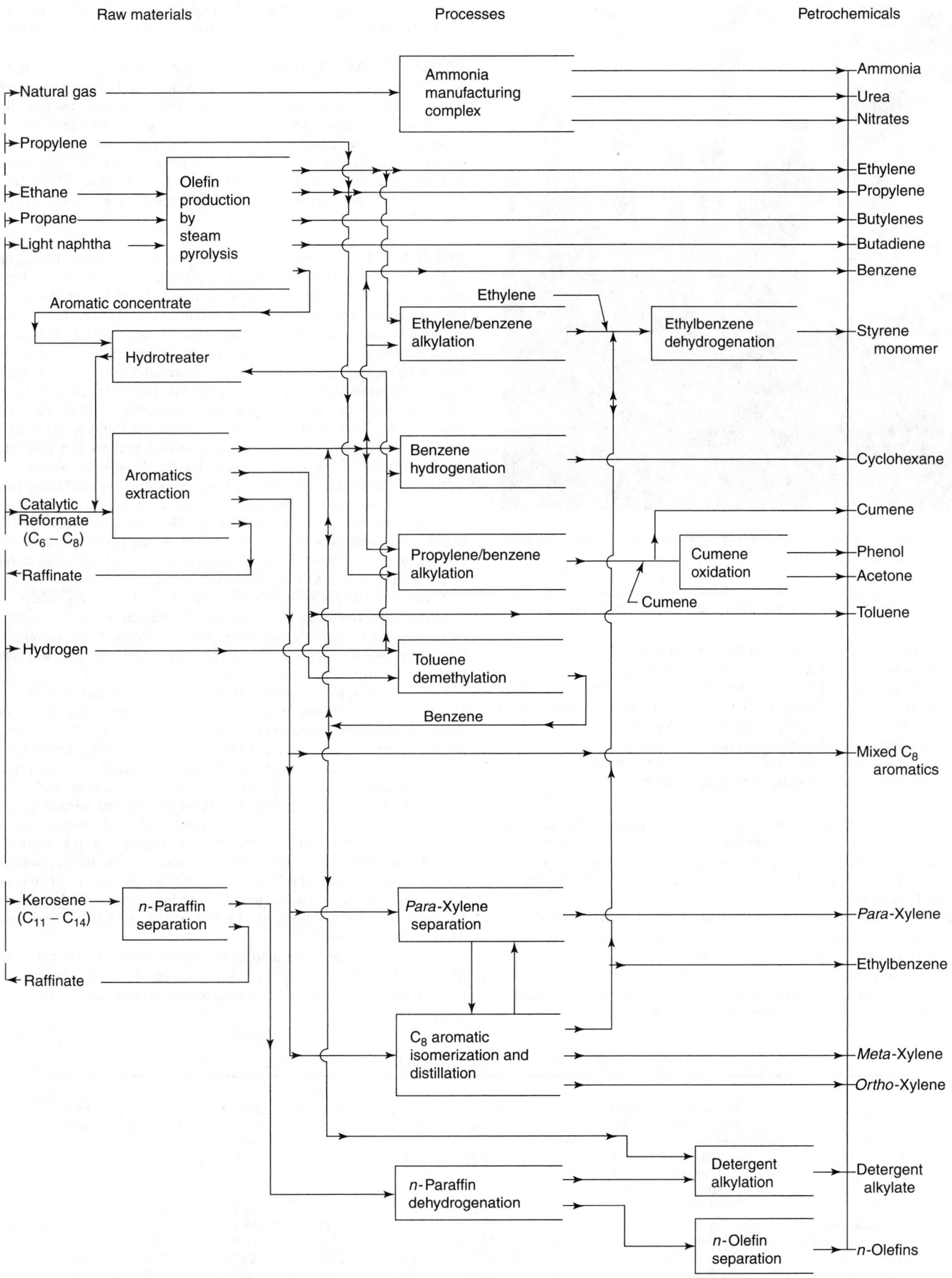

Raw materials	Processes	Petrochemicals

Fig. 1.   Interlocking processes and flow of materials in a representative petrochemical complex. (*UOP, Inc*)

Fig. 2.   Portions of a representative solvent-producing plant, using petrochemicals as raw materials

Among the most important petrochemicals manufactured are:

Acetic acid	Ethylene dichloride	Phenol
Acetone	Ethylene glycol	Polyethylene
Acrylonitrile	Ethylene oxide	Polypropylene
Benzene	Formaldehyde	Polyvinyl chloride
Cumene	Isopropyl alcohol	Styrene
Cyclohexane	Maleic anhydride	Toluene
Ethylbenzene	Methanol	Vinyl chloride
Ethylene	Phthalic anhydride	Xylenes

The chemical unit operations, such as distillation, extraction, and various separation operations, and the chemical unit processes, such as alkylation, dehydrogenation, hydrogenation, and isomerization, are essentially identical to those operations used in the manufacture of chemicals from other sources.

In order to save materials transportation costs, a petrochemical plant frequently will be located adjacent to a petroleum refinery or gas processing plant. Short pipelines can be used in place of leasing long pipelines or having to depend upon tank car shipments by rail or truck. This also contributes to the overall safety of production. A representative petrochemical plant adjacent to a refinery is shown in Fig. 2 on.

An excellent report of the petrochemical industry is prepared annually by the *Oil and Gas Journal* and *Hydrocarbon Processing* magazine.

PETROLATUM.   A semisolid or liquid mixture of hydrocarbons derived by distillation of paraffin-base petroleum fractions. The solid form (mineral jelly) may be either water-white or pale yellow. Its chief uses are in mild ointments, cosmetics, softener in rubber mixtures and food processing (release agent in bakery products, dehydrated fruits and vegetables), defoaming agent (beet sugar, yeast). The liquid form (white mineral oil) is used as a laxative, textile lubricant and dispersing agent. There are three grades of both solid and liquid types with various specifications (USP, NF, and FCC).

PETROLEUM.   A natural oil, ranging in color through black, brown, and green, to a light amber shade. It is often termed crude oil, and consists principally of hydrocarbons, that is, compounds of carbon and hydrogen, but varying amounts of oxygen-, nitrogen-, and sulfur-bearing compounds are almost invariably present. The term *mineral oil*, which was and is often used as a synonym for petroleum, is inadequate, for most geologists believe that it was derived from organic material resulting from reactions of organic materials such as plants and animals buried in sedimentary rocks. The more important of these geologic formations in which petroleum is found are the Tertiary period of the Cenozoic era (50% of the world's oil production comes from these rocks, including regions in California and the Gulf Coast of the United States. Russia, Venezuela, Malaysia, Iran, and Iraq); the Cretaceous period of the Mesozoic era (including the East Texas, Kuwait, and Bahrein fields); the Jurassic period of the Mesozoic era (including the Arkansas and Rocky Mountain regions of the United States, and Saudi Arabia); and the Mississippian period of the Paleozoic era (including the West Texas, Pennsylvania, and Mid-Continent regions of the United States, and the Alberta, Canada, fields).

Petroleum oils vary considerably in composition, even when closely associated geographically. In some areas of the United States, for example, crude oils near the surface may have quite a different chemical composition from those found in deeper strata. Depth alone, however, does not correlate significantly with composition.

Analysis of typical crude oils found in representative areas of the United States are given in Table 1. It may be generalized that crudes found in the eastern and midwestern sections of the United States are predominantly sweet and paraffinic; those found along the Gulf Coast usually are naphthenic; those occurring in the inland southwest are sour and naphthenic; and those found along the west coast are asphaltic. The waxy, sweet paraffinic oils found in Pennsylvania first became prominent because of the high quality of lubricating oils and greases that could be made from them. The severe stresses imposed by the bearings and close-fitting reciprocating surfaces of machinery led to the development of refining processes and the discovery of additive materials whereby many other crude oils also can be transformed into excellent lubricants. Even Pennsylvania oils require special refining and additives to meet present quality specifications.

Analyses of some crude petroleums found outside the United States are given in Table 2, which illustrates the variety of crudes existent, but is not intended to give a full representation of worldwide petroleum source compositions.

TABLE 1.   ANALYSIS OF REPRESENTATIVE U.S. CRUDE OILS

Property	McComb, Mississippi	Southwest Texas	East Texas	Wyoming (Sour)	New Mexico	N. Kenia Peninsula, Alaska	San Ardo, Calif.	Ospelousas, Louisiana	Velma, Okla.
Total sulfur, wt%	0.07	0.45	0.2	3.33	1.0	1.04	1.93	0.08	1.13
Pour point, °C	15.6	−1.1	12.8	−20	−3.9			4.4	
°F	60	30	55	−5	25			40	<−30
Gasoline, vol%	35.5	32.0	29.0	6.3	37.8	14.4	1.9	26.1	22.3
Kerosene, vol%	18.1	12.1	10.1	9.1		18.0	16.1	18.9	17.3
Diesel fuel, vol%	14.6	38.0	13.8	14.0		18.4	10.6	22.9	8.5
Gas oil, vol%	28.1	12.6		30.7	41.2	22.3	23.3	27.9	31.9
Asphalt bottoms, vol%	3.7	5.3	47.1	39.9	20.8	26.9	48.1	4.2	20.0
Metals in gas oils, ppm									
Nickel	0.06						0.15		
Vanadium	0.08						<0.1		
Salt, lb/1000 bbl	4	<0.5	31	0.6	14	76		5	78

TABLE 2. ANALYSES OF REPRESENTATIVE WORLD CRUDE OILS

Property	Arabian	Minas, Central Sumatra Topped	Putomayo, Colombia	Gulf Nigeria	Zulia, Venezuela	Iran	Kuwait
Total sulfur, wt%	3.05	0.2	0.49	0.16	1.69	1.12	2.62
Pour point, °C	−36.1	−17.8	7.2	−6.7	<−15	15	<−15
°F	−33	0	45	20	<5	5	<5
Gasoline, vol%	29.1	11	34.1	24.9	18.9	32.2	25.5
Kerosene, vol%	16.0	16	9.3	26.5	14.1	18.3	13.7
Gas oils, vol%	12.5	14	40.7	19.3			
Residuum, vol%	42.4	59	15.9	29.3			
Metals in gas oils, ppm							
Vanadium	0		25	7			
Nickel	0		11	5			
Iron	3						
Salt, lb/1000 lb	12		trace	5			

API Gravity. This parameter (API stands for American Petroleum Institute), expressed in "degrees," is mathematically related to specific gravity and can be determined with a hydrometer. The specific gravity of water (arbitrarily defined as unity) is 10.00 when expressed as degrees API. API gravity usually, although not infallibly, indicates the gasoline and kerosine contents of the crude. As an example, the Mississippi, Texas, New Mexico, and Louisiana crudes have API gravities between approximately 35 and 40; as do the Arabian, Iranian, and Colombian crudes. The gasoline content (that fraction boiling below about 400°F (204°C) of these crudes ranges from about 25% to over 35% by volume. The kerosene portions of such "light" crudes also are usually high. In contrast, Wyoming sour crude with an API gravity of 17.9 contains but 6% gasoline and about 40% asphalt. California crude has an even greater content of residuum and almost no gasoline.

Sulfur Content. The amount of sulfur in crude is important in terms of handling the crude within the refinery and the undesirable effects of sulfur in finished products. High-sulfur crudes require special materials of construction for refinery equipment because of their corrosiveness. Certain refinery processes require desulfurization of sour charge stocks prior to use as a feedstock, not only because of their corrosiveness, but also because of the effect of sulfur-bearing compounds on expensive catalysts. From the standpoint of the consumer, sulfurous gasoline has an unforgettably offensive odor unless specially sweetened and it may corrode the fuel system and engine parts, as well as pollute the atmosphere after it has been burned.

Other factors indicated in the data of Tables 1 and 2 include: **Pour Point**—defined as the lowest temperature at which the material will pour and a function of the composition of the oil in terms of waxiness and bitumen content; **Salt Content**—which is not confined to sodium chloride, but usually is interpreted in terms of NaCl. Salt is undesirable because of the tendency to obstruct fluid flow, to accumulate as an undesirable constituent of residual oils and asphalts, and a tendency of certain salt compounds to decompose when heated, causing corrosion of refining equipment; **Metals Content**—heavy metals, such as vanadium, nickel, and iron, tend to accumulate in the heavier gas oil and residuum fractions where the metals may interfere with refining operations, particularly by poisoning catalysts. The heavy metals also contribute to the formation of deposits on heated surfaces in furnaces and boiler fireboxes, leading to permanent failure of equipment, interference with heat-transfer efficiency, and increased maintenance.

Natural Gas, Oil Shales, and Tar Sands. Natural gas is not formally defined as a component of crude petroleum, although natural gas commonly exists in the same geological formations, often directly in contact with crude petroleum. However, a large percentage of natural gas wells are not associated with producing oil wells. See also **Natural Gas**.

The oils derived from oil shales are not true petroleum, although they are petroleumlike products after being subjected to specialized chemical processing. Shales are sedimentary rocks that have a relatively high content of a bituminous substance called *kerogen* and 30–60% organic matter and fixed carbon. Kerogen, although not a definite chemical compound, yields an oily substance when heated (retorted) in the absence of air. Extraction of oil shale with ordinary solvents produces no oil, and their solubility in solvents is low. This evidence supports the conclusion that the "oil" is the result of a chemical change, i.e., the thermal cracking or fragmenting (pyrolysis) of the molecules that make up kerogen. See also **Oil Shale**.

Tar sands is an expression commonly used in the petroleum industry to describe sandstone reservoirs impregnated with a very heavy viscous crude oil which cannot be produced through a well by conventional production techniques. Two other terms, *bituminous sands* and *oil sands*, are gaining favor. The heavy viscous petroleum substances impregnating the "tar sands" are called asphaltic oils. See also **Tar Sands**.

Petroleum processing and petroleum end-products are described in the article on **Petroleum Refining** immediately following this article.

Origin and Geology of Petroleum

Among the general theories for explaining the origin of petroleum, the most widely accepted is the *organic theory*, which can be quickly summarized. Over millions of years, rivers flowed to the seas, carrying large volumes of mud and sand to be spread out by currents and tides over the sea bottoms near the gradually changing shorelines. New deposits were distributed, layer upon layer, over the floors of the seas. Because of the increasing weight of these accumulations, the sea floors slowly sank, building up a thick series of mud and sand layers. High pressure and chemical forces ultimately converted these layers into sedimentary rocks of the type that often contain petroleum—the sandstones, shales, limestones, and dolomites. The organic theory further stipulates most importantly that tiny marine organisms were buried with the silt. In an airless environment and under high pressures and elevated temperatures, these carbon- and hydrogen-containing minuscule life-forms were converted over an extremely long time span into hydrocarbons. This theory, of course, requires acceptance of the concept of drastically altered shorelines, because obviously oil deposits are found in many parts of the world long distances from the present coastlines.[1]

Geologists find it particularly difficult to trace the history of a given hydrocarbon deposit, because the oil and gas may have moved as the result of numerous seismic events, again occurring over a very long time span. A past requisite for commercially exploitable hydrocarbon deposits has been prior movement and concentration of large quantities of hydrocarbons in various forms of *traps*. In contrast with oil shales and tar sands, natural gas and petroleum flow relatively easily in permeable underground structures and, consequently, tend to concentrate, greatly assisting the economic exploitation of these materials.

The movement of petroleum from the place of its origin to the traps where accumulations are found is believed to have occurred in an upward direction. This movement took place as the result of the tendency for oil and gas to rise through the ancient seawater with which the pore spaces of the sedimentary formations were filled when originally laid down. An underground porous formation or series of rocks which occur in some shape favorable to the trapping of oil and gas must also be covered or adjoined by a layer or rock that provides a covering or seal for the trap. A seal of this type, frequently called a *cap rock*, stops further upward movement of petroleum through the pore spaces.

As oil and gas gathered in the upper part of a trap, because of differences in weight of gas, oil, and salt water, these fluids also separated vertically, much in the same manner as if these materials were all present in a bottle.

---

[1] In a more recent, alternate theory, Thomas Gold (Cornell University) suggests that, in contrast with the organic sediments theory, the prime source of natural gas is primordial, abiotic methane rising from deep within the earth's mantle. This is discussed in further detail in the article on **Natural Gas**.

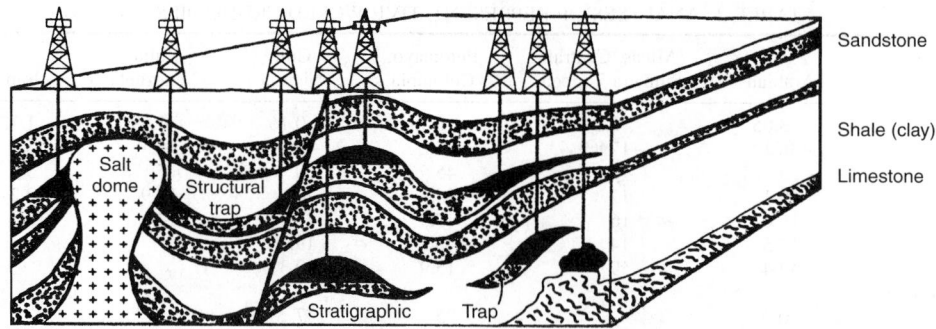

Fig. 1. Composite diagram showing different types of oil and gas accumulations (shown in solid black). *Structural traps* where petroleum deposits may have accumulated are often found along the edges of salt domes or along fault lines. *Stratigraphic traps* may exist where reservoir rock is "pinched off" by denser strata. These accumulations are the "pools" or "reservoirs" which, singly or in groups, compose an oil or gas "field." The pores of the reservoir rock contain oil or gas or both, always accompanied by briny water. The fluids tend to be layered, with gas at the top of the trap, oil in the middle, and water underneath. Most of the petroleum which ever existed has been obliterated, either by attenuation in the earth's crust, or by exposure to heat and pressure high enough to break down its chemical bonds. The accumulations that do exist have endured against long odds. Additional detail on oil and gas reservoirs will be found in Figs. 2 through 7

Thus gas, if any is present, is found in the highest parts of the trap, followed by oil (and oil with gas) below the gas, and finally salt water below the oil. Experience has indicated that the salt water seldom was completely displaced by oil or gas from the pore spaces, even within the trap. Even in the midst of oil and gas accumulation, pore spaces within the trap may contain from 10 to 50% or more of salt water. It appears that the remaining water (termed *connate water*) fills the smaller pores and also exists as a coating or film, covering the rock surfaces of the larger pore spaces; thus oil and/or gas are apparently contained in water-jacketed pore spaces. The geological structures called traps are petroleum reservoirs, i.e., they are the oil and gas fields that are explored and produced. All oil fields contain some gas, but the quantity may range widely. See also *Natural Gas*.

Types of Oil Accumulations

A composite diagram showing different types of oil and gas accumulations is given in Fig. 1.

Structural Traps. The attitude of the rocks, whether they are folded, fractured, displaced, or otherwise disturbed, is called their geologic structure. Traps that are due to geologic structure are known as *structural traps*.

A common structural trap, the *anticline*, is an upward bulge in the rock layers which forms an arch capable of holding oil under its apex. The buoyancy of oil and gas carries them upward through porous rock layers into the apex until they are trapped by an impermeable layer. Anticlinal type of folded structure is shown in Fig. 2. Reservoirs formed by folding of the rock layers or strata usually have the shapes shown in Fig. 2(a) and (b). These traps were filled by upward migration of oil and/or gas through the porous strata or beds to the location of the trap. Further movement was arrested by a combination of the forms of the structure and the seal or cap rock provided by the formation covering the structure.

Examples of domal structures are the Conroe Oil Field in Montgomery County, Texas and the Old Ocean Gas Field in Brazoria County, Texas. Another example of a reservoir formed by an anticlinal structure is the Ventura Oil Field in California.

Another type of structural trap is the *fault trap*. A fault is a fracture in the earth's crust along which movement has occurred such that a porous rock layer is offset by a nonporous layer. The oil moving along a porous stratum is dammed or blocked by an impermeable shale or limestone. See Fig. 3. Examples of fields of this type occur along the Mexia fault zone of East-Central Texas.

The *salt dome* is another interesting form of structural trap. This type of trap is found along the Gulf Coasts of Texas and Louisiana and in western Colorado and Utah. This type of structure resulted from the upward thrust of a great mass of salt far below the earth's surface. When a salt dome rose through a layer of oil-bearing sedimentary rock, oil may have been trapped in anticlines above the dome, or in structures similar to faults along its flanks. One famous example of a salt-dome reservoir is the Spindletop field, near Beaumont, Texas. It was "brought in" in 1901 by a 100,000-barrel-a-day gusher, giving birth to the modern petroleum industry. Another example of a salt-dome field is the Sugarland Oil Field in Fort Bend County, Texas. See Fig. 4.

Stratigraphic Traps. Petroleum geologists also seek another kind of trap, the *stratigraphic trap*, which results when a porous layer is "pinched"

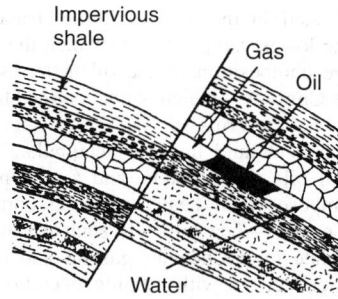

Fig. 3. Example of a fault structural trap. The oil is confined in traps like this because of the tilt of the rock layers and faulting

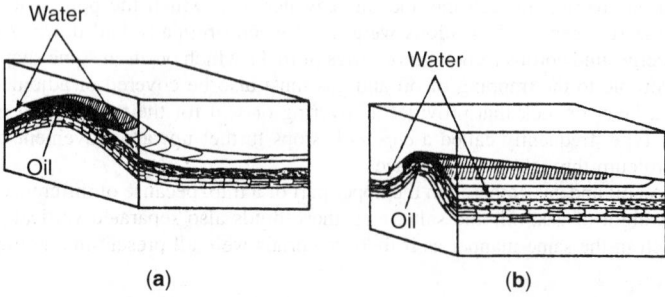

Fig. 2. Examples of anticline structural traps: (**a**) Oil accumulates in a dome-shaped structure. The dome is circular in outline. (**b**) Anticlinal trap that is long and narrow, differing from the dome configuration

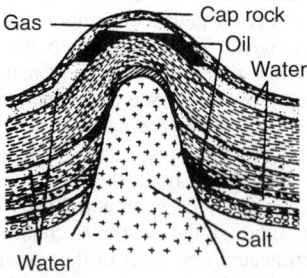

Fig. 4. Example of a salt-dome structural trap. One of the earliest and greatest reservoirs of this type was Spindletop, brought into production in 1901 and located near Beaumont, Texas

or phased out between two nonporous layers. Caught in an underground envelope of impermeable rock, the oil accumulates to form a reservoir. This type of trap may have formed from buried beaches or sandbars. The famous East Texas field is a "strat" trap. See Fig. 5. Because structural features such as anticlines and faults are often more obvious and easier for the geologist and geophysicist to detect, more fields thus far have been found in structural traps than in their stratigraphic counterparts. Stratigraphic traps are usually discovered only after exhaustive studies have been made of rock samples from outcrops and from core samples from wells drilled over large areas.

The serpentine plug, shown in Fig. 6, is an interesting type of trap, an example of which is the Hilbig Field in Bastrop County, Texas. As illustrated, a porous serpentine plug has formed a reservoir within itself by intruding into nonporous surrounding formations.

Lens-Type Traps. These form in limestone and sand. In this type of trap the reservoir is sealed in its upper regions by abrupt changes in the amount of connected pore space within a formation. A trap formed in sand is shown in Fig. 7(a). An example is the Burbank Field in Osage County, Oklahoma. This type of trap may occur in sandstones where irregular deposition of sand and shale occurred at the time the formation was laid down. In these cases, oil is confined within the porous parts of the rock by the nonporous parts of rock surrounding it. A lens-type trap formed in limestone is shown in Fig. 7(b). In limestone formations there are frequent areas of high porosity with a tendency to form traps. Examples of limestone reservoirs of this type are found in the limestone fields of West Texas.

Reef-Type Traps. These have accounted for some of the most important production in recent years. The reef is generally considered a type of stratigraphic trap. The reef was formed under the right combination of conditions by the remains of millions of small underwater animals. Building their limestone residences on top of those built by their ancestors, the tiny

creatures produced columns or mounds, often reaching several hundred feet high. If eventually surrounded by impermeable rock layers, the reef could become a trap for oil and gas.

Petroleum Production—Geophysics

In the petroleum industry, the word *production* is generally interpreted as the obtaining of crude oil, natural gas, and other related natural hydrocarbons, whether located in reservoirs under the land or under the oceans. Production generally does not relate to the manufacture of final end-products, such as naphtha, gasoline, diesel fuel, lubricating oils, etc. Products obtained by the processing of natural hydrocarbons are commonly referred to as *refined*. In a sense, then, petroleum production is analogous to other natural raw material mining, such as coal and metals mining.

The fundamental phases of petroleum production include: (1) the initial *exploration* required to find heretofore undiscovered oil and gas reservoirs; (2) *primary* and *secondary recovery* methods, which make use of both naturally occurring (or *primary*) reservoir energy and the application of *secondary* energy sources, such as the injection of gas or water; and (3) *enhanced oil recovery* used to increase ultimate oil production beyond that achievable with primary and secondary methods. Enhanced oil recovery (EOR) methods increase the proportion of the reservoir by improving the "sweep" efficiency, reducing the amount of residual oil in the swept zones (increasing the displacement efficiency), and reducing the viscosity of thick oils.

Petroleum Exploration

The exploration required to identify previously undiscovered oil and gas reservoirs is grossly affected by the economics and politics of the worldwide oil markets (supply and demand). The experience of one year cannot easily be extrapolated to that of subsequent years.

Although the petroleum geologist has for several years employed highly sophisticated instrumentation and computer technology, the fact remains that the geologist, in many respects, is still a *detective*, often commencing with the barest of clues and following the search hopefully to a successful conclusion. Decisions based on uncertain information are normal in petroleum exploration. How to capitalize on the creativity of the geological and geophysical scientists further complicates decision making.

Modern petroleum exploration geophysics is based on three important earth properties: (1) density, (2) magnetization; and (3) acoustic response.

Gravity Surveys. Gravitational pull is an effective way to measure the density of the strata lying far below the earth's surface. The *gravimeter* came into oil exploration use in about 1900. If there are dense rocks near the surface, a spring-loaded weight in the instrument will weigh more—obviously since denser rocks exert greater gravitational pull than those which are less dense. Usually, the denser rocks are older and their presence near or far from the surface indicates an underlying shallow or deep basement. Thus, a structural "high," such as an anticline, will often appear as a high reading on the gravimeter. Conversely, a structural "low," such as a syncline, will yield opposite effects. The low density of a salt dome will be reflected on the gravimeter as having a lesser gravity value. When using a gravity meter offshore, it is necessary to place the instrument so that it remains level, regardless of the movement of the vessel. The rise and fall of the ship and other effects of its travel must be measured by other equipment and taken into account when the gravity data are processed.

Magnetic Surveys. The natural magnetic properties of rocks can be useful in the early stages of petroleum exploration. More than three centuries ago, it was found that a freely suspended magnetic needle will move away from a level position in response to the presence of nearby iron. In 1879, this principle was applied in the first *magnetometer*, a simple device combining a magnetized needle with a compass. Both types and depths of rocks far underground affect how much and in which direction the needle will move.

Using a magnetometer, earth scientists can measure the depths at which basement rocks lie. These rocks are usually from the Precambrian period, dating back over 600 million years. They contain large amounts of magnetite, a naturally occurring iron oxide having magnetic properties. Where petroleum-bearing sedimentary rocks overlie these basement rock formations, the magnetometer can be used to indicate variations in the thickness of these sedimentary layers. This information can provide the geologist with a valuable clue as to the probability that those sediments, bent by the underlying basement rocks, may provide the right conditions for the accumulation of recoverable quantities of oil or natural gas.

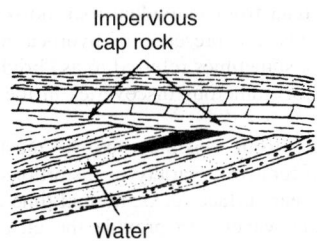

Fig. 5. Example of a stratigraphic trap. This unconformity represents the condition where upward movement of oil has been halted by the impermeable cap rock laid down across the cutoff (possibly by water or wind erosion) surfaces of the lower beds. This type of reservoir is found in the great East Texas field

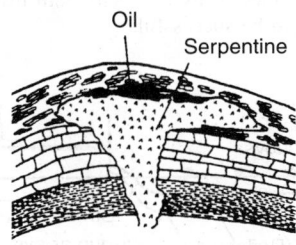

Fig. 6. Example of a serpentine plug as found in the Hilbig Field in Bastrop County, Texas

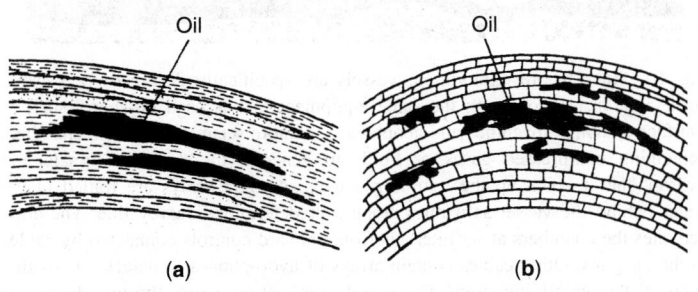

Fig. 7.   Lens-type traps: **(a)** sandstone; **(b)** limestone

It is not always necessary to place instruments on the ground to measure magnetic attraction of the subterranean rock layers. Magnetometers can also be used from aircraft flying high above the surface to determine quickly areas of general interest. Use of airborne magnetometers also eliminates the need for and the problems sometimes associated with personnel actually getting on the site.

Seismic Surveys. Gravity and magnetic measurements are considered preliminary. The principal detailing tool used is the seismograph.[2]

The type of experiment used to make a seismic survey in petroleum exploration will be determined by the type of information required, the types of rock strata anticipated, environmental considerations, the type of terrain, and, of course, the economic costs involved.

In petroleum exploration, reflection seismic surveying is the method predominantly used. The seismograph records variations in the way rocks reflect sound waves sent downward from a surface source. The reflected sound waves vary with the type, depth, density, and dip of the rocks encountered. The returning sound waves made from a series of points along the survey path can then be displayed graphically to form a seismic record for interpretation by earth scientists. These principles are diagrammed and explained in Fig. 8.

Four sources of energy may be used to generate the sound waves: (1) controlled explosives; (2) vibrators; (3) weight dropping (where heavy weights are dropped to create the waves); and (4) compressed gases (producing bursts of energy using compressed air or propane). In exploring onshore sources, controlled explosives and vibrators are most often used. In offshore work, compressed air guns are most frequently used. These sources have been shown to cause virtually no damage to marine life.

When *controlled explosives* are used, the most common method is to place carefully measured charges in shallow "shot holes" a few inches in diameter drilled from a truck-mounted drill. These charges are then detonated to produce the sound waves needed. In less accessible areas, a portable drill may be used and, in certain environmentally sensitive areas, the charges may be mounted on stakes above the ground to minimize plant disturbance.

Truck-mounted vibrators are frequently used. Metal plates, used singly or in groups, are pressed against the ground and vibrated briefly creating a frequency sweep similar to that used in radar, to send a brief acoustic signal into the ground. The vibrators cause little disturbance to the surrounding area and are often used in sensitive locations, in cities and along highways, where dynamite charges would be less acceptable. See Fig. 9.

The principles of seismic surveying are the same regardless of the type of equipment used. As the waves strike the various strata, some are

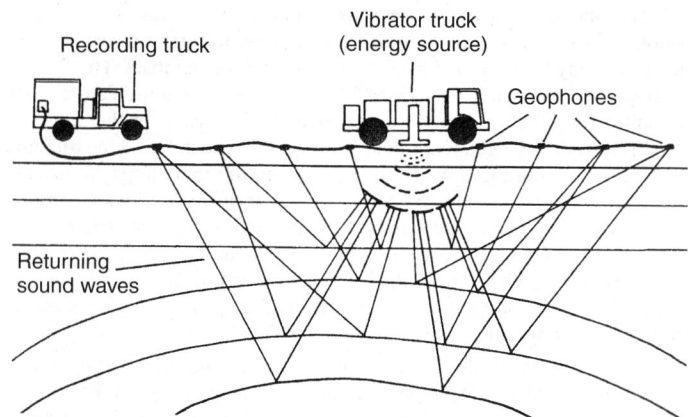

Fig. 9.   Seismic survey using a truck-mounted vibrator as the energy source for creating reflected acoustic sounds. This method is used in sensitive areas near cities and along highways because disturbance to the environment is minimal. (*American Petroleum Institute*)

reflected back to the surface, where they are picked up by sensing devices called *geophones* (on land) and *hydrophones* (offshore). See Fig. 10. These sensors convert acoustic information into electrical signals, whence the data are recorded on magnetic tape for processing at a computer center.

Seismic Data Processing. Continuing advances in technology, such as fiber optics, have enabled the capabilities of seismic surveys to be markedly increased so that, where required, a 3-dimensional picture of the rocks in the subsurface can be obtained. Advanced processing has also enabled discrete anomalies in seismic data from individual rock horizons to be analyzed. Under certain conditions, the presence of hydrocarbons can be directly detected. This analysis, sometimes referred to as "bright spot" technology, has been responsible for numerous discoveries in the Gulf of Mexico in recent years.

Raw data gathered from seismic surveys must be processed to compensate for and to remove a variety of distortions—unwanted noises created by weathered near-surface rocks, normal time delays, and echoing by rebounding acoustic waves—to provide the clearest possible image of the strata below. Computers can restore these distortions in a fraction of the time that was formerly required to adjust the data painstakingly by hand. Advanced techniques not only permit presentations in three dimensions, but also in color, and to create contour maps and models of subterranean features. However, even with the use of sophisticated tools, there remains a large measure of uncertainty. History has shown repeatedly that a prospective area rejected by one petroleum firm has been accepted by another and proved to be successful.

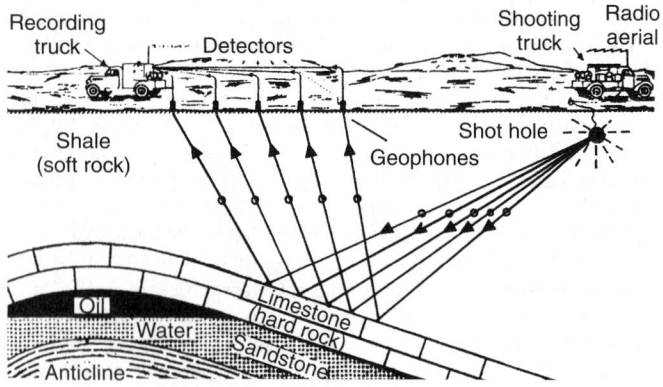

Fig. 8.   Onshore seismic field operation, using the shot hole energy method. Energy from a controlled explosion is directed toward underlying rock structures and is reflected to indicate shape of formation below—in this case an anticline. Geophones placed on the ground surface by the surveying crew measure and record the reflected acoustic energy. With the survey truck safely away from the shot hole, the explosives are detonated by an assigned radio frequency from the truck

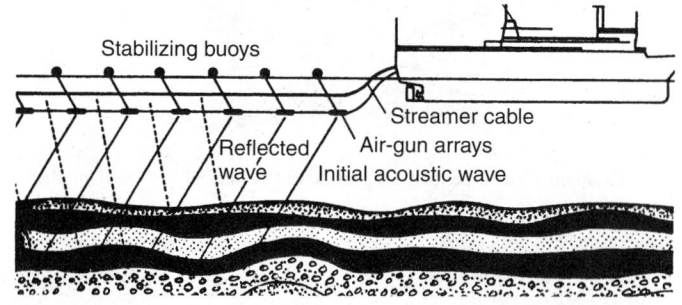

Fig. 10.   In offshore exploration, vessels are specifically designed for seismic surveying. Instead of using the same equipment that is used on-shore, such as "surface shaking" machines and geophones, offshore seismic crews use chambers containing compressed gases or fluids to generate the acoustic signals and hydrophones to pick up the returning sounds. Air-gun arrays are trailed in the water behind the vessel as it plies along a predetermined survey line. The crew activates the chambers at set intervals from onboard controls connected by cables to the air guns. Other cables contain arrays of hydrophones to detect sounds that echo off the underlying strata. The vessels are kept on course through the use of radar, loran, and satellite navigation equipment. (*After Exxon*)

[2] The first "seismograph" is recorded to have been an inverted bronze urn with a pendulum inside and used by the Chinese in about 160 A.D. to announce earthquakes. Seismographs were "reinverted," so to speak, during World War I to locate heavy German artillery. In 1920, seismography for use in petroleum exploration was first demonstrated.

After thorough analysis of seismic and other exploration data, the next steps are management decision making and approval to proceed with exploratory drilling. The ultimate exploratory tool is the *drill bit*.

Exploratory Drilling

Onshore Drilling. When Col. Edwin Drake brought in the first commercial oil well in 1859, he struck oil at a depth of 59 feet, 8 inches (18.2 meters). See Fig. 11. Most early wells were less than 400 feet (122 meters) deep. Shallow oil and gas wells were fully exploited many years ago. Deep producing wells today often exceed depths of 25,000 feet (7,620 meters) and dry holes have been drilled to a depth in excess of 31,000 feet (9,449 meters). In an average year, wildcat wells reach a depth of about 6,000 feet (1,829 meters). Depth, however, is only one of the factors that makes the search for petroleum difficult in modern times. Increasingly, drilling must be done in remote places where it is costly to bring in materials and labor. For example, onshore locations, such as those found on the North Slope of Alaska, can result in drilling costs that are 10 times as high as they would be in the lower 48 states.

The Rotary Drill. This is the most commonly used method and consists of a rotary drill, a power source, a derrick and lifting and lowering devices, and a bit attached to a length ("string") of tubular high-tensile-strength steel. See Fig. 12. The drill string passes through a rotary table that turns it and thus provides the torque needed for the drilling operations. The weight applied to the formation is also a critical factor.

Shown in the upper right of this view is the *swivel* (front and side view) which permits the drill pipe to rotate while mud is pumped down to clean the hole. Shown in the lower right is a three-cone drill bit, with cutaway showing the bearings on which it rotates; and at the extreme lower right, the face of a diamond bit revealing openings through which fluids may pass. (*American Petroleum Institute; Exxon Corp.*)

During the drilling operation, a special *drilling mud* (mixture of clay, water, and chemical additives) is pumped down through the hollow drill string and bit into the borehole. The fluid is forced up the borehole and through the area between the drill string and the casing (the "annulus") to the surface. There it is cleaned and recirculated into the well. The fluid helps to cool and lubricate the bit, control the pressures within the well, provide a protective and stabilizing coating to some permeable formations, and brings the rock cuttings up the borehole to the surface. The consistency

Fig. 11.   Colonel Edwin Drake (right) and Peter Wilson, a druggist who endorsed a $500 bank loan for Drake, confer in front of the world's first commercial oil well near Titusville, Pennsylvania. Initially, Drake rigged a large wheel powered by steam to raise and lower a cable and iron bit. Later connected to a crude drill pipe and pump, this well produced about 35 barrels a day. (*ca. 1861*)

of the fluid is carefully monitored and adjusted to compensate for pressure changes within the well, as the bit penetrates the various rock strata.

"Spudding" is the actual start of drilling a well and is akin to the first shovel of dirt at groundbreaking. A large bit, frequently from 18 to 38 inches (46–97 cm) in diameter, is used to drill a hole to a depth of from 10 to 100 feet (3–30 meters). The hole is then lined with a conductor pipe ("casing"). The space between the casing and the drilled hole (the "borehole") is filled with cement.

Drilling is then resumed using a smaller bit and, after the borehole reaches several hundred feet (meters), the bit and drill string are hoisted out of the well and another length of pipe ("surface casing") is lowered into the borehole and cemented in place. Besides preventing the generally unconsolidated surface formation from sloughing into the hole, the casing also protects the freshwater strata ("aquifers") from being contaminated by the drilling mud.

As the drilling proceeds, additional casings of concentrically smaller diameter are lowered into the well and sealed in place until the final depth ("target zone") is reached. During the drilling, the drill bit and string must be removed from the well whenever the bit becomes dull and requires changing or cores are taken from the well. The coring process involves a special cylindrical rock bit, generally with a diamond-encrusted face and a cylinder ("core barrel") into which the core passes and is retained for recovery at the surface. These cores are analyzed to determine the type of rocks penetrated and their porosity, permeability, chemical analysis and possible hydrocarbon content. See Fig. 13.

Drilling Geometry. Deviation surveying was introduced into oil-well drilling technology in 1929. Before that time, it was generally assumed that a hole properly started as a vertical hole would remain essentially vertical. In many instances, this was not a realistic assumption because many "vertical" holes were found to be quite crooked. Crooked holes not only caused operational problems, but also resulted in false indications of depth. Since the early 1930s, drilling contracts usually have specified a maximum deviation of 3 to 5 degrees. The problem of drilling a straight hole usually is simpler with uniform materials, such as limestone, and more difficult when laminar formations of sandstone and shale are encountered. Often of even greater concern than a crooked hole is an irregular, "jagged" hole that does not have a graceful bending contour in the vertical. The presence of abrupt changes in angle interferes with the casing program and ultimately with production. Although the mechanics involved in causing nonvertical drilling are not fully understood, much has been learned through experience and great improvements in drilling have been made. For one thing, deviation results from flexibility of the drill string (drill collars), the forces acting upon the string causing it to bend. A relatively simple change to square collars, as shown in Fig. 14, has brought about marked improvement.

Of course, it is frequently desirable to utilize a controlled directional drilling technique. There are several reasons, as indicated by Fig. 15. The three most commonly patterned directional holes are shown in Fig. 16. The planned course of direction depends upon several factors, including rig capacity, hole size, mud program, types of formation, and the casing program. Meticulous surveying is required to achieve the desired results. Several types of drilling tools may be required.

Offshore Drilling. When exploration moves offshore, standard drilling equipment obviously must be supplemented by some sort of structure that provides a stable platform for operations. The structure also must be movable, given the odds against a single wildcat finding commercial quantities of petroleum.

The first offshore exploration in the 1930s in the Gulf of Mexico was conducted from rigs on barges which could be towed to drilling sites and submerged to rest on the bottom during operations. These were forerunners of the twin-hulled submersible rig, which has an upper hull housing crew quarters and working spaces and a lower hull providing the buoyancy needed to move the unit. See Fig. 17(a) on p. 2695.

The jack-up or self-elevated rig, introduced in the 1950s, is a barge with movable legs, which can be lowered to the sea floor and the barge jacked into drilling position above the water. Jack-ups are used in water depths up to about 300 feet (91 meters). See Fig. 17(b) on p. 2695.

In deeper waters, exploration is conducted from floating rigs, including submersibles and drill ships. Drill ships with conventionally shaped hulls of seagoing vessels are not so stable as semisubmersibles (semis) in rough waters, but can be moved from location to location much faster.

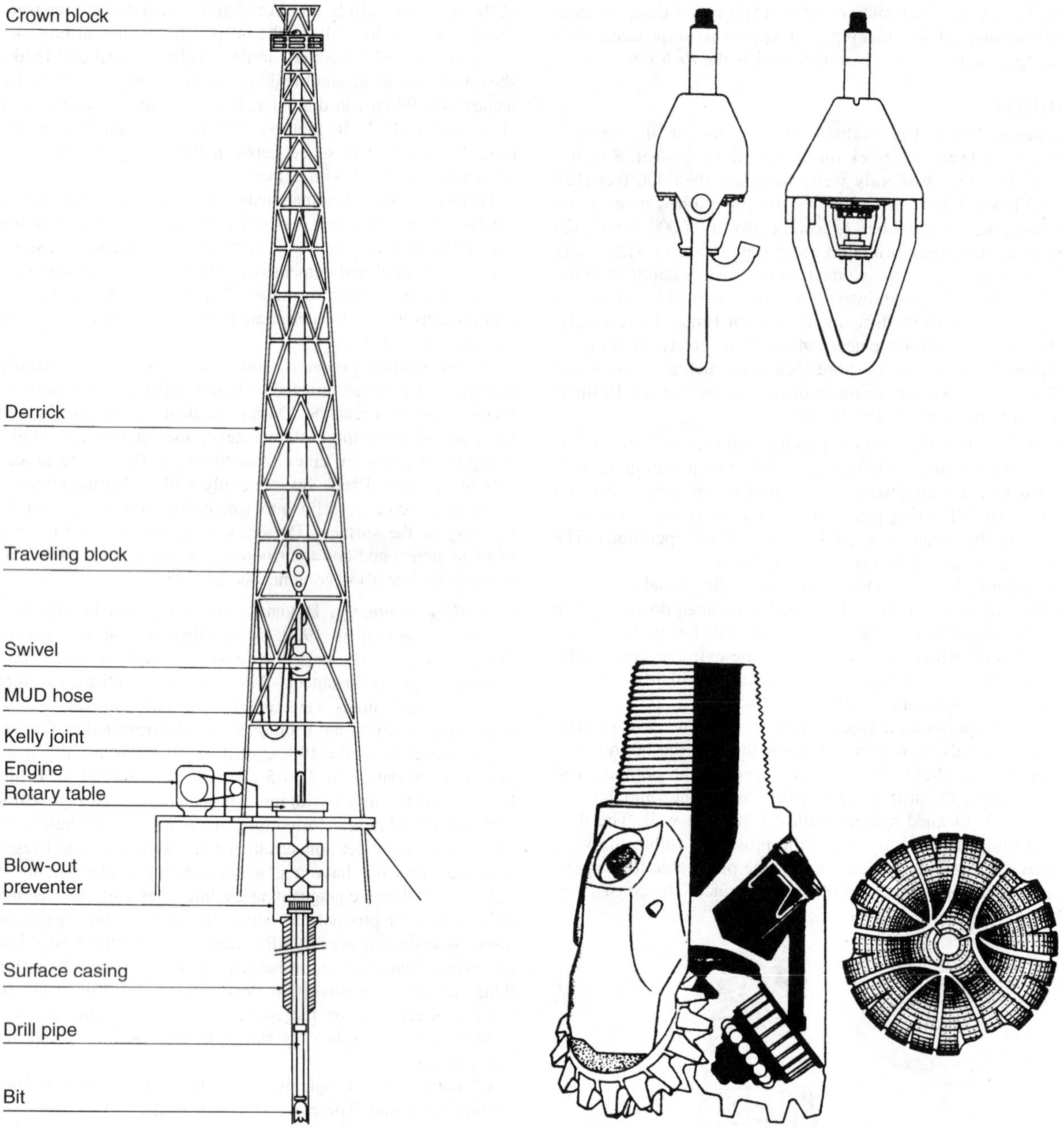

Crown block

Derrick

Traveling block

Swivel

MUD hose

Kelly joint

Engine

Rotary table

Blow-out preventer

Surface casing

Drill pipe

Bit

Fig. 12.   A rotary rig has four systems. The *rotary system* consists of a turntable, a swivel, a square or hexagonal pipe length called a "kelly," which transmits rotary motion from the turntable to the drill pipe, and the drill "string" itself. A *circulating system* of pumps, hoses and other apparatus keeps mud circulating through the well. The *hoisting system* includes the derrick, a drawworks, hoisting blocks, and other equipment needed to lift and lower heavy pipe joints and casing. The *power system* usually consists of diesel engines and generators, set apart from the rig, which provide power for the electric motors that drive the rotary, hoisting, and pumping equipment. The elevated floor allows installation of blowout preventer beneath the platform

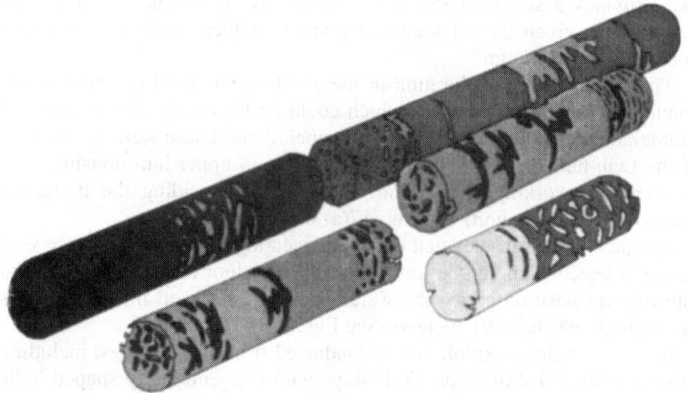

Fig. 13.   Rock core samples cut by a diamond-faced core bit reveal underlying structure and the possible presence of hydrocarbons. (*Exxon Corp*)

A critical requirement of all floating rigs is the ability to maintain position over the wellhead while drilling proceeds. Semis and drill ships use either multiple anchors or "dynamic positioning" systems to keep on station. A dynamic positioning system uses thruster engines which, responding to signals from acoustic beacons on the sea floor, automatically make the adjustments required to maintain the rig in position. Hydraulic devices keep a constant tension on the drill string to prevent the up-and-down motion of the sea from being transmitted to the drill bit.

Semis and drill ships find limited use in arctic waters where ice covers the sea most of the year.

In offshore operations, exploration wells are almost always plugged and abandoned even when they strike petroleum. Their sole function is to find oil or gas and to delineate the reservoir. The operator uses this information to pick a location for a permanent production platform from which development wells will be drilled to recover as much petroleum as economically possible. In onshore operations, however, successful exploration wells also become producers.

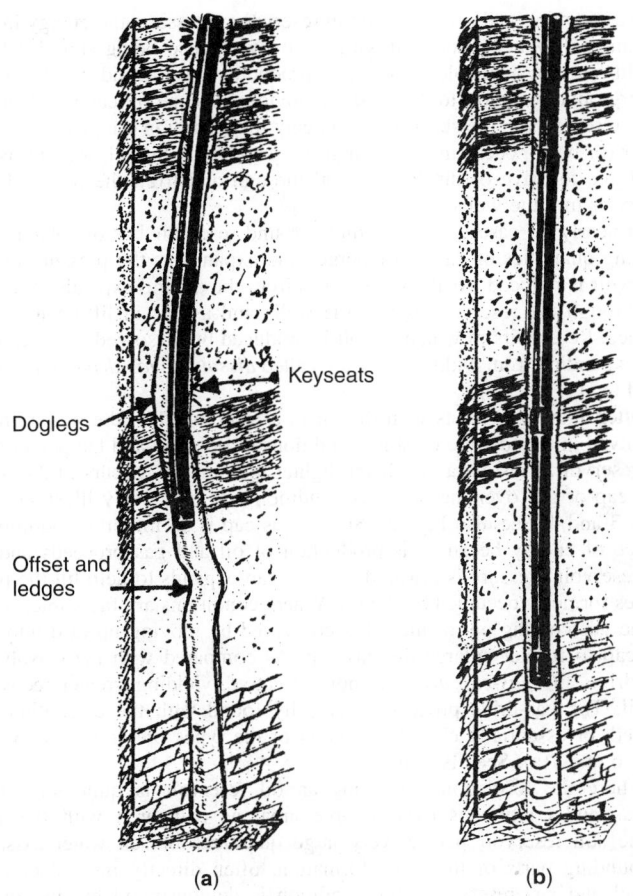

Fig. 14. Example of a crooked hole (**a**) drilled without a square collar, and a relatively straight hole (**b**) drilled with a square collar. (*Drilco*)

Measuring Well Characteristics

At selected intervals during the drilling, generally before the casing is run or when formations with hydrocarbon indications are encountered, measurements may be taken of the characteristics of the borehole and surrounding strata. Wire line logging tools are used.

In the early days of the industry, little was known about "downhole" geophysics, that is, the physical characteristics of the subsurface strata, how they might be measured, and what could be learned from such measurement. Since the Drake well, more than 3 million wells have been drilled in the United States. An estimated 27,000 fields have been found. With each new discovery, additional data become available and patterns begin to emerge. It was not until the late 1920s that technological changes were introduced that would have a lasting impact on "logging" (recording) the characteristics within the well during exploratory drilling. The first well-logging device (electrical resistivity log), invented by Conrad Schlumberger (France), was introduced into the United States. In 1934, a second development, the *spontaneous potential* (SP) *curve*, was introduced.

The *resistivity log* was lowered by cable into the borehole, with the drill bit and drill stem removed, thus enabling the recording of the electrical resistivity of the rock layers that the bit had penetrated. The record helped to identify the hydrocarbon content of the reservoir rock (since both oil and gas have different resistivities than does salt water). It was also used to correlate rock horizons between wells, which proved to be an invaluable tool in subsurface mapping.

The *SP Curve* recorded the differences in natural electrical potential between the fluids in the adjacent formations and those within the uncased borehole. This curve was soon accepted as an indicator of the porosity of the rock strata and as a means of locating the boundaries of rock beds. See Fig. 18.

Since then, a number of downhole measurements have been devised that use radioactive, acoustic, and electrical methods.

Presently, televiewers are used to look at rock features in boreholes and computers are programmed to compare, synthesize, and integrate the new range of measurements, thus providing more reliable information and

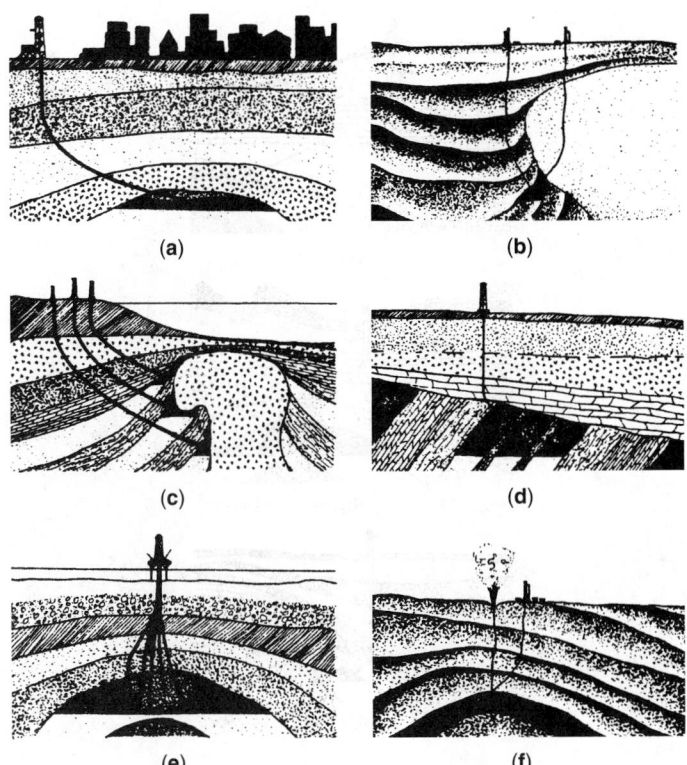

Fig. 15. Some applications for controlled directional drilling: (**a**) Reaching formations which lie below inaccessible locations, such as towns, rivers, and lakes. (**b**) Formations sometimes are found below the overhanging cap of a salt dome. A well may be drilled around this cap, or through the salt and deflected into the productive formation. (**c**) Formations below harbors or the ocean floors sometimes can be reached from rigs located on the shore. (**d**) Directional drilling into the intersection of several oil sands from a single wellbore. Obviously, a straight hole would be less effective in this type of situation, (**e**) Offshore drilling is usually most economic when several directional wells can be drilled from a single platform. As many as 20 or more wells can be drilled from a small area. (**f**) Drilling of a relief well to intersect a wild, cratered well near the source of pressure. Mud and water can be pumped in to kill the blowout. This technique, first used in 1934, helped to establish the importance of directional drilling. (*Petroleum Extension Service, The University of Texas at Austin*)

definition of rock properties and formation fluids thousands of feet (meters) below the surface.

Until recently, these measurements could be taken only when the drill bit and string were removed from the borehole. However, in the mid-1980s, a new dimension was added, namely, *measurement while drilling* (MWD) instruments. These devices are mounted above the drill bit and around the drill string to provide a continuing source of data on downhole characteristics. This advancement reduces the drilling downtime previously required when measurements were taken.

Testing. Modern wire line logs will indicate with a good degree of accuracy the potential of a hydrocarbon zone. If the zone is sufficiently promising to warrant further study, a formation test will be undertaken.

Generally, a drill stem test is carried out—either in the open hole or after the hole has been cased. However, the case hole test is the most reliable.

Basically, the drill stem test involves attaching a tubing assembly to the end of the drill pipe, isolating the test zone with rubber packers, and perforating the zone. The tool is then opened so that the fluids or gas in the formation can flow up the drill pipe for metering at the surface. During this process, extensive pressure measurements are taken, which can help to indicate the extent of the reservoir and the rate at which the hydrocarbons could be recovered. Prior to describing how a well is finally completed (if the hole is not dry!), it is in order to describe the forces utilized to transfer the oil from the reservoir to the surface.

Well Drive Systems

It is convenient to classify oil and gas reservoirs in terms of the type of natural energy and forces available to produce the oil and gas. At the time

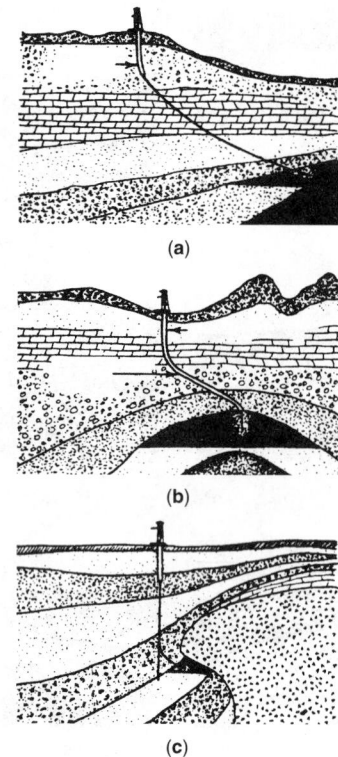

(a)

(b)

(c)

Fig. 16.   Principal types of directional drilling patterns: (**a**) The most widely used directional drilling pattern is one in which the initial deflection is obtained at relatively shallow depth. Then, surface casing is set and cemented through the deviated section of hole. From that point, the angle is maintained as a straight line to the target zone. (**b**) This pattern is also initially deflected at shallow depth. Surface casing is set and cemented. Drilling continues on a straight line to a point where the hole is gradually returned to vertical. After intermediate casing is set, drilling is continued to final depth. This type hole is used when undesirable formations must be penetrated and isolated with an intermediate casing string. (**c**) In this pattern, deflection commences at a greater depth. Drift angle is maintained on a straight line to the target. This type hole may be used for exploratory drilling from a dry hole. Normally, the deflected part of this hole is not protected by casing during drilling operations. (*Petroleum Extension Service, The University of Texas at Austin*)

oil was forming and accumulating in reservoirs, pressure and energy in the gas and salt water associated with the oil were also being stored, which would later be available to assist in producing the oil and gas from the underground reservoir to the surface. Obviously, since oil cannot lift itself from reservoirs to the surface, it is largely the energy in the gas or the salt water (or both), occurring under high pressures with the oil, that furnishes the force to drive or displace the oil through and from the pores of the reservoir into the wells.

In nearly all cases, oil in an underground reservoir has dissolved in it varying quantities of gas that emerges and expands as the pressure in the reservoir is reduced. As the gas escapes from the oil and expands, it drives oil through the reservoir toward the wells and assists in lifting it to the surface. Reservoirs in which the oil is produced by dissolved gas escaping and expanding from within the oil are called *dissolved-gas-drive reservoirs*. See Fig. 19.

Often more gas exists with the oil in a reservoir than the oil can hold dissolved in it under the existing conditions of pressure and temperature in the reservoir. This extra gas, being lighter than the oil, occurs in the form of a cap of gas over the oil. This condition was previously illustrated by Figs. 3 and 4. See also Fig. 20. Such a gas cap is an important additional source of energy because, as production of oil and gas proceeds and as the reservoir pressure is lowered, the gas cap expands to help fill the pore spaces formerly occupied by the oil. Where conditions are favorable, some of the gas coming out of the oil is conserved by moving upward into the gas cap to further enlarge the gas cap. As compared with the dissolved-gas drive, the *gas-cap drive* is more effective, yielding greater recovery of oil. The gas-drive process is typically found with the discontinuous, limited, or essentially closed reservoirs of the types previously shown in Figs. 6 and 7(a). See also Fig. 20.

Where the formation containing an oil reservoir is quite uniformly porous and continuous over a large area, as compared with the size of the soil reservoir per se, very large quantities of salt water exist in surrounding parts of the same formation, often directly in contact with the oil and gas reservoir. This condition is demonstrated by previously shown Figs. 2, 3, 4, and 5. These large quantities of salt water occur under pressure and provide a large additional store of energy to assist in producing oil and gas. A situation like this is termed *water-drive reservoir* and is shown in Fig. 21. The energy supplied by the salt water comes from expansion of the water as pressure in the petroleum reservoir is reduced by production of oil and gas. Water will compress, or expand, to the extent of about one part in 2500 per 100 psi (6.8 atmospheres) change in pressure. Although this effect is slight with reference to small quantities, the phenomenon becomes of importance when changes in reservoir pressure

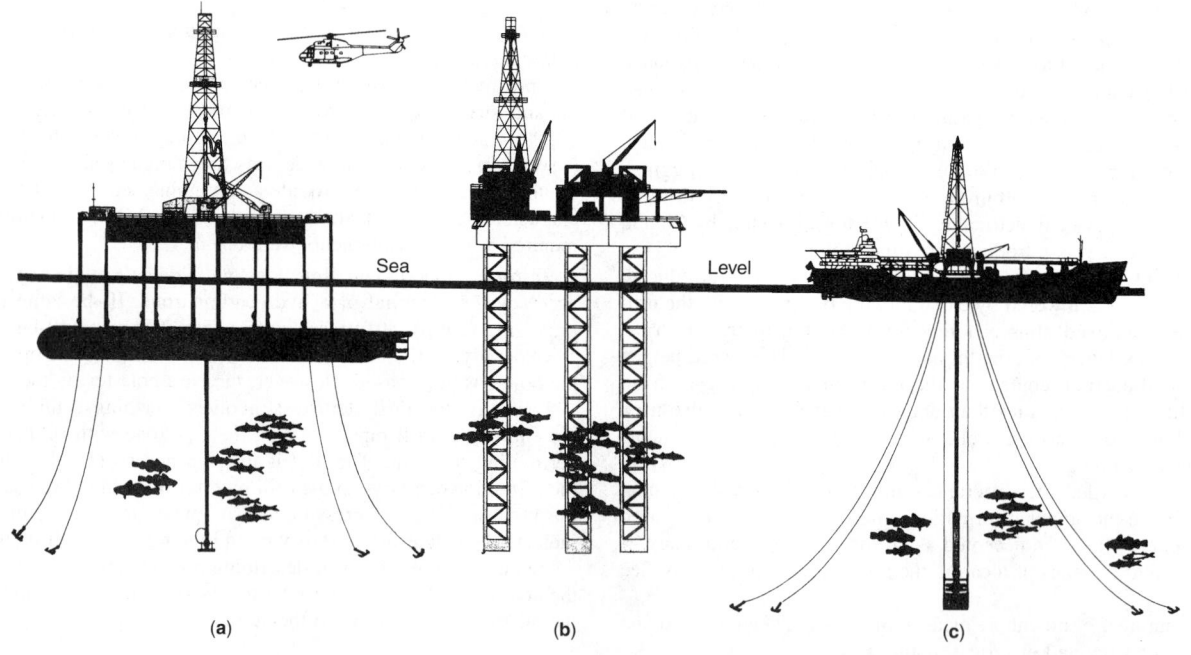

(a)   (b)   (c)

Fig. 17.   Offshore drilling schemes: (**a**) Big, pontoon-mounted "semisubmersible" rig that is indispensable for exploratory drilling in rough water; (**b**) self-elevating or *jack-up* drilling rig widely used in water depth of less than 300 feet (91 meters); (**c**) turret mooring allows a drill ship to head into prevailing winds and currents while positioned over the well. Helicopters and boats are required to transport personnel to and from offshore well sites. (*Exxon Corp*)

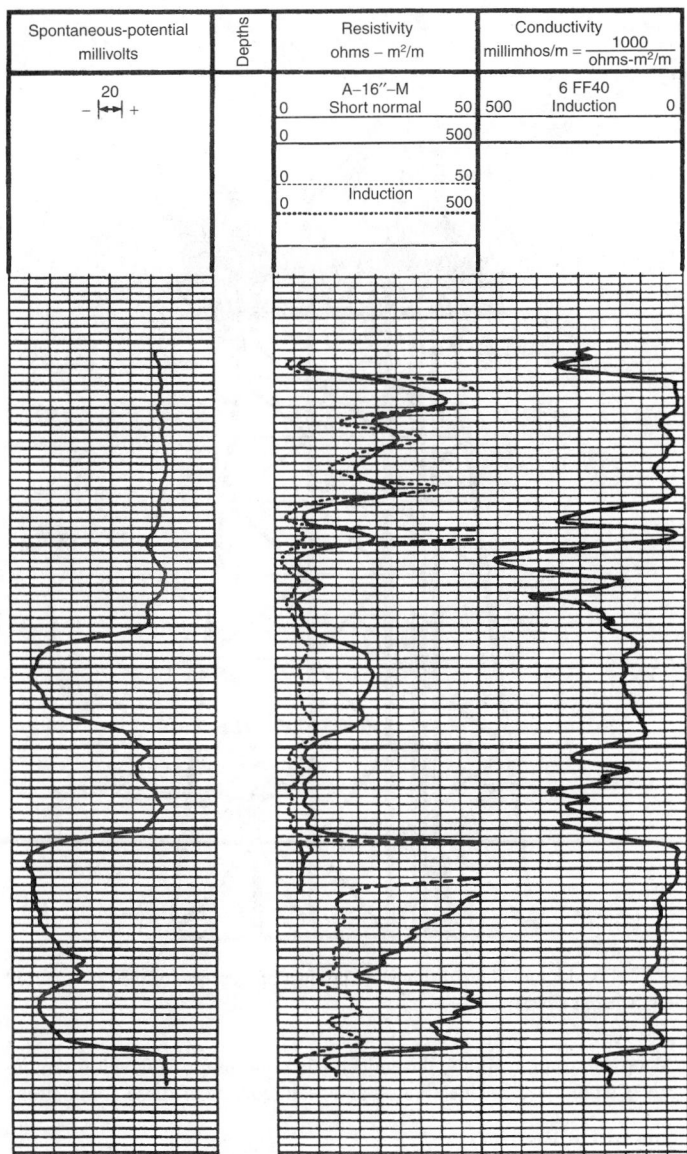

Spontaneous-potential millivolts	Depths	Resistivity ohms – m²/m	Conductivity millimhos/m = $\frac{1000}{\text{ohms-m}^2/\text{m}}$
20 − ↔ +		A–16″–M Short normal 0 — 50	6 FF40 Induction 500 — 0
		0 — 500	
		0 — 50 Induction	
		0 — 500	

Fig. 18. Oil well logging provides valuable information on the "down-hole" characteristics of oil and gas wells. Spontaneous potential (SP), electrical resistivity/conductivity logs are frequently recorded simultaneously

Fig. 19. Dissolved-gas drive. (*Texas Mid-Continent Oil and Gas Association*)

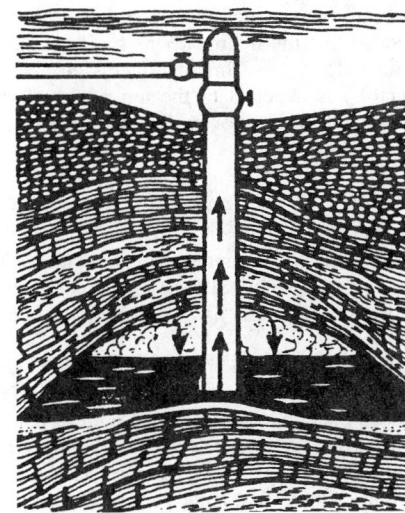

Fig. 20. Gas-cap drive. (*Texas Mid-Continent Oil and Gas Association*)

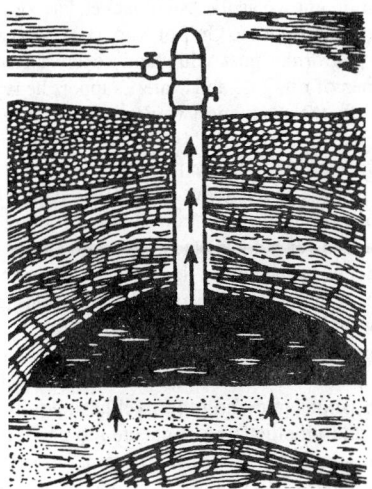

Fig. 21. Water drive. (*Texas Mid-Continent Oil and Gas Association*)

affect large volumes of salt water that are often contained in the same porous formation adjoining or surrounding a petroleum reservoir.

The expanding water moves into the regions of lowered pressure in the oil- and gas-saturated portions of the reservoir caused by production of oil and gas, and retards the decline in pressure. In this way, the expansive energy in the oil and gas is conserved. As shown in Fig. 4, the expanding water also moves and displaces the oil and gas in an upward direction out of the lower parts of the reservoir. By this natural process, the pore spaces vacated by oil and gas produced are filled with water, and oil and gas are progressively moved toward the wells.

The water drive is generally the most efficient oil-production process. Oil fields in which water drive is effective are capable of yielding recoveries ranging up to 50% of the oil originally in place, if (1) the physical nature of the reservoir rock and of the oil are conducive to the process, (2) care is exercised in completing and producing the wells, and (3) the rate of withdrawal of products is optimal.

When pressures in an oil reservoir have fallen to the point where a well will not produce by natural energy, some method of artificial lift must be used. Oil-well pumps are of three general types: (1) pumps located at the bottom of the hole run by a string of rods, (2) pumps at the bottom of the hole run by high-pressure liquids, and (3) bottom-hole centrifugal pumps. Another method involves the use of high-pressure gas to lift the oil from the reservoir.

Well Completion

Production casing must be set through which the oil and/or gas can be brought safely to the surface. The "pay zone" (productive area) is then sealed off with cement. With the production casing in place, hollow charges are fired through it into the production formation and the drilling mud is gradually displaced, so that the hydrocarbons can flow into the well-bore and up to the surface. There, a "Christmas tree" (an assembly of valves and special connectors) is attached to the top of the production casing. This device controls the flow of oil or gas into the gathering pipelines. See Fig. 22.

Deepwater Production

Two basic types of platforms may be used in deepwater production—*fixed leg* and *compliant*. Each has its advantages and limitations. These facilities can provide all the functions required for drilling, completing, producing, and maintaining conventional wells or a combination of conventional and subsea wells.

Nearly all offshore fields have been developed with fixed-leg platforms. In 1947, for example, a fixed-leg platform weighing 1200 tons was in operation in 20 feet (6 meters) of water out of sight of land. Twenty years later, such platforms weighing in excess of 6500 tons were in use in 340 feet (104 meters) of water. See Fig. 23. To date (1986), the tallest fixed-leg platform, weighing 58,000 tons, is located in 1025 feet (312 meters) of water in the Gulf of Mexico. This platform, completed in 1979, has a total height from seabed to top of the derricks of 1265 feet (386 meters)—taller than the Empire State Building in New York City.

This type of design, known as the "steel jacket," accounts for most of the hundreds of platforms that dot the Gulf of Mexico. For larger oil fields, such as the North Sea, platforms must withstand severe environmental forces, handle large volumes of oil, gas, and water, support heavy equipment, and accommodate 200 to 300 production workers. Here, a favored type is the concrete "gravity platform," so called because its own immense weight pins it to the sea bottom and no piles are needed to secure it. Rigid platforms are impractical in waters much more than 1000 feet (305 meters) deep. An alternative for deeper water is a "compliant" structure, such as a guyed tower, a slender steel tower held in place by a radial array of anchor cables. Heavy weights attached to the cables lie on the bottom some distance away; these keep the cables taut under normal sea conditions and lift gradually in storms, to allow the tower to tilt slightly to absorb wave forces.

When a platform is not a practical way to develop an offshore field, the operator may "complete" the well using a submerged production system, in which case, the Christmas tree and other wellhead equipment are installed on the sea bottom and pipelines are connected to carry off the petroleum, either to shore, to a nearby platform, or to a vessel or storage buoy moored in the area. Divers can be used to make the necessary connections.

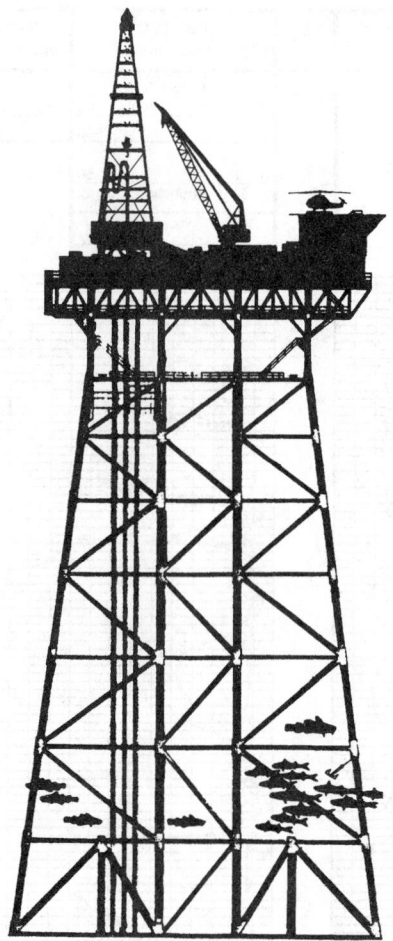

Fig. 23.   Fixed-leg offshore drilling and production platform of the "steel jacket" variety. (*Exxon Corp*)

For deepwater applications, the industry is continuing to develop remote, diverless techniques for installation and maintenance of these completions.

Enhanced Oil Recovery

Enhanced oil recovery (EOR) methods increase ultimate oil production beyond that achievable with primary and secondary methods. This is accomplished by increasing the proportion of the reservoir affected. EOR methods are of three broad groups: (1) thermal, (2) miscible, and (3) chemical.

Thermally Enhanced Recovery. Because oil becomes thinner and flows more easily when it is heated, considerable effort has been devoted to the development of techniques that introduce heat into a reservoir to improve recovery of the heavier, more viscous crude oils. Hot water flooding has been tried, but it is seldom used today because it contains too little heat energy and is very slow to warm the oil and rock surrounding an injection well. More heat is needed for efficiency.

Steam contains the extra heat energy that is required and it has been widely used by the petroleum industry since the mid-1960s to stimulate the production of thick oils. Two techniques, steam stimulation and steam flooding, are currently used.

Steam stimulation or steam soaking uses a well as both injector and producer. High-pressure steam is injected directly into the production zone for several days to weeks. After this period, the reservoir area around the well is allowed to soak in the new heat energy for an additional period. During this time, most of the steam condenses to hot water. After the soak period, the well is brought back into production to recover the heated (thinner) oil and hot water near the wellbore. Because natural driving forces are relied upon to move the oil through the reservoir during the production phases, steam stimulation generally increases the *rate* of recovery rather than the *amount* of oil that ultimately may be recovered. This technique is particularly adapted to certain California fields containing heavy crude oils, as well as fields in the Orinoco oil belt of Venezuela and the Cold Lake area of Alberta, Canada.

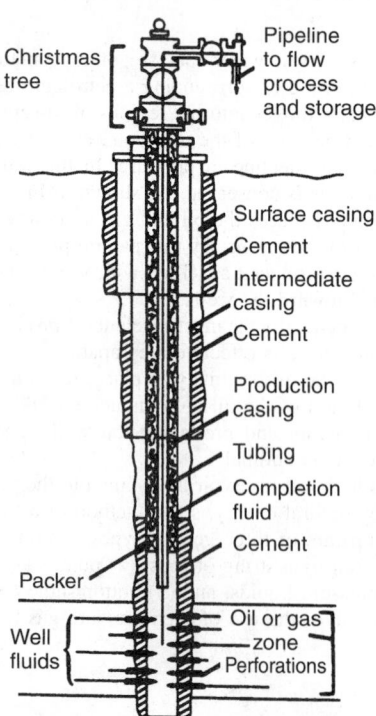

Fig. 22.   Completed oil well showing the flow of oil into and up the well to the pipeline connection at the Christmas tree. (*American Petroleum Institute*)

Steam flooding is more sophisticated and difficult than steam stimulation. This technique uses separate injection and production wells to improve both *rate* and *amount* of production.

Miscible Recovery. Oil and water do not mix and they do not flow with equal facility through a porous rock. Over the years, many miscible flood processes have been tested, the most successful of which have been: (1) hydrocarbon miscible recovery; (2) carbon dioxide miscible flooding; and (3) chemically enhanced recovery.

Depending on the composition of the oil and the reservoir temperature and pressure, light hydrocarbons in liquid form, such as liquefied petroleum gas (LPG) and including propane, butane, and ethane, may be miscible with crude oil. Where conditions are right, natural gas can be used to drive a bank of injected light hydrocarbon liquids through the reservoir to form a miscible flood. The disadvantage of this method is that it involves the prolonged use of valuable hydrocarbon liquids, some of which may never be recovered. The use of natural gas alone has received consideration in special situations where high pressure, combined with low reservoir temperature, may result in miscibility.

Carbon dioxide miscible recovery is a preferred method. Carbon dioxide may not be initially miscible with crude oil, but when it is forced into an oil reservoir, some of the smaller, lighter hydrocarbon molecules in the contacted crude will vaporize and mix with the CO_2, forming a wall of enriched gas (CO_2 plus light hydrocarbons). If the temperature and pressure of the reservoir are suitable, this wall of enriched gas will mix with more of the crude, forming a "bank" of miscible solvents capable of efficiently displacing large volumes of crude oil. Carbon dioxide is found in underground deposits and can be produced through wells similar to gas wells. But its production and transportation to the oil reservoir can add significantly to the cost. A project has been proposed that would involve the construction of a long CO_2 pipeline from a Colorado CO_2 well to a large oil field in Texas. The economic feasibility of the project remains to be proved.

The use of chemicals to coax more oil out of the ground has been investigated for many years. Chemically enhanced methods are of three major types: (1) polymer flooding; (2) surfactant flooding; and (3) alkaline flooding.

Experts estimate that prudent but aggressive application of enhanced recovery technology to known reservoirs in the United States could result in the production of 20 to 30 billion barrels of oil that might otherwise be lost forever. The amount of oil that might be recovered from known oil fields worldwide is estimated to be in the range of 100 to 200 billion barrels. See Fig. 24.

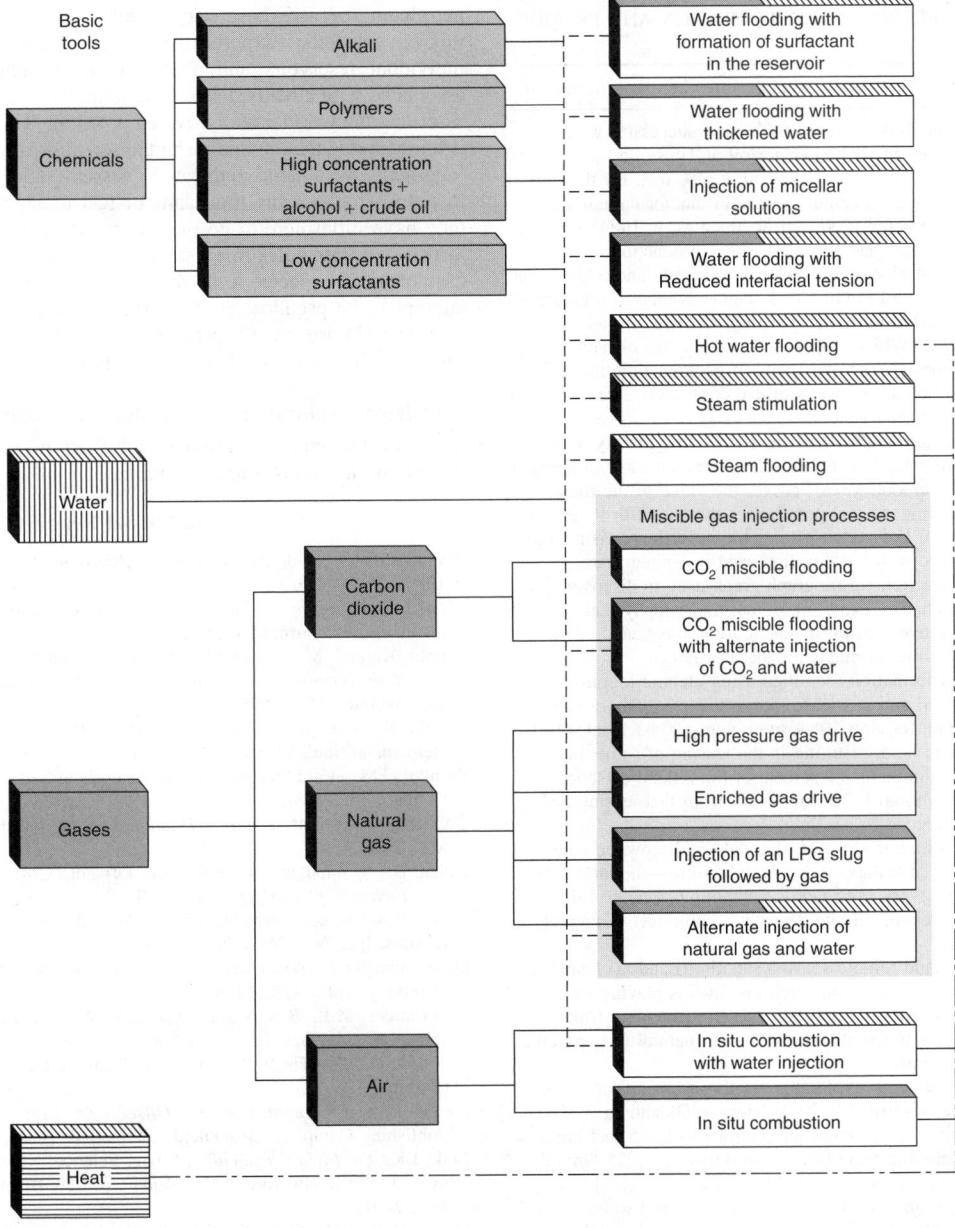

Fig. 24.   Enhanced oil recovery techniques employ heat, gases, chemicals, and water—singly or in combinations—to reduce the factors that inhibit oil recovery and to augment reservoir energy. (*Exxon Corp*)

Petroleum Reserves

In 1988, the U.S. Geological Survey reduced its prior estimate of the oil and gas remaining to be discovered in the United States by 40%. As pointed out by R.A. Kerr (reference listed), "If the new estimates hold up, it would solidify a new realism in the agency's view of energy resources. In 1972, the agency claimed that there were 450 billion barrels of oil and 2100 trillion cubic feet of gas left to be found—Figures that the USGS itself soon characterized as four times too high." A revised estimate in 1981 showed 83 billion barrels and 594 trillion cubic feet left undiscovered. Even this estimate was challenged as too high by some industry experts.

The foregoing is exemplary of how assumptions of remaining oil and gas reserves are made and of the lack of confidence that *any* figures appeared to enjoy as of the 1990s. Generally, there is a consensus that, given current rates of consumption, oil reserves should be visualized as becoming exhausted in terms of decades rather than centuries!

Everyone recognizes that crude oil exists in finite amounts, but no one really knows how much of it remains in the earth. Since E.L. Drake drilled his first well in 1859, a vast industry has been established and oil became the predominant worldwide fuel. But petroleum geologists freely admit that definitive knowledge about the resource base that made this growth possible remains *elusive*. At any given time, the amount of oil available for consumption depends primarily on two factors: (1) the producibility of

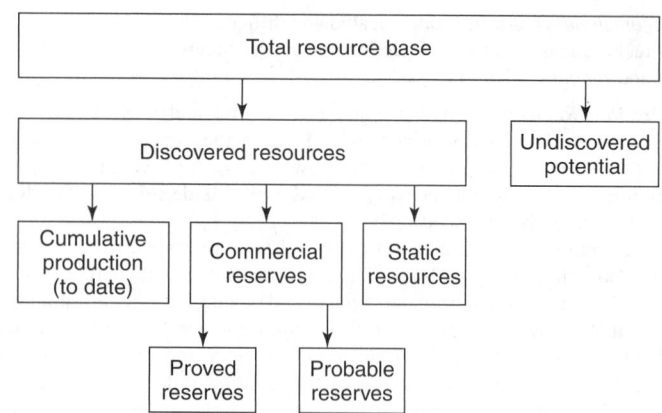

Fig. 25. Commonly used approach for handling statistics in connection with estimating oil reserves

already discovered reserves; and (2) the production policies of governments in countries where those reserves exist. Over the long run, however, it is the amount of recoverable oil left in the ground, including those volumes yet undiscovered, that will be decisive. Estimates of the remaining petroleum resource base vary widely—in fact, too widely to record in this encyclopedia. Experts disagree about the size and producibility of individual reservoirs and about the total national and world reserves associated with already discovered fields. They are even further apart when assessing the world's undiscovered potential. This is not surprising when it is remembered that, in this current unscientific arena, the experts are making judgments—frequently educated guesses—about hydrocarbons contained in porous rocks many thousands of feet under the earth's surface. Often, they have little more to go on than the data from a few widely dispersed 8-inch-diameter holes in existing fields—and still less information in the case of fields expected to be discovered in the future. Consequently, in an attempt to be pseudoscientific, experts generally refer to three classes of reserves: (1) proved; (2) probable; and (3) potential. The chart shown in Fig. 25 represents a well-accepted approach to reserves analysis.

Petroleum Exploration and Production Progress

Some of the major milestones achieved in petroleum exploration and production technology are summarized in Table 3.

TABLE 3. MILESTONES IN PETROLEUM EXPLORATION AND PRODUCTION TECHNOLOGY

1853	Dr. Albert Gesner manufactures kerosene from coal.
1859	Edwin Drake completes the first successful well drilled in the search for oil at Titusville, Pennsylvania, striking oil at $69\frac{1}{2}$ feet. By the start of the 20th century, crude oil and/or natural gas were being produced in 20 states; in 1984, 33 of the 50 states had some oil or gas production.
1865	First oil pipeline, 2 inches in diameter and 32,000 feet long, laid at Oil Creek, Pennsylvania, to transport oil from the field to the Oil Creek Railroad.
1883	Dr. I.C. White proposes the theory that oil and gas deposits could be found in geological anticlines.
1896	First "offshore" wells drilled from piers extending into California waters.
1899	Threllfall and Pollock devise the first gravity meter.
1901	"Spindletop" oil field is discovered on a salt dome near Beaumont, Texas; proves the value of the rotary drilling rig and popularizes the use of drilling mud.
1914	Reginal Fessenden patents the reflections seismograph.
1924	Electric well logging first used in United States; refraction seismograph graph used to discover Orchard, Texas, salt dome; first geophysical discovery using magnetic torsion-balance.
1939	First airborne magnetometer developed.
1942	Fluid formation identified using electric logging.
1954	First oil and gas lease sale in federal offshore area held; through 1984, more than 100 such sales had been held, resulting in the leasing of some 38 million acres (4 percent of the federal offshore area); and federal revenues from that leasing had exceeded $77 billion.
1968	The Prudhoe Bay, Alaska, field is discovered some 250 miles above the Arctic Circle—the largest U.S. discovery evermade—containing some 10 billion barrels of oil and 26 trillion cubic feet of natural gas.
1972	First land remote sensing satellite (Landsat) launched; information from such satellites is playing an increasingly important role in identifying from space potential deposits of oil, natural gas and other minerals.
1974	Record-depth exploratory well—a natural gas well—drilled to 31,441 feet in Oklahoma.
1979	World's tallest fixed-leg platform—1,265 feet tall and weighing 59,000 tons—installed in 1,025 feet of water in the Gulf of Mexico.
1984	Exploratory well drilled in world record water depth—6,942 feet—off the coast of New England.

Source: American Petroleum Institute.

Additional Reading

Abelson, P.H.: "Hydrocarbon Energy Revisited," *Science*, 1433 (September 29, 1989).

Ahmed, T.H.: *Reservoir Engineering Handbook,* 2nd Edition, Butterworth-Heinemann, Inc., Woburn, MA, 2001.

Arnold, K. and M. Stewart, Jr.: *Surface Production Operations: Design of Oil-Handling Systems and Facilities,* Vol. 1, 2nd Edition, Butterworth-Heinemann, Inc., Woburn, MA, 1998.

Arnold, K. and M. Stewart, Jr.: *Surface Production Operations,* Butterworth-Heinemann, Inc., Woburn, MA, 1986.

Bethke, C.M., et al.: "Supercomputer Analysis of Sedimentary Basins," *Science*, 261 (January 15, 1988).

Chaudhry, A.: *Oil Well Testing Handbook,* Butterworth-Heinemann, Inc., Woburn, MA, 2003.

Dalen, B.: "Computer Control Reduces Offshore Costs, Safety Risks," *Instrumentation Technology*, 22 (December 1989).

Dawe, R.A.: *Modern Petroleum Technology: 2 Volume Set,* 6th Edition, John Wiley & Sons, Inc., New York, NY, 2000.

Devereaux, S.: *Practical Well Planning and Drilling Manual,* PennWell Publishing Company, Tulsa, OK, 1998.

Economides, M.J., B.N. Murali, and L.T. Watters: *Petroleum Well Construction,* John Wiley & Sons, Inc., New York, NY, 1998.

Elliott, D.H.: "Is There Any Oil and Natural Gas (Antarctica)?" *Oceanus*, 32 (Summer 1988).

Esmaeili, H.: *The Legal Regime of Offshore Oil Rigs in International Law,* Ashgate Publishing Company, Brookfield, VT, 2001.

Fink, J.K.: *Oil Field Chemicals,* Elsevier Science, New York, NY, 2003.

Gluyas, J. and R. Swarbrick: *Petroleum Geology,* Blackwell Science, Inc., Malden, MA, 2001.

Hapgood, F.: "The Quest for Oil," *National Geographic*, 226 (August 1989).

Hyne, N.J.: *Nontechnical Guide to Petroleum Geology, Exploration, Drilling, and Production,* PennWell Publishing Company, Tulsa, OK, 1995.

Kerr, R.A.: "Oil and Gas Estimates Plummet," *Science*, 1330 (September 22, 1989).

Lee, D.B.: "Oil in the Wilderness—An Arctic Dilemma," *National Geographic*, 858 (December 1988).

Leffler, W.L., G. Sterling, and R. Pattarozzi: *Deepwater Petroleum Exploration and Production: A Nontechnical Guide,* PennWell Corporation, Tulsa, OK, 2003.

Longwen, W.: "China's Exploration and Development of Offshore Oil and Gas," *Oceanus*, 32 (Winter 1989/1990).

Lynch, M.C.: "Preparing for the Next Oil Crisis," *Chem. Eng. Progress*, 20 (March 1988).

Lyons, W.C.: *Standard Handbook of Petroleum and Natural Gas Engineering,* Vol. 1, 6th Edition, Butterworth-Heinemann, Inc., Woburn, MA, 2001.

Lyons, W.C.: *Standard Handbook of Petroleum and Natural Gas Engineering,* Vol. 2, 6th Edition, Butterworth-Heinemann, Inc., Woburn, MA, 1996.

Miall, A.D.: *Principles of Sedimentary Basin Analysis,* 3rd Edition, Springer-Verlag, Inc., New York, NY, 1999.

McCain, W.D.: *The Properties of Petroleum Fluids,* 2nd Edition, PennWell Publishing Company, Tulsa, OK, 1990.

Nelson, R.C.: "Chemically Enhanced Oil Recovery: The State of the Art," *Chem. Eng. Progress*, 50 (March 1989).

Olah, G.A. and A. Molnar: *Hydrocarbon Chemistry,* 2nd Edition, John Wiley & Sons, Inc., New York, NY, 2003.

Pate-Cornell, M.E.: "Organizational Aspects of Engineering System Safety: The Case of Offshore Platforms," *Science*, 1210 (November 30, 1990).

Rippee, B. and R.L. Busby: *2003 International Petroleum Encyclopedia,* PennWell Corporation, Tulsa OK, 2003.

Rutledge, G.: "Arctic Oil," *Chem. Eng. Progress*, 6 (October 1989).

Schmidt, R.L.: "Thermal Enhanced Oil Recovery: Current Status and Future Needs," *Chem. Eng. Progress*, 47 (January 1990).

Selly, R.C.: *Elements of Petroleum Geology,* 2nd Edition, Morgan Kaufmann Publishers, Orlando, FL, 1997.

Short, J.A.: *Introduction to Directional and Horizontal Drilling,* PennWell Publishing Company, Tulsa, OK, 1993.

Staff: "U.S. Oil and Gas Outlook Brightens," *Chem. Eng. Progress*, 6 (December 1989).

Staff: *International Petroleum Encyclopedia,* PennWell Publishing Company, Tulsa, OK, 2000.

Staff: "Mobil's Arnold Stancell and the Pursuit of Oil," *Chem. Eng. Progress*, 70 (April 1990).

Tearpock, D.J. and R. Bischke: *Applied Subsurface Geological Mapping,* 2nd Edition, Prentice Hall, Inc., Upper Saddle River, NJ, 2002.

Twiss, R.J. and E.M. Moores: *Structural Geology,* W. H. Freeman Company, New York, NY, 1995.

Van Der Pluijm, B.A. and S. Marshak: *Earth Structure: An Introduction to Structural Geology and Tectonics,* The McGraw-Hill Companies, Inc., New York, NY, 1997.

Web Reference

American Association of Petroleum Geologists: http://www.aapg.org/

PETROLEUM REFINING. Because crude oils exhibit important differences from the standpoint of processing and final end-products from one geographic resource to the next (see Tables 1 and 2 of the preceding article on **Petroleum**), very few petroleum refineries operate on exactly the same basis. The principles of the operations performed are remarkably similar, but there are basic differences in the amount of throughputs for given products. The latter varies from one geographic area to the next and with the season of the year. Modern petroleum refineries are designed with considerable built-in flexibility, so that the production of petroleum products can be customized for specific market demands. For example, the production of greater amounts of heating fuels during the fall and winter season as contrasted with larger production of gasolines and diesel fuels during summer months, when the demand is greatest. In contrast with a few decades ago, modern refineries are integrated to produce numerous products and often will also serve nearby petrochemical plants where petroleum products become starting ingredients for a vast variety of chemicals used in many industries, such as plastics, solvents, polymers, fibers, among many others.

A representative integrated petroleum refinery is shown in Fig. 1.

The major processing units fundamental to the manufacture of fuel products from crude oil include: (1) crude distillation; (2) catalytic reforming; (3) catalytic cracking; (4) catalytic hydrocracking; (5) alkylation; (6) thermal cracking; (7) hydrotreating; and (8) gas concentration. Refineries also will use numerous auxiliary processes, such as treating units to purify both liquid and gas streams, waste-management and pollution-control systems, cooling-water systems, units to recover hydrogen sulfide (or elemental sulfur) from gas streams, desalters, electric-power stations, steam-producing facilities, and provisions for storage of crude oil and products.

Petroleum refining and petrochemical production is a 24-hour, 365-day operation with a very minimum of time planned for downtime. Unless one has visited a refinery firsthand, it is very difficult to comprehend the size and complexity of the equipment used. See Figures 2, 3, and 4.

Crude Distillation. To minimize corrosion of refining equipment, a crude-oil distillation unit generally is preceded by a *desalter*, which reduces the inorganic salt content of raw crudes. Salt concentrations vary widely (from nearly zero to several hundred pounds, expressed as NaCl per

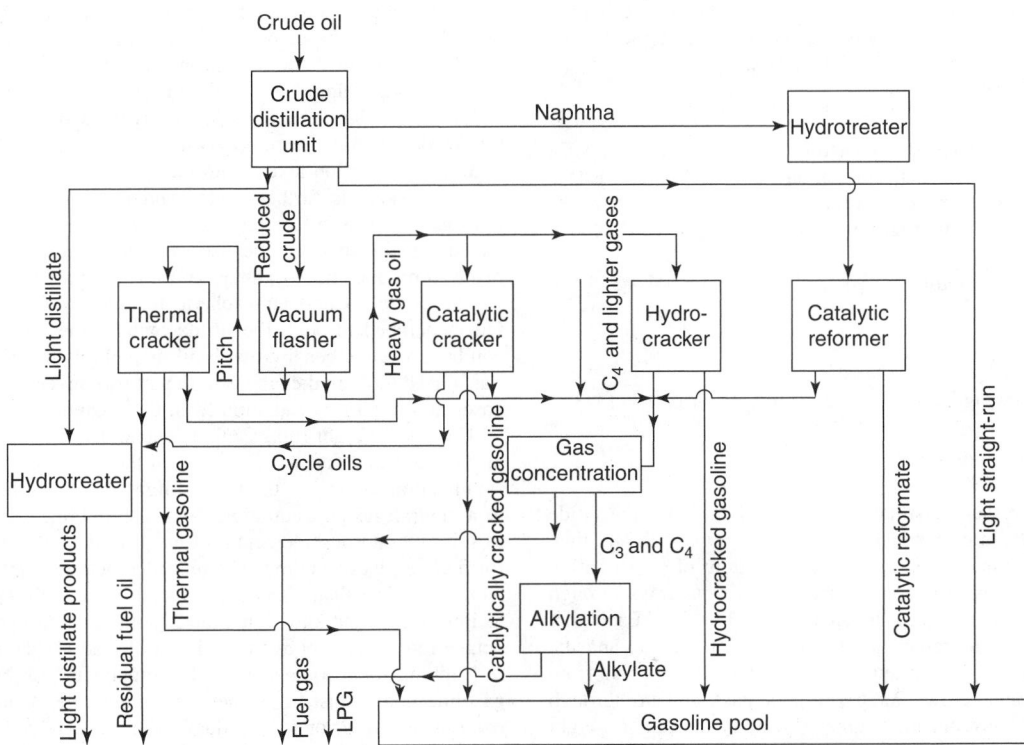

Fig. 1.   Principal flow of materials in an integrated petroleum refinery for producing various fuels and raw materials for petrochemical plants

Fig. 2.   Aerial view of small portion of a Texas petroleum refinery, showing the hydrocracking unit just left of center of view

1,000 barrels). The crude unit functions simply to separate the crude oil physically, by fractional distillation, into components of such boiling ranges that they can be processed by appropriately selected equipment in a long train of processing operations which follow. Although the boiling ranges of components (or fractions) vary between refineries, a typical crude distillation unit will resolve the crude into the following fractions:

By distillation at atmospheric pressure,

1. A light straight-run fraction, consisting primarily of C_5 and C_6 hydrocarbons. These also will contain any C_4 and lighter gaseous hydrocarbons that are dissolved in the crude.
2. A naphtha fraction having a nominal boiling range of $200°-400°F$ $(93°-204°C)$.
3. A light distillate with boiling range of $400°-540°F$ $(204°-343°C)$.

By vacuum flashing

1. Heavy gas oil, having a boiling range of $650°-1,050°F$ $(343°-566°C)$.
2. A nondistillable residual pitch.

In the atmospheric-pressure distillation section of the unit, the crude oil is heated to a temperature at which it is partially vaporized and then introduced near, but at some distance above, the bottom of a distillation column. This cylindrical vessel is equipped with numerous trays through which hydrocarbon vapors can pass in an upward direction. Each tray contains a layer of liquid through which the vapors can bubble, and the liquid can flow continuously by gravity in a downward direction from one tray to the next one below. As the vapors pass upward through the succession of trays, they become lighter (lower in molecular weight and more volatile with lower boiling temperature). The liquid flowing downward becomes progressively heavier (higher in molecular weight and

less volatile with higher boiling temperature). The countercurrent action results in fractional distillation or separation of hydrocarbons based upon their boiling points. A liquid can be withdrawn from any preselected tray as a net product. Thus, the lighter liquids, such as naphtha, exit from trays near the top of the column, whereas heavier liquids, such as diesel oil, exit from trays near the bottom of the column. Thus, the boiling range of the net product liquid depends upon the tray from which it is taken. The vapors containing the C_6 and lighter hydrocarbons are withdrawn from the top, while a liquid stream boiling at about $650°F$ $(343°C)$ is taken from the bottom. The portion taken from the bottom is called *atmospheric residue*.

This residue is further heated and introduced into a vacuum column operated at an absolute pressure of about 50 millimeters of mercury, a vacuum maintained by the use of steam ejectors. A flash separation is made to produce heavy gas oil and nondistillable pitch.

The crude oil and atmospheric residue are heated in tubular heaters. Oil is pumped through the inside of the tubes contained in a refractory combustion chamber fired with oil or fuel gas in such manner that heat is transferred through the tube wall in part by convection from hot combustion gases and in part by radiation from the incandescent refractory surfaces.

The light straight-run gasoline contains all hydrocarbons lighter than C_7 in the crude and consists primarily of the native C_5 and C_6 families. After stabilization to remove the C_4 and lighter hydrocarbons (which are routed to a central gas-concentration unit), the stabilized C_5/C_6 blend is treated to remove odorous mercaptans and passed to the refinery gasoline pool for final product blending. The unleaded octane number (Research Method Number) is less than 70, and thus blending or further processing is required to improve its antiknock qualities. Isomerization can be used to improve octane rating, as well as the addition of lead alkyls.

Naphtha to become a suitable component for blending into finished gasoline pools must be further processed. The octane number will range from 40 to 50. Prior to introduction into a catalytic-reforming unit, most naphtha feedstocks are hydrotreated in the interest of prolonging the life of the reforming catalyst.

Fig. 3. Erection of 9.5-meter (31-foot) diameter, 263-ton vacuum tower at a petroleum refinery in Kuwait. (*The Fluor Corp*)

Fig. 4. Erection of a 675-ton hydrocracking reactor at Shauaiba, Kuwait, refinery, the world's first all-hydrogen refinery. (*The Fluor Corp*)

Gas oil separated from the crude by vacuum distillation, plus portions of light distillates, is the feedstock to catalytic cracking units. The main function of catalytic cracking is to convert into gasoline those fractions having boiling ranges higher than that of gasoline. Remaining uncracked distillates (cycle oils) are used as components for domestic heating fuels (generally after hydrotreating) and to blend with residual fractions to reduce their viscosity to make acceptable heavy fuel oils. In some refineries, cycle oils are hydrocracked to complete their conversion to gasoline.

The aforementioned processes are described in more detail in separate entries: **Alkylation**; **Cracking Process**; **Distillation**; and **Hydrotreating**.

Petroleum Terminology

An abridged glossary of terms used to describe petroleum products and processes would include:

Additives, Diesel Fuel—Chemicals for reducing smoke emissions and for cold weather conditioning.

Additive, Gasoline—In some instances, several functions can be combined in one chemical compound to provide "multifunctional" additives. Increasingly stringent control of automotive emissions in several countries has stimulated the development and use of "extended-range" detergents which are designed to promote peak engine performance by maintaining engine cleanliness. Some gasoline additives include:

Antiknock compounds to increase octane number.

Antioxidants to provide gasoline storage stability.

Antirust agents to prevent corrosion in gasoline-handling systems.

Detergents to control carburetor and induction system cleanliness.

Dyes to indicate kind of antiknock compounds used and to identify brand and grade of gasoline.

Viscosity index improves make lube oil more effective over a wide range of temperatures.

Alkylation—A refinery process for chemically combining isoparaffin with olefin hydrocarbons. The product, *alkylate*, has a high octane value and is blended with motor and aviation gasoline to improve the antiknock value of the fuel.

Base Oil—A refined or untreated oil used in combination with other oils and additives to produce lubricants.

Blending—The process of mixing two or more oils having different properties to obtain a final blend having the desired characteristics. This can be accomplished by off-line batch processes or by in-line operations as part of continuous-flow operations.

Bright Stock—High-viscosity, fully refined and dewaxed lubricating oils produced by the treatment of residual stocks and used to compound motor oils.

Catalytic Cracking—A refinery process that converts a high-boiling range fraction of petroleum (gas oil) to gasoline, olefin feed for alkylation, distillate, fuel oil, and fuel gas by use of a catalyst and heat.

Catalytic Reforming—A catalytic process to improve the antiknock quality of low-grade naphthas and virgin gasolines by the conversion of naphthenes (such as cyclohexane) and paraffins into higher-octane aromatics (such as benzene, toluene, and xylenes). There are approximately ten commercially licensed catalytic reforming processes, including fully regenerative and continuously regenerative designs.

Cetane Number—The cetane number (C.N.) of a fuel is the percentage by volume of normal cetane in a mixture of cetane and alpha-methylnaphthalene which matches the unknown fuel in ignition quality when compared with a standard diesel engine under specified conditions. The C.N. scale ranges from 0 to 100 C.N. for fuels equivalent in ignition quality to alpha-methylnaphthalene and cetane, respectively. For routine-testing, secondary reference fuels having cetane values of about 25 and 74 are blended in any desired proportion.

Clear Octane—The octane number of a gasoline before the addition of antiknock additives.

Cloud Point—The aniline cloud point is a measure of the paraffinicity of a fuel oil, a high value indicating a straight-run paraffinic oil and a low value indicating an aromatic, a naphthenic, or a highly cracked oil.

Coking—Distillation to dryness of a product containing complex hydrocarbons, which break down in structure during distillation, such as tar or crude petroleum. The residue is called coke.

Cracking—A process carried out in a refinery reactor in which the large molecules in the charge stock are broken up into smaller, lower-boiling, stable hydrocarbon molecules, which leave the vessel as overhead (unfinished cracked gasoline, kerosenes, and gas oils). At the same time, certain of the unstable or reactive molecules in the charge stock combine to form tar or coke bottoms. The cracking reaction may be carried out with heat and pressure (thermal cracking) or in the presence of a catalyst (catalytic cracking).

Cycle Stock—Unfinished product taken from a stage of a refinery process and recharged to the process at an earlier period in the operation.

Deasphalting—Process for removing asphalt from petroleum fractions, such as reduced crude. A common deasphalting process introduces liquid propane, in which the nonasphaltic compounds are soluble while the asphalt settles out.

Desulfurization—The removal of sulfur or sulfur-bearing compounds from a hydrocarbon by any one of a number of processes, such as hydrotreating.

Distillate—That portion of a liquid which is removed as a vapor and condensed during a distillation process. As fuel, distillates are generally within the 400° to 650°F (204° to 343°C) boiling range and include Nos. 1 and 2 fuel, diesel, and kerosene.

End Point—The temperature at which the last portion of oil has been vaporized in ASTM or Engler distillation. Also called the final boiling point.

Equilibrium Volatility of a Gasoline—The volatility of a gasoline is determined by the Reid vapor pressure and the ASTM distillation data. The Reid vapor pressure is the vapor pressure of a gasoline at 100°F (37.8°C) under specified conditions. The distillation curve of a fuel indicates the temperatures at which the various amounts of a given sample are distilled under specified test conditions. However, gasoline will completely evaporate in the presence of air at a temperature much lower than the end-point of the distillation curve. According to O.C. Bridgeman (U.S. National Bureau of Standards, Research Paper 694), the volatility of a gasoline is the temperature at which a given air-vapor mixture is formed under equilibrium conditions at a pressure of one atmosphere, when a given percentage is evaporated. According to this definition, one gasoline is more volatile than another for any given percentage evaporated if it forms the given air-vapor mixture at a lower temperature. Distillation temperature curves, for a given text sample, plot amount of sample distilled over (percentage of sample) at the time a given temperature has been reached.

Fire Point—The lowest temperature at which a fuel ignites and burns for at least 5 seconds under specified test conditions.

Flare—A device for disposing of gases by burning.

Flash Point—The lowest temperature at which a flash appears on the fuel surface when a test flame is applied under specified test conditions. This property is an approximate indication of the tendency of the fuel to vaporize.

Flue Gas Expander—A turbine used to recover energy where combustion gases are discharged under pressure to the atmosphere. The pressure reduction drives the impeller of the turbine.

Fractions—Refiner's term for the portions of oils containing a number of hydrocarbon compounds but within certain boiling ranges, separated from other portions in fractional distillation. They are distinguished from pure compounds which have specified boiling temperatures, not a range.

Fuel Oils—Any liquid or liquifiable petroleum product burned for the generation of heat in a furnace or firebox or for the generation of power in an engine. Typical fuels include clean distillate fuel for home heating and higher-viscosity residual fuels for industrial furnaces.

Gas Oil—A fraction derived in refining petroleum with a boiling range between kerosene and lubricating oil.

Heating Oils—A trade term for the group of distillate fuel oils used in heating homes and buildings as distinguished from residual fuel oils used in heating and power installations. Both are burned-fuel oils.

Heavy Ends—The highest-boiling portion of a gasoline or other petroleum oil.

Hydrocracking—The cracking of a distillate or gas oil in the presence of catalyst and hydrogen to form high-octane gasoline blending stock.

Hydrogenation—A refinery process in which hydrogen is added to the molecules of unsaturated (hydrogen-deficient) hydrocarbon fractions. It plays an important part in the manufacture of high-octane blending stocks for aviation gasoline and in the quality improvement of various petroleum products.

Hydrotreating—A treating process for the removal of sulfur and nitrogen from feedstocks by replacement with hydrogen.

Isomerization—A refining process which alters the fundamental arrangement of atoms in the molecule. Used to convert normal butane into isobutane, as alkylation process feedstock, and normal pentane and hexane into isopentane and isohexane, high-octane gasoline components.

Kinematic Viscosity—The absolute viscosity of a liquid (in centipoises) divided by its specific gravity at the temperature at which the viscosity is measured.

Knock—The sound or "ping" associated with the autoignition in the combustion chamber of an automobile engine of a portion of the fuel-air mixture ahead of the advancing flame front.

Lead Susceptibility—The increase in octane number of gasoline imparted by the addition of a specified amount of tetraethyl lead.

Low-Sulfur Crude Oil—Crude oil containing low concentrations of sulfur-bearing compounds. Crude is usually considered to be in the low-sulfur category if it contains less than 0.5% (weight) sulfur. Examples of low-sulfur crudes are offshore Louisiana, Libyan, and Nigerian crudes.

Lube Stock—Refinery term for fraction of crude petroleum suitable in terms of boiling range and viscosity to yield lubricating oils when further processed and treated.

Mercaptans—Compounds of sulfur having a strong, repulsive, garlic-like odor. A contaminant of "sour" crude oil and products.

Octane Number—The octane rating of a motor fuel is defined in terms of its knocking characteristics relative to those of blends of isooctane (2,3,4-trimethylpentane) and n-heptane, and a rating of 100 to isooctane. The octane number of an unknown fuel is numerically equal to the volume percent of isooctane in a blend with *n*-heptane which has the same knocking tendency as the unknown fuel when both the unknown and the reference blend are run in a standard single-cylinder engine operated at specified conditions. Motor Method octane numbers are measured at more severe engine conditions and are numerically lower than those determined by the milder Research Method. The difference between the two numbers is termed *sensitivity*.

Polymer—A product of polymerization of normally gaseous olefin hydrocarbons to form high-octane hydrocarbons in the gasoline boiling range. Polymerization is the process of combining two or more simple molecules of the same type, called monomers, to form a single molecule having the same elements in the same proportions as in the original molecule, but having different molecular weights. The combination of two or more dissimilar molecules is known as copolymerization—and the product is called a *copolymer*.

Pour Point—This property is defined as the lowest temperature at which the fuel will pour and is a function of the composition of the fuel. Normally, the pour point of a fuel should be at least 10 to 15 degrees below the anticipated minimum use temperature.

Presulfide—A step in the catalyst regeneration procedure which treats the catalyst with a sulfur-bearing material such as hydrogen sulfide or carbon bisulfide to convert the metallic constituents of the catalyst to the sulfide form in order to enhance its catalytic activity and stability.

Process Unit—A separate facility within a refinery, consisting of many types of equipment, such as heaters, fractionating columns, heat exchangers, vessels, and pumps, designed to accomplish a particular function within the refinery complex. For example, the crude processing unit is designed to separate the crude into several fractions, while the catalytic reforming unit is designed to convert a specific crude fraction into a usable gasoline blending stock.

Raffinate—In solvent refining, that portion of the oil that remains undissolved and is not removed by the selective solvent.

Refinery Pool—An expression for the mixture obtained if all blending stocks for a given type of product were blended together in production ratio. Usually used in reference to motor gasoline octane rating.

Refluxing—In fractional distillation, the return of part of the condensed vapor to the fractionating column to assist in making a more complete separation of the desired fractions. The material returned is called *reflux*.

Residual Fuel Oils—Topped crude petroleum or viscous residuums obtained in refinery operations. Commercial grades of burner-fuel oils Nos. 5, and 6 are residual oils and include Bunker fuels.

Riser Cracking—Applied to fluid catalytic cracking units where the mixture of feed oil and hot catalyst is continuously fed into one end of a pipe (riser) and discharges at the other end where catalyst separation is accomplished after the discharge from the pipe. There is no dense phase bed through which the oil must pass because all the cracking occurs in the inlet pipe (riser).

Road Octane—A numerical value based upon the relative antiknock performance of an automobile with a test gasoline as compared with specified reference fuels. Road octanes are determined by operating a car over a stretch of level road or on a chassis dynamometer under conditions simulating those encountered on the highway.

SAE Numbers—A classification of motor, transmission, and differential lubricants to indicate viscosities, standardized by the Society of Automotive Engineers. They do not connote quality of the lubricant.

Smoke Point—The smoking tendency of a fuel is indicated by this value, which is the maximum height of a specified type of flame in a given wick lamp that results in no visible smoke.

Solvent Extraction—The process of mixing a petroleum stock with a selected solvent, which preferentially dissolves undesired constituents, separating the resulting two layers, and recovering the solvent from the raffinate (the purified fraction) and from the extract by distillation.

Sour Crude—Crude oil which (1) is corrosive when heated, (2) evolves significant amounts of hydrogen sulfide on distillation, or (3) produces light fractions which require sweetening. Sour crudes usually, but not necessarily, have high sulfur content. Examples are most West Texas and Middle East crudes.

Specific Gravity—The specific gravity of a petroleum fuel is the ratio of the weight of a given volume of the product at 60°F to the weight of an equal volume of distilled water at the same temperature, both weights corrected for air buoyancy. The relation between API gravity scale and specific gravity is: $°API = 141.5/(sp\ gr\ 60/60°F) - 131.5$.

Stability—In petroleum products, the resistance to chemical change. Gum stability in gasoline means resistance to gum formation while in storage. Oxidation stability in lubricating oils and other products means resistance to oxidation to form sludge or gum in use.

Sweet Crude—Crude oil that (1) is not corrosive when heated, (2) does not evolve significant amounts of hydrogen sulfide on distillation, and (3) produces light fractions which do not require sweetening. Examples are offshore Louisiana, Libyan, and Nigerian crudes.

Tricresyl Phosphate (TCP)—Colorless to yellow liquid used as a gasoline and lubricant additive and plasticizer. Formula, $PO(OC_6H_4Ch_3)_3$.

Tetraethyl Lead (TEL)—A volatile lead compound which is added to motor and aviation gasoline to increase the antiknock properties of the fuel. $Pb(C_2H_5)_4$. The use of this compound has diminished in recent years because of pollution regulations.

Thermal Cracking—A refining process which decomposes, rearranges, or combines hydrocarbon molecules by the application of heat without the aid of a catalyst.

Topped Crude—A residual product remaining after the removal, by distillation or other processing means, of an appreciable quantity of the more volatile components of crude petroleum.

Unsaturates—Hydrocarbon compounds of such molecular structure that they readily pick up additional hydrogen atoms. Olefins and diolefins, which occur in cracking, are of this type.

Vacuum Distillation—Distillation under reduced pressure, which reduces the boiling temperature of the material being distilled sufficiently to prevent decomposition or cracking.

Vapor Lock—The displacement of liquid fuel in the feed line and the interruption of normal motor operation, caused by vaporization of light ends in the gasoline. Vaporization occurs when the temperature at some point in the fuel system exceeds the boiling points of the volatile light ends.

Virgin Stock—Oil processed from crude oil which contains no cracked material. Also called straight-run stock.

Visbreaking—Lowering or breaking the viscosity of residuum by cracking at relatively low temperatures.

Viscosity—This is generally expressed in terms of the time required for a given quantity of fuel to flow through a capillary tube under specified conditions. Kinematic viscosity v is viscosity divided by mass density, or $v = \mu\rho$. The unit in cgs units is called the *stoke*; a customary unit is the *centistoke* ($\frac{1}{100}$ of a stoke). The value of the kinematic viscosity in (cm²/seconds) can be obtained from the indications in seconds t of various viscometers by:

$$Saybolt\ Universal,\ when\ 32 < t < 100 = 0.00226t - 1.95/t$$

$$Saybolt\ Universal,\ when\ t > 100 = 0.00220t - 1.35/t$$

$$Saybolt\ Furol,\ when\ 25 < t < 40 = 0.0224t - 1.84/t$$

$$Saybolt\ Furol,\ when\ t > 40 = 0.0215t - 0.50/t$$

Yield—In petroleum refining, the percentage of product or intermediate fractions based on the amount charged to the processing operation.

Zeolitic Catalyst—Since the early 1960s, modern cracking catalysts contain a silica-alumina crystalline structured material called zeolite. This zeolite is commonly called a molecular sieve. The admixture of a molecular sieve in with the base clay matrix imparts desirable cracking selectivities.

Crude Oil Pipelines

Transportation of crude to petroleum refineries is essentially a part of the refining operation. Petroleum movement can be measured in barrels, tons

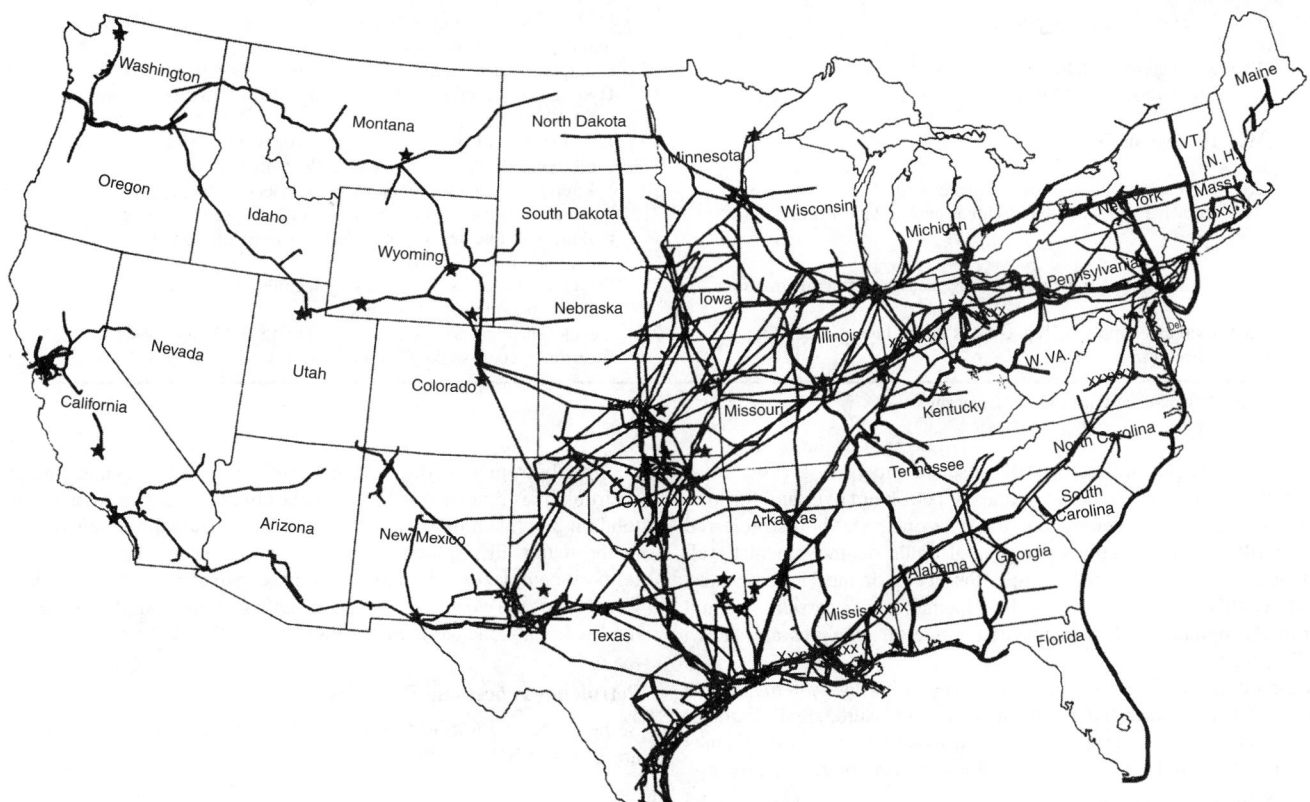

Fig. 5. Major interstate pipelines and waterways in the United States. Pipelines in Alaska not shown. Asterisks indicate principal refineries. (*American Petroleum Institute*)

TABLE 1. CHRONOLOGY OF MAJOR PETROLEUM INDUSTRY ADVANCEMENTS

Challenge	Action
1. Better gasoline quality and greater quantity needed	1. Thermal cracking processes (about 1910)
2. Need to improve odor and stability of gasoline and kerosene	2. Refining with chemical solutions and synthesis and use of oxidation inhibitors; started in late 1920s
3. Better gasoline quality and greater quality needed	3a. 19. Development of processes of dehydrogenate n-paraffins to n-olefins and alkylate benzene with them (mid-1960s) Discovery of tetraethyl lead (1921)
	b. Polymerization of light olefins to make "poly" gasoline by catalysis (mid-1930s)
	c. Catalytic cracking invented and improved (late 1930s)
4. Combat grade aviation gasoline testing above 100 octane needed for World War II	4. Alkylation of light olefins with light isoparaffins by catalysis; discovered in 1932; commercialized in early 1940s
5. More aromatic hydrocarbons needed, especially toulene for TNT; benzene, toluene, and other high-octane aromatics needed for combat grade aviation gasoline and for chemical synthesis	5a. Catalytic reforming to make toluene from petroleum naphthas (early 1940s), using non-noble metal catalyst
	b. Extractive distillation of toluene from *reformate* with phenol and other materials (early 1940s)
	c. Extraction with SO_2, suggested in 1907 to purify *kerosene*, applied to secure aromatics from reformate (early 1940s)
	d. Alkylation of propylene with benzene *using* solid H_3PO_4 catalyst to make cumene (early to mid-1940s)
6. Butadiene needed for synthetic rubber in wartime	6. *Thermal* and catalytic processing applied to petroleum distillates, "quickly" butadiene program (early 1940s)
7. More isobutane for alkylation in wartime aviation-gasoline program	7. *Isomerization* of n-butane (early 1940s)
8. Improve quality of straight-run gasoline	8. Catalytic reforming using noble-metal *catalyst* (1949)
9. Remove catalyst poisons and sulfur compounds from gasoline and naphtha	9. Catalytic hydrotreating (early 1950s)
10. Increase supplies of pure aromatic hydrocarbons and aromatic concentrates	10. Liquid-liquid solvent extraction *processes* using aqueous glycols and improved contacting means (1952)
11. Purify kerosenes and light and heavy distillates	11. *Modified* catalytic hydrotreating (mid-1950s)
12. Improve quality of light hydrocarbons used in gasoline	12. New catalytic isomerization processes using noble-metal catalysts, converting C_4, C_5, and C_6 n-paraffins to isoparaffins (mid-1950s)
13. Increase production of light fuels and gasoline; reduce production of heavy fuels	13. Development of catalytic hydrocracking processes having great flexibility (1959–1960)
14. Ethylbenzene for styrene manufacture	14. Catalytic alkylation process developed, uniting benzene directly with dilute ethylene in refinery gases (1958)
15. Separation of normal paraffins from mixtures with isoparaffins	15. Molecular sieves used as solid adsorbants (1959, but not commercialized until late 1960s)
16. Increase benzene supply and decrease toluene	16. Hydrodealkylation of toluene; produce naphthalene from alkyl naphthalenes (early 1960s)
17. Synthesize cyclohexane for nylon	17. Catalytic hydrogeneration of benzene (early 1960s)
18. Improve quality of heavy fuel oils	18. Hydrodesulfurization of heavy fuels, also by hydrocracking (mid-1960s)
19. Biodegradable synthetic detergents	19. Development of processes of dehydrogenate n-paraffins to n-olefins and alkylate benzene with them (mid-1960s)
20. Increase production of p-xylene	20. Isomerization of C_8 aromatics to p-xylene (late 1950s)
21. Improve supplies of individual pure xylene isomers	21. Adsorptive separation of p-xylene in high yield and purity, making possible separation of other isomers by precise fractionation (early 1970s)
22. Utilization of metals-containing heavy petroleum fractions	22. Development of hydroprocessing techniques to effectively convert to synthetic crude oils
23. Increase supply of liquid fuels in the face of declining petroleum reserves	23. Development of processes for producing liquid and gaseous fuels from coal, shale oil, and tar sands

Source: Universal Oil Products Company.

moved, barrel-miles, ton-miles, etc. Because oil pipelines are the most economical means of moving large volumes of petroleum overland for long distances, ton-miles (movement of one ton over one mile) is the preferred unit. About 740 billion ton-miles per year of crude oil movement by all transportation modes is required to meet the needs of the United States. About half of this quantity is moved by pipeline; water carriers account for most of the remainder. Major interstate crude oil pipelines are shown in map (Fig. 5).

As domestic crude oil is produced, pipeline gathering systems collect and move it to central locations by means of low-pressure, small-diameter pipelines. Usually, these gathering lines feed into pipeline working tanks where the oil is held until it is ready for shipment by a crude oil *trunk line*. Gathering systems include pumping stations, meters, and samplers. About 90% of the gathering lines in the United States are 6 inches (diameter) or smaller. About 34% of all oil pipeline mileage in the United States consists of gathering lines. Crude oil trunk lines, larger in diameter, receive crude

oil directly from gathering systems and also from barges and tankers. Crude oil trunk lines range from 8 to 56 inches in diameter and account for 30% of total oil pipeline mileage in the nation. These trunk lines generally originate in the major oil producing areas of the nation and terminate at the main refinery complexes. The largest of these refinery complexes is in the Gulf Coast area of Texas and Louisiana. Other major complexes are located in the St. Louis-Chicago area, northern Ohio, the East and West Coasts.

Petroleum Processing Progress

Some of the major milestones achieved in petroleum processing technology are summarized in Table 1.

Additional Reading

Abraham, O.C. and F.G. Prescott: "Make Isobutylene from Tertiary Butyl Alcohol," *Hydrocarbon Processing*, 51 (February 1992).

Ansari, R.M. and M.O. Tade: *Nonlinear Model-Based Process Control: Applications in Petroleum Refining,* Springer-Verlag, Inc., New York, NY, 2000.

Chang, E.J. and S.M. Leiby: "Ethers Help Gasoline Quality," *Hydrocarbon Processing,* 41 (February 1992).

Chaput, G., et al.: "Pretreat Alkylation Feed," *Hydrocarbon Processing,* 51 (September 1992).

Dawe, R.A.: *Modern Petroleum Technology: 2 Volume Set,* 6th Edition, John Wiley & Sons, Inc., New York, NY, 2000.

Desai, P.H., et al.: "Enhance Gasoline Yield and Quality," *Hydrocarbon Processing,* 51 (November 1992).

Devlin, J.F., L.A. Edwards, and J.E. Crosby: "Fluid Coker Benefits from Advanced Control," *Hydrocarbon Processing,* 55 (June 1992).

Elliott, J.D.: "Maximize Distillate Liquid Products," *Hydrocarbon Processing,* 75 (January 1992).

Gary, J.H. and G.E. Handwerk: *Petroleum Refining: Technology and Economics,* 4th Edition, Marcel Dekker, Inc., New York, NY, 2001.

Johansen, T., K.S. Raghuraman, and L.A. Hackett: "Trends in Hydrogen Plant Management," *Hydrocarbon Processing,* 119 (August 1992).

Jones, D.S.: *Elements of Petroleum Processing,* John Wiley & Sons, Inc., New York, NY, 1995.

Leffler, W.L.: *Petroleum Refining in Nontechnical Language,* 3rd Edition, PennWell Corporation, Tulsa, OK, 2003.

Magee, J.S. and G.E. Dolbear: *Petroleum Catalysis in Nontechnical Language,* PennWell Publishing Company, Tulsa, OK, 1998.

McKetta, J.J.: *Petroleum Processing Handbook,* Marcel Dekker, Inc., New York, NY, 1992.

Meyers, R.A.: *Handbook of Petroleum Refining Processes,* 3rd Edition, The McGraw-Hill Companies, Inc., New York, NY, 2003.

Monfils, J.L., et al.: "Upgrade Isobutane to Isobutylene," *Hydrocarbon Processing,* 47 (February 1992).

Nierlich, F.: "Oligomerize for Better Gasoline," *Hydrocarbon Processing,* 45 (February 1992).

Parkash, S., Ph. D: *Refining Processes Handbook,* Butterworth-Heinemann, Inc., Woburn, MA, 2003.

Reynolds, B.E., E.C. Brown, and A. Silverman: "Clean Gasoline Via Vacuum Residuum Hydrotreating and Residuum Fluid Catalytic Cracking," *Hydrocarbon Processing,* 43 (April 1992).

Rippee, B. and R.L. Busby: *2003 International Petroleum Encyclopedia,* PennWell Corporation, Tulsa, OK, 2003.

Speight, J.G.: *Petroleum Chemistry and Refining,* Taylor & Francis, Inc., Philadelphia, PA, 1997.

Speight, J.G. and B. Ozum: *Petroleum Refining Processes,* Vol. 85, Marcel Dekker, Inc., New York, NY, 2001.

Speight, J.G.: *The Chemistry and Technology of Petroleum,* 3rd Edition, Marcel Dekker, Inc., New York, NY, 1999.

Staff: "Refining Handbook, '92," *Hydrocarbon Processing,* 133 (November 1992).

Staff: "What's Ahead in 1993?" *Hydrocarbon Processing,* 33 (December 1992).

Wagner, E.S. and G.F. Froment: "Steam Reforming Analyzed," *Hydrocarbon Processing,* 69 (July 1992).

Wauquier, J.P.: *Petroleum Refining: Crude Oil, Petroleum Products, Process Flowsheets,* Gulf Publishing Company, Houston, TX, 2000.

Wheatcroft, G.: "Present and Future Refinery Information Systems," *Hydrocarbon Processing,* 101 (May 1992).

Web Reference

American Petroleum Institute: http://api-ec.api.org/intro/index_noflash.htm

pH (Abrasion). The term *abrasion pH* was originated by Sevens and Carron in 1948 to designate the pH values obtained by grinding materials in water as a useful aid in the field identification of minerals. The pH value ranges from 1 for ferric sulfate minerals, such as coquimbite, konelite, and rhomboclase, to 12 for calcium-sodium carbonates, such as gaylussite, pirssonite, and shortite.

The recommended technique for determination of abrasion pH is to grind, in a nonreactive mortar, a small amount of the mineral in a few drops of water for about one minute. Usually, a pH test paper is used. Values obtained in this manner are given in the left-hand column of the accompanying table. Another method, proposed by Keller et al. in 1963 involves the grinding of 10 grams of crushed mineral in 100 milliliters of water and noting the pH of the resulting slurry electronically.

pH (Hydrogen Ion Concentration). A measure of the effective acidity or alkalinity of a solution. It is expressed as the negative logarithm of the hydrogen-ion concentration. Pure water has a hydrogen ion concentration equal to 10^{-7} moles per liter at standard conditions. The negative logarithm of this quantity is 7. Thus, pure water has a pH value of 7. The pH scale usually is considered as extending from 0 to 14. When a strong acid fully dissociates (or ionizes) in water, a 1 N solution of this acid will have a pH value of 0.0. Conversely, a 1 N base fully ionized in water will have a pH value of 14. Both hydrochloric acid and sodium hydroxide come close to meeting these stipulations. Because of the logarithmic nature of the pH scale, there is a tenfold change in hydrogen- and hydroxyl-ion concentration per unit change of pH. Thus, a slightly acidic solution having a pH of 6 will contain ten times as many active hydrogen ions as a solution of pH 7. See also **pK**.

Effective acidity or alkalinity is stressed in pH measurement—not the total hydrogen present. Sulfuric acid and boric acid both contain significant amounts of hydrogen. Nearly all the hydrogen in sulfuric acid dissociates in the presence of sufficient water to become free hydrogen ions. On the other hand, when boric acid is added to water, it dissociates very little into free hydrogen ions. The pH of a 0.1 N sulfuric acid solution will be about 1.3 whereas for the same concentration, boric acid will have a pH of about 5.3. Thus, sulfuric acid is called a strong acid; boric acid a weak acid. In all materials, of course, dissociation increases with temperature, thus the same solution will have a somewhat different pH at a lower temperature than at a higher temperature. Pure water is neutral at a temperature of 25°C, having a concentration of 1×10^{-7} hydrogen ions and 1×10^{-7} hydroxyl ions and, consequently, a pH of 7. Dissociation is less at 0°C, at which temperature the hydrogen-ion concentration is 0.34×10^{-7}, or a pH of 7.47 (slightly basic rather than neutral). But, at a temperature of 100°C, dissociation is greater. The hydrogen ion concentration is 8×10^{-7} and the pH is 6.10 (or slightly acid). The pH of various substances is given in Table 1.

Buffer solutions can be added to resist changes in pH despite the addition of acid or base to the solution. This is explained under **Buffer (Chemical).**

pH is measured in two basic ways: (1) colorimetrically, usually where high accuracy is not required and manual methods suffice; and (2) electrometrically. Color changes are based upon various organic dyes which alter their color within a relatively narrow range of pH values. Numerous dyes are required to cover the full pH range. Electrometric methods are used both in the laboratory and on-line for process control. They are continuous and easily adapted to automatic control systems. The possibility that a thin glass membrane of special composition could develop a potential in relation to hydrogen ion concentration was described as early as 1909 by the German chemist, Fritz Haber. Little progress was made until the middle-1920s. Glass electrodes are now the standard approach to electrometric pH measurement, after periods of trial with quinhydrone and antimony electrodes. The glass electrode responds in

TABLE 1. pH VALUES OF VARIOUS SUBSTANCES (AT 25°C)

Material	pH
Seawater	7.75 to 8.25
Soils	3 to 10
Plant tissues and fluids	About 5.2
Animal tissues and fluids	About 7.0 to 7.5
Blood	7.35–7.5
Urine	5.0–7.0
Milk	6.5–7.0
Gastric juice	1.7
Pancreatic juice	7.8
Intestinal juice	7.7
Internal tissue fluids:	
Minimum, below which acidosis ensues	7.0
Maximum, above which tetany ensues	7.8
Hydrochloric acid ($1N$)	0.1
Hydrochloric acid ($0.1N$)	1.08
Hydrochloric acid ($0.001N$)	3.00
Sulfuric acid ($1.0N$)	0.32
Sulfuric acid ($0.1N$)	1.17
Acetic acid ($1N$)	2.37
Lemon juice	2.0–2.2
Acid fruits	3.0–4.5
Fruit jellies	3.0–3.5
Sodium hydroxide ($1N$)	13.73
Sodium hydroxide ($0.1N$)	12.84
Ammonia (10% NH_3)	11.8
Limewater, $Ca(OH)_2$ saturated	12.4
Trisodium phosphate, 2%	11.95

a predictable fashion throughout the 0 to 14 pH range, developing 59.2 millivolts per pH unit at 25°C, values which are consistent with the classical Nernst equation. Contrary to earlier pH electrodes, the glass electrode is not influenced by oxidants or reductants in solution. With suitable temperature compensation, pH measurements can be made up to 100°C and higher. In pH measurement, a second or reference electrode is required to complete the circuit. After trials with numerous electrodes (the hydrogen electrode is the standard) for practical plant and laboratory applications, the mercury-mercurous chloride (calomel) electrode is widely used. There is also some use of the silver-silver chloride reference electrode.

pH control systems are widely used in waste control and neutralization systems, in pulp and paper manufacture, in food processing, and in the manufacture of numerous organic chemicals. pH measurement is very important in the medical field.

PHARMACEUTICALS. Pharmaceuticals are best viewed as drug-containing products in dosage forms. These forms are designed and manufactured to deliver safe and effective therapeutic responses each time administered within appropriate regimens and even after storage under well-documented conditions in scientifically designed packaging for designated time periods. Thus, pharmaceuticals are actual drug delivery systems.

Various technologies are required to produce drug products. Both federal and state laws and regulations exist in the United States to control the manufacture and distribution of pharmaceuticals. The U.S. drug distribution system is multifaceted including drug usage within the community and hospitals, under long- or short-term home health care or pharmacy practice. Individual pharmaceuticals are covered elsewhere.

In the United States, there is no national qualifying or licensing body for pharmacists. Licensure requirements are promulgated by State boards of pharmacy that administer examinations, issue internship requirements, and oversee the practice of pharmacy. The National Association of Boards of Pharmacy serves the collective needs of the state boards. This organization has no licensure authority. However, it has developed a standardized licensure examination (NABPLEX), which as of this writing is used by 48 states (see **Licensing**).

Several national organizations serve the professional needs of U.S. pharmacists. The American Pharmaceutical Association (APhA), founded in 1852, is composed of the Academy of Pharmaceutical Research and Science, Academy of Pharmaceutical Practice and Management, and the Academy of Students of Pharmacy. Other organizations include the American Society of Health-Systems Pharmacists (ASHP), National Association of Chain Drug Stores (NACDS), and National Association of Retail Druggists (NARD).

The American College of Apothecaries represents pharmacists whose practices can best be described as emphasizing prescription and related products.

The pharmaceutical industry is represented by several organizations. Examples are the Pharmaceutical Research and Manufacturers of America, the Non-Prescription Drug Manufacturers Association, and the National Pharmaceutical Council. The schools and colleges of pharmacy are organized as the American Association of Colleges of Pharmacy, representing both schools and colleges, and faculty members.

Each state has a professional pharmacy organization, some of which are affiliated with the American Pharmaceutical Association. Similarly, state organizations of hospital pharmacists exist in affiliation with the ASHP. Likewise, local or county associations exist in most instances. Each national association publishes a journal as do most state organizations. The *Federal Register* reports proposed and enacted federal regulatory occurrences several times a week. Each state has a similar publication to report its legislation and regulatory developments, e.g., *The Pennsylvania Bulletin*.

Drugs and Drug Products

The U.S. Food and Drug Administration (FDA) approved 22 new drugs and one biotech medicine during 1994. These new drug entities had an adjusted average review time of 19.7 months, from filing of the New Drug Application (NDA) at the FDA to time of approval. This was down from the 25.6 months for the 26 new entities approval in 1993. In the total drug development and approval process it takes approximately 12 years for an experimental drug to go from the lab to the medicine chest. Only about 5 in 5000 new chemical entities that enter preclinical (lab and animal studies) testing reach human clinical testing (Phase I, II, and III) and only one of the five tested clinically is approved. On average a pharmaceutical manufacturer invests ~$360 million to get one new drug to the consumer or patient.

In the United States, through the NDA review process, pharmaceutical companies that seek FDA approval for new drug products are assessed user fees by FDA to gain faster approval, by virtue of the U.S. Prescription Drug User Fee Act of 1992. In 1962, amendments to the U.S. Federal Food, Drug and Cosmetic Act promulgated regulations concerning the requirements for premarketing approval by the FDA. This legislation established requirements of proof of both safety and therapeutic efficacy and strict control of human clinical testing, for example, which have extended the time and cost to market a new drug.

The increase in time to bring a new drug to the point of FDA approval, that the 1962 amendments generated, reduced the length of the effectiveness of the patent period. During the same period, the 1960s, availability of generic drug products began to increase significantly. The FDA at that time utilized the Abbreviated New Drug Approval (ANDA) process developed in the 1962 amendments for review and approval of generic drug products that were to be marketed after FDA approval and patent expiration of the originator new drug entity. Some legal questions, however, arose as to the use of the ANDA procedure for generic approval for both pre-1992 and post-1962 new drug approvals.

The world trade agreement, the General Agreement on Tariffs and Trade (GATT), resulted in a U.S. federal law, the Uruguay Round Agreement Act (URAA), that became effective in June 1995. Under this Act, numerous drugs are projected to gain months or even years of additional patent protection, depending on current patent expiration dates. The GATT provides new prescription drugs with 20 years of patent protection from the patent application date.

The 1962 Amendments also mandated a review of safety and therapeutic efficacy for U.S. nonprescription, i.e., over-the-counter or proprietary, products. There are an estimated 125,000–300,000 U.S. OTC products covering a variety of sizes, dosage form types, and dosage form strengths. The FDA has increased its approval rate for the switch of prescription drugs to nonprescription status in the 1990s. This procedure has gained impetus as more than 450 OTC products in 1994–1995 used ingredients and dosages only available by prescription in 1974–1975.

The principal OTC pharmaceutical products include cold remedies, vitamins and mineral preparations, antacids, analgesics, topical antibiotics, antifungals and antiseptics, and laxatives. Others include suntan products, ophthalmic solutions, hemorrhoidal products, sleep aids, and dermatological products for treatment of acne, dandruff, insect parasites, burns, dry skin, warts, and foot care products. More recent prescription-to-OTC switches have included hydrocortisone, antihistamine and decongestant products, antifungal agents, and, as of 1995, several histamine H_2-receptor antagonists.

Personnel. A large number of personnel trained in a wide range of special skills are needed for the development of a new drug. Skills include organic synthesis, medical and analytical chemistry, microbiology and immunology, biochemistry, physiology, pharmacology, toxicology, and pathology. Likewise, in the development of safe, stable, and therapeutically effective drug products various physical chemistry principles apply and specialists trained in this phase of development, pharmaceutics, assume such responsibility. These people become involved in the preformulation studies that investigate the properties of the new drug for inclusion in dosage forms, in the scale-up procedures that are needed to transfer dosage form preparation from laboratory batch sizes to manufacture batch sizes, and in the actual manufacture of the product. These specialists work closely with chemical engineers, especially during the scale-up phase.

Concepts and Processes. Contemporary dosage forms are drug delivery systems, designed and manufactured to achieve safe and effective therapeutic responses each time the forms are used as part of an appropriate regimen. Each drug product involves several interrelated concepts that must be considered in its design and manufacture. Examples include the following:

Component/concept	Requirement
drug (active ingredient)	purity, stability, accuracy in measurement
nontherapeutic ingredients (excipients)	needed for safe and effective delivery of the active ingredient
unit process/manufacturing technology	procedures needed to ensure batch-to-batch, dose-to-dose reliability of safe and effective response
packaging/labeling	designed for patient compliance and product stability
quality assurance procedures	to protect the drug product throughout its projected shelf-life
storage	to ensure stability and safety/efficacy

Attention to various physiochemical parameters of the drug moiety, such as particle size, crystalline form, and solubility, is vital to the design of a dosage form, as are its purity and accurate measurement. Nontherapeutic or excipient ingredients are selected to ensure stability (buffers, chelating agents, antioxidants, antimicrobial preservatives), and accuracy and precision of dosage (diluents, vehicles). Various types of excipients are used for specific types of dosage forms in order to permit their manufacture and desired therapeutic performances. Other excipients function as processing aids. Lubricating agents are solids used in tablet compression to lubricate the diewalls and punch faces to prevent sticking, capping, and/or excessive die-wall wear. Polymers find wide excipient use in dosage form design as viscosity-building agents in suspensions and emulsions and in the control of drug release in products prepared to achieve longer (8–12 h) than usual therapeutic periods. Various excipients are used to provide drug palatability for patients, e.g., colorants and flavoring agents.

The selection of excipient ingredients is important. These must be both chemically and physically compatible with the drug moiety and cannot negatively affect product stability or therapeutic performance, i.e., bioavailability.

The various preparation processes and technologies used in drug product manufacture also can effect product safety, stability, and performance, e.g., compression during tablet manufacture. The principal processes used in dosage form manufacture are as follows.

Dosage form types	Processes
liquid solutions	dissolution and filtration
parenterals	sterilization, lyophilization
liquid dispersion (suspensions, emulsions)	dispersion/wetting of solids, homogenization
semisolid dispersions (ointments, creams)	levigation, melting
liquid/semisolid capsules	soft gelatin encapsulation
suppositories	molding
solids (granules, capsules, tablets)	comminution, blending, granulation, compression, coating
aerosols	specialized packaging under pressure
general	heating, cooling, mixing

The therapeutically active drug can be extracted from plant or animal tissue, or be a product of fermentation, as in the case of antibiotics.

Biological characterization includes toxicological studies, dose relationships, routes of administration, identification of side effects, and absorption, distribution, metabolism, and excretion patterns. If the results are still acceptable, product formulation and dosage form are developed.

Application for discovery and product patents must be made early in the process. Appropriate labels are designed and the product is submitted to the FDA for approval to begin human testing in the form of an Investigational New Drug Application (INDA). When such approval is granted, a clinical evaluation is developed which includes general testing for human pharmacology in healthy volunteers; clinical studies for therapeutic safety and efficacy in volunteer patients who are suffering from the disease for which the drug has therapeutic promise; and drug samples are made available to select clinicians for use on large numbers of patients.

Manufacturing, analytical, and quality control procedures are established. Specifications for raw and in-process materials, as well as for final products per USP/NF and in-house standards are also determined. Process and formula validation assures that each technological procedure in manufacture accomplishes its purpose most efficiently, e.g., blending times for powdered mixtures in tableting, and that each formula ingredient is present in optimal concentrations. Thus, it serves to ensure process control, reproducibility, and content uniformity.

Stability studies are developed to assure a desirable shelf-life period. These also establish limits of acceptability for impurities and degradation compounds, when present, and determine acceptable storage conditions for raw materials and the manufactured products. Stability studies are thus important to the determination of expiration dates for drug products.

Finally, all data, including the results of the clinical investigation, are collected in a New Drug Application (NDA) and sent to the FDA. Once approved, the new drug goes into production. After manufacturing begins, the new drug products must be monitored in clinical use in the marketplace for reports of untoward reactions. This amounts to post-approval surveillance known as Phase IV. All such reports must be submitted to the FDA in a timely manner.

Bioavailability, Bioequivalence, and Pharmacokinetics. Bioavailability can be defined as the amount and rate of absorption of a drug into the body from an administered drug product. It is affected by the excipient ingredients in the product, the manufacturing technologies employed, and physical and chemical properties of the drug itself, e.g., particle size and polymorphic form. Two drug products of the same type, e.g., compressed tablets, that contain the same amount of the same drug are pharmaceutical equivalents, but may have different degrees of bioavailability. These are chemical equivalents but are not necessarily bioequivalents. For two pharmaceutically equivalent drug products to be bioequivalent, they must achieve the same plasma concentration in the same amount of time, i.e., have equivalent bioavailabilities.

Bioavailability, important to the design and preparation of drug products, can be affected adversely by the selection of excipients and/or the manufacturing processes used. Excessive pressure used in the compression of tablets, for example, could cause a tablet to pass through the gastrointestinal tract with no therapeutic effect.

Pharmacokinetics is the study of how the body affects an administered drug. It measures the kinetic relationships between the absorption, distribution, metabolism, and excretion of a drug. To be a safe and effective drug product, the drug must reach the desired site of therapeutic activity and exist there for the desired time period in the concentration needed to achieve the desired effect. Too little of the drug at such sites yields no positive effect (<MEC); too much (>MTC) leads to toxicity. For intravenous administration there is no absorption factor. Total body elimination includes both metabolic processing and excretion.

In cases of all but intravenous administration, dosage forms must make the active moiety available for absorption, i.e., for drug release. This influences the bioavailability and the drug's pharmacokinetic profile. Ideally the drug is made available to the blood for distribution and elimination at a rate equal to those processes. Through technological developments drug product design can achieve release, absorption, and elimination rates resulting in durations of activity of 8–12 hours, i.e., prolonged action/controlled release drug products.

Manufacturing

Compressed Tablets. This popular type of dosage form offers convenience, stability, accuracy and precision, and good bioavailability of active ingredients. After the best formulation has been established, compressed tablets can be manufactured at high rates of speed on advanced equipment. Tablets can be made to achieve rapid drug release or to produce delayed, repeat, or prolonged therapeutic action. Tablets are produced directly by compression of powder blends or granulations, which include a small percentage of fine, particle-sized powders.

Granulation. Granulation methods can be wet or dry. Wet granulation cannot be used for drugs that are sensitive to moisture and heat. The powered drug and diluent are blended with a dispersion of the binder excipient, e.g., gelatin, to a consistency that can be screened to 840–1800-μm granules (10–20 mesh). These granules are dried on trays in hot-air ovens or fluid-bed dryers. Dry granulation is used when the drug is not stable under the conditions of wet granulation and when the combined powders of a formulation cannot be compressed directly.

Direct Compression. This process is relatively simple and time saving. All the ingredients are blended and then compressed into the final tablet. This is an excellent method, but encumbered by a number of problems. Not all substances can be compressed directly, necessitating a granulation step. Likewise, the flow properties of many blends of fine, particle-sized powders are not such as to ensure even filling of the die cavities of tablet presses. In addition, air entrapment can occur.

The availability of spray-dried lactose, microcrystalline cellulose, and other excipients allows for the use of granular rather than powdered phases. This eliminates some of the problems of particle segregation according to size (demixing) and even flow to the die. Direct compression eventually may be the preferred method of tablet preparation.

Tablet Press. The main components of a tablet compression machine (press) are the dies, which hold a measured volume of material to be compressed (granulation), the upper punches which exert pressure on the down stroke, and the lower punches which move upward after compaction to eject the tablets from the dies. Mechanical components deliver the necessary pressure. The granulation is fed from a hopper with a feed-frame on rotary-type presses and a feeding shoe on single-punch presses. A smooth and even flow ensures good weight and compression uniformity. Using the proper formulation, demixing in the hopper is minimized.

Compressed tablets that are composed of several layers require specially adapted presses designed with several fed hoppers. For a two-layer tablet, one granulation is first fed to a die and partially compressed into a soft tablet. The second granulation is added, and the total die components then are compressed fully. Such procedures are used when the tablet ingredients may be incompatible, which requires separate granulations. If needed, a layer of inert ingredient, e.g., lactose, is inserted between the two.

Layered tablets are also used for a prolonged or sustained therapeutic effect. In this case, one layer disintegrates and dissolves rapidly to provide the initial dosing, whereas the other is designed for controlled release.

Formulation. Compressed tablet formulations contain several types of inert, adjuvant ingredients necessary for proper preparation and therapeutic performance. Tablets designed to be swallowed need diluent, disintegrating, binding (adhesive), and lubricating inert ingredients, whereas troches or lozenges intended to be dissolved slowly in the mouth should not disintegrate quickly, need more binder, and no disintegrant. Lactose or dicalcium phosphate are common diluents, whereas starch and cellulose derivatives are used as disintegrating agents.

Glidants are needed to facilitate the flow of granulation from the hopper. Lubricants ensure the release of the compressed mass from the punch surfaces and the release/ejection of the tablet from the die. Combinations of silicas, corn starch, talc, magnesium stearate, and high molecular weight poly(ethylene glycols) are used. Most lubricants are hydrophobic and may slow down disintegration and drug dissolution.

Colors and flavors increase the elegance and acceptability of the product. Sometimes colors are used for identification.

Effervescent tablets disintegrate by virtue of the chemical reaction occurring in water between component ingredients, such as sodium bicarbonate and citric or tartaric acid, to achieve release of carbon dioxide.

Coating. Sugar or film coatings offer protection from moisture, oxygen, or light and mask unpleasant taste or appearance. Enteric coatings delay the release of active ingredients in the stomach and may prolong the onset of therapeutic activity. The latter are used for drugs that are unstable to gastric pH or enzymes, cause nausea and vomiting or irritation to the stomach, or should be present in high concentration in the intestines, e.g., preoperative sterilization of the gut or as anthelmintics. Effectiveness depends on the varying pH patterns of the gastrointestinal tract and the enzymes present for dissolution and aqueous solubility.

Enteric coating is also used for repeat-action tablets, which contain an enteric-coated core tablet and a sugar or film-coated second dose, permitting the administration of two doses simultaneously. The core dose is released several hours after the initial, outer dose.

Some tablets that provide a sustained period (up to 8–12 h) of therapy may be coated during processing. A portion is released first to bring the drug to the desired blood concentration (onset of activity), whereas a sustained-release portion maintains an effective level for a prolonged

period of time (duration of activity), e.g., by coating erosion or diffusion of drug through it.

A more recent development in tablet coating involves the use of gelation as the coating material to produce geltabs. If a tablet is compressed as a capsule-shaped unit prior to gelatin coating it is called a gelcap.

Capsules. Capsules are made in two types. In hard-gelatin capsules, powders or granules are enclosed in rigid gelatin shells. Soft-gelatin capsules contain glycerol as well as gelatin and maintain plastically even when dried. Hard-gelatin capsules are made in two sections, cap and body, which are then filled, whereas soft-gelatin capsules are formed and filled in succession in one manufacturing procedure. Soft-gelatin capsules are generally filled using nonaqueous solutions, although powders can also be used. Most drug companies buy the hard-gelatin shells from external sources. These are made by dipping precisely tooled pins into controlled solutions of gelatin. A film of gelatin adheres to the pins. Upon drying, the units are trimmed to specified length, removed from the pins, and the cap and body portions are joined. Various colors can be incorporated. See also **Gelatin**.

The formulations of filled, hard-gelatin capsules are generally less complex than those of compressed tablets, and require no binders or disintegrators. Upon swallowing, the capsule shell dissolves quickly and the powder ingredients are available for dissolution. Because no initial disintegration step is needed, bioavailability of drugs in capsule formulations is generally better than that of compressed tablets. The capsules are filled by various high speed machines. Occasionally the pharmacist has to perform this procedure manually.

Prolonged Action/Controlled Release, Orally Administered Solid Dosage Forms. The therapeutic purpose of prolonged action and controlled release solid, oral drug products is to maintain safe and effective concentrations of the drug in the blood for 2–4 times longer than those times achieved using regular compressed tablets or capsules. This is accomplished by releasing one portion of the drug quickly, whereas the remaining portion is released at a rate that approaches the elimination rate. Ideally, the second portion should be released at a zero-order rate to achieve this profile. The technologies used for such controlled release only approach such a rate, but do accomplish the increased therapeutic period. These oral products mainly use diffusion-controlled or dissolution-controlled release profiles. The more recognized technologies used to achieve these methods include ion-exchange resins, coated micropellets, barrier coatings, drug embedment in either slowly eroding or plastic matrices, swelling hydrogels of various polymer resins, drug complexation, and osmotic pressure controlled tablets. Other technologies that have been attempted or tested include altered density micropellets, prodrugs, and bioadhesives. See also **Barrier Polymers**.

The best drug candidates for incorporation into prolonged action systems are uniformly absorbed throughout the gastrointestinal (GI) tract, have medium (2–8 h) biological half-lives, and are prescribed for chronic maintenance use. Drugs in large doses are difficult to formulate into such products.

Liquid Dosage Forms. Simple aqueous solutions, syrups, elixirs, and tinctures are prepared by dissolution of solutes in the appropriate solvent systems. Adjunct formulation ingredients include certified dyes, flavors, sweeteners, and antimicrobial preservatives. These solutions are filtered under pressure, often using selected filtering aid materials. The products are stored in large tanks, ready for filling into containers. Quality control analysis is then performed.

Dosage forms of naturally occurring materials having therapeutic activity are prepared by extractive processes, especially percolation and maceration. Examples of such dosage forms have included certain tinctures, syrups, fluid extracts, and powdered extracts.

Solutions for external or oral use do not require sterilization but generally contain antimicrobial preservatives. Ophthalmic solutions and parenteral solutions require sterilization.

For the preparation of suspensions and emulsions, colloid mills and homogenizers, respectively, are used. Ultrasonic mills that utilize vibrating reeds in restricted chambers to reduce the particle size of the dispersed ingredients can also be employed. See also **Colloid Systems**.

Semisolid Dosage Forms. The ingredients that constitute the base of ointments, e.g., petrolatum and waxes, are melted together, powdered drug components are added, and the mass stirred with cooling. Generally, the product then is passed through a roller mill to achieve the particle-size range desired for the dispersed solid. Pastes are ointments having relatively large, dispersed solid content, and are prepared similarly.

Creams are semisolid emulsions either water-in-oil (w/o) or oil-in-water (o/w).

Suppositories are semi-rigid, plastic dosage forms are designed to deliver a unit dose of medication to body cavities, i.e., rectum, vagina, or urethra. Depending on the base, suppositories either melt (cocoa butter) at body temperature or dissolve (poly(ethylene glycol)s, glycerogelatin) in the fluids of the cavity. They can be used for systemic therapy (rectal suppositories) or for localized treatment. Rectal suppositories are a route of administration in comatose conditions or after gastrointestinal surgery, and for pediatric patients. On a large scale, suppositories are produced by molding.

Parenteral Dosage Forms. The most commonly used forms for drug products designed and manufactured for injection through the skin include those meant for subcutaneous, intramuscular, and intravenous administration.

Intravenous aqueous injections provide an excellent means of achieving a rapid therapeutic response. Parenteral product design, e.g., vehicle and other excipient selection, as well as choice of route of administration, can prolong therapeutic activity and increase onset times. Thus, oily solutions, suspensions, or emulsions can be administered by subcutaneous or intramuscular routes to create prolonged effect, i.e., depot injection.

Several factors of design and manufacture are of great importance: sterility, absence of pyrogens and foreign particulate matter, and tonicity. The last, when adjusted to the osmotic pressure of body fluids in the case of aqueous solutions, reduces the risk of tissue irritation and pain.

Lyophilization. Lyophilization is essentially a drying technology. Some drugs and biologicals are thermolabile and/or unstable in aqueous solution. Utilization of freeze drying permits the production of granules or powders that can be reconstituted by the addition of water, buffered solution, or mixed hydrophilic solvents just prior to use, e.g., certain antibiotic suspensions.

Ophthalmic Dosage Forms. Ophthalmic preparations can be solutions, e.g., eye drops, eyewashes, ointments, or aqueous suspensions. They must be sterile and any suspended drug particles must be of a very fine particle size. Solutions must be particle free and isotonic with tears. Thus, the osmotic pressure must equal that of normal saline (0.9% sodium chloride) solution. Hypotonic solutions are adjusted to be isotonic by addition of calculated amounts of tonicity adjusters, e.g., sodium chloride, boric acid, or sodium nitrate.

Radiopharmaceuticals. Radioactive isotopes for human use in the diagnosis and treatment of disease states are called radiopharmaceuticals. Whereas the dosage form types used, e.g., solutions or injections, are traditional, special handling of these products during compounding, transport, and use is vital. Most are administered intravenously and shortly after preparation. Specialized pharmacies prepare these products overnight and transport them to hospitals for early administration by members of nuclear medicine departments.

Aerosols. Pressurized containers to deliver aerosolized drug products through appropriate systems of valves and actuators have been available since the 1950s. See also **Aerosols**. Such dosage forms are used as external applications of lotions and creams, for oral inhalation, or for treatment of the vaginal cavity, e.g., contraceptive foams. Aerosols contain two- or three-phase systems, wherein a volatile liquid or ad-mixture of liquids is sealed in a container in equilibrium with a vapor phase (propellant). Upon actuation and delivery of the product, the propellant evaporates quickly, and fine dispersion of the drug settles on the area of application. For aerosol products that need accurate dosing, metered valves are used with the valve chamber being recharged between each actuation or dose.

The popularity of aerosols has been declining. A widely used group of propellants, the fluorinated hydrocarbons, have been restricted in use since it was found that they can harm the environment by reducing the ozone layer of the upper atmosphere. See also **Pollution (Air)**; and **Ozone**.

Biotechnology and Dosage Forms. In drug development, biotechnology generally is recognized as a term that identifies those technologies that utilize living organisms in the production and/or alteration of chemical entities that have potential therapeutic activity. Besides the production of pharmacologically or biochemically active moieties, these technologies also have been used to produce food ingredients, vaccines, diagnostic testing reagents, and agricultural products. See also **Fermentation**; and **Vaccine Technology**.

Packaging. The packaging components of pharmaceutical products are vital to their safe and effective use. Besides serving the patient as a convenient unit of use, the composite package (unit container, labeling, and shipping components) must provide appropriate identification and necessary information for proper use (including warnings and cautions) and preservation of the product's chemical and physical integrity.

Labeling. Labeling, controlled by FDA regulations, includes not only the affixed labels, but also the package inserts that provide more detailed information. Trade, generic, or common name, dose, number of dose units present, and name and address of manufacturer and distributor are required. For nonprescription products, adequate directions for use are required. Prescription products must bear the phrase, "Caution: Federal law prohibits use without a prescription" on their labels.

All drug labels must include batch or lot numbers. The nature of the drug product may require special cautionary phrases, e.g., "store in cool place or refrigerator," "protect from light," and "shake well before using." In the 1990s, labels also carry the expiration date, i.e., shelf-life. This information is expected to become mandatory.

Labeling information also includes warnings as to possible side effects, e.g., drowsiness, and potential harm if used with other drugs or certain foods (drug–drug or drug–food interactions). Inserts are generally intended for use by physicians or pharmacists and give name and description of the product, mode of administration, dosage regimen, therapeutic indications and contraindications, precautions and side effects, units of supply, and literature citations. All labeling must be approved by the FDA as part of the New Drug Application.

Containers. The USPXXIII–NFXVIII lists container requirements such as well-closed, tight, or light-resistant. Most containers are light-resistant (amber) glass or plastic. The latter is break-resistant and lightweight, which reduces shipping costs and increases safety.

In hospitals and long-term care units, unit-dose packages are used more and more. This system allows better control of the dispensed drugs in institutional settings and precludes the dispensing of larger numbers of doses than needed.

Quality Control and Quality Assurance

Quality control (QC) involves the regular, daily assessment and/or analysis, according to established protocols and standards, of all ingredients, processes, and finished products. Official USP/NF monographs, for example, provide various chemical, physical, and biological tests and specifications for assurance of purity, potency, and stability of component ingredients used to prepare and package drug products. The FDA requires process validation procedures as QC constituents. The FDA also monitors QC standards through the requirements of the Current Good Manufacturing Procedures regulations.

PAUL ZANOWIAK
Temple University

Additional Reading

Ansel, H.C. and N.G. Popovich: *Pharmaceutical Dosage Forms and Drug Delivery Systems,* 5th Edition, Lea & Febiger, Philadelphia, PA, 1990, pp. 92–133.

Banker, G.S.: in G.S. Banker and C.T. Rhodes, eds., *Modern Pharmaceutics,* 2nd Edition, Marcel Dekker, New York, NY, 1990, pp. 15–20.

Kayser, O. and R.H. Muller: *Pharmaceutical Biotechnology: Drug Discovery and Clinical Applications,* John Wiley & Sons, Inc., Hoboken, NJ, 2004.

Lee, D. and M. Webb: *Pharmaceutical Analysis,* CRC Press LLC., Boca Raton, FL, 2003.

Trends in U.S. Pharmaceutical Sales and R & D: 1990–93 PMA Annual Survey Report, Pharmaceutical Manufacturers Association, Washington, DC, 1993.

Walsh, G.: *Biopharmaceuticals: Biochemistry and Biotechnology,* 2nd Edition, John Wiley & Sons, Inc., New York, NY, 2003.

Zanowiak, P.: in *Ullmann's Encyclopedia of Industrial Chemistry,* VA19, VCH Verlagsgesellschaft, MbH, Weinheim, Germany, 1991, pp. 241–271.

Web References

American Pharmaceutical Association (AphA): http://www.aphanet.org/

American Pharmaceutical Association Academy of Pharmacy Practice: http://www.aphanet.org/APPM/APPMpig.html

Pharmaceutical Research and Manufacturers of America, (PhRMA) Home: http://www.phrma.org/

U.S. Food and Drug Administration: http://www.fda.gov/

PHARMACEUTICALS, CHIRAL.

Stereoisomers are compounds which have the same molecular formula but differ in the arrangement of their atoms in space. Chiral compounds are compounds which have nonsuperimposable mirror images. Enantiomers are pairs of stereoisomers which are nonsuperimposable mirror images; they possess identical physical and chemical properties within an achiral environment. Stereoisomers other than enantiomers, i.e., diastereomers, are identified by distinct physical and chemical properties including melting points, spectral characteristics, and rates of reaction with both chiral and achiral reactants. Enantiomers, however, are only distinguished when in the presence of a homochiral environment such as polarized light, chiral solvents, chiral reagents, or chiral molecules such as biomolecules, e.g., nucleic acids, proteins, and carbohydrates. The two molecules in a pair of enantiomers rotate a plane of polarized light with equal intensities, but in opposite directions. The dextrorotatory isomer (+ or d) rotates the plane of polarized light clockwise; the levorotatory isomer (− or l) rotates the plane of polarized light counterclockwise. An equal mixture of (+) and (−)-enantiomers is a racemic mixture or racemic compound and does not rotate a plane of polarized light. Optical rotation, an intrinsic property of the substance, has no bearing on drug-macromolecule interactions. It is the absolute configuration of the homochiral compound that is important for its interaction with biomolecules.

Absorption, metabolism, and biological activities of organic compounds are influenced by molecular interactions with asymmetric biomolecules. These interactions, which involve hydrophobic, electrostatic, inductive, dipole–dipole, hydrogen bonding, van der Waals forces, steric hindrance, and inclusion complex formation give rise to enantioselective differentiation. Within a series of similar structures, substantial differences in biological effects, molecular mechanism of action, distribution, or metabolic events may be observed. For example, (R)-carvone (1) has the odor of spearmint whereas (S)-carvone (2) has the odor of caraway.

The amino acids L-leucine, L-phenylalanine, L-tyrosine, and L-tryptophan all taste bitter, whereas their D-enantiomers taste sweet. See also **Amino Acids**.

The importance of optical isomers with regard to biological effect has a long history beginning with the observations of Pasteur. The U.S. FDA requires that both enantiomers of a drug be individually tested when associated toxicities occur near the effective dose of the racemic substance. It has been suggested that the use of racemic drugs in human subjects cannot be justified until both enantiomers are tested thoroughly, both individually and in composite mixtures. Often, side effects of therapeutics are not discovered until after large-scale marketing. The distomer (therapeutically inactive enantiomer) may be at best a nontoxic impurity, but is often associated with dangerous side effects as exemplified by the thalidomide problem.

The thrust toward homochiral drugs by leading researchers and organizations such as the FDA, the rapidly expanding technology of asymmetric syntheses and chiral separations, the decreased side effects found with homochiral drugs, and the potential financial benefits are expected to ensure that the majority of chiral synthetic drugs will, in the future, be available in enantiomerically pure form. See also **Pharmaceuticals**.

Background

Nomenclature. Compounds which have tetrahedral atoms having four different substituents are often chiral. These tetrahedral atoms are referred to as stereocenters or stereogenic atoms; the terms asymmetric atom or asymmetric center are considered misnomers. The letters D and L are used to denote the absolute configurations of amino acids and sugars according to Fischer-Rosanoff nomenclature (See also **Sugar**). In this system, dextrorotatory glyceraldehyde (3) is arbitrarily assigned an absolute configuration of D. In a Fischer projection, the most highly oxidized carbon is placed on top and the last stereocenter determines the absolute configuration L or D. Examples of Fischer projections are shown for D- and L-glyceraldehyde (4) and D- (5) and L-glucose (6) where the arrow denotes

the determining stereocenter.

CHO
H—OH
CH2OH
(3)

CHO
HO—H
CH2OH
(4)

CHO
H—OH
H—OH
HO—H
H—OH
CH2OH
D-glucose [492-62-6]
(5)

CHO
HO—H
HO—H
H—OH
HO—H
CH2OH
L-glucose [921-60-8]
(6)

If the heteroatom attached to the last stereocenter projects to the right, the compound is of the D-configuration; if the heteroatom points to the left, the compound is of the L-configuration. There is no simple relationship between sign of rotation (d (+) or l (−)) and the absolute configuration D or L. However, optical activity may be related empirically to absolute configuration by observing changes in optical rotation with varying wavelength, i.e., optical rotatory dispersion (ord) and circular dichroism (cd).

Because of ambiguities involved in this nomenclature, the Cahn-Ingold-Prelog rules were introduced and are widely used to designate the absolute configuration of stereocenters. Groups attached to the stereogenic atom are assigned priorities according to the atomic number of the atom attached to the stereogenic center. Highest priority is given to the atom with the highest atomic number. The molecule is drawn such that the function of lowest priority (d) is directed away from the viewer:

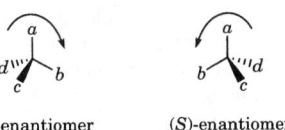

(R)-enantiomer   (S)-enantiomer

If the observed order of priority of the remaining three functions ($a > b > c$) is in a clockwise direction, the absolute configuration is designated R (rectus or right); if counterclockwise, the configuration is S (sinister or left). Diastereomers which differ at a single stereocenter are called epimers.

Enantiomeric purity, measured as the enantiomeric excess (ee) of an isomer, is determined by the formula (% major isomer)—(% minor isomer). Thus, if a chiral drug is said to be of 50% ee, the composite mixture contains 75% of one enantiomer and 25% of the other. Enantioselectivity refers to the greater activity of one enantiomer over its mirror image. Enantiospecificity is rarely observed and implies that one enantiomer possesses 100% of the observed activity; in most cases it is more accurate to use the term highly enantioselective. The pharmacologically more active enantiomer is termed the eutomer and the less active enantiomer is referred to as the distomer.

The therapeutic efficacy of a drug is generally measured in terms of ED_{50} or ID_{50} which represent the concentration of drug which produces 50% of the maximum effect or 50% of maximum inhibition.

Role of Homochiral Molecular Building Blocks. Generally L-amino acids and D-sugars are found in biological systems.

The formation of D-amino acids in polypeptides and in monomeric form during processing of proteinaceous foods has raised considerable concern about associated nutritional and toxicity effects. Racemization of amino acids occurs under strongly basic or acidic conditions, which are conditions used in food processing. The presence of aldehydic contaminants enhances the rate of amino acid racemization through formation of stereogenically labile α-imino acids. D-Amino acids may be utilized in a nutritional manner if they are converted to L-amino acids. For the most part, D-amino acids are generally no more toxic than their L-enantiomers. Thus, foods containing proteins with high concentrations of D-amino acid residues may be useful for weight management.

Modeling of Drug-Receptor Interactions. The identification of molecular interactions between drugs and their receptor or enzyme targets and the three-dimensional spatial requirements of the macromolecular binding pocket are important for the rational design of new, more selective, and potent pharmaceuticals. Methods used to explore such interactions include nmr spectroscopy of receptor-ligand complexes, molecular modeling, point mutation analysis, and binding assays of conformationally constrained, stereochemically defined small molecules. See also **Molecular Modeling**. X-ray crystal structure analysis of macromolecule-ligand complexes provides information concerning important molecular interactions which give rise to the observed affinity between the macromolecule and ligand. Computers are used to graphically display calculated crystal structures. Numerous computer programs have been developed and refined which are capable of determining the energy minimized structures of such complexes.

Molecular modeling techniques, although aesthetically pleasing, are far from reliable owing to the large number of associated variables.

Methods for the Preparation of Homochiral Drugs

Resolution Methods. Chiral pharmaceuticals of high enantiomeric purity may be produced by resolution methodologies, asymmetric synthesis, or the use of commercially available optically pure starting materials. Resolution refers to the separation of a racemic mixture. Classical resolutions involve the construction of a diastereomer by reaction of the racemic substrate with an enantiomerically pure compound. The two diastereomers formed possess different physical properties and may be separated by crystallization, chromatography, or distillation. A disadvantage of the use of resolutions is that the best yield obtainable is 50%, which is rarely approached. However, the yield may be improved by repeated racemization of the undesired enantiomer and subsequent resolution of the racemate. Resolutions are commonly used in industrial preparations of homochiral compounds.

Three general methods exist for the resolution of enantiomers by liquid chromatography. Conversion of the enantiomers to diastereomers and subsequent column chromatography on an achiral stationary phase with an achiral eluant represents a classical method of resolution. Diastereomeric derivatization is problematic in that conversion back to the desired enantiomers can result in partial racemization. Direct resolution of the enantiomers without derivatization is performed by use of an achiral stationary phase with a chiral mobile phase or, more commonly, by use of a chiral stationary phase and an achiral mobile phase. Ligand-exchange chromatography, using Cu^{2+} and proline, and chiral ion pair chromatography, which involves use of a chiral counterion in the mobile phase, exemplify the former method. Chromatographic resolution of enantiomers using chiral stationary phases is advantageous over the previously described methods in that no derivatization is required, there is no need to employ expensive mobile phases, and the method does not require complicated product analysis. Stationary phases include cyclodextrins, protein bonded supports, chiral polymers, and the Pirkle type. See also **Polymers**. The three-point rule for chiral recognition is used to rationalize the separation of enantiomers by use of a homochiral stationary phase and an achiral eluent. The (*S*)-enantiomer possesses three favorable interactions with the stationary phase and is therefore expected to traverse the column at a slower rate than its (*R*)-enantiomer, which maintains only two favorable interactions.

(*S*)-enantiomer   (*R*)-enantiomer

Methods Employing Enantiomerically Pure Starting Materials. A large number of optically pure natural products are commercially available and relatively inexpensive. Amino acids, carbohydrates, and terpenes are some of the homochiral building blocks used routinely in enantiomeric syntheses. The stereocenter(s) in such building blocks are used as either chiral synthons (chirons) or as chiral auxiliaries. Chirons are an inexpensive source of chirality and their use in organic synthesis generally produces enantiomerically pure compounds of known absolute configuration. The commercially available polyether antibiotic monensin, which contains 17 stereocenters, is synthesized by the preparation and coupling of three fragments, each prepared from commercially available optically active starting materials.

Many notable examples of the synthesis of complex natural products from optically pure starting materials have been reported. One synthesis of considerable interest is that of taxol, a potent antitumor agent used clinically. The starting material used in the first total synthesis of taxol

is produced in enantiomerically pure form from inexpensive and readily available *l*-camphor.

Asymmetric Induction Methodologies. Asymmetric synthesis is defined as the construction of new chiral centers within a prochiral molecule, with the condition that one optical isomer is formed to a greater extent than the other. The most common type of asymmetric induction involves the conversion of a trigonal carbon atom to a tetrahedral carbon atom by use of a reagent that is biased toward preferential attack from one side or face of the prochiral molecule. Generally, asymmetric induction involves diastereotopic transition states wherein one transition state is favored due to steric and electronic effects which govern the selective formation of one enantiomer. The primary advantage of asymmetric synthesis resides in the stereoselective production, in general, of either enantiomer of a compound; synthesis of both enantiomers is not always possible using chiral synthons or auxiliaries. Asymmetric syntheses avoid the use of inefficient resolutions, and often the reagent or catalyst is recyclable making the synthetic process both material and cost efficient. Asymmetric reduction of ketones, epoxidation of allylic alcohols, hydroboration, hydrogenation, and dihydroxylation reactions, as well as asymmetric cycloaddition reactions, allyl borations, and aldol condensations represent several classes of the numerous enantioselective reactions developed since the 1960s.

Analysis of Synthetic Homochiral Drugs

Determination of Absolute Configuration. X-irradiation of a crystal produces a diffraction pattern from which the relative spatial orientation of the atoms that make up the molecule may be determined. If the crystal is made of homochiral molecules, the absolute configuration of the compound may be deduced.

Chemical conversion of compounds to intermediates of known absolute configuration is a method routinely used to determine absolute configuration. This is necessary because x-ray analysis is not always possible; suitable crystals are required and determination of the absolute configuration of many crystalline molecules cannot be done because of poor resolution. Such poor resolution is usually a function of either molecular instability or the complex nature of the molecule.

ORD and CD also provide a basis by which the absolute configuration of a compound may be correlated with that of a known compound of similar structure by observing changes in degree of rotation with wavelength.

Determination of Enantiomeric Purity. In order to analyze the biological properties of a single enantiomer, the optical purity of the compound should be enantiomerically pure, i.e., 100% ee. Contrasting reports on the differences in pharmacological activity of single enantiomers, as well as the misinterpretation of data, are often a result of unknowingly testing enantiomerically impure material. The oldest and perhaps easiest method for determining optical purity is by measuring optical rotation and comparing the value with that reported for the enantiopure compound. There are several drawbacks to this method. The assumption must be made that the reported literature value is without error, and truly represents the optically pure compound. Numerous examples exist in which unambiguous methods, i.e., chiral gc, hplc, nmr, for determination of optical purity reveal that the previously reported values for optically pure compounds were in error. Variables such as temperature, solvent, concentration, purity of the compound, type of cell, and even differences between polarimeters employed in the measurement influence the observed degree of rotation. Therefore, polarimetry measurements for determination of optical purity deviate by at least ±4%.

[1]H-nmr is commonly used to determine enantiomeric purity and is reliable to above 98% ee.

Chiral Pharmaceuticals

Enantiomeric Pairs. Enantioselective differences in absorption, metabolism, clearance, drug–macromolecule binding affinity, and other factors, which culminate in the observed enantioselective efficacy chiral drugs, are considered.

Antihypertensive Agents. Hypertension (high blood pressure) is a significant risk factor for cardiovascular diseases such as angina heart attacks, and strokes. β-Adrenoceptor (adrenergic nervous system receptors of the β-type) antagonists (β-blockers), calcium channel blockers, angiotensin-converting enzyme (ACE) inhibitors, and potassium channel activators

(KCAs) are among the numerous classes of drugs developed to control hypertension. See also **Enzyme Inhibitors**. β-Adrenoceptor antagonists exemplified by the phenoxypropanolamine derivatives propranolol and alprenolol and or the phenethanolamine drugs such as sotalol and require for activity both the ethanolamine portion and an aromatic ring. Furthermore, the correct spatial arrangement of the phenyl, ethylamine, and hydroxyl moieties is critical for β-blockade.

It has been demonstrated that the β_1-selectivity is due to the para-substituents of β-adrenoceptors. In contrast, ($-$)-erythro-isoetharine, a bronchodilator, is 80 times more selective for β_2-adrenergic receptors than for β_1-receptors. Isoetharine contains an α-alkyl substituent, thus producing four isomeric compounds. The ($-$)-erythro isomer is 100-fold more active than the ($-$)-threo isomer and has more than 500 times the activity of either of the ($+$)-isomers and in blocking electrically stimulated spasms. In general, introduction of α-alkyl substituents on both β-blockers and agonists provides diastereomers with increased β_2-selectivity, but often with compromised potency.

(7)

Cromakalim (**7**) is a potassium channel activator commonly used as an antihypertensive agent. The rationale for the design of cromakalim is based on β-blockers such as propranolol and atenolol. Conformational restriction of the propanolamine side chain as observed in the cromakalim chroman nucleus provides compounds with desired antihypertensive activity free of the side effects commonly associated with β-blockers. Enantiomerically pure cromakalim is produced by resolution of the diastereomeric (S)-α-methylbenzylcarbamate derivatives. X-ray crystallographic analysis of this diastereomer provides the absolute stereochemistry of cromakalim. Biological activity resides primarily in the ($-$)-(3S, 4R)-enantiomer. In spontaneously hypertensive rats, the ($-$)-(3S, 4R)-enantiomer, at dosages of 0.3 mg/kg, lowers the systolic pressure 47%, whereas the ($+$)-(3R, 4S)-enantiomer only decreases the systolic pressure by 14% at a dose of 3.0 mg/kg.

Nonsteroidal Antiinflammatory Drugs. Nonsteroidal antiinflammatory drugs (NSAIDs) include, among the numerous agents of this class, aspirin (acetylsalicylic acid), the arylacetic acids indomethacin and sulindac, and the arylpropionic acids, (S)-(**8**) and (R)-(**9**) ibuprofen, (S)-(**10**) and (R)-(**11**), flurbiprofen naproxen, and fenoprofen. See also **Analgesics, Antipyretics, and Antiinflammatory Agents; and Salicylic Acid and Related Compounds**.

(8) R = CH$_3$; R′ = H
(9) R = H; R′ = CH$_3$

(10) R = CH$_3$; R′ = H
(11) R = H; R′ = CH$_3$

Although the arylpropionic acids contain a stereogenic center they are generally marketed as racemic mixtures. The only exception is naproxen, which is marketed as its (S)-enantiomer. NSAIDs produce their antiinflammatory effects by inhibiting cyclooxygenase (COX), the enzyme which catalyzes the first transformation in the biosynthetic conversion of arachidonic acid to the 20 carbon prostaglandins.

CNS Depressant Drugs. Central nervous system (CNS) depressant drugs including antianxiety agents (benzodiazepines), sedative-hypnotics, general anesthetics, and certain spasticity agents all demonstrate high degrees of enantioselective activity. Barbiturates are commonly prescribed for their sedative-hypnotic activities. In general, the (S)-($-$)-enantiomers possess CNS depressant activities, whereas the (R)-($+$)-isomers (R)-($+$)-isomers often produce an excitatory effect. In humans, (R)-($+$)-pentobarbital is found bound to human plasma proteins to a lesser extent than the (S)-($-$)-isomer (36.6% free vs 26.5% free) and is subsequently

cleared 14% faster. This increased rate of clearance is not sufficient to account for the two- to threefold greater duration of action of (S)-($-$)-pentobarbital, and suggests that the difference in activity between the enantiomers is due to the pharmacodynamics of the more potent (S)-isomer. (S)-($+$)-Hexobarbital, the eutomer, is eliminated about 2.5 times more slowly than the inactive (R)-($-$)-isomer, a result of differences in hepatic metabolism. Diazepam (**12**), an achiral benzodiazepine, undergoes stereoselective metabolism to (S)-($+$)-oxazepam (**13**) in the liver. (S)-($+$)-Oxazepam produces antianxiety effects to a greater degree than the mirror image isomer (**14**).

(12) X, Y = H
(13) X = H; Y = OH
(14) X = OH; Y = H

Antibiotic and Antimicrobial Drugs. The antimicrobial agents flumequine and methylflumequine (S-25930) effectively eliminate a number of microbial pathogens via inhibition of the topoisomerase II enzyme of c-DNA containing bacteria. (see **Antibacterial Agents, synthetic**. The (S)-enantiomers of both drugs are much more potent than the (R)-enantiomers. The potent analogue (S)-($-$)-ofloxacin is 8–125 times more potent than its enantiomer although it is sold only as the racemate. In humans the disposition of (R)- and (S)-enantiomers of ofloxacin is stereoselective due to differences in renal clearance rates. This difference, however, does not fully explain the large enantioselective difference in antibacterial potency. β-Lactam antibiotics. See also **Antibiotics**, such as the penicillins and cephalosporins, require the (3S, 5R, 6R)-configuration of the β-lactam functionality combined with a D-amine in either the 6-position (penicillins) or 7-position (cephalosporins) to produce optimal activity.

Opioid Analgesic Drugs. (5R, 6S, 9R, 13S, 14R)-($-$)-Morphine (**15**) and its closely related relatives ($-$)-codeine (**16**) and ($-$)-heroin (**17**) are potent analgesics, while their ($+$)-isomers possess no analgesic effects, α-Dextropropoxyphene (DARVON) (**18**) is a marketed analgesic whereas its enantiomer, α-levopropoxyphene (NOVRAD) (**19**) is sold as an antitussive devoid of analgesic activity. These analgesics produce their biological effects via stimulation of the opioid receptor subclasses mu-, delta-, kappa-, and sigma.

(15) R = R′ = OH
(16) R = —OCH$_3$; R′ = OH
(17) R, R′ = —OC—CH$_3$

(18) R = —phenyl—; R′ = OCCH$_2$CH$_3$;
(19) R = OCCH$_2$CH$_3$; R′ = phenyl

Anticoagulant Drugs. Warfarin, a potent anticoagulant, was first isolated from spoiled clover hay and identified as the agent responsible for the hemorrhagic symptoms associated with the death of livestock in the 1930s. This functionalized coumarin derivative exerts its effect via competitive inhibition of vitamin K-dependent carboxylation of blood clotting factors. Warfarin is generally administered as the racemate, even though (S)-warfarin is fivefold more active than (R)-warfarin in both rats and humans. The (S)-enantiomer is eliminated at a higher rate than its antipode in humans, but in rats, (R)-warfarin is more rapidly eliminated.

Neurotransmitters. Histamine receptors are found in at least two subtypes designated H$_1$ and H$_2$. See also **Histamine and Histamine Antagonists**. H$_1$-receptor antagonists produce vasoconstriction, while H$_2$-receptor antagonists inhibit gastric secretion. Neobenodine (**20**) and (**21**), the chiral p-methylphenyl analogue of benadryl (**22**), is an antihistamine marketed as the racemate. The (R)-($+$)-isomer (**20**) is 65 times more potent than its ($-$)-enantiomer (**21**) when tested in guinea pig ileum. Chlorpheniramine (**23**) and (**24**) is also an enantioselective H$_1$-antagonist, wherein the (S)-enantiomer (23) is most potent. It has been demonstrated that the more potent enantiomer of diphenhydramine and pheniramine drugs is the one in which the aryl moiety, alkylamine group, and the p-substituted aryl functionality occur in a clockwise orientation.

(20) R = H; R' = —⟨ ⟩—CH₃

(21) R = —⟨ ⟩—CH₃ ; R' = H

(22) R = H; R' = —⟨ ⟩

(23) R = H; R' = —⟨ ⟩—Cl

(24) R = —⟨ ⟩—Cl; R' = H

Antineoplastic Drugs. Cyclophosphamide (25) produces antineoplastic effects via biochemical conversion to a highly reactive phosphoramide mustard (26); it is chiral owing to the tetrahedral phosphorus atom. The therapeutic index of the (S)-(−)-cyclophosphamide (25) is twice that of the (+)-enantiomer due to increased antitumor activity; the enantiomers are equally toxic. The effectiveness of the DNA intercalator drugs adriamycin (27) and daunomycin (28) is affected by changes in stereochemistry within the aglycon portions of these compounds. Inversion of the carbohydrate C-1 stereocenter provides compounds without activity. The carbohydrate C-4 epimer of adriamycin, epirubicin, is as potent as its parent molecule, but is significantly less toxic. (R)-3-Ethyl-3(4-pyridyl)piperidine-2,6-dione (29), useful in the treatment of certain breast cancers, is a 20-fold more potent aromatase inhibitor (IC₅₀ = 10 μM) than is its (S)-enantiomer.

(25)   **(26)**

(27) R = —COCH₃
(28) R = —COCH₂OH

(29)

Peptidomimetics. Many drugs mimic natural small peptides. For example, morphine (15) is believed to be a natural peptidomimetic for the enkephalins. Similarly, FK-506 mimics the binding of peptidal FK-506 to the intracellular receptor, FKBP12. Numerous small endogenous peptides have been characterized which possess potent cellular signaling and homeostatic regulating activities. The regulation of glycolysis, growth, mitosis, and apoptosis, as well as the maintenance of blood pressure and the natural relief of pain, exemplify a few of the regulatory actions of peptide hormones. Exogenous control, through the use of synthetic compounds, of the activities of these regulatory elements is highly desirable as demonstrated by the use of drugs such as the ACE inhibitor captopril. The design of peptidomimetics is complicated owing to the flexibility and stereochemical complexity of such hormones. Numerous small peptides have been synthesized which possess tremendous enzyme inhibitory and receptor binding activities *in vitro*; HIV protease inhibitors are one example. Unfortunately, the use of such peptides *in vivo* generally is not successful, as peptidase enzymes rapidly degrade synthetic peptides.

Several methods are being studied to enhance the stability of peptide mimics and improve their stereochemical similarity to the endogenous peptides.

Economic Aspects

Drugs classified as either natural or semisynthetic in origin accounted for ~22% of the market share in 1991. Nearly 94% of the agents are chiral compounds and are sold as single enantiomers. Chiral synthetic drugs make up 38% of the market share and 43% of these are sold as single enantiomers, a two- to threefold increase since 1981. Achiral or symmetrical synthetic drugs make up 40% of the drug market. The vast number of marketed racemic drugs are being reinvestigated and newer pharmacological data as well as production technology are being patented.

A steady increase in the number of homochiral drugs on world markets has created an increased demand for enantiomerically pure intermediates as well as for enantioselective technologies. Many pharmaceutical companies are pursuing new financial opportunities and gaining improved bargaining positions by producing patent protected and more expensive enantiomerically pure drugs from unprotected racemic pharmaceuticals.

DONALD T. WITLAK
ALLEN T. HOPPER
University of Wisconsin-Madison

Additional Reading

Hyneck, M., J. Dent, and J.B. Hook: in C. Brown, ed., *Chirality in Drug Design and Synthesis,* Academic Press, Inc., San Diego, CA, 1990, pp. 1–28.

Kayser, O., and R.H. Muller: *Pharmaceutical Biotechnology: Drug Discovery and Clinical Applications,* John Wiley & Sons, Inc., Hoboken, NJ, 2004.

Lee, D., and M. Webb: *Pharmaceutical Analysis,* CRC Press LLC., Boca Raton, FL, 2003.

Levin, S., and S. Abu-Lafi: in P.R. Brown and E. Grushka, eds., *Advances in Chromatography,* Vol. 33, Marcel Dekker, Inc., New York, NY, 1993, pp. 233–266.

Ott, R.J., and K.M. Giacomini: in I.W. Wainer, ed., *Drug Stereochemistry Analytical Methods and Pharmacology,* Second Edition, Revised and Expanded, Marcel Dekker, Inc., New York, NY, 1993, pp. 281–314.

Sheldon, R.A.: *Chirotechnology: Industrial Synthesis of Optically Active Compounds,* Marcel Dekker, Inc., New York, NY, 1993, pp. 271–341.

Walsh, G.: *Biopharmaceuticals: Biochemistry and Biotechnology,* 2nd Edition, John Wiley & Sons, Inc., New York, NY, 2003.

PHARMACODYNAMICS. Pharmacodynamics is the study of drug action primarily in terms of drug structure, site of action, and the biochemical and physiological consequences of the drug action. The availability of a drug at its site of action is determined by several processes, including absorption, metabolism, distribution, and excretion. These processes constitute the pharmacokinetic aspects of drug action. The onset, intensity, and duration of drug action are determined by these factors as well as by the availability of the drug at its receptor site(s) and the events initiated by receptor activation.

Both pharmacokinetic and pharmacodynamic processes are involved in mediating nonconstant expressions of drug action. Thus, resistance to the actions of a drug, e.g., in the development of antibiotic-resistant bacteria or of barbiturate tolerance, can arise from changes in drug metabolism and/or alterations in the receptor target site. Factors controlling drug resistance may be whole-body, cellular, or individual events. Decreased absorption, increased metabolism, or increased elimination reduce circulating drug levels and affect the whole body. Increased drug metabolism, increased concentration of an agent that antagonizes drug action, decreased affinity or concentration of a drug receptor, and depletion of an agent that mediates drug action are examples of cellular events; and genetic factors controlling metabolism, receptor alterations, and disease states are examples of individual events. Individual variation in the susceptibility to a particular drug or class of drugs also may arise from genetically based pharmacokinetic factors as well as from specific receptor-linked changes.

For a large number of drugs, including neurotransmitters, peptide and protein hormones, and their analogues and antagonists, the cell membrane is the principal locus of action. Concepts of cell membrane structure are derived from the original Davson-Danielli lipid bilayer hypothesis. More specifically, the membrane is viewed as a dynamic fluid mosaic or a matrix of fluid bilayer in which there are asymmetrically inserted proteins and glycoproteins. Phospholipids and proteins diffuse laterally and the resultant protein-protein communication is of considerable importance to the understanding of membrane-receptor function. Despite the dynamic nature of the membrane and the absence of global organization, local organization is possible through the local assembly of individual protein components and the attachment of membrane proteins to the subcellular structure of contractile proteins. However, the cell membrane is not the site of action of all drugs. A number of drugs, including steroid and thyroid hormones, exert their effects intracellularly at the level of the genetic material as well as at the plasma membrane. Other agents, including polypeptide growth factors, exert their effects not only at the plasma membrane through tyrosine kinase receptors, but also on cell growth and differentiation at the genetic level.

Drug Discovery and Regulation

In the United States and elsewhere, the introduction of a new drug is subject to a sequence of well-defined stages of development and approval. Each stage involves either scientific testing or submission and preparation of data and analysis review (Fig. 1).

An investigational new drug (IND) application usually initiates the process for drug approval. The IND derives from the concept that a specific molecule or molecules may have a particular therapeutic benefit. Preclinical data are analyzed to determine the implications of such molecules for human pharmacology, chemical composition, manufacturing processes, and the protocols for subsequent clinical work. Clinical trials are usually carried out in at least three phases. Phase one involves a small number of individuals and is designed to find information about basic safety and response issues. In phase two studies, the drug is employed on a larger number of individuals (100–200) who suffer from the condition that the drug is designed to treat. Phase three studies involve a much larger group of patients and are designed to assess safety, efficacy, and dosage regimens in a broad range of patients across lines of age, race, and gender. Phase three studies may involve several thousand patients and be carried out at several sites.

New drug application (NDA) is the process through which the U.S. Food and Drug Administration (FDA) authorizes the marketing of a new drug. In the NDA, the data are intended to demonstrate the safety and efficacy of the drug in its intended application. After approval, the drug becomes available to the public. Subsequently, dosage amounts and forms may be modified according to experience, new indications may be added, and contraindications may be noted. All of the changes require regulatory approval. A drug in human use is subject to constant surveillance.

The Receptor Concept

Drug receptors are chemical entities which are typically, but not exclusively, small molecules that interact with cellular components, frequently at the plasma membrane level. There are many types of receptors; heat, light, immune, hormone, ion channel, toxin, and virus are but a few that can excite a cell. The receptor concept can be applied generally to signal recognition processes where a chemical or physical signal is recognized. This recognition is translated into response and the process can be seen as a flow of information.

Elucidation of the structural requirements for drug interaction at the recognition site is by the study of structure-activity relationships (SAR), in which, according to a specific biologic response, the effects of systematic molecular modification of a parent drug structure are determined. Such studies have permitted the classification of discrete classes of pharmacological receptors.

The demonstration of the existence of strictly defined SARs, which is perhaps the most important criterion of drug action at a specific receptor site, has made possible the most important pharmacologic discoveries. For example, the analgesic actions of morphine and related agents, which are indicative of specific receptors, led to the discovery of endogenous opiate peptides, i.e., the leucine and methionine enkephalins and endorphins.

Pharmacokinetic Aspects of Drug Action

The receptor represents the locus of drug action. However, the pharmacokinetic processes of absorption (drug entry), distribution, metabolism, and excretion play principal roles in determining *in vivo* time courses and concentrations of drugs and thus modify actions initiated at receptors.

Drug Entry. Drugs enter the body by one of two routes. In enteral administration (sublingual, oral, rectal), the drug enters directly the gastrointestinal tract. In the parenteral route, the drug bypasses the gastrointestinal tract by, among others, subcutaneous (sc), intramuscular, intravascular (iv), inhalational, intraperitoneal (ip), intravaginal, and intranasal routes. Each route has a particular set of advantages and disadvantages. Patient convenience is high in the oral route; speed of action and ability to control concentrations are high in the iv route; and nonoral routes are best for unstable or insoluble drugs.

In light of the recognized importance of achieving stable, reproducible plasma concentrations of drugs, particular attention is given to pathways and devices, including sustained-release formulations, pumps, and transdermal entry processes that ensure such properties.

Drug Distribution. After administration, a drug may be distributed either generally or selectively in the body. The distribution pattern depends on many factors, including the pattern and time-course of blood flow, diffusion of drugs into tissues, binding of drugs to plasma proteins and cellular compartments, and elimination kinetics and mechanisms.

Drug Metabolism. Generally, metabolism (biotransformation) of drugs increases their water solubility as well as the rate and ease of elimination, but reduces their volume of distribution. Many drug-metabolizing pathways have arisen during evolution to deal with foreign compounds present in food materials. Although metabolism generally leads to more polar and less active compounds, there are exceptions. Metabolic pathways have also been exploited to design prodrugs, materials that are converted to active species through biotransformation.

Biotransformation reactions can be classified as phase I and phase II. In phase I reactions, drugs are converted to product by processes of functionalization, including oxidation, reduction, dealkylation, and hydrolysis. Phase II or synthetic reactions involve coupling the drug or its polar metabolite to endogenous substrates and include methylation, acetylation, and glucuronidation (Table 1).

The biotransforming pathways are subject to manipulation and modification in a variety of ways. Drug metabolism also depends on age and sex. Drug metabolism may also produce toxic materials.

Drug Elimination. Drugs are removed from their sites of action through metabolism, storage, and excretion. These processes are not necessarily independent and drugs are frequently metabolized prior to excretion. Indeed, for lipophilic drugs this is virtually a necessity. Drugs are excreted via the kidneys, biliary systems, intestines, and lungs.

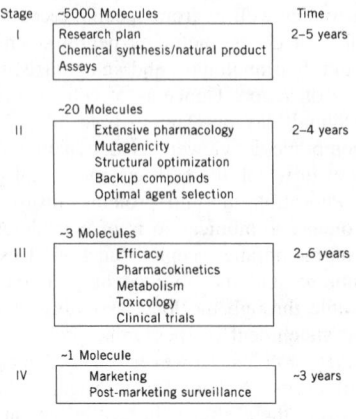

Fig. 1.   Pathway for drug development

TABLE 1. BIOTRANSFORMATION REACTIONS

Pathways	Reactions types	Examples
	PHASE I REACTIONS	
oxidative	aliphatic and aromatic oxidation	phenobarbital, phenytoin
	N- and O-dealkylation	desipramine, phenacetin
	N-oxidation	guanethidine
	oxidative deamination	amphetamine
	desulfuration	thiobarbitol
	dehalogenation	chloroform
hydrolytic	esters and amides	procaine, lidocaine
reductive	azo reduction	prontosil
	nitro reduction	chloramphenicol
nonmicrosomal oxidative	alcohol and aldehyde oxidation	ethanol
	purine oxidation	6-mercaptopurine
	oxidative deamination (monoamine oxidase)	serotonin
	PHASE II REACTIONS	
coupling	glucuronidation	acetaminophen
	acetylation	isoniazid
	glycine conjugation	salicylic acid
	sulfate conjugation	steroids, phenols
	methylation	norepinephrine

The process of reabsorption depends on the lipophilic-hydrophilic balance of the molecule. Charged and ionized molecules are reabsorbed slowly or not at all. Reabsorption of acidic and basic metabolites is pH-dependent, an important property in detoxification processes in drug poisoning. Both passive and active carrier-mediated mechanisms contribute to tubular drug reabsorption.

Clinical Pharmacokinetics. Clinical pharmacokinetics attempts to define the relationship between drug concentration and therapeutic response. The underlying assumption is that response is proportional to drug concentration at the site of action. This concentration is dependent on many factors that are frequently pharmacokinetic determinants. The most important factors are defined as clearance, bioavailability, and volume of distribution.

Clearance, CL, is defined by $DL = CL \cdot C_{ss}$, where DR represents dosing rate and C_{ss} the steady-state concentration of the drug.

Once the steady-state concentration is known, the rate of drug clearance determines how frequently the drug must be administered. Because most drug elimination systems do not achieve saturation under therapeutic dosing regimens, clearance is independent of plasma concentration of the drug.

The half-life, $t_{1/2}$, for a drug in plasma, i.e., the time it takes for the concentration of a drug to be reduced by 50%, is determined by both volume of distribution, V, and clearance: $t_{1/2} = 0.693 \cdot V/CL$. The bioavailability of a drug can be defined as the fraction of a does, F, that reaches the systemic circulation. When $F < 1$, $F \cdot DR = CL \cdot C_{ss}$.

Pharmacodynamic Aspects of Drug Action

Although the same general principles of chemical specificity apply to all ligand-macromolecular interactions, the term receptor is generally applied to those cellular macromolecules and macromolecular complexes with which ligands, physiological or synthetic, interact both to complex and to initiate a physiological response. Receptors are conveniently viewed as existing in several principal classes, i.e., G-protein-coupled receptors, ligand-gated ion channels, voltage-gated ion channels, tyrosine kinase receptors, guanylyl cyclase receptors, and steroid hormone receptors. All of these receptors form homologous classes according to structure and mechanisms of action. G-protein-coupled receptors form a homologous class of membrane proteins characterized by seven transmembrane domains and the ability to couple to guanine (G) nucleotide-binding proteins.

Ligand-gated ion channels represent a significant family of ion channels that feature as an integral component of their multimeric subunit organization a receptor site for either acetylcholine (nicotine acetylcholine receptor (AChR)), amino acids including glycine and γ-aminobutyric acid (GABA) (inhibitory transmitters), or glutamic acid (excitatory transmitter). The interaction of the ligand with the endogenous receptor site causes channel opening or closing.

An important characteristic of both classes of ion channel is that they possess multiple drug binding sites. Many of the channel-active drugs have achieved particular therapeutic importance, including, for example, the Ca^{2+} antagonists, widely used for a number of cardiovascular disorders, such as hypertension.

Structure–Activity Relationships. Until the mid-1980s, the attempted correlation of chemical structure and biological activity was the only available approach to the definition of receptor site structures. The basic assumption in the analysis of structure-activity relationships (SAR) is the existence of a definable mutual complementarity between the structure of the drug and its corresponding binding site. This application is limited when applied in empirical fashion. Many drug molecules are flexible structures and, although conformations in the solution and solid states can be determined by spectroscopic and crystallographic methods, these bear no necessary relationship to those adopted at the receptor site. The possibility of mutual conformational adaptation of both the drug and the receptor site during the binding process adds a further complication. Furthermore, there may exist multiple drug-binding modes at the receptor such that transitions in binding modes occur at some point in a structurally related series. An additional problem in the quantitative interpretation of SAR is that of the relationship between biological response and drug-receptor interaction. Despite these limitations, SARs have been of great value in providing qualitative concepts of binding site geometry, classifying receptors, furnishing evidence for the existence of new classes of receptor-specific drugs, and generating new and therapeutically effective compounds.

The simplest SARs occur in homologous series of compounds. Thus a linear relationship exists between carbon chain length and biological activity in 1-alkanol-mediated anesthesia (see **Anesthetics**). The activity can be related to the water:cell partition coefficient. For other homologous series, however, such linear relationships may not be observed; for example, in the antagonistic activity of α, ω-bistrimethylammonium alkanes at acetylcholine receptors where binding to sites of defined anionic site geometry probably is involved.

Relatively unambiguous monotonic SARs also occur where activity depends on the ionization of a particular functional group. A classic example is that of the antibacterial sulfonamides where activity is exerted by competitive inhibition of the incorporation of p-aminobenzoic acid into folic acid. The bell-shaped relationship is consistent with the sulfonamide acting as the anion but permeating into the cell as the neutral species.

The SAR is also determined at the level of stereochemistry of interaction. In principle, three limiting situations can apply to the stereochemistry of drug-receptor interactions: the enantiomers may not differ in activity; the species may differ quantitatively; or they may differ qualitatively.

The issue of drug stereoselectivity has become one of both developmental and regulatory significance. In principle, a racemic drug possesses only 50% of the active ingredient, and the rest may have other or interacting pharmacologic activities, which may contribute a metabolic burden or be inert. Over 50% of clinically available drugs have chiral centers and only about 10% of synthetic chiral drugs are marketed in homochiral (enantiomerically pure) form. In contrast, drugs that are naturally occurring substances, obtained from or related to naturally occurring molecules, are frequently homochiral.

There is increasing pressure to develop homochiral drugs.

Often pharmacologic agonist activity decreases and is lost with progressive structural change.

Increasing attention has been paid to the generation of quantitative structure—activity relationships in which the effects of molecular substitution on pharmacologic activity can be interpreted in terms of the physico-chemical properties of the substituents. These approaches are based on the extrathermodynamic analysis of substituent effects.

Advancing technology permits increasing attention to the definition of the three-dimensional structure of the ligand in its bioactive conformation as it binds to the receptor or active site. This bioactive conformation is not necessarily the solution or the crystal structure of the ligand, which is often the most experimentally accessible structure. It is of critical importance to define the three-dimensional structure of the ligand complexed with its target. This resolution permits not only the understanding of a particular ligand–macromolecule, but also the *in vivo* design of ligand homologues that may have tighter or more selective affinities for the site.

Considerable effort must be applied to obtaining adequate quantities of the protein target and its structural solution, together with the structural solution of the complexed ligand, either by x-ray or solution nmr techniques. Alternatively, homology modeling may be possible when the structure of a homologue protein is already available. Although many examples of ligand–protein structure determinations are available, some of the most interesting targets, e.g., membranebound receptors, defy structural solution at the necessary resolution. The examination of the real structure of ligand–receptor complexes should be an increasingly important and integral part of the drug discovery process.

Quantitative Aspects of Drug–Receptor Interactions. As a general rule, pharmacological responses are graded and a defined relationship exists between the concentration of a drug and the receptor response. This usually is expressed as a concentration–response (A–R) relationship in linear or semilogarithmic coordinates and usually is referred to as a dose–response curve. The shape of these curves offers a clear analogy to processes of physical adsorption but, because of the complexity of the sequence of events between drug–receptor interaction and the response, the interpretation of dose–response curves is not simple. A quantitative understanding of drug–receptor interactions is crucial both to the nontrivial interpretation of structure–activity relationships and to the determination of the mechanisms by which drug–receptor complexes initiate pharmacological response.

Nonreceptor-Mediated Drug Action. At least one important class of drugs, the general anesthetics, has been assumed not to owe its therapeutic activities to a specific receptor process. Anesthetic potency shows an excellent linear correlation with partition coefficient and this has been extrapolated to a definition of action at a lipid site. The phospholipids of cell membranes, particularly nerve cells, have been considered as principal targets for general anesthetic action. It has been hypothesized that anesthetics may disrupt phospholipid structure by fluidizing or expanding the cell membrane or by altering the phase relationships of the phospholipids. However, it is possible that anesthetics bind to hydrophobic sites on proteins and thus affect directly excitable cell behavior. This latter proposal is consistent both with the activity of the gaseous general anesthetics and with the activity of structurally more complex agents, e.g., 3α-hydroxy-5α-pregnane-11,20-dione, 3α-hydroxy-5-pregn-16-ene-11,20-dione, and 1,5-desmethyl-5-cyclohexenylbarbituric acid.

Although most anesthetics are achiral or are administered as racemic mixture, the anesthetic actions are stereoselective. This property can define a specific, rather than a nonspecific, site of action. Stereoselectivity is observed for such barbiturates as thiopental, pentobarbital, and secobarbital. The (S)-enantiomer is modestly more potent. Additionally, the volatile anesthetic isoflurane also shows stereoselectivity. The (S)-enantiomer is the more active. Further evidence that proteins might serve as appropriate targets for general anesthetics come from observations that anesthetics inhibit the activity of the enzyme luciferase. The potencies parallel the anesthetic activities closely.

It is likely that a principal target of the general anesthetics is neuronal ion channels of both voltage-gated and ligand-gated classes. Interactions at GABA-mediated inhibitory channels is a significant, but not exclusive, target. Thus, a general anesthetic may have specific but multiple, rather than nonspecific, sites of action.

Receptor–Effector Coupling. The informational signal initiated by drug–receptor interaction must be translated to biological response. This is activated by a variety of effector-coupling processes that lead to ionic or biochemical changes, including ion channel opening and closing; the formation of second messengers such as cyclic adenosine-3'-5'-monophosphate (cAMP) and inositol-1,4,5-triphosphate (IP$_3$); and protein phosphorylation through protein kinase A (cAMP-dependent) and protein kinase C (CA^{2+}-dependent), or through autophosphorylation (tyrosine kinase receptors). In these systems, it is increasingly clear that the individual components of a receptor system may be linked in multiple ways. The virtue of this organization lies in the multiple coupling processes permitted beyond a set of components.

These cascades serve as operational amplifiers of the initial ligand–receptor interaction. In each step of the process, amplification by several powers of 10 may occur so that an original signal may be multiplied several millionfold.

G-Protein Coupling. The heterotrimeric guanosine triphosphate (GTP) binding proteins, known as G-proteins, are a principal family of proteins serving to couple membrane receptors of the G-protein family to ionic and biochemical processes. The G-proteins are heterotrimers made of three families of subunits, α, β, and γ, which can interact specifically with discrete regions on G-protein-coupled receptors. This includes most receptors for neurotransmitters and polypeptide hormones (see **Neuroregulators**). G-protein-coupled receptors also embrace the odorant receptor family and the rhodopsin-linked visual cascade.

The underlying coupling mechanisms are defined by the enzymatic activity of the G-protein, that of hydrolyzing GTP, i.e., GTPase activity. In the inactive state, the heterotrimeric G-protein is liganded to the diphosphate GDP. Receptor activation reduces the affinity of the α-subunit for GDP and increases the affinity for GTP. The GTP-liganded complex then dissociates to the GTP-bound activated α-subunit and the β- and γ-subunits. These dissociated subunits then interact with the corresponding effectors. The effectors include adenylyl cyclase, phospholipase C, cGMP phosphodiesterase, some ion channels (K$^+$, Ca^{2+}), and receptor kinases. These signals may be excitatory or inhibitory according to the class of G-protein, some of which are listed for G-protein-linked adenylyl cyclase:

A critical component of the G-protein effector cascade is the hydrolysis of GTP by the activated α-submit (GTPase). This provides not only a component of the amplification process of the G-protein cascade but also serves to provide further measures of drug efficacy. The coupling process also depends on the stoichiometry of receptors and G-proteins. A reduction in receptor number should diminish the efficacy of coupling and thus reduce drug efficacy.

The ability of receptors to couple to G-proteins and initiate GTPase activity may also be independent of ligand.

The principal intracellular messengers derived from activation of G-protein-coupled receptors are cAMP and IP$_3$. cAMP may be degraded by phosphodiesterase (PDE) or it may activate cAMP-dependent protein kinase (PKA). The activation of this enzyme involves dissociation of the inactive form. (R$_2$C$_2$) into the active form which subsequently phosphorylates specific proteins. In contrast, IP$_3$, one of the products of receptor-mediated phospholipase C breakdown of phosphatidylinositol (PI), acts on specific receptors in the endoplasmic reticulum to release Ca^{2+} from intracellular sources. The other product of PI turnover is a 1,2-diacylglycerol that activates protein kinase C (PKC). This is also the receptor for the tumor-promoting phorbol esters. These diacylglycerols can be cleaved by monoacyl- or diacylglycerol kinases to yield arachidonic acid, a precursor to the prostaglandins and thromboxanes.

Ion Channels. The excitable cell maintains an asymmetric distribution across both the plasma membrane, defining the extracellular and intracellular environments, as well as the intracellular membranes which define the cellular organelles. This maintained asymmetric distribution of ions serves two principal objectives. It contributes to the generation and maintenance of a potential gradient and the subsequent generation of electrical currents following appropriate stimulation. Moreover, it permits the ions themselves to serve as cellular messengers to link membrane excitation and cellular response. In some instances, the current itself may be the response, as, for example, in the electric organ of electric fishes. In most instances, however, the current serves to initiate or modulate another cellular response, including propagation of impulses in nerve fibers, and alteration of the sensitivity of membranes to other stimuli or coupling to cellular responses such as contraction and secretion. In the latter examples, a role for calcium is particularly prominent because Ca^{2+} can serve as both a current-carrying and a messenger species.

Regulation of ion channels by drugs may have excitatory or inhibitory effects according to the channels affected.

Channels may be regulated exclusively by electrical or chemical signals corresponding to purely voltage-gated or ligand-gated channels, respectively. Regardless of regulatory mechanism, ion channels may be regarded as allosteric enzymes. The function is to accelerate the transit of ions across an essentially impermeable barrier and to be responsive to a variety of heterotropic signals.

Ion channels may be regarded as pharmacological receptors frequently possessing a multiplicity of drug binding sites. These sites may be for endogenous physiological regulators or for endogenous or synthetic agents.

Tyrosine Kinase Receptors. The polypeptide growth factors control cell proliferation, differentiation, and survival. Several distinct subfamilies of receptor tyrosine kinases exist and at least nine have been characterized. These include families for epidermal growth factor, insulin and insulin-related factors, fibroblast growth factors, and neurotrophin receptors such as nerve growth factor and brain-derived neurotrophic factor. All of these receptors have kinetics that share certain fundamental signaling properties. Ligand binding to the extracellular domain activates a tyrosine kinase of the cytoplasmic domain. Subsequently, a variety of downstream signaling molecules are activated. These include phospholipase C, GTPase activating factor (GAP), *Ras*, and MAP kinases.

Guanylyl Cyclase Receptors. Cyclic GMP concentrations (cGMP) rise in response to a number of cell signals. Membrane-associated guanylyl cyclase catalyzes the conversion of guanosine triphosphate (GTP) to cGMP. This enzyme resembles in organization the tyrosine kinases having an intracellular protein kinase-like domain and a cyclase catalytic domain. The enzymes are activated by several distinct species that include atrial natriuretic peptide (ANF) and peptides related to the heart-stable enterotoxins.

In contrast, the soluble guanylyl cyclases are regulated by nitric oxide and NO-forming drugs through the Ca^{2+} calmodulin-dependent nitric oxide synthase.

Stimulation (Gs)	Inhibition (Gi, Go)
β-adrenergic	opiate
H$_2$-histamine	muscarinic
dopamine	α$_1$-adrenergic
polypeptide hormones (glucagon, ACTH, etc)	adenosine (fat cells) A$_1$, prostaglandins (fat cells)
adenosine (platelets, lymphocytes) A$_2$	
prostaglandins (platelets)	polypeptide hormones
serotonin (5-HT$_{1\alpha}$)	somatostatin, neuropeptide Y, atriopeptin

Receptor Regulation and Defects. Specific recognition and the initiation of response are the accepted attributes of the drug–receptor interaction. However, target cells can alter on both short- and long-term time scales their sensitivity to drugs. Such regulation, achieved by altering the number and/or affinity of receptors, is well established for all receptor systems and can be viewed as an integral component of the drug–receptor interaction. In this view, subsequent to the formation of the drug–receptor complex with agonist, the continued existence of the drug–receptor complex may lead to one or more phases of desensitization, according to which there may occur initially transient and subsequently prolonged phases of reduced or lost sensitivity. Occupancy by antagonist, in contrast, leads to an increased number of receptors and increased drug sensitivity. This phenomenon may contribute to clinical rebound during abrupt withdrawal from drugs, including β-blockers. Additional to this homologous regulation, receptor sensitivity may be controlled through heterologous influences, whereby hormones, including thyroid and corticosteroids, regulate other receptors. These regulatory events are made possible because pharmacologic receptors, in common with other cellular components, are in dynamic balance between synthesis and degradation. This balance is sensitive to a number of influences that include agonist and antagonist presence.

There are probably several processes that contribute to the total desensitization process and these may be directed homologously (to own receptor) or heterologously (to other receptor). Additionally, the influences may be directed at the receptor itself and affect only that receptor, i.e., specific desensitization, or may affect other receptor processes as well, i.e., nonspecific desensitization.

An increasing number of diseases are known to be linked to defects in receptor structure, function, or coupling. The defects may lie at several locations: in the structure of the receptor, which may alter its ability either to bind drugs, to be inserted into the membrane, or to couple to effectors (including G-proteins); in the coupling protein; or in the presence of autoantibodies, which can proceed to activate, block, or lyse the receptors and its components.

Components of Drug Action and Responses to Drugs. The response to a drug can vary among race, gender, and age groups. It may vary according to disease state and age, and it may vary according to the time of administration. These factors may have several origins, including (*1*) compliance, the ability or desire of the subject to take a drug according to a specific regimen; (2) pharmacokinetic, disease-, age-, race-, and gender-based factors that contribute to variable absorption, distribution, metabolism, and excretion of a drug; and (*3*) pharmacodynamic, disease-, age-, race-, and gender-based factors that contribute to variable drug–receptor interactions.

DAVID J. TRIGGLE
State University of New York at Buffalo

Additional Reading

Gilman, A.G. and co-workers, eds.: *The Pharmacological Basis of Therapeutics,* 8th Edition, Pergamon Press, New York, NY, 1990.

Pratt, W.B. and P. Taylor, eds.: *Principles of Drug Action: The Basis of Pharmacology,* 3rd Edition, Churchill Livingstone, New York, NY, 1990.

Wermuth, C.G. ed.: *The Practice of Medicinal Chemistry,* Academic Press, San Diego, CA, London, UK, and New York, NY, 1996.

Wolff, M.E. ed.: *Burger's Medicinal Chemistry and Drug Discovery,* Vol. I, Principles and Practice, Wiley-Interscience, New York, NY, 1995.

PHASE DIAGRAM (Metallurgy).

A graphical representation defining the phase fields of a multiphase system, such as an alloy, in a coordinate system using the temperature and the compositions of the phases as coordinates. A phase diagram may be an equilibrium diagram, but it may also sometimes show the boundaries of the phase field under nonequilibrium conditions corresponding to specific conditions of heating or cooling. See iron carbon equilibrium diagram in entry **Iron Metals, Alloys, and Steels**.

Additional Reading

Frick, J.P.: *Woldman's Engineering Alloys,* 9th Edition, ASM International, Materials Park, OH, 2000.

Gupta, K.P.: *Phase Diagrams of Ternary Nickel Alloys: Part I and II,* ASM International, Materials Park, OH, 1990.

Kassner, M.E. and D.E. Peterson: *Phase Diagrams of Binary Actinide Alloys,* ASM International, Materials Park, OH, 1995.

Massalski, T.B.: *Binary Alloy Phase Diagrams,* 2nd Edition, ASM International, Materials Park, OH, 1990.

Massalski, T.B.: *Binary Alloy Phase Diagrams Materials Network User,* ASM International, Materials Park, OH, 1996.

Rogi, P. and J.C. Schuster: *Phase Diagrams of Ternary Boron Nitride and Silicon Nitride Systems,* ASM International, Materials Park, OH, 1992.

Subramanian, P.R., D.J. Chakrabarti, and D.E. Laughlin: *Phase Diagrams of Binary Copper Alloys,* ASM International, Materials Park, OH, 1994.

Staff: *ASM Handbook, Vol. 3, Alloy Phase Diagram,* 10th Edition, ASM International, Materials Park, OH, 1992.

Staff: ASM International *Superalloys: A Technical Guide,* 2nd Edition, ASM International, Materials Park, OH, 2002.

Villars, P., A. Prince, and H. Okamoto: *Handbook of Ternary Alloy Phase Diagrams,* ASM International, Materials Park, OH, 1995.

PHASE RULE.

The phase rule, due to Gibbs, gives the number F of intensive variables which can be fixed arbitrary in a system in equilibrium. This number is also called the variance or the number of degrees of freedom of the system. It is given by

$$F = 2 + (C' - R) - P$$

where C' is the number of components, R, the number of independent chemical reactions, and P, the number of phases.

In terms of the number of independent components, C, Equation (1) may also be written

$$F = 2 + C - P$$

If $F = 0$ the system is invariant. We cannot fix either temperature or pressure arbitrarily. Equilibrium can only be established at isolated points. An example is the *triple point* at which pure solid, liquid and vapor are in equilibrium ($F = 2 + 1 - 3 = 0$).

If $F = 1$ the system is *monovariant*. We can, for example, fix the temperature, but the equilibrium pressure is then fixed. This is the situation for a system containing one component and two phases.

If $F = 2$ the system is *bivariant*. Within certain limits both pressure and temperature can be given arbitrarily. This is the situation for $C = 1$ and $P = 1$, or $C = 2$ and $P = 2$.

PHENACITE.

The mineral phenacite is a beryllium silicate corresponding to the formula Be_2SiO_4. It is hexagonal but the crystals are usually rhombohedral in habit. It has a conchoidal fracture; is brittle, hardness, 7.5–8; specific gravity, 3; luster, vitreous; colorless to yellowish or reddish, sometimes brown; transparent to translucent. Phenacite is found in pegmatites with topaz, quartz and microcline, and occurs also in emerald-bearing mica schists of the Ural Mountains. It is found also in France, Norway, Switzerland, Africa, Brazil, and Mexico; and in the United States, in Oxford County, Maine; Carroll County, New Hampshire; and in Chaffee and El Paso Counties in Colorado. It derives its name from the Greek meaning deceiver, as it resembles quartz and topaz with which it is associated. It is sometimes spelled phenakite. It has been used as a gem.

PHENOCLAST.

A textural term proposed by R.M. Field in 1916 for coarsely graded clastic sedimentary rocks in which the largest or "show" particles or fragments are referred to as phenoclasts, regardless of their shape or composition. The term implies that the larger constituents of the glomerate have been derived from prelithified rock. Rounded fragments are called pebbles or spheroclasts, which when lithified by means of matrix (sand and clay) and cement form a conglomerate. Angular fragments are called anguclasts (Field), which when lithified by means of matrix (sand and clay) and cement form a breccia.

PHENOCRYST.

A textural term proposed by Iddings in 1892 for macroscopic crystals which are relatively much larger than the crystalline matrix of the igneous rock in which they occur. Rocks which have phenocrysts are called porphyritic. The term phenocryst is derived from the Greek, meaning show, and crystal.

PHENOL.

1. A class of aromatic organic compounds in which one or more hydroxy groups is attached directly to the benzene ring. Examples are phenol itself (benzophenol), the cresols, xylenols, resorcinol, naphthols. Although technically alcohols, their properties are quite distinctive.

2. Phenol (carbolic acid; phenylic acid; benzophenol; hydroxybenzene), C_6H_5OH. Phenol is a white, crystalline substance that turns pink or red if not perfectly pure, or if under influence of light; absorbs water from the air

and liquefies. It has a distinctive odor and a sharp burning taste. It is toxic by ingestion, inhalation, and skin absorption, and is a strong irritant to tissue. When in a very weak solution, phenol has a sweetish taste; specific gravity 1.07; mp 42.5–43°C; bp 182°C; flash point 77+ °C. Soluble in alcohol, water, ether, chloroform, fixed or volatile oils, and alkalies.

Most of the phenol used in the United States is made by the oxidation of cumene, yielding acetone as a byproduct. The first step in the reaction yields cumene hydroperoxide, which decomposes with dilute sulfuric acid to the primary products, plus acetophenone and phenyl dimethyl carbinol. Other processes include sulfonation, chlorination of benzene, and oxidation of benzene. The compound is purified by rectification.

Major uses of phenol include production of phenolic resins, epoxy resins, and 2,4-D (regulated in many countries); as a selective solvent for refining lubricating oils; in the manufacture of adipic acid, salicylic acid, phenophthalein, pentachlorophenol, acetophenetidine, picric acid germicidal paints, and pharmaceuticals; as well as use as a laboratory reagent. Special uses include dyes and indicators, and slimicides.

High-boiling phenols are mixtures containing predominantly meta substituted alkyl phenols. Their boiling point ranges from 238–288°C they set to a glass below −30°C. They are used in phenolic resins, as fuel-oil sludge inhibitors, as solvents and as rubber chemicals.

Phenol is regarded as a dangerous chemical. Refer to *Dangerous Properties of Industrial Materials*, 10th Edition, Sax and R.J. Lewis, Editors, Wiley, New York, 1999.

PHENOLATE PROCESS. A process for removing hydrogen sulfide from gas by the use of sodium phenolate, which reacts with the hydrogen sulfide to give sodium hydrosulfide and phenol. This can be reversed by steam heat to regenerate the sodium phenolate.

PHENOL COEFFICIENT. In determining the effectiveness of a disinfectant using phenol as a standard of comparison, the phenol coefficient is a value obtained by dividing the highest dilution of the test disinfectant by the highest dilution of phenol that sterilizes a given culture of bacteria under standard conditions of time and temperature.

See also **Disinfectant**.

PHENOL-FURFURAL RESIN. A phenolic resin that has a somewhat sharper transition than phenol-formaldehyde from the soft, thermoplastic stage to the cured, infusible state and can be fabricated by injection molding since it has little tendency to harden before curing conditions are reached.

See also **Phenolic Resins**.

PHENOLIC RESINS. These resins have been known since the 1870s when Baeyer first investigated the reactions of phenols and aldehydes. However, his findings were not commercially utilized until Dr. L. H. Baekeland disclosed his classic work in 1907. Through his use of high-pressure molding, he provided a solution to the problem of making quick-curing moldings which did not blister or crack. Contemporary with Dr. Baekeland was the work of Lebech and Aylsworth who provided the key to the application of large commercial quantities of phenolic Novolacs by suggesting the use of hexamethylenetetramine as a curing agent. Since that time, and because of their desirable price-to-property relationship, phenolic resins have enjoyed steady growth despite the encroachment into their areas of application by a few thermoplastics and other thermosetting materials.

Although in the pure state phenolic resins are quite weak and brittle, they are highly regarded among the plastic materials as being capable of producing very strong physical bonds with a large variety of materials at very low concentration. Consequently, phenolics have found use in many applications as binders. In addition to the strength they impart as bonding agents in matrixes, phenolic resins also possess resistance to chemical attack by all but the most polar organic solvents. The only inorganic reagents that have a deleterious effect on them are the strongest and most oxidizing of the acids and the strongest bases.

For many years prior to the development of high-temperature thermoplastics and thermosets, such as the polyimides, polysulfones, and epoxies, phenolic molding material dominated the high temperature-resistant market. This emphasizes their ability to resist temperature degradation in the 400–500°F (204–260°C) range. Because phenolics were found to possess excellent ablative properties, it has been reported that both the American and Soviet space efforts used them in combination with certain other polymeric compounds in heat shield materials. Phenolics, like other aromatic hydrocarbon-based resins, possess excellent resistance to high-energy radiation degradation.

Unlike most thermoplastics, phenolic resins and moldings are characterized by high flexural modulus and good tensile strength while having relatively low impact resistance. In addition to their good physical strength properties, phenolic resins are used in the manufacture of many electrical devices where high dielectric breakdown strength and electrical resistance, in combination with excellent dimensional stability, are required.

Chemistry of Phenolics

Phenolic compounds are capable of chemically combining with a large number of aldehydes and other compounds to yield an almost infinite spectrum of modified polymers. However, the reaction of a phenol with an aldehyde (most commonly encountered is that between phenol and formaldehyde) leads to the formation of only two classes of phenolic resins. These are Novolacs and resols. In general, these two classes of resins may be differentiated by the fact that Novolacs are prepared with an acid catalyst and substantially less than one mole of aldehyde per mole of phenol and require the addition of a curing catalyst to become thermosetting; while resols, or single-stage resins as they are commonly called, are prepared with from 1 to 3 moles of aldehyde per mole of phenol and employ a basic condensation catalyst, and are inherently thermosetting.

Cured single-stage and two-stage resins.

Novolacs

The aldehyde content of Novolac resins is insufficient to render the resin thermosetting, hence, they are true thermoplastics provided that a curing agent is not added. Novolacs may be stored indefinitely in the pulverized state at moderate temperatures even mixed with curing agent.

The final cure speed of simple Novolacs may be accurately controlled by the use of the proper condensation catalyst in the initial phase of the reaction. Structurally, a Novolac consists of a series of phenol nuclei joined by methylene

$$
(-\overset{\displaystyle H}{\underset{\displaystyle H}{C}}-)
$$

links at the *o* and *p* positions. Only two of the three possible *o* and *p* positions on each ring within the polymeric chain are substituted with a methylene group. Only one position on each of the two terminal phenol groups is substituted with a methylene group, hence, two positions on each terminal ring and one on each internal ring is available for future reactions, including the curing reactions. When the unsubstituted positions are predominantly the para positions, very fast-curing resins result. Slower-setting resins are obtained when the unoccupied positions are the ortho positions. When hexa is used as the hardener at approximately the 10% level, the reactive sites are joined by a

$$
-\overset{\displaystyle H}{\underset{\displaystyle H}{C}}-N-\overset{\displaystyle H}{\underset{\displaystyle H}{C}}-
$$

linkage. One mole of ammonia is liberated for approximately every three of the above links formed.

Unaltered Novolacs and two-stage resins* find application in grinding wheel bonding, molding material, brake linings and clutch faces, foundry sand binding, premix and wood fiber bonding, thermal insulation. Modified phenolics are found in adhesives, coatings, and aerospace applications.

Resols

On the other hand, resols contain sufficient aldehyde to make them thermosetting without a curing agent. Consequently, they have only finite storage stability and care must be exercised to minimize both the length and the temperature of storage. The high aldehyde-to-phenol ratios used in the preparation of resols insure that a high percentage of the reactive *o* and *p* positions are utilized in either methylene links or are substituted by a hydroxymethyl group. It is these hydroxymethyl groups which function as crosslinking sites in the final curing reaction. In phenolformaldehyde resols, the ratio of methylene groups to hydroxymethyl groups is an important factor in determining the solubility of these resins. Low ratios insure water solubility in almost infinite proportions, while resins with high ratios can be dissolved in only low molecular weight alcohols, ketones, ethers, and esters. Solubility of resols in nonpolar solvents, such as hydrocarbons, is always very low.

In addition to acting as a crosslinking site, the hydroxymethyl group and unsubstituted *o* and *p* positions may be used as reactive sites to join numerous other compounds to the phenolic polymer. These modified resins often possess many properties normally not attributable to phenolics in the unaltered state.

Resols find application as impregnating resins, in laminating paper, cloth, glass and asbestos; as a pickup agent in grinding wheels, exterior and marine plywood, premix and granular molding material, adhesives, wood waste and particle board manufacture, and in coatings.

The final curing step of both classes of phenolics is accomplished by exposing the resin or the resin containing matrix to temperatures in the 190–450°F (88–232°C) range for an appropriate length of time to render the resin infusible. In most applications high pressure is also applied concurrently with the heating cycle to eliminate blistering, which would normally occur if the trapped gases (ammonia in the case of Novolacs, and water in the case of resols) generated in the cure were allowed to escape unrestrained.

Comparative infrared analysis of the two classes of resins in their pure state show that only resols have strong absorptions in the 1,000 and 880 cm^{-1} range. As Novolacs require a curing agent, their presence may be inferred by a strong, sharp absorption at 510 cm^{-1} which is indicative of the most common curing agent used, hexamethylenetetramine. Unfortunately, hexa also has strong absorptions at or near 1,000 and 880 cm^{-1} which makes differentiation between resols and two-stage

---

* By convention, two-stage resins are defined as a mixture of Novolac and curing agent, which is capable of thermosetting.

resins, by infrared spectroscopy, possible only for the experienced. In the cured state, it is very difficult to determine the class identity of an unknown sample.

PHILLIP A. WAITKUS
Plastics Engineering Company
Sheboygan, Wisconsin

PHENOLICS. These are products of the condensation reaction of phenol and formaldehyde.[1] Water is the byproduct of this reaction. Substituted phenols and higher aldehydes may be incorporated to achieve specific resin properties, e.g., flexibility, reactivity, or compatibility with elastomers and other polymers. A variety of phenolic resins can be produced by adjusting the formaldehyde:phenol molar ratio and the resinification temperature and catalyst. Single-stage (*resole*) resins are produced with an alkaline catalyst and a molar excess of formaldehyde. The reaction is carefully controlled to allow the production of low-molecular-weight, noncrosslinked resins. Single-stage resins complete the curing reaction in a heated mold with no additional catalyst. A three-dimensionally crosslinked, insoluble, and infusible polymer is formed.

Two-stage resins (*novolacs*) are produced by the acid-catalyzed reaction of phenol and a portion of the required formaldehyde. The resin product is brittle at room temperature. It can be melted, but it will not crosslink. Novolacs can only be cured by the addition of a hardener, almost always formaldehyde supplied as hexamethylene tetramine. Upon heating, the latter compound decomposes to yield ammonia and formaldehyde.

Phenolic resins are available in flake, powder, and liquid forms. A wide variety of industrial applications has been developed, including foundry molds and cores; plywood and particle board; brake and clutch linings; glass, cellulose, and foam insulation; grinding wheels and coated abrasives; adhesives and glues, rubber tackifiers; coatings; varnishes; and electrical and decorative laminates. Resole and novolac resins are combined with a variety of fillers, reinforcements, and additives to produce phenolic engineering plastics. See also **Paints and Coatings**.

Application development for phenolics has been spurred by weight and cost savings inherent in metal replacement and parts consolidation. Thermoplastics have been replaced by phenolics where creep resistance and thermal stability are required in downsized parts or applications in hostile environments.

See also **Phenolic Resins**.

PHENYLALANINE. See **Amino Acids**.

PHENYLBUTAZONE. See **Analgesics, Antipyretics, and Antiinflammatory Agents**.

PHILLIPSITE. The mineral phillipsite is a zeolite, a hydrous silicate of potassium, calcium, and aluminum, corresponding to formula $(K, Na_2Ca)(Al_2Si_4)O_{12} \cdot 4-5H_2O$. It is monoclinic, forming penetration twins, and sometimes crosses resembling orthorhombic or tetragonal forms. It also may occur in radial groups. Phillipsite is a brittle mineral; hardness, 4–4.5; specific gravity, 2.2; luster, vitreous; color, white to light red; translucent to opaque. Like other zeolites, it is found in veins and cavities in basalts, and sometimes in more acidic rocks. It is believed to be a low-temperature mineral. Phillipsite is found in Italy, especially in the lavas of Vesuvius and Monte Somma, and in the basalts of Germany, Ireland, and Australia. It has been reported from Greenland. This mineral was named in honor of the British mineralogist William Phillips.

PHLOGOPITE. The mineral phlogopite is a magnesium-bearing mica, with but little iron, corresponding essentially to the formula $K(Mg, Fe)_3(AlSi_3)O_{10}(F, OH)_2$. Fluorine is sometimes present. This mica is monoclinic like muscovite, biotite and lepidolite, forming prismatic crystals, occasionally very large, and occurring also in scales and plates. Its cleavage is basal and highly perfect with elastic laminae; hardness, 2–2.5; specific gravity, 2.76–2.90; luster, pearly to submetallic; color, yellowish-brown, green, white and colorless; transparent to translucent; may exhibit asterism, probably due to minute inclusions. Phlogopite is more nearly a characteristic of metamorphic than igneous rocks although occasionally occurring in the latter if they are rich in magnesia and with but little iron. Phlogopite is found especially in Rumania, Switzerland, Italy, Finland,

---

[1] Data furnished by *Occidental Chemical Corp.*

Sweden and Madagascar where it occurs in the crystalline limestones in huge crystals. In the United States it occurs in New York State at Edwards, Hammond, DeKalb, Monroe and, in New Jersey, at Franklin. In Canada it is found at many places in Ontario and Quebec.

The name phlogopite comes from the Greek word meaning like fire, referring to the copper-like reflections often observed in the reddish-brown varieties.

Phlogopite is in demand commercially by the electrical industry for use as an insulator.

PHONONS. Many of the thermal and vibrational properties of solids can be explained by considering the material to be a volume made up of a gas of particles called *phonons*. This particle description is a method of taking into account the actual motion of the atoms and molecules in the solid. Since each atom possesses energy due to its thermal environment, and since there are forces between the atoms that keep the solid together, each atom tends to oscillate about its equilibrium position. The formal mathematical development, obtained through solving the equations of motion of the array of individual atoms and molecules, indicates that the thermal energy of the solid is contained in certain combinations of particle vibrations which are equivalent to standing elastic waves in the sample and are called normal modes. Each normal mode contains a number of discreet quanta of energy $E = \hbar$ where ω is the frequency of the mode (or wave) and $\hbar$ is Planck's constant divided by 2π. Each of these quanta is called a phonon (in analogy with the light quanta or photon whose energy-frequency relationship is identical). Phonons are considered only as particles, each having an energy $E = \hbar$, a momentum q, and a velocity $v = \partial \omega / \partial q \sim \omega / q$. Analogous to the energy levels of electrons in a solid, phonons can have only certain allowed energies.

The phonon is of importance to many phenomena: electron mobility, optical absorption, electron spin resonance, electron tunneling, and superconductivity. The phonon spectrum represents a detailed picture of the forces that hold solids together. Thus, it is clear why the phonon has been and will continue to be of fundamental importance in solid-state physics.

PHOSGENE. See **Carbonyls**; **Chlorinated Organics**.

PHOSGENITE. This mineral is a chlorocarbonate of lead, $Pb_2(CO_3)$ Cl_3, crystallizing in the tetragonal system, associated with other lead minerals of secondary origin, e.g., cerussite and anglesite; hardness, 2–3; specific gravity 6.133; prismatic to tabular crystals, also massive and granular, adamantine luster; color, white, gray, brown, green or pink; transparent to translucent. Some specimens show yellowish fluorescence under ultraviolet light.

Found in the United States in California, Colorado, Arizona, and New Mexico. Magnificent crystals up to 5 inches (12.5 centimeters) in diameter have been found at Monte Poni, Sicily; as fine crystals in England at Derbyshire and Matlock; and in Poland, Russia, Tasmania, Australia, and Namibia.

PHOSPHATE FERTILIZERS. See **Fertilizer**.

PHOSPHATE ROCK. A natural rock consisting largely of calcium phosphate and used as a raw material for manufacture of phosphate fertilizers, phosphoric acid, phosphorus, and animal feeds. Recovery of uranium from the manufacture of phosphoric acid and other phosphate chemicals is expected to become an important source of this metal. Phosphate rock is the primary source of superphosphate, prepared by treatment of the pulverized rock with sulfuric acid (superphosphate having 16–18% P_2O_5) or by acidifying with phosphoric acid (triple superphosphate having 40–48% P_2O_5). Nitric acid is sometimes used, i.e., nitrophosphate. Defluorinated phosphate rock is the source of phosphate used in animal feeds and feed concentrations. Important deposits are in the U.S. (Florida, North Carolina, Tennessee, California, Wyoming, Montana, Utah, and Idaho), North Africa (Morocco, Libya, and Algeria), the former U.S.S.R., and various islands in the Pacific.

See also **Fertilizer**; **Phosphoric Acid**; and **Phosphorus**.

PHOSPHATES. See **Phosphorus**.

PHOSPHATING. See **Conversion Coatings**.

PHOSPHAZENE (Phosphonitrile). A ring or chain polymer that contains alternating phosphorus and nitrogen atoms with two substituents on each phosphorus atom. Characteristic structures are cyclic trimers, cyclic tetramers, and high polymers. The substituent can be any of a wide variety of organic groups, halogen, amino, etc. Most cyclic trimers are crystalline, solids, organosoluble, and stable to weather conditions; the high polymers (polyphosphazenes) are elastomeric or thermoplastic. A copolymer of phosphazene and styrene has been investigated for use as a flame-retardant.

PHOSPHINE. See **Phosphorus**.

PHOSPHOLIPIDS. These compounds belong to a group of fatty acid compounds sometimes referred to as complex lipids. The simplest are esters of fatty acids with glycerol phosphate and are called *phosphatidic acids*. There are also phosphatidylcholines or lecithins, phosphatidylethanolamines, phosphatidylserines, and phosphatidylinositols. The latter may have one or more additional phosphate groups attached to the inositol. A similar series also exists containing an aldehyde attached to the 1-position of the glycerol, in the form of an α, β-unsaturated ether. These are commonly referred to as *plasmalogens*.

The percentage of phospholipid content of tissues varies little under normal physiological conditions, thus giving rise to the term *element constant*, in contrast to the triglycerides, which have been called the *element variable*.

Phospholipids are considered to be involved in the transport of triglycerides through the liver, especially during mobilization from adipose tissue. Conditions which could be interpreted as interfering with phosphatidylcholine formation, such as deficiency of choline or its precursors, result in a pronounced increase in liver triglycerides.

Mitochondrial phospholipids play a role in electron transport and oxidative phosphorylation, two mechanisms by which the cell accomplishes the final oxidation of the metabolites to produce energy. Phospholipids also are linked in the transport of ions, especially sodium, across membranes.

In summary, phospholipids (phosphatides) comprise a group of lipid compounds that yield, upon hydrolysis, phosphoric acid, an alcohol, fatty acid, and a nitrogenous base. They are widely distributed throughout nature.

PHOSPHOR BRONZE. See **Copper**.

PHOSPHORESCENCE. See **Luminescence**.

PHOSPHORIC ACID. [CAS: 7664-38-2]. Generally the term *phosphoric acid* refers to orthophosphoric acid, H_3PO_4. Anhydrous orthophosphoric acid is a white, crystalline solid, which melts at 42.35°C. It forms a hemihydrate, $2H_3PO_4 \cdot H_2O$, which melts at 29.32°C. Although it is possible to produce almost any desired concentration, it is common practice to supply the material as a solution containing from 75% H_3PO_4 (melting point, 17.5°C) to 85% H_3PO_4 (melting point, 21.1°C). When phosphoric acid is heated to temperatures above about 200°C, water of constitution is lost. A series of acids is formed by the dehydration, ranging from pyrophosphoric acid, $H_4P_2O_7$, to metaphosphoric acid, $(H_3PO_4)_n$. Salts of the dehydrated acids are used for the preparation of certain types of liquid fertilizers and have been used in some detergents. However, to counter the effects of "phosphate pollution," there has been a serious cutback in this latter use of the phosphates. See also **Fertilizer**.

One, two, or three of the hydrogens in phosphoric acid may be neutralized, leading to a series of products which range widely in their hydrogen ion concentration (pH): monosodium phosphate, NaH_2PO_4, with a pH of 4.0; disodium phosphate, Na_2HPO_4, with a pH of 8.3 (approximate); and trisodium phosphate, Na_3PO_4, with a pH of 12.0. Other phosphorous acids of little commercial importance are hypophosphorous acid, H_3PO_2; orthophosphorous acid, H_3PO_3; and pyrophosphorous acid, $H_4P_2O_5$.

Manufacture

The major sources of H_3PO_4 traditionally have been mineral deposits of phosphate rock. Mining operations are extensive in a number of locations, including the United States (Florida), the Mediterranean area, and Russia, among others. The major constituent of most phosphate rocks is fluorapatite, $3Ca_3(PO_4)_2 \cdot CaF_2$. The supply of high-grade phosphates, the raw material of choice for producing high-purity phosphoric acid by the wet process, is rapidly decreasing in some areas.

Two major methods are utilized for the production of phosphoric acid from phosphate rock. The *wet process* involves the reaction of phosphate rock with sulfuric acid to produce phosphoric acid and insoluble calcium sulfates. Many of the impurities present in the phosphate rock are also solubilized and retained in the acid so produced. While they are of no serious disadvantage when the acid is to be used for fertilizer manufacture, their presence makes the product unsuitable for the preparation of phosphatic chemicals.

In the other method, the *furnace process*, phosphate rock is combined with coke and silica and reduced at high temperature in an electric furnace, followed by condensation of elemental phosphorus. Phosphoric acid is produced by burning the elemental phosphorus with air and absorbing the P_2O_5 in water. The acid produced by this method is of high purity and suitable for nearly all uses with little or no further treatment.

Basic reactions of the wet process are

$$3\ Ca_3(PO_4)_2 \cdot CaF + 10\ H_2SO_4 + 20\ H_2O$$

$$\longrightarrow 10\ CaSO_4 \cdot 2\ H_2O + 6\ H_3PO_4 + 2\ HF$$

Numerous side reactions also occur. Phosphate rock and sulfuric acid, together with recycled weak liquors, are carefully metered to a large, stirred reactor, providing retention for 4–8 hours. Conditions in the reaction are carefully controlled to maintain preselected conditions. Temperatures (77–83°C) are controlled by removing excess heat of reaction with a vacuum cooler, or by blowing air through the phosphoric acid slurry. The slurry contains precipitated gypsum and is sent to a filter. The gypsum is washed with water in several countercurrent steps, and weak liquor is returned to the reaction stage. For most uses, the acid requires further concentration, normally done in vacuum evaporators. Merchant-grade acid is generally concentrated to about 54% P_2O_5 (75% H_3PO_4). See Fig. 1.

Effluents and gypsum disposal pose problems. Fluorine is evolved at various steps in the process and scrubbers are required to reduce release to the atmosphere. Gypsum is frequently piled in diked areas or dumped into abandoned mines. Wastewater from these plants is heavily contaminated with fluorine, phosphates, sulfates, and other compounds. It is commonly impounded in large ponds, where a portion of the contaminants may precipitate or be lost by other processes. The cooled effluent from the

Fig. 1. View of three-stage evaporation process used for concentrating wet-process phosphoric acid. (*Swenson*)

ponds is recycled to the production unit. Any excess water must be treated with lime before it can be allowed to enter streams.

Developments of recent years include plants designed to precipitate the calcium sulfate in the form of the hemihydrate instead of gypsum. In special cases, hydrochloric acid is used instead of sulfuric acid for rock digestion, the phosphoric acid being recovered in quite pure form by solvent extraction. Solvent-extraction methods have also been developed for the purification of merchant-grade acid, which normally contains impurities amounting to 12–18% of the phosphoric acid content. Processes for recovering part of the fluorine in the phosphate rock are in commercial use.

Although more costly to operate, the electric-furnace process produces phosphoric acid of high purity. A mixture of coke, silica, and phosphate rock is formed into nodules by heating in a nodulizing kiln, and the resulting lump material is transferred to the electric furnace, where it is heated with an electric current introduced by means of graphite electrodes. The entire charge is melted, and elemental phosphorus is volatilized. The slag is tapped off intermittently while the phosphorus vapor is condensed. The phosphorus is then burned in air and the P_2O_5 is absorbed in water. Reactions are

$$2\ Ca_3(PO_4)_2 + 6\ SiO_2 + 10\ C \longrightarrow P_4 + 10\ CO + 6\ CaSiO_3$$

$$P_4 + 5\ O_2 \longrightarrow 2\ P_2O_5$$

$$P_2O_5 + 3\ H_2PO \longrightarrow 2\ H_3PO_4$$

PHOSPHORIC ACID (Fuel Cell).   See **Fuel Cells**.

PHOSPHORS AND PHOSPHORESCENCE.   A large variety of substances become luminescent when stimulated or excited by suitable radiation, or by emissions, such as cathode rays or beta-rays. This phenomenon is complex and exhibited in various aspects. In some cases, the light is emitted only so long as the exciting emission is maintained, in which case it is called *fluorescence*. In other cases, the luminescence persists after the excitation is removed and it is then called *phosphorescence*. It has long been known, for example, that zinc sulfide, under certain conditions, glows brightly for a time after exposure to daylight or lamplight, but the luminosity decays rapidly and disappears, usually within a few minutes. The electroluminescent phosphor of zinc sulfide-zinc selenide-copper has the property that the wavelength of the emitted radiation increases with increasing selenium content. The white luminescence of some television tubes is obtained from a combination of cadmium-zinc sulfide phosphors, one that is blue-emitting and the other yellow-emitting. Also, in some color television tubes, the blue-emitting and green-emitting phosphors are of the sulfide type, but earlier use of sulfides for the red-emitting "dots" on the tube surface were replaced by rare-earth red-emitting phosphors. One composition used is prepared by combining about 4% europium oxide and 65% yttrium oxide, with various vanadium compounds and calcining the mixture. Rare-earths also have been used in producing phosphors for high-pressure mercury-arc lamps. These phosphors increase the proportion of red light emitted by reducing the green, blue, and ultraviolet portions.

Quantitatively, phosphorescence may be defined as luminescence that is delayed by more than 10^{-8} seconds after excitation. It may be associated with transitions from a higher excited state to a lower one, the energy going into a radiationless rearrangement of the system. If the lower state is metastable, its lifetime may be considerable before it finally decays by a highly forbidden radiative transition to the ground state. In the case of zinc sulfide, the process depends upon the ionization of activator atoms, the freed electrons being trapped and only released slowly for recombination.

See also **Luminescence**.

PHOSPHORUS.   [CAS: 7723-14-0]. Chemical element, symbol P, at. no. 15, at. wt. 30.9738, periodic table group 15, mp 44.1°C (α-white), bp 280°C (α-white), sp gr 1.82 (white), 2.20 (red).

Four allotropes of phosphorus are known, the hexagonal β-white, stable only below −77°C, the cubic α-white (mp 44.1°C), the violet, and the black (which is thermodynamically the most stable). The α-white form is usually taken as the standard state. The violet is obtained by continued heating at 500°C of a solution of phosphorus in lead. When α-white phosphorus is heated to 250°C in the absence of air, a red variety (mp 590°C) is obtained which is believed to consist of a mixture of the α-white and violet allotropes, although the studies of the violet component in the mixture have shown that at least four polymorphic forms of red (violet) phosphorus exist.

White phosphorus is considered to be made up largely of P_4 molecules, as is the liquid and vapor up to 800°C, where dissociation becomes appreciable. The P_4 molecule is a tetrahedron, with single covalent bonds between the P atoms, and each having an unshared pair of electrons. White phosphorus is much more reactive than red or violet.

Black phosphorus has a graphite-like structure and has a similar electrical conductivity.

There is one stable nuclide, ^{31}P. Six radioactive isotopes have been identified, ^{28}P through ^{30}P and ^{32}P through ^{34}P, all with short half-lives, measured in terms of seconds, minutes, or days. See also **Radioactivity**. In terms of terrestrial abundance, phosphorus ranks 10th with an estimated average content of igneous rocks being 0.13% phosphorus. The element ranks 19th in abundance in seawater, there being an estimated 325 tons of phosphorus per cubic mile (70 metric tons per cubic kilometer) of seawater. In terms of cosmic abundance, phosphorus is ranked 15th among the elements. The element was first identified by Hennig Brandt in Germany in 1669 during an experiment in which he was distilling urine with sand and coal. White phosphorus is very toxic.

First ionization potential 11.0 eV; second, 19.81 eV; third, 30.04 eV; fourth, 51.1 eV; fifth, 64.698 eV. Oxidation potentials $H_3PO_2 + H_2O \longrightarrow H_3PO_3 + 2H^+ + 2e^-$, 0.59 V; $P + 3H_2O \longrightarrow H_3PO_3 + 3H^+ + 3e^-$, 0.49 V; $P + 2H_2O \longrightarrow H_3PO_2 + H^+ + e^-$, 0.29 V; $H_3PO_3 + H_2O \longrightarrow H_3PO_4 + 2H^+ + 2e^-$, 0.20 V; $PH_3(g) \longrightarrow P + 3H^+ + 3e^-$, 0.04 V; $P + 2OH^- \longrightarrow H_2PO_2^- + e^-$, 1.82 V; $P + 5OH^- \longrightarrow HPO_3^{2-} + 2H_2O + 3e^-$, 1.71 V; $H_2PO_2^- + 3OH^- \longrightarrow HPO_3^{2-} + 2H_2O + 2e^-$, 1.65 V; $HPO_3^{2-} + 3OH^- \longrightarrow PO_4^{3-} + 2H_2O + 2e^-$, 1.05 V; $PH_3(g) + 3OH^- \longrightarrow P + 3H_2O + 3e^-$, 0.87 V.

Other physical characteristics of phosphorus are given under **Chemical Elements**.

Because of its reactivity, phosphorus does not occur in nature in the elemental form. Phosphate rock is the principal source of phosphorus and phosphorus compounds. Very large deposits of phosphate rock occur and are worked in the Bone Valley area of Florida, as well as deposits in Tennessee, Idaho, and South Carolina. Large deposits are mined in Northern Africa (Morocco and Tunisia). Very significant reserves have been found on several of the Pacific Islands, the reserves on Christmas Island estimated at some 30 million tons (27 million metric tons) and those on Nauru Island in excess of 100 million tons (90 million metric tons). There also are large active mining operations in the Mediterranean area as well as in the former Soviet Union. Known reserves assure a supply for several centuries. The mineral apatite, $Ca_3(PO_4)_2 \cdot CaCl_2$ or $\cdot CaF_2$, found in Quebec, Virginia, Brazil, and the South Pacific also contains high percentages of phosphorus, up to 20% P_2O_5. The main constituents of most phosphate rocks is fluorapatite, $3Ca_3(PO_4)_2 \cdot CaF_2$. These rocks contain 30–37% P_2O_5.

Most phosphorus raw materials are converted into phosphorus and phosphorus compounds, such as phosphoric acid, on an extremely high-tonnage basis. Percentagewise, relatively little elemental phosphorus is produced for consumption as an end product. See also **Fertilizer**. Phosphorus is important both to plant and animal nutrition. Traditionally, phosphorus compounds have been key components of cleaning compounds and detergents although there have been trends to reduce or eliminate phosphates from high-consumption items. See also **Detergents**.

Production of Elemental Phosphorus

The tricalcium phosphate in phosphate rock, mixed with coke and silica, is thermally reduced to yield P_2 vapor. The phosphorus vapors condense to a liquid and the carbon monoxide produced is returned for burning in the furnace. The process requires much heat and, in addition to the heat provided by the combustion of the coke and the heating value of the recycled carbon monoxide, an electric arc also is used. The reaction takes place in very large furnaces at a temperature of 1,300–1,500°C and at atmospheric pressure. A 70-MW furnace will produce 44,000 short tons (39,600 metric tons) of P_4 per year, equivalent to 100,000 tons (90,000 metric tons) of P_2O_5 (if converted to acid). Although there are numerous intermediate and side reactions, the overall reaction is: $Ca_3(PO_4)_2 + 5C + 3SiO_2 \longrightarrow P_2 + 5CO + 3Ca \cdot SiO_3$. Byproduct ferrophosphorus alloy and calcium silicate slag are tapped from the furnace periodically. Maximum furnace efficiency occurs when the SiO_2/CaO weight ratio is about 0.8. This ratio also assures a minimum melting-point eutectic for the melt and thus lengthens furnace life. This process was originally developed by Readman in England in 1888. The first 1,500-kW furnace in the United States was installed at Niagara Falls, N.Y. in 1896 because of the availability of low-cost energy. For many years the proximity of Tennessee brown stone (a phosphate rock) to the low-cost power of the Tennessee Valley Authority made a good economic combination. Worldwide production of phosphorus by this process is about $\frac{3}{4}$-billion short tons (0.675 billion metric tons) per year (installed capacity). In the United States, about 80% of the phosphorus produced is immediately converted to the oxide and thence to phosphoric acid. The remaining 20% has gone into alloys, organic intermediates for oil and fuel additives, pesticides, plasticizers, and pyrotechnics. In addition to use in detergents, cleaning compounds, and degreasing formulations, phosphoric acid has been consumed in the preparation of liquid fertilizers, water-treatment, pharmaceutical, and chemical products. Phosphorus-containing fertilizers, such as single superphosphate, wet-process orthophosphoric acid, triple superphosphate, ammonium phosphate, and nitrophosphates do not require elemental phosphorus (or the resulting pure P_2O_5) in their preparation, but are manufactured by directly reacting phosphate rock with requisite chemicals, such as H_2SO_4 or HNO_3. See also **Fertilizer**.

Chemistry and Compounds

Like carbon, phosphorus is covalently bound to its neighboring atoms in all of its compounds, except perhaps for some metallic phosphides. Indeed, the chemistry of carbon and that of phosphorus are somewhat similar as might be expected from the diagonal relationship of these elements in the periodic table.

Probably the major difference between carbon and phosphorus is that the former element is quite closely restricted to the use of s- and p-orbitals, because of the relatively high energy of d-orbitals in the case of second-period elements; whereas, phosphorus, being a third-period element, can use d-orbitals in bonding. For both carbon and phosphorus, the most common hybridization for σ-bonding is approximately the tetrahedral sp^3. However, in order to form π-bonds, carbon must go to lower hybrids: sp^2 and sp. Phosphorus, on the other hand, does not do this but can employ d-orbitals for π-bonding. This difference between carbon and phosphorus in sigma bond strength, and the ease with which phosphorus uses its d-orbitals for attachment of attacking nucleophilic groups, can be used to explain why catenation is common in carbon compounds, while at the same time phosphorus compounds containing long chains of connected phosphorus atoms have not yet been synthesized.

The known coordination numbers exhibited by phosphorus within the molecule-ions containing this element are 1, 3, 4, 5, and 6 which, to at least a first approximation, exhibit the symmetry of p, p^3, sp^3, sp^3d and sp^3d^2 hybridization, respectively. A very large number (several thousand in each case) of triply- and quadruply connected phosphorus compounds are known; but there are only a few compounds of higher coordination number in which d-orbitals are involved in the σ-bond base structure. These are the halogen compounds PF_5, PCl_5, PBr_5, PCl_2F_3, PBr_2F_3 and the pentaphenyl compound $(C_6H_5)_5P$, in which the phosphorus is quintuply connected to its neighboring atoms, and the PF_6^- and PCl_6^- anions, in which the phosphorus has a six-fold coordination. The singly connected phosphorus atoms appear only in compounds occurring at very high temperatures. Although singly connected phosphorus is not known under ordinary conditions, interpretation of diatomic spectra has given considerable information.

Several generalities can be stated concerning the phosphorus compounds that are stable under normal conditions:

1. In those compounds in which phosphorus shares electrons with three neighboring atoms, there are three σ-bonds, with little or no π-character, from the phosphorus.
2. In those compounds in which phosphorus shares electrons with four neighboring atoms there are four σ-bonds, with an average of about one π-bond per P atom.
3. When electrons are shared with five or six neighboring atoms, there is less than one full σ-bond for each connection between the phosphorus and a neighboring atom with apparently very little π-bonding.

These generalities are obviously dependent to a considerable extent upon the specific atoms connected to the phosphorus and, indeed, it is possible that the observed differences between the triply and quadruply connected phosphorus atoms may be attributed primarily to the individual ligands. Fluorine appears to contribute nearly as much shortening (assuming that the tabulated values for the fluorine bond length are correct) to the P—F connection in the triply connected as in the quadruply connected phosphorus compounds. On the other hand, chlorine shows essentially no shortening, whether attached to either triply or quadruply connected phosphorus.

Phosphine. [CAS: 7803-51-2]. PH_3, and its substitution products, have a pyramidal structure. The P—H bond length is 1.42 Å, and the H—P—H angle is 93°. Hypophosphites, containing the radical

are produced by alkaline hydrolysis of white phosphorus. The barium salt yields hypophosphorous acid, H_3PO_2, upon acidification with sulfuric acid. In this acid, only one H is capable of ionization, suggesting the experimentally confirmed formula $H_2P(O)OH$. Hypophosphorous acid and its salts are reducing agents, although their reaction rates are somewhat low, which is usually explained by an equilibrium between H_3PO_2 and its hydrate form, H_5PO_3.

Phosphorus Halides. PX_3, P_2X_4 and PX_5 are formed by direct reaction of the elements, though the pure substances require special methods. Mixed halides are also known. The trihalides are covalent pyramidal compounds, the X—P—X bond angles being generally between 98° and 104°. They all undergo hydrolysis, the rate being roughly inversely proportional to the sum of the atomic numbers of the halogen atoms.

The halogen derivatives of pentavalent phosphorus may be grouped on the basis of their structure into three classes, the pentahalides, oxyhalides and related compounds, and the fluorophosphoric acids. The pentahalides (except the pentaiodide, which is unknown) are produced by reaction of the elements, or, in the case of mixed halides, by reaction of a halogen and a phosphorus trihalide in correct proportions. Their structure in the vapor state has been determined to be a trigonal bipyramid and their bonding is covalent. In the solid, however, phosphorus pentachloride, [CAS: 10026-13-8], PCl_5, is $[PCl_4^+]$ $[PCl_6^-]$ and phosphorus pentabromide, [CAS: 7789-69-7], PBr_5, is $[PBr_4^+]Br^-$. Various mixed halides are known. The five halogen atoms are not in equivalent positions. One may be ionized, with the other four forming sp^3d orbitals or there may be a transition state, to explain the nonequivalence of the exchange between the three equatorial chlorine atoms and the two apical ones in PCl_5 in carbon tetrachloride solution. The pentahalides react with excess water to yield phosphoric acid and hydrohalic acids, but with less water to form phosphorus oxyhalides instead of phosphoric acid.

Phosphorus oxyhalides have the tetrahedral structure

$$:\!\overset{\textstyle :\ddot{X}:}{\underset{\textstyle :\ddot{X}:}{\ddot{X}\!:\!P\!:\!\ddot{O}:}}$$

These compounds, particularly $POCl_3$ and $POFCl_2$, readily form complexes with metal halides. Closely analogous to the oxyhalides are the thiohalides of general formula PSX_3 and the phosphorus nitrilic halides, $(PNX_2)_n$, the chloride of the latter being obtained by partial ammonolysis of PCl_5, and existing as cyclic or polymeric structures of alternate nitrogen and phosphorus atoms.

Phosphorus Oxides and Oxyacids. The principal oxides are related to the acids which they yield when dissolved in H_2O in the following manner:

Trioxide, P_2O_3	Hypophosphorous acid, H_3PO_2
Tetroxide, P_2O_4	Phosphorous acid, H_3PO_3
Pentoxide, P_2O_5	Hypophosphoric acid, $H_4P_2O_6$
plus $3H_2O$	Orthophosphoric acid, $2H_3PO_4$
plus $2H_2O$	Pyrophosphoric acid, $H_4P_2O_7$
plus $1H_2O$	Metaphosphoric acid, $2HPO_3$

Normally when the term "phosphoric acid" is used, it is with reference to orthophosphoric acid, H_3PO_4. Anhydrous orthophosphoric acid is a white crystalline solid that melts at 42.35°C. It forms a hemihydrate, $2H_3PO_4 \cdot H_2O$, which melts at 29.32°C. Although practically any desired concentration can be produced, it is common to supply the material as a solution containing from 75% H_3PO_4 (mp −17.5°C) to 85% H_3PO_4 (mp 21.1°C). When phosphoric acid is heated to above 200°C, the water of constitution is lost. Thus, a series of acids is formed by the dehydration, ranging from pyrophosphoric acid, $H_4P_2O_7$, to metaphosphoric acid, $(HPO_3)_n$. Salts of the dehydrated acids are used for the preparation of certain kinds of liquid fertilizers and are present in numerous cleaning

compounds. The dehydrated acids can form water-soluble complexes with many metals, such as calcium. One, or two, or three of the hydrogens of phosphoric acid may be neutralized. When one hydrogen is replaced with sodium, for example, the product is slightly acidic; while replacement of all three hydrogens yields a highly alkaline product. The acidity of the solutions is: NaH_2PO_4, a pH of 4.0; Na_2HPO_4, a pH of about 8.3; Na_2PO_4, a pH of 12.0. Although of interest scientifically, the other acids of phosphorus, hypophosphorous acid, H_3PO_2, orthophosphorous acid, H_3PO_3, and pyrophosphorous acid, $H_4P_2O_5$, are not important commercially.

There are two main processes for the industrial production of phosphoric acid, H_3PO_4, from phosphate rock: (1) the *wet process* which involves the reaction of phosphate rock with H_2SO_4 to yield phosphoric acid and insoluble calcium sulfates. Several of the impurities present in the rock dissolve and remain with the product acid. These are not important when the acid is used for fertilizer manufacture. However, the impurities are deleterious to the manufacture of phosphorus chemicals. For a purer product, (2) the *furnace process* is used, wherein the phosphate rock is combined with coke and silica, producing elemental phosphorus as previously described. Oxidation of the phosphorus produces P_2O_5 which, when combined with H_2O, yields H_3PO_4.

Phosphorus sesquioxide, P_4O_6, produced by controlled oxidation of white phosphorus, is hydrolyzed in the cold to produce phosphorous acid, H_3PO_3, a colorless solid (mp 73°C). Only two of its H atoms are capable of ionization, thus compounds such as M_3PO_3 do not exist. This fact leads to the formula $HP(O)(OH)_2$. Phosphorous acid is a somewhat stronger acid than phosphoric, and both it and the phosphite ion (HPO_3^{2-}) are strong reducing agents.

Hypophosphorous acid was discussed earlier in this entry.

Metaphosphorous acid, HPO_2, is produced by atmospheric combustion of PH_3, but in aqueous solution it hydrates to H_3PO_3.

Phosphorous acid is used in solution, and is usually a reducing agent. That is, in air it changes to phosphoric acid, with hot concentrated sulfuric acid it yields phosphoric acid plus SO_2, with copper sulfate it yields finely divided copper metal, with silver nitrate it yields finely divided silver metal, with permanganate after some time it yields manganous. Occasionally it is an oxidizing agent, e.g., with zinc plus dilute H_2SO_4 it yields phosphine.

Phosphorus tetroxide, P_2O_4, is obtained along with red phosphorus by heating P_4O_6 at 290°C in a closed tube. It is believed to have the formula, P_8O_{16}, and to consist of trivalent and pentavalent phosphorus. Hypophosphoric acid, $H_4P_2O_6 \cdot 2H_2O$, cannot be produced directly from the tetroxide. The acid decomposes into phosphorous and phosphoric acids on heating, and must be prepared by indirect methods, such as treatment of white phosphorus with an HNO_3 solution of $Cu(NO_3)_2$.

Phosphorus(V) oxide is the chief product of atmospheric oxidation of phosphorus, whence it is obtained as the β-allotrope, of formula P_4O_{10}. Several other allotropes are obtained by various thermal treatments of the β-form, differing in structure and physical properties. The compound hydrates rapidly to form various phosphoric acids. With an excess of H_2O, orthophosphoric acid, $(HO)_3PO$, is formed, mp 42.3°C. It is a triprotic acid, yielding $H_2PO_4^-$, HPO_4^{2-} and PO_4^{3-} ions. The other crystalline phosphoric acid, pyrophosphoric acid, is believed to have the formula $(HO)_2P(O)-O-P(O)-(OH)_2$. Its acid solution undergoes hydrolysis to the orthoacid. It is tetraprotic, yielding the ions $H_3P_2O_7^-$, $H_2P_2O_7^{2-}$, $HP_2O_7^{3-}$ and $P_2O_7^{4-}$.

Classification of Phosphates. Audrieth and Hill have proposed a classification on the basis of structure, that is, to divide them into glassy phosphates and crystalline phosphates, the latter class being subdivided into (1) linear phosphates and polyphosphates, and (2) cyclic phosphates. Three important members of class (1) are the orthophosphates, containing the PO_4^{3-} ion, the pyrophosphates, having the $P_2O_7^{4-}$ ion, and the triphosphates, having the $P_3O_{10}^{5-}$ ion. The general structural unit of the linear phosphates is the tetrahedron, containing a phosphorus atom surrounded by four oxygen atoms covalently linked to it. Such tetrahedra are linked through a common oxygen atom to form linear polyphosphates, the $P_2O_7^{4-}$ ion having two such tetrahedra, and the $P_3O_{10}^{5-}$ ion, three. Moreover, the tetrahedra may also be double linked through oxygen atoms to form cyclic structures, as in the trimetaphosphate ion, $P_3O_9^{3-}$, which has three such tetrahedra and the tetrametaphosphate ion, $P_4O_{12}^{4-}$, which has four tetrahedra. In general, heating acid salts of simpler phosphates produces polyphosphates (loss of H_2O), while alkaline hydrolysis reverses the process. The glassy phosphates, produced by fusion and rapid cooling of metaphosphates, appear to be true glasses, containing anions with molecular weights well into the thousands.

The widely diverse functionality of phosphates makes them of exceptional importance to technologists, particularly in the food processing field. Phosphoric acid finds many direct uses as an acidulant. It has three available hydrogens which can be replaced one by one with alkali metals, forming a series of *orthophosphate salts* with pH levels ranging from moderately acid (pH = 4) to strongly alkaline (pH = 12). This wide pH range makes phosphates very useful for adjusting the pH of food and chemical systems to almost any desired level. Heating orthophosphates converts them to condensed phosphates containing two, three, or more phosphorus atoms per molecule. The *condensed phosphates*, or *polyphosphates*, have many properties that the orthophosphates do not enjoy. They are polyelectrolytes, and have dispersing or emulsifying properties. They can sequester or chelate metals, such as calcium, magnesium, iron, and copper, rendering these metals nonreactive. This functionality is useful for controlling oxidative rancidity and color formation, as both are catalyzed by metal ions.

Condensed phosphates containing two atoms of phosphorus are *pyrophosphates*. Sodium acid pyrophosphate (SAPP) is used as a leavening acid in baking, and is particularly useful because of the way it can be modified to give different rates of reaction. Pyrophosphates are good sequestrants for iron and copper, which often catalyze oxidation in fruits and vegetables. Thus, the use of a pyrophosphate effectively prevents the discoloration of such foods during preparation and storage.

Condensed phosphates containing three atoms of phosphorus are *tripolyphosphates*, the most important of which is sodium tripolyphosphate (STPP). This compound reacts with the protein in meat, fish, and poultry to prevent denaturing or loss of fluids. This property is sometimes called "moisture binding." STPP also solubilizes protein, which aids in binding diced cured meat, fish, and poultry. It also emulsifies fat to prevent separation.

The chain length of phosphates can be increased further by melting and chilling to form a glass. Glassy sodium phosphates are generally called *sodium hexametaphosphates* (SHMP). SHMPs have excellent sequestering power toward calcium and magnesium. They are used in meat treatment as a partial replacement for STPP to improve solubility in strong pickling brine or to prevent hardness precipitation in very hard water. Considerably more information on the use of phosphates in food processing will be found in the *Foods and Food Production Encyclopedia*, D.M. Considine (editor), Van Nostrand Reinhold, New York, 1981. See also **Fertilizer**.

Peroxyphosphoric Acids. Two are known—peroxymonophosphoric acid, $HOOP(O)(OH)_2$, prepared by treatment of P_4O_{10} with hydrogen peroxide, and peroxydiphosphoric acid, $(HO)_2(O)POOP(O)(OH)_2$, prepared from metaphosphoric acid and peroxide or by electrolysis of an alkali hydrogen phosphate solution (cf. preparation of peroxydisulfuric acid). They and their salts are strong oxidants.

Fluorophosphoric Acids. [CAS: 13537-32-1]. H_2PO_3F, HPO_2F_2, (POF_3 is not a protic acid), "HPF_6" are obtained by replacing one or more hydroxyl groups with fluorine. They are strong fuming acids, like H_2SO_4 in most properties except its oxidizing power. Their salts are also known, extending as far as completely fluorinated MPF_6 where M is an alkali metal. The solubilities of the monofluorophosphates parallel those of the sulfates; while those of the di- and hexafluorophosphates parallel those of the perchlorates.

Upon acidification or neutralization of solutions containing phosphate anions with other anions, such as those of molybdenum and tungsten, complexes are formed which can readily be crystallized as salts, called phosphomolybdates, phosphotungstates, etc.

Oxygen-nitrogen Compounds of Phosphorus. These may be classified as aquo, aquo ammono, or ammono derivatives. The first group includes the various acids and oxides already discussed. The second group includes the amido phosphoric acids in which one or more of the hydroxy groups of the acid are substituted by amino groups. Thus there is amidophosphoric acid (*orthophosphoric acid* is understood when the substituted acid is not specified), amidopyrophosphoric acid, diamidophosphoric acid, and triamidophosphoric acid. The substitution of all—OH groups of phosphoric acid gives phosphoryl triamide, $OP(NH_2)_3$. Related compounds are phosphoryl amide imide, $OP(NH_2):NH$ and phosphoryl nitride, $(OPN)_x$.

The second group also includes imidodiphosphoric acid,

$$HN \begin{matrix} PO(OH)_2 \\ \\ PO(OH)_2 \end{matrix}$$

diamidotriphosphoric acid,

$$\begin{matrix} PO(OH)_2 \\ HN \\ PO(OH)_2 \\ HN \\ PO(OH)_2 \end{matrix}$$

and still longer chain acids. Finally, many other derivatives are possible because

$$N\equiv P\diagdown$$

of the stability of the arrangement. This gives rise to the phosphonitrilic acids, $[NP(OH)_2]_x$, as well as to many ammono derivatives containing only the two elements. The latter include phosphonitrilamide, $[NP(NH_2)]_x$, phospham $[NPNH]_x$, and phosphoric nitride, $[P_3N_5]_x$. The phosphonitrilic chlorides, already discussed, are derivatives of phosphonitrilamide.

Organophosphorus Compounds. Most of the industrially important organic compounds of phosphorus commence with one of the basic inorganic phosphorus compounds, such as PCl_3, $POCl_3$, P_2S_5, and P_2O_5, reacted with an appropriate organic intermediate. Ester intermediates, such as alkyl phosphoryl chlorides, are made by the addition of primary alcohols to $POCl_3$. Triaryl phosphate plasticizers and gasoline additives, such as tricresyl phosphate (TCP), can be prepared from PCl_5 and an appropriate phenolic compound. Alkyl diaryl phosphates can be made from $POCl_3$ and corresponding phenols. A number of thiophosphate esters contain PS plus an ethyl or methyl group and a substituted aryl group and are based on $PSCl_3$ or P_2S_5. These compounds are finding use for pesticide control. Dialkyl dithiophosphates may be prepared from P_2S_5 and appropriate intermediates. They are finding application as flotation-agents, oil-additives, and insecticides. It is much more difficult to prepare organophosphorus compounds containing a C—P bond than it is to form the esters. Numerous organic phosphorus compounds are found in nearly all life processes and remain to be better understood before they can be synthesized. Classes of compounds of this type include the phosphoglycerides required for fermentation, the adenosine phosphates needed in photosynthesis and muscle activity, and the very complex phosphorus-containing groups identified in the nucleotides. Of structural interest are the catenation compounds, as illustrated below, which contain many cyclic phosphates and oxygen-linked chains. Compounds of this type include tetrachlorodiphosphine, Cl_2PPCl_2, tetraphenyldiphosphine $(C_6H_5)_2PP(C_6H_5)_2$, diphosphobenzene, $C_6H_5PPC_6H_5$, and tetramethyl hypophosphate, $(CH_3O)_2(O)PP(O)(OCH_3)_2$.

Inorganic Macromolecules. After the accelerated activity in the development of new polymers that took place during the past 30 or 40 years, some researchers observed some lessening of polymer research and polymer achievements during the 1970s. Not all authorities agreed, but most agreed that the time had arrived when polymer chemistry and applications deserved evaluation both in terms of the past and the future. Synthetic polymers generally have a number of relatively negative features—flammability (derived from their organic nature); a tendency to melt, oxidize, and char at high temperatures in regular atmospheric conditions (again, a result of their organic nature); and a tendency to become stiff and brittle at low temperatures. Many also have a tendency to soften, swell and dissolve in a number of common substances, such as gasoline, jet fuel, hot oil, and numerous other hydrocarbons. In terms of medical applications, most organic polymers tend to initiate a clotting reaction of the blood and many tend to cause toxic, irritant, and sometimes carcinogenic responses. Observers also noted that most polymer research in some way initiated with petrochemicals.

In the early 1970s, a number of investigators decided to shift emphasis and to look at a number of inorganic elements including silicon, phosphorus, sulfur, boron, and some metal atoms that might make up the backbone of a polymer. It was reasoned that the presence of some of these materials in the backbone might remedy some of the aforementioned shortcomings. It should be pointed out that as early as the 1940s, silicone, or poly(organosiloxane) polymers were developed and have proved highly satisfactory in many applications. Peters et al. (1976) reported on a new class of thermally stable polymers, which are based upon alternating siloxane and carborane units. In 1965, Allcock et al. (Pennsylvania State University) synthesized the first poly(organophosphazenes). Since that time, well over 60 new polymers have been made and presently constitute a substantial class of new elastomers. They appear to solve some of the biomedical problems previously mentioned. As pointed out by Allcock (1976), all linear, high polymeric polyphosphazenes have the general structure shown by (a), below. It is interesting to note that over a century ago researchers in Germany and Britain found that phosphorus pentachloride will react with ammonia or ammonium chloride to yield a volatile, white solid. It is now known that this product has the form of (b). Later experimentation showed that the compound would melt under strong heating to form a transparent rubbery material. Involving a ring-opening polymerization of the cyclic trimer, a poly(dichlorophosphazene), shown in (c), is formed. Mainly for the reason that the compound hydrolyzes slowly in the presence of atmospheric moisture to form a crusty mixture of ammonium phosphate and phosphoric acid, the substance was not given serious thought for many years.

Allcock and associates, during the 1960s, subjected the cyclic trimer to new procedures and, after considerable research, were successful in developing polymers that have molecular weights up to and sometimes exceeding 3 to 4 million. The investigators found that the introduction of different substituent groups had a marked effect on the properties of the polymers. See (d). The further detailed research is beyond the scope of this book, but interesting details can be found in the Allcock reference. In summarizing one of their reports, the researchers stated: "Polyphosphazenes are emerging as a new class of macromolecules that have an obvious future as technological elastomers, films, fibers, and textile treatment agents. However, they also possess almost unique attributes for use in biomedicine as reconstructive plastics or as drug-carrier molecules. Moreover, their possible value as 'pseudo-protein' model polymers is an exciting prospect."

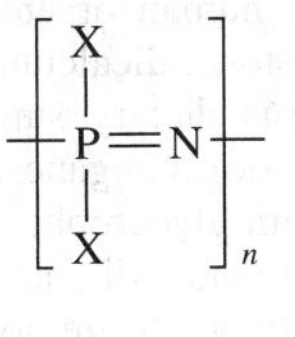

(a) 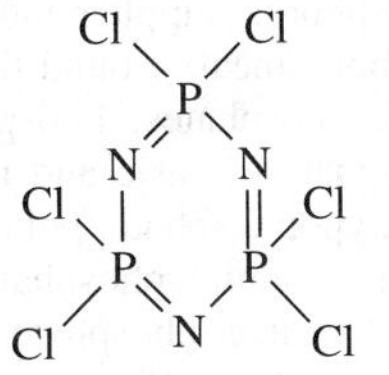   (b)

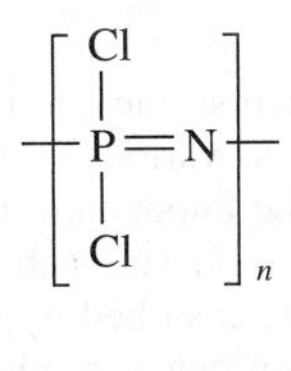

(c) 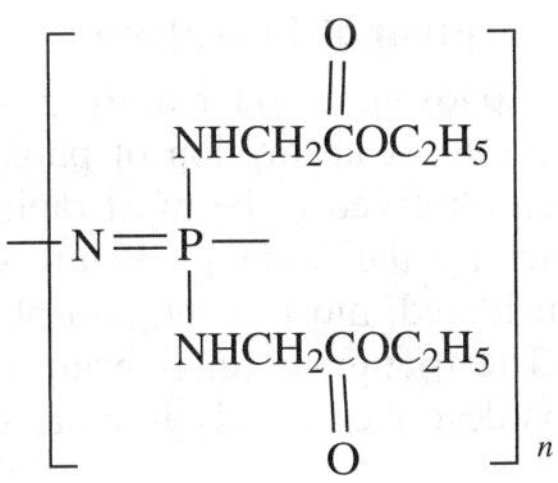   (d)

Phosphorus Ylides. In 1979, G. Wittig received a share of the Nobel Prize for Chemistry in recognition of his development of the use of phosphorus-containing compounds as important reagents used in organic synthesis. According to the selection committee, Wittig's most important achievement was "the discovery of the rearrangement reaction that bears his name. In the Wittig reaction an organic phosphorus compound with a formal double bond between phosphorus and carbon is reacted with a carbonyl compound. The oxygen of the carbonyl compound is exchanged for carbon, the product being an olefin. This method of making olefins has opened up new possibilities, not the least of which is the synthesis of biologically active substances containing carbon-to-carbon double bonds. For example, vitamin A is synthesized industrially using the Wittig reaction." As early as 1919, some 30 years prior to Wittig's work, the first phosphorus ylide was described by Staudinger. This was diphenylmethylenetriphenylphosphorane, formed by pyrolysis of a phosphazine precursor. During that period, Staudinger conceived the possibility of olefin synthesis by condensation of this ylide with a carbonyl compound and visualized a four-membered phosphorus-oxygen heterocyclic (oxaphosphetane) as the intermediate. Staudinger's work was accomplished during a period when practical application of the concept was in doubt. At that time, the Lewis theory of electronic structure was new. The exact bonding of phosphonium salts was somewhat veiled and controversial. Attempts of an olefin synthesis were put aside. For more detail, see **Wittig Reaction**.

Health and Safety

At ambient temperatures white phosphorus spontaneously ignites when exposed to air. It has an autoignition temperature of 30°C. As a result, any human exposure to white phosphorus can cause severe thermal burns to the skin and eyes. The vapor from phosphorus can cause severe lung irritation, followed by a build-up of fluid in the lungs. Continuous long-term inhalation of white phosphorus vapor (>0.1 mg/m3) can result in bone loss to the jawbone structure causing loosening of teeth and severe pain and swelling of the jaw. This condition is commonly referred to as phossy jaw. Some evidence exists that increased infant mortality can result when pregnant women are exposed to P_4 vapor in excess of 0.075 mg/kg/day. Ingestion of white phosphorus is potentially fatal. The lowest reported fatal dose is 1 mg/kg for humans. Absorption through the skin is also possible but is only considered to be a moderate hazard compared to the other routes of exposure.

When exposed to air, white phosphorus oxidizes to phosphorus pentoxide forming copious quantities of white smoke. This smoke may be irritating but is not considered to be toxic. White phosphorus solid, liquid, or vapor is also extremely reactive with oxidizers such as strong acids, alkaline hydroxides, halogens, and nitrates. Contact of phosphorus with water or oxidizers also generates phosphine CAS: 7803-51-2, PH_3, a highly toxic and flammable gas. Phosphine has an 8-h time-weighted average exposure limit of 0.3 ppm. Under alkaline conditions the rate of PH_3 formation is high. At neutral or acidic pH, the PH_3 generation is slow but still very hazardous if the PH_3 is allowed to accumulate in a confined vapor space. The safest commercial handling conditions for molten phosphorus are generally considered to be from pH 6 to 8 at 45–65°C.

Phosphorus production plants and users should ensure that processing of the material is contained and that potential high exposure areas are well ventilated. Workers must wear aluminized fiber glass or Kevlar flame-retardant full protective clothing, face shield with hard hat, rubber boots, and heavy rubber gloves when handling or transferring the product. The phosphorus should always be kept under neutral pH water at temperatures less than 65°C or under an inert atmosphere to avoid oxidation and exposure hazards. High potential exposure areas should also be equipped with well-maintained, water-filled safety tubs, deluge systems, and water-spray extinguishing systems as a precautionary measure. If high exposure levels to phosphorus vapors or phosphine are anticipated, then self-contained breathing apparatus units should be utilized. Individuals exposed to phosphorus through skin or eye contact should have the exposed area flushed immediately with large amounts of water. The affected area should be kept wet until all of the phosphorus is removed or flushed away. Victims of phosphorus inhalation should immediately be removed to an area with fresh air and have artificial respiration administered, if necessary. Workers who have had dental surgery and pregnant women should be kept away from phosphorus exposure areas completely. Anyone who ingests phosphorus should drink a large volume of water and be induced to vomit. Medical assistance should be obtained as soon as possible after any instance of phosphorus exposure.

Additional Reading

Allcock, H.R.: "Polyphosphazenes: New Polymers with Inorganic Backbone Atoms," *Science*, **193**, 1214–1219 (1976).

Burges, R.J.: "Choose the Right Alloys for Fertilizer Acids," *Chem. Eng. Progress,* 82 (November 1992).

Considine, D.M. and G.D. Considine: *Foods and Food Production Encyclopedia,* Van Nostrand Reinhold, New York, NY, 1982.

Corbridge, D.E.: *Phosphorus: An Outline of Its Chemistry, Biochemistry and Technology,* 5th Edition, Elsevier Science, New York, NY, 1995.

Corbridge, D.E.: *Phosphorus 2000: Chemistry, Biochemistry and Technology,* Elsevier Science, New York, NY, 2000.

Dillon, K.B., F. Mathey, and J.F. Nixon: *Phosphorus: The Carbon Copy,* John Wiley & Sons, Inc., New York, NY, 1998.

Emsley, J.: *The 13th Element: The Sordid Tale of Murder, Fire and Phosphorus,* John Wiley & Sons, Inc., New York, NY, 2000.

Greenwood, N.N. and A. Earnshaw: *Chemistry of the Elements,* 2nd Edition, Butterworth-Heinemann, Inc., Woburn, MA, 1997.

Kent, J.A.: *Riegel's Handbook of Industrial Chemistry,* 9th Edition, Chapman & Hall, New York, NY, 1992.

Krebs, R.E.: *The History and Use of Our Earth's Chemical Elements: A Reference Guide,* Greenwood Publishing Group, Inc., Westport, CT, 1998.

Lewis, R.J. and N.I. Sax: *Sax's Dangerous Properties of Industrial Materials,* 10th Edition, John Wiley & Sons, Inc., New York, NY, 2000.

Lide, D.R.: *CRC Handbook of Chemistry and Physics,* 84th Edition, CRC Press, LLC., Boca Raton, FL, 2003.

Peters, E.N., et al.: *Rubber Chemistry,* 1976.

Somerville, R.L.: "Reduce Risks of Handling Liquefied Toxic Gas (Phosgene)," *Chem. Eng. Progress,* 64 (December 1990).

Stevenson, F.J. and M.A. Cole: *Cycles of Soils: Carbon, Nitrogen, Phosphorus, Sulfur, Micronutrients,* 2nd Edition, John Wiley & Sons, Inc., New York, NY, 1999.

Vedejs, E.: "1979 Nobel Prize for Chemistry," *Science,* **207,** 42–44 (1980).

PHOSPHORUS (In Biological Systems).

Phosphorus is required by every living plant and animal cell. Deficiencies of available phosphorus in soils are a major cause of limited crop production. Phosphorus deficiency is probably the most critical mineral deficiency in grazing livestock. Phosphorus, as orthophosphate or as the phosphoric acid ester of organic compounds, has many functions in the animal body. As such, phosphorus is an essential dietary nutrient.

The biological roles of phosphorus include: (1) anabolic and catabolic reactions, as exemplified by its essentiality in high-energy bond formation, e.g., ATP (adenosine triphosphate), ADP (adenosine diphosphate), etc., and the formation of phosphorylated intermediates in carbohydrate metabolism; (2) the formation of other biologically significant compounds, such as the phospholipids, important in the synthesis of cell membranes; (3) the synthesis of genetically significant substances, such as DNA (deoxyribonucleic acid) and RNA (ribonucleic acid); (4) contributing to the buffering capacity of body fluids, cells, and urine; and (5) the formation of bones and teeth. Like calcium, the majority of the phosphorus in the vertebrate body is contained in the hard tissues; in the adult, approximately 80–86% of the total body phosphorus is contained in the bones and teeth, with the balance found in the soft tissues and body fluids.

In a very interesting dissertation by F.H. Westheimer (*Science,* **235,** 1173–1177, 1987), the role of phosphates in living substances is described. As pointed out by Westheimer, phosphate esters and anhydride dominate the living world but are seldom used as intermediates by organic chemists. Phosphoric acid is specially adapted for its role in nucleic acids because it can link two nucleotides and still ionize; the resulting negative charge serves both to stabilize the diesters against hydrolysis and to retain the molecules within a lipid membrane. A similar explanation for stability and retention also holds for phosphates that are intermediary metabolites and for phosphates that serve as energy sources. Phosphates with multiple negative charges can react by way of the monomeric metaphosphate ion PO_3^- as an intermediate. No other residue appears to fulfill the multiple roles of phosphate in biochemistry. Stable negatively charged phosphates react under catalysis by enzymes; organic chemists, who can only rarely use enzymatic catalysis for their reactions, because they need more highly reactive intermediates than phosphates.

Most of the coenzymes are esters of phosphoric or pyrophosphoric acid. The main reservoirs of biochemical energy, adenosine triphosphate (ATP), creatine phosphate, and phosphoenolpyruvate are phosphates. Many intermediary metabolites are phosphate esters, and phosphates or pyrophosphates are essential intermediates in biochemical syntheses and degradations. The genetic materials DNA and RNA are phosphodiesters.

Phosphorus deficiencies are not common in humans and most other species, but they have been observed in ruminants. Symptoms of the deficiency are loss of appetite and a depraved appetite (termed "pica") where the animal chews and consumes extraneous items, such as wood, clothing, bones, etc. Vitamin D deficiency may accentuate a marginal lack of phosphorus in the diet.

Cereals and meats are the major sources of phosphorus in human diets. Phosphorus deficiencies in most regions have not been a serious problem in human nutrition. Insofar as food is concerned, the primary value of phosphorus fertilizers is that they generally increase the total food production; not the content of phosphorus in the food per se.

Experimental phosphorus deficiency can be induced by feeding diets low in this element and by including excesses of calcium, strontium, barium, beryllium, and other cations that precipitate phosphates in the intestinal tract. In this situation, bone formation ceases, and the following histological bone changes have been noted in experimental animals: (1) a thickening of the epiphyseal plate and the formation of a typical rachitic metaphysis; (2) wide osteoid borders of trabecular bone and a considerable rarefaction of the shaft; and (3) irregular or complete cessation of calcification of the zone of provisional calcification of the cartilage matrix. Rickets can be produced in the laboratory by feeding a diet high in calcium and low in phosphorus, and containing little or no vitamin D.

The nutrient requirement for phosphorus depends upon the particular species and the physiological status of the animal. During growth, lactation, gestation, and egg laying, a higher phosphorus content of the diet is generally required in poultry than for the maintained adult. The availability of phosphorus in the diet varies with its chemical form and the animal species in question. Diets high in foods of plant origin may contain a considerable portion of phosphorus in the form of phytic acid, which is the hexaphosphoric acid ester of inositol. When the acid occurs as salts of calcium, magnesium, sodium, etc., it is referred to as phytin. Phytate phosphorus usually is less available than inorganic phosphate to such species as rat, chicken, dog, pig, and human. However, a phytase has been shown to be present in the intestine and intestinal secretions of some animals, and the formation of this enzyme is dependent, in part, on the presence of vitamin D. Through the action of phytase, some of the phytate phosphorus would be made available for absorption.

Experimentation has indicated that, under normal dietary conditions and calcium intake, food phytate is of no nutritional concern in humans. The microbial population of the ruminant also elaborates a phytase enzyme that makes phytate phosphorus readily available in this class of animals. Phytates may be of nutritional consequence for another reason—dietary calcium can be bound in an unavailable, insoluble complex, thereby decreasing the absorption of this element.

Many studies have involved determination of the availability of phosphorus from other organic and inorganic sources. In chicks, orthophosphates, superphosphates, and phosphate rock products are good sources of phosphorus, whereas metaphosphate and pyrophosphate are relatively unavailable to the species. Most organic phosphorus sources, such as casein, pork liver, and egg phospholipid are found to be as available as inorganic phosphorus. Commonly used phosphorus supplements in human or animal nutrition or both are steamed bone meal, ground limestone, dicalcium phosphate, and defluorinated rock phosphates. Phosphorus dietary supplements include magnesium phosphate (dibasic and tribasic) manganese glycerophosphate and manganese hypophosphate, potassium glycerophosphate, sodium ferric pyrophosphate, sodium phosphate (mono-, di-, and tri-), and sodium pyrophosphate. Of course, phosphate compounds are not always added in the interest of augmenting phosphorus, but for the other elements which may be contained in the compound.

Absorption of Phosphate

The phosphate ion readily passes across the gastrointestinal membrane. The rate of absorption of phosphate at various intestinal sites in rats has been observed to be most rapid in the duodenum, followed in decreasing order by the jejunum, ileum, colon, and stomach. When transit time is considered, most of the phosphorus is absorbed by the ileum.

The triangular relationship between calcium, phosphorus, and vitamin D is described briefly in entry on **Calcium (In Biological Systems)**.

Plasma Phosphate

Once absorbed, phosphorus enters the blood and the majority is present therein as orthophosphate ions. At an ionic strength of 0.165 and at 37°C (98.6°F), calculations show that the proportional concentration of the orthophosphate ions in plasma for $H_2PO_4^-$ is 18.6×10^{-30}; for HPO_4^{2-} is 81.4×10^{-30}; and for PO_4^{3-} is 8×10^{-30}. About 12% of the phosphorus present is bound to proteins. During egg laying in birds, the concentration of nonionized phosphorus compounds in plasma is greatly increased. The administration of diethylstilbestrol (regulated in some countries) to cockerels results in the formation of a plasma phosphoprotein, which forms relatively firm complexes with calcium. The function of the phosphoprotein appears to be one of phosphorus transport; in laying birds, the phosphoprotein is incorporated in egg yolk.

The approximate average plasma phosphorus levels for several species, in milligrams per 100 milliliters of plasma, are: pigs, 8.0; sheep, cattle, and goats, 6.0; horse, 2.3. Erythrocytes contain considerably more phosphorus than plasma, mostly in the form of organic esters. Some of the latter are acid soluble and hydrolyzable by intracellular enzymes.

Plasma phosphate appears to be homeostatically controlled. The primary organ concerned appears to be the kidney, although the skeleton also may play a role. Parathyroid hormone, by way of its direct action on the kidney and bone, is a significant hormonal factor.

Phosphate Excretion

The excretion of body phosphorus occurs via the kidney and intestinal tract, the distribution between these pathways varying with species. For example, relatively small amounts of phosphorus are endogenously excreted into the feces of rat, pig, and human, but in the bovine, perhaps 50% or more of the fecal phosphorus may be from endogenous sources.

The amount of phosphorus excreted in the urine varies with the level of ingested phosphorus and factors influencing phosphorus availability and utilization. It has been shown that in the dog, when plasma phosphate is normal or low, over 99% of the filtered ion is reabsorbed, presumably in the upper part of the proximal tubule. Increased plasma concentrations of alanine, glycine, and glucose depress phosphate reabsorption.

Phosphate of Hard Tissues

Body phosphorus contained in the intracellular matrix of bone and teeth is of the general form of hydroxyapatite, $Ca_{10}(PO_4)_6(OH_2)$, this calcium phosphate salt providing the characteristic hardness of ossified tissue. Phosphate ions are also adsorbed onto the surface of bone crystals and exist in the hydration layers of the crystals. Early theories of calcification placed special emphasis on the role of alkaline phosphatase and organic esters of phosphoric acid. As part of the theory, it was stipulated that, with the hydrolysis of phosphate esters at the site of calcification, the K_{sp} for bone salt would be exceeded. Although phosphatase may have a function in bone formation, as in the synthesis of organic matrix, its role as earlier depicted has been revised. Later research emphasized the specific and characteristic properties of collagen and other substances, such as chondroitin sulfate. This is related to the local mechanism of calcification; the other component of calcification is the humoral mechanism whereby an adequate supply of calcium, phosphate, and other ions is made available to the calcifying site. A later theory proposed that either the functional groups on collagen are anionic, initially binding Ca^{2+}, or that the first reaction is with phosphate or phosphorylated intermediates. The first-held moiety of bone salt (Ca^{2+} or phosphate) subsequently attracts or binds the other component, providing the aggregation or "seed" for subsequent crystal growth. Since an ATPase-type enzyme has been demonstrated in cartilage, suggesting that ATP may be intimately involved in the calcification mechanism, another proposal is along the line that pyrophosphate is transferred from ATP to free amino groups of collagen, leading to nucleation and followed by combination with calcium and bone salt formation. Or, the ATP provides energy which increases the calcification mechanism.

Dietary inorganic phosphates have been shown to protect experimental animals against dental caries. Orthophosphates were effective cariostats, but $Na_4P_2O_7$ and $Na_5P_3O_{10}$ were not. Dicalcium phosphate, $CaHPO_4$, did not decrease dental caries unless a high level of NaCl was also included in the diet.

Toxicity

Although many phosphorus-containing compounds are vital to life processes, as previously described, there are also many phosphorus compounds that are quite toxic—elemental phosphorus, for example. While the elemental form is dangerous because of its low combustion temperature, its absorption also has an acute effect on the liver. The long and continued absorption of small amounts of phosphorus can result in necrosis of the mandible or jaw bone (sometimes called "phossyjaw"). Chronic phosphorus poisoning occurs particularly through the lungs and

gastrointestinal tract. The most common symptom is the necrosis of the jaw already mentioned, but this is also usually accompanied by anemia, loss of appetite, gastrointestinal weakness, and pallor. Other bones and teeth may be adversely affected.

Phosphine is a very toxic gas. Inhalation of phosphine causes restlessness, followed by tremors, fatigue, slight drowsiness, nausea, vomiting, and, frequently severe gastric pain and diarrhea. Although most cases recover without after-effects, in some cases, coma or convulsions may precede death.

Phosphorus-halogen compounds are quite toxic.

Phosphorus in Soils

When phosphorus fertilizers are added to soils deficient in available forms of the element, increased crop and pasture yields ordinarily follow. Sometimes the phosphorus concentration in the crop is increased, and this increase may help to prevent phosphorus deficiency in the animals consuming the crop, but this is not always so. Some soils convert phosphorus added in fertilizers to forms that are not available to plants. On these soils, very heavy applications of phosphorus fertilizer may be required. Some plants always contain low concentrations of phosphorus even though phosphorus availability from the soil may be good. See also **Fertilizer**.

PHOSPHORYLATION (Oxidative). This is an enzymic process whereby energy released from oxidation-reduction reactions during the passage of electrons from substrate to oxygen over the electron transfer chain is conserved by the synthesis of adenosine triphosphate (ATP) from adenosine diphosphate (ADP) and inorganic orthophosphate. Since ATP is the major source of energy for biological work, and since most of the net gain of ATP in the animal cell derives from oxidative phosphorylation, research in the area has been very active.

Oxidative phosphorylation was discovered simultaneously and independently in 1939 by Kalckar (Denmark) and by Belitzer (the former U.S.S.R.). It was recognized by these workers that aerobic phosphorylation was different from and independent of phosphorylation supported by glycolysis. In addition, they found that the stoichiometry of phosphate esterification (ATP synthesis) and oxygen utilized was two or more, or that the reduction of one atom of oxygen to form water may be accompanied by the "activation" of two or more molecules of phosphorus (P_i), thus leading to an expression of the efficiency of the energy-conserving system. The efficiency expression is known as the P/O ratio, i.e., the ratio of molecules of P_i esterified per atom of oxygen utilized.

The quantitative importance of the ATP synthesized at the expense of energy liberated during electron transfer in the mitochondrion is realized when one follows the conservation of energy during the metabolism of a molecule, such as glucose in the cell. The oxidation of one mole of glucose to carbon dioxide and water is accompanied by the release of 673,000 calories. In order to degrade the glucose molecule to a form that can be metabolized further by mitochondrial enzymes, the glycolytic enzymes consume two molecules at ATP and also synthesize two molecules of ATP in the presence of oxygen, a net energy conservation of zero. The mitochondrion may then degrade the pyruvate supplied by glycolysis to carbon dioxide and water, yielding a net total of 38 molecules of ATP, mostly at the level of the electron transfer process. Thirty-eight molecules of ATP per molecule of glucose results in between 260,000 and 380,000 calories conserved, between 39 and 56% of the total energy released in the complete oxidation of glucose, the remainder being released directly to heat. Inasmuch as the mitochondrion is approximately 50% effective in conserving energy from its major substrate, it is indeed an efficient machine.

PHOSPHORYLATION (Photosynthetic). Photosynthetic conversion of light energy into the potential energy of chemical bonds involves an electron transport chain, and the phosphorylation of ADP (adenosine diphosphate)

$$ADP + P \xrightarrow[\text{Chlorophyll}]{\text{Light}} ATP$$

as intermediate stages. The process of phosphorylation defined by the foregoing equation was discovered simultaneously by Arnon and coworkers for green plant chloroplasts and by Frenkel, working with Geller and Lipmann, for bacteria in 1954. For both systems, the heart of the mechanism is the creation of a very oxidizing and a very reducing component, utilizing the energy of the photoexcited stage of one of the pigment (chlorophyll) molecules. This process will be designated a photoact. The redox components are both members of a photosynthetic electron transport chain, bound to the membranes of the chloroplasts (for green plants) or chromatophores (for bacteria). The photoact can be considered as electron transport against the thermochemical gradient, i.e., away from the member which is a better electron acceptor (high oxidation-reduction potential), through the excited chlorophyll, then to the member which is a better electron donor (low oxidation-reduction potential). Subsequent steps consist of ordinary, dark electron transport with the thermochemical gradient. The energy in at least one of these redox reactions is conserved as ATP by a phosphorylation reaction analogous to that found in oxidative phosphorylation by mitochondria. See also **Phosphorylation (Oxidative)**.

In bacteria, the photoact proper is accomplished by a special kind of bacteriochlorophyll, amounting to only 3% of the total present. It is unique in having a peak in absorption at 870–890 micrometers, or further into the infrared than the remaining 97% of the chlorophyll molecules. It is unique not by virtue of a difference in structure, but because of its environment—most probably in close association or complexing with cytochrome molecules. Since its absorption extends to longer wavelengths, it is an energy trap, and the function of the bulk of the bacterio-chlorophyll is that of capturing light and transmitting it to this active center.

Components of the electron transport chain in bacteria have been shown to include b- and c-type cytochromes, ubiquinone (fat-soluble substitute quinone, also found in mitochondria), ferredox (an enzyme containing nonheme iron, bound to sulfide, and having the lowest potential of any known electron-carrying enzyme) and one or more flavin enzymes. Of these a cytochrome (in some bacteria, with absorption maximum at 423.5 micrometers, probably c_2) has been shown to be closely associated with the initial photoact. Some investigators were able to demonstrate, in chromatium, the oxidation of the cytochrome at liquid nitrogen temperatures, due to illumination of the chlorophyll. At the very least this implies that the two are bound very closely and no collisions are needed for electron transfers to occur.

In both bacterial chromatophores and green plant chloroplasts, the existence of photo-induced high-energy intermediates or states leads to reversible confirmational changes in the structures of the membranes, and to gross swelling and shrinking. These are observed by changes in light scattering, viscosity, and sedimentation properties, and by electron microscope studies. The mechanisms may include ion transport, followed by water diffusion, internal pH changes leading to conformation changes of proteins, or possibly something resembling a contractile protein.

PHOTOCHEMISTRY AND PHOTOLYSIS.

When certain substances are subjected to light, a chemical change results. Such reactions comprise *photochemistry*. The production of an image on a photographic plate is an example. Photosynthesis in the green leaf of a plant is another. Where the change involves chemical decomposition of the radiated material, the process is termed *photolysis*. As used in this context, the term light includes both visible light and ultraviolet radiation. One of the better known and most extensive examples of photolysis is the production of ozone, O_3, in the upper atmosphere, a reaction critical to life on earth because ozone acts as a filter of the middle- and far-ultraviolet radiations which destroy living organisms. The oxygen molecule, O_2, absorbs solar ultraviolet radiation with a wavelength of 190 nanometers, and the energy absorbed breaks the molecule to the atomic state. The released oxygen atoms may combine with oxygen molecules present to form ozone, or the freed oxygen atoms may recombine to form O_2. Thus, there is a continuing combination of processes in dynamic equilibrium, that is, the synthesis and the photolysis of ozone.

Similarly, oceanic nitrite (NO_2^-) photolysis by natural light produces detectable concentrations of nitric oxide (NO). This latter forms in the oceans during daylight and disappears rapidly at sunset when recombination occurs: $NO_2 \longrightarrow NO + O \longrightarrow NO_2$.

Isomerism can also be induced photochemically, although such processes are less well understood and probably require the presence of additional free radicals. The cytotoxic metabolite bilirubin can cause brain damage in infants with neonatal jaundice; this is prevented by exposing the child to intense blue light. The bilirubin is photochemically converted in the skin to metastable geometric isomers, which can be transported in the blood and excreted in bile.

Two major instances of photochemical reactions that have reached deeply into modern civilization are the photosensitive silver and uranium salts and dyes which are the basis of photography and the manufacture of Vitamin D by the ultraviolet irradiation of ergosterol.

Photochemical reactions are highly specific and their products are quite different from those of thermochemical reaction processes.

Sunlight in the near infrared, visible, and near ultraviolet regions possesses considerable energy; utilization of this through photochemical reactions could make a considerable contribution to energy resources. Since biosynthesis itself is relatively inefficient in conversion of solar energy, emphasis has been placed upon the fabrication of *artificial* photochemical systems. One of the more promising approaches has involved application of photoelectric chemical cells or catalysts of semiconductor materials.

The absorption of light by semiconductors creates electron-hole pairs ($e^- h^+$) which can be separated because their components diffuse in different directions. The energies of these moieties can be stored by several mechanisms or used in photocatalysis or photosynthesis for nitrogen fixation, formation of amino acids, methanol, etc. The efficiencies of such conversions depend almost entirely upon the semiconductor material, and as yet these efficiencies are too low for significant application. Currently the most promise is demonstrated by the use of titania on a platinum substrate or single crystals of strontium titanate. See also **Photoelectric Effect**.

Fundamental Considerations. In photochemical reactions, light supplies the energy necessary for the activation of the reacting molecules (Grotthus, 1818, and Draper, 1839). Sometimes the light waves absorbed by a body produce only an increase in temperature, sometimes fluorescence as in the cases of eosin and fluorescein, and sometimes chemical change. The reaction of hydrogen and chlorine in light was studied by Bunsen and Roscoe (1862), and they discovered that the amount of chemical change is proportional to the intensity of the light and to the length of time of exposure to the light. The first law of photochemistry (Draper-Grotthus) states that light that is absorbed causes chemical change. The energy of light is measured in quanta, and according to the Stark-Einstein law,

$$E = Nhc/\lambda$$

where N is Avogadro's constant, h is Planck's constant, c is velocity of light, λ is wavelength of light; that is, each molecule that takes part in a chemical reaction induced by exposure to light absorbs one quantum of radiation causing the reaction. Photochemical processes are of two kinds: primary and secondary. The primary process in a photochemical reaction is limited by the Einstein law to the absorption of one quantum by a molecule or atom. A knowledge of the spectrum of the reactants is necessary to determine what happens in this process. The molecule may be disrupted into fragments or an electron may be excited from a lower orbit to a higher one. Which of these events takes place can often be determined by spectroscopic studies. The secondary process deals with the fate of the molecular fragments or of the excited molecules. The excited molecule may emit its extra energy as light, causing fluorescence; it may lose it by transferring it to other molecules as thermal energy; or it may cause a chemical reaction. On the other hand, the molecular fragments may either recombine to give the original reactant or cause further chemical reactions. The study of the quantum yield (which is the number of molecules reacting divided by the number of quanta absorbed), is used as a means of formulating the secondary processes. If the quantum yield is less than one, fluorescence, deactivation or recombination of fragments must take place. If the quantum yield is unity every photon absorbed decomposes one molecule. When the quantum yield is greater than unity (and in some reactions it may be as high as a million) chain reactions are involved. The classical example of such a reaction is the combination of hydrogen and chlorine. The primary reaction is Cl_2 and light $\longrightarrow$ 2Cl. The chain propagation reactions are

$$Cl_2 + h\upsilon \longrightarrow 2Cl$$

$$Cl + H_2 \longrightarrow HCl + H$$

$$H + Cl_2 \longrightarrow HCl + Cl$$

creating a cycle which is only stopped by

$$Cl + Cl \longrightarrow Cl_2$$

$$H + H \longrightarrow H_2$$

Since the last two processes are slow compared to the two before them, one quantum of light can bring about a combination of a million molecules of hydrogen and chlorine.

See also **Photosynthesis**.

The existence of microbes capable of utilizing light energy to drive the synthesis of cellular components was firmly established by Winogradsky and Molisch by the late 19th century, but it was largely through the work of Van Neil in the 1930s on the physiology of the purple sulfur bacteria that a clearer picture of their photosynthetic processes started to emerge. The photochemical reaction center (RC) is a now well-defined physical entity which exists as an energy sink able to convert the energy of excitation into an electron transfer event. From the use of this RC, three broad patterns of microbiological photosynthesis are known. See Table 1.

The chlorophylls are the pigments responsible for initiating primary photochemistry in the cyanobacteria and higher plants and absorb light energy at 870 nm converting appropriate species to the single excited state and rapidly transferring an electron to a single molecule. Chlorophyll is remarkably similar to hemoglobin in that iron is replaced by magnesium as the chelated metal. Any other transition metals in that position would cause quenching of the initial photochemical reaction.

The carotenoids are largely responsible for the color of the green and purple bacteria and function in both light absorption and energy transfer processes as well as extending the usable light wavelengths and protecting against harmful photooxidation.

In green and purple photosynthetic bacteria, most of the bulk bacterial chlorophyll is inactive in photochemistry and does not undergo photooxidation when excited. Energy is transferred randomly between adjacent pigment molecules until trapped by the RC or lost as fluorescence or heat.

But the photochemistry of bacterial activity must also be associated with photoeffects in the surrounding media, and in this context a rapid increase in the concentration of hydrogen peroxide has been observed when natural surface and ground waters are exposed to sunlight. The hydrogen peroxide is photochemically generated from organic constituents present in the water. Humic materials are believed to photochemically reduce oxygen to give the superoxide anion, which subsequently disproportionates to hydrogen peroxide. Since both hydrogen peroxide and peroxide radical are known to affect biological systems, they may be important factors in the photochemical reactions of photosensitive bacteria and other natural life forms.

Laser Chemistry

Lasers generate a high-intensity output of monochromatic photon energy, and studies of the photochemical reactions induced by this have created a virtual subdivision of photochemistry known as "laser chemistry." While the output of a laser can heat, anneal, burn, cut, or be used instrumentally as a spectral source, we are concerned here only with those chemical effects attributable to the photon output at wavelengths between near infrared and near ultraviolet, i.e., between about 12 and 0.2 microns.

When atoms or molecules are excited conventionally by elevated temperatures or pressure, they can follow several reaction paths yielding a variety of byproducts in addition to the desired substance. Since the basis of a chemical reaction is to weaken or break or make specific chemical bonds to yield the final product, energy ideally should be selectively introduced at the particular level necessary to accomplish this. The high energy and monochromaticity of laser output are ideal for imposition of the specific energy changes that induce or catalyze chemical changes.

The absorption of a quantum of energy by an atom or a molecule takes it from a low energy state to a higher one, and the jump will affect the different properties of the atom or molecule depending upon the amount of energy in the quantum. When absorbed, a quantum of visible or ultraviolet radiation raises an electron to a higher orbit; on the other hand, a quantum of infrared radiation will alter energy levels on an atomic basis.

A laser can supply a precise amount of energy to an atom or molecule, thus effecting a transition from one excited state to a higher excited one. Once known of the energy level displacement required to effect a chemical reaction is available, the laser can provide the specific energy for the specific excitation required. However, the energy input must be related to the total energy dissipation, for, if excess energy leads to ionization or dissociation, a continuum of allowed energy levels will be developed rather than the required discrete levels. If the energy is thus fragmented, the required reaction will proceed only weakly, if at all. Excess energy may be redistributed in two ways. It is either transferred from the excited vibrational state to one or more other vibrational states of the molecule, or it is transferred directly into rotational and translational states. The first mode of energy translation proceeds appreciably more rapidly than the second. Time is a further controlling factor in laser chemistry. The reaction must proceed in time either shorter than, or equal to, that required for transfer of vibrational energy from one state to another in the same molecule; molecule dissociation or atom ionization must take place before there is any depletion of energy by molecular or atomic collisions. Where a reaction proceeds within the lengthy period required for transfer of energy from the initial vibrational state to the much lower rotational and translational states, one cannot hope for laser action to effect a significant degree of reaction specificity, since the effect is basically a thermal one.

A goal of early laser-induced photochemistry was the initiation of specific chemical reactions, to fabricate specific chemicals, or to separate isotopes. Although the specificity of laser-induced photochemistry is important, equally significant is the ability of the laser to confine excited regions to microscopic areas. Thus, one of the better known capabilities of the laser beam is that even a low-powered laser can produce highly intense spots of light of submicrometer dimensions.

In many photochemical reactions, a specific excitation wavelength leads to a specific set of molecular fragments and ultimately products. This has enabled the dissociation of a large variety of molecular gases which can then deposit on, dope, or etch a semiconductor wafer. Further, laser excitation can photodeposit metals, insulators, and semiconductors by decomposing one or more photosensitive gases. Direct writing with lasers through such reactions as:

$$CF_3Br \xrightarrow{h\nu} CF_3 + Br; \text{ or } Al(CH_3)_3 \xrightarrow{h\nu} Al + 3 CH_3$$

is just beginning to be applied to the fabrication of complete microelectronic devices and the modification of actual circuits.

Operation of visible-light and ultraviolet lasers costs more per photon produced than does infrared laser operation. Partly because of this, appreciable interest has centered over the past several years on unimolecular reactions driven by infrared lasers. But absorption of a single infrared photon will raise a molecule only one step in the energy ladder, and, to be dissociated, the molecule will require the absorption of many infrared photons in sequence. The carbon dioxide laser can supply this requirement cheaply and efficiently. A mole of photons (6.02×10^{23}) costs only a few cents in the infrared, but several dollars in the visible and near ultraviolet ranges. This has aided continued study of multiple-photon infrared laser excitation. Much study has gone into an exciting and fundamental reaction and its implication. When sulfur hexafluoride (SF_6) is irradiated by infrared laser light, it decomposes to the pentafluoride (SF_5) and fluorine (F). When the laser is tuned to the vibrational absorption of $^{32}SF_6$ in a mixture with $^{34}SF_6$, only the $^{32}SF_6$ decomposes, leaving the residual gas enriched some 3000-fold in $^{34}SF_6$. Changing the frequency of the irradiating light slightly from emission at 10.61 μ to emission at 10.82 μ selectively decomposes the $^{34}SF_6$ molecule. This method of isotope separation by lasers is being extensively studied for the separation of fissionable U^{235} from nonfissionable U^{238}.

Laser-induced processes are expected to increase in number and expand in application, but the principal obstacle to large-scale introduction of the

TABLE 1. BROAD PATTERNS OF MICROBIOLOGICAL PHOTOSYNTHESIS*

Bacterial group	Type of photosynthesis	Pigment in primary photoactivation	Electron donors	Products	Carbon sources
Eubacteria	Anoxygenation	Bacterial chlorophyll	H_2, H_2S, S, Organics	ATP + NO, D, P(H)	CO_2 + Organics
Eubacteria	Oxygenation	Chlorophylls	H_2O, H_2S	ATP + NO, D, P(H)	CO_2 + Organics
Archeobacteria	Halogenation	Bacterial rhodopsin	Do not participate	ATP	Organics

(* *After Keely and Dow.*)

laser into the chemical industry is an economic one. Laser photos are still much more expensive than those from thermal sources, and the initial application will undoubtedly be directed to those specialty chemicals and isotopes whose current cost far exceeds that of large volume chemicals.

A typical example of this is in the preparation of extraordinary divalent carbon intermediates (carbenes) by application of ultrafast laser techniques. Thus, photoexcitement of diphenyldiazomethane (DPDM) to an excited singlet state breaks the $C=N_2$ bond, releasing diphenylcarbene (DPC) in a singlet state and ultimately allows its stabilization in the triplet ground state.

The increasing number of known photochemical reactions is still very small in comparison with those in ground state chemistry, and our understanding of all the factors controlling photochemical reactions is quite primitive. In some cases it is the ease of conversion to the ground state that is significant, while in others it is the energy hypersurface surrounding the excited state that dictates the energy pathways through which the free electrons will move back toward the ground state, and hence the nature of the photochemical products.

Laser Femtochemistry

As explained by A. H. Zewail (California Institute of Technology), "Femtochemistry is concerned with the very act of the molecular motion that brings about chemistry, chemical bond breaking, or bond formation on the femtosecond (10^{-15} second) time scale. With lasers it is possible to record snapshots of chemical reactions with sub-angstrom resolution. This strobing of the transition-state region between reagents and products provides real time observations that are fundamental to understanding the dynamics of the chemical bond."

A longstanding problem in chemistry has been the development of a better understanding of the transition state between reagents and products. Several investigative approaches, including thermodynamics, kinetics, and synthesis, have been used to systematize masses of experimental data to ascertain the rates and mechanisms of reactions. Over the past few decades, advanced understanding of molecular reactions have stemmed from the use of molecular beams, chemiluminescence, and, more recently, laser techniques. In laser-molecular beam research, a laser is used (1) to excite one of the reagent molecules, thus influencing reaction probability, or (2) to initiate a unimolecular process simply by providing energy to a molecule. In what has been called a "half-collision" unimolecular process, the fragmentation of the excited molecule can be determined. As pointed out by Zewail, "During the last three decades many reactions have been studied and these methods, with the help of theory, have become the main source of information for deducing the nature of the potential surface of a reaction."

Unimolecular reactions, as typified by

$$ABC^* \longrightarrow [A \cdot BC]^{\ddagger *} \longrightarrow A + BC,$$

and bimolecular reactions, as shown by

$$A + B \longrightarrow [ABC]^{\ddagger} \longrightarrow AB + C,$$

have been studied in terms of their *time scale*.

In an excellent paper, commenting on the progress made from the "picosecond" era to the "femto-second" era of research, Zewail observes, "Prior to femtochemistry, molecular beams and picosecond lasers were combined, which led to studies of collision-free energy redistribution in molecules and state-to-state rates of reaction, but the time resolution was still not sufficient to directly view the process of bond breaking or bond formation. However, in femtochemistry, the 'shutter speed' has reached the 10^{-15} regime, so it is now possible to observe chemistry as it happens—the transition-state region between reagents and products. · · · The strobing of these ultrafast molecular reactions, stemming from the happy marriage between ultrafast lasers and chemistry, is what forms the central theme in real time femtochemistry."

In an extension of this technology, A. S. Moffat reported in 1992 two additional goals of development: (1) the use of lasers to *control* chemical reactions, and (2) development of "tunable" lasers that can vary the wavelength of the light they generate. Thus, the light source can be tailored exactly to the vibrational frequency of the bond targeted.

Time-Resolved Photoacoustic Calorimetry

As defined by K. S. Peters and G. J. Snyder (University of Colorado), time-resolved photoacoustic calorimetry is an experimental technique

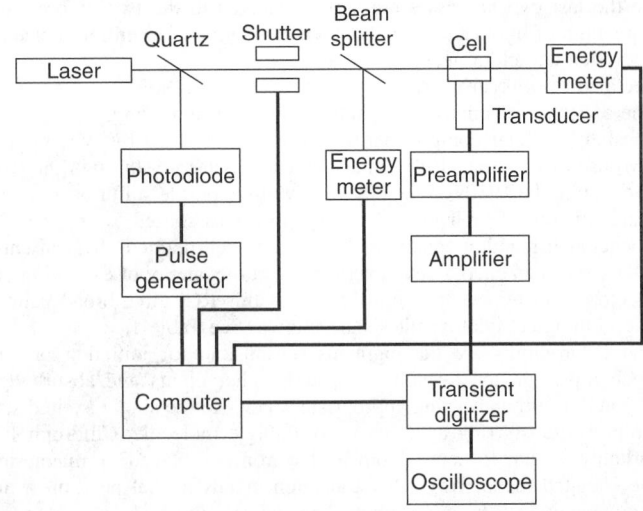

Fig. 1.   Schematic representation of the time-resolved photoacoustic calorimeter. Solid lines indicate light path; heavy solid lines represent signal paths. (*After Peters and Snyder*)

that measures the dynamics of enthalpy changes on the time scale of nanoseconds to microseconds for reactions initiated by absorption of light. As pointed out, "When the reaction is carried out in water, it is also possible to obtain the dynamics of the corresponding volume changes. This method has been applied to a variety of biochemical, organic, and organometallic reactions."

Although pulsed time-resolved photoacoustic calorimetry is in an early phase of development (1994), it is proving to be a powerful technique for understanding the dynamics of enthalpy and volume changes for ground- and excited-state species. Peters and Snyder, developers of the technique, observe, "For reactions in water, the problem is directly approached through temperature-dependent studies. For reactions in organic solvent, there must be further investigations into the magnitude of the effect. At some future date, time-resolved photoacoustic calorimetry will be extended onto the 1 ns time scale . . . . The technique should find wide ranging applications to problems in chemistry and biochemistry that include solid-state reactions and dynamics of proteins in membranes."

A schematic representation of the instrumental technique used is shown in Fig. 1.

R. C. VICKERY, D.Sc.,
Blanton/Dade City, Florida

Additional Reading

Andrews, D.L.: *Lasers in Chemistry,* 3rd Edition, Springer-Verlag, Inc., New York, NY, 1997.

Dunning, T.H., Jr., et al.: "Theoretical Studies of the Energetics and Dynamics of Chemical Reactions," *Science,* **453** (April 22, 1988).

Eisenthal, K.B., et al.: "Divalent Carbon Intermediates: Laser Photolysis," *Science,* **225,** 1439–1445 (1984).

Gust, D. and T.A. Moore: "Mimicking Photosynthesis," *Science,* **35** (April 7, 1989).

Horspool, W.M., and F. Lenci: *CRC Handbook of Organic Photochemistry and Photobiology,* CRC Press LLC., Boca Raton, FL, 2003.

Murov, S.L., I. Carmichael, and G.L. Hug: *Handbook of Photochemistry,* 2nd Edition, Marcel Dekker, Inc., New York, NY, 1993.

Neckers, D.C., D.H. Volman, and G. Von Bunau: *Advances in Photochemistry,* Vol. 24, John Wiley and Sons, Inc., New York, NY, 1999.

Neckers, D.C., D.H. Volman, and G. Von Bunau: *Advances in Photochemistry,* Vol. 25, John Wiley and Sons, Inc., New York, NY, 1999.

Neckers, D.C., D.H. Volman, and G. Von Bunau: *Advances in Photochemistry,* Vol. 26, John Wiley & Sons, Inc., New York, NY, 2001.

Neckers, D.C., T. Wolff, G. Von Bunau, and W.S. Jenks: *Advances in Photochemistry,* Vol. 27, John Wiley & Sons, Inc., New York, NY, 2002.

Osgood, R.M. and T.F. Deutsch: "Laser Induced Chemistry for Microelectronics," *Science,* **227,** 709–714 (1985).

Peters, K.S. and G.J. Snyder, "Time-Resolved Photoacoustic Calorimetry: Probing the Energetics and Dynamics of Fast Chemical and Biochemical Reactions," *Science,* 1053 (August 26, 1988).

Ramamurthy, V. and K.S. Schanze: *Organic, Physical and Materials Photochemistry,* Marcel Dekker, Inc., New York, NY, 2000.

Staehelin, L.A. and P. Aentzer: "Photosynthetic Membranes and Light Harvesting Systems," in *Encyclopedia of Plant Physiology*, Vol. 19, Springer-Verlag, New York, NY, 1986.

Truhlar, D.G. and M.S. Gordon: "From Force Fields to Dynamics: Classical and Quantal Paths," *Science*, **491** (August 3, 1990).

Wayne, C.E. and R.P. Wayne: *Photochemistry,* Oxford University Press, Inc., New York, NY, 1996.

Zewail, A.H.: "Laser Femtochemistry," *Science*, 1645 (December 23, 1988).

PHOTOELECTRIC CONSTANT.   A quantity equal to h/e where h is the Planck constant, and e, the electronic charge, and which multiplied by the frequency of any radiation exciting photoemission gives the potential difference corresponding to the quantum energy absorbed by the escaping photoelectron.

$$h/e = 4.1349 \times 10^{-7} \text{erg} \cdot \text{sec} \cdot \text{emu}^{-1}$$

$$= 1.3793 \times 10^{-17} \text{erg} \cdot \text{sec} \cdot \text{esu}^{-1}$$

PHOTOELECTRIC EFFECT.   Changes in electrical characteristics of substances due to radiation, generally in the form of light. Radiation of sufficiently high frequency (short wavelength), impinging on certain substances, particularly, but not exclusively, metals, causes bound electrons to be given off with a maximum velocity proportional to the frequency of the radiation, i.e., to the entire energy of the photon. The Einstein photoelectric law, first verified by Millikan, states:

$$E_k = hv - \omega$$

where E_k is the maximum kinetic energy of an emitted electron, h is the Planck constant, v is the frequency of the radiation (frequency associated with the absorbed photon), and ω is the energy necessary to remove the electron from the system, i.e., the photoelectric work function for the surface of the emitting substance. An inverse photoelectric effect results from the transfer of energy from electrons to radiation. For example, in an x-ray tube, there is observed the transfer of energy from electrons accelerated by the anode voltage to radiation emitted by the target. This radiation exhibits a continuous spectrum at lower voltages, upon which are superimposed, at higher voltages, intense lines characteristic of the anode material.

Two principal aspects of the photoelectric effect are described here: (1) Photoconductivity; and (2) photovoltage.

Photoconductivity is the phenomenon evidenced by the increase in electrical conductivity of a material by the absorption of light or other electromagnetic radiation. Although insulating or semiconducting materials exhibit this effect to some degree, there are relatively few materials that give sufficiently large changes of conductivity with illumination for application of the principle in useful devices. The principle can be explained briefly by using a cadmium sulfide photoconductor as an example. As in the case of luminescence, the band-type of energy level diagram is useful. See Fig. 1. Transition 1 represents absorption of a photon of energy at least equal to that of the band gap, giving rise to a free electron and a free hole. Transition 2 represents absorption at a local crystalline imperfection (defect or impurity), also producing a free electron, but with a hole trapped in the vicinity of the imperfection. While these carriers are "free" in the crystal, the conductivity can be greatly enhanced, so that the conductivity in the light can be a million times that in the dark. Recombination of the carriers may occur via transition (3), which is a "direct" electron-hole recombination across the band-gap, or via step 4, an electron recombining with a center containing a hole, so that they no longer contribute to the conductivity.

For a material in which one type of carrier predominates (i.e., electrons), the change in conductivity with illumination can be given as:

$$\Delta\sigma = \Delta n e \mu + e n \Delta\mu \qquad (1)$$

where $\Delta\sigma$ = conductivity change, Δn is the change in free carrier density, e is the electronic charge, μ is the carrier mobility, and $\Delta\mu$ is the change in carrier mobility. Usually, the first term in Equation (1) predominates.

Photoconductivity gain, G, may be defined as the number of interelectrode transits that can be made by an electron until the photogenerated hole is eliminated by recombination. For the case treated here, namely where one type of carrier predominates, the gain, G, can be stated as:

$$G = \tau\mu^{VL-2} \qquad (2)$$

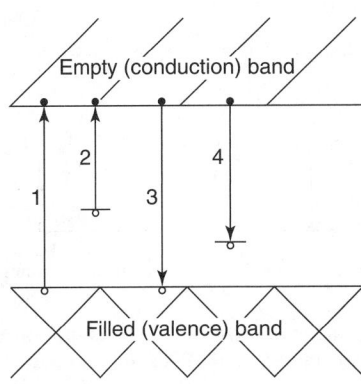

Fig. 1.   Simplified band model for photoconduction processes

where τ is the carrier lifetime, μ is the mobility, V is the applied voltage, and L is the spacing between electrodes. Since the specific sensitivity, S, varies as the product of carrier lifetime and mobility, $S\alpha\mu\tau$, we can state that

$$G\alpha(VL^{-2}S) \qquad (3)$$

Commercially, photoconductive devices are used as (1) Detectors of radiation; (2) switches which are sensitive to light and which can actuate relays; and (3) in combination with other photoelectronic materials, such as electroluminescent materials, as image intensifiers. Germanium and silicon devices of the $p-n$ junction phototransistor type have long been used in computer detectors; lead sulfide has been used in photocells for infrared detection; cadmium sulfide or cadmium selenide have been used in photocells for detection of light in the visible range; zinc oxide and selenium devices have been used in photocopying machines; antimony sulfide has been used in television pickup tubes. There are numerous other applications.

Photovoltage or the *photovoltaic effect* may be defined as the conversion of light photons to electrical voltage by a material. Becquerel, in 1839, was the first to discover that a photovoltage was developed when light was shining on an electrode in an electrolyte solution. Nearly half a century elapsed before this effect was observed in a solid, namely, selenium. Again, many years passed before successful devices, such as the photoelectric exposure meter, were developed. Radiation is absorbed in the neighborhood of a potential barrier, usually a *p-n* junction, or a metal-semiconductor contact, giving rise to separated electron-hole pairs, which create a potential. An equivalent circuit for a photovoltaic cell is shown in Fig. 2, where R_{SH} and R_S are the internal shunting and series impedances; I_J is the junction current; R_L is the load resistance; I_S is a constant-current generator; and V_L and I_L are the voltage and current developed across the load. With R_L optimum, the maximum conversion efficiency, η_{max}; can be given by:

$$\eta_{max} = \frac{100 \, V_{mp} I_{mp}}{P_{in}}$$

in which case, V_{mp} and I_{mp} are the voltage and current across R_L, and P_{in} is the radiant input power.

Photovoltaic cells have found numerous applications in electronic and aerospace applications, notably in satellites (solar cells) for instrument power. See also **Solar Energy**. Materials used, in order of decreasing theoretical efficiency, include gallium arsenide (24%); indium phosphide (23%); cadmium telluride (21%); silicon (20%); gallium phosphide (17%); and cadmium sulfide (16%). See Fig. 3. In the past, disadvantages of photovoltaic cells have included: (1) high susceptibility to radiation damage; (2) high cost; and (3) requirement for auxiliary battery power

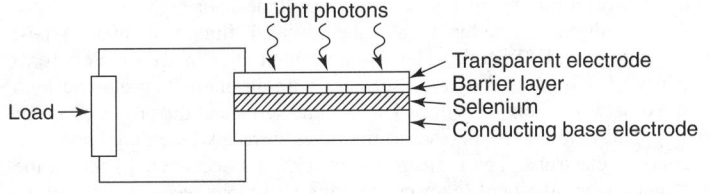

Fig. 2.   Diagram of photovoltaic cell

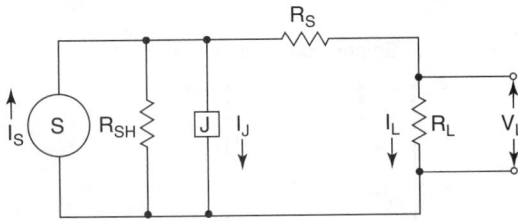

Fig. 3.   Photovoltaic cell equivalent circuit

when a source of radiation for the cells is not available. Intensive research is underway in the mid-1970s to improve the production of known materials, i.e., upping the practical efficiency to approach the theoretical efficiency—as well as decreasing production costs; and to find new materials and combinations. Such findings are extremely important to certain of the proposed approaches for utilizing solar energy on a large scale. Much of this research is currently of a proprietary nature. The scope of the problem can be visualized from published Figures giving ranges of 0.2 to 0.5 volts per cell with current output of a cell with an area of two square centimeters amounting to only about 0.05 ampere. Consequently, without significant materials and efficiency improvements, extremely large numbers of cells are needed for any large-scale, practical solar energy application. Also, for maximum utility, a means must be found to orient the cells wherever possible so that the incident light will be perpendicular to the face of the cells.

See also **Photoemission and Photomultipliers**; **Photometers**; **Photon and Photonics**; and **Photovoltaic Cells**.

PHOTOEMISSION AND PHOTOMULTIPLIERS.

Photoemission is the ejection of electrons from a substance as a result of radiation falling on it. Photomultipliers make use of the phenomena of photoemission and secondary-electron emission in order to detect very low light levels. The electrons released from the photocathode by incident light are accelerated and focused onto a secondary-emission surface (called a dynode). Several electrons are emitted from the dynode for each incident primary electron. These secondary electrons are then directed onto a second dynode where more electrons are released. The whole process is repeated a number of times depending upon the number of dynodes used. In this manner, it is possible to amplify the initial photocurrent by a factor of 10^8 or more in practical photomultipliers. Thus, the photomultiplier is a very sensitive detector of light.

The major characteristics of the photomultiplier with which the user is generally most concerned include: (1) sensitivity, spectral response, and thermal emission of photocathodes; (2) amplification factor; and (3) noise characteristics and the signal-to-noise ratio.

Many different types of photocathodes are used in photomultipliers. With a selection of various cathodes, it is possible to cover the range of response from the soft x-ray region (approximately 5 to 500 Å) to the near infrared (approximately 12,000 Å). Materials and combinations used include cesium-oxygen-silver; cesium-antimony; cesium-antimony-bismuth; sodium-potassium antimony; sodium-potassium-cesium-antimony; copper-iodine; and cesium-iodine. The thermal emission at 25°C of copper iodide and cesium iodide tends to run less than the other materials.

The amplification factor in a photomultiplier depends upon the secondary emission characteristics of the dynode and to some extent on the design of the multiplier structure. Important secondary-emission surfaces used in commercial photomultipliers are of two types: (1) alkali metal compounds, e.g., cesium-antimony; and (2) metal oxide layers, e.g., magnesium oxide on silver-magnesium alloy. The alkali metal compounds have higher gain at low primary electron energy (of the order of 75 volts). The metal oxide layers show less fatigue at high current density of emission (i.e., at several microamperes per square centimeter or higher).

The multiplier structures may be divided into two main types: (1) dynamic; and (2) static. The dynamic multiplier in its simplest form consists of two parallel dynode surfaces with an alternating electric field applied between them. Electrons leaving one surface at the proper phase of the applied field are accelerated to the other surface where they knock out secondary electrons. These electrons, in turn, are accelerated back to the first plate when the field reverses, creating still more secondary electrons. Eventually, the secondary electrons are collected by an anode placed in

the tube; if they are not, a self-maintained discharge occurs. In practice, dynamic multipliers have been replaced by static ones mainly because the latter have better stability and are easier to operate.

Static multipliers may be either magnetically or electrostatically focused. In a magnetic type, primary electrons impinging on one side of a dynode cause the emission of secondary electrons from the opposite side. These electrons are then focused onto the next dynode by means of the axial magnetic field. The more common types of electrostatic multiplier structures use focusing from one stage to the next. Deposition of thin semiconductor secondary emission surfaces onto insulating substrates has been used in designing rugged miniature multiplier structures. In unfocused electrostatic structures, there is less sensitivity to stray electric and magnetic fields.

Dark noise in photomultipliers is caused by: (1) leakage current across insulating supports; (2) field emission from electrodes; (3) thermal emission from the photocathode and dynodes; (4) positive ion feedback to the photocathode; and (5) fluorescence from dynodes and insulator supports. Careful design can eliminate all but item (3). Associated with the photocurrent from the photocathode is shot noise. There is also shot noise from secondary emission in the multiplier structure.

A major use for photomultipliers has been in the scintillation counter where in combination with a fluorescent material, it is used to detect nuclear radiation. They also have been used in star and planet tracking for guidance systems as well as in star photometry and quantitative measurements of soft x-rays in outer space. Additional uses include facsimile transmission, spectral analysis, process control, and wherever extremely low-light levels must be detected. For applications in photometers, see also **Photometers**.

PHOTOGRAPHIC CHEMISTRY.

In photographic films and papers the sensitive surface usually consists of microscopic grains of a silver halide, suspended in gelatin. Exposure to light renders the halide particles susceptible to reduction to metallic silver by developing agents containing a reducing agent, as well as an accelerator, preservative, and restrainer. The accelerator increases the activity of the reducing agent (due principally to ionization of the phenolic agents to their active form) and is usually an alkaline compound. The preservative, usually sodium sulfite, minimizes air oxidation. The restrainer helps to prevent "fog" (reduction of silver halide gains that have not been exposed to light) and is almost always potassium bromide.

Color sensitizers are dyes added to silver halide emulsions to broaden their response to various wavelengths. Unsensitized emulsions are most responsive in the blue region of the spectrum and thus do not correctly represent the light spectrum striking them. Widely used sensitizers include the cyanine dyes, the merocyanines, the benzooxazoles, and the benzothiazoles. Cryptocyanine sensitizes the extreme red and infrared.

In color photography diethyl-p-phenylenediamine is an important developer because its oxidation product readily couples with a large number of phenol and reactive methylene compounds to form indophenol and indoaniline dyes, which are the basis of most of the current color processes.

See also **Holography**; and **Photography and Imagery**.

PHOTOGRAPHIC GRADE.

A chemical in which impurities known to be photographically harmful are limited to safe levels and inert impurities are restricted to levels that will not reduce the required assay strength.

PHOTOGRAPHY AND IMAGERY.

The unique light-sensing properties of silver halide crystals have been recognized since the 1500s. In spite of many technical advances in nonsilver halide (e.g., electronic) technologies, chemically based silver halide systems continue to dominate in the ability to record images of superb image quality and archival characteristics. Photochemical reduction in which the silver ion, Ag+, in the ionic silver halide crystal is reduced to elemental silver, Ag0 was first observed by the alchemist Fabricius in 1556. As photochemical reduction continues, elemental silver atoms aggregate and grow into clusters of a colloidal size sufficient to scatter light and produce hue shifts. The science of photography uses this photochemical property of silver halide to form images and record scenes. One of the earliest researchers to produce such a photochemical image was Schultze in 1727.

Imagery is the representation (pictorial, graphical, etc.) of a subject by sensing quantitatively the patterns of electromagnetic radiation emitted by, reflected from, or transmitted through a subject of interest (object, body, scene, etc.). Imagery is not wavelength-limited, but is achievable (theoretically if not practically) with all bands of the electromagnetic spectrum—gamma rays, x-rays, ultraviolet radiation, visible light, infrared radiation, radar and radio waves.

Chemical imagery or *traditional photography*, as initially conceived and as commonly practiced, depends upon visible light and uses an optical light-gathering and focusing system (camera) and a light-sensitive medium (film emulsion) to record (store) the image—a *photo-image*. The subsequent availability of infrared, ultraviolet, and x-ray sensitive films extended the capabilities of traditional photography well beyond its dependence upon visible light. The word *photography* derives from the Greek roots *photos* (light) and *graphos* (to draw). Coining of the term is usually attributed to Herschel, although this has not been proved conclusively. Herschel did use the term in a memo dated January 17, 1839 and in a technical paper given on March 14, 1839.

Electronic Imagery, instead of using chemical means (emulsions), takes advantage of the sensitivity of various electronic detectors to different bands of the electromagnetic spectrum. The energy received is transduced by these sensors into an electronic or electrical effect (change of resistance, current, emf, the emission of electrons, etc.), from which effects an option of ways to process and display the information is available. The most common form of electronic imagery is found in television. Image orthicons, vidicons, and the more recent TV cameras using charge-coupled devices are among recent developments in electronic imagery.

Electronic imagery is particularly attractive for situations where image information must be transmitted over long distances where digitized signals offer greater accuracy and reliability—and where the incoming information is immediately compatible with digital data processing and computing equipment. Electronic imagery also has made certain imaging tasks possible, such as radar imaging, where traditional photographic means do not suffice.

While techniques are available in traditional photography to enhance raw information, these methods are largely of a qualitative, aesthetic nature rather than of a quantitative, scientific nature. In handling tiny pixels (one of the dots or resolution elements making up a digitized picture) of information, it becomes possible to computer program the processing of image information as it is received or after it is retrieved from tape or other electronic storage medium. The pixels can be measured one at a time at a rapid rate for brightness, sense (detection of obviously bad information), and other quantities, and over a wide scale of selections (for example, black = 0; medium gray = 32; white = 63) so that groupings of input information can be made to provide better contrasts (in pattern and blackness or color) when the information is all regrouped and reassembled for display. Accomplishments of this nature are probably best exemplified by the image intensification and color enhancement of pictures returned to earth from various space explorations. This is discussed further in this entry, but also see entries on specific planets.

With the flexibility of modern data processing and computing equipment, a vast array of programming techniques can be applied to the handling of pixels similar to the handling of any other kind of information. Also, with electronic imagery, a full reconstitution of a scene need not always be a primary objective. For example, as color may be related to chemical composition (discussed later) an astrochemist may call for the proportion of certain "colored" pixels in an entire scene or part of a scene and thus make at least a preliminary judgment as to the composition of rocks or soil in a scene without having to see the entire scene reconstituted.

Whereas the final results of traditional photography are usually in the form of prints or transparencies (with attendant problems), electronic imagery can be projected at will on cathode ray tubes and, if desired, combined with computer graphics—or conventional photos can be made from digitized information.

Early History of Photography. The concept of a camera obscura (dark chamber) was first described by Giovanni Battista Porta in 1558 when he put a lens in a hole in the shutter of an otherwise fully darkened room. His objective was to drawn an image of the outside by means of tracing a pattern projected by the lens onto the screen rather than attempting to draw the scene simply by looking at it. Because there was a great desire to capture scenes and subjects on paper, canvas, and other media, but a scarcity of artistic and drawing abilities among many of the populace, the principle of the camera obscura persisted for some 250 years, with improvements and refinements of the optical system used.

During the sixteenth and seventeenth century, it became known to technically curious persons that a number of substances would change color when exposed to light. These observations provided the first hint that perhaps an image could be captured permanently and thus save a lot of time and labor and also provide a means of relatively quickly producing duplicates of any given subject. There were two main problems: (1) finding a suitable medium that would respond within a reasonably short time span to the projected image; and (2) finding some way to hold or fix an image without permitting the medium to follow a full course of development and thus completely obliterate the captured image. The properties of silver salts were discovered in 1725 by Johann Heinrich Schulze at the University of Altdorf. He found that chalk, when moistened with a solution of nitric acid and silver nitrate, became darker upon exposure to light. Early experiments involved contacting objects with the silver medium to produce silhouettes. The first image obtained through use of a camera-like technique was achieved in France by Joseph NicÉphore Niepce. His first success came in 1813, using paper which he had soaked in sodium chloride, followed by immersion in a silver nitrate solution, upon which silver chloride was precipitated throughout the paper. The crude image obtained was a negative and persisted for only a short period because he had not developed a required fixative. Niepce then turned his efforts to an asphalt process, known as *heliogravure*, which he used for copying prints. The process was quite insensitive, but an image (exposure of about 8 hours in direct sunlight) of some buildings is regarded as the *first photograph*. This old photograph is now part of the Gernsheim collection in the United States.

Daguerreotype. The first practical process of photography was invented by Louis J.M. Daguerre of Paris in 1837, although the details of the process were not published until 1839. The process was used chiefly for portraiture and became obsolete within a few years after the introduction of the wet collodion process in 1851. Although the daguerreotype process was the original, modern photography is based on the negative-positive methods introduced the same year by William H. Fox-Talbot of England. This was known as the *calotype* process. In the daguerreotype process a light-sensitive layer of silver iodide is formed on a silver plate by contact with iodine. After exposure in the camera, a positive image is produced when the image is exposed to mercury and heated. The mercury, by attaching itself to the unexposed portions, forms a positive image. The silver iodide remaining was removed at first with a solution of sodium chloride (salt) which was soon replaced, however, with sodium thiosulfate (hypo), the properties of which had been discovered by Herschel in 1819. The daguerreotype image so produced is very weak. In 1840 Fizeau described a process of toning with gold which greatly increased the strength of the image and was generally adopted.

At first, from 5 to 10 minutes' exposure was required on open landscapes and street scenes. The invention of a fast, large-aperture portrait lens by Petzval in 1841 and the discovery by Goddard in London (1840) of the superior sensitivity of silver bromide reduced the time of exposure to a few seconds.

Problems were encountered in preparing positive prints from the calotype negatives because of reproduction of the grain of the paper that contained the negative. Attempts were made to wax or oil paper negatives, but these were essentially unsuccessful. De Saint-Victor attempted to coat plates with albumin (egg white). Upon hardening of the albumin, the plates were bathed in silver nitrate, causing precipitation of silver iodide within the film of albumin. This was not successful because the sensitivity of the plates was greatly lowered.

Early Emulsions. The use of collodion in photographic emulsions dates from 1851 when Frederick Scott Archer published details of his wet collodion process. Although this process is no longer in general use, it can be used in making the half-tone negatives required in photoengraving. In the collodion process, a clean glass plate is first coated with collodion containing potassium iodide and potassium bromide. It is next sensitized by immersion in a solution of silver nitrate. It is then placed in a plate holder—specially designed for the handling of the wet plate—and the exposure made. After exposure it is developed in a solution of ferrous sulfate and fixed in potassium cyanide, or in hypo, washed and dried. The wet collodion process, as it is used by the photoengraver, results in a negative of high density and extreme contrast, high resolution and with an extremely fine grain. These characteristics render wet collodion well adapted to the requirements of photoengraving. Much later, the wet collodion process was essentially replaced by the gelatino-bromide emulsions of similar characteristics.

Collodion printing-out paper was introduced by Obernetter of Munich in 1867 and was for many years the favorite printing process of the portrait and professional photographer. It was in general use until the early years of the present century when it was gradually replaced by developing-out paper.

Gelatin Emulsions. Not true emulsions, but suspensions of minute silver halide crystals dispersed in a protective colloid medium (gelatin), the suggestion of replacing collodion with gelatin was first made by R.L. Maddox in 1871. The first plates made by Maddox were not very sensitive, but their advantages far outweighed their defects, leading to further developments by Charles Bennett in England in the late 1870s, and the first mass production attempts by George Eastman in 1880. One of the several contributions of Eastman to photography was his early recognition of making and marketing gelatin dry plates on a large scale, eliminating the need for the photographer to prepare his own plates, as well as the need for developing and fixing the plates immediately thereafter. There soon followed the concept of strip film, making it unnecessary to change plates after each exposure. Eastman avoided the grain problem by using a coating that enabled the stripping of the thin layer of gelatin from the paper support. Later, in 1889, he replaced the paper support with a transparent plastic support (nitrocellulose), thus making it possible to produce prints without the need of stripping the gelatin layer from the support. Eastman's goals were to make it easy for the masses of people to take photographs in a simplified manner and, through mass production, market equipment at a price within grasp of the public.

Gelatin is a preferred photographic colloid because the sensitizing bodies in the gelatin make possible emulsions with great sensitivity and speed. Gelatin is an excellent emulsifying agent and is readily transformed, from gel to a liquid or the reverse, by changes in temperature. The latter property makes coating of supports and emulsion processing and working feasible. The strong protective action of gelatin lowers the rate of reduction of unexposed silver halide crystals in developers so that image formation is readily obtained.

Silver halides employed in emulsions are the chloride, the bromide and the iodide. Negative emulsions are composed of silver bromide with a small amount of silver iodide. Positive emulsions for films and paper contain silver chloride, or mixtures of silver chloride and silver bromide in varying amounts, according to the tone, speed, and contrast desired.

In photomicrographs of negative emulsions, the crystals of silver bromide appear as flat triangular or hexagonal plates with rounded corners. Some globular, needle-shaped or diamond shaped crystals may also be observed, as in Fig. 1. The thickness of the flat plates is approximately one-tenth of their diameter. The size of silver bromide crystals range from less than one to four micrometers. Crystals of silver bromide, as used in positive emulsions, are quite uniform and seldom exceed 0.5 micrometer in diameter. Multi-layered emulsions contain approximately 1 billion (10^9) crystals per square centimeter for low-speed emulsions to 1.0×10^{-8} centimeter for high-speed negative emulsions.

The characteristics of an individual emulsion are primarily dependent on two factors, the size-frequency distribution of the crystals and the composition of the silver halide crystals. For instance, Fig. 2 illustrates silver bromide grains with slightly rounded corners due to the presence of a silver complexing agent. The chief problems of the emulsion-maker are the production of uniform suspensions of silver halide crystals with proper size-frequency distribution and the correct composition in gelatin, and the ability to reproduce results. In an attempt to meet these needs along with increased sharpness for high-speed negative color films, *Eastman Kodak* has developed T-Grain type emulsions of silver bromide which also contain some iodide. Illustrated in Fig. 3, these new emulsions exhibit flat grains with relatively sharper edges. As a result, when compared to traditional high-speed color negative films they are capable of producing more clearly defined images.

Classification of Emulsions

1. *Printing-out emulsions.* These emulsions produce images on exposure without development. They are used largely for making portrait proofs which are distinguished by their red or purplish color. Emulsions of this type differ from others in that they usually contain silver nitrate, some free silver, silver salt of an organic acid and a weak free acid. These are known as P.O.P. Proof Papers.

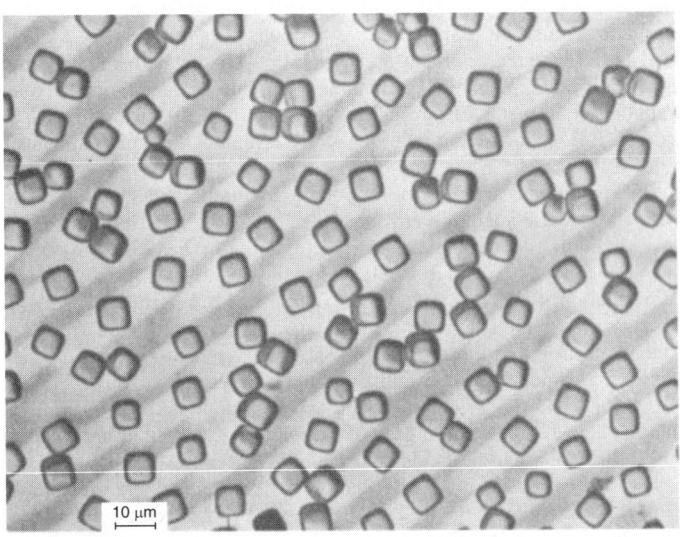

Fig. 2. Using the carbon replica technique, this is an electron micrograph of cubic silver bromide grains in which the corners have been slightly rounded due to the presence of a silver complexing agent. (*Photo by Dr. Donald L. Black, Eastman Kodak Company*)

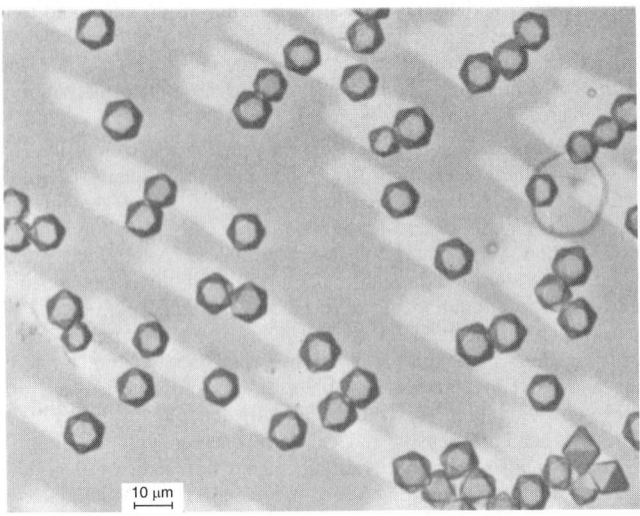

Fig. 1. Using the carbon replica technique, this is an electron micrograph of octahedral silver bromide grains. (*Photo by Dr. Donald L. Black, Eastman Kodak Company*)

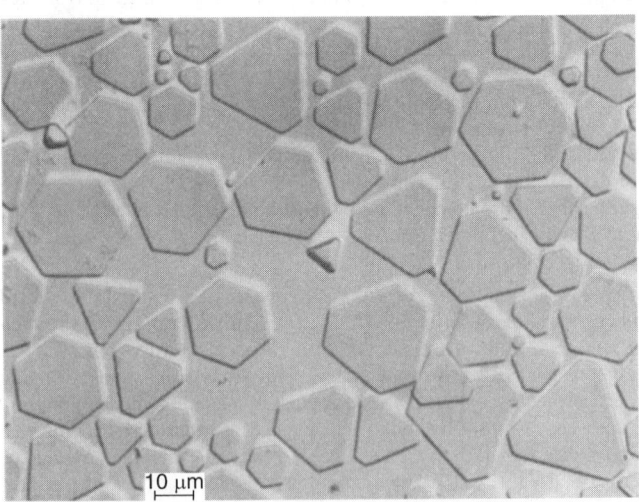

Fig. 3. Using the carbon replica technique, this is an electron micrograph of an Eastman Kodak T-Grain type emulsion of silver bromide that contains some iodide. (*Photo by Dr. Donald L. Black, Eastman Kodak Company*)

2. *Developing-out emulsion.* Emulsions for development have an excess of alkaline halides. By varying the composition of the silver halide and treatment, developing-out emulsions may be prepared which are suitable for either negative or positive purposes.

   a. *Negative emulsions.* Negative emulsions are prepared by adding a small amount of a soluble iodide to the bromide used in making the silver halide. The mixed crystals of silver-bromiodide formed are more sensitive to light and produce emulsions with greater speed than silver bromide alone. Negative emulsions are referred to as neutral emulsions if precipitation of the silver halide is carried out in a gelatin solution with an excess of soluble bromide. They are referred to as ammonia emulsions if the precipitation takes place in a gelatin solution with an excess of soluble bromide in the presence of ammonia or ammoniacal silver. The latter method produces emulsions with coarser grains, which have the highest sensitivity.

   b. *Positive emulsions.* Positive emulsions are prepared by precipitating silver halides containing chloride or mixtures of chloride and bromide in gelatin. The size of the crystals formed are smaller than those of negative emulsions and have a lower sensitivity. Positive emulsions are divided into four classes, according to the composition of the silver halides and their properties.

- *Chloride Emulsions.* Because of their slow speed chloride emulsions are used largely for contact printing.
- *Bromide Emulsions.* Bromide emulsions are very sensitive and fast. They are used for projection printing exclusively.
- *Chlor-Bromide Emulsions.* In chlor-bromide emulsions the amount of silver chloride is greater than that of silver bromide. These emulsions are somewhat faster than chloride emulsions and used for contact or slow projection printing. Chlor-bromide emulsions produce warm-toned silver images with a brown or brown-black color.
- *Brom-Chloride Emulsions.* Brom-chloride emulsions contain more silver bromide than silver chloride. They are faster than chlor-bromide emulsions and used for projection printing where black images and speed printing are desired. Image tones of brom-chloride emulsions are not as warm as chlor-bromide images nor as cold as bromide images.

Other Emulsion Additives. In addition to chemical and spectral sensitizers, several other classes of chemical compounds are added to emulsions before coating. Additives are used to facilitate coating operations, e.g., surfactants (qv) and viscosity enhancers; to reduce spontaneous development in unexposed regions, e.g., tetraazaindenes and mercaptotetrazoles; and to reduce abrasion and permit high temperature processing, e.g., aldehydes.

For certain component compositions the viscosity and surface tension of the melted emulsion may not allow adequate emulsion spreading on the support during the coating procedures. For these situations, various surfactants that act as spreading agents are available to control the surface tension. The sulfiding reaction is a thermally activated process with an activation energy near 126 kJ/mol (30 kcal/mol). Therefore, quenching the reaction by cooling the emulsion from 60 to 25°C does not eliminate the reaction but rather reduces the rate by about two orders of magnitude. After long storage of the emulsion, continued sensitization may produce a catalytic activity in the silver halide grains and unwanted photographic fog upon development. This can be controlled partly by additions of such stabilizers as halide ions, acid, benzimidazoles, benzotriazoles, benzothiazolium salts, and mercaptotetrazoles. Many of these compounds adsorb to silver and complex with silver ions. Specifically as a result of these interactions, phenylmercaptotetrazole restrains development and enhances sensitivity even in freshly coated samples. Quantum-mechanical analyses coupled with photographic data suggest that the best stabilizers not only bind with silver ions but are also poor reducing agents. Azaindene compounds satisfy both of these properties and are effective stabilizers and development antifoggants. Fog control for Au-sensitized emulsions can be, in part, achieved using thiocyanate. Gelatin cross-linking agents (hardeners) represent another class of materials that may be added before coating. These compounds render the coated emulsion layers more resistant to abrasion during handling and improve the thermal stability of the gelatin.

The desire for reduced development times required to produce an image has necessitated the use of solutions with increased activities. Temperature increases, pH increases, and increased oxidizability are all variations directed toward shortening process times. Unfortunately, high temperature processing tends to soften and dissolve the gelatin emulsions; therefore the gelatin must be hardened before development. The enhanced thermal stability and improved mechanical durability produced by hardeners result from the formation of three-dimensional bridging of various sites within the gelatin molecules. Both inorganic, e.g., chromium salts, and organic, e.g., aldehydes, compounds have been used as hardeners. Chromium appears to complex with carbonyl groups, whereas many of the organic hardeners seem to cross-link between the amino groups in the gelatin molecules.

In most color photographic products, organic compounds such as couplers or redox dye releasers are added to the melted emulsions before coating. These compounds are essential to the development reactions that produce the dye molecules composing color images.

Manufacture of Commercial Emulsions. Although the details are proprietary, the basic procedures of manufacturing commercial emulsions are known. A portion of the gelatin in the formula is swelled by soaking in water and later dissolved with heat. Mixtures of soluble bromides and iodides, or chlorides, are placed in water solution and added to the gelatin solution. Precipitation of silver halides is accomplished by slowly adding a solution of silver nitrate, while stirring, to the mixture. The relative concentration of the solutions, the rate of addition and temperature during mixing, are factors which control the formation, size and dispersion of the crystals in gelatin. The emulsion is then heated or "ripened" at 40–80°C to recrystallize the silver halides and readjust the size-frequency distribution. Following ripening, more gelatin is added and the emulsion is chilled so it will set quickly. The emulsion is then placed in a press and forced through a screen to break it into shreds or noodles, which are washed, in cold running water to remove the potassium nitrate formed, the excess soluble halides, and certain soluble byproducts of the reaction. Chloride emulsions are often prepared without washing or with only a limited washing. After washing, the emulsion is drained, remelted, and additional gelatin and certain agents, such as fog preventatives, are added. The emulsion is then heated, or "after-ripened," to form sensitizing nuclei on the silver halide crystals. This operation increases the sensitivity and contrast of the emulsion and is necessary for the preparation of high-speed negative emulsions. Certain preservatives, or stabilizers, are added so the emulsion can be stored in refrigerated rooms until needed. Before coating the emulsion is melted and sensitizing dyes, hardening agents, wetting agents, etc., are added. After thorough mixing, filtering and heating to coating temperature, it is placed in a coating machine. Supports, as film, paper, or glass, with substratum coatings are fed through machines at proper rates so they become coated with emulsions in uniform layers of desired thickness. The coated supports pass over chill boxes to set the emulsion and then through a series of drying compartments where the rate of drying is carefully controlled so as not to change the sensitivity on the surface. Following drying, the coatings are inspected under proper safelights and the film or paper is cut to desired size and packaged.

Numerous variations in the manufacturing process make possible a wide range of film characteristics, including film speed and spectral sensitivity. Film, unlike the human eye, can extend beyond the visible region of the spectrum. High-speed film can capture the details of a fast-moving object, seen only as a blur by the eye. By extended exposure, film can capture images entirely too faint to be seen by the eye. The three main types of film emulsions for black-and-white photography are: (1) *ordinary* (color-blind; sensitive to blue light only); (2) *orthochromatic* (sensitive to all but red light); and (3) *panchromatic* (sensitive to light of all colors). Ordinary and orthochromatic films generally offer greater contrast than most panchromatic emulsions. However, the response of panchromatic emulsions can be modified by use of color filters. Film is available in several sizes and formats. Obviously, a delineation of film specifications is beyond the scope of this encyclopedia. Some excellent references are listed at the end of this entry.

Color Films. The trichromatic theory of vision was first proposed by Thomas Young, a British physicist in 1801. He was the first to propose that the retina of the eye incorporates three different types of receptors, responding to blue, green, and red light, respectively. The theory was elaborated upon to the extent that color perception is based upon the stimulation of two receptors, with light stimulating both red and green receptors seen as yellow light; light equally stimulating all three types of receptors seen as white, etc. Young concluded that it should be feasible to match any color of the spectrum through the proper mixing of blue, green, and red light. Although not essentially interested in color photography, Maxwell effectively demonstrated the principle by way of specially-prepared lantern slides before the Royal Institution in London in 1861. Maxwell had demonstrated the *additive color principle* (mixing of blue, green, and red light).

Practical color photography on a massive amateur scale, of course, could not depend upon the preparation of three separate photographs and the use of three projectors, but rather dictated a process that would combine the three records on one plate. In 1907, the LumiÈre Autochrome plate was developed. This was comprised of a very coarse mosaic of potato starch grains, one third of which was dyed blue; another third, green, and the remaining third, red. An emulsion layer was exposed, with the light first passing through the mosaic. In the *Kodacolor* system of 1928, filters in the camera were used instead of color mosaics. A major problem of the mosaic and filter approaches was that of loss of light as it passed through one or the other media, greatly reducing sensitivity and loss of brightness of a projected image. See also **Additive Color Process**.

In the *subtractive color system*, the phenomenon of absorption is involved. A dye that will absorb red light will, in turn, reflect green and blue light, thus appears a greenish-blue (cyan); a dye that will absorb green light appears a bluish-red (magenta); and a dye that absorbs blue light appears yellow. Thus, cyan, magenta, and yellow are the three primary subtractive colors. A mixture of all three dyes in proper portion will absorb all primary light and thus appear black. Most processes of color photography make use of a subtractive synthesis to yield prints or transparencies. See also **Subtractive Color Process**.

Color-separation negatives are photographic negatives that record the relative intensities of the primary colors used in the analysis necessary to reproduce a subject by means of color photography. In three-color photography, for example, the separation negatives are records, in terms of silver densities, of the amounts of red, green and blue light received at the camera from the subject.

A set of color-separation negatives may be prepared by photographing the subject three times on separate color-sensitive emulsions so that each is a record of one of the primary colors. A panchromatic emulsion is generally employed with a set of tricolor filters, the colors of the primaries. It is only necessary, however, to obtain the color records on separate negatives so it is also possible to use for each record any combination of color filter and emulsion sensitivity that will record one of the primary colors. A set of color-separation negatives may be made by exposing (1) each one in turn in a camera, (2) by the use of a color camera that will expose them simultaneously, or (3) in a tripack.

It is common practice to balance a set of color-separation negatives, by altering the exposure and development times, so that a gray scale will be recorded equally on each negative. The particular densities desired are dependent on the method of color synthesis to be employed.

The majority of color is by use of integral tripacks. There are three layers of photographic emulsion in the tripack, one layer sensitive to red light, another layer to green light, and another to blue light. They are coated, one on top of the other. Since silver iodobromide emulsions usually selected for film emulsions are sensitive to blue light, sensitivity to the green and red light must be conferred by sensitizing dyes. Although this sensitivity can be obtained, the dyes do not negate the emulsion's natural sensitivity to blue light. Thus, those layers that are sensitive to green and red light must be protected from blue light. This is accomplished by inserting a yellow filter layer that will absorb the blue light. Chloride emulsions on the other hand are sensitive only to ultraviolet light. Whereas they do not require a yellow-filter layer, they have to incorporate a filter for exclusion of ultraviolet light. There are a number of dyes that may be used in dye-transfer systems, but for tripacks it is necessary to select only those dyes that will be formed during the development process. In 1912, the German scientist, Rudolf Fischer, discovered the role of couplers. In his early version of a color film, he placed three layers of emulsion one atop another as previously described and he also incorporated a coupler in each layer to cause the development of a particular color. Fischer's concept was brilliant, but the actual process failed because the couplers and sensitizers tended to wander from layer to layer.

In 1931, Leopold Godowsky, a violinist, and Leopold Mannes, a pianist, and both avid amateur photographers made crude experiments in a home laboratory on a type of color film that ultimately became *Kodachrome*, released by Eastman in 1935. In the *Kodachrome* process, the couplers are laced in the developers instead of in the emulsions. Phenols are usually the couplers that form cyan dyes; nitriles or pyrazolones form magenta dyes; and esters, ketones, or amides form yellow dyes. There are many hundreds of couplers and, consequently, there is continuing improvement in color film. Space here does not permit a detailed description of such important matters as the negative-positive system, reversal systems for transparencies, color corrections, etc., but these areas are well covered in some of the listed references.

Direct Positive Images. Even in the early days of black-and-white photography and the early work of Daguerre and Fox-Talbot, it was realized that there would be a great advantage gained from a system that would initially produce a positive rather than a negative image. As early as the late 1930s, Hippolyte Bayard and Robert Hunt proposed systems, but these did not produce satisfactory results. The *chemical transfer* process was developed by A. Rott in Belgium in 1939 and found application in the document-copying field. In 1947, E.H. Land demonstrated a camera which produced a finished black-and-white print without need for a negative and one that was available to the photographer within a very short period, approximately one minute. This was the first model of the *Polaroid* camera. In chemical transfer, a normal emulsion is used. Immediately after exposure and while within the camera, it is developed in a solution containing combined developer-fixer agents. The emulsion is in contact with a special positive white paper, not light sensitive, on which the finished image is printed. The developing reagent is of a jellylike consistency and in early models was contained in pouches or pods, one for each picture. The exposed grains develop in the normal fashion. The unexposed grains are dissolved by the fixing agent. Thus, in the unexposed areas, the dissolved halide is silver which forms on the nuclei in the receiving sheet. In connection with partially exposed areas, the developing grains and the receiving sheet nuclei compete for the silver. Thus, a negative image is formed on the original film or paper, whereas a positive image appears on the receiving sheet. Subsequent to the first Polaroid camera, models were developed to provide a permanent negative as well as print, with the processing time reduced to seconds.

To achieve an instant color film that could rival 35 mm color quality, current generation *Polaroid Spectra* system film utilizes two different color chemistries for greater control of the self-developing image formation process. Composed of 18 microscopically thin coated layers in a rectangular format for both horizontal and vertical composition, this film is able to produce photographs of improved color separation, saturation and brilliance. This is a result of combining images created in three chemical sandwiches in the film negative. Each sandwich is sensitive to red, green, or blue light and consists of a photosensitive emulsion and a related image dye.

For the *Spectra* film, the blue-light sensitive sandwich has been radically altered by the utilization of thiazolidine dye release. This required the creation of new molecules and a new dye-release mechanism involving only a minute quantity of silver. By utilizing this hybrid imaging system, chemical interaction and molecular cross talk between the red-, green-, and blue-sensitive sandwiches have been reduced, which results in greater color definition.

This new material also affords substantial improvement in yellow dye saturation and in recording pastels. By controlling chemical crosstalk between the red and green layers, *Spectra* film is further able to produce more brilliant greens, which is a difficult photographic accomplishment because of the low reflectivity of green in nature. This material also reproduces a broader range of hues and tints when compared with earlier-generation self-developing color films. The transparent support through which the image is viewed is thinner than previous *Polaroid* films and enhances image quality.

Special Films

Infrared Films. The first photographs by infrared radiation appear to have been made about 1880 by Sir William Abney, using a specially prepared collodion emulsion. Abney is reported to have photographed a boiling tea-kettle, but efforts by others to repeat his work were not particularly successful. In 1903, the first real infrared sensitizer, Dicyanine, was discovered. While the sensitizing action extended to a wavelength of 960 nanometers in the infrared, the exposure was too long for general photography and Dicyanine-sensitized plates were used chiefly in infrared spectroscopy. The discovery of more efficient sensitizers in the early twenties, beginning with Kryptocyanine and neocyanine and followed by the penta- and tetra-carbocyanines, has made it possible to prepare films and plates whose sensitivity in the infrared is such that they can be used for general photography, including aerial and motion-picture photography.

Infrared-sensitive films and plates may be divided into two classes: (1) materials of relatively high speed to the extreme red and infrared, i.e., from approximately 700 to 900 nanometers (nm), and (2) materials sensitive to much longer wavelengths but of lower sensitiveness. The former are used for general photography, for aerial photography and cinematography; the latter for spectroscopy in the infrared and other scientific applications requiring sensitivity to wavelengths longer than about 900 nm. All infrared-sensitive materials are sensitive to violet and blue and to the extreme visible red, as well. Photographs made without a filter resemble those made on an ordinary blue-sensitive material. For most purposes it is sufficient to use an orange or light-red filter which will absorb blue and violet light. In this case, the picture is made partly by infrared and partly by the extreme red. The result, however, is generally only slightly different from that obtained with infrared radiation alone. For true infrared photographs, a visually opaque filter transmitting the infrared only must be used. No filter, however, is required when photographing hot bodies such as an electric flatiron, hot castings, or high-pressure boilers, provided that these show no visible glow.

All infrared-sensitive materials must be loaded and developed in total darkness, as safelight screens, even those for panchromatic films and plates, transmit the infrared freely.

Certain precautions are necessary when making pictures with infrared-sensitive materials. The bellows and shutter blades of some cameras, although perfectly safe for ordinary photographic films and plates, transmit the infrared and fog infrared-sensitive films. The slides of some film and plate holders transmit the infrared sufficiently to cause fog. Although some modern lenses are corrected for the infrared, with most, the focal distances for the visible and for the infrared are different. Usually, it is sufficient to extend the lens a distance equal to 2% of the focal length beyond the visual focus. Even this may often be ignored when using a lens of short focal length or a small diaphragm.

Infrared photographs of landscapes are quite different from those made in the usual way. Green foliage is reproduced light and blue sky and water almost black. The shadow portions of the subject are dark and without detail. The general effect is that of a photograph made by moonlight, particularly if the print is made rather dark. As a matter of fact, most night scenes in motion pictures are really infrared photographs.

Since infrared radiation is not scattered by atmospheric haze, as is light, distant objects are rendered sharper and more distinctly in infrared photographs. Objects invisible to the eye because of the intervening haze are often reproduced sharply in an infrared photograph. In fact, one of the most important applications of infrared photography is in photographing distant objects, whether from the ground or the air. Infrared photographs, however, cannot be made through dense fog.

The scientific applications of infrared photography are numerous and important.

Aerial Photography in the Infrared. Extensive use of color infrared film (CIR) has been made in the field of aerial photography for such applications as crop sensing and inventorying, flood assessments, etc. Both normal color film and CIR consist of three separate layers of emulsion on a clear base material. It will be recalled that in normal color film one emulsion layer is sensitive to blue light, one to green light, and one to red light. The images recorded on the three emulsion layers of normal color film combine in the final image to form colors which closely match those of the original subject. CIR film, sometimes referred to as "false color film," also produces combinations of blue, green, and red in the final image; but the blue color results from exposure by green light; the green color from exposure by red light; and red color by exposure of the infrared sensitive layer by infrared energy. Therefore, the images are called false color images. See Fig. 4. Actually, all three layers of CIR film are also sensitive to blue light. For this reason, the film is always exposed through a minus-blue (yellow) filter which eliminates blue light before it reaches the film. The infrared energy needed to expose the infrared sensitive layer

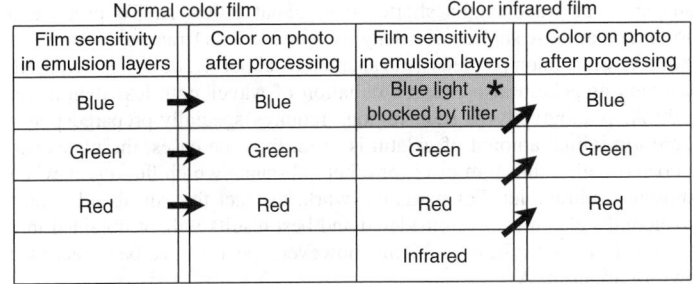

Normal color film		Color infrared film	
Film sensitivity in emulsion layers	Color on photo after processing	Film sensitivity in emulsion layers	Color on photo after processing
Blue	Blue	Blue light ★ blocked by filter	Blue
Green	Green	Green	Green
Red	Red	Red	Red
		Infrared	

Fig. 4.   Film sensitivities and final image color of normal color and color-infrared films. *Blue light absorbed by yellow filter. (*U.S. Geological Survey*)

is reflected energy, not heat energy. Heat energy does not enter into the image forming process of CIR film.

One of the most important features of CIR film is the manner in which vegetation is recorded. Healthy green plants appear in shades of red, because healthy plants reflect sunlight strongly in the photographic infrared region (therefore strongly exposing the infrared sensitive layer) while simultaneously reflecting relatively little energy in the visible region (therefore offering little exposure to the green and red sensitive layers). For all practical purposes, living healthy vegetation is the only natural source of high-infrared reflection coupled with low visible reflection. Because of the unique reflectance characteristic of healthy vegetation, the film was originally used by the military to differentiate between real vegetation and painted camouflage material.

Another unique and variable aspect of CIR photography results from the fact that plants do not reflect strongly in the photographic infrared when they are severely stressed or have died, and as a result, no longer appear red on the photographs, in contrast with normal, healthy vegetation. The reasons behind this phenomenon are complex, yet the ability to distinguish between healthy and stressed or dead vegetation by using CIR is very important for vegetation analysis. This characteristic is particularly useful in determining crop damage due to flooding.

Generalizations about the photographic appearance of other features commonly found on the agricultural landscape can also be made. Clear water usually appears very dark blue or black, but muddy or turbid water appears light to medium blue. This is useful for satellite tracking of pollution situations. Fresh grain stubble appears very light or almost white, whereas clean plowed fields of dark soil usually appear dark blue.

Other Applications of Infrared Film. Among several other scientific uses are:

1. *In Medical Photography.* For the study of the following: diseases and conditions affecting the venous pattern not revealed by light; the progress of healing beneath certain scabs; the eye, to determine atrophy; histological specimens, to reveal structures below the surface and invisible to the eye. Thermography has been used to detect tumors, the skin temperature often being as much as 1 to 2°C higher than that of the surrounding skin.
2. *In Industry.* For the study of irregularities in the dyeing and weaving of textile fibers, the interior of furnaces, the detection of carbon in lubricating oils, infrared spectroscopy of metals and alloys.
3. *In Astronomy.* For detection of nebulae and stars otherwise invisible because of astronomical haze or because their radiation lies chiefly in the infrared; in infrared spectroscopy, for the determination of the composition, the temperature, and the movement of stars and nebulae.
4. *In Criminology.* For deciphering writing or printing that has been crossed out with other inks to render it illegible; for obtaining copies of charred documents, detecting erasures, revealing finger prints, identifying blood and other stains, uncovering secret writings, etc.

Ultraviolet Films. In the near-ultraviolet region, photography is the same as with visible light. However, at shorter wavelengths, many materials are not transparent to ultraviolet radiation. For example, glass is not transparent at wavelengths shorter than about 3,000 Å. To produce a photograph at these shorter wavelengths, a quartz lens (transparent to about 1,800 Å) or a fluorspar lens (transparent to about 1,200 Å) is required. Inasmuch as gelatin also absorbs radiation of wavelength less than about 2,200 Å, photography in such regions requires specially-prepared plates where a minimal amount of gelatin is used. In some cases, the plates can be coated with a thin film of oil or other substance which fluoresces when exposed to ultraviolet. For particular work, the fact that air absorbs short wavelengths also must be considered and best results will be obtained in a vacuum. Even with these problems, however, spectra have been recorded down to about 50 Å.

Ultraviolet photography and spectroscopy have found particular usefulness in the study of combustion processes. Some of the very short-lived chemical species occurring during combustion can be observed in the near-ultraviolet region. Along with infrared, ultraviolet techniques also have been used for detecting the retouching of paintings.

Descriptions of cameras, other photographic equipment and the many thousands of applications of photography are beyond the scope of this volume.

Additional Reading

NOTE: Because of their large numbers, general books on photography are not included here. References that support some of the more specialized areas of photography and imagery, as reported in this entry, are included.

Alper, J.: "Echo-Plannar MRI: Learning to Read Minds," *Science*, 556 (July 30, 1993).

Barger, M.S. and W.B. White: *The Daguerreotype,* Smithsonian Institution Press, Washington, DC, 1991.

Barger, M.S. and W.B. White: *Daguerreotype: Nineteenth-Century Technology and Modern Science,* Johns Hopkins University Press, Baltimore, MD, 2000.

Barinaga, M.: "Biology Goes to the Movies," *Science*, 1204 (November 30, 1990).

Bentley, J.: "Coloring the Invisible World," *Technology Review (MIT)*, 54 (July 1991).

Beardsley, T.: "Sharper Image: Picosecond Photography May Reveal Tumors," *Sci. Amer.*, 32 (October 1991).

Becher, P.: *Emulsions: Theory and Practice,* 3rd Edition, Oxford University Press, Inc., New York, NY, 2001.

Benaron, D.A.: "Optical Time-of-Flight Absorbance Imaging of Biological Media," *Science*, 1463 (January 22, 1993).

Booth, S.A.: "Video To Go: Camcorders," *Popular Mechanics*, 38 (January 1991).

Cipra, B.A.: "Image Capture by Computer," *Science*, 1288 (March 10, 1989).

Corcoran, E.: "Not Just a Pretty Face: Compressing Pictures with Fractals," *Sci. Amer.*, 77 (March 1990).

Corcoran, E.: "Body Heat: Quantum-Well Infrared Photodetectors," *Sci. Amer.*, 123 (October 1991).

Cornwell, T.J.: "The Applications of Closure Phase to Astronomical Imaging," *Science*, 263 (July 21, 1989).

Crease, R.P.: "Biomedicine in the Age of Imaging," *Science*, 554 (July 30, 1993).

Doan, V.V. and M.J. Sailor: "Luminescent Color Image Generation on Porous Silicon," *Science*, 1791 (June 26, 1992).

Drury, S.A.: *Guide to Remote Sensing: Interpreting Images of the Earth,* Oxford University Press, Inc., New York, NY, 1990.

Grimm, T. and M. Grimm: *The Basic Book of Photography,* 4th Edition, Penguin USA, New York, NY, 1998.

Hedgecoe, J.: *The Photographer's Handbook,* 3rd Edition, Alfred A Knopf, Inc., Westminster, MD, 1992.

Huang, et al.: "Optical Coherence Tomography," *Science*, 1178 (November 22, 1991).

Izatt, J.A., et al.: "Ophthalmic Diagnostics Using Optical Coherence Tomography," *SPDIE Proceedings*, 1877 (1993).

Jenkins, F.A., Jr., K.P. Dial, and G.E. Goslow, Jr.: "A Cineradiographic Analysis of Bird Flight," *Science*, 1495 (September 16, 1988).

Lam, D., Man-Kit, and B.W. Rossiter: "Chromoskedasic Painting," *Sci. Amer.*, 80 (November 1991).

Lillesand, T.M.M. and R.W. Kiefer: *Remote Sensing and Image Interpretation,* 4th Edition, John Wiley & Sons, Inc., New York, NY, 1999.

London, B. and J. Upton: *Photography,* 6th Edition, Addison Wesley Longman, Inc., Redding, MA, 1997.

Mollet, H. and A. Grubenmann: *Formulation Technology: Emulsions, Suspensions, Solid Forms,* John Wiley & Sons, Inc., New York, NY, 2001.

Newhall, B.: *Daguerreotype in America,* 3rd Edition, Dover Publications, Inc., Mineola, NY, 1999.

Ourmazd, A., et al.: "Quantifying the Information Content of Lattice Images," *Science*, 1571 (December 22, 1989).

Pappas, D.L., et al.: "Atom Counting at Surfaces," *Science*, 64 (January 6, 1989).

Peterson, I.: "Needle Imaged in Animal-Tissue Haystack," *Science News*, 325 (May 25, 1991).

Pool, R.: "Molecular Photography with an X-ray Flash," *Science*, 295 (July 15, 1988).

Pool, R.: "Making 3-D Movies of the Heart," *Science*, 28 (January 4, 1991).

Richards J.A. and D.E. Ricken: *Remote Sensing Digital Image Analysis,* 3rd Edition, Springer-Verlag, Inc., New York, NY, 1999.

Richelson, J.T.: "The Future of Space Reconnaissance," *Sci. Amer.*, 38 (January 1991).

Roberts, L.: "Mapping by Color and X-rays," *Science*, 425 (April 28, 1989).

Romer, G.B., J. Delamoir: "The First Color Photographs," *Sci. Amer.*, 88 (December 1989).

Silverman, J., J.M. Mooney, and F.D. Shepherd: "Infrared Video Cameras," *Sci. Amer.*, 78 (March 1992).

Staff: "Odyssey (Reviews of Photos in First 100 Years of National Geographic Magazine)," *Natl. Geographic*, 322 (September 1988).

Staff: "New Projectors," *Hughesnews*, 1, Culver City, California (February 21, 1992).

Staff: "Imaging Technologies, Inscribing Science," *Camera Obscura*, 28 (1992).

Vager, Z., R. Naaman, and E.P. Kanter: "Coulomb Explosion Imaging of Small Molecules," *Science*, 426 (April 28, 1989).

Stroebel, L.D., J. Compton, and I. Current: *Basic Photographic Materials and Processes,* 2nd Edition, Butterworth-Heinemann, Inc., Woburn, MA, 2000.

Vander Voort, G.F.: "Metallography," *Advanced Materials & Processes*, 71 (January 1990).

Van Sant, T., et al.: *The Earth—From Space: A Satellite View of the World,* Spaceshots, Inc., Manhattan Beach, CA, 1990.

Waters, A.J., M.J. Bader, J.R. Grant, G.S. Forbes, et al.: *Images in Weather Forecasting: A Practical Guide for Interpreting Satellite and Radar Imagery,* Cambridge University Press, New York, NY, 1997.

Wilkie, D.S. and J.T. Finn: *Remote Sensing Imagery for Natural Resources Monitoring: A Guide for First-Time Users,* Columbia University Press, New York, NY, 1996.

Zwingle, E., H.E. Edgerton, and B. Dale: "'Doc' Edgerton—The Man Who Made Time Stand Still," *Natl. Geographic*, 464 (October 1987).

PHOTOIONIZATION.   This process, which is also called the atomic *photoelectric effect*, is the ejection of a bound electron from an atom by an incident photon whose entire energy is absorbed by the ejected electron. This statement means that photoionization cannot occur unless the energy of the photon is at least equal at the ionization energy of the particular electron in the particular atom; any excess of energy in the photon above this value appears as kinetic energy of the ejected electron.

PHOTOLUMINESCENCE.   See **Luminescence**.

PHOTOLYSIS.   See **Free Radical**; **Photochemistry and Photolysis**.

PHOTOMETERS.   Instruments for the measurement of luminous intensity, luminous flux density, and illumination. In usual terminology, only instruments that respond to the central portion of the electromagnetic spectrum, i.e., the ultraviolet, visible, and infrared regions, are called photometers. Essentially, a photometer is comprised of a transducer, which transforms electromagnetic waves (photons) into an electric current, and a current-measuring readout device. In the simplest form, the instrument could be a voltage-generating photocell connected to a microammeter. Photographic exposure meters and light meters that measure ambient illumination are usually of this type. The latter are furnished with green filters, which correspond to the relative spectral sensitivity of the human eye. Photoresistors also are used for this purpose. Photoresistors require a voltage source (battery) in the circuit. The microammeter reads out the change in resistance caused by the illumination. These devices are more sensitive, but not of high precision because of fatigue effects of photoresistors. For precision work, photomultiplier tubes are usually the transducer selected. Some specific types of photometers and spectrophotometers include:

Atomic-Absorption Photometer.   This instrument operates on the very specific spectral absorption of an atomized sample rather than emission. The equipment is comprised of a stabilized hollow-cathode lamp (one for each element to be analyzed), a flame with sample nebulizer, a monochromator, and a photometer.

Brightness Meter.   A special type of reflection meter for evaluating the brightness of paper and similar products by measuring the diffuse reflectance in the blue range of the spectrum. Actually, these meters quantify the yellow characteristics of the paper.

Circular Dichrograph. An instrument similar to a spectropolarimeter. Instead of a change in angle of optical rotation versus wavelength, the instrument records the difference in dichroic absorption versus wavelength.

Color-Difference Meter. A specially designed reflection meter for assessing small color variations.

Colorimeter. An instrument for routine chemical analysis. Compounds or ions which absorb light in the visible part of the spectrum (400 to 800 nanometers) or which are convertible by specific reagents to such compounds can be analyzed with a colorimeter. The instrument typically incorporates an incandescent light bulb as light source, filters to separate the spectral region, a cuvette to contain the sample solution, and a photometer. See also **Colorimetry.**

Densitometer. An instrument used to measure the attenuation of a beam passing through, or reflected from the surface of solid samples.

Ellipsometer. An instrument for determining the thickness of very thin films of monomolecular dimensions. Essentially, the instrument is a polarization interferometer that utilizes a photometer as a read-out device.

Flame Photometer. See *Atomic-Absorption Photometer* in this entry.

Fluorimeter (also Fluorometer or Fluorophotometer). In this instrument, the sample is excited by a light beam of suitable short wavelength. The remitted fluorescent light is picked up by a photometer, usually placed 90 degrees from incidence. A filter or monochromator is provided which excludes the exciting waveband and transmits the fluorescent light. See also **Fluorometers.**

Footcandle Meter. A color-corrected illumination meter calibrated in footcandles.

Glossmeter. An instrument for measuring specularly reflected light from the surface of a flat sample. The angle of incidence and the angle of light pickup are identical and opposed from the normal to the surface. A typical glossmeter consists of a light source and simple optics to direct a defined beam onto the sample. In the opposite direction, there is a light detector connected to a readout meter.

Hemoglobinometer. A specialized colorimeter for determining hemo globin in blood.

Light-Scattering Photometer. In one type, suspended particles are determined (counted). Another type is used to determine the molecular weight of macromolecules dispersed in solution. The former type operates on the basis of a nephelometer; the latter is of much higher precision and uses monochromatic light.

Lux Meter. Essentially a footcandle meter calibrated in international lux units. (1 footcandle = 10.8 lux.)

Nephelometer. An instrument for determining particle size or particle concentration by measuring the amount of light transmitted or scattered by the suspended particles. Quantitative determinations are made by comparing a given sample with a known standard. See also **Nephelometry.**

Opacimeter. A reflection meter specifically designed to evaluate the opacity of thin sheets, such as paper, by measuring the diffuse reflectance over a white and a black surface in turn.

Optical-Emission Spectrometer. Similar to flame photometer (atomic-absorption photometer) except that an electric spark rather than a flame is used to vaporize (atomize) unknown samples.

Polarimeter. An instrument for determining the concentration of optically active compounds in solution by determining the angle of rotation of plane-polarized light passing through the sample. See also **Polarimetry.**

Reflection Meter. A photometer arranged to pick up diffusely reflected light from the surface of a flat sample. The spectral evaluation of the reflected light permits quantitative color evaluation as seen by the human observer.

Refractometer. An instrument for determining the refractive index of solutes. Most of these instruments use photometric readout systems.

Saccharimeter. A polarimeter calibrated in "sugar degrees" for analyzing the concentration of sugar solutions.

Spectrofluorimeter. A fluorimeter with two separate monochromators. One serves to scan through the spectrum of the exciting light source; the other scans the emitted fluorescent light.

Spectrophotometer. An instrument comprising a light source, means of monochromatizing the light, a sample space, and a photometer. These instruments normally determine concentration of a solute by measurement of light attenuation, the logarithm of absorption being proportional to the concentration. If the instrument is designed to operate in the infrared region, it is known as an infrared spectrophotometer. If in the ultraviolet region, an ultraviolet spectrophotometer, etc.

Additional Reading

Decusatis, C.: *Handbook of Applied Photometry,* Springer-Verlag, Inc., New York, NY, 1998.

Heranshaw, J.B.: *The Measurement of Starlight: Two Centuries of Astronomical Photometry,* Cambridge University Press, New York, NY, 1996.

Swatland, H.J.: *Computer Operation for Microscope Photometry,* CRC Press, LLC., Boca Raton, FL, 1997.

PHOTOMETRIC ANALYSIS. Chemical analysis by means of absorption or emission of radiation, primarily in the near UV, visible, and infrared portions of the electromagnetic spectrum. It includes such techniques as spectrophotometry, spectrochemical analysis, Raman spectroscopy, colorimetry, and fluorescence measurements.

See also **Colorimetry**; and **Raman Spectrometry.**

PHOTON AND PHOTONICS. In common usage, a photon is a quantum of electromagnetic energy. The energy of a photon is hv, where h is the Planck constant, and v is the frequency associated with the photon. The term photon usually refers to a plane-wave quantum of electromagnetic energy, for which the momentum is hv/c, and the component of angular momentum in the direction of the momentum is $\pm\hbar$, where c is the velocity of light and $\hbar$ is $h/2\pi$.

The word *photonics* entered the scientific vocabulary in the mid-1980s to describe a communications transmission system that converted digital information into pulses of light that traveled over an optical fiber cable (fiber optics/light wave communication). The first crude cables were used in the mid-1970s. An exploratory system was established in a network between three buildings in downtown Chicago. Since then, the growth of optical fiber networks worldwide has been no less than dramatic. However, photonics also has a broader connotation and parallels in its hardware and systems aspects the well-established microwave technology, thus, a *photonics technology*. See also **Optical Fiber Systems.**

The existence of the photon was first suggested by Planck's famous research, about 1900, into the distribution in frequency of blackbody radiation. Planck arrived at agreement with the experimental distribution only by making the drastic (for that period in science) assumption that the radiation exists in discrete amounts with energy $E = hf$, where f is the frequency of the radiation and h is Planck's constant, $6.626 (10)^{-27}$ erg sec. Confirmation of the existence of these quanta of electromagnetic energy was provided by Einstein's interpretation of the photoelectric effect (1905). Einstein made it clear that electrons in a solid absorb light energy in the discrete amounts hf. The full realization that the photon is a particle with energy and momentum was provided by the Compton effect (1922), an aspect of the scattering of light by free electrons. Compton showed that features of the scattering are understood by balancing energy and momentum in the collision in the usual way, the light considered as a beam of photons each with energy hf and momentum hf/c.

The modern point of view is that, for every particle that exists, there is a corresponding field with wave properties. In the development of this viewpoint, the particle aspects of electrons and nuclei were evident at the beginning and the field or wave aspects were found later (this was the development of quantum mechanics). In contrast, the wave aspects of the photon were understood first (this was the classical electromagnetic theory of Maxwell) and its particle aspects only discovered later. From this modern viewpoint, the photon is the particle corresponding to the electromagnetic field. It is a particle with zero rest mass and spin one.

For a photon moving in a specific direction, the energy E and the momentum q of the particle are related to the frequency f and wavelength λ of the field by Planck's equation $E = hf$ and the de Broglie equation $q = h/\lambda$. As for all massless particles, the energy and momentum are related by $E = cq$ and the photon can only exist moving at light speed c. Another property of all massless particles is that, given the momentum, the particle can exist in just two states of spin orientation. The spin can be parallel or antiparallel to the momentum, but no other directions are possible. The photon state with the spin and momentum parallel (antiparallel) is said to be right- (left-) handed and is a right- (left-) hand circularly polarized wave. In analogy with the neutrino, one can say that the state has positive (negative) helicity and can call the right-handed particle the antiphoton, the left-handed particle the photon. There is an operation, CP conjunction, that converts a photon state into an antiphoton state and vice versa. It is possible to superpose photon and antiphoton states in such a way that the superposition is unchanged by CP conjugation and so gives a type of photon that is its own antiparticle. The photons produced by transitions between states of definite parities in atoms or nuclei are their own antiparticles in this sense. As for all particles with integer spin, the photon follows Bose-Einstein statistics. This means that a large number of photons may be accumulated into a single state. Macroscopically observable electromagnetic waves, such as those resonating in a microwave cavity, for example, are understood to be large numbers of photons all in the same state. The photon, among all particles, is unique in having its states be macroscopically observable in this way.

Additional Reading

Ackerman, E., et al.: "A 3 to 6 GHz Microwave/Photonic Transceiver for Phased-Array Interconnects," *Microwave J.*, 60 (April 1992).

Cusack, J.: "Photonics at Rome Laboratory," *Microwave J.*, 72 (February 1992).

Fujimoto, J.G. and M.S. Patterson: *Advances in Optical Imaging and Photon Migration,* Optical Society of America, Washington, DC, 1998.

Howe, H.: "Let There Be Light," *Microwave J.*, 24 (January 1992).

Joannopoulos, J.D., R.D. Meade, and J.N. Winn: *Photonic Crystals,* Princeton University Press, Princeton, NJ, 1995.

Kaminow, I.P. and T.L. Koch: *Optical Fiber Telecommunications IIIA,* Morgan Kaufmann Publishers, Orlando, FL, 1997.

Polifko, D. and H. Ogawa: "The Merging of Photonic and Microwave Technologies," *Microwave J.*, 75 (March 1992).

Pradhan, T.: *The Photon,* Nova Science Publishers, Inc., Huntington, NY, 2001.

Render, D.J.: "Photonics—Fast Track for Tomorrow's Communications," *AT&T Technology,* **II**, (1), 1987. *A classic reference.*

Sakoda K.: *Optical Properties of Photonic Crystals,* Springer-Verlag, Inc., New York, NY, 2001.

Soukoulis, C.M.: *Photonic Band Gaps and Localization,* Kluwer Academic Publishers, Norwell, MA, 1993.

Zmuda, H. and E.N. Toughlian: "Adaptive Microwave Signal Processing: A Photonic Solution," *Microwave J.*, 58 (February 1992).

Web Reference

Optical Society of America: http://www.osa.org/

PHOTONEUTRON.   A neutron emitted from a nucleus in a photonuclear reaction.

PHOTONUCLEAR REACTION.   A nuclear reaction induced by a photon. In some cases the reaction probably takes place via a compound nucleus formed by absorption of the photon followed by distribution of its energy among the nuclear constituents. One or more nuclear particles then "evaporate" from the nuclear surface, or occasionally the nucleus undergoes photofission. In other cases the photon apparently interacts directly with a single nucleon, which is ejected as a photoneutron or photoproton without appreciable excitation of the rest of the nucleus.

PHOTOPOLYMER.   A polymer or plastic so made that it undergoes a change on exposure to light. Such materials can be used for printing and lithography plates, photographic prints, and microfilm copying. The light may cause further polymerization or cross-linking, or it may cause degradation. One application involves the use of esters of polyvinyl alcohol that cross-link and so become insoluble, whereas unexposed portions of the material remain soluble.

PHOTOSENSITIVE GLASS.   Certain clear silicate glass containing ingredients capable of forming permanent photographic images when subjected to action of X-rays or ultraviolet light and subsequent heat treatment.

See also **Glass**.

PHOTOSYNTHESIS.   This is the most important of all biological processes. With negligible exceptions the existence of the entire biological world hinges upon this process. From a few simple inorganic compounds and from the sugar made in photosynthesis are erected all of the complex kinds of molecules essential to the construction of the bodies of plants and animals or to maintenance of their existence. Some of these subsequent synthetic processes occur in the plant body, others in the bodies of animals after they have ingested plant materials as foods. Likewise, the energy used by plants and animals represents sunlight energy that has been entrapped in sugar molecules during photosynthesis. The entire organic world runs by the gradual expenditure of the energy capital accumulated in photosynthesis.

Under suitable conditions of temperature and water supply, the green parts of plants, when exposed to light, abstract and use carbon dioxide from the atmosphere and release oxygen to it. These gaseous exchanges are the opposite of those occurring in respiration and are the external manifestation of the process of photosynthesis by which carbohydrates are synthesized from carbon dioxide and water by the chloroplasts of the living plant cells in the presence of light. For each molecule of carbon dioxide used, one molecule of oxygen is released. A summary chemical equation for photosynthesis is:

$$6\,CO_2 + 6\,H_2O \xrightarrow{\text{light}} C_6H_{12}O_6 + 6\,O_2$$

In this process, the radiant energy of sunlight is stored as chemical energy in the molecules of carbohydrates and other compounds that are derived from them.

All photosynthetic organisms, except bacteria, use water as the electron or hydrogen donor to reduce various electron acceptors, and from the water they evolve molecular oxygen. Anaerobic bacteria cannot endure such oxygen, but derive their sustenance through slightly different photosynthetic routes:

$$2\,H_2S + CO_2 \xrightarrow{\text{light}} (CH_2O) + H_2O + 2S$$

or

$$2\,CH_3CHOHCH_3 + CO_2 \xrightarrow{\text{light}} (CH_2O) + CH_3COCH_3 + H_2O$$

Photosynthesis takes place in chlorophyll-containing cells only when carbon dioxide, water, and light are available, and when a suitable temperature prevails. Although carbon dioxide constitutes, on the average, only 0.03% of the atmosphere, land plants are entirely dependent upon this source for the carbon dioxide used in photosynthesis. It has been shown experimentally that an increase in the carbon dioxide concentration of the atmosphere results in an increased rate of photosynthesis. On the other hand, a deficiency of water results in a reduced rate of photosynthesis. In nature, sunlight is the source of radiant energy used in photosynthesis, although plants will also photosynthesize under artificial light sources of suitable quality and intensity.

The total radiant energy received at the earth's surface is 1–2 gram calories/square centimeter/minute, depending upon altitude, or approximately 1 hp/10–20 square feet. For crop plants in the field, a maximum of 2–3% of this energy remains stored in the plants at the end of the growing season. During that time about 20% more is actually used in photosynthesis and lost by respiration of the plant, the remainder of the energy being dissipated by re-radiation, transmission through the leaves, and evaporation of water from the plant.

The intensity, quality and daily duration of illumination all have influence on the amount of photosynthesis accomplished per day. Clearly, the longer the daily period of illumination, the more photosynthesis will be accomplished by a plant in the course of a day. The minimum light intensity at which a measurable rate of photosynthesis occurs varies according to species, but is seldom less than 1% of full midday summer sunlight. Under natural conditions, maximum rates of photosynthesis are attained in single leaves of many species at 25–35% of full sunlight intensity, and in some shade species at even lower intensities. For equal intensities, more photosynthesis appears to occur in the orange–short red and blue parts of the spectrum than in the green and yellow. This is because the chlorophyll pigments of the leaves absorb light energy at wavelengths of 6600 and 4250 micrometers. Radiation is most intense in the green and, if this radiation were absorbed, the plant could not utilize it and would overheat.

The range of temperatures most suitable for relatively rapid rates of photosynthesis is not the same for all kinds of plants. In general, it is higher in tropical than in temperate species, and higher in temperate species than in those of subarctic regions. Increase in temperature results in an increase in the rate of photosynthesis up to an optimum which varies with the variety of plant, but which, for most temperate zone species, lies within the range of 20–30°C. With increase above the optimum, the rate of photosynthesis progressively decreases.

In the vascular plants, photosynthesis occurs chiefly in the leaves. Carbon dioxide diffuses into the intercellular spaces of the leaf from the atmosphere via the stomates, and then dissolves in the moist walls of the mesophyll cells. In solution, the carbon dioxide diffuses to the surface of the chloroplasts, which are the actual seat of the photosynthetic process. The first major step in photosynthesis is the absorption of radiant energy by the plant pigments in the chloroplasts, with the generation of electrons. The plant pigment consists of two closely similar pigments, chlorophyll a and chlorophyll b, which are porphyrin-derived complexes of magnesium and which, upon excitation by radiant energy, become electron donors. See also **Chlorophylls**. The chloroplast is a complex, self-replicating organelle that possesses its own DNA and is able to synthesize at least a few of the proteins needed for its own functioning. It is filled with membranous thylakoid sacs which are specifically designed to harness the energy available in the excited electron and to carry out the light phase of photosynthesis. In this, the light energy captured is converted into the chemical energy of adenosine triphosphate (ATP) and nicotinamide-adenine dinucleotide phosphate (NADPH). See also **Adenosine Phosphates**; and **Coenzymes**. Hydrogen atoms are removed from water and used to reduce NADP, leaving behind molecular oxygen. Simultaneously, adenosine diphosphate (ADP) is phosphorylated to ATP:

In the second, or dark, reaction phase, NADPH and ATP provide the

$$Water + NADP^+ + PO_4 + ADP \xrightarrow{light} Oxygen + NADPH + H^+ + ADP$$

energy to reduce carbon dioxide to glucose and are themselves oxidized or decomposed:

$$CO_2 + NADPH + H^+ + ATP \longrightarrow Glucose + NADP^+ + ADP + PO_4$$

Peter Mitchell (Nobel Prize, 1978) of Great Britain was the first to realize, and to propose in his chemi-osmotic theory, that the energy required for the ADP-ATP reaction could be derived by an accretion of protons in the thylakoid sac to the point at which the electrochemical gradient across the membrane could effect the proton transport required as the driving force for this reaction. See also **Phosphorylation (Photosynthetic)**.

In most plants, the water used in photosynthesis is absorbed by the roots from the soil, whence it is translocated to the leaves. Except for a small portion used in respiration, the oxygen liberated in the process diffuses out of the leaf into the atmosphere, mostly through the stomates. Carbohydrates other than hexoses are synthesized in the leaves, apparently as a result of secondary reactions following photosynthesis. Sucrose invariably accumulates in actively photosynthesizing leaf cells. This more complex sugar is built up from the molecules of the simpler hexoses. In most plants, insoluble starch also accumulates in leaf cells during photosynthesis. This carbohydrate is synthesized by the condensation of numerous glucose molecules. The sucrose and starch contents of leaves decrease at night as a result of the continued translocation from the leaves to other parts of the plant. The sucrose is probably translocated as such, but the starch must first be converted into simpler, soluble sugars before it can move out of the leaves. Synthesis of starch is not restricted to the green parts of plants; a familiar example of this is the accumulation of

starch in potato tubers. Starch in the nongreen cells is made from glucose, which comes from the leaves or other photosynthetic organs. Starch occurs in cells in the form of small grains, the type of grain formed in each kind of plant being more or less characteristic of that species.

For many years, the nature and location of the complex of proteins (sometimes referred to as the "engine" of photosynthesis) were poorly understood. During the 1980s, much more was learned as the result of research carried out by Johann Deisenhofer (Howard Hughes Medical Institute), Robert Huber, and Harmut Michel (Max Planck Institute), and for this work the investigators were awarded the 1988 Nobel Prize for chemistry. The protein complex, called the membrane-bound proteins, are difficult to define structurally because they do not crystallize readily and thus could not be subjected to x-ray crystallography. However, over a period of three years, the researchers were able to create crystals and thus were able to determine precisely the position of some 10,000 atoms in the protein complex.

With this better understanding, researchers are able to depict how plants, algae, and rhodopseudomonads carry out synthesis. Also, it has been hypothesized that such membrane-bound proteins may have a functional role in some diseases, such as cancer and diabetes.

Finally, it must be realized that photosynthesis is not the sole prerogative of the higher plants. More than half the photosynthesis on the earth's surface is carried out in the oceans by phytoplankton.

See also references list at the end of entry on **Photochemistry and Photolysis**.

R. C. VICKERY, D.Sc.,
Blanton/Dade City, Florida

Additional Reading

Amato, J.: "A Shady Strategy for Photosynthesis," *Science News*, 246 (October 20, 1990).

Anderson, B., J. Barber, and H. Salter: *Molecular Genetics of Photosynthesis*, Oxford University Press, Inc., New York, NY, 1996.

Blankenship, R.E.: *Molecular Mechanisms of Photosynthesis*, Blackwell Science, Inc., Malden, MA, 2001.

Blaxter, J.H.S. and A.J. Southward: *Advances in Marine Biology*, Academic Press, Inc., San Diego, CA, 1993.

Bogorad, L. and I.K. Vasil: *The Photosynthetic Apparatus*, Academic Press, Inc., San Diego, CA, 1991.

Charles-Edwards, D.A.: *Mathematics of Photosynthesis and Productivity*, Academic Press, Inc., San Diego, CA, 1981.

Coleman, G. and W.J. Coleman: "How Plants Make Oxygen," *Sci. Amer.*, 50 (February 1990).

Darnell, J., H. Lodish, and D. Baltimore: *Molecular Cell Biology*, 4th Edition, W. H. Freeman and Company, New York, NY, 1999.

Falkowski, P.G. and J.A. Raven: *Aquatic Photosynthesis*, Blackwell Science, Inc., Malden, MA, 1996.

Hall, D.O. and K. Rao: *Photosynthesis*, 6th Edition, Cambridge University Press, New York, NY, 1999.

Herring, P.J., et al.: *Light and Life in the Sea*, Cambridge University Press, New York, NY, 1990.

Hogan, J.: "1988 Nobel Prize for Chemistry," *Sci. Amer.*, 33 (December 1988).

Holden, C.: "Picture-Perfect Plankton," *Science*, 681 (February 7, 1992).

Kirk, J.T.O.: *Light and Photosynthesis in Aquatic Ecosystems*, 2nd Edition, Cambridge University Press, New York, NY, 1994.

Metzler, D.: *Biochemistry*, 2nd Edition, Academic Press, Inc., San Diego, CA, 2002.

Miller, K.R.: "A Particle Spanning the Photosynthetic Membrane," *J. Ultrastruct., Res.*, **54**, 1, 159–167 (1976).

Miller, K.R.: "The Photosynthetic Membrane," *Sci. Amer.*, **241**, 4, 102–113 (1979).

Ort, D.R.: *Oxygenic Photosynthesis: The Light Reactions*, Kluwer Academic Publishers, Norwell, MA, 1996.

Pessarakli, M.: *Handbook of Photosynthesis*, Marcel Dekker, Inc., New York, NY, 1996.

Raghavendra, A.S.: *Photosynthesis*, Cambridge University Press, New York, NY, 2000.

Sherman, K., L. Alexander, and B. Gold: *Large Marine Ecosystems: Patterns, Processes, and Yields*, AAAS Books, Waldorf, MD, 1992.

Stoecker, D.K.: "Photosynthesis Found in Some Single-Cell Marine Animals," *Oceanus*, 49 (Fall 1987).

Stryer, L. and J.L. Tymoczko: *Biochemistry Extended, Chapters 1–34*, 5th Edition, W. H. Freeman and Company, New York, NY, 2002.

Yunus, M. and Dr. U. Pathre: *Probing Photosynthesis: Mechanisms, Regulation, and Adaptation*, Taylor & Francis, Inc., Philadelphia, PA, 2000.

Zilsnov, V.K.: "Living Marine Resources," *Oceanus*, 29 (Summer 1991).

PHOTOVOLTAIC CELLS.   A photovoltaic (PV) solar power system is a complete electrical source that uses solar cells to directly convert light energy into electricity. The system can be self-contained and completely autonomous or it can work in tandem with other conventional fuel-based sources of power to offer robust power availability.

A solar cell is a semiconductor device that can convert light instantaneously into direct-current (d-c) electricity. A number of cells are typically connected together in series in a weather-resistant package such that enough voltage is generated to recharge a 12-volt lead–acid storage battery, the most common storage device used in conjunction with solar power. Such a package of cells is designated a PV module, which is often constructed of an external sheet of strengthened glass and polymeric encapsulation. The most common size module is $0.5–1\ m^2$ in area and delivers between 25 and 150 watts of power. See also **Batteries**.

The advantages of photovoltaic cells as a source of electric power over alternative power sources may be characterized as follows: solar cells capture sunlight, an essentially inexhaustible and nonpolluting energy source which is freely distributed, and directly convert that light into electricity; photovoltaic generation of electricity requires no machinery with moving parts and produces no noise, waste, or polluting by-products; photovoltaic systems are modular and therefore can be adapted for a variety of applications. Solar power systems are particularly useful in areas where power lines cannot be readily or inexpensively routed.

Solar cells have been used extensively and successfully to power satellites in space since the late 1950s, where their high power-to-weight ratio and demonstrated reliability are especially desirable characteristics. On earth, where electrical systems typically provide large amounts of power at reasonable costs, three principal technical limitations have thus far impeded the widespread use of photovoltaic products: solar cells are expensive, sunlight has a relatively low power density, and commercially available solar cells convert sunlight to electricity with limited efficiency. Clearly, terrestrial solar cells must be reasonably efficient, affordable, and durable. International efforts are dedicated to obtaining such devices.

The power density of sunlight is about $1350\ W/m^2$ at elevations just above the earth's atmosphere. Less than $1000\ W/m^2$ is typically incident on earth after filtering through the atmosphere. Due to the low power density of sunlight and limited conversion efficiencies, the most efficient solar modules can generate about $250\ W/m^2$ in peak sunlight conditions. The maximum power output of a solar cell or module is defined in peak watts (W_{peak}), a rating based on a standard measurement method established by international consensus. A solar panel of one square meter area nominally produces one kilowatt hour of electricity per day. For most large-scale, power-producing applications, solar modules have conversion efficiencies above 10% in order to minimize the total cost of a generating system.

Chemistry

Crystalline silicon $p-n$ junction solar cells are the principal commercially available type and are used here to illustrate the operation of a solar cell. When sunlight falls on a solar cell, a voltage is induced and an electric current flows in an external circuit that is connected to the cell. Each atom in the silicon crystal lattice is surrounded by and bound to four equidistant neighboring atoms. The outermost shell of electrons of each silicon atom contains four valence electrons, and each of the four valence electrons in the crystal lattice is shared in a bonding orbital with an electron from one of its four nearest neighbors. This electron pair or covalent bond firmly binds the crystal. If all the valence electrons were inexorably bound, as they would be at 0 degrees kelvin, the silicon crystal would be an insulator because no free electrons would be available, and conduction would be precluded. See Fig. 1(a). However, the covalent bonds can be broken, e.g., by thermal excitation. See Fig. 1(b). The energy required to break a covalent bond is the bond energy or energy gap, E_g. In silicon, E_g is ca 1.1 eV.

The absence of an electron from a covalent bond leaves a hole and the neighboring valence electron can vacate its covalent bond to fill the hole, thereby creating a hole in a new location. The new hole can, in turn, be filled by a valence electron from another covalent bond, and so on. Hence, a mechanism is established for electrical conduction that involves the motion of valence electrons but not free electrons. Although a hole is a conceptual artifact, it can be described as a concrete physical entity to keep track of the motion of the valence electrons. Because holes and electrons move in opposite directions under the influence of an electric field, a hole has the same magnitude of charge as an electron but is opposite in sign.

The energy in light also can break the bonds of silicon valence electrons. Each photon has energy equal to the product of Planck's constant and the frequency of the light, i.e., $E = h\upsilon$, where E is photon energy, h is Planck's constant, and υ is the frequency of light. Solar photons range in energy from 0.5 eV for infrared to 4 eV for ultraviolet.

When a photon having energy equal to or greater than E_g is absorbed by the silicon crystal, the photon breaks a covalent bond, thereby freeing an electron and forming a hole. An electron is excited by a photon from a valence-energy band in a covalent bond into a conduction-energy band. The electron, which is transformed into a mobile, negatively charged carrier, leaves behind a mobile hole and consequently the photon has formed a free electron–hole pair.

If the hole and electron are not kept apart, they recombine to produce a small amount of thermal energy within the crystal and no net current flow. When the holes and electrons are kept apart, collected, and made to flow in a circuit outside the crystal, they produce electric current in that circuit. Solar cells are equipped with a barrier or a junction which provides an internal electric field that segregates photogenerated electrons and holes. Thus, although unmodified silicon has an equal number of holes and electrons, a $p-n$ junction silicon solar cell consists of two charge-dissimilar regions which are separated by a junction: one region is rich in holes (positive), i.e., p-type silicon, and the other is rich in electrons (negative), i.e., n-type silicon. Such regions do not occur naturally; they are fabricated by doping, i.e., replacing some silicon atoms in the lattice with atoms having a valence other than four. Replacement of a few silicon atoms, i.e., ca one in several million, causes large increases in the electrical conductivity of the resultant doped crystal.

Atoms of elements that are characterized by a valence greater than four, e.g., phosphorus or arsenic (valence = 5), are one type of dopant. These high valence dopants contribute free electrons to the crystal and are called donor dopants. If one donor atom is incorporated in the lattice, four of the five valence electrons of donor dopants are covalently bonded, but the fifth electron is very weakly bound and can be detached by only ca 0.03 eV of energy. Once it is detached, it is available as a free electron, i.e., a carrier of electric current. A silicon crystal with added donor dopants has excess electron carriers and is called n-type (negative) silicon. See Fig. 1(c).

When a silicon crystal is doped with atoms of elements having a valence of less than four, e.g., boron or gallium (valence = 3), only three of the four covalent bonds of the adjacent silicon atoms are occupied. The vacancy at an unoccupied covalent bond constitutes a hole. Dopants that contribute holes, which in turn act like positive charge carriers, are acceptor dopants and the resulting crystal is p-type (positive) silicon. See Fig. 1(d).

Conductivity in doped silicon crystals is determined by the properties of the added charge carriers or majority carriers. In n-type silicon, electrons are majority carriers and holes are minority carriers. There are fewer holes in n-type silicon than in undoped silicon because the large number of electrons causes some recombination with preexisting holes. In p-type silicon, holes are the majority carriers and electrons are the minority carriers. Fewer electrons are present in p-type silicon than in undoped silicon because of the recombination of some electrons with the enhanced population of holes.

Junctions

Four different types of junctions can be used to separate the charge carriers in solar cells: (1) a homojunction joins semiconductor materials of the same substance, e.g., the homojunction of a $p-n$ silicon solar cell separates two oppositely doped layers of silicon; (2) a heterojunction is formed between two dissimilar semiconductor substances, e.g., copper sulfide, Cu_xS, and cadmium sulfide, CdS, in Cu_xS–CdS solar cells; (3) a Schottky junction is formed when a metal and semiconductor material are joined; and (4) in a metal–insulator–semiconductor junction (MIS), a thin insulator layer, generally less than 0.003-μm thick, is sandwiched between a metal and semiconductor material.

Fabrication methods that are generally used to make these junctions are diffusion, ion implantation, chemical vapor deposition (CVD), vacuum deposition, and liquid-phase deposition for homojunctions; CVD, vacuum deposition, and liquid-phase deposition for heterojunctions; and vacuum deposition for Schottky and MIS junctions.

Efficiency

The most efficient silicon cells produced are based on $p-n$ homojunctions and convert 23.1% of the energy in incident light set to simulate the global air mass (AM) 1.5 spectrum, an artificial reference spectrum used to standardize measurement of PV power, with an intensity of $1000\ W/m^2$

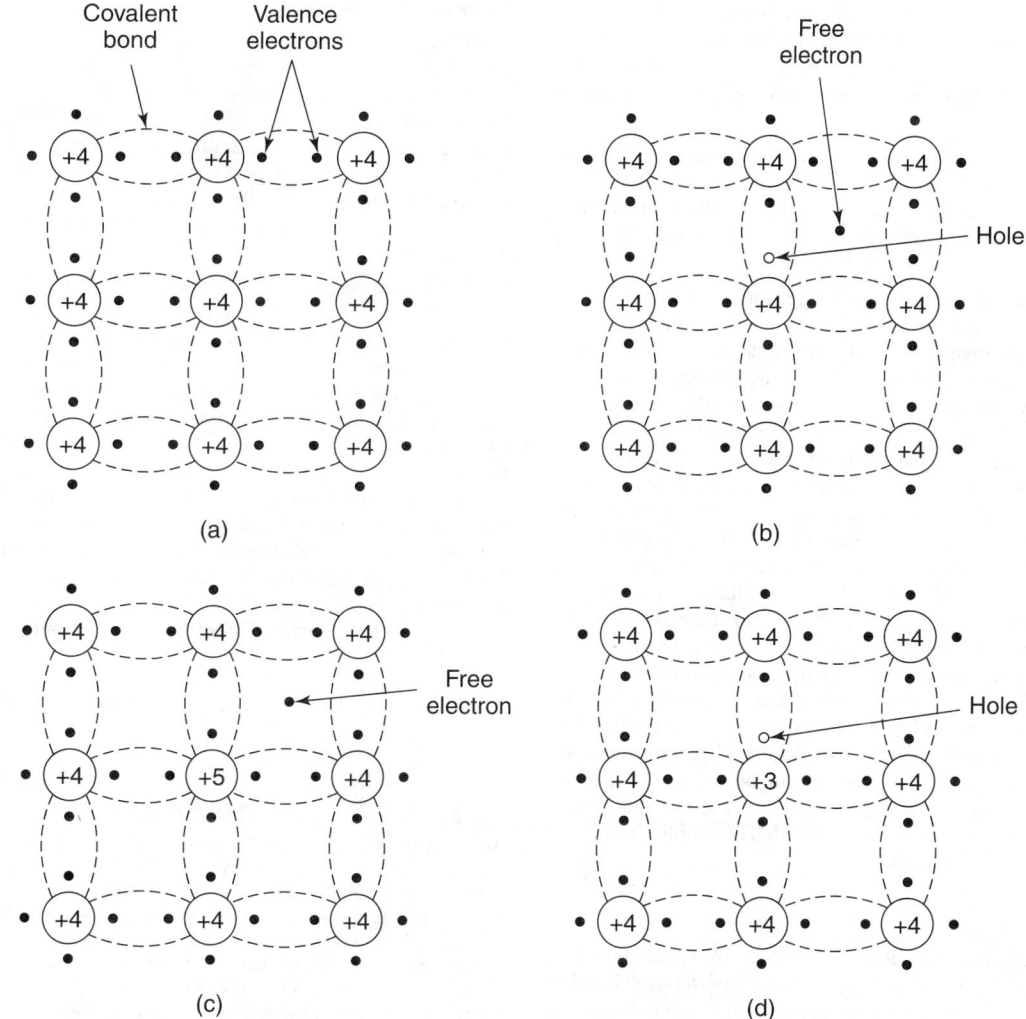

Fig. 1. (a) Silicon (valence = 4) crystal lattice shown in two dimensions with no broken bonds, $T = 0$ K; (b) silicon crystal lattice with a broken bond; (c) silicon crystal lattice with a silicon atom displaced by a donor dopant, i.e., n-doped (valence = 5); and (d) silicon crystal lattice with a silicon atom displaced by an acceptor dopant, i.e., p-doped (valence = 3)

at 25°C. This is the definition of peak sunlight test conditions. In theory, silicon $p - n$ junction solar cells can convert a maximum approaching 26% of the energy in AM 1.5 sunlight to electricity. Approximately 75% of the energy in sunlight is lost to factors intrinsic to the silicon material.

In comparison, $p - n$ homojunction cells made of more costly semiconductor materials, e.g., indium phosphide, InP, and gallium arsenide, GaAs, which have energy gaps of 1.2–1.4 eV, and maximum theoretical conversion efficiencies of ca 28–30%, depending on the device construction and layering of junctions.

Commercial Silicon Solar Cells

Silicon cells are hundreds of micrometers (μm) thick in order to facilitate handling with minimal breakage, although most solar radiation is absorbed in the first 20–30 μm. The junction in a silicon cell usually is ca 0.2–0.5 μm from the surface of the cell. The crystal surface has many broken bonds that act as recombination centers. In conventional silicon cells, a comb or narrow metal grid lattice is connected to a current-carrying bus to collect charge carriers from the side of the cell facing the sun. The fingers are small enough in total area so that minimal cell area is in their shadow.

Antireflection coatings are used over the silicon surface which, without the coating, reflects ca 35% of incident sunlight. Materials such as titanium dioxide, TiO_2, tantalum pentoxide, Ta_2O_5, or silicon nitride, Si_3N_4, ca 0.08-μm thick are common.

Types of Solar Cells. There are three basic technology options for making solar cells with dozens of variations on each. These approaches are conveniently grouped as follows: thick ($\sim$300 μm) crystalline materials, concentrator cells, and thin ($\sim$1 μm) semiconductor films.

Thick Crystalline Materials. Crystalline silicon technology is the worldwide industry standard. The total cost of solar cells made from ingots reflects the costs of the silicon raw material used in forming an ingot, cutting and etching thin silicon wafers from the ingot, fabricating and encapsulating the cells, and assembling them into modules. An attractive cost-reducing approach is to grow good quality crystalline sheets directly from molten silicon. Smoothly grown sheets ca 100-μm thick require little or no cutting and polishing and incur little waste.

Gallium arsenide is a promising material for gaining the advantages of high efficiency. It is superior to silicon in several respects. The Eg of GaAs, ca 1.4 eV, is higher than that of silicon and is in the range that provides the highest calculated conversion efficiency for a single-junction cell. Because of this high efficiency and the fact that it does not decline as rapidly as that of silicon cells with increasing temperature, GaAs single-crystal cells are attractive for use as concentrator cells.

Gallium arsenide solar cells advanced in the 1980s for space use because they weighed much less than silicon cells of similar output, since GaAs absorbs sunlight much more strongly than silicon.

Concentrator Cells and Systems. Concentrators circumvent the problem of high semiconductor material cost by using mirrors or lenses to concentrate sunlight on small surface areas of more expensive solar cells. Concentration allows more power to be produced from a given amount of photosensitive material.

Concentrator optics vary from low ratio designs, e.g., concentration of sunlight of an order of magnitude by Winston collectors, which do not require elaborate tracking of the sun, to much higher ratio systems based on parabolic mirrors or Fresnel lenses and which require precise, two-axis tracking. Three types of concentrator systems are being developed which operate at low level (<30 times), mid-level (100–400 times), and high level (>400 times) sunlight concentrations. The cell specifications and engineering requirements for each of these types of systems are quite different. Specially designed silicon has shown potential for use in concentrator systems.

Thin Film. In the thin-film approach, raw material usage is generally more than two orders of magnitude less and patterning is more direct.

Good solar cell results have been obtained from cells of materials, including polycrystalline silicon, amorphous silicon–hydrogen (α-Si:H) alloys, Cu_xS–CdS, $CuInSe_2$–CdS, and CdTe.

Electrochemical Photovoltaic Cells. The application of photoelectrochemistry in solar energy conversion technologies includes biomass conversion, photoelectrolysis, photogalvanic cells, electrochemical photovoltaic cells, etc. In electrochemical photovoltaic cells, electric energy is converted directly from sunlight by absorption of light in a semiconductor electrode. In many respects, these cells closely resemble conventional solid-state cells, except that the charge-separating barrier layer is formed at the interface between a semiconductor surface with a liquid electrolyte. When sunlight is incident on the semiconductor electrode, free holes and electrons are created. The relevant minority carriers must migrate to the interface and be separated; these carriers then react with the electrolyte either through oxidation or reduction. The counterelectrode reverses the reaction, thereby maintaining the electrolyte balance. The semiconductor electrode material may be either polycrystalline or amorphous material because in some cases the poorer material properties cause relatively little degradation of conversion efficiencies. In addition, incorporation of a third electrode may make possible *in situ* storage. The main disadvantage of these cells is the instability of the semiconductor electrode, especially under sunlight, for extended periods of operation. Electrochemical cells could be inexpensive, since the electrode–electrolyte barriers usually are easy to form, but appropriate deployment strategies have not yet been identified. The stability problems encountered to date have been extensive.

Balance of Systems

A solar photovoltaic system contains, in addition to solar cells and module(s), an array structure to support the modules, power-conditioning circuitry for control and modification of the output, and a means of storing energy if required. All elements beyond the module are referred to as balance-of-system (BOS) components. The cost of BOS items is nominally about equal to the cost of the PV module. However, the BOS fractional cost contribution can vary from one- to two-thirds of the total installed cost of a system, depending on application.

Material Availability and Environmental Impact

Photovoltaic systems must satisfy four principal requirements before solar photovoltaic conversion can provide a significant portion of general energy needs. The system costs must be low enough to be competitive with other means of energy generation, the amount of energy generated during the life cycle of a photovoltaic system must be substantially greater than the energy required to fabricate the system to meet the criteria of a sustainable technology, the materials used in the cells must be available to generate a substantial portion, i.e., at least a few percent of world energy needs, and the fabrication and utilization of the conversion systems should not cause more environmental problems than other competing energy systems.

Silicon is the second most abundant element in the world and is not toxic. Inherent in the use of materials other than silicon for solar cells are challenges of material availability and environmental safety. In terms of production of CdS-based cells, sulfur is abundant, but the world's resources of cadmium, tellurium, selenium, and indium are much less than those of silicon. However, these resources are several orders of magnitude greater than the amount needed to provide photovoltaic power production of 50,000 MW/yr. Similarly, although arsenic is plentiful, the supply of gallium for GaAs cells is limited. However, studies have concluded that the gallium supply also is sufficient for substantial manufacturing scale.

Although photovoltaic conversion is nonpolluting, environmental, health, and safety aspects must be considered, especially with regard to harmful emission and waste products resulting from the production of the solar cell modules. It has been shown that, with proper encapsulation and a proactive recycling program, it should be possible to minimize environmental concerns.

Photovoltaic Markets

In the mid-1990s, utility applications have once again begun receiving a great deal of attention due to a profound paradigm shift that appears to be taking place in the utility industry. Rather than replacing or adding large central fossil-fueled or nuclear generation facilities, small PV systems deployed at the outer extremities of the grid can be cost-effectively used to manage demand profiles, defer transmission hardware upgrades, and support electrical service quality (voltage, power factor, etc) during periods of peak demand in locations where the utility grid transmission is unidirectional.

PV Market Segment Categories. Solar modules are used to provide power to a broad range of industrial, commercial, and consumer systems and products. Most participants in the PV industry use the following categories to describe the various market segments, which group applications by functional product requirement, system type, sales channel, and client base. These include the following: specialties, e.g., spacecraft circuits, calculator chips, automobile sunroofs, and building facades; industrial power, i.e., telecommunications, warning/signal lights, and remote data gathering; rural and off-grid electrification, e.g., lighting, water pumping and purification, refrigeration, and recreational travel and boating; consumer convenience, e.g., garden and security lighting and small battery charging; and grid-connected power, i.e., distributed grid support and peaking power augmentation.

See also **Photoelectric Effect**; and **Solar Energy**.

Charles F. Gay
National Renewable Energy Laboratory
Chris Eberspacher
UNISUN

Additional Reading

Annan, R.H., W.L. Wallace, T. Surek, E. Boes, and L.O. Herwig: *Department of Energy Review of the U.S. Photovoltaic Industry,* Report ST-211-3488, Solar Energy Research Institute, Golden, CO, 1989.

Cody, G.D. and T. Tiedje, in B. Abeles, A. Jacobson, and P. Sheng: *Energy and the Environment,* World Scientific, Teaneck, N.J., 1994.

Day, J. and R.O. Johnson: *Distributed PV Applications, Report PM-36,* Strategies Unlimited, Mountain View, CA, 1992.

Hamakawa, Y.: *Thin-Film Solar Cells,* Springer-Verlag New York, Inc., New York, NY, 2004.

Luque, A. and S. Hegedus: *Handbook of Photovoltaic Science and Engineering,* John Wiley & Sons, Inc., New York, NY, 2003.

Markvart, T.: *Practical Handbook of Photovoltaics: Fundamentals and Applications,* Elsevier Science, New York, NY, 2003.

Markvart, T.: *Solar Electricity,* 2nd Edition, John Wiley & Sons, Inc., New York, NY, 2000.

Marti, A. and A. Luque: *Next Generation Photovoltaics,* Institute of Physics Publishing, Philadelphia, PA, 2003.

Smith, K.: *Survey of U.S. Line-Connected Photovoltaic Systems,* EPRI GS-6306, Palo Alto, CA, 1989.

Staff: *Maintenance and Operations of Stand-Alone Photovoltaic Systems,* Naval Facilities Engineering Command, Southern Division, rev. 1991.

PHTHALIC ACID. [CAS: 88-99-3]. $C_6H_4(COOH)_2$, formula weight 166.13, mp 208°C (ortho), 330°C (meta and iso), the ortho form sublimes and the meta and iso forms decompose with heat, sp gr 1.593 (ortho). Phthalic acid is very slightly soluble in H_2O, soluble in alcohol, and slightly soluble in ether. The solid form is colorless, crystalline. Because of their chemical reactivity and versatility, phthalic acid derivatives find wide use as starting and intermediate materials in important industrial organic syntheses. A common starting material is phthalic anhydride which is formed when phthalic acid loses water upon heating. See also **Phthalic Anhydride**; and **Terephthalic Acid**.

Orthophthalic acid is made by the oxidation of naphthalene (1) with H_2SO_4 fuming heated, in the presence of mercuric sulfate $-SO_2$ is also formed and recovered; (2) with air in the presence of vanadium pentoxide at 450 to 520°C. Orthophthalic acid also is formed when benzene compounds containing carbon ortho-substituted groups are oxidized. Orthophthalic acid is used in the manufacture of indigo and other dyes.

PHTHALIC ANHYDRIDE. [CAS: 85-44-9]. $C_6H_4(CO)_2O$, formula weight 148.11, mp 130.8°C, bp 284.5°C, sp gr 1.527. Phthalic anhydride is very slightly soluble in H_2O, soluble in alcohol, and slightly soluble in ether. The compound is a high-tonnage chemical and is widely used in a variety of industrial organic syntheses. Although phthalic anhydride may be derived directly from phthalic acid by heating and dehydration, it usually is prepared on a large scale by (1) oxidizing naphthalene, or (2) from the petroleum derivative, orthoxylene. Phthalic anhydride, in addition to its use as a raw and intermediate material for syntheses, finds wide application in the chlorinated form as a compounding ingredient for plastics. The chlorine content is approximately 50%. The compound provides increased stability and improved resistance of plastics to high temperatures.

Representative reactions of phthalic anhydride include: (1) phthalic anhydride reacts with phosphorus pentachloride to form phthalyl chloride which, upon rearrangement, can be transformed to unsymmetrical phthalyl chloride; (2) both forms of phthalyl chloride react with zinc plus acetic acid to form unsymmetrical phthalide, or with benzene plus aluminum chloride to form unsymmetrical-diphenylphthalide (phthalophenone); (3) phthalic anhydride reacts with NH_3 to form phthalimide $C_6H_4(CO)_2NH$; (4) phthalimide reacts with KOH in alcohol to form potassium phthalimide; (5) treatment of potassium phthalimide with an alkyl halide (e.g., ethyl chloride) forms an alkyl phthalimide (e.g., ethyl phthalimide); (6) ethyl phthalimide, when heated with fuming HCl, yields the primary amine $C_2H_5NH_2$ (ethyl amine) in a reaction used for the production of many primary amines and known as Gabriel's synthesis; (7) ethyl phthalimide, when treated with sodium hypochlorite, forms sodium anthranilate which upon treatment with an acid yields anthranilic acid; (8) phthalic anhydride reacts with phenol to form phthaleins, such as phenolphthalein, when in the presence of concentrated H_2SO_4; (9) phthalic anhydride reacts with resorcinol to form resorcinolphthalein (fluorescein); (10) fluorescein reacts with bromine to form tetrabromo-fluorescein, the potassium salt of which is eosin (a red dye for wool and silk); (11) phthalic anhydride reacts with N-diethyl-meta-aminophenol to form N-diethyl-meta-aminophenolphthalein (rhodamine) which is a red dye.

Health and Safety Factors

Phthalic anhydride is a severe irritant to the eyes, respiratory tract, and skin, especially to moist tissue. The solid may burn skin tissue if it is in contact with it for a significant amount of time. Repeated exposure may result in asthma, irritation of mucous membranes, and diseases of the respiratory tract and digestive organs. Contact with skin or the eyes should be followed immediately by washing with large quantities of water.

There are explosion hazards with phthalic anhydride, both as a dust or vapor in air and as a reactant. Water, carbon dioxide, dry chemical, or foam may be used to extinguish the burning anhydride.

See also **Phthalic Acid**; and **Terephthalic Acid**.

PHTHALOCYANINE COMPOUNDS. Phthalocyanine, $C_{32}H_{18}N_8$, compounds have found widespread acceptance in a variety of applications. The discovery of iron phthalocyanine and the elucidation of its structure led to the commercial application of copper phthalocyanine.

Copper phthalocyanine (**1**) was developed in the 1930s and is the most commonly used blue organic pigment in the coatings, paint, and printing inks industry. Phthalocyanine forms complexes with numerous metals. Various complexes with 66 chemical elements are known. Phthalocyanines are structurally related to naturally occurring dyes such as hemoglobin and chlorophyll A.

(1)

Physical Properties

The density of β-phthalocyanine, H_2Pc, is 1.43 g/cm^3; β-copper phthalocyanine, CuPc, 1.61 g/cm^3; and polychloro-copper phthalocyanine, 2.14 g/cm^3. The color of most phthalocyanines ranges from blue-black to a metallic bronze, depending on the manufacturing process and the chemical and crystalline form of the material. The colors of the finely divided pigment forms vary from dark blue to green, as phthalocyanines absorb in the visible region at 600–700 μm. Most compounds do not melt but sublime above 200°C. CuPc can be sublimed without decomposition at

500–580°C under an inert gas and normal pressure and at 900°C under vacuum. It decomposes vigorously, however, at 405–420°C in air and in nitrogen between 460–630°C. The thermodynamic stability of the five crystalline forms of CuPc increases in the sequence $\alpha = \gamma < \delta < \varepsilon < \beta$. The solubility of most phthalocyanines in water and organic solvents is very low. The α-form, however, is slightly soluble in polar solvents and converts rapidly to the β-form.

Chemical Properties

The chemical properties of phthalocyanines depend mostly on the nature of the central atom. Phthalocyanines are stable to atmospheric oxygen up to approximately 100°C. Mild oxidation may lead to the formation of oxidation intermediates that can be reduced to the original products. In aqueous solutions of strong oxidants, the phthalocyanine ring is completely destroyed and oxidized to phthalimide. Oxidation in the presence of ceric sulfate can be used to determine the amount of copper phthalocyanine quantitatively.

Phthalocyanine compounds exhibit favorable catalytic properties which makes them interesting for applications in dehydrogenation, oxidation, electrocatalysis, gas-phase reactions, and fuel cells.

Manufacturing and Processing

Phthalocyanine compounds have been synthesized with various metals. The most important metal phthalocyanines are derived from phthalodinitrile, phthalic anhydride, Pc derivatives, or alkali metal Pc salts.

The route from o-phthalodinitrile can be represented 4 $C_8H_4N_2$ + M $\longrightarrow$ MPc, where M is a bivalent metal, metal halide, metal alcoholate, or an equivalent amount of metal of valence other than two in a 4:1 molar ratio. If a solvent, e.g., trichlorobenzene, benzophenol, pyridine, nitrobenzene, or quinoline, is used, the reaction takes place at approximately 180°C. Without a solvent the dry mixture must be heated to ca 300°C to initiate the exothermic reaction.

The synthesis from phthalimide derivatives, e.g., diimidophthalamide (or phthalimide) is usually carried out in a solvent such as formamide. Metal phthalocyanines may also be prepared using alkali metal salts or from metal-free phthalocyanine by boiling the latter in quinoline with metal salt.

Industrial production of copper phthalocyanine usually favors either the phthalic anhydride–urea process or the o-phthalodinitrile process. Both can be carried out continuously or batchwise in a solvent or bake process of the solid reactants.

Crude copper phthalocyanine must be treated to obtain a satisfactory pigment in regard to the crystal modification and optimal particle size. See also **Pigments (Organic)**. The particle size of crude phthalocyanine can be reduced by chemical or mechanical methods.

The second process to finish phthalocyanine, which is more important for β-copper phthalocyanine, involves grinding the dry or aqueous form in a ball mill or a kneader. Agents such as sodium chloride, which have to be removed by boiling with water after the grinding, are used. Solvents like aromatic hydrocarbons, xylene, nitrobenzene or chlorobenzene, alcohols, ketones, or esters can be used.

Incorporation of less than a stoichiometric amount of alkyl sulfonamides of copper phthalocyanine into copper phthalocyanine improves the pigment's properties in rotogravure inks.

Performance in ink and coatings can be improved by addition of surfactants, dispersants, resins, or copper phthalocyanine derivatives with long aliphatic chains, $CuPc(CH_2-NHR)_3$, to stabilize the pigment in the binder system. Another possibility is wet-milling of aqueous pigment dispersions incorporating an organic medium, e.g., glycols, polyethers, or surfactants.

Some references cover direct preparation of the different crystal modifications of phthalocyanines in pigment form from both the nitrile–urea and phthalic anhydride–urea process. Metal-free phthalocyanine can be manufactured by reaction of o-phthalodinitrile with sodium amylate and alcoholysis of the resulting disodium phthalocyanine. The phthalic anhydride–urea process can also be used. Other sodium compounds or an electrochemical process have been described. Production of the different crystal modifications has also been discussed.

Perchloro- and perchlorobromo copper phthalocyanine are important organic green pigments. They are accessible through direct chlorination of copper phthalocyanine in a eutectic melt of aluminum and sodium chloride or in a chlorosulfonic acid medium. Bromine can be used instead of chlorine in the $AlCl_3-NaCl$ melt to obtain polybromochloro copper phthalocyanine.

Phthalocyanine sulfonic acids, which can be used as direct cotton dyes, are obtained by heating the metal phthalocyanines in oleum.

Polymeric phthalocyanines, which possess a higher stability compared to the monomers, can be obtained by combining a phthalocyanine with a polymer. The linking of the polymeric chain can occur at the central metal atom, the phenyl rings, through bridging or attachment to a polymeric chain.

Uses

Approximately 90% of the phthalocyanines (predominantly copper phthalocyanine) are used as pigments. In addition, they have found acceptance in many types of dyestuffs, e.g., direct and reactive dyes, water-soluble and solvent-soluble dyes with physical and chemical binding, azo-reactive dyes, azo nonreactive dyes, sulfur dyes, and vat dyes.

Available Forms. Phthalocyanines are available as powders, in paste, or liquid forms. They can be dispersed in various media suitable for aqueous, nonaqueous, or multipurpose systems, e.g., polyethylene, polyamide, or nitrocellulose. Inert materials like clay, barium sulfate, calcium carbonates, or aluminum hydrate are the most common solid extenders. Predispersed concentrates of the pigments, like flushes, are interesting for manufacturers of paints and inks, who do not own grinding or dispersing equipment. Pigment–water pastes, i.e., presscakes, containing 50–75% weight of water, are also available.

Colorants. The pigmentary forms of copper phthalocyanine are by far the most important commercial products of that class. They provide excellent color properties, excellent resistance to heat and light, acid and alkali, and are extremely insoluble in most solvents. They are less expensive than other organic pigments and color practically every type of printing ink, paint, plastic, and textile. Other uses include the coloring of roofing granules, cements and plasters, fine art paint, soaps, detergents, and other cleaning products. The two principal classes of copper phthalocyanine pigments are the blues and the greens. The blues may be further classified as the α- and β-crystal types, and the greens as the chlorinated and brominated derivatives.

Phthalocyanines have interesting properties as catalysts, lasers, semiconductors, lubricants, or as photographic components.

Health and Safety Factors

Phthalocyanines do not pose any significant risk to human health in the environment or the workplace. In several studies, no carcinogenic risk or toxicity to humans was revealed. The FDA approved the use of CuPc in general and ophthalmolic surgery, for contact lenses, and food packaging. Phthalocyanine Blue may be used as a colorant for coatings that are used in manufacturing, packing, processing, preparing, treatment, packaging, transporting, or holding food. The TLV value for CuPc is 10 mg/m^3.

Polychlorinated biphenyls (PCBs) have been detected in pigments manufactured in trichlorobenzene, but not in those made with nonchlorinated solvents. High boiling hydrocarbons or esters are suitable replacements.

GERD LOEBBERT
BASF Corporation

Additional Reading

Booth, G.: in K. Venkataraman, ed., *The Chemistry of Synthetic Dyes,* Vol. V, Academic Press, Inc., New York, 1971, p. 241.

Lever, A.B.P.: in H.J. Emeleus and A.G. Sharpe, eds., *Advances in Inorganic Chemistry and Radiochemistry,* Vol. 7, Academic Press, Inc., New York, 1965, pp. 27–113.

Leznoff, C.C. and A.B.P. Lever, eds.: *Phthalocyanines: Properties and Applications,* VCH Verlagsgesellschaft, Weinheim, Germany, Vol. 1, 1989; Vols. 2 and 3, 1993.

Lide, D.R.: *CRC Handbook of Chemistry and Physics,* 84th Edition, CRC Press LLC., Boca Raton, FL, 2003.

Moser, F.H. and A.L. Thomas: *Phthalocyanine Compounds,* Reinhold Publishing Co., New York, 1963.

Nalwa, H.S.: *Supramolecular Photosensitive and Electroactive Material,* Elsevier Science, New York, NY, 2001.

PHYCOCOLLOID.   One of several carbohydrate polymers (polysaccharides) occurring in algae (seaweed). They are hydrophilic colloids having a tendency to absorb water, with swelling, and to form gels of varying strength and consistency. The chief types of phycocoloid are carrageenan from Irish moss, algin from brown algae, and agar from red algae. They contain complex galactose and mannose sugars and are sometimes considered seaweed mucilages.

See also **Carbohydrates**.

PHYSICAL CHEMISTRY.   Application of the concepts and laws of physics to chemical phenomena in order to describe in quantitative (mathematical) terms a vast amount of empirical (observational) information. A selection of only the most important concepts of physical chemistry would include the electron wave equation and the quantum mechanical interpretation of atomic and molecular structure, the study of the subatomic fundamental particles of matter. Application of thermodynamics to heats of formation of compounds and the heats of chemical reaction, the theory of rate processes and chemical equilibria, orbital theory and chemical bonding, surface chemistry (including catalysis and finely divided particles) the principles of electrochemistry and ionization. Although physical chemistry is closely related to both inorganic and organic chemistry, it is considered a separate discipline. See also **Inorganic Chemistry** and **Organic Chemistry**.

Additional Reading

Atkins, P.W.: *Physical Chemistry,* Oxford University Press, New York, NY, 1978.

Hiemenz, P.C. and R. Rajagopalan: *Principles of Colloid and Surface Chemistry,* Marcel Dekker Inc., New York, NY, 1997.

Hunter, R.J.: *Introduction to Modern Colloid Science,* Oxford University Press, New York, NY, 1993.

Monk, P.M.S.: *Physical Chemistry: Understanding Our Chemical World,* John Wiley & Sons, Inc., Hoboken, NJ, 2004.

Silbey, R.J., M.G. Bawendi, and R.A. Alberty: *Physical Chemistry,* 4th Edition, John Wiley & Sons, Inc., Hoboken, NJ, 2004.

Sun, S.F.: *Physical Chemistry of Macromolecules: Basic Principles and Issues,* John Wiley & Sons, Inc., Hoboken, NJ, 2004.

PHYTOCHEMISTRY.   That branch of chemistry dealing with (1) plant growth and metabolism and (2) plant products. The former includes the absorption of inorganic nutrients (nitrogen, phosphorus, potassium, carbon dioxide, water, etc.) to form sugars, starches, proteins, fats, vitamins, etc. and is closely associated with photosynthesis. Plant products comprise a vast group of natural materials and chemicals; besides those used directly as foods, these include alkaloids, cellulose, lignin, dyes, glucosides, essential oils, resins, gums, tannins, rubbers, terpene hydrocarbons, and glycerides (fats and oils). Some of these are basic raw materials for industry (paper, pharmaceuticals, food, paint, perfume, flavoring, leather, rubber); there are also many miscible plant products such as drugs, poisons, and pigments. Phytochemistry also embraces the study of plant hormones or growth regulators (auxin, gibberellin, synthetic types).

Additional Reading

Conn, E.E.: *Opportunities for Phytochemistry in Plant Biotechnology,* Perseus Publishing, Bolder, CO, 1988.

Robins, R.J. and F.A. Tomas-Barberan: *Phytochemistry of Fruits and Vegetables,* Oxford University Press, New York, NY, 1997.

Romeo, J.T.: *Phytochemical Signals and Plant-Microbe Interactions: Recent Advances in Phytochemistry,* Kluwer Academic Publishers, Norwell, MA, 1998.

pi BOND.   A covalent bond formed between atoms by electrons moving in orbitals that extend above and below the plane of an organic molecule containing double bonds. A double bond consists of one pi and one sigma bond, and a triple bond consists of one sigma and two pi bonds.

See also **Metallocenes**; and **Orbitals**.

PICKLING.   (1) Removal of scale, oxides, and other impurities from metal surfaces by immersion in an inorganic acid, usually sulfuric, hydrochloric, or phosphoric. Rate of scale removal varies inversely with concentration and temperature; the usual concentration is 15% at or above 100°C. The rate is also increased by electrolysis. (2) A method of food preservation involving use of salt, sugar, spices, and organic acids (acetic). (3) Preserving or preparing hides for tanning by immersion in a 6–12% salt solution, together with enough acid to maintain pH at 2.5 or less.

PICOLINES.   See **Pyridine and Derivatives**.

PIDGEON PROCESS.   Also referred to as ferrosilicon process or silicothermic process. Process for the production of high-purity magnesium metal from dolomite or magnesium oxide by reduction with ferrosilicon at 1150°C under high vacuum.

PIEZOCHEMISTRY.   Study of reactions occurring at very high pressures, e.g., in the interior of the earth's crust.

PIEZOELECTRIC EFFECT.   The interaction of mechanical and electrical stress-strain variables in a medium. Thus, compression of a crystal of quartz or Rochelle salt generates an electrostatic voltage across it, and conversely, application of an electric field may cause the crystal to expand or contract in certain directions. Piezoelectricity is only possible in crystal classes which do not possess a center of symmetry. Unlike electrostriction, the effect is linear in the field strength.

The directions in which tension or compression develop polarization parallel to the strain are called the piezoelectric axes of the crystal. Thus the axis of a hexagonal quartz crystal indicated by the arrows in Fig. 1 is known as an "X-axis," and a plate cut, as shown, with its faces perpendicular to this direction is an "X-cut"; while one cut with its faces parallel to the lateral faces of the crystal is a "Y-cut."

The magnitude of the piezoelectric polarization is proportional to the strain and to the corresponding stress, and its direction is reversed when the strain changes from compression to tension. The principal piezoelectric constants of a crystal are the polarizations per unit stress along the piezoelectric axes. While these constants are much greater for Rochelle salt than for quartz, the latter is better adapted to some purposes because of its greater mechanical strength. It is also stable at temperatures over 100°C.

If a quartz plate is subjected to a rapidly alternating electric field, the inverse piezoelectric property causes it to expand and contract alternately. As an elastic body, the plate has a certain natural frequency of expansion and contraction in the direction of the field, and if the field is made to alternate with the same frequency, the plate responds with a vigorous resonant vibration. This reacts, through the direct piezoelectric property, to augment the electric oscillations. A circuit arranged for this purpose, as in Fig. 2, is known as a piezoelectric or crystal oscillator, the crystal itself, P, being the piezoelectric resonator; T is the oscillation transformer, and C a variable condenser. This device has been much used as a frequency control in radio transmitters. Both X-cut and Y-cut quartz plates are subject to changes of frequency with temperature, due to change of elastic modulus; but certain planes in the crystal have been found, oblique to both X and Y, such that plates cut parallel to them are nearly free from the temperature effect.

In addition to natural quartz, Rochelle salts, and tourmaline, synthetic crystals, such as ethylenediamine tartrate (EDT), dipotassium tartrate (DKT), and ammonium dihydrogen phosphate (ADP) have varying suitability as piezoelectric elements. While Rochelle salt has a greater piezoelectric effect than any other crystal, it has the disadvantage of a greater sensitivity to temperature change than quartz. EDT has an advantage over quartz when used in frequency-modulated oscillators because of

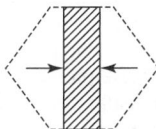

Fig. 1.   Hexagonal quartz crystal showing X-axis

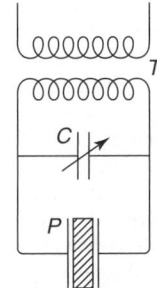

Fig. 2.   Piezoelectric oscillator circuit

the wide gap between its resonant and antiresonant frequencies. See also **Quartz**.

PIEZOELECTRICITY.   Electric energy created by application of pressure to ceramics of plastics. Devices utilizing this phenomenon are gas flame igniters, ultrasonic welding tools, and sonar navigation aids.
See also **Piezoelectric Effect**.

PIG IRON.   Product of blast-furnace reduction of iron oxide in the presence of limestone. About half the ore is converted to iron. Average analysis is 1% silicon, 0.03% sulfur, 0.27% phosphorus, 2.4% manganese, 4.6% carbon, balance iron. Pig iron is the basic raw material for steel and cast iron. In metal terminology a "pig" is a bar or ingot of cooled metal.

PIGMENT DISPERSIONS.   A pigment dispersion in a concentrated form is a uniform distribution of very fine color pigment particles in a suitable medium or carrier. Such a dispersion is normally used for applying color to the surface of a substrate, such as an ink film on paper or a paint film on a steel surface. It is also used for mass coloring, as in the case of plastics. Considering the high cost and specialized equipment in its preparation, a dispersion is manufactured in relatively small batches in highest concentration of pigment. The concentrate made in such a manner is usually diluted, reduced, or extended to produce the finished product.

Dispersion

Organic and inorganic pigment powders are finely divided crystalline solids that are essentially insoluble in application media such as ink or paint. The carrier used for dispersion of a pigment is usually a liquid or solid, such as a polymer, that is deformable at the processing conditions of high temperature and/or shear. The color strength of the dispersed pigment increases markedly with decrease in particle size. Optimum color strength from a given pigment in practice requires a mean particle size of the order of 0.1 μm or less, which is half the wavelength of the light involved. Therefore, the dispersion process involves size reduction of the pigment particle to the smallest practical size, reasonably complete wetting of its solid surfaces by the carrier, and stabilization of the resulting dispersion.

Because the intensity and color strength of pigments are largely dependent on the exposed surface, it is desirable to reduce the particles to primary particle size. This is the size of the solid pigment crystals as they are precipitated in their synthesis. In practice, the size reduction processes are limited by the nature of pigment, dispersion system, constraints of the processing equipment, the requirements imposed by the product application, and the overall economics. The maximum aggregate size permissible in a given dispersion system depends on the thickness of the film or the coating. For example, the dispersion used for architectural coatings can tolerate a much larger pigment aggregate than a similar dispersion used for automotive finishes, which requires finer particles. Any dispersion system, however, is expected to contain a very small number of these largest aggregates. Generally, it is important to reduce most aggregates to the smaller size to achieve color strength, gloss, film integrity, and durability.

In a dispersed pigment system, a primary pigment particle refers to an individual crystal and a loosely formed association of the pigment crystals from the manufacturing process. Size reduction beyond primary particle size requires excessive energy, but it also has an adverse effect on the visual properties of the pigment. Generally, the particle size of most organic pigments is much smaller initially by precipitation than optimum primary particles, but the particles tend to grow to a much larger size when their formation is complete.

Organic pigments, such as the azo red and yellow pigments, in the process of striking the color undergo definite crystal growth following their precipitation from the aqueous media. The individual crystals are joined together due to forces on the crystal surfaces to form the aggregate. These are held together as static systems by van der Waals forces. Subsequent processing to recover the pigment product results in the formation of agglomerates, which are large associations of pigment crystals and aggregates. The agglomerates are held together by forces that are much weaker than those present within the aggregates. Typically, agglomerates are joined at the edges and corners in a loose matrix form. It is possible to generate an even larger association of pigment agglomerates or flocculates during further processing. These formations are loosely held together and are usually easy to break down by application of shear. Various surface treatments are used to suppress the formation of large aggregates and, thereby, ease the dispersion process. These treatments range from the classical approach of rosination to additions of a variety of surface-active agents at the synthesis step. However, occasionally large agglomerates, several millimeters in diameter, form during the initial stages of dispersion in a highly viscous system. The commercial processes used in dispersion manufacturing may not fully eliminate the aggregates. However, the design and operation of pigment dispersion equipment is aimed at application of mechanical forces to break down the agglomerates and even some less tightly held aggregates. Ideally, an excellent dispersion should consist mainly of primary pigment particles and few loosely held aggregates.

Wetting of the pigment surface constitutes a critical step in achieving a stable and uniform pigment dispersion. Wetting refers to displacement of adsorbed gases (usually air) on the surface of pigment particles, followed by attachment of a vehicle system to the pigment surface. Since the vehicles used for many dispersion systems are viscous, it follows that the penetration of vehicles to the pigment surface is slow and, hence, aided by external mechanical forces. Thus, the grinding (size reduction) and wetting of pigment are frequently carried out simultaneously. The adsorbed gases are displaced on application of shear, and the action also provides smearing of vehicle on the pigment surface and exposes a new surface for wetting. The system of wetted fine primary pigment particles, must be stabilized to prevent reversal of the dispersion process. It is usually done by surrounding the particles with a protective colloid or buffer, which blocks the reagglomeration action of particles. In some cases, the stabilization is attained by addition of ions to establish similar charges on all particles.

Flushing

Flushing processes are used extensively in preparing organic pigment dispersion concentrates for color printing ink applications. The process can be described as a direct transfer of pigment from an aqueous phase to an oil or nonaqueous phase without drying. When the pigment presscake is mixed with an oil-based vehicle or a carrier, water is separated from the pigment surface and replaced by the vehicle. Most organic pigments demonstrate an affinity for hydrocarbon oils and lend themselves to easy dispersion in oil by the process of flushing. See also **Pigments (Organic)**. Inorganic pigments, on the other hand, have to be treated with cationic surfactants to make their surface lipophilic. The majority of inorganic pigments are usually dried and dispersed as dry powders in the carrier, as opposed to being flushed. Techniques used for dispersion of these pigments are different and should be treated as special cases. See also **Pigments (Inorganic)**.

The process of flushing typically consists of the following sequence: phase transfer; separation of aqueous phase; vacuum dehydration of water trapped in the dispersed phase; dispersion of the pigment in the oil phase by continued application of shear; thinning the heavy mass by addition of one or more vehicles to reduce the viscosity of dispersion; and standardization of the finished dispersion to adjust the color and rheological properties to match the quality to the previously established standard.

Flushing is frequently used for the manufacture of large quantities of a dispersion having a specific pigment in a compatible vehicle system. The flushed products, typically containing 28–40% pigment, offer sufficient flexibility to the formulator to produce the finished offset ink. The flushed products exhibit superior gloss, transparency, and strengths, compared to those produced by dispersing the dried pigment. Flushing is particularly important to dispersions of organic pigments, such as Diarylide Yellow (CI Pigment Yellow 12, CI 21090) and Alkali Blue (Pigment Blue 61, CI 42765) because the drying process is detrimental to the product quality of these pigments.

Equipment

Various types of equipment are used commercially to manufacture dispersed pigment concentrates or finished dispersion products used by printing ink, and the coatings and plastics industry. These include kneaders or internal mixers; close tolerance mills; high speed fluid energy mills; ball and pebble mills; san, bead, and shot mills.

Uses

The formulation of dispersed pigment concentrates is influenced by the manufacturing process, as well as the performance parameters desired in the final application. The finished product in many cases is significantly different in formulation than the concentrate to achieve desired properties.

One of the principal factors to be considered is the concentration of pigment in the dispersion concentrate. Compatibility of the carrier (solvent additives, etc) used in the preparation of concentrated dispersion and that used in the finished color product also plays an important role. In some cases this can be difficult because the carriers having the best performance, from the standpoint of processing, could be poor in the application systems. However, in the majority of the applications, particularly in coatings and colored plastics, the concentration of the pigment in the finished product is quite low, and the incompatibility problem is easily overcome.

Generally, the pigment dispersion concentrates are formulated for specific end use. They can be supplied as flushed pigments, dispersions or pastes for offset inks, chip dispersions for solvent and aqueous inks, and color concentrates for coloring large quantities of plastics. Although it is feasible for the end user to prepare the pigment dispersion concentrates, it is usually more cost effective and technologically advantageous to manufacture these dispersions by the pigment manufacturers of specialty dispersion houses. Three significant areas of application for concentrated dispersions, i.e., printing inks, coatings, and plastics.

GERD LOEBBERT
ANAND S.G. SHARANGPANI
BASF Corporation

Additional Reading

Lewis, P.A.: *Pigment Handbook,* Vol. 1, 3rd Edition, John Wiley & Sons, Inc., New York, NY, 1988.

Patton, T.C.: *Paint Flow and Pigment Dispersion: A Rheological Approach to Coating and Ink Technology,* 2nd Edition, John Wiley & Sons, Inc., New York, NY, 1979.

Smith, H. MacDonald: *High Performance Pigments,* John Wiley & Sons, Inc., New York, NY, 2001.

PIGMENTATION (Plants).   The distinctive green color of leaves and other plant organs results from the presence in such organs of two pigments called chlorophyll *a* and chlorophyll *b*. In the higher plants these pigments occur only in the chloroplasts. These pigments play so important a role in the fundamental process of plant life, photosynthesis, that their chemical reactions are discussed at length in that entry. The chlorophylls are not water-soluble but can be readily dissolved out of leaf tissues with alcohol, acetone, ether, or other organic solvents. The resulting solutions exhibit the phenomenon of fluorescence; they are deep green when held between an observer and the light, but deep red when viewed in reflected light. By suitable treatments it is possible to obtain pure crystals of chlorophyll from such solutions. Most leaves contain considerably more chlorophyll *a* than chlorophyll *b*, often two to three times as much. In the organs of the higher seed plants, with rare exceptions, chlorophyll is synthesized only upon exposure to light. Leaves of grass that develop under a board, for example, contain no chlorophyll. In the leaves of mosses, ferns, and gymnosperms, however, chlorophyll develops in the dark as well as in the light.

Invariably associated with the chlorophylls in the chloroplasts are the yellow pigments, the carotenes and the xanthophylls. These pigments are not, however, restricted in their occurrence to the chloroplasts, but may also be present in nongreen parts of the plant where they commonly occur in chromoplasts. Collectively, these pigments, together with certain others which are closely related chemically, are called the carotenoids. Carotene refers to a class of orange-yellow pigments. They are especially abundant in the roots of carrots. These compounds are of considerable importance because they are the precursors of vitamin A, one molecule of β-carotene being split into two molecules of vitamin A by a simple hydrolytic reaction.

Lycopene, a red pigment of this class, is responsible for the red color of the fruits of tomato, pepper, rose, and some other species. The commonest xanthophylls found in leaves are lutein and zeaxanthin, although others also occur. Another xanthophyll is fucoxanthin, which imparts to brown algae their distinctive color. None of the carotenoids is water soluble, but all of them can be extracted from plant tissues with suitable organic solvents.

Most of the red, blue, and purple pigments of plants belong to the group of anthocyanins. In general, the anthocyanins are red in an acid solution and change in color through purple to blue as the solution becomes more alkaline. Red pigmentation resulting from the presence of the anthocyanins is found in flowers, fruits, bud scales, young leaves and stems, and sometimes even mature leaves as in those of the red cabbage. Blue and purple pigmentation due to the presence of anthocyanins occur principally in flowers and fruits. The anthocyanins are diglucosides of the compounds pelargonidin, cyanidin, delphinidin and apigenidin. These compounds are closely similar in structure, all having the double ring benzopyrylium.

Another group of cell sap water-soluble pigments is the *anthoxanthins*. These pigments are also chemically related to the glucosides. Anthoxanthins often occur in the plant in a colorless form but under suitable conditions their typical yellow or orange color becomes apparent. Some yellow flowers, such as yellow snapdragons, owe their color to the presence of anthoxanthins, but the color of the majority of kinds of yellow or orange flowers is due to carotenoid pigments.

The autumnal coloration of leaves in temperate regions is one of the most spectacular accompaniments of the march of the seasons. Both carotenoid and anthocyanin pigments play an important role in autumnal leaf coloration which is not, contrary to popular opinion, a result of the action of frost. Brilliant development of the anthocyanin pigments in the fall is, however, favored by dry weather during which cool, but not frosty, nights alternate with clear days. During the late summer and early fall the chlorophyll in the leaves gradually decomposes. In many species this simply results in unmasking the yellow carotenoid pigments already present, accounting for the yellow autumnal pigmentation of such species as birch, sycamore, aspen, and tulip trees. In other species synthesis of anthocyanins occurs more or less concomitantly with the disintegration of the chlorophyll; this accounts for the reds or purplish reds characteristic in the autumnal coloration of such species as many oaks, maples, sumacs, and dogwood.

Except in flowers, white is an uncommon color in the externally visible parts of plants, and results from the complete absence of pigments. In some species white streaks or other markings are of common or regular occurrence in leaves, and in the leaves of some species such as roses completely white leaves or even entire branches bearing only white leaves sometimes occur. Such branches cannot be propagated because no photosynthesis can take place in the absence of chlorophyll. As long as such branches remain attached to a plant bearing green leaves they can obtain necessary food from the branches bearing normally pigmented leaves.

See also **Annatto Food Colors**; **Carotenoids**; **Chlorophylls**; **Colorants (Foods)**; and **Photosynthesis**.

PIGMENTS (Inorganic).

Inorganic pigments, black, white, or colored inorganic substances produced and marketed as fine powders, are an integral part of many decorative and protective coatings and are used for the mass coloration of plastics, fibers, paper, rubber, glass, cement, glazes, and porcelain enamels. See also **Enamels, Porcelain or Vitreous**. These materials are colorants in printing inks, cosmetics, and markers, e.g., crayons. In all these applications the pigments are dispersed, i.e., they do not dissolve, in the media forming a heterogeneous mixture. See also **Pigment Dispersions**. In nature, inorganic pigments contribute to the color of some rocks and minerals. See also **Colorants (Foods)**; and **Paints and Coatings**.

Originally, only fine powders used for coloring various media were defined as pigments. This definition has been expanded to include many powdery materials, e.g., metallic powders and powders having magnetic or anticorrosive properties, which are intentionally dispersed (not dissolved) into media to increase value and/or impart some special properties.

Chemically, inorganic pigments are quite simple materials and include elements, their oxides, mixed oxides, sulfides, chromates, silicates, phosphates, and carbonates. The application usefulness of inorganic pigments is determined by physical as well as chemical properties.

Properties

The value of pigments results from their physical–optical properties. These are primarily determined by the pigments' physical characteristics (crystal structure, particle size and distribution, particle shape, agglomeration, etc.) and chemical properties (chemical composition, purity, stability, etc.). The two most important physical–optical assets of pigments are the ability to color the environment in which they are dispersed and to make it opaque.

The opacity of a pigment lies in its ability to prevent a transmission of light through the medium. White pigments disperse the whole visible light spectrum more effectively than they absorb it; black pigments do the opposite. Color results when pigment particles absorb only certain portions of the visible light spectrum while dispersing the rest of it.

The opacity of pigments is a function of the pigment particle size and the difference between the pigment's refractive index and that of the medium in which pigment particles are dispersed. The multiple light dispersion in the pigment–medium interface results in the appearance that the light is transmitted through a much thicker layer than it actually is. A pigment having a particle size between 0.16–0.28 μm gives the maximum dispersion of the visible light.

The most common measurements of pigment properties comprise elemental analysis, impurity content, crystal structure, particle size and shape, particle size distribution, density, and surface area. These parameters are measured so that pigments' producers can better control production, and set up meaningful physical and chemical pigments' specifications. Measurements of these properties are not specific only to pigments. The techniques applied are commonly used to characterize powders and solid materials, and the measuring methods have been standardized in various industries.

Coloristic properties of pigments are best evaluated by dispersing them into the media they were developed to color, e.g., plastics, glass enamels, glazes, etc. The measured characteristics include color, color strength, opacity, lightfastness, weathering, heat stability, chemical stability, and rheological properties. The dispersing media and the processing conditions can strongly influence the results. Because pigments, as any fine powders, have a tendency to segregate by size during transportation and handling, the use of proper sampling methods is critical for getting meaningful physical and chemical data.

Chemical Properties. Elemental profile, impurity content, and stoichiometry are determined by chemical or instrumental analysis. Instrumental analytical methods are usually faster, can be automated, and can be used to determine very small concentrations of elements. Atomic absorption spectroscopy and x-ray fluorescence methods are the most useful instrumental techniques in determining chemical compositions of inorganic pigments. Chemical analysis of principal components is carried out to determine pigment stoichiometry.

Crystal Structure. Crystal structure, the information about compounds such as impurities or unreacted materials present in the pigment, the presence of various crystal phases, and the degree of the crystallinity, can be resolved using an x-ray diffractometer. In most cases, analyzed pigments are not completely unknown, and the powder diffraction pattern is sufficient in finding present phases or unreacted starting materials in the pigment. The x-ray analysis has become an indispensable tool of the inorganic pigments' development and production.

Physical Properties. Particle size and distribution are the most fundamental measured properties of powders (see **Size Measurement of Particles**). These properties impact a number of pigment characteristics. Those affected the most are the color, color strength, hiding power, and rheological properties. Actual powders consist of a population of particles of many different shapes. To permit a good description of powder population, a representative sample of the powder must be collected, measured, and the results interpreted using statistical methods. For inorganic pigments to be useful in most applications, they must have an average particle size between 0.1 and 10 μm. A character, i.e., color or pattern, of a substrate becomes obscured when coated with a pigment containing film such as a paint or a ceramic glaze. The degree of the obscuration (opacity) depends on the amount and type of pigment used and the thickness of the applied film.

The ability of a coating to hide the substrate is called its hiding power. Hiding power of a uniform coating is expressed as the area of substrate that can be hidden by a unit volume of the coating (ft^2/gal or m^2/L).

The ability of a pigment to change the color of an opaque film is known as its tinting strength. Both the hiding power and tinting strength are the fundamental pigment properties. Hiding power and tinting strength can be determined visually or instrumentally.

Color matching is a process in which a technician prepares a formulation, i.e., a mixture of pigments in a desired medium, that has the desired color effects. A good color match in one medium, e.g., plastic, is not always a good match in another medium, e.g., ceramic glaze.

Experienced color matchers can achieve a good color match by trial and error without using any instrumentation. In some cases, however, this technique can be a lengthy process. To get the most cost-effective match in the shortest possible time, the use of a computer color matching system is preferable.

Many pigments, when exposed to high intensity light such as direct sunlight or uv lamp, can get darker, change their shade or lose the color saturation. The color and its saturation change mainly for organic pigments. Inorganic pigments, particularly those containing ions that can exist in several oxidation states, usually get darker. Some color changes can be reversible; others are permanent.

Lightfastness is measured by exposing pigmented film to an artificial or natural light for a predetermined time. It is a relative term where the color of a sample exposed to a known light source is compared to its original color values.

Weathering is the ability of the colored system, i.e., the coating, paint, etc., not the pigment alone, to resist light and environmental conditions. Changes in color and gloss are two main factors that are evaluated in weathering tests.

Heat stability is measured as a change in the hue of the colored system and a degree of yellowing of the white system after exposure to a desired high temperature for a certain time. This property can also be expressed as the maximum temperature at which the color of the system does not change.

In determining the chemical resistance, color changes of pigmented binder surfaces are measured after their exposure to various chemicals, such as water–sulfur dioxide or water–sodium chloride systems. These systems imitate the environment to which the colored articles could become exposed.

The surfaces of pigment particles can have different properties and composition than the particle centers. This disparity can be caused by the absorption of ions during wet milling, e.g., the $-OH$ groups, on the surface.

Most inorganic pigments are hydrophilic and therefore can be readily wetted only by polar solvents, e.g., water. The wettability and dispersion of inorganic pigments in an organic matrix (polymer, solvent) can be improved by the physical or chemical absorption of surface-active compounds containing polar groups, such as $-NH_2$, $-OH$, or longer aliphatic chains on pigment particles. The absorption of these compounds makes the pigment surface hydrophobic. Compounds that help to form a bridge between inorganic particles and an organic polymeric matrix are called coupling agents, the most common being tetrafunctional organometallic compounds.

Specifications, Standards, and Quality Control

Production and product quality of most pigment producers are controlled through pigments' standards. Whenever a pigment is developed or significantly improved, a new standard is set that represents the average production results, not necessarily the best ones. If all production processes are under control, the properties of the produced pigment's lots are evenly distributed around those of the standard.

White Pigments

The most common white pigments are titanium dioxide, zinc oxide, leaded zinc oxide, zinc sulfide, and lithopone, a mixture of zinc sulfide and barium sulfate. The use of lead whites and antimony oxides has been decreasing steadily for environmental reasons.

Titanium Dioxide. Chemically, titanium white is titanium dioxide either in an anatase or rutile form.

Properties. Crystals of titanium dioxide, TiO_2, can exist in one of the three crystal forms: rutile, anatase, and brookite. Only anatase and rutile forms have good pigmentary properties, and rutile is more thermally stable. Compared to other white pigments, titanium dioxide has the highest refractive index, giving white paints formulated with these pigments the highest coverage, ca 38 m^2/g.

Titanium whites resist various atmospheric contaminants such as sulfur dioxide, carbon dioxide, and hydrogen sulfide. Owing to its chemical inertness, titanium dioxide is a nontoxic, environmentally preferred white pigment.

Titanium is the seventh most common metallic element in the earth's crust. Titanium minerals are plentiful in nature. The most common mineral/raw materials used for the production of titanium dioxide pigments are shown in Table 1.

Use. Titanium dioxide is used mainly in the production of paints and lacquers, plastics, and paper. Other applications include the pigmentation of printing inks, rubber, textiles, leather, synthetic fibers, ceramics, white cement, and cosmetics.

Other White Pigments.

Zinc Oxide. By volume, zinc oxide is the second most significant white pigment. Its pigmentary properties are good, providing good coverage. It has a good lightfastness and is well miscible with other pigments. With the

TABLE 1. MINERAL/RAW MATERIALS FOR TiO₂ PRODUCTION

Mineral/raw material	Main composition	TiO₂, %
ilmenite	$FeO-TiO_2$	35–65
leucoxene	$Fe_2O_3-TiO_2(+TiO_2)$	60–90
rutile	TiO_2	90–98
rutile, synthetic	TiO_2	85–96
anatase	TiO_2	80–90
titanium slag	$TiO_2(Fe)$	70–85

increasing popularity of titanium dioxide white pigment in the twentieth century, the pigmentary use of zinc oxide has been declining.

Whereas zinc oxide was originally used as a pigment, its most important modern application is to aid in vulcanizing synthetic and natural rubber. Paint and coating industries use zinc white mainly as an additive to improve anticorrosion properties, mildew resistance, and durability of external coatings.

Zinc Sulfide. Whereas zinc sulfide is important mainly as a component of the composite white pigment lithopone, it also has a limited use as a single pigment. After titanium white, it has the second highest refractive index of all the white pigments. However, its chemical and thermal resistances are inferior to those of TiO_2.

Zinc sulfide is used in applications where white color shade and low abrasivity are required. In printing inks and paints it also contributes to stability and good rheological and printing properties.

Lithopone. Lithopone is a mixture of ZnS and BaSO₄. The pigmentary properties of the mixture are determined by zinc sulfide, and therefore lithopone pigments are characterized by the amount of ZnS present in the mixture. The amount of ZnS in commercial lithopones varies from 15 to 60%.

Lithopones are used in water-based paints because of their excellent alkali resistance, in paper manufacturing as a filler and opacifying pigment, and in rubber and plastics as a whitener and reinforcing agent.

Lead Whites. Basic lead carbonate, sulfate, silicosulfate, and dibasic lead phosphite are commonly referred to as lead whites. Usage is limited because of environmental restrictions placed on the use of lead-containing compounds.

Colored Pigments

Iron Oxide Pigments. In general, all iron pigments are characterized by low chroma and excellent lightfastness. They are nontoxic, nonbleeding, and inexpensive. They do not react with weak acids and alkalies, and if they are not contaminated with manganese, do not react with organic solvents. However, properties vary from one oxide to another.

Natural Iron Oxides. The earth's crust contains about 7 wt % iron oxides, but only a few deposits are rich enough in iron to be suitable for mining pigmentary-quality iron oxides. Deposits that are a suitable source of natural iron oxide pigments are usually hydrated aluminum silicates that contain various amounts and forms of iron oxide.

Iron oxides are supplied to the market as red, ocher, sienna, and umber natural pigments. The hue of the natural iron oxide pigments is determined by raw material composition and processing.

About 60% of the natural iron oxide pigments is used to color cement and other building materials. About 30% is consumed in the production of paints. For coloring plastics and rubber, synthetic iron oxide pigments are preferred. The main advantage of the natural iron oxide pigments, as compared to the synthetic ones, is cost. However, the quality is inferior, and in most cases, they are consumed in close proximity to the mines.

Synthetic Iron Oxides. Advantages of synthetic iron oxides over their natural counterparts include chemical purity, more uniform particle size and size distribution, and in the case of precipitated oxides the ability to prepare the pigment in predispersed vehicle systems by flushing techniques.

Iron Oxide Reds. From a chemical point of view, red iron oxides are based on the structure of hematite, $\alpha\text{-Fe}_2O_3$, and can be prepared in various shades, from orange through pure red to violet. Different shades are controlled primarily by the oxide's particle size, shape, and surface properties.

Synthetic red iron oxides are prepared in a variety of grades from light to dark. These are sold under a variety of names, e.g., Indian red, Turkey red, and Venetian red.

Iron Oxide Yellows. From a chemical point of view, synthetic iron oxide yellows, also known as iron gelbs, are based on the iron(III) oxide–hydroxide, α-FeO(OH), known as goethite. Color varies from light yellows to dark buffs and is primarily determined by particle size, which is usually between 0.1 and 0.8 μm. Because of their resistance to alkalies, these are used by the building industry to color cement.

Iron Blacks. Chemically, iron blacks are based on the binary iron oxide, $FeO \cdot Fe_2O_3$. Most of the black iron oxide pigments contain iron(III) oxide impurities, giving a higher ratio of iron(III) than would be expected from the theoretical formula.

Iron Browns. Iron browns are often prepared by blending red, yellow, and black synthetic iron oxides to the desired shade. The most effective mixing can be achieved by blending iron oxide pastes, rather than dry powders.

Complex Inorganic Color Pigments. Based on the crystal structure, the Color Pigments Manufacturers' Association (CPMA) has classified 53 key inorganic pigments into 14 categories; these inorganic colorants are known as complex inorganic color pigments. The original name, mixed-metal oxide are pigments (MMO), did not accurately describe the chemical nature of all the classified pigments.

Mixed-Metal Oxide Pigments. Mixed-metal oxide pigments can be considered a subcategory of complex inorganic color pigments. In reality these pigments are not mixtures but rather solid solutions or compounds consisting of two or more metal oxides. Structurally, mixed-metal oxide pigments belong to one of 14 structure types. The most common ones are rutile and spinel. The commercial significance is in their thermal, chemical, and light stability, combined with low toxicity. When these are employed for coloring glass enamels and ceramics, they are sometimes referred to as colors or stains; when used to color paints and plastics, they are known as pigments.

The color of mixed-metal oxide pigments results from the incorporation of chromophores, into the structure of stable host oxides.

Pigments having a spinel structure are widely used by the ceramic and plastic industries. They cover a wide range of colors and many are thermally stable up to 1400°C and are resistant to molten glass. Another advantage is their intermiscibility, allowing the user a choice of creating many intermediate colors.

Spinel compounds have a common chemical formula, AB_2X_4. Structurally they have a cubic symmetry and are derived from magnesium aluminate, $MgAl_2O_4$, a naturally occurring mineral.

Structurally, all rutile pigments are derived from the most stable titanium dioxide structure, i.e., rutile. The crystal structure of rutile is very common for AX_2-type compounds such as the oxides of four-valent metals, as well as halides of divalent elements.

Zircon pigments are derived from the tetragonal zirconium silicate, $ZrSiO_4$. Because of the high temperature (up to 1600°C) and chemical stability of zirconium silicate, zircon pigments can be used in the formulations of high temperature (1300–1400°C) glazes. Zirconium silicate is also used as an opacifier in porcelain and vitreous enamels.

In pigments, zirconium silicate serves as the host lattice for various chromophores, such as vanadium, praseodymium, iron, etc.

Bismuth Vanadate. The use of bismuth vanadate, $BiVO_4$, as a nontoxic, yellow pigment with good hiding strength and lightfastness was patented by DuPont in 1978. At least two pigment producers, Ciba and BASF, are marketing this pigment primarily for plastic and paint applications. Some users have already replaced the toxic pigment lead chromate in their paint formulations with a combination of organic pigments and bismuth vanadate.

Chromium(III) Pigments. There are two green pigments based on chromium in the +3 oxidation state. The first one is chromium oxide, Cr_2O_3; the second is hydrated chromium oxide, $Cr_2O_3 \cdot xH_2O$.

Chromium(III) Green Pigment. Chromium oxide green is characterized by outstanding lightfastness and has excellent resistance to acids, alkalies, and high temperatures. Because it weathers extremely well, chromium oxide green is applied as a colorant for roofing granules, cement, concrete, and outdoor industrial coatings. It is also used in ceramic applications.

Hydrated Chromium(III) Green Pigment. Hydrated chromium oxide has a brilliant green color and is referred to as Gingnet's green. It exhibits a limited hue range, is semitransparent, and has a low opacity, but provides excellent lightfastness and alkali resistance.

Ultramarine Pigments. Ultramarines are derived from lazurite (Lapis Lazuli), a semiprecious stone; they can be prepared in many shades.

Chemically, ultramarines are complex sodium aluminates having a zeolite structure. Composition varies within certain wt % ranges, i.e., Na_2O, 19–23; Al_2O_3, 23–29; SiO_2, 37–50; and S, 8–14.

Ultramarine pigments are used in printing inks, textiles, rubber, artists' colors, cosmetics, and laundry bluing. Because of their thermal stability they are also used to color roofing granules.

Cyanide Iron Blues. Cyanide iron blue, also known as Prussian blue, is one of the oldest industrially produced, inorganic pigments. Chemically, cyanide iron blues are based on the $\{Fe^{2+}[Fe^{3+}(CN)_6]\}^-$ anion. The charge is balanced by sodium, potassium, or ammonium cations.

Iron blues are mainly used by the printing industry for coloring printing inks. In Europe, cyanide blues are used for coloring fungicides.

Cadmium Pigments. Historically, cadmium pigments have been very important, providing a range of clean, bright shades of yellow, orange, red, and maroon colors. This importance, however, has decreased because of environmental issues. Only a few pigment producers are willing to continue cadmium pigment production.

Lead Chromate Pigments. Lead chromate, $PbCrO_4$, occurs in nature as the orange-red mineral crocoite. Synthetically prepared lead chromate and its solid-state solutions with lead sulfate, $PbSO_4$, or lead molybdate, $PbMoO_4$, are known to have excellent pigmentary properties. The usage of these pigments has been steadily decreasing because of environmental regulations restricting the production and the use of lead-containing products.

Black Pigments

Black pigments can be divided into two basic groups. The first group is represented by carbon blacks. Many other inorganic black pigments, called noncarbon blacks, also are available. These belong chemically to the colored pigment category. Examples are spinel and rutile blacks, iron blacks, and some inclusion zircon pigments.

Carbon Blacks. Carbon black is one of the oldest pigments known. More than 90% of the production of this pigment is consumed by the rubber industries, in particular, by the tire industry as a reinforcing agent. The rest is used for coloring plastics, printing inks, and paints. Particle size of carbon blacks varies from 5 to 500 μm and can be controlled by the process conditions and feedstock (see **Carbon Black**).

Environmentally, carbon blacks are relatively stable and unreactive. There is no evidence that these materials are toxic to humans or animals.

Extenders and Opacifiers

Extender pigments are low-cost, generally colorless or white pigments with a refractive index less than 1.7. Sometimes these pigments are also referred to as fillers. Many extenders are derived from natural sources and display many diverse properties. They are added to various formulations to improve technical and application properties and to reduce costs. Like pigments, extenders are dispersed in media in which they do not dissolve, but compared to pigments they do not have any significant coloristic properties.

In coating applications, extender pigments control gloss, viscosity, texture, suspension, and durability. Extender pigments also enhance the opacity of white hiding pigments, e.g., TiO_2. In plastics applications, extenders influence numerous properties of the resin including melt viscosity, thermal conductivity, and electrical properties, tensile strength, and moisture resistance.

Opacifiers are fine inorganic powders, usually white, that are used to reduce the transparency of ceramic glazes and porcelain enamels. The coating becomes opaque because the particles of the opacifier scatter and reflect the incident light. When inorganic pigments are combined with white opacifiers, pastel colors are obtained.

Commercially, the most important opacifiers for glazes are ZrO_2, $ZrSiO_4$, and SnO_2.

Miscellaneous Pigments

Luminescent Pigments. Luminescence is the ability of matter to emit light after it absorbs energy. Materials that have luminescent properties are known as phosphors, or luminescent pigments. If the light emission ceases shortly after the excitation source is removed ($>10^{-8}$ s), the process is fluorescence. The process with longer decay times is referred to as phosphorescence.

Semiconducting sulfides that can be represented by the formula $nZnS(1 - n)CdS:A$, where A stands for an activator, and $n = 0.15–1$, are typical of fluorescence pigments. Phosphorescence pigments can be expressed by the general formula $nZnS(1 - n)CdS:Cu$, where $n = 0.78–1.0$ and the amount of the Cu+ activator is only a few hundredths of a percent.

Phosphorescent pigments are used in military applications, plastics, and paints. Zinc sulfide doped with Ag+ (blue) cations, or with Cu+ (green) cations are important pigments for the production of color television screens.

Metal Effect Pigments. Some metals, when prepared as small flakes, impart a special metallic appearance to the coatings and plastics in which they are dispersed. Metals most often used in these applications are aluminum (aluminum bronzes), copper and copper-zinc alloys (gold bronzes), and in smaller amounts zinc, tin, nickel, gold, silver, and stainless steel.

Nacreous Pigments. Nacreous, i.e., pearlescent pigments are used for creating special decorative effects typical of natural pearls. Nacreous pigments are fine, thin, plate-like transparent particles having a high refractive index. Because of these physical characteristics, when dispersed in a transparent film, they produce a silky appearance.

Manufacture of the most popular nacreous pigments involves coating mica with 50–300-nm films of TiO_2, Fe_2O_3, or Cr_2O_3. The mica, which alone does not have a high enough refractive index for creating nacreous luster, provides the required transparent platelet base. The oxide coating provides the necessary high refractive index.

Transparent Pigments. Pigments having chemical composition corresponding to colored or white opaque pigments can, under certain circumstances, appear transparent in a medium. This happens when the particle size of these pigments becomes very small (2–15 nm), and if the particle refractive index is comparable to the refractive index of the media in which the particles are dispersed. Because of the very small particle size, the preparation of these pigments is much more complicated than the preparation of their nontransparent analogues. Their large surface area makes their dispersion difficult and they have a strong tendency to agglomerate.

Environmental Aspects

Some inorganic pigments contain heavy metals. Thus production, use, and disposal are becoming more and more regulated. In the United States there are several federal regulations that control the use and disposal of heavy metals.

The Resource Conservation and Recovery Act (RCRA) controls the disposal of hazardous waste. SARA Title III governs the toxic inventory and emission reporting; the Clean Water Act (CWA) sets the limits for metals that can be present in water discharge; and the Clear Air Act (CAA) Amendments of 1990 control the abatements of all materials in the air.

TABLE 2. ELEMENTS POTENTIALLY PRESENT IN INORGANIC PIGMENTS

Element	Federal regulation[a]				
	RCRA	SARA	OSHA	CWA	CAA
Al				+	
Ag	+	+			
As	+	+		+	+
Ba	+	+		+	
Be		+		+	
Cd	+	+	+	+	+
Co		+		+	+
Cr	+	+		+	+
Cu		+		+	
Hg	+	+		+	+
Mn		+			+
Mo				+	
Ni		+		+	
Pb	+	+	+	+	+
Sb	+	+		+	
Se	+	+		+	
Ti				+	
Tl		+			
Zn				+	

[a] A + indicates that the element is regulated by the particular act.

The Occupational Safety and Health Administration (OSHA) regulates the exposure to chemicals in the workplace. From the point of view of the inorganic pigments industry, the limits established for lead and cadmium exposure are particularly important. A comprehensive lead standard adopted by OSHA in 1978 has been successful in reducing the potential for lead contamination in the workplace.

Table 2 lists those metals regulated by federal law that are or might be present in inorganic pigments.

To assure the future of inorganic pigments, research efforts are directed toward the development of environmentally acceptable pigments, pigments that when produced under well-controlled conditions do not release any toxic materials into the environment whether during production, use, or disposal.

MIREK NOVOTNY
Cerdec Corporation
Z. SOLC
M. TROJAN
University of Pardubice

Additional Reading

Lewis, P.A.: *Pigment Handbook,* Vol. 1, 3rd Edition, John Wiley & Sons, Inc., New York, NY, 1988.
Patton, T.C.: *Paint Flow and Pigment Dispersion: A Rheological Approach to Coating and Ink Technology,* 2nd Edition, John Wiley & Sons, Inc., New York, NY, 1979.
Smith, H. MacDonald: *High Performance Pigments,* John Wiley & Sons, Inc., New York, NY, 2001.

PIGMENTS (Organic). Pigments are colored, colorless, or fluorescent particulate organic or inorganic finely divided solids which are usually insoluble in, and essentially physically and chemically unaffected by, the vehicle or medium in which they are incorporated. They alter appearance either by selective absorption and/or scattering of light. They are usually incorporated by dispersion in a variety of systems and retain their crystal or particulate nature throughout the pigmentation process. The large number of systems vary widely from paints to plastics to inks and fibers.

Dyes, on the other hand, are colored substances which are soluble or go into solution during the application process and impart color by selective absorption of light. In contrast to dyes, whose coloristic properties are almost exclusively defined by their chemical structure, the properties of pigments also depend on the physical characteristics of its particles.

The description of colored organic pigments excludes consideration of inorganic pigments, as well as black pigments which consist of specially treated forms of carbon and white pigments which are entirely of inorganic origin.

Pigments are categorized according to their generic name and chemical constitution in the *Color Index* (CI), published by the Society of Dyers and Colourists, and the American Association of Textile Chemists and Colorists.

Significant pigment attributes are tinctorial strength, durability (photochemical stability), hiding power, transparency, and heat and solvent resistance. Other properties include brightness (saturation), gloss, rheology, crystal stability, bleed resistance, flocculation resistance, and other properties associated with specialized applications.

The development of modern organic pigments started with the synthesis of dyestuffs for the textile industry. The period up to 1900 was characterized by the discovery and development of many dyes derived from coal-tar intermediates. Rapid advances in color chemistry were initiated after the discovery of diazo compounds and azo derivatives (shown to be largely hydrazone derivatives). The wide color potential of this class of pigments and their relative ease of preparation led to the development of azo colors, which represent the largest fraction of manufactured organic pigments.

The most important advance in pigment technology after World War I was the discovery of the relatively complex structure but easily synthesized copper phthalocyanines, which were characterized by excellent brightness, strength, bleed resistance, and lightfastness. See also **Phthalocyanine Compounds**.

After World War II the most important discovery was the family of red-violet quinacridone pigments, followed by the mostly yellow-orange benzimidazolone, the isoindolinone pigments, and the red diaryl pyrrolopyrroles.

Color and Constitution

The term chromophore is used to designate π-electron-containing moieties (conjugated double bonds) which contribute to the selective absorption of visible light. Generally, organic compounds absorb light in the ultraviolet (210–400 nm) and visible (400–750 nm) region of the spectrum at characteristic wavelengths. The intensities of these absorptions vary due to the excitation of the more loosely held electrons in the molecule. All unsaturated groups have remarkably similar $\pi - \pi^*$ transitions regardless of the atoms contained in the common chromophores.

The presence of $\pi - \pi$ or $n - \pi$ conjugated systems does not assure absorption of visible light or generation of color. However, all colored organic compounds, including pigments, possess extended conjugated resonance systems. Thus, whereas 1,4-diphenylbutadiene is colorless, 1,6-diphenylhexatriene is colored.

colorless   colored

All absorbed light is complementary to reflected light which produces observed color. All colors are a function of the wavelength of absorbed light as shown in Table 1. Thus, if a pigment absorbs only blue light it imparts an orange color, whereas when it absorbs orange light the observed color is blue.

Properties of Pigments

The physical and chemical characteristics that control and define the performance of a commercial pigment in a vehicle system include its chemical composition, chemical and physical stability, solubility, particle size, shape and particle size distribution, degree of dispersion, crystal morphology including polymorphic forms, refractive index, specific gravity, electronic spectra with particular emphasis on extinction coefficients in the visible spectrum, surface area, and the presence of impurities, extenders, and surface modifying agents. Invariably a pigment is used in a vehicle system, therefore its ultimate performance in use derives from both physical and chemical pigment-vehicle interaction. Performance of most pigments is system dependent.

Unlike inorganic pigments, organic pigments are relatively strong and bright (saturated), but their fastness properties, though adequate for the purposes for which they are used, vary widely from poor to outstanding.

Strength. The inherent strength of a pigment depends on its light-absorbing characteristics, which are related to its molecular and crystalline structure. In addition, strength is a function of particle size or surface area. The ability of a pigment to absorb light increases with decreasing particle size or increasing surface area, until the particles become entirely translucent or transparent to incident light. Being finely divided pigment particles have a great tendency to aggregate and agglomerate into crystal assemblies. To obtain the inherent strength of a pigment the aggregates must be completely broken down to individual crystals by application of work and their reagglomeration or flocculation prevented. Total breakdown to single crystals in practical systems seldom happens.

Pigment strength in a vehicle also depends on the typical character of other components in a pigmented system insofar as they absorb or scatter light. Strength comparisons are usually made with a series of samples featuring varying amounts of a pigment incorporated in a vehicle with a corresponding series in the same vehicle containing a reference pigment. Instrumental comparisons are commonly practiced.

A good absolute theoretical comparison of strength is represented by the area under the absorption bands in the visible spectrum, or less accurately by the molecular extinction coefficients at the maximum wavelength of absorption.

Brightness or Saturation. The saturation of a colored pigment is a measure of its brightness or cleanliness as opposed to dullness of hue. Generally, if a pigment absorbs light over a wide range of wavelengths, i.e., shows broad absorption bands, or contains more than one chromophore, the pigment is likely to be duller than a pigment with sharp absorption bands due to a single chromophore. Because pigments are frequently used in combinations or blends, the brightness is determined by the selective absorption of the individual pigments and this significantly affects the brightness of the reflected color.

TABLE 1. COLORS OF ABSORBED LIGHT AND THE CORRESPONDING COMPLEMENTARY COLORS AS A FUNCTION OF WAVELENGTH

Wavelength, nm	Color of absorbed light	Complementary color
400–420	violet	yellow-green
420–450	indigo blue	yellow
450–490	blue	orange
490–510	blue-green	red
510–530	green	purple
530–545	yellow-green	violet
545–580	yellow	indigo blue
580–630	orange	blue
630–720	red	blue-green

Fastness. Fastness describes the characteristics of a pigment in terms of its color stability in a pigmented system upon exposure to light, weather, heat, solvents, or various chemical agents. Ideally, a pigment should be insoluble and chemically and photochemically inert. Only a few organic pigments approach such perfection.

The development of resins, plastics, fibers, elastomers, etc. that are processed at progressively higher operating and curing temperatures has created a need for pigments that stand up for relatively long periods of time to a hostile environment. They must remain essentially unaltered when incorporated into plastics such as polypropylene, ABS, or nylon at relatively high temperatures.

Once a pigment is incorporated into a system, it is expected to be durable and withstand the combined chemical and physical stresses of weather, solar radiation, heat, water, and industrial pollutants. Because a pigment is totally enveloped by the medium which is itself not inert, various pigments perform differently in different systems. Thus, a pigment may be lightfast or weatherfast in one system and fail in another.

Dispersibility. The dispersibility of a pigment is measured by the effort required to develop the full tinctorial potential of a pigment in a vehicle system. Dispersibility differs from system to system depending on pigment–medium interaction and compatibility.

Small particle size pigments, especially the very small crystals, seldom exist as individual entities, but as strongly coherent aggregates or less firmly bound agglomerates. A wide variety of additives are being used to reduce aggregation or agglomeration and flocculation to improve dispersibility and color strength of organic pigments. These include resins, especially those related to abietic acid, aliphatic amines, amides, substituted derivatives of pigments themselves, and various combinations thereof. The additives are most effective if they are present when the pigment crystals are being generated.

Hiding Power and Transparency. Hiding power of a pigment is a function of its strength, that is, its absorption coefficients, its particle size, or light-scattering coefficient, and relative refractive indexes of pigment and vehicle. Light scattering has a powerful influence on opacity and goes through a maximum as a function of particle size. The maximum occurs at a particle size which is approximately half the wavelength of absorbed visible light.

Similarly, a pigment which absorbs much light increases hiding even when light scattering is insufficient, and the higher the refractive index the greater the hiding power of a pigment.

Conversely, to increase transparency light scattering must be minimized by particle size reduction. The smaller the particle size and the better the dispersion the greater the transparency.

Other Working Properties. Other properties which facilitate pigment incorporation and use include compatibility with a system, oil absorption, rheological characteristics, gloss, distinctness of image, wettability, migration fastness, plate-out, polymer distortion, etc. Most of these properties are controlled during pigment manufacture or formulation, others require special treatments to overcome undesirable effects.

Types of Pigments

All organic pigments have to be synthesized and nearly all have to be conditioned or finished. The physical conditioning is as important as its chemical constitution, and it has become an important separate process step in the manufacture of organic pigments.

Azo Pigments. Azo pigments provide good examples of materials in which conditioning is an integral part of the synthesis process. The coupling process for azo components is simple. An aromatic amine is diazotized by treatment with nitrous acid under conditions which vary from dilute mineral acid to concentrated sulfuric acid, depending on the basicity of the amine. The simplest method, i.e., direct coupling, involves running the diazo solution directly into the solution or suspension of the coupling component. In inverse coupling, the coupling component is run into the diazo solution. The most elegant technique, simultaneous coupling, entails running both the diazo and coupling component simultaneously into water or a dilute buffer.

The monoazo and disazo pigments contain one or more chromophoric groups usually referred to as the azo $-(N=N)-$ group. However, it has been shown by x-ray diffraction analysis and nuclear magnetic resonance (nmr) techniques that azo pigments exist in the hydrazone rather than the azo tautomeric form. The hydrazone form, which has three intramolecular hydrogen bonds, renders the molecule planar (with the exception of the aniline moiety) which is a stabilizing influence.

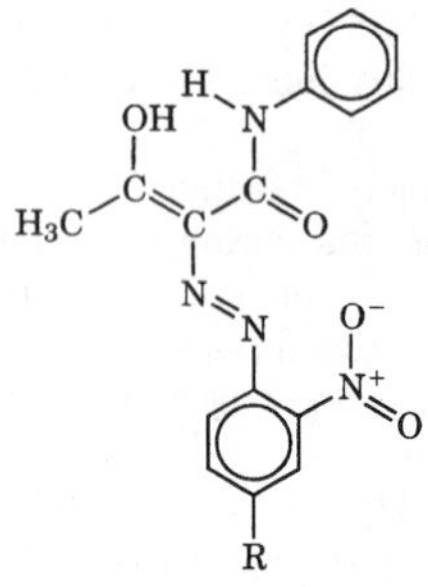

azo form 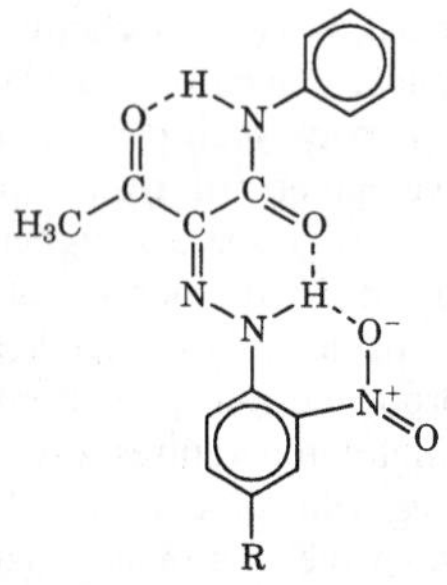   hydrazone form

Azo pigments, one of the oldest and most diverse group of pigments, comprise two types. One type consists of pigments that are insoluble in the aqueous reaction medium in which they are synthesized. Most simple azo pigments show poor bleed characteristics, but relatively good acid and alkali resistance. They show acceptable lightfastness in deep shades but poor tint lightfastness. The second type are laked or precipitated azo pigments derived from components substituted with sulfonic and/or carboxylic acid groups. These pigments are characterized by good to excellent bleed resistance, poor acid and alkali resistance, fair to good lightfastness in deep shades, and poor tint lightfastness. Also available are special azo pigments which show very good overall properties and therefore find applications in fairly demanding systems.

Monoazo Pigments. Monoazo yellow pigments are represented by the following general formula:

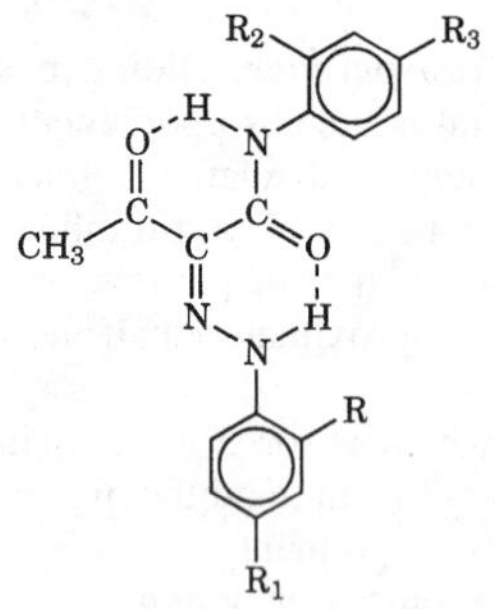

Many of the pigments carry a nitro group in the diazonium component, usually in the ortho position ($R = NO_2$). Among the acetoacetarylide components the *o*-methoxy derivative ($R_2 = OCH_3$, $R_3 = H$) is one of the most important in the production of azo pigments. The colors of these pigments range from red to green-shade yellows.

These pigments are sensitive to heat and bleed in most paint solvents. They are, however, resistant to acids and bases. They are used extensively in emulsion paints, paper coating compositions, inks, and, depending on particle size, can in some cases be used outdoors because of excellent lightfastness in full shades.

Benzimidazolones. This class of pigments derives its name from 5-aminobenzimidazolone, which upon reaction with diketene or 2-hydroxy-3-naphthoyl chloride leads to compounds that can be coupled with a variety of diazotized amines.

The acetoacetarylides yield yellow to orange pigments and the naphthoic acid amides yield red and brown pigments.

Diarylide Yellows. Diarylide or disazo yellow pigments are represented by the following general structural formula:

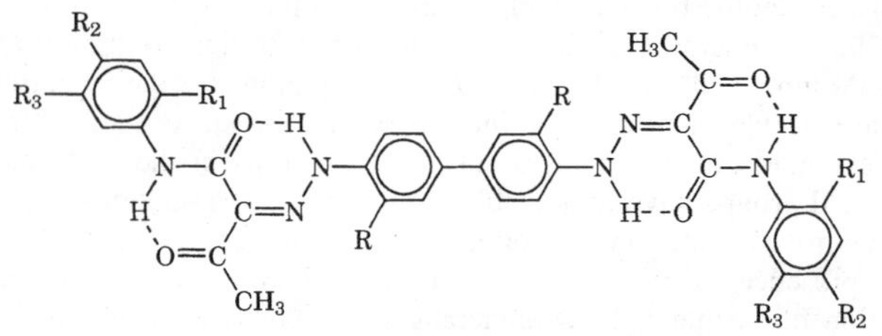

The chemistry and process of manufacture are very similar to the monazo pigments. Diarylides show significantly greater tinctorial strength and superior bleed and heat resistance than the conventional monoazo pigments. However, they are generally inferior to the monoazo pigments in lightfastness.

Based on high strength and versatile transparency most of the diarylides are used in a variety of printing inks, and in some plastics where temperature restrictions of 200°C have been imposed.

Monoazo Yellow Salts. Several monoazo yellow salts have gained popularity since the 200°C temperature restriction on the use of diarylide yellows has been imposed. An example Pigment Yellow 168, the calcium salt of diazotized 3-nitro-4-aminobenzenesulfonic acid coupled with acetoaceto-2-chloroanilide, provides a clean, somewhat greenish yellow color which shows good migration resistance but relatively poor tinctorial strength. It is used in polyethylene and inexpensive industrial finishes where the durability requirements are not high.

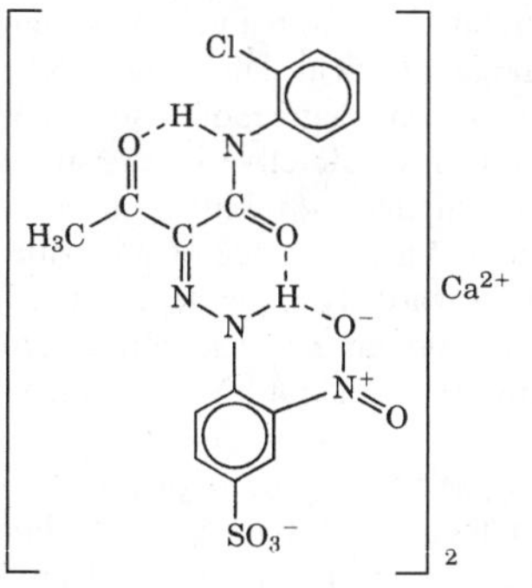

PY 168

Dinitraniline Orange. Dinitraniline Orange or Pigment Orange 5 is a strong and bright orange pigment with relatively low hiding power and good lightfastness in full shades, but poor tint lightfastness. It shows poor bleed but acceptable base resistance and finds principal application in air-drying systems, including a variety of printing inks.

Pyrazolone Orange. Pyrazolone Orange or Pigment Orange 13 is a disazo pigment of high strength and bright color with good lightfastness in full shades and poor tint lightfastness. It is characterized by fair base and chemical resistance and is used primarily in inks, with limited application in paints and plastics.

Azo Reds and Maroons. Pigment Red 3 is one of the most popular organic red pigments used in industrial finishes. Its hue varies considerably with particle size and therefore several shades are commercially available. Its principal application is in air drying paints, and to a limited extent in printing inks.

Para Red or Pigment Red 1 is an intense, reasonably opaque red which shows poor lightfastness, particularly in tints. The related pigment Parachlor Red (Pigment Red 4) is an intense yellowish red. Both pigments show poor bleed and bake resistance and a tendency to bloom in enamels. The pigments are used in some inks and in low cost articles such as detergents, floor polishes, colored pencils, etc. The use of these pigments has declined markedly as a result of greater quality demands by the coatings industry.

Lithol Red or Pigment Red 49:1 is one of the most important of the precipitated salt pigments. A family of sodium (PR 49), barium (PR 49:1), calcium (PR 49:2), and strontium (PR 49:3) salts of diazotized Tobias acid or 2-naphthylamine-1-sulfonic acid coupled with 2-naphthol. These reds are used where brightness, bleed resistance, and low cost are of primary importance.

PR 49:1

The BON or BONA Reds and Maroons derive their name from β-hydroxynaphthoic acid, also known as 3-hydroxy-2-naphthoic acid. BON is used as a general coupling component for the entire group with various diazotized amines containing salt-forming groups.

Lithol Rubine (Pigment Red 57) is the calcium salt of diazotized 2-amino-5-methylbenzenesulfonic acid coupled with 3-hydroxy-2-naphthoic acid. It ranks high among organic pigments in production volume and use.

Lithol Rubine is characterized by high tinctorial strength, good bleed, and bake resistance but poor alkali, soap, and acid resistance. Its lightfastness is considered fair and varies within a wide range of shades obtained by inclusion of auxiliary agents.

Red 2B defines the important barium (PR 48:1), strontium (PR 48:3), calcium (PR 48:2), and manganese (PR 48:4) salts of diazotized 2-amino-4-chloro-5-methylbenzenesulfonic acid coupled with 3-hydroxy-2-naphthoic acid, bright red pigments. They exhibit high strength, good bleed, and bake resistance, but poor resistance to alkali, soap, and acids, and fair lightfastness. The main fields of application are printing inks, plastics, and inexpensive industrial paints.

The manganese salt, i.e., manganese 2B (PR 48:4) is a bluish red, characterized by superior masstone lightfastness and outdoor durability. It finds use in some automotive and other high quality industrial finishes, as well as in some plastics and a variety of printing inks.

The pyrazolone reds are disazo pigments which provide high color strength and reasonable lightfastness in full shades but poor tint lightfastness, good bake, bleed, and chemical resistance. Some find application in plastics such as poly(vinyl chloride) where they show good dielectric properties, making them useful for cable insulations, in rubber, and specialized printing inks.

Naphthol Reds and Maroons. Naphthol Reds and Maroons are monoazo pigments which provide a wide range of colors from yellowish and medium red to bordeaux, maroon, and violet, and are characterized by high strength but marginal migration resistance. Depending on the substitution pattern some are strongly migrating and others are more or less resistant to migration. Lightfastness is generally marginal to good. Pigment Red 112 is a brilliant medium red pigment, approaching the shade of Toluidine Red, which is used in a variety of printing inks, air drying, and emulsion paints.

Another important pigment in this class is Pigment Red 170 which provides medium shades of red, and when particle-grown produces an opaque modification which shows improved migration resistance and lightfastness. It is used in high grade industrial paints and, in combination with high performance pigments, in automotive finishes. The transparent type which is tinctorially strong finds applications in a variety of printing inks.

Azo Condensation Pigments. A further improvement in heat stability of azo pigments was achieved by the condensation disazo pigments due to an enlarged molecular framework and higher molecular weight. Formally they are composed of two monoazo or more accurately two monohydrazone units, which are attached to each other by an aromatic dicarbonamide bridge.

The pigments are used primarily in plastics, including polypropylene fibers, because of very good bleed resistance, heat stability, and lightfastness. The reds also find use in printing inks, primarily for high quality products.

Lakes. Lakes are either dry toner pigments that are extended with a solid diluent, or an organic pigment obtained by precipitation of a water-soluble dye, frequently a sulfonic acid, by an inorganic cation or an inorganic substrate such as aluminum hydrate.

Basic dyes are characterized by bright shades and high strength but poor lightfastness. However, when laked by precipitation with soluble salts of organic acids such as tannic acid, or inorganic heteropoly-acids like phosphotungstic (PTA; M = W) and phosphomolybdic (PMA; M = Mo), and the combined phosphotungstomolybdic acid (PTMA), the resulting

pigments retain the dyes' tinctorial attributes, but become insoluble and show improved lightfastness.

Copper Phthalocyanines. Copper phthalocyanine (CPC) approximates an ideal pigment (Pigment Blue 15). This class of pigments offers extreme brightness, tinctorial strength, bleed and chemical resistance, stability to heat, and migration. The pigments show excellent weatherfastness but are restricted to the blue and green regions of the spectrum. Phthalocyanine blue and green are among the most important organic pigments on the worldwide market.

Copper Phthalocyanine Blue. CPC blue exists in several polymorphic modifications, two of which, the red-shade blue alpha and greenshade blue beta form, are of great commercial significance. Beta is the thermodynamically more stable phase and is the product resulting from manufacture by the two basic processes using either phthalonitrile or phthalic anhydride as starting materials. The alpha form is usually obtained by conversion from the beta form and has to be stabilized to prevent phase reconversion.

Copper Phthalocyanine Green. CPC green is obtained by electrophilic substitution of CPC blue with chlorine. The typical polychloro-CPCs are blue-shade green pigments. To provide yellower shades of green, bromine is substituted for chlorine. Like CPC blue the green pigments show outstanding pigmentary properties, but are lower in tinting strength with progressive halogen substitution, particularly with bromine. See also **Phthalocyanine Compounds**.

Quinacridones. Quinacridone pigments offer generally outstanding fastness properties across the visible spectrum from red-shade yellows to scarlet, maroon, red, magenta, and violet color ranges. The pigments are practically insoluble in most common solvents and therefore show excellent migration resistance in most application media. The low solubility is attributed to effective intermolecular hydrogen bonding which is also responsible for the photochemical stability. The various available colors are the result of polymorphism and various substitution patterns. The parent compound Pigment Violet 19 exists in three polymorphic modifications. The red gamma and violet beta forms are commercial pigments, whereas the red alpha form is metastable.

Due to the excellent pigmentary properties, quinacridones are used in many industries but particularly in automotive finishes, emulsion paints, plastics, and fibers.

Pigment Violet 19

Diaryl Pyrrolopyrroles. The 1,4-diketo-3,6-diarylpyrrolo(3,4-c)pyrroles are the most recently discovered class of pigments ranging in color from orange to bluish reds. These pigments are synthesized by base-catalyzed condensation of higher diakyl esters of succinic acid with aromatic nitriles. One important member of this class is Pigment Red 254, which is a very opaque yellowish red pigment of outstanding durability, brightness, and chemical resistance. Another is the parent compound Pigment Red 255, which is a high performance orange pigment. Both are used in automotive finishes, and a higher strength variation of PR 254 is used in plastics applications.

PR 254

Vat Dye Pigments. Vat dyes have been used for a long time for coloring textile fibers. As pigment technology evolved, new methods of particle size reduction have been successfully applied to largely insoluble dyes. Only a few of the very large number of vat dyes have found application in the pigment field.

Perylenes. Perylene pigments are either the 3,4,9,10-tetracarboxylic dianhydride or more often *N, N'*-substituted diimides.

The pigments are manufactured either by reaction of the dianhydride with an amine or *N, N'*-dialkylation of the diimide. They are characterized by high tinctorial strength, excellent solvent stability, very good weatherfastness, moderate brightness, and range in color from red to violet.

Most applications are in high grade industrial paints, especially automotive finishes. Some types are used primarily in plastics and fibers.

Perinones. The most important pigment in this family is the orange perinone, Pigment Orange 43; is fairly weatherfast and heat stable, and is used primarily in plastics and fiber applications.

Thioindigo. The most important thioindigo pigment is the redviolet Pigment Red 88. Although still used in some paint and plastic systems, it is being replaced by pigments of higher quality.

Aminoanthraquinone Pigments. Pigment Red 177 has the chemical structure of 4, 4'-diamino-1, 1'-dianthraquinonyl. It is the only known pigment with unsubstituted amino groups which are involved in both intra- and intermolecular hydrogen bonding. The bluish red pigment is used in plastics, industrial and automotive paints, and specialized inks. See also **Dyes: Anthraquinone.**

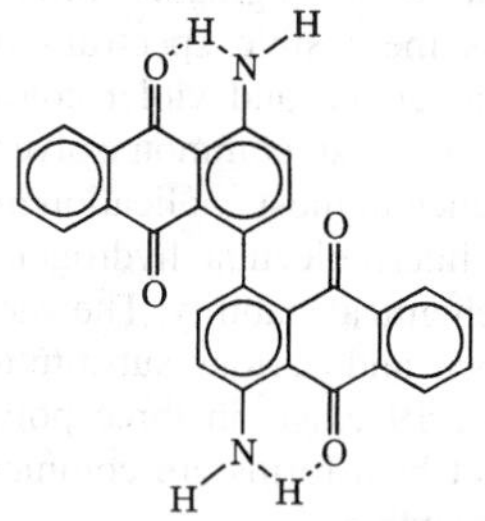

PR 177

Pigment Yellow 147 is a reddish shade yellow pigment used primarily in certain plastics and in polyester and polypropylene fibers.

Indanthrone. Pigment Blue 60 is a very red-shade blue pigment that shows outstanding weatherfastness in full shade as well as in light white reductions. It is used primarily in metallized automotive finishes where it is sometimes more weatherfast than some copper phthalocyanine blues.

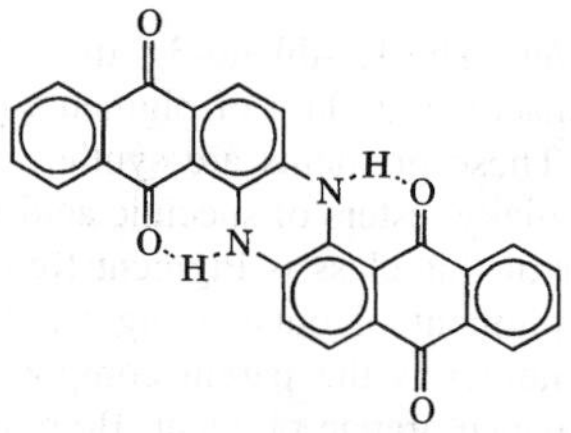

PB 60

Dioxazine. Carbazole Violet (Pigment Violet 23) is a bluish violet pigment that is uncommonly strong, resistant to solvents, and shows fair weatherfastness. It is used primarily as a shading pigment with copper phthalocyanines and for toning whites in a variety of systems.

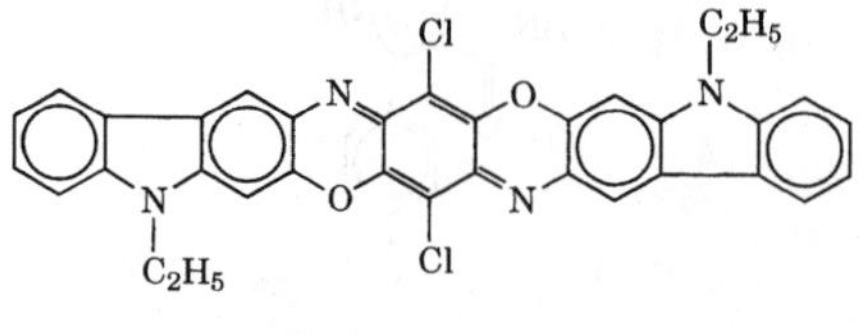

Isoindolinones and Isoindolines. Tetrachloroisoindolinone pigments are characterized by very good lightfastness, heat stability, migration resistance, and chemical inertness. Although Pigment Yellow 110, a red-shade yellow, is relatively weak, it finds extensive use in automotive and

other high grade finishes and in a variety of plastics and ink applications.

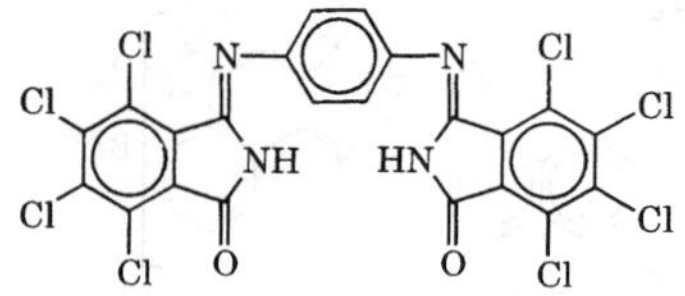

PY 110

Among the isoindoline pigments is Pigment Yellow 139, a reddish yellow pigment which differs in color as a function of particle size. The opaque version is the reddest. Although marginal in chemical resistance the pigment is used in the paint and plastics industries.

Quinophthalones. The quinophthalone pigments are prepared by condensation of quinaldines with a variety of aromatic anhydrides. One pigment in this series, Pigment Yellow 138, is a reasonably weatherfast greenish yellow pigment of good heat stability. The main field of application is paints and plastics.

Uses

Organic pigments are used for decorative and/or functional effects. In paints, for example, pigments provide color and contribute to exposure durability of the systems, which is particularly true for high performance pigments. Other functional effects include hiding power and high visibility, such as is displayed with daylight fluorescent pigments. They are used in various printing processes for textiles, plastics, and safety markings of various types.

The most important and established use for pigments is the imparting of color to a variety of materials and compositions. Examples are surface coatings for exteriors and interiors of automobiles and houses with oil- or water-based paints; wood stains, leather and artificial leather finishes, printing inks and many other applications.

Testing and Standardization

Pigments are subjected to a number of tests before they are released to customers. Testing is complicated because of the great diversity of pigment types and uses. A given pigment may be dispersible in one system but poorly dispersible in another, and can exhibit different durability depending on the system; performance is system dependent. Standardization is carried out against a standard sample for coloristics and a variety of working properties. Among the tests, depending on the pigment type, may be thermal stability, hiding power, rheology, migration, chemical stability, gloss, distinctness of image, durability, etc.

In the process of testing, color deviations are expressed in the CIELAB system or the equivalent polar LHC system. In either case tested samples must fall within acceptable ranges or limits established versus a standard by the pigment manufacturer and accepted by the pigment user.

In dispersing a pigment by an established method, acceptable pigment strength vs a standard must be achieved, even though an ideal dispersion normally is rarely realized.

Health and Safety Factors

Since pigments are generally insoluble, unlike most dyes, they are usually not bioavailable and consequently are generally not absorbed or metabolized. Nevertheless many health-related studies have been carried out and reported in the literature.

Acute toxicity of organic pigments has been studied extensively. The most common measure of toxicity is LD_{50} expressed in mg/kg of body weight which has a lethal effect on 50% of test animals after a single (oral, dermal, etc.) administration. These tests assess toxicity vs other known compounds. A large LD_{50} value represents a low degree of toxicity. Pigments in general have very low levels of acute toxicity.

Chronic toxicity defines a specific dose or exposure level that will produce measurable, long-term toxic effects, including carcinogenicity.

One area which requires special comment is a study which showed that certain diarylide pigments processed in polymers above 200°C and particularly above 240°C decompose to give off 3,3-dichlorobenzidine, an animal carcinogen. As a consequence diarylide pigments (not, however, condensation disazo pigments) are not recommended for use in any applications where they might be exposed to temperatures exceeding 200°C.

The hazards associated with handling pigments are specified by an OSHA Hazards Communication Standard, which also requires labeling and employee information and training.

Ecological Effects. The starting materials for manufacture of organic pigments are as diverse as the pigments themselves. However, most starting materials are derived from petroleum or natural gas sources. Although many pigments are synthesized in water, a variety of organic solvents are also employed by the industry. The effective utilization of all starting materials and solvents, and reduction of undesirable by-products, is a primary objective of the organic pigment industry.

E. E. JAFFE
Ciba-Geigy Corporation

Additional Reading

Clark, E.A. and P. Anliker: in O. Hutzinger, ed., *Organic Dyes and Pigments, The Handbook of Environmental Chemistry,* Vol. 3, Springer-Verlag, Berlin, 1980.

Color Index, 3rd Edition, Vol. 4, The Society of Dyers and Colourists, Bradford, Yorkshire, England; American Association of Textile Chemists and Colorists, Research Triangle Park, NC.

Herbst, W. and K. Hunger, *Industrial Organic Pigments,* VCH Publishers, Inc., New York, NY, 1993.

Lewis, P.A.: *Pigment Handbook,* Vol. 1, 3rd Edition, John Wiley & Sons, Inc., New York, NY, 1988.

Patton, T.C. ed.: *Pigment Handbook,* Vols. I, II, and III, John Wiley and Sons, Inc., New York, NY, 1973.

Patton, T.C.: *Paint Flow and Pigment Dispersion: A Rheological Approach to Coating and Ink Technology,* 2nd Edition, John Wiley & Sons, Inc., New York, NY, 1979.

Smith, H. MacDonald: *High Performance Pigments,* John Wiley & Sons, Inc., New York, NY, 2001.

PILOT PLANT. A trial assembly of small-scale reaction and processing equipment that is the intermediate stage between laboratory experiment and full-scale operation in the production of a new product. The functions of this stage are (1) to furnish chemical engineers with design data needed to construct a large-scale plane, (2) to resolve the many problems inherent in conversion from batch to continuous production, (3) to eliminate the differences that accompany change from constant laboratory conditions to a less closely controlled environment, and (4) to provide management with a basis for cost evaluation and estimation of the capital requirements of the new product. Because the size of the pilot plant varies with the nature of the product, it must be determined on a individual basis.

PINACOL REARRANGEMENT. See **Rearrangement (Organic Chemistry).**

PINNER REACTION. Formation of imino esters (alkyl imidates) by addition of dry hydrogen chloride to a mixture of a nitrile and an alcohol. Treatment of alkyl imidates with ammonia or primary or secondary amines affords amidines, while treatment with alcohols yields ortho-esters.

PIPETTE. A slender glass tube open at both ends and having an expanded area at or near the center designed to contain a specific volume of liquid, e.g., 5 ml. Liquid is drawn into the tube by oral or, for the sake of safety, some other form of suction.

PIRIA REACTION. Formation of arylsulfamic acids or sulfonation products or both by refluxing aromatic nitro compounds with a metal sulfite and boiling the mixture with dilute acid to yield the amines and sulfamic acids.

PITCH. (1) A carbonaceous, tacky residue resulting from distillation of coal tar, petroleum, pine tar, and fatty acids. Some types, such as glance pitch, occur naturally. They are used chiefly as sealants, roofing compounds, and wood preservatives. Synthetic carbon fibers are made from petroleum pitch. (2) In papermakers' terminology, a mixture of calcium carbonate, calcium soaps from wood components, and miscellaneous residues from materials used in paper manufacture. Pitch of this type is a production nuisance that requires close control. (3) The degree of slope of an inclined plane as in a screw auger, as measured by the distance between the flights or treads.

See also **Coal Tar and Derivatives.**

PITCHBLENDE. A massive variety of uraninite or uranium oxide found in metallic veins. Contains 55–75% UO_2, up to 30% UO_3, usually a little water, and varying amounts of other elements. Thorium and the rare earths are generally absent.

See also **Uranium.**

PITZER EQUATION. Equation for the approximation of data for heats of vaporization for organic and simple inorganic compounds. It is derived from temperature and reduced temperature relationships.

pK. A measurement of the completeness of an incomplete chemical reaction. It is defined as the negative logarithm (to the base 10) of the equilibrium constant K for the reaction in question. The pK is most frequently used to express the extent of dissociation or the strength of weak acids, particularly fatty acids, amino acids, and also complex ions, or similar substances. The weaker an electrolyte, the larger its pK. Thus, at 25°C for sulfuric acid (strong acid), pK is about -3.0; acetic acid (weak acid), p$K = 4.76$; boric acid (very weak acid), p$K = 9.24$. In a solution of a weak acid, if the concentration of undissociated acid is equal to the concentration of the anion of the acid, the pK will be equal to the pH.

PLANT GROWTH MODIFICATION AND REGULATION. Those chemical substances having the most to do with plant growth and form are given the general term *plant hormones.* A plant hormone, or *phytohormone,* may be defined as an organic compound produced naturally in plants, which controls growth or other functions at a site remote from its place of production, and which is very active in minute amounts. Three chemically quite different types of compounds apparently act as plant hormones: the *auxins,* the *gibberellins,* and the *kinetins.* In addition, the growth of roots is dependent upon vitamins of the B group which are synthesized in leaves and transported thence to the roots, thus qualifying as hormones.

Auxins

The best-studied hormones are those belonging to the class of auxins. These are defined as organic substances which promote growth along the longitudinal axis, when applied in low concentrations to shoots of plants freed as far as practical from their own inherent growth-promoting substances. Auxins generally have additional properties, but this one is critical.

Natural auxins have been identified in a number of instances. Indole-3-acetaldehyde occurs in a number of etiolated seedlings and in pineapple leaves; indole-3-acetonitrile has been isolated from cabbage and its presence indicated in a number of plants. One of the most widely occurring auxins is indole-3-acetic acid, which has been isolated in pure form from fungi and from corn (maize) grains. Its presence has been conclusively demonstrated by biochemical and chromatographic tests in a wide variety of flowering plants, including mono- and dicotyledons.

Many synthetic auxins have been produced, including 2,4-dichloro-phenoxyacetic acid or 2,4-D; naphthalene-1-acetic acid; and 2,3,6-trichlorobenzoic acid, among others. Used as a herbicide, 2,4-D is described in **Herbicides.**

By definition, these synthetic compounds are not hormones, although they are sometimes loosely referred to as hormone-type compounds.

An auxin is formed in fruits, seeds, pollen, root tips, coleoptile tips, young leaves, and especially in developing buds. The auxin travels away from the site of production in shoots by a special transporting system, depending on oxygen, which moves it in a predominantly polar direction from apex toward base. Movement in the opposite direction, i.e., from base toward apex, takes place to a variable extent depending upon the tissue and the plant. In the course of the polar transport, a large part of the auxin becomes bound and is no longer transportable. The transport is rather specifically inhibited by related compounds, particularly 2,3,5-triiodobenzoic acid, 2,4-D, and other synthetic auxins, which are transported either more slowly or to a much lesser extent in the polar system. Auxin applied artificially to intact plants can travel rapidly upward by penetrating into the conducting tissues of the wood, where it is carried upward in the transpiration stream.

In its normal polar, downward movement, the auxin stimulates the cells below the tip to elongate and sometimes to divide. Specific tissues, notably the cambium, are caused to divide laterally by auxin coming from the developing buds, which accounts for the wave of cell division occurring in tree trunks in the spring. Stimulation of other stem cells to divide leads to the production of root initials, which grow out as lateral roots. Cells of the young ovary are commonly caused to multiply and enlarge so that an apparently normal fruit is produced without requiring pollination (*parthenocarpic fruit*). This latter phenomenon indicates that the growing seeds normally secrete an auxin to which enlargement of the fruit is due, a conclusion which has been directly confirmed by bioassay in several fruit types.

Gibberellins can also cause enlargement of fruit. On reaching the lateral buds, however, auxin inhibits their elongation into shoots, and this accounts for *apical dominance*, i.e., suppression of the growth of lateral buds by the terminal bud of a shoot. Auxin also inhibits the falling off of leaves or fruits, which normally occurs when they are mature or aged by the formation of an *abscission layer* of special cells whose walls come apart. That the leaves or fruits do not absciss earlier is due to their steady production of auxin, which prevents formation of these cells. In the root, auxin inhibits elongation except in very low concentrations, but its level therein is usually low. Auxin can be transported for a short distance from the root apex toward the base, but the transport is not fully polar and in the more basal parts of the root the transport is slight.

When the shoot is placed horizontal, auxin is transported toward the lower side, causing accelerated growth there and hence upward curvature (*geotropism*); in the root, this causes decreased growth on the lower side and hence downward curvature. However, in the downward geotropic curvature of roots, other phenomena appear to enter in, and the complexities are not yet fully resolved. When shoots are illuminated from one side, auxins accumulate on the shaded side and, therefore, the plant curves toward the light (*phototropism*). Both geotropic and phototropic auxin movements have been confirmed with carboxyl-^{14}C-labeled compounds. The first observed effect when auxin is applied is the acceleration of the streaming of cytoplasm, but acceleration of growth begins in 7 to 14 minutes at about 23°C.

In plants that flower on short days, auxin may inhibit flowering; in plants that flower on long days, however, if close to the transition from the vegetative to the flowering state, auxin may promote flowering. In hemp and some of the squashes, auxin modifies the sexuality of the flowers toward femaleness. In the special case of pineapple, auxin directly causes flowering in an unusually clear-cut and quantitative response.

The principal uses of synthetic auxins are to promote the formation of roots on stem cuttings, to prevent abscission, especially of apples and pears, to induce flowering in pineapples, and occasionally to produce seedless fruits. The largest use, however, is that of weed killing. This action depends upon the fact that, at concentrations from 100 to 1000 times those concentrations occurring naturally, the auxins are highly toxic. Monocotyledonous plants, however, are usually resistant. In years past, 2,4-D has been favored in North and South America, whereas 2-methyl-4-chlorophenoxyacetic acid (*methoxone*) has been popular in Europe. However, as of the early 1980s, the regulatory status of these compounds in various countries is under study and may be subject to change.

Some chemically related compounds antagonize the action of auxins, for example, relieving the inhibition of root growth caused by 2,4-D. In contrast, 2,3,5-triiodbenzoic acid synergizes the action.

Gibberellins

These compounds were originally isolated from a parasitic fungus that causes excessive leaf elongation in rice plants. The mechanisms and applications of this group of compounds are described in **Gibberellic Acid and Gibberellin Plant Growth Hormones**.

Kinetins

Considerably less is known about this class of compounds. The first one to be discovered, produced by autoclaving yeast nucleic acid, was 6-furfurylaminopurine. Somewhat later, zeatin was isolated from immature corn (maize) kernels. The kinetins promote cytokinesis and protein synthesis, thus causing amino acids to accumulate where kinetins are synthesized (or externally applied) and maintaining the chlorophyll content of yellowing leaves. The kinetins antagonize auxin in apical dominance, releasing lateral buds from inhibition by a terminal bud or by applied auxin. It is believed that through the same mechanism, the kinetins promote the development of buds and leaves on tissue cultures. Their action is primarily local, and if there is transport *in vivo*, it probably occurs mainly in the transpiration stream (where amino acids are also often found).

Ethylene

The production of ethylene in fruit tissue and in small amounts in leaves may justify its consideration as a hormone, functioning in the gaseous state. Cherimoyas and some varieties of pear produce 1000 times the effective physiological concentration. Ethylene formation is closely linked to oxidation and may be centered in the mitochondria. Its effects are to promote cell-wall softening, starch hydrolysis, and organic acid disappearance in fruits—the syndrome known as *ripening*. Ethylene also decreases the geotropic responses of stems and petioles.

Daminozide Growth Modifier

Daminozide, the chemical name of which is 2,2-dimethyldrazide, was developed in the early 1960s as a modifying or regulating agent for the growth process of several food plants. The action varies with each plant. For example, on apple, the compound accelerates the start of flower budding, restricts nonproductive vegetative growth, and assist in fruit drop control. It is also claimed that the compound accelerates fruit coloring and helps to retain the firmness of the fruit. For some of these and other similar reasons, the compound has been used effectively for certain varieties of grape (particularly Concord), for peanuts (groundnuts), for tomatoes, nectarines (except Cherokee), and peaches. Other commercial designations are Alar, B-Nine, Kylar, and Sadh.

Ethephon Growth Modifier

This compound, (2-chloroethyl)phosphonic acid was developed in the United States in the mid-1960s and is used effectively on a number of fruit and vegetable crops for controlling a variety of factors. These include loosening fruit and causing earlier ripening of the fruit (apple, blackberry, blueberry, cherry, cranberry, filbert, tangerine, and walnuts); for encouraging uniform ripening and increasing yield (pepper and tomato); to improve color as well as accelerate maturity (cranberry); and to decrease time required for degreening in citrus fruits, particularly lemon. Other commercial designations for this compound include Cepha, Ethrel, and Florel.

Maleic Hydrazide

This compound, 1,2-dihydro-3,6-pyridazinedione, is also used as a growth regulator, herbicide, and plant modifier. It is used in the treatment of tobacco plants; as a post-harvest sprouting inhibitor; and as a sugar content stabilizer in sugar beets.

PLANT HORMONES (Gibberellins).   See **Gibberellic Acid and Gibberellin Plant Growth Hormones**.

PLASMA.   The portion of the blood remaining after removal of the white and red cells and the platelets; it differs from serum in that it contains fibrinogen, which induces clotting by conversion into fibrin by activity of the enzyme thrombin. Plasma is made up of more than 40 proteins and also contains acids, lipids, and metal ions. It is an amber, opalescent solution in which the proteins are in colloidal suspension and the solutes (electrolytes and nonelectrolytes) are either emulsified or in true solution. The proteins can be separated from each other and from the other solutes by ultrafiltration, ultracentrifugation, electrophoresis, and immuno-chemical techniques. See also **Blood**.

PLASMA FREQUENCY.   The oscillation frequency of plasma electrons about an equilibrium charge distribution is called the plasma or Langmuir frequency and is

$$\omega_p = \sqrt{\frac{4\pi n_e q^2}{m}} = 5.7 \times 10^5 \, n_e^{1/2} \text{ (radians/sec)}$$

where n_e is the electron density, m is the electron mass, and q is the electronic charge in esu. Its value is

$$(5.7 \times 10^{-5} \text{ radian cm}^{-3/2} \text{ sec}^{-1})n_e^{1/2}$$

PLASMA (Particle).   1. An assembly of ions, electrons, neutral atoms and molecules in which the motion of the particles is dominated by electromagnetic interactions. This condition occurs when the macroscopic electrostatic shielding distance (Debye length) is small compared to the dimensions of the plasma. Because of the large electrostatic potentials which would result from an inhomogeneous distribution of unlike charges, a plasma is effectively neutral. Thus there are equal numbers of positive and negative charges in every macroscopic volume of a plasma. Also, because a plasma is a conductor, it interacts with electromagnetic fields. The study of these interactions is called *hydromagnetics* or *magnetohydrodynamics*.

2. A collection of electrons and ions, usually at a high enough temperature so that the ionization level is about 5% and at densities such that the Debye shielding distance is much smaller than the macroscopic dimensions of the system. See also **Fusion Energy**.

PLASMID. A strand or fragment of genetic material existing outside the chromosomes in certain types of bacteria. R-type plasmids, which are present in *E. coli*, impart resistance to antibiotics in organisms that are exposed to them. The plasmids can be transferred from animals to humans, as well as to other, harmful bacteria that also become resistant to antibiotics. Feeding of traces of antibiotics to animals is believed to promote the growth of E. *coli* and, thus, to produce strains of pathogenic bacteria that are not amenable to antibiotic treatment. For this reason FDA has recommended elimination of certain antibiotics from animal feeds, e.g., penicillin, oxytetracycline, and chlortetracycline. Synthetic plasmids have been used successfully in recombinant DNA research.

PLASTIC DEFORMATION. When a metal or other solid is plastically deformed it suffers a permanent change of shape. The theory of plastic deformation in crystalline solids such as metals is complicated but well advanced. Metals are unique among solids in their ability to undergo severe plastic deformation. The observed yield stresses of single crystals are often 10^{-4} times smaller than the theoretical strengths of perfect crystals. The fact that actual metal crystals are so easily deformed has been attributed to the presence of lattice defects inside the crystals. The most important type of defect is the dislocation. See also **Creep (Metals)**; **Crystal**; and **Hot Working**.

PLASTIC FILM. A thermoplastic film less than 0.022 cm (0.010 inch) in thickness.

PLASTIC FLOW. A type of rheological behavior in which a given material shows no deformation until the applied stress reaches a critical value called the *yield value*. Most of the so-called plastics do not exhibit plastic flow. Common putty is an example of a material having plastic flow.

PLASTICITY. A rheological property of solid or semisolid materials expressed as the degree to which they will flow or deform under applied stress and retain the shape so induced, either permanently of for a definite time interval. It may be considered the reverse of elasticity. Application of heat and/or special additives is usually required for optimum results.

See also **Plasticizers**; and **Thermoplastic**.

PLASTICIZERS. High-boiling solvents or softening agents, usually liquid, added to a polymer to facilitate processing or to increase flexibility or toughness. (Where these effects are achieved by chemical modification of the polymer molecule, e.g., through copolymerization, the resin is said to be "internally plasticized.")

Thermoplastic polymers are composed of long-chain molecules held together by secondary valence bonds. When incorporated in the polymer with the aid of heat or a volatile solvent, plasticizers replace some of these polymer-to-polymer bonds with plasticizer-to-polymer bonds, thereby facilitating movement of the polymer chain segments and producing the physical changes described. Thermosetting polymers, which consist of three-dimensional networks connected through primary valence bonds, are not usually amenable to such softening by external plasticizers. Not all the thermoplastics can be plasticized satisfactorily. Polyvinyl chloride polymers and copolymers, and cellulose esters respond particularly well to plasticizing and represent the major outlets for plasticizers.

The results obtained by addition of plasticizer vary with different polymers. In polyvinyl chloride, for example, plasticizer concentrations of 30–50% convert the hard, rigid resin to rubber-like products having remarkably high elastic recovery, while similar plasticizer concentrations in cellulose acetate produce tough but essentially rigid products.

The plasticizer field has grown tremendously since camphor was patented as plasticizer for nitrocellulose in 1870. The reported U.S. production of plasticizers in 1968 exceeded 1.3 billion pounds. Close to

two-thirds of this production volume is used in vinyl chloride polymers and copolymers. Plasticized polyvinyl chloride is fabricated at elevated temperature into film and sheeting, and into molded and extruded articles; it is also processed in the form of dispersions, which may be applied as coatings and subsequently baked to form continuous films. These dispersions include plastisols, in which the plasticizer is the continuous phase; organosols, in which a volatile solvent is added; and water dispersions. Plastisols are also used in molding.

Compatibility

Where the polymer-to-plasticizer attraction is strong, the plasticizer has high compatibility with the resin and is said to be of the "primary" or "solvent" type. With polyvinyl chloride this attraction is furnished particularly well by ester groups. Where the polymer-to-plasticizer attraction is low the plasticizer is of the "secondary" or "nonsolvent" type; it functions as a spacer between polymer chains but cannot be used alone because of limited compatibility. This is manifested by exudation of the plasticizer from the resin. Secondary plasticizers are often employed to take advantage of other desirable properties which they may impart. Where they are used merely to cheapen the formulation, they are usually referred to as "extenders."

Plasticizing Efficiency

It is customary when evaluating plasticizers in polyvinyl chloride to compare them at concentrations which produce a standard apparent modulus in tension, as measured at room temperature. Since the stress-strain relationship is generally nonlinear it is necessary to specify a given point on the stress-strain curve as well as the rate of loading or straining. The efficiency may be expressed as the concentration of a given plasticizer necessary to produce this standard modulus. Other properties, e.g., indentation hardness, may take the place of tensile modulus.

Such tests, while they constitute an adequate basis for routine evaluation of plasticizers, furnish only a rudimentary picture of the elastic properties of the plasticized resin. More complete studies supply valuable information. For example, tensile creep tests have shown that polyvinyl chloride resin plasticized with trioctyl phosphate will deform more in response to stresses of short duration than will resin plasticized with tricresyl phosphate; the reverse is true for stresses of long duration.

For many applications low-temperature flexibility of the plasticized composition is also important. Plasticizers of low viscosity and low viscosity-temperature gradient are usually effective at low temperature. There is also a close relationship between rate of oil extraction and low-temperature flexibility: plasticizers effective at low temperature are usually rather readily extracted from the resin. Plasticizers containing linear alkyl chains are generally more effective at low temperature than those containing rings. Low-temperature performance is evaluated by measurement of stiffness in flexure or torsion or by measurement of second-order transition point, brittle point or peak dielectric loss factor.

Permanence

Where a plasticizer is used in thin films it is important that it have low vapor pressure. For polyvinyl chloride a plasticizer vapor pressure of no more than 4 mm at 225°C has been suggested as a rough criterion, although much higher vapor pressures are tolerated in plasticizers for cellulose acetate. Volatile losses are determined, not only by the plasticizer vapor pressure, but also by the plasticizer-resin interaction; plasticizers of limited compatibility may exhibit unexpectedly high volatile loss. Comparisons are usually made by heating a plasticized sample in contact with activated carbon and measuring the weight loss of the sample. Resistance of the plasticizer to migration determines whether the plasticized composition will mar or soften varnished surfaces with which it comes in contact. Stability and water-and oil resistance are further factors in plasticizer permanence. Plasticizers which have poor oil resistance are usually also the worst offenders with respect to marring.

Commercial Plasticizers

Phthalates. These esters, prepared from *o*-phthalic anhydride, constitute the most important group of plasticizers from the stand-point of production and sales volume. Among these the dioctyl phthalates are the most widely used in vinyl chloride resins, where they are preferred because they offer a good compromise with respect to a wide range of properties: satisfactory volatility, good compatibility, fair low-temperature flexibility, and moderately low cost. Most popular of the group is the 2-ethylhexyl ester, known as DOP, but the isoctyl, n-octyl, and capryl esters also find use. Increased emphasis on low volatility and plastisol viscosity stability has in recent years led to use of octyl decyl and didecyl phthalates. The lower alkyl phthalates are generally employed in resins other than polyvinyl chloride. Dibutyl phthalate is used chiefly in nitrocellulose lacquers and polyvinyl acetate adhesives; diethyl, dimethyl, and di(methoxyethyl) phthalates are used in cellulose acetate. For low toxicity, methyl and ethyl phthalyl ethyl glycolates are favored for cellulose acetate, and butyl phthalyl butyl glycolate for vinyl chloride polymers and nitrocellulose.

Phosphates. The phosphates, second only to phthalates in production volume, are favored for flame resistance and low volatility. Tricresyl phosphate (mixed meta and para isomers) is the most popular; it is used in polyvinyl chloride and in nitrocellulose lacquers. Resins plasticized with tricresyl phosphate are deficient in low-temperature flexibility. Diphenyl cresyl phosphate and triphenyl phosphate are other examples, the former for polyvinyl chloride, the latter for cellulose acetate. Diphenyl-2-ethylhexylphosphate is preferred to tricresyl phosphate in polyvinyl chloride where its low toxicity and improved low-temperature flexibility are required. Tri(2-ethylhexyl)-phosphate is outstanding among phosphates used in polyvinyl chloride with respect to low-temperature flexibility; in flame- and oil resistance, however, it is inferior to tricresyl phosphate. Tri(butoxyethyl)phosphate finds some use in synthetic rubber.

Esters of Aliphatic Dibasic Acids. This group consists of the adipates, sebacates, and azelates. These esters lead the field for low-temperature flexibility and efficiency in vinyls: they are useful in the preparation of low-viscosity stable plastisols. The principal disadvantages are high cost and poor solvent resistance. They are generally incompatible with cellulose acetate. Cheapest but most volatile are the adipates. Among the sebacates, the 2-ethylhexyl ester is outstanding in polyvinyl chlorides for low-temperature flexibility and low volatility; the butyl ester is used in polyvinyl butyral. The 2-ethylexyl ester of azelaic acid falls between the corresponding adipate and sebacate in price and properties.

Fatty Acid Esters. Monohydric alcohol esters of fatty acids usually have limited compatibility with vinyl chloride polymers. However, small amounts of these esters are useful for imparting low-temperature flexibility and softness and for their lubricant action during processing. Butyl oleate and stearate are leaders in this group. Methyl and butyl acetyl ricinoleates are used in nitrocellulose. Various glycol and glycerol esters of fatty acids have properties similar to the monohydric alcohol esters. Castor oil (glyceryl triricinoleate) and its derivatives are used in nitrocellulose. Triethylene glycol (di(2-ethylbutyrate) is the leading polyvinyl butyral plasticizer.

Epoxidized esters of fatty acids are similar in their plasticizing properties to other fatty acid esters, with the added advantage that they improve heat and light stability of vinyl resins by acting as hydrogen chloride scavengers. Although generally used together with conventional stabilizers, they can also be used as the sole stabilizer.

Polymeric Plasticizers. Polyesters prepared from dicarboxylic acids and diols enjoy some use in polyvinyl chloride resins for specialized applications. Ranging in properties from somewhat viscous liquids to soft resins, these plasticizers have outstanding permanence: they are of very low volatility and generally show good resistance to extraction and migration. The acids used commercially in preparing them include adipic, sebacic and azelaic; among the diols employed are propylene glycol and 2-ethyl-1,3-hexanediol; monobasic acids may be included in the preparation to limit molecular weight. Reported molecular weights of commercial polyester plasticizers range from 850 to 8000. Those of highest molecular weight are the most permanent but are somewhat difficult to process because of high viscosity and slow solvation of the resin. Other disadvantages include poor low-temperature properties and high cost. Most polyesters are rated as secondary plasticizers and are blended with the more efficient monomeric types.

Butadiene-acrylonitrile copolymers may also be used in polyvinyl chlorides and are rated as primary plasticizers for these resins. Like polyesters, they are favored for permanence, particularly oil resistance, and because they permit high loading with filler without serious impairment of physical properties. They are difficult to incorporate in vinyl resin and are often used together with conventional plasticizers. GRS rubber and polyisobutylene are also blended with certain polymers for plasticizer-like action. Other addition polymers have been studied as plasticizers, among them a series of promising liquid polymers of dibutyl itaconate.

Other Plasticizers. Acetyl tributyl citrate is an outstanding nontoxic plasticizer for polyvinyl chloride in food packaging. Plastisol formulations containing this ester have exceptional viscosity stability. Acetyl triethyl citrate is a good plasticizer for the cellulosics and acetyl trioctyl citrate shows promise with vinylidene chloride polymers. Other compounds of diverse nature find application as plasticizers. These include tetra-n-butyl thiodisuccinate, camphor, o-nitrobiphenyl and partially hydrogenated isomeric terphenyls.

CHARLES J. KNUTH
Pfizer, Inc.
New York

PLASTIC PIPE. Tubes, cylinders, conduits, and continuous length piping made (1) from thermoplastic polymers unreinforced (polyethylene, polyvinyl chloride, ABS polymers, polypropylene) or (2) from thermosetting polymers (polyesters, phenolics, epoxies) blended with 60–80% of such reinforcing materials as chopped asbestos or glass fibers to increase strength. The latter type is a reinforced plastic. In general the properties of plastic tubing or pipe are those of the polymers that comprise it. Most have good resistance to chemicals, corrosion, weathering, etc., combined with flexibility, light weight, and high strength. They are combustible but generally slow burning. The reinforced type is widely used as underground conduit for transportation of gases and fluids, including city water services, sewage disposal systems, etc. Its use in buildings is subject to local building codes. For additional information contact Plastic Pipe Institute, 1825 Connecticut Ave., NW, Suite 680, Washington, DC 20009. http://www.plasticpipe.org/applications/productinfo03_1_1.php

PLASTICS. These are materials formed from resins through the application of heat, pressure, or both. Most starting materials prior to the final fabrication of plastic products exhibit more or less plasticity—hence the term plastic. However, the great majority of plastic end-products are quite *nonplastic*, i.e., they are nonflowing, relatively stable dimensionally, and are hard. There are scores of different kinds of plastics. They fall into two broad categories: (1) *thermoplastic resins*, which can be heated and softened innumerable times without suffering any basic alteration in characteristics; and (2) *thermosetting resins*, which once set at a temperature critical to a given material cannot be resoftened and reworked. Since most plastic fabrication methods, such as casting, molding, or extruding, involve heat, the thermosetting materials must be properly and accurately formed during any thermal cycling that exceeds the critical temperature.

The principal kinds of thermoplastic resins include: (1) acrylonitrile-butadiene-styrene (ABS) resins; (2) acetals; (3) acrylics; (4) cellulosics; (5) chlorinated polyethers; (6) fluorocarbons, such as polytetra-fluorethylene (TFE), polychlorotrifluoroethylene (CTFE), and fluorinated ethylene propylene (FEP); (7) nylons (polyamides); (8) polycarbonates; (9) polyethylenes (including copolymers); (10) polypropylenes (including copolymers); (11) polystyrenes; and (12) vinyls (polyvinyl chloride). The principal kinds of thermosetting resins include: (1) alkyds; (2) allylics; (3) the aminos (melamine and urea); (4) epoxies; (5) phenolics; (6) polyesters; (7) silicones; and (8) urethanes.

The development of electrically conductive plastics is reviewed by R.B. Kaner and A.G. MacDiarmid (*Sci. Amer.*, 106–111, February 1988).

Numerous plastics are described in terms of manufacture and chemical and physical properties throughout this volume. Consult the alphabetical index.

Additional Reading

Elias, Hans-Georg: *An Introduction to Plastics,* 2nd Edition, John Wiley & Sons, Inc., New York, NY, 2003.

Harper, C.A. and E.M. Petrie: *Plastics Materials and Processes: A Concise Encyclopedia,* John Wiley & Sons, Inc., New York, NY, 2003.

Richardson, T.L. and E. Lokensgard: *Industrial Plastics: Theory and Application,* Delmar Learning, Albany, NY, 2003.

Rosato, D.V.: *Plastic Product Material and Process Selection Handbook,* Elsevier Science, New York, NY, 2004.

Wright, R.E.: *Product Design for Thermosetting Plastics,* CRC Press LLC., Boca Raton, FL, 2004.

Web References

Society of Plastics Engineers—SPE: http://www.4spe.org/
Plastics Technology Online: http://www.plasticstechnology.com/

PLASTIDS. Pigments in plants are often located in special bodies called plastids. There are many kinds of plastids: leucoplasts, those which contain no pigment and which are therefore colorless; chloroplasts, those which contain chlorophyll (by far the commonest kind); and chromoplasts, colored plastids that do not contain chlorophyll.

Leucoplasts occur in parts of stems and roots where light fails to penetrate. They absorb glucose and change it to starch.

Chloroplasts occur in cells exposed to light. They are indispensable to photosynthesis. In the algae the shapes of these bodies are many; in a large number of cases the plastid is a thick cup-shaped body occupying the greater part of the volume of the cell; in other algae the plastids have a central mass from which radiating plates of arms extend outward to the cell wall; spiral, net-shaped and ring-shaped plastids are not uncommon in this group of plants. In some algae and in nearly all higher plants, the chloroplasts are small subspherical or lens-shaped bodies, varying in number from one to many in a single cell. Always the chloroplasts are found embedded in the cytoplasm of the cell. In many plants the continuous movement of the cytoplasm in the cell carries the plastids along with it; in others these bodies have a fixed position. In certain algae and in many cells in higher plants, as for example in the palisade layer of leaves, the chloroplasts may change their position so that they will receive the most favorable amount of light. If the light intensity is low, they will present their flat surface to it; whereas if the light intensity is high, the plastid rotates so that it is placed edgewise to the light. Chloroplasts contain chlorophyll and other pigments. See also **Pigmentation (Plants)**.

PLASTISOL. A dispersion of finely divided resin in a plasticizer. Atypical composition is 100 parts resin and 50 parts plasticizer, forming a paste that gels when heated to 150C as a result of solvation of the resin particles by the plasticizer. If a volatile solvent is included, the plastisol is called an *organosol*.

Plastisols are used for molding thermoplastic resins, chiefly polyvinyl chloride. See also **Plasticizers**.

PLATFORMING. The process in which octane ratings of gasoline are raised by dehydrogenating naphthenes to aromatics, cracking high-boiling paraffins, and isomerizing paraffins to form products of greater chain branching. Desulfurization also takes place in this process.

PLATINUM AND PLATINUM GROUP. [CAS: 7440-06-4]. Chemical element, symbol Pt, at. no. 78, at. wt. 195.09, periodic table group 10 (formerly 8—transition metals), mp 1,772°C, bp 3,727 to 3,927°C, density 21.37 g/cm^3 (solid), 21.5 g/cm^3 (single crystal) (20°C). Elemental platinum has a face-centered cubic crystal structure. The five stable isotopes of platinum are ^{192}Pt, ^{194}Pt through ^{196}Pt, and ^{198}Pt. The seven unstable isotopes are ^{188}Pt through ^{191}Pt, ^{193}Pt, ^{197}Pt, and ^{199}Pt. In terms of earthly abundance, platinum is one of the scarce elements. Also, in terms of cosmic abundance, the investigations by Harold C. Urey (1952), using a figure of 10,000 for silicon, estimated the figure for platinum at 0.016. No notable presence of platinum in seawater has been found.

Electronic configuration

$$1s^2 2s^2 2p^6 3s^2 3p^6 3d^{10} 4s^2 4p^6 4d^{10} 4f^{14} 5s^2 5p^6 5d^9 6s^1$$

Ionic radius Pt^{2+} 0.52 Å. Metallic radius 1.3873 Å. First ionization potential 8.96 eV. Oxidation potentials Pt $\longrightarrow$ Pt^{2+} + 2e$^-$, ca. −1.2 V; Pt + 2OH$^-$ $\longrightarrow$ Pt(OH)$_2$ + 2e$^-$, −0.16 V.

Platinum is one member of a family of six elements, called the *platinum metals*, which almost always occur together. Before the discovery of the sister elements, the term *platinum* was applied to an alloy with Pt as the dominant metal, a practice that persists to some degree even today. The major properties of the platinum metals are given in Table 1. See also **Iridium Osmium**; **Palladium**; **Rhodium**; and **Ruthenium**.

Occurrence

These metals occur in both primary and secondary deposits. The primary deposits are generally associated with Ni-Cu sulfide ores. The Sudbury ores of Canada and the deposits of the Bushveld complex of South Africa are of this type. Native platinum occurs as a primary deposit in the Ural Mountains of the former U.S.S.R. and also in the Choco district of Colombia. Weathering and erosion of these deposits have resulted in the formation of secondary, or placer, deposits of native Pt in riverbeds and streams. One nugget of Pt found in the Urals weighed over 25 pounds (11.3 kilograms). Most of the world's platinum comes from Canada, the former

TABLE 1. REPRESENTATIVE PROPERTIES OF PLATINUM GROUP METALS

Property	Iridium	Osmium	Palladium	Platinum	Rhodium	Ruthenium
Atomic volume, cm^3/g-atom	8.54	8.43	8.88	9.09	8.27	8.29
Atomic radius	1.355	1.350	1.373	1.335	1.342	1.336
Crystalline form	fcc	hcp	fcc	fcc	fcc	hcp
Lattice parameters, Å, a	3.8389	2.7341	3.8902	3.9310	3.804	2.7041
B	—	4.3197	—	—	—	4.2814
Thermal conductivity at 20°C, (cal)(cm)/(s)(cm^3)(°C)	0.14	—	0.168	0.166	0.21	—
Electrical resistivity at 0°C, micro-ohm-cm	5.3	9.5	10.8	10.6	4.5	7.2
Thermal expansivity, °C × 10^6 at 20°C	6.6	6.6	12.4	9.0	8.3	9.6
Hardness, Mohs scale	6.5	7.0	4.8	4.3	—	6.5
Specific heat, cal/g-atom at 20°C	0.031	0.031	0.0584	0.031	0.059	0.057
Heat of fusion, kcal/mole	6.3	7.0	4.0	4.7	5.2	6.1
Heat of vaporization, kcal/mole	134.7	150	90	122	118.4	135.7

fcc = face-centered cubic
hcp = hexagonal close-packed

U.S.S.R., and South Africa. Minor amounts have been found in Alaska, Colombia, Ethiopia, Japan, Australia, and Sierra Leone.

Because of their unique properties and in spite of their high initial cost, the platinum metals find many applications in industry. Since used platinum metals retain a large portion of their initial value, many scrap materials are a major source of recoverable platinum metals. Practically every application of platinum generates scrap in some form, which is eventually returned to the platinum refiner for recycling. Although there are ample mine reserves, they soon would be depleted without constant scrap recycling.

Refining Processes

The refining procedures are a good introduction to the complex chemistry of the platinum metals. Some of these methods are still the best analytical techniques available for the separation of the metals. South African ore is smelted to form a Cu-Ni matte containing small amounts of the platinum metals (0.18%). The matte is melted, cast into anodes, and electrolytically dissolved. The contained Cu is deposited at the cathode, the Ni remains in the H$_2$SO$_4$ electrolyte, and the Pt metals are contained in the anode slimes. The resulting Cu is refined and the NiSO$_4$ solution purified and crystallized. The anode slimes are treated by roasting to remove sulfur and leached with dilute H$_2$SO$_4$ and air to remove Cu and Ni. The leached slimes are treated with aqua regia. The aqua regia solution is evaporated to concentrate the solutions and expel the excess HNO$_3$. The residue from this treatment contains Rh, Ir, Ru, Os, and Ag. The solution contains Pt, Pd, and Au.

Platinum is first removed by precipitating as ammonium hexachloroplatinate(IV) [(NH$_4$)$_2$PtCl$_6$] by the addition of a saturated solution of NH$_4$Cl. The precipitate is washed, dried, and calcined to form platinum sponge about 98% pure. The sponge is purified by redissolving in aqua regia and evaporating the solution to dryness with NaCl. The resulting sodium hexachloroplatinate is dissolved in H$_2$O and boiled with NaBrO$_3$ to convert impurities, such as Ir, Rh, Pd, and base metals, to valence states which produce readily filterable hydroxides. The Pt left in solution is free of impurities. It is then treated with NH$_4$Cl, and the pure ammonium hexachloroplatinate precipitate is calcined at 1000°C to pure Pt sponge.

The first aqua regia solution is treated with $FeSO_4$ to precipitate the gold. Pd is precipitated by oxidizing the solution with HNO_3 and adding NH_4Cl. Ammonium hexachloropalladate(IV) is formed (analogous to the Pt compound). This salt is purified by dissolving in NH_4OH, filtering off the impurities, and reprecipitating the Pd by the addition of HCl. The insoluble complex $Pd(NH_3)_2Cl_2$ is formed, which when calcined and reduced in H_2 yields pure Pd sponge.

The insolubles from the first aqua regia treatment are fused with a flux of litharge, soda ash, borax, and carbon in a gas-fired furnace at 1000°C for 1 hour. This procedure converts silica, alumina, and some base metals to slag. The precious metals are retained in the lead phase. The lead portion is heated with HNO_3, which dissolves the Pb and Ag. The Pb is precipitated as a sulfate and then the Ag as a chloride. The residue is treated with concentrated H_2SO_4 at 300°C. Rh will dissolve, leaving Ir, Ru, and Os as insolubles. The Rh solution is treated with Zn powder, precipitating an impure Rh. The impure Rh is heated in an atmosphere of Cl_2. Many impurities form volatile chlorides at this temperature and are expelled. Rh forms a polymeric trichloride, which is insoluble in aqua regia. The rhodium trichloride is digested in aqua regia for several hours, then filtered, dried, and calcined, yielding a commercial grade of Rh sponge.

The residue insoluble in H_2SO_4 is fused with Na_2O_2, poured into thin slabs and cooled. Ir is oxidized in the fusion of IrO_2, which is insoluble in H_2O. Ru and Os form soluble sodium salts and are separated from the Ir by filtration. The insoluble IrO_2 is dissolved in aqua regia, and ammonium hexachloroiridate(IV) is precipitated by the addition of NH_4Cl. Calcining yields pure Ir sponge.

The filtrate from the dissolution of the Na_2O_2 fusion contains $NaRuO_4$ and $NaOsO_4$. Ethyl alcohol is added to the solution, causing the precipitation of RuO_2, which is separated by filtration.

The Ru is purified by distilling with Cl_2. Volatile ruthenium tetroxide is collected. A saturated solution of NH_4Cl is added, causing the precipitation of ammonium hexachlororuthenate(III). The precipitated salt is calcined in H_2, yielding commercial Ru sponge.

The filtrate from the alcohol precipitation of Ru contains the Os. The solution is neutralized with HCl and is treated with powdered Zn, reducing the Os to the metallic state. Osmium tetroxide is formed by roasting the impure Zn in a current of O_2. The volatile OsO_4 is trapped in an aqueous solution of KOH. Ethanol is added to the solution, precipitating potassium osmate(VI), which is mixed with an excess of NH_4Cl and calcined in an atmosphere of H_2. The resulting Os sponge is leached to remove KCl, leaving a commercial-grade Os sponge.

The refining of secondary scrap follows much the same procedures with minor variations. For example, solid metallic Pt and especially the Rh and Ir alloys of Pt are very difficult to dissolve in aqua regia. Therefore the scrap generally is alloyed with Cu, Ni, Pb, or Zn before dissolution with acids.

Uses

These metals, in various forms, currently are used as catalysts for a wide variety of reactions. Products include high-octane gasoline, HNO_3, H_2SO_4, HCN, vitamins, antibiotics, H_2O_2, cortisone, alkaloids, and fuel-cell chemicals. These catalysts also are used to remove trace impurities, e.g., acetylene in ethylene or O_2 in H_2, or noxious constituents of partial combustion, e.g., automobile exhausts. Although substitutes are being sought, Pt is by far the best catalyst for pollution control of auto exhausts. In the future, the catalytic converters currently installed in automobiles will become a significant source of platinum metals. See also **Catalysis**.

The corrosion resistance of the Pt metals has made the Pt crucible and the Pt electrodes commonplace laboratory tools. The glass industry makes use of large amounts of Pt and its alloys for manufacturing very pure glass. Synthetic fibers often are extruded through spinnerettes made of Pt alloys. The large use of Pt metals in dental and medical devices, in jewelry, and for decorative purposes is based on the corrosion resistance and general appearance of these metals.

Because of their high melting points and stability, Pt alloys have found applications in thermocouples, resistance thermometers, potentiometer windings, electrodes, insoluble anodes, high-temperature furnace winding, crucibles that can withstand corrosive materials at high temperature, and generally as materials of construction that will not contaminate products at very high temperatures. Often Pt and Pd are alloyed with Rh, Ir, Ru, or Os to increase their strength, hardness, and corrosion resistance.

Platinum metals, in particular Pd, find extensive use in the electrical industry. Most of these metals are used as contacts, particularly in telephone relays, where their resistance to oxidation and sulfidization results in

circuits of reliability and stability. Alloys of Pt find use as grids for electronic tubes, in electrodes for aircraft spark plugs, for contact metal in printed and solid-state circuits, and in pressure-rupture disks.

In the medical field, *cis*-dichlorodiammineplatinum(II) has been available for cancer therapy (*Science*, **192**, 774–775, 1976).

Platinum Compounds

Platinum forms many di- and tetravalent compounds. The latter valence is more common and more stable. Pt in compact form is inert to all mineral acids except aqua regia. Under oxidizing conditions, fused alkalies will attack Pt to some extent. Molten halides, carbonates, and sulfates have little effect on the metal. Concentrated boiling H_2SO_4, fused cyanides, and fused alkaline sulfides will attack the finely divided metal. Pt is vigorously attacked by Cl_2 at elevated temperatures. In hot aqua regia or HCl containing chlorate ion or H_2O_2, the metal slowly dissolves, yielding a solution of hexachloroplatinic acid, H_2PtCl_6.

Platinum(II) hydroxide is made by adding KOH to a solution of platinum(II) chloride. The unstable black powder is easily oxidized by air and must therefore be handled in an inert atmosphere. In hot alkali or HCl, it disproportionates into the platinum(IV) compound and the metal. Very careful dehydration results in the formation of a gray powder that approaches the composition of platinum(II) oxide. Platinum(II) oxide can also be made by combining the elements at 420–440°C at an O_2 pressure of 8 atm.

When a solution of hexachloroplatinic(IV) acid is boiled for some time with NaOH, all the chloride ions are replaced by hydroxide ions. The resulting sodium hexahydroxoplatinate(IV), $Na_2Pt(OH)_6$, is soluble in the basic solution, but it can be precipitated as hexahydroxoplatinic(IV) acid, $H_2Pt(OH)_6$, by the addition of acetic acid. The hydroxide ions of the salt are replaced by the corresponding ions of mineral acids when the compound is dissolved in acid. Hexahydroxoplatinic acid can be dehydrated to yield compounds corresponding to the tri-, di-, and monohydrate of platinum(IV) oxide. The last water molecule cannot be removed without some destruction of the dioxide.

Brown-black, insoluble, anhydrous, platinum(IV) oxide is made by fusing hexachloroplatinic(IV) acid with $NaNO_3$ at about 500°C. The alkali salts are washed out with H_2O to free the fine insoluble residue of platinum(IV) oxide. This compound is known as *Adam's catalyst*.

When Pt is heated to 500°C in the presence of Cl_2, yellow-green, insoluble platinum(II) chloride is formed. At a pressure of 1 atm of Cl_2, the compound is stable from 435 to 581°C. It can also be made by heating hexachloroplatinic(IV) acid in Cl_2 at about 500°C. Platinum(II) chloride is soluble in HCl as tetrachloroplatinic(II) acid. It forms many salts that are water-soluble. These salts can be made by reducing a hot sol7ution of the corresponding hexachloroplatinate(IV) with oxalic acid or SO_2. Platinum(III) chloride has a narrow range of stability. It can be made by contacting Pt or a platinum chloride with 1 atm of Cl_2 at 364–374°C. This dark-green to black compound is practically insoluble in cold concentrated HCl but does dissolve on warming, forming a mixture of tetrachloroplatinic(II) and hexachloroplatinic(IV) acids. Anhydrous platinum(IV) chloride is very difficult to prepare. This brown soluble solid can be made by heating hexachloroplatinic(IV) acid in Cl_2 at 360°C. The most common Pt compound, hexachloroplatinic(IV) acid, is readily made by dissolving Pt in aqua regia, followed by several evaporations with additional HCl to destroy nitrosyl compounds. The acid crystallizes as a hexahydrate. It is difficult to stop the evaporation at just this point, and slight local overheating causes excess loss of water. The sodium salt is quite soluble, and the compound is resistant to hydrolysis in basic solution, allowing the bromate hydrolysis to precipitate base metals and other Pt metals as their hydroxides. The Pt remains in solution. The insolubility of ammonium hexachloroplatinate(IV) often is used in refining Pt. Its slight solubility can be overcome sufficiently by mass action to allow its use as a gravimetric procedure for the determination of Pt. This yellow compound decomposes at red heat, yielding pure Pt sponge. The insolubility of the potassium salt is used for the gravimetric determination of potassium.

A series of di-, tri-, and tetrabromides is well known. Platinum(II) iodide is precipitated as a black insoluble compound by the addition of 2 equiv of iodide to a hot solution of platinum(II) chloride. The black, insoluble, graphitelike substance, platinum(III) iodide, is made by combining the elements in a sealed tube at 350°C.

In contrast with Pd, Pt does form a Pt(IV) iodide. When a concentrated solution of hexachloroplatinic(IV) acid is treated with a hot solution of KI, this brown-black substance is precipitated. The compound is somewhat unstable and light-sensitive. It dissolves in excess KI to form the complex salt, rather unstable.

Pt forms a nonvolatile tetrafluoride, a pentafluoride, and a volatile hexafluoride. The dark-red PtF_6 melts at $56.7°C$ and is very reactive. It even reacts with O_2 at $21°C$ to form dioxygenylhexafluoroplatinate(V), O_2PtF_6.

When sulfur and Pt sponge are ignited, some platinum(II) sulfide is formed. The naturally occurring mineral is called cooperite. When heated in air or H_2, the products are metallic Pt and S. Platinum(IV) sulfide can be made by heating ammonium hexachloroplatinate(IV) or Pt and S at $650°C$. When precipitated by H_2S from chloroplatinic acid, the compound may exist as $PtS_2 \cdot H_2S$.

Divalent and tetravalent Pt probably form as many complexes as any other metal. The platinum(II) complexes are numerous with N_2, S, halogens, and C. The tetranitritoplatinum complexes are soluble in basic solution. Tetranitritoplatinum(II) ion is formed when a solution of platinum(II) chloride is boiled, at about neutral pH, with an excess of $NaNO_3$. The ammonium salt may explode when heated. Generally, platinum-metal nitrites should be destroyed in solution. They never should be heated in the dry form. Platinum(II) complexes most often have a coordination number of 4. Many compounds have been prepared with olefins, cyanides, nitriles, halides, isonitriles, amines, phosphines, arsines, and nitro compounds.

Platinum(IV) has a coordination number of 6. It forms complexes with halides, nitrogen and sulfur compounds, and other donors but to a lesser extent than platinum(II).

Health and Safety Factors

In bulk metallic form, the PGMs are not hazardous to health. However, like many other metals in finely divided form, PGM powders can be hazardous to handle. Powdered platinum is a powerful catalyst and is liable to ignite combustible materials. Powdered iridium can ignite in air and palladium dust is combustible.

LINTON LIBBY
Chief Chemist, Simmons Refining Company
Chicago, Illinois

Additional Reading

Carter, G.F. and D.E. Paul: *Materials Science and Engineering,* ASM International, Materials Park, OH, 1991.

Greenwood, N.N. and A. Earnshaw: *Chemistry of the Elements,* 2nd Edition, Butterworth-Heinemann, Inc., Woburn, MA, 1997.

Krebs, R.E.: *The History and Use of Our Earth's Chemical Elements: A Reference Guide,* Greenwood Publishing Group, Inc., Westport, CT, 1998.

Lide, D.R.: *CRC Handbook of Chemistry and Physics,* 84th Edition, CRC Press, LLC, Boca Raton, FL, 2003.

Meyers, R.A.: *Handbook of Chemicals Production,* The McGraw-Hill Companies, Inc., New York, NY, 1986.

Parker, P.: *McGraw-Hill Encyclopedia of Chemistry,* 2nd Edition, The McGraw-Hill Companies, Inc., New York, NY, 1993.

Schweitzer, P.A.: *Corrosion Resistance Tables,* 3rd Edition, ASM International, Materials Park, OH, 1991.

Staff: *ASM Handbook—Properties and Selection: Nonferrous Alloys and Pure Metals,* ASM International, Materials Park, OH, 1990.

Stwertka, A. and E. Stwertka: *A Guide to the Elements,* Oxford University Press, Inc., New York, NY, 1998.

PLUTONIUM. [CAS 7440-07-5]. Actinide radioactive metal. Atomic number 94. Symbol Pu. This element does not occur in nature except in minute quantities as a result of the thermal neutron capture and subsequent beta decay of ^{238}U; all isotopes are radioactive; atomic weight tables list the atomic weight as [242]; the mass number of the second-most-stable isotope ($t_{1/2} = 3.8 \times 10^5$ years). The most stable isotope is ^{244}Pu ($t_{1/2} = 7.6 \times 10^7$ years). Electronic configuration $1s^2 2s^2 2p^6 3s^2 3p^6 3d^{10} 4s^2 4p^6 4d^{10} 4f^{14} 5s^2 5p^6 5d^{10} 5f^6 6s^2 6p^6 7s^2$. Ionic radii Pu^{4+} 0.86 Å; Pu^{3+} 1.01 Å (Zachariasen). Oxidation potentials in acid solution $Pu \longrightarrow Pu^{3+} + 3e^-$, 2.03 V; $Pu^{3+} \longrightarrow Pu^{4+} + e^-$, −0.982 V; $Pu^{4+} + O_2 + 3e^- \longrightarrow PuO_2^+$, −1.17 V; $PuO_2^+ \longrightarrow PuO_2^{2+} + e^-$, −0.91 V. Oxidation potential in alkaline solution $Pu^{3+} + 4H_2O \longrightarrow Pu(OH)_4 + 4H^+ + e^-$, 0.4 V; $Pu(OH)_4 \longrightarrow PuO_2^+ + 2H_2O + e^-$, −1.0 V; $PuO_2^+ + 2OH^- \longrightarrow PuO_2(OH)_2 + e^-$, −0.8 V.

The isotope of major importance is ^{239}Pu ($t_{1/2} = 2.44 \times 10^4$ years). The importance of this isotope stems from its property of being fissionable with slow neutrons, together with the fact that the problem of its mass production has been solved. The first isotope to be produced was ^{238}Pu ($t_{1/2} = 86.4$ years).

Processes for the isolation and purification of plutonium, including the enrichment of spent nuclear reactor fuels, are described in the entry on **Nuclear Power Technology**. These processes take advantage of Pu's several oxidation states, each of which has different chemical properties. The processes may involve carrier precipitation, solvent extraction, and ion exchange.

Plutonium is of major importance because of its successful use as an explosive ingredient in nuclear weapons and the role it plays in the industrial applications of nuclear power. Exemplary of the energy available from plutonium: (1) One pound (0.45 kilogram) $\cong$ 10 million kilowatts; (2) one kilogram (2.2 pounds) = 22 million kilowatts; (3) one kilogram (2.2 pounds) = 20,000 tons of chemical explosive. Plutonium has the important nuclear property of being readily fissionable with neutrons. ^{238}Pu was used in the Apollo lunar missions to power seismic and other experimental instruments placed on the lunar surface. Because comparatively large quantities of plutonium are produced in reactors, the amount available for various applications has increased considerably during recent years. It is estimated that as of the late 1980s, nuclear reactors throughout the world are producing in excess of 20,000 kilograms of Pu per year. Within a few years, there will be an accumulation of some 300,000 kilograms of Pu or more. The element is available for purchase by qualified potential users.

In a typical fast breeder nuclear reactor, most of the fuel is ^{238}U (90 to 93%). The remainder of the fuel is in the form of fissile isotopes, which sustain the fission process. The majority of these fissile isotopes are in the form of ^{239}Pu and ^{241}Pu, although a small portion of ^{235}U can also be present. Because the fast breeder converts the fertile isotope ^{238}U into the fissile isotope ^{239}Pu, no enrichment plant is necessary. The fast breeder serves as its own enrichment plant. The need for electricity for supplemental uses in the fuel cycle process is thus reduced. Several of the early liquid-metal-cooled fast reactors used plutonium fuels. The reactor "Clementine," first operated in the United States in 1949, utilized plutonium metal, as did the BR-1 and BR-2 reactors in the former Soviet Union in 1955 and 1956, respectively. The BR-5 in the former Soviet Union, put into operation in 1959, utilized plutonium oxide and carbide. The reactor "Rapsodie" first operated in France in 1967 utilized uranium and plutonium oxides.

Plutonium was the second transuranium element to be discovered. The isotope ^{238}Pu was produced in 1940 by Seaborg, McMillan, Kennedy, and Wahl at Berkeley, California by deuteron bombardment of uranium in a 150-cm cyclotron. Plutonium exists in trace quantities in naturally occurring uranium ores. The metal is silvery in appearance, but tarnishes to a yellow color when only slightly oxidized. A relatively large piece will give off sensible heat as the result of alpha decay. Large pieces are capable of boiling water.

Chemical Properties

Plutonium has the oxidation states (III), (IV), (V), and (VI), and a complex chemistry in aqueous solutions, as can be judged from such a multiplicity of states. A large number of solid compounds corresponding to these states have been made, and they are in general similar in formulas and properties to the corresponding compounds of uranium and neptunium. An important difference, especially as regards ranges of stability of these compounds, arises as a result of the much greater stability of the (III) and (IV) states of plutonium. This also leads to differences in the aqueous solution chemistry of plutonium as compared to uranium and neptunium. The pentavalent state, like that of uranium, but unlike that of neptunium, is unstable in aqueous solution with respect to disproportionation.

The ionic species corresponding to the four oxidation states of plutonium vary with the acidity of the solution. In moderately strong (one-molar) acid the species are Pu^{3+}, Pu^{4+}, PuO_2^+, and PuO_2^{2+}. The ions are hydrated but it is not possible at present to assign a definite hydration to each ion. The potential scheme of these ions in one-molar perchloric acid is the following:

$$Pu \xrightarrow{+2.03\text{ V}} Pu^{3+} \xrightarrow{-0.982\text{ V}} Pu^{4+} \xrightarrow{-1.17\text{ V}} PuO_2^+ \xrightarrow{-0.91\text{ V}} PuO_2^{2+}$$

(with -1.043 V from Pu^{4+} to PuO_2^{2+} and -1.023 V from Pu^{3+} to PuO_2^+)

The potentials are in volts relative to the hydrogen–hydrogen-ion couple as zero.

The values given for the potential scheme in one-molar acid may be altered extensively by a change in hydrogen ion concentration (pH) or as a result of the addition of substances capable of forming complex ions with the plutonium species. Among such substances are sulfate, phosphate, fluoride, and oxalate ions, and various organic compounds, especially those known as chelating agents. The tetrapositive and hexapositive ions are complexed appreciably even by nitrate and chloride ions. The stability of the complex formed with a specified anion increases in the order: PuO_2^+, Pu^{3+}, PuO_2^{2+}, Pu^{4+}.

The hydrolysis of the ions follows a similar order; Pu^{4+} begins to hydrolyze even in tenth-molar acid and in hundredth-molar acid forms partly the hydroxide, $Pu(OH)_4$, and partly a colloidal polymer of variable but approximate composition $Pu(OH)_{3.85}X_{0.15}$, where X is an anion present in the solution. Further reduction of the acidity results in the hydrolysis of PuO_2^{2+} near pH 5, of Pu^{3+} at about pH 7, and of PuO_2^+ at about pH 9.

The plutonium ions in aqueous solution possess characteristic colors: blue-lavender for Pu^{3+}, yellow-brown to green for Pu^{4+}, and pink-orange for PuO_2^{2+}.

Plutonium monoxide occasionally appears on the surface of metal exposed to atmospheric oxidation, but is prepared more conveniently by treating the oxychloride with barium vapor at about 1,250°C. The oxide is classified with the interstitial compounds rather than with the typical metal oxides.

The so-called sesquioxide ($PuO_{1.5-1.75}$) is a typical mixed oxidation state oxide, similar to those formed by uranium, praseodymium, terbium, titanium, and many other metals. Its composition shows continuous variation with changes in temperature and pressure of oxygen above the oxide.

Plutonium dioxide (yellow-green to brown, cubic) is the most important oxide of the element. Almost all compounds of plutonium are converted to the dioxide upon ignition in air at about 1,000°C.

The important halides and oxyhalides of plutonium are PuF_3 (purple, hexagonal), PuF_4 (brown, monoclinic), PuF_6 (red-brown, orthorhombic), $PuCl_3$ (green, hexagonal), $PuCl_4$ (green-yellow, tetragonal), $PuBr_3$ (green, orthorhombic), PuI_3, $PuOF$, $PuOCl$, $PuOBr$, $PuOI$.

All of the halides except the hexafluoride and the triiodide may be prepared by the hydrohalogenation of the dioxide or of the oxalate of plutonium(III) at a temperature of about 700°C. With hydrogen fluoride the reaction product is PuF_4, unless hydrogen is added to the gas stream, in which case the trifluoride is produced. With hydrogen iodide the reaction product is $PuOI$, and the other oxyhalides may be formed by the addition of appropriate quantities of water vapor to the hydrogen halide gas. Plutonium triiodide is produced by the reaction of the metal with hydrogen iodide at about 400°C. The hexafluoride is produced by direct combination of the elements or by the reaction $2PuF_4 + O_2 \longrightarrow PuF_6 + PuO_2F_2$ at high temperature. The hydrides of plutonium include PuH_2 (black, cubic) and PuH_3 (black, hexagonal).

Plutonium forms several binary compounds that are of interest because of their refractory character and stability at high temperatures. These include the carbide, nitride, silicide, and sulfide of the element.

The monocarbide is formed by reacting the dioxide in intimate mixture with carbon at about 1,600°C. The mononitride may be obtained by heating the trichloride in a stream of anhydrous ammonia at 900°C; it is prepared more easily, however, by reacting finely divided metal with ammonia at 650°C. Although the lower temperatures are favorable to the production of higher nitrides, none are obtained, in contrast to the uranium-nitrogen system in which compositions up to $UN_{1.75}$ are easily realized.

The disilicide is formed when a slight stoichiometric excess of calcium disilicide is heated with plutonium dioxide in vacuum at about 1,550°C. The disilicide is only moderately stable in air and burns slowly to the dioxide when heated to about 700°C.

Plutonium "sesquisulfide" may be prepared by prolonged treatment of the dioxide in a graphite crucible with anhydrous hydrogen sulfide at 1,340°–1,400°C, or by the reaction of the trichloride with hydrogen sulfide at 900°C.

Handling Precautions. Care must be taken in the handling of plutonium to avoid unintentional formation of a critical mass. Plutonium in liquid solutions is more apt to become critical than solid plutonium. The shape of the mass also determines criticality. Plutonium's chemical properties also increase handling difficulty. Metallic plutonium is pyrophoric, particularly in finely divided form. Because of the high rate of emission of alpha particles, and the physiological fact that the element is specifically absorbed by bone marrow, plutonium, like all of the transuranium elements, is a radiological poison and must be handled with special equipment and precautions. To assure the safety of personnel, plutonium operations are normally handled in an essentially closed system, such as a *glovebox*. In addition, shielding is required when certain isotopes, including ^{240}Pu and ^{241}Pu, are present in appreciable quantity. Because research continues on the hazards and toxicity of plutonium, specific toxicity data should be sought from current authoritative literature, including government (U.S., UK, France, etc.) publications. As of the early 1980s, permissible body burden was established at 0.6 microgram; lung burden at 0.25 microgram. Chemical toxicity is trivial compared with radiation effects. The permissible levels for plutonium are the lowest for any of the radioactive elements.

The behavior of actinides in natural waters has great relevancy to the safe long-term storage of radioactive wastes. The enhanced solubility of plutonium and other actinides in the water of Mono Lake, California was studied by a group of scientists with the U.S. Geological Survey (Denver, Colorado). J.M. Cleveland and associates found that the solubility of plutonium in Mono Lake water is enhanced by the presence of large concentrations of indigenous carbonate ions and moderate concentrations of fluoride ions. In spite of the complex chemical composition of this water, only a few ions govern the behavior of plutonium, as demonstrated by the fact that it was possible to duplicate plutonium speciation in a synthetic water containing only the principal components of Mono Lake water. See reference listed.

Practical Utilization. Since the potential reserves of ^{235}U are limited, some point will be reached where this power source no longer will be competitive with fossil fuels, synthetic fuels, solar power plants, etc.—unless the development of means for the practical utilization of plutonium can be achieved. An important element of nuclear fuel cost is the credit received from the sale or future utilization of plutonium after its recovery from spent fuel. The plutonium credit is realistic only if the plutonium is used for power production, since, at present, there are few commercial uses envisioned where it would yield a similar economic return.

See also **Chemical Elements**.

NOTE: References pertaining to plutonium in nuclear reactors and nuclear wastes are listed at end of entry on **Nuclear Power Technology**.

Additional Reading

Albright, D. and F. Berkhout: *Plutonium and Highly Enriched Uranium, 1996: World Inventories, Capabilities, and Policies,* Oxford University Press, Inc., New York, NY, 1996.

Cleveland, J.M., T.F. Rees, and K.L. Nash: "Plutonium Speciation in Water from Mono Lake, California," *Science,* **222**, 1323–1325 (1983).

Lewis, R.J. and N.I. Sax: "Sax's Dangerous Properties of Industrial Materials," 10th Edition, John Wiley & Sons, Inc., New York, NY, 1999.

Lide, D.R.: *CRC Handbook of Chemistry and Physics,* 84th Edition, CRC Press, LLC, Boca Raton, FL, 2003.

Kent, J.A.: *Reigel's Handbook of Industrial Chemistry,* 9th Edition, Chapman Hall, New York, NY, 1992.

Krebs, R.E.: *The History and Use of Our Earth's Chemical Elements: A Reference Guide,* Greenwood Publishing Group, Inc., Westport, CT, 1998.

Parker, P.: *McGraw-Hill Encyclopedia of Chemistry,* 2nd Edition, The McGraw-Hill Companies, Inc., New York, NY, 1993.

Classical References

Kennedy, J.W., G.T. Seaborg, E. Segrè, and A.C. Wahl: "Properties of 94(239)," *Phys. Rev.,* **70**, 7/8, 555–556 (1946).

Seaborg, G.T.: "The Chemical and Radioactive Properties of the Heavy Elements," *Chem. Eng. News,* **23**, 2190–2193 (1945).

Seaborg, G.T., E.M. McMillan, J.W. Kennedy, and A.C. Wahl: "Radioactive Element 94 from Deuterons on Uranium," *Phys. Rev.,* **69** (7/8), 366–367 (1946).

Seaborg, G.T. and A.C. Wahl: "The Chemical Properties of Elements 94 and 93," *J. Amer. Chem. Soc.,* **70**, 1128–1134 (1948).

Seaborg, G.T. (editor): *Transuranium Elements,* Dowden, Hutchinson & Ross, Stroudsburg, Pennsylvania, 1978.

PNEUMOKONIOSES. Diseases of the lungs produced by inhalation of dusts, particularly those containing silica, asbestos and other inorganic material, or certain vegetable substances, notably sugar cane waste and raw cotton dust (brown lung).

Silicosis occurs in industries in which the air is polluted by silica dust, e.g., pottery, metal grinding, sandblasting and mining in rock. The inhaled silica gives rise to the production of diffuse fibrosis in the lungs; moreover it facilitates the growth of the tubercle bacillus so that tuberculosis is a possible complication. A special form of silicosis, called anthracosis (black lung), occurs in coal miners who are exposed to a mixed dust, mainly of coal, with a small proportion of silica.

Asbestosis is much less common but more serious than silicosis, since once contracted it is more rapidly fatal, and is associated with a liability to lung cancer.

Another form of pneumokoniosis is an acute and often fatal form which results from inhalation of beryllium, much used in the manufacture of fluorescent lamps.

Corticosteroid therapy has had encouraging results but the primary consideration in these diseases is the environmental protection of exposed workers.

POISON. (1) Any substance that is harmful to living tissues when applied in relatively small doses. The most important factors involved in effective dosage are (a) quantity or concentration, (b) duration of exposure, (c) particle size or physical state of the substance, (d) its affinity for living tissue, (e) its solubility in tissue fluids, and (f) the sensitivity of the tissues or organs. Sharp distinction between poisons and non-poisons is not always possible, because many variables must be taken into consideration in each case. Poisons are divided into four classes by the shipping regulatory agencies, as follows:

- *Poison A:* A gas or liquid so toxic that and extremely small amount of the gas of the vapor formed by the liquid is dangerous to life.
- *Poison B:* Less toxic liquids and solids that are hazardous either by contact with the body (skin adsorption) or by ingestion.
- *Poison C:* Liquids or solids that evolve toxic or strongly irritating fumes heated or when exposed to air (excluding class A poisons).
- *Poison D:* Radioactive materials.

See also **Toxicity**; and **Toxic Substances**.

Note: A computerized poison information center is operated by the FDA in Washington, DC. The National Clearinghouse for Poison Information Centers is located at 5401 Westbard Ave., Bethesda, MD 20016.

(2) In nuclear technology, any material with a high capture probability for neutrons that may divert an undesirable number of neutron from the fission chain reaction. (3) A substance that reduces or destroys the activity of a catalyst. Carbon monoxide and phosphorus arsenic, or sulfur compounds have this effect on the formation of ammonia from hydrogen and nitrogen gases, and the gases must be highly purified to avoid this. Another example is the poisoning of the platinum catalyst used in emission-control devices by organic lead compounds.

POLANYI, JOHN C. (1929–). Awarded the Nobel prize in chemistry in 1986 jointly with Dudley R. Herschbach and Yuan T. Lee. Herschbach reported that the energies of reactions of colliding beams of isolated alkali metal atoms and alkyl halide molecules appeared mostly as vibrational excited states of products. Polanyi characterized the excited states by the infrared light emitted by product molecules. His work also led to the development of lasers. Born in Germany, Polanyi studied in England and later became a Canadian citizen. Doctorate awarded by Manchester University, England, in 1952.

POLAR. Descriptive of a molecule in which the positive and negative electrical charges are permanently separated, as opposed to non-polar molecules in which the charges coincide. Polar molecules ionize in solution and impart electrical conductivity. Water, alcohol, and sulfuric acid are polar in nature; most hydrocarbon liquids are not. Carboxyl and hydroxyl groups often exhibit an electric charge. The formation of emulsions and the action of detergents are dependent on this behavior.

See also **Dipole Moment**.

POLAR COMPOUND. See **Compound (Chemical)**.

POLARIMETER. An instrument for measuring the amount by which the plane of polarization of plane polarized light is rotated in passing through a medium (usually a liquid).

See also **Polarimetry**.

POLARIMETRY. The basic principles of polarimetry as a method of quantitative chemical analysis were established over 150 years ago. The method is simple and nondestructive. A polarimeter measures the angle of rotation of linearly polarized light upon passage of the light from the unknown sample. Saccharimetry represents the polarimetric analysis of sugar and is a specialized area with its own form of instrumentation and well-established procedures of international acceptance. Polarimetry in other fields is less standardized, but is extensively used for the qualitative determination of numerous alkaloids, steroids, pharmaceutical, and organic chemical products. See also **Saccharimeter**.

Polarimetric instruments are operative with asymmetric molecules in the direct measurement of circular dichrosm (i.e., the difference of absorption of the left and right circularly polarized light as it passes through the sample). The technique is analogous to absorptiometry.

When plane polarized light passes through an anisotropic medium, the refractive indices of the two beams which emerge, which are right-hand and left-hand polarized, respectively, are not the same. This causes a phase difference between the two component beams and the resultant beam is rotated in its plane of polarization as it emerges from the medium.

Molecules of inherent structural asymmetry are anisotropic; they are *optically active* and exhibit *optical rotation* in solution. The typical optically active center is a carbon atom with four different substituents. In addition, any structural dissymmetry that results in a spatial left- and right-handedness will cause optical activity. Compounds of these types of come in a right-hand (R) and left-hand (L) form. When equal amounts of these two forms are mixed (racemic mixtures) there is no optical rotation because the activity of the two forms exactly cancel. Internal compensation of optically active centers in complex molecules is also found. Left- and right-handed optical isomers were first studied by Pasteur well over 100 years ago, and extensive surveys are found in most organic chemical texts.

Visual Polarimeters

A typical visual polarimeter is shown in Fig. 1. Light source may be a sodium or a mercury arc (less usual is the cadmium arc for the 509-millimicron and 644-millimicron lines). A filter isolates the emission line for monochromatic illumination. (While an instrument does not produce absolutely monochromatic light, the term is used by spectroscopists to describe light within a very narrow wavelength range, such as 0.2 millimicron.) The light then passes a polarizer prism system. This is usually a Nicol prism (a prism made of calcite that is cut and recemented in such a way that the incident light is split into a linear polarized beam which is transmitted, while the second beam is reflected and absorbed). The polarized beam is then passed through the analyzer, which is essentially identical with the polarizer. One of these two elements (usually the analyzer) can be rotated, and it is provided with a graduated circle for the precise read-out in angular degrees. By using a large circular scale and a vernier, a precision of $0.002°$ can be obtained in research-type polarimeters.

The principle of measurement is straightforward. If the two "Nicols" are oriented identically with respect to their optic axes, maximum light is passed. When they are crossed ($90°$), the intensity is at minimum (following a $\sin^2$ law). A refinement in all commercial visual polarimeters is that the observation of the crossed analyzer position is made easier by a half-shade field. See Fig. 2. Because the human eye is a comparative, rather than absolute, light-measuring device, very much better precision can be obtained by comparing two adjacent fields, rather than attempting to evaluate the brightness of a single field. The half-shade fields are created by an auxiliary prism, and the details of the optical arrangement can be found in the literature. Here we are only concerned with the operational features. The zero position of the instrument is that angle at which the two (or three) segments of the observed field are *equally* dim. Between the polarizer and the analyzer, a space is provided to accept the sample. The sample is placed in a tube that has precisely ground ends corresponding to the light path. End windows are held to the tube by gasketed fittings.

Routine polarimetric determinations are simple enough. First the polarimeter is balanced to zero degrees with the solvent. Then the solution is placed into the instrument, the instrument is rebalanced, and the angle α read off the scale. Nevertheless, when many measurements are taken, this becomes somewhat tedious. For the assessment of the half-shade field, the operator's eyes must be dark-adapted. Extended work in a darkened room peering through the eyepiece at an almost black field is tiring. The precision of visual polarimetric measurements will tend to increase rapidly at first, as the observer's eyes become adapted, but then it will decrease gradually

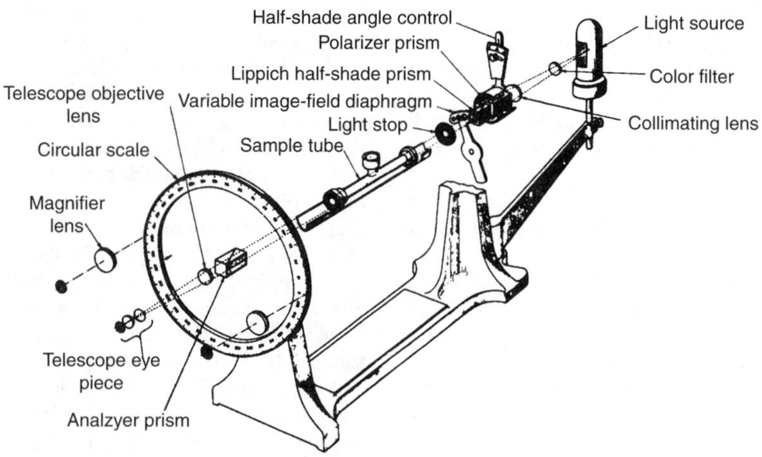

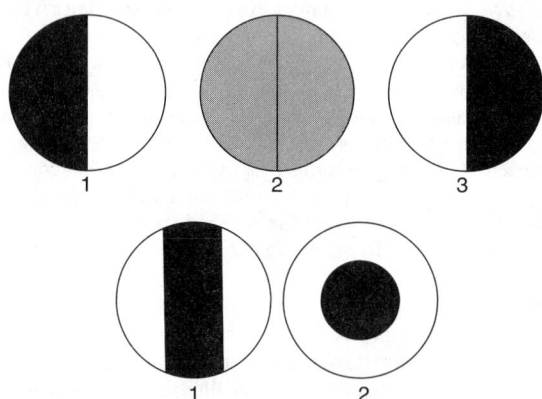

Fig. 1.   Schematic diagram of a visual polarimeter

Fig. 2.   Aspect of half-shade fields. Top diagram shows conventional split fields with (2) at the balance point. Bottom diagrams show special field configurations when at the off-balance point

because of fatigue. These facts justified the introduction of photoelectric polarimeters.

Photoelectric Polarimeters

In one design, there is no split half-shade field, but rather the analyzer is mechanically flip-flopped over an adjustable angle. At balance (and only at balance), the two extreme positions of the analyzer yield equal (low) intensity. A sensitive photomultiplier-photometer serves as a null indicator. The analyzer prism is manually rotated until a minimum deflection results on the large photometer scale and then the angle of rotation is read visually. In another design, Faraday cells are used. In the Faraday effect, a magnetic field induces optical rotatory power in liquids and solids by its influence on the atomic electron configuration. By using an electromagnet surrounding a glass rod, or a suitable crystal or solution, an alternating optical rotation can be introduced. This is analogous to the flip-flop just described. In the second design, a rotating polarizer is used, followed by the sample cell and the Faraday modulator. The emerging light passes through the analyzer and a photomultiplier pickup. The latter actuates a servosystem which drives the polarizer to balance. When the polarizer is exactly crossed with the analyzer, then, and only then will the alternating polarization introduced by the Faraday modulator have equal magnitude. This is the null point for the servosystem which thus establishes the balance automatically. The operator places the sample into the instrument (after it has been set to zero with the solvent) and then reads off the value of rotation from a magnified scale which allows estimation to 0.0025°. Several other designs have been developed for various measurements in optical rotatory dispersion. A number of spectropolarimeters, both automatic and recording, are available.

Kinetic Polarimetry

Polarimetry is particularly well suited for kinetic studies. The reason lies in the cyclic nature of the phenomenon, which allows the measurement of small changes in the angle of rotation with equal precision in the presence and absence of large background values. Moreover, subtle changes in structure, which are common in enzyme reactions, are often strongly reflected in rotatory power.

Spectropolarimetry

The determination of a rotatory dispersion curve is a prerequisite to the establishment of a sensitive polarimetric technique. This is analogous to colorimetry, where a complete spectrophotometric curve is required to establish the absorption peak best suited for routine work at a fixed wavelength. Similarly, in polarimetry, the gain in sensitivity may be enormous when working at an extremum (peak or trough). Thus, spectropolarimetry plays a role even though the analytical technique is simply polarimetry at a fixed wavelength. It is important to make a judicious choice in wavelength in an industrial assay.

POLARIZATION.   1. The process of bringing about a partial separation of electrical charges of opposite sign in a body by the superposition of an external field.

2. A vector quantity representing the dipole moment per unit volume of a dielectric medium. In rationalized units, the electric induction in a dielectric is given by $\mathbf{D} = \varepsilon\mathbf{E}$, which can be written

$$\mathbf{D} = \varepsilon_r \varepsilon_0 \mathbf{E} = \varepsilon_0 E + (\varepsilon_r - 1)\varepsilon_0 \mathbf{E}$$

where ε_r is the relative permittivity or dielectric constant (κ) of the medium. The term

$$(\varepsilon_r - 1)\varepsilon_0 \mathbf{E}$$

is the additional induction attributable to the matter of the dielectric, and is called the polarization of the dielectric. The coefficient $(\varepsilon_r - 1)$ is the "electric susceptibility" of the dielectric, and is often written as χ_e. In unrationalized systems,

$$\chi_e = \frac{\varepsilon_r - 1}{4\pi}$$

3. The process of confining the vibrations of the magnetic (or electric) field vector of light or other radiation to one plane.

4. The formation of localized regions near the electrodes of an electric cell during electrolysis, of products which modify (usually adversely) the further flow of current through the cell.

5. In electrochemistry, the increase of solution resistance due to gas accumulation at the electrode or chemical depletion in part of the solution.

POLARIZED LIGHT.   Whenever ordinary light is reflected from a glass plate, a varnished table-top, or other polished dielectric surface, we find upon suitable examination that a much larger part of the reflected beam is vibrating at right angles to the plane of incidence than in that plane; whereas in the incident beam there was no evidence of any preferential direction of vibration. A little experimenting shows that at a certain angle of incidence (the "polarizing angle," different for different dielectrics), the component vibrating in the plane of reflection is practically extinguished, all vibration being confined to the plane at right angles to this. The light is then said to be plane-polarized. The effect is more conveniently produced by a Nicol prism or by one of the polarizing films that polarize by transmission with less loss of light.

When the light passes through two such polarizers in succession, as in a polariscope, the fraction of it finally emerging depends upon the angle between the transmission planes of the polarizers, and varies all the way from nearly 100% to zero. The same effect may be produced by two reflections at glass plates turned to reflect in different planes but at the same (polarizing) angle. It seems probable that when the polarizing films above mentioned have been further perfected and cheapened, this intensity-reducing effect will be turned to account in reducing automobile headlight glare.

Metallic reflectors do not produce plane-polarization, but when plane-polarized light falls on a polished metal, its vibration is in general changed from a rectilinear to an elliptic one, and the light is said to be elliptically polarized.

When plane-polarized light traverses a crystal exhibiting double refraction, such as calcite, at right angles to its axis, it is transformed into elliptically polarized, or even circularly polarized, light.

If plane-polarized light is passed through quartz along its axis, or in any direction through one of the many optically active liquids such as turpentine or sugar solution, it undergoes optical rotation. That is, its vibration plane is twisted around through an angle that steadily increases with the distance traversed in the substance. Different substances have very different rotatory power, and some rotate one way and some the other. The sacharimeter is especially adapted to the study of this effect in liquids, especially sugar solutions.

POLAROGRAPHIC ANALYZERS. Polarography is an electrometric method of chemical analysis that is based on the current-voltage relationship at a special type of electrode. In most electrometric processes, it is desirable to use electrodes of relatively large surface area and, in some cases, to stir the solution, and thus avoid an effect termed "concentration polarization." The cause of this effect is the formation of a region in the solution that differs in composition from the main body of the solution. There is a depletion layer in cases where the ions of the solution are discharging on the electrode, and an excess layer in cases where the electrode is ionizing. Under such conditions, the value of the current at the electrode will not be determined solely by the usual factors: impressed electromotive force, electrode potential, and concentration of the ions in question in the body of the solution. It will also be affected by the rate of diffusion of these ions, from the body of the solution into the depletion layer. Since this rate of diffusion depends in turn upon the concentration of the ions in question in the body of the solution, it can be used as a measure of that concentration by constructing a cell in such a way as to maximize the effect of "concentration polarization." This is done by using a *microelectrode* (an electrode having a very small area of contact with the solution) as the electrode at which the ionic reaction to be measured is to occur, and a large or normal size electrode as the other one in the cell. As a further step, there is added to the solution an excess of an electrolyte that is inert to the electrochemical reaction, so that the diffusion effect on the ion under analysis will not be masked by its migration effect, that is, by its role in carrying current through the cell.

The conditions described for the polarized electrode are met by the dropping mercury electrode, shown in Fig. 1. The size of the polarized

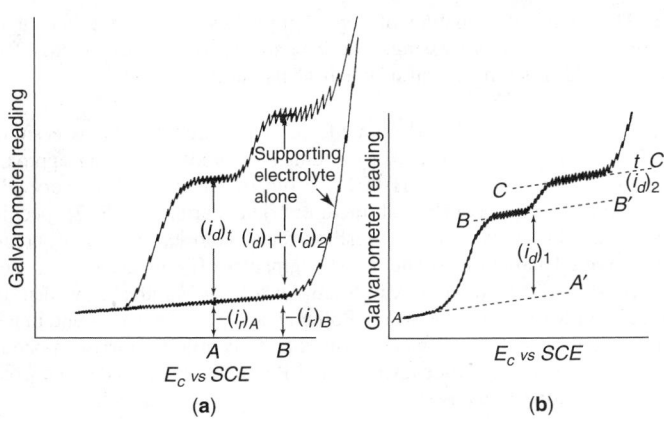

Fig. 2. Measuring a diffusion current: (**a**) exact method; (**b**) extrapolation method

electrode, which is merely a forming drop of mercury, is certainly small. It has the further advantage that it is constantly being replaced so that no solid deposit can form on its surface and so contribute an unwanted polarization effect to the one being measured. (Since the polarographic process involves the reduction of the ions, there is a deposition of metal on the mercury drop in all cases in which they are reduced to the metal, the effect of which is overcome by the constant replacement of the drop with fresh mercury.) This advantage is not processed by the other electrode, the pool of mercury in the bottom of the cell. For that reason, among others, the mercury pool electrode is not used as the other electrode in many instruments, being replaced by a standard calomel electrode connected electrically to the solution by a salt bridge, as described earlier in this entry. Moreover, the simple galvanometer and slidewire arrangement for measuring current and potential difference, respectively, is replaced and supplemented by more sensitive and easier-operated measuring and control devices, often automatic in operation and provided with recorders for producing the graphs of the current-potential difference directly. In fact, this instrumentation is so sensitive that it records the small fluctuations in the graph that occur during the formation of the drop of mercury and the beginning of the next one, the characteristic polarographic waves shown in Fig. 2. As can be seen from this figure, the polarographic method makes it possible to measure more than one concentration at a time, as represented by the two waves shown.

In addition to its advantage of permitting the determination of more than one ionic concentration in the same run, polarography also is most useful in determining very small ionic concentrations, of the order of millimoles per liter or lower.

POLAR MOLECULE. 1. A molecule with a positive charge on one end and a negative charge on its other end. Its vector sum of its bond dipoles is not zero.

2. Molecule in which the electrons forming the valency bond are not symmetrically arranged.

POLE FIGURE. A diagram used in metallurgy to show the preferred orientation of crystals in a metal. A pole figure is prepared by plotting, on a statistical basis, the positions in space of the poles of a specific crystallographic plane using a stereographic projection as the basis of the representation. The data for a pole figure is normally obtained using x-ray diffraction techniques.

POLISH. 1. A solid powder or a liquid or semi-liquid mixture that imparts smoothness, surface protection, or a decorative finish. The most widely used solid polishing agent is fine-ground red iron oxide (rouge), applied to the surface of plate glass, backs of mirrors, and optical glass. A wide variety of liquid and pastelike polishes are based on vegetable waxes (carnauba and candelilla), combined with softeners, fillers, and pigments or emulsified in alcohol or other solvent. Furniture polishes often contain red oil, lemon oil, and petroleum solvent; most types of metal and wood polish contain organic solvents and, hence, are flammable liquids. Nail polishes are nitrocellulose lacquers, usually with amylacetate solvent.

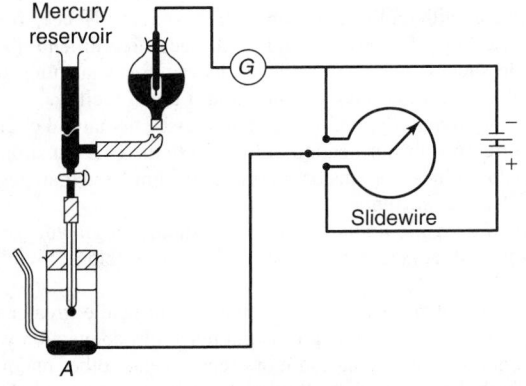

Fig. 1. Dropping mercury cathode assembly

2. The hard outer coating of cereal grains, especially rice, which is usually removed in processing. These coatings are rich in vitamin B_1. Their removal robs the cereal of much of its nutritive value.

POLLUCITE. The mineral pollucite is rather rare. It contains cesium, aluminum, silicon, and oxygen, its chemical composition being approximately $(Cs, Na)_2(Al_2Si_4)O_{12} \cdot H_2O$. It is isometric, usually in cubic crystals or crystalline masses; conchoidal fracture; brittle; hardness, 6.5–7; specific gravity, 2.9; luster, vitreous on fresh surfaces; colorless and transparent. Found on the Island of Elba and in the pegmatites of Maine, and as masses 3–4 feet (0.9–1.2 meters) thick in South Dakota; at Varutrask, Sweden; in Italy; and in Kazakhastan, Russian. Pollucite and petalite were found in the granites of Elba and at first named pollux and castorite for the two famous brothers of Roman mythology, Castor and Pollux. Pollucite is derived from the Latin genitive *Pollucis.*

POLLUTION (Air). Prior to the Industrial Revolution (circa 1840s), the composition of "pure" air making up Earth's surrounding atmospheres essentially remained constant for several thousand years. True, certain natural phenomena, such as volcanic eruptions, may have altered the atmospheric composition over relatively short time spans.

By many orders of magnitude, the greatest alteration of the atmosphere occurred during the middle Precambrian period, between 2.9 and 1.8 billion years ago. It is generally accepted that prior to that time, the terrestrial atmosphere was chemically of a reducing nature—as contrasted with an oxidative environment of the present general composition required to support humans and other mammals and life forms which abound on the earth today. See also **Air.** Brought about by greatly accelerated plant growth, that earlier natural change represented the most dramatic pollution effect ever suffered by the earth's environment. During that period, the oxygen liberated by plant activity proved to be a very toxic substance for anaerobic life forms and eradicated most of the biotic community existing at that time. New types of life had to develop which were capable of survival in an oxidative environment. Geochemical processes took on new characteristics, based upon the slow oxidative degradation of both organic and inorganic materials.

Alteration of the earth's atmosphere as the result of human (anthropogenic) activities is extremely recent on the life scale of the earth. This altering process was essentially commenced when humans first discovered and started to use fire as a means of heating, cooking, etc. It is the *combustion* of organic fuels today that is the principal contribution to anthropogenic air pollution. For centuries the pollutants added to the atmosphere by humans were essentially insignificant in terms of the mass and the dynamics of the earth's atmosphere. Except on a local and sometimes regional basis, air pollution was no problem prior to the invention of the steam engine and, of course, the later invention of the internal combustion engine. Traditionally, air pollution has a direct relationship with increasing population and the growing sophistication of the population, which demands ever increasing quantities of energy and the manufacture of goods by processes which yield byproducts that require removal to some kind of sink, the earth's atmosphere being one of these sinks.

With few exceptions, air pollutants ultimately fall by gravity to the surface of the earth. On land, pollution of the soil and freshwater lakes and rivers and ultimately the groundwater occurs. Fallout on the seas and oceans also occurs, but unless radioactive, the effects are less easy to discern except on the long term. It is indeed difficult to separate air and water pollution. The relationship is explored in the article on **Wastes and Pollution.** The winds contribute both to the spread and, in some instances, to the contribution of air pollutants. Frequently, as in the case of "acid rain," the precipitation of water (an excellent solvent) in the form of rain, snow, sleet, ice pellets, etc. causes entrainment of pollutants (gases, mists, particles, etc.). Thus the soils, rocks, lakes, and rivers are subject to the corrosive and biodestructive processes brought about by the presence of alien substances. Acid rain is described later in this article.

During the last few decades and including the early 1990s, the total amount of pollutants in the atmosphere has increased exponentially. Currently, many hundreds of different pollutants, largely from anthropogenic sources, rise into the atmosphere. Even though many of these substances are measured on a scale of parts per million (ppm) or even parts per billion (ppb), a majority of these substances cause ill effects on the health and well-being of air-breathing creatures as well as corroding and eroding structures.

The following *incomplete* list is given here to dramatize the complexity of the current air pollution problem, to demonstrate the probable impracticality of finding a few simple solutions, and to illuminate why air pollution measurement and control presently requires the skills of many hundreds of scientists and engineers. To accomplish these tasks, many excellent instrumental techniques are required, as exemplified by chromatography, laser radar, various forms of spectrometers, and particle analyzers, among other advanced analytical tools. The use of radioactive isotopes has been effective in many instances of pollution source tracing.

Partial List of Air Pollutants

Common Gases—carbon monoxide (CO), carbon dioxide (CO_2), nitrogen oxides (NO_x), ammonia (NH_3), hydrogen sulfide (H_2S), chlorine (Cl_2).

Volatile Inorganics—sulfuric acid (H_2SO_4), hydrochloric acid (HCl), nitric acid (HNO_3), hydrogen peroxide (H_2O_2).

Volatile Organics—hydrocarbons*, fluoroalkenes, alcohols, polychlorinated biphenyls (PCBs), ketones, aldehydes.

Free Radicals—hydroxyl, sulfate.

Solid Particles—carbonaceous and metal particles (lead, zinc, manganese, cadmium, chromium).

Formulated Commercial Products and Byproducts—insecticides, pesticides, fungicides, herbicides, solvents, coatings, chlorinated fluorocarbons, petroleum and petrochemical products, plastics, fibers.

Radioactive Substances—from weapon testing, radon in some enclosed spaces.

Pollution also affects the manufacture of certain materials and products. This is evidenced by the need for "clean rooms" in metrology standards laboratories and in the production of certain electronics materials and of component assembly operations. In addition to elaborate filtering systems, such rooms are held at a slightly positive pressure (above outside atmospheric pressure) to prevent the entry of raw air from the outside.

Air Pollution Settings

In developing preventive and remedial technologies for air pollution abatement, it is helpful to consider the fundamental settings in which pollution occurs.

(1) *Workplace pollution* usually represents the highest concentration and length of exposure to specific pollutants. It usually is the most obvious mode of air pollution and, consequently, the easiest to correct. In this setting, workers breathe specific pollutants on a day-to-day basis. Well-known and publicized examples would include: miners; chemical plant workers; farmers; textile workers; metal production workers; transportation-related personnel (who are exposed to high concentrations of carbon monoxide and other gaseous products of fossil fuel combustion as well as coal dust and other carbonaceous particles and volatile hydrocarbons); and insulation installers (who are exposed to airborne tiny particles and strands of glass, plastics, and natural minerals, such as asbestos and mica). In the industrial and manufacturing complex, which has continued to expand markedly during the last several decades, hundreds of specific examples of exposure to dangerous air pollutants could be recited.

(2) *Point-source pollution* extends beyond a specific entity within a plant, such as a particular machine or process, and comprises a source of pollution that may emanate from only one or a few particular facilities within a small area. Point-source pollutants become mixed in the atmosphere with pollutants from other sources. Hence, beyond the immediate vicinity of a given facility, sources are difficult to identify. Considerable success has been achieved by isotopic matching to specific coals and other fuels that may be burned at a given facility.

(3) *Confined-area pollution*, where there may be no natural circulation of air and no effective air purification effluent system. This situation is found in factories or mines and inside improperly ventilated garages, service stations, and vehicular tunnels.

(4) *Limited-area pollution* occurs on a small geographic scale, such as strips of land adjacent to major highways or close to a pollution point source.

(5) *Regional pollution* is pollution that occurs in the greater part of a city, valley, or basin and frequently is publicized in connection with cities like Los Angeles, London, and, in more recent years, other major cities of the world, including New York. Regional pollution is particularly affected by weather conditions, such as, for example, an inversion layer hanging

over a natural basin. When pollution occurs on this scale, the beginnings of ecological damage (to trees, plants, natural life, etc.) are seen.

(6) *International pollution* or wide-area pollution is the occurrence of massive air pollution, usually extending over many years, to the extent that the atmosphere becomes severely overloaded with pollutants—that is, the atmosphere's ability to contain pollutants is exceeded. After holding pollutants for long periods, during which time prevailing winds transport the pollutants over long distances (from one country to the next, for example), the point is reached where pollutants "drop out" and contaminate the topography below. This is the type of pollution that has been the subject of debate, for example, between Canada and the United States—where pollutants from coal-burning power plants in the Midwestern states of the United States are transported to northeastern Canadian provinces (and northeastern U.S. states as well). Such pollution damages forests and lakes. See also "Acid Rain" later in this entry.

Other examples of international pollution occur in other parts of the world, but are less well understood at this time.

(7) *Worldwide pollution* is simply an extension of wide-area pollution and encompasses emissions that essentially become mixed with the entire atmosphere of the Earth. Even though the mixing time may be quite long, it is reasonable to assume that, over time, the ultimate effects of almost *any* pollution ultimately will affect the atmosphere on a worldwide basis. For example, as the result of nuclear events, particularly those that occurred prior to the ban on nuclear weapons testing, radioactive particles could be discerned over extensive regions of Earth. Depletion of Earth's ozone layer also is exemplary of how destruction of ozone can be caused by chlorine molecules, essentially without regard to where the pollutants are released into the atmosphere. **The Energy vs. Environment Conflict**

Just as it is difficult to separate the topics of air and water pollution, so is it hard to separate the problems of pollution from energy generation and consumption. With exception of some of the nontraditional sources of energy, such as nuclear energy and the more direct utilization of solar energy (as contrasted with combustion), the needs of the earth's population for energy tend to follow a collision course with concerns over the environment. For example, until the nontraditional energy sources can be reduced to practical usage (of which economics is an important, if not scientific factor in the equation), coal, wood, biomass, and other organic fuels when burned are air polluters unless very costly measures are taken to treat the effluents. Even when the numerous chemical and electroprecipitation measures, among others, are taken, there remains the problem of increasing the carbon dioxide content of the atmosphere.

Energy has an impact on the environment by tending to worsen the environment. The environment has an impact on energy by requiring considerable energy to alleviate the degradation of environmental quality. Environmental concerns also tend to limit the energy options available. It is generally agreed that the standard of living of the technologically sophisticated and developed nations is at least partially the result of inexpensive energy. Generally, the societies in these countries have not readied themselves to very high energy costs, there being a realization that these increased costs will, by and large, come out of so-called discretionary income and, consequently, impact the standard of living in a negative way. The significance of the incremental addition of energy required to effect environmental protection is shown in Table 1.

Principal Air Pollutants

The major air pollutants as identified by a number of countries in recent years in connection with pollutant regulatory programs are: (1) Particulate matter; (2) nitrogen oxides (NO_x); (3) sulfur oxides (SO_x); (4) hydrocarbons; and (5) carbon monoxide (CO).

Particulates and Aerosols. These may be comprised of numerous mineral and organic materials and frequently result from such operations as milling, crushing, screening, grinding, and demolition operations—as well as quarries and cement plants. Soot and fly ash as well as heavy carbonaceous smoke, arising from fuel-burning operations and smudge pots, also may fall into this category of pollutants. Aerosols generally are considered to be very tiny spherical droplets of a liquid that may be as small as 0.01 micrometer in diameter. These small liquid particles and the larger liquid particles, including mists and sprays, along with dusts, permit numerous physical separating and isolating means that do not apply to gases and vapors. Recent investigations of particle size distribution of atmospheric aerosols have revealed a multimodal character, usually with a bimodal mass, volume, or surface area distribution and frequently trimodal

TABLE 1. AIR POLLUTION COSTS FOR REPRESENTATIVE TECHNOLOGIES (10^6 BTU/TON OF PRODUCT)*

Process option	Primary energy source	Process energy (10^6 Btu)	Air pollution control energy (10^6 Btu)	Percent of total for air pollution control
Glassmaking				
side port regenerative furnace	natural gas	7.0	0.57	7.5
side port regenerative furnace with preheat of charge	natural gas	5.7	0.37	6.0
electric furnace	electric power	8.2	0.03	0.3
coal gasification	coal	8.6	0.9	9.5
direct coal firing	coal	7.0	0.65	8.5
Cement				
long kiln (conventional)	oil	5.6	0.07	1.2
suspension preheater with long kiln	oil	4.2	0.05	1.2
fluid bed	oil	5.0	0.1	2.0
Copper Production				
roast-reverb smelting (conventional)	gas, oil, or coal	22.0	5.3	19.4
flash smelting (90–95% sulfur recovery)	oil	10.0	7.8	43.8

* 10^6 Btu = 252×10^3 Calories

surface area distribution near sources of fresh combustion aerosols. These modes are attributed to the following factors: (a) the course mode (2 micrometers and greater) is formed by relatively large particles generated mechanically or by evaporation of liquid from droplets containing dissolved substances; (b) the nuclei mode (0.03 micrometer and smaller) is formed by condensation of vapors from high-temperature processes or by gaseous reaction products; and (c) the intermediate or accumulation mode (0.1 to 1.0 micrometer) is formed by coagulation of nuclei. This evidence indicates that atmospheric particles tend to form a stable aerosol having a size distribution ranging from 0.1 to 1.0 micrometer in general. However, larger and smaller particles occur. The larger particles (greater than 1.0 micrometer) settle out, and the very fine particles (smaller than 0.1 micrometer) tend to agglomerate to form larger particles which remain suspended. The nuclei mode tends to be highly transient and is concentration-limited by coagulation with both other nuclei and also particles in the accumulation mode. Therefore, the particulate content of a source emission and the ambient air can be viewed as composed of two portions, i.e., the settleable and the suspended.

Control of emissions in both size ranges is required because both settleable and suspended atmospheric particulates have deleterious effects upon the environment. Significantly, it is the suspended particles from an upper level of about 2 to 5 micrometers and smaller that health experts consider most harmful to humans because particles of this size have been found to penetrate the body's natural defense mechanisms and reach most deeply into the lungs. Efforts to control particulate emissions to the atmosphere have historically been geared to maximizing the efficiency of control (by weight) of the overall particulate loading emanating from the generating process. This work has led to the empirical understanding that present systems can perform with high control efficiencies down to a particle size of about 2–3 micrometers. But, below this size, the control efficiency appears to decrease with decreasing particle size to a minimum between 1.0 and 0.1 micrometer; and then increases again. This relationship of control efficiency and particle size is highly significant to any strategy for controlling particulate air pollution, and serves to underscore the need to adequately measure and evaluate both ambient particulate air pollution and source emissions.

The particle diameters of some substances commonly found in the atmosphere or important in various manufacturing operations are given in

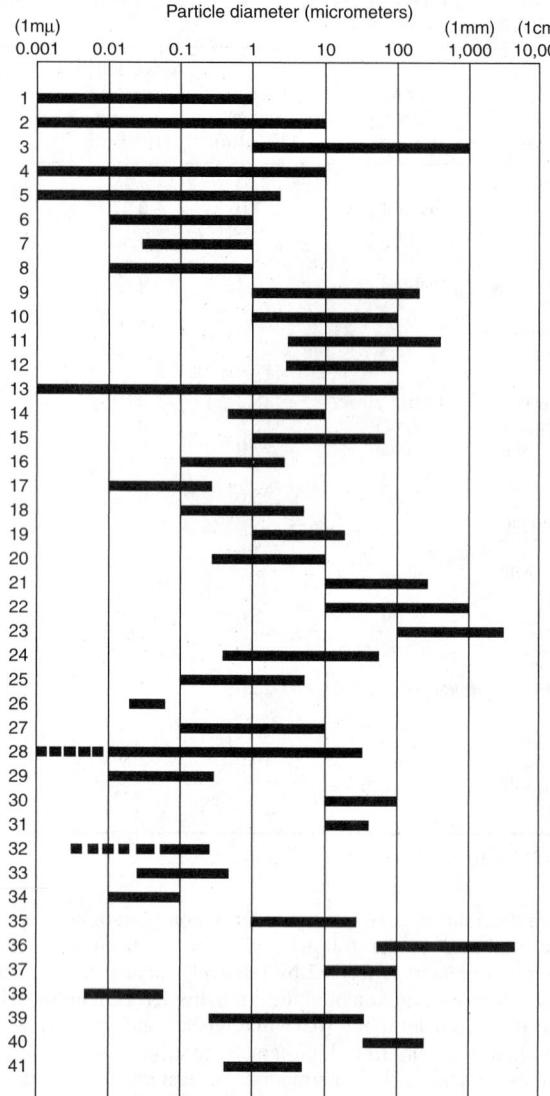

Fig. 1.   Range of diameters of particles and particle dispersoids of substances commonly encountered in the atmosphere or associated with various manufacturing operations. (*Stanford Research Institute*)

Legend:

1. Gas dispersoids—solid fumes
2. Gas dispersoids—liquid mists
3. Gas dispersoids—solid dusts
4. Gas dispersoids—liquid sprays
5. Common atmospheric dispersoids
6. Smoke—rosin
7. Smoke—oil
8. Smoke—tobacco
9. Fly ash
10. Coal dust
11. Pulverized coal
12. Cement dust
13. Metallurgical dusts and fumes
14. Insecticide dusts
15. Milled flour
16. Fumes—ammonium chloride
17. Fumes—zinc oxide
18. Fumes—alkali
19. Sulfuric acid concentrator mist
20. Contact sulfuric acid mist
21. Ore flotation mist

22. Ground limestone fertilizer
23. Beach sand
24. Ground talc
25. Paint pigments
26. Colloidal silica
27. Spray dried milk
28. General atmospheric dust
29. Carbon black
30. Plant pollen
31. Plant spores
32. Nuclei (Aitken)
33. Nuclei (Sea salt)
34. Nuclei (Combustion)
35. Nebulizer drops
36. Hydraulic nozzle drops
37. Pneumatic nozzle drops
38. Viruses
39. Bacteria
40. Human hair (diameter)
41. Most severely lung-damaging Dust

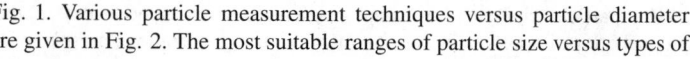

Fig. 1. Various particle measurement techniques versus particle diameter are given in Fig. 2. The most suitable ranges of particle size versus types of

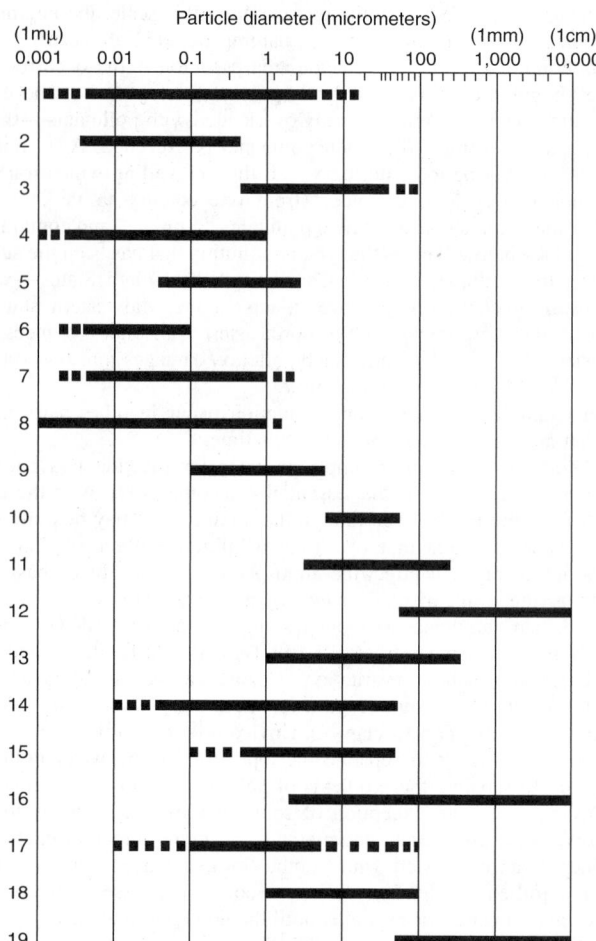

Fig. 2.   Various particle measurement techniques versus particle diameter. (*Stanford Research Institute*)

Legend:

1. Electron microscope
2. Ultramicroscope
3. Microscope
4. Ultracentrifuge
5. Centrifuge
6. X-ray diffraction
7. Adsorption
8. Nuclei counter
9. Impingers
10. Electroformed sieves

11. Elutriation
12. Sieving
13. Sedimentation
14. Turbidimetry
15. Permeability
16. Scanners
17. Light scattering
18. Electrical conductivity
19. Visible to eye

gas cleaning equipment are given in Fig. 3. Modern instrumental techniques are shown in Fig. 4.

Source Identification of Airborne Particles. In recent years, ingenious methods of identifying the point source of airborne particles have been developed. If not specific point sources of pollution, rather small regional areas can often be identified. Source of particles can be important for numerous reasons, including the enforcement of regulation and also in sorting out, for example, the various distant sources that contribute to acid rain pollution.

As reported by Olmez and Gordon (University of Maryland), the concentration pattern of rare earth elements on fine airborne particles (less than 2.5 micrometers in diameter) is distorted from the crustal abundance pattern in areas influenced by emissions from oil-fired plants and refineries. The ratio of lanthanum (La) to samarium (Sm) is often greater than 20 (crustal ratio is less than 6). The unusual pattern apparently results from the distribution of rare earths in zeolite catalysts used in refining oil. Oil industry emissions have been found to perturb the rare earth pattern even in very remote locations, such as the Mauna Loa Observatory in Hawaii.

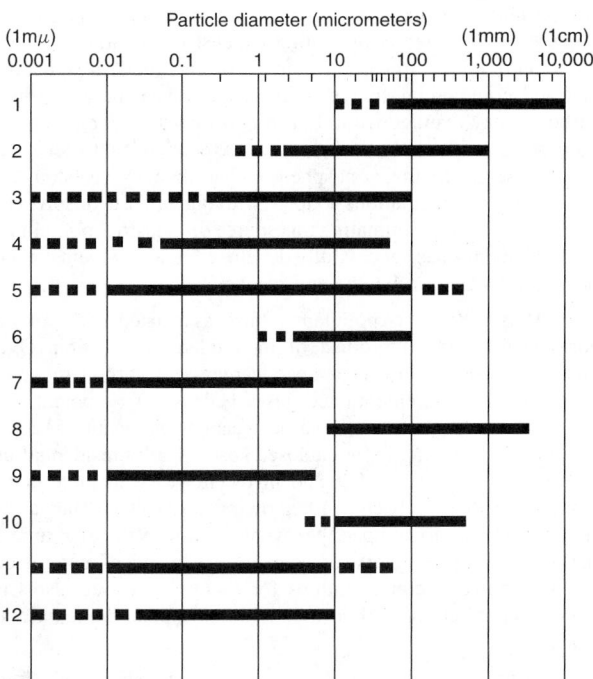

Fig. 3.   Types of gas cleaning equipment versus particle size diameters. (*Stanford Research Institute.*)

Legend:

1. Settling chambers
2. Centrifugal separators
3. Liquid scrubbers
4. Cloth collectors
5. Packed beds
6. Plain air filters
7. High-efficiency air filters
8. Impingement separators
9. Thermal precipitators
10. Mechanical separators
11. Electrical precipitators
12. Ultrasonic methods

Rare earth ratios are probably better for long-range tracing of oil emissions than vanadium (V) and nickel (Ni) concentrations because the ratios of rare earths on fine particles are probably not influenced by deposition and other fractionating processes. Emissions from oil-fired plants can be differentiated from those of refineries on an urban scale by the much smaller amounts of V in the latter. Pb in urban areas originates mainly from combustion of leaded gasoline. Arsenic (As), selenium (Se), and other chalcophile elements are usually associated with coal-fired plants or sulfide ore smelters (or both). In addition to the use of receptor models on an urban scale, they also have been used on a global scale to identify sources of Arctic haze, based upon manganese (Mn) and V ratios.

R.W. Shaw (U.S. Army Research Office, Research Triangle Park, North Carolina) commented that in analyzing particle samples, one must always consider that a particle sample could be produced by various sources. Chemical-element balance copes with this complication by positing (on the basis of the known ratios) several different elemental concentrations that could be generated by the suspected sources. It then finds the "mix" that best fits the actual concentrations on the collected samples. Such manipulations might reveal, for example, that 80% of the fine fraction of a sample is a byproduct of coal combustion and 15% comes from motor vehicles. Or it might show that 94% of the lead in a specimen is from motor vehicles, 4% from the burning of refuse, and 1% from the burning of coal.

Smoky Mountain Haze. A rather thorough study of what contributes to the increasing haze that exists in the Great Smoky Mountains National Park in Tennessee was reported by Shaw in 1987. The overall purpose of the study was to determine the contributions of natural emissions, as from vegetation, and motor vehicle and industrial emissions. The researchers also conducted surveys in the Georgia region (former Soviet block), the latter highly regarded for clear air and massive areas of evergreen forests. The Smoky Mountains haze is ascribed to sulfate particles in the air. It was tentatively concluded that the sulfate is a byproduct of coal burned at distant power plants. Surprisingly, the concentration of coal-derived particles in the Smoky Mountains was not much lower than concentrations found at several

sites in an industrialized city, St. Louis, Missouri. Sulfate particles also were found in Georgia, but considered to be much older and the particles contained thirty times more sulfur than in sulfur dioxide gas. These studies, however, did not make a direct link with remote coal-burning plants. This was accomplished later in a 16-month study in the Ohio River valley, with sampling stations located in rural Kentucky, Indiana, and Ohio, far away from cities, roads, and power plant smoke plumes. About 50% of the fine-particle mass was found to be sulfate. The total concentration nearly equaled that found in the industrialized cities of the regions. The link was confirmed by a consistent association between sulfate concentration and the trace element selenium. No other probable sources of selenium exist in or near the Ohio River valley. The investigators (R.W. Shaw, et al.) concluded in their report that there is little doubt that fuel combustion is the main source of acid deposition and particulate fallout.

Shaw also describes two techniques that now make it possible to analyze particles without removing them from collection filters. To determine the mass of a sample, technicians insert the particle-laden filter between a source that emits beta particles and a detector that counts them. As the mass increases, the number of particles that can penetrate the sample decreases. To determine the atomic elements in a specimen, laboratory workers may also separately carry out x-ray fluorescence spectroscopy. X-rays passed through the sample cause each element to emit characteristic x-rays. The energy levels of the rays reveal the identity of the elements: the intensity of the x-rays (number emitted) reflects the concentrations.

Sulfur Oxides. (Sulfur dioxide, SO_2, and sulfur trioxide, SO_3). The primary sources of these oxides SO_x are sulfur-bearing fuels—as used for heat and power, both industrially and residentially. Chemical and metallurgical plants of various kinds also emit SO_x as the result of processing activities, such as the manufacture of sulfuric acid, the roasting of ores, etc. In order of decreasing pollution, the fossil fuels contributing to SO_x pollution are: (a) Untreated coal; (b) untreated petroleum fuels, particularly those originating from so-called sour crude oils; and (c) natural gas. Thus, the preference for natural gas by many large fuel users, such as power plants. With only small variations in the cost of raw fossil fuels, there was an advantage in burning a naturally low-sulfur fuel as contrasted with installing elaborate SO_x removal or reduction systems. But, with a rapidly lessening natural gas supply and accompanying higher costs, it has become economically attractive to pay more for desulfurized coal and petroleum fuels, as well as to install SO_x abatement equipment. The allowable sulfur content of oil and coal fuels varies from one community to the next, ranging from 0.50% by weight or less, up to 4% and slightly higher. Such regulations usually take into consideration new versus old fuel-burning equipment, the incidence of serious pollution in a given area, as well as economic impact and practicability. Logically, for some years to come, such regulations must represent a compromise of social, economic, and technological factors. The chemical nature of the oxides of sulfur is given in the entry on **Sulfur**. Treatment of SO_x effluents is described later in this article.

Nitrogen Oxides. These compounds result from all fossil-fuel combustion processes where air is used as the oxidant. Oxygen from the air and nitrogen combine at combustion flame temperatures to form nitric oxide, NO, according to $N_2 + O_2 \longleftrightarrow 2NO$. The rate at which NO is formed and decomposed depends largely upon temperature. For the majority of stationary combustion processes, there is too short a residence time for the full oxidation of NO to NO_2, an estimated average of only 5 to 10% of this reaction occurring. Thus, it is important to observe that although NO_x emissions generally are given as "equivalent NO_2," the predominant NO_x in combustion gases is NO. Several factors affect the generation of NO_x pollutants. Factors which tend to decrease NO_x emissions are: (a) decrease in excess air for combustion; (b) decrease in preheat temperature; (c) decrease in the heat-release rate; (d) increase in the heat-removal rate; (e) increase in back-mixing; and (f) decrease in fuel nitrogen content. With exception of very large installations, coal appears to generate more NO_x than oil; and oil generates more NO_x than natural gas. Thus, as with SO_x, natural gas is the preferred fuel when properly burned to minimize NO_x.

The major sources of NO_x are the large fuel-burning operations as previously mentioned, automotive vehicles, and certain chemical plants, notably nitric acid manufacturing facilities. Research to date indicates that effective steps toward reducing the overall emission of NO_x can be effected from stationary combustion sources by: (a) using low excess air firing; (b) providing for two-stage combustion; (c) utilizing flue-gas recirculation; and (d) using water injection. These objectives, when reduced to terms of hardware, mean changes in the configuration, location, and spacing of burners, and the kinds of firing and combustion techniques used. Two-stage combustion is defined as firing all fuel below stoichiometric amounts of primary air in a first stage of combustion, followed by injecting air in a second stage, whereupon burnout of the fuel is completed. There is removal of heat between the two stages. The formation of NO in the first stage is limited because the available oxygen for combustion with nitrogen is limited. The removal of heat between stages kinetically limits the formation of NO when excess air is added to the second stage. Experience shows that a 90% reduction in NO_x emission can be achieved in this manner. By recirculating flue gas, both the peak flame temperature and oxygen content are lowered. Injecting low-temperature steam or water also provides a diluting effect. Although probably of limited value for electric utility boilers (because thermal efficiency is lowered), the water-injection technique may be one of the better ways to reduce NO_x emissions in connection with internal-combustion engines of the stationary type. The situation in the case of internal combustion engines for automotive vehicles is considerably more complicated—there is a wide range of loads on such engines, high performance is required at all loading conditions, and the combustion process from fuel to exhaust must be kept simple and hence low-cost. As of the early 1980s, there remained some differences in opinion as regards the use of catalytic converters to remove NO_x from automobile exhausts. See also **Catalytic Converter (Internal Combustion Engine)**; **Combustion (Fuels)**; and **Petroleum**. The chemical characteristics of NO_x are described under **Nitrogen**.

Hydrocarbons. Extensive pollution of air occurs from the introduction of hydrocarbons either from (a) the incomplete combustion of hydrocarbon fuels in both stationary and vehicular engines; or (b) from paint spraying, solvent cleaning, printing, chemical and metallurgical, and other plants that use various fluids that have a high hydrocarbon content. Engine design and tuning are major factors in abating exhaust hydrocarbons. The intent is to fully burn the hydrocarbon content of the fuel. In a major city, industrial and commercial sources of organic solvent fumes (principally hydrocarbons) may average from 300 to 600 tons/day. For years, without legal restrictions, some operators found it more economical to permit vapor-laden air to escape to the atmosphere rather than to invest in solvent recovery equipment. Regulations coupled with higher costs of solvents have gone a long way toward eliminating this source of industrial pollution. Also, the chemical industry has successfully developed newer solvents, which are less volatile, and easier to handle and recover.

Carbon Monoxide. This pollutant is also associated with combustion operations, again being a product of incomplete combustion. Over the years, there has been a much greater awareness of carbon monoxide as a pollutant than the aforementioned gases because of its potent toxicity, dramatized by numerous deaths in earlier years as the result of keeping an automobile engine running in an enclosed space. Faulty residential heaters continue to take their toll of life and in recent years an important killer is the outdoor grill or hibachi with glowing coals taken into a camper or cabin as a means to temper the evening chill. Vehicular tunnel and large parking garage designers, of course, have practiced careful control over carbon monoxide concentrations for many years. See also **Carbon Monoxide**. The effects of carbon monoxide on human beings are shown in Fig. 5.

Other Pollutants. Some of these are gases; others fall into the particulate category. Of considerable importance are beryllium dust—very toxic and arising from ore preparation and metalworking operations, but of relatively limited extent because this metal is not a common structural

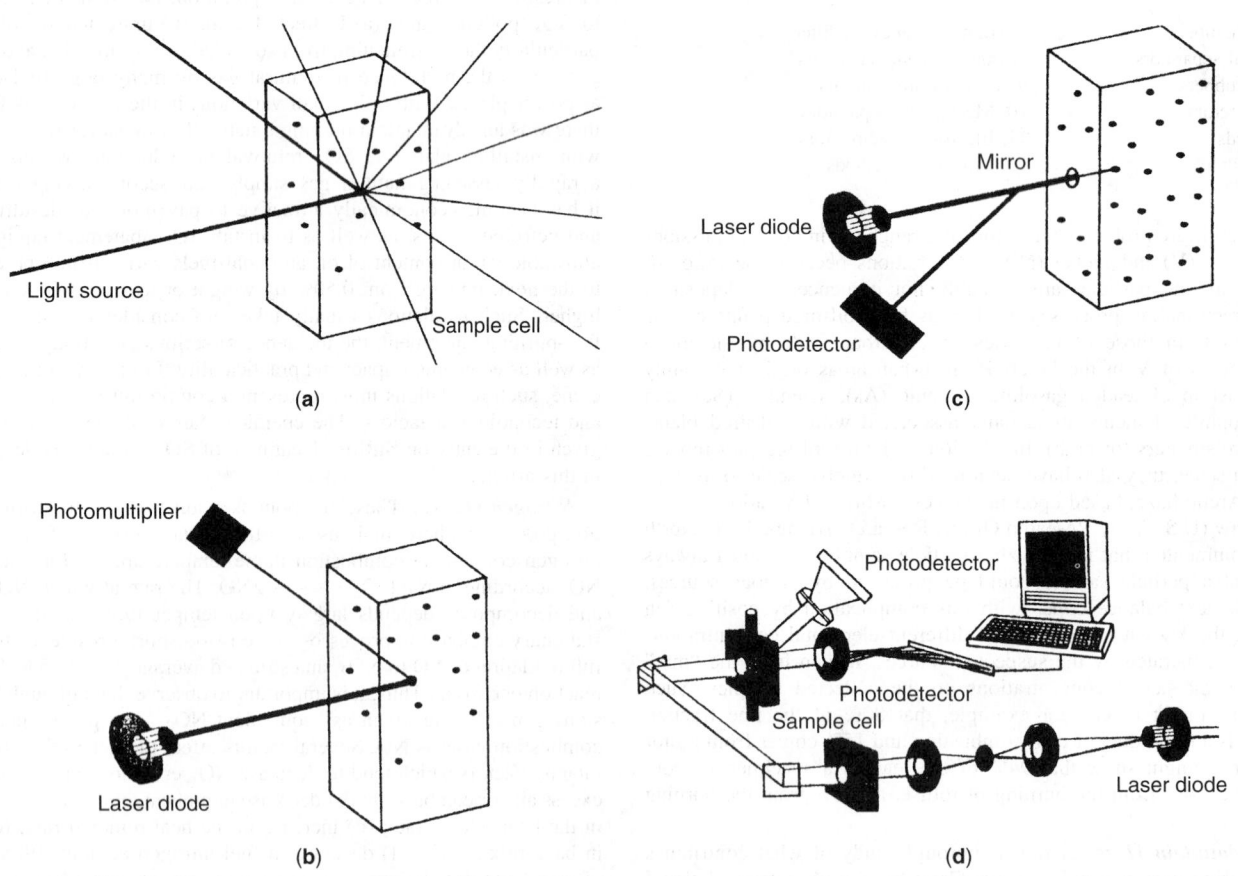

Fig. 4.   Instrumentation for measuring air pollution particle dimensions. (**a**) Some fine-particle-size analyzers determine particle size by measuring the Doppler shift of light as it is scattered by moving particles. Smaller particles move faster, causing a greater Doppler shift in the light that they scatter. (**b**) In some conventional analyzers, light passes completely through an extremely dilute suspension and scatters in all directions. The detector measuring the Doppler shifts of the scattered light has no high-level signal to reference against the resulting low-level signal. The resulting low-level signals require amplification from photomultipliers, which can introduce noise errors. (**c**) In instrument shown (*Microtrac—Leeds & Northrup*), light travels to the sample via an optical wave guide. A mirror reflects some of the light, creating a high-level reference signal. Moving particles backscatter the light penetrating the mirror. The instrument combines the reflected and backscattered light to create a high-level signal sufficiently strong to be fed directly to a solid-state photodetector with a requirement for amplification. (**d**) Infrared light, emitted from a long-lived solid-state laser diode, scatters when it hits particles in the sample cell. Multi-train optics direct all scattered light onto solid-state photodetectors, which measure scattered-light angles and send signals to a computer control module. (*Leeds & Northrup*)

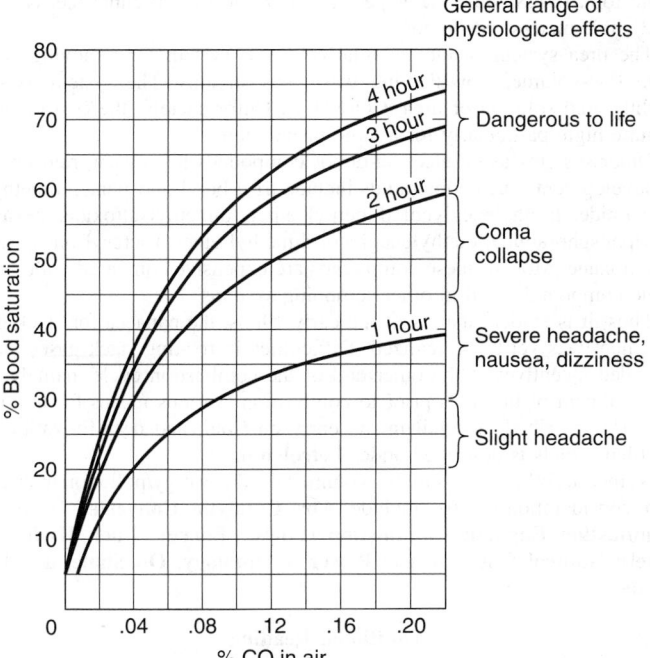

Fig. 5. Effects of carbon monoxide on human beings. This chart can be considered only as a general guide because the percent of CO blood saturation will vary with exertion, excitement, fear, depth of respiration, anemia, and general physical condition of the individual

material. Other contaminants include fluorides; metal fumes, such as arsenic, lead, and zinc; organic phosphates, notably from crop dusting and spraying; numerous kinds of organic vapors, including chlorinated hydrocarbons and hydrofluorocarbons (used in aerosol containers but suspect in connection with altering the ozone content of the upper atmosphere—see also **Aerosols**); radioactive fallout, such as ^{14}C, ^{137}Ce, and ^{90}Sr, arising from nuclear-device testing–no longer the major concern it was prior to the nuclear test ban accepted by most nations; and uranium dust. See also **Asbestos**; **Beryllium**; and **Pneumokonioses**.

Acid Rain

This term is used almost exclusively in connection with the effect of airborne pollutants on the natural health of forests and lakes. The term was coined by Angus Smith several years ago, when he referred to the effect of industrial emissions on precipitation over the British midlands. Unfortunately, the term does *not* fully or accurately describe what may be occurring in certain forests and lakes.

The topic of acid rain during the 1980s and early 1990s was one of controversy and of incomplete answers in terms of official policy and science—after an expenditure of many millions of dollars. In 1980, the National Acid Precipitation Assessment Program (NAPAP) was established and subsequently consumed thousands of scientific research hours and costly field investigations, including the use of numerous helicopter expeditions to northwestern mountain and lake areas of the United States and Canada. Thousands of hours of computer power were consumed.

Segments of the jigsaw puzzle were worked out in excruciating detail, but only a few of the larger pieces of the puzzle were put together. Bottom-line conclusions and recommendations needed for governmental and private forestry operators were not forthcoming. A 6,000-page report was generated by NAPAP. But, in terms of a summary, the findings can be condensed to the few following points:

1. Acid rain has adversely affected aquatic life in about 10 percent of eastern lakes and streams.
2. Acid rain has contributed to the decline of red spruce at high elevations by reducing that species' cold balance.
3. Acid rain has contributed to erosion and corrosion of buildings and materials.

4. Acid rain and related pollutants, especially fine sulfate particles, have reduced visibility throughout the Northeast and in parts of the West.

Ellis Cowling (Ecologist, North Carolina State University) observed in early 1991, "You can say no symptoms were found, and we looked hard"—but that is different from saying no problem exists. A. Johnson (University of Pennsylvania) has observed, "It is the marginal ecosystems, the tenuous ones on the edge, that are affected by acid rain at current levels. I think control measures are justified. They will probably go a small way toward restoring some of the ecosystems that have been altered. And they will largely prevent more harm from occurring." See also **Acid Rain**.

Air Treatment Methodologies

Numerous techniques have been used to reduce, if not fully eliminate, various forms of air pollution over the years. For example, in connection with particulates, see entry on **Electrostatic Precipitator**. Most methods, however, fall into what might be termed the category of wet-scrubbing processes. As of the early 1980s, there were nearly 100 such processes on the market. Progress in this field has been steady, but not characterized by major breakthroughs. The concentration of sulfur dioxide in stack gases emitted by steam generation plants is usually in the range of only 400 to 2,000 parts per million (ppm). However, the volume of gases produced by the utility industry results, for example, in the liberation of large tonnages of sulfur dioxide into the atmosphere.

Chemical scrubbing systems for SO_2 absorption fall into two broad categories: (a) Disposable systems; and (b) regenerative systems. Typical of systems in use for a number of years are those that use an aqueous slurry of an insoluble calcium compound, which can be discarded after use. Disposable SO_2-removal systems use aqueous slurries of finely ground materials, such as lime, limestone or dolomite, to produce a mixture of insoluble sulfites and sulfates. On passing through the scrubber, SO_2 from the waste gas dissolves to form sulfurous acid: $SO_2 + H_2O \longrightarrow H_2SO_3$. The dissolved SO_2 reacts with the lime, $Ca(OH)_2$ or limestone, $CaCO_3$, to form insoluble calcium sulfite, $CaSO_3$: $Ca(OH)_2 + H_2SO_3 \longrightarrow CaSO_3 + 2H_2O$; $CaCO_3 + H_2SO_3 \longrightarrow CaSO_3 + H_2O + CO_2$. Unfortunately, SO_2 is less soluble (and hence less easily removed by scrubbing) in slightly acid solutions, so that it is extremely difficult in practice to operate a calcium-based system in such a manner that SO_2 removal is maximized while the quantities of calcium chemicals are minimized in order to approach stoichiometric conditions. As calcium-based slurry systems are usually operated at pH 6–10, disposal of the very large masses of used slurry presents a major problem. A typical power station using a calcium-based SO_2-removal slurry system will produce several hundred tons of spent slurry per day. A further disadvantage of lime or limestone systems is their marked tendency to precipitate insoluble calcium salts inside the scrubber. Unless the scale is removed, the scrubber shortly becomes inoperable.

Although chemically analogous to calcium-based systems, magnesium-based scrubbing systems possess several advantages. A slurry of finely divided magnesium hydroxide, $Mg(OH)_2$, is pumped through the scrubber to remove SO_2 from stack gases. Insoluble magnesium sulfite, $MgSO_3$, is formed: $Mg(OH)_2 + SO_2 \longrightarrow MgSO_3 + H_2O$. Hydrated magnesium sulfite, $MgSO_3 \cdot 6H_2O$, can be disposed of as such, although it is usually heated to produce a rich stream of SO_2 and regenerate MgO. The SO_2 is compressed, liquefied, and stored in tanks for market; or catalytically oxidized to sulfur trioxide, SO_3, and treated with water to produce sulfuric acid, H_2SO_4. Alternatively, the SO_2 is mixed with hydrogen sulfide, H_2S, to produce elemental sulfur by the Claus process: $SO_2 + 2H_2S \longrightarrow 3S + 2H_2O$. Absorption efficiency of SO_2 attainable in a magnesium system is good, and removal efficiencies from 90 to 95% have been claimed without difficulty at reasonable liquor recirculation and MgO feed rates. As with calcium systems, serious scaling occurs due to build-up of insoluble $MgSO_3$.

Scrubbing solutions containing sodium (or other alkali metals) compounds have been extensively studied for removal of SO_2. Justification for the use of sodium compounds includes: (a) complete solubility in water with no formation of scale; and (b) simple reactions with SO_2 : $Na_2CO_3 + SO_2 \longrightarrow Na_2SO_3 + CO_2$; $2NaHCO_3 + SO_2 \longrightarrow Na_2SO_3 + 2CO_2 + H_2O$; $2NaOH + SO_2 \longrightarrow Na_2SO_3 + H_2O$. In one commercial process, a scrubbing solution of sodium sulfite is used, which readily absorbs SO_2 to form the bisulfite: $Na_2SO_3 + H_2O + SO_2 \longrightarrow 2NaHSO_3$. In practice, only a portion of the Na_2SO_3 is converted to $NaHSO_3$ because the SO_2 absorption efficiency diminishes as the bisulfite concentration increases. The resulting solution is heated to decompose the bisulfite and thermally regenerate the sulfite. The gaseous SO_2 is compressed, liquefied and handled as previously mentioned under the magnesium system.

Ammonia-based chemicals appear to have some advantages over sodium systems. They are less costly, and regeneration by conventional means is possible, with the byproduct, ammonium sulfate, a marketable commodity for fertilizer.

Solutions containing ammonium sulfate, with or without the addition of ammonium hydroxide, have been widely used. The ammonium system can operate effectively only within a pH range of 4.0 to 7.0. As the pH value increases above 7.0, progressively more gaseous ammonia is liberated and this reacts in the gaseous phase with water vapor and SO_2 to produce a dense aerosol (white plume) which is difficult for scrubbers to remove. In an ammonia system, in order to regenerate the scrubbing solution, the ammonium bisulfite and sulfite mixture is heated to drive off gaseous SO_2 : $2NH_4HSO_3 \longrightarrow (NH_4)_2SO_3 + H_2O + SO_2$. Alternatively, the ammonium bisulfite/sulfite mixture can be treated with calcium hydroxide. Gaseous ammonia is evolved and trapped in water, which is then recirculated to the scrubber.

Sodium citrate also is used in an SO_2 removal system. The solution is buffered at a pH 3.0–3.7 by the citrate ion, sulfur dioxide is absorbed, and an equilibrium mixture of sodium bisulfite and citric acid is produced.

$$
\begin{array}{c}
CH_2COONa \\
| \\
HO-C-COONa \\
| \\
CH_2COONa
\end{array}
+ 3\,SO_2 + 3\,H_2O \longrightarrow
\begin{array}{c}
3\,NaHSO_3 \\
+ \\
CH_2COOH \\
| \\
HO-C-COOH \\
| \\
CH_2COOH
\end{array}
$$

The bisulfite leaving the scrubber is then reduced with gaseous hydrogen sulfide, which precipitates elemental sulfur by a modified Claus reaction:

$$
3\,NaHSO_3 +
\begin{array}{c}
CH_2COOH \\
| \\
HO-C-COOH \\
| \\
CH_2COOH
\end{array}
+ 6\,H_2S
$$

$$
\longrightarrow
\begin{array}{c}
CH_2COONa \\
| \\
HO-C-COONa \\
| \\
CH_2COONa
\end{array}
+ 9\,H_2O + 9\,S
$$

A formate system uses two reactions involving potassium formate, HCOOK, which is regenerated after recovery of elemental sulfur. This method has the advantage over other wet scrubbing methods in that no precipitation of insoluble intermediates occurs at any stage of the process. Disadvantages include the need to heat K_2CO_3 solution, at high temperature and pressures, with carbon monoxide to regenerate the potassium formate. The energy requirements thus are high.

While it has been demonstrated that solutions of NaOH, $NaHCO_3$, and Na_2CO_3 are effective for SO_2 removal, these solutions are not effective for removal of mixtures of NO and NO_2, particularly when the gas stream velocities are reasonably high. Under conditions where from 95 to 99% SO_2 may be removed, the solutions may only be effective in removing from 5 to 15% NO_x. The fundamental difference between SO_2 and NO_x removal is that NO_x gases (mixtures of NO, N_2O_3, NO_2) are approximately 1,000 to 2,000 times less soluble in water than SO_2 at any given temperature. It has been found that conventionally designed wet scrubbers often do not provide sufficient liquid-to-gas contact surface areas or residence times to permit the NO_x to dissolve in the scrubbing solution. Consequently several stages may be required. Concentrations of NO_x in the range of 20,000 to 40,000 ppm require from 6 to 12 stages.

If no SO_2 is present, a sodium-based process may be used to remove NO_x efficiently. In the Neville-Krebs process (patent applied for), removal efficiencies of 60 to 90% have been achieved from gas streams containing up to 1,500 to 2,000 ppm of NO_x passing through a 3-stage scrubber (1 cubic foot; 0.028 cubic meter per stage) at 150 to 500 cubic feet (4.2 to 14.2 cubic meters) per minute.

The urea system is another relatively new system for removing NO_x from low-volume, slow-flowing waste gas streams. The system uses a slightly acid solution of urea, $CO(NH_2)_2$. Unfortunately, the cost of urea is quite high, particularly for a large installation.

Other systems using electron-donor compounds have been tried or are in development. Such compounds include tri-*n*-butyl phosphate, dimethylformamide, triethyleneglycol dimethyl ether, dimethylsulfoxide, hexamethylphosphoramide, diethyleneglycol dimethyl ether, tricresyl phosphate, and dioxane. Most of these compounds are expensive compared with inorganic compounds used in other scrubbing systems.

Thus, it is evident that as of the early 1980s, the panacea for stack gas treatment was yet to be realized. Difficulties in treating stack gases have provided incentives at the other end of the combustion cycle, namely, in the treatment of the fuels prior to combustion. Various means for treating coal are described in detail in the entry on **Coal**; and desulfurization of petroleum fuels is described under **Petroleum**.

Numerous other entries in this volume take the energy/pollution interface into consideration. These include: **Air**; **Catalytic Converter (Internal Combustion Engine)**; **Combustion (Fuels)**; **Energy**; **Fuel**; **Hydrogen (Fuel)**; **Natural Gas**; **Nuclear Power Technology**; **Oil Shale**; and **Tar Sands**.

Additional Reading

Abelson, P.H.: "Asbestos Fiasco," *Science*, 1017 (March 2, 1990).

Abelson, P.H.: "New Technology for Cleaner Air," *Science*, 793 (May 18, 1990).

Abelson, P.H.: "Excessive Fear of PCBs," *Science*, 361 (July 26, 1991).

Alley, E.R., L.B. Stevens, and W.L. Cleland: *Air Quality Control Handbook*, The McGraw-Hill Companies, Inc., New York, NY, 1998.

Arya, S.P.: *Air Pollution Meteorology and Dispersion*, Oxford University Press, Inc., New York, NY, 1998.

Ashmore, M., L. Emberson, and F. Murray: *Air Pollution Impacts on Crops and Forests: A Global Assessment*, World Scientific Publishing Company, Inc., River Edge, NJ, 2001.

Ayres, J., R. Richards, and R. Maynard: *Air Pollution and Health*, Vol. 3, World Scientific Publishing Company, Inc., River Edge, NJ, 2002.

Barner, R.A. and A.C. Lasaga: "Modeling the Geochemical Carbon Cycle," *Sci. Amer.*, 74 (March 1989).

Baron, P.A. and K. Willeke: *Aerosol Measurement: Principles, Techniques, and Applications*, 2nd Edition, John Wiley & Sons, Inc., New York, NY, 2001.

Barth, H.G.: "Particle Size Analysis," *Analytical Chemistry*, (June 15, 1991).

Bell, N. and M. Treshow: *Air Pollution and Plant Life*, 2nd Edition, John Wiley & Sons, Inc., New York, NY, 2002.

Benedick, R.E.: *Ozone Diplomacy: New Directions in Safeguarding the Planet*, Harvard University Press, Cambridge, MA, 1997.

Bohn, H.: "Consider Biofiltration for Decontaminating Gases," *Chem. Eng. Progress*, 34 (April 1992).

Boss, M.J. and D.W. Day: *Air Sampling and Industrial Hygiene Engineering*, Lewis Publishers, Boca Raton, FL, 2000.

Boubel, R.W., D.L. Fox, and D.B. Turner: *Fundamentals of Air Pollution*, 3rd Edition, Academic Press, San Diego, CA, 1994.

Brimblecombe, P.: *Air Composition and Chemistry*, 2nd Edition, Cambridge University Press, New York, NY, 1995.

Clement, R.E.: "Environmental Analysis," *Analytical Chemistry*, 270T (June 15, 1991).

Colls, J.: *Air Pollution: An Introduction*, Chapman Hall, New York, NY, 1996.

Conrad, J.: "An Acid-Rain Trilogy," *American Forests*, 21 (November–December 1987).

Cordasco, E.M., C. Zenz, and S.L. Demeter: *Environmental Respiratory Diseases*, John Wiley & Sons, Inc., New York, NY, 1997.

Crawford, M.: "Scientists Battle Over Grand Canyon Pollution," *Science*, 911 (February 23, 1990).

Davenport, G.B.: "Understand the Air-Pollution Laws that Affect Chemical Process Industries Plants," *Chem. Eng. Progress*, 40 (April 1992).

de Nevers, N.: *Air Pollution Control Engineering*, 2nd Edition, The McGraw-Hill Companies, Inc., New York, NY, 1999.

Downing, T.M.: "Preparing for New Smokestack Monitoring Regulations," *Instruments and Control Systems*, 47 (February 1992).

Ebert, L.B.: "Is Soot Composed Predominantly of Carbon Clusters?" *Science*, 1469 (March 23, 1990).

Fulkerson, W., R.R. Judkins, and J.K. Sanghvi: "Energy from Fossil Fuels," *Sci. Amer.*, 128 (September 1990).

Graedel, T.E. and P.J. Crutzon: "The Changing Atmosphere," *Sci. Amer.*, 58 (September 1989).

Hall, J.V., et al.: "Valuing the Health Benefits of Clean Air," *Science*, 812 (February 14, 1992).

Hesketh, H.E.: *Air Pollution Control: Traditional and Hazardous Pollutants,* CRC Press, LLC., Boca Raton, FL, 1996.

Hobbs, P.V. and L.F. Radke: "Airborne Studies of the Smoke from the Kuwait Oil Fires," *Science,* 987 (May 15, 1992).

Hoffman, D.J.: "Increase in the Stratospheric Background Sulfuric Acid Aerosol Mass in the Past 10 Years," *Science,* 996 (May 25, 1990).

Holdren, J.P.: "Energy in Transition," *Sci. Amer.,* 156–163 (September 1990).

Holgate, S.T., J.M. Samet, R.L. Maynard, and H.S. Koren: *Air Pollution and Health,* Academic Press, Inc., San Diego, CA, 1999.

Hutterman, A. and D. Godbold: *Effects of Acid Rain on Forest Processes,* John Wiley & Sons, Inc., New York, NY, 1994.

Kennedy, I.R.: *Acid Soil and Acid Rain: Research Studies in Botany and Relate Applied Fields,* 2nd Edition, John Wiley & Sons, Inc., New York, NY, 1992.

Koenig, J.Q.: *Health Effects of Ambient Air Pollution: How Safe Is the Air We Breathe?* Kluwer Academic Publishers, Norwell, MA, 2000.

Little, C.E.: "The California X-Disease," *Amer. Forests,* 32 (July 8, 1992).

Liu, D.H. and B.G. Liptbak: *Air Pollution,* Lewis Publishers, Boca Raton, FL, 1999.

Lyons, C.E.: "Environmental Problem Solving: The 1987–88 Denver Brown Cloud Study," *Chem. Eng. Progress,* 6171 (May 1990).

Majewski, M.S., P.D. Capel, and R.J. Gilliom: *Pesticides in the Atmosphere: Distribution, Trends, and Governing Factors,* CRC Press, LLC., Boca Raton, FL, 1999.

Matthews, S.W. and J.A. Sugar: "Is Our World Warming? Under the Sun," *National Geographic,* 66 (October 1990).

Mohnen, V.A.: "The Challenge of Acid Rain," *Sci. Amer.,* 30 (August 1988).

Nazaroff, W.W. and L. Alvarez-Cohen: *Environmental Engineering Science,* John Wiley & Sons, Inc., New York, NY, 2000.

Nierenberg, W.A.: "Atmospheric Carbon Dioxide: Causes, Effects, and Options," *Chem. Eng. Progress,* 27 (August 1989).

Ondov, J.M. and W.R. Kelly: "Tracing Aerosol Pollutants with Rare Earth Isotopes," *Analytical Chemistry,* 691A (July 1, 1991).

Patrick, D.R.: *Toxic Air Pollution Handbook,* Van Nostrand Reinhold Company, Inc., New York, NY, 1997.

Regens, J.L. and R.W. Rycroft: *The Acid Rain Controversy,* University of Pittsburgh Press, Pittsburgh, PA, 1988.

Roberts, L.: "Learning from the Acid Rain Program," *Science,* 1302 (March 15, 1991).

Schifftner, K.C.: *Air Pollution Control Equipment Selection Guide,* CRC Press, LLC., Boca Raton, FL, 2002.

Schneider, T.: *Air Pollution in the 21st Century: Priority Issues and Policy,* Elsevier Science, New York, NY, 1998.

Schnelle, K.B., C.A. Brown, C. Carelli, and F. Kreith: *Air Pollution Control Technology Handbook: A Handbook Series for Mechanical Engineering,* CRC Press, LLC., Boca Raton, FL, 2001.

Sher, E.: *Handbook of Air Pollution from Internal Combustion Engines: Pollutant Formation and Control,* Academic Press, Inc., San Diego, CA, 1998.

Snow, R.H. and T. Allen: "Effectively Measure Particle-Size Classifier Performance," *Chem. Eng. Progress,* 29 (January 1993).

Staff: "ICI Plans U.S. Plant for CFC Substitute," *Chem. Eng. Progress,* 11 (February 1990).

Stradling, D.: *Smokestacks and Progressives: Environmentalists, Engineers, and Air Quality in America 1881–1951,* Johns Hopkins University Press, Baltimore, MD, 1999.

Turco, R.P.: *Earth under Siege: From Air Pollution to Global Change,* 2nd Edition, Oxford University Press, Inc., New York, NY, 2001.

Van Wormer, M.B.: "Use Air Quality Auditing as an Environmental Management Tool," *Chem. Eng. Progress,* 62 (November 1991).

Wallich, P.: "Dark Days: Eastern Europe Brings to Mind the West's Polluted Past," *Sci. Amer.,* 16 (August 1990).

Wark, K., C.F. Warner, and W.T. Davis: *Air Pollution: Its Origin and Control,* 3rd Edition, Addison Wesley Longman, Inc., Redding, MA, 1997.

Wettestad, J.: *Clearing the Air: European Advances in Tackling Acid Rain and Atmospheric Pollution,* Ashgate Publishing Company, Brookfield, VT, 2002.

POLLUTION (Nuclear). See **Nuclear Reactor**.

POLLUTION (Petroleum Fuels). See **Petroleum**.

POLLUTION (Water and Soil). See **Wastes and Pollution**.

POLONIUM. [CAS 7440-02-06]. Chemical element, symbol Po, at. no. 84, at. wt. 210 (mass number of the most stable isotope), mp 252°C, bp 960°C, sp gr 9.4. The element was first identified as an ingredient of pitchblende by Marie Curie in 1898. The element occurs in nature only as a decay product of thorium and uranium. Because of limited availability and high cost, relatively few practical uses for the element have been found. Meteorological instruments for measuring the electrical potential of air have used small quantities of the metal. It is interesting to note that when Mme. Curie first identified polonium, she found that an electroscope was

a far better instrument for detecting the metal than spectroscopic means. Polonium-plated metal rods and strips have been used as static dissipators in textile coating equipment and in various electrical equipment. The alpha rays from the polonium ionize the air, causing it to conduct and draw off accumulations of static electrical charges. Polonium is a member of periodic group 16 (formerly 6a).

Three isotopes of polonium occur in the uranium $(4n + 2)$ radioactive series: ^{218}Po (radium A), $t_{1/2}$ 3.05 min; ^{214}Po (radium C'), $t_{1/2}$1.6 × 10^{-4} s; and ^{210}Po (radium F), $t_{1/2}$ 138.4 days, and the most stable isotope of polonium. It is used as a source of a-radiation. The thorium $(4n)$ series has two isotopes, ^{216}Po (thorium A), $t_{1/2}$ 0.16 s, and ^{212}Po (thorium C'), $t_{1/2}$3 × 10^{-7} s. The actinium $(4n + 3)$ series also has two isotopes, ^{215}Po (actinium A), $t_{1/2}$1.83 × 10^{-3} s, and ^{211}Po (actinium C'), $t_{1/2}$ 0.52 s, which occurs in a 0.3% branched chain disintegration of ^{211}Bi (actinium C). Several other isotopes of polonium have been prepared, one of which occurs in the neptunium $(4n + 1)$ series as ^{213}Po, $t_{1/2}$4.2 × 10^{-6} s.

Polonium exhibits the allotropy of the lower members of the chalcogen group, having a low-temperature, cubic form, α-polonium, and a high-temperature, rhombohedral form, β-polonium.

The tendency of the chalcogens to show increasing metallic character as one moves down the periodic table is quite marked for polonium; in fact, it resembles lead more than it does tellurium. Its compounds have a more ionic character in its lower oxidation states than do the tellurium compounds. The stability of the 6+ state is low, the existence of polonate(VI) ion being doubtful. The common oxidation states of the element are 2+ and 4+.

The halides, consisting of both dihalides and tetrahalides, are covalent and volatile, and they are not well characterized. The fluorides have not been established. The complex $PoCl_6{}^{2-}$ is known. Polonium compounds are usually colored, a fact that is useful in following their reactions. Thus, polonium(II) chloride, $PoCl_2$, formed by dissolving polonium(IV) oxide, PoO_2, in HCl is pink, and an oxidation by heating or treatment with chlorine yields yellow $PoCl_4$. Polonium(IV) bromide, $PoBr_4$, dark red, gives purple polonium(II) bromide, $PoBr_2$, on heating. $PoBr_4$ also gives ammonium polonium bromide, $(NH_4)_2[PoBr_6]$ with ammonia. Complex iodides $M_2[PoI_6]$ have been prepared.

Metallic polonium reacts with air readily on heating, to form PoO_2, which exists in a yellow face-centered form having fluorite structure at low temperatures, and a red tetragonal one on heating. Polonium(IV) hydroxide, $Po(OH)_4$, precipitated from polonium(IV) solutions by ammonia, exhibits only slight activity, and is thus not amphiprotic. On reaction of polonium with HNO_3, $Po(NO_3)_4$ is formed, and on treatment of polonium(IV) chloride, $PoCl_4$, with H_2SO_4, polonium(IV) sulfate, $Po(SO_4)_2$, is formed, both being ionic-type salts, as indeed are other oxyacid compounds. The sulfate, however, is quite reactive, being hydrated in solution, dehydrated on removal from solution, and forming a basic compound, $2PoO_2 \cdot SO_3$, on heating. H_2S precipitates black polonium(II) sulfide, PoS.

Additional Reading

Greenwood, N.N. and A. Earnshaw: *Chemistry of the Elements,* 2nd Edition, Butterworth-Heinemann, Inc., Woburn, MA, 1997.

Krebs R.E.: *The History and Use of Our Earth's Chemical Elements: A Reference Guide,* Greenwood Publishing Group, Inc., Westport, CT, 1998.

Lide, D.R.: *CRC Handbook of Chemistry and Physics,* 84th Edition, CRC Press, LLC., Boca Raton, FL, 2003.

Parker, P.: *McGraw-Hill Encyclopedia of Chemistry,* 2nd Edition, The McGraw-Hill Companies, Inc., New York, NY, 1993.

Stwertka, A. and E. Stwertka: *A Guide to the Elements,* Oxford University Press, Inc., New York, NY, 1998.

POLONOVSKI REACTION. Demethylation of tertiary (or Heterocyclic) amine *N*-oxides on treatment with acetyl chloride or acetic anhydride to give *N*-acylated secondary amines and formaldehyde, along with *O*-acylated aminophenols as a result of a side reaction.

POLYACETYLENE. A linear polymer of acetylene having alternate single and double bonds, developed in 1978. It is electrically conductive, but this property can be varied in either direction by appropriate doping either with electron acceptors (arsenic pentafluoride or a halogen) or with electron donors (lithium, sodium). Thus, it can be made to have a wide range of conductivity from insulators to *n*- or *p*-type semiconductors to strongly conductive forms. Polyacetylene can be made in both *cis* and *trans* modifications in the form of fibers and thin films, the conductivity of the fibers increasing with their degree of orientation. Films can be applied on glass or metal substrates. Though still in an experimental stage, these polymers have significant possibilities for industrial application, e.g., in batteries.

POLYALLOMER RESINS.

These are block copolymers prepared by polymerizing monomers in the presence of anionic coordination catalysts. The polymer chains in polyallomers are composed of homopolymerized segments of each of the monomers employed. The structure of a typical polyallomer can be represented as:

$$(PPPPPPPP\cdots)_x(EEEE\cdots)_y$$

where P represents a propylene molecule and E an ethylene molecule. The number and length of the individual segments can be varied within wide limits depending on the process conditions, monomer concentrations, and catalyst systems employed. The copolymers exhibit the crystallinity normally associated only with the stereoregular homopolymers of these monomers.

The word *polyallomer* is derived from the Greek words *allos, meros*, and *poly*. *Allos* means "other" and denotes a differentiation from the normal. The word *meros* means "parts" and the prefix *poly* is added to show that these materials are polymeric. Since allomerism is defined as a constancy of crystalline form with a variation in chemical composition, the polyallomers are examples of allomerism in polymer chemistry.

Polyallomers can be synthesized in slurries by contacting the parent monomers with anionic coordination catalysts of the Ziegler-Natta type at temperatures of 70–80°C. Polyallomers are also synthesized in solution at 140–200°C by contacting the parent monomers with hydrogen-reduced alpha-titanium trichloride and a lithium-containing cocatalyst. In both the slurry and solution processes the polyallomers are formed by alternate polymerization of the monomers employed. In the synthesis of polyallomers from monomers which differ widely in polymerization rates it is sometimes desirable to alternately polymerize a single monomer and then a mixture of monomers to obtain the most desirable properties.

Evidence for the crystalline nature of the polyallomers includes x-ray diffraction patterns, infrared spectra, and for olefin polyallomers, low solubility in hydrocarbon solvents.

The physical properties of polyallomers are generally intermediate between those of the homopolymers prepared from the same monomers, but frequently represent a better balance of properties than blends of the homopolymers. This is illustrated by comparing the properties of a propylene-ethylene polyallomer containing 2.5% ethylene with polypropylene, high-density polyethylene, and a blend of 5% high-density polyethylene and 95% polypropylene.

Compared to polypropylene, a propylene-ethylene polyallomer has a lower brittleness temperature, higher impact strength, and less notch sensitivity. Compared to high-density polyethylene, a propylene-ethylene polyallomer is harder, higher-melting, and of higher impact strength. Other advantages of this polyallomer over high-density polyethylene includes its excellent resistance to environmental stress cracking and its low and uniform mold shrinkage, which minimizes sinks and voids in molded parts. A propylene-ethylene polyallomer thus overcomes the most serious property deficiencies of polypropylene (poor impact and low-temperature properties) and of high-density polyethylene (poor stress-crack resistance and excessive mold shrinkage).

Propylene-ethylene polyallomer is used in vacuum forming, blow molding, injection molding, film and sheeting, wire covering and pipe. Propylene-ethylene polyallomer is the easiest of all polyolefins to vacuum form. The excellent melt strength and broad processing range permit a deep draw. Reproduction of mold detail and surface finish is very good. In film and sheeting, propylene-ethylene polyallomer has optical properties equal to polypropylene but much higher impact strength, particularly at low temperatures. In wire covering, blow molding, and pipe extrusion operations, the combination of good processability and excellent environmental stress crack resistance is important.

H. J. HAGEMEYER, JR.
M. B. EDWARDS
Longview, Texas

POLYAMIDE RESINS.

These are synthetic polymers that contain an amide group, $-CONH-$, as a recurring part of the chain. Poly-alpha-aminoacids, i.e., proteins, whether natural or synthetic, are not normally included in this classification.

The polyamides trace their origin to the studies of W. H. Carothers, begun in 1928, on condensation polymerization, a process that involves the repetition many times of a reaction known to the organic chemist as a condensation reaction because it links two molecules together with the loss of a small molecule. Esterification and amidation are examples:

$$CH_3COOH + C_2H_5OH \rightleftharpoons CH_3COOC_2H_5 + H_2O$$

acetic acid   ethyl alcohol   ethyl acetate or
ethyl ester of
acetic acid

$$CH_3COOCH_3 + C_4H_9 NH_2 \rightleftharpoons CH_3CONHC_4H_9 + CH_3OH$$

methyl acetate   butyl amine   N-butylacetamide

Polymerization requires at least two reactive groups per molecule as in the following example:

$$n\ H_2N(CH_2)_6 NH_2\ +n\ HOOC(CH_2)_4COOH \rightleftharpoons$$

hexamethylenediamine   adipic acid

$$H[NH(CH_2)_6NHCO(CH_2)_4CO(CH_2)_4CO]nOH+$$

poly(hexamethylene adipamide)

$$(2n-1)H_2O$$

The first truly high-molecular-weight polyamide was made this way in 1935 and lead to the development of the first wholly man-made fiber which became popularly known as *nylon*, a name coined by Du Pont for fiber-forming polyamides. Numerals representing the number of carbon atoms in first the diamine and then the diacid are used to identify the nylon. Thus, poly(hexamethylene adipamide) is nylon-66 (six-six, not sixty-six), and poly(hexamethylene sebacamide) made from the 10-carbon sebacic acid is nylon-610 (six-ten).

Both reacting species may be present in the same molecule:

$$n\ H_2N(CH_2)_{10}COOH \rightleftharpoons$$

11-aminoundecanoic acid

$$H[NH(CH_2)_{10}CO]_nOH + (n-1)H_2O$$

poly(11-aminoundecanoic acid)

Here a single number is used to indicate the number of carbon atoms in the original nomomer, i.e., nylon-11 ("eleven" not "one-one"). In some instances the cyclic analogue or lactam is more accessible than the amino acid and is polymerized by a ring-opening rather than condensation mechanism:

$$n\ CH_2 \begin{array}{c} CH_2-CH_2-C=O \\ | \\ CH_2-CH_2-NH \end{array} + H_2O \rightleftharpoons$$

epsilon-caprolactam

$$H[NH(CH_2)_5CO]_n OH$$

polycaprolactam or polycaproamide
nylon-6

$$n\ (CH_2)_{11} \begin{array}{c} C=O \\ | \\ NH \end{array} + H_2O \rightleftharpoons H[NH(CH_2)_{11}CO]_n OH$$

laurolactam   nylon-12

Some of the monomers commonly used to prepare the nylon resins are shown in the accompanying table. Both petrochemical and vegetable products provide the source materials that are transformed into the reactive intermediates. The table correctly suggests that there is a wider choice in diacids than in diamines. The most important commercial polyamide resins are nylons-66 and -6. Other commercial nylons include 610, 612, 11, and 12.

The above equations illustrate via the double arrows an important facet of polyamides—the equilibrium nature of the polymerization reactions. Achieving and maintaining useful molecular weights (about 10,000 or more) for plastics applications require low moisture contents in order to avoid the reverse reaction of hydrolysis. Most commercial nylons are processed at melt temperatures in excess of 200°C, and molecular weight stability requires a water content below 0.3 weight percent. At a molecular weight of about 11,300, nylon-66 and nylon-6 have, on the average, 99 amide groups linking together 100 monomer units with one unreacted amine group and one unreacted acid group at the ends of the 700-atom long chain. This corresponds to 99% reaction. Because nylon plastics have average molecular weights in the 11,000 to 40,000 range, the need for pure materials and freedom from side reactions is seen to be essential for successful polymerization.

An average molecular weight (number average) is used because any one nylon comprises a broad range of molecular weights. Nylons typically have a "most probable distribution" in which the weight average is twice the number average.

As made, nylon-6 contains up to ten weight percent of lactam monomer that has to be extracted for most applications. Nylons made from larger lactam rings contain only about one percent monomer.

Not all polyamide resins are nylons, and not all nylons are polyamide resins. Some nylons such as nylon-4 or those with a high content of relatively inflexible rings are too unstable or have too high a melt viscosity to be melt processible and are not normally included in the "polyamide resin" category. However, spinning or casting from solution permits some such polymers to be converted into useful fibers or films. A class of polyamide resins distinct from the nylons is based upon polymerization or dimerized vegetable oil acids and polyalkylene polyamines (e.g., ethylenediamine or diethylenetriamine). These are relatively low in molecular weight (2000–10,000), vary from liquids to low melting solids, and are more soluble and flexible than the higher molecular weight, more crystalline nylons.

Nylons are semicrystalline polymers with fairly sharp melting points varying from 180°C for nylon-2 to 270° for nylon-66. The melting point increases in zigzag fashion with increasing concentration of amide groups, being higher where there is an even number of chain atoms between the amide groups. For example, nylon-66 averages five CH_2-groups per CONH but has either four or six chain atoms between amide groups; nylon-6 also has five CH_2/CONH but always has five chain atoms between amides and melts 40°C lower than nylon-66. The degree of crystallinity and the morphology of nylons are more readily controlled by choice of processing conditions than other crystalline polymers such as polyethylene or polyacetal.

The nylons are typically tough and strong. Nylon-66 was the first thermoplastic to provide a combination of stiffness and toughness suitable for mechanical applications. Nylons were therefore the first members of the family of engineering thermoplastics that now include newer materials such as the polyacetals, polysulfones, and polycarbonates. Nylons are outstanding in withstanding repeated impact, in resistance to organic solvents, and in water resistance. A low coefficient of friction, as ASTM self-extinguishing rating, reasonable electrical properties, adequate creep resistance, good fatigue properties, and good barrier properties, particularly to oxygen, are also characteristic of the nylons. Properties change with temperature but are often acceptable in the interval of about −60 to 110°C. Properties may change also with relative humidity because nylons characteristically absorb moisture and water acts as a plasticizer, that is, stiffness decreases and toughness increases. Water absorption decreases as the amide group concentration in the nylon decreases; for example, nylon-6 at saturation contains 9.5% water and nylon-11, 1.9%. Nylons are somewhat notch sensitive, and parts are typically designed to avoid sharp corners. Nylons are attacked by strong acids, oxidizing agents, and a few specific salt solutions such as aqueous potassium thiocyanate or methanolic lithium chloride.

Control of crystallinity and morphology provides one tool for changing properties as desired with any given nylon. But there are many other tools for modification of nylons, and this viability has been an important factor in meeting specific market demands in the face of increased competition from newer materials. Copolymerization and plasticization are alternatives that provide lower melting and tougher compositions with a somewhat different balance of properties. Lubricants as processing aids, nucleating agents to accelerate crystallization and give a little stiffer product, molybdenum disulfide or graphite to improve lubricity in gears or bearings, colorants, fire retardants, carbon black for weather resistance, and antioxidants for better resistance to thermal oxidation are examples of additives employed to achieve specific effects. Combination with other polymers is another

technique. Reacting nylon-66 with formaldehyde in alcohol solution has yielded a low melting derivative, that can be cross-linked to provide a thermoset. Glass fiber reinforcement has proved to be particularly effective in nylons and has yielded products of exceptional strength, stiffness, and heat resistance.

The nylon resins are converted into useful shapes principally by injection molding or extrusion. These shapes are most often in finished form, but forming is sometimes employed to impart added strength via orientation. Machining of extruded stock shapes is often appropriate where the number of parts does not justify manufacture of a costly mold. Nylon-6 castings are made by the base catalyzed, anhydrous polymerization of monomer in molds that are relatively inexpensive because they do not have to sustain high pressures. Rotational molding, fluidized bed coating, and electrostatic spray coating are also used, especially with low melting nylons such as nylon-11.

Football face guards, hammer handle gears, sprockets, journal bearings, bristles, filaments, refrigerant tubing, film as a cooking pouch, coil forms, casters, package strapping, loom parts, automobile dome lights, power tool housings, and virtually thousands of other diverse articles attest to the performance of nylon resins in electrical appliances, automotive parts, business equipment, consumer products, and other industrial and home applications.

The non-nylon polyamide resins include relatively low-melting solids that are used with or without modifiers in hot melt cements, heat seal and barrier coatings, inks for flexographic printing of plastic film, and other specialty adhesives and coatings such as varnishes to provide a glossy, transparent, protective layer over print. The large hydrocarbon side chains that are present in the dimerized vegetable oil acids used to make these polyamide resins contribute to their low crystallinity, low water absorption, and flexibility. The fluid, "reactive" polymides are lowest in molecular weight and contain an excess of amine groups. Monomers with more than two amine groups per molecule such as diethylenetriamine ($H_2NCH_2CH_2NHCH_2CH_2 NH_2$) are used, and care must be taken during polymerization to avoid premature gelation. The excess amine groups permit interaction with epoxy, aldehyde, hydroxmethyl, anhydride, acrylic, and other groups in other resins to produce a variety of thermosetting formulations. Combinations with epoxy or phenol-formaldehyde polymers

MONOMERS USED TO PREPARE NYLON RESINS

Monomer	Formula	Source(s)
hexamethylene diamine	$H_2N(CH_2)_6 NH_3$	butadiene, furfural, or propylene
adipic acid	$HOOC(CH_2)_4COOH$	butadiene or cyclohexane
suberic acid	$HOOC(CH_2)_6COOH$	butadiene or acetylene
azelaic acid	$HOOC(CH_2)_7COOH$	oleic acid
sebacic acid	$HOOC(CH_2)_8COOH$	castor oil
dodecanedioic acid	$HOOC(CH_2)_{10}COOH$	butadiene
"dimer" acid	$HOOC-C_{34}H_{60}-COOH$	oleic and linoleic acids
caprolactam	$(CH_2)_5$, with CO and NH (lactam ring)	toluene, benzene, or cyclohexane
7-aminoheptanoic acid	$H_2N(CH_2)_6COOH$	cyclohexane or ethylene
capryllactam	$(CH_2)_7$, with CO and NH (lactam ring)	butadiene or acetylene
9-aminononanoic acid	$H_2N(CH_2)_8COOH$	soybean oil
11-aminoundecanoic acid	$H_2N(CH_2)_{10}COOH$	castor oil
laurolactam	$(CH_2)_{11}$, with CO and NH (lactam ring)	butadiene

with or without added fillers are the most common. The epoxy compositions are cured at relatively low temperatures and are useful for bonding aluminum in aircraft, steel in automobiles, wood, leather, glass, ceramics, and other materials. The resin combination broadens the capacity of the adhesive to wet a variety of surfaces and enhances bond strengths. The phenolic-polyamide compositions require higher curing temperatures but withstand higher temperatures in use. They tend to form bonds less tough than the epoxy mixes but bond strongly to copper and other common components of printed circuits. The phenolic blends also produce acid and alkali resistant coatings useful as container linings or wire coatings.

M. I. KOHAN
E. I. Du Pont de Nemours & Co., Inc.
Wilmington, Delaware

POLYAMIDE-IMIDE RESINS. An injection-moldable, high-performance engineering thermoplastic, polyamide-imide is the condensation polymer of trimellitic anhydride and various aromatic diamines with the general structure:

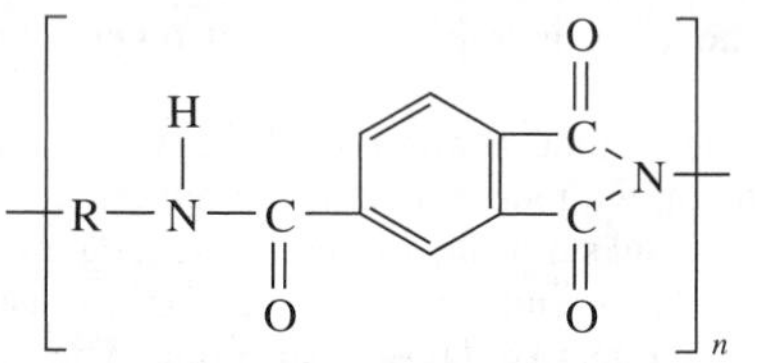

Polyamide-imides are available[1] in unfilled; 30% glass fiber-filled; 30% graphite fiber-filled; and modified 40% glass-filled grades. The unfilled grade has the highest impact resistance, while the graphite fiber-filled grade has the highest modulus or stiffness. The modified version offers the lowest cost while still maintaining an impressive slate of properties. The resins can be molded into complex precision parts and also can be extruded and machined to close tolerances. The resins are used extensively in the aerospace industry, offering significant weight reduction by replacing metal parts. Aircraft usage includes jet engine components, compressor and generator parts, and electronic/electrical devices. Polyamide-imide resins are also used in the hydraulic/pneumatic equipment industry for wear surfaces, bushings, seals, vanes, and flow control devices. The automative and heavy equipment industries use this material as parts in transmissions, universal joints, and power-assisted devices. Internal combustion engines use many polyamide-imide structural-mechanical parts, such as valve train components.

The material is opaque and characterized by good dimensional stability, creep resistance, impact resistance, and superior mechanical properties that persist up to temperatures of about 500°F (260°C). Unfilled polyamide-imide has a tensile strength of about 27,000 psi (186 mPa); flexural strength of about 35,000 psi (241 mPa); compressive strength of about 32,000 psi (221 mPa); and an elastic modulus of about 750,000 psi (5172 mPa). Mechanical properties at 450°F (232°C) exceed those of many polymers at room temperature. At cryogenic temperature, the unfilled polymer has a tensile strength of about 31,500 psi (241 mPa) with 6% elongation at −196°C (liquid nitrogen). Heat deflection is high (525°F; 273°C), while the coefficient of linear thermal expansion is low. When burned, polyamide-imide produces a char rather than drip and produces very little smoke. Electrical properties are attractive. Radiation resistance is good. The resin is virtually unaffected by aliphatic and aromatic hydrocarbons, halogenated solvents, and most acid and base solutions. It is attacked by high-temperature caustic, steam, and some acids. At 50% relative humidity, the material (70°F; 21°C) will absorb about 1% moisture in 1000 hours.

POLYARYLATES. These are clear, amorphous thermoplastics that combine clarity, high heat deflection temperatures, high impact strength, good surface hardness, and good electrical properties with inherent ultraviolet stability and flame retardance. No additives or stabilizers are required to provide these properties. Polyarylates are aromatic polyesters that are manufactured from various ratios of iso- and terephthalic acids with bisphenol A.[1] The resultant products are free-flowing pellets which can be processed by a variety of thermoplastic techniques in transparent and

opaque colors. Because polyarylate's weatherability is obtained through polymer chemistry rather than additives (as with most UV-resistant polymers), the properties of polyarylate do not deteriorate significantly with time. (Over 5000 hours of accelerated weathering and actual Florida and Arizona aging resulted in virtually no change in performance with respect to luminous light transmittance, haze, gloss, yellowness, and impact.) The flammability characteristics are inherent. Properties include a high oxygen index, low smoke density, low flame spread, and low toxic gas formation. Because the flammability properties are achieved without additives, the resultant products of combustion are essentially limited to Commercially available as CO_2, CO, and water.

Polyarylates are offered in several glass-reinforced versions with loading available up to 40%. The glass fibers provide higher stiffness, improved tensile strength, and higher heat deflection temperatures. Polyarylates may also be mineral-filled and reinforced with other fibers, such as carbon. Alloys/blends with other polymers are also available.

These materials are useful in outdoor applications, such as high-intensity discharge lighting (traffic signals), automobile halogen headlamp lenses and bodies, and rear-end elevated automobile stop lights. High-temperature lighting and microwave cookware are suitable applications. Electronic/electrical connectors and housings are also important applications for the polymer.

Polyarylates have good optical properties. Luminous light transmission can range from 84% to 88% with only 1% to 2% haze. Refractive index is 1.61. An important feature of the polyarylate family is high heat resistance demonstrated by a 340°F (171°C) heat deflection temperature at 264 psi (1.8 mPa). The material exhibits good retention of properties at high temperature exposures: 270,000 psi (1380 mPa) at 300°F (149°C); and over 200,000 psi at 350°F (177°C).

Polyarylates are injection molded, using standard screw machines, as well as extruded, foam molded, and blow molded. Melt temperatures range from 600 to 680°F (316 to 360°C). Mold temperatures should be maintained between 200 and 300°F (93 and 149°C).

POLYBASITE. A mineral antimony sulfide of silver $(Ag,Cu)_{16}Sb_2S_{11}$, in which copper substitutes for silver to approximately 30 atomic percent. It crystallizes in the monoclinic system; hardness, 2–3; specific gravity, 6.3; color, black, dark ruby red in thin splinters with metallic luster; nearly opaque. From the Greek, meaning *many*, suggesting the many-metal basis.

Occurs in low-temperature silver deposits commonly associated with silver and lead minerals. Found in various Western States in the United States, and as superb crystals at Arizpe and Las Chiapas, Mexico; in Chile, Peru, Sardinia, Germany, and Australia.

POLYBENZIMIDAZOLES. These are heterocyclic polymers that have outstanding high thermal characteristics, the highest obtainable in commercial polymers. These materials also have superior ablative and hydrolytic stability as well as high compressive and dimensional stability. Polybenzimidazoles essentially are unaffected by solvents, acids, and bases. They are marketed in stock shapes and as finished parts. The materials are not available in resin form. Hoechst Celanese markets the products under the tradename *Celazole*.®

Parts are produced by a high-pressure sintering process wherein the melt polycondensation resin is densified and then coalesced. Metallurgical pressures at temperatures exceeding over 400°C (750°F) are required. Polybenzimidazoles have repeating benzimidazole groups in the polymer backbone. These materials were synthesized first in 1961. Currently, the products result from a melt polycondensation reaction of aromatic, bis-ortho-diamines (diphenylisophthalate). By way of compounding, fabrication and end-use performance characteristics can be customized to specific needs. Often the materials are preferred for tribological applications inasmuch as they have the desirable characteristics of low coefficient of friction, low abrasion, and good high-temperature dimensional stability.

Polybenzimidazoles have been used for seals, mechanical components, electrical connectors, valve seats, and as components of materials-handling equipment in the petrochemical, geothermal, chemical process, aerospace, defense, automotive, and electrical products industries. These materials frequently are procured as replacement parts for other materials in an effort to improve equipment performance.

---

[1] Commercially available as *Torlon*™. (*Amoco Chemicals Co.*)
[1] Data furnished by *Celanese Engineering Resins*.

POLYBUTYLENE RESINS. These materials (PBs) are semicrystalline polyolefin thermoplastics based on poly(1-butene) and include homopolymers and a series of poly(1-butene-ethylene) copolymers. The resins available commercially[1] are manufactured via stereospecific Ziegler-Natta polymerization of 1-butene monomer. The commercial products are based on isotactic (98% to 99.5%), high-molecular-weight (230,000 to 750,000) polymer. Five crystalline modifications of poly(1-butene) are known. Of these, the glass transition temperature ranges from about $-4°F$ ($-20°C$) for the homopolymers to about $-30°F$ ($-34:18C$) for the high-ethylene copolymers.

PB resins generally are resistant to acids, bases, solvents, paraffinic and naphthenic oils, detergents, and various chemicals. Resistance decreases, however, at elevated temperatures. They have good moisture barrier and electrical insulation properties. They exhibit a broad range of flexibility: tensile moduli vary from 41,500 psi (286 mPa) for homopolymers to 7500 psi (52 mPa) for the high-ethylene polymers. The resins are particularly resistant to creep, environmental stress cracking, chemicals, and abrasion. PB resins are offered in a special pipe grade, film grades, and five general-purpose grades.

PB pipe can be fabricated by conventional single-screw extrusion technology using vacuum or pressure sizing for dimensional control. The pipes can be joined by thermal fusion or mechanical fittings. Applications include cold and hot water plumbing. Other uses include well, heat pump, and fire-sprinkler piping as well as specialty hoses. Large-diameter PB pipe finds uses in the transport of abrasive or corrosive materials at high temperature, as found in the mining, chemical, and power generation industries.

PB film is usually made by the blown film process, but also can be cast on chill rolls. Film applications include food and meat packaging, compression wraps, and hot-fill containers. The material can be formulated to provide a wide range of seal strengths for peelable or easy-opening packaging.

The ability of PB to accept high filler loading (up to 80%) has resulted in its use as a color, mineral filler, and flame-retardant concentrate carrier. Polybutylene is compatible with polypropylene in all proportions and as a modifier it provides enhancement of processibility, impact, and weld line strength in injection molding and extrusion. It also provides improved impact strength and heat stability in films and improved hand and bondability in fibers of polypropylene. A comparatively recent use of PBs is for hot-melt adhesives and sealants where high strength, high-shear adhesion failure temperature, and a long open time are needed. They are particularly suited for use with aliphatic tackifying resins.

POLYBUTYLENE TEREPHTHALATE POLYSTERS. A semicrystalline thermoplastic polyester. Because of its rapid crystallization, injection molding is the preferred method of processing. The material has been used for many years in the connector industry because of its good chemical resistance, high-temperature capabilities, good electrical properties, and long flow lengths in thin sections. This material (PBT) typically is formed in a transesterification reaction between 1,4-butanediol and dimethylterephthalate. Unmodified PBT is translucent in thin sections and opaque white in thicker sections.

The glass transition temperature of PPBT is about 52°C (125°F). Melting point is about 230°C (440°F). Unreinforced PBT is obtainable in several molecular weights. Compounded resins are available with numerous types and levels of fillers and reinforcements. Glass fiber reinforcement has a wide spectrum of physical properties. These materials can be made flame-retardant through the use of additives.

Exceptional electrical properties and temperature resistance qualify the material for numerous electrical parts—connectors, coil bobbins, light sockets, terminal blocks, fuse holders, and motor parts. PBT provides weight reduction of final parts. PBT has replaced a number of thermoset materials and is particularly popular for appliance housings and fibers for paint brushes.

POLYCARBONATES. Polycarbonates are an unusual and extremely useful class of polymers. The vast majority of polycarbonates are based on bisphenol A (BPA) and sold under various trade names. BPA polycarbonates, having glass-transition temperatures, in the range of 145–155°C, are widely regarded for optical clarity and exceptional

impact resistance and ductility at room temperature and below. Other properties, such as modulus, dielectric strength, or tensile strength are comparable to other amorphous thermoplastics at similar temperatures below their respective glass-transition temperatures. Whereas below their T_gs most amorphous polymers are stiff and brittle, polycarbonates retain their ductility.

Important products are based on polycarbonate in blends with other materials, copolymers, branched resins, flame-retardant compositions, and foams.

Properties and Characterization

Solubility and Solvent Resistance. The majority of polycarbonates are prepared in methylene chloride solution. Chloroform, *cis*-1,2-dichloroethylene, *sym*-tetrachloroethane, and methylene chloride are the preferred solvents for polycarbonates. Hydrocarbons and aliphatic alcohols, esters, or ketones do not dissolve polycarbonates. Acetone promotes rapid crystallization of the normally amorphous polymer, and causes catastrophic failure of stressed polycarbonate parts.

In general, polycarbonate resins have fair chemical resistance to aqueous solutions of acids or bases, as well as to fats and oils. Chemical attack by amines or ammonium hydroxide occurs, however, and aliphatic and aromatic hydrocarbons promote crazing of stressed molded samples. BPA polycarbonate has excellent resistance to hydrolysis.

Certain blends and copolymers of polycarbonate demonstrate dramatically improved solvent resistance. The blend of polycarbonate and poly(butylene terephthalate) combines the toughness of polycarbonate with the solvent resistance of the semicrystalline polyester.

Copolycarbonates of BPA and hydroquinone (HQ) can be prepared via the intermediacy of cyclic oligomeric cocyclics. Although hydroquinone linear oligomers having degrees of polymerization greater than two are insoluble in CH_2Cl_2, the cyclic analogues remain soluble when randomly cyclized with BPA.

Molecular Weight and Viscosity. BPA polycarbonates are commercially available in a wide range of molecular weights. As the molecular weight increases, melt and solution viscosities increase proportionally. Molecular weights may be determined or inferred by several means, including gel-permeation chromatography, light-scattering chromatography, measurement of intrinsic or inherent viscosity, and measurement of melt viscosity and flow. Correlation of intrinsic viscosity (IV), [η], with weight-average mol wt (M_w) has been carried out on carefully characterized polycarbonate samples. The following relationship exists when [η] is in mL/g.

$$[\eta] = 41.2 \times 10^{-3} \cdot M_w^{0.69}.$$

The mechanical properties of polycarbonate, e.g., tensile strength, impact resistance, flexural strength, elongation, etc., improve dramatically with increasing polymer intrinsic viscosity up to a value of about 0.45 dL/g. After that point, slight increases in mechanical properties are seen with increasing molecular weight, but melt viscosity continues to climb. At IV values greater than 0.6 dL/g, the melt viscosity becomes so high that processing is very difficult.

Spectroscopy and Analysis. Polycarbonates have a strong C=O stretching-band at 1770 cm^{-1}, and strong C—O stretching bands at 1220 and 1235 cm^{-1}, distinguishing them from polyesters.

Differential scanning calorimetry reveals a T_g at around 154°C, shifting slightly with molecular weight or level of branching. End group and impurity analysis is best revealed by hydrolysis of the polycarbonate using KOH-methanol in tetrahydrofuran under nitrogen, followed by reversed-phase hplc analysis or by spectroscopic techniques. Trace levels of impurities, are determined by standard analytical techniques.

Structure and Crystallinity. The mechanical–optical properties of polycarbonates are those common to amorphous polymers. The polymer may be crystallized to some degree by prolonged heating at elevated temperature (8 d at 180°C), or by immersion in acetone. Powdered amorphous powder appears to dissolve partially in acetone, initially becoming sticky, then hardening and becoming much less soluble as it crystallizes. Enhanced crystallization of polycarbonate can also be caused by the presence of sodium phenoxide end groups.

Film or fibers derived from low molecular weight polymer tend to embrittle on immersion in acetone; those based on higher molecular weight polymer (>0.60 dL/g) become opaque, dilated, and elastomeric.

Glass-Transition Temperature and Melt-Behavior. The T_g of BPA polycarbonate is around 150°C, which is unusually high compared to other

---

[1] Commercially available as *Duraflex™*. (*Shell Chemical Co.*)

thermoplastics. The high glass-transition temperature can be attributed to the bulky structure of the polymer, which restricts conformational changes, and to the fact that the monomer has a higher molecular weight than the monomer of most polymers. The high T_g is important for the utility of polycarbonate in many applications, because, as the point which marks the onset of molecular mobility, it determines many of the polymer's properties such as dimensional stability, resistance to creep, and ultimate use temperature. Polycarbonates of different structures may have significantly higher or lower glass-transition temperatures.

BPA polycarbonate becomes plastic at temperatures around 220°C. The viscosity decreases as the temperature increases, exhibiting Newtonian behavior, with the melt viscosity essentially independent of the shear rate. At the normal injection molding temperature of 270–315°C, the melt viscosity drops from 1,100 to 360 Pa · s (11,000 to 3,600 poise). Because the viscosity of polycarbonate can only be reduced by increasing the temperature, the ultimate limit on molecular weight is controlled by the processing conditions and the thermal stability of the polymer.

Thermal, Flame-Retardant, and Hydrolytic Behavior. BPA polycarbonate exhibits excellent thermal stability, especially in the absence of oxygen and water. At temperatures above 400°C, rapid decomposition and cracking occur. BPA has an oxygen index of 26; this indicates that under test conditions, an atmosphere of 26% oxygen is required for combustion. Owing to thermal–oxidative stability, polycarbonate has some inherent flame resistant properties and can be classified as V-2 according to UL94 of the Underwriters Laboratory. Several polycarbonate grades have additives to increase the flame-retardant properties, and to decrease smoke.

Because of the low solubility of water in the resin, BPA polycarbonates are inherently resistant to aqueous acid and base, although strong nucleophilic bases can cause hydrolysis.

Optical Properties. Polycarbonate is a transparent colorless polymer, making it attractive for glass replacement. Visible light transmission is about 90%, and haze is minimal (1–2%). Absorption in the ultraviolet region is essentially complete. Polycarbonate's high (1.584) refractive index and light weight relative to glass make it attractive for eyewear.

Special polycarbonate grades have been developed for the optical information storage market e.g., compact disks.

Mechanical Properties. The room temperature modulus and tensile strength are similar to those of other amorphous thermoplastics, but the impact strength and ductility are unusually high. Whereas most amorphous polymers are glass-like and brittle below their glass-transition temperatures, polycarbonate remains ductile to about −10°C. The stress–strain curve for polycarbonate typical of ductile materials, places it in an ideal position for use as a metal replacement. Weight savings as a metal replacement are substantial, because polycarbonate is only 44% as dense as aluminum and one-sixth as dense as steel.

Impact strength can be measured by a variety of methods, including notched Izod, tensile impact, and falling dart impact. Polycarbonates are among the highest rated engineering polymers for impact resistance, and are the toughest transparent materials known.

Glass-reinforced polycarbonates are sold as high modulus materials having properties approaching those of metals, while retaining the basic plastic attributes of low cost processing, dielectric character, resistance to corrosion, and inherent color.

Preparation

Interfacial Polymerization. Most BPA polycarbonate is produced by an interfacial polymerization process utilizing phosgene. The interfacial process for polycarbonate preparation involves stirring a slurry or solution of BPA and 1–3% of a chain stopper, such as phenol, p-t-butylphenol, or p-cumylphenol, in a mixture of methylene chloride and water, while adding phosgene in the presence of a tertiary amine catalyst.

Transesterification. The transesterification process is an environmentally friendly process that utilizes no solvent during polymerization, producing neat polymer directly.

Processing. Polycarbonates may be fabricated by all conventional thermoplastic processing operations, of which injection molding is the most common.

Injection blow molding of polycarbonates produces an assortment of containers from 20-L water bottles and 0.25-L milk bottles to outdoor lighting protective globes.

Conventional thermoforming of sheet and film is applicable to the production of skylights, radomes, signs, curved windshields, prototype production of body parts for automobiles, skimobiles, boats, etc. Because BPA polycarbonate is malleable, it can be cold-formed like metal, and may be cold-rolled, stamped, or forged.

Health and Safety Factors

Polycarbonate is considered a slight or nonexistent fire hazard. Odor and volatiles are negligible. Processing fumes, which include water, carbon dioxide, diphenyl carbonate, methylene chloride, and phenol, are not formed in levels considered to be hazardous. Polycarbonate has very low acute oral and dermal toxicity, is not a primary skin irritant, and does not cause systemic or local sensitization. Polycarbonate does not degrade during storage, and no heating or cooling requirements are necessary.

Uses

Extreme toughness, transparency, low color, resistance to burning, and maintenance of engineering properties over a wide thermal range are the outstanding properties of polycarbonate that make it useful for a variety of applications. Glazing and sheet are the largest markets for polycarbonate resins. Other uses for polycarbonates include automotive components; packaging; electrical, electronic, and technical applications; and medical and health-care related applications.

Polycarbonate is popular because of its clarity, impact strength, and low level of extractable impurities.

Other Polycarbonates, Blends, and Copolymers

Blends of polycarbonate with ABS and with poly(butylene terephthalate) (PBT) have shown significant growth since the mid-1980s.

Copolymers. The copolymer of tetrabromoBPA and BPA was one of the first commercially successful copolymers. Low levels of brominated comonomer lead to increased flame resistance.

Polyester carbonates can be prepared by the copolymerization of BPA with diacyl chlorides, leading to high heat materials with $T_g \sim 190°C$.

Some of these block copolymers have improved low temperature impact strength and higher stress–crack resistance than neat BPA polycarbonate.

Blends. The concept of blending two or more commercially available materials to create a new material having properties different from either starting material has generated a great deal of interest. Polycarbonate blends are used to tailor performance and price to specific markets.

Fundamental studies of blends of polycarbonate with acrylonitrile–butadiene–styrene (ABS) indicate that the presence of ABS greatly decreases the melt viscosity in the blend, enhancing processibility. A synergistic improvement of the notched impact strength at low temperature is also seen for polycarbonate–ABS blends.

Polycarbonate–polyester blends are used on exterior parts for the automotive industry. Such blends combine the toughness and impact strength of polycarbonate with the crystallinity and inherent solvent resistance of PBT, PET, and other polyesters.

The most significant blends are with polyurethanes, polyetherimides, acrylate–styrene–acrylonitrile (ASA), acrylonitrile–ethylene–styrene (AES), and styrene–maleic anhydride (SMA).

DANIEL J. BRUNELLE
General Electric

Additional Reading

LeGrand, D.G., and J.T. Bendler: *Handbook of Polycarbonate Science and Technology,* Marcel Dekker, Inc., New York, NY, 1999.
Read, C.S.: *CEH Marketing Research Report,* SRI International, 1993.
Schnell, H.: *Ang. Chem.* **68**, 633 (1956).
Sperling, L.H.: *Introduction to Physical Polymer Science,* 3rd Edition, John Wiley & Sons, Inc., New York, NY, 2001.
Stevens, M.P.: *Polymer Chemistry: An Introduction,* 3rd Edition, Oxford University Press, New York, NY, 1998.
U.S. Pat. 3,028,365 (1962), H. Schnell, L. Bottenbruch, and G. Grimm (to Bayer AG).
U.S. Pat. 3,153,008 (1964), D.W. Fox (to General Electric).

POLYCYCLO-HEXYLENE-DIMETHYLENE TEREPHTHALATE. PCT is 1,4-cyclohexylene-dimethylene terephthalate and is a high-temperature, semicrystalline thermoplastic polyester. PCT possesses excellent thermal properties. Injection molding is the predominant method of processing glass fiber-reinforced grades of PCT. It is widely used for products where excellent thermal (heat-resistant) properties are needed, as exemplified by surface-mountable electronic components, automotive parts, and dual-ovenable cookware. The material also is used for flexible electronic circuitry. PCT-based polycarbonate is a polymeter that provides melt blends that exhibit excellent clarity, toughness, chemical resistance, flow, and gloss.

PCT is differentiated from other thermoplastic polyesters by its higher heat deflection temperature. Continuous use temperatures of up to 150°C (300°F) are possible.

Principal uses for this material are found in the electrical/electronics industries; automotive parts, such as alternator armatures and pressure sensors; optical uses, such as safety goggles; and garden vehicles, such as mower decks and shrouds, tractor hoods, grills, and fenders.

POLYDIMETHYLSILOXANE.

A silicone polymer developed for use as a dielectric coolant and in solar energy installations. It also may have a number of other uses. It is stated to be highly resistant to oxidation and biodegradation by microorganisms. It is degradable when exposed to a soil environment by chemical reaction with clays and water, by which it is decomposed to silicic acid, carbon dioxide, and water.

POLYELECTROLYTES.

These are macromolecules with incorporated ionic constituents. Polyelectrolytes may be cationic or anionic, depending on whether the fixed ionic constituents are positive or negative. Examples of cationic polyelectrolytes are polyvinyl-ammonium chloride and poly-4-vinyl-N-methyl-pyridinium bromide. Examples of anionic polyelectrolytes are potassium polyacrylate, polyvinylsulfonic acid, and sodium polyphosphate. If a polyelectrolyte contains both fixed positive and negative ionic groups, it is called a polyampholyte. Polyelectrolytes may be synthesized by polymerization of a monomer containing the ionic substituent, as for instance the polymerization of acrylic acid to polyacrylic acid, or by attaching the ionic constituent by chemical means to an already existing macromolecule, as for instance in the quaternization of poly-4-vinyl-pyridine with methyl bromide, or in the preparation of sodium carboxymethylcellulose from natural cellulose. Many macromolecules occurring in nature are polyelectrolytes. Examples are gum arabic, which carries carboxylate groups; carrageenin, which contains sulfate groups; proteins, which carry both negative carboxylate and positive ammonium groups; and nucleic acids, which contain negative phosphate groups and basic purine and pyrimidine groups, which acquire positive charges at low pH. Inorganic long-chain polyphosphates have also been isolated from biological materials.

A solution of a polyelectrolyte in water or other suitable solvent conducts an electric current, indicating that the polyelectrolyte is ionized. Transference and electrophoresis measurements show that both the macroion and the counterions (gegenions) contribute to the conductance. Because the counterions are osmotically active, polyelectrolytes show much higher osmotic pressures and diffusion rates than do nonionogenic macromolecules. The osmotic pressure of a polyelectrolyte solution is greatly reduced by the addition of a simple electrolyte which distributes itself among the two sides of the membrane according to the thermodynamic theory of Donnan equilibrium. Polyelectrolytes are called weak if they carry weakly ionized groups such as −COOH, and strong if they carry strongly ionized groups such as −COONa. On titrating polyacrylic acid with sodium hydroxide, the pH increases much more slowly than it does in a corresponding titration of a monocarboxylic acid, thus indicating a pronounced buffering capacity. Even in the case of strong polyelectrolytes, the full osmotic activity of the counterions is not realized; as counterions leave the macroion, the electrostatic potential on the latter builds up making it increasing difficult for additional counterions to escape. This binding effect becomes especially strong with multivalent counterions, whose effective concentration may be rendered several orders of magnitude smaller by a polyelectrolyte than their stoichiometric concentration.

The electric charge on the macroion has several important secondary effects. If the macroion is a flexible chain, intramolecular repulsion between charged segments will stretch out the macroion from a coiled to a more rod-like structure, resulting in much larger solution viscosities than are usually obtained with uncharged polymers under corresponding conditions. Intermolecular repulsion causes the macroions to arrange themselves so that they are as far from each other as is possible. With this ordering, the light scattering which is characteristic of solutions of ordinary macromolecules is greatly diminished, often to the vanishing point, as a result of destructive interference. These secondary effects of charge may be reduced by the addition of simple electrolytes which screen the charged elements from each other, and in some cases also lower the charge by specific counterion binding. At high enough concentrations of added salt, the light scattering of the polyelectrolyte may become sufficiently pronounced to allow its use for the determination of the molecular weight.

An interesting class of polyelectrolytes, denoted by polysoaps, is obtained by attaching soap-like molecules to the polymer chain. Such a polysoap is for instance produced by the quaternization of polyvinyl-pyridine with n-dodecyl bromide. The polysoap molecules differ from ordinary polyelectrolytes in that they may reach protein-like compactness in solution. They behave like prefabricated soap micelles and solubilize hydrocarbons and other compounds insoluble in water.

While the applications of polyelectrolytes for practical purposes depend on their general ionic properties, nevertheless large differences appear among individual members of the class in their applicability to a specific use. When polyelectrolytes are absorbed at interfaces, they affect the zeta-potential and a suspending action may result. Adsorption at growing crystal surfaces is also believed to be the reason for the high effectiveness of small amounts of certain polyelectrolytes in preventing or retarding the precipitation of calcium carbonate. The dispersion of clays by polyelectrolytes is applied in oil-well drilling. The ability of long-chain polyelectrolytes to bind together small particles has found uses in soil conditioning and in the flocculation of phosphate slimes. Because of their effect on the solution viscosity, certain polyelectrolytes are used as thickening agents. Because of their ability to bind di- and trivalent cations, some anionic polyelectrolytes are used in water softening and as enzyme inhibitors. When polyelectrolytes are adsorbed or otherwise incorporated into membranes, they make the latter permselective, hindering small ions of the same charge as the macroion from passing through the membrane while allowing free passage to small ions of opposite charge. The well-known ion-exchange resins are polyelectrolytes which have been cross-linked to prevent them from dissolving.

The most important and widespread use of polyelectrolytes is to aid in the removal of small suspended solids from waste water in the primary, secondary and dewatering stages of treatment.

ULRICH P. STRAUSS
Rutgers University
New Brunswick, New Jersey

POLYENE.

Any unsaturated aliphatic or alicyclic compound containing more than four carbon atoms in the chain and having at least two double bonds. Examples are pentadiene, cyclooctatriene.

POLYESTER FIBERS.

The principal characteristics of these fibers are described in the entry on **Fibers**. Polyester fibers are defined as synthetic fibers containing at least 80% of a long-chain polymer compound of an ester of a dihydric alcohol and terephthalic acid. The first polyester fiber to be commercialized was prepared from the ester in which the dihydric alcohol was ethylene glycol; this fiber is the material used in the largest quantity by the textile industry. For some other commercial uses, the ester 1,4-dimethyldicyclohexyl terephthalate is also used.

The original process, still in use for making the polymer, employs dimethyl terephthalate (DMT) and ethylene glycol as raw materials. A later process, using direct esterification of terephthalic acid (TPA) with ethylene glycol, also gained acceptance after the increased availability of highly purified TPA. With either process, the first step is the preparation of the intermediate diester, *bis*-hydroxyethyl terephthalate (bisHET), which then is further condensed to the polymer.

The basic process for making polyester fibers from the polymer is called *melt spinning*, i.e., heating the polymer above its melting point, forcing it through small holes in a metal plate, and then quenching the molten stream as it issues from the holes by means of a current of cool air. The spun yarn is weak and highly extensible because the polymer molecules are randomly oriented. To impart strength and dimensional stability the yarns must be drawn at temperatures above the glass-transition temperature of the material by pulling the yarn between two *godet wheels*, the second of which is rotating at a speed three to six times as fast as the first. The higher the draw ratio, i.e., the ratio of the two speeds, the more oriented the molecules become and the stronger the yarn.

The two main classes of polyester fibers are continuous-filament yarns and short-cut fibers, called staple. A wide range of deniers is available in continuous-filament yarns, varying from very fine deniers of about 20 up to 2000 for heavy industrial yarns. (Denier is the weight in grams of 9000 meters of yarn). The number of filaments in these yarns ranges from about 7 for the 20-denier yarns up to 384 for the heavy material. Staple fiber is produced in sizes ranging from 0.5 to 1.5 denier per filament. The finer deniers are used in making blends with cotton and rayon for apparel, while the coarser-denier yarns generally are used for carpets. Staple lengths vary from $1\frac{1}{4}$ to 6 inches (3.1 to 15.2 centimeters).

Fiber with no added delustrant is designated as *clear*. *Bright* fiber has about 0.1% titanium dioxide (TiO_2); *semidull* fiber has about 0.25% TiO_2; and *dull* fiber has up to 2% TiO_2. Other variations in physical properties and dyeing characteristics include optically brightened, high-modulus, high-shrink, high-tenacity, low-pilling, deep-dyeable, and cationic-dyeable fibers.

POLYESTER FILM.

Continuously extruded polyester sheet of various thickness, especially useful in electrical equipment because of its high resistivity. Its tensile strength of 25,000 psi is much greater than that of other plastic films. Sensitized polyester film is used in magnetic tapes, in the photocopying technique known as repography.

POLYESTER RESINS.

Any of a group of synthetic resins, which have polycondensation products of dicarboxylic acids with dihydroxy alcohols. They are thus a special type of alkyd resin but, unlike other types, are not usually modified with fatty acids or drying oils. The outstanding characteristics of these resins are their ability, when catalyzed, to cure or harden at room temperature under little or no pressure. Most polyesters now produced contain ethylenic unsaturation, generally introduced by unsaturated acids. The unsaturated polyesters are usually cross-linked through their double bonds with a compatible monomer, also containing ethylenic unsaturation, and thus become thermosetting. Flame resistance is imparted by using either acid or glycol ingredients having a high content of halogens, e.g., HET acid.

The principal unsaturated acids used are maleic and fumaric. Saturated acids, usually phthalic and adipic, may also be included. The function of these acids is to reduce the amount of unsaturation in the final resin, making it tougher and more flexible. The acid anhydrides are often used if available and applicable. The dihydroxy alcohols most generally used are ethylene, propylene, diethylene, and dipropylene glycols. Styrene and diallyl phthalate are the most common cross-linking agents. Polyesters are resistant to corrosion, chemicals, solvents, etc.

Common uses for polyester resins are: reinforced plastics; automotive parts; boat hulls; foams; encapsulation of electrical equipment; protective coatings; ducts; flues; and other structural applications; low pressure laminates; magnetic tapes; piping; bottles; non-woven disposable filters and low-temperature mortars.

See also **Alkyd Resins**; and **Polyester Fibers**.

Additional Reading

Deligny, P., and N. Tuck: *Resins for Surface Coatings, Alkyds & Polyesters*, Vol. 22, 2nd Edition, John Wiley & Sons, Inc., New York, NY, 2001.

Scheirs, J., and T.E. Long: *Modern Polymers*, John Wiley & Sons, Inc., New York, NY, 2003.

POLYETHER-ETHERKETONE.

Abbreviated PEEK, polyether-ether ketone is a high-temperature resistant thermoplastic suitable for wire coating, injection molding, film, and advanced composite fabrication. The wholly aromatic structure of PEEK contributes to its high-temperature performance. Its crystalline character gives it important advantages, including resistance to organic solvents and dynamic fatigue, and retention of ductility on short-term heat aging. The material is available as dry, free-flowing granules and exhibits very low water absorption. Continuous service at temperatures up to 470°F (243°C) and intermittent use up to 600°F (316°C) are possible. PEEK has good resistance to aqueous reagents, with long-term performance in super-heated water at 500°F (260°C). It resists attacks over a wide pH range, from 60% sulfuric acid to 40% sodium hydroxide at elevated temperatures. Attack can occur with some concentrated acids. No organic solvent attack has been observed on molded parts, although a limited range of solvents will stress-craze highly stressed PEEK-coated wire. Radiation resistance is excellent. Typical applications include wire and cable, automotive engine parts, aerospace components, valve plates, valve linings, oil well data logging tools, bearings, woven monofilament, and film.

POLYETHERIMIDE.

This is an amorphous, high-performance thermoplastic. The material is characterized by high strength, rigidity, heat resistance, dimensional stability, and electrical properties, combined with broad chemical resistance and processibility. Unmodified polyetherimide is amber-transparent in color and exhibits inherent flame resistance and low smoke evolution without the use of additives. The material is commercially available in several grades—unreinforced and in 10, 20, 30, and 40% glass fiber reinforced formulations for general-purpose molding and extrusion. Also available are easy-flow and release grades, wear-resistant grades, carbon fiber-reinforced grades for high-strength and static dissipation, along with a family of high-heat grades. A relatively new family of polyetherimide blends is available for use in vapor-phase soldering environments and for high-impact applications.

Polyetherimide has a chemical structure based on repeating aromatic imide and ether units. High performance strength characteristics at high temperatures are provided by rigid imide units, while the ether linkages confer the chain flexibility required for good melt processibility and flow. Polyetherimide is resistant to a wide range of chemicals, including most hydrocarbons, alcohols, and fully halogenated solvents. It is resistant to mineral acids and tolerates short-term exposure to mild bases. These resins are rated for 170–180°C (338–356°F) in continuous-use applications. Intermittent use at 200°C (392°F) is possible.

Resins are used in electronic/electrical applications (connectors, circuit boards that are vapor and wave solderable, microwave transparent radomes, integrated circuit chip carriers, miniature switches, explosion proof enclosures, lamp reflectors, and high-precision fiber optic components). Polyetherimide is used for medical components that require all forms of sterilization. Other uses are found in the transportation field, dual-ovenable cookware, as well as bearings, fasteners, and advanced composites.

POLYETHYLENE.

A thermoplastic molding and extrusion material available in a wide range of flow rates (commonly referred to as melt index) and densities. Polyethylene offers useful properties, such as toughness at temperatures ranging from −76 to +93°C, stiffness, ranging from flexible to rigid, and excellent chemical resistance. The plastic can be fabricated by all thermoplastic processes.

Polyethylenes are classified primarily on the basis of two characteristics, namely, density and melt index. The former is the criterion used to distinguished the type; and the latter for the designation as to category (ASTM-D-1248). ASTM type I polyethylene (sp. gr. 0.910–0.925) is commonly referred to as low-density, conventional, or high-pressure polyethylene. ASTM type II polyethylene (sp. gr. 0.926–0.940) is commonly referred to as medium-density or intermediate-density polyethylene. ASTM type III polyethylene (sp. gr. 0.941–0.965) is commonly called high-density, linear, or low-pressure polyethylene. High-density type III polyethylene has been divided into two ranges of density: 0.941–0.959 (considered type III); and 0.960 and higher, commonly considered type IV. Within each density classification, products with different melt indexes are categorized numerically as follows: category 1 has a melt index (MI) greater than 25; category 2 has an MI greater than 10 to 25; category 3, MI > 1.0 to 10; category 4, MI > 0.4 to 1.0; category 5 has a 0.4 maximum.

Chemical Composition. Polyethylene is formed from the polymerization of ethylene under specific conditions of temperature and pressure and in the presence of a catalyst, according to:

The reaction is exothermic and may form polymer from a molecular weight of 1000 to well over 1 million. The high-pressure process, which normally produces types I and II, uses oxygen, peroxide, or other strong oxidizers as catalyst. Pressure of reaction ranges from 15,000 to 50,000 psi (~1,020–3,400 atmospheres). The polymer formed in this process is highly branched, with side branches occurring every 15–40 carbon atoms on the chain backbone. Crystallinity of this polyethylene is approximately 40–60%. Amorphous content of the polymer increases as the density is reduced.

The low-pressure processes, such as slurry, solution, or gas phase, can produce types I, II, III, and IV polyethylenes. Catalysts used in these process vary widely, but the most frequently used are metal alkyls in combination with metal halides or activated metal oxides. Reaction pressures normally fall within 50 to 500 psi (~3.4–34 atmospheres). Polymer produced by this process is more linear in nature, with branching occurring about every 1000 carbon atoms. Linear polyethylene of types I and II is approximately 50% crystalline and types III and IV are as high as 85% crystalline.

Ethylene has been polymerized with other monomers, e.g., propylene, butene-1, hexene, ethyl acrylate, vinyl acetate, and acrylic acid, to develop such specific properties as environmental stress crack resistance, low-temperature toughness, and improved flexibility and toughness. High-molecular-weight (HDPE) and chlorinated polyethylenes have been developed to extend the property range of polyethylenes from extremely rigid to elastomeric.

Applications

Polyethylene products include extruded films for food packaging (baked goods, frozen foods, produce); nonfood packaging (heavy-duty sacks, industrial liners, shrink and stretch pallet wrap); nonpackaging (agricultural, diaper liners, industrial sheeting, trash bags); extrusion coating of films, foils, paper, and paperboard; blow molding of bottles, drums, tanks, toys, and pails; injection molding of industrial containers, closures, housewares, toys; extrusion of electrical cable jacketing, pipe, sheet, and tubing; and rotational molding of tanks, drums, toys, and sporting goods.

Properties

Tensile strength, hardness, chemical resistance, surface appearance, and flexural modulus increase with an increase in density (from type I through type IV).

Polyethylene is translucent to opaque white in thick sections, opacity increasing with density. Relatively clear film can be extruded from polyethylene, especially if it is quenched rapidly. The plastic accepts pigmentation readily. Most coloring is performed using dry-blend techniques. Color dispersion devices are required to ensure thorough mixing of resin and pigment.

Mechanical properties of polyethylenes vary with density and melt index. Low-density polyethylenes are flexible and tough; high-density products are quite rigid and have creep resistance under load. Toughness is the primary mechanical property affected by melt index, with lower-melt-index polyethylenes having greater toughness. Under loads, polyethylene is subject to creep, stress relaxation, or a combination of both.

Excellent dielectric characteristics at all frequencies and high electrical resistivity have made polyethylene one of the most important insulating materials for wire and cable.

At no-load conditions, polyethylene has good heat resistance. However, small loads can cause distortion at relatively low temperatures. Dimensional stability of polyethylene is fair to good. Dimensional changes caused by crystallization during cooling usually occur in a non-uniform pattern, resulting in warpage. Narrower molecular weight distribution resins within given families result in less warpage. Types I and II polyethylenes produced by the low-pressure process offer significant improvement in heat distortion temperatures. This property is directly related to melting point and is much higher for low-pressure, low-density resins than for conventional LDPE resins. This allows molded parts to be exposed to significantly higher service temperatures, e.g., dishwasher parts, without undergoing distortion or warpage. Most shrinkage occurs within 48 hours after fabrication and for type I and type II materials is 0.01–0.03 inch/inch (centimeter/centimeter).

Rupture of molecular bonds by external and internal stress in the presence of certain compounds is referred to as environmental stress cracking. Small molecular fractures in the amorphous regions propagate until visible cracks appear. In time, the part may fail. Chemical agents which accelerate stress cracking in polyethylene include detergents; aliphatic and aromatic hydrocarbons; soaps; animal, vegetable, and mineral oils; ester-type plasticizers; organic acids; and aldehydes, ketones, and alcohols. There is no adequate test for stress cracking.

Deterioration occurs in uncolored polyethylene exposed to weather. Ultraviolet light causes photoactivated oxidation. Satisfactory weathering formulations contain 2–2.5% well-dispersed carbon black and stabilizers. The carbon black prevents ultraviolet light penetration.

Unmodified polyethylenes are flammable and are classified in the slow-burning category by the National Board of Fire Underwriters. Burning rate is approximately 1–1.5 inches (2.5–3.8 centimeters) per minute. The flammability of polyethylene may be retarded significantly by the addition of flame retardant compounds, such as antimony trioxide along with halogenated compounds.

At room temperature, polyethylene is insoluble in practically all organic solvents, although softening, swelling, and environmental stress cracking can occur. At high temperatures, some concentrated acids and oxidizing agents chemically attack polyethylene. Above 60°C, the material becomes increasingly soluble in aliphatic and chlorinated hydrocarbons. Chemical resistance increases slightly as density is increased.

Polyethylene is water-resistant and is a good water vapor barrier. Less than 0.1% water is absorbed in a 2-inch (5-centimeter), 1/8-inch (3-millimeter) thick disk of polyethylene in 24 hours. Transmission of other gases is high when compared with that of most other plastics. Polyethylene is not satisfactory for retention of vacuum.

Fabrication

Polyethylene is readily fabricated by all methods of thermoplastic processing. The principal methods used are film and sheet extrusion, extrusion coating, injection molding, blow molding, pipe extrusion, wire and cable extrusion coating, rotomolding, and hot melt and powder coatings.

Decorating

Polyethylene parts are decorated by silk screening, hot stamping, or dry offset printing. For satisfactory printing, the surface must be oxidized by hot air, flame, chlorination, sulfuric acid-dichromate solution, or electronic bombardment. Hot air or flame methods are used with molded parts; flame or electronic methods with films. Inks specially made for polyethylene give best results. Roll-leaf hot stamping does not require pretreatment of the surface.

Design

Because of high mold shrinkage, parts must be carefully designed to minimize warpage. Wall cross-sectional thicknesses should be uniform throughout the part. Large flat areas should be avoided. Corners should be curved rather than square. Stiffening ribs should be less than 80% of the thickness of the wall to which they are attached. Thermoformed parts require liberal radii and draft angles. Slight undercuts can be incorporated when a female mold is used. Dimensional variations in a part made of polyethylene are difficult to predict. In general, greater tolerances should be allowed than with more rigid plastics.

B. W. Heinemeyer
The Dow Chemical Company
Freeport, Texas

Additional Reading

Gsell, R.A., H.L. Stein, and J.J. Ploskonka: "Characterization and Properties of Ultra-High Molecular Weight Polyethylene," *American Society for Testing & Materials*, West Conshohocken, PA, 1998.

Harris, J.M.: *Poly(Ethylene Glycol): Chemistry and Biological Applications*, American Chemical Society, Washington, DC, 1997.

Peacock, A.J.: *Handbook of Polyethylene: Structures, Properties, and Applications*, Marcel Dekker, Inc., New York, NY, 2000.

POLYHALITE.   Polyhalite, K_2Ca, $Mg(SO_4)_4 \cdot 2H_2O$ is a late evaporate mineral associated with halite, sylvite and carnallite from the famous oceanic salt deposits at Stassfurt, Germany, and near Carlsbad, New Mexico. It is of triclinic crystallization, with color grading from gray to brick-red; hardness, 3–3.5; specific gravity 2.78; translucent with vitreous luster; very bitter taste. It is a source of potassium.

POLYIMIDES.   These are heat-resistant polymers which have an imide group ($-CONHCO-$) in the polymer chain. Polyimides, poly(amide-imides), and poly(esterimides) are commercially available.

Poly(amide-imides) are prepared by the thermal degradation of a soluble poly[amide-(amic acid)]. The latter may be produced by the condensation of an aliphatic diamine with less than a molar equivalent of pyromellitic dianhydride or with a molar equivalent of a derivative of trimellitic anhydride, such as the acyl chloride in dimethylacetamide as shown in the following equation:

poly[amide(amic acid)]

$$\xrightarrow[(-H_2O)]{\Delta} \text{poly(amide-imide)}$$

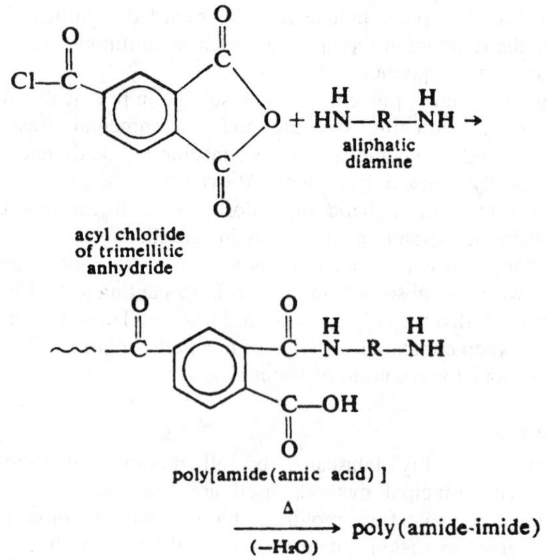

The poly(amide-imides) are soluble in dimethylacetamide but are insoluble in less polar solvents such as toluene and perchloroethylene. They are used for wire enamels, high-temperature adhesives, laminates and molded articles.

The poly(ester-imides) are produced by the thermal decomposition of the soluble poly(amic acids) which are obtained by the condensation of an aromatic diamine and the bis-(ester anhydride) of trimellitic anhydride as shown in the following equation:

These poly(ester-imides) have good electrical properties. Their tensile-modulus is about 400,000 psi (2759 MPa) at 25°C and approximately 50 percent of this modulus is retained at 200°C. Poly(ester-imide) films fail when heated at 240°C for 1000 hrs.

Polyimides are produced by the thermal dehydration of the soluble poly(amic -acid) which is obtained by the condensation of a diamine, such as 4, 4'-diaminophenyl ether and a dianhydride, such as pyromellitic dianhydride called PMDA as shown in the following equation:

It is customary to apply these polymers as the poly(amic -acids) and to dehydrate the film, coating, fiber or molded forms by heating to produce the polyimides. Polyimides are insoluble in most solvents but are attacked by alkalies, ammonia and amines. These heat resistant polymers are used without fillers and with a graphite filler.

Polyimide films have excellent electrical properties and a tensile modulus of over 400,000 psi at 25°C. Over 60 percent of this modulis is retained at 200°C. Polyimide wire enamels are stable for up to 100 thousand hours at 200°C. Polyimide fibers have a tenacity of 7 g/denier at 25°C and over 1000 hrs at 283°C is required to reduce the value to 1 g/denier.

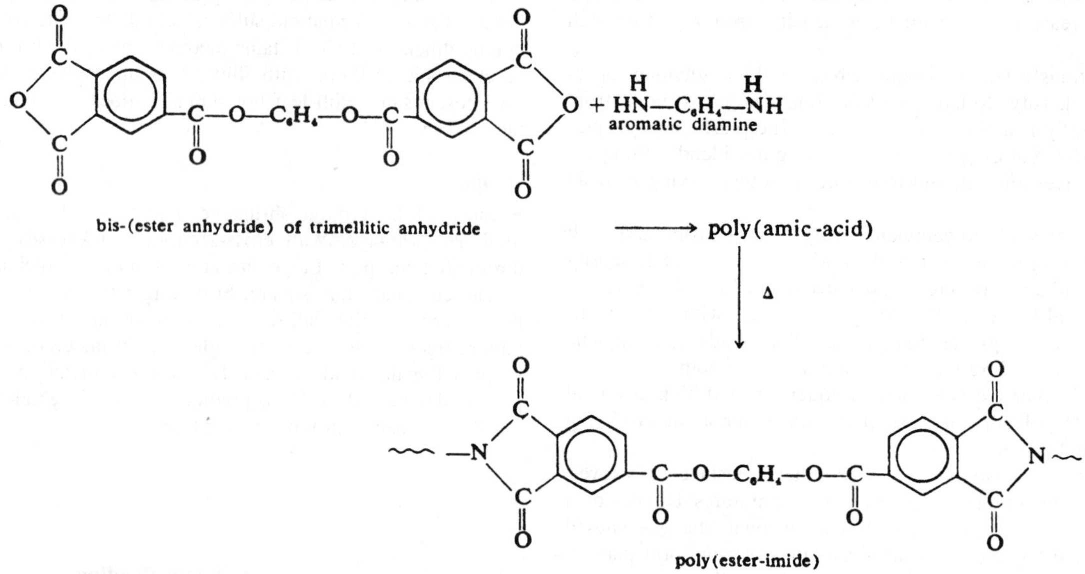

bis-(ester anhydride) of trimellitic anhydride $\longrightarrow$ poly(amic -acid)

poly(ester-imide)

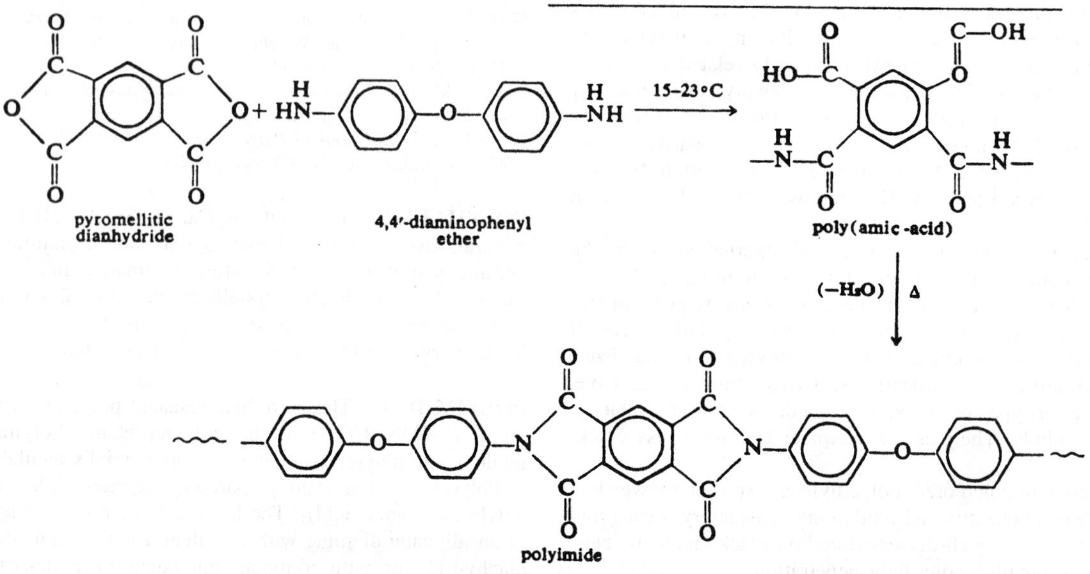

pyromellitic dianhydride   4,4'-diaminophenyl ether   poly(amic -acid)

polyimide

The coefficient of linear expansion of polyimides is $4.0–5.0 \times 10^{-5}$ in./in./°C. The heat deflection is 680°F (360°C). Polyimides have been used as binders for abrasive wheels, high-temperature laminates, wire coatings, insulating varnishes and in aerospace applications.

RAYMOND B. SEYMOUR
University of Houston
Houston, Texas

POLYISOPRENE.   See **Rubber (Natural).**

POLYMERIZATION.   Chemical reaction, particularly in organic chemistry, which provides very large molecules by a process of repetitive addition. These are of great practical importance in the field of rubbers, plastics, coatings, adhesives and synthetic fibers. The initial materials which give rise to such reactions are called *monomers*; they have molecular weights between 50 and 250 and have certain reactive or functional groups which enable them to undergo *polymerization*. The large molecules, which are formed by a polymerization reaction are called *high polymers, macromolecules*, or simply *polymers*; they usually consist of several hundred and in many cases even of several thousand monomeric units and, consequently, have molecular weights of many hundred thousands and even of several millions. The number of monomers contained in a polymer molecule determines its *degree of polymerization* (D.P.).

Polymerization processes never lead to macromolecules of uniform character but always to a more or less broad mixture of species with different molecular weights, which can be described by a *molecular weight distribution function*. The individual macromolecules of such a system belong to a *polymer-homologous series*; the molecular weight and the degree of polymerization of a given material have, therefore, always the character of *average* values.

There exist many ways to assemble small molecules to give large ones and, hence, there exist several types of polymerization reactions. The most important are the following:

(1) *Vinyl-type addition polymerization.* Many olefins and diolefins polymerize under the influence of heat and light or in the presence of catalysts, such as free radicals, carbonium ions or carbanions. Free radicals are particularly efficient in starting polymerization of such important monomers as styrene, vinylchloride, vinylacetate, methylacrylate or acrylonitrile. The first step of this process—the so-called *initiation* step—consists in the thermal or photochemical *dissociation* of the catalyst, and results in the formation of two *free radicals*:

$$
\underset{\text{Catalyst molecule}}{R-R} \xrightarrow[\text{or light}]{\text{heat}} \underset{\substack{\text{two free radical} \\ \text{type]fragments}}}{2R} \cdot \qquad (1)
$$

The most commonly used catalysts are peroxides, hydroperoxides and aliphatic azocompounds, which need activation energies between 25 and 30 kcal for decomposition.

The free radicals R· attack the monomer and react with its double bond by adding to it on one side and reproducing a new free electron on the other side:

$$
R \cdot + CH_2 = CHX \longrightarrow R - CH_2 - CHX \cdot \qquad (2)
$$

This step is called *propagation* reaction; it adds more and more monomer units to the growing chain and builds up the macromolecules while the free radical character of the chain end is maintained. Each single addition represents the reaction of a free radical with a monomer molecule—a process which requires an activation energy of 8–10 kcal.

Whenever two free radical chain ends collide with each other they can react in such a manner that the resulting products have lost their free radical character and are converted into normal stable molecules. One way is a process of *recombination*:

$$
R-(-CH_2-CHX-)-_xCH_2-CHX\cdot
$$
$$
+ R-(-CH_2-CHX-)-_yCH_2-CHX\cdot
$$
$$
\longrightarrow R-(-CH_2-CHX)-CH_2-CHX
$$
$$
-CHX-CH_2-(CH_2-CHX-)-_yR
$$

where *one* macromolecule of the degree of polymerization $(X + y + 2)$ is formed. The other is a process of *disproportionation*:

$$
R-(-CH_2-CHX)_x-CH_2-CHX\cdot
$$
$$
+ R-(-CH_2CHX-)-_yCH_2-CHX\cdot
$$
$$
\longrightarrow R-(-CH_2-CHX-)-_x-CH_2-CHX\cdot \qquad (4)
$$
$$
+ R-(-CH_2-CHX-)-_yCH_2=CH_2X
$$

where a hydrogen atom moves from one molecule to the other so that one of the two resulting molecules—the $(x + 1)$ mer—has a double bond at its end, whereas the other one—the $(y + 1)$ mer—has a saturated chain end. Reactions in the course of which free radicals are destroyed are called *termination* or *cessation* steps; they convert the transient reactive intermediates into stable polymer molecules.

Vinyl-type addition polymerization can also be carried out with *acidic catalysts* such as boron trifluoride or tin tetrachloride and with *basic catalysts* such as alkali metals or alkali alkyls. An example of the first case is the low-temperature polymerization of isobutene, which gives "Vistanex" and butyl rubber; an example of the second type is the polymerization of butadiene with sodium, which leads to buna rubber.

(2) Another important kind of addition polymerization is the formation of polyethers by the opening of epoxy ring compounds. Polyoxyethylene ("Carbowax") is produced by a sequence of additions of ethylene oxide to an alcohol or amine, as initiator:

$$
CH_3-CH_2OH + CH_2\underset{O}{\overset{\diagdown\diagup}{-}}CH_3 \rightarrow
$$
$$
CH_3-CH_2-O-CH_2-CH_3-OH +
$$
$$
CH_2\underset{O}{\overset{\diagdown\diagup}{-}}CH_2 \rightarrow
$$
$$
CH_3-CH_2-O-CH_2-CH_3-O\cdot
$$
$$
-CH_2-CH_2OH, \text{ and so on}
$$

No termination reaction occurs in this case; the reaction proceeds until all the monomer is used. This process is catalytically accelerated by the presence of alkali. A similar addition polymerization involving the opening of a ring compound is the conversion of caprolactam into polycaprolactam ("Perlon" or 6-nylon) under the influence of acidic or basic catalysts. All addition polymerizations are typical *chain reactions* with at least two or three different elementary steps cooperating in building up the resulting macromolecules.

(3) There exist other, different classes of reactions which form large molecules, namely processes in the course of which a *small fragment*, usually H_2O, is *split out* of two reacting monomers and where the monomers are chosen in such a manner that the removal of the fragment can be repeated many times. Multi step reactions of this type are called *polycondensations*; they involve the use of at least a pair of bifunctional monomers and proceed by a sequence of identical condensation steps. One important process of this type is the formation of polyesters from glycols and dicarboxylic acids. Thus the progressive removal of water from ethylene glycol and adipic acid leads to a soft, rubbery polyester ("Paracon")

$$
HOCH_2-CH_2OH
$$
$$
+ HOOC-(-CH_2-)-_4COOH
$$
$$
\longrightarrow HOCH_2CH_2-O-CO-(-CH_2-)-_4COOH
$$
$$
+ HOCH_2-CH_2OH \longrightarrow HOCH_2-CH_2OCO
$$
$$
= -(-CH_2-)_4-COOCH_2-CH_2OH, \text{ and so on.}
$$

As long as in processes of this type only *bifunctional* monomers are used, the resulting macromolecules are *linear* and, as a consequence, are of the soluble and fusible type. They can be used as fiber formers, rubbers or thermoplastic resins. If, however, some of the monomers are tri- or tetra-methylolurea, the reaction leads to three-dimensional polymeric networks which are hard and brittle thermosetting resins, such as "Bakelite" or "Glyptal."

The preceeding classification of polymerization reactions concentrates essentially on the organic chemical character of the involved monomers and on the mechanism of their interaction. There exists, however, another classification which is concerned about the manner in which polymerization reactions are carried out in practice and which is of interest and importance whenever industrial application is contemplated. We shall, therefore, briefly enumerate here the most important *polymerization techniques.*

(1) Polymerization in the *gas phase* is usually carried out under pressure (several thousand psi) and at elevated temperatures (around 200°C); the most important example is the polymerization of ethylene to form polythene.

(2) Polymerization in *solution*, essentially under normal pressure and at temperatures from $-70°C$ to $70°C$; important examples are the production of butyl rubber with boron trifluoride and the synthesis of the various "Vinylites" with benzoyl peroxide.

(3) Polymerization in *bulk* (or in block) under normal pressure in the temperature range from room temperature to about 150°C. The batch polymerization of methylmethacrylate to give "Lucite" or "Plexiglass" and the continuous polymerization of styrene to give the various types of polystyrene can be quoted as examples.

(4) Polymerization in *suspension* (bead or pearl polymerization) under normal pressure in the range from 60 to 80°C operates with a suspension of globules of an oil-soluble monomer in water and uses a monomer soluble catalyst. Substantial quantities of polystyrene and polyvinyl acetate are made by this method.

(5) Polymerization in *emulsion* under normal pressure and in the temperature range from $-20°C$ to 60°C uses a fine emulsion of oil-soluble monomers in water and initiates the reaction with a system of water-soluble catalysts. This method is probably the most important of all, because it is used in very large scale in the copolymerization of butadiene and styrene and in the polymerization of many other monomers, such as chloroprene and vinyl chloride, to produce latices of the various synthetic rubbers.

See also **Molecule**; and several articles which follow.

(Reprinted from the 4th Edition because the article by Herman F. Mark is such an apt review of the fundamentals of polymerization)

Additional Reading

Mark, H.F., N. Bikales, and J.I. Kroschwitz: *Encyclopedia of Polymer Science and Technology,* 3rd Edition, Volumes 1–4, Part 1, John Wiley & Sons, Inc., New York, NY, 2003.
Matyjaszewski, K., and T.P. Davis: *Handbook of Radical Polymerization,* John Wiley & Sons, Inc., New York, NY, 2002.
Odian, G.: *Principles of Polymerization,* 4th Edition, John Wiley & Sons, Inc., Hobokon, NJ, 2004.
Stevens, M.P.: *Polymer Chemistry: An Introduction,* 3rd Edition, Oxford University Press, New York, NY, 1998.

POLYMERIZATION (Emulsion).

Since an aqueous system provides a medium for dissipation of the heat from exothermic addition polymerization processes, many commercial elastomers and vinyl polymers are produced by the emulsion process. This two-phase (water-hydrophobic monomer) system employs soap or other emulsifiers to reduce the interfacial tension and disperse the monomers in the water phase. Aliphatic alcohols may be used as surface tension regulators.

Formulas for emulsion polymerization also include buffers, free radical initiators, such as potassium persulfate ($K_2S_2O_8$), chain transfer agents, such as dodecyl mercaptan ($C_{12}H_{25}SH$). The system is agitated continuously at temperatures below 100°C until polymerization is essentially complete or is terminated by the addition of compounds such as dimethyl dithiocarbamate to prevent the formation of undesirable products such as cross-linked polymers. Stabilizers such as phenyl Beta-naphthylamine are added to latices of elastomers.

The final product in latex form may be used for water-type paints or coatings or the water may be removed from the finely divided high-molecular weight polymer. Separation may be brought about by the addition of electrolytes, freezing or spray drying.

It is believed that polymerization of hydrophobic monomers is initiated by free radicals in the aqueous phase and that the surface-active oligomers produced migrate to the interior of the emulsifier micelles where propagation continues. Monomer molecules dispersed in the water phase also solubilize by diffusing —to the expanding lamellar micelles. These micelles disappear as the polymerization continues and the rate may be measured by noting the increase in surface tension of the system.

RAYMOND B. SEYMOUR
University of Houston, Houston, Texas

Additional Reading

Mark, H.F., N. Bikales, and J.I. Kroschwitz: *Encyclopedia of Polymer Science and Technology,* 3rd Edition, Volumes 1–4, Part 1, John Wiley & Sons, Inc., New York, NY, 2003.
Matyjaszewski, K., and T.P. Davis: *Handbook of Radical Polymerization,* John Wiley & Sons, Inc., New York, NY, 2002.
Odian, G.: *Principles of Polymerization,* 4th Edition, John Wiley & Sons, Inc., Hobokon, NJ, 2004.
Stevens, M.P.: *Polymer Chemistry : An Introduction,* 3rd Edition, Oxford University Press, New York, NY, 1998.

POLYMERIZATION (Oxidative-Coupling).

A technique for preparation of high-molecular-weight linear polymers. Schematically the reaction is represented below and involves the oxidative coupling of certain organic compounds containing two active hydrogen atoms to give a linear polymer. The hydrogens ultimately, with oxygen, form water. High-molecular-weight polymers have been prepared in this manner from phenols, diacetylenes, and dithiols.

$$n\text{HRH} \xrightarrow[\text{catalyst}]{\frac{n}{2}\ O_2} \left(R\right)_n + n\text{H}_2\text{O}$$

When 2,6-dimethylphenol is oxidized with oxygen in the presence of an amine complex of a copper salt as catalyst a high-molecular weight polyether (PPO) is formed.

The reaction is exothermic and proceeds rapidly at room temperature. The polymerization is generally performed by passing oxygen or air through a stirred solution of the catalyst and monomer in an appropriate solvent. When the desired molecular weight is attained, the polymer is isolated by dilution of the reaction mixture with a nonsolvent for the polymer. The precipitated polymer is then removed by filtration, washed thoroughly and dried. The polymer is soluble in most aromatic hydrocarbons and chlorinated hydrocarbons and insoluble in alcohols, ketones and aliphatic hydrocarbons.

A large number of other 2,6-disubstituted phenols have been oxidatively coupled. A representative list of the results is presented below.

Polymer formation readily occurs if the substituent groups are relatively small and not too electro-negative. When the substituents are bulky, the predominant product is the diphenoquinone formed by a tail-to-tail coupling. No appreciable reaction occurs when 2,6-dinitrophenol is oxidized even at 100°C.

R_1	R_2	Principal Product
methyl	methyl	polymer
methyl	ethyl	polymer
methyl	*i*-propyl	polymer
methyl	*t*-butyl	diphenoquinone
methyl	phenyl	polymer
methyl	chloro	polymer
methyl	methoxy	polymer
ethyl	ethyl	polymer
i-propyl	*i*-propyl	diphenoquinone
t-butyl	*t*-butyl	diphenoquinone
methoxy	methoxy	diphenoquinone
nitro	nitro	no reaction

A family of engineering thermoplastics based on the above technology includes PPO polyphenylene oxide, Noryl® thermoplastic resins (modified phenylene oxide) and glass reinforced varieties of each. The phenylene oxide resins are characterized by: (1) outstanding hydrolytic stability; (2) excellent dielectric properties over a wide range of temperatures and frequencies; and (3) outstanding dimensional stability at elevated temperatures. Because of these properties modified phenylene oxides are finding major application in the areas of business machine housings, appliances, automotive, TV and communications, electrical/electronic, and water distribution.

Oxidative polymerization of 2,6-diphenylphenol yields a crystallizable polymer that is characterized by a very high melting point ($\sim$480°C) and excellent electrical properties. It can be spun into a fiber with excellent thermal, oxidative and hydrolytic stability. It is marketed under the trademark Tenax®.

By performing the oxidation at elevated temperatures the phenols which would ordinarily yield polymers are converted instead to diphenoquinones. These quinones are readily reduced to the corresponding hydroquinones, compounds which promise to be useful as antioxidants and polymer intermediates.

Oxidative coupling of diacetylenes yields another unusual class of polymers. From *m*-diethynylbenzene, for example, is obtained a high-molecular weight polymer that can be cast into a tough, flexible film.

The polymer contains 96.75% carbon and on heating to about 350°F (177°C) or above it spontaneously rearranges to an insoluble and infusible material. When ignited the hydrogen in the polymer burns leaving a carbon residue.

In the same manner dithiols can be converted to polydisulfides.

ALLAN S. HAY
General Electric Company
Schenectady, New York

Additional Reading

Mark, H.F., N. Bikales, and J.I. Kroschwitz: *Encyclopedia of Polymer Science and Technology,* 3rd Edition, Volumes 1–4, Part 1, John Wiley & Sons, Inc., New York, NY, 2003.
Matyjaszewski, K., and T.P. Davis: *Handbook of Radical Polymerization,* John Wiley & Sons, Inc., New York, NY, 2002.
Odian, G.: *Principles of Polymerization,* 4th Edition, John Wiley & Sons, Inc., Hobokon, NJ, 2004.
Stevens, M.P.: *Polymer Chemistry : An Introduction,* 3rd Edition, Oxford University Press, New York, NY, 1998.

POLYMERIZATION (Radical). In addition polymerization, polymer is the sole product of the reaction so that the monomer and polymer have essentially the same chemical composition—for example, monomeric styrene and polystyrene. In a polymerization of this type, polymer is formed by a stepwise reaction in which molecules of monomer are added one at a time to a reactive center; the center grows in size while retaining its reactivity.

In a radical polymerization, the reactive centers are free radicals and the process is a typical chain reaction. The monomers in radical polymerizations normally contain carbon-carbon double bonds in their molecules; styrene is typical. Usually radical polymerization is performed in the liquid phase. The chain reaction can be divided into the following steps:

1. *Initiation.* Formation of a reactive free radical and its capture by monomer to form a center
2. *Propagation.* Reaction of a center with a molecule of monomer to form a larger center
3. *Termination.* Deactivation of a center so that it becomes incapable of further growth

4. *Transfer.* Reaction of a center with another molecule so that further growth of that particular center is prevented but a new center, capable of growth, is formed.

Commonly in radical polymerizations, initiation occurs continuously at a steady rate and is balanced by termination so that a steady concentration of growing centers (usually in the region of 10^{-8} mole/1) is established. The number of propagation reactions greatly exceeds the number of reactions of other types so that *macromolecules* are built up. The life-time of an active center is very much less than the duration of the whole process of polymerization and so the macromolecules are produced even in the earliest stages; there is not a continuous rise in the molecular weight of the polymeric product as found in polymerizations of certain other types. It is instructive to consider in some detail the component reactions in the overall process of radical polymerization.

Initiation

In principle, the simplest method for initiation is to add to the purified monomer a small amount of a substance which dissociates to fairly reactive free radicals. This initiator (or sensitizer) is chosen so that its decomposition occurs at a suitable rate at the working temperature; thus az*oiso*-butyronitrile is commonly used at about 60°C dissociating according to the equation:

$$(CH_3)_2C(CH) \cdot N:N \cdot C(CN)(CH_3)_2$$
$$\longrightarrow 2(CH_3)_2C(CN) \cdot + N_2$$

The radical adds to monomer thus

$$(CH_3)_2C(CN) \cdot + CH_2 : CHX$$
$$\longrightarrow (CH_3)_2C(CN) \cdot CH_2 \cdot CHX \cdot$$

forming the starting point of a polymer chain, i.e., an *end-group*; this reaction is the real initiation of polymerization. Initiators of other types are also used, notably peroxides, both organic and inorganic. In some cases, the initiator is chosen to give free radicals under the influence of light; this process can be useful for initiating polymerizations at comparatively low temperatures. Two-component initiating systems are widely used in this connection, an example being

$$H_2O_2 + Fe^{2+} \longrightarrow Fe^{3+} + OH^- + \cdot OH$$

which clearly would be selected for aqueous systems. At elevated temperatures or under the influence of external sources of energy (light, high-energy radiations, ultrasonics, mechanical work), many monomers polymerize apparently spontaneously without deliberate addition of sensitizer; the mechanisms of initiation under such circumstances are not completely understood.

Propagation

The propagation reaction in a radical polymerization can be represented by the general equation

$$P \cdot CH_2 \cdot CHX \cdot + CH_2 : CHX$$
$$\longrightarrow P \cdot CH_2 \cdot CHX \cdot CH_2 \cdot CHX \cdot$$

which corresponds to the conversion of a carbon-carbon double bond to two carbon-carbon single bonds. The group $-CH_2 \cdot CHX-$ in the polymer chain is referred to as the *monomer* unit. If the growing center includes more than a few monomer units, the characteristics of the growth reaction are reasonably supposed to be independent of the size of the center.

The growth reaction is exothermic (in the region of 20 kcal/mole, i.e., about 80 kj/mole); under some circumstances, polymerizations may become self-heating and difficult to control. The growth reaction involves a decrease in entropy since a free molecule of monomer becomes organized in a polymer chain. The opposing effects of changes in enthalpy and entropy indicate that, for every polymerizing system, there is a *ceiling temperature* below which the growth reaction is favored thermodynamically but above which the reverse process is favored; the value of the ceiling temperature depends on the nature of the monomer and on its concentration in the system. Certain monomers, e.g., α-methyl styrene, were once thought not to polymerize by a radical mechanism but it is now clear that they will do so provided that the experiment is performed below the ceiling temperature.

The growth reaction shown above represents *heat-to-tail* addition; the CHX groups occur at alternate sites along the main polymer chain and the unpaired electron is sited on a substituted carbon atom. Head-to-head addition, to give a polymer radical $P \cdot CH_2 \cdot CHX \cdot CHX \cdot CH_2\cdot$, may occur occasionally but is likely to be followed by tail-to-tail addition to give $P \cdot CH_2 \cdot CHX \cdot CHX \cdot CH_2 \cdot CH_2 \cdot CHX\cdot$ which can be regarded as the normal growing radical. Head-to-head groupings may well be sites of instability in the polymer.

The substituted carbon atoms in the polymer chain are asymmetric. *Stereoregular polymers* are produced if all these carbon atoms have the same configuration (all *d* or all *l*) or if the *d* and *l* configurations occur alternately; pronounced stereo-regularity is seldom achieved in radical polymerizations except perhaps at very low temperatures. When dienes are polymerized by a radical mechanism, the resulting polymers contain several distinct types of monomer unit, thus butadiene can give rise to $-CH_2 \cdot C(CH:CH_2)-$, $-CH_2 \cdot CH:CH \cdot CH_2-$ *cis*, and $-CH_2 \cdot CH:CH \cdot CH_2-$ *trans*.

Termination

In many radical polymerizations, termination occurs by interaction of pairs of growing radicals, either by combination to give $P \cdot CH_2 \cdot CHX \cdot CHX \cdot CH_2 \cdot P$ or by disproportionation to give $(P \cdot CH:CHX + P \cdot CH_2 \cdot CH_2X)$. The relative importances of these alternative processes depend upon the chemical nature of the monomer and, to a lesser extent, upon the temperature in the sense that the chance of disproportionation rises as the temperature is increased. Combination gives rise to a head-to-head grouping in the chain, and disproportionation to some unsaturated end-groups for molecules; both structural features may give rise to instability.

Termination can occur for polymer radicals of any size and so there is inevitably a wide *distribution of sizes* among the final molecules. The distribution can be predicted by application of kinetic principles and can be determined experimentally by fractionation of the whole polymer, e.g., by gel permeation chromatography. It is possible to quote only *average molecular weights* for polymers; they can be determined by several experimental methods, e.g., osmometry and viscometry.

The average *chain length* or *degree of polymerization* (DP) of the molecules in a sample of polymer is the average number of monomer units contained in them. The average *kinetic chain length* (v) in a polymerization is the number of growth reactions which, on average, occur between an initiation step and the corresponding termination process. The relationship between degree of polymerization and kinetic chain length depends on the relative frequencies of combination and disproportionation (for 100% combination DP = $2v$; for 100% disproportionation DP = v) but may also be affected by the occurrence of transfer reactions (see later).

Termination is commonly *diffusion-controlled*, i.e., it is governed by the rate at which the reactive sites in growing radicals can come together rather than by chemical factors. In viscous media, termination may be so seriously impeded that both the overall rate of polymerization and the degree of polymerization increase markedly. In systems where the polymer is insoluble in the reaction medium, polymer radicals may be trapped in the precipitated material and be able to grow but unable to participate in termination processes.

Transfer

The average molecular weight of a polymer produced in a particular system may be substantially reduced by occurrence of some types of transfer reactions. If the system contains certain substances, e.g., mercaptans, a growing polymer radical may abstract hydrogen thus

$$P \cdot + R \cdot SH \longrightarrow P \cdot H + RS\cdot$$

giving a dead polymer molecular and a new radical which can react with monomer to reinitiate polymerization. If reinitiation is 100% efficient, the effect of transfer of this type is to reduce the average degree of polymerization without affecting the rate of polymerization or the kinetic chain length. In practice, transfer is commonly accompanied by retardation since some of the new radicals are consumed in side-reactions instead of reacting with monomer; this type of transfer is said to be degradative.

Other components of the polymerization mixture, including monomer and initiator, may engage in transfer reactions. They are particularly significant for allyl monomers for which *degradative transfer* to monomer

is of such importance that rates and degrees of polymerization are very low. The radical produced in the reaction

$$P \cdot CH_2 : CH \cdot CH_2 \cdot O \cdot CO \cdot CH_3$$

$$\longrightarrow P \cdot H + CH_2 : CH \cdot CH(O \cdot CH \cdot CH_3)\cdot$$

is so stabilized by resonance that it is not reactive enough to initiate efficiently.

Transfer to polymer, causing reactivation of a polymer molecule at some point along its length, leads to the growth of *branches*. The process can occur intermolecularly and also intramolecularly; the latter process is particularly important in the free radical polymerization of ethylene at high pressure where it leads to the production of numerous short branches which considerably affect the properties of the polymer.

Transfer to polymer, the subsequent growth of branches and termination of their growth by combination lead to *cross-linking* whereby the separate polymer molecules are united to form an insoluble three-dimensional network. Cross-linking is however much more likely to occur during the polymerization of those monomers which contain more than one carbon-carbon double bond per molecule. The monomer unit in the polymer first formed still possesses an unsaturated grouping which can participate in another polymerization chain. Certain monomers of this type however engage in a special type of reaction so that reaction of one double bond in a monomer is immediately followed by reaction of the second double bond; this type of growth is shown by, for example, methacrylic anhydride.

$$P \cdot + CH_2 : C(CH_2) \cdot CO \cdot O \cdot CO \cdot C(CH_3) : CH_2 \rightarrow$$

$$P \cdot CH_2 \cdot \overset{\cdot}{C}(CH_3) \cdot CO \cdot O \cdot CO \cdot C(CH_3) : CH_2 \rightarrow$$

$$\begin{array}{c} P \cdot CH_2 \cdot C(CH_3) \cdot CH_3 \cdot \overset{\cdot}{C}(CH_3) \cdot \\ \quad | \qquad\qquad\qquad / \\ \quad CO{-}O{-}CO \end{array}$$

Inhibitors and Retarders

Various substances can reduce the rate at which a monomer is converted to polymer. Inhibitors completely suppress polymerizations whereas retarders only reduce the rate. The former deactivate very readily the primary radicals so that growth of polymer chains cannot begin; the latter deactivate growing polymer radicals so causing premature termination. Inhibitors are commonly used to stabilize monomers during storage. Many nitro compounds and quinones act as inhibitors and retarders.

Copolymerization

A process known as copolymerization can occur if reactive radicals are generated in a mixture of monomers; the resulting polymer molecules contain monomer units of more than one type. Copolymerization is of great significance academically, where it leads to information about the reactivities of monomers and radicals, and also industrially where it is used for the production of materials with special properties. Usually the composition of a copolymer is different from that of the mixture of monomers from which it is derived. For this reason, the average compositions of feed and copolymer drift during the course of a copolymerization. There are useful analogies between copolymerization and fractional distillation; special mixtures of monomers producing copolymers without change of composition are said to give azeotropic copolymerizations.

Extensive tables of so-called *monomer reactivity ratios* are available and make it possible to predict the compositions of copolymers formed from particular mixtures of monomers.

In many binary copolymerizations, there is a pronounced tendency for the two types of monomer unit to alternate along the copolymer chain. In extreme cases, there is almost perfect alteration, notably for pairs of monomers, e.g., maleic anhydride and stilbene, which do not polymerize on their own. Ternary copolymerizations are of practical importance; the kinetic treatments developed for binary copolymerizations can be extended to these systems.

J. C. Bevington
University of Lancaster
Lancaster, England

Additional Reading

Mark, H.F., N. Bikales, and J.I. Kroschwitz: *Encyclopedia of Polymer Science and Technology,* 3rd Edition, Volumes 1–4, Part 1, John Wiley & Sons, Inc., New York, NY, 2003.

Matyjaszewski, K., and T.P. Davis: *Handbook of Radical Polymerization,* John Wiley & Sons, Inc., New York, NY, 2002.

Odian, G.: *Principles of Polymerization,* 4th Edition, John Wiley & Sons, Inc., Hobokon, NJ, 2004.

Stevens, M.P.: *Polymer Chemistry : An Introduction,* 3rd Edition, Oxford University Press, New York, NY, 1998.

POLYMERS.

Polymers are very large molecules made by covalently binding many smaller molecules. The word polymer is derived from the Greek *poly* (many) and *meros* (part). The size of polymer molecules imparts many interesting and useful properties not shared by low molecular weight materials. Polymers are the fundamental materials of plastics, rubbers and most fibers, and surface coatings and adhesives, and as such are essential to modern society. Also, many important constituents of living organisms, e.g., proteins and cellulose, are biopolymers.

Classification and Nomenclature

Polymers were initially classified according to their response to temperature. Those that are softened (plasticized) reversibly by heat are known as thermoplastics. Others, though they might initially be liquid or soften once upon heating, undergo a curing (setting) reaction that solidifies them, and further heating leads only to degradation. These are known as thermosets. The ability of polymers to soften and flow at least once is one of their most valuable assets, as it allows them to be formed into complex shapes easily and inexpensively.

In general, polymers are formed by two types of reactions: condensation and addition. The formation of a polyester by polycondensation may be illustrated as follows.

$$x\,\text{HOROH} + x\,\text{HOOCR'COOH} \longrightarrow \text{H}\!\!-\!\!\!\left(\text{ORO}-\overset{\overset{\text{O}}{\|}}{\text{C}}\text{R}'\overset{\overset{\text{O}}{\|}}{\text{C}}\right)_{\!\!x}\!\!\text{OH} + (2x-1)\,\text{H}_2\text{O}$$

diol   diacid   polyester

In the polyester formula shown, parentheses enclose the repeating unit. The quantity x is the degree of polymerization, sometimes also called the chain length, the number of repeating units strung together like identical beads on a string. Neglecting the ends of the molecule, which is usually justified for large x, the molecular weight M of the polymer molecule is given by $M = mx$, where m is the molecular weight of the repeating unit. Since x can easily be in the thousands, the term macromolecules is also used to describe these materials.

Addition or chain-growth polymerization involves the opening of a double bond to form new bonds with adjacent monomers, as typified by the polymerization of ethylene to polyethylene:

$$x\,\text{H}_2\text{C}\!=\!\text{CH}_2 \longrightarrow \!\!-\!\!(\text{CH}_2-\text{CH}_2)\!\!-\!\!_x$$

Because no molecule is split out, the molecular weight of the repeating unit is identical to that of the monomer.

In terms of molecular structure, there are three principal categories of polymers, illustrated schematically in Figure 1. If each monomer is difunctional, that is, can react with other monomers at two points, a linear polymer is formed. Polymers that contain two different repeating units, say A and B, are known as copolymers. A linear polymer with a random (AABBABAAABABB) arrangement of the repeating units is a random or statistical copolymer, or just copolymer. It is termed poly(A-*co*-B), with the primary constituent listed first. A molecule in which the two repeating units are arranged in long, contiguous blocks $([A]_x[B])_y$ is a block (*b*) copolymer, poly (A-*b*-B).

A few points of tri- or higher functionality introduced along the polymer chains, either intentionally or through side reactions, give a branched polymer. Branches may grow from a linear backbone. A branched structure with the backbone consisting of one repeating unit (A) and the branches of another (B), is a graft (*g*) copolymer, poly(A-*g*-B).

As the length and frequency of branches increase, they may ultimately reach from chain to chain. If all the chains are connected together, a cross-linked or network polymer is formed. Cross-links may be built in during the polymerization reaction or may be created chemically or by radiation between previously formed linear or branched molecules (curing or vulcanization).

Structure and Properties

Various levels of structure ultimately determine the properties of a polymer.

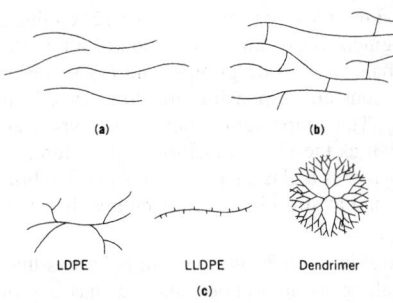

Fig. 1. Schematic diagram of polymer structures: (**a**) linear; (**b**) cross-linked; and (**c**) branched, where LDPE = low density polyethylene and LLDPE = linear low density polyethylene.

Molecular Weights. With the exception of some naturally occurring polymers, all linear and branched polymers consist of molecules with a distribution of molecular weights. Two average molecular weights are commonly defined; the number-average, $\overline{M}_n$, and the weight-average, $\overline{M}_v$.

It may be shown that $\overline{M}_w \geq \overline{M}_n$. The two are equal only for a monodisperse material, in which all molecules are the same size. The ratio $\overline{M}_w/\overline{M}_n$ is known as the polydispersity index and is a measure of the breadth of the molecular weight distribution.

Most molecular weight characterization now is done by size-exclusion chromatography (sec), also known as gel-permeation chromatography (gpc).

Size-exclusion chromatography easily and rapidly gives the complete molecular weight distribution and any desired average.

Secondary Bonding. The atoms in a polymer molecule are held together by primary covalent bonds. Linear and branched chains are held together by secondary bonds: hydrogen bonds, dipole interactions, and dispersion or van der Waal's forces. By copolymerization with minor amounts of acrylic ($\text{CH}_2\!=\!\text{CHCOOH}$) or methacrylic acid followed by neutralization, ionic bonding can also be introduced between chains. Such polymers are known as ionomers.

$$\sim \text{COOH} + \text{M(OH)}_2 + \text{HOOC} \sim \longrightarrow \sim \text{COO}^{-+}[\text{M}]^{+-}\text{OOC} \sim +2\,\text{H}_2\text{O}$$

Secondary bonds are considerably weaker than the primary covalent bonds. When a linear or branched polymer is heated, the dissociation energies of the secondary bonds are exceeded before the primary covalent bonds are broken, freeing up the individual chains to flow under stress. When the material is cooled, the secondary bonds reform. Thus, linear and branched polymers are generally thermoplastic. On the other hand, cross-links contain primary covalent bonds like those that bond the atoms in the main chains. When a cross-linked polymer is heated sufficiently, these primary covalent bonds fail randomly, and the material degrades. Therefore, cross-linked polymers are thermosets.

Stereoisomerism. Vinyl monomers, $\text{CH}_2\!=\!\text{CHR}$, generally polymerize in a head-to-tail fashion, placing the R group on every other carbon atom in the chain backbone. If a chain is conceptually stretched out, the carbon atoms in the backbone will lie in a plane. The arrangement in which the R groups are all on one side of that plane is the isotactic stereoisomer. Regular alternation of the R groups from side to side is the syndiotactic form. Random placement of the R groups is the atactic (without order) polymer. Stereoisomers are formed during polymerization, and cannot be altered subsequently by rotation about the bonds.

Crystallinity. Crystals are an ordered, regular arrangement of units in a repeating, three-dimensional lattice structure. Small molecules, which in the liquid state have three-dimensional mobility, crystallize readily when cooled. It is not so easy for polymers, because a repeating unit cannot move independently of its neighbors in the chain. Nevertheless, some polymers can and do crystallize, though never completely.

Liquid-crystal polymers exhibit considerable order in the liquid state, either in solution (lyotropic) or melt (thermotropic). When crystallized from solution or melt, they have a high degree of extended-chain crystallinity, and thus have superior mechanical properties.

The Amorphous Phase and T_g. Not all polymers crystallize, and even those that do are not completely crystalline. Noncrystalline polymer is termed amorphous. Four types of molecular motion have been identified in amorphous polymers. Listed in order of decreasing activation energy, they are *(1)* translational motion of entire molecules, *(2)* coiling and uncoiling of 40–50 C-atom segments of chains, *(3)* motion of a few (five to six) atoms along the main chain or on side groups, and *(4)* vibrations of individual atoms. Type 1 motions are responsible for flow. Type 2 motions give rise to rubber elasticity. The temperature below which type 1 and 2 motions are frozen out is known as the glass-transition temperature, T_g. Below its T_g, an amorphous polymer is a glass; hard, rigid, and often brittle. Above T_g, it becomes rubbery, and at still higher temperatures, if it is not cross-linked, it flows easily.

Effects of Crystallinity on Properties. In polymers that can crystallize, the ratio of crystalline to amorphous material has a profound effect on properties. Because the chains are packed more tightly and efficiently in the crystalline areas than in the amorphous, the crystalline phase has a higher density and greater mechanical strength. In fact, density is a common measure of degree of crystallinity.

The stiffest polymers are both crystalline and have a glassy amorphous phase. They are often useful as engineering (structural) plastics.

Solubility. Cross-linking eliminates polymer solubility. Crystallinity sometimes acts like cross-linking because it ties individual chains together. Thus, there are no solvents for linear polyethylene at room temperature, but as it is heated toward its crystalline melting point, T_m (135°C), it dissolves in a variety of aliphatic, aromatic, and chlorinated hydrocarbons. A rough guide to solubility is that like dissolves like, i.e., polar solvents tend to dissolve polar polymers and nonpolar solvent dissolve nonpolar polymers.

Polymer Synthesis

Step-Growth Polymerization. Step-growth polymerization is characterized by the fact that chains always maintain their terminal reactivity and continue to react together to form longer chains as the reaction proceeds, i.e., $x-\text{mer} + y-\text{mer} \rightarrow (x+y)\text{-mer}$. Because there are reactions that follow this mechanism but do not produce a molecule of condensation, the terms step-growth and polycondensation are not exactly synonymous.

Chain-Growth Polymerization. Chain-growth polymerizations are characterized by chains that propagate by adding one monomer molecule at a time, i.e., $x-\text{mer} + \text{monomer} \rightarrow (x+1)\text{-mer}$. There are, however, several mechanisms by which this occurs.

Free-Radical Addition. In free-radical addition polymerization, the propagating species is a free radical. The free radicals are most commonly generated by the thermal decomposition of a peroxide or azo initiator. See also **Initiators (Free-Radical)**.

Unlike step-growth polymerization, free-radical chains do not continue to grow as the reaction proceeds. The average lifetime of a growing chain, from initiation to termination, is typically less than a second. Thus, high molecular weight polymer is produced right from the beginning.

Polymerization Processes. Free-radical polymerization is carried out in a variety of ways.

Bulk polymerization involves only monomer and initiator. It gives the greatest polymer yield per unit of reactor volume and a very pure polymer.

In solution polymerization an inert solvent is added to the reaction mass. The solvent adds its heat capacity and reduces the viscosity, facilitating convective heat transfer.

In suspension polymerization, the organic reaction mass is dispersed in the form of droplets 0.01–1 mm in diameter in a continuous aqueous phase. Each droplet is a tiny bulk reactor. Heat is readily transferred from the droplets to the water, which has a large heat capacity and a low viscosity, facilitating heat removal through a cooling jacket.

In emulsion polymerization the organic monomer is emulsified with soap in an aqueous continuous phase.

Ionic Polymerization. Addition polymerization may also be initiated and propagated by anions. Ionic polymerizations are almost exclusively solution processes.

There are some important differences between anionic and free-radical addition. First, unlike free-radical initiators, which decompose and start chains randomly throughout the course of the reaction, anionic initiators ionize readily in fairly polar organic solvents or at low concentrations in hydrocarbons, and chains are started immediately, one for each molecule of initiator. Second, in the absence of impurities, there is no termination.

When the initial monomer supply is exhausted, the anionic chain ends retain their activity. Thus, these anionic chains have been termed living polymers. If more monomer is added, they resume propagation. If it is a second monomer, the result is a block copolymer.

Cationic polymerization has been used commercially to polymerize isobutylene and alkyl vinyl ethers, which do not respond to free-radical or anionic addition. See also **Elastomers**; and **Rubber (Synthetic)**.

Stereospecific Polymerization. In the early 1950s, Ziegler observed that certain heterogeneous catalysts based on transition metals polymerized ethylene to a linear, high density material at modest pressures and temperatures. Natta showed that these catalysts also could produce highly stereospecific poly-α-olefins, notably isotactic polypropylene, and polydienes. They shared the 1963 Nobel Prize in chemistry for their work. More recently, metallocene catalysts that provide even greater control of molecular structure have been introduced.

STEPHEN L. ROSEN
University of Missouri-Rolla

Additional Reading

Allcock, H., F. Lampe, and J. Mark: *Contemporary Polymer Chemistry,* 3rd Edition, Prentice Hall, Inc., Upper Saddle River, NJ, 2003.

Bahadur, P., and N.V. Sastry: *Principles of Polymer Science,* CRC Press LLC., Boca Raton, FL, 2002.

Bower, D.I.: *Introduction to Polymer Physics,* Cambridge University Press, New York, NY, 2002.

Brandrup, J., and E.H. Immergut, eds.: *Polymer Handbook,* 3rd Edition, Wiley-Interscience, New York, NY, 1989.

Brandrup, J.D.R. Bloch, E.A. Grulke, E.H. Immergut, and A. Abe: *Polymer Handbook,* 2 Vol., 4th Edition, John Wiley & Sons, Inc., New York, NY, 2003.

Cheremisinoff, N.P.: *Condensed Encyclopedia of Polymer Engineering Terms.* Elsevier Science & Technology Books, New York, NY, 2001.

Flory, P.J.: *Principles of Polymer Chemistry,* Cornell UP, Ithaca, N.Y., 1953.

Fried, J.: *Polymer Science and Technology,* 2nd Edition, Prentice Hall Professional Technical Reference, Upper Saddle River, NJ, 2003.

Kroschwitz, J.I.: *Encyclopedia of Polymer Science and Technology,* 12 Volume Set, 3rd Edition, John Wiley and Sons, Inc., Hoboken, NJ, 2004.

Morawetz, H.: *Polymers: The Origins and Growth of a Science,* Dover Publications, Inc., Mineola, NY, 2002.

Odian, G.: *Principles of Polymerization,* 3rd Edition, Wiley-Interscience, New York, NY, 1991.

Rosen, S.L.: *Fundamental Principles of Polymeric Materials,* 2nd Edition, Wiley-Interscience, New York, NY, 1993.

Rubinstein, R., and R.H. Colby: *Polymer Physics,* Oxford University Press, New York, NY, 2003.

Solomons, T.W. Graham, C.B. Fryhle, and M.M. Shenkman: *Organic Chemistry,* 8th Edition, John Wiley & Sons, Inc., New York, NY, 2003.

Scheirs, J., and T.E. Long: *Modern Polymers,* John Wiley & Sons, Inc., New York, NY, 2003.

Walton, D.J., and J.P. Lorimer: *Polymers,* Oxford University Press, New York, NY, 2001.

POLYMERS (Electroconductive). Most polymers are electrical insulators and have conductivities of 10^{-15} ohm^{-1} cm^{-1} or less. However, there are several ways to arrive at compositions of polymeric nature that have higher conductivity. A simple way to obtain such a system is to use electrically conductive fillers such as metal powders or special types of carbon black. In these physical mixtures the polymer itself does not become conductive but acts only as an inert matrix to keep the conducting filler particles together. Conduction then occurs through chains of touching, conducting particles. Control of conductivity is limited in that it tends to be high as long as continuous chains of conducting particles are present. When fewer particles are present and an insufficient number of contacts between them can be established, the conductivity drops sharply. With metal fillers, conductivities of 10^2 ohm^{-1} cm^{-1} can be realized. The particle size and the effectiveness of dispersion are important. In polymer-filler systems the conducting particles may rearrange under the influence of thermal or mechanical cycles, and the bulk conductivity tends to change as a result of such cycles.

Ionic conductivity can be found in polyelectrolytes such as the salts of polyacrylic acid, sulfonated polystyrene or quaternized polyamines (ion-exchange resins). When dry, these materials have low conductivities. However, in the presence of small amounts of polar solvents or water—some of these polyelectrolytes are somewhat hygroscopic—electrical conductivity can be observed. The currents are carried by ions (protons, for instance). Such systems can only be used in cases where very small currents are expected. Large currents would result in observable electrochemical changes of the materials. In applications as antistatic electricity coatings, conductivities of 10^{-8} ohm^{-1} cm^{-1} are sufficient.

Thermal decomposition of a large number of organic solids yield carbonaceous materials which are electrically conductive. It is believed that the conductive pyrolysis products are of polymeric nature, and that at high temperatures a carbon skeleton similar to graphite is formed. Since these products are insoluble, infusible mixtures, very little is known about their structure. Variation of the pyrolysis conditions leads to products with different conductivities. A well-studied example is polyacrylonitrile. Upon pyrolysis, an originally colorless piece of polyacrylonitrile (Orlon) fabric turns black with remarkable retention of its structure and becomes electrically conductive. Depending on the pyrolysis conditions, conductivities up to 10^{-1} ohm^{-1} cm^{-1} have been obtained. It is believed that an aromatic system of condensed six-membered rings analogous to graphite is formed.

A number of polymers of more defined chemical structure exhibit *electronic* electrical conduction. According to one of the early concepts, long conjugated unsaturated chains would make good electronic conductors, assuming that resonance would render a fraction of the electrons in the molecules mobile, and thus give rise to electrical conductivity. Synthesis of long conjugated chains has been attempted by polymerization of acetylene derivatives (phenylacetylene), by dehydration or dehydrohalogenation of polyalcohols (polyvinylalcohol) or polyhalides (polyvinylchloride) and by polycondensations of suitable monomeric reaction partners, for instance, diamines with dialdehydes. In addition to the conjugated systems with only carbon in the chain and those with carbon and nitrogen, polymeric chelates have also been reported. Here the d-orbitals of the transition elements are supposed to form a part of the conjugated system.

Problems associated with the study and fabrication of these polymeric materials arise from the fact that many of them cannot be purified because crosslinking renders them infusible or insoluble, or both. Consequently the molecular weights and other structural details cannot be determined. Some noncrosslinked polymers of these types have been described with low molecular weight and low conductivities.

In another approach, the fact that crystalline monomeric charge transfer complexes exhibit electrical conductivity led to preparation of polymeric charge transfer complexes. These can be obtained from a polymeric electron donor and a monomeric electron acceptor or from a polymeric acceptor and a monomeric donor, the former type being the more common. These polymers are not crosslinked and some are soluble, but their conductivities are generally low.

Another example of an extension of the properties of monomeric compounds into the realm of polymers is the case of the 7,7,8,8-tetracyanoquinodemethan (TCNQ) compounds. Some monomeric, salt-like derivatives of TCNQ have conductivities of the order of 1 ohm^{-1} cm^{-1}. Apparently stacks of TCNQ$^-$ ions and neutral TCNQ are responsible for these high conductivities. The polymeric TCNQ compounds consist of polycations, TCNQ$^-$ ions and neutral TCNQ, and have conductivities ranging from 10^{-10} to 10^{-3} ohm^{-1} cm^{-1}. These polymeric materials are soluble in organic solvents, can have high molecular weights (several million) and can be cast as films from solutions. Although the compounds are polyelectrolytes, they exhibit electronic conduction when dry. Among the many types of electrically conducting polymeric compositions the TCNQ derivatives seem to have an advantage because of an attractive combination of properties, namely controllable molecular weight, solubility, known chemical structure, fair chemical and thermal stability and electronic conduction controllable over several orders of magnitude.

The possibility of synthesizing polymeric superconductors has been proposed, but at the present time these ideas have not been confirmed by successful experiments.

JOHN H. LUPINSKI
General Electric Company
Schenectady, New York
KENNETH D. KOPPLE
Illinois Institute of Technology
Chicago, Illinois

Additional Reading

Allcock, H., F. Lampe, and J. Mark: *Contemporary Polymer Chemistry,* 3rd Edition, Prentice Hall, Inc., Upper Saddle River, NJ, 2003.
Bahadur, P., and N.V. Sastry: *Principles of Polymer Science,* CRC Press LLC., Boca Raton, FL, 2002.
Bower, D.I.: *Introduction to Polymer Physics,* Cambridge University Press, New York, NY, 2002.
Brandrup, J.D.R. Bloch, E.A. Grulke, E.H. Immergut, and A. Abe: *Polymer Handbook,* 2 Vol., 4th Edition, John Wiley & Sons, Inc., New York, NY, 2003.
Cheremisinoff, N.P.: *Condensed Encyclopedia of Polymer Engineering Terms.* Elsevier Science & Technology Books, New York, NY, 2001.
Fried, J.: *Polymer Science and Technology,* 2nd Edition, Prentice Hall Professional Technical Reference, Upper Saddle River, NJ, 2003.
Kroschwitz, J.I.: *Encyclopedia of Polymer Science and Technology,* 12 Volume Set, 3rd Edition, John Wiley and Sons, Inc., Hoboken, NJ, 2004.
Morawetz, H.: *Polymers: The Origins and Growth of a Science,* Dover Publications, Inc., Mineola, NY, 2002.
Rubinstein, R., and R.H. Colby: *Polymer Physics,* Oxford University Press, New York, NY, 2003.
Solomons, T.W. Graham, C.B. Fryhle, and M.M. Shenkman: *Organic Chemistry,* 8th Edition, John Wiley & Sons, Inc., New York, NY, 2003.
Scheirs, J., and T.E. Long: *Modern Polymers,* John Wiley & Sons, Inc., New York, NY, 2003.
Walton, D.J., and J.P. Lorimer: *Polymers,* Oxford University Press, New York, NY, 2001.

POLYMERS (Inorganic). Most inorganic materials can be considered polymeric since they are built up of a relatively simple atomic grouping repeated a very large number of times. Metals and simple ionic materials are easily excluded, but there still remains a large group of covalently bonded, regularly repeating materials. For example, many mineral silicates are based on the monomer $[SiO_4]^{4-}$ which is covalently bonded to form the large, two-dimensional sheets from which these materials are built. Still, we do not normally think of most of these inorganic, covalently bonded polymeric materials as polymers, because their behavior is so different from what we have come to expect of organic polymers. Such properties as high viscosity in the melt and in solution, rubbery elasticity, moldability, ability to form fibers, films, and so on, are not possessed by most of these materials. In a few cases, enough of them are present to suggest the underlying similarity in structure, for example, in the silicate minerals, crysotile asbestos forms fibers of excellent textile quality. Such samples show the possibility of obtaining useful inorganic polymers.

In the light of this discussion inorganic polymers will be considered to be those materials in which the main polymiric chain contains no organic carbon and in which behavior similar to that of organic polymers can be developed.

The question "Why is there such a difference in behavior between the usual inorganic and organic polymeric materials?" is helpful in guiding such a development. The contrast must be due to differences in molecular structure. For example, in the case of quartz the $[SiO_4]^{4-}$ tetrahedra are covalently bonded together. The high regularity of the structure and the large number of cross-links per $[SiO_4]^{4-}$ unit lead to a material which is strong and dimensionally stable, but brittle. The same situation of over-crosslinking can be found with organic polymers. If the number of cross-links in quartz is reduced by substituting organic groups, such as methyl, for some of the oxygen-silicon linkages, the silicone polymers are produced. These polymers, the only commercial inorganic ones, show that inorganic materials which behave as organic polymers can be made. However, a number of obstacles are found which are not as troublesome with organic polymers. For example, six to eight membered rings are more stable than long-chains. In the case of organic materials, if chains can be formed initially, they have considerable stability. With inorganic materials, the bonds are much more labile (constantly forming and breaking) and the long chain may break down to a collection of smaller rings.

Other factors which influence the properties of polymers can be illustrated by examining the bond energies or bond strengths and the ionic character of bonds based on Si as contrasted to similar ones based on C.

From the bond energies we would expect the homo-atomic silane polymers with Si—Si bonds to be considerably less stable than the more familiar C—C chain polymers. This expectation fits the observed facts. On the other hand, one could expect little gain in stability in the carbon series by going to an ether linked chain (—C—O—C—), while in the silicon series a silicon-oxygen linkage is stronger than any of the others. This is reflected in the very good stability of the silicone polymers. From bond energies one might also expect that a chain of alternating Si and N atoms would have good stability.

Bond Energies			
Si–Si	53 kcal	C–C	83 kcal
Si–O	106	C–O	86
Si–N	82	C–N	73
Si–C	78		

Ionic Character			
Si–Si	0	C–C	0
Si–O	51%	C–O	22%
Si–N	30	C–N	7
Si–C	12		

In addition to pure thermal stability, if the polymer is to be heated in air, one must also consider oxidative stability. In the carbon series oxidation always leads to more stable species and tends to occur, but in the silicon series there is a much higher tendency towards reaction with oxygen. This is the principal reason for the low utility of the silane polymer. Finally, a third factor in polymer stability is the ease of attack by solvents, acids, bases, etc. This is largely determined by the ionic nature of the bonds involved. The silica based polymers should be more susceptible to such attack than carbon, since they have a higher percent of ionic nature.

We do find that acidic or basic water solutions attack silicones when they are heated together under pressure. Their resistance is still high, however, because of other details of the way the polymer molecules are bound together.

The polymers which have been used to illustrate problems of inorganic polymer formation have been heteroatomic, that is, their chains are built from different atoms alternating with each other. The other structure mentioned has been homoatomic—all the atoms in the chain are the same. There are only a few homoatomic polymers of any promise. Most elements will form only cyclic materials of low molecular weight if they polymerize at all. In addition to the silane polymers, black phosphorus, a high-pressure modification of the element, forms in polymeric sheets.

Boron has similar tendencies in its compounds. The outstanding member of this class is sulfur. A transition from S_8 rings to long sulfur chains takes place over a narrow temperature range around 159°C. An increase in the viscosity of the liquid by 2000 times or more, within a range of 25°, is the tangible evidence of polymerization. The material also forms rubbery, plastic and fibrous forms when chilled to room temperature. However, it has a strong tendency to revert to the cyclic form unless stable groups are placed at the end of the chain or copolymerization hinders the process. Attempts to improve the stability of polymeric sulfur have met with some success. This is the only homoatomic inorganic polymer which appears technically interesting at present.

The class of heteroatomic polymers, besides containing the silicones, offers more promise for useful materials. Most nonmetals and many of the less positive metals form heteroatomic compositions. In many cases they are high polymers. The silicones themselves behave quite the same as organic polymers and are used as oils, rubbers and resins. The rubber is vulcanized either by the reaction of organic peroxides with the methyl groups on the chain or by incorporating groups such as Si–OH or Si–OR which crosslink on exposure to moisture in the atmosphere. The properties in which they excel over organic polymers are high thermal stability, resistance to oxidation and inertness to organic reagents. These are usually the special properties one hopes to get from inorganic polymers. The polymers may be modified by substitution of other groups for the methyl groups on the side chain and by copolymerization with other heteroatoms in which B–O–Si, Al–O–Si, Sn–O–Si, Ti–O–Al and other combinations are produced.

A similar class is the titanates. Three-dimensional Ti–O chains form pigments and pigment binders for paints and water-proofing compounds for use on cloth. Their properties can be modified by substituting monofunctional groups for some of the oxygen, for example, by forming esters to interrupt the chains.

The polyphosphates have also been widely studied. Here the phosphate ion is found as a high polymer. The molecular weight of the polymer ranges from 250,000 to 2,000,000. The polyphosphates are water soluble and form fibers. No uses have been found for this class of materials. They hydrolyze slowly in atmospheric moisture and also embrittle on standing.

Other attempts to base a polymer on B–N heteroatomic chains are being vigorously pursued although the B–O bond with an energy of 130 kilocalories is thermally more stable than the B–N bond at 100 kilocalories. The borates formed with B–O chain links, however, are too hydrolytically unstable and too thoroughly cyclized to be useful. B–N compounds are also plagued with the same weakness. However, the very high thermal stability of low molecular weight materials has encouraged the search for high polymers with the same basic structure. The combination of boron and nitrogen approximates that of carbon with carbon due to its location in the Periodic Table. See **Boron**.

One other area of materials deserves mention here. Coordination polymers are found when metal atoms are joined together by coordinating bonding involving some bridging group, e.g.

In view of the high thermal stability of monomeric chelation compounds, coordination polymers were expected to be promising for use at high temperatures. This has not proved to be the case. Thermal stabilities are usually lower than for low molecular weight materials. In addition, if the polymerization goes beyond a few monomer units, the materials tend to become insoluble and infusible so they cannot be fabricated into useful items. See **Chelation Compounds**.

T. E. FERRINGTON
Clarksville, Maryland

Additional Reading

Allcock, H., F. Lampe, and J. Mark: *Contemporary Polymer Chemistry,* 3rd Edition, Prentice Hall, Inc., Upper Saddle River, NJ, 2003.

Bahadur, P., and N.V. Sastry: *Principles of Polymer Science,* CRC Press LLC., Boca Raton, FL, 2002.

Bower, D.I.: *Introduction to Polymer Physics,* Cambridge University Press, New York, NY, 2002.

Brandrup, J.D.R. Bloch, E.A. Grulke, E.H. Immergut, and A. Abe: *Polymer Handbook,* 2 Vol., 4th Edition, John Wiley & Sons, Inc., New York, NY, 2003.

Cheremisinoff, N.P.: *Condensed Encyclopedia of Polymer Engineering Terms.* Elsevier Science & Technology Books, New York, NY, 2001.

Fried, J.: *Polymer Science and Technology,* 2nd Edition, Prentice Hall Professional Technical Reference, Upper Saddle River, NJ, 2003.

Kroschwitz, J.I.: *Encyclopedia of Polymer Science and Technology,* 12 Volume Set, 3rd Edition, John Wiley and Sons, Inc., Hoboken, NJ, 2004.

Morawetz, H.: *Polymers: The Origins and Growth of a Science,* Dover Publications, Inc., Mineola, NY, 2002.

Rubinstein, R., and R.H. Colby: *Polymer Physics,* Oxford University Press, New York, NY, 2003.

Solomons, T.W. Graham, C.B. Fryhle, and M.M. Shenkman: *Organic Chemistry,* 8th Edition, John Wiley & Sons, Inc., New York, NY, 2003.

Scheirs, J., and T.E. Long: *Modern Polymers,* John Wiley & Sons, Inc., New York, NY, 2003.

Walton, D.J., and J.P. Lorimer: *Polymers,* Oxford University Press, New York, NY, 2001.

POLYMERS (Organic). Organic high polymers have a great number of different chemical structures, ranging from completely nonpolar to very polar and even ionic materials. They all clearly resemble each other, however. The basis for this resemblance is that many of their properties are governed by their high molecular weights, which range from 5,000 to tens of millions. For example, as the molecular weight increases in a given polymer family, the tensile strength of the polymer increases markedly. In some cases it approaches that of steel on a weight-for-weight basis, especially when oriented fibers are fabricated.

In a similar way, the viscosity of the molten material changes from a free flowing liquid at low molecular weights to the highly viscous polymeric liquid where flow may be observed only over a long period of time or under a considerable applied pressure. A property which shows up only in the case of high polymers is rubbery elasticity. Here again the development of a sufficiently long and flexible molecule is necessary before rubbery behavior develops.

Because of this striking dependence on molecular size, the measurement of molecular weight and dimensions is very important. Some of the most significant early work of Staudinger was the demonstration of the existence of large molecules joined by covalent bonds. (Others felt that such large molecules were not possible). In order to determine these molecular properties the molecules must be dissolved. Thus each molecule can be separated from its neighbors and its effect measured independently. Solution properties such as osmotic pressure, light scattering and viscosity are used to measure the molecular weight of polymers. In the case of many natural polymers the ultracentrifuge has proved uniquely useful.

The osmotic pressure determination of molecular weights is based on the thermodynamic interaction of solvent and solute to lower the activity of the solvent. Experimentally, the solution is separated from the solvent by a semipermeable membrane. The solvent tends to pass through the membrane to dilute the solution and bring the activity of the solvent in both phases to equilibrium. The quantitative measurement of this tendency is obtained by allowing the liquid solution to rise in a vertical capillary connected to the solution compartment. The equilibrium height it achieves or the rate at which it rises can be measured.

The measurements are converted to effective pressure (π) at zero polymer concentrations (c) and the average molecular weight ($\overline{M}_n$) gotten from the following relation:

$$\lim_{c \to 0} \frac{\pi}{c} = \frac{RT}{\overline{M}_n}$$

(R = gas constant; T = absolute temperature).

The light-scattering method is based on similar thermodynamic inter-actions. In any solution there are random variations in concentration and refractive index. These scatter some light out of a beam passing through the liquid. In a polymer solution the nature of the fluctuations and thus the amount of scattered light (τ) depend on the attractive forces between polymer and solvent molecules. This, in turn, depends on the polymer molecular weight ($\overline{M}_w$). The following equation describes the behavior:

$$\lim_{c \to 0} \frac{Hc}{\tau} = \frac{l}{\overline{M}_w}$$

(H = a constant)

This method was developed by Peter J. W. Debye in 1944. Its evolution has been one of the most stimulating chapters of polymer physics.

In contrast to these thermodynamic methods, the viscosity molecular weight determination depends on the interference in the flow of the solvent caused by the dissolved molecules. In contrast to osmometry and light scattering, it has not been possible to develop the viscosity effect into an absolute measure of molecular weight. Rather, it must be calibrated, preferably by light scattering measurements.

The relationship between the measured limiting specific viscosity [η] and the molecular weight ($\overline{M}_v$) is as follows:

$$\lim_{c \to 0} \frac{\eta_{sp}}{c} = [\eta] = K\overline{M}_v^a$$

K and a are determined by calibration for a given polymersolvent system.

$$\eta_{sp} = \frac{\eta_{\text{solution}}}{\eta_{\text{solvent}}} - 1$$

In each of the equations above a different symbol has been used for the molecular weight. Most polymers are heterodisperse, i.e., have many molecular weight species of the same chemical nature, thus the experiments yield an average molecular weight. Osmotic pressure gives a lower average ($\overline{M}_n$ or number average) because it emphasizes the effect of small molecules, while light scattering emphasizes the larger molecules and gives a higher average, ($\overline{M}_w$ or weight average). The viscosity average ($\overline{M}_v$) is between the two and closer to the weight.

Many natural polymers are monodisperse (all molecules have the same molecular weight). In this case the ultracentrifuge which separates materials according to their effective density in solutions is a most powerful tool for molecular weight determination. With poly-disperse materials, the interpretation of ultracentrifuge results becomes more complex and widespread application of this method to synthetic polymer molecular weight determination has not yet been achieved.

In addition to the primary effect of the great length of the molecule, the details of the distribution of functional groups along the polymer chain modify the behavior of these materials. This leads to differences in their applications. For example, natural rubber exists in the rubbery state at room temperature. If cooled below zero degrees Celsius it becomes a hard, inflexible material, brittle and easily broken. On the other hand, if heated too far above room temperature, it begins to flow quite rapidly and behaves more like a fluid than a rubber. The same pattern is observed with other materials. For example, polystyrene, hard and brittle at room temperature, becomes rubbery when heated up sufficiently. The study of the mechanical behavior of polymers at various temperature is called rheology. The temperature at which the rubbery material becomes glassy is called the glass transition temperature (T_g). This transition temperature depends on the nature of the backbone and the substituent groups on the polymer chain. Rubbery materials, e.g., polyisoprene, polychloroprene, polybutadiene, the copolymer of butadiene and styrene, etc., have molecular chains with considerable flexibility. Usually small side groupings and irregularities in the chain prevent them from coming together in a regular structure. Instead the molecules stay in an amorphous random packing much like a pile of cooked spaghetti. As with cooked spaghetti, there is a tendency for the whole mass to flow, by the movement of chains past one another. With natural rubber this flow at room temperature and above had to be inhibited by tying the chains together with chemical bonds before a useful product was obtained. This cross-linking is called vulcanization in the case of rubber. The process was discovered by Charles Goodyear in 1839. A similar cross-linking to inhibit the motion of the chains is necessary to make useful products from the newer synthetic rubbers also.

Other polymers are not rubbery at room temperature, instead they exist in the glassy state. They are amorphous, but because of more bulky substituents on the polymer chain the molecule is less flexible. There is less ease of molecular motion under applied stress at room temperature. Examples of such polymers are polystyrene and polymethylmethacrylate, which are transparent due to their amorphous, homogeneous nature. When heated they first become rubbery and then, at higher temperatures, show viscous flows so that they may easily be molded. When they are cooled to room temperature the rate of flow is vanishingly small due to the stiff chains; thus items made in this way can be used at ordinary temperatures if they are not required to bear too large a load. These common organic glasses are brittle and easily broken on impact. One of the interesting problems for polymer development is to obtain impact resistance without losing transparency and without increasing the cost by an excessive amount.

A related class of polymers is the crystalline, thermoplastic materials. These also are fabricated by heating to a high temperature so that they flow; but when they are cooled ordered regions develop within them, which makes them translucent. They have much tendency to flow because of these mechanical "crosslinks" and have good dimensional stability. Polyethylene and polypropylene belong to this class. Here the chain is simple and regular so that different polymer molecules, or different parts of the same molecule, can pack next to each other. The same situation exists with "Teflon" (polytetrafluoroethylene).

As might be expected, there are polymers intermediate between the crystalline and glassy ones. For example, polyvinylchloride shows enough order to prevent its classification as glassy, but not enough to be considered crystalline.

Many crystalline polymers form part of another class of materials, the fiber-forming polymers. The formation of fibers of significant strength depends on the growth of ordered structures when the fiber is stretched. Thus crystalline materials, such as polypropylene, whose crystallites can grow on elongation, form strong fibers. In some cases fiber formation is aided by polar groups on the polymer chain. These interact with each other to give strong attractive forces which aid the molecular alignment that is needed. For example, polyacrylonitrile, the base polymer of "Orlon" and "Acrilan," has a −CN dipole in each monomer unit. The cooperative attraction of hundreds of these units along a chain gives a very strong cumulative effect. The fibers are strong even though they are not crystalline. The intermediate degree of order they develop is referred to as paracrystallinity.

In the case of other common fibers the polar groups are found in the polymer chain itself. In nylons the amide group,

$$\begin{array}{c} \text{H} \quad \text{O} \qquad \text{O} \quad \text{H} \\ | \quad\ \ || \qquad\ \ || \quad\ | \\ -\text{R}-\text{N}-\text{C}-\text{R}-\text{C}-\text{N}- \end{array}$$

offers the possibility of both dipolar attraction and hydrogen bonding. The ester group in "Dacron" and similar polyesters contributes the −C=O dipole. See **Fibers**.

The introduction of oxygen and nitrogen into the chain changes its flexibility, stability toward chemical reaction, resistance to solvents, strength and other properties. In this way quite extensive changes in behavior are obtained. Most of these heterochain polymers are prepared by condensation or ring opening reactions. The first important work in this field was the classic investigation of W. H. Carothers in the late 1920's on polyesters and polyamides. See **Polyimides**.

The toughness, high melting points and high tensile strength of many of these polymers have since led to their widespread use of fibers, films and molded objects. In general these polymers are rather high cost materials, which are used, because of their unusual properties, in places where ordinary polymers are inadequate. For example, polysulfide rubbers show outstanding solvent resistance, while the silicones are a unique class of materials inert to many environmental conditions and unwettable by most liquids. Polycarbonates and acetals are so dimensionally stable that they can be used in place of metals in molded items. Many of the interesting developments in polymers over the last few years have involved new syntheses and new variations in structure of such heterochain materials.

The materials described above are thermoplastic resins, i.e., they all melt on being heated to sufficiently high temperatures and can be molded while molten. This characteristic is associated with molecular chains which are long and stringlike with few branches on them. If, however, the polymer chains have many covalent bonds linking them together into a network, a thermosetting resin develops which may flow at an early stage of its history, but is insoluble and infusible after the full crosslinking reaction has taken place. Any of the previous chain compositions can be used in making thermosetting resins if provision for crosslinking is made by using multifunctional monomers. Some are used more commonly, e.g., the epoxy resins, phenolformaldehyde, urea formaldehyde, melamine, etc. Separate articles describing the properties of many of these plastics are found in other parts of this encyclopedia. In general they are useful because of their inertness to solvents, resistance to dimensional change on heating, rigid dimensions, physical strength, chemical resistance and abrasion resistance.

Many of the varieties of polymers which have been discussed have analogs in polymers isolated from natural systems, for example, natural rubber is a polyisoprene with a purely carbon chain. Other natural polymers are based on the C—O—C bond. The cellulose and starch polymers, which are found in plants are composed of chains of six-membered carbon-oxygen rings joined through an oxygen linkage. Cellulose and starches differ from each other in the spatial orientation of the links joining the six membered rings.

Products from different sources in each class differ in degree of branching of the molecule and in amount of crosslinking.

In unmodified cellulose the hydroxyl groups give a large amount of hydrogen bonding which leads to insolubility in most solvents. On the other hand if these are changed by chemical reactions to ether or ester groups a much more tractable material results. Cellulose acetate, butyrate and nitrate; methyl and ethyl ether and carboxy methyl ether are widely used modified celluloses. Starches also are modified, but much less commercial success has been had with them.

The polymers described above have been chemically pure, although physically heterodisperse. It is often possible to combine two or more of these monomers in the same molecule to form a copolymer. This process produces still further modification of molecular properties and, in turn, modification of the physical properties of the product. Many commercial polymers are copolymers because of the blending of properties achieved in this way. For example, one of the important new polymers of the past ten years has been the family of copolymers of acrylonitrile, butadiene and styrene, commonly called ABS resins. The production of these materials has grown rapidly in a short period of time because of their combination of dimensional stability and high impact resistance. These properties are related to the impact resistance of acrylonitrile-butadiene rubber and the dimensional stability of polystyrene, which are joined in the same molecule.

Since they are organic materials most polymers are not water soluble, however, water solubility can be obtained by substituting the proper side groups on the polymer chain. Such polymers include the nonionic materials polyvinyl alcohol, polyethylene glycol, etc., where the strong dipolar and hydrogen bonding interactions cause the solubility; and the polyelectrolytes where ionizable groups such as the carboxylate, sulfonate, quaternary ammonium, etc., cause the solubility. Many of these water soluble polymers are used to increase the viscosity of water based systems. As little as 0.1

or 0.2% of the polymer is needed to produce a very viscous solution. They also are of wide biological interest.

T. E. FERRINGTON
Clarkesville, Maryland

Additional Reading

Allcock, H., F. Lampe, and J. Mark: *Contemporary Polymer Chemistry,* 3rd Edition, Prentice Hall, Inc., Upper Saddle River, NJ, 2003.

Bahadur, P., and N.V. Sastry: *Principles of Polymer Science,* CRC Press LLC., Boca Raton, FL, 2002.

Bower, D.I.: *Introduction to Polymer Physics,* Cambridge University Press, New York, NY, 2002.

Brandrup, J.D.R. Bloch, E.A. Grulke, E.H. Immergut, and A. Abe: "Polymer Handbook, 2 Vol., 4th Edition, John Wiley & Sons, Inc., New York, NY, 2003.

Cheremisinoff, N.P.: *Condensed Encyclopedia of Polymer Engineering Terms.* Elsevier Science & Technology Books, New York, NY, 2001.

Fried, J.: *Polymer Science and Technology,* 2nd Edition, Prentice Hall Professional Technical Reference, Upper Saddle River, NJ, 2003.

Kroschwitz, J.I.: *Encyclopedia of Polymer Science and Technology,* 12 Volume Set, 3rd Edition, John Wiley and Sons, Inc., Hoboken, NJ, 2004.

Morawetz, H.: *Polymers: The Origins and Growth of a Science,* Dover Publications, Inc., Mineola, NY, 2002.

Rubinstein, R., and R.H. Colby: *Polymer Physics,* Oxford University Press, New York, NY, 2003.

Solomons, T.W. Graham, C.B. Fryhle, and M.M. Shenkman: *Organic Chemistry,* 8th Edition, John Wiley & Sons, Inc., New York, NY, 2003.

Scheirs, J., and T.E. Long: *Modern Polymers,* John Wiley & Sons, Inc., New York, NY, 2003.

Walton, D.J., and J.P. Lorimer: *Polymers,* Oxford University Press, New York, NY, 2001.

POLYMERS, WATER-SOLUBLE.

Any substance of high molecular weight which swells or dissolves in water at normal temperature. These fall into several groups, including natural, semisynthetic, and synthetic products. Their common property of water solubility makes them valuable for a wide variety of applications as thickeners, adhesives, coatings, fooe additives, textile sizing, etc.

Natural

This type is principally composed of gums, which are complex carbohydrates of the sugar group. They occur as exudations of hardened sap on the bark of various tropical species of trees. All are strongly hydrophilic. Examples are arabic, tragacanth, karaya.

Semisynthetic

This group (sometimes called water-soluble resins) includes such chemically treated natural polymers as carboxymethylcelluose, methylcellulose, and other cellulose esters, as well as various kinds of modified starches (esters and acetates).

Synthetic

The principal members of this class are polyvinyl alcohol, ethylene oxide polymers, polyvinyl pyrrolidone, polyethyleneimine.

POLYMETHINE DYES.

Polymethine dyes (PMD) represent a large class of organic colored compounds that contain a chain of methine groups $(-CH=)$ as the basic constitutive elements. According to S. Daehne's triad theory, polymethines, together with polyenes and aromatics, are the three main types of conjugated systems. The term polymethine dyes was introduced by W. Koenig in 1922. The formula of PMDs having the stable closed electron shell, in its general form, can be written as two resonance structures, where n is the number of vinylene groups in the polymethine chain,

$$[G_1^+ \!\!\leftarrow\!\! CH\!\!=\!\!CH\!\!\overline{)_n}\!\!-\!\!CH\!\!=\!\!G_2]X^\mp \leftrightarrow [G_1\!\!=\!\!CH\!\!\leftarrow\!\!CH\!\!=\!\!CH\!\!\overline{)_n}\!\!-\!\!G_2^\pm]X^\mp$$

(1)

and G_1 and G_2 are terminal or end groups of various chemical structure, i.e., acyclic, carbo-, or heterocyclic residues. Polymethine chains of symmetrical dyes $(G_1 = G_2)$ consist of an odd number of carbon atoms, and these systems carry charges, i.e., they exist as either cations or anions. In formula **(1)**, end groups differ by a number of π-electrons, and thus one is written in electron donor, and the other in electron acceptor, form.

The simplest PMDs include the polymethines (**2**), streptocyanines (**3**), oxonols (**4**), and merocyanines (**5**).

$$H_2C=CH \fbox{\!} CH=CH \fbox{\!}_n CH_2^{\pm} \qquad R_2N^+=CH \fbox{\!} CH=CH \fbox{\!}_n NR_2$$

(2)   (3)

$$O=CH \fbox{\!} CH=CH \fbox{\!}_n O^- \qquad R_2N \fbox{\!} CH=CH \fbox{\!}_n CH=O$$

(4)   (5)

A great number of different heterocyclic residues have been used as the terminal groups of PMDs. PMDs containing residues with quaternary nitrogen atoms are traditionally called cyanine dyes.

Polymethines with branched polymethine chains also exist. Among these, PMDs with symmetrically branched chains are the best known; they are referred to as trinuclear polymethine dyes (TPMD) (**6**).

(9a)   (9b)

PMDs demonstrate pronounced absorption and contain fluorescence bands that are relatively narrow and highly intense, which arise from electron transitions occurring within the polymethine chromophore

$$\fbox{\!} (CH—CH)_n CH \fbox{\!}$$

. These spectral properties give rise to a wide range of applications of PMDs such as silver halide sensitizers, laser media components, polymerization initiators, etc.

According to one classification, symmetrical dinuclear PMDs can be divided into two classes, A and B, with respect to the symmetry of the frontier molecular orbital (MO). Thus, the lowest unoccupied MO (LUMO) of class-A dyes is antisymmetrical and the highest occupied MO (HOMO) is symmetrical, and the π-system contains an odd number of π-electron pairs. On the other hand, the frontier MO symmetry of class-B dyes is the opposite, and the molecule has an even number of π-electron pairs.

For convenience, unsymmetrical PMDs should be considered as the derivatives of the corresponding symmetrical polymethines, commonly called parent dyes. In contrast to symmetrical compounds, unsymmetrical PMDs can contain an even number of methine groups in the polymethine chains, for example, styryls (**7**), where X = S, O, NCH$_3$, C(CH$_3$)$_2$, or CH=CH; and (**8**), where Y = O, S, Se or NCH$_3$.

(6)

Unsubstituted PMDs (**2**) or dyes containing odd alternate hydrocarbon residues as end groups can exist in two relatively stable forms distinguished by a π-electron pair, e.g., α, ω-diphenylpolymethines (**9**).

(7)   (8)

Electron Structure

A considerable number of experiments have shown that symmetrical PMDs in the ground state have an all-trans configuration and are nearly planar with practically equalized carbon-carbon bonds and slightly alternating valence angles within the polymethine chain.

Electron Transitions

PMD color or the nature of the electron transitions produces the widest application for PMDs. Depending on the polymethine chain length, the end-group topology, and the electron shell occupation, polymethines can absorb light in uv, visible, and near-ir spectral regions.

Chemical Reactivity

As conjugated systems with alternating π-charges, the polymethine dyes are comparatively highly reactive compounds. Substitution rather than addition occurs to the equalized π-bond. If the nucleophilic and electrophilic reactions are charge-controlled, reactants can attack regiospecifically.

Protonation. As expected from π-electron distribution, the proton attacks the negatively charged odd positions in the polymethine chain e.g.,

$$H^+$$

$$G^+ - CH=CH-CH=CH-CH=G \longrightarrow G^+ - CH_2$$

$$-CH=CH-CH=CH-G^+$$

The destruction of the total π-system causes the color to vanish the protonated molecule absorbs light in the uv region. Protonation has been proved to be reversible.

Other Electrophilic Reactants. Reversibility of the electrophilic reactions enables substituted dye derivatives to be obtained. Halogen atoms are mobile in the polymethine chain, and the derivatives themselves can function as halogenation reagents.

The dye can be formulated by means of a phosgene, Chloromethylation leads to the alkylated polymethines; nitration of cyanines results in mononitro-substituted compounds.

Nucleophilic Reagents. In contrast to electrophilic reactions, nucleophiles attack positively charged, even carbons in the chain. The reactions lead to the exchanging of substituents or terminal residues. Thus, SR and OR groups, or halogen atoms can be exchanged by other suitable nucleophiles.

If the dye contains no mobile substituents in the chain, nucleophiles attack primarily the end carbon atoms (changing of terminal residues). Nucleophilic reactions with the methylene bases of the corresponding heterocycles result in polymethines containing new end groups.

The asymmetrical polymethines appear to be ambivalent systems, and the number of possible reaction paths increases considerably as a result.

Reactions with Parting of Radicals. The one-electron oxidation of cationic dyes yields a corresponding radical dication. The stability of the radicals depends on the molecular structure and concentration of the radical particles. They are susceptible to radical-radical dimerization at unsubstituted, even-membered methine carbon atoms.

Photochemistry. The most important photochemical processes that proceed from the excited state are geometrical isomerization and photochromic reactions. Photoisomerization of polymethines is a reversible trans-cis transfer. The cis-isomer absorbs at longer wavelength with a smaller intensity than the trans-isomer.

Applications of Polymethines. The most important reason for the large number of technical applications of polymethine dyes is their relatively low electron-transition energies and their highly intense and narrow spectral bands. Indeed, polymethines display strong light absorption and emission, between 300 and 1600 nm. These dyes have been used as photographic sensitizers and desensitizers, as laser dyes, as probes of membrane potentials, and in other applications where the theoretical aspects of polymethines are useful.

Spectral Sensitization. Photographic silver halide emulsions are active with light only up to about 500 nm. However, their sensitivity can be extended within the whole visible and near-ir spectral region up to about 1200–1300 nm. This is reached by the addition of deeply colored dyes that transfer excited electrons.

According to the electron-transfer mechanism of spectral sensitization, the transfer of an electron from the excited sensitizer molecule to the silver halide and the injection of photoelectrons into the conduction band are the primary processes. Thus, the lowest vacant level of the sensitizer dye is situated higher than the bottom of the conduction band. The regeneration of the sensitizer is possible by reactions of the positive hole to form radical dications. If the highest filled level of the dye is situated below the top of the valence band, desensitization occurs because of hole production.

Based on correlations between energy level positions and electrochemical redox potentials, it has been established that polymethine dyes with reduction potentials less than −1.0 V (vs SCE) can provide good spectral sensitization. On the other hand, dyes with oxidation potentials lower than +0.2 V are strong desensitizers.

Improvement of spectral sensitization can be accomplished by dye combinations. The effect has been found to often be greater than the predicted additive sensitivity increase. This phenomenon is called supersensitization, which is applied most effectively to polymethine aggregates. The opposite phenomenon, a decrease of sensitivity, is known as desensitization. The main reasons for desensitization are the results of relative electron level positions as well as the secondary processes of the photoelectrons.

Quantum Electronics and Laser Dyes. In quantum electronics, PMDs are usually applied as mode-locking compounds in passive mode-locked lasers as well as active laser media. The required characteristics of dyes used as passive mode-locking agents and as active laser media differ in essential ways. For passive mode-locking dyes, short excited-state relaxation times are needed; dyes of this kind are characterized by low fluorescence quantum efficiencies caused by the highly probable nonradiant processes. On the other hand, the polymethines to be applied as active laser media are supposed to have much higher quantum efficiencies, approximating a value of one.

Photopolymerization. In many cases polymerization is initiated by irradiation of a sensitizer with ultraviolet or visible light. The excited state of the sensitizer may dissociate directly to form active free radicals, or it may first undergo a bimolecular electron-transfer reaction, the products of which initiate polymerization.

Synthesis

By varying the molecular structure, it is possible to synthesize dye initiators with the required characteristics. Polymethine dyes with different chain length, end groups, and substituents, or with other variations of the chromophore, have been synthesized.

Polymethine dyes consist of three main structural elements: two identical or different end groups and a conjugated chain containing an odd number of methine groups. There are many possibilities for changing the chromophore constitution: using new heterocyclic systems for the end groups, introducing specific substituents in either the chain or in the residues, branching of the polymethine chromophore, replacement of the methine groups by heteroatoms, and cyclization of the chain by conjugated or unconjugated bridges.

General Aspects. As a rule, the end-group synthones have the following reactive centers: an activated methyl or methylene group with high CH acidity, a functional group (OR, SR, X(halide), NR_2) leaving as an anion in the reaction, and a carbonyl or heteroanalogous group as a leaving group. Complementary reactive centers are needed in the chain synthones in the α- and ω-positions. In particular, derivatives of formic acid are used to prepare monomethine dyes; for dyes with longer chromophores, the application of vinylogous aminals or ω-methylpolyenals are preferred.

Molecular Design

Extension of the applications of polymethine dyes has required special spectral and other characteristics. As a rule, the search for and synthesis of promising new compounds having desired properties imply the preliminary estimation of their most important parameters on the basis of elaborated theoretical conceptions. Thus, an effective way of governing electron properties consists of the variation of molecular topology of polymethines.

The first encouraging results for engineered dyes having desired ground- and excited-state properties were reported in 1992. To design effective spectral sensitizers, it is necessary to engineer dyes having special positions of their frontier levels, desired wavelengths of the absorption band, and special thermodynamic stability after light excitation.

Also, using dyes as laser media or passive mode-locked compounds requires numerous special parameters, the most important of which are the band position and bandwidth of absorption and fluorescence, the luminescence quantum efficiency, the Stokes shift, the possibility of photoisomerization, chemical stability, and photostability. Applications of PMDs in other technical or scientific areas have additional special requirements.

ALEXY D. KACHKOVSKI
Institute of Organic Chemistry
National Academy of Sciences of the Ukraine

Additional Reading

Fabian, J. and S. Daehne: *J. Mol. Structure (Theochem.)* **92**, 217 (1983).
Hamer, F.M.: *The Cyanine Dyes and Related Compounds,* Interscience Publishers, New York, NY, 1964.
Mason, S.F.: in K. Venkataraman, ed., *The Chemistry of Synthetic Dyes,* Vol.3, Academic Press, Inc., New York, NY, 1970, p. 169.
Tyutyulkov, N. and co-workers: *Polymethine Dyes: Structure and Properties,* St. Kliment Ohridski University Press, Sofia, Bulgaria, 1991.

POLYMETHYLBENZENES.

Polymethylbenzenes (PMBs) are aromatic compounds that contain a benzene ring and three to six methyl group substituents (for the lower homologues see **Benzene; Toluene; Xylenes and Ethylbenzene**). Included are the trimethylbenzenes, C_9H_{12} (mesitylene (**1**), pseudocumene (**2**), and hemimellitene (**3**)), the tetramethylbenzenes, $C_{10}H_{14}$ (durene (**4**), isodurene (**5**), and prehnitene (**6**)), pentamethylbenzene, $C_{11}H_{16}$ (**7**), and hexamethylbenzene, $C_{12}H_{18}$ (**8**). The PMBs are primarily basic building blocks for more complex chemical intermediates.

Physical Properties

The structures of the eight PMBs are shown here and their physical and thermodynamic properties are given in Table 1.

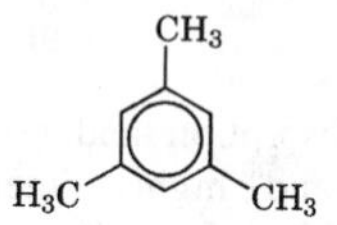

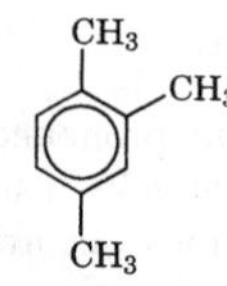

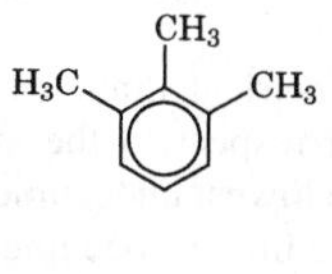

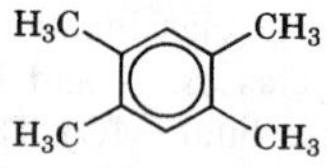

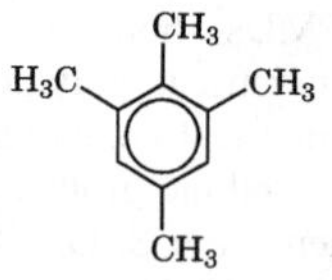

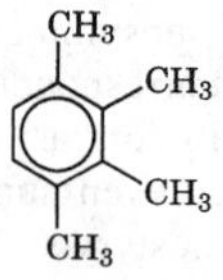

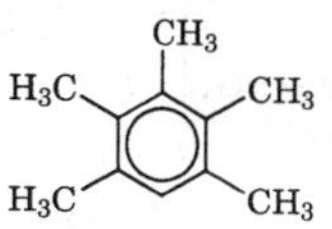

Manufacture

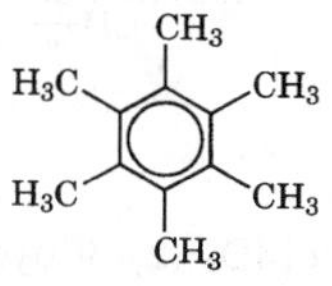

High purity mesitylene, hemimellitene, and durene are often produced synthetically, whereas pseudocumene is obtained from extracted C_9 reformate by superfractionation.

Koch Chemical Company is the only U.S. supplier of all PMBs (except hexamethylbenzene).

Health and Safety Factors

The PMBs, as higher homologues of toluene and xylenes, are handled in a similar manner, even though their flash points are higher (see Table 1). Containers are tightly closed and use areas should be ventilated. Breathing vapors and contact with the skin should be avoided.

Uses

Pseudocumene is used as a component in liquid scintillation cocktails for clinical analyses. Pseudocumene and durene are oxidized to trimellitic anhydride and pyromellitic dianhydride, respectively. Mesitylene is a key building block for important antioxidants and agricultural chemicals. Prehnitene, isodurene, pentamethylbenzene, and hexamethylbenzene

TABLE 1. PHYSICAL AND THERMODYNAMIC PROPERTIES OF POLYMETHYLBENZENES

Property	Systematic (benzene) name							
	1,3,5-Trimethyl-benzene	1,2,4-Trimethyl-benzene	1,2,3-Trimethyl-benzene	1,2,4,5-Tetramethyl-benzene	1,2,3,5-Tetramethyl-benzene	1,2,3,4-Tetramethyl-benzene	Pentamethyl-benzene	Hexamethyl-benzene
mol wt	120.194	120.194	120.194	134.221	134.221	134.221	148.248	162.275
bp, °C	164.74	169.38	176.12	196.80	198.00	205.04	231.9	263.8
flash point, °C	43.0	46.0	51.0	67.0	68.0	73.0		
density, g/cm^3								
at 20°C	0.8651	0.8758	0.8944	0.8875[a]	0.8903	0.9052	0.917[b]	solid
25°C	0.8611	0.8718	0.8905	0.8837[a]	0.8865	0.9015	0.913[b]	solid
freezing point, °C in air at 101.3 kPa[b]	−44.694	−43.881	−25.344	79.240	−23.689	−6.229	54.35	165.7
refractive index, n_D at 25°C	1.49684	1.50237	1.51150	1.5093[a]	1.5107	1.5181	1.525[a]	solid
surface tension, mN/m (= dyn/cm), at 20°C	28.84	29.72	31.28	solid	33.51	35.81	solid	solid
critical temperature, °C	364.20	376.02	391.32	401.85	405.85	416.55		
critical pressure, kPa[b]	3127	3232	3454	2940	2860	2860		
critical volume, cm^3/mol	427	427	427	482	482	482		
heat of vaporization at bp, kJ/mol[c]	39.0	39.2	40.0	45.52	43.81	45.02	45.1	48.2
heat of formation at 25°C, liquid, kJ/mol[c]	−63.4	−61.8	−58.5	−119.87[d]	−96.35	−90.20	−135.1[d]	−171.5[b]
heat of combustion, kJ/mol[c] at 25°C	5193.1	5194.8	5198.0	5816.0[a]	5839.6	5845.7	6490.8[a]	
dielectric constant, at 20°C		2.383	2.636					
specific heat, C_p, liquid, at 25°C, J/mol·K)[c]	200.5	214.9	216.4		240.7	238.3		

[a] Supercooled liquid.
[b] To convert kPa to atm, divide by 101.3.
[c] To convert J to cal, divide by 4.184.
[d] Crystal.

have no significant commercial uses. The higher polymethylbenzenes show potential as highly regiospecific methylation agents for methylation of 4-alkylbiphenyls to form 4,4′-alkylmethylbiphenyls which can be oxidized to the monomer 4,4′-biphenyldicarboxylic acid (see **Liquid Crystalline Materials**).

H. W. EARHART
Consultant
ANDREW P. KOMIN
Koch Chemical Company

Additional Reading

H.W. Earhart, *The Polymethylbenzenes*, Noyes Development Corp., Park Ridge, N.J., 1969.
U.S. Pat. 3,542,890 (Nov. 24, 1970), H.W. Earhart and G. Sugerman (Sun to Koch Industries Inc.).

POLYMORPHISM. 1. A phenomenon in which a substance exhibits different forms. Dimorphic substances appear in two solid forms, whereas trimorphic exist in three, as sulfur, carbon, tin, silver iodide, and calcium carbonate. Polymorphism is usually restricted to the solid state. Polymorphs yield identical solutions and vapors (if vaporizable). The relation between them has been termed "physical isomerism." See **Allotropes** under **Chemical Elements**. See also **Mineralogy**.

2. The occurrence of individuals of distinctly different structure or appearance within a species. In many cases two such forms occur and the species is said to be dimorphic rather than polymorphic.

Polymorphism depends upon many different conditions in various groups of animals. The various forms may be adapted for different places in a life cycle, for special parts in a colonial or social organization, or for special stages in a metamorphosis. They may also result from the incidence of different environmental conditions due to seasons or to unusual climatic conditions.

POLYOL. A polyhydric alcohol, i.e., one containing three or more hydroxyl groups. Those having three hydroxyl groups (trihydric) are glycerols; those with more than three are called sugar alcohols, with general formula $CH_2OH(CHOH)_nCH_2OH$, where n may be from 2 to 5. These react with aldehydes and ketones to form acetals and ketals.

See also **Alcohols**; and **Glycerol**.

POLYOLEFIN. A class or group name for thermoplastic polymers derived from simple olefins; among the more important are polyethylene, polyproplene, polybutenes, polyisoprene, and their copolymers. Many are produced in the form of fibers. This group comprises the largest tonnage of all thermoplastics produced.

See also **Elastomers**; **Polyethylene**; **Polypropylene**; and **Thermoplastic**.

POLYORGANOSILICATE GRAFT POLYMER. An organoclay to which a monomer or an active polymer has been chemically bonded, often by the use of ionizing radiation. An example is the bonding of styrene to a polysilicate containing vinyl radicals, resulting in the growth of polystyrene chains from the surface of the silicate. Such complexes are stable to organic solvents. They have considerable use potential in the ion-exchange field, as ablative agents, reinforcing agents, and hydraulic fluids.

POLYPEPTIDE. A compound composed of two or more amino acids, similar in many properties to the natural peptones. The amino acids are joined by peptide groups

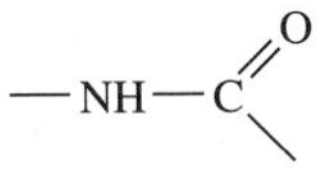

formed by the reaction between an $-NH_2$ group and a

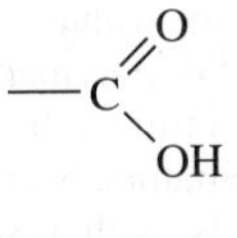

group, whereby there is elimination of a molecule of water, and formation of a valence bond. They may be termed di-, tri-, tetra-, etc., peptide according to the number of amino acids present in the molecule.

The sequence of amino acids in the chain of a protein is of critical importance in the biological functioning of the protein, and its determination is very difficult. The chains may be relatively straight, or they may be coiled or helical. In the case of certain types of polypeptides, such as the keratins, they are crosslinked by the disulfide bonds of cystine. Linear polypeptides can be regarded as proteins. See also **Amino Acids**; and **Proteins**.

R.C.V.

POLYPROPYLENE. [CAS: 9003-07-0]. A synthetic crystalline thermoplastic polymer, $(C_3H_5)_n$, with molecular weight of 40,000 or more. Low-molecular-weight polymers are also known which are amorphous in structure and used as gasoline additives, detergent intermediates, greases, sealant, and lubricating oil additives. They are also available as high-melting-point waxes.

Polypropylenes are derived by the polymerization of propylene with stereospecific catalyst, such as aluminum alkyl. These polymers are translucent, white solids, insoluble in cold organic solvents, softened by hot solvents. They maintain their strength after repeated flexing. They are degraded by heat and light unless protected by antioxidants. Polypropylenes are readily colored, exhibit good electrical resistance, low water absorption and moisture permeability. They have rather poor impact strength below 15°F (-9.5°C). They are not attacked by fungi or bacteria, resist strong acids and alkalies up to about 140°F (60°C), but are attacked by chlorine, fuming nitric acid, and other strong oxidizing agents. They are combustible, but slow burning.

Polypropylenes are available as molding powder, extruded sheet, cast film, textile staple, and continuous-filament yarn. They find use in packaging film; molded parts for automobiles, appliances, and housewares; wire and cable coating; food container closures; bottles; printing plates; carpet and upholstery fibers; storage battery cases; crates for soft-drink bottles; laboratory ware; trays; fish nets; surgical casts; and a variety of other applications.

See also **Fibers**; and **Polymers**.

POLYSTYRENE. [CAS: 9003-53-6]. General purpose (or crystal) polystyrene is a clear, water-white, glassy polymer commonly derived from coal tar and petroleum gas. Physical properties of this material can be altered by addition of modifying agents, such as rubber (for increased toughness), methyl or α-methyl styrene (for heat resistance, methyl methacrylate (for improved light stability), and acrylonitrile (for chemical resistance). In general, varying the level of modifying agent (e.g., comonomer) will alter the level of desired property improvement.

Special grades of polystyrene include impact polystyrene modified with ignition-resistant chemical additives. These were developed because of increased emphasis on product safety and used in many electrical and electronic appliances. The addition of flame-retardant chemicals does not make the polymer noncombustible, but increases its resistance to ignition and decreases the rate of burning when exposed to a minor fire source.

Chemistry

The polymerization of styrene is an exothermic chain reaction which proceeds by all known polymerization techniques. This reaction can be shown schematically as:

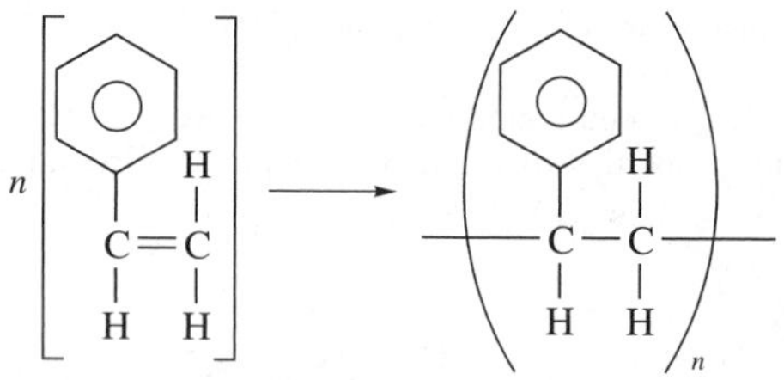

The exact nature of the beginning and end of such a polymer chain is not certain. In general, the polymer can be characterized by its average degree of polymerization, i.e., the value of n, or more precisely by the distribution of n values. The heat of polymerization is 17.4 ± 0.2 kcal/mole at 26.9°C. The reaction may be initiated by heat or by means of catalysts. Organic peroxides are typical initiators. Styrene also will polymerize in the presence of various inert materials, such as solvents, fillers, dyes, pigments, plasticizers, rubbers, and resins. Moreover, it forms a variety of copolymers with other mono- and polyvinyl monomers.

It is a matter of general observation that with styrene, the polymerization-rate curves will exhibit three distinct phases, the nature of which can be determined by the polymerization conditions and the purity of the monomer: (1) an initial slow period at the beginning of the reaction, known as the *induction period*, which appears to be associated with the presence of an inhibitor or other impurity in the monomer; (2) a period of relatively rapid polymerization, which persists almost to the end of the reaction, and for which the rate is exponentially dependent upon temperature; and (3) a final slowing down in rate as the reaction approaches completion and the

monomer becomes exhausted. This effect is particularly apparent at low temperatures with relatively impure monomers.

General Properties

The *specific gravity* of general purpose and impact polystyrene is 1.05. It can vary for copolymers. It is higher for some specialty grades. Density varies slightly with pressure, but for practical purposes, the polymer is noncompressible.

In terms of *heat-resistance*, deflection temperatures range from about 66 to 99°C (170 to 215°F), depending upon the formulation. Continuous resistance to heat for polystyrene is usually 60 to 80°C (140 to 175°F). Time and load have a significant influence on the useful service temperature of a part.

Polystyrene is nontoxic when free from additives and residuals. It has no nutritive value and does not support fungus or bacterial growth.

Dimensional stability of polystyrene resins is excellent. Mold shrinkage is small. The low moisture absorption (about 0.02%) allows fabricated parts to maintain dimensions and strength in humid environments.

General-purpose polystyrene is water white, and transmission of visible light is about 90%. Modifiers reduce this property, and translucence results. The refractive index is about 1.59; critical angle about 39. Polystyrene molecules do not have the same optical properties in all directions. When molecules become oriented in a given direction during fabrication, a double refraction occurs and a birefringence effect can be observed if the part is examined through a polarized lens under a polarized light source. Injection moldings often exhibit birefringence in a random pattern. This can be beneficial if the birefringence is in the direction of load.

In terms of *weatherability*, polystyrene does not exhibit ultraviolet stability and is not considered weather-resistant as a clear material. Continuous, long-term exposure results in discoloration and reduction of strength. Improvement in weatherability can be obtained by the addition of ultraviolet absorbers, or by incorporating pigments. The best pigmenting results are obtained with finely dispersed carbon black.

In terms of *chemical resistance*, polystyrene has a high resistance to water, acids, bases, alcohols, and detergents. Chlorinated solvents will mar the surface and, in the presence of an external load or high internal stresses, will cause failure. Aliphatic and aromatic hydrocarbons, in general, will dissolve polystyrene. Such foodstuffs as butter and coconut oil should be avoided. The chemical resistance depends upon chemical concentration, time, and stress.

Typical mechanical properties of polystyrene are given in the accompanying table. The long-term load-bearing strength of most polystyrene materials is about one-third of the typical tensile strength given in Table 1.

Uses

Packaging applications are the most extensive. Meat, poultry, and egg containers are thermoformed from extruded foamed polystyrene sheet. The fast-food market also accounts for a substantial amount of polystyrene for takeout containers where the insulation value of a foamed container is an advantage. Containers, tubs, and trays formed from extruded impact polystyrene sheets are used for packaging a large variety of food. Biaxially oriented polystyrene film is thermoformed into blister packs, meat trays, container lids, and cookie, candy, pastry, and other food packages where clarity is required.

TABLE 1. COMPRESSION MOLDED PROPERTIES OF POLYSTYRENE

Property	General purpose psi (MPa)	Impact psi (MPa)
Tensile strength	5500–8000 (38–55)	2500–5000 (17–35)
Compressive strength	21,000–16,000 (145–110)	4500–9000 (31–62)
Flexural strength	9000–15,000 (62–104)	5000–10,000 (35–69)
Tensile (Young's) modulus	400,000–500,000 (2760–3450)	200,000–400,000 (1880–2760)
Impact strength, Izod, foot-pounds/inch	0.3–0.5	1–4
Hardness, Rockwell M	65–80	60
Elongation, %	0.8–2.0	5–50

Housewares is another large segment of the use of polystyrenes. Refrigerator door liners and furniture panels are typical thermoformed impact polystyrene applications. Extruded profiles of solid or foamed impact polystyrene are used for mirror or picture frames, and moldings for construction applications.

General-purpose polystyrene is extruded either clear or embossed for room dividers, shower doors, glazings, and lighting applications. Injection molding of impact polystyrene is used for household items, such as flower pots, personal care products, and toys. General-purpose polystyrene is used for cutlery, bottles, combs, disposable tumblers, dishes, and trays.

Injection blow molding can be used to convert polystyrene into bottles, jars, and other types of open containers.

Impact polystyrene with ignition-resistant additives is used for appliance housings, such as those for television and small appliances. Structural foam impact polystyrene modified with flame-retardant additives is used for business machine housings and in furniture because of its decorability and ease of processing. Consumer electronics, such as cassettes, reels, and housings, is a fast growing area for use of polystyrenes. Medical applications include sample collectors, petri dishes, and test tubes.

In an effort to make homes and other buildings more energy efficient, the use of polystyrenes in extruded foam board with flame-retardant additives for walls and under slabs has experienced exceptional growth in recent years. Used as a sheeting material, extruded foam board complies with the requirements of the major building codes as well as federal and military specifications.

In general, polystyrene is used in applications where ease of fabrication and decorability are required. Polystyrene has excellent electrical properties, good thermal and dimensional stability, resistance to staining, and low cost. General purpose polystyrene is preferred where clarity is also of prime concern. Impact polystyrene is preferred where toughness is needed.)

POLYSULFONE. This is a transparent, heat-resistant, ultrastable high-performance engineering thermoplastic. It is amorphous in nature and has low flammability and smoke emission. It possesses good electrical properties that remain relatively unchanged up to temperatures near its glass transition temperature of 374°F (190°C). The molecular structure of polysulfone features the diaryl sulfone group. This group tends to attract electrons from the phenyl rings. Oxygen atoms para to the sulfone group enhance resonance and produce oxidation resistance. High resonance also strengthens the bonds spatially, fixing the grouping into a planar configuration. The polymer consequently has good thermal stability and rigidity at high temperatures. Ether linkages provide chain flexibility, thereby imparting good impact strength. The polymer resists hydrolysis and aqueous acid and alkaline environments, because the linkages connecting the benzene rings are hydrolytically stable.

Polysulfone is available in transparent and opaque colors in both molding and extrusion grades (unfilled). A special medical grade meets U.S.P. criteria. Two mineral-filled grades are available; one is designed specifically for plating using conventional techniques; the other is a combination of polysulfone compounds with glass fiber or beads as well as other fillers, such as *Teflon.*

Polysulfone is widely used in medical instrumentation and trays to hold instruments during sterilization. It is also used in food processing equipment, including piping, scraper blades, milking machines, steam Tables, microwave oven cookware, coffee makers, coffee decanters, and beverage dispensing tanks. Electrical/electronic uses include connectors, fuse and switch housings, coil bobbins and cores, TV components, capacitor film, and structural circuit boards. In chemical processing equipment, uses are found in corrosion-resistant piping, both transparent and glass fiber-bonded, tower packing, pumps, filter modules, and membranes.

Polysulfone has high resistance to acids, alkalies, and salt solutions, and good resistance to detergents, oils, and alcohols, even at elevated temperatures under moderate stress. It is attacked by polar organic solvents, such as ketones, chlorinated hydrocarbons, and aromatic hydrocarbons. Polysulfone can be used continuously in steam at temperatures up to 300°F (149°C). Maximum stress in water at 180°F (82°C) is about 2000 psi (14 mPa) for steady loads and up to 2500 psi (17 mPa) for intermittent loads. Polysulfone offers a good combination of electrical properties—dielectric strength and volume resistivity are high, while dielectric constant and dissipation factor are low. The latter two properties (which determine lossiness) remain relatively constant over a wide range of temperatures and frequencies (including microwave). Polysulfone can be plated by an electroless nickel or copper process.

POLYTROPIC PROCESSES. The expansion or compression of a constant weight of gas may assume a variety of forms, depending on the extent to which heat is added to or rejected from the gas during the process, and also on the work done. There are, theoretically, an infinite number of ways possible in which a gas may expand from an initial pressure p_1, and volume v_1 to a final volume v_2. All these expansions may be grouped generically as polytropic expansions, and all could be represented graphically on the PV plane by the family of curves $pv^n = C$. They are all, in theory, perfectly reversible. n may have any positive value, 0 to ∞, and having been selected numerically it defines the type of expansion. From the infinite number of possible polytropic expansions, it is worthwhile to isolate four that deserve special attention. When one of the four physical characteristics, to wit, pressure, temperature, entropy, or volume, remains constant, expansions of more than ordinary interest are denoted, since they are frequently employed in a practical way, in situations which can be subjected to thermodynamic analysis. The value of the exponent n of the polytropic family for each of these is:

isobaric $\qquad n = 0$
isothermal $\qquad n = 1$
isentropic $\qquad n = \gamma$ (γ = ratio of specific heat at constant pressure to that at constant volume)
isometric $\qquad n = \infty$

Note, however, that the first and last are limiting cases, since in the first the pressure remains constant, and in the fourth it approaches zero. Note also that the second applies strictly only to ideal gases.

These thermodynamic processes, as they occur in useful machines, are not often of the exact polytropic form desired. For example, an isentropic process, which is exemplified, at least theoretically, by expansion of the burned gases after the explosive combustion in the gasoline engine, is modified slightly by the interchanging of heat between gases and cylinder wall, whereas a true isentropic has no heat either added or rejected in this way. The particular polytropic curve that would suit these conditions of expansion would depart somewhat from the adiabatic form.

During a polytropic process conditions of the working medium are constantly varying, and analysis may be aimed at determining one of the following: the work done, the heat added, the variation of temperature, and the change of entropy. Some information may be obtained merely by comparing the value of the exponent n with certain other data. For example, if n lies between 0 and 1, the temperature rises during an expansion and falls during a compression; when n is greater than 1, the temperature falls during expansion and rises during compression. Also, when n is less than γ, heat must be added to obtain an expansion, whereas when it is greater than γ, heat must be expelled. From the above it will be noted that there is a certain range of polytropic expansion in which, although heat is added, the temperature falls. This may seem to some to be paradoxical, but it is readily explained. During these expansions work is being done by the gas at a rate greater than that at which heat is being added, with the result that the deficiency must be made up from within the gas. The only way that this may be accomplished is for the gas to cool and give up some of its internal energy.

The equations for work done and for heat added in the case of the general polytropic expansions are:

$$W = \frac{p_1 v_1 - p_2 v_2}{n - 1}$$

$$Q = (p_1 v_1 - p_2 v_2)\left(\frac{1}{n-1} - \frac{1}{\gamma - 1}\right)$$

Both of these are expressed in foot-pounds. Sometimes a substitution of a definite value of n in one or the other of these equations leads to an indeterminate; for example, with the isothermal,

$$W = \frac{p_1 v_1 - p_2 v_2}{1 - 1}$$

But since the equation of the isothermal for an ideal gas is

$$pv = C$$

$$p_1 v_1 = p_2 v_2$$

and the work equation becomes indeterminate,

$$W = \frac{0}{0}$$

By approaching the isothermal from a different angle, however, the equation

$$W = pv \log_e \frac{v_2}{v_1}$$

may be deduced for work done.

POLYURETHANES.

[CAS: 9009-54-5]. These materials comprise a conglomerate family of polymers in which formation of the urethane group

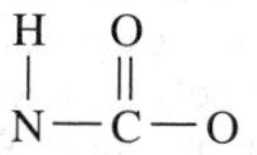

$$
\begin{array}{cc}
H & O \\
| & \| \\
N & - C - O
\end{array}
$$

is an important step in polymerization. Because the urethane linkage usually is formed by reaction of hydroxyl and isocyanate groups, urethane chemistry is the chemistry of isocyanates. The high reactivity of isocyanates and knowledge of the catalysis of isocyanate reactions have made possible the simple production of diverse polymers from low- to moderate-molecular-weight liquid starting materials. Several isocyanates (tolylene diisocyanate, hexamethylene diisocyanate, dicyclohexylmethane diisocyanate, etc.) are used in preparing polyurethanes. All are low-viscosity liquids at room temperature with the exception of 4, 4'-diphenylmethane diisocyanate (MDI), which is a crystalline solid. The aromatic isocyanates are more reactive than the aliphatic isocyanates and are widely used in urethane foams, coatings, and elastomers. The cyclic structure of aromatic and alicyclic isocyanates contributes to molecular stiffness in polyurethanes.

Flexible and rigid urethane foams, probably the most familiar of the polyurethanes, are produced in very large quantities. Foam formulations contain isocyanates and polyols with suitable catalysts, surfactants for stabilization of foam structure, and blowing agents, which produce gas for expansion. The largest volume of flexible urethane foam is used as a cushioning material. Expanding uses for flexible foam include carpet underlays and bedding. Weight reduction programs in the transportation field also take advantage of polyurethane forams for seating and trim. Rigid foams find application in insulation for appliances. Thermoplastic urethane elastomers form a widely used family of engineering materials, which appear to combine the best properties of elastomers and thermoplastics. They are tough, have high load-bearing capacity, low-temperature flexibility, and resistance to oils, fuels, oxygen, ozone, abrasion, and mechanical abuse. Possible carcinogenic properties are being studied. See also **Elastomers**; and **Urethane Polymers**.

POLYVINYL ALKYL ETHERS.

These products have properties which range from sticky resins to elastic solids. They are obtained by the low-temperature cationic polymerization of alkyl vinyl ethers having the general formula $ROCH=CH_2$. These monomers are prepared by the addition of the selected alkanol to acetylene in the presence of sodium alkoxide or mercury(II) catalyst. As shown by the following equations, the latter yields an acetal which must be thermally decomposed to produce the alkyl vinyl ether.

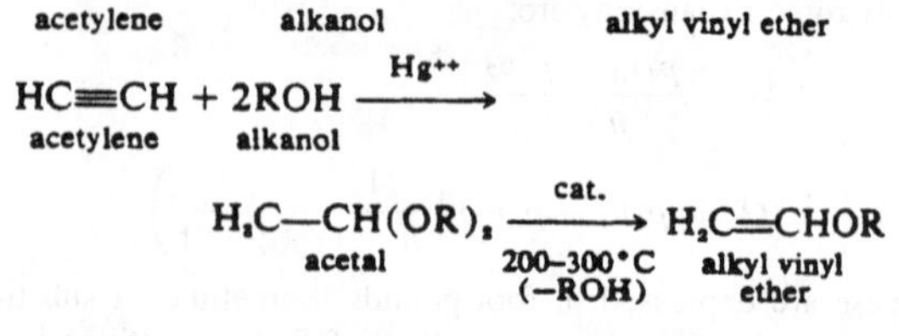

$$\underset{\text{acetylene}}{HC\equiv CH} + \underset{\text{alkanol}}{ROH} \xrightarrow[\text{130-180}^\circ C]{Na^+, OR^-} \underset{\text{alkyl vinyl ether}}{H_2C=CHOR}$$

$$\underset{\text{acetylene}}{HC\equiv CH} + \underset{\text{alkanol}}{2ROH} \xrightarrow{Hg^{++}}$$

$$\underset{\text{acetal}}{H_2C-CH(OR)_2} \xrightarrow[\substack{200-300^\circ C \\ (-ROH)}]{cat.} \underset{\substack{\text{alkyl vinyl} \\ \text{ether}}}{H_2C=CHOR}$$

These monomers are also produced by an oxidative process in which the alkanols are added directly to ethylene and the alkyl ethers are thermally decomposed to produce hydrogen and the alkyl vinyl ethers.

Commercial polymers have been produced from methyl, ethyl, iso-propyl, n-butyl, isotubtyl, t-butyl, stearyl, benzyl and trimethylsilyl vinyl ethers. The poly(methyl vinyl ether) called PVM or Resyn is produced by the polymerization of the monomer by boron trifluoride in propane at $-40^\circ C$ in the presence of traces of an alkyl phenyl sulfide. The polymer may have isotactic, syndiotactic or stereoblock configurations depending on the solvent and catalyst used.

Nonpolar solvents favor the formation of ion pairs between the polymer cation and the counteranion and favor the production of isotactic polymers.

Soluble catalysts, such as diethyl aluminum chloride and ethyl aluminum dichloride, also affect the stereoregularity of the polymer chains. The tendency for the formation of stereoregular polymers is decreased as the size of the alkyl group is increased. Typical structures of these polymers are shown below:

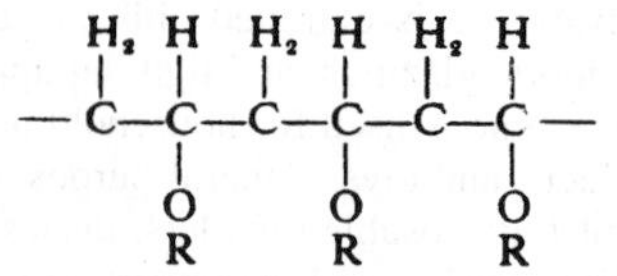

isotactic polymer

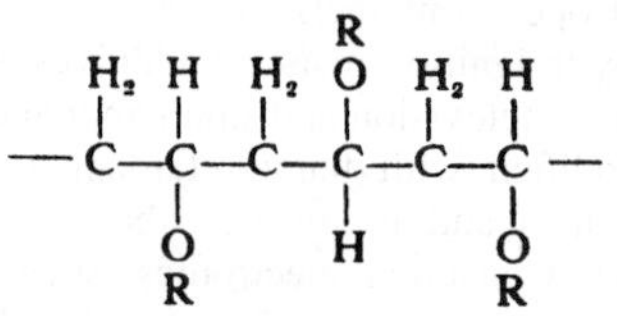

Syndiotactic polymer

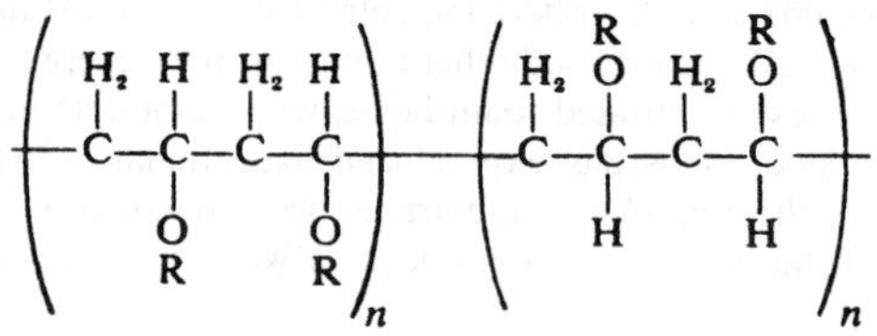

stereoblock polymer

Poly(methyl vinyl ether) is soluble in cold water but becomes insoluble in a reversible process when the temperature is raised to 35°C. This sticky polymer has a glass transition temperature of $-20^\circ C$. It has been used as an adhesive and as a heat sensitizer for polymer latices.

Poly(vinyl ethyl ether) is soluble in ethanol, acetone and benzene. It is a rubbery product which may be cross-linked by heating with dicumyl peroxide. Poly(vinyl isobutyl ether), has a glass transition temperature of $-5^\circ C$. It has been used as an adhesive for upholstery, cellophane and adhesive tape.

The processing properties of poly(vinyl chloride) has been improved by copolymerizing vinyl chloride with a small amount of vinyl alkyl ether. Copolymers of vinyl alkyl ethers and maleic anhydride are used as water soluble thickeners, paper additives, textile assistants and in cleaning formulations.

RAYMOND B. SEYMOUR
University of Houston
Houston, Texas

POLYVINYL CHLORIDE (PVC).

[CAS: 9002-86-2]. The manufacture of polyvinyl chloride resins commences with the monomer, vinyl chloride, which is a gas, shipped and stored under pressure to keep it in a liquid state; bp $-14^\circ C$, fp $-160^\circ C$, density ($20^\circ C$), 0.91. The monomer is produced by the reaction of hydrochloric acid with acetylene. This reaction can be carried out in either a liquid or gaseous state. In another technique, ethylene is reacted with chlorine to produce ethylene dichloride. This is then catalytically dehydrohalogenated to produce vinyl chloride. The by-product is hydrogen chloride. A later process, oxychlorination, permits the regeneration of chlorine from HCl for recycle to the process.

Polymerization may be carried out in any of the following manners:

(1) *Suspension*: a large particle size dispersion or suspension of vinyl chloride is made in water by addition of a small quantity of emulsifying agent. The product after polymerization and drying consists of granules.

(2) *Emulsion*: a larger quantity of emulsifier is employed, resulting in a fine particle size emulsion. The polymer after spray drying, is a finely divided powder suitable for use in organosols and plastisols.

(3) *Solution*: vinyl chloride is dissolved in a suitable solvent for polymerization. The resultant polymer may be sold in solution form, or dried and pelletized.

Emulsions may be polymerized by use of a water-soluble catalyst (initiator), such as potassium persulfate, or a monomer-soluble catalyst, such as benzoyl peroxide, lauroyl peroxide or azobisisobutyronitrile. Suspension and solution polymerizations employ the monomer soluble catalysts only. In addition to the above-mentioned initiators, diisopropyl peroxydi-carbonate may also be employed, where lower-temperature polymerization may be desired, e.g., to reduce branching and minimize degradation.

Because of the low level of emulsifiers and protective colloids, the suspension polymer types are most suitable for electrical applications and end uses requiring clarity. This form is also employed in the bulk of extrusion and molding applications. Cost is lower than for emulsion and solution forms. The emulsion or dispersion resins are employed mainly for organosol and plastisol applications where fast fusion with plasticizer at elevated temperature will occur as a result of the fine particle size of the resin.

Monomers such as vinyl acetate or vinylidene chloride may be copolymerized with vinyl chloride. Up to 15% of the comonomer may be employed. Vinyl acetate increases the solubility, film formation and adhesion. Processing or forming temperatures are generally lowered. Chemical resistance and tensile strength decrease with increasing amount of vinyl acetate.

Rigid Vinyls

These have been separated into two categories according to ASTM:

Type I is rigid PVC with excellent chemical resistance, physical properties and weathering resistance such as obtained from unplasticized high molecular weight PVC.

Type II has the added feature of high impact resistance but with slightly lower chemical and physical requirements.

Perhaps the most important applications for rigid PVC will be in building. This is a rapidly growing market. Fabrication is via extrusion. Examples of applications are pipe, siding, roofing shingles, panels, glazing, window and door frames, rain gutters and downspouts.

Blow-molded bottles, which exhibit excellent product resistance, and good clarity, are also expected to become an important outlet for rigid PVC.

Formulations for extrusion generally include light and heat stabilizers, lubricants, which facilitate molding, and colorants. These materials are generally purchased in a compounded ready to use cube form, in order to minimize irregularities in blending, etc.

The outstanding characteristics of these rigid vinyls are chemical, solvent and water resistance; resistance to weathering when properly stabilized, therefore permitting long-term outdoor exposure; and low cost. Abrasion and impact resistance are satisfactory.

A major deficiency is heat sensitivity. Here, degradation begins with the split-off of HCL. The resultant unsaturation leads to cross-linking and chain cission, causing a degradation of the physical properties. Maximum service temperature for continuous exposure should not exceed 150–175°C. Cold flow or creep is another deficiency, which leads to dimensional changes in materials under constant load, e.g., water pipe under constant service pressures will tend to enlarge in diameter, resulting in decreased strength; long spans of pipe or siding may sag. Temperature accelerates this effect.

PVC has a high coefficient of expansion, one of the highest for all plastic materials, and substantially higher than metals and wood. Therefore, design allowances must be made to provide for movement in order to avoid buckling, breakage, etc.

Flexible PVC

An unusually wide variety of products and usages are possible with plasticized vinyls. Typical applications include floor and wall coverings, boots, rainwear, jackets, upholstery, garden hose, electrical insulation, film and sheeting, foams and many others.

The primary processing techniques are by means of extrusion, calendering and molding. Special techniques involve organosols and plastisols.

Plasticizers used to develop the desired flexibility and performance are selected on the basis of cost and application requirements, e.g., temperature; service life; exposure to solvents, chemicals, water, UV, food; tensile strength; abrasion resistance; flexibility; tear strength, etc.

Plasticizers must be classed as *primary*, where high compatibility is limited, thus restricting the amount that can be tolerated. The addition of secondary plasticizers may import special properties or simply reduce cost (extender plasticizer).

Primary plasticizers may be further subdivided. The *phthalate* types are by far the most popular due to cost and ease of incorporation. Dioctyl phthalate and diisooctyl phthalate are typical of this class. They exhibit good general-purpose properties. *Phosphate* plasticizers are also important for general-purpose use. Typical of these are tritolyl phosphate and trixylenyl phosphate. These plasticizers also impart fire retardant properties. *Low-temperature* plasticizers, such as dibutyl sebacate, are used where good low-temperature flexibility is required. For maximum compatibility and minimum cost, a typical plasticizer combination would be a blend of 50% DOP and 50% dibutyl sebacate.

Polymeric plasticizers are generally polyesters with a relatively low molecular weight. They are used where resistance to high temperatures and freedom from migration and extraction are required. Polymerics are more difficult to incorporate, have poor low-temperature properties, and are expensive.

Epoxy plasticizers are epoxidized oils and esters. These are generally classed with the polymerics. However, molecular weight is lower. Therefore, resistance to extraction and heat are slightly inferior. Low-temperature properties are better and epoxies are more easily incorporated.

Extender plasticizers, which are used mainly to reduce cost, consist of chlorinated waxes, petroleum residues, etc. Incorporation of excessive amounts may result in exudation on aging. The chlorinated types decrease flammability.

Organosols and Plastisols

Plastisols are dispersions of powdered PVC resin in plasticizer. A typical composition would consist of 100 parts of PVC resin dispersed in 50 parts of DOP. The resultant paste when heated to 300°F (149°C) fuses or "fluxes" into a solid plastic mass. Stability of this plastisol at room temperature may range from several weeks to several months depending on the plasticizers and resins employed.

An organosol is the same mixture as described above, with the addition of solvent to reduce viscosity. These find their major applications in coatings. The solvent is evaporated before fusion of the film. Various pigments, colorants, stabilizers and fillers may be added, depending on the desired properties. Emulsion polymerization resins are generally employed because of their fast fusion rates. Coarser particle sized PVC resins would require extended time at the elevated temperature.

Plastisols allow the use of inexpensive manufacturing techniques, such as slush and rotational molding, casting, dipping, etc. They are employed for the manufacture of a large variety of parts, e.g., toys, floor mats, handles and many others.

Foams are made by the addition of blowing agents to the plastisol. These may be continuously applied to a moving substrate which includes a pass at an elevated temperature where foaming occurs, followed by fusion of the plastisol.

Organosols find their major application in coatings, which may be applied by spray, dip, knife, roller, etc. Typical products are coated aluminum siding, fabrics, paper, industrial coatings, etc.

An important development was the use of plasticizers which crosslink upon application of heat and thus produce a more rigid end product. This extends the range of products obtainable by plastisol techniques into rigids. By varying the amount of crosslinking plasticizer incorporated, various levels of flexibility are obtained.

HAROLD A. SARVETNICK
Westfield, New Jersey

POLYVINYLIDENE CHLORIDE. [CAS: 9002-86-2]. A stereoregular, thermoplastic polymer is produced by the free-radical chain polymerization of vinylidene chloride ($H_2C=CCl_2$) using suspension or emulsion techniques. The monomer has a bp of 31.6°C and was first synthesized in 1838 by Regnault, who dehydrochlorinated 1,1,2-trichloroethane which he obtained by the chlorination of ethylene. The copolymer product has been produced under various names, including *Saran*. As shown by the following equation, the product, in production since the late 1930s, is produced by a reaction similar to that used by Regnault nearly a century earlier:

$$H_2ClCCHCl_2 \; + \; Ca(OH)_2 \xrightarrow[-2H_2O]{90°C}$$

1,1,2-tri-chloroethane · calcium hydroxide

$$CaCl_2 \; + \; 2H_2C=CCl_2$$

calcium chloride · vinylidene chloride

Since this monomer readily forms an explosive peroxide

$$\left(H_2C\!-\!CCl_2 \right) \overset{O_2}{}$$

it must be kept under a nitrogen atmosphere at -10°C in the absence of sunlight.

The copolymers were patented by Wiley, Scott, and Seymour in the early 1940s. A typical formulation for emulsion copolymerization contains vinylidene (85 g), vinyl chloride (15 g), methylhydroxypropylcellulose (0.05 g), lauroyl peroxide (0.3 g) and water (200 g). More than 95 per cent of these monomers are converted to copolymer when this aqueous suspension is agitated in an oxygen-free atmosphere for 40 hrs at 60°C. The glass transition temperature of the homopolymer is −17°C. It has a specific gravity of 1.875 and a solubility parameter of 9.8.

Because of its high crystallinity, the homopolymer (PVDV) is insoluble in most solvents at room temperature. However, since the regularity of repeating units in the chain is decreased by copolymerization, Saran is soluble in cyclic ethers and aromatic ketones. This copolymer (100 g) is plasticized by the addition of α-methyl-benzyl ether (5 g), stabilized against ultraviolet light degradation by 5-chloro-2-hydroxybenzophenone (2.0 g) and heat stabilized by phenoxypropylene oxide (2.0 g).

The poly(vinylidene chloride-co-vinylchloride) may be injection molded and extruded. Extruded pipe and molded fittings which were produced in large quantity in the 1940s have been replaced to some extent by less expensive thermoplastics. A flat extruded filament is used for scouring pads and continuous extruded circular filament is used for the production of insect screening, filter clothes, fishing nets and automotive seat covers.

A large quantity of this copolymer is extruded as a thin tubing which is biaxally stretched by inflating with air at moderate temperatures before slitting. This product, called Saran Wrap, has a tensile strength of 15,000 psi (103 MPa). Since it has a high degree of transparency to light and a high coefficient of static friction (0.95) it is widely used for the protection of foods in the household. It has a low permeability value for gases such as oxygen and nitrogen.

Poly(vinylidene chloride-co-acrylonitrile) is widely used as a latex coating for cellophane, polyethylene and paper. Since this copolymer is soluble in organic solvents, it is also used as a solution coating. The resistance to vapor permeability and the ease of printing on polyethylene and cellophane is increased by coating with this vinylidene chloride copolymer.

The tensile strength of both film and fiber is increased tremendously by cold drawing 400–500 per cent. Thus, tensile strengths as high as 40,000 psi (276 MPa) in the direction of draw have been obtained by cold drawing.

RAYMOND B. SEYMOUR
University of Houston
Houston, Texas

POLVINYLIDENE FLUORIDE. This product is made by the free-radical chain polymerization of vinylidene fluoride ($H_2C=CF_2$). This odorless gas which has a boiling point of −82°C is produced by the thermal dehydrochlorination of 1,1,1-chlorodifluoroethane or by the dechlorination of 1,2-dichloro-1,1-difluoro-ethane. As shown by the following equations, 1,1,1-chlorodifluoroethane may be obtained by the hydrofluorination and

chlorination of acetylene and by the hydrofluorination of vinylidene chloride or of 1,1,1-trichloroethane.

Polyvinylidene fluoride is polymerized under pressure at 25–150°C in an emulsion using a fluorinated surfactant to minimize chain transfer with the emulsifying agent. Ammonium persulfate is used as the initiator. The homopolymer is highly crystalline and melts at 170°C. It can be injection molded to produce articles with a tensile strength of 7000 psi (48 MPa), a modulus of elasticity in tension of 1.2×10^5 psi and a heat deflection of 300°F (149°C).

Poly(vinylidene fluoride) is resistant to most acids and alkalies but it is attacked by fuming sulfuric acid. It is soluble in dimethylacetamide but is insoluble in less polar solvents. Copolymers have been produced with ethylene, tetrafluoroethylene, chlorotrifluoroethylene and hexafluoroethylene. The latter is an elastomer called Viton or Fluorel.

The homopolymer is used as a chemical resistant coating for steel, for tank linings, hose, and pump impellors. The elastomeric copolymer with hexafluoroethylene when cured with hexamethylenediamine is used as a seal, gasket, 0-ring, tubing, coating and lining.

POMERANZ-FRITSCH REACTION. Formation of isoquinolines by the acid-catalyzed cyclization of benzalaminoacetals prepared from aromatic aldehydes and aminoacetal.

PORCELAIN. ($4K_2O•AI_2O_3•3SiO_2$), high impact strength; impermeable to liquids and gases; resistant to chemicals except hydrogen fluoride and hot, strong caustic solutions; usable up to 1093°C but subject to heat shock. D 2.41, Mohs hardness 6–7, compression strength 100,000 psi. Porcelain is a mixture of clays, quartz, and feldspar usually containing at least 25% alumina. Ball and china clays are ordinarily used. A slip or slurry is formed with water to form a plastic, moldable mass, which is then glazed and fired to hard, smooth solid.

Uses

Reaction vessels, spark plugs, electrical resistors, electron tubes, corrosion-resistant equipment, ball mills and grinders, food-processing equipment, piping, valves, pumps, and laboratory ware.

See also **Ceramics**; **Porcelain Enamel**; and **Porcelain, Zircon**.

PORCELAIN ENAMEL. A substantially vitreous inorganic coating bonded to metal by fusion above 426°C (ASTM). Composed of various blends of low-sodium frit, clay, feldspar, and other silicates; ground in a ball mill; and sprayed onto a metal surface (steel, iron, or aluminum), to which it bonds firmly after firing, giving a glasslike fire-polished surface.

PORCELAIN, ZIRCON. ($ZrO_2•SiO_2$). A special high-temperature porcelain used for spark plugs and furnace trays because of its high mechanical strength and heat-shock resistance. Usable up to 1700°C with high dielectric strength but rather lower power factor at high frequencies.

PORE. 1. A minute cavity in epidermal tissue as in skin, leaves, or leather, having a capillary channel to the surface that permits transport of water vapor from within outward but not the reverse. 2. A void of interstice between particles of a solid such as sand minerals or powdered metals, that permits passage of liquids or gases through the material in either direction. In some structures, such as gaseous diffusion barriers and molecular sieves, the pores are of molecular dimensions, i.e., 4–10 Å units. Such microporous structures are useful for filtration and molecular separation purposes in various industrial operations. 3. A cell in a spongy structure made by gas formation (foamed plastic) that absorbs water on immersion but releases it when stressed.

See also **Molecular Sieves**; and **Semipermeable Membrane (or Semipermeable Diaphragm)**.

POROMERIC. A term coined to describe the microporosity, air permeability, and water and abrasion resistance of natural and synthetic leather. The pores decrease in diameter from the inner surface to the outer and thus permit air and water vapor to leave the material while excluding water from the outside. Polyester-reinforced urethane resins have been used as leather substitutes with some success, primarily for shoe uppers.

PORPHYRIN. Any of several physiologically active nitrogenous compounds occurring widely in nature. The parent structure is comprised of four pyrrole rings, shown in I, II, III, and IV in Fig. 1, together with four nitrogen atoms and two replaceable hydrogens, for which various metal atoms can be readily substituted. A metal-free porphyrin molecule has the structure shown in the diagram. Porphyrins of this type have been made synthetically by passing an electric current through a mixture of ammonia, methane, and water vapor. Some biochemists suggest that this phenomenon may account for the early formation of chlorophyll and other porphyrins which have been essential factors in the development of life.

The most important porphyrin derivatives are characterized by a central metal atom; hemin is the iron-containing porphyrin essential to mammalian blood, and chlorophyll is the magnesium-containing porphyrin that catalyzes photosynthesis. Other derivatives include the cytochromes, which function in cellular metabolism, and the phthalocyanine group of dyes. Porphyrins are described in considerable detail in a 20-volume set of books, The Porphyrin Handbook, Academic Press, New York, NY, 2003.

PORPHYRY. Porphyry is a textural term applied to igneous rocks in which one or more of the mineral constituents present exists as well crystallized individuals in a ground mass that is relatively of much finer grain. The derivation of the word presents an interesting study. The gasteropods of the genus *Murex* were much used for obtaining a purple dye; the Greek name for both the animal and the dye is the same. A certain Egyptian rock which was once much used for building and ornamental purposes displays very prominent crystals in a purplish ground-mass and so the same Greek word was applied to it, then later came to mean all rocks of this general appearance. Modern use now restricts the term porphyry to the description of texture alone as in the case of the Egyptian rock.

PORTER, GEORGE (1920–2002). An English chemist who won the Nobel prize for chemistry in 1967 with Manfred Eigen and Ronald George Wreyford Norrish. His research concerned fast chemical reactions wand the chemistry of photosynthesis. He was educated at Cambridge University and taught there before going on to other posts.

PORTLAND CEMENT. See **Cement**; **Gypsum**.

POSITRON. The positron is one of many fundamental bits of matter. Its rest mass (9.109×10^{-31} kilogram) is the same as the mass of the electron, and its charge ($+1.602 \times 10^{-19}$ coulomb) is the same magnitude, but opposite in sign to that of the electron. The positron and electron are antiparticles for each other. The positron has spin 1/2 and is described by Fermi-Dirac statistics, as is the electron.

The positron was discovered in 1932 by C.D. Anderson at the California Institute of Technology while doing cloud chamber experiments on cosmic rays. The cloud chamber tracks of some particles were observed to curve in such a direction in a magnetic field that the charge had to be positive. In all other respects, the tracks resembled those of high-energy electrons. The discovery of the positron was in accord with the theoretical work of Dirac on the negative energy of electrons. These negative energy states were interpreted as predicting the existence of a positively charged particle.

Positrons can be produced by either nuclear decay or the transformation of the energy of a gamma ray into an electron-positron pair. In nuclei that are proton-rich, a mode of decay that permits a reduction in the number of protons with a small expenditure of energy is positron emission. The reaction taking place during decay is

$$p^+ \to n^0 + e^+ + v$$

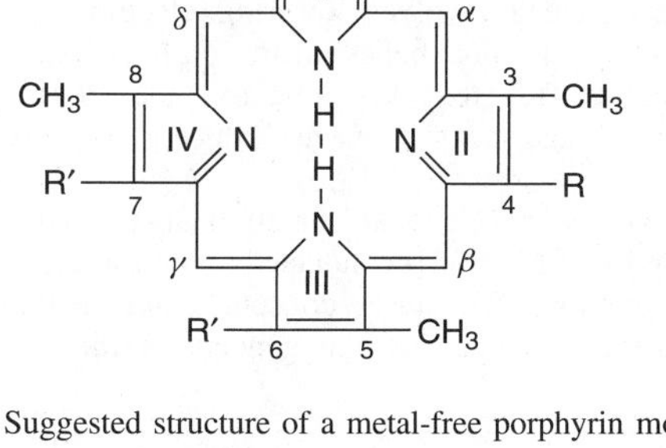

Fig. 1.   Suggested structure of a metal-free porphyrin molecule

where p^+ represents the proton, n^0 the neutron, e^+ the positron, and v a massless, chargeless entity called a neutrino. See also **Neutrino**. The positron and neutrino are emitted from the nucleus while the neutron remains bound within the nucleus. Although none of the naturally occurring radioactive nuclides are positron emitters, many artificial radioisotopes that decay by positron emission have been produced. The first observed case of positron decay of nuclei was also the first observed case of artificial radioactivity. An example of such a nuclear decay is

$$_{11}Na^{22} \longrightarrow {}_{10}Ne^{22} + e^+ + v \text{ (half-life 2.6 years)}$$

This decay provides a practical, usable source of positrons for experimental purposes.

The process of pair production occurs when a high-energy gamma ray interacts in the electromagnetic field of a nucleus to create a pair of particles—a positron and an electron. Pair production is an excellent example of the fact that the rest mass of a particle represents a fixed amount of energy. Since the rest energy ($E_{rest} = m_{rest} - c^2$) of the positron plus electron is 1.022 MeV, this energy is the gamma energy threshold and no pair production can take place for lower-energy gammas. In general, the cross section for pair production increases with increasing gamma energy and also with increasing Z number of the nucleus in whose electromagnetic field the interaction takes place.

The positron is a stable particle (i.e., it does not decay itself), but when it is combined with its antiparticle, the electron, the two annihilate each other and the total energy of the particles appears in the form of gamma rays. Before annihilation with an electron, most positrons come to thermal equilibrium with their surroundings. In the process of losing energy and becoming thermalized, a high-energy positron interacts with its surroundings in almost the same way as does the electron. Thus, for positrons, curves of distance traversed in a medium as a function of initial particle energy are almost identical with those of electrons.

It is energetically possible for a positron and an electron to form a bound system similar to the hydrogen atom, with the positron taking the place of the proton. This bound system has been called "positronium" and the chemical symbol Ps has been assigned. Although the possibility of positronium formation was predicted as early as 1934, the first experimental demonstration of its existence came in 1951 during an investigation of positron annihilation rates in gases as a function of pressure. The energy levels of positronium are about one-half those of the hydrogen atom, since the reduced mass of positronium is about one-half that of the hydrogen atom. This also causes the radius of the positronium system to be about twice that of the hydrogen atom.

In principle, positronium can be observed through the emission of its characteristic spectral lines, which should be similar to hydrogen's except that the wavelengths of all corresponding lines are doubled. Positronium is also the ideal system in which the calculations of quantum electrodynamics can be compared with experimental results. Measurement of the fine-structure splitting of the positronium ground state has served as an important confirmation of the theory of quantum electrodynamics.

It is possible for a positron–electron system to annihilate with the emission of one, two, three, or more gamma rays. However, not all processes are equally probable.

See also **Particles (Subatomic)**.

Additional Reading

Ali, A.: *High Energy Electron Positron Physics,* World Scientific Publishing Company, Inc., Riveredge, NJ, 1988.
Jean, Y.C., D.M. Schrader, and P.E. Mallon: *Principles and Applications of Positron and Positronium Chemistry,* World Scientific Publishing Company, Inc., Riveredge, NJ, 2003.
Krause-Rehberg, R., and H.S. Leipner: *Positron Annihilation in Semiconductors: Defect Studies,* Vol. 127, Springer-Verlag New York, Inc., New York, NY, 1998.

POSITRONIUM. A quasi-stable system consisting of a positron and a negatron bound together. Its set of energy levels is similar to that of the hydrogen atom (electron and proton). However, because of the different reduced mass, the frequencies associated with the spectral lines are less than half of those of the corresponding hydrogen lines. The mean life of positronium is at most about 10^{-7} seconds, its existence being terminated by negatron-positron annihilation. See also **Positron**.

POTASSIUM. [CAS: 7440-09-7]. Chemical element, symbol K, at. no. 19, at. wt. 39.098, periodic table group 1 (alkali metals), mp 63.3°C, bp 760°C, density 0.86 g/cm^3 (20°C). Elemental potassium has a body-centered cubic crystal structure. Potassium is a silver-white metal, can be readily molded, and cut by a knife, oxidizes instantly on exposure to air, and reacts violently with H_2O, yielding potassium hydroxide and hydrogen gas, which burns spontaneously in air with a violet flame due to volatilized potassium element, is preserved under kerosene, burns in air at a red heat with a violet flame. Discovered by Davy in 1807.

There are three naturally occurring isotopes, ^{39}K through ^{41}K, of which ^{40}K is radioactive with a half-life of 1.3×10^9 years. In ordinary potassium, this isotope represents only 0.0119% of the content. There are four other known isotopes, all radioactive, ^{38}K and ^{42}K through ^{44}K, all with relatively short half-lives measured in minutes and hours. In terms of abundance, potassium ranks seventh among the elements occurring in the earth's crust. In terms of content in seawater, the element ranks eighth, with an estimated 1,800,000 tons of potassium per cubic mile (388,000 metric tons per cubic kilometer) of seawater. First ionization potential 4.339 eV; second, 31.66 eV. Oxidation potential $K \longrightarrow K^+ + e^-$, 2.924 V. Other important physical properties of potassium are given under **Chemical Elements**.

Potassium does not occur in nature in the free state because of its great chemical reactivity. The major basic potash chemical used as a source of potassium is potassium chloride, KCl. The potassium content of all potash sources generally is given in terms of the oxide K_2O. The majority of potash produced comes from mineral deposits that were formed by the evaporation of prehistoric lakes and seas which had become enriched in potassium salts leached from the soil. In addition to natural deposits of potassium salts, large concentrations of potassium also are found in some bodies of water, including the Great Salt Lake and the Salduro Marsh in Utah, the Dead Sea between Israel and Jordan, and Searles Lake in California. All of these brines are used for the commercial production of potash.

The main potassium minerals are sylvite, KCl, sylvinite, KCl/NaCl, carnallite, $KCl \cdot MgCl_2 \cdot 6H_2O$, kainite, $MgSO_4 \cdot KCl \cdot 3H_2O$, polyhalite, $K_2SO_4 \cdot MgSO_4 \cdot 2CaSO_4 \cdot 2H_2O$, langbeinite, $K_2SO_4 \cdot 2MgSO_4$, jarosite, $K_2Fe_6 (OH)_{12}(SO_4)_4$, leucite, $K_2O \cdot Al_2O_3 \cdot 4SiO_2$, alunite, $K_2Al_6(OH)_{12}(SO_4)_4$, microcline, $K_2O \cdot Al_2O_3 \cdot 6SiO_2$, muscovite, $K_2O \cdot 3Al_2O_3 \cdot 6SiO_2 \cdot 2H_2O$, biotite, $H_2K(Mg, Fe)_3(Al, Fe)(SiO_4)_3$, and orthoclase, $K_2O \cdot Al_2O_3 \cdot 6SiO_2$. See also **Alunite**; **Biotite**; and **Polyhalite**. The principal workable mineral deposits are in Stassfurt, Germany, Alsace, New Mexico, Saskatchewan, the former Soviet Union, Spain, Poland, Italy, the Atlantic Seaboard of the United States, and Utah. There are significant potassium reserves in many other parts of the world, notably in Canada and the former Soviet Union. World consumption of potash is about 18 million tons annually. Potassium metal is obtained by electrolysis of fused potassium hydroxide or chloride fluoride mixture in a specially designed cell.

Uses

Like so many of the chemical elements, the compounds of potassium are far more important than elemental potassium—by several orders of magnitude. The uses for metallic potassium are extremely limited, mainly because metallic sodium serves about the same needs and is much less costly. Sodium production, for example, exceeds potassium production by a factor of at least 1,000. A large amount of elemental potassium is used to produce the superoxide, KO_2 which finds application in gas-mask canisters. The compound also goes into the production of a sodium-potassium alloy, which is used as a heat-exchange medium. This alloy also has been used in magnetohydrodynamic power generation and as a catalyst for the removal of CO_2, H_2O, and oxygen from inert-gas systems. The handling precautions for potassium metal are similar to those for sodium metal. See also **Sodium**.

Chemistry and Compounds

Potassium is more electropositive than sodium in many of its reactions, as is consistent with its position in group 1. Its reaction with H_2O is more vigorous and it reacts violently with liquid bromine, and readily on heating with solid iodine.

Because of the ease of removal of its single 4s electron (4.339 eV) and the difficulty of removing a second electron (31.66 eV) potassium is exclusively monovalent in its compounds, which are electrovalent. (Some experimental work indicates that the potassium alkyls may be covalent, but even they form conducting solutions in other metal alkyls.)

Potassium solutions in liquid NH_3 react readily with the elements on the further right side of the periodic table to produce normal and poly compounds such as potassium sulfide, K_2S, and tetrapotassium plumbide, K_4Pb, in the first instance and K_2S_6 and K_4Pb_9 in the second. Ammoniates are not formed by potassium as readily as by sodium or lithium and solubility of salts exhibits a minimum at the cation: anion radius ratio of 0.75 (potassium fluoride, KF, 16 moles per kilogram, potassium chloride, KCl, 0.0177 moles per kilogram, potassium bromide, KBr, 2.26 moles per kilogram, potassium iodide, KI, 11.09 moles per kilogram). Potassium nitrate reacts in liquid ammonia with potassium amide, KNH_2 to form the azide, KN_3.

Like the other alkali metals, potassium forms compounds with virtually all the anions, organic as well as inorganic. Like sodium bicarbonate, the reactivity of potassium bicarbonate with many metallic oxides permits of the preparation of many compounds (such as the meta- and pyroarsenates) which are unstable in aqueous solution. For a general discussion of these reactions, and for a general picture of the inorganic salts of potassium, see the discussion of the compounds of sodium, which differ principally in their greater degree of hydration and greater number of hydrates. However, potassium, rubidium, and cesium coordinate with large organic molecules even though they do not with water. Potassium, like the others, coordinates with salicylaldehyde. It is believed to have two coordination numbers, 4 and 6. The tetracoordinate compounds of potassium (and sodium) are the most stable. The following reasons are given: (1) Increasing atomic number carries with it increasing electropositiveness and ease of ionization, which diminishes the tendency to coordinate. (2) The increasing distance of the nucleus from the coordinating electrons with increasing atomic volume makes it less likely that additional electrons will be held with ease. (3) On the other hand, there is an increase in the maximum coordination number with the elements of higher atomic number. These factors are in keeping with a maximum stability for the tetracoordinate compounds occurring with potassium.

Charge Density Waves in Potassium

Frequently, because of their relatively simple electronic structure, the alkali metals are selected as a basis for the study of the behavior of electrons in solids. As early as 1964, Overhauser (Purdue University) predicted the existence of "charge density waves," a phrase coined by Overhauser, in the potassium atom. This conclusion was the result of calculations made by Overhauser to the effect that K, in its lowest energy or ground state, does not exhibit a uniform distribution of its free electrons (which cause K to behave as a metal), but rather the electron density varies sinusoidally with a characteristic wavelength—and that this usually is not an integral multiple of the crystal lattice constant. This concept, of course, was not in agreement with the traditional conclusion that free electrons are uniformly distributed. The reasoning—the sinusoidal clumping lowers the electron energy, which in turn causes the lattice to distort and, as explained by Robinson (1986), this distortion is an attempt to reduce the huge electric fields generated by the separation between the positive charge of the K ions and the negative charge of the electrons.

At the time of Overhauser's work in 1964, experimental examples were not available and the concept was generally considered academic. Several years later, however, investigators working with layered materials (electrons essentially move in only two directions) and with linear conductors (motion is essentially in one direction) attributed a charge density wave phenomenon to what has been termed the Pierls instability. The latter effect, which involves lowering electron energy and lattice distortion, currently is not believed to apply to the simple, three-dimensional metals (K etc.). In summary, the Pierls instability and the Overhauser charge density waves concept appear to be similar, but different.

In 1985, Giebultowicz (National Bureau of Standards), Overhauser, and Werner (University of Missouri) conducted a neutron diffraction study and tentatively proved the Overhauser concept. Some solid-state physicists are seeking further evidence. If the concept is fully confirmed, some modifications in the thinking of how electrons behave in solids may be required.

Salt-Forming Properties. One major difference between potassium and sodium in their salt-forming properties is the much greater ability of potassium to form alums, although potassium does not form quite as many types of these compounds as do the higher alkali metals, or ammonium or monovalent thallium.

Potassium also differs from sodium, and especially from lithium, in the greater stability of its salts of polarizable polyatomic anions, such as peroxide, superoxide, azide, polysulfide, polyhalides, etc. The rubidium and cesium salts, on the other hand, are even more stable.

Among the other inorganic compounds of potassium are the following:

Bromate. Potassium bromate, [CAS: 7758-01-2], $KBrO_3$, white solid, soluble, mp 434°C, upon heating oxygen is evolved and the residue is potassium bromide; formed by electrolysis of potassium bromide solution under proper conditions. Used as a source of bromate and bromic acid.

Carbonate. Potassium carbonate, [CAS: 584-08-7], potash, pearl ash, K_2CO_3, white solid, soluble, formed (1) in the ash when plant materials are burned, (2) by reaction of potassium hydroxide solution and the requisite amount of CO_2. Used (1) in making special glasses, (2) in the making of soft soap, (3) in the preparation of other potassium salts (a) in solution, (b) upon fusion; potassium hydrogen carbonate, potassium bicarbonate, potassium acid carbonate, $KHCO_3$, white solid, soluble, (4) in vat dyeing and textile printing, (5) in titanium enamels, (6) in boiler water treating compounds, (7) in photographic chemical formulations, (8) in electroplating baths, and (9) as an important absorbent for CO_2 in the process industries.

Chlorate. Potassium chlorate, [CAS: 3811-04-9], chlorate of potash, $KClO_3$, white solid, soluble, mp about 350°C, powerful oxidizing agent, and consequently a fire hazard with dry organic materials, such as clothes, and with sulfur; upon heating oxygen is liberated and the residue is potassium chloride; formed by electrolysis of potassium chloride solution under proper conditions. Used (1) in matches, (2) in pyrotechnics, (3) as disinfectant, (4) as a source of oxygen upon heating. (Hazardous! Use of potassium perchlorate is recommended instead.)

Chloride. Potassium chloride, [CAS: 7447-40-7], KCl, colorless or white crystals; strong saline taste. Occurs naturally as sylvite. Soluble in water; slightly soluble in alcohol. Sp. gr. 1.987; mp 772°C; sublimes at 1500°C; noncombustible; low toxicity. Used in fertilizers, as a source of potassium salts; pharmaceutical preparations; photography; spectroscopy; plant nutrient; salt substitute; laboratory reagent. See also **Fertilizer**.

Chloroplatinate. Potassium chloroplatinate, [CAS: 16921-30-5]. K_2PtCl_6, yellow solid, insoluble, formed by reaction of soluble potassium salt solution and chloroplatinic acid. Used in the quantitative determination of potassium.

Chromate. Potassium chromate, [CAS: 7789-00-6], K_2CrO_4, yellow solid, soluble, formed by reaction of potassium carbonate and chromite at a high temperature in a current of air, and then extracting with water and evaporating the solution. Used (1) as a source of chromate, (2) in leather tanning, (3) in textile dyeing, (4) in inks.

Cobaltinitrite. Dipotassium sodium cobaltinitrite, $K_2NaCo(NO_2)_6 \bullet H_2O$, golden yellow precipitate, formed by reaction of sodium cobaltinitrite solution in acetic acid with soluble potassium salt solution. Used in the detection of potassium.

Cyanate. Potassium cyanate, [CAS: 590-28-3], KCNO, white solid, soluble, formed along with lead metal by reaction of potassium cyanide and lead monoxide solids upon heating. Source of cyanate.

Cyanide. Potassium cyanide, [CAS: 151-50-8], cyanide of potash, KCN, white solid, soluble, very poisonous, formed by reaction of calcium cyanamide and potassium chloride at high temperature. Used as a source of cyanide and for hydrocyanic acid, but usually replaced by the cheaper sodium cyanide. Also used in metallurgy, electroplating,

extraction of gold from ores, as a pesticide and fumigant, in photography and analytical chemistry. Upon acidification, produces dangerous HCN gas.

Dichromate. Potassium dichromate, [CAS: 7778-50-9], chromate of potash, $K_2Cr_2O_7$, red solid, soluble, powerful oxidizing agent, formed by acidifying potassium chromate solution and then evaporating. Used (1) in matches, (2) in leather tanning and in the textile industry, (3) as a source of chromate, (4) in pyrotechnics, (5) in colored glass, (6) as an important laboratory reagent, (7) in blueprint developing, and (8) in wood preservation formulations.

Hydroxide. Potassium hydroxide, [CAS: 1310-58-3], caustic potash, potassium hydrate, KOH, white solid, soluble, mp 380°C, formed (1) by reaction of potassium carbonate and calcium hydroxide in H_2O, and then separation of the solution and evaporation, (2) by electrolysis of potassium chloride under the proper conditions, and evaporation. Used in the preparation of potassium salts (1) in solution, and (2) upon fusion. Also used in the manufacture of (3) soaps, (4) drugs, (5) dyes, (6) alkaline batteries, (7) adhesives, (8) fertilizers, (9) alkylates, (10) for purifying industrial gases, (11) for scrubbing out traces of hydrofluoric acid in processing equipment, (12) as a drain-pipe cleaner, and (13) in asphalt emulsions.

Hypophosphite. Potassium hypophosphite, [CAS: 7782-87-8], KH_2PO_2, white solid, soluble, formed (1) by reaction of hypophosphorous acid and potassium carbonate solution, and then evaporating, (2) by reaction of potassium hydroxide solution and phosphorus on heating (poisonous phosphine gas evolved).

Iodate. Potassium iodate, [CAS: 7758-05-6], KIO_3, white solid, soluble, melting point 560°C, formed (1) by electrolysis of potassium iodide under proper conditions, (2) by reaction of iodine and potassium hydroxide solution, and the fractional crystallization of iodate from iodide. Used as a source of iodate and iodic acid.

Manganate. Potassium manganate, K_2MnO_4, green solid, soluble, permanent in alkali, formed by heating to high temperature manganese dioxide and potassium carbonate, and then extracting with water, and evaporating the solution. The first step in the preparation of potassium manganate and permanganate from pyrolusite.

Nitrate. Potassium nitrate, [CAS: 7757-79-1], saltpeter, niter, KNO_3, white solid, soluble, mp 333°C, formed by fractional crystallization of sodium nitrate and potassium chloride solutions. Used (1) in matches, explosives, pyrotechnics, (2) in the pickling of meat, (3) in glass, (4) in medicines, (5) as a rocket-fuel oxidizer, and (6) in the heat treatment of steel. See also **Fertilizer**.

Nitrite. Potassium nitrite, [CAS: 7758-09-0], KNO_2, yellowish-white solid, soluble, for med (1) by reaction of nitric oxide plus nitrogen tetroxide and potassium carbonate or hydroxide, and then evaporating, (2) by heating potassium nitrate and lead to a high temperature and then extracting the soluble portion (lead monoxide insoluble) with H_2O, and evaporating. Used as a reagent (diazotizing) in organic chemistry.

Oxides. See discussion later in entry.

Perchlorate. Potassium perchlorate, [CAS: 7778-74-7], $KClO_4$, white solid, very slightly soluble, mp 610°C, but above 400°C decomposes with evolution of oxygen gas and formation of potassium chloride residue; formed (1) by electrolysis of potassium chlorate under proper conditions, (2) by heating potassium chlorate at 480°C and then fractional crystallization. Used (1) as a convenient and safe (preferred to use of potassium chlorate) method of preparing oxygen by heating, (2) in the determination of potassium in soluble salt solution.

Periodate. Potassium periodate, [CAS: 7790-21-8], KIO_4, white solid, very slightly soluble, mp 582°C, formed by electrolysis of potassium iodate under proper conditions.

Permanganate. Potassium permanganate, [CAS: 7722-64-7], permanganate of potash $KMnO_4$, purple solid, soluble, formed by oxidation of acidified potassium manganate solution with chlorine, and then evaporating. Used (1) as disinfectant and bactericide, (2) in medicine, (3) as an important oxidizing agent in many chemical reactions.

Persulfate. Potassium persulfate, [CAS: 7727-21-1], $K_2S_2O_8$, white solid, slightly soluble, formed by electrolysis of potassium sulfate under proper conditions. Used (1) as a bleaching and oxidizing agent, (2) as an antiseptic.

Silicate. Potassium silicate, K_2SiO_3, colorless (when pure) glass, soluble, mp 976°C, formed by reaction of silicon oxide and potassium carbonate at high temperature, similar in properties and uses to the more common sodium silicate.

Sulfates. Potassium sulfate, [CAS: 7778-80-5], sulfate of potash, K_2SO_4, white solid, soluble. Common constituent of potassium salt minerals. Used (1) as an important potassium fertilizer, (2) in the preparation of potassium or potash alums; potassium hydrogen sulfate, $KHSO_4$, white solid, soluble; potassium pyrosulfate, $K_2S_2O_7$, white solid, soluble, formed by heating potassium hydrogen sulfate to complete loss of H_2O. See also **Fertilizer**.

Sulfides. Potassium sulfide, [CAS: 1312-73-8], K_2S, yellowish to reddish solid, soluble, formed by heating potassium sulfate and carbon to a high temperature; potassium hydrogen sulfide, potassium bisulfide, potassium acid sulfide KHS, formed in solution by reaction of potassium hydroxide or carbonate solution and excess H_2S.

Sulfite. Potassium sulfite, [CAS: 10117-38-1], $K_2SO_3 \cdot 2H_2O$; potassium hydrogen sulfite, $KHSO_3$; white solids, similar in properties and formation to the corresponding sodium sulfites.

Thiocarbonate. Potassium thiocarbonate, K_2CS_3, yellow solid, soluble, formed by reaction of potassium sulfide and CS_2.

Thiocyanate. Potassium thiocyanate, [CAS: 333-20-0], potassium sulfocyanide, potassium sulfocyanate, potassium rhodanate, KCNS, white solid, soluble, mp about 170°C, formed by fusing potassium cyanide and sulfur, and then crystallizing. Used as a source of thiocyanate.

In addition to the inorganic salts, potassium forms such binary compounds as a phosphide, K_3P, by direct union with phosphorus, a boride, KB_6, by electrolysis of fused fluorides and borates in the presence of a metal boride, a nitride, and the oxides. Of the latter, direction reaction of potassium and oxygen yields the superoxide, KO_2, a paramagnetic, orange-colored substance. The likelihood of KO_2 having a monomeric structure is supported by these properties, since the O_2^- ion would have an odd electron, which would confer paramagnetism and color upon the compound. The lower oxides of potassium, K_2O and K_2O_2, which are less stable in air than the superoxide, have been prepared, as have their hydrates. K_2O unites explosively with the oxygen of the air. One other oxide, K_2O_3, has been reported, but this appears to be a double salt of KO_2 and K_2O_2. The properties of potassium hydroxide are in keeping with its position in Group 1; thus its heat of solution is somewhat lower than that of rubidium hydroxide, RbOH, or cesium hydroxide, CsOH, and much higher than that of lithium hydroxide, LiOH, and NaOH.

The organic compounds of potassium include many oxycompounds, such as salts of organic acids, alcohols and phenols (alkoxides, phenoxides, etc.). A few potassium-carbon linked compounds have been reported, such as a phenylisopropyl potassium, $C_6H_5C_3H_7K$, and a carbonyl compound of unknown composition, $K_x(CO)_x$. The adduct of ethyl potassium and diethyl-zinc is a true salt, $K_2[Zn(C_2H_5)_4]$, potassium tetraethylzincate.

Health and Safety Factors

Reactions of potassium with water and oxygen are hazardous and safe handling is a concern. Potassium oxidizes slowly in air at room temperature, and it usually ignites if it sprays hot into the air. The peroxide and superoxide products may explode in contact with free potassium metal or organic materials including hydrocarbons. Thus, packaging (qv) under oils is less desirable than packaging under an inert cover gas or in a vacuum. Potassium can react with entrapped air in oils to form the superoxide. The encrustation of potassium with superoxide (as a yellow crust) developed during storage has been known to detonate by friction from cutting. Potassium encrusted with a peroxide and superoxide layer should be destroyed immediately by careful, controlled disposal.

Potassium forms corrosive potassium hydroxide and liberates explosive hydrogen gas upon reaction with water and moisture. Airborne potassium dusts or potassium combustion products attack mucous membranes and skin causing burns and skin cauterization. Inhalation and skin contact must be avoided. Safety goggles, full face shields, respirators, leather gloves, fire-resistant clothing, and a leather apron are considered minimum safety equipment.

See also **Potassium and Sodium (In Biological Systems)**.

Additional Reading

Giebultowicz, T.M., A.S. Overhauser, and S.A. Werner: *Phys. Rev. Lett.,* **56**, 1485 (1986).
Greenwood, N.N. and A. Earnshaw: *Chemistry of the Elements,* 2nd Edition, Butterworth-Heinemann, Inc., Woburn, MA, 1997.
Krebs, R.E.: *The History and Use of Our Earth's Chemical Elements: A Reference Guide,* Greenwood Publishing Group, Inc., Westport, CT, 1998.
Lide, D.R.: *CRC Handbook of Chemistry and Physics,* 84th Edition, CRC Press, LLC., Boca Raton, FL, 2003.
Parker, P.: *McGraw-Hill Encyclopedia of Chemistry,* 2nd Edition, The McGraw-Hill Companies, Inc., New York, NY, 1993.
Robinson, A.L.: "Charge Density Waves Seen in Potassium," *Science,* **232**, 713 (1986).
Stwertka, A. and E. Stwertka: *A Guide to the Elements,* Oxford University Press, Inc., New York, NY, 1998.

POTASSIUM AND SODIUM (In Biological Systems).   Potassium and sodium play major roles in biological processes. Because of the numerous parallels between these two elements in metabolism, they are treated in a single entry, with appropriate distinctions made.

Potassium is required by both plants and animals. Although the total amount of potassium in most soils is usually rather high, the level of available or soluble forms of the element is frequently too low to meet the needs of growing plants. Deficiencies of plant-available potassium are more frequent in the soils of the eastern rather than of the western United States. See also **Soil**. Potassium in the form of soluble potassium salts is a very common constituent of fertilizers. See also **Fertilizer**.

Many plants will not grow at normal rates unless the plant tissues, especially the leaves, contain as much as 1 or 2% potassium and, for some plants, even higher concentrations are required. Therefore, if a plant grows at all, it will nearly always contain sufficient potassium to meet the requirements of the people or animals that consume the plant. Potassium deficiencies do occur in humans and animals, but these are largely due to metabolic upsets and illnesses that interfere with the utilization of potassium in the body, or via excessive losses of potassium from the body, rather than due to inadequate levels of dietary potassium.

The general role of potassium fertilizers in improving human and animal nutrition is to help increase food and feed supplies rather than to improve the nutritional quality of the crops produced. Excessive use of potassium fertilizers may decrease the concentration of magnesium in crops. Sodium is essential to higher animals that regulate the composition of their body fluids and to some marine organisms, but it is dispensable for many bacteria and most plants except for the blue-green algae. Potassium, on the other hand, is essential for all, or nearly all forms of life. The importance of these cations for all forms of life has been related to the predominance of sodium and potassium in the ocean where primitive forms of life are thought to have originated and developed. During most of the period of evolvement of living organisms, there has been little change in the sodium and potassium content of seawater, either as to proportion or total amount. The body fluids of sea animals are, in most instances, similar to seawater in sodium and potassium level and ratio. In freshwater and terrestrial animals, the sodium and potassium level of body fluids is usually somewhat lower, and the ratio is likely to vary from the 40:1 ratio of seawater. Most fresh waters contain small and variable amounts of sodium and potassium, usually in a ratio of from 1:1 to 4:1.

Despite the higher level of sodium in natural water, potassium is universally the characteristic cation found within both plant and animal cells. Although sodium is not an absolute requirement for most plants and bacteria, it is found in these organisms and is essential to higher animals where it is the principal cation of the extracellular fluids. Sodium and potassium are important constituents of both intra- and extracellular fluids. Generally, the best external and internal medium for function of cells not adjusted to low salt levels is a medium involving a balance of sodium and potassium.

Beyond the osmotic effects depending on the sum of the concentration of the ions in the solution, Ringer found in 1882 that to maintain the contractility of an isolated frog heart, it was necessary to perfuse it with a medium containing sodium, potassium, and calcium ions in the proportion of seawater. It has since been recognized that the normal life activities of tissues and cells may depend on a proper balance among the inorganic cations to which they are exposed. Sodium is required for the sustained contractility of mammalian muscle, while potassium has a paralyzing effect. Thus, a balance is necessary for normal function. Other investigators have found that the antagonism among univalent and divalent cations observed by Ringer is demonstrable with various simpler or more complicated organisms or biological systems.

Excessive salt in soil, such as soils recently soaked with seawater, is toxic to most plants, although there are many plants, e.g., those of the salt marshes and the sea, which are adapted to a high salt concentration. Ingestion of seawater by man as the only source of water is eventually fatal because of the inability of the body to eliminate salt at a concentration comparable to that of seawater. This results in accumulation of salt, with severe toxic effects and eventually fatal results.

It is probable that potassium is absorbed by the plant roots from the soil by an active transport mechanism which carries it through the cell wall structure. Similarly, potassium and sodium if required, are accumulated by animals also by active transport. The actual cellular content of potassium and sodium is likewise controlled by transport mechanisms that specifically move potassium in and sodium out of the cell against the concentration gradient. The energy for this is derived from the metabolic processes of the cell. The nature of these transport mechanisms has not been fully determined.

Ions and Transport Mechanisms

Potassium differs from most other essential constituents of plant and animal cells in that it is not built into the cell as a part of an organic compound, but is rather an ion from a soluble inorganic or organic salt. Potassium ions may chelate with cellular constituents, such as polyphosphates. The ion is of the correct size to fit into the water lattice adsorbed by the protein in the cell. In general, the potassium and sodium ions are attracted to protein or other colloidal or structural units having a negative charge. Mucopoly-saccharides within the cell, on the cell surfaces and of the intercellular structures, are of particular importance in holding cations, such as potassium and sodium. Active centers of other configurational features of the proteins in the cell may be affected or altered by the potassium held by electrostatic or covalent binding. There are several enzyme systems activated by potassium.

In general, most of the sodium and potassium in the animal is in a dynamic state, being exchanged between different parts of the cell, between the cell and the extracellular fluid, and intermixing with ingested sodium and potassium in body fluids.

Most cellular constituents do not selectively bind potassium in preference to sodium. Myosin of muscle fibers, for example, will bind either. But, in contrast, the mitochondria and ribosomes are organized cellular organelles able to selectively take up or extrude potassium. This accounts for only a part of the potassium held in the cell.

In blue-green algae and some yeasts, sodium may in part replace cellular potassium. While potassium is usually the principal cation concerned with the maintenance of the osmotic pressure within the cell, sodium contributes appreciably to the total, and amino acids and other organic compounds may help make up any deficit, particularly in marine invertebrates.

The sodium content of the body extracellular fluids of marine invertebrates from the coelenterate through the arthropod phyla is approximately that of seawater. In freshwater and terrestrial invertebrates, the sodium of body fluids varies over a wide range and there is considerable variation among vertebrates. There are both fish and crustaceans so highly adaptable that they are able to live in either fresh or salt water.

Osmotic Pressure Regulation

The regulation of osmotic pressure within the cell and the control of the passage of water into or out of the cell is dependent to a considerable extent on the control of the potassium and sodium in the cell by the transport systems of the cell wall. The cell wall itself is of protein-lipid composition and is in general impermeable to the passage of water and inorganic salts. Recent studies of the cell walls with electron microscopes and with the use of other investigative techniques indicate that the cell wall contains pores connecting the cell contents with the extracellular fluid, or in some plants, with other cells. In cells having an endoplasmic reticulum, the intracellular vacuolar system may have openings through the cell wall communicating with the extracellular fluid. The ease with which water passes in or out of the cell in response to changes in external or internal osmotic pressure varies over an extreme range, from easy passage to rigid control, depending on the cell and its functions.

Phagocytosis and pinocytosis may bring salts and water, as well as other substances, into the cell.

In some unicellular organisms, osmotic equilibrium may be maintained by a contractile vacuole, which collects water; in other organisms, water may be excreted through the cell wall. The kidney and sweat glands of higher animals, gills of fish and salt glands of birds serve to excrete salt. Most animals, through control of sodium and potassium excretion and loss, are able to adapt to a wide range of intake.

The importance of sodium chloride in nutrition has been recognized from the beginning of history. Agricultural populations that lived on cereal grains, nuts, berries, and other vegetable foods poor in sodium, experienced a hunger for salt which led them to go to great lengths to obtain the mineral. This was particularly true if they lived in a hot climate with the attendant increased loss of salt in perspiration. Similarly, herbivorous animals will travel long distances to supply their need for additional salt. In contrast, peoples or animals subsisting on meat, milk and other foods receive quite appreciable amounts of sodium salts in the diet, and experience no special desire or hunger for salt. See also **Sodium Chloride**.

In plants, the meristematic tissues in general are particularly rich in potassium, as are other metabolically active regions, such as buds, young leaves, and root tips. Potassium deficiency may produce both gross and microscopic changes in the structure of plants. Effects of deficiency reported include leaf damage, high or low water content of leaves, decreased photosynthesis, disturbed carbohydrate metabolism, low protein content and other abnormalities.

Since potassium is found abundantly in most natural foods consumed by animals, deficiency is ordinarily no problem. With prolonged maintenance through parenteral (intravenous) feeding when normal oral feeding is not possible, potassium must be supplied.

Role of Kidney

Experimental potassium deficiency in rats results in stunted growth, loss of chloride with hypochloremic acidosis, loss of potassium and increase of sodium in muscle. In man, disease of the gastrointestinal tract, involving loss of secretions through vomiting or diarrhea, may result in serious loss of both sodium and potassium. Trauma, surgery, anoxia, ischemia, shock and any damage to or wasting away of tissues may result in loss of cellular potassium to the extracellular fluid and plasma, and the loss from the body through kidney excretion. Recovery with rapid uptake or potassium by the tissues may result in low plasma levels. Low extracellular potassium concentration may cause muscular weakness, changes in cardiac and kidney function, lethargy, and even coma in severe cases. There are no reserve stores of either sodium or potassium in the animal body, so any loss beyond the amount of intake comes from the functional supply of cells and tissues. The kidney is the key regulator of the sodium and potassium content of higher animals and makes possible adaptation to wide variations of intake. In the glomerulus of the kidney nephron (or individual unit), an ultrafiltrate containing the smaller molecules of plasma is normally produced. As this ultrafiltrate passes down the kidney tubule, 97.5% or more of the sodium is actively resorbed, along with nearly all of the potassium. The remaining 2.5% of the sodium is sufficient to account for even the maximum sodium excretion. Potassium is added to the filtrate in the distal tubule through exchange for sodium. Control of this exchange appears to be the principal mode of action of aldosterone, which thus exerts a final control over sodium excretion. Aldosterone is a steroid hormone from the adrenal cortex, secretion of which seems to result from lowering of the Na/K ratio in the blood. Water is passively resorbed with the electrolytes along the length of the tubule.

Water excretion is further controlled by the antidiuretic hormone from the posterior pituitary gland which acts to increase water resorption in the kidney through making the collecting tubule permeable to water for additional resorption beyond what took place in the tubule. The posterior pituitary gland secretes the hormone as a rapid and sensitive response to a rise in the osmotic pressure of the extracellular fluid. The osmotic pressure of the extracellular fluid is, of course, principally due to its sodium chloride content.

With low intake of sodium, excretion is reduced to a very low level to conserve the supply in the body. Potassium is not so efficiently conserved.

The kidney regulates the acid-base balance of the body by control over resorption of sodium ions, which may exchange for hydrogen ions in the kidney tubule. Since most dietaries are of acid-ash, the urine is usually more acid than the original plasma filtrate and much of the phosphate excreted is thus changed to the acid monosodium salt. Within the range of normal variability, with an alkaline ash diet, the urine may become alkaline, and in extreme instances, some sodium bicarbonate may be excreted.

The salts of the buffer pairs responsible for control of the pH of plasma and extracellular fluid involve sodium as the principal cation, while the cellular buffers involve potassium salts. See also **Acid-Base Regulation (Blood)**; and **Diuretic Agents**.

Additional Reading

Benos, D.J. and D.M. Fambrough: *Amiloride-Sensitive Sodium Channels: Physiology and Functional Diversity,* Academic Press, Inc., San Diego, CA, 1999.

Evans, J.M., T.C. Hamilton, S.D. Longman, and G. Stemp: *Potassium Channels and Their Modulators: From Synthesis to Clinical Experien,* Taylor & Francis, Inc., Philadelphia, PA, 1997.

Young, D.B.: *Role of Potassium in Preventive Cardiovascular Medicine,* Kluwer Academic Publishers, Norwell, MA, 2001.

POTENTIATOR.   A term used in the flavor and food industries to characterize a substance that intensifies the taste of a food product to a far greater extent than does an enhancer. The most important of these are the 5′-nucleotides. They are approved by the FDA. Their effective concentration is measured in parts per billion, whereas that of an enhancer such as MSG is in parts per thousand. The effect is thought to be due to synergism. Potentiators do not add any taste of their own, but intensify the taste response to substances already present in the food.

POUND, ROBERT (1919–).   Pound is a Canadian-born American physicist who pioneered many fruitful ideas and is especially remembered for co-discovering, with Purcell, nuclear magnetic resonance (NMR) and establishing it as one of physics' most valuable analytical techniques. NMR is used as an analytical technique in chemical research, medical diagnosis, and a number of other fields.

Pound worked with his associate, Glen A. Rebka, Jr., carrying out an experiment using the Mossbauer effect to measure the gravitational effects of electromagnetic radiation and to test the predictions of Einstein's theory of general relativity. Pound's experiments continued and results predicted the Red Shift discovery.

During WW II, Pound worked at the Submarine Signal Company and then at MIT's radiation laboratory helping to develop radar and microwave technology. After the war, he became a professor at Harvard in 1948 and stayed until his retirement in 1989. Among his many awards have been the Thompson Memorial Award of the Institute of Radio Engineers in 1948, The Eddington Medal of the Royal Astronomical Society in 1965, and the National Medal of Science in 1990.

See also **Nuclear Magnetic Resonance (NMR) and Magnetic Resonance Imaging (MRI)**.

J.M.I.

POUR POINT.   1. The lowest temperature at which a liquid will flow when a test container is inverted. 2. The temperature at which an alloy is cast.

POUR POINT DEPRESSANT.   An additive for lubricating and automotive oils that lowers the pour point (or increases the flow point) by 11.0°C. The agents now generally used are polymerized higher esters of acrylic acid derivatives. They are most effective with low-viscosity oils.

See also **Petroleum**.

POWDER.   Any solid, dry material of extremely small particle size ranging down to colloidal dimensions, prepared either by comminuting larger units (mechanical grinding), combustion (carbon black, lampblack), or precipitation via a chemical reaction (calcium carbonate, etc.). Powders that are so fine that the particles cannot be detected by rubbing between thumb and forefinger are called *impalpable*. Typical materials used in powder form are cosmetics, inorganic pigments, metals, plastics (molding powders), dehydrated dairy products, pharmaceuticals, and explosives. Metal powders are used to make specialized equipment by sintering and pressing (powder metallurgy), as well as sprayed coatings and paint pigments (aluminum, bronze). Thermoplastic polymers in powder form are used in a technology known as powder molding. Thermosetting polymers are used in the sprayed coatings field for autos, machinery, and other industrial applications in which they have many advantages over sprayed solvent coatings.

See also **Carbon Black**; and **Powder Metallurgy**.

POWDER METALLURGY.   Powder metallurgy (PM) embraces the production of finely divided metal powders and their union through the use of pressure and heat into useful articles. The temperatures required are below the fusion point of the principal constituent, and bonding depends on interdiffusion of the metal particles in the solid state. It is necessary to provide intimate contact between particles, hence reducing atmospheres are provided in the sintering process to prevent formation of oxide films. Readily oxidized powders such as aluminum require special technique.

Probably the most important applications of powder metallurgy are those in which a product is made which cannot be duplicated by other methods. There are many examples of this kind. The melting point of tungsten, 6,100°F (3.371°C), is much too high for ordinary melting and casting methods and the only way in which filaments for electric lights can be made is to draw them from rods of compacted and sintered tungsten powder. The cemented carbide cutting tools are another important product of refractory nature readily made by powder metallurgy.

Self-lubricating bronze bearings having controlled porosity are products that can be made only by powder metallurgy. The pores are impregnated with oil, and flow to the bearing surface is maintained by capillary action. Graphite is incorporated with the metal powder in one type of oil-less bearing. A material made from powdered copper and graphite is used for electric-current collector brushes, and tungsten-copper or tungsten-silver combinations are used for electric contact points. In contrast to these high-conductivity materials, a high-resistance element is produced from a mixture of copper and porcelain powders, combining a metal with a nonmetallic substance.

Advances in PM Technology

Particularly during the past decade, remarkable progress was made in PM technology. Major trends in the early 1990s included: (1) rapid solidification processing (RSP), (2) liquid-dynamic compaction (LDC); (3) self-propagating high-temperature synthesis (SHS); (4) greater use of intermetallics and additives in PM products; (5) advancements in PM injection molding; and (6) improvements in heat treating PM parts—not to mention the appearance of PM in products and structures traditionally made by other metallurgical processes, such as seamless tubing.

Rapid Solidification Processing.   RSP holds high promise for producing engineering alloys with refined microstructures, improved chemical homogeneity, extended solute solubility, and possible retention of metastable phases. RSP usually involves cooling rates greater than 100°C/second (212°F/s). For high cooling rates, RSP products must have a large surface-to-volume ratio, and thus are commonly in the form of powder, flakes, or ribbon. To be commercially acceptable, such rapidly solidified particulates must be consolidated into fully dense, metallurgically bonded forms suitable for engineering applications. RSP properties are quite sensitive to heat treatment and the desired properties can easily be lost without careful control over the consolidation process. Among the consolidation methods currently in commercial or near-commercial use include hot extrusion, hot isostatic pressing, vacuum or inert-atmosphere pressing or sintering, and powder forging. Unfortunately, these processes require elevated temperatures for relatively long times, which may destroy the benefits achieved by RSP. A major problem involves the tenacious oxide that forms on the surface of many RSP materials, particularly aluminum, nickel, and stainless steels.

A shock wave moving through the medium at velocities in excess of that of sound appears to be one solution to this problem. The shock wave can greatly exceed the yield stress. Passage of the shock wave causes plastic flow, interparticle melting and bonding, and can produce a fully dense, metallurgically bonded product. Three methodologies have evolved for introducing a shock wave: (1) use of a gas gun incorporating propellants or compressed gas; (2) direct application of explosives; and (3) impact of a projectile accelerated by explosives.

Guns are available of several designs. In one configuration, a high-pressure burst of gas launches a projectile down an evacuated tube where the projectile imparts a shock wave by driving a punch into the powder bed. As pointed out by Wright, the gun may be in the form of a high-impact press in which a reusable piston is accelerated in an evacuated chamber by introducing a rapid burst of gas into the breach. The impact of the ram produces a pressure pulse.

Hitchcox (1986) describes a process being developed at the Massachusetts Institute of Technology, which uses high-velocity pulses of an inert gas to atomize a stream of molten metal. Semisolid droplets of the metal are collected as rapidly solidified "splats" on a chilled metallic substrate. (This liquid dynamic compaction (LDC) process is attractive from a cost standpoint.) Substrates can be flat surfaces, molds, or shaped containers. The splats build up rapidly, forming high-density bodies suitable for further processing. Because the splats are thin, they cool at relatively high rates (1000°C/second; 1800°F/s). It is claimed that the LDC process improves ductility and fracture toughness because oxides and powder particle boundaries are minimized. Although in an early stage of development, materials such as high-strength aluminum and superalloys and (FeCo)-Nd-B have been produced with the process. Grant (MIT) reports that rapidly solidified material may exhibit grain sizes as fine as 0.2 micrometer (8 microinches) after crystallization of glasses. The fine grain size allows superplastic forming of aluminum alloys, stainless steels, and other materials.

Self-Propagating High-Temperature Synthesis. SHS usually involves an exothermic reaction producing temperatures in excess of 2500°C (4532°F). In essence, a mixture of compressed powders is ignited with a heat source in air or an inert atmosphere and in an instant, a refractory compound or multicomponent material results. SHS eliminates the need for high-temperature furnaces as required by conventional processes. Processing time is shortened to seconds or minutes versus hours and days as required with normal sintering. The products are usually of a higher purity, some having less than 0.2% (wt) of unreacted elements. This is the result of vaporizing volatile contaminants during the "explosion." SHS has been used to produce borides, carbides, and other difficult materials and is considered to have much potential for making ceramic matrix composites with unique microstructures.

In SHS, there are fundamentally two types of reactions: (1) *thermite*, where oxidation-reduction produces multiphase products, such as cermets; and (2) *compound formation*, as resulting from the starting elements, such as $Ti + 2B = TiB_2$. A combination of the two types of reaction also can be used. SHS requires a strong exothermic reaction where the heat of reaction is at least 40 kcal/mole (168,000 Joules/mole). The adiabatic temperature must be greater than the melting point of the product in order to produce a liquid phase for enhancing diffusion. Sheppard also breaks the reactions into (1) propagating, and (2) bulk. *Propagating reactions* are initiated locally, so that a synthesis wave of reactants, or, conversely, chemical activators can be added to accelerate the reaction. Also, if a higher reaction temperature required, preheating of the reactants is practiced.

Examples of products made by the SHS process include borides, carbides, chalcogenides, hydrides, intermetallic compounds, nitrides, silicides, carbonitrides, sulfides, cemented carbides (cermets), and various heterogeneous mixtures (microcomposites).

PM Intermetallics and Additives. An example of improved materials for which PM technology may solve past metallurgical processing problems is found in turbine parts, where high-temperature performance and oxidation resistance is mandatory. Aluminides of iron, nickel, and titanium have received consideration for a number of years, not only because they appear to meet the two foregoing criteria, but also because of their relatively low density, high strength, and corrosion resistance. Conventional casting of these materials results in unacceptable inhomogeneities. This has led to the evaluation of several PM methodologies, including hot isostatic pressing (HIP), vacuum hot pressing (VHP), injection molding, transient liquid-phase sintering, reactive sintering, and hot extrusion.

Of considerable promise, reflecting research at Rensselaer Polytechnic Institute, is *reactive sintering*. This process involves a transient liquid phase. The reaction takes place above the lowest eutectic temperature in the system, but still at a temperature at which the compound remains in the solid phase. Research has shown that a transient liquid forms at the lowest eutectic temperature and spreads through the compact during heating. Actually, the reaction is approximately spontaneous because heat is liberated due to the thermodynamic stability of the compound's high melting temperature. In terms of the reaction of nickel and aluminum powders, a temperature over 550°C (1020°F) is the optimum. The time required for processing is relatively short (about one-half hour). Densities over 97% (of theoretical) are obtained. Even with the presence of some residual porosity, the ductility and strength of the product are good, which properties are retained after subsequent high-temperature exposure.

Researchers at Case Western Reserve University and the NASA Lewis Research Center, both located in Cleveland, Ohio, have evaluated *hot extrusion* as a candidate process. In essence, the process consists of canning the powder (prealloyed aluminide powders [FeAl, NiAl, and Ni₃Al]) and then extruding the material at a temperature and area-reduction ratio sufficiently high to produce satisfactory material flow and efficient filling of interparticle spaces, the latter for eliminating porosity and to encourage grains to recrystallize dynamically.

A basic advantage of PM technology has been that of minimizing or eliminating machining in making a final part. Nevertheless, some machining operations may be required. Traditionally, the machinability of sintered PM steels, for example, is poor, mainly due to porosity, hardness, and low thermal conductivity. Porosity causes an interrupted cut and causes tool wear—with the possible results of both higher tool costs and poorer surface finish. In recent years, PM techniques have been improved by the incorporation of additive, notably manganese sulfide (MnS), to enhance machinability.

Powder metallurgy also is playing a major role in the pioneering but rapidly developing technology of *nanofabrication*. Melding the technologies of PM and electronic components manufacture are reducing operational minute machine parts to submicron levels. See Fig. 1.

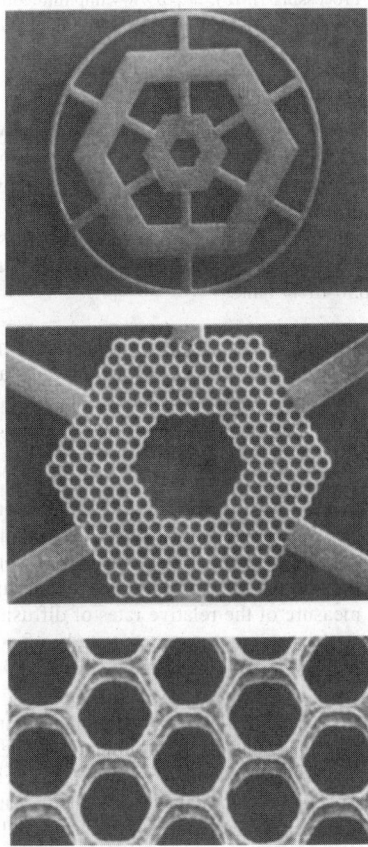

Fig. 1.   Representative cross sections of tiny (submicrometer) mechanical parts that can be produced by nanofabrication technology. (*Cornell University*)

Additional Reading

Alman, D.E. and J. Newkirk: *Powder Metallurgy Alloys and Particulate Materials for Industrial Application,* The Minerals, Metals & Materials Society, Warrendale, PA, 2000.

Anderson, I.E.: "Boost in Atomizer Pressure Shaves Powder-Particle Sizes," *Advanced Materials and Processes,* 30 (July 1991).

Craighead, H.G.: *The National Nanofabrication Facility at Cornell University,* Cornell University, Ithaca, New York, NY, October 1990.

Froes, F.H.: "Powder Metallurgy," *Advanced Materials and Processes,* 55 (January 1990).

German, R.M.: *Powder Metallurgy of Iron and Steel,* John Wiley & Sons, Inc., New York, NY, 1998.

Keishi Gotoh, K. and H. Masuda: *Powder Technology Handbook,* 2nd Edition, Marcel Dekker, Inc., New York, NY, 1997.

Hitchcox, A.L.: "Advances in Powder Metallurgy Cover Many Fields," *Advanced Materials and Processes,* 63–65 (December 1986).

Jenkins, I. and J.V. Wood: *Powder Metallurgy: An Overview,* Ashgate Publishing Company, Brookfield, VT, 1991.

Kloecker, C.J.: "Hammers Take on Presses for Forging PM Steel," *Advanced Materials and Processes,* 37 (July 1991).

Marquis F.D.S.: *Powder Materials: Current Research and Industrial Practices,* The Minerals, Metals & Materials Society, Warrendale, PA, 1999.

Scott, W.W., Jr.: "Engineering the Part," *Advanced Materials and Processes,* 4 (July 1991).

Staff: *Properties and Selection: Nonferrous Alloys and Special-Purpose Materials,* ASM International, Materials Park, OH, 1991.

Staff: *ASM Handbook: Powder Metal Technologies and Applications,* Vol. 7, ASM International, Materials Park, OH, 1998.

Staff: "Top Powder Metallurgy Parts Honored," *Advanced Materials and Processes,* 8 (August 1991).

Staff: "Forecast for Metals," *Advanced Materials and Processes,* 17 (January 1991); 17 (January 1992); 18 (January 1993).

Staff: *Metallic and Inorganic Coatings, Metal Powders, and Sintered P/M Structural Parts,* American Society for Testing & Materials, West Conshohocken, PA, 2001.

Suslick, K.S.: "Ultrasound 'Makes a Hit' with Metal Powder," *Advanced Materials and Processes,* 10 (September 1990).

Thummler, F. and R. Oberacker: *An Introduction To Powder Metallurgy,* Ashgate Publishing Company, Brookfield, VT, 1994.

Web References

Institute of Materials Processing (IMP): http://www.imp.mtu.edu/
The Minerals, Metals, Materials Society: http://members.tms.org/Staff.asp

POWER (Nuclear). See **Nuclear Reactor**.

PPB. Parts per billion. One part per billion is a frequently used dimension for expressing the composition and analysis of substances—as found in air, water, food substances, etc. Instrument developments and other assay techniques perfected during the past decade or so have made the determination of such minute quantities a practical possibility for many materials. One part per billion is approximately equivalent to 1 drop in a 10,000-gallon (37,850-liter) tank.

PPM. Parts per million. One part per million is a common dimension for expressing the composition and analysis of substances—as found in air, water, raw materials, food substances, etc. One part per million is approximately equivalent to about 1/32 ounce (1 gram) in 1 ton of substance. One gram is exactly one-millionth of a metric ton.

PRANDTL NUMBER. A dimensionless number equal to the ratio of the kinematic viscosity to the thermometric conductivity (or thermal diffusivity). For gases, it is rather under one and is nearly independent of pressure and temperature, but for liquids the variation is rapid. Its significance is as a measure of the relative rates of diffusion of momentum and heat in a flow and it is important in the study of compressible flow and heat convection. See also **Heat Transfer**.

PRASEODYMIUM. [CAS: 7440-10-0]. Chemical element symbol Pr, at. no. 59, at. wt. 140.91, second in the Lanthanide Series in the periodic table, mp 934°C, bp 3,512°C, density 6.769 g/cm³ (20°C). Elemental praseo dymium has a close-packed hexagonal crystal structure at 25°C. The pure metallic praseodymium is silver-gray in color, the luster dulling rapidly upon exposure to air and forming a nonadherent oxide which hastens the process of oxidation. When pure, the metal is soft and workable with ordinary tools. Processing and handling require storage under a nonreactive liquid or inert atmosphere or vacuum. Finely-divided praseodymium is pyrophoric, burning at a red heat. There is only one isotope of the element in nature ^{141}Pr. It is not radioactive and has a low acute-toxicity rating. Fourteen artificial isotopes have been produced. Of the light (or cerium-group) rare-earth metals, praseodymium is the fourth most plentiful and ranks 59th in abundance of the elements in the earth's crust, exceeding tantalum, mercury, bismuth, and the precious metals, excepting silver. The element was first identified by C.A. von Welsbach in 1885. Electronic configuration

$$1s^2 2s^2 2p^6 3s^2 3p^6 3d^{10} 4s^2 4p^6 4d^{10} 4f^2 5s^2 5p^6 5d^1 6s^2$$

Ionic radius, Pr^{3+} 1.01 Å, Pr^{4+} 0.90 Å. Metallic radius, 1.828 Å. First ionization potential, 5.42 eV; second, 10.55 eV. Other important physical properties of praseodymium are given under **Rare-Earth Elements and Metals**.

Primary sources of the element are bastnasite and monazite, which contain from 4 to 8% praseodymium. Plant capacity involving liquid-liquid or solid-liquid organic ion-exchange processes for recovering the element is in excess of 100,000 pounds Pr_6O_{11} annually. Metallic praseodymium is obtained by electrolysis of Pr_6O_{11} in a molten fluoride electrolyte, or by a calcium reduction of PrF_3 or $PrCl_3$ in a sealed-bomb reaction.

For many years, praseodymium has been a component of light rare-earth mixtures used in mischmetal, a pyrophoric alloy used in cigarette-lighter "flints." Mixtures of cerium, lanthanum, neodymium, and praseodymium, as oxides and fluorides, are used in the cores of arc carbons for the production of light of greater intensity. Similar mixtures of rare-earth oxides, including praseodymium, are used in optical glass polishing formulations. Mixtures of the lanthanide compounds, including about 5% praseodymium, find application as catalysts in petroleum cracking processes. A mixture containing 10% Pr, 30% Nd, and 60% La is used for cracking crude oil and comprises the largest single use of the element as well as of all other Lanthanide elements. Use of elemental praseodymium as a colorant for glass was one of the early applications. The color ranges from clear yellow to green and finds use in sunglasses, protective glasses for industry, art objects of glass, tableware, and optical filters. In the manufacture of ceramic tile, a praseodymia-zirconia yellow stain is used. Metallurgically, the most important intermetallic compound is $PrCo_5$, which has unsurpassed permanent magnetic properties. The compound has a very high resistance to demagnetization and has a high magnetic saturation value. PrNi5 has been used for adiabatic magnetization cooling of samples down to the milli-Kelvin range for low-temperature research. Investigations continue into further electronic and optical uses of the element and its compounds.

Additional Reading

Greenwood, N.N. and A. Earnshaw: *Chemistry of the Elements,* 2nd Edition, Butterworth-Heinemann, Inc., Woburn, MA, 1997.

Krebs, R.E.: *The History and Use of Our Earth's Chemical Elements: A Reference Guide,* Greenwood Publishing Group, Inc., Westport, CT, 1998.

Lide, D.R.: *CRC Handbook of Chemistry and Physics,* 84th Edition, CRC Press, LLC., Boca Raton, FL, 2000.

Parker, P.: *McGraw-Hill Encyclopedia of Chemistry,* 2nd Edition, The McGraw-Hill Companies, Inc., New York, NY, 1993.

Stwertka, A. and E. Stwertka: *A Guide to the Elements,* Oxford University Press, Inc., New York, NY, 1998.

PRECIPITATE. ($\downarrow$, ppt). Small particles that have settled out of a liquid or gaseous suspension by gravity, or that result from a chemical reaction. Precipitated compounds, such as blanc fixe (barium sulfate, are prepared in this way, for example, by the reaction $BaCl_2 + Na_2SO_4 \longrightarrow NaCl + BaSO_4$. In formulas, a downward vertical arrow, $\downarrow$, or "ppt" is sometimes used to indicate a precipitate. A class of organic pigments called *lakes* are made by precipitating an organic dye onto an inorganic substrate. Colloidal particles dispersed in a gas, as flue dust in industrial stacks, can be precipitated by introducing an electric charge opposite to that which sustains the particles.

See also **Sedimentation**.

PRECIPITATION HARDENING. A large number of alloys are hardenable by a heat treating procedure known as precipitation hardening. Hardening is accomplished by the controlled precipitation of many minute particles of a second crystalline phase (or phases) inside the crystals of the primary metal. In order that the precipitation may be effected, the hardening constituent must be more soluble at higher temperatures than it is at lower temperatures, so that heating of the solid metal at an elevated temperature causes the second phase to dissolve into the matrix. If a precipitation hardening alloy is heated and held at an elevated temperature so as to dissolve the hardening phase and then is quenched to room temperature, a supersaturated solid solution is obtained. This heating and quenching operation is known as the solution treatment. The second phase of precipitation hardening is known as the aging treatment wherein the second phase is precipitated out of the supersaturated solid solution by holding the metal either at room temperature or some intermediate temperature well below the temperature employed in the solution treatment. The various stages involved in the formation of the nuclei of the precipitation particles may be very complex. In general, however, the aim of the aging process is to obtain a distribution of the precipitated particles that produces maximum hardness. This will usually occur when the particles are submicroscopic in size and extremely numerous. Their hardening effect on the crystal lattice of the matrix crystals is believed to result from local strains that they produce in the matrix. These latter hinder the normal easy motion of dislocations, thereby hardening the metal. The term age hardening is synonymous with precipitation hardening, but when so used generally refers to metals aged at room temperature.

PRECURSOR.   In biological systems, an intermediate compound or molecular complex present in a living organism which, when activated physiochemically, is converted to a specific functional substance. Sometimes the prefix "pro" is used to indicate that a compound in question plays the role of a precursor. Examples from the history of vitamin and other essential chemical developments include: ergosterol (pro-vitamin D2), which is activated by ultraviolet radiation to form vitamin D; carotene (pro-vitamin A) is a precursor of vitamin A; prothrombin forms thrombin upon activation in the blood-clotting mechanism.

PREFERENTIAL.   Descriptive of the selectivity of action, either chemical or physiochemical, exhibited by a substance when in contact with two other substances; it may be due either to chemical affinity or to surface phenomena. An example of a preferential chemical combination is that of hemoglobin with carbon monoxide, with which it unites 200 times as readily as it does with oxygen when expose to a mixture of the two. Such phenomena as adsorption, corrosion, and the wetting of dry powders by liquids are other examples.

PREGESTOGENS AND PROGESTINS.   See **Steroids**.

PREGEL, FRITZ (1869–1930).   An Austrian chemist who won the Nobel prize in 1923. He was also a medical doctor who worked in micromechanical analysis and developed determinations for hydrogen, carbon, nitrogen, and organic groups using micromethods. He was educated at Tubingen, Leipzig, and Berlin.

PREHNITE.   Prehnite is a hydrous silicate of calcium and aluminum, $Ca_2Al_2Si_3O_{10}(OH)_2$, crystallizing in the orthorhombic system. Usual occurrence as intergrown crystals of reniform, stalactitic character, and as rounded groups of such crystals; hardness, 6–6.5; specific gravity 2.90–2.95; luster, vitreous to pearly; color, various shades of light green to gray or white; translucent. Though not a zeolite it is found associated with them and with datolite and calcite, in veins and cavities of basic rocks, sometimes in granites, syenites, or gneisses. It is found in Austria, Italy, the Harz Mountains, France, Scotland, and the Republic of South Africa, where it was originally discovered. Magnificent crystal casts after an unknown mineral have been found in a single large cavity in the basaltic rocks near Bombay, India. In the United States well-known localities are Somerville, Massachusetts; Farmington, Connecticut; Paterson, New Jersey; and Keweenas County, Michigan. Named for Colonel Prehn, its discoverer, who was an early Dutch Governor of the Cape of Good Hope colony.

PRELOG, VLADIMIR (1906–1998).   A Swiss organic chemist who won the Nobel prize for Chemistry in 1975 along with John W. Cornforth for his research into the stereochemistry of organic molecules and reactions. Although educated in Yugoslavia, he spent many years in Zurich.

PREMIX MOLDING.   A mixture of plastic ingredients prepared in advance of the molding or extruding operation and stored in bags or bins

until required. It is made by mixing the components (resin, filler, fibrous materials such as glass and necessary curatives) in a dough blender. Storage life may be from a few days to a year or more, depending on formulation. Such mixtures are then calendered or extruded after warming to suitable temperature.

PRENENOLONE.   See **Steroids**.

PREPOLYMER.   An adduct or reaction intermediate of a polyol and a monomeric isocyanate, in which either component is in considerable excess of the other. A polymer of medium molecular weight having reactive hydroxyl and $-NCO$ groups.

See also **Polymers**.

PRESERVATIVE.   Any agent that prolongs the useful life of a material. Food products are preserved by; (1) low temperature, (2) ionizing radiation (X and Y rays), (3) antioxidants, (4) fungicides, (5) aldehydes, (6) paints and others.

See also **Food Additives**.

PRESSURE.   If a body of fluid is at rest, the forces are in equilibrium or the fluid is in static equilibrium. The types of force that may act on a body are shear or tangential force, tensile force, and compressive force. Fluids move continuously under the action of shear or tangential forces. Thus, a fluid at rest is free in each part from shear forces; one fluid layer does not slide relative to an adjacent layer. Fluids can be subjected to a compressive stress, which is commonly called *pressure*. The term may be defined as force per unit area. The pressure units may be dynes per square centimeter, pounds per square foot, torr, mega-Pascals, etc. Atmospheric pressure is the force acting upon a unit area due to the weight of the atmosphere. Gage pressure is the difference between the pressure of the fluid measured (at some point) and atmospheric pressure. Absolute pressure, which can be measured by a mercury barometer, is the sum of gage pressure plus atmospheric pressure.

Pascal's law states that the pressure in a static fluid is the same in all directions. This condition is different from that for a stressed solid in static equilibrium. In such a solid, the stress on a plane depends upon the orientation of that plane. A liquid in contact with the atmosphere is sometimes called a free surface. A static liquid has a horizontal free surface if gravity is the only type of force acting.

Imagine a body of static fluid in a gravitational field. The mass of the fluid is m (in grams) and the weight of the fluid is mg (as dynes) where g is the local gravitational acceleration. Figure 1 shows a large region of any static fluid with a very small or infinitesimal element. Figure 2 indicates the element in detail. The vertical distance z is measured positively in the direction of decreasing pressure (up); dA is an infinitesimal area; p is the pressure acting on the top surface; and $(p + dp)$ is the pressure acting on the bottom surface. The pressure difference is due only to the weight of the fluid element. Let r represent density, which is mass per unit volume (as grams per cubic centimeter). Thus the weight of the element is $\rho\, g\, dz\, dA$. Considering the element as a free body, an accounting of forces in the vertical direction gives:

$$dp\,dA = -\rho g\,dz\,dA; \quad dp = -\rho g\,dz \tag{1}$$

As z is measured positively upward, the minus sign indicates that the pressure increases with an increase in height. This fundamental equation

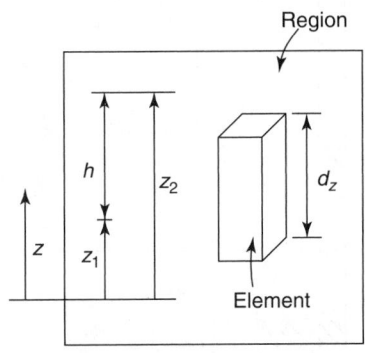

Fig. 1.   Large region of any static fluid

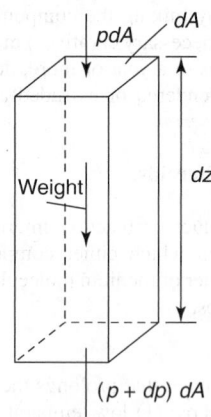

Fig. 2.   Vertical forces on infinitesimal element

of fluid statics can be applied to all fluids. In integral form, Equation (1) becomes:

$$\int_1^2 \frac{dp}{g} = \int_1^2 dz = -(z_2 - z_1) \tag{2}$$

where 1 refers to one level and 2 refers to another level. The functional relation between pressure p and the combination ρg must be established before Equation (2) can be integrated. There are two major cases: (a) incompressible fluids, in which the density ρ is a constant; and (b) compressible fluids, in which the density ρ varies.

Liquids can be considered as incompressible in many cases. For small differences in height, a gas might be regarded as incompressible. For an incompressible fluid, with constant g, Equation (2) becomes:

$$p_2 - p_1 = -\rho g(z_2 - z_1) \tag{3}$$

The term $(z_2 - z_1)$ may be called a static "pressure head," and it can be expressed in feet or inches of water, or some height of any liquid. For example, barometric pressure can be expressed in inches of mercury.

A manometer is a device that measures a static pressure by balancing the pressure with a column of liquid in static equilibrium. Many types of manometers are used. The common mercury barometer is essentially a manometer for measuring atmospheric pressure; a mercury column in a glass tube balances the weight of the air above the mercury. Figure 3 illustrates a manometer in which the left leg is open to the atmosphere; the liquid has a specific weight (weight per unit volume) $\rho_2 g$. In the other leg is a liquid of specific weight $\rho_1 g$. Starting with the left leg, the gage pressure p_A is:

$$p_A = h_2 \rho_2 g$$

Since the fluid is in static equilibrium, the pressure p_B at point B equals the pressure at point A. Thus:

$$p_A = p_B = h_2 \rho_2 g$$

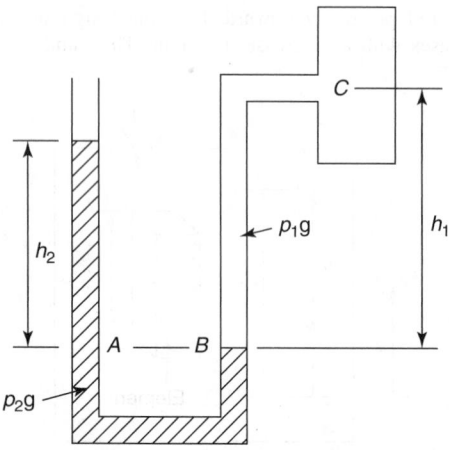

Fig. 3.   Manometer

The pressure p_C at point C is less than that at B. Thus:

$$p_B - p_C = h_1 \rho_1 g$$

Then the gage pressure at point C is:

$$p_C = g(h_2 \rho_2 - h_1 \rho_1)$$

When a body of any kind is partly or fully immersed in a static fluid, every part of the body surface in contact with the fluid is pressed on by the fluid. The pressure is greater on the areas more deeply immersed. The resultant of all these fluid pressure forces is an upward or buoyant force. The pressure on each part of the body is independent of the body material. Archimedes' principle states that the buoyant force equals the weight of the displaced fluid.

Equation (3) is for the special case of an incompressible fluid. As an example of a compressible fluid, consider an isothermal or constant-temperature layer of gas. The equation of state for such a gas can be written:

$$p = \rho R T_1 \tag{4}$$

where T_1 is the given absolute temperature and R is a gas constant or gas factor depending upon the gas. Assuming a constant g, Equation (2) gives:

$$\frac{RT_1}{g} \int_1^2 \frac{dp}{p} = -(z_2 - z_1)$$

$$z_2 - z_1 = \frac{RT_1}{g} \log_e \frac{p_1}{p_2} \tag{5}$$

Equation (5) is sometimes called a "barometric height" relation. For an isothermal atmosphere, a measurement of the temperature T_1 and the static pressure (as with a barometer) at two different levels will provide data for the calculation of the height difference.

Other pressure designations include:

Vacuum. A gage pressure below atmospheric.
Hydrostatic Pressure. The pressure at a point below a liquid surface due to the height of fluid above it.
Tons-on-Ram. The force that acts over a given area as in various types of hydraulic machinery.
Partial Pressure. The pressure exerted by one component in a system, usually one gas or vapor in a mixture.
Internal Pressure. The effect of the attractive forces of the molecules of a substance, which is called pressure because its result is the same as that of an added external pressure. In liquids, its effect appears as the ability of liquids to stand substantial negative pressures without rupture.
Cohesion Pressure. A term in Van der Waal's equation introduced to take care of the effect of molecular attraction. It is usually expressed as a/V^2, where a is a constant and V is the volume of the gas.
Pressure Measurement. Liquid-column elements, such as the manometer, are commonly used for pressure measurement. A variety of diaphragm and other elastic elements is used to measure pressure. A metallic diaphragm element is primarily a device for measuring relatively low pressures. It consists of a single diaphragm or of one or more capsules connected together, so that upon pressure application, each capsule deflects. The total deflection is the sum of the deflections of all capsules. A variety of bellows elements is similarly used in pressure gages. One of the most common forms of pressure gage makes use of a bourdon-spring element. Gages for medium-to-high vacuums usually incorporate an electronic type transducer. See also **Vacuum Gages.** Electrical transducers, such as strain gages, moving-contact resistance elements, inductance, reluctance, capacitative, and piezoelectric devices also are used in pressure detection systems. Pressure not only is important as a key variable for direct measurement, but differential pressures are commonly measured in connection with various flowmeters that use a differential-producing element, such as an orifice plate, to measure flow. Manometers and other pressure sensors are also used in liquid-level measuring devices.
High-Pressure Technology. Until the mid-1970s, the limit to most high-pressure experimentation was confined to about 300 kilobars.

As of 1988, the maximum pressure created in the laboratory by the diamond anvil pressure cell approximates 5 million atmospheres.

Theoretical estimates, however, forecast that diamond is stable up to 23 million atmospheres with respect to any phase transition. Although plastic deformation would limit its capability, predictions for the diamond anvil cell are for pressures somewhere between 5 and 23 million atmospheres.

PRIESTLEY, JOSEPH (1733–1804).   Priestley was an English chemist who researched relationships among plants, air, and animals. After meeting Benjamin Franklin he became interested in science and the two men became lifelong friends. Priestley started doing chemical experiments as a hobby, but it soon became a passion. He had little scientific education but his observations were very keen.

Priestley lived near a brewery and his curiosity about how it operated and about the gases involved lead him to discover a gas (carbon dioxide) was heavier than air. He found water and this heavy "air" made a great drink and in 1773 he was awarded a medal by the Royal Society for his invention of soda water. In 1774, he announced the results of his experiment, which described the unusual properties of a new "air", this was in fact, the discovery of oxygen. His experiments with "air" and gases were important for leading to the first ballooning flights.

Priestley also researched relationships among plants, air, and animals. He observed the respiration of plants, by which they take in carbon dioxide and produce oxygen. His observation helped others understand the process. He observed "green matter", which now we know as photosynthesis.

He was a strong religious and political leader and was persecuted for his support of the American Revolution. He came to America in 1794 and spent his last years experimenting in his laboratory. His research in America resulted in the discovery of carbon monoxide (1799).

See also **Oxygen.**

J.M.I.

PRIGOGINE, IIYA (1917–2003).   A Belgian chemist who won the Nobel prize for chemistry in 1977 for his contributions to nonequilibrium thermodynamics particularly the theory of dissipative structures. The main theme of the scientific work of Ilya Prigogine has been a better understanding of the role of time in the physical sciences and in biology. He has contributed significantly to the understanding of irreversible processes, particularly in systems far from equilibrium. His education was at the University of Brussels. The Center for Statistical Mechanics and Thermodynamics at the University of Texas bears his name. order.ph.utexas.edu/research/glimpse.html

PRILLS.   Small, round, or acicular aggregates of a material, usually a fertilizer, that are artificially prepared. In the explosives field, prills-and-oil consists of 94% coarse, porous ammonium nitrate prills and 6% fuel oil.

PROGESTERONE.   See Hormones; Steroids.

PROMETHIUM.   [CAS: 7440-12-2]. Chemical element symbol Pm, at. no. 61, at. wt. 145 (mass number of the most stable isotope), fourth in the Lanthanide Series in the periodic table, mp 1042°C, bp 3000°C (estimated), density 7.26 g/cm^3 (20°C). Elemental promethium has a double hexagonal closepacked crystal structure at 25°C. The pure metallic promethium is silverwhite in color, is soft, and can be cast or machined. The naturally occurring isotope ^{147}Pm is radioactive with a half-life of 2.52 years. Consequently, the element must be handled within a shielded area. Eighteen artificially produced isotopes, ranging from ^{140}Pm to ^{146}Pm and from ^{148}Pm to ^{158}Pm have been identified, all with very short half-lives. Many of the properties of promethium remain classified by the United States Atomic Energy Commission, or are known by other proprietary sources. Although first identified as an element by J.A. Marinsky, L.E. Glendenin, and C.D. Coryell in 1947, the element was not available on more than a gram-scale for several years. Electronic configuration

$$1s^2 2s^2 2p^6 3s^2 3p^6 3d^{10} 4s^2 4p^6 4d^{10} 4f^5 5s^2 5p^6 5d^1 6s^2$$

Ionic radius 0.98 Å. Other important physical properties of promethium are given under **Rare-Earth Elements and Metals.**

^{147}Pm is extracted from the wastes of uranium or plutonium reactors, the most important source of the element. ^{146}Pm and ^{148}Pm also are derived from reactor wastes. In 1970, ^{147}Pm became available in kilogram quantities. ^{147}Pm has been under intensive study as a heat and power source; however, before it can be used for this, ^{146}Pm and ^{148}Pm, which

produce penetrating gamma radiation, must be eliminated. The desirable property of ^{147}Pm is that it decays by beta emission only, at a low energy level compared with most fission products, and thus requires only light to moderate shielding. ^{147}Pm has been used to activate luminescent phosphors. Beads (Microspheres®, 3 M Company) containing ^{147}Pm mixed with a phosphor provide a long-lived, reliable green light and were used by astronauts to assist in docking and other maneuvers in outer space. Commercial applications of ^{147}Pm as a power source include beta-voltaic cells for surgical implant with heart pumps and pacemakers.

Additional Reading

Greenwood, N.N. and A. Earnshaw: *Chemistry of the Elements,* 2nd Edition, Butterworth-Heinemann, Inc., Woburn, MA, 1997.
Krebs, R.E.: *The History and Use of Our Earth's Chemical Elements: A Reference Guide,* Greenwood Publishing Group, Inc., Westport, CT, 1998.
Lide, D.R.: *CRC Handbook of Chemistry and Physics,* 84th Edition, CRC Press, LLC., Boca Raton, FL, 2003.
Parker, P.: *McGraw-Hill Encyclopedia of Chemistry,* 2nd Edition, The McGraw-Hill Companies, Inc., New York, NY, 1993.
Stwertka, A. and E. Stwertka: *A Guide to the Elements,* Oxford University Press, Inc., New York, NY, 1998.

PROMOTER.   1. A substance that, when added in relatively small quantities to a catalyst, increases its activity, e.g., aluminum and potassium oxide are added as promoters to the iron catalyst used in facilitating a combination of hydrogen and nitrogen to form ammonia.

2. In ore flotation, a substance that provides the minerals to be floated with a water-repellent surface that will adhere to air bubbles. Such reagents are generally more or less selective toward minerals of certain classes.

PROOF.   The ethanol content of a liquid at 15.5 C, stated as two times the percentage of ethanol by volume. One gallon of 95% alcohol is therefore equivalent to 1.9 gallons of proof alcohol. In the U.S., the alcohol tax is based on the number of proof gallons.

PROPANE.   [CAS: 74-98-6], $CH_3 \cdot CH_2 \cdot CH_3$, formula weight 44.09, colorless gas, mp -187.1°C, by -42.2°C, sp gr 0.585 (at -45°C). The gas is slightly soluble in H_2O, moderately soluble in alcohol, and very soluble in ether. Although a number of organic compounds which are important industrially may be considered to be derivatives of propane, it is not a common starting ingredient. The content of propane in natural gas varies with the source of the natural gas, but on the average is about 6%. Propane also is obtainable from petroleum sources.

Liquefied propane is marketed as a fuel for outlying areas where other fuels may not be readily available and for portable cook stoves. In this form, the propane may be marketed as LPG (liquefied petroleum gas) or mixed with butane and pentane, the latter also constituents of natural gas (1.7% and 0.6%, respectively). LPG also is transported via pipelines in certain areas. The heating value of pure propane is 2520 Btu/ft^3 (283 Calories/m^3); butane 3260 Btu/ft^3 (366 Calories/m^3); and pentane 4025 Btu/ft^3 (452 Calories/m^3). Propane and the other liquefied gases are clean and appropriate for most heating purposes, making them very attractive where they are competitively priced.

PROPELLANTS (Rocket and Missile).   See Rocket Propellants.

PROPIONATE PLASTICS.   See Cellulose Ester Plastics (Organic).

PROPIONIC ACID AND PROPIONATES.   See Antimicrobial Agents (Foods).

PROSTAGLANDINS.   A group of physiologically active compounds (PGs) derived from fatty acids with 20 carbon atoms (approximate formula, $C_{20}H_{36}O_5$). The compounds originally were isolated as lipid-soluble extracts from sheep and human prostates. Later studies have shown that prostaglandins are found in most mammalian tissues. There are numerous prostaglandins, individually named by the substituents present on the cyclopentane ring that is part of the parent molecule, prostanoic acid. Thus, they are identified as PGA$_1$, PGE$_1$, PGI$_2$ (prostacyclin), etc. The chemical structure and metabolic functions of the prostaglandins have been established, in most cases, with considerable accuracy. Some have been synthesized. Each prostaglandin has specific effects. The compounds participate in pulmonary circulation and hypertension, with varying vasodilator and vasoconstrictor effects. Prostaglandins of the E series have been implicated as a cause of hypercalcemia—they resorb fetal bone in vitro, urinary prostaglandin metabolites are elevated in certain hypercalcemic patients with malignancy, and clinically very important, in certain cancer patients with hypercalcemia. Chemical improvement has been seen after treatment with indomethacin, which inhibits prostaglandin synthesis. Prostaglandins also are implicated in systemic mastocytosis, due partly to marked overproduction of prostaglandin D_2. Prostacyclin plays an important role in platelet function, acting as an effective antiaggregating agent. The prostaglandins are involved in the biochemical pathways that participate in bronchial asthma. PGs are synthesized ubiquitously in the body from unsaturated fatty acid precursors with high rates of production by the seminal vesicles and renal medulla. The metabolism of prostaglandins occurs mainly in the lungs, renal cortex, and liver, with the metabolites excreted in the urine. The most prolific source of natural prostaglandins is a marine organism (gorgonian sea whip) found in great numbers in coral reefs, notably in the Caribbean area. Intermediates and chemical analogs derived from this organism are sometimes referred to as *syntons*.

Additional Reading

Champe, P.C. and R.A. Harvey: *Lippincott's Illustrated Reviews Biochemistry*, 2nd Edition, Lippincott Williams & Wilkins, Philadelphia, PA, 1994.

Marks, F. and G. Furstenberger: *Prostaglandins, Leukotrienes, and Other Eicosanoids: From Biogenesis to Clinical Application*, John Wiley & Sons, Inc., New York, NY, 1999.

Yazici, Z. and G.C. Folco: *Advances in Prostaglandin, Leukotriene, and Other Bioactive Lipid Research: Basic Science and Clinical Applications*, Kluwer Academic Publishers, Norwell, MA, 2003.

PROTACTINIUM. [CAS: 7440-13-13]. Chemical element, symbol Pa, at. no. 91, at. wt. 231.036, radioactive metal of the Actinide Series, mp is estimated at less than 1600°C. All isotopes are radioactive. The most stable isotope is ^{231}Pa with a half-life of 3.43×10^4 years. The latter is a second-generation daughter of ^{235}U and a member of the actinium $(2n + 3)$ decay series. See also **Radioactivity**. Electronic configuration

$$1s^2 2s^2 2p^6 3s^2 3p^6 3d^{10} 4s^2 4p^6 4d^{10} 4f^{14} 5s^2 5p^6 5d^{10} 5f^2 6s^2 6p^6 6d^1 7s^2$$

Ionic radii Pa^{4+} 0.91 Å; Pa^{3+} 1.06 Å. See also **Chemical Elements**.

The probable existence of protactinium was predicted as early as 1871 by Mendeleev to fill up the space on his periodic table between thorium (at. no. 90) and uranium (at. no. 92). He termed the unconfirmed element *ekatantalum*. In 1926, O. Hahn predicted the properties of the element in considerable detail, including descriptions of its compounds. In 1930, Aristid v. Grosse isolated 2 milligrams of what then was termed ekatantalum pentoxide and showed that element 91 differed in all reactions with comparable amounts of tantalum compounds with exception of precipitation by NH_3. However, credit for the discovery of protactinium generally is attributed to Lise Meitner and Otto Hahn in 1917.

Protactinium-231 yields actinium-227 by α-particle emission and has a half-life of 3.43×10^4 years. Its other isotopes include two isomers of mass number 234: uranium X_2 with a half-life of 1.17 minutes, and uranium Z with a half-life of 6.7 hours, the former being an excited state which undergoes de-excitation to give the latter. Other nuclear species have mass numbers 225–230, 232, 233, 235 and 237.

Protactinium (of mass number 231) is found in nature in all uranium ores, since it is a long-lived member of the uranium series. It occurs in such ores to the extent of about $\frac{1}{4}$ part per million parts of uranium. An efficient method for the separation of protactinium is by a carrier technique using zirconium phosphate which, when precipitated from strongly acid solutions, coprecipitates protactinium nearly quantitatively. Then the protactinium is separated from the carrier by fractional crystallization of zirconium oxychloride.

Isotopes of protactinium can also be produced artificially, i.e., by the nuclear reactions of other elements with such particles as deuterons, neutrons, and alpha-particles. Thus, when thorium is bombarded with deuterons of various high energies, five of the reactions are: ^{232}Th(d, 4n)^{230}Pa, ^{232}Th(d, 6n)^{228}Pa, ^{232}Th(d, 7n)^{227}Pa, ^{232}Th(d, 8n)^{226}Pa, and ^{230}Th(d, 3n)^{229}Pa.

Quantitative methods of obtaining protactinium start from the carbonate precipitate from the treatment of the acid extract of certain uranium ores. After this carbonate precipitate is dissolved, the protactinium remains in the silica gel residue, from the solution of which it is obtained on a manganese dioxide carrier. An alternate method effects final separation

of the protactinium by formation of a complex compound, protactinium-cupferron, and its extraction with amyl acetate.

The methods of purification include the use of ion exchange resins, the precipitation of protactinium peroxide and the extraction of aqueous solutions of protactinium salts by various organic solvents.

Protactinium metal is prepared: (1) by reducing the tetrafluoride with metallic barium at about 1,500°C; (2) by heating the halide, usually the iodide, under a high vacuum; and (3) by bombardment of the oxide under high vacuum with 35-keV electrons for hours at a current strength of 0.005–0.010 Amperes.

As early as 1965, investigators at Los Alamos (Fowler et al., 1965) reported that protactinium metal is superconductive below 1.4 K. In 1972, researchers at Harwell (Mortimer, 1972) reported no superconductivity of the metal down to approximately 0.9 K. An exchange of information to resolve the differences in data was conducted over the next few years (Fowler, 1974; Hall et al., 1977). Smith, Spirlet, and Mueller (1979) reported that differences in experimental research were due to problems with the crystal structure of the metal and sample purity that arise when dealing with radioactive material. These investigators observed very-high-purity protactinium, produced by the Van Arkel procedure, and observed an extremely steep superconductivity transition at 0.42 K in protactinium in the presence of rather high self-heating. The superconducting transition temperature and upper critical magnetic field of protactinium were measured by alternating-current susceptibility techniques. Inasmuch as the superconducting behavior of protactinium is affected by its $5f$ electron character, it has been further confirmed that protactinium is a true actinide element.

The predominant oxidation state of the element is (V). There is some evidence that the (IV) state is obtained under certain reduction conditions. When the pentapositive form is not in the form of a complex ion it may exist in solution as PaO_2^+. The compounds are very readily hydrolyzed in aqueous solution yielding aggregates of colloidal dimensions, thus showing marked similarity to niobium and tantalum in this respect. These properties play a dominant role in the chemical properties of aqueous solution, because the element is so easily removed from solution by hydrolysis and adsorption. Protactinium coprecipitates with a wide variety of substances, and it seems likely that the explanation for this lies in the hydrolytic and adsorptive behavior.

The element is difficult to maintain in aqueous solution in the form of simple salts. Solubility data seem to indicate that such amounts as can be dissolved probably do so entirely by formation of complex ions. Fluoride ion strongly complexes protactinium, and it is due to this that protactinium compounds are in general soluble in hydrofluoric acid.

Protactinium oxide may be prepared from the hydrated oxide or the oxalate by ignition. The product is a dense white powder with a very high melting point; the ignited material is not hygroscopic and maintains a constant weight upon exposure to the air. The formula Pa_2O_5 has been determined indirectly, and there is evidence for the existence of $PaO_{2.25}$ (air oxidation) and PaO_2 (reduction of P_2O_5 by H_2).

Volatile protactinium pentachloride has been prepared in a vacuum by reaction of the oxide with phosgene at 550°C or with carbon tetrachloride at 200°C. Reduction of this at 600°C with hydrogen leads to protactinium(IV) tetrachloride, $PaCl_4$, which is isostructural with uranium(IV) tetrachloride, UCl_4. The pentachloride can be converted into the bromide or iodide by heating with the corresponding hydrogen halide or alkali halide.

The volatile fluoride protactinium(V) fluoride, PaF_5, or possibly protactinium(V) oxyfluoride, $PaOF_3$, is formed at relatively low temperatures such as 200°C from the action of agents such as bromine tri- or pentafluoride, BrF_3 or BrF_5, on one of the protactinium oxides. At higher temperatures, treatment of Pa_2O_5 with hydrofluoric acid and hydrogen yields PaF_4.

The reduction of protactinium to the (IV) state in aqueous solution can be accomplished by reducing agents, such as zinc amalgam, and polarographically.

Additional Reading

Fowler, R.D., et al.: *Phys. Rev. Lett.*, **15**, 860 (1965).

Fowler, R.D., et al.: *Proceedings of the 13th International Conference on Low Temperature Physics* (K.D. Timmerhaus, et al.), Plenum, New York, NY, 1974.

Greenwood, N.N. and A. Earnshaw: *Chemistry of the Elements*, 2nd Edition, Butterworth-Heinemann, Inc., Woburn, MA, 1997.

Hall, R.O.A., J.A. Lee, and M.J. Mortimer: *J. Low Temp. Phys.*, **27**, 305 (1977).

Krebs R.E.: *The History and Use of Our Earth's Chemical Elements: A Reference Guide,* Greenwood Publishing Group, Inc., Westport, CT, 1998.

Lide, D.R.: *CRC Handbook of Chemistry and Physics,* 84th Edition, CRC Press, LLC., Boca Raton, FL, 2003.

Mortimer, J.J.: Harwell Report AERE-R 7030 (1972).

Parker, P.: *McGraw-Hill Encyclopedia of Chemistry,* 2nd Edition, The McGraw-Hill Companies, Inc., New York, NY, 1993.

Smith, J.L., J.C. Spirlet, and W.C. Miller: "Superconducting Properties of Protactinium," *Science,* **205,** 188–190 (1979).

Stwertka, A. and E. Stwertka: *A Guide to the Elements,* Oxford University Press, Inc., New York, NY, 1998.

PROTEASE.   A proteolytic enzyme that weakens or breaks the peptide linkages in proteins. They include some of the more widely known enzymes such as pepsin, trypsin, ficin, bromelin, papain, and rennin. Being water soluble they solubilize proteins and are commercially used for meat tenderizers, bread baking, and digestive aids.

See also **Enzyme**; and **Proteins**.

PROTECTIVE COATING.   A film or thin layer of metal glass of paint applied to a substrate primarily to inhibit corrosion, and secondarily for decorative purposes. Metals such as nickel, chromium, copper, and tin are electrodeposited on the base metal; paints may be sprayed or brushed on. Vitreous enamel coatings are also used; these require baking. Zinc coating are applied by continuous bath process in which a strip of ferrous metal is passed through molten zinc.

See also **Corrosion**; **Electroplating**; and **Paints and Coatings**.

PROTEIN CALORIE MALNUTRITION (PCM).   See **Proteins**

PROTEIN HYDROLYSATE.   Solutions of protein hydrolyzed into its constituent amino acids.

PROTEINS.   Along with the carbohydrate and lipid[1] components of the animal diet, protein substances are a major source of nutrition and energy for the living system. Because of his high regard for the proteins, but well before they were really understood, the Dutch chemist Gerardus Mulder (1802–1880) pioneered the use of the term *protein,* derived from the Greek word meaning "to come first." Although proteins furnish energy to the body and thus can be considered body fuels, as are the carbohydrates and fats, the major nutritional roles of the proteins reside in other functions, usually of a highly specific nature. Thus, there are structural, contractile, process-activating, and transport proteins, among others, which essentially are responsible for the chemical workability of the animal system.

Considering the research tools available, the amount of qualitative and quantitative information pertaining to proteins collected over several decades of effort has been tremendous. The data amassed have been highly beneficial to the medical and health sciences, notably in terms of dietary requirements and protein deficiency diseases, to biologists, and, of course, to organic chemists. Past protein research has led to the development of many useful protein substances for industry and commerce. Scientists stretched the limitations of their available instrumental techniques (crystallography, electron microscopy, chromatography, electrophoresis) in their efforts to better understand protein structure and protein function. With the advent of molecular biology (studying proteins at the molecular level), the potential for learning more pertaining to structure and of what proteins do and how they behave in living organisms increased, conservatively speaking, by an order of magnitude or more. As of the later 1980s, protein science has progressed just a little beyond the initial efforts to reduce protein studies to the molecular level. Highlights are summarized in the latter portion of this article. There are several keys to expanding protein knowledge, two of the most important of which are continued mapping of organism genomes, notably mapping the human genome; and the continuing development of improved instrumental and procedural techniques. In using this newly acquired knowledge to manipulate protein structures, the term "protein engineering" is sometimes used. Protein engineering largely lies in the future.

Protein Requirements

In the growing animal body, a significant portion of proteins consumed is required for the creation of new tissue. This results in an increasing

[1] Fats, oils, fatty acids, phospholipids, and sterols.

requirement for proteins in the diet of humans, for example, up to about the age of 20 years, at which time the protein requirement tends to level off to a fairly stable figure. After body maturity, the portion of proteins needed for tissue maintenance is greater than the need for new tissue building. It must be emphasized, however, that immediately at the commencement of life both new tissue building and tissue maintenance take place and even as the body grows older, the two needs continue—only the proportions between the two roles change.

Proteins, on a weight basis, are second only to water in their presence in the human body. If the factor of water is discounted, then about 50% of the body's dry weight is made up of numerous protein substances, distributed about as follows: 33% in muscles; 20% in bones and cartilage; 10% in skin; the remaining 37% in numerous other body tissues. With exception of the urine and bile in the normal healthy individual, all other body fluids contain from small to relatively large portions of protein substances.

Chemically, proteins are distinguished from other body substances in that all proteins contain nitrogen. Some contain sulfur, phosphorus, iron, iodine, cobalt, and other elements, some of which are generally not thought of as components of the life process, but which nevertheless do play extremely important roles (e.g., as catalysts), even if present only in very minute quantities.

In considering the importance of proteins to building and maintaining body functions, it must be emphasized that proteins consumed essentially are raw materials that contain the building blocks for the creation of different proteins. These building blocks are the amino acids of which the protein molecules consumed are constructed and of which the proteins restructured in the body (after consuming or metabolizing the raw materials) are also constructed. Thus, the desirability of proteins for the diet is based upon the best combination of amino acids present. Therefore, some foods are desirable from a protein nutrition standpoint not only because, with relation to their carbohydrate and fat content, they contain a high percentage of protein, but also because they contain most or all of the amino acids needed to form new proteins within the body. See also **Amino Acids**.

Examples of this situation (desirable versus less desirable proteins) popularly cited are the soybean proteins and the grain proteins. With exception of the sulfur-bearing amino acids, notably methionine, the amino acid balance of soybean proteins is reasonably good. With exception of the amino acid lysine, the amino acid balance of grain proteins is reasonably good. By mixing protein substances from these two sources, an excellent source of protein for the human diet is obtained, this explaining growing trends toward fortification of wheat and other cereal flours with soy flour. There are scores of examples of this type which are representative of the trend toward so-called *fabricated foods*.

From years of experience in studying the dietary needs of humans, nutritionists and biologists established the hen egg as having the most perfect balance of amino acids in a natural protein substance. Against this standard, other foods can be rated in their performance. In naming the following food substances in order of their diminishing chemical score, it should be stressed that these foods are arranged only in terms of this one nutritional criterion: fish (70), beef (69), cow's milk, whole (60), brown rice (57), polished white rice (56), soybeans (47), green leaves (45), brewer's yeast (44), groundnuts (peanuts) (43), whole grain maize (corn) (41), cassava (manioc) (41), common dry beans (34), white potato (34), white wheat flour (32). The foregoing food items were selected *randomly* to provide a sense of the spectrum of foods from this one particular standpoint. The Figures represent only the chemical balance of amino acids present and not the total amount of protein available as a weight percentage of food intake, or from the standpoint of protein utilization, once ingested.

In looking at a number of food substances, again a random selection, from the standpoint of total protein (with no regard to quality) in an average serving, the following amounts of protein (grams) are present: fried chicken breasts (27.8); canned tunafish (24), cooked round roast of beef (24), roasted leg of lamb (22), oven-cooked pork loin (21), dry cooked soybeans (13), whole milk (1 cup) (9), canned red beans (7.5), cheddar cheese (1 ounce = 28 grams) (7), fresh cooked lima beans (6.5), egg (medium size) (6), vanilla ice cream (6), fried crisp bacon (5), baked potato (3), cooked broccoli (2.5), cooked oatmeal (2.5), enriched white bread (1 slice) (2), cooked green snap beans (1), lettuce (1/4 head) (1), and reconstituted frozen orange juice (1).

Consequences of Protein Deficiency. Because proteins are so important to numerous and very complex bodily functions, years of research have just commenced to provide some understanding of most of the mechanisms involved. As would be expected, recognition of the extreme manifestations of protein deficiencies has taken place, at least to the extent of providing new guidelines for assisting millions of inadequately fed people in several regions of the world. As further experience is gained in researching the gross problem, the important subtleties of protein performance within the body will become more apparent.

Exemplary of a better understanding and appreciation of protein nutrition is a comparison of the 1945 report of the Food and Agriculture Organization (United Nations) with more recent findings, recommendations, and nomenclature used. In the first *World Food Survey*, the terms *undernourishment* and *malnourishment* were used throughout the report. The general interpretation of undernourishment was taken to mean an inadequate caloric intake, i.e., insufficient energy input to support normal body functions and activities, with body weight loss the inevitable result. Similarly, *malnourishment* was taken to mean a deficiency of one or all of the protective nutrients, such as proteins, vitamins, and minerals. During the last few years, inasmuch as these two problems are so interrelated, the term *protein-calorie malnutrition* (PCM) has come into wide use. PCM of early childhood, particularly in regions that are a part of some of the less developed countries, is quite widespread. PCM apparently is manifested in minor ways at first, but when prolonged very severe syndromes become evident. These include the conditions known as *kwashiorkor* and *marasmus*.

Kwashiorkor usually occurs in the second or third year in the life of a child. Edema is the principal symptom. The condition arises from a combination of circumstances, but the primary cause appears to be a weaning diet that is both inadequate and indigestible and, notably, is lacking of protein. The principal calories are supplied by carbohydrate. The condition is accelerated by repeated infections of a bacterial, parasitic, or vital nature. Without treatment, the disease is fatal in most cases.

Nutritional marasmus is a severe manifestation of PCM and is a condition that usually occurs during the first year of life. Again, it arises from a combination of conditions, frequently widespread in many regions, of feeding an overly diluted formula of cow's milk, thus reducing the protein input well below minimum needs. The condition is accelerated by filthy surroundings and contaminated bottles. Characteristic of the syndrome are a wasting of muscle and subcutaneous fat, a body weight that may be only 60% of standard, and diarrhea. Children who have access to human milk usually are protected against marasmus and diarrheal disease.

A more recent finding and term now used for a protein deficiency syndrome is *PCM-plus*, or *infantile obesity*. This is a condition that occurs among the more affluent populations where an infant is bottlefed, where hygiene is adequate, and where funds are adequate. Overfeeding of an improperly balanced formula can cause the condition. The condition does not occur with breast feeding because the volume of intake is regulated by the infant's appetite and thirst.

Sources of Proteins

The two basic categories of protein sources for the animal diet are other *animals* (living or dead) and *plants*. Thus, in the animal category as a source of human and pet protein foods, there are what might be called terminal sources or nonreplenishing sources, in which the living animal is killed and disassembled into its protein-containing parts. The most common examples including the meaty flesh and organs of beef cattle, pigs, sheep, and horses and goats, as well as the more occasional sources of meat, such as deer, elephant, hippopotamus, etc., depending upon availability and regional eating preferences. To these sources are added the flesh and organs of birds (chickens, ducks, turkeys, pheasants, etc.) and of fish caught in saline and fresh waters. In the overall animal protein category, one also would include those less conventional and essentially unexplored categories, such as earthworms and single-cell proteins (produced by microorganisms) and algae. Renewable or repeating protein sources from living animals, of course, include the milk from dairy cows and buffaloes and the eggs from hens, from which hundreds of high-protein foods (cheese, for example) are prepared. And, to this category, must be added the excellent source of protein provided by human milk to the nursing infant.

Plants, of course, also require protein to build and maintain their life processes and, consequently, are protein sources for the animal diet. In the case of herbivores, plants are essentially the exclusive source of proteins, energy, and all other dietary elements.

In terms of percentage of protein content of basic sources, the animal sources far excel the plant sources. For example, the protein content of some typical unfortified foods is as follows: 20–30% for cooked poultry and meats; 19–30% for cooked or canned fish; 25% for cheese; 13–17% 17% for cottage cheese; 16% for nuts; 13% for whole eggs; 7–14% for dry cereals; 8.5–9% for white bread; 7–8% for cooked legumes; and about 2% for cooked cereals.

Of course, in achieving the higher protein contents of meat from poultry and cattle, a rather costly two-step production process is involved, wherein the animal first converts plant proteins (as from grasses) into animal protein. In a sense, the animal both converts and concentrates the protein source for humans. Several economic factors enter into the picture—the utilization of land, the costs of labor, the additional costs of feed materials, and the costs related to a greater time span of production, among others. As a case in point, an animal must be fed between 3 and 10 pounds (1.4 and 4.5 kilograms) of grain to produce 1 pound (0.45 kilogram) of meat. All of these factors in recent years, particularly in consideration of protein shortages in many regions of the world, have given rise to conflicting opinions pertaining to the ever-increasing production and consumption of meat, not only in several of the western nations of the world, but in the developed nations of the Orient as well. A few authorities have suggested that the western countries should cut back on meat production, thus making more land, skills, etc. available to increasing vegetable protein production to the level where a generous excess supply would be available to underdeveloped countries as well as amply supplying the protein needs of the developed countries. Quickly, these arguments penetrate not only into technological and economic factors, but psychological considerations as well—because any moves of this type necessarily require drastic changes in eating habits, and to bring them about successfully would require much more governmental regulation and policing than any system of private enterprise is likely to tolerate. Further, attitudes tend to swing rather widely from times of grain surplus to times of grain shortage.

Fortunately, as of the early 1980s, it appeared that protein-processing techniques were providing a very satisfactory compromise, even though the industry is just getting underway toward a large-scale operation. Protein meat extenders, for example, wherein meat and vegetable protein are blended to produce an edible product that retains much of what is desired of meats, including their good protein content, are finding acceptance. The wide acceptance of vegetable protein in analogue meat products has many hurdles to overcome, but it appears that a solid start has been made. The hurdles not only include acceptability in the marketplace, but also some justifiable resistance on the part of cattle and poultry producers. For many reasons, the transition, if it ultimately takes place, will occur over quite a long period of time. Because of continuing economic inflation, the earlier cost advantages that tended to favor blends of meat and vegetable proteins have become less significant.

An early impetus to soy protein foods was given when the United States introduced soy protein products into its overseas donation program in 1966 as a component of foods formulated to meet special needs of certain population groups. Chief among these were children in developing nations, especially the weanling infant and preschool child whose requirements for growth put special demands on diet composition. Pregnant and lactating mothers also had dietary needs frequently not met in countries where food supplies were marginal. Beyond these needs, there were nutrient deficiencies in large population groups, which could be best overcome by enrichment or fortification of commonly eaten foods.

Shortages in the domestic supply of nonfat dry milk, which developed in 1965, stimulated the development of high-protein formulated foods which would serve as supplements in the diets of the children or in the emergency feeding of adults. These formulations had to pass rigid specifications, one of the principal criteria being the recommended daily dietary allowance for protein, vitamins, and minerals. The U.S. Department of Agriculture and the U.S. Agency for International Development developed the guidelines and designed various formulated foods. Among these formulations were Corn-Soy Milk (CSM), Corn-Soy Blend (CSB), and Wheat-Soy Blend (WSB).

Further impetus was given to protein blends in foods when such products were introduced into the domestic food assistance program in the United States. Soy protein foods were introduced into school lunch and breakfast programs for which federal assistance has been given in the form of a subsidy administered by the federal government. Soy-fortified foods also were distributed to needy families through a family food distribution program.

Textured soy protein products in their use as meat alternatives have become increasingly popular in school lunch programs since their introduction in 1971. A soy-modified macaroni was introduced into the family food assistance program a number of years ago.

Less Conventional Sources of Protein. In addition to the traditional animal sources of protein already described and the very large amounts of vegetable protein derived from the soybean, other sources of protein on a large scale for the future are under intense study. Among these are (1) oil-seed crops, such as rapeseed and cottonseed; (2) leaf proteins; (3) algae; and (4) single-cell protein.

Rapeseed, one of the five most widely produced oilseeds, is cultivated mainly in India, Canada, Pakistan, France, Poland, Sweden, and Germany. Past objections to using rapeseed as a source of edible protein has been its content of deleterious glucosinolates. Considerable research has been conducted in Sweden to develop a rapeseed protein concentrate. The first full-scale production plant using a new process was installed in Alberta, Canada. The plant, with a capacity of 5000 tons/year produces a material containing 65% protein. Rapeseed is rich in essential amino acids, with exception of methionine, which soybeans also lack.

Cottonseed offers an attractive source of protein provided that certain objectionable ingredients can be removed. One of these is gossypol, a substance in cottonseed gland that is harmful to humans. A process developed by the U.S. Department of Agriculture has been designed to turn out a satisfactory edible cottonseed protein product. Employing solvent-extraction techniques, the first plant was built in Texas. Cottonseed flour extrudes easily and can be water-extracted to produce a nearly 100% protein isolate. The product has been used as a bland extender and fortifier for processed meats, baked goods, candies, and cereals. Research of a different approach has been used in Central America. In this approach, iron compounds are used to tie up the gossypol in nontoxic form without having to remove it.

Leaf Protein Concentrates. Laboratories in Hungary, Japan, the United Kingdom, and the United States, among other countries, have been engaged in perfection of a leaf protein concentrate process, with emphasis upon increasing yields and palatability and reducing flavor problems and cost. To date, alfalfa appears to be most attractive as a source of leaf protein. Alfalfa will produce more protein per unit of land than most other crops—up to 2800–4000 pounds/acre (3136–4480 kilograms/hectare). It has been estimated that the raw material costs for edible protein from alfalfa would be about 50% that for soybean meal. Several processes have been worked out, ranging from a green curd containing 52% protein to a white powder containing about 90% protein.

Single-Cell Protein. The advantages of single-cell protein (SCP) made from growing microorganisms are several: (1) SCP is independent of agricultural or climatic conditions; (2) SCP doubles in mass rapidly for high production rates and fast genetic experimentation; (3) the crop is free of surface-area limitations, and (4) the protein in microbial cells is generally of a high nutritional quality. Many of the processes proposed and tested, some with limited operating experience, commence with hydrocarbon feedstocks—gas oil and normal paraffin substrates. Two objections have been raised. The first is the possibility that carcinogenic polyaromatic materials present in gas oil may be passed along to the final protein product. The second is an adverse public reaction. A more recent, third objection is the proposition that perhaps technology should be concentrating on manufacturing fuels from farm products rather than food from petroleum products.

Some of the more recent SCP process concepts start with other materials, such as ethanol, acetic acid, starches, sugars, and cellulosic products that may be more available and particularly so in the protein-needy developing countries.

Algae have the highest intrinsic rates of photosynthesis and growth found among green plants. Human food and animal feed are being produced from algae. In Japan, a full plant-scale production harvests algae from open ponds to yield green powder extract that can be used for animal or human consumption. The genus *Chlorella* has perhaps received the most research to date.

Conservation Sources of Protein. Tightening pollution restrictions have forced cheese makers in many regions to end a long-time practice of dumping whey (with its high biological oxygen demand) as a liquid waste. Although many of these manufacturers are now evaporating or spray-drying whey to produce a whole-solids product, several fractionation techniques have been devised to separate a concentrated protein. In the United States,

whey as a byproduct of cheese making totals well over 30 billion pounds (13.6 billion kilograms) per year. From 6.5 to 7% of the whey is solids, of which 0.9% is protein. Some authorities believe that whey and other milk-based protein ingredients offer a high growth potential among all of the non-soybean sources.

Fish protein concentrate is regarded by some authorities as having a high long-term potential. A major restraint is competition for the whole fish. As fish food sources become increasingly competitive, fishes currently considered "trash" fishes from a fresh marketing viewpoint may ultimately become more desirable for table use. Animal-feed fish meal also will be a strong contender for available fish. In terms of processes required for preparing fish-protein concentrate, extraction processes using single or mixed solvents of isopropanol, ethylene dichloride, ethanol, and hexane already have been developed. Experiments with enzymatic processing also are underway.

Chemical Nature of Proteins

In defining a protein structurally, it is first necessary to define a peptide. Peptides are compounds made up of two or more amino acids covalently bound in an amide linkage. The characteristic amide linkage, in which the carboxyl group of one amino acid joins with the amino group of the next amino acid, is called a peptide bond. A peptide is a chain of amino acid residues. Provided that the chain is not circular or blocked at either of the ends, the peptide has an N-terminal amino acid, bearing a free amino group, and a C-terminal amino acid, bearing a free carboxyl group. This is illustrated as follows:

$$H - NHCHR'CO - OH$$
$$H - NHCHRCO - OH + H - NHCHR''CO - OH$$
$$\downarrow -2H_2O$$
$$\underset{\text{N-terminal}}{H - NHCHRCO} - \underset{\text{Nonterminal}}{NHCHR'CO} - \underset{\text{C-terminal}}{NHCHR''CO - OH}$$

Usually a form of shorthand is used to represent the structure of a peptide. For example, H-Val-Gly-Ala-OH, represents a peptide where abbreviation for each amino acid is given in terms of three letters each (Val = valine; Gly = glycine; Ala = alanine). Abbreviations for other amino acids are given in entry on **Amino Acids.** The H denotes the amino terminal (N-terminal) and the suffix OH denotes the carboxyl terminal (C-terminal). Peptides may consist of from two to eight amino acid residues and thus are known as dipeptides, tripeptides, or oligopeptides (eight), depending upon the number of residues contained. A peptide consisting of ten or more amino acid residues and with a molecular weight in the range of $1-5 \times 10^3$ is called a polypeptide. Emil Fischer, father of protein chemistry, proposed early in the twentieth century that proteins are peptide in nature. Actually, no sharp demarcation exists between large polypeptides and small proteins. Examples of small proteins include insulin (hormone protein), protamine, and some components of histone (basic proteins of chromosomes).

Almost all proteins are comprised of amino acid residues, more than 100 in number, and their molecular weight may range from 10^4 to 10^7. A few examples include: Insulin (6×10^3); ribonuclease (13×10^3); lysozyme (eggwhite) (15×10^3); chymotrypsinogen (21×10^3); ovalbumin (43×10^3); serum albumin (66×10^3)—all of the foregoing being single peptide chains. Multiple chains include: Hemoglobin (68×10^3); gamma globulin (IgG) (160×10^3); fibrinogen (340×10^3); urease (460×10^3); thyroglobulin (640×10^3); myosin (850×10^3); hemocyanin (octopus) $(2,800 \times 10^3)$; hemocyanin (snail) $(8,900 \times 10^3)$; and tobacco mosaic virus $(40,000 \times 10^3)$. Proteins of huge molecular weight (millions) are enormous aggregates of protein subunits, each of which may be so large (molecular weight = $1.5-10 \times 10^4$) in most instances. The independent peptide chains that constitute a protein molecule are often held by the disulfide bridges of cystine residues. From the diagram below, it will be seen that in a single chain the bridges may hold together two quite distant points in terms of the linear amino acid sequence, forming a large loop structure:

$$
\begin{array}{c}
\text{H} \text{---} \text{---} \text{NHCHCO} \text{---} \text{---} \\
\mid \\
\text{CH}_2 \\
\mid \\
\text{S} \\
\mid \\
\text{S} \\
\mid \\
\text{CH}_2 \\
\mid \\
\text{HO} \text{---} \text{---} \text{COCHNH} \text{---} \text{---}
\end{array}
$$

Although more than 200 amino acids have been found in living organisms, only 20 alpha-amino acids of the L configuration have been found serving as the building units for proteins and related peptides. These 20 amino acids occur in varying proportions in different proteins. Some proteins are fully lacking in one or more of them. Some amino acids occur only in some of the proteins. For example, hydroxyproline has been found only in collagen and elastin (proteins of animal connective tissue) and in gelatin derived from collagen.

Numerous classifications of proteins have been proposed over the years. In terms of function, there are:

a. *Structural Proteins.* Proteins that support the skeletal structures, maintain the form and position of organs, impart the structural rigidity to walls of containers for biological fluids, and often form part of the external tissues. In keeping with their functions, they are insoluble in many liquids, especially body fluids, and are otherwise relatively resistant to biochemical reactions. The proteins of nails, horn, hoofs, and hair are familiar examples.

b. *Contractile Proteins.* Those substances that have the property of undergoing a change in configuration, which results in a change in length or shape. Thus they give the organism the power to move itself, its parts, or other objects. The proteins of muscles are prominent examples.

c. *Process-Activating Proteins.* As used here, the term *process* includes the biochemical reactions, which are catalyzed by enzymes, and in some of which the cytochromes play an intermediate role; it also includes the endocrine reactions activated by the hormones, some of which are proteins.

d. *Transport Proteins.* Proteins which transport an essential substance or factor, from that part of the organism where it becomes available from a source external to the organism to the point where it is used. Examples are many of the chromoproteins, such as hemoglobin, or the blue hemocyanins (from mollusks) which contain copper instead of iron as does hemoglobin, or the chlorophyll-protein complexes of plants.

Another basis of classification is that of solubility, which has been applied to proteins from all sources, plant and animal. (a) Thus the albumins were soluble in water and coagulable by heat. They included serum albumin, egg albumin, lactalbumin (from milk), leucosin (from wheat), and legumelin (from legumes, chiefly peas). (b) The globulins are soluble in neutral salt solutions and in strong acids and alkalies. They include blood globulin (which has been separated by electrophoresis into alpha, beta, and gamma fractions, and is further discussed later in this entry), ovoglobulin (from egg yolk), edestin (from hempseed), phaseolin (from beans), arachin (from peanuts), and amandin (from almonds). (c) The glutelins, such as glutenin from wheat, are soluble in dilute acids and alkalies, and insoluble in neutral salt solutions. (d) The scleroproteins are quite insoluble, and the structural proteins (group I mentioned above) belong to this group. All these groups, and several others not included here, are simple proteins, i.e., they consist only of polypeptide chains of amino acids. The many conjugated proteins must then be classified upon the basis of their nonprotein portions: glycoproteins which contain carbohydrate groups, lipoproteins which contain lipid groups, chromoproteins which contain metal-containing complexes that are usually colored, as hemoglobin contains heme.

Still another classification places proteins into three major categories: (a) Simple proteins; (b) conjugated proteins; and (c) derived proteins. The last classification embraces all denatured proteins and hydrolytic products of protein breakdown and no longer is considered a general class.

A relatively simplistic concept of a protein structure is indicated in Fig. 1. The molecular weight for the hemoglobins is on the order of 68,000. They are conjugated proteins and consist of four heme groups and the

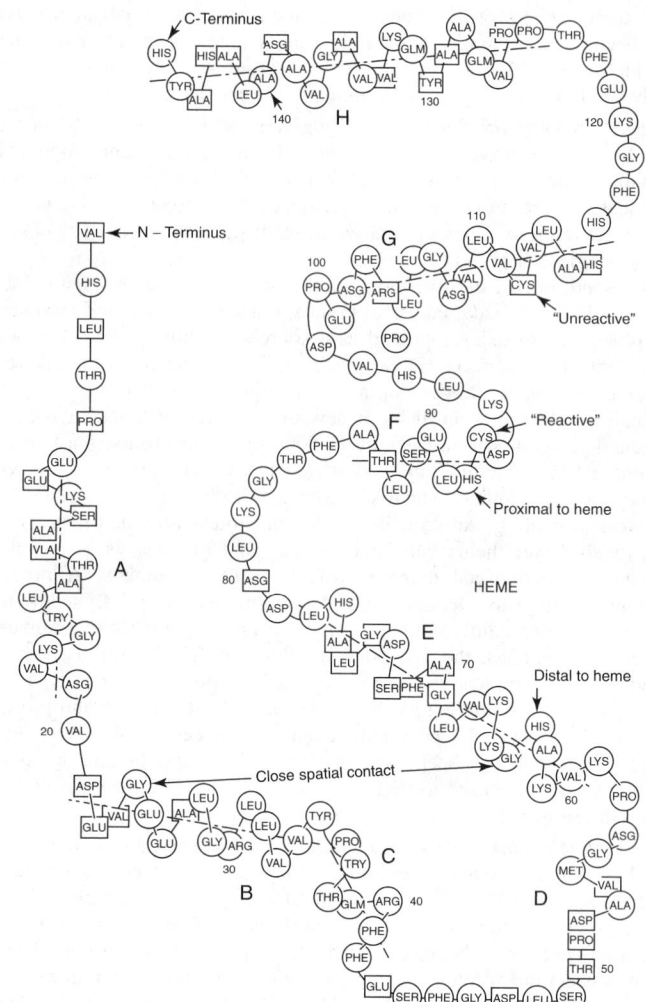

Fig. 1.   Simplified representation of the beta chain of human hemoglobin A

globin portion. The heme group is a porphyrin in which the metal ion coordinated is iron, which may be Fe^{3+} or Fe^{2+}, but only in the latter case (ferrohemoglobin) can the molecule bond molecular oxygen and be effective in respiration, i.e., by forming oxyhemoglobin. The globin portion of the molecule consists of four polypeptide chains. These chains are designated as alpha, beta, gamma, etc. according to their amino acid composition. Normal adult hemoglobin consists of two alpha chains and two beta chains. The composition and conformation of the beta chain are shown in the diagram, together with the point of attachment of the heme groups: Note that they are attached to histidine groups. It has been learned that the central iron atom in heme, which is chelated to the porphyrin ring by four bonds, is attached to the polypeptide chain in adult human hemoglobin by three imidazole ligands of the globin chain, which belong to the histidines at positions 58, 87, and 89 of the alpha chain. See also **Hemoglobin**.

Besides hemoglobin, other proteins of blood are of considerable importance. They are the plasma proteins, serum albumin and fibrinogen, and the globulins. Serum albumin is responsible for the major part of the osmotic pressure of human plasma. Its molecular weight is on the order of 68,000. It is a typical globular protein, having nearly one-half helical character. Although not as nearly symmetrical as hemoglobin or myoglobin, it has a symmetry indicated by its molecular dimensions of 150 Å long and 38 Å wide. It is the smallest, and most abundant, of the plasma proteins; for this reason, and also because of its relatively low isoelectric point, it undergoes migration rapidly in an electric field. See also **Electrophoresis**. By this method it may be separated into two types of molecules, similar in composition except for the presence of a single cysteine residue in one and not the other. However, it contains cystine residues, which form seventeen disulfide bridges cross-linking the polypeptide chain, i.e., the molecule of serum albumin consists of a single polypeptide chain.

Another plasma protein to be discussed here is fibrinogen, which is the chief substance involved in the process of blood clotting. Its molecular weight is on the order of 330,000. It contains all twenty of the amino acids described in that entry as the most general in proteins, although it is relatively low in cysteine, and highest in the acidic amino acids (aspartic and glutamic acids). The process of clotting occurs in three major steps. In the first the substance prothrombin, a blood glycoprotein containing about 5% carbohydrate as glucosamine and a hexose sugar, is converted to the clotting enzyme thrombin. (The latter is unstable, and hence must be formed when needed.) The conversion process is catalyzed by the calcium ion and a group of substances known as thromboplastins. In the second step the enzyme thrombin catalyzes the transformation of fibrinogen to an activated form, called profibrin, with an altered pattern of electric charge. This change is considered to be due to the liberation of two short-chain polypeptides (one bearing 18 amino acid residues and the other 20), and a corresponding change in the character of the remainder of the fibrinogen molecule, which collectively constitute the substance profibrin. In the third step, this mixture of substances undergoes spontaneous polymerization to form the substance fibrin, which has been shown in electron microscope photographs to consist of a network of striated fibers. This polymerization occurs in stages, and in some views of the process they are divided into two steps, polymerization and clotting, the former being regarded as the formation of linear polymers, and the latter as their cross-linking by an enzymatic reaction whereby disulfide bonds are formed.

Animal organisms generally require effective assistance of intestinal flora, as in ruminants to assimilate inorganic nitrogen into a very wide variety of foreign substances, called antigens.

The life span of individual proteins in living organisms is relatively short—about 4 months for hemoglobin and but a week or two for serum albumin. The aged proteins are digested by proteolytic enzymes of tissues, such as cathepsin. A significant portion of the recovered amino acids may be available for the biosynthesis of new proteins, but another part is catabolized and the nitrogen is excreted as urea in mammals, uric acid in birds, reptiles, and insects, or ammonia in organisms of lower classes. For the maintenance of nitrogen balance and for growth, the human organism requires a daily intake of from 70 to 80 grams of proteins. The present world population thus requires 3×10^7 tons of animal proteins and 10×10^7 tons of plant proteins annually.

Living organisms can synthesize their own proteins from amino acids. In terms of the ability to carry out the de novo synthesis of amino acids, however, there are wide variations among different organisms. Plants, for example, can synthesize amino acids from nitrogen in the form of ammonium salts or nitrate and other simple compounds. The annual production of cereal and vegetable proteins so assimilated from inorganic nitrogen over the world is estimated to be about 10×10^7 tons. Of this, 4×10^7 tons are provided by wheat; 2×10^7 tons by rice; and 1×10^7 tons by corn and other sources. Lactic acid and some other microorganisms require preformed amino acids for growth, lacking own ability to synthesize. However, some microorganisms perform well with ammonium sulfate and carbohydrate as the sole sources of nitrogen, sulfur, and carbon. In some cases, they accumulate particular amino acids in a process referred to as amino acid fermentation.

Animal organisms generally require effective assistance of intestinal flora, as in ruminants, to assimilate inorganic nitrogen into body protein. This accounts for the human needs of a daily requirement of 70–80 grams of protein. However, over half of the protein-constituent amino acids can be derived from other amino acids by their own enzymic reactions. Thus, amino acids are classified as essential or nonessential. Amino acid requirements vary with the physiological state of the animal, age, and possibly with the nature of the intestinal flora.

The Food and Agricultural Organization (FAO) established the following essential amino acids in the ratios indicated:

	Percent of Crude Protein
Isoleucine	4.2
Leucine	6.2
Lysine	4.2
Methionine	2.2
Phenylalanine	2.8
Threonine	2.8
Tryptophan	1.6
Valine	5.0

A distribution of amino acids in dietary proteins can be obtained accordingly by taking both animal and plant proteins at a ratio of 1:3–4. Although plant proteins are lower cost, they are markedly deficient in some essential amino acids. Their protein efficiency is low without addition of deficient amino acids. Enrichment of human and animal diets with free amino acids, such as lysine, methionine, threonine, and tryptophan, as a substitute for animal proteins, has proved successful.

Excesses of amino acids are not harmful—with few exceptions. An imbalance of amino acids can result in a few instances. For example, a rat that feeds on eggwhite proteins with threonine or isoleucine added in high concentrations can experience an undesirable imbalance.

Several industries are based upon proteins as exemplified by the keratins of wool, feather, or horn; the fibroin of silk; the collagenous tissues as leather; proteins in milk, wheat, soybean, egg, and numerous other natural substances. Making cheese from milk casein and flavor seasonings from plant or fish proteins are old processes. Gelatin derived from collagen has been used widely in processed foods and as an adhesive material and in photography. Gluten in cereal is a protein. Major proteins in meat are *myosin*; in egg, *ovalbumin*; in rice, *oryzenin*; in soybean, *glycinin*; and in corn, *zein*.

Protein Quality and Evaluation. Protein quality relates to the efficiency with which various food proteins are used for synthesis and maintenance of tissue protein. Food industry evaluators of protein nutritional quality must operate on several levels of awareness. In particular, manufacturers of processed foods must measure the biological value of the protein content of a variety of processed foods for several reasons: (1) to comply with various governmental regulations; (2) to satisfy nutrition labeling regulations; and (3) an accurate knowledge of protein effectiveness is required in developing new food products and in controlling sources of protein ingredients. Protein quality is also very important in the formulation of animal feedstuffs.

In a number of countries, including the United States, the stipulated measurement of protein quality is the so-called protein efficiency ratio (PER), which may be defined as the *gain in weight* divided by the *weight of protein consumed* by experimental laboratory animals. As of the early 1980s, the AOAC (Association of Official Analytical Chemists) method, defined in 1975, is only one of the codifications of PER work since the concept was first proposed in 1919. More specifically, the PER is the ratio of the weight gained by a group of ten weanling rats fed a diet containing about 10% protein, to the weight of protein consumed over a 28-day period. No sample to be studied should contain less than 1.8% nitrogen according to the AOAC method, and the diet should supply 1.6% nitrogen. Since the samples are not analyzed for protein, but rather for nitrogen, and since protein efficiency ratio rather than nitrogen efficiency ratio is reported, it is important to be clear about whether one of the specific nitrogen factors or the conventional 6.25 figure is used to calculate the protein in the final diet.

Advances in Protein Chemistry

For many years, research was directed to a better understanding of the structure of proteins, notably based upon X-ray crystallography. Remarkable structural details were evidenced. But this avenue of research tended to regard proteins as being static in nature, whereas more recent findings show that indeed proteins are dynamic and that, if they were rigid, they simply could not function. The internal motions that underlie their workings are best explored in computer simulations. As pointed out by Karplus and McCammon, it is now recognized that the atoms in a protein molecule are in a state of constant motion. Thus, what the crystallographer finds is at best a representation of a protein's average structure. The chemical bonds between the atoms along the polypeptide chain in a protein act much like springs. There are also weaker forces between unbonded atoms, including forces that prevent more than one atom from occupying the same point in space at any given time. Thus, in a protein consisting of many atoms, the total force acting on any one atom at any given time depends upon the positions of all the others. Not surprising, the solution of Newton's equations of motion for determining the positions and velocities of all the atoms in a protein requires a high-speed computer. Such calculations constitute what is called a *molecular-dynamics simulation*.

In summarizing their recent research, Karplus (Harvard University) and McCammon (University of Houston) observe that, from future research, much will be learned regarding how to calculate the rates of enzymatic reactions, and the binding of small molecules to large ones, as well as the role of flexibility and fluctuations in the function of macromolecules. For example, it should become possible to determine how particular solvent conditions and amino acid sequences produce certain patterns of protein fluctuations. Such information will become useful in applying new genetic technologies in a practical way.

Protein research of this kind is important because all enzymes are proteins. They catalyze the speed of essential reactions in living systems, including the synthesis of proteins themselves. Knowledge of the dynamics of proteins will assist in better understanding those proteins that transport small molecules, electrons and energy to specific parts of an organism where they are needed. Those proteins of a structural nature, which make up fibrous tissue and muscle, also will be better understood.

As pointed out by Phillips, the level of understanding of enzyme (protein) action has been achieved for many enzymes through the use of chemical, crystallographic, and spectroscopic methods. Gene science, however, has enormously advanced protein studies. By using cellular machinery for protein synthesis, proteins can be manufactured with any primary structure and then introducing whatever changes seem useful in the chemical constitution of naturally occurring proteins. With further knowledge, at some future date it most likely will be possible to design and manufacture fully novel proteins with new and useful properties.

Currently, the most useful advances are being made by the detailed modification of existing protein structures. A very small change, often involving only a single base, is made of the DNA coding for the protein. This is followed by use of natural cellular machinery (frequently bacteria) to synthesize the modified protein. The method is known as *site-directed mutagenesis*, which was first used in 1982. Classical chemical modification of protein structure still is used, but the site-directed mutagenesis approach is usually more straightforward and reliable.

Phillips has projected an imaginary oligopeptide with side chains grouped in accordance with their properties to illustrate intricacies of structure and regions of specializing functions. See Fig. 2.

Much progress is being made in connection with *fibronectins*, those adhesive proteins that act as biological organizers by holding cells in position and guiding their migration. Studies are now revealing the molecular bases for the functions of fibronectins. As observed by Hynes, within the complex architecture of a multicellular organism most normal cells remain reasonably stationary. They are anchored to basement membranes and connective tissue, which is made up mainly of a fibrous mesh of proteins and other substances. In the adults of most species, only

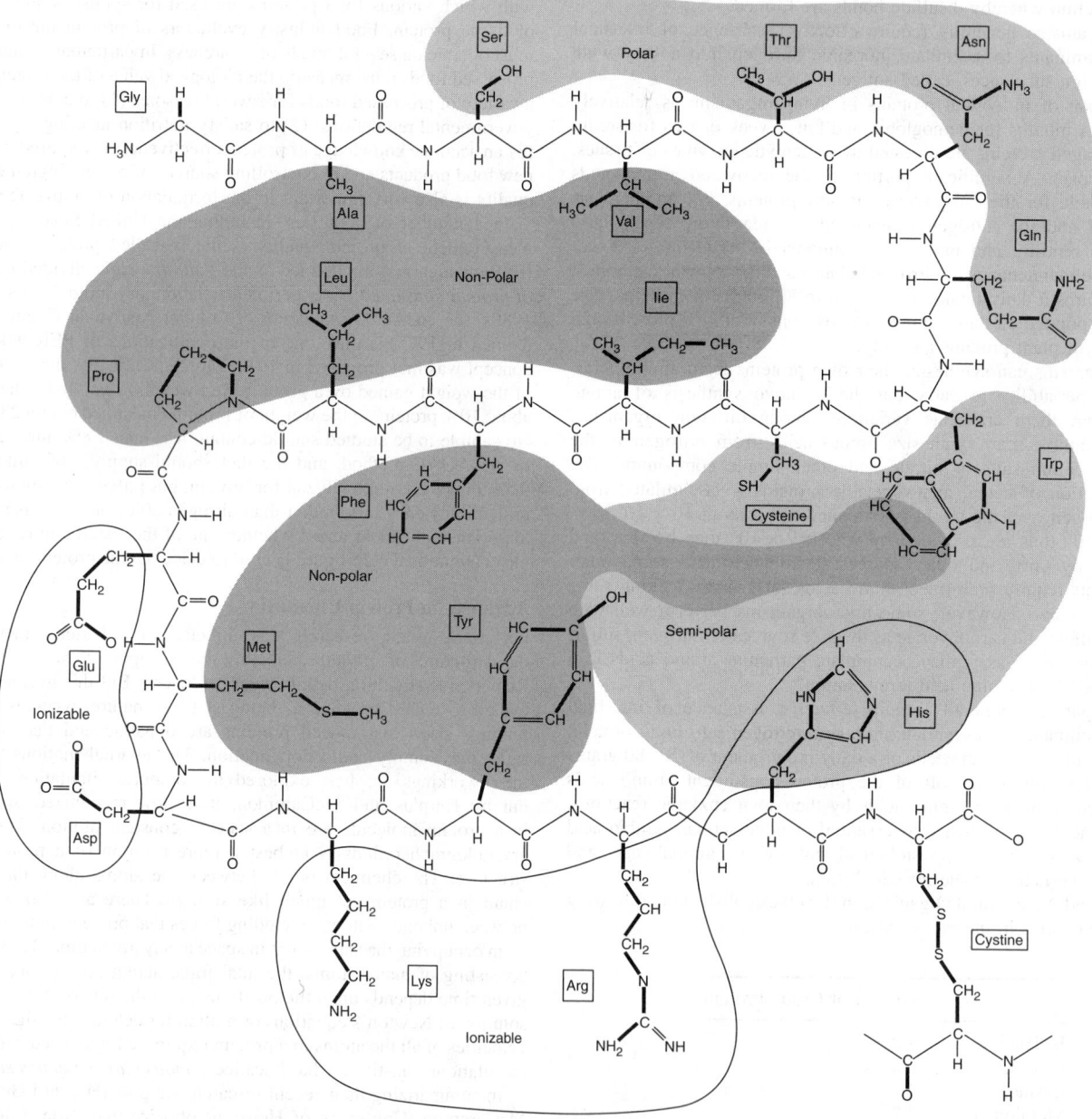

Fig. 2. Facsimile depiction of an imaginary oligopeptide with side chains grouped in accordance with their properties as proposed by Phillips (1987) in an excellent summary of "Protein Engineering" in the new and exceptional publication, *Scientific & Technology Review (The University of Wales)*. All twenty amino acids are represented as shown by three-letter abbreviations in the boxes on the diagram. Polar, semipolar, nonpolar, and ionizable portions of the hypothetical oligopeptide are indicated by shaded and dotted areas. Also, note disulfide bridge shown. (*After Phillips*)

a few cell types will routinely move through this extracellular matrix. It is known that during embryonic development and wound healing, some cells migrate extensively and usually unerringly. The question is asked—how can the organization of these cells be both fixed and dynamic? Glycoproteins (those with attached sugars) may be part of the answer. Of these glycoproteins, the fibronectins are currently the best understood. These molecules have several functions—they can assemble into fibrils, bind to cells, and link cells to other kinds of fibrils in the extracellular matrix. Fibronectin, of course, is a critical component of the blood clotting function. Several lines of research are now being followed in fibronectin studies. These are well described and illustrated in the Hynes reference. It has been suggested that, inasmuch as cancer most frequently involves metastasis (migration of tumor cells to unrelated tissues elsewhere in the body), there may be some connection with fibronectins, because their currently best understood role is that of keeping cells in place and when they move they control their migration.

W.R. Schaffer (University of California, Berkeley) and colleagues have been investigating what are known as *isoprenoids*. These compounds are structurally related lipophilic molecules that perform a wide variety of essential cellular functions. These lipids include such functionally diverse molecules as cholesterol, ubiquinone, dolichols, and chlorophyll, yet isoprenoids are derived from a common precursor, mevalonic acid. These studies may lead to a better understanding of the Ras oncogenic proteins.

In recent years, molecular biologists have found that proteins, in their various roles (binding of receptors, assembling into cellular structures, catalyzing metabolic reactions, etal) depend largely on their three-dimensional structure. For quite some time, biologists have been successful with their techniques for sequencing the amino acids that make up a given protein. In contrast, progress was slow toward determining how the protein chain of components folds into a three-dimensional structure.

It was not until quite recently that neural computers have been used to solve the protein-folding problem. Some proteins have been described in the past as being contorted into "tangled" structures, sometimes likened to a twisting telephone cord. Attempts to predict a folding pattern would require the computation of each part of a chain and its effect on adjacent parts of the chain, a computation process of great magnitude. Further, protein crystals are difficult to develop, thus eliminating or at least reducing the effectiveness of x-ray crystallography. Nuclear magnetic resonance (NMR) also has been used, but unfortunately tends to be limited to the smaller proteins and requires much computer time. Although several thousand proteins have been amino acid sequenced, only a few hundred structures have been determined.

In 1988, researchers T.J. Sejnowski and N. Olan (Johns Hopkins University) reasoned that a computer (NETtalk), that had been designed to pronounce written English words might be applied to the protein structure problem—because NETtalk also depended upon deciphering that occurs at the junctions of numerous separations in a word (as it may appear hyphenated) and that this analysis may be similar to the occurrence of a multi-hyphenated structure exhibited by proteins. The researchers explain that a learning rule modifies the network so that eventually the network will produce the correct phoneme a large percentage of the time. Further work along these lines resulted in a network that could correctly predict over 64% of a test sequence.

S. Brunak and R.M.J. Cotterill (Technical University of Denmark) pursued the approach further, based upon data inputs from NMR and x-ray diffraction. The neural network approach remains very active so that encoding of the intricacy of interconnections may be achievable.

In addition to studying the structure (folding) of proteins for fundamental knowledge, the study of enterotoxins has the additional incentive where life-threatening diseases are concerned, particularly toward the development of improved vaccines.

In research conducted at the University of Groningen (Netherlands) over a 14-year period, scientists succeeded in developing a pure crystal of the cholera toxin. Over 25,000 diffraction measurements of pure crystals, it became possible to generate a computer image of the cholera toxin. It has been observed that all bacterial toxins act in the same manner—one component is an enzyme that performs the invasive function and another component performs destruction once it enters the cell.

Research also has indicated that *E. coli* and diphtheria toxin perform in a similar manner. Active research programs currently are being conducted at Harvard University and the University of California, Los Angeles.

Similar structural determination studies are going forward to determine enzyme structures. In 1991, S. Taylor, D. Knighton, J. Sowadski, and colleagues (University of California, San Diego) announced their development of the three-dimensional structure of a protein kinase.

As aptly put by Doolittle (1985)—"If DNA is the blueprint of life, then proteins are the bricks and mortar."

Additional Reading

Abbott, N.L. and T.A. Hatton: "Liquid-Liquid Extraction for Protein Separations," *Chem. Eng. Progress*, 31 (August 1988).

Angeletti, R.H.: *Proteins: Analysis and Design,* Academic Press, Inc., San Diego, CA, 1998.

Barton, G.J.: *Protein Structure and Prediction,* Blackwell Science, Inc., Malden, MA, 2002.

Bollag, D.M., S.J. Edelstein, and M.D. Rozycki: *Protein Methods,* 2nd Edition, John Wiley & Sons, Inc., New York, NY, 1996.

Bohr, H.G.: *Neural Network Prediction of Protein Structures,* Springer-Verlag, Inc., New York, NY, 2001.

Bowie, J.U., et al.: "Deciphering the Message in Protein Sequences: Tolerance to Amino Acid Substitutions," *Science*, 1306 (1990).

Branden, C. and J. Tooze: *Introduction to Protein Structure,* 2nd Edition, Garland Publishing, Inc., New York, NY, 1998.

Brown, W.E. and G.C. Howard: *Modern Protein Chemistry: Practical Aspects,* CRC Press, LLC., Boca Raton, FL, 2001.

Builder, S.E. and W.S. Hancock: "Analytical and Process Chromatography in Pharmaceutical Protein Production," *Chem. Eng. Progress*, 42 (August 1988).

Clore, G.M. and A.M. Gronenborn: "Structures of Larger Proteins in Solution: Three- and Four-Dimensional Heteronuclear NMR Spectroscopy," *Science*, 1390 (June 7, 1991).

Considine, D.M. and G.D. Considine: *Foods and Food Production Encyclopedia,* Van Nostrand Reinhold Company, Inc., New York, NY, 1982.

Copeland, R.A.: "Proteins: Masterpieces of Polymer Chemistry," *Today's Chemist*, 53 (June 1992).

Creighton, T.E.: *Protein Function: A Practical Approach,* 2nd Edition, Oxford University Press, Inc., New York, NY, 1997.

DeGrado, W.F., Z.R. Wasserman, and J.D. Lear: "Protein Design, a Minimalist Approach," *Science*, 622 (1989).

Deutscher, M.P. and J.N. Abelson: *Guide to Protein Purification,* Vol. 182, Academic Press, Inc., San Diego, CA, 1990.

Fersht, A.: *Structure and Mechanism in Protein Science: A Guide to Enzyme Catalysis and Protein Folding,* W. H. Freeman Company, New York, NY, 1999.

Gennadios, A. and C.L. Weller: "Edible Films and Coatings from Wheat and Corn Proteins," *Food Techy.*, 63 (October 1990).

Gierasch, L.M. and J. King: "Protein Folding," *Amer. Assn. for the Adv. of Science*, Waldorf, MD, 1990.

Hall, A.: "The Cellular Function of Small GTP-Binding Proteins," *Science*, 635 (August 10, 1990).

Hoffman, M.: "New 3-D Protein Structures Revealed," *Science*, 382 (July 26, 1991).

Hoffman, M.: "New Role Found for a Common Protein 'Motif'," *Science*, 742 (August 16, 1991).

Hoffman, M.: "Playing Tag with Membrane Proteins," *Science*, 650 (November 1, 1992).

Hynes, R.O. and K.M. Yamada: "Fibronectins: Multifunctional Modular Glycoproteins," *J. of Cell Biology*, **95**(2), Part I, 369–377 (November 1982).

Hynes, R.O.: "Molecular Biology of Fibronectin," *Ann. Rev. of Cell Biology*, **1**, 67–90 (1985).

Hynes, R.O.: "Fibronectins," *Sci. Amer.*, 42–51 (June 1986).

Karplus, M. and J.A. McCammon: "Dynamics of Proteins: Elements and Function," *Ann. Rev. of Biochemistry*, **52**, 263–300 (1983).

Karplus, M. and J.A. McCammon: "The Dynamics of Proteins," *Sci. Amer.*, 42–51 (April 1986).

Kinoshita, J.: "Net Result: Folded Protein," *Sci. Amer.*, 24 (April 1990).

Knighton, D.R., et al.: "Crystal Structure of the Catalytic Subunit of Cyclic Adenosine Monophosphate-Dependent Protein Kinase," *Science*, 407 (July 26, 1991).

Lesk, A.M.: *Introduction to Protein Architecture: The Structural Biology of Proteins,* Oxford University Press, Inc., New York, NY, 2000.

Linder, M.E. and A.G. Filman: "G Proteins," *Sci. Amer.*, 56 (July 1992).

Marx, J.L.: "New Family of Adhesion Proteins Discovered," *Science*, 1144 (March 3, 1989).

Nakai, S. and H.W. Modler: *Food Proteins: Processing Applications,* Vol. 2, John Wiley & Sons, Inc., New York, NY, 1999.

Neurath, H.: *Protein Science,* Cambridge University Press, New York, NY, 1991.

Otting, G., E. Liepinsh, and K. Wuthrich: "Protein Hydration in Aqueous Solution," *Science*, 974 (November 15, 1991).

Patthy, L.: *Protein Evolution,* Blackwell Science, Inc., Malden, MA, 1999.

Phillips, D.C.: "Protein Engineering," *Review (Univ. of Wales)*, 46 (March 1987).

Richards, F.M.: "The Protein Folding Problem," *Sci. Amer.*, 54 (January 1991).

Radousky, H.B., G. Hammond, Z. Xu, et al.: *Gene Families: Studies of DNA, RNA, Enzymes and Proteins,* World Scientific Publishing Company, Inc., River Edge, NJ, 2001.

Schaffer, W.R., et al.: "Enzymatic Coupling of Cholesterol Intermediates to a Mating Pheromone Precursor and to the Ras Protein," *Science*, 1133 (September 7, 1990).

Sikorski, Z.E.: *Chemical and Functional Properties of Food Proteins*, CRC Press, LLC., Boca Raton, FL, 2001.

Skolnick, J. and A. Kolinski: "Simulations of the Folding of a Globular Protein," *Science*, 1121 (November 23, 1990).

Smith, D.M.: "Meat Proteins," *Food Techy.*, 116 (March 1988).

Utermann, G.: "The Mysteries of Lipoprotein (a)," *Science*, 904 (1989).

Villafranca, J.J.: *Current Research in Protein Chemistry: Techniques, Structure, and Function*, Academic Press, Inc., San Diego, CA, 1990.

Walker, J.M.: *Protein Protocols Handbook*, 2nd Edition, Humana Press, Totowa, NJ, 2002.

Walsh, G.: *Proteins: Biochemistry and Biotechnology*, 2nd Edition, John Wiley & Sons, Inc., New York, NY, 2002.

Whiting, R.C.: "Ingredients and Processing Factors that Control Muscle Protein Functionality," *Food Techy.*, 104 (April 1988).

Wuthrich, K.: "Protein Structure Determination in Solution by Nuclear Magnetic Resonance Spectroscopy," *Science*, 45 (1989).

PROTHROMBIN.   See **Anticoagulants**; **Blood**.

PROTIUM.   The lighter isotope of hydrogen, with a single proton and electron, and constituting 98.51% of ordinary hydrogen is termed protium.

PROTON.   The proton is the atomic nucleus of the element hydrogen, the second most abundant element on earth. Positively charged hydrogen atoms or "protons" were identified by J.J. Thomson in a series of experiments initiated in 1906. Although the structure of the hydrogen atom was not correctly understood at that time, several properties of the proton were determined. The electric charge on the proton was found to be equal but opposite in sign to that of an electron. The traditionally accepted proton mass is 1836 times the electron rest mass, or 1.672×10^{-24} grams.[1]

An estimate of the size of the proton and an understanding of the structure of the hydrogen atom resulted from two major developments in atomic physics: the Rutherford scattering experiment (1911) and the Bohr model of the atom (1913). Rutherford showed that the nucleus is vanishingly small compared to the size of an atom. The radius of a proton is on the order of 10^{-13} centimeter as compared with atomic radii of 10^{-8} centimeter. Thus, the size of a hydrogen atom is determined by the radius of the electron orbits, but the mass is essentially that of the proton.

In the Bohr model of the hydrogen atom, the proton is a massive positive point charge about which the electron moves. By placing quantum mechanical conditions upon an otherwise classical planetary motion of the electron, Bohr explained the lines observed in optical spectra as transitions between discrete quantum mechanical energy states. Except for hyperfine splitting, which is a minute decomposition of spectrum lines into a group of closely spaced lines, the proton plays a passive role in the mechanics of the hydrogen atom. It simply provides the attractive central force field for the electron.

The proton is the lightest nucleus, with atomic number one. Other singly charged nuclei are the deuteron and the triton, which are nearly two and three times as heavy as the proton, respectively, and are the nuclei of the hydrogen isotopes deuterium (stable) and tritium (radioactive). The difference in the nuclear masses of the isotopes accounts for a part of the hyperfine structure called the isotope shift.

In 1924, difficulties in explaining certain hyperfine structures prompted Pauli to suggest that a nucleus possesses an intrinsic angular momentum or "spin" and an associated magnetic moment. The proton spin quantum number I is $\frac{1}{2}$, and the angular momentum is given by $[I(I + 1)h^2/(2\pi)^2]^{1/2}$, where h is Planck's constant. The intrinsic magnetic moment is 2.793 in units of nuclear magnetons (0.50504×10^{-23} erg/gauss), which is about a factor of 660 less than the magnetic moment of the electron.

Two types of hydrogen molecule result from the two possible couplings of the proton spins. At room temperature, hydrogen gas is made up of 75% orthohydrogen (proton spins parallel) and 25% parahydrogen (proton spins antiparallel). Several gross properties, such as specific heat, strongly depend upon the ortho or para character of the gas.

See also **Particles (Subatomic)**.

PROTON-PROTON REACTION.   A thermonuclear reaction in which two protons collide at very high velocities and combine to form a deuteron. The resultant deuteron may capture another proton to form tritium and the latter may undergo proton capture to form helium. The proton–proton reaction is now believed to be the principal source of energy within the sun and other stars of its class. A temperature of the order of five million degrees Kelvin and high hydrogen (proton) concentrations are required for this reaction to proceed at rates compatible with energy emission by such stars.

PROUSTITE.   This ruby-silver mineral crystallizes in the hexagonal system; its name is a product of its scarlet-to-vermilion color when first mined. It is a silver arsenic sulfide. Ag_3AsS, of adamantine luster. Hardness of 2–2.5; specific gravity of 5.55–5.64. Usual crystal habit is prismatic to rhombohedral; more commonly occurs massive. Conchoidal to uneven fracture; transparent to translucent; color, scarlet to vermilion red. Light sensitive; must be kept in dark environment to maintain its primary character. A product of low-temperature formation in most silver deposits. Notable world occurrences include the Czech Republic and Slovakia, Saxony, Chile and Mexico. Found in minor quantities in the United States; the most exceptional occurrence at the Poorman Mine, Silver City District, Idaho where a crystalline mass of some 500 pounds (227 kilograms) was recovered in 1865. It was named for the famous French chemist, Louis Joseph Proust.

PROVITAMIN.   The precursor of a vitamin. Examples are carotene and ergosterol, which upon activation become Vitamin A and Vitamin D, respectively.

See also **Vitamin**; **Vitamin A**; and **Vitamin D**.

PSEUDOMORPH.   In mineralogy and geology, a mineral, having the crystal form of one species and the chemical composition of another. Typical pseudomorphs are malachite in the form of cuprite, barite in the form of quartz, limonite in the form of pyrite. In such cases of pseudomorphism the evidence seems to be that there has been a complete chemical and molecular change but without any change of the original outward form. See also **Mineralogy**.

PSEUDOPLASTIC SUBSTANCES.   See **Rheology**.

PSILOMELANE.   Psilomelane is a massive black mineral, essentially a basic oxide of barium with divalent and quadrivalent manganese, corresponding to the formula $BaMn^{2+}Mn^{4+}O_{16}(OH)_4$. It crystallizes in the monoclinic system, but is found only in massive, botryoidal or reniform to earthy habits; hardness, 5–6, less in earthy varieties; specific gravity, 6.45; color, black to gray; opaque; submetallic to dull luster. It is a product of secondary weathering of manganese carbonates and silicates. Of widespread occurrence, usually associated with pyrolusite. Major world occurrences include Michigan in the United States, Scotland, Sweden, France, Germany, and India. It is a major source of manganese. The word psilomelane is derived from the Greek words meaning smooth and black, in reference to the smooth black surfaces so often exhibited.

PSI PARTICLE.   Discovery of this subatomic particle in 1974 was announced independently by Ting (Brookhaven National Laboratory) who named it the *J particle* and by B.D. Richter (Stanford) who named it the *psi particle*. The discovery of this particle resolved a number of important problems in particle physics. Intensive research on the psi particle was carried out by Richter and the Stanford group during 1975 and 1976 and is reported firsthand by Richter (*Science*, **196**, 1286–1297,1977). As pointed out by Richter, the four-quark theoretical model became much more compelling with the discovery of the psi particles. The long life of the psi is explained by the fact that the decay of the psi into ordinary hadrons requires the conversion of both c and $\bar{c}$ into other quarks and antiquarks. See also **Particles (Subatomic)**.

PSYCHOACTIVE DRUGS.   See **Enkephalins and Endorphins**.

---

[1] Particularly since the early 1970s, physicists have been seeking a grand unification theory to explain all the elementary particles of matter and all the forces acting between them. Although this goal continues to be elusive, work toward that end is producing many new findings and revised concepts. In the main part of this entry, the traditional viewpoints on the proton are described. Some of the more recent postulations are given toward the end of the entry.

PTOMAINE. A group of highly toxic substances (derivatives of ethers of polyhydric alcohols) resulting from the putrefaction or metabolic decomposition of animal proteins. Examples that have been isolated and prepared synthetically are cadaverine (1,5-diaminopentane), muscarine (hydroxyethyltrimethylammonium hydroxide), putrescine (tetraethylenediamine), and neurine (trimethylvinyl-ammonium hydroxide).

Note: The term *ptomaine poisoning* is usually a misnomer for other types of food poisoning.

PULP (Wood) PRODUCTION AND PROCESSING.

Pulps can be defined as fibrous products derived from cellulosic fiber-containing materials and used in the production of hardboard, fiberboard, paperboard, paper, and molded-pulp products. With suitable chemical modification, pulps can be used in the manufacture of rayon, cellulose acetate, and other familiar products. Pulps can be produced from any material containing cellulosic fiber; but in North America and several other regions of the world, wood is the predominant source of pulp. This description is confined to the production and processing of wood pulp.

Wood is a cellular substance chemically composed of roughly 70% holo cellulose, 25% lignin, and 5% water and ethyl alcohol-benzene soluble extractives. These percentages are based on oven-dry wood.

The chemical composition and physical character of wood vary from species to species, within species grown in different geographical locations, and within a given tree, depending upon the location of the fiber cell in the tree. Both lignin (noncarbohydrate) and holocellulose (carbohydrate) are polymeric substances. Holocellulose is composed of approximately 70% alpha cellulose and 30% hemicellulose, the long-chained alpha cellulose being characterized by nonsolubility in alkali; whereas the shorter-chained hemicellulose is alkali-soluble, the degree depending upon the alkali concentration. Lignin concentration in wood substance is greatest in the middle lamella (the zone around each individual fiber cell), decreasing in concentration through the cross section of the fiber, and reaching a concentration of about 12% at the inner layer of the fiber adjacent to the fiber cavity, or lumen. It is the middle-lamella material (lignin and hemicellulose) that cements the fiber cells together, thus giving rigidity to the fibrous wood structure.

The objective of wood pulping is to separate the cellulose fibers one from another in a manner that preserves the inherent fiber strength while removing as much of the lignin, extractives, the hemicellulose materials as required by pulp end-use considerations. Wood pulp to be used for the manufacture of hardboard, for example, requires only the removal of water-soluble wood sugars and sufficient fiberization, i.e., separation of fibers, to permit effective felting of the fibers in a sheet-forming operation. In a subsequent operation in which the felted fiber sheet is subjected to high pressure and heat, the lignin in the fiber mass softens and flows, ultimately acting as a bonding agent cementing the fibers together into a coherent hardboard. At the other extreme, wood pulp to be used for rayon manufacture must be of a high alpha-cellulose content (~ 88–93%), have extremely low amounts of noncarbohydrate material, and be well fiberized to permit uniform reactions during chemical processing.

Pulping Processes

Wood is converted to pulp by mechanical and chemical actions, which constitute the pulping process. Their selection depends upon the type of wood supply available and the pulp qualities desired. Pulps can be characterized on the basis of the unbleached pulp yields achieved by the pulping process used, i.e., the yield of oven-dry (OD) pulp obtained from oven-dry debarked wood.

Five major types, or classes, of pulps, related to pulp yield ranges normally considered to define each class of pulp, are shown in Fig. 1. Pulp yield is a direct indication of degree of chemical action (delignification and chemical attack on carbohydrate and other nonligneous material). Also shown in this figure are the degrees of defibration effected by chemical and mechanical action utilized to produce the pulp, although this representation is not strictly correct. For example, in producing a full chemical pulp, wood chips are subjected to chemical action (digestion or cooking) in a pressure vessel. When digestion is completed, the cooked and softened chips retain the same physical form as the raw chips originally charged to the digester. But they separate into essentially discrete fibers as a result of mechanical action occurring upon sudden release of the chips from the pressure vessel into a receiving tank, which ordinarily is at atmospheric pressure.

At the other extreme, no chemicals are used in the production of mechanical pulp, and defibration is effected by subjecting wood to a

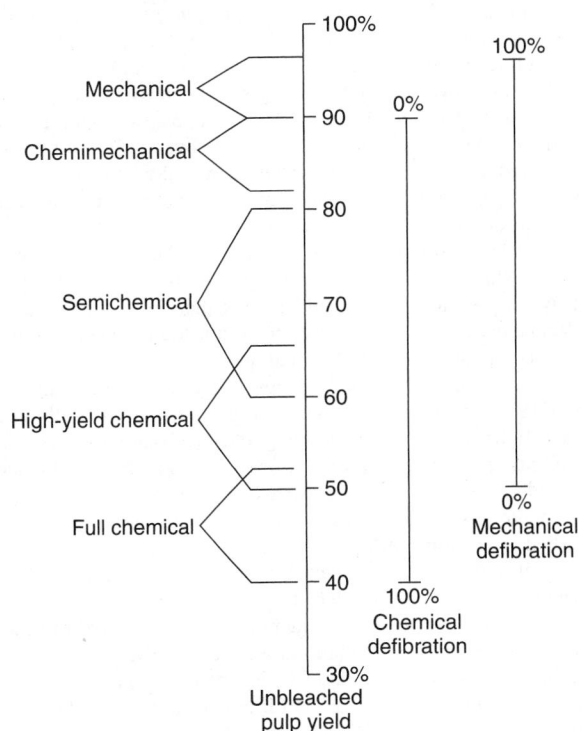

Fig. 1.   Wood pulp characterized on basis of yield

mechanical grinding or attrition action. In this instance, the defibration is aided by some small degree of chemical change and solubilization of wood substance occasioned by heat generated by the grinding operation.

The pulps listed in Fig. 1 are characterized on an unbleached basis as produced by processes conventionally called *pulping processes*. In many instances, these pulps must be further treated chemically to remove residual lignin, hemicellulose, and color bodies before they can be considered suitable for use in specific applications. This further treatment is called *bleaching*, and the bleaching operation is actually an extension of the pulping process.

Customarily, pulping processes and bleaching processes are considered separately, although the choice of bleaching process is highly dependent upon the pulping process used. With this distinction between pulping and bleaching in mind, it will be understood that the pulping processes that are briefly described here pertain only to the production of *unbleached* pulps.

The soda, kraft, and sulfite pulping processes are used to prepare full chemical pulps. The soda process, which uses sodium hydroxide as the cooking chemical for delignification purposes, has largely been superseded by the kraft process, which is characterized by its use of sodium hydroxide and sodium sulfide as active delignification agents in the chip-cooking phase of the process.

Chip-digestion parameters are digester pressure and temperature, digestion time to and at maximum temperature, amount of active alkali used per unit weight of OD wood (percent active alkali), percentage ratio of sulfide to active alkali (percent sulfidity), and weight ratio of cooking liquor (including chip moisture) to OD wood weight. No two kraft pulp mills use the same set of parameter values. Such values must be frequently adjusted, even within a given mill, because of variations in incoming wood and pulp-quality requirements.

Kraft processes are applicable to nearly all species of wood, and effective means of recovering spent cooking chemicals for recycle in the process have been developed. Some sodium and sulfur losses do occur and are replenished in the cooking-liquor system by adding sodium sulfate at the recovery boiler, where it is converted to sodium carbonate and sulfide. In order to maintain a proper sulfur-to-sodium ratio in the recovered chemicals, other chemicals, such as sodium carbonate, sodium sulfite, and sulfur, are sometimes used for chemical makeup.

In contrast to the highly alkaline (pH 11–13) kraft processes, sulfite pulping processes are acidic in nature and are of two general types: (1) the *acid sulfite processes* utilize calcium, sodium, magnesium, or ammonium bisulfite in combination with free or excess sulfur dioxide as cooking chemicals (pH 1.7–2.3). (2) The *bisulfite processes* use sodium, magnesium, or ammonium bisulfite (pH 3.5–5.5) for chip digestion.

Several sulfite processes are multistage and use various combinations of acid sulfite and bisulfite cooking stages and can even use the alkaline kraft cook as one of the multistages. Although spent calcium acid sulfite cooking liquor can be incinerated, there is no recovery of calcium or sulfur. The sodium and magnesium bases can be recovered with or without sulfur recovery, and spent ammonium base liquor can be burned with recovery of sulfur as an option.

High-yield chemical pulps can be produced by the soda, kraft, or sulfite processes, in which chemical use and digestion time and/or temperature are suitably reduced to effect a milder cook than used for full chemical pulps. Mechanical defibrators are used to complete the separation of wood fibers not accomplished by the chemical action.

Semichemical pulps are usually prepared by the neutral sulfite semi-chemical (NSSC) process, although modifications of the full chemical processes can be used. Active pulping chemicals are (in the sodium-base NSSC process) sodium sulfite buffered with sodium bicarbonate (pH 7.0–9.0) and (in the ammonium-base NSSC process) ammonium sulfite with ammonium hydroxide used as a buffer. Defiberization is usually accomplished by attrition mills of the disk type.

Mechanical pulps are produced by two basic processes: (1) *stone groundwood pulp* (SGW) is produced by the defibration action of natural or artificial grindstones rotated at moderate speeds (200–300 rpm) against bark-free bolts of roundwood axially aligned across the peripheral face of the stone in the presence of water. By air-pressurization of the grinder, a *pressurized groundwood pulp* (PGW) of improved quality can be produced. (2) *Refiner mechanical pulp* (RMP) is produced by the attrition action upon raw wood chips of an open (atmospheric) discharge disk refiner. By preheating the chips in a pressurized vessel via direct steaming at temperatures of 120°C or higher and fiberizing the heated chips in either a pressurized or atmospheric disk refiner, *thermomechanical pulp* (TMP) is produced.

Chemimechanical pulps (CMP) are produced by processes in which roundwood or chips are treated with weak solutions of pulping chemicals, such as sulfur dioxide, sodium sulfite, sodium bisulfite or sodium hydrosulfite, followed by mechanical defibration. By presteaming chemically treated chips before attrition, *chemithermo-mechanical pulps* (CTMP) are produced. The mild chemical action, augmented by heat, softens wood lignin and promotes easier defibering with less fiber damage than achieved by the purely mechanical processes.

Wood Pulping Operations. The preceding description of pulps and pulping processes were given as a background to the following descriptions of the various operations involved in the preparation of wood pulp. The pulping system of a typical kraft linerboard mill, as indicated in the simplified flow diagram in Fig. 2, is illustrative of that required for the preparation of both full and high-yield chemical pulps. Linerboard normally is two-layered. The base, or primary sheet, is formed from a high-yield chemical pulp (50–54% yield) and the top, or secondary sheet, is formed from a full chemical pulp, either unbleached (48–50% yield) or bleached (46–48% unbleached yield), laid upon the wet primary sheet on the sheet-forming wire.

Pulp, paper, and paperboard mills are characterized by high capital investment costs and use of high tonnage and rugged but precisely engineered machinery capable of continuous operation with minimum maintenance. A modern kraft linerboard mill with a capacity of 1000 short tons per day (900 metric tons) will have an installed cost, excluding woodlands, of from $275,000 to $325,000 (1986 dollars) per daily on of board produced. Indication of machinery sizes will be given in the following paragraphs.

Wood-Chip Preparation. As indicated in Fig. 2, pulping operations begin with receipt of wood at the mill site. Pulpwood is supplied in log form (roundwood) or chips in accordance with specifications set by the pulp mill. Roundwood is usually received with bark on and in lengths and diameters suitable for proper handling in the wood-preparation equipment at the mill. It has been customary for mills to specify multiple lengths of pulpwood, i.e., 4 feet (1.2 meters) and 8 feet (2.4 meters) as standard receipts, but there is a trend to the procurement of tree-length logs, up to 70 feet (21 meters) in length, either exclusively or in combination with short logs. Another trend has been to the use of chips already prepared, except perhaps for final screening, by independent suppliers or by satellite wood yards operated by the pulp mill itself.

Although linerboard mills formerly used only softwoods (coniferous) for pulping, continued improvements in pulping and board-making technology have permitted the inclusion of up to 20% or more of hardwoods (deciduous) in the wood furnished to the mill, with improved utilization of woodlands as a beneficial result. Softwood and hardwood species are processed through the chipping operation and stored separately; they are either blended into the digester and cooked together, or they are processed separately and the respective pulps blended just ahead of the linerboard machine.

Former practice was to store pulpwood receipts in either a debarked or unbarked condition in stacks or random piles in the wood yard and to reclaim the yard wood for processing into chips just a few hours in advance of chip needs at the digester. A common practice today is to convert the wood into chips immediately after pulpwood receipt and to place the chips, usually by belt or air conveyance, in chip piles built up on concrete or asphalt pads. Separate piles are provided for softwood and hardwood chips, and storage capacities of 40,000 cords or greater can be maintained.

Pulp logs are conveyed to the debarking area, where they are cut to proper length, if necessary, and sorted. Accepted logs are mechanically fed into one end of a large horizontal, cylindrical drum, usually consisting of one or more sections constructed of spaced steel plates, channels, or bars mounted in carrying rings and supported on trunnions and driven by ring gears or suspended from an overhead structure by heavy chains, one or more of which are motor driven. This *barking drum rotates* at a speed of 5–8 rpm, and as the logs tumble about in passing from the intake to the discharge end of the drum, bark removal is effected by the logs rubbing against each other or against the bars or plates constituting the drum shell. Provision can be made for introduction of steam into the feed-end for log de-icing when needed.

Bark removal also can be accomplished by use of a *ring barker* or *hydraulic barkers* which employ high-pressure water jets for bark stripping. Bark removed is collected, shredded in a hog or hammer mill, and used as fuel in steam boilers, where it contributes about 3863 kJ/kg (9000 Btu/pound) of dry solids.

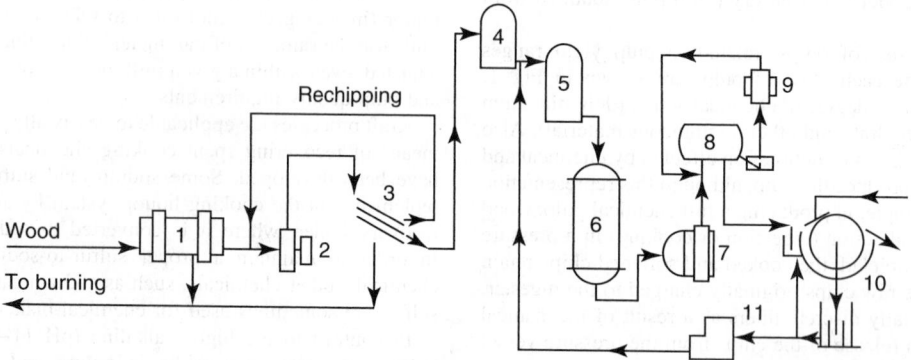

Fig. 2. Flow diagram of kraft pulp mill: (1) debarking, (2) chipping, (3) screening, (4) steaming, (5) impregnating, (6) digesting, (7) fibrilizing, (8) screening, (9) fiberizing, (10) washing, (11) chemical recovery

Debarked wood is conveyed to a chipper for conversion into chips of proper length for chemical treatment in a subsequent cooking operation. Chip length of 0.5–1.0 inch (12.7 to 25.4 millimeters) is conventional.

Chip Digestion. This cooking operation is accomplished in either a batch or continuous digester. A chip digester is essentially a large pressure vessel provided with suitable raw-chip and cooking liquor feed ports and a cooked-pulp discharge port. It is equipped with means for heating and maintaining its contents to and at a specified temperature for the required periods of time.

Batch digesters are vertical, stationary, cylindrical pressure vessels into which chips and cooking liquor are charged under atmospheric conditions. Heating of the digester, after sealing of the feed ports, is effected by direct steam addition or by continual withdrawal of liquor through screened ports and reintroduction of the liquor, after passage through external heat exchangers, onto the top (and sometimes into the bottom) of the chip mass within the vessel. Often, a combination of the direct and indirect heating methods is used. Modern batch digesters are typically 4000–6000 cubic feet (113–170 cubic meters) in volume, with height-to-diameter ratios of 3.5–5.5, and pre-cook pulp capacities of 10–12 tons (9–18 metric tons).

Continuous digesters have been developed as part of the highly successful effort to convert pulp and papermaking from a series of strictly batch operations into an integrated series of continuous operations. A number of successful types of continuous digesters range from horizontal and inclined tube (single or multiple) designs, in which the chip charge is moved through the digester by mechanical screw or bucket conveyors, to vertical digesters, in which chip movement is effected by gravity. See Fig. 3.

Screened chips are conveyed from storage to a chip supply bin in the digester house. The chip bin is designed so that low pressure steam recovered from the hot, spent cooking liquor can contact the chips, preheat them and expel most of the air from the chip interior. If hardwood and softwood chips are to be cooked together, they are blended by weight proportion during the transfer to the chip bin. The chips drop by gravity from the bin to a chip meter, either a twin-screw or a multi-pocket rotary feeder, the speed of which determines chip and cooking liquor flow rate to the digester and pulp discharge rate.

Metered chips drop into a low-pressure rotary feeder valve, through which they are introduced to a steaming vessel maintained at a pressure of about 15–55 psi gage (1 to 3.5 atmospheres). There the chips are further preheated, the remaining air expelled from the chip interior, and chip moisture is leveled in preparation for impregnation of cooking liquor. Since cooked chips are continuously removed from the bottom of the digester, chips pass downward in the digester, replacing those discharged. The time of passage through the cooking zone is normally 90–120 minutes. As cooked chips reach the bottom zone of the digester, the hot, spent cooking liquor is displaced with cooler filtrate from the pulp washers, and removed via extraction strainers to a heat recovery system in which steam is generated to be used to precondition the chips being fed to the digester. As the partially cooled chips move further down the digester, they are plowed to a central well in the bottom of the digester, and are mixed with more filtrate from the pulp washers for dilution and final cooling. Mechanical forces exerted in the transfer of chips from the digester to the blow tank effect fiberization of the chips, the degree of which depends upon cooking conditions. The fibrous material in the blow tank is called pulp, and separate blow tanks are normally used to collect the several types of pulp produced alternately in the digester.

Pulp Screening and Washing. Pulp (brown stock) discharged to the blow tank is in admixture with black liquor, a water solution of spent and residual cooking chemicals and dissolved wood substance, and is at a consistency of from 10 to 18%. The term *consistency* has a meaning peculiar to the pulp and paper industry and refers to the percentage ratio of washed, dry (either oven- or air-dried) fiber to total fiber slurry weight. The fiber bundles left in the pulp after blowing must be fiberized, i.e., separated into discrete fibers, and the black liquor removed in order for the pulp to be refined (a conditioning of individual fibers) and formed into a fiber sheet on the linerboard machines.

Pulp is diluted with filtrate from the pulp washer to a consistency of about 4.5% in the lower portion of the blow tank and fed to fibrilizers, which serve the purposes of metal trapping, fiber-bundle breaking, rough screening, and pumping.

Removal of the black liquor from screened brown stock is usually accomplished on rotary-drum vacuum filters, arranged for multistage countercurrent washing, as shown in Fig. 4.

Refining is accomplished by disk mills, equipped with different plate designs or patterns than those used for defibration. During the refining operation, cellulose fibrils, which wind spirally around the fiber at various positions in its cell wall, are loosened, the cell wall swells due to water absorption, and the fiber is conditioned for sheet formation and inter-fiber binding in the paper- or board-making operation.

Chemical Recovery. Economic and environmental control factors dictate that chemical and heat values of black liquor solids be carefully

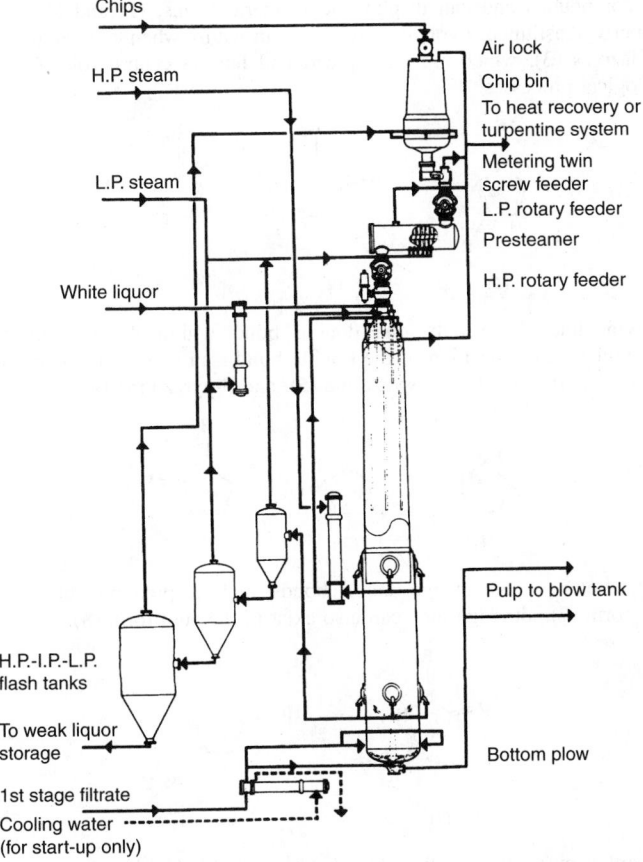

Chips
Air lock
H.P. steam
Chip bin
To heat recovery or turpentine system
Metering twin screw feeder
L.P. steam
L.P. rotary feeder
Presteamer
White liquor
H.P. rotary feeder

Pulp to blow tank

H.P.-I.P.-L.P. flash tanks

To weak liquor storage

Bottom plow

1st stage filtrate
Cooling water (for start-up only)

Fig. 3.   Continuous digester system (*Ingersoll Rand Co.*)

Fig. 4.   Line of three brown stock washers, 9.5 feet (3 meters) in diameter and 16 feet (3 meters × 5 meters) long, equipped with multiport circumferential valve. First stage washer is shown in foreground. (*Ingersoll-Rand Co.*)

conserved, and the recovery system of the modern kraft pulp mill has developed into a highly sophisticated system with still more improvement in efficiency continually being sought. See also **Papermaking and Finishing**.

HENRY F. SZEPAN (retired)
DUNBAR G. TERRY (retired)
Ingersoll-Rand Co., Impco Division
Nashua, New Hampshire

PULTRUSION. A technique for making certain products from glass-reinforced plastics, such as rods, electrical insulators, etc. It involves passage of continuous bundles of glass fiber, which have been impregnated with liquid resin through an oven at the rate of 18 inches per minute at 140°C (285°F).

PUMICE. A highly porous igneous rock, usually containing 67–75% SiO_2, and 10–20% Al_2O_3, with a glassy texture. Potassium, sodium, and calcium are generally present; insoluble in water; not attacked by acids. High gas content, when suddenly discharged by volcanic action; congeal in the form of a highly vesicular natural glass called pumice. When ground, mixed with an appropriate binder and pressed into cakes it is the "pumice stone" of commerce which is used as a light abrasive.

Uses. Concrete aggregate, heat and sound insulation, filtration, finishing glass and plastics, road construction, scouring preparations, paint fillers, absorbents, support for catalysts, and dental abrasive.

PURINE BASES. See **Nucleoproteins and Nucleic Acids**.

PURINES. [CAS: 120-73-0]. Derivatives of the dicyclodiureide of malonic and oxalic acids. The dicyclodiureide is uric acid and the parent compound is purine: so that uric acid is 2,6,8-trioxypurine or the keto form of 2,6,8-trihydroxypurine. Caffeine, theobromine, and theophylline are other important purine compounds.

Uric acid CAS: 69-93-2, ($C_5H_4O_3N_4$) is a white solid, insoluble in cold water, alcohol or ether, sparingly soluble in hot water. Uric acid is a weak dibasic acid thus forming two series of salts, most of which are very slightly soluble in water (lithium urate soluble).

Uric acid is found in the urine, blood, and muscle juices of carnivorous animals (herbivorous animals secrete hippuric acid), in the excrement of birds, serpents and insects, and is an oxidation product of the complex nitrogenous compounds of the animal organism.

Purine Metabolism

Purines are major building blocks for the nucleic acids, DNA and RNA. Adenine, also a purine, plays several important roles—as a cofactor component in energy metabolisms and in enzymatic reactions in which the coenzymes NAD^+ and $NADP^+$ are involved. The end product of purine metabolism is uric acid. It has been well established for many years that biochemical shortcomings in purine metabolism are the principal cause of gout. An average adult male will excrete between 200 and 600 milligrams of uric acid in the urine per day, representing about two-thirds of the total uric acid production in the body. Less than 10–20% of uric acid can be accounted for directly as dietary intake. When insufficient uric acid is excreted, a condition known as *hyperuricemia* will result. When the concentration of uric acid nears the saturation threshold, precipitation in tissues commences. Increased amounts of uric acid may be produced as the result of faulty enzyme activity or other abnormal factors that may occur in the purine metabolism system. Hyperuricemia may be evidenced by the development of an acute, extremely painful, swollen, inflamed joint, frequently at the base of the great toe (podagra). This condition is most commonly encountered in obese, overindulgent people. Usually this condition persists for several days to several weeks without treatment. The condition may recur periodically. The condition responds well to the administration of colchicine. See also **Alkaloids**. Treatment also includes removal of carbohydrates from the diet for a few days, as well as deprivation of alcohol and certain medications, such as thiazide diuretics. Abnormalities in purine metabolism also may create a purine nucleoside phosphorylase (PNP) deficiency, which ultimately may surface as *hypoplastic anemia*.

Purine, uric acid, and other associated compounds play a role in organic synthesis of industrial products.

PYCNOMETER. A device for measuring densities of liquids. It is a container, usually in the form of a bottle or a pipette-like tube, the capacity of which is accurately known and which may be completely filled with the liquid. The difference in weight when filled and when empty, together with the known volume of the liquid, gives the density. The pipette form has a mark to show how far to fill it, and is bent into a V-shape to facilitate immersion in a temperature bath. A familiar design is the "specific gravity bottle," a small flask with a ground and perforated stopper, and sometimes provided with a thermometer. In one of the most precise forms the stopper has a conical top with the capillary leading to the apex, and both neck and stopper are covered by a tight-fitting ground-glass cap to prevent evaporation.

A preliminary step necessary to precise work with the pycnometer is the determination of its two volume constants; that is, the constants of the linear equation expressing the capacity as a function of the temperature. This is done by filling with distilled water and weighing accurately several times at each of two temperatures near the ends of the range for which the pycnometer is to be used. The bottle form is also adapted to the precise measurement of densities of solids. See also **Specific Gravity**.

PYRARGYRITE. An antimony-bearing silver mineral corresponding to the formula Ag_3SbS_3. It crystallizes in the hexagonal system, commonly in rhombic prismatic forms. It displays a rhombohedral cleavage; fracture, conchoidal to uneven; brittle; hardness, 2.5; specific gravity, 5.24; luster, adamantine to submetallic; color, deep red, but being light sensitive alters readily to black. In thin fragments deep red by transmitted light, otherwise practically opaque; streak, purplish red. Pyragyrite occurs with proustite, other silver minerals, and galena, and sphalerite. It is found in the Harz Mountains, in the Czech Republic and Slovakia, Bolivia, Chile, Mexico, and in the United States in Colorado, Idaho, and Nevada. In Canada it is found in the Cobalt region of the Province of Ontario. It derives its name from the Greek words meaning fire and silver.

PYRAZOLES, PYRAZOLINES, AND PYRAZOLONES. The compounds of this article, i.e., five-membered heterocycles containing two adjacent nitrogen atoms, can best be discussed according to the number of double bonds present. Pyrazoles contain two double bonds within the nucleus, imparting an aromatic character to these molecules. They are stable compounds and can display the isomeric forms, (**1**) and (**2**), when properly substituted. Pyrazoles are scarce in nature when compared to the imidazoles (**3**), which are widespread and have a central role in many biological processes.

(1)   (2)   (3)

Pyrazolines have only one double bond within the nucleus and, depending on the position of the double bond, can exist in three separate forms: 1-pyrazoline (**4**), 2-pyrazoline (**5**), and 3-pyrazoline (**6**).

(4)   (5)   (6)

Pyrazolones, contain two double bonds, and are predominantly in the keto form (**7**), although they can also exist in the enol form (**8**).

(7)   (8)

Neither pyrazolidines (**9**), which have no double bonds, nor pyrazoline diones (**10**), with two double bonds, and pyrazolidine triones (**11**), which have three double bonds, are covered in this article.

(9)   (10)   (11)

Despite their scarcity in nature, the title compounds have found use in many applications, including pharmaceuticals, agricultural chemicals, and dyes.

Theoretical Methods

A number of theoretical studies on the reactivity of pyrazoles have been published. However, due to the difficulties involving these calculations, the studies often only approximate the actual reactions occurring in the laboratory.

Structural Elucidation

Among the modern procedures utilized to establish the chemical structure of a molecule, nuclear magnetic resonance (nmr) is the most widely used technique. Mass spectrometry is distinguished by its ability to determine molecular formulas on minute amounts, but provides no information on stereochemistry. The third most important technique is x-ray diffraction crystallography, used to establish the relative and absolute configuration of any molecule that forms suitable crystals. Other physical techniques, although useful, provide less information on structural problems.

Nuclear Magnetic Resonance Spectroscopy. The main application of nmr in the field of pyrazolines is to determine the stereochemistry of the substituents and the conformation of the ring. For pyrazolones, nmr is useful in establishing the structure of the various tautomeric forms.

X-Ray Diffraction. Because of the rapid advancement of computer technology, this technique has become almost routine and the structures of moderately complex molecules can be established sometimes in as little as 24 hours.

Miscellaneous Techniques. The use of ultraviolet (uv) and infrared (ir) spectroscopy has diminished drastically as newer and more powerful procedures have been introduced. However, uv is still useful in studying the tautomeric structures and ionization constants of pyrazoles.

Physical Properties

Pyrazoles in general are stable compounds, as demonstrated by pyrazole itself, which distills at 186°C at atmospheric pressure. The boiling point (bp) increases with an increase in the number of alkyl substituents on carbon. *N*-Methylation decreases both the bp and the melting point (mp) as a result of the elimination of hydrogen bonding. Pyrazoles with substituents at C_3 (C_5) are tautomeric mixtures and form azeotropes. The solubility of pyrazole in H_2O is about 1 g/mL, but it is much less soluble in organic solvents. Pyrazole is a weak base ($pK_a = 2.5$) and can be protonated by strong acids; strong bases yield metal salts. The pyrazolines resemble the pyrazoles in their physical properties. They are liquids with a high bp or low mp. Pyrazolines are basic and the ease of protonation is dependent on the position of the double bond. Most pyrazolones are solids and the mp usually decreases in the presence of substituents at N_1. Simple low molecular weight pyrazolones are soluble in hot water and the higher mol wt materials are soluble in most organic solvents. Hydrogen bonding has strong influence on the predominant tautomeric form. 3-Pyrazolones are more basic than the isomeric 5-pyrazolones.

Chemical Reactivity

Pyrazoles. The chemical reactivity of the pyrazole molecule can be explained by the effect of individual atoms. The N-atom at position 2 with two electrons is basic and therefore reacts with electrophiles. The N-atom at position 1 is unreactive, but loses its proton in the presence of base. The combined two N-atoms reduce the charge density at C_3 and C_5, making C_4 available for electrophilic attack. Deprotonation at C_3 can occur in the presence of strong base, leading to ring opening. Protonation of pyrazoles leads to pyrazolium cations that are less likely to undergo electrophilic attack at C_4, but attack at C_3 is facilitated. The pyrazole anion is much less reactive toward nucleophiles, but the reactivity to electrophiles is increased. Chlorination of pyrazole yields 4-chloropyrazole (12) and bromination

can produce mono-, di-, or tribromo pyrazoles (13). 3-Methylpyrazole on treatment with chlorine in acetic acid yields the pentachloropyrazole derivative (14).

(12)   (13)   (14)

The pyrazole ring is resistant to oxidation and reduction. Only ozonolysis, electrolytic oxidations, or strong base can cause ring fission. On photolysis, pyrazoles undergo an unusual rearrangement to yield imidazoles via cleavage of the N_1-N_2 bond, followed by cyclization of the radical intermediate to azirine.

Oxidation of N_1-substituted pyrazoles to 2-substituted pyrazole-1-oxides using various peracids facilitates the introduction of halogen at C_3, followed by selective nitration at C_4. The halogen atom at C_3 or C_5 is easily removed by sodium sulfite and acts as a protecting group. Formaldehyde was used to direct the selective introduction of electrophiles at C_5 in a simple one-pot procedure.

Pyrazolines. The chemical properties of pyrazolines are governed by their relative instability. They readily undergo ring cleavage, and are easily reduced and oxidized. Loss of nitrogen occurs in pyrazolines lacking a substituent at N_1 to give a mixture of olefins and cyclopropanes, the latter being predominant. This elimination occurs near the map and can be catalyzed by uv light, aluminum oxide, and many other substances. Mild reduction of pyrazolines leads to pyrazolidines. Sodium–alcohol, tin–HCl, or Raney nickel cause ring cleavage, yielding diamines or aminonitrile derivatives. Pyrazolines are easily oxidized to pyrazoles by many reagents, such as bromine, permanganate, and lead tetraacetate. Besides pyrazole formation, rearrangements or side-chain oxidations may also occur. Oxidation with peracids produce N-oxides. Pyrazolines lacking a substituent at N_1 undergo reactions typical of secondary amines, such as acylation, benzoylation, nitrosation, carbamate, and urea formation.

Pyrazolones. The oxo derivatives of pyrazolines, known as pyrazolones, are best classified as follows: 5-pyrazolone, also called 2-pyrazolin-5-one (15); 4-pyrazolone, also called 2-pyrazolin-4-one (16); and 3-pyrazolone, also called 3-pyrazolin-5-one (17). Within each class of pyrazolones many tautomeric forms are possible; for simplicity only one form is shown.

(15)   (16)   (17)

Substitution at N_1 decreases the possible number of tautomers: for 3-pyrazolones, two tautomeric forms are possible, (18) and (19), which in nonpolar solvents are both present in about the same ratio. 5-Pyrazolones exhibit similar behavior.

(18)   (19)

In 4-pyrazolones, the enol form predominates, although the keto form has also been observed.

enol   keto

The tautomeric character of the pyrazolones is also illustrated by the mixture of products isolated after certain reactions. Thus alkylation normally takes place at C_4, but on occasion it is accompanied by alkylation on O and N. Similar problems can arise during acylation and carbamoylation reactions, which also favor C_4. Pyrazolones react with aldehydes and ketones at C_4 to form a carbon–carbon double bond, eg (**20**). Coupling takes place when pyrazolones react with diazonium salts to produce azo compounds, e.g. (**21**).

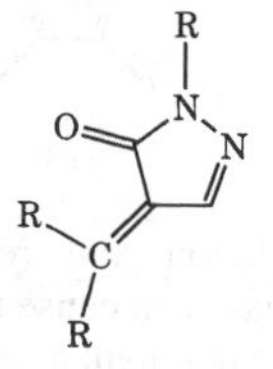

(**20**) 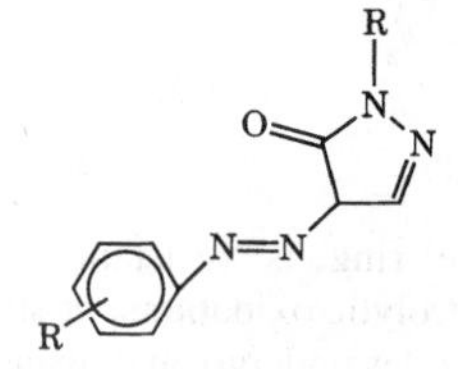   (**21**)

Compounds of type (**21**) are widely used in the dye industry. See also **Azo Dyes**.

Synthesis

In general, the synthesis of pyrazoles and related compounds can be classified into one of four principal categories, with the first two classes being by far the most important: (*1*) from the reaction of hydrazine or its derivatives with β-bifunctional compounds, or compounds that give rise to such functionality (eq. 1) (*2*) by 1,3-dipolar cycloaddition, usually involving diazo compounds (eq. 2) (*3*) by ring-opening of more complex systems already containing the pyrazole nucleus; and (*4*) by chemical, thermal, or photochemical rearrangement of other monocyclic heterocycles. Examples from each class follow.

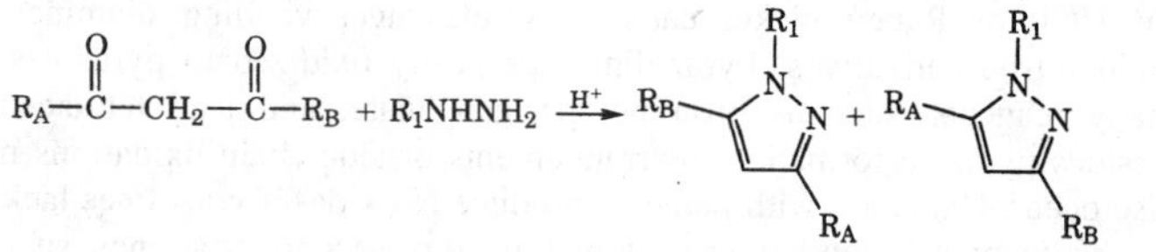

Health Factors

Pyrazole is considered a toxic material because in rats it causes hepatomegaly, anemia, and atrophy of the testis. It also inhibits the enzyme alcohol dehydrogenase, leading to severe hepatotoxic effects and liver necrosis when administered in combination with alcohol. Pyrazolones with a free NH group are easily nitrosated and give rise to nitrosamines, which cause tumors in the liver of test animals. The analgesics antipyrine (**22**) and aminopyrine (two pyrazolones), (**23**), if admixed with nitrites, are mutagenic when tested *in vitro*; however, when tested in the absence of nitrites, negative results are obtained.

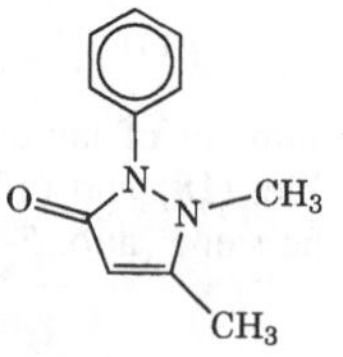

(**22**) 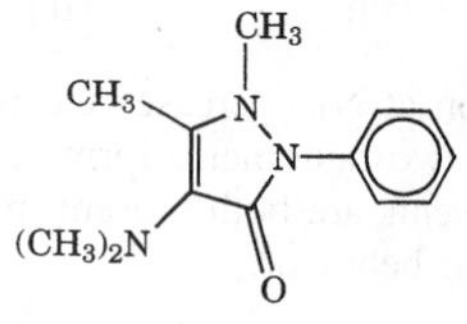   (**23**)

Pyrazolone-type drugs, such as phenylbutazone and sulfinpyrazone, are metabolized in the liver by microsomal enzymes, forming glucuronide metabolites that are easily excreted because of enhanced water solubility.

Applications

Pyrazoles, pyrazolines, and pyrazolones have all found wide use in many fields. Their greatest utility resides in pharmaceuticals, agrochemicals, dyes (textile and photography), and to a lesser extent in plastics. The main uses of the pharmaceuticals that incorporate the pyrazole nucleus are as antipyretic, antiinflammatory, and analgesic agents. To a lesser extent, they have shown efficacy as antibacterial/antimicrobial, antipsychotic, antiemetic, and diuretic agents.

Compounds containing the pyrazole nucleus have also found utility in agriculture. The organophosphate and carbamoyl functionalities, which impart insecticidal activity through linkage to many organic molecules. These compounds act by interfering with acetyl-cholinesterase in the cholinergic synapses.

Pyrazole derivatives have also considerable herbicidal activity.

GABE I. KORNIS
Pharmacia & Upjohn Inc.

Additional Reading

Behr, L. C., R. Fusco, and C. H. Jarboe: in R. H. Wiley, ed., *Pyrazoles, Pyrazolines, Pyrazolidines, Indazoles and Condensed Rings*, Vol. 22 of A. Weissberger, ed., *The Chemistry of Heterocyclic Compounds*, Wiley-Interscience, New York, NY, 1967.
Eicher, T., and S. Hauptmann: *The Chemistry of Heterocycles: Structure, Reactions, Syntheses, and Applications*, 2nd Edition, John Wiley & Sons, Inc., New York, NY, 2003.
Elguero, J.: in A. R. Katritzky and C. W. Rees, eds., *Comprehensive Heterocyclic Chemistry*, Vol. 4, Pergamon Press, Oxford, U.K., 1984.
Jacobs, T. L.: in R. C. Elderfield, ed., *Heterocyclic Compounds*, Vol. 5, John Wiley & Sons, Inc., New York, NY, 1957.
Katritzky, A. R.: *Handbook of Heterocyclic Chemistry*, Pergamon Press, Oxford, U.K., 1985.

PYRIDINE AND DERIVATIVES. Pyridine, [CAS: 110-86-1], is a slightly yellow or colorless liquid; hygroscopic; bp, $115.5°C$; fp, $-41.7°C$; unpleasant odor; burning taste; slightly alkaline in reaction; soluble in water, alcohol, ether, benzene, and fatty oils; specific gravity, 0.978; flash point (closed cup), $20°C$; autoignition temperature, $482°C$. Pyridine, a tertiary amine, is a somewhat stronger base than aniline and readily forms quaternary ammonium salts.

Pyridine and derivatives of pyridine occur widely in nature as components of alkaloids, vitamins, and coenzymes. These compounds are of continuing interest to theoretical physical, organic, and biochemistry and to industrial chemistry. Pyridine and derivatives have many uses, e.g., herbicides and pesticides, pharmaceuticals, feed supplements, solvents and reagents, and chemicals for the polymer and textile industries.

Structure and Nomenclature. The pyridine group consists of a six-membered, heterocyclic, aromatic compound with one nitrogen atom in the ring. The parent compound of this group is pyridine I with ring positions numbered as shown. Alternative denotations of the 1, 2, and 4 positions in the ring are alpha, beta, and gamma, respectively.

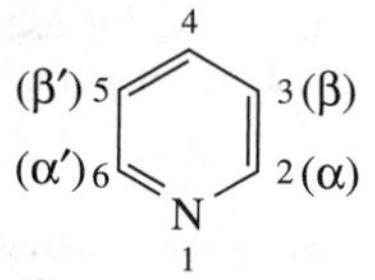

I 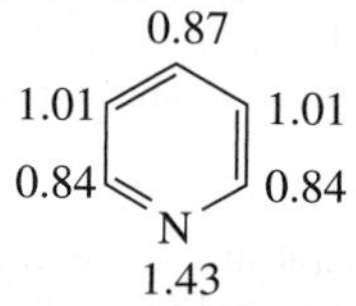   II

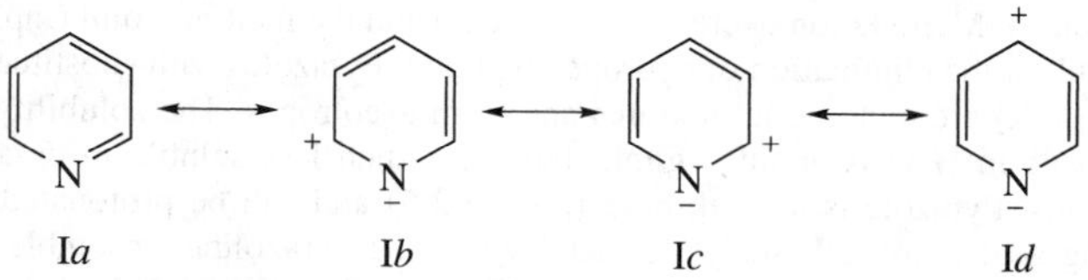

The behavior of pyridine in substitution reactions can be understood on the basis of its resonance structures (Ia–d) and on the basis of the electron-density distribution at the various ring positions as derived from molecular-orbital-theoretical calculations. An example of the published pi-electron density distribution is shown in II. The resonance energy of pyridine is 35 kcal/mole (versus 39 kcal/mole for benzene).

Electrophilic substitution occurs at the 3 and 5 positions, but usually requires drastic conditions because the species actually being attacked is a pyridinium ion. For example, nitration of pyridine with KNO_3 and concentrated H_2SO_4 at $300°C$ gives a 15% yield of 3-nitropyridine. Electrophilic substitution in the pyridine ring is facilitated by the presence of electron-donating substituents.

Nucleophilic substitution occurs in the 2, 4, and 6 positions of pyridine under relatively mild conditions. As an example, amination of pyridine with sodium amide in *N*, *N*-dimethylaniline at $180°C$ gives 2-aminopyridine in good yield.

Homolytic (free-radical) substitution may occur in any of the 2 to 6 positions of pyridine. Thus, the reaction of pyridine with benzene-diazonium salts gives a mixture of 2-, 3-, and 4-phenylpyridine.

Many pyridine derivatives difficult to make directly from pyridine are readily accessible starting from pyridine N-oxide, made by oxidation of pyridine with hydrogen peroxide in acetic acid. As but one example, the nitration of pyridine N-oxide gives 4-nitropyridine N-oxide in high yield. Reduction of the D-oxide to the parent pyridine nucleus is readily effected by hydrogenation or reagents, such as PCl$_3$ or triphenyl phosphine.

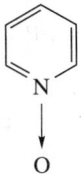

Pyridine N-oxide

Trivial names for the methylpyridines are the *picolines*; the dimethyl-pyridines are the *lutidines*; and the trimethylpyridines (and in older literature the ethyldimethylpyridines) are the *collidines*. The refractive indices for these alkyl pyridines and for pyridine itself fall in the range: $n_D^{20} \sim 1.50-1.51$.

Production of Pyridine and Homologues

Coke Manufacture By-products. In United States practice, coking of coal is done almost exclusively by the high-temperature (900–1200°C) process. For many years, the major source of the pyridines was the chemical-recovery coke oven. The volatiles produced in the coke oven are only partially condensed. The noncondensed gases are passed through a scrubber (the ammonia saturator) containing sulfuric acid. After removal of crystals (ammonium sulfate), a solution of ammonium sulfate and pyridinium sulfates is obtained and treated with ammonia to liberate and contained pyridine bases (~ 70% is pyridine itself). See also **Coal Tar and Derivatives.** The balance of the pyridine bases is extracted from the crude coal tar, i.e., the condensed, main portion of the volatilization products from coking. The crude tar contains approximately 0.1–0.2% pyridine bases. Further separation of the pyridines involves a rather complex series of extractions, distillations, and crystallizations.

Synthetic Methods of Manufacture. Due to rising demand, production of the pyridine bases by large-scale synthesis passed the volume of tar bases extracted from coal tar in the 1960s. By the early 1970s, capacity in the United States for the synthetic manufacture of pyridine, the picolines, and 2-methyl-5-ethylpyridine (MEP) was in the tens of millions of pounds. All of these products can be made by condensation reactions of aldehydes and ammonia. MEP is no longer made in the United States.

When acetaldehyde and ammonia in a 3:1 mole ratio are fed over dehydration-dehydrogenation catalysts, such as PbO or CuO on alumina, ThO$_2$, or ZnO or CdO on silica-alumina, or CdF$_2$ on silicamagnesia at 400–500°C and atmospheric pressure, an equimolar mixture of 2- and 4-picolines can be obtained in 40–60% yields. When a mixture of acetaldehyde, formaldehyde, and ammonia in about 2:1:1 mole ratio is passed over such catalysts, pyridine and 3-picoline are produced; their ratios are usually 1:0.8, but the amounts of pyridine can be increased by changes in the feed.

The lowest-cost synthetic pyridine base, 2-methyl-5-ethylpyridine, is made in a liquid-phase process from paraldehyde (derived from acetaldehyde) and aqueous ammonia in the presence of ammonium acetate at approximately 102–190 atmospheres and 220–280°C in 70–80% yield. Minor byproducts include 2- and 4-picoline.

A new synthetic method for preparing 2-methylpyridine has been commercialized. This process involves the acid/base-catalyzed condensation of acetone with acrylonitrile to make 5-oxo-hexanonitrile. Then the nitrile is converted to a 2-methylpyridine by catalytic cyclization/dehydrogenation:

$$CH_3COCH_3 + CH_2{:}CHCN \longrightarrow$$
acetone   acrylonitrile

$$CH_3COCH_2CH_2CH_2CN \longrightarrow$$

5-oxo-hexanonitrile   (2-methylpyridine)

Much recent work has been done on the synthesis of pyridines from alkynes and nitriles over cobalt catalysts. For example, 2-vinylpyridine has been obtained in good yield from acetylene and acrylonitrile using a cyclopentadienyl-cobalt catalyst. Pyridine has also been obtained from cyclopentadiene and ammonia over a silica/alumina catalyst.

In the synthetic processes, mixtures of products are often obtained. Variation in the supply/demand balance of the alkyl pyridine isomers has led to much research on processes which may alleviate such imbalances, including development of the catalytic hydrodealkylation of alkyl pyridines to pyridine as well as the alkylation of pyridine.

Major Uses of Pyridine Derivatives

The applications of these compounds are wide-ranging and new uses are proliferating. The following examples are a selection of important commercial products, but hardly a complete listing.

Herbicides. A major outlet for pyridine (20–30 million lb/yr world-wide) is in the manufacture of the desiccant herbicides and aquatic weed killers, such as 1,1'-ethylene-2,2'-dipyridilium dibromide, known as Diquat;® and 1,1'-dimethyl-4,4'-dipyridilium dichloride (or dibromide or dimethylsulfate), known as Paraquat.®

1,1'-ethylene-2,2'-dipyridilium dibromide

1,1'-dimethyl-4,4'-dipyridilium dichloride

4-Amino-3,5-dichloro-6-fluoro-2-pyridyloxyacetic acid, tradenamed Sta rane,® is a herbicide used to control broadleafed weeds and brush species, and certain deep-rooted perennial weeds. 3,5,6-Trichloro-2-pyridyloxyacetic acid, tradenamed Garlon,® is a herbicide used for vegetation management, such as in rights-of-way.

4-amino-3,5-dichloro-6-fluoro-2-pyridyloxyacetic acid

3,5,6-trichloro-2-pyridyloxyacetic acid

A new class of herbicides, the pyridylosy-phenoxyalkanoic acids, is typified by *n*-butyl 2-[4-(5-trifluoromethyl-2-pyridyloxy)phenoxy] propionate, tradenamed Fusillade,® active against annual and perennial grasses:

n-butyl-2-[4-(5-trifluoromethyl-2-pyridyloxy)phenoxy]propionate

The newest class of pyridine herbicides with pre- and post-emergent activity, the pyridinesulfoneamides, is typified by *N*-(2-chloro-3-pyridine-sulfonyl)-*N'*-[2-(4-chloro-5,6-dimethylpyrimidyl] urea.

N-(2-chloro-3-pyridinesulfonyl)-*N'*-{2-4-chloro-5,6-dimethylpyrimidyl]urea

2-picoline (2-methyl pyridine) is the source of 2-chloro-6-trichloro methylpyridine, known as N-Serve,® which is useful as a fertilizer additive for reduction of nitrogen losses in the soil due to bacterial oxidation. 2-Picoline also is the starting material for the production of 4-amino-3,5,6-trichloropicolinic acid, a powerful broad spectrum herbicide for broad-leaved plants, known as Tordon.® 3,6-Dichloropicolinic acid, tradenamed Lontrel® and Format® in different formulations, is used for the postemergence control of broadleafed weeds.

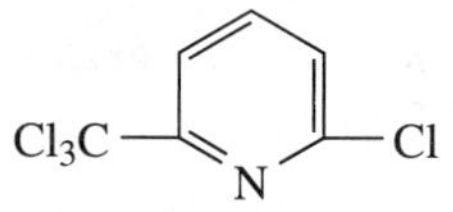

2-chloro-6-trichloromethylpyridine

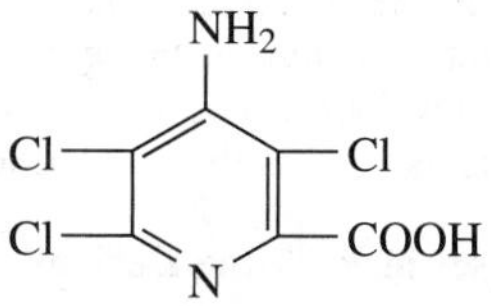

4-amino-3,5,6-trichloropicolinic acid

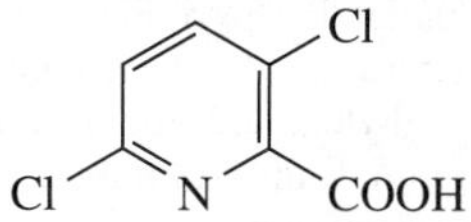

3,6-Dichloropico-linic acid

2,3-Lutidine (2,3-dimethylpyridine) is the starting material for the herbicide 2-[4,5-dihydro-4-methyl-4-(1-methylethyl)-5-oxo-1H-imidazol-2-yl]-3-pyridinecarboxylate (compound with 2-propanamine), tradenamed Arsenal.®

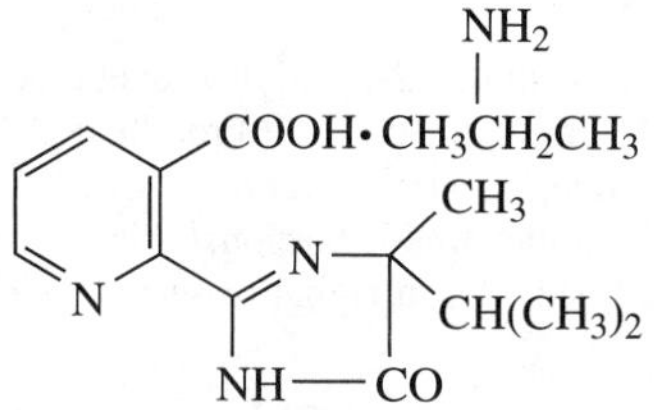

2-[4,5-dihydro-4-methyl-4-(1-methylethyl)-5-oxo-1H-imidazol-2-y1]3-pyridinecarboxylate (compound with 2-propanamine)

Pesticides. The compound 2-picoline is a component of 1-[(4'-amino−2'-n-propyl-5'-pyrimidinyl)methyl]-2-picolinium chloride hydrochloride, known as Amprolium,® a broad-spectrum coccidiostat. A newer coccidiostat is 3,5-dichloro-4-hydroxy-2,6-lutidine and known as Clopidol.®

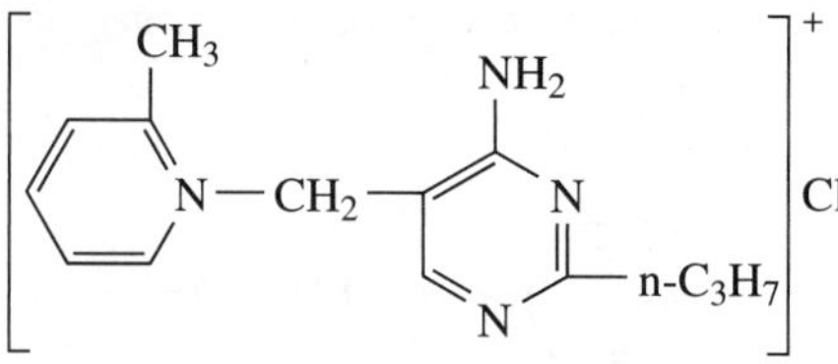

1-[(4'-amino-2'-n-propyl-5'-pyrimidinyl)methyl]-2-picolinium chloride hydrochloride

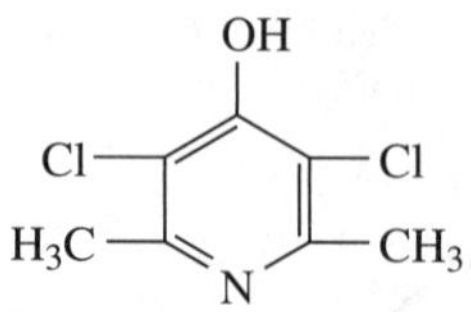

3,5-dichloro-4-hydroxy-2,6-lutidine

The acaricide O,O-diethyl-O-(3,5,6-trichloro-2-pyridyl) thiophosphate, known as Dursban,® is used to control ectoparasites. The similar, O,O-dimethyl-O-3,5,6-trichloro-2-pyridyl) thiophosphate, tradenamed Reldan® and Tumar,® is a nonsystemic insecticide/acaricide. Di(n-propyl)isocinchomerate, known as MGK Repellent 326® is used in fly repellents and is

made by oxidation of 2-methyl-5-ethylpyridine and esterification of the isocinchomeronic acid obtained. Nicotine (sulfate) (Black Leaf 40®) is used as an agricultural insecticide, as an external parasiticide, and as an anthelminthic, and is obtained by extraction of tobacco wastes (not by synthesis).

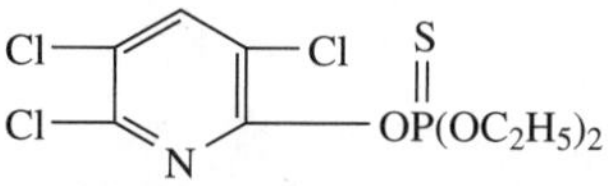

O,O-diethyl-O-(3,5,6-trichloro-2-pyridyl) thiophosphate

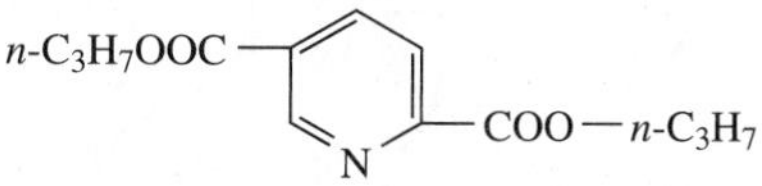

Di (n-propyl) isocinchomerate

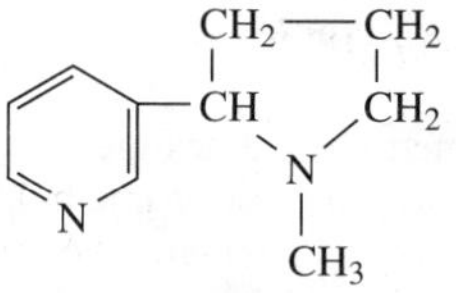

Nicotine

3-(2-Methylpiperidino)propyl-3,4-dichlorobenzoate, tradenamed Pi-pron,® is a foliar fungicide for the control of powdery mildew:

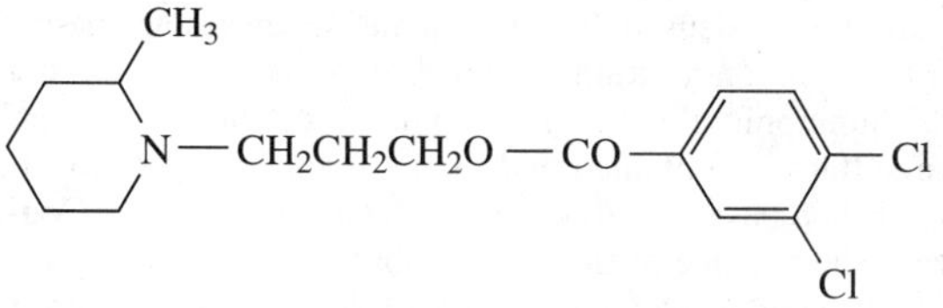

3-(2-methylpiperidino)propyl   3,4-dichlorobenzoate

Dimethyl 3,5,6-trichloro-2-pyridyl phosphate, known as Fospirate® or Dowco® 217, is an insecticide useful in antiflea collars for dogs and cats. The compound 4-aminopyridine, known as Avitrol® 100, and 4-nitropyridine-N-oxide, known as Avitrol® 200, are useful as bird repellents.

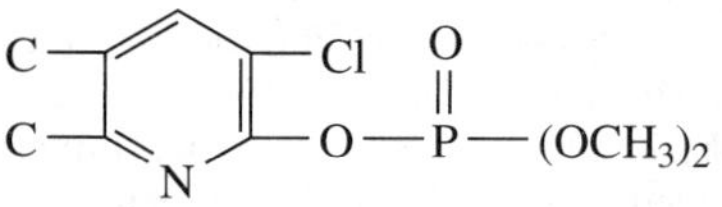

Dimethyl-3,5,6-trichloro-2-pyridyl phosphate

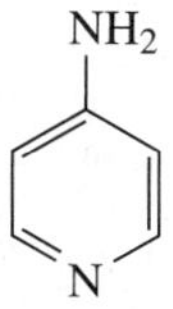

4-aminopyridine

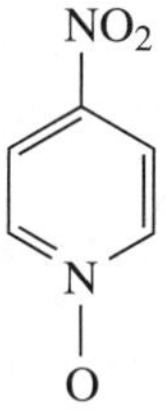

4-nitropyridine-N-oxide

Pharmaceuticals. A wide variety of pyridine compounds, with varying, and often multiple, drug action are used commercially. A few examples are given. A number of antihistamines contain the pyridine moiety in their structure, as exemplified by chlorpheniramine maleate (2-[p-chloro-α-(2-dimethylaminoethyl)benzyl] pyridine acid maleate); doxylamine succinate (2-[α-(2-dimethylamino)ethoxy-α-methylbenzyl]-pyridine acid succinate); and pyrilamine maleate (2-(2-dimethyl-aminoethyl-2-p-methoxybenzyl) aminopyridine acid maleate). These products are synthesized, e.g., from the appropriate benzylpyridines or aminopyridines.

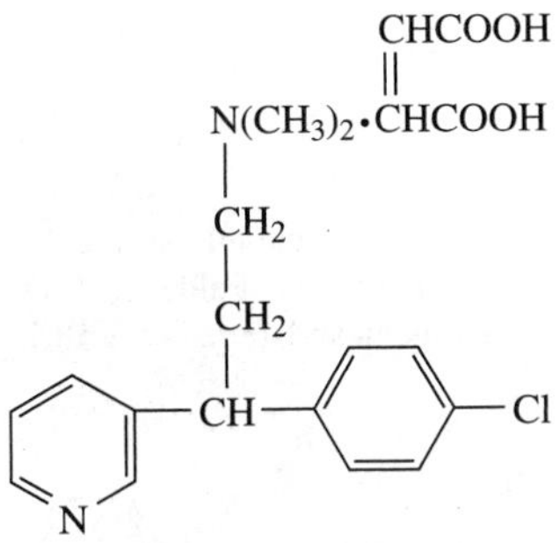

Chlorpheniramine maleate

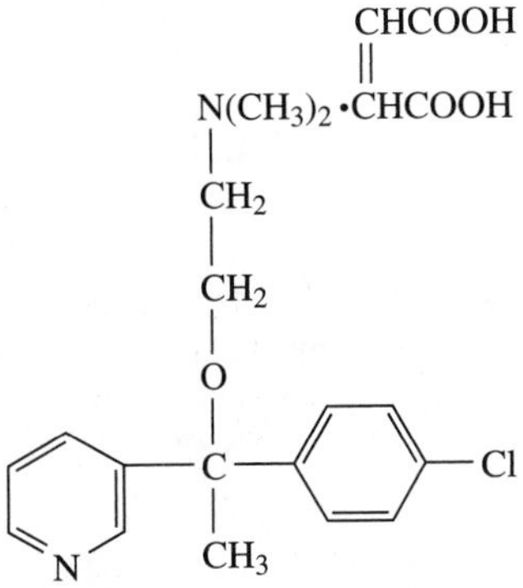

Doxylamine succinate

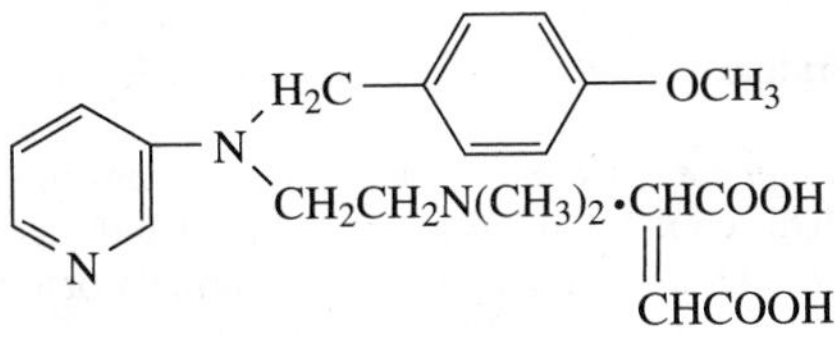

Pyrilamine maleate

Cetylpyridinium chloride is used as a germicide and antiseptic, e.g., in mouthwashes; it is made by quaternization of pyridine with cetyl chloride.

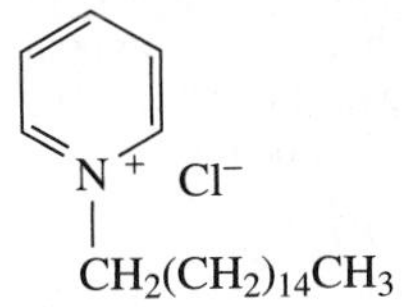

Cetylpyridinium chloride

Isonicotinehydrazide, also known as isoniazid, is an important antitubercular drug made by oxidation of 4-alkylpyridine (or 2,4-lutidine) or by hydrolysis of 4-cyanopyridine to isonicotinic acid (pyridine 4-carboxylic acid) and reaction of an ester or the acid chloride of the latter with hydrazine.

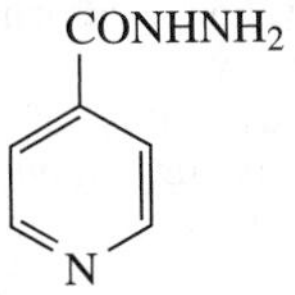

Isonicotinehydrazide

Meperidine hydrochloride (1-methyl-4-carbethoxy-4-phenylpiperidine), also known as Demerol,® is an important narcotic and analgesic. It is not made from piperidine, but rather by ring-closure reactions of appropriate precursors.

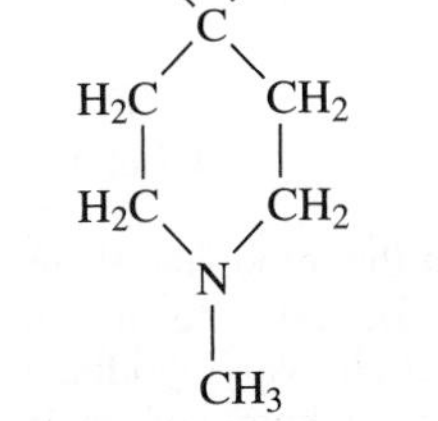

Meperidine hydrochloride

Cephapirin sodium, tradenamed Bristocef,® Cefadyl,® Today® (and others) is a cephalosporin C antibiotic:

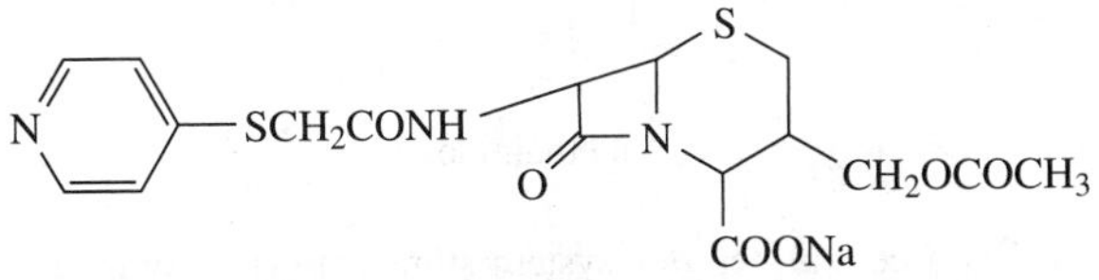

Cephapirin sodium

Nalidixic acid (1-ethyl-7-methyl-1.8-naphthridine-4-one-3-carboxylic acid), many tradenames (e.g., Nalidicron®), is an antibacterial. *Bisacodyl* [4,4′-(2-pyridylmethylene)diphenol diacetate], tradename Dulcolax,® is a laxative.

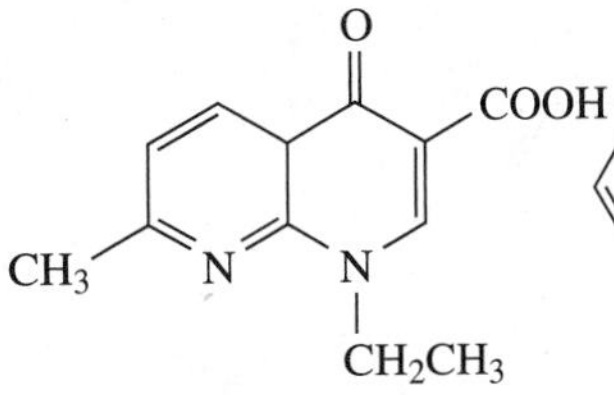

Nalidixic

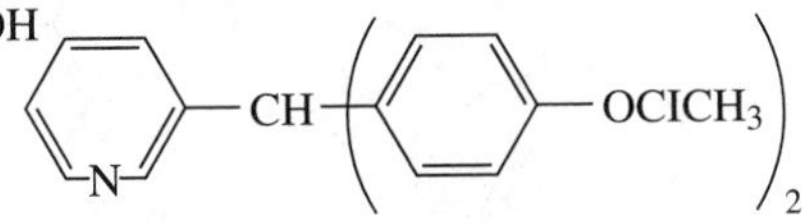

Bisacodyl

Nifedipine [1,4-dihydro-2,5-dimethyl-3,5-dicarbmethoxy-4-(2-nitrophenyl)pyridine], tradename Procardia,® is used in the treatment of angina.

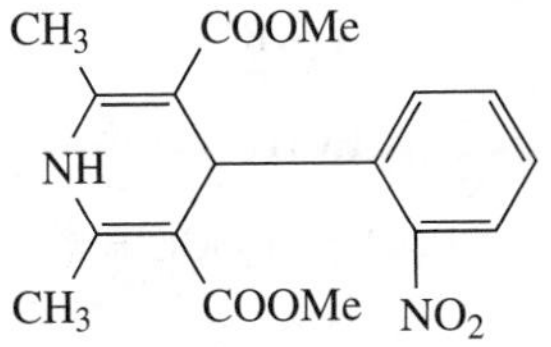

Nifedipine

Nicotinic acid and nicotinamide, members of the vitamin B group and used as additives for flour and bread enrichment, and as animal feed additive among other applications, are made to the extent of 24 million pounds (nearly 11 million kilograms) per year throughout the world. Nicotinic acid (pyridine-3-carboxylic acid), also called *niacin*, has many uses. See also **Niacin**. Nicotinic acid is made by the oxidation of 3-picoline or 2-methyl-5-ethylpyridine (the isocinchomeric acid produced is partially decarboxylated). Alternatively, quinoline (the intermediate quinolinic acid) is partially decarboxylated with sulfuric acid in the presence of selenium dioxide at about 300°C, or with nitric acid, or by electrochemical oxidation. Nicotinic acid also can be made from 3-picoline by catalytic ammoxidation to 3-cyanopyridine, followed by hydrolysis.

Nicotinamide is prepared by partial hydrolysis of the nitrile, or by amination of nicotinic acid chloride or its esters. Some of the compounds mentioned in the foregoing are shown below.

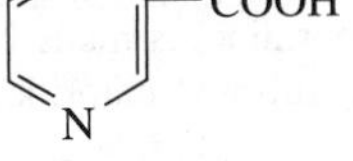

Nicotinic acid

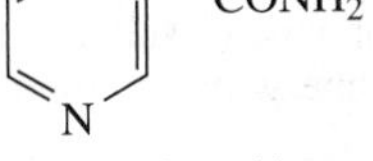

Niacinamide

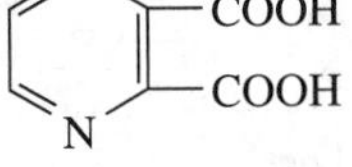

Quinolinic acid

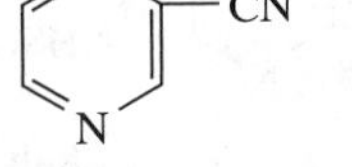

3-cyanopyridine

Several esters of nicotinic acid are used as vasodilators.

Nikethamide is a respiratory and heart stimulant, used beneficially against overdoses of barbiturates and morphine. Also known as Coramine,® this compound (*N*, *N*-diethylnicotinamide) is made by reaction of nicotinic acid esters or the acid chloride with diethylamine. Its formula is shown below.

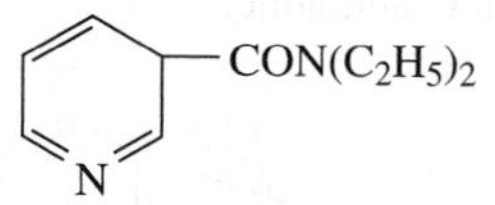

Nikethamide

Pipadrol is a central nervous system stimulant. This compound, α,α-diphenyl-2-piperidinemethanol, is made by condensation of 2-pyridyl-magnesium chloride with benzophenone and catalytic hydrogenation of the pyridine ring of the resultant carbinol. Its formula is shown below.

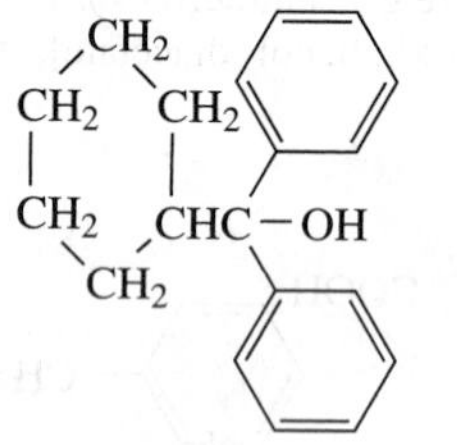

Pipadrol

Piperocaine hydrochloride is used as a local anesthetic. This compound (d,l-(2-methylpiperidino)propyl benzoate hydrochloride) is made by reaction of 2-methylpiperidine with 3-chloropropyl benzoate. Its formula is shown below.

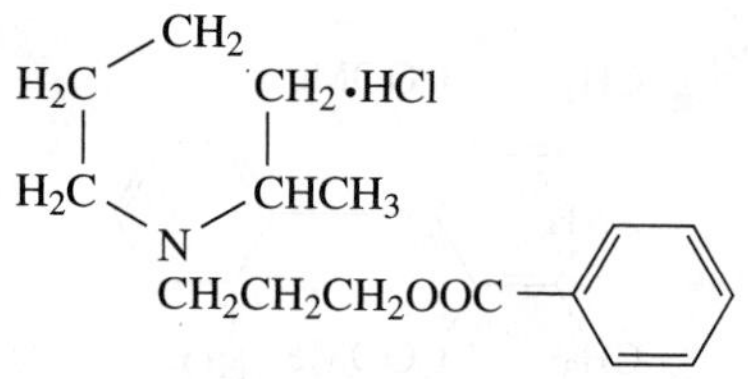

Piperocaine hydrochloride

Pyrithione (zinc salt of) is used as a component of antidandruff shampoos and as a bactericide in soap and detergent formulations. This compound (2-mercaptopyridine N-oxide) exists in equilibrium with N-hydroxy-2-pyridinethione and is a fungicide and bactericide, prepared by reaction of 2-chloropyridine N-oxide with sodium hydrosulfide and sodium sulfide. This compound is also known as Omadine.® Its formula is shown below.

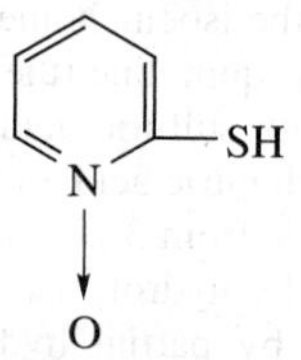

Pyrithione

Sulfapyridine is used to treat dermatitis herpetiformis and also has been used by veterinarians against pneumonia, shipping fever, and foot rot of cattle. This compound (2-sulfanylamidopyridine) is made by condensation of 2-aminopyridine with the appropriate sulfonyl chloride. Its formula is shown below.

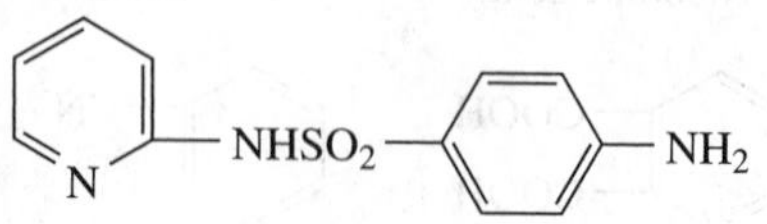

Sulfapyridine

Vitamin B6 is described in detail under **Vitamin B₆ (Pyridoxine)**. This is 2-methyl-3-hydroxy-4,5-di(hydroxymethyl)pyridine or pyridoxol. World demand of this compound is estimated at about 5 million pounds (about 2.3 million kilograms) per year. Commercial production is by synthesis, starting, for example, with the base-catalyzed condensation of cyanoacetamide and ethoxyacetylacetone. The formula for pyridoxol is

shown below.

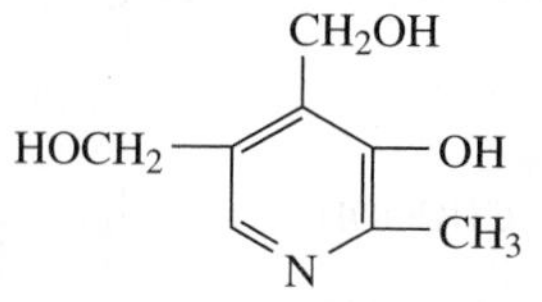

Pyridoxol

Methyridine or 2-(2-methoxyethyl)pyridine, also called Mintic,® is used as an anthelmintic. *Piroxicam*, also known as Feldene,® is a relatively new anti-inflammatory for the treatment and relief of arthritis. See formulas below.

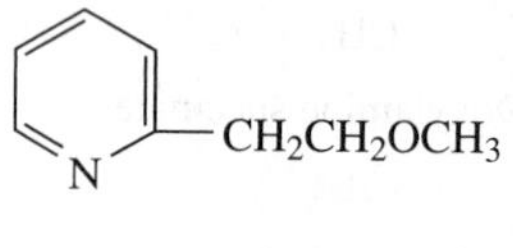

Methyridine 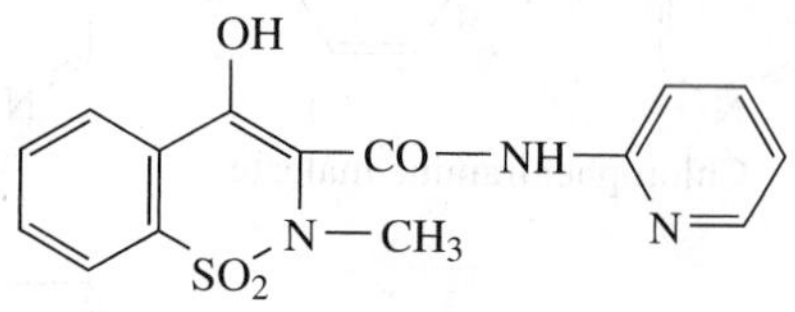   Piroxicam

Pyridinol carbamate has been used as an anti-inflammatory/anti-arteriosclerotic. This compound 2,6-pyridinedimethanol-bis-(N-methyl carbamate) is also known as Anginin.® See the formula below.

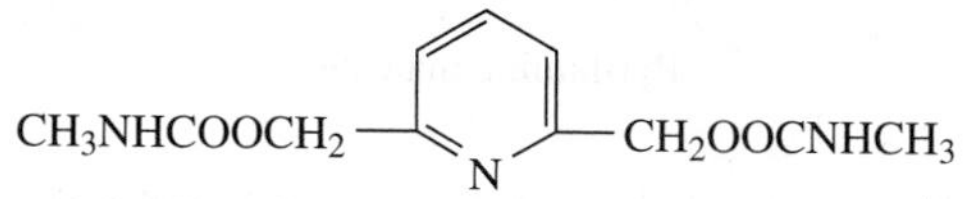

Pyridinol carbamate

Pyrithioxin is a neurotropic agent that reduces the permeability of the blood-brain barrier to phosphate. This compound, 3,3′-dithio-dimethylene-bis-(5-hydroxy-6-methyl-4-pyridinemethanol), is also known as Life® and Bonifen.® Its formula is shown below.

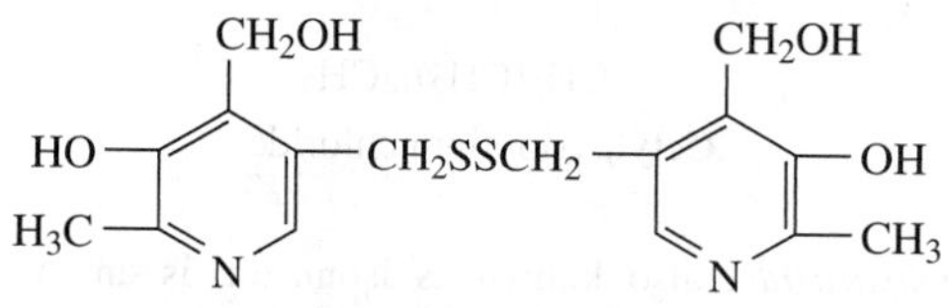

Pyrithioxin

Textile Chemicals. Pyridine derivatives find a number of quite different applications in the textile and related fields.

Stearamidomethylpyridinium chloride is used in waterproofing textiles. It is made by reacting pyridine hydrochloride with stearamide and formaldehyde. *Vinylpyridines* are used as components of acrylonitrile copolymers to improve the dyeability of polyacrylonitrile fibers. The commercially important products are 2-vinylpyridine; 4-vinylpyridine; and 2-methyl-5-vinylpyridine. Formulas are shown below.

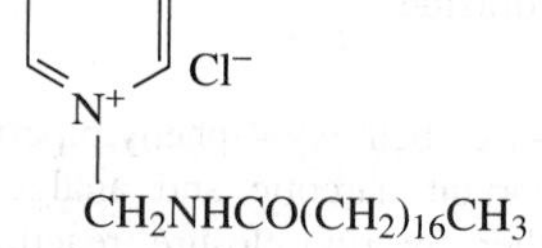

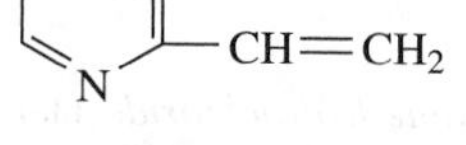

Stearamidomethylpyridinium   2-vinylpyridine
chloride

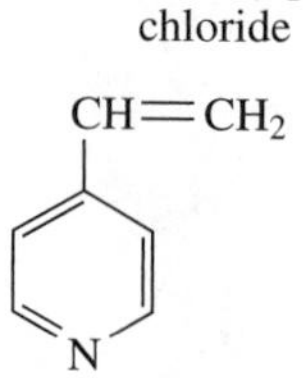

4-vinylpyridine 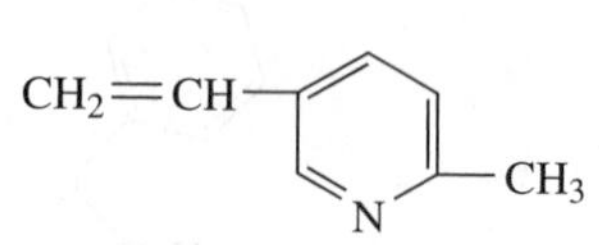   2-methyl-5-vinylpyridine

2-Vinylpyridine is used in the terpolymer latex component of tire cord dips to improve the bonding of textile to rubber. Rubber tires built with steel cord, however, do not require vinylpyridine latex-based adhesives for the steel belt. Therefore, the consumption of vinylpyridines may be affected in the future.

Other. The pyridines and methylpyridines and their mixtures are used as chemical processing aids (e.g., acid acceptors, solvents) and as industrial corrosion inhibitors.

Piperidine, the hydrogenation product of pyridine, is used as an intermediate for drugs and for making rubber-vulcanization accelerators, e.g., piperidinium pentamethylenedithiocarbamate (also known as Accelerator 552®). On a commercial scale, piperidine (hexahydropyridine) is prepared by the catalytic hydrogenation of pyridine, e.g., with nickel catalysts at from 68 to 136 atmospheres pressure and at 150–200°C, or under milder conditions with noble-metal catalysts. Pyridine derivatives can be similarly reduced to substitute piperidines. See formulas below.

Piperidine   Piperidinium pentamethylendithiocarbamate

4-*N*,*N*-Dialkylaminopyridines have found use as catalysts for acylation reactions.

There are developing applications for linear and crosslinked poly vinylpyridines in photovoltaic cells and batteries, electron beam resists, as catalysts and reagents (e.g., in pollution control).

The hindered-amine light stabilizers for polymers are piperidine derivatives. An example of these products is bis-(2,2,6,6-tetramethyl-4-piperidinyl)sebacate, tradenamed Tinuvin 770,® useful as a light stabilizer for polyolefins and styrenics.

Bis-(2,2,6,6-tetramethyl-4-piperidinyl)sebacate

There is growing evidence of developing high-technology uses of pyridines, particularly as quaternary salts, as components of electrolytic capacitors, photoconductors, rechargeable batteries, complex-coated electrodes for photosensors, electrochromic display elements and photoresist matrix resins.

Health and Safety Factors

Pyridine Acute Toxicology. Pyridine causes gastrointestinal upset and central nervous system (CNS) depression at high levels of exposure. The odor of pyridine can be detected at extremely low concentrations (12 ppb).

Acute Toxicology of Pyridine Derivatives. In general, many pyridines are reasonably safe to handle and do not represent a serious hazard. However, some types of aminopyridines are poisons. Quaternary salts of pyridines can also be toxic. Chloropyridines, especially polychloropyridines, can potentially be mutagenic, teratogenic, and carcinogenic.

Safety Aspects in Handling and Exposure. Pyridine compounds are ubiquitous in the natural environment, and are often found in foods as minor flavor and fragrance components. Some synthetic pyridines are used as food additives. A high proportion of pyridine compounds shows some type of bioactivity, albeit mostly minor, such as herbicidal, insecticidal, or medicinal activity. Therefore, all the normal precautions should be exercised when handling pyridines that would be used when handling other organic products that are potentially bioactive.

Pyridine and alkylpyridines are excellent solvents for many materials, a property that must be taken into account when selecting O-rings, gaskets, and other sealants that are in contact with liquids.

HANS DRESSLER
Koppers Company, Inc.
Monroeville, Pennsylvania

Additional Reading

Coffey, S.: *Six Membered Heterocyclic Compounds with a Single Atom in the Rind, Pyridine, Polymethyl-Epyridines, Quinoline, Isoquinoline and Their Derivatives,* Elsevier Science, New York, NY, 1977.

Eicher, T., and S. Hauptmann: *The Chemistry of Heterocycles: Structure, Reactions, Syntheses, and Applications,* 2nd Edition, John Wiley & Sons, Inc., New York, NY, 2003.

Katritzky, A.R., and C.W. Rees: *Comprehensive Heterocyclic Chemistry: Six-Membered Rings with One Nitrogen Atom,* Elsevier Science, New York, NY, 1984.

PYRITE. The mineral pyrite or iron pyrites is iron disulfide, FeS_2, its isometric crystals usually appearing as cubes or pyritohedrons. It has a slightly conchoidal to uneven fracture; brittle; hardness, 6–6.5; specific gravity, 5; metallic luster; color, pale to normal brass-yellow; streak, greenish-black; opaque. Arsenic, nickel, cobalt, copper, and gold may be found in small quantities in pyrite, auriferous pyrite being sometimes a very valuable ore. Pyrite is the commonest of the sulfide minerals, and is of worldwide occurrence. It is found associated with other sulfides, or with oxides, in quartz veins, in sedimentary and metamorphic rocks, in coal beds, and as the replacement material in fossils. There are many well-known pyrite localities, among which are the Rio Tinto mines in Spain, where copper-bearing pyrite is obtained from huge deposits. Magnificent crystals and crystal groups occur at Ambasaguas (Logrono) in Spain; Quirivulca, Peru; and from the Island of Elba. In the United Stated pyrite is found in California, New York, and Virginia in workable deposits. The name pyrite is derived from the Greek word meaning fire, because of the sparks that result when pyrite is struck with steel.

PYROGENETIC MINERALS. A term for the primary magmatic minerals of igneous rocks as distinguished from those minerals which are the result of special and later processes such as come under the head of pneumatolytic, hydrothermal, etc.

PYROLUSITE. The mineral pyrolusite, manganese dioxide (MnO_2), crystallizes in the tetragonal system, but may be only pseudomorphous after manganite. It is found massive or in indistinct crystalline aggregates, often acicular, and as dendritic growths on fractured rock surfaces and as inclusions within moss agates and other chalcedony varieties of quartz. Hardness, 6–6.5 (crystals), 2–6 (massive); specific gravity, 5.06; luster, metallic; color, steel gray to black; streak black; opaque. Pyrolusite is found as replacement deposits and as residual and sedimentary masses. Psilomelane is its usual associate. European localities for pyrolusite are in Bohemia, Saxony, the Harz Mountains, England, and elsewhere. Other deposits occur in India and Brazil. In the United States it is found in Arkansas and Michigan. It is an ore of manganese. It is from this latter use that it derives the name pyrolusite, from the Greek words meaning *fire* and *to wash.*

PYROLYSIS. Transformation of a compound into one or more other substances by heat alone, i.e., without oxidation. It is thus similar to destructive distillation. Although the term implies decomposition into smaller fragments, pyrolytic change may also involve isomerization and formation of higher-molecular-weight compounds. Hydrocarbons are subject to pyrolysis, e.g., formation of carbon black and hydrogen from methane at 1300°C and the decomposition of gaseous alkanes at 500–600°C. The latter is the basis of thermal cracking (pyrolysis) in the production of gasoline. An application of pyrolysis is the conversion of acetone into ketone by decomposition at about 700°C; the reaction is $CH_3COCH_3 \rightarrow H_2C=C=O + CH_4$. Pyrolysis of natural gas or methane at about 2000°C and 100 mm mercury pressure produces a unique form of graphite. Synthetic crude oil can be made by pyrolysis of coal, followed by hydrogenation of the resulting tar. Large-scale pyrolysis of solid wastes has been considered in connection with several synfuel projects.

PYROMETER. An instrument for measuring temperatures of 1800°C or higher, for example, molten steels, hot springs, volcanoes, etc. There are three kinds: (1) thermocouples of the graphite to silicone carbide type; (2) optical, in which the indications depend on the brightness at some one wavelength of the hot body whose temperature is being measures; and (3) radiation, in which the indications depend on the radiance of a source of radiant energy.

See also **Thermocouple.**

PYROMETRIC CONES. Small cones that differ in the temperatures at which they soften on heating. They are made of clay and other ceramic materials and are used in the ceramic industries to show furnace temperatures within ranges. In practice, three or four of the cones which have softening points at consecutive temperature ranges are used, and the increase in kiln temperature is judged from the progressive deformation of the cones.

PYROMORPHITE.

The mineral pyromorphite is lead chlorophosphate with a formula corresponding to $Pb_5(PO_4)_3Cl$. The phosphorus is sometimes replaced by arsenic and the lead by calcium. It occurs in prismatic, sometimes hollow, hexagonal crystals or may appear in massive forms. It is brittle; hardness, 3.5–4; specific gravity, 7.04; luster, resinous; color, green, yellow-green, yellow, brown, and less often gray or white; translucent to opaque.

Pyromorphite is a secondary mineral associated with other lead minerals, but is seldom found in large quantities. It has probably resulted from the action of waters bearing phosphoric acid upon the preexisting lead minerals. Localities for pyromorphite are in the Ural Mountains, Saxony, France, Spain, Cornwall and Cumberland, England; in Scotland, Zaire, and Australia. In the United States pyromorphite has been found in Chester and Montgomery Counties, Pennsylvania; in Davidson County, North Carolina, and in the Coeur d'Alene mining district of Idaho. The name is derived from the Greek words meaning fire and form.

PYROPHORIC MATERIAL.

Any liquid or solid that will ignite spontaneously in air at about 130F (54.4C). Titanium dichloride and phosphorus are examples of pyrophoric solids; tributylaluminum and related compounds are pyrophoric liquids. Sodium, butyllithium, and lithium hydride are spontaneously flammable in moist air because they react exothermically with water. Such materials must be stored in an atmosphere of inert gas or under kerosene. Some alloys (barium, misch metal) are called pyrophoric because they spark when slight friction is applied.

PYROPHYLLITE.

The mineral pyrophyllite is a hydrous silicate of aluminium corresponding to the formula $Al_2Si_4O_{10}(OH)_2$. Monoclinic with a basal cleavage, it is usually, however, in foliated, radiated lamellar, or fibrous masses, sometimes compact. It is a soft mineral with a greasy feel; hardness, 1–2; specific gravity, 2.65–2.9; luster, pearly to dull; color, white, greenish, grayish, yellowish, and brownish; translucent to opaque. It is found making up schists or in foliated masses in the Ural Mountains, in Switzerland, Sweden, Brazil, and in the United States in Pennsylvania, North Carolina, Georgia, and California. It is used to some extent for the same purpose as is the mineral talc, and also for making slate pencils, hence the name pencil stone sometimes applied to pyrophyllite.

PYROTECHNICS.

Pyrotechnics involves the combination of science and art to chemically generate heat, and from that heat create light, color, audible effects, and gas pressure for entertainment, emergency signaling, and military applications. The civilian side of pyrotechnics includes fireworks, highway flares (fusees), air bag inflators, and special effects devices for the entertainment industry. Military and aerospace pyrotechnics include a wide range of devices for illumination, signaling, obscuration, and gas generation. A pyrotechnic mixture typically contains one or more oxygen-rich oxidizers and one or more fuels, which undergo an exothermic reaction when heated to the ignition temperature of the mixture. The heat that is produced then creates the desired pyrotechnic effect. The selection of the chemicals used in a pyrotechnic composition, as well as the particle sizes of the chemicals and the degree of intimacy to which the composition is blended, determine in large part the speed of the pyrotechnic reaction. Safety in all aspects of manufacturing and using pyrotechnic mixtures and devices is important.

The industry is professionally represented by the Pyrotechnic Guild International, Inc. http://www.pgi.org/. The primary ingredients of pyrotechnic products are as follows:

1. *Oxidizers:* potassium nitrate, potassium chlorate, or potassium perchlorate; ammonium perchlorate; barium chlorate and nitrate; strontium nitrate.
2. *Fuels:* aluminum, magnesium, antimony sulfate, dextrin, sulfur, and titanium.
3. *Binders:* dextrin and various polymers.

Colored flames are produced by strontium compounds (red); barium compounds (green); copper carbonate, sulfate, and oxide (blue); sodium oxalate and cryolite (yellow); and magnesium, titanium, or aluminum (white). Black powder is used as the propellant.

PYROXENE.

This is the name given to a closely related group of minerals, all of which show a distinct cleavage angle of 87° or 93° parallel to the fundamental prism. Chemically the pyroxenes are metasilicates corresponding to the formula $RSiO_3$, where R may be calcium, magnesium, iron, or less commonly manganese, zinc, sodium, or potassium. Rarely titanium, zirconium, or fluorine may be present. A general formula is $ABSi_2O_6$, where A is Ca, Na, Mg, or Fe^{2+}, and B is Mg, Fe^{3+}, or Al. Sometimes the Si is replaced by Al. The pyroxenes crystallize in the orthorhombic, and monoclinic systems, like the amphiboles, the chief difference between the two groups being the cleavage angles, which for amphibole are 56° and 124°. Pyroxene crystals tend to be short, stout, complex prisms as opposed to the long, slender, and simpler amphiboles.

The pyroxenes are common in the more basic igneous rocks, both intrusive and extrusive, and may be developed by the metamorphic processes in gneisses, schists, and marbles.

For descriptions of members of the pyroxene group, see also **Acmite-Aegerine**; **Augite**; **Diallage**; **Diopside**; **Enstatite**; **Hypersthene**; **Jadeite**; and **Spodumene**.

PYRRHOTITE.

The mineral pyrrhotite, sometimes called magnetic pyrites, is a sulfide of iron with varying amounts of sulfur. Analyses indicate formulae $Fe_{1-x}S$. Pyrrhotite exists in two modifications: it is monoclinic below, and hexagonal above 138°C (280°F). It is a brittle mineral; hardness, 3.5–4.5; specific gravity, 4.53–4.97; luster, metallic; color, reddish bronze-yellow when fresh, otherwise tarnished; streak, grayish-black; magnetic. It may carry nickel, generally as pentlandite, when it becomes a valuable nickel ore as at Sudbury, Ontario. Pyrrhotite is commonly associated with the basic igneous rocks like gabbro, and norite, and occurs with chalcopyrite, magnetite, and pyrite. Besides being apparently of magmatic origin, it has been found as contact metamorphic and as vein deposits. Austria, Italy, Saxony, Bavaria, Switzerland, Norway, Sweden, and Brazil have deposits of more or less importance, and in the United States it has been found associated with andalusite crystals at Standish, Maine; also at Brewster, New York; Lancaster County, Pennsylvania, and elsewhere. At Ducktown, Tennessee, it is found together with copper and zinc minerals. It is mined for its nickel content, in the form of admixed pentlandite, in Sudbury, Ontario.

Pyrrhotite derives its name from the Greek word *pyrrhos*, meaning reddish, in reference to the color of the fresh ore.

PYRROLE AND RELATED COMPOUNDS.

[CAS:109-97-7]. Pyrrole (monoazole, C_4H_5N or C_4H_4NH), contains a ring of 1 nitrogen and 4 carbons, with 1 hydrogen attached to nitrogen and to each carbon:

Beta prime HC — CH Beta } C-compounds
Alpha prime HC — CH Alpha }
NH } N-compounds

Pyrrole is a colorless liquid, boiling point 131°C, insoluble in water, soluble in alcohol or ether. Pyrrole dissolves slowly in dilute acids, being itself a very weak base; resinification takes place readily, especially with more concentrated solutions of acids; and on warming with acid a red precipitate is formed. Pyrrole vapor produces a pale red coloration on pine wood moistened with hydrochloric acid, which color rapidly changes to intense carmine red. Pyrrole may be made (1) by reaction of succinimide

H_2C — CO
                NH
H_2C — CO

with zinc and acetic acid, or with hydrogen in the presence of finely divided platinum heated, (2) by reaction of ammonium saccharate or mucate $COONH_4 \cdot (CHOH)_4 \cdot COONH_4$ with glycerol at 200°C by loss of carbon dioxide, ammonia, and water.

When pyrrole is treated with potassium (but not with sodium) or boiled with solid potassium hydroxide, potassium pyrrole C_4H_4NK is formed, which is the starting point for *N*-derivatives of pyrrole, since reaction of the potassium with halogen of organic compound and with carbon dioxide, readily occurs. When pyrrole is treated with magnesium metal and ethyl bromide in ether, pyrrole magnesium bromide plus ethane is formed, which may be used as the starting point for C-derivatives of pyrrole, since reaction with sodium alcoholates readily occurs (with separation of magnesium oxybromide).

The pyrrole nucleus has been shown to be present in the complex substances chlorophyll (the green coloring matter of plants), hematin (the red coloring matter of blood), and in the coloring matter of bile.

PYRUVIC ACID.   See **Carbohydrates**; **Coenzymes**; **Vitamin**.

Q

QUAD. An energy unit that has come into use in recent years in predicting future energy requirements on a national basis. One quad equals 10^{15} Btus (British thermal units), which is the energy equivalent of 10^{12} cu ft natural gas, or 182 million barrels of oil, or 42 million tons of coal, or 293 billion kilowatt-hours of electricity.

QUADRUPLE POINT. The temperature at which four phases are in equilibrium, such as ice, saturated salt solution, water vapor, and salt.

QUANTUM CHEMISTRY. The use of the principles of quantum mechanics for the resolution of problems in chemistry, notably in connection with the electronic structure of molecules. Some authorities attribute the beginnings of this field to James and Coolidge who, as early as 1933, theorized on the molecular structure of hydrogen and, in these efforts, demonstrated that the Schrödinger equation (primarily proposed for atoms) could be applied to molecules. Over a period of years, studies by other researchers followed. In 1968, for example, Kolos and Wolniewicz carefully investigated the dissociation energy of the hydrogen molecule. But, as early as 1960, some investigators in quantum chemistry turned their attention to methylene (CH_2) as the molecular target of choice. The predictive powers of computational quantum chemistry have since been demonstrated in connection with other molecules. Schaefer (1986) suggests, however, that methylene is the paradigm for computational quantum chemistry. In his paper, three important roles for quantitative theory are outlined: (1) theory precedes experiment, (2) theory overturns experiment, as resolved by later experiments, and (3) theory and experiment work together to gain insight that is afforded independently to neither. See also **Quantum Mechanics**.

Additional Reading

Goddard, W.A.: "Theoretical Chemistry Comes Alive: Full Partner with Experiment," *Science,* **227**, 917–923 (1985).
Haken, H., and H.C. Wolf: *Molecular Physics and Elements of Quantum Chemistry,* 2nd Edition, Springer-Verlag, New York, Inc., New York, NY, 2004.
Hayward, D.O.: *Quantum Mechanics for Chemists,* John Wiley & Sons, Inc., New York, NY, 2003.
House, J.E.: *Fundamentals of Quantum Chemistry,* 2nd Edition, Elsevier Science, New York, NY, 2003.
Levine, I.N.: *Quantum Chemistry,* Prentice-Hall, Inc., Upper Saddle River, NJ, 1999.
Lowdin, Per-Olov, E. Brandas, J. Sabin, and Mike Zerner : *Advances in Quantum Chemistry,* Vol. 39, Academic Press, Inc., San Diego, CA, 2001.
Roberts, M.W.: "Chemistry in Two Dimensions," *Review (University of Wales),* 58 (Autumn 1987).
Schaefer, H.F.: "Methylene: A Paradigm for Computational Quantum Chemistry," *Science,* **231**, 1100–1107 (1986).
Stucky, G.D. and J.E. MacDougall: "Quantum Confinement and Host/Guest Chemistry: Probing a New Dimension," *Science,* 669 (February 9, 1990).
Veszpremi, T. and M. Feher: *Quantum Chemistry: Fundamentals to Applications,* Kluwer Academic Publishers, Norwell, MA, 1999.
Warren, W.S., Rabiz, H., and M. Dahleh: "Coherent Control of Quantum Dynamics: The Dream is Alive," *Science,* 1581 (March 12, 1993).
Wasserman, E. and Schaefer H.F.: "Letters—Methylene Geometry," *Science,* **233**, 829–830 (1986).
Wilson, S., and P.F. Bernath: *Handbook of Molecular Physics and Quantum Chemistry,* John Wiley & Sons, Inc., New York, NY, 2003.

QUANTUM EFFICIENCY. A measure of the efficiency of conversion or utilization of light or other energy, being in general the ratio of the number of distinct events produced in a radiation sensitized process to the number of quanta absorbed (the intensity-distribution of the radiation in frequency or wavelength should be specified). In the photoelectric and photoconductive effects, the quantum efficiency is the number of electronic charges released for each photon absorbed. For a phototube, the quantum efficiency is defined as the average number of electrons photometrically emitted from the photocathode per incident photon of a given wavelength. In photochemistry, the quantum efficiency or yield is the ratio of the number of molecules transformed to the number of quanta of radiation absorbed.

QUANTUM ELECTRODYAMICS. A quantized field theory of the interaction between electrons, positrons and radiation based on the quantized form of the Maxwell equations and the Dirac electron theory. The theory is characterized by its remarkably accurate predictions (see also **Positronium**) and its meaningless results. The latter arise from divergent integrals that appear in the development of the theory by perturbation techniques based on expansion in powers of the fine structure constant. These divergences may be pictured in terms of the model of a vacuum as consisting of an infinite sea of negative energy electrons, since the introduction of a charge into this distribution causes infinite currents to be induced.

In 1948, techniques introduced by Schwinger and Feynman enabled these difficulties to be avoided, without being removed. Their relativistically covariant development of the theory allowed such infinite terms to be treated unambiguously, and in particular terms which are to be understood as electrodynamic contributions to the charge and mass of a particle were put in a form which is invariant under Lorentz transformations. The program of charge renormalization and renormalization of mass then enabled such terms to be related to the experimentally observed charge and mass of the particle. See also **Quantum Mechanics**.

QUANTUM MECHANICS. The wave theory of light as originally developed by Maxwell in the 1860s became well established, but it did not accommodate certain phenomena. For example, experiments on thermal radiation uncovered gross disagreement or contradiction with classical theories. The equilibrium distribution of electromagnetic radiation (i.e., emission and absorption of radiation at constant temperature) in a hollow cavity could not be explained on the basis of classical electrodynamics (Maxwell's equations plus the laws of motion of particles). Thermal radiation is a certain function of the temperature (T) of the emitting body. When dispersed by a prism, thermal radiation forms a continuous spectrum. It was found that the energy distribution of the radiation had a regular dependence on its wavelength. Furthermore, the energy E_v as a function of the temperature of the material did not depend upon the structure of the cavity or its shape. On these bases, it was shown that the energy E_v should have a functional dependence upon frequency v, at temperature T, in the form:

$$E_v = v^3 F\left(\frac{cT}{v}\right)$$

All attempts to find the correct form of the function F on the basis of classical theory failed. The classical theory led to the now well-known "ultraviolet catastrophe," since the contribution of high frequencies caused the energy to assume an infinite value. The difficulty was removed by a hypothesis of Planck, according to which the energy of a monochromatic wave with frequency n can only assume those values which are integral multiples of energy hv, i.e., $E_n = nhv$, where n is an integer referring to the number of "photons." Thus the energy of a single photon of frequency v is:

$$E = hv \qquad (1)$$

The finiteness of Planck's constant h and its resulting implications laid the foundations of quantum theory. Quantum theory, like the special theory of relativity, was discovered through experiments on electromagnetic phenomena and their theoretical interpretations.

The fundamental equation of quantum mechanics, Eq. (1), implies, on the one hand, that energy of radiation stays concentrated in limited regions of space in amounts of $h\nu$ and, therefore, behaves like the energy of particles. On the other hand, it establishes a definite relationship between the frequency ν and the energy E of an electromagnetic wave. This dual behavior of light corresponds, in one way, to experimental situations of the interference properties of radiation, for the description of which one uses the wave theory of light. In another way, it corresponds to the properties of exchange of energy and momentum between radiation and matter, which require for their explanation the particle picture of light. Thus, the dual behavior of light has necessitated the quantum description (quantization) of the electromagnetic field. A unified point of view was formulated quantitatively by de Broglie, according to which all forms of energy and momentum related to matter will manifest a dual behavior of belonging to a wave or particle description of the physical system, depending upon the type of experiment performed.

The most interesting example of a quantum mechanical object is the photon itself. By using the relativistic and quantum mechanical definition of the photon energy, we can obtain a quantitative formulation of the concepts just described. The relativistic form of the total energy of a particle with rest mass m and momentum ρ is:

$$E = c(\rho^2 + m^2c^2)^{1/2} \qquad (2)$$

We set $m = 0$ and obtain the relativistic definition of the energy of a photon:

$$E = c\rho \qquad (3)$$

Hence the first unification of relativity and quantum theory originated from the combination of Eqs. (1) and (3) in the form

$$c\rho = h\nu \qquad (4)$$

By using $\nu\lambda = c$ for the plane electromagnetic wave, we obtain the fundamental statement of quantum mechanics:

$$\lambda\rho = h \qquad (5)$$

valid for all particles with or without mass, where

$$\lambda = \frac{\lambda}{2\pi} ; = \frac{h}{2\pi} \qquad (6)$$

These assumptions of quantum theory have laid the foundations of new physical and philosophical concepts for the process of measurement in physics and the definition of physical reality.

It is necessary to develop a dynamic theory to describe the wave character of material particles. In the case of particles with mass, one has the possibility of comparing their kinetic energies with their rest masses. If the kinetic energy is small compared to rest energy, then we can formulate a nonrelativistic theory. However, with the photon there exists no possibility for the formulation of a nonrelativistic theory. There are important advantages in entering quantum mechanics via the photon:

1. The energy of a photon is a quantum mechanical quantity, $E = h\nu$.
2. It has provided a natural basis to postulate the wave-particle relation, $\lambda = h\rho$.
3. The wave aspects of the photon are completely described by charge-free Maxwell equations. Therefore, it is natural to try to reconcile Planck's hypothesis with the wave theory of light.

During 1979, scores of distinguished scientists reviewed the accomplishments of Albert Einstein who was born a century earlier (March 14, 1879). Many papers describing and reviewing the works of Einstein were prepared in honor of the centenary of Einstein's birth. Among Einstein's accomplishments, three were cited by Viktor Weisskopf, one of the speakers at the Pontifical Academy of Science (Vatican City) on November 10, 1979 at a special session devoted to Einstein. The relation between waves and particles was given as one of these, the other two, topically related, were special and general relativity. In commenting on Einstein's interest in wave-particle duality, Weisskopf made the following observations.

... Einstein's idea started a truly revolutionary development in physics: quantum mechanics. It opened up wide new horizons and clarified many outstanding problems in our view of the structure of matter. Quantum mechanics is based on the idea of wave-particle duality. Einstein first applied this idea to the nature of light, but it was soon applied to the nature of elementary entities such as electrons and other constituents of matter.

The idea was that all these entities exhibit both wave and particle properties. This double nature taxes our imagination: few things differ as much as a beam of particles and a running wave. In a beam of particles, matter is concentrated in small units, whereas a wave spreads continuously over space. Still, wave and particle properties are observed for electrons and other fundamental entities.

The wave nature of electrons explains so many previously unexplained facts for the following reason. If waves are confined to a finite region of space, they form characteristic shapes and patterns that are specific to the nature of the confinement. [Figure 8 in the entry on **Chemical Elements** shows waves in space confined to the neighborhood of a central point.] Only those and no other patterns can develop in this sort of confinement. But this is just the confinement that electrons suffer when they are confined around the atomic nucleus by electric attraction. The electron waves in atoms must assume some of these patterns. The simple patterns are "lower" than the more complex ones; they are lower in energy. Indeed, the electrons in an atom assume the lowest possible patterns.

This is the explanation of the stability of atoms—it takes energy to change to the next higher pattern. For example, the energy of molecular collisions in air is not sufficient to change the electron patterns in oxygen. Thus oxygen survives unchanged the many millions of collisions in air.

The typical shapes of the electron patterns determine the specific properties of atoms. For example, in the oxygen atom the electrons fill the lowest patterns up to the fourth one. The resulting pattern combination is characteristic for oxygen and is responsible for its properties; it determines how oxygen combines with other atoms (forming water with hydrogen, for example) and how the atoms fall into a symmetrical crystalline order when they form solids, such as ice crystals.

The electron patterns are the primal shapes of nature. Fundamentally, all of nature's shapes can be traced to such patterns. Even the properties of living substances are based on them—in particular, the properties of the molecules that carry the hereditary code. In the final scientific analysis, the stability of electron wave patterns causes the same flowers to bloom every spring and makes children similar to their parents.

Einstein started this great development as early as 1905 by an almost unimaginable act of vision, when he concluded that the concept of such an electromagnetic wave does not suffice to explain important properties of light. He drew the revolutionary conclusion that there must exist light-particles, the photons. The particle-wave duality was born. Einstein recognized the fertility of his idea, but he was never completely satisfied with the conceptual basis of quantum mechanics. The lack of complete causality and the frequent use of probability instead of certainty were always a matter of deep concern for him.

The next great development in physics was again an outgrowth of Einstein's ideas. Dirac was not satisfied with the fact that early quantum mechanics did not fit into the framework of relativity theory. The velocities of electrons in ordinary atoms are so small compared to the speed of light that the neglect of relativity theory did not matter much. But what about wave mechanics of particles that move much faster? Dirac was able in 1927 to unite relativity with quantum mechanics.

In so doing, Dirac discovered a new symmetry in nature, the matter-antimatter symmetry. He discovered it, not by experimenting, but solely by putting together the two great ideas of Einstein: the space-time unity of relativity and the wave-particle duality of quantum mechanics. Dirac saw that for every particle there must be an antiparticle with opposite charge. Although in our own environment we find only negatively charged electrons and protons with positive charge, which is ordinary matter, Dirac concluded that nature must also admit the opposite side. Such anti-matter, he predicted, would not be stable in the presence of ordinary matter; it would annihilate when in touch with it; in a sort of explosion where the masses would be transformed into energy—a direct manifestation of Einstein's equivalence of mass and energy.

A few years later the antielectron was found, and almost 30 years later, the antiproton. Antimatter indeed exists in nature, as Dirac predicted from Einstein's work. This theoretical prediction was one of the greatest intellectual achievements of science. Today, beams of antimatter are produced in many laboratories; they run in carefully evacuated tubes in order not to hit any ordinary matter until they reach their target, where they annihilate with the target substance.

Also, at the aforementioned Pontifical Academy Session on Einstein, P.A.M. Dirac observed:

> By 1905, the wave theory of light based on Maxwell's equations was well established, but certain phenomena would not fit in. It seemed that emission and absorption of light occur discontinuously. This led Einstein to the view that the energy is concentrated in discrete particles. It was a revolutionary idea, very hard to understand, as the successes of the wave theory were undeniable. It seemed that light had to be understood sometimes as waves, sometimes as particles, and physicists had to get used to it. The idea was incorporated into Bohr's theory of the hydrogen atom and forms an essential part of it.
>
> The statistics of an assembly of light particles was studied by Bose, who found that ordinary statistics was not applicable. The laws for the new statistics were formulated jointly by Bose and Einstein. By studying an atom in statistical equilibrium, Einstein saw the necessity for the phenomenon of stimulated emission of radiation. This effect is, in the first place, extremely small, but it can be very much enhanced with a suitable apparatus, because of the new statistics. This led to the laser, a useful tool in present-day technology, which we owe to Einstein.
>
> The appearance of waves connected with particles was shown by de Broglie to be applicable to all particles, not just those having the velocity of light. De Broglie worked out the mathematical relations between waves and particles, using only the requirements of special relativity. He found that the waves move faster than light. However, they cannot be used to transmit signals faster than light, which is an important feature of special relativity.
>
> De Broglie's theory was extended by Schrödinger and led to wave mechanics, which is fundamental for modern atomic theory. Here again, we have a long line of development of physics, originated by Einstein.

In 1926, the Schrödinger equation described the motion of the de Broglie phase waves under the influence of an externally applied potential, and the physical significance of the phase wave ψ was recognized particularly by Born by identifying $\psi^*(q_k)\psi(q_k)d\tau$ with the probability of finding the system in the element of configuration space $d\tau$ between q_1 and $q_1 + dq_1$, etc. Independently in 1925 Heisenberg developed a calculus of observable quantities, representing dynamical variables such as momentum, position, etc., by means of matrices, the time rate of change of a variable X being given by $iX = XH - HX$ where H is the Hamiltonian of the system. This formulation (matrix mechanics) of quantum theory is equivalent to the Schrödinger formulation (wave mechanics). However, it emphasizes the role played by the observer in the measurement of a physical quantity, and the fact that natural limits imposed on measurements he makes must be incorporated into a theory which purports to describe such measurements. Thus in particular to specify the momentum p and corresponding position x of a particle is strictly speaking not legitimate since the very measurement of the one will lead to an unpredictability of the other given by the Heisenberg indeterminacy relation $\Delta x \Delta p <$. Dynamical variables which cannot be measured simultaneously with arbitrary precision are thus represented by matrices, or, more generally, operators, which may not commute, while a system in the state ψ has a definite value for the dynamical variable A if ψ is an eigenfunction of the operator A, i.e., $A\psi = a\psi$ (a = number). Thus if A and B do not commute ψ cannot be at once an eigenfunction of both A and B. A system in an eigenstate of energy is thus described by the equation $H\psi = E\psi$ where H is the Hamiltonian of the system. In the Schrödinger representation (wave mechanics) ψ is regarded as a function of position and time and the momentum p appearing in the Hamiltonian is represented by the operator $-i$ grad, which automatically yields the commutation relation $p_i x_j - x_j p_i = -i\delta_{ij}$. In the Heisenberg representation (matrix mechanics) the position and momentum are represented by matrices which satisfy this commutation relation, and ψ by a constant vector in Hilbert space, the eigenvalues E being the same in two cases.

The Hamiltonian of a particular system is formally identical with that of the classical theory, the simplest, for one particle of mass m moving in a potential V, being

$$H = \frac{p^2}{2m} + V = E$$

which in the Schrödinger representation gives the Schrödinger equation

$$\left(-\frac{\hbar^2}{2m}\nabla^2 + V\right)\psi = E\psi$$

In the presence of a magnetic field B derived from the vector potential A it is necessary to replace

$$\mathbf{p} = -i\hbar\nabla$$

by

$$\mathbf{p} - \frac{e}{c}\mathbf{A} = -i\hbar\left[\nabla - \frac{ie}{hc}\mathbf{A}\right]$$

and in addition to note the contribution $-\mu B$ to the energy arising from the magnetic moment μ of the particle. The vector μ is itself an operator, being, for an electron, $(e/mc)\mathbf{S}$ where $\mathbf{S}$ is the electron spin, $\mathbf{S} = (h/2)\sigma$ the components of σ being the Pauli spin operators.

The value e/mc for the electron gyromagnetic ratio was first postulated by Uhlenbeck and Goudsmit and later shown to be a consequence of the Dirac electron theory.

Nonrelativistic quantum mechanics, extended by the theory of electron spin and by the Pauli exclusion principle, provides a reliable theory for the computation of atomic spectral frequencies and intensities, of cross sections for scattering or capture of electrons by atomic systems, of chemical bonds and many properties of solids, including magnetic properties, although with much more complicated systems it has not always proved possible to develop with adequate accuracy the consequences of the theory. Quantum mechanics has also had a limited success in nuclear theory although in this field it is possible that a more fundamental system of mechanics is required.

Relativistic quantum mechanics is a generalization of nonrelativistic quantum mechanics in which the quantum equations of motion satisfy the principle of relativity. In its simplest form the equation of motion of a particle is the Klein-Gordon equation, but since this neglects the spin properties of the particle, the best verified form of relativistic quantum mechanics is provided by the Dirac electron theory. In general the relativistic quantum theory of a system may always be derived, in accordance with present knowledge, from a Lagrangian, the relativistically covariant properties of the resulting equations of motion being automatically assured if the Lagrangian is invariant under Lorentz transformations.

See also **Quantum Chemistry**; and **Quantum Number**.

Additional Reading

Adar, R.K.: "A Flaw in a Universal Mirror," *Sci. Amer.*, 50 (February 1988).

Ahimony, A.: "The Reality of the Quantum World," *Sci. Amer.*, 46 (January 1988).

Batalin, I.A., Isham, C.J., and G.A. Vileovisky, Editors: *Quantum Field Theory and Quantum Statistics,* Hilger, Bristol, U.K., 1987.

Canright, G.S. and S.M. Girvin: "Fractional Statistics: Quantum Possibilities in Two Dimensions," *Science*, 1197 (March 9, 1990).

Chakaraborty, T. and P. Pietilainen: *The Fractional Quantum Hall Effect,* Springer-Verlag, New York, NY, 1988.

Clarke, J. et al. : "Quantum Mechanics of a Macroscopic Variable: The Phase Difference of a Josephson Junction," *Science*, 992 (February 26, 1988).

Cohen-Tannoudii, C., Dupont-Roc, J., and G. Grynberg: *Photons and Atoms: Quantum Electrodynamics,* Wiley, New York, NY, 1989.

Davies, P., Editor: *The New Physics,* Cambridge University Press, New York, NY, 1989.

Dirac, P.A.M.: "Einstein," Einstein Session of the Pontifical Academy, Vatican City (November 10, 1979). Reprinted in *Science*, **207**, 1161–1162 (1980).

Eisenstein, J.F. and H.L. Stormer: "The Fractional Quantum Hall Effect," *Science*, 1510 (June 22, 1990).

Ellis, P.M. and Y.C. Tang, Editors: *Trends in Theoretical Physics,* Addison—Wesley, Redwood City, CA, 1990.

Fayer, M.D.: *Elements of Quantum Mechanics,* Oxford University Press, Inc., New York, NY, 2000.

Freedman, D.H.: "A Chaotic Cat Takes a Swipe at Quantum Mechanics," *Science*, 626 (August 9, 1991).

Greiner, W.: *Quantum Mechanics: An Introduction,* 2nd Edition, Springer-Verlag, Inc., New York, NY, 2001.

Haken, H., and H. C. Wolf: *Molecular Physics and Elements of Quantum Chemistry*, 2nd Edition, Springer-Verlag New York, Inc., New York, NY, 2004.

Hameka, H. F.: *Quantum Mechanics: A Conceptual Approach,* John Wiley & Sons, Inc., Hoboken, NJ, 2004.

Hayward, D. O.: *Quantum Mechanics for Chemists,* John Wiley & Sons, Inc., New York, NY, 2003.

House, J. E.: *Fundamentals of Quantum Chemistry,* 2nd Edition, Elsevier Science, New York, NY, 2003.

Imry, Y. and R.A. Webb: "Quantum Interference and the Aharonov-Bohm Effect," *Sci. Amer.*, 56 (April 1989).

Mehra, J. and H. Rechenberg: *The Fundamental Equations of Quantum Mechanics 1925–1926: The Reception of the New Quantum Mechanics 1925–1926*, Springer-Verlag, Inc., New York, NY, 2000.

Mehra, J. and H. Rechenberg: *The Historical Development of Quantum Theory: The Completion of Extensions of Quantum Mechanics—1926–1941*, Springer-Verlag, Inc., New York, NY, 2001.

Penrose, R. and C.J. Isham, Editors: *Quantum Concepts in Space and Time*, Oxford University Press, New York, NY, 1986.

Phillips, A. C.: *Introduction to Quantum Mechanics: (Manchester Physics Series)*, John Wiley & Sons, Inc., New York, NY, 2003.

Pool, R.: "Quantum Chaos: Enigma Wrapped in a Mystery," *Science*, 803 (February 17, 1989).

Pope John Paul II: "Einstein," Einstein Session of the Pontifical Academy, Vatican City (November 10, 1979). Reprinted in *Science*, **207**, 1165–1167 (1980).

Powell, C.S.: "Can't Get There from Here: Quantum Physics Puts a New Twist on Zeno's Paradox," *Sci. Amer.*, 24 (May 1990).

Ruthen, R.: "Quantum Pinball: A Quantum System Can Be Observed Without an Observer," *Sci. Amer.*, 36 (November 1991).

Schwinger, J.S. and C. Clarice Schwinger: *Quantum Mechanics: Symbolism of Atomic Measurements*, Springer-Verlag, Inc., New York, NY, 2001.

Styer, D.F.: *The Strange World of Quantum Mechanics*, Cambridge University Press, New York, NY, 2000.

Trefil, J.: "Quantum Physics' World," *Smithsonian*, 66 (August 1987).

Waldrop, M.M.: "Viewing the Universe as a Coat of Chain Mail: New Calculations Have Pointed the Way to Quantum Gravity and Suggested a Novel Structure for the Sub-sub-Microscopic World," *Science*, 1510 (December 14, 1990).

Weisskipf, V.F.: "Einstein," Einstein Session of the Pontifical Academy, Vatican City (November 10, 1979). Reprinted in *Science*, **207**, 1163–1164 (1980).

Wilson, S., and P. F. Bernath: *Handbook of Molecular Physics and Quantum Chemistry*, John Wiley & Sons, Inc., New York, NY, 2003.

Ziock, K. and D. Pocanic: *Introduction to Quantum Mechanics*, Cambridge University Press, New York, NY, 2002.

QUANTUM NUMBER. A number assigned to one of the various quantities that describe a particle or state. Many different characteristics of atomic and nuclear systems, as well as of those entities that are introduced as a part of particle physics, are described by means of quantum numbers. The quantum numbers arise from the mathematics of the eigenvalue problem and may be related to the number of nodes in the eigenfunction. Any state may be described by giving a sufficient set of compatible quantum numbers. In the customary formulations, each quantum number is either an integer (which may be positive, negative, or zero) or an odd half-integer.

Quantum Number (Magnetic). A quantum number that describes the component of the angular momentum vector of an atomic electron or group of electrons in the direction of an externally applied magnetic field. The values of these components are restricted, i.e., quantized. The symbol for the magnetic quantum number is m.

Quantum Number (Orbital). A quantum number characterizing the orbital angular momentum of an electron in an atom or of a nucleon in the shell-model description of the atomic nucleus. The symbol for the orbital quantum number is l.

Quantum Number (Principal). A quantum number that, in the old Bohr model of the atom, determined the energy of an electron in one of the allowed orbits around the nucleus. In the theory of quantum mechanics, the principal quantum number is used most commonly to describe the atomic shell in which the electrons are located. In a somewhat general way, it is related to the energy of the electronic states of an atom. The symbol for the principal quantum number is n. In x-ray spectral terminology, a K-shell is identical to an $n = 1$ shell, and an L-shell to an $n = 2$ shell, etc.

Quantum Number (Spin). A number that describes that part of the total angular momentum of the electron that is due to the rotation of the electron on its own axis. This contribution is quantized, having only a single value $\frac{1}{2}$ in terms of $h/2\pi$ units of angular momentum (h = Planck's constant). The magnetic quantum number associated with the spinning electron can have two values, either $+\frac{1}{2}$ or $-\frac{1}{2}$. The spin angular momentum of the electron can then couple in more than one way with its orbital angular momentum to provide a basis for the occurrence of many multiplet lines in atomic spectra. The symbol for the spin quantum number is s.

See also **Electron**; **Orbitals**; **Proton**; and **Pauli Exclusion Principle**.

QUANTUM THEORY OF SPECTRA. The present theory of spectra, which is based on an idea that there exist in each atom or molecule certain permitted energy levels. An atom or molecule absorbs or radiates energy as it moves from one energy level to another. The frequency (v) of the radiation associated with such change of energy level is given by

$$E_1 - E_2 = hv$$

E_1, E_2 are the energy levels and h is the Planck constant.

QUARKS. Quarks are fundamental matter particles that are constituents of neutrons and protons and other hadrons. There are six different types of quarks, (physicists call them 'flavors'), each of which have a unique mass. The two lightest, unimaginatively called 'up' and 'down' quarks, combine to form protons and neutrons. The heavier quarks aren't found in nature and have so far only been observed in particle accelerators. No one has ever seen a quark. Yet physicists seem to know quite a lot about the properties and behavior of these ubiquitous elementary particles.

History

The quark hypothesis was first proposed independently in 1964 by M. Gell-Mann and G. Zweig, both at the California Institute of Technology. These researchers pointed out that all the known hadrons (i.e., particles that interact via the strong nuclear force) of that time could be constructed out of simple combinations of three particles (and their three antiparticles). These hypothetical particles had to have slightly peculiar properties (the most peculiar being a fractional electric charge). Gell-Mann called them quarks and they were designated p, n, and λ because they somewhat resemble the proton, neutron, and Λ^0 hyperon. The theory supposed that three quarks bind together to form a baryon, while a quark and an antiquark bind together to form a meson. If it is supposed that the binding is such that the internal motion of the quarks is nonrelativistic (which requires that the quarks be massive and sit in a broad potential well), then many quite detailed properties of the hadrons can be explained. One notable exception—the *omega minus particle* discovered in 1964.

A brief review of the complexities to which the quark theory is addressed is in order. Particles which can interact via the strong nuclear force are called hadrons. Hadrons can be divided into two main classes—the mesons (with baryon number zero) and the baryons (with nonzero baryon number). Within each of the classes there are small subclasses. The subclass of baryons which has been known the longest consists of those particles with spin $\frac{1}{2}$ and even parity. The members of this class are the proton, the neutron, the Λ^0 hyperon, the three Σ hyperons and the two Ξ hyperons. There are no baryons with spin $\frac{1}{2}$ and even parity (or, to the usual notation, $J^P = \frac{1}{2}^+$). The next 'family' of baryons has ten members, each with $J^P = \frac{3}{2}^+$. The mesons can be grouped into similar families. One of the first successes of the quark model was to explain just why there should be eight baryons with $J^P = \frac{1}{2}^+$, ten with $\frac{3}{2}^+$, etc., and why the various members of these families have the particular quantum numbers observed.

The chief drawback of the quark model, as voiced soon after the theory was described, has been the failure to produce a beam of individual quarks, much as one can produce a beam of pions, kaons, negative protons, and so on. As of the early 1980s, this objection remains to be satisfied, but because of much theoretical work and new discoveries since 1964, a majority of the community of particle physicists now accept the theory (with modifications and refinements thereof). Explanations offered for the failure to date to isolate quarks is that quarks exist and have already been seen in numerous experiments, but without recognition. Or, that quarks do not exist, but that the hadrons behave as if they did. A somewhat similar belief was once held by some scientists concerning the neutrino. Some theories have yielded quark-like models in which the particles are not fractionally charged, but they were found to have a number of deficiencies. A compelling argument is that the mass of quarks is greater than the energy available and many scientists are looking forward to more powerful accelerators now underway.

The initial quark model was formulated to explain the diversity of the hadrons and not to explicitly describe the internal structure of any particle. It was inevitable, however, that with further research there was a tendency to identify new findings with the hypothetical quarks. A number of properties of the partons, such as their intrinsic spin angular momentum, have been measured and have proved to be consistent with the predictions of the quark model.

During the course of developing several theories concerning the nature of quarks, physicists required a new nomenclature. To persons outside the field, the words used seem as peculiar as the word quark itself. Some writers have referred to the whimsical character of the names used. The kinds of quarks are called *flavors*.[1] The three kinds of flavors initially proposed are *up*, *down*, and *strangeness*. It was proposed a bit later that a fourth flavor, *charm*, should be added to describe a new property of matter. Discovery of the *psi* or *J particle* in 1974 provided strong evidence for the charmed quark. More recent research has pointed toward the need for a fifth quark, called *bottom*. Inasmuch as the other four quarks appear to be organized in pairs, it is generally assumed that there is also a sixth flavor, named *top*.

Quarks also possess another distinctive property that governs their binding together to form hadrons. This property, called *color*, is of three varieties (*red, blue,* and *yellow*) and thus numerous combinations are available, but only certain combinations of them seem possible at this juncture. The property of color[2] plays a role in binding the quarks together in a hadron. Some physicists point out that this is analogous to the role of electric charge in binding together the particles that make up an atom, described by a precise and well-tested theory called quantum electrodynamics. It will be recalled that this theory allows the attraction or repulsion between two charged particles to be communicated by exchange of photons (quanta of electromagnetic radiation).

A theory modeled on the quantum electrodynamics theory has been proposed which describes the interactions between colored quarks and it is known as *quantum chromodynamics*. Mediation of these forces is by hypothetical particles called *gluons*, whose role it is to "glue" the quarks together. Eight kinds of gluon are required. Gluons are themselves colored particles. This means that they are subject to the same forces they transmit. This contrasts with the photon, which communicates electromagnetic forces between charged particles, but in itself carries no electric charge, so that a particle can emit or accept a photon without changing its charge. A number of successful predictions have been made by quantum chromodynamics, but as yet they do not compare with quantum electrodynamics in terms of precision.

In the quantum chromodynamics theory, it is suggested that the effective strength of the force between quarks is small when at close range, but much greater when the quarks are separated by a distance comparable to the diameter of a proton. This concept, in itself, is counter to the much better understood forces of electromagnetism and gravitation, where forces become greater as bodies become closer and weaker as bodies recede from one another. It would appear that, within a hadron, the quarks are constrained very little, whereas the energy required to extract them increases at a rapid rate. If it is simply a matter of mounting sufficient energy, then perhaps quarks have not been found in past experiments simply because too little energy was applied. Or, as some physicists have proposed, perhaps the quarks are permanently confined in some manner. For example, K. A. Johnson and a team at the Massachusetts Institute of Technology proposed a *bag model* of quark confinement in 1979. Simply stated, it is hypothesized that quarks are confined in bags analogous to the bubbles in a liquid. This is an interesting concept and is well described by K. A. Johnson in *Sci. Amer.*, **241**, 1, 112–121, 1979. A number of other concepts have been proposed to explain the peculiar behavior of quarks.

Among other experimental objectives in connection with quarks, physicists have been looking for evidence of gluons in the jets of debris which result from particle collisions, notably between electrons and positrons. It is reasoned that some events should produce two oppositely directed jets that become increasingly thin and pencil-like as the collision energy increases. According to some investigators, double-jet events have occurred during the last few years. Some jets have been described as having the shape of a tennis racket. Findings of this kind have been reported by Deutsches Elektronen-Synchrotron (DESY) in Hamburg. In an August 1979 conference held at Fermilab (Fermi National Accelerator Laboratory), researchers from DESY reported events in which three jets form a pattern like the *Mercedes* 3-pronged star. They appeared in early experiments using a new electron-positron colliding beam storage ring called PETRA. For the time being, these results were expressed as "mimicking the expectations of the quantum chromodynamics theory."

---

[1] One definition of flavor is: "the characteristic quality of something—distinctive nature."

[2] Color, as used here, has nothing to do with visual color. The manner in which different colored quarks combine in quantum mechanics is suggestive of the way in which visual colors combine. Hadrons are "colorless" since they are averages of the three colors.

See also **Hadrons** and **Particles (Subatomic)**.

Additional Reading

Brambilla, N., and G. Prosperi: *Quark Confinement and the Hadron Spectrum: Proceedings of the 5th International Conference,* World Scientific Publishing Company, Inc., Riveredge, NJ, 2003.

Letessier, J., and J. Rafelski: *Hadrons and Quark Gluon Plasma,* Cambridge University Press, New York, NY, 2002.

Smith, T.P.: *Hidden Worlds: Hunting for Quarks in Ordinary Matter,* Princeton University Press, Princeton, NJ, 2003.

Zichichi, A.: *From Quarks and Gluons to Quantum Gravity: Proceedings of the International School of Subnuclear Physics,* World Scientific Publishing Company, Inc., Riveredge, NJ, 2003.

Web References

Quarks: http://hyperphysics.phy-astr.gsu.edu/hbase/particles/quark.html
Quark—Wikipedia: http://en.wikipedia.org/wiki/Quark
Theory: Quarks http://www2.slac.stanford.edu/vvc/theory/quarks.html

QUARTZ. [CAS: 14808-60-7]. The mineral quartz, oxide of the nonmetallic element silicon, is the commonest of minerals, and appears in a greater number of forms than any other. Its formula is SiO_2. Quartz commonly occurs in prismatic hexagonal crystals terminated by a pyramid. This pyramid is due to the equal development of two rhombohedrons, and may be observed in cases where one rhombohedron predominates. Cleavage is not observed; the fracture is typically conchoidal; hardness is 7; specific gravity, 2.65; luster, vitreous to greasy or dull; colorless to white, pink, purple, yellow, blue, green, smoky brown to nearly black; transparent to opaque.

There are two distinct modifications of quartz, depending upon the temperature at which they were formed. The low-temperature variety is formed below 573°C and is the more common sort, being found in veins, geodes, etc. It is called low-quartz. The high-temperature modification is formed between 573°C and 870°C, and is found chiefly in granites and granite or rhyolite porphyries. This is called high-quartz. Above 870°C tridymite is the stable form of SiO_2. The differences between high- and low-quartz are entirely crystallographic, low-quartz having a vertical axis of threefold symmetry and three horizontal axes of twofold symmetry, while high-quartz has a vertical axis of six-fold symmetry and six horizontal axes of twofold symmetry. It is usual to separate the many kinds of quartz into (1) crystalline or vitreous varieties, actual crystals or vitreous crystalline masses, and (2) cryptocrystalline varieties, mostly compact nonvitreous sorts, but which may show a crystalline structure under the microscope.

1. Crystalline or Vitreous: Rock crystal, colorless crystals or masses. Amethyst, clear violet or purple, either crystals or masses. Rose quartz, usually massive but rarely in crystals, delicate shades of pink or rose, sometimes red. Citrine or yellow quartz, sometimes called false or Spanish topaz, light to deep yellow. Smoky quartz, smoky brown to almost black, often called cairngorm stone from Cairngorm, Scotland. Milky quartz, often showing delicate opalescence, transparent to nearly opaque, often with a greasy luster. Aventurine quartz incloses glistening scales of mica or hematite. Rutilated quartz incloses needle-like prisms of rutile called "fleches d'amour." Other acicular minerals such as actinolite, tourmaline, and epidote, may also be thus inclosed; Cat's Eye shows a peculiar opalescence, probably due to inclosed masses of some fibrous mineral. Tiger's Eye is a siliceous pseudomorph after crocidolite of a golden yellow brown color.

2. Cryptocrystalline: The following cryptocrystalline varieties of quartz are treated under their own headings: agate, basanite, bloodstone, carnelian, chalcedony, chert, chrysoprase, flint, heliotrope, jasper, moss agate, onyx, plasma, prase, sard, and sardonyx. Quartz readily forms pseudomorphs after various minerals or structures. Silicified wood is a quartz pseudomorph after the organic material of which it originally consisted. Quartz is often pseudomorphic after calcite, barite, and fluorite. Quartz is an essential constituent of many igneous rocks, for example, granites, granite porphyries, and felsites, as well as quartz diorites and their surface equivalents, the dacites. In the metamorphic rocks quartz Figures very largely in the gneisses and schists, and, of course, in quartzite. In the sedimentary rocks most sandstones are composed chiefly of grains of quartz, and quartz forms veins and nodules in limestones.

Of the many places that have yielded fine specimens of quartz, a few include: the Swiss Alps, the Piedmont of Italy, the Island of Elba, Dauphiné in France, Cumberland in England, Banffshire in Scotland, and Madagascar, Uruguay, Mexico and Brazil. Magnificent rose quartz crystals occur at the Arassuahy-Jequitinhonha District, Minas Gerais, Brazil. In the United States the following localities are well known: Paris, Maine, especially for rose quartz; Herkimer County, New York, for small but very brilliant crystals found in the Cambrian dolomites or in the soil. Amethyst County, Virginia, furnishes amethysts, as do Lincoln and Alexander Counties, North Carolina. Other localities for amethyst and smoky quartz are South Dakota, in the Black Hills; the Pikes Peak district, Colorado; Yellowstone National Park, Wyoming; Jefferson County, Montana; and in Canada in the Province of Ontario in the Thunder Bay region. The word *quartz* is believed to have been originally of German origin. Besides the use of the different varieties of quartz for jewelry and other ornamental purposes, this mineral has extensive industrial uses in the ceramic arts, optical and other sorts of scientific instruments, abrasive, scouring, polishing materials, and for refractories.

Certain mineral classes of low symmetry possess no center of symmetry, and their axes, known as polar axes, have different properties at their terminal ends. Quartz belongs to one of those classes. When quartz is exposed to an exerted compressive or mechanical stress along one of these polar axes, electrical charges are developed on that axis; a negative charge is produced at one end, a positive charge at the opposing end. Conversely, when quartz crystals are subjected to an applied electric field along a polar axis, mechanical strains will be developed in those crystals. This phenomenon is known as piezoelectricity. Plates or disks cut perpendicular to such polar axes and properly oriented with established specifications are subject to mechanical vibrations (oscillations) at predetermined frequencies under an applied electric field. Those frequencies are designed to coincide with and stabilize the circuit frequency of radio transmitters and receivers. This property is utilized extensively in the control of frequency oscillations in the field of radio telemetry. Between January 1942 and V-J Day over 70 million such units were manufactured for the United States armed forces, and consumed over 4 million pounds (1.8 million kilograms) of radio grade quartz. The excessive demand for natural quartz of required quality to produce those wafers resulted in the development of a new industry, synthesizing quartz to meet the demand.

See also **Piezoelectric Effect**.

ELMER B. ROWLEY
F.M.S.A., formerly Mineral Curator
Department of Civil Engineering, Union College
Schenectady, New York

QUARTZITE. A hard, tough, and compact metamorphic rock composed almost wholly of quartz sand grains that have been recrystallized to form a particularly massive siliceous rock. The term is also used for non-metamorphosed quartzose sandstones and grits whose clastic grains have been firmly cemented by silica that has grown in optical continuity around each grain.

QUARTZ PORPHYRY. One of the hypabyssal or effusive rocks chemically related to the granite or alkali family but rich in silica, which occurs as quartz phenocrysts in a crypto- or microcrystalline ground mass.

See also **Porphyry**.

QUATERNARY AMMONIUM COMPOUNDS. There are a vast number of quaternary ammonium compounds (quaternaries). Many are naturally occurring and have been found to be crucial in biochemical reactions necessary for sustaining life. Many quaternaries are also produced synthetically and are commercially available.

Most quaternary ammonium compounds have the general formula $R_4N^+X^-$ and are a type of cationic organic nitrogen compound. The nitrogen atom, covalently bonded to four organic groups, bears a positive charge that is balanced by a negative counterion. Heterocyclics, in which the nitrogen is bonded to two carbon atoms by single bonds and to one carbon by a double bond, are also considered quaternary ammonium compounds. The R group may either be equivalent or correspond to two to four distinctly different moieties. These groups may be any type of hydrocarbon: saturated, unsaturated, aromatic, aliphatic, branched chain, or normal chain. They may also contain additional

functionality and heteroatoms. Examples include methylpyridinium iodide (**1**); benzyldimethyloctadecylammonium chloride (**2**); and di(hydrogenated tallow)alkyldimethylammonium chloride (**3**), where $R = C_{14} - C_{18}$.

(1)   **(2)**   **(3)**

Nomenclature

Quaternary ammonium compounds are usually named as the substituted ammonium salt. The anion is listed last. Substituent names can be either common (stearyl) or IUPAC (octadecyl). If the long chain in the compound is from a natural mixture, the chain is named after that mixture, e.g., tallowalkyl. Prefixes such as di- and tri- are used if an alkyl group is repeated. Complex compounds usually have the substituents listed in alphabetical order.

Properties

Physical Properties. Most quaternary compounds are solid materials that have indefinite melting points and decompose on heating. Physical properties are determined by the chemical structure of the quaternary ammonium compound as well as any additives such as solvents. The simplest quaternary ammonium compound, tetramethylammonium chloride, is very soluble in water and insoluble in nonpolar solvents. As the molecular weight of the quaternary compound increases, solubility in polar solvents decreases and solubility in nonpolar solvents increases.

The ability to form aqueous dispersions is a property that gives many quaternary compounds useful applications.

Higher order aliphatic quaternary compounds, where one of the alkyl groups contains ~10 carbon atoms, exhibit surface-active properties. These compounds compose a subclass of a more general class of compounds known as cationic surfactants.

Chemical Properties. Reactions of quaternaries can be categorized into three types: Hoffman eliminations, displacements, and rearrangements. Thermal decomposition of a quaternary ammonium hydroxide to an alkene, tertiary amine, and water is known as the Hoffman elimination (eq. 1a).

$$CH_3CH_2CH_2\overset{+}{N}(CH_3)_3 + {}^-OH \xrightarrow[displacement]{elimination}$$

$$\begin{aligned} &CH_3CH_2\!=\!\!=\!CH_2 + N(CH_3)_3 + H_2O \quad (1a)\\ &CH_3CH_2CH_2N(CH_3)_2 + CH_3OH \quad (1b) \end{aligned}$$

Displacement of a tertiary amine from a quaternary (eq. 1b) involves the attack of a nucleophile on the α-carbon of a quaternary and usually competes with the Hoffman elimination.

The Stevens rearrangement (eq. 2) is a base-promoted 1,2-migration of an alkyl group from a quaternary nitrogen to carbon. The Sommelet-Hauser rearrangement (eq. 3) is a base-promoted 1,2-migration of a benzyl group to the *ortho*-position of that benzyl group.

$$C_6H_5COCH_2\overset{+}{\underset{CH_2C_6H_5}{N}}(CH_3)_2 \xrightarrow{{}^-OH} C_6H_5COCH\overset{}{\underset{CH_2C_6H_5}{N}}(CH_3)_2 + H_2O \quad (2)$$

$$\quad (3)$$

Naturally Occurring Quaternaries

Many types of aliphatic, heterocyclic, and aromatic derived quaternary ammonium compounds are produced both in plants and invertebrates. Examples include thiamine (vitamin B$_1$) (**4**). See also **Vitamin**; choline

(**5**); and acetylcholine (**6**). These have numerous biochemical functions.

(**4**)

(**5**)   (**6**)

Biochemically, most quaternary ammonium compounds function as receptor-specific mediators; they also function biochemically as messengers.

Analytical Test Methods

There are no universally accepted wet analytical methods for the characterization of quaternary ammonium compounds. The American Oil Chemists' Society (AOCS) has established, however, a number of applicable tests.

The chain length composition of quaternaries can be determined by gas chromatography.

Mass spectral analysis of quaternary ammonium compounds can be achieved by fast-atom bombardment (fab) ms.

Liquid chromatography has been widely applied for analysis of quaternaries. Modified reverse-phase columns can provide chain length information, whereas normal-phase chromatography results in groupings of alkyl distributions.

Nuclear magnetic resonance (nmr) spectroscopy is useful for determining quaternary structure.

Toxicology and Environmental Fate

Some quaternary ammonium compounds are potent germicides, toxic in small (mg/L range) quantities to a wide range of microorganisms. Bactericidal, algicidal, and fungicidal properties are exhibited. Many quaternaries are considered to be moderately to severely irritating to the skin and eyes.

Most uses of quaternary ammonium compounds can be expected to lead to these compounds' eventual release into wastewater treatment systems except for those used in drilling muds. Useful properties of the quaternaries as germicides can make these compounds potentially toxic to sewer treatment systems. It appears, however, that quaternary ammonium compounds are rapidly degraded in the environment and strongly sorbed by a wide variety of materials.

The threat of accidental misuse of quaternary ammonium compounds coupled with potential harmful effects to sensitive species of fish and invertebrates has prompted some concern. Industry has responded with an effort to replace the questionable compounds with those of a more environmentally friendly nature. A newer class of compounds containing an ester linkage has been developed which are more readily biodegraded.

Synthesis and Manufacture of Quaternaries

Quaternary ammonium compounds are usually prepared by reaction of a tertiary amine and an alkylating agent (eq. 4). Some alkylating reagents pose significant health concerns and require special handling techniques.

$$R-\underset{\underset{}{|}}{\overset{\overset{R'}{|}}{N}}-R'' + R'''X \longrightarrow R-\overset{+}{\underset{\underset{R'''}{|}}{\overset{\overset{R'}{|}}{N}}}-R'' \quad X^- \qquad (4)$$

Synthesis and Manufacture of Amines. The chemical and business segments of amines and quaternaries are closely linked. The majority of commercially produced amines originate from three amine raw materials: natural fats and oils, α-olefins, and fatty alcohols. The amines are then used to produce a wide array of commercially available quaternary ammonium compounds. Some individual quaternary ammonium compounds can be produced by more than one synthetic route. See also **Amines**.

Uses

Uses of quaternary ammonium compounds range from surfactants to germicides and encompass a number of diverse industries.

Fabric Softening. The single largest market for quaternary ammonium compounds is as fabric softeners. The use of quaternary surfactants as fabric softeners and static control agents can be broken down into three main household product types: rinse cycle softeners; tumble dryer sheets; and detergents containing softeners, also known as softergents.

Hair Care. Quaternary ammonium compounds are the active ingredients in hair conditioners. Quaternaries are highly substantive to human hair because the hair fiber has anionic binding sites at normal pH ranges.

Other Uses. An important market for quaternaries is sanitation. Quaternaries find use as disinfectants and sanitizers in hospitals, building maintenance, and food processing; in secondary oil recovery for drilling fluids; and in cooling water applications. See also **Petroleum**. Additional applications of quaternaries include the manufacture of organo-modified clays and use as phase-transfer catalysts.

Other important classes of quaternaries are the polyamine-based (polyquats) and the perfluorinated quaternaries, both of which have a number of applications.

Maurice Dery
Akzo Nobel Chemicals Inc.

Additional Reading

Jungermann, E. ed.: *Cationic Surfactants*, Marcel Dekker, Inc., New York, NY, 1969.
Salamone, J. and W. Rice: in J. I. Kroschwitz, ed., *Encyclopedia of Polymer Science and Engineering*, 2nd Edition, John Wiley & Sons, Inc., New York, NY, 1988.
Specialty Surfactants Worldwide in Specialty Chemicals, SRI International, Menlo Park, CA, 1989, pp. 81–94.

QUELET REACTION. Passage of dry hydrogen chloride through a solution in inligroin of a phenolic ether and an aliphatic aldehyde in the presence or absence of a dehydration catalyst to yield α-chloroalkyl derivatives by substitution in the *para* position to the ether group or in the *ortho* position in *para*-substituted phenolic ethers.

QUENCHING. Immersion of hot metals in liquid baths in order to effect rapid cooling. In steel heat-treating practice quenching oils give slower and brine solutions give faster cooling rates than water. Dilute caustic solutions are sometimes used for rapid cooling rates comparable to brine. These baths are maintained at or near room temperature. For special quenching procedures requiring baths held at moderately elevated temperatures, molten metals, such as lead, and fused salts may be used.

Quenching of ordinary steel is for the purpose of hardening it. Certain non-hardenable steels, for example austenitic stainless steel, and many non-ferrous metals may be cooled rapidly from elevated temperatures for other reasons.

QUICKSILVER. See **Mercury**.

QUINOLINES AND ISOQUINOLINES. The isomeric heterocycles quinoline (**1**) and isoquinoline (**2**) possess structures that occur frequently in alkaloids and pharmaceuticals for example, quinine and morphine. See also **Alkaloids**.

(**1**)   (**2**)

Quinoline and isoquinoline are aromatic, but less intensely than benzene.

Comparative Properties

Physical Properties. Both (**1**) and (**2**) are weak bases, showing pK_a 4.94 and 5.40, respectively. Selected physical data for quinoline and isoquinoline are given in Table 1.

Chemical Properties. The presence of both a carbocyclic and a heterocyclic ring facilitates a broad range of chemical reactions for (**1**) and (**2**). Quaternary alkylation on nitrogen takes place readily, but unlike pyridine both quinoline and isoquinoline show addition by subsequent

TABLE 1. PHYSICAL PROPERTIES OF QUINOLINE AND ISOQUINOLINE

	Value	
Property	Quinoline	Isoquinoline
mp, °C	−15.6	26.5
bp, °C	238	243
ΔH_{vap}, kJ/mol[a]	46.4	49.0
n_D^{20}	1.6268	1.6148
d^{20}, g/cm^3	1.0929	1.0986
K_a	1.25×10^{-5}	3.80×10^{-6}
viscosity at 30°C, mPa·s(= cP)	2.997	3.2528
T_c	509	530

[a] To convert J to cal, divide by 4.186.

reaction with nucleophiles. Nucleophilic substitution is promoted by the heterocyclic nitrogen. Electrophilic substitution takes place much more easily than in pyridine, and the substituents are generally located in the carbocyclic ring. Their facile formation of crystalline salts with either inorganic or organic acids and complexes with Lewis acids is in each case of considerable interest.

Quinoline

Reactions. Quinoline exhibits the reactivity of benzene and pyridine dine rings, as well as its own unique reactions.

As an aromatic system, (**1**), shows important synthetic and mechanistic nistic nitro group chemistry. See also **Nitration**. The experimental conditions employed usually determine the product structure.

The main sulfonation product of quinoline at 220°C quinolinesulfonic acid; at 300°C it rearranges to 6-quinolinesul acid.

Unlike pyridine, quinoline undergoes facile addition to the nitrogen-containing ring. Allylmagnesium chloride reacts with quinoline in deoxygenated tetrahydrofuran to produce 80% 2-allyl-1,2-dihydroquinoline. Similar results are observed with vinyl Grignard reagents and with alkyllithium reagents.

Treatment of quinoline with cyanogen bromide, the von Braun reaction, in methanol with sodium bicarbonate produces a high yield of 1-cyano-2-methoxy-1,2-dihydroquinoline (**3**).

(**3**)

2-Aminoquinoline is obtained from quinoline in 80% yield by treatment with barium amide in liquid ammonia.

As with nitration, halogenation under acidic conditions favors reaction in the benzenoid ring, whereas reaction at the 3-position takes place in the neutral molecule.

The synthesis of quinolinic acid and its subsequent decarboxylation to nicotinic acid has been accomplished directly in 79% yield using a nitric–sulfuric acid mixture above 220°C. A wide variety of oxidants have been used in the preparation of quinoline N-oxide.

The ring nitrogen of quinoline reacts with a wide variety of alkylating and acylating agents to produce useful intermediates, for example, N-benzoylquinolinium chloride (**4**).

(**4**)

The direct introduction of carbon–carbon bonds in quinoline rings takes place in low yield and with little selectivity.

Quinoline may be reduced rather selectively, depending on the reaction conditions. Catalytic reduction with platinum oxide in strongly acidic solution at ambient temperature and moderate pressure gives a 70% yield of 5,6,7,8-tetrahydroquinoline. Further reduction of this material with sodium–ethanol produces 90% of *trans*-decahydroquinoline.

Manufacture From Coal Tar. Commercially, quinoline is isolated from coaltar distillates. Tar acids are removed by caustic extraction, and the oil is distilled to produce the methylnaphthalene fraction (230–280°C).

Syntheses of Quinolines

Skraup Synthesis. This general method, used for many quinolines, consists of heating a primary aniline with glycerol, concentrated sulfuric acid, and an oxidizing agent. Often the nitrobenzene corresponding to the aniline employed is used as the oxidant, and iron(II) sulfate is added to moderate the often violently exothermic process. The use of compounds related to acrolein, such as crotonaldehyde and methyl vinyl ketone, allow substituents to be placed in the heterocyclic ring. With ortho- and para-substituted anilines, a single product is usually found; meta-derivatives produce mixtures.

(1)

Döbner-von Miller Synthesis. A much less violent synthetic pathway, the Döbner-von, Miller, uses hydrochloric acid or zinc chloride as the catalyst. α, β-Unsaturated aldehydes and ketones make the dehydration of glycerol unnecessary, and allow a wider variety of substitution patterns. No added oxidant is required.

Combes Synthesis. When aniline reacts with a 1,3-diketone under acidic conditions a 2,4-disubstituted quinoline results, e.g., 2,4-dimethylquinoline from 2,4-pentadione.

Conrad-Limpach-Knorr Synthesis. When a β-keto ester is the carbonyl component of these pathways, two products are possible. Aniline reacts with ethyl acetoacetate below 100°C to form 3-anilinocrotonate, which is converted to 4-hydroxy-2-methylquinoline by placing it in a preheated environment at 250°C. If the initial reaction takes place at 160°C, acetoacetanilide forms and can be cyclized with concentrated sulfuric acid to 2-hydroxy-4-methylquinoline. This example of kinetic vs thermodynamic control has been employed in the synthesis of many quinoline derivatives. They are useful as intermediates for the synthesis of chemotherapeutic agents.

Pfitzinger Reaction. Quinoline-4-carboxylic acids are easily prepared by the condensation of isatin with carbonyl compounds. The products may be decarboxylated to the corresponding quinolines.

Frieländer Synthesis. The methods cited thus far all suffer from the mixtures which usually result with meta-substituted anilines. The use of an ortho-disubstituted benzene for the subsequent construction of the quinoline avoids the problem. In the Frieländer synthesis a starting material like 2-aminobenzaldehyde reacts with an α-methyleneketone in the presence of base. The difficulty of preparing the required anilines is a limitation in this approach.

New Synthetic Approaches. There have been a number of efforts to prepare quinolines by routes quite different from the traditional methods. In one, the cyclization of 3-amino-3-phenyl-2-alkenimines using alkali metals leads to modest yields of various 4-arylaminoquinolines. Because this structure is found in many natural products and few syntheses of it exist, the method merits further investigation.

The importance of quinolinium salts to dye chemistry accounts for the long, productive history of their synthesis. The reaction of N-methylformanilide with ketones, aldehydes, ketone enamines, or enol acetates in phosphoryl chloride leads to high yields of N-methylquinolinium salts.

Toxicology. Quinoline is a poison when it enters the body by any of the normal routes, i.e., ingestion, or subcutaneous or intraperitoneal injection. Even contact with the skin produces a moderate toxic reaction, and can result in severe irritation. There is evidence that quinoline is mutagenic, and long exposure can produce lung problems.

Uses

Antioxidants. The 1,2-dihydroquinolines have been used in a variety of ways as antioxidants. These compounds react with aldehydes, and the products are useful as food antioxidants.

Corrosion Inhibitors. Steel-reinforcing wire and rods embedded in concrete containing quinoline or quinoline chromate are less susceptible to corrosion. Treating the surface of metals with 8-hydroxyquinoline makes them resistant to tarnishing and corrosion. Ethylene glycol-type antifreeze may contain quinoline or its derivative to prevent corrosion.

Agricultural Chemicals. A herbicide possessing activity comparable to 2,4-D is found in compounds like quinolyl esters of N-substituted dithiocarbamic acids. A wide variety of compounds containing the quinoline system are herbicides. Derivatives and salts of 8-quinolinecarboxylic acid as well as quinolyl carbamates are each useful insecticides. The copper salt of 8-hydroxyquinoline is an effective fungicide.

Polymers. Quinoline and its derivatives may be added to or incorporated in polymers to introduce ion-exchange properties.

Platinum-group metals form complexes with chelating polymers with various 8-mercaptoquinoline derivatives.

Metallurgy. The extraction and separation of metals and plating baths have involved quinoline and certain derivatives see also **Electroplating**. The extraction of metal ions depends on the chelating ability of 8-hydroxyquinoline. Dilute solutions of heavy metals such as mercury, cadmium, copper, lead, and zinc can be purified using quinoline-8-carboxylic acid adsorbed on various substrates.

Catalysts. Acrolein and methacrolein 1,4-addition polymerization is catalyzed by lithium complexes of quinoline. The peracetic acid epoxidation of a wide range of alkenes is catalyzed by 8-hydroxyquinoline.

Medicine. In addition to the naturally occurring compounds, a large number of synthetic quinolines have been prepared and studied for use in medicine.

Quinoline Dyes. The reaction of 2-methylquinoline with phthalic anhydride produces a 2:1 mixture of 2-(2-quinolinyl)-1,3-indandione ((**5**), R=H) and 2-(6-methyl-2-quinolinyl)-1,3-indandione ((**5**), R = CH$_3$).

(5)

This mixture is known as Quinoline Yellow A and is most widely used with polyester fibers. Several other quinoline dyes are commercially available and find applications as pigments, biological stains, and analytical reagents.

Isoquinoline

The widespread occurrence of the isoquinoline (**2**) structure in such important alkaloids as those found in cactus, opium, and curare has created a long-standing interest in its synthesis and properties.

Reactions. In general, isoquinoline undergoes electrophilic substitution reactions at the 5-position and nucleophilic reactions at the 1-position. Nitration with mixed acids produces a 9:1 mixture of 5-nitroisoquinoline and 8-nitroisoquinoline. Sulfonation of isoquinoline gives a mixture with 5-isoquinolinesulfonic acid as the principal product.

Amination of isoquinoline with sodamide in neutral solvents gives 1-aminoisoquinoline.

Direct bromination of isoquinoline hydrochloride in a solvent like nitrobenzene gives an 81% yield of 4-bromoisoquinoline.

The oxidation of isoquinoline has also been examined using ruthenium tetroxide. In this instance, the observation that phthalic acid is the only significant product (58%) was made; this fact is both important and difficult to explain. Isoquinoline is also oxidized to its N-oxide by peracids. The N-oxides of isoquinolines have proved to be excellent intermediates for the preparation of many compounds.

Isoquinoline can be reduced quantitatively over platinum in acidic media to a mixture of *cis*-decahydroisoquinoline and *trans*-decahydroisoquinoline.

Synthesis of Isoquinoline and Isoquinoline Derivatives.

Bischler–Napieralski Reaction. This synthetic method involves the cyclodehydration of N-acyl derivatives of β-phenethylamines to 3,4-dihydroisoquinolines, such as 1-methyl-3,4-dihydroisoquinoline.

Pictet-Spengler Synthesis. An acidic catalyst results in the condensation of β-phenethylamines with carbonyl compounds to give 1,2,3,4-tetrahydroisoquinolines.

Pomeranz-Fritsch Synthesis. Isoquinolines are available from the cyclization of benzalaminoacetals under acidic conditions.

Miscellaneous Synthetic Reactions. A number of *o*-disubstituted benzenes have been used to prepare isoquinolines. For example, the Radziszewski method and subsequent dehydration converts *o*-cyanomethylbenzoic acid to homophthalimide in 90% yield.

Toxicology. Isoquinoline is a poison when ingested or injected intraperitoneally. Even in cases of skin contact it is moderately toxic. Its vapors are irritating to the eyes, nose, and throat.

Uses. Isoquinoline and isoquinoline derivatives are useful as corrosion inhibitors, antioxidants, pesticides, and catalysts. They are used in plating baths and miscellaneous applications, such as in photography, polymers, and azo dyes. Numerous derivatives have been prepared and evaluated as pharmaceuticals. Isoquinoline is a main component in quinoline still residue bases, which are sold as corrosion inhibitors and acid inhibitors for pickling iron and steel.

4-Aminoisoquinoline is a component of an ethylene glycol-based corrosion inhibiting antifreeze agent.

A great many alkaloids and synthetic medicinal compounds are isoquinoline derivatives.

K. Thomas Finley
State University of New York, Brockport

Additional Reading

Jones, G.: "Quinolines," Part I, 1977; Part II, 1982; Part III, 1990; G. Grenthe, "Isoquinolines," Part I, 1981; F. G. Kathawala, G. M. Coppola, and H. F. Schuster, Part II, 1990; and G. M. Coppola and H. F. Schuster, Part III, 1995, in A. Weissberger and E. C. Taylor, eds., *The Chemistry of Heterocyclic Compounds*, Vols. 32 and 38, Wiley-Interscience, New York.

QUINONES. These useful compounds have played a central role in both theoretical and practical organic chemistry since the 1840s. The compound 1,4-benzoquinone (**1**)

(1)

provides the generic name quinone. This simple, descriptive nomenclature has been abandoned by *Chemical Abstracts*, but remains widely used. The systematic name for (**1**) is 2,5-cyclohexadiene-1,4-dione. Several examples of quinone synonyms are given in Table 1. Common names are used in this article.

Simple quinones have two notable physical properties: odor and color. The 1,4-benzo- and 1,4-naphthoquinones and many of their derivatives have high vapor pressures and pungent, irritating odors. The single-ring compounds are often found as constituents of insects chemical defense against predators. In general, the 1,2-quinones are vibrant in color, ranging from orange to red, whereas the 1,4-quinones are usually lighter, i.e., yellow to orange.

The quinones have excellent redox properties and are thus important oxidants in laboratory and biological synthons. The presence of an extensive array of conjugated systems, especially the α, β-unsaturated ketone arrangement, allows the quinones to participate in a variety of reactions. Characteristics of quinone reactions include nucleophilic substitution; electrophilic, radical, and cycloaddition reactions; photochemistry; and normal and unusual carbonyl chemistry.

Physical Properties

Selected physical data for various quinones are given in Table 2.

Chemical Properties

Biochemical Reactions. The quinones in biological systems play varied and important roles. In insects they are used for defense purposes, and the vitamin K family members, which are based on 2-methyl-1,4-naphthoquinone, are blood-clotting agents. See also **Vitamin**; and **Vitamin K**.

Two groups of substituted 1,4-benzoquinones are associated with photosynthetic and respiratory pathways; the plastoquinone, e.g., plastoquinone

(2), and the ubiquinones, e.g., ubiquinone (3), are involved in these processes. Although they are found in all living tissue and are central to life itself, a vast amount remains to be learned about their biological roles.

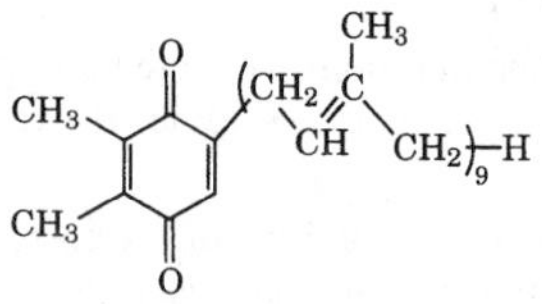

(2) 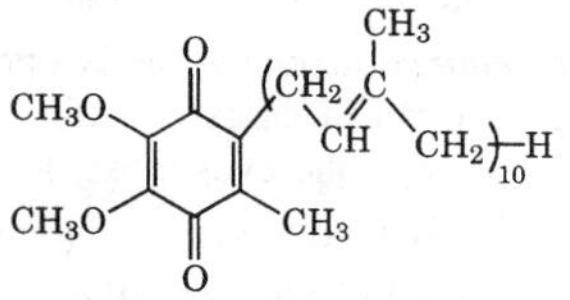   (3)

Quinones of various degrees of complexity have antibiotic, antimicrobial, and anticancer activities.

Dehydrogenation. The oldest and still important synthetic use of quinones is in the removal of hydrogen, especially for aromatization. This method has often been applied to the preparation of polycyclic aromatic compounds. Quinones are used extensively in the dehydrogenation of steroidal ketones. Such reactions are marked by high yield and selectivity. Generally, the results when using nonsteroidal ketones are disappointing.

Oxidation. The use of 1,4-benzoquinone (1) in combination with palladium(II) chloride converts terminal alkenes such as 1-hexene to alkyl methyl ketones in high yield (81%). The quinone appears to reoxidize the palladium.

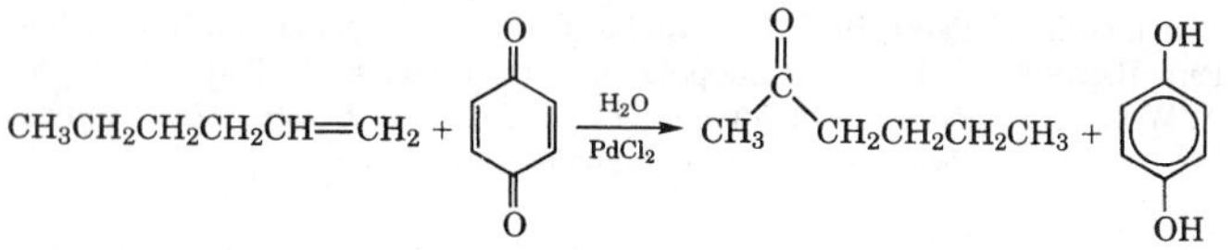

(1)

Photochemical Reactions. Increased knowledge of the centrality of quinone chemistry in photosynthesis has stimulated renewed interest in their photochemical behavior. Synthetically interesting work has centered on the 1,4-quinones and the two reaction types most frequently observed, ie [2 + 2] cycloaddition and hydrogen abstraction.

TABLE 1. QUINONE NOMENCLATURE

Common name	Synonym
1,4-benzoquinone	*p*-benzoquinone
1,2-benzoquinone	*o*-benzoquinone
1,4-napthoquinone	*α*-naphthoquinone
4,4′-diphenoquinone[a]	diphenoquinone
o-chloranil[b]	tetrachloro-*o*-quinone
p-chloranil[c]	tetrachloro-*p*-quinone
2,3-dichloro-5,6-dicyano-1,4-benzoquinone[d]	DDQ

[a] Bis-2,5-cyclohexadien-1-ylidene,4,4′-dione; 4-(4-oxo-2,5-cyclohexadiene-1-ylidene)-2,5-cyclohexadien-1-one is also used.

[b] 3,4,5,6-Tetrachloro-3,5-cyclohexadiene-1,2-dione.

[c] 2,3,5,6-Tetrachloro-2,5-cyclohexadiene-1,4-dione.

[d] 4,5-Dichloro-3,6-dioxo-1,4-cyclohexadiene-1,2-dicarbonitrile.

Addition Reactions. The addition of nucleophiles to quinones is often an acid-catalyzed, Michael-type reductive process. The addition of benzenethiol to 1,4-benzoquinone (1) was studied by A. Michael for a better understanding of valence in organic chemistry. The presence of the reduced product thiophenylhydroquinone, the cross-oxidation product 2-thiophenyl-1,4-benzoquinone, and multiple-addition products such as 2,5-bis(thiophenyl)-1,4-benzoquinone and 2,6-bis(thiophenyl)-1,4-benzoquinone, is typical of many such transformations.

Nucleophilic Substitution Reactions. Many of the transformations realized through Michael additions to quinones can also be achieved using nucleophilic substitution chemistry. In some instances the stereoselectivity can be markedly improved in this fashion, e.g., in the reaction of benzenethiol with esters ($R^3 = CH_3C=O$) and ethers ($R^3 = CH_3$) of 1,4-naphthoquinones. 2-Bromo-5-acetyloxy-1,4-naphthoquinone, $R^1 = Br$, yields 75% of 2-thiophenyl-5-acetyloxy-1,4-naphthoquinone, $R^1 = SC_6H_5$. 3-Bromo-5-methoxy-1,4-naphthoquinone, $R^2 = Br$, yields 82% of 3-thiophenyl-5-methoxy-1,4-naphthoquinone $R^2 = SC_6H_5$.

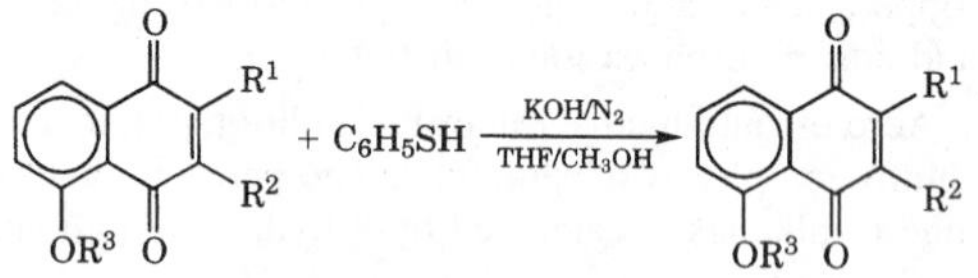

Syntheses

Syntheses of quinones often involve oxidation, because this is the only completely general method. Thus, in several instances, quinones are the reagents of choice for the preparation of other quinones. Oxidation has been especially useful with catechols and hydroquinones as starting materials. The preparative utility of these reactions depends largely on the relative oxidation potentials of the quinones.

For the preparation of ≤10 g of a quinone, the oxidation of a phenol with Fremy's salt in the Teuber reaction is the method of choice. A wide range of phenols has been used, including some having 4-substituents.

In small-scale syntheses, a wide variety of oxidants has been employed in the preparation of quinones from phenols. Of these reagents, chromic acid, ferric ion, and silver oxide show outstanding usefulness in the oxidation of hydroquinones. Thallium(III) trifluoroacetate converts 4-halo- or 4-*tert*-butylphenols to 1,4-benzoquinones in high yield. For example, 2-bromo-3-methyl-5-*t*-butyl-1,4–benzoquinone (4) has been made by this route.

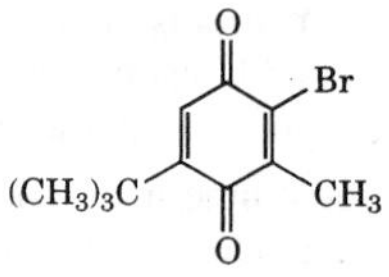

(4)

Thallium trinitrate oxidizes naphthols and hydroquinone monoethers, respectively, to quinones and 4,4-dialkoxycyclohexa-2,5-dienones, e.g., 4,4-dimethoxy-2-methyl-2,5-cyclohexadienone (5). The yield of (5) is 89%.

TABLE 2. PHYSICAL PROPERTIES OF SELECTED QUINONES

Name	Color	Mp, °C	Crystalline form	Solubility Soluble	Insoluble
1,4-benzoquinone	yellow	113, 116	monoclinic prisms	alcohol, ether	water, pentane
1,2-benzoquinone	red	60–70 dec	plates or prisms	ether, benzene	pentane
1,2-naphthoquinone	yellow-red orange	145–147	needles	water, alcohol	ligroin
1,4-naphthoquinone	bright yellow	125, 128.5	needles	alcohol, benzene	water, ligroin
3,4,5,6-tetrachloro-1,2-benzoquinone	orange-red	133, 122–127			
2,3,5,6-tetrachloro-1,4-benzoquinone	yellow	290, 294	monolinic prisms	ether	water, ligroin
2,3-dichloro-5,6-dicyano-1,4-benzoquinone	bright yellow	201–202 dec	plates		
2-chloro-1,4-benzoquinone	yellow-red	57	rhombic-hexagonal	water, alcohol	
2,5-dichloro-1,4-benzoquinone	pale yellow	161–162	monoclinic prisms	ether, chloroform	water, alcohol
2,5-dimethyl-1,4-benzoquinone	yellow	125		ether, alcohol	water, alcohol
2-methyl-1,4-benzoquinone	yellow	69	plates or needles	ether, alcohol	water
3-chloro-1,2-naphthoquinone	red	172 dec	needles	alcohol, benzene	water
2,3-dichloro-1,4-naphthoquinone	yellow	193, 195	needles	benzene, chloroform	water, alcohol
2-methyl-1,4-naphthoquinone	yellow	105–107	needles	ether, benzene	water, alcohol

(5)

The oxidation of 4-bromophenols to quinones can also be accomplished using periodic acid.

The anodic oxidation of hydroquinone ethers to quinone ketals yields synthetically useful intermediates that can be hydrolyzed to quinones at the desired stage of a sequence.

Manufacture

With the exceptions of 1,4-benzoquinone and 9,10-anthraquinone, quinones are not produced on a large scale, but a few of these are commercially available. See also **Anthraquinone**. The few large-scale preparations involve oxidation-of aniline, phenol, or aminonaphthols, e.g., (**6**), from which 1,2-napthoquinone (**7**) is obtained in 93% yield.

In the case of 1,4-benzoquinone (**1**), the product is steam-distilled, chilled, and obtained in high yield and purity. Direct oxidation of the appropriate unoxygenated hydrocarbon has been described for a large number of ring systems, but is generally utilized only for the polynuclear quinones without side chains. A representative sample of quinone uses is given in Table 3.

TABLE 3. USES OF SELECTED QUINONES

Quinone	Use
1,4-benzoquinone	oxidant, amino acid determination
2-chloro-, 2,5-dichloro-, and 2,6-dichloro-1,4-benzoquinone	bactericides
2,3-dichloro-5,6-dicyano-1,4-benzoquinone	oxidation and dehydrogenation agent
2-methyl-and 2,3-dimethyl-1,4-naphthoquinone	vitamin K substitutes, antihemorrhagic agents

Health and Safety Factors

Because of the high vapor pressure of the simple quinones and their penetrating odor, adequate ventilation must be provided in areas where these quinones are handled or stored. Quinone vapor can harm the eyes. In either solid or solution form quinone can cause severe local damage to the skin and mucous membranes. Swallowing benzoquinones may be fatal. The higher quinones are less of a problem because of their decreased volatility.

K. Thomas Finley
State University of New York, Brockport

Additional Reading

Patai, S. ed.: *The Chemistry of Quinonoid Compounds,* John Wiley & Sons, Inc., New York, NY, 1974.

Patai, S. and Z. Rappoport, eds.: *The Chemistry of Quinonoid Compounds,* Vol. 2, John Wiley & Sons, Inc., New York, NY, 1988.

Thomson, R.H.: *Naturally Occurring Quinones,* 3rd Edition, Chapman and Hall, London, UK, 1987.

R

RACEMIZATION. The conversion of an optically active compound i.e., one that rotates the plane of polarized light, into its racemic or optically inactive form is known as *racemization*. In this process half of the optically active compound is converted into its mirror image. The resultant mixture of equal quantities of the dextro- and levo-rotatory isomers is without effect on plane-polarized light due to external compensation (meso forms are internally compensated). Racemization of compounds possessing more than one asymmetric atom may yield products with residual optical activity due to the presence of unchanged centers of asymmetry. An example of this is found in the mutarotation of either α- or β-D-glucose in which only 1 of the 5 asymmetric centers is affected by the opening and closing of the hemiacetal ring.

Racemization may occur in molecules in which structural changes, such as those due to resonance, enolization, substitution or elimination of groups, temporarily destroy the asymmetry needed to maintain the optical activity. Also, Walden inversion of half of an optically active isomer can yield a racemate without the destruction of the center of asymmetry; this phenomenon is observed in the reaction of D-butanol-2 with $HClO_4$.

In the case of the optically active acids, racemization is postulated as the result of the enolization mechanism:

$$\underset{R'}{\overset{R}{\diagdown}}C^*H-COOH \rightleftharpoons \underset{R'}{\overset{R}{\diagdown}}C=C\underset{OH}{\overset{OH}{\diagup}}$$

In this case it is seen that the hydrogen attached to the alpha carbon migrates to the oxygen of the carbonyl group forming a carbon-to-carbon double bond and thereby destroying the asymmetry previously found at the alpha carbon atom. Reversion of the enol to the acid reforms the center of asymmetry on a random basis (i.e., giving the same quantities of dextro and levo forms of the acid). Variations of pH, temperature, solvents and catalysts are likely to change the rates of racemization. The progress of the reaction is conveniently followed by examination of the reaction mixture with plane-polarized light in a polarimeter. A similar racemization of disodium L-cysteine in liquid ammonia in the presence of $NaNH_2$ is reported to occur through abstraction of the α-proton to form a carbanion followed by random recombination of the proton.

Although most racemization reactions have been observed to occur in solution, L-leucine on the surface of silicates has been completely racemized by heating to 200°C for 6 hours. Racemization has also been observed in optically active compounds which possess no asymmetric atom but owe their activity to hindered rotation around a single C—C bond. For example, 8'-methyl-1,1'binaphthyl-8-carboxylic acid has been resolved into its optically active forms and racemized by heating in dimethylformamide.

One of the earliest known examples of racemization was described by Pasteur in his studies of tartaric acid. By heating D-tartaric acid to 165°C in water he partially converted it into a mixture of D-, L- and *meso*-tartaric acids. The occurrence of some DL-tartaric acid along with the natural D-tartaric acid in the wine industry is explained on the basis of partial racemization.

The process of racemization has a number of practical application in the laboratory and in industry. Thus, in the synthesis of an optical isomer it is frequently possible to racemize the unwanted isomer and to separate additional quantities of the desired isomer. By repeating this process a number of times it is theoretically possible to approach a 100% yield of synthetic product consisting of only one optical isomer. An example of the utilization of such a process is found in the production of pantothenic acid and its salts. In this process the mixture of D- and L-2-hydroxy-3,3-butyrolactones are separated. The D-lactone is condensed with the salt of beta-alanine to give the biologically active salt of pantothenic acid. The remaining L-lactone is racemized and recycled.

The process of racemization is important in the survival and growth of living cells and is catalyzed by a group of enzymes called racemases. Alanine racemase, for example, is able to convert D-alanine to DL-alanine if a suitable alpha keto acid is also present. In this reaction the asymmetry of the alpha-carbon atom of alanine is lost as the amino acid is converted to the keto acid and back. This process is analogous to the well-known process of transamination in which racemization seldom occurs.

See also **Amino Acids**; and **Radioactivity**.

LOUIS H. GOODSON
Midwest Research Institute
Kansas City, Missouri

RACKING. Experimental cold-stretching of unvulcanized rubber, whose behavior under stress is unique among natural materials. A thin narrow strip stretched at, for example, 500–600% at 0°C will retain that extension indefinitely after release of stress as long as the low temperature persists. In this state it loses its elasticity and has virtually 100% permanent set. It also displays a crystalline X-ray pattern similar to that of a fiber, in contrast to the amorphous structure of the unstretched state. On exposure to room temperature it slowly retracts to its original length; higher temperature it slowly retracts its rate of recovery. Tests made on racked rubber have shown that crude rubber can be exposed to any degree of low temperature, for any length of time without impairment of its properties.

RADIATION. 1. The emission and propagation of energy through space or through a material medium in the form of waves; for instance, the emission and propagation of electromagnetic waves, or of sound and elastic waves.

2. The energy propagated through space or through a material medium as waves; for example, energy in the form of electromagnetic waves or of elastic waves. The term radiation, or radiant energy, when unqualified, usually refers to electromagnetic radiation; such radiation commonly is classified, according to frequency, as radio-frequency, microwave, infrared, visible (light), ultraviolet, x-rays, and γ-rays. Radiation may also be designated as monochromatic, when it has, ideally, one wavelength, or actually a narrow band of wavelengths; or as heterogeneous, when it has two or more narrow bands of wavelengths (or particles of two or more narrow energy ranges); or as homogeneous, when it has only one narrow band of wavelengths (or consists of essentially monoenergetic particles).

3. Corpuscular emissions, such as alpha- and beta-radiation or rays of mixed or unknown type.

4. In a looser sense the term *radiation* also includes energy emitted in the form of particles that possess mass and may or may not be electrically charger, (i.e., α [positive] and β [negative] and also neutrons. Beams of such particles may be considered as "rays". The charged particles may all be accelerated and the high energy imparted to "beams" in particle accelerators such as cyclotrons, betatrons, synchrotrons, and linear accelerators.

Radiation is used in medicine in the form of X-rays and radioactive isotopes. It is used in industry in many ways, e.g., as vitamin activator, sterilizing agents, and polymerization initiator; it is also the basis of all types of spectroscopic analysis.)

RADIATION (Biochemistry). The study of substances having the ability to protect cells and body tissue against the deleterious effects of ionizing radiation. See also **Radiation (Ionizing)**. Because one of these effects is to deprive proteins of sulfhydryl (^-SH) groups necessary for cell division, the injection of compounds rich in this radical (notably cysteine) has been successfully tried with laboratory animals. Thiourea has been found to protect DNA from depolymerization by X-rays; enzymes containing ^-SH groups inactivated by radiation are reactivated by addition of glutathione.

Some of the other radiochemically induced reactions that adversely affect biochemical activity are:

1. Formation of hydrogen peroxide (a biological poison) by free radical mechanism.
2. Denaturation of proteins.
3. Change in substituent groups of amino acids.
4. Oxidation of hemoglobin.
5. Depolymerization of DNA.

RADIATION (Industrial).

Chemical or physiochemical changes induced by exposure to various types and intensities of radiation are as follows:

1. Synthesis of ethyl bromide from hydrogen bromide and ethylene, using α-radiation from cobalt-60.
2. Cross-linking of such polymers as polystyrene and polyethylene with either β- or γ-radiation.
3. Vulcanization of rubber with ionizing radiation.
4. Polymerization of methyl methacrylate monomer with cobalt-60 as source of γ-rays. Free radical formation is involved in both cross-linking and polymerization reactions. This technique is also being applied in the textile finishing field for grafting and cross-linking fibers with chemical agents for durable-press fabrics.
5. Processing of various foods (cooking, drying, pasteurizing, etc.) by electromagnetic energy in the microwave region of the spectrum; preservation and sterilization of food products by ionizing radiation (γ- and X-rays). The dosage of radiation is strictly controlled, and FDA approval is required. Irradiation is also effective in inhibiting sprouting and preventing insect infestation of stored grains.
6. Curing or hardening of organic protective coatings (paints, inks) by exposure to infrared, UV, or electron-beam radiation. Required are a monomer or oligomer and a photoninitiator, which induces polymerization by free radical formation.

See also **Radioactivity**.

RADIATION (Ionizing).

Extremely short-wave-length, highly energetic, penetrating rays of the following types:

1. γ-rays emitted by radioactive elements and isotopes (decay of atomic nuclei).
2. X-rays generated by sudden stoppage of fast-moving electrons.
3. Subatomic charged particles (electrons, protons, deuterons) when accelerated in acyclotron or betatron.

The term is restricted to electromagnetic radiation at least as energetic as X-rays, and to charged particles of similar energies. Neutrons also may induce ionization.

Such radiation is strong enough to remove electrons from any atoms in its path, leading to the formation of free radicals. These short-lived but highly reactive particles initiate decomposition of many organic compounds. Thus ionizing radiation can cause mutations in DNA and in cell nuclei; adversely affect protein and amino acid mechanisms; impair of destroy body tissue; and attack bone marrow, the source of red blood cells. Exposure to ionizing radiation for even a short period is highly dangerous, and for an extended period may be lethal. The study of the chemical effects of such radiation is called radiation chemistry or (in the case of body reactions) radiation biochemistry.

See also **Radiation (Industrial)**; and **Radioactivity**.

RADIOACTIVE WASTE.

Disposal of waste containing radioisotopes and of spent nuclear reactor fuel presents a serious problem for which there is as yet no completely satisfactory solution. Such wastes may remain radioactive for thousands of years and can constitute a long-term hazard that is restraining the development of nuclear-generated electric power. Safe disposal techniques are being intensively studied. Ocean dumping, practiced some years ago, is no longer permissible. Small amounts of low-level wastes containing radioisotopes can be diluted sufficiently with an inert material to reduce its activity to an acceptable point. High-level reactor waster, for example, at Hanford are stored in concrete tanks lined with steel and buried under a foot of concrete and 5 or 6 ft of soil.

Containers of compressed alumina (corundum) have been recommended, as this material remains impervious to water indefinitely. Storage in the form of calcine (granular solid) and in borosilicate glass is a promising possibility under active investigation. The DOE has recommended disposal in deep geologic formations, although the heat generated by the radioactivity could cause fracturing of the surrounding rock structures; this would admit water that would eventually rise to the surface after being contaminated. A test program involving storage in basalt is being conducted by the DOE. Storage is salt formations is under serious consideration because they are self-sealing and free from water.

See also **Nuclear Power Technology**.

RADIOACTIVITY.

The spontaneous disintegration of the nucleus of an atom with the emission of radiation. This phenomenon was discovered by Becquerel in 1896 by the exposure-producing effect on a photographic plate by pitchblende (uranium-containing mineral) while wrapped in black paper in the dark. Soon after this, it was found that uranium minerals and uranium chemicals showed more radioactivity than could be accounted for by the uranium content. About the same time, radioactivity of thorium minerals and thorium chemicals was also discovered.

The excess radioactivity of mineral over chemical uranium led Pierre and Marie Curie to experiment with the mineral. To detect the presence of radioactivity the discharge of a charged gold-leaf electroscope was used. A quantitative estimation of the amount of radioactivity was made by observing the rate of drop of the gold leaf. By chemically separating the uranium mineral into fractions and examining each fraction by the electroscope, they found in the bismuth element fraction the first new radioactive element to be discovered. It was named polonium (1898). They found that polonium disappeared rapidly, half of its radioactivity vanishing in about six months. The fraction containing barium element was also found by them to be radioactive. Repeated fractional crystallizations of the chloride and bromide solutions made possible the recovery by them of practically pure salt of the second new radioactive element. It was named radium (1898).

Radium is chemically similar to barium; it displays a characteristic optical spectrum; its salts exhibit phosphorescence in the dark, a continual evolution of heat taking place sufficient in amount to raise the temperature of 100 times its own weight of water 1°C every hour; and many remarkable physical and physiological changes have been produced. Radium shows radioactivity a million times greater than an equal weight of uranium and, unlike polonium, suffers no measurable loss of radioactivity over a short period of time (its half life is 1620 years). From solutions of radium salts, there is separable a radioactive gas; radium emanation, radon, which is a chemically inert gas similar to xenon and disintegrates with a half life of 3.82 days, with the simultaneous formation of another radioactive element, Radium A (polonium-218).

Definitions[1]

Decay. The diminution of a radioactive substance due to nuclear emission of alpha or beta particles, gamma rays or positrons.

Decay Constant. A constant λ that relates the instant rate of radioactive decay of a radioactive species to the number of atoms N present at a *given time t*:

$$-(\partial N/\partial) = \lambda N$$

Where N is the number of atoms present at time *zero*:

$$N = N_o e^{-\lambda t}$$

Decay Product. A *nuclide* that results from the radioactive disintegration of a radionuclide that is formed either directly or as the result of progressive transformations in a radioactive series. The nuclide thus produced is sometimes called the daughter or *daughter* element.

Half-Life (Radioactive Element). The average time required for one-half of the atoms in a sample of radioactive element to decay. The half-life $t\frac{1}{2}$ is given by

$$t_{\frac{1}{2}} = (\ln 2)/\lambda$$

Types of Radioactivity

Beginning in 1899 and continuing through the next two decades, E. Rutherford and his associates conducted a rather thorough study of

---

[1] Data pertaining to radioisotopes is provided in quantitative detail in the "Handbook of Chemistry and Physics," 80th Edition, CRC Press, LLC., Boca Raton, FL. 2000.

the radiations emitted by radioactive substances. During this study the radiations were found to be of three types, called alpha, beta, and gamma radiations. In kind, they resemble anode rays, cathode rays, and x-rays, respectively. In this behavior toward electrical and magnetic fields, the resemblance is qualitatively complete: (1) Alpha rays are positively charged particles of mass number 4 and slightly deflected by electrical and magnetic fields. (2) Beta rays are negatively charged electrons, and strongly deflected by electrical and magnetic fields. (3) Gamma rays are undeflected by electrical and magnetic fields, and of wavelength of the order of 10^{-8} to 10^{-9} centimeters.

Alpha Ray Emission. Alpha rays have a definite velocity and a definite range for each radioactive nuclide. The velocity is from 5–7% that of light. *Range* is defined as the distance traversed in a homogeneous medium before absorption. The penetrating power of alpha rays is the smallest of the three kinds of rays, the beta rays being of the order of 100 times, and the gamma rays 10,000 more penetrating. The alpha rays are twice-ionized nuclei of helium (He^{2+}). Ramsay and Royds (1909) experimentally demonstrated that accumulated alpha particles, quite independently of the matter from which they have been expelled, consist of helium. They sealed radon in a glass tube with a wall so thin that the alpha particles passed through the wall into a surrounding vessel and after six days the optical spectrum of helium was observed. Helium itself does not diffuse through such a wall. Therefore, alpha particles on losing their positive charge become ordinary helium. This is the first instance of the production of a known element during radioactive transformation. The loss of a single alpha particle by an atom leaves the residual atom four units less in mass number, and two units less in atomic number. The shooting of alpha particles was visibly registered by Crooke's spinthariscope in which the tip of a wire, coated by a tiny amount of radium salts, was placed near a screen coated with zinc blende. Viewed in the dark with a magnifying eyepiece, each alpha particle striking the zinc blende target was observed to produce a visible scintillation. The detection and counting of single alpha particles was accomplished by Rutherford and Geiger (1908), by the deflection of an electrometer needle upon the arrival of each alpha particle in a gas at low pressure in an electric field somewhat below the sparking point.

Beta Radiation. Beta rays are electrons. They have varying velocities almost up to that of light. The loss of a single negatron by an atom leaves the residual atomic nucleus the same in mass number and one unit greater in atomic number, while the loss of a positron or an orbital electron capture leaves the residual atomic nucleus the same in mass number, and one unit less in atomic number.

Double-Beta Decay. In a scholarly paper, M. Moe (University of California, Irvine) and S. Rosen (Los Alamos National Laboratory) describe what is considered to be a very rare radioactive event, namely that of double-beta decay. Searching since 1971, Moe and colleagues observed a double-beta event directly in 1987. In an event of this type, using a selenium atom, which contains 48 neutrons and 34 protons, as an example, two of the neutrons decay simultaneously into two protons. During the process, two beta rays and two antineutrinos are generated. Application of an external magnetic field will cause the paths of the ejected electrons to spiral, but in different directions. A double spiral of this type, when observed, is indicative of a double-beta event. The atom remaining has gained two additional protons and lost two neutrons, as compared with its original state (i.e., prior to the double-beta decay event). The selenium has been transformed to krypton.

The neutrino problem is described in the article on **Particles (Subatomic)**. The double-beta decay event may contribute to the solution of that problem. In their introductory to the aforementioned article, the authors observe, "The future of fundamental theories that account for everything from the building blocks of the atom to the architecture of the cosmos hinges on studies of this rarest of all observed radioactive events."

Gamma Radiation. Gamma rays are photons of electromagnetic radiation. This radiation is much more penetrating than alpha or beta particles. The presence of gamma rays from 30 milligrams of radium can be observed in an electroscope after passing through 30 centimeters of iron (Rutherford). For the protection of the operator, radium is kept in lead outer containers or screened by lead sheets.

The naturally occurring radioactive elements at the upper end of the periodic table of elements form a number of series, the elements of each series existing in radioactive equilibrium, unless individual elements are separated chemically away from the series. These series include

TABLE 1. THE URANIUM SERIES

Radioelement	Corresponding element (2)	Symbol	Radiation	Half-life
Uranium I	Uranium (92)	^{238}U	α	4.51×10^9 yr
Uranium X_1	Thorium (90)	^{234}Th	β	24.1 days
Uranium X_2^*	Protactinium (91)	^{234}Pa	β and I.T.	1.17 min
99.87% | 0.13%				
Uranium II	Uranium (92)	^{234}U	α	2.48×10^5 yr
Uranium Z	Protactinium (91)	^{234}Pa	β	6.66 hr
Ionium	Thorium (90)	^{230}Th	α	7.5×10^4 yr
Radium	Radium (88)	^{226}Ra	α	$1.62^3 10^3$ yr
Ra Emanation	Radon (86)	^{222}Rn	α	3.82 days
Radium A	Polonium (84)	^{218}Po	α and β	3.05 min
99.96% | 0.04%				
Radium B	Lead (82)	^{214}Pb	β	26.8 min
Astatine-218	Astatine (85)	^{218}At	α	2 sec
Radium C	Bismuth (83)	^{214}Bi	β and α	19.7 min
99.96% | 0.04%				
Radium C	Polonium (84)	^{214}Po	α	1.5×10^{-4} sec
Radium C''	Thallium (81)	^{210}Tl	β	1.32 min
Radium D	Lead (82)	^{210}Pd	β	19.4 yr
Radium E	Bismuth (83)	^{210}Bi	β and α	2.6×10^6 yr
~100% | ~10^{-5}%				
Radium F	Polonium (84)	^{210}Po	α	138.4 days
Thallium-206	Thallium (81)	^{206}Tl	β	4.23 min
Radium G (end product)	lead (82)	^{206}Pb	None	Stable

* Undergoes ismeric transition (I.T.) to form uranium Z(^{234}Pa); the latter has a half life of 6.66 hr, emitting β radiation and forming Uranium II ^{234}U.

the Uranium Series, the Thorium Series, and the Actinium Series. See Tables 1, 2, and 3. These arrangements are useful in showing the decay chain (i.e., the parent-daughter) relationships of radioactive elements, including such concepts as radioactive equilibrium. Other naturally occurring radioactive elements also exist, including, for example, ^{40}K, ^{87}Rb, and ^{148}Sm.

Ionizing Radiation. This type of radiation is of major importance because it represents a biological and environmental hazard. Radioactive isotopes contribute to this potential danger. The extent of damage varies immensely with the dose (exposure over a long or short period of time) and with the source material. The principal ionizing radiations are summarized in Table 4.

Artificially Induced Radioactivity

In addition to the radionuclides already discussed, there are also the great numbers of artificially produced radioactive elements. They are represented

TABLE 2. THE THORIUM SERIES

Radioelement	Corresponding element	Symbol	Radiation	Half-life
Thorium	Thorium	^{232}Th	α	1.39×10^{10} yr
Mesothorium I	Radium	^{228}Ra	β	6.7 yr
Mesothorium II	Actinium	^{228}Ac	β	6.13 hr
Radiothorium	Thorium	^{228}Th	α	1.90 yr
Thorium X	Radium	^{224}Ra	α	3.64 days
Th Emanation	Radon	^{220}Rn	α	54.5 sec
Thorium A	Polonium	^{216}Po	α	0.16 sec

~100% | 0.014%

Thorium B	Lead	^{212}Pb	β and α	10.6 hr
Astatine-216	Astatine	^{216}At	α	3×10^{-4} sec
Thorium C	Bismuth	^{212}Bi	β and α	60.5 min

66.3% | 33.7%

Thorium C	Polonium	^{212}Po	α	3×10^{-7} sec
Thorium C″	Thallium	^{208}Tl	β	3.1 min
Thorium D (end product)	Lead	^{208}Pb	None	Stable

TABLE 3. THE ACTINIUM SERIES

Radioelement	Corresponding element	Symbol	Radiation	Half-life
Actinouranium	Uranium	^{235}U	α	7.07×10^{8} yr
Uranium Y	Thorium	^{231}Th	β	25.6 hr
Protactinium	Protactinium	^{231}Pa	α	3.25×10^{4} yr
Actinium	Actinium	^{227}Ac	β and α	21.7 yr

98.8% | 1.2%

Radioactinium	Thorium	^{227}Th	α	18.2 days
Actinium K	Francium	^{223}Fr	β	21 min
Actinium X	Radium	^{223}Ra	α	11.7 days
Ac Emanation	Radon	^{219}Rn	α	3.92 sec
Actinium A	Polonium	^{215}Po	α and β	1.83×10^{-3} sec

~100% | ~5×10^{-4}%

Actinium B	Lead	^{211}Pb	β	36.1 min
Astatine-215	Astatine	^{215}At	α	~10^{-4} sec
Actinium C	Bismuth	^{211}Bi	β and α	2.16 min

99.96% | 0.32%

Actinium C	Polonium	^{211}Po	α	0.52 sec
Actinium C″	Thallium	^{207}Tl	β	4.76 min
Actinium D (end product)	Lead	^{207}Pb	None	Stable

in the Neptunium Series and in various collateral series, because, in addition to the three main natural, and the one artificial, disintegration series of radioelements, each has been found to have at least one parallel or collateral series. The main series and the collateral series have different parents, but they become identical when, in the course of disintegration, they have a member in common. Collateral with the natural uranium series is an artificial series discovered in the United States by M.H. Studier and E.K. Hyde. Its parent is ^{230}Pa formed by the bombardment of thorium with alpha particles or deuterons of high energy. The decay scheme of the series has been found to be

$$^{230}\text{Pa} \xrightarrow{\beta-} {}^{230}\text{U} \xrightarrow{\alpha} {}^{226}\text{Th} \xrightarrow{\alpha} {}^{222}\text{Ra} \xrightarrow{\alpha} {}^{218}\text{Rn} \xrightarrow{\alpha} {}^{214}\text{Po}$$
$$\longrightarrow \text{Uranium Series}$$

The loss of the alpha particle by the emanation, ^{218}Rn, leads to the formation of ^{214}Po, which is identical with radium C′ of the Uranium Series; the subsequent decay of the collateral series thus becomes identical with that of the main Uranium Series at this point.

Another collateral Uranium Series has for its progenitor ^{226}Pa which is found among the products of bombardment of thorium with 150-MeV deuterons. The decay scheme is represented by:

$$^{226}\text{Pa} \xrightarrow{\alpha} {}^{222}\text{U} \xrightarrow{\alpha} {}^{218}\text{Th} \xrightarrow{\alpha} {}^{214}\text{Ra} \xrightarrow{\alpha} {}^{210}\text{Bi}$$
$$\longrightarrow \text{Uranium Series}$$

Still other collateral series are the following:

$$^{228}\text{Pa} \xrightarrow{\alpha} {}^{224}\text{Ac} \xrightarrow{\alpha} {}^{220}\text{Fr} \xrightarrow{\alpha} {}^{216}\text{At} \xrightarrow{\alpha} {}^{212}\text{Bi}$$
$$\longrightarrow \text{Thorium Series}$$

$$^{232}\text{Pa} \xrightarrow{\alpha} {}^{228}\text{U} \xrightarrow{\alpha} {}^{224}\text{Th} \xrightarrow{\alpha} {}^{220}\text{Ra} \xrightarrow{\alpha} {}^{216}\text{Rn} \xrightarrow{\alpha} {}^{212}\text{Po}$$
$$\longrightarrow \text{Thorium Series}$$

$$^{227}\text{Pa} \xrightarrow{\alpha} {}^{223}\text{Ac} \xrightarrow{\alpha} {}^{219}\text{Fr} \xrightarrow{\alpha} {}^{215}\text{At} \xrightarrow{\alpha} {}^{211}\text{Bi}$$
$$\longrightarrow \text{Actinium Series}$$

$$^{239}\text{U} \xrightarrow{\beta} {}^{239}\text{Np} \xrightarrow{\beta} {}^{239}\text{Pu} \xrightarrow{\alpha} {}^{235}\text{U} \longrightarrow \text{Actinium Series}$$

$$^{239}\text{U} \xrightarrow{\alpha} {}^{225}\text{Th} \xrightarrow{\alpha} {}^{221}\text{Ra} \xrightarrow{\alpha} {}^{217}\text{Rn} \xrightarrow{\alpha} {}^{213}\text{Po}$$
$$\longrightarrow \text{Neptunium Series}$$

See Table 5.

Frederic and Irene Joliot-Curie found in 1933 that boron, magnesium, or aluminum, when bombarded with α-particles from polonium, emit neutrons, proton, and positrons, and that when the source of bombarding particles was removed, the emission of protons and neutrons ceased, but that of positrons continued. The targets remained radioactive, and the emission of radiation fell off exponentially just as it would for a naturally occurring radioelement. The results of this work may be stated in two equations as follows:

$$^{4}\text{He} + {}^{27}\text{Al} \longrightarrow {}^{30}\text{P} + n$$

$$^{30}\text{P} \xrightarrow{\beta+} {}^{30}\text{Si}$$

The first of these equations shows that the result of the nuclear reaction in which aluminum is bombarded with α-particles is the emission of a neutron and the production of a radioactive isotope of phosphorus. The second equation shows the radioactive disintegrations of the latter to yield a stable silicon atom and a positron. Continuation of this line of investigation by several research groups confirmed that radioactive nuclides are formed in many nuclear reactions.

Generally, if any two isobars differ in charge by $\pm e$, one had a higher ground-state energy than the other and is beta radioactive. Any nuclide that

TABLE 4. PRINCIPAL TYPES OF IONIZING RADIATION

Name	Symbol	Location in aton	Relative rest mass	Charge
Proton $(H^1)^+$	p	Nucleus	1	+1
Neutron	n	Nucleus	1	0
Electron	e	Outer shells	0.00055	−1
Beta⁻ (electron)	β	Emitted during decay processes	0.00055	−1
Beta⁺ (positron)	β⁺		0.00055	+1
Alpha $(He^4)^{++}$	α		4	+2
Gamma[a] (photon)	γ	Emitted during decay processes	0.0	0

[a] X-rays of equal energy are identical, but of extranuclear origin. Although only the gamma or x-rays are electromagnetic in character and thus "radiations" (particles) is often not made. X-rays are distinguished from gamma rays only with respect to their origins. Gamma rays result from nuclear interactions or decays. X-rays result from transitions of atomic or free electrons, produced artificially by bombarding metallic targets with energetic electrons. As pointed out by Mel and Todd (see references), it is sometimes difficult to make a clear distinction between ionizing and nonionizing electromagnetic radiations, particularly in condensed phases. The ionization potential of gaseous elements, that is, the energy required for removal of the first, electron, ranges from 3.9 eV (cesium) to 24.6 eV (helium). Although ultraviolet and even visible light in special cases can ionization, the general assumption is that the more energetic x- or gamma radiation is needed to insure ionization. Neutrons also lead to ionization, but for other reasons. Ionization, of course, is not the only interaction of high-energy particles and radiations with matter. The excitation of atomic electrons into higher-energy states always accompanies ionization as well.

can be formed from a nuclear reaction and is not one of the known stable nuclides is radioactive. Nuclides having higher atomic number (Z) than the nearest stable isobar decay to it through positron emission (β^+ decay) or orbital electron capture. Nuclides with lower Z than the nearest stable isobar decay to it through negatron emission (β^- decay). Occasionally, as for ^{64}Cu, a radioactive nuclide is located between two stable isobars and can decay to either of them, in this case either ^{64}Ni or ^{64}Zn. The simplest radioactive nuclide is the neutron, which has a half life of 12 minutes, and decays into a proton, a negatron, and a neutrino.

Energy-level diagrams for nuclear transformations are usually drawn to show the relative energies of levels of an entire neutrally charged atomic system. Since the nuclear charge increases in magnitude by e if a beta radioactive nucleus emits a negatron, one additional external electron must be added to maintain a neutral atom. On the other hand, the nuclear charge changes by $-e$ during positron emission; therefore, in order to maintain a neutral atom, an electron must also be lost by one of the atomic shells. Thus, for negatron decay the total energy difference between initial and final energy states is only the sum of the negatron kinetic energy and the neutrino energy. For positron emission, however, the atom loses a minimum energy equal to twice the rest energy of an electron, $2m_0c^2$. These energy relationships are shown schematically in Fig. 1. During orbital electron capture the nucleus loses a single positive charge merely by taking an electron from one of its own atomic shells. The only energy loss is that energy emitted as X radiation during rearrangement of the atomic shells following electron capture and the energy carried away by the neutrino. See Fig. 1 on p. 2925.

A table of nuclides showing mass number and isotopic abundance is given in the entry on **Chemical Elements**.

Exotic Nuclei and Their Decay. As reported by J.C. Hardy (Chalk River Nuclear Laboratories, Atomic Energy of Canada, Ltd.), recent advances in nuclear accelerators and experimental techniques have led to an increasing ability to synthesize new isotopes. As isotopes are produced with more and more extreme combinations of neutrons and protons in their nuclei, new phenomena are observed, and the versatility of the nucleus is increased as a laboratory for studying fundamental forces. Hardy reports that, among the newly discovered decay modes are: (1) proton radioactivity, (2) triton, two-proton, two-neutron, and three-neutron decays that are beta-delayed, and (3) ^{14}C emission in radioactive decay. Precise tests of the properties of the weak force have also been achieved.

The fundamental usefulness of exotic nuclei and their decay assures a continuing interest in the field. New heavy-ion accelerators will ensure that

TABLE 5. THE NEPTUNIUM SERIES

Element (2)	Symbol	Radiation	Half-life
Curium (96)	^{245}Cm	α	9300 yr
Plutonium (94)	^{241}Pu	β	13.2 yr
Americium (95)	^{241}Am	α	458 yr
Neptunium (93)	^{237}Np	α	2.20×10^6 yr
Protactinium (91)	^{233}Pa	β	27.4 days
Uranium (92)	^{233}U	α	1.62×10^5 yr
Thorium (90)	^{239}Th	α	7340 yr
Radium (88)	^{225}Ra	β	14.8 days
Actinium (89)	^{225}Ac	α	10.0 days
Francium (87)	^{221}Fr	α	4.8 min
Astatine (85)	^{217}At	α	1.8×10^{-2} sec
Bismuth (83)	^{213}Bi	β and α	47 min
Polonium (84)	^{213}Po	α	4.2×10^{-6} sec
Thallium (81)	^{209}Tl	β	2.2 min
Lead (82)	^{209}Pb	β	3.3 hr
Bismuth (83) (end product)	^{209}Bi	None	Stable

96% | 4%

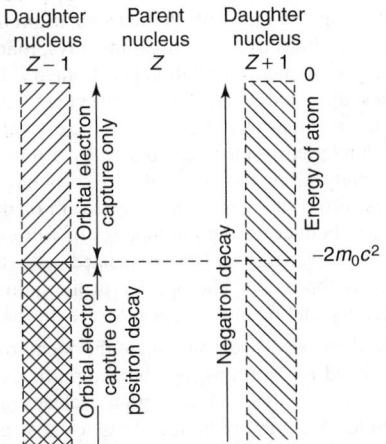

Fig. 1.   The energy regions for which negatron emission, positron emission, and orbital electron capture are energetically possible

this interest is matched by an ever-increasing capability to synthesize new isotopes and provide the nuclear laboratory with renewed flexibility. Hardy further emphasizes that applications uniquely suited to the decay modes of exotic nuclei, are starting to appear and indeed are sophisticated. Many of the new forms provide greater detail than can be obtained with stable nuclei. As an example, beta-delayed proton decay has been used to time the life of excited nuclear states. The technique is sensitive to lifetimes in the range of 10^{-16} second, a span in which it has few competitors, and has been applied to a number of different nuclei.

Useful Applications for Isotopes

Although care must always be exercised to avoid undue exposure to various radioisotopes, these materials have found wide acceptance in analytical chemistry, medicine, radiocarbon and other radioelement dating (geology, archaeology, etc.), and other special situations—for example, as a fuel source in spacecraft.

The radionuclides commercially available and most commonly used for a number of the foregoing applications include: antimony-125; barium-133, 207; bismuth-207; bromine-82; cadmium-109, 115 m; calcium-45; carbon-14; cerium-141; cesium-134, 137; chlorine-36; chromium-51; cobalt-57, 58, 60; copper-64; gadolinium-153; germanium-68; gold-195, 198; hydrogen-3 (tritium); indium-111, 114 m; iodine-125, 129, 131; iron-55, 59; krypton-85; manganese-54; mercury-203; molybdenum-99; nickel-63; phosphorus-32, 33; potassium-42; promethium-147; rubidium-86; ruthenium-103; samarium-151; scandium-46; selenium-75; silver-110 m; sodium-22, strontium-85; sulfur-35; technetium-99; thallium-204; thulium-171; tin-113, 119 m, 121 m; titanium-44; ytterbium-169; and zinc-65.

Radioisotopes in Chemical Analysis

There is a wide range of applications for methods of analysis that are based upon the energies and intensities of the radiations emitted by radioactive nuclides. These techniques sometimes are termed *radiometric methods of analysis*. The methods are not restricted to the determination of substances initially radioactive, since there is wide use of methods involving the irradiation of stable nuclides to produce radioactive ones, followed by measurement of their radiations, from which the composition of the original stable substance can be inferred. This method is *radioactivation analysis*. Another method for the use of measurements of radioactivity in the analysis of stable substances is that of *tracer techniques*, that is, by the addition to them of radioactive nuclides, which can then be used to follow the course of various reactions or processes. There are various ways of introducing the radioactive nuclides, which are discussed later in this entry.

All methods of radiometric analysis involve, of course, the use of various radiation detection devices. The devices available for measuring radioactivity will vary with the types of radiations emitted by the radioisotope and the kinds of radioactive material. Ionization chambers are used for gases; Geiger-Müller and proportional counters for solids; liquid scintillation counters for liquids and solutions; and solid crystal or semi-conductor detector scintillation counters for liquids and solids emitting high-energy radiations. Each device can be adopted to detect and measure radioactive material in another state, e.g., solids can be assayed in an ionization chamber. The radiations interact with the detector to produce a signal.

Since many radionuclides decay with gamma rays, many measurements are being made by gamma-ray scintillation spectrometry. Usually, a crystal detector, such as a sodium iodide crystal, is connected to a spectrometer. As described above, the gamma-rays interact with the crystal to produce light pulses which are converted to electrical pulses by a multiplier phototube. The pulse height analyzer of the spectrometer sorts out the gamma-rays of various energies. From this operation, a spectrum of the radionuclide's gamma-rays can be obtained to the *photopeaks* of full-energy pulses and the continuum of lower-energy pulses associated with the decay of the radionuclide. The photopeak, or photopeaks, in the gamma-ray spectrum can be used to identify and quantitatively measure the radionuclide.

Radioactivation Analysis. The principle of this technique is that a stable isotope when irradiated by neutrons, by charged particles such as protons or deuterons or by gamma rays, can undergo a nuclear reaction to produce a radioactive nuclide. After the radionuclide is formed, and its radiations have been characterized by radiation detection devices, calculations can be made of the elements contained in the sample before irradiation.

An important reaction used quite widely for this purpose is irradiation by neutrons and measurement of the energies of radiations emitted. The source of the neutrons may be a nuclear reactor, a particle accelerator, or an isotopic source, that is, a sealed container in which neutrons are produced by alpha rays emitted by a source such as radium, sodium-24(^{24}Na), yttrium-88(^{88}Y), etc., and arranged so that the alpha rays react-with a substance such as beryllium which in turn emits neutrons. The neutrons react with stable nuclides in the sample to produce radioactive ones. Thus

ordinary sodium undergoes a nuclear reaction with neutrons as follows:

$$\text{Na}^{23} + n \longrightarrow \text{Na}^{24} + \gamma$$

The ^{24}Na decays with a half-life of 15 hours to yield gamma rays and β-particles

$$\text{Na}^{24} \longrightarrow \text{Mg}^{24} + e^- + \gamma$$

Moreover the energies of these β-particles (electrons) are known to be 1.39 MeV and that of the gamma-rays 1.38 MeV so that the measured values of these magnitudes are characteristic of substances containing sodium. (Measurement of the γ-radiation is the usual procedure.) At least 70 of the elements can be activated in this way, by the capture of thermal neutrons, i.e., by *neutron activation analysis*. An activation analysis follows a procedure similar to that shown in Fig. 2. In almost all analyses, the sample materials are not treated before the bombardment, but are placed directly into the bombardment capsule or container. The length of the bombardment interval is usually determined by the half-life of the radionuclide used for the element of interest and the flux of nuclear particles.

The post-bombardment processing of the activated sample may follow either a nondestructive assay of the radioactivity in the sample (gamma-ray scintillation spectrometry is used most often for this) or a chemical processing of the sample prior to the radioactivity assay. Techniques involving either precipitation, electrodeposition, solvent extraction, and ion exchange or some combination of these form the basis of the radio-chemical separation techniques used in activation analysis.

Neutron activation has been successfully applied to a great variety of determinations of small concentrations of elements present in alloys, for example, vanadium and manganese in iron. Other metallurgical applications include the determination of some 70 elements: including the metals, aluminum, antimony, arsenic, barium, bismuth, cadmium, calcium, cerium, cesium, and so on alphabetically down the list. Minerals and soils have also been extensively analyzed by the method. However, it is not restricted to trace quantities or to inorganic substances, one interesting application being the determination of phosphorus, oxygen and nitrogen in organic phosphorus compounds. Sodium has been determined in blood plasma, and numerous other biochemical determinations have been made accurately. See also **Neutron**.

Tracer Analysis. This method is readily performed with radioactive isotopes because their ease of detection by measurement of their

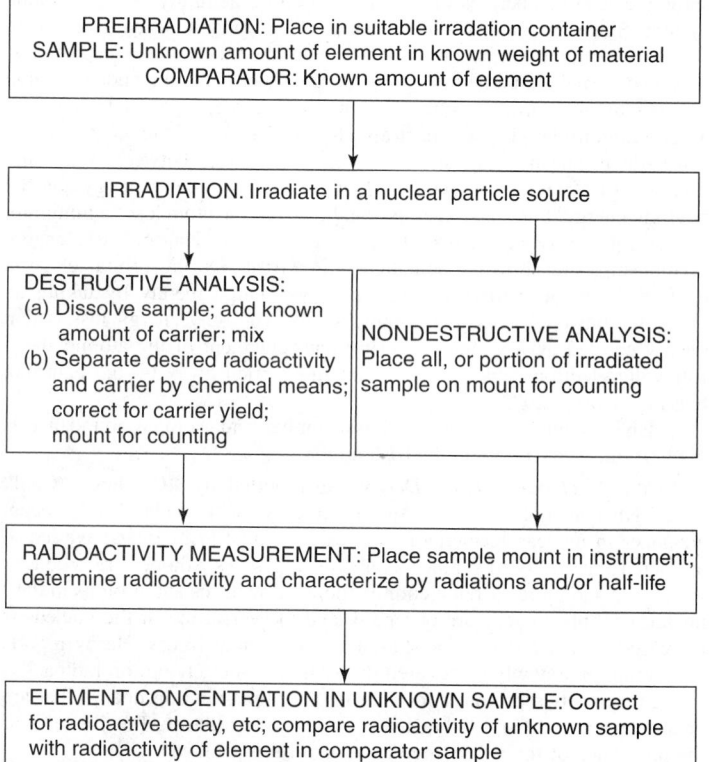

Fig. 2.   Representative program followed in activation analysis

radioactivity makes them effective means of "tagging" their stable isotope counterparts (e.g., ^{24}Na(sodium-24) to tag ordinary sodium). Since compounds are usually involved, rather than elements, one merely synthesizes enough of the radioactive compound to tag the compound under analysis. The tagged compound may be followed through any analytical scheme, industrial system, or biological process. It is essential that a compound be tagged with an atom, however, which is not readily exchangeable with similar atoms in other compounds under normal conditions. For example, tritium could not be used to trace an acid if it were inserted on the carboxyl group where it is readily exchanged by ionization with the solvent.

Radiometric methods employing reagent solutions or solids tagged with a radionuclide have been used to determine the solubility of numerous organic and inorganic precipitates, or as a radioreagent for titrations involving the formation of a precipitate. In this type of application it is necessary to establish the ratio between radioactivity and weight of radionuclide plus carrier present. This may be established by evaporating an aliquot to dryness, weighing the residue, and measuring the radioactivity.

Closely related to tracer analysis is the method of *isotopic dilution analysis*. Here, instead of checking the effectiveness of a method from known amounts of an element in the sample, and of its radioactive isotope, one knows only the amount of radioactive isotope added, and by precipitating or otherwise separating the total amount of that element present, and then measuring its radioactivity, one determines its amount, and hence the amount present in the original sample.

Radioactive tracer methods lend themselves well to research applications in studying entire processes in science and industry, and in the biological as well as the physical sciences.

Industrial Applications for Radioisotopes

Radioisotopes are widely used in the measurement of process variables, including the level of liquids and solids in tanks, silos, and other vessels, the density and specific gravity of fluids and solids, the thickness of sheets and coatings, the moisture content of soils and other solids, the mass flow of materials in pipelines or on belts, and the determination of chemical composition of raw materials, in-process materials, and end-products. Representative examples of these applications are given in Table 6.

Density, level, and thickness measurements all depend upon the determination of the number of radiations per unit time penetrating the sample and producing a measurable signal in the radiation detector. When the amount of matter between the source and the detector increases, there usually is a decrease in the signal. The following relationship demonstrates the exponential nature of the attenuation of beta or gamma radiation:

$$\frac{I}{I_0} = Be^{-\mu\rho t}$$

TABLE 6. REPRESENTATIVE RADIOISOTOPE PROCESS INSTRUMENTATION APPLICATIONS

Variable measured	Representative applications	Radioisotope frequently used
Mass flow in pipes	Slurries and viscous or corrosive materials	^{60}Co, ^{124}Sb
Mass flow on belts	Conveyance of solids, such as coal, woodchips, gravel, etc.	varies with material
Level	Slurries, viscous materials, solids, crushed rock, aircraft and rocket fuels, etc.	^{137}Cs, ^{60}Co, ^{226}Ra and daughter
Density—fluids	Coal slurries, sewage sludges, granular materials, black and green liquor (paper pulp processing), chemicals	^{137}Cs, ^{60}Co, ^{226}Ra and daughter
Density—solids	Compacted soil density for roads, footings, dams, asphaltic concrete	^{137}Cs, ^{60}Co, ^{226}Ra and daughter
Thickness	Thin plastic sheets, films	^{14}C
	Light paper and plastics	^{85}Kr
	Heavy paper, thin metal, rubber	^{90}Sr-^{90}Y
Moisture	Soils and solids	Neutron sources: ^{226}Ra-Be, ^{210}Po-Be, ^{124}Sb-Be, ^{239}Pu-Be, ^{252}Cf, ^{241}Am-Be

where I_0 = initial radiation intensity

I = radiation intensity through absorbing material

B = a buildup factor dependent on the energy and collimation of the source, and on ρ and t. B accounts for radiation which has been "scattered" or changed in direction by interactions that do not stop the radiation.

μ = absorption coefficient, dependent on composition of absorbing material, energy of radiation, and source detector geometry

ρ = specific gravity of absorbing material

t = thickness of absorbing material

Beta radiation has a finite range, whereas in theory x- and gamma rays are exponentially attenuated. It should be noted that x-radiation of each energy exhibits sharp changes in absorption coefficients for certain absorbers.

Where density and specific gravity measurements are made of liquids and slurries, major signal generation changes can result from temperature changes. Thus, compensation for temperature changes should be built into the instrumentation. Major errors also may result from the presence of air bubbles, grease deposits, and pipewall corrosion.

In a typical level-measuring gage, a gamma-emitting radioactive source is mounted in a shielded holder on one side of the vessel and a suitable detector is mounted on the other side. Several radiation receivers, mounted at vertical increments, are required to provide a range of levels. Greater accuracy and flexibility can be obtained with a motor-driven level gage. Here the radiation source and detector cell move up and down in unison as controlled by a motor-driven drum. Potentiometers provide output voltages; one indicates full range in feet; a second potentiometers indicates inches within range. A digital readout also may be provided.

Moisture measurements can be made by using a source of fast neutrons. If such a source (Sb-Be, Pu-Be, ^{252}Cf) is located on or in a medium containing hydrogen, some of the neutrons will be slowed (moderated or thermalized) by collisions with the hydrogen atoms. The number of slow neutrons per unit area per unit time can be determined with a detector selectively sensitive only to slow neutrons. A representative detector of this type is a ^{3}He or BF$_3$-filled proportional or Geiger-Mueller detector. The relative response of the detector will provide a measure or the hydrogen content of the medium surrounding or adjacent to the source. Thus, if water is the only or principal hydrogen-containing variable constituent of the medium, the technique can be used to measure moisture content.

Pressure measurements involve the interaction of alpha radiations with a gas, which results in the formation of positive and negative ions. The latter can be collected and measured as electric current. The number of ions produced in a gas by alpha particles depends upon the density and composition of the gas. Where either of these factors is known the other can be inferred from these measurements. Several vacuum gages employ this principle.

For the measurement of thickness and coverage (coatings) of very thin materials, the absorption of alpha radiation may be used as a measure of weight/unit area. For moderately thick materials, the beta radiation emitted by ^{90}Sr (strontium-90), ^{85}Kr (krypton-85), or ^{14}C (carbon-14) is used. For greater thicknesses, bremsstrahlung or gamma sources are used.

The uniformity of mixing may be determined by mixing radioactive ^{24}Na in sodium chloride as the tagged compound and, after mixing, determining the uniformity of the presence of the isotope in the samplings. The mixing patterns in fluid catalytic cracking units have been ascertained by tagging catalysts with ^{51}Cr (chromium-51), ^{46}Sc (scandium-46), and ^{144}Ce (cerium-144).

Radioisotope techniques also find application in determining diffusion rates in the study of metals, porous bodies, liquids, and gases. Tritiated water is used in studies of the permeability of thin, flexible plastic sheets. The thin sample sheets are placed over a dish containing tritiated water (^{3}H, heavy hydrogen in the water) solidified by gelatin. Then methane is passed over the upper surface of the septum and into a counter. Pulses of the counter resulting from the tritium diffused in the methane are measured. Similar techniques have been used for studying flow patterns of underground water supplies of for tracing cross flow between oil wells. In one application, water tagged with ^{131}I (iodine-131) is used in the study of the subsurface flow of water in secondary oil recovery—to determine the path, velocity, and carrying strata of the water.

Three main techniques are used for applying radioisotopes to the measurement of flow rate: (1) peak timing, (2) dilution, and (3) total count. In peak timing, a gamma emitter (^{60}Co or ^{124}Sb) is injected quickly at a point close to the section of the pipe in which the velocity is to be determined. The time of passage of the peak of the tracer wave is determined by two detectors located at a known distance apart and external to the pipe. The dilution technique is based on the fact that the concentration of mixed tracer in a line resulting from the continuous bleeding of a tracer at a known rate into the line will be inversely proportional to the relative flow rates of the line and the bleeder, assuming that the tracer will uniformly mix with the flowing liquid. In the total-count method, flow rate is based upon measurement of the total counts from a radioactive tracer that has been added to the flowing stream. The total count bears a simple inverse relation to the flow rate.

Radioisotopes also are used for detection of interfaces in pipelines. Since some pipelines handle many different stocks, ranging from crude oil to finished petroleum products to chemicals, etc., effective control requires a knowledge of the precise instant when the interface between two materials passes the control point. This is obtained by adding to the interface a radioactive nuclide, which emits a strong beam of radiation, such as ^{124}Sb. The half-life of the radioisotope used must be quite short to insure that the activity will have decayed to a safe level before it reaches the ultimate user of the transported material.

Metallurgical Applications. Some of the outstanding industrial applications of radioactivity have been in metallurgy. The wear of piston rings in internal combustion engines is important in the selection of alloys for this service. Since the rate of wear is relatively slow, a long period of test would be required to give effects measurable by ordinary methods. However, if the piston ring is made radioactive by irradiation, and the lubricating oil is then tested for the radioactivity of particles of metal abraded from the ring, the more sensitive radiation counters will show measurable and comparable results after relatively short test periods.

Another metallurgical application of radioactivity is in determining the distribution of metals in alloys. A case in point is the molybdenum alloys, where the distribution of the molybdenum between the various solid phases is important to the properties of the alloy. Such studies can be made by polishing a sample of the alloy, etching its surface with properly chosen reagents, and examining its grain structure under the microscope. This process can often be shortened by enriching the molybdenum with one of its radioactive isotopes before adding this metal to the melting furnace. Alloys containing such radioactive elements can be photographed directly through the microscope, whereupon the radioactive areas of the metal show dark on the film.

A related metallurgical application is the use of radioactive nuclides, which emit high energy gamma rays and have a reasonably long half-life, such as ^{60}Co, 113-day ^{182}Ta or 100-day ^{88}Y, to take radiographs of metallic articles. This replaces the x-ray technique for the detection of holes, inclusions and other internal defects in castings and other parts. A similar technique is employed in the beta-ray gauges used in checking the uniformity of the thickness of sheets of material and various articles, such as paper, plastics, steel, textiles, rubber, glass, roofing, flooring, and cigarettes. Here the highly penetrating gamma rays would not be appreciably absorbed by the thin, nonmetallic material; therefore, radioactive nuclides emitting beta rays are used instead. This type of application constitutes an extensive use of radioactive radiations for testing purposes.

Radioisotopes in Medicine

Radiopharmaceuticals are almost ideal diagnostic tools because radioisotope tracers do not alter body physiology, and they permit external monitoring with minimal instrumentation. Presently, there are three major areas of nuclear medicine: (1) physiological function studies, (2) radionuclide imaging procedures, and (3) therapeutic techniques.

Physiologic Function Testing. An example of this application is the assay of thyroid hormone levels in the blood which, in turn, can aid in the assessment of thyroid function. The radioactive iodine uptake test, which involves the administration of a dose of ^{131}I (iodine-131) to the patient, is also a valuable procedure in assessing thyroid function. At present, the technique is best reserved for problem cases rather than used as a primary screening test. The main disadvantage of this test is the effect of the dietary intake of iodine, which reacts in various ways in different individuals.

The gamma camera, with computer-assisted data analysis, is used together with ^{131}I-hippuran to measure renal function. The renogram is of most clinical value in the assessment of ureter impairment in pre-and postoperative patients with carcinoma of the cervix and other pelvic and gynecological tumors.

Radionuclides also are useful in assessment of hematological status to detect anemia and iron deficiency, and in studying radioactivity in feces in order to detect significant blood loss through the gastrointestinal tract. Although considerable development remains, radioisotopes show promise of facilitating differentiation between well-vascularized and ischemic tumors and organs.

Radionuclide Imaging Procedures. Brain tumors can be detected by external counting of radionuclides, a procedure introduced into general clinical use in the late 1950s. A significant advance in brain tumor imaging was the introduction of the gamma camera, which permitted more rapid studies with multiple views, as well as dynamic cerebral blood flow assessment.

The ^{85}Sr (strontium-85) scanning of metastatic bone disease is another important tumor localization technique. Since metastatic lesions of bone are frequently associated with new bone formation, there is usually a significant localization of several radioisotopes in the general vicinity of the metastasis. Early metastatic lesions often go undetected on roentgenographic examination because a 30–50% change in bone density is required to produce visible changes on X-ray examination. Bone scans, however, are generally positive quite early in the development of metastasis. Patients with prostatic carcinoma and carcinoma of the breast are most often candidates for study with this technique.

The liver scan, using radioactive colloids, utilizes a slightly different approach to tumor detection. In this scan, the radioisotope concentrates in the normal tissue, and the tumor appears as a nonradioactive, or "cold" area. This procedure is often an indicated procedure in the cancer patient because of the frequency of liver metastases, and because the liver is not easily visualized using routine radiographic techniques. There are limitations to the approach. Lesions that are smaller than two centimeters in diameter generally go undetected because of limitations of resolution of scanning devices. A number of other disease conditions also may interfere with localization of radiocolloids, producing defects on the liver scan that are indistinguishable from neoplastic disease.

Lung scans also are useful in checking for changes before and after radiation treatment of carcinoma of the lung. Although not widely used at present, a technique for detecting bronchial obstruction has been developed using inhalation of radioactive aerosols. A liver-pancreas scan also can be performed, although interpretation of pancreatic scans is often difficult because of normal variation in size and shape and in trace concentration. When the scan appears to be within normal limit range, however, the presence of disease is unlikely.

Thyroid scans with ^{131}I are useful in determining the activity of thyroid nodules in the intact thyroid gland. A nonradioactive, "cold" nodule indicates a higher risk of thyroid carcinoma, but the scan alone is not recommended as a technique of selecting patients for surgery. After removal of a thyroid carcinoma, a scan of the neck may demonstrate areas of increased activity in the cervical lymph nodes and other organs, indicating metastatic disease.

Scintigram techniques of the kidney can be helpful in distinguishing between cysts and neoplasms, and salivary gland scanning can be useful in confirming abnormality in the salivary gland where tumor is suspected. Lymph node scans with the radiocolloid injected subcutaneously on the dorsum of the feet can be used as screening procedures for lymph flow.

The search continues for a general tumor-scanning agent. Although several radionuclides have been found to localize tumors of widely different types and regions of the body, current interest is in the use of ^{67}Ga-citrate, which is undergoing a wide clinical trial and may prove to be useful in the localization of lymphomas as well as some adenocarcinomas. Medical imaging techniques are discussed in several articles of this encyclopedia. Consult alphabetical index.

Therapeutic Techniques. Probably the most prominent therapeutic use of radiopharmaceuticals is radioactive iodine in the treatment of metastatic thyroid cancer. ^{131}I has a half-life of about 8 days and emits gamma and beta rays. When iodine salts are taken into the body, most of the dose is concentrated in the thyroid gland. A dose of radioactive iodine salt similarly concentrates in the thyroid gland. When there is a cancer in the thyroid gland, or the gland is overactive (hyperthyroidism), the excessive tissue may be destroyed by the radiation from the radioactive iodine that has been administered. Although removal of metastatic thyroid cancer is not always achieved with ^{131}I therapy, significant palliation can occur. In some instances of lung metastasis and lymph node metastasis in the neck, patients may show no evidence of recurrence, even many years after treatment.

Radiophosphorus is used in the treatment of patients with a number of diseases. This element has a half-life of about 14 days and emits beta rays. It is taken up by the body in the greatest quantity by those tissues which manufacture blood cells. In polycythemia vera, a condition in which too many red blood cells are formed, the radiation from this isotope often brings about a sufficient suppression of the blood cell-making tissues to alleviate some of the symptoms of the disease. Leukemia patients, in whom there is an excessive production of white cells, are offered added comfort and, in some instances, prolongation of life by the use of radiophosphorus. This element also may be used in treatment of metastatic cancer to the bone and, although the treatment is not used in an attempt to eradicate cancer, it can result in significant palliation of pain in some patients.

^{198}Au (gold-198), in the form of a suspension in water, has found increasing use in the treatment of certain types of cancer. Isotopes of gallium, sodium, arsenic, and other elements have been tested for possible uses in medicine and show some promise. Other methods for using nonsealed sources include arterial therapy of liver cancer, endolymphatic therapy of lymph node cancer, and intracavitary therapy of pleural and peritoneal cancer. The basic principle back of the internal use of all radioactive isotopes depends upon the concentration of the isotope in some particular tissue. The search for elements that are concentrated in each of the organs by the selective abilities of the tissues, or elements that concentrate in tumor tissue as contrasted with normal tissues, is the key to all techniques in radiodiagnosis.

^{60}Co (cobalt-60), a gamma and beta ray emitter with a half-life of about 5.3 years, is used in cancer therapy. Small pieces of radioactive metallic cobalt, made radioactive in a nuclear reactor, are placed into a proper shielding device and the radiation from them used in place of a high-powered x-ray machine or radium implantations in treating patients with localized cancer. Thousands of ^{60}Co irradiators are in use. Among the most significant advances in radiotherapy since 1925 has been the development of supervoltage equipment and the ^{60}Co megavoltage units. The latter are the most suitable, since they are compact, have high activity source in a small volume, are flexible and adapt to many geometric patterns for therapeutic use, and are easy and economical to maintain. With the ^{60}Co isotope, supervoltage is now made available throughout the world, and at a very low cost when compared with the cost of radium per se, or of x-ray generators. ^{137}Cs (cesium-137) units also are in use, but do not have the same therapeutic usefulness as the ^{60}Co, and the activity cannot be concentrated as easily as with cobalt.

See also **X-Ray Scan and Other Medical Imagery**.

Relative Biological Effectiveness (RBE). The roentgen is a measure of the intensity of ionizing radiation in air. One roentgen corresponds to the creation of 1 esu of charge in 1 ml of standard air. The roentgen can be considered as a measure of energy dissipation in air, and its definition has been extended to cover ionization or energy dissipation in other media. One roentgen corresponds to a radiation field dissipating 83.8 ergs/g of air, which dissipates approximately 93.8 ergs/g of body tissue. Ionization in tissue is a measure of physical damage. Thus, allowable radiation exposures for human tissue are expressed in roentgens. Since the same number of roentgens from various types of radiation produce different amounts of body damage, a term called roentgen equivalent man (rem) is used in stating allowable radiation exposure values. The rem = $R \times$ rbe, where rbe, known as *relative biological effectiveness*, has the values given in Table 7.

Radioactive Dating Techniques

Age determinations using radioactive nuclides may be looked upon as processes that are the inverse of half-life measurements. If a radionuclide of known half life exists within an object, the age of that object can be determined either by measuring the number of radionuclides that remain or the number of product nuclides of the radioactive decay. In these determinations it is assumed that, if we know the half life of the radionuclide, an elapsed time t, or age, for the object can be found by using the formula $t = (\ln N_1/N_2)/\lambda$, where λ is the decay constant of the

TABLE 7. VALUES OF RELATIVE BIOLOGICAL EFFECTIVENESS

Type of radiation	rbe
X- and gamma radiation	1
Beta rays	1
Alpha rays	20
Fast neutrons[a]	10
Thermal or slow neutrons	5

[a]Having energies in the range 0.1 to 10 MeV. Above 10 MeV, the rbe increases rapidly.

radionuclide and N_1 and N_2 are the amounts of the radionuclide present at the beginning and the end of the interval spanning the time t.

In any use of radioactive dating or age determining processes, a basic assumption is, in general, that the concentration of the radioactive element is changed during the life of the sample only by its natural decay process, and that the accuracy of the determination depends primarily, therefore, upon the accuracy with which the half-life of that radionuclide is known.

Ages of specimens may sometimes be determined by other methods that the measurement of radioactivity, as by combination of radioactive measurements with mass spectroscopic determinations.

D.Q. Bowen (The University College of Wales, Aberystwyth) stresses that, in studies of paleoclimatology (global warming periods, etc.), "The last 130,000 years is especially important because it includes geological analogues which may be useful for predicting future changes in climate, and against which the predicted trace-gas induced global warming may be evaluated."[2]

Preliminary climate modeling suggests that natural trends will eventually overcome the predicted global warming, but better dating of past changes is required to refine such models. Two such dating methods are: (1) *thermoluminescence dating* of sediments, and (2) *amino acid geochronology* of fossil mollusks. All physical and biological sciences involved in research into the Quaternary Period[3] of the last 2.4 million years were revolutionized by the reinterpretation of Emoiliani's (1955) classic work on oxygen isotope variability in marine microfossils. For the reader with a scholarly interest in this topic, the Bowen reference listed contains a depth of understanding of radioactive decay dating in this area.

Age of Rocks. In the table of nuclides given under **Chemical Elements**, there are listed a number of naturally occurring radionuclides with long half-lives. From these known half-lives, the geological age of a rock may be calculated. One method of making this estimate is based upon the amount of radionuclide and its daughter nuclide contained in the rock. This method is based upon various assumptions, which may be stated as follows:

1. Since the rock was formed, the parent nuclidic content of the mineral has been changed only by radioactive decay.
2. All the decay products produced by the parent nuclide have been retained since the mineral was formed.
3. The geological separation of the parent and daughter elements at the time of formation of the mineral was sufficient to make the determination of the decay products unambiguous. For example, if a uranium mineral does not exclude all lead at the time it is formed, the isotopic abundance of the lead at the time of formation cannot be calculated with certainty.
4. The radioactive decay scheme of the parent nuclide is well known.

Another method uses the decay of cosmogenic isotopes that are produced in the atmosphere and then incorporated into terrestrial reservoirs. Examples of this approach include standard ^{14}C and ^{10}Be dating.

The contributions of modern chemistry, including the availability of separated isotopes, the extension of the range of mass spectrometers, and the developments of new chemical methods, which make possible the determination of microgram quantities, have extended the range of application of radioactive age measurements. This extension has been either

[2] The most appropriate past analogue for predicated global warming is from the mid-12. Global temperatures at that time simulate the predicted warming at high latitudes, whereas analogues based on the warmest interglaciations of the past 2 to 4 million years only give appropriate warming in middle latitudes.

[3] The Quaternary Period is subdivided into the Pleistocene Epoch and, commencing at 10,000 years ago, the Holocene Epoch. The Holocene is synonymous with the "Postglacial or present interglaciation."

to minerals that contain relatively little of the parent element, but maintain a good separation of the parent and daughter elements when they are formed; or to minerals containing radioactive elements that have a very low natural abundance, such as ^{40}K, or a very long half-life, such as ^{87}Rb. Although these extensions have in turn introduced certain new problems and forced some compromises, they have made possible certain conclusions about geological questions and have opened new avenues for research.

A number of possible radioactive dating methods exist, but each method is practical, of course, only if the appropriate radionuclide exists in the mineral. One series of possible dating methods is based on the decay of natural uranium and natural thorium. If the rock has retained the helium produced by the decay of ^{238}U, for example, 8 helium atoms should exist for each nuclide of ^{238}U that has decayed through its complete chain to ^{206}Pb, since 8 alpha particles result from this chain. From a measure of the ratio of the amount of helium to the amount of ^{238}U in the rock, a calculation may then be possible of the age of the rock. In this method, corrections must be made for the decay of ^{235}U and of ^{232}Th, both of which are the initiating nuclides for a natural chain of radioactive nuclides. Because the half lives of ^{232}Th and ^{238}U are different, another method for determining the age of a rock containing both these nuclides is the measurement of the ratio of the amount of ^{206}Pb to the amount of ^{208}Pb, which are the ultimate decay products of the ^{238}U and ^{232}Th chains, provided neither of these isotopes of lead existed in appreciable quantity prior to formation of the rock. A related measurement is the ratio of radiogenic lead (either ^{206}Pb or ^{208}Pb) to nonradiogenic lead (^{204}Pb), which can be assumed to have been of primordial origin. Another correction that may be necessary, especially if the rock comes from a high altitude, is a determination of the amount of helium that has been produced as a result of spallation reactions caused by very high-energy cosmic radiation. Other radioactive age-dating systems are those of potassium-argon (which consists of the decay of ^{40}K to ^{40}Ar, by electron capture, a process with a half-life of 1.27×10^{10} years) and rubidium-strontium which consists of the decay of ^{87}Rb to ^{87}Sr, by electron emission, a process having a half-life of 4.7×10^{10} years.

One conclusion drawn from radioactive measurements is that the pre-Cambrian history of the earth's crust extends beyond 2,700 million years. The pegmatites that have been found to be this old are located in North America and Australia, and they probably exist on all the continents. The oldest rocks in the United States that have been measured are on the south rim of the Bridger Mountains near the Wind River Canyon in Wyoming. These ancient pegmatites intrude geologic formations of sedimentary and volcanic rocks that, themselves, are the result of even more ancient processes than those in which they were formed. Thus, a period of the order of 3,000 million years or more is available for geologic processes that have formed the crust seen today.

Next, the facility to measure the absolute age of micas in igneous intrusives of pre-Cambrian sediments provides a method of correlating these sediments wherever they occur in much the same fashion that fossil correlation of more recent sedimentary formations is possible. A method that is independent of the lithologic characteristics and the general structure of the sediments will provide a crucial test of the validity of these criteria, which have been all that was available to the geologist. Further, any attempts to look for more subtle evidence of such things as changes in the composition of the atmosphere or origins of life itself must be fitted into a time scale of the pre-Cambrian.

Radiocarbon Dating. This is a method of estimating the age of carbon-containing materials by measuring the radioactivity of the carbon in them. The validity of this method rests upon certain observations and assumptions, of which the following statement is a brief summary. The cosmic rays entering the atmosphere undergo various transformations, one of which results in the formation of neutrons, which in turn, induce nuclear reactions in the nuclei of individual atoms of the atmosphere. The dominant reaction is

$$n + {}^{14}N \longrightarrow {}^{14}C + p$$

in which the neutrons react with the nuclei of nitrogen atoms of mass number 14 (which make up the nitrogen molecules that constitute nearly $\frac{4}{5}$ of the atmosphere) to form carbon atoms of mass number 14 and protons (p). The ^{14}C atoms are radioactive, having a half life of about 5730 years. The largest rate of formation of ^{14}C atoms from cosmic rays is at 30,000–50,000 feet (9,144–15,240 meters) above sea level and at higher geomagnetic latitudes, although formation occurs at varying rates throughout the entire atmosphere. The ^{14}C atoms react with oxygen in the

atmosphere to form carbon dioxide, which is mixed with the nonradioactive carbon dioxide in the atmosphere, and with it gains worldwide distribution by various processes. The radioactive $^{14}CO_2$ enters the carbon cycle in which plants take up carbon dioxide from the atmosphere to form carbohydrates, which enter through plant foods into the composition of animals. In another world-wide process, also of exchange nature, carbon dioxide is dissolved in seawater and then, under changing conditions of acidity and temperature, is partially evolved from the seawater again. As a result of these and other processes, the ^{14}C formed in the atmosphere by cosmic rays tends to become distributed throughout all the nonradioactive carbon, not only in the atmosphere, but in the biosphere, the hydrosphere, and even the upper levels of the lithosphere (there are many carbonate-containing minerals).

Obviously this wide distribution of the ^{14}C formed in the atmosphere takes time; it is believed to require a period of 500–1000 years. This time is not, however, a deterrent to radiocarbon dating because of two factors; the long half-life of ^{14}C and the relatively constant rate of cosmic-ray formation of ^{14}C in the earth's atmosphere over the most recent several thousands of years. These considerations lead to the conclusion that the proportion of ^{14}C in the carbon reservoir of the earth is constant, and that the addition by cosmic ray production is in balance with the loss by radioactive decay. If this conclusion is warranted, then the carbon dioxide on earth many centuries ago had the same content of radioactive carbon as the carbon dioxide on earth today. Thus, radioactive carbon in the wood of a tree growing centuries ago had the same content as that in carbon on earth today. Therefore, if we wish to determine how long ago a tree was cut down to build an ancient fire, all we need to do is to determine the relative ^{14}C content of the carbon in the charcoal remaining, using the value we have determined for the half life of ^{14}C. If the carbon from the charcoal in an ancient cave has only $\frac{1}{2}$ as much ^{14}C radioactivity as does carbon on earth today, then we can conclude that the tree which furnished the firewood grew 5730 ± 30 years ago.

As pointed out by Muller (1978), there are well-documented differences between the ages of materials determined by dating with radioisotopes and the ages determined by other means, such as tree-ring counting. In addition to systematic effects, there are statistical errors due to the limited number of atoms observed. Both types of errors can be considered to be fluctuations in n, the number of atoms observed. A relationship can be derived between the magnitude of these fluctuations and the resulting error in the estimation of age of the sample:

$$n = ke^{-t/\tau} \text{ or } t = \tau \ln(k/n)$$

where τ is the mean life of the isotope, t is the age of the sample, and k is the initial number of radioactive atoms in the sample multiplied by the efficiency for detecting them. If n has errors associated with it of $+\delta n_1$ and $-\delta n_2$, then the corresponding values of t will be:

$$t = \tau \ln \frac{k}{n_{\delta n_2}^{\delta n_1}} = \tau \ln(k/n) \frac{+}{-} \left| \begin{array}{c} \ln(1 - \delta n_2/n) \\ \ln(1 + \delta n_1/n) \end{array} \right|$$

Muller has shown that for $n = 1$, inverse Poisson statistics gives $n_1 = 1.36$ and $n_2 = 0.62$ and thus the foregoing equation becomes:

$$t = \tau \ln(k)^{+0.96\tau}_{-0.86\tau}$$

Further details of this method can be found in aforementioned reference.

Determination of the ratio of two oxygen isotopes has been effective in fixing the age of fossil sediments and can provide information about ice formation and, possibly, water temperatures. The lighter isotope ^{16}O evaporates preferentially and thus precipitation and hence ice in glaciers and polar caps should be enriched with ^{16}O relative to seawater. Thus, fluctuations in the amount of water locked up as ice can be determined from variations in the oxygen isotope ratio of fossils locked up in deep-sea sediments. And, because this ratio also varies with water temperature, thermal information also can be gleaned. Kennett (University of Rhode Island) has employed this technique in determining when significant amounts of ice first formed at the poles. This research has indicated that the Antarctic ice cap formed only about 16 million years ago, after Australia had split off and moved away from Antarctica, leaving the latter continent isolated at the pole and surrounded by the fast-moving circumpolar current.

In 1986, J.I. Hedges and colleagues (see reference) reported on how dissolved and particulate organic material transported by rivers can provide a continuous record of physical and biological processes at work within the drainage basin. Rivers also contribute a potentially important quantity of organic matter to the ocean, where the dissolved component may exhibit an appreciable residence time. Although the magnitude of the global river contribution is known, the dynamics of organic materials within terrestrial ecosystems and their effects on the composition of the corresponding marine reservoirs are poorly understood. This is particularly true for rivers draining topical rain forests, which account for about 40% of the total riverine organic carbon discharged into the ocean. Hedges and a team studied the Amazon River System using organic carbon-14 dating methods. They found that coarse and fine suspended particulate organic materials and dissolved humic and fulvic acids transported by the Amazon River all contain bomb-produced carbon-14, indicating relatively rapid turnover of the parent carbon pools. However, the carbon-14 contents of these coexisting carbon forms are measurably different and may reflect varying degrees of retention of soils in the drainage basin.

Dating by Accelerator Mass Spectrometry. The cyclotron is mainly used as a source of energetic particles. The cyclotron also can be used as a very sensitive mass spectrometer. Alvarez and Cornog (1939) were the first researchers to use a cyclotron in this manner. This was in connection with their discovery of the true nuclear properties of ^{3}He and tritium. Within the last few years, Muller and associates at the Lawrence Berkeley Laboratory have used this method in a search for integrally charged quarks in terrestrial material. For radioisotope dating, the cyclotron is tuned to accelerate the isotope of interest and the sample is introduced into the ion source, preferably as a gas. For radioisotope dating, the greatest gains over radioactive counting techniques apply to the longer-lived species, which have lower decay rates. It has been estimated that the cyclotron can be used to detect atoms or simple molecules that are present at the 10^{-16} level or greater. For ^{14}C dating, the Berkeley investigators indicate that one should be able to go back 40,000 to 100,000 years with 1-to-100 microgram carbon samples; for ^{10}Be dating, 10–30 million years with from 1 cubic millimeter to 10 cubic centimeter rock samples; and for tritium dating, 160 years with a 1-liter water sample. Over 50 cyclotrons are in operation today that could perform radioisotope dating and, although the instruments are costly, the cost for a dating determination experiment may not be much higher than for decay dating technology.

Other isotopes with which an accelerator mass spectrometer may be effective include ^{26}Al, ^{36}Cl, ^{53}Mn, ^{81}K, and ^{129}I. Chlorine-36 has a half-life of 300,000 years and may be used for dating water in underground reservoirs. ^{10}Be is produced in the atmosphere at the rate of 1.5×10^{-2} atom cm^{-2} sec^{-1} by cosmic rays that break up oxygen and nitrogen nuclei. ^{10}Be has been used in studies of both seafloor spreading and manganese nodule formation. Although tritium (^{3}H) has a short mean life of 17.8 years, tritium dating has been important in cosmic-ray physics, hydrology, meteorology, and oceanography. For example, if one desires to know how long an underground water reservoir may require for refilling, the age of the water can be determined by tritium dating methods.

In 1986, researchers at the Research Laboratory for Archaeology and the History of Art, University of Oxford, reported on how the radioactive carbon-14 isotope can be separated from other atoms in a sample by use of accelerator mass spectrometry, thus making it possible to derive more accurate chronologies from much smaller archaeological or anthropological specimens. For details, consult Hedges/Gowlett reference listed.

In 1992, K.R. Ludwig and colleagues of the U.S. Geological Survey reported on their use of mass-spectrometric ^{230}Th-^{234}U-^{238}U dating of the Devil's Hole calcite vein (Derbyshire, England), which contains a long-term climatic record, but requires accurate chronological control for its interpretation. Mass-spectrometric U-series ages for samples from core DH = 11 yielded ^{230}Th ages, with precisions ranging from less than 1000 years to less than 50,000 years for the oldest samples. The ^{234}U/^{238}U ages could be determined to a precision of approximately 20,000 years for all ages. Tentatively, the researchers have concluded, "Overall, the U-series ages form a remarkably self-consistent suite of age determinations. Because this consistency is both internal (from replicate samples) and external (from the stability of the overall age-distance trend), it seems highly unlikely that the dates have been significantly corrupted by open-system processes, such as uranium gain or loss or alpha-recoil phenomena. The apparent ideality of the U-Th system in the vein material is probably the result of continuous submergence in water that showed limited secular variations of its physical and chemical properties."

Non-Radioactivity Dating Techniques

Several methods in addition to those involving radioactivity have been used to estimate the ages of various materials and objects.

Obsidian Hydration Rate. Obsidian (rhyolitic volcanic glass) can be used as a key to age determinations for both archeological and geological purposes. As pointed out by Friedman and Long (1976), the method depends upon the fact that obsidian absorbs water from the atmosphere to form a hydrated layer, which thickens with time as the water slowly diffuses into the glass. The hydrated layer can be observed and measured under a microscope on thin sections cut normal to the surface. To convert the measured hydration thickness to an age, the equation relating thickness to time must be known. This requires not only the form of the equation (functional dependence), but also the constants in it. Prior to the early 1960s, age could be related to hydration thickness only if combined with known history of a region or through the use of carbon-14 techniques. In the mid-1960s, Friedman and associates conducted actual experimental hydration experiments on obsidian, exposing the materials (taken from the Valles Mountains in New Mexico) to a temperature of 100°C and steam at a pressure of 1 atmosphere over a 4-year period. An equation of the form, $T = kt^{1/2}$ was developed, where T = thickness of hydration layer, t = time, and k is a constant. Investigators have developed a procedure for calculating hydration rate of a sample from its silica content, refractive index, or chemical index and a knowledge of the effective temperature at which the hydration occurred. The effective hydration temperature (EHT) can either be measured or approximated from weather records. The investigators concluded that if the EHT can be determined and measured for the hydration of a particular obsidian, it should be possible to carry out absolute dating to ±10% of the true age over periods as short as several years and as long as millions of years.

Manufactured Glass Objects. Other investigators (Lanford, 1977) have extended the principles applying to obsidian to manufactured glass, which extends back for thousands of years and thus can be useful to archeologists. However, as observed by Lanford, one cannot use the same optical method for measuring the thickness of hydration layers as used with obsidian. The hydration of the two materials differs. Also, glass that is less than a few hundred years old would generally have hydration layers thinner than the wavelength of visible light. The optical method is destructive in that it requires removal of a slice of glass from an object, something much discouraged by art historians and dealers. The Lanford method involves a resonant nuclear reaction between ^{15}N and ^{1}H for measuring the distribution of hydrogen in solids. With this technique, complete depth profiles of the surface hydration layer can be obtained in a fully nondestructive manner. Lanford summarizes by observing that this method of hydration dating need not be limited to glass. Since most silicates are unstable against slow reactions with atmospheric water, many may develop surface hydration layers suitable for dating and authenticating. The glazes on pottery are chemically similar to glass, and it may be possible that a dating method for glazed pottery based upon these procedures can be developed.

Amino Acid Racemization. This dating method is based upon the incorporation of L-amino acids, exclusively, into proteins by living organisms. As pointed out by the researchers Masters and Zimmerman (1978), given sufficient periods of time over which proteins are preserved after synthesis, a number of spontaneous chemical reactions take place. Among these is racemization, which converts L-amino acids into their enantiomers, the D-amino acids. The different amino acids racemize at various rates, and these rates (as with all chemical reactions) are proportional to temperature. One of the fastest racemization rates known is that of aspartic acid, with a half-life of 15,000 years at 20°C. It follows, then, that the older a fossilized material may be, the higher will be its D-aspartic acid content or D/L Asp ratio. Once the k_{asp} is known for a given fossil locality, the age of a specimen can be calculated from the D/L ratio.

This method was used in the examination of an Eskimo who died 1600 years ago. The body was discovered in a frozen state on St. Lawrence Island, Alaska in 1972 and remained frozen until it was brought to Fairbanks in 1973. Examination of the female individual revealed that she had a skull fracture, probably resulting from instant burial caused by a landslide. Aspartic acid racemization analysis of a tooth from the mummy yielded an age at death of 53 ± 5 years, which correlated well with earlier estimates based upon morphological features. This method is an example of the need to preserve mummies (Alaskan, Egyptian, and Peruvian, among others) for application of new dating techniques as they develop.

The racemization dating technique fell out of favor with many paleoanthropologists in the 1980s, but staged a comeback of acceptance in the early 1990s. This resurgence was the result of analyzing samples of African ostrich egg fragments collected from the Border Cave located on the east coast of South Africa. These or close-by sites also contain reputedly human bones. If the age of these bones is determined to be in the range of 100,000 years, this would provide an additional bit of independent evidence for the theory that modern humans came from Africa. In the Border Cave determination, researchers found that amino acid racemization (AAR) time dating compared relatively closely with electron-spin resonance dating carried out at Cambridge University. AAR techniques revealed an age of 80,000 to 100,000 years for the egg-shells, whereas electron-spin resonance dating of human bones shows 60,000 years. See also **Racemization**.

Geochemical Methods. These methods usually involve a combination of chemical analysis of materials coupled with the curiosity of a detective. This is not a singular methodology, but incorporates numerous disciplines and a lot of past experience with the materials in question. The authentication of ancient marble sculptures prior to their procurement by the J. Paul Getty Museum is an example. In this case, an expert geochemist with past experience with dolomite marble studied a phenomenon known as *dedolomitization,* wherein after many centuries the exposed surface of the marble changes into calcite, or calcium carbonate. Thus, this provides a key to age as well as to source of the original material. See Margolis reference listed.

After using numerous dating techniques over a span of several years, by the late 1980s, the much publicized "Shroud of Turin" was finally explained to the satisfaction of most investigators. The answer was put forth by W.C. McCrone, a microscopist who specializes in authenticating art objects. McCrone examined fibers and other materials lifted from the surface of the cloth with adhesive tape. He determined that light-colored portions of the figure were comprised of a gelatin-based medium speckled with particles of red ocher and that fibers from the dark areas (representing blood) contained stains not of blood, but rather were made up of particles of vermilion. The vermilion was found to be of the type developed for artists in the Middle Ages. Also, the practice of painting linen with gelatin-based temperas was common during the late 13th and the 14th century. Final conclusion: The "shroud" had been forged by some unknown 14th century artist.

The use of thermoluminescence (i.e., heating small particles to a high temperature and then analyzing the emanations spectroscopically) can be a key to an object's age. In a paper by P. Lang presented at the 1988 Pittsburgh Conference and Exposition on Analytical Chemistry and Applied Spectroscopy, modern methods involving Fourier transform spectrometry or Raman spectrometry are described. As early as 1818, Sir Humphrey Davey presented a paper along these lines to the Royal Society on the colors used by the ancients in painting.

Tree-ring counting as a measure of age is described under Pine Trees. However, it is of note to mention an August 1990 report on how tree rings were used to reveal the age of the "oldest road," which was found in a peat bog in 1970. The road is located in southwestern England. It dates back 4000 years.

Additional Reading

Abelson, P.H.: "Isotopes in Earth Science," *Science,* 1357 (December 9, 1988).

Alpen, E.L.: *Radiation Biophysics,* 2nd Edition, Morgan Kaufmann Publishers, Orlando, FL, 1997.

Appenzeller, T.: "Roving Stones: A Landmass Was Wandering Over Three Billion Years Ago," *Sci. Amer.,* **19** (February 1990).

Avignone, F.T., III and R.I. Brodzinski: *A Review of Recent Developments in Double-Beta Decay,* 21 (A. Faessler, Editor) Pergamon Press, 1988.

Badash, L.: "The Age-of-the-Earth Debate," *Sci. Amer.,* 90 (August 1989).

Barnes, D.M.: "Probing the Authenticity of Antiquities with High-Tech Attacks on a Microscale," *Science,* 1374 (March 18, 1988).

Beardsley, T.: "Fallout: New Radiation Risk Estimates Prompt Calls for Tighter Controls," *Sci. Amer.,* **35** (March 1990).

Bowen, D.Q.: "The Last 130,000 Years," *Review (University of Wales),* **39** (Spring 1989).

Bower, B.: "Eggshells Help Date Ancient Human Sites," *Science News,* 215 (April 7, 1990).

Brooks, A.S. et al.: "Dating Pleistocene Archeological Sites by Protein Diagenesis in Ostrich Eggshell," *Science,* 60 (April 6, 1990).

Chesley, J.T., A.N. Halliday, and R.C. Scrivener: "Samarium-Neodymium Direct Dating of Fluorite Mineralization," *Science,* **949** (May 17, 1991).

Cobb, C.E., Jr.: "Living with Radiation," *National Geographic,* **403** (April 1989).

Curie, M.: *Radioactive Substances,* Dover Publications, Inc., Mineola, NY, 2002.

Eisenbud, M. and T.F. Gesell: *Environmental Radioactivity from Natural, Industrial, and Military Sources,* 4th Edition, Morgan Kaufmann Publishers, Orlando, FL, 1997.

Elliott, S.R., A.A. Hahn, and K.M. Moe: "Direct Evidence for Two-Neutrino Double-Beta-Decay in $_{82}$Se," *Physical Review Letters,* **59**, 18, pp. 2020–2023 (November 2, 1987).

Friedman, I. and W. Long: "Hydration Rate of Obsidian," *Science,* **191**, 347–352 (1976).

Gaisser, T.K.: "Gamma Rays and Neutrinos as Clues to the Origin of High Energy Cosmic Rays," *Science,* 1049 (March 2, 1990).

Greiner, W. and A. Sandulescu: "New Radioactivities," *Sci. Amer.,* 58 (March 1990).

Hamilton, D.P.: "U.S. Faces Uncertain Medical Isotope Supply," *Science,* 603 (July 31, 1992).

Hardy, J.C.: "Exotic Nuclei and Their Decay," *Science,* **227**, 993–999 (1985).

Hedges, J.I., et al.: "Organic Carbon-14 in the Amazon River Systems," *Science,* **231**, 1129–1131 (1986).

Hedges, R.E.M. and J.A.J. Gowlett: "Radiocarbon Dating by Accelerator Mass Spectrometry," *Sci. Amer.,* 100–107 (January 1986).

Horgan, J.: "The Shroud of Turin," *Sci. Amer.,* 18 (November 1988).

Huntley, B. and I.C. Prentice: "July Temperatures in Europe from Pollen Data 6000 Years Before Present," *Science,* **687** (August 5, 1988).

Knoll, G.F.: *Radiation Detection and Measurement,* 3rd Edition, John Wiley & Sons, Inc., New York, NY, 1999.

Lanford, W.A.: "Glass Hydration: A Method of Dating Glass Objects," *Science,* **196**, 975–976 (1977).

L'Annunziata, M.: *Handbook of Radioactivity Analysis,* 2nd Edition, Academic Press, Inc., San Diego, CA, 2003.

Lowenthal, G. and P. Alrey: *Practical Applications of Radioactivity and Nuclear Radiations,* Cambridge University Press, New York, NY, 2001.

Ludwig, K.R., et al.: "Mass-Spectrometric ^{230}Th- ^{234}U-^{238}U Dating of the Devil's Hole Calcite Vein," *Science,* **284** (October 9, 1992).

Margolis, S.V.: "Authenticating Ancient Marble Sculpture," *Sci. Amer.,* **104** (June 1989).

Marshall, E.: "Racemization Dating: Great Expectations," *Science,* **790** (February 16, 1990).

Masters, P.M. and M.R. Zimmerman: "Age Determination of an Alaskan Mummy: Morphological and Biochemical Correlation," *Science,* 201, 811–812 (1978).

Moe, M.K. and S.P. Rosen: "Double-Beta Decay," *Sci. Amer.,* **48** (November 1989).

Monastersky, R.: "Coral Corrects Carbon Dating Problems," *Science News,* **356** (June 9, 1990).

Muller, R.A., Stephenson, E.J., and T.S. Mast: "Radioisotope Dating with an Accelerator: A Blind Measurement," *Science,* **201**, 347–348 (1978).

Pasachoff, N.: *Marie Curie and the Science of Radioactivity,* Oxford University Press, Inc., 1997.

Pszczola, D.W.: "Food Irradiation," *Food Technology,* **92** (June 1990).

Rusting, R.: "A Clock in the Trees: Tree Rings Reveal the Precise Age of the Oldest Road," *Sci. Amer.,* **30** (April 1990).

Smith, F.A.A.: *A Primer in Applied Radiation Physics,* World Scientific Publishing Company, Inc., River Edge, NJ, 1999.

Staff: "Rocks of Ages," *Technology Review (MIT),* **80** (April 1990).

Strauss, S.: "Archeology's Dating Game," *Technology Review (MIT),* **8** (October 1987).

Swisher, C.C., III and D.R. Prothero: "Single-Crystal ^{40}Ar/^{39}Ar Dating of the Eocene-Oligocene Transition in North America," *Science,* **760** (August 17, 1990).

Theodorsson, P.: *Measurement of Weak Radioactivity,* World Scientific Publishing Company, Inc., River Edge, NJ, 1996.

RADIOACTIVITY (Mineral). See **Mineralogy**.

RADIOCHEMISTRY. The subdivision of chemistry that deals with the properties and uses of radioactive materials in industry; biology, and medicine, including tracer research and radioactive waste disposal.

RADIOISOTOPES. See **Radioactivity**.

RADIUM. [CAS: 7440-14-4]. Chemical element symbol Ra, at. no. 88, at. wt. 226.025, periodic table group 2 (alkaline earths), mp 700°C, bp 1,140°C, density 5 g/cm³ (20°C). Radium metal is white, rapidly oxidized in air, decomposes H_2O, and evolves heat continuously at the rate of approximately 0.132 calorie per hour per mg when the decomposition products are retained, and the temperature of radium salts remains about 1.5°C above the surrounding environment. Radium is formed by radioactive transformation of uranium, about 3 million parts of uranium being accompanied in nature by 1 part radium. Radium spontaneously generates radon gas at approximately the rate of 100 mm³ per day per gram of radium, at standard conditions. Radium usually is handled as the chloride or bromide, either as solid or in solution. The radioactivity of the material decreases at a rate of about 1% each 25 years. All isotopes of radium are radioactive. See also **Radioactivity**. The first ionization potential of radium is 5.227 eV; second, 10.099 eV. Other important physical properties of radium are given under **Chemical Elements**.

One year after the discovery of X-rays by Röntgen (1895), Henri Becquerel investigated the relationship between the phosphorescence of various salts after their exposure to sunlight and the fluorescence in an operating x-ray tube. One of the salts under investigation was potassium-uranium sulfate. After exposure to sunlight, Becquerel noted that the salt not only emitted visible light, but also rays similar to x-rays that were able to penetrate the heavy black paper and thin metal foils within which his photographic plates were wrapped. During a period of cloudy weather, Becquerel stored the salt and photographic plates in a closet, awaiting further sunny days. Later, when he inspected the package, he noted that a very intense image had been developed on the photographic plate even though it had not received much prior exposure to sunlight. By further experiments, Becquerel confirmed that the intense image was derived directly from the presence of the salt, regardless of any exposure to sunlight. This constituted the first demonstration of radioactivity. Through further investigations, Ernest Rutherford demonstrated that both alpha and beta radiations were emitted by the salt. Rutherford learned that the alpha rays were easily absorbed by thin sheets of paper, whereas the beta rays acted in the same manner as observed by Becquerel. Later Mme. Marie Curie found that thorium produced about the same intensity of radioactivity as uranium. Further tests disclosed that the uranium ore with which she was working (pitchblende) exhibited more radioactivity than could be accounted for by its uranium content alone. Subsequently, Mme. Curie and her husband, Pierre Curie, successfully separated two previously unknown elements, radium and polonium. Thus, both radium and polonium were identified as chemical elements in 1898. It was found that each of these elements was over a million times more radioactive than uranium.

Radium gained prominence not only from its scientific interest, heralding a whole new area of physics and chemistry, but from its wide use in therapeutic medicine, as an ingredient (very dangerous) of luminous paints, and in various instruments for inspecting structures, such as metal castings. Commercially, radium generally is marketed as the bromide or sulfate and is extremely radioactive in these forms. Use of radium in medical technology has largely been replaced by other sources of radioactivity.

Radium occurs in pitchblende, and in carnotite along with uranium. Radium was first obtained from the uranium residues of pitchblende of Joachimsthal, the Czech Republic and Slovakia, later from carnotite of southwestern Colorado and eastern Utah. Richer ores have been found in Republic of Congo and in the Great Bear region of northwestern Canada.

In an interesting narrative, Landa (1979) describes and depicts what the author calls the "first nuclear industry," one that flourished during the first third of the 20th century. At that time, the prized element was *not* uranium, but rather it was radium. Radium initiated the concept of radiotherapy for the treatment of cancerous tumors. Even at the peak of radium production no more than a few hundred grams per year were purified. At one time the price approached $180,000 per gram (in early 1900s dollars). In one large extraction plant located in Denver, Colorado (National Radium Institute), carnotite ore was processed by a direct-dissolution method. During the three years (1914–1916) this plant was in operation, only 8.5 grams of Ra had been purified at an average cost of about $38,000 per gram. Pitchblende deposits from the Haut Katanga district of the Belgian Congo were discovered, but because of World War I, were not commercially exploited until 1921. Whereas it previously had required between 300 and 400 tons of a typical American carnotite ore to produce one gram of Ra, less than ten tons of the Kantanga ore were required. In 1931, an extraction plant at Port Hope, Ontario, designed to process uranium found in outcrops along the shores of Great Bear Lake in northwestern Canada, was commissioned. The hazards of radium production and handling were slow to surface and were not seriously recognized until the 1920s after a report had been prepared which implicated the ingestion of Ra in jaw necrosis in radium dial painters and lung cancer in uranium miners. Several workers, possibly including Marie Curie, died of conditions that were probably the result of long-term radiation exposure. By the 1930s, numerous precautions were initiated, including the design of tunnels at the Great Bear Lake mines to minimize radon hazards to miners.

The radium isotope of mass number 226 occurs in the uranium ($2n + 2$) alpha-decay series. Its half-life is 1,620 years, and it yields radon-222 by α-disintegration. Other naturally occurring isotopes of radium are ^{228}Ra in

the thorium series, half-life 6.7 years, producing actinium-228 by β-decay, which yields by β-decay thorium-228, which in turn yields ^{224}Ra, half-life 3.64 days, giving radon-220 by α-decay. Another naturally, occurring isotope of radium is found in the actinium series; it is ^{223}Ra, half-life 11.7 days, giving radon-219 by α-decay. In the neptunium series there is ^{225}Ra, half-life 14.8 days, undergoing β-decay to actinium-225. Other isotopes of radium include those of mass numbers 219, 221, 225, 227, 229, and 230.

Chemically related to barium, radium is recovered from its ores by addition of barium salt, followed by treatment as for recovery of barium, usually as the sulfate. The sulfates of barium and of radium are insoluble in most chemicals, so they are transformed into carbonate or sulfide, both of which are readily soluble in HCl. Separation from barium is accomplished by fractional crystallization of the chlorides (or bromides, or hydroxides). Dry, concentrated radium salts are preserved in sealed glass tubes, which are periodically opened by experienced workers to relieve the pressure. The glass tubes are kept in lead shields.

In many of its chemical properties, radium is like the elements magnesium, calcium, strontium and barium, and it is placed in group 2, as is consistent with its $6s^26p^67s^2$ electron configuration. Its sulfate ($K_{sp} = 4.2 \times 10^{-15}$) is even more insoluble in water than barium sulfate, with which it is conveniently coprecipitated. Like barium and other alkaline earth metals, it forms a soluble chloride ($K_{sp} = 0.4$) and bromide, which can also be obtained as dihydrates. Radium also resembles the other group 2 elements in forming an insoluble carbonate and a very slightly soluble iodate ($K_{sp} = 8.8 \times 10^{-10}$).

Additional Reading

Greenwood, N.N. and A. Earnshaw: *Chemistry of the Elements,* 2nd Edition, Butterworth-Heinemann, Inc., Woburn, MA, 1997.

Hawley, G.G. and R.J. Lewis: *Hawley's Condensed Chemical Dictionary,* 13th Edition, John Wiley & Sons, Inc., New York, NY, 1999.

Krebs, R.E.: *The History and Use of Our Earth's Chemical Elements: A Reference Guide,* Greenwood Publishing Group, Inc., Westport, CT, 1998.

Landa, E.R.: "The First Nuclear Industry," *Sci. Amer.,* **180** (November 1982).

Lewis, R.J. and N.I. Sax: *Sax's Dangerous Properties of Industrial Materials,* 10th Edition, John Wiley & Sons, Inc., New York, NY, 1999.

Lide, D.R.: *CRC Handbook of Chemistry and Physics,* 84th Edition, CRC Press, LLC., Boca Raton, FL, 2003.

Williams, P.L. and J.L. Burson: *Industrial Toxicology: Safety and Health Applications in the Work Place,* John Wiley & Sons, Inc., New York, NY, 1997.

RADON. [CAS: 10043-92-2]. Chemical element symbol Rn, at. no. 86, at. wt. 222 (mass number of the most stable isotope), periodic table group 18 (inert gases), mp $-71°C$, bp $-61.8°C$. First ionization potential, 10.745 eV. Density 9.72 g/l($0°C$, 760 torr), $7.5\times$ more dense than air. The gas has been liquefied at $-65°C$ and solidified at $-110°C$. Radon was first isolated by Ramsay and Gray in 1908. Prior to acceptance of the present designation, radon was called niton or radium emanation. See also **Radioactivity**.

^{222}Rn is formed by the alpha disintegration of ^{226}Ra. Actinon, its isotope of mass number 219, is produced by alpha disintegration of ^{223}Ra (AcX) and is a member of the Actinium Series. Similarly, thoron, its isotope of mass number 220, is a member of the thorium series. Since the name "radon" may be considered to be specific for the isotope of mass number 222 (from the radium series), the term "emanation" is sometimes used for element number 86 in general. Other isotopes of radon include those of mass numbers 209–218 and 221.

A fluorine compound of radon has been formed by reaction of the elements under higher temperature and pressure, similar to the conditions for forming xenon fluorides. Radon forms a hydrate at atmospheric pressure at $0°C$. It forms a compound with phenol, $Rn \cdot 2C_6H_5OH$ that is stable enough to give a sharply defined melting point at $50°C$. At low temperatures and pressures, HCl, hydrobromic acid, H_2S, SO_2, and CO_2 all add considerable percentages of radon; the HCl product, although possibly not a compound in the classical sense, being stable enough for its use as a method for separating radon from other gases.

Ionizing Radiation from a Natural Source

Since the mid-1980s, there has been a growing concern among some pollution experts with the topic of *indoor air pollution*, and as a major part of this problem, the presence in homes and other structures of ionizing radiation in the form of radon gas. It is beginning to appear that radon's earlier designation as radium emanation was not inappropriate. Most health

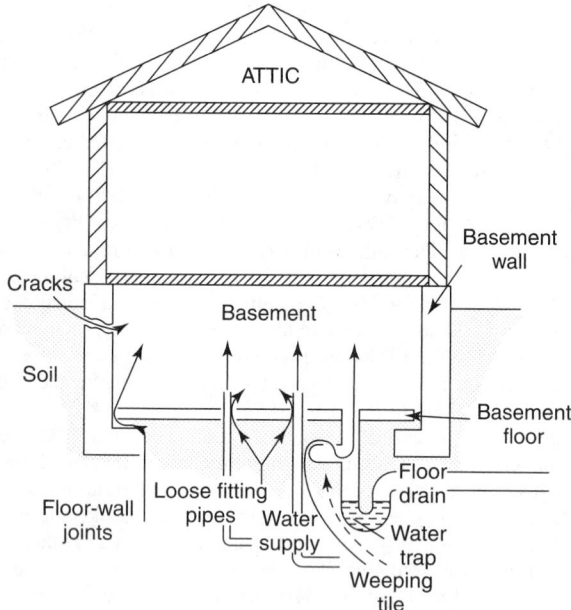

Fig. 1. Ports of entry for radon gas into average residence. (*National Indoor Environmental Institute, Plymouth Meeting, Pennsylvania*)

hazards faced by the citizens of industrialized nations are anthropogenic. Not so in the case of radon. Numerous experts have concluded that the largest dose of ionizing radiation an average person may receive during a lifetime is the radioactive radon gas that emanates from rock formations as the uranium they contain decays. Some epidemiologists and environmental scientists believe that the gas can cause lung cancer in people who have done nothing more hazardous than to live in houses built over such rock formations. Others do not agree and caution has been exercised to avoid the tremendous costs that could be involved in taking remedial actions.

Radon issues continuously from the ground since uranium is present in virtually all rocks and soils. Radon dissipates quickly in open air, but when trapped inside a building, the gas can accumulate in concentrations tens, hundreds, and even thousands of times higher than out of doors.

It is important to note that radon per se is not the direct radiation hazard, but rather it is certain daughters (radioactive-decay products of the radon—mainly isotopes of polonium) that contribute the major radiation dose to lung tissue. These isotopes are chemically reactive. They can stick, either in elemental form or adsorbed onto minute airborne particles, to the lining of the bronchial passageways, whence they eradicate the surrounding tissue.

Dangerous radon concentrations were first noted in Sweden. Radiation measurements were made in well-insulated houses and immediately from that experience it was postulated that energy-efficient houses, designed to minimize the ventilation rate and thus conserve heating or cooling losses, were in effect "radon traps." Later investigations have discounted the importance of ventilation. Attention was shifted from "tight" houses to ways in which radon may enter a building—in an effort to explain the wide disparity of data collected from thousands of structures. In these studies, building materials per se were rarely implicated as major sources of radon. Pathways for radon to enter buildings from underneath are illustrated in Fig. 1.

Because there are so many unanswered questions, most experts are attempting to quell any alarm complex that might arise among the public and to continue to collect data before drawing tentative conclusions. The U.S. Environmental Protection Agency is conducting a comprehensive survey of domestic radon levels. Radon concentrations are being measured in a random nationwide sampling of residential buildings, during which survey consistent instrumentation techniques will be used.

Additional Reading

Abelson, P.H.: "Uncertainties About Health Effects of Radon," *Science,* **353** (October 19, 1990).
Abelson, P.H.: "Mineral Dusts and Radon in Uranium Mines," *Science,* 777 (November 8, 1991).

Web References

National Safety Council Radon Information Page: http://www.nsc.org/ehc/ radon.htm
NEHA—National Radon Proficiency Program: http://www.radongas.org/
Radon information from the U.S. EPA: http://www.epa.gov/iaq/radon/
The High Radon Project-Lawrence Berkeley National Laboratory: http://eande.lbl.gov/IEP/high-radon/hr.html
USGS Radon Information: http://sedwww.cr.usgs.gov:8080/radon/radonhome.html

RAMAN SPECTROMETRY.   This form of spectrometry is based upon the Raman effect which may be described as the scattering of light from a gas, liquid, or solid with a shift in wavelength from that of the usually monochromatic incident radiation. Discovered by the Indian physicist, C.V. Raman in 1928, it has also been called the Smekal-Raman effect, the former investigator having made some earlier theoretical predictions about it. If the polarizability of a molecule changes as it rotates or vibrates, incident radiation of frequency v, according to classical theory, should produce scattered radiation, the most intense part of which has unchanged frequency. This is Rayleigh scattering. In addition, there should be Stokes and anti-Stokes lines of much lesser intensity and of frequencies $v \pm v_k$, respectively, where v_k is a molecular frequency of rotation or vibration. The anti-Stokes line is always many times less intense than the Stokes line and this fact is satisfactorily explained by the quantum mechanical theory of the effect. The vibrational Raman effect is especially useful in studying the structure of the polyatomic molecule. If such a molecule contains N atoms it can be shown that there will be $3N-6$ fundamental vibrational modes of motion only ($3N-5$ if the molecule is a linear one). Those accompanied by a change in electric moment can be observed experimentally in the infrared. The remaining ones, if occurring with a change in polarizability, will be observable in the Raman effect. Thus both kinds of spectroscopic measurements are usually required in a complete study of a given molecule.

Like infrared spectrometry, Raman spectrometry is a method of determining modes of molecular motion, especially the vibrations, and their use in analysis is based on the specificity of these vibrations. The methods are predominantly applicable to the qualitative and quantitative analysis of covalently bonded molecules rather than to ionic structures. Nevertheless, they can give information about the lattice structure of ionic molecules in the crystalline state and about the internal covalent structure of complex ions and the ligand structure of coordination compounds both in the solid state and in solution.

Both the Raman and the infrared spectrum yield a partial description of the internal vibrational motion of the molecule in terms of the normal vibrations of the constituent atoms. Neither type of spectrum alone gives a complete description of the pattern of molecular vibration, and, by analysis of the difference between the Raman and the infrared spectrum, additional information about the molecular structure can sometimes be inferred. Physical chemists have made extremely effective use of such comparisons in the elucidation of the finer structural details of small symmetrical molecules, such as methane and benzene. But the mathematical techniques of vibrational analysis are not yet sufficiently developed to permit the extension of these differential studies to the Raman and infrared spectra of the more complex molecules that constitute the main body of both organic and inorganic chemistry.

The analytical chemist can use Raman and infrared spectra in two ways. At the purely empirical level, they provide "fingerprints" of the molecular structure and, as such, permit the qualitative analysis of individual compounds, either by direct comparison of the spectra of the known and unknown materials run consecutively, or by comparison of the spectrum of the unknown compound with catalogs of reference spectra.

By comparisons among the spectra of large numbers of compounds of known structure, it has been possible to recognize, at specific positions in the spectrum, bands which can be identified as "characteristic group frequencies" associated with the presence of localized units of molecular structure in the molecule, such as methyl, carbonyl, or hydroxyl groups. Many of these group frequencies differ in the Raman and infrared spectra.

When a transparent medium was irradiated with an intense source of monochromatic light, and the scattered radiation was examined spectroscopically, not only is light of the exciting frequency, v, observed (Rayleigh scattering), but also some weaker bands of shifted frequency are detected. Moreover, while most of the shifted bands are of lower frequency, $v - \Delta v_1$, there are some at higher frequency, $v + \Delta v_1$. By analogy to fluorescence spectrometry (see below), the former are called *Stokes bands* and the latter *anti-Stokes bands*. The Stokes and anti-Stokes

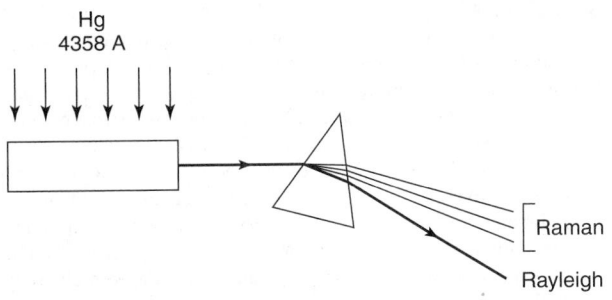

Fig. 1.   Optical system to observe a Raman spectrum. The Rayleigh scattering is at the wavelength of the exciting line

bands are equally displaced about the Rayleigh band; however, the intensity of the anti-Stokes bands is much weaker than the Stokes bands and they are seldom observed. This article deals only with the more intense Stokes bands. The geometric arrangement for observing the Raman effect is shown diagrammatically in Fig. 1.

Additional Reading

Long, Fayer, M.D.: *Ultrafast Infrared and Raman Spectroscopy,* Marcel Dekker, Inc., New York, NY, 2001.

Laserna, J.J.: *Modern Techniques in Raman Spectroscopy,* John Wiley & Sons, Inc., New York, NY, 1996.

Lindon, J.C., G.E. Tranter, J. John, and L. Holmes: *Encyclopedia of Spectroscopy and Spectrometry,* Academic Press, Incorporated, San Diego, CA, 2000.

McCreery, R.L.: *Raman Spectroscopy for Chemical Analysis,* John Wiley & Sons, Inc., New York, NY, 2000.

Nakamoto, K.: *Infrared and Raman Spectra of Inorganic and Coordination Compounds,* Vol. 2, 5th Edition, John Wiley & Sons, Inc., New York, NY, 1997.

Nakamoto, K. and J.R. Ferraro: *Introductory Raman Spectroscopy,* Academic Press, Inc., 1994.

Ogilvie, J.F.: *The Vibrational and Rotational Spectrometry of Diatomic Molecules,* Academic Press, Inc., San Diego, CA, 1998.

Pelletier, M.J.: *Analytical Applications of Raman Spectroscopy,* Blackwell Science, Inc., Malden, MA, 1999.

Schrader, B.: *Infrared and Raman Spectroscopy: Methods and Applications,* John Wiley & Sons, Inc., New York, NY, 1995.

Socrates, G.: *Infrared and Raman Characteristic Group Frequencies: Tables and Charts.* 3rd Edition, John Wiley & Sons, Inc., New York, NY, 2001.

RAMSAY, SIR WILLIAM (1852–1916).   A British chemist born in Scotland who received the Nobel prize for chemistry in 1904. He participated in the discovery of helium, argon, neon, xenon, and krypton. Much of his work concerned investigations of inert gases. He was also known for studies in organic, physical, and inorganic chemistry. Ramsay was educated at the Universities of Glasgow, Heidelberg, and Tubingen, and was a Professor at the Universities of Bristol and London.

RANKINE.   A scale of absolute temperature based on Fahrenheit degrees. Temperatures on the Rankine scale are 9/5 (or 1.8) times those on the Kelvin scale.

   See also **Absolute Temperature**.

RAOULT'S LAW.   The vapor pressure of a substance in equilibrium with a solution containing the substance is equal to the product of the mole fraction of the substance in the solution and the vapor pressure of the pure substance at the temperature of the solution. The law is not applicable to most solutions, but is often approximately applicable to a mixture of closely similar substances, particularly the substance present in high concentration.

   See also **Vapor Pressure**.

RAPESEED OIL.   See **Vegetable Oils (Edible)**.

RARE-EARTH ELEMENTS AND METALS.   Sometimes referred to as the "fraternal fifteen," because of similarities in physical and chemical properties, the rare-earth elements actually are not so rare. This is attested by Fig. 1, which shows a dry lake bed in California that alone contains well in excess of one million pounds of two of the elements, neodymium and praseodymium. The world's largest rare earth body and mine near Baotou, Inner Mongolia, China is shown in Fig. 2. It contains 25 million tons of rare earth oxides (about one quarter of the world's human reserves. The term *rare* arises from the fact that these elements were discovered in scarce materials. The term *earth* stems from the fact that the elements were first isolated from their ores in the chemical form of oxides and that the old chemical terminology for oxide is earth. The rare-earth elements, also termed Lanthanides, are similar in that they share a valence of 3 and are treated as a separate side branch of the periodic table, much like the Actinides. See also **Actinide Contraction**; **Chemical Elements**; **Lanthanide Series**; and **Periodic Table of the Elements**.

   The properties of the Rare-earth elements are given in Tables 1 and 2. Pronunciation of the elements is as follows: Cerium (*sear' ium*),

Fig. 1.   Dry lake mineral bed near Mountain Pass, California contains over one million pounds of neodymium and nearly one-half million pounds of praseodymium, both elements once regarded as "rare earths" and of limited scientific curiosity. During recent years, the rare earths have become significant materials in the electronic, chemical, metallurgical, glass, cryogenic, nuclear, and ceramic refractory industries. Lanthanum, another rare-earth element, is more abundant than lead

Fig. 2. Open-pit operation at Baiyunebo mine. (*The Chinese Society of Rare Earth*)

dysprosium (*dis pröz' ium*), erbium (*ur' bium*), europium (*yoo rö pium*), gadolinium (*gado lin' ium*), holmium (*hol' mium*), lanthanum (*Ian' tha num*), lutetium (*loo tee' shium*), neodymium (*neo dim' ium*), praseodymium (*pra zee o dim' ium*), promethium (*pro mee' thium*), samarium (*sa mar' ium*), scandium (*scan de'ium*) terbium (*tur' bium*), thulium (*thoo' lium*), ytterbium (*i tur' bium*), and yttrium (*it' rium*). The lanthanides are further described by individual alphabetical entries for each element.

C.A. Arrhenius, in 1787, noted an unusual black mineral in a quarry near Ytterby, Sweden. This was identified later as containing yttrium and rare-earth oxides. With the exception of promethium, all members of the Lanthanide Series had been discovered by 1907, when lutetium was isolated. In 1947, scientists at the Atomic Energy Commission at Oak Ridge National Laboratory (Tennessee) produced atomic number 61 from uranium fission products and named it promethium. No stable isotopes of promethium have been found in the earth's crust.

Natural mixtures of these elements have been used commercially since the early 1900s. Mischmetal is the source of the hot spark in cigarette lighter flints. The mixed rare-earth fluorides are burned in the cores of carbon electrodes to create the intense sunlike illumination required by motion-picture projectors and searchlights. The mixed rare-earth oxides, which contain primarily cerium dioxide, are used to grind and polish almost all optical lenses and television faceplates. In the late-1940s, it was discovered that the rare-earth metals effectively control the shape of carbon in normally brittle cast iron, resulting in ductile or nodular iron. During the 1950s, interest in several of the pure elements (europium, gadolinium, dysprosium, samarium, and erbium) was stimulated because these elements have the highest thermal-neutron-absorption properties among the elements. These elements have found application in control rods and as burnable poisons. Yttrium metal was fabricated into tubing and mill products because it is almost transparent to thermal neutrons and has a unique stability at high temperature in contact with liquid uranium, potassium, and sodium. Nuclear aircraft and submarine propulsion programs were the main impetus for these efforts. Radioactive promethium has been used as a power source for pacemakers.

Early in the 1960s, mixtures of the rare-earth elements were incorporated with synthetic molecular-sieve catalysts, resulting in increased petroleum refining efficiency. Various rare-earth compounds have been found to act as catalysts in several chemical processes, such as hydrogenation. Rare-earth mixed oxides and especially CeO_2 are being used in auto exhaust catalysts. In 1964, a new red phosphor for color television was discovered. Relatively large quantities of highly purified europium and yttrium oxides were needed as commercial color television production started. Rare-earth phosphors are also being used in color monitors for computers, X-ray screens, fluorescent lamps, UV-conversion phosphors, dental and surgical lasers, electro- and thermoluminescent devices, and fiber optics.

Permanent magnets having properties several times superior to any other known materials were developed in 1967. Praseodymium, yttrium, samarium, lanthanum, and cerium are alloyed with cobalt in the range

RCo_5 to R_2Co_{17}, where $R = a$ rare-earth element. The new family of permanent-magnet materials is bringing about improvements in power generation and electronic communications. Conventional applications now include watches, electric motors, computer printers, automotive devices, frictionless bearings, and loudspeakers. Novel applications for the powerful magnets include magnetic earrings and use in medical treatments.

About 15 years later a new family of iron-neodymium-based permanent magnets were discovered. In 1983 several research groups in the United States and Japan announced the discovery of a new compound with the probable composition, $R_2Fe_{14}B$ ($R = a$ light rare earth lanthanide, predominantly neodymium). These materials exhibit extremely powerful magnetic qualities as compared with traditional magnet materials and about 10–20% stronger than the Sm-Co magnets. One shortcoming, is the loss of desirable magnetic qualities at elevated temperatures above about 200°F. Applications for the new magnetic materials span a wide range from the very sophisticated applications, such as found in nuclear magnetic resonance imaging systems, down to the inexpensive magnets used in toys and around the home. The $Nd_2Fe_{14}B$ materials are made either by rapid solidification or by powder metallurgy techniques. Some investigators have found that the addition of 6% cobalt increases the Curie temperature 100 K while others found that small additions of Dy (less than a 10% substitution of the Nd) are added to raise the Curie temperature. The crystal structure of $Nd_2Fe_{14}B_1$ is tetragonal, an anisotropic structure that contributes to the high coercivity. The relatively low concentrations of light lanthanides and boron, and the fact that the magnetic moments in iron and the lanthanides align parallel (ferromagnetically) to each other, allow the magnetization to remain high. See also **Magnetism**.

Metallurgical Uses

During the 1960s, the rare-earth metals were established as reactive and refining metals in the iron and steel industry. As alloying elements, lanthanum and yttrium improve the high-temperature oxidation and corrosion properties of superalloys. Other metallurgical applications of the rareearths include welding solders, brazing alloys, nonferrous alloys (such as magnesium and aluminum), and dispersion hardening of complex alloys (eg. Y_2O_3 dispersed in nickel based alloys). In the 1990s complex LaNi$_5$-based alloys were developed for use in rechargeable metal hydride batteries, which are slowly replacing NiCd batteries in many applications because of their superior performance and because the LaNi$_5$-based materials are more environmentally friendly. Another fairly recent development is the use of mixed rare earths as fertilizers in agriculture. The Chinese have been quite successful in improving various crop yields, but this application has seen only limited use in other countries.

Miscellaneous applications of rare-earth compounds, complexes and alloys include: MRI (magnetic resonance imaging) contrasting agents in the medical field; high temperature oxide superconductors; magnetic recording alloys and magnetic bubble devices; electronic components, capacitors and semiconductors; optical glasses and fiber optics; magnetic cooling and refrigeration; corrosion inhibitors; dying and printing textiles; cosmetics; oxidizer in self-cleaning ovens; fuel cell cathodes; and Y_2O_3 (and other R_2O_3)-stabilized ZrO_2 as oxygen sensors, electrolytes, structural ceramics, and synthetic jewelry.

The Institute for Physical Research and Technology sponsors a Rare-Earth Information Center at Iowa State University. Ames, Iowa, which provides a comprehensive service to science and industry by cataloging the vast amount of technical information generated about these elements each year.

Occurrence

Rare-earth minerals exist in many parts of the world; the overall potential supply is essentially unlimited. As a group, these elements rank fifteenth in abundance, somewhat more plentiful than zinc. Rare-earth minerals generally are classified as sources for *light* (La through Gd) or *heavy* (Y plus Tb through Lu). Typical mineral distributions are given in Table 3.

Until 1964, monazite, a thorium-rare-earth phosphate, REPO$_4$Th$_3$ (PO$_4$)$_4$, was the main source for the rare-earth elements. Australia, India, Brazil, Malaysia, and the United States are active sources. India and Brazil supply a mixed rare-earth chloride compound after thorium is removed chemically from monazite. Bastnasite, a rare-earth fluocarbonate mineral; REFCO$_3$, is a primary source for light rare earths. From 1965 to about 1985, an open-pit resource at Mountain Pass, California, has furnished about two-thirds of world requirements for rare-earth oxides. In the early

TABLE 1. ATOMIC AND THERMAL PROPERTIES OF RARE-EARTH ELEMENTS

Property	21 Sc Scandium	39 Y Yttrium	57 La Lanthanum	58 Ce Cerium	59 Pr Praseodymium	60 Nd Neodymium	61 Pm Promethium	62 Sm Samarium	63 Eu Europium	64 Gd Gadolinium	65 Tb Terbium	66 Dy Dysprosium	67 Ho Holmium	68 Er Erbium	69 Tm Thulium	70 Yb Ytterbium	71 Lu Lutetium
Estimated abundance: ppm	6	33	30	60	8.2	28	0	6.0	1.2	5.4	0.9	3.0	1.2	2.8	0.5	3.0	0.5
g/ton	—	28–70	5–18	20–46	3.5–5.5	12–24	0	4.5–7	0.14–1.1	4.5–6.4	0.7–1	4.5–7.5	0.7–1.2	2.5–6.5	0.2–1	2.7–8	0.8–1.7
Atomic constants: Atomic weight	44.96	88.91	138.91	140.12	140.91	144.24	(145)	150.36	151.96	157.25	158.93	162.50	164.93	167.26	168.93	173.04	174.97
Metallic radius, Å, (CN = 12)	1.641	1.801	1.879	(+3)1.846 (+4)1.672	1.828	1.821	1.811	1.804	(+2)2.042 (+3)1.798	1.801	1.783	1.774	1.766	1.757	1.746	(+2)1.939 (+3)1.741	1.735
Å, (CN = 12) Volume, cm³/g	15.04	19.89	22.60	(+3)21.43 (+4)15.92	20.80	20.58	(20.24)	20.00	(+2)28.98	19.90	19.31	19.00	18.75	18.45	18.12	(+2)24.84 (+3)17.98	17.79
Atom Density, g/cm³	2.989	4.469	6.146	6.770	6.773	7.008	7.264	7.520	5.244	7.901	8.230	8.551	8.795	9.066	9.321	6.966	9.841
lb/in.³	0.108	0.161	0.222	0.244	0.244	0.253	0.262	0.271	0.189	0.285	0.297	0.308	0.317	0.327	0.336	0.251	0.355
Crystal structure at 25°C	hcp	hcp	dhcp	fcc	dhcp	dhcp	dhcp	rhom	bcc	hcp	hcp	hcp	hcp	hcp	hcp	fcc	hcp
Unpaired 4f electrons	0	0	0	1	2	3	4	5	6	7	6	5	4	3	2	1	0
Number of isotopes: Natural	1	1	2	4	1	7	0	7	2	7	1	7	1	6	1	7	2
Artificial	11	14	19	15	14	7	15–18	11	16	11	17	12	18	12	17	10	14
Lattice constants, A: a	3.309	3.648	3.774	5.161	3.672	3.658	3.65	3.629	4.583	3.634	3.605	3.592	3.578	3.559	3.538	5.485	3.505
c	5.268	5.732	12.171		11.833	11.797	11.656	26.207		5.781	5.697	5.650	5.618	5.585	5.554	—	5.549
Ionic radius, Å: +2								1.19	1.17							1.00	
+3	0.745	0.900	1.045	1.010	0.997	0.983	0.97	0.958	0.947	0.938	0.923	0.912	0.901	0.890	0.880	0.868	0.861
+4				0.80	0.78						0.76						
Color of 3⁺ ion (in solution)	Colorless	Colorless	Colorless	Colorless	Green	Reddish violet	Pink	Yellow	Pale pink	Colorless	Almost colorless	Yellow	Pink	Reddish violet	Green	Colorless	Colorless
Electronegativity	1.28	1.177	1.117	(+3)1.123 (+4)1.43	1.130	1.134	1.139	1.145	(+2)0.98 (+3)1.152	1.160	1.168	1.176	1.184	1.192	1.200	(+2)1.02 (+3)1.208	1.216
Absorption bands, 3⁺ ion, Å	None	None 2380 2520	None 4822 5885	2105 2220 5745 7395	4445 4690 7025 7355	3540 5218 4020	5485 5680	3625 3745 2754 2756	3755 3941 4875	2729 2733 9100	3694 3780 4508 5370	3504 3650 4870 5228	2870 3611 7800	3642 3792 7420 7975	3600 6825 6404	9750 6525	None 8030 8680
Thermal properties: Melting point °C	1541	1522	918	798	931	1021	1042	1074	phantom 822	1313	1365	1412	1474	1529	1545	phantom 819	1663
°F	2806	2772	1684	1468	1708	1868	1908	1965	1512	2395	2489	2574	2685	2784	2813	1506	3025
Boiling point at 1 atm. °C	2836	3345	3464	3443	3520	3074	3000	1794	1529	3273	3230	2567	2700	2868	1950	1196	3402
°F	5137	6053	6267	6229	6368	5565	5432	3261	2784	5923	5846	4653	4892	5194	3542	2185	6156
Heat of fusion ΔH_f kcal/g atom	3.370	2.724	1.482	1.305	1.646	1.706	1.84	2.060	2.201	2.390	2.579	2.643	4.063	4.756	4.015	1.831	5.26
Heat of sublimation ΔH_s at 25°C, kcal/g atom	90.30	101.5	103.0	101.0	84.99	78.30	83	49.40	41.90	95.00	92.90	69.41	71.89	75.79	55.50	36.35	102.20
Heat capacity ΔC_p at 25°C, cal/(g atom)	6.09	6.34	6.48	6.43	6.55	6.55	6.52	7.05	6.62	8.87	6.91	6.62	6.50	6.72	6.45	6.38	6.41
Coefficient of expansion, per °C ×10⁻⁶	10.2	10.6	12.1	6.3	6.7	9.6	11	12.7	35.0	9.4	10.3	9.9	11.2	12.2	13.3	26.3	9.9
Nuclear properties: Thermal neutron capture, barns/atom	17	1.31	8.9	0.73	11.6	50		5,600	4,300	40,000	46	1100	64	170	125	37	108

*Table compiled by Molybdenum Corporation of America, White Plains, N.Y. (Joseph G. Cannon); edited by Rare-Earth Information Center, Energy and Mineral Resources Research Institute, Iowa State University, Ames, Iowa (Karl A. Gschneidner, Jr. and N. Kippenhan). Data from S.R. Taylor, Abundance of Chemical Elements in the Continental Crust: A New Table, Geochim. Cosmochim. Acta, vol. 28, pp. 1273–1285, 1964; E.T. Teatum, et al., Compilation of Calculated Data Useful in Predicting Metallurgical Behavior of Elements in Binary Alloy Systems, Univ. Calif., Los Alamos Sci. Lab. Rep. LA-4003, pp. 11–12, Dec. 24, 1968; Clifford A. Hampel, "Rare Metals Handbook," 2d ed., chaps. 1 and 35, Van Nostrand Reinhold Company, New York, 1961; O.A. Songina, "Rare Metals: Scandium, Yttrium, Lanthanide and Actinides," chap. 6, trans. from Russian (1970), 3d ed (1964), U.S. Dept. of Interior and The National Science Foundation, Washington, DC.; Karl A. Gschneidner, Jr., "Solid State Physics," vol. 16, "Physical Properties and Interrelationships of Metallic and Semimetallic Elements," pp. 275–426, Academic, New York, 1964; Clifford A. Hampel, "The Encyclopedia of the Chemical Elements," Van Nostrand Reinhold Company, New York, 1968; R. Hultgren, R.L. Orr, and K.K. Kelley, supplement to "Selected Values of Thermodynamic Properties of Metals and Alloys," Wiley, New York, 1963; Data from of the Department of Mineral Technology and Lawrence Radiation Laboratory, The University of California, Berkeley, Calif. (data and revision published periodically). Data from Karl A. Gschneidner, Jr. and Leroy Eyring, eds., "Handbook on the Physics and Chemistry of the Rare Earths, Vol. 1," North-Holland, Amsterdam, (1979).

TABLE 2. MECHANICAL, ELECTRICAL, AND OXIDE PROPERTIES OF RARE-EARTH ELEMENTS

Atomic number	21	39	57	58	59	60	61	62	63	64	65	66	67	68	69	70	71
symbol	Sc	Y	La	Ce	Pr	Nd	Pm	Sm	Eu	Gd	Tb	Dy	Ho	Er	Tm	Yb	Lu
element	Scandium	Yttrium	Lanthanum	Cerium	Praseodymium	Neodymium	Promethium	Samarium	Europium	Gadolinium	Terbium	Dysprosium	Holmium	Erbium	Thulium	Ytterbium	Lutetium
Mechanical properties:[†]																	
Yield strength: kg/mm²	17.6	4.3	12.8	2.9	7.4	7.2	N.A.	6.9	N.A.	1.5	N.A.	4.4	22.6	6.1	N.A.	0.7	N.A.
1,000 psi	28.0	6.1	18.2	4.1	10.5	10.2	N.A.	9.8	N.A.	2.1	N.A.	6.3	32.1	8.7	N.A.	1.0	N.A.
Elongation, %	5.0	34	7.9	22	15.4	25	N.A.	17	N.A.	37	N.A.	30	5	11.5	N.A.	43	N.A.
Tensile strength: kg/mm²	26.0	13.2	13.3	11.9	15.0	16.7	N.A.	15.9	N.A.	12	N.A.	14.2	26.4	13.9	N.A.	5.9	N.A.
1,000 psi	37.0	18.8	18.9	16.9	21.3	23.8	N.A.	22.6	N.A.	17.1	N.A.	20.2	37.5	19.8	N.A.	8.4	N.A.
Vickers hardness, 10-kg load, kg/mm²	—	41	38	29	37	35	63	40	17	42	38	44	46	42	48	17	44
Elastic properties (values in parentheses estimated): Compressibility, $cm^2/kg \times 10^{-6}$	1.73	3.98	3.23	4.96	3.39	3.09	(2.96)	2.60	11.76	2.59	2.52	2.44	2.37	2.23	2.21	7.26	2.06
Shear modulus, $kg/cm^2 \times 10^6$	0.297	0.260	0.152	0.122	0.150	0.169	(0.183)	0.199	(0.079)	0.226	0.232	0.259	0.269	0.289	(0.310)	0.101	0.276
Young's modulus, $kg/cm^2 \times 10^6$	0.759	0.648	0.392	0.306	0.387	0.431	(0.471)	0.510	0.186	0.569	0.582	0.643	0.665	0.672	0.754	0.314	0.697
Poisson's ratio	0.279	0.246	0.288	0.248	0.289	0.279	(0.278)	0.282	0.167	0.254	0.255	0.238	0.237	0.250	0.217	0.207	0.261
Electrical properties at 25°C: Resistivity, $\mu\Omega\text{-cm}$	56.2	59.6	61.5	74.4	70.0	64.3	75	94.0	90.0	131	115	92.6	81.4	86	67.6	25	58.2
Hall coefficient, $V\text{-cm}/(A)(Oe) \times 10^{12}$	−0.13	−0.77	−0.35	+1.81	+0.71	+0.97	N.A.	−0.2	+24.4	−4.48	−4.3	−2.7	−2.3	−0.34	−1.8	+3.77	−0.54
Work function, eV	3.5	3.1	3.5	2.9	2.7	3.2	3.1	2.7	2.5	3.1	(3.1)	(3.1)	(3.1)	(3.1)	(3.1)	(2.6)	(3.1)
phantom	phantom	phantom	phantom	phantom	phantom	phantom	phantom	phantom	phantom	phantom	phantom	phantom	phantom	phantom	phantom	phantom	phantom
Magnetic properties: Moment, theoretical for 3+ ion, Bohr magnetons	0	0	0	2.5	3.6	3.6	N.A.	1.6	3.5	7.95	9.7	10.6	10.6	9.6	7.6	4.5	0
Susceptibility, emu/g atom $\times 10^6$	295	191	101	2430	5320	5650	N.A.	1275	33,100	356,000	193,000	99,800	70,200	44,100	26,100	71	17.9
Curie temperature, °C	None	None	None	None	None	None	N.A.	None	None	+20	−53	−185	−254	−253	−248	None	None
Neel temperature, °C	None	None	None	−260.6	None	−253	N.A.	−258	−184	None	−43	−97	−143	−188	−215	None	None
Metal oxide: Formula	Sc_2O_3	Y_2O_3	La_2O_3	CeO_2	Pr_6O_{11}	Nd_2O_3	Pm_2O_3	Sm_2O_3	Eu_2O_3	Gd_2O_3	Tb_4O_7	Dy_2O_3	Ho_2O_3	Er_2O_3	Tm_2O_3	Yb_2O_3	Lu_2O_3
Color	White	White	White	Buff	Black	Light blue	White	Cream	Pale pink	White	Dark brown	Cream	Cream	Rose	Light green	White	White
Molecular weight	137.92	225.81	325.82	172.12	1021.79	336.48	342	348.70	351.92	362.50	747.69	373.00	377.86	382.52	385.87	394.08	397.94
Melting point: °C	2403	2410	2300	2210	2183	2233	2320	2269	2291	2339	2303	2228	2330	2344	2341	2355	2427
°F	4357	4370	4172	4010	3961	4051	4208	4116	4156	4242	4117	4042	4226	4251	4246	4271	4401
Density g/cm³	3.88	5.03	6.58	7.22	6.83	7.31	7.60	7.11	7.29	7.61	7.87 (Tb_2O_3)	8.16	8.41	8.65	8.90	9.21	9.41

* Table compiled by Molybdenum Corporation of America, White Plains, N.Y. (Joseph G. Cannon); edited by Rare-Earth Information Center, Energy and Mineral Resources Research Institute, Iowa State University, Ames, Iowa (Karl A. Gschneidner, Jr. and N. Kippenhan). Data from S.R. Taylor, Abundance of Chemical Elements in the Continental Crust: A New Table, *Geochim. Cosmochim. Acta*, vol. 28, pp. 1273–1285, 1964; E.T. Teatum, et al., Compilation of Calculated Data Useful in Predicting Metallurgical Behavior of Elements in Binary Alloy Systems, *Univ. Calif., Los Alamos Sci. Lab. Rep.* LA-4003, pp. 11–12, Dec. 24, 1968; Clifford A. Hampel, "Rare Metals Handbook," 2d ed., chaps. 1 and 35, Van Nostrand Reinhold Company, New York, 1961; O.A. Songina, "Rare Metals: Scandium, Yttrium, Lanthanide and Actinides," chap. 6, trans. from Russian (1970), 3d ed. (1964), U.S. Dept. of Interior and The National Science Foundation, Washington, DC.; Karl A. Gschneidner, Jr., "Solid State Physics," vol. 16, "Physical Properties and Interrelationships of Metallic and Semimetallic Elements," pp. 275–426, Academic, New York, 1964; Clifford A. Hampel, "The Encyclopedia of the Chemical Elements," Van Nostrand Reinhold Company, New York, 1968; R. Hultgren, R.L. Orr, and K.K. Kelley, supplement to "Selected Values of Thermodynamic Properties of Metals and Alloys," Wiley, New York, 1963; Data from Department of Mineral Technology and Lawrence Radiation Laboratory, The University of California, Berkeley, Calif. (data and revisions published periodically). Data from Karl A. Gschneidner, Jr. and Leroy Eyring, eds., in "Handbook on the Physics and Chemistry of the Rare Earths, Vol. 1 (metals) & 3 (oxides)," North-Holland, Amsterdam, 1978, 1979.
† Highest reported value for metal at room temperature after 10–50% reduction in area or annealed or as-cast; purity unknown.
N.A.—not available.

TABLE 3. RARE EARTH CONTENT OF SEVERAL PRIMARY SOURCE MINERALS

Rare earth element	Bastnasite		Monazite			Xenotime Malaysia (%)	Uranium residues Ontario, Canada (%)	Ion-adsorption Clays (China)		Loparite, Russia (%)
	Calif. (%)	China[a] (%)	China[b] (%)	Australia[c] (%)	Brazil and India (%)			Longman (%)	Xunwu (%)	
La	32.0	22.8	23.4	23.9	22.8	0.5	0.8	2.2	29.8	25.0
Ce	49.0	49.8	45.7	46.0	45.7	5.0	3.7	1.1	7.2	50.5
Pr	4.4	6.2	4.2	5.0	5.0	0.7	1.0	1.1	7.1	5.0
Nd	13.5	18.5	15.7	17.4	18.9	2.2	4.1	3.5	30.2	15.0
Sm	0.5	1.0	3.0	2.5	3.0	1.9	4.5	2.3	6.3	0.7
Eu	0.1	0.2	0.1	0.05	0.1	0.2	0.1	0.1	0.5	0.1
Gd	0.3	0.7	2.0	1.5	1.7	4.0	8.5	5.7	4.2	0.6
Tb		0.1	0.1	0.04	0.2	1.0	1.2	1.1	0.5	—
Dy		0.1	1.0	1.2	0.5	8.7	11.2	7.5	1.8	0.6
Ho		0.1	0.05	0.1	2.1	2.1	2.6	1.6	0.3	0.7
Er	0.1		0.5	0.2	0.1	5.4	5.5	4.3	0.8	0.8
Tm		0.1	0.5	0.01	—	0.9	0.9	0.6	0.1	0.1
Yb			0.5	0.1	0.1	6.2	4.0	3.3	0.6	0.2
Lu		0.1	0.1	0.04	—	0.4	0.4	0.5	0.1	0.2
Y	0.1	0.5	3.0	2.4	2.0	60.8	51.4	64.1	10.1	1.3

[a] Baiyunebo iron ore mine, Inner Mongolia
[b] Guangdong/Guangxi Provinces
[c] Western Australia

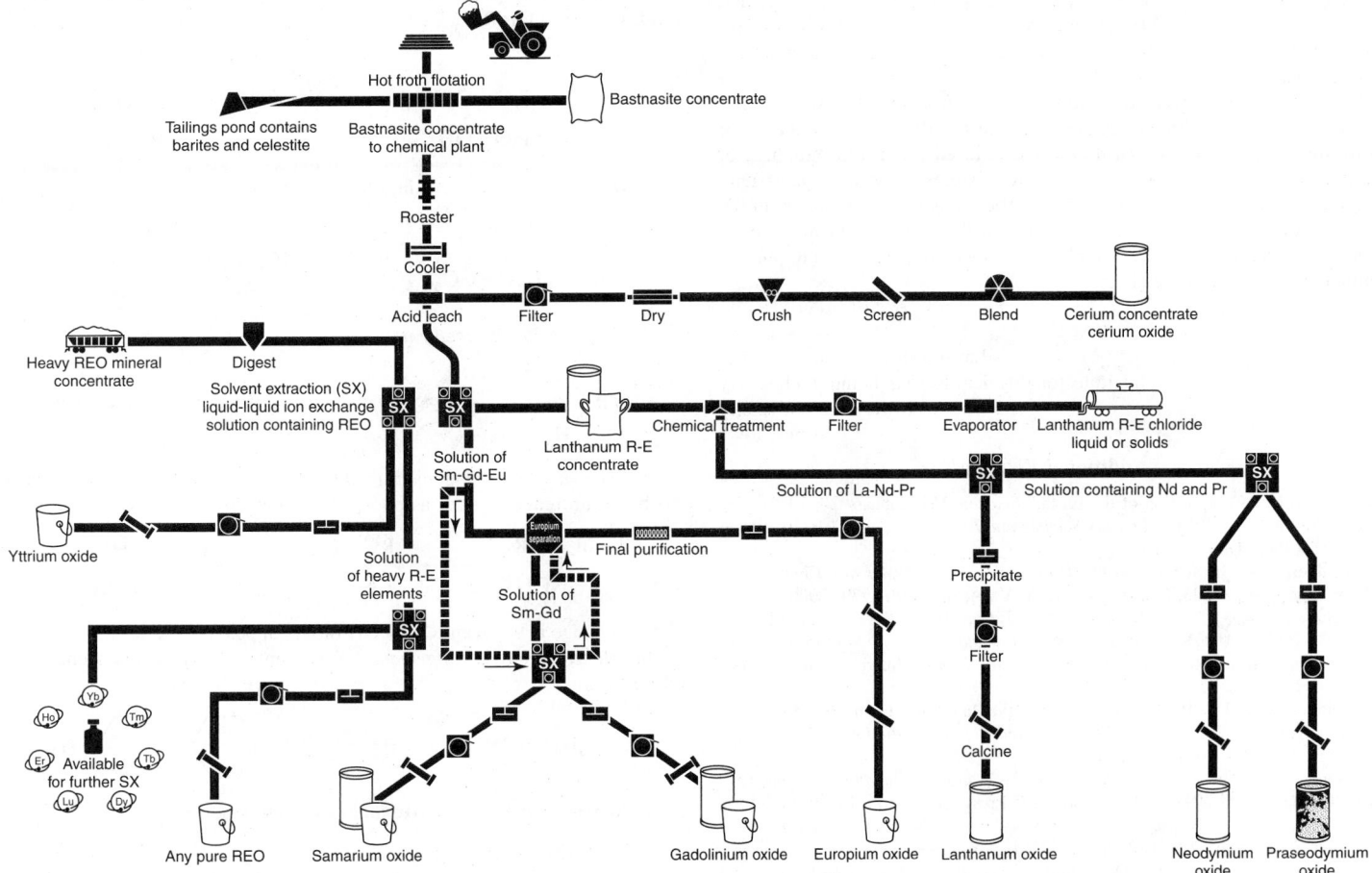

Fig. 3. Separation and purification of rare-earth elements. REO = rare-earth oxides; SX = solvent extraction

1980s the Chinese started to produce rare earths from their Baiyunebo mine, which contains both bastnasite and monazite, located about 100 miles from Baotou, Inner Mongolia. In the early 1990s the rare earth production from this mine exceeded that mined in Mountain Pass. In 1998 the Mountain Pass operation was temporarily closed because of a blocked wastewater pipe. Once the governmental delays due to environmental issues are approved, production will resume. The main source for yttrium and heavy rare-earths is a by-product of uranium mining in the Elliott Lake Region, Ontario. Some xenotime, found in Malaysia, is processed in Japan and Europe.

A highly generalized flowsheet of the production of some of the rare-earth oxides is shown in Fig. 3. Crushed and finely ground bastnasite contain about 70% rare-earth oxides is roasted under oxidizing conditions to convert soluble trivalent cerium compounds to insoluble tetravalent CeO_2. The roasted product is leached with HCl, which dissolves the remaining rare earths (La, Pr, Nd, Sm, Eu, Gd), leaving behind a concentrated cerium product. The solution is passed through liquid-liquid organic solvent extraction (SX) cells, resulting in a primary separation of La-Nd-Pr from Sm-Eu-Gd. Further SX separates a pure lanthanum solution and a concentrated Nd-Pr solution, which another SX circuit separates. Europium is reduced to a divalent state in solution and precipitated. A final SX system separates and purifies gadolinium and samarium. Pure elements are usually precipitated as oxalates and calcined to oxides.

In connection with production of the heavy rare earths, monazite, containing about 55% rare-earth oxides and 5% thorium, is treated in one of two ways: (1) finely ground particles are leached with hot H_2SO_4, which dissolves thorium and the rare earths, leaving an insoluble residue; or (2) finely ground particles are reacted with hot caustic (NaOH), which dissolves the phosphate, creating a solution of trisodium phosphate, which may be recovered as a by-product. The thorium and rare-earth hydrate cake is then dissolved in H_2SO_4. Thorium sulfate is selectively precipitated by pH adjustment. Separation of the other rare earths in solution is usually completed by selective absorption on ion-exchange resins and elution from ion-exchange columns. After thorium is removed from the H_2SO_4 solution, the rare earths remaining are precipitated, using NaOH, forming a double salt, $NaRESO_4 \cdot xH_2O$, known as pink salt. This salt is dissolved in HCl, treated to remove impurities, and evaporated until the hydrated $RECl_3 \cdot 6H_2O$ can be cast.

In the case of the Canadian yttrium-heavy rare earth concentrate, this is leached with HNO_3, causing all rare earths to go into solution. Solvent extraction separates yttrium from the other heavy rare earths, each of which can eventually be separated by further solvent extraction. In the case of xenotime, this is leached with hot H_2SO_4 and separation of yttrium and the heavy rare earths completed in ion-exchange columns. The liquid-liquid organic solvent extraction cycle is complete within 5–10 days and is a continuous process. The resin ion-exchange cycle requires 60–90 days and is a batch process. Both processes result in pure rare earth oxides and chemicals.

Mischmetal is produced commercially by electrolysis. The usual starting ingredient is the dehydrated rare earth chloride produced from monazite or bastnasite. The mixed rare earth chloride is fused in an iron, graphite, or ceramic crucible with the aid of electrolyte mixtures made up of potassium, barium, sodium, or calcium chlorides. Carbon anodes are immersed in the molten salt. As direct current flows through the cell, molten mischmetal builds up in the bottom of the crucible. This method is also used to prepare lanthanum and cerium metals.

K. A. GSCHNEIDNER, JR.
B. EVANS
Rare-Earth Information Center
Institute for Physical Research and Technology
Iowa State University
Ames, Iowa

Additional Reading

Anderson, D.L.: "Composition of the Earth," *Science*, **367** (January 20, 1989).

Carter, G.F. and D.E. Paul: *Materials Science and Engineering,* ASM International, Materials Park, Ohio, 1991.

Gschneidner, K.A., Jr. and L. Eyring: *Handbook on the Physics and Chemistry of Rare Earths,* Vol. 1–28, Elsevier Science B.V., Amsterdam, 1979–2000.

Gschneidner, K.A., Jr., B.J. Beaudry, and J. Capellen: "Rare Earth Metals," pp. 720–732 in *Metals Handbook, Properties and Selection: Nonferrous Alloys and Special Purpose Materials,* 10th Edition, Vol. 2, ASM International, Materials Park, OH, 1990.

Gschneidner, K.A., Jr.: "Physical Properties of the Rare Earth Elements," pp. 4–112 to 4–121 in *CRC Handbook of Chemistry and Physics 1996–1997,* 77th Edition, CRC Press, LLC., Boca Raton, FL, 1997.

Hammond, C.R.: "The Elements," pp 4–1 to 4–34 in *CRC Handbook of Chemistry and Physics, 1996–1997,* 77th Edition, CRC Press, LLC., Boca Raton, FL, 1997.

Lewis, R.J. and G.G. Hawley: *Hawley's Condensed Chemical Dictionary,* 13th Edition, John Wiley & Sons, Inc., New York, NY, 1999.

Lewis, R.J. and N.I. Sax: *Sax's Dangerous Properties of Industrial Materials,* 10th Edition, John Wiley & Sons, Inc., New York, NY, 1999.

Lide, D.R.: *CRC Handbook of Chemistry and Physics,* 84th Edition, CRC Press, LLC., Boca Raton, FL, 2003.

Mayers, R.A.: *Handbook of Chemicals Production Processes,* The McGraw-Hill Companies, Inc., New York, NY, 1986.

White, R.M.: "Opportunities in Magnetic Materials," *Science,* **229**, 11 (1985).

RARE GAS. Any of the six gases composing the extreme right-hand group of the periodic table, namely helium, neon, argon, krypton, xenon, and radon. They are preferably called *noble gases* or (less accurately) *inert gases.* The first three have a valence of 0 and are truly inert, but the others can form compounds to a limited extent.

See also **Periodic Table of the Elements**.

REACTOR (Nuclear).   See Nuclear Power Technology.

REALGAR.   The mineral realgar is a monosulfide of arsenic corresponding to the formula AsS. It is monoclinic, showing short prismatic crystals, or may be in granular or compact masses. It is a soft sectile mineral; hardness, 1.5–2; specific gravity, 3.5; luster, resinous; color, red to orange-yellow; transparent to translucent. Realgar occurs associated with other arsenic minerals and with gold, silver, and lead ores, although not in great quantities. It has been found as a hot-spring deposit and in volcanic sublimations. Realgar has been found in Macedonia, Japan, Switzerland; and in the United States, in Yellowstone National Park, as a hot-spring deposit, and in Utah and Nevada. The name realgar is derived from the Arabic words *rahj al ghar*, which means the powder of the mine.

REARRANGEMENT (Organic Chemistry).   These are reactions involving the transfer of an atom or group from one part of the molecule to another. Tautomerism is a special case of rearrangements in which the two forms are in dynamic equilibrium. See "Tautomerism" in entry on **Isomerism**. When such reactions take place, the establishment of structural formulas becomes complex. Some of the better known rearrangements include:

1. *Allyl Rearrangement.*

$$CH_3CH:CHCH_2Br \longrightarrow CH_2 \cdot CHCHBrCH_3.$$

2. *Pinacol Rearrangement* takes place when pinacol is heated with dilute acid, pinacolin formed:

3. *Benzil Rearrangement* into benzylic acid:

4. *Hoffman Rearrangement.*

5. *Beckmann Rearrangement* results when oximes of ketones are treated with certain reagents such as phosphorus pentachloride.

The exchange is between R′ and OH on the opposite sides of the CN bond.

6. *Benzidine Rearrangement.* Treatment of hydrazobenzene with strong acids.

7. *Rearrangement of hydroxylamine derivatives.*

8. *Rearrangement in reaction of nitrous acid and alkyl amines.*

9. *Walden Inversion.* A rearrangement takes place in the reaction of optically active substances. The structural formula does not change but the configuration changes due to the interchange of two groups attached to the asymmetrical carbon atom during the course of the reaction.

Additional Reading

Buncel, E., and C.C. Lee: *Isotopes in Molecular Rearrangements,* Elsevier Science, New York, NY, 1975.

Harwood, L.M.: *Polar Rearrangements,* Oxford University Press, New York, NY, 1991.

Manion, J., and J. Jeffers: *Molecular Rearrangement Reactions: Azobenzene and Benzilic Acid,* Chemical Education Resources, Inc., American Chemical Society, Washington, DC, 1999.

Stutz, A.E.: *Glycoscience: Epimerisation. Isomerisation and Rearrangement Reactions of Carbohydrates,* Springer-Verlag New York, Inc., New York, NY, 2001.

REAUMUR. A temperature scale in which 0 is the freezing point of water and 80 is its boiling point.

RECALESCENCE. A phenomenon exhibited by iron and some other ferromagnetic metals. If iron is heated white hot and allowed to cool, it will, at a certain temperature, suddenly evolve enough heat to halt the cooling and even produce a momentary heating. This is easily exhibited by stretching an iron wire against the tension of a spring and arranging a lever index to show slight changes in length. The wire is first heated by an electric current. As it cools and contracts, the index will at a certain point give a perceptible jerk, and then resume its steady motion of contraction. The effect is due to an exothermic change in the crystalline structure. The reverse phenomenon, exhibited on heating, is called "decalescence." For cast iron, the recalescence point is a little below $700°C$. Pure iron has two such points, at $780°C$ and $880°C$.

A somewhat analogous effect is exhibited by some amorphous solids upon devitrification, which takes place when the temperature becomes high enough for the substance to crystallize. Noncrystalline sodium silicate, for example, has such a transition point near $500°C$, where it suddenly begins to glow.

RECLAIMING. Recovery and reuse of scrap materials, either in low percentage in new product manufacture or in larger proportions in products in which the highest quality is not essential. Among the materials widely reclaimed in industry are aluminum, steel, paper, rubber, glass, crankcase oil, greases, etc. Solid materials are comminuted, the contaminants being removed with organic solvents or strong alkali solutions (paint from metals, ink from paper, fabric and metals from tires). In the case of cross-linked elastomers, more intensive solvent and heat treatment is necessary. The resulting product is used as an adulterant in low-quality scrap rubber to oil, with recovery of carbon black and metal inserts, indicates that substantial value may be obtained by this method. Devulcanization by means of microwave radiation is also under development. The term *reprocessing* refers specifically to the recovery of nuclear fuels from reactor waste.

RECOMBINANT DNA.   See **Immune System and Immunochemistry**.

RECOMBINATION. The process by which a positive and a negative ion join to form a neutral molecule or other neutral particle. In the literature of atmospheric electricity, this term is applied both to the simple case of the capture of free electrons by positive atomic or molecular ions, and also to the more complex case of the neutralization of a positive small ion by a negative small ion, or a similar (but much more rare) neutralization of large ions.

Recombination is, in general, a process accompanied by emission of radiation. The light emitted from the channel of a lightning stroke is recombination radiation. The much less concentrated recombinations steadily occurring in all parts of the atmosphere where ions are forming and disappearing does not yield observable radiation. The intermediate ions are forming, and disappearing does not yield observable radiation.

RED TIDE. Yellow-to-reddish discoloration of seawater due to rapid multiplication of various species of plantlike microorganisms, called dinoflagellates, which occurs seasonally in areas of warm water, especially off the coast of or California, Florida and occasionally as far north as New England. Some, though by no means all, of the species are poisonous. Concentration of these organisms (phytoplankton) may be as high as 10^8 units/L. The shellfish are lethal to free-swimming fish. Shellfish are unharmed by them but are able to store and concentrate the toxin, which causes paralytic poisoning when they are eaten by humans. So potent is this poison, that death may result from ingestion of milligram amounts.

REFINING. Essentially a separation process whereby undesirable components are removed from various types of mixture to give concentrated and purified products. Such separation may be effected as follows:

- Mechanically, by pressing, centrifuging, filtering, etc.
- By electrolysis.
- By distillation, solvent extraction, or evaporation.
- By chemical reaction.

One or more of these operations is applied to:

- Food products.
- Petroleum.
- Lubricating oils.
- Metals.

As regards petroleum, refining is generally understood to include not only fractional distillation of crude oils to naphthas, low-octane gasoline, kerosene, fuel oil, and asphaltic residues, but also the processes involved in thermal and catalytic cracking (hydroforming, reforming, etc.) for production of high-octane gasoline.

See also **Petroleum**.

REFINING (Petroleum).   See **Petroleum**.

REFLUX. In distillation processes in which a fractionating column is used, the term *reflux* refers to the liquid that has condensed from the rising vapor and been allowed to flow back down the column toward the still. As it does so, it comes into intimate contact with the rising vapor, resulting in improved separation of the components. The separation resulting from contact of the countercurrent streams of vapor and liquid is called rectification or fractionation.

REFORMATSKY REACTION. Dating back to 1887, this reaction depends on interaction between a carbonyl compound, an α-halo ester, and activated zinc in the presence of anhydrous ether or ether-benzene, followed by hydrolysis. The halogen component for example ethyl bromoacetate, combines with zinc to form an organozinc bromide that adds to the carbonyl group of the second component to give a complex readily hydrolyzed to carbinol. The reaction

$$Zn + BrCH_2COOC_2H_5 \rightarrow$$
$$BrZnCH_2COOC_2H_5 \quad \overset{}{\searrow}C=O$$

is conducted by the usual Grignard technique except that the carbonyl component is added at the start. Magnesium has been used in a few reactions in place of zinc but with poor results, for the more reactive organometallic reagent tends to attack the ester group; with zinc this side reaction is not appreciable, and the reactivity is sufficient for addition to the carbonyl group of aldehydes and ketones of both the aliphatic and aromatic series. The product of the reaction is a β-hydroxy ester and can be dehydrated to the α,β-unsaturated ester; thus the product from benzaldehyde and ethyl bromoacetate yields ethyl cinnamate. α-Bromo esters of the types $RCHBrCO_2C_2H_5$ and $RR'CBrCO_2C_2H_5$ react satisfactorily, but

Ethyl β-phenyl-β-hydroxy-propionate (b.p. $130°/6$ mm.)

β- and γ-bromo derivatives of saturated esters do not have adequate reactivity. Methyl γ-bromocrotonate (BrCH$_2$CH=CHCO$_2$CH$_3$), however, has a reactive, allylic bromine atom and enters into the Reformatsky reaction.

L. F. and M. FIESER
Harvard University
Cambridge, Massachusetts

REFORMING. Decomposition (cracking) of hydrocarbon gases or low-octane petroleum fractions by heat and pressure, either without a catalyst (thermoforming), or with a specific catalyst. The latter is most extensively used. The principal cracking reactions are (1) dehydrogenation of cyclohexanes to aromatic hydrocarbons; (2) dehydrocyclicization of certain paraffins to aromatics; (3) isomerization, i.e., conversion of straight-chain to branched-chain structures, as octane to isooctane. These result in substantial increase in octane number. Steam reforming of natural gas is an important method of producing hydrogen; steam reforming of naphtha is used to produce substitute natural gas. See also **Ammonia**; **Cracking Process**; **Isomerization**; and **Petroleum**.

REFRACTION. The term refraction properly applies to the change of direction which radiation, especially light, or sound experiences on passing obliquely from one medium to another in which its velocity of propagation is different. The physical nature of the effect can be visualized by considering a regiment marching in columns of platoons across a boundary between smooth turf and freshly plowed ground. If the line of march is perpendicular to the boundary, the platoons are simply slowed up and thus crowded more closely together; but if it is oblique, one end of each platoon is retarded sooner than the other, and the file swings around to a direction nearer the normal. See Fig. 1. A train of waves is similarly affected as it passes into a new medium with change of velocity.

It is easy to show that if i is the angle of incidence and r the angle of refraction at such a boundary (Fig. 2), the refraction is governed by a simple relation known as *Snell's law*:

$$\sin i = n \sin r$$

in which n has the same value for various angles of incidence and refraction. This constant n, known as the refractive index, depends upon the character of the wave train and of the two media. Physically it represents the ratio of the velocity of the disturbance in the first medium to that in the second. For light passing from one medium to another in which its velocity is greater, so that $n < 1$, we may, for a sufficiently large angle of incidence, encounter the curious phenomenon known as total reflection.

Refraction often occurs in a single medium due to variations in its properties resulting from changes in conditions through the portion of the medium traversed by the radiation or sound. The twinkling of stars is caused by differences in refractive index in the atmosphere resulting from differences in temperature. Sound also exhibits this temperature refraction.

Specific refraction is a relationship between the refractive index of a medium at any definite wavelength and its density, of the form

$$r = \left(\frac{n^2 - 1}{n^2 + 2}\right)\left(\frac{1}{\rho}\right)$$

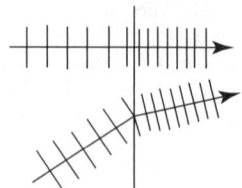

Fig. 1.   Change of wavelength and (in general) of direction of wave train upon entering new medium

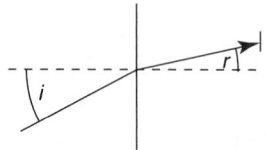

Fig. 2.   Angles of incidence (i) and refraction (r)

in which r is the specific refraction of the medium, n is its index of refraction at any definite wavelength, and ρ is its density. The relation does not always give a constant value of r as the density is varied, and hence must be considered as an approximation.

Molar refraction is the product of the specific refraction by the molecular weight. The form of this relationship is

$$R = Mr$$

in which R is the molar refraction, M is the molecular weight, and r is the specific refraction. The direct form of this relationship is

$$R = \left(\frac{n^2 - 1}{n^2 + 2}\right)\left(\frac{M}{\rho}\right)$$

in which R is the molar refraction, n is the index of refraction for any chosen wavelength, M is the molecular weight, and ρ is the density.

An empirical relationship between molar refraction, density, and molar volume, that applies to many liquids over a considerable range of temperatures is that of Eykiman:

$$R = \frac{M(n^2 - 1)}{\rho(n + 0.4)} = \frac{V(n^2 - 1)}{(n + 0.4)}$$

in which R is the molar refraction at a given wavelength, n is the index of refraction at that wavelength, M is the molecular weight, ρ is the density, and V is the molar volume.

Atomic refraction is the product of the specific refraction of an element by its atomic weight.

Standard refraction is the refraction that would occur in an idealized atmosphere in which the refractive index decreases uniformly with height at the rate of 39×10^{-6} per kilometer. Standard refraction may be included in ground wave calculations by use of an effective earth radius of 8.5×10^6 meters, or $\frac{4}{3}$ the geometrical radius of the earth.

REFRACTIVE INDEX. The phase velocity of radiation in free space divided by the phase velocity of the same radiation in a specified medium. Because of the Snell law (see also **Refraction**) the refractive index may also be defined as the ratio of the sine of the angle of incidence to the sine of the angle of refraction.

The absolute index for all ordinary transparent substances is greater than 1 (see Table 1); but there are some special cases (X-rays and light in metal films, which are discussed below) for which the index of refraction is less than unity. Since the absolute index for air exceeds unity by less than 0.0003, the relative indices for solids and liquids in air are very nearly equal to their absolute indices. It should be noted that since the refractive index varies with the wavelength, any exact statement of its value must specify the wavelength to which it refers; in Tables it is usually given for sodium light of frequency 5,893A. See also **Dispersion**.

Various relationships have been used to express the refractive index. Thus there are several semi-empirical relationships expressing the refractive index of a medium as a function of wavelength:

$$n^2 = 1 + \frac{A_1 \lambda^2}{\lambda^2 - \lambda_1^2}$$

where A_1 is a constant characteristic of the material and λ_1 is an idealized absorption wavelength of the medium. When $\lambda_1 \ll \lambda$, this reduces to

$$n^2 = 1 + A_1 + \frac{A_1 \lambda_1^2}{\lambda^2}$$

TABLE 1. ABSOLUTE INDEX

Substance	Absolute Index
Air	1.0002926
Bromine	1.661
Carbon dioxide	1.00045
Diamond	2.419
Glass	1.5 to 1.9
Glycerine	1.4729
Helium	1.000036
Ice	1.31
Rock salt	1.516
Water (20°C)	1.333

or

$$n = A + \frac{B}{\lambda^2} \text{(Cauchy Formula)}$$

A better approximation is

$$n = A + \frac{B}{\lambda^2} + \frac{C}{\lambda^4} + \cdots$$

For equations for specific and molar refraction, see also **Refraction**.

As a consequence of his electromagnetic theory of light, Clerk Maxwell obtained the relation between the dielectric constant of a medium and its refractive index n:

$$\varepsilon = n^2$$

This relation holds only under rather restrictive conditions, such as measurement with light of long wavelength, absence of permanent dipoles in the substance, etc.

For strongly absorbing media, such as metals, the customary refractive index must be replaced by the complex refraction index $n(1 - i\kappa)$ where κ is called the absorption index. Then the reflectivity for normal incidence is

$$R = \frac{(n-1)^2 + n^2\kappa^2}{(n+1)^2 - n^2\kappa^2}$$

The refractive index for metals varies over a much larger range than for conventional dielectrics, e.g., sodium at $\lambda = 0.546$ micrometer, $n = 0.052$, while for silicon at $\lambda = 0.589$ micrometer, $n = 4.24$. The refractive index is sometimes defined as the relative dielectric constant of a medium, $\sqrt{\varepsilon/\varepsilon_0}$. This expression is invariably a function of the wavelength of the radiation.

REFRACTOMETERS. Several types of instruments, called refractometers, have been devised for measuring the refractive index of any substance. See also **Photometers**. Special forms are used for solids, for liquids, and for gases. Solid and liquid refractometers usually depend upon the principle of total reflection and the fact that the sine of the critical angle is equal to the refractive index for light passing from the more to the less refractive medium. The critical angle is what is measured, or deduced from other measured angles.

Suppose that a specimen of the solid or liquid to be tested is brought into optical contact with one face of a glass prism (or "block") of known, higher refractive index and known angle, and that a slightly convergent pencil of light, entering the test substance, is directed at grazing incidence upon the interface between it and the prism. Those rays incident at less than 90° to the normal of the interface enter the prism; the others do not, and the boundary between is sharply defined. The resulting half-pencil traverses the prism and emerges from the other face where the direction of its cutoff edge can be observed (Fig. 1). The angle between the cutoff boundary of the pencil and the first prism face, inside the prism, is the critical angle, and can be easily calculated from the observations and the known data. This is the principle of the Pulfrich refractometer.

Another form, due to Abbe, is used for liquids. A film of the liquid is enclosed between two similar glass prisms (Fig. 2), and the total reflection

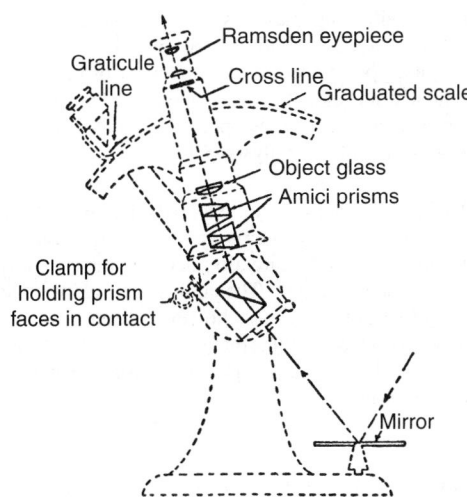

Fig. 2.   Optical system of Abbe refractometer

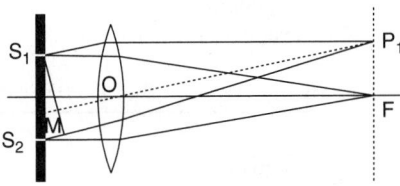

Fig. 3.   When a double slit $S_1 S_2$, illuminated by light from a distant narrow source, is placed in front of the objective of a telescope, interference fringes are observed in the focal plane $P_1 F$

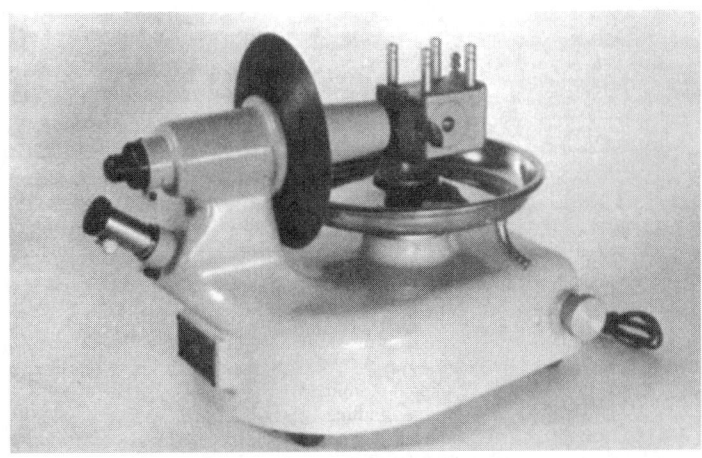

Fig. 4.   Abbe refractometer equipped with sodium are for studies of turbid liquids. (*Gaertner Scientific Corp*)

at the interface observed. Any spectrometer, with a pair of good prisms (preferably right-angled) mounted on the prism table, can be used in this way.

Rayleigh utilized an interference method for measuring the indices of gases. Using a collimator to render the rays parallel (Fig. 3), the stream of light entering one of the slits of the apparatus for Young's experiment is passed through a tube of the gas to be tested. The resulting retardation in phase causes a shift of the interference fringes, the amount of which gives the retardation and hence the refractive index of the gas relative to the air outside the tube. See also **Refraction**; and **Refractive Index**.

A chemical analytical technique is based upon measurement of refractive index. The velocity of light in a material, and therefore the refractive index, depends upon several physical properties of the sample. Theoretical studies have indicated that the refractive index is related to the number, charge, and mass of vibrating particles in the material through which the light is passing. Further, it has been possible to relate refractive index to density and molecular weight for classes of compounds that have a relatively

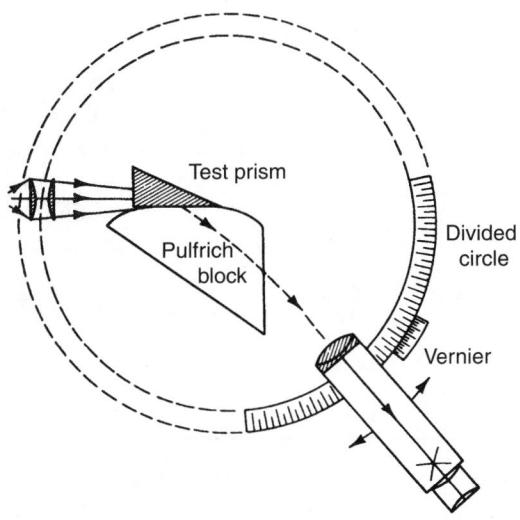

Fig. 1.   Pulfrich refractometer

constant number of vibrating particles per unit weight. The number of vibrating particles in a compound is determined by the atoms in the structure and by the type of electronic bonding. Correlations of this sort have been particularly successful for the analysis of hydrocarbon mixtures. Several techniques have been developed and applied in the petroleum industry. See Fig. 4. See also **Analysis (Chemical)**.

REFRACTORIES. Refractories are materials that resist the action of hot environments by containing heat energy and hot or molten materials. There is no well-established line of demarcation between those materials that are and those that are not refractory. The ability to withstand temperatures above 1100°C without softening has, however, been cited as a practical requirement of industrial refractory materials. (See also **Ceramics**). The type of refractories used in any particular application depends on the critical requirements of the process.

Physical Forms

Refractories may be preformed (shaped) or formed and installed on-site. Castables, gunning mixes, and plastic and ramming mixes are used either for repair or for complete new construction of what is known as monolithic linings. The use of monolithics instead of constructions using shaped products has increased; monolithic installations have become as common as conventional shaped product construction.

Brick. The standard dimensions of a refractory brick are 229 mm (9 in.) length, 114 mm (4.5 in.) width, and 64 mm (2.5 in.) thickness. This is known as a standard straight. Quantities of bricks are given in brick equivalents, that is, the number of standard straight bricks that have a volume equal to that of the particular installation.

Other common refractory forms include setter tile and kiln furniture, fusion-cast shapes, cast and hand-molded refractories, insulating refractories, castables and gunning mixes, plastic refractories and ramming mixes, mortars, composite refractories, and refractory coatings.

Raw Materials

In the past, refractory raw materials were used essentially as mined minerals. Selective mining yielded materials of the desired properties and only in cases of expensive raw materials, such as magnesite, was a beneficiation process required. However, demand for high purity natural raw materials and synthetically prepared refractory grain made from combinations of high purity and beneficiated raw materials has increased (Table 1). The material produced upon firing raw as-mined minerals or synthetic blends is called grain, clinker, co-clinker, or grog. Recycled materials produced by the manufacturers and recovered from users are also used to reduce waste. Use of recycled materials is limited in the United States. It is more common in Europe and is expected to increase.

Silica. The most common refractory raw materials are ganister, which is a dense quartzite, and silica gravels.

TABLE 1. PHYSICAL PROPERTIES OF REFRACTORY RAW MATERIALS

Material	Main crystalline phases	Specific gravity, g/cm³	
		Bulk	True
SILICA			
ganister	quartz		2.66
gravel	quartz		2.61
CLAYS			
flint clays	kaolinite, quartz, illite	2.55	
plastic clays	kaolinite, quartz, illite		
kaolin	kaolinite		
fireclay	kaolinite		
plastic	kaolinite, illite		
HIGH ALUMINA			
natural siliceous bauxite,[a] ca 70% Al_2O_3	mullite	2.85–2.95	3.1–3.2
ca 60% Al_2O_3	mullite	2.75–2.85[b]	2.95–3.95
ca 50% Al_2O_3		2.65	
South American bauxite[a]	corundum, mullite	3.1	3.6–3.7
Chinese bauxite[a]		3.20	
kyanite[b]	kyanite		3.5–3.7
sillimanite[c]	sillimanite		3.23
synthetic fused alumina, black	α-alumina	3.87	4.01
gray	α-alumina	3.95	3.98
sintered alumina	α-alumina	3.45–3.6	3.65–3.80
sintered mullite		2.85	
sintered magnesium aluminate	spinel, periclase	3.33	
fused mullite	mullite	3.1	3.45
calcium aluminate cement, low purity	calcium monoaluminate		
high purity	α-alumina, calcium monoaluminate		
ZIRCONIUM			
zircon	zircon		4.2–4.6
baddeleyite, ZrO_2	baddeleyite		5.5–6.5
BASIC RAW MATERIALS			
calcined magnesias			
natural magnesite	periclase	3.2	
	periclase	3.4	
	magnesite, dolomite, calcite	3.40	
	d	3.39	
seawater	periclase	3.44	
brine	periclase	3.41	
dolomite[e]	periclase + CaO		
chrome ore	chromite spinel	4.2	

[a] Calcined.

[b] Another type of ca 60% Al_2O_3 natural siliceous bauxite has a bulk density of 2.70 g/cm³.

[c] Raw.

[d] Unclassified.

[e] Both regular and low flux dolomite.

Fireclay. Fireclays consist mainly of the mineral kaolinite, $Al_2O_3 \cdot 2SiO_2 \cdot 2H_2O$, with small amounts of other clay minerals, quartzite, iron oxide, titania, and alkali impurities.

Alumina. The naturally occurring raw materials are bauxites, sillimanite group minerals, and diaspore clays (see **Aluminum**).

Mullite. Although mullite is found in nature, for example, as inclusions in lava deposits on the island of Mull, Scotland, no commercial natural deposits are known. It is made by burning pure sillimanite minerals or sillimanite–alumina mixtures.

Zirconia. Zircon (zirconium silicate), the most widely occurring zirconium-bearing mineral, is dispersed in various igneous rocks and in zircon sands. Zircon can be used as such in zircon refractories or as a raw material to produce zirconia, ZrO_2.

Basic Raw Materials. Basic raw materials include magnesite dolomite, forsterite, chrome ore, silicon carbide, beryllia, thoria, and carbon (graphite).

General Properties

Oxides. Beryllium and magnesium oxides are stable to very high temperatures in oxidizing environments. Beryllia has good electrical insulating properties and high thermal conductivities; however, its high toxicity restricts its use. Stabilized cubic or partially stabilized ZrO_2 is the most useful simple oxide for operations above 1900°C.

Carbon, Carbides, and Nitrides. Carbon (graphite) is a good thermal and electrical conductor. It is not easily wetted by chemical action, which is an important consideration for corrosion resistance. As an important structural material at high temperature, pyrolytic graphite has shown a strength of 280 MPa (40,600 psi). It tends to oxidize at high temperatures, but can be used up to 2760°C for short periods in neutral or reducing conditions. When heated under oxidizing conditions, silicon carbide and silicon nitride, Si_3N_4, form protective layers of SiO_2 and can be used up to ca 1700°C. Silicon carbide has very high thermal conductivity and can withstand thermal shock cycling without damage. It also is an electrical conductor and is used for electrical heating elements.

Borides and Silicides. These materials do not show good resistance to oxidation.

Metals. The highest melting refractory metals are tungsten (3400°C), tantalum (2995°C), and molybdenum (2620°C). All show poor resistance to oxidation at high temperatures.

Phase Equilibria. Phase diagrams represent the chemical equilibria that exist among one, two, or three components of a system under the influence of temperature and pressure. Reference to a phase diagram permits the determination of the amount and composition of solid and liquid phases that coexist under certain specified conditions of temperature and pressure for a particular system. Using such information, the occurrence of physical and chemical changes within a system or between systems at high temperatures can be predicted.

Physical Properties. Brick bulk density depends on the specific gravity of the constituents and the porosity. Usually the highest density possible is desired. Upon firing, the grains and matrix form glassy, direct, or solid-state ceramic bonds. Sintering is generally accompanied by shrinkage. Particle size distribution, forming method, and firing process contribute to texture, whereas permeability is related to porosity, which in turn is dependent upon texture.

Mechanical Properties. The transverse strength (modulus of rupture) at room temperature is related to the degree of bonding. Fine-grained refractories generally are stronger than coarse-grained types and those having a low porosity are stronger than those of high porosity.

Generally, high temperature strength is lower than room temperature strength. The development of solid-state or direct-bonded basic brick requires high firing temperatures, and is impeded by glassy phases. By referring to phase diagrams, refractory compositions may be designed that avoid the development of such phases.

The modulus of elasticity (MOE) is related to the strength and can be used as a nondestructive quality control test on high cost special refractory shapes.

Thermal Properties. Refractories, like most other solids, expand upon heating, but much less than most metals. The degree of expansion depends on the chemical composition.

Reheat Change. Most refractory bricks are not chemically in equilibrium before use. During the prolonged heating in service, additional reactions occur that may cause the brick to shrink or expand. Considerable expansion may be caused by gas formation during heating. In general, basic brick exhibits good volume stability at high temperatures.

Thermal Conductivity. The refractory thermal conductivity depends on the chemical and mineral composition of the material and increases with decreasing porosity.

Thermal Spalling. The susceptibility to thermal cracking and spalling depends on certain characteristics of the raw material and the macrostructure of the particular refractory. Spalling resistance may be increased by either preventing cracks from forming or preventing cracks from growing. The approach used to effect spall resistance determines which properties of the refractory are optimized.

Refractoriness. Refractoriness is the resistance to physical deformation under the influence of temperature. It is determined by the pyrometric cone equivalent (PCE) test for aluminosilicates and resistance to creep or shear at high temperature.

Manufacture

Processing. Some materials can be used without further processing, although many must be subjected to heat treatment. In the case of synthetic grain, the selected and beneficiated raw materials are blended in the desired proportions and formed into suitable shapes for calcination by briquetting, pelletizing, or extrusion. The term *Calcination* is used to indicate heat treatment to sinter or burn (dead burn) the refractory grain to a stable dense material as well as to decompose minerals. Calcination may be carried out to rotary kilns, shaft kilns, multiple-hearth furnaces, or fluidized-bed reactors. Both raw and processed materials can be fused or melted in electric-arc furnaces.

Crushing and Grinding. Some raw materials, such as hard clay and quartzite, must first be crushed to grains small enough for the grinding equipment. Almost all raw materials require grinding after primary crushing.

Screening. To obtain a high density product, the mix is made from materials that have been sized into classes by means of standard screens. In a continuous screening operation, the ground raw materials are generally fed to vibrating high capacity screens that may be heated. Material that does not pass the screen is returned to the grinding system for further size reduction. Coarse and medium fine-grain sizing is accomplished by the aforementioned methods, whereas fine-sized materials generated in rod mills, ball mills, ring-roll mills, etc., are classified by air separators.

Mixing. As in other ceramic processes, more than one type of raw material is often required for a refractory product. The purpose of mixing is to uniformly distribute the various ingredients (see **Mixing and Blending**). Although the specific steps and equipment involved in the mixing of batches for fireclay, high alumina, and basic refractories are somewhat different, the general principles are similar. Mixes that are to be dry-pressed contain 2–6% binding liquid, depending on the plasticity of the raw material bond system and the fineness of the mix. The ingredients may be blended in a pug mill, dry pan, or other type mixer and tempered with the bonding ingredients. Tempering, in the sense used here, denotes the kneading action produced on the mix, usually in a muller mixer.

Forming. Most refractory shapes are formed by mechanical equipment, but some very large or intricate shapes require hand molding in wooden, steel-lined molds with loose liners to permit easy removal of plaster of Paris molds.

Refractory shapes are generally produced on a mechanical toggle press, screw press, or hydraulic press. Some special shapes are produced by air-ramming which is similar to hand molding, except that reinforced steel molds are required. Special shapes can also be formed by slip casting and hot pressing.

Drying. The drying step for large shapes is critical. Extremely large fireclay and silica shapes are sometimes allowed to dry on a temperature-controlled floor heated by steam or air ducts embedded in the concrete. Smaller shapes are generally dried in a tunnel dryer. The ware is placed on cars that enter the cold end and exit at the hot end.

Curing. Some chemically bonded bricks require some elevated heat treatment that is typically higher than the tempering process, but at a lower temperature than that required to form ceramic bonds.

Bricks are fired or burned in kilns to develop a ceramic bond within the refractory and attain certain desired properties. This step does not apply to chemically or organically bonded products. Burned brick may be impregnated with tar or pitch to improve corrosion resistance.

Specialty Refractories. Bulk refractory products include gunning, ramming, or plastic mixes, granular materials, and hydraulic setting castables and mortars. These products are generally made from the same raw materials as their brick counterparts.

Economic Aspects

The principal consumers of refractories are the iron and steel industries. There has been a decrease in refractories consumption that coincided with technological changes in the manufacture of steel. Steady improvements in basic oxygen furnace (BOF) practice and improvements in refractory composition and design led to improved refractory performance. More sophisticated ladle metallurgical practice has been employed, leading to improved steel quality and improved refractory performance.

Analytical and Test Methods

The test methods applicable to refractories are summarized in Table 2.

Health and Safety Factors

Industrial refractories are by their very nature stable materials and usually do not constitute a physiological hazard. This is not so, however, for unusual refractories that might contain heavy metals or radioactive oxides, such as thoria and urania, or to binders or additives that may be toxic. Inhalation of certain fine dusts may constitute a health hazard. For example, exposure to silica, asbestos, and beryllium oxide dusts over a period of time results in the potential risk of lung disease.

Selection and Uses

Any manufacturing process requiring refractories depends on proper selection and installation. When selecting refractories, environmental conditions are evaluated first, then the functions to be served, and finally the expected length of service. All factors pertaining to the operation, service design, and construction of equipment must be related to the physical and chemical properties of the various classes of refractories.

By far the most common industrial refractories are those composed of single or mixed oxides of Al, Ca, Cr, Mg, Si, and Zr. These oxides exhibit relatively high degrees of stability under both reducing and oxidizing conditions. Carbon, graphite, and silicon carbide have been used both alone and in combination with the oxides. Refractories made from these materials are used in ton-lot quantities, whereas silicides are used in relatively small quantities for specialty application in the nuclear, electronic, and aerospace industries.

The common industrial refractories are classified into acid, SiO_2 and ZrO_2; basic, CaO and MgO; and neutral, Al_2O_3 and Cr_2O_3. Oxides within

each group are generally compatible with each other, whereas mixtures of acid and basic oxides often give low melting products. Neutral oxides are generally compatible with both acidic and basic oxides.

Reactions Between Refractories and Liquids. The response of a refractory to a chemical environment generally depends on its slag resistance which, in turn, depends on the compositions and properties of slag and refractory. Other factors include temperature, severity of thermal cycling or shock of the process, velocity and agitation of the slag in contact with the refractory, and the abrasion to which the refractory is subjected. Thus similar refractories placed in similar furnaces can wear at vastly different rates under different operation practices.

Reactions Between Refractories and Gases. Reactions of refractories and gases can be quite destructive. The gases generally penetrate the pores of the refractory destroying its structure. An example is the disintegration of aluminosilicates in blast furnaces caused by the deposition of carbon from carbon monoxide. The growth of the carbon deposit causes the brick to rupture. Therefore, a brick of low iron and alkali content having a dense, low permeability is preferred.

Reactions Between Refractories. Dissimilar refractories can react vigorously with each other at high temperatures.

H. DAVID LEIGH III
Clemson University

Additional Reading

Arpe, Hans-Jurgen, W. Russey, G. Schulz, and S. Hawkins: *Ullmann's Encyclopedia of Industrial Chemistry, Refractory Ceramics to Silicon Carbide,* John Wiley & Sons, Inc., New York, NY, 1997.

Banerjee, S.: *Monolithic Refractories,* World Scientific Publishing Company, Inc., Riveredge, NJ, 1998.

Carniglia, S.C. and G.L. Barna: *Handbook of Industrial Refractories Technology—Principles, Types, Properties and Applications,* Noyes Publications, Park Ridge, NJ, 1992.

Janeway, P.A. ed.: *Bull. Am. Ceram. Soc.* **73**(10), 46–55 (1994).

Manual of ASTM Standards on Refractory Materials, 8th Edition, American Society for Testing and Materials, Philadelphia, PA, 1957.

Pierson, H.O.: *Handbook of Refractory Carbides and Nitrides: Properties, Characteristics, Processing, and Applications,* Noyes Publications, Park Ridge, NJ, 1996.

Singer, F. and S.S. Singer: *Industrial Ceramics,* Chemical Publishing Co., Inc., New York, NY, 1964.

Staff: *Nonferrous Metals–Nickel, Cobalt, Lead, Tin, Zinc, Cadmium, Precious, Reactive, Refractory Metals and Alloys: Materials for Thermostats, Electrical Heating and Resistance Contacts, and Connectors,* American Society for Testing & Materials, West Conshohocken, PA, 2004.

REFRIGERANT. Any substance that by undergoing a change of phase (solid to liquid to vapor) lowers the temperature of its environment because of its latent heat. Melting ice, with latent heat of 80 calories per gram removes heat and exerts a considerable cooling effect. Most commercial refrigerants are liquids whose latent heat of vaporization results in cooling.

The specific fluid used to produce refrigeration is a very important consideration. The refrigeration industry uses a variety of compounds that fall into three general categories: halocarbons, hydrocarbons, and inorganic fluids. For saturated and unsaturated aliphatic hydrocarbons and aliphatic halogenated hydrocarbons designated as refrigerants, a numerical coding system has been developed which describes the refrigerant molecular structure by the use of a general designation having the form ABCD, in which A is the number of double bonds, B the number of carbon atoms less one, C the number of hydrogen atoms plus one, and D the number of fluorine atoms.

Inorganic refrigerants are described somewhat differently. They are assigned three-digit numbers the first of which is 7, with the following two numbers comprising the molecular weight.

Refrigerant mixtures are divided into two categories, azeotropes and zeotropes. In zeotropes, the equilibrium vapor and liquid compositions are different, and composition shifting can occur in a refrigeration cycle.

Within the family of halocarbon refrigerants are compounds that contain chlorine, fluorine, and carbon, the chlorofluorocarbons (CFCs). Hydrochlorofluorocarbons (HFCs) contain chlorine, fluorine, carbon and hydrogen. Halocarbons that contain only carbon, fluorine, and hydrogen are called hydrofluorocarbons (HFCs).

See also **Refrigeration.**

TABLE 2. ASTM TEST METHODS FOR REFRACTORIES

Material	Test identification	Properties
burned brick	C20, C830	apparent porosity, water adsorption, bulk density
brick, various shapes	C133, C607, C93	crushing strength, modulus of rupture
basic brick	C456	hydration resistance
brick and tile	C154	warpage
granules	C357, C493	bulk density
periclase grains	C544	hydration
mortar	C198	cold-bonding strength
air-setting plastics	C491	modulus of rupture
castables	C298	modulus of rupture
granular dead-burned dolomite	C492	hydration
fireclay plastics	C181	workability index
castables	C417	thermal conductivity
plastics	C438	thermal conductivity
general refractories	C288	disintegration in CO atmosphere
	C135	true specific gravity
	C201	thermal conductivity
	C92	sieve analysis and water content

REFRIGERATION. A process of cooling or freezing a substance to a temperature lower than that of its surroundings and maintaining that substance in a cold state. Refrigeration can be accomplished by arranging heat transfer from a warm body to a colder body through processes such as convection or thermal conduction. Other, more erotic methods include the exploitation of thermoelectric properties of semiconductors, the magnetothermoelectric effects in semimetals, or the diffusion of ^{3}He atoms across the interface between distinct phases of liquid helium having high and low concentrations of ^{3}He in ^{4}He, among other methods. See also **Thermoelectric Cooling.**

Mainly because of two factors, (1) energy conservation and cost and (2) the desirability of phasing out the use of chlorofluorocarbons (CFCs) in an attempt to slow the deterioration of Earth's ozone layer, refrigeration technology is under intense scrutiny as of the early 1990s.

In an effort to accelerate refrigeration technology, a group of U.S. electric utilities established in 1992 the "Super Efficient Refrigerator Program" (SERP), which will award approximately $30 million to that U.S. manufacturer who first succeeds in producing the most efficient home refrigerator. Targets that must be met include:

1. Consumes as little as 400 kWh per year, compared with 1993 federal efficiency standards of 704 kWh for comparable refrigerators.
2. The refrigerator will be moderately priced.
3. The system will require no chlorofluorocarbons.

In this article, after a review of the common contemporary means of refrigeration, some new or revised innovative methods for refrigeration and cooling systems will be described.

Contemporary Refrigeration Systems

Most commercial refrigeration systems operate on a cyclic basis. A refrigerator operating in this manner may be considered a heat pump, for it continuously extracts heat from a low-temperature region and delivers it to a high-temperature region. It is rated by its *coefficient of performance*, which may be defined as the ratio of the heat removed from the cold region per unit of time to the net input power for operating the device, in symbols $K = Q_t/P$. Vapor-absorption and thermoelectric refrigeration systems have lower coefficients of performance than vapor-compression refrigerators, but they have other characteristics that are superior, such as quietness of operation and compactness.

A *vapor-compression refrigerator* consists of a compressor, a condenser, a storage tank, a throttling valve, and an evaporator connected by suitable conduits with intake and outlet valves. See Fig. 1. The refrigerant is a liquid that partly vaporizes and cools as it passes through the throttling valve. Among the common refrigerants are ammonia, sulfur dioxide, and various halides of methane and ethane.

CFCs have been used widely in all kinds of cooling systems, including automobile air-conditioning systems. But for their atmospheric pollution problems, they are excellent refrigerants.

Nearly constant pressures are maintained on either side of the throttling valve by means of the compressor. The mixed liquid and vapor entering

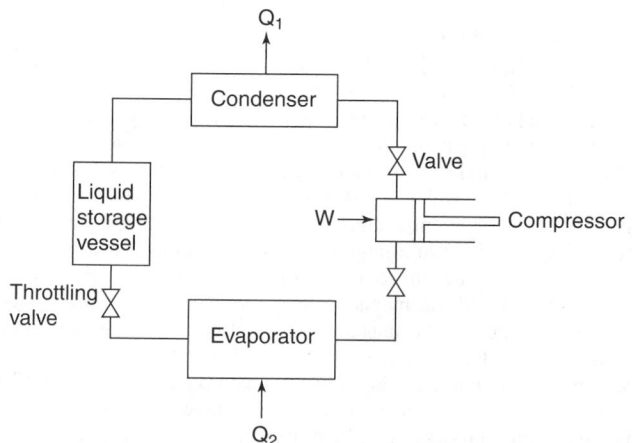

Fig. 1.   Vapor-compression refrigeration system

the evaporator is colder than the near-surround; it absorbs heat from the interior of the refrigerator box or cold room and completely vaporizes. The vapor is then forced into the compressor, where its temperature and pressure increase as the result of compression. The compressed vapor then pours into the condenser, where it cools down and liquifies as the heat is transferred to cold air, water, or other fluid medium in the cooling coils. Comparative tests have shown that the coefficient of performance of vapor-compression refrigerators depends very little on the nature of the refrigerant. Because of mechanical inefficiencies, the actual value may be well below an ideal value, which ordinarily lies between 2 and 3.

In a *vapor-absorption refrigeration system*, there are no moving parts. The added energy comes from a gas or liquid fuel burner or from an electrical heater, as *heat*, rather than from a compressor, as *work*. See Fig. 2. The refrigerant used in this example is ammonia gas, which is liberated from a water solution and transported from one region to another by the aid of hydrogen. The total pressure throughout the system is constant and therefore no valves are needed.

Heat from the external source is supplied to the generator, where a mixture of ammonia and water vapor with drops of ammoniated water is raised to the separator in the same manner as water is raised to the coffee in a percolator. Ammonia vapor escapes from the liquid in the separator and rises to the condenser, where it cools and liquefies. Before the liquefied ammonia enters the evaporator, hydrogen, rising from the absorber, mixes with it and aids in the evaporation process. Finally, the mixture of hydrogen and ammonia vapor enters the absorber, where water from the separator dissolves the ammonia. The ammonia water returns to the generator to complete the cycle. In this cycle, heat enters the system not only at the generator, but also at the evaporator, and heat leaves the system at both the condenser and the absorber to enter the atmosphere by means of radiating fins.

No external work is done, and the change in internal energy of the refrigerant during a complete cycle is zero. The total heat $Q_a + Q_c$ released to the atmosphere per unit of time by the absorber and the condenser equals the total heat $Q_g + Q_c$ absorbed per unit of time from the heater at the generator and from the cold box at the evaporator. Thus $Q_e = Q_a + Q_c - Q_g$, and therefore, the coefficient of performance is $K = Q_e/Q_g = [(Q_a + Q_c)/Q_g] - 1$.

The vapor-absorption refrigerator is free from intermittent noises, but it requires a continuous supply of heat. Once very popular for households, refrigeration systems of this type are now most frequently found in camping facilities and some rural areas where commercial electric power may not be easily available.

In a *dilution refrigeration system*, the properties of helium are used advantageously for attaining very low temperatures. Below a temperature of $0.87°$K, liquid mixtures of ^{3}He and ^{4}He at certain concentrations separate into two distinct phases. One is a concentrated (^{3}He-rich) phase floating on the other, denser (^{4}He-rich) phase with a visible interface between them. The concentrations of ^{3}He in the two phases are functions of temperature, approaching 100% in the concentrated phase and about 6% in the dilute phase at $0°$K. The transfer of ^{3}He atoms from the concentrated to the dilute phase, like an evaporation process, entails a latent heat, an increase in entropy, and a lowering of temperature. The main features of a recirculating dilution refrigeration system are shown in Fig. 3. The pump forces helium vapor (primarily ^{3}He) from the still into the condenser, where it is liquefied at a temperature near $1°$K in a bath of rapidly evaporating ^{4}He, through a flow controller that consists of a narrow tube of suitable diameter to obtain an optimum rate of flow, and then through the still where its temperature is further reduced to about $0.6°$K. The liquefied ^{3}He next passes through a heat exchanger so as to reduce its temperature to nearly that of the dilution chamber, by giving up thermal energy to the counter-flowing dilute phase, before entering the concentrated phase therein.

The diffusion of ^{3}He atoms from the concentrated into the dilute phase within this chamber can produce steady temperatures of very low values ($0.01°$K or less). Liquid ^{3}He from the dilute phase then passes through the heat exchanger to the still, where it is warmed to transform the liquid to the vapor phase that goes to the pump, thus completing the cycle. Modified versions of this system have been constructed, sometimes with an added single-cycle process for temporarily producing temperatures lower than previously mentioned. The low-temperature limit in any system of this type is governed largely by two important sources of inefficiency that cannot be completely eliminated—heat leakage, especially severe because of the extreme range of temperatures, and recirculation of some ^{4}He with ^{3}He.

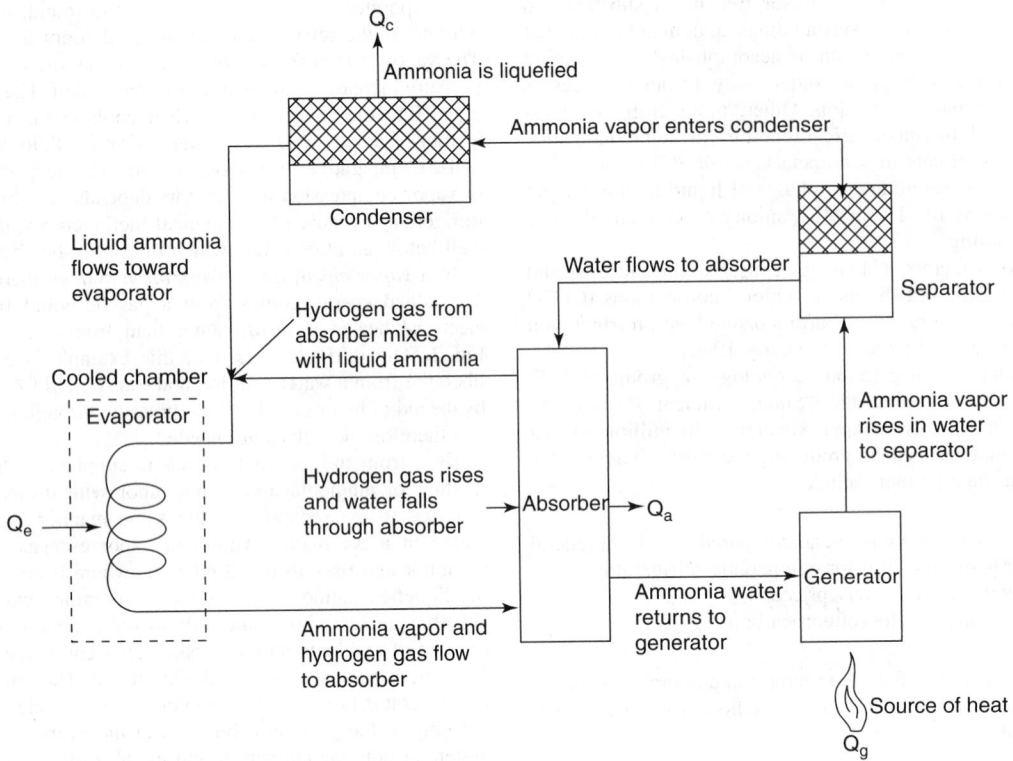

Fig. 2.   Vapor-absorption refrigeration system

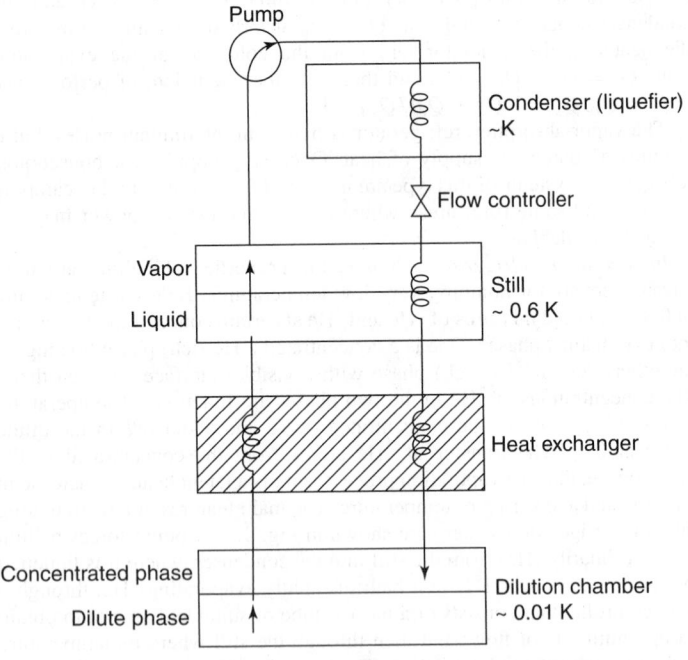

Fig. 3.   Dilution refrigeration system

See also **Helium**; and **Thermodynamics**.

Cool-Storage Concept

In a majority of buildings and nearly all residences, automobiles, and other forms of transportation, comfort cooling systems operate on command—that is, the air conditioner is "on" or "off" in obedience to the thermostat. Thus, the air conditioner starts and stops a number of times during a 24-hour period. Studies have shown that the performance of cooling equipment operates at highest efficiency when used continuously. It has been estimated by authorities that about 25% of the total power consumption of an air-conditioning unit can be attributed to discontinuous operation.

Such inefficient operation essentially can be alleviated by converting to a "cool-storage" system. In this concept, a large pool of cooling medium is created during off-peak hours. The medium is stored in fluid or semi-fluid (slushy) form in a holding tank, which becomes the source of very cold water that can be circulated through a structure. There are definite parallels with circulating hot water or steam heat.

The principle of cool storage was used many years ago when motion picture theaters introduced air-conditioning to the general public. Precooling of the theater prior to opening and during off-peak load hours saved on electricity costs and also cut down on the investment in air-conditioning equipment.

In establishing a cool storage system, there is a choice of cooling media that may be used: (1) cold water just above the freezing point; (2) ice (commonly used with blowers in theaters, shops, and railway cars in the earlier days of air conditioning); (3) so-called "slippery ice," which is comparable to the mixtures containing calcium magnesium acetate used for deicing aircraft; and (4) eutectic salt mixtures. Slippery ice has the desirable property of a flowing slush that does not cling to metal parts.

Many years ago, groups of owners of large adjacent buildings would buy their steam from a district steam plant for heating purposes. This same concept also can be applied to a district cooling system. See Figs. 4 and 5.

Magnetic Refrigeration

The basic physical principle behind magnetic refrigeration (or cooling) is the ability of a ferromagnetic material near its magnetic ordering temperature (called the Curie temperature) to heat-up when magnetized (i.e., placed in a magnetic field), and to cool-off when demagnetized (i.e., the magnetic field is removed). This phenomenon is known as the magnetocaloric effect, which was discovered in 1881. It was first utilized in the mid-1930s to reach temperature below $1°K$ by cooling a paramagnetic salt $[Gd_2(SO_4)_3 \cdot 7H_2O]$ in a large magnetic field to as low a temperature as possible by pumping on liquid He (about $1.5°K$), and when salt was thermally isolated and the magnetic field removed the sample cooled to well below $1°K$ due to the magnetocaloric effect. This one step process is known as adiabatic demagnetization, and it is still used today to reach temperatures in the microkelvin range (the record stands at $0.000025°K$).

In order to make magnetic refrigeration practical one must use a continuous cyclic process in which heat must be rejected when the ferromagnet experiences a magnetic field increase (just as the heat is rejected during the compression of a gas in a conventional gas cycle

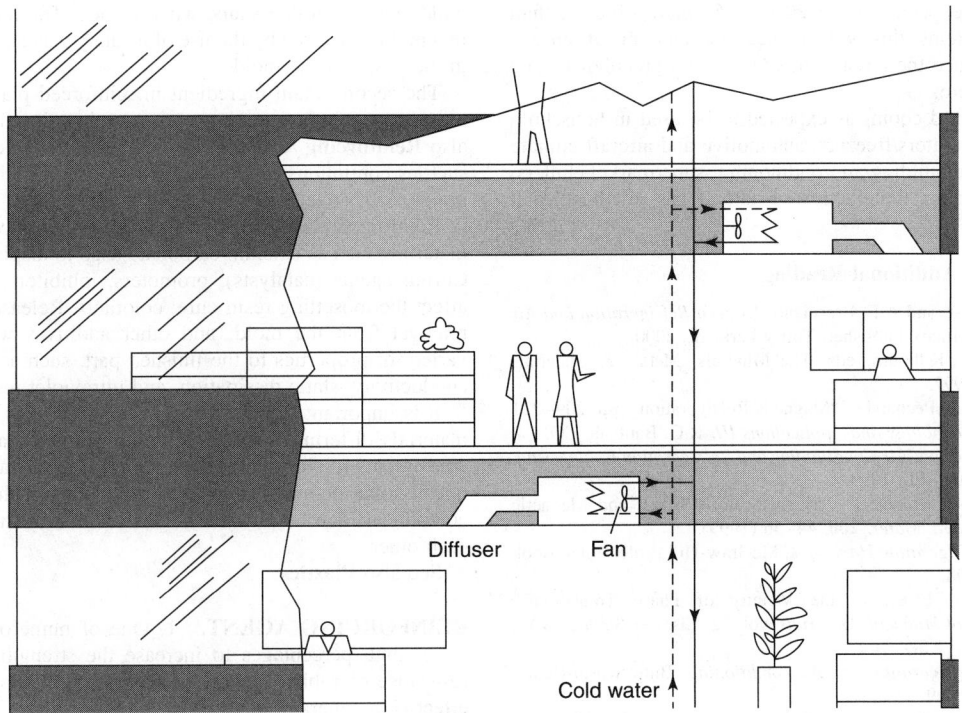

Fig. 4. The cool storage system shown here differs from conventional air conditioners mainly by the inclusion of a storage tank. The tank contains a thermal medium (water, ice, or eutectic salt) that stores cooling generated by the refrigeration unit. When cooling is required, a water solution from the storage tank is circulated in a pipe system that runs throughout the building. The storage capacity effectively decouples the refrigeration process from the building load, allowing building operators to generate cooling during "off-peak" hours of the local electric utility at a time when rates are lower. (*Source: Electric Power Research Institute*)

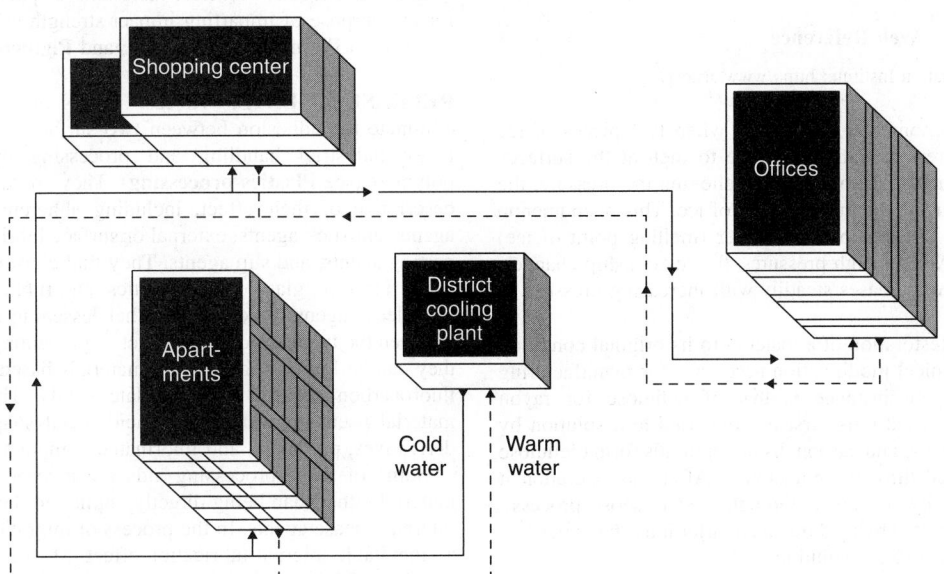

Fig. 5. District cooling. The basic components of cool storage systems also exist in district cooling systems, which use a central plant to cool nearby buildings. Although some systems of this kind are already in use in the United States, they are much more common in Europe. Some authorities believe that district cooling would be a profitable business for many utilities because they could sell cooling while tapping generation capacity that typically goes unused in the off-peak hours. (*Electric Power Research Institute*)

refrigerator). The cooling is achieved by removing the magnetic field experienced by the magnetic refrigerant (this is equivalent to the gas expansion step in the gas cycle refrigerator). One of the first continuously operating magnetic refrigerators, which was used to cool from 1 to 0.2°K, was built and tested in the mid-1950s. Over the next 40 years progress was slow, but in 1997 Zimm and co-authors announced the successful testing of a proof-of-principle continuous near room temperature magnetic refrigerator, which showed for the first time that magnetic refrigeration is a viable technology and competitive with vapor compression cooling. They used gadolinium metal as the magnetic refrigerant and water as the heat transfer medium and were able to cool down to almost 0°C (273°K), while the heat was rejected at a temperature of 38°C from the hot heat exchanger. They achieved a maximum coefficient of performance, COP, (the cooling power divided by the power required to operate the apparatus) of 16 using a maximum field change of 0 to 5 T at a frequency of 0.17 Hz. For smaller field changes the COP was less, about 4 for a 0 to 1.5 T magnetic field change. Gas compression refrigeration has COPs ranging from 2 to 4.

Since magnetic refrigeration uses a non-toxic solid refrigerant, it is an environmentally friendly technology because it replaces the commonly used gaseous refrigerants such as ozone-depleting chemicals (chloro-fluorocarbons), hazardous chemicals (ammonia), and greenhouse gases (hydro-chloro-fluorocarbons and hydro-fluorocarbons). Furthermore, since magnetic refrigeration is expected to be 20 to 30% more efficient than conventional cooling systems this will reduce the amount of energy consumed and in turn reduce the amount of CO_2 (also a greenhouse gas) released into the atmosphere.

Magnetic refrigeration and cooing is expected to be used in household air conditioning and refrigerators/freezers, automotive and aircraft climate control systems, large scale building air conditioning, supermarket chillers, frozen food processing plants, liquefaction of natural gas (methane) and hydrogen gas.

Additional Reading

Althouse, A.D., C.H. Turnquist, and A.F. Bracciano: *Modern Refrigeration and Air Conditioning,* Goodheart-Willcox Publisher, Tinley Park, IL, 2000.

Bounds, T.W. Ellis, and B.T. Kilbourn, eds, The Minerals, Metals & Materials Society, Warrendale, PA, 1997.

Gschneidner, K.A., Jr. and V.K. Pecharsky, "Magnetic Refrigeration," pp. 209–221 in *Rare Earths: Science, Technology and Applications III,* R.G. Bautista, CO.

Kreith, F., Z. Lavan, and P. Norton: *Air Conditioning and Refrigeration Engineering,* CRC Press LLC., Boca Raton, FL, 1999.

Pecharsky, V.K. and K.A. Gschneidner, Jr. "Magnetocaloric Effect and Magnetic Refrigeration," *J. Magn. Magn. Mater.* **200**, 44–56 (1999).

Stoecker, W.F.: *Industrial Refrigeration Handbook,* McGraw-Hill Professional Book Group, New York, NY, 1998.

Tishin, A.M. "Magnetocalorid Effect in the Vicinity of Phase Transitions," pp. 395–524 in *Handbook of Magnetic Materials,* vol. 12, Elsevier Science B.V., Amsterdam, 1999.

Trott, A.R. and T. Welch: *Refrigeration and Air-conditioning,* Butterworth-Heinemann, Inc., Woburn, MA, 1999.

Wang, S.K.: *Handbook of Air Conditioning and Refrigeration,* 2nd Edition, McGraw-Hill Professional Book Group, New York, NY, 2000.

Whitman, B.C., B. Johnson, and J. Tomczyk: *Refrigeration and Air Conditioning Technology,* 4th Edition, Delmar Publishers, Albany, NY, 1999.

Zimm, C., A. Jastrab, A. Stenberg, V. Pecharsky, K. Gschneidner, Jr., M. Osborne, and I. Anderson, "Description and Performance of a Near-room Temperature Magnetic Refrigerator," *Adv. Cryo. Engin.* **43**, 1759–1766 (1998).

Web Reference

Air-Conditioning and Refrigeration Institute: http://www.ari.org/

REGELATION. The phenomenon that occurs when two pieces of ice are rubbed together, the pressure causing the ice to melt at the surfaces of contact while the temperature drops, and, on relieving the pressure, the two surfaces freeze together, producing one mass of ice. This phenomenon is due to reduction of the freezing point of water (melting point of ice) under increased pressure. At very high pressures, the relationship changes, and the melting point of ice increases steadily with increasing pressure.

REGENERATION. 1. Restoration of a material to its original condition after it has undergone chemical modification necessary for manufacturing purposes. The most common instance is that of cellulose for rayon production. The wood pulp used must first be converted to a solution by reaction with sodium hydroxide and carbon disulfide; in this form (cellulose xanthate) it can be extruded through spinnerettes. After this operation it is regenerated to cellulose by passing it through acid (viscose process). Collagen can also be regenerated by acid treatment after it has been purified for use in food products by alkaline solution.

2. Renewal or reactivation of a catalyst that has accumulated reaction residues such as coke, usually accomplished by passage of steam or reducing gases over the catalyst bed.

3. Replenishing the sodium ions of a zeolite of similar ion-exchange agent by treatment with sodium chloride solution. Molecular sieves are regenerated by heat removal of the water ($200°C$), followed by treatment with an inert gas.

REINFORCED PLASTICS. Reinforced plastics are commonly referred to as composites or, more specifically, polymer composites. Not all composites are reinforced plastics; ceramic/metal-matrix composites and concrete are good examples of nonpolymeric composites. Reinforced plastics are also referred to RP, FRP (fiberglass-reinforced plastic), and GRP (glass-reinforced plastic) interchangeably.

Common to all reinforced plastics are two ingredients, resin and reinforcement. Resin is an inorganic material, usually of high molecular weight, that can be molded and set into a final shape. Resins are of two basic types. Thermoplastic resins; soften upon heating, are shaped in a mold, and retain that shape when cooled. Thermosetting resins are placed in a mold and cured by the use of a catalyst, heat, or both, until they harden in the shape of the mold.

The second main ingredient in reinforced plastic is the reinforcement, e.g., fibers of glass, carbon, boron, mineral, cellulose, or polymers. See also **Reinforcing Agent**. Reinforcements can be configured in many ways, such as continuous or chopped strands, milled fibers, rovings, tows, mats, braids, and woven fabrics.

Reinforced plastics may also include fillers (qv), which are inexpensive materials such as calcium carbonate used to displace resin and reduce cost. Curing agents (catalysts), promoters, inhibitors, and accelerators, which affect thermosetting resin cure; colorants. Release agents (qv) to facilitate removal from the mold; and other additives which can impart a wide variety of properties to the finished part, such as fire resistance, electrical conductivity, static dissipation, and ultraviolet resistance.

It is important to note that reinforced plastics remain a combination of materials differing in form or composition on a macro scale. The main constituents (resin, reinforcement, and filler) retain their identities and do not dissolve or merge into each other; rather, they act in concert. These components can be physically identified and exhibit an interface between each other.

See also **Plastics**.

REINFORCING AGENT. 1. One of numerous fine powders used in rather high percentages to increase the strength, hardness, and abrasion resistance of rubber, plastics, and flooring compositions. The reinforcing effect is in general a function of the particle size of the powder. The finest of all is channel carbon black, whose surface area may be as great an 18 acres per pound. Other widely used reinforcing agents are thermal and furnace blacks, magnesium carbonate, zinc oxide, hard clay (kaolin), and hydrated silicas. Though some reinforcing agents have positive coloring properties, the term should not be used as a general synonym for pigment.

2. Fibers, fabric, or metal insertions in plastics, rubber, flooring, etc., for the purpose of imparting impact strength and tear resistance.

See also **Pigments (Inorganic)**; and **Pigments (Organic)**.

RELEASE AGENTS. Release agents are substances that control or eliminate the adhesion between two surfaces. They are used to expedite many industrial handling and processing operations, particularly of polymers (see **Plastics processing**). They are known by a variety of terms descriptive of their effect, including abherents, abhesives, antiblocking agents, antistick agents, external or surface lubricants, mold-release agents, parting agents, and slip agents. They find considerable use in the adhesive, food, furniture, glass, metal, plastics, and rubber industries.

Release agents function by either lessening intermolecular interactions between the two surfaces in contact or preventing such close contact. Thus, they can be low surface-tension materials based on aliphatic hydrocarbon, fluorocarbon groups, or particulate solids. The principal categories of material used are waxes, fatty acid metal soaps, other long-chain alkyl derivatives, polymers, and fluorinated compounds.

Some of these processing aids are incorporated into the bulk of the material rather than being directly applied to the surface and are known as internal release agents. In the process of migration to the surface they have an inevitable internal lubrication effect. Most internal lubricants also have an external lubricant effect, particularly at higher concentrations because usually some material finds its way to the exterior.

Product Types and Requirements

Release agents are available in a wide variety of forms and are formulated for numerous modes of application. Product types include neat liquids, solutions, powders, flakes, pastes, emulsions, dispersions, sprays, and films. Some are general-purpose inert products intended for a broad range of applications, including home consumer uses, whereas others are highly specific, designed to react *in situ* with particular substrates. There is a trend away from products containing volatile organic compounds (VOCs) and ozone-depleting substances, and toward water-based and high solids formulations. Users are also switching from the heavier metal soaps. Many products serve more than one processing function and contain other additives; such products are usually proprietary.

The choice of a release agent depends on the process conditions involved and the nature of the contacting substrates. Apart from the obvious ease of release, other important requirements are minimal buildup of residues on mold substrate, minimal effect on the molded article, adequate film-forming ability, compatibility with secondary operations and other processing parameters, health and safety requirements, and cost.

Classification of Release Agents

The diversity of release products and the wide range of release problems make classification difficult. One approach is by product form, with subdivisions such as emulsions, films, powders, reactive or inert sprays, reactive coatings, and so on. Another approach is by application, e.g., metal casting, rubber processing, thermoplastic injection molding, and food preparation and packaging.

A classification by chemical type is given in Table 1. It does not attempt to be either rigorous or complete. The broad classes of release materials available are given in the chemical class column, the principal types in the chemical subdivision column, and one or two important selections in the specific examples column.

Mechanism of Release

Release agents are used to reduce the adhesion between materials. The main mechanisms of adhesion are interdiffusion, electrostatic attraction, surface energetics and wettability, and mechanical interlocking. The important surface and interfacial properties include surface topography (composition, structure, and roughness), surface tension or energy and its effect on substrate and adhesive wettability, the thermodynamic work of adhesion, and chemical reactivity at the interface. Nonsurface properties of the thin-film interfacial phase include its ability to set to a cohesive solid after wetting the substrate and its viscoelastic response to deformation controlled by factors such as degree of crystallinity, molecular weight and distribution, number of cross-links, and presence of fillers. Surface treatments that enhance adhesion do so by removing weak boundary layers, changing surface topography, or changing the chemical nature of the surface and by introducing strongly attracting functional groups.

An inversion of these arguments indicates that release agents should exhibit several of the following features: (1) act as a barrier to mechanical interlocking; (2) prevent interdiffusion; (3) exhibit poor adsorption and lack of reaction with at least one material at the interface; (4) have low surface tension, resulting in poor wettability, i.e., negative spreading coefficient, of the release substrate by the adhesive; (5) low thermodynamic work of adhesion; (6) low intermolecular forces across the interface, e.g., an absence of electrostatic or polar attractions; (7) display nonsetting or

TABLE 1. CHEMICAL CLASSIFICATION OF RELEASE AGENTS

Chemical class	Chemical subdivisions	Specific examples
waxes	petroleum waxes	paraffin wax, microcrystalline wax
	vegetable waxes	carnauba wax
	animal waxes	lanolin
	synthetic waxes	polyethylene wax, Fischer-Tropsch wax
fatty acid metal soaps	metal stearates	magnesium stearate, zinc stearate
	other	calcium ricinoleate
other long-chain alkyl derivatives	fatty esters	diethylene glycol monostearate, hydrogenated castor oil
	fatty amides and amines	ethylenebis(stearamide), oleyl palmitamide
	fatty acids and alcohols	palmitic acid, oleic acid
polymers	polyolefins	polypropylene
	silicones	polydimethylsiloxane, polymethyl(nonafluorohexyl)-siloxane
	fluoropolymers	polytetrafluoroethylene, poly(fluoroethers)
	natural polymers	cellophane
	other	polyoxalkylenes, poly(vinyl alcohol)
fluorinated compounds	fluorinated fatty acids	perfluorolauric acid
inorganic materials	silicates	talc
	clays	kaolin, mica
	other	silica, graphite

low cohesive interactions within the release phase; and (8) provide a weak boundary layer.

Many of these features are interrelated. Finely divided solids such as talc are excellent barriers to mechanical interlocking and interdiffusion. They also reduce the area of contact over which short-range intermolecular forces can interact. Because compatibility of different polymers is the exception rather than the rule, preformed sheets of a different polymer usually prevent interdiffusion and are an effective way of controlling adhesion, provided no new strong interfacial interactions are thereby introduced. Surface tension and thermodynamic work of adhesion are interrelated, as shown in equations 1, 2, and 3, and are a direct consequence of the intermolecular forces that also control adsorption and chemical reactivity.

The work of adhesion, W_A, is the change in energy per unit surface area when two interfaces come into contact, as given in equation 1 where σ_1 and σ_2 are the surface energy of each phase and σ_{12} the interfacial energy between them.

$$W_A = \sigma_1 + \sigma_2 - \sigma_{12} \tag{1}$$

For liquid systems these surface energies expressed in mJ/m^2 are numerically equivalent to the surface tensions in mN/m(=dyn/cm). If the adhesive is phase 1 and the release coating is phase 2, then the spreading coefficient, S, of 1 on 2 is as given in equation 2.

$$S = \sigma_2 - \sigma_1 - \sigma_{12} \tag{2}$$

Because the work of cohesion, W_C, of the adhesive is $2\,\sigma_1$, equation 3 follows.

$$W_A - W_C = S \tag{3}$$

This simple yet fundamental relationship states that when the forces of adhesion are less than those of cohesion, i.e., when W_A is less than W_C, the spreading coefficient is negative and failure will be at the interface between the adhesive and release coating, the desired situation in a release coating application. Conversely, when the spreading coefficient is positive, undesirable cohesive failure in the adhesive or release coating will be the case. These equations clearly demonstrate the importance of the release coating surface tension being lower than that of the adhesive.

The intermolecular forces of adhesion and cohesion can be loosely classified into three categories: quantum mechanical forces, pure electrostatic forces, and polarization forces. Quantum mechanical forces give rise both to covalent bonding and to the exchange interactions that balance the attractive forces when matter is compressed to the point where outer electron orbits interpenetrate. Pure electrostatic interactions include Coulomb forces between charged ions, permanent dipoles, and quadrupoles. Polarization forces arise from the dipole moments induced in atoms and molecules by the electric fields of nearby charges and other permanent and induced dipoles.

The forces involved in the interaction at a good release interface must be as weak as possible. They cannot be the strong primary bonds associated with ionic, covalent, and metallic bonding; neither are they the stronger of the electrostatic and polarization forces that contribute to secondary van der Waals interactions. Rather, they are the weakest of these types of forces, the so-called London or dispersion forces that arise from interactions of temporary dipoles caused by fluctuations in electron density. They are common to all matter. The surfaces that are solid at room temperature and have the lowest dispersion-force interactions are those comprised of aliphatic hydrocarbons and fluorocarbons.

Among the hydrocarbons, the lowest surface tension values are found for surfaces comprising closely packed methyl groups. This is a characteristic of waxes and oriented long-chain alkyl derivatives.

Because the forces of attraction prevail when molecules are brought into sufficiently close proximity under normal conditions, release is best effected if both the strength of the interaction and the degree of contact are minimized. Aliphatic hydrocarbons and fluorocarbons achieve the former effect, finely divided solids the latter. Materials such as microcrystalline wax and hydrophobic silica combine both effects.

MICHAEL J. OWEN
Dow Corning Corporation

Additional Reading

Lammerting, H.: "Release Agents," in W. Gerhartz, ed., *Ullmann's Encyclopedia of Industrial Chemistry,* Vol. A23, VCH Publishers Inc., New York, NY, 1993.

Owen, M.J. and J.D. Jones: "Silicone Release Coatings," in J.C. Salamone, ed., *Polymeric Materials Encyclopedia,* Vol. 10, CRC Press, Boca Raton, FL, 1996.

Satas, D.: *Coatings Technology Handbook,* Marcel Dekker, Inc., New York, NY, 1991.

Stepek, J. and H. Daoust: *Additives for Plastics,* Springer-Verlag, New York, NY, 1983.

REMSEN, IRA (1846–1927). An American chemist born in New York. He began his career in medical practice but abandoned it for chemistry and went to Germany to study. He received his doctorate in chemistry from the University of Gottingen in 1870. Returning to the U.S. he taught physics and chemistry at Williams. He was later invited to join the staff of John Hopkins University where he became head of the department of chemistry. There he established the first graduate curriculum in chemistry in the U.S. based on the system then in use in Germany. In 1879, saccharin was discovered in his research laboratory. He wrote several widely used textbooks and founded the *American Journal of Chemistry,* which later merged with JACS. Remsen served as President of Johns Hopkins from 1901–1912. He was also recognized as one of the great teachers of chemistry.

REPELLENT. 1. A substance that causes and insect of animal to turn away from it or reject it as food. Repellents may be in the form of gases (olfactory), liquids, or solids (gustatory). Standard repellents for mosquitos, ticks, etc., are citronella oil, dimethyl phthalate, *n*-butylmesityl oxide oxalate, DEET, and 2-ethyl hexanediol-1,3. Actidione is the most effective rodent repellent, but is too toxic and too costly to use. Copper naphthenate and lime/sulfur mixtures protect vegetation against rabbits and deer. Shark repellents are copper acetate or formic acid mixed with ground asbestos. Bird repellents are chiefly based on taste, but this sense varies widely with the type of bird so that generalization is impossible. α-Naphthol, naphthalene, sandalwood oil, quinine, and ammonium compounds have been used, with no uniformity or result.

2. A substance that, because of its physicochemical nature, will not mix or blend with another substance. All hydrophobic materials have water-repellent properties due largely to differences in surface tension or electric charges, e.g., oils, fats, waxes, and certain types of plastics. Silicone resin coatings can keep water from penetrating masonry by lining the pores, not by filling them; they will not exclude water under pressure.

REPPE PROCESS. Any of several processes involving reaction of acetylene (1) with formaldehyde to produce 2-butyne-1,4-diol which can be converted to butadiene; (2) with formaldehyde under different conditions to produce propargyl alcohol and, form this, allyl alcohol; (3) with hydrogen cyanide to yield acrylonitrile; (4) with alcohols to give vinyl ethers; (5) with amines or phenols to give vinyl derivatives; (6) with carbon monoxide and alcohols to give esters of acrylic acid; (7) by polymerization to produce cyclooctatetraene; and (8) with phenols to make resins. The use of catalysts, pressures up to 30 atm, and special techniques to avoid or contain explosions are important factors in these processes.

See also **Acetylene**.

RESINS (Acetal). These are thermoplastic resins, obtainable both as homopolymers and copolymers, and produced principally from formaldehyde or formaldehyde derivative. Acetal resins have the highest fatigue endurance of commercial thermoplastics. A variety of ionic initiators, such as tertiary amines and quaternary ammonium salts, are used to effect polymerization of formaldehyde. Chain transfer, shown by the following reactions, controls the molecular weight of resulting resins:

Step 1, initiation of new chain:

$$-CH_2OCH_2O^- + H_2O \longrightarrow -CH_2OCH_2OH + OH^-$$

Step 2, reaction of growing chain with H_2O, releasing hydroxyl ion:

$$OH^- + CH_2O \longrightarrow HOCH_2O^-$$

Step 3, end-capping of high-molecular-weight polyoxymethylene glycol to provide stable commercial resin:

$$HOCH_2O^- + nCH_2O \longrightarrow HOCH_2O(CH_2O)_{n-1}CH_2O^-$$

Starting ingredients may be formaldehyde or the cyclic trimer trioxane, $CH_2OCH_2OCH_2O$. Both form polymers of similar properties. Boron trifluoride of other Lewis acids are used to promote polymerization where trioxane is the raw material.

Acetals provide excellent resistance to most organic compounds except when exposed for long periods at elevated temperatures. The resins have limited resistance to strong acids and oxidizing agents. The copolymers and some of the homopolymers are resistant to the action of weak bases. Normally, where resistance to burning, weathering, and radiation are required, acetals are not specified. The resins are used for cams, gears, bearings, springs, sprockets, and other mechanical parts, as well as for electrical parts, housings, and hardware.

RESINS (Acrylonitrile-Butadiene-Styrene). Commonly referred to as ABS resins, these materials are thermoplastic resins which are produced by grafting styrene and acrylonitrile onto a diene-rubber backbone. The usually preferred substrate is polybutadiene because of its low glass-transition temperature (approximately $-80°C$). Where ABS resin is prepared by suspension or mass polymerization methods, stereospecific diene rubber made by solution polymerization is the preferred diene. Otherwise, the diene used is a high-gel or cross-linked latex made by a hot emulsion process.

ABS resins possess an attractive balance of impact resistance, hardness, tensile strength, and elastic modulus properties. The temperature range is wide—from -40 to $107°C$ (-40 to $225°F$). Other advantages include chemical resistance, high gloss, and nonstaining properites. The dimensional stability of ABS is good and creep resistance is excellent. The resins exhibit low water absorption or volume change at varying humidities.

Commercial ABS is in the form of custom color-matched compounded pellets, or granular resin for compounding or alloying with other plastics. A representative alloying ingredient is polyvinyl chloride (PVC). Almost all standard thermoplastic converting processes can be used with ABS plastics. Injection-molded parts include telephone sets, refrigerator parts, plumbing fixtures, fittings, radio, television, and appliance housings, and auto parts. ABS can be extruded into sheet, pipe, and various cross sections.

Thermoforming of large surface areas and deep draws from sheet stock are possible. Examples of parts which involve extrusion and subsequent thermoforming include lawnmower housings, refrigerator liners, pipe and conduit, vehicle bodies, snow-mobile shrouds, and camper bodies. The various thermoforming techniques applicable include plug and air assist, vacuum snapback, vacuum-plug forming, and drape forming.

The compatibility of ABS with other plastics makes them useful as impact modifiers and as processing additives with many other polymers to achieve a variety of final product specifications. Substitution of α-methyl styrene for styrene increases heat-distortion temperatures; or of methacrylonitrile for acrylonitrile improves barrier properties to gases such as carbon dioxide in connection with carbonated beverage containers. Over 75 grades of ABS are commercially available, including self-extinguishing, electroplating, antistatic expandable, glass-rein-forced, high-heat, cold-forming, and low-gloss sheet grades.

Three polymerization steps are involved in ABS manufacture. The process is shown in block diagram format in Fig. 1.

Step 1—Polybutadiene rubber is formulated by feeding butadiene, water, an emulsifier, and catalyst into a glass-lined reactor. This is an exothermic reaction. About 80% conversion is achieved in a period of about 50 hours. The residual butadiene monomer is recovered by steam-stripping and recycled.

Step 2—Polybutadiene rubber is further polymerized, but in the presence of styrene and acrylonitrile monomers. This is done in low-pressure reactors under a nitrogen atmosphere. In this operation, the monomers are grafted onto the rubber backbone through the residual unsaturation remaining from the first step.

Step 3—In a separate step, styrene-acrylonitrile (SAN) resin is prepared by emulsion, suspension, or mass polymerization by free-radical techniques. The operation is carried out in stainless-steel reactors operated at about $75°C$ ($167°F$) and moderate pressure for about 7 hours. The final chemical operation is the blending of the ABS graft phase with the SAN resin, plus adding various antioxidants, lubricants, stabilizers, and pigments. Final operations involve preparation of a slurry of fine resin particles (via chemical flocculation), filtering, and drying in a standard fluid-bed dryer at $121–132°C$ ($250–270°F$) inlet air temperature.

Assistance of the Marbon Division, Borg-Warner Corporation in preparation of this entry is appreciated.

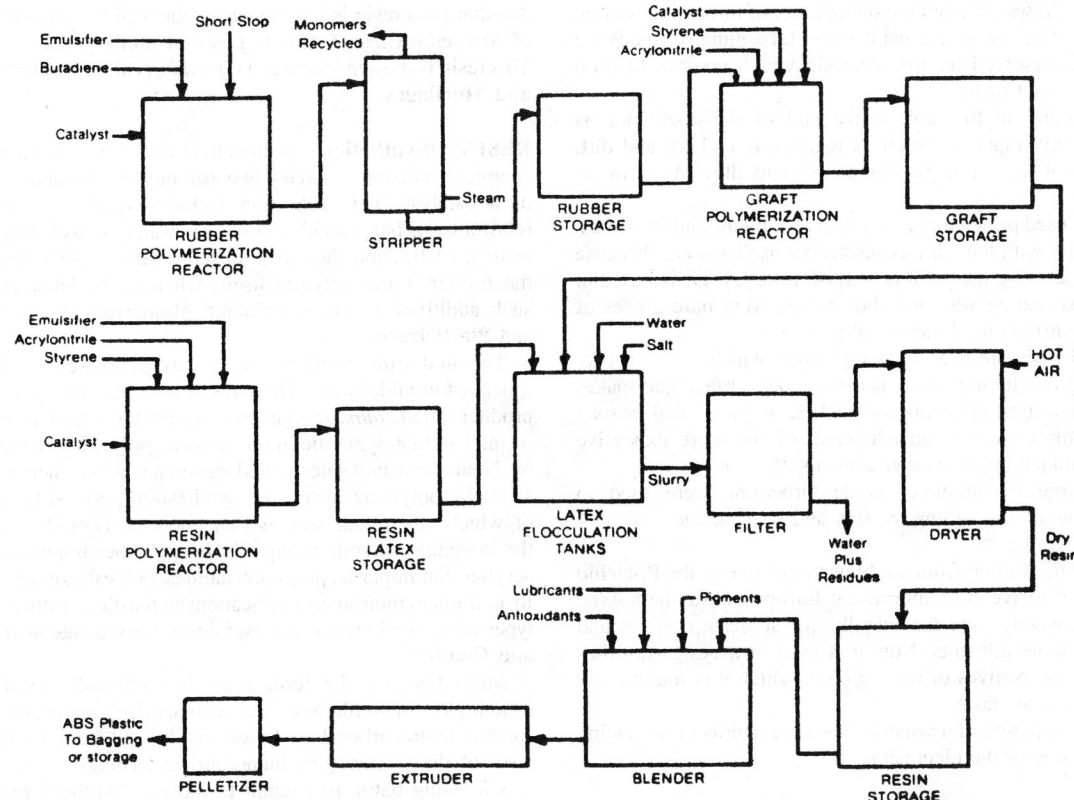

Fig. 1.   Process for manufacturing acrylonitrile-butadiene-styrene (ABS) resins. (*Marbon Div., Borg-Warner Corp.*)

RESINS (Natural).

RESINS (Natural).   Complex compounds composed of carbon, hydrogen, and relatively small amounts of oxygen, which are secreted in various tissues of many plants. In the pine family, where resins are very common, they are secreted as oleoresins in resin canal cells, which break down finally, producing resin canals. These canals appear as longitudinal ducts in the sapwood and inner bark, connected laterally by resin canals in the compound wood rays, thus forming an extensive network. A common name given to the oleoresin in this group is pitch, the sticky juice that exudes from the plant wherever it is wounded. On exposure to the air the volatile oil in this pitch (oil of turpentine) gradually evaporates, leaving a clear hard glassy substance, the resin, which forms a protective coating over the wound.

Most resins have the same physical properties, being clear, translucent, and of a yellow or brownish color. Amber, a fossil resin, is a more or less familiar example. Resins are insoluble in water, but soluble in common organic solvents such as ether and alcohol. All resins burn with a sooty flame. Resins seem to be mainly of value to the plan in that they form protective coverings against the entrance of disease-producing organisms and also prevent excessive loss of water from the thin-walled tissues exposed in the wound.

Resins are separated into several classes. Many of the resins contain almost no volatile oil and are hard, without taste or odor. These are the varnish or hard resins. Other resins, when removed from the plant in which they are formed, and dissolved in volatile oils, form a thick semi-solid mass: these are the oleoresins. In still other cases the resin occurs in combination with a gum, forming a gum resin.

Hard Resins

Several of the hard resins, used mainly for making varnishes, are called copals. Most of them come from Africa and are either found in fossil form or obtained from living plants. Other copals come from Australia, New Zealand and East Indian Islands. The plants that form them are members of the legume and pine families. The African copals are products of several species of *Trachylobium*, fairly large trees growing in east Africa and Madagascar. The best resin from these trees occurs in a fossil form, often deeply buried in the ground—sometimes in regions where the trees no longer grow. These resins dissolve slowly and are used in making varnishes, which are very durable. A South American tree of large size, *Hymenaea*

courbaril, also of the Leguminoseae, yields a very similar resin, which is also found in lumps in the ground around the trees, and used in varnishes.

Another copal is obtained from *Agathis australis*, a very large coniferous tree native in Australia and New Zealand, where it is known as the Kauri pine. Like the other copals, that from the Kauri pine is found in lumps buried in the ground. Most of these lumps are 1 or 2 inches in diameter, but some are much larger, weighing up to 100 pounds. Nearly all of this resin comes from the northern part of North Island of New Zealand. It is frequently called Kauri gum, though it is not a gum, but a true copal resin. Another group of hard resins, known as dammar resins, is obtained from many different trees growing in southern Asia and the East Indian Islands. These resins dissolve readily in alcohol, forming spirit-varnishes.

One of the commonest and most important of the hard resins is rosin, obtained by distilling the pitch, or turpentine, which is a product of several of the native pines of the southeastern United States. This rosin, also known as colophony, is a very important product of that region. Originally the turpentine was obtained by chopping a deep hollow in the base of the trunk of the tree and allowing it to fill up with the turpentine, which was then scooped out. This method was very destructive and wasteful, since much of the oleoresin, turpentine, was lost during the process. The weakened trees were easily blown down.

Now turpentine is obtained by cutting V-shaped gouges in the bark and inserting metal gutters beneath the gouges. These gutters carry the turpentine to a cup placed underneath. As soon as the cut is made, turpentine begins to flow and continues to do so for two or three days, gradually slowing as the drying turpentine allows resin to accumulate and plug the wounds. A new flow is obtained by cutting off a narrow strip of bark from the upper edge of the cut. The process is continued as long as the pitch will flow, which is usually all summer and well along into the late fall. Each tree may be turpentined for 6 to 7 successive years or even longer before it ceases to be profitable.

The crude turpentine collected in the cups and the product that has dried on the wound of the tree are removed and carried to the still. Here the turpentine, to which a little water is added, is carefully heated to drive off the oil of turpentine present, together with the water added. The distillate is condensed by passing it through a coil around which cold water is flowing. It is collected in a barrel or any suitable container. The two substances, water and oil of turpentine, which make up the distillate, are immiscible and soon separate, the lighter oil of turpentine rising to the top and floating on the water, which is drawn off from the bottom. Oil of turpentine is often called spirits of turpentine, or, in the paint trade, turpentine. In medicine the word turpentine is reserved for the oleoresin which upon distillation yields oil of turpentine and rosin.

The residue remaining in the tank at the end of the distillation is skimmed to remove any impurities such as twigs, bits of bark and dirt, and run into vats to cool. Then it is put into barrels and allowed to harden, forming rosin.

Oil of turpentine is used principally as a solvent for paints and varnishes, because it mixes readily with the various substances used and also because it evaporates quickly, causing the paint or varnish to dry. It is also used in making such things as sealing wax and shoe polish. Very pure grades of turpentine (the oleoresin) are used medicinally.

Large quantities of rosin are used in sizing paper, which makes it take ink without spreading or blotting, gives it a smoother surface and makes it heavier. Rosin is also used in cheaper varnishes, in paints, and in soap making. It is furthermore used as an adulterant of the more expensive resins. Linoleum manufacturers use large amounts of rosin.

In early times, large quantities of crude turpentine were used to waterproof the rigging of the sailing vessels and to calk the seams of the hull.

Mastic is a hard resin exuding from the branches of one of the Pistachio trees, *Pistacia lentiscus*, native of Mediterranean Europe and southern Asia. Formerly, it was extensively used medicinally, for stomach troubles and dysentery, as well as other ailments. Now it is used in making varnishes and in lithographic work. Natives of the region in which it is found chew mastic, which has a pleasant taste.

Since turpentine is a mixture of a volatile substance, spirits of turpentine and a hard resin, it is one of the oleoresins.

Oleoresins

Canada balsam is one of the oleoresins. It is obtained from the bark of *Abies balsamea*, the common balsam fir of northern North America. Canada balsam, because its refractive index is so near that of glass, is much used in optical work and in preparing materials for examination with a microscope.

Little used today is Dragon's blood, an oleoresin obtained from the fruits of *Daemonorops draco*, a native palm of southeastern Asia and the Molucca Islands. The resins exudes from the surface of the ripening fruits. It is removed from them by boiling in water. The resin is then moulded into balls or long sticks. It is sometimes used in making varnishes and lacquers.

True lacquer, obtained from the juice of *Rhus verniciflua*, a sumac tree of southeastern Asia, is another oleoresin. To obtain the juice lateral cuts are made in the bark. The exuded sap is collected not only from these cuts but from small branches which are cut off and soaked in water. The juice is cleaned of any foreign substances by straining it through hemp cloth. By slow heating, either artificial or by the sun, the juice is evaporated and stored until used. Lacquer is a poisonous substance, causing intense irritation of the skin in many people. Others seem to be immune. Lacquer is usually applied over some soft wood, commonly soft pine, the pores of which have first been filled by rubbing in a paste of rice and resin, followed by a paste of soft clay and resin. The surface is then covered with cloth and layer after layer of lacquer put over that. Each layer is allowed to dry and rubbed down very smooth before the next layer is added. Any color to be added is mixed with the lacquer, with each colored layer covered by a clear layer before another is put on. The final product is a thick covering composed of many thin layers of lacquers. If this is carved the edges of the carving, on careful examination, will show the fine lines separating the different layers. Lacquering is a very old industry, having been carried on in China since the sixth century.

Certain resins occur in combination with fragrant volatile oils. One of these is benzoin, obtained from *Styrax benzoin* by cutting notches in the bark and allowing the resin to collect in them. It is used in making perfumes, in incense, and as a source of benzoic acid, used medicinally.

Another fragrant oleoresin is storax, obtained from *Liquidambar orientalis*, a medium-sized tree growing in southwestern Asia. The resin is obtained by boiling the bark and wood of young branches. It is used medicinally and also in incense.

Gum Resins

Gum resins include myrrh, which exudes from the trunk and branches of *Commiphora myrrha*, a tree growing in the region around the Red Sea. The lumps of resin are used medicinally, and also in making incense.

Another gum resin is frankincense, obtained by cutting notches in the stem of *Boswellia carterii*, which grows in northeastern Africa and in Arabia. This resin is used in incense. *Asafoetida* is also a gum resin. See also **Gums and Mucilages**.

RESINS (Synthetic). A manufactured high polymer resulting from a chemical reaction between two (or more) substances, usually with heat or a catalyst. This definition includes synthetic rubbers and silicones (elastomers), but excludes modified, water soluble polymers (often called resins). Distinction should be made between a synthetic resin and a plastic; the former is the polymer itself, whereas the latter is the polymer plus such additives as filters, colorant, plasticizers etc. See also **Elastomers**; and **Plasticizers**.

The first truly synthetic resin was developed by Baikeland in 1911 (phenol-formaldehyde). This was soon followed by a petroleum-derived product called *coumarone-indene*, which did indeed have the properties of a resin. The first synthetic elastomer was polychloroprene (1931) originated by Nieuwland and later called neoprene. Since then many new types of synthetic polymers have been synthesized perhaps the most sophisticated of which, are nylon and its congeners (polyamides, by Carothers), and the inorganic silicone group (Kipping). Other important types are alkyds, acrylics, aminoplasts, polyvinyl halides, polyester, epoxies, and polyolefins. In addition to their many applications in plastics, textiles, and paints, special types of synthetic resins are useful as ion-exchange media. See also **Paints and Coatings**.

Note: Because the term *resin* is so broadly used as to be almost meaningless, it would be desirable to restrict it application to natural, organsoluble, hydrocarbon-based products derived from trees and shrubs. But in view of the tendency of inappropriate terminology to "gel" irreversibly, it is a losing battle to attempt to replace "synthetic resin" with the more precise "synthetic polymer." See also **Gums and Mucilages**; and **Resins (Natural)**.

RESONANCE. 1. In chemistry, resonance (or mesomerism) is a mathematical concept based on quantum mechanical considerations (i.e., the wave functions of electrons). It is also used to describe or express the true chemical structure of certain compounds that cannot be accurately represented by any one valence-bond structure. It was originally applied to aromatic compounds such as benzene, for which there are many possible approximate structures, none of which is completely satisfactory. See also **Benzene**.

The resonance concept indicates that the actual molecular structure lies somewhere between these various approximations, but is not capable of objective representation. The idea can be applied to any molecule, organic or inorganic in which and electron pair bond is present. The term *resonance hybrid* denotes a molecule that has this property. Such molecules do not vibrate back and forth between two or more structures, nor are they isotopes or mixtures. The resonance phenomenon is rather an idealized expression of an actual molecule that cannot be accurately pictured by any graphic device.

2. In the terminology of spectroscopy, resonance is the condition in which the energy state of the incident radiation is identical with that of the absorbing atoms, molecules or other chemical entities. Resonance is applied in various types of instrumental analysis such as nuclear resonance absorption and nuclear magnetic resonance. See also **Absorption (Process)**; and **Nuclear Magnetic Resonance (NMR) and Magnetic Resonance Imaging (MRI)**.

Note: The multiple meanings of *nucleus* and *resonance* can be a source of confusion, especially when these terms are closely associated, as in *nuclear magnetic resonance* and *resonance of molecular nucleus*. In the first of these expressions, *nucleus* is used in sense (1) under "nucleus," and *resonance* in sense (2) under "resonance." In the second expression, *nucleus* is used in sense (3) under "nucleus" and resonance in sense (1) under *resonance*.

RETIFICATION. The enrichment or purification of the vapor during the distillation process by contact and interaction with a countercurrent stream of liquid condensed from the vapor.

See also **Reflux**.

RETORTING. 1. A process much used in the early years of chemistry for destructive distillation of heavy organic liquids and for laboratory separations. It involves the use of a cylindrical vessel made of glass (for laboratory work), fireclay, or metal, with a neck bent at a downward angle to facilitate distillation. For gas manufacture, the equipment is built on a heavier scale to handle destructive distillation of coal. 2. Processing shale oil. 3. Heating canned or pouched foods with steam to stop bacterial growth. 4. Volatilization of mercury from gold and silver amalgams.

REVEGETATION. Generally refers to the purposeful seeding and planting of an area that once was covered with grass, trees, shrubs, forbs, etc., but which was denuded of vegetation because of a natural disturbance, such as a lightning-caused fire, or because of an artificial disturbance, such as mining and construction projects. A great deal of attention has been given to seeding and planting programs in connection with large areas that have become barren as the result of unplanned strip mining of the coal fields, notably in Appalachia.

Extraction of coal by surface mining has disturbed about two million acres of land in 26 states of the United States, approximately 50% of this disturbance taking place between 1965 and 1975. Ninety percent is on land owned by mining companies, farmers, and other private interests. Even though there has been extensive damage, most of the mining was done under then existing state laws and regulations. Regulations are becoming more rigid. Land disturbed by surface mining is generally an undeveloped land resource. Evidence indicates, however, that all but a small percentage of such land is capable of producing some tangible or intangible societal benefit. Mined sites, appropriately reshaped, can be developed for many uses. In the Midwestern and Appalachian coal fields, they have been developed, in relatively few instances to date, into successful and profitable production of agricultural and horticultural crops. Conversion of such areas into recreational and wildlife developments is a popular target. Industrial and residential sites have been established on areas disturbed by surface mining. Sites planted to trees yield various forest products. As of the early 1980s, some progressive mining companies have identified opportunities and have developed areas that provide social and economic benefits. Such reclamation and revegetation programs, coupled with superior surface mining methods (See also **Coal**), can yield vitally needed coal without extensive spoiling of the land.

Spoil Evaluation

In evaluating old sites that were mined by conventional methods, prior to the use of such techniques as haulback methods, (valley fill or head-of-the-hollow method, mountaintop leveling, etc.), it is necessary to determine both chemical and physical characteristics of the spoil. Chemical factors include toxic ion concentrations and availability of nutrients. The important physical features include particle size, texture, and color of the spoil. Spoil evaluations are often complicated by changes that occur in weathering. Each rock stratum has distinct chemical and physical characteristics. Many of these properties change when fragments of rock are exposed to moisture, light, air, and temperature variations. Release of iron, sulfur, manganese, and aluminum compounds may result in off-site pollution and failure of vegetation. The release of essential nutrients, such as phosphorus, potassium, calcium, and magnesium may, on the other hand, improve the growth of vegetation, reduce acidity problems, and increase the possibilities of reclamation.

Physically, the spoil is a heterogeneous mass of rock fragments derived from the rock strata above the coal. The size of rock fragments after mining and reshaping depends on the size of the basic particles and the strength of the cementing materials, which naturally hold small mineral particles together. Fine-grained shales strongly bonded together may break into hard, platy fragments that resist weathering. Coarse-grained sandstones weakly cemented together may disintegrate rapidly into sand. Spoil may be a dynamic material, more reactive than natural soils that have been exposed to the process of soil genesis for centuries. The chemical and physical properties of a spoil may change rapidly for several years. The rate of change will gradually lessen as characteristics of natural soils develop. Needless to say, however, time required for natural reclamation processes is exceedingly long and provides no relief to the immediate problem.

Evaluations of the chemical properties of a soil could be perplexing if an attempt were made to include all components that affect plant growth. Methods in use today rely on indicators for judging the interaction of many factors at one time. Soil reaction or hydrogen ion concentration (expressed as pH) is the most widely used indicator. Acidity alone may or may not affect the establishment of plants. Changes in acidity determine the concentration of toxic ions and nutrients in the soil solution. At pH levels below 4.0, two chemical changes may occur: Toxic ions, such as manganese, aluminum, and iron, become more available to plants and some essential nutrients become less so. When the pH of a spoil is between 5.5 and 7.0, the concentration of toxic ions in the soil solution decreases and more essential nutrients become available. A soil classification system for acidic spoils is given in Table 1. No system has been proposed for the alkaline or sodic spoils in the western U.S. coal fields. The same principles can be applied, but the toxic ions or cations may differ.

Other chemical analyses have been considered for identifying characteristics that result in specific nutrient deficiencies or toxicities. Phosphorus is often deficient (0 to 7 parts per million) in spoil material. Several laboratory analyses provide estimates of plant-available phosphorus. Total soluble salts in the soil solution may be important for the very strongly acid or alkaline spoils. This does not necessarily mean high concentrations of sodium salts, but salts of all anions and cations in the soil solution. Some consideration has been given to laboratory analyses that give concentrations of exchangeable aluminum, manganese, and hydrogen. These analyses can identify specific toxicities and thus permit the selection of tolerant plant materials. Research has been initiated to determine whether the heavy metals, such as copper, zinc, nickel, mercury, and cobalt, occur in concentrations toxic to plants.

Texture and clay mineralogy influence the degree of compaction that results from heavy equipment passing over the surface during mining and reshaping. There is evidence that compaction may limit plant growth by reducing water infiltration and nutrient release. Texture may also determine the rate of release of toxic ions or nutrients from soil particles. Smaller particles expose a larger surface area per unit volume to the forces of weathering than coarse fragments. This results in a more rapid release of chemicals. Surface color is important because it influences heat exchange at the spoil surface. Dark materials absorb solar energy so the spoil may attain temperatures lethal to plants. The degree of risk depends on the season of the year, the slope of the exposed surface, and the exposure of the site. The highest risk may occur during summer months on steep slopes of black or dark gray material facing south or west.

Preparation of the Site

Treatments to prepare the site for seeding or planting are determined after evaluation of site data and establishment of the land-management options and objectives. Grass and legumes often require more surface preparation than trees and shrubs. Uncompacted fill slopes and freshly reshaped surfaces may make acceptable seedbeds. Hard crusts form on many spoils; fine clay-size particles are consolidated by the drying action of wind and sun. This crust may be broken by rainfall, frost, or mechanical scarification. In many cases, ground cover will be denser if it is seeded into fresh spoil or where spoil surfaces have been broken by natural or mechanical scarification.

TABLE 1. SYSTEM FOR CLASSIFYING SPOILS (ACIDIC)

		Acidity	
Class number	Description	pH value	Extent on area sampled
1	Toxic	Less than 4.0	More than 75%
2	Marginal	Less than 4.0	50 to 75%
3	Acid[a]	4.0 to 6.9	50 to 75%
4	Calcareous	7.0 or more	More than 50%
5	Mixed	(Too varied to be classified as any above)	—

	Texture
Group	Description of texture
A	Chiefly sand, sandstones, or sandy shales
B	Chiefly loamy materials and silty shales
C	Chiefly clay and clay shales

Combine acidity and textural classes to describe spoil type.

[a] Acid spoils may be subdivided into two classes: pH 4.1 to 5.4; and pH 5.5 to 6.9.
Source: Northeastern Forest Experiment Station, Forest Service, U.S. Department of Agriculture.

Treatment to create special surface configurations may be required on toxic spoils, on steep slopes, or in geographical locations where precipitation is low or unfavorably distributed throughout the year. Rows or depressions, made by machinery, trap precipitation, moderate wind velocities, reduce evapotranspiration rates, and moderate spoil temperature extremes. The orientation of the furrows with respect to direction of slope, exposure to sun, or prevailing winds may determine the effectiveness of the treatment. Amendments to modify acidity should be applied weeks or months before seeding. Scarification will mix the ameliorating material into the top 6 to 8 inches of spoil and create a neutralized layer for root development. On extremely rocky spoils where scarification is not practical, frost action and infiltrating water may carry the neutralizing materials below the spoil surface. The rate of application of neutralizing material depends on the type of vegetation to be seeded, the present and predicted acidity of the surface spoil, and the neutralizing capacity of the material used. Agricultural limestone is preferred for most treatments, but it can be used only in areas accessible to application equipment. No practical system has been developed for applying large amounts of agricultural limestone to steeply sloping land. Small quantities of lime can be applied with a hydroseeder to steep slopes. Finely ground limestone or hydrated lime may be mixed with water to form a slurry. Repeated applications may be necessary to achieve high rates per acre. The cost of repeated treatments could make this procedure impractical. Alkaline power plant fly ash and bottom ash are neutralizing materials, but they are usually not as effective as limestone. Fly ash contains various quantities of essential plant nutrients, such as phosphorus, zinc, molybdenum, and boron. However symptoms of boron toxicity have been observed on plants growing on sites treated with large amounts of fly ash. Rock phosphate provides a slowly available source of phosphorus and helps to neutralize acidity. Spoil acidity may react with rock phosphate to slowly release plant-available phosphorus.

Fertilization is generally recommended for grass and legume crops. Nitrogen and phosphorus have been used to accelerate the growth of trees on spoils. Tests show that most spoils are deficient in nitrogen. Phosphorus occurs in various concentrations, but deficiencies often limit plant growth. Potassium is usually adequate. There is little information about the concentrations of other nutrients. For most land-management objectives, the formulation of the fertilizer makes little difference when equivalent rates are applied. High-analysis fertilizers, ammonium nitrate, triple superphosphate, and ammonium phosphate are often preferred. Using low-analysis fertilizers on sites that require high rates of application may increase the total soluble-salt concentration to levels toxic to some plants.

Rates of fertilization vary with the seeded crop, the inherent fertility of the spoil, and the land-management objective. Nitrogen and phosphorus at 50 and 22 pounds per acre (56 and 246 kilograms per hectare) respectively, are sufficient to establish grass on many spoils. Higher rates of phosphate are important for legumes. More consideration is being given to retreatment at regular intervals. Multiple treatments are attractive because they reduce the chance of unacceptable cover and increase the probability of achieving land use objectives within the shortest time.

Much research on fertilizer application remains to be done. Research results thus far show that placing selected fertilizers in or near the planting hole increases the growth of black locust. This leguminous species responded to additions of nitrogen and phosphorus placed near seedlings on extremely acid spoil. In other trials, direct-seeded black locust made more rapid growth after surface applications of nitrogen and phosphorus. Five tons of lime per acre per foot of soil increased the growth of loblolly, shortleaf, Virginia, and a hybrid (pitch/loblolly) pine planted in extremely acid spoil. Pitch pine did not respond to the liming treatments. Ten tons of lime per acre-foot reduced the growth of the pines.

Municipal waste products have been considered for surface-mine reclamation. Waste applications could improve soil texture, add essential plant nutrients, and provide mulch for seed and seedlings. Shredded or composted waste could be applied to the surface, and mechanical methods could be used to incorporate it into the spoil. Sewage sludge can be applied in water slurry or in dried form. Mixtures of shredded or composted waste and sewage sludge offer another possibility. Thus far, use of these materials has been restricted to relatively small demonstration areas for public health reasons.

Seeding and Planting

For each of the major coal-producing regions in the United States, there is a group of preferred plant species. These may be grouped under four major categories: (1) grasses; (2) forbs; (3) trees; and (4) shrubs. The grasses and forbs may be further classified as (a) temporary; (b) semipermanent; and (c) permanent, depending on life expectancy of the plant. Temporary species give prompt and effective site protection by reason of quick growth, fibrous roots, and ability to endure unfavorable site conditions. Semipermanent species are perennials that will be ultimately replaced by permanent vegetation. Under favorable conditions, permanent species will persist for many years.

The grasses are a varied group of plant species well suited to surface-mine restoration. Experience has shown that many grasses are adapted to wide ranges of climate, spoil texture, nutrient regimes, and toxic ion concentrations. Germination is usually rapid, and growth often produces a crop, or at least site protection, during the first growing season. Annual grasses often grown as agricultural crops may be used to provide quick site protection. These temporary quick-cover crops also may serve as nurse crops for slower-developing perennials. Some species are better adapted to summer seeding; others should be seeded in the fall.

Leguminous forbs are considered by many to be essential components of ground-cover mixtures. The fixation of atmospheric nitrogen by the legumes benefits associated plants. This assists in maintaining a vigorous ground cover and may reduce the need for retreatment. The leguminous forbs generally are less tolerant of toxic ion concentrations and require more phosphorus than the grasses.

Forbs not classified as legumes are being evaluated in the western United States. The emphasis is on species that are components of natural vegetative cover and are common invaders of disturbed areas. Initial evaluations indicate that some species may be useful for surface-mine revegetation.

Inoculation with specific strains of rhizobium bacteria stimulates nodulation on leguminous forbs. Commercial inoculants are available for the important legume species. Rhizobium bacteria may not survive or produce effective nodules in acidic spoils with pH below 5.0.

Grass and legume ground covers reduce tree growth, but this is often unavoidable because some state laws and regulations require a herbaceous cover. Manipulation of the species composition and reduction of ground-cover density may minimize the adverse effects.

Trees are often planted by hand, using one of several planting tools. On selected spoils, machine planting is possible. Most planting stock is small, 1 to 2 years old, and bare-rooted. Spacing between trees varies by species, site characteristics, and end-objectives. Plantings may be mixtures of several species, or pure plantings of one species. The arrangement may be random or designed to protect seedlings from environmental extremes. There is increasing interest in the establishment of trees by direct seeding. Success has been achieved with several pine species in Alabama. Black locust is used in West Virginia and Kentucky, The planting of shrub species has not been emphasized in surface-mine revegetation. Shrubs have little tangible value; benefits accrue from site protection, wildlife food and cover, and aesthetics.

Costs of Refuse Disposal and Land Reclamation

A predominant advantage of so-called strip mining over underground mining over the years and particularly with reference to smaller, previously unestablished mining operations, has been one of economics. Obviously, as costs of reclamation of stripped areas and of improved surface mining techniques are added to strip mining costs, the differential becomes much less. Past despoilage of the land thus has resulted from comparative economics. While some progressive operators have reduced their profits in order to lessen the damage to the land, such voluntary action is simply more than can be expected on a massive scale. Concerns with land spoilage prevention and reclamation of spoiled lands, unfortunately, are coincident timewise with a growing demand for coal. This, in the early 1980s, hardly felt the impact of the numerous coal conversion programs most of which were still under test and development for the gasification and liquefaction of coal in terms of the new coal technology for supplanting waning supplies of natural gas and petroleum energy sources. The practical, interim solution to the energy/environment problem must be one of compromise.

Note: See references listed at end of article on **Coal**.

The cooperation of W. T. Plass of the U.S. Dept. of Agriculture Forest Service, Northeastern Forest Experimentation, Princeton, West Virginia in making information available for this summary is gratefully acknowledged.

Additional Reading

Holzworth, L.K. and R.W. Brown: *Revegetation with Native Species: Proceedings, 1997 Society for Ecological Restoration Annual Meeting,* DIANE Publishing Company, Collingdale, PA, 2000.

Munshower, F.F.F.: *Practical Handbook of Disturbed Land Revegetation,* Lewis Publishers, Boca Raton, FL, 1994.

Vogel, W.G.: *Manual for Training Reclamation Inspectors in the Fundamentals of Soils and Revegetation,* DIANE Publishing Company, Collingdale, PA, 1998.

REVERSE OSMOSIS. A method for separating inorganic salts and simple organic compounds under pressure. The size of the species usually is considered in terms of molecular weight. Solvent transport through reverse-osmosis membranes is substantially diffusive in nature. The membranes used are anisotropic (thin-skinned, overlaying a porous substructure). In all probability, the skin contains no pores. If water is the solvent phase, the water passes through the membrane by true diffusion. At some point in its passage through the membrane, the solvent water actually becomes a part of the membrane water structure. As low-molecular-weight solutes exhibit osmotic pressure when concentrated, system pressure in excess of the osmotic pressure of the concentrated solutes must be applied to create a pressure driving force. Depending upon materials being separated and membrane used, reverse osmosis system pressures may range from 200 to nearly 1500 pounds per square inch (13.6 to 102 atmospheres), but the lower values are most frequently encountered.

It has been well appreciated for many years that reverse osmosis concentration has considerable potential in the chemical and food processing field. Reverse osmosis is used for processing whey. A new approach to the production of protein isolates and concentrates from oilseed flour combines ultrafiltration and reverse osmosis. See also **Ultrafiltration**. In 1977, this process was optimized for use with soy flour. The process uses semipermeable membranes to directly process protein extracts from defatted soy flour. Some researchers have found that essentially all the solubilized protein is recovered and the generation of wheylike products is avoided. Other desirable features of the combined ultrafiltration/reverse osmosis process include its shorter time span, increased yield of isolate (since whey proteins are harvested along with proteins normally precipitated), and the enhanced nitrogen solubility of the products. For the ultrafiltration part of the system, second-generation, noncellulosic membranes have been used. The membranes used in reverse osmosis are cellulose-based because noncellulosic reverse osmosis membranes have not been fully developed. Cellulose acetate membranes are generally restricted to operating temperatures below about 50°C and a pH of less than 9.

Japanese researchers have developed a method for washing membranes and overcoming the adhesive effects of mandarin orange juice fouling on reverse osmosis membranes. The importance of keeping fluxes high by controlling concentration polarization and membrane fouling has been observed by several researchers. Other observers have noted that high axial velocities and turbulence promoters minimize the flux decrease caused by concentration polarization and fouling. Considerable research has been undertaken as regards the energy requirements for reverse-osmosis process and these are reviewed in some of the references listed.

Molecular diffusion in reverse osmosis is represented schematically in the accompanying diagram.

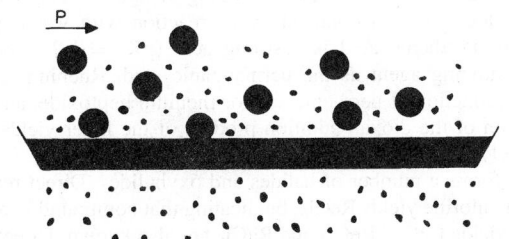

Schematic representation of molecular diffusion in reverse osmosis.

See also **Desalination**; and **Membrane Separations Technology**.

Additional Reading

Amjad, Z.: *Reverse Osmosis,* Chapman & Hall, New York, NY, 1992.

Baker, R.W.: *Membrane Technology and Applications,* 2nd Edition, John Wiley & Sons, Inc., Hoboken, NJ, 2004.

Ebara, K. et al.: "The Advanced Waste Water Treatment by Reverse Osmosis Method," *Kagaku Kojo,* **20**, 4, 69 (1976).

Lawhon, J.T., et al.: "Optimization of Protein Isolate Production from Soy Flour Using Industrial Membrane Systems," *Jrnl. of Food Science,* **43**, 2, 361–364 (1978).

Nomura, D., and I. Hayakawa: "Studies on Concentration of Orange Juice by Reverse-osmosis Process," *Jrnl. of Food Science Technology* (Japan), **23**, 404 (1976).

Parekh, B.S.: *Reverse Osmosis Technology: Applications of High-Purity-Water Production,* Marcel Dekker, Inc., New York, NY, 1988.

Rautenbach, R., and R. Albrecht: *Membrane Processes,* John Wiley & Sons, Inc., New York, NY, 1989.

Schwartzberg, H.C.: "Energy Requirements for Liquid Food Concentration," *Food Technology,* **31**, 3, 68–69 (1977).

Watanabe, A., Kimura, S., and S. Kimura: "Flux Restoration of Reverse-Osmosis Membranes by Intermittent Lateral Surface Flushing for Orange Juice Processing," *Jrnl. of Food Science,* **43**, 3, 985–988 (1978).

REYNOLDS NUMBER. A dimensionless number that establishes the proportionality between fluid inertia and the sheer stress due to viscosity. The work of Osborne Reynolds has shown that the flow profile of fluid in a closed conduit depends upon the conduit diameter, the density and viscosity of the flowing fluid, and the flow velocity.

Pipe Reynolds number, R_D, is the dimensionless ratio, $VD\gamma/g\mu_e$,

where V = Velocity in any units consistent with the rest of the equation

D = Inside diameter of the conduit (pipe)

γ = Specific weight in any units consistent with the rest of the equation

g = Acceleration of gravity, feet/second^2

μ_e = Absolute viscosity, pound-second/feet^2

The foregoing equation is inconvenient for commercial use because commonly used units of measurement are rarely consistent. Two alternative equation forms are:

$$R_D = \frac{6.32 \times \text{Rate of Flow (pounds/hour)}}{\text{Pipe Diameter (inches)} \times \text{Absolute Viscosity (centipoise)}} \quad (1)$$

$$R_D = \frac{52.77 \times \text{Rate of Flow (gallons/hour at flowing temperature)}}{\text{Pipe Diameter (inches)} \times \text{Kinetic Viscosity (centistokes)}} \quad (2)$$

Flow generally is considered to be fully laminar at Reynolds numbers below 1500; in transition between 1500 and 4000; and turbulent above 4000. Thus, the Reynolds number is a most useful tool in constructing piping system designs and in sizing flowmeters. Orifice or throat Reynolds number, $R_d = R_D/d/D$, where d = diameter of orifice bore (inches); D = inside diameter of conduit or pipe (inches).

Another general way of expressing Reynolds number is by the ratio

$$\frac{\text{Scale Velocity} \times \text{Scale length}}{\text{Kinematic Viscosity}}$$

This also is referred to as the Reynolds criterion. Reynolds numbers are only comparable when they refer to geometrically similar flows; and, provided that all the boundary conditions can be described by the scale velocity and scale length, flows of the same Reynolds number are dynamically similar. For an airfoil, it is

$$\frac{\text{Air Velocity} \times \text{Chord of the Airfoil}}{\text{Kinematic Viscosity of the Air}}$$

Reynolds numbers are of value in the various fields because tests of models are directly comparable to full-scale results of geometrically similar shapes if the Reynolds ratio for the model equals that of the actual or full-scale project. This has its practical applications in the field of hydrodynamics, in the study of water resistance of hulls or floats, and in the study of water velocities, levee problems, etc., of large rivers. It is used also to establish the best proportions of hydraulic turbines through the use of models. Much of the science of aeronautics rests upon experimental data obtained in wind tunnels. Dangerous inaccuracies might exist in drawing conclusions for actual construction from model tests, unless either the model were tested at a Reynolds number equal to that of the completed project, or due corrections and allowances were made for the Reynolds number. See also **Heat Transfer**.

RHENIUM. [CAS: 7440-15-5]. Chemical element, symbol Re, at. no. 75, at. wt. 186.2, periodic table group 7, mp 3180°C, bp 5627°C, density

21.04 g/cm^3 (20°C). Elemental rhenium has a close-packed hexagonal crystal structure.

Rhenium is a platinum-white, very hard metal; stable in air below 600°C (at this temperature, the metal begins to generate a white, nonpoisonous, vaporous oxide, Re$_2$O$_7$); practically insoluble in HCl or hydrofluoric acid, but soluble in HNO$_3$ with the formation of perrhenic acid; forms sodium rhenate when fused with NaOH and nitrate. Discovered by Noddack and Tacke in 1925 in tantalite, wolframite, and columbite by the Moseley X-ray spectrographic method of analysis, and later found present in molybdenite, from which rhenium is obtained. Predicted by Mendeleev, in 1871, as an element to be discovered with properties resembling manganese, and named by him dvi-manganese.

There are two natural isotopes, ^{185}Re and ^{187}Re, of which the latter is radioactive with respect to beta decay, having a half-life of 5×10^{10} years reverting to ^{187}Os. Other radioactive isotopes include ^{177}Re, ^{178}Re, ^{180}Re, ^{183}Re, ^{184}Re, ^{186}Re, ^{188}Re, and ^{189}Re. The latter isotope has a long half-life, something less than 10^3 years; the half-lives of the remaining isotopes are comparatively short, measured in minutes, hours, or days.

Because Re and Os are highly refractory and siderophilic elements, the Re-Os isotope system is important in studies concerned with metal phases and high-temperature inclusions of meteorites. R.J. Walker (Carnegie Institute of Washington) and J.W. Morgan (U.S. Geological Survey) observe, "Potential applications of the system include dating meteorites, especially with respect to the chronology of the assembly and subsequent metamorphism of genetically disparate components, and providing estimates of the initial Os isotopic composition and Re/Os ratio of the early earth. This ratio is an important chemical tracer for understanding the formation of the earth's core and the chemical evolution of the mantle and crust."

In terms of abundance, rhenium ranks 75th among the elements, based upon estimated contents of the universe. Rhenium is not plentiful in the earth's crust, being essentially confined to association with the mineral molybdenite. First ionization potential, 7.87 eV. Oxidation potentials Re + 4H$_2$O → ReO$_4^-$ + 8H$^+$ + 7e$^-$, −0.15 V; Re + 8OH$^-$ → ReO$_4^-$ + 4H$_2$O + 7e$^-$, 0.81 V. Other important physical properties of rhenium are given under **Chemical Elements**.

Rhenium is a minor constituent (100 ppb) of molybdenite-bearing porphyry copper ores, of which there is extensive mining in the United States and South America. Commercial rhenium is recovered from by-product molybdenum. As the result of roasting MoS$_2$ and MoO$_3$, the rhenium is concentrated to levels of 300–1,000 ppm. At the high process temperatures, the rhenium is oxidized and volatilized as rhenium heptoxide, Re$_2$O$_7$. This compound is recovered from flue gases by way of wet scrubbing and chemical separation techniques as the relatively crude ammonium perrhenate. The latter compound is reduced with hydrogen to produce rhenium metal.

Uses

Rhenium finds rather wide use as a catalyst in selective hydrogenation and other chemical reactions. It sometimes is used in conjunction with platinum in reforming operations. Additional processes for which rhenium has been tested and used as a catalyst include alkylation, de-alkylation, dehydrochlorination, dehydrogenation, dehydroisomerization, enrichment of water, hydrocracking, and oxidation. The outstanding feature of rhenium catalysts is their high selectivity, very important in hydrogenation reactions. Rhenium catalysts also resist such catalyst poisons as nitrogen, sulfur, and phosphorus. In terms of activity, rhenium commonly surpasses cobalt, molybdenum, and tungsten type catalysts and approximates palladium, nickel, and platinum catalysts.

Because of the very heavy ionic weight (250) of the perrhenate ion, it is one of the heaviest simple anions obtainable in readily soluble salts. It has found use as a precipitant for potassium and some other heavy univalent ions; also as a precipitant for such complex ions as Co(NH$_3$)$_6^{2+}$, and for the separation of alkaloids and organic bases. Perrhenate also is used in the fractional crystallization of the rare-earth elements.

When rhenium is added to other refractory metals, such as molybdenum and tungsten, ductility and tensile strength are improved. These improvements persist even after heating above the recrystallization temperature. An excellent example is the complete ductility shown by a molybdenum-rhenium fusion weld. Rhenium and rhenium alloys have gained some acceptance in semiconductor, thermocouple, and nuclear reactor applications. The alloys also are used in gyroscopes, miniature rockets, electrical contacts, electronic-tube components, and thermionic converters.

Because rhenium is very difficult to machine with carbide tools and other conventional methods, electrical-discharge machining (EDM), electrochemical machining (ECM), abrasive cutting, or grinding is recommended.

Rhenium can be consolidated by powder metallurgy techniques, inert-atmosphere arc melting, and thermal decomposition of volatile halides. In the powder metallurgy process, bars are pressed at 200 MPa, and this is followed by vacuum sintering at 1200°C and hydrogen sintering at 2700°C. Rhenium is usually fabricated from sintered bar by cold working, following by annealing. Reductions of 10–20% can be taken with intermediate anneals for one or two hours at 1700°C. Primary working is by rolling, swaging, or forging. Wire drawing is possible down to 2 mils for strip and wire. Because of its excellent ductility at room temperature, rhenium is suitable for forming of complex shapes (Knipple, 1979).

The cost and scarcity of pure rhenium preclude its extensive use for large structural components. Typical applications include wear-resistant electrical contacts and mass spectrometer cathodes. Rhenium also is being studied for use in the reactor core of space nuclear power systems as, for example, fuel-pin liners that prevent interaction with the fuel and also provide high neutron capture under some operating conditions. Rhenium coatings are used to enhance the heat resistance of carbon and graphite parts in low-oxygen environments. Tungsten and molybdenum alloys containing rhenium have been used for heating elements, compact electromagnetic coils, high-temperature thermocouples, anti-friction and wear parts, and high-temperature elastic elements. Rhenium and rhenium-containing alloys are principally produced by powder metallurgy. The use of vacuum or inert-gas atmosphere melting or chemical vapor deposition processes tend to be cost prohibitive.

As observed by B.D. Bryskin (Sandvik Rhenium Alloys Inc., Elyria, Ohio), "Rhenium has poor oxidation resistance, but is chemically inert in most oxygen-free atmospheres. When it oxidizes in air, Re$_2$O$_7$ is emitted as a white smoke. When hot worked in air, rhenium is embrittled by grain boundary penetration of liquid-phase oxide. Consequently, hot deformation of Re is practical only in a nonoxidizing protective atmosphere (hydrogen or vacuum), or if the workpiece is encapsulated by an oxidation-resistant material. Rhenium is unique among refractory metals in that it does not form a stable carbide. The solubility for carbon is rather high, resulting in a eutectic melting point at about 2773 K (2500°C; 4530°F) and 0.085% (wt) carbon." Further excellent application information is given in the Bryskin reference listed.

Chemistry and Compounds

Rhenium has a $5d^5 6s^2$ electron configuration and all oxidation states from 0 to 7+ are known, although the heptavalent is the most stable state.

The supposed compounds of Re(−I), formulated as M^I Re(H$_2$O)$_4$, have been shown actually to be tetrahydrorhenates(III), e.g., potassium tetrahydrorhenate, KReH$_4$. These compounds are obtained by reducing potassium perrhenate, in a solution of water and ethanolamine, with the alkali metal. In addition to these compounds, however, rhenium differs from its congener manganese in that it forms more compounds that are in the higher valence group. Instead of the larger number of divalent compounds formed by manganese, the stable compounds of rhenium begin with the trivalent ones. The oxides include Re$_2$O$_3$, ReO$_2$, ReO$_3$, and Re$_2$O$_7$. The heptoxide differs from the corresponding manganese compound in its stability, as does the acid obtained by its reaction with water, perrhenic acid, HReO$_4$. Perrhenic acid is a strong acid (pK$_A$ = 1.25), but not as strong an oxidizing agent as the permanganic acid. Rhenium dioxide is obtained by reduction of perrhenic acid or rhenium heptoxide, and thermal decomposition of the dioxan addition product of the latter yields rhenium trioxide, ReO$_3$.

Rhenium forms a number of halides and oxyhalides. Direct reaction by heating with chlorine yields ReCl$_5$, but heating that compound in a nitrogen atmosphere yields ReCl$_3$. ReCl$_4$ and ReCl$_6$ are also known. Direct reaction by heating with fluorine carries the halogenation further, to ReF$_6$ and ReF$_7$, which may be reduced to ReF$_4$. Direct reaction by heating with bromine yields ReBr$_3$. Reaction of the halides with oxygen, or the oxides or oxyacid salts with halogen-containing substances yields a considerable number of oxyhalogen compounds, e.g., ReOF$_4$, ReO$_2$F$_2$, ReO$_2$Cl$_2$, ReO$_2$Br$_2$, ReO$_3$F and ReO$_3$Cl · ReO$_3$F may also be prepared by the action of liquid HF on KReO$_4$ and ReO$_3$Cl may be prepared by reaction of ReO$_3$ and Cl$_2$.

The complexes of rhenium have not been studied as extensively as those of manganese. Among the known complex ions are Re(NH$_3$)$_6^{3+}$,

$Re(CN)_6^{3-}$, $Re(CN)_8^{4-}$, $Re(CN)_8^{3-}$, $ReO_2(CN)_4^{2-}$, $ReCl_4^-$, ReF_6^{2-}, $ReCl_6^{2-}$, $ReBr_6^{2-}$ and ReI_6^{2-}.

The sulfides include ReS_2 and Re_2S_7.

Unlike manganese, rhenium is reported to form an alkyl compound, trimethyl rhenium, $Re(CH_3)_3$. Like manganese, it forms a dirhenium compound with carbon monoxide, $(CO)_5ReRe(CO)_5$, as well as hydrogen-halogen and alkyl-carbonyl compounds. It also forms a dicyclopentadienyl compound, $(C_5H_5)ReH$.

Additional Reading

Bryskin, B.D.: "Rhenium and Its Alloys," *Advanced Materials and Processes*, **22** (September 1992).

Greenwood, N.N. and A. Earnshaw: *Chemistry of the Elements,* 2nd Edition, Butterworth-Heinemann, Inc., Woburn, MA, 1997.

Horan, M.F. et al.: "Rhenium-Osmium Isotope Constraints on the Age of Iron Meteorites," *Science,* 1118 (February 28, 1992).

Knipple, W.R.: "Rhenium," in *Metals Handbook,* 9th Edition, Vol. 2, American Society for Metals, Metals Park, OH, 1979.

Lewis, R.J. and J.I. Sax: *Sax's Dangerous Properties of Industrial Materials,* 10th Edition, John Wiley & Sons, Inc., New York, NY, 1999.

Lide, D.R.: *CRC Handbook of Chemistry and Physics,* 84th Edition, CRC Press, LLC., Boca Raton, FL, 2003.

Meyer, C.: "Ore Metals Through Geologic History," *Science,* **227,** 1421–1428 (1985).

Savittskii, E.M. and M.A. Tylkina (editors): *The Study and Use of Rhenium Alloys* (translated from the Russian), Amerind Publishing Co., Pvt. Ltd. Available through U.S. Bureau of the Interior, or National Science Foundation, Washington, DC. (1978).

Sinfelt, J.H.: "Bimetallic Catalysts," *Sci. Amer.,* 90–98 (September 1985).

Staff: *RE Rhenium: Organorhenium Compounds: Binuclear Compounds,* Springer-Verlag, New York, NY, 1996.

Staff: *ASM Handbook—Properties and Selection: Nonferrous Alloys and Pure Metals,* ASM International, Materials Park, OH, 1990.

Staff: *Forecast for Metals,* Advanced Materials and Processes, (Published annually in January issue.)

Walker, R.J. and J.W. Morgan: "Rhenium-Osmium Isotope Systematics of Carbonaceous Chondrites," *Science,* **519** (January 27, 1989).

RHEOLOGY.   The study of the response of materials to an applied force. Rheology deals with the deformation and flow of matter.

Heraclitus, a pre-Socratic metaphysician, recognized in the fifth century B.C. that πάντα ρεί, or "everything flows." Long before Heraclitus, the prophetess Deborah, fourth judge of the Israelites, had sung that "the mountains flowed before the Lord" in celebrating the victory of Barak over the Canaanites (*Judges* 5:5). Reiner (1964) proposed the dimensionless quantity D (for *Deborah*), where:

$$D = \frac{\text{time of relaxation}}{\text{time of observation}} = \frac{\tau}{t} \tag{1}$$

The difference between solids and liquids is found in the magnitude of D. Liquids, which relax in small fractions of a second, have small D. Solids have a large D. A sufficient time span can reduce the Deborah number of a solid to unity, and impact loading can increase D of a liquid. Viscoelastic materials are best characterized under conditions in which D lies within a few decades of unity.

Force Balance Equation

When a force f is applied to a body, four things may happen. The body may be accelerated, strained, made to flow, or slide along another body. If these four responses are added, one can write an expression for motion in one direction:

$$f = m\ddot{x} + r\dot{x} + sx - f_0 \tag{2}$$

where m is the mass, r is a damping parameter related to viscosity, s to elasticity, and f_0 to the yield value. Evaluation of the coefficients $m, r,$ and s involves the measurement of displacements x and their time derivatives in a manner which links these kinematic variables via an equation of state, such as given in Eq. (2), to stress σ (force per unit area) and its time derivatives.

Scope of Rheology

In contrast to the discipline of mechanics, wherein the responses of bodies to unbalanced forces are of concern, rheology concerns balanced forces which do not change the center of gravity of the body. Since rheology involves deformation and flow, it is concerned primarily with the evaluation

of the coefficients r and s of Eq. (2). The coefficients account for most of the energy dissipated and stored, respectively, during the process of distorting a body. Most rheological systems lie between the two extremes of ideality—the Hookean solid and the Newtonian liquid.

Measurements of Viscosity and Elasticity in Shear (Simple Shear)

Shear viscosity η and shear elasticity G are determined by evaluating the coefficients of the variables $\dot{x}$ and x, respectively, which result when the geometry of the system has been taken into account. The resulting equation of state balances stress against shear rate $\dot{\gamma}$ (reciprocal seconds) and shear γ (dimensionless) as the kinematic variables. For a purely elastic, or Hookean, response:

$$\sigma = G\dot{\gamma} \tag{3}$$

and for a purely viscous, or Newtonian response:

$$\sigma = \eta\dot{\gamma} \tag{4}$$

As a consequence, G can be measured from stress-strain measurements, and η from stress-shear rate measurements.

Elasticoviscous behavior is described in terms of the additivity of shear rates:

$$\dot{\gamma} = \frac{\sigma}{\eta} + \frac{\dot{\sigma}}{G} \tag{5}$$

whereas viscoelastic behavior is characterized by the additivity of stress, according to Eq. (2);

$$\sigma = G\gamma + \eta\dot{\gamma} \tag{6}$$

Relaxation

Numerous attempts have been made to fit simplified mechanical models to the two behavior patterns described by Eq. (6). One can picture the elastic element as a spring-arrayed network parallel with the viscous element to give essentially a (Kelvin) solid with retarded elastic behavior, wherein:

$$\frac{\eta_k}{G_k} = \text{retardation time (sec)} \tag{7}$$

or as a (Maxwellian) series network which flows when stressed or relaxes, under constant strain:

$$\frac{\eta_m}{G_m} = \text{relaxation time (sec)} \tag{8}$$

and transient experiments may be designed to measure these parameters singly. In real systems, a single relaxation (or retardation) time fails to account for experimental results. A distribution of relaxation time exists.

Dynamic Studies

When Eq. (2) is written in the form:

$$\ddot{x} + 2k\dot{x} + \omega_1^2 x = 0 \tag{2b}$$

the equation suggests that the variation in stress should be cyclic. Rheometers are designed so that the system may oscillate in free vibration of natural resonant frequency ω_1, or else so that a cyclic shearing stress of the form $f_0 \cos \omega t$ is impressed on the sample over a frequency range which spans ω_1. In neither case is the material strained beyond its range of linearity. Equation (2b) represents a damped harmonic oscillator, providing that the coefficients are constant (i.e., providing that they do not depend on the strain magnitude). Not all systems meet this requirement in the strict sense, with the result that one of the first checks the experimenter makes is for linearity. Doubling the amplitude of oscillation should double the stress and should not change the phase relationships between the cyclic stress and the deformation.

Time-Temperature Equivalence (Steady-State Phenomena)

The creep of a viscoelastic body or the stress relaxation of an elasticoviscous one is employed in the evaluation of η and G. In such studies, the long-time behavior of a material at low temperatures resembles the short-time response at high temperatures. A means of superimposing data over a wide range of temperatures has resulted which permits the mechanical behavior of viscoelastic materials to be expressed as a master curve over a reduced time scale covering as much as twenty decades (powers of ten).

Polymeric materials generally display large G values (10^{10} dynes/cm^2 or greater) at low temperatures or at short times of measurement. As either of these variables is increased, the modulus drops slowly at first, then attains a steady rate of roughly one decade drop per decade increase in time. If the material possesses a yield value, this steady drop is arrested at a level of G which ranges from 10^7 downward.

Dynamic Behavior

The application of sinusoidal stress to a body leads inevitably to the complex modulus G^*, where

$$G^* = G' + iG'' = G' + i\omega\eta' \qquad (9)$$

where G' is the in-phase modulus (σ/γ) which represents the stored energy, and G'' is the out-of-phase modulus $(\sigma/\dot\gamma)$ representing dissipated energy (as its relation to η suggests); the variable against which G' and η' are determined is the circular frequency ω. Superposition of variable temperature data or variable frequency data provides a master curve of the type previously described for steady-state parameters.

Problems in Three Dimensions (State of Stress)

The forces and stresses applied to a body may be resolved in three vectors, one normal to an arbitrarily selected element of area and two tangential. For the yz plane, the stress vectors are σ_{xx} and σ_{xy}, σ_{xz}, respectively. Six analogous stresses exist for the other orthogonal orientations, giving a total of nine quantities, of which three exist as commutative pairs $(\sigma_{rs} = \sigma_{sr})$. The state of stress, therefore, is defined by three tensile or normal components $(\sigma_{xx}, \sigma_{yy}, \sigma_{zz})$ and three shear or tangential components $(\sigma_{xy}, \sigma_{xz}, \sigma_{yz})$. The shear components are most readily applicable to the determination of η and G.

Strain Components

For each stress component σ there exists a corresponding strain component γ. Even for an ideally elastic body, however, a pure tension does not produce a pure γ_{xx} strain; γ components exist which constrict the body in the y and z directions.

The complete stress-strain relation requires the six σs to be written in terms of the six γ components. The result is a 6×6 matrix with 36 coefficients k_{rs} in place of the single constant. Twenty-one of these coefficients (the diagonal elements and half of the cross elements) are needed to express the deformation of a completely anisotropic material. Only three are necessary for a cubic crystal, and two for an amorphous isotropic body. Similar considerations prevail for viscous flow, in which the kinematic variable is $\dot\gamma$.

Applications

The study of rheology is important to many sciences and technologies. Numerous instruments have been constructed to make manual, semiautomatic, and automatic measurements (and control) of such variables as viscosity, and consistency (flowability). These variables are particularly important in the food processing, chemical, and plastics industries, where many of the products and intermediates lie between Hookean solids and Newtonian liquids. Consistency measurement is of large importance in the paper manufacturing industry. See also **Papermaking and Finishing**; and **Pulp (Wood) Production and Processing**.

Some terms commonly used in industry include:

Newtonian Substance. Fundamentally, liquids or suspensions in liquids when subjected to a shear stress behave in two ways: (1) A Newtonian substance undergoes deformation, the ratio of shear rate (flow) to shear stress (force) is constant. See Fig. 1.

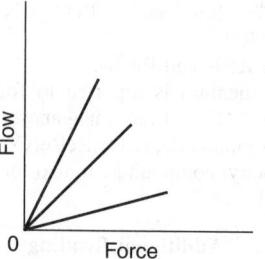

Fig. 1.   Behavior of Newtonian substances

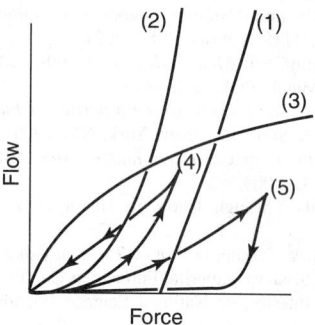

Fig. 2.   Behavior of non-Newtonian substances: (1) true plastic (sometimes called a Bingham body); (2) pseudo plastic; (3) dilatant; (4) thixotropic; and (5) rheopectic

Non-Newtonian Substance. (2) In a non-Newtonian substance, the ratio of shear rate to flow is not constant. See Fig. 2.

Dilatant. The initial flow of a dilatant substance under a low shear stress is at a high rate; further increases in shear stress result in a lower flow rate. A dilatant substance is sometimes called an inverted plastic or inverted pseudoplastic.

Pseudoplastic. A material of this type appears to have a yield stress beyond which flow commences and increases sharply with increase in stress. In practice, such substances are found to exhibit flow at all shear stresses, although the ratio of flow to force increases negligibly until the force exceeds the apparent yield stress.

Thixotropic. The flow rate of a thixotropic substance increases with increasing duration of agitation, as well as with increased shear stress. When agitation is stopped, internal shear stress exhibits hysteresis. Upon reagitation, generally less force is required to create a given flow than is required for first agitation.

Rheopectic. If certain thixotropic suspensions are rhythmically shaken or tapped, they will "set" or build up very rapidly, a phenomenon termed rheopexy. Apparent viscosity of a rheopectic substance increases with time (duration of agitation) at any constant shear rate.

See Table 1. See also **Colloid Systems**; **Fluid and Fluid Flow**; **Stoke's Law**; **Viscoelasticity**; and **Viscosity**.

Additional Reading

Acierno, D. and A.A. Collyer: *Rheology and Processing of Liquid Crystal Polymers,* Chapman & Hall, New York, NY, 1996.

TABLE 1. EXAMPLES OF NON-NEWTONIAN FLUIDS EXHIBITING DIVERSE RHEOLOGIC PROPERTIES

Pseudoplastic	Plastic	Trixotropic*	Rheopectic	Dilatant*
Catsup	Chewing gum	Silica gel	Bentonite sols	Quicksand
Printers ink	Tar	Most paints	Gypsum in water	Peanut butter
Paper pulp	Various slurries	Glue		Many candy compounds
		Molasses		
		Lard		
		Fruit juice concentrates		
		Asphalts		

* Some liquids may change from thixotropic to dilatannt or vice versa as the temperature or concentration changes.

Cheremisinoff, N.P.: *Introduction to Polymer Rheology and Processing,* CRC Press, LLC., Boca Raton, FL, 1999.

Goodwin, J.W., and R.W. Hughes: *Rheology for Chemists: An Introduction,* The Royal Society of Chemistry, London, UK, 2000.

Gupta, R.K.K.: *Polymer and Composite Rheology,* Marcel Dekker, Inc., New York, NY.

Larson, R.G., and K.E. Gubbins: *Structure and Rheology of Complex Fluids,* Oxford University Press, New York, NY, 1998.

Morrison, F.A.: *Understanding Rheology,* Oxford University Press, Inc., New York, NY, 2000.

Munson, B.R., D.F. Young, T.H. Okiishi, and B.G. Young: *Fundamentals of Fluid Mechanics,* 4th Edition, John Wiley & Sons, Inc., New York, NY, 2003.

Tanner, R.I.: *Engineering Rheology,* 2nd Edition, Oxford University Press, Inc., New York, NY, 2000.

Tanner, R.I. and K. Walters: *Rheology: An Historical Perspective,* Elsevier Science, New York, NY, 1998.

RHEOPECTIC SUBSTANCES.   See Rheology.

RHODIUM.   [CAS: 7440-16-6]. Chemical element, symbol Rh, at. no. 45, at. wt. 102.906, periodic table group 9, mp. 1,963 to 1,969°C, bp. 3,627 to 3,827°C, density 12.44 g/cm^3 for solid (20°C). Elemental rhodium has a face-centered cubic crystal structure. The one stable isotope is ^{103}Rh. The seven unstable isotopes are ^{99}Rh through ^{101}Rh and ^{104}Rh through ^{107}Rh. In terms of earthly abundance, rhodium is one of the scarce elements. Also, in terms of cosmic abundance, the investigation by Harold C. Urey (1952), using a figure of 10,000 for silicon, estimated the figure for rhodium at 0.0067. No notable presence of rhodium in seawater has been found.

Electronic configuration is $1s^22s^22p^63s^23p^63d^{10}4s^24p^64d^85s^1$. Ionic radii Rh^{3+} 0.75 Å, Rh^{4+} 0.65 Å. Metallic radius 1.345 Å. First ionization potential 7.7 eV. Other physical properties of rhodium will be found under **Platinum and Platinum Group**. See also **Chemical Elements**.

Rhodium was discovered by Wollaston (England) in 1803. Compact Rh is almost insoluble in all acids at 100°C, including aqua regia. Hot concentrated H$_2$SO$_4$ will slowly dissolve the finely divided metal. When alloyed with 90% or more of Pt, it is soluble in aqua regia. The metal is attacked by fused bisulfates. Rh is soluble in molten Pb. This is the basis of the classic separation of Rh and Ir.

Rh compounds exhibit valences of 2, 3, 4, and 6. The trivalent form is by far the most stable. When Rh is heated in air, it becomes coated with a film of oxide. Rhodium(III) oxide, Rh$_2$O$_3$, can be prepared by heating the finely divided metal or its nitrate in air or O$_2$. The rhodium(IV) oxide is also known. Rhodium trihydroxide may be precipitated as a yellow compound by adding the stoichiometric amount of KOH to a solution of RhCl$_3$. The hydroxide is soluble in acids and excess base. When the freshly precipitated Rh(OH)$_3$ is dissolved in HCl at a controlled pH, a yellow solution is first obtained in which the aquochloro complex of Rh behaves as a cation. The hexachlororhodate(III) anion is formed when the solution is boiled for 1 hour with excess HCl. The solution chemistry of RhCl$_3$ is often very complex. Two trichlorides of Rh are known. The trichloride formed by high-temperature combination of the elements is a red, crystalline, nonvolatile compound, insoluble in all acids. When Rh is heated in molten NaCl and treated with Cl$_2$, Na$_3$RhCl$_6$ is formed, a soluble salt that forms a hydrate in solution. Rhodium(III) iodide is formed by the addition of KI to a hot solution of trivalent Rh.

Rhodium(III) sulfate exists in yellow and red forms. If Rh(OH)$_3$ is dissolved in cold H$_2$SO$_4$, the product is the yellow form, in which the sulfate is ionic. If this solution is evaporated in hot H$_2$SO$_4$, the product is a red, nonionic sulfate. When Rh is treated with F$_2$ at 500–600°C, RhF$_3$ is slowly formed. This compound is practically insoluble in water, concentrated HCl, HNO$_3$, H$_2$SO$_4$, HF, or NaOH.

If a solution of RhCl$_3$ is treated with NaNO$_2$, the very soluble sodium hexanitritorhodate(III), Na$_3$Rh(NO$_2$)$_6$, is formed. The solubility of this compound in alkaline solution makes it useful for refining, as many base metals are precipitated as their hydroxides under these conditions. The analogous ammonium and potassium salts are relatively insoluble.

When H$_2$S is passed into a solution of a trivalent Rh salt at 100°C, the hydrosulfide, Rh(SH)$_3$, is formed. This black precipitate is insoluble in (NH$_4$)$_2$S. Rh forms many complexes with NH$_3$, amines, cyanide, chloride, bromide, and numerous polynitrogen and polyoxygen chelating agents.

EDITOR'S NOTE: In 1982, J. Halpern (University of Chicago) reported that rhodium complexes containing chiral phosphine ligands catalyze the hydrogenation of olefinic substrates such as alpha-aminoacrylic acid derivatives, producing chiral products with very high optical yields.

Elucidation of the mechanisms of such reactions leads to the conclusion that the stereoselection is dictated not by the preferred initial binding of the substrate to the chiral catalyst, but rather by the much higher reactivity of the minor diastereomer of the catalyst-substrate adduct corresponding to the less favored binding mode. In the Halpern 1982 reference listed, a relatively restricted class of asymmetric catalytic reactions, namely, the hydrogenation of alpha-acylaminoacrylic acid derivatives and related substrates, catalyzed by rhodium complexes containing chiral phosphine ligands, is discussed.

LINTON LIBBY
Chief Chemist, Simmons Refining Company
Chicago, Illinois

Additional Reading

Bauccio, M.L.: *ASM Metals Reference Book,* 3rd Edition, ASM International, Materials Park, OH, 1993.

Carter, G.F. and D.E. Paul: *Materials Science and Engineering,* ASM International, Materials Park, OH, 1991.

Davis, J.R.: *Metals Handbook,* 2nd Edition, ASM International, Materials Park, OH, 1998.

Greenwood, N.N. and A. Earnshaw: *Chemistry of the Elements,* 2nd Edition, Butterworth-Heinemann, Inc., Woburn, MA, 1997.

Halpern, J.: "Mechanism and Stereoselectivity of Asymmetric Hydrogenation," *Science,* **217,** 401–407 (1982).

Krebs, R.E.: *The History and Use of Our Earth's Chemical Elements: A Reference Guide,* Greenwood Publishing Group, Inc., Westport, CT, 1998.

Lewis, R.J. and N.I. Sax: *Sax's Dangerous Properties of Industrial Materials,* 10th Edition, John Wiley & Sons, Inc., New York, NY, 1999.

Lide, D.R.: *CRC Handbook of Chemistry and Physics,* 84th Edition, CRC Press, LLC., Boca Raton, FL, 2003.

Meyers, R.A.: *Handbook of Chemicals Production Processes,* The McGraw-Hill Companies, Inc., New York, NY, 1986.

Staff: *ASM Handbook—Properties and Selection: Nonferrous Alloys and Pure Metals,* ASM International, Materials Park, OH, 1990.

RHODOCHROSITE.   Rhodochrosite, manganese carbonate, MnCO$_3$, is a rose-pink to red hexagonal mineral, occurring as small crystals, in cleavable masses, granular or compact. It is a brittle mineral; hardness, 3.5–4; specific gravity, 3.7 (pure MnCO$_3$, usually 3.4–3.6); luster, vitreous to pearly; color, various shades of pink, red and reddish-brown; transparent to opaque; streak, white. Rhodochrosite has a perfect rhombohedral cleavage. Rhodochrosite is a product of high-temperature metamorphic deposits; as a gangue mineral in ore veins of hydrothermal origin, and as secondary residual deposits from bodies of manganese or iron oxides, and in sedimentary deposits precipitated like siderite by organic matter acting, in the absence of oxygen, upon bicarbonates.

Spectacular, gemmy red rhombohedrons, up to 3 inches (7.5 centimeters) on an edge occur at the Sweet Home Mine, Alma, Colorado, and the John Reed Mine, Alicante, Colorado. Exceptionally fine stalactitic formations are found in Catamarca Province, Argentina. Recently the most beautiful, large transparent gem red scalenohedron crystals ever found have come from Hotazel and the Kalahari manganese field in northern Cape Province, South Africa. This region encompasses one of the largest and richest known manganese deposits in the world. Other localities for this mineral are in Rumania, Saxony, Westphalia, and Cornwall, England. In the United States rhodochrosite is found at Franklin, New Jersey; Butte Montana; and in various localities in Colorado and Nevada. The name rhodochrosite is derived from the Greek meaning rose, and color.

RHODONITE.   The mineral rhodonite, manganese metasilicate, (Mn,Fe, Mg)SiO$_3$, crystallizes in the triclinic system forming large, irregular tabular crystals but usually occurring massive. Prismatic and basal cleavages excellent; fracture conchoidal to uneven; hardness, 5.5–5.6; specific gravity, 3.57–3.76; luster, vitreous to pearly on cleavage faces; color, red to brownish-red; rarely yellow to gray; streak, white; transparent to translucent. A variety containing much calcium is called bustamite. Zinc may replace the manganese in rhodonite; it is then known as fowlerite. Rhodonite is found in the Harz Mountains, Germany; in the Urals of the former U.S.S.R.; in Hungary, Italy, and Sweden. Bustamite from Mexico, Franklin and Sterling Hill, New Jersey, occurs with fowlerite.

Rhodonite has been occasionally used for an ornamental stone. Its name is derived from the Greek meaning a rose, because of the color.

RIBONUCLEASE.   An enzyme that causes splitting of ribonucleic acid. Pancreatic ribonuclease for example, cleaves only phosphodiester bonds that are linked to pyrimidine-3′-phosphates. It is a critical regulator of life processes in the cell. The first enzyme to be synthesized (1969), it is composed of 124 amino acid residues. It is one of the proteins for which the sequence of amino acids has been elucidated (the order of sequence of amino acids is of critical importance in the functioning of enzymes, genes, and nucleotides).

See also **Genetic Code**; and **Genetics and Gene Science (Classical)**.

RIBONUCLEIC ACID (RNA). See **Genetics and Gene Science (Classical)**.

RIBOSOME. A ribonucleoprotein, the smallest organized structure in the cell. Ribosomes occur in all cells including bacteria, fungi, algae, and protozoa. They are the central point of protein synthesis. They contain from 45 to 60% of ribonucleic acid (RNA), the balance being proteins. Ribosome crystals have been produced; in the electron microscope these appear as sheets of black dots, each sheet (one ribosome thick) containing hundreds of ribosomes in recurring groups of four.

See also **Genetics and Gene Science (Classical)**.

RICHARDS, THEODORE W. (1868–1928). An American chemist born in Germantown, PA. He was the first American to receive the Nobel prize in chemistry (in recognition of his accurate determinations of the atomic weight of a large number of chemical elements). He studied chemistry at Haverford and Harvard, with a doctorate in chemistry from Harvard where he later became Erving Professor of Chemistry. An outstanding experimental chemist, his major interests were atomic weights, thermochemistry, and thermodynamics. He was also a brilliant teacher. He was president of the ACS in 1914, and the recipient of many honorary awards, including the Davy, Faraday, and Gibbs medals.

RIEBECKITE. The mineral riebeckite, essentially sodium iron silicate, $Na_2(Fe_3{}^{2+}, Fe_2{}^{3+})_5Si_8O_{22}(OH)_2$, is a monoclinic member of the amphibole group, usually in prismatic crystals. It has a prismatic cleavage; hardness, 5; specific gravity, 3.32–3.382; vitreous luster; color dark bluish to black. It occurs in granites and syenites chiefly. It is found in Greenland, Portugal, Madagascar, and South Africa (crocidolite), and in the United States at Quincy, Massachusetts; near Pikes Peak, Colorado, and the San Francisco Mountains, Arizona.

RNA. See **Genetics and Gene Science (Classical)**; **Nucleoproteins and Nucleic Acids**.

ROBINSON, SIR ROBERT (1886–1975). An English chemist who won the Nobel prize for chemistry in 1947. His work began on plants with biological significance, particularly those containing alkaloids. He synthesized brazilin and haematoxylin. In organic chemistry he discovered an important qualitative electronic theory. He explicated the penicillin structure as well. His education, which began at the University of Manchester, took him throughout Europe.

ROCKET PROPELLANTS. Chemical rocket propellant can be classified in several ways, including liquid or solid, monopropellant, bipropellant or tripropellant, cryogenic or storable, hypergolic or nonhypergolic, double-base, composite or composite double-base. Propellant classification frequently influences the classification of rocket engines—for example, monopropellant rocket, liquid rocket, solid rocket, and hybrid rocket (usually using a liquid oxidizer and a solid fuel). Propellants for chemical rockets serve two primary functions as contrasted to one function for nuclear, solar, electrical or laser heated rockets. In chemical propellant rockets, the propellant is both the energy source and the ejected mass or "working fluid."

Compared with the almost limitless number of chemical compounds that exist or can be formed, the number of chemical propellants in common use is relatively few. This situation arises from criteria including costs, source availability, toxicity, resistance to shock, and other requirements imposed by the vehicle application and the propulsion system design. Another practical reason is that extensive overlap of physical, chemical, and economic properties are displayed by many of the theoretically possible propellants. During the 1960s, the universities, industry, and government in the United States pursued extensive research programs for synthesizing new chemical compounds viewed as candidate propellants,

which would increase the performance capability of chemical rockets. Although dozens of compounds were synthesized, few results reached the production line. This does not mean, however, that a scientific breakthrough in increasing the molecular energy of a propellant may not be ultimately obtainable.

The characteristics desired of a rocket propellant are several in number and can be divided into economic, safety, materials compatibility, engine-cycle needs and vehicle requirements.

In general terms, the engine-cycle needs ideally are: (1) a propellant or propellant combination that has a high heat of reaction per unit weight (also called heat of combustion). Most vehicles add a requirement for high heat of reaction per unit volume of propellant to minimize the vehicle size. (2) reaction products that are all gaseous, that have a very low molecular weight and that have a very high temperature of dissociation.

In addition to specific impulse, the vehicle requirements usually influence propellant selection in terms of storability, density, toxicity, and other hazards, and other application-sensitive factors, including exhaust plume properties and radar cross section and radiation emissions. Other factors being essentially equal, the higher the heat of reaction of a propellant (or combination), the more attractive the propellant. Sharp exceptions to this rule occur in some missiles because of volume limitations, the need for smokeless exhaust or similar restraints.

The heat released by a propellant is the difference in heat between the constituents and the end-products of combustion.

$$\Delta H_r^\circ = \left[\sum_{k,\ products} \eta_k (\Delta H_f^\circ)_k - \sum_{j,\ reactants} \eta_j (\Delta H_f^\circ)_j \right]$$

where ΔH_r° is the heat generated; ΔH_f° is the standard heat of formation of the constituent at reference temperature (298K); and η is the number of moles of each j reactant or k product. Large heat release is afforded by reaction products having large negative values, while the reactants should have positive, or at least small negative values, if possible. The heat of reaction is often noted in energy/weight units, such as kilocalories/gram.

Specific impulse, I_{sp}, the universally accepted measure of rocket engine performance, can also be used to indicate the performance of propellants. The most commonly stated expansion ration is $1,000 \rightarrow 14.7$ giving "sea-level specific impulse at 1,000 psi chamber pressure." Sometimes the expansion ratio is $1,000 \rightarrow 0.2$ to indicate specific impulse for high-altitude or space flight.

By definition, specific impulse I_{sp} is:

$$I_{sp} = \frac{F}{\dot{W}}$$

with the I_{sp} units being seconds; the short designation for units of thrust (force) per units of propellant mass flow per second.

For an ideal rocket with the nozzle exhaust pressure being the ambient pressure, the thrust, recognizing Newton's second law of motion, is:

$$F = \frac{\dot{W} c_e}{g}$$

where W is propellant flow rate in pounds per second; c_e is exhaust velocity in feet per second; and g is the gravitational constant in feet/second/second.

In practice, only about 10% of the elements on the periodic chart are adaptable to chemical rocket propellants. Propellants have made little use of elements other than hydrogen, carbon, nitrogen, oxygen, chlorine, fluorine, aluminum, boron, and beryllium.

Liquid propellants fall into two broad classes: (1) earth storable (monopropellants and bipropellants), and (2) cryogenic, depending upon whether they can be kept in the vehicle tankage for months and years, or must be used in a few hours or days. The theoretical performances of storable and cryogenic bipropellants combusting ideally at 1,000 psia chamber pressure and expanding to sea-level pressure without loss, assuming shifting chemical equilibrium of the combustion products during expansion in the engine exhaust nozzle, are listed in Tables 1 and 2. For comparison purposes, a few properties of the more common monopropellants are listed in Table 3. Water (not listed) as a source of hydrogen and oxygen via electrolysis has merit as a propellant in long-life satellites (5 years plus), equipped with solar electric cells.

Solid propellants fall into three general types: (1) double-base, (2) composite; and (3) composite double-base. Double-base propellants form a

TABLE 1. STORABLE LIQUID BIPROPELLANT COMBINATIONS

Oxidizer	Fuel	Oxygen-fuel ratio by weight for maximum I_{sp}	Bulk specific gravity	Theoretical I_{sp}[a]
Nitrogen tetroxide	50/50 Hydrazine/ Unsymmetrical dimethylhydrazine	2.0	1.19	289
Nitrogen tetroxide	Unsymmetrical dimethylhydrazine	2.6	1.17	286
Nitrogen tetroxide	Hydrazine	1.3	1.21	292
Red fuming nitric acid (15% NO$_2$)	Unsymmetrical dimethylhydrazine	3.4	1.28	266
Red fuming nitric acid (15% NO$_2$)	Kerosene-type fuel	5.6	1.37	257
Maximum density red fuming nitric acid (mixture of red fuming nitric acid and N$_2$O$_4$)	Unsymmetrical dimethyl-hydrazine	2.9	1.29	278
N$_2$O$_4$ with 15% NO	Unsymmetrical dimethylhydrazine	2.6	1.15	288
Hydrogen peroxide	Hydrazine	2.0	1.24	287
Hydrogen peroxide	Unsymmetrical dimethylhydrazine	4.2	1.22	284
Chlorine trifluoride	Hydrazine	2.8	1.64	295
Chlorine trifluoride	Unsymmetrical dimethyl-hydrazine	3.0	1.39	280
Chlorine pentafluoride	Hydrazine	2.7	1.47	313
Chlorine pentafluoride	Unsymmetrical dimethyl-hydrazine	2.9	1.34	297
Hydrazine	Pentaborane	1.3	0.80	328

[a] $1,000 \rightarrow 14.7$ psia, shifting equilibrium (chemical composition of exhaust gases changes during nozzle flow).

TABLE 2. CRYOGENIC LIQUID BIPROPELLANT COMBINATIONS (AT LEAST ONE PROPELLANT IS CRYOGENIC)

Oxidizer	Fuel	Pounds oxidizer/pound fuel stotchiometric	Maximum I_{sp}	Bulk specific gravity maximum I_{sp}	Theoretical I_{sp}[a]
Liquid oxygen	Kerosene-type fuel	3.41	2.6	1.02	300
Liquid oxygen	Hydrazine	3.0	0.9	1.07	313
Liquid oxygen	Unsymmetrical dimethylhydrazine	—	1.7	0.98	310
Liquid oxygen	Ammonia	2.37	1.4	0.89	294
Liquid oxygen	Ethyl alcohol	2.09	1.8	0.99	290
Liquid oxygen	Methane	—	3.3	0.82	311
Liquid oxygen	Liquid hydrogen	7.95	4.2	0.29	290
Liquid fluorine	Liquid hydrogen	19.0	8.0	0.46	412
Liquid fluorine	Hydrazine	2.71	2.3	1.31	365
Liquid fluorine	Kerosene-type fuel	4.07	2.6	1.21	322

[a] $1,000 \rightarrow 14.7$ psia, shifting equilibrium (chemical composition of exhaust changes during nozzle flow).

TABLE 3. LIQUID MONOPROPELLANTS

Propellant	Specific gravity at 68°F	Theoretical I_{sp}[a]	Exhaust (Average molecular weight)
Hydrogen peroxide	1.39	165	22.68
Hydrazine	1.01	199	12.77
Nitromethane	1.12	245	20.34
Ethylene oxide	0.89	199	20.50

[a] $1,000 \rightarrow 14.7$ psia, shifting equilibrium (chemical composition of exhaust gases changes during nozzle flow).

homogeneous cured propellant, usually a nitrocellulose-type of gunpowder dissolved in nitroglycerin plus minor percentages of additives. Both the major ingredients are explosives and both contribute to the functions of fuel, oxidizer, and binder. Composite propellants form a heterogeneous propellant grain with the oxidizer crystals and a powdered fuel (usually aluminum) held together in a matrix of synthetic rubber (or plastic) binder such as polybutadiene. Normally, composite propellants are less hazardous to manufacture and handle than double-base propellants. Composite double-base propellants are a combination of the two aforementioned types—usually a crystalline oxidizer (ammonium perchlorate) and powdered aluminum fuel held together in a matrix of nitrocellulose-nitroglycerin. The hazards of processing and handling this type of propellant are similar to those experienced with the double-base propellants. The characteristics of several common solid propellants are given in Table 4.

Ingredients are generally classified according to their function, e.g., fuel, oxidizer, binder, curing agent, burn-rate catalyst, etc. Ingredients used in small amounts are called additives and usually have functions other than the fuel, oxidizer, or binder. For example, an additive can reduce the viscosity of the propellant during mixing and casting (pouring) of the propellant, increase the burning rate of the propellant, or improve the storage stability. Often an ingredient serves or affects more than one function, the most diffused situation relating to composite double-base ingredients where the binder is a nitrocellulose-nitroglycerine complex with each of these two ingredients having its own fuel and oxidizer chemical elements. The binder contributes also as a fuel and in some propellant formulations, such as asphalt-base nonmetallized propellants, the binder is the fuel.

Ammonium perchlorate, NH_4ClO_4, is the most widely used crystalline oxidizer in solid propellants. Because of its characteristics, including compatibility with other propellant materials, specific impulse performance, quality uniformity and availability, it dominates the solid oxidizer field. Both ammonium and potassium perchlorate are only slightly soluble in water, a favorable trait for propellant use. Nitronium perchlorate is objectionably hygroscopic, is relatively incompatible with available binders, and detonates easily. All of the perchlorate oxidizers produce hydrogen chloride in their reaction with fuels. Their exhaust bases are toxic and corrosive to the extent that care is required in firing rockets, particularly the very large rockets, to safeguard operating personnel and communities in the path of exhaust clouds. Ammonium perchlorate is available in the form of small white crystals and close control of the size range and percentage of several sizes present in a given quantity or batch is required, since particle size influences propellant processing and the physical and ballistic properties of the finished propellant.

Inorganic nitrates are relatively low-performance oxidizers as compared with the perchlorates. However, ammonium nitrate is used in some applications for economy and because of its smokeless and relatively nontoxic exhaust. Its main use is in low-burning rate, low-performance applications, such as gas generators for turbine pumps.

TABLE 4. CHARACTERISTICS REPRESENTATIVE SOLID PROPELLANTS

Propellant type	I_{sp} (seconds)	Flame temperature (°F)	Flame temperature (°C)	Density Pounds/ cubic inch	Density Grams/cubic centimeter	Burning rate Metal content (Weight %)	Burning rate inches/ second	Centimeters/ second	Hazards class (military)	Stress strain (psi) −60°F (−51°C)	Stress strain (%) 150°F (66°F)	Processing method
DB	255	5,340	2,449	0.057	1.58	0	0.45	1.1	7	4,600/2	490/60	Extruded
DB/AP/Al[a]	258	6,990	3,866	0.069	1.91	25	0.78	2.0	2	2,750/5	120/50	Extruded
DB/AP-HMX/Al[a]	272	6,630	3,666	0.067	1.85	20	0.55	1.4	7	2,375/3	50/33	Solvent Cast
PVC/AP[b]	239	4,810	2,654	0.065	1.80	0	0.45	1.1	2	369/150	38/220	Solvent Cast
PVC/AP/Al[c]	253	6,120	3,382	0.069	1.91	20	0.45	1.1	2	259/150	38/220	Solvent Cast
PBAN/AP/Al[c]	265	5,600	3,093	0.063	1.74	19	0.55	1.4	2	520/16 (at −10°F)	71/28	Cast
PU/AP/Al[c]	263	6,000	3,316	0.065	1.80	23	0.27	0.7	2	1,170/6	75/33	Cast
CTPB/AP/Al[c]	265	5,540	3,060	0.063	1.74	19	0.45	1.1	2	325/26	88/75	Cast
HTPB/AP/AL[c]	264	5,540	3,060	0.063	1.74	19	0.40	1.0	2	910/50	90/33	Cast
PBAA/AP/AL[c]	265	5,660	3,127	0.063	1.74	20	0.32	0.8	2	500/13	41/31	Cast

Al Aluminum
HTPB Hydroxy-terminated polybutadiene
AP Ammonium perchlorate
PBAA Polybutadiene-acrylic acid polymer
CTPB Carboxy-terminated polybutadiene
PBAN Polybutadiene-acrylic acid-acrylonitrile terpolymer
DB Double base
PU Polyurethane
HMX Cyclotetramethylene tetranitramine
PVC Polyvinyl chloride
Notes:
[a] AP/Al optimized with 40% DB as binder.
[b] AP/Al optimized with 20% binder.
[c] AP/Al optimized with 15% binder.

One or two crystalline high explosives, such as HMX (cyclotetramethylene tetranitramine) and RDX (cyclotrimethylene trinitramine), are sometimes included in a propellant formulation to achieve a specific performance characteristic. Depending upon the objectives, the percent can range from 5 to 50%.

The one prominent solid fuel is powdered aluminum, and it is used in a wide variety of composite and composite double-base propellant formulations, usually being between 14 and 22% of the propellant by weight.

Boron, even though it appears as one of the high-energy fuels and is lighter than aluminum, has not proven to be a practical fuel because it is so difficult to burn with high efficiency in combustion chambers of reasonable length. Beryllium burns much more easily than boron and improves the specific impulse of a solid propellant motor, usually by about 15 seconds, but as a powder or dust it is highly toxic to animals and humans. The technology with composite propellants using powdered beryllium fuel is sufficiently advanced for vehicle application, with space travel being the most likely appli cation.

Theoretically, both aluminum hydride, AlH_3, and beryllium hydride, BeH_2, are attractive fuels because of their high heat release and gas volume contribution. Both are difficult to manufacture and both deteriorate chemically during storage due to loss of hydrogen. Because of these difficulties, coupled with relatively modest I_{sp} gains, these compounds remain experimental.

Hybrid rocket propellants are various combinations of solid and liquid propellants, usually a solid fuel and a liquid oxidizer. Sometimes, a third propellant, liquid hydrogen, is added, not for energy release, but as a low-molecular-weight working fluid. The main advantages of a hybrid rocket are: (1) use of liquid and solid propellant combinations offering the highest performance attainable with chemical rockets; (2) simplicity of a solid grain (usually fuel); (3) a liquid for nozzle cooling and thrust modulation (compared with a solid rocket); (4) restart capabilities; and (5) good storability and safe storage characteristics.

The chemical bond energy present in propellant molecules is the energy source used by chemical rocket engines to date. This source affords energy densities of approximately 3 kilocalories/gram in the liquid hydrogen/liquid oxygen combination, and up to about 5.7 kilocalories/gram with the lithium/fluorine combination. Theoretically, supplemental energy can be added to molecules or molecular fragments that, upon recombination or relaxation to their normal energy state, release significant amounts of energy. For example, 52 kilocalories/gram is theoretically released when

two hydrogen atoms (free radicals) recombine to form hydrogen. Even higher energy densities, as much as 100 kilocalories/gram, are theoretically available from lightweight molecules, such as helium, that are in an excited state.

Metastable, in the sense of propellant ingredients, means that the "energized" molecule, atom, or molecular fragment, tends to promptly return to its normal state. Some molecular species distinctly assume a metastable state upon excitation with the lifetime at room temperature being 10^{-3} to 10^{-2} second as compared with less than 10^{-6} for nonmetastable excited species. Atoms subjected to excitation move into a more energetic state of translational motion of vibration or into a high-energy electron orbital state; diatomic molecules do likewise. Molecules containing more than two atoms can experience higher translational rotational motion, as well as higher electron orbital state.

Most of the research to date on metastable propellants has been with gaseous atoms and molecules. Obviously, energized ingredients in a condensed phase, solid or liquid, would be needed for most rockets. The primary objectives to be reached, if metastable ingredients are to benefit rocket propulsion, are: (1) an efficient process for energizing the ingredients; and (2) a means of storing the ingredients for days at a time without appreciable energy loss. Actual use of metastable ingredients in a rocket is envisioned in the company of liquid hydrogen, or other low-molecular-weight working fluid.

Limited research has been conducted on two approaches to generating and storing (stabilizing) metastable propellant ingredients: (1) free radicals, specifically, atomic hydrogen; and (2) helides, which are excited states of helium. In the late-1950s, the U.S. National Bureau of Standards produced low concentrations of free radicals and stored them in inert matrices at very low temperatures. More recently, an approach has been taken to generate hydrogen atoms, immediately condensed at liquid-helium temperature, in the presence of a high density (70 to 100 kilogauss) magnetic field for the purpose of stabilizing the hydrogen atoms. Theoretically, the high-strength magnetic field is capable of aligning the spin of the electron of the hydrogen atom so as to prevent recombination into the hydrogen molecule.

Triplet helium has a theoretical energy level of 114 kilocalories/gram above the ground state. Assuming release of this energy and subsequent expansion through a rocket nozzle gives a specific impulse of 2,800 seconds. Techniques for generating activated helium and other noble gases are well known, but concentrating and storing these metastable species is quite another matter inasmuch as they revert to their ground state by collision processes. Experimental approaches to activating helium and trapping the helium molecules in a hydrocarbon wax have been reported.

The creation and use of metallic hydrogen (hydrogen derived from normal hydrogen subjected to about 2 megabars pressure) should release about 52 kilocalories/gram upon transitioning from the metallic to the normal solid form. The concept dates back to 1935, but interest has been renewed because some scientists believe that metallic hydrogen exists in some large planets.

Antimatter Rockets

Sufficient atomic particle research has been accomplished to warrant discussion of possible methods of applying energy available from particle mass annihilation to rocket propulsion. Complete conversion of matter to energy would allow exhaust velocities near that of light to be obtained from a propulsion device. Antimatter, by definition is matter made up of antiparticles, such as antineutrons, negatrons (antiprotons), and positrons (antielectrons). An annihilation property is known to exist between particles with one particle termed the antiparticle of the other.

Rocket design concepts envisioned for utilizing the reaction between atomic particles and antiparticles (matter and antimatter) are based upon the following postulations: (1) annihilation products can be accelerated using electrical and magnetic forces (consider the annihilation reaction of a neutrino with an antineutrino, yielding a proton and an electron); (2) annihilation products can be used indirectly to heat a working fluid for thermal expansion through a nozzle (consider the annihilation reaction of hydrogen and antihydrogen, leaving high-energy gamma rays); (3) antimatter possesses negative gravitational mass although its inertial mass may be positive. This could give rise to antigravity propulsion; and (4) annihilation products of ordinary quanta give rise to the possibility of a photon-expelling beam for the direct generation of thrust.

Before any form of antimatter rocket can exist, a lightweight method must be developed for producing antiparticles at a flow rate of grams/second in contrast with the few dozen of antiparticles produced in research laboratory generators. Also, a practical storage or containment method must arise inasmuch as antiparticles explode violently upon contact with normal matter. Reference 5 gives a performance estimate of an I_p of 3.06×10^7 seconds for a rocket propelled vehicle with a thrust/weight ratio of 10^{-7}.

Multiple Uses of Propellants

Propellants, both solid and liquid, are used in many secondary propulsion applications, including crew capsule ejection, attitude control and station-keeping of satellites, braking of reentry vehicles, extravehicular space operations—as well as being essential in rocket engine igniters, signal and illumination flares, and fuel-cell type electric generators. New developments, such as the high-powered gas dynamic laser[6] continue to broaden the field of applications.

Acknowledgment

Information for this entry as furnished by Mr. Donald M. Ross, Consulting Engineer, Lancaster, California and for confirmation of tabular performance data by Mr. Curtis C. Selph, Propellant Research Engineer, U.S. Air Force Rocket Propulsion Laboratory, Edwards, California, are gratefully acknowledged.

Additional Reading

Baryakhtar, V.H. and T. Rosendorfer: *Demilitarisation of Munitions: Reuse and Recycling Concepts for Conventional Munitions and Rocket Propellants,* Kluwer Academic Publishers, Norwell, MA, 1997.

Cohen, W.: "New Horizons in Chemical Propulsion," *Astronautics and Aeronautics,* 2, **12,** 46–51 (1973).

Davenas, A.: *Solid Rocket Propulsion Technology,* Elsevier Science, New York, NY, 1992.

Kent, J.A.: *Riegel's Handbook of Industrial Chemistry,* 9th Edition, Chapman & Hall, New York, NY, 1992.

Quinn, L.P., et al.: *High Energy Storage Investigations,* Rept. AFRPL TR-71-36, U.S. Air Force Rocket Propulsion Laboratory, Edwards, California, 1971.

Ross, D.M.: *Propellants,* in "Energy Technology Handbook," (D.M. Considine, editor), McGraw-Hill, New York, NY, 1977.

ROENTGEN. A unit of radiation, that quantity of X-rays or gamma rays which will produce, as a consequence of ionization, 1 electrostatic unit of electricity in 1 cubic centimeter of dry air measured at 0 degrees C and standard atmospheric pressure. See also **Radioactivity.**

RONTGEN, WILHELM CONRAD (1845–1923). Rontgen became one of the outstanding experimental physicists of his time. In 1869, he received his doctorate in mechanical engineering from the University of Zurich. He held various positions at higher institutions throughout Germany until, in 1888, he became the chair for the physics department at the University of Wurzburg.

In 1895, Rontgen was the first scientist to observe and record X-rays. While experimenting with a cathode ray discharge tube that was completely enclosed in cardboard so no light could escape, he observed that a sheet of barium platino-cyanide coated paper lying several feet away from the discharge tube was glowing in the dark. Rontgen realized the importance of this accidental discovery and he devoted six straight weeks of working in his laboratory investigating the X-ray. While investigating, he actually saw the image of his bones within his fingers and deduced that his X-rays must be a very short wavelength form of electromagnetic radiation. He published his work in a series of classic papers and the phenomenon of his X-rays had immediate use in hospitals in the diagnosis and treatment of fractures. His discovery could have been patented, but Rontgen gave it to humanity. Later, when he received a cash prize for receiving the first Nobel Prize (1901) in Physics, Rontgen gave the money to the University of Wurzburg.

Rontgen was truly a great scientist and great humanitarian. His discovery of the X-ray changed the world and today it has universal usage.

See also **X-Ray.**

J.M.I.

ROTATION AXIS. A symmetry element possessed by certain crystals, whereby the crystal can be brought into a physically equivalent position by rotation about an axis which can be onefold, twofold, threefold, fourfold, or sixfold, according to whether the crystal can be brought into self-coincidence by the operations of rotation through 360, 180, 120, 90, or 60 degrees about the rotation axis. See also **Mineralogy.**

ROWLAND, FRANK SHERWOOD (1927–). An American who won the Nobel prize for chemistry along with Paul Crutzen and Mario Molina in 1995 for their work in atmospheric chemistry, particularly concerning the formation and decomposition of ozone.

ROXEL PROCESS. A process for increasing the flame resistance of cotton by chemically modifying the fiber, i.e., by applying a formulation consisting of tetrakis(hydroxymethyl)phosphonium chloride (THPC), triethylolamine, and urea in aqueous solution. These substances cross-link with the hydroxyl groups of the cellulose to form an efficient and durable protective medium. This process is used in the manufacture of blankets, curtains, bed linen, and the industrial safety garments.

RUBBER (Natural). *Natural rubber* is the name applied to the polymer *cis*-polyisoprene obtained chiefly from the *Hevea brasiliensis* tree.[1] Originally, the tree grew wild in the Amazon valley, but during the last part of the 19th century it was planted in well-organized plantations in tropical lands of the Far East and later in Africa. See Table 1. The average rubber trees stand about 40–50 feet (12–15 meters) high. For optimum growth, a tropical climate having 80 inches (203 centimeters) or more of annual rainfall is required.

Rubber comes from the tree as a milky white fluid, which is a colloidal suspension of rubber in a liquid consisting mostly of water. The tree is tapped by well-trained workers, who use a sharp-edged tool, and the cutting action goes at an angle of 30° from top left to bottom right. It is important that the rubber latex-bearing cells be cut, but that the blade not wound the inner cambium layer, as this would harm the tree. A cup is hung below the cut to collect the white, milklike latex, which contains about 35% rubber,

---

[1] It is alleged that English scientist Joseph Priestly observed that the material could be used for rubbing out lead pencil marks and thus gave the material the name rubber. ("Introduction to the Theory of Perspective," Joseph Priestly, 1835.)

TABLE 1. NATURAL RUBBER EXPORTED

Producing area	Quantity produced (1000 long tons)				
	1940	1960	1970	1975	1979
Malaysia	547	775	1304	1424	1595
Indonesia	543	587	755	788	866
Other Asian countries, including Oceania	283	400	484	525	704
Africa	16	149	207	197	141
Tropical America	26	15	10	12	19

Note: 1 long ton is slightly more than 1 metric ton (1016 kilograms).

the remainder being water, protein, resins, organic materials, and other plant substances.

The yield of the *Hevea* tree can be increased by applying chemicals to the bark. These include 2-chloroethylphosphonic acid, which supplies small quantities of ethylene gas. This type of chemical is applied in a high-viscosity liquid form, usually mixed with palm oil or other diluent. The function of it is to stabilize the latex so that it continues to flow for a longer time and thus increases rubber yield. Because of the higher rubber yield per tapping operation, the cost of tapping labor is reduced. When stimulants are used, the tree is given a longer rest period between tappings to avoid diseases that would eventually kill it. See also **Plant Growth Modification and Regulation**.

During the 1980s, the annual yield for Malaysian plantations was about 1010 pounds/acre (1133 kilograms/hectare), which includes high-yielding trees which produce 1500 pounds/acre (1680 kilograms/hectare), as well as older, lower-yielding clones.

Properties of Natural Rubber

Chemically, natural rubber or *cis*-polyisoprene, has a broad molecular-weight distribution, ranging from several million to about one hundred thousand.

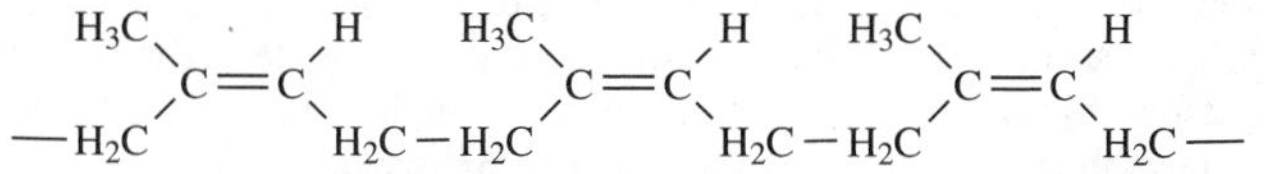

Natural rubber is soluble in practically all aromatic and aliphatic hydrocarbons and particularly in halogenated hydrocarbons. When cements and solvent adhesives are made using natural rubber, methylethylketone (MEK) frequently is used to reduce viscosity. Although MEK is not a solvent, it tends to disperse large molecular particles, resulting in lower-viscosity dilution. Crude rubber is decomposed by heat and can be cyclized at 250°C. It can easily be hydrogenized and reacts readily with halogens. The stress-strain properties of natural rubber are the best of all the elastomeric polymers. In vulcanized films made by the latex process, the tensile strength may exceed 6000 pounds per square inch (41 mPa), and ultimate elongation is as high as 700% or more.

Natural rubber is readily attacked by oxygen. Copper and manganese, if present in amounts greater than the specified 0.001%, greatly accelerate oxidation. There are, however, naturally occurring antioxidants in natural rubber that help preserve it until vulcanization. All vulcanized natural-rubber products contain added antioxidants to ensure satisfactory life.

Rubber burns quite readily and generates more than 10,000 cal/g. The specific gravity of rubber is 0.934, a property utilized in concentrating natural-rubber latex by the centrifuge process. The serum, which is mostly water and has a specific gravity of about 1.0, tends to separate readily from the rubber. The liquid concentrated latex is used in making foam rubber, dipped goods, adhesives, and carpet backing for nonwoven carpets. An industry has developed around this application, which involves

spreading foamed latex onto the underside of carpeting, making an integral carpet-foam system.

Compounding and Vulcanization

Crude rubber in the raw state has few applications with the exception of crepe soles for shoes. To make commercial rubber products, the material must be mixed with a variety of chemicals and vulcanized into desirable end shapes. Charles Goodyear discovered in 1839 that adding sulfur to rubber and heating the mixture greatly enhances the physical properties of rubber. The material no longer becomes tacky in warm weather and in cold weather it does not become brittle. The material is much tougher, and the quality of products made this way results in service for a much longer period of time. In addition to sulfur, which crosslinks the large rubber molecules and makes it a giant organic molecule, zinc oxide, organic accelerators, antioxidants, reinforcing pigments, and other processing aids are used in compounding rubber for useful vulcanized products.

The function of the antioxidant is to improve service life of the product against such well-known degrading agents as oxygen, light, and nitroso compounds. One theory is that the antioxidant selectively reacts with the degrader, slowing down its reaction with the rubber molecule, which would result in scission and eventually poorer physical properties. During recent years, considerably aggravated by air pollution, degradation of vulcanized rubber by small quantities of ozone (a few parts per million) in the air has become a serious problem. Ozone has little noticeable effect on unstretched rubber, but even under slight stretch it causes cracks in the surface, which grow perpendicularly to the direction of extension. Hundreds of different antioxidants and antiozonants are employed, amine and phenol complexes being the basis of most. See also **Antioxidants**.

Accelerators act as catalysts of vulcanization, but, unlike most catalysts, they undergo chemical change during the reaction. Benzothiazyl disulfide is one of the oldest types, dating back to 1925, but it still accounts for the greatest use in the industry today. Besides the thiazole types, other popular accelerators are sulfenamides, aryl guanidines, dithiocarbames (extremely fast accelerators used mostly in latex compounds), and thiurams (also very fast and often used as a secondary accelerator to hasten the vulcanization rate). Accelerators also contribute to improved ageing properties of the end product.

Stearic acid is an activator of vulcanization, as is zinc oxide, both reacting to form zinc stearate, which enhances the activity of the organic accelerators. Zinc stearate is impractical to add directly to the rubber because its slippery, lubricating nature makes it difficult to mix in the batch (see Fig. 1).

With an accelerated system, a simple network structure with dialkenyl mono- and disulfide crosslinks and conjugated triene units as main-chain modifications is obtained:

With an unaccelerated sulfur-natural-rubber system, the poor crosslinking efficiency results in sulfur being incorporated into the rubber network as long polysulfide crosslinks, cyclic monosulfides, and vicinal crosslinks, which are very close together and act physically as a single cross-link (see Fig. 2).

It is theorized that between the complex network structure of the unaccelerated system and the simpler network structure of the accelerated system, structures made up of the two models represent natural-rubber vulcanizates made at various times and temperatures of cures, with different reactant concentrations, and showing the effects of other variants.

At any given degree of crosslinking, the tensile strength is highest with polysulfide bonds. High elongation at break is obtained by slightly decreasing the crosslinking action. If lower elongation is required, slightly excessive crosslinking is used, usually accompanied by higher tensile strength. Vulcanization of rubber decreases its solubility in solvents, and this property frequently is used as a qualitative measure of cure.

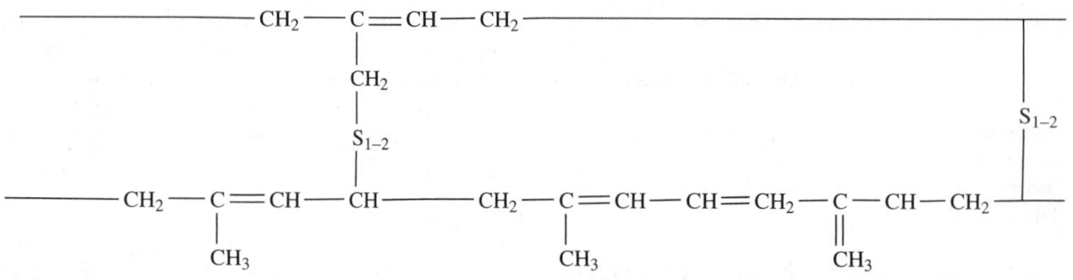

Fig. 1.   Accelerated system

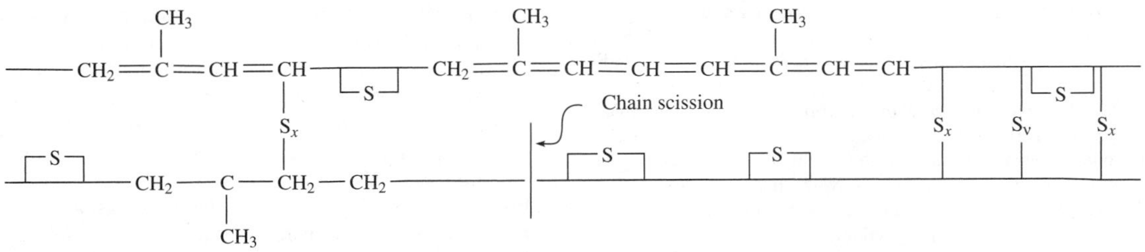

Fig. 2.   Unaccelerated system

Vulcanization by sulfur accounts for practically all the commercial products. However, peroxide types of curing systems may be used, especially for some of the synthetic rubbers.

Ultrahigh-frequency (UHF) energy may be used for preheating and precuring rubber compounds for continuous vulcanization (CV) of rubber, containing carbon black, for such applications as weather stripping, tubing, hose, and, in some instances, tire tread compounds.

Carbon black is the major reinforcing pigment used, not only for natural rubber, but for practically all the synthetic rubbers. As much as 40–50 parts by weight, based upon 100 parts of rubber, is used in all tire-tread compounds. Carbon black greatly increases tensile strength at low elongations (modulus) and results in longer-wearing tires. Colloidal silica contributes some reinforcing properties to rubber, but not to the same degree as carbon black. See also **Carbon Black**.

Uses of Natural Rubber

Thousands of flexible products requiring top performance characteristics are made of natural rubber, e.g., huge earthmover tires, truck tires, tires for large aircraft, bridge supports, and surgeons' gloves. The treads of most passenger car tires in the United States consist mainly of styrenebutadiene synthetic rubber because of lower cost and lower temperature buildup during use.

The use of natural rubber in passenger car tires has increased in recent years due to the industry going from bias to radial types which, in North America, now account for 75% of the total. Higher degree of tack or cohesive bonding during the building of the radial tire, as compared with that of styrene-butadiene rubber, is largely responsible for this.

Because of its excellent high- and low-temperature properties, many products used in the arctic and tropical areas of the world are made from natural rubber. However, it is not suitable for applications where there is contact with naphtha, e.g., gasoline hoses, because the solvent swells the material. Almost all elastic bands are made from natural rubber. Because of its excellent tack properties, the material is used in solvent and latex form as the base for adhesives.

With the dependence of synthetic rubber on petroleum, natural rubber, which is produced by solar energy, may look increasingly attractive over the years ahead.

Processing Raw Materials

Field latex is bulked in large tanks at a factory adjacent to the rubber estate. If a high-solids latex is desired, the field latex is strained, stabilized with ammonia or other chemicals, such as soap and bactericide, and either centrifuged or creamed to 62–68% total solids.

Smoked Sheet. For making ribbed smoked sheet, the field latex is immediately mixed with dilute formic acid in long horizontal tanks. Because fresh latex is somewhat protected by a protein surface layer, it does not coagulate or gel immediately on addition of the acid. Within a few hours, however, the rubber particles in the latex gel form a spongy mass which is then run through a series of smooth metal rolls with clearance decreased from one set to the next, an arrangement that squeezes out the serum and densifies the wet rubber. Water is run over the wet coagulum to wash out non-rubber materials and dirt. The last unit consists of ribbed rolls, which imprint ribbed markings on the sheet. After drying in air for a few hours, the sheets are hung in a drying shed at 40–50°C until dry. Modern installations use efficient drying tunnels. Sheets are inspected by holding them over a strong light to determine clarity, color, presence of dirt and other factors. The rubber is classified by various grades. Sheets then are piled up and squeezed in a baling machine to form 250-pound (~113-kilogram) bales that measure 19 × 19 × 24 inches (~ 48 × 48 × 61 centimeters).

Crepe. Another popular type of commercial rubber, known as crepe, consists of two major classes—*pale crepe* and *thick blanket crepe*. Pale crepe is made by adding sodium hydrogen sulfite, $NaHSO_3$, to field latex to inhibit discoloration and softening during processing. Formic acid is used as the coagulant. The wet coagulum is passed through rolls with longitudinal grooves, which give the rubber a crepe-like appearance. Water running over the surface cleans out dirt and other nonrubber ingredients. Sheets are hung up to dry in circulating warm air. The quality of pale crepe is assessed on its whiteness and how good the finished rubber appears.

Blanket crepes are of lower quality and are made from wet slabs obtained usually from small landholders. These are creped, dried, and baled. Other types of crepe are made from coagulum left in collection cups and from dried skin remaining from the tapping incision. In addition to collecting latex, a tapper collects all dried and coagulated rubber that remains from the previous round, usually as skin in the cup or on the tapping panel.

Grading of Rubber. Commercial grades of natural rubber are classified into two main groups: (1) "Green Book International Grades," and (2) "Technically Specified Forms." The former depends on a visual grading system, the source of the rubber and the method of preparation. This system, dating back many years and kept current by the International Rubber Quality and Packing Conference Committee, consists of 35 grades under 8 major types, such as Ribbed Smoked Sheets, White and Pale Crepe, Estate Brown Crepes, Compo Crepes, Thin Brown Crepes (Remills), Thick Blanket Crepes, and Pure Smoked Blanket Crepe. Publisher of the "Green Book" is the Rubber Manufacturers Association, Inc., New York.

Technically Specified Rubbers (TSR), originated by the Malaysian Rubber Producers Association, classifies rubber not only on the basis of the source of rubber, but on its physical properties, such as dirt content, ash, and nitrogen content, volatile matter, plasticity, and, with the higher-quality grades, cure rate and color are standardized. This type of rubber is packaged in 75-pound (34-kilogram) bales, wrapped in transparent plastic, and the color of the printing on the bale identification strips indicate whether the source is latex grade, sheet material grade, blended grades, or field grades. An additional convenience of this rubber is that the bales can be charged into the mixing machine (Banbury) without removing the wrapper. This is in contrast with the Green Book grades, which are bonded together by a press, with the outside layer treated with soapstone to keep the bales from sticking together. These larger bales require cutting before they can be charged into the mixer.

Guayule

During the past few years, natural rubber from the desert shrub *Parthenium argentatum* has been under intensive study by scientists in the United States and Mexico as a possible domestic source of natural rubber. This plant grows wild in the arid areas of Mexico and the United States. In 1910, guayule produced 10% of the world's rubber, but lower-cost Hevea rubber from the Far East displaced it from the market. Rubber in the guayule plant is present in the roots and branches of the shrub and must be separated and purified by a flotation and solvent system. The purified product is equivalent in chemical properties to the Hevea rubber. An advantage of guayule is that it can be grown on semiarid land that is not suitable for other crops. Presently, agricultural experimentation on increasing rubber yield of the plant is underway. The U.S. government has passed legislation providing funds to help in developing an American-based guayule industry. Several large rubber product manufacturers have experimental plots planted with the shrub. The National Research Council, Washington, DC., published a report, "Guayule: An Alternative Source of Natural Rubber," in 1977.

Additional Reading

Bhowmick, A.K., M.M. Hall, and H.A. Benarey: *Rubber Products Manufacturing Technology,* Marcel Dekker, Inc., New York, NY, 1994.

Hepburn, C.: *Rubber Technology,* 3rd Edition, Butterworth-Heinemann, Inc., Woburn, MA, 1998.

Loadman, M.J.: *Analysis of Rubber and Rubber-like Polymers,* Kluwer Academic Publishers, Norwell, MA, 1998.

Mark, J.E., F.R. Eirich, and B. Erman: *Science and Technology of Rubber,* 2nd Edition, Academic Press, Inc., San Diego, CA, 1994.

Sethuraj, M.R. and N.M. Mathew: *Natural Rubber: Biology, Cultivation, and Technology,* Elsevier Science, New York, NY, 1992.

Tinker, A.J. and K.P. Jones: *Blends of Natural Rubber: Novel Techniques for Blending with Specialty Polymers,* Chapman & Hall, New York, NY, 1998.

Thomas H. Rogers, Consultant (Rubber and Plastics Industries), formerly Research Manager, Goodyear Tire and Rubber Company, Akron, OH.

RUBBER (Synthetic).   Any of a group of manufactured elastomers that approximate one or more of the properties of natural rubber. Some of these are: sodium polysulfide ("Thiokol"), polychloroprene (neoprene), butadiene-styrene copolymers (SBR), acrylonitrilebutadiene copolymers (nitril rubber), ethylenepropylene-diene (EPDM) rubbers, synthetic polyisoprene ("Coral," "Natsyn"), butyl rubber (copolymer of isobutylene and isoprene), polyacrylonitrile ("Hycar"), silicone (polysilorane), epichlorohydrin, polyurethane ("Vulkollan").

The properties of these elastomers are widely different. All require vulcanization. In general, sulfur is used only for unsaturated polymers, peroxides, quinones, metallic oxides, or diisocyanates effect vulcanization with saturated types. Many are special-purpose rubbers, some can be used in tires when loaded with carbon black, others have high resistance to attack by heat and hydrocarbon oils and thus are superior to natural rubber for steam hose, gasoline and oil-loading hose. Most are available in latex form.

See also **Elastomers.**

RUBIDIUM.   [CAS: 7440-17-7]. Chemical element symbol Rb, at. no. 37, at. wt. 85.468, periodic table group 1, mp 38.9°C, bp 686°C, density 1.53 g/cm^3 (20°C). Elemental rubidium has a body-centered cubic crystal structure.

Rubidium is a silver-white, very soft metal; tarnishes instantly on exposure to air, soon ignites spontaneously with flame to form oxide; best preserved in an atmosphere of hydrogen rather than in naphtha; reacts vigorously with H$_2$O forming rubidium hydroxide solution and hydrogen gas. Discovered by Bunsen and Kirchhoff in 1860 by means of the spectroscope.

There are two naturally occurring isotopes ^{85}Rb and ^{87}Rb, of which the latter is unstable with respect to beta decay ($t_{12} = 5 \times 10^{10}$ years) into ^{87}Sr. There are eight other known radioactive isotopes ^{81}Rb through ^{84}Rb, ^{86}Rb, and ^{88}Rb through ^{90}Rb, all with comparatively short half-lives, measured in terms of minutes, hours, or days. In terms of abundance, rubidium ranks 34th among the elements in the earth's crust. In terms of content in seawater, the element ranks higher (18th) with an estimated 570 tons of rubidium per cubic mile of seawater. First ionization potential 4.176 eV; second, 27.36 eV. Oxidation potential Rb → Rh$^+$ + e$^-$, 2.99 V. Other important physical properties of rubidium are given under **Chemical Elements.**

Rubidium occurs in lepidolite (lithium aluminosilicate, in amount up to 1% Rb), in certain mineral waters and rare minerals. Rubidium salts may be recovered from the mother liquor upon crystallization of (1) lithium salts, (2) potassium salts. Rubidium metal is obtained by electrolysis of the fused chloride out of contact with air.

Uses

The main uses of rubidium are in photocathodes and photo-electric cells. However, rubidium cells are inferior to cesium cells in their sensitivity and range. Although very small quantities are involved, rubidium gas cells now perform as secondary time standards, on the order of quartz crystal oscillators, inasmuch as they must be referenced to more accurate systems. The rubidium systems have a characteristic resonance at 6,835 MHz and, unlike other atomic frequency standards, require little power and are relatively compact. Portable rubidium atomic clocks were introduced by the U.S. Army in 1963. They weight as little as 44 pounds (20 kilograms) and occupy a volume of only about 1 cubic foot (0.028 cubic meter). The units operate on 110-V current, on the 24-V output of military vehicles, or both. Clocks of this type are used to synchronize radar nets, to assist in the accurate tracking of missiles and satellites, and to set precise radio broadcasting frequencies. Rubidium-vapor instruments also were developed as absolute-type magnetometers and introduced in 1958 by U.S. government scientists. The rubidium-vapor magnetometer uses a rubidium lamp, mounted in the tank coil of a radio-frequency oscillator.

After collimating and filtering, the rubidium light is circularly polarized and then passed through a rubidium-vapor cell, after which it is focused on a sensitive photocell. Numerous combinations of amplifier parameters and various rubidium isotopes permit considerable range in the measurement of ambient magnetic fields. Inasmuch as the total world range is from 15,000 to 80,000 gammas, a system capable of this span finds use anywhere in the world.

Potential uses of rubidium include use as a fuel for ion-propulsion engines and as a heat-transfer medium.

Rubidium alloys easily with potassium, sodium, silver, and gold, and forms amalgams with mercury. Rubidium and potassium are completely miscible in the solid state. Cesium and rubidium form an uninterrupted series of solid solutions. These alloys, in various combinations, are used mainly as getters for removing the last traces of air in high-vacuum devices and systems.

Small quantities of rubidium are found in certain foods, including coffee, tea, tobacco, and several other plants. There is evidence indicating that trace quantities of the element are required by living organisms.

Chemistry and Compounds

Rubidium is more electropositive than potassium (or the lower alkali metals) as is consistent with its position in main group 1. It reacts more vigorously with H$_2$O, and ignites on exposure to oxygen.

Because of the ease of removal of its single 5s electron (4.159 eV) and the difficulty (27.36 eV) of removing a second electron, rubidium is exclusively monovalent in its compounds, which are electrovalent.

In its solutions in liquid NH$_3$, rubidium is, like the other alkali metals, a powerful reducing agent, so that in such solutions titrations of rubidium polysulfide with rubidium are made by electrometric methods. The solubility of rubidium salts in liquid NH$_3$ increases markedly with the radius of the anion (rubidium chloride, RbCl, 0.024 moles per kilogram, rubidium bromide, RbBr, 1.35 moles per kilogram, and rubidium iodide, RbI, 10.08 moles per kilogram). However, in water they exhibit minimum solubility at cation: anion radius ratio of 0.75 (rubidium fluoride, RbF, 12.5 moles/kilogram, RbCl 6.8 moles/kilogram, RbBr, 6.6 moles/kilogram, RbI 7.2 moles/kilogram).

As in the case of the other alkali metals, rubidium forms compounds generally with the inorganic and organic anions; for a general discussion of these compounds, see the entry on **Sodium,** because the sodium compounds differ principally in their greater extent of hydration and greater number of hydrates. However, rubidium coordinates with large organic molecules, such as salicylaldehyde, even though it does not with H$_2$O.

One respect in which rubidium and cesium are outstanding among the alkali metals is the readiness with which they form alums. Rubidium alums are known for all of the trivalent cations that form alums, Al^{3+}, Cr^{3+}, Fe^{3+}, Mn^{3+}, V^{3+}, Ti^{3+}, Co^{3+}, Ga^{3+}, Rh^{3+}, Ir^{3+}, and In^{3+}.

As in the case of potassium and cesium, rubidium forms a superoxide on reaction of the metal with oxygen. The compound is dark brown in color and paramagnetic, and hence believed to contain the O$_2^-$ ion with an odd electron, and to have the formula RbO$_2$. On heating, it loses oxygen to form Rb$_2$O$_3$. Rubidium also forms a peroxide Rb$_2$O$_2$, and a normal oxide, Rb$_2$O, which is prepared by heating rubidium nitrite with metallic rubidium.

Rubidium hydroxide, [CAS: 1310-82-3]. RbOH, is the strongest, except for cesium hydroxide, CsOH (and francium hydroxide, FrOH), of the alkali hydroxides, as would be expected from its position in the periodic table. For the same reason, it has the next smallest lattice energy (146.6 kilocalories per mole).

The most numerous organic compounds of rubidium are those of oxy compounds, such as the salts of organic acids, the alcohols and phenols (alkoxides, phenoxides, etc.). An ethyl rubidium-zinc diethyl adduct has been reported, RbZn(C$_2$H$_5$)$_3$, which is certainly the true salt, rubidium triethylzincate, Rb[Zn(C$_2$H$_5$)$_3$].

Additional Reading

Christensen, J.N., J.L. Rosenfeld, and D.J. DePaolo: "Rates of Tectonometamorphic Processes from Rubidium and Strontium Isotopes in Garnet," *Science,* 1465 (June 23, 1989).

Greenwood, N.N. and A. Earnshaw: *Chemistry of the Elements,* 2nd Edition, Butterworth-Heinemann, Inc., Woburn, MA, 1997.

Krebs, R.E.: *The History and Use of Our Earth's Chemical Elements: A Reference Guide,* Greenwood Publishing Group, Inc., Westport, CT, 1998.

Lewis, R.J. and N.R. Sax: *Sax's Dangerous Properties of Industrial Materials,* 10th Edition, John Wiley & Sons, Inc., New York, NY, 1999.

Lide, D.R.: *CRC Handbook of Chemistry and Physics,* 84th Edition, CRC Press, LLC., Boca Raton, FL, 2003.

Staff: *ASM Handbook—Properties and Selection: Nonferrous Alloys and Special-Purpose Materials,* American International, Materials, Park, OH, 1990.

Zhu, O., et al.: "X-ray Diffraction Evidence for Nonstoichometric Rubidium-C60 Intercalation Compounds," *Science,* 545 (October 25, 1991).

RUTHENIUM.

[CAS: 7440-18-8]. Chemical element, symbol Ru, at. no. 44, at. wt. 101.07, periodic table group 8 (platinum metals), mp 2,310°C, bp 3,900°C, specific gravity 12.41 (20°C). Elemental ruthenium has a close-packed hexagonal crystal structure. The seven stable isotopes are 96Ru, 98Ru through 102Ru, and 104Ru. The five unstable isotopes are 95Ru, 97Ru, 103Ru, 105Ru, and 106Ru. In terms of earthly abundance, ruthenium is one of the scarce elements. Also, in terms of cosmic abundance, the investigation by Harold C. Urey (1952), using a figure of 10,000 for silicon, estimated the figure for ruthenium at 0.019. No notable presence of ruthenium in seawater has been found. Ruthenium was discovered by Claus (Germany) in 1844.

Electronic configuration $1s^2 2s^2 2p^6 3s^2 3p^6 3d^{10} 4s^2 4p^6 4d^7 5s^1$. Ionic radius Ru⁴⁺ 0.60 Å. Metallic radius 1.3251 Å. First ionization potential 7.5 eV. Other physical properties of ruthenium will be found under **Platinum and Platinum Group**. See also **Chemical Elements**.

The chemistry of Ru is still poorly understood. The existence of at least eight valence states, coupled with the tendency to complex with many ions, often results in the presence of several different complexes in a given solution.

Ru metal is quite refractory. It is not significantly soluble in any single acid; even aqua regia has little effect. At room temperature, the metal does not react with O_2, but, when heated in air, a film of the dioxide appears. The metal is insoluble in fused sulfates. Molten alkali slowly dissolves the metal. The rate of attack is rapid under oxidizing conditions, and a molten mixture of NaOH and Na_2O_2 will readily dissolve the metal.

The finely divided metal is soluble in hypohalites if an excess of alkali is present. At red heat, the metal combines with Cl_2 to form the dichloride. Ruthenium(VIII) oxide is formed when an alkaline ruthenium solution is treated with a strong oxidant, such as chlorine, or bromate ion when the Ru is in acid solution.

Ruthenium(III) hydroxide is formed by the action of alkali on a solution of ruthenium(III) chloride. It is easily oxidized by air to the tetravalent state. The dioxide, RuO_2, forms when the metal is heated in air. Hydrous ruthenium(IV) oxide can be precipitated by adding alcohol to a less than 3-M NaOH solution of ruthenium(VIII) oxide, followed by boiling. Above 3-M NaOH, complete reduction is not obtained. The hydrous oxide that is soluble in concentrated HCl tends to occlude impurities.

The only known octavalent Ru compound is the tetroxide, RuO_4, which exists in a yellow and a brown form. The volatile and poisonous tetroxide melts at about 25°C and sublimes readily. It may explode in contact with oxidizable substances or when heated above 100°C. It is formed by distillation from either an alkaline or acid solution under strongly oxidizing conditions. The tetroxide is moderately water-soluble. When dissolved in alkali, it initially forms a green solution of heptavalent perruthenate of the form $MRuO_4$, which further reduces to the orange ruthenate M_2RuO_4. The reduction to the hexavalent state is quicker in strong alkali. The ruthenates also are made by fusing finely divided metal with a mixture of alkali hydroxide and nitrate or peroxide.

Anhydrous ruthenium(III) chloride, $RuCl_3$, is made by direct chlorination of the metal at 700°C. Two allotropic forms result. The trihydrate is made by evaporating an HCl solution of ruthenium(III) hydroxide to dryness or reducing ruthenium(VIII) oxide in a HCl solution. The trihydrate, $RuCl_3 \cdot 3H_2O$, is the usual commercial form. Aqueous solutions of the trihydrate are a straw color in dilute solution and red-brown in concentrated solution. Ruthenium(III) chloride in solution apparently forms a variety of aquo- and hydroxy complexes. The analogous bromide, $RuBr_3$, is made by the same solution techniques as the chloride, using HBr instead of HCl.

Ruthenium(III) iodide, RuI_3, is a black, insoluble compound precipitated by the addition of iodide ion to a solution of $RuCl_3$.

Tetravalent ruthenium chloride, $RuCl_4$, and the hydroxychloride, $Ru(OH)Cl_3$, are intermediate products when $RuCl_3$ is prepared by evaporating the tetroxide in HCl. When the hydroxychloride in hot HCl is treated with Cl_2, it is converted to the tetrachloride. The anhydrous tetrachloride also is known. The tetrabromide and tetraiodide have not been isolated; attempts to prepare these compounds result in the formation of the respective trihalides.

The only pentavalent Ru compounds known are the fluorides; RuF_5 is made by combining the elements. The compound melts at 107°C and boils at 313°C. The salt $NaRuF_6$ was recently made by mixing $RuCl_3$, with NaCl and treating the mixture with BrF_3.

Ru forms many complex ions. The nitrosyl compounds are frequently encountered by accident due to the great affinity of Ru for the nitrosyl group. Ruthenium(III) nitrosylchloride, $Ru(NO)Cl_3 \cdot 4H_2O$, is a by-product of most solutions of $RuCl_3$ in aqua regia or solutions containing HNO_3. It also is present in HCl solutions resulting from a KOH and nitrate fusion of the metal. The chloride and bromide are respectively raspberry and violet in solution. Alkaline chlorides form complex salts of the type $M_2Ru(NO)Cl_5$, which can be crystallized from solution. A black gelatinous precipitate of the nitrosylhydroxide, $RuNO(OH)_3$, is slowly formed when a solution of the nitrosylchloride is heated with a strong base. A series of nitrato- and nitro-derivatives of nitrosylruthenium also have been described and separated.

It is generally accepted that the disulfide is the only certain sulfide of Ru. It is formed by the action of H_2S on a solution of Ru or from the elements at about 1000°C. When ruthenium(IV) sulfide is treated with HNO_3, the sulfate is formed.

Dichlorodicarbonylruthenium(II), $Ru(CO)_2Cl_2$, is formed when $RuCl_3$ is heated above 210°C in the presence of CO. It is a yellow, insoluble, volatile compound. The bromine and iodine analogs are similarly formed.

When finely divided Ru metal is heated at 180°C under 200 atm of CO, pentacarbonylruthenium(0), $Ru(CO)_5$, is formed.

Ruthenium forms a large number of complex ions with amines.

Recently, a new group of organometallic sandwich compounds, called *metallocenes*, has been discovered. Ruthenocene is made in about 50% yield by reacting RuCl3 with cyclopentadienylsodium in tetrahydrofuran. After refluxing and distilling the solvent, the light-yellow crystals of ruthenocene are sublimed. The compound, $Ru(C_5H_5)_2$, undergoes a large number of substitution reactions typical of aromatic systems.

Ruthenium is commonly used with other platinum metals as a catalyst for oxidations, hydrogenations, isomerizations, and reforming reactions. The synergetic effect of mixing ruthenium with catalysts of platinum, palladium, and rhodium has been found for the hydrogenations of aromatic and aliphatic nitro compounds, ketones, pyridine, and nitriles.

LINTON LIBBY
Chief Chemist, Simmons Refining Company
Chicago, Illinois

Additional Reading

Coles, D.G. and L.D. Ramspott: "Migration of Ruthenium-106 in a Nevada Test Site Aquifer," *Science,* **215**, 1235–1237 (1982).

Davis, J.R.: *Metals Handbook,* 2nd Edition, ASM International, Materials Park, OH, 1998.

Greenwood, N.N. and A. Earnshaw: *Chemistry of the Elements,* 2nd Edition, Butterworth-Heinemann, Inc., Woburn, MA, 1997.

Krebs, R.E.: *The History and Use of Our Earth's Chemical Elements: A Reference Guide,* Greenwood Publishing Group, Inc., Westport, CT, 1998.

Lewis, R.J. and N.I. Sax: *Sax's Dangerous Properties of Industrial Materials,* 10th Edition, John Wiley & Sons, Inc., New York, NY, 1999.

Lide, D.R.: *CRC Handbook of Chemistry and Physics,* 84th Edition, CRC Press, LLC., Boca Raton, FL, 2003.

Seddon, E.A. and K.R. Seddon: *The Chemistry of Ruthenium,* Elsevier Science, New York, NY, 1984.

Sinfelt, J.H.: "Bimetallic Catalysts," *Sci. Amer.,* 90–98 (September 1985).

Staff: *ASM Handbook—Properties and Selection: Nonferrous Alloys and Special-Purpose Materials,* American International, Materials Park, OH, 1990.

RUTHERFORD, ERNEST (1871–1937).

Rutherford was a British physicist who was born in the South Island of New Zealand and is famous for his pioneering work in nuclear physics and for his theory of the structure of the atom.

Rutherford was awarded a scholarship to be a research student at the University of Cambridge and began research under J.J. Thomson. He soon abandoned research on his radio wave detector to work on the power of X-rays to confer electric charge on gases but soon turned to researching the problem of the rays emitted by thorium. Rutherford found three kinds of radiation, which he named alpha, beta, and gamma. In collaboration with Frederick Soddy, he was able to isolate a substance, thorium X, and identify the phenomenon of radioactive half-life and formulated an explanation of radioactivity. Rutherford was awarded the 1908 Nobel Prize for chemistry for his work in radioactivity.

In 1907, Rutherford moved to the University of Manchester and in 1909 he discovered the atomic nucleus. In 1911 he announced his revolutionary idea on the nature of the atom and he developed a model of the atom showing it similar to the solar system. He proposed the idea that almost all the mass and all the positive electricity in an atom was densely concentrated in a tiny nucleus and the electrons circled around it like planets around the sun.

While at Manchester, Rutherford produced the first human "nuclear reaction" with the disintegration of a non-radioactive atom, dislodging a single particle. He became famous as the man who "split the atom." In 1919, Rutherford succeeded J.J. Thomson as Cavendish Professor of Physics at Cambridge. He was a leader of a research team encouraging others in the investigation of the nucleus.

See also **Radioactivity**; and **Proton**.

J.M.I.

RUTHERFORDIUM.   See Chemical Elements.

RUTILE.   A mineral, composed of titanium dioxide, which occurs in three distinct forms: as rutile, a tetragonal mineral usually of prismatic habit, often twinned; as octahedrite (anatase), a tetragonal mineral of pseudo-octahedral habit; and as brookite, an orthorhombic mineral. Both octahedrite (anatase) and brookite are relatively rare minerals.

Rutile has a sub-conchoidal fracture; is brittle; luster, metallic-adamantine; color, commonly reddish-brown but sometimes yellowish, bluish or violet; streak, brown; transparent to opaque. Rutile may contain up to 10% of iron.

Experiments in the artificial preparation of titanium dioxide appear to show that rutile is the most stable form and produced at the highest temperature, brookite at a lower temperature, and octahedrite (anatase) at a still lower temperature.

Rutile is found as an accessory mineral in many kinds of igneous rocks, and to some extent in gneisses and schists. In groups of acicular crystals it is frequently seen penetrating quartz as the "flèches d'amour" from Grisons, Switzerland, and Brazil. Rutile is found also in Austria, Italy, Norway, South Australia, and Brazil. In the United States it occurs in Vermont, Massachusetts, Connecticut, New York, Pennsylvania, Virginia, Georgia, North Carolina, and Arkansas.

Rutile derives its name from the Latin *rutilus*, red, in reference to the deep red color observed in some specimens when viewed by transmitted light.

RUZICKA, LEOPOLD (1887–1976).   A chemist who won the Nobel prize in 1939 with Adolf Friedrich Johann Butenandt. His work involved research in organic synthesis including polymethylenes and higher terpenes. He was the first chemist to synthesize musk, androsterone, and testosterone from cholesterol. His medical degree was awarded at the University of Basel, Switzerland, although he was born in Croatia and educated partially in Germany.

S

SABATIER, PAUL (1854–1941). A French chemist who received the Nobel prize in Chemistry in 1912 along with Victor Grignard. His work involved the behavior of oxides as oxidizing catalysts and as agents for dehydrating and dehydrogenating. He received his PhD. In Nimes, France, and went on to become lecturer and faculty member in Toulouse, France.

SABATIER-SENDERENS REDUCTION. Catalytic hydrogenation of organic compounds in the vapor phase by passage over hot, finely divided nickel (the oldest of all hydrogenation methods).

SACCHARIDE. See **Carbohydrates**.

SACCHARIMETER. An instrument for the measurement of sucrose solutions. A saccharimeter differs from a polarimeter in that the saccharimeter uses white light, whereas a polarimeter is operated with sharply monochromatic light. Consequently, a saccharimeter can only be used with sugar solutions (in which case the quartz compensates for the rotatory dispersion of sucrose). Conversely, a polarimeter is suitable for the measurement of optical rotation of any solution, including sugar. However, saccharimetric sugar determinations are the basis of internationally accepted *sugar degrees* ($°S$), and saccharimeters are appropriately calibrated. When a polarimeter is employed that utilizes monochromatic light rather than a quartz wedge, deviations from sugar degrees will be found in some solutions, and they may not be inconsequential. The International Sugar Scale assigns $100°S$ to a pure sucrose solution of normal weight (26 grams in 100 milliliters of pure water) at $20°C$ and a 200-millimeter light path, measured in a saccharimeter with white light and a dichromate filter. Of course, an exact numerical conversion from sugar degrees to angular degrees is possible with pure sucrose ($100°S$ corresponds to $\alpha = 34.6°$), but in practical, more or less impure solutions, the relationship is not exactly predictable, and the sugar scale is conventionally and legally binding. It is not surprising, then, that virtually all sugar laboratories use visual saccharimeters (also called polariscopes). They differ in vintage, the half-shade presentation, and in construction features, but they all use the quartz-wedge compensating principle, and they read out in sugar degrees. A typical instrument will have a split half-shade field or a triple field for observation. The quartz wedge is equipped with a fine scale that is read off a second observation tube. It is graduated in $°S$ (e.g., -30 to $+110°S$) and comes with a vernier, readable to $1/10°S$.

See also **Photometers**; **Polarimetry**; and **Polarized Light**.

SACCHARIN. See **Sweeteners**.

SACCHAROMYCES. See **Carbohydrates**.

SAFFLOWER SEED OIL. See **Vegetable Oils (Edible)**.

SALICYLIC ACID AND RELATED COMPOUNDS. [CAS: 69-72-7]. Salicylic acid or $C_6H_4(OH)(COOH)$ is a white solid, melting points $159°C$, sublimes at $76°C$, insoluble in cold water, soluble in hot water, alcohol, or ether. With ferric chloride solution, salicylic acid solutions are colored violet (distinction from benzoic acid).

Salicylic acid may be obtained (1) from oil of wintergreen, which contains methyl salicylate, or (2) by heating dry sodium phenate C_6H_5ONa plus carbon dioxide under pressure at $130°C$ and recovering from the resulting sodium salicylate by adding dilute sulfuric acid.

Uses of Salicylic Acid

Approximately 60% of the salicylic acid produced in the United States has been consumed in the manufacture of aspirin. However, several applications of the acid are growing. Salicylic acid USP, EP, and other pharmacopeial grades are used medically as antiseptic, disinfectant, antifungal, and keratolytic agents. Carbonless copypaper using salicylic acid and alkyl salicylic acid derivatives has become an active application area. The salicylic acid is incorporated into a resin or is encapsulated as one of the agents for pressure or thermally activated imaging.

Salicylic acid derivatives form surfactants that can influence the rheologic properties of solutions. They impart controllable and useful viscous and elastic properties to aqueous liquids.

Salts of Salicylic Acid. A large number of salts of salicylic acid have been prepared and evaluated for therapeutic or other commercial use. Sodium salicylate has analgesic, antiinflammatory, and antipyretic activities. Magnesium salicylate, an analgesic and antiinflammatory agent, appears to have exceptional ability to relieve backaches. It is also used for the symptomatic relief of arthritis. Bismuth subsalicylate is taken orally in combination with other ingredients for protective, antacid action as well as antidiarrheal and antiseptic effects.

Esters of Salicylic Acid. The esters of salicylic acid are commercially produced by esterification of salicylic acid with the appropriate alcohol using a strong mineral acid such as sulfuric as a catalyst. To complete the esterification, the excess alcohol and water are distilled away and recovered. The main commercial applications for salicylate esters are as uv sunscreen agents and as flavor and fragrance agents. Several have application as topical analgesics.

Other Derivatives of Salicylic Acid. p-Aminosalicylic acid and its salts have been used in the treatment of tuberculosis. p-Aminosalicylic acid can be prepared by the carboxylation of m-aminophenol. Methylene-5,5-disalicylic acid, produced by heating two parts salicylic acid with 1–1.5 parts of 30–40 wt % formaldehyde in the presence of an acid catalyst, is used as an intermediate in the production of bacitracin methylenedisalicylate.

Salicylamide, prepared by the reaction of methyl salicylate with ammonia, has mild analgesic, antiinflammatory, and antipyretic properties. Salicylanilide, prepared by heating salicylic acid and aniline in the presence of phosphorus trichloride, is used as an intermediate in the production of other chemicals and as a slimicide, fungicide, and medicament.

Acetylsalicylic Acid (Aspirin). Acetylsalicylic acid, (o-acetyoxybenzoic acid) was first synthesized in 1853 by reaction of acetyl chloride with sodium salicylate. As a drug, acetylsalicylic acid was introduced in Germany in 1899 and into the United States in 1900. Aspirin is a registered trademark of Bayer in many nations, but in the United States and the United Kingdom, aspirin is accepted as the generic name for acetylsalicylic acid. See **Acetylsalicylic Acid**.

m-Hydroxybenzoic Acid. Of the three hydroxybenzoic acids, the metaisomer is of least commercial importance.

Unlike salicylic acid, m-hydroxybenzoic acid does not undergo the Friedel-Crafts reaction. It can be converted in 80% yield to m-aminophenol by the Schmidt reaction, which involves treating the acid with hydrazoic acid in trichloroethylene in the presence of sulfuric acid at $40°C$. m-Hydroxybenzoic acid is reported as an intermediate in the manufacture of germicides, preservatives, pharmaceuticals, and plasticizer.

p-Hydroxybenzoic Acid. p-Hydroxybenzoic acid is of significant commercial importance. The most familiar application is the use of several of its esters as preservatives, known as parabens. See also Antimicrobial Agents (Foods). Also of interest is the use in liquid crystal polymer applications.

Salicyl Alcohol. Salicyl alcohol [CAS: 90-01-7]: (saligenin, o-hydroxybenzyl alcohol) crystallizes from water in the form of needles or white rhombic crystals. It occurs in nature as the bitter glycoside, salicin [CAS: 138-52-3], which is isolated from the bark of *Salix helix, S. pentandra, S. praecos*, some other species of willow trees, and the bark of a number of species of poplar trees such as *Polpulus balsamifera, P. candicans,* and *P. nigra.*

The alcohol, which sublimes readily and is very soluble in alcohol and ether, has the following properties: melting point, 86°C; density 13°/25°, 1.161 g/cm³; heat of combustion, 3.542 mJ/mol (846.6 kcal/mol); and solubility in 100 mL water at 22°C, 6.7 g.

Saligenin has been used medically as an antipyretic and appears to possess marked topical analgesic powers in concentrations of 4–10%. Saligenin's taste is pungent at first and then numbing. Unsymmetrical diphenylolmethanes prepared from salicyl alcohol and substituted phenol at 160–170°C in the presence of alkaline catalysts have been claimed as resin components and resin-hardening agents.

Thiosalicylic Acid. Thiosalicylic acid, [CAS: 147-93-3], (*o*-mercapto-benzoic acid), a sulfur-yellow solid that softens at 158°C, has a melting point of 164°C. It sublimes, is slightly soluble in hot water but freely soluble in glacial acetic acid and alcohol, and yields dithiosalicylic acid, CAS 527-89-9, upon exposure to air.

Thiosalicylic acid has been used as an anthelmintic, bactericide, and fungicide. It has also been used as a rust remover, a corrosion inhibitor for steel, and a polymerization inhibitor. In photography, it has application in print-out emulsions and as an activator for photographic emulsions.

SALT. A compound formed by replacement of part or all of the hydrogen of an acid by one (or more) element(s) or radical(s) that are essentially inorganic. Alkaloids, amines, pyridines, and other basic organic substances may be regarded as substituted ammonias in this connection. The characteristic properties of salts are the ionic lattice in the solid state and the ability to dissociate completely in solution. The halogen derivatives of hydrocarbon radicals and esters are not regarded as salts in the strict definition of the term.

In the classical concept of the process of neutralization, whereby an acid and a base in solution react to form a salt, the proton of the acid and hydroxyl ion of the base react to form water, leaving the cation of the base and the anion of the salt by recombination.

Upon evaporation of the solvent, the salt is obtained as such, frequently as crystals, sometimes with and sometimes without water of crystallization. A salt, when dissolved in an ionizing solvent, or fused (e.g., sodium chloride in water), is a good conductor of electricity and when in the solid state forms a crystal lattice (e.g., sodium chloride crystals possess a definite lattice structure for both sodium cations (Na^+) and chloride anions (Cl^-), determinable by examination with x-rays).

A broader definition than that confined to solutions is demanded in some fields of chemistry (e.g., in high temperature reactions of acids, bases, and salts). In the formation of metallurgical slags, at furnace temperatures, calcium oxide is used as base and silicon oxide and aluminum oxide as acids; calcium aluminosilicate is produced as a fused salt. Sodium carbonate and silicon oxide when fused react to form the salt sodium silicate with the evolution of carbon dioxide. In this sense:

$$\begin{bmatrix} \text{Oxide of any} \\ \text{element functioning} \\ \text{as a metal-that is,} \\ \text{as a base} \end{bmatrix} \text{plus} \begin{bmatrix} \text{Oxide of any} \\ \text{element functioning} \\ \text{as a nonmetal-that} \\ \text{is, as an acid} \end{bmatrix} \text{yields[Salt]}$$

Iron and sulfur when heated react to form the salt ferrous sulfide. In this sense:

$$\text{metal plus nonmetal yields salt}$$

Salts therefore, are prepared (1) from solutions of acids and bases by neutralization and separation by evaporation and crystallization; (2) from solutions of two salts by precipitation where the solubility of the salt formed is slight (e.g., silver nitrate solution plus sodium chloride solution yields silver chloride precipitate [almost all as solid], and sodium nitrate present in solution as sodium cations and nitrate anions [recoverable as sodium nitrate, solid by separation of silver chloride and subsequent evaporation of the solution]); (3) from fusion of a basic oxide (or its suitable compound—sodium carbonate above) and an acidic oxide (or its suitable compound—ammonium phosphate), since ammonium and hydroxyl are volatilized as ammonia and water. Thus, sodium ammonium hydrogen phosphate

$$\begin{array}{c} NH_4 \\ | \\ Na - PO_4 \\ | \\ H \end{array}$$

yields sodium metaphosphate, $NaPO_3$, upon heating. (4) Salts also are prepared from reaction of a metal and a nonmetal.

Reactions of salts as such in solution, without decomposition of cation or anion, are dependent upon the presence of the cation and the anion of salt.

An *acid salt* is a salt in which all of the replaceable hydrogen of the acid has not been substituted by a radical or element. These salts, in ionizing, yield hydrogen ions and react like the acids (e.g., $NaHSO_4$, $KHCO_3$, Na_2HPO_4).

An *amphiprotic* (also called *amphoteric*) *salt* is a salt that may ionize in solution either as an acid or a base, and react either with bases or acids, according to the conditions.

A *basic salt* is a salt contains combined base as $Pb(OH)_2 \; Pb(C_2H_3O_2)_2$, a basic acetate of lead. These salts may be regarded as formed from the basic hydroxides by partial replacement of hydroxyl (e.g., $HO-Zn-Cl$). They react like bases and, when soluble, ionize to yield hydroxyl ions.

A *complex salt* is a saline compound having the structure of a combination of two or more salts and that is regarded as the normal salt of a complex acid. Complex salts do not split into a mixture of the constituent salts in solution, but furnish a complex ion that contains one of the bases (e.g., potassium molybdophosphate and potassium platinochloride).

A *double salt* is a substance consisting of two simple salts that crystallize together in definite proportions and exist independently in solution (distinction from complex salts). The alums are representative double salts.

An *inner salt* is a member of a special class of internal salts in which an acid group and a neutral group coordinate with metals to form a cyclic complex. These salts occur widely in analytical chemistry, (where they are formed between metallic ions and organic reagents) in dyestuffs, in life processes (chlorophyll and hematin belong to this class of compounds), and in many other fields.

An *internal salt* is a compound in which the acidic or basic groups that react to produce the salt linkage (which may or may not entail the formation of water) are in the same molecule. This particular salt linkage may consist of a polar or a nonpolar bond.

A *mixed salt* is a salt of a polybasic acid in which the hydrogen atoms are replaced by different metallic atoms or positive radicals.

A *pseudo salt* is a compound that has some of the normal characteristics of a salt, but lacks certain others, notably the ionic lattice in the solid state and the property of ionizing completely in solution. The absence of these properties is due to the fact that the bonds between the metallic and nonmetallic radicals are covalent or semicovalent instead of polar. Because these salts do not ionize completely, they are also called *weak salts.*

SALT BATH. A molten mixture of sodium, potassium, barium, and calcium chlorides or nitrates to which sodium carbonate and sodium cyanide are sometimes added. Used for hardening and temperature of metals and for annealing both ferrous and nonferrous metals. Temperatures used may be as high as 1,315°C for hardening high-speed steels. Commercial mixtures are available for a variety of specifications.

SALT (NaCI). See **Sodium Chloride.**

SALTING OUT. Reduction in the water solubility of an organic solid or liquid by adding a salt (usually sodium chloride) to an aqueous solution of the substance. Ions of the dissolved salt attract and hold water molecules, thus making them less free to react with the solute. The result of this is to decrease the solubility of the solute molecules with consequent separation or precipitation. Colloidal suspensions of proteins, soaps, and similar substances are precipitated in this way.

SALTPETER. Potassium nitrate. Sodium nitrate is often called Chile saltpeter, and calcium nitrate is sometimes called Norway saltpeter.

SAMARIUM. [CAS: 7440-19-9]. Chemical element symbol Sm, at. no. 62, at. wt. 150.35, fifth in the Lanthanide Series in the periodic table, mp 1,073°C, bp 1,791°C, density 7.520 g/cm³ (20°C). Elemental samarium has a rhombohedral crystal structure at 25°C. The pure metallic samarium is silver-gray in color, retaining a luster in dry air, but only moderately stable in moist air, with formation of an adherent oxide. When pure, the metal is soft and malleable, but must be worked and fabricated under an inert gas atmosphere. Finely divided samarium as well as chips from working are pyrophoric and ignite spontaneously in air, burning at 150–180°C. There are seven natural isotopes of samarium ^{144}Sm, ^{147}Sm through ^{150}Sm, ^{152}Sm, and ^{154}Sm. Eleven artificial isotopes have been identified. The natural ^{147}Sm isotope is weakly radioactive with a half-life of 2.5×10^{11} years. The samarium isotope mixture is the second highest (after gadolinium) of all elements in terms of its thermal-neutron-absorption cross-section (5,800 barns at 0.025 eV). The cross-section of ^{149}Sm is about 40,000 barns. Samarium ranks 62nd in abundance of the elements in the earth's crust, exceeding tantalum, mercury, bismuth, and the precious metals, excepting silver. The element was first identified by Lecoq de Boisbaudran in 1879. Electronic configuration

$$1s^2 2s^2 2p^6 3s^2 3p^6 3d^{10} 4s^2 4p^6 4d^{10} 4f^5 5s^2 5p^6 5d^1 6s^2$$

Ionic radius Sm^{2+} 1.11 Å, Sm^{3+} 0.964 Å. First ionization potential 5.6 eV; second 11.1 eV. Other important physical properties of samarium are given under **Rare-Earth Elements and Metals**.

The principal sources of samarium are monazite (4.5% Sm$_2$O$_3$) and bastnasite (0.5% Sm$_2$O$_3$). Current demands for the element are met by the coproduction with europium and gadolinium from these minerals. The residues of uranium mining (Canada) also contain about 4.5% Sm$_2$O$_3$. Unlike the other light rare-earth metals, the salts and oxide of samarium do not reduce to metal using barium, calcium, or lithium, nor can electrolytic processes be used. The most effective reducing agent is lanthanum, which is mixed with Sm$_2$O$_3$ and heated under vacuum in a tantalum crucible. The samarium metal volatilizes and is condensed as powder or sponge on coiled tantalum or copper condenser plates. Subsequently, the samarium must be remelted under an argon or inert atmosphere before it is cast into graphite or tantalum molds.

Samarium has been alloyed with gadolinium and aluminum to produce nuclear reactor hardware that will absorb neutrons for short periods. The use of samarium in intermetallics, cermets, and other chemical forms for use in nuclear applications holds promise. Small quantities of Sm$_2$O$_3$ are used in optical-glass filters and to encase lanthanum borate glass rods, which then are drawn into fine fibers for fiberoptics applications. The element has been used as a coding agent for inks used in data handling systems. Small amounts also have been used for activating phosphate-type phosphors. The addition of samarium oxide produces a strong narrow emission in the near-infrared spectral region. The most significant use of samarium is in the permanent-magnet alloys SmCo$_5$ and Sm$_2$Co$_{17}$. The strength of these magnets Second to that of Nd$_2$Fe$_{14}$B by only a small margin. But, because of costs the utilization of the SmCo$_5$–Sm$_2$Co$_{17}$ alloys is much less than that of the Nd–based permanent magnets. The Sm–base permanent magnets have a much higher magnetic ordering temperature and are used in high temperature greater than 250°F (or 120°C) applications. See also **Cobalt**.

See references listed at ends of entries on Chemical Elements; and Rare-Earth Elements and Metals.

Note: This entry was revised and updated by K. A. Gschneidner, Jr, Director, and B. Evans, Assistant Chemist, Rare-earth Information Center, Institute for Physical Research and Technology, Iowa State University, Ames, Iowa.

SANDSTONE. Sand grains cemented by such substances as silica, carbonate of lime or iron oxide, so as to form a solid rock is called sandstone. It occurs usually in beds of varying thickness, depending upon the conditions under which the original sediments were laid down. Because it is normally well-jointed and easy to work, sandstone has been much used for building purposes. Unfortunately, however, as most sandstones are quite porous, the weathering action of the atmospheric agencies may have a very deleterious effect upon them.

SANGER, FREDERICK (1918–). An English biochemist who won the Nobel prize for chemistry in 1958. His research was on protein structure. He identified the amino acid sequence of the protein insulin. His PhD. was awarded from Cambridge University.

SAPONIFICATION. A special case of hydrolysis in which an ester is converted into an alcohol and a salt of the appropriate acid by reaction with an alkali. Though the operation has numerous applications throughout the chemical industry, it is noteworthy because some 80% of standard soap is prepared by this method. The esters may be of mono- or polybasic acids and mono- or polyhydric alcohols, the physical conditions under

which the reaction occurs being suitably varied to secure an adequate rate. The alkali most commonly used is sodium hydroxide, because of cost and water solubility, but other appropriate alkaline materials are suitable.

Since the preparation of soap is typical of a fairly complex reaction, chemically, and since it is common, it serves as a useful example of the saponification operation. The complication, of course, occurs because the usual esters used for soap are the glycerol esters of fatty acids, saturated and unsaturated. Thus the saponification of stearin (glycerol tristearate) is commonly shown as follows:

$$C_{17}H_{35}COO—CH_2$$
$$C_{17}H_{35}COO—CH + 3NaOH \rightarrow$$
$$C_{17}H_{35}COO—CH_2$$

$$3C_{17}H_{35}COONa + HO—CH_2$$
$$HO—CH$$
$$HO—CH_2$$

Actually, the saponification appears to progress stepwise, the first hydrolytic reaction taking place as follows:

$$C_{17}H_{35}COO—CH_2$$
$$C_{17}H_{35}COO—CH + NaOH \rightarrow$$
$$C_{17}H_{35}COO—CH_2$$

$$C_{17}H_{35}COO—CH_2$$
$$C_{17}H_{35}COO—CH + C_{17}H_{35}COONa$$
$$HO—CH_2$$

The diglyceride formed is subsequently split to the monoglyceride, which finally is converted to glycerol, if sufficient alkali is present. Thus, the reaction is a bimolecular one rather than quadrimolecular, as is commonly indicated. In actual practice, the fats used are complex glycerides of a number of saturated and unsaturated acids, rather than the stearin shown here.

Technologically, the saponification operation varies in degree of difficulty depending on the ester. The reaction rate differs for different esters, for one thing, bu another determining factor is the contact area possible between the alkali and the ester. In the case mentioned above, the fat at the start is insoluble and immiscible in water, so that reaction in a nonagitated vessel would be very slow, occurring only at the limited interface.

Though saponification is the dominant reaction when the techniques described above are used, in many instances side reactions may occur which may profoundly modify the products. Oxidation, of course, is one of the more obvious things to guard against, since many times the esters being treated are unsaturated. Both isomerization and polymerization, however, may occur under the alkaline conditions obtaining, the unconjugated polyethenoid acids becoming conjugated during treatment. This especially appears to be true in the case of highly unsaturated compounds. However, by properly controlling reactants and conditions, saponification remains a very flexible and useful industrial and laboratory operation possible of wide application and at low cost.

See also **Soaps**.

STANLEY B. ELLIOTT
Bedford, Ohio

SAPONIFICATION NUMBER. See **Vegetable Oils (Edible)**.

SATURATED COMPOUND. See **Organic Chemistry**.

SATURATION. 1. The state in which all available valence bonds of an atom (especially carbon) are attached to other atoms. The straight-chain paraffins are typical saturated compounds.

2. The state of a solution when it holds the maximum equilibrium quantity of dissolved matter at a given temperature.

SAYBOLT UNIVERSAL VISCOSITY. The efflux time in seconds (SUS) of 60 mL of sample flowing through a calibrated Universal orifice in a Saybolt viscometer under specified conditions.

See also **Viscosity**; and **Viscometer (or Viscosimeter)**.

SCALE. 1. A calcareous deposit in water tubes of steam boilers resulting from deposition of mineral compounds present in the water, e.g., calcium carbonate.

2. A type of paraffin or petroleum was from which all but a few percent of oil has been removed by hydraulic pressing and subsequent processing.

3. A graduated standard of measurement in which the units (degrees) are defined in relation to some property of what is measured, e.g., temperature scale, Brix scale, and Baume scale.

4. The markings indication such units as on a thermometer or graduate.

5. A weighing device that may be of the beam type in which weights (poises) and lever systems are used, or the direct-reading spring type in which the gravitational pull of the object being weighed is counterbalanced by a known constant spring force.

SCANDIUM. [CAS: 7440-20-2]. Chemical element, symbol Sc, at. no. 21, at. wt. 44.956, periodic table group 3, is a member of the rare earth group of elements along with yttrium and the lanthanides, mp 1540°C, bp 2850°C, density 2.989 g/cm^3 (alpha form), 3.73 g/cm^3 (beta form). The alpha form is close-packed hexagonal and beta is the face-centered cubic allotrope. Scandium is a relatively soft metal with a silvery luster. The metal is stable in air. Scandium combines readily with acids and at elevated temperatures with oxygen, halogens, and chalcogenides. ^{45}Sc occurs in nature and is not radioactive. Nine radioactive isotopes have been identified ^{40}Sc through ^{44}Sc and ^{46}Sc through ^{51}Sc, all with relatively short half-lives, ranging from a fraction of a second up to 84 days. Scandium occurs widely throughout nature, but in reasonably concentrated forms only in a few uncommon minerals. Abundance in the Earth's crust is estimated at approximately $5-6 \times 10^{-4}\%$, ranking it ahead of such elements as antimony, bismuth, silver, and gold. It is estimated that a cubic mile of seawater contains about 375 pounds of the element. Scandium was predicted by Mendeleev in 1869, at which time he called it *ekaboron* and foretold accurately a number of its properties. A small amount of scandium oxide was extracted from euxenite and gadolinite by Nilson in 1879, a material that Nilson called *scandia*. In the same year, Cleve isolated a greater quantity of the oxide, from which several compounds were prepared and favorably compared with Mendeleev's predictions for ekaboron. Ionic radius Sc^{3+} 0.745. First ionization potential, 6.54 eV; second, 12.80 eV; third, 24.76 eV. Other physical properties are given under **Chemical Elements**.

Scandium occurs in some ores with the Rare Earth Series elements. It is easily separated from the Lanthanides, as well as yttrium, by taking advantage of the greater solubility of its thiocyanate in ether. The three recognized scandium minerals are thortveilite, a silicate; and sterrettite and kolbeckite, both phosphates. Wiikite and bazzite, complex niobates and silicates, are known to contain more than 1% scandium. The recovery of scandium from uranium and tungsten from mining tailings, however is the major source of the element. Scandium has not been found without the other Rare earth elements. The element usually is separated from ore extracts and concentrates by precipitation as the oxalate. Scandium metal with a purity of 99.9% has been produced.

In water solutions, the scandium ion has a triple positive charge. Studies show, however, that the simple Sc^{3+} ion seldom exists. Rather, the form is highly polymerized and hydrolyzed—with hydroxy-bonded structures. In forming compounds, scandium parallels aluminum, yttrium, gallium, indium, and tellurium. Several carbides of scandium have been reported, the most stable carbide being ScC, including scandium clusters in fullerene cages.

Like the hydroxides of the Rare earth, scandium hydroxide, $Sc(OH)_3$, is precipitated by addition of alkalies to solutions of scandium salts; however, the latter is precipitated at pH 4.9, while the former require pH 6.3 or more, a property which is utilized in one method of separation. Upon heating the hydroxide (or certain oxyacid salts), scandium oxide, Sc_2O_3 is produced. Scandium hydroxide is less acidic than aluminum hydroxide, requiring boiling KOH solution to form the complex potassium compound, $K_2[Sc(OH)_5 \cdot H_2O] \cdot 3H_2O$.

All four trihalides of scandium are known. The trifluoride is very slightly soluble in H_2O, and is precipitated from scandium nitrate, $Sc(NO_3)_3$,

solutions by hydrofluoric acid. It dissolves in alkali fluorides to yield the complex ion $[ScF_6]^{3-}$. The chloride is formed in solution by treating the hydroxide or oxide with HCl, yielding hydrated crystals on concentration, which give hydroxychlorides on heating. The bromide is also prepared from the oxide or hydroxide and hydrobromic acid, or in anhydrous form from the oxide, carbon, and bromine, on heating. The iodide is also prepared by the latter method.

The thiocyanate is prepared in solution by adding ammonium thiocyanate, NH_4SCN to HCl solutions of the chloride. Both basic and double carbonates are known. The former is precipitated from Sc^{3+} solutions by adding carbonate solutions, and is probably $Sc(OH)CO_3 \cdot H_2O$. The latter are obtained by the use of an excess of the soluble carbonate. Normal, basic, and double sulfates are known. The first exists in several degrees of hydration; the second is obtained as $Sc(OH)SO_4 \cdot 2H_2O$, by treating the normal sulfate tetrahydrate with the hydroxide. The alkali double sulfates and alums are obtained by treating the sulfate solution with an excess of the alkali (or ammonium) sulfate solution.

The nitrate is readily obtained by action of dilute HNO_3 on the hydroxide. In aqueous solution, the anhydrous nitrate yields a monobasic nitrate on heating.

To date, the applications for scandium and its compounds have been limited, mainly because of its high cost. The main uses of scandium are in sporting goods equipment and in lighting, including automotive headlights. In former application scandium is added to improve the strength of aluminum, while in the latter ScI_3 is used in metal halide arc lamps to increase the intensity and to improve the color rendition to match that of natural sunlight. Minor applications include: scandium substitution for yttrium in yttrium garnets for electronic uses, analytical standards, semiconductor applications, welding wire, and metallurgical research.

Note: This entry was revised and updated by K. A. Gschneidner, Jr., Director, and B. Evans, Assistant Chemist, Rare-earth Information Center, Institute for Physical Research and Technology, Iowa State University, Ames, IA.

Additional Reading

Carter, G.F. and D.E. Paul: *Materials Science and Engineering,* ASM International, Materials Park, Ohio, 1991.

Gschneidner, K.A., Jr., B.J. Beaudry, and J. Capellen: *Rare Earth Metals,'* pp/720–732 in "*Metals Handbook 10th Edition, Vol. 2, Properties and Selection: Nonferrous Alloys and Special Purpose Materials,* ASM International, Metals Park, OH, 1990.

Gschenidner, K.A., Jr.: *Physical Properties of the Rare Earth Elements,* pp. 4–112 to 4–121 in *CRC Handbook of Chemistry and Physics,* 77th Edition, CRC Press, LLC., Boca Raton, FL, 1997.

Hammond, C.R.: *The Elements,* pp. 4–1 to 4–34 in *CRC Handbook of Chemistry and Physics,* 77th Edition, CRC Press, LLC., Boca Raton, FL, 1997.

Horovitz, C.T., K.A. Gschneidner, Jr., G.A. Melson, D.H. Youngblood, and H.H. Schock: *Scandium. Its Occurrence, Chemistry, Physics, Metallurgy, Biology and Technology,* Academic Press, London, UK, 1975.

Lewis, R.J. and N.I. Sax: *Dangerous Properties of Industrial Materials,* 10th Edition, John Wiley & Sons, Inc., New York, NY, 1992.

SCANNING TUNNELING MICROSCOPE. This microscopic technique (STM), invented by Gerd K. Binnig and Heinrich Rohrer (IBM Research Laboratory, Zurich, Switzerland) in 1981, generates topographic images of surfaces with atomic resolution. With the STM, scientists have obtained previously unseen images of gold, silicon, nickel, oxygen, and carbon atoms. STM views can reveal flaws and contaminants in atomic surface structure. Detailed views of three-dimensional atomic "landscapes" are improving and will continue to improve the knowledge of surface physics and chemistry and thus be of exceptional value in technical fields as varied as metallurgy, magnetism, semiconductor technology, and biology. Before development of the STM technique, scientists had been puzzled about the exact surface structure of silicon, the basic material from which computer chips are made. Many models of the silicon surface have been constructed. The STM has enabled researchers to sort out prior assumptions and hypotheses. In connection with gold, the STM has revealed a surface structure created by the spontaneous formation of ribbon-like facets, features that previously had not been distinguished with such clarity. The renowned physicist, Wolfgang Pauli, many years ago observed, "The surface was invented by the devil," with reference to the fact that the surface of a solid is its boundary or interface with the environment outside and beyond the solid. Whereas atoms within a solid interact with other atoms within the solid, those atoms on the surface can interact only with those atoms directly underneath and those atoms in the environment beyond the surface. Thus, surface atoms behave with different rules.

Instrumental means for investigating the atomic pattern of surfaces have included electron microscopy and later (1950s), the field-ion microscope (invented by Edwin W. Müller). These techniques continue, but are generously abetted by the STM. In 1984, the inventors of the STM received the Hewlett-Packard Europhysics Prize for outstanding achievements in solid state physics and the King Faisal International Prize in Science. In 1986, Binnig and Rohrer shared in the Nobel Prize in Physics.

Expanded Applications of STM

Since its introduction for practical uses in the early 1980s, the scanning tunneling microscope has found scores of applications that were not contemplated at the outset of STM technology. These applications include the manufacture of optical grating masters, the manufacture of recording thin-film magnetic recording heads, the manufacture of compact disk stampers, and the repair of costly masks used for integrated circuit manufacture—just to mention a few examples of practical usage in the electronics field. STM can be operated under water and other fluids, permitting the examination of biological materials in a more natural setting. The versatility of the STM is demonstrated by the appearance of hundreds of technical papers in the literature over a 1-year period. The STM can achieve lateral resolution of 1 to 2 angstroms and, in the vertical dimension, better than 0.05 angstrom. To overcome the relatively few limitations of STM, the technology has spawned several other microscopic techniques, including the atomic force microscope, the friction force microscope, the magnetic force microscope, the electrostatic force microscope, the attractive mode force microscope, the scanning thermal microscope, the optical absorption microscope, the scanning ion-conductance microscope, the scanning near-field optical microscope, the scanning acoustic microscope, and the molecular dipstick microscope—all resulting from the character of innovative thinking that scientists have come to apply to the application of new forms of energy to microscopy, once wholly dependent upon materials interactions with visible light. STM has played an invaluable role, not only in the development of new materials, but in understanding how surface atoms differ from the permanently embedded atoms of a material.

STM Fundamentals

As put forth in their excellent 1985 paper (reference listed), Binnig and Rohrer credit *electron tunneling* as the phenomenon that underlies the operation of the STM. As indicated by Fig. 1, an electron cloud occupies the space between the surface of the sample and the needle tip used. The cloud is a consequence of the *indeterminacy*[1] of the electron's location (a result of its wavelike properties). Because the electron is "smeared out" so to speak, there is a probability that it can lie beyond the surface boundary of a conductor. The density of the electron cloud decreases exponentially with distance. A voltage-induced flow of electrons through the cloud thus is extremely sensitive to the distance between the surface and the tip. As the STM tip is swept across the surface, a feedback mechanism senses the flow (the *tunneling current*) and holds constant the height of the tip above the surface atoms. In this way, the tip follows the contours of the surface. The motion of the tip is read and processed by computer and displayed on a screen or plotter. By sweeping the tip through a pattern of parallel lines, a high-resolution, three-dimensional image of the surface is obtained.

Views of the STM are shown in Figs. 2 and 3. Binnig and Rohrer describe the STM as having two stages, suspended from springs, that operate within a cylindrical steel frame. The microscope mechanism is contained by the innermost stage. Vibration is a severe problem, requiring special preventive measures. To obtain images with high resolution (Fig. 4) of surface structures, obviously the microscope must be protected from even very small vibrations as may be caused by sound and footsteps and other disturbances within the laboratory. To subdue vibration, copper plates are attached to the bottom of the stainless-steel frame and magnets are attached to the bottom of the inner and outer stages. Any disturbance causes the copper plates to move up and down in the field generated by the magnets. The movement induces eddy currents in the plates. Interaction of the eddy currents with the magnetic field retards the motion of the plates and thus the motion of the microscope stages. Where investigations

---

[1] The roots of atomic structure extend back over 50 years, to the development of the concept of quantum mechanics and experiments in 1927 by Davisson and Germer, the researchers who confirmed the wave nature of the electron. The first experimental verification of tunneling was made about three decades ago by Ivar Giaever.

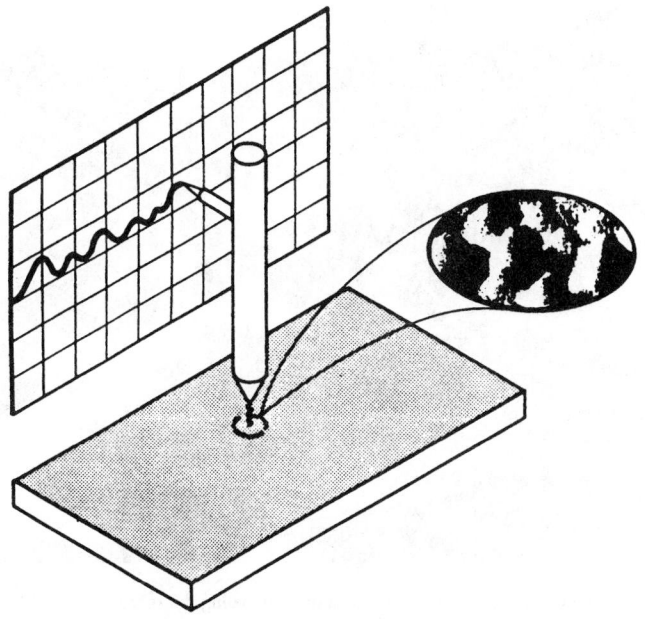

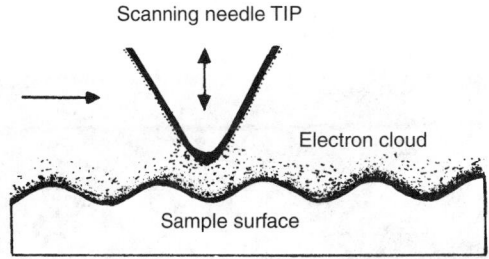

Fig. 1. As the probe tip of the STM is scanned across the microscopic "hills and valleys" of a surface, the vertical position of the probe is precisely adjusted to maintain a constant tip-to-surface distance. This is accomplished by keeping the tunneling current constant. Consequently, the probe follows the surface contour as it moves, yielding a 2-dimensional, enlarged representation of the surface contour for each such scan. A full 3-dimensional image is obtained by assembling an entire sequence of scans. (*IBM Corporation*)

Fig. 2. Inventors Rohrer (left) and Binnig (right) shown adjusting the sample in the chamber of an early (1981) version of the scanning tunneling microscope. Later (1986), these scientists shared in the Nobel Prize in Physics for developing the STM instrument and technique. (*IBM Corporation*)

require vacuum conditions, a steel cover is placed over the outer frame of the microscope.

The microscopic device incorporates a sample and a scanning needle. Piezoelectric materials (expand or contract with applied voltage) make it

Fig. 3.   Miniaturized version of the scanning tunneling microscope. The STM has become a very important tool for materials research and, in particular, to gaining new knowledge of computer chips. STMs are being used or are under construction by some fifty research groups worldwide in a broad range of physical, chemical, biological, and materials studies. (*IBM Corporation*)

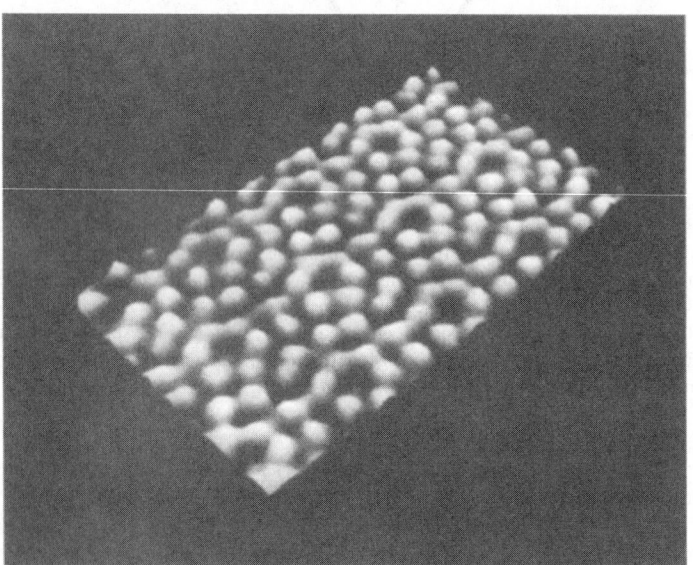

Fig. 4.   Magnified millions of times are surface atoms of silicon. The image is computer-generated from data produced by a scanning tunneling microscope. (*IBM Corporation*)

possible for the device to resolve features that are about a hundredth the size of an atom. A piezoelectric drive positions the sample on a horizontal metal plate and a piezoelectric tripod then sweeps the scanning needle over the surface of the sample, simultaneously achieving high stability and precision. See also Figs. 5 through 8.

Not only does the STM portray atomic topography, but it also reveals atomic composition. The inventors observe that the tunneling current depends both on the tunnel distance and the electronic structure of the surface and on the fact that each atomic element has an electronic structure unique to itself. The ability of the STM to resolve both topography and electronic structure assures wide use of the technique in numerous fields. Unlike other imaging techniques, the STM does not alter or partially destroy the sample. The STM already has demonstrated its utility in biology even though lateral resolutions of only 10 angstroms can be achieved. In this instance, the relatively poor resolving power of the microscope is more than compensated for by its ability to provide a direct and nondestructive method of viewing biological samples. Researchers

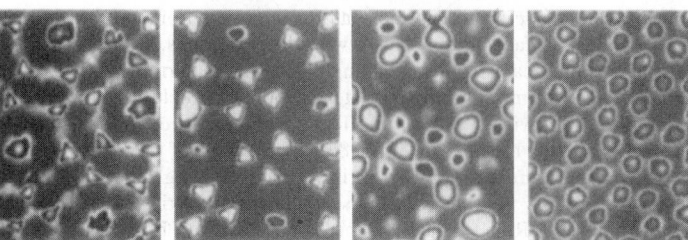

Fig. 5.   Looking successively deeper into the atomic structure of silicon. Sequence shows surface magnified some ten million times, to a depth of about nine angstroms (36 billionths of an inch). Shown from left to right are the geometric positions of the atoms and three different classes of electronic bonds. (**a**) shows the position of the top atoms; (**b**) the dangling bonds that reach up from those atoms; (**c**) the dangling bonds that reach up from other atoms in the second layer in the surface; and (**d**) bonds (called "back bonds") that reach out sideways from the atoms in the second layer in the surface. (*IBM Corporation*)

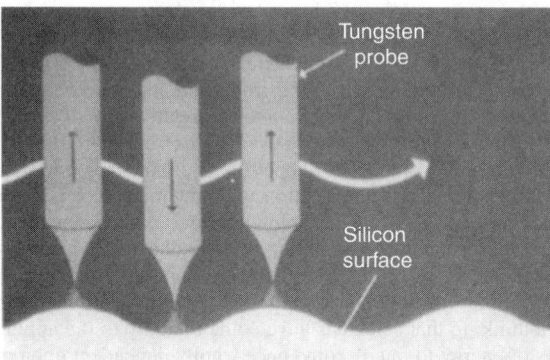

Fig. 6.   As probe is scanned across silicon surface, its height above the surface is adjusted to keep the tunneling current constant. The monitoring of those height changes provides the desired topographic information. (*IBM Corporation*)

Fig. 7.   Tungsten probe tip for scanning tunneling microscope as seen by a scanning electron microscope. Such probes, produced by field-ion microscopy, can be constructed with tips only one atom in width. (*IBM Corporation*)

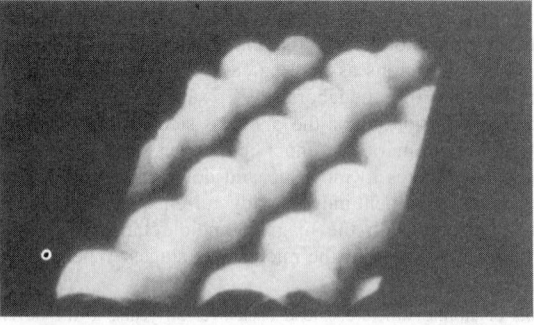

Fig. 8.   Scanning tunneling microscope image of gallium arsenide (GaAs). Atoms to the left of each row are gallium; others are arsenic. (*IBM Corporation*)

at the IBM Zurich Research Laboratory and the Swiss Federal Institute of Technology have scanned the surface of DNA and observed a series of zigzags which correspond to its helical structure. Researchers at the Autonomous University of Madrid have made detailed examinations of viruses, notably phi 29 which measures $40 \times 300 \times 200$ angstroms. It is currently visualized that the STM also will be very useful for testing electronic circuits, particularly as further reductions in size are achieved.

Additional Reading

Abelson, P.H.: "Phenomena at Interfaces," *Science*, 1357 (March 18, 1988).

Bai, C.: *Scanning Tunneling Microscopy and Its Application,* 2nd Edition, Springer-Verlag Inc., New York, NY, 1999.

Bard, A.J. et al.: "Chemical Imaging of Surfaces with the Scanning Electrochemical Microscope," *Science*, 68 (October 4, 1991).

Baro, A.M., Binnig, G., Rohrer, H., Gerber, C., et al.: "Real-Space Observation of 2×1 Structure of Chemisorbed Oxygen on NI(110) by Scanning Tunneling Microscopy," *Physical Review Letters*, **52**(15), 1304–1307 (April 9, 1984).

Beebe, T.P., Jr., et al.: "Direct Observation of Native DNA Structures with the Scanning Tunneling Microscope," *Science*, 370 (January 20, 1989).

Bindell, J.B.: "Elements of Scanning Electron Microscopy," *Advanced Materials & Processes*, 20 (March 1993).

Binnig, G., Rohrer, H., Gerber, C., and E. Weibel: "Facets as the Origin of Reconstructed Au(110) Surfaces," *Surface Science*, **131**, L379–L384 (1983).

Binnig, G. and H. Rohrer: "The Scanning Tunneling Microscope," *Sci. Amer.*, 50–56 (August 1985).

Binnig, G. and H. Rohrer: *The Scanning Tunneling Microscope*, in The Laurates' Anthology, 72, Scientific American, Inc., New York, NY, 1990.

Birdi, K.S.: *Scanning Tunneling Microscopy (STM) and Atomic Force Microscopy (AFM): Applications in Surface and Colloid Chemistry*, CRC Press, LLC., Boca Raton, FL, 2001.

Bonnell, D.A.: *Scanning Tunneling Microscopy and Spectroscopy: Theory, Techniques, and Applications*, John Wiley & Sons, Inc., New York, NY, 1993.

Clemmer, C.R. and T.P. Beebe, Jr.: "Graphite: A Mimic for DNA and Other Biomolecules in Scanning Tunneling Microscope Studies," *Science*, 640 (February 8, 1991).

Chen, J.: *Introduction to Scanning Tunneling Microscopy*, Oxford University Press, Inc., New York, NY, 1993.

Epstein, A.W.: "A Tunneling Microscope Can Cleave Molecules," *Sci. Amer.*, 26 (April 1988).

Flam, F.: "Scopes with a Light," *Science*, 30 (April 5, 1991).

Giaever, I.: "Energy Gap in Superconductors Measured by Electron Tunneling," *Physical Review Letters*, **5**(4), 147–148 (August 15, 1960).

Hansma, P.K. et al.: "Scanning Tunneling Microscopy and Atomic Force Microscopy: Application to Biology and Technology," *Science*, 209 (1988).

Hansma, P.K. et al.: "The Scanning Ion-Conductance Microscope," *Science*, 641 (February 3, 1989).

Kinoshita, J.: "Scanning Tunneling Microscope Spawns Diverse Applications," *Sci. Amer.*, 33 (July 1988).

Pomerantz, M. et al.: "Rectification of STM Current to Graphite Covered with Phthalocyanine Molecules," *Science*, 1115 (February 28, 1992).

Pool, R.: "The Children of the STM," *Science*, 634 (February 9, 1990).

Pool, R.: "A New Role for the STM," *Science*, 130 (December 7, 1990).

Schardt, B.C., Yau, Shueh-Lin, and F. Rinaldi: "Atomic Resolution Imaging of Adsorbates on Metal Surfaces in Air: Iodine Adsorption on Pt(III)," *Science*, 1050 (February 24, 1989).

Smith, D.P.E. et al.: "Smectic Liquid Crystal Monolayers on Graphite Observed by Scanning Tunneling Microscopy," *Science*, 43 (July 7, 1989).

Takayanagi, K. et al.: "Structure Analysis of the Silicon(111) 7×7 Reconstructed Surface by Transmission Electron Diffraction," *Surface Science*, 164, 367 (1985).

Tromp, R.M. and E.J. van Loenen: "Ion-Beam Crystallography on Silicon Surfaces III. Si(111)," *Surface Science*, 155, 441 (1985).

Whitman, L.J. et al.: "Manipulation of Adsorbed Atoms and Creation of New Structures on Room-Temperature Surfaces with the Scanning Tunneling Microscope," *Science*, 1206 (March 8, 1991).

Wiesendanger, R. and H.J. Guntherodt: *Scanning Tunneling Microscopy III: Theory of STM and Related Scanning Probe Methods,* 2nd Edition, Springer-Verlag Inc., New York, NY, 1998.

SCAPOLITE. The mineral scapolite is a silicate of calcium and aluminum that contains also some potassium, sodium, and chlorine. The name identifies all intermediate members of a series with the following end members: Marialite $3Na(AlSi^3)O^8 \cdot NaCl$, Meinoite $3Ca(Al^2Si^2)O^8 \cdot CaCO_3$. Its tetragonal crystals are coarse and thick, often very large. It occurs also in massive forms. It has a distinct prismatic cleavage; subconchoidal fracture; is brittle; hardness, 5.5–6; specific gravity, marialite, 2.5–2.62, meionite, 2.72–2.78; luster, vitreous to rather dull, color, white to gray, red, green, blue, or yellow; translucent to opaque, rarely transparent.

Scapolite is found in the metamorphic rocks, particularly those rich in calcium; also in contact metamorphic deposits in limestones. It has been found in basic igneous rocks, probably as a secondary mineral. Notable localities are Lake Baikal, Siberia; Arendal, Norway; and Madagascar. In the United States, it is found in Massachusetts, New York, and New Jersey. Greenville, in the Province of Quebec, Canada is an important locality. Superb transparent yellow gem crystals have recently been found in Brazil and Tanzania. Wernerite (scapolite) was named in honor of A.O. Werner, a famous German mineralogist (1749–1817).

SCAVENGING. 1. The use of an unspecific precipitate to remove from solution by adsorption or coprecipitation a large fraction of one or more undesirable radionuclides. Voluminous gelatinous precipitates are usually used as scavengers, e.g., $Fe(OH)_3$. 2. The removal of impurities from molten metal by addition of substances to form slags, or other compounds that can readily be removed. 3. The removal of unwanted gases from systems, e.g., of products of combustion from an internal combustion engine, or residual gases from an evacuated tube.

SCHEELE, C. W. (1742–1786). One of the outstanding early chemical thinkers and experimenters, Scheele was a Swedish scientist who discovered a number of previously unknown substances, among which are tartaric acid, chlorine, manganese salts, arsine, and copper arsenite (Scheele's green). He also noted the oxidation states of various metals, observed the nature of oxygen two years before Priestley's discovery, and discovered the chemical action of light on silver compounds, thus laying the foundation of photochemistry and photography. Scheelite, or natural calcium tungstate, is named after him.

See also **Scheelite.**

SCHEELITE. The mineral scheelite is calcium tungstate, $CaWO_4$, with molybdenum substituting for tungsten up to 25% in the molybdian scheelite variety. It is a tetragonal with an octahedral habit although also at times tabular, and may occur massive. It displays an octahedral cleavage; is brittle; hardness, 4.5–5; specific gravity, 6.1; luster, vitreous; color, white to yellowish, reddish, greenish and brownish; white streak; transparent to translucent. Scheelite is found in pegmatite and ore veins associated with granites, also as a contact metamorphic mineral. It is known from the Czech Republic and Slovakia, Saxony, Italy, Alsace, Finland; Cumberland and Cornwall in England; and Mexico. Crystals of exceptional length (6–10 inches; 15–25 centimeters) are found at various localities in Korea and Japan; and in the United States, in Connecticut, Colorado, South Dakota, Arizona, Nevada, and California. The Swedish chemist, Karl Wilhelm Scheele, discovered tungsten in this mineral, which later was named for him.

The mineral fluoresces vivid bluish white to white; or yellowish white with increasing molybdenum content under exposure to short-wave ultraviolet light.

SCHIST. The schists form a great group of metamorphic rocks chiefly notable for the preponderance of the lamellar minerals such as the micas, chlorite, talc, hornblende, graphite, etc. Quartz often occurs in drawn out grains to such an extent that a quartz schist is produced. Most schists have in all probability been derived from clays and muds which have passed through a series of metamorphic processes involving the production of shales, slates and phyllites as intermediate steps. Certain schists have been derived from fine-grained igneous rocks such as lavas and tuffs. Most schists are mica schists, but graphite and chlorite schists are common. Schists are named for the prominent or perhaps unusual mineral constituent, as garnet schist, tourmaline schist, glaucophane schist, etc. The word schist is derived from the Greek meaning to split, with reference to the easy separation of these rocks in a direction parallel to that in which the platy minerals lie.

SCHMIDLIN KETENE SYNTHESIS. Formation of ketene by thermal decomposition of acetone over electrically heated wire at 500–750 degrees by a reaction involving radical formation with generation of methane and carbon monoxide.

SCHMIDT REACTION. Acid catalyzed addition of hydrazonic acid to carboxylic acids, aldehydes, and ketones to give amines, nitriles, and amides, respectively.

SCHORIGIN (or Shorygin) REACTION. Organometallic reactions of the Grignard type, employing sodium in place of magnesium; the reaction of alkyl sodium compounds with carbon dioxide to give monobasic acids is sometimes known as the Wanklyn reaction.

SCHOTTEN-BAUMANN REACTION. Acylation of alcohols with acyl halides in aqueous alkaline solution.

SCHWEITZER'S REAGENT. A solution of copper hydroxide in strong ammonia used in analytical chemistry as a test for wool. It dissolves cotton, silk, and linen.

SCLEROMETER. An apparatus for determining the hardness of a material by measuring the pressure on a standard point that is required to scratch the material. A scleroscope is a similar apparatus, which measures hardness by determining the rebound of a standard ball dropped on the subject material from a fixed height.

SCOLECITE. This mineral is a zeolite, a hydrous calcium-aluminum silicate, $CaAl_2Si_3O_{10} \cdot 3H_2O$. It occurs in slender monoclinic prisms and in fibrous and nodular masses. Hardness is 5; specific gravity, 2.27; luster vitreous to silky; transparent to translucent. When heated, some specimens of scolecite curl up like worms; hence its name, derived from the Greek meaning a worm. This mineral occurs with other zeolites, at Baden, Switzerland; Iceland; Greenland; the Deccan region of India; and in the United States at Golden, Colorado, and Paterson, New Jersey. Single crystals up to 12 inches (30 centimeters) in length have recently been found in a single large cavity in the basaltic trap rocks near Nasik, India.

SCORODITE. This hydrated arsenate of ferric iron and aluminum $(Fe_3+, Mg_3+)AsO_4 \cdot 2H_2O$, crystallizing in the orthorhombic system, is the iron-rich isomorphous end member of a complete series extending to the aluminum-rich mineral Mansfieldite. Crystals usually occur as drusy crusts. Also occurs as massive, compact, and earthy material. Hardness of 3.5–4, with specific gravity of 3.278. Vitreous to subadamantine luster, of pale green to liver-brown color.

Scorodite occurs as a secondary alteration mineral in the oxidized zone of metallic arsenic-containing veins. The mineral also may be a product of deposition from certain hot springs. World localities of note include Siberia; Laurium, Greece; Carinthia; Cornwall, England; and Nevada, Utah in the United States. Currently being deposited by hot springs at Yellowstone National Park in Wyoming.

SEABOARD PROCESS. Method of removing hydrogen sulfide from a gas by adsorption in sodium carbonate solution. Sodium bicarbonate and sodium hydrosulfide are formed. Blowing air through the solution, hydrogen sulfide is released and carried off, and the sodium carbonate is regenerated.

SEABORG, GLENN T. (1912–1999). An American atomic scientist, and one of the first scientists to produce a transuranium element. He received is A. B. degree at the University of California, Los Angeles in 1934, and his doctorate at the University of California, Berkeley in 1937. As a graduate student in the 1930s doing wet chemistry research for his advisor Gilbert Newton Lewis, Seaborg devoured the text *Applied Radiochemistry* by Otto Hahn, of the Kaiser Wilhelm Institute for Chemistry in Berlin. For several years, Seaborg conducted important research in artificial radioactivity using the Lawrence cyclotron at Cal Berkeley. He was excited to learn from others that nuclear fission was possible, but also chagrined, as his own research might have led him to the same discovery. In 1939 he became an instructor in chemistry at UC Berkeley.

In the same year in which he produced plutonium, 1941, he also discovered that the isotope U^{235} undergoes fission under appropriate conditions. He therefore was responsible for two different approaches to the development of nuclear weapons. At this time he was transferred to the Manhattan Project and was part of Enrico Fermi's team, which achieved the first nuclear chain reaction in 1942.

In 1951 along with Edwin McMillan he received the Nobel prize in chemistry for the creation of the first transuranium elements. He created plutonium, americium, curium, berkelium, and californium at Berkeley.

The element seaborgium was named for him in honor of his accomplishments. It was so named while he was still alive, which proved extremely controversial. For the remainder of his life, Seaborg was the only person in the world who could write his address in chemical elements: Seaborgium, Lawrencium, Berkelium, Californium, Americium (Glenn Seaborg, Lawrence Berkeley National Laboratory, Berkeley, California, United States of America).

See also **Chemical Elements**.

SEABORGIUM. See **Chemical Elements**.

SEALANTS. Any organic substance that is soft enough to pour or extrude and is capable of subsequent hardening to form a permanent bond with the substrate. Most sealants are synthetic polymers (silicones, urethanes, acrylics, polychloroprene) that are semisolid before application and later become elastomeric.

Performance Characterists

Movement Capabilities. The movement capability of a sealant is the amount of displacement the sealant can endure in extension or compression without failing. In general, sealants with lower modulus of elasticity can handle higher movements.

Sealants can be broadly divided into classes according to the amount of movement they can successfully handle. High performance sealants such as silicones, urethanes, and polysulfides can typically handle movements of 25% or higher. Medium performance sealants such as some acrylics can handle movements of 10–25%. Low performance sealants such as butyls, putties, and caulks accommodate movements under 10%.

Modulus. The measure of the stress of a sealant at a specific strain is referred to as the modulus of elasticity, sometimes called the *secant modulus*. This important sealant property describes the force exerted by a sealant as it is stressed. Because a primary function of sealants is to adhere to the substrates it is in contact with, the forces generated by a joint opening or closing are transmitted by the sealant to the substrate–sealant bond line. For this reason, it is important to know the modulus of the sealant and also the strength of the substrate.

Durability. A primary factor in sealant durability is its ability to resist decay from environmental elements. For most typical applications this includes extremes of high and low temperature, water, oxidation, and sunlight.

Sealant Types and Formulations

Silicones. Commercially available silicone sealants are typically one of three curing types: moisture-reactive (curing) sealants, moisture-releasing (latex) sealants, and addition-curing sealants. Of these three types, moisture-curing silicones make up the vast majority of silicone sealants sold.

The formulation of moisture-curing silicones includes a silicone polymer, filler, a moisture-reactive cross-linker, and sometimes a catalyst. A newer class of silicone sealants are known as the silicone latex sealants. These sealants are silicone-in-water emulsions that cure by evaporation of the emulsifying water. Addition-curing silicones in general are two-part systems that cure by the platinum-catalyzed reaction of a silicone hydride with typically a vinyl group attached to silicon. Because no by-products are generated by the cure, there are few volatiles and no shrink in thick sections.

Urethanes. The basis for urethane chemistry is the reaction of an isocyanate group with a component containing an active hydrogen. The first step in formulating a urethane sealant is to prepare what is commonly called the prepolymer, typically by reaction of a hydroxy-terminated polyether with a stoichiometric amount of a diisocyanate.

Although polyethers are the main building blocks of the prepolymer, other materials such as polyesters, polythioethers, and polybutadienes are also used. Most urethanes use blends of polymers to achieve desired properties. Urethane sealants have good inherent adhesion to most substrates, but silane adhesion promoters are often used to improve this adhesion.

See also **Urethane**; and **Urethane Polymers**.

Polysulfides. Polysulfide sealants were the first high performance synthetic elastomeric sealants produced in the United States. The basic polymers are mercaptan-terminated (HS–R–SH), with molecular weights ranging from 1000 to ca 8000. Curing occurs through the mercaptan groups by oxidation and results in an S–S linkage. In addition to the terminal SH groups, pendent SH groups are located along the polymer chain and contribute to cross-linking. Polysulfide polymers provide inherent resistance to fluel and quite good resistance to alkali. In contrast to the silicone polymers, they have low gas permeability. Different curing agents have different effects on the properties of the polysulfide sealant. Therefore, the choice of curing agent depends on the sealant application. Adhesion promoting silanes are often added to improve adhesion to various substrates. As is the case with urethane sealants, silanes with a dual-reactive nature are typically used.

Acrylics. There are two principal classes of acrylic sealants: latex acrylics and solvent-release acrylics. High molecular weight latex acrylic polymers are prepared by emulsion polymerization of alkyl esters of acrylic acid. The emulsion polymers are compounded into sealants by adding fillers, plasticizers, freeze–thaw stabilizers, thickeners, and adhesion promoters. As is true of the silicone latex sealants, the acrylic latex sealants are easy to apply and clean with water.

Another class of acrylic sealants are the solvent-releasing acrylics. Acrylic monomers are polymerized in a solvent. The natural adhesion of most of the solvent-releasing acrylics produces some of the best unprimed adhesion in the sealant industry. However, slow, continual cure generally produces large compression sets and limits their use to low movement applications. Also, the relatively high amounts of solvent and traces of acrylic monomer in these formulations limits their use to outdoor applications, usually in construction.

Butyls. Butyl-based materials are sold in the form of preformed tapes, thermoplastic hot melts, and one-part solvent-releasing sealants. Butyl polymers are made by the copolymerization of 97–98 mol % isobutylene with 2–3% isoprene. Another butyl-based polymer, polyisobutylene, is produced by the polymerization of isobutylene. Formulations of butyl-based sealants also include plasticizer, filler, and tackifier resins. Polybutenes are common plasticizers for butyl sealants. Solvents, such as mineral spirits, are used for the one-part solvent-releasing formulations. As the solvent leaves the typical one-part butyl, the sealant hardens and loses its elastomeric ability. This limits the use of solvents to low movement applications where durability is not of high concern.

Manufacture and Processing

Almost all sealants contain a mixture of a powdered filler incorporated into a viscous liquid, which results in a viscous liquid sealant having a paste-like consistency. Processing conditions can have a dramatic effect on sealant rheology, cure time, and physical properties. Typical processing variables are mixer speed (rpm), time, temperature, and vacuum. Order of ingredient addition is also important.

Health and Safety Factors

Most sealants use mineral-based fillers, which may contain small amounts of crystalline silica. Crystalline silica is a known cause of silicosis, a debilitating disease of the lung. Some but not all sealants use flammable ingredients, but for those that do, proper inserting and grounding are needed to prevent potential explosions. In general, sealants should be used in areas that have good ventilation. A careful review of the material safety data sheet is fundamental before use of any sealant.

Uses

Each class of sealants has certain attributes inherent to the polymer on which it is based. These attributes often define the sealant's applications and limitations. Future developments are likely to feature the production of more silicone sealants that do not pick up dirt, more latex acrylic sealants that have high performance properties. Urethanes that have improved uv stability, and high performance polysulfides that are made in the United States.

RICHARD A. PALMER
JEROME M. KLOSOWSKI
Dow Corning Corporation

Additional Reading

Amstock, J.S.: *Handbook of Adhesives and Sealants in Construction,* The McGraw-Hill Companies, Inc., New York, NY, 2000.

Petrie, E.M.: *Handbook of Adhesives & Sealants,* The McGraw-Hill Companies, Inc., New York, NY, 1999.

SEASON CRACKING. Spontaneous cracking of brass and other metals on standing. Intergranular cracks result from the action of residual internal stresses from cold-working operations aided by surface corrosion. Cold-worked high-zinc brasses sometimes fail during storage in ordinary atmosphere. Ammonia salts and other specific reagents greatly accelerate cracking in brasses subject to this defect. Many other metals, including stainless steels, are subject to cracking under certain conditions of stress and corrosion.

SEAWATER (Desalination). See **Desalination**.

SEAWATER (Elements in). See **Chemical Elements**.

SEDIMENTATION. The settling out by gravity of solid particles suspended in a liquid, the rate of settling being defined by Stoke's law. See also **Stoke's Law**. This method is used industrially in water purification. It is also an analytical procedure for separation of solids of different particle size, as well as for molecular weight determination. The sedimentation of large molecules in a strong centrifugal field permits determination of both average molecular weights and the distribution of molecular weights in certain systems. When a solution containing polymer or other large molecules is centrifuged at forces up to 250,000 times gravity, the molecules begin to settle, leaving pure solvent above a boundary that progressively moves toward the bottom of the cell. An optical system is provided for viewing this boundary, and a study as a function of the time of centrifuging yields the rate of sedimentation for the single component, or for each of many components of a polydisperse system. These sedimentation rates may then be related to the corresponding molecular weights of the species present after the diffusion coefficients for each species are determined by independent experiments. The result of this work is the distribution of molecular weights in the sample, which is attainable by few other methods.

See also **Precipitate**.

SEGER CONE. A series of substances having different fusion temperatures might serve roughly to measure the temperature of high-temperature regions such as furnaces, since, with a series of substances having progressively increasing fusing temperatures, the temperature naturally lies between the fusion temperature of the last substance fused, and that of the next not yet fused. A series of artificially prepared mixtures, mostly of the oxides such as clays, lime, feldspar, have been designed to form a series of "Seger cones." There are 60 mixtures covering a temperature range from 590 to 2,000°C.

SELECTION RULES (Energy Levels). It was found early in the study of atomic spectra that radiative transitions between certain pairs of energy levels seldom or never occur. A set of rules which are expressed in terms of the differences of the quantum numbers of the two states involved allow a prediction of allowed transitions and forbidden transitions. The conditions for allowed transitions are:

$$\Delta L(\text{orbital angular momentum}) = \pm 1$$

$$\Delta J(\text{total angular momentum}) = 0 \text{ or } \pm 1$$

$$\Delta M(\text{magnetic orientation}) = 0 \text{ or } \pm 1$$

The selection rules are not rigorously obeyed. In atoms that do not exhibit Russell-Saunders coupling, the quantum numbers L and S are not defined. Even in atoms that do have this type of coupling, forbidden transitions are merely of lower probability than allowed ones, and they may occur from a state from which no transitions are allowed by the rules, if conditions are such that collisions of the second kind do not remove the atom from the initial state before it radiates (e.g., at extremely low pressures).

Similar selection rules hold for molecular spectra. In fact, let ψ_i and ψ_j be wave functions for two levels in any quantum mechanical system. Then if P is the appropriate operator, a transition between levels i and j is permitted if the matrix element

$$\int \psi_i^* P \psi_j d\tau$$

does not vanish. Here ψ_i^* is the complex conjugate of ψ_i, $d\tau$ is a volume element including all of the variables involved in the two wave functions, and the operator may refer to electric or magnetic dipole radiation, quadrupole radiation, polarizability, etc. If the integral vanishes, the transition is forbidden. Frequently, symmetry properties and group theory may be used to determine whether the matrix element does or does not vanish. This is very helpful since evaluation of the integral itself may be difficult or impossible.

SELECTION RULES (Nuclear). A set of statements that serves to classify transitions of a given type (emission or absorption of radiation, beta decay, and so forth) in terms of the spin and parity (I and π) quantum numbers of the initial and final states of the systems involved in the transitions, in such a way that transitions of a given order of inherent probability (after making allowance for the influence of varying energy, charge and size of system, and so forth) are grouped together. The group having highest probability of taking place per unit time is said to consist of allowed transitions; all others are called forbidden transitions. Table 1 lists the selection rules for radiative transitions: each entry gives the character (E = electric, M = magnetic) and the multipole order (1 for dipole, 2 for quadrupole, 3 for octopole, ...) of the predominant radiation mechanism for the indicated spin change ΔI and parity change $\Delta \pi$; the entry "none" means that radiative transitions are strictly forbidden.

TABLE 1. SELECTIVE RULES FOR RADIATIVE TRANSITIONS

			ΔI				
	0	0					
$\Delta \pi$	$I = 0$	$I \neq 0$	1	2	3	4	5
No	None	$M1$	$M1$	$E2$	$M3$	$E4$	$M5$
Yes	None	$E1$	$E1$	$M2$	$E3$	$M4$	$E5$

Fermi selection rules and Gamow-Teller (GT) selection rules are alternative sets of rules for allowed beta transitions; both are currently believed to be valid, so that a transition allowed according to either set is actually allowed.

Table 2 lists the selection rules for beta decay: the entry A means that for the indicated spin and parity change the transition is allowed; I, means that it is first forbidden; II, second forbidden ...

TABLE 2. SELECTION RULES FOR BETA DECAY

				ΔI			
$\Delta \pi$	0	1	2	3	4	5	6
No	A	A	II	II	IV	IV	VI
Yes	I	I	I	III	III	V	V

SELENIUM. [CAS: 7782-49-2]. Chemical element, symbol Se, at no. 34, at. wt. 78.96, periodic table group 16, mp 217°C, bp 685°C, density 4.82 g/cm^3 (solid), 4.86 (single crystal). Selenium has a large number of allotropes, some of which have not been fully investigated. On heating selenium above its melting point and cooling it, a red vitreous mass is formed, probably a mixture of allotropes. A red, amorphous allotrope is precipitated by SO$_2$ from selenous acid solutions. On heating at above 150°C, the red vitreous form changes to a gray hexagonal form, the stable form at ordinary temperatures, with metallic properties, one of which is photo-conductivity. By evaporation of a CS$_2$ solution of the red vitreous form below 72°C, a red α-monoclinic form is obtained; evaporation above 72°C gives β-monoclinic selenium. Black hexagonal selenium, believed to have a ring structure, is produced by heating amorphous selenium to near its melting point. Unlike sulfur, liquid selenium apparently has only one form.

There are six natural occurring isotopes: ^{74}Se, ^{76}Se through ^{78}Se, ^{80}Se, and ^{82}Se, and seven known radioactive isotopes ^{72}Se, ^{73}Se, ^{75}Se, ^{79}Se, ^{81}Se, ^{83}Se, and ^{84}Se. With exception of ^{79}Se which has a half-life of something less than 6×10^4 years, the half-lives of the other isotopes are comparatively short, measured in minutes, hours, or days. In terms of abundance, selenium ranks 34th among the elements occurring in the earth's crust. It is estimated that a cubic mile of seawater contains about 14 tons (3 metric tons per cubic kilometer) of selenium. First ionization potential 9.75 eV; second 21.3 eV; third, 33.9 eV; fourth, 42.72 eV;

fifth, 72.8 eV. Oxidation potentials H$_2$Se(aq) $\longrightarrow$ Se + 2H$^+$ + 2e$^-$, 0.36 V; Se + 3H$_2$O $\longrightarrow$ H$_2$SeO$_3$ + 4H$^+$ + 4e$^-$, $-.740$ V; H$_2$SeO$_3$ + H$_2$O $\longrightarrow$ SeO$_4^{2-}$ + 4H$^+$ + 2e$^-$, -1.15 V; Se^{2-} $\longrightarrow$ Se + 2e$^-$, 0.78 V; Se + 6OH$^-$ $\longrightarrow$ SeO$_3^{2-}$ + 3H$_2$O + 4e$^-$, 0.36 V; SeO$_3^{2-}$ + 2OH$^-$ $\longrightarrow$ SeO$_4^{2-}$ + H$_2$O + 2e$^-$, -0.03 V. Other important physical properties of selenium are given under **Chemical Elements**.

Selenium was first identified by Berzelius in 1817. The element is found associated with volcanic activity, as for example in cavities of Vesuvian lavas and in the volcanic tuff of Wyoming (about 150 parts per million).

Selenium occurs as selenide in many sulfide ores, especially those of copper, silver, lead, and iron, and is obtained as a by-product from the anode mud of copper refineries. The mud is (1) fused with sodium nitrate and silica, or (2) oxidized with HNO$_3$, and the H$_2$O extract is then treated with HCl and SO$_2$, whereupon free selenium is separated.

Uses

Selenium is widely used in photoelectric cells. The element alters its electrical resistance upon exposure to light. The response is proportional to the square root of incident energy. Selenium cells are most sensitive in the red portion of the spectrum. Although an external emf must be applied, the resistance is low and amplification is easy. In the selenium photovoltaic cell configuration, a thin film of vitreous or metallic selenium is coated onto a metal surface. Then, a transparent film of another metal, often platinum, is placed over the selenium. A cell of this type generates its own emf, with a decrease in internal resistance with increasing irradiation. The response essentially is proportional to incident energy. The cells are not importantly sensitive to small temperature changes.

Advantage of the unipolar conduction characteristic of selenium is taken in arc rectifiers. In a typical unit, a nickel or nickel-plated steel or an aluminum disk with a thin layer of selenium applied to one side is used. Selenium also is added to copper alloys and to stainless steel to increase machinability. Advantages claimed for selenium copper are high machinability, combined with hot-working properties and high electrical conductivity. As a decolorizer in glass, selenium counteracts green shades arising from ferrous ingredients. Sodium selenite is used in the production of red enamels and in the manufacture of clear red glass. Addition of from 1 to 3% selenium to vulcanized rubber increases abrasion resistance. The element also is used in photographic and printing reproduction chemicals.

Selenium is also used as an additive to lead-antimony battery grid metal and as a vulcanizing agent to improve temperature and abrasion resistance of rubber.

Chemistry and Compounds

Due to its $4s^24p^4$ electron configuration, selenium, like sulfur, forms many divalent compounds with two covalent bonds and two lone pairs, and d hybridization is quite common, to form compounds with Se oxidation states of 4+ and 6+.

While selenium dioxide, [CAS: 7446-08-4], SeO$_2$, can be produced by direct reaction of the element with oxygen activated by passage through HNO$_3$, the compound is easily made by heating selenious acid, H$_2$SeO$_3$. Selenium dioxide sublimes at 315–317°C, and is readily reduced by SO$_2$ to elemental selenium. Selenium trioxide, SeO$_3$, is not prepared from the dioxide by oxidation, although selenium does react with oxygen to form SeO$_3$ and SeO$_2$ in an electric discharge. Preferred method of preparing SeO$_3$ is by refluxing potassium selenate with sulfur trioxide. The reverse reaction, hydration of SeO$_3$ to selenic acid, [CAS: 7783-08-6], H$_2$SeO$_4$, occurs easily. Selenous acid, H$_2$SeO$_3$, produced by hydration of SeO$_2$, is a stronger oxidizing agent than sulfurous acid as judged by its quantitative oxidation of iodide ion in acid solution, but is a weaker acid (ionization constants 2.4×10^{-3} and 4.8×10^{-9} at 25°C). It forms salts, the selenites, many of which, especially those of the heavy metals, are reduced to selenides by hydrazine. Many of the selenites, e.g., those of nickel, mercury, and ferric ion, are very slightly soluble in H$_2$O. Selenous acid is readily oxidized by halogens in the presence of silver ion or 30% H$_2$O$_2$ to selenic acid, H$_2$SeO$_4$. Selenic acid is as strong an acid as H$_2$SO$_4$, and it is more readily reduced, reacting with hydrobromic acid and hydriodic acid to form selenous acid or (at high concentration) elemental selenium. Like sulfate ion, SO$_4^{2-}$, SeO$_4^{2-}$ is tetrahedral in crystals.

Hydrogen selenide, [CAS: 7783-07-5], H$_2$Se, is a stronger acid than H$_2$S (ionization constants of H$_2$Se, 1.88×10^{-4} and about 10^{-10}) and is less readily obtained from selenides than H$_2$S from sulfides (the selenides of aluminum, iron and magnesium, Al$_2$Se$_3$, FeSe, and MgSe, require heating with H_2O or dilute acids). In general, the metal selenides are prepared by direct combination of the elements. Those of transition groups 3–8, 1 and 2 and main groups 3 and 4 exhibit many instances of well-defined compounds, berthollide compounds, and substitutional solid solutions. Thus, four intermediate phases are found in the palladium-selenium system, Pd_4Se, $Pd_{2.8}Se$, $Pd_{1.1}Se$, and $PdSe_2$.

Selenium hexafluoride, SeF_6, the only clearly defined hexahalide, is formed by reaction of fluorine with molten selenium. It is more reactive than the corresponding sulfur compound, SF_6, undergoing slow hydrolysis. Selenium forms tetrahalides with fluorine, chlorine, and bromine, and dihalides with chlorine and bromine. However, other halides can be found in complexes, e.g., treatment of the pyridine complex of SeF_4 in ether solution with HBr yields $(py)_2SeBr_6$. Selenium tetrafluoride also forms complexes with metal fluorides, giving $MSeF_5$ complexes with the alkali metals.

Selenium forms several oxyhalides, e.g., $SeOF_2$, $SeOCl_2$, and $SeOBr_2$, the first two being liquids and the last a crystalline solid, mp 41.6°C. Selenium also forms tetraselenium tetranitride, Se_4N_4.

Selenocyanates, $M^I SeCN$, corresponding to the thiocyanates, are prepared by addition of selenium to soluble cyanides. They are similar to the thiocyanates except that HSeCN immediately decomposes in acid to selenium and hydrogen cyanide. The heavy metal selenocyanates are less soluble than the corresponding thiocyanates.

Selenium forms "thio"-type compounds, such as $SeSO_3$ by reaction of selenium and sulfur trioxide, $SeSO_3^{2-}$ (selenosulfates) by reaction of selenium and sulfites, SeS^{2-} (selenosulfides) by reaction of selenium with sulfides, as well as diselenides, Se_2^{2-}, and polyselenides, Se_x^{2-}.

Carbon diselenide is an evil-smelling liquid, and COSe and CSSe are also known.

Biological Role of Selenium

Some very interesting examples of the effect of soils on the nutritional quality of plants are associated with selenium. The element has not been found to be required by plants, but it is required in very small amounts by warm-blooded animals and probably by humans. However, selenium in larger quantities can be very toxic to animals and humans.

In large areas of the world, the soils contain very little selenium in forms that can be taken up by plants. Crops produced in these areas are, therefore, very low in selenium. A selenium deficiency in livestock is a serious problem. A deficiency causes a form of muscular dystrophy in younger animals and poor reproductive qualities in the adult animals. For prevention, sodium selenate or sodium selenite, sometimes augmented with vitamin E, is added in proper proportions to feedstuffs. Some areas, including the Plains and Rocky Mountain states in the United States have soils that are rich in available selenium. In regions like these, selenium toxicity is a problem. The situation is particularly serious in Arizona, California, Montana, Nevada, New Mexico, and South Dakota.

An interesting feature of selenium is that it occurs naturally in several compounds and these vary greatly in their toxicity and in their value in preventing selenium-deficiency diseases. In its elemental form, selenium is essentially insoluble and biologically inactive. Inorganic selenates or selenites and some of the selenoamino acids in plants are very active biologically, whereas some of their metabolites that are excreted by animals are not biologically active. In well-drained alkaline soils, selenium tends to be oxidized to selenates and these are readily taken up by plants, even to levels that may be toxic to the animals that eat them. In acid and neutral soils, selenium tends to form selenites and these are insoluble and unavailable to plants. Selenium deficiency in livestock is most often found in areas with acid soils and especially soils formed from rocks low in selenium.

In 1934, the mysterious livestock maladies on certain farms and ranches of the Plains and Rocky Mountain states were discovered to be due to plants with so much selenium that they were poisoning grazing animals. Affected animals had sore feet and lost some of their hair; many died. Over the next 20 years, researchers found that the high levels of selenium occurred only in soils derived from certain geological formations of high selenium content. They also found that a group of plants, called *selenium accumulators*, had an extraordinary ability to extract selenium from the soil. These accumulators were mainly shrubs or weeds native to semiarid and desert range lands. They usually contained about 50 parts per million (ppm) or more of selenium, whereas range grasses and field crops growing nearby contained less than 5 ppm selenium. These findings helped ranchers to avoid the most dangerous areas when grazing livestock.

In 1957, selenium was found to be essential in preventing liver degeneration of laboratory rats. Since then, research workers have found that certain selenium compounds, either added to the diet or injected into the animal, would prevent some serious disease of lambs, calves, and chicks. That selenium is an essential nutrient element for birds and animals has been established.

In most diets used in livestock production, from 0.04 to 0.10 ppm of selenium protects the animal from deficiency diseases. If the diet is very high in vitamin E, the required level of selenium may be lower.

In terms of human dietary requirements, much of the wheat for breadmaking in the United States is produced in selenium-adequate sections of the country. Bread is generally a good source of dietary selenium.

Selenomethionine decomposes lipid peroxides and inhibits *in vivo* lipid peroxidation in tissues of vitamin-E-deficient chicks. Selenocystine catalyzes the decomposition of organic hydroperoxides. Selenoproteins show a high degree of inhibition of lipid peroxidation in livers of sheep, chickens, and rats. Thus, some forms of selenium exhibit *in vivo* antioxidant behavior.

Health and Safety Factors

Commercial elemental selenium along with the stable metallic selenides are considered relatively nontoxic. However, other selenium compounds which include the reactive selenides; the gaseous, volatile, and soluble compounds; and particularly hydrogen selenide, the halides, oxides, and the organics are highly toxic and must be handled with care. Selenium can enter the body through inhalation, ingestion, or absorption through the skin where it accumulates primarily in the liver and kidney. Symptoms of selenium poisoning include bronchial irritation, gastrointestinal distress, nasopharyngeal irritation, and garlic on the breath.

Selenium plays a dual role in a living organism, depending on the compound and the amount adsorbed. Controlled small doses of some compounds are used in medicine and as diet supplements, for example, ca 0.1 ppm of diet dry matter for livestock. Larger amounts can be toxic.

Additional Reading

Carter, G.F. and D.E. Paul: *Materials Science and Engineering,* ASM International, Materials Park, Ohio, 1991.
Frankenberger, W.T. Jr. and R.A. Enberg: *Environmental Chemistry of Selenium,* Marcel Dekker, Inc., New York, NY, 1998.
Greenwood, N.N. and A. Earnshaw: *Chemistry of the Elements,* 2nd Edition, Butterworth-Heinemann, Inc., Wodurn, MA, 1997.
Hatfield, D.L.: *Selenium: Its Molecular Biology and Role in Human Health,* Kluwer Academic Publishers, Norwell, MA, 2001.
Lewis, R.J. and N.I. Sax: *Sax's Dangerous Properties of Industrial Materials,* 10th Edition, John Wiley & Sons, Inc., New York, NY, 1999.
Lide, D.R.: *CRC Handbook of Chemistry and Physics,* 84th Edition, CRC Press, LLC., Boca Raton, FL, 2003.
Liotta, D. and R. Monahan III: "Selenium in Organic Synthesis," *Science,* **221** 356–361 (1986).
Marshall, E.: "High Selenium Levels Confirmed in Six States," *Science,* **231,** 111 (1986).
Meyers, R.A.: *Handbook of Chemicals Production Processes,* McGraw-Hill Companies, Inc., New York, NY, 1986.
Reamer, D.C. and W.H. Zoller: "Selenium Biomethylation Products from Soil and Sewage Sludge," *Science,* **208,** 500–502 (1980).
Staff: "Plants Can Eat Toxic Soil," *National Food Review,* **42** (October–December, 1989).
Staff: *ASM Handbook—Properties and Selection: Nonferrous Alloys and Pure Metals,* ASM International, Materials Park, Ohio, 1990.

SEMENOV, NIKOLAI N. (1896–1986). A Russian chemist and physicist who won the Nobel Prize in 1956. He authored books on the chain reaction and problems of chemical kinetics and reactivity, as well as many articles. His work concerning thermal combustion and explosion is utilized in rockets and jet engines. He received his doctorate at Leningrad State University.

SEMICARBAZONES. The products of the reaction between an aldehyde or a ketone with semicarbazide are termed *semicarbazones.*

$$CH_3 \cdot CHO + NH_2 \cdot CO \cdot NHNH_2 \longrightarrow$$
(acetaldehyde)   (semicarbazide)

$$CH_3 \cdot CH{:}N \cdot NH \cdot CONH_2 + H_2O$$
(acetaldehyde semicarbazone)

$$(CH_3)_2CO + NH_2 \cdot CO \cdot NH \cdot NH_2 \longrightarrow$$
(acetone)

$$(CH_3)_2C{:}N \cdot NH \cdot CO \cdot NH_2 + H_2O$$
(acetone semicarbazone)

SEMICONDUCTORS

SEMICONDUCTORS. Materials and devices known as *semiconductors* have been the backbone of the electronics industry for many years. Semiconductors did not enter the industry in a major way, however, until several years after the vacuum tube (valve) had been well established. In terms of perspective, it is interesting to note that at least one semiconductor device predated the vacuum tube in the early days of radio communication. This was the then familiar galena crystal and accompanying whisker used in early crystal set radio receivers.

With the continuing attention to developing reliable and efficient vacuum tubes, which occurred over a long time span, interest in semiconductors stagnated—with the exception of the emerging radar technology of the World War II era. Massachusetts Institute of Technology's Radiation Laboratory became active in the investigation of crystal rectifiers and engaged in exploratory studies and in the development of very pure semiconducting materials, notably silicon and germanium. Several other research institutions during this same period became interested in the theoretical aspects of solid state and semiconductors, including the energy levels of solids and the charged carrier transport in silicon and a few other materials. Out of this early phase of semiconductor technology came, in 1947, the invention of the transistor (contraction for transfer resistor). Inventors Shockley, Bardeen, and Brattain (Bell Laboratories) received the Nobel Prize in physics in 1956 for their accomplishment.

The transistor had a tremendous impact, constituting the birth of modern electronics. The transistor led to the development of almost innumerable semiconductor device configurations and ultimately to the phasing out of the vacuum tube, except for rather special and limited applications.

Innovations to fabricate semiconductors with improved performance and smaller size (microelectronics) for many years has been a continuing process without apparent end. As of the very late 1980s, the enhancement of semiconductors continues apace. Circuit integration has gone through several major phases—from IC (integrated circuit) to LCI (large-scale integration) to VLSI (very large-scale integration) to VHSIC (very high-speed integrated circuits) to ULSI (ultralarge-scale integration)—to the point where it is difficult to find appropriate adjectives to describe developments in ICs.

With few exceptions, it is only within recent years that serious concern has been expressed by experts in what the physical limits on size and performance may be. Fortunately, thus far the advancements in fabrication (chip-making processes), such as electron beam and molecular beam lithography, have allayed these concerns in the short term. Although some interest in nonsilicon materials has always been present, there has been a recent reawakening of interest in the use of gallium, indium, arsenic, phosphorus, and antimony—this interest intensifying because of the opportunities such elements offer over silicon. Although silicon which for years has been the basis for volume-produced devices, is not seriously threatened at this juncture, some authorities feel that, in the long term, the emphasis on silicon will be heavily shared with these other materials.

Even with the aforementioned materials, the microelectronics and optoelectronics industries have relied almost exclusively on *inorganic* materials. This is expected to change. Some authorities now forecast that in the early 2000s, there will be a major shift, namely, to *molecular electronics*. Even today, considerable attention for the long term is being paid to the *organic* solid state. The richness of the variety of organic molecular materials available offers enormous potential compared with the relative paucity of structures achievable with inorganic compounds, even when due allowance is made for the exciting developments in inorganic quantum well semiconductors. A hint of what may be achieved along these lines is given by the progress thus far made in liquid crystals, piezoelectric and photoconducting polymers. This implies, of course, that organics no longer will be confined to their traditional applications in electronics, such as for insulation, adhesion, or encapsulation.

Molecular electronics currently is defined as the use of organic molecular materials to perform an active function in the processing of information and its transmission and storage. An alternative definition has been suggested, namely, the achievement of *switching* on a molecular scale. As observed by G.G. Roberts (University of Oxford), "It is interesting to note that only a modest diminution in the size of electronic circuit components is required before the scale of individual molecules is reached; in fact many existing circuit elements could already be accommodated within the area occupied by a leukemia virus."

Some investigators forecast that during the first quarter of the 21st Century, molecular electronics will lead to *supermolecular electronics* where signal transport and control will be effected by nanometer-scale assemblies.

Numerous applications of contemporary semiconductors are described throughout this encyclopedia. See also **Molecular and Supermolecular Electronics**.

Nature of Semiconductors

From the standpoint of their use in electronics, semiconductors are distinguished from other classes of materials by their characteristic electrical conductivity σ. The electrical conductivities of materials vary by many orders of magnitude, and consequently can be classified as: (1) the perfectly conducting superconductors; (2) the highly conducting metals ($\sigma \approx 10^6$ mho/centimeter); (3) the somewhat less conducting semimetals ($\sigma \approx 10^4$ mho/centimeter); (4) the semiconductors covering a wide range of conductivities ($10^3 \gtrsim \sigma \gtrsim 10^{-7}$ mho/centimeter); and (5) the insulators, also covering a wide range ($10^{-10} \gtrsim \sigma \gtrsim 10^{-20}$ mho/centimeter).

These low-conductivity materials are characterized by the great sensitivity of their electrical conductivities to sample purity, crystal perfection, and external parameters, such as temperature, pressure, and frequency of the applied electric field. For example, the addition of less than 0.01% of a particular type of impurity can increase the electrical conductivity of a typical semiconductor like silicon or germanium by six or seven orders of magnitude. In contrast, the addition of impurities to typical metals and semimetals tends to decrease the electrical conductivity, but this decrease is usually small. Furthermore, the conductivity of semiconductors and insulators characteristically decreases by many orders of magnitude as the temperature is lowered from room temperature to 1 K. On the other hand, the conductivity of metals and semimetals characteristically increases in going to low temperatures, and the relative magnitude of this increase is much smaller than are the characteristic changes for semiconductors. The principal conduction mechanism in metals, semimetals, and semiconductors is electronic, whereas both electrons and the heavier charged ions may participate in the conduction processes of insulators.

Classification

It is customary to classify a semiconductor according to the sign of the majority of its charged carriers, so that a semiconductor with an excess of negatively charged carriers is termed *n*-type. A semiconductor with an excess of positively charged carriers is called *p*-type, while a material with no excess of charged carriers is considered to be perfectly compensated. Many of the important semiconductor devices depend upon fabricating a sharp discontinuity between the *n*- and *p*- type materials, the discontinuity being called a *p-n* junction.

Other Characteristics

Even though most semiconductors exhibit a metallic luster when inspected visually, this does not provide a reliable criterion for their classification, since the electrical conductivity of all materials is frequency dependent. Visual inspection tends to be sensitive to the conductivity properties at visible frequencies ($\sim 10^{15}$ Hz). Although materials with a high optical reflectivity tend also to exhibit high D.C. conductivity, these two properties are not necessarily correlated in semiconductors and metals. An example of a metal without metallic luster is ReO_3 (rhenium trioxide), a semitransparent reddish solid. On the other hand, most of the common semiconductors do exhibit metallic luster primarily because electronic excitation across their fundamental energy gaps can be achieved at infrared frequencies. At low frequencies, the principal conduction mechanism is free carrier conduction, which is important in metals and is present to some extent in semiconductors that contain impurities or are found at elevated temperatures. In contrast, interband transitions dominate the conduction process at very high frequencies. Interband transitions contribute to the conductivity by about the same order of magnitude in semiconductors, metals, and insulators.

Since the D.C. conductivity due to free carriers is characteristically low in semiconductors and insulators, the generation of free carriers by exposure to light at infrared, visible, and ultraviolet frequencies can lead to a large increase in the D.C. conductivity. This photoconductive effect, which is not observed in metals or semimetals, can be enormous in low-conductivity semiconductors (an increase in the D.C. conductivity of CdS (cadmium sulfide) by 8 orders of magnitude is observed). The effects of ultraviolet light, for example have been used successfully in reprogramming EPROM (electrically programmable read-only memory) devices, exposure to UV light causing trapped charges to leak off.

Because of the extreme sensitivity of semiconductors to impurities, temperature, pressure, light exposure, and certain other factors, these materials can be exploited in the fabrication of useful devices, such as the crystal diode, the transistor, integrated circuits, photodetectors, and light switches.

Flow of Current in Semiconductors

The flow of electric current depends upon the acceleration of charges by an externally applied electric field. Only those charges that resist collisions or scattering events are effective in the conduction process. Because of collisions, charged particles in a solid are not accelerated indefinitely by the applied field, but rather, after every scattering event, the velocity of a charged particle tends to be randomized. Thus the acceleration process must start anew after each scattering event and charged particles achieve only a finite velocity along the electric field $\mathbf{E}$, the average value of the velocity being denoted by v_D, the drift velocity. The effectiveness of the charge transport by a particular charged particle is expressed by the mobility μ, which is defined as $\mu = v_D/E$. The mobility of a particle with charge e and mass m can be related directly to the mean time between scattering events (also called the relaxation time) by the expressed $\mu = e\tau/m$. The electrical conductivity s depends upon the mobility of the charged carriers as well as on their concentration n, and is simply written as $\sigma = ne\mu$, where e is the charge of the carriers. The advantage of expressing the conductivity in this form is the explicit separation into a factor n that is highly sensitive to external parameters, such as temperature, pressure, optical excitation, irradiation, and into another factor μ, which depends characteristically on scattering mechanisms and on the electronic structure of the semiconductor.

The classical theory for electronic conduction in solids was developed by Drude in 1900. This theory has since been reinterpreted to explain why all contributions to the conductivity are made by electrons which can be excited into unoccupied states (Pauli principle) and why electrons moving through a perfectly periodic lattice are not scattered (wave-particle duality in quantum mechanics). Because of the wavelike character of an electron in quantum mechanics, the electron is subject to diffraction by the periodic array, yielding diffraction maxima in certain crystalline directions and diffraction minima in other directions. Although the periodic lattice does not scatter the electrons, it nevertheless modifies the mobility of the electrons. The cyclotron resonance technique is used in making detailed investigations in this field.

The origin of the energy barrier[1] for carrier generation is directly connected with the energy levels for electrons in a solid. Considering electrons in a solid from a tight-binding point of view, the discrete energy levels of the free atom broaden in the solid to form energy bands. For materials that are well described by the tight-binding approximation, the width of the energy bands is sufficiently small so that an energy gap is formed between the energy bands; in the forbidden energy gap there are no bound states. Of particular importance to the conduction properties of a solid is the fact that *all* of the available states in each band would be filled if each atom were to contribute exactly two electrons, thereby causing every solid with an odd number of electrons per atom to be metallic; while solids with an even number of electrons per atom would be insulating or semiconducting. The occurrence of energy bandgaps is also a consequence of the weak binding approximation, whereby the periodic potential itself is responsible for creating bandgaps through the mixing of states separated by a reciprocal lattice vector. See also **Solid-State Physics**.

For semiconductors, the excitation energy lies in the range 0.1 to about 2 eV. Thermal fluctuations are sufficient to excite a small, but significant, fraction of electrons from the occupied levels (the valence band) into the unoccupied levels (the conduction band). Both the excited electrons and the empty states in the valence band (aptly called *holes*) may move under the influence of an electric field, providing a means for conduction of current. (A hole acts like an electron with a positive charge.) Such electron-hole pairs may be produced not only by thermal energy, but also by incident light, providing photo-effects.

Crystallographic defects, in general, are also electronic defects. In metals, they provide scattering centers for electrons, increasing the

resistance to charge flow. The resistance wire in many electric heaters, in fact, consists of an ordinary metal, such as iron, with additional alloying elements, such as nickel or chromium, providing scattering centers for electrons. In semiconductors and insulators, alloying elements and defects provide an even greater variety of effects, since they can change the electron-hole concentrations drastically in addition to providing scattering centers. The semiconductor industry has been built on the alloying of silicon, and a few other elements, including germanium, with *selected impurities* in carefully controlled concentration and geometry. Some of the other emerging semiconductor materials are described later.

Doping

This is a process for purposely adding impurities to a semiconductor (or production of a deviation from stoichiometric composition in order to achieve a desired characteristic). Doped material thus is no longer intrinsic but is impure and called extrinsic. If a trivalent impurity is introduced into silicon or germanium, *holes* are created and the material is said to be *p*-type. Introduction of a pentavalent element into silicon or germanium, on the other hand, creates *free electrons* and the material is said to be *n*-type. Because of thermal effects, free electrons and holes are always being produced in silicon and germanium (intrinsic generation of electron-hole pairs). Consequently, there will be some electrons in *p*-type material and some holes in the *n*-type material. These carriers are referred to as *minority carriers*. Electrons in *n*-type material and holes in *p*-type material are termed *majority carriers.*

The process of placing impurities in the near-surface region of solids is accomplished by a procedure known as *implanting*. A commonly used implanting procedure is to accelerate impurity ions in an electrostatic field with the energy sufficient to impinge with the desired force on the solid target. Known as *ion implantation*, this carefully controlled and reproducible procedure has been widely used to dope semiconductors to create *p-n* junction formations. A certain amount of damage, however, occurs in the semiconductor material in some cases. The surface may become amorphous, or because the implanted dopants may not reach substitutional spots in the crystal lattice the ions may not become electrically active. Thus, it is necessary to anneal the solid for electrical activation of the implanted ions as well as to remove any damage.

Semiconductor Device Configurations

Representative of semiconductor devices is the diode, a two-terminal device which has the property of permitting current to flow with practically no resistance in one direction and offering nearly infinite resistance to current flow in the opposite direction. Applications of the diode are numerous, as in gating circuits used in digital computers.

A widely used semiconductor diode is the *p-n junction diode*. Imagine a crystal (single) of silicon doped so half the material is *p*-type and the other half is *n*-type. The internal boundary between the two extrinsic regions is a *p-n junction*, and the resulting device is a *diode*. See Fig. 1. Three possible configurations of the *p-n* junction are shown in Fig. 2. The energy diagrams for these three configurations are also shown in Fig. 2. Similar diagrams can be generated for holes. When the diode is unbiased, no net flow of electrons takes place across the junction. Assuming that some electrons on the *n*-side have sufficient energy to overcome the potential hill, electrons on the *p*-side (minority carriers) "slide down" the hill, making the net current flow zero. For the reverse biased example, the potential hill is raised and only the few minority carriers from the *p*-side slide down. This results in a minute reverse saturation current. When the diode is forward biased, the potential hill is lowered. This enables electrons to climb over the hill and current flow occurs. The same considerations apply to holes. In fact, the total diode current is equal to the sum of the electrons and holes flowing across the junction.

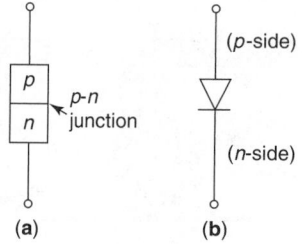

Fig. 1.   (**a**) Configuration of *p-n* junction diode. (**b**) electrical symbol

[1] The sets of discrete but closely adjacent energy levels, equal in number to the number of atoms, that arise from each of the quantum states of the atoms of substance when the atoms condense to a solid from a nondegenerate gaseous condition, make up the energy band (also called the Bloch band). For a semiconductor, the highest energy level is the conduction band, containing only the excess electrons resulting from crystal impurities. The next highest level is the valence band, and it is usually completely filled with electrons. In between these bands is the forbidden band, which is wider for an insulating material than for a semiconductor and which vanishes in a conducting material.

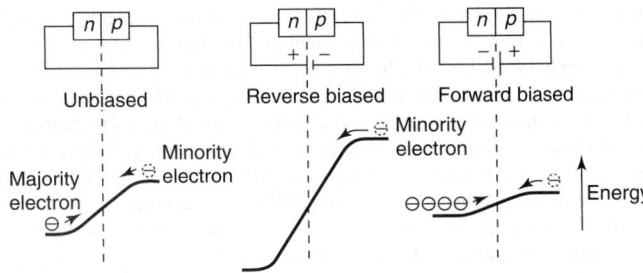

Fig. 2.   Three possible configurations of a p-n junction diode

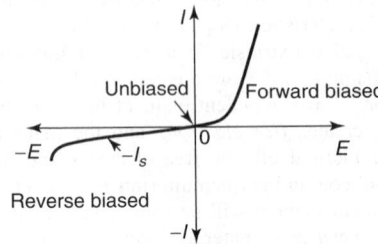

Fig. 3.   Characteristic curve of a semiconductor diode

The characteristic curve of a semiconductor diode is shown in Fig. 3. An equation for this curve, called the *rectifier equation*, is expressed as:

$$I = I_s(e^{-11600E/T} - 1)$$

where I = diode current, amperes

    I_s = reverse saturated current (temperature dependent), amperes

    E = diode biasing voltage (+E for forward bias;—E for reverse bias), volts

    T = absolute temperature (0°C + 273°), degrees Kelvin

At room temperature (300 K) and $E > 0.1$ volt,

$$I \cong I_s^{39E}$$

Where E is more negative than 0.1 volt,

$$I \cong -I_s$$

An example of a simple rectifier employing a p-n junction diode is given in Fig. 4. During the positive half-cycle (0° to 180°) of the A.C. sinusoidal waveform v_s, the diode is forward-biased and conducts. The voltage v_L across load resistance R_L is, therefore, nearly identical to that of v_s for the positive half-cycle. For the negative half-cycle (180° to 360°), the diode is reverse biased and does not conduct. No current flows in R_L, and $v_L = 0$ during the negative half-cycle. Because the diode conducts for only one-half cycle, the circuit of Fig. 4 is called a *half-wave rectifier*. The waveform of v_L is only unidirectional. To obtain steady D.C., like that from a battery, a filter is required. An example of an elementary filter is a large-valued capacitor placed across the load resistor.

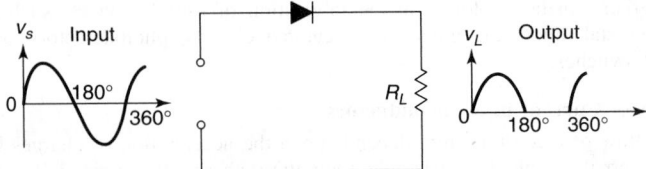

Fig. 4.   Simple rectifier employing a p-n junction diode

The circuit of Fig. 4 can also be used as a detector of amplitude-modulated (AM) radio waves. Fig. 5(a) illustrates the components of an AM wave. If this is applied to the input of Fig. 4, the wave is rectified and the output appears as shown in Fig. 5(b). Placing a small-valued capacitor across R_L filters out the carrier frequency and the desired modulating signal is obtained, as shown in Fig. 5(c).

A bipolar transistor on a single silicon crystal is diagrammed in Fig. 6.

Metal Oxide Semiconductors

The metal oxide semiconductor field effect transistor (MOSFET) is representative of another class of semiconductors. In n-MOS device, two islands of n-type silicon are created in a p-type silicon substrate. A thin layer of nonconducting SiO₂ lies on top of the silicon substrate. Direct connections on a *source* and *drain* are made to the two islands, while a metal *gate* is coupled to the silicon substrate by capacitance. Usually the source and substrate are electrically connected and held at a potential of zero volts. The drain is held at a positive voltage. In this condition, no current flows into the MOS device. When a positive potential is applied to the gate, the electric field attracts a majority of electrons to the thin layer at the surface of the crystal under the gate. Since this region is normally p-type, the surface becomes "inverted" creating a continuous n-type channel between source and drain, thus allowing large currents to flow. This creates a current amplification as in a bipolar transistor. An advantage of MOSFET over bipolar transistors is that they require no isolation islands and thus can be packed more closely on a silicon chip.

Complementary MOS devices (CMOS) have been widely used in recent years. See Fig. 7. The CMOS is made up of a n-MOS and a p-MOS. A main advantage of the CMOS is its low power consumption.

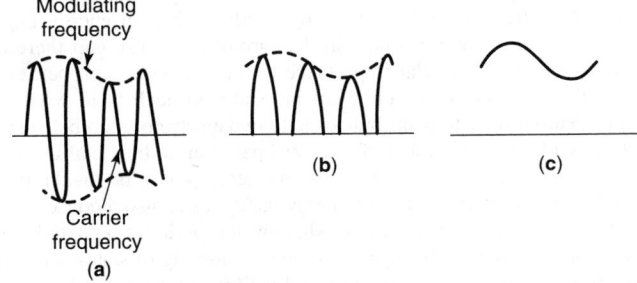

Fig. 5.   Use of p-n junction diode as AM radio detector

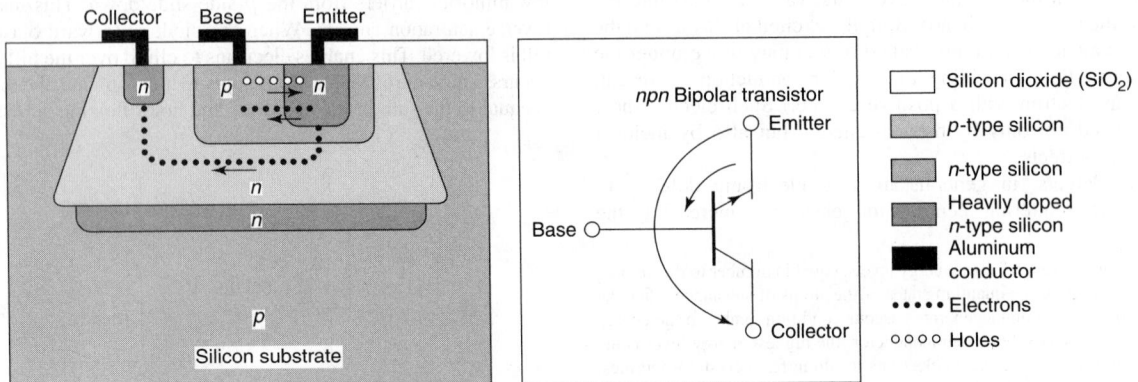

Fig. 6.   A bipolar transistor on a single silicon crystal. (*R.T. Kurnik*, "Chemical vapor deposition in microelectronics," *Chemical Engineering Progress, vol. 81, pp. 30–35, May, 1985*)

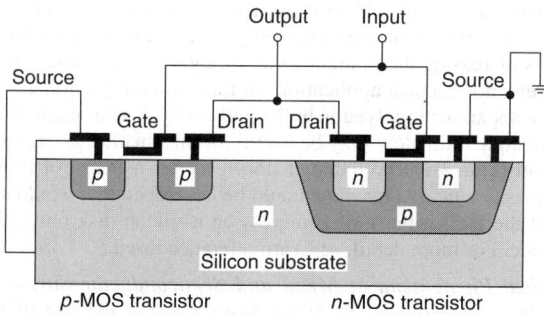

Fig. 7. Complementary MOS device (CMOS) on a single silicon crystal. (*R.T. Kurnik, "Chemical vapor deposition in microelectronics," Chemical Engineering Progress, vol. 81, pp. 30–35, May, 1985*)

Other Materials for Semiconductors

Although silicon (and germanium at one time) is the unquestioned principal semiconductor material to date, silicon does have limitations. For example, it is not easy to integrate electronic and photonic devices in the same microchip. Silicon has a relatively narrow range of temperature tolerance, is susceptible to radiation damage and has "slow" electrons compared with some other materials. Elements in Periodic Table Groups III and IV (now officially called Groups 13 and 15) have fast electrons. For example, the differences between electrons in gallium arsenide and silicon stem basically from the differing chemical characteristics. Electrons in gallium arsenide at low electric fields behave like very light particles which can move easily through the vibrating (and obstructing) crystal lattice of atoms. In contrast, the electrons in silicon behave like heavy particles that move sluggishly under the influence of an applied voltage. The result is significantly faster operating times in microchip operation. See Fig. 8. Fast electrons translate into fast switches. Such switches, when multiplied by thousands or even hundreds of thousands, comprise the basic building blocks of a digital integrated circuit (commonly called a microchip). As pointed out by Allyn, Flahive, and Wemple (1986), there are two classes of speed: (1) Maximum speed achievable, no matter how much "push" is provided by the applied voltage. This is known as *saturated voltage*. Gallium arsenide materials have an advantage in saturated voltage over silicon of 1.5; and with indium—gallium arsenide compounds, the advantage reaches 2.5. (2) The second speed relates to the ease with which electrons can be brought up to full speed (*low-field electron mobility*). Higher mobility in the Groups 13–15 (III–V) semiconductors means that the electrons reach full speed at lower operating voltages. These speed advantages are particularly important in terms of interdevice wiring, which tends to dominate the speed of high-density microchips. Major emphasis on these newer semiconductor materials is directed on the Schottky gate field effect transistor. See Fig. 9.

In the 1950s and 1960s, considerable investigation was made of amorphous *chalcogenide glasses* for possible use in semiconductor devices. The glasses are named for the chalcogens (Group 16, formerly Group VI in the Periodic Table). Early in their consideration, these materials created a considerable controversy among solid-state physicists. Claims were made

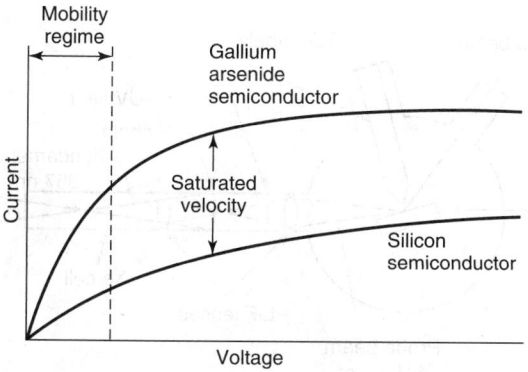

Fig. 8. Current-voltage characteristics of two hypothetical devices of identical physical size. The gallium arsenide curve rises faster and reaches peak velocity faster than the silicon. This means that the group III–V (13–15) electrons produce significantly faster operating times in microchips. (*AT&T Technology*)

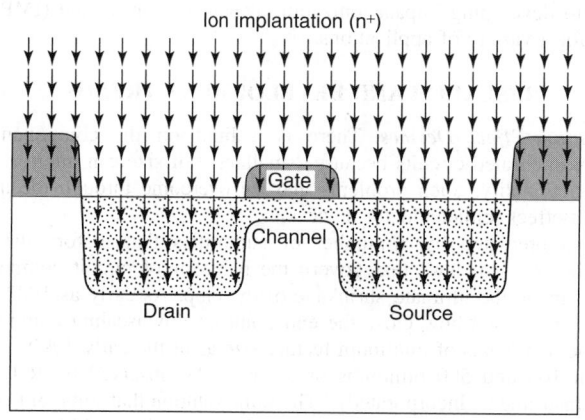

Fig. 9. A Schottky barrier gate used in the metal-semiconductor field-effect transistor (MESFET) in AT&T gallium arsenide microchips. The tiny gate is only one micrometer wide (1/25,400 inch). The gate electrode is deposited before the ion-implantation process so that the gate material will "shade" the channel under it from the "ion rain" that doses the exposed material. (*AT&T Technology*)

and challenged as regards their possible impact on further revolutionizing the semiconductor industry. However, it has been shown that chalcogenide glasses can "switch," but some scientists observe that almost any material will switch under the right conditions. Compositions proposed for memory switches are exemplified by $Te_{81}Ge_{15}Sb_2S_2$, and for non-memory switch materials, $Te_{40}As_{35}Si_{18}Ge_7$. It has also been shown that transitions occur in these glasses when they are exposed to intense light and thus possible photographic uses have been proposed.

Semiconductors used in solar cells are described under **Solar Energy**.

Gallium Arsenide Power Sources

GaAs was first synthesized in 1929 by V.M. Goldschmidt. Its semiconducting properties were not studied until 1952 by H. Welker. The first GaAs p-n junction used for power generation at microwave frequencies was the tunnel diode. Later, GaAs varactor diodes were used in harmonic frequency multipliers and parametric amplifiers at microwave and mm-wave frequencies because of the inherent higher cut-off frequencies possible with gallium arsenide. In 1963, J.B. Gunn discovered the negative resistance property of GaAs, after which GaAs diodes and field effect transistors (FETs) were developed.

The idea for using diodes for generation and amplification of power at microwave frequencies was suggested by A. Uhlir, Jr. Frequency multipliers have been used for power generation since 1958. These devices depend on the nonlinear reactance or resistance characteristics of semiconductor diodes. Generally, there are three types of multiplier diodes—step recovery diodes, variable resistance multiplier diodes, and variable capacitance multiplier diodes.

T.B. Ramachandran (Microwave Device Technology Corporation) notes that there are two inherent major disadvantages for current GaAs FET power devices:

1. The devices are surface oriented. Since the active region is close to the surface, the surface effects tend to affect the device performance. This may be seen in the noise performance of GaAs FETS close to the carrier.
2. To increase the power output, the breakdown voltage must be increased. Active channel doping has to be decreased in order to increase the breakdown voltage. This reduction in doping density decreases the maximum current density, and this tends to reduce the total power output.

J.B. Gunn (International Business Machines) noticed in 1963 the current instabilities in GaAs at high electric fields. Known as the Gunn effect, or the transferred electron effect, Gunn diodes have been used as a low-cost source for microwaves since 1968. These components are comparatively easy to manufacture and hence the cost is low. GaAs Gunn diodes are used from C-band (4 GHz) through W band (100 GHz).

W.T. Read (AT&T Bell Laboratories) first reported microwave oscillations in silicon p-n junctions in 1965. During the interim, much research has gone into developing impact ionization avalanche transit time (IMPATT) diodes for a variety of applications.

RESEARCH AND DEVELOPMENT TRENDS

Quantum-Effect Devices. There is a limit on the components of ordinary integrated circuits because "smallness" of size can interfere with their functionality. Such problems may be overcome through the use of quantum-effect semiconductor devices.

It was predicted by a number of authorities that, before the year 2000, the physical laws that govern the behavior of circuit components would impede the ultimate shrinkage of the chip. As early as 1982, P.K. Chatterjee stressed how close the end point on downscaling components might be. Estimates of minimum feature size as of the early 1990s ranged between 100 and 500 billionths of a meter. As observed by R.T. Bate (Texas Instruments Incorporated), "The same solution that some of the very phenomena that impose size limits on ordinary circuits could be exploited in a new generation of vastly more efficient devices. The functional bases for these devices are quantum-mechanical effects that carry semiconductor technology into a realm of physics where subatomic particles behave like waves and pass through formerly impenetrable barriers. With the so-called quantum semiconductor device, I believe it will be possible to put the circuitry of a supercomputer on a single chip."

Doped silicon, doped and undoped gallium arsenide, and aluminum gallium arsenide have been used as the basis for quantum devices. Of course, size reduction of these proportions pose difficult production tasks. In addition to shrinking size, quantum devices can be expected to be faster and more efficient. A prototype quantum chip, with features one-hundredth of the size of an ordinary chip may appear as shown in Fig. 10. An operational semiconductor device based upon the quantum effect should appear prior to the year 2000. Aggressive research currently is being carried out by AT&T Bell Laboratories, IBM Corporation, the Massachusetts Institute of Technology, Hughes Research Laboratories, Texas Instruments Corporation, the University of Cambridge, Philips Research Laboratory, and the University of Glasgow, among others. As stressed by R.T. Bate, "The commitment of so many research teams to a problematic technology attests to the tremendous potential of these devices and to the faith that they will take the lead in the next semiconductor revolution. The costs and risks involved must be borne in order to revitalize a rapidly maturing electronics industry; the results can only benefit a society that has learned to depend on integrated circuits in many ways."

Atom Switch. By employing the technique of the Scanning Tunneling Microscope, D.M. Eigler and a research team at the IBM Almaden Research Division, San Jose, California, have improvised an "atom switch." Through careful movement of a single xenon atom between the microscope's tip or a nickel surface, the researchers have altered the amount of tunneling current between tip and sample. When the xenon rests on the surface, this is tantamount to the switch's off position. The switch is turned on by applying a 64-millisecond (0.8 V pulse) to the tip. This causes the xenon to jump to the tip, thus increasing the tunneling current by a factor of about seven. As of the present, no practical applications of this switching action are planned because the apparatus involved is bulky and costly. Some scientists believe that the principle ultimately may be useful for information storage systems. Other scientists have observed that, if storing a bit in a cluster of 1000 atoms ever becomes practical, a machine could be developed that would store the contents of the U.S. Library of Congress on a silicon disk only 12 inches (30 cm) wide. For more detail, see Yam reference listed.

Dynamical Phenomena at Metal and Semiconductor Surfaces. This topic has been investigated in recent years through the use of ultrafast measuring techniques involving lasers and nonlinear optics. As reported by J. Bokor (AT&T Bell Laboratories), "Understanding of the rates and mechanisms for relaxation of optical excitation of the surface itself as well as those of adsorbates on the surface is providing new insight into surface chemistry, surface phase transitions, and surface recombination of charge carriers in semiconductors." The combination of lasers and nonlinear optical techniques is now being brought to bear on the next frontier in surface physics, namely surface dynamics. Ultrafast lasers allow for the study of picosecond and femtosecond processes directly in the time domain, circumventing the ambiguities attendant on linewidth measurements for the determination of lifetimes. One may anticipate continued growth in the diversity of applications of these techniques to the understanding of the complexities of surface dynamics. See Fig. 11.

Amorphous Silicon. According to P.G. LeComber (University of Dundee), the most important difference between crystalline silicon and an amorphous semiconductor is that in the latter there is a continuous distribution of localized states within the forbidden energy gap. Another important difference concerns the mobility of the electrons or holes. LeComber observes, "In an amorphous material the periodicity of the lattice only extends over a few atomic spacings. Under these conditions, the electron transport may no longer be considered as band motion with occasional scattering, as in crystalline theory. In this case, the electron motion is essentially a diffusive process that can be considered to be similar to the Brownian motion of small particles in liquids. Properties of particular importance in the application of amorphous silicon films include:

- Thin films (about 1 micrometer thick).
- Low deposition temperature.
- Large area growth on many substrates, such as glass, metals, and flexible plastics.
- Mechanically very hard.
- Chemically very stable.
- Inert material.
- Extremely photoconductive.
- Room temperature electrical conductivity can be controlled over ten orders of magnitude by doping for both *n*-type and *p*-type material.
- Ease of sequentially producing *p*-type and *n*-type material by switching from one gas mixture to another.
- Easy to pattern arrays of devices using conventional photolithographic techniques developed for crystalline silicon.

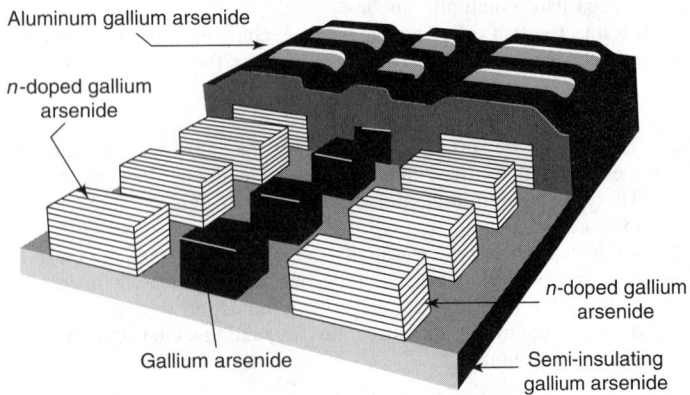

Fig. 10. Quantum chip consisting of four materials. Final product is about $\frac{1}{100}$th size of conventional chip. Current flows from one negatively doped (n-doped) gallium arsenide block to another by way of a layer of aluminum gallium arsenide, a gallium arsenide cube, and thence to an other aluminum gallium arsenide layer. Current conductivity of a quantum device is extremely sensitivity and thus capable of exacting control. (*This idealized model is suggested by R.T. Bate in the scholarly reference cited*)

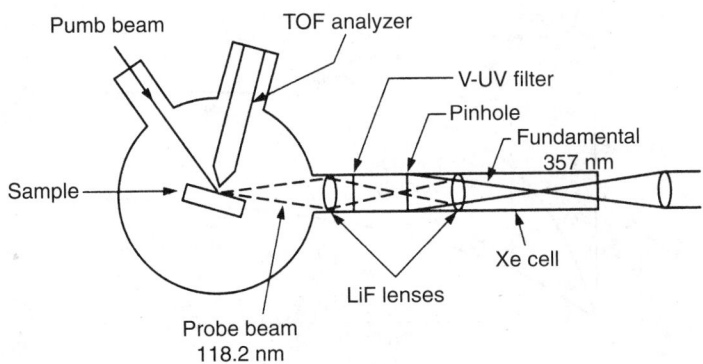

Fig. 11. Experimental arrangement used for picosecond time- and angle-resolved photoemission spectroscopy. TOF = time of flight; V-UV = visible-ultraviolet; LiF = lithium fluoride Xe = xenon. (*source: AT&T Bell Laboratories*)

Hybrid Ferromagnetic-Semiconductor Structures. G.A. Prinz and researchers at the Materials Science and Technology Division of the Naval Research Laboratory, Washington, DC, have been studying hybrid ferromagnetic-semiconductor materials through the use of modern thin-film techniques. Thus far, the team has researched and demonstrated combinations of Fe/Ge, Fe/GaAs, Fe/ZnSe, and Co/GaAs. The researchers observe, "Ultrahigh-vacuum growth techniques are being used to grow single-crystal films of magnetic materials. These growth procedures, carried out in the same molecular beam epitaxy systems commonly used for the growth of semiconductor films, have yielded a variety of new materials and structures that may prove useful for integrated electronics and integrated optical device applications."

Useful characteristics of hybrid ferromagnetic-semiconductor structures include:

1. Produce significant changes in the electrical and optical properties.
2. Coupling of devices to a radiation field, particularly in the microwave range.
3. Such devices provide a source of spin-polarized carriers.

Details of this research can be found in the Prinz reference listed.

Microclusters. These may be defined as small aggregates of atoms that make up a distinct phase of matter. The chemistry of clusters is highly reactive and selective. Their principle area of future application is catalysis. However, clusters also hold some promise for electronic applications. As observed by M.A. Duncan and D.H. Rouvray (University of Georgia), "Thin films of clusters possessing desirable electronic qualities could be of great interest in microelectronics. It is possible to envision applications in optical memories, image processing and superconductivity. Given the potential for construction of parts from networks of clusters, it may eventually be possible to make electronic devices on a molecular scale. Ultimately a machine might be designed that could serve as a link between solid-state electronics and biological systems, such as systems of neurons. Such a link might convey data from a television camera to the brain of a blind person."

See also separate article on **Molecular and Supermolecular Electronics**.

Additional Reading

Allison, J.: *Electronic Engineering Semiconductors and Devices,* 2nd Edition, McGraw-Hill Companies, Inc., New York, NY, 1990.

Allyn, C.L., Flahive, P.G., and S.H. Wemple: "Choosing from Column III and Column IV," *Record (AT&T Bell Laboratories),* 4–11 (January 1986).

Bate, R.T.: "The Quantum-Effect Device: Tomorrow's Transistor?" *Sci. Amer.,* 96 (March 1988).

Bierman, H.: "Material Advances Pave the Way for Device and System Improvements," *Microwave J.* 26 (October 1990).

Bokor, J.: "Ultrafast Dynamics at Semiconductor and Metal Surfaces," *Science,* 1130 (December 1, 1989).

Brennan, K.F.: *The Physics of Semiconductors: With Applications to Optoelectronic Devices,* Cambridge University Press, New York, NY, 1999.

Brennan, K.F.: *Theory of Modern Electronic Semiconductor Devices,* John Wiley & Sons, Inc., New York, NY, 2002.

Brodsky, M.H.: "Progress in Gallium Arsenide Semiconductors," *Sci. Amer.,* 68 (February 1990).

Brophy, J.J.: *Basic Electronics for Scientists,* 5th Edition, McGraw-Hill Companies, Inc., New York, NY, 1990.

Dimitrijev, S.: *Understanding Semiconductor Devices,* Oxford University Press, Inc., New York, NY, 2000.

DiSalvo, F.J.: "Solid-State Chemistry: A Rediscovered Chemical Frontier," *Science,* 649 (February 9, 1990).

Duncan, M.A. and D.H. Rouvray: "Microclusters," *Sci. Amer.,* 110 (December 1989).

Ellowitz, H.I.: "1991 U.S. GaAs Foundry Update," *Microwave J.,* 42 (August 1991).

Fink, D.G. and D. Christiansen: *Electronics Engineers' Handbook,* 3rd Edition, McGraw-Hill Companies, Inc., New York, NY, 1989.

Fisk, Z. et al.: "Heavy-Electron Metals: New Highly Correlated States of Matter," *Science,* 33 (January 1, 1988).

Geinovatch, V.G.: "Prognostications from the Edge," *Microwave J.,* 26 (April 1991).

Geis, M.W. and J.C. Angus: "Diamond Film Semiconductors," *Sci. Amer.,* 84 (October 1992).

Goldstein, A.N., Echer, C.M., and A.P. Alivisatos: "Melting in Semiconductor Nanocrystals," *Science,* 1425 (June 5, 1992).

Kemerley, R.T. and D.F. Fayette: "Affordable MMICs for Air Force Systems," *Microwave J.,* 172 (May 1991).

LeComber, P.G.: "Amorphous Silicon—Electronics Into the 21st Century," University of Wales Review, 31 (Spring 1988).

Mouthaan, T.J.: *Semiconductor Devices Explained: Using Active Simulation,* John Wiley & Sons, Inc., New York, NY, 1999.

Neamen, D.A.: *Semiconductor Physics and Devices: Basic Principles,* 2nd Edition, McGraw-Hill Higher Education, New York, NY, 1997.

Pool, R.: "Clusters: Strange Morsels of Matter," *Science,* 1186 (June 8, 1990).

Prinz, G.A.: "Hybrid Ferromagnetic Semiconductor Structures," *Science,* 1092 (November 23, 1990).

Ramachandran, T.B.: "Gallium Arsenide Power Sources," *Microwave J.,* 91 (January 1990).

Schroder, D.K.: *Semiconductor Material and Device Characterization,* 2nd Edition, John Wiley & Sons, Inc., New York, NY, 1998.

Singh, J.: *Semiconductor Devices: Basic Principles,* John Wiley & Sons, Inc., New York, NY, 2000.

Soref, R.: "Silicon-Based Optical-Microwave Integrated Circuits," *Microwave J.,* 230 (May 1992).

Sze, S.M.M.: *Semiconductor Devices: Physics and Technology,* 2nd Edition, John Wiley & Sons, Inc., New York, NY, 2001.

Van Zant, P.: *Microchip Fabrication,* 4th Edition, McGraw-Hill Companies, Inc., New York, NY, 2000.

Whitaker, J.C.: *Semiconductor Devices and Circuits,* CRC Press, LLC., Boca Raton, FL, 1999.

Wiley, J.B. and R.B. Kaner: "Rapid Solid-State Precursor Synthesis of Materials," *Science,* 1093 (February 28, 1992).

Yablonovitch, E.: "The Chemistry of Solid-State Electronics," *Science,* 347 (October 20, 1989).

Yam, P.: "Atomic Turn-On: First Atom Switch," *Sci. Amer.,* 20 (November 1991).

SEMIPERMEABLE MEMBRANE (or Semipermeable Diaphragm).

A membrane or septum through which one (or more) of the substances composing a mixture or solution may pass, but not all.

In osmotic pressure determinations, semipermeable membranes permit the passage of a solvent but not of certain colloidal or dissolved substances. Many natural membranes are semipermeable, e.g., cell walls; other membranes may be made artificially, e.g., by precipitating copper cyanoferrate(II) in the interstices of a porous cup, the cup serving as a frame to give the membrane stability.

Semipermeable membranes are also used in the separation of gases. See Fig. 1. When a semipermeable membrane is placed in a gas mixture, being impermeable to gas 2 and allowing gas 1 to pass, the force exerted on it will equal the area times the partial pressure of gas 2 only. While there are no ideal semipermeable membranes for gases, there exist in practice reasonable approximations to them, such as incandescent platinum or palladium sheets, which can be penetrated by hydrogen but not by other gases. A film of water also acts as a semipermeable membrane for gases, since it is pervious to NH_3 or SO_2 because of their solubility in water, but gases which are not easily soluble are held back.

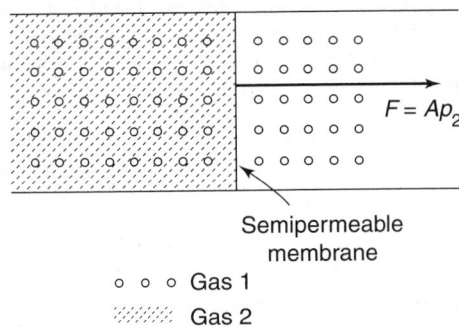

Fig. 1. Separation of gases by semipermeable membrane

See also **Desalination**.

SEPIOLITE.

The mineral sepiolite or meerschaum is soft, white, light in weight, and occurs in clay-like nodular masses. It is a complex, hydrous magnesium silicate corresponding to the formula $Mg_4Si_6O_{15}(OH)_2 \cdot 6H_2O$. It crystallizes in the orthorhombic system; hardness, 2–2.5; specific gravity, 2; color, white, grayish white, sometimes a yellowish- or bluish-green; opaque. It is capable of floating on water, hence the name meerschaum or sea foam. It occurs in Asia Minor associated with serpentine and magnesite, and may be derived from the latter. Other deposits are in the Czech Republic and Slovakia, Morocco, and Spain; and in the United States in Pennsylvania and New Mexico. The name meerschaum is from the German. Sepiolite is from the Greek, meaning cuttlefish, referring to the similarity of the bone of that animal to the light, porous sepiolite. The material is used in the manufacture of smoking pipes.

SEQUENTIAL ANALYSIS. The analysis of material derived by a sequential method of sampling, that is to say, it is the data, not the analysis, which are sequential.

In sequential sampling the members are drawn one by one (or in groups) in order, and the results of the drawing at any stage decide whether sampling is to continue. The sample size is thus not fixed in advance but depends on the actual results and varies from one sample to another. The sampling terminates according to predetermined rules which are decided by the degree of precision required.

SEQUESTERING AGENTS.   See Chelates and Chelation.

SERANDITE.   The mineral serandite is a hydrated manganese-sodium silicate corresponding to the formula $Mn_2NaSi_3O_8(OH)$, crystallizing in the triclinic system, of pseudo-monoclinic character. Color, rose-red, pink; transparent; brittle, and uneven fracture. Prominent basal and prismatic cleavage; vitreous to pearly luster. Crystals thick tabular or prismatic, and as intergrown aggregates. Occurs as superb crystals in a carbonatite zone in a host body of nepheline-syenite in association with analcime, aegerine, and other rare minerals at Mt. St. Hilaire, Quebec, Canada. Its only other known world occurrence is on the Island of Rouma, Los Islands, Guinea.

SERPENTINE.   This is a group name for minerals encompassing two principal polymorphic forms: *chrysotile* and *antigorite*. This monoclinic mineral of hydrous magnesium silicate composition $Mg_3Si_2O_5(OH)_4$ is essentially a product of metamorphic alteration of ultrabasic rocks rich in olivine, pyroxene, and amphibole. Serpentine crystals are unknown except as pseudomorphic replacements of other minerals, e.g., after clinochlore crystals at the Tilly Foster Mine, Brewster, New York, Antigorite occurs as platy masses; *chrysotile* as silky fibers. Most massive serpentine rocks are composed essentially of antigorite. The hardness is 2–5, specific gravity ranges from 2.2 (fibrous varieties) to 2.65 (massive varieties). Color usually mottled green. The name *serpentine* stems from the mottled character, somewhat resembling the skin of a serpent. There is a greasy to wax-like luster in massive material; silky in fibrous material. The minerals are translucent.

Chrysotile fibers are the source of commercial asbestos, although fibrous amphiboles also contribute to similar usage. Asbestos is economically valuable for its incombustibility and low conductivity of heat, thus as fireproofing and insulating material. See also **Asbestos**.

Chrysotile deposits of economic value are found in Quebec, Canada, in the former U.S.S.R., and in South Africa. Minor occurrences are found in the United States in Vermont, New York, New Jersey, and Arizona. *Verd antique* marble (serpentine marble) is quarried extensively near West Rutland, Vermont.

ELMER B. ROWLEY
F.M.S.A. Union College
Schenectady, New York

SESAME SEED OIL.   See Vegetable Oils (Edible).

SET (Permanent).   When a solid has been strained beyond the elastic limit and the deforming stress is completely removed, in general the strain does not decrease ultimately to zero but to some nonvanishing value, known as a permanent set.

SEWAGE SLUDGE (Energy Source).   See Wastes as Energy Sources; Sludge.

SHALE.   A fine-grained sedimentary rock whose original constituents were clays or muds. It is characterized by thin laminae breaking with an irregular curving fracture, often splintery, and parallel to the often indistinguishable bedding planes.

SHELL.   1. In physical chemistry, this term is applied to any of the several sets of, or orbits of, the electrons in an atom as they revolve around the nucleus. They constitute a number of principal quantum path representing successively higher energy levels. There may be from one to seven shells, depending on the atomic number of the element and corresponding to the seven periods of the periodic table. The shells are usually designated by number, though letter symbols have been used, i.e., K, L, M, N, O, P, Q. The laws of physics limit the number of electrons in the various shells as follows: two in the first (K), eight in the second (L), 18 in the third (M), and 32 in the fourth (N). With the exception of hydrogen and helium, each shell contains two or more orbital, each of which is capable of holding a maximum of two electrons.

See also **Orbitals**; **Quantum Number**; and **Pauli Exclusion Principle**.

2. The hard integument of mollusks and crustaceans, consisting mostly of calcium carbonate, chitin, etc.

3. The brittle covering of avian eggs, chiefly calcium carbonate, lime, etc. The formation of proper shell structures in certain species of birds is said to be adversely affected by DDT and similar insecticidal contaminants of their food.

4. The shells of nuts are cellulosic in character. Some contain industrially useful oils.

SHELL (Atomic).   See Chemical Elements.

SHELLAC.   A secretion or excretion of the lac insect, *Coccus lacca*, found in the forests of Assam and Siam. Freed from wood it is called "seed lac." It is soluble in alkaline solutions such as ammonia, sodium borate, sodium carbonate and sodium hydroxide, and also in various organic chemicals. When dissolved in acetone or alcohol, shellac yields the familiar shellac varnish of superior gloss and hardness. Orange shellac is bleached with sodium hypochlorite solution to form white shellac. See also **Paints and Coatings**.

SHERARDIZING.   The process for applying an adherent protective coating of zinc to steel parts by heating at 700°F (371°C) in contact with zinc dust in a rotating-drum container.

SHIFT REACTION (Water Gas).   See Coal; Substitute Natural Gas (SNG).

SIDERITE.   This mineral is a carbonate of iron, $FeCO_3$. It is hexagonal with rhombohedral crystals, and also occurs in various massive forms. It has a rhombohedral cleavage; uneven fracture; is brittle; hardness, 3.75–4.25; specific gravity, 3.96; luster, vitreous to pearly; color, gray, yellowish- or greenish-gray, green, reddish-brown and brown. Siderite is found as concretionary masses in the sedimentary rocks; as a replacement mineral from the action of iron solutions upon limestones; and in metalliferous veins as a gangue mineral. It is relatively common. Siderite is found in Austria, Saxony, the Czech Republic and Slovakia, France, England, Italy, Greenland, Australia, Brazil and Bolivia. In the United States important localities are in Connecticut, Pennsylvania, New Jersey, Ohio, and Washington. It is an iron ore. The mineral was at one time called chalybite.

SILICATES (Soluble).   The most common and commercially used soluble silicates are those of sodium and potassium. Soluble silicates are systems containing varying proportions of an alkali metal or quaternary ammonium ion and silica. The soluble silicates can be produced over a wide range of stoichiometric and nonstoichiometric composition and are distinguished by the ratio of *silica* to *alkali*. This ratio is generally expressed as the *weight percent ratio* of silica to alkali-metal oxide (SiO_2/M_2O). Particularly with lithium and quaternary ammonium silicates, the molar ratio is used.

Sodium silicates find wide application in many types of detergents and cleaning compounds and have been used for many years as adhesives and cements. Both sodium and potassium silicates are important bonding agents in a large variety of ceramic cement and refractory applications, notably because of their heat stability and resistance to chemicals. Alkali-metal silicate bonds are used in high-temperature ceramic products in the fabrication of electrical components. Soluble silicates find wide application for pelletizing, granulating, and briquetting finely divided particles, such as clays, fertilizers, and ores. Sodium silicates also are used as bonding materials for foundry mold and core compositions. Because of their adherence properties, soluble sodium and potassium silicates are widely used as coatings. Frequently, sodium silicates are used to protect against water-line corrosion in tanks. The ability to form sols and gels is an interesting and very useful characteristic of soluble silicates. Silica gels are used in a major way as desiccants and as carriers for the production of petroleum-cracking catalysts, as well as raw materials in the manufacture of zeolites. Activated sols are used in water clarification.

Generally, sodium and potassium silicates are made by fusion of pure sand with alkali-metal carbonate or alkali-metal sulfate and carbon. This operation is carried out in large open-hearth furnaces heated to a temperature range of 1300–1500°C. The resulting glasses may be used in this form, or dissolved in water to produce silicate solutions. Sodium and potassium silicate solutions also can be made by dissolving sand in sodium or potassium hydroxide solution at elevated temperatures and pressures. Lithium silicate glasses, although insoluble in water, can be made by dissolving silica gel in, or mixing silica sols with lithium hydroxide solutions. Anhydrous sodium metasilicate is made from the anhydrous melt. This salt crystallizes rapidly from its aqueous solution at temperatures in the range of 80–85°C.

The most important property of sodium and potassium silicate glasses and hydrated amorphous powders is their solubility in water. The dissolution of vitreous alkali is a two-stage process. In an ion-exchange process between the alkali-metal ions in the glass and the hydrogen ions in the aqueous phase, the aqueous phase becomes alkaline, due to the excess of hydroxyl ions produced while a protective layer of silanol groups is formed in the surface of the glass. In the second phase, a nucleophilic depolymerization similar to the base-catalyzed depolymerization of silicate micelles in water takes place.

When sodium silicate solutions of intermediate ratios are concentrated to a thick gum, they become very sticky and tacky. This property is important to many of the adhesive applications. It is related to high cohesion and low surface tension rather than primarily to viscosity.

The stability of soluble silicate solutions depends strongly on pH and concentration. The addition of acids and acid-forming compounds gives rise to the formation of silica gels. Soluble alkali-metal silicate solutions are not compatible with most organic water-miscible solvents. The addition of alcohols and ketones causes phase separation into liquid layers. A few organic systems, however, particularly polyols, such as glycols, glycerins, sugars, and polyethylene glycols, are compatible and miscible with alkali-metal silicate solutions. See also **Adhesives**; and **Glass**.

SILICIC.   1. Containing or pertaining to silicon. 2. Containing silicic acid (ortho) H_4SiO_4; or silicic acid (meta) H_2SiO_3; or silicic acids of a higher degree of hydration (disilicic acids, trisilicic acids, etc.).

SILICIFICATION.   An important geochemical process by which certain sedimentary rocks such as limestones and dolomites, or calcareous fossils are partially or entirely replaced by silica, SiO_2. See also **Chert**; and **Flint**.

SILICON.   [CAS: 7440-21-3] Chemical element, symbol Si, at. no. 14, at. wt. 28.086, periodic table group 14, mp 1408–1,412°C, bp 2,355°C, density 2.242 g/cm³ (solid crystalline, 20°C), 2.32 g/cm³ (single crystal, 20°C). Elemental silicon has a face-centered cubic crystal structure (diamond structure). The existence of a hexagonal form of silicon with a wurtzite-type structure and with lattice parameters $a = 3.80$ Å and $c = 6.28$ Å was established in 1963 (Wentorf-Kasper). Claims to different parameters were made by Jennings-Richman (1976). These differences are discussed by Kasper-Wentorf (1977). Much new knowledge concerning the crystalline structures and phase transitions of silicon has been gained during the mid-1980s, notably from research under immensely high pressures and investigations involving the tunneling microscope, as described shortly.

The common form of silicon is a dark-gray, hard solid. It can be obtained as a brown microcrystalline powder, which is not an allotrope of the gray form. Both forms are unaffected by air at ordinary temperatures, but when heated in air to high temperatures a protective layer of oxide is formed. Silicon reacts with nitrogen at high temperatures to form the nitride; with chlorine to form the chloride, with several metals to form silicides. Crystalline silicon is unattacked by HCl or HNO_3, or H_2SO_4, but is attacked by hydrofluoric acid to form silicon tetrafluoride gas. Silicon is soluble in NaOH solution forming sodium silicate and hydrogen gas. Silicon reacts with dry chlorine to form silicon tetrachloride.

There are three naturally occurring isotopes, ^{28}Si through ^{30}Si, and three radioactive isotopes have been identified, ^{27}Si, ^{31}Si, and ^{32}Si. The latter isotope has a half-life of approximately 700 years, while the half-lives of the other two are short, measured in terms of seconds and hours.

Lavoisier showed in 1787 that SiO_2 was not a single element and indicated that it was the oxide of a hitherto unknown element. In the early 1800s, Scheele, Davy, Gay-Lussac, and Thénard attempted to isolate the element, but were not successful. In 1871, Berzelius discovered silicon in a cast-iron melt and, in 1823, succeeded in isolating the element by reduction of potassium fluorosilicate with potassium. Small laboratory amounts were produced by H.E. Sainte-Claire Deville in 1854 and by C. Winkler in 1864. It was not until 1900 that the effective properties of silicon as a deoxidizing agent for steel production were observed. Shortly thereafter, ferrosilicon alloys, using quartzite, coke, and iron pellets, were produced in electric refining furnaces of the type already in use for making calcium carbide. With this technique, it was possible to produce silicon of about 98% purity. It remained for the rigid purity requirements of semiconductors many years later before silicon of higher purities was produced.

Silicon is ranked second in the order of chemical elements appearing in the earth's crust, an average of 27.72% occurring in igneous rocks. In terms of seawater, it is estimated that a cubic mile of seawater contains about 15,000 tons of silicon (3240 metric tons per cubic kilometer). In terms of abundance throughout the universe, silicon is ranked seventh. First ionization potential 8.149 eV; second, 16.27 eV; third, 33.30 eV; fourth, 44.95 eV. Oxidation potentials $Si + 2H_2O \longrightarrow SiO_2 + 4H^+ + 4e^-$, 0.86 V; $Si + 6OH^- \longrightarrow SiO_3^{2-} + 3H_2O + 4e^-$, 1.73 V.

Other important physical properties of silicon are given under **Chemical Elements**.

Because of its chemical reactivity, silicon does not occur in elemental form in nature. The element is present in igneous rocks and clays as alumino-silicate; as the oxide SiO_2 in quartz, sand. (Fig. 1), flint, and the gems amethyst, jasper, chalcedony, agate, onyx, tridymite, opal, crystobalite; as silicates in zircon (zirconium silicate, $ZrSiO_4$), in willemite (zinc silicate, Zn_2SiO_4), in wollastinite (calcium silicate, $CaSiO_3$), in serpentine (magnesium silicate, $Mg_3Si_2O_7$). Impure (up to 98% Si) silicon is obtained from the oxide (1) by igniting with aluminum powder, or (2) by reduction with carbon in an electric furnace. See also **Cancrinite**.

Silicon Production for Alloys: production of raw steel requires about 1.6–1.7 kilograms of silicon per metric ton of steel. The silicon is used in the form of ferrosilicon, which contains about 20% silicon. It is estimated that about 3 million metric tons of ferrosilicon are consumed annually in steelmaking. The 20%-silicon-content ferrosilicon can be made in a conventional blast furnace. Ferrosilicons with higher silicon contents (45, 75, 90, and 98%) must be produced in electric furnaces. The raw materials are pure quartzites. The presence of impurities, such as Al_2O_3 and CaO, interfere with the melting process because of the formation of dross. The reducing agent used is chemical coke. For the very high concentrations of silicon

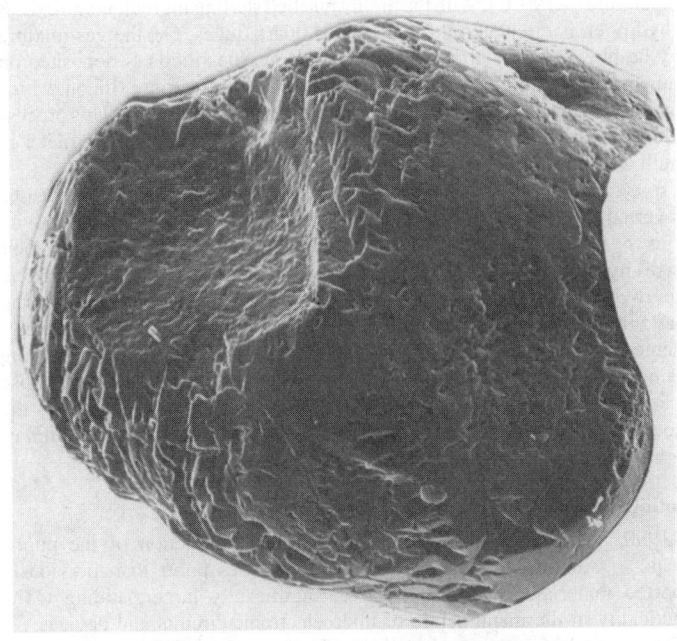

Fig. 1.   Grain of sand, originally magnified 100×

(90–98%), ash-free petroleum coke or charcoal are used. Iron is added in the form of small pellets or chips in the production of the 45–75%-silicon alloys.

For certain metal alloys, a calcium silicon alloy is required. This alloy also is used as a steel deoxidizer and is favored because it forms a low-melting-point calcium silicate product. A representative composition of the alloy is: 30–33% Ca, 60–64% Si, 3–5% Fe, 1–2% Al, 0.3–0.6% C, and less than 0.15% S and P.

Silicon is used in the primary and secondary aluminum industry. The purity of silicon for metallurgical purposes ranges from 96.7 to 98.5% silicon; 0.10 to 0.75% aluminum; 0.03 to 0.04% calcium; with the remainder being principally iron.

Silicon Carbide

This compound is an important industrial abrasive, having a hardness of 9.5 on the Mohs scale. In this compound, each silicon atom is surrounded tetrahedrally by four carbon atoms, and similarly, each carbon atom is surrounded by four silicon atoms. Silicon carbide is made by reducing pure quartz (glass-sand) with petroleum coke in an electric-resistance type furnace, known as the Acheson process. The product is hexagonal crystals ranging from light-green to black. It is used as a ceramic raw material for dross-repellent linings as well as for many abrasive applications.

Silicon carbide also has been recognized for many years because of its having a unique set of electronic material advantages over silicon and gallium arsenide. Not only can SiC withstand higher device operating temperatures (approximately 650°C compared with silicon's 150°C), but SiC devices can operate with ten times the voltage capability and three times the thermal conductance capability. And, they are mechanically much more robust than traditional semiconductors.

The foregoing characteristics enable the configuration of a whole new family of high-power microwave and high-temperature electronics that can withstand high radiation for military and commercial systems.

Only recently has it been possible to produce uniform, centimeter-sized crystals. A high-purity vapor transport growth process has been developed (Westinghouse) to produce 1.5-inch (3.8-cm) device-grade SiC crystals and wafers. These comprised the building blocks for a demonstration of microwave transistors in the early 1990s. See Fig. 2.

Super- or Hyperpure Silicon

For semiconductor use, there can be only one atom of impurity for every 100,000 silicon atoms! The starting material for the manufacture of hyperpure silicon is silicon tetrachloride, $SiCl_4$, or trichlorosilane, $SiCl_3H$. Both of these materials can be reduced with hydrogen to yield a compact deposition of silicon on hot surfaces, ranging from 800–1,200°C. The starting compounds are purified of boron and phosphorus by fractional distillation and absorption techniques. The process hydrogen is purified by passing it through molecular sieves under high pressure, followed by absorption techniques at a low temperature (−190°C). With the highly purified starting ingredients an excess of hydrogen is circulated through heated quartz tubes. Or, the gas mixture may be blown into quartz bell jars, whereupon the silicon is deposited on filaments of tantalum or tungsten or on thin rods of hyperpure silicon, which may be heated by electrical resistance or radiofrequency energy. This process yields polycrystalline rods of silicon which range up to about one meter in length and 150 millimeters in diameter.

To be used in seimconductor devices, the polycrystalline silicon must be converted to single crystals of a defined, predetermined type of conductivity (*n* or *p* type). The crystals must be rigidly controlled as regards their resistivity, and possess the highest degree of crystallographic perfection. The two crystal-growing techniques used are: (1) crucible free vertical float zoning which removes all residual impurities, including phosphorus, arsenic, and oxygen, but boron is essentially irremovable by floating zoning, or (2) crucible pulling in which the crystals, particularly those of lower resistivity, are drawn out of a melt in a process known as the Czochralski technique. Both processes must be conducted under helium or argon, often under a vacuum of 10^{-5} torr.

Production of Ultrapure Silicon Crystal

In 1990, Westinghouse engineers reported the production of the purest crystal of silicon ever made—namely, four times purer than previously reported material. The crystal also is significantly larger, adding to its practicality in the manufacture of microelectronic circuits and devices.

The cylindrical structure, called a *boule*, weighs 22 pounds (10 kg) and is over a yard (meter) long, with a diameter of just over 3 inches

Fig. 2.   Researcher Dan Barrett (Westinghouse Science & Technology Center) checks the hot (2400°C) crystal growth furnace that he designed for physical vapor transport growth of single crystals of silicon carbide

(8 cm). Impurities are a few parts in 100 billion, compared with more than 10 parts in 100 billion previously reported for 1-inch (2.5-cm) diameter ultrapure crystals.

Crystal boules are sliced into wafers on which microelectronic circuits and power semiconductor devices are fabricated. An important use of the wafers is for infrared dectors for space, defense, and environmental applications.

Liquid-Solution Synthesis of Silicon Crystals

In late 1992, J.R. Heath (IBM Watson Research Laboratory) reported on a liquid-solution phase technique for preparing submicrometer-sized silicon single crystals. The synthesis is based on the reduction of $SiCl_4$ and $RSiCl_3$ (R = H, octyl) by sodium metal in a nonpolar organic solvent at high temperatures (385°C) and high pressure (above 100 atmospheres). For R = H, the synthesis produces hexagonal silicon single crystals ranging from 5 to 2000 nanometers. For *R* = octyl, the synthesis also produces hexagonal-shaped silicon single crystals.

Light Emission from Silicon

Because of silicon's successes in the electronic components field, research has been going on to find a form of Si that will produce luminous radiation. Because of former failures, numbers of scientists have given up this research. However, independently in French and British laboratories during 1990, some success has been achieved. These researchers have found that, if one etches Si into structures so tiny that the electronic behavior of the material is transformed, full-color emission from what are termed "silicon quantum wires." A British researcher L. Canham (Royal Signals and Radar Establishment, Malvern, England) has observed that, to make silicon quantum wires, a process for sculpting silicon, known for some 30 years, is the basis. A silicon wafer is immersed in an acid electrochemical bath, which bores into the disk to produce extremely small so-called "wormholes." The latter are etched chemically, enlarging them until they meet one another. The result is a columnar structure of silicon. The latter are about a micron high, and 50 of them, stacked end to end, would span an area about equivalent to a cross-section of a human hair and are only a few nanometers thick ($\frac{1}{15,000}$), smaller than a hair. The researchers have found that, when such a structure is bathed in ultraviolet light, light emission occurs. The emitted wavelength is determined by the porosity of the Si layer.

J.P. Harbison (Bellcore, Redbank, New Jersey) observes, "This is not the moment of the breakthrough for light-emitting silicon, but it is the moment when a lot of people are realizing its potential."

See also **Crystal**; and **Semiconductors**.

Research on Silicon Structure and Surface Properties

The rather unusual properties of silicon have intrigued scientists for many years. It possesses the physical properties of a *metalloid* (exhibits properties of both a metal and nonmetal). In several ways, silicon resembles germanium and, to a lesser extent, it resembles arsenic and boron. Silicon is a semiconductor of electricity, the conductivity rising with temperature. Silicon, in pure form, is intrinsically a semiconductor. The presence of impurities in very minute amounts markedly increases its conductivity. By introducing elements of group 13 (such as boron), which have a deficiency of electrons, the *p*-type silicon results. Therein, electricity is conducted by migration of electron vacancies or holes. On the other hand, introduction of elements of group 15, such as arsenic or phosphorus, in which there is no deficiency of electrons, the *n*-type silicon results, in which extra electrons carry an increased current because of their migration. Scientists have not been satisfied with oversimplified explanations such as that just given. As a key material in the microelectronics field, where the processing of silicon into chips and other configurations for electronic components is essentially effected at the surface of the silicon, particular interest concerns those crystalline structural details that play a role in the electronic nature of the element. However, prior to the emergence of solid-state technology, scientists were puzzled by what appeared to be crystal structure and surface anomalies and, consequently, research dates back many years, with progress largely determined by the instrumentation available to investigators. The *tunneling microscope*, the invention of which is accredited to G.K. Binnig and H. Rohrer and partially to E.W. Müller (who also invented the field-ion microscope in the 1950s), has contributed much toward an understanding of the surface of the silicon crystal, an understanding which is expected to be translated ultimately in manufacturing improvements and better final properties of silicon-based electronic components. See also **Scanning Tunneling Microscope**.

The relatively recent availability of means to create extremely high pressures (see also **Diamond Anvil High Pressure Cell**) has made it possible to gain further insights into the character of silicon. It has been learned from such experimentation that at a pressure of 110 kilobars (about 1.6 million pounds per square inch), silicon enters *truly metallic* phases. At the pressure stated, silicon abruptly assumes a structure similar to the beta form of tin. At this pressure and at a temperature of 6 degrees Kelvin (six Celsius degrees above absolute zero) the metal becomes superconducting, that is, it offers no resistance to the passage of electrons. At a pressure of 130 kilobars, the beta-tin form of silicon transforms into what has been designated as the *primitive hexagonal* phase, a phase first discovered in 1984. This research was conducted by Cohen, Chang, and Dacorogna (University of California, Berkeley). Prior to this experiment, it was not thought that such a phase would exist in the crystal of any chemical element. The researchers entered into theoretical calculations after the experiment to better understand the properties of the primitive hexagonal phase. The calculations were lengthy and required a CRAY/X-MP computer. A major finding was that the bonds linking the atoms in each of the planes defined by the hexagons should be weaker than the bonds linking the atoms in adjacent planes. This indicates that the electronic charge distribution should be inhomogeneous along one dimension. This inhomogeneity is an indication of a good superconductor because, in effect, it provides corridors through which electrons can move.

In their investigation, the researchers turned back to the much earlier hypotheses of quantum mechanics, including the properties of phonons (in quantum mechanics, a phonon can be treated as a particle, one that interacts with electrons). A strong interaction improves the opportunities for superconductivity. However, where the coupling is too strong, the integrity of the lattice may collapse and a structural phase transition may occur. Testing for superconductivity at such high pressures will be difficult. The Berkeley group predicts that the superconducting temperature will rise to a value greater than 10 degrees as the pressure nears the value required for the transition from the simple hexagonal phase to the hexagonal closed-packed phase. If expectations are proved, silicon could be the best superconductor of all chemical elements. Translating this to practical application, of course, may or may not be feasible at some future date. See also **Superconductivity**.

Silicon Chemistry and Compounds

Like carbon, silicon forms chiefly covalent bonds, but its greater atomic radius enables it to form positive ions more readily. Unlike carbon and tin, silicon is not allotropic, having only one elemental form, the diamond structure in which each atom is surrounded tetrahedrally by four others to which it is covalently bonded. An apparently amorphous brown powder, produced by combustion of silane, SiH_4, has been found to be a microcrystalline variety of this covalently-bonded structure. Much research has been conducted during the 1970s and 1980s pertaining to the more exotic silicon compounds, such as the *disilenes*, and to silicon-mediated organic synthesis. These topics are discussed later in this article. The following several paragraphs are devoted to the large number of traditional silicon compounds whose constitution and characteristics have been well established over the years.

Silicon Dioxide, (SiO_2)

This compound exists in at least eleven distinct crystalline forms. Several of them are obtained by heating α-quartz, which has a number of transition points, to produce β-quartz, and to give various forms of tridymite and crystobalite. The unit of structure is the tetrahedron in which each silicon atom is covalently bonded to four oxygen atoms, and the variation is in the ways these tetrahedra are interconnected (by oxygen atoms) to form a three-dimensional system.

Silicon dioxide is converted by hydrofluoric acid into silicon tetrafluoride, SiF_4, a gas. SiF_4 can also be produced directly from the elements, as can the other tetrahalides, silicon tetrachloride, $SiCl_4$ (a liquid), silicon tetrabromide, $SiBr_4$ (a liquid), and silicon tetraiodide, SiI_4 (a solid). The silicon halides hydrolyze much more readily than the carbon halides, because the unoccupied silicon $3d$ orbitals are energetically not far above its $3s$ and $3p$ orbitals. This fact also permits the formation of the sp^3d^2 hybrid bonds of the fluorosilicate ion, SiF_6^{2-}, and additional compounds of the halides, e.g., $SiX_4 \cdot 2$ pyridine. Silicon also is intermediate between carbon and the higher members of main group 4 of the periodic table in forming a dichloride, $SiCl_3$, by strong heating of silicon with silicon tetrachloride.

Quartz and other forms of silica react very slightly with water to form monosilicic acid, $(SiO_2)_n + 2nH_2O \longrightarrow nSi(OH)_4$. As shown, this reaction is a depolymerization followed by a hydrolysis, and proceeds rapidly with hot alkalis or fused alkali metal carbonates, yielding soluble silicates containing the SiO_4^{4-} and $(SiO_3^{2-})_n$ ions. The hydrolysis reaction is geologically important, because it is considered to be the starting point in the formation of the innumerable silicate minerals that occur so widely in nature, just as many of the silica minerals may have originated by the reverse reaction. Many of the more complex silicic acids are considered to form by polymerization of $Si(OH)_4$ molecules by sharing of OH ions between two silicon ions (octahedrally coordinated by six hydroxyl ions) followed by condensation with the loss of water to produce

$$-\overset{|}{\underset{|}{Si}}-O-\overset{|}{\underset{|}{Si}}-$$

linkages. The polymerization of silicic acid is carried out industrially to produce silica gel, a stable sol of colloidal particles. The various methods involve careful removal of H_2O, the catalytic effect of acid or alkali (or fluoride ion) and controlled pH. Many varieties of silica gel have been made, including the zerogels and aerogels, in which the aqueous phase is displaced by a gaseous one.

In 1992, Yeganeh-Haeri, Weidner, and Parise (Center for High Pressure Research, State University of New York, Stony Brook) used laser Brillouin spectroscopy to determine the adiabatic single-crystal elastic stiffness coefficients of silicon dioxide in the alpha-cristobalite structure. This SiO_2 polymorph, unlike other silicas and silicates, was found to exhibit a negative Poisson's ratio. Alpha-cristobalite contracts laterally when compressed and expands laterally when stretched. Tensorial analysis of the elastic coefficients showed that Poisson's ratio reached a maximum value of -0.5 in some directions, whereas averaged values for the single-phased aggregate yielded a Poisson's ratio of -0.16.

Silicon Dioxide as a Chemical Intermediate

In 1992, R.M. Laine (University of Michigan, Ann Arbor) announced the development of a process that transforms sand and other forms of silica into reactive silicates that can be used to synthesize unusual silicon-based chemicals, polymers, glasses, and ceramics. The Laine procedure produces pentacoordinate silicates directly from low-cost raw materials—silicon dioxide, ethylene glycol, and an alkali base. The mixture is approximately a 60:1 ratio of silica gel, fused silica (or sand) to metal hydroxide and ethylene glycol. Heating the mixture slowly, the ethylene glycol and water (used to put the materials in solution) boils off. The resulting glycolatosilicates, unlike the hexa- and tetracoordinate forms, are reactive and offer potential for synthesizing a wide range of materials. Laine observes, "The new silicon chemistry could produce alternatives to many petrochemical-based products and could be competitive with or superior to present carbon-based materials." Thus far, a number of materials have been produced by the process:

1. A clear polymer capable of conducting electric current when spread in a thin layer across a flat surface. Potential includes applications in batteries, heated windshields, and electrochromic windshields.
2. A fire retardant polymer that is easily impregnated into wood to "petrify" the material, making it stronger and nonflammable.
3. Liquid-crystal polymers stable to about 425°C (800°F), with potential for uses in watch displays and aerospace instrumentation.
4. Silicate glasses capable of withstanding high temperatures.

Silicates

The great number of naturally occurring silicates result, as just indicated, from the polymerization and dehydration of monosilicic acid to form, ultimately, such groups and ions as $(Si_2O_7)^{6-}$, $(Si_3O_9)^{6-}$, $(Si_4O_{12})^{8-}$, and $(Si_6O_{18})^{12-}$. Various cations, such as those of boron, B^{3+}, aluminum, Al^{3+}, etc., in the structure lie at the centers of anionic polyhedra having as anions the O^{2-} ions of neighboring SiO_4 tetrahedra, in which each Si O bond has an electrostatic bond strength of 1. Cations of lower charge density, on the other hand, like sodium, Na^+, potassium, K^+, calcium, Ca^{2+}, etc., are located interstitially. The great variety of the silicates is due to the considerable degree of isomorphism, exhibited not only by elements of the same group, but by elements of different groups, whereby they partly replace each other in the complex silicates, and by no means necessarily in stoichiometric proportions. Thus troosite may be represented by the formula $(Zn, Mn)_2SiO_4$, chrysolite by $6Mg_2SiO_4 \cdot Fe_2SiO_4$, and vermiculite by $(Mg, Fe)_3(AlSi)_4O_{10} \cdot (OH)_2 \cdot 4H_2O$, even the silicon in vermiculite being partly replaced (by aluminum). One plane of classification of the silicates is upon the basis of the linking of the SiO_4 tetrahedra:

A. *Discrete silicate radicals.*

1. Single tetrahedral (SiO_4^{4-}), e.g., phenacite, Be_2SiO_4.
2. Two tetrahedra (Si_2O_7)$^{6-}$, e.g., hardystonite, $Ca_2ZnSi_2O_7$.
3. Three tetrahedra (Si_3O_9)$^{6-}$, e.g., benitoite, $BaTiSi_3O_9$.
4. Four tetrahedra (Si_4O_{12})$^{18-}$, e.g., axinite, $(Fe, Mn) Ca_2Al_2BO_3 Si_4O_{12}$.
5. Six tetrahedra (Si_6O_{18})$^{12-}$ e.g., beryl, $Be_3Al_2Si_6O_{18}$.

B. *Silicon-oxygen chains of indefinite length.*

1. Single chains with one silicon atom to three oxygen atoms, e.g., diopside, $CaMg(SiO_3)_2$.
2. Double chains (Si:O = 4:11), e.g., tremolite, $Ca_2Mg_5(Si_4O_{11})_2 (OH)_2$.

C. Silicon-oxygen sheets. (Si:O = 2:5), e.g., talc, $Mg_3(SiO_5)_2(OH)_2$.
D. *Silicon-oxygen spatial networks.*

1. Composition SiO_2 (composed of interlinked SiO_4 tetrahedral), e.g., quartz, SiO_2.
2. Composition $M_n(Si, Al)_nO_{2n}$, e.g., feldspar, KSi_3AlO_8. These are probably based upon silicon and aluminum tetrahedra, variously linked.

Silanes

The increasingly large number of silicon compounds produced by industrial processes may be systematized about the silanes and their substitution products, just as the silicates are about the SiO_4 tetrahedron. Silicon, like carbon, forms a number of hydrides, though their number is much more limited. The silane series, analogous to the paraffin hydrocarbons, has at least six members, silane (SiH_4), disilane ($H_3Si-SiH_3$), ... hexasilane ($H_3Si-SiH_2-SiH_2-SiH_2-SiH_2-SiH_3$). They are increasingly unstable, hexasilane dissociating at room temperature. They are halogenated with free halogens to form substituted silanes, and catalytically with the hydrogen halides. The halosilanes react with NH_3 to form silylamines or silazanes and are hydrolyzed by water to form siloxanes. Prosiloxane, H_2SiO, polymerizes readily but disiloxane, $H_3Si-O-SiO_3$, and the higher siloxanes, although

they polymerize, can readily be studied. They have properties like the ethers and other analogous carbon compounds. Hydrogen-containing siloxanes, such as $HO_2Si-SiO_2H$ are also known and polymerize readily. There are also ring siloxanes, such as siloxen, which has a polymerized structure of epoxy form (a powerful reducing agent)

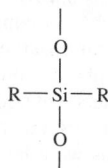

Silyl and polysilyl radicals also combine with nitrogen, arsenic, and other main group 5 elements, as with sulfur and selenium.

The silazanes are such compounds of silicon, nitrogen and hydrogen of the general formula $H_3Si(NHSiH_2)nNHSiH_3$, being called disilazane, trisilazane, etc., according to the number of silicon atoms present. (In disilazane, n in the above formula has a value of 0, in trisilazane it is 1, etc.)

The silthianes are sulfur compounds having the general formula $H_3Si(SSiH_2)_n SSiH_3$ which are called disilthiane, trisilthiane, etc., according to the number of silicon atoms present. They have the generic name *silthianes*. (In disilthiane, n in the above formula has a value of 0, in trisilthiane, a value of 1, etc.)

Silicones

These are semiorganic polymers with a quartz-like structure in which various organic groups are attached to the silicon atom. By varying the kind and number of organic groups, a variety of materials ranging from liquids through gels and elastomers to rigid solids (resins) can be produced.

The organosilicon compounds may be regarded as substituted silanes, although of course their preparation is not usually in this way. Thus, ethyl silicate, $Si(OC_2H_5)_4$, is prepared from silicon tetrachloride and ethyl alcohol, and tetraethyl silane, $Si(C_2H_5)_4$, is prepared from silicon tetrachloride and diethylzinc. The silicon-carbon bond, unlike the carbon-carbon bond, has about 12% of ionic character, varying somewhat with the atoms or groups attached to the two atoms. Other types of organosilicon compounds include the esters, the alkoxyhalosilanes, the higher tetra-alkylsilanes (prepared from silicon tetrachloride and Grignard reagents), the alkylsilanes (H partly replaced by R), the alkylhalosilanes, the alkylalkoxysilanes, the alkylsilylamines, some aryl compounds of the foregoing types, and many related derivatives of disilane and the polysilanes. Other types of compounds are those having silicon-carbon chains and the organosiloxane compounds, for which the name "silicones" is often used. These are essentially chains or networks of groups

$$R-\underset{\underset{O}{|}}{\overset{\overset{O}{|}}{Si}}-R$$

joined by oxygen atoms attached to the silicon atoms as shown. There are many other groups of silicon compounds, as well as individual ones.

Aluminates

Many complex silico-aluminates or aluminosilicates are formed in nature. Of these, clay in more or less pure form (pure clay, kaolinite; kaolin, china clay, $H_4Si_2Al_2O_9$ or $Al_2O_3 \cdot 2SiO_2 \cdot 2H_2O$) is of great importance. Clay is formed by the weathering of igneous rocks, and is used in the manufacture of bricks, pottery, porcelain, Portland cement. Sodium aluminosilicate is used in water purification to remove dissolved calcium compounds.

Fluosilicate

Sodium fluosilicate, Na_2SiF_6, white solid slightly soluble; magnesium fluosilicate, $MgSiF_6$, white solid, soluble.

Sulfides

Silicon monosulfide, SiS, yellow solid, somewhat volatile, formed by heating to redness crystalline silicon in sulfur vapor, reactive with water;

silicon disulfide, SiS_2, white crystals, formed by heating amorphous silicon and sulfur, and then subliming, reactive with water.

Nitrides

Trisilicon tetranitride, Si_3N_4, by heating silicon oxide plus carbon to 1,500°C in a current of nitrogen gas.

Silicates

See also **Adhesives**.

Silicon-Silicon Double Bond

Of the chemical elements, Si is closest to carbon in terms of its chemical properties. Multiple bonds pervade carbon chemistry and thus it is no surprise that investigators, over a period of many years, have been seeking evidence of multiple bonding in silicon. As early as 1911, Kipping reported compounds exhibiting this bonding, but these substances were later shown to be polymers or cyclic oligomers. It was not until the 1960s that good evidence was reported for the existence of $Si = C$ (silene), $Si = Si$ (disilene), and $Si = O$ (silanone) compounds. The full reality of such compounds, however, was not reported until 1981. At that time, a silene and a disilene, each of which is stable at room temperature, were reported by two separate groups. Brook, et al. reported on a silene; West, et al. reported on a disilene. It has since been concluded that many disilenes can be prepared, including compounds that are unexpectedly stable. Molecules containing $Si = Si$ bonds can be synthesized by several routes. The key to stabilization of these compounds is to provide large substituents bonded to the Si atoms so that polymerization is blocked. It has been determined that disilenes react chemically by addition across the double bond, as do alkenes. Tetramesityldisilene, as reported by West, also undergoes a wide variety of addition reactions previously unestablished in organic chemistry. The result is several "new" and unusual types of molecules, the details of which are reported in the West paper listed under references.

Silicon-Mediated Organic Synthesis

As reported by Paquette (reference listed), since the late 1960s, organic chemists have used the chemical properties of tetracovalent silicon to achieve a variety of new synthetic transformations. Paquette (Ohio State University) summarizes, "In carbon-functional silanes, exceptional stabilization is provided to a carbocation center in the beta position when the carbon-silicon bond lies in plane. This phenomenon directs electrophilic attack to the silicon-substituted carbon in aryl-, vinyl-, and alkynylsilanes and to carbon-3 in allylsilanes. For different reasons, silicon also stabilizes a carbon-metal bond in the alpha position. Consequently, access to many silicon-containing organometallics is readily available. The exceptional strength of silicon-oxygen and silicon-fluorine bonds is yet another factor that controls the chemical reactivity of silicon reagents. In recent developments, preparative chemists have taken advantage of these properties in imaginative and useful ways." In the Paquette paper, these observations are developed in exceptional and illustrated detail.

Reactions of Elemental Silicon

In the late 1980s, E.A. Pugar and P.E.D. Morgan and a team of researchers at the Rockwell International Science Center, Thousand Oaks, California, conducted a thorough effort to understand "Low Temperature Direct Reactions Between Elemental Silicon and Liquid Ammonia or Amines for Ceramics and Chemical Intermediates." Details are given in reference cited.

Because of the important potential applications of silicon nitride, the use of low-cost starting materials, such as elemental silicon and liquid ammonia or amines, may be more effective than the existing chloride method. In earlier work, this process was found to form silicon di-imide $(Si(NH)_2)$, but required purification steps to remove chloride.

Pugar and Morgan elucidate their research and include a summary of the work of other researchers over the years. The report concludes: "Through the use of modern sensitive probes, direct elemental silicon reactions with liquid ammonia, silicon-hydrazine and silicon-organic amines have been discovered. The reaction of elemental silicon with nitrogen-containing reagents, under rather benign conditions, can produce ceramic precursors and with further chemical treatments can produce fibers, films, and other commercial and industrial products."

Nomenclature of Silicon Compounds

The name of the compound SiH_4 is *silane*. Compounds having the general formula $H_3Si \cdot [SiH_2]_n \cdot SiH_3$ are called disilane, trisilane, etc., according

to the number of silicon atoms present. Compounds of the general formula Si_nH_{2n+2} have the generic name *silanes*. Example: Trisilane, $H_3Si \cdot SiH_2 \cdot SiH_3$.

Compounds having the formula $H_3Si \cdot [NH \cdot SiH_2]_n \cdot NH \cdot SiH_3$ are called disilazane, trisilazane, etc., according to the number of silicon atoms present; they have the generic name *silazanes*. Example: Trisilazaine, $H_3Si \cdot NH \cdot SiH_2 \cdot NH \cdot SiH_3$.

Compounds having the formula $H_3Si \cdot [S \cdot SiH_2]_n \cdot S \cdot SiH_3$ are called disilthiane, trisilthiane, etc., according to the number of silicon atoms present; they have the generic name *silthianes*. Example: Trisilthiane, $H_3Si \cdot S \cdot SiH_2 \cdot S \cdot SiH_3$.

Compounds having the formula $H_3Si \cdot [O \cdot SiH_2]_nO \cdot SiH_3$ are called disiloxane, trisiloxane, etc., according to the number of silicon atoms present; they have the generic name *siloxanes*. Example: Trisiloxane, $H_3Si \cdot O \cdot SiH_2 \cdot O \cdot SiH_3$.

For designating the positions of substituents on compounds named as silanes, silazanes, silthianes, and siloxanes, each member of the fundamental chain is numbered from one terminal silicon atom to the other. When two or more possibilities for numbering occur, the same principles are followed as for carbon compounds. Examples:

1-Butyl-2,3-dichloro-2-pentyltrisilane
$Cl \cdot SiH_2 \cdot SiCl(C_5H_{11}) \cdot SiH_2 \cdot C_4H_9$
2-Methyl-3-pentyloxytrisilazane
$SiH_3 \cdot N(CH_3) \cdot SiH(OC_5H_{11}) \cdot$
$H \cdot SiH_3$
1-Methoxytrisiloxane
$CH_3O \cdot SiH_2 \cdot O \cdot SiH_2 \cdot O \cdot SiH_3$

The names of representative radicals containing silicon are shown below. These illustrate the principles on which any further radical names should be formed.

Silicon, hydrogen	
silyl	H_3Si-
silylene	$H_3Si=$
silylidyne	$HSi \equiv$
disilanyl	$H_3Si \cdot SiH_2-$
trisilanyl	$H_3Si \cdot SiH_2SiH_2-$
disilanylene	$-SiH_2 \cdot SiH_2-$
trisilanylene	$-SiH_2 \cdot SiH_2 \cdot SiH_2-$
cyclohexasilanyl	$\begin{array}{c} SiH_2 \cdot SiH_2 \cdot SiH_2 \\ \vert \qquad\qquad \vert \\ SiH_2 \cdot SiH_2 \cdot SiH_2 \end{array}$
Silicon, hydrogen, oxygen	
siloxy	$H_3Si \cdot O-$
disiloxanyl	$H_3Si \cdot O \cdot SiH_2-$
disilanoxy	$H_3Si \cdot SiH_2 \cdot O-$
disiloxanoxy	$H_3Si \cdot O \cdot SiH_2 \cdot O-$
Silicon, hydrogen, sulfur	
silylthio	$H_3Si \cdot S-$
disilanylthio	$H_3Si \cdot SiH_2 \cdot S-$
disilthianyl	$H_3Si \cdot S \cdot SiH_2-$
disilthianylthio	$H_3Si \cdot S \cdot SiH_2 \cdot S-$
Silicon, hydrogen, sulfur, oxygen	
disilthianoxy	$H_3Si \cdot S \cdot SiH_2 \cdot O-$
disiloxanylthio	$H_3Si \cdot O \cdot SiH_2 \cdot S-$
Silicon, hydrogen, nitrogen	
silylamino	$H_3Si \cdot NH-$
disilanylamino	$H_3Si \cdot SiH_2 \cdot NH-$
disilazanyl	$H_3Si \cdot NH \cdot SiH_2-$
disilazanylamino	$H_3Si \cdot NH \cdot SiH_2 \cdot NH-$
Silicon, hydrogen, nitrogen, oxygen	
disilazanoxy	$H_3Si \cdot NH \cdot SiH_2 \cdot O-$
disiloxanylamino	$H_3Si \cdot O \cdot SiH_2 \cdot NH-$

Compound radical names may be formed in the usual manner. Examples:

silyldisilanyl	$(H_2Si)_2SiH-$
disilyldisilanyl	$(H_3Si)_3Si-$
triphenylsilyl	$(C_6H_5)_3Si-$

Open-chain compounds which have the requirements for more than one of the structures already defined are named, if possible, in terms of silane, silazane, silthiane, or siloxane containing the largest number of silicon atoms. Examples:

3-Siloxytrisilthiane
$H_3Si \cdot S \cdot SiH \cdot SiH_3$

$| $
$O \cdot SiH_3$

1-Siloxy-3-(disilthianoxy)trisilthiane
$H_3Si \cdot S \cdot SiH \cdot S \cdot SiH_2 \cdot OSiH_3$

$|$
$O \cdot SiH_2 \cdot S \cdot SiH_3$

When there is a choice between two parent compounds possessing the same number of silicon atoms, the order of precedence is siloxanes, silthianes, silazanes, and silanes. Examples:

1-Silylthiodisiloxane
$SiH_3 \cdot O \cdot SiH_2 \cdot S \cdot SiH_3$
1-Silylaminodisilthiane
$SiH_3 \cdot S \cdot SiH_2 \cdot NH \cdot SiH_3$
1-Phenyl-3-silyldisiloxane
$SiH_3 \cdot SiH_2 \cdot O \cdot SiH_2 \cdot C_6H_5$

Cyclic silicon compounds having the formula $[SiH_2]_n$ are called cyclotrisilane, cyclotetrasilane, etc., according to the number of members in the ring; they have the generic name *cyclosilanes*. Example:

Cyclotrisilane $SiH_2 \cdot SiH_2 \cdot SiH_2$

Cyclic compounds having the formula $[SiH_2 \cdot NH]_n$ are called cyclodisilazane, cyclotrisilazane, etc., according to the number of silicon atoms in the ring. They have the generic name *cyclosilazanes*. Example:

Cyclotrisilazane
$HN \cdot SiH_2 \cdot NH \cdot SiH_2 \cdot NH \cdot$ SiH_2

Cyclic compounds having the formula $[SiH_2 \cdot S]_n$ have the generic name *cyclosilthianes* and are named similarly to the cyclosilazanes. Example:

Cyclotrisilthiane
$S \cdot SiH_2 \cdot S \cdot SiH_2 \cdot O \cdot$ SiH_2

Cyclic compounds having the formula $[SiH_2 \cdot O]_n$ have the generic name *cyclosiloxanes* and are named similarly to the cyclosilazanes. Example:

Cyclotrisiloxane
$O \cdot SiH_2 \cdot O \cdot SiH_2 \cdot O \cdot$ SiH_2

Cyclosilanes, cyclosilazanes, cyclosilthianes, and cyclosiloxanes are numbered in the same way as carbon compounds of similar nature. Examples:

2-Methoxycyclotrisilazane
$HN \cdot SiH_2 \cdot NH \cdot SiH_2 \cdot NH \cdot SiH \cdot OCH_3$

2-Methoxycyclotrisilthiane
$S \cdot SiH_2 \cdot S \cdot SiH_2 \cdot S \cdot SiH \cdot OCH_3$

2-Methoxycyclotrisiloxane
$O \cdot SiH_2 \cdot O \cdot SiH_2 \cdot O \cdot SiH \cdot OCH_3$

Polycyclic siloxanes (polycyclic compounds whose members consist entirely of alternating silicon and oxygen atoms) are named as bicyclosiloxanes, tricyclosiloxanes, etc., or as spirosiloxanes, and are numbered according to methods in use for carbon compounds of similar nature. Polycyclic

silthianes, silazanes, and silanes are treated similarly. Examples:

3,3,5,5,9,9-Hexamethyl-1,7-diphenylbicyclo[5,3,1]pentasiloxane

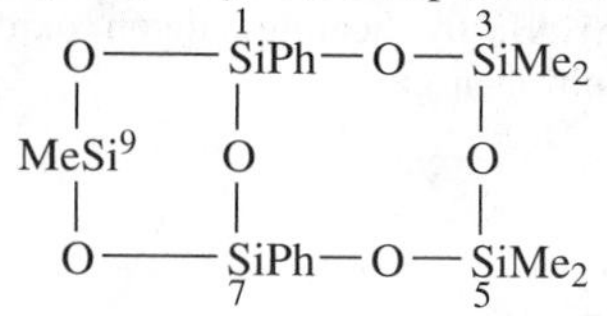

Tetramethyltricyclo[3,3,1,1]tetrasiloxane

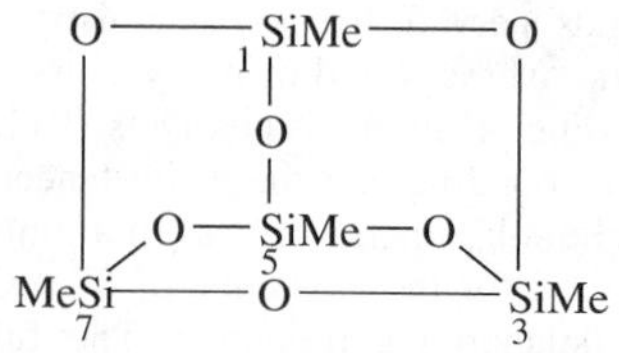

The names of compounds containing silicon atoms as heteromembers (with or without other heteromembers) but not classifiable as (linear or cyclic) silanes, silazanes, silthianes or siloxanes are derived with the aid of the oxa-aza convention. Examples:

2,2,4,4,6,6-Hexamethyl-2,4,6-trisilaheptane
$(CH_3)_3Si \cdot CH_2 \cdot Si(CH_3)_2 \cdot CH_2 \cdot Si(CH_3)_3$
2,4,6,8,-Tetraoxa-5-carbonsoilane
$SiH_3 \cdot O \cdot SiH_2 \cdot O \cdot CH_2 \cdot O \cdot SiH_2 \cdot O \cdot SiH_3$

Octaphenyloxacyclopentasilane

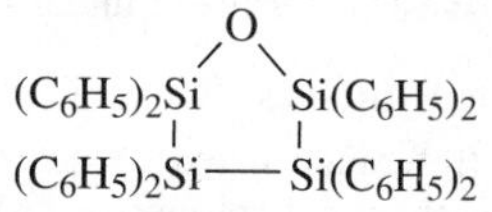

Hydroxy-derivatives in which the hydroxyl groups are attached to a silicon atom are named by adding the suffixes *ol, diol, triol*, etc., to the name of the parent compound. Examples:

Silanol	$H_3Si \cdot OH$
Silanediol	$H_2Si(OH)_2$
Silanetriol	$HSi(OH)_3$
Disilanehexaol	$(HO)_3Si \cdot Si(OH)_3$
Disiloxanol	$H_3Si \cdot O \cdot SiH_2 \cdot OH$
	$SiH_2 \cdot SiH_2 \cdot SiH \cdot OH$
Cyclohexasilanol	$\quad|\qquad\qquad\quad|$
	$SiH_2 \cdot SiH_2 \cdot SiH_2$

Polyhydroxy-derivatives in which hydroxyl group is attached to a silicon atom are named wherever possible in accordance with the principle of treating like things alike. Example:

1,13,5,5-Pentamethyltrisiloxane-1,3,5-triol

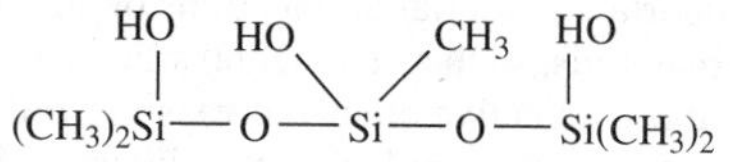

Otherwise they are named in accordance with the principle of the largest parent compound. Example:

2-Hydroxysilyltetrasilane-1,4-dio
$SiH_2 \cdot OH$

$|$
$HO \cdot SiH_2 \cdot SiH_2 \cdot SiH \cdot SiH_2 \cdot OH$

Substituents other than hydroxyl groups (functional atoms or groups and hydrocarbon radicals) attached to silicon are expressed by appropriate prefixes or suffixes. Examples:

Ethyldisilane

$CH_3 \cdot CH_2 \cdot SiH_2 \cdot SiH_3$

Hexachlorodisiloxane

$Cl_3Si \cdot O \cdot SiCl_3$

Dibutyldichlorosilane

$(CH_3 \cdot CH_2 \cdot CH_2 \cdot CH_2)_2SiCl_2$

Silylamine

$H_3Si \cdot NH_2$

Silanediamine

$H_2Si(NH_2)_2$

Silanetriamine

$HSi(NH_2)_3$

N-Methylsilylamine

$H_3Si \cdot NH \cdot CH_3$

N,N-Dimethylsilylamine

$H_3Si \cdot N(CH_3)_2$

N,N'-Dimethylsilanediamine

$H_2Si(NH \cdot CH_3)_2$

N,N',N''-Trimethylsilanetriamine

$HSi(NH \cdot CH_3)_3$

Acetoxytrimethylsilane

$(CH_3)_3Si \cdot O \cdot OC \cdot CH_3$

Diacetoxydimethylsilane

$(CH_3)_2Si(O \cdot OC \cdot CH_3)_2$

Compounds containing carbon as well as silicon and in which there is a "reactive group" in the carbon-containing portion of the molecule not shared by a silicon atom are named in terms of the organic parent compound wherever feasible. Examples:

α-Trimethylsilylacetanilide

$(CH_3)_3Si \cdot CH_2 \cdot NH \cdot C_6H_5$

1-Trichlorosilylethanol

$Cl_3Si \cdot CH(OH) \cdot CH_3$

2-Trimethylsilylethanol

$(CH_3)_3Si \cdot CH_2 \cdot CH_2OH$

(Hydroxydimethylsilyl)methanol

$(CH_3)_2Si \cdot CH_2OH$
$\qquad\quad |$
$\qquad\quad OH$

α-(Hydroxydimethylsilyl)acetanilide

$(CH_3)_2Si \cdot CH2 \cdot CO \cdot NH \cdot C_6H_3$
$\qquad\quad |$
$\qquad\quad OH$

(Silylmethyl)amine

$H_3Si \cdot CH_2 \cdot NH_2$

But by rules 70.16 and 70.17:

(Methoxymethyl)silanol

$CH_3O \cdot CH_2 \cdot SiH_2 \cdot OH$

N-Methylsilylamine

$H_3Si \cdot NH \cdot CH_3$

Compounds in which metals are combined directly with silicon are, in general, named as derivatives of the metal. Example:

(Triphenylsilyl)lithium

$(C_6H_5)_3SiLi$

However, in exceptional cases, the metal may be named as a substituent. Example:

Sodium p-(sodiosilyl)benzoate

$p-NaO_2C \cdot C_6H_4 \cdot SiH_2Na$

Metallic salts of hydroxy-derivatives may be named in the customary manner. Example:

Sodium salt of triphenylsilanol

$(C_6H_5)_3Si \cdot ONa$

ESSENTIALLY NONCHEMICAL PROPERTIES OF SILICON

Were it not for the firm establishment of silicon as an indispensable material for modern electronics, the other exceptionally attractive properties of Si may have been overlooked for many years. Over the past few decades, electronics components manufacturers have mastered the skills required for manufacturing microminiature components, and this experience has given Si a head start for use in other subminiature structures. Si has been recognized as an outstanding material for making micromachined subminiature structures essentially just within the past decade, and it has become one of the key materials in the comparatively new field of *nanotechnology*.

As a mechanical material, silicon is stronger than steel, it does not show mechanical hysteresis, and it is highly sensitive to stress. This combination of properties qualifies Si as an excellent sensor for detecting acceleration, pressure, force, and other variables encountered in processing and manufacturing. One method of measuring fluid flow, for example, traditionally has depended upon sensing pressure differentials, as in the case of an orifice-type flowmeter. Thus, silicon sensors can be used. Silicon accelerometers can employ the same piezoresistive sensing technique used in pressure sensors.

In addition to sensors, Si can be used at the subminiature scale for the production of tiny pipe, nozzles, and valves required by automatic control systems. Thus, the weight and bulk of future control systems may be reduced by several orders of magnitude.

As of the early 1990s, engineers are working at the "edge" of a new kind of robotics—that is, subminiature handling devices that can master the handling requirements of the new nanomanufacturing technology. Such robots would be miniature, fully integrated silicon systems drawing heavily on the technologies of silicon-integrated electronics and micromachining. Semi-intelligent robots could be used in many manufacturing and control tasks. According to some researchers, such robots could have intelligence at the lowest possible system level, thus allowing them to function semiautonomously, with occasional input from a central control system.

Biological Applications of Silicon Technology

H.M. McConnell (Stanford University) and a team of researchers have developed a silicon-based device called a cytosensor (microphysiometer) that can be used to detect and monitor the response of cells to a variety of chemical substances, particularly ligands for specific plasma membrane receptors. As pointed out by McConnell, "The microphysiometer measures the rate of proton excretion from 10^4 to 10^6 cells. The instruments serves two distinct functions. In terms of detecting specific molecules, selected biological cells in this instrument serve as detectors and amplifiers. The microphysiometer can also investigate cell function and biochemistry. A major application of this instrument may prove to be screening for new receptor ligands. In this respect, the instrument appears to offer significant advantages over other techniques." More detail is given in the McConnell reference listed.

Additional Reading

Amato, I.: "Shine On, Holey Silicon," *Science*, 922 (May 17, 1991).

Aufderhaar, H.C.: *Silicon*, in "Metals Handbook," 9th edition, Vol. 2, ASM International, Metals Park, OH, 1989.

Binnig, G. and H. Rohrer: "The Scanning Tunneling Microscope," *Sci. Amer.*, **253**(2), 50–56 (August 1985).

Boland, J.J. and G.N. Parsons: "Bond Selectivity in Silicon Film Growth," *Science*, 1304 (May 29, 1992).

Bryzek, J., Mallon, J.R., Jr., and R.H. Grace: "Silicon's Synthesis: Sensors to Systems," *Instrumentation technology*, 40 (January 1989).

Carter, G.F. and D.E. Paul: *Materials Science and Engineering*, ASM International, Materials Park, OH, 1991.

Chabal, Y.J.: *Fundamental Aspects of Silicon Oxidation*, Springer-Verlag Inc., New York, NY, 2001.

Connally, J.A. and S.B. Brown: "Slow Crack Growth in Single-Crystal Silicon," *Science*, 1537 (June 12, 1992).

Corcoran, E.: "Holey Silicon," *Sci. Amer.*, 102 (March 1992).

Dunn, W.: "Micromachined Sensors for Automotive Applications," *Sensors*, 54 (September 1991).

Feng, Z.C. and R. Tsu: *Porous Silicon*, World Scientific Publishing Company, Inc., Riveredge, NJ, 1994.

Golovchenko, J.A.: "The Tunneling Microscope: A New Look at the Atomic World," *Science*, **232**, 48–53 (1986).

Greenwood, N.N. and A. Earnshaw: *Chemistry of the Elements*, 2nd Edition, Butterworth-Heinemann, Woburn, MA, 1997.

Heath, J.R.: "A Liquid-Solution-Phase Synthesis of Crystalline Silicon," *Science*, 1131 (November 13, 1992).

Henkel, S.: "Silicon Microvalves Fabricated on Bimetallic Diaphragms," *Sensors,* 4 (December 1991).

Iyer, S.S. and Y-H Xie: "Light Emission from Silicon," *Science,* 40 (April 2, 1993).

Jackson, K.: *Silicon Devices: Structures and Processing,* John Wiley & Sons, Inc., New York, NY, 1998.

Jennings, H.M. and M.H. Richman: *Science,* **193,** 1242 (1976).

Kasper, J.S. and R.H. Wentorf, Jr.: "Hexagonal (Wurtzite) Silicon," *Science,* **197,** 599 (1977).

Laine, R.M.: "Beach Sand: Material of the Future?" *Advanced Materials & Processes,* 6 (February 1992).

LeComber, P.G.: "Amorphous Silicon—Electronics Into the 21st Century," University of Wales Review, 31 (Spring 1988).

Lide, D.R.: *CRC Handbook of Chemistry and Physics,* 84th Edition, CRC Press, LLC., Boca Raton, FL, 2003.

Link, B.: "Field-Qualified Silicon Accelerometers," *Sensors,* 28 (March 1993).

Maugh, T.H., II: "A New Route to Intermetallics (Metal Silicides)," *Science,* **225,** 403 (1984).

McConnell et al.: "The Cytosensor Microphysiometer: Biological Applications of Silicon Technology," *Science,* 1906 (September 25, 1992).

Meyers, R.A.: *Handbook of Chemicals Production Processes,* McGraw-Hill Companies, Inc., New York, NY, 1986.

Nalwa, H.S.: *Silicon-Based Materials and Devices,* Academic Press, Inc., San Diego, CA, 2001.

Paquette, L.A.: "Silicon-Mediated Organic Synthesis," *Science,* **217,** 793–800 (1982).

Pensl, G. and H. Matsunami: *Silicon Carbide: A Review of Fundamental Questions and Applications to Current Device Technology,* John Wiley & Sons, Inc., New York, NY, 1997.

Pugar, E.A. and P.E.D. Morgan: "Low Temperature Direct Reactions Between Elemental Silicon and Liquid Ammonia or Amines for Ceramics and Chemical Intermediates," in Report issued by Rockwell International Science Center, Thousand Oaks, California (September 1988).

Rappoport, Z. and Y. Apeloig: *The Chemistry of Organic Silicon Compounds,* John Wiley & Sons, Inc., New York, NY, 2001.

Robinson, A.L.: "Consensus on Silicon Surface Structure Near," *Science,* **232,** 451–453 (1986).

Schubert, U.: *Silicon Chemistry,* Springer-Verlag Inc., New York, NY, 1999.

Simpson, T.L. and B.E. Volcani: *Silicon and Siliceous Structures in Biological Systems,* Springer-Verlag Inc., New York, NY, 1981.

Staff: "Grace with Pressure (Silicon)," *Sci. Amer.,* **253**(2), 62–64 (August 1985).

Staff: *ASM Handbook—Properties and Selection: Nonferrous Alloys and Pure Metals,* ASM International, Materials Park, OH, 1990.

Staff: "Silicon Atoms 'See the Light'," *Advanced Materials & Processes,* 6 (November 1990).

Staff: "Tough MoSi₂ Composites also Combat Oxidation," *Advanced Materials & Processes,* 26 (January 1991).

Strausser, Y.E.: *Characterization in Silicon Processing,* Butterworth-Heinemann, Inc., Woburn, MA, 1993.

Street, R.A.: *Technology and Applications of Amorphous Silicon,* Springer-Verlag Inc., New York, NY, 2001.

Tanaka, K. and H. Okamoto: *Amorphous Silicon,* John Wiley & Sons, Inc., New York, NY, 1999.

Travis, J.: "Building a Silicon Surface, Atom by Atom," *Science,* 1354 (March 13, 1992).

Wentorf, R.H., Jr. and J.S. Kasper: *Science,* **139,** 338 (1963).

West, R.: "Isolable Compounds Containing a Silicon-Silicon Double Bond," *Science,* **225,** 1109–1114 (1984).

Yeganeh-Haeri, A., Weidner, D.J., and J.B. Parise: "Elasticity of Alpha-Cristobalite: A Silicon Dioxide with a Negative Poisson's Ratio," *Science,* 650 (July 31, 1992).

Yun, W. and R.T. Howe: "Recent Developments in Silicon Micro-accelerometers," *Sensors,* 31 (October 1992).

Yun, W. and R.T. Howe: "Sigma-Delta Modulator Interfacing with Silicon Micro sensors," *Sensors,* 11 (May 1993).

Zdebick, M.: "A Revolutionary Actuator for Microstructures," *Sensors,* 26 (February 1993).

SILICON CHIP (Fabrication).   See **Semiconductors**.

SILICONE RESINS.

The chemistry of the silicones is based on the hydrides, or silanes, the halides, the esters, and the alkyls or aryls. The silicon oxides are composed of networks of alternate atoms of silicon and oxygen so arranged that each silicon atom is surrounded by four oxygen atoms and each oxygen atom is attached to two independent silicon atoms:

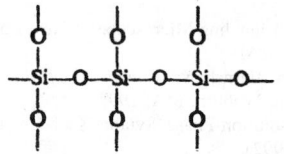

Such a network can be described as a series of spiral silicon-oxygen chains crosslinked with each other by oxygen bonds. If some of the oxygen atoms are replaced with organic substituents, a linear polymer will result:

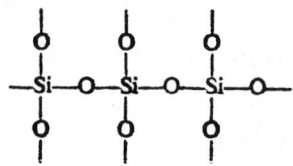

Taking into consideration the stability of structures involving C−Si bonds, it is evident that the basic chain itself must be comparable in its stability to that of silica and the silicate minerals, and if the R substituents contain no carbon-to-carbon bonds, as for example with methyl groups, the combination should have excellent thermal stability and chemical resistance.

Among the most efficient of silicone monomers in early use as building blocks in the preparation of silicone resins are the halogen alkyl or aryl silanes. Compounds of the type represented by the formula R_2SiCl_2 are capable of undergoing hydrolysis to form long-chain polymers of varying consistencies and viscosities with a predetermined number of molecules of R_3SiCl as chain stoppers. If cross links are desired, tri-functional compounds such as $RSiCl_3$ can be used.

Many silicone resins have been prepared through the use of silazine monomers, that is, compounds with amino groups attached directly to silicon. Di-2-pyridyldichlorosilane has also been described as an intermediate in the preparation of oils, emulsifying agents and resins. Fluorinated aromatic rings are found in many silicone resins. The esterification of dimethylbis-(*p*-carboxylatophenyl)-silane with glycol or glycerol yields thermoplastic materials. Trimethyl-*p*-hydroxyphenylsilane is acceptable as a monomer in resin formation when compounded with hexamethylenetetramine. Trichlorosilane is often a constituent of cohydrolysis monomeric mixtures.

Emulsifying agents have been prepared from quaternary ammonium salts with silicon in the cation. There is a large number of alkyd-silicone resins.

Water-soluble metal salts of alkyltrisilanols are efficient in the reduction of surface tension. A silicone putty is made by compounding a benzene-soluble silicone polymer with silica powder and an inorganic filler.

Chlorinated alkyl or aryl groups are often found in polysiloxane resins. One of the chief advantages of this type of halogenated product lies in its reduced tendency to burn. Whereas diphenyloxosilane polymer burns readily in a flame, the introduction of one, two or three chlorines in each benzene ring progressively reduces this tendency. Physically, these resins appear in many different forms, from horny through sticky-resinous to rubber-like, depending on the conditions of combination and composition.

Some of the most important types of silicone resins, useful as coating preparations, are analogous to the alkyds. Glycerol, for instance, is allowed to react with trialkylethoxysilanes, by which reaction one or more hydroxyls of the glycerol are replaced by R_3SiO. Polysiloxane resins with terminal ethoxyl groups can be used. Synthesis of silicone resins with terminal halogen or hydroxyl is also possible, though generally the halogen disappears in hydrolysis of further processing.

Ethyl silicate (tetraethoxysilane) is often used without modification as a water-repellent material for concrete and masonry in general. All, or nearly all, the ethoxyl groups are hydrolyzed by the moisture of the air to form cross-linked water-repellent polymers. The material is applied in desirable thickness, dissolved in some volatile solvent which soon evaporates. Silicone resins which are partially condensed before application, or even fully condensed, can also be used here. In the latter case, hardness is achieved on evaporation of the solvent. Certain silicone resins are useful as hydrophobic agents for the impregnation of paper and fabrics.

The simplest silicone resins are formed by the almost simultaneous hydrolysis and condensation (by dehydration) of various mixtures of methylchlorosilanes. Ice water often suffices for the first step, but advanced condensation to resinous materials of satisfactory thermosetting properties generally comes about on heating. As far as solvents are concerned, water alone has its disadvantages in that the organic materials are so slightly soluble therein. Mixed solvents are commonly used, generally water with such compounds as dioxane, one of the amyl alcohols, dibutyl ether or even an aromatic hydrocarbon. Warm water hydrolysis of di-*t*-butyldiaminosilane forms noncrystallizable liquids or resinous products, and this resinification can be controlled. Among catalysts for these condensations may be found ferric chloride, the hydroxyl ion, triethyl borate, stannic chloride and sulfuric acid.

A water-methylene dichloride mixture is satisfactory as a hydrolyzing agent on groups of compounds such as phenyltrichlorosilane, dimethyldichlorosilane and methyltrichlorosilane. The value of higher-boiling ethers lies in their ability to provide higher-boiling reaction systems.

Ferric chloride is sometimes an important constituent of the hydrolyzing mixture. The formation of gels during polymerization can be controlled. Patents are in existence covering the hydrolysis of chlorosilanes by pouring their solutions onto the surface of a swirling solution of the active electrolyte.

Alkaline hydrolysis of dialkyldialkoxysilanes can sometimes be used for the purpose of preparing silicone resins.

Triethyl orthoborate affects polymerization by dehydration, probably reacting with the water which it abstracts, to form boric acid and alcohol. This principle is commonly used. Antimony pentachloride is used on occasion and also sulfuric acid, but there is always danger that the latter will split off an alkyl or an aryl group. Sulfuric acid also sometimes induces equilibration. This *tendency* on the part of the acid *reacts* sometimes with the opposite effect.

Some polysiloxanes are curable with lead monoxide, with a consequent reduction in both curing time and temperature. High-frequency electrical energy vulcanizes in one case at least. Zirconium naphthenate imparts improved resistance to high temperatures. Barium salts are said to prevent "blooming." Sulfur dichloride is also used. Some resins are solidified by pressure vulcanization, using di-*t*-butyl peroxide. Improvements are to be found in lower condensation temperatures and shorter times of treatment.

Viscosity is often regulated by bubbling air through solutions of polysiloxanes or the liquid material itself. In this manner, alkyl side chains are oxidized and oxygen bridges set up between silicon atoms. Obviously, the greater the number of such cross links, other influences constant, the greater will be the viscosity.

Addition of glycerol, phthalic anhydride and "butylated melamine formaldehyde resins" is sometimes found to improve the thermosetting properties of silicone resins. Methylsilyl triacetate has the same effect in certain cases. Some silicone resins can be advantageously modified by the addition of polyvinyl acetyl resins or nitroparaffins.

A solution of cellulose nitrate in butyl acetate, diluted with toluene, can be plasticized with dibutyl phthalate and tetraethoxysilane. After application to glass, the lacquer cannot be stripped from the surface even after soaking in boiling water. It has been stated, however, that tetraethoxysilane sometimes decreases the tensile strength of a lacquer in spite of its effect in increasing adhesive properties. A uniformly lustrous appearance is imparted to compounds containing plasticized ethylcellulose, cellulose fibers and pigment, by applying liquid polymerized alkyl polysiloxanes with an average radical/silicon ratio of between 1185 and 2.20. Several resinous materials are known which contain sulfur connected to carbon but not directly to silicon.

Inorganic fillers include titanium dioxide, "Celite" and zinc oxide. Lithium or lead salts of acetic acid, stearic acid or phenol are sometimes used as fillers. Silica and alumina are also feasible. Trimethyl-β-hydroxyethylammonium bicarbonate has been used as a curing agent.

There are any number of review articles and patents covering the increase in serviceability of paints and varnishes which are admixed with silicone resins. Products suitable for use as plasticizers, paint vehicles, etc., are sometimes prepared from mixtures which include phthalic anhydride. Silicone paints, in general, show high adhesion and permit greater retention of color and tint. Of special importance here is the absence of color in the resin and the freedom from discoloration during baking and curing. At high temperatures, silicone varnishes show much higher electrical resistance than others. Paints admixed with silicone resins generally show increased resistance to alkalies and to the elevated temperatures involved in baking processes.

Treating spinnarets with silicon resins eliminates much of the plugging. Aluminum alkoxides are successful as hardening agents.

Copolymers of adipic acid, glycerol, and 1,3-diacetoxy-methyltetra-methyldisiloxane, or similar compound, are sometimes used. The silicon adaptation of the alkyd resin possesses, in general, increased hardness and flexibility. In addition, there is greater stability at higher temperatures. Copolymers can also be prepared using amyldibutoxyboron. Antiknock properties are claimed for this type of product as well as increased heat resistance.

The most important development in the chemistry of silicone resins embodies the preparation of monomeric silanes with at least one alkenyl group attached to silicon. Hydrolyzable groups are also present, so that polymerization can take place in two ways—by the conventional hydrolytic processes followed by condensation and by addition polymerization on the double bond, usually through the catalytic activity of benzoyl peroxide or similar agent. Some of the products find use as textile finishes, lubricating oils, additives or molding compounds. Vinyl and allyl groups are most common, with an occasional methallyl.

Triethylsilyl acrylate can be induced to undergo hydrolysis of the ethoxyl radicals to a desired extent forming linear or cross linked polymers. Addition polymerization will also take place on the double bond of the acrylate radical. More stable monomers result from the use of allyl or vinyl groups instead of acrylates. The latter contain a silicon-oxygen-carbon linkage which is always more or less susceptible to hydrolysis.

Other copolymers of this type with vinyl acetate or vinyl butyral resins have been found satisfactory for use in the lamination of wood, glass and metals. Glass-resin adhesives are also known.

General resistance to external influences constitutes the most outstanding property of silicone resins. Ultimate failure and breakdown is probably attributable, more than do anything else, to eventual oxidation of the radicals.

Among the more recent developments in this field may be mentioned the use of certain silicone resins, particularly those containing vinyl groups, as adhesives. Others find value as woodsealing products. A silicone-glycol copolymer has been reported with curing properties, and alkyd resins are now modified with silicones. Combination epoxide-silicone resins have been investigated. A harder type of silicone resins sometimes results from the processing of monomers containing H or olefinic group attached to silicon. Dental impression materials are coming more to the front from this source as well. Finally, more attention is being devoted to studies of the relation between chemical structure and thermodynamic properties.

Resistance to γ-radiation is especially valuable. "Mouth-tissue-simulating molding compositions" for dentures are on the market. Optical lenses prepared from silicone resins are not affected by hot climates. Mold release agents are still the subject of research, as are resins with long storage capabilities. Coating bananas with silicone resins reduces the possibility of bruising. Heat-transfer compositions must have high molecular weights. In oil wells, a new use has come to the fore in the prevention of sand flow. Transparent resins are still in demand, some of which are porous.

Compounding silicone resins with phenyl formaldehyde or melamine polymers seems to produce good results. Sulfur in the resin makes a product suitable for use as fuel hose, gaskets and gas tanks. Phosphorus is found in silicone resin coatings but the product is liable to be toxic. Titanium is found in waterproof coatings. Tin compositions have low toxicity. Some resins of high viscosity contain Si—O—N units. Crosslinking, however, is not recommended. The tendency to become brittle is ever present although a few instances are known to be contrary.

HOWARD W. POST
Williamsville, New York

SILLIMANITE. The mineral sillimanite is an aluminum silicate, the formula Al_2SiO_5 being like that of andalusite and kyanite. It is orthorhombic, usually in slender prisms, but may be fibrous or massive. Its hardness is 6.5–7.5; specific gravity, 3.23–3.27; luster, vitreous to silky; color, various shades of gray, grayish-green, and grayish-brown; transparent to translucent. It occurs in granites and gneisses as tiny prisms and aggregates, and is often associated with andalusite, cordierite and corundum. Sillimanite has been found in Bavaria, the Czech Republic and Slovakia, France, India, the Malagasy Republic, Myanmar and Ceylon, the latter two localities furnishing transparent sapphire-blue gem stones. In the United States sillimanite has been found in Connecticut, New York, Pennsylvania, Delaware, North Carolina, and in California, where in Inyo County is the largest deposit in the world. This mineral was named in honor of Benjamin Silliman for many years professor of chemistry and natural science at Yale University. Sillimanite is used in the manufacture of spark plug "porcelains" and laboratory ware.

SILOXANES.   See **Silicon.**

SILTHIANES.   See **Silicon.**

SILVER. [CAS: 7440-22-4]. Chemical element, symbol Ag (from Latin *argentum*), at. no. 47, at. wt. 107.868 ± 0.003, periodic table group 11, mp 961.93°C, bp approximately 2,212°C, density 10.50 g/cm³ (20°C). Elemental silver has a face-centered cubic crystal structure.

Silver is a white metal, softer than copper and harder than gold. When molten, silver is luminescent and occludes oxygen, but the oxygen is released upon solidification. As a conductor of heat and electricity, silver is superior to all other metals. Silver is soluble in HNO_3 containing a trace of nitrate; soluble in hot 80% H_2SO_4; insoluble in HCl or acetic acid; tarnished by H_2S, soluble sulfides and many sulfur-containing organic substances (e.g., proteins); not affected by air or H_2O at ordinary temperatures, but at 200°C, a slight film of silver oxide is formed; not affected by alkalis, either in solution or fused. There are two stable, naturally occurring isotopes, ^{107}Ag and ^{109}Ag. In addition, there are reported to be 25 less stable isotopes, ranging in half-life from 5 seconds to 253 days.

In terms of cosmic abundance, the estimate of Harold C. Urey (1952), using silicon as a base with a figure of 10,000, silver was assigned an abundance figure of 0.023. In terms of abundance in sea water, silver is ranked number 43 among the elements, with an estimated content of 1.5 tons per cubic mile (0.324 metric ton per cubic kilometer) of sea water.

Electronic configuration is $1s^2 2s^2 2p^6 3s^2 3p^6 3d^{10} 4s^2 4p^6 4d^{10} 5s^1$. First ionization potential is 7.574 eV; second, 21.4 eV; third, 35.9 eV. Oxidation potentials: $Ag \longrightarrow Ag^+ + e^-$, $E^0 = -0.7995$ V; $Ag^+ \longrightarrow Ag^{2+} + e^-$, -1.98 V; $2Ag^+ OH^- \longrightarrow Ag_2O + H_2O + 2e^-$, -0.344 V; $Ag_2O + 2OH^- \longrightarrow 2AgO + H_2O + 2e^-$, -0.57 V; $2AgO + 2OH^- \longrightarrow Ag_2O_3 + H_2O + 2e^-$, -0.74 V. Other important physical properties of silver are given under **Chemical Elements**.

Occurrence and Processing

Silver is widely distributed throughout the world. It rarely occurs in native form, but is found in ore bodies as silver chloride, or more frequently, as simple and complex sulfides. In former years, simple silver and gold-silver ores were processed by amalgamation or cyanidation processes. The availability of ores amenable to treatment by these means has declined. Most silver is now obtained as a byproduct or coproduct from base metal ores, particularly those of copper, lead, and zinc. Although these ores are different in mineral complexity and grade, processing is similar.

All the ores are concentrated in complex mills by selective froth flotation to produce individual copper, zinc, lead, and, infrequently, silver concentrates. The copper and lead concentrates are smelted to produce lead and copper bullions from which silver is recovered by electrolytic or fine refining. The silver bearing zinc concentrates are commonly processed by leaching and electrolytic methods. Silver is ultimately recovered as a byproduct from zinc plant residues. Canada is a leading silver mining country. Other important sources of silver are Mexico, the United States, Peru, the former U.S.S.R., and Australia. See also **Mineralogy**.

A substantial portion of the total world silver supply is obtained from recycled scrap. Much of this scrap comes from photographic film, jewelry and the electrical field. The high value of the scrap dictates accurate sampling and careful feed preparation. Efficient and fast processing is required to minimize metal losses and a tie-up of high-value materials. The highly complex nature of plant feed, with respect to physical form, chemical composition, and grade, requires use of complex and highly flexible processing procedures.

Uses of Silver

Silver in the twentieth century can be classified an industrial commodity. For most of the 19th century, silver was a monetary metal. Industrial consumption of silver is principally in photographic film, electrical contacts, batteries and brazing alloys. Sterling silver and silver plated copper alloys are used extensively for tableware and for jewelry and other decorative art. Recently, the field of commemorative and collector arts has become a substantial market for silver alloys, particularly sterling silver.

The predominant place of silver salts as photographic receptors is not the result of any unusual primary sensitivity to illumination, but is due to the fact that they undergo an unusual secondary amplification process called "development." Silver salts, like the salts of many metals, when immersed in solutions of many reducing agents, are changed to metallic silver. The photographic system depends upon the fact that when certain mild reducing agents (called "photographic developers") are chosen, the rate of reduction is increased many fold if the silver salt crystals carry very small amounts of metallic silver at the developer-crystal interface. The effect produced by the original light exposure is amplified in the development process by a factor of 100 billion. Whereas new photographic or recording devices are being developed not involving silver, none yet approach the packing density of a fine-grained image possible using silver. Thus, it appears that silver will be used in photographic recording for many years to come.

Among the electrical uses for silver are electrical contacts, printed circuits, and batteries. By far, the primary use is in electrical contacts where the high electrical and thermal conductivities, as well as corrosion and oxidation resistances, of silver are major reasons for its selection. Although silver has a strong tendency to weld under heavy currents, this is counteracted by alloying or by adding nonmetallic substances (such as cadmium oxide) to the silver matrix. The use of silver-cadmium oxide and silver-tungsten materials in electrical contact applications is widespread. The alloys used to improve the wear resistance and to reduce the sticking tendency of silver include silver-gold, silver-copper, silver-palladium, and silver-platinum. More complex alloys include silver-copper-nickel, silver-magnesium-nickel, silver-gold-cadmium-copper, and silver-cadmium-copper-nickel. Silver-cadmium oxide alloys are unique materials and are prepared either by combining silver and cadmium oxide by powder metal techniques or by the internal oxidation of a silver-cadmium alloy. Electrical alloys, which are impossible to combine by conventional melting, lend themselves to powder metal fabrication. Such composite structures as silver-graphite, silver-iron, and silver-tungsten are good examples of these types of materials.

In silver batteries, the silver oxide-zinc secondary battery has found its place in applications where energy delivered per unit of weight and space is of prime importance. The major disadvantages lie in their high cost and relatively short life. Consequently, a large part of the silver battery market is concerned with defense and space components. See also **Batteries**.

Prior to World War II, consumption of silver in silverware and jewelry was the largest industrial use of silver. Competition from stainless steel in flatware and holloware has contributed to a decline in overall use. Most consumption of silver in silverware and jewelry is in the form of sterling silver, an alloy of silver with approximately 7.5 weight percent copper. Silver plate, which is silver electroplated on a base metal, varies widely in specification. The thickness, expressed for example in penny-weights of pure silver per gross of teaspoons can range from a low of 1 to as high as 200.

In the 1920s and 1930s, low-temperature silver-copper brazing alloys were found to be useful on copper and its alloys and iron and its alloys (including stainless steel). Silver and copper form a simple eutectic system with limited solid solubility. This system can absorb elements such as zinc, cadmium, tin, and indium. These additions lower its melting temperature. It also can absorb higher melting elements such as nickel or palladium. These raise its melting temperature, but may improve its wetting characteristics, corrosion resistance, and strength at elevated temperatures. Silver solders or brazing alloys have the ability of making joints far stronger and more durable than common soft-solder (such as lead-tin) alloys. They are used in most refrigeration systems to join copper tubing. Also, extensive use is found in the assembly of automotive parts, military components, aircraft assemblies, and other hard goods manufacture. The nominal composition of a popular brazing alloy, ASTM Classification BAg-1 is silver 45%, copper-15%, zinc-16%, cadmium-24%.

One silver alloy containing about 70% silver, 26% tin, 3% copper, and 1% zinc is unique in that it is used extensively by dentists in combination with mercury to fill cavities in teeth. The "amalgam" manufacturers supply dentists with the alloy in the form of powder (filed, or more recently, atomized). This is mixed with mercury, using from 8 to 5 parts of mercury to 5 of alloy, and the cavity is packed. In the cavity, a metallurgical reaction takes place in which the silver-tin compound in the alloy becomes a durable silver-tin-mercury compound.

Silver, its oxides, halides and other salts play important roles in chemistry. Silver is an excellent catalyst in oxidizing reactions such as in the production of formaldehyde from methanol and oxygen, ethylene oxide from ethylene and oxygen, and glyoxal from ethylene glycol and oxygen. Silver has oligodynamic properties, that is, the ability of minute amounts of silver in solution to kill bacteria. Modern technology has made use of this property in various ways, mainly as a means of purifying water.

Small amounts of silver are used annually in such diverse applications as a backing for mirrors, and in control rods for pressurized water nuclear reactors. Miscellaneous uses like this account for only a small fraction of total silver consumption.

Chemistry of Silver

Silver(I) oxide, [CAS: 20667-12-3], Ag_2O, is made by action of oxygen under pressure on silver at $300°C$, or by precipitation of a silver salt with carbonate-free alkali metal hydroxide; it is covalent, each silver atom (in solid Ag_2O) having two collinear bonds and each oxygen atom four tetrahedral ones; two such interpenetrating lattices constitute the structure. Silver(I) oxide is the normal oxide of silver. Silver(II) oxide, AgO, is formed when ozone reacts with silver, and thus was once considered to be a peroxide. Silver(III) oxide, Ag_2O_3, has been obtained in impure state by anodic oxidation of silver.

All of the silver(I) halides of the four common halogens are well known. The fluoride may be prepared from the elements, the chloride by action of hydrogen chloride gas at $150°C$, upon silver, and the bromide and iodide by ionic reactions in solution. The chloride, bromide, and iodide are essentially insoluble in H_2O, but the fluoride is soluble. There is also a subfluoride, Ag_2F, which may be prepared as a cathodic deposit by electrolysis of silver(I) fluoride AgF, or by evaporation of finely divided silver with silver(I) fluoride in dilute hydrofluoric acid. It is an anisotropically conducting solid and is considered to be made up in the solid state of two silver layers, metallic-bonded to each other, and ionic-covalent bonded to a single fluorine layer. It has reverse cadmium iodide structure. Silver subchloride, Ag_2Cl is made by reaction of Ag_2F and phosphorus trichloride. Silver(II) fluoride, AgF_2, made by action of fluorine upon a silver(I) halide, is a fluorinating agent or catalyst for fluorinations. The silver(I) halides vary markedly in ionicity, the values given by Pauling being AgF 70%, AgCl 30%, AgBr 23% and AgI 11%. This is reflected in their crystal structures and in their solubility in water (or rather, their relative insolubility). The first three have sodium chloride structure, AgI has wurtzite structure; AgF has a molal solubility of 14, and the pK_{sp} values of the others are 9.75, 12.27 and 16.08, respectively.

Silver differs markedly from copper in forming few oxy compounds. One of these is silver oxynitrate or silver(II, III) nitrate which has the empirical formula $AgO_{1.148}(NO_3)_{0.453}$, in which the average oxidation number of silver is 2.448. It is prepared by action of fluorine upon aqueous silver nitrate or is obtained as an anodic deposit by electrolysis of silver nitrate in dilute HNO_3.

Silver enters into complex formation with many ions and molecules. With halogens, the silver complexes are fewer than the copper ones. Silver chloride dissolves in HCl with the formation of such chloroargentate ions as $(AgCl_2)^-$, $(AgCl_3)^{2-}$, and possibly $(AgCl_4)^{3-}$. Complex ions with bromide, $(AgBr_2)^-$ and $(AgBr_3)^{2-}$ are more stable, as are those with iodide, than those with chloride. Complexes of the type Ag_2Cl^+, Ag_3Cl^{2+}, Ag_2Br^+, Ag_3Br^{2+}, Ag_4Br^{3+}, $Ag_2Br_6^{2-}$, Ag_2I^+, Ag_3I^{2+}, Ag_4I^{3+}, $Ag_2I_6^{4-}$, $Ag_2I_7^{5-}$ and $Ag_3I_8^{5-}$ are also known. With ammonia the ions $(Ag(NH_3)_2)^+$ and $(Ag(NH_3)_3)^+$ are definitely known and others may exist. Similar complexes are formed with amines and diamines. With cyanides, silver forms very stable complexes, the number of CN−ions in the complex depending somewhat upon the excess of cyanide, so that $(Ag(CN)_2)^-$, $(Ag(CN)_3)^{2-}$, and $(Ag(CN)_4)^{3-}$ are definitely known. With thiosulfates, silver forms various complexes. In dilute solution, $(Ag_2(S_2O_3)_2)^{2-}$ exists, while in high concentration of $S_2O_3^{2-}$ ion, the complex $(Ag_2(S_2O_3)_6)^{10-}$ has been identified. In HNO_3 solution Ag^+ is easily oxidized to Ag^{2+} by peroxydisulfate. From this solution complex compounds of dipositive silver can be prepared, which are stable because coordination radically alters the oxidation potential of Ag(I) to Ag(II). They include pyridine complexes such as $(Ag(py)_4) \times (NO_3)_2$. 8-Hydroxyquinoline complexes containing the ions $(Ag(oxin)_2)^{2+}$, and o-phenanthroline complexes containing the ion $(Ag(o-phen)_2)^{2+}$. Silver(III) is known in the square, planar complex AgF_4^-, which has been prepared as $KAgF_4$ by direct fluorination of a mixture of potassium chloride and silver chloride. Silver(III), like Cu(III), also occurs in tellurate and periodate complexes.

Other silver compounds include: Silver chromate [CAS: 7784-01-2] (Ag_2CrO_4), yellow to red to brown precipitate by reaction of silver nitrate solution and potassium chromate solution.

Silver dichromate $(Ag_2Cr_2O_7)$, red precipitate by reaction of silver nitrate solution and potassium dichromate solution, changing to silver chromate upon boiling with H_2O.

Silver phosphate (Ag_3PO_4), [CAS: 7784-09-0] yellow precipitate, by reaction of silver nitrate solution and disodium hydrogen phosphate solution, soluble in HNO_3 and in NH_4OH, turns dark on exposure to light.

Silver sulfate [CAS: 10294-26-5] (Ag_2SO_4), white precipitate, by the action of silver nitrate solution and potassium sodium or ammonium sulfate solution or H_2SO_4, mp of silver sulfate $652°C$.

Silver sulfide [CAS: 21548-73-2] (Ag_2S), black precipitate, by the reaction of silver nitrate solution and hydrogen sulfide.

Silver forms several compounds or complexes with proteins by the action of silver oxide with gelatin in alkali solution, or with albumin, or by suspension in casein solution and by other methods. Such silver-protein complexes containing from 19 to 23% of silver are known as "mild silver protein" and are used as antiseptic solutions. They are readily soluble in H_2O.

DONALD A. CORRIGAN
Handy & Harman
Fairfield, Connecticut

Additional Reading

Carapella, S.C., Jr. and D.A. Corrigan: *Properties of Pure Silver,* Metals Handbook, 9th Edition, Vol. 2, ASM International, Metals Park, OH, 1979.

Coxe, C.D., McDonald, A.S., and G.H. Sistare, Jr.: *Properties of Silver and Silver Alloys,* Metals Handbook, 9th Edition, Vol. 2, ASM International, Metals Park, OH, 1979.

Coxe, C.D., McDonald, A.S., and G.H. Sistare, Jr.: *Silver-Base Brazing Alloys,* Metals Handbook, 9th Edition, Vol. 2, ASM International, Metals Park, OH, 1979.

Davis, J.R.: *Metals Handbook,* 2nd Edition, ASM International, Metals Park, OH, 1998.

Friend, W.Z.: *Corrosion Resistance of Precious Metals,* Metals Handbook, 9th Edition, Vol. 2, ASM International, Metals Park, OH, 1979.

Gale, N.H. and Z. Stos-Gale: "Lead and Silver in the Ancient Aegean," *Sci. Amer.,* 176−192 (June 1981).

Greener, E.H.: *Dental Materials,* Encyclopedia of Materials Science and Engineering, MIT Press, Cambridge, MA, 1986.

Greenwood, N.N. and A. Earnshaw: *Chemistry of the Elements,* 2nd Edition, Butterworth-Heinemann, Inc., Woburn, MA, 1997.

Lechtman, H.: "Pre-Columbian Surface Metallurgy," *Sci. Amer.,* 56−53 (June 1984).

Lide, D.R.: *CRC Handbook of Chemistry and Physics,* 84th Edition, CRC Press, LLC., Boca Raton, FL, 2003.

Parker, P.: *McGraw-Hill Encyclopedia of Chemistry,* 2nd Edition, The McGraw-Hill Companies, Inc., New York, NY, 1993.

Sinfelt, J.H.: "Bimetallic Catalysts," *Sci. Amer.,* 90−98 (September 1985).

Stwertka, A. and E. Stwertka: *A Guide to the Elements,* Oxford University Press, Inc., New York, NY, 1998.

Waterstrat, R.M. and G. Dickson: *Dental Amalgam (Hg, Ag, Sn, Cu, Zn),* Metals Handbook, 9th Edition, Vol. 2, ASM International, Metals Park, OH, 1979.

Zysk, E.D.: *Precious Metals and Their Use,* Metals Handbook, 9th Edition, Vol. 2, ASM International, Metals Park, OH, 1979.

SIMPLE DISTILLATION. Distillation in which no appreciable rectification of the vapor occurs, i.e., the vapor formed from the liquid in the still is completely condensed in the distillate receiver and does not undergo change in composition due to partial condensation or contact with previously condensed vapor.

SIZING COMPOUND. 1. A material such as starch, gelatin, casein, gums, oils, waxes, asphalt emulsions, silicones, rosin, and water-soluble polymers applied to yarns, fabrics, paper, leather, and other products to improve or increase their stillness, strength, smoothness, or weight.

2. A material used to modify the cooked starch solutions applied to warp ends prior to weaving.

SKUTTERUDITE. This mineral includes an isomorphous series with *smaltite-chloanthite,* essentially cobalt/nickel arsenides, (Co, Ni) $As_{2−3}$, crystallizing in the isometric system. The usual habit is cubic, octahedral, or cubo-octahedral. The mineral also occurs in massive and granular forms. Skutterudite has a metallic luster; hardness of 5.5 to 6.0, a specific gravity of 6.5. The mineral is opaque with tin-white to silver-gray color. The nickel-rich material alters surficially to annabergite (green color); the cobalt-rich material to erythrite (rose color). The streak is black. The mineral is an essential ore of cobalt and nickel.

Skutterudite is found in moderate-temperature veins, commonly associated with other cobalt/nickel minerals, e.g., cobaltite and nickeline. The mineral was named for its occurrence at Skutterud, Norway. Important ore sources are Norway, Bohemia, Saxony, Spain, France, and New South Wales, Australia. Notable occurrences are in Ontario, Canada, mainly Sudbury, South Lorrain, and Gowganda.

SLACK. 1. Descriptive of a soft paraffin was resulting from the incomplete pressing of the settlings from the petroleum distillate. Though it has some applications in this form, it is actually an intermediate product between the liquid distillate and the scale wax made by expressing more of the oil.

2. Specifically, to react calcium oxide (lime) with water to form calcium hydroxide (slaked or hydrated lime), the reaction is $CaO + H_2O \longrightarrow Ca(OH)_2 + heat$. The alternate spelling "slake" has the same meaning.

SLAG. Slag is a fused product occurring in connection with metallurgical and combustion processes. It is composed of the oxidized impurities in a metal, and of a fluxing substance, and of ash. In the steel industry, slag is the neutralized product of anhydrous compounds entering into the process. Slag is of great importance to the operator of a steel furnace or a cupola, in that, through the slag, impurities are separated and removed from the metal. By floating as a molten covering on the pool of metal, slag protects it from oxidation and serves to keep it clean. By controlling the character of slag, and continuous observation, the metallurgist insures that the metal is of the quality desired.

Molten ash is one of the products of combustion of coal in certain high-capacity boiler furnaces. It is also called slag. In some plants, the ash is removed from the furnace in this fluid form. Such furnaces are known as slag tap furnaces. Slag has some commercial value as ballast, coarse aggregate for concrete, road metal, etc.

SLATE. A fine-grained homogeneous sedimentary rock composed of clay or volcanic ash which has been metamorphosed (foliated) so as to develop a high degree of fissility or salty cleavage which is usually at a high angle to the planes of stratification. This high degree of fissility makes the better grades of slates an extremely useful roofing material which, however, has been somewhat replaced in recent years by synthetic and manufactured substitutes. The finest slates in the world come from Wales, Britain.

SLUDGE. When fresh sewage is admitted to settling tanks a certain amount of the solid matter in suspension will settle out, 50% more or less for sedimentation periods of an hour and a half or so. This collection of solids is known as fresh sludge. Such sludge will become actively putrescent in a short time and in modern treatment plants must be passed on from the sedimentation tank before this stage is reached. This may be done in two common ways. The fresh sludge may be passed through the slot in an Imhoff tank to the lower story or digestion chamber. Here, decomposition by anaerobic bacteria takes place with considerable liquefaction and reduction in volume. After the decomposition process has run its course (in 6–9 months) the resulting sludge is called "digested" sludge and is relatively inoffensive in character. It may be disposed of by drying on sludge drying beds and spreading on the land. It has little, if any, fertilizing value, being in the nature of humus. The sludge digestion chamber is operated on a periodic schedule of sludge withdrawals.

Alternatively, plain sedimentation basins with mechanical equipment for continuous collection of the fresh sludge may be used. The fresh sludge, so collected, is discharged into separate sludge digestion tanks which operate on the principle of the lower story of the Imhoff tank except that by means of higher and better temperature control the digestion cycle is much more rapid and efficient than for the Imhoff tank.

SMALLEY, RICHARD E. (1943–). An American who won the Nobel prize for chemistry along with Robert. F. Curl, Jr. and Sir Harold W. Kroto in 1996, the 100th anniversary of Alfred Nobel's death. The trio won for the discovery of the C_{60} compound called buckminsterfullerene. He graduated from the University of Michigan and earned a Ph.D. from Princeton University.

See also **Buckminsterfullerene (Buckyballs).**

SMART MATERIALS. From a technical and simple point of view, a smart material is a material that responds to its environment in a timely manner. To expand on this definition, a smart material is one that receives, transmits, or processes a stimulus and responds by producing a useful effect, which may include a signal that the material is acting upon it. Stimuli may include strain, stress, temperature, chemicals, an electric field, a magnetic field, hydrostatic pressures, different types of radiation, and other forms of stimuli. Transmission or processing of the stimulus may be in the form of an absorption of a photon, of a chemical reaction, of an integration of a series of events, of a translation or rotation of segments within the molecular structure, of a creation and motion of crystallographic defects or other localized conformations, of an alteration of localized stress and strain fields, and of others. The useful effects produced could be a change in color, index of refraction, stress or strain distribution, or volume. Also, incorporated within the definition of smart materials is the ability to be reversible.

Under the proper set of environments and circumstances all materials are smart and depict smart behavior at some point during their life cycle. Some examples of technically smart behaving materials are piezoelectric materials, electrostrictive materials, magnetostrictive materials, electrorheological materials, magnetorheological materials, thermoresponsive materials, pH-sensitive materials, uv-sensitive materials, smart polymers, smart gels (hydrogels), smart catalysts, and shape memory alloys.

Smart structures are structures that incorporate at least one smart material within itself and from the effort produced by the smart material causes an action.

Piezoelectric Materials

Piezoelectric materials are materials that exhibit a linear relationship between electric and mechanical variables. The direct piezoelectric effect can be described as the ability of materials to convert mechanical stress into an electric field; and the reverse, to convert an electric field into a mechanical stress. The use of the piezoelectric effect in sensors is based upon the latter property.

There are two principal types of materials that can function as piezoelectrics: the ceramics and polymers. The piezoelectric materials most widely used are the piezoceramics based upon the lead zirconate titanate, PZT. The advantages of these piezoceramics are that they have a high piezoelectric activity and they can be fabricated in many different shapes.

A newer class of materials called smart tagged composites has been developed for structural health monitoring applications. These composites consist of PZT-5A particles embedded into the matrix resin (unsaturated polyester) of the composite.

Electrostrictive Materials

Electrostrictive materials are materials that exhibit a quadratic relationship between mechanical stress and the square of the electric polarization. Electrostriction can occur in any material. Whenever an electric field is applied, the induced charges attract each other, thus, causing a compressive force. This attraction is independent of the sign of the electric field. Typical electrostrictive materials include such compounds as lead manganese niobate, lead titanate (PMN:PT), and lead lanthanium zirconate titanate (PLZT).

Magnetostrictive Materials

As materials show mechanical deformation induced by electric fields, the same type of material response can be observed when the stimulus is a magnetic field. Shape changes are the largest in ferromagnetic and ferrimagnetic solids. The repositioning of domain walls that occur when these solids are placed in a magnetic field leads to hysteresis between magnetization and an applied magnetic field.

Materials that have shown a response to magnetic stimuli have primarily been inorganic in chemical composition, alloys of iron, nickel, and cobalt doped with rare earths. However, there has been a great interest in the development of organic and organometallic magnets.

In comparing organic magnets with organometallic magnets there are several key differences between the two types. The first is that the organic-based magnets do not contain metal atoms. The second difference involves the fact that in organic-based magnets, the coupled spins residue entirely in the p orbitals; whereas in the organometallic-based magnets, they are either in the p or d orbitals, or a combination of the two.

Electrorheological Materials

Electrorheological materials are fluids whose viscous properties are modified by applying an electric field. There are many electrorheological fluids, which are usually a uniform dispersion or suspension of particles within a fluid. In an applied electric field the particles orient themselves in fiberlike structures (fibrils). When the electric field is off, the fibrils disorient themselves. The damping characteristics of the system can be changed (flexible to rigid). Electrorheological fluids are non-Newtonian fluids, that is, the relationship between shear stress and shear strain rate is nonlinear. The changes in viscous properties of electrorheological fluids are obtained only at relatively high electric fields in the order of 1 kV/mm.

Magnetorheological Materials

Magnetorheological materials (fluids) are the magnetic equivalent of electrorheological fluids. In this case, the particles are either ferromagnetic or ferrimagnetic solids that are either dispersed or suspended within a liquid and the applied field is magnetic.

An adaptation of magnetorheological fluids is a series of elastomeric matrix composites embedded with magnetic particles such as iron. During the thermal cure of the elastomer, a strong magnetic field was applied to align the iron particles into chains. These chains of iron particles were locked into place within the composite through the cross-linked structure of the cured elastomer. The resistance of the composite to changes in modulus or deformation was controlled by an external magnetic field. When stimulated by a compressive force, the composite was 60% more resistant to deformation in a magnetic field.

Thermoresponsive Materials

Polymeric materials are unique because of the presence of a glass-transition temperature. At the glass-transition temperatures, the specific volume of the material and its rate of change changes, thus, affecting a multitude of physical properties.

Materials that typify thermoresponsive behavior are polyethylene–poly-(ethylene glycol) copolymers that are used to functionalize the surfaces of polyethylene films (smart surfaces). When the copolymer is immersed in water, the poly(ethylene glycol) functionalities at the surfaces have solvation behavior similar to poly(ethylene glycol) itself. The ability to design a smart surface in these cases is based on the observed behavior of inverse temperature-dependent solubility of poly(alkene oxide)s in water.

pH-Sensitive Materials

By far the most widely known classes of pH-sensitive materials are those classes of chemical compounds that include the acids, bases, and indicators. The most interesting of these are the indicators. These materials change colors as a function of pH and usually are totally reversible.

In addition to acting as a means of observing changes in pH in titrations and in chemical reactions, indicators have been used in the development of novel chemical indicating devices.

Other examples of pH-sensitive materials are the smart hydrogels and smart polymers.

Light-Sensitive Materials

There are several different types of material families that exhibit different kinds of responses to a light stimuli. One type comprises materials that exhibit electrochromism. This is a change in color as a function of an electrical field. Other types of behaviors include thermochromism (color change with heat), photochromic material (reversible light-sensitive materials), photographic materials (irreversible light-sensitive materials), and photostrictive materials (shape changes due to light usually caused by changes in electronic structure).

Smart Polymers

Even though smart polymers have been used in all types of applications and can exhibit all types of stimuli–response behaviors, the term, smart polymers, has been used as a separate category of smart materials. In medicine and biotechnology, smart polymer systems usually involve aqueous polymer solutions, interfaces, and hydrogels. These are polymeric systems that are capable of responding strongly to slight changes in the external medium; a first-order transition accompanied by a sharp decrease in the specific volume of the system. The presence of a poor solvent is one of the main conditions for this phenomenon in swollen polymer networks or linear polymers to occur. A poor solvent causes the forces of attraction between the polymer chain segments to overcome the repulsion forces associated with the extended volume, thus, leading to the collapse of the polymer chain.

Smart polymers can respond to environmental stimuli such as temperature, pH, ions, solvents, reactants, light or uv radiation stress, recognition, electric fields, and magnetic fields. These stimuli once acted upon, result in changes in phases, shape, optics, mechanics, electric fields, surface energies, recognition, reaction rates, and permeation rates. The polymers that fit into this category include the naturally occurring polymers, acrylic polymers and copolymers, and polymers based on combining acid monomers with basic monomers.

Smart Gels (Hydrogels)

Smart (intelligent) gels (or hydrogels) are not new. They are finally reaching commercialization after thirty years of research and development. The concept of smart gels is also more complex than the simple concept of solvent-swollen polymer networks. It is the behavior of the solvent-swollen polymer networks in conjunction with the material being able to respond to other types of stimuli; such as temperature, pH, and concentrations of solvents. An example of a smart gel chemical composition consists of an entangled network of two polymers; one is a poly(acrylic acid) (PAA), and the second, a tri-block copolymer containing poly(propylene oxide) (PPO) and poly(ethylene oxide) (PEO) in a PEO–PPO–PEO sequence. The PPA portion of the smart gel system is a bioadhesive and is pH responsive. The PPO segments are hydrophobic that help solubilize lipophilic substances in medical applications, and the PPO segments tend to aggregate, thus resulting in gelation at body temperatures.

Smart Catalysts

One class of smart catalysts is based on homogeneous rhodium-based poly(alkene oxide)s, in particular those with a poly(ethylene oxide) backbone. Traditionally chemical catalyzed reactions proceed in a manner in which the catalysts become more soluble and active as the temperature is raised. This can lead to exothermal runaways, thus, posing both safety and yield problems. The behavior of these smart catalysts is different from that of traditional catalysts. As the temperature increases, they become less soluble, thus precipitating out of solution and becoming inactive. As the reaction mixture cools down, a smart catalyst redissolves and becomes active again.

Smart Memory Alloys

Shape-memory alloys undergo thermomechanical changes as they pass from one phase to another. The crystalline structure of such alloys based on nickel and titanium enters the martensitic phase as the alloy is cooled below a critical temperature. In this stage, the alloy is easily manipulated through large strains with a little change in stress. As the temperature of the alloy is increased above the critical (transformation) temperature it changes into the austentic phase. In the austentic phase, the alloy regains its high strength and high modulus. It behaves like a "normal" metal. The alloy shrinks during the transformation from the martensitic to austentic phase.

The use of shape-memory alloys as actuators depends on their use in the plastic martensitic phase that has been constrained within the structural device. Shape-memory alloys (SMAs) can be divided into three functional groups; one-way SMAs, two-way SMAs, and magnetically controlled SMAs. The magnetically controlled SMAs show great potential as actuator materials for smart structures because they could provide rapid strokes with large amplitudes under precise control. The most extensively used conventional shape-memory alloys are the nickel–titanium- and copper-based alloys (see **Shape-Memory Alloys**).

Elastorestrictive Materials

This class of smart materials is the mechanical equivalent of electrostrictive and magnetostrictive materials. Elastorestrictive materials exhibit high hysteresis between strain and stress. This hysteresis can be caused by motion of ferroelastic domain walls. This behavior is more complicated and complex near a martensitic phase transformation.

Materials with Unusual Behaviors or Unusual Materials

Only a few materials fit into this category; they seldom can be categorized into one of the above material classes. Water fits into the category of materials with unusual behavior. Water is one of the few materials that expands upon freezing. It changes volume by approximately 8% transiting from the liquid to the solid state.

Fullerene and its derivatives can be included in the unusual material category. One interesting application of fullerenes as smart materials has been in the area of embedding fullerenes into sol–gel matrices for the purpose of enhancing optical limiting properties. A semiconducting material with a magnetic ordering at 16.1 K was produced from the reaction of the fullerene C_{60} with tetra(dimethylamino)ethylene.

JAMES A. HARVEY
Hewlett-Packard Company
Oregon Graduate Institute of Science & Technology

Additional Reading

Committee on New Sensor Technologies: Materials and Applications, National Materials Advisory Board, Commission on Engineering and Technical Systems, *National Research Council Report: Expanding the Vision of Sensor Materials,* National Academy Press, Washington, DC 1995.

Miller, J. S. and A. J. Epstein: *Chem. Eng. News,* 30–41 (Oct. 2, 1995).

Rogers, C. A.: *Scientific American* **273**(3), 122–126 (Sept. 1995).

Udd, E.: *Fiber Optic Smart Structures,* John Wiley & Sons, Inc., New York, NY, 1995.

SMECTIC LIQUID CRYSTALS. See **Liquid Crystals**.

SMELTING. The process of heating ores to a high temperature in the presence of a reducing agent, such as carbon (coke), and of a fluxing agent to remove the accompanying rock gangue is termed smelting. Iron ore is the most abundantly smelted ore. It contains about 20% gangue (clay and sand). The ore is heated in an air blast furnace with coke and limestone (fluxing agent) at a temperature above the melting point of iron and slag (fusion mixture of impurities and flux). The molten iron (the more dense material) and molten slag (the less dense material) are removed separately from the furnace. See also **Arsenic**; **Cadmium**; **Cobalt**; **Copper**; **Indium**; **Iron Metals, Alloys, and Steels**; **Lead**; **Silver**; and **Tin**.

SMITHSONITE. Smithsonite is zinc carbonate, $ZnCO_3$, a hexagonal mineral with a rhombohedral cleavage. It is a brittle mineral; hardness, 4–4.5; specific gravity, 4.3–4.5; luster, vitreous to dull; color, usually white, but may be colored yellowish or brownish or perhaps blue or green due to impurities. It is translucent to opaque. Smithsonite is a secondary mineral after sphalerite or may replace limestone or dolomite. It is sometimes called calamine (but true calamine is a zinc silicate) and often associated with it. Smithsonite occurs in Siberia, Greece, Rumania, Austria, Sardinia, Cumberland and Derbyshire, England; New South Wales, Australia; South West Africa, and Mexico. In the United States, it is found in Pennsylvania, Wisconsin, Missouri, Arkansas, and Utah. This mineral was named in honor of James Smithson, whose legacy founded the Smithsonian Institution at Washington, DC.

SMOG. A coined word denoting a persistent combination of smoke and fog occurring under appropriate meteorological conditions in large metropolitan or heavy industrial areas. The discomfort and danger of smog is increased by the action of sunlight on the combustion products in the air, especially sulfur dioxide, nitric oxide, and exhaust gases (photochemical smog). Strongly irritant and even toxic substances may be present e.g., peroxybenzoyl nitrate. Fatalities have resulted from exposure from exposure to particularly severe photochemical smogs.

See also **Pollution (Air)**.

SMOKE. A colloidal or microscopic dispersion of a solid in gas, and aerosol. (1) Coal smoke: A suspension of carbon particles in hydrocarbon gases or in air, generated by combustion. The larger particles can be removed by electrostatic precipitation in the stack (Cottrell). Dark color, nauseating odor.

See also **Cottrell, Frederick G. (1877–1948)**; **Pollution (Air)**; and **Smog**.

(2) Wood smoke: Light-colored particles of cellulose ash, pleasant aromatic odor. Smoke from special kinds of wood (e.g., hickory, maple) is used to cure ham, fish, etc., also to preserve crude rubber.

(3) Chemical smoke: Generated by chemical means for military purposes (concealment, signaling, etc.).

(4) Metallic smoke (fume): An emanation from heated metals or metallic ores, the particles being of specific geometric shapes. Such smoke is particularly damaging to vegetation in the neighborhood of zinc and tin smelters.

(5) Cigarette smoke: There is conclusive evidence that the tars occurring in cigarette smoke can lead to lung cancer; chief factors are age of individual at initiation of smoking, extent of inhalation, an amount smoked per day. Polonium, a radioactive element, is known to occur in cigarette smoke; more than 100 compounds have been identified including nicotine, cresol, carbon monoxide, pyridene, and benzopyrene, the latter a carcinogen.

SMOKELESS POWDER. Nitrocellulose containing about 13.1% nitrogen, produced by blending material of somewhat lower (12.6%) and slightly higher (13.2%) nitrogen content, converting to a dough with alcohol-ether mixture, extruding, cutting, and drying to a hard, horny product. Small amounts of stabilizers (amines) and plasticizers are usually present, as well as various modifying agents (nitrotoluene, nitroglycerin salts).

SNG. See **Substitute Natural Gas (SNG)**.

SOAPS. Chemically, a soap is defined as any salt of a fatty acid containing 8 or more carbon atoms. Structurally a soap consists of a hydrophilic (water compatible) carboxylic acid which is attached to a hydrophobic (water repellent) hydrocarbon. Soap molecules thus combine two types of behavior in one structure; part of the molecule is attracted to water and the other part is attracted to oil. This feature underlies the function of these materials as surface active agents, or surfactants. Soaps are one class of surfactants. The other classes generally are called detergents. See also **Colloid Systems** and **Detergents**.

All surfactants, including soaps, demonstrate a common physical property—when they dissolve they preferentially concentrate at solution surfaces. These surfaces are known as the *interfacial regions* or regions where one continuous phase, such as water, stops and another, such as oil, begins. By their presence at the interface, surfactants lower the total energy associated with maintaining that boundary and thereby stabilize it. Without surfactants, a mixture of oil and water will soon separate into two distinct phases where the total surface area across which water and oil contact each other will be minimal. Adding soap to the water reduces its surface tension—the energy needed to maintain contact between the oil and the water. The oil then can be broken into microscopic droplets, which are dispersed in the water. Creation of these droplets, however, is accompanied by a huge increase in the interfacial contact area between oil and water. The dispersion of the oil in water is only possible, and only can be maintained over a period of time, because the surfactant reduces the energy associated with the large surface over which oil and water are in contact with each other. This phenomenon is the basis for the cleansing action of soaps and other detergents. Stabilization of the interface between the water used to cleanse and oils and other water-insoluble soils facilitates the dispersion of these materials into the water.

Although soaps and synthetic detergents have similar physical properties, several factors distinguish between them. Soap is generally made from natural fats and oils (oleochemicals). Some important synthetic detergents are also derived from oleochemicals, but almost no ordinary soaps are produced from petrochemicals. Fats and oils are triglycerides which contain three fatty acids, the basic structural unit of soaps, chemically linked to a glycerine backbone. As the "soap" chemical structure basically exists in natural triglycerides, with relatively straightforward processing operations, soap can be obtained from fats and oils.

Another important distinguishing feature of soaps is that they form a curdy, insoluble compound in hard water due to interaction between the carboxylate soap structure and calcium and magnesium ions in the water. Synthetic detergents, which generally are based on sulfate or sulfonate chemical structures for the water-attracting portion of the molecule, have less affinity for these metals and thus work well in all types of water. In addition, since these synthetics maintain their surfactancy, they also function to disperse objectionable curd. For these reasons, the synthetic detergents have generally replaced soaps in heavy-duty cleaning (laundry, floors, woodwork). Soaps, however, remain popular for mild cleaning and particularly for personal cleansing.

Personal Cleansing Soap Products

The major soap-based products which one commonly encounters are soap bars. Two broad categories of bar soaps may be defined: *basic cleaning bars*, which are natural soaps without extra ingredients and comprise about 20% of the market; and bars with special ingredients to provide a benefit beyond fundamental cleansing. The latter category may be further subdivided into *deodorant soaps* and *skin care bars*. Generally most of these bars command a higher retail price than basic cleaning bars, with skin care bars priced above deodorant soaps.

Deodorant soaps add fragrances that are partially substantive to the skin and that mask body odors, and antimicrobial agents. The antimicrobials, such as *Triclocarban®*, are deposited on the skin and inhibit bacterial growth and associated malodors.

Skin care bars are formulated with ingredients for which specific skin benefits are claimed. Consumers generally recognize and are concerned that personal cleansing products can dry the skin, leaving it feeling rough, itchy, and tight, and looking powdery and scaly. To counter these effects, particularly during the dry winter months, they may elect to use a cleansing bar containing a moisturizer, as well as increasing their use of body oils and hand and body lotions. Skin care claims for these products are based on the inclusion of moisturizers such as glycerin, cocoa butter, lanolin, cold cream and vitamins to the soap.

The mildness of soap bars toward the skin can also be enhanced by the process of *superfatting*. In superfatting, excess fatty acid is added to the soap during processing. This water-insoluble material functions as an emollient, significantly improving the mildness and the lathering of the bar.

Manufacture of Soap

Ingredients. The primary materials used in the manufacture of bar soaps are natural fats and oils. The performance and physical properties of soap bars can be varied by altering the blend of fats and oils used to make the neat soap. The most common materials used are top-quality animal tallows and coconut oil with blends ranging from 50% to 85% tallow. Generally it is found that bars containing higher proportions of coconut soap are physically harder, more brittle, lather more, and are more expensive to produce due to the higher cost of coconut oil. It is therefore common practice to vary the blend of tallows and coconut oil to meet the desired properties and price of each product.

These basic materials eventually are converted to their neutral salts by use of some alkaline material, such as sodium hydroxide. Additional, minor ingredients are added, e.g. sodium silicate or magnesium sulfate, to control alkalinity, odor, and aging stability.

The basic process is that of reacting fat stocks with alkali to form soap (direct saponification) and glycerin, followed by washing to remove the glycerin. Two methods of direct saponification are in common use (*kettle method* and *continuous saponification*). An alternative method is splitting fat stocks with water (hydrolysis) to form fatty acids and glycerine, followed by neutralization of the fatty acids with alkali.

Kettle Method. The pioneers used a simplified kettle process when they boiled animal fat and wood ashes (for alkalinity) for several hours in a large pot. The modern soap kettle has a capacity of 60,000–300,000 pounds (27,216–136,080 kg) and is equipped for heating, settling, and blending the fats, alkali, salt, and water.

The kettle first is charged with fat and a sodium hydroxide solution. Then follows a sequence of heating, separating, and washing to convert the raw materials to *finished base soap* and to separate the impurities and byproducts. The process normally takes several days for any single kettle. Although there have been improvements in handling and purification such as continuous centrifugation, the basic kettle process of saponifying fats directly with caustic remains unchanged.

Continuous Saponification. Fat stocks, plus caustic and salt solutions, are fed continuously into an autoclave operating under pressure at typically about 250°F (120°C). A recycle stream provides sufficient soap concentration to solubilize the fat stream for good contacting with the caustic. The soap-lye-glycerin mix moves to a mixer/cooler to complete saponification. The cooler temperature reduces the solubility of soap in the lye and aids separation. See also **Saponification**.

Glycerin and excess caustic are removed by several stages of countercurrent washing with fresh washing solution. The washing and separation stages usually take the form of a series of mixers and centrifugal separators or a continuous countercurrent contactor, such as a rotating disk contractor (RDC) in a vertical column. The mix from the saponifier is fed near the bottom of the RDC and washing solution near the top. The lower-density soap rises through the falling wash solution. Washed soap exits at the top while spent lye (glycerin plus lye solution) exits out the bottom of the RDC column. Spent lye is processed to recover the glycerin.

The washed soap is converted to finished base soap (neat soap) by a final composition adjustment called *fitting*. Fitting is accomplished by adding water (plus salt as needed), which causes a phase separation. Depending on the salt concentration the separated phase is either a lye or niger phase. A centrifuge or kettles can be used to separate the two phases.

Hydrolyzer Process. The development of continuous hydrolysis provides basic improvements in the processing of fats into soap. There are several advantages over the kettle process: (1) better quality soaps can be made from darker fats; (2) glycerin recovery is simplified, because no salt is needed and the resulting finished glycerin is of higher quality; (3) a single hydrolyzer unit produces about the same quantity of soap as 10 kettles, thus effecting savings in manufacturing space and a reduction of in-process inventory; and (4) greater flexibility is possible in controlling the chemical and physical properties of the finished soap. The hydrolyzing process consists essentially of (1) hydrolysis, (2) fatty acid distillation, (3) post-hardening (optional), (4) neutralization, and (5) glycerin recovery. The basic hydrolyzing process is shown in Fig. 1.

Hydrolysis. Development of continuous hydrolyzing was the key step toward this continuous soap making process. In this reaction, fat and water react to form fatty acid and glycerin:

$$(RCOO)_3C_3H_5O + H_2O \rightleftharpoons 3RCOOH + C_3H_5(OH)_3$$

where *R* is an alkyl of C_8 or larger. This equation represents the complete hydrolysis. Actually, the reaction takes place in a stepwise fashion, forming intermediate diglyceride and monoglyceride.

The reaction can be accomplished only through intimate contact between water and fat molecules. High temperature makes it possible to dissolve an appreciable quantity of water in the fat phase and to obtain this intimate contact. At room temperature, water and fat are essentially insoluble. At elevated temperature, the solubility of water increases to 12–25%, depending upon the type of fat. At the higher temperatures, high pressures also are necessary to keep the water from flashing into steam.

The reaction is reversible. In order to make it proceed to the right, the proportion of water to fat can be increased or the glycerin can be removed. Removal of glycerin is used as the reaction-forcing method. The required combination of high temperature, high pressure, and continuous glycerin removal is accomplished in a countercurrent hydrolyzer column. Fat stocks, blended in the proper formula, are mixed with dry zinc oxide catalyst. The mixture is maintained at about 212°F (100°) to ensure dryness and to keep the catalyst in solution. Hot water for the hydrolysis reactions is put under high pressure by piston-type feed pumps with adjustable drives so that the rates and proportions of fat to water can be accurately controlled. The fat and water are heated to the hydrolyzing temperature by direct steam injection or by heat exchangers. The fats are pumped into the column near the bottom, and the water enters near the top. Thus, a countercurrent flow of water downward through rising fatty material is obtained.

The hydrolysis occurs in a two-phase reaction system. The fats and fatty acids flow continuously with droplets of water falling through them. Glycerin from the hydrolysis is dissolved in the excess water falling through the column. The rate-limiting factor is the transfer of glycerin into the water droplets. Zinc oxide catalyzes the reaction of forming zinc soap, which increases the glycerin transfer across the oil-water interface. Fresh water entering the column at the top reduces the glycerin to the lowest possible point, while a glycerin-water seat maintained at the bottom of the column (where the glycerin content is highest) prevents fat from washing out.

The fatty material passes upward through the column with about 99% completeness in splitting. The fatty acids, saturated with water,

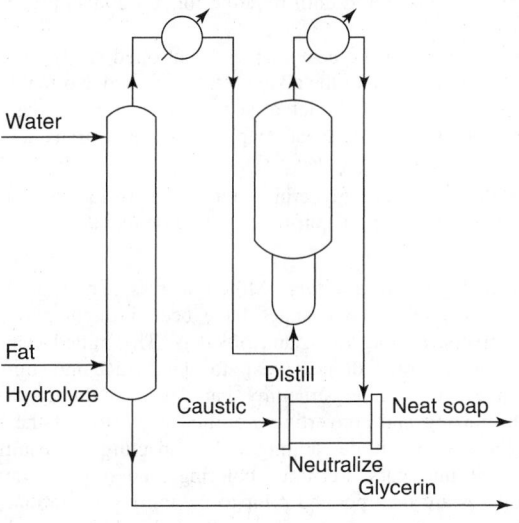

Fig. 1.   Basic hydrolyzer process used in soap manufacture

are discharged through an orifice into a flash tank. The dissolved water vaporizes, cooling the fatty acids and blanketing them with steam. The fatty acid contains the zinc soap catalyst and the remaining unsplit fat.

The column, pumps, and piping in contact with the hot fatty acid are made from corrosion-resistant stainless steel. The column is a hollow vessel, containing no baffles, trays, or packing material of any kind. The quality of the hydrolyzing operation is determined by the degree of split obtained on the fat. The fatty acid stream contains very little free glycerin, if any.

Distillation. The second key step in continuous soapmaking is distillation. Originally, fatty acids made in hydrolyzers were acid washed to split out the zinc soap and then bleached to improve color, but continuous distillation of the hydrolyzer fatty acids results in lighter soap from darker stocks at lower cost.

The fatty acids from the hydrolyzer are collected in the still feed tank and vacuum-dried to reduce moisture to low levels. Then they are flash-distilled at an absolute pressure of 2–5 mm Hg. The still bottoms are recirculated through heat exchangers back to the still to carry the heat necessary for vaporizing the fatty acids. The still bottoms, which contain the zinc soap catalyst and unsplit fat, are removed from the system, acidulated to remove the zinc, and frequently used in animal feeds. The fatty acid vapors from the still pass to several water condensers in series. The condensed fatty acids drop to a surge tank for posthardening or directly for neutralization.

The two prime objectives of this process, maintenance of good odor and color in the distillate and proper bottoms yield, are achieved by effective control over vacuum, temperature, and distillation rate.

Posthardening. Not shown in the figure is an optional further treatment of the fatty acids known as *posthardening.* This operation involves hydrogenation of some of the unsaturated carbon-carbon bonds on the fatty acid molecules. Originally, the purpose of this step was to improve color and odor. As such, the hardening was intended only to eliminate polyunsaturates, leaving the majority of the monounsaturates unaffected. A greater amount of hardening can be performed, however, to tailor some of the physical properties of the finished bar characteristics. The fatty acids from distillation are heated and passed with a metered hydrogen supply through hardening tubes which contain a fixed bed of granular nickel catalyst where the hydrogenation takes place. The hardened fatty acids flow through a filter to remove traces of catalyst. The filtered stock drops to a flash tank, where excess hydrogen is removed. Hardening is controlled by temperature, pressure, hydrogen flow, residence time, and catalyst age. The fatty acids then are cooled for neutralization.

Neutralization. The saponification reaction between alkaline solutions and fatty acids is almost instantaneous:

$$RCOOH + NaOH \longrightarrow RCOONa + H_2O$$

Each reactant is metered accurately into the neutralizer, where intimate mixing occurs and the reaction takes place. Soap from the neutralizer is discharged at about 200°F (93°C) to a blend tank equipped with agitation and recirculation to ensure uniform composition of the soap. This base soap (or *neat soap*) is stored until required for subsequent processing into finished bars.

The characteristics of the neat soap are controlled easily by accurately governing the composition of the alkaline solution used. Normal hydrolyzer neat soap contains about 69% actual soap, 30% water, and less than 1% NaCl, plus other stabilizers. Neat soap is a uniform, translucent, white, viscous fluid at 180–200°F (82–93°C).

Glycerin Recovery. The glycerin water stream from the hydrolyzer is concentrated by evaporation, purified, and subsequently sold or used in other processes.

Milled Bar Soap Manufacture. Milled soap is a high-grade soap in which critical crystal-phase changes have been brought about through the use of mixers, milling rolls, and plodders. The milled soap is made by drying a good grade of neat soap to about 15% moisture content, breaking up the crystalline structure that develops during drying and cooling, plasticizing and converting a sufficient portion of the soap to a desirable phase condition, de-aerating and compacting the resulting mass, and forming it into bars. Perfume, coloring matter, preservatives, and special additives are incorporated prior to the milling operation. A milled bar is particularly hard, dense, and smooth, and it lathers freely without forming excessive soft soap on the surface of the bar.

Drying. Liquid base soap is dried from a 30% water liquid form to a solid of about 15% water content. If desired, some minor ingredients may be blended into the soap stream prior to drying. Methods of drying used in common practice are (1) chip drying, (2) atmospheric flash drying, and (3) vacuum flash drying.

Chip Drying. Sometimes called *ribbon drying,* this process involves spreading a thin layer of hot base soap on a large chilled drum, which cools and firms up the soap. Drying is promoted primarily by the difference in water vapor pressure between the soap chips and the air surrounding them. No attempt is made to increase drying rate by heating the soap itself.

Atmospheric Flash Drying. A tower similar to a synthetic-granules spray-drying tower is used. The heat for drying, however, is put into the soap by heating it under high pressure before flashing it into the tower. During flashing, the pressure on the soap is abruptly relieved and soap moisture flashes to steam. Air to the tower is used for cooling. The soap temperature as it enters the flashing nozzles determines the final moisture of the dried soap.

An alternative method involves flashing the soap from the nozzle onto the surface of a chilled drum. The resultant solid soap is scraped off in flake form. This process called *chill flake drying,* is the method of choice for drying sticky soap/synthetic combination formulas.

Vacuum Flash Drying. In this most recent technique, drying takes place in a vacuum vessel similar to an atmospheric tower but smaller. The soap is similarly heated before flashing but under less pressure, so that boiling (actually drying of the soap) occurs in the heat exchangers. Since there is boiling in the heaters, the moisture of the dried soap depends primarily upon soap flow rate, soap pressure, and steam pressure to the heater and to a minor extent on the absolute pressure in the vacuum chamber. The final temperature of the soap depends entirely upon the absolute pressure in the vacuum chamber.

Mixing. After drying, the soap noodles or flakes are mixed with all additional ingredients required by the final product formula. Mixing is done in batch processes or continuously. These ingredients include dye, perfume, preservatives, deodorants, opacifier, and special purpose items. The type and proportion of these materials is largely what makes one brand of milled soap different from another.

In batch mixing, dried soap and additives are measured and dumped into a dry blender, where macro-mixing occurs. The batch process is cumbersome and slow, and it is difficult to maintain uniform quality. Continuous mixing operations for improving economy and efficiency of mixing include precision metering devices to measure the additives into the soap noodles as they are pulverized and conveyed through the mixer. Although these ingredients constitute but a small portion of the total product, their effect on the physical properties, e.g., softness, resistance to cracking, lathering, and resistance to dissolving, are considerable.

Milling. The three objectives of milling are: (1) thorough and intimate final mixing of the soap, perfume, and other ingredients without overheating; (2) crushing lumps of overdried soap and pulverizing them into pieces too small to appear as lumps or hard specks in the finished bar; and (3) conversion of a sufficient portion of the soap into the waxy, plastic phase of cold working. Soap is milled by forcing it through a series of rolls, thus subjecting it to a strong shearing action. This cold working at the proper moisture content changes the crystalline structure or phase of the soap. Temperature control during milling is important. If the temperature is too low, the wrong crystal structure will be formed, resulting in soft soap or a hard, brittle structure prone to cracking. If the temperature is too high, the soap will become sticky and difficult to process further.

Another method for complete mixing and working uses multiple plodding and screening, in which the soap and additives are pushed together through finer and finer mesh screens.

Plodding. After milling, it is necessary to form the soap into a shape for making the final bar. This usually is accomplished with a plodder, which essentially is a large-size meat grinder with a barrel that terminates in a cone. The plodder functions to compact the pellets or flakes of soap into a solid mass, squeeze out any pockets of entrapped air, and extrude it as a firm, uniform, and continuous strip.

Operations that follow include cutting, stamping, wrapping, and packing.

Transparent Bar Manufacture. Most milled bars are opaque and contain a whitening agent (titanium dioxide) to create a uniform appearance. By eliminating this whitener and carefully controlling processing conditions, a bar that is transparent can be produced. This transparency results when the soap crystals are reduced to microscopic size which then allows light to pass through the structure. It is also important to achieve the correct soap phase. The control of soap phase is a function of the ratio of tallow to coconut soaps, the milling temperature, and soap moisture which must be maintained within the very rigid limits.

Floating-Bar Manufacture. Base soap made from the desired blends of fat and oil first is flash-dried to a moisture content of about 22%. It then enters a mechanical mixer called a *crutcher*, where it is thoroughly mixed with perfume, preservatives, and air. The amount of air controls the density of the final product, giving the bar a density of less than one and making it floatable.

From the crutcher, the mix goes to a freezer to reduce the temperature of the soap to the point where it will hold its shape when extruded. In the earlier steps, the soap mix is in liquid form. Rapid chilling is required to put it into a solid state. The machine is similar to a commercial ice-cream freezer, consisting of a horizontal cylinder surrounded by a jacket and housing a rotating shaft (mutator) on which scraping blades are mounted. The liquid soap mix from the crutcher is pumped into one end of the cylinder. A refrigerated brine solution is circulated through the jacket to chill the soap. The scraping blades on the mutator remove the chilled soap from the cylinder walls and maintain uniformity of the mix. The nose of the freezer is equipped with an oblong orifice through which the soap is extruded, after chilling, in the form of a continuous ribbon, which has the same cross section as the final bar. There follows a series of cooling, storing, stamping, and packaging operations.

R. MARC DAHLGREN and JOHN N. KALBERG
Ivory Technical Center,
The Procter & Gamble Company
Cincinnati, Ohio

Additional Reading

Bailey, A.E. and Y.H. Hui: *Bailey's Industrial Oil and Fat Products,* 5th Edition, John Wiley & Sons, Inc., New York, NY, 1996.

Basta, N.: *Shreve's Chemical Process Industries Handbook,* 6th Edition, The McGraw-Hill Companies, Inc., New York, NY, 1993.

Woolatt, E.: *The Manufacture of Soaps, Other Detergents, and Glycerin,* John Wiley & Sons, Inc., New York, NY, 1985.

Web References

Detergent Chemistry: History: http://www.chemistry.co.nz/deterghistory.htm

The Procter & Gamble Company: http://www.pg.com/sitesearch/google.jhtml?
_DARGS=%2Fsitesearch%2Fgoogle.jhtml

SODALITE. An isometric mineral, a sodium aluminum silicate containing sodium chloride, with the chemical composition $Na_4Al_3 (SiO_4)_3Cl$, potassium sometimes replacing a small amount of sodium. It is commonly found as dodecahedrons or simply massive. When observed sodalite has a dodecahedral cleavage; conchoidal to uneven fracture; brittle; hardness, 5.5–6; specific gravity, 2.14–2.30; luster, vitreous to greasy; color grayish to greenish or yellowish, may be white. It is often a beautiful blue and may sometimes be red. It is transparent to translucent; streak, white. Sodalite is found in igneous rocks of nephelite-syenite type which have been produced from soda rich magmas. Sodalite also has been found in the lavas of Vesuvius. Common minerals associated with it are nephelite and cancrinite. It occurs in the Ilmen Mountains of the former U.S.S.R.; at Vesuvius and Monte Somma, Italy; in Norway and Greenland. In Canada, in British Columbia and in Ontario, beautiful blue sodalite is found; and in the United States similar material comes from Kennebec County, Maine. The mineral derives its name from the fact of its soda content.

SODA NITRE. The mineral soda nitre or Chile saltpeter is naturally occurring sodium nitrate, $NaNO_3$. Its hexagonal crystals are rare, this mineral usually being found in crystalline aggregates, crusts or masses. It is soft; hardness, 1.5–2; specific gravity, 2.266; vitreous luster; colorless or white to yellow or gray; transparent to opaque. Soda nitre is a most important mineral commercially, being used in the manufacture of nitric acid, other nitrates and fertilizers. The chief soda nitre deposits of the world are those found in the Atacama and Tarapaca deserts of northern Chile, although others exist in the Argentine and Bolivia. Some small deposits have been found in California, New Mexico and Nevada. The origin of these nitrate deposits is far from being well understood. They have been regarded as nitrates formed originally by oxidation of organic matter and subsequently leached out. Guano, the excrement of birds, might be the original source of the nitrates. Ground water and ancient marine deposits have been suggested as well as the possibility of derivation from nitric acid produced in the atmosphere during electrical storms. Some investigators consider that the nitrates may have come from volcanic sources.

SODA PULP PROCESS.   See **Pulp (Wood) Production and Processing**.

SODDY, FREDERICK (1877–1965).   A British physicist who won the Nobel prize in chemistry in 1921. His work was concerned with radioactive elements and atomic energy. His concept of isotopes and the displacement law of radioactive change is basic to nuclear physics. His education was at Oxford and Glasgow. He later worked in Canada and Australia.

SODIUM.   [CAS: 7440-23-5]. Chemical element, symbol Na, at. no. 11, at. wt. 22.9898, periodic table group 1 (alkali metals), mp 97.82°C, bp 882.9°C, density 0.971 g/cm³ (solid at 0°C), 0.9268 g/cm³ (liquid at mp). Elemental sodium has a face-centered cubic crystal structure.

Sodium is a silvery-white metal. It can be readily molded and cut by knife. It oxidizes instantly on exposure to air, and reacts with water violently, yielding sodium hydroxide and hydrogen gas, consequently is preserved under kerosene, and burns in air at a red heat with yellow flame. Discovered by Davy in 1807.

There is only one naturally occurring isotope, ^{23}Na. There are five known radioactive isotopes, ^{20}Na through ^{22}Na, and ^{25}Na, all with short half-lives except ^{22}Na with a half-life of 2.6 years. See also **Radioactivity**. In terms of abundance, sodium ranks sixth among the elements occurring in the earth's crust, with an average of 2.9% sodium in igneous rocks. In terms of content in seawater, the element ranks fourth (due mainly to excellent solubility of its compounds), with an estimated 50,000,000 tons of sodium per cubic mile of seawater. First ionization potential 5.138 eV. Oxidation potential $Na \rightarrow Na^+ + e^-$, 2.712 V.

Other important physical properties of sodium are given under **Chemical Elements**.

Sodium does not occur in nature in the free state because of its great chemical reactivity. Sodium occurs as sodium chloride in the ocean (1.14% Na); in salt deposits (salt, halite, NaCl), e.g., in Michigan, New York, Louisiana, in Great Britain, and in Germany; in salt lakes, e.g., the Dead Sea (3% Na), Great Salt Lake; in common rocks (average of the solid shell of the earth 2.75% Na) as sodium nitrate (Chile saltpeter, $NaNO_3$) in Chile; as sodium borate (rasorite, kernite, $Na_2B_3O_7 \cdot 4H_2O$, in California; tinkal, $Na_2B_4O_7 \cdot 10H_2O$, in Tibet); and as sodium carbonate Na_2CO_2 and sulfate Na_2SO_4 in certain salt lake areas. See also **Sodium Chloride**.

Although sodium metal was isolated in 1807, it remained a laboratory curiosity until Oersted discovered in 1824 that sodium metal will reduce aluminum chloride to produce pure aluminum metal. This discovery led to the development of a commercial process for the manufacture of sodium. The first cell was designed by Castner in 1886 and a plant was built in Niagara Falls, N.Y., because of availability of low-cost electric power, for the electrolysis of fused NaOH. This process was made obsolete in 1921 by introduction of the Downs process in which a mixture of fused sodium chloride and calcium chloride is electrolyzed to produce metallic sodium. The modern cells have four anodes (graphite) surrounded by a steel cathode. Wire mesh diaphragms extend down into the electrolysis zone to prevent recombination of product sodium and chlorine. The use of calcium chloride in the cell significantly lowers the melting point of the mix. Sodium chloride has a mp 800°C, calcium chloride, mp 772°C, the two-salt eutectic, mp 505°C. Calcium has limited solubility in sodium. The excess calcium reacts with the sodium chloride present, $Ca + 2NaCl \rightarrow 2Na + CaCl_2$, and thus does not contaminate the sodium metal to a large degree. The sodium, which is saturated with calcium, is cooled in a riser pipe. This reduces the solubility of Ca in Na, precipitating Ca, which falls back into the cell, where it reacts to form more Na. The Na that overflows at the top of the riser pipe contains 1% or less of Ca. The Na is further purified by filtration at a temperature near its melting point, reducing the Ca content to about 0.05%. The cells operate at about 8 V, with groups of 25 to 40 cells connected in series.

Uses

Like so many of the chemical elements, the compounds of sodium are far more important than elemental sodium—by several orders of magnitude.

Among the attractions of molten sodium metal as a heat-transfer medium are: (1) low density compared with other metals and combinations of salts, contributing to low cost per unit volume and thus relative ease of pumping, sodium being about one-half that of the more commonly applied nitrate-nitrite heat-transfer salts; (2) relatively low vapor pressure even at temperatures as high as 550°C; (3) greater heat capacity than most common metals in liquid form, the thermal conductivity being 5 to 10× greater than the conductivities of lead or mercury and 50× higher than for most organic heat-transfer media; and (4) the viscosity of molten sodium is quite low. Despite these fine qualifications, however, the use of sodium as a heat-transfer medium has enjoyed a mixed reception over the years, partially attributable to a lack of marketing thrust in its behalf. Sodium is fifth among the metals in terms of electrical conductivity—hence bus bars are constructed from steel pipe filled with sodium. The characteristic yellow sodium light, created by the passage of an electric current through sodium vapor, is used for commercial and industrial lighting. Sodium is used to modify aluminum-silicon alloys. Normally coarse and brittle, such alloys can be transformed into fine-grained alloys with good casting properties through the addition of a fraction of 1% of sodium. Sodium also has been used as a hardening agent in bearing metals. When added with an alkaline-earth metal, such as calcium, sodium increases the hardness of lead. The German alloy "Bahnmetal" is an alloy of this type.

Generally, plain carbon steel containers are sufficient for handling metallic sodium at temperatures not in excess of the metal's boiling point. All-welded pipeline construction and bellows-sealed packless valves are usually used. Because of the metal's violent reactivity with H_2O, conventional fire extinguishers, including CO_2 and chlorinated hydrocarbons, should not be used. The preferred fire-retarding agents are salt, graphite, and soda ash, but they must be dry. Sand usually is not recommended because it is difficult to obtain perfectly dry sand in an emergency. In manufacturing operations involving sodium, particularly at reasonably high temperatures, an apron, leggings, and a complete face covering should be used. At normal temperatures, or where only small quantities of the metal are required, as may be the case in a research laboratory, conventional protective gear and goggles and gloves usually suffice.

Chemistry and Compounds

Sodium metal is obtained by electrolysis of fused sodium chloride or hydroxide out of contact with air. Its uses are limited in extent, but important in particular cases, as in the liberation of a metal from its chloride by reaction of sodium to form sodium chloride, and in certain reactions of organic chemistry.

The ionization potential of sodium (5.138 eV) is second to that of lithium and higher than those of the other alkali metals. However, the measured value of its oxidation potential against a normal aqueous solution of its ion is 2.712 V, the lowest of the group. Potassium is more electropositive in many of its reactions, even with water; though both react vigorously to produce the hydroxide and hydrogen, the reaction of potassium is more vigorous. With bromine, sodium reacts only slowly without heating and with iodine scarcely at all even on heating; potassium reacts violently with bromine, and with iodine on heating.

Because of the ease of removal of its single $3s$ electron (5.138 eV) and the great difficulty of removing a second electron (47.29 eV), sodium is exclusively monovalent in its compounds, which are electrovalent. Some experimental work indicates that the sodium alkyls may be covalent, but even they form conducting solutions in other metal alkyls.

Sodium Atoms Confined. In an interesting experiment conducted at the National Bureau of Standards in 1985, Migdall and colleagues trapped slow-moving neutral Na atoms in a magnetic field that created an energy well for the atoms. Robinson (1985) reported that approximately 10^5 sodium atoms in a trap volume of 20 cubic centimeters were stopped for a brief instant of time. For many years, spectroscopists have visualized the ideal sample where a collection of atoms or molecules would reside motionless in space for a period of time. Because of this experiment, researchers are closer to this goal. In the experiment, it was found that the particles gradually leak out, with time constants ranging from 0.1 to 1 second. Similar experiments have been conducted at AT&T Bell Laboratories (Holmdel, New Jersey). The theoretical trapping time in a perfect vacuum has been estimated as greater than 1000 seconds. The importance of these experiments is explored in considerable detail by Robinson in the reference listed.

Like lithium, sodium and its compounds have been studied extensively in solution in liquid NH_3. Sodium metal in such solutions slowly or with catalysis forms the amide, $NaNH_2$. The solution of the metal is a powerful reducing agent, reacting with metallic salts to free the metal, with which it may form an intermetallic compound

$$Na + AgCl \longrightarrow NaCl + Ag$$

$$9Na + 4Zn(CN)_2 \longrightarrow 8NaCN + NaZn_4$$

Sodium chloride also forms the amide, or at low temperatures the pentammoniate, $NaCl \cdot 5NH_3$.

Like the other alkali metals, sodium forms compounds with virtually all the anions, organic as well as inorganic. These compounds are remarkable for their great variety and for the fact that the reactivity of sodium bicarbonate with many metallic oxides permits preparation of many compounds that are unstable in aqueous solution. While other alkali bicarbonates react similarly, the general discussion of these compounds, and of the inorganic alkali salts generally, is appropriately given in this book under this entry for sodium, from which such a great number of inorganic (as well as organic) salts has been prepared.

Thus, normal (ortho) sodium arsenates $Na_3AsO_4 \cdot xH_2O$ and acid arsenates exist both in solution and in the solid state, whereas the meta- and pyroarsenates exist only as solids, but are readily prepared by heating arsenic pentoxide, As_2O_5, and sodium bicarbonate in correct proportions to produce the primary and secondary sodium arsenates, whence the meta- and pyroarsenates are obtained by heating

$$NaH_2AsO_4 \longrightarrow Heat\ NaAsO_3 + H_2O$$

$$2Na_2HAsO_4 \longrightarrow Heat\ Na_4AsO_2O_7 + H_2O$$

Similarly, the boron salts include metaborates, $NaBO_2 \cdot xH_2O$, tetrab-orates, $Na_2B_4O_7 \cdot xH_2O$, other polyborates, $Na_2B_{10}O_{16} \cdot xH_2O$, at least one orthoborate, Na_3BO_3, and peroxyborates, such as $NaBO_3 \cdot H_2O$. See also **Boron**. Other important sodium salts include the carbonates, cyanides, cyanates, hexacyanoferrates, $Na_4Fe(CN)_4$ and $Na_3Fe(CN)_6$, halides, poly-halides, hypohalites, halites, halates, perhalates, permanganates, ortho-, pyro-, meta-, fluoro-, and peroxyphosphates, hyposulfites, sulfites, sul-fates, thiosulfates, peroxysulfates, polythionates, tungstates, vanadates, uranates, etc.

In addition to the simple compounds, sodium forms double salts of various types, although because of the relatively small size of the Na^+ ion, the number of sodium alums (see also **Alum**) is relatively small.

In addition to the inorganic salts, sodium forms such binary compounds as a phosphide, Na_3P, by direct union with phosphorus, a nitride, Na_3N, by direct union with nitrogen when activated electrically (which decomposes partly to give sodium amide, NaN_3, also obtained by heating sodium nitrate with sodium amide) and the oxides. Sodium monoxide, Na_2O, is obtained by heating the nitrite with the metal, displacing the nitrogen. Sodium peroxide, Na_2O_2, is the most stable oxide, obtained by reaction of the elements. Sodium superoxide is known, NaO_2, and one other oxide, Na_2O_3, has been reported. Sodium hydroxide, $NaOH$, is very soluble in H_2O and soluble in alcohol. It is almost completely ionized in water at ordinary concentrations, although its basic character is less than those of the higher elements in the group ($pK_B = -0.70$).

The detailed chemistry and applications of some of the more important compounds, other than those already discussed, follow.

Aluminate: Sodium aluminate, [CAS: 11138-49-1], $NaAlO_2$, white solid, (1) by reaction of aluminum hydroxide and NaOH solution, (2) by fusion of aluminum oxide and sodium carbonate, the solution reacts with CO_2 to form aluminum hydroxide. Used as a mordant, and in water purification. See also **Aluminum**.

Aluminosilicate: Sodium aluminosilicate is used as a water softener for the removal of dissolved calcium compounds.

Amide: Sodamide, sodamine, $NaNH_2$, white solid, formed by reaction of sodium metal and dry NH_3 gas at 350°C, or by solution of the metal in liquid ammonia. Reacts with carbon upon heating, to form sodium cyanide, and with nitrous oxide to form sodium azide, NaN_3.

Bromide: Sodium bromide, [CAS: 7647-15-6], $NaBr$, white solid, soluble, mp 755°C. Used in photography and in medicine. See also **Bromine**.

Carbonates: Sodium carbonate (anhydrous), soda ash, [CAS: 497-19-8], Na_2CO_3, sodium carbonate decahydrate, washing soda, sal soda, $Na_2CO_3 \cdot 10H_2O$, white solid, soluble, mp 851°C, formed by heating sodium hydrogen carbonate, either dry or in solution. Commonly bought and sold in quantity on the basis of oxide Na_2O determined by analysis (58.5% Na_2O equivalent to 100.0% Na_2CO_3).

Soda ash is a very-high-tonnage chemical raw material and approaches a production rate of 10 million tons/year in the United States. About 40% of soda ash is used in glassmaking; approximately 35% goes into the production of sodium chemicals, such as sodium chromates, phosphates, and silicates; nearly 10% is used by the pulp and paper industry; the remainder going into the production of soaps and detergents and in nonferrous metals refining. The first process for preparing soda ash was developed by Leblanc during the first French Revolution. In the Leblanc process, sodium chloride first is converted to sodium sulfate and subsequently the sulfate is heated with limestone and coke: (1) $Na_2SO_4 + 2C \rightarrow Na_2S + 2CO_2$; (2) $Na_2S + CaCO_3 \rightarrow Na_2CO_3 + CaS$. During the mid-1800s, the Solvay process was introduced. In this process, CO_2 is passed through an NH_3-saturated sodium chloride solution to form sodium bicarbonate, then followed by calcination of the bicarbonate: (1) $NH_3 + CO_2 + NaCl + H_2O \rightarrow HNaCO_3 + NH_4Cl$; (2) $2HNaCO_3 + heat + Na_2CO_3 + CO_2 + H_2O$. A large proportion of soda ash now is derived from the natural mineral trona, which occurs in great abundance near Green River, Wyoming. Chemically trona is sodium sesquicarbonate, $Na_2CO_3 \cdot NaHCO_3 \cdot 2H_2O$. After crushing, the natural ore is dissolved in agitated tanks to form a concentrated solution. Most of the impurities (boron oxides, calcium carbonate silica, sodium silicate, and shale rock) are insoluble in hot H_2O and separate out upon settling. Upon cooling, the filtered sesquicarbonate solution forms fine needle-like crystals in a vacuum crystallizer. After centrifuging, the sesquicarbonate crystals are heated to about 240°C in rotary calciners whereupon CO_2 and bound H_2O are released to form natural soda ash. The crystals have a purity of 99.88% or more and handle easily without abrading or forming dust and thus assisting glassmakers and other users in obtaining uniform and homogeneous mixes.

Chlorate. Sodium chlorate, chlorate of soda, [CAS: 7775-09-9], $NaClO_3$, white solid, soluble, mp 260°C, powerful oxidizing agent and consequently a fire hazard with dry organic materials, such as clothes, and with sulfur; upon heating oxygen is liberated and the residue is sodium chloride; formed by electrolysis of sodium chloride solution under proper conditions. Used (1) as a weedkiller (above hazard), (2) in matches, and explosives, (3) in the textile and leather industries.

Chloride. Sodium chloride, common salt, rock salt, halite, NaCl, white solid, soluble, mp 804°C. See also **Sodium Chloride**.

Chromate. Sodium chromate [CAS: 7775-11-3]. $Na_2CrO_4 \cdot 10H_2O$, yellow solid, soluble, formed by reaction of sodium carbonate and chromite at high temperatures in a current of air, and then extracting with water and evaporating the solution. Used (1) as a source of chromate, (2) in leather tanning, (3) in textile dyeing, (4) in inks.

Citrate. Sodium citrate, $Na_3C_6H_5O_7 \cdot 5\frac{1}{2}H_2O$ white solid, soluble, formed (1) by reaction of sodium carbonate or hydroxide and citric acid, (2) by reaction of calcium citrate and sodium sulfate or carbonate solution, and then filtering and evaporating the filtrate. Used in soft drinks and in medicine.

Cyanide. Sodium cyanide, [CAS: 143-33-9], NaCN, white solid, soluble, very poisonous, formed (1) by reaction of sodamide and carbon at high temperature, (2) by reaction of calcium cyanamide and sodium chloride at high temperature, reacts in dilute solution in air with gold or silver to form soluble sodium gold or silver cyanide, and used for this purpose in the cyanide process for recovery of gold. The percentage of available cyanide is greater than in potassium cyanide previously used. Used as a source of cyanide, and for hydrocyanic acid.

Dichromate. Sodium dichromate, [CAS: 10588-01-9], $Na_2Cr_2O_7 \cdot 2H_2O$, red solid, soluble, powerful oxidizing agent, and consequently a fire hazard with dry carbonaceous materials. Formed by acidifying sodium chromate solution, and then evaporating. Used (1) in matches and pyrotechnics, (2) in leather tanning and in the textile industry, (3) as a source of chromate, cheaper than potassium dichromate.

Dithionate. Sodium dithionate, "sodium hyposulfate," [CAS: 7775-14-6], $Na_2S_2O_6 \cdot 2H_2O$, white solid, soluble, formed from manganese dithionate solution and sodium carbonate solution, and then filtering and evaporating the filtrate.

Fluorides. Sodium fluoride [CAS: 7681-49-4], NaF, white solid, soluble, formed by reaction of sodium carbonate and hydrofluoric acid, and then evaporating. Used (1) as an antiseptic and antifermentative in alcohol distilleries, (2) as a food preservative, (3) as a poison for rats and roaches, (4) as a constituent of ceramic enamels and fluxes; sodium hydrogen fluoride, sodium difluoride, sodium acid fluoride, $NaHF_2$, white solid, soluble, formed by reaction of sodium carbonate and excess hydrofluoric acid, and then evaporating. Used (1) as an antiseptic, (2) for etching glass, (3) as a food preservative, (4) for preserving zoological specimens.

Fluosilicate. Sodium fluosilicate, Na_2SiF_6, white solid, very slightly soluble in cold H_2O, formed by reaction of sodium carbonate and hydrofluosilicic acid. Used (1) in ceramic glazes and opal glass, (2) in laundering, (3) as an antiseptic.

Formate. Sodium formate, [CAS: 141-53-7], $NaCHO_2$, white solid, soluble, formed by reaction of NaOH and carbon monoxide under pressure at about 200°C. Used (1) as a source of formate and formic acid, (2) as a reducing agent in organic chemistry, (3) as a mordant in dyeing, (4) in medicine.

Hydride. Sodium hydride, [CAS: 7646-69-7], NaH, white solid, reactive with water yielding hydrogen gas and NaOH solution, formed by reaction of sodium and hydrogen at about 360°C. Used as a powerful reducing agent.

Hydroxide. Sodium hydroxide, caustic soda, sodium hydrate, "lye," [CAS: 1310-73-2], NaOH, white solid, soluble, mp 318°C, an important strong alkali, not as cheap as calcium oxide (a strong alkali) nor sodium carbonate (a mild alkali), but of wide use. Formed (1) by reaction of sodium carbonate and calcium hydroxide in H_2O, and then separation of the solution and evaporation, (2) by electrolysis of sodium chloride solution under the proper conditions, and evaporation. Commonly bought and sold in quantity on the basis of oxide Na_2O determined by analysis (77.5% Na_2O equivalent to 100.0% NaOH). Used (1) in the manufacture of soap, rayon, paper ("soda process"), (2) in petroleum and vegetable oil refining, (3) in the rubber industry, in the textile and tanning industries, (4) in the preparation of sodium salts, (a) in solution, (b) upon fusion. See Fig. 1.

Hypochlorite. Sodium hypochlorite, [CAS: 7681-52-9], NaOCl, commonly in solution by (1) electrolysis of sodium chloride solution under proper conditions, (2) reaction of calcium hypochlorite suspension in water and sodium carbonate solution, and then filtering. Used (1) as a bleaching agent for textiles and paper pulp, (2) as a disinfectant, especially for water, (3) as an oxidizing reagent.

Hypophosphite. Sodium hypophosphite, [CAS: 7681-53-0], $NaH_2PO_2 \cdot H_2O$, white solid, soluble, formed (1) by reaction of hypophosphorous acid and sodium carbonate solution, and then evaporating, (2) by reaction of NaOH solution and phosphorous on heating (poisonous phosphine gas evolved).

Hyposulfite. Sodium hyposulfite, sodium hydrosulfite (not sodium thio sulfate), $Na_2S_2O_4$, white solid, soluble, formed by reaction of sodium hydrogen sulfite and zinc metal powder, and then precipitating sodium hyposulfite by sodium chloride in concentrated solution. Used as an important reducing agent in the textile industry, e.g., bleaching, color discharge.

Iodide. Sodium iodide, [CAS: 7681-82-5], NaI, white solid, soluble, mp 651°C, formed by reaction of sodium carbonate or hydroxide and hydriodic

Fig. 1. Triple-effect evaporator used in concentrating soda solutions and preparation of solid sodium hydroxide

acid, and then evaporating. Used in photography, in medicine and as a source of iodide.

Manganate. Sodium manganate, Na_2MnO_4, green solid, soluble, permanent in alkali, formed by heating to high temperature manganese dioxide and sodium carbonate, and then extracting with water and evaporating the solution. The first step in the preparation of sodium permanganate from pyrolusite.

Nitrate. Sodium nitrate, nitrate of soda, Chile saltpeter, "caliche," [CAS: 7631-99-4], $NaNO3$, white solid, soluble, mp 308°C, source in nature is Chile, in the fixation of atmospheric nitrogen HNO_3 is frequently transformed by sodium carbonate into sodium nitrate, and the solution evaporated. Used (1) as an important nitrogenous fertilizer, (2) as a source of nitrate and HNO_3, (3) in pyrotechnics, (4) in fluxes.

Nitrite. Sodium nitrite, [CAS: 7632-00-0], $NaNO_2$, yellowish-white solid, soluble, formed (1) by reaction of nitric oxide plus nitrogen dioxide and sodium carbonate or hydroxide, and then evaporating, (2) by heating sodium nitrate and lead to a high temperature, and then extracting the soluble portion (lead monoxide insoluble) with H_2O and evaporating. Used as an important reagent (diazotizing) in organic chemistry.

Oleate. Sodium oleate, [CAS: 143-19-1], $NaC_{18}H_{33}O_2$, white solid, soluble, froth or foam upon shaking the H_2O solution (soap), formed by reaction of NaOH and oleic acid (in alcoholic solution) and evaporating. Used as a source of oleate.

Oxalates. Sodium oxalate, [CAS: 62-76-0], $Na_2C_2O_4$, white solid, moderately soluble, formed (1) by reaction of sodium carbonate or hydroxide and oxalic acid, and then evaporating, (2) by heating sodium formate rapidly, with loss of hydrogen. Used as a source of oxalate; sodium hydrogen oxalate, sodium binoxalate, sodium acid oxalate, $NaHC_2O_4 \cdot H_2O$, white solid, moderately soluble.

Palmitate. Sodium palmitate, $NaC_{16}H_{31}O_2$, white solid, soluble, froth or foam upon shaking the H_2O solution (soap), formed by reaction of NaOH and palmitic acid (in alcoholic solution) and evaporating. Used as a source of palmitate.

Permanganate. Sodium permanganate, permanganate of soda, [CAS: 10101-50-5], $NaMnO_4 \bullet 3H_2O$, purple solid, soluble, formed by oxidation of acidified sodium manganate solution with chlorine, and then evaporating. Used (1) as disinfectant and bactericide, (2) in medicine.

Phenate. Sodium phenate, sodium phenoxide, sodium phenolate, [CAS: 139-02-6], $NaOC_6H_5$, white solid, soluble, formed by reaction of sodium hydroxide (not carbonate) solution and phenol, and then evaporating. Used in the preparation of sodium salicylate.

Phosphates. Trisodium phosphate, tribasic sodium phosphate, [CAS: 7601-54-9], $Na_3PO_4 \bullet 12H_2O$, white solid, soluble, formed (1) by reaction of sodium hydroxide and the requisite amount of phosphoric acid, and then evaporating, (2) by reaction of disodium hydrogen phosphate plus sodium hydroxide, and then evaporating. Used (1) as a cleansing and laundering agent, (2) as a water softener, (3) in photography, (4) in tanning, (5) in the purification of sugar solutions; disodium hydrogen phosphate, dibasic sodium phosphate, $Na_2HPO_4 \cdot 12H_2O$, white solid, soluble, formed (1) by reaction of dicalcium hydrogen phosphate and sodium carbonate solution, and then evaporating the solution, (2) by reaction of sodium carbonate and the requisite amount of phosphoric acid, and then evaporating. Used (1) in weighting silk, (2) in dyeing and printing textiles, (3) in fireproofing wood, paper, fabrics, (4) in ceramic glazes, (5) in baking powders, (6) to prepare sodium pyrophosphate; sodium dihydrogen phosphate, monobasic sodium phosphate, $NaH_2PO_4 \cdot H_2O$, white solid, soluble, formed (1) by reaction of sodium carbonate and the requisite amount of phosphoric acid, and then evaporating, (2) by reaction of calcium monohydrogen phosphate and sodium carbonate solution, and then evaporating the solution. Used (1) in baking powders, (2) in medicine, (3) to prepare sodium metaphosphate; sodium pyrophosphate, $Na_4P_2O_7 \cdot 10H_2O$, white solid, soluble, mp about 900°C, formed by heating disodium hydrogen phosphate to complete loss of water, followed by crystallization from water solution. Used in electroanalysis; sodium metaphosphate, $NaPO_3$, white solid, soluble, mp 617°C, formed by heating sodium dihydrogen phosphate or sodium ammonium phosphate to complete loss of water, is an easily fusible phosphate forming colored phosphates with many metallic oxides, e.g., cobalt oxide. The hexametaphosphate, $(NaPO_3)_6$, is an important water-conditioning agent forming soluble complex compounds with many cations, e.g., Ca^{2+}, Mg^{2+}. Many polyphosphate compounds are known; their various uses include water softening and ion exchange. They are widely formulated in detergents, as are several of the simpler phosphates.

Phosphites. Disodium hydrogen phosphite, $Na_2HPO_3 \cdot 5H_2O$, white solid, soluble, formed by reaction of phosphorous acid and sodium carbonate, and then evaporating at a low temperature, mp of anhydrous salt is 53°C, at higher temperatures yields sodium phosphate and phosphine gas; sodium dihydrogen phosphite, $NaH_2PO_3 \cdot 2\frac{1}{2}H_2O$, white solid, soluble, formed by reaction of phosphorous acid and NaOH cooled to −23°C when the crystalline salt separates.

Salicylate. Sodium salicylate, [CAS: 54-21-7], $NaC_7H_5O_3$, white solid, soluble, formed by reaction of sodium phenate and CO_2 under pressure. Used as a source of salicylate and for salicylic acid.

Silicate. Sodium silicate, sodium metasilicate, "water glass," [CAS: 6834-92-0], Na_2SiO_3, colorless (when pure) glass, soluble, mp 1,088°C, formed by reaction of silicon oxide and sodium carbonate at high temperature; solution reacts with CO_2 of the air, or with sodium carbonate solution or ammonium chloride solution, yielding silicic acid, gelatinous precipitate. Sodium silicate solution is used (1) in soaps, (2) for preserving eggs, (3) for treating wood against decay, (4) for rendering cloth, paper, wood noninflammable, (5) in dyeing and printing textiles, (6) as an adhesive (e.g., for paper boxes) and cement. Sold as granular, crystals, or 40° Baumé solution.

Silicoaluminate. (See aluminosilicate, above.)

Silicofluoride. (See fluosilicate, above.)

Stearate. Sodium stearate, [CAS: 822-16-2], $NaC_{18}H_{35}O_2$, white solid, soluble, froth or foam upon shaking the water solution (soap), formed by reaction of NaOH and stearic acid (in alcoholic solution) and evaporating. Used as a source of stearate.

Sulfates. Sodium sulfate (anhydrous), "salt cake," [CAS: 7757-82-6], Na_2SO_4, sodium sulfate, decahydrate, "Glauber's salt," $Na_2SO_4 \cdot 10H_2O$, white solid, soluble, formed by reaction of sodium chloride and H_2SO_4 upon heating with evolution of hydrogen chloride gas. Used (1) in dyeing, (2) along with carbon in the manufacture of glass, (3) as a source of sulfate, (4) to prepare sodium sulfide; sodium hydrogen sulfate, sodium bisulfate, sodium acid sulfate, "nitre cake," $NaHSO_4$, white solid, soluble, formed by reaction of sodium nitrate and H_2SO_4, upon heating, with evolution of HNO_3. Used (1) as a cheap substitute for H_2SO_4. (2) in dyeing, (3) as a flux in metallurgy; sodium pyrosulfate, $Na_2S_2O_7$, white solid, soluble, formed by heating sodium hydrogen sulfate to complete loss of H_2O.

Sulfides. Sodium sulfide, [CAS: 1313-82-2], Na_2S, yellowish to reddish solid, soluble, formed (1) by heating sodium sulfate and carbon to a high temperature. Used (1) as the cooking liquor reagent (along with sodium hydroxide) in the "sulfate" or "kraft" process of converting wood into paper pulp, (2) as a depilatory, (3) in sheep dips, (4) in photography, engraving and lithography, (5) in organic reactions, (6) as a source of sulfide, (7) as a reducing agent; sodium hydrogen sulfide, sodium bisulfide, sodium acid sulfide, NaHS, formed in solution by reaction of NaOH or carbonate solution and excess H_2S.

Sulfites. Sodium sulfite, [CAS: 7757-83-7], Na_2SO_3, white solid, soluble, dilute solution readily oxidized in air, but retarded by mannitol (carbohydrates), formed by reaction of sodium carbonate or hydroxide solution and the requisite amount of SO_2, at high temperature yields sodium sulfate and sodium sulfide. Used (1) as a source of sulfite, (2) as a reducing agent, (3) to prepare sodium thiosulfate, (4) as a food preservative, (5) as a photographic developer, (6) as a bleaching agent and antichlor in the textile industry; sodium hydrogen sulfite, sodium bisulfite, sodium acid sulfite, $NaHSO_3$, white solid, soluble, formed by reaction of sodium carbonate solution and excess sulfurous acid. Uses similar to those of sodium sulfite.

Tartrate. Sodium tartrate, [CAS: 868-18-8], $Na_2C_4H_4O_6 \cdot 2H_2O$, white solid, soluble, formed by reaction of sodium carbonate solution and tartaric acid. Used in medicine; sodium potassium tartrate, Rochelle salt, $NaKC_4H_4O_6 \cdot 4H_2O$, white solid, soluble. Used (1) in medicine, (2) as a source of tartrate.

Thiosulfate. Sodium thiosulfate, "Hypo" [CAS: 7772-98-7], $Na_2S_2O_3 \cdot 5H_2O$, white solid, soluble, formed by reaction of sodium sulfite and sulfur upon boiling, and then evaporating. Used (1) in photography as fixing agent to dissolve unchanged silver salt, (2) as a reducing agent and antichlor. See also **Sodium Thiosulfate.**

Tungstate. Sodium tungstate, (sodium wolframate), [CAS: 13472-45-2], $Na_2WO_4 \cdot 2H_2O$, white solid, soluble, by reaction of NaOH solution and tungsten trioxide upon boiling, and then evaporating. Used (1) in fireproofing fabrics, (2) as a source of tungsten for chemical reactions.

Uranate. Sodium uranate, uranium yellow, Na_2UO_4, yellow solid, insoluble, formed by reaction of soluble uranyl salt solution and excess sodium carbonate solution. Used (1) in the manufacture of yellowish-green fluorescent glass, (2) in ceramic enamels, (3) as a source of uranium for chemical reactions.

Vanadate. Sodium vanadate, sodium orthovanadate, Na_3VO_4, white solid, soluble, formed by fusion of vanadium pentoxide and sodium carbonate. Used (1) in inks, (2) in photography, (3) in dyeing of furs, (4) in inoculation of plant life.

The larger number of organic compounds of sodium are for great part derivatives of oxygen-containing compounds such as salts of organic acids (several of which are discussed above), alcoholic and phenolic compounds (carboxylates, alkoxides, phenoxides, etc.). However, in some cases, sodium derivatives of nitrogen-containing compounds, as sodium benzamide, $C_6H_5C(O)NHNa$, and sodium anilide, C_6H_5NHNa, contain sodium-nitrogen bonds, while even sodium-boron bonds exist in certain boron-containing compounds, as sodium triphenylborene, $NaB(C_6H_5)_3$, and others; and in a number of compounds sodium is carbon-connected, as in methylsodium, CH_3Na, ethylsodium, C_2H_5Na, cyclopentadienylsodium, C_5H_5Na, and sodium triphenylmethane, $NaC(C_6H_5)_3$.

The organometallic compounds of sodium may be divided into two groups, differing in properties. One group, e.g., ethylsodium, consists of compounds that are colorless, insoluble in organic solvents, and that electrolyze readily in diethylzinc solution. Another group, e.g., benzylsodium, $C_6H_5CH_2Na$, are colored, and soluble in organic solvents. Like all the alkali metals, sodium coordinates with salicylaldehyde. Its tetracovalent compounds, with those of potassium, are the more stable of the group, for the following reasons: (1) Increasing ionic size carries with it increasing electropositiveness and ease of ionization, which diminishes the tendency to coordinate. (2) The increasing distance of the nucleus from the coordinating electrons with increasing atomic volume makes it less likely that additional electrons will be held with ease. (3) On the other hand, there is an increase in the maximum coordination number with the elements of higher atomic number. These factors are in keeping with a maximum stability for the tetracovalent compounds occurring with sodium.

Sodium in Biological Systems. Sodium is essential to higher animals which regulate the composition of their body fluids and to some marine organisms. The several important roles played by the sodium cation in biological systems, frequently in concert with the potassium cation are described in the entry on **Potassium and Sodium (In Biological Systems)**.

Additional Reading

Emsley, J. and J. Neruda: *The Elements,* Oxford University Press, Inc., New York, NY, 1996.

Garrett, D.E.: *Sodium Sulfate: Handbook of Deposits, Processing and Use,* Academic Press, Inc., San Diego, CA, 2001.

Greenwood, N.N. and A. Earnshaw: *Chemistry of the Elements,* 2nd Edition, Butterworth-Heinemann, Inc., Woburn, MA, 1997.

Kent, J.A.: *Riegel's Handbook of Industrial Chemistry,* 9th Edition, Chapman & Hall, New York, NY, 1992.

Krebs, R.E.: *The History and Use of Our Earth's Chemical Elements: A Reference Guide,* Greenwood Publishing Group, Inc., Westport, CT, 1998.

Lagowski, J.J.: *MacMillan Encyclopedia of Chemistry,* Vol. 1, MacMillan Library Reference, New York, NY, 1997.

Lewis, R.J. and N.I. Sax: *Sax's Dangerous Properties of Industrial Materials,* 10th Edition, John Wiley & Sons, Inc., New York, NY, 2000.

Lide, D.R.: *CRC Handbook of Chemistry and Physics,* 84th Edition, CRC Press, LLC., Boca Raton, FL, 2003.

Meyers, R.A.: *Handbook of Chemicals Production Processes,* The McGraw-Hill, Companies, Inc., New York, NY, 1986.

Parker, S.P.: *McGraw-Hill Encyclopedia of Chemistry,* 2nd Edition, The McGraw-Hill Companies, Inc., New York, NY, 1993.

Robinson, A.L.: "Sodium Atoms Stopped and Confined," *Science,* **229,** 39–41 (1985).

Staff: *ASM Handbook—Properties and Selection: Nonferrous Alloys and Pure Metals,* ASM International, Materials Park, OH, 1990.

Stwertka, A. and E. Stwertka: *A Guide to the Elements,* Oxford University Press, Inc., New York, NY, 1998.

SODIUM CARBONATE.   See **Leavening Agents**.

SODIUM CHLORIDE.   NaCl, formula weight 58.44, white solid, cubic crystal structure, mp 800.6°C, bp 1,413°C, sp gr 2.165. Commonly called "salt," the mineral name for rock salt is *halite*. See also **Halite (Rock Salt)**. The compound is soluble in H_2O (35.7 g/100 g H_2O at 0°C; 39.8 g/100 g H_2O at 100°C), only slightly soluble in alcohol, and insoluble in HCl. Sodium chloride is produced in nearly all nations of the world, but some only have a sufficient supply for local needs. The leading salt-producing nations include the United States, China, the former U.S.S.R., West Germany, France, the United Kingdom, India, Italy, Canada, and Mexico. In 28 states of the United States and in several provinces of Canada, salt occurs as bedded or domed deposits. Most of the rock salt produced in the United States comes from Michigan, New York, Texas, Ohio, Louisiana, and Kansas. Purity ranges from 97% NaCl for Kansas salt to 99% purity and higher for Louisiana salt. The main impurities are calcium sulfate (0.5–2%), dolomite, quartz, calcite, and traces of iron oxides. Natural rock salt is mined much as coal and usually marketed without purification, after crushing and screening. For most industrial and consumer requirements, the impurities are harmless. There is no evidence that bacteria exist in rock salt. Additionally, there is some solar salt production in the Great Salt Lake area of Utah and on the west coast. Salt deposits date back to past geologic ages and are believed to be the results of evaporated impounded sea water.

Purified salt for table and industrial processing requirements of a special nature is made by dissolving raw sodium chloride in H_2O and then evaporating the H_2O to form a final product. There are several types of evaporated salt, including *granulated salt* in which each crystal is a tiny cube, and *grainer* or *flake salt*, made up of irregularly shaped crystals, often thin and flaky and unusually soft. A process for producing evaporated salt is shown in Fig. 1. Holes are drilled into the salt deposits, after which H_2O is pumped into the beds to create a brine which then is brought to the surface for refining. In this method, all insolubles are left in the bed. After some pretreatment to remove hardness and dissolved gases, the semipure brine is evaporated in multiple-effect vacuum pans. The salt crystallizes as perfect cubes of NaCl. In the system shown, each vacuum pan performs not only as an evaporator, but also as a boiler. The vapors from a preceding pan are used to heat the contents of the following pan. This system of heat economizing is possible because each succeeding pan in the series is under less pressure—hence the contents boil at a lower temperature. See Fig. 2. The lower pressure in succeeding pans results from condensation of the vapors as well as assistance from vacuum pumps. Crystal size is controlled by evaporation rate, the latter depending on the degree of vacuum, temperature, and agitation maintained. When grown to proper size, the crystals drop to the bottom of the pans and fall into the salt legs, from which they are drawn continuously in the form of a slurry. After washing, filtering, cooling, and screening, they are packaged. See also **Evaporation**.

Grainer salt is made by surface evaporation of brine in flat pans open to the atmosphere. Heat usually is furnished by steam pipes located a few inches below the tank bottom. Crystals form at the surface of the brine and are held there temporarily by surface tension. Thus, they grow laterally for awhile and form thin flakes. But, as they grow, they tend to sink and this process imparts a peculiar, hollow pyramid-like structure to them. Such crystals are called *hopper crystals*. Ultimately, the crystals sink to the bottom where they are scraped to one end of the pan. The crystals are fragile and during handling they break up, finally assuming a flake-like shape. Thus, the term *flake salt.*

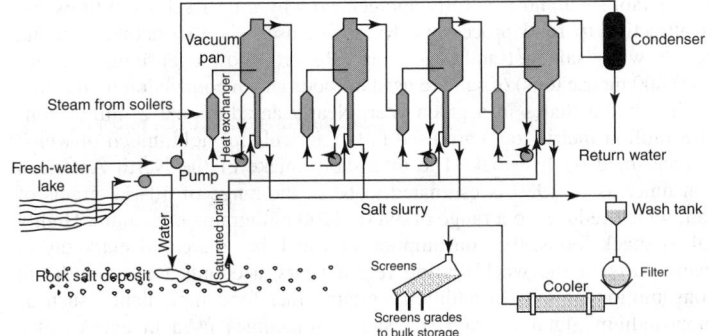

Fig. 1.   Multiple-effect vacuum pans used in production of sodium chloride from brine. The saturated brine is formed by pumping fresh water directly into the rock salt deposit, leaving insoluble materials in the deposit

Fig. 2. Portion of train of evaporator bodies in a multiple-effect vacuum evaporation system used in production of sodium chloride

In the *recrystallizer process* for making salt, advantage is taken of the fact that the solubility of NaCl increases with temperature whereas the solubility of the principal impurity, $CaSO_4$, decreases with temperature. In *solar* facilities, the raw brine is pumped into concentrating ponds where most of the H_2O is evaporated. Some of the impurities are precipitated out in this stage, after which the saturated brine is transferred to crystallization ponds where the salt crystallizes out at a high degree of purity. Since evaporation occurs at the surface of the ponds, hopper crystals are formed as in the grainer process, with flake salt being the final product.

Uses. Sodium chloride is a very high-tonnage material. In addition to its familiar use in the diets of man and animals, representing a small part of total production, large quantities are used by highway departments to control icy road conditions, in agriculture, and as a basic chemical raw material. The chemical industry consumes about two-thirds of the salt produced, the majority of it going to electrolytic plants. Some of the basic inorganic chemicals that require salt as a starting material include soda ash, calcium chloride, caustic soda, sodium sulfate, sodium bisulfate, HCl, sodium cyanide, sodium hypochlorite, and chlorine. See also **Chlorine**; and **Sodium**.

Salt and the Diet

In food processing, the preservative and organoleptic qualities of salt are well established and it is fully appreciated why use of salt even to excess is attractive to food processors. Excessive usage is also habitual among people who "salt first and taste later." Reports show over 1 million tons (900,000 metric tons) of salt are used in foods and in connection with eating in the United States in a given year. Nearly an additional 2 million tons (1.8 million metric tons) are used in the agricultural field, much of which is consumed by livestock. The total daily intake of the North American consumer as of 1980 is estimated to be in the range of 10–12 grams of salt, which reduces to a range of 3900–4700 milligrams of sodium. Highly salted snack foods, the consumption of which has increased markedly in many parts of the world during recent years, accounts for a significant consumption of salt. In addition, certain other food ingredients, such as monosodium glutamate and soy sauce, sometimes used in excess, also contribute to the average intake of sodium.

Sodium and chloride are not normally retained in the body even when there is a high intake. See also **Chloride (Biological Aspects)**. Amounts consumed in excess of need are excreted, so that the level in the body is maintained within very narrow limits, as is also the chloride, regardless of intake. The primary route of excretion is via the urine, with substantial amounts also lost in sweat and feces. About 50% of the sodium in the human body is located in the extracellular body fluids; 10% inside the cells; and 40% in the bones. Chloride is found mainly in the gastric juice and other body fluids.

Essential though sodium is to the normal functioning of the human body, there has been considerable concern over the last few years, about the amount of salt in the diet. This concern centers mainly on possible relationship between salt and hypertension (high blood pressure).

Hypertension afflicts more than 20% of the world population, with an estimated 24 million cases in the United States as of 1980. In 1976, Marx reported that, in about 90% of these cases, the actual causes of hypertension cannot be pinpointed.—This was in face of the fact that research on the possible role of sodium in essential hypertension had been underway for 60 years or longer. Tests of unmedicated persons with essential hypertension have been found to indicate a lowering of blood pressure when sodium intake is restricted below one gram per day—and that the blood pressure rises again if additional sodium is taken. However, in other studies, some persons have retained a normal blood pressure level even when fed substantially increased amounts of salt (or other sodium-ion-furnishing substances). In 1976, Freis reported positive correlations between estimated average salt consumption of various ethnic populations and their incidence of hypertension. But such studies are complicated by many factors, including the inability to control or eliminate other possible causes of hypertension, such as obesity, genetic predisposition, general nutritional status, and potassium intake. It also has been generally proved extremely difficult to determine differences between individuals within these cultures.

Nevertheless, the concern remains on the part of a large number of professional people who feel that someday a definitive correlation will be made. And, with considerable awareness of the lay public in this regard, very definite pressures are being exerted on food processors to reduce salt usage and to more accurately label their merchandise in this regard.

The physiology of the sodium-potassium relationship is explained in some detail in the entry on **Potassium and Sodium (In Biological Systems)**.

Concerning the sodium content (much of which is derived from salt), the following composition data may be of interest. The figures in parentheses are milligrams of sodium per 100 grams of food.

Meats: Canadian bacon (2555), bacon (1077), cured ham (860), beef liver (136), pork chops (60), ground beef (48).
Cheeses: Parmesan (1848), process (1421), blue (1396), brick (557), cream cheese (294).
Other dairy products: Ice cream (83), whole milk (50), sherbet (45).
Miscellaneous foods: Pretzels (7800), soy sauce (regular) (6082), dill pickles (4000–5000), soy sauce (mild) (3569), green olives (2400), soda crackers (1100), salted peanuts (groundnuts) (418), eggs (122).
Vegetables: Beet greens (130), celery (126), dandelion greens (76), kale (75), spinach (60), beets (60), watercress (52), turnips (49), carrots (47), artichokes (43), collards (43), mustard greens (32), Chinese cabbage (20). Other common vegetables range between (10) and (18).

Sodium Chloride and Energy

As pointed out by Wick (*Oceanus*, **22**, 4, 28, 1980), most of the energy in the oceans is bound in thermal and chemical forms. Although thermal energy is presently commanding the most attention, within the past few years another, rather unusual, form has received notice. Where rivers flow into the oceans a completely untapped source of energy exists—represented by a large osmotic pressure difference between fresh and salt water. If economical ways to tap these salinity gradients could be developed, large quantities of energy would be available. See also **Solar Energy**.

SODIUM HYPOCHLORITE. See **Bleaching Agents**.

SODIUM THIOSULFATE. [CAS: 7772-98-7], $Na_2S_2O_3 \cdot 5H_2O$, formula weight 248.19, white crystalline solid, decomposes above 48°C, sp gr 1.685. Also known as "hypo" and sometimes misnamed "hyposulfite," sodium thiosulfate is very soluble in H_2O (301.8 parts in 100 parts H_2O at 60°C), soluble in ammonia solutions, and very slightly soluble in alcohol. When sodium thiosulfate is added to an acid, thiosulfuric acid $H_2S_2O_3$ may be formed, but only for an instant, immediately decomposing into sulfur and SO_2.

Sodium thiosulfate is used: (1) to dissolve silver chloride, bromide, iodide in the photographic "fixing" bath, soluble sodium silver thiosulfate being formed plus sodium chloride, bromide, iodide; (2) in reaction with iodine in solution, sodium tetrathionate and sodium iodide being simultaneously formed, or with ferric salt solution, sodium tetrathionate and ferrous being simultaneously formed; and, (3) in reaction with chlorine as an "antichlor" forming sulfate and chloride. Sodium thiosulfate reacts with silver nitrate solutions yielding silver sulfide, brown precipitate, and with permanganate yielding manganous. Sodium amalgam changes sodium thiosulfate to sodium sulfide plus sodium sulfite.

Sodium thiosulfate is formed: (1) by reaction of sodium sulfite solution and sulfur upon warming; (2) by reaction of sodium sulfite solid and sulfur upon heating; and, (3) by complex reaction of sulfur and sodium hydroxide solution upon warming. Sulfur yields sodium sulfide plus sodium sulfite, and the latter reacts with excess sulfur, forming sodium thiosulfate. The sodium sulfide present may be converted into sodium thiosulfate by passing in SO_2 until the solution changes from yellow to colorless.

There are numerous other thiosulfates, including potassium, magnesium, calcium, barium, mercury, lead, and silver. All are soluble in H_2O except Ba, Pb, and Ag thiosulfates.

Thiosulfates are commonly identified as follows:

1. Dilute acids precipitate sulfur from thiosulfates (difference from sulfides and sulfites).

2. Zinc sulfate and sodium hexacyanoferrate(II) give no color (difference from sulfites).

SOFT. A nontechnical word used by chemists in several senses, it describes the following: (1) an acid having little or no positive charge and whose valence electrons are easily excited; (2) water that is relatively free from calcium compounds. See also **Water (Hard)**; (3) wood from coniferous trees.

SOFTENER. 1. A substance used when dry powders are added to a polymeric material (e.g., rubber or plastic) to reduce the friction of mechanical mixing and to facilitate subsequent processing. It exerts both lubricating and dispersing action, often by means of emulsification. Examples are vegetable oils, asphaltic materials, and stearic acid, the latter being especially effective with carbon black. It is difficult to distinguish precisely between softeners and plasticizers; in general, softeners do not enter into chemical combination with the polymer, and their softening effect tends to be temporary.

2. A fatliquoring agent used to soften leather.

3. A sulfonate oil, fatty alcohol, or quanternary ammonium compound used in textile finishing to impart superior "hand" to the fabric and facilitate mechanical processing.

4. A substance that reduces the hardness of water by removing or sequestering calcium and magnesium ions; among those used are various sodium phosphates and zeolites.

SOFTENING (Water). See Water Conditioning; **Water Treatment (Boiler)**.

SOIL. All consolidated earth material over bedrock. Soil is approximately equivalent to regolith.[1] Agriculturally, soil is any one of many varied natural media that support or can support land plant growth outdoors; or, when in containers, indoors. The lower limit of *topsoil* is normally the lower limit of biologic activity, which usually coincides with the common rooting of native perennial land plants. The word *soil* is derived from the Latin *solum* for "ground."

The upper part of the regolith is divided into topsoil and subsoil. The topsoil is usually a relatively thin layer or zone of the more highly decomposed mineral constituents of the regolith and contains a varying proportion of organic material called *humus*. This soil zone is the habitat

of the shallow-rooted plants, such as most grasses. The topsoil usually passes gradationally into the subsoil, which supplies some of the moisture and food for the deeply rooted plants and trees. The subsoil may or may not pass gradationally into the underlying bedrock. Topsoil is easily destroyed by erosion when not protected by a mantle of vegetation.

Soil serves: (1) As a foundation for holding plants in place, whether tiny grasses or huge trees; (2) as a protective covering for the root structures of plants; (3) as a source and/or medium of exchange for supplying plants with nutrients; (4) and as a reservoir for moisture upon which growing plants can draw. Soil also must be capable of allowing excessive moisture to pass through its pores and drain to a lower level so that the soil will not remain excessively wet. Properties such as permeability and strength are not only of importance to the agriculturist, but to civil engineers and construction people who excavate, drill through, and handle soil in connection with buildings, roads, tunnels, etc.

Soil is a subsystem that interacts as part of a four-element system: (1) The *climatic or environmental subsystem* prevails immediately above the ground level and thus is the microclimate for a particular location. The principal variables of this subsystem are temperature, humidity, precipitation, and solar radiation—all of which interact constantly with soil. (2) The characteristics and patterns of the *hydrologic subsystem* determine essentially how water reaches the soil, both from above and below, and how water is carried away or drained from the soil. (3) The *plant subsystem* reduces the nutrient and moisture content of the soil, depending upon the particular uptake characteristics of a given plant. The plant subsystem also contributes in a major way to hold the soil in place and to protect it from disintegration and destruction by water and wind erosion. The plant subsystem also distributes moisture over the top of the soil so that all porous paths of the soil can be used to transport water rather than overloading and hence enlarging only some of the pores. The plant subsystem also protects the top of the soil against drying into a hard crust during periods of drought. Once disturbed, the characteristics of soil are difficult to replicate—a problem that arises when large projects, such as strip-mining, remove vast amounts of soil. (4) The *soil subsystem*, the properties of which are described briefly in this entry.

Lack of sufficient attention to the long-term protection of soil has caused innumerable problems and losses over the years. Although warnings of gross problems arising from soil destruction and land mismanagement were given in North America as early as the latter part of the 1600s, it required the rudest of awakenings to precipitate national interest and action in soil conservation. This came in the early 1930s in the form of the Great Dust Bowl, a national disaster that affected some 96 million acres (38.4 million hectares) of farmland in the southern part of the Great Plains region, involving parts of Kansas, Oklahoma, Texas, New Mexico, and Colorado. During just a few years of severe droughts, accompanied by frequent high winds, literally billions of tons of soil were lost. Organic matter, clay, and silt were lifted and carried for great distances. There were times when the heavily laden skies as far east as the Atlantic coast were darkened. Sand and silt dunes from 4 to 10 feet (1.2 to 3 meters) in height were formed in many locations. In some parts of the Dust Bowl, as much as 80% of the land suffered from wind erosion. Parallel situations have occurred in several other areas of the world.

The Soil Conservation Service of the U.S. Department of Agriculture was established in 1935. Concentrated and participative programs with land users in the Great Plains region from the mid-1950s to the present time have provided impressive improvements: (1) 2.4 million acres (1 million hectares) of permanent vegetative cover have been established; (2) 1.0 million acres (0.4 million hectares) of field and wind stripcropping have been introduced; (3) 169 thousand (68 thousand hectares) of grasslands have been reestablished; (4) 41 thousand acres (16.8 thousand hectares) of trees or shrubs have been placed as windbreaks; (5) 81 thousand miles (150 thousand kilometers) of terraces have been constructed; (6) 5.4 million acres (2.2 million hectares) of brush control have been provided; and (7) 9 thousand miles (16.7 thousand kilometers) of pipelines to provide water for livestock grazing lands have been installed.

Soil Characteristics and Classification

A soil is a naturally occurring three-dimensional body with morphology and properties resulting from effects of climate, flora and fauna, parent rock materials, topography, and time. A soil occupies a portion of the land surface, is mappable and is composed of horizons that parallel the land surface. A vertical section downward through all the horizons of the soil is called a *soil profile*. See Fig. 1.

---

[1] A general term for the entire layer or mantle of fragmental and loose, incoherent, or unconsolidated rock material, of whatever origin (residual or transported) and of a much varied character, that nearly everywhere forms the surface of the land and overlies or covers the more coherent bedrock. It includes rock debris (weathered in place) of all kinds, volcanic ash, glacial drift, alluvium, loess and aeolian deposits, vegetal accumulations, and soils.

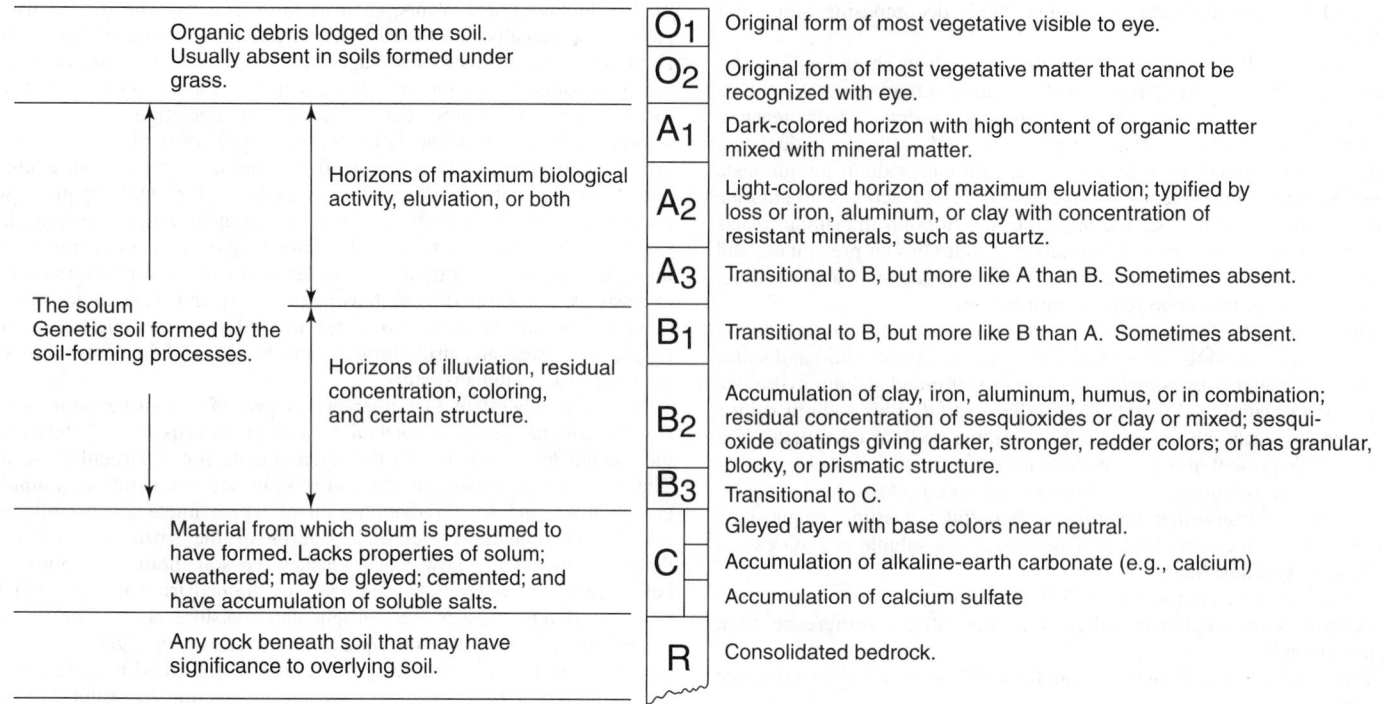

Fig. 1.   Hypothetical soil profile that has all principal horizons. Not all horizons shown are present in any given profile, but every profile has some of them. Terms used in diagram: *Eluviation* is the downward movement of soluble or suspended material in a soil from the A horizon to the B horizon by groundwater percolation. The term refers especially, but not exclusively, to the movement of colloids, whereas the term *leaching* refers to the complete removal of soluble materials. *Illuviation* is the accumulation of soluble or suspended material in a lower soil horizon that was transported from an upper horizon by the process of eluviation. *Gleying* is soil mottling, caused by partial oxidation and reduction of its constituent ferric iron compounds, due to conditions of intermittent water saturation. Process is also called *gleization*. (*Adapted from USDA diagram*)

Characterization of a soil requires selection of a representative profile that is described as quantitatively as possible, utilizing comparative charts for color, structure, and other properties, and accurately measuring soil horizons. Soils are collected from horizons and analyzed for particle-size distribution, pH, organic carbon, nitrogen, free iron oxide, calcium carbonate equivalent, moisture tension, cation-exchange capacity, extractable cations (calcium, magnesium, hydrogen, sodium, potassium), base saturation, and bulk density, among other factors.

Soil classification has been oriented to soil properties in recent years, but still is tempered with concepts of soil genesis, with external associations, and with the use of the soil. The first systematic classification was by Dokuchaiev in Russia in 1882. Based upon field and laboratory characteristics, soils were grouped into three categories—*normal soils* of the dry-land vegetative zones and moors, *transitional soils* of washed or dry land sediments; and *abnormal soils*. The system involved properties of the soil with external associations of climate and vegetation. Later, an associate (Sibirtsev) renamed the highest classes *zonal, intrazonal,* and *azonal*.

A traditional classification of soils includes three categories: (1) *Young soils*. These usually show their relationship to the parent material and are typical flood plain and hilly land deposits, when the soil surfaces are constantly being replenished or disturbed. (2) *Mature Soils*. These usually cover relatively flat lands where there are good drainage conditions but relatively little erosion. The development of these soils has gone so far in some cases, particularly in semi-arid regions, that little relation is shown to the parent material and their nature has therefore been principally determined by climatic and organic factors. (3) *Old Soils*. These usually cover old flat surfaces that have not been disturbed by erosion or sedimentation for a long time. Such soils, due to the dominance of climatic factors in their formation, have lost many of their original characteristics and have, therefore, developed abnormal features. When soils are intensively cultivated their mineral and organic constituents are rapidly depleted and must be replenished by rotation of crops and the application of natural fertilizers. The method of allowing the land to remain fallow is now known to be inefficient. The complete removal of vegetable cover, such as may result from overgrazing, deforestation, or dry farming, exposes the soil to rapid erosion and destruction.

A more scientific classification of soils, adapted by the U.S. Department of Agriculture, is given in Table 1. Systematic classification of soils in the United States began with the work of Coffey in 1912 and resulted in the first comprehensive system by Marbut in 1936. Considering the size of the earth and the large number of soils represented, the detailed cataloging of soils for any country is a tremendous task. To simplify the task to some extent and to make findings more meaningful from a practical viewpoint, the European Commission on Agriculture (Working Party on Soil Classification and Survey) in 1966 correlated types of soils (soil units) with several regions designated by geography, geology, and climatology, and, to some extent, by the traditional use of the soils. Mixed criteria enter into soil classification schemes simply because the various physical or chemical characteristics, considered separately or together, do not fully identify a soil. The principal categories adopted by the European Commission include: Lowlands, Mountainous Areas and Highlands, Volcanic Areas, Zones of the Tundras and Fields, Zones of the Boreal Forests of Conifers and Birch, Zones of Mixed Forests of Conifers and Broadleaved Trees, Zones of the Central European Beech Forests and Oak Forests, Zones of the Oak Forests and of the Atlantic Heaths, Zones of the Continental Oak Forest, Zones of the European Grassland, Zones of the Mediterranean Sclerophyll Forests, Zones of the Juniper Forests and the Mediterranean Steppes, Zones of the Montane Mediterranean Forests, Zones of the Subalpine Mediterranean Forests, and Zones of the Arabo-Caspian Steppes.

Soil Genesis. The origin and processes of soil formation usually are inferred, by relating measured morphological, physical, and chemical properties of a part to other parts of a given soil. And, during the last several decades, laboratory experimentation has revealed a better understanding of many of these processes. A factor to be stressed is that, in general, these processes occur over very long periods of time and frequently under multivariate conditions—conditions that are extremely difficult to duplicate and speed up in the laboratory. In the late 1950s, an interesting group of experiments revealed information concerning the formation of something similar to podzolic soil. Organic and distilled water leachates from tree leaves were passed through columns of different soil materials. Bleached surface layers and subjacent layers of stronger color formed in the columns. Effluent solutions from the base of the columns contained detectable amounts of calcium, magnesium, iron, manganese, phosphorus,

TABLE 1. SOIL CLASSIFICATION SYSTEM (U.S.D.A)

Order and suborders	Definitions and properties
Entisols	Weakly developed soils on freshly exposed rock or recent alluvium without genetic horizons. While alluvium may be rich in plant nutrients, entisols often are too shallow, too wet, or too dry for agricultural purposes.
E1	*Aquents.* Seasonally or perennially wet.
E2	*Orthents.* Loam or clay texture, often shallow to bedrock.
E3	*Psamments.* Sand or loamy sand texture.
Vertisols	Clay soils that have deep wide cracks during periods of moisture deficiency. During rainfall, vertisols swell, slide and produce warping.
V1	*Uderts.* Usually moist, with cracks open less than 90 days/year.
V2	*Usterts.* Dry and cracked more than 90 days/year.
Inceptisols	Soils that are beginning to show development of genetic horizons. Inceptisols lack evidence of weathering and usually are found in humid climates where leaching is active.
I1	*Andepts.* Soils containing amorphous or allophanic clay, often associated with volcanic ash and/or pumice.
I2	*Aquepts.* Seasonally or perennially wet.
I3	*Ochrepts.* Soil with thin, light colored surface horizons.
14	*Tropepts.* Continuously warm or hot.
15	*Umbrepts.* Dark surface horizons; medium to low base supply.
Aridisols	Soils which contain little organic matter or nitrogen. They are usually dry for more than 6 months/year. In numerous areas, salts accumulate on or near the soil surface. Since the nutrient content, except nitrogen, of aridisols is often high, these soils can be productive with irrigation and nitrogen application. Salt accumulation can be a problem with some crops.
D1	*Undifferentiated aridisols.*
D2	*Argids.* Soils with horizons of clay accumulation.
Mollisols	Soils with dark, thick, organic-rich surface horizon, high base supply. Mollisols are highly fertile and can support a variety of crops.
M1	*Albolls.* Soils with seasonally high water tables.
M2	*Borolls.* Cool or cold soils.
M3	*Rendolls.* Soils with subsurface accumulations of calcium carbonate, but no clay.
M4	*Udolls.* Temperate or warm, usually moist.
M5	*Ustolls.* Temperate or hot. Dry more than 90 days/year.
M6	*Xerolls.* Cool to warm. Moist in winter and continuously dry more than 60 days/year.
Spodosols	Soils found primarily in cool and humid forested regions. Spodosols have subsurface accumulations of amorphous materials, mainly iron and aluminum oxides. These soils are usually strongly leached, but can be used for crop support with addition of lime and fertilizer.
S1	*Undifferentiated spodosols.*
S2	*Aquods.* Seasonally wet.
S3	*Humods.* Soils with subsurface accumulations of organic matter.
S4	*Orthods.* Soils with subsurface accumulations of organic matter, iron, and aluminum.
Alfisols	Soils of middle latitudes and degraded grasslands soils. Alfisols are strongly weathered, with gray to brown surface horizons, a subsurface clay accumulation, and a medium-to-high base supply. With adequate lime and fertilizer, the alfisols will continue to produce a variety of crops.
A1	*Boralfs.* Cool soils.
A2	*Udalfs.* Temperate to hot. Usually moist.
A3	*Ustalfs.* Temperate to hot. Dry more than 90 days/year.
A4	*Xeralfs.* Temperate to warm. Moist in winter and continuously dry more than 60 days in summer.
Ultisols	Strongly weathered soils of the middle and low latitudes. Ultisols are usually moist and low in organic matter. These soils have experienced a high degree of mineral alteration and extensive leaching. With fertilizer additions and good management, ultisols can support crops.
U1	*Aqults.* Seasonally wet.
U2	*Humults.* Temperate or warm. Moist all year. High content of organic matter.
U3	*Udults.* Temperate to hot. Usually moist.

Order and suborders	Definitions and properties
U4	*Ustults.* Warm or hot. Dry more than 90 days/year.
Oxisols	The predominant soils of the Tropics. Oxisols have experienced the greatest degree of mineral alteration and horizon development of any soil. The humus breakdown is rapid and the soils are usually deep and porous. Oxisols require fertilization to support continued crop production.
O1	*Orthox.* Hot and nearly always moist.
O2	*Ustox.* Warm or hot. Dry for long periods, but moist for at least 90 days/year.
Histosols	Bog or peat soils composed primarily of vegetative debris in various stages of decomposition.
Mountain	Soils with various temperature and moisture parameters. Altitude, aspect, steepness of slope, and relief cause these soils to vary greatly within short distances. In many places, soil will be entirely absent.
Soil-absent	Rugged mountains and icefields.

Note: Further details can be obtained from "Soil Classification, A Comprehensive System, 7th Approximation," Soil Conservation Service, U.S. Department of Agriculture, Washington, DC. (Published periodically).

potassium, and sodium. Very fine silicate clays, e.g., illite, montomorillonite, vermiculite, and chlorite, also were suspended in the effluent. Removal, transfer, and transformation were demonstrable experimentally. Examination of the columns showed that clay was partially removed from the bleached layers and was deposited in voids in the lower layers. The experiments showed removal (*eluviation*) and addition (*illuviation*) actually occurring and at a much accelerated pace.

Organic matter probably best illustrates additions to a soil and is formed in the biological decomposition of plant and animal residues by soil microorganisms. Plants supply most of the organic matter as a dry material added to the soil surface and as roots in the subsurface. It has been estimated that short grass prairie in semiarid regions may annually add 0.7 ton/acre (1.6 metric tons/hectare) of dry matter; tall grass prairie in subhumid regions, 0.8 to 1.7 tons/acre (1.8 to 3.8 tons/hectare); pine forest in more humid areas, 2.1 tons/acre (4.7 metric tons/hectare); and tropical rain forest, from 45 to 90 tons/acre (101 to 202 metric tons/hectare). Under bluegrass roots, additions may amount to 2.4 tons/acre (5.4 metric tons/hectare) in the top 4 inches (10 centimeters) of soil.

During decomposition, plant materials are converted to carbon dioxide, water, mineral elements, and other chemically altered substances. Less-resistant materials are consumed first by soil microbes—so that more resistant plant materials remain with the new organic compounds that are synthesized by the organisms. At any time, the organic matter at a place in the soil reflects an equilibrium state of the addition of new material to the system, removal of more readily decomposable materials, and transformation to other forms by microorganisms and other agents. Organic matter also may be transferred within the soil by physical and physicochemical processes. Burrowing animals, worms, and insects turn over the soil and physically mix adjacent portions. Freezing and thawing and wetting and drying also assist in the process. Colloidal organic matter may be flushed downward or laterally and coagulate as coatings on structural aggregates in the soil.

The more unstable organic compounds are rapidly oxidized to carbon dioxide and water by various biochemical processes, while the more stable fractions accumulate. Conjugated ring compounds containing carbon, hydrogen, oxygen, nitrogen, phosphorus, and sulfur and other elements in small quantities accumulate in relatively stable organic and organomineral colloidal complexes. Lignin-like, phytin-like, and nucleo-protein-like compounds are included. Sorption of the organic matter on mineral colloid surfaces, particularly layer silicates, such as montmorillonite, helps to stabilize the organic matter against biochemical oxidation. In tropical soils, high stability of soil organic matter is imparted by coatings of aluminum hydroxide and red ferric oxide. Organic and iron oxide colloids, when fairly abundant, stabilize the soil into porous aggregates through which ample air and water can circulate.

Localized spots of decomposing organic matter are important in reducing small but important quantities of iron to ferrous form and manganese to divalent form so that they become available to plants. Moderately to highly alkaline soils sometimes have inadequate activity of the reduced forms of iron and manganese, particularly in the absence of sufficient organic matter.

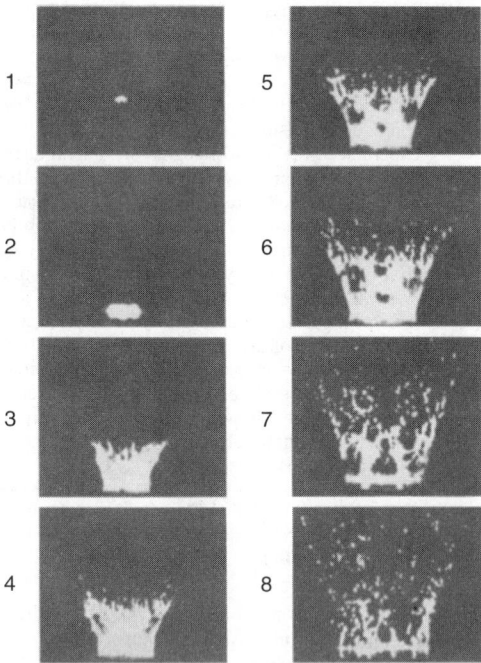

Fig. 2. Interaction of raindrops with the soil surface is an important component of the erosion process. Frames shown here were made by a 16-millimeter movie camera capable of speeds from 150 to 8000 frames per second. In the experiment, water drops are released from a tower 40 feet (12 meters) high and strike a plate glass target table. Drops from 5 to 6 millimeters in diameter are produced. Target plate is covered with water approximately 0.5 millimeter deep. (*North Central Soil Conservation Research Center, U.S. Department of Agriculture, Morris, Minnesota*)

Radiocarbon dates of organic matter from the surface horizons of soils not only reflect the equilibrium status, but point out the turnover of the system. One example of research, for example, has shown that in the Edina soil in southern Iowa, organic matter is 410 ± 110 years old in the 6-inch (15-centimeter) top layer. In the next subjacent layer, the age is 840 ± 220 years. At depths of 23 to 25 inches (58.4 to 63.5 centimeters), the organic carbon is 1545 ± 110 years old. The entire soil has been estimated as 14,000 years old.

The four kinds of changes that develop soil horizons are dependent upon many basic processes, such as hydration, oxidation, reduction, solution, precipitation, freezing, thawing, wetting, drying, among others. These processes, in turn, are dependent upon the four fundamental factors of soil formation: (1) nature of the parent material; (2) topography; (3) climate; and (4) biological activity that occurs in the upper strata of the soil. To

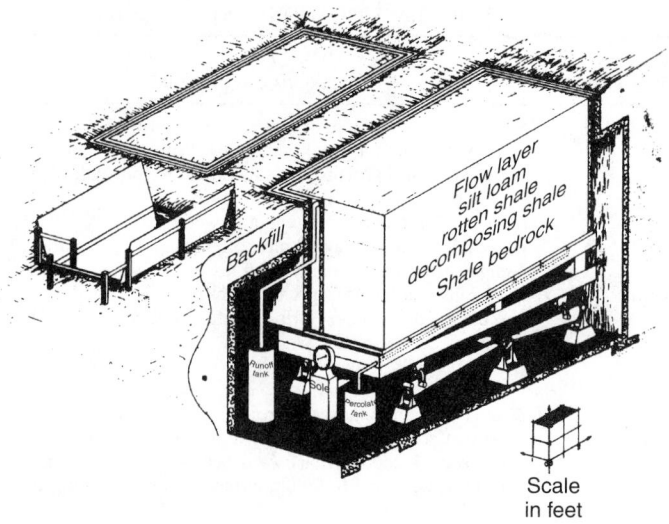

Fig. 3. A lysimeter, which provides a means for isolating soil masses and recording weight changes and water percolation. Such instruments provide accurate assessments of moisture behavior in soil. The lysimeter shown here represents $\frac{1}{500}$ acre (0.0008 hectare) and is 8 feet (2.4 meters) deep. The soil weighs 65 tons (58.5 metric tons), yet can be weighed to a precision of 5 pounds (2.3 kilograms). Soil scientists use lysimeters to study evapotranspiration, moisture consumption by crops, precipitation, water movement, and pollution. (*USDA diagram*)

these factors must be added time and imposed manual and mechanical manipulation (as by tilling, planting, etc.) and chemical manipulation (as by fertilizing and use of various control chemicals that seep into the soil).

Soil is destroyed by two principal processes—*water erosion* and *wind erosion*. The word *erosion* is the physical removal of all or part of established soil by washing or blowing away. Erosion, in some instances, also brings soil to convenient locations, but usually in so doing, unless carried out over long periods as in the development of bottomlands and deltas where crops can be grown, the new muddy, fine, highly unconsolidated and disintegrated soil causes more problems than immediate benefits. The bringing in or transfer of soil by water is commonly referred to as *sedimentation*. Much research has been carried out in connection with water and wind erosion. Typical of fundamental research is the study of splash patterns as shown in Fig. 2. The lysimeter, as shown in Fig. 3, also has been effectively used. The effects of wind erosion have been extensively studied, as exemplified by Figs. 4 and 5.

Additional Reading

Angers, D.A., E.G. Gregorich, L.W. Turchenek, et al.: *Soil and Environmental Science Dictionary*, CRC Press, LLC., Boca Raton, FL, 2001.

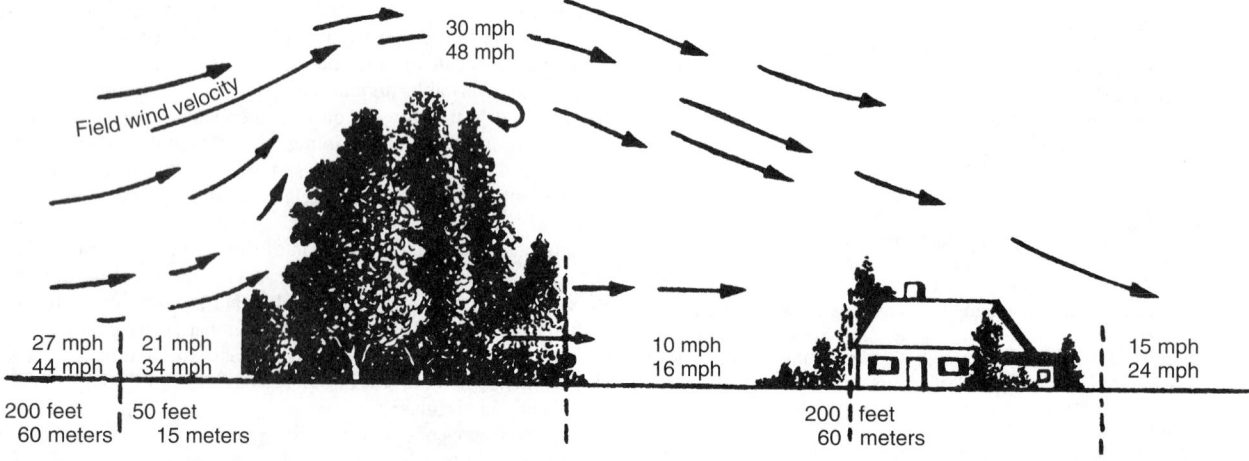

Fig. 4. Windbreaks reduce wind currents. Part of the air current is diverted over the top of the trees and part of it filters through the trees. Breaks like this reduce wind erosion of soil. Farmstead, livestock, and wildlife windbreaks should be relatively dense and wide to permit maximum protection close to the trees. Field, orchard, and garden-type windbreaks need not be so wide and dense. (*USDA diagram*)

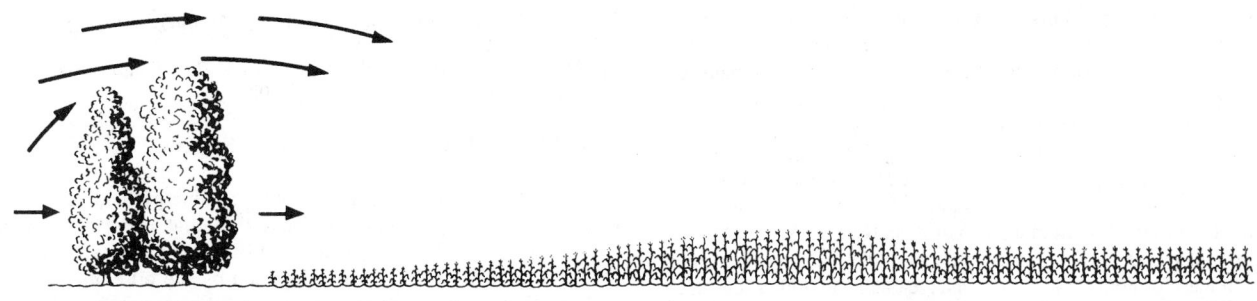

Fig. 5.   Sometimes crop yields are lowest next to a tree windbreak. A common error made by some growers is to observe only the immediately adjacent area. The greatest gains are from a distance of 30 to 45 feet (9 to 13.5 meters) from the windbreak. (*USDA diagram*)

Bohn, H.L., B.L. McNeal, and G.A. O'Connor: *Soil Chemistry,* 3rd Edition, John Wiley & Sons, Inc., New York, NY, 2001.

Bowles, J.E.: *Foundation Analysis and Design,* 5th Edition, The McGraw-Hill Companies, Inc., New York, NY, 1995.

Brady, N.C. and R.R. Weil: *The Nature and Properties of Soils,* 13th Edition, Prentice Hall, Inc., Upper Saddle River, NJ, 2001.

Carroll, R.C. and J.H. Vandermeer: *Agroecology,* The McGraw-Hill Companies, Inc., New York, NY, 1990.

Das, B.M.: *Soil Mechanics,* Oxford University Press, Inc., New York, NY, 2001.

FAO: *Soil Maps of the World,* Food and Agriculture Organization of the United Nations, Rome, Italy. (Revised periodically.)

Fisher, R.F. and W.L. Prtichett: *Ecology and Management of Forest Soils,* 3rd Edition, John Wiley & Sons, Inc., New York, NY, 1999.

Franklin, J.A. and M.B. Dusseaultz: *Rock Engineering Applications,* The McGraw-Hill Companies, Inc., New York, NY, 1991.

Frenkel, H. and A. Meiri: *Soil Salinity,* John Wiley & Sons, Inc., New York, NY, 1985.

Goldman, S. et al.: *Erosion and Sediment Control Handbook,* The McGraw-Hill Companies, Inc., New York, NY, 1986.

Hausmann, M.R.: *Engineering Principles of Ground Modification,* The McGraw-Hill Companies, Inc., New York, NY, 1990.

Lal, R.: *Soil Erosion in the Tropics,* The McGraw-Hill Companies, Inc., New York, NY, 1990.

Levy, R.: *Chemistry of Irrigated Soils,* John Wiley & Sons, Inc., New York, NY, 1984.

Pierzynski, G.M., G.F. Vance, and J.T. Sims: *Soils and Environmental Quality,* 2nd Edition, CRC Press, LLC., Boca Raton, FL, 2000.

Rendig, V.V. and H.M. Taylor: *Principles of Soil-Plant Interrelationships,* The McGraw-Hill Companies, Inc., New York, NY, 1989.

Singer, M.J. and D.N. Munns: *Soils: An Introduction,* 5th Edition, Prentice Hall, Inc., Upper Saddle River, NJ, 2001.

USDA: *Soil Conservation Reports,* National Soil Survey Laboratory, U.S. Department of Agriculture, Washington, DC, (Published periodically.)

Warrick, A.W.: *Soil Physics Companion,* CRC Press, LLC., Boca Raton, FL, 2001.

SOIL CHEMISTRY. This field includes all aspects of the study of soil as a chemical system. The eight chemical elements in soils which generally surpass 1% by weight are oxygen, silicon, aluminum, iron, calcium, magnesium, potassium, and sodium; the eleven elements making up 0.2 to 1% include titanium, hydrogen, phosphorus, manganese, fluorine, sulfur, strontium, barium, carbon, chlorine, and chromium. The most abundant minerals present in less-weathered soils are quartz, feldspars, micas and colloidal layer silicates including vermiculite, chlorite, and montmorillonite. Calcareous soils contain calcite and dolomite. More-weathered soils contain larger amounts of more resistant minerals such as kaolinite, halloysite allophane, hematite, goethite, gibbsite, anatase, pyrolusite, tourmaline, and zircon. The organic matter, or humus, content of soils varies from less than 1% to over 80%. Generally, upland soils range from 1 to 8% organic matter, while less well-drained soils are frequently higher.

Soils developed under coniferous forests often accumulate acid organic matter at the surface; the resulting leaching through the soil of chelating organic acids bleaches (*podzolizes*) the mineral soil beneath. These soils are gray when plowed. Organic matter from hardwood trees and grasses which are high in bases, particularly calcium, accumulates in the soil and causes a dark color in the surface horizon. Poor drainage leads to the development of light-colored gray horizon within the soil column (*profile*), owing in part to the reduction of iron oxides to ferrous form. A bluish color is sometimes present, particularly when vivianite, $(Fe)_3(PO_4)_2 \cdot 8H_2O$, forms. Soluble soil salts, mainly chlorides and sulfates of sodium, calcium, and magnesium, when present in quantities over 0.1 to 0.7% cause a condition known as salinity or soil alkali. If much Na_2CO_3 is present, some organic matter is mobilized and together with FeS, colors the soil black, giving rise to the name black alkali.

The most reactive portion of the soil resides in colloidal organic matter, layer silicates, and hydrous oxides of iron, aluminum, and occasionally manganese and titanium. The colloids of soil have a negative electrostatic charge arising through carboxyls of organic compounds and through excess negative charge of oxygen in the silicate structure. The negative charge is neutralized by exchangeable cations, giving systems known as colloidal electrolytes. When these exchangeable ions, i.e., counterions, are hydrogen or aluminum, the colloids act as a moderately strong acid. Different colloids range in the strength of acidity as evidenced by the shapes of the titration curves, which are analogous to the shapes of those of weak and strong soluble acids. The colloids are hydrophilic and subject to flocculation in the presence of dilute salt solutions, owing to repression of the charge developed by dissociated cations. The flocculation is reversible.

Important chemical characteristics of the soil include the total exchange capacity for cations, expressed as total meq of cations per 100 gm of soil, and the base status, which is the percentage saturation of the negative charge with cations such as calcium, magnesium and sodium. The more productive soils are about 80% saturated with calcium and magnesium. Excessive hydrogen and aluminum saturation (much over 15%) is termed soil acidity. Excess sodium saturation (12% or more) leads to dispersiveness of the soil and poor productivity.

There are also positive charges associated with aluminum and iron colloids of soils. These charges give rise to phenomenon known as anion exchange capacity, which is mainly concerned with phosphorus chemistry; the usual soluble anions such as nitrate, chloride, and sulfate are little held. Synthetic organic soil conditioners are long-chain organic molecules with carboxyl charges along the chain which react with the positive charges of the soil particles. These colloidal molecules can bind the soil particles into aggregates. Natural humus of soil acts in a similar way until oxidized by soil organisms.

Analytical methods employed in soil chemistry include the standard quantitative methods for the analysis of gases, solutions, and solids, including colorimetric, titrimetric, gravimetric, and instrumental methods. The flame emission spectrophotometric method is widely employed for potassium, sodium, calcium, and magnesium; barium, copper and other elements are determined in cation exchange studies. Occasionally arc and spark spectrographic methods are employed.

The most commonly made chemical determination is that of soil pH measurement, as an indicator of soil acidity. The glass electrode has proved the most satisfactory method for soil pH measurement because the moistened soil rapidly equilibrates in contact with the glass surface, no reagents are added to the soil, and the soil CO_2 tension is not disturbed by bubbling through of gases. Colorimetric indicators are also employed. Soils of pH 4.3 to 5 are highly acid, of pH of 5 to 6 are moderately acid, of pH 6.3 to 6.6 are very slightly acid, of pH 6.7 to 7.3 are considered neutral; soils of pH 8 to 9 are moderately alkaline, and of pH 9 to 11 are very alkaline. For acid soils, pH measurement serves as a guide to agricultural liming practices. For many crops, such as alfalfa, the soil is adjusted to pH 6.5 to 7 by the addition of ground limestone, the active ingredients of which are $CaCO_3$ in calcic limestone and $CaCO_3$ $MgCO_3$ in dolomitic limestone. A soil colloidal acid may be represented as HX. Then the liming reaction, by which the exchangeable calcium is increased, is

$$CaCO_3 + 2HX \longrightarrow CaX_2 + CO_2 + H_2O$$

The proton donor, X, represents a variety of organic and inorganic donors, including the reaction $Al(OH_2)_6 \longrightarrow Al(OH)(OH_2)_5 + H^+$. The reaction is hastened by fine grinding of the limestone, and liming materials are graded on the basis of fineness and $CaCO_3$ equivalence. Burned lime (CaO), marl $(CaCO_3)$, and sugar refinery wastes $[Ca(OH)_2]$ are also used in liming. For some crops, owing to disease susceptibility and preferences, soils are kept more acid, as low as pH 5.3. Calcium and magnesium, necessary to plant growth, are furnished to plants from exchangeable form.

Fine grains (finer than 20 μ in diameter) of mica $[KAlSi_3Al_2O_{10}(OH)_2]$, and potassium feldspar $(KAlSi_3O_8)$ slowly undergo chemical weathering in soils with the release of potassium into exchangeable form. Other ions such as calcium, sodium, and iron are also released by mineral weathering. In subhumid and arid regions the release of potassium is fast enough for crop production but must be supplemented by the addition of potash fertilizer salts in more leached soils of humid regions.

Soil chemists have rapid chemical tests for measurement of the amounts of plant-available K, P, and N, as well as other elements in soils which are essential to plants. When the quantity of an element is too low for efficient crop production, it is added as fertilizer, such as KCl, $Ca(H_2PO_4)_2$, or ammonium or nitrate salts. Large chemical fertilizer industries are required for mining, refining, and preparation of chemical salts for soil application as fertilizers.

Extraction of soils for analysis of the readily available nutrients include replacement of exchangeable cations by salt solutions, dilute acids, and dilute alkalies such as $NaHCO_3$. Fluoride solutions are employed to repress iron, aluminum, and calcium activity during the extraction of phosphorus. Extraction of the soil solution is effected by displacement in a soil column, often through the application of pressure across a pressure membrane. The soil solution is analyzed by conductance and elemental analysis methods. Also, the total elemental analysis of soils is made by Na_2CO_3 fusion of the soil followed by classical geochemical analysis methods.

Organic compounds of great variety have accumulated in soils as residues from plant and animal life of the soil. The more unstable compounds of these residues are rapidly oxidized to CO_2 and H_2O by biochemical processes, while the more stable fractions accumulate. Conjugated ring compounds containing the elements C, H, O, N, P, S, and several other elements in small quantities accumulate in relatively stable organic and organomineral colloidal complexes. Lignin-like, phytin-like, and nucleoprotein-like compounds are included. Sorption of the organic matter on mineral colloid surfaces, particularly on layer silicates, such as montmorillonite, helps to stabilize the organic matter against biochemical oxidation. In tropical soils, high stability of soil organic matter is imparted by coatings of aluminum hydroxide and red ferric oxide. Organic and iron oxide colloids, when fairly abundant, stabilize the soil into porous aggregates through which ample air and water can circulate. Decomposition of soil organic matter, especially when hastened by tillage, gradually releases HNO_3, H_2SO_4, and H_3PO_4 in amounts which are highly significant in nutrition of crops. Much of the nitrogen, sulfur, and phosphorus required by crops is furnished in this way.

The oxidation potential of well-aerated soils is low $(-0.5$ V$)$ and of reduced soils is high $(+0.30$ V$)$. These relationships are sometimes expressed by soil scientists as reduction potentials or redox potentials, in which case the algebraic signs are the opposite. The oxidation potential is advantageously measured with a platinum-blackened electrode in the soil in place in the field. Moderately good aeration is a requirment of a productive soil. The oxidation status may also be tested in the field by rapid spot tests for ferric and ferrous iron in soils. Most of the dilute acid-soluble iron is in ferric form in well drained soils. Localized spots of decomposing organic matter are important in reducing small but important quantities of iron to ferrous form and manganese to divalent form so as to be available to plants. Moderately to highly alkaline soils sometimes have inadequate activity of the reduced forms of iron and manganese, particularly in the absence of sufficient organic matter. Small quantities of Cu, Zn, B, and Mo must be present in productive soils in forms which have enough activity to be available to growing plants.

M. L. JACKSON
University of Wisconsin
Madison, Wisconsin

Additional Reading

Andrews, J. E., T. Jickells, and P. Brimblecombe: *An Introduction to Environmental Chemistry,* Blackwell Publishers, Malden, MA, 2004.
Essington, M. E.: *Soil and Water Chemistry: An Integrative Approach,* CRC Press LLC., Boca Raton, FL, 2003.
Sparks, D. L.: *Environmental Soil Chemistry,* 2nd Edition, Elsevier Science & Technology Books, New York, NY, 2002.

SOL AND SOLATION. See **Colloid Systems**.

SOLAR ENERGY. The vast quantity of energy received by Earth from the sun and the potential for converting that energy into more useful forms for society has intrigued scientists, engineers, and social planners for decades. This interest was sharpened by the oil embargo of the 1970s and, for about a decade after that, tremendous interest was displayed, by the scientific and lay community alike, in alternative energy sources, including a turn to solar energy. Energy from the sun was considered by many people as a relatively low-cost and essentially pollution-free source, particularly in contrast with polluting, nonrenewable, so-called fossil fuels and with nuclear fuels, which many people consider in a negative light. During the 1970s, but tapering off in the 1980s, many, many millions of dollars were invested by governments worldwide and by private institutions, architectural and solar equipment firms, and energy supply firms toward the development of practical, economically competitive solar energy systems.

As a result of that activity, progress in designing passive solar energy systems into office and factory buildings has been impressive, but not nearly so extensive as once estimated. Active solar energy systems, in which solar radiation is converted into another energy form (usually electrical) has also progressed, but the number of outstandingly successful installations is relatively limited, and essentially these are presently regarded as still in an experimental phase. In contrast, considerable research continues to be directed toward solar cells (solid-state devices that convert solar radiation into electric power), but it should be immediately stressed that solar cells for communication and other satellites and space vehicles are vitally needed, because they provide an energy source difficult to obtain in other ways. Cost, in this instance, is not supercritical, but one finds that solar cells for building, etc. heating and power still are essentially noncompetitive. Some relatively low-cost, small solar-powered devices designed mainly for public consumption have appeared in recent years.

To put solar energy into perspective, one should review other energy resources as well. Fortunately, throughout this encyclopedia, such energy information is available. See list of articles at end of this description and also consult alphabetical index.

Availability of Solar Energy

Not to be confused with insulation, the word *insolation* (acronym for "incoming solar radiation") defines the rate at which direct solar radiation is incident upon a unit horizontal surface at any point on or above the surface of the earth. The unit of insolation is the Langley, named after Samuel Pierpoint Langley (1834–1906), an American astronomer, physicist and pioneer in the utilization of solar energy:

$$1 \text{ langley} = 1 \text{ gram calorie per square centimeter}$$
$$= 3.687 \text{ Btu per square foot}$$

Fundamental to the practical application of real-time solar energy systems is the amount of energy received from the sun at any given location at any particular time. The energy received varies with the geometry of the sun-earth system—and thus varies with latitude, season of the year, time of day, as well as with local weather conditions. On a typical day in June, anywhere from 500 to 700 langleys of solar energy can be expected in most parts of the United States, whereas in December, with shorter days and more inclement weather, only from 100 to 300 langleys can be expected. In June, for example, both Saskatoon, Canada and Tampa, Florida receive about 600 langleys of solar energy daily, but in December, the amount of such energy received in Saskatoon drops to 75 langleys per day (only 12% of the June value at that location), whereas, in Tampa, more than 50% of the June amount is received in December. In terms of equipment costs, this translates into needing four times the solar-energy collector surface in Saskatoon as that required in Tampa in order to supply the same amount of power year-round. Maps of the type shown in Fig. 1 can be helpful in this regard.

The solar constant (intensity of solar radiation outside the Earth's atmosphere at the mean distance between the earth and the sun) has been determined by measurements from satellites and high-altitude aircraft and is 1.353 kilowatts per square meter. This extraterrestrial radiation,

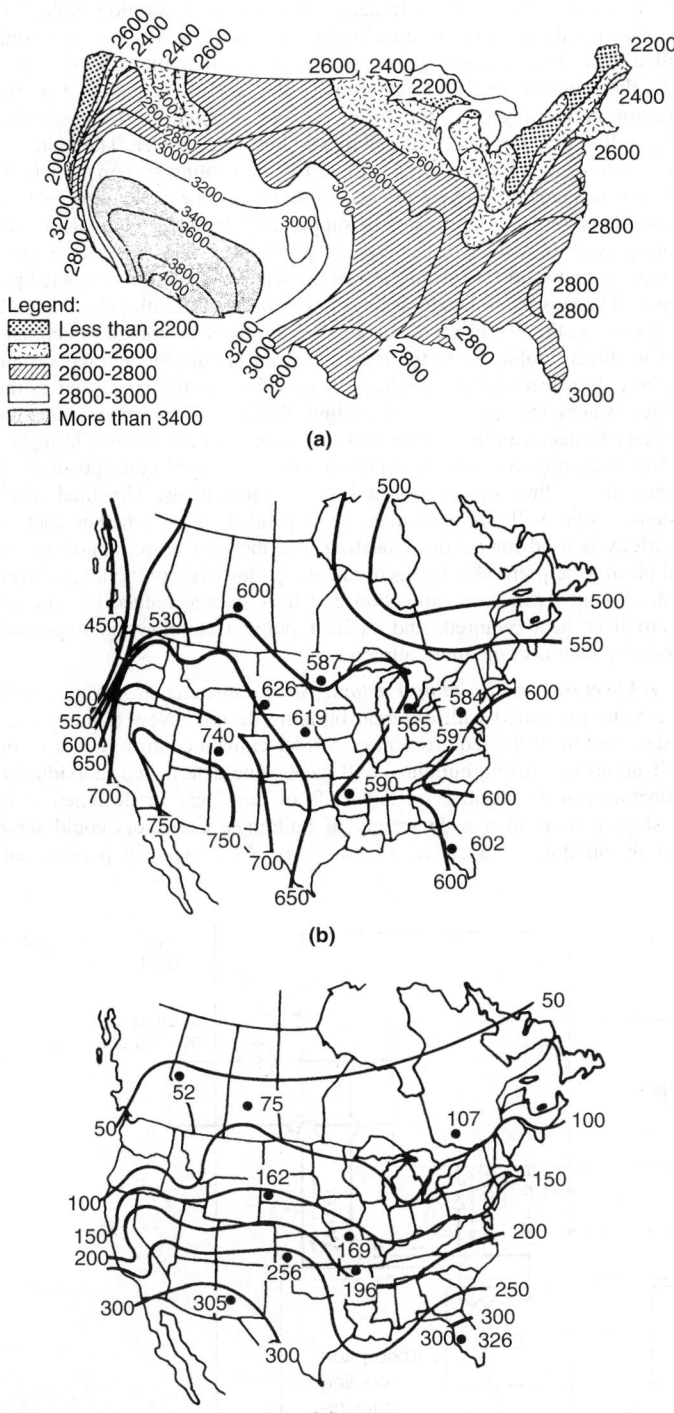

Fig. 1.   Availability of solar energy (insolation): (**a**) Average number of hours of sunshine per year (United States); (**b**) median daily insolation in langleys (North America in June); (**c**) in December. (*National Oceanic and Atmospheric Administration*)

which corresponds closely to that of a blackbody at 5762°K, is 7% in the ultraviolet range (wavelength less than 0.39 micrometer) and 47% in the visible range (wavelengths from 0.38 to 0.78 micrometer), with the balance in the near-infrared (largely with wavelengths of less than 3 micrometers). Radiation is depleted as it passes through the atmosphere by a combination of scattering and absorption; the radiation that reaches the ground—the raw material of this energy source—can vary from almost none, under heavy cloud cover, to 85–90% of the solar constant under *very clear* skies.

SOLAR ENERGY FOR BUILDING AND RESIDENCE COMFORT

The application of solar energy for residences and commercial and public buildings tends to fall into three categories of increasing complexity: (1) heating only; (2) cooling only; and (3) combined heating and cooling.

Simple Heating System with Hot Storage. As shown in Fig. 2, the basic elements of this system, exclusive of pumps, valves, and controls, are: (a) a solar collector; (b) an auxiliary heating device; (c) hot storage system; and (d) heater element fan and air duct system. The solar collector absorbs heat energy from the sun and transfers it to a heat-transfer fluid, which conveys the heat to a hot storage system. From the hot storage, heat is withdrawn from storage through a heater coil, where an air-circulating fan carries heat from the coils into an air duct system. When the solar collector cannot provide an adequate amount of heat to maintain the hot storage at a minimum temperature, the auxiliary heating device, such as a fuel burner, electric resistance heating, or an electric driven heat pump, comes on. This auxiliary heat could be added to hot storage as shown in Fig. 2, or used to directly heat the room air. For the case of electric heating, heat addition to storage will provide the opportunity to limit auxiliary heat addition to nonpeak hours.

Solar Cooling System. As shown in Fig. 3, the basic elements of this system are: (a) a solar collector; (b) an auxiliary heating device; (c) a cold storage system; and (d) a heat-actuated refrigeration loop (absorption cycle). The solar collector absorbs heat energy and transfers it to a heat-transfer fluid, which in turn conveys the solar heat to the generator or boiler of the heat-actuated refrigeration loop. This loop can also be driven by auxiliary heating when the solar heat input is not adequate. If the heat-actuated refrigeration loop were of the Rankine-cycle type, rather than an absorption cycle, it might also be possible to drive the refrigeration loop with auxiliary power rather than auxiliary heating—a more favorable situation if the auxiliary is electric rather than fuel. The refrigeration loop cools the cold storage reservoir from which home cooling is supplied upon demand.

Combined Solar Heating and Cooling System. A system of this type is shown in Fig. 4. This is only one of a variety of possible systems. The major elements of this system are: (a) a solar collector; (b) an auxiliary furnace with heating coils; (c) a storage system (hot in winter; cold in summer); (d) absorption refrigeration cycle; and (e) necessary valving and controls. The system is designed to provide both heating and cooling upon demand. The heat energy generated by the solar collector is directed to either the hot storage tank or the refrigeration cycle generator, according to the seasonal mode of operation. When solar energy (either direct or stored) cannot supply the required heating or cooling load, auxiliary heating or cooling can be used.

In the winter mode of operation, solar energy is gathered at the collector and is pumped directly to the storage system. From the storage system, heat is extracted according to household needs through the heating/cooling coil in the main air duct. This coil is controlled by 3-way valves which are open to heating and shut to cooling. Heat is then extracted from the coil by air fans and carried into the house. At certain times, the solar collector and storage system will not be able to provide enough heat to maintain the heat needs of the household. In these cases, the auxiliary furnace will assume the heating load until the solar system is able to provide heating.

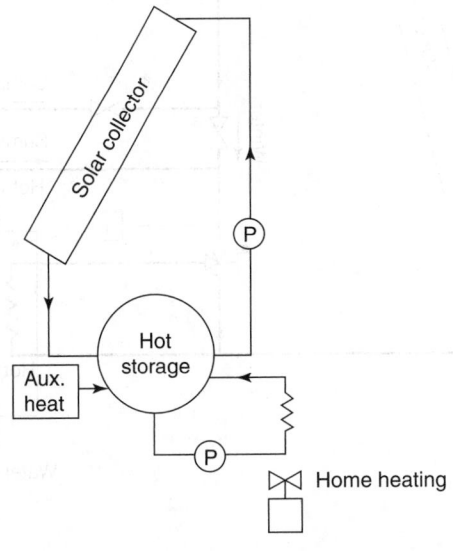

Fig. 2.   Basic elements of solar heating system

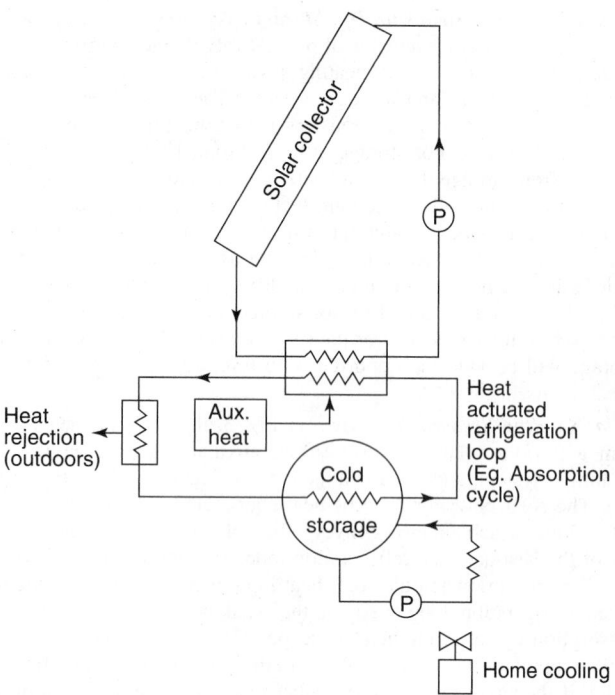

Fig. 3.   Basic elements of solar cooling system. Cold storage only

condition exists, the auxiliary furnace will drive the absorption cycle. The two 3-way valves on the heating/cooling coil will be actuated to permit chilled water from the absorption machine to circulate through this coil.

In the summer mode of operation, solar energy is gathered at the collector and pumped in the form of heat directly to the absorption refrigeration machine and to the domestic hot water heater. The collected energy serves as the main driving force for the cooling system. When the collector is unable to provide the necessary energy for the cooling system, the auxiliary furnace is activated to supplement the energy load. Once the cooling cycle is activated, the cooling produced is directed to the same storage system used for storing heat in the winter. Cooling is extracted per household needs from the storage system through the heating/cooling coil. In this mode, the 3-way valves of the coil are open to storage system and shut to direct cooling from the cooling cycle. Should the storage system not be able to provide the cooling, these valves would be closed to the storage system and open to direct cooling. While on the summer mode, the auxiliary furnace can be used to heat the house on occasional cold nights.

The foregoing systems are representative of general concepts and not necessarily of final designs or optimum arrangements. The final detail system design will, in particular, be dependent upon whether fuel or electricity is used for auxiliary heating. For the near term, natural gas or fuel oil may be preferable to electric heating. However, in the longer term, as developments in solar collection and heat storage reduce the amount of auxiliary heat required, and as heat pump technology is improved, electricity may become more attractive.

Architectural and Building Factors. Solar climate control systems will have to be integrated with different building designs. New buildings can be designed to fit the requirement of a solar climate control system while applications to existing buildings will have to be determined individually. Collectors can be installed on flat roofs of buildings, or designed to fit the sloping roofs of a wide variety of buildings. Collectors could serve a single building or a cluster of buildings. As previously pointed out,

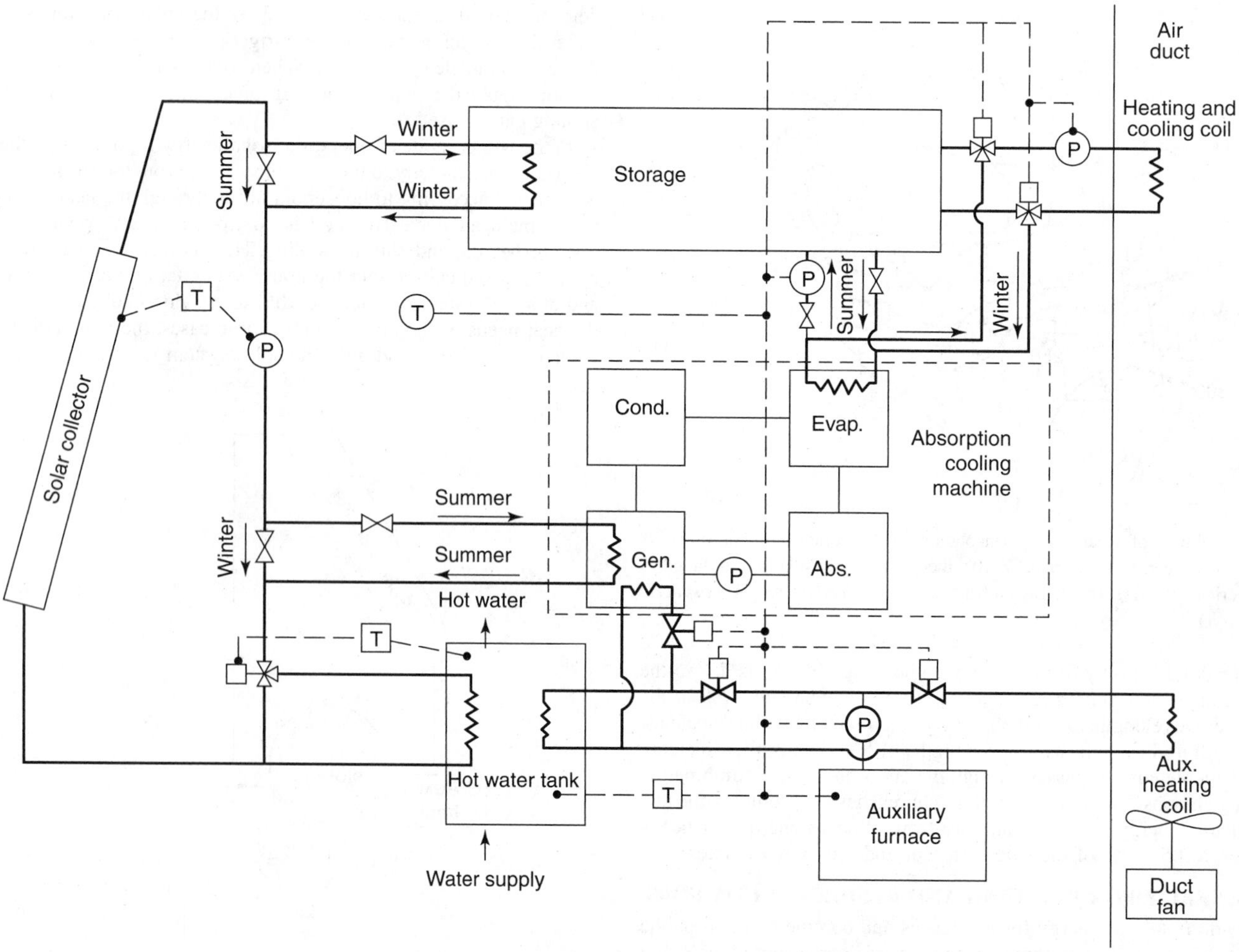

This furnace will also provide auxiliary heat for domestic hot water when solar energy cannot provide this function. At times, while on winter mode operation, there will be days when cooling may be required. When this

Fig. 4.   Combined solar climate heating and cooling system

the geometry of the collector installation varies with location latitude. Economics to a large degree will be determined by availability of reasonably sustained sunshine.

Flat-Plate Collectors. Essentially since the outset of solar energy technology for environmental heating and cooling of living and working enclosures, the flat-plate collector has dominated the field. Within recent years, however, some of the initial needs for collectors of this nature have been obviated by more attention being given to the design of passive solar collectors, described a bit later. Also, for very large commercial installations, there has been a trend toward the use of nonfocusing or trough or line-focusing concentrators, also described later.

The essential features of a flat-plate solar collector are shown in Fig. 5. A blackened receiver surface covered by one or more special glass plates is used. Since the glass is transparent to the incident solar radiation, but opaque to the reradiated energy, the solar collector, like a greenhouse, serves to trap solar energy. The working fluid used to remove the heat from the collector can be either air flowing between the blackened surface and the glass plate or water (or some other liquid) flowing in tubes attached to the blackened plate. Solar collection efficiency is defined as the ratio of usable energy collected per unit time to the incident solar flux. Efficiency, η, may be calculated as:

$$\eta = \alpha\tau - \frac{q_L}{q_{in}}$$

where: α is the absorptivity for sunlight; τ is the transmittance of the glass plate; q_L is the heat loss from the collector; and q_{in} is the incident solar flux. The heat loss q_L is, in turn, dependent upon the emissivity, ϵ, for low-temperature radiation. Typical performance characteristics for a solar energy collector are shown in Fig. 6, which is a plot of collector efficiency versus temperature of the absorber plate for an incident radiation of 300 Btu/(hour) (square feet); 814 kcal/(hour) (square meter). Note that the efficiency falls off as absorber temperature rises. Efficiency, of course, also drops off rapidly as the incident radiation is reduced, since the heat loss term is a function of absorber temperature only.

A problem that must be faced by architects and engineers is the need to integrate collectors into building and residence design in a way to maximize thermal performance and, at the same time, provide an esthetically satisfactory structure. A major variable, depending upon energy requirements and the average solar insolation over a year, is the amount of collector area needed. Obviously, the larger the energy requirements and the less favorable the insolation, the greater the problem. Because collectors

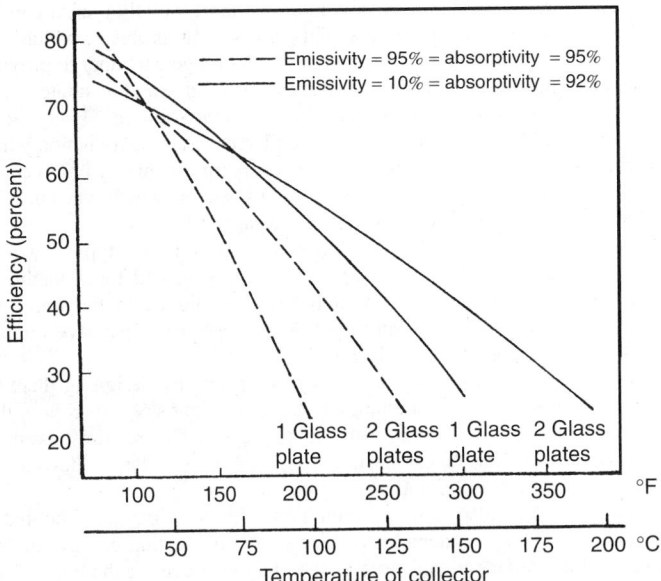

Fig. 6. Relative solar collection efficiency of plate-type solar collectors with different emissivities

must be oriented within rather narrow limits if they are to maximize their capture of solar radiation, the problem of retrofitting many existing structures sometimes renders a project impractical. Thus, the general pattern has been one of concentrating on new construction, particularly for low-rise, flat-roofed buildings, such as schools and shopping centers, where cooling may be more important than heating. Where surrounding area is available, as, for example, an adjacent parking area or facility, the collectors can be installed apart from the structure to be heated and/or cooled. Also, by turning to advanced collector designs, which provide greater efficiency, even at greater initial cost, architects and builders can better cope with the problem of collector area required. Passive heating of buildings is also a possibility in many cases.

Evacuated-Tube Collectors. In this type of collector, an inner glass cylinder, blackened to absorb solar radiation, is enclosed within an outer protective cylinder. The space between the two cylinders is evacuated. The inner cylinder is usually coated with material that reduces energy loss through reradiation. Transfer of heat is accomplished by a fluid (air or liquid) that flows through the inner cylinder. These collectors are similar to flat-plate collectors in that they can use both direct and diffuse light. However, the evacuated-tube collectors operate better during the early and late parts of the day. The vacuum provides such excellent insulation that they are less affected by high winds and cold weather than the flat-plate collectors. The output of the evacuated-tube collectors is essentially independent of ambient temperature and their efficiency is generally 40–50%. Ordinarily these collectors operate at about 180°F (82°C) for space-heating applications and certain process heating uses. Equipped with reflectors, they can operate up to 240°F (116°C), which is sufficient to drive absorption air conditioners. Cylindrical evacuated tube collectors can absorb radiation coming from any direction (360° aperture). Usually, they are mounted in arrays with a spacing of about one cylinder diameter between tubes and with a reflective material behind them.

Large-scale collectors, with or without focusing (radiation concentrators) are described a bit later.

Heat Storage. A comparison of heat-storage capacity on a volumetric basis between various storage media shows that water can store 62.5 Btu per cubic foot per degree Fahrenheit (311.5 kcal per cubic meter per degree Celsius. Rocks, bricks and gravel can store about 36 Btu/cu. ft/degree F (179.4 kcal/cu. meter/degree C). In addition to fluid media and solid media, advantage can be taken of the latent heat of a phase transition. Some salts, which melt in the desired temperature range, can store about 60 Btu/cu. ft/degree F (299 kcal/cu. meter/degree C) as sensible heat, and 9500 Btu/cu. ft./degree F (47,348 kcal/cu. meter/degree C) at the melting point as heat of fusion. However, these salts tend to undercool rather than crystallize during the cooling cycle. While undercooling can be prevented by use of nucleating agents, the fixed rate of crystal growth is very slow in

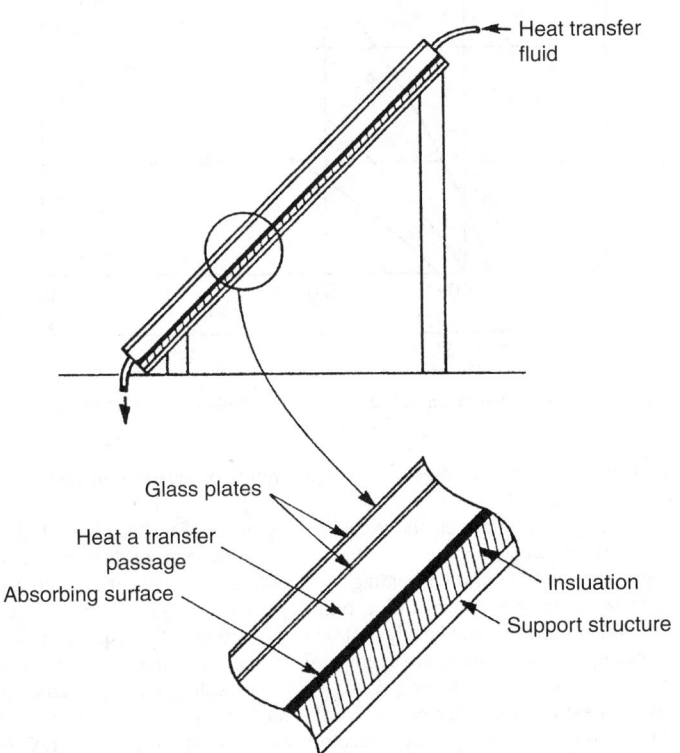

Fig. 5. Basic elements of a flat-plate solar collector

most salt hydrates. Heat cannot be withdrawn more rapidly than it can be supplied by the growth of crystals. This is a serious problem, which can only be partially overcome by the design of the storage container to provide a large heat-transfer area. Based upon these alternatives, an insulated tank of water to store heat is one of the most efficient solutions. Early work concentrated on $Na_2SO_4 \cdot 10H_2O$ which undergoes a phase transition when heated at 32°C (89.6°F). Because phase separation of this hydrate occurs on cycling, other chemical systems are being sought which can undergo thousands of cycles without loss of storage capacity.

Typically optimum storage capacity varies from 1 to 3 typical winter days' solar energy supply, depending upon the site, and for a medium-size residence or small commercial building would be in the range of 1000–2000 gallons (38–76 hectoliters). A section of a comparatively small solar-heated building is shown in Fig. 7.

Collectors also can be used as energy dissipaters by designing them to lose heat by convection and radiation to the clear night sky. The role of the collector is thus fully reversed for the cooling cycle. This requires a system for moving insulation, unless design compromises in collector design are made for both the heating and cooling cycles.

Solar collectors also can be combined with heat pumps. The latter can serve as an independent (auxiliary) source of heating energy, or the collector-storage system can serve as the energy source for the evaporator of the heat pump. The latter system has apparent advantages of lowering mean collector temperature and raising the mean evaporator temperature of the heat pump, thus improving the performance of each. See Fig. 8.

In addition to close cycle absorption cooling, open cycles are of potential interest. Desiccants can be used to absorb water vapor from room air, which then can be evaporatively cooled. The desiccant is regenerated and recycled. Löf has suggested the use of triethylene glycol as a desiccant, with solar-heated air for regeneration. Lithium chloride also has been proposed as a desiccant.

Heat-Actuated Refrigeration. A variety of heat-actuated refrigeration cycles has been proposed for solar air conditioning. These can be divided into heat engine types, such as the Rankine and Stirling cycles, and the absorption machines. Most successful to date have been the lithium bromide-water and the ammonia-water absorption cycles. Regardless of type, operating temperature is a tradeoff when coupling a solar collector to a heat-actuated refrigeration machine. The efficiency of solar collectors decreases with temperature. On the other hand, the coefficient of heat-actuated refrigerators increases with generator temperature. Figure 9 shows Carnot coefficient of performance as a function of generator temperature for an evaporator temperature of 45°F (7°C) and for heat rejection temperatures of 120°F (49°C) (typical of air cooling) and 100°F (38°C) (warm water cooling, such as might be achieved with well water). For the case of absorption machines, it is assumed that absorber and condenser operate at a common heat-rejection temperature. Using these plots, the collection efficiency–Carnot coefficiency of performance product for the simple single-pane black collector reaches a maximum rate at about 175°F (80°C) and falls off rapidly at higher temperatures. For a more refined collector, such as a single pane with selective surface collector, the efficiency

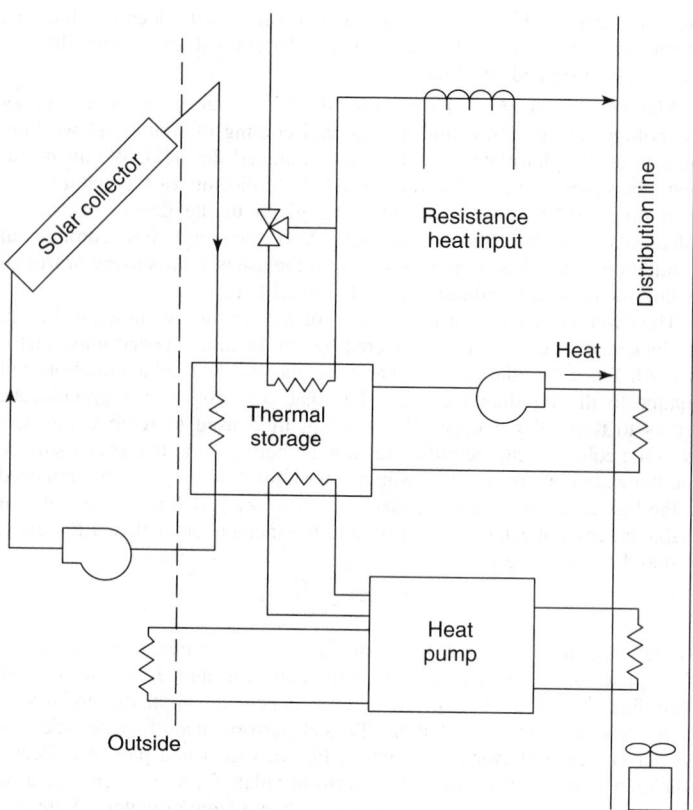

Fig. 8.   Solar heating system with heat pump auxiliary

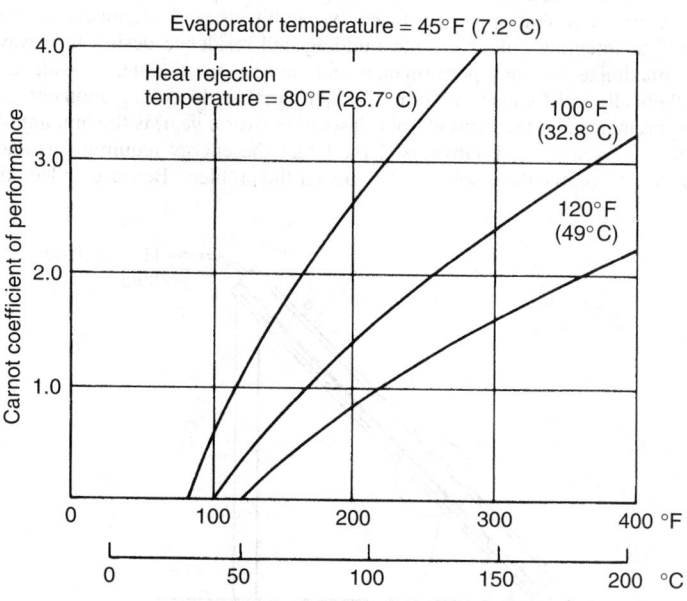

Fig. 9.   Carnot coefficient of performance of heat-actuated refrigerators

coefficient of performance-product is maximum and nearly constant in the range of 200 to 300°F (93 to 149°C).

To achieve temperatures above the boiling point of water, concentrators are required as well as collectors. One of the largest high-temperature solar energy systems for building heating and cooling commenced operation in late 1978, at an eight-story office building (Minneapolis), which houses about 500 employees, and has over 100,000 square feet (929 square meters) of working space. During an average year, the system was designed to generate about 50% of heating needs, 80% of cooling energy needs, and 100% of heat for hot water needs. On the roof of a parking ramp adjacent to the building, 252 trough-like collectors track the sun to focus its rays on liquid-filled pipes. The liquid, which may reach 177°C (350°F), is pumped to an isolation heat exchanger that allows different fluids, pressures, and

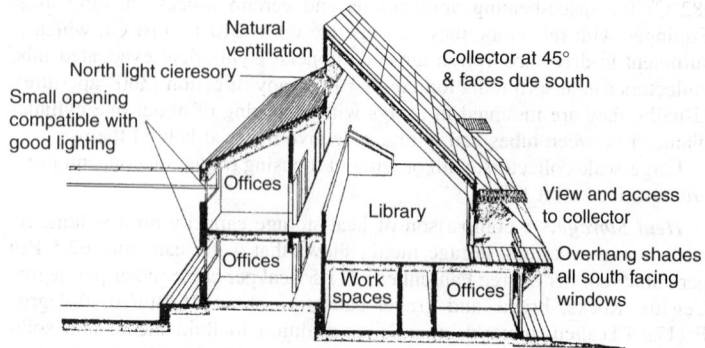

Fig. 7.   Section of solar-heated building. Solar collector has an area of 3500 square feet (325 square meters) facing south at an angle of 45 degrees. There are about 8000 square feet (743 square meters) of working space. Estimates of heat loss indicate heat demand is in range of 40,000–70,000 Btu (10,080–17,640 kcal) per day. Located in the northeastern United States, the building was designed to furnish between 65 and 75% of total seasonal heating load

flow rates to be used in the two-loop system, dictated because of the cold winter temperatures. Excess heat is stored underground in two 18,000-gallon (681 hectoliters) tanks until required. Each row of solar collectors is under the control of a local system that uses a photosensitive sun tracker and bidirectional electric drive A.C. motor. Wind, solar isolation and other safety control circuits are continuously monitored to protect the collector field from damage by weather excesses.

Passive Solar Heating and Cooling. Although not always the practical solution, one of the most sensible approaches to the utilization of solar energy does not require pumps, fans, etc., but utilizes the building or residence structure per se as the solar radiation absorber (or insulator). An example of the modern approach to passive approaches is, ironically, incorporated in Montezuma's Castle, built around 700 A.D. by the cliffdwelling Indians of Arizona. The basic philosophy is to design the structure to capture and retain heat during winter and to remain cool during summer. For example, the windows that face north are made small or largely eliminated. These windows do not contribute much to heat collection during summer or winter and thus, if small, diminish heat leakage in either direction during all seasons. South-facing windows are made large because they are required as radiation collectors during winter. However, they are protected during the summer season by an overhanging roof. These features alone, of course, are common-sense ideas that have been used by some designers for many years. Passive systems also take advantage of heavy masonry walls (or other sources of thermal mass) which can absorb solar radiation during the day and reradiate some of it at night. Such an arrangement could be called a "concrete collector." The designer can improve the effectiveness of extensive south-facing windows by constructing an interior masonry wall adjacent to the windows (lighting becomes a problem for special design). In an experimental building (Wallasey School, Liverpool, England), the south wall of the two-story concrete structure is made up essentially of double-glass windows with a heat-storage (or insulating) wall. The only supplementary heat required is derived from body heat of the students and heat radiated by electric lights. A structure of this type, of course, is subject to wider interior temperature variations than those to which much of society has become accustomed. Temperature swings can be reduced by partially decoupling thermal storage from the living and working space. In this concept, solar radiation entering the south-facing windows is absorbed by a masonry wall (sometimes called a Trombe wall) or by a water wall in which water-filled drums are placed. This wall insulates the building from high temperatures during daytime and transmits stored energy to the structure for warming during nighttime. There is an office building and warehouse in northern New Mexico which incorporates a water wall passive system and which provides 95% of the energy required for heating.

Some designers use a "roof pond," in which plastic bags filled with liquid are exposed to the sun during the day. They are covered with an insulating panel at night so that they can radiate stored heat downward to the structure. The cycle can be reversed to provide cooling during summer.

Investigations indicate that the optimum thickness for concrete thermal storage walls is about 30–40 centimeters (1–1.3 feet). Innovations in passive systems are appearing at a rapid rate. Some of these include movable insulation for shielding glass areas at night, and more compact thermal storage systems, such as ceiling tiles which have been developed by the Massachusetts Institute of Technology. These tiles contain a material that undergoes a phase change at 75°F (24°C), storing heat as the material melts and later releasing the heat to the room as the material solidifies. It is expected that, as passive systems improve, there will be a considerable impact on the traditional flat-plate collector.

Large-Scale, High-Temperature Solar Energy Systems

Systems in this category require concentration of solar radiation prior to its collection and utilization. Concentrators fall into three categories:

(1) *Nonfocusing concentrators* have the advantage that they do not have to continuously track the sun and thus do not require optical precision. Also, they can utilize both diffuse and direct radiation and thus are partially operable on cloudy days. They do not, however, operate as efficiently as focusing types, particularly during the early and late periods of the day. A simply designed nonfocusing concentrator essentially will consist of a stationary mirror or reflector located next to a flat-plate collector. In another approach, placing reflectors behind evacuated-tube collectors also accomplishes a modest degree of concentration. An advanced nonfocusing concentrator, known as the *compound parabolic concentrator* (CPC) was developed by the Argonne National Laboratory as the result of experience gained from designing light-concentrating devices for use in high-energy physics experiments. The device incorporates a parabolic surface designed to provide the maximum amount of radiation to an absorber for a given concentration ratio. In one configuration, the radiation is concentrated 1.8 times and, when operated in conjunction with a stationary collector, will reach temperatures as high as 250°F (121°C). Units with even higher concentration ratios (3× up to 6×) have been developed for use with evacuated-tube collectors and can achieve temperatures between 300 and 450°F (150 and 232°C). At a concentration ratio of 6, it is necessary to reorient the collector once each month. Currently, manufacturing costs are relatively high, but the CPC holds promise for a number of future applications.

(2) *Trough or line-focusing concentrators* track the sun by focusing in one direction only. Concentrations in this category of device range from 10× to 100× and they are capable of achieving temperatures between 200 and 600°F (93 and 316°C). On average, these devices deliver a minimum of 50% of the solar energy available to the heat-transfer medium in the absorber. In one configuration, mirrors form a parabolic trough that focuses radiation onto a linear absorber. Usually the mirrors are constructed of polished metal or coated plastic; the absorbers are blackened metallic pipe or evacuated-glass tubes. See Fig. 10. The entire assembly or array tracks the sun. Although normally considered in terms of relatively large thermal capacities, versions have been offered for residence and small building applications. Other, more sophisticated versions, operating at the high-temperature range, are used to drive Rankine-cycle heat engines for pumping irrigation water in a number of locations in the southwestern United States. Other installations include water heating for industrial processes. In another type of line-focusing concentrator, the optics are altered to utilize plastic Fresnel lenses for focusing the radiation onto the absorbers. Apparently a similar thermal result can be

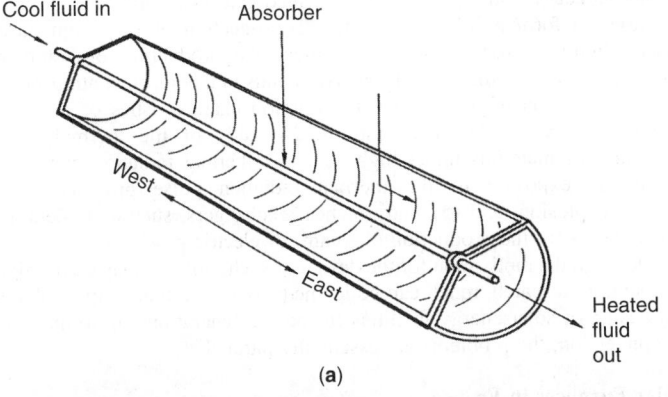

(a)

(b)

Fig. 10. (a) One configuration of an optical-type concentrator with axial absorber; (b) parabolic troughs under test for various uses, including irrigation pumping for croplands

accomplished, as compared with the parabolic mirror approach, with a smaller optical surface.

(3) In another one-axis tracking concentrator, a fixed trough is made up of flat mirrors and tracking is accomplished by moving the absorber. Known as the *Russell collector*, this unit apparently can reach temperatures up to 900°F (482°C).

Two-axis focusing systems will be described a bit later.

High-Temperature Solar Energy

Prior to the serious consideration of high-temperature solar energy as a source of electric power, much was learned from the design of solar furnaces, the objective of which was the production of extremely high temperatures for materials testing, a very useful research function that continues. Much knowledge was and is continuing to be learned from the operation of solar furnaces—knowledge that is helpful in the design of solar power plants. It is fitting here, as a backdrop to describing the solar power tower concept, to present information on solar furnaces. Historically, solar furnaces have been selected for high-temperature research and development activities where a highly concentrated source of nonpolluting radiant energy is required. Generally such activities can be categorized as: (1) high-temperature chemistry involving the formation of very pure or otherwise unique materials; (2) high-temperature processing by which a material is fused, purified or otherwise improved; (3) high temperature property measurements involving the determination of the behavior of a material under conditions which require a noncontaminating environment; (4) determination of the thermal shock resistance or other behavior of materials in a high-temperature, high-heat flux radiant energy environment; and (5) study of high temperature solar-thermal conversion systems. Certain of these applications may be refined further by conducting the operation in an optically transparent vessel or one containing a transparent window such as fused quartz through which the radiant energy may pass and in which the composition and pressure of the atmosphere can be controlled.

A few examples of the types of high temperature studies which have been conducted in the previously described categories are: (1) gas phase reactions to form pyrolytic graphite; (2) production of very high purity fused aluminum oxide and fused silica, the production of stabilized zirconia and the purification of reactive metals in a controlled atmosphere; (3) determination of microwave transmission characteristics of dielectric materials at very high temperatures; (4) study of the thermal shock resistance of materials under high heat flux thermal radiation conditions simulating exposure to the thermal radiation pulse provided by a nuclear explosion; and (5) study of heat exchangers, such as boilers and superheaters for the production of steam for electric power generation.

Although the motivation for the design of such furnaces may be for high-temperature research, much can be learned from them that is applicable to the design of solar energy facilities for power generation. Up to the point of conversion, the problems are essentially parallel.

Solar Furnaces in France

In 1948, under the leadership of Professor F. Trombe, the Centre National de la Recherche Scientifique (CNRS) in Paris undertook the design, construction, and development of the world's first large solar furnace at Montlouis in the French Pyrenees mountains. This furnace was completed in 1952, and provided 50 kilowatts of thermal energy. The Montlouis solar furnace became the prototype design for other large high-temperature solar furnaces. Basically, this design utilized a single large heliostat (array of numerous flat mirror elements) which continuously tracked the sun to direct the sun's rays onto a concentrating reflector (parabolic or spherical) consisting of many smaller mirror elements each of which was contoured to concentrate the incident radiation at a common focal point. In the case of the Montlouis furnace, the heliostat was 43 feet (13.1 meters) wide and 34 feet (10.4 meters) tall and contained 540 flat mirrors each 50×50 centimeters. The concentrating reflector was made up of 3,500 mirrors 16×16 centimeters arranged in a parabolic configuration 36 feet (11 meters) wide and 30 feet (9.1 meters) high with a focal length of 6 meters. Each of the 3,500 flat mirror elements in the parabolic concentrator was mechanically contoured and aligned to focus the radiation received from the heliostat onto the focal point of the parabola.

The successful performance of Montlouis solar furnace led to the use of its design as the prototype for the next three large single heliostat-concentrator solar furnaces which were to be built during the next twenty years. All three of these furnaces were similar to the Montlouis furnace

in size, operation and thermal power level and were constructed by: (1) U.S. Army Quartermaster Corps, Natick, Massachusetts; (2) Tohoku University, Sendai, Japan; and (3) the French Army's Laboratoire, Central de L'Armement, Odeillo, Font-Romeu, France. In 1973 the U.S. Army's solar furnace was moved to the Nuclear Weapon Effects Laboratory, White Sands Missile Range, New Mexico, where it became operational in 1974.

Although the Montlouis solar furnace played a major role in developing applications for high-temperature solar energy, and in providing design information for the three other large solar furnaces, its most valuable contribution to the field of high-temperature solar energy was the experience and background it provided the CNRS Solar Energy Laboratory. This led them to design and construct the CNRS 1,000-kilowatt solar furnace.

The CNRS 1,000 kilowatt solar furnace is located at Odeillo, Font-Romeu, altitude of 5,900 feet (1798 meters) about 25 miles (40 kilometers) east of Andorra and 5 miles (8 kilometers) west of Montlouis. At this location, the sun shines as many as 180 days a year and solar intensities as high as 1,000 watts per square meter are common. The solar furnace was completed on October 1, 1970, after more than 10 years of construction.

Figure 11 is a schematic of the CNRS 1,000-kilowatt solar furnace. This furnace utilizes 63 heliostats to direct the sun's rays onto the surface of the giant parabolic concentrator.

The 63 heliostats are each 7.5 meters wide by 6 meters high and contain 180 single flat mirror elements 50×50 centimeters. The total area of mirror surface in the 63 heliostats is 2,835 square meters or over one-half the playing area of a football field.

The heliostats are located directly north of the parabola and are arranged on eight terraces. Each terrace corresponds in elevation to one of the floors of the building supporting the concentrating parabola. A solar beam of constant energy is thus directed horizontally and southward from the heliostats to the mirrors that make up the concentrating parabola.

Each heliostat is designed to illuminate a specific area on the parabola and is equipped with a dual optical control system, which maintains the proper orientation for each heliostat by means of a dual hydraulic system. This dual system permits each heliostat to be operated in either a "search" or "track" mode. In both cases, the optical guidance system uses an optical tube, which contains four photodiodes that control the heliostat motion in east-west and up-down direction.

When operating in the "search" mode a short (10-centimeter) optical tube with a 40 degree acceptance angle is used to activate the "fast" hydraulic system, which operates in an on-off mode to quickly bring the heliostat to within the operating range of the "track" system. In the "track" mode a 100-centimeter optical tube is used to control a slower acting hydraulic system which operates in a proportional control mode. The size of the sun's image at the base of the 100-centimeter tube is $\frac{1}{2}$-inch (13 mm) in diameter and the accuracy of the control is one minute of arc.

The concentrating parabola has a focal length of 18 meters, is 40 meters high and 54 meters wide, and the focal axis is 13 meters above the first floor. The parabola consists of 9,500 initially flat glass mirrors that were mechanically curved and adjusted to provide a solar image of minimum diameter at the focal point. Almost two years were required to accomplish these two precise adjustments which were completed on 1

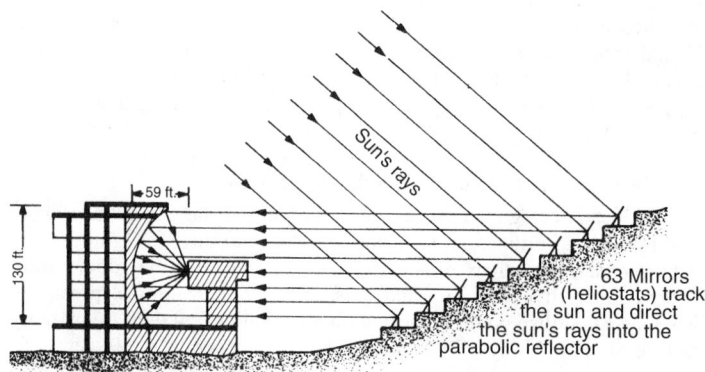

Parabolic reflector concentrates sun's rays onto target area

Fig. 11. Schematic representation of the 1000-kilowatt solar furnace at Odeillo, Font-Romeu, France. (*Centre National de la Recherche Scientifique*)

Fig. 12. Large parabolic reflector and focal building in foreground. Concentrated energy is directed at the solar furnace located within the focal building. Installation is at Odeillo, Font Romeu, France. (*Photo by Glenn D. Considine*)

Fig. 13. Field of heliostat-controlled collector-reflector mirrors, which direct their energy to the parabolic reflector at the solar furnace installation at Odeillo, Font-Romeu, France. (*Photo by Glenn D. Considine*)

October 1970. Figure 12 shows the parabola and the focal building into which the concentrated solar energy is directed. See also Fig. 13.

The solar energy incident on an area of about 2,000 square meters is concentrated by the parabolic reflector onto an area of less than 0.3 square meters. Sixty percent of the total thermal energy (about 600 kilowatts) is concentrated in an area of less than 0.08 square meter at the center of the focal plane of the parabola.

The diameter of the image of the sun at the focal point is 17 centimeters and 27% of the thermal energy (about 270 kilowatts) is concentrated in this area. Heat flux data in watts per square centimeter in the focal area are presented graphically in Fig. 14. Curve 0 represents the heat flux at the focal plane. Curve d/2 shows the heat flux and temperature at a plane removed one-half the diameter of the solar image (8.5 centimeters) behind the focal plane. Curve d presents the same data on a vertical plane removed one diameter of the solar image (17 centimeters) behind the focal plane.

Solar Tower Energy Collector

Authorities have observed that solar energy can be usefully collected optically from one square mile (2.6 square kilometers) of surface area, or even larger, and concentrated onto a central receiver by an array of heliostats, i.e., independently steered mirrors. By judiciously spacing mirrors over 35% of the area, such a system in the desert southwest of the United States, for example, could collect 2800 megawatt-hours thermal per day in midwinter and almost twice that amount of energy in midsummer. In order that the reflected radiation from this field be efficiently intercepted, the central receiver would have to be several hundred meters high.

Unlike the Odeillo installation, previously described, where a field of heliostats finally focus their energy to a small aperture by way of a huge parabolic reflector, in the solar tower approach, the energy from each mirror is directed to a central receiving tower, located high above the field, as

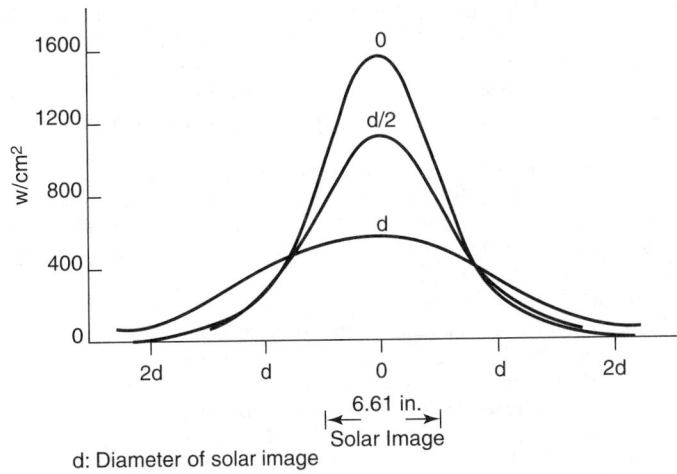

Fig. 14. Solar energy versus distance from focal point in Odeillo solar furnace. (*Georgia Institute of Technology*)

shown schematically in Fig. 15. Shown is a large array of heliostats by which essentially flat mirrors are automatically steered to reflect or redirect the incident solar radiation to a high tower. It is assumed that the terrain of the heliostat field is flat. However, a gentle slope southward would be advantageous. After reflection from the mirrors, the redirected solar energy can be absorbed and converted to heat by a black body receiver placed in the focal region. The heat can be transported down the supporting tower by way of liquid metal and/or steam lines and can be stored or used to operate a conventional turbine generating station. Alternate uses would be direct conversion to electricity by way of high-power-density solar cells placed in the focal region, or use of the heat to produce a fuel thermodynamically. Two-axis control can be obtained by either hydraulically or electrically operated servo-mechanisms that derive a signal from a simple position-sensing element.

It requires energy collected from an area like a square mile to be of interest to power utilities. Energy collection from hundreds of such installations would be required to make a significant impact on the energy supply. If one replaces the 300-meter tower with geometrically identical systems using 100-meter towers, it is found that 9 such towers would be required. Although the cost of nine smaller towers would compare with the cost of a single, large tower, additional costs and losses would be incurred in connecting heat-transfer lines to a central generator to handle 9 collectors. Also, heliostats smaller than about 20 square meters are not economical because the cost per heliostat of the support, actuator, and steering systems have a substantial fixed component.

The first choice of heat-transfer fluid would be steam because it would appear not to require any new technology. However, because the flux density that must be absorbed and transferred to the fluid can be appreciably higher than in conventional steam plants, efficient operation may require some new technology. Also, because of the large daily and seasonal heat flux variations, the design of the receiver is not trivial and may ultimately be best accomplished by utilization of liquid metal, such as sodium, for heat transfer from the receiver surface to a steam line. Liquid sodium technology has been developed for nuclear reactors and operating temperatures of 550 to 650°C appear reasonable. Sodium also presents a promising high-temperature thermal storage medium. In general, the

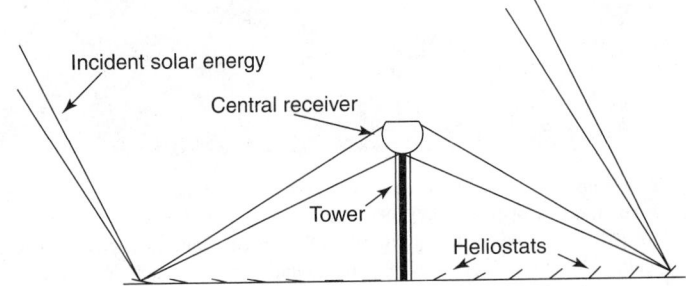

Fig. 15. Schematic diagram of solar tower energy concept

thermal cycling due to the intermittent nature of solar energy of this high-temperature system will have to be considered in any detailed design. The black body boiler surface should not deteriorate at high temperature in the presence of air. If deterioration is a problem and if convective losses are appreciable an inverted cavity design can be used in order to avoid the use of vacuum jacketing.

Solar energy may best be utilized at first by a steam generator operating in a solar-only mode with short-term storage of an hour or less to provide operational stability. Because utilities require very high reliability, little new plant capacity credit would be given such a plant because of possible cloudiness. New plant credit could be given if there were possibilities of using liquid fuels, such as liquid petroleum gas or oil for backup. This could be accomplished by adding a simple low-cost, but possibly inefficient burner to add heat. Although liquid fuels may be in short supply, they are easily stored and afford an ideal way of giving a solar plant high reliability as an intermediate plant. Depending on actual operating conditions, it may be that very little liquid fuel is burned. The comparison of such a hybrid plant should be made with a solar plant that has a gas turbine as a backup. An alternate approach might be to store solar energy as sensible heat, perhaps in underground cavities, or as latent heat.

Some authorities believe that close ties between solar power and conventional electric power plants—so-called solar thermal electric designs—represent an ideal approach to the large-scale use of solar energy. In this concept, instead of a solar-powered facility being linked to traditional power-generating facilities by way of the electric grid, a fossil fuel backup for a solar system would be located at the same site as the solar electric plant. Savings could be realized through the common usage of certain equipment items. In one design, the oil-fired backup would use the same turbine as a power tower system. Some designers have estimated

that this additional capability would add only about 0.26% of the total cost of the solar electric facility.

There are other authorities who believe that adapting solar energy to electric utilities will limit the economic potential of solar energy. The basic problem, is that both technologies are very capital intensive and that the electric utility, because of the high fixed costs of generation, transmission, and distribution capacity, represents a poor backup for solar energy systems. On the other hand, the solar collection system, because it represents pure, high-cost capital and because of its outage problems, cannot be considered as a part-load source of auxiliary energy for the electric utility system.

Solar Energy Plant at Electric Utility Level

Construction commenced in 1975 on an experimental 10 MWe central receiver pilot plant in a combined effort by two electric utilities in the southwestern United States, Southern California Edison and the Los Angeles Department of Water & Power, who worked in cooperation with the U.S. Department of Energy and the California Energy Commission. The start of continuous electric power production commenced in August 1984 and the plant is now up to its design capacity of 10 MWe. The plant, known as *Solar One*, is located in Daggett, California just off Highway 40 and east of Barstow. A panoramic view of the facility, clearly showing nearly 2000 heliostat-controlled mirrors focusing their collected energy on a boiler atop a 300-foot (91-meter) tower, is given in Fig. 16.

Although large and very impressive, *Solar One* is regarded as an experimental pilot plant for proving and testing technological improvements that can be incorporated in future commercial-size plants. The Daggett plant is a scale model of a 100 MWe generating plant. On its own, *Solar One* is currently furnishing the electricity requirements for a community of about

Fig. 16.   Panoramic view of the world's largest solar thermal electric power plant, located on the Mojave Desert, near Daggett, California. The 10 megawatt (electric) facility commenced operation in 1982 and achieved design capacity in 1984. The collector tower (receiver upon which solar energy is reflected) is located atop a 300-foot (91-meter) tower. The north field of heliostats (mirrors kept in synchronism with the movement of the sun), 1240 in number, is shown in background; the south field (578 heliostats) is shown in foreground. Operating facilities, turbine generators, and storage tank are shown in circular middle section under the tower. The facility is operated by Southern California Edison and the Los Angeles Department of Water and Power, in conjunction with the U.S. Department of Energy and the California Energy Commission

6000 people. *Solar One* relies on a combination of both old and new solar technology. Certain features not found in typical commercial generating plants allow great flexibility in plant operation. Several different types of solar central receiver plants can be simulated within this one project.

The Basic Concept of Solar One. Computer-controlled mirrors (helio stats) totaling 1818 in number form a circular array around a central tower. Within the receiver, the solar energy is transformed into high-temperature thermal energy in a water-steam heat transport fluid. The thermal energy can be converted to electric power immediately or stored to extend plant operation. See Fig. 17. The collected solar energy is most efficiently put to work as receiver steam to power a turbine-generator (Path A). If the energy is to be stored, receiver steam follows path B and heats oil that is routed to and from the thermal storage tank. Energy is discharged from storage by using hot oil from the tank (path C) to generate steam, which is then sent to the turbine along path D.

The thermal storage system uses oil as both a thermal storage medium and a heat transport fluid. The maximum operating temperature of the storage system is 575°F (300°C). As a result, electricity is generated less efficiently than when 960°F (515°C) receiver-supplied steam is used directly in the turbine.

The operating temperature of the storage system simulates steam generation conditions in industrial plants and the chemical processing industry. Furthermore, because storage-supplied heat can supplement solar-supplied energy, *Solar One* can simulate a plant that uses both conventional fuels and solar energy.

Heliostats. The facility receives 3600 to 4000 hours of sunlight/year (9.8 to 10.9 hours/day). Construction of the 1818 heliostats for the pilot plant demonstrated that prototype designs can be successfully produced in volume quantities with conventional manufacturing techniques. Each heliostat has a reflective area of 430 square feet (39.3 square meters). The heliostat glass is specially formulated to contain a minimum amount of impurities. As a result, 91% of the incident sunlight can be reflected when the mirror surface is clean. A close-up of a set of mirrors (in a vertical position for demonstration purposes) is given in Fig. 18.

The vertical and horizontal movement of the heliostats is directed by a control system—a microprocessor in each heliostat, a controller to regulate groups of up to 32 heliostats, and a central computer. Over 97% of the heliostats are available more than 98% of the time. Operation of the heliostats has suggested areas for further research and development; for example, rain water may be sufficient to maintain the cleanliness of the mirrors, and mechanical rinsing may be required only in dry months.

The heliostats, as shown in Fig. 19, are distributed in a south field (578) and a north field (1240). The mirrors are slightly concave (approximately 1000 foot focal length with $\frac{1}{6}$-inch curvature in a 10-foot length). The total weight of a heliostat, as previously shown in Fig. 18, is 4312 pounds (1956 kg). The heliostats are normally stowed in a vertical position except when high wind conditions exist. During daylight hours, of course, the mirrors are rotated by a drive mechanism to follow the direct solar rays as closely as possible. It is interesting to note that the sun's position is calculated rather than sensed—so that even when clouds briefly cover the sun, maximum energy is reflected.

Receiver. On top of the steel tower rests the cylindrical receiver, a superheated steam boiler that is 14 meters (46 feet) tall and 7 meters (23 feet) in diameter. The receiver weighs almost 50 tons and is positioned over 20 stories above the ground. Feedwater is pumped to the bottom of the receiver, where it is vaporized to superheated steam in a single pass to the receiver's top. The steam is then piped to the turbine-generator at the foot of the tower. This steam can also provide heat to the thermal storage system.

Thermal Storage System. On a clear day, the receiver can generate sufficient steam to simultaneously operate the turbine and also deliver

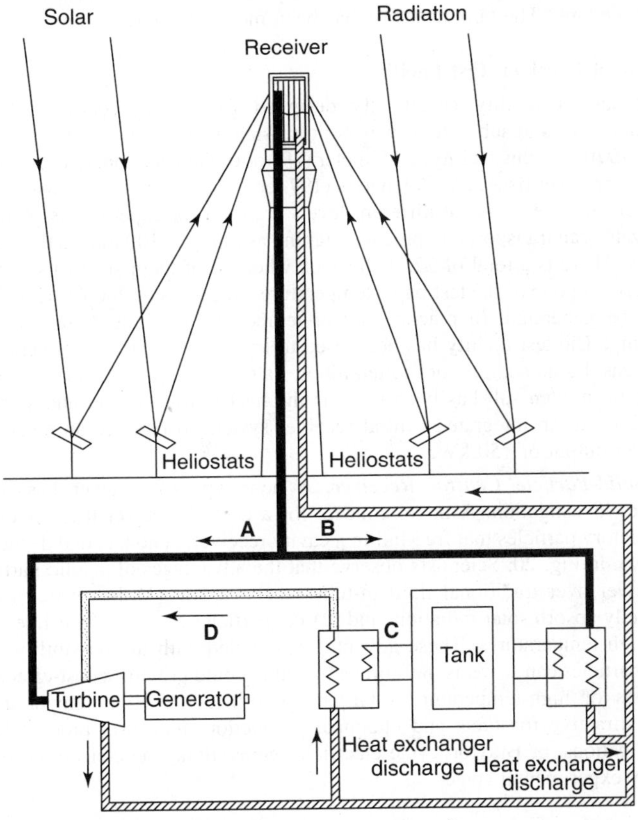

Fig. 17. General concept of *Solar One*. Within the receiver, the solar energy is transformed into high-temperature thermal energy in a water-steam heat transport fluid. The thermal energy can be converted to electric power immediately or stored to extend plant operation. The collected solar energy is most efficiently used as *receiver steam* to power a turbine-generator (Path A). If the energy is to be stored, receiver steam follows Path B and heats oil that is routed to and from the thermal storage tank (Path C) to generate steam, which is then sent to the turbine along Path D. The thermal storage system uses oil as both a thermal storage medium and as a heat-transport fluid

Fig. 18. Close-up of a heliostat rack assembly shown in vertical position for demonstration purposes. Note reflection of tower in mirrors. *Solar One* has a total of 1818 heliostats

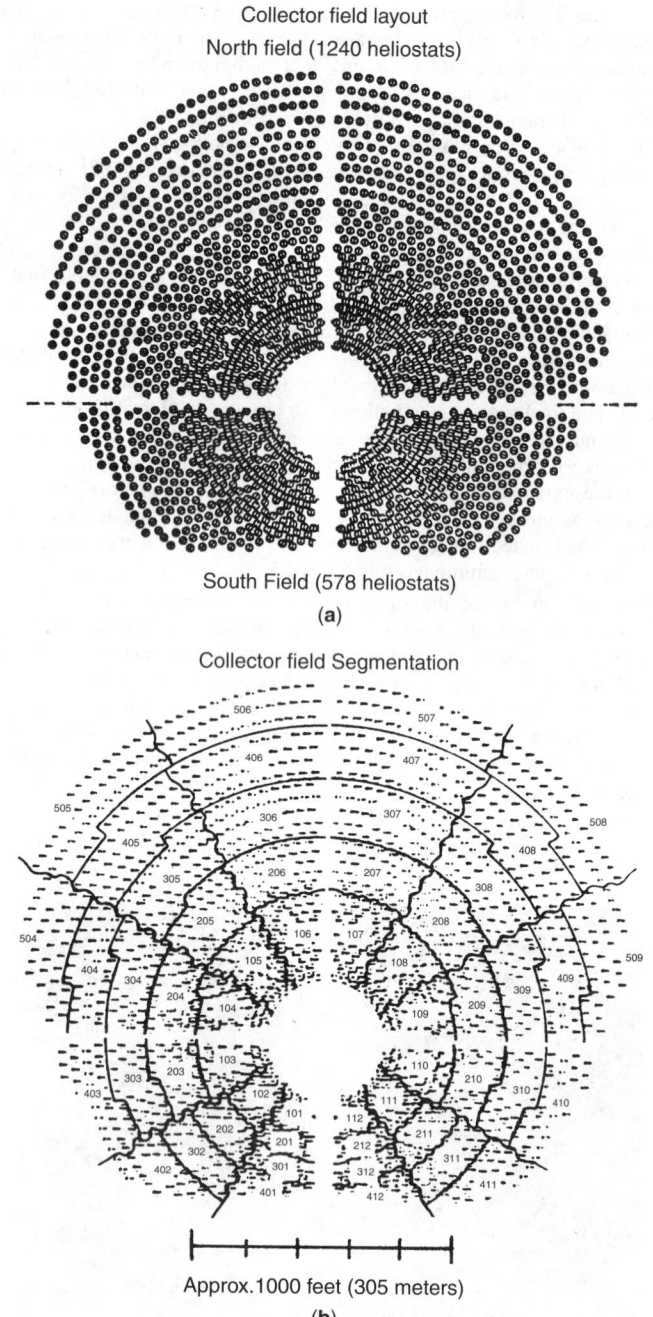

Collector field layout

North field (1240 heliostats)

South Field (578 heliostats)

(a)

Collector field Segmentation

Approx.1000 feet (305 meters)

(b)

Fig. 19. (a) The heliostats are distributed in two fields—North and South; (b) for control purposes, the heliostats are segmented

heat to the storage system. The thermal storage can generate power in the evening or during periods of cloud cover. It also provides steam for start-up in the morning and for keeping selected portions of the plant warm when the plant is not operating. The steel-walled insulated storage tank has a capacity of 3.5 million liters and sits on lightweight insulating concrete. The tank is filled with sand, rocks, and a high-temperature thermal oil. Steam from the receiver is routed through heat exchangers to heat the thermal oil, which is then pumped into the tank to heat the rock and sand. This stored thermal energy can then be transferred to the turbine generator for electrical power production.

Power Generation Station. The turbine-generator is rated at 12.5 megawatts and is sized to handle the full plant system output plus all internal plantloads. The dual-admission turbine has a high-pressure steam inlet for steam produced by the receiver and a low-pressure inlet for steam produced from thermal storage. The rated turbine thermal-to-electric efficiency from receiver to steam is 35%. The efficiency is 25% from the lower quality thermal storage system.

Master Control System. In the morning the operator, through keyboard commands, positions the heliostats at standby operating points, begins water circulation in the receiver, and then issues a command to the system to start up the plant. At this point, a computer takes over and automatically directs heliostats to track the receiver. When receiver steam conditions are correct, steam is routed to the turbine. The operator then synchronizes the turbine to the electric grid, after which the minimal manual attention is needed. If conditions change, such as a cloud passing over, the control system automatically makes adjustments to keep the plant in the best operating state. If some abnormal event occurs, alarm messages tell the operator which parameters are out of normal operating range. The operator can, at any time, make changes in any plant operating condition. While the pilot plant control system was designed for controlling a water-steam central receiver solar plant, the basic functions and operating philosophy are readily adaptable to other power plants.

Performance. The requirement for production of 10 MWe was exceeded by a peak production of 12.1 MWe. Similarly, the required 7 MWe net generation from storage was exceeded by an output of 7.3 MWe. The plant also has successfully operated down to 0.5 MWe, which is considerably lower than the designed minimum operating production level of 2 MWe. The minimum sunlight threshold for operation was designed as 450 W per square meter, yet the plant has operated in direct solar radiation levels as low as 300 W per square meter. In an endurance test, the receiver and storage system kept the turbine continuously on-line for 33.6 hours and generated 127 MWe net.

Solar One was designed to have 95% of the heliostats available at any one time. Between April 1982 and April 1983, 98% of the heliostats were available for operation. This percentage later increased to 99%. The establishment of the sharp thermal gradient (thermocline) needed for the storage system has been verified. Gradients of 49°C/meter have been measured. Equally important is the very low rate of heat loss from the storage tank. The tank heat loss has been measured at 1.3% day.

Central Receiver Test Facility

The largest facility specifically designed for testing central receiver components and subsystems was built in Albuquerque, New Mexico in the late 1970s. At this facility, a 15-meter (49-foot) diameter concrete-and-steel receiver tower rises some 60 meters (197 feet) above the ground. Within the tower, three test bays at different levels are used for experiments. A huge elevator can transport equipment weighing as much as 100 tons to these test bays. There is a total of 222 heliostats; when all of them are focused on a receiver in one of the test bays, temperatures in excess of 2000°C (3632°F) can be generated. In practice, lower temperatures are used for receiver testing. The test facility has been used to try out innovative receivers that use gas, liquid sodium, or molten nitrate salts for thermal transport. In one system, molten salt has been used as the heat transport fluid and storage medium in an integrated central receiver system to produce an electrical power output of 750 kWe.

Solid-Particle Central Receiver. A new type of receiver has been under investigation. A novel concept for a central receiver uses sandsize refractory particles that free-fall in a cavity receiver. A conceptual design is shown in Fig. 20. Scientists observe that the advantages of a solid particle receiver over traditional fluid in-tube receivers are: (1) the particles can directly absorb solar radiation, and (2) the particles maintain their integrity at high temperatures. These advantages, coupled with the possibility that the particles can serve as the storage medium, could provide a cost-effective means of high-temperature solar energy utilization. High temperatures are attractive for fuels and chemical production, industrial process heat applications, or Brayton cycle electricity generation. The concept is in an early experimental stage.

Heat Engine Cycles for Solar Power

Heat engines for conversion of solar energy to electric power ideally should have the following attributes: (1) low cost per kilowatt output capacity; (2) long life and reliable operation with minimal maintenance; (3) safe and environmentally acceptable operation; (4) characteristics compatible with cycle top temperatures up to 1,000 K; and (5) efficiency approaching Carnot values.

Heat engines that are potential candidates for coupling a solar heat source include thermoelectric, thermionic, thermochemical, magnetohydrodynamic, Rankine, Brayton (simple or recuperated), and cascaded cycles.

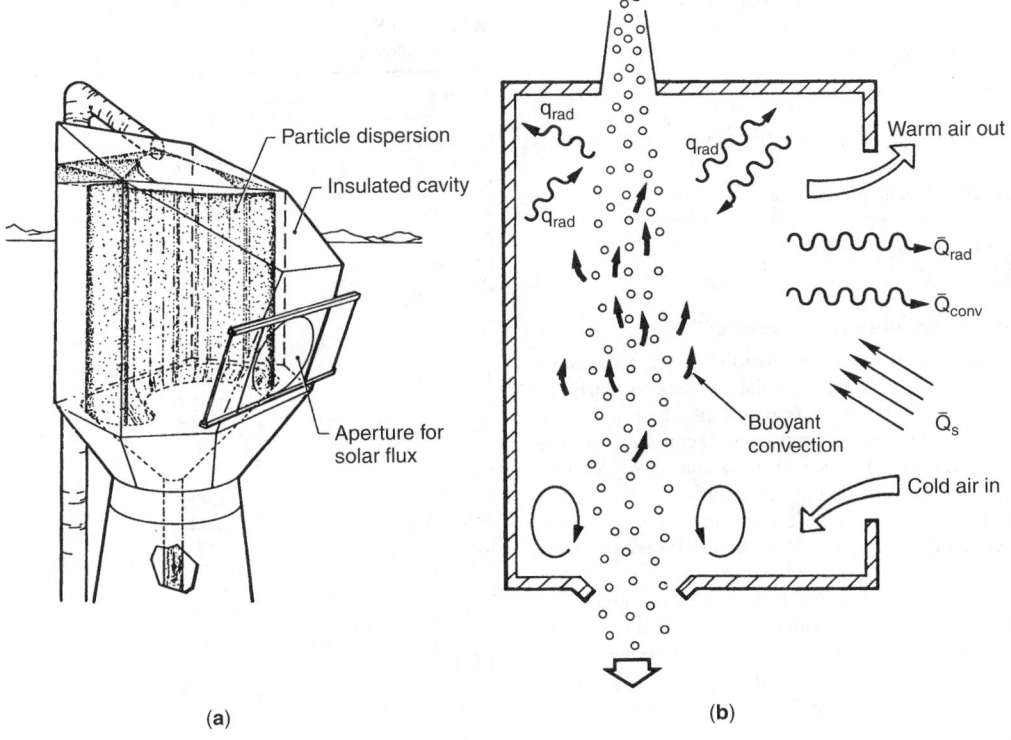

Fig. 20.   Solid-particle central solar energy receiver: (a) conceptual design; (b) thermal phenomena in a solid particle receiver. (*Sandia National Laboratories*)

Rankine Cycle. The steam-Rankine cycle employing steam turbines has been the mainstay of utility thermal electric power generation for many years. The cycle, as developed over the years, is sophisticated and efficient. The equipment is dependable and readily available. A typical cycle (Fig. 21) uses superheat, reheat, and regeneration. Heat exchange between flue gas and inlet air adds several percentage points to boiler efficiency in fossil-fueled plants. Modern steam Rankine systems operate at a cycle top temperature of about 800 K with efficiencies of about 40%. All characteristics of this cycle are well suited to use in solar plants.

Brayton Cycle. In recent years, attention has been drawn to the Brayton cycle as a potential and practical alternative to the steam Rankine cycle for solar power and for high-temperature gas-cooled nuclear reactors. The Brayton cycle is most familiar in its open form as used in aircraft gas turbines. The open Brayton cycle cannot compete with steam-Rankine in efficiency. In a power-generation application, cycle efficiencies on the order of 20% would be expected. However, the Brayton cycle can achieve higher efficiency through recuperation, sometimes called regeneration. A representative cycle diagram is given in Fig. 22. The working fluid is an

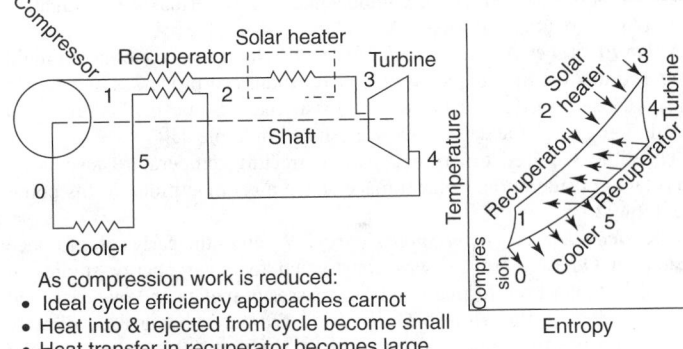

As compression work is reduced:
- Ideal cycle efficiency approaches carnot
- Heat into & rejected from cycle become small
- Heat transfer in recuperator becomes large
- Cycle pressure ratio approaches zero
- Cycle mass flow per unit shaft power approaches ∞

Fig. 22.   Recuperated Brayton cycle diagram

inert gas, typically helium. Inert gas mixtures, such as helium-xenon, have been studied and have potential advantages.

The recuperated Brayton cycle approaches Carnot efficiency in the ideal limit. As compressor and turbine work are reduced, the average temperatures for heat addition and rejection approach the cycle limit temperature. The limit is reached as compressor and turbine work (and cycle pressure ratio) approach zero and fluid mass flow per unit power output approaches infinity. It can be expected from this that practical recuperated Brayton cycles would operate at relatively low pressure ratios, but be very sensitive to pressure drop. With the assumption of constant gas specific heat over the cycle temperature range, a good assumption for helium, the cycle efficiency of a recuperated Brayton cycle may be expressed:

$$\eta e = 1 - \left[\frac{\dfrac{r_{pc}\zeta - 1}{\eta_b} + \dfrac{\Delta T_r}{T_0}}{\dfrac{T_3}{T_0}\eta_T \left[1 - \left(\dfrac{G}{r_{pc}}\right)^{\zeta} \right] + \dfrac{\Delta T_r}{T_0}} \right]$$

where r_{pc} is compressor pressure ratio ($r > 1$)
         η_b is compressor efficiency

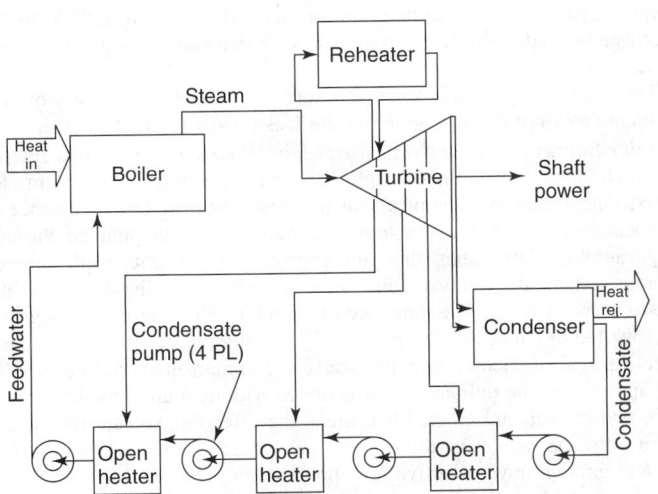

Fig. 21.   Schematic diagram of regenerative-reheat steam Rankine cycle

ΔT_r is temperature difference across recuperator ($T_4 - T_2$ in Fig. 21)

T_0 is cycle lower limit temperature

T_3 is cycle top temperature

η_T is turbine efficiency

ζ is specific heat factor, $(\gamma - 1)/\gamma = 0.4$ for $\gamma = 1.67$ as for helium

G is the product of the four pressure drop factors,

$$\left(\frac{P_1}{P_2}\right)\left(\frac{P_2}{P_3}\right)\left(\frac{P_4}{P_5}\right)\left(\frac{P_5}{P_0}\right)$$

Solar Energy for Industrial/Metallurgical Processes

It is interesting to note that experiments on melting large masses of metals were conducted at the French Odeillo solar furnace as early as the mid-1970s. Since the mid-1980s, solar furnaces in other locations have researched what promises to be an important technology in the near future—namely, Solar Induced Surface Transformation of Materials (SISTM). In the past, numerous heat-treating methodologies have been practiced widely. Less traditional methods have been developed in recent years, including cladding/coating, self-propagating high-temperature synthesis, thin-film deposition, and ion-, laser-, and electron-beam processes.

Researchers have found that, for delivering large fluxes on target, the solar furnace is less capital intensive than competing methods that require an intermediate energy-conversion step. For example, a 1-square-meter (11 ft^2) solar dish can deliver about 1 kW of optical power to a target. To deliver the same amount of radiant energy, an arc lamp would have to be powered by the electrical output of an 11-square-meter (120 ft^2) dish, and a carbon dioxide laser would require the energy supplied by 26-square-meter (280 ft^2) disk.

The automotive industry, which turned to laser transformation hardening of engine and other drivetrain components, is now seriously investigating the "solar hardening" process.

Solar Furnaces No Longer Uncommon. The past decade has brought increased interest in using solar energy for industrial processes, not just to conserve energy or avoid the use of fossil fuels. Some of the important installations, as of the early 1990s, are listed in Table 1.

The time required to bring a part to treating temperature also is an important consideration. Solar furnaces have no competition on this point. See Table 2.

Chemicals Directly from Solar Energy. Whereas the early solar furnace located in Odeillo, France, was constructed for solar energy studies in general and later used for melting huge masses of metals and materials and the solar plant in the Mojave Desert in California is being used directly to generate electric power, scientists in the early 1990s were taking a somewhat different approach.

Researchers have recognized that the regions where ample sunshine is available for harvesting seldom coincide with the population and industrial centers of the world. Further, solar energy is of an intermittent nature because it depends upon clear skies with no cloud cover. I. Dostrovsky (Weizmann Institute of Science) observes that most of the limited amount

TABLE 2. SOLAR FURNACE TIME TO REACH MELTING POINT OF MATERIALS
(When exposed to absorbed solar flux of 20 MW/m^2)

Material	Melting point, T_m, °C	Time to reach T_m, [1] sec
Carbides		
TiC	3,200	9.14
NbC	3,500	0.86
ZrC	3,540	0.112
SiC	3,830	0.56
Metals		
Al	660	0.42
Cu	1,083	3.10
Ni	1,453	1.03
Steel	1,535	0.79
Ti	1,670	0.23
Cr	1,857	1.46
Mo	2,617	4.70
W	3,407	9.80
Nitrides		
Si$_3$N$_4$	1,900	0.059
AlN	2,200	1.320
BN	3,000	0.545
TiN	3,200	0.611
Oxides		
SiO$_2$	1,720	0.014
TiO$_2$	1,870	0.044
Al$_2$O$_3$	2,050	1.00
V$_2$O$_3$	2,410	0.107
CaO	2,580	1.66
HfO$_2$	2,780	0.188
MgO	2,800	2.59
ZrO$_2$	2,900	0.089

Source: Solar Energy Research Institute.

of research on solar energy focuses on converting sunlight to electricity, mainly by photovoltaic and thermal methods, or, in the case of the Daggett, California, utility installation, to furnish heat to supply steam turbines.

An alternative approach is that of using solar energy to produce chemicals that can be stored and transported much as present fossil fuels. One installation along these lines is now undergoing demonstration operation in Saudi Arabia. This project, known as *HYSOLAR*, is a joint venture of Saudi Arabia and Germany. In essence, the facility consists of a plant that produces hydrogen. See also **Hydrogen (Fuel)**.

One process involves the high-temperature decomposition of sulfuric acid, using recoverable iodine as an intermediate reactant. In a first stage, sulfuric acid yields water and sulfur dioxide. In a second stage, the sulfur dioxide plus water and iodine yield hydrogen iodide and sulfuric acid. In the third stage, the hydrogen iodide yields hydrogen (the desired product) and recoverable iodine.

In an electrolytic process, hydrogen is produced by electrolyzing a mixture of sulfur dioxide and water to produce sulfuric acid and hydrogen. In still another electrolytic process, bromine, sulfur dioxide, and water react to form hydrogen bromide and sulfuric acid. By applying 0.62 V to the hydrogen bromide molecules, hydrogen is yielded and the original bromine is recovered.

This type of approach is somewhat reminiscent of the chemistry of coal gasification. More detail is given in the Dostrovsky reference listed.

Solar Energy for Detoxifying Hazardous Chemicals. The Solar Energy Research Institute, Golden, Colorado, has developed a system for detoxifying hazardous chemicals in polluted groundwater. In essence, a photocatalyst is added to the polluted water and then pumped through long, narrow glass tubes that are exposed to sunlight. High-energy photons activate the catalyst, which in turn breaks the pollutants down into nontoxic components. The tubes are mounted in reflecting glass troughs to improve the efficiency of the process. The system has proved particularly effective against trichloroethylene, once a common industrial cleaner. In this application, the polluted water is mixed with titanium dioxide catalyst. Hydroxyl radicals are created that break the offending solvent into water, carbon dioxide, and very dilute hydrochloric acid. The next step is that of determining how effective the process may be in removing other chlorinated hydrocarbons, as well as such substances as benzene, various pesticides, and textile dyes.

TABLE 1. COMPARISON OF HIGH-FLUX SOLAR FACILITIES

Location	Total power Kw	Peak flux MW/m^2
Albuquerque, New Mexico		
Central receiver test facility	5000	2.4
Furnace	22	3.0
Atlanta, Georgia		
Furnace	1.3	9.5
Golden, Colorado	10.0	2.5
(measured using nonimaging		
secondary concentrator)		20.0
White Sands, New Mexico	30	3.6
Odeillo, France		
Horizontal furnace	1000	16.0
Vertical furnace	6.5	15.0
Rehovot, Israel		
Central receiver test facility	2900	
Furnace	16	11.0
Uzbek, Russia	1000	17.0

Source: Solar Energy Research Institute.

Photovoltaic Conversion (Solar Cells)

Photovoltaic devices made of selenium have been known since the 19th Century. Pioneering research in semiconductors, which led to the invention of the transistor in 1947, formed the basis of the modern theory of photovoltaic performance. From this research, the silicon solar cell was the first known photovoltaic device that could convert a sufficient amount of the sun's energy to power complex electronic circuits. The conventional silicon cell is a solid-state device in which a junction is formed between single crystals of silicon separately doped with impurity atoms in order to create n (negative) regions and p (positive) regions which respectively are receptors to electrons and to "holes" (absence of electrons). See also **Semiconductors**. The first solar cell to be demonstrated occurred at Bell Laboratories (now AT&T Bell Laboratories) in Murray Hill, New Jersey in 1954.

In a photovoltaic device, the energy in light is transferred to electrons in the semiconductor when a photon collides with an atom in the material with enough energy to dislodge an electron from a fixed position in the material. A common technique for producing a voltage is by creating an abrupt discontinuity in the conductivity of the cell material (typically silicon) through the addition of dopants. A basic limit on the performance of these devices stems from the fact that light photons lacking the energy needed to lift electrons from the valence to the conduction bands ("band gap" energy) cannot contribute to photovoltaic current, and from the fact that the energy given to electrons which exceeds the minimum excitation threshold cannot be recovered as useful electric current. Most of the photon energy not recovered as electricity is converted to thermal energy in the cell.

Photon energies in the visible light spectrum vary from 1.8 eV (deep red) to 3 eV (violet). In silicon, about 1.1 eV is needed to produce a photovoltaic electron; in gallium arsenide (GaAs), this is about 1.4 eV. Silicon is a comparatively poor absorber of light and consequently silicon cells must be from 100 to 200 micrometers thick to capture an acceptable fraction of the incident light. This places limitations on crystal grain size and thus, with present technology, single crystals must be used. Polycrystalline materials may alter this problem favorably and much research is being directed toward developing polycrystalline materials and, in general, for finding methods to minimize the impact of grain boundaries.

Thin films of gallium arsenide (GaAs) and cadmium sulfide/copper sulfide (CdS/Cu_2S) show potential because they are better absorbers of light and can be made thinner than crystalline silicon. Smaller crystal grains can be tolerated better than with crystalline silicon. See also **Thin Films**. These can be spray- or vapor-deposited, thus simplifying manufacturing. One possible drawback of the CdS and GaAs materials is their toxicity, particularly hazardous during manufacturing operations.

Where solar cells are used in *concentrated* sunlight, efficiency becomes of particular importance because of its effect upon total collector area needed, this being a major cost component of a solar energy system. A number of ingenious collector configurations have been developed. Further, there is the concept of the *thermophotovoltaic cell*, which may be able to achieve efficiencies as high as 30–50% through shifting the spectrum of light reaching the cell to a range where most of the photons are close to the minimum excitation threshold for silicon cells. High efficiencies in intense radiation can be achieved, for example, with GaAs cells by covering them with a layer of $Ga_xAl_{1-x}As$, a material that reduces surface and contact losses. Clearly the interface between cell and solar radiation is of as great importance as development of new cell materials per se.

So-called *wet solar cells* show promise, particularly because of their relative ease of fabrication. In this type of photovoltaic cell, the junction is formed between a semiconductor and a liquid electrolyte. No doping is required because a junction forms spontaneously when a suitable semiconductor, such as GaAs, is contacted with a suitable electrolyte. Three knotty problems (accelerated oxidation of surface of semiconductor; exchange of ions between semiconductor and electrolyte forming a blocking layer; and deposition of ions of impurities on the surface of the semiconductor) all have been solved and thus the concept now appears technically viable.

Over a number of years, the photovoltaic cell developers received large financial incentives from the U.S. government. For example, the National Photovoltaics Act of 1978 was passed by the U.S. Congress, which authorized an expenditure of $1.5 billion for research, development, and demonstration of solar cell systems for converting sunlight into electric power. Also, in connection with the Federal Non-Nuclear Energy Research and Development Act of 1974, which established the concept of "net energy"—that is, the effect of new devices and systems on the overall energy balance. Projects were evaluated on the basis of their "potential for production of net energy." Although there have been some breakthroughs of particular significance to scientists and a gradually expanding market for photovoltaics in addition to use in space, particularly in various consumer products, the long awaited and ultimate application (generation of electric power in impressive amounts at competitive prices) has remained elusive. Scores of analyses have been made and forecasts range from very pessimistic to quite optimistic. The era of practically achieving this goal on the part of the photovoltaic cell community tends to be progressively shifted outward into the future. Forecasts usually are based upon numerous assumptions that are subject to periodic change and their reporting is best left to the periodicals and thus are not detailed here.

It is in order, however, to sum up the observations of the Electric Power Research Institute (EPRI): Photovoltaics need significant additional research to reduce cost and increase efficiency of the cells as well as their support systems (tracking and D.C.-to-A.C. power conversion) before they can be competitive with conventional electricity supply technologies. Current manufacturing costs for flat-panel arrays of interconnected, encapsulated cells are approximately $5000 per peak kW, and balance-of-plant costs double the effective system cost to $10,000 per peak kW. This compares roughly with $300 per kW for combustion turbines; $1400/kW for pulverized coal plants; and $2500/kW for nuclear plants. Two classes of photovoltaic converters that appear to show the most promise for producing large amounts of power are (1) the inexpensive, flat-plate, thin-film devices with target prices of less than $1500 per peak kW and efficiencies of 15%, compared with their current costs of $5000 per peak kW and efficiencies of about 10%; and (2) very high-efficiency, high-concentration devices with target prices less than $1500 per peak kW and efficiencies of 25%, compared with their current costs of $7000 per peak kW and efficiencies to utilities (largely subsidized or experimental programs).

Some authorities estimate that photovoltaic utility capacity could range from 0.6 to 16 GW by the year 2010, provided that needed technical performance is achieved. See also **Photovoltaic Cells**.

Satellite Energy Collectors

Having proven their value in connection with relatively small space satellites, probes, etc., a huge satellite energy collector was first proposed in the late 1960s and, largely on the basis of national concerns with energy supplies precipitated by the oil embargo of the 1970s, considerable attention was given at the design level and in the literature to a solar power satellite (SPS). One proposal called for a space-based array requiring about 90 square kilometers (55 square miles)! That is about the size of Manhattan Island. The satellite would be in a geosynchronous orbit some 36,000 kilometers (22,000 miles) above Earth. Because nearly all authorities now consider such a project very "futuristic," no further details are reported here.

Additional Reading

Asbury, J.G. and R.O. Mueller: "Solar Energy and Electric Utilities," *Science*, **195**, 445–450 (1977).

Asbury, J.G., Maslowski, C., and R.O. Mueller: "Solar Availability for Winter Space Heating," *Science*, **206**, 679–681 (1979).

Beattie, D.A.: *History and Overview of Solar Heat Technologies*, MIT Press, Cambridge, MA, 1997.

Becker, M.: *Solar Thermal Central Receiver Systems*, Springer-Verlag, Inc., New York, NY, 1987.

Considine, D.M.: *Solar Absorption Coating and Heat-Pipe System*, in "Energy Technology Handbook: (D.M. Considine, editor), The McGraw-Hill Companies, Inc., New York, NY, 1977.

Dostrovsky, I.: *Energy and the Missing Resource*, Cambridge University Press, New York, NY, 1988.

Dostrovsky, I.: "Chemical Fuels from the Sun," *Sci. Amer.*, 102 (December 1991).

Flood, D.J.: "Space Solar Cell Research," *Chem. Eng. Progress*, 62 (April 1989).

Goswami, D.Y., F. Kreith, and J.F. Kreider: *Principles of Solar Engineering*, 2nd Edition, Taylor & Francis, Inc., Philadelphia, PA, 1999.

Gupta, B.P.: "Solar Thermal Technology: Research and Development and Applications," Proceedings of the Fourth International Symposium, Albuquerque, NM, 1990.

Holden, C.: "Sunlight Breaks Down Hazardous Chemicals," *Science*, 1215 (September 13, 1991).

Hubbard, H.M.: "Photovoltaics Today and Tomorrow," *Science*, 297 (April 21, 1989).

Laird, F.N.: *Solar Energy, Technology Policy, and Institutional Values*, Cambridge University Press, New York, NY, 2001.

Stanley, J.T., Fields, C.L., and J.R. Pitts: "Surface Treating with Sunbeams," *Advanced Materials & Processes*, 16 (December 1990).

Waterbury, R.C.: "Solar Pump Delivers Remote Power," *InTech*, 74 (January 1990).

Wieder, S.: *An Introduction to Solar Energy for Scientists and Engineers*, Krieger Publishing Company, Melbourne, FL, 1992.

Wilson, H.G., MacCready, P.B., and C.R. Kyle: "Lessons of Sunnyracer," *Sci. Amer.*, 90 (March 1989).

Winter, Carl-Jochen and J. Nitsch: *Hydrogen as an Energy Carrier: Technologies, Systems, Economy*, Springer-Verlag, Inc., New York, NY, 1988.

Note: For earlier references on solar energy, see prior edition of this encyclopedia.

Web References

American Solar Energy Society: http://www.ases.org/

Electric Power Research Institute: http://www.epri.com/

Georgia Institute of Technology University Center of Excellence for Photovoltaics Research and Education: http://www.ece.gatech.edu/research/UCEP/

National Renewable Energy Laboratory: http://www.nrel.gov/

National Solar Thermal Test Facility: http://www.sandia.gov/Renewable_Energy/solarthermal/nsttf.html

Office of Power Technologies: http://www.eren.doe.gov/power/

University of Florida Solar Energy and Energy Conversion Laboratory: http://www.me.ufl.edu/SOLAR/

University of Massachusetts Renewable Energy Research Laboratory: http://www.ecs.umass.edu/mie/labs/rerl/

US Environmental Protection Agency Clean Energy: http://www.epa.gov/global warming/actions/cleanenergy/index.html

100 Top Energy Sites: http://www.100topenergysites.com/

SOL–GEL TECHNOLOGY.

The goal of sol–gel technology is to use low temperature chemical processes to produce net-shape, net-surface objects, films, fibers, particulates, or composites that can be used commercially after a minimum of additional processing steps. See also **Thin Films**. Sol–gel processing can provide control of microstructures in the nanometer size range, i.e., 1–100 nm (0.001–0.1 μm), which approaches the molecular level. These materials often have unique physical and chemical characteristics. See also **Nanotechnology (Molecular)**.

Sols are dispersions of colloidal particles in a liquid. *Colloids* are nanoscaled entities dispersed in a fluid. *Gels* are viscoelastic bodies that have interconnected pores of submicrometric dimensions. A gel typically consists of at least two phases, a solid network that entraps a liquid phase. *Sol–gel technology* is the preparation of ceramic, glass, or composite materials by the preparation of a sol, gelation of the sol, and removal of the solvent.

Compositions

Nucleation of particles in a very short time followed by growth without supersaturation yields monodispersed colloidal particles that resist agglomeration. A large range of colloidal powders having controlled size and morphologies have been produced using these concepts. Materials include oxides, hydroxides, carbonates, sulfides, as well as various mixed phases or composites and coated particles. Controlled hydrolysis of alkoxides has also been used to produce submicrometer TiO_2, doped TiO_2, ZrO_2, doped ZrO_2, doped SiO_2, $SrTiO_3$, and even cordierite powders. Emulsions have been employed to produce spherical powders of mixed cation oxides, such as yttrium aluminum garnets (YAG), and many other systems. Sol–gel powder processes have also been applied to fissile elements. Spray-formed sols of UO_2 and $UO_2–PuO_2$ were formed as rigid gel spheres during passage through a column of heated liquid. Abrasive grains based on sol–gel-derived mixed alumina are important commercial products. Powders for superconductors and magnetic ceramics were also developed using the sol–gel technology. See also **Magnetic Materials**.

Glass and polycrystalline ceramic fibers have been prepared using the sol–gel method.

Sol–Gel Process Steps

Overview. Three approaches are used to make most sol–gel products: method 1 involves gelation of a dispersion of colloidal particles; method 2 employs hydrolysis and polycondensation of alkoxide or metal salts precursors followed by supercritical drying of gels; and method 3 involves hydrolysis and polycondensation of alkoxide precursors followed by aging and drying under ambient atmospheres.

Production of net-shape silica components serves as an example of sol–gel processing methods. A silica gel may be formed by network growth from an array of discrete colloidal particles (method 1) or by formation of an interconnected three-dimensional network by the simultaneous hydrolysis and polycondensation of a chemical precursor (methods 2 and 3). When the pore liquid is removed as a gas phase from the interconnected solid gel network under supercritical conditions (critical-point drying, method 2), the solid network does not collapse and a low density aerogel is produced. Aerogels can have pore volumes as large as 98% and densities as low as 80 kg/m^3.

When the pore liquid is removed at or near ambient pressure by thermal evaporation, i.e., by drying (methods 1 and 3), shrinkage occurs and the monolith is termed a xerogel. If the pore liquid is primarily alcohol-based, the monolith is often termed an alcogel. The generic term gel is usually applied to either xerogels or alcogels, whereas aerogels are usually specified as such. A gel is defined as dried when the physically adsorbed solvent is completely evacuated. This occurs between 100 and 180°C.

A dried gel still contains a very large concentration of chemisorbed hydroxyls on the surface of the pores. Thermal treatment in the range of 500–800°C desorbs the hydroxyls and thereby decreases the contact angle and the sensitivity of the gel to rehydration stresses, resulting in a stabilized gel. Heat treatment of a gel at elevated temperatures substantially reduces the number of pores and their connectivity owing to viscous phase sintering. This is termed *densification*. The density of the material increases and the volume fraction of porosity decreases during sintering. The porous gel is transformed to a dense glass when all pores are eliminated. Densification is complete at 1250–1500°C for silica gels made by method 1 and as low as 1000°C for gels made by method 3. The densification temperature decreases as the pore radius decreases and surface area of the gels increases. Silica glass made by densification of porous silica gel is amorphous and nearly equivalent in structure and density to vitreous silica made by fusing quartz crystals or sintering of SiO_2 powders made by chemical vapor deposition (CVD) of $SiCl_4$.

Seven processing steps are involved to various degrees in making sol–gel-derived silica monoliths by methods 1, 2, and 3. The emphasis herein is primarily on net-shape sol–gel-derived silica monoliths made by the alkoxide process (method 3) prepared under ambient pressures.

In method 1, a suspension of colloidal powders, or sol, is formed by mechanical *mixing* of colloidal particles in water at a pH that prevents precipitation. In method 2 or 3, a liquid alkoxide precursor such as $(SiOR)_4$, where R is CH_3 (TMOS), C_2H_5 (TEOS), or C_3H_7, is hydrolyzed by mixing with water (eq. 1).

Hydrolysis

$$H_3C-O-\underset{\underset{OCH_3}{|}}{\overset{\overset{OCH_3}{|}}{Si}}-O-CH_3 + 4\,H_2O \longrightarrow HO-\underset{\underset{OH}{|}}{\overset{\overset{OH}{|}}{Si}}-OH + 4\,CH_3OH \qquad (1)$$

As soon as any hydrolyzed species is present, condensation proceeds. The hydrated silica tetrahedra interact in a condensation reaction (eq. 2), forming $\equiv Si-O-Si \equiv$ bonds.

Condensation

$$HO-\underset{\underset{OH}{|}}{\overset{\overset{OH}{|}}{Si}}-OH + HO-\underset{\underset{OH}{|}}{\overset{\overset{OH}{|}}{Si}}-OH \longrightarrow HO-\underset{\underset{OH}{|}}{\overset{\overset{OH}{|}}{Si}}-O-\underset{\underset{OH}{|}}{\overset{\overset{OH}{|}}{Si}}-OH + H_2O \qquad (2)$$

Linkage of additional $\equiv Si-OH$ tetrahedra occurs as a polycondensation reaction (eq. 3) and eventually results in a SiO_2 network. The H_2O and alcohol expelled from the reaction remain in the pores of the network.

Polycondensation

$$HO-\underset{\underset{OH}{|}}{\overset{\overset{OH}{|}}{Si}}-O-\underset{\underset{OH}{|}}{\overset{\overset{OH}{|}}{Si}}-OH + 6\,Si(OH)_4 \longrightarrow \qquad (3)$$

The hydrolysis and polycondensation reactions initiate at numerous sites within the TMOS/H_2O solution as mixing occurs. When sufficient interconnected Si–O–Si bonds are formed in a region, the material responds cooperatively as colloidal (submicrometer) particles or a sol. The size of the sol particles and the cross-linking within the particles, i.e., the density, depends on the pH and R ratio, where $R = [H_2O]/[Si(OR)_4]$.

Because the sol is a low viscosity liquid, it can be *cast* into a mold.

After some time the colloidal particles and condensed silica species link together to become a three-dimensional network. The physical characteristics of the gel network depend greatly on the size of particles and extent of cross-linking prior to *gelation*. At gelation, the viscosity increases sharply, and a solid object results in the shape of the mold.

The process that involves a continuous change in structure and properties of a completely immersed gel in liquid after the gel point is called *aging*. The shrinkage of the gel and the resulting expulsion of liquid from the pores during aging is called syneresis. During aging, polycondensation continues along with localized solution and reprecipitation of the gel network, which increases the thickness of interparticle necks and decreases the porosity.

During *drying*, the liquid is removed from the interconnected pore network. Large capillary stresses can develop during drying when the pores are small (< 20 nm). These stresses can cause gels to crack catastrophically unless the drying process is controlled by either decreasing the liquid surface energy by addition of surfactants, elimination of very small pores (method 1), supercritical evaporation which avoids the vapor–liquid interface (method 2), or producing a homogenous structure free of defects by controlling the rates of hydrolysis and condensation (method 3).

Dehydration or chemical stabilization, the removal of surface silanol (Si–OH) bonds from the pore network, results in a chemically stable ultraporous solid. Porous gel–silica made in this manner by method 3 is optically transparent, having both interconnected porosity and sufficient strength to be used as unique optical components when impregnated with optically active polymers, such as fluors, wavelength shifters, dyes, or nonlinear polymers.

Heating the porous gel at high temperatures causes *densification*. The pores are eliminated and the density ultimately becomes equivalent to quartz or fused silica.

Hydrolysis and Polycondensation. At gel time, events related to the growth of polymeric chains and interaction between colloids slow down considerably and the structure of the material is frozen. Post-gelation treatments (aging, drying, stabilization, and densification) alter the structure of the original gel, but the resultant structures all depend on the initial structure. Relative rates of hydrolysis (eq. 1) and condensation (eq. 2) determine the structure of the gel. Many factors influence the kinetics of hydrolysis and condensation, because both processes often occur simultaneously. The most important variables are temperature, nature and concentration of electrolyte, nature of solvents, and type of alkoxide precursor. Pressure also influences the gelation process. Condensation may result in a spectrum of structures ranging from molecular networks to colloidal particles. Under acidic conditions, more linear structures are formed prior to gelation. Under basic conditions, the distribution of polysilicate species is much broader and characteristic of branched polymers having a high degree of cross-linking, whereas for acidic conditions there is a lower degree cross-linking. Thus, the shape and size of polymeric structural units are determined by the relative values of the rate constants for hydrolysis and polycondensation reactions.

Gelation. A sol becomes a gel when it can support a stress elastically, defined as the gelation point or gelation time, t_g. A sharp increase in viscosity accompanies gelation. A sol freezes in a particular polymer structure at the gelation point. This frozen-in structure may change appreciably with time, depending on the temperature, solvent, and pH conditions or on removal of solvent.

The time of gelation changes significantly with sol–gel chemistry. One method of measuring t_g determines the viscoelastic response of the gel as a function of shear rate.

The system evolves from a sol, where individual particles interact more or less weakly with each other, to a gel, which is a continuous network occupying the entire volume. The techniques available to follow structural evolution at the nanometer scale of sol–gel networks include small-angle x-ray scattering (saxs), neutron scattering, light scattering, and transmission electron microscopy. Scattering studies show that acid-catalyzed sols develop a more linear structure with less branching, whereas base-catalyzed systems have highly ramified structures.

The classical or mean field theory of polymerization is useful for visualizing the conditions for gelation. This model yields a degree of reaction, p_c, of one-third at the time of gelation for chemical species having functionality equal to four. Two-thirds of the possible connections are still available and therefore may play a role in subsequent processing. This value is lower than the experimental evidence but represents the minimum degree of reaction before gelation can occur. Experimental results indicate that $0.6 < p_c \le 0.84$ for silica sol–gel systems. Percolation theory is also used to represent gelation.

Aging. When a gel is maintained in its pore liquid, the structure and properties continue to change long after the gel point. This process is called aging. Four aging mechanisms can occur, singly or simultaneously: polycondensation, syneresis, coarsening, and phase transformation. Polycondensation reactions (eqs. 2 and 3), continue to occur within the gel network as long as neighboring silanols are close enough to react. This increases the connectivity of the network and its fractal dimension. Syneresis is the spontaneous shrinkage of the gel and resulting expulsion of liquid from the pores. Coarsening is the irreversible decrease in surface area through dissolution and reprecipitation processes.

During aging, there are changes in most textural and physical properties of the gel. Inorganic gels are viscoelastic materials responding to a load with an instantaneous elastic strain and a continuous viscous deformation. Because the condensation reaction creates additional bridging bonds, the stiffness of the gel network increases, as does the elastic modulus, the viscosity, and the modulus of rupture.

Drying. For porous systems, there are three stages of drying. During the first stage of drying the decrease in volume of the gel is equal to the volume of liquid lost by evaporation. The compliant gel network is deformed by the large capillary forces that cause shrinkage of the object. Changes in pore size during drying as well as a shift in composition of pore liquid can affect the rate of drying in stage 1. For large-or small-pore gels the greatest changes in volume, weight, density, and structure occur during stage-1 drying. Stage 1 ends when shrinkage ceases. The second stage begins when the critical point is reached. The critical point occurs when the strength of the network has increased owing to the greater packing density of the solid phase, which is sufficient to resist further shrinkage. In stage 2, liquid transport occurs by flow through the surface films that cover partially empty pores. The liquid flows to the surface where evaporation takes place. The flow is driven by the gradient in capillary stress. Because the rate of evaporation decreases in stage 2, this is termed the first falling rate period. The third stage of drying is reached when the pores have substantially emptied and surface films along the pores cannot be sustained. The remaining liquid can escape only by evaporation from within the pores and diffusion of vapor to the surface. During this stage, called the second falling rate period, there are no further dimensional changes, only a slow progressive loss of weight until equilibrium is reached, which is determined by the ambient temperature and partial pressure of water.

When gels crack, they do so at distinct points within the drying sequence. Cracking during stage 1 is rare but can occur when the gel has had insufficient aging and strength, and therefore does not possess the dimensional stability to withstand the increasing compressive stress. Most failures occur during the early part of stage 2, the point at which the gel stops shrinking. Cracking during stage 3 seldom occurs.

Stabilization. A critical step in preparing sol–gel products and especially Type VI silica optical components is stabilization of the porous structure. Both thermal and chemical stabilization is required in order for the material to be used in an ambient environment. The reason for the stabilization treatment is the large concentration of hydroxyls on the surface of the pores of these high (> 400 m^2/g) surface area materials. *Chemical stabilization* involves removing the concentration of surface hydroxyls and surface defects, such as metastable three-membered rings, below a critical level so that the surface is not stressed by rehydroxylation in use. Thermal stabilization involves reducing the surface area sufficiently to enable the material to be used at a given temperature without reversible structural changes. The mechanisms of thermal and chemical stabilization are interrelated because of the extreme effects that surface hydroxyls and chemisorbed water have on structural changes. Full densification of gels, such as the transformation of gel–silica to a glass, is nearly impossible without dehydration of the surface prior to pore closure. *Dehydration* of a gel requires removal of two forms of water: free water within the ultraporous gel structure, i.e., physisorbed water, and hydroxyl groups associated with the gel surface, i.e., chemisorbed water. The amount of physisorbed water adsorbed to the silica particles is directly related to the number of hydroxyl groups existing on the surface. When a silica gel has been completely dehydrated, there are no surface hydroxyl groups to adsorb the free water, and the surface is hydrophobic. It is the realization of this critical point that is the focus for making stable gel products.

Densification. Densification, the final treatment process of gels, occurs between 1000–1700°C, depending on the radii of the pores and the surface area. Controlling the gel–glass or gel–ceramic transition to retain the initial shape of the starting material is difficult. It is essential to eliminate volatile species prior to pore closure and density gradients owing to nonuniform thermal or atmosphere gradients Using appropriate successful stabilization treatments, it is possible to produce monolithic, dense, gel-derived glasses without pressure or heating to temperatures above the melting point. There are at least four mechanisms responsible for the shrinkage and densification of gels: capillary contraction, condensation, structural relaxation, and viscous sintering. It is likely that several mechanisms operate at the same time, e.g., condensation and viscous sintering.

Alumina Derived from Sol–Gel

Aluminum oxide (alumina), Al_2O_3, has high technological value. Sol–gel processing of alumina has created novel applications and improved some of its properties. Products such as catalyst carriers, abrasives, fibers, films for electronic applications, aerogels, and membranes for molecular filtration have been developed based on sol–gel processing. See also **Alumina Adsorption (Process)**; and **Bauxite**; **Aluminum**; and **Aluminum Alloys and Engineered Materials**.

Hydrolysis and Condensation Reactions of Aluminum Alkoxides. Aluminum is less electronegative than silicon, causing it to be more electrophilic and thus less stable toward hydrolysis. These features are responsible for the greater rates of hydrolysis and condensation of aluminum alkoxides when compared to the rates of silicon alkoxide reactions. Hydrolysis and condensation reactions probably occur by nucleophilic addition, followed by proton transfer and elimination of either alcohol or water under neutral conditions. Both reactions are catalyzed by the addition of acid or base. When acids are added, they protonate organic or hydroxyl groups, creating reactive species and eliminating the requirement for proton transfer as an intermediate step. Bases deprotonate water or OH groups, leading to the formation of strong nucleophiles.

Peptization. Aggregation of small inorganic polymeric chains and clusters produced from hydrolysis and condensation reactions of aluminum alkoxides forms macroparticles and nonuniform aggregate. These aggregates that are broken during the peptization step effect the formation of a clear sol having a narrow distribution of particle size. Another mechanism associated with peptization is the production of surface charges on the colloids that eventually leads to either gelation or dispersion.

Gelation. Mechanisms of gelation of alumina sols derived from alkoxides differ from gelation of silicon alkoxide sols. Whereas the sol–gel transition for silica sols is basically a consequence of interactions between long inorganic chains, the gel transition in alumina sols results from colloidal growth by dissolution and reprecipitation processes (Ostwald ripening), followed by formation of linkages between particles. These linkages are initialized by physical–chemical interactions between surface-charged colloids that eventually produce a three-dimensional network, formed by interconnected colloids. The initial contact points between colloids are responsible for neck formation by a coarsening process, followed by particle reshaping and densification. Gelation can be induced either by eliminating excess of water added during hydrolysis or by adding electrolytes to the peptize sol prepared in temperatures above 60°C that leads to sol flocculation.

Drying of Gels. Drying of alumina gels prepared using high temperatures of hydrolysis consists of two steps: sol concentration and pore liquor removal. The rate of sol concentration can dictate some gel properties. High rates of sol concentration lead to less efficiency in packing of colloids. Transparent monolit's can be prepared by drying high density gels.

Applications

Sol–Gel Processing of Thin Films. The sol–gel method enables the production of ceramic films having thickness from 10–1000 nm. See also **Thin Films**. The rheological characteristics of the sol allow the deposition of a film by several procedures: dip coating, spin coating, electrophoresis, thermophoresis, and settling. Dip and spin coating are the most frequently used procedures. Dip coating can be divided into five stages: immersion, startup, deposition, drainage, and evaporation. The fluid

mechanical boundary layer, which is pulled with the substrate, splits into two. The inner layer moves upward with the substrate, while the outer layer is returned to the bath. The thickness where the split occurs is responsible for the thickness of the film. The spin coating process can be divided into four stages: deposition, spinup, spinoff, and evaporation. In the first stage, an excess of liquid is distributed along the surface that is to undergo deposition. The spinup stage is related to flow of liquid along all the surface, driven by centrifugal force. In the spinoff stage, the excess of liquid flows to the perimeter and is eliminated as droplets. The solvent is eliminated in the fourth stage by evaporation, which leads to the thinning of the film.

Sol–Gel Fibers. During sol-to-gel evolution, changes in the rheology of the sol can be used to allow fiber pulling. Formation of elongated polymers in a solution is a requirement for spinnability, i.e., the ability to form fibers. Reduced viscosity for solutions of chain-like or spherical polymers is independent of concentration, whereas linear polymers give a direct relation between reduced viscosity and concentration. Acidic pHs and low values for the molar ratio between water and alkoxide result in the production of linear polymers that exhibit spinnability.

Organic–Inorganic Hybrids. Ceramics and polymers have been combined into high performance composites. The association between high modulus and high strength ceramic fibers, such as glass, carbon, and boron fibers, having the inherent ductility and toughness of some polymers, enables the fabrication of materials having special properties. The integration of different types of materials is restricted by the high temperature processing conditions usually employed in ceramic fabrication. The sol–gel method enables preparation of ceramic materials in a temperature range compatible with organic polymer stability, and involves mechanisms of network formation such as hydrolysis and polycondensation reactions, that are similar to the polymerization reactions of polymers. Thus, sol–gel can be used to produce new types of composites involving ceramics and polymers. These are called organic–inorganic hybrids or ceramers.

Several types of organic–inorganic hybrids have been prepared by using different polymers coupled with TEOS. The basic procedure involves dissolution of the polymer in THF (20 wt %) followed by addition of TEOS with an acidic water solution. Some of the polymers, i.e., poly(methyl methacrylate), poly(vinyl acetate), poly(vinyl pyrrolidone), and poly(N,N-dimethylamide), yielded transparent films, this demonstrating the absence of macrophase separation. Polycarbonate, poly(acrylic acid), and Nylon trogamid lead to the production of opaque films. Extensive work has been done in terms of combining nanometric clay particles using either Nylon or polyimide. Montmorillonite has been modified by cation exchange using aminolauric acid, and the new groups attached on the surface of the clay bonded to the polymer by initiation of polymerization.

Sol–Gel Bioactive Glasses. Bioactive glasses and ceramics bond to both soft and hard tissue. The chemical bond formed between the implant and tissue can provide the desired adhesion required in many medical and dental applications. Two types of sol–gel processing yield bioactive materials in the $SiO_2–CaO–P_2O_5$ system having high bioactivity index. The bioactivity of the gel-derived materials is equivalent or greater than melt-derived glasses. See also **Glass**

LARRY L. HENCH
RODRIGO OREFICE
University of Florida

Additional Reading

Brinker, C. J. and G. W. Scherer: *Sol–Gel Science,* Academic Press, New York, NY, 1990.

Brinker, C. and co-workers: in L. L. Hench and J. K. West, eds., *Chemical Processing of Advanced Materials,* John Wiley & Sons, Inc., New York, NY, 1992.

Iler, R. K.: *The Chemistry of Silica,* John Wiley & Sons, Inc., New York, NY, 1979.

Mackenzie, J. D.: in L. L. Hench and D. R. Ulrich, eds., *Ultrastructure Processing of Ceramics, Glasses and Composites,* John Wiley & Sons, Inc., New York, NY, 1984.

SOLID. Matter in its most highly concentrated form, i.e., the atoms or molecules are much more closely packed than in gases or liquids and thus more resistant to deformation. The normal condition of the solid state is crystalline structure—the orderly arrangement of the constituent atoms of a substance in a frame work called a lattice. See also **Crystal**. Crystals are of many types and normally have defects and impurities that profoundly affect their applications, as in semiconductors. The geometric structure of solids is determined by X-rays that are reflected at characteristic angles from the crystalline lattices, which act as diffraction gratings.

Some materials that are physically rigid, such as glass, are regarded as highly viscous liquids because they lack crystalline structure. All solids can be melted (i.e., the attractive forces acting between the crystals are disrupted) by heat and are thus converted to liquids. For ice, this occurs at 0°C; for some metals the melting point may be as high as 3300°C. Some solids convert by sublimation directly to a gas.

SOLID-STATE CHEMISTRY. Study of the exact arrangement of atoms in solids, especially crystals, with particular emphasis on imperfections and irregularities in the electronic and atomic patterns in a crystal and the effects of these on electrical and chemical properties.

See also **Crystal**; and **Semiconductors**.

SOLID-STATE PHYSICS. The study of the physical properties (crystallographic, electrical and electronic, magnetic, acoustic, optical, thermal, mechanical, etc.) of substances in the solid phase.

In years past, much emphasis has been given to crystalline solids and this continues, but there has been a growing shift of interest to polymeric and amorphous substances as well. Much attention in the past has been given to metals and this also continues apace, but other substances are now under very serious investigation, including the ceramics, glasses, and organics. Interest in the solid state, of course, was given a tremendous boost by the discovery of semiconductors in the 1940s. During the intervening years, this interest has been spurred by other electronic and electrical materials, including dielectrics, piezoelectrics, ferroelectrics, conductors and superconductors, electrodes, insulators, contacts, and polymers and macromolecular materials, notably those that are electroactive. Interest outside the electronics field, notably in the science of ceramics, glasses, and entirely new materials, such as composites, also has been adding to the body of knowledge of the solid state. However, because of the great need for solid materials with special properties for a host of applications, solid-state theory has tended to lag practice.

Nevertheless, solid-state theory has made excellent progress during the past decade. Just a few examples would include:

Excitonic Matter

The interaction of light with solid matter is a phenomenon of fundamental importance for exploring the quantum mechanics of materials. This field dates back to Einstein's finding that light energy is carried by quantized packets of radiation (photons). More recently, it has been found that a conduction electron can combine with a positively charged "hole" in a semiconductor to create an *exciton*, which, in turn, can form molecules and liquids. Some authorities consider the exciton as a new phase of matter. It was learned several years ago that the energy of incident photons can be converted inside a crystal into what might be termed short-lived neutral entities, i.e., excitons. As reported in an excellent paper by Wolfe and Mysyrowicz (1984), the exciton resembles the hydrogen atom. It consists of two oppositely charged carriers bound together by electrostatic attraction. In the hydrogen atom, the positive charge is a proton, which is surrounded by the negatively charged electron. In the exciton, the positive charge has a mass of an estimated $\frac{1}{1000}$th that of the proton. In the Wolfe/Mysyrowicz paper (details far beyond the scope of this encyclopedia), the investigators address several interesting questions. Can the exciton propagate freely through the crystal like a free hydrogen atom in a gas? Can two or more excitons combine to form a molecule? Can the excitonic "atoms" or the molecules made up of them form liquid or solid phases? Can more exotic phases of condensed excitonic matter come into being? How are excitons created by light in a crystal? Why does a crystal absorb light at all?

Electron Transport in Solids

It is well established (elucidated in several articles in this encyclopedia) that the production of integrated circuits (ICs) requires manufacturing techniques of extreme precision and sophistication. The purity of materials used is also far higher than experienced by most other materials-processing industries. It has been observed by Howard, Jackel, Mankiewich, and Skocpol (AT&T Bell Laboratories), in a 1986 paper, that a single-crystal silicon wafer 15 cm or more in diameter can be obtained with concentrations of undesired dopants at less than 1 part in 10 billion and with only about one defect per square centimeter. Accuracy in recent years is in terms of a few nanometers, and feature sizes in commercial circuits are down to 1 micrometer (micron) and getting smaller. Thus, it is no surprise

that the silicon transistor can serve as a model for investigating numerous areas of the solid state. Using new patterning techniques, devices almost $\frac{1}{100}$th the size of commercial ICs can be made, making it possible to study transport physics in microstructures only a few hundred atoms across.

In 1985, two research institutions (IBM and AT&T Bell Laboratories) reported that electrons can travel through a semiconductor without being slowed by collisions (*ballistically*). The report was based upon experimental data showing a ballistic peak in the electron energy spectra of gallium arsenide (GaAs) test devices. This is reported in more detail in article on **Arsenic**.

Electroactive Polymers and Macromolecular Electronics

Electro active polymers are of particular interest in connection with their use in fabricating improved electronic microstructures. Scientists at AT&T Bell Laboratories have been active in the investigation and development of electroactive polymers notably for electrodes. As reported by Chidsey and Murray (1986), electrodes can be coated with electrochemically reactive polymers in several microstructural formats called sandwich, array, bilayer, micro-, and ion-gate electrodes. These microstructures can be used to study the transport of electrons and ions through the polymers as a function of the polymer oxidation state, which is essential for understanding the conductivity properties of these new chemical materials. The microstructures also exhibit potentially useful electrical and optical responses, including current rectification, charge storage and amplification, electron-hole pair separation, and gates for ion flow. In their well-illustrated paper, the investigators explore the three broad categories of electroactive polymers: (1) pi-conjugated, electronically conducting polymers; (2) polymers with covalently linked redox groups (redox polymers); and (3) ion-exchange polymers. In summary, the authors observe that although macromolecular electronics is still at a rudimentary level, the concepts involved are quite novel and with continued development may lead to practical applications. See also entry in this encyclopedia, **Molecular and Supermolecular Electronics**.

Quantized Hall Effect

In 1980, at the Max Planck Institute, Klaus von Klitzing discovered the quantized Hall effect, a phenomenon that occurs in certain semiconductor devices at low temperatures in very strong magnetic fields. As pointed out by Halperin (1986), the quantized Hall effect is observed in artificial structures known as two-dimensional electron systems. The conduction electrons in these systems are trapped in a very thin layer, such that the electronic motion perpendicular to the layer is frozen into its lowest quantum mechanical stage and thus plays no role in the conductivity of the device. In his experiment, Klitzing worked with a silicon field effect transistor (MOSFET). Electrons are trapped in what is called an inversion layer near the surface of a silicon crystal that is covered with a film of insulating silicon oxide, on top of which is deposited a metal gate electrode, used to control the density of conduction electrons in the inversion layer. This effect had been predicted as early as 1975 by Japanese investigators. Considerable detail pertaining to von Klitzing's experimental apparatus is given in the Halperin paper.

Surface Physics

Closely allied with solid state physics is the discipline of surface science. Investigations in this area have been quite intense during the past decade, notably in connection with catalysts. A catalyst is a species that changes the rate of a reaction and yet is regenerated by that reaction so that it seems to be unchanged in the net reaction. Although there are enzyme catalysts, for example, the majority of industrially interesting catalysts are found among the metals, the surfaces of which serve to catalyze reactions. The first catalytic phenomena were observed as early as 1835 by Berzelius and later better quantified by Ostwald in 1894. Aided by the great volume of catalysts used industrially ($ billions/year), the incentive for research is large. See articles on **Catalysis**; **Scanning Tunneling Microscope**; and **Silicon**.

Extremely High-Pressure Research

The invention and refinement of the modern diamond anvil cell (Carnegie Institution) occurred in the mid-1980s. This is a tool par excellence for optical, infrared and Raman spectroscopy and enables the researcher to study the changes in the electronic structure and chemical binding caused by the application of high pressure. Phase transitions, which involve changes in the atomic architecture can be determined with the diamond cell using the x-ray diffraction technique. Studies with the diamond anvil cell have been particularly valuable for obtaining geophysical information—for example, the state of silicate minerals and oxides in the mantle region right up to the core-mantle boundary to provide a view of the earth's interior, where high pressure and high temperature conditions exist. In solid-state physics, there is the fascinating challenge of making metallic hydrogen under ultrahigh pressure. This extraordinary change from a very good insulating to a metallic state in hydrogen is predicted to occur near 3 to 4 million atmospheres. See article on **Diamond Anvil High Pressure Cell**.

The foregoing examples are but a few to indicate the continuing vigorous research into the nature of the solid state. See also **Superconductivity**.

Concepts of Solids Simplified

The atoms that comprise a solid can be considered for many purposes to be hard balls which rest against each other in a regular repetitive pattern called the crystal structure. Most elements have relatively simple crystal structures of high symmetry, but many compounds have complex crystal structures of low symmetry. The determination of crystal structures, of atom location in the crystal, and of the dependence of many physical properties upon the inherent characteristics of the perfect solid is an absorbing study, one that has occupied the lives of numerous geologists, mineralogists, physicists, and other scientists for many years.

The rigid, hard-ball model is not adequate to explain many properties of solids. To begin with, solids can be deformed by finite forces, thus solids must not be completely rigid. Furthermore, atoms in a solid possess vibrational energy, so the atoms must not be precisely fixed to mathematically defined lattice points. This deformability of solids is built into the model by the assignment of deformable bonds (springs) between nearest atom neighbors. This ball-and-spring model has many successes; one important early use was that of Einstein to devise a reasonably successful theory of specific heats. Later incorporation by Debye of coupled motion of groups of atoms led to an even more successful theory.

Several measures exist of the strength of these bonds. One is the size of the elastic constants—for most solids, Young's modulus is about 10^{11} newtons per square meter. The other is the frequency of vibration of the atoms—values around 10^{13} to 10^{14} Hz are found.

The lack of perfection occasioned by elastic deformation of solids is but one of many kinds of crystalline imperfections. Defects are frequently found in crystals, produced in nature and in the laboratory. These defects may be characterized by three principal parameters—their geometry, size, and energy of formation.

All real crystals have atoms which occupy external surface sites and which do not possess the correct number of nearest neighbors as a consequence. Thus, a surface is a seat of energy and is characterized by surface tension. Furthermore, internal surfaces exist, grain boundaries and twin boundaries across which atoms are incorrectly positioned. In a crystal of reasonable size—say 1 cubic centimeter, these two-dimensional defects, called *surface defects*, contain only about 1 atom in 106, a rather small fraction. Even so, surfaces are important attributes of solids.

Some defects have extent in only one dimension—*line defects*. The most prominent of these, the dislocation, is a line in the crystal along which atoms have either an incorrect number of neighbors or neighbors which have not the correct distance or angle. In 1 cubic centimeter of a real crystal, one might find a wide variation of length of dislocations present—from near zero to perhaps 10^{11} centimeters.

Defects which have extent of only about an atomic diameter also exist in crystals—the *point defects*. Vacant lattice sites may occur—*vacancies*. Extra atoms—*interstitials*—may be inserted between regular crystal atoms. Atoms of the wrong chemical species—*impurities*—also may be present.

The properties of defects are intimately related to their energy of formation. A standard against which this energy can be compared is provided by the energy of sublimation—the energy necessary to separate the ions of a solid into neutral, noninteracting atoms. This energy is about 81,000 calories per mole for a typical metal, copper, at room temperature, about 3.5 eV per atom. Energies of surfaces, both free surfaces and grain boundaries, are about 1000 ergs per square centimeter, about 1 eV per surface atom. Dislocation energies are of similar size per atom length of dislocation, about 1 to 5 eV, so the energy of a dislocation is about 10^{-4} erg per centimeter of length. Point defects, too, possess an inherent energy of about 1 eV each. Vacancies in copper have an energy of about 1 eV; self-interstitials, 2 or 3 eV.

The energies per atom of these various defects, surface, line, and point, are all much larger than the average thermal energy per atom in a solid at reasonable temperatures. This thermal energy kT is only about $\frac{1}{40}$ eV

at room temperature. Thus, defects can be produced only by conditions which exist during manufacturing (artificial and natural) by external means, such as plastic deformation or particle bombardment; or by large local fluctuations in thermal energy away from the average.

The total amount of energy bound up in ordinary concentrations of these defects is not large as compared to the total thermal energy of a solid at normal temperatures. All the vacancies in equilibrium in copper, even at the melting point, comprise less than 10 calories of energy per mole, much less than the enthalpy at 1357 K (the melting point) of more than 7000 calories per mole. In a material with very heavy dislocation density, 10^{12} centimeters per cubic centimeter, the total dislocation energy is only a few calories per mole. And the total energy of a free surface of a compact block of 1 mole of copper is even less: about 10^{-3} calorie. Thus, the inherent energy of these defects is not large; even so they are immensely important in controlling many phenomena in crystals—as in the case of semiconductor devices.

Crystallographic defects need not remain stationary in the crystal; they may move about with time. Some of these movements may reduce the overall free energy of the solid; others (these are chiefly movement of the point defects) may simply be the wandering of random walk. Since these movements require larger than kT, the motion of defects depends upon rather large local fluctuations in energy. Consequently, their rate of motion depends upon temperature through a Boltzmann factor exp $(-\Delta H/RT)$, where ΔH is the enthalpy increase necessary to move the defect from the lowest-energy site to the top of the barrier.

A convenient description of the crystalline structure of solids is thus seen to consist of successive stages of approximation. First, the mathematically perfect geometrical model is described; then departures from this perfect regularity are permitted. The deformability of solids is allowed for by letting the force constants between adjacent atoms be finite, not infinite. Then, misplacement of atoms is permitted and a variety of crystalline irregularities, called defects, is described. Some of these defects have intrinsic features which affect properties of the crystal; other affect the properties by their motion from site to site in the crystal. In spite of their relatively small number, defects are of immense importance.

Electronic Structure of Solids

In principle, the electronic structure of solids is determined by the electronic structure of the *free atoms* of which the solid is composed. Since the free atom structure is known rather well, especially for atoms of lower atomic number, the electronic structure of solids should be subject to determination by calculation. This is not the case. A wide variety of interactions occur between the electrons on adjacent atoms as they approach the equilibrium distance characteristic of solids. These interactions are of such complex nature that they tend to defy concise definition and involve such a host of charged particles, electrons, and ion cores, that only approximate calculations can usually be made. Nevertheless, the use of approximate models allows many general features of the electronic structure to be deduced, especially when close interplay between theory and experiment is established. As for the crystalline structure of solids, two stages are useful in understanding the electronic structure. First, the perfect electronic structure is defined. Then, irregularities in this structure, again termed defects, are described. Although both the geometry and energy of crystalline defects are defined, description of the geometry of the charge distribution of many of the electron defects is difficult, and one must generally be content with description of the formation energy of the defect.

The nuclei of the atoms in a solid and the inner electrons form ion cores with energy levels little different from corresponding levels in free atoms. The characteristics of the valence electrons are modified greatly, however. The state functions of these outer electrons greatly overlap those of neighboring atoms. Restrictions of the Pauli Exclusion Principle and the Uncertainty Principle force modification of the state functions, and the development of a set of split energy levels becomes a quasi-continuous band of levels of width, which are several electron volts for most solids. Importantly, unoccupied levels of the atoms are also split into bands. The electronic characteristics of solids are determined by the relative position in energy of the occupied and unoccupied levels as well as by the characteristics of the electrons within a band.

Metals. The solid is called a *metal* if excitation of electrons from the highest filled levels to the lowest unoccupied levels can occur with infinitesimal expenditure of energy. Thus, excitation can occur by means of many external forces, such as electric fields, heat, light, radio waves. Metals are, therefore, good conductors of electricity and of heat; they are opaque to light and they reflect radio waves.

Insulators. Some solids have wide spacing between the occupied and the unoccupied energy states—2 eV or more. Such solids are called *insulators* since normal electric fields cannot cause extensive motion of the electrons. Examples are diamond, sodium chloride, sulfur, quartz, mica. They are poor conductors of electricity and heat and are usually transparent to light (when not filled with impurities or defects).

Semiconductors. Solids with conductivity properties intermediate bet ween those of metals and insulators are called *semiconductors.* For them, the excitation energy lies in the range 0.1 to about 2 eV. Thermal fluctuations are sufficient to excite a small, but significant, fraction of electrons from the occupied levels (the valence band) into the unoccupied levels (the conductance band). Both the excited electrons and the empty states in the valence band (aptly called *holes*) may move under the influence of an electric field, providing a means for conduction of current. Such electron-hole pairs may be produced not only by thermal energy, but also by incident light, providing photo-effects. The inverse process, emission of light by annihilation of electrons and holes in suitably prepared materials, provides a highly efficient light source (example, light-emitting diodes).

Crystallographic defects, in general, are also electronic defects. In metals, they provide scattering centers for electrons, increasing the resistance to charge flow. The resistance wire in many electric heaters consists of an ordinary metal, such as iron with additional alloying elements such as nickel or chromium providing scattering centers for electrons. In semiconductors and insulators, however, alloying elements and defects provide an even greater variety of effects, since they can change the electron-hole concentrations drastically in addition to providing scattering centers. This is the basis of semiconductor technology. See also **Semiconductors.**

Interactions of Solids with Light. Solids are useful because of their interaction with external forces or stimuli, such as electric and magnetic fields, heat, and mechanical forces. Yet among these interactions, probably the most important I s the interrelation between matter and light. This interaction, important to all photosynthetic phenomena and the production of food; to the artificial generation of light; to the use of phosphors in cathode-ray tubes—is also the basis of spectroscopy and its use in the study of solids. In this field, first came investigation of emission and absorption of radiation from free atoms. Later investigations included emission and absorption of radiation by atoms in solids—giving rise to maser and laser phenomena, Mossbauer spectroscopy, nuclear magnetic resonance, X-ray diffraction, infrared spectroscopy, fluorescence, the Raman effect, microwave emission and absorption, among many other useful effects.

Band Theory of Solids

The success of the simple free electron theory of metals was so striking that it was natural to ask how the same ideas could be applied to other types of solids, such as semiconductors and insulators. The basic assumption of the free electron theory is that the atoms may be stripped of their outer electrons, the resulting ions arranged in the crystalline lattice, and the electrons then poured into the space between.

The free electron model results from the neglect of the interaction of the various atoms and of the periodic variation of the potential in which the electrons move, i.e., as their distance from the nearest metallic ion changes. When the former is taken into account, it is found that each energy eigenstate of an isolated atom is split into N non-degenerate states, where N is the number of atoms in the crystal. The group of levels that result from a single atomic state form an *allowed band*. If we start from the free electron picture and consider the effect of the periodic variations of potential, the Bloch theorem leads to the conclusion that there will be discontinuities in the plot of energy vs. momentum whenever the wave vector $\mathbf{k}$ has magnitude and direction such that it satisfies the Bragg law for reflection, in which λ may be set equal to $1/\mathbf{k}$ to give $\mathbf{k} \cdot \mathbf{d} = n$. Here $\mathbf{k}$ is the wave vector, $\mathbf{d}$ is the vector separation of two atomic planes in the crystal, and n is an integer equal to the scalar product. As with the atomic interaction model, the number of eigenstates between two energy breaks is equal to the number of atoms in the crystal. Thus, either approach leads to the existence of a manifold of energy levels occurring in groups of N closely spaced levels, the groups being separated by energies that are often very large compared with the spacing of levels within a group, somewhat as shown in Fig. 1. Each group of levels is known as an *allowed band*; the energies between groups are said to be in a *forbidden band*. Because these levels depend on the properties of the body as a whole, the entire macroscopic crystal may be considered to be a single giant molecule.

The electrical, mechanical, and thermal properties of the crystal are then largely determined by the electrons in the energy levels within the highest occupied bands.

Because electrons obey the Pauli Exclusion Principle, not more than two of them (with oppositely directed spin) can exist in any single energy level. In thermal equilibrium at the zero of absolute temperature, than, all of the levels up to some particular energy, determined by the number of electrons present, will be occupied and all above this energy will be vacant. This highest level is known as the *Fermi level*. At higher temperatures there will not be a sharp discontinuity in occupancy—some of the levels below the Fermi level will be vacant and some above it will be occupied. The *Fermi level* is then defined as the energy of the state that has a 50% chance of being occupied.

The Fermi level is determined by the number of electrons present, and the properties of the material are therefore dependent on whether this energy falls near the bottom, top, or middle of an allowed band. If the number of electrons is such as to exactly fill certain bands, with a wide gap above them, the material will be an insulator (m). If the gap is very narrow, or if there are impurities present to create extra levels, the substance may be semiconducting (o). See Fig. 2. In these cases, it is difficult to supply sufficient thermal energy to an electron to promote it into the conduction band above the gap, where alone it is free to carry an electric current. In a metal (n), however, there is always a partially filled band, in which the electrons behave in many respects as if they were free. The existence of the partially filled band may be due either to the fact that each atom contains an odd number of electrons or to the overlapping of two allowed bands, each of which will be partly filled. Direct evidence for the existence of bands is provided by the soft X-ray emission spectra, but the importance of the theory is not so much its correctness in detail as the simplicity of the band scheme by which the energy relations between various phenomena may be shown on a single diagram.

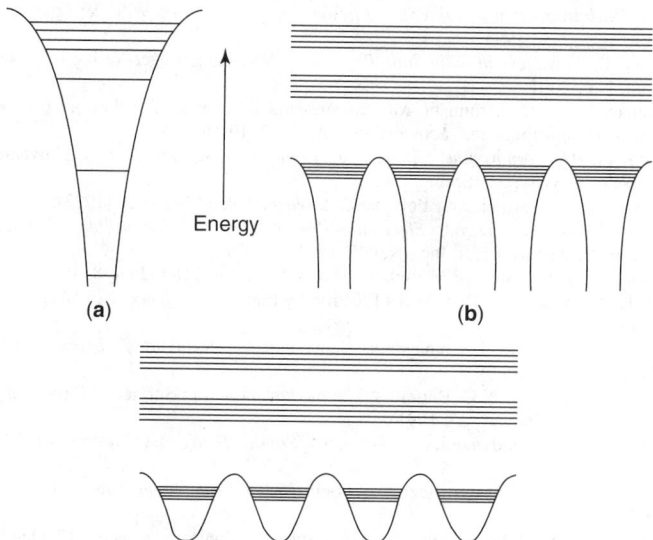

Fig. 1. Origin of the energy levels in a crystalline solid. The curves represent potential energy versus distance. At (**a**), the potential energy is that of an isolated ion; the energy levels, represented by the horizontal lines, are sharp. At (**b**), the overlap of the fields of the ions lowers the potential energy curve between the atomic positions and results in a splitting of each atomic level into a band of allowed levels. At (**c**), the model is derived from one in which the electrons are free, subject only to a periodic potential resulting from the ionic fields

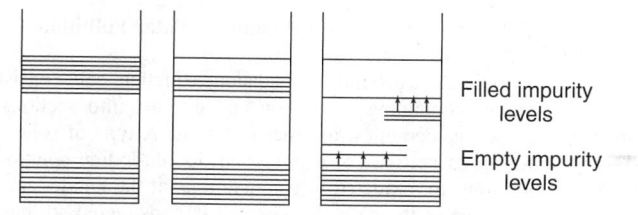

Fig. 2. Band diagrams of *m*, insulator; *n*, metal; and *o*, semiconductor

Additional Reading

Ashcroft, N.W.: *Solid State Physics,* 2nd Edition, Harcourt Brace College Publishers, San Diego, CA, 2001.

Bate, R.T.: "The Quantum-Effect Device: Tomorrow's Transistor?" *Sci. Amer.,* 96 (March 1988).

Blakely, J.M.: *Surfaces and Interfaces,* in Encyclopedia of Materials Science and Engineering (M.B. Bever, Ed.), MIT Press, Cambridge, MA, 1988.

Bokor, J.: "Ultrafast Dynamics at Semiconductor and Metal Surfaces," *Science,* 1130 (December 1, 1989).

Brodsky, M.H.: "Progress in Gallium Arsenide Semiconductors," *Sci. Amer.,* 68 (February 1990).

Caruana, C.M.: "The Interdisciplinary Approach to Surface Science," *Chem. Eng. Progress,* 64 (July 1987).

Chidsey, C.E.D. and R.W. Murray: "Electroactive Polymers and Macromolecular Electronics," *Science,* **231,** 25–31 (1986).

Chin, G.Y.: *Magnetic Materials,* in Encyclopedia of Materials Science and Engineering (M.B. Bever, Ed.), MIT Press, Cambridge, MA, 1988.

DeShazer, L.G.: *Optical Materials,* in Encyclopedia of Materials Science and Engineering (M.B. Bever, Ed.), MIT Press, Cambridge, MA, 1988.

DiSalvo, F.J.: "Solid-State Chemistry: A Rediscovered Chemical Frontier," *Science,* 649 (February 9, 1990).

Ehrenreich, H. and F. Spaepen: *Solid State Physics: Fullerene Fundamentals,* Vol. 48, Academic Press, Inc., San Diego, CA, 1997.

Ehrenreich, H. and F. Saepen: *Solid State Physics: Advances in Research and Applications,* Vol. 55, Academic Press, Inc., San Diego, CA, 2000.

Fisk, Z. et al.: "Heavy-Electron Metals: New Highly Correlated States of Matter," *Science,* 33 (January 1, 1988).

Halperin, B.I.: "The 1985 Noble Prize in Physics (Quantized Hall Effect)," *Science,* **231,** 820–822 (1986).

Heiblum, M. and L.F. Eastman: "Ballistic Electrons in Semiconductors," *Sci. Amer.,* 102–111 (February 1987).

Howard, R.E. et al.: "Electrons in Silicon Microstructures," *Science,* **231,** 346–349 (1986).

Karasz, F.E. and T.S. Ellis: *Polymers: Structure, Properties, and Structure-Property Relations,* in Encyclopedia of Materials Science and Engineering (M.B. Bever, Ed.), MIT Press, Cambridge, MA, 1988.

Kittel, C.: *Introduction to Solid State Physics,* 7th Edition, John Wiley & Sons, Inc., New York, NY, 1995.

Kramer, B.: *Advances in Solid State Physics 41,* Vol. **41,** Springer-Verlag Inc., New York, NY, 2001.

Landman, U. et al.: "Atomistic Mechanisms and Dynamics of Adhesion, Nanoindentation, and Fracture," *Science,* 454 (April 27, 1990).

LeComber, P.G.: *Amorphous Silicon—Electronics into the 21st Century,* University of Wales Review, 31 (Spring 1988).

Lovinger, A.J.: "Ferroelectric Polymers," *Science,* **220,** 1116–1121 (1983).

Mott, N.F. and E.A. Davis: *Electronic Processes in Non-Crystalline Materials,* Oxford University Press, Inc., New York, NY, 1979.

Pool R.: "Clusters: Strange Morsels of Matter," *Science,* 1184 (June 8, 1990).

Pool, R.: "A Transistor That Works Electron by Electron," *Science,* 629 (August 10, 1990).

Prinz, G.A.: "Hybrid Ferromagnetic Semiconductor Structures," *Science,* 1092 (November 23, 1990).

Williams, E.D. and N.C. Bartelt: "Thermodynamics of Surface Morphology," *Science,* 393 (January 25, 1991).

Wolfe, J.P.: "Thermodynamics of Excitons," *Physics Today,* **35**(12), 46–54 (March 1982).

Wolfe, J.P. and A. Mysyrowicz: "Excitonic Matter," *Sci. Amer.,* 98–107 (March 1984).

Yablonovitch, E.: "The Chemistry of Solid-State Electronics," *Science,* 347 (October 20, 1989).

SOLIDUS CURVE. A curve representing the equilibrium between the solid phase and the liquid phase in a condensed system of two components. The relationship is reduced to a two-dimensional curve by disregarding the influence of the vapor phase. The points on the solidus curve are obtained by plotting the temperature at which the last of the liquid phase solidifies, against the composition, usually in terms of the percentage composition of one of the two components.

SOLID WASTES. See **Wastes and Pollution; Water Pollution**

SOLION. A small electrochemical oxidation-reduction cell consisting of a small cylinder containing a solution and divided into sections by platinum gauze, porous ceramics, or other materials. A type of solion for detecting sound waves consists of a potassium iodide-iodine solution in which the iodide ions are oxidized to triiodide ions at the anode, and the reverse process occurs at the cathode. The cell is constructed so that the sound waves cause agitation of the solution between the electrodes, and

thus change the current. In addition to detection of sound, solions can be designed to detect changes in other conditions, such as temperature, pressure, and acceleration.

SOLUBILITY. A property of a substance by virtue of which it forms mixtures with other substances which are chemically and physically homogeneous throughout. The degree of solubility is the concentration of a solute in a saturated solution at any given temperature. The degree of solubility of most substances increases with a rise in temperature, but there are cases (notably the organic salts of calcium) where a substance is more soluble in cold than in hot solvents.

SOLUBILITY PRODUCT. A numerical quantity dependent upon the temperature and the solvent, characteristic of electrolytes. It is the product of the concentrations of ions in a saturated solution and defines the degree of solubility of the substance. When the product of the ion concentrations exceeds the solubility product, precipitation commonly results. Strictly speaking, the product of the activities of the ions should be used to determine the solubility product, but in many cases the results obtained using concentrations, as suggested by Nernst, are correct.

SOLUBILIZATION. Defined loosely, solubilization is the enhancement of the solubility of one substance, the solubilizate, by another substance, the solubilizing agent or solubilizer. More strictly, it is a process occurring in the presence of a solvent, whereby one species, the solubilizing agent, diminishes the activity coefficient of another species, the solubilizate, and both species are soluble thereafter, J. W. McBain, who coined this term, used it to denote the dissolution of an otherwise insoluble material brought about by interaction with micelles, a type of colloid, present in the solvent. The definition given here, however, is more inclusive than his original concept, and could be extended logically to systems whose characteristics are remote or completely apart from colloidal behavior. Practice nevertheless limits the term to usage in which there is either a close or a marginal relationship to micelles, and the literature of solubilization refers chiefly to systems in which the solubilizers are micelle formers. For example, potassium laurate solubilizes hydrocarbons in water, and calcium xenylstearate solubilizes water in hydrocarbons because of the micelle-forming nature of the respective solubilizing agents. However, the striking similarity among interactions between various agents and both soluble and insoluble species makes it undesirably arbitrary to restrict the term solubilization rigidly to its original usage.

Because absolute insolubility does not exist in nature, insolubility must be considered a matter of degree. Consequently, if an *apparently* insoluble species, in unlimited excess, is in contact with a solvent it must have a finite concentration and activity in the solvent at equilibrium. A solubilizing agent added to the system may interact with this species by coordination, hydrogen bonding, dipole interaction, complex formation, or in some other manner. In any case, the interaction results in a decrease of the effective concentration, or activity, of this species. Accordingly, more of the solubilizate progressively dissolves until its activity returns to the initial equilibrium value in the pure solvent, whereupon the activity coefficient is correspondingly less. If a species is freely soluble, or even infinitely miscible with a solvent, an interaction causes no *apparent* increase in the solubility of the species, but its activity, as evidenced by its osmotic behavior, nevertheless similarly decreases. The activity represents the tendency of the species to escape from the solution. Since solubility depends upon a balance between the opposing tendencies to enter and to leave the solution, the decreased activity is in effect equivalent to increased solubility. Solubilization is said to occur then, regardless of the independent solubility or insolubility in the pure solvent.

The salts of high-molecular weight organic acids are particularly important solubilizing agents. In nonpolar solvents such as hydrocarbons, they form colloidal aggregates known as association micelles. Most frequently such a micelle constitutes a limited number of salt monomers associated into a spheroidal cluster, with the polar ends of the salt monomers oriented toward the interior, and the nonpolar hydrocarbon ends at the periphery. Other polar species such as water, alcohol, acids, and dyes can be solubilized by these micelles in a variety of ways. In benzene solution, for example, zinc dinonylnaphthalene sulfonate can solubilize at least six moles of water for each equivalent weight of the salt present. The solution remains transparent, and no phase separation is observed. During the progressive addition of six moles of water per gram-equivalent of salt, the micelles expand to aggregations containing ten acid residues per unit, whereas the water-free micelles contain only seven. The water molecules are believed to be held in the polar core of the micelle where the environment is favorable to their retention.

Methanol, on the other hand, decreases the size of magnesium phenylstearate micelles. As methanol is solubilized by this salt in toluene solution, the micelle size decreases progressively from 23 salt monomers per aggregate to as little as 2 at a methanol concentration of 2% by weight. Each of these dimers is then associated with ten molecules of the alcohol. The partial pressure of methanol over the solution is demonstrably less than that over the salt-free methanol-toluene solution of equal methanol concentration. Rhodamine B dissolves very sparingly in pure benzene as the colorless and nonfluorescent base form. It is converted to the brilliantly fluorescent colored form by the addition of any of numerous micelle-forming solubilizers. No major changes in micelle size are believed to result from this solubilization, and it is postulated that a dye molecule replaces a monomer of the solubilizing agent in the matrix of the micelle.

In aqueous solutions, salts of high-molecular weight acids from micelles whose orientations are the reverse of those in nonpolar solvents. The hydrocarbon portions of the monomers, being insoluble in water, are oriented inward, whereas the ionic, or polar ends are oriented outward. Solubilization by these agents is complicated by dissociation of cations from surfaces of the aggregates and by the resulting surface charges developed. Both polar and nonpolar species such as hydrocarbons, dyes, alcohols, fats, organic acids, and a wide variety of soluble and insoluble species are solubilized in aqueous micellar solutions. Micelle enlargement is frequently said to follow from solubilization by these salts in aqueous solution, although the possibility of reduction of micelle size should not be excluded from consideration.

A nonpolar solubilizate such as hexane penetrates deeply into such a micelle, and is held in the nonpolar interior hydrocarbon environment, while a solubilizate such as an alcohol, which has both polar and nonpolar ends, usually penetrates less, with its polar end at or near the polar surface of the micelle. The vapor pressure of hexane in aqueous solution is diminished by the presence of sodium oleate in a manner analogous to that cited above for systems in nonpolar solvents. A 5% aqueous solution of potassium oleate dissolves more than twice the volume of propylene at a given pressure than does pure water. Dimethylaminoazobenzene, a water-insoluble dye, is solubilized to the extent of 125 mg per liter by a 0.05 M aqueous solution of potassium myristate. Bile salts solubilize fatty acids, and this fact is considered important physiologically. Cetyl pyridinium chloride, a cationic salt, is also a solubilizing agent, and 100 ml of its N/10 solution solubilizes about 1 g of methyl ethyl-butyl either in aqueous solution.

Among other species that are good solubilizing agents are the nonionic compounds such as the polyethylene oxide-fatty acid condensates and the fatty esters of polyalcohols. A wide variety of nonionic solubilizing agents is possible, but most of those available are of variable composition. They can be effective in both aqueous and nonaqueous solutions.

The colloidal nature of some systems can disappear completely as solubilization proceeds. For instance, when methanol is solubilized by magnesium and sodium dinonylnaphthalene sulfonates, the aggregates decrease in size to a degree beyond which they can be considered micelles. In toluene solutions, micelles of these salts dissociate progressively on the addition of methanol increments until each of the particles in these solutions contains only one salt monomer when the methanol concentration reaches about 2% by weight. Probably the properties of some species which cause them to aggregate are those which make them good solubilizing agents, but it is evident that micelles are not a necessary condition for solubilization.

Accordingly, it is logical for solubilization to occur in systems which show no colloidal behavior, although frequently the effect in these cases is described by other proper terminology. Usually the term "solubilization" is applied in cases where the solubilizing agent is effective in small quantities, but arbitrary limitations of quantity might confuse the basic concept of solubilization. The terms cosolvency, hydrotropy, and "salting in" are used sometimes to describe effects which may be considered within the broad general scope of solubilization.

Applications of solubilization, although not always completely understood, range widely. A solubilizing agent can be used to bring an otherwise insoluble substance into solution where it is needed for a specific use, or it can be incorporated in a formulation to suppress the activity of an unwanted species which otherwise cannot be eliminated or prevented from occurring.

In the pharmaceutical industry, drugs which are insoluble in pure water are solubilized by suitable agents to form homogeneous solutions. Dyes are solubilized for more efficient penetration and uniform coloring of fabrics. Soaps and detergents in aqueous solution are effective cleansing agents because they solubilize oily and greasy residues which may be flushed away from contaminated surfaces, although other effects may be equally important in the process. Removal of silver halides from photographic papers and films by aqueous fixing solutions may be considered solubilization by noncolloidal solubilizers. Certain oil-soluble salts dissolved in dry cleaning fluids can solubilize water. The water, which is solubilized in the micelles can in turn solubilize inorganic salts. The salts are then retained in the polar cores of the micelles where the water is held. This effect is referred to as secondary solubilization. In automotive fuels and lubricating oils, nonaqueous detergents are used to maintain engine cleanliness by solubilizing products of oxidation and combustion which tend to form sludges and gums, and to suppress the destructive effects of acids and other species generated in operation. Other solubilizing agents are used in these fluids to incorporate otherwise insoluble additives for oxidation and corrosion inhibition.

SAMUEL KAUFMAN
Naval Research Laboratory
Washington, District of Columbia

SOLUTIONS. The equilibrium of a saturated solution represents a balance between the potentials and entropies of the molecules present in the two phases. These depend upon pressure, temperature, and the kind and strength of the attractions between the molecules. The attractions may be classified as interactions between ions, dipoles, metallic atoms, and the "electron clouds" of nonpolar molecules, differing among themselves in kind, in range, and in strength. The potential energy of the molecules of a nonpolar liquid is measured appropriately for the purpose of solubility relations by its energy of vaporization per cc, called its "cohesive energy density." The square root of this quantity will be used below as a "solubility parameter" δ.

We consider, first, the mutual solubility of two nonpolar liquids, whose molecules have practically equal sizes, and equal attractive and repulsive forces. When they are brought into contact, thermal agitation will cause mutual diffusion until the two species are uniformly distributed. The mixing process has produced maximum molecular disorder, and therefore entropy, which is given by the expression, for 1 mole of solution,

$$\Delta S^M = -R(x \ln x_1 + x_2 \ln x_2), \tag{1}$$

where R is the gas constant and x_1 and x_2 the respective mole fractions. The partial molal entropies of transfer of 1 mole from pure liquid to solution are

$$\bar{s}_1 - \bar{s}_1^0 = -R \ln x_1, \tag{2}$$

for component 1, and with subscript 2, for the other component.

The partial molal free energies of transfer are related to the *fugacities* in pure liquid, f^0 (vapor pressure corrected for deviation from the perfect gas law), and in solution, by the equations,

$$\overline{F}_1 - \overline{F}_1^0 = -R \ln(f_1/f_1^0). \tag{3}$$

and its counterpart.

Liquids such as are here postulated mix with no heat effect; therefore $F_1 - F_1^0 = -T(s_1 - s^0)$, etc.; therefore

$$f_1/f_1^0 = x_1 \quad \text{and} \quad f_2/f_2^0 = x_2 \tag{4}$$

which is Raoult's law, and defines the *ideal* solution.

If one of the components of an ideal solution, e.g., component 2, is a solid, its fugacity, f_2^s, is less than the fugacity of the pure, supercooled liquid, and limits the amount that can dissolve to $x_2 = f_2^s/f_2^0$. The ratio f_2^s/f_2^0, can be calculated from its melting point and heat of fusion.

Most solutions deviate from Raoult's law. The curved lines in Fig. 1 represent positive deviations, with $f_1/f_1^0 > x$. The ratio f_1/f_1^0 is called *activity*, and

$$f_1/f_1^0 = a, \quad \text{and} \quad a_1/x_1 = \gamma, \tag{5}$$

the *activity coefficient*.

Regular Solutions

The internal forces of a pair of liquids are seldom so nearly alike as to permit their mixture to obey Raoult's law very closely throughout the whole range of composition. In the absence of chemical interaction, the attraction between two different molecular species, provided their dipole moments are zero or small, is approximately the geometric mean of the attractions between the like molecules. Since a geometric mean is less than an arithmetic mean, the mixing is accompanied by expansion and absorption of heat. The partial molal heat of transfer per mole from pure liquid to solution is given with fair accuracy for many systems by the equation,

$$H_2 - H_2^0 = v_2 \phi_1^2 (\delta_2 - \delta_1)^2 \qquad (6)$$

and its cognate. where $v \equiv$ molal volume, δ is a *solubility parameter*, the square root of the energy of vaporization per cm³, and ϕ_2 is volume fraction.

Thermal agitation, except in the liquid-liquid critical region, suffices to give essentially maximum randomness of mixing, especially when one component is dilute, so that the entropy of mixing may be practically ideal, although the heat of mixing is not, and the partial molal free energy can be computed by combining the entropy and the heat terms, Eqs. (2), (5), and (6),

$$RT \ln a_2^s/x_2 = v_2 \phi_1^2 (\delta_1 - \delta_2)^2 \qquad (7)$$

This equation neglects the effects of expansion upon both the heat and the entropy, but the errors largely cancel when combined in Eq. (7).

A plot of a_2 vs. x_2 for symmetrical systems (i.e., $v_1 \approx v_2$) is shown in Fig. 1 for a series of values of the heat term. It shows how the partial vapor pressure of a component of a binary solution deviates positively from Raoult's law more and more as the components become more unlike in their molecular attractive forces. Second, the place of T in the equation shows that the deviation is less the higher the temperature. Third, when the heat term becomes sufficiently large, there are three values of x_2 for the same value of a_2. This is like the three roots of the van der Waals equation, and corresponds to two liquid phases in equilibrium with each other. The criterion is that at the critical point the first and second partial differentials of a_2 and a_1 are all zero.

The presence of a dipole in one component adds a temperature-dependent component to its self-attraction and also induces a dipole in the other component. The effect can often by allowed for, for practical purposes, by an empirical adjustment of its solubility parameter.

If the dipole is hydrogen bonding, then this component is "associated," and it mixes less readily with a nonpolar second component.

If the components are, respectively, electron-donor and acceptor, or basic and acidic in the generalized sense of Gilbert Lewis, negative deviations from Raoult's law occur, with enhancement of solubility.

The effects of these various factors are well illustrated by solutions of iodine, I_2. In Fig. 2 are plotted the saturation values of $\log x_2$ for iodine against $\log T$. The slopes of the lines, when multiplied by R, give the entropy of transfer of iodine from solid to saturated solution. The solid lines are for violet solutions, from which chemical equilibria are absent. The positions of the lines are determined by the solubility parameters: how well is seen in the accompanying table, where δ-values are given for iodine

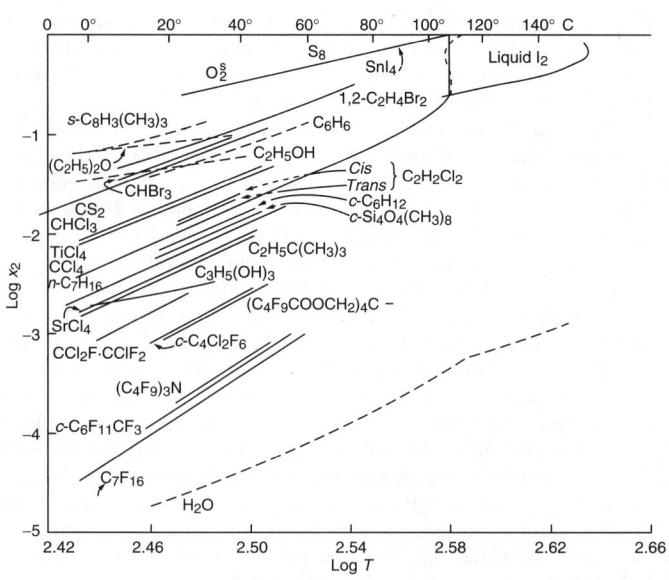

Fig. 2.   Solubility of iodine

in a spread of solvents calculated by means of Eq. (7) from the measured values of x_2. The broken lines indicate nonviolet solutions.

The factors that cause solutions of iodine to deviate from the behavior of regular solutions are illustrated in Fig. 3, in which values of the left hand member of Eq. (7) are plotted against those of the right for iodine solutions at 25°C; a_2^s is the activity of solid iodine; x_2 denotes measured solubility; v_2 is the extrapolated molal volume of liquid iodine, 59 cm³; ϕ_1 is the volume fraction of the solvent, ~ 1.0; $\delta_2 = 14.1$; δ_1 is the solubility parameter of the solvent. Illustrative values of x_2 and δ_1 are given in accompanying table.

The points on line A are all for regular solutions, conforming to Eq. (7) over large ranges of x_2. Line B starts with a point for iodine in cycyohexane, next a point for methylcyclohexane, followed by one for dimethylcyclohexane. The point below is for ethylcyclohexane. Line C is for normal alkanes, from $C_{16}H_{34}$ to C_5H_{12}; groups D and E are for branched alkanes. Displacements from line A increase with increasing ratios of $-CH_3$ to $-CH_2$. The reason for this is not clear.

Line F contains points for aromatics, from benzene at the top to mesitylene at the bottom. All complex with iodine, altering its color. Group G consists of CH_2Cl_2 and 1,1- and 1,2- $C_2H_4Cl_2$, with strong dipoles, which enhance energy of vaporization without increasing solvent power for iodine.

Gases

Gas solubilities may be expressed as (1) volume of gas dissolved in unit volume of solvent, known as the *Ostwald coefficient*, designated by γ; (2) the volume of gas reduced to 0°C and 1 atmosphere dissolved in unit volume of solvent, known as the Bunsen coefficient, designated α; (3) the mole fraction, x; or (4) the moles per liter, c, dissolved at 1 atmosphere partial pressure. Henry's law, that the amount of gas dissolved is proportional to its partial pressure, holds rather well at moderate pressures in the basence of a chemical equilibrium. The fact that a substance is a gas at 1 atmosphere and ordinary temperatures indicates that its attractive forces are low and that consequently its solubility will be greater in solvents with low δ-values; also that solubility of different gases in the same solvent will be higher the higher the critical temperature of the gas.

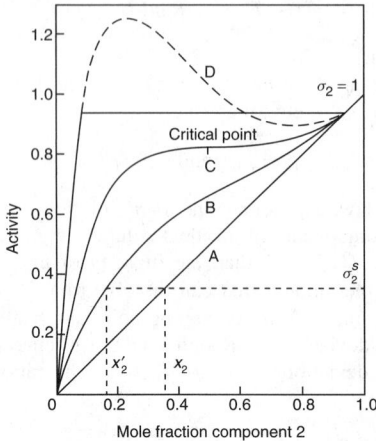

Fig. 1.   Activity versus mole fraction for varying deviations from Raoult's law

δ-VALUES FOR I₂, 25°C

Solvent	Molal Vol. CC.	δ_1	100_{x_2}
n-C_7F_{16}	227.0	5.7	.0185
$SiCl_4$	115.3	7.6	.499
Cyclo-C_6H_{12}	109.	8.2	.918
CCl_4	97.1	8.6	1.147
$TiCl_4$	110.5	9.0	2.15
CS_2	60.6	9.9	5.46
$CHBr_3$	87.8	10.5	6.16

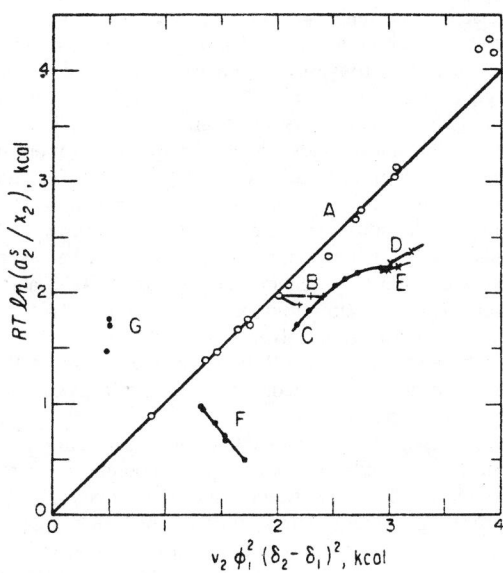

Fig. 3. Relation between energy of solution of iodine derived from measured solubility, x_2, and that calculated from solubility parameters

Line A (beginning at lower left)
CS_2, $CHCl_3$, $TiCl_4$, cis-$C_{10}H_{18}$, $trans$-$C_{10}H_{18}$, $CCl_4 c$-C_6H_{12}, c-C_5H_{10}, $SiCl_4$, CCl_3CF_3, $CCl_2F \cdot CClF_2$, $C_4Cl_3F_7$, c-$C_4Cl_2F_6$, C_7F_{16}.
Line B (left to right)
c-C_6H_{12} (on line A), c-$C_6H_{11}C_2H_5$ (below), c-$C_6H_{11}CH_3$, c-$C_6H_{10}(CH_3)_2$.
Line C (left to right, normal paraffins)
$C_{16}H_{34}$, $C_{12}H_{26}$, C_8H_{18}, C_7H_{16}, C_6H_{14}, C_5H_{12}.
Line D (left to right)
2,3-$(CH_3)_2C_4H_8$, 2,2-$(CH_3)_2C_4H_8$.
Line E (left to right)
2,2,3-$(CH_3)_3C_4H_7$, 2,2,4-$(CH_3)_3C_5H_9$.
Line F (top to bottom)
C_6H_6, $C_6H_5CH_3$, p-$C_6H_4(CH_3)_2$, m-$C_6H_4(CH_3)_2$, 1,3,5-$C_6H_3(CH_3)_3$.
Group G (from top)
1,2-$C_2H_4Cl_2$, CH_2Cl_2, 1,1-$C_2H_4Cl_2$.

The solubility of a number of gases at 1 atmosphere partial pressure and 25°C expressed as $RT \ln x_2$ is plotted in Fig. 4 against the squares of the solubility parameters of a number of solvents. A high amount of regularity is evident for all except the gases SF_6 and CF_4, whose molecules attract molecules of the solvents very selectively. Similar irregularity is evident in the case of the solvent $(C_4F_9)_3N$. In all other cases the positions of missing points could be predicted with confidence.

Variations of solubility with temperature are illustrated in Fig. 4 for 10 gases in cyclohexane. The slopes of the lines times the gas constant R give values for the entropy of solution. In decending from C_2H_6 to He the

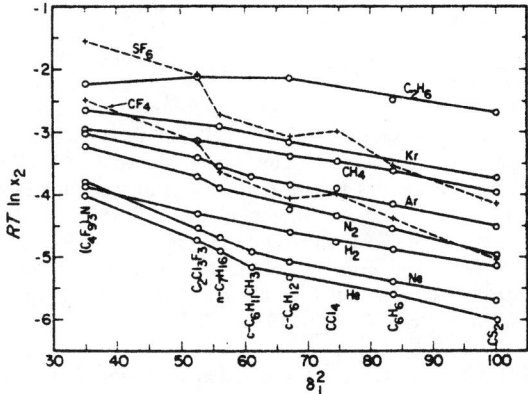

Fig. 4. Solubility of gases, $\log x_2$ at 25°C and 1 atm versus square of solubility parameter of solvents

entropy increases from −8.7 cal/deg mole to +8.1 partly from increases in entropy of dilution, −$R \ln x_2$, but also because the successive gases attract the surrounding solvent molecules less and less strongly, but since they have the same kinetic energy they finally almost blow bubbles permitting more freedom of motion to adjacent molecules of solvent.

The foregoing interesting phenomena are treated at length in "Regular and Related Solutions," by J. H. Hildebrand, J. M. Prausnitz, and R. L. Scott, Van Nostrand Reinhold, New York, 1970.

Solid Solutions

The formation of a solid solution requires not only attractive forces which are not too different, but also identical crystal structures. The latter condition is found most frequently among solids whose molecules are rotating, giving highly symmetrical crystals. See **Crystal**.

Metallic Solutions

In the absence of compounds, these follow the foregoing rules to a fair extent, but with added complications on account of the states of their electrons. The metals have a wide range of solubility parameters and exhibit many cases of incomplete miscibility in the liquid state.

Salt Solutions

The most obvious requirement necessary in a solvent for a salt is that it shall have a high dielectric constant, as is the case with water, liquid ammonia, hydrogen fluoride, and, in a smaller degree, methyl alcohol, in order to weaken the coulombic attraction of its ions for one another. It is possible to formulate the equilibrium between a solid salt and a solution of its ions by considering the changes in energy and entropy involved in vaporization of the solid to gaseous ions, and hydration of the ions. This would be relatively simple if the lattice energy of the solid and the hydration of the ions were solely electrostatic; but the process involves also van der Waals forces, polarization, covalent forces, hydrogen bonding, and entropy changes, which, in the case of water, are considerable, by reason of the ice structure persisting in water and the different structure of water of hydration. Consequently, such a breakdown of the problem, while it may serve to suggest comparisons, is better for explanation than for prediction.

The Periodic System offers the most useful guide by virtue of the trends it reveals; e.g., the decreasing solubility in water of the sulfates and the increasing solubility of the hydroxides of the elements of Group II in descending the group.

Liquid ammonia, because of its lower dielectric constant, is in general a much poorer solvent for salts than water; but this is offset to some extent toward salts of electron-acceptor, Lewis acid cations by its greater basic, electron-donor character.

Insight into the nature of electrolytes in water solutions is afforded by their effects in varying concentration upon the freezing point of water, Δt, at varying concentrations, m moles per 1000 grams. In Fig. 5 $\Delta t/m$ is plotted against m on a logarithmic scale.

The molal lowering of nonelectrolytes is illustrated by sucrose and H_2O_2. These enter so easily into the hydrogen bonded structure of water that they give the theoretical lowering, 1.86° up to 0.1 M in the case of sucrose and to 10 M by H_2O_2.

Binary electrolytes, such as KCl, although completely ionized, even in the solid state, lower the freezing point less than $2 \times 1.86°$, even when as dilute as $10^{-3} M$. This was at first attributed to incomplete ionization but is now explained by the long range of electrostatic forces. Note that Mg^{++} and SO_4^- are less independent than K^+ and $Cl^- \cdot AgNO_3$, unlike KCl, etc., is a weak salt, and undissociated molecules increase rapidly with concentration. The ions nearer to an ion of one sign are those of opposite sign, therefore electric conductivity is less than the sum of ionic conductivities extrapolated to zero concentration.

This effect has been formalized in the concept of *ionic strength*, expressed as

$$I = \frac{1}{2}(m_1 z_1^2 + m_2 z_2^2 + m_3 z_3^2 + \cdots)$$

where z is the ionic charge. Applied to solutions of KCl, K_2SO_4 and $MgSO_4$ the values of I are respectively, 0.01, 0.03, and 0.04. Ionic strength is significant for dealing with the equilibrium and kinetic properties of an ion in mixtures of electrolytes.

Concentrated solutions are strongly affected by ionic hydration. Its strength depends upon ionic radius and charge, therefore it is in general stronger for cations than anions. K_2SO_4 and $MgSO_4$ both yield 3 ions,

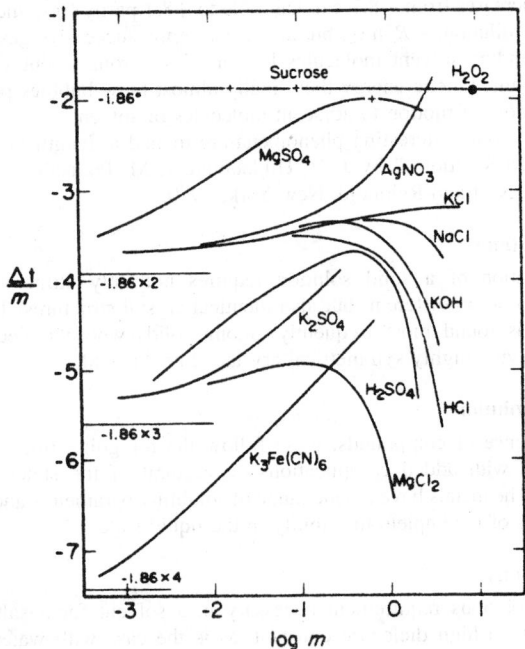

Fig. 5.   Molal lowering of freezing points at different concentrations

but the hydration is stronger for Mg^{++} than K^+, Na^+ than K^+, Na^+ than Ag^+. The line for K_2SO_4 ascends whereas the one for $MgSO_4$ plunges downward, (a) because the strong hydration of Mg^{++} diminishes the coulombic attraction of SO_4^- and (b) because it ties up molecules of water, decreasing the amount of solvent.

JOEL H. HILDEBRAND
University of California
Berkeley, California

SOLVENT.   The term solvent generally denotes a liquid that dissolves another compound to form a homogeneous liquid mixture in one phase. More broadly, the term is used to mean that component of a liquid, gaseous, or solid mixture which is present in excess over all other components of the system. A *chemical solvent* is the term used for solvents in those instances where the process of solution is attended by a chemical reaction between the solvent and the solute. In contrast, a *physical solvent* is one that does not react with the solute. A *dissociating solvent* is one in which solutes that associate in many other solvents enter into solution as single molecules. For instance, various carboxylic acids associate and thus give abnormal elevations of the boiling point, abnormal depressions of the freezing point, etc., in many organic solvents; but in water, however, they do not associate. For this reason water is called a dissociating solvent for such solutes. A liquid that dissolves or extracts a substance from solution in another solvent without itself being very soluble in that other solvent is termed an *immiscible solvent*. A solvent whose constituent molecules do not possess permanent dipole moments and do not form ionized solutions, is termed a *nonpolar solvent*. *Polar solvents*, on the other hand, consist of polar molecules, that is, molecules that exert local electrical forces. In such solvents, acids, bases, and salts, that is, electrolytes, in general, dissociate into ions and form electrically conducting solutions. Water, ammonia, and sulfur dioxide are typical polar solvents. A *normal solvent* is one that does not undergo chemical association, namely, the formation of complexes between its molecules.

A *leveling solvent* is a solvent in which the acidity or basicity of a solute is limited (or leveled) by the acidity or basicity of the solvent itself. For example, the strongest acid that can exist in water is oxonium ion, H_3O^+. Consequently, even though HCl (for example) is intrinsically a much stronger acid than H_3O^+, its acidity in aqueous solution is "leveled" to that of H_3O^+ through the reaction $HCl + H_2O \rightleftharpoons H_3O^+ + Cl^-$. Likewise the very strong base KNH_2 is leveled in water to the basicity of OH^-

$$KNH_2 + H_2O \rightleftharpoons K^+ + OH^- + NH_3$$

The solvents that are leveling to both acids and bases are self-ionized solvents, e.g., water, ammonia, alcohols, carboxylic acids, nitric

acid, etc. Basic non-protonic solvents are leveling to acids, but not to bases (i.e., they are differentiating toward bases), e.g., pyridine, ethers, ketones, etc., since the strongest acid attainable is the protonated solvent molecule (e.g., $C_5H_5N + HCl \rightleftharpoons C_5H_5NH^+ + Cl^-$), whereas there is no corresponding basic species derived from the solvent. Though solvents leveling to bases but not to acids are in principle much more difficult to find, in practice, very strong acids like H_2SO_4 and $HClO_4^-$ are limiting to bases because the species HSO_4^- and ClO_4^-, which will be formed by almost any basic substance, are the strongest bases attainable in these solvents—$B^- + HClO_4 \rightleftharpoons HB + ClO_4^-$—whereas practically no other acid is capable of producing the cations $H_3SO_4^+$ and $H_2ClO_4^+$ in these solvents (i.e., they are differentiating toward acids).

Differentiating solvents are solvents in which neither the acidity of acids nor the basicity of bases is limited by the nature of the solvent. These solvents are not self-ionized. The aliphatic hydrocarbons and the halogenated hydrocarbons are such solvents.

In industry it is generally understood that solvents are simple or complex, pure or impure, compounds or mixtures of compounds (either natural or synthetic), which dissolve many water-insoluble products like fats, waxes, resins, etc., forming homogeneous solutions; that such organic solvents dissolve these water-insoluble products in various proportions depending on the solvent power of the solvent, the degree of solubility of the solute, and the temperature; and that the solute can be recovered with its original properties by the removal of the solvent from the solution. It is also understood in industry that there is a much more limited number of solvents which do not have the properties given above but which nevertheless are of considerable importance; they are the inorganic solvents like water, liquid ammonia, liquid metals, and the like.

Solvents have been classified on various arbitrary bases: (1) boiling point, (2) evaporation rate, (3) polarity, (4) industrial applications, (5) chemical composition, (6) proton donor and proton acceptor relationships, and (7) behavior toward a dye, Magdala Red. Thus on the basis of industrial application one can classify solvents as those for (1) acetyl-cellulose, (2) pyroxylin, (3) resins and rubber, (4) cellulose ether, (5) chlorinated rubber, (6) synthetic resins, and (7) solvents and blending agents for cellulose ester lacquers. Solvents classified according to chemical composition are noted below.

The term *solvent action* is understood to mean any process of making substances water-soluble; but in a broader interpretation the term is understood to be the phenomenon of making a substance soluble in a solvent. Solvent power, diluting power, solvency and similar expressions indicate the property of solvents to disperse the molecules of a solute or vehicle thereby causing a decrease in viscosity.

The most common solvent is water. Water dissolves a great many gases, liquids, and solids, and is much used for this purpose. Other liquids similarly dissolve many substances without reacting chemically with them. Important considerations in connection with the choice of solvent for a given case are (1) vapor pressure and boiling point, (2) solvent power under stated conditions of temperature, (3) ease and completeness of recoverability by evaporation and condensation, and completeness of separation from dissolved material by evaporation, (4) heat of vaporization, (5) miscibility with water or other liquid, if present, (6) inertness to chemical reaction with the materials present, and with the apparatus, (7) inflammability and explosiveness, (8) odor and toxicity; (9) cost of solvent, loss in process, cost of recovering.

See also **Pollution (Air)**.

Colligative Properties of Solutions

When solute is added to a pure solvent, thus forming a solution, properties of the solvent are altered, including (1) osmotic pressure; (2) vapor pressure (lowered); (3) melting point (lowered); and (4) boiling point (elevated). These properties bear a relationship to the number of solute molecules in solution and not to the nature of the molecules. These phenomena are explained by enhanced tension in the solvent. Complete explanation of these changes is beyond the scope of this book, but reference is suggested to H.T. Hammel's article on "Colligative Properties of a Solution" (*Science*, **192**, 748–756, 1976).

SOLVENT EXTRACTION.   A separation operation that may involve three types of mixture:

- A mixture composed of two or more solids, such as a metallic ore.

- A mixture composed of a solid and a liquid.
- A mixture of two or more liquids.

One or more components of such mixture are removed (extracted) by exposing the mixture to the action of a solvent in which the component to be removed is soluble. If the mixture consists of two or more solids, extraction is performed by percolation of an appropriate solvent through it. This procedure is also called leaching, especially if the solvent is water; coffee making is and example. Synthetic fuels can be made from coal by extraction with a coal-derived solvent followed by hydrogenation.

In liquid-liquid extraction one or more components are removed from a liquid mixture by intimate contact with a second liquid that is itself nearly insoluble in the first liquid and dissolves the impurities and not the substance that is to be purified. In other cases, the second liquid may dissolve i.e., extract from the first liquid, the component that is to be purified, and leave associated impurities in the first liquid. Liquid-liquid extraction may be carried out by simply mixing the two liquids with agitation and then allowing them to separate by standing. It is often economical to use counter-current extraction, in which the two immiscible liquids are caused to flow past or through one another in opposite directions. Thus fine droplets of heavier liquid can be caused to pass downward through the higher liquid in a vertical tube or tower.

The solvents used vary with the nature of the products involved. Widely used are water, hexane, acetone, isopropyl alcohol, furfural, xylene, liquid sulfur dioxide, and tributyl phosphate. Solvent extraction is an important method of both producing and purifying such products as lubrication and vegetable oils, pharmaceuticals, and nonferrous metals.

SOLVOLYSIS. A generalized conception of the relation between a solvent and a solute (i.e., a relation between two components of a single-phase homogeneous system) whereby new compounds are produced. In most instances, the solvent molecule donates a proton to, or accepts a proton from a molecule of solute, or both, forming one or more different molecules. A particular case of special interest occurs when water is used as solvent, in which case the interaction between solute and solvent is called *hydrolysis*.

SOMMELET REACTION. Preparation of aldehydes from aralkyl or alkyl halides by reaction with hexamethylenetetramine followed by mild hydrolysis of the formed quaternary salt.

SONN-MULLER METHOED. Preparation of aromatic aldehydes from anilides by conversion of an acid anilide with phosphorus pentachloride to an imido chloride, reduction of the imido chloride with stannous chloride, and hydrolysis of the obtained anil.

SONOCHEMISTRY. Ultrasonic irradiation of liquids causes high energy chemical reactions to occur, often with the emission of light. The origin of sonochemistry and sonoluminescence is acoustic cavitation: the formation, growth, and implosive collapse of bubbles in liquids irradiated with high intensity sound. The collapse of bubbles caused by cavitation produces intense local heating and high pressures, with very short lifetimes. In clouds of cavitating bubbles, these hot-spots have equivalent temperatures of roughly 5000 K, pressures of about 1000 atmospheres, and heating and cooling rates above 10^{10} K/s. In single-bubble cavitation, conditions may be even more extreme. Thus, cavitation can create extraordinary physical and chemical conditions in otherwise cold liquids. Sonoluminescence in general may be considered a special case of homogeneous sonochemistry; however, recent discoveries in this field have heightened interest in the phenomenon in and by itself. See also **Ultrasonics**.

Acoustic Cavitation

The chemical effects of ultrasound do not arise from a direct interaction with molecular species. Ultrasound spans the frequencies of roughly 15 kHz to 1 GHz. With sound velocities in liquids typically about 1500 m/s, acoustic wavelengths range from roughly 10 to 10^{-4} cm. These are not molecular dimensions. Consequently, no direct coupling of the acoustic field with chemical species on a molecular level can account for sonochemistry or sonoluminescence. Instead, sonochemistry and sonoluminescence derive principally from acoustic cavitation, which serves as an effective means of concentrating the diffuse energy of sound.

Compression of a gas generates heat. The compression of bubbles during cavitation is more rapid than thermal transport, which generates a short-lived, localized hot-spot.

If the acoustic pressure amplitude of a propagating acoustic wave is relatively large (greater than ≈0.5 MPa), local inhomogeneities in the liquid (e.g., gas-filled crevices in particulates) can give rise to the explosive growth of a nucleation site into a cavity of macroscopic dimensions, primarily filled with vapor. Such a bubble is inherently unstable, and its subsequent collapse can result in an enormous concentration of energy. This violent cavitation event has been termed "transient cavitation". A normal consequence of this unstable growth and subsequent collapse is that the cavitation bubble itself is destroyed. Gas-filled remnants from the collapse, however, may give rise to reinitiation of the process. The generally accepted explanation for the origin of sonochemistry and sonoluminescence is the hot-spot theory, in which the potential energy given the bubble as it expands to maximum size is concentrated into a heated gas core as the bubble implodes.

Two-Site Model of Sonochemical Reactivity

The transient nature of the cavitation event precludes conventional measurement of the conditions generated during bubble collapse. Chemical reactions themselves, however, can be used to probe reaction conditions. The effective temperature realized by the collapse of clouds of cavitating bubbles can be determined by the use of competing unimolecular reactions whose rate dependencies on temperature have already been measured. The sonochemical ligand substitutions of volatile metal carbonyls were used as these comparative rate probes (eq. 1), where the symbol $\xrightarrow{)))}$ represents ultrasonic irradiation of a solution, and L represents a substituting ligand. These kinetic studies revealed that there were in fact

$$M(CO)_x \xrightarrow{)))} M(CO)_{x-n} + n\ CO \xrightarrow{L} M(CO)_{x-n}(L)_n$$

$$\text{where M = Fe, Cr, Mo, W} \tag{1}$$

two sonochemical reaction sites: the first (and dominant site) is the bubble's interior gas-phase while the second is an *initially* liquid phase. The latter corresponds either to heating of a shell of liquid around the collapsing bubble or to droplets of liquid ejected into the hot-spot by surface wave distortions of the collapsing bubble.

Microjet Formation during Cavitation at Liquid–Solid Interfaces

A very different phenomenon arises when cavitation occurs near extended liquid–solid interfaces. There are two proposed mechanisms for the effects of cavitation near surfaces: microjet impact and shockwave damage. Whenever a cavitation bubble is produced near a boundary, the asymmetry of the liquid particle motion during cavity collapse can induce a strong deformation in the cavity. The potential energy of the expanded bubble is converted into kinetic energy of a liquid jet that extends through the bubble's interior and penetrates the opposite bubble wall. Because most of the available energy is transferred to the accelerating jet, rather than the bubble wall itself, this jet can reach velocities of hundreds of meters per second. Because of the induced asymmetry, the jet often impacts the solid boundary and can deposit enormous energy densities at the site of impact. Such energy concentration can result in severe damage to the boundary surface. The second mechanism of cavitation-induced surface damage invokes shockwaves created by cavity collapse in the liquid. The impingement of microjets and shockwaves on the surface creates the localized erosion responsible for much of ultrasonic cleaning and many of the sonochemical effects on heterogeneous reactions. In this process, the erosion of metals by cavitation generates newly exposed, highly heated surfaces that are highly reactive.

Sonoluminescence

In addition to driving chemical reactions, ultrasonic irradiation of liquids can also produce light. As with sonochemistry, sonoluminescence derives from acoustic cavitation. There are two separate forms of sonoluminescence: multiple-bubble sonoluminescence (MBSL) and single-bubble sonoluminescence (SBSL). Since cavitation is a nucleated process and liquids generally contain large numbers particulates that serve as nuclei, the cavitation field generated by a propagating or standing acoustic wave typically consists of very large numbers of interacting bubbles, distributed over an extended region of the liquid. If this cavitation is sufficiently intense to produce sonoluminescence, then this phenomenon is called multiple-bubble sonoluminescence (MBSL).

Under the appropriate conditions, the acoustic force on a bubble can be used to balance against its buoyancy, holding the single bubble isolated in the liquid by acoustic levitation. This permits examination of the dynamic characteristics of the bubble in considerable detail, from both a theoretical and an experimental perspective. Such a bubble is typically quite small, compared to an acoustic wavelength (e.g., at 20 kHz, the resonance size is approximately 150 μm). For rather specialized but easily obtainable conditions, a single, stable, oscillating gas bubble can be forced into such large amplitude pulsations that it produces sonoluminescence emissions on each (and every) acoustic cycle. This phenomenon is called single-bubble sonoluminescence.

Sonochemistry

In a fundamental sense, chemistry is the interaction of energy and matter. In large part, the properties of a specific energy source determine the course of a chemical reaction. Ultrasonic irradiation differs from traditional energy sources (such as heat, light, or ionizing radiation) in duration, pressure, and energy per molecule. The immense local temperatures and pressures and the extraordinary heating and cooling rates generated by cavitation bubble collapse mean that ultrasound provides an unusual mechanism for generating high energy chemistry. Furthermore, sonochemistry has a high-pressure component, which suggests that one might be able to produce on a microscopic scale the same macroscopic conditions of high temperature–pressure "bomb" reactions or explosive shockwave synthesis in solids.

Experimental Design. A variety of devices have been used for ultrasonic irradiation of solutions. There are three general designs in use presently: the ultrasonic cleaning bath, the direct immersion ultrasonic horn, and flow reactors. The originating source of the ultrasound is generally a piezoelectric material, usually a lead zirconate titanate ceramic (PZT), which is subjected to a high a-c voltage with an ultrasonic frequency (typically 15 to 50 kHz).

The ultrasonic cleaning bath has been used successfully for a variety of liquid–solid heterogeneous sonochemical studies. Lower acoustic intensities can often be used in liquid–solid heterogeneous systems, because of the reduced liquid tensile strength at the liquid–solid interface. The low intensity available in these devices (≈ 1 W/cm^2), however, can prove limiting. The most intense and reliable source of ultrasound generally used in the chemical laboratory is the direct immersion ultrasonic horn (50 to 500 W/cm^2) which can be used for work under either inert or reactive atmospheres or at moderate pressures (<10 atmospheres). Commercially available flow-through reaction chambers which will attach to these horns allow the processing of multiliter volumes. Homogeneous sonochemistry typically is not a very energy efficient process (although it can be more efficient than photochemistry), whereas heterogeneous sonochemistry is several orders of magnitude better. Unlike photochemistry, whose energy inefficiency is inherent in the production of photons, ultrasound can be produced with nearly perfect efficiency from electric power. A primary limitation of sonochemistry remains the small fraction of the acoustic power actually involved in the cavitation events. Sonochemistry is strongly affected by a variety of external variables, including acoustic frequency, acoustic intensity, bulk temperature, static pressure, ambient gas, and solvent. The frequency of the sound field is not a commonly altered variable in most sonochemistry. Changing sonic frequency alters the resonant size of the cavitation event and to some extent, the lifetime of the bubble collapse, but the overall process remains unchanged.

Acoustic intensity has a dramatic influence on the observed rates of sonochemical reactions. Below a threshold value, the amplitude of the sound field is too small to induce nucleation or bubble growth. Above the cavitation threshold, increased intensity of irradiation (from an immersion horn, for example) will increase the effective volume of the zone of liquid which will cavitate, and thus, increase the observed sonochemical rate.

The effect of the bulk solution temperature lies primarily in its influence on the bubble content before collapse. With increasing temperature, in general, sonochemical reaction rates are *slower*. This reflects the dramatic influence which solvent vapor pressure has on the cavitation event: the greater the solvent vapor pressure found within a bubble prior to collapse, the less effective the collapse. Increases in the applied static pressure increase the acoustic intensity necessary for cavitation, but if equal numbers of cavitation events occur, the collapse should be more intense. In contrast, as the ambient pressure is reduced, eventually the gas-filled crevices of particulate matter which serve as nucleation sites for the formation of cavitation in even "pure" liquids, will be deactivated, and therefore the observed sonochemistry will be diminished.

The choice of ambient gas will also have a major impact on sonochemical reactivity. Monatomic gases give much more heating than diatomic, which are much better than polyatomic gases (including solvent vapor). The choice of the solvent also has a profound influence on the observed sonochemistry. In addition to vapor pressure, other liquid properties, such as surface tension and viscosity, will alter the threshold of cavitation. The chemical reactivity of the solvent is often much more important. No solvent is inert under the high temperature conditions of cavitation.

Homogeneous Sonochemistry: Bond Breaking and Radical Formation. The primary products of ultrasound in aqueous solutions are H$_2$ and H$_2$O$_2$; there is strong evidence for various high-energy intermediates, including HO$_2$, H·, OH·, and perhaps $e^-_{(aq)}$. The sonolysis of water, which produces both strong reductants and oxidants, is capable of causing secondary oxidation and reduction reactions.

Suslick and co-workers established that virtually all organic liquids will generate free radicals upon ultrasonic irradiation, as long as the total vapor pressure is low enough to allow effective bubble collapse. The sonalysis of sample hydrocarbons, (e.g., *n*-alkanes) creates the same kinds of products associated with very high temperature pyrolysis.

Applications of Sonochemistry to Materials Synthesis. Of special interest is the development of sonochemistry as a synthetic tool for the creation of unusual inorganic materials. More generally, ultrasound has proved extremely useful in the synthesis of a wide range of nanostructured materials, including high surface area transition metals, alloys, carbides, oxides and colloids. Sonochemical decomposition of volatile organometallic precursors in high boiling solvents produces nanostructured materials in various forms with high catalytic activities. Nanometer colloids, nanoporous high surface area aggregates, and nanostructured oxide supported catalysts can all be prepared by this general route. Sonochemistry has important applications with polymeric materials. Substantial work has been accomplished in the sonochemical initiation of polymerization and in the modification of polymers after synthesis.

Another important application has been the sonochemical preparation of biomaterials, most notably protein microspheres. Using high intensity ultrasound and simple protein solutions, a remarkably easy method to make both air-filled microbubbles and nonaqueous liquid-filled microcapsules has been developed. These microspheres are stable for months, and being slightly smaller than erythrocytes, can be intravenously injected to pass unimpeded through the circulatory system. The mechanism responsible for microsphere formation is a combination of *two* acoustic phenomena: emulsification and cavitation. Ultrasonic emulsification creates the microscopic dispersion of the protein solution necessary to form the proteinaceous microspheres. The long life of these microspheres comes from a sonochemical crosslinking of the protein shell. These protein microspheres, have a wide range of biomedical applications.

Heterogeneous Sonochemistry: Reactions of Solids and Liquids. The use of ultrasound to accelerate chemical reactions in heterogeneous systems has become increasingly widespread. The use of high intensity ultrasound to enhance the reactivity of reactive metals as stoichiometric reagents has become an especially routine synthetic technique for many heterogeneous organic and organometallic reactions. Applications of ultrasound to electrochemistry have also seen substantial recent progress. Another important application for sonoelectrochemistry is the electroreductive synthesis of sub-micrometer powders of transition metals.

Sonocatalysis. Ultrasound has potentially important applications in both homogeneous and heterogeneous catalytic systems. The inherent advantages of sonocatalysis include *(1)* the use of low ambient temperatures to preserve thermally sensitive substrates and to enhance selectivity; *(2)* the ability to generate high energy species difficult to obtain from photolysis or simple pyrolysis; and *(3)* the mimicry of high temperature and pressure conditions on a microscopic scale. Homogeneous catalysis of various reactions often uses organometallic compounds that are often catalytically inactive until loss of metal-bonded ligands (such as carbon monoxide) from the metal. Ultrasound can induce ligand dissociation, permitting initiation of homogeneous catalysis by ultrasound. Heterogeneous catalysts often require rare and expensive metals. The use of ultrasound may permit activating less reactive, but also less costly, metals.

KENNETH S. SUSLICK
University of Illinois at Urbana-Champaign

Additional Reading

Crum, L. A., J. L. Reisse, and K. S. Suslick: *Sonochemistry and Sonoluminescence,* Kluwer Academic Publishers, Norwell, MA, 1998.

Luche, Jean-Louis, and C. Bianchi: *Synthetic Organic Sonochemistry,* Kluwer Academic Publishers, Norwell, MA, 1998.

Mason, T. J. ed.: *Advances in Sonochemistry,* vols. 1–4, JAI Press, New York, NY, 1990, 1991, 1993, 1996.

Mason, T. J. and J. P. Lorimer: *Sonochemistry: Theory, Applications and Uses of Ultrasound in Chemistry,* Ellis, Horwood, Ltd., Chichester, U.K., 1988.

Mason, T. J., and J. P. Lorimer: *Applied Sonochemistry: Uses of Power Ultrasound in Chemistry and Processing,* John Wiley & Sons, Inc., New York, NY, 2002.

Price, G. J. ed.: *Current Trends in Sonochemistry,* Royal Society of Chemistry, Cambridge, 1992.

Suslick, K. S. "Sonochemistry of Transition Metal Compounds," in R. B. King, ed., *Encyclopedia of Inorganic Chemistry,* John Wiley & Sons, Inc., New York, NY, Vol. 7.

SONOLYSIS. The breaking up (molecular fragmentation) of molecules by ultrasonic radiation. Examples: sonolysis in pure water produces hydrogen atoms, hydroxyl radicals, molecular hydrogen, oxygen peroxide; acetonitrile in an argon atmosphere produces molecular hydrogen, nitrogen, and methane.

SORBITOL. See **Sweeteners**.

SORPTION. A generalized term for the many phenomena commonly included under the terms adsorption and absorption when the nature of the phenomenon involved in a particular case is unknown or indefinite.

SOYBEAN OIL. See **Vegetable Oils (Edible)**.

SOY PROTEIN. See **Proteins**.

SPACE CHEMISTRY. Experiments carried out on the space shuttle in the early 1980s indicate that unique types of chemical reactions occur in outer space, and the actual products may result that are not achievable under the terrestrial environment. Several factors are believed to account for this, primarily zero gravity, through absence of oxygen and enhanced magnetic effects may also play a part. Several encouraging results have already been obtained. Among projects that have that have been carried out or are contemplated are the following:

- Uniform polymer microspheres that are over twice as large as possible on earth have been made due to zero gravity.
- More effective electrophoresis reactions for making biological materials have been discovered, probably also because of zero gravity.
- Possibilities exist for (a) making unique alloys in space that are not possible on earth, for example lead-copper, lead-zinc, and aluminum indium (b) purer crystals for microelectronics; (c) better glass for fiber optics; (d) new drugs and pharmaceuticals.

Future experiments will involve human cells, enzymes, and hormones. See also **Space Processing**.

SPACE PROCESSING. Because the residual accelerations resulting from atmospheric drag and gravity gradient effects in low earth orbit are typically on the order of 10^{-6} times earth-gravity, the term microgravity generally is used to describe this acceleration environment. Because by the mid-1990s the space shuttle had been operational for more than a decade, a large number of microgravity experiments were conducted on various flights. Several microgravity-emphasis missions, in which the shuttle was flown in an attitude that minimized acceleration disturbances, were also flown to accommodate experiments that were exceptionally sensitive to accelerations. The more academically oriented experiments designed to address fundamental issues of materials processing are sponsored by the NASA Office of Microgravity Science and Applications; experiments designed to address issues of more direct interest to industrial research are sponsored by the NASA Office of Space Access and Technology (formerly the Office of Commercial Programs).

Materials Experiments in Space

Protein Crystal Growth. By the mid-1990s, the protein crystal growth experiments produced the most spectacular results of all the space processing experiments. The importance of x-ray crystallography as a mechanism for determining three-dimensional structure of complex macromolecules has placed new demands on the ability to grow large (ca 0.5 mm on a side), highly ordered crystals of a vast variety of biological macromolecules in order to obtain high resolution x-ray diffraction data. There are numerous difficulties encountered in attempts to grow macromolecular crystals of biological interest, and the ability to grow such crystals of sufficient size and quality has become an important step for advancement in this field. The difficulties encountered prompted some investigators to consider growing protein crystals in reduced gravity. The first protein growth experiment in reduced gravity was carried out in 1983 on Spacelab 1; crystals of lysozyme and β-galactosidase grew substantially larger in space than in ground control experiments.

The first U.S. protein crystal growth experiment under reasonably well-controlled conditions was carried out in September 1988 on a space transportation system (STS-26), the first shuttle flight after the Challenger accident. The γ-interferon sample and the porcine elastase sample grew much larger than the ground control samples. The isocitrate lyase sample grew as discrete prisms, whereas the ground control crystals always grew dendritically. Crystals of all three of these proteins exhibited significantly higher x-ray diffraction resolution than any produced on Earth. This was true even when some of the smaller space-grown crystals were compared with larger Earth-grown crystals.

Following STS-26, there have been many other attempts to grow a variety of protein crystals in space. Considered as a whole, these experiments have produced a mixed set of results. In some cases the space experiments yielded no crystals or produced crystals that were inferior to those grown on Earth. However, there have been a number of cases in which the space-grown crystals were larger and better ordered than the best ever grown on the earth.

Solution Growth of Small-Molecule Crystals. At least some of the advantages obtained from growth of macromolecular crystals in microgravity appear to carry over to the growth of small-molecule crystals. Triglycine sulfate (TGS) crystals were grown from solution on Spacelab-3 and again on IML-1 using a novel cooled sting method. Supersaturation was maintained by extracting heat through the seed, which was mounted on a small heat pipe, and in turn was attached to a thermoelectric device. By growing under diffusion-controlled transport conditions, it was hoped it would be possible to avoid liquid–vapor inclusions. These inclusions are the most common types of defect in solution-grown crystals and are believed to be caused by unsteady growth conditions resulting from convective flows.

Typically, TGS crystals are grown on ⟨001⟩ oriented seeds, because growth on the ⟨010⟩ face tends to be nonuniform and multifaceted. However, in the absence of convection, growth on the ⟨001⟩ seeds on Spacelab-3 was mostly around the periphery of the seed. Therefore, seeds with a natural ⟨010⟩ face were cut from a polyhedral TGS crystal for the experiments on IML-1. The crystal was grown in space with a 4°C undercooling, which produced a growth rate of 1.6 mm/d. Even though this is somewhat larger than typical growth rates (because of the limited time available) the quality of the space-grown crystal was extremely good. Growth was very uniform and the usual growth defects in the vicinity of the seed that form during the transition from dissolution to growth, known as ghost of the seed, were notably absent.

Vapor Crystal Growth. For materials that lend themselves to physical or chemical vapor transport, growth from the vapor offers some attractive alternatives to growth from the melt. Growth can take place at temperatures considerably lower than the melting point, thus avoiding some of the higher temperature problems associated with melt growth. Several crystal growth experiments on the shuttle have produced provocative results that are not at all understood, e.g., growth of unseeded GeSe crystals by physical vapor transport using an inert noble gas as a buffer in a closed tube on STS-7. In the ground control experiment, many small crystallites formed a crust inside the growth ampul at the cold end. The flight experiment produced dramatically different results; the crystals apparently nucleated away from the walls and grew as thin platelets that eventually became entwined with one another, forming a web that was loosely contained by the ampul. Even more striking was the appearance of the surfaces of the space-grown crystals. These were mirror-like and almost featureless, exhibiting only a few widely spaced growth terraces. By contrast, the crystallites in the ground control experiments conducted under identical thermal conditions had many pits and irregular closely spaced growth terraces.

Test of Dendritic Growth Models. The microgravity environment provides an excellent opportunity to carry out critical tests of fundamental theories of solidification without the complicating effects introduced by buoyancy-driven flows. This advantage was used to carry out a series of experiments to elucidate dendrite growth kinetics under well-characterized diffusion-controlled conditions in pure succinonitrile (SCN). Dendrite tip velocities were measured as a function of undercooling over a range from 0.05–1.5 K. Comparing these measurements with ground-based measurements, it was possible to show that effects of convection are more significant at the smaller undercoolings and are still important up to undercoolings as large as 1.3 K. Even in microgravity, a slight departure in the data was noted at the smallest undercooling, which was attributed to the residual acceleration of the spacecraft. These data also allow the determination of the scaling constant important in the selection of the dynamic operating state, which the present theories have been unable to provide.

Electrodeposition. Electrodeposition experiments in reduced gravity have produced some intriguing results. An early experiment on the German TEXUS rocket, using higher current densities than can normally be used on Earth, reported the deposition of amorphous Ni on Au substrates. In a series of experiments on the Consort Rocket, it was possible to repeat this result.

ROBERT J. NAUMANN
University of Alabama in Huntsville

Additional Reading

DeLucas, L. J. and co-workers: *Science* **246**, 651 (1989).
DeLucas, L. J. and co-workers: *J. Cryst. Growth* **135**, 172 (1994).
Glicksman, M. E. M. B. Koss, and E. A. Winsa: *Phys. Rev. Lett.* **73**, 573 (July 25, 1994).
Riley, C. H. Abi-Akar, B. Benson, and G. Maybee: *J. Spacecraft Rockets* **29**, 386 (1990).

Web References

NASA/Marshall Space Flight Center: http://stp.msfc.nasa.gov/
NASA Life and Microgravity Science: http://www.ssl.umd.edu/space/microgravity.html
NASA - Space Research - Fundamental Microgravity Research in the Physical Sciences: http://spaceresearch.nasa.gov/research_projects/microgravity.html

SPECIFIC GRAVITY. For a given liquid, the specific gravity may be defined as the ratio of the density of the liquid to the density of water. Because the density of water varies, particularly with changes in temperature, the temperature of the water to which a specific gravity measurement is referred should be stated. In exacting, scientific observations, the reference may be to pure (double-distilled) water at 4°C (39.2°F). In engineering practice, the reference frequently is to pure water at 15.6°C (60°F). A value of unity is established for water. Thus, liquids with a specific gravity less than 1 are lighter than water; those with a specific gravity greater than 1 are heavier than water. From a practical standpoint, it usually is more meaningful to express the specific gravity of gases with reference to pure air rather than to pure water. Thus, for a given gas, the specific gravity may be defined as the ratio of the density of the gas to the density of air. Since the density of air varies markedly with both temperature and pressure, exacting observations should reflect both conditions. Common reference conditions are 0°C and 1 atmosphere pressure (760 torr; 760 millimeters Hg; 29.92 inches Hg).

Specific Gravity Scales

Arising essentially from a lack of communication between various scientific and industrial communities, a number of different specific gravity scales were formulated in earlier times. Because so much data and experience have been accumulated in terms of these scales, several methods of expressing specific gravity persist in common use. The most important of these scales are defined here.

API Scale—This scale was selected in 1921 by the American Petroleum Institute, the U.S. Bureau of Mines, and the National Bureau of Standards (Washington, DC) as the standard for petroleum products in the United States.

$$\text{Degrees hydrometer scale (at } 15°C; 60°F) = \frac{141.5}{\text{sp gr}} - 131.5$$

Balling Scale—This scale is used mainly in the brewing industry to estimate percent wort but also is used to indicate percent by weight of either dissolved solids or sugar liquors. Hydrometers are graduated in percent weight at 60°F or 17.5°C.

Barkometer Scale—This scale is used essentially in the tanning and tanning-extract industry. Water equals zero. Each scale degree equals a change of 0.001 in specific gravity. The following formula applies:

$$\text{Sp gr} = 1.000 \pm 0.001 \times (\text{degrees Barkometer})$$

Baumé Scale—This scale is used widely in connection with the measurement of acids and light and heavy liquids, such as syrups. The scale originally was proposed by Antoine Baumé, a French chemist, in 1768. The scale has been widely accepted because of the simplicity of the numbers which represent liquid specific gravity. Two scales are in use:

$$\text{For light liquids,}° \text{Be} = \frac{140}{\text{sp gr}} - 130$$

$$\text{For heavy liquids,}° \text{Be} = 145 - \frac{145}{\text{sp gr}}$$

The standard temperature for these formulas is 15.6°C (60°F).

To calibrate his instrument for heavy liquids, Baumé prepared a solution of 15 parts by weight of sodium chloride in water. On his hydrometer, Baumé marked zero at the point to which the float submerged in pure water; and he marked the scale 15 at the point to which the float submerged in the salt solution. He then divided the distance between the two marks into 15 equal spaces (or degrees as he termed them). In connection with liquids lighter than water, Baumé prepared a 10% sodium chloride solution. In this case, he marked the scale zero at the point to which the float submerged in the salt solution; and he marked the scale 10 at the point to which the float submerged in pure water. Thus, he created a scale which provided increasing numbers with decreases in density.

Users of the Baumé method found that the scale generally read 66 when the float was submerged in oil of vitriol. Thus, early manufacturers of hydrometers calibrated the instruments by this method. There were variations in the Baumé scale, however, because of lack of standardization in hydrometer calibration. Consequently, in 1904, the National Bureau of Standards made a careful survey and finally adopted the scales previously given for light and for heavy liquids.

Brix Scale—This scale is used almost exclusively by the sugar industry. Degrees on the scale represent percent pure sucrose by weight at 17.5°C (63.5°F).

Quevenne Scale—This scale is used for milk testing and essentially represents an abbreviation of specific gravity. For example, 20° Quevenne indicates a specific gravity of 1.020; 40° Quevenne, a specific gravity of 1.040, and so on. One lactometer unit approximates 0.29° Quevenne.

Richter, Sikes, and Tralles Scales—These are alcoholometer scales, which indicate directly in percent ethyl alcohol by weight in water.

Twaddle Scale—This scale is the result of attempting to simplify the measurement of industrial liquids heavier than water. The range of specific gravity from 1.000 to 2.000 is divided into 200 equal parts. Thus, 1° Twaddle equals 0.005 sp gr.

An abridged compilation of specific gravity conversions is given in Table 1. The specific gravities of numerous materials are given throughout this volume.

Determination of Specific Gravity. The principal means for measuring specific gravity (and density) of liquids and gases are listed in Table 2.

Hand Hydrometer. This instrument consists essentially of a long, slender glass float weighted at the lower end and provided with a scale so graduated that the depth to which the instrument sinks in the liquid indicates the specific gravity by direct reading of the scale. See Fig. 1. The numbering of the scale increases from the top downward. The instrument sometimes is proportioned so that the numbering begins with unity at the top, being applicable only to liquids heavier than water; in others, it increases from the top to unity at the lower end and is for use with liquids lighter than water; in still other designs, unity is marked at the middle of the scale and thus the instrument may be used for both light and heavy liquids. To be sensitive, the stem carrying the scale must be slender. It may be observed that the scale intervals corresponding to equal increments

TABLE 1. SPECIFIC GRAVITY SCALE EQUIVALENTS

Specific gravity 60°/60°F	°Baume	°API	Specific gravity 60°/60°F	°Baume	°FAPI
0.600	103.33	104.33	0.800	45.00	45.38
0.620	95.81	96.73	0.820	40.73	41.06
0.640	88.75	89.59	0.840	36.67	36.95
0.660	82.12	82.89	0.860	32.79	33.03
0.680	75.88	76.59	0.880	29.09	29.30
0.700	70.00	70.64	0.900	25.56	25.72
0.720	64.44	65.03	0.920	22.17	22.30
0.740	59.19	59.72	0.940	18.94	19.03
0.760	54.21	54.68	0.960	15.83	15.90
0.780	49.49	49.91	0.980	12.86	12.89
			1.000	10.00	10.00

Specific gravity 60°/60°F	°Baume	°Twaddle	Specific gravity 60°/60°F	°Baume	°Twaddle
1.020	2.84	4	1.500	48.33	100
1.040	5.58	8	1.520	49.61	104
1.060	8.21	12	1.540	50.84	108
1.080	10.74	16	1.560	52.05	112
1.100	13.18	20	1.580	53.23	116
1.120	15.54	24	1.600	54.38	120
1.140	17.81	28	1.620	55.49	124
1.160	20.00	32	1.640	56.59	128
1.180	22.12	36	1.660	57.65	132
1.200	24.17	40	1.680	58.69	136
1.220	26.14	44	1.700	59.71	140
1.240	28.06	48	1.720	60.70	144
1.260	29.92	52	1.740	61.67	148
1.280	31.72	56	1.760	62.61	152
1.300	33.46	60	1.780	63.54	156
1.320	35.15	64	1.800	64.66	160
1.340	36.79	68	1.820	65.33	164
1.360	38.38	72	1.840	66.20	168
1.380	39.93	76	1.860	67.04	172
1.400	41.43	80	1.880	67.87	176
1.420	42.89	84	1.900	68.68	180
1.440	44.31	88	1.920	64.98	184
1.460	45.68	92	1.940	70.26	188
1.480	47.03	96	1.960	71.02	192
			1.980	71.77	196
			2.000	72.50	200

Note: 60°F = 15.6°C

TABLE 2. SPECIFIC GRAVITY AND DENSITY INSTRUMENTATION

Instrument	Liquids	Gases	Solids
Hydrometers			
Nicholson's hydrometer			x
Hand type	x		
Photoelectric type	x		
Inductance-bridge type	x		
Specific-gravity balance		x	
Fixed-volume methods			
Balanced-flow vessel	x		
Displacement meter	x		
Chain-balanced plummet	x		
Buoyancy gas balance		x	
Differential-pressure methods			
Liquid-purge systems	x		
Air-bubbler systems	x		
Viscous-drag method		x	
Boiling-point rise system	x		
Radiation gages	x		
Pycnometer[a]	x		x

[a] See separate editorial entry under **Pycnometer**.

of density cannot be equal where the stem is of uniform diameter. These intervals, in fact, are inversely proportional to the square of the density, being much smaller at the lower than at the upper end of the scale. To

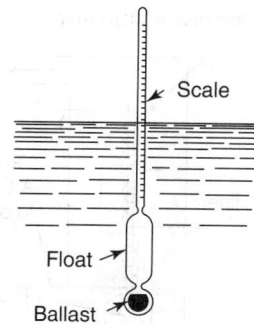

Fig. 1.   Hydrometer for liquids

avoid this, some hydrometers are graduated with an arbitrary scale having uniform spacing, as in the case of the Baumé scale, the readings of which may be converted into density by reference to tables. Nicholson devised a hydrometer for measuring the densities of small solids, the specimen being placed on the hydrometer, first above and then below the surface of the water in which the instrument floats, and its volume deduced from the resulting alteration in buoyant force. See Fig. 2. Hand hydrometers are used extensively where automatic, continuous, and remote readings of specific gravity or density are not required. However, a simple hydrometer can be used in a standpipe, equipped with an overflow at reading level, thus allowing for visual observations of a continuously flowing liquid.

Automated Hydrometers. One form of hydrometer utilizes an opaque stem which, as the stem rises and falls, effects the amount of light which passes through a slit to a photocell. In this way, the photocell output is proportional to specific gravity and may be recorded by an electric instrument. In another industrial version, designed for remote transmission, the hydrometer is contained within a metal cylinder. A rod connects the bottom portion of the float to an armature, which moves vertically between inductance coils. Changes in inductance are transmitted by cable to a central instrument panel receiving instrument.

Chain-balanced Float. In this device, a submerged plummet, which is self-centering and operates essentially without friction, moves up or down with changes in specific gravity. A section of chain is fastened to the bottom of the plummet to provide a counter-buoyancy force. The effective chain weight varies as the plummet moves up and down. For each value of density or specific gravity, the plummet assumes a specific point of equilibrium. By means of a differential transformer, readings may be transmitted to a receiving instrument. A resistance thermometer bridge may be used where compensation for density changes with temperature is required. See Fig. 3.

Balanced-flow Vessel. In this method of specific gravity measurement, a vessel with a fixed-volume is weighed automatically by means of a scale, spring, or use of a pneumatic force-balance system. The liquid being measured flows continuously into and out of the vessel by way of flexible connections. Accuracy of the system is very good.

Displacement-type Meter. In this instrument, the liquid being measured flows continuously through a chamber. A displacer element, usually a hollow metal sphere or cylinder containing air, is submerged fully in the liquid. The buoyant force on the displacer is measured, often by a pneumatic force-balance system, and any variations in this force reflect changes in the specific gravity or density of the liquid. The system can be compensated for changes in liquid temperature by thermostatically heating the chamber.

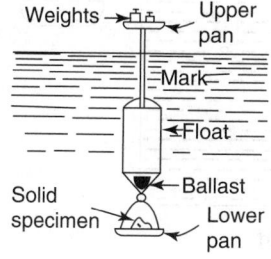

Fig. 2.   Nicholson's hydrometer for small solids

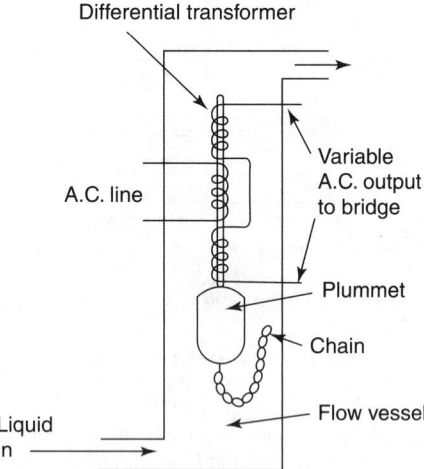

Fig. 3.   Chain-balanced float or plummet-type specific gravity or density meter

Differential-pressure Method

As shown in Fig. 4, two bubbler tubes, the exit of one being lower than the other, are installed in a vessel containing the liquid being measured. Air is bubbled into the liquid through these tubes. The difference in pressure required by each tube represents the weight of a constant-height column of the liquid equal to the difference in level of the ends of the two tubes. Thus, the instrument can be calibrated directly in terms of specific gravity or density. The accuracy ranges from 0.3 to 1% and provides a good approach for liquids that do not tend to crystallize in the measuring pipes. The system is used extensively in the pulp and paper industry for the measurement of white liquor, light black liquor, and bleach. There are several variations of the air-bubbler method, including the use of a reference column and a system with range suppression.

Boiling-point Rise System

For certain liquids, the temperature of a boiling solution of the unknown may be compared with that of boiling water at the same pressure. For a given solution, the boiling-point elevation may be calibrated in terms of specific gravity at standard temperature. Usually two resistance thermometers are used. The system finds use in the control of evaporators to determine the endpoint of evaporation. Good accuracy is achieved in the determination of one dissolved component, or of mixtures of fixed composition.

Nuclear-type Density Meters

As illustrated in Fig. 5, a radioisotope source is placed on one side of a pipeline while a detector is placed on the opposite side. Transmitted

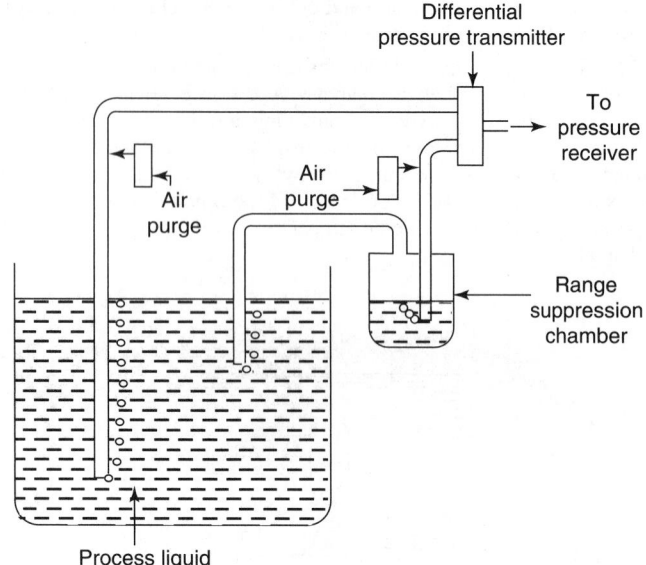

Fig. 4.   Differential-pressure method for specific gravity measurement

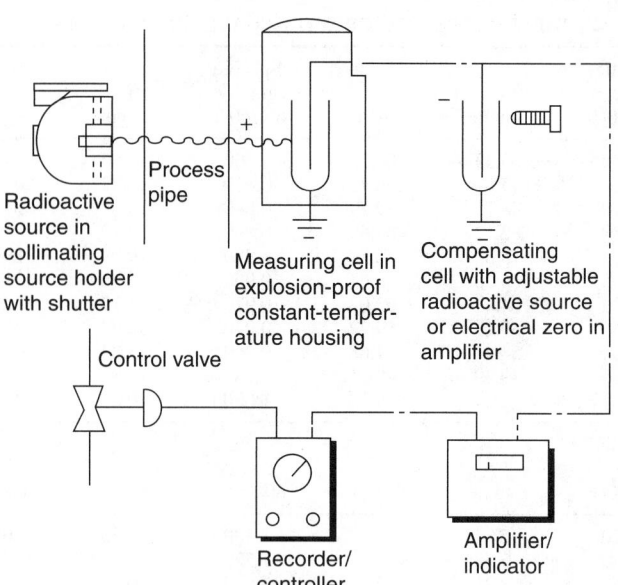

Fig. 5.   Nuclear radiation-type density meter

radiation is in proportion to the density of the material within the pipeline. Standard radiation detectors and associated electrical instrumentation can be used. An A.C. type amplifier is used for accurate determinations when the measurement spans are narrow. For wider spans, D.C. amplifiers may be used. A compensating cell may be used for zero suppression. This essentially eliminates zero drift rate that results from radioisotope source decay. The method is particularly attractive for the density and specific gravity determination of slurries. See also **Radioactivity**.

Gas Density Measurements

In the gas specific gravity balance, a tall column of gas is weighed by the floating bottom of the vessel. This weight measurement is converted to the motion of an indicating pointer or recording pen over a graduated scale, calibrated in units of density or specific gravity. In a viscous-drag type of instrument, one set of impellers is driven in a chamber containing a standard reference gas, and the power required to achieve rotation is measured. A matching set of nonrotating impellers is located in a chamber containing the gas under test. The sets of impellers are connected by a linkage and measure the relative drag shown by the tendency of the impellers to rotate in the test gas. The balance point is a function of relative gas density. In the buoyancy gas balance, a displacer is mounted on a balance beam in the vessel containing the test gas. A reading of the air pressure required to maintain the displacer at a perfectly horizontal position (observed through a window in the chamber) is read from a manometer. Then, the air is displaced in the chamber by the gas under test. The foregoing procedure is repeated. The ratio of pressure with air to pressure with the gas provides a measure of the density of the gas relative to air. This is primarily a laboratory type measurement.

SPECIFIC HEAT.   Sometimes called specific heat capacity. The quantity of heat required to raise the temperature of unit mass of a substance by one degree of temperature. The units commonly used for its expression are the unit mass of one gram, the unit quantity of heat in terms of the calorie. See also **Heat**.

Specific Heat at Constant Pressure

The amount of heat required to raise unit mass of a substance through one degree of temperature without change of pressure. Usually denoted by C_p, when the mole is the unit of mass, and c_p when the gram is the unit of mass.

Specific Heat at Constant Volume

The amount of heat required to raise unit mass of a substance through one degree of temperature without change of volume. Usually denoted by C_v, when the mole is the unit of mass, and c_v when the gram is the unit of mass.

SPECIFIC VOLUME. The volume of a substance or entity per unit mass, obtained by dividing the volume by the mass; and expressed in units of length to the third power and reciprocal units of mass. Reciprocal of the density.

See also **Density**.

SPECTROCHEMICAL ANALYSIS (Visible). Chemical systems that exhibit a selective light absorptive capacity are colored. Hence, the terms *colorimetric analysis* and *colorimetry* often are used to designate the measurement of such systems when the objective is to determine the concentration of the constituent responsible for the color. The use of the term *colorimetry* in this respect is not to be confused with the use of the same term in physics where the term refers strictly to the measurement of color. See also **Colorimetry**.

Visible spectrometry may be used to determine a constituent constituting the major part of a sample, but it also may be applied to the determination of trace quantities. Some methods are applicable to amounts of a few parts per million, with some tests sensitive to 0.01 ppm or less.

Like other areas of spectroscopy, visible spectrometry has a wide range of applications. Included are most of the elements, many anions, functional groups, and innumerable compounds.

The main practical problems in the methodology of visible spectrometry are: (1) to prepare a suitable colored solution; and (2) to measure the light absorptive capacity of this solution, or to compare it with that of a colored solution of known concentration.

Visual spectrochemical analysis has largely been displaced by other, more automated instrumental methods.

SPECTRO INSTRUMENTS. *Spectro* is used as a prefix for a wide assortment of analytical instruments. *Spectro* is derived from *spectrum*, which originally referred to the component colors that make up visible light, the so-called rainbow colors of violet, indigo, blue, green, yellow, orange, and red. A very simple device made up of a glass prism to break up sunlight into color bands is referred to as a *spectroscope*. Much more sophisticated instruments are available for manual manipulation and observation, which still rely on this basic, simple principle; these are termed visual spectroscopes, and the field is called visual spectroscopy.

Over the years, the term *spectrum* has increased in application and meaning and now embraces the total electromagnetic radiation span—no longer being confined to the visible portion. Concurrently, the term *spectroscope* has broadened in meaning so that mention of spectroscope no longer signifies an instrument which operates in the visible region. This situation gave rise to the need for modifying words for use with the term *spectroscope* to signify the portion of the electromagnetic spectrum with which the instrument is concerned. Thus, there are infrared spectroscopes, x-ray spectroscopes, ultraviolet spectroscopes, microwave spectroscopes, and so on. Further, the term *spectrum* has widened to include practically any orderly array of phenomena. For example, a mass spectrometer sorts atoms or radicals by their atomic weight—over a spectrum of values. Other spectrometers analyze the decay "spectra" of radioactive isotopes.

In terms of what is measured or observed, there are (1) portions of the electromagnetic spectrum: gamma-ray, cosmic ray, x-ray, ultraviolet, infrared, far-infrared, microwave, and radiowave instruments; (2) regions pertaining to the energies of particles: beta ray (electrons), protons, neutrons, and mass associated instruments; and (3) instruments dealing with other "spectra" such as radioactive decay and Mossbauer effects.

The suffix *scope* was ample when spectro instruments were essentially manual and confined to visual observations with the unaided eye. As the functions of these instruments increased, new suffixes were required to completely describe the instruments. Thus, if the instrument provides a record of the measurement, either by means of photographic film or by pen recording on a chart, the suffix *graph* is used. Where a meter is utilized to display the information detected by a device such as a bolometer, thermocouple, or thermistor, and so on, the suffix *meter* may be used. In cases where an instrument is designed to measure the intensity of various portions of spectra (again not confined to visible light), the suffix *photometer* may be used. Thus, there are spectrographs (spectrography), spectrometers (spectrometry), and spectrophotometers (spectrophotometry).

But, with increasing complications of instrument design and flexibility of use, the foregoing terms still are not sufficient to fully describe many instruments. Although some instruments display a continuous spectrum of what is being measured, other instruments filter out, for example, certain incident radiation. In other words, certain radiation may be absorbed, giving rise to the terms *absorption spectrometer* or *absorption spectrograph*. In some cases, the energy level of the specimen under analysis must be raised, as by means of flame heating or a spark discharge. Instruments in this category are of the emission type—and hence such terms as *flame photometer* and *optical emission spectrometer*. The names of spectro instruments can be complicated further where some special function may be incorporated in the title. A few examples are:

Spectroheliograph: a spectrograph designed for use with a telescope and for making photographs of the sun, in which the radiation of a particular element in the radiation of the sun may be recorded.

Ultraviolet-visible Spectropolarimeter: an instrument for measuring the rotation of the plane of polarization in accordance with wavelength and light intensity.

Spectrophotofluorometer: an instrument which provides means for both controlling the exciting wavelength and for identifying and measuring light output of a fluorescing sample. If the term *fluorometer* alone were used, it could indicate an instrument with a filtered light input that will measure the fluorescent light from a sample, usually by wavelength.

See also **Spectroscopy (Instrumental Analysis)**.

SPECTROSCOPE. Several types of instruments for producing 0and viewing spectra are included under this term. Variations in form are due, not only to differences in principle, but also to the type of radiation or phenomena to be examined. Terminology employed in this field of instrumentation is described under **Spectro Instruments**.

The earlier spectroscopes, developed by Fraunhofer, Ångström, and others in the early part of the last century, adapted Newton's discovery of the dispersion of light by a prism. The essential features shown in Fig. 1 are a slit, S, a collimator, C, for rendering the light from the slit parallel before entering the prism, one or more dispersing prisms, P, and a telescope, T (or a camera), for forming images of the slit in the various wavelengths and thus providing a method for viewing or photographing the spectrum. The light passes through these in the order named, being deviated by the prism through various angles according to the wavelength. When a spectroscope is provided with a graduated circle for measuring deviations, it is called a "spectrometer." The "direct-vision" or nondeviation spectroscope, employing an Amici prism, is a compact instrument for qualitative purposes. The photographic spectroscope, known as a "spectrograph," is now almost universally used in spectral research.

Many modern spectroscopes employ the diffraction grating instead of the prism. In the concave grating spectroscope, developed by Rowland, the collimating lens and telescope or camera objective are unnecessary because of the focusing effect of the grating itself. See Fig. 2.

The growth of the importance of infrared spectrography and spectrophotometry in determining the structure of compounds and the composition of substances has led to the development of many infrared spectroscopes and other instruments. Most infrared spectroscopes and spectrophotometers employ front-surface mirrors instead of lenses. This eliminates the necessity for energy to pass through glass, quartz, or similar material. Furthermore,

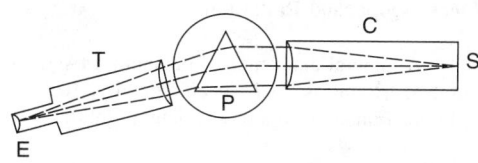

Fig. 1.   Simple prism spectroscope: Collimator, C; slit, S; prism, P; telescope, T; and eyepiece for viewing spectrum, E

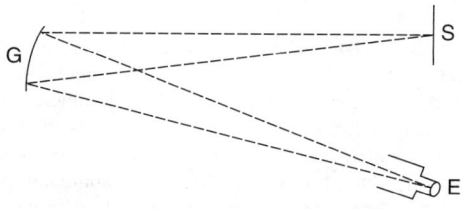

Fig. 2.   Concave grating spectroscope: Slit, S; grating, G, eyepiece or plate-holder, E

it would be difficult or impossible to make lenses that would bring rays of the widely varying wavelengths in the infrared region to focus at one point. Occasionally a rock salt lens will be found in an infrared spectrophotometer, but such lenses are usually not essential optical components. Parabolic mirrors bring energy of all wavelengths to focus at one point. Reflection from most metallic surfaces is generally very efficient in the infrared region. Both gratings and prisms can be used for dispersing the energy, but prisms seem to be more common, perhaps because energy in the infrared region is at a premium and none can be wasted in higher order spectra. The materials most suitable in the infrared region are quartz, calcium fluoride, sodium chloride, and sodium bromide, all in the form of single crystals.

The energy source for an infrared spectroscope or spectrophotometer is a Nernst glower or a Globar. The receiving element for the infrared radiation must be a bolometer, Golay cell or thermistor since photocells are not sensitive in this region of the spectrum. Indeed, the thermoelectric element chosen must be extremely sensitive, since the average energy in the dispersed beam is very small.

Most available infrared instruments use the Littrow mount for the prism, the beam being reflected from a plane mirror behind the prism and thus returning it through the prism a second time. This doubles the dispersion produced. Actually, a double-pass system is also used so that the beam goes through the prism four times. Other design modifications include those with single beam and double monochromator, double beam and double monochromator, and related combinations. See also **Infrared Radiation**.

Check alphabetical index for a wide range of spectroscopic instruments that are based upon numerous materials-energy relations. Also see **Analysis (Chemical)**.

SPECTROSCOPY (Instrumental Analysis).

A branch of analytical chemistry devoted to identification of elements and elucidation of atomic and molecular structure by measurement of the radiant energy absorbed or emitted by a substance in any of the wavelength of the electromagnetic spectrum in response to excitation by an external energy source. The types of absorption and emission spectroscopy are usually identified by the wavelength involved, γ-ray, X-ray, UV, visible infrared, microwave, and radiofrequency. The technique or spectroscopic analysis was originated by Fraunhofer who in 1814 discovered certain dark (D) lines in the solar spectrum, which were later identified as characterizing the element sodium. In 8161 Kirchoff and Bunsen produced emission spectra and showed their relationship to Fraunhofer lines X-ray spectroscopy was utilized by Moseley (1912) to determine the precise location of elements in the periodic system. Since then, a number of sophisticated and highly specialized techniques have been developed including the Raman spectroscopy, nuclear magnetic resonance, nuclear quadrupole resonance, dynamic reflectance spectroscopy, laser, microwave, and γ-ray spectroscopy, and electron paramagnetic resonance.

See also **Spectrochemical Analysis (Visible)**; **Spectro Instruments**; **Spectroscope**; and **Raman Spectrometry**.

SPECTRUM.

The radiant energy emitted by a substance as a characteristic band of wavelengths by which it can be identified.

See also **Spectroscope**; and **Radiation**.

SPERRYLITE.

A mineral diarsenide of platinum, $PtAs_2$. Crystallizes in the isometric system. Hardness, 6–7; specific gravity, 10.58; color, white; opaque. Named after Francis L. Sperry, Sudbury, Ontario.

SPERRY PROCESS.

An electrolytic process for the manufacture of lead carbonate, basic (white lead) from desiliverized lead containing, some bismuth. The impure lead forms the anode. A diaphragm separates anode and cathode compartments, and carbon dioxide is passed into the solution. Impurities, including bismuth, remain on the anode as a slime blanket.

SPHALERITE BLENDE.

Also known as zinc blende, this mineral is zinc sulfide, (Zn, Fe)S, practically always containing some iron, crystallizing in the isometric system frequently as tetrahedrons, sometimes as cubes or dodecahedrons, but usually massive with easy cleavage, which is dodecahedral. It is a brittle mineral with a conchoidal fracture; hardness, 2.5–4; specific gravity, 3.9–4.1; luster, adamantine to resinous, commonly the latter. It is usually some shade of yellow brown or brownish-black, less often red, green, whitish, or colorless; streak, yellowish or brownish, sometimes white; transparent to translucent. Certain varieties are phosphorescent or fluorescent. Sphalerite is the commonest of the zinc-bearing minerals, and is found associated with galena, chalcopyrite, tetrahedrite, barite, and fluorite, as a result of contact metamorphism, and as replacements and vein deposits.

There are very many European localities, including Saxony; Bohemia; Switzerland; Cornwall, in England; Spain; Sweden; Japan; and elsewhere. In the United States, sphalerite is found in Arkansas, Iowa, Wisconsin, Illinois, Colorado, New Jersey, Pennsylvania, Ohio, and especially in the area that includes parts of Kansas, Missouri, and Oklahoma. The word sphalerite is derived from the Greek, meaning treacherous, and its older name, blende, meaning blind or deceiving, refers to the fact that it was often mistaken for lead ore.

SPHENE.

This mineral occurs as a yellow, green, gray, or brown calcium titanosilicate, corresponding to formula $CaTiSiO_2$, crystallizing in the monoclinic system. Fracture conchoidal to uneven; brittle; habit usually wedge-shaped and flattened crystals, also massive and lamellar; luster, resinous to adamantine; transparent to opaque; hardness, 5–5.5; specific gravity, 3.45–3.55.

Sphene is an accessory mineral of widespread occurrence in igneous rocks, and calcium-rich schists and gneisses of metamorphic origin, and very common in nepheline-syenites. In the United States sphene is found in Arkansas, California, New Jersey, New York; in Ontario and Quebec in Canada; and from Greenland, Brazil, Norway, France, Austria, Finland, Russia, Madagascar and New Zealand, as well as many other world localities.

SPINEL.

The mineral spinel is one of a group of minerals which crystallize in the isometric system with an octahedral habit, and whose chemical compositions are analogous. These minerals are combinations of bivalent and trivalent oxides of magnesium, zinc, iron, manganese, aluminum, and chromium, the general formula being represented as $R''O \cdot R_2''O_3$. The bivalent oxides may be MgO, ZnO, FeO, and MnO, and the trivalent oxides Al_2O_3, Fe_2O_3, Mn_2O_3, and Cr_2O_3. The more important members of the spinel group are spinel, $MgAl_2O_4$; gahnite, zinc spinel, $ZnAl_2O_4$, franklinite $(Zn, Mn^{2+}, Fe^{2+})(Fe^{3+}, Mn^{3+})_2O_4$, and chromite, $FeCr_2O_4$. True spinel has long been found in the gem-bearing gravels of Sri Lanka and in limestones of Burma and Thailand.

Spinel usually occurs in isometric crystals, octahedrons, often twinned. It has an imperfect octahedral cleavage; conchoidal fracture; is brittle; hardness, 7.5–8; specific gravity, 3.58; luster, vitreous to dull; transparent to opaque; streak white; may be colorless, rarely through various shades of red, blue, green, yellow, brown, or black. These colors are doubtless due to small amounts of impurities. The clear red spinels are called spinel-rubies or balas-rubies and were often confused with genuine rubies in times past. Rubicelle is a yellow spinel. A violet-colored manganese-bearing spinel is called almandine spinel.

Spinel is found as a metamorphic mineral, and also as a primary mineral in basic rocks, because in such magmas the absence of alkalies prevents the formation of feldspars, and any aluminum oxide present will form corundum or combine with magnesia to form spinel. This fact accounts for the finding of both ruby and spinel together. In addition to the localities mentioned above, which yield beautiful specimens, spinel is found in Italy and Sweden and in Madagascar. Also in the United States in Orange County, New York, and in Sussex County, New Jersey, are many well-known spinel localities. Spinel is found also in Macon County, North Carolina, and in Canada in Quebec and Ontario.

The name spinel is derived from the Greek, meaning a spark, in reference to the fire-red color of the sort much used for gems. Balas ruby is derived from Balascia, the ancient name for Badakhshan, a region of central Asia situated in the upper valley of the Kokcha River, one of the principal tributaries of the Oxus.

ELMER B. ROWLEY
F.M.S.A., formerly Mineral Curator
Department of Civil Engineering
Union College
Schenectady, New York

SPIRO COMPOUND.   See **Compound (Chemical)**.

SPODUMENE.

The mineral spodumene is a lithium aluminum silicate corresponding to the formula $LiAlSi_2O_6$ and occurs in monoclinic prismatic crystals, occasionally of very large size. It also occurs massive. Spodumene has a perfect prismatic cleavage often very noticeable; uneven to splintery fracture; brittle; hardness, 6.5–7.5; specific gravity, 3–3.2; luster, vitreous to dull; color, grayish- to greenish-white, green, yellow and purple. Its streak is white; it is transparent to translucent. Spodumene is characteristically a mineral of the pegmatites, and it is found in Sweden, Ireland, Madagascar and Brazil. In the United States it is found especially in the pegmatites of Oxford County, Maine; in the towns of Goshen, Huntington and Chesterfield in western Massachusetts; at Branchville, Connecticut; in North Carolina; in South Dakota in huge crystals and in San Diego and Riverside Counties in California.

The name spodumene is derived from the Greek meaning ash-colored, particularly appropriate for the slightly weathered varieties. Hiddenite, the beautiful emerald-green or yellow-green spodumene that is used as a gem, was named for W.E. Hidden. Kunzite, named in honor of George F. Kunz, is a transparent lilac to rose-colored spodumene from Madagascar and California, and recently as magnificent, large gem crystals of both purple and yellow color from the Hindu-Kush Mountains, Nuristan Province in Afghanistan. Beautiful gem stones are cut from such crystals, but its easy cleavage discourages its use as a wearable gem. Spodumene alters rather readily to a mass of albite and muscovite. The commercial use of spodumene is chiefly as a source of lithium compounds.

See also **Lithium**.

SPRAY DRYING. A process used in the production of numerous chemical and food products. It is widely used in connection with the production of powdered milk and instant coffee preparations. The spray drying is unique among dryers in that it dries a finely divided droplet by direct contact with the drying medium (usually air) in an extremely short retention time (3 to 30 seconds). This short contact time results in minimum heat degradation of the dried product, a feature that led to the popularity of the spray dryer in the food and dairy industries during its early development. In the case of coffee extract, water in the feed will range from 50 to 70%.

Atomization

In as much as the spray dryer operates by drying a finely divided droplet, the feed to the dryer must be capable of being atomized sufficiently to ensure that the largest droplet produced will be dried within the retention time provided. There are different requirements on the degree of

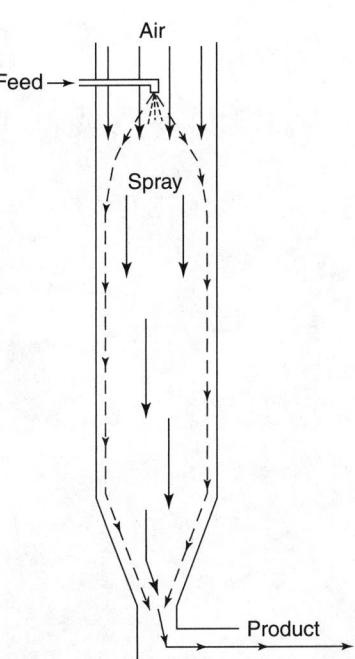

Fig. 2. Spray dryer featuring parallel flow

atomization needed to result in the desired product. These factors including minimizing the fine and/or coarse fractions, controlling particle dryness, and controlling bulk density. All commercial atomizers, whether of the centrifugal-wheel, pressure-nozzle, or other types, will produce a particle-size distribution that follows a probability curve. As the total energy input increases, the average particle size will decrease and the particle-size distribution will improve, i.e., the spread between the largest and smallest particles will be less.

Centrifugal-wheel Atomizers

The wheel consists of a disk, which is rotated at very high speed (1700–50,000 revolutions per minute). See Fig. 1. Feed generally is introduced to the center, with centrifugal force dispersing the feed and throwing out a thin film to the periphery. As the film leaves the disk, it breaks up into a thread, which in turn forms droplets. The disk is located in the hot-air stream so that even though droplets are thrown toward the wall of the dryer, the hot air travels cocurrently and dries the particle sufficiently to prevent wall build-up upon contact. A spray dryer with a wheel atomizer must be relatively large in diameter and shorter than a dryer with pressure nozzles.

Pressure-Nozzle Atomizers

This system consists of an orifice placed after a fixed mechanism, called a core, swirl chamber, or whizzer, depending upon the manufacturer. A

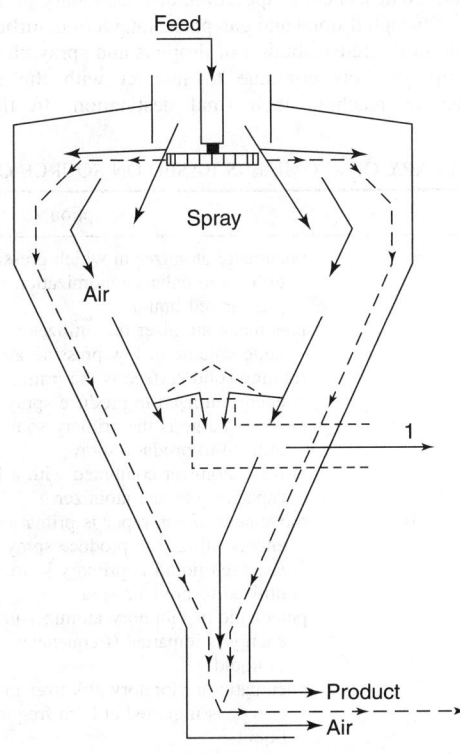

Fig. 1. Spray dryer with wheel atomizer: (1) Air outlet when drying chamber is used for initial separation

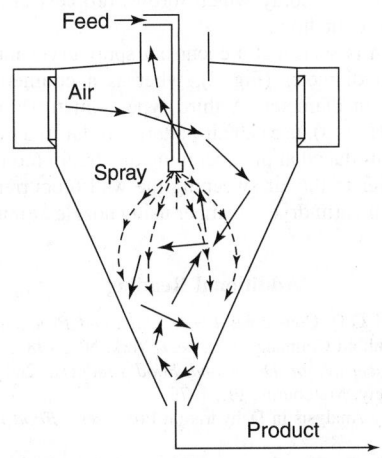

Fig. 3. Mixed-flow spray dryer

Fig. 4.   Tall-form chamber employing nozzle atomization. System is particularly suited for dense particles requiring high-pressure atomization. (*Stork-Bowen*)

high-pressure pump moves the feed to the nozzle body at a pressure of from 250 to 8000 pounds per square inch (17 to 544 atmospheres). The feed slurry is pumped through high-pressure piping to the whizzer, where a spin is imparted to the fluid before it enters the nozzle orifices. This results in a hollow-cone spray which throws droplets either cocurrent or countercurrent to the air flow.

The flow pattern is such that a cocurrent spray dryer must be relatively long and small in diameter (Fig. 2), whereas a countercurrent dryer is shorter and larger in diameter. A third type, sometimes referred to as a mixed-flow dryer (Fig. 3), uses an air pattern similar to a cyclone collector, i.e., the spray is introduced at the upcoming air stream (countercurrent) and the particles transfer to the air sweeping the wall (cocurrent).

A multistory, tall-form dryer chamber using nozzle atomization is shown in Fig. 4.

Additional Reading

Considine, D.M. and G.D. Considine: *Foods and Food Production Encyclopedia,* Van Nostrand Reinhold Company, Inc., New York, NY, 1982.

Farrall, A.W.: *Engineering for Dairy and Food Products,* 2nd Edition, Krieger Publishing Company, Melbourne, FL, 1979.

Flink, M.M.: "Energy Analysis in Dehydration Processes," *Food Technology,* **31**(3), 77–84 (1977).

Masters, K.: *Spray Drying Handbook,* 5th Edition, Halsted Press, Washington, DC, 1991.

Toledo, R.T.: *Fundamentals of Food Process Engineering,* 3rd Edition, Aspen Publishers, Inc., Gaithersburg, MD, 1999.

Web Reference

Pulse Combustion Spray Drying Systems: http://www.pulsedry.com/

SPRAYS. A spray comprises a cloud of liquid droplets randomly dispersed in a gas phase. Depending on the application, sprays may be produced in many different ways. The purposes of most sprays are *(1)* creation of a spectrum of droplet sizes to increase the liquid surface-to-volume ratio, *(2)* metering or control of the liquid throughput, *(3)* dispersion of the liquid in a certain pattern, or *(4)* generation of droplet velocity and momentum. The mechanical devices designed to generate sprays are commonly called atomizers. In the past, the design of atomizers and spray processes was based on traditional fluid dynamic principles and empirical methods. The technology of spraying has advanced rapidly through computer-aided design, mathematical modeling, and sophisticated instrumentation.

Liquid Atomizers

The transformation of bulk liquid to sprays can be achieved in many different ways. Basic techniques include applying hydraulic pressure, electrical, acoustic, or mechanical energy to overcome the cohesive forces within the liquid. Atomizers can be classified according to the energy source used to achieve liquid breakup. Table 1 provides a summary of various atomizers.

Physics of Liquid Atomization

Liquid atomization involves a series of complicated physical processes. These processes can generally be divided into three different flow regimes: internal flow, breakup, and droplet dispersion. The internal flow regime extends from the atomizer inlets to the discharge orifice where liquid emerges. The liquid breakup regime starts at the atomizer exit plane and ends at a certain distance downstream where primary atomization is complete. The final process of atomization is the dispersion regime where spherical droplets gradually evolve into a particular spray pattern.

Droplet Dispersion. The primary feature of the dispersed flow regime is that the spray contains generally spherical droplets. In most practical sprays, the volume fraction of the liquid droplets in the dispersed region is relatively small compared with the continuous gas phase. Depending on the gas-phase conditions, liquid droplets can encounter acceleration, deceleration, collision, coalescence, evaporation, and secondary breakup during their evolution. Through droplet and gas-phase interaction, turbulence plays a significant role in the redistribution of droplets and spray characteristics.

After breakup, droplets continue to interact with the surrounding environment before reaching their final destination. In theory, each

TABLE 1. SUMMARY OF ATOMIZERS BASED ON SOURCE OF ENERGY

Atomizer	Description
air-assisted	pneumatic atomizer in which pressurized air is utilized to enhance atomization produced by pressurized liquid
airblast	pneumatic atomizer that utilizes a relatively large volume of low pressure air
centrifugal	rotating solid surface is the primary source of energy utilized to produce spray
electrostatic	electric charge is the primary source of energy utilized to produce spray
piloted airblast	airblast atomizer combined with a lower capacity pressure atomizer
pneumatic (twin-fluid)	movement of gas/vapor is primary source of energy utilized to produce spray
pressure	pressurized liquid is primary source of energy utilized to produce spray
sonic	pneumatic or vibratory atomizer in which energy is imparted (frequencies > 20 KHz) to liquid
ultrasonic	pneumatic or vibratory atomizer in which energy is imparted at high frequency to liquid
vibratory (piezoelectric)	oscillating solid surface is primary source of energy

droplet group produced during primary breakup can be traced by using a Lagrangian calculation procedure. Droplet size and velocity can be determined as a function of spatial locations.

Spray Characteristics

Spray characteristics are those fluid dynamic parameters that can be observed or measured during liquid breakup and dispersal. They are used to identify and quantify the features of sprays for the purpose of evaluating atomizer and system performance, for establishing practical correlations, and for verifying computer model predictions. Spray characteristics provide information that is of value in understanding the fundamental physical laws that govern liquid atomization.

Spray Parameters. There are several common spray parameters.

Droplet Size Distribution. Most sprays comprise a wide range of droplet sizes. Some knowledge of the size distribution is usually required, particularly when evaluating the overall atomizer performance.

Mean Diameters. Several mean diameters are frequently used to represent the statistical properties of droplets produced by liquid automizers. These mean diameters include volume mean and Sauter mean diameters.

Median Diameter. The median droplet diameter is the diameter that divides the spray into two equal portions by number, length, surface area, or volume. Median diameters may be easily determined from cumulative distribution curves.

Number Density and Volume Flux. The determination of number density and volume flux requires accurate information on the sample volume cross-sectional area, droplet size and velocity, as well as the number of droplets passing through the sample volume at any given instant of time. Volume flux is the volume contained by the droplets passing through a unit cross-sectional area per unit interval of time.

Cone Angle. The spray cone angle is one of the most important parameters in the specification of atomizers. A common method of defining the spray cone angle is to draw two tangent lines originating at the orifice and extending to the outermost spray edges at a specified axial distance.

Patternation. The spray pattern provides important information for many spray applications. It is directly related to the atomizer performance. The pattern information must be able to reveal characteristics such as skewness, degree of pattern hollowness, and the uniformity of liquid flux over the entire cross-sectional area.

Spray Dynamic Structure. Detailed measurements of spray dynamic parameters are necessary to understand the process of droplet dispersion. Improvements in phase Doppler particle analyzers (PDPA) permit *in situ* measurements of droplet size, velocity, number density, and liquid flux, as well as detailed turbulence characteristics for very small regions within the spray.

Spray Correlations. One of the most important aspects of spray characterization is the development of meaningful correlations between spray parameters and atomizer performance. The parameters can be presented as mathematical expressions that involve liquid properties, physical dimensions of the atomizer, as well as operating and ambient conditions that are likely to affect the nature of the dispersion. Empirical correlations provide useful information for designing and assessing the performance of atomizers. Dimensional analysis has been widely used to determine nondimensional parameters that are useful in describing sprays.

Spray Instrumentation

An ideal droplet measurement instrument should (*1*) not interfere with the spray pattern or breakup process, (*2*) provide for large representative samples, (*3*) permit rapid sampling or counting, (*4*) have adequate resolution and accuracy over a wide range of droplet sizes, and (*5*) accommodate variations in the liquid and ambient gas properties. Significant advances have been made in the development of laser diagnostic techniques for measuring sprays. Prior to selecting such an instrument, users should have a thorough understanding of its capabilities and limitations. Existing droplet measurement techniques may be classified into three broad categories: (*1*) optical nonimaging techniques; (*2*) imaging techniques; and (*3*) nonoptical methods.

Industrial Applications

Although atomizers are usually small components in many industrial spray applications, they play an important role in determining the

TABLE 2. SUMMARY OF ATOMIZER SPRAYS FOR SPECIFIC APPLICATIONS

Atomizer spray	Special application
cone spray, hollow or solid	aerating water, brine sprays, chemical processing, coil defrosting, dust control, evaporative condensers, evaporative coolers, industrial washers, roof cooling, spray ponds, spray coating, spray drying, gas scrubbing and washing, humidification, gas cooling, cooling towers, coal washing, degreasing, gravel washing, dish washing, foam control, suspensions and slurries for food and chemical products, pollution control, and oil heating
flat spray	asphalt or tar laying, bottle washing, coal and gravel washing, foam control, degreasing, metal cleaning—rinsing, spray coating, vehicle washing and water misting, descaling, roll cooling, quenching, and agricultural spraying
plain jet spray	rocket engines, diesel engines, agitation, mixing of liquids, cataphoresis plants, and cutting
air atomizing spray	chemical processing, continuous casting, cooling casting and molds, curing concrete products, evaporative coolers, foam control, incineration, quenching, spray coating, spray painting, spray drying, flue gas desulfurization, pollution control, gas turbine engines, and medical spray

performance and efficiency of the entire process. It has long been recognized that atomizers must be properly selected to achieve optimum performance. More recently it has become necessary to comply wit stringent environmental regulations to reduce waste and pollution. Though spray requirements differ from one application to another, the spray pattern or shape appears to be a sensible criterion for selectiry liquid atomizers for certain processes. Table 2 lists a variety of applications that are based on the pattern of the spray.

CHIEN-PEI MAO
ROGER TATE
Delavan Inc.

Additional Reading

Bachalo, W.D. and M.J. Houser: *Opt. Eng.* **23**(5), 583 (1984).

Bayvel, L. and Z. Orzechowski: *Liquid Atomization,* Taylor & Francis Ltd, London, 1993.

Ghavami-Nasr, G.: *Industrial Sprays and Atomization: Design, Analysis, and Applications* Springer-Verlag New York, LLC., New York, NY, 2002.

Giffen, E. and A. Muraszew: *The Atomization of Liquid Fuels,* John Wiley & Sons Inc., New York, NY, 1953.

Lavernia, E.J. and Yue Wu: *Spray Atomization and Deposition,* John Wiley & Sons, Inc., New York, NY, 1996.

Lefebvre, A.H.: *Atomization and Sprays,* Hemisphere Publishing Corp, New York, NY, 1989.

Sirignano, W.A.: *Fluid Dynamics and Transport of Droplets and Sprays,* Cambridge University Press, New York, NY, 1999.

SPUTTERING. In a gas discharge, material is removed, as though by evaporation, from the electrodes, even though they remain cold. This phenomenon is known as sputtering. 2. The term is also used for the corresponding phenomenon when the discharge is through a liquid. In the first case, sputtering is a nuisance that limits the life of a device; in the second case, it is put to work to make colloidal solutions of metals. 3. A result of the disintegration of the metal cathode in a vacuum tube due to bombardment by positive ions. Atoms of the metal are ejected in various directions, leaving the cathode surface in an abraded and roughened condition. The ejected atoms alight upon and cling firmly to the tube walls and other adjacent surfaces, forming a blackish or lustrous metallic film. This effect is often utilized to form very fine-grained coatings of metal upon surfaces of glass, quartz, etc., purposely exposed to the sputtering. Films of different metals can be obtained by using cathodes made of these metals. Glass plates may be thus silvered, or suspension fibers of spun quartz rendered conducting for use in electrometers, etc.

STABLIZER. Any substance that tends to keep a compound, mixture, or solution from changing its form or chemical nature. Stabilizers may retard a reaction rate, preserve a chemical equilibrium, act as antioxidants, keep pigments and other components in emulsion form, or prevent the particles in a colloidal suspension from precipitating.

See also **Inhibitor**.

STACHYOSE.   See Sweeteners.

STAINLESS STEEL.   See Iron Metals, Alloys, and Steels.

STALACTITE AND STALAGMITE. A *stalactite* is a deposit of a mineralized solution, commonly calcium carbonate, which hangs like an icicle from the roof or wall of a limestone cavern. See Fig. 1. The formation of the stalactite usually is quite a slow process. Corresponding columnar structures built upward from the floors of caves beneath the stalactites are called *stalagmites*. Stalactite is derived from the Greek, meaning to fall in drops. Stalagmite, also derived from the Greek, means that which drops. When a stalactite from the top of a cave and a stalagmite from the floor of the cave join, the resulting singular structure is called a *column*.

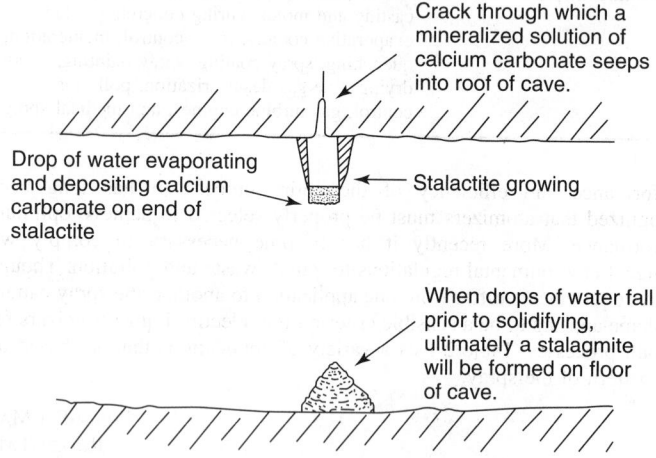

Fig. 1.   Formation of stalactite and stalagmite

STANDARD CONDITIONS.   Many physical and chemical phenomena and substances are defined in terms of *standard* conditions. In some instances, a temperature commonly prevailing in a chemical laboratory may be selected. Thus, when comparing a number of substances, such as their index of refraction, one may find lists in handbooks that give these values as measured by a given temperature and pressure. The Smithsonian Tables, for example, include such data for scores of substances. The researcher then can seek formulas for converting such values to other temperatures/pressures. Interpolations in some instances can be linear over a wide range of values, or the relationships may be nonlinear.

In the case of gases, properties may be tabulated in terms of their existence at 0°C and 760 mm pressure. To determine the volume of a gas at some different temperature and pressure, corrections derived from known relationships (Charles', Amonton's, Gay-Lussac's, and other laws) must be applied as appropriate. In the case of pH values given at some measured value (standard for comparison), the same situation applies. Commonly, lists of pH values are based upon measurements taken at 25°C. The pH of pure water at 22°C is 7.00; at 25°C, 6.998; and at 100°C, 6.13. Modern pH instruments compensate for temperature differences through application of the Nernst equation.

Standard conditions are not necessarily consistent with standards definitions. The careful researcher will always take note of the conditions stated for determining values in a tabulated list.

STANDARD STATE.   The stable form of a substance at unit activity. The stable state for each substance of a gaseous system is the ideal gas at 1 atmosphere pressure; for a solution it is taken at unit mole fraction; and for a solid or liquid element it is taken at 1 atmosphere pressure and ordinary temperature.

STANLEY, WENDELL M. (1904–1971).   An American biochemist who won the Nobel prize for chemistry in 1946 along with John H.

Northrop and James B. Sumner. His work on virus research resulted in isolation of crystals proving the virus to be proteinaceous. In the 1930s, he was concerned with isolating nucleic acid from crystallized virus, and the reproduction of influenza virus. His doctorate was from the University of Illinois. His many accomplishments included membership in the National Advisory Cancer Council of the United States Public Health Service in the 1950s.

STANNITE (Mineral).   This mineral is a sulfo-stannate of copper and iron, sometimes with some zinc, corresponding to the formula, Cu_2FeSnS_4. It is tetragonal; brittle with uneven fracture; hardness, 4; specific gravity, 4.3–4.5; metallic luster; color, gray to black, sometimes tarnished by chalcopyrite; streak, black; opaque. The mineral occurs associated with cassiterite, chalcopyrite, tetrahedrite, and pyrite, probably the result of deposition by hot alkaline solution. Stannite occurs in Bohemia; Cornwall, England; Tasmania; Bolivia; and in the United States in South Dakota. It derives its name from the Latin word for "tin," *stannum*.

STARCH.   [CAS: 9005-25-8]. Chemically, starch is a homopolymer of α-D-glucopyranoside of two distinct types. The linear polysaccharide, amylose, has a degree of polymerization on the order of several hundred glucose residues connected by alpha-D-(1 → 4)-glucosidic linkages. The branched polymer, amylopectin, has a DP (degree of polymerization) on the order of several hundred thousand glucose residues. The segments between the branched points average about 25 glucose residues linked by alpha-D-(1 → 4)-glucosidic bonds, while the branched points are linked by alpha-D-(1 → 6)-bonds. See Fig. 1.

Most cereal starches are made up of about 75% amylopectin and 25% amylose molecules. However, root starches are slightly higher in amylopectin, while waxy corn*. and waxy milo starch contain almost 100% amylopectin. At the other extreme, high amylose corn starch and wrinkled pea starches contain 60–80% amylose. The molecules of amylose and amylopectin are synthesized by enzymes inside the living cell in plastids known as amyloplasts and are deposited as starch granules. These granules are microscopic in size, ranging from 3–8 micrometers in diameter for rice starch up to 100 micrometers for the larger potato starch granules. Corn starch usually falls in a range of 5–25 micrometers. An experienced observer usually can identify the genetic origin of a sample of starch by the size and shape of the granules. The granules are insoluble in cold water, but swell rapidly when heated to the gelatinization temperature range for the particular starch involved. As the granules swell, they lose their characteristic cross under polarized light and imbibe water rapidly until they are many times their original size. Upon continued heating or mechanical shear, the swollen granules begin to disintegrate and the viscosity, having reached a maximum, begins to decrease. However, there usually are some granules and some segments of granules that do not completely disperse in aqueous systems even under the most stringent conditions.

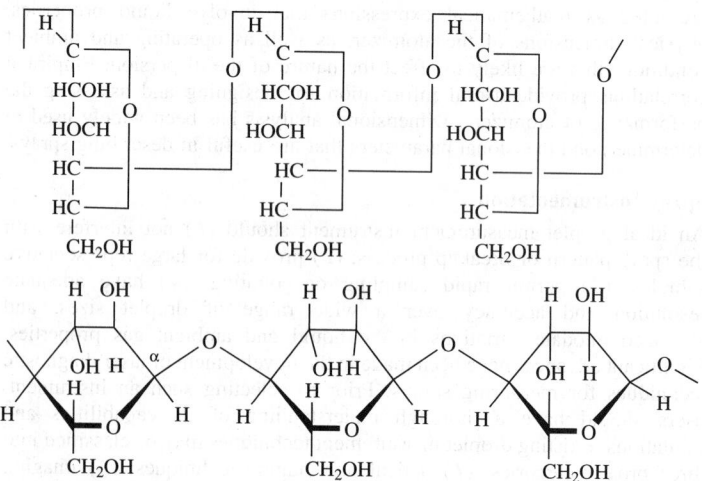

Fig. 1.   A segment of the starch molecule

* With exception of North America, where the plant is called *corn*, other English-speaking people call it *maize*. French = *mais*; Spanish = *maiz*

As the partially dissolved paste is cooled, the hydrated molecules and segments of granules begin to precipitate. In a dilute system (approximately 1%), the segments and molecules retrograde or precipitate. At higher concentrations, sufficient intermolecular and intersegment bonds form to fix the entire system into three-dimensional gel. The rigidity of this gel is affected by many factors, but the amylose content is perhaps the most significant. High amylose starches, when thoroughly cooked, form very rigid gels. Waxy corn or waxy milo starch paste form little, if any, gel structure when cooled.

While some wheat and potatoes are processed in the United States, over 90% of all starch is produced from corn in what is called the *corn wet milling industry*. Close to one-quarter of a billion bushels of corn, representing about 5% of the total corn crop, is converted into wet-process products. The corn refining process is illustrated in Fig. 2. Shelled corn is delivered to the wet-milling plant in boxcars containing an average of 2,000 bushels (50.8 metric tons) per car, and unloaded into a grated pit. The corn is elevated to temporary storage bins, and then to scale hoppers for weighing and sampling. The corn passes through mechanical cleaners designed to separate unwanted substances, such as pieces of cobs, sticks, and husks, as well as metal and stones. The cleaners agitate the kernels over a series of perforated metal sheets; the smaller foreign materials drop through the perforations, while a blast of air blows away chaff and dust, and electromagnets draw out nails and bits of metal. Coming out of the storage bins, the corn is given a second cleaning before going into very large "steep" tanks.

At this point, the use of water becomes an essential part of the corn refining process. The cleaned corn is typically moved into large wooden or metal tanks holding 2,000 to 6,000 bushels (50.8 to 152.4 metric tons), and soaked for 36 to 48 hours in circulating warm water 49°C (120°F) containing a small amount of sulfur dioxide to control fermentation and to facilitate softening. At the end of the steeping process, the steepwater contains much of the soluble protein, carbohydrates, and minerals of the corn kernel, and is drawn off as the first by-product of the process. Steepwater, unmodified and modified, is an essential nutrient for production of antibiotic drugs, vitamins, amino acids, and fermentation chemicals. See Fig. 3. It is also an effective growth supplement for animal feeds.

Fig. 3.   Triple-effect steepwater evaporator. The third effect (forced circulation) is shown in background; second effect (falling-film, recirculating) is middle unit. The first effect vapor head is shown in foreground. (*Swenson, Whiting Corp*)

From the steeps, the softened kernels go through degerminating mills, which are designed not for fine grinding, but rather for tearing the soft kernels apart into coarse particles, freeing the rubbery oil-bearing germ without crushing it, and loosening the bran. The wet, macerated kernels then are sluiced into flotation tanks, called germ separators, or centrifugal hydrocyclones. The germs, lighter than the other components of the kernel, float to the surface, and are skimmed off. By oil expellers or extractors (heat and pressure) and by means of solvents, practically all of the oil is removed as another byproduct to be settled, filtered, refined, and otherwise processed into clear, edible oil for salad dressing and frying, and "corn oil foods" or "soap stock" for soap manufacture. The residue of the germ, after oil-extraction, is ground and marketed as corn germ meal, or may become a part of corn gluten feed or meal.

The remaining mixture of starch, gluten, and bran (hull), which is finely ground, is washed through a series of screens to sieve the bran from the starch and gluten. The hull becomes part of corn gluten feed.

The remaining mixture of gluten and starch is pumped from the shakers to high speed centrifugal machines, which, because of the difference in specific gravity, separate the relatively heavier starch from the lighter gluten. After further processing, the protein-rich gluten is marketed as such, or becomes corn gluten meal, or may be mixed with steepwater, corn oil meal, and hulls to become corn gluten feed. Gluten may also be made to yield a highly versatile protein, *zein*; amino acids, such as glumatic acid, leucine, and tyrosine; and xanthophyll oil, for poultry rations.

Having been separated from the kernels, the starch is now ready for washing, drying, or further processing into numerous dry-starch products, or into dextrin, or for conversion into syrup and sugar. From a 56-pound (25 kilograms) bushel of corn, approximately 32 pounds (14.5 kilograms) of starch result, about 14.5 pounds (6.6 kilograms) of feed and feed products, about 2 pounds (0.9 kilogram) of oil, the remainder being water.

Starch Conversion

More than half of the total production of starch is converted into syrup dextrins or dextrose by acid hydrolysis and/or enzyme action or heat treatment.

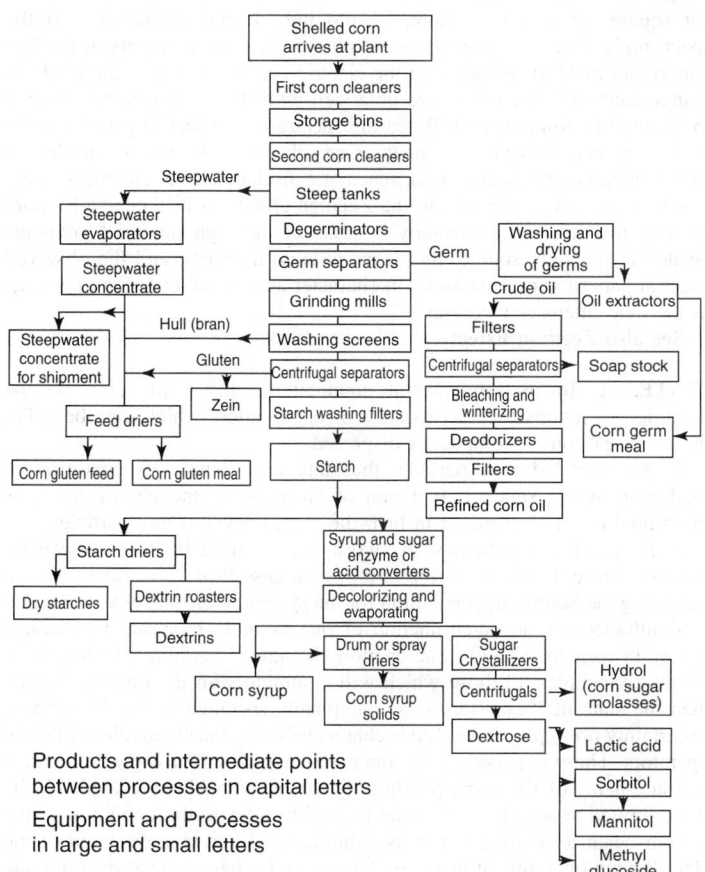

Fig. 2.   The corn (maize) refining process. (*Corn Refiners Association, Inc*)

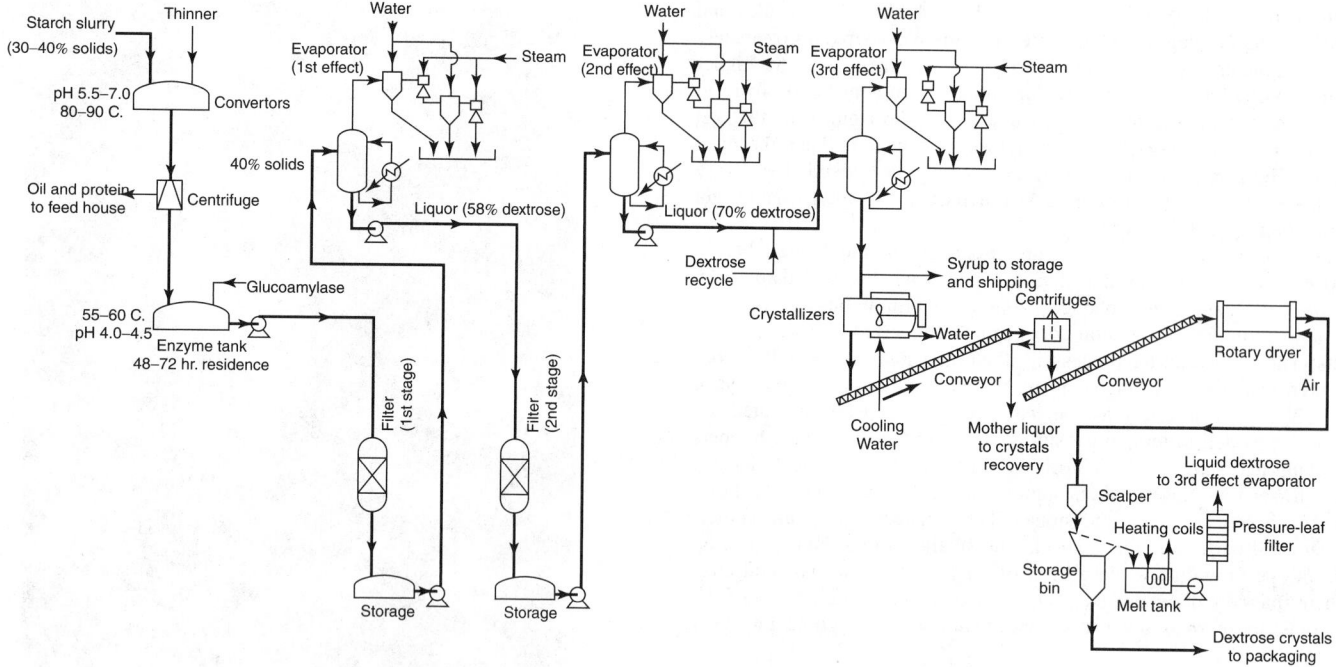

Fig. 4.   Enzyme process for converting starch into dextrose. (*A.E. Staley Mfg. Co.*)

Starch, mixed with water, and heated in the presence of weak hydrochloric acid, breaks down chemically by hydrolysis. If the hydrolysis or conversion of corn starch is interrupted before final conversion, a noncrystallizing corn syrup is obtained. Many varieties may be made by supplemental use of enzymes to meet specific functional requirements. The solids content is varied to suit the requirements of the users. Corn syrup is used in a wide variety of food products, including baby foods, breakfast foods, cheese spreads, chewing gum, chocolate products, confectionary, cordials, frostings and icings, peanut butter, sausage, and for numerous industrial products, including adhesives, dyes and inks, explosives, metal plating, plasticizers, polishes, textile finishes, and in leather tanning.

A process for converting starch to dextrose is shown in Fig. 4. The enzyme process shown overcomes flavor and color difficulties of the hydrochloric acid method. The enzyme is obtained by growing a mold (*Aspergillus phoenicis*, a member of the *Aspergillus niger* group). The mold yields the key glucoamylase as well as transglucosylase. The latter must be eliminated because it catalyzes the formation of undesirable glucosidic linkages. Through a special process, almost pure glucoamylase is obtained.

Purified starch slurry (30–40% solids), made from dent corn, is received in the converters from basic processing at the corn plant. A preliminary conversion using alpha-amylase enzyme or acid is carried out at 80–90°C (176–194°F), during which 15–25% of the starch is converted into dextrose. This thins the starch slurry, allowing easier addition of the glucoamylase enzyme. It also prevents formation of unhydrolyzable gelatinous material during the main conversion, and results in increased dextrose yields of from 3–4%. Thinning also reduces evaporation costs because starch concentrations of 30–40% can be handled compared with the 12–20% limit for the acid process. Before the main conversion, the starch-dextrose slurry is centrifuged to remove oil and protein by-products, which are processed for animal feed.

The slurry then goes to a 25,000-gallon (946-hectoliter) enzyme tank where, at pH 4.0–4.5 and 60°C, the major reaction with the glucoamylase takes place. It is a batch operation requiring about 72 hours. When conversion is complete, the batch (97–98.5% dextrose on a dry basis) is passed through a preliminary decolorizing filter of powdered carbon and then pumped on to the first of three evaporators. The remaining operations are evaporation and crystallization, followed by centrifuging, and rotary drying for dextrose crystals; and by a remelting and filtering process for handling of outsize crystals, the resulting liquid being returned to the third effect evaporator for reprocessing.

Additional Reading

Bourne, G.H.: *Nutritional Value of Cereal Products, Beans and Starches,* S. Karger Publishers, Inc., Farmington, CT, 1989.

Galliard, T.: *Starch: Properties and Potential,* John Wiley & Sons, Inc., New York, NY, 1987.

Preiss, J., M.N. Sival, and S.L. Taylor: *Starch: Basic Science to Biotechnology,* Vol. 41, Academic Press, Inc., San Diego, CA, 1998.

Schenck, F.W., R.E. Hebeda: *Starch Hydrolysis Products: Worldwide Technology, Production, and Applications,* John Wiley & Sons, Inc., New York, NY, 1992.

Whistler, R.L.: *Starch,* 3rd Edition, Academic Press, Inc., San Diego, CA, 2001.

STARK EFFECT.   In 1913, Stark showed that every line in the Balmer series of hydrogen, when excited in a strong electric field of 100 kilovolts per square centimeter or more, is split into several components. If the spectrum is observed perpendicular to the field, some members of the line pattern are plane-polarized with the electric vector parallel to the field (*p*-components) and the others are polarized with the electric vector normal to the field (*s*-components). When the spectrum is observed parallel to the field, only unpolarized *s*-components are observed. A similar splitting of lines is noted in the cathode dark space of a discharge tube. The Stark effect is similar, in many respects, to the Zeeman effect but it is generally more difficult to study experimentally because of the high potential gradients needed in the light source. Its theory is quite different, and the observed spectral pattern varies markedly in character and in number of components as the field intensity increases.

See also **Zeeman Effect**.

STATE.   1. In its fundamental connotation, this term refers to the condition of a substance, as its state of aggregation, which may be solid, liquid, or gaseous—compact or dispersed.

2. As extended to a particle, the state may denote its condition of oxidation, as the state of oxidation of an atom, or the energy level, as the orbital of an electron, or in fact, the energy level of any particle.

3. In quantum mechanics, the word state is used in its most general context to refer to the condition of a system described by a wave function satisfying the Schrödinger equation for the system, when this wave function is simultaneously an eigenfunction of one or more quantum mechanical operators corresponding to one or more dynamical variables. If this set of operators includes all those which will commute with the ones in the set, then the state of the system is as completely specified as the Heisenberg uncertainty principle allows and is characterized by the eigenvalues of these operators. These eigenvalues are the results which will always be found if measurements of the corresponding dynamical variables are made. In its more limited sense, the word state is used to refer to the condition of the system when its wave function is simultaneously an eigenfunction of the Hamiltonian operator. In this case the system is characterized by a definite value of the energy, i.e., the eigenvalue of the Hamiltonian operator, and is said to be an *energy eigenstate*, a *stationary state*, or a definite *energy state*.

STATISTICAL MECHANICS. One major problem of physics involves the prediction of the macroscopic properties of matter in terms of the properties of the molecules of which it is composed. According to the ideas of classical physics, this could have been accomplished by a determination of the detailed motion of each molecule and by a subsequent superposition or summation of their effects. The Heisenberg indeterminacy principle now indicates that this process is impossible, since we cannot acquire sufficient information about the initial state of the molecules. Even if this were not so, the problem would be practically insoluble because of the extremely large numbers of molecules involved in nearly all observations. Many successful predictions can be made, however, by considering only the average, or most probable, behavior of the molecules, rather than the behavior of individuals. This is the method used in statistical mechanics.

In the general approach to classical statistical mechanics, each particle is considered to occupy a point in phase space, i.e., to have a definite position and momentum, at a given instant. The probability that the point corresponding to a particle will fall in any small volume of the phase space is taken proportional to the volume. The probability of a specific arrangement of points is proportional to the number of ways that the total ensemble of molecules could be permuted to achieve the arrangement. When this is done, and it is further required that the number of molecules and their total energy remain constant, one can obtain a description of the most probable distribution of the molecules in phase space. The Maxwell-Boltzmann distribution law results.

When the ideas of symmetry and of microscopic reversibility are combined with those of probability, statistical mechanics can deal with many stationary state nonequilibrium problems as well as with equilibrium distributions. Equations for such properties as viscosity, thermal conductivity, diffusion, and others are derived in this way.

The development of quantum theory, particularly of quantum mechanics, forced certain changes in statistical mechanics. In the development of the resulting quantum statistics, the phase space is divided into cells of volume h^f, where h is the Planck constant and f is the number of degrees of freedom. In considering the permutations of the molecules, it is recognized that the interchange of two identical particles does not lead to a new state. With these two new ideas, one arrives at the Bose-Einstein statistics. These statistics must be further modified for particles, such as electrons, to which the Pauli exclusion principle applies, and the Fermi-Dirac statistics follow.

It is often possible to obtain similar or identical results from statistical mechanics and from thermodynamics, and the assumption that a system will be in a state of maximal probability in equilibrium is equivalent to the law of entropy. The major difference between the two approaches is that thermodynamics starts with macroscopic laws of great generality and its results are independent of any particular molecular model of the system, while statistical methods always depend on some such model.

STAUROLITE. The mineral staurolite is a complex silicate of iron and aluminum corresponding to the formula $(Fe, Mg, Zn)_2Al_9Si_4O_{23}(OH)$ but somewhat varying and may carry magnesium or zinc. It is orthorhombic, prismatic, twins common, often producing cruciform crystals. It is a brittle mineral; fracture, subconchoidal; hardness, 7–7.5; specific gravity, 3.65–3.83; luster, subvitreous to resinous; color, dark brown, sometimes reddish to nearly black; grayish streak; translucent to opaque. Staurolite is a metamorphic mineral usually the result of regional rather than contact metamorphism, and is common in schists, phyllites and gneisses together with garnet, kyanite, and tourmaline.

Well-known European localities are in Switzerland and Brittany; and in the United States this mineral is common in the schists of New England, and those of the southern Alleghenies. Frequently the crystals are found loose in the soil after the disintegration of the country rock. The name staurolite is derived from the Greek meaning a cross, in reference to the twin crystals, the more nearly perfect crosses being somewhat in demand as curios.

STEAM. Steam, gaseous H_2O, is the most important industrially used vapor and, after water, the most common and important fluid used in chemical technology. Steam is generated from water by boiling, flash evaporation, and throttling from high to low pressure. The phase change occurs along the saturation line such that the specific volume of steam is larger than that of the boiling water. Thermal energy, i.e., the heat of evaporation, is absorbed during the process. At the critical and supercritical pressures, the water–steam distinction disappears, and the fluid can go from water-like properties to steam-like properties without an abrupt change in density or enthalpy. Properties of steam can be divided into thermodynamic, transport, physical, and chemical properties.

Physical Properties

Official Properties. The International Association for Properties of Water and Steam (IAPWS), an association of national committees that maintains the official standard properties of steam and water for power cycle use, maintains two formulations of the properties of water and steam. The first is an industrial formulation, the official properties for the calculation of steam power plant cycles. This formulation is appropriate from 0.001 to 100 MPa (0.12–1450 psia) and from 0 to 800°C (32–1472°F) and also from 0.001 to 10 MPa (0.12–145 psia) between 800 and 2000°C (1472–3632°F). This formulation is used in the design of steam turbines and power cycles. IAPWS maintains a second formulation of the properties of water and steam for scientific and general use from 0.01 MPa (extrapolating to ideal gas) at 0°C (1.45 psia at 32°F) to the highest temperatures and pressures for which reliable information is available.

Thermodynamic Properties. Ordinary water contains three isotopes of hydrogen (qv), i.e., $^1H, ^2H$, and ^3H, and three of oxygen (qv), i.e., $^{16}O, ^{17}O$, and ^{18}O. The bulk of water is composed of ^1H and ^{16}O. Tritium, ^3H, and ^{17}O are present only in extremely minute concentrations, but there is about 200-ppm deuterium, ^2H, and 1000-ppm ^{18}O in water and steam. See **Deuterium** and **Tritium**. The thermodynamic properties of heavy water are subtly different from those of ordinary water. The properties given herein are for ordinary water having the usual mix of isotopes.

Vapor pressure is one of the most fundamental properties of steam. Figure 1 shows the vapor pressure as a function of temperature for temperatures between the melting point of water and the critical point. This line is called the saturation line. Liquid at the saturation line is called saturated liquid; liquid below the saturation line is called subcooled. Similarly, steam at the saturation line is saturated steam; steam at higher temperature is superheated. Properties of the liquid and vapor converge at the critical point, such that at temperatures above the critical point, there is only one fluid. Along the saturation line, the fraction of the fluid that is vapor is defined by its quality, which ranges from 0 to 100% steam.

The density of saturated water and steam is a function of temperature. As the temperature approaches the critical point, the densities of the liquid and vapor phase approach each other. This fact is crucial to boiler construction and steam purity, because the efficiency of separation of water from steam depends on the density difference.

The enthalpies and internal energies of steam and water also converge at the critical point. The heat capacity at constant pressure, C_p, is defined as the derivative of enthalpy with respect to temperature. The value of C_p becomes very large in the vicinity of the critical point. The variation is much smaller for the heat capacity at constant volume, C_v.

Transport Properties. Viscosity, thermal conductivity, the speed of sound, and various combinations of these with other properties are called steam transport properties, which are important in engineering calculations. The speed of sound is important to choking phenomena, where the flow of steam is no longer simply related to the difference in pressure. Thermal conductivity is important to the design of heat-transfer apparatus. See **Heat-exchange Technology**. Sharp declines in each of these properties occur at the transition from liquid to gas phase, i.e., from water to steam.

Miscellaneous Properties. The dielectric constant is a physical property having great importance to the chemical properties of hot water and steam.

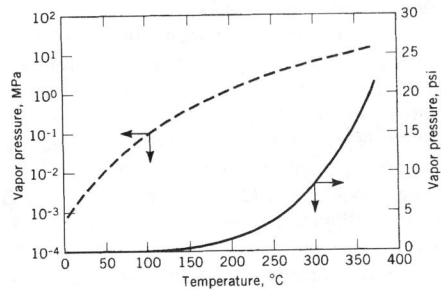

Fig. 1. Vapor pressure of ordinary water, where (——) represents linear and (————) logarithmic scale. To convert MPa to psi, multiply

Along the saturation line, the steam and water values converge at the critical point. The ability of water to dissolve salts results from the high dielectric constant. The precipitous drop in water dielectric constant in the region of the critical point is very important to the solubility of salts in water near the critical temperature. Many salts exhibit declining solubilities as the critical temperature is approached and then exceeded. The drop in dielectric constant is largely a result of the decline in density.

The ion product of water is the product of the molality of the hydrogen and hydroxide ions, $K_w = m_{H+} m_{OH^-}$. The ion product increases with temperature to 250°C and then declines. The initial increase is the temperature effect, and the later decline is on account of the decline in dielectric constant of water. This variation means that neutral pH, which is the square root of the ion product, varies with temperature.

Chemical Properties

Molecular Nature of Steam. The molecular structure of steam is not as well known as that of ice or water. There are indications that in the steam phase some H_2O molecules are associated in small clusters of two or more molecules.

Solvent. The solvent properties of water and steam are a consequence of the dielectric constant. At 25°C, the dielectric constant of water is 78.4, which enables ready dissolution of salts. As the temperature increases, the dielectric constant decreases. The solubility of many salts declines at high temperatures. As a consequence, steam is a poor solvent for salts. Although the solubility of salts in steam is small, it has great significance to corrosion of steam system components, particularly steam turbines. At the critical point and above, water is a good solvent for organic molecules.

Reactant. Steam can behave as an oxidant. Steam reacts with salts so that the salts dissociate into the respective hydroxide and acid. For sodium salts, the sodium hydroxide is largely in a liquid solution and the acid is volatile. See also **Coal Conversion (Clean Coal) Processes**; **Nuclear Power Technology**; and **Petroleum Refining**

JAMES BELLOWS
Westinghouse Electric Corporation

Additional Reading

Cohen P. *ASME Handbook on Water Technology for Thermal Power Systems,* ASME, New York, NY, 1989.
Meyer C. A. and co-workers: *Steam Tables,* 6th Edition, ASME, New York, NY, 1993.
Steam: Its Generation and Use, 40th Edition, Babcock and Wilcox Co., New York, NY, 1992.
White H. J. and co-workers: "Proceedings of the 12th International Conference on the Properties of Water and Steam, Orlando, Fla.," 1994, Begell House, New York, NY, 1995.

STEAM DISTILLATION. See **Distillation**.

STEAM REFORMING. See **Ammonia**; **Substitute Natural Gas (SNG)**.

STEARIC ACID AND STEARATES. [CAS: 57-11-4]. Stearic acid $H \cdot C_{18}H_{35}O_2$ or $C_{17}H_{35} \cdot COOH$ or $CH_3(CH_2)_{16} \cdot COOH$ is a white solid, melting point 69°C, boiling point 383°C, insoluble in water, slightly soluble in alcohol, soluble in ether. Stearic acid may be obtained from glyceryl tristearate, present in many solid fats such as tallow, and in smaller percentage in semisolid fats (lard) and liquid vegetable oils (cottonseed oil, corn oil), by hydrolysis. The crude stearic acid, after separation of the water solution of glycerol, is cooled to fractionally crystallize the stearic and palmitic acids, which are then separated by filtration (oleic acid in the liquid), and fractional distillation under diminished pressure. With sodium hydroxide, stearic acid forms sodium stearate, a soap. Most soaps are mixtures of sodium stearate, palmitate and oleate.

The following are representative esters of stearic acid: Methyl stearate $C_{17}H_{35}COOCH_3$, melting point 38°C, boiling point 215°C at 15 millimeters pressure; ethyl stearate $C_{17}H_{35}COOC_2H_5$, melting point 35°C, boiling point 200°C at 10 millimeters pressure; glyceryl tristearate [tristearin $C_3H_5(COOC_{17}H_{35})_3$], melting point 70°C approximately.

Stearic acid is used (1) in the preparation of metallic stearates, such as aluminum stearate for thickening lubricating oils, for waterproofing materials, and for varnish driers, (2) in the manufacture of "stearin" candles, and is added in small amounts to paraffin wax candles. As the glyceryl ester, stearic acid is one of the constituents of many vegetable and animal oils and fats.

See also **Rubber (Natural)**.

STEARONE. An aliphatic ketone, insoluble in water, stable to high temperatures, acids, and alkalies; compatible with high-melting vegetable waxes, paraffins, and fatty acids; incompatible with resins, polymers and organic solvents at room temperature but compatible with them at high temperature.

STEELS AND STEELMAKING. See **Iron Metals, Alloys, and Steels**.

STENGEL PROCESS. A method of making ammonium nitrate fertilizer from anhydrous ammonia and nitric acid. The fertilizer particles can be made in different sizes according to use.

STEPHANITE. The mineral stephanite, silver antimony sulfide, Ag_5SbS_4, is found in short prismatic or tabular orthorhombic crystals. It is a brittle mineral; hardness, 2–2.5; specific gravity, 6.25; metallic luster; color, black; streak, black; opaque.

Stephanite occurs associated with other silver minerals and is believed to be primary in character. Localities are in the Czech Republic and Slovakia, Saxony, the Harz Mountains, Sardinia; Cornwall, England; Chile and Mexico. In the United States it is found in Nevada, where it is an important silver ore. It was named for the Archduke Stephan of Austria, mining director of that country at the time this mineral was first described.

STEPHEN ALDEHYDE SYNTHESIS. Preparation of aldehydes from nitriles by reduction with stannous chloride in ether saturated with hydrochloric acid. The intermediate aldimine salts have to be hydrolyzed. The best results are obtained in the aromatic series.

STEREOCHEMISTRY. Two molecules are said to be sterioisomers if they possess identical chemical formulas with the same atoms bonded one to another, but differ in the manner these atoms are arranged in space. Thus *sec*-butanol can exist in two forms, I and II, which cannot be superimposed on each other.

This particular example represents one class of stereoisomers known as *enantiomers*, which may be defined as two molecules that are mirror images but are nonetheless nonsuperimposable. Such molecules are said to possess opposite configuration. If these isomers are separated (*resolved*), the separate enantiomers have been found to rotate the plane of plane-polarized light. This phenomenon of *optical activity* has been known for well over a century. A 50–50 mixture of two enantiomers is optically inactive or *racemic*, since the rotation of light by one enantiomer is precisely compensated by the rotation of light in the opposite direction by the other enantiomer.

Physiological activity is closely related to configuration. Thus the left-rotating, or *levo*, form of adrenalin is over ten times more active in raising the blood pressure than is the right-rotating, or *dextro* form. Many organic chemicals essential to plants and animals are optically active. Enzymes, which catalyze chemical reactions in the body, are frequently programmed to accept only one enantiomer. All the essential amino acids, generally of the formula III, are of the *levo* type, although important

exceptions exist. Recently, several amino acids were found by NASA in a meteorite that presumably originated from the asteroid belt between Mars and Jupiter. The proof that the amino acids were extraterrestrial came from the fact that they were racemic. Any terrestrial contaminants from laboratory handling would have been optically active and *levo*.

Many examples of optically active molecules contain an asymmetric carbon atom, that is, one with four different groups attached, as in I-III. A wide variety of other atoms may also be asymmetric (IV-VI). An asymmetric center is by no means

a necessary condition for enantiomerism. Well-known examples of nonsuperimposable mirror images without asymmetric atoms are allenes (VII), spiranes (VIII), biphenyls (IX), and various

inorganic complexes. Molecules that can support optical activity are said to be *chiral*, and to possess *chirality* (meaning *handedness*, since the human hand is chiral).

The process of converting optically active materials into equal amounts of the enantiomers is called *racemization*. Ordinarily this process requires breaking of bonds to form a symmetrical (*achiral*) intermediate, and reforming the bonds to generate the racemic material. For special cases such as phosphines (IV) and sulfoxides (VI), in which one "substituent," is a nonbonding electron pair, configurational inversion may occur without breaking any bonds. For such molecules the tetrahedral enantiomers may interconvert through a metastable planar intermediate.

Stereoisomers that are not enantiomers are called *diastereoisomers*. Three classes may be distinguished: configurational, geometrical, and conformational isomers. Configurational diastereomers include molecules with more than one chiral center. Thus 2,3-dichlorobutane can exist in three configurationally

different forms, X-XII. Although forms X and XI are enantiomers, XII is a stereoisomer that is not a mirror image of X or XI. It is therefore termed a diastereomer of X and XI. Even though there are two asymmetric centers in XII, it is superimposable on its mirror image. The molecule is therefore achiral. The term *meso* is applied to molecules that contain chiral centers but are achiral as a whole. The molecules of nature frequently have many asymmetric centers. If a molecule has n centers, there can be 2^n stereoisomers, although this number may be reduced if some of the diastereoisomers are *meso* or if certain ring constraints are present. Glucose is one of the aldohexose sugars, which contain four chiral centers (disregarding the phenomenon known as anomerism). The naturally occurring *dextro*-glucose is enantiomeric to *levo*-glucose, and diastereomeric to the other fourteen isomers.

Geometrical isomers differ in the arrangement of groups about certain bonds, rather than about a chiral center. 2-Butene may exist in *cis* (XIII) or *trans* (XIV) forms. In the former case the methyl groups lie on the same side of the double bond, and in the latter on opposite sides. Chirality is not

important in geometrical isomerism. The isomers may be interconverted

if the double bond is broken to leave a residual single bond about which rotation may occur, followed by reformation of the double bond.

Geometrical isomerism may occur not only in alkenes (XIII, XIV), but also in oximes (XV), azo compounds (XVI), and many other doubly bonded systems. More importantly, cyclic

molecules exhibit this type of isomerism; the average plane of the ring serves as the reference. Molecule XVII is therefore named *cis*-1,2-dimethylcyclopropane, and XVIII is the *trans* isomer. Interconversion of XVII and XVIII would require breaking and reforming a ring bond.

Conformational isomers differ only in the arrangements of atoms obtainable by rotations about one or more single bonds. *meso*-2,3-Dichlorobutane can exist not only in the form XII, but also as the representation XIX. To take a simpler case,

n-butane may exist in two conformational forms, known as the *gauche* (XX) and the *anti* (XXI). These isomers may inter-convert by rotation about the C−C single bond, a process that requires an energy of only 3−5 kilocalories/mole. No bonds are broken, in contrast to the manner by which geometrical isomers are interconverted. In a more complicated but very common case, substituents on the six-membered cyclohexane ring may assume either the equatorial (XXII) or axial (XXIII) positions. These isomers may interconvert by *ring reversal*, Eq. (1),

which consists of a sequence of single-bond rotations. β-D-(+)-Glucose has the specific conformation of XXIV, in which all the substituents are equatorial.

Stereochemistry is one of the most important characteristics of a reaction mechanism. In the nucleophilic displacement of iodide ion on *sec*-butyl bromide, Eq. (2), the reaction is known to occur with inversion. If one

disregards the identity of the

halogen, then the starting material and product have opposite configurations. Thus the iodide ion must attack the C–Br bond from the backside, thereby effecting an inversion of the chiral center.

Important mechanistic consequences may also derive from geometrical isomerism. *cis*-3,4-Dimethylcyclobutene may ring-open to form either *cis, trans*- or *trans,trans*-2,4-hexadiene. The methyl groups may rotate away from each other (disrotation, Eq. (3)) to form the *trans, trans* isomer. Alternatively, they

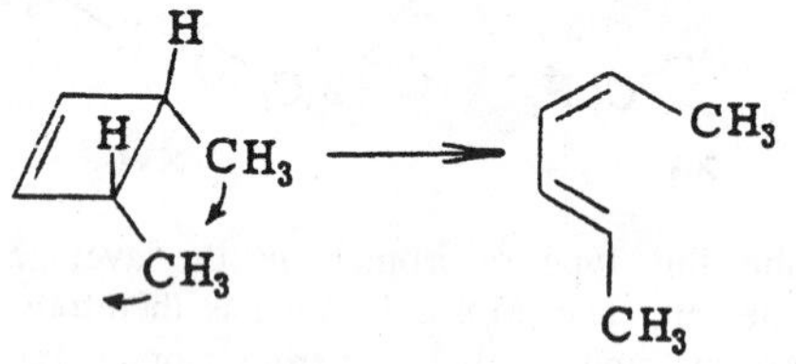

may rotate in the same direction (conrotation, Eq. (4)) to form the *cis,trans* isomer. An alternative disrotatory mode to form

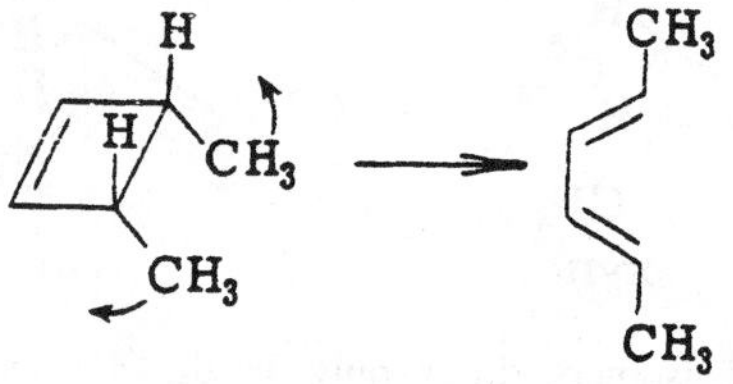

a *cis,cis* isomer by rotation of the methyl groups toward each other need not be considered because the methyl groups cannot pass by each other. Recent experimental and theoretical consideration of this problem has demonstrated that only the conrotatory mode is permitted for this ring opening. Thus *cis*-dimethylcyclobutene always forms only *cis,trans*-hexene.

JOSEPH B. LAMBERT
Northwestern University
Evanston, Illinois

Additional Reading

Eames, J., and J. M. Peach: *Stereochemistry at a Glance,* Blackwell Publishers, Malden, MA, 2002.

Green, M. M., E. W. Meijer, and R. J. M. Nolte: *Topics in Stereochemistry, Materials - Chirality: A Special Volume in the Topics in Stereochemistry Series,* John Wiley & Sons, Inc., New York, NY, 2003.

Mislow, K.: *Introduction to Stereochemistry,* Dover Publications, Inc., Mineola, NY, 2003.

Morris, D. G., and E. Abel: *Stereochemistry,* John Wiley & Sons, Inc., New York, NY, 2002.

Robinson, M. J.: *Organic Stereochemistry,* Oxford University Press, New York, NY, 2001.

STEREOISOMERISM. If one considers a molecular unit consisting of three unlike atoms A, B, and C which are connected (*bonded*) to each other to form a linear system, three arrangements are possible: A–B–C, A–C–B, and B–A–C. By employing the simple test of superimposability it is seen that these arrangements are nonidentical and are termed *structural isomers* (Gr. *isos*, equal). Structural isomers are thus chemical species that have the same molecular formula (the same number and types of atoms) but differ in the sequence in which the atoms are bonded. Of great significance is the fact that these isomers are separated by an energy barrier: in order to convert, for example, A–B–C into A–C–B an input of energy would be necessary to break the existing A–B and B–C bonds, followed by a rearrangement of the sequence of the atoms to yield A–C–B. The rearrangement process is termed an *isomerization* and the magnitude of the energy required (the barrier) for the conversion has important consequences regarding the number of arrangements that may exist for structural isomers. This will be discussed shortly.

Turning to a four-atom arrangement consisting of A_2B_2, four structural isomers are possible if linearity of the system is again assumed: A–A–B–B; A–B–A–B; A–B–B–A; B–A–A–B. Each arrangement can be characterized in terms of interatomic distances (A to B, B to B, and A to A) and such distances would be distinctive for each isomer. It is not mandatory, however, that any one of these A_2B_2 combinations exists in a linear form and the consequences of nonlinearity will be viewed. If (for the ABBA case) planarity of the unit still exists but the internuclear angles A–B–B are set at, say, 120°, two arrangements are apparent:

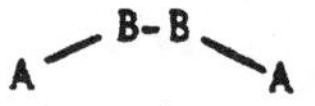

Arrangements **1** and **2** are clearly different forms of the same structural isomer and are called *stereoisomers* (Gr. *stereo*, solid or space). The relationship between **1** and **2** may be expressed in terms of the geometry of the molecule and hence these forms have also been called *geometrical isomers.* (Recalling the provisos set down for this system—planarity and angles—the difference between **1** and **2** is merely the location of one A with respect to the other in the unit.)

The statement made earlier about the energy requirement for the conversion of one structural isomer into another will now be examined. Conversion of **2** into **1** may be viewed in the simplest manner is involving a rotation of 180° about the B–B internuclear axis:

This operation involves no reshuffling of the atomic arrangement (or constitution) within the unit but does represent an isomerization. How readily such a conversion occurs then depends upon the size of the energy barrier to rotation. Looking at specific examples, the compound $CH_3-N=N-CH_3$, exists in two different and stable stereochemical arrangements (**3**, *cis* and **4**, *trans*) corresponding to **1** and **2** above:

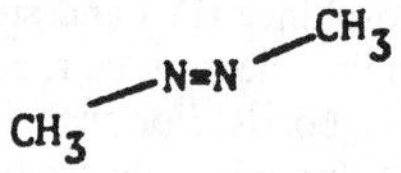

The C–N–N–C atoms are coplanar in each form and the interconversion of **3** and **4** requires a relatively high amount of energy. On the other hand, hydrogen peroxide (H_2O_2), which may be thought of as existing in similar *cis* and *trans* stereochemical arrangements (**5** and **6**), does *not* exhibit stereoisomerism

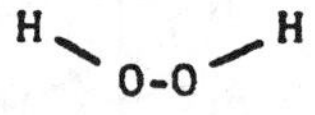

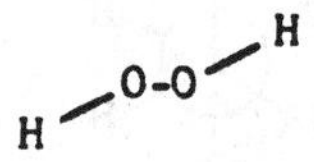

and only one isolable H_2O_2 is known. The difference in the two systems lies in the fact that rotation about the O–O single bond in H_2O_2 is relatively "free" (i.e., requires little energy) whereas the interconversion of **3** and **4** is a higher energy process that necessitates breaking a π bond before rotation may occur.

We now consider the case of four atoms (called *ligands*) that are attached to a center atom. If the general case consists of grouping A_{abcc}, square planar and tetrahedral structures, among others, may result:

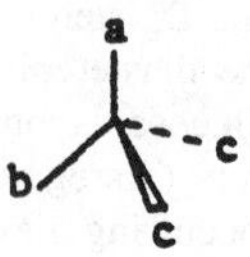

The square planar forms **7** and **8** are stereoisomers which are nonequivalent to the single nonplanar tetrahedral arrangement **9**. Now of grouping A_{abcd} is examined one predicts three square planar stereoisomers and the following *two* tetrahedral stereoisomers (**10** and **11**):

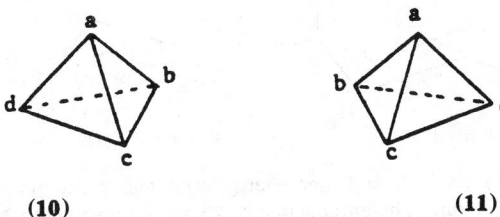

(10)   **(11)**

The relationship of **10** to **11** is that of the right hand to the left hand and these nonsuperimposable stereoisomers are mirror images or *enantiomers* (Gr. *enantio-*, opposite). Thus, *chiral* (Gr. *cheir*, hand) molecules are those which possess mirror images and arise when appropriate conditions of geometry and number and types of ligands are present in a system.

A molecule that has a mirror image is also said to be dissymmetric while one that does not (an achiral molecule) have an enantiomer is nondissymmetric. The classification of a given structure as dissymmetric or nondissymmetric is based upon the presence (or lack) of symmetry elements (axes, planes) in the structure.

It is important to note that in either **10** or **11** the magnitude of any internuclear angle (e.g., a—A—c), or any bond length (e.g., A—b), or the distance between any two ligands is exactly the same. This is not true in the case of the achiral square planar isomers that may be written for the A_{abcd} system.

The single most important physical property that differentiates enantiomers is their ability to rotate the plane of plane polarized light. This property is called *optical activity* and is displayed only by chiral molecules. Thus, stereoisomers which are also chiral are known as *optical isomers*. Chiral molecules that rotate polarized light in a clockwise fashion are termed *dextrorotatory* (*d*) while those that rotate the beam counterclockwise are *levorotatory* (*l*). Enantiomers have optical rotations of the same magnitude but of different signs (*d* or *l*).

The structures **10** and **11** denoted above contain a single chiral center **A**, the atom to which the ligands are attached. If two different such centers, A and B, are in a molecule the number of optical isomers is increased to four:

$$A\pm \qquad\qquad A\pm$$
$$B\pm$$

one chiral center   two different chiral centers

where + and − refer to the handedness or *configuration* at the chiral center. In the AB system, each optical isomer will have a mirror image whose configurations are opposite at the

$$A + \quad A- \qquad\qquad A+ \quad A-$$
$$B+ \quad B- \qquad\qquad B- \quad B+$$

mirror images   mirror images

chiral centers. The relationship of A + B+ or A − B− to A + B− (or A − B+) cannot be an enantiomeric one for the obvious reason that a given optical isomer may have only one mirror image. Instead, the relationship is said to be *diastereomeric* (Gr. *dia*, apart). Any given AB optical isomer will therefore have one enantiomer and two diastereomers.

We again examine a specific case. In 3-chloro-2-butanol, $CH_3CH-CHCH_3$

     | | , and A ± B ± situation—exists whose four isomers

     Cl   OH

(**12–15**) are shown in three dimensions so that the mirror image relationship of **12** to **13** and **14** to **15** is readily apparent.

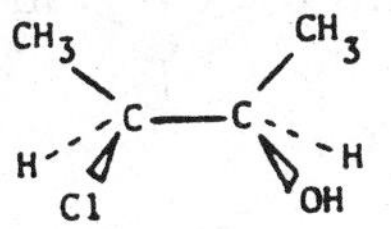

(The carbon atoms in the middle of the carbon chain are the chiral centers A and B.) Taking **12** (which is assumed to be the A + B+ combination) if one views down the bond axis of the chiral centers from the right a projection of the molecule results which shows the orientation of ligands to one another.

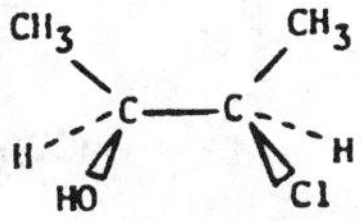

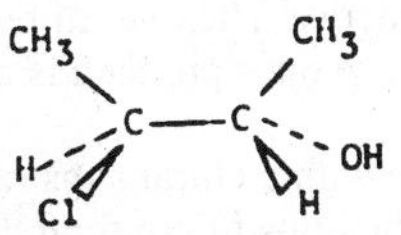

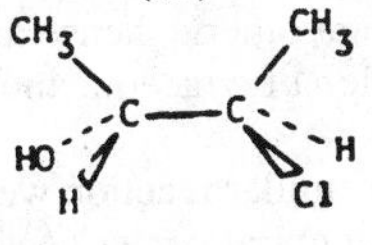

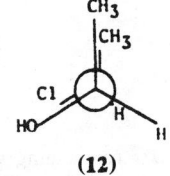

(12)

As noted earlier rotation may occur about single bonds in molecules and if a clockwise rotation of 180° is made about the bond axis of the chiral centers, a different form (*conformation*), **16**, results.

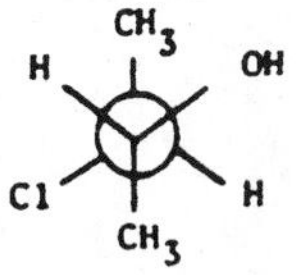

(16)

Conformations **16** and the *infinite number* of others that are obtained by rotation about the single bond in **12** are all nonidentical but the energy barrier separating them is small hence only one chiral compound having configurations ++ at the chiral centers may be isolated under normal conditions for **12**.

If two *identical* chiral centers are present in a molecular unit the number of stereoisomers is reduced to three: A + A+ and A − A− represent a pair of enantiomers but A + A− and A − A+ are identical arrangements. The +− form is said to be a *meso* or optically inactive diastereomer of the active forms A + A+ and A − A−.

The three tartaric acids may be used to illustrate the method employed in depicting three-dimensional molecules in two-dimension projections.

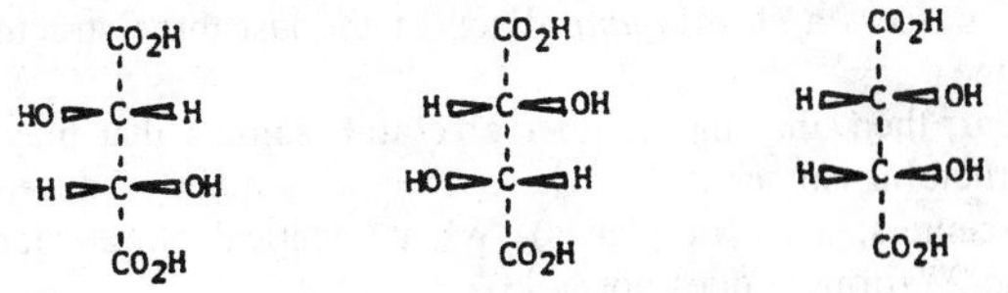

enantiomers   meso

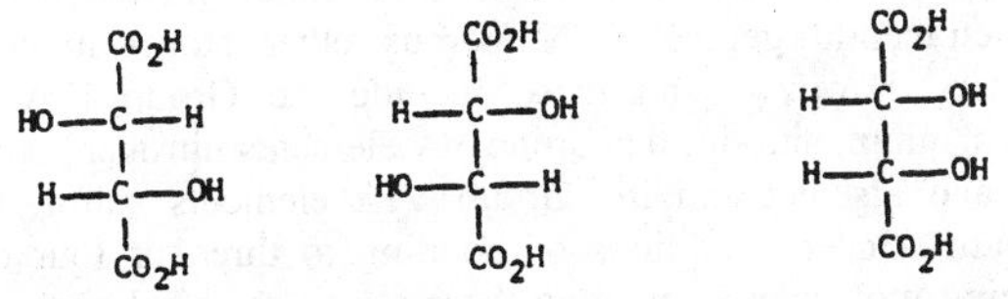

D   enantiomers   L   meso

The top formulas show the three-dimensional relation of the groups along the main carbon chain (dashed groups lie below the plane of the paper, bold above) while the bottom formulas correspond to projections in two dimensions obtained by lifting the dashed substituents into the plane of reference and pushing the bold groups down into the plane. In this process a unique projection is obtained for each three-dimensional molecule. The symbols D and L under the formulas for the enantiomeric tartaric acids are notations used to relate the configuration at the bottom chiral center to that of a standard compound, glyceraldehyde.

The presence of a chiral center is a sufficient, but not a necessary condition, for the existence of chirality in a molecule. For example, numerous biphenyl derivatives may exist in chiral pairs

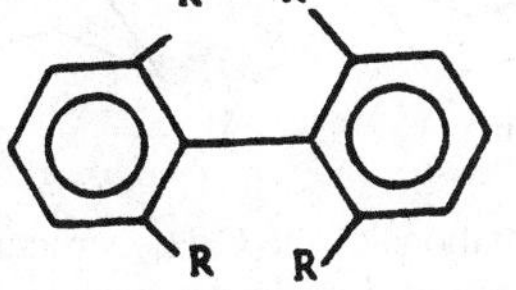

if the size of the R groups is large enough to restrict rotation about the single bond connecting the two rings. The restriction causes the rings to adopt a nonplanar orientation and raises the energy barrier to rotation about the connecting bond. No chiral center is present and the resulting enantiomers in the example following

are said to possess a *chiral axis* (coinciding with the connecting bond). Axial chirality is also found in the compounds known as allenes

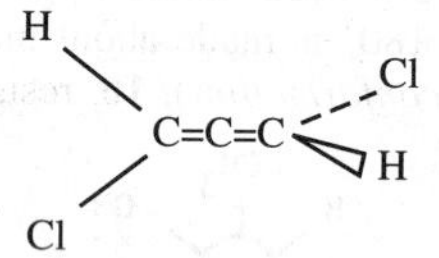

and in spiranes

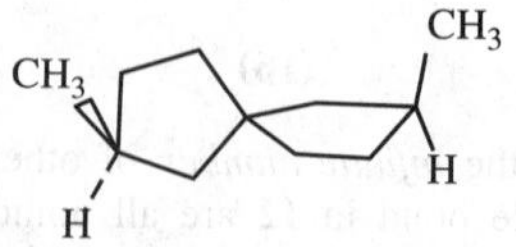

while chiral structures such as *trans*-cyclooctene

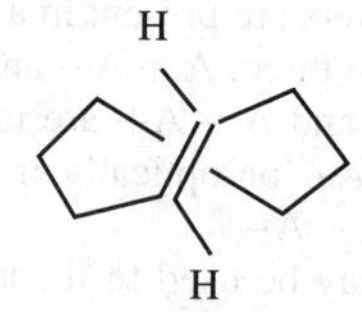

are said to possess a *chiral plane*. Each of the last three structures has a mirror image.

It is clear, then, that the number of chiral isomers that may exist for a given structural isomer is 2^n, where n = the number of different chiral elements (centers, axes, or planes). When identical chiral elements are present, the 2^n formula does not hold.

The examples used above to illustrate centers of chirality included carbon atoms which were attached to four unlike ligands, resulting in localized tetrahedral geometry. Numerous other atoms may serve as chiral centers, however, and these include the Group IVA elements silicon, germanium, and tin; the Group VA elements nitrogen, phosphorus, antimony, and arsenic; and the Group VIA elements sulfur, selenium, and tellurium. Under conditions of bonding to three or four dissimilar ligands, chiral molecules containing these atoms as chiral centers may be isolated. Also, the geometric form about the chiral center need not be tetrahedral, for octahedral complexes of the transition metals or their ions (Co^{3+}, Cr^{3+}, etc.) may be chiral when substituted by the proper number and type of ligands.

Conformations in Six-Membered Rings. The nonplanar ring compound, cyclohexane, which contains six contiguous $-CH_2-$ units, may exist in chair or boat forms (hydrogens are not shown):

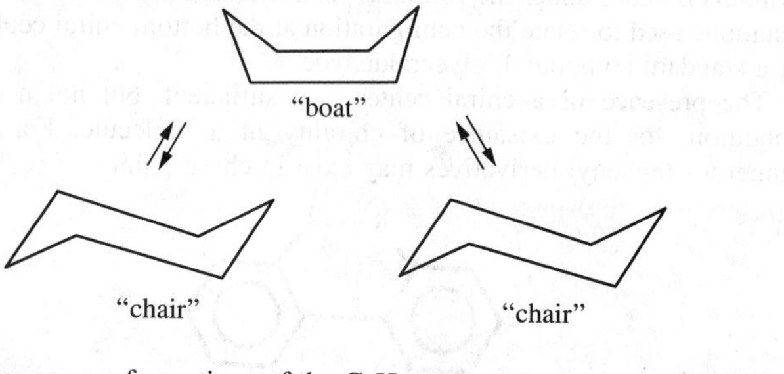

These forms are conformations of the C_6H_{12} structure and are separated by energy barriers which restrict, but do not prohibit, interconversion of the conformations. Substitution of a group on the ring yields a single nonchrial structure that exists in two main conformations, one with the substituent R

axial and the other *equatorial* to the main plane of the ring.

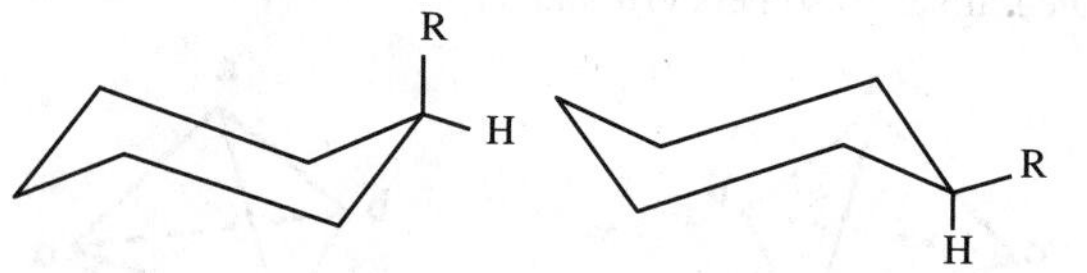

R *axial* (**17**)   R *equatorial* (**18**)

Conformation **18** is in a lower energy state and predominates in the equilibrium mixture. The introduction of a second substituent into the ring gives rise to three structural isomers (shown here in projection):

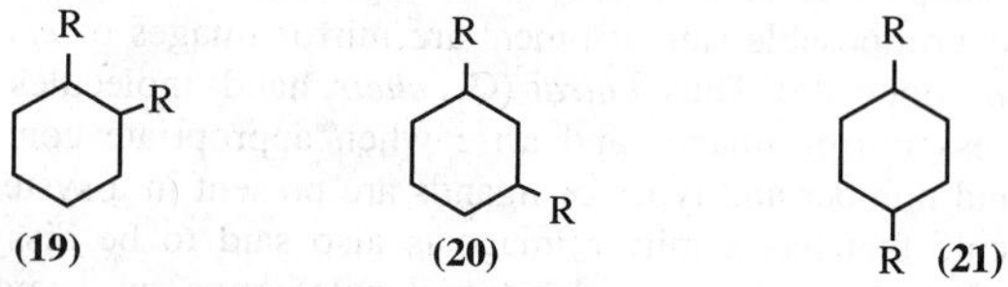

(**19**)   (**20**)   (**21**)

In **19** and **20**, both carbon atoms in the ring to which the R groups are bonded are identical chiral centers and hence one pair of enantiomers and one meso diastereomer exist for each structure. In **21** no chiral center is present but the isomers that are possible in this case (shown below in the preferred conformations)

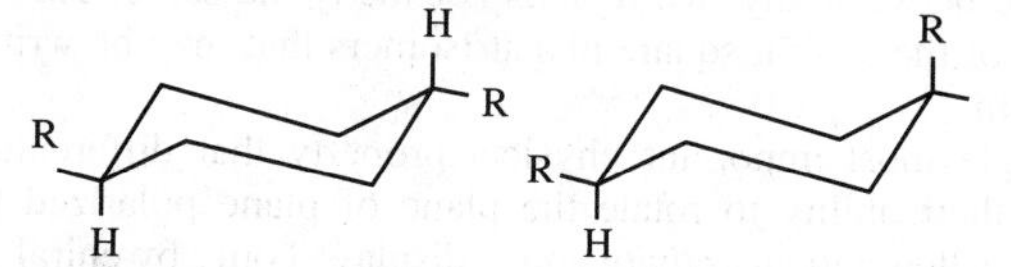

trans   *cis*

may be viewed as having a diastereomeric relationship.

Consequences of Molecular Chirality. A mixture containing an equal number of molecules of enantiomers is known as a *racemic modification*. The preparation and reactions of these modifications (as well as the individual enantiomers themselves) represent important aspects of the study of stereochemistry.

If a molecule
$$CH_3-\underset{\overset{\|}{O}}{C}-CH_2CH_3$$
is converted into

$$CH_3-\underset{\overset{|}{OH}}{CH}-CH_2CH_3$$
by some achiral reagent (e.g., hydrogen gas and a

catalyst) the chiral center in the product molecule may be considered as being generated by approach of a hydrogen from the top *or* bottom "face" of the planar $\underset{\overset{|}{C}}{C-C-O}$ grouping in reactant molecule

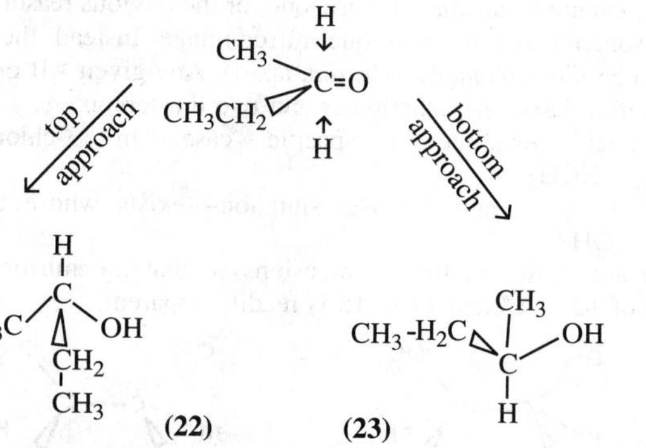

(**22**)   (**23**)

The molecule (**22**) produced by "top" approach is the enantiomer of that (**23**) resulting from "bottom" approach. In fact, the pathways leading to each are enantiomeric, hence are of equal energy. The overall result is thus the production of a racemic modification, since one approach is as probable as another.

Now if a similar reaction were conducted with a chiral substrate that has a preexisting chiral center (A+), the combinations of configurations at the centers in the product molecules would be A + B+ and A + B−. These are diastereomers, the pathways involving their information are diastereomeric (unequal energy!), and hence they are produced in unequal amounts. Such a case is illustrated as follows:

$$CH_3-\overset{\displaystyle H}{\underset{\displaystyle C=O}{\overset{|}{\underset{|}{C}}}}-OH \longrightarrow$$

$$H-\overset{CH_3}{\underset{\underset{CH_3}{|}}{\overset{|}{\underset{|}{\overset{\displaystyle C-OH}{C-OH}}}}} \quad + \quad HO-\overset{CH_3}{\underset{\underset{CH_3}{|}}{\overset{|}{\underset{|}{\overset{\displaystyle C-OH}{C-H}}}}}$$

diastereomers

Diastereomeric reaction pathways may be obtained in numerous other ways. An interesting case is represented by the biological reduction of acetaldehyde-1-D, $CH_3-\overset{\underset{\displaystyle D}{|}}{C}=O$. The product is the chiral structure

$CH_3-\overset{\underset{\displaystyle D}{|}}{C}H-OH$ and a racemic modification might be expected from the

reduction. In fact, the transfer of a hydrogen during the enzymatic reduction (an enzyme is a large chiral molecule) to one face of the acetaldehyde is diastereomeric with the transfer to the opposite face, hence (very) unequal amounts of the enantiomers are formed. These interactions may be described as $E+\ldots A+$ and $E+\ldots A-$, where $E+$ represents the chirality of the enzyme and $A+$ or $A-$ represent the incipient chirality in the reduced acetaldehyde molecules.

The last example reflects in a modest way the importance of the study of stereoisomerism. Biological conversions represent a glorious array of diastereomeric reactions and interactions: a given chiral amino acid is metabolized but its enantiomer is not; a certain complex drug (often a chiral molecule) alleviates pain but its enantiomer is inactive; and subtle changes in structure alter a given chiral compound's action completely in the human body.

ALEX T. ROWLAND
Gettysburg College
Gettysburg, Pennsylvania

STEREOREGULAR POLYMERS. The properties of natural and synthetic high polymers and their applications in plastics, fibers, elastomers, adhesives and coatings are determined in large part by (a) their average molecular weights, (b) the forces between the long chain molecules and (c) geometrical considerations, especially the degree of regularity of repeating units. Molecular weight characteristics and forces have received attention for many years. In contrast, the importance of stereoregularity vs. irregularity of chemical substituents became fully appreciated only around 1950. In general, strong interchain forces and regularity promote normal crystallinity, along with high strength, high softening temperatures, hardness and insolubility in common solvents.

Some polymers have regular structure free of diastereoisomerism because of the symmetry of the monomers, such as vinylidene chloride, $CH_2=CCl_2$ and isobutene $CH_2=C(CH_3)_2$. However, the term stereoregular polymer is generally reserved for stereoregular polymer structures derived from unsymmetrical monomers which can be obtained by special ionic methods of polymerization (usually from heterogeneous systems). These polymerization processes often using complex catalysts such as Ziegler-Natta activated transition metal catalysts, e.g., from AlR_3 and polymeric $TiCl_3$, have been called stereoregulated, stereospecific or oriented polymerizations. In 1964 the writer suggested the word *stereopolymerization* to describe those special ionic polymerization systems for treating unsymmetrical ethylenic monomers to obtain either normally crystalline polymers with DDDDD regularity, permanently amorphous polymers with irregular DLDDL sequences or intermediate structures according to conditions chosen. Such control of polymer stereoisomerism was achieved first with vinyl alkyl ethers, but the first stereoregular polymers to become the basis of a major industry were the stereoregular or isotactic propylene polymers (heat-resistant molding plastics and fibers).

Crystallinity has been one of the principal effects by which stereoregularity or tacticity has been studied in polymers. However, as expected, not all stereoregular polymers are equally crystallizable. Differences in chemical reactivity, nuclear magnetic resonance and infrared have given useful information about stereoregularity. However, for comparing polymers from a given monomer x-ray diffraction and solubility data are most reliable for estimating tacticity.

Interest in controlling steric configurations and stereoisomerism in polymers by polymerization conditions developed only slowly. Staudinger and Schwalbach in 1931 suggested that invariable low crystallinity in polyvinyl acetate might be caused by diastereoisomerism, that is randomness in D and L positions of the acetate groups along the chain molecules[1]. Branching and deviations from head to tail addition were studied mean-while as types of isomerism. However not until 1948 were examples of stereoregulated polymerizations of a vinyl-type monomer disclosed by Schildknecht and coworkers. In both early types of stereopolymerizations vinyl isobutyl ether diluted by liquid propane could be treated at low temperatures. Addition of gaseous boron fluoride gave very rapid polymerizations to rubber-like substantially amorphous high polymers. Careful addition of cold boron fluoride etherate, immiscible with liquid propane at $-78°C$ or above produced a slow growth or proliferous polymerization to form normally crystalline polymers.

This suggested that it might be possible to prepare normally crystalline polymers from other unsymmetrical monomers of the type $CH_2=CHY$ in ionic heterogeneous systems.

The discovery of stereopolymerizations of 1-alkenes by use of Ziegler-Natta catalysts in heterogeneous systems in 1954, and subsequent studies of polymer structure, attracted world-wide attention to this field. Propylene and 1-butene, which are monoallylic compounds, had not been homopolymerized by conventional ionic or free radical conditions to give linear high polymers suitable as plastics or other synthetic materials. Short branches in polyethylenes had been shown to reduce crystallinity and hardness. Natta and coworkers demonstrated the ability of catalysts such as those from reaction of aluminum alkyls with titanium halides to form normally crystalline, surprisingly high softening polymers from propylene. A helix of three monomer units explained the regularity required for crystallization and the identity period observed from x-ray diffraction. Stereoregular isotactic polymers of crystal melting ranges shown in Table 1 were prepared by slow heterogeneous ionic polymerizations using special catalysts at moderate pressures and temperatures.

Stereoregular propylene polymer plastics have outstanding utility, for example, heat resistance superior to that of polyethylenes in sterilizable hospital devices. By copolymerization and control of the degree of stereoregularity brittleness at low temperatures can be avoided. Stereoregular 1-butene and isobutylethylene polymers are also manufactured, but the isotactic polymers from styrene and from methyl methacrylate are too brittle for much use. Crystallizable polystyrenes also have been prepared by heterogeneous anionic polymerizations (Lewis basic catalysts) and crystalline methyl methacrylate polymers can be prepared using Grignard catalysts.

Soluble catalysts derived from organoaluminum compounds and vanadium halides promote formation of atactic elastomeric propylene polymers and copolymers such as ethylene-propylenediene terpolymer rubbers (EPDM). Amorphous adhesive propylene homopolymers have some commercial use. Syndiotactic or DLDLD propylene polymers have been reported but their structures and properties have not been completely established.

Isomeric isoprene polymers are formed biologically as natural rubber (cis-1,4) and balata or gutta-percha (largely *trans*-1,4). Modified Ziegler-Natta type catalysts, colloidal lithium or lithium alkyls (in absence of ethers) were found to give predominantly cis-1,4 polymer rubbers from isoprene. *Cis*-1,4-polybutadiene, so-called synthetic natural rubbers, have become important in tires and in graft copolymerization with styrene for high-impact plastics. Different conditions of polymerization give rigid *trans*-1,4-diene polymers resembling balata.

TABLE 1. CRYSTAL MELTING RANGES OF SOME STEREOREGULAR POLYMERS

Isotactic stereopolymers	Melting range	Stereopolymers	Approximate maximum melting point
Propylene	165–176°C	Isotactic 1,2-butadiene	120°C
1-Butylene	120–136	Syndiotactic 1,2-butadiene	154
1-Amylene	60–70	*Cis*-1,4-butadiene	+1
Isopropylethylene	300–310	*Trans*-1,4-butadiene	148
Isobutylethylene	235–250	*Cis*-1,4-isoprene	22
Isoamylethylene	about 110	*Trans*-1,4-isoprene	65
4,4-Dimethyl-1-pentene	>380	*Cis*-1,4-(2,3-dimethyl butadiene)	190
Styrene	230–250	*Trans*-1,4-(2,3-dimethyl butadiene)	260
Isobutyl vinyl ether	100–130		
Methyl methacrylate	160		

Although precise mechanisms of the stereopolymerizations are yet uncertain, several characteristics become evident. The reactions are predominantly heterogeneous, ionic reactions at low or moderate temperatures. The more stereoregular polymers grow as a separate phase upon the solid or immiscible liquid catalyst. However, some monomers such as vinyl isobutyl ether can form somewhat isotactic polymer fractions even from homogeneous solutions of Lewis acid and monomer. In contrast, the polymers from vinyl isopropyl ether, which apparently are stereoregular when obtained by slow growth polymerization using boron fluoride etherate catalysts, nevertheless do not crystallize readily. BF_3 can be used to form isotactic vinyl isobutyl ether polymers if it is applied in a separate phase of methylene chloride immiscible with liquid propane. Relatively polar solvents which favor separation of gegen ions from their growing macroions generally impair stereospecificity.

Although the Ziegler-Natta catalyst systems for stereopolymerization of 1-olefins were regarded by Natta, Mark, and others as examples of anionic polymerizations, the writer considers them as a special type of cationic polymerization. A consistent system relating monomer structure to response to catalyst types is only possible if propylene containing an electron repelling methyl group attached to the ethylene nucleus polymerizes with Lewis acid catalysts (cationic polymerization). Propylene as an allyl compound lacks sufficient electron withdrawal from the ethylene group to homopolymerize by free radical initiation (peroxide, azo catalysts or ultraviolet light) and it also lacks sufficient electron donation (as in isobutene) for homopolymerization by conventional cationic system. Ziegler-Natta catalysts have been observed to homopolymerize some other monoallyl compounds. An intensive study of the literature by the writer and Mabel D. Reiner showed no well-characterized homopolymers of high molecular weight obtained from monoallyl compounds by free radical or conventional ionic catalyst systems.

Transition metal catalysts for polymerization of 1-alkenes similar to those of Ziegler were developed in DuPont laboratories and have been called coordination catalysts.

Outside of vinyl addition polymerizations some crystallizable stereoregular polymers also have been prepared. An example is isotactic polymer from propylene oxide

$$-CH\ CH_2O-$$
$$|$$
$$CH_3$$

made by using ferric chloride complex catalysts for proliferous type reactions.

After the demonstrations of preparation of stereoregular polymers having novel properties by means of special ionic methods, the possibilities of free radical methods were examined extensively. It must be concluded that in free radical systems the structures of homopolymers and copolymers can be little influenced by specific catalysts and other reaction conditions, but are determined largely by monomer structure. This is consistent with the relative uniformity of comonomer reactivity ratios in radical copolymerizations. However, it has been found possible to obtain somewhat more syndiotactic structure, DLDL, than normally obtained by radical reactions, at low temperatures and by selecting solvents. Examples are polyvinyl chlorides of higher than usual crystallinity from polymerizations at low temperature e.g., −50°C under ultraviolet light.

Although they do not crystallize, polyvinyl acetates prepared at low temperatures apparently are more syndiotactic since they yield more than usually crystalline polyvinyl alcohols by saponification. Monomers of high polarity such as vinyl trifluoracetate by radical polymerization can form relatively syndiotactic polymers from which more crystalline polyvinyl alcohols can be prepared by saponification.

C. E. SCHILDKNECHT
Gettysburg College
Gettysburg, Pennsylvania

STERLING SILVER. Silver alloy, usually with copper, containing at least 92.5% silver.

STEROIDS. Steroids are members of a large class of lipid compounds called terpenes that are biogenically derived from the same parent compound, isoprene, C_5H_8. Steroids contain or are derived from the perhydro-1,2-cyclopentenophenanthrene ring system (**1**) and are found in a variety of different marine, terrestrial, and synthetic sources. The vast diversity of the natural and synthetic members of this class depends on variations in side-chain substitution (primarily at C17), degree of unsaturation, degree and nature of oxidation, and the stereochemical relationships at the ring junctions.

(1)

There are many classes of natural and synthetic steroids best known for their wide array of biological activity. The naturally occurring steroids can be subdivided into several categories that include (*1*) nonhormonal, mammalian steroids; (*2*) vitamin D; (*3*) hormonal steroids; and (*4*) other naturally occurring steroids. See also **Hormones**; **Vitamin**; and **Vitamin D**.

Structure and Nomenclature

The position-numbering and ring-lettering conventions for steroids are shown in (**1**). Positions 18 and 19 are often angular methyl groups; in addition, position 19 can be a hydrogen and is not substituted when the A-ring is aromatic. Position 17 can be substituted, unsubstituted, and/or oxygenated. Compounds are systematically named as derivatives of the parent hydrocarbons shown in Figure 1. Substituents that extend below the plane of the steroid are referred to as α and are designated by a broken line; those attached to the plane of the steroid from above are called β and are shown by a bold or solid line. Substituents of unknown configuration are indicated by a wavy line. Generally, the ring junctions have an all-trans relationship with the hydrogen attached to C9 on the α-face, unless otherwise indicated. Changes in steroid nomenclature that have been introduced since 1972 include a wider use of the (*R*), (*S*)-system for designating the stereochemistry in the side chain. Although the systematic nomenclature for steroids has been firmly established, the most common and most important steroids are often designated by trivial names.

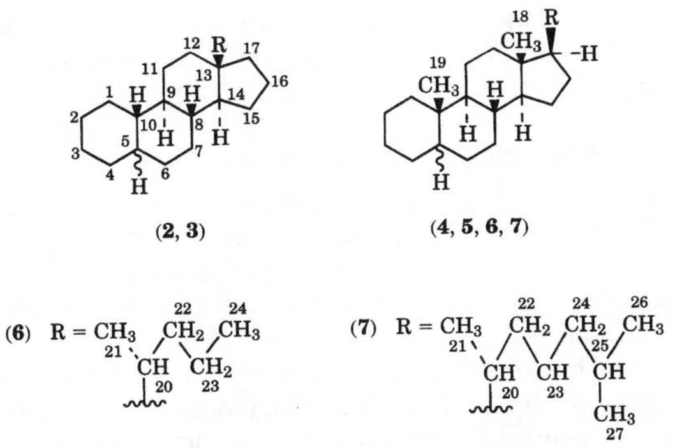

Fig. 1. Nomenclature of the parent hydrocarbon ring skeletons. Gonane (**2**) R = H; estrane (**3**) R = CH₃; androstane (**4**) R = H; pregnane (**5**) R = C₂H₅; cholane (**6**) R as shown; and cholestane (**7**) R as shown.

Classification of Biologically Active, Natural Steroids

Nonhormonal Mammalian Steroids

Sterols and Cholesterol. Natural sterols are crystalline C_{26}–C_{30} steroid alcohols containing an aliphatic side chain at C17. Sterols were first isolated as nonsaponifiable fractions of lipids from various plant and animal sources and have been identified in almost all types of living organisms. By far, the most common sterol in vertebrates is cholesterol (**8**). Cholesterol serves two principal functions in mammals. First, cholesterol plays a role in the structure and function of biological membranes. Secondly, cholesterol serves as a central intermediate in the biosynthesis of many biologically active steroids, including bile acids, corticosteroids, and sex hormones.

Bile Acids and Alcohols. Bile acids have been detected in all vertebrates that have been examined and are a result of cholesterol metabolism. The C_{24} acid, 5β-cholanic acid (**9**) is the structural derivative of the majority of bile acids in vertebrates. Most mammalian bile acids have a cis-fused A–B ring junction resulting in a nonplanar steroid nucleus. Bile acids, like sterols, typically contain a C3α-hydroxyl group (lithocholic acid: 3α-hyroxycholanic acid.

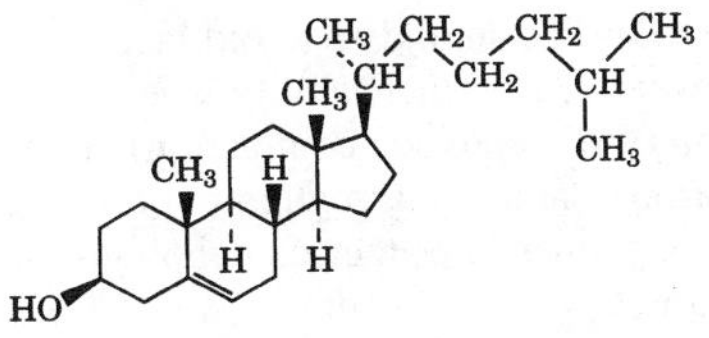

(**8**)

Along with the C3α-hydroxyl group, bile acids may contain a hydroxyl at C7α, at C12α, and at other positions. Bile salts, cholesterol, phospholipids, and other minor components are secreted by the liver.

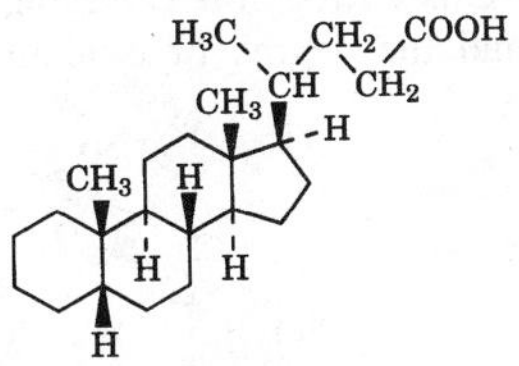

(**9**)

Vitamin D. The term vitamin D refers to a group of seco-steroids that possess a common conjugated triene system of double bonds. Vitamin D₃ (**10a**) and vitamin D₂ (**10b**) are the best-known examples (Fig. 2). Vitamin D₃ (**10a**) is found primarily in vertebrates, whereas vitamin D₂ (**10b**) is found primarily in plants. The term *vitamin* is a misnomer. Vitamin D₃ is a prohormone that is converted into physiologically active form, primarily 1,25-dihydroxyvitamin D₃ (**11**), by successive hydroxylations in the liver and kidney. This active form is part of a hormonal system that regulates calcium and phosphate metabolism in the target tissues.

Steroid Hormones. Generally, steroid hormones are metabolically short-lived steroids produced in small amounts by various endocrine glands. They serve as chemical messengers that regulate a variety of physiological and metabolic activities in vertebrates. Steroid hormones bind to soluble, intracellular receptor molecules. In the nucleus of the cell, dimeric steroid receptors bind to DNA and together with a heteromeric complex of proteins, regulate gene transcription. Molecules that interfere with steroid hormone gene regulation are called antagonists or antihormones. Steroid hormones can be subdivided into sex hormones (androgens, estrogens, and progestins) and corticosteroids (glucocorticoids and mineralocorticoids).

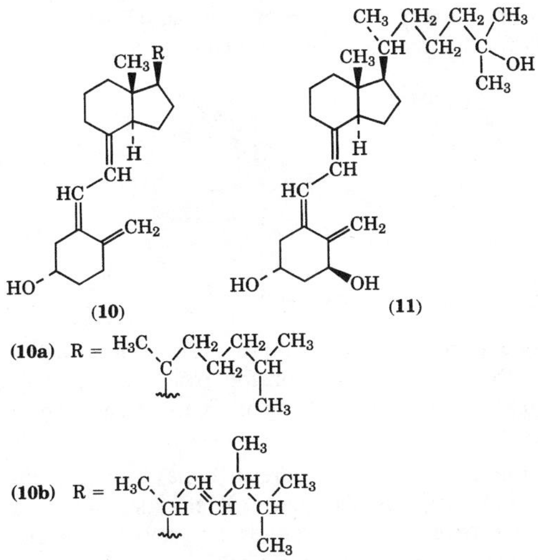

Fig. 2. Vitamin D: prohormones (**10**), and active hormone (**11**)

Sex Hormones. Androgens, estrogens, and progestins are steroids that are secreted primarily by the genital glands. From a chemical point of view, the division of the sex hormones into these three groups is convenient; however, they may possess common physiological properties. Therefore, the sex hormones are organ-specific rather than sex-specific.

Androgens are C_{19} steroids that contain the basic perhydro-1,2-cyclopentenophenanthrene ring system with the C18 and C19 angular methyl group. A primary function of androgens is to maintain the male sex organs and secondary sex characteristics. Examples of androgens are testosterone (**12**) and dehydroepiandrosterone (DHEA) (**13**). DHEA is one of the most abundant steroids in human males; however, it is not a potent androgen.

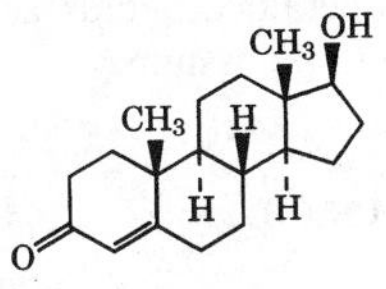

(**12**)

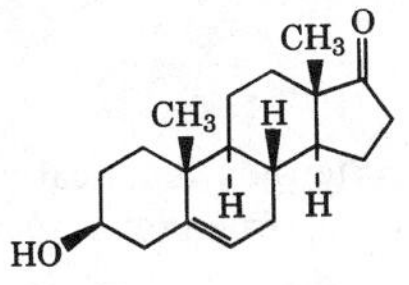

(**13**)

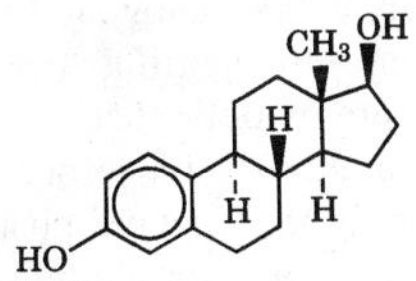

(**14**)

Estrogens. Estrogens are characterized by having an aromatic A-ring and thus having a phenolic character. Estrogens stimulate the growth and development of the female reproductive organs and the secondary sex characteristics. Another primary function of estrogens along with progesterone, is to regulate the ovulatory cycle. Estrogens, as with all the steroid hormones, are important for healthy growth and development in women and men. The main production site of estradiol (**14**) is the female ovary; however, small amounts of estrogens are produced in testes and the adrenal cortex. Synthetic and natural estrogens play an important role in the treatment of osteoporosis in post-menopausal women. Antiestrogens are important for the treatment of breast cancer.

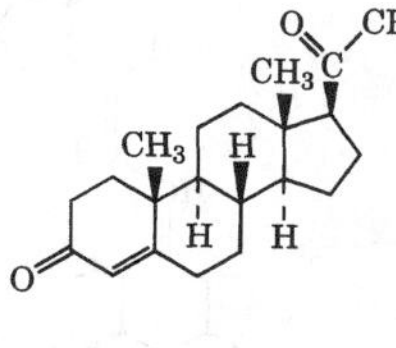

(15)

Progesterone (**15**), the principal progestin in mammals, is secreted primarily by the corpus luteum of the ovary. A main responsibility of progesterone, together with estrogen, is to prepare the endometrium for pregnancy.

Corticosteroids. Although the adrenal cortex secretes small amounts of androgens and estrogens, the major secretory steroids from this gland are called corticosteroids. Corticosteroids have several biological activities, including the regulation of electrolyte balance by mineralocorticoids and carbohydrate and protein metabolism by glucocorticoids.

Natural, potent glucocorticoids possess a Δ^4-3-one group, an oxygen substituent at C11β (necessary for agonism), and a C17β-2-hydroxyethan-1-one sidechain. Atypical example is cortisol (**16**). The principal effects of glucocorticoids are to mobilize fat and protein from tissues, utilize these nutrients to supply energy for the body, and decrease the rate of carbohydrate utilization. Thus, they are diabetogenic and act as functional insulin antagonists. Glucocorticoids are also potent inhibitors of inflammation, and they are used as therapeutics.

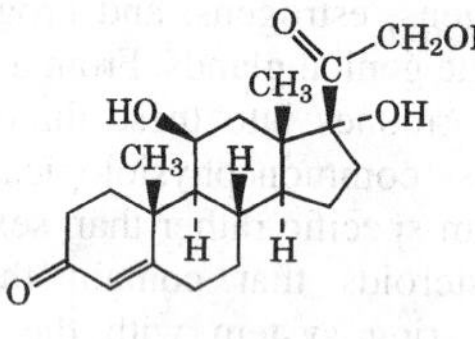

(16)

Aldosterone (**17**), the most potent natural mineralocorticoid, also possesses a Δ^4-3-one group, an oxygen substituent at C11β, and a C17β-2-hydroxyethan-1-one side chain. In addition, the C18 of aldosterone is oxidized to an aldehyde. Mineralocorticoids act to retain sodium and to prevent the retention of excess potassium.

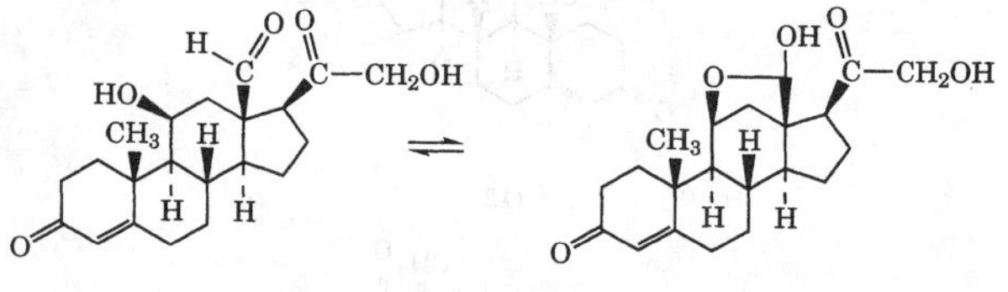

(17)

Other Natural Steroids. Steroids are nearly ubiquitous to all living organisms and have a variety of structural variations.

Sapogenins and Saponins. Steroids isolated from a variety of plant sources that contain a spiroketal between hydroxyl moieties at C16 and C26 and a carbonyl at C22 are called sapogenins (**18**). Sapogenin aglycones have been an important source of starting materials for the commercial steroid industry. Saponins are widely distributed in plants and marine organisms and consist of a steroid or terpene skeleton attached to a saccharide. Because of diversity in structure, pharmacology, and biological activities, saponins have been studied for a number of different commercial applications. Saponins have been used as detergents, foaming agents, and fish toxins. Although toxic to fish, saponins are nontoxic when ingested by humans. Another commercial application of saponins is in food flavoring.

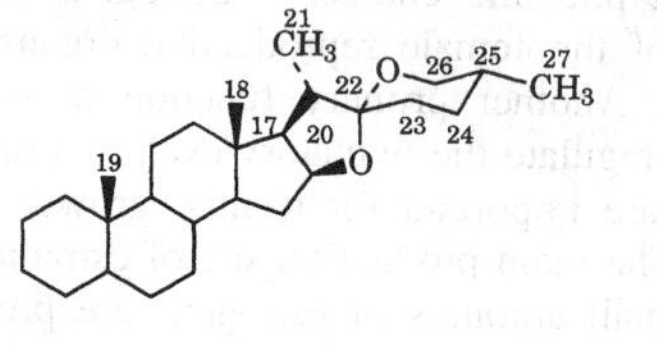

(18)

Plant Sterols. Sterols have been identified in almost all types of living organisms and can be isolated, in varying quantities, from many different plants. Similar to cholesterol, plant sterols have a structural and functional role in biological systems and serve as intermediates in the biosynthesis of an assortment of biologically active steroids.

Steroid Alkaloids. Steroid alkaloids are compounds isolated from plants and some higher animals that possess the basic steroidal skeleton with nitrogen(s) incorporated as an integral part of the molecule. The nitrogen can be located within the perhydro-1,2-cyclopentenophenanthrene ring system or in a side chain.

Steroid alkaloids have been isolated from four families of terrestrial plant sources (*Solanaceae, Liliaceae, Apocynaceae,* and *Buxaceae*), two animal sources (*Salamandra* and *Phyllobates*), and several marine sources. Steroid alkaloids can be classified based on structure and fall into a variety of categories. The spirosolanes contain a C_{27} cholestane skeleton with a C_{20} spiroaminoketal moiety. Solanidine-type steroidal alkaloids are a small subclass. The largest subclass of steroidal alkaloids is the secosoline bases; (**19**) is a general secosolanidine. The pregnane-type alkaloids have one or more nitrogens attached to a pregnane skeleton. The buxus alkaloids, isolated from evergreen shrubs, contain carbon substitution at C4 and C14 and either a cyclopropane moiety between C9, C10, and C19 or the B-ring expanded diene. Buxus alkaloids have been used as folk remedies for a variety of disorders, including venereal disease, tuberculosis, cancer, and malaria. The samanine, jerveratrum, and ceveratrum-type compounds all have a structurally altered C_{27} steroid skeleton. The samanine alkaloids have an expanded A-ring with the formation of an isoxazoline ring system and a cis-A–B ring junction. Ritterazines and cephalostatins are among steroid alkaloids recently isolated from marine sources. When assayed *in vitro*, cephalostatins are among the most potent cytotoxins ever screened by the National Cancer Institute.

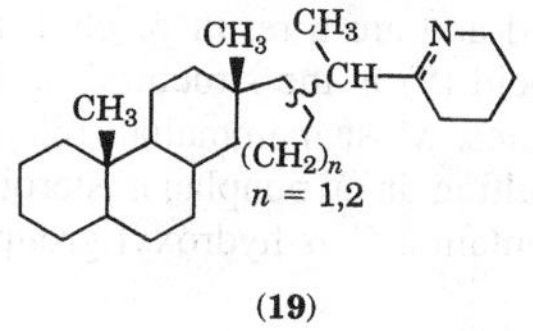

(19)

Cardiac Steroids. Cardiac steroids (steroid lactones) and corresponding glycosides are characterized by their ability to exert a powerful inotropic (increasing the force of cardiac contraction) effect, and are used both for their inotropic and antiarrhythmic properties. The two most prevalent cardiac aglycones are the cardenolides and bufadienolides. The cardenolides are C_{23} steroids that have a C17β-substituted five-membered lactone that is generally α,β-unsaturated, an unusual β-faced oxygen on C14, and a bile acid-like cis-A–B ring junction. Cardenolides are exemplified by digitoxigenin (**20**) which is an active ingredient in digitalis. The bufadienolides differ in that they are C_{24}-steroids that possess a C17β-substituted six-membered lactone ring that generally has two degrees of unsaturation. Other structural variations in both series are the stereochemistry at C3 and the degree of oxidation on the nucleus and side chains.

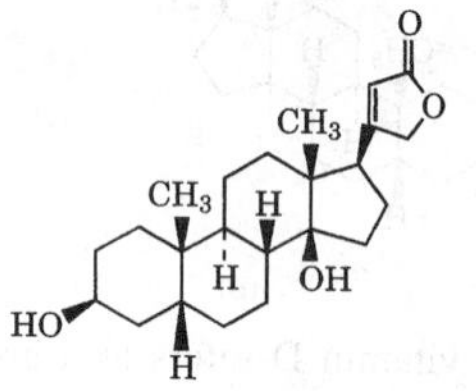

(20)

Withanolides. Withanolides are C_{28}-steroidal lactones that are isolated from the Solanaceae plant family. Withanolides are characterized by an ergostane-type skeleton, the C17-side chain of which is transformed into a six-member lactone ring. The withanolides and the related ergostanes are the only known natural steroids obtained from the same family that have representatives with both α- and β-orientations of the C17 side chain.

Ecdysteroids. Ecdysteroids can be isolated from many species of the animal kingdom that belong to the phyla Protomia, e.g., insects, worms, and arthropods, as well as a variety of different plant species. Ecdysteroids include the molting hormones; however, not all the over 60 ecdysteroids that have been isolated are active hormones. Ecdysteroids from animals are referred to as zooecdysteroids and from plants are referred to as phytoecdysteroids.

Marine Sterols. Several hundred unique sterol structures have been elucidated from a variety of marine invertebrates. A single nucleus can be used to describe most terrestrial sterols, but no single template suffices for marine sterols. Similar to cholesterol, marine sterols play a critical role in both the physiology and biochemistry of biological systems.

Steroid Antibiotics. The steroid antibiotics are a structurally diverse class of steroids that have a common biological function, i.e., antibacterial, antifungal, antiviral, or antitumor activities. This group of compounds can overlap with other steroid classes listed above. Fusidic acid, helvolic acid, and cephalosporin P_1 (**21**) exemplify a set of antibacterial steroids that contain a prolanostane skeleton with an unique trans–syn–trans–antitrans stereochemistry. These compounds inhibit the growth of gram-positive bacteria by inhibiting protein synthesis, but have little activity against gram-negative bacteria. An antibiotic isolated from the tissues of the dogfish shark is the steroid alkaloid squalamine, a broad-spectrum antibiotic that exhibits potent antimicrobial activity against fungi, protozoa, viruses, and both gram-negative and gram-positive bacteria.

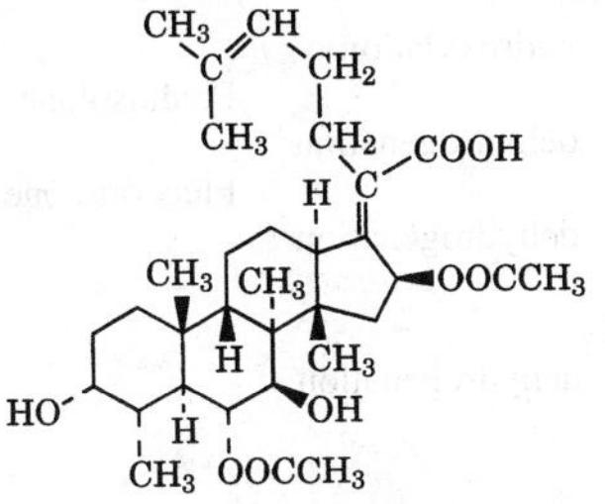

(21)

Biosynthesis

Steroids are members of a large class of lipid compounds called terpenes. Using acetate as a starting material, a variety of organisms produce terpenes by essentially the same biosynthetic scheme (Fig. 3). The self-condensation of two molecules of acetyl coenzyme A (CoA) forms acetoacetyl CoA. Condensation of acetoacetyl CoA with a third molecule of acetyl CoA, then followed by an NADPH-mediated reduction of the thioester moiety produces mevalonic acid (**22**). Phosphorylation of (**22**) followed by concomitant decarboxylation and dehydration processes

produce isopentenyl pyrophosphate. Isopentenyl pyrophosphate isomerase establishes an equilibrium between isopentenyl pyrophosphate and 3,3-dimethylallyl pyrophosphate (**23**). The head-to-tail addition of these isoprene units forms geranyl pyrophosphate. The addition of another isopentenyl pyrophosphate unit results in the sesquiterpene (C_{15}) farnesyl pyrophosphate (**24**). Both of these head-to-tail additions are catalyzed by prenyl transferase. Squalene synthetase catalyzes the head-to-head addition of two achiral molecules of farnesyl pyrophosphate, through a chiral cyclopropane intermediate, to form the achiral triterpene, squalene (**25**).

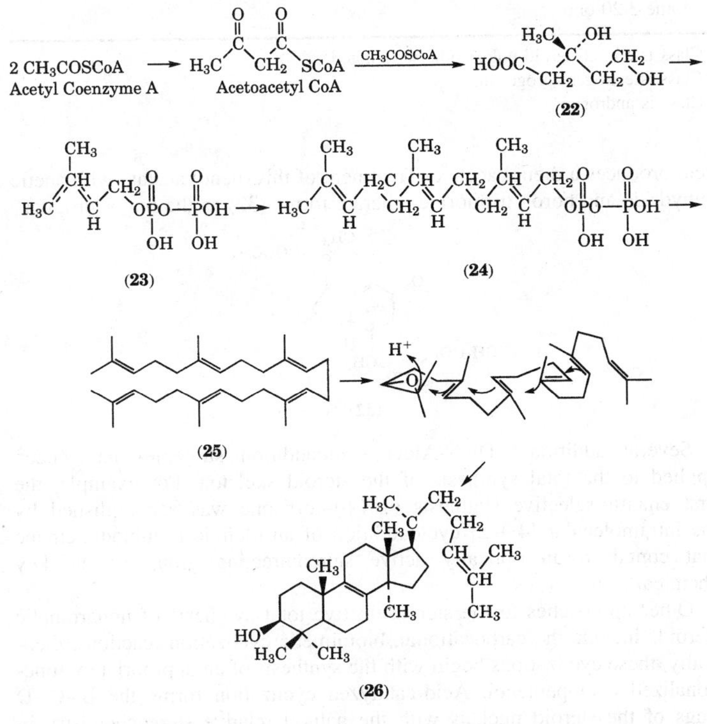

Stereospecific 2,3-epoxidation of squalene, followed by a non-concerted carbocationic cyclization and a series of carbocationic rearrangements, forms lanosterol (**26**) in the first steps dedicated solely toward steroid synthesis. Cholesterol is the principal starting material for steroid hormone biosynthesis in animals. The cholesterol biosynthetic pathway is composed of at least 30 enzymatic reactions. Lanosterol and squalene appear to be normal constituents, in trace amounts, in tissues that are actively synthesizing cholesterol.

Manufacture and Synthesis

There are three general processes for steroid production: (*1*) direct isolation from natural sources, (*2*) partial synthesis from steroid raw materials that have been isolated from plants and animals, and (*3*) total synthesis from nonsteroidal starting materials.

Direct Isolation. The two most important classes of steroid pharmaceuticals that are isolated directly from natural products are some estrogens and most cardiac steroids. Compounds with estrogenic activity have been isolated from different sources, including urine from pregnant women and from pregnant mares. Cardiac steroids occur in small amounts in various plants with a wide geographical distribution.

Partial Syntheses.
Raw Materials and Extraction. The variety of natural sources of steroid raw materials is vast, and the exact details of manufacturing processes are ambiguous closely held industrial secrets. However, the most widely utilized raw materials for the partial synthesis of steroids appear to be the following: (*1*) the sapogenins, for example, diosgenin (**27**), (*2*) the structurally related steroid alkaloids, (*3*) sterols, such as cholesterol (**8**), and (*4*) bile acids.

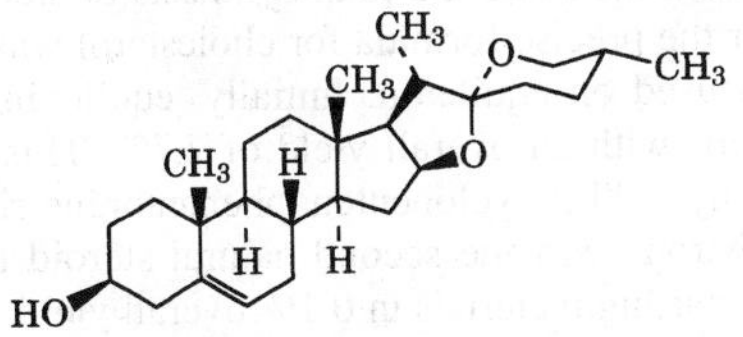

(27)

Plants of the genus *Dioscorea* are the most common source of diosgenin. This genus occurs abundantly in tropical and subtropical regions throughout the world.

Owing to periodic fluctuations in the price of diosgenin, alternative raw materials such as solasodine have been used for the synthesis of steroid drugs. In the U.S., the plant sterols stigmasterol and β-sitosterol are a significant raw material for the synthesis of antiinflammatory glucocorticoids and other steroid hormones.

Methods of Partial Synthesis. Partial syntheses are done typically by chemical degradation or fermentation/biotransformation.

An important commercial method for the commercial synthesis of steroids is the chemical degradation of diosgenin. The Marker degradation became the principal method for commercial steroid synthesis in the 1940s and 1950s, and modifications of this process are still in use. When diosgenin is heated to approximately 200°C in acetic anhydride, elimination and acetylation of the oxygen in the F-ring produce the bis-acetylated enol ether. Oxidative cleavage of the enol ether with chromium trioxide followed by elimination of the C16-acyl-oxygen results in steroid. Selective hydrogenation of the α, β-unsaturated ketone in the D-ring from the sterically less hindered α-face forms pregnenolone (**28**). Pregnenolone is readily converted into progesterone (**15**) under oxidative conditions.

This process was improved and expanded to provide starting materials for the C19-sex hormones that include estrogens and androgens. Another commercial method that has been used for the production of progesterone

2 CH$_3$COSCoA   Acetyl Coenzyme A   Acetoacetyl CoA   (22)

(23)   (24)

(25)

(26)

Fig. 3.   Abbreviated terpene biosynthesis

is the chemical degradation of the side chain of stigmasterol.

Fermentation/Biotransformation. Commercial biotechnology operations have focused on microbial agents for specific transformations of individual steroid substrates. The regio- and stereoselective hydroxylation of every site on virtually every steroid nucleus is possible. Many of these hydroxylation steps are of commercial importance. For example, the 9α-, 11α-, 11β-, and 16α-hydroxylations are key steps in the industrial synthesis of synthetic corticosteroid antiinflammatory drugs. These steps are accomplished almost exclusively by microbial transformations. In addition to hydroxylations, other useful microbial oxidations of steroids include alcohol oxidations, epoxidations, oxidative cleavage of carbon–carbon bonds, introduction of double bonds, peroxidations, and heteroatom oxidations. Other invaluable microbial steroid transformations include reductions, degradations, A-ring aromatization, resolutions, isomerizations, conjugations, hydrolyses, heteroatom introduction, and sequential reactions.

There are two principal biotechnological applications dealing with steroids. Microbial agents are used for processing raw materials into useful intermediates for general steroid production and for specific transformations of steroids to advanced intermediates or finished products (Table 1).

Processing Raw Materials. Along with the aforementioned chemical methods of processing steroid raw materials, microbial transformations have been and are used in a number of commercial degradation processes. The microbial degradation of the C17 side chain of the two most common sterols, cholesterol (**8**) and β-sitosterol, is a principal commercial method for the preparation of starting materials in Japan and the U.S.

Representative Partial Syntheses. The synthesis of 19-nor-steroids was stimulated by the development of orally active progestins as birth control agents.

Total Synthesis.

Estranes. Investigations into the total synthesis of steroids began in the 1930s shortly after the precise formula for cholesterol was established. The earliest studies focused on equilenin. Initially, equilenin was synthesized in 20 chemical steps with an overall yield of 2.7%. This synthesis helped to confirm the perhydro-1,2-cyclopentenophenanthrene ring system of the steroid nucleus. Estrone was the second natural steroid to be synthesized from nonsteroidal starting materials in 0.1% overall yield in 18 steps. Since these original processes, a vast number of total syntheses of aromatic A-ring steroids have appeared.

An asymmetric synthesis of estrone begins with an asymmetric Michael addition of lithium enolate (**29**) to the scalemic sulfoxide (**30**). Direct treatment of the crude Michael adduct with *meta*-chloroperbenzoic acid to oxidize the sulfoxide to a sulfone, followed by reductive removal of the bromine affords (**31**) X = α and βH; R = H in over 90% yield.

The most recent, and probably most elegant, process for the asymmetric synthesis of (+)-estrone applies a tandem Claisen rearrangement and intramolecular ene-reaction. Most 19-norsteroid contraceptive agents are produced by total synthesis from nonsteroidal starting materials.

Androstanes and Pregnanes. The first total syntheses of nonaromatic steroids that contain the C19-angular methyl substituent were accomplished in the early 1950s. These syntheses all began with starting materials containing a two-ring system. A more recent ring annulation strategy for the total synthesis of steroids begins with the formation of the C–D-ring system as a suitably functionalized indane. Condensation of the pyrrolidine enamine of cyclopentanone with ene-one results in the bicyclic keto-ester in 60–70% yield. Treatment of the latter with isopropenyl acetate and sulfuric

TABLE 1. COMMERCIAL MICROBIAL TRANSFORMATIONS USED TO PRODUCE ADVANCED INTERMEDIATES OR FINISHED PRODUCTS

Substrate[a]	Transformation	Product	Organism
progesterone	11α-hydroxylation oxidation/lactonization		*Rhizopus nigricans* *Cylindrocarpon radicicola*
11-deoxycortisol (17α-derivatives)	11β-hydroxylation	cortisol/derivatives	*Curvularia lunata*
6α-fluoro-16α-methyl-21-hydroxypregn-4-ene-3,20-dione	11β-hydroxylation	Paramethasone	*Curvularia lunata*
11-deoxy-16-methylene-cortisol	11β-hydroxylation	Prednylidene	*Curvularia lunata*
9α-fluorohydro-cortisone	1-dehydrogenation 16α-hydroxylation	Triamcinolone	*Arthrobacter simplex*
hydrocortisone	1-dehydrogenation	Prednisolone	*Arthrobacter simplex* or *Bacillus lentus*
6α-fluoro-16α-methyl corticosterone	1-dehydrogenation	Fluocortolone	
11β,21-dihydroxy-pregna-4,17(20)-dien-3-one	1-dehydrogenation		*Septomyxa affinis*
rac-3-methoxy-8,14-secoestra-1,3,5-(10),9(11)-tetra-ene-14,17-dione (Secosteroid)[b]	17-ketone reduction		*Saccharomyces uvarum*
androst-4-ene-3,17-dione[c]	17-ketone reduction		*Saccharomyces* sp.
21-acetoxy-17α-hydroxypregne-nolone	Δ^5-3β-alcohol dehydrogenase		*Flavobacterium dehydrogenans*
6α-fluoro-21-hydroxy-16α-methyl-pregn-4-ene-3,20-one	9α-hydroxylation		*Curvularia lunata*

[a] Class is corticosteroid unless otherwise noted.
[b] Class is estrogen–progestin.
[c] Class is androgen.

acid produces a dienol acetate. Treatment of this dienol acetate with acetic anhydride and boron trifluoride etherate forms (**32**) as the major product.

Several additional Diels-Alder cycloaddition strategies have been applied to the total synthesis of the steroid skeleton. For example, the first enantio-selective synthesis of (+)-cortisone was accomplished by the intramolecular [4 + 2] cycloaddition of an olefinic *o*-quinodimethane that contained an optically active stereodirecting group as the key chemical step.

Other approaches to the stereoselective total synthesis of nonaromatic steroids include the carbocationic, biomimetic cyclization reactions. Generally, these cyclizations begin with the synthesis of an appropriately functionalized cyclopentenol. Acid-catalyzed cyclization forms the B–C–D rings of the steroid nucleus with the natural relative stereochemistry in a single step.

Removable cation-stabilizing auxiliaries have been investigated for polyene cyclizations. For example, a silyl-assisted carbocation cyclization has been used in an efficient total synthesis of lanosterol. Other conditions for the cyclization of polyenes and of ene-ynes to steroids have been investigated. Oxidative free-radical cyclizations of polyenes produce steroid nuclei with exquisite stereocontrol. Besides the aforementioned A-ring aromatic steroids and contraceptive agents, partial synthesis from steroid raw materials has also accounted for the vast majority of industrial-scale steroid synthesis.

An interesting breakthrough in steroid endocrinology occurred with the discovery of a novel class of steroid antihormones. Several 11β-substituted 19-norsteroids display potent antiprogestinal activity. For example, RU-486 (**33**) is marketed in Europe as a contragestive agent. The synthesis of RU-486 demonstrates a unique method for functionalization of the 11β-position of a steroid nucleus.

(33)

Uses: Therapeutics and Toxicology

Steroid Hormones

Sex Hormones. The largest economic impact of synthetic estrogen and progestin production has been for use as contraceptive agents and for treatment and prevention of osteoporosis. Mixtures of estrogens and progestins have been used as contraceptive agents since the early 1960s. The principal mode of steroid contraceptive action is exerted at the hypothalamic–pituitary–ovarian and uterine sites. Thus, contraceptive steroid mixtures have been used to treat a variety of related abnormal states including endometriosis, dysmenorrhea, hirsutism, polycystic ovarian disease, dysfunctional uterine bleeding, benign breast disease, and ovarian cyst suppression.

Estrogens are routinely prescribed to post-menopausal women to prevent the development and exacerbation of osteoporosis, because it can increase bone density and reduce fractures. Estradiol (**14**) or conjugated estrogens are typical agents used for the prevention and treatment of osteoporosis.

Antiprogestins, such as RU-486 (17β-hydroxy-11β-(4-dimethylamino-phenyl-1)-17α-(prop-1-ynyl)-estra-4,9-diene-3-one) (**33**) and ZK98299 11β-(4-dimethylaminophenyl)-17α-hydroxy-17β-(3-hydroxypropyl-13α-methyl-4,9-gonadien-3-one) represent a new class of drugs for fertility regulation. Also, these drugs have potential applications in the treatment of uterine cancer.

During the 1960s and 1970s a wide range of estrogens and antiestrogens were synthesized primarily to study reproductive endocrinology. The focus of clinical applications of many of these antiestrogens has shifted to breast cancer therapy. Although structurally different, these antiestrogens bind to the estrogen receptor in the breast cancer cell and exert a profound influence on cell replication.

Testosterone, alkylated testosterone, or testosterone esters are the primary anabolic–androgenic steroid drugs. The medicinal uses for these drugs include treatment of certain types of anemias, hereditary angioedema, certain gynecological conditions, protein anabolism, certain allergic reactions, and use in replacement therapy in gonadal failure states. However, anabolic–androgenic steroids are best known for their nonmedical, and illegal, use to aid in body-building or to increase skeletal muscle size, strength, and endurance.

Corticosteroids. The greatest portion of steroid drug production is aimed at the synthesis of glucocorticoids, which are highly effective agents for the treatment of chronic inflammation. Glucocorticoids exert their effects by binding to the cytoplasmic glucocorticoid receptor within the target cell and thus either increase or decrease transcription of a number of genes involved in the inflammatory process. Glucocorticoids are used to treat a variety of different diseases that are exacerbated by inflammation, such as arthritis, asthma, rhinitis, and skin irritations.

Corticosteroids are the most efficacious treatment available for the long-term treatment of asthma, and inhaled corticosteroids are considered to be a first-line therapy for asthma. Rhinitis is characterized by nasal stuffiness with partial or full obstruction, and itching of the nose, eyes, palate,

or pharynx, sneezing, and rhinorrhoea. If left untreated it can lead to more serious respiratory diseases such as sinusitis or asthma. Nasal spray topical corticosteroids are widely regarded as the reference standard in rhinitis therapy.

Other Therapeutics Steroids.

Saponins. Synthetic steroids that are structurally related to saponins have been shown to lower plasma cholesterol in a variety of different species.

Heterocyclic Steroids. Steroid 5α-reductase (types 1 and 2) converts testosterone (**12**) to the physiologically more potent androgen dihydrotestosterone (DHT) (**34**). The type 1 isoform occurs in nongenital skin, whereas the type 2 isoform is the predominant form in the prostate (the type 1 isoform is present in a lesser extent) and genital skin fibroblasts. There has been much interest in developing inhibitors of steroid 5α-reductase as a therapy for a variety of disorders associated with elevated levels of DHT, including benign prostatic hyperplasia (BPH), some prostatic cancers, certain skin disorders, and male pattern baldness.

(34)

Analytical Methods

The field of steroid analysis includes identification of steroids in biological samples, analysis of pharmaceutical formulations, and elucidation of steroid structures. Many different analytical methods, such as ultraviolet (uv) spectroscopy, infrared (ir) spectroscopy, nuclear magnetic resonance (nmr) spectroscopy, x-ray crystallography, and mass spectroscopy, are used for steroid analysis.

Generally, the most powerful method for structural elucidation of steroids is nuclear magnetic resonance (nmr) spectroscopy. A definitive method for structural determination is x-ray crystallography. Extensive x-ray crystal structure determinations have been done on a wide variety of steroids. In addition, other analytical methods for steroid quantification or structure determination include, mass spectrometry, polarography, fluorimetry, radioimmunoassay, and various chromatographic techniques.

BRADLEY P. MORGAN
MELINDA S. MOYNIHAN
Pfizer, Inc.

Additional Reading

Briggs, M.H. and J. Brotherton: *Steroid Biochemistry and Pharmacology,* Academic Press, London, UK, 1970.

Connolly, S.: *Steroids,* Heinemann Library, Woburn, MA, 2000.

Duax, W.L. and D.A. Norton: *Atlas of Steroid Structure,* IFI/Plenum Data, New York, NY, 1975.

Fieser, L.F. and M. Fieser: *Steroids,* Reinhold Publishing Corp., New York, NY, 1959.

Genazzani, A.R., F. Petraglia, and R.H. Purdy: *The Brain: Source and Target for Sex Steroid Hormones,* CRC Press, LLC, Boca Raton, FL, 1996.

Handa, R.J., S. Hayashi, E. Terasawa, and M. Kawata: *Neuroplasticity, Development, and Steroid Hormone Action,* CRC Press, LLC, Boca Raton, FL, 2001.

Heftmann, E.: *Steroid Biochemistry,* Academic Press, Inc., New York, NY, 1970.

Karch, S.B.: *The Pathology of Drug Abuse,* 2nd Edition, CRC Press, LLC., Boca Raton, FL, 1996.

Kirk, D.N., B. Hill, H.L. Makin, and G.M. Murphy: *Dictionary of Steroids: Chemical Data Structure,* CRC Press, LLC, Boca Raton, FL, 1999.

Lukas, S.E.: *Steroids,* Enslow Publishers, Inc., Berkeley Heights, NJ, 2001.

Milne G.W.A.: *Ashgate Handbook of Endocrine Agents and Steroids,* Ashgate Publishing Company, Brookfield, VT, 2000.

Monroe, J.: *Steroid Drug Dangers,* Enslow Publishers, Inc., Berkeley Heights, NJ, 1999.

Veldhuis, J.D. and A. Giustina: *Sex-Steroid Interactions with Growth Hormone,* Springer-Verlag, Inc., New York, NY, 1999.

Web References

AboutSteroids.com: http://www.aboutsteroids.com/
AmericanAcademy of Pediatrics: Steroids: http://www.aap.org/family/steroids.htm
Anabolic Steroid Abuse: http://www.steroidabuse.org/
SteroidsInfo.com: http://www.steroidsinfo.com/

STEVENS REARRANGEMENT.   Migration of an alkyl group from a quaternary ammonium salt to an adjacent carbanionic center on treatment with strong base. The product is a rearranged tertiary amine, sulfonium, or sulfide.

STIBNITE.   The mineral stibnite, antimony sulfide, Sb_2S_3, is found in radiated groups of acicular orthorhombic crystals or in other sorts of aggregates, as well as blades, also as columnar or granular masses. It shows a highly perfect pinacoidal cleavage; conchoidal fracture; hardness, 2; specific gravity, 4.63–4.66; luster, metallic and very brilliant on cleavage faces or freshly fractured surfaces. Its color is a steely gray; the streak very similar in color, may be covered with a black, sometimes iridescent tarnish.

Stibnite is the most common antimony mineral known and is the chief ore of that metal. It is a primary ore mineral and occurs with other antimony minerals and galena, sphalerite, and silver ores. It is found in Germany, Rumania, the Balkans, Italy, Borneo, Peru, Japan, China, Mexico; and in the United States in California and Nevada.

The name stibnite is derived from the Latin word for antimony, *stibium*.

STILBITE.   The mineral stilbite, $NaCa_2(Al_5Si_{13})O_{36} \cdot 14H_2O$, is a zeolite, the compound monoclinic crystals of which are usually grouped in approximately parallel positions, forming sheaf-like aggregates, which have a soft pearly luster, whence the name stilbite from the Greek, meaning luster. The less commonly used term desmine is likewise from the Greek, meaning a bundle. Stilbite has one perfect cleavage; uneven fracture; is brittle; hardness, 3.5–4; specific gravity, 2–2.2; luster, vitreous to pearly; color, usually white but may be brownish, yellowish, red or pink. Its streak is white, and it is transparent to translucent. Like the other zeolites stilbite occurs in cavities in basalts and traps, rarely in granites and gneisses. Of the many localities may be mentioned Trentino, Italy; the Harz Mountains; Valais, Switzerland; Arendal, Norway; the Ghats Mountains of India; and Mexico. The Triassic traps of New Jersey and Pennsylvania furnish specimens as do also rocks of the same age in Nova Scotia. This mineral sometimes is called *desmine*.

STOICHIOMETRY.   The mathematics of chemical reactions and processes. It relates to all the quantitative aspects of chemical changes, both mass and energy. Stoichiometry is based on the absolute laws of conversion of mass and of energy and on the chemical law of combining weights. This basis makes stoichiometry as exact as any other branch of mathematics.

The law of conservation of mass dictates that, regardless of the nature of the changes undergone in a physical or chemical process, the total mass of all the materials in the system remains the same, even though the physical states and chemical compositions of the materials may change. Likewise, the law of conservation of energy is based upon the fact that the total energy in a reacting system remains constant even though the level or form of the energy may change. In radioactive transformations, however, a slight correction must be applied to the law of the conservation of mass. Mass and energy have been found to be interconvertible, so that in general the total energy of the system remains constant even though there may be small mass changes.

The above concepts form the basis for weight and heat balance calculations. Such calculations are of great significance in engineering practice for the purpose of evaluating performance of existing operations or designing new manufacturing facilities and equipment.

The basic laws of conservation specifically state that matter or energy in a given system cannot be created or destroyed, and accordingly this requires that the following equality holds true:

$$Input = output + accumulation$$

For continuous, steady flow systems the change in in-process inventory is zero during any interval of time. In this case, therefore, the above expression reduces to the simplified form of input = output.

In making material weight balances the above relation may be applied to a single unit of the operation, or to the over-all operation with reference to the separate elements and/or the total mass entering and leaving the system. This method of analysis can best be exemplified by means of a synthetic problem: Let it be assumed that consideration is being given to a continuous, steady flow system, to which X pounds of material is fed per minute and from which Y pounds of useful product are analyzed to contain $a\%$ and $b\%$ by weight of a certain constituent, respectively. It is desired to determine the extent of unmeasured loss, Z pounds per minute, incurred

from the system and the average concentration, c, of the said constituent in the waste stream.

First, it may be written that input = output. By dividing each side of this equality by an element of time, this relationship can then be transformed into the following expression:

$$Rate\ of\ input = rate\ of\ output$$

Then, a total weight balance can be written to express this statement of equality in terms of the quantities specified in the problem:

$$X = Y + Z \qquad (1)$$

A similar balance may also be written in terms of the constituent in question:

$$\frac{a}{100}X = \frac{b}{100}Y + \frac{c}{100}Z \qquad (2)$$

Finally, by algebraic solution of the two simultaneous equations it follows that

$$Z = X - Y \qquad (3)$$

and

$$c = \frac{aX - bY}{X - Y} \qquad (4)$$

The above example is only a simple illustration of a weight balance. Similarly, the reaction between elements and compounds may be symbolically expressed to portray the principle of conservation of matter. For example, if hydrogen is completely burned to water, the reaction between it and oxygen can be represented as follows:

$$(Hydrogen) + (Oxygen) \longrightarrow (Water)$$

$$H_2 + \tfrac{1}{2}O_2 \longrightarrow H_2O$$

$$(2.02\ Wgt.\ units) + (16.00\ Wgt.\ units) \longrightarrow (18.02\ Wgt.\ units)$$

It would be found that these materials would always react in the same relative proportions to form water in an amount equal to the total weight of reactants. The relative weights indicated are equal to the molecular weights of the materials in question. Even if a reaction does not go to completion, the quantities which did react would be proportional to the combining weights expressed in the balanced chemical equation.

Since the element of time is usually involved as the basis of a stoichiometric calculation, proper quantitative deductions often depend on adequate knowledge of other laws or principles, such as those governing rates of reaction and those pertaining to chemical equilibria. When materials in the gaseous state are involved, the general gas laws are of great utility.

Another independent relation for a system is obtainable by applying the law of conservation of energy, which requires that energy input equals energy output. A valid equality of this type must include all forms of energy such as potential energy, kinetic energy, internal energy, flow work, electrical energy, etc. This type of equality results in the so-called "total energy balance." Another very useful but similar expression is the Bernoulli mechanical energy balance for steady mass flow of fluids. However, heat energy is very frequently the only primary effect in a process so that, in such cases, the total energy balance can be simplified to the very advantageous expression of heat input equals heat output. This constitutes the basis for heat balances which, together with weight balances, are the most useful tools in any stoichiometric calculations.

Since chemical reactions involve combination of atoms or molecules to form new compounds or decomposition of compounds to form simpler ones, it is most convenient in stoichiometric calculations to employ molecular units rather than weight units. This particular kind of unit is called a "mole" and represents the quantity of substance numerically equal to its molecular weight. This weight quantity may be based on any system of weight units desired, and it is thus necessary to designate this basis by referring to pound moles, gram moles, etc.

A particular chemical reaction may be written to embody both laws of conservation of mass and energy as demonstrated below:

$$FeS + \tfrac{7}{4}O_2 \longrightarrow \tfrac{1}{2}Fe_2O_3 + SO_2 + 268,000\ Btu$$

This equation states that 1 pound mole of ferrous sulfide reacts with 7/4 moles of oxygen to form $\tfrac{1}{2}$ mole of ferric oxide and 1 mole of sulfur dioxide, accompanied by a release of heat amounting to 268,000 Btu.

However, to assign a specific meaning to the numerical value for this heat release, it is customary to specify a reference temperature and pressure for the reaction, these being 25°C and 1 atmosphere in the example cited. In making heat balance calculations, it is then convenient to choose these conditions as the datum level and then calculate the heat input and heat output quantities above or below the reference state.

WALTER C. LAPPLE
Alliance, Ohio

STOKE'S LAW.

(1) The rate at which a spherical particle will rise or fall when suspended in a liquid medium varies as the square of its radius; the density of the particle and the density and viscosity of the liquid are essential factors. Stoke's law is used in determining sedimentation of solids, creaming rate of fat particles in milk, etc. (2) In atomic processes, the wavelength of fluorescent radiation is always longer than that of the exciting radiation.

STORAGE BATTERY.

A secondary battery so called because the conversion of chemical to electrical energy is reversible and the battery is thus rechargeable. An automobile battery usually consists of 12–17 cells with plates (electrodes) made of sponge lead (negative plate or anode) and lead dioxide (positive plate or cathode) that is in the form of a paste. The electrolyte is sulfuric acid. The chemical reaction that yields electric current is $Pb + PbO_2 + 2H_2SO_4 \leftrightarrow 2PbSO_4 + 2H_2O + 2e$. More complicated and expensive types have nickel-iron, nickel-cadmium, silver-zinc, and silver-cadmium as electrode materials. A sodium-liquid sulfur battery for high-temperature operation as well as a chlorine-zinc type using titanium electrodes have also been developed.

As part of the U.S. effort to replace gasoline with another form of energy, DOE is supporting short-and long-term research on batteries for electric vehicles at Argonne National Laboratories. These types intended to deliver 20–30 kwh are in the short-term program: improved lead-acid, nickel-zinc and nickel-iron. The long-term program includes lithium-metal sulfide, sodium-sulfur (β-battery), zinc-chlorine, and metal-air. Independent research indicates that a zinc-nickel oxide system has encouraging possibilities. A lead-acid battery for storing energy from solar cells has been reported to have a life of 5–7 years.

STRAIN THEORY.

A theory first proposed by von Baeyer to explain the relative stability of various carbon compounds. It may be stated in the form: The regular tetrahedral-symmetric position is the most stable of all possible positions of neighboring carbon compounds; variations from this position produces increased energy content, and hence strain. Since the angle at the vertex of a regular tetrahedron is 109° 28; this theory ascribes minimum strain to cyclopentane, of the polymethylenes. The theory is borne out by the lesser stability of cyclobutane and cyclopropane, but not to the degree that might be expected by the stability of some of the higher-membered rings. In that case, the lesser strain is often due to a spatial or three-dimensional structure.

Extensions of the strain theory have been made, with varying success, to other hydrocarbon ring structures, saturated and unsaturated, to ring compounds in which the hydrogen atoms have been variously substituted, and to rings containing atoms other than carbon, as well as to bicyclic and polycyclic systems.

STREAMING (Molecular).

Application of kinetic theory to the flow of gas through a tube at low pressures, such that the mean free path is large compared with the diameter of the tube. In this case, the streaming of the gas is due to the random motion of the molecules, and to the density gradient down the tube, so that the numbers of molecules traversing a given cross section in opposite directions is different. For a tube of circular cross section, the mass flowing per second is proportional to the pressure difference and the cube of the radius.

STRIPPING COLUMN.   See Distillation.

STROMATOLITE.

A term that has been generally applied to variously shaped (often domal), laminated, calcareous sedimentary structures formed in a shallow-water environment under the influence of a mat or assemblage of sediment-binding blue-green algae that trap fine (silty) detritus and precipitate calcium carbonate and that commonly develop colonies or irregular accumulations of a constant shape, but with little or no microstructure. It has a variety of gross forms, from near-horizontal to markedly convex, columnar, and subspherical. Stromatolites were originally considered animal fossils, and although they are still regarded as fossils because they are the products of organic growth, they are not fossils of any specific organism, but rather consist of associations of different genera and species of organisms that can no longer be recognized and named or that are without organic structures. An excellent treatise on stromatolites is *Stromatolites* (M.R. Walter, editor), Elsevier, New York, 1976.

STRONG INTERACTION.   See Particles (Subatomic).

STRONTIANITE.

The mineral strontianite is strontium carbonate, $SrCO_3$, usually occurring in whitish-yellow or whitish-green masses of radiated acicular crystals, or in fibrous or granular form. When distinctly crystallized it is obviously orthorhombic, but such crystals are rare. It has a nearly perfect prismatic cleavage; uneven fracture; brittle; hardness, 3.5; specific gravity, 3.785; luster, vitreous; color, as above, also green, gray and colorless; streak, white; transparent to translucent. Strontianite occurs in veins, chiefly in limestones, occasionally in the crystalline rocks, and it is usually associated with calcite and celestite. It is found in the metalliferous veins in the Harz Mountains and Saxony. It is commercially important in Westphalia where it is mined for use in the beet sugar industry. In the United States, crystalline masses and gorges of strontianite are found in Schoharie County, New York, long a famous locality for this mineral.

STRONTIUM.

[CAS: 7440-24-6], Chemical element, symbol Sr, at. no. 38, at. wt. 87.62, periodic table group 2, mp 769°C, bp 1384°C, density 2.54 g/cm³ (20°C). Below 215°C, elemental strontium has a face-centered cubic crystal structure; between 215–605°C, a hexagonal close-packed crystal structure; and above 605°C, a body-centered cubic crystal structure.

Strontium is a silver-white metal, soft as lead, malleable, ductile, oxidizes rapidly on exposure to air, burns when heated in air emitting a brilliant light and forming oxide and nitride, reacts with H_2O yielding strontium hydroxide and hydrogen gas. Discovered by Hope and by Klaproth in 1793, and isolated by Davy in 1808.

There are four stable isotopes, ^{84}Sr and ^{86}Sr through ^{88}Sr, and seven known radioactive isotopes, ^{82}Sr, ^{83}Sr, ^{85}Sr, and ^{89}Sr through ^{92}Sr, all with relatively short half-lives measurable in hours or days except 90Sr which has a half-life of about 26 years. The latter isotope represents a hazard from nuclear blasting activities because of its long half-life, tendency to contaminate food products, such as milk, and retention in the body. See also **Radioactivity**. In terms of abundance, strontium is 21st among the elements occurring in the rocks of the earth's crust. In terms of the content of sea water, the element ranks 11th, with an estimated 38,000 tons of strontium per cubic mile (9120 tons/cubic kilometer) of seawater. First ionization potential 5.692 eV; second, 10.98 eV. Oxidation potentials $Sr \longrightarrow Sr^{2+} + 2e^-$, 2.89 V; $Sr + 2OH^- + 8H_2O \longrightarrow Sr(OH)_2 \cdot 8H_2O + 2e^-$, 2.99 V. Other important physical properties of strontium are given under **Chemical Elements**.

Occurrence and Characteristics

Strontium occurs chiefly as sulfate (celestite, $SrSO_4$) and carbonate (strontianite, $SrCO_3$) although widely distributed in small concentration. The commercially exploited deposits are mainly in England. The sulfate or carbonate is transformed into chloride, and the electrolysis of the fused chloride yields strontium metal.

As is to be expected from its high oxidation potential (2.89 V) strontium, like calcium and barium, reacts readily with all halogens, oxygen and sulfur to form halides, oxide and sulfide. See also **Celestite**; and **Strontianite**. In all its compounds it is divalent. It reacts vigorously with H_2O to form the hydroxide, displacing hydrogen and it forms a hydride with hydrogen. Strontium hydroxide forms a peroxide on treatment with H_2O_2 in the cold. Strontium exhibits little tendency to form complexes; the amines formed with NH_3 are unstable, the β-diketones and alcoholates are not well characterized, and the chelates formed with ethylenediamine and related compounds are the only representatives of the type.

Strontium Compounds

Strontium acetate, [CAS: 543-94-2], $Sr(C_2H_3O_2)_2$, white crystals, soluble, formed by reaction of strontium carbonate or hydroxide and acetic acid.

Strontium carbide (acetylide), SrC_2, black solid, formed by reaction of strontium oxide and carbon at electric furnace temperature; the carbide reacts with water yielding acetylene gas and strontium hydroxide.

Strontium carbonate, [CAS: 1633-05-2], $SrCO_3$, white solid, insoluble ($K_{sp} = 9.4 \times 10^{-10}$), formed (1) by reaction of strontium salt solution and sodium carbonate or bicarbonate solution, (2) by reaction of strontium hydroxide solution and CO_2. Strontium carbonate decomposes at 1,200°C to form strontium oxide and CO_2, and is dissolved by excess CO_2, forming strontium bicarbonate, $Sr(HCO_3)_2$, solution.

Strontium chloride, [CAS: 10476-85-4], $SrCl_2 \cdot 6H_2O$, white crystals, soluble, formed by reaction of strontium carbonate or hydroxide and HCl. Anhydrous strontium chloride, $SrCl_2$, absorbs dry NH_3 gas.

Strontium chromate, [CAS: 7789-06-2], $SrCrO_4$, yellow precipitate ($K_{sp} = 3.75 \times 10^{-5}$) formed by reaction of strontium salt solution and potassium chromate solution.

Strontium cyanamide, $SrCN_2$, formed with the cyanide, $Sr(CN)_2$, by heating strontium carbide at 1,200°C with nitrogen.

Strontium hydride, SrH_2, white solid, formed by heating strontium metal or amalgam in hydrogen gas at 250°C. Is reactive with H_2O, yielding strontium hydroxide and hydrogen gas.

Strontium nitrate, [CAS: 10042-76-9], $Sr(NO_3)_2$, white crystals, soluble, formed by reaction of strontium carbonate or hydroxide and HNO_3.

Strontium oxide, [CAS: 1314-11-0], SrO, white solid, mp about 2,400°C, reactive with H_2O to form strontium hydroxide ($K_{sp} = 3.2 \times 10^{-4}$); strontium peroxide, $SrO_2 \cdot 8H_2O$, white precipitate, by reaction of strontium salt solution and hydrogen or sodium peroxide, yields anhydrous strontium peroxide SrO_2, upon heating at 130°C in a current of dry air.

Strontium oxalate, [CAS: 814-95-9], SrC_2O_4, white precipitate ($K_{sp} = 5.6 \times 10^{-8}$) formed by reaction of strontium salt solution and ammonium oxalate solution.

Strontium sulfate, [CAS: 7759-02-6], $SrSO_4$, white precipitate ($K_{sp} = 3.2 \times 10^{-7}$), formed by reaction of strontium salt solution and H_2SO_4 or sodium sulfate solution, insoluble in acids. On heating with carbon strontium sulfate yields strontium sulfide, SrS, while on boiling with sodium carbonate solution, $SrSO_4$ yields strontium carbonate.

Strontium sulfide, SrS, [CAS: 1314-96-1], grayish-white solid (thermodynamic K_{sp} 500) reactive with water to form strontium hydrosulfide, $Sr(SH)_2$, solution. Strontium hydrosulfide is formed (1) by reaction of strontium sulfide and H_2O, (2) by saturation of strontium hydroxide solution with H_2S. Strontium polysulfides are formed by boiling strontium hydrosulfide with sulfur.

Editor's Note re Strontium Isotope Research

At any given time, the Sr isotope composition in seawater is uniform throughout the ocean because the oceanic residence time of Sr (5 million years) exceeds the mixing time of the oceans (~1000 years). However, over geologic time, the $^{87}Sr/^{86}Sr$ ratio in seawater has varied as the result of fluxes of Sr to the oceans from various sources. These would include submarine hydrothermal activity, fluxes from rivers, and submarine recycling, the latter occurring by limestone recrystallization and erosion of ancient sedimentary carbonate. J. Hess and colleagues (University of Rhode Island) reported in 1986 that the seawater Sr isotope composition appears to be a smoothly varying function of time and can be useful for high-precision correlations of oceanic sediments for certain periods of time. These researchers prepared a detailed record of the Sr isotope ratio during the last 100 million years by measuring this ratio in well over a hundred foraminifera samples. Sample preservation was evaluated from scanning electron microscopy studies, measured Sr/Ca ratios, and pore water Sr isotope ratios. Results show that the marine Sr isotope composition can be used for correlating and dating well-preserved authigenic marine sediments throughout much of the Cenozoic to a precision of ±1 mil years. See also **Cordierite**.

In 1990, R.C. Capo and D.J. DePaolo (University of California, Los Angeles and Berkeley, respectively) reported that "marine carbonate samples indicate that during the past 2.5 million years the $^{87}Sr/^{86}Sr$ ratio of seawater has increased by 14×10^{-7}. The high average rate of increase of this ratio indicates that continental weathering rates were exceptionally high. Nonuniformity in the rate of increase suggests that weathering rates fluctuated by as much as ±30 percent of present-day values. Some of the observed shifts in weathering rate are contemporaneous with climatic changes inferred from records of oxygen isotopes and carbonate preservation in deep sea sediments."

Studies of Metamorphism. As reported by J.N. Christensen, J.L. Rosenfield, and D.J. DePaolo (University of California, Berkeley), "Measurement of the radial variation of the $^{87}Sr/^{86}Sr$ ratio in a single crystal

from a metamorphic rock can be used to determine the crystal's growth rate. Such variation records the accumulation of ^{87}Sr from radioactive decay of ^{87}Rb (rubidium) in the rock matrix from which the crystal grew. This method can be used to study the rates of petrological processes associated with mountain building." This methodology has been applied by the researchers mentioned to the study of the rates of tectonometamorphic processes from rubidium and strontium isotopes in garnet."

Isotopic Tests for Upwelling Water. In studies of the Yucca Mountain, Nevada, area as a potential site for a high-level nuclear waste repository, the area has been aggressively scrutinized geologically for possible upwelling of deep-seated waters. Strontium and uranium isotopic compositions of hydrogenic materials were used by scientists J.S. Stuckless, Z.E. Peterman, and D.R. Muhs (U.S. Geological Survey, Denver, Colorado) to assist in confirming other geological methods. Their findings indicated in 1991 that the vein deposits are isotopically distinct from groundwater in the two aquifers that underline Yucca Mountain, thus indicating that the calcite could not have precipitated from groundwater and thus providing evidence against upwelling water at the site.

STEPHEN E. HLUCHAN
Business Manager, Calcium Metal
Products, Minerals
Pigments & Metals Division, Pfizer Inc.
Wallingford, Connecticut

Additional Reading

Capo, R.C. and D.J. DePaolo: "Seawater Strontium Isotopic Variations from 2.5 Million Years Ago to the Present," *Science*, **51** (July 6, 1990).

Christensen, J.N., J.L. Rosenfeld, and D.J. DePaolo: "Rates of Tectonometamorphic Processes from Rubidium and Strontium Isotopes in Garnet," *Science*, 1405 (June 21, 1989).

Greenwood, N.N. and A. Earnshaw: *Chemistry of the Elements,* 2nd Edition, Butterworth-Heinemann, Inc., Woburn, MA, 1997.

Hess, J., M.L. Bender, and J.-G. Schilling: "Evolution of the Ratio of Strontium-87 to Strontium-86 in Seawater from Cretaceous to Present," *Science*, **231**, 979–983 (1986).

Krebs, R.E.: *The History and Use of Our Earth's Chemical Elements: A Reference Guide,* Greenwood Publishing Group, Inc., Westport, CT, 1998.

Lewis, R.J. and N.I. Sax: *Sax's Dangerous Properties of Industrial Materials,* 10th Edition, John Wiley & Sons, Inc., New York, NY, 1999.

Lide, D.R.: *CRC Handbook of Chemistry and Physics,* 84th Edition, CRC Press, LLC, Boca Raton, FL, 2003.

Macdougall, J.D.: "Seawater Strontium Isotopes, Acid Rain, and the Cretaceous-Tertiary Boundary," *Science*, 485 (January 29, 1988).

Meyers, R.A.: *Handbook of Chemicals Production,* The McGraw-Hill Companies, Inc., New York, NY, 1986.

Parker, S.P.: *McGraw-Hill Encyclopedia of Chemistry,* 2nd Edition, The McGraw-Hill Companies, Inc., New York, NY, 1993.

Staff: *ASM Handbook—Properties and Selection: Nonferrous Alloys and Pure Metals,* ASM International, Materials Park, OH, 1990.

Stuckless, J.S., Z.E. Peterman, and D.R. Muhs: "U and Sr Isotopes in Ground Water and Calcite, Yucca Mountain, Nevada: Evidence Against Upwelling Water," *Science*, 551 (October 25, 1991).

Stwertka, A. and E. Stwertka: *A Guide to the Elements,* Oxford University Press, Inc., New York, NY, 1998.

STYRENE. Styrene, $C_6H_5CH=CH_2$, is the simplest and by far the most important member of a series of aromatic monomers. Also known commercially as styrene monomer (SM), styrene is produced in large quantities for polymerization. It is a versatile monomer extensively used for the manufacture of plastics, including crystalline polystyrene, rubber-modified impact polystyrene, expandable polystyrene, acrylonitrile–butadiene–styrene copolymer (ABS), styrene–acrylonitrile resins (SAN), styrene–butadiene latex, styrene–butadiene rubber (SBR), and unsaturated polyester resins. See also **Acrylonitrile Polymers**.

Properties

Styrene is a colorless liquid with an aromatic odor. Important physical properties of styrene are shown in Table 1. Styrene is infinitely soluble in acetone, carbon tetrachloride, benzene, ether, *n*-heptane, and ethanol. Polymerization generally takes place by free-radical reactions initiated thermally or catalytically. Styrene undergoes many reactions of an unsaturated compound, such as addition, and of an aromatic compound, such as substitution.

TABLE 1. PHYSICAL PROPERTIES OF STYRENE MONOMER

Property	Value				
boiling point (at 101.3 kPa = 1 atm, °C)	145.0				
freezing point, °C	−30.6				
flash point (fire point), °C					
Tag open-cup	34.4 (34.4)				
Cleveland open-cup	31.4 (34.4)				
autoignition temperature, °C	490.0				
explosive limits in air, %	1.1–6.1				
refractive index, n_D^{20}	1.5467				
	at 0°C	20°C	60°C	100°C	140°C
viscosity, mPa·s(= cP)	1.040	0.763	0.470	0.326	0.243
surface tension, mN/m (= dyn/cm)	31.80	30.86	29.01	27.15	25.30
density, g/cm³ ᵃ	0.9237	0.9059	0.8702	0.8346	
heat of formation (liquid) at 25°C, ΔH_f, kJ/molᵇ	147.36				
heat of polymerization, kJ/molᵇ	74.48				

ᵃ Density at 150°C is 0.7900 g/cm³.
ᵇ To convert J to cal, divide by 4.184.

Ethylbenzene Manufacture

Styrene is manufactured from ethylbenzene. Ethylbenzene is produced by alkylation of benzene with ethylene. The reaction takes place on acidic catalysts and can be carried out either in the liquid or vapor phase.

Commercial ethylbenzene is manufactured almost exclusively for captive use to produce styrene. Most of the ethylbenzene plants built before 1980 are based on use of aluminum chloride catalysts. Aluminum chloride is an effective alkylation catalyst but is corrosive. The newer plants are based on zeolite catalysts.

Zeolite-Based Alkylation. Zeolites have the advantage of being noncorrosive and environmentally benign. The Mobil-Badger vapor-phase ethylbenzene process was the first zeolite-based process to achieve commercial success. It is based on a synthetic zeolite catalyst, ZSM-5, and has the desirable characteristics of high activity, low oligomerization, and low coke formation. See also **Molecular Sieves**.

In the Mobil-Badger vapor-phase process, fresh and recycled benzene are vaporized and preheated to the desired temperature and fed to a multistage fixed-bed reactor. Ethylene is distributed to the individual stages. Alkylation takes place in the vapor phase. Separately, the polyethylbenzene stream from the distillation section is mixed with benzene, vaporized and heated, and fed to the transalkylator, where polyethylbenzenes react with benzene to form additional ethylbenzene. The combined reactor effluent is distilled in the benzene column. Benzene is condensed in the overhead for recycle to the reactors. The bottoms from the benzene column are distilled in the ethylbenzene column to recover the ethylbenzene product in the overhead. The bottoms stream from the ethylbenzene column is further distilled in the polyethylbenzene column to remove a small quantity of residue. The overhead polyethylbenzene stream is recycled to the reactor section for transalkylation to ethylbenzene.

A liquid-phase process based on an ultraselective Y (USY)-type zeolite catalyst, called the Lummus-UOP process, is similar to the Mobil-Badger vapor-phase process. The differences are primarily in the catalysts, reaction conditions, reactor sizes, yields, and product specifications. The zeolite-based processes require more benzene recycle than the aluminum chloride-based processes. The EBMax technology, based on a Mobil zeolite catalyst called MCM-22, overcomes the oligomerization problem that plagues other liquid-phase alkylation processes. The catalyst is highly active for alkylation but inactive for oligomerization and cracking.

Aluminum Chloride-Based Alkylation. An improved aluminum chloride-based process was developed by Monsanto in the 1970s. Using a presynthesized aluminum chloride complex and operating the reactor at higher temperature and pressure, the catalyst inventory is reduced to below its solubility in the reaction mixture. The reactants and the catalyst complex are mixed in the reactor to form a homogeneous liquid. The transalkylation

of polyethylbenzenes is carried out separately. These improvements result in a higher yield.

Other Technologies. Ethylbenzene can be recovered from mixed C_8 aromatics by superfractionation. The Alkar process, commercialized in 1960, uses boron trifluoride on alumina support as the catalyst. It has been used for polymer-grade as well as dilute ethylene feeds.

Styrene Manufacture

Styrene manufacture by dehydrogenation of ethylbenzene is used for nearly 90% of the worldwide styrene production. The rest is obtained from the coproduction of propylene oxide (PO) and styrene (SM).

Dehydrogenation. The dehydrogenation of ethylbenzene to styrene takes

$$C_6H_5CH_2CH_3 \rightleftharpoons C_6H_5CH=CH_2 + H_2$$

place on a promoted iron oxide–potassium oxide catalyst in a fixed-bed reactor at the 550–680°C temperature range in the presence of steam. The reaction is limited by thermodynamic equilibrium. Low pressure favors the forward reaction. Dehydrogenation is an endothermic reaction. High temperature favors dehydrogenation both kinetically and thermodynamically but also increases by-products from side reactions and decreases the styrene selectivity.

The main by-products in the dehydrogenation reactor are toluene and benzene. The formation of toluene accounts for the biggest yield loss. Other by-products include carbon dioxide and various hydrocarbons.

Dehydrogenation catalysts usually contain 40–90% Fe_2O_3, 5–30% K_2O, and promoters such as chromium, cerium, molybdenum, calcium, and magnesium oxides. Dehydrogenation is carried out either isothermally or adiabatically. In principle, isothermal dehydrogenation has the dual advantage of avoiding a very high temperature at the reactor inlet and maintaining a sufficiently high temperature at the reactor outlet. In practice, these advantages are negated by formidable heat-transfer problems. In an adiabatic reactor, the endothermic heat of reaction is supplied by the preheated steam that is mixed with ethylbenzene upstream of the reactor. As the reaction progresses, the temperature decreases. To obtain a high conversion of ethylbenzene to styrene, usually two, and occasionally three, reactors are used in series with a reheater between the reactors to raise the temperature of the reaction mixture.

Other than the reactor system, the distillation column that separates the unconverted ethylbenzene from the crude styrene is the most important and expensive equipment in a styrene plant. To minimize yield losses and to prevent equipment fouling by polymer formation, polymerization inhibitors are used in the distillation train, product storage, and in vent gas compressors.

The qualities of the styrene product and toluene by-product depend primarily on three factors: the impurities in the ethylbenzene feed-stock, the catalyst used, and the design and operation of the dehydrogenation and distillation units. Other than benzene and toluene, the presence of which is usually inconsequential, possible impurities in ethylbenzene are C_7–C_{10} nonaromatics and C_8–C_{10} aromatics. The condensed reactor effluent is separated in the settling drum into vent gas (mostly hydrogen), process water, and organic phase. The organic phase with polymerization inhibitor added is pumped to the distillation train.

Benzene and toluene by-products are recovered in the overhead of the benzene–toluene distillation column. The bottoms from the benzene–toluene column are distilled in the ethylbenzene recycle column, where the separation of ethylbenzene and styrene is effected. The bottoms, are pumped to the styrene finishing column. The overhead product from this column is purified styrene. The bottoms are further processed in a residue-finishing system to recover additional styrene from the residue.

PO–SM Coproduction. The coproduction of propylene oxide and styrene includes three reaction steps: (*1*) oxidation of ethylbenzene to ethylbenzene hydroperoxide, (*2*) epoxidation of ethylbenzene hydroperoxide with propylene to form α-phenylethanol and propylene oxide, and (*3*) dehydration of α-phenylethanol to styrene.

$$C_6H_5CH_2CH_3 + O_2 \longrightarrow C_6H_5CH(CH_3)OOH$$

$$C_6H_5CH(CH_3)OOH + CH_2=CHCH_3 \longrightarrow C_6H_5CH(CH_3)OH + \underset{O}{CH_2CHCH_3}$$

$$C_6H_5CH(CH_3)OH \longrightarrow C_6H_5CH=CH_2 + H_2O$$

The recovery and purification facilities in such a process are complex. One reason is that oxygenated by-products are made in the reactors. Oxygenates hinder polymerization of styrene and cause color instability. Elaborate purification is required to remove the oxygenates.

Specifications and Analysis

The freezing point measurement, standard method for the determination of styrene assay until the 1970s, has been largely replaced by gas chromatography. Color is measured spectrophotometrically and registered on the APHA or the platinum–cobalt scale.

Health and Safety Factors

Styrene is mildly toxic, flammable, and can be made to polymerize violently under certain conditions. However, handled according to proper procedures, it is a relatively safe organic chemical.

While styrene is not confirmed as a carcinogen, it is considered a suspect carcinogen. Styrene liquid is inflammable and has sufficient vapor pressure at slightly elevated temperatures to form explosive mixtures with air. Properly inhibited and attended, styrene can be stored for an extended period of time.

Uses

Commercial styrene is used almost entirely for the manufacture of polymers.

Common applications for polystyrene include packaging, food containers, and disposable tableware; toys; furniture, appliances, television cabinets, and sports goods; and audio and video cassettes. Expandable polystyrene is widely used in construction for thermal insulation.

Uses for ABS are in sewer pipes, vehicle parts, appliance parts, business machine casings, sports goods, luggage, and toys.

SB latex is used in coatings, carpet backing, paper adhesives, cement additives, and latex paint.

SBR is used primarily in tires, vehicle parts, and electrical components.

The principal uses for UPR are in putty, coatings, and adhesives. Glass-reinforced UPR is used for marine, construction, and vehicle materials, as well as for electrical parts.

Derivatives

A large number of compounds related to styrene have been reported in the literature. Those having the vinyl group $CH_2=CH-$ attached to the aromatic ring are referred to as styrenic monomers. Several of them have been used for manufacturing small-volume specialty polymers. The specialty styrenic monomers that are manufactured in commercial quantities are vinyltoluene, *para*-methylstyrene, α-methylstyrene, and divinylbenzene. In addition, 4-*tert*-butylstyrene (TBS) is a specialty monomer that is superior to vinyltoluene and *para*-methylstyrene in many applications. Other styrenic monomers produced in small quantities include chlorostyrene and vinylbenzene chloride. With the exception of α-methylstyrene, which is a by-product of the phenol–acetone process, these specialty monomers are more difficult and expensive to manufacture than styrene.

Vinyltoluene. Vinyltoluene (VT) is a mixture of *meta*- and *para*-vinyltoluenes, typically in the ratio of 60:40. This isomer ratio results from the ratio of the corresponding ethyltoluenes in thermodynamic equilibrium. Vinyltoluene is produced for special applications. Its copolymers are more heat-resistant than the corresponding styrene copolymers, and it is used as a specialty monomer for paint, varnish, and polyester applications.

para-Methylstyrene. PMS is the para isomer of vinyltoluene in high purity. PMS is made by alkylation of toluene with ethylene to *p*-ethyltoluene, followed by dehydrogenation of *p*-ethyltoluene.

Divinylbenzene. This is a specialty monomer used primarily to make cross-linked polystyrene resins. The largest use of divinylbenzene (DVB) is in ion-exchange resins for domestic and industrial water softening. Ion-exchange resins are also used as solid acid catalysts for certain reactions, such as esterification. Divinylbenzene is manufactured by dehydrogenation of diethylbenzene, which is an internal product in the alkylation plant for ethylbenzene production.

α-Methylstyrene. This compound is not a styrenic monomer in the strict sense. The methyl substitution on the side chain, rather than the aromatic ring, moderates its reactivity in polymerization. It is used as a specialty monomer in ABS resins, coatings, polyester resins, and hot-melt adhesives. As a copolymer in ABS and polystyrene, it increases the heat-distortion

resistance of the product. In coatings and resins, it moderates reaction rates and improves clarity. α-Methylstyrene (AMS) is produced as a by-product in the production of phenol and acetone from cumene.

SHIOU-SHAN CHEN
Raytheon Engineers & Constructors

Additional Reading

Chem. Mark. Rep. **248**(5), 41 (July 24, 1995).

Maerz, B., S.S. Chen, C.R. Venkat, and D. Mazzone: "EBMax: Leading Edge Ethylbenzene Technology from Mobil/Badger," 1996 DeWitt Petrochemical Review, Houston, TX, Mar. 19–21, 1996.

Styrene/Ethylbenzene, PERP report 94/95-8, Chem Systems, Tarrytown, NY, Mar. 1996.

U.S. Pat. 4,066,706 (Jan. 3, 1978), J.P. Schmidt (to Halcon International, Inc.).

STYRENE-BUTADIENE RUBBER.

Styrene–butadiene rubber (SBR), an elastomer, is a copolymer of three parts 1,3-butadiene and one part styrene. It is a synthetic rubber used mainly in the manufacture of automobile tires.

In the late 1920s Bayer & Company began studies of the emulsion polymerization process of polybutadiene for producing synthetic rubber. Incorporation of styrene as a comonomer produced a superior polymer compared to polybutadiene. The product, Buna S, was the precursor of the single largest-volume polymer produced in the 1990s, emulsion styrene–butadiene rubber (ESBR).

In the mid-1950s, the Nobel Prize-winning work of K. Ziegler and G. Natta introduced anionic initiators which allowed the stereospecific polymerization of isoprene to yield high cis-1,4 structure, much like natural rubber. At almost the same time, another route to stereospecific polymer architecture by organometallic compounds was announced.

In the 1960s, anionic polymerized solution SBR (SSBR) began to challenge emulsion SBR in the automotive tire market. Organolithium compounds allow control of the butadiene microstructure, not possible with ESBR. Because this type of chain polymerization takes place without a termination step, an easy synthesis of block polymers is available, whereby glassy (polystyrene) and rubbery (polybutadiene) segments can be combined in the same molecule. These thermoplastic elastomers (TPE) have found use in nontire applications.

Physical Properties

Desirable properties of elastomers include elasticity, abrasion resistance, tensile strength, elongation, modulus, and processibility. These properties are related to and dependent on the average molecular weight and mol wt distribution, polymer macro- and microstructure, branching, gel (cross-linking), and glass-transition temperature (T_g).

Emulsion polymerization gives SBR polymer of high molecular weight. Because it is a free-radical-initiated process, the composition of the resultant chains is heterogeneous, with units of styrene and butadiene randomly spaced throughout. Unlike natural rubber, which is polyisoprene of essentially all cis-1,4 configuration, giving an ordered structure and hence crystallinity, ESBR is amorphous. Unlike SSBR, the microstructure of which can be modified to change the polymer's T_g, the T_g of ESBR can be changed only by a change in ratio of the monomers. Glass-transition temperature is that temperature where a polymer experiences the onset of segmental motion.

The glass-transition temperatures for solution-polymerized SBR as well as ESBR are routinely determined by nuclear magnetic resonance (nmr), differential thermal analysis (dta), or differential scanning calorimetry (dsc).

For routine analysis of SBR polymers, gpc is widely accepted.

Advantages of natural rubber and isoprene rubber (NR/IR) are high resilience and strength, and abrasion-resistance. BR shows low heat buildup in flexing, good resilience, and abrasion-resistance. Random SBR is low in price, wears well, and bonds easily. Block SBR is easily injection-molded, and is not cross-linked. Applications of NR/IR include tires, tubes, belts, bumpers, tubing, gaskets, seals, foamed mattresses, and padding. BR is used in tire treads and mechanical goods, as is random SBR. Block SBR is used in toys, rubber bands, and mechanical goods.

Raw Materials

The monomers butadiene and styrene, are the most important ingredients in the manufacture of SBR polymers. For ESBR, the largest single material is water; for solution SBR, the solvent.

The quality of the water used in emulsion polymerization affects the manufacture of ESBR. Water hardness and other ionic content can directly affect the chemical and mechanical stability of the polymer emulsion (latex). Solution polymerization can use various solvents, primarily aliphatic and aromatic hydrocarbons. SSBR polymerization depends on recovery and reuse of the solvent for economical operation as well as operation under the air-quality permitting of the local, state, and federal mandates involved.

Styrene. Commercial manufacture of this commodity monomer depends on ethylbenzene, which is converted by several means to a low purity styrene, subsequently distilled to the pure form. A small percentage of styrene is made from the oxidative process, whereby ethylbenzene is oxidized to a hydroperoxide or alcohol and then dehydrated to styrene. A popular commercial route has been the alkylation of benzene to ethylbenzene, with ethylene, after which the crude ethylbenzene is distilled to give high purity ethylbenzene. See also **Styrene**.

Butadiene. Economic considerations favor recovering butadiene from by-products in the manufacture of ethylene. Butadiene is a by-product in the C4 streams from the cracking process. For use in polymerization, the butadiene must be purified to 99 + %. Crude butadiene is separated from C_3 and C_5 components by distillation. Separation of butadiene from other C_4 constituents is accomplished by salt complexing/solvent extraction. See also **Butadiene**.

Soap. A critical ingredient for emulsion polymerization is the soap, which performs a number of key roles, including production of oil (monomer) in water emulsion, provision of the loci for polymerization (micelle), stabilization of the latex particle, and impartation of characteristics to the finished polymer. Both fatty acid and rosin acid soaps, mamly derived from tall oil, are used in ESBR.

Polymerization

ESBR and SSBR are made from two different addition polymerization techniques: one radical and one ionic. ESBR polymerization is based on free radicals that attack the unsaturation of the monomers, causing addition of monomer units to the end of the polymer chain, whereas the basis for SSBR is by use of ionic initiators.

Free-radical initiation of emulsion copolymers produces a random polymerization in which the trans/cis ratio cannot be controlled. The nature of ESBR free-radical polymerization results in the polymer being heterogeneous, with a broad molecular weight distribution and random copolymer composition. The microstructure is not amenable to manipulation, although the temperature of the polymerization affects the ratio of trans to cis somewhat.

In solution-based polymerization, use of the initiating anionic species allows control over the trans/cis microstructure of the diene portion of the copolymer. In solution SBR, the alkyllithium catalyst allows the 1,2 content to be changed with certain modifying agents such as ethers or amines. Anionic initiators are used to control the molecular weight, molecular weight distribution, and the microstructure of the copolymer.

SBR Compounding and Processing

The art of compounding requires extensive experience and knowledge of the many compound ingredients. A typical rubber compound in addition to polymer contains one or more ingredients from the following general classes: vulcanizing agents, accelerators, accelerator activators, antioxidants, pigments, and softeners.

The vulcanizing agent, which supplies the bridge between the polymer chains, is furnished predominantly by the sulfur molecule in commercial formulations. Peroxide vulcanizers that produce carbon-to-carbon cross-links are also important.

Accelerators are chemical compounds that increase the rate of cure and improve the physical properties of the compound. Accelerator activators are chemicals required to initiate the acceleration of the curing process.

Antioxidants are routinely added to the compounds over and above those contained in the polymer at manufacture. Antiozonants prevent or reduce polymer degradation by the active ozone molecule. Some antioxidant compounds, such as the *para*-phenylene-diamines, are excellent as antiozonants.

Pigments improve or change polymer properties as well as lower product costs. Reinforcement of SBR by carbon blacks allows this family of polymers to compete with natural rubber. See also **Carbon Black**. It is the most important attribute of the pigment in SBR processing. Softeners,

i.e., plasticizers, reinforcing agents, extenders, lubricants, tackifiers, and dispersing aids, are used as processing aids to enhance mixing of uncured stocks and soften cured compounds.

Economic Aspects and Uses

Styrene–butadiene elastomers, emulsion and solution types combined, are reported to be the largest-volume synthetic rubber. The actual percentage has decreased steadily since 1973. The decline has been attributed to the switch to radial tires (longer milage) and the growth of other synthetic polymers. SBR is forecast to remain the dominant elastomer of all synthetic polymers. In the late 1990s, use of SBR has encompassed the following: tires and tire-related products, including tread rubber, 80%; mechanical goods, 11%; other automotive uses, 6%; and adhesives, chewing gum base, shoe products, flooring, etc, for the remaining 3%.

Health and Safety Factors

Air quality and plant effluent have been monitored and more or less regulated from the inception of SBR manufacture. Most local and state governments have strict discharge permits that limit what kind of chemicals and how much of it can be emitted into the environment. Both styrene and butadiene are considered suspect carcinogens.

There is an industry trend to supply SBR certifiably free of volatile nitrosamines or nitrosatable compounds. Of primary concern to local, state, and federal governments is the growing stockpile of scrap tires. The threat of huge piles of scrap tires catching fire is cited as a principal concern. Such fires pollute the air and threaten groundwater as the large quantities of oil released in the incomplete burning become a serious runoff problem. Although use of scrap tires is projected to increase rapidly, the only economically feasible use has been as a fuel or fuel supplement in utility and industrial applications.

RICHARD R. LATTIME
The Goodyear Tire & Rubber Company

Additional Reading

The Vanderbilt Rubber Handbook, 13th ed., R. T. Vanderbilt Co., Inc., Norwalk, CT, 1990.

Whitby, G.S. ed.: *Synthetic Rubber*, John Wiley & Sons, Inc. New York, NY, 1954.

STYRENE-MALEIC ANHYDRIDE. A thermoplastic copolymer made by the copolymerization of styrene and maleic anhydride. Two types of polymers are available—impact-modified SMA terpolymer alloys (*Cadon*®) and SMA copolymers, with and without rubber impact modifiers (*Dylark*®). These products are distinguished by higher heat resistance than the parent styrenic and ABS families. The MA functionality also provides improved adhesion to glass fiber reinforcement systems. Recent developments include terpolymer alloy systems with high-speed impact performance and low-temperature ductile fail characteristics required by automotive instrument panel usage.

Copolymers show chemical resistance generally similar to that of polystyrene and terpolymers similar to that of ABS (acrylonitrile-butadiene-styrene). Neither type is recommended for use in strongly alkaline environments. All impact versions have good natural color and products are available in a wide range of colors. Copolymer crystal grades have good clarity and gloss.

Glass-reinforced SMA polymers are used as electrical connectors, consoles, top pads, and as supports for urethane-padded instrument panels. There are several additional automotive uses. SMA are also found in coffee makers, steam curlers, power tools, audio cassette components, business machines, vacuum cleaners, solar heat collectors, electrical housing, and fan blades, among others.

SUBATOMIC PARTICLES. See **Particles (Subatomic)**.

SUBLIMATION. The direct transition, under suitable conditions, between the vapor and the solid state of a substance. If solid iodine is placed in a tube and slightly warmed, it vaporizes and the vapor reforms into crystals on the cooler parts of the tube. Many crystalline substances, both metallic and nonmetallic, may be similarly sublimated in a vacuum; fairly large crystals of selenium have been thus prepared. The most familiar sublimates are frost and snow. As in the case of other changes of state, sublimation is accompanied by the absorption or evolution of heat, the quantity of which per unit mass is called the heat of sublimation of the substance. At pressures near the triple point the heat of sublimation is approximately equal to the sum of the heats of fusion and vaporization. In physical and chemical literature, it is customary to regard as sublimation only the transition from solid to vapor, not from vapor to solid; but meteorologists do not make this distinction.

Sublimation plays a major role in the freeze-drying of foods. See also **Freeze-Drying**.

SUBLIMATION (Heat of).   The quantity of heat required at constant temperature (and pressure) to evaporate unit mass of a solid. In sublimation, the change is directly from solid to vapor, without appearance of the liquid phase.

SUBSTITUTE NATURAL GAS (SNG).   A rather general term for describing an artificially produced relatively-high Btu gas that compares favorably with natural gas as a fuel. SNG also may refer to synthetic natural gas; or, on some occasions, to synthesis gas. The latter generally connotes a specially constituted gas to be used as the raw material for a chemical process, such as an ammonia synthesis—thus ammonia synthesis gas, etc. The Btu value of natural gas typically lies within the range of 975 to 1,180 Btu per standard cubic foot (110–133 Calories/cubic meter). Thus, to substitute for and to compete with natural gas (where available), the artificially-produced gas must have a Btu content within this general range. Substitute or synthetic gases generally fall into two categories: (1) low-Btu-value gases with a Btu content of 400 to 600 Btu per cubic foot (50–67 Calories/cubic meter) or lower, sometimes suitable for combined-cycle power generation schemes or for subsequent enrichment to increase the Btu content; and (2) high-Btu-value gases (sometimes referred to as pipeline gases) which have a Btu content generally within the 950 to 1,050 Btu per cubic foot (107–118 Calories/cubic meter) range. Such gases, properly treated to remove traces of unwanted impurities and corrosives, can be introduced into transcontinental pipelines and handled essentially in the same manner as natural gas.

SNG is derived from coal, various petroleum fractions, and waste products. Gases produced from coal are described under **Coal**.

The most practical and economic source of raw material for producing SNG varies with the proximity to raw materials, the relative cost of raw materials, and by numerous other factors which affect the complex energy balance of a given nation and geographical location. Naphtha may make a logical choice of starting material in one area, whereas coal would be most logical in another area. Also, for some years to come—until SNG processes become better proved on a day-to-day operating basis—a somewhat more costly raw material, if available, may be the only practical answer. With proven processes, waste materials as a source of SNG is a very sensible approach, but in some areas the costs of collecting wastes (or the availability of sufficient waste products) may prohibit this approach. As proponents of various schemes and concepts have found upon undertaking detailed, practical development of concepts, a process will not necessarily be successful even though initial gross statistics "prove" the wisdom of the concept.

Catalytic Rich Gas (CRG) Process

This process was developed from the work of a team at the Gas Council's Midlands Research Station (MRS) (England) led by Dr. F. J. Dent. In the late 1950s, it became apparent that, due to the postwar increase in refining capacity in Europe, naphtha was becoming available as a potential feedstock for gas making and that its use would be more economical than coal carbonization, which was then the major source of fuel gas in the United Kingdom. The first semicommercial plant, producing 4 million standard cubic feet per day of rich gas, was commissioned in 1964. Within the next five years, nearly 40 units were installed in the United Kingdom for production of rich gas and town gas (470–500 Btu per standard cubic foot; 53–56 Calories/cubic meter). Plants were also installed in Japan, Italy, Brazil, and the United States.

The overall reactions which occur in the steam reforming of naphtha are:

(1) $4C_6H_{14} + 10H_2O \longrightarrow 19CH_4 + 5CO_2$ Exothermic
(2) $CH_4 + H_2O \rightleftharpoons CO + 3H_2$ Endothermic
(3) $CO + H_2O \rightleftharpoons CO_2 + H_2$ Slightly Exothermic

At all practical temperatures, reaction (1) proceeds almost to completion; no significant quantities of higher hydrocarbons exist at the outlet of the CRG reactor.

Reactions (2) and (3) are reversible; the concentrations of the five components CH_4, H_2O, CO_2, CO and H_2 which result are governed by thermodynamic equilibrium. Raising the reaction temperature shifts the equilibrium for both reactions to the right. Thus at low temperatures the exothermic reaction (1) predominates, while at high temperatures the overall reaction is endothermic. At approximately 500–550°C the reaction is thermally neutral.

Naphthas boiling up to 185°C can be reformed at pressures up to 600 psig. Naphthas with final boiling point up to 240°C may be reformed at lower pressures. Higher olefin contents may be accepted provided that sufficient hydrogen is available in the recycle gas to saturate the feed in the desulfurization section. Higher aromatic contents may be accepted but the catalyst life will be reduced.

A typical rich gas leaving the CRG reactor has the following composition:

CO_2	23.0 mol.% (dry)
CO	0.7
H_2	12.8
CH_4	63.5
	100.0%
Calorific Value (Btu/standard bic foot)	675

Higher hydrocarbons are present in negligible quantities.

The calorific value of this gas is too high for direct use as town gas (470–500 Btu/standard cubic foot in the United Kingdom) and too low for SNG. However, by removing the CO_2 the calorific value is increased to about 870 Btu/standard cubic foot, which is useful for enriching lean gas (e.g. from an Imperial Chemical Industries (ICI) naphtha reformer) to town gas quality. Long has suggested that by enriching this gas with LPG (liquefied petroleum gas) a satisfactory SNG may be obtained.

Alternatively, the calorific value may be changed by bringing the components to a new equilibrium at a different temperature. In the Series "A" Process, part of the rich gas is further reformed at high temperature and remixed with the remaining rich gas. After water gas shift and partial CO_2 removal a 500 Btu/standard cubic foot product is obtained which is fully interchangeable with the town gas distributed in the United Kingdom.

If the subsequent stage is at a lower temperature, carbon oxides and hydrogen recombine to methane, increasing the calorific value. Following CO_2 removal, very little enrichment is required to achieve a product fully interchangeable with natural gas.

CRG catalyst is deactivated by low concentrations of sulfur and chlorine compounds. To achieve removal of sulfur to very low concentrations (less than 0.2 ppm), British Gas developed their own process which is always used in association with CRG catalyst. Organic sulfur compounds are hydrogenated to H_2S over nickelmolybdenum catalyst at about 380°C. The hydrogen is usually generated by reforming rich gas from the CRG reactor with added steam in a tubular reformer. Alternatively, gas from the reactor may be used directly, while in some plants it is normal to use CO_2-free town gas. The H_2S is then absorbed on zinc oxide or, in the case of many United Kingdom town gas plants, Luxmasse (hydrated ferric oxide).

Because of the large stream sizes involved, many of the SNG plants in the United States incorporate a bulk sulfur removal stage using a hydrofining process. The H_2S produced is commonly recovered as elemental sulfur by a Stretford plant, in which the hydrofiner off gas is washed with an aqueous alkaline solution which is regenerated by oxidation with air.

Chlorine compounds, if present in concentration higher than 1 ppm, not only deactivate the CRG catalyst but also interfere with the absorption of H_2S. Therefore they are removed by hydrogenation to HCl which is absorbed on a proprietary absorbent.

Reforming in the CRG process occurs adiabatically at 450–550°C at pressures up to about 600 psig (41 atmospheres). The reactor is a vertical cylindrical pressure vessel containing a bed of the special high-nickel catalyst which is supported on a grid or on inert ceramic balls. The gas flow is downwards through the bed and distributors are provided at inlet and outlet. A layer of ceramic balls on top of the bed prevents disturbance of the catalyst by the entering gas.

Normal practice is to install two reactors in parallel, of which one is working at any time. The catalyst charge in each vessel is designed for 3–6 months operation at full load. This system avoids unnecessary exposure of catalyst to high temperatures, minimizes the catalyst loss in

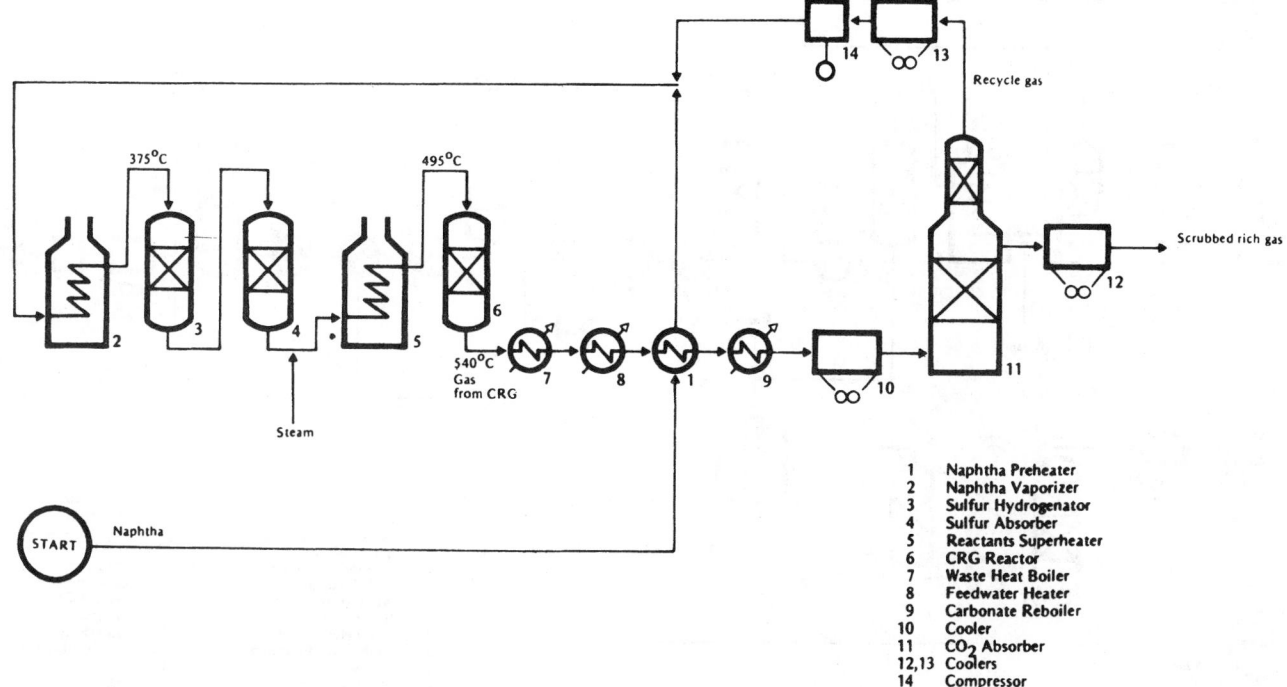

Fig. 1.   Typical flowsheet for rich gas plant. (*Woodall-Duckham Limited.*)

1	Naphtha Preheater
2	Naphtha Vaporizer
3	Sulfur Hydrogenator
4	Sulfur Absorber
5	Reactants Superheater
6	CRG Reactor
7	Waste Heat Boiler
8	Feedwater Heater
9	Carbonate Reboiler
10	Cooler
11	CO₂ Absorber
12,13	Coolers
14	Compressor

the event of damage by maloperation and provides instant standby if such damage occurs.

The flow diagram for a rich gas plant producing gas with a calorific value of 710 Btu/standard cubic foot is shown in Fig. 1. The product is used to enrich lean gas from an ICI (Imperial Chemical Industries) naphtha reformer which has a calorific value of about 320 Btu/standard cubic foot to the town gas standard of 500 Btu/standard cubic foot (56 Calories/cubic meter). Typical gas analyses are given in Table 1.

In what is termed a Series "A" Process, part of the gas from the CRG reactor is reformed with additional steam and the resulting lean gas is reblended with the remaining rich gas. The mixed gas is then subjected to water gas shift and partial carbon dioxide removal, yielding a product with a calorific value of 470–500 Btu/standard cubic foot (53–56 Calories/cubic meter). Typical gas analyses are shown in Table 2. By varying the proportion of gas which flows to the tubular reformer and the degree of CO_2 removal, the characteristics of the product gas can be made interchangeable with any of the different standards employed by the United Kingdom Area Boards and Japanese and European gas companies. A variation of this process has been used in Italy. The calorific value of important Libyan LNG (1395 Btu/standard cubic foot; 157 Calories/cubic meter) was too high for direct use as pipeline gas. The LNG was, therefore, fractionated and the heavy ends (C_2H_6—C_6H_{14}) subjected to processing.

If the rich gas from the CRG reactor is passed over another bed of high-nickel catalyst at a lower temperature, the equilibrium of the five components is reestablished. Carbon oxides react with hydrogen to form methane and the calorific value of the gas is increased. It should be noted that this methanation step differs from that encountered in ammonia synthesis gas production; because of the high steam content the temperature rise is reduced and there is no possibility of temperature "runaway" as the

exit temperature can never rise above the temperature corresponding to equilibrium at the inlet composition, i.e. the CRG exit temperature.

In order to minimize cold enrichment and to achieve a very low carbon monoxide content in the product, a second methanation stage is frequently employed. To achieve sufficient "driving force" to make the reaction proceed, the water vapor content is reduced by cooling the gas, rejecting condensate, and reheating to the required reaction temperature.

Table 3 shows the effect of the second methanation stage on product calorific value. While consumption of LPG is minimized, the capital cost is increased and the overall thermal efficiency is slightly reduced.

If part of the purified naphtha vapor from desulfurization is allowed to bypass the CRG reactor, it can be fully gasified by reaction with

TABLE 1. GAS ANALYSES IN RICH GAS PLANT

	Recycle gas	Gas from CRGR	Scrubbed rich gas
CO_2 (mol %)	0.9	20.3	13.5
CO	1.8	1.4	1.5
H_2	23.3	19.1	20.7
CH_4	74.0	59.2	64.3
	100.0	100.0	100.0
Calorific Value			
(Btu/standard cubic foot)	816	654	710
(Calories/cubic meter)	91.7	73.5	79.8

TABLE 2. GAS ANALYSES IN SERIES "A" PLANT

	Recycle gas	Rich gas	Reformed gas	Mixed gas	Converted gas	Product gas
CO_2 (mol %)	1.0	21.6	13.4	16.1	21.3	13.5
CO	3.4	0.9	13.7	9.6	2.7	3.0
H_2	60.9	15.3	59.0	44.9	48.4	53.2
CH_4	34.7	62.2	13.9	29.4	27.6	30.3
	100.0	100.0	100.0	100.0	100.0	100.0
Calorific Value						
(Btu/standard cubic foot)	550	670	370	466	438	480
(Calories/cubic meter)	61.8	75.3	41.6	52.4	49.2	53.9

TABLE 3. GAS ANALYSES—SNG PRODUCTION BY DOUBLE METHANATION

	Recycle gas	1st Stage gas	2nd stage gas	3rd stage gas	Scrubbed gas	Product gas
CO_2 (mol %)	0.5	21.7	22.0	21.9	0.5	0.50
CO	1.5	0.8	0.1	0.1	0.1	0.04
H_2	86.5	12.8	3.7	0.4	0.5	0.53
CH_4	11.5	64.7	74.2	77.6	98.9	97.98
C_3H_8	—	—	—	—	—	0.95
	100.0	100.0	100.0	100.0	100.0	100
Calorific Value						
(Btu/standard cubic foot)	395	687	750	773	986	1000
(Calories/cubic meter)	44.3	77.2	84.3	86.9	110.8	112.4

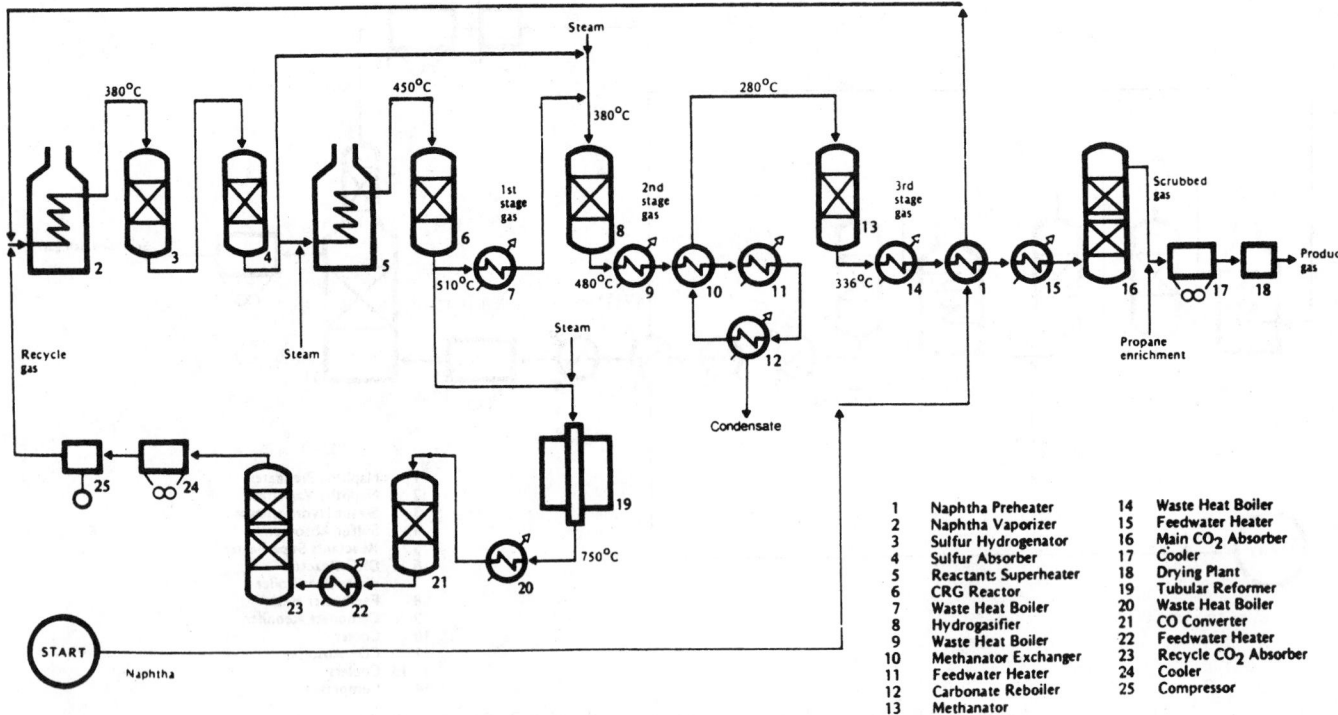

Fig. 2. Typical flowsheet for SNG plant—hydrogasification route. (*Woodall-Duckham Limited.*)

1	Naphtha Preheater	14	Waste Heat Boiler
2	Naphtha Vaporizer	15	Feedwater Heater
3	Sulfur Hydrogenator	16	Main CO_2 Absorber
4	Sulfur Absorber	17	Cooler
5	Reactants Superheater	18	Drying Plant
6	CRG Reactor	19	Tubular Reformer
7	Waste Heat Boiler	20	Waste Heat Boiler
8	Hydrogasifier	21	CO Converter
9	Waste Heat Boiler	22	Feedwater Heater
10	Methanator Exchanger	23	Recycle CO_2 Absorber
11	Feedwater Heater	24	Cooler
12	Carbonate Reboiler	25	Compressor
13	Methanator		

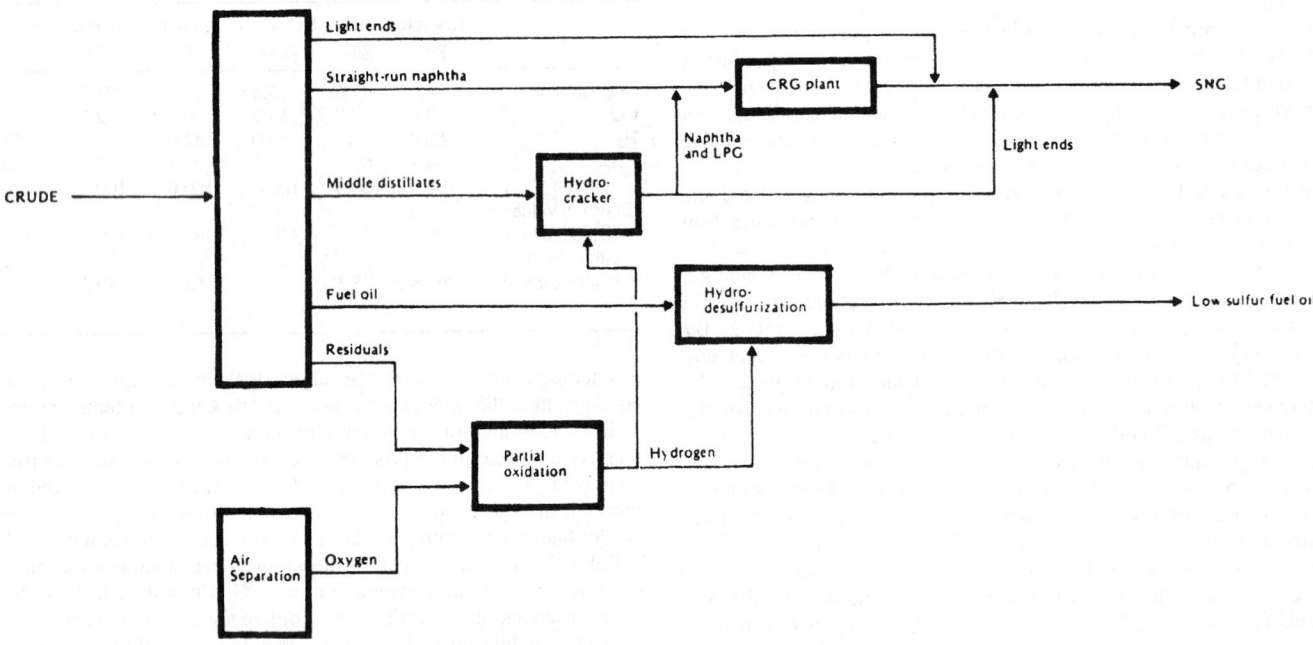

Fig. 3. SNG and low-sulfur fuel oil from crude. (*Woodall-Duckham Limited.*)

the hydrogen and steam in the rich gas. This reaction occurs at lower temperatures than the CRG reaction, and is known as hydrogasification. It has the advantage that the total steam requirement for the process is reduced, although with heavier feedstocks it may be necessary to add a little steam to the hydrogasifier in order to ensure that carbon is not formed by the Boudouard reaction. Since less makeup steam has to be generated from fired boilers, the overall efficiency is improved by 1–2%. The capital cost is slightly lower than that of the methanation route.

The calorific value of the product from hydrogasification is lower than that from single methanation, particularly with high carbon/hydrogen feedstocks because of the additional steam required. However, by adding a final methanator, the calorific value can be increased to that obtained from double methanation, again with increased capital cost and reduced efficiency. This process (Fig. 2) is used in the first operational SNG plant in the United States at Harrison, N.J. Typical gas analyses are given in Table 4.

TABLE 4. GAS ANALYSES—SNG PRODUCTION BY HYDROGASIFICATION

	Recycle gas	1st Stage gas	2nd Stage gas	3rd Stage gas	Scrubbed gas	Product gas
CO_2 (mol %)	0.5	21.8	21.7	21.9	0.5	~0.50
CO	1.6	0.7	0.6	0.1	0.1	~0.70
H_2	86.5	13.3	6.2	0.6	0.8	~0.79
CH_4	11.4	64.2	71.5	77.4	98.6	~97.56
C_3H_8	—	—	—	—	—	~1.08
	100.0	100.0	100.0	100.0	100.0	
Calorific Value (Btu/standard cubic foot)	395	683	733	772	984	1000
(Calories/cubic meter)	44.4	76.7	82.4	86.7	110.6	112.4

Because of the lower temperature, the catalyst in the hydrogasifier, which is the same as that in the CRG reactor, is slowly deactivated by polymer formation. The activity may be recovered "*in situ*" by heating in hydrogen. Two hydrogasifiers are therefore provided in parallel so that regeneration can be carried out without interrupting production.

The CRG process is one of a range of processes developed by the British Gas Corporation for production of fuel gases. The range and application of these processes and their impact has been described by Hebden, illustrating the effect on capital cost of increasing the carbon/hydrogen ratio of the feedstock.

An alternative method of handling crude oil is the "energy refinery" in which crude is split into a number of fractions which can be treated by proven processes to yield two products, SNG and low sulfur fuel oil.

One such scheme is shown in Fig. 3. The advantages of using the CRG process as the final stage in the production of SNG are high efficiency and low capital cost, the predictable quality of the product gas, and the absence of by-products.

Methane-Rich Gas (MRG) Process. A process which produces methane gas from feedstock hydrocarbons, such as naphtha, liquefied petroleum gases (LPG), and refinery gas, developed by Japan Gasoline Co., Ltd., in collaboration with its affiliate, Nikki Chemical Co., Ltd., for application to town-gas facilities. The MRG process had its origin in the high-temperature hydrocarbon steam reforming technology. First efforts culminated in a successful installation in 1956. In the tendency to employ heavier hydrocarbons as feedstock, carbon formation in the low-temperature range posed a problem. The Japan Gasoline Co. took this up as a main research subject and continued the study of low-temperature-range reaction, concentrating especially on the difference of product gas properties and carbon formation according to reaction conditions and catalyst specifications. As a result, a new catalyst was developed which converts butane or naphtha to a gas consisting mainly of methane, hydrogen, and carbon dioxide, with a negligible amount of carbon monoxide. This was the first stage of development of the present MRG process.

In 1964, while developing practical applications for the town-gas industry, a pilot plant with a daily capacity of 15,000 cubic meters was built. Continuous test runs were conducted over a long term, in cooperation with Osaka Gas Co., Ltd., thus starting commercial production of equipment for the process. Based upon results of the aforementioned test runs, a commercial-size town-gas plant (200,000 cubic meters daily capacity) was constructed at the Hokkoh plant of Osaka Gas Co., Ltd. This was followed by two plants each of 500,000 cubic meters daily capacity in 1967 and 1969. Additional plants followed not only for town-gas uses, but also for petrochemical needs. Late in 1971, an MRG plant incorporating a wet methanation system went into operation at Keiyo Gas Co., Ltd., near Tokyo, with a capacity of 105,000 cubic meters per day. In late 1972, a complete MRG-based SNG plant, consisting of gasification, methanation, and CO_2 removal sections was completed for the same firm, with a capacity of 200,000 cubic meters per day. In early 1974, Boston Gas Co. (U.S.) started up a 1,070,000 cubic meters per day SNG plant which employs a two-stage MRG gasification system.

The basic reactions of the MRG process consist of three stages: (1) hydrodesulfurization of sulfur compounds in the hydrocarbon feedstock; (2) low-temperature steam reforming (gasification) of desulfurized hydrocarbons; and (3) methanation reaction between hydrogen and carbon dioxide in methane gas available by gassification.

Sulfur compounds contained in hydrocarbon feedstock vary, depending on the types of crudes and their boiling points. Naphtha, for example, contains mainly mercaptans, disulfides, and thiophenes. Such sulfur compounds deteriorate the activity of the low-temperature steam-reforming MRG catalyst. They should be removed to some degree before the feedstock enters the system. Major reactions of the hydrodesulfurization step are:

$$RSH + H_2 \longrightarrow RH + H_2S$$
$$R-S-R' + 2H_2 \longrightarrow RH + R'H + H_2S$$
$$R-S-S-R' + 3H_2 \longrightarrow RH + R'H + 2H_2S$$

$$\boxed{}_S \!-R + 3H_2 \longrightarrow RC_4H_9 + H_2S \quad \text{(alkyl group R not fixed to a specific carbon)}$$

In as much as these are all exothermic reactions, low ambient temperatures are favorable from the standpoint of equilibrium theory, but in

consideration of reaction rates, general processes are operated in a range of 350–400°C with the aid of a highly active catalyst (like Co-Mo or Ni-Mo), involving the side reactions:

$$CO_2 + 4H_2 \rightleftharpoons CH_4 + 2H_2O$$
$$CO + 3H_2 \rightleftharpoons CH_4 + H_2O$$

The foregoing reactions are highly exothermic and significantly raise reaction temperatures. The MRG process, however, does not involve such adverse side reactions with use of a special, selective hydrodesulfurizing catalyst (developed by Japan Gasoline Co. and Nikki Chemical). The MRG process uses part of product gas for hydrodesulfurization, and even if it contains only 20–25% hydrogen and as high as 20–23% carbon oxides, only the proper hydrodesulfurization reactions take place. The MRG process features a recycle use of product gas for hydrodesulfurization purposes without any special treatment.

To eliminate hydrogen sulfide formed in hydrodesulfurization reactions, two solutions are available: (1) fixation by H_2S contact with an adsorbent (zinc oxide) via the reaction: $ZnO + H_2S \longrightarrow ZnS + H_2O$; and (2) physical removal by stripping. The hydrodesulfurization system is most economically practical with feedstocks containing less than 200 to 500 ppm sulfur. The removal of H_2S by stripping after hydrodesulfurization with an external hydrogen supply may be applied to naphtha stocks contaminated by trace metals as well as those high in sulfur.

Gasification by low-temperature steam-reforming reactions, the heart of the MRG process, is carried out between liquid hydrocarbons and steam over catalyst to form methane, hydrogen, and carbon oxides. In order to increase the calorific value of product gas to the values similar to natural gas, methanation reactions are required. Hydrogen in product gas is reacted with CO_2 and CO to form methane, with only a small portion unconverted. Methanation reactions are:

$$CO + 3H_2 \rightleftharpoons CH_4 + H_2O \quad 49.3 \text{ kcal/g-mole at } 25°C$$
$$CO_2 + 4H_2 \rightleftharpoons CH_4 + 2H_2O \quad 9.8 \text{ kcal/g-mole at } 25°C$$

After methanation, the gas goes to a scrubber to remove CO_2 for further purification.

Because the MRG process is mainly based on steam reforming, the success of the process hinges on the reliable availability of steam. Steam should be controlled at a constant level somewhat above the projected requirements to assure continuous, effective reforming. Adequate steam must be on hand at all times to assure effective control of the steam/naphtha ratio. In summary, for the proper feedstocks and economic situations, the MRG process offers the following: (1) a wide variety of feedstocks can be used; (2) broad selection of calorific value of final product gas. Product gas is available in a range of calorific values from 5,500 to 9,400 kcal/cubic meter; (3) high-pressure operation. Many conventional town-gas plants which operate at near atmospheric pressure require an additional compressor to convey product gas through pipelines. The MRG process does not require any additional compressor, but does permit operation at high pressures—up to approximately 1,140 pounds per square inch gage (77.6 atmospheres), enabling high-pressure, long-distance transportation of product gas. (4) Noncomplex equipment is used. The use of a drum-type reactor makes the reactor design quite simple, resulting in a compact design of the overall system. In terms of product gas calorific value, the MRG reactor requires only $\frac{1}{6}$ to $\frac{1}{8}$ the area of a general coke oven and about $\frac{1}{3}$ that of a conventional high-temperature steam-reforming plant; (5) sulfur-resistant catalyst; and (6) high thermal efficiency. Operation at low temperatures and high pressures permits thermal efficiency as high as 92–96%, depending on desulfurized feedstock used.

Hydrocracking-Hydrogasification Process. In a continuous, two-step process developed by the Institute of Gas Technology, crude oil can be hydrocracked to approximately diesel oil weight and then made to react noncatalytically with hydrogen at elevated pressures (500–1,500 psig; 34–102 atmospheres) and temperatures 593–760°C to produce methane-rich gas containing about 30% (volume) hydrogen and 10% ethane. This gas can be desulfurized and methanated to yield pipeline gas. By adjusting the conditions of hydrocracking, enough heavy fuel oil can be produced to provide feedstock for hydrogen production by partial oxidation, or a low-sulfur fuel oil product can be made if desired.

Interest has centered on plants to produce substitute natural gas from light distillate feedstocks, such as naphtha. However, when naphtha is in

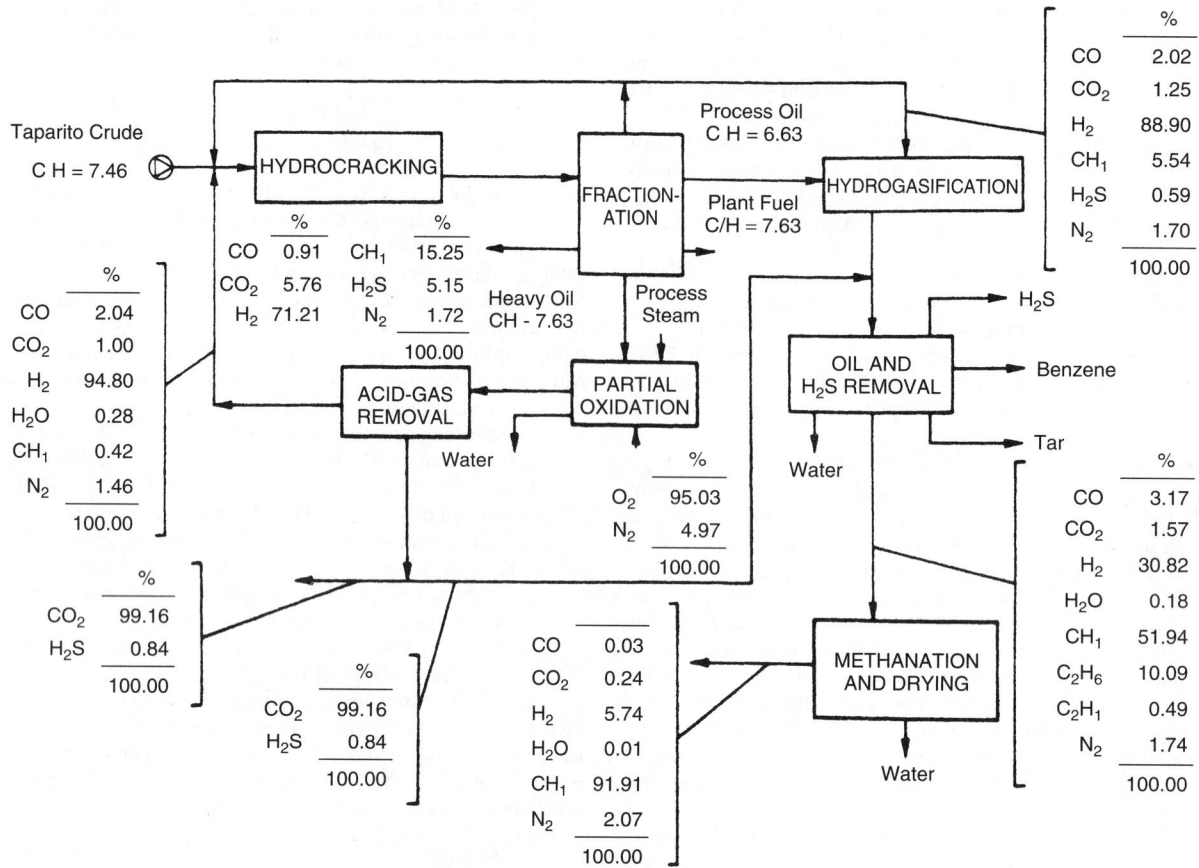

Fig. 4.   Block flow diagram for pipeline gas production. (*Institute of Gas Technology.*)

short supply, a process capable of converting the more plentiful crude and residual oils to pipeline gas is needed. A number of processes were developed over 20 years ago to provide supplemental gas in winter periods when demand is high. Because the supplemental gas was required for only 20 to 30 days during the year, only cyclic thermal cracking at atmospheric pressure was used in order to avoid the high capital cost of more complex continuous processes. However, with the need for base-load gas, a continuous process would be feasible.

The major difficulty associated with the production of pipeline gas from crude and residual oils is carbon deposition during the gasification step, which leads to reactor vessel plugging. One approach to this problem has been to conduct pressure gasification with hydrogen in a fluidized bed of coke, allowing carbon to deposit on the coke particles. To avoid accumulation of these deposits, a small amount of coke is continuously withdrawn from the bed. This technique, developed by the British Gas Council, gasifies the oil in a single reaction step. An approach to the problem of carbon deposition taken by the Institute of Gas Technology is to eliminate the carbon-forming materials in the heavy oil prior to hydrogasification by catalytic hydrocracking. In developing this concept, experiments were conducted on both the hydrocracking and the hydrogasification operations for a variety of feedstocks, ranging from kerosine to Bunker-C fuel oil. Distillate feeds required no hydrocracking.

A simplified flowsheet based on this concept (hydrocracking, separation, and hydrogasification) is shown in Fig. 4, where 250 billion Btu/day (63 billion Calories/day) of pipeline-quality gas is produced from 59,350 barrels/day of Taparito crude. The overall fuel efficiency of the process is 67%, allowing for all utility requirements, including oxygen production. The design of hydrogen plants based on partial oxidation of residual oils and the design of hydrocracking operations are well established arts. The only unusual component of the process, therefore, is the hydrogasification reactor itself. Offsites not shown on the flowsheet include an oxygen plant, sulfur-recovery facilities (Claus plant), power and steam generation equipment, and water treatment facilities.

At the hydrogasification conditions used, about 90% of the 360°C endpoint feed oil is gasified, yielding a raw gas containing about 52% methane and 10% ethane, with the remainder principally hydrogen. About 60% of the liquid products is benzene. Total liquid products are removed

first by separation in a knockout drum and then by straw oil scrubbing. Benzene, the last traces of which are removed from the gas by activated carbon, is recovered and sold. Very heavy oil is used for plant fuel. Since the excess hydrogen in the gas is to be methanated with carbon dioxide, a small amount of carbon dioxide from the hydrogen plant is added to the gas prior to the removal of hydrogen sulfide. Most of the hydrogen in the gas at this point will react with ethane during methanation. Because carbon dioxide is used for methanation, hydrogen sulfide must be selectively removed from the gas prior to methanation. Approximately 48,400 metric tons/year of elemental sulfur are recovered for disposal from waste gas streams in the plant. The waste gas streams from the acid-gas removal unit upstream of the methanator and from the first stage of the acid-gas removal unit in the hydrogen plant are sent to a Claus plant.

Noncatalytic Partial-Oxidation Gasification. Designed for the partial combustion or oxidation of hydrocarbons, a process of this type is particularly suitable for converting heavy, sulfur-containing residual fuels and heavy crude oils into a mixture of hydrogen and carbon monoxide in inert gases. The process is carried out by injecting oil and air (or oxygen) through a specially-designed burner assembly into a closed combustion vessel, where partial oxidation occurs at about 1,316°C. The term partial oxidation describes the net effect of a number of component reactions that occur in a flame supplied with less than stoichiometric oxygen.

In the fuel injection region of the reactor, hydrocarbons leaving the atomizer at about preheat temperature are intimately mixed with air or oxygen. The atomized hydrocarbon is heated and vaporized by back radiation from the flame and reactor walls. Some cracking of the hydrocarbons to carbon, methane, and hydrocarbon radicals may occur during this brief phase. When the fuel and air or oxygen reach the ignition temperature, part of the hydrocarbons reacts with oxygen in a highly exothermic reaction to produce carbon dioxide and water. Practically all available oxygen is consumed in this phase. The remaining hydrocarbons which have not been oxidized react with steam and the combustion products from reaction to form carbon monoxide and hydrogen. The carbon produced during gasification is recovered as a soot-in-water slurry.

Depending upon the desired heating value of the product gas, either oxygen or air may be used as oxidant. Nitrogen present in the air acts as a moderator for temperature control in the reactor and does not enter into the reactions. When either oxygen or air enriched with oxygen is used, a quantity of steam must be injected into the reactor for temperature moderation. Air oxidation alone requires no steam. The latter method produces a low heating-value fuel gas (approximately 120 Btu/standard cubic foot; 13.5 Calories/cubic meter) due to the presence of nitrogen. Oxygen feed produces a medium heating-value gas (approximately 300 Btu/standard cubic foot; 33.7 Calories/cubic meter).

The net products of the process are high-pressure steam, clean wastewater, and carbon-free fuel gas. While the high-pressure steam is saturated, pressures over 1,100 psig (74.8 atmospheres) have been commercially demonstrated and increases to substantially higher pressures in commercial practice are anticipated. Under any condition, using superheating, this steam is easily converted into an attractive feed for steam turbines. With appropriate design, oxygen-based oxidation units can be made almost entirely energy self-sufficient.

Gas from Solid Wastes. In one process, municipal refuse is charged at the top of a shaft furnace and is pyrolyzed as it passes downward through the furnace. Oxygen enters the furnace through tuyeres near the furnace bottom and passes upward through a 1,425–1,650°C combustion zone. The products of combustion then pass through a pyrolysis zone and exit at about 93°C. The offgas then passes through an electrostatic precipitator to remove flyash and oil formed during pyrolysis, both of which are recycled to the furnace combustion zone. The gas then passes through an acid absorber and a condenser. The clean fuel gas has a heating value of about 300 Btu/cubic foot (33.7 Calories/cubic meter) and a flame temperature equivalent to that of natural gas. As the solid waste passes downward through the furnace, it contacts the exiting pyrolysis products and traps a portion of the oil and flyash while itself losing moisture. After passing through the pyrolysis and combustion zones, the remaining solid waste is removed as a slag from the furnace bottom. The system has a net thermal efficiency of about 65% in converting solid waste to fuel gas. Process losses include energy losses in the conversion process and energy required for the operation of the onsite cryogenic gas separation unit for production of 95% oxygen needed by the system. The clean fuel gas is low in sulfur (about 15 ppm) and is essentially free of nitrogen oxides.

Another process involves the anaerobic digestion of a solid waste and water or sewage slurry at 60°C for 5 days to produce a methane-rich gas. Solid waste is prepared by shredding and air classification prior to being blended with water or sewage sludge to a 10 to 20% solids concentration. The slurry is heated and placed in a mixed digester for 5 days detention. The digestor gas is drawn off and separated into carbon dioxide and methane. The spent slurry from the digester is pumped through a heat exchanger to partially heat the incoming slurry prior to filtration. The filtrate is returned to the blender and the sludge is used as landfill. Heat addition to the refuse slurry is required to maintain the required digester temperature. This process is suited for use on sewage sludge, animal manures, and other high-moisture-content solid wastes. It is estimated that the process reduces the volume of volatile solids by 75% while producing about 3,000 cubic feet (85 cubic meters) of methane per metric ton of incoming solid waste. The major residue is a sludge that requires landfilling or incineration. About 10% of the methane is consumed in heating the digester feed.

Methanol as Source of SNG. Methanol can be produced from a large range of feedstocks by a variety of processes. Natural gas, liquefied petroleum gas (LPG), naphthas, residual oils, asphalt, oil shale, and coal are in the forefront as feedstocks to produce methanol, with wood and waste products from farms and municipalities possible additional feedstock sources. In order to synthesize methanol, the main feedstocks are converted to a mixture of hydrogen and carbon oxides (synthesis gas) by steam reforming, partial oxidation, or gasification. The hydrogen and carbon oxides are then converted to methanol over a catalyst.

The concept of utilizing associated or natural gas for production of methanol which could be transported more economically than LNG from areas of surplus to areas of shortage was examined in the mid-1960s. At that time, the largest single-stream plant designed had a capacity of 900 metric tons of methanol per day. A fuel plant which might need to produce—say 22,500 metric tons per day of methanol was assessed on the economics basis of 25 times the small plant. It was quickly ascertained that methanol fuel delivered, for example, to the United States from the Middle East could not compete with local natural gas supplies which were than available at low cost within the United States. With shorter local supplies accompanied by much greater costs, the possibilities of economically feasible large-scale fuel methanol production now appear much more promising.

Conversion of natural gas (or other petroleum components) at the source to methanol, shipment as methanol, and reconversion of methanol into pipeline gas at point of use—versus the concept of liquefying natural gas and shipping LNG for regasification at point of use—probably will be a problem of some controversy for a number of years, pending assessment of actual system operating costs for both systems on a large-scale. See also **Natural Gas**.

SUBTRACTIVE COLOR PROCESS. A method of photographic color synthesis using two or more superimposed colorants, which selectively absorb their complementary colors from white light.

Most modern processes of color photography make use of a subtractive synthesis to yield prints or transparencies. In a three-color process, the colorants cyan, magenta, and yellow are used to control the amounts of red, green and blue in a beam of white light. See Fig. 1. This beam of white light may be either that of a projector with its color transparency, or the light reflected from a white support, such as paper, on which the color reproduction is printed. In the first case, the light passes through the colorants once, while in the print viewed by reflection the light must traverse the colorants twice.

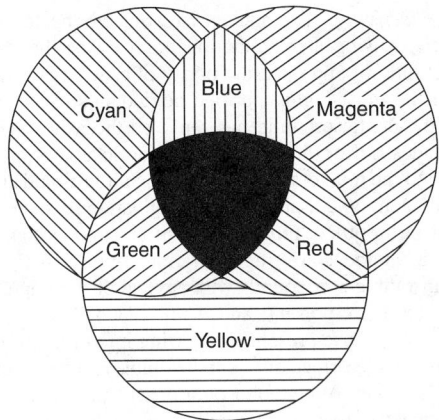

Fig. 1. Superimposed color filters

The colorants are positive or negative images. A cyan positive image (a cyan colorant controls red light), for example, may be prepared from the negative that recorded the red present in the subject. The magenta and yellow colorant images are likewise made from green and blue record negatives. These colorant images are superimposed in register to yield the final reproduction. The three colorants may be in separate removable layers or they may be physically inseparable as in the modern integral tripacks. The contrast of the colorant images must be approximately double for a picture to be viewed by transmitted light as compared to one to be viewed by reflection. The accuracy of color reproduction by a subtractive synthesis as compared to an additive is chiefly dependent on how satisfactorily the three colorants cyan, magenta, and yellow fulfill their role, as red, green and blue absorbers respectively. Color correction is often adopted to improve the accuracy of reproduction when using the colorants generally available.

See also **Photography and Imagery**.

SUCROSE. See **Carbohydrates**; **Sugar**; and **Sweeteners**.

SUGAR. The two principal sources of sucrose (table sugar, saccharose) are sugarcane, a tropical perennial grass (*Saccharum officinarium*), accounting for slightly over 60% of world sugar production; and the sugar beet, a biennial plant (*Beta vulgaris*), accounting for nearly 40% of world sugar production. Relatively minor commercial sources of saccharose include sorghum and the sugar maple tree (sap). Sucrose also occurs in honey. The basic chemical and physical properties of sucrose, $C_{12}H_{22}O_{11}$, are described in the entry on **Carbohydrates**.

Cane Sugar Manufacture

The amount of sucrose in the natural juice of the cane ranges from 10% to nearly 17% (weight), depending upon the variety, the nature of the growing season, and the time of harvesting. In addition to sucrose, cane juice contains from 1 to 2.5% glucose or reducing sugars. Various nonsugars range from 1 to 3% and are made up of carbohydrate polymers, such as gums, and polysaccharides, such as pectins—plus a number of other substances in small quantities.

Crushing the Cane. Upon receipt of cane at the mill, the stalks are washed and cut into several smaller pieces, after which they are fed to a series (frequently three) roller mills. Three heavy, serrated roller crushers are used for each of these milling operations. Two of the rollers turn in opposite direction, while a third roller guides the flow of the stalks through the crushing operation. Where three sets of mills are used, the adjustment of the spacing between the crushing rollers will be wider for the first mill than the second mill, with the narrowest spacing for the third set of crushing rollers. The fibers are sprayed during crushing with a small amount of maceration water (from 5 to 20% of the weight of the cane). This facilitates extraction of sucrose. In some installations, the juice from the third mill is returned to the first and second mills as maceration water. The concentration of sucrose in juice from the first mill will usually be about 0.2% greater than that from the second mill; that of the third mill will be about 0.5% less than that of the first mill. However, the juice of the third mill will contain greater concentrations of gummy matter and some of the other impurities. The result of the total macerating action, which fully ruptures the plant cells, is a gray- to dark-green, cloudy juice that must be treated to effect a separation of impurities.

Liming and Clarifying. The ancient sugarmakers heated the raw juice and added ashes, causing a precipitation of many of the impurities, but the final product did not approach the purity of cane sugar marketed today which is one of the most highly purified compounds found in commerce. Over the past few centuries, lime has replaced the ashes and, during the past several decades, sulfur dioxide and phosphoric acid or phosphates also have been added to the total clarifying process. Sulfur dioxide bleach acts as an antimicrobial agent, and assists in the coagulation of such substances as albumin present in the juice. Further, the sulfur dioxide makes it possible to use more lime in the clarifying operation. Lime functions in several ways, forming insoluble compounds with several of the impurities present, neutralizes organic acids present, and when added in sufficient quantity, also reacts with the glucose present, converting it to organic acids. Most of the calcium compounds formed are quite insoluble and thus can be removed by settling or filtration. When phosphoric acid is added, the insoluble tricalcium phosphate is formed. A number of researchers have suggested that phosphates other than phosphoric acid may be preferable. Various sodium, ammonium, potassium, and calcium orthophosphates have been proposed as additives to the sugar solution, along with lime. The control of pH is critical if all lime is to be removed from the juice. Jung (U.S. Patent 3,347,705 issued in 1967) developed the use of polyphosphoric acid in combination with a dicarboxylic acid for the clarification of sugar juices. The primary objection to any excess lime in the solution is later scaling that will be caused in processing equipment.

Upon leaving the crushing mill, the juice is first treated with sulfur dioxide. The design of clarifying equipment has changed much during the past few decades. In earlier installations, the clarifiers were rectangular or circular metal pans, each with a capacity up to 1200 gallons (45 hectoliters).

A modern cane juice clarifier or proprietary design is shown in Fig. 1. As shown, the unit is equipped with separate provisions in each compartment for feed, overflow takeoff and mud withdrawal which allows the unit to operate essentially as four totally independent clarifiers enclosed in a common housing. Juice is introduced as the top-center of each compartment through a hollow rotating center tube. This tube is fitted with a series of ports and scalpers that serve as feed introduction points. Located directly below each port, and attached to the center tube, are feed deflection baffles which insure uniform feeding and impede the natural tendency of the incoming juice to mix with the settled muds. Also attached to the center tube are the various sets of rake arms.

As the feed enters each compartment, it first strikes the deflection baffle, then flows outward at a decreasing velocity creating minimum turbulence. The various sets of rotating rake arms move the settled muds to the mud discharge boot located at the center of each tray. The mud is then withdrawn from each compartment separately. Overflow piping removes the clarified juice from each compartment independently at multiple points around the periphery of the clarifier, through a single overflow box where accurate flow distribution is easily maintained and controlled at one point. Standard capacities of the units range from 10,800 gallons (409 hectoliters) and a mud-thickening area of 312 square feet (29 square meters) to 140,800 gallons (5329 hectoliters) and a mud-thickening area of 4068 square feet (378 square meters).

Evaporation. After clarification and filtration, the juice goes to evaporators (vacuum pans), where upon concentration of the solution, small crystals grain out. Continuous evaporation produces a very thick mixture (*masscuite*), which is a mixture of sugar grains that are suspended in thick molasses. This mixture is centrifuged which throws off most of the molasses, leaving raw sugar, sometimes referred to as *centrifugal sugar*. At this point, the sugar is from 96 to 97% pure. The molasses may be reworked 2 or 3 times more to increase the yield of sugar. Although the remaining molasses may contain up to 50% sugar, the impurities present prevent any further formation of crystals. At this point the residue is called blackstrap and is further treated for use in animal feedstuffs. Molasses is described further a bit later.

Cane Sugar Refining. Raw cane sugar mills, as just described, produce the raw sugar. Refiners then further process the raw sugar into the more familiar white crystalline sugar. This 2-stage sugar production process for cane sugar stems from the economics of processing raw sugar in relatively small cane-producing regions and then refining the sugar on a much larger scale, usually thousands of miles closer to the markets. Traditionally, much of the raw sugar was imported from tropical, underdeveloped countries that lacked resources for constructing complete refineries. There are cases, however, where both processing and refining are performed at a single location or where the raw mill and refinery are adjacent and operated under one ownership.

The raw sugar as received at the refinery is mixed with sugar syrup for the purpose of dissolving the molasses residuals which still stick to the crystals. The heavy mixture resulting is sometimes called *magma*. This mix is centrifuged, after which the crystals are steam-treated and at this point are almost white. Again, the sugar crystals are dissolved in sugar syrup and, once again, are treated with lime and phosphoric acid in order to precipitate impurities present. From this operation, the effluent is filtered through bone char to yield a purified solution. Again, the solution is evaporated, crystallized, centrifuged, the final moisture content adjusted, after which the product is packaged.

Beet Sugar Production

The German chemist Marggraf discovered the presence of sugar in beets as early as 1747. Early laboratory methods to extract sugar from the beet proved overwhelmingly costly as contrasted with processing the traditional source, sugar cane. Little progress was made for over a half-century, when in 1802, another German chemist, Achard, found a way to extract sugar from the beet root on a relatively large scale. For a few years, a small manufacturing operation in Silesia prospered, mainly because of political factors that drove up the price of cane sugar. In 1812, Napoleon ordered an establishment of the beet sugar industry in France. Early attempts toward beet sugar extraction were made in the United States (Massachusetts) in 1838, followed by efforts over the subsequent 30 years in Illinois, Wisconsin, and California. The first real success in the United States was achieved in the late 1870s by a factory in Alvarado, California.

The principal operations in sugar beet processing today include thorough washing of the beets, after which whirling knives slice the beets into

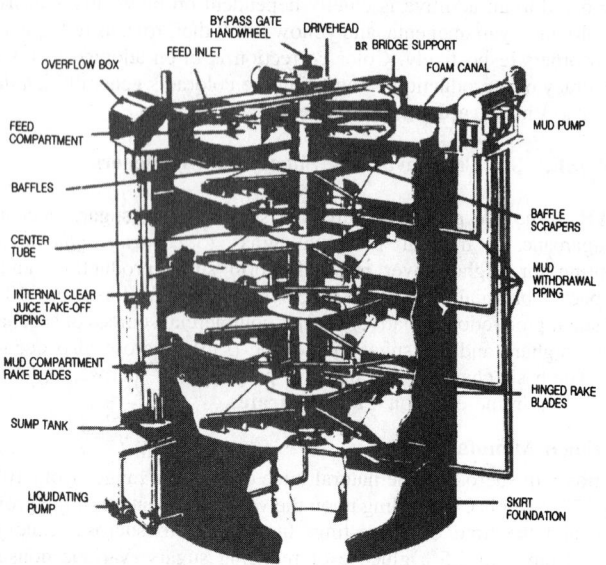

Fig. 1.   Modern cane juice clarifier. (*Dorr-Oliver Rapidorr 444*^{TM})

thin strips, called *cossettes*. These are immersed in hot water where the sugar is removed from the beets by diffusion. The resulting solution is *raw juice*. This juice is purified in a process (carbonation), wherein lime and carbon dioxide are added to cause undesired impurities in the raw juice to precipitate out of the solution (as in the case of cane sugar previously described). This resulting, purified liquid is *thin juice*. Filtering and settling operations remove solid particles and impurities from the thin juice. This juice is concentrated by boiling off water to form *thick juice*. Further filtering ensures that all solid particles are eliminated. Sugar crystals are formed by boiling the thick juice under vacuum. The resulting mixture of crystals and liquid is known as *fillmass*. This mixture is spun and washed in high-speed centrifugals to separate the sugar crystals from the liquid. These crystals are now pure white sugar (sucrose). After further crystallization of the separated liquid, additional sugar and an important by-product (molasses) is obtained. The white sugar crystals are dried by tumbling in warm air in long rotating drums (granulators), after which the sugar is ready for market. The residue of the beets (*pulp*) is sold for livestock feed in either wet or dried form. Some molasses may be added to the pulp prior to drying.

Additional Reading

Considine, D.M., and G.D. Considine, Editors: *Foods and Food Production Encyclopedia,* Van Nostrand Reinhold, New York, NY, 1982.

Junk, W.R., and H.M. Pancoast: *Handbook of Sugars for Processors, Chemists, and Technologists,* Avi, Westport, Connecticut, 1973.

Shallenberger, R.S., and G.G. Birch: *Sugar Chemistry,* Avi, Westport, Connecticut, 1975.

SULFITE PULP PROCESS. See **Pulp (Wood) Production and Processing**.

SULFONAMIDE DRUGS. In 1935, Domagk, a German researcher, was the first to observe the clinical value of *prontosil*, a red compound derived from azo dyes. Paraaminobenzenesulfonamide was shown to be the effective portion of the prontosil molecule. This substance was given the name *sulfanilamide*. This was the first of a group of related drugs to receive wide clinical trial. It was found to be effective in the treatment of hemolytic streptococcal and staphylococcal infections. Within a short span of years, related drugs were synthesized and given clinical trials. These included *sulfapyridine, sulfathiazole, sulfaguanidine, sulfadiazine,* and *sulfamerazine*. These drugs acted by inhibiting the growth of bacteria rather than by killing organisms.

Even though numerous adverse side effects were observed over a period of time, the sulfonamides played an important role in medicine prior to the advent of the antibiotics. In recent years, the importance of the so-called *sulfa drugs* has diminished considerably, but for certain situations they are still considered important antimicrobials. Presently the sulfonamides are mainly used to treat uncomplicated urinary tract infections, including prostatitis, due to *E. coli*. They are also used to treat a number of noncardial infections. At one time the sulfa drugs were widely used in the treatment of meningococcal meningitis and bacillary dysentery. Unfortunately, the bacilli responsible for these diseases developed, over the years, a resistance to the drugs, severely reducing their efficacy.

Within the last few years, some new sulfa drugs have been introduced, including trimethoprim-sulfamethoxazole. This drug has broadened the scope in treatment of urinary tract infections derived from species in addition to *E. coli*, namely, *Klebsiella, Enterobacter,* and *Porteus* species. This drug also is used for the treatment of acute otitis media in children, particularly those instances where strains of *H. influenzae* and *streptococcus pneumoniae* may be suspected. The drug is also used to treat systemic infections that may arise from chloramphenicol- and ampicillin-resistant *Salmonella*; as well as infections attributed to *Pneumocystis carinii*.

Also, the nature of sulfonamide compounds (relatively short duration of action, capability of entering into synergism with other drugs, poor absorption, and topical effectiveness, not to mention relatively low cost) is taken advantage of in what is sometimes called short-acting sulfonamides. Short-acting sulfonamides include sulfisoxazole, sulfadiazine, and trisulfapyrimidines. An intermediate-acting sulfonamide in current use is sulfamethoxazole. This drug does tend to cause renal damage arising from sulfonamide crystalluria.

Sulfacetamide eyedrops continue to be used for treatment of superficial ocular infections. Sometimes silver-sulfadiazine cream is applied to burn surfaces to minimize or prevent bacterial growth, as well as preventing invasive infection.

The adverse effects of sulfonamides include hypersensitivity reactions, as manifested by rashes, photodermatitis (allergic reaction to light), so-called drug fever, nausea, and vomiting. These reactions occur with some frequency when sulfonamides are administered. Less frequently encountered is crystalluria, previously mentioned, but with the risk lessened in the case of sulfisoxazole. Sulfa drugs also occasionally cause hemolytic anemia, agranulocytosis, and kernicterus (in infants) when the drugs are given to nursing mothers. In rare instances, sulfa drugs may precipitate hepatitis, aplastic anemia, renal tubular necrosis, and certain blood disorders.

SULFONATION AND SULFATION. Sulfonation and sulfation, chemical methods for introducing the SO_3 group into organic entities, are related and usually treated jointly.

In sulfonation, an SO_3 group is introduced into an organic molecule to give a product having a sulfonate, CSO_3, moiety. The compound may be a sulfonic acid, a salt, or a sulfonyl halide requiring subsequent alkaline hydrolysis. Aromatic hydrocarbons are generally directly sulfonated using sulfur trioxide, oleum, or sulfuric acid. Sulfonation of unsaturated hydrocarbons may utilize sulfur trioxide, metal sulfites, or bisulfites. The latter two reagents produce the corresponding hydrocarbon metal sulfonate salts in processes referred to as sulfitation and bisulfitation, respectively. Organic halides react with aqueous sodium sulfite to produce the corresponding organic sodium sulfonate. In instances where the sulfur atom at a lower valance is attached to a carbon atom, the sulfonation process entails oxidation. Thus the reaction of a paraffin hydrocarbon with sulfur dioxide and oxygen is referred to as sulfoxidation; the reaction of sulfur dioxide and chlorine is called chlorosulfonation. The sulfonate group may also be introduced into an organic molecule by indirect methods through a primary reaction, e.g., esterification, with another organic molecule already having an attached sulfonate group.

Sulfation is defined as any process of introducing an SO_3 group into an organic compound to produce the characteristic $C-OSO_3$ configuration. Typically, sulfation of alcohols utilizes chlorosulfuric acid or sulfur trioxide reagents. Unlike the sulfonates, which show remarkable stability even after prolonged heat, sulfated products are unstable toward acid hydrolysis. Hence, alcohol sulfuric esters are immediately neutralized after sulfation in order to preserve a high sulfation yield.

In sulfamation, also termed *N*-sulfonation, compounds of the general structure R_2NSO_3H are formed as well as their corresponding salts, acid halides, and esters. The reagents are sulfamic acid (amido-sulfuric acid), SO_3–pyridine complex, SO_3–tertiary amine complexes, aliphatic amine–SO_3 adducts, and chlorine isocyanate–SO_3 complexes.

Uses for Derived Products and Sulfonation Technology

Sulfonation and sulfation processes are utilized in the production of water-soluble anionic surfactants as principal ingredients in formulated light-duty and heavy-duty detergents, liquid hand cleansers, general household and personal care products, and dental care products. Other commercially significant product applications include emulsifiers, lube additives, sweeteners, pesticides, medicinals, ion-exchange resins, dyes and pigments.

Sulfonation and sulfation processes are important tools for organic synthesis of specific molecules and positional isomers.

Application chemists are most interested in physical and functional properties contributed by the sulfonate moiety, such as solubility, emulsification, wetting, foaming, and detersive properties. Products can be designed to meet various criteria including water solubility, water dispersibility, and oil solubility. The polar SO_3 moiety contributes detersive properties to lube oil sulfonates and dry-cleaning sulfonates.

Process Selection and Options. Because of the diversity of feedstocks, no one process fits all needs. An acceptable sulfonation/sulfation process requires (*1*) the proper reagent for the chemistry involved and the ability to obtain high product yields; (*2*) consistency with environmental regulations such that minimal and disposable byproducts are formed; (*3*) an adequate cooling system to control the reaction and to remove significant heat of reaction; (*4*) intimate mixing or agitation of often highly viscous reactants to provide adequate contact time; (*5*) products of satisfactory yields and marketable quality; and (*6*) acceptable economics. Viscosity constraints may play a significant role not only dictating agitation/mixing requirements but also seriously affecting heat-exchange efficiency.

Reagents

Reagents for direct sulfonation and sulfation reactions are listed in Table 1. Unlike sulfuric acid reactions which usually require 3–4 moles per mole of organic feedstock resulting in substantial "spent acid" requiring disposal, SO_3 generally reacts essentially stoichiometrically thus producing high-purity products directly.

By 1987, sulfur trioxide reagent use in the United States exceeded that of oleum for sulfonation. Sulfur trioxide source is divided between liquid SO_3 and *in situ* sulfur burning. The latter is integrated into sulfonation production facilitates.

Liquid SO_3 is commercially available as both unstabilized and stabilized liquids. Unstabilized liquid SO_3 can be utilized without problem as long as moisture is excluded, and it is maintained at ca 27–32°C. Stabilized liquid SO_3 has an average in that should the liquid freeze (16.8°C), in the absence of moisture pickup, the SO_3 remains in the gamma-isomer form and is readily remeltable. Gaseous SO_3 can also be obtained by stripping 70% oleum (70%SO_3 : 30%H_2SO_4) or by utilizing SO_3 converter gas (6–8% SO_3) from H_2SO_4 production, or by vaporizing liquid SO_3 which is then generally diluted with moisture free air.

Sulfur trioxide is an extremely strong electrophile that rapidly seeks to enter into transient or permanent relationships or reactions with organics containing electron donor elements, such as oxygen, nitrogen, halogen, and phosphorus. In some instances, SO_3 may first form a transient intermediate adduct at some moderate temperature, which at some higher temperature becomes unstable, liberating SO_3. This subsequently may react to produce a stable sulfonated product, often accompanied by a difficult to control strong or violent exotherm. The reactivity of SO_3 can be moderated by the use of solvents (such as liquid SO_2, or halogenated hydrocarbons), or by the use of SO_3 adducts, (such as SO_3–Trimethylamine or SO_3–Pyridine).

Sulfonation

All sulfonation is concerned with generating a carbon sulfur(VI) bond in the most controlled manner possible using some form of the sulfur trioxide moiety. Sulfonation can be carried out in a number of ways using the reagents listed in Table 1. Sulfur trioxide is a much more reactive sulfonating reagent than any of its derivatives. Care should be taken with all sulfonating reagents owing to the general exothermic nature of the reaction.

The variety of reagents available makes possible the conversion of a wide range of aromatics into sulfonic acids. The reactivity of compounds that are activated toward electrophilic attack are so high that often alternative reagents are used in order to minimize undesirable by-products largely formed owing to excessive heating.

Aromatic Compounds. The accepted general mechanism for the reaction of an aromatic compound with sulfur trioxide involves an activated intermediate as shown in equation 1.

$$R{-}C_6H_5 + SO_3 \longrightarrow [R{-}C_6H_5SO_3]^* \longrightarrow R{-}C_6H_4SO_3H \qquad (1)$$

The reaction of sulfur trioxide and benzene in an inert solvent is very fast at low temperatures. Yields of 90% benzenesulfonic acid can be expected. Increased yields of about 95% can be realized when the solvent is sulfur dioxide.

Several thousand different synthetic dyes are known, having a total worldwide consumption of 298 million kg/yr. Many dyes contain some form of sulfonate as $-SO_3H$, $-SO_3Na$, or $-SO_2NH_2$.

The world's largest volume synthetic surfactant is linear alkylbenzene sulfonate (LAS), which was developed as a biodegradable replacement for nonlinear alkylbenzene sulfonates (BAB). LAS is derived from the sulfonation of linear alkylbenzene (LAB). Detergent sulfonates use LAB in the 236 to 262 molecular weight range, having a C_{11}–C_{13} alkyl group. The simplest sulfonation route uses 100% sulfuric acid. Continuous falling film SO_3 sulfonation systems utilizing either vaporized and dry air diluted gaseous SO_3 (3–8% SO_3) or in-situ sulfur burning integrated air diluted SO_3 generating systems (3–8% SO_3) have become the method of choice for the sulfonation of most aromatics, as well as for the sulfation of alcohols.

Sulfonated toluene, xylene, and cumene, neutralized to the corresponding ammonium or sodium salts, are important industrially as hydrotropes or coupling agents in the manufacture of liquid cleaners and other surfactant compositions.

Sulfitation and Bisulfitation of Unsaturated Hydrocarbons. Sulfites and bisulfites react with compounds such as olefins, epoxides, aldehydes, ketones, alkynes, aziridines, and episulfides to give aliphatic sulfonates or hydroxysulfonates.

Sulfosuccinates and Sulfosuccinamates. The principal sulfonating reagent in these cases is the bisulfite molecule which readily attacks electron-deficient carbon centers. Variations in the choice of starting material can give a broad spectrum of products of widely varying chemical and physical properties.

TABLE 1. REAGENTS FOR DIRECT SULFONATION AND SULFATION REACTIONS[a]

Reagent	Formula	Physical form	Advantages	Disadvantages	Applications
sulfur trioxide liquid	SO_3	liquid	low cost, concentrated reagent	extremely reactive; charring	very few
gas	SO_3	gas, 3–8% SO_3	low cost, stoichiometric reactions; preferred reagent	requires significant dry diluent gas; mole ratio sensitive; liquid storage	most every sulfonation and sulfation reaction
sulfur burning	SO_3	gas *in situ*, 3–8% SO_3	lowest cost SO_3 produced *in situ*; preferred reagent	catalyst requires startup time; higher investment cost	most every sulfonation and sulfation reaction
chlorosulfuric acid	$ClSO_3H$	liquid	stoichiometric reactions	expensive; produces HCl gas, disposal problem	alcohol sulfation, dyes, etc.
oleum	$H_2SO_4 \cdot SO_3$	liquid	low cost	reactions not stoichiometric; 3–4 mol generally required	dyes, alkylated aromatic sulfonation; continuous sulfation of alcohols
sulfuric acid	H_2SO_4[b]	liquid	low cost, easily handable	reactions not stoichiometric; generally requires 3–4 mol	hydrotrope sulfonation of aromatics using azeotropic water removal, etc.
sodium bisulfite	$NaHSO_3$	solid, 38% liquid	simple processing	higher cost, except for sulfur burning	sulfosuccinates, lignin, olefins, Streker reaction
sodium sulfite	Na_2SO_3	solid, 38% liquid	simple processing	higher cost	Streker reaction, etc.
sodium bisulfite, hydroperoxide catalyst	$NaHSO_3$, O_2	solid, 38% liquid	sulfonation of olefins	requires hydroperoxide catalyst; costly	sulfonation of olefinic hydrocarbons producing primary paraffin sulfonation
sulfamic acid	H_2NSO_3H	solid	stoichiometric reaction, mild, simple	high cost; limited to NH_4 salt; heating to ca 150°C	small specialties, sulfations
sulfuryl chloride	SO_2Cl_2	liquid	few	expensive, usually required catalyst	chlorosulfonation reactions; mostly research
sulfur dioxide and chlorine	SO_2, Cl_2	gases	few, relatively inexpensive	not generally stoichiometric; need catalyst	chlorosulfonation of paraffins, produces HCl
sulfur dioxide and oxygen	SO_2, O_2	gases	few, inexpensive	not stoichiometric; requires catalyst	sulfoxidation of paraffins

[a] In order of descending reactivity.
[b] 93–100%.

Unsaturated Hydrocarbons. The reaction of long-chain, i.e., $C_{12}-C_{18}$, α-olefins with strong sulfonating agents leads to surface-active materials. The overall product of continuous falling film SO_3 sulfonation of α-olefins, termed α-olefin sulfonate (AOS), is really a mixture containing both alkenesulfonates (65–70%) and hydroxyalkanesulfonates (20–25%), along with small amounts of disulfonated products (7–10%). The composition of the final product varies as a result of manufacturing conditions.

AOS prepared from α-olefins in the $C_{12}-C_{18}$ range are most suitable for detergent applications.

Fatty Acid Esters. Fatty acid ester sulfonates are manufactured by reaction of the corresponding hydrogenated (usually methyl) ester and a strong sulfonating agent, such as sulfur trioxide, in order to sulfonate on the alpha-position of the ester. The procedure for the reaction and equipment requirements are very similar to those for the production of LAS. Sodium fatty acid ester sulfonates are known to be highly attractive as surfactants, because they are produced from renewable natural resources and their biodegradability is almost as good as alkyl sulfates.

Petroleum and Related Feedstocks. Petroleum sulfonate by-products were the first petrochemical product. Since that time, By-product petroleum sulfonates have gradually found utilization in a great many applications, including as lubricant additives for high performance engines; as emulsifiers, flotation agents, and corrosion inhibitors; and for enhanced oil recovery. The importance of petroleum sulfonates has grown to the point where these compounds are produced as coproducts, or even as primary petrochemicals.

Factors impacting petroleum sulfonation operations since the late 1970s include the many significant changes and modernizations petroleum refineries have undergone leading to the closing of many refineries practicing oil sulfonation processes; white oil manufacturing technology has eliminated sulfonation and thus sludge disposal to utilize the more cost-efficient hydrogenation process; a principal shift has developed in the use of first-intent oil-soluble synthetic alkylated aromatic sulfonates in place of the traditional petroleum sulfonates for lube additives, and the synthetic sulfonates are made by continuous SO_3 sulfonation processes; and the large projected need for petroleum sulfonates for enhanced oil recovery processes has ceased owing to a significant and prolonged drop in crude oil market prices. Hence there has been a significant drop in the production of natural petroleum sulfonates.

Lignin. Lignosulfonates are complex polymeric materials obtained as by-products of wood pulping where lignin is treated with sulfite reagents under various conditions See also **Pulp (Wood) Production and Processing**. Lignin polymers contain substantial amounts of guaiacyl units, followed by *p*-hydroxyphenyl and syringyl units. Two principal wood pulping processes are utilized: the sulfite process and the kraft process. Sulfonation of lignin mainly occurs on the substituted phenyl–propene precursors at the alphacarbon next to the aromatic ring.

Styrene and Vinyl Monomer, Polymer, and Copolymer Sulfonates. The incorporation of sulfonates into polymeric material can occur either after polymerization or at the monomer stage. The sulfonic acid group is strongly acidic and can therefore be used to functionalize the polymer backbone to the desired degree. The ability of sulfonic acids to exchange counterions has made these polymers prominent in industrial water treatment applications, separators in electrochemical cells, and selective membranes of many types.

The simplest monomer, ethylenesulfonic acid, is made by elimination from sodium hydroxyethyl sulfonate and polyphosphoric acid. Ethylenesulfonic acid is readily polymerized alone or can be incorporated as a copolymer using such monomers as acrylamide, allyl acrylamide, sodium acrylate, acrylonitrile, methylacrylic acid, and vinyl acetate. Styrene and isobutene fail to copolymerize with ethylene sulfonic acid.

Sulfation

Sulfation is the generation of an oxygen sulfur(IV) bond, where the oxygen is attached to the carbon backbone, in the most controlled manner possible, using some form of sulfur trioxide moiety. When sulfating alcohols, the reaction is strongly exothermic. Examples of feedstocks for such a process include alkenes, alcohols, or phenols. Unlike the sulfonates, which exhibit excellent stability to hydrolysis, the alcohol sulfates are readily susceptible to hydrolysis in acidic media. The sulfation of fatty alcohols and fatty polyalkoxylates has produced a substantial body of commercial detergents and emulsifiers.

Linear ethoxylates are the preferred raw materials for production of ether sulfates used in detergent formulations because of uniformity, high purity, and biodegradability. The alkyl chain is usually in the C_{12} to C_{13} range having a molar ethylene oxide: alcohol ratio of anywhere from 1:1 to 7:1. Propoxylates, ethoxylates, and mixed alkoxylates of aliphatic alcohols or alkyl phenols are sulfated for use in specialty applications.

Alcohols and Alkoxylates. The preferred method of sulfation uses some form of a continuous thin-film SO_3 reactor.

Sulfamation

Sulfamation is the formation of a nitrogen sulfur(VI) bond by the reaction of an amine and sulfur trioxide, or one of the many adduct forms of SO_3. Heating an amine with sulfamic acid is an alternative method. A practical example of sulfamation is the artificial sweetener sodium cyclohexylsulfamate, produced from the reaction of cyclohexylamine and sulfur trioxide See also **Sweeteners**. Sulfamic acid is prepared from urea and oleum. Whereas sulfamation is not greatly used commercially, sulfamic acid has various applications.

Industrial Processes

A wide array of industrial processes is suitable for the manufacture of sulfated and sulfonated products. Process selection is dependent on the specific chemistry involved, choice and cost of reagents, physical properties of feedstocks and derived products, product volume requirements, operational mode (batch, continuous), quality of derived products, and possible generation and disposal of by-products as well as operating and equipment investment costs. Another important consideration is the location of the sulfonation plant relative to raw material suppliers, particularly for the more limited liquid SO_3 supplier's plants. On the other hand, molten sulfur used for *in situ* sulfur burning and gaseous SO_3 generation is readily available throughout the United States and worldwide. Another consideration for process selection is plant versatility in sulfonating a variety of feedstocks.

The handling of highly acidic sulfonation reagents and the actual sulfonation processing conditions for the production of acidic reaction products and by-products present a number of corrosion problems which must be carefully addressed. Special stainless steel alloys or glass-lined equipment are often used, although the latter generally has poorer heat-exchange properties. All environment regulations or restrictions must also be met. For example, in utilizing $ClSO_3H$ reagent, HCl gaseous by-product is generated requiring its recovery by adsorption or neutralization.

The viscosity of sulfonation and sulfation reaction mixtures increases with conversion, often producing extremely high viscosities. Sulfonation process design must accommodate such viscosities.

Batch processes are currently used for the manufacture of small volume specialty sulfonates based on H_2SO_4, oleum, $ClSO_3H$, sulfite, or SO_3 reagents. Production of large volume sulfonates or alcohol sulfates generally utilize continuous SO_3 falling-film processes based on multitubular or concentric designed reactor systems.

EDWARD A. KNAGGS
Consultant
MARSHALL J. NEPRAS
Stepan Company

Additional Reading

Andersen, K.K.: in D.N. Jones, ed., *Sulphonic Acids and Their Derivatives,* Vol. 3, Pergamon Press, Oxford, U.K., 1991.

deGroot, W.H.: *Sulphonation Technology in the Detergent Industry,* Kluwer Academic Publishers, Dorrecht, the Netherlands, 1991.

Gilbert, E.E.: *Sulfonation and Related Reactions,* Interscience Publishers, New York, 1965; reprinted by R. E. Kreiger Publ. Co., Melbourne, FL.

Patai, S. and Rappoport, Z. eds.: *The Chemistry of Sulphonic Acids, Esters and Their Derivatives,* John Wiley & Sons, Ltd., Chichester, U.K., 1991.

SULFONE POLYMERS. Polysulfone is a transparent, heat-resistant, ultrastable and high-performance engineering thermoplastic. It is amorphous and has low flammability and smoke emission. Electrical properties are good; the material remains essentially unchanged up to near its glass transition temperature, 190°C (374°F). The molecular structure of polysulfone features the diaryl sulfone group, a group that tends to attract electrons from the phenyl rings. Oxygen atoms para to the sulfone group enhance resonance and produce oxidation resistance. High resonance also strengthens the bonds spatially, fixing the grouping into a planar configuration. Thus, the polymer has good thermal stability and rigidity at high temperatures. Ether linkages provide chain flexibility, thus imparting good impact strength.

The resistance to acids, alkalies, and salt solutions is high and also good in terms of detergents, oils, and alcohols even at elevated temperatures under moderate stress. Polysulfones, however, are attacked by polar organic solvents, such as ketones, chlorinated hydrocarbons, and aromatic hydrocarbons. The material can be used continuously in steam up to temperatures of 93°C (300°F). Maximum stress in water at about 82°C (180°F) is 2000 psi (steady loads) and 2500 psi (intermittent loads). In long-term performance at 150°C (300°F), polysulfone increases about 10% in strength and modulus values, retaining 90% of its dielectric strength and 70% of its impact strength.

Polysulfone is widely used in medical instrumentation and trays for holding instruments during sterilization. Food processing applications, such as piping, scraper blades, steam tables, microwave oven cookware, and beverage dispensing tanks, are numerous. Electrical/electronic applications include connectors, automotive fuses and switch housings, soil bobbins and cores, television components, capacitor film, and structural circuit boards. In chemical processing equipment, uses include corrosion-resistant piping, tower packing, pump parts, filter modules, and membranes. Polysulfone is available in both molding and extrusion grades. A special medical grade is available. Also available are polysulfone compounds with glass fiber or beads, as well as fillers, such as Teflon®.

SULFONIC ACIDS. Sulfonic acids are classically defined as a group of organic acids which contain one or more sulfonic, $-SO_3H$, groups, The general formula of organic sulfonic acids RSO_3H, where the R-group may be derived from many different sources. Typical R-groups are alkane, alkene, alkyne, and arene. The R-group may contain a wide variety of secondary functionalities such as amine, amide, carboxylic acid, ester, ether, ketone, nitrile, phenol, etc. Sulfonic acid derivatives, where the R-group is derived from an inorganic source such as a halide, oxygen (i.e., sulfate), or amine (i.e., sulfamic acid), are often referred to as sulfuric acid derivatives.

Physical Properties

The physical properties of sulfonic acids vary greatly depending on the nature of the R-group. Sulfonic acids can be described as having similar acidity characteristics to sulfuric acid. Sulfonic acids are prone to thermal decomposition, i.e., desulfonation, at elevated temperatures. However, several of the alkane-derived sulfonic acids show excellent thermal stability, as shown in Table 1. Arene-based sulfonic acids are thermally unstable.

Sulfonic acids are such strong acids that in general they can be considered greater than 99% ionized.

Chemical Properties

Sulfonic acids are prepared on a commercial scale by the sulfonation of organic substrates using a variety of sulfonating agents, including sulfur trioxide (diluted in air), sulfur trioxide (in sulfur dioxide), sulfuric acid, oleum (fuming sulfuric acid), chlorosulfuric acid, sulfamic acid, trialkylamine−sulfur trioxide complexes, and sulfite ions. Other methods of sulfonic acid production, practiced on an industrial scale, include the oxidation of thiols, sulfide, disulfides, sulfoxides, sulfones, and sulfinic acids. See also **Sulfonation and Sulfation**.

TABLE 1. PHYSICAL PROPERTIES OF SULFONIC ACIDS

Acid	Mp,°C	Bp,[a]°C	Density d_4^{25}, g/cm^3
methanesulfonic acid	20	122	1.48
ethanesulfonic acid	−17	123	1.33
propanesulfonic acid	−37	159	1.19
butanesulfonic acid	−15	149	1.19
pentanesulfonic acid	−16	163	1.12
hexanesulfonic acid	16	174	1.10
benzenesulfonic acid	44	172[b]	
p-toluenesulfonic acid	106	182[b]	
1-naphthalenesulfonic acid	78	dec	
2-naphthalenesulfonic acid	91	dec	1.44
trifluoromethanesulfonic acid	none	162[c]	1.70

[a] At 133 Pa (1 mm Hg) unless otherwise noted.
[b] At 13.3 Pa (0.1 mm Hg).
[c] At 101.3 kPa = 760 mm Hg.

General Reaction Chemistry of Sulfonic Acids. Sulfonic acids may be used to produce sulfonic acid esters, which are derived from epoxides, olefins, alkynes, allenes, and ketenes, as shown in Figure 1. Phosphorus pentachloride and phosphorus pentabromide can be used to convert sulfonic acids to the corresponding sulfonyl halides.

Halogenation of sulfonic acids, which avoids production of a sulfonyl halide, can be achieved under oxidative halogenation conditions.

Sulfonic acids may be subjected to a variety of transformation conditions. Sulfonic acids may be hydrolytically cleaved, using high temperatures and pressures, to drive the reaction to completion.

Aromatic sulfonic acid derivatives can be nitrated using nitric acid, in H_2SO_4. Sulfones may be treated with hydrazine derivatives to give the corresponding ring-opened sulfonic acid.

Production

At the end of the 1990s, there were four primary methods of sulfonic acid production in the United States: falling film sulfonation; oleum sulfonation; chlorosulfuric acid sulfonation; and SO_3 solvent-based sulfonation.

The vast majority of sulfonic acids were produced using continuous falling film sulfonation technology, which utilizes vaporized SO_3 mixed with air. This technology dominates the sulfonation industry owing to the capability of high product throughput and low by-product waste streams.

Analytical and Test Methods

Modern analytical techniques have been developed for complete characterization and evaluation of a wide variety of sulfonic acids and sulfonates. Titration is the most straightforward method of evaluating sulfonic acids. Spectroscopic methods for sulfonic acid analysis include ultraviolet spectroscopy, infrared spectroscopy, and ^1H and ^{13}C nmr spectroscopy. Modern separation techniques of sulfonates include liquid chromatography and ion chromatography. See also **Chromatography**.

Health and Safety Factors

In general, unneutralized sulfonic acids are regarded as moderate to highly toxic substances. However, slight detoxification, via the introduction of a sulfonic acid moiety, is observed for nitrobenzene and aminobenzene. Sulfonic acids emit toxic SO_x fumes upon heating to decomposition. Halogenated sulfonic acids, such as trifluoromethane sulfonic acid, also release toxic halogen-containing fumes when heated to decomposition.

Sulfonic acids have essentially the same corrosive characteristics as does concentrated sulfuric acid. Detergent-based sulfonic acids pose a contact hazard, as they are very corrosive to the skin.

When sulfonic acids are neutralized to sulfonic acid salts, the materials become relatively innocuous and low in toxicity, as compared to the parent sulfonic acid.

Environmental Issues

Linear alkylbenzenesulfonic acid is the largest intermediate used for surfactant production in the world. Owing to the large volumes of production and consumption of linear alkylbenzenesulfonate, much attention has been paid to its biodegradation and a series of evaluations have been performed to thoroughly study its behavior in the environment. Much less

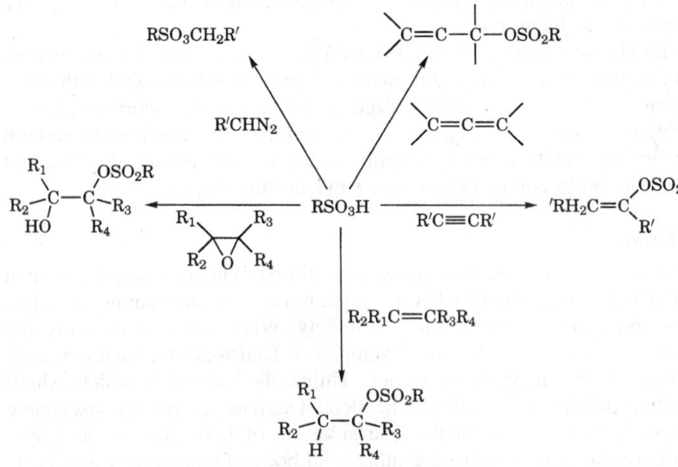

Fig. 1. Reaction chemistry of sulfonic acids

attention has been paid to the environmental impact of other sulfonic acid-based materials.

Linear alkylbenzenesulfonate showed no deleterious effect on agricultural crops exposed to this material. Kinetics of biodegradation have been studied in both wastewater treatment systems and natural degradation systems. Studies have concluded that linear alkylbenzenesulfonate does not pose a risk to the environment. Linear alkylbenzenesulfonate has a half-life of approximately one day in sewage sludge and natural water sources and a half-life of one to three weeks in soils. Aquatic environmental safety assessment has also shown that the material does not pose a hazard to the aquatic environment.

Uses

Surfactants and Detergents Uses. Perhaps the largest use of sulfonic acids is the manufacture of surfactants and surfactant formulations. In almost all cases, the parent sulfonic acid is an intermediate which is converted to a sulfonate prior to use. The largest volume uses for sulfonic acid intermediates are the manufacture of heavy-duty liquid and powder detergents, light-duty liquid detergents, hand soaps (see **Soaps**), and shampoos.

Lignosulfates, a complex mixture containing sulfonated lignin, are used as dispersing agents, wetting agents, binding agents, and sequestering agents. Dry forms of the materials are used as road binders, concrete additives, animal feed additives, and in vanillin production.

Naphthalenic, lignin, and melamine-based sulfonic acids are used as dispersion and wetting agents in industry. The sulfonate (**1**) is also widely used as a dispersing agent in dyestuff manufacture and high temperature dyeing of polyester fibers. A derivative of (**1**) based on 4-aminobenzene sulfonic acid has also been produced.

(**1**)

Other commercial naphthalene-based sulfonic acids, such as dinonylnaphthalene sulfonic acid, are used as phase-transfer catalysts and acid reaction catalysts in organic solvents.

Sulfonic Acid-Based Dyestuffs. Sulfonic acid-derived dyes are utilized industrially in the areas of textiles, paper, cosmetics, foods, detergents, soaps, leather, and inks, both as reactive and disperse dyes. Of the principal classes of dyes, sulfonic acid derivatives find utility in the areas of acid, azoic, direct, disperse, and fiber-reactive dyes. Sulfonic acid-based azo dyes and intermediates are characterized by the presence of one or more azo, RN=NR, groups.

Amide-Based Sulfonic Acids. The most important amide-based sulfonic acids are the alkenylamidoalkanesulfonic acids. These include 2-acrylamidopropanesulfonic acid, 2-acrylamido-2-methylpropanesulfonic acid, 3-acrylamido-2,4,4-trimethylpentanesulfonic acid, 2-acrylamido-2-(*p*-tolyl)ethanesulfonic acid, and 2-acrylamido-2-pyridylethanesulfonic acid.

Biological Uses

Taurine (2-aminoethanesulfonic acid), is the only known naturally occurring sulfonic acid. The material is an essential amino acid for cats and is used extensively by Ralston Purina Company as a food supplement in cat food manufacture.

Sulfonic acids have found greatly expanded usage in biological applications. Whereas the toxicity of sulfonic acids is in general rather high, several sulfonic acids are beneficially utilized *in vivo*. Taurocholic acid is an important bile component, aiding in the digestion of fat.

Potent inhibition of the herpes simplex virus has been observed using biphenyl disulfonic acid urea copolymers. Sulfonic acid derivatives have been shown to be potent antihuman immunodeficiency virus (anti-HIV) agents.

Other Applications. Hydroxylamine-*O*-sulfonic acid has many applications in the area of organic synthesis. The acid has found application in the preparation of hydrazines from amines, aliphatic amines from activated methylene compounds, aromatic amines from activated aromatic compounds, amides from esters, and oximes.

Petroleum sulfonates have found wide usage in enhanced oil recovery technology.

A variety of barium sulfonates have found use in antifriction lubricants for high speed bearing applications. Calcium and sodium salts of sulfonated olefins, esters, or oils are used for the enhancement of extreme pressure properties of grease and gear lubricants.

PAUL S. TULLY
Stepan Company

Additional Reading

Bank, R.E. and R.N. Hazeldine: *The Chemistry of Organic Sulfur Compounds,* Vol. 2, Pergamon Press, Inc., New York, 1966.

Knaggs, E.A.: *CHEMTECH*, 436–445 (July 1992), for a review of major surfactant sulfonic acids.

Sandler, S.R. and W. Karo: *Organic Functional Group Preparation,* Vol. I, Academic Press, Inc., New York, 1983.

United States International Trade Commission, "Synthetic Organic Chemicals, United States Production and Sales, 1991," USITC Publication 2607, Washington, D.C., Feb. 1993, pp. 12–3, 12–11–12-14.

SULFOXIDES. Sulfoxides are compounds that contain a sulfinyl group covalently bonded at the sulfur atom to two carbon atoms. They have the general formula RS(O)R′, ArS(O)Ar′, and ArS(O)R, where Ar and Ar′ = aryl. Sulfoxides represent an intermediate oxidation level between sulfides and sulfones. The naturally occurring sulfoxides often are accompanied by the corresponding sulfides or sulfones. The only commercially important sulfoxide is the simplest member, dimethyl sulfoxide (DMSO) or sulfinylbismethane.

Sulfoxides occur widely in small concentrations in plant and animal tissues.

Properties

For the most part, sulfoxides are crystalline, colorless substances, although the lower aliphatic sulfoxides melt at relatively low temperatures. The lower aliphatic sulfoxides are water soluble; but as a class the sulfoxides are not soluble in water. They are soluble in dilute acids and a few are soluble in alkaline solution. DMSO is a colorless liquid; selected properties are listed in Table 1. Dimethyl sulfoxide generally undergoes typical sulfoxide reactions. It is used herein as an illustrative example.

Thermal Stability. Dimethyl sulfoxide decomposes slowly at 189°C to a mixture of products that includes methanethiol, formaldehyde, water, bis(methylthio)methane, dimethyl disulfide, dimethyl sulfone, and dimethyl sulfide. The decomposition is accelerated by acids, glycols, or amides. Sulfoxides undergo oxidation, reduction, carbonsulfide cleavage, and Pummerer reactions.

Methylsulfinyl Carbanion. Strong bases, e.g., sodium hydride or sodium amide, react with DMSO producing solutions of methylsulfinyl carbanion, known as the dimsyl ion, which are synthetically useful. The solutions also provide a strongly basic reagent for generating other carbanions.

TABLE 1. SELECTED PROPERTIES OF DIMETHYL SULFOXIDE

Property	Value		
boiling point, °C	189.0		
conductivity, at 20°C, S/cm	3×10^{-8}		
dielectric constant, at 25°C, 10 MHz	46.7		
dipole moment, C·m[a]	1.4×10^{-29}		
entropy of fusion, J/(mol·K)[b]	45.12		
free energy of formation gas, C_{graph}, S_2(g), at 25°C, kJ/mol[b]	115.7		
freezing point, °C	18.55		
refractive index, n^{25}D	1.4768		
flash point, open cup, °C	95		
density, g/cm³, at 25°C	1.0955		
viscosity, mPa·s(= cP)	1.996^{25}	1.396^{45}	0.68^{100}

[a] To convert C·m to debye, divide by 3.336×10^{-30}.
[b] To convert J to cal, divide by 4.184.

Methoxydimethylsulfonium and Trimethylsulfoxonium Salts. Alkylating agents react with DMSO at the oxygen. For example, methyl iodide gives methoxydimethylsulfonium iodide as the initial product. The alkoxysulfonium salts are quite reactive and, upon continued heating, either decompose to give carbonyl compounds or rearrange to the more stable trimethylsulfoxonium salts.

Complexes. The sulfoxides have a high (ca 4) dipole moment, which is characteristic of the sulfinyl group, and a basicity about the same as that of alcohols. They are strong hydrogen-bond acceptors. They would be expected, therefore, to solvate ions with electrophilic character, and a large number of DMSO complexes of metal ions have been reported.

Synthesis and Manufacture

The sulfoxides are most frequently synthesized by oxidation of the sulfides.

Dimethyl Sulfoxide. Dimethyl sulfoxide is manufactured from dimethyl sulfide (DMS), which is obtained either by processing spent liquors from the kraft pulping process or by the reaction of methanol or dimethyl ether with hydrogen sulfide.

Health and Safety Factors

Dimethyl sulfoxide is a relatively stable solvent of low toxicity. However, DMSO can penetrate the skin and may carry with it certain chemicals with which it is combined under certain conditions. Dimethyl sulfoxide has received considerable attention as a useful agent in medicine. In veterinary medicine, DMSO is used for horses and dogs as a topical application to reduce swelling resulting from injury or trauma.

Uses of Dimethyl Sulfoxide

Polymerization and Spinning Solvent. Dimethyl sulfoxide is used as a solvent for the polymerization of acrylonitrile and other vinyl monomers, and as a reaction solvent for other polymerizations. It is also used as a solvent for displacement reactions, solvent for base-catalyzed reactions, extraction solvent, solvent for electrolytic reactions, cellulose solvent, pesticide solvent, and clean-up solvent.

Additional Reading

Epstein, W.W. and F.W. Sweat: *Chem. Rev.* **67**(3), 247 (1967).

Martin, D. and H.G. Hauthal: *Dimethyl Sulfoxide,* Halsted Press, a division of John Wiley & Sons, Inc., New York, NY, 1975.

Thyagarajan, B.S. and N. Kharasch: *Intrascience Sulfur Reports,* Vol. 1, The Chemistry of DMSO, Intrascience Research Foundation, Santa Monica, CA, 1966.

SULFUR. [CAS: 7704-34-9]. Chemical element, symbol S, at. no. 16, at. wt. 32.064, periodic table group 16, mp 112.8°C (rhombic), 119.0°C (monoclinic), 120.0°C (amorphous), bp 444.7°C (all forms), sp gr 2.07 (rhombic), 1.96 (monoclinic), 2.046 (amorphous). Atomic weight varies slightly because of naturally occurring isotopes 32, 33, 34, and 36, the total possible variation amounting to ±0.003.

The stable isotopes of sulfur are ^{32}S, ^{33}S, ^{34}S, and ^{36}S. There are three known radioactive isotopes, ^{31}S, ^{35}S, and ^{37}S, with ^{35}S having the longest half-life (87.1 days). See also **Radioactivity**. Electronic configuration $1s^2 2s^2 2p^6 3s^2 3p^4$. Ionic radius S^{2-} 1.855 Å, S^{6+} 0.29 Å (Pauling). Covalent radius 1.07 Å. In terms of abundance, sulfur ranks fourteenth among the elements occurring in the earth's crust, with an estimated 520 grams per metric ton. In seawater, the element ranks fifth, with an estimated 894 grams per metric ton.

First ionization potential 10.357 eV; second, 23.3 eV; third, 34.9 eV; fourth, 47.08 eV; fifth, 63.0 eV; sixth 87.67 eV. Oxidation potentials $H_2S(aq) \longrightarrow S + 2H^+ + 2e^-$, -0.141 V; $H_2SO_3 + H_2O \longrightarrow SO_4^{2-} + 4H^+ + 2e^-$, -0.20 V; $S + 3H_2O \longrightarrow H_2SO_3 + 4H^+ + 4e^-$, -0.45 V; $SO_3^{2-} + 2OH^- \longrightarrow SO_4^{2-} + H_2O + 2e^-$, 0.90 V; $S^{2-} \longrightarrow S + 2e^-$, 0.508 V; $HS^- + OH^- \longrightarrow S + H_2O + 2e^-$, 0.478 V. Other important physical properties of sulfur are given under **Chemical Elements**.

Sulfur has a large number of allotropes. The ordinary form, α-sulfur, is rhombic having a crystal unit cell composed of sixteen S_8 molecules. At 95.5°C it undergoes transition to β-sulfur, which is monoclinic and also has a molecular weight (in solution in carbon disulfide) corresponding to S_8. Four other monoclinic forms have been identified microscopically: γ-sulfur, prepared by heating α-sulfur to 150°C, cooling to 90°C, and inducing crystallization by friction, ρ-sulfur, S_6, prepared by extracting

an acidulated sodium thiosulfate solution with toluene, as well as δ-sulfur, and λ-sulfur. There is also a tetrahedral form, θ-sulfur, crystallized from a carbon disulfide solution of rhombic sulfur treated with balsam. The first liquid form to appear is λ-sulfur, a pale yellow liquid, obtained on heating sulfur to 120°C. Above 160°C, this form changes to a viscous, dark-brown liquid consisting mainly of μ-sulfur. A third liquid allotrope, π-sulfur is considered to exist in molten sulfur, in equilibrium with the other two forms, having its greatest concentration at about 180°C. Sulfur vapor has been shown to contain S_8, S_6, S_4, and S_2 molecules. Several other allotropes of sulfur have been produced, including two paramagnetic forms, purple and green in color, by low-temperature processing.

Sulfur occurs as free sulfur in many volcanic districts, and may have been formed in part by sublimation, by decomposition of hydrogen sulfide, or metallic sulfides, or by organic agencies. It is often associated with limestones and gypsum. Sulfur is found in Spain, Iceland, Japan, Mexico, and Italy. It occurs especially in Sicily, which was the producer for the world until about the beginning of the twentieth century, when Herman Frasch, by inventing the superheated-water method of mining sulfur, made available the great Louisiana and Texas deposits. This method of mining is at the same time a method of purifying sulfur, because in the process of heating, accompanying materials remain unmelted at the temperature at which sulfur melts and is drawn off. In the Louisiana and Texas deposits the sulfur is associated with gypsum, occurring in the caprock overlying the salt plugs that have pierced the strata underlying the Gulf coastal plain. In the United States, sulfur is also found in California, Colorado, Nevada, and Wyoming. Sulfur also occurs as (1) sulfides, e.g. cobaltite, iron disulfide, pyrite, FeS_2, lead sulfide, galenite, PbS, copper iron sulfide, copper pyrite, $CuFeS_2$, zinc sulfide, zinc blende, ZnS, mercury sulfide, cinnabar, HgS; and (2) as sulfates, e.g., calcium sulfate, gypsum, $CaSO_4 \cdot 2H_2O$, barium sulfate, barite, $BaSO_4$. Several of these minerals are described under separate alphabetical entries.

Sulfur Production and Use

The manufacture of H_2SO_4 accounts for nearly 90% of all sulfur consumed. Of this, about 50% of the H_2SO_4 goes into fertilizer production, nearly 20% into chemical manufacture, 5% into pigments, about 3% each for iron and steel production and the manufacture of rayon and synthetic films, and about 2% for various petroleum processes. The balance of over 15% of H_2SO_4 is consumed by a large number of other industries, this all giving credence to the use of H_2SO_4 production figures as an overall economic index. The 10% of the sulfur not going into H_2SO_4 is converted into numerous chemicals that are consumed by a variety of industries, the largest among these being pulp and paper production and the manufacture of carbon disulfide.

Sulfur Compounds

In addition to the compounds described in the following paragraphs, see also **Hydrogen Sulfide**; **Mercaptans**; **Sodium Thiosulfate**; **Sulfuric Acid**; **Sulfurous Acid**; **Thiocyanic Acid**; **Thioethers**; **Thiophene**; and **Thiourea**.

Sulfur-Oxygen Compounds. Due to its $3s^2 3p^4$ electron configuration sulfur, like oxygen, forms many divalent compounds with two covalent bonds and two lone electron pairs, but d-hybridization is quite common, to form compounds with oxidation of 4+ and 6+.

A number of suboxides of sulfur have been reported, but in general their composition has not been clearly established. Polysulfur oxides of formula $S_{8-16}O_2$ are formed by reaction of hydrogen sulfide and sulfur dioxide. Also, when sulfur is burned with oxygen in very limited supply disulfur monoxide, S_2O is formed. This has the structure

$$\overset{\cdot\cdot}{\underset{\displaystyle \cdot\overset{\cdot\cdot}{\underset{\cdot\cdot}{S}}\cdot \quad \cdot\overset{\cdot\cdot}{\underset{\cdot\cdot}{O}}}{S}}$$

A mixture of sulfur dioxide, SO_2, and sulfur vapor, at low pressure and with an electric discharge, forms sulfur monoxide, SO. Its presence is shown from its absorption spectrum, but upon separation it disproportionates at once to sulfur and SO_2. Sulfur sesquioxide, S_2O_3, is formed by reaction of powdered sulfur with anhydrous SO_3; S_2O also disproportionates (at 20°C in nitrogen) to sulfur and SO_2. Sulfur dioxide, SO_2, is formed by the combustion in air or oxygen of sulfur and sulfur compounds generally, except those in which sulfur is in a higher state of oxidation. Sulfur dioxide has an O—S—O bond angle of 119.5°. The sigma bonds utilize essentially sulfur p orbitals, with dp hybridization for the pi bonds. Its oxidation to sulfur trioxide, SO_3, by atmospheric oxygen attains a significant rate only at higher temperatures, but can be materially increased by catalysts. Sulfur trioxide is also evolved from oleum on heating. It exists in the vapor state chiefly as the planar monomer, in which the oxygen atoms are spaced symmetrically (120° angles) about the sulfur atom, and it has S—O bond lengths of 1.43 Å. Liquid SO_3 is partly trimerized, and exists in three physical forms.

Sulfur tetroxide is formed by reaction of pure oxygen and sulfur dioxide under the silent electric discharge. It is not obtained pure, but in a variable SO_3/SO_4 ratio, and as a polymerized white solid. Another peroxide, $(SO_2OOSO_2O)_x$, which is written as S_2O_7, is known.

Of the 16 oxyacids of sulfur that are recognized, only four have been isolated. The more important oxyacids of sulfur are: (1) Thiosulfurous acid, $H_2S_2O_2$, structure not established, existing only in compounds, an oxidizing agent for Fe^{2+}, H_2S and HI; (2) Sulfoxylic acid, H_2SO_2, existing only in salts and other compounds, e.g., $ZnSO_2$, SCl_2, $S(OR)_2$, structure probably

$$\text{H}:\ddot{\text{O}}:\text{S}:\ddot{\text{O}}:\text{H}$$

(3) Dithionous acid (or hydrosulfurous acid), $H_2S_2O_4$, existing only in compounds, widely used reducing agent, chiefly as the sodium salt, for organic substances, also reduces Sb^{3+}, Ag^+, Pb^{2+}, Cu^{2+} to the elements, structure

$$\text{H}:\ddot{\text{O}}:\ddot{\text{S}} \quad \ddot{\text{S}}:\ddot{\text{O}}:\text{H}$$

(4) Sulfurous acid, H_2SO_3, produced by hydration of SO_2, not isolated but existing in many salts, the sulfites and acid sulfites, and many organic compounds, including the dialkyl or diaryl sulfites and the alkyl or aryl sulfonic acid esters, which suggests two possible structures, $(HO)_2SO$ and $H(HO)SO_2$, although the acid dissociation constants (first, 1.25×10^{-2}, and second, 5.6×10^{-8}) suggest the structure with only one unhydrogenated oxygen atom. Sulfurous acid and sulfites are fairly strong reducing agents, but the HSO_3^- ion may act as an oxidizing agent, as for formates and related compounds. Other compounds of SO_2 are the metabisulfites or pyrosulfites, containing the ion

$$^-:\ddot{\text{O}}:\ddot{\text{S}}:\ddot{\text{O}}:\ddot{\text{S}}:\ddot{\text{O}}:^-$$

which enters into equilibrium with water to form acid sulfite. (5) Thiosulfuric acid, $H_2S_2O_3$, existing only in compounds, the anion having the structure

$$^-:\ddot{\text{O}}:\ddot{\text{S}}:\ddot{\text{O}}:^-$$

and widely used as a coordinating ion for forming complexes with metals; it also is an oxidizing agent, and is used in iodometric titrations. (6) Dithionic acid, $H_2S_2O_6$, existing only in compounds but stable in dilute solution at room temperature, and differing in its stability to hydrolysis and oxidation from the polythionates,

$$^-:\ddot{\text{O}}:\ddot{\text{S}} \quad \ddot{\text{S}}:\ddot{\text{O}}:^-$$

(7) Polythionic acids, $H_2S_nO_6$, in which n has values of 3, 4, 5, 6 and others, some of which have been reported to have values indefinitely high (20–80), structure not established, though there is evidence that they consist of two sulfonic acid groups connected by a linear chain of sulfur atoms. An interesting property of the polythionates that are very rich in sulfur ($n > 20$) is their slight tendency to decompose to give free S. (8) Sulfuric acid, H_2SO_4, structure

$$\text{H}:\ddot{\text{O}}:\ddot{\text{S}}:\ddot{\text{O}}:\text{H}$$

strong acid, formed by hydration of sulfur trioxide, completely dissociated (first ionization) in aqueous solutions up to 40%; above that concentration dissociation decreases and hydrate formation occurs. Both normal and acid sulfates are formed by metallic elements, though the products of their direct reaction with the acid vary with temperature. (9) Sulfuric acid dissolves SO_3, the product of a 1:1 ratio being pyrosulfuric or disulfuric acid. $H_2S_2O_7$, which forms the pyrosulfates, also obtainable by heating acid sulfates, structure HO(O)(O)SOS(O)(O)OH. Two series of alkali metal pyrosulfates are known: those formed from SO_3 and the metal sulfates and those formed from H_2SO_4 and the metal sulfates, which have the pyrosulfuric acid structure. (10) Peroxymonosulfuric acid is produced by addition of SO_3 to concentrated H_2O_2, its salts are fairly stable, and it has the structure HOS(O)(O)OOH. (11) Peroxydisulfuric acid is produced by reaction of concentrated H_2O_2 on H_2SO_4 or by electrolysis of acid sulfate solutions; its salts are fairly stable and it has the structure HOS(O)(O)OOS(O)(O)OH.

Hydrogen Sulfide. H_2S is a weak acid ($pK_{A1} = 7.00$), ($pK_{A2} = 12.92$) stronger than water but weaker than H_2Se, as expected from its position in the periodic system; its reducing strength exhibits the same relation. Its long use in analytical chemistry is due to the differential solubility of many sulfides with variation of the pH of an aqueous solution. Hydrogen persulfide, H_2S_2, structure HSSH, with an S—S bond distance of 2.05 Å, formed from an alkali metal polysulfide solution and HCl at low temperatures, is the first of a group of hydrogen polysulfides of the general formula H_2S_x.

Sulfur Halides. Many are known. Those that have been identified and whose properties have been determined include the fluorine compounds, S_2F_2, SF_4, SF_6, S_2F_{10}, the chlorine compounds, S_2Cl_2, SCl_2, SCl_4 and the bromine compound, S_2Br_2. Sulfur chlorides of general formula S_nCl_2 are known up to $n \approx 20$. A similar series of cyanides, $S_n(CN)_2$, is known. Derivatives of SCl_4, e.g., SCl_3CN, have been prepared and the list of derivatives of SF_6 is rapidly growing, including S_2F_{10}, $(SF_5)_2O$, $(SF_5)_2O_2$, SF_5Cl, SF_5Cl_3, $SF_4(CF_3)_2$, $(SF_5)_2CF_2$, SF_5OF, SF_5OSO_2F, etc. Derivatives of SF_4 include $C_6H_5SF_3$, $(SF_3)_2CF_2$, etc. All of them except the higher fluorides hydrolyze readily, and they are essentially covalent in character. The simple compounds can be prepared directly from the elements, the activity of the halogen determining the product obtained: fluorine yielding SF_6 and S_2F_{10} and the other fluorides being prepared from those; chlorine and bromine yielding the monohalides, from which the others are obtained by continued halogenation.

Sulfur Oxyhalides. Four general compositions of oxyhalides of sulfur have been known for many years. In one of these, sulfur has a 4+ oxidation state, the thionyl halides, SOX_2, and in three of which it has a 6+ oxidation state, thionyl tetrafluoride, the sulfuryl and pyrosulfuryl halides, SOF_4, SO_3X_2 and $S_2O_5X_2$, respectively. As is the case for the simple halides, no iodine compounds are known, but polyhalogen ones, such as SOFCl and SO_2FCl exist.

Isolable Oxysulfuranes. Sulfuranes, as described by Musher (1969), are compounds of sulfur(IV) in which four ligands are attached to sulfur and have in common with rare-gas compounds such as XeF_2 an electronic structure involving a formal expansion of the valence shell of the central atom from 8 to 10 electrons. Martin and Perozzi (1976) pointed out that the incorporation of oxygen ligands makes possible a wide range of new structural types that illustrate structure-reactivity relationships in a particularly illuminating way.

For many years, it was postulated that most types of sulfuranes were intermediate (not isolable) compounds. However, the isolable halosulfuranes have been well established for many years. The first known of these, SCl_4, was prepared by Michaelis and Schifferdecker in 1873. In 1911, it was found that SF_4, while highly reactive, was thermally stable. However, the compound was not fully described until 1929. Development of SF_4 led to the creation of a family of stable fluorosulfuranes and their derivatives. It was found that the fluorines in these compounds can be replaced by aryl or perfluoroalkyl groups (Tyczkowski, 1953). Kimura and Bauer (1963) described the geometry of SF_4 as a distorted trigonal bipyramid with two fluorines and a lone pair of electrons occupying equatorial positions, with the other two fluorines in apical positions. The postulated structures of SF_4 (a), and of a derivative (b) are shown below:

(a)   (b)

In the early 1970s, Sheppard, by reacting SF_4 with pentafluorophenyllithium, prepared an isolated sulfurane with four carbon-centered ligands, namely, *tetrakis*-(pentafluorophenyl)sulfurane, $(C_6F_5)_4S$. Martin and Perozzi (1976) prepared the first isolable diaryldialkyloxysulfurane. If it is protected against moisture, the researchers found the compound to be stable over an indefinite period at room temperature. The research in this interesting area continues, some of the details of which are well described by Martin–Perozzi (1976). Summarizing the situation, the researchers observe that the development of synthetic methods for oxysulfuranes has made a wide range of isolable compounds of hypervalent sulfur available for study. Structure-reactivity correlations are now becoming evident as a result of such study. The fact that oxygen is dicoordinate makes it possible to sythesize cyclic oxysulfuranes and to use the pronounced changes of reactivity which accompany cyclization to design new, potentially useful sulfurane reagents stable enough to allow isolation.

Sulfur-Nitrogen Compounds. Many of the sulfur-nitrogen compounds are sulfuric acid derivatives. Three of these compounds correspond to replacement of the hydrogen atoms of ammonia with one, two and three—SO_3H radicals, the monosubstituted compound being aminesulfonic (sulfamic) acid, and being readily separated, the others known only in their salts, the aminedisulfonates (imidodisulfonates) and aminetrisulfonates (nitrilotrisulfonates). Other amines, such as hydroxylamine and hydrazine have similarly related compounds. See also **Hydrazine**; and **Hydroxylamine**. Diamino derivatives of the sulfoxy acids are also known, such as sulfamide, $H_2NSO_2NH_2$. Imidosulfinamide, $HN (SONH_2)_2$, has been prepared by reaction of $SOCl_2$ and ammonia (also directly from SO_2 and ammonia), and a trimer of sulfimide, $(O_2SNH)_3$, by ammoniation of SO_2Cl_2. It is cyclic in structure, composed of alternate $>NH$ and $>SO_2$ groups. Nitrosulfonates, containing the ion $SO_3 NO^-$ and dinitrososulfonates, containing $SO_3N_2O_2^{2-}$, are also known.

The most important sulfur-nitrogen compound is tetrasulfur tetranitride, S_4N_4, prepared in many ways, including the direct reaction of ammonia and sulfur. All data on its structure are in accord with a puckered eight-member ring, or a cage with N—S connections. The question as to whether there are also transannular N—N or S—S bonds has not been clearly settled. On hydrogenation it adds 4 H atoms, on fluorination it forms $S_4N_4F_4$, structure

and SN_2F_2 the latter reacting with SNF to form SNF_3, structure F_2SNF. Other thiazyl compounds, prepared from S_4N_4 and the halogens or sulfur halides, include $(ClSN)_3$, S_4N_3Cl, S_4N_3Br, S_4N_3I. These last are salts, i.e., $[N_4S_3]X$, and salts of other anions can also be prepared. Other sulfur-nitrogen compounds known are SN_2, S_4N_2, S_5N_2, and S_2N_2, the last being formed by heating S_4N_4.

Thiocyanogen, $(SCN)_2$, is formed by treatment of a metal thiocyanate with bromine in an organic solvent. It reacts with organic compounds in a manner completely analogous to the free halogens, lying between bromine and iodine in oxidizing power. The alkali metal and alkaline earth metal thiocyanates are prepared by fusing the cyanides with sulfur, and the other metal thiocyanates, as well as the organic ones, are usually prepared from the alkali metal thiocyanates. Many selenium analogs of thio compounds can be made, including $SeSO_3$, SO_3Se^{2-}, SSe^{2-}, etc.

In addition to carbon disulfide (odorless when pure), carbon subsulfide, $S=C=C=C=S$, an evil-smelling red oil and carbon monosulfide, (CS_x), are known as well as COS, CSSe and CSTe. Because of its similarity to oxygen, and the reactivity of its acids, sulfur enters widely into organic compounds.

For biological aspects of sulfur. See **Sulfur (In Biological Systems)**.

Additional Reading

Dalrymple, D.A. and T.W. Trofe: "An Overview of Liquid Redox Sulfur Recovery," *Chem. Eng. Progress,* 43 (March 1989).

Greenwood, N.N. and A. Earnshaw: *Chemistry of the Elements,* 2nd Edition, Butterworth-Heinemann, Inc., Woburn, MA, 1997.

Kent, J.A.: *Riegel's Handbook of Industrial Chemistry,* 9th Edition, Chapman & Hall, New York, NY, 1992.

Kimura, K. and S.H. Bauer: *J. Chem. Phys.,* **39**, 3172 (1963).

Krebs, R.E.: *The History and Use of Our Earth's Chemical Elements: A Reference Guide,* Greenwood Publishing Group, Inc., Westport, CT, 1998.

Lewis, R.J. and N.I. Sax: *Sax's Dangerous Properties of Industrial Materials,* 10th Edition, John Wiley & Sons, Inc., New York, NY, 2000.

Lide, D.R.: *CRC Handbook of Chemistry and Physics,* 84th Edition, CRC Press, LLC, Boca Raton, FL, 2003.

Loretta, J. and P.W. Atkins: *Chemistry: Molecules, Matter and Change,* W.H. Freeman and Company, New York, NY, 1999.

Martin, J.C. and E.F. Perozzi: "Isolable Oxysulfurances in Organic Chemistry," *Science,* **191**, 154–159 (1976).

Meyers, R.A.: *Handbook of Chemicals Production Processes,* The McGraw-Hill Companies, Inc., New York, NY, 1986.

Mollare, P.D.: "From Calcasieu to Caminada: A Brief History of the Louisiana Sulfur Industry," *Chem. Eng. Progress,* 73 (March 1989).

Parker, S.P.: *McGraw-Hill Encyclopedia of Chemistry,* 2nd Edition, The McGraw-Hill Companies, Inc., New York, NY, 1993.

Stwertka, A. and E. Stwertka: *A Guide to the Elements,* Oxford University Press, Inc., New York, NY, 1998.

Trofe, T.W., D.A. Dalrymple, and F.A. Scheffel: *Stetford Process Status and R&D Needs,* Topical Report GRI-87/0021, Gas Research Institute, Chicago, IL, 1987.

Tyczkowski, E.A. and L.A. Bigelow: *J. Amer. Chem. Soc.,* **75**, 3523 (1953).

SULFURIC ACID. Infrequently termed "oil of vitriol," sulfuric acid, [CAS: 7664-93-9], H_2SO_4, is a colorless, oily liquid, dense, highly reactive, and miscible with water in all proportions. Much heat is evolved when concentrated sulfuric acid is mixed with water and, as a safety precaution to prevent spluttering, the acid is poured into the water rather than vice versa. Sulfuric acid will dissolve most metals. The concentrated acid oxidizes, dehydrates, or sulfonates most organic compounds, sometimes causing charring. There are numerous commercial and industrial uses for H_2SO_4 and these include the manufacture of fertilizers, chemicals, inorganic pigments, petroleum refining, etching, as a catalyst in alkylation processes, in electroplating baths, for pickling and other operations in iron and steel production, in rayon and film manufacture, in the making of explosives, and in nonferrous metallurgy, to mention only some of its numerous uses. Because of its wide use industrially, some economists over the years have included sulfuric acid consumption among their economic indicators.

Most countries with significant industrial activity and particularly in chemicals production will have significant capacities for making sulfuric acid. In some countries, H_2SO_4 is the leading chemical in terms of tonnage production. Depending upon suppliers, H_2SO_4 is commercially available in a number of strengths, ranging from 77.7% H_2SO_4 (60° Baumé, sp. gr. 1.71) through 93.2% H_2SO_4 (66° Baumé), 98% H_2SO_4, 99% H_2SO_4, and 100% H_2SO_4 (sp. gr. 1.84).

Fundamentally, there are two kinds of sulfuric acid plants: (1) those that use the dry gas (sulfur burning) process; and (2) those that use the wet gas process. In the first type, the raw materials are elemental sulfur and water. In the second type, the sulfur dioxide feed may come from a variety of sources, including metallurgical smelters (copper, zinc, lead, etc.), pyrite roasters, waste acid decomposition furnaces, and hydrogen sulfide burners. In these plants, the SO_2 gas stream enters the acid plant containing a large amount of water vapor. The gas is usually hot (260–430°C) and dusty, and also may contain a number of impurities, such as fluorides, that could harm the catalyst in the contact section of the plant. These incoming gases thus require cooling and purification in the series of scrubbers and electrostatic precipitators, followed by drying prior to entering the contact section of the plant.

In either type of plant, sulfur dioxide is converted to sulfur trioxide in the contact portion of the plant. The reaction $SO_2 + \frac{1}{2}O_2 \longrightarrow SO_3$ is effected by passing the SO_2 over a catalyst, usually vanadium pentoxide (V_2O_5). The catalyst in the converter vessel is usually in the form of small pellets and typically arranged in four layers. Provision is made for removal of the heat of reaction after each layer or stage. The catalyst may be used for a number of years with only a very moderate decrease in activity.

From this fundamental point, the sulfuric acid plant designer has a number of alternatives and options to consider. Two factors are of major import in sulfuric acid plant design today, namely, recovery

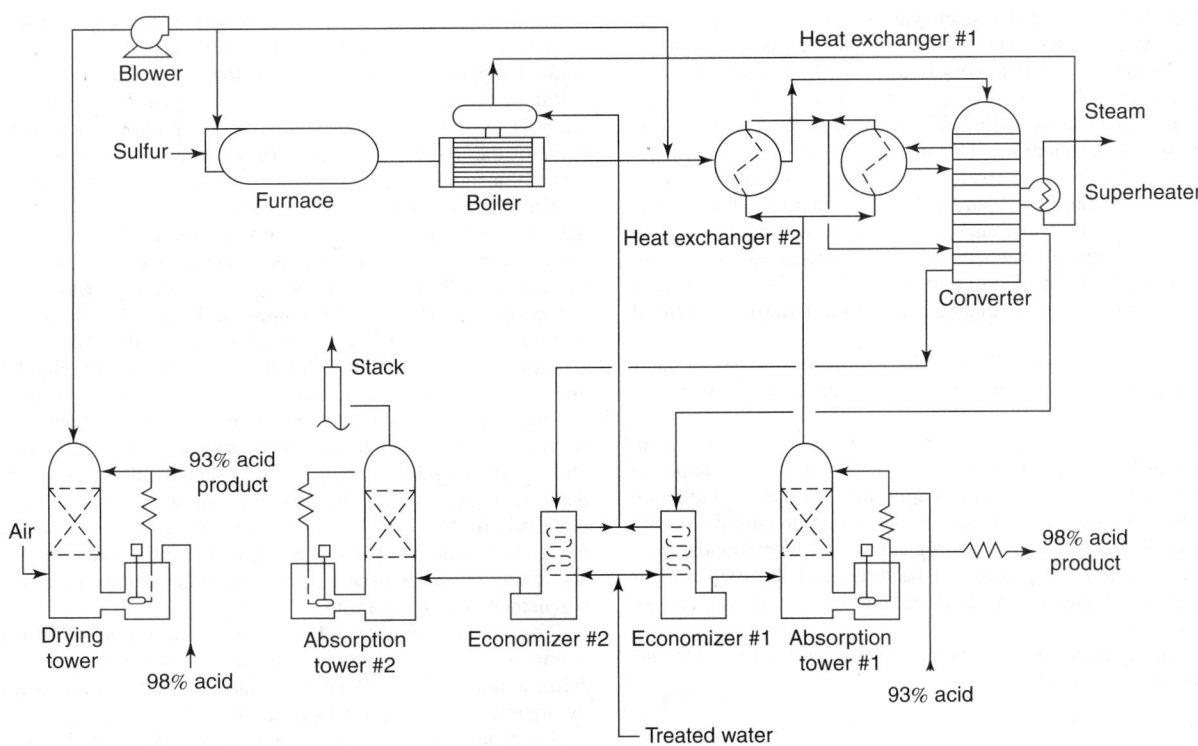

Fig. 1.   Representative sulfuric acid plant of the sulfur-burning, double contact (DC), double absorption (DA) type

and conservation of energy, and minimizing environmental impact. For example, in the relatively simple plants of a few years age, the SO_2 need contact the catalyst but once and the absorption of the resulting SO_3 in water (a solution of sulfuric acid) could be handled in a single absorption tower. Recycling could be kept to a minimum. In the modern sulfuric acid plant, double contact (DC) of the gases with catalyst and double absorption (DA) of the gases is commonly practiced. Designs are available in numerous configurations, each offering various advantages in terms of energy conservation, pollution minimization, initial and operating costs. A typical sulfur-burning DC/DA sulfuric acid plant is shown in Fig. 1.

Of the approximately 40 million tons (36 million metric tons) of sulfuric acid manufactured in the United States per year, about 90% is used in the production of fertilizers and other inorganic chemicals. Much of the remaining 10% of H_2SO_4 is used by the petroleum, petrochemical, and organic chemicals industries. Much of this latter acid is involved in recycling kinds of processes. As pollution regulations in various countries become more restrictive, spent acid may become a much more attractive raw material than has been the case in the past.

As pointed out by Sander and Daradimos (1978), a regeneration of sulfuric acid of high quality can only be attained by thermal decomposition back to sulfur dioxide at high temperatures, where all organic impurities are completely burned—followed by reprocessing the SO_2 gases by the contact process to concentrated acid or oleum.

Reactivity of Sulfuric Acid

Dilute sulfuric acid reacts: (1) with many hydroxides, e.g., sodium hydroxide, to yield two series of sulfates (the acid is dibasic), e.g., sodium sulfate or sodium hydrogen sulfate, depending upon the ratio of acid to base reacting, (2) with many ordinary oxides, e.g., magnesium oxide, to yield the corresponding sulfate, e.g., magnesium sulfate solution, (3) with some carbonates, e.g., zinc carbonate, to yield the corresponding sulfate, e.g., zinc sulfate solution plus carbon dioxide gas (calcium carbonate is soon coated by a layer of calcium sulfate, which prevents further reaction), (4) with some sulfides, e.g., ferrous sulfide, to yield the corresponding sulfate, e.g., ferrous sulfate plus hydrogen sulfide gas, (5) with many metals, e.g., zinc, if not too pure (but not copper), to yield the corresponding sulfate, e.g., zinc sulfate solution plus hydrogen gas, (6) with solutions of some salts to yield the corresponding sulfate, e.g., barium chloride, changed to barium sulfate precipitate, calcium citrate, malate, tartrate to calcium sulfate precipitate and the free organic acid in solution.

Higher strengths of sulfuric acid react similarly in kind to the cases of (1), (2), (3), (6) above, but not, in general, as in cases (4) and (5) above. Copper and concentrated sulfuric acid yield copper sulfate and sulfur dioxide gas. Iron reacts similarly, yielding ferric sulfate in the place of copper sulfate.

A number of other reactions of sulfuric acid are characteristic of its higher strengths. Concentrated sulfuric acid is thus (7) an oxidizing agent, and a further example is the oxidation of sulfur to sulfur dioxide (the reacting sulfuric acid is reduced to sulfur dioxide), (8) a sulfonating agent, e.g., naphthalene sulfonated to naphthalene-sulfonic acids (mono-, alpha or beta, di- several), (9) an esterification agent, e.g., methyl alcohol esterified to dimethyl sulfate $(CH_3O)_2SO_2$, melting point $-32°C$, boiling point $189°C$, or methyl hydrogen sulfate $CH_3O \cdot SO_2OH$, ethyl alcohol esterified to diethyl sulfate $(C_2H_5O)_2SO_2$, melting point $-26°C$, boiling point $208°C$, or ethyl hydrogen sulfate $C_2H_5O \cdot SO_2OH$, (10) a dehydration agent, e.g., formic acid into carbon monoxide, sugar blackened with separation of carbon, (11) an addition agent, e.g., ethylene into ethyl hydrogen sulfate, (12) a nonvolatile acid upon heating, e.g., with sodium chlorite or nitrate, hydrogen chloride or nitric acid, respectively, is volatilized and sodium sulfate or sodium hydrogen sulfate remains as a residue.

Additional Reading

Behrens, D.: *DECHEMA Corrosion Handbook: Corrosive Agents and Their Interaction with Materials, Sulfuric Acid,* Vol. 8, John Wiley & Sons, Inc., New York, NY, 1991.

Kent, J.A.: *Riegel's Handbook of Industrial Chemistry,* 9th Edition, Chapman & Hall, New York, NY, 1992.

Lewis, R.J., N.I. Sax: *Sax's Dangerous Properties of Industrial Materials,* 10th Edition, John Wiley & Sons, Inc., New York, NY, 2000.

Lide, D.R.: *CRC Handbook of Chemistry and Physics,* 84th Edition, CRC Press, LLC, Boca Raton, FL, 2003.

Parker, S.P.: *McGraw-Hill Encyclopedia of Chemistry,* 2nd Edition, The McGraw-Hill Companies, Inc., New York, NY, 1993.

Sander, U., G. Daradimos: "Regenerating Spent Acid," *Chem. Eng. Progress,* **74,** 57–67 (1978).

SULFUR (In Biological Systems).   Sulfur, in some form, is required by all living organisms. It is utilized in various oxidation states, including sulfide, elemental sulfur, sulfite, sulfate, and thiosulfate by lower forms and in organic combinations by all. The more important sulfur-containing organic compounds include the amino acids (cysteine, cystine, and methionine, which are components of proteins); the vitamins thiamine and biotin; the cofactors lipoic acid and coenzyme A; certain complex lipids of nerve tissues, the sulfatides; components of mucopolysaccharides, the sulfated polysaccharides; various low-molecular-weight compounds, such as glutathione and the hormones vasopressin and oxytocin; and many therapeutic agents, such as the sulfonamides and penicillins, as well as oral hypoglycemic agents sometimes used in treatment of diabetes mellitus. Sulfhydryl groups of the cysteine residues in enzyme proteins and related compounds, such as hemoglobin, play a key role in many biocatalytic processes; sulfhydryl-disulfide interchange reactions involving the cysteine residues of proteins are critical events in the immune processes, in transport across cell membranes, and in blood clotting. The S—S bridges between these residues are important in the maintenance of the tertiary structure of most proteins.

The electronic structure of sulfur is such that a variety of oxidation states are readily obtainable. It can be said that a sulfur cycle exists in nature, as noted in Fig. 1.

The oxidation and reduction of elemental sulfur and sulfide occur in different species of bacteria, e.g., the oxidation of sulfides via elemental sulfur to sulfate takes place in *Chromatia*, the alternative oxidation to sulfate in *Thiobacillus*. The reduction of sulfate to sulfide occurs in *Desulfovibrio*. The biosynthesis of organic sulfur compounds from sulfate takes place mainly in plants and bacteria, and the oxidation of these compounds to sulfate is characteristic of animal species and of heterotrophic bacteria.

The amino acids cysteine and cystine are interconverted by oxidation-reduction reactions, as shown by

$$
\begin{array}{ccc}
\text{S}\!\!-\!\!\!-\!\!\!-\!\!\!-\!\!\text{S} & & \text{S} \\
| \quad\quad | & & | \\
\text{CH}_2 \quad \text{CH}_2 & & \text{CH}_2 \\
| \quad\quad | & & | \\
\text{CNNH}_2 \quad \text{CNNH}_2 + 2\,\text{H} \rightleftharpoons & 2\text{CNNH}_2 \\
| \quad\quad | & & | \\
\text{COOH} \quad \text{COOH} & & \text{COOH} \\
\text{(Cysteine)} & & \text{(Cystine)}
\end{array}
$$

Cystine was first isolated from a urinary calculus by Wollaston in 1805. It was shown to be a component of protein by Morner in 1899 and independently by Embden in 1900. Proof of its structure was given by Friedman in 1902. See also **Amino Acids**; **Coenzymes**; **Proteins**; and **Vitamin**.

In the chain from soils to plants to humans, inorganic sulfur, or more accurately, the sulfate ion (SO_4^{2-}), is taken up by plants and converted within the plant to organic compounds (the sulfur amino acids). These amino acids combine with other amino acids to make up plant protein. When the plant is eaten by a human or by livestock animals, the protein is broken down and the amino acids are absorbed from the digestive tract and recombined in the proteins of the animal body. The most important feature of sulfur in the food chain is that plants use inorganic sulfur compounds to make sulfur amino acids, whereas animals and humans use the sulfur amino acids for their own processes and excrete inorganic sulfur compounds resulting from the metabolism of the sulfur amino acids.

Ruminants, such as cattle, sheep, and goats, can use inorganic sulfur in their diets because the microorganisms in the rumen convert the inorganic sulfur into sulfur amino acids and these are then absorbed farther along in the digestive tract.

Soils very low in available sulfur are common in a number of regions of the world. In the United States, low-sulfur soils are frequently found in the Pacific Northwest and in some parts of the Great Lakes states. For many years, sulfur in the form of calcium sulfate was an accessory part of most commercial phosphate fertilizers, and this probably helped to prevent development of widespread sulfur deficiency in crops grown

where these fertilizers were used. Volatile sulfur compounds from smoke, particularly before tight pollution controls, were an important source of sulfur for plants growing near industrial centers. In some cases, excessive sulfur in the air can cause injury to the plants. The trend toward high-analysis fertilizers without sulfur and air pollution abatement diminishes some of the inadvertent sources of sulfur for plants and crops and creates a need for more deliberate use of sulfur-containing fertilizers.

The extent to which any plant will convert inorganic sulfur taken up from the soil into amino acids and incorporate these into protein is controlled by the genetics of the plant. Increasing the available sulfur in soils to levels in excess of those needed for optimum plant growth will not increase the concentration of sulfur amino acids in plant tissues. To meet the requirements for sulfur amino acids in human diets, the use of food plant species with the inherited ability to build proteins with high levels of sulfur amino acids is required in addition to that supplied by way of the soil.

Since animals tend to concentrate in their own proteins the sulfur amino acids contained in the plants they eat, such animal products (meat, eggs, and cheese) are valuable sources of the essential sulfur amino acids in human diets. In regions where the diet is composed almost entirely of foods of plant origin, deficiencies of sulfur amino acids may be critical in human nutrition. Frequently, persons in such areas (also voluntary vegetarians) are also likely to suffer from a number of other dietary insufficiencies unless supplemental sources are used.

Diets of corn (maize) and soybean meal are usually fortified with sulfur amino acids for pigs and chickens. Sometimes fishmeal, a good source of sulfur amino acids, is added to the diets, or sulfur amino acids synthesized by organic chemical processes may be used.

Since ruminants can utilize a wide variety of sulfur compounds, any practice to increase the sulfur in plants may help to meet the requirements of these animals. Sheep appear to have a higher requirement for sulfur than most other animals, perhaps because wool contains a fairly high level of sulfur. Adding sulfur fertilizers to soils used to produce forage for sheep may improve growth and wool production, even though no increased yield of the forage crop per se may be noted.

Sulfate and Organic Sulfates. Inorganic sulfate ion (SO_4^{2-}) occurs widely in nature. Thus, it is not surprising that this ion can be used in a number of ways in biological systems. These uses can be divided primarily into two categories: (1) formation of sulfate esters and the reduction of sulfate to a form that will serve as a precursor of the amino acids cysteine and methionine; and (2) certain specialized bacteria use sulfate to oxidize carbon compounds and thus reduce sulfate to sulfide, while other specialized bacterial species derive energy from the oxidation of inorganic sulfur compounds to sulfate.

Among the variety of sulfate esters formed by living cells are the sulfate esters of phenolic and steroid compounds excreted by animals, sulfate polysaccharides, and simple esters, such as choline sulfate. The key intermediate in the formation of all of these compounds has been shown to be 3'-phosphoadenosine-5'-phosphosulfate (PAPS). This nucleotide also serves as an intermediate in sulfate reduction.

In organisms that utilize sulfate as a source of sulfur for synthesis of cysteine and methionine, the first step in the reduction process is the formation of PAPS. This is not surprising since the direct reduction of sulfate ion itself is an extremely difficult chemical process. It is known that the reduction of esters and anhydrides occurs much more readily than the reduction of corresponding anions. Following activation, the sulfuryl group of PAPS is reduced to sulfite ion (SO_3^{2-}) by reduced triphosphopyridine nucleotide (TPNH) and a complex enzyme system. Following the reduction of PAPS to sulfite, additional reduction steps readily produce hydrogen sulfide, which appears to be a direct precursor of the amino acid cysteine.

Sulfur Compounds in Onion and Garlic

Dating back to antiquity, there have claims made for the curative and preventative physiological powers of onion and garlic. Dr. Eric Block (State University of New York at Albany), a specialist in the organic chemistry of sulfur, and colleagues, have investigated the chemistry of onion and garlic over a period of years. Some of the results were reported in the Block (1985) reference listed. As pointed out by Block, the cutting of an onion or a garlic bulb releases a number of low-molecular-weight organic molecules that incorporate sulfur atoms in bonding forms rarely encountered in nature. These molecules are highly reactive and they change spontaneously into other organic sulfur compounds, which in turn participate in further transformations. Researchers have cataloged a

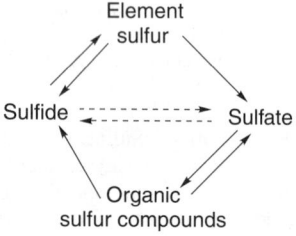

Fig. 1.   Sulfur cycle

number of biological effects of the extracts from these bulbs, including antibacterial and antifungal properties. Other extracts act as antithrombotic agents (inhibit blood platelets). As early as 1721, a drink consisting of wine and macerated garlic (*vinaigre des quatre voleurs*) was used as an antibiotic in France and is still available today! Pasteur (1858) reported on the antibacterial properties of garlic. Albert Schweitzer is reported to have used garlic in the treatment of amoebic dysentery in Africa. As reported by Block, laboratory investigations have shown that garlic juice diluted in one part in 125,000 inhibits the growth of bacteria of the genera *Staphylococcus, Streptococcus, Vibrio* (including *V. cholerae*) and *Bacillus* (including *B. typhosus, B. dysenteriae,* and *B. Enteritidis*). Lacrimatory factors contained in these bulbs are well known.

Serious research commenced in 1844 by Theodor Wertheim, a German chemist. He attributed some of the properties of garlic, "*mainly to the presence of a sulfur-containing, liquid body, the so-called garlic oil. All that is known about the material is limited to some meager facts about the pure product, which is obtained by steam distillation of bulbs of Allium sativum. Since sulfur bonding has been little investigated so far, a study of this material promises to supply useful results for science.*" Wertheim suggested the name *allyl* for the oil. Today, allyl is used for chemicals in the C_3H_5 series ($CH_2=CHCH_2$).

Another German investigator (Semmler, 1892) also produced garlic oil via steam distillation. The oil yielded diallyl disulfide, $CH_2=CHCH_2SSCH_2$ $CH \equiv CH_2$, with minor amounts of diallyl trisulfide and diallyl tetrasulfide present. The oil yielded by similar experimentation with onions was different, containing essentially propionaldehyde, C_2H_5CHO, plus a number of sulfur compounds, of which dipropyl disulfide, $C_6H_{12}S_2$, was one.

Using less harsh methods, Cavallito (1944) produced an oil, $C_6H_{10}S_2O$, by extracting the garlic with ethyl alcohol (room temperature). This oil was found to be more powerful than penicillin or sulfaguanidine against *B. typhosus.* The exact formula of Cavallito's oil was found to be allyl-2-propenethiosulfinate, $CH_2=CHCH_2S(O)SCH_2CH=CH_2$. Cavallito gave this substance the common name, *allicin.* Precursor molecules for allicin have been identified and it has been established that allicin is not developed in garlic until it is initiated by an enzyme, termed *allinase.*

As pointed out by Block, allinin is the "first natural substance found to display optical isomerism due to mirror-image forms at sulfur as well as at carbon." The research by Block and others is well delineated in the Block (1985) reference. Although much remains to be learned, there is now a long line of hard scientific evidence that garlic and onion have beneficial physiological properties. Most scientists to date suggest that the properties of these bulbs are best exploited by consuming the fresh products, rather than from extracts, particularly when the latter are derived from harsh methodologies, such as steam distillation.

Preservatives. Sulfur compounds, such as sulfur dioxide and sodium bisulfite, are used commercially to preserve the color of various food products, such as orange juice, dehydrated fruits and vegetables, such as apricots, carrots, peaches, pears, potatoes, and many others. Concentrated sulfur dioxide is used in wine-making to destroy certain bacteria. The color preservation of canned green beans and peas is enhanced by dipping the produce in a sulfite solution prior to canning. In 1986, some of these compounds and uses were put under closer regulation in the United States.

The sulfatases are a widely distributed group of enzymes that hydrolyze simple sulfate esters to inorganic sulfate.

Sulfur-Based Pesticides. Sulfur (elemental) has been used as an effective acaricide, fungicide, and insecticide. For ease of use, a number of special formulations are available, ranging from sulfur dusts (up to 95% sulfur); a wettable powder (30 to 90%); and paste-like solutions in which the sulfur is ground to a fine colloidal form. Such formulations may contain up to 50% sulfur. Target plant diseases of sulfur when used as a fungicide include: apple scab, brown rot, downy and powdery mildew, and peach scab. Against insects, sulfur is effective for mite, scale, and thrip. Most formulations are not injurious to honeybees.

Specific sulfur control chemicals include: (1) Calcium polysulfide (lime-sulfur), dating back to the 1850s and available as a solution (up to 31% sulfur) or as a dry powder (up to 70% sulfur). The compound is effective against anthracnose, apple scab, brown rot, powdery mildew, and peach leaf curl—and against mite and scale insects. (2) Sodium polysulfide and sodium thisulfate mixtures—used for spraying and dipping fruit, adding color to the product, and prolonging the period during which the fruit can be picked. (3) Sodium thiosulfate pentahydrate, which prevents discoloration of some green vegetables (use is regulated).

Role of Sulfur in Tidal Wetlands. As pointed out by Luther, et al. (1985 reference listed), the biogeochemical role of sulfur in tidal wetlands presently is subject to considerable research. Sulfur is an important redox element under natural aquatic conditions and is responsible for several important biogeochemical processes, including (1) sulfate reduction, (2) pyrite formation, (3) metal cycling, (4) salt-marsh ecosystem energetics, and (5) atmospheric sulfur emissions. These processes depend upon the formation of one or more sulfur intermediates, which may have any oxidation state between +6 and −2. The intermediate oxidation states may be organic or inorganic. In their study, Luther and colleagues analyzed sulfur species in pore waters of the Great Marsh, Delaware. Anticipated findings reported were bisulfide increases with depth due to sulfate reduction and subsurface sulfate excesses and pH minima, the result of a seasonal redox cycle. Not expected was the pervasive presence of thiols, such as glutathione, particularly during periods of biological production. It appears that salt marshes may be unique among marine systems in producing high concentrations of thiols. Polysulfides, thiosulfate, and tetrathionate also showed seasonal subsurface maxima. The findings suggest a dynamic seasonal cycling of sulfur in salt marshes involving abiological and biological reactions and dissolved and solid sulfur species. The researchers suggest that the chemosynthetic turnover of pyrite to organic sulfur is the likely pathway for this sulfur cycling. It follows that the material, chemical, and energy cycles in wetlands appear to be optimally synergistic.

Additional Reading

Block, E.: "The Chemistry (Sulfur Compounds) of Garlic and Onions," *Sci. American,* **252**(3), 114–119 (1985).

Considine, D.M. and G.D. Considine: *Food and Food Production Encyclopedia,* Van Nostrand Reinhold Company, Inc., New York, NY, 1982.

Greyson, J.C.: *Carbon, Nitrogen and Sulfur Pollutants and Their Determination in Air and Water,* Marcel Dekker, Inc., New York, NY, 1990.

Lide, D.R.: *CRC Handbook of Chemistry & Physics,* 84th Edition, CRC Press, LLC., Boca Raton, FL, 2003.

Luther, G.W., III et al.: "Inorganic and Organic Sulfur Cycling in Salt-Marsh Pore Waters," *Science,* **232**, 746–749 (1986).

Maynard, D.G.: *Sulfur in the Environment,* Marcel Dekker, Inc., New York, NY, 1998.

Mitchell, S.C.: *Biological Interactions of Sulfur Compounds,* Taylor & Francis, Inc., Philadelphia, PA, 1996.

Mudahar, M.S., J.S. Kanwar: *Fertilizer Sulfur and Food Production,* Kluwer Academic Publishers, Norwell, MA, 1986.

Stevenson, F.J., M.A. Cole: *Cycles of Soils: Carbon, Nitrogen, Phosphorus, Sulfur, Micronutrients,* 2nd Edition, John Wiley & Sons, Inc., New York, NY, 1999.

SULFUROUS ACID. [CAS: 7782-99-2], H_2SO_3, formula weight 82.08, colorless liquid, prepared by dissolving SO_2 in H_2O. Reagent grade H_2SO_3 contains approximately 6% SO_2 in solution. As a bleaching agent, sulfurous acid is used for whitening wool, silk, feathers, sponge, straw, wood, and other natural products. In some areas, its use is permitted for bleaching and preserving dried fruits. The salts of sulfurous acid are sulfites.

Sulfurous acid is a strong reducing agent, being oxidized to H_2SO_4: (1) on standing in contact with air; (2) by chlorine, bromine, iodine, yielding HCl, HBr, or HI, respectively; (3) by HNO_3 or nitrous acid yielding nitric oxide; and (4) by permanganate. Sulfurous acid is itself reduced by zinc and dilute H_2SO_4 to H_2S. Sulfurous acid also may be formed by the reaction of a sulfite or bisulfite solution and an acid.

Sodium sulfite, [CAS: 7757-83-7], Na_2SO_3 and sodium hydrogen sulfite, [CAS: 7631-90-5], $NaHSO_3$ are formed by the reaction of sulfurous acid and NaOH or sodium carbonate in the proper proportions and concentrations. Sodium sulfite, when dry and upon heating, yields sodium sulfate and sodium sulfide. Sodium pyrosulfite (sodium metabisulfite), [CAS: 7681-57-4], $Na_2S_2O_5$ is a common sulfite. Crystalline sulfites are obtained by warming the corresponding bisulfite solutions. Calcium hydrogen sulfite $Ca(HSO_3)_2$ is used in conjunction with excess sulfurous acid in converting wood to paper pulp. Sodium sulfite and silver nitrate solutions react to yield silver sulfite, a white precipitate, which upon boiling decomposes forming silver sulfide, a brown precipitate.

An esterification agent, sulfurous acid forms dimethyl sulfite $(CH_3O)_2$ SO, bp 126°C and diethyl sulfite $(C_2H_5O)_2SO$, bp 161°C. Sulfites give a white precipitate with barium chloride, soluble in HCl with evolution of SO_2. Sulfites decolorize iodine in acid solution.

SULFUROUS ACID DERIVATIVES. See **Herbicides.**

SULFUR OXIDES (Pollution).   See **Pollution (Air).**

SULFUR (Vulcanization).   See **Rubber (Natural).**

SUNFLOWER OIL.   See **Vegetable Oils (Edible).**

SUPERACIDS.   See **Acids and Bases.**

SUPERCONDUCTIVITY.   A property of a material that is characterized by zero electric resistivity and, ideally, zero permeability. The phenomenon of superconductivity was discovered in 1911 by Heike Kamerlingh Onnes (University of Leiden) as the outcome of a remarkable achievement in those years—the liquefaction of helium for the first time. Helium condenses at atmospheric pressure at $4.2°K$. Onnes, using the newly available very low temperature substance, proceeded to investigate the electrical resistance of various metals at low temperatures. Even prior to quantum mechanics, it had been predicted that if absolute zero could be achieved in a metal having a perfectly regular interatomic structure, the electrical resistance of the metal would be zero. Onnes found that the resistance of a mercury wire suddenly dropped to zero at $4.2°K$, which indeed is a very low temperature, but still well above absolute zero. Onnes and other investigators at Leiden researched superconductivity in other metals, such as lead, which was found to become superconductive at $7.2°K$. It should be mentioned at this point that scientists of that period were biased in their thinking of conductivity and superconductivity in terms of metals. The first stable conducting organic material was not synthesized until 1960 and it was not until 1979 that a superconducting organic material was isolated. The so-called "hottest" superconductor was announced in mid-1993 by a research team at Eidgenossische Technische Hochschule in Zurich. The new material, considered toxic, is made of two distinct compounds that commence to be superconductive at $133°K$. Prior record was claimed for $Tl_2Ca_2Ba_2Cu_3O_{10}$ at $127°K$.

Rediscovery. The topic and potential of superconductivity essentially was rediscovered in the late 1970s and early 1980s, during which period the research activity safely can be described as zealous. This was fueled by the scores of probable applications of superconductors, but which to date largely remain as promises. Research continues at a good pace, and the topic is much better understood as compared with a decade ago. As pointed out later in this article, much excellent technological fallout has occurred from many millions of dollars invested in research. The search for the ultimately practical superconductor, although still elusive, has reinforced the multidisciplinary sciences involved, notably physics and chemistry.

Application Targets. Among the ultimately practical applications for superconductivity, especially those of significant future commercial values, are: (1) magnetic shielding, (2) magnetic resonance imaging magnets for medicine and research, (3) electric utility transmission lines and load-leveling storage coils, (4) magnetic separators for materials processing, (5) higher-speed ($10\times$) switching and signal transmission for computers, (6) more compact electronics with finer interconnect lines, (7) extremely compact electric motors and actuators, (8) no-loss portable electrical storage "batteries," and (9) noncontact bearings and magnetically levitated vehicles. Special areas of interest by the military in superconductors include (1) infrared optical detector elements, (2) high-speed millimeter and submillimeter-wave electronics for advanced radar countermeasures, (3) magnetic anomaly detection of submarines, and (4) free-electron laser components.

Considerable research of a contemplative or assumptive nature continues to go forward—that is, studying how the "ideal" superconductor can be applied, once developed. Examples of progress are shown in Figs. 1, 2, and 3.

Fundamental Research Findings

For several years following the previously mentioned work of the Louden scholars, investigators concentrated principally on metallic elements and alloys, including indium, tin, vanadium, molybdenum, niobium-zirconium, and niobium-tin, among many others. A number of basic discoveries were made. For example, finding that the property of superconductivity could be destroyed by the application of a magnetic field equal to or greater than a critical field H_c. This H_c, for a given superconductor, is a function of the temperature given approximately by

$$H_c = H_0(1 - T^2/T_c^2) \qquad (1)$$

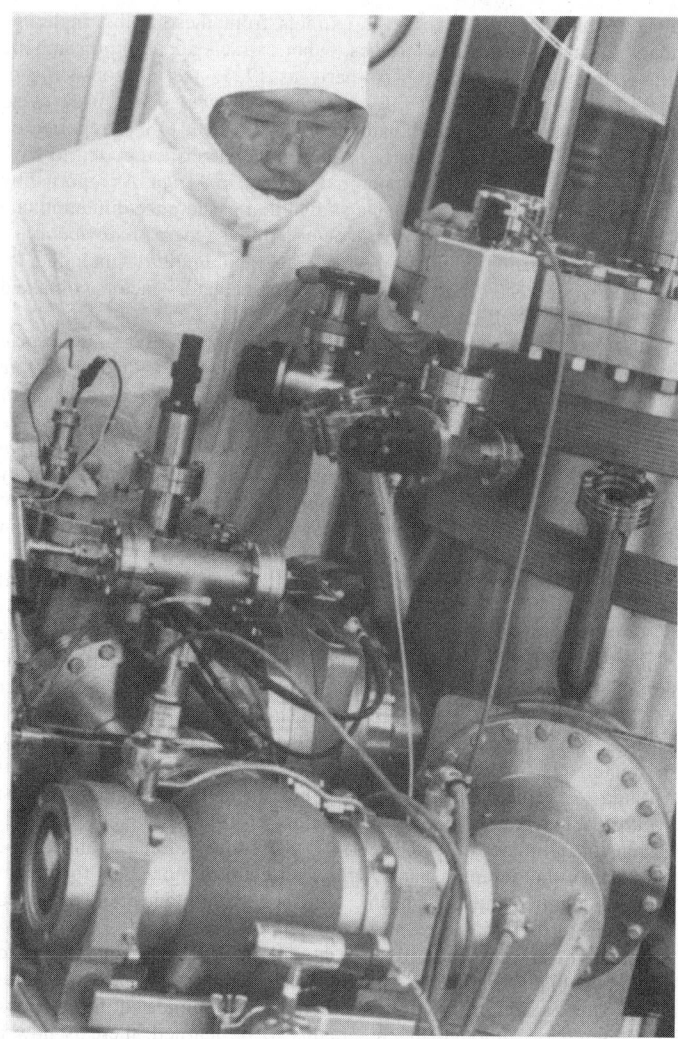

Fig. 1.   Special equipment required to fabricate low-temperature superconducting junctions. Josephson junctions are comprised of aluminum oxide sandwiched between layers of niobium. These trilayer devices are considered vital to the very-high-speed signal processing demands of next-generation computers, radar, and communication systems. Shown in illustration is scientist Dr. Joonhee Kang. (Westinghouse Electric Corporation)

where H_0, the critical field at $0°K$, is in general different for different superconductors and has values from a few gauss to a couple of thousand gauss. For applied magnetic fields less than H_c, the flux is excluded from the bulk of the superconducting sample, penetrating only to a small depth λ into the surface. The value of λ (called the penetration depth) is in the range 10^{-5} to 10^{-6} centimeter. Thus the magnetization curve for a superconductor is

$$B(\text{inside}) = 0 \qquad \text{for } H < H_c$$
$$B(\text{inside}) = B(\text{outside}) \qquad \text{for } H > H_c$$

This magnetization behavior is reversible and cannot therefore be explained entirely on the basis of the zero resistance. The reversible magnetization behavior is called the Meissner effect.

The existence of the penetration depth λ suggests that a sample having at least one dimension less than λ should have unusual superconducting properties, and such is indeed the case. Thin superconducting films, of thickness d less than λ, have critical fields higher than the bulk critical field, approximately in the ratio of λ to d. This result follows qualitatively from the thermodynamics of the Meissner effect: the metal in the superconducting state has a lower free energy than in the normal state, and the transition to the normal state occurs when the energy needed to keep the flux out becomes equal to this free energy difference. But in the case of a thin film with $d < \lambda$, there is partial penetration of the flux into the film, and thus one must go to a higher applied field before the free energy difference is compensated by the magnetic energy.

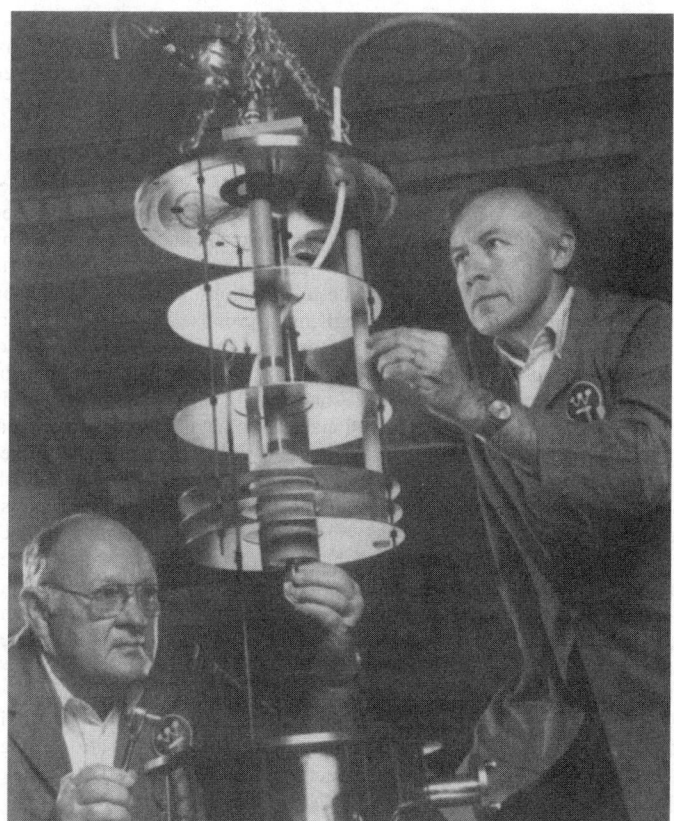

Fig. 2. Electrical lead comprised of a high-temperature superconductor can carry a current of 2000 amperes. A variety of uses include magnetic resonance imaging and superconducting magnetic energy storage. (*Westinghouse Electric Corporation*)

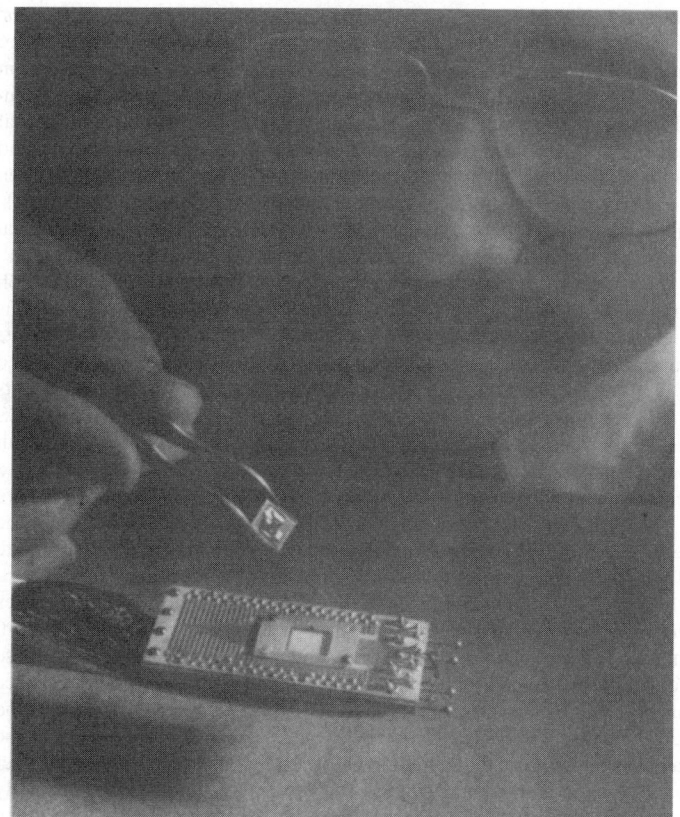

Fig. 3. Scientist Donald L. Miller holds an integrated circuit chip comprising a high-resolution superconducting analog-to-digital converter. The one-square-centimeter chip, known as a counting converter, holds promise as an unprecedented combination of high resolution and low power consumption, as needed in future air traffic control radar and infrared space-tracking applications. The 12-bit circuit (Josephson junction) has a resolution of 1 part in 4000. (*Westinghouse Electric Corporation*)

It is clear that the existence of the critical field also implies the existence of a critical transport electrical current in a superconducting wire, i.e., that current I_c, which produces the critical field H_c at the surface of the wire. For example, in a cylindrical wire of radius r, $I_c = \frac{1}{2} r H_c$. This result is called the Silsbee rule.

All of the above properties distinguish superconductors from "normal" metals. There is another very important distinction, which contains a clue to understanding some of the properties of superconductors. In a normal metal at $0°K$, the electrons, which obey Fermi statistics, occupy all available states of energy below a certain maximum energy called the Fermi energy ζ. Raising the temperature of the metal causes electrons to be singly excited to states just above the Fermi energy. There is for all practical purposes a continuum of such excited energy states available above the Fermi energy. The situation is quite different in a superconductor; it turns out that in a superconductor, the lowest excited state for an electron is separated by an energy gap ϵ from the ground state. The existence of this gap in the excitation spectrum has been confirmed by a wide range of measurements: electronic heat capacity, thermal conductivity, ultrasonic attenuation, far infrared and microwave absorption, and tunneling. The energy gap is a monotonically decreasing function of temperature, having a value ~ 3.5 kT_c at $0°K$ (where k is the Boltzmann constant) and vanishing at T_c.

The superconducting state has a lower entropy than the normal state, and therefore one concludes that superconducting electrons are in a more ordered state. Without, for the present, inquiring more deeply into the nature of this ordering, one can state that a spatial change in this order produced say by a magnetic field will occur, not discontinuously, but over a finite distance ζ, which is called the *coherence length*. The coherence length represents the range of order in the superconducting state and is typically about 10^{-4} centimeter, though we shall see later that it can in some superconductors take much lower values and lead to some remarkable properties.

Measurements of the transition temperature on different isotopes of the same superconductor showed that T_c is proportional to $M^{-1/2}$, where M is the isotopic mass. This isotope effect suggests that the mechanism underlying superconductivity must involve the properties of the lattice, in addition to those of the electrons. Another indication of this is given by the behavior of allotropic modifications of the same element: white tin is superconducting, while grey tin is not, and the hexagonal and face-centered cubic phases of lanthanum have different transition temperatures. A third, and most striking, indication is that the current vs voltage characteristic of a superconducting tunneling junction shows a structure which is intimately related to the phonon spectrum of the superconductor.

The superconducting properties of alloys present a bewildering variety of phenomena. They show a great deal of magnetic hysteresis, with little indication of a perfect Meissner effect. The Silsbee rule is inapplicable, and the resistive transition occurs at fields generally very much higher than in pure superconductors. For example, a wire of Nb_3Sn can carry a current of 10^5 amperes/cubic centimeter in an applied field of 100 kilogauss, while a similar wire of lead would carry about 10^3 amperes/cubic centimeter in a field of only 100 gauss. When experiments are done using well-annealed (preferably single-crystal) alloys, it is found that the critical currents drop considerably, and the magnetic behavior becomes reversible but still quite unlike that of pure superconductors. The flux is excluded from the interior of the sample up to a well-defined field H_{c1}. When the applied field is raised further, flux begins to penetrate, even though the resistance remains zero, until a second critical field H_{c2} is reached, at which the flux penetration is complete, and normal resistance is abruptly restored.

Superconductivity Theory

The theory of superconductivity has developed along two lines, the phenomenological and the microscopic. The phenomenological treatment was initiated by F. London, who modified the Maxwell electromagnetic equations so as to allow for the Meissner effect. His theory explained the existence and order of magnitude of the penetration depth, and gave a qualitative account of some of the electrodynamic properties. The treatment was extended by V.L. Ginzburg and L.D. Landau, and by A.B. Pippard, who in particular emphasized the concept of the range of coherence. A.A.

Abrikosov used these ideas to develop a model for alloy superconductors. He showed that if the electronic structure of the superconductor were such that the coherence length ζ becomes smaller than the penetration depth λ, one would get magnetic behavior similar to that observed in alloys, with two critical fields H_{c1} and H_{c2}. The problem of high critical currents in unannealed (or otherwise metallurgically imperfect) alloys and compounds is more complicated because it involves the interaction between the microscopic metallurgical structure and the superconducting properties. This is an area of great research activity because of the technological implication to be mentioned later.

The microscopic theory of superconductivity was initiated by H. Fröhlich, who first recognized the importance of the interactions of electrons with lattice vibrations and in fact predicted the isotope effect before its experimental observation. The detailed microscopic theory was developed by J. Bardeen, L.N. Cooper and J.R. Schrieffer in 1957, and represents one of the outstanding landmarks in the modern theory of solids. The BCS theory, as it is called, considers a system of electrons interacting with the phonons, which are the quantized vibrations of the lattice. There is a screened coulomb repulsion between pairs of electrons, but in addition there is also an attraction between them via the electron-phonon interaction. If the net effect of these two interactions is attractive, then the lowest energy state of the electron system has a strong correlation between pairs of electrons with equal and opposite momenta and opposite spin and having energies within the range $k\theta$ (where θ is the Debye temperature) about the Fermi energy. This correlation causes a lowering of the energy of each of these Cooper pairs (named after L.N. Cooper who first pointed out their existence on the basis of some general arguments) by an amount ϵ relative to the Fermi energy. The energy ϵ may be regarded as the binding energy of the pair, and is therefore the minimum energy which must be supplied in order to raise an electron to an excited state. We see thus that the experimentally observed energy gap follows from the theory. The magnitude ε_0 of the gap at $0°$K is

$$\varepsilon_0 \approx 4k\theta \exp\left(-\frac{1}{NV}\right)$$

where N is the density of electronic states at the Fermi energy and V is the net electron-electron interaction energy. The superconducting transition temperature T_c is given by

$$3.5kT_c \approx \varepsilon_0$$

It has been shown that the BCS theory does lead to the phenomenological equations of London, Pippard and Ginzburg and Landau, and one may therefore state that the basic phenomena of superconductivity are now understood from a microscopic point of view, i.e., in terms of the atomic and electronic structure of solids. It is true, however, that we cannot yet, *ab initio*, calculate V for a given metal and therefore predict whether it will be superconducting or not. The difficulty here is our ignorance of the exact wave functions to be used in describing the electrons and phonons in a specific metal, and their interactions. However, we believe that the problem is soluble in principle at least.

The range of coherence follows naturally from the BCS theory, and we see now why it becomes short in alloys. The electron mean free path is much shorter in an alloy than in a pure metal, and electron scattering tends to break up the correlated pairs, so that for very short mean free paths one would expect the coherence length to become comparable to the mean free path. Then the ratio $\kappa \approx \lambda/\zeta$ (called the Ginzburg-Landau order parameter) becomes greater than unity, and the observed magnetic properties of alloy superconductors can be derived. The two kinds of superconductors, namely those with $k < 1/\sqrt{2}$ and those with $k > 1/\sqrt{2}$ (the inequalities follow from the detailed theory) are called respectively type I and type II superconductors.

Challenges to Established Theories. It is interesting to note that some theoreticians struggle with describing how superconductivity occurs at high temperatures in the newer, ceramic superconductors. This is understandable because the classic theory of superconductivity is tied to metals. Most ceramic superconductors discovered to date incorporate distinctive layers of copper and oxygen atoms. One question posed by some researchers, "Is the mechanism of high-temperature superconductivity the same in hole superconductors as it is in electron superconductors?"

Researchers at the Brookhaven National Laboratory, in applying X-ray techniques to a cerium-doped electron superconductor developed at the University of Tokyo, found that the holes of a hole superconductor are linked to oxygen atoms in the copper-oxygen layers, whereas in an electron superconductor the electrons are associated with copper atoms. This is exemplary of how easy it is for former theories to become outdated when new material combinations are tested for their superconductivity.

Superconductivity Research

In 1962, B. Josephson recognized the implications of the complex order parameter for the dynamics of the superconductor, and in particular when one considers a system consisting of two bulk conductors connected by a "weak link." This research led to the development of a series of weak link devices commonly called Josephson junctions. See also **Josephson Tunnel-Junction**. These devices hold much promise for achieving ultra high-speed computers where switching time is of the order of 10^{-11} second.

Good success also has been achieved in the use of certain type II superconductors, such as Nb-Zr and Nb-Ti alloys, and Nb_3Sn, in making electromagnets. In a conventional electromagnet employing normal conductors, the entire electric power applied to the magnet is consumed as Joule heating. For a magnet to produce 100 kilogauss in a reasonable volume, the power requirement can run into megawatts. In striking contrast, a superconducting magnet develops no Joule heat because its resistance is zero. Indeed, if such a magnet has a superconducting shunt placed across it after it is energized, the external power supply can be removed, and the current continues to flow indefinitely through the magnet and shunt, maintaining the field constant. Superconducting magnets have been constructed producing very strong fields in usable volumes. There is a natural upper limit to the critical field possible in such superconductors, given by the paramagnetic energy of the electrons (due to their spin moment) in the normal state becoming equal to the condensation energy of the Cooper pairs in the superconducting state. This leads to a limit of about 360 kilogauss for a superconductor with a T_c of $20°$K.

As investigators accumulated data upon data, many emphasized the practical as well as theoretical aspects of superconductors. The ultimate superconductor, of course, would be one that operated at room temperature or above. The materials must be manufacturable in a useful form, such as strong ductile wires for high-field magnets, electrical machinery, and power transmission lines, situations which could be even more important in commercial and industrial application than their value to science per se. (Traditionally, superconducting materials have been hard, brittle, and difficult to process.) Although superconductors that would operate at room temperature and above present a long-range target, lesser targets, including practical ways to cool them with liquid nitrogen instead of liquid helium and possibly, even better, operate them within a closed-cycle refrigeration system is the goal in the shortrange. Useful superconductors in large-scale applications must retain their properties not only at high temperatures, but also in the presence of high magnetic fields and while carrying large electrical currents. Praveen Chaudhari (IBM) has observed that new superconductors will enter the marketplace rapidly when intensive materials engineering produces easily cooled, mechanically robust conductor configurations that can handle high current densities ($100,000 +$ A/cm^2) under powerful magnetic fields ($10 + T$), while maintaining stable superconductivity.

Johannes Georg Bednorz and Karl Alexander Mueller (IBM Zurich Research Laboratory) after several years devoted to a study of oxide compounds (not in terms of superconductivity) proceeded with the working hypothesis that an increase in the density of charge carriers in a material (either as electrons or as positively charged "holes") possibly would lead to a rise in transition temperature. They commenced a search for nickel- and copper-containing oxides. Early in 1986, they found a certain form of barium lanthanum copper oxide that evidenced the onset of superconductivity at temperatures as high as $35°$K (12 degrees over the previous record). They encountered skepticism, because the facts did not square with accepted theory that limits the phenomenon to well below $35°$K. Shortly thereafter, however, researchers at the University of Tokyo, the University of Houston, and AT&T Bell Laboratories confirmed the Bednorz-Mueller findings. On October 14, 1987, the Nobel prize in physics was awarded to these two researchers and a speaker for the Royal Swedish Academy of Sciences observed that their work inspired "the explosive development in which hundreds of laboratories the world over commenced work on similar material." [It should be observed that Ching-Wu Chu and colleagues (University of Houston) did announce in February 1987 that a related class of ceramics (a certain form of yttrium barium copper oxide) remained superconducting up to $94°$K, proclaiming that to be the first superconductor which could be cooled by liquid nitrogen (bp = 77°K) instead of requiring helium.] That announcement in itself also precipitated a "rush" of researchers to the ceramics.

Technological Fallout of Superconductivity Research

While in the course of finding viable superconductors for commercial applications, researchers have produced valuable ancillary information.

Quantization of Energy. In a scholarly paper, D.G. McDonald (U.S. National Institute of Standards and Technology, Boulder, Colorado) observes, "Ideas about quantized energy levels originated in atomic physics, but research in superconductivity has led to unparalleled precision in the measurement of energy levels. Microscopic things can be identical; macroscopic things cannot. This proposition is so imbued in the minds of physicists that it is interesting to see that it is false in the following sense. In the past, physicists believed that only atoms and molecules could have identical states of energy, but recent experiments have shown that much larger bodies, superconductors in macroscopic quantum states, have equally well-defined energies." In the paper, McDonald uses the novelty of the Josephson effect to illustrate the primary point of the technical paper. See McDonald reference listed.

Structural Chemistry. In an enlightening paper, R.J. Cava (AT&T Bell Laboratories) asserts, "The discovery of high-temperature superconductivity in oxides based on copper and rare and alkaline earths at first caught the solid-state physics and materials science communities completely by surprise. Since the earliest 30 to 40°K superconductors based on $La_{2-x}(Ca,Sr,Ba)_xCuO_4$, many new superconducting copper oxides have been discovered, with ever-increasing chemical and structural complexity. The current record transition temperature is held[1] by $Tl^2Ba^2Ca^2Cu^3O^{10}$, a material whose processing requires the stoichiometric control of five elements, each with considerably different chemical characteristics." In the Cava paper, the crystal structures of the known copper oxide superconductors are described, with particular emphasis on the manner in which they fall into structural families. The local charge picture—a framework for understanding the influence of chemical composition, stoichiometry, and doping on the electrical properties of complex structures—is also described.

This probing of complex and previously unattended solid materials typifies technical fallout from superconductor research.

Impact on Materials Processing and Chemical Engineering. In an interesting paper, R. Kumar (Indian Institute of Science, Bangalore) points out how processing considerations for achieving high-temperature superconductors has introduced new process engineering problems not contemplated heretofore. In a paper (reference listed), Kumar observes that processes involve multicomponent solid-solid reactions; mixing of fine powders; simultaneous precipitation of many ions from solutions, emulsions, microemulsions, and liquid membranes; the flow of cohesive powders with and without binders; the flow of thin films over partially wetted particles; grain boundary growth and composition; quick evaporation using pulsed lasers; mixing of molecules during their flight paths and the influence of oxygen jets; deposition of particles on substrates; and other relatively unfamiliar processing techniques.

Additional Reading

Amato, I.: "Finally, a Hotter Superconductor," *Science*, 755 (May 7, 1993).

Beardsley, T.M.: "Unsuperconductivity," *Sci. Amer.*, 22D (April 1989).

Bishop, D.J., P.L. Gammel, and D.A. Huse: "Resistance in High-Temperature Superconductors," *Sci. Amer.*, 48 (February 1993).

Brosha, E.L. et al.: "Metastability of Superconducting Compounds in the Y-Ba-Cu-O System," *Science*, 196 (April 9, 1993).

Caruana, C.M.: "Superconductivity: The Near and Long Term Outlook," *Chem. Eng. Progress*, 72 (May 1988).

Cava, R.J.: "Structural Chemistry and the Local Charge Picture of Copper Oxide Superconductors," *Science*, 656 (February 9, 1990).

Conradson, S.D., I.D. Raistrick, and A.R. Bishop: "Axial-Oxygen-Centered Lattice Instabilities and High-Temperature Superconductivity," *Science*, 1394 (June 15, 1990).

Erwin, S.C. and W.E. Pickett: "Theoretical Fermi-Surface Properties and Superconducting Parameters for K3C60," *Science*, 842 (November 8, 1991).

Fisk, Z. et al.: "Heavy-Electron Metals: New Highly Correlated States of Matter," *Science*, 33 (January 1, 1988).

[1] As of 1990.

Fisk, Z., G. Aeppli: "Superstructures and Superconductivity," *Science*, 38 (April 2, 1993).

Foner, S., T.P. Orlando: "Superconductors: The Long Road Ahead," *Technology Review (MIT)*, 36 (February 1988).

Gabelle, T.H., J.K. Hulm: "Superconductivity—The State that Came in from the Cold," *Science*, 367 (January 22, 1988).

Haroche, S., J.-M. Raimond: "Cavity Quantum Electrodynamics," *Sci. Amer.*, 54 (April 1993).

Hazen, R.M.: "Perovskites," *Sci. Amer.*, 74 (June 1988).

Iqbal, Z. et al.: "Superconductivity at 45°K in Rb/Tl Codoped C60 and C60/C70 Mixtures," *Science*, 826 (November 8, 1991).

Ishiguiro, T., K. Yamaji, and G. Saito: *Organic Superconductors,* 2nd Edition, Springer-Verlag, Inc., New York, NY, 1997.

Ketterson, J.B. and S. Song: *Superconductivity,* Cambridge University Press, New York, NY, 1999.

Kumar, R.: "Chemical Engineering and the Development of Hot Superconductors," *Chem. Eng. Progress*, 17 (April 1990).

Laughlin, R.B.: "The Relationship Between High-Temperature Superconductivity and the Fractional Quantum Hall Effect," *Science*, 525 (October 28, 1988).

Lee, P.J.: *Engineering Superconductivity,* Wiley-IEEE Press, New York, NY, 2001.

Little, W.A.: "Experimental Constraints on Theories of High-Transition Temperature Superconductors," *Science*, 1390 (December 9, 1988).

Luss, D. et al.: "Processing High-Temperature Superconductors," *Chem. Eng. Progress*, 40 (September 1989).

McDonald, D.G.: "Superconductivity and the Quantization of Energy," *Science*, 177 (January 12, 1990).

Murphy, D.W. et al.: "Processing Techniques for the 93°K Superconductor $Ba_2YCu_3O_7$," *Science*, 922 (August 19, 1988).

Pool, R.: "Superconductor Patents: Four Groups Duke It Out," *Science*, 931 (September 1, 1989).

Pool, R.: "Superconductivity Stars React to the Market," *Science*, 373 (January 25, 1991).

Poole, C.P., H.A. Farach, and R.J. Creswick: *Superconductivity,* Academic Press, Inc., San Diego, CA, 1996.

Poole, C.P., H.A. Farach, and R.J. Creswick: *Handbook of Superconductivity,* Academic Press, Inc., San Diego, CA, 1999.

Ross, P. and R. Ruthen: "Squeezed Hydrogen Forms Metal with Superconducting Potential," *Sci. Amer.*, **26** (November 1989).

Schrieffer, J.R.: *The Theory of Superconductivity,* Perseus Publishing, Boulder, CO, 1999.

Shrivastava, K.N.: *Superconductivity,* World Scientific Publishing Company, Inc., Riveredge, NJ, 2000.

Shumay, W.C. Jr.: *Superconductor Materials Engineering,* Advanced Materials & Processes, 49 (November 1988).

Sleight, A.W.: "Chemistry of High-Temperature Superconductors," *Science*, 1519 (December 16, 1988).

Staff: "Trying to Cooperate in Order to Compete," *Technology Review (MIT)*, 13 (February/March 1991).

Stix, G.: "Superconducting SQUIDS," *Sci. Amer.*, 112 (March 1991).

Sun, J.Z. et al.: *Elimination of Current Dissipation in High Transition Temperature Superconductors, Science,* 307 (January 19, 1990).

Tinkham, M.: *Introduction to Superconductivity,* 2nd Edition, The McGraw-Hill Companies, Inc., New York, NY, 1995.

Wolsky, A.M., R.F. Giese, and E.J. Daniels: "The New Superconductors: Prospects for Applications," *Sci. Amer.*, 60 (February 1989).

SUPERCOOLING. The cooling of a liquid below its freezing point without the separation of the solid phase. This is a condition of metastable equilibrium, as is shown by solidification of the supercooled liquid upon the addition of the solid phase, or the application of certain stresses, or simply upon prolonged standing.

SUPERFLUIDITY. The term used to describe a property of condensed matter in which a resistance-less flow of current occurs. The mass-four isotope of helium in the liquid state, plus over 20 metallic elements, are known to exhibit this phenomenon. In the case of liquid helium, these currents are hydrodynamic. For the metallic elements, they consist of electron streams. The effect occurs only at very low temperatures in the vicinity of the absolute zero (−273.16°C or 0 K). In the case of helium, the maximum temperature at which the effect occurs is about 2.2 K. For metals, the highest temperature is in the vicinity of 20 K.

If one of the metals (commonly referred to as superconductors) is cast in the form of a ring and an external magnetic field is applied perpendicularly to its plane and then removed, a current will flow round the ring induced by Faraday induction. This current will produce a magnetic field, proportional to the current, and the size of the current may be observed by measuring this field. Were the ring (e.g., one made of lead) at a temperature above 7.2 K, this current and field would decay to zero in a fraction of a second.

But with the metal at a temperature below 7.2 K before the external field is removed, this current shows no sign of decay even when observations extend over a period of a year. As a result of such measurements, it has been estimated that it would require 10^{99} years for the supercurrent to decay. Such persistent or "frictionless" currents in superconductors were observed in the early 1900s—hence they are not a recent discovery.

In the case of liquid helium, these currents are hydrodynamic, i.e., they consist of streams of neutral (uncharged) helium atoms flowing in rings. Since, unlike electrons, the helium atoms carry no charge, there is no resulting magnetic field. This makes such currents much more difficult to create and detect. Nevertheless, as a result of research carried out in England and the United States during the late 1950s and early 1960s, the existence of supercurrents in liquid helium has been established.

SUPERMOLECULAR CHEMISTRY. See **Molecular Recognition**.

SUPEROXIDES.
These compounds are characterized by the presence in their structure of the O_2^- ion. The O_2^- ion has an odd number of electrons (13) and, as a result, all superoxide compounds are paramagnetic. At room temperature all superoxides have a yellowish color. At low temperature many of them undergo reversible phase transitions which are accompanied by a color change to white. Superoxide compounds known to be stable at room temperature are:

Sodium superoxide	NaO_2
Potassium superoxide	KO_2
Rubidium superoxide	RbO_2
Cesium superoxide	CsO_2
Calcium superoxide	$Ca(O_2)_2$
Strontium superoxide	$Sr(O_2)_2$
Barium superoxide	$Ba(O_2)_2$
Tetramethylammonium superoxide	$(CH_3)_4NO_2$

The superoxides are generally prepared by one of three methods:

(1) Direct oxidation of the metal, metal oxide, or metal peroxide with pure oxygen or air. All alkali metal superoxides, with the exception of lithium have been prepared in this manner. The superoxides of potassium, rubidium, and cesium form quite readily upon direct oxidation of the molten metal in air or oxygen at atmospheric pressure. Attempts to prepare sodium superoxide under the same conditions result in the formation of sodium peroxide, Na_2O_2. As a result, it was generally felt, prior to 1949, that sodium superoxide was not stable enough to be synthesized. However, in 1949 this superoxide was prepared for the first time, in good yield and purity, by the direct oxidation of sodium peroxide at 490°C under an oxygen pressure of 298 atm. Sodium superoxide is now commercially available and is prepared by a high-temperature, high-pressure, direct oxidation of the peroxide. It is now known that the pale yellow color common in commercial grade sodium peroxide is due to the presence of 5 to 10% sodium superoxide.

(2) Oxidation of an alkali metal dissolved in liquid ammonia with oxygen. All the alkali metal superoxides have been prepared by this method. Although lithium superoxide (LiO_2) has not been isolated in a room temperature-stable form, it has been demonstrated that when lithium is oxidized in liquid ammonia at −78°C the superoxide does form and is stable at that temperature.

(3) Reaction of hydrogen peroxide with strong bases. Hydrogen peroxide can be caused to react with strong inorganic bases to form intermediate peroxide compounds which disproportionate to yield superoxides. The alkaline earth metal superoxides, and sodium, potassium, rubidium, cesium, and tetramethylammonium superoxide have been obtained via this process. Claims have also been made for the synthesis of lithium superoxide via this method; however, such claims have not been adequately substantiated.

Using the formation of potassium superoxide as an example, the reactions involved in this process are:

$$2KOH + 3H_2O_2 \longrightarrow K_2O_2 \cdot 2H_2O_2 + 2H_2O$$

followed by

$$K_2O_2 \cdot 2H_2O_2 \longrightarrow 2KO_2 + 2H_2O.$$

From the commercial point of view the most important of the superoxides is KO_2. This compound has been in large scale commercial production for many years. It is manufactured in very good yield and purity by air oxidation of the molten metal. This compound is utilized in self-contained breathing devices which are widely used in fire fighting operations and in mine rescue work. The function of the superoxide is to provide oxygen and to remove exhaled carbon dioxide. This unique capability of superoxides is explained by the following chemical reactions:

$$2KO_2(s) + HOH(v,l) \longrightarrow 2KOH(s,soln) + 3/2O_2(g)$$

and

$$2KOH(s,soln) + CO_2(g) \longrightarrow K_2CO_3(s,soln) + H_2O$$

where s = solid, v = vapor, l = liquid, g = gas, and soln = solution.

Up to 34% of the weight of potassium superoxide is available as breathing oxygen. The lower molecular weight NaO_2 is capable of supplying up to 43% of its weight as oxygen. Thus, sodium superoxide is a better oxygen storage compound. However, it has not been widely used due to its relatively high cost. The cost of KO_2 is much less. The use of superoxides for maintaining proper oxygen and carbon dioxide levels in the atmospheres of space vehicles, space stations, and submarines has been of some interest.

The handling and storage of superoxides requires care and caution. Chemically they are powerful oxidizing agents and strong bases and as a result, they react vigorously with acids and organic materials. All superoxides are extremely hydroscopic, thus their safe storage requires the use of tightly sealed, clean, dry containers.

The chemical bond between the superoxide ion, O_2^-, and the metal ion is ionic in nature. Melting points of potassium, rubidium and cesium superoxide have been determined, and in keeping with the ionic nature of the compounds, the melting temperatures are high, in the order of 400°C.

The most reliable technique for the analysis of superoxides is that developed by Seyb and Kleinberg. In this method the superoxide sample is treated with a mixture of glacial acetic acid and diethyl or dibutyl phthalate. The superoxide reacts with the acetic acid to yield oxygen, hydrogen peroxide, and potassium acetate. The amount of superoxide in the sample is related to the amount of oxygen evolved which is measured with a gas buret. The stoichiometry of the analytical reaction is:

$$2KO_2 + 2HC_2H_3O_2 \longrightarrow 2KC_2H_3O_2 + H_2O_2 + O_2$$

It is important that a sufficiently dilute glacial acetic acid-diethyl phthalate mixture be used. Contact of undiluted glacial acetic acid with the superoxide will result in a violent and uncontrollable reaction.

As a result of the paramagnetic nature of superoxides, it is possible to determine their purity by means of paramagnetic susceptibility measurements. The use of this method is limited by its poor accuracy.

A. W. PETROCELLI
Westerley, Rhode Island

SUPER POSITION (Nernst Principle of).
The potential difference between junctions in similar pairs of solutions which have the same ratio of concentrations are the same even if the absolute concentrations are different, e.g., the same potential difference exists between normal solutions of HCl and KCl as exists between tenth-normal solutions of HCl and KCl.

SUPERSATURATED VAPOR.
A vapor that remains dry, although its heat content is less than that of dry and saturated vapor at the pressure. Supersaturation is an unstable condition, and is found in the steam emerging from the nozzles of a steam turbine. The abnormality of the phenomenon is similar to that of supercooling. See also **Supercooling**. Supersaturation of the steam probably results from the very rapid expansion of steam in the nozzle, permitting the traverse of a short distance before the condensation of moisture is completed. At a certain definite point, however, known as the Williams limit, the supersaturation vanishes, and the steam regains the wet state which would be normal in view of the pressure and the heat content. Supersaturation of vapor is impossible in the presence of numerous charged ions or dust particles.

SUPERSATURATION (Chemical).
The condition existing in a solution when it contains more solute than is needed to cause saturation. Thermodynamically, this type of supersaturation is closely allied to supersaturation of a vapor, since the solute cannot crystallize out in solutions free from impurities or seed crystals of the solute. See also **Supersaturated Vapor**.

SURFACE. In physical chemistry the area of contact between two different phases or states of matter, e.g., finely divided solid particles and air or other gas (solid-gas); liquids and air (liquid-gas); insoluble particles and liquid (solid-liquid). Surfaces are the sites of the physiochemical activity between the phases that is responsible for such phenomena as adsorption, reactivity, and catalysis. The depth of a surface is of molecular order of magnitude. The term *interface* is approximately synonymous with *surface*, but it also includes dispersions involving only one phase of matter, i.e., solid-solid or liquid-liquid.

SURFACE ACTIVE AGENTS. See **Detergents**.

SURFACE CHEMISTRY. This topic deals with the behavior of matter, where such behavior is determined largely by forces acting at surfaces. Since only condensed phases, i.e., liquids and solids, have surfaces, studies in surface chemistry require that at least one condensed phase be present in the system under consideration. The condensed phase may be of any size ranging from colloidal dimensions to a mass as large as an ocean. Interactions between solids, immiscible liquids, liquids and solids, gases and liquids, gases and solids, and different gases on a surface fall within the province of surface chemistry.

Surface forces determine whether one material will wet and spread on a substrate, e.g., whether a liquid will wet a solid and spread into crevices and pores to displace air. This seemingly simple phenomenon is of cardinal importance in determining the strength of adhesive joints and of reinforced plastics; it establishes the printing and writing qualities of inks; lubricants will wet and spread over entire surfaces or be confined to limited working areas depending upon built-in wetting or nonwetting properties; ores are floated if the surrounding liquid is readily displaced by air bubbles; the dispersion of pigments in paints depends upon wetting of the individual particles by the liquid; the action of a foam breaker frequently depends upon its ability to spread on the foam; secondary oil recovery often involves displacement of oil from sand by water; wetting is also a factor in detergency; water and soil repellancy depend upon nonwetting.

Wetting or nonwetting often depends upon the adsorption of a solute at a surface or interface. The bulk liquid phase either advances or recedes, depending upon the nature of the solute and the condensed phase. However, there are many phenomena where adsorption is essential to the process but wetting is not a factor. For example, toxic gases and cigarette tars are removed by adsorption on suitable substrates; color bodies are removed from vegetable oils by adsorption on activated clays; heterogeneous catalysis requires the adsorption of reactants on the catalytic surface; dyeing of fabrics is an adsorption process; dispersions and emulsions are stabilized by the adsorption of suitable solutes and flocculated by the adsorption of other solutes; foaming depends upon adsorption; chromatography is a preferential adsorption process; the action of many corrosion inhibitors depends upon their adsorption on metal surfaces.

The spreading of an insoluble monolayer is a process analogous to adsorption with a number of specialized applications. Thus, cetyl alcohol is spread as a monoloayer on reservoirs to retard the evaporation of water. Some antifoaming agents act by spreading as monolayers.

Because of the widespread applications of surface chemistry, practically all industries, knowingly or otherwise, make use of the principles of surface chemistry. Countless cosmetic and pharmaceutical products are emulsions—lotions, creams, ointments, suppositories, etc. Food emulsions include milk, margarine, salad dressings and sauces. Adhesive emulsions, emulsion paints, self-polishing waxes, waterless hand cleaners and emulsifiable insecticide concentrates are commonplace examples of emulsions, which fall within the province of surface chemistry. Other products which function in accordance with the principles of surface chemistry include detergents of every variety, fabric softeners, antistatic agents, mold releases, dispersants and flocculants.

Surface forces are merely an extension of the forces acting within the body of a material. A molecule in the center of a liquid drop is attracted equally from all sides, while at the surface the attractive forces acting between adjacent molecules results in a net attraction into the bulk phase in a direction normal to the surface. Because of unbalanced attraction at the surface, the tendency is for these molecules to be pulled from the surface into the interior, and for the surface to shrink to the smallest area that can enclose the liquid. The work required to expand a surface by 1 sq cm in opposition to these attractive forces is called the surface tension.

This concept applies equally well to solids. Molecules in a solid surface are also in an unbalanced attractive field and possess a surface tension or surface free energy. While the surface tension of a liquid is easily measured, this is much more difficult to do for a solid, since to increase the surface extraneous work must be done to deform the solid.

In the case of solid or liquid solutions it is frequently observed that one component of the solution is present at a greater concentration in the surface region than in the bulk of the solution. Thus, for an ethanol-water system, the surface region will contain an excess of ethanol. The concentration of water will be higher at the surface than in the bulk, if the solute is sulfuric acid. Molybdenum oxide dissolved in glass will concentrate at the surface of the glass. The concentrating of solute molecules at a surface is called adsorption.

If a clean solid is exposed to the atmosphere, molecules of one or more species present in the atmosphere will deposit on the surface. If the clean solid is immersed in a solution, molecules of one or another species present in the solution will be apt to concentrate at the solid-liquid interface. These phenomena are also referred to as *adsorption*.

All adsorption processes result from the attraction between like and unlike molecules. For the ethanol-water example given above, the attraction between water molecules is greater than between molecules of water and ethanol. As a consequence, there is a tendency for the ethanol molecules to be expelled from the bulk of the solution and to concentrate at the surface. This tendency increases with the hydrocarbon chain-length of the alcohol. Gas molecules adsorb on a solid surface because of the attraction between unlike molecules. The attraction between like and unlike molecules arises from a variety of intermolecular forces. London dispersion forces exist in all types of matter and always act as an attractive force between adjacent atoms and molecules, no matter how dissimilar they are. Many other attractive forces depend upon the specific chemical nature of the neighboring molecules. These include dipole interactions, the hydrogen bond and the metallic bond.

There is an additional explanation for the tendency of a solute such as ethanol to concentrate at the surface of a liquid, which originates with Langmuir. According to his "principle of independent surface action" each portion of a molecule behaves independently of other portions of the molecule in its attraction to other molecules or functional groups on a molecule. The attraction between the CH_3CH_2 portion of the ethanol molecule and water arises from relatively weak London dispersion forces, as compared with the additional attraction of strong hydrogen-bonding forces acting between the hydroxyl group of the alcohol and water. Hydrogen bonding is also responsible for the strong attraction between water molecules. As a consequence, not only is the alcohol concentrated at the surface, it is also oriented with the hydroxyl group toward the water and the hydrocarbon chain directed outward. Since the attraction between adjacent hydrocarbon molecules is less than that between adjacent water molecules, hydrocarbon liquids have lower surface tensions than water, and the surface tension of an aqueous alcohol solution is intermediate between that of liquid hydrocarbons and water.

As noted earlier, the phenomenon of adsorption is encountered in diverse applications. Medical applications are often the most complex and the least understood. For example, replacement hearts and kidney machines require plastics that can be kept in contact with human blood for long periods of time. However, foreign material in contact with blood results in clotting. The material first becomes coated with adsorbed protein. Some time later the clotting process begins, apparently due to activation of the Hageman factor, one of the proteins in blood, at the blood-material interface. The activation initiates a chain reaction that results in the conversion of fibrinogen to fibrin. It has been suggested that the Hageman factor is helical in form and that adsorption results in an unfolding of the protein helix with exposure of certain active sites which then initiate the clotting of blood. Other proteins are also adsorbed and their biological function may be altered, but little is known about this.

The ideal surface for contact with human blood is the surface of blood vessels, and the immediate surface contains heparinoid complexes. Heparin, a negatively charged polysaccharide, has been bonded to silicon rubber and other polymers. In one procedure, a quaternary ammonium compound is first adsorbed on the polymer substrate and heparin is in turn adsorbed on the positively charged surface. Chemical bonding of heparin has also been achieved. Such surfaces do not cause clotting of contacted blood.

As noted earlier, the phenomenon of adsorption is encountered in such diverse applications as the separation of components in chromatography, the removal of toxic gases by activated charcoal, heterogeneous catalytic reactions and the dyeing of fabrics. The surface area of solids is most commonly determined by the adsorption of nitrogen on the surface of the solid at −915°C. Nitrogen is assigned an area of 16.2 Å^2 per molecule. The method is due to Brunauer, Emmett and Teller and is referred to as the BET method for determining surface area.

The fundamental adsorption equation is due to Gibbs. In the case of a solution containing a single solute,

$$\Gamma = \frac{1}{RT}\left(\frac{\partial \dot{\gamma}}{\partial \ln a}\right)_T$$

where Γ is the excess of solute at the surface as compared with the concentration of solute in the bulk liquid expressed in moles per sq cm, R is the gas constant, T is the absolute temperature γ is the surface tension of the solution, and a is the activity of the solute. Where the solution is sufficiently dilute, the concentration of the solute may be substituted for its activity.

The equation also applies to the adsorption of a gas on a solid. At low gas pressures, p, the equilibrium pressure of the gas can be substituted for a, the activity of the solute. The amount of gas adsorbed v/V is equivalent to the surface excess Γ, where v is equal to the volume of gas adsorbed per gram of solid and V is the molar volume of the gas. The total free energy change at constant pressure is $\Sigma \delta \gamma$, where Σ is the area per gram of solid.

When a drop of liquid is placed on the surface of a solid, it may spread to cover the entire surface, or it may remain as a stable drop on the solid. There is a solid-liquid interface between the two phases. In the case of liquids that do not spread on the solid, the bare surface of the solid adsorbs the vapor of the liquid until the fugacity of the adsorbed material is equal to that of the vapor and the liquid.

The equation relating contact angle to surface tension, generally ascribed to Young or Dupre, is

$$\gamma_{Se} = \gamma_{SL} + \gamma_{L}\cos\theta$$

where γ_{Se} is the surface tension of the solid covered with adsorbed vapor, γ_{SL} is the solid-liquid interfacial tension, γ_{L} is the surface tension of the liquid and θ is the contact angle.

As a general rule, organic liquids and aqueous solutions will spread on high-energy surfaces, such as the clean surfaces of metals and oxides. The rule has a number of exceptions. For one, certain organic liquids will deposit a low-energy film by adsorption on higher-energy surfaces over which the bulk liquid will not spread.

Zisman discovered that there is a critical surface tension characteristic of low-energy solids, such as plastics and waxes. Liquids that have a lower surface tension than the solid will spread on that solid, while liquids with a higher surface tension will not spread. Examples of critical surface tension values for plastic solids in dynes per cm are: "Teflon," 18; polyethylene, 31; polyethylene terephthalate, 43; and nylon, 42–46. As one indication of the way this information can be used in practical applications, one can consider the bonding of nylon to polyethylene. If nylon were applied as a melt to polyethylene, it would not wet the lower-energy polyethylene surface and adhesion would be poor. However, molten polyethylene would spread readily over solid nylon to provide a strong bond.

There are a large number of materials that exhibit a pronounced tendency to concentrate at surfaces and interfaces and thus alter the surface properties of matter. These materials are called surface-active agents or surfactants. Depending upon the manner in which they are used or the purpose they serve in specific applications, they may be referred to as detergents, emulsifying agents, foaming agents or foam stabilizers, antibacterial agents, fabric softeners, flotation reagents, antistatic agents, corrosion inhibitors, or by other names. There are two general ways of classifying surfactants. According to solubility, they are classified as water or oil soluble. The other classification is according to change type. Those that do not ionize are called nonionic surfactants. If they ionize and the surface-active ion is anionic, the material is an anionic surfactant. If the surface-active ion carries a positive charge, it is called a cationic surfactant.

Only molecules with certain specific types of configurations exhibit surface activity. In general, these molecules are composed of two segregated portions, one of which has low affinity for the solvent and tends to be rejected by the solvent. The other portion has sufficient affinity for the solvent to bring the entire molecule into solution. Water-soluble soaps are probably the oldest surfactants. The long hydrocarbon chain has a low affinity for water and is referred to as the hydrophobic or nonpolar

portion of the molecule. The carboxylate group has a high affinity for water and is called the hydrophilic or polar portion.

LLOYD OSIPOW
New York

Additional Reading

Birdi, K.S.: *Handbook of Surface and Colloid Chemistry,* 2nd Edition, CRC Press LLC., Boca Raton, FL, 2002.
Carley, A.F., G.J. Hutchings, M.S. Spencer, and P.R. Davies: *Surface Chemistry and Catalysis,* Kluwer Academic Publishers, Norwell, MA, 2002.
McCash, E.M.: *Surface Chemistry,* Oxford University Press, New York, NY, 2001.
Sposito, G.: *Surface Chemistry of Natural Particles,* Oxford University Press, New York, NY, 2004.

SURFACE TENSION.   Fluid surfaces exhibit certain features resembling the properties of a stretched elastic membrane; hence the term surface tension. Thus, one may lay a needle or a safety-razor blade upon the surface of water, and it will lie at rest in a shallow depression caused by its weight, much as if it were on a rubber air-cushion. A soap bubble, likewise, tends to contract, and actually creates a pressure inside, somewhat after the manner of a rubber balloon. The analogy is imperfect, however, since the tension in the rubber increases with the radius of the balloon, and the pressure inside, which would otherwise decrease, remains approximately constant; while the liquid "film tension" remains constant and the pressure in the bubble falls off as the bubble is blown.

Surface tension results from the tendency of a liquid surface to contract. It is given by the tension σ across a unit length of a line on the surface of the liquid. The surface tension of a liquid depends on the temperature; it diminishes as temperature increases and becomes 0 at the critical temperature. For water σ is 0.073 newtons/meter at 20°C, and for mercury, it is 0.47 newtons/meter at 18°C.

Surface tension is intimately connected with capillarity, that is, rise or depression of liquid inside a tube of small bore when the tube is dipped into the liquid. Another factor related to this phenomenon is the angle of contact. If a liquid is in contact with a solid and with air along a line, the angle θ between the solid-liquid interface and the liquid-air interface is called the angle of contact. See Fig. 1. If $\theta = 0$, the liquid is said to wet the tube thoroughly. If θ is less than 90°, the liquid rises in the capillary; and if more than 90°, the liquid does not wet the solid, but is depressed in the tube. For mercury on glass, the angle of contact is 140°, so that mercury is depressed when a glass capillary is dipped into mercury. The rise h of the liquid in the capillary is given by $h = 2\sigma\cos\theta/r\rho g$, where r is the radius of the tube, ρ the density of the liquid, and g is the acceleration due to gravity.

Surface tension can be explained on the basis of molecular theory. If the surface area of liquid is expanded, some of the molecules inside the liquid rise to the surface. Because a molecule inside a mass of liquid is under the forces of the surrounding molecules, while a molecule on the surface is only partly surrounded by other molecules, work is necessary to bring molecules from the inside to the surface. This indicates that force must be

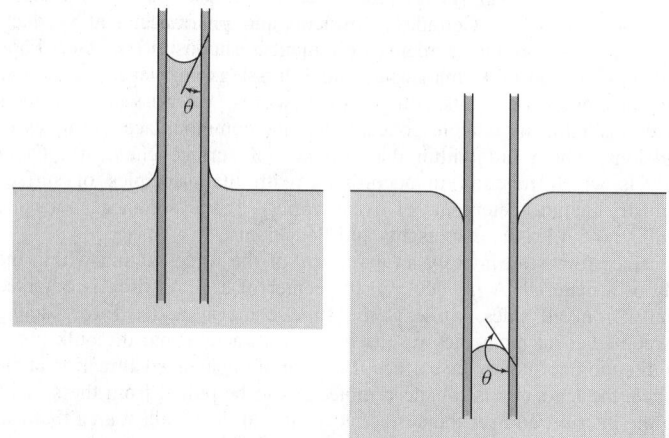

Fig. 1.   Interrelationship between surface tension and capillarity: (*Left*) Case where angle theta is less than 90° (water); (*Right*) case where angle theta is greater than 90° (mercury)

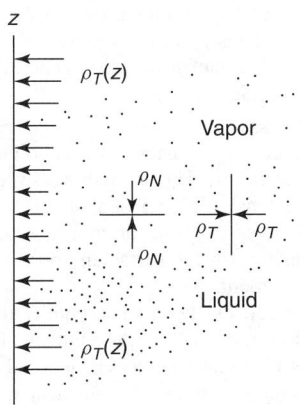

Fig. 2.   Stress relationships in surface tension

applied along the surface in order to increase the area of the surface. This force appears as tension on the surface and when expressed as tension per unit length of a line lying on the surface, it is called the surface tension of the liquid.

The molecular theory of surface tension was dealt with by Laplace (1749–1827). But, as a result of the clarification of the nature of intermolecular forces by quantum mechanics and of the more recent developments in the study of molecular distribution in liquids, the nature and value of surface tension have been better understood from a molecular viewpoint. Surface tension is closely associated with a sudden, but continuous change in the density from the value for bulk liquid to the value for the gaseous state in traversing the surface. See Fig. 2. As a result of this inhomogeneity, the stress across a strip parallel to the boundary—ρ_N per unit area—is different from that across a strip perpendicular to the boundary—ρ_T per unit area. This is in contrast with the case of homogeneous fluid in which the stress across any elementary plane has the same value regardless of the direction of the plane.

The stress ρ_T is a function of the coordinate z, the z-axis being taken normal to the surface and directed from liquid to vapor. The stress ρ_N is constant throughout the liquid and the vapor. Figure 2 shows the stress ρ_N and ρ_T. The stress $\rho_T(z)$ as a function of z is also shown on the left side of the figure.

SURFACTANTS.   The term surfactant, contraction of surface-active agent, is used to describe organic substances having certain characteristics in structure and properties. The term detergent is often used interchangeably with surfactant. As a designation for a substance capable of cleaning, detergent can also encompass inorganic substances when these do in fact perform a cleaning function. More often, however, detergent refers to a combination of surfactants and other substances, organic or inorganic, formulated to enhance functional performance, specifically cleaning, over that of the surfactant alone. It is so used herein.

Surfactants are characterized by the following features. Amphipathic structure: surfactant molecules are composed of groups of opposing solubility tendencies, typically an oil-soluble hydrocarbon chain and a water-soluble ionic group; solubility: a surfactant is soluble in at least one phase of a liquid system; adsorption at interfaces: at equilibrium, the concentration of a surfactant solute at a phase interface is greater than its concentration in the bulk of the solution; orientation at interfaces: surfactant molecules and ions form oriented monolayers at phase interfaces; micelle formation: surfactants form aggregates of molecules or ions called micelles when the concentration of the surfactant solute in the bulk of the solution exceeds a limiting value, the so-called critical micelle concentration (CMC), which is a fundamental characteristic of each solute–solvent system; and functional properties: surfactant solutions exhibit combinations of cleaning (detergency), foaming, wetting, emulsifying, solubilizing, and dispersing properties.

The presence of two structurally dissimilar groups within a single molecule is the most fundamental characteristic of surfactants. The surface behavior (surface activity) of the surfactant molecule is determined by the makeup of the individual groups, solubility properties, relative size, and location within the surfactant molecule.

Different designations describe the opposing groups within the surfactant molecules, e.g., hydrophobic (water hating) and hydrophilic (water liking),

lipophobic (fat hating) and lipophilic (fat liking), oleophobic (fat (oil) hating) and oleophilic (fat (oil) liking), and lyophobic (solvent hating) and lyophilic (solvent liking). The terms polar and nonpolar are also used to designate water-soluble and water-insoluble groups, respectively.

Surface activity is not limited to aqueous systems; however, because water is present as the solvent phase in the overwhelming proportion of commercially important surfactant systems, its presence is assumed in much of the common terminology of industry. Thus, the water-soluble amphipathic groups are often referred to as solubilizing groups.

Surfactants are classified depending on the charge of the surface-active moiety. In anionic surfactants, this moiety carries a negative charge. In cationic surfactants, the charge is positive. In nonionic surfactants, there is no charge on the molecule, the solubilizing contribution can be supplied by side groups. Finally, in amphoteric surfactants, solubilization is provided by the presence of positive and negative charges in the molecule. In general, the hydrophobic group consists of a hydrocarbon chain containing ca 10–20 carbon atoms. The chain may be interrupted by oxygen atoms, a benzene ring, amides, esters, other functional groups, and/or double bonds. A propylene oxide hydrophobe can be considered a hydrocarbon chain in which every third methylene group is replaced by an oxygen atom. In some cases, the chain may carry substituents, most often halogens. Siloxane chains have also served as the hydrophobe in some surfactants.

Hydrophilic, solubilizing groups for anionic surfactants include carboxylates, sulfonates, sulfates, and phosphates. Cationics are solubilized by amine and ammonium groups. Ethylene oxide chains and hydroxyl groups are the solubilizing groups in nonionic surfactants. Amphoteric surfactants are solubilized by combinations of anionic and cationic solubilizing groups.

The molecular weight of surfactants may be as low as ca 200 up to the thousands for polymeric structures. A surfactant with a straight-chain C_{12}-hydrophobe and a solubilizing group is generally an effective structure. The optimum can be higher by several carbon atoms or even slightly lower than 12 depending on the nature of the polar group and the desired function of the surfactant.

In the application of surfactants, physical and use properties, precisely specified, are of primary concern. Chemical homogeneity is of little significance in practice. In fact, surfactants are generally polydisperse mixtures, such as the natural fats as precursors of fatty acid-derived surfactant structures; e.g., coconut oil contains glycerol esters of C_6–C_{18} fatty acids. Nonionic surfactants of the alcohol ethoxylate type are polydisperse not only with respect to the hydrophobe but also in the number of ethylene oxide units attached.

Commercial surfactants are complicated mixtures exceedingly difficult to separate into pure molecular species.

Physical Chemistry of Interfaces

The usefulness of surfactants stems from the effects that they exert on the surface, interfacial, and bulk properties of their solutions and the materials their solutions come in contact with.

Phenomena at Liquid Interfaces.   The area of contact between two phases is called the interface; three phases can have only a line of contact, and only a point of mutual contact is possible between four or more phases. Combinations of phases encountered in surfactant systems are L–G, L–L–G, L–S–G, L–S–S–G, L–L, L–L–L, L–S–S, L–L–S–S–G, L–S, L–L–S, and L–L–S–G, where G = gas, L = liquid, and S = solid. An example of an L–L–S–G system is an aqueous surfactant solution containing an emulsified oil, suspended solid, and entrained air (see **Emulsions; Foam**). This embodies several conditions common to practical surfactant systems. First, because the surface area of a phase increases as particle size decreases, the emulsion, suspension, and entrained gas each have large areas of contact with the surfactant solution. Next, because interfaces can exist only between two phases, analysis of phenomena in the L–L–S–G system breaks down into a series of analyses, i.e., surfactant solution to the emulsion, solid, and gas. It is also apparent that the surfactant must be stabilizing the system by preventing contact between the emulsified oil and dispersed solid. Finally, the dispersed phases are in equilibrium with each other through their common equilibrium with the surfactant solution.

Figures 1a and 1b represent typical gas–liquid and liquid–liquid interfaces at equilibrium. Assuming that gas, G, consists of air and vapor of the liquid, L, at equilibrium, there is continuous movement of liquid molecules through the gaseous interfacial region R_G because rates of evaporation and condensation at the interface I_G are equal (Fig. 1). Liquid molecules are also moving continuously into and out of I_G through the liquid interfacial region, R_L. R_G and R_L represent nonhomogeneous transitional regions between the homogeneous phases, G and L. Systems are known in which R_G and R_L have thicknesses equivalent to two or more layers of molecules, but for most analyses the interface I_G can be considered as consisting of a single layer of molecules.

For thermodynamic treatment of surface phenomena, the thickness of the boundary regions can often be ignored or their effect eliminated by selection of a convenient location for the interface I_{GL}. The liquid–liquid interface, I_{LL} (Fig. **1b**) is similarly associated with interfacial regions, R_A and R_B, which can be treated like the gas–liquid interface in most analyses. Because few liquids are completely immiscible, mutual saturation is taken as the equilibrium condition.

Energy of Adhesion. The interfacial energy between two mutually insoluble saturated liquids, A and B, is equal to the difference in the separately measured surface energies of each phase: $\gamma_{AB} = \gamma_A - \gamma_B$, where γ is free-surface or interfacial energy. The term γ_{AB} represents the energy that must be added to the system to separate the liquids.

Contact Angle. The line of contact between the three phases of a G–L–S system is the locus of all points from which the angle of contact between the liquid and the solid can be measured. The drop of liquid, L, is resting on the solid, S, and both phases are exposed to the gas, G, at equilibrium saturation of the liquid in air (gas). The drop is assumed to be small enough for the flattening pressure of gravity to be negligible. The vector X_G is tangent to the liquid at its contact with the solid. The angle between the tangent and the surface of the solid is called the contact angle, θ. The equilibrium value of θ is an indicator of the energy relationships between liquid–liquid and liquid–solid interfaces.

Effects of Surfactants on Solutions. A surfactant changes the properties of a solvent in which it is dissolved to a much greater extent than is expected from its concentration effects. This marked effect is the result of adsorption at the solution's interfaces, orientation of the adsorbed surfactant ions or molecules, micelle formation in the bulk of the solution, and orientation of the surfactant ions or molecules in the micelles, which are caused by the amphipathic structure of a surfactant molecule. The magnitude of these effects depends to a large extent on the solubility balance of the molecule. An efficient surfactant is usually relatively insoluble as individual ions or molecules in the bulk of a solution.

Positive adsorption, the concentration of one component of a solution at a phase boundary, results in a lowering of the free-surface energy of the solution. Accumulation of a surfactant at a solution interface means that the attractive forces between surfactant and solvent are less than the attraction between solvent molecules. As thermal diffusion brings surfactant molecules into the surface, accumulation occurs because the solute molecules cannot re-enter the solution against the stronger mutual attraction of the solvent molecules. Negative adsorption occurs when the attraction between solute and solvent molecules is greater than that between solvent molecules, and exists in concentrated aqueous solutions of inorganic compounds such as NaOH. It is associated with a surface tension slightly higher than for pure water.

Practical applications of surfactants usually involve some manner of surfactant adsorption on a solid surface. This adsorption is always associated with a decrease in free-surface energy, the magnitude of which must be determined indirectly. The force with which the adsorbate is held on the adsorbent may be roughly classified as physical, ionic, or chemical. Physical adsorption is a weak attraction caused primarily by van der Waals forces. Ionic adsorption occurs between charged sites on the substrate and oppositely charged surfactant ions, and is usually a strong attractive force. The term chemisorption is applied when the adsorbate is joined to the adsorbent by covalent bonds or forces of comparable strength.

Physical and ionic adsorption may be either monolayer or multilayer. Capillary structures in which the diameters of the capillaries are small, i.e., one to two molecular diameters, exhibit a marked hysteresis effect

on desorption. Sorbed surfactant solutes do not necessarily cover all of a solid interface and their presence does not preclude adsorption of solvent molecules. The strength of surfactant sorption generally follows the order cationic > anionic > nonionic.

Micelles. Surfactant molecules or ions at concentrations above a minimum value characteristic of each solvent-solute system associate into aggregates called micelles. The formation, structure, and behavior of micelles have been extensively investigated. The term critical micelle concentration (CMC) denotes the concentration at which micelles start to form in a system comprising solvent, surfactant, possibly other solutes, and a defined physical environment.

Micelle size is expressed as the micellar molecular weight or, more generally, the aggregation number, i.e., the number of monomers making up the micelles. Micellar aggregation numbers generally lie between 20 and 100, for single-chain anionic and cationic surfactants. Large aggregation numbers (>1000) have been reported for nonionic micelles, especially as the cloud point is approached.

Small micelles in dilute solution close to the CMC are generally believed to be spherical. Under other conditions, micellar materials can assume structures such as oblate and prolate spheroids, vesicles (double layers), rods, and lamellae.

Micellar properties are affected by changes in the environment, e.g., temperature, solvents, electrolytes, and solubilized components. These changes include complicated phase changes, viscosity effects, gel formation, and liquefication of liquid crystals.

Measurement of Surface Activity. Each surface-active property can be measured in a variety of ways and the method of choice depends on the characteristics of the substance to be tested. The most frequently determined properties are surface tension (γ_{SG}, γ_{LG}), interfacial tension (γ_{LL}, γ_{LG}), contact angle (θ), and CMC.

Anionic Surfactants

Carboxylate, sulfonate, sulfate, and phosphate are the polar, solubilizing groups found in most anionic surfactants. In dilute solutions of soft water, these groups are combined with a 12–15 carbon chain hydrophobe for best surfactant properties. In neutral or acidic media, or in the presence of heavy-metal salts, e.g., Ca, the carboxylate group loses most of its solubilizing power.

Of the cations (counterions) associated with polar groups, sodium and potassium impart water solubility, whereas calcium, barium, and magnesium promote oil solubility. Ammonium and substituted ammonium ions provide both water and oil solubility. Triethanolammonium is a commercially important example. Salts (anionic surfactants) of these ions are often used in emulsification. Higher ionic strength of the medium depresses surfactant solubility. To compensate for the loss of solubility, shorter hydrophobes are used for application in high ionic-strength media.

Carboxylates. Soaps represent most of the commercial carboxylates. The general structure of soap is $RCOO^- M^+$, where R is a straight hydrocarbon chain in the $C_9 - C_{21}$ range and M^+ is a metal or ammonium ion. Interruption of the chain by amino or amido linkages leads to other structures which account for the small volumes of the remaining commercial carboxylates.

Large volumes of soap are used in industrial applications as gelling agents for kerosene, paint driers, and as surfactants in emulsion polymerization. See also **Soaps**. Concern over water eutrophication resulted in a ban of phosphorus in laundry detergents. Phosphates have been effectively replaced by combinations of zeolite, citrate, and polymers, coupled with rebalanced synthetic active systems. Soap itself is generally present only as a minor component of surfactants.

Polyalkoxycarboxylates surfactants are produced either by the reaction of sodium chloroacetate with an alcohol ethoxylate or from an acrylic ester and an alcohol alkoxylate. Because of the presence of the ethylene oxide linkages, these products possess a higher aqueous solubility which manifests itself in greater compatibility with cationic surfactants and polyvalent cations.

N-Acylsarcosinates. Sodium *N*-lauroylsarcosinate is a good soaplike surfactant. The amido group in the hydrophobe chain lessens the interaction with hardness ions. *N*-Acylosarcosinates are prepared from a fatty acid chloride and sarcosine.

Acylated Protein Hydrolysates. These surfactants are prepared by acylation of protein hydrolysates with fatty acids or acid chlorides.

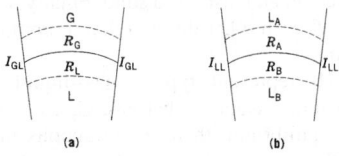

Fig. 1.   (a) Gas–liquid (GL) interface; **(b)** liquid–liquid (LL) interface

Acylated protein hydrolysates are mild surfactants recommended for personal-care products.

Sulfonates. The sulfonate group, $-SO_3M$, attached to an alkyl, aryl, or alkylaryl hydrophobe, is a highly effective solubilizing group. Sulfonic acid surfactants are strong, their salts are relatively unaffected by pH, they are stable to oxidation, and because of the strength of the C—S bond also stable to hydrolysis. Sulfonates interact moderately with the hardness, Ca^{2+} and Mg^{2+}, but significantly less so than carboxylates. Sulfates can be tailored for specific applications by introduction of double bonds or ester or amide groups, either into the hydrocarbon chain or as substituents. Because the introduction of the SO_3H function is inherently inexpensive, e.g., by oleum, SO_3, SO_2Cl_2, or $NaHSO_3$, sulfonates are heavily represented among high volume surfactants. See also **Sulfonation and Sulfation**.

Sulfonates include alkylbenzenesufonates (ABS), the most widely used of the non-soap surfactants; short-chain alkylarenesulfonates; lignosulfonates; napthalenesulfonates; α-olefinsulfonates; petroleum sulfonates; sulfonates with ester, amide, and ether linkages; and fatty acid ester sulfonates.

Sulfates and Sulfated Products. The sulfate group, $-OSO_3M$, where M is a cation, represents the sulfuric acid half-ester of an alcohol and is more hydrophilic than the sulfonate group because of the presence of an additional oxygen atom. Attachment of the sulfate group to a carbon atom of the hydrophobe through the C—O—S linkage limits hydrolytic stability, particularly under acidic conditions. Usage of sulfated alcohols and sulfated alcohol ethoxylates has expanded dramatically since the 1970s as the detergent industry reformulates consumer products to improve biodegradability, lower phosphate content, and move from powder to liquid.

Sulfates and sulfated products include alcohol sulfates, ethoxylated and sulfated alcohols; ethoxylated and sulfated alkylphenols; and sulfated natural oils and fats.

Phosphate Esters. Mono and diesters of orthophosphoric acid:

$$RO-\underset{\underset{HO}{|}}{\overset{\overset{HO}{|}}{P}}=O \text{ and } HO-\underset{\underset{RO}{|}}{\overset{\overset{RO}{|}}{P}}=O$$

and their salts are useful surfactants. In contrast to sulfonates and sulfates, the resistance of alkyl phosphate esters to acids and hard water is poor. Calcium and magnesium salts are insoluble. In the acid form, the esters show limited water solubility, although their alkali metal salts are more soluble. The surface activity of phosphate esters is good, although in general it is somewhat lower than that of the corresponding phosphate-free precursors. Thus, a phosphated nonylphenol ethoxylated with 9 mol of ethylene oxide is less effective as a detergent in hard water than its nonionic precursor. At higher temperatures, however, the phosphate surfactant is significantly more effective.

Because of high costs and the limitations noted above, phosphate surfactants find application in specialty situations where such limitations are of no concern. As specialty surfactants, phosphate esters and their salts are remarkably versatile. Applications include emulsion polymerization of vinyl acetate and acrylates; dry-cleaning compositions where solubility in hydrocarbon solvents is a particular advantage; textile mill processing where stability and emulsifying power for oil and wax under highly alkaline conditions is necessary; and industrial cleaning compositions where tolerance for high concentrations of electrolyte and alkalinity is required. In addition, phosphate surfactants are used as corrosion inhibitors, in pesticide formulations, in papermaking, and as wetting and dispersing agents in drilling mud fluids.

Nonionic Surfactants

Unlike anionic or cationic surfactants, nonionic surfactants carry no discrete charge when dissolved in aqueous media. Hydrophilicity in nonionic surfactants is provided by hydrogen bonding with water molecules. Oxygen atoms and hydroxyl groups readily form strong hydrogen bonds, whereas ester and amide groups form hydrogen bonds less readily. Hydrogen bonding provides solubilization in neutral and alkaline media. In a strongly acid environment, oxygen atoms are protonated, providing a quasi-cationic character. Each oxygen atom makes a small contribution to water solubility; more than a single oxygen atom is therefore needed to solubilize a nonionic surfactant in water. Nonionic surfactants are compatible with ionic and amphoteric surfactants. Because a polyoxyethylene group can easily

be introduced by reaction of ethylene oxide with any organic molecule containing an active hydrogen atom, a wide variety of hydrophobic structures can be solubilized by ethoxylation.

Polyoxyethylene Surfactants. Polyoxyethylene-solubilized nonionics (ethoxylates) are moderate foamers and do not respond to conventional foam boosters. Foaming shows a maximum as a function of ethylene oxide content. Low foaming nonionic surfactants are prepared by terminating the polyoxyethylene chain with less soluble groups such as polyoxypropylene and methyl groups. Ethoxylates can be prepared to attain almost any hydrophilic–hydrophobic balance. For incorporation into powdered products, they suffer from the disadvantage of being liquids or low melting waxes, which complicates the manufacture of free-flowing, crisp powders. Solid products are manufactured with ethoxylates of high ethylene oxide content. The latter, however, are too water soluble to provide good surface activity.

Base-catalyzed ethoxylation of aliphatic alcohols, alkylphenols, and fatty acids can be broken down into two stages: formation of a monoethoxy adduct and addition of ethylene oxide to the monoadduct to form the polyoxyethylene chain. Polyoxyethylene surfactants include alcohol ethoxylates and akylphenol ethoxylates.

Carboxylic Acid Esters. In the carboxylic acid ester series of surfactants, the hydrophobe, a naturally occurring fatty acid, is solubilized with the hydroxyl groups of polyols or the ether and terminal hydroxyl groups of ethylene oxide chains. Included in this group are glycerol esters; polyoxyethylene esters; and hydrosorbitol esters; ethoxylated anhydrosorbitol esters; natural ethoxylated fats, oils, and waxes; and glycol esters of fatty acids.

Carboxylic Amides. Carboxylic amide nonionic surfactants are condensation products of fatty acids and hydroxyalkyl amines. They include diethanolamine condensates, monoalkanolamine condensates, and polyoxyethylene fatty acid amides.

Fatty Acid Glucamides. Fatty acyl glucamides (FAGA) or polyhydroxyamides (PHA) have been adopted by detergent manufacturers in the United States and Europe. FAGA is produced via reaction of fatty acid methyl ester with N-methyl glucamine and attendant elimination of methanol. The methyl ester would be produced via the standard route of transesterification with fatty triglycerides; the glucamine, via reaction between glucose and methylamine with attendant hydrogenation and elimination of water.

Fatty acid glucamides are used in dishwashing liquids and heavy-duty liquids. Benefits include improved mildness for dishwashing liquids and improved enzyme stability in fabric washing detergents.

Polyalkylene Oxide Block Copolymers. The higher alkylene oxides derived from propylene, butylene, styrene, and cyclohexene react with active oxygens in a manner analogous to the reaction of ethylene oxide. Because the hydrophilic oxygen constitutes a smaller proportion of these molecules, the net effect is that the oxides, unlike ethylene oxide, are hydrophobic. The higher oxides are not used commercially as surfactant raw materials except for minor quantities that are employed as chain terminators in polyoxyethylene surfactants to lower the foaming tendency. The hydrophobic nature of propylene oxide units, $-CH(CH_3)CH_2O-$, has been utilized in several ways in the manufacture of surfactants.

Block polymer nonionic surfactants are not strongly surface-active but exhibit commercially useful surfactant properties. Aqueous solutions characteristically foam less than those of other surfactant types. They act as detergents, wetting and rinsing agents, demulsifiers and emulsifiers, dispersants, and solubilizers. They are used in automatic dishwashing detergent compositions, cosmetic preparations, spin finishing compositions for textile processing, metal-cleaning formulations, papermaking, and other technologies.

Cationic Surfactants. The hydrophobic moiety of a cationic surfactant carries a positive charge when dissolved in aqueous media, which resides on an amino or quaternary nitrogen. A single amino nitrogen is sufficiently hydrophilic to solubilize a detergent-range hydrophobe when protonated in dilute acidic solution; e.g., laurylamine is soluble in dilute hydrochloric acid. For increased water solubility, additional primary, secondary, or tertiary amino groups can be introduced or the amino nitrogen can be quaternized with low molecular weight alkyl groups such as methyl or hydroxyethyl. Quaternary nitrogen compounds are strong bases that form essentially neutral salts with hydrochloric and sulfuric acids. Most quaternary nitrogen surfactants are soluble even in alkaline aqueous solutions. Polyoxyethylated amino surfactants behave like nonionic surfactants in alkaline solutions and like cationic surfactants in acid solutions.

Cationic surfactants are widely used in acidic aqueous and non-aqueous systems as textile softeners, conditioning agents, dispersants, emulsifiers, wetting agents, sanitizers, dye-fixing agents, foam stabilizers, and corrosion inhibitors. To some extent, the usage pattern mirrors that of the anionic surfactants in neutral and alkaline solutions. The positively charged cationic surfactants are more strongly adsorbed than anionic or nonionic surfactants on a variety of substrates including textiles, metal, glass, plastics, minerals, and animal and human tissue, which can often carry a negative surface charge. Substantivity of cationic surfactants is the key property in many applications. In general, they are incompatible with anionic surfactants. Reaction of the two large, oppositely charged ions gives a salt insoluble in water. Ethoxylation moderates the tendency to form insoluble products with anionic surfactants.

Many benzenoid quaternary cationic surfactants possess germicidal, fungicidal, or algicidal activity. Solutions of such compounds, alone or in combination with nonionic surfactants, are used as detergent sanitizers in hospital maintenance. Classified as biocidal products, their labeling is regulated by the U.S. EPA.

Amines. Aliphatic mono-, di-, and polyamines derived from fatty and main acids make up this class of surfactants. Primary, secondary, and tertiary monoamines with C_{18} alkyl or alkenyl chains constitute the bulk of this class. The products are sold as acetates, naphthenates, or oleates. Principal uses are as ore-flotation agents, corrosion inhibitors, dispersing agents, wetting agents for asphalt, and as intermediates for the production of more highly substituted derivatives.

In addition to the mono- and dialkylamines, representative structures of this class of surfactants include *N*-alkyltrimethylene diamine, $RNH(CH_2)_3NH_2$, where the alkyl group is derived from coconut, tallow, and soybean oils; or is 9-octadecenyl, 2-alkyl-2-imidazoline, where R is heptadecyl, heptadecenyl, or mixed alkyl, and 1-(2-aminoethyl)-2-alkyl-2-imidazoline, where R is heptadecyl, 8-heptadecenyl, or mixed alkyl.

This group includes amine oxides, ethoxylated alkylamines, 1-(2-hydroxyethyl)-2-imidazolines, and alkoxylates of ethylenediamine.

Amine oxides have attracted widespread interest as replacements for alkanolamides as foam builders in liquid hand-dishwashing compositions.

2-Alkyl-1-(2-hydroxyethyl)-2-imidazolines are used in hydrocarbon and aqueous systems as antistatic agents, corrosion inhibitors, detergents, emulsifiers, softeners, and viscosity builders. They are prepared by heating the salt of a carboxylic acid with (2-hydroxyethyl) ethylenediamine at 150–160°C to form a substituted amide; 1 mol water is eliminated to form the substituted imidazoline with further heating at 180–200°C. Substituted imidazolines yield three series of cationic surfactants: by ethoxylation to form more hydrophilic products; quaternization with benzyl chloride, dimethyl sulfate, and other alkyl halides; and oxidation with hydrogen peroxide to amine oxides.

Quaternary Ammonium Salts. The quaternary ammonium ion is a much stronger hydrophile than primary, secondary, or tertiary amino groups, strong enough to carry a hydrophobe into solution in the surfactant molecular weight range, even in alkaline media. The discrete positive charge on the quaternary ammonium ion promotes strong adsorption on negatively charged substrates, such as fabrics, and is the basis for the widespread use of these surfactants in domestic fabric-softening compositions. See also **Quaternary Ammonium Compounds**.

Amphoteric Surfactants

Amphoteric surfactants contain both an acidic and basic hydrophilic group. Ether or hydroxyl groups may also be present to enhance the hydrophilicity of the surfactant molecule. Examples of amphoteric surfactants include amino acids and their derivatives in which the nitrogen atom tends to become protonated with decreasing pH of the solution. Amino acid salts, under these conditions, contain both a positive and a negative charge on the same molecule.

Amphoteric surfactants are generally considered specialty surfactants, however, usage has expanded significantly. They do not irritate skin and eyes, exhibit good surfactant properties over a wide pH range, and are compatible with anionic and cationic surfactants. A basic nitrogen and an acidic carboxylate group are the predominant functional groups.

Imidazolinium Derivatives. Amphoteric imidazolinium derivatives are prepared from the 2-alkyl-1-(2-hydroxyethyl)-2-imidazolines and from sodium chloroacetate.

Imidazolinium derivatives are recommended as detergents, emulsifiers, wetting and hair conditioning agents, foaming agents, fabric softeners, and antistatic agents. There is some evidence that in cosmetic formulations certain imidazolinium derivatives reduce eye irritation caused by sulfate and sulfonate surfactants present in these products.

Uses

Detergency, i.e., cleaning, is the primary function of household and personal products. More recently, a secondary function, such as softening in combination with detergency in laundry detergents or conditioning in combination with detergency in shampoos, has been offered as an additional product benefit. In general, products have tended toward functional specialization.

Surfactants are widely used outside the household for a variety of cleaning and other purposes. Often the volume or cost of the surfactant consumed in industrial applications is small compared to benefit.

JESSE L. LYNN, JR.
BARBARA H. BORY
Lever Company

Additional Reading

Fainerman, V.B., R. Miller, and D. Mobius: *Surfactants: Chemistry, Interfacial Properties, Applications,* Elsevier Science, New York, NY, 2001.

Hummel, D.O.: *Handbook of Surfactant Analysis: Chemical, Physico-Chemical and Physical Methods,* John Wiley & Sons, Inc., New York, NY, 2000.

Rosen, M.J.: *Surfactants and Interfacial Phenomena,* 3rd Edition, John Wiley & Sons, Inc., Hoboken, NJ, 2004.

Schmitt, T.M.: *Analysis of Surfactants,* 2nd Edition, Marcel Dekker, Inc., New York, NY, 2001.

Spitz, L. ed.: *Soaps and Detergents,* AOCS Press, Champaign, IL, 1996.

Swisher, R.D. ed.: *Surfactant Biodegradation, Surfactant Science Series,* 2nd Edition, Vol. 18, Marcel Dekker, Inc., New York, NY, 1987.

Swisher, R.D. ed.: *Surfactant Biodegradation, Surfactant Science Series,* Vol. 3, Marcel Dekker, Inc., New York, NY, 1970.

Tadros, T.F.: *Surfactants,* Academic Press, London, 1984.

Witten, T.A.: *Structured Fluids: Polymers, Colloids, Surfactants,* Oxford University Press, New York, NY, 2004.

Zana, R. ed.: *Surfactant Solutions: New Methods of Investigation, Surfactant Science Series,* Vol. 22, Marcel Dekker, Inc., New York, NY, 1986.

SUSPENSION. A system in which very small particles (solid, semisolid, or liquid) are more or less uniformly dispersed in a liquid or gaseous medium. If the particles are small enough to pass through filter membranes, the system is a colloidal suspension (or solution). Examples of solid-in-liquid suspensions are comminuted wood pulp in water, which becomes paper on filtration; the fat particles in milk; and the red corpuscles in blood. A liquid-in-gas suspension is represented by fog or by an aerosol spray. If the particles are larger than colloidal dimensions they will tend to precipitate if heavier than the suspending medium, or to agglomerate and rise to the surface if lighter. This can be prevented by incorporation of protective colloids. Polymerization is often carried out in suspension, the product being in the form of spheres or beads.

See also **Colloid Systems**.

SVEDBERG, THEODOR (1884–1971). A Swedish chemist who won the Nobel prize in 1926. Author of *Die Methoden zur Herstellung Kolloider Losungen anorganischer Stoffe*. His work included research in colloidal chemistry, molecular size determination, and methods of electrophoresis, as well as the development of the ultracentrifuge, for separation of colloidal particles in solution. His education was in Sweden with later work done at the University of Wisconsin before returning to Uppsalla.

SWARTS REACTION. Fluorination of organic polyhalides with antimony trifluoride (or zinc and mercury fluorides) in the presence of a trace of a pentavalent antimony salt.

SWEETENERS. Drawings in Egyptian tombs depicting beekeeping practices and honey production attest that the demand for sweet-tasting

TABLE 1. SUGAR CONSUMPTION PER CAPITA PER YEAR

Country	Refined sugar	
	Pounds	Kilograms
Israel	150	68.0
Bulgaria	130	59.0
Australia	119	54.0
New Zealand	110	49.9
Costa Rica	108	49.0
Cuba	107	48.5
Switzerland	106	48.1
United States	102	46.3
Hungary	99	44.9
Iceland	98	44.9
Poland	95	43.1
Sweden	94	43.1
Austria	92	41.7
Czechoslovakia	92	41.7
European Economic Community	89	40.4
Norway	87	39.5

Source: International Sugar Organization.

substances dates back to 2600 B.C. Sugar consumption varies considerably from one country to the next as shown in Table 1.

In terms of sugar consumption in the United States, until the early 1940s, sucrose from sugarcane and sugar beets accounted for a very high volume of the fundamental sweeteners. Since that time, there has been continuously increasing consumption of corn sweeteners and other caloric sweeteners, notably high-fructose corn syrup (HFCS). Of course, a marked impact on sucrose consumption occurred with the introduction of artificial sweeteners, particularly of saccharin and aspartame. Sweeteners fall into two general categories—*nutritive* and *nonnutritive*.

Nutritive Sweeteners

In addition to their sweetening power, nutritive sweeteners are effective preservatives in numerous foods. Sweeteners tie up water, essential for microorganism growth, thus preventing or inhibiting spoilage. Nutritive sweeteners also serve as food for yeasts and other fermenting agents, so important in many processes, including baking. The principal functional properties of sucrose are (1) browning reactions, (2) fermentability, (3) flavor enhancement, (4) freezing-point depression, (5) nutritive solids source, (6) osmotic pressure, (7) sweetness, (8) texture tenderizer, and (9) viscosity/bodying agent.

Among the principal *natural sugars* are fructose, glucose (also called dextrose), honey, invert sugar, lactose, maltose, raffinose and stachyose, sucrose, sugar alcohols, and xylitol.

Dextrose Equivalent. A means for comparing one sugar with another. The total amount of reducing sugars, expressed as dextrose (glucose), that is present in a given sugar syrup is calculated as a percentage of the total dry substance. More technically, the dextrose equivalent (DE) is the number of reducing ends of sugar that will react with copper. The DE can be measured in several ways.

Fructose. Also called *levulose* or *fruit sugar*, $C_6H_{12}O_6$. It is the sweetest of the common sugars, being from 1.1 to 2.0 times as sweet as sucrose. Fructose is generally found in fruits and honey. An apple is 4% sucrose, 6% fructose, and 1% glucose (by weight). A grape (*Vitis labrusca*) is about 2% sucrose, 8% fructose, 7% glucose, and 2% maltose (by weight) (Shallenberger). Commercially processed fructose is available as white crystals, soluble in water, alcohol, and ether, with a melting point between 103 and 105°C (217.4 and 221°F) (decomposition). Fructose can be derived by the hydrolysis of insulin; by the hydrolysis of beet sugar followed by lime separation; and from cornstarch by enzymic or microbial action.

Dry crystalline fructose is reported to have a sweetness level of 180 on a scale in which sucrose is represented at 100 (Andres, 1977). In cool, weak solutions and at lower pH, sweetness value is reported to be 140–150. At neutral pH or higher temperatures, the sweetness level drops, and at 50°C (122°F) sweetness equals that of a corresponding sucrose solution. A synergistic sweetness effect is reported between sucrose and fructose. A 40–60% fructose/sucrose mixture in a 100% water solution is sweeter than either component under comparable conditions (Unpublished report, University of Helsinki, 1972).

Glucose. Also known as *grape sugar* or *dextrose*, this is the main compound into which other sugars and carbohydrates are converted in the human body and thus is the major sugar found in blood. Glucose is naturally present in many fruits and is the basic "repeating" unit of the starches found in many vegetables, such as potato. Purified glucose takes the form of colorless crystals or white granular powder, odorless, with a sweet taste. Soluble in water, slightly soluble in alcohol. Melting point is 146°F (294.8°F). Glucose finds many uses—confectionery, infant foods, brewing and winemaking, caramel coloring, baking, and canning. Glucose is derived from the hydrolysis of corn starch with acids or enzymes. Glucose is a component of invert sugar and glucose syrup. Glucose was first obtained (1974) from cellulose by enzyme hydrolysis.

Corn (maize) syrup is a sweetener derived from corn starch by a process that was first commercialized in the 1920s. Corn syrup is composed of glucose and a variety of sugars described as the "maltose series of oligosaccharides." These syrups are not as sweet as sucrose, but are very often used in conjunction with sugar in confections and other food products.

Five types of corn sweeteners are commercially available: (1) *Corn syrup* (glucose syrup), with a DE of 20 or more, is a purified and concentrated aqueous solution of mono-, di-, and oligosaccharides. High fructose corn syrup (HFCS) is prepared by enzymatically converting glucose to fructose with glucose isomerase. (2) *Maltodextrin*, concentrated solutions or dried powders of disaccharides, characterized with a DE of less than 20. The manufacturing process is similar to that of corn syrup except that the conversion process is stopped at an earlier stage. (3) *Dried corn syrup* is a granular, crystalline, or powder product, from which a portion of the water has been removed. (4) *Dextrose monohydrate* is a purified and crystallized form of D-glucose, and contains one molecule of water of crystallization per molecule of D-glucose. (5) *Dextrose anhydrous* is primarily D-glucose with no water of crystallization.

Galactose. A monosaccharide commonly occurring in milk sugar or lactose. Formula, $C_6H_{12}O_6$.

Honey. A natural syrup which varies in composition and flavor, depending upon the plant source from which the nectar was collected by the honeybee, the amount of processing, and the duration of storage. The principal sugars contained in honey are fructose and glucose, the same components as in table sugar. There are minute amounts of vitamins and minerals in honey, but these are not usually considered in terms of calculating minimum requirements.

Invert Sugar. A mixture of 50% glucose and 50% fructose obtained by the hydrolysis of sucrose. Invert sugar absorbs water readily, and is usually only handled as a syrup. Because of its fructose content, invert sugar is levorotatory in solution, and sweeter than sucrose. Invert sugar is often incorporated in products where loss of water must be minimized. Commercially, invert sugar is obtained from the inversion of a 96% cane sugar solution. This sugar is used in various foods, in the brewing industry, confectionery field, and in tobacco curing.

Lactose. *Milk sugar* or *saccharum lactis*, $C_{12}H_{22}O_{11} \cdot H_2O$. Purified lactose is a white, hard, crystalline mass or white powder with a sweet taste, odorless. It is stable in air, soluble in water, insoluble in ether and chloroform, very slightly soluble in alcohol. The compound decomposes at 203.5°C (398.3°F). Lactose is derived from whey, by concentration and crystallization. Cow's milk contains about 5% lactose. Because of its relative lack of sweetening power, lactose is not considered a sweetener in the usual sense. It is used as a bulking agent in numerous food products. Lactose can be used effectively as a carrier for artificial sweeteners to give a free-flowing powder that is easily handled. There has been interest in the hydrolysis of lactose into glucose and galactose, both enzymatically and chemically. It has been reported that glucose and galactose are known to be sweeter than lactose itself. The relative sweetness of sugars is not a constant relationship, but depends upon many factors, including pH, temperature, and presence of other constituents. Mixtures of sugars can make a different sweetness impression than that of individual sugars alone. Synergistic sweetness often results from a combination of sugars.

Maltose. Also known as *malt sugar*, maltose is a product of the fermentation of starches by enzymes or yeast. Barley malt, which is used as an adjunct in brewing, enhances the flavor and color of beer because of its maltose content. Maltose also is formed by yeast during breadmaking. Maltose is the most common reducing disaccharide, $C_{12}H_{22}O_{11} \cdot H_2O$, composed of two molecules of glucose. It is found in starch and glycogen. Purified maltose takes the form of colorless crystals, melting point, 102–103°C (215.6–217.4°F). Soluble in alcohol; insoluble in ether. Combustible. Maltose is used as a nutrient, sweetener, and culture medium.

Raffinose and Stachyose. These are sugars found in significant amounts in some foods, such as beans. These sugars are not digested in the stomach and upper intestine as are other disaccharides. They are fermented by bacteria in the lower digestive tract, producing gases and sometimes causing discomfort from flatulence. Raffinose is a trisaccharide composed of one molecule each of D(+)-galactose, D(+)-glucose, and D(−)-fructose, $C_{18}H_{32}O_{16} \cdot 5H_2O$. Raffinose is sometimes used in the preparation of other saccharides.

Sucrose. Table sugar, also known as *saccharose*. Sucrose is a disaccharide, composed of two simple sugars, glucose and fructose, chemically bound together, $C_{12}H_{22}O_{11}$. Hard, white, dry crystals, lumps, or powder. Sweet taste, odorless. Soluble in water; slightly soluble in alcohol. Solutions are neutral to litmus. Decomposes in range of 160–186°C (320–366.8°F). Combustible. Optical rotation = +33.6°. Derived from sugarcane or sugar beets and also obtainable from sorghum. Sucrose is the most abundant free sugar in the plant kingdom and has been used since antiquity (Mead and Chem, 1977).

Turbinado sugar is raw sugar that has been refined to remove impurities and most of the molasses. It is edible when produced under sanitary conditions and has a molasses flavor. *Brown sugar* consists of sucrose crystals covered with a film of molasses syrup that give the characteristic color and flavor. The sucrose content varies from 91 to 96%. *Confectioner's* or *powdered sugar* is another form of sucrose made by grinding the sugar crystals. It is usually mixed with about 3% starch to prevent clumping. It is used for household baking, canning, and table use, or industrially where rapid solution in cold liquids is desirable.

Sugar Alcohols. These are polyols, chemically reduced carbohydrates. Important in this group are *sorbitol, mannitol, maltitol,* and *xylitol.* Xylitol is described later.

Polyols are frequently used sugar substitutes and are particularly suited to situations where their different sensory and functional properties are attractive. In addition to sweetness, some of the polyols have other useful properties. For example, although it contains the same number of calories/gram as other sweeteners, sorbitol is absorbed more slowly from the digestive tract than is sucrose. It is, therefore, useful in making foods intended for special diets. When consumed in large quantities (1–2 oz; 25,059 g)/day, sorbitol can have a laxative effect, apparently because of its comparatively slow intestinal absorption.

When sugar alcohols are ingested, the body converts them first to fructose, which does not require insulin to facilitate its entry into the cells. For this reason, ingesting these sweeteners (including fructose itself) does not cause the immediate increase in blood sugar level which occurs upon eating glucose or sucrose. Within the body, however, the fructose is rapidly converted to other compounds, which *do* require insulin in their metabolism. One effect of this stepwise metabolism is to "damp out" the peaks in blood sugar levels which occur immediately after ingesting sucrose, but which are absent after ingesting fructose, even if the eventual insulin requirements are the same. Thus, individuals with metabolic problems should not make the assumption that fruit sugars are perfectly all right to consume, but first should consult their physicians. In fact, some health scientists are dubious about pursuing the apparent claims for substituting fructose and sugar alcohols for sucrose as a major sweetener, particularly for diabetics, until more research is done on their long-range nutritional and biophysiological consequences. Research interest has also focused on these sweeteners because of their relatively low potential for causing dental caries. Studies have shown about a 30% reduction in dental caries in laboratory animals on sorbitol and mannitol diets, and virtually complete elimination of caries in the animals when on xylitol diets.

Xylitol. This is a 5-carbon sugar alcohol that occurs widely in nature—raspberries, strawberries, yellow plums, cauliflower, spinach, and many other plants. Although widely distributed in nature, it is present in low concentrations and this makes it uneconomic to extract the substance directly from plants. Thus, commercial xylitol must be produced from xylan or xylose-rich precursors through the use of chemical, enzymatic, and other bioprocessing conversions. A frequently used source has been birch tree chips. Other appropriate starting materials include beech and other hardwood chips, almond and pecan shells, cottonseed hulls, straw, cornstalks (maize), and corn cobs. The base source in the aforementioned agricultural waste materials is hemicellulose xylan. The hemicellulose is

acid hydrolyzed to yield xylose which, followed by hydrogenation and chromatographic separation, yields xylitol.

Xylitol is equally as sweet as sucrose. This property is of advantage to food processors because in reformulating a product from sucrose to xylitol, approximately the same amounts of xylitol can be used. Because xylitol has a negative heat of solution, the substance cools the saliva, producing a perceived sensation of coolness, quite desirable in some food products, notably beverages. Recently, this property has been used in an iced-tea-flavored candy distributed in the European market. As of the late 1980s, 28 countries have ruled positively in terms of xylitol for use in commercial products. Xylitol has been found particularly attractive for use in chewing gum, mint and hard candies, and as a coating for pharmaceutical products. Xylitol has the structural formula shown below, with a molecular weight of 152.1. It is a crystalline, white, sweet, odorless powder, soluble in water and slightly soluble in ethanol and methanol. It has no optical activity.

Isomalt. Developed in Germany, isomalt is described as an energy-reduced bulk sweetener and marketed in Europe under the tradename *Palatinit*™*mark*. The compound is produced from sucrose in a two-step process, as shown in Fig. 1.

$$\text{HOCH}_2 - \overset{\overset{\displaystyle H}{|}}{\underset{\underset{\displaystyle OH}{|}}{C}} - \overset{\overset{\displaystyle OH}{|}}{\underset{\underset{\displaystyle H}{|}}{C}} - \overset{\overset{\displaystyle H}{|}}{\underset{\underset{\displaystyle OH}{|}}{C}}\text{CH}_2\text{OH}$$

In the first step, the easily hydrolyzable 1–2 glucoside linkage between the glucose and fructose moieties of sucrose are catalyzed by immobilized enzymes to produce isomaltulose, *Palatinos.*™ *mark* After crystallization, the isomaltulose is hydrogenated in a neutral aqueous solution using a nickel catalyst.

It is claimed that isomalt is odorless, white, crystalline, and sweet tasting without the accompanying taste or aftertaste. Sweetening power is from 0.45 to 0.6 that of sucrose. A synergistic effect is achieved when isomalt is combined with other artificial sweeteners and sugar substitutes. Principal applications are in confections, pan-coated goods, and chewing gum. The substance was approved for use in most European countries in 1985. Classification of isomalt as a GRAS substance was petitioned in the United States. (GRAS = generally regarded as safe.)

Aspartame. This synthetic sweetener is included with the nutritive sweeteners because it does have some caloric value (when metabolized as a protein, it releases 4 kcal/g). The relationship between sweetness of aspartame and sucrose is almost linear when plotted on a log-log scale. Aspartame is 182 times sweeter than a 2% sucrose solution, but only 43 times sweeter than a 30% solution. The clean, full sweetness of aspartame is similar to that of sucrose and complements other flavors.

The full name of aspartame is *aspartylphenylalanine*, a dipeptide that degrades to a simple amino acid. It has been reported as easily metabolized by humans. Aspartame was accidentally discovered in 1965 with the synthesis of a product for ulcer therapy. Aspartame is metabolized by the same biochemical pathway as proteins, yielding phenylalanine, aspartic acid, and methanol. Because of the byproduct phenylalanine, which some individuals are unable to metabolize, appropriate labeling is required. This is a concern for individuals with phenylketonuria (PKU). Aspartame was first approved in the United States in 1974, then banned in 1975. In July of 1981, it was approved for use in various foods, dry beverage mixes, and in tabletop sweeteners. Approval for use in carbonated beverages was granted in July 1983.

Currently, aspartame is used in tabletop sweeteners (*Equal* in the U.S.; *Egal* in Quebec, Canada; and *Canderal* in Europe and the U.K.). Aspartame currently is incorporated as the exclusive sweetening ingredient in nearly all diet soft drinks in the United States. In other countries, it may be blended with saccharin at a level close to 50% of the saccharin level. Soft-drink manufacturers have taken some measures to enhance stability by raising pH slightly and by more closely controlling the inventory for carbonated soft drinks. Notable differences in sweetness are perceived at a 40% loss in aspartame level.

Crystalline Maltitol. Classified as a bulk sweetener with taste and mouthfeel similar to sucrose, crystalline maltitol contains maltitol as the major component (88+%), with small amounts of sorbitol, maltotriitrol, and hydrogenated oligosaccharides. Its use is in tabletop sweeteners, chocolate, candy, and baked goods. Maltitol has been a major component of hydrogenated glucose syrup in the United States since 1977 and has been

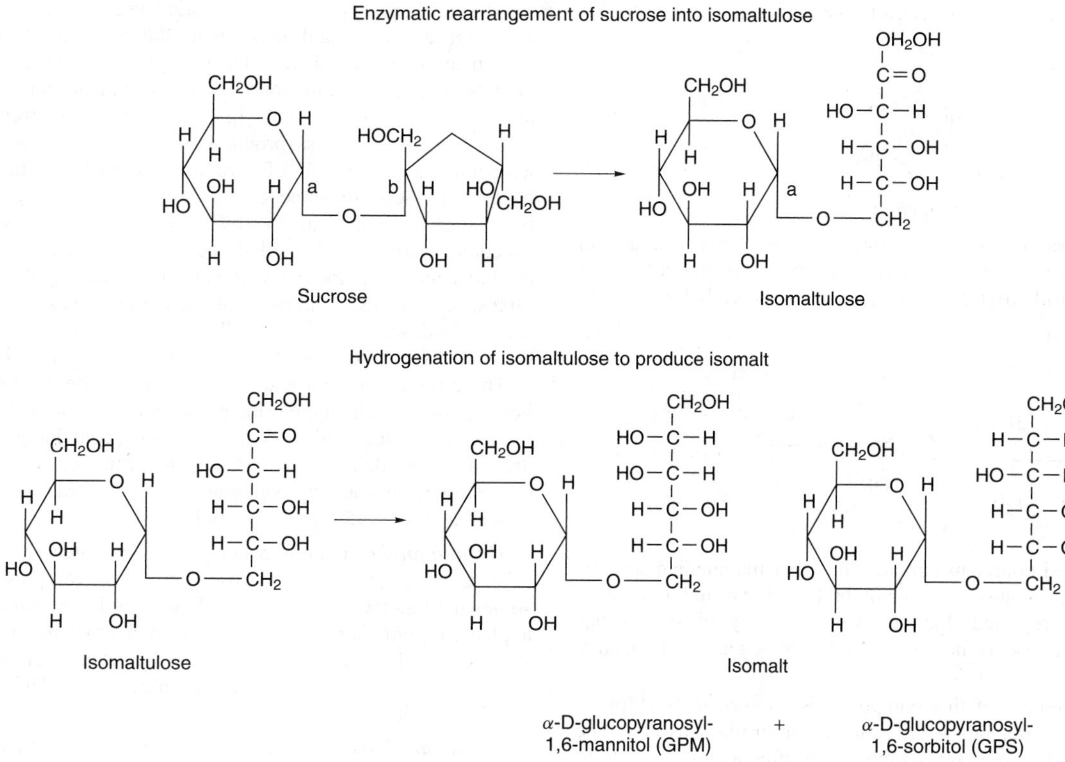

Enzymatic rearrangement of sucrose into isomaltulose

Sucrose → Isomaltulose

Hydrogenation of isomaltulose to produce isomalt

Isomaltulose → Isomalt

α-D-glucopyranosyl-1,6-mannitol (GPM)   +   α-D-glucopyranosyl-1,6-sorbitol (GPS)

Fig. 1.   Isomalt

used in Japan since 1963. The product was introduced in Europe in 1984. Classification of crystalline maltitol as a GRAS substance was petitioned in the United States in 1986.

Nonnutritive Sweeteners

There are several currently used and a number of potential noncaloric sweeteners, including saccharin, cyclamate (banned in the U.S., but permitted in approximately 40 other countries), acesulfame K, monellin (from the serendipity berry), stevioside, glycyrrhizin, hernandulcin, neosugar, miraculin (from miracle fruit), and a sweetener-enhancer, thaumatin, are being investigated.

Saccharin. A noncaloric sweetener that is about 300 times as sweet as sugar. The compound is manufactured on a large scale in several countries. It is made as saccharin, sodium saccharin, and calcium saccharin, as shown by formulas below.

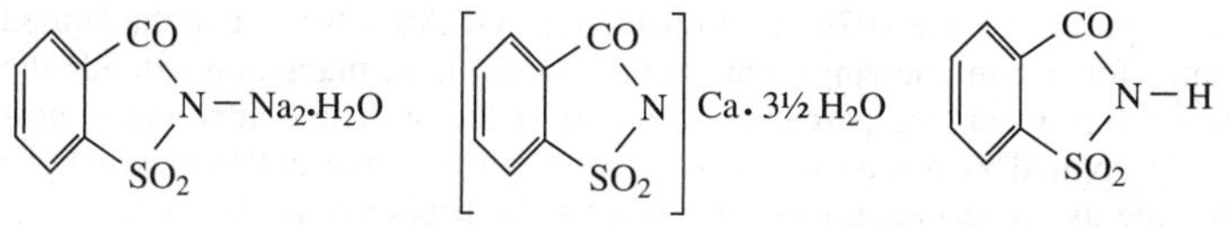

Sodium saccharin   Calcium saccharin   Saccharin

Saccharin (ortho-benzosulfimide) was discovered in 1879 by I. Remsen and C. Fahlberg when they were researching the oxidation products of toluene sulfone amide. The most common forms of saccharin are sodium and calcium saccharin, although ammonium and other salts have been prepared and used to a very limited extent. The saccharins are white, crystalline powders, with melting points between 226 and 230°C (438.8 and 446°F). Soluble in amyl acetate, ethyl acetate, benzene, and alcohol; slightly soluble in water, chloroform, and ether. Saccharin is derived from a mixture of toluenesulfonic acids. They are converted into the sodium salts, then distilled with phosphorus trichloride and chlorine to obtain the ortho-toluene sulfonyl chloride, which by means of ammonia is converted into ortho-toluenesulfamide. This is oxidized with permanganate, then treated with acid, and saccharin is crystallized out. In food formulations, saccharin is used mainly in the form of its sodium and calcium salts. Sodium bicarbonate may be added to provide improved water solubility.

Saccharin is used in conjunction with aspartame in carbonated beverages. Other uses include tabletop sweeteners, dry beverage blends, canned fruits, gelatin desserts, cooked and instant puddings, salad dressings, jams, jellies, preserves, and baked goods.

For many years, saccharin has been under investigation by a number of countries. As of the late 1900s, some questions remained unresolved.

Cyclamate. Group name for synthetic, nonnutritive sweeteners derived from cyclohexylamine or cyclamic acid. The series includes sodium, potassium, and calcium cyclamates. Cyclamates occur as white crystals, or as white crystalline powders. They are odorless and in dilute solution are about 30 times as sweet as sucrose. The purity of commercially available compounds is approximately 98%.

Discovered in 1937 and patented in 1940, cyclamate is a derivative of cyclohexylamine, specifically, cyclohexane sulfonic acid. The sodium salt form is normally used, but the calcium salt may be substituted in low-sodium diets. See structural formulas below.

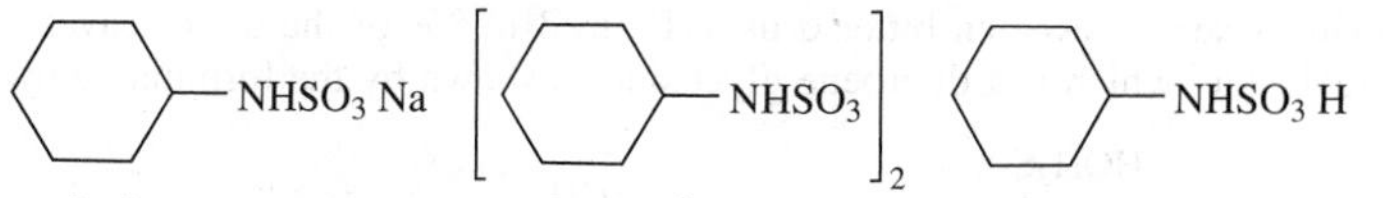

Sodium cyclamate   Calcium cyclamate   Cyclamic acid

Once widely used, cyclamate was prohibited in the United States in 1970. Although used in many other countries, reapproval in the United States has not yet been established. An independent review of the possible carcinogenicity of cyclamate was conducted in April 1985 by the National Academy of Sciences/National Research Council at the request of the Food and Drug Administration. The review concluded that cyclamate itself is not a carcinogen, although it may serve as a promotor or cocarcinogen in the presence of other substances.

Acesulfame-K. This substance (potassium salt of the cyclic sulfanomide), 6-methyl-1,2,3-oxathiazine-4(3H)-1,2,2-dioxide, shown below, was developed by Karl Clauss (Hoechst Celanese Corporation, Somerville, New Jersey) in 1967. The compound is a white, odorless, crystalline substance with a sweetening power 200 times that of sucrose. A synergistic effect is produced when the substance is combined with a number of other sweeteners. The substance is calorie-free and not metabolized in the human body. Approval of the use of Acesulfame-K was given by the Food and Drug Administration (FDA) in the United States in 1983 and it is found in scores of popular retail products, including yogurt, rice pudding, and soft drinks.

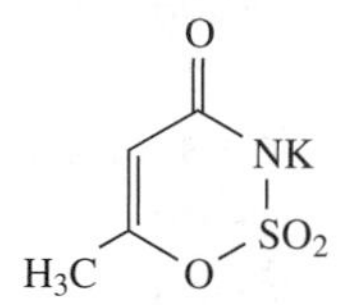

Acesulfame-K

Sucralose. Developed in England during the mid-1980s, testing and evaluation commenced in 1988. The structural formula of the compound (a chlorinated disaccharide derived from sucrose) is shown below.

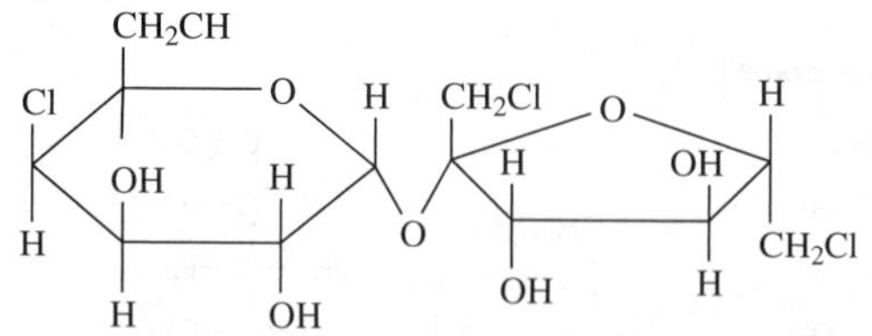

Sucralose

Sucralose is absorbed poorly in humans and other mammalian species. The small portion that is absorbed is not broken down and is quickly excreted. It has been reported that an extensive array of studies has demonstrated that sucralose is nontoxic—not carcinogenic, teratogenic, mutagenic, or caloric.

Monellin. The sweetness of this compound is claimed to be 1500 to 3000 times that of sucrose, but a different flavor profile prevails. The detection of a sweet taste is slow, commencing after a few seconds in contact with the taste buds, then gradually increasing to its peak intensity. The sweet taste can persist for up to an hour. The source is the relatively rare serendipity berry, the fruit of a noncultivated West African vine. Extraction of the sweet component is effected by treating the berry with a series of enzymes (pectinase and bromelain), followed by dialysis and chromatographic separation. The compound resulting contains the protein *monoellin* with a molecular weight of about 10,700 and composed of two nonidentical polypeptide chains of 50 and 43 amino acids. Neither of the individual chains imparts sweetness. Regulatory measures have not been instituted because of the compound's apparent instability and limited raw resources for processing. However, to date, tests with mice have shown no evidence of toxicity.

Stevioside. Derived from the roots of the herb *Stevia rebaudiana*, this compound has found limited use in Japan and a few other countries as a low-calorie sweetener having about 300 times the sweetening power of sucrose. The compound has not been investigated thoroughly by a number of countries with strong regulatory agencies and, therefore, is not on the immediate horizon for wide consideration as a sweetener.

The dried leaves of *S. rebaudiana* have been used in Paraguay for many years to sweeten bitter drinks. From 3 to 8% of the dried leaves is stevioside, which is a diterpene glycoside as shown by the formula below.

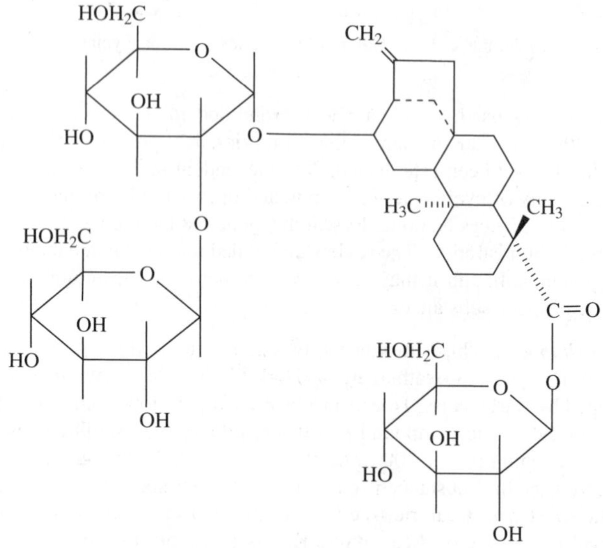

Stevioside

Glycyrrhizins. These are noncaloric sweeteners approximately 50 times as sweet as sugar and used as a flavor enhancer under the GRAS classification in the United States. Glycyrrhizins, which have a pronounced licorice taste, are used in tobacco, pharmaceuticals, and some confectionary products. They are available in powder or liquid form and with color, or as odorless, colorless products. These compounds are stable at high temperatures (132°C; 270°F) for a short time and thus can be used in bakery products. In some chocolate-based products, the sweetener has been used to replace up to 20% of the cocoa. The sweetener also has excellent foaming and emulsifying action in aqueous solutions. Typical products in which these sweeteners may have application include cake mixes, ice creams, candies, cookies, desserts, beverages, meat products, sauces, and seasonings, as well as some fruit and vegetable products. Generally available as malted and ammoniated glycyrrhizin.

The basic compound is a triterpene glycoside. It is extracted from the licorice root, of which the principal sources are China, Russia, Spain, Italy, France, Iran, Iraq, and Turkey. The roots, containing 10% moisture, are dried and shredded, after which they are extracted with aqueous ammonia, concentrated in vacuum evaporators, precipitated with sulfuric acid, and crystallized with 95% ethyl alcohol.

Hernandulcin. Tasting panels have estimated that this substance is 1000 times sweeter than sucrose, but the flavor profile is described as somewhat less pleasant than that of sucrose. Hernandulcin is derived from a plant, *Lippia dulcis* Trev, commonly known as "sweet herb" by the Aztecs as early as the 1570s. It has been categorized as noncarcinogenic, based upon standard bacterial mutagenicity tests. The economic potential is being studied.

Neosugar. This is another substance in early stages of development and testing. The compound is composed of sucrose attached in a beta(2-1) linkage to 2, 3, or 4 fructose units.

Miraculin. Rather than a sweet-tasting substance, miraculin is described as a taste-modifying substance that elicits a sweet taste to tart foods. The product has been reported as used by African cultures for over a century. The compound is derived from a shrub (*Synsepalum dulcificum*) which grows in West Africa. Miraculin is a glycoprotein with a molecular weight ranging from 42,000 to 44,000. Approval of the Food and Drug Administration has thus far been denied, awaiting further tests. A GRAS category was denied in 1974.

Thaumatin. This is a protein extracted and purified from *Thaumatococcus danielli*, a plant that is found in West Africa. The leaves of the plant have been used for many years in Africa for wrapping food during cooking. Claims have been made that thaumatin is from 2000 to 2500 times sweeter than 8–10% solution of sucrose. The final product is odorless, cream-colored and imparts a lingering licorice-like aftertaste. The substance synergizes well with monosodium glutamate (MSG) and is used in typical Japanese seasonings as well as in chewing gum, pet foods, and certain pharmaceuticals (to mask unpleasant flavor notes). Use in Japan has been approved since 1979. It is considered a GRAS substance in the United States for use in chewing gum. In this application, thaumatin extends the flavor and boosts the perceived duration of flavor. The compound is normally applied as a dust to the surface of gum. Some authorities believe that the use of thaumatin in pet foods has high potential.

Sweeteners in Formulating and Processing

In using low-calorie sweeteners in various food products, the problems are not limited to flavor, but often much more importantly involve texture, acidity, storage stability, and preservability, among others. Acceptable nonnutritively sweetened products cannot be developed by the simple substitution of artificial sweeteners for sugars. Rather, the new product must be completely reformulated from the beginning. Three examples follow.

Jams, Jellies, and Preserves. Traditional products in this category contain 65% or more soluble solids. In low-calorie analogs, soluble solids range from 15% to 20%. Under these circumstances, commonly used pectins (high methoxyl content) do not suffice. Thus, special LM (low methoxyl) pectins must be used, along with additional gelling agents, such as locust bean gum, guar gum, and other gums and mucilagenous substances, some of which may require some masking. In the absence of sugar, a preservative, such as ascorbic acid, sorbic acid, sorbate salts, propionate salts, and benzoates, usually is required to the extent of about 0.1% (weight).

Soft Drinks. In addition to providing sweetness, sugar also functions to provide mouthfeel and to stabilize the carbon dioxide of soft drinks. To contribute to mouthfeel, the use of hydrocolloids and sorbitol has been attempted with limited success. Hydrocolloids also help to some degree with the problem of carbonation retention, but the principal solutions to this problem involve avoiding all factors which contribute to carbonation loss. Thus, the requirement for very well filtered water to eliminate particulates as possible nucleation points; any substances that promote foaming must be avoided; any emulsifying agents used in connection with flavoring agents must be handled carefully to avoid foaming; carbonation should be carried out at low temperature (34°F; 1.1°C); and trace quantities of metals must be absent from the water.

Bakery Products. These foods are among the most difficult as regards the use of artificial sweeteners. A listing of the functions of sugar in baked goods beyond that of providing sweetness is indicative of these problems. Sugar contributes to texture in forming structures, in providing moist and tender crumbs by counteracting the toughening characteristics of flour, milk, and egg solids. In the emulsification process required to retain gas during leavening, sugar is an effective accessory agent. Ingredients frequently used in bakery products to compensate for the absence of sugar include carboxymethylcellulose, mannitol, sorbitol, and dextrins, but, generally, these have not been very satisfactory—either to processor or consumer. This remains a large area of challenge for the food processors and ingredient manufacturers.

Evaluating Synthetic Sweeteners. Evaluation of new sweeteners, unlike that of most functional food ingredients, is not possible using totally objective means. There are no general rules leading to structure/function relationships for all classes of sweeteners. The principal judgments must rely on human sensory panel tests. The training and administration of sensory panels for sweeteners are beyond the scope of this volume.

Additional Reading

Andres, C.: "Alternate Sweeteners," *Food Processing*, **38**(5), 50–52 (1977).

Barndt, R.L. and G. Jackson: "Stability of Sucralose in Baked Goods," *Food Technology*, 62 (January 1990).

Bartoshuk, L.M.: "Sweetness: History, Preference, and Genetic Variability," *Food Technology*, 108 (November 1991).

Birch, G.G.: "Chemical and Biochemical Mechanisms of Sweetness," *Food Technology*, 121 (November 1991).

Chen, J.C.P. and Chung-Chi Chou: *Chen-Chou Cane Sugar Handbook: A Manual for Cane Sugar Manufacturers and Their Chemists,* 12th Edition, John Wiley & Sons, Inc., New York, NY, 1993.

Corti, A.: *Low-Calorie Sweeteners: Present and Future,* S. Karger Publishers, Inc., Farmington, CT, 1999.

DeMan, J.M.: *Principles of Food Chemistry,* 3rd Edition, Aspen Publishers, Inc., Gaithersburg, MD, 1999.

Farber, S.A.: "The Price of Sweetness," *Technology Review (MIT)*, 46 (January 1990).

Fennema, O.R.: *Food Chemistry,* 3rd Edition, Marcel Dekker, Inc., New York, NY, 1998.

Grenby, T.H.: *Advances in Sweeteners,* Blackie Academic & Professional, New York, NY, 1999.

Igoe, R.S. and Y.H. Hui: *Dictionary of Food Ingredients,* 4th Edition, Aspen Publishers, Inc., Gaithersburg, MD, 2001.

Keller, W.E. et al.: "Formulation of Aspartame-Sweetened Frozen Dairy Dessert without Bulking Agents," *Food Technology*, 102 (February 1991).

Kretchmer, N. and C. Hollenbeck: *Sugars and Sweeteners,* CRC Press, LLC., Boca Raton, FL, 1991.

Lindley, M.G.: "From Basic Research on Sweetness to the Development of Sweeteners," *Food Technology*, 134 (November 1991).

Nabors, L.O'Brien: *Alternative Sweeteners,* 3rd Edition, Marcel Dekker, Inc., New York, NY, 2001.

Noble, A.C., N.L. Matysiak, and S. Bonnans: "Factors Affecting the Time- Intensity Parameters of Sweetness," *Food Technology*, 128 (November 1991).

O'Mahony, M.: "Techniques and Problems in Measuring Sweet Taste," *Food Technology*, 128 (November 1991).

Pepper, T. and P.M. Olinger: "Xylitol in Sugar-Free Confections," *Food Technology*, 98 (October 1988).

Read, N.W. and J. Donelly: *Food and Nutritional Supplements: Their Role in Health and Disease,* Springer-Verlag, Inc., New York, NY, 2001.

Shallenberger, R.S.: "Predicting Sweetness from Chemical Structure and Knowledge of the Chemoreception Mechanism of Sweetness," *Institute of Food Technologists Symposium,* Saint Louis, MO, 1979.

Staff: "Applications of Aspartame in Baking," *Food Technology*, 56 (January 1988).

Staff: "Evaluation of Advanced Sweeteners," *Food Technology*, 60 (January 1988).

Staff: "FDA Clears Hoechst's Non-Caloric Sweetener for Use in Dry Foods," *Food Technology*, 108 (October 1988).

Welti-Chanes, J. and G.V. Barbosa-Canovas: *Engineering and Food for the 21st Century,* CRC Press, LLC, Boca Raton, FL, 2002.

Wnnia, S.M.: "Modeling the Sweet Taste of Mixtures," *Food Technology*, 140 (November 1991).

Wong, D.W.S.: *Mechanism and Theory in Food Chemistry,* Chapman & Hall, New York, NY, 1999.

SYENITE. A coarse-grained, granular, therefore intrusive, igneous rock of the general composition of granite except that quartz is either absent or present in a relatively small amount. The feldspars are alkaline in character and the dark mineral is usually hornblende. Soda-lime feldspars may be present in small quantities. The term syenite was originally applied to hornblende granite like that of Syene in Egypt from whence the name is derived. Syenite is not a common rock, some of the more important occurrences being, in the United States, in New England, Arkansas, Montana, and New York State (syenite gneisses), and elsewhere, in Switzerland, Germany, and Norway.

SYLVANITE. A mineral, a telluride of gold and silver approximating the formula $AgAuTe_4$. Sylvanite is monoclinic, occurring in bladed, columnar, and granular forms as well as arborescent and branching. It is a brittle mineral; hardness, 1.5–2; specific gravity, 8.16; luster, metallic; color and streak, steel gray to yellowish-gray. This mineral is found associated with gold and tellurides of gold and silver or with sulfides such as pyrite. It is found in Rumania, Australia, Colorado and California. It was named for Rumanian Transylvania where it was first found.

Krennerite is another telluride of gold and silver with a similar composition to sylvanite, but crystallizing in the orthorhombic system. Calaverite is a gold telluride with only a small silver content.

SYLVITEA. A mineral, potassium chloride, KCl, occurring in cubes, or as cubes modified by octahedra. Sylvite is therefore isometric. It has a perfect cubic cleavage; uneven fracture; is brittle; hardness, 2; specific gravity, 1.9; luster, vitreous; colorless when pure but may be white, bluish, yellowish or reddish due to impurities. It is soluble in water. It is much rarer than halite and has been found as sublimates at Mt. Vesuvius and as bedded deposits at Stassfurt, Germany. Extensive deposits occur in sedimentary deposits in the Permian basin of southwestern New Mexico, near Carlsbad, in the United States.

It is used as a source of potash salts. Potassium chloride was called by the early chemists *sal digestivus Sylvii,* whence the name of the mineral.

SYMMETRY. Arrangement of the constituents of molecule in a definite and continuously repeated space pattern or coordinate system. It is described in terms of three parameters called elements of symmetry. (1) The center of symmetry, around which the constituent atoms are located in an ordered arrangement; there is only one such center in a molecule, which may or may not be an atom. (2) Planes of symmetry, which represent division of a molecule into mirror-image segments. (3) Axes of symmetry, represented by lines passing through the center of symmetry; if the molecule is rotated it will have the same position in space more than once in a complete 360-degree turn, e.g., the benzene molecule with 6 axes of symmetry requires 60-degree rotation to return to its identical position.

See also **Stereochemistry**.

SYNDETS. See **Detergents**.

SYNERESIS. The contraction of a gel with accompanying pressing out of the interstitial solution or serum. Observed in the clotting of blood, with silicic acid gels, etc. See also **Colloid Systems**.

SYNGE, RICHARD L. M. (1914–1994). An Irish mathematician and physicist who won the Nobel prize for chemistry in 1952 along with Archer J. P. Martin for their invention of partition chromatography. His research was on the application of methods of physical chemistry to isolate and analyze proteins, with special attention to antibiotic peptides and higher plants. He received his doctorate from Cambridge.

SYNTHESIS (Chemical). The process of building chemical compounds through a planned series of steps (reactions, separations, etc.). Synthesis usually is the method of choice: (1) when the desired compound is not present in natural materials from which it can be isolated: (2) when the compound cannot be easily obtained from reacting readily available materials in a few simple steps; and (3) although a compound may be available within a natural complex, the economic separation and purification are prohibitive, or often in the case of biochemicals, too little natural raw material is available to meet the demand.

Even more important, synthesis plays a key role in developing new, untried chemical structures which, on paper, appear to have properties that may be of great value, e.g., a new synthetic material, a new drug, or a new fuel. Chemicals by design from prior knowledge of related materials generally are created via the route of synthesis. Further, synthesis is fundamental to broadening the base of chemical knowledge. Sometimes unexpected results occur, i.e., compounds with unusual, unexpected, and often desirable practical chemical and/or physical properties.

Because of the hundreds of thousands of organic substances already established, but many yet remaining to be "built," organic synthesis predominates. Most of the synthetics (elastomers, fibers, and other polymers, coatings, films, adhesives, and numerous other products) that have appeared during the last 30 to 40 years resulted from research involving organic synthesis. Some of the early work in organic synthesis dealt with the creation of certain fatty acids and ketones. A few examples are given to provide an insight into the workings of synthesis.

In the following examples, only the main starting ingredients and products are shown. No attempt is made to indicate byproducts or the conditions of the reactions involved:

(a)   Target compound: Ethylpropylacetic acid, $(C_2H_5)(C_3H_7)CH:COOH$

   (1)   Acetic anhydride → ethyl acetate
           (+ alcohol)
   (2)   Ethyl acetate → ethyl acetoacetate
           (sodium + dilute acids)
   (3)   Ethyl acetoacetate → sodium derivative of ethyl acetoacetate
           (+ sodium ethoxide)
   (4)   Sodium derivative of ethyl acetoacetate → ethyl ethylpropyl acetoacetate
           (+ propyl iodide)
   (5)   Ethyl ethylpropyl acetoacetate → ethylpropylacetic acid
           (concentrated alcohol and potash)

(b)   Target compound: Butyl acetone, $CH_3 \cdot CO \cdot CH_2 \cdot C_2H_4$
   (1) through (3), same as given in example (a)
   (4) Sodium derivative of ethyl acetoacetate → ethylbutylpropyl acetoacetate
           (+ butyl iodide)
(5) Ethylbutylpropyl acetoacetate → butyl acetone
           (+ dilute alcohol and potash)
(c)   Target compound: n-valeric acid, $CH3 \cdot CH_2 \cdot CH_2CH_2COOH$

   (1)   Potassium chloroacetate → potassium cyanoacetate
           (+ potassium cyanide)
   (2)   Potassium cyanoacetate → ethyl malonate
           (+ alcohol and hydrogen chloride)
   (3)   Ethyl malonate → sodium derivative of ethyl malonate
           (+ sodium ethoxide)
   (4)   Sodium derivative of ethyl malonate → ethylpropyl malonate
           (+ propyl iodide)

The compounds on the right-hand side of intermediate reactions are often called *intermediates*. See also **Intermediate (Chemical)**.

Some of the notable syntheses from the early history of the technique include:

Inorganic Syntheses

1746	Sulfuric acid (chamber process)
1800	Soda ash (Le Blanc process)
1861	Soda ash (Solvay process)
1890	Sulfuric acid (contact process)
1912	Ammonia (Haber-Bosch process)

Organic Syntheses

1828	Urea (Wohler)
1857	Mauveine (Perkin)
1869	Celluloid (Hyatt)
1877	Ethylbenzene (Friedel-Crafts)
1884	Rayon (Chardonnet)
1910	Phenolic resins (Baekeland)
1910	Neoarsphenamine (Ehrlich)
1920	Aldehydes, alcohols (Oxo synthesis)
1925	Insulin (Banting)
1927	Methanol
1930	Neoprene (Nieuwland)
1935	Nylon (Carothers)
1940	Styrene-butadiene rubber
1950	Polyisoprene

SYNTHESIS GAS.   For a number of industrial organic syntheses that proceed in the gaseous phase, it is advantageous to prepare a chargestock to specification. When a mixture of gases is so prepared, the term *synthesis gas* is often used. Thus, there are several mixtures which qualify under this definition: (1) a mixture of H_2 and N_2 used for NH_3 synthesis; (2) a mixture of CO and H_2 for methyl alcohol synthesis; and (3) a mixture of CO, H_2, and olefins for the synthesis of oxo-alcohols. Ammonia synthesis gas is described briefly here.

The hydrogen required for NH_3 synthesis gas may be obtained in commercial quantities from coke oven water gas; from steam reforming of hydrocarbons; from the partial oxidation of hydrocarbon chargestocks; or from the electrolysis of H_2O. The nitrogen required may come from the introduction of air to the process, or where specifically required, pure nitrogen may be obtained from an air separation plant. Since NH_3 synthesis occurs under high pressure, it is advantageous to generate the synthesis gas at high pressure and thus avoid additional high compression costs. For this and other economic situations, coke oven gas and hydrogen from electrolysis are eliminated. This leaves hydrocarbons as the logical choice.

In the steam-hydrocarbon reforming process, steam at temperatures up to 850°C and pressures up to 30 atmospheres reacts with the desulfurized hydrocarbon feed, in the presence of a nickel catalyst, to produce H_2, CO, CO_2, CH_4, and some undecomposed steam. In a second process stage, these product gases are further reformed. Air also is added at this stage to introduce nitrogen into the gas mixture. The exit gases from this stage are further purified to provide the desired 3 parts H_3 to 1 part N_2 which is the correct empirical ratio for NH_3 synthesis. See also **Ammonia**.

T

TACHYLYTE (or Tachylite). Pure tachylite is a natural, basic black glass, which may form along the chilled contacts of dikes or sills. It also occurs as a rind on basic pillow lavas that have been suddenly chilled by plunging into water. Occasionally it forms entire flows from certain Hawaiian volcanoes.

TACONITE. A low-grade iron ore consisting essentially of a mixture of hematite and silica. It contains 25% iron. Found in the Lake Superior district and western states.

TACTICITY. The regularity of symmetry in the molecular arrangement of structure of a polymer molecule. Contrasts with random positioning of substituent groups along the polymer backbone, or random position with respect to one another of successive atoms in the backbone chain of a polymer molecule.

TAFEL REARRANGEMENT. Rearrangement of the carbon skeleton of substituted acetoacetic esters to hydrocarbons with the same number of carbon atoms by electrolytic reduction to a lead cathode in alcoholic sulfuric acid.

TALC. [CAS: 14807-96-6]. The mineral talc is a magnesium silicate corresponding to the formula $Mg_3Si_4O_{10}(OH)_2$ which occurs as foliated to fibrous masses, its monoclinic crystals being so rare as to be almost unknown. It has a perfect basal cleavage, the folia nonelastic although slightly flexible; it is sectile and very soft; hardness, 1; specific gravity, 2.5–2.8; luster, waxlike or pearly; color, white to gray or green; translucent to opaque. It has a distinctly greasy feel. Talc is a metamorphic mineral resulting from the alteration of silicates of magnesium like pyroxenes, amphiboles, olivine and similar minerals.

It is found chiefly in the metamorphic rocks, often those of a more basic type due to the alteration of the minerals above mentioned. Some localities are the Austrian Tyrol, the St. Gotthard district of Switzerland, Bavaria and Cornwall, England. In Canada, talc is found in Brome County, Quebec and Hastings County, Ontario. In the United States, well-known localities are to be found in Vermont, New Hampshire, Massachusetts, Rhode Island, New York, Pennsylvania, Maryland, and North Carolina.

A coarse grayish-green talc rock has been called soapstone or steatite and was formerly much used for stoves, sinks, electrical switchboards, etc. Talc finds use as a cosmetic, for lubricants and as a filler in paper manufacturing. Most tailor's "chalk" consists of talc. The origin of the word talc is not definitely known.

See also terms listed under **Mineralogy**.

TAMM LEVELS. Surface states; the extra electron energy levels found at crystal surfaces.

TANNIN. Substances found in many plants; generally related to one of the phenols, pyrogallol or catechol. By their action on animal skins, they cause changes that make the skins resistant to decomposition and at the same time leave them flexible and very strong, greatly improved in wearing qualities. Skins so treated are said to be tanned, and are called leather. Tanning is a very old art, having been practiced in China since long before the Christian era. It was also known to the American Indians before the arrival of the Europeans.

Tannins are found in various parts of the plant, appearing frequently in leaves, and in the cortical tissues of stems. Tannins may be found in the walls of cells or in the vacuoles; often their presence causes the cell to appear dark-colored. Many fruits, such as the persimmon, contain large amounts of tannin, especially before they are ripe. Wound tissues, and especially the hypertrophied tissues known as galls, which result from the

bits of certain insects, are particularly rich in tannins. Tannins appear to be by-products of the metabolism of the plant. When present in the epidermal cells, tannins are seen as a deterrent to snails, which might injure the leaf by feeding on it, to parasitic fungi, which might otherwise enter the leaf tissue, and as a protection against desiccation, since they form substances impervious to water.

An important source of tannin is the bark of various trees, especially that of the hemlock and several species of oaks. The bark is removed from the tree in sheets approximately 4 feet long. Stripping from the tree is usually done in the spring, when the cambial cells are most active and the bark separates easily. To remove the bark, two rings are cut completely through the bark and around the tree. A longitudinal slit is made through the bark from one ring to the other. With the use of a blunt, long-handled implement, the bark is then pried loose from the tree and allowed to dry. By felling the tree, the entire trunk may be stripped of its bark in this way. The dried bark is shipped to mills, where the tannin is extracted. Tannins from these barks are used to tan leather for shoe-soles and other heavy leathers. The wood of the chestnut tree yields a tannin similarly used.

Trees of the genus *Schinopsis*, native to the southern part of South America, including southern Brazil, Bolivia and other southern countries are very important source of tannin. These trees are known by the name "quebracho," which means "ax-breaker," because of their very hard, dense, heavy, dark-red wood, which is cut with difficulty. The heartwood of the tree contains 20–27% tannin, which is obtained by cutting the wood into small chips and extracting with water. This tannin is often used in combination with tannins from other plants.

The bark of many other trees yields large amounts of tannins. Among these are the mangrove, and several species of Acacia, known as wattles, natives of Australia. Fruits also may be a source of tannin. The fruits of *Terminalia chebula*, called *myrobalans*, are an important tannin source. The tree is a native of tropical Asia. Another fruit rich in tannin is *divi-divi*, the pods of a legume, *Caesalpinia coriaria*, which is native in tropical America and the West Indies. Sumac leaves, especially those of *Rhus coriaria*, a shrub or small tree native in Mediterranean Europe, are rich in tannins. To obtain the tannin, the plants are cut down and spread out to dry. The leaves are then removed from the stems and packed into bags, which are shipped to the mills. There the leaves are first cleaned and then ground up. The tannins from this source are used in manufacturing fine leathers, like glove leathers. Leaves of other species of sumac, including the various American sumacs, also contain tannins which, however, are not so valuable and are little used.

Tannins are solids, soluble in water or alcohol, usually extracted by hot water, insoluble in ether, chloroform, carbon disulfide, benzene, soluble in alcohol-ether mixture, and in ethyl acetate, possessing a bitter astringent taste. Tannins (1) yield precipitates with gelatin, proteins (connected with the property of making leather from hides), alkaline salt solutions of many heavy metals, e.g., lead acetate, copper acetate (precipitate brown), antimonyl tartrate, concentrated dichromate solution, also by chromic acid (1% CrO_3); (2) yield dark blue or green coloration with ferric salt solutions; (3) in alkaline solution, absorb oxygen and yield dark colored solution; (4) with iodine in potassium iodide plus small proportion of ammonium hydroxide, yield red color (5) with dilute solution potassium hexacyanoferrate(II) in ammonium hydroxide, yield a red to brown coloration (care not to use excess reagent).

While tannins probably vary in composition, the type generally termed tannic acid is a pentadigalloylglucose for hydrolysis yields diagallic acid and glucose.

Additional Reading

Hemingway, R.W. and J.J. Karchesy: *Chemistry and Significance of Condensed Tannins*, Perseus Books, Boulder, CO, 1989.

Hemingway, R.W. and P.E. Laks: *Plant Polyphenols: Synthesis, Properties and Significance,* Kluwer Academic Publishers, Norwell, MA, 1992.

Lemmens, R.H. and N. Wulijarni-Soetjipto: *Dye and Tannin-Producing Plants,* Balogh Scientific Books, Champaign, IL, 1991.

Salunkhe, D.K., J.K. Chavan, and S.S. Kadam: *Dietary Tannins: Consequences and Remedies,* CRC Press, LLC., Boca Raton, FL, 1990.

TANTALITE. This black mineral, $(Fe, Mn)(Ta, Nb)_2O_6$, is isomorphous with columbite and dimorphous with tapiolite. Tantalite occurs in pegmatites and is a principal ore of tantalum.

TANTALUM. [CAS: 7440-25-7]. Chemical element symbol Ta, at. no. 73, at. wt. 180.948, periodic table group 5, mp 2,996°C, bp 5,427°C, density 16.65 g/cm³ (solid at 20°C), 17.1 (single crystal). Elemental tantalum has a body-centered cubic crystal structure. Because of high mp, it is considered a refractory metal.

Tantalum is a slightly bluish metal; ductile, malleable, and when polished resembles platinum; burns upon being heated in air; insoluble in HCl or HNO_3, but soluble in hydrofluoric acid or a mixture of hydrofluoric and HNO_3. The tough, impermeable oxide film formed on the metal when exposed to air makes tantalum the most resistant of all metals to atmospheric corrosion. Tantalum was first identified by Ekeberg as a new element in yttrium minerals in 1802 and was first obtained in pure form by Berzelius in 1820 by heating potassium tantalofluoride with potassium. There is one, naturally occurring stable isotope ^{181}Ta. ^{180}Ta also occurs naturally (isotopic abundance 0.012%), with a half-life of something greater than 10^7 years. At least nine other radioactive isotopes have been identified ^{176}Ta through ^{179}Ta and ^{182}Ta through ^{186}Ta. With exception of ^{179}Ta (half-life of about 600 days), the remaining half-lives are expressed in minutes, hours, or days. ^{182}Ta has been used as a source of gamma rays. See also **Radioactivity**. In terms of abundance, tantalum does not appear on the list of the first 36 elements that occur in the earth's crust and hence is relatively scarce. Also, tantalum does not appear on the list of the first 65 elements that are found in seawater. First ionization potential, 7.7 eV. Oxidation potential $2Ta + 5H_2O \leftarrow Ta_2O_5 + 10H^+ + 10e^-$, 0.71V. Other important physical properties of tantalum are given under **Chemical Elements**.

Tantalum is found in a number of oxide minerals, which almost invariably also contain niobium (columbium). The most important tantalum-bearing minerals are tantalite and columbite, which are variations of the same natural compound $(Fe, Mn)(Ta, Nb)_2O_6$. Much of the tantalum concentrates has been obtained as a byproduct from tin mining; in recent years, tin slags, which are a byproduct of the smelting of cassiterite ores, such as those found in the Republic of Congo. Nigeria, Portugal, Malaya, and Thailand have been an important raw material source for tantalum.

The first successful industrial process used to extract tantalum and niobium from the tantalite-columbite-containing minerals employed alkali fusion to decompose the ore, acid treatment to remove most of the impurities, and the historic Marignac fractional-crystallization method to separate the tantalum from the niobium and to purify the resulting K_2TaF_7. Most tantalum production now employs recovery of the tantalum and niobium values by dissolution of the ore or ore concentrate in hydrofluoric acid. Then the dissolved tantalum and niobium values are selectively stripped from the appropriately acidified aqueous solution and separated from each other in a liquid-liquid extraction process using methyl isobutyl ketone (MIBK) or other suitable organic solvent. The resulting purified tantalum-bearing solution is generally treated with potassium fluoride or hydroxide to recover the tantalum in the form of potassium tantalum fluoride, K_2TaF_7, or with ammonium hydroxide to precipitate tantalum hydroxide, which is subsequently calcined to obtain tantalum pentoxide, Ta_2O_5. Tantalum metal is generally obtained by sodium reduction of K_2TaF_7, although electrolysis of K_2TaF_7 and carbon reduction of Ta_2O_5 in an electric furnace have also been used. Tantalum metal can absorb large volumes of hydrogen during heating in a hydrogen-bearing atmosphere at an intermediate temperature range (450–700°C). The hydrogen is readily removed by heating in vacuum at higher temperatures.

Uses

Tantalum is used widely, although in small quantities in the electronics industry in electrolytic capacitors, emitters, and getters. The corrosion resistance of tantalum has been compared with that of glass. Additionally, the metal has a high heat-transfer coefficient and is easy to fabricate. Consequently, it finds use in equipment that must resist strong corrosive attack, as in the manufacture of HCl, hydrogen peroxide, in chromium plating baths, in bromine heaters and stills, and in the preparation of corrosive fine chemicals, such as ethyl bromide. The metal also has been used in resistance heaters in very high-temperature furnaces and for some nuclear reactor parts.

Alloys

Tantalum is added to nickel and nickel-cobalt superalloys for gas-turbine and jet-engine parts. Several surgical applications for tantalum have developed because of the inertness of the metal to body fluids and the tolerance of the body for the metal. Tantalum may be placed in the skull or other body parts without rejection. Strips and screws made of tantalum are used for holding broken pieces of bone and tantalum wire mesh is used for surgical staples, braid for sutures, and reinforcements. Tantalum-base alloys are used for aerospace structures and space power systems, principally because of the high-temperature stability and strength of these alloys. They operate satisfactorily at temperatures in excess of 1,600°C. Tantalum alloys are used in heat exchangers. See Fig. 1. Small additions of zirconium to tantalum increases its tensile strength at normal temperatures and up to approximately 1,200°C. Also, when added in amounts of about 5%, hafnium, molybdenum, rhenium, tungsten, and vanadium also increase the strength of tantalum. The tensile strength of ternary alloys of tantalum (Ta with 30% Nb and 5% Zr of V) at room temperature is about 3X that of tantalum alone. A tantalum-tungsten alloy is used for fabricating springs for high temperature and high vacuum applications.

The trend in the chemical industry is to use increasingly high processing temperatures and pressures, requiring stronger materials with better corrosion resistance. With these objectives in mind, researchers have been studying the influence of the alloying elements tungsten, molybdenum, niobium (columbium), hafnium, zirconium, and rhenium on both the mechanical properties and corrosion behavior of pure tantalum. They have found that additions of only 1% to 3% molybdenum to tantalum, for example, has a marked effect in decreasing the susceptibility of pure tantalum to hydrogen embrittlement in severely corrosive conditions. Not only is the corrosion rate of tantalum decreased, but mechanical properties, such as strength and room temperature workability, are also improved.

Tantalum-tungsten alloys have been successfully developed, which exhibit at least four key advantages: (1) tungsten causes a considerable solid solution hardening (SSH) effect in tantalum; (2) tungsten shows almost no evaporation during electron beam melting; (3) tungsten is less costly than tantalum; and (4) the corrosion rate of tantalum is but slightly influenced by the addition of tungsten up to about 10% (wt).

Tantalum-hafnium and tantalum-zirconium alloys are less suitable for aggressive acid environments. Compared with tantalum-tungsten alloys, tantalum rhenium alloys are superior in corrosion resistance and to hydrogen embrittlement, but the major disadvantage is the high cost of rhenium.

Fig. 1. High-heat-transfer bayonet-style exchangers employing tantalum alloy tubes. Each exchanger uses 104 tubes. (*Fansteel*)

TABLE 1. REPRESENTATIVE PROPERTIES OF TANTALUM ALLOYS

Alloy additions (Weight, %)	Forms commercially available	Code name	Typical high-temperature strength			
			Temperature, °C	Tensile, Mpa	Temperature, °C	10-h rupture, Mpa
None	All	Unalloyed Ta	1315	59	1315	7
7.5 W (P/M alloy)	Wire, strip	FS61	25	1140	—	—
2.5 W, 0.15 Nb	All	FS63	95	315	—	—
25 W	All	KBI 6	95	315	—	—
0 W	All	Ta-10 W	1315	345	1315	140
8 W, 2 Hf	All	T-111	1315	255	—	—
8 W, 1 Re, 1 Hf, 0.025°C	All	Astar 811C	1315	275	—	—
40 Nb	All	KBI 40	260	290	—	—
37.5 Nb, 2.5 W, 2 Mo	All	KBI 41	260	515	—	—

(After Advanced Materials & Processes).

Some of the properties of tantalum and its alloys are given in Table 1.

Chemistry and Compounds

As might be expected from its 5d 36s 2 electron configuration, tantalum forms pentavalent compounds. In fact, they constitute the great majority of tantalum compounds, although the valences 2, 3, and 4 are known. However, the existence of the Ta^{5+} ion is very brief, since it readily coordinates with H_2O, OH^-, and other anions or molecules. Tantalum is extremely resistant to chemical action, not being attacked by acids other than hydrofluoric acid, and by alkalies only upon fusion. Even fluorine and oxygen react only on heating.

Tantalum pentoxide, formed by heating the metal with oxygen, reacts with hydrofluoric acid, alkali bisulfates or alkali hydroxides, forming tantalates with the latter. It reacts with a number of halogen compounds to give tantalum pentafluoride, pentachloride and pentabromide, TaF_5, $TaCl_5$, and $TaBr_5$. (Carbon tetrachloride is often used in this preparation of $TaCl_5$.) These compounds readily undergo hydrolysis, and may form oxyhalides, such as TaO_2F and $TaOBr_3$. They may be reduced, but with difficulty, $TaCl_5$ when heated with aluminum yielding the tetrachloride, $TaCl_4$. The trihalides, $TaCl_3$ and $TaBr_3$ have also been prepared. Tantalum(V) fluoride combines with other fluorides, notably the alkali metal fluorides, to yield complexes, such as K_2TaF_7 and Na_3TaF_8.

Other complexes of tantalum(V) are formed with oxygen-function compounds, such as o-dihydroxybenzene and acetylacetone.

In addition to Ta_2O_5, another oxide is known, TaO_2, which may be formed by active-metal reduction (as is the tetrachloride), except that the pentoxide is heated with magnesium rather than aluminum. It forms with alkali metals the metatantalates, $MTaO_3$, the orthotantalates, M_3TaO_4, and pyrotantalates, $M_4Ta_2O_7$, as well as such polytantalates as $M_8Ta_6O_{19}$, the latter requiring fusion with the alkali hydroxides.

The only known sulfide, which is produced by heating with carbon disulfide, is TaS_2, but at least two nitrides are known, TaN and Ta_3N_5, the latter being unstable.

The organometallic compounds of tantalum all involve oxygen bonding, with the exception of a dicyclopentadienyl compound, $(C_5H_5)_2TaBr_3$. The others are alkoxy compounds, such as $(C_2H_5O)_3TaCl_2$, $Ta(OC_2H_5)_5$, $Ta(OCH(C_2H_5)CH_3)_5$, etc., with the exception of bis(fluorosulfonyloxy) trichlorotantalane, $Cl_3Ta(OS(O_2)F)_2$.

A major portion of this article was furnished by M. SCHUSSLER FANSTEEL North Chicago, Illinois

Additional Reading

Cardonne, S.M. et al.: "Tantalum and Its Alloys," *Advanced Materials & Processing*, 16 (September 1992).

Carter, G.F. and D.E. Paul: *Materials Science and Engineering*, ASM International, Materials Park, OH, 1991.

Davis, J.R.: *Metals Handbook*, 2nd Edition, ASM International, Materials Park, OH, 1998.

Greenwood, N.N. and A. Earnshaw: *Chemistry of the Elements*, 2nd Edition, Butterworth-Heinemann, Inc., Woburn, MA, 1997.

Gypen, L.A. and A. Deruyttere: "New Tantalum Base Alloys for Chemical Industry Applications," *Metal Progress*, **127**(2), 27–34 (February 1985).

Hala, J.: *Halides, Oxyhalides and Salts of Halogen Complexes of Titanium, Zirconium, Hafnium, Vanadium, Niobium and Tantalum*, Vol. 40, Elsevier Science, New York, NY, 1989.

Hawley, G.G. and R.J. Lewis: *Hawley's Condensed Chemical Dictionary*, 13th Edition, John Wiley & Sons, Inc., New York, NY, 1999.

Krebs, R.E.: *The History and Use of Our Earth's Chemical Elements: A Reference Guide*, Greenwood Publishing Group, Inc., Westport, CT, 1998.

Lide, D.R.: *CRC Handbook of Chemistry and Physics*, 84th Edition, CRC Press, LLC., Boca Raton, FL, 2003.

Staff: *Properties and Selection: Nonferrous Alloys and Pure Metals*, ASM International, Materials Park, OH, 1990.

Yau, Te-Lin and K.W. Bird: "Know Which Reactive and Refractory Metals Work for You," *Chem. Eng. Progress*, **65** (February 1992).

TAR ACID. Any mixture of phenols present in tars or tar distillates and extractable by caustic soda solutions. Usually refers to tar acids from coal tar and includes phenol, cresols, and xylenols. When applied to the products from other tars it should be qualified by the appropriate prefix, e.g., wood tar acid, lignite tar acid, etc. See also **Coal Tar and Derivatives**.

TARNISH. A reaction that occurs readily at room temperature between metallic silver and sulfur in any form. The well-known black film that appears on the silverware results from reaction between atmospheric sulfur dioxide and metallic silver, forming silver sulfide. It is easily removable with a cleaning compound and is not a true form of corrosion. Plating with a mixture of silver and indium will increase tarnish resistance. Gold will also tarnish in the presence of a high concentration of sulfur in the environment.

TAR SANDS. Also called *bituminous sands* and *oil sands*, tar sands represent a vast potential of petroleum like energy reserves and a reservoir of materials for the preparation of syncrudes. In 1988, the processing of tar sands is steadily approaching a state of economic viability. Although there are major technological problems remaining in the recovery and processing of tar sands into practical fuels, the overriding factor affecting progress in this field is a combination of economics and technology.

The heavy, viscous petroleum substances impregnating the tar sands are called *asphaltic oils*. Other names used to describe these oils include *maltha*, *brea*, and *chapapote*. Asphaltic petroleums are most commonly confused with, but are *not* related to *asphaltites* (gilsonite, glance pitch, and grahamite); the *asphaltic pyrobitumens* (elaterite, wurtzilite, albertite, and impsonite); the native *mineral wax* (ozokerite); and the *pyrogenous distillates* of bituminous substances (tar and pitch).

Tar sands are composed of a mixture of 84–88% sand and mineral-rich clays, 4% water, and 8–12% bitumen. Bitumen is a dense, sticky, semisolid that is about 83% carbon. The substance does not flow at room temperature and is heavier than water. At higher temperatures, it flows freely and floats on water. Characteristics of tar sands important to mining, recovery, and processing include grain size, composition, sortability, porosity, permeability, and microscopic habitat.

Tar Sand Resources

The presence of tar sands in North America was noted by American Indians several centuries ago. Pitch recovered from surface deposits was used for waterproofing canoes. It is reported that Columbus observed asphalt from Pitch Lake in Trinidad and used the material for repairing his ships on his third voyage to the West Indies in 1498. The same bitumen deposit was reported by Sir Walter Raleigh in 1595. For several centuries the material was used for repairing vessels.

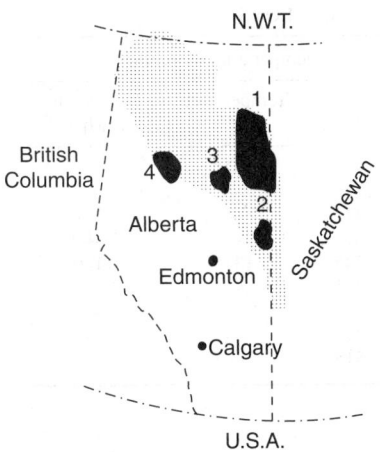

Fig. 1.   Location of major tar (oil) sand deposits in Alberta, Canada: (1) Athabasca, (2) Cold Lake, (3) Wabasca, and (4) Peace River, N.W.T. (Northwest Territories)

Fig. 2.   General view of upgrading plant for processing bitumen mined from Athabasca deposit. In the foreground is the sulfur stockpile; in the background is a tailings pond. Right background shows the extraction plant; left is the coke pile. (*Alberta Government Photographic Service*)

In 1962, reports were received of tar sands in the Bjorne Formation of Triassic age on Melville Island in the Canadian Arctic Archipelago near the southern margin of the Sverdrup sedimentary basin. Subsequent investigations of these deposits by officers of the Geological Survey of Canada revealed them to be a seepage derived from the oxidation and polymerization of 19–31° API gravity oil, with a total in-place reserves of only 30 million barrels. The largest find of tar sands in Canada (and possibly in the world) was located in the subsurface of northern Alberta in the valley of the lower Athabasca River, along a distance of 160 kilometers (100 miles). This is now known as the Athabasca deposit. As indicated by Mossop (1980), McMurray Formation deposition began in the Athabasca region in Early Cretaceous time. The surface on which the initial sediments were laid down was an exposed landscape of Devonian limestone. It is envisioned that, during the McMurray period, the region underwent gradual subsidence, with the Boreal Sea[1] slowly transgressing across it from the north. The McMurray sand deposition

---

[1] Boreal is a term that pertains to the north, or to things located in northern regions. The Boreal region is characterized by tundra and taiga—a climatic zone having a definite winter that experiences snow and a short summer that is generally hot, characterized by a large annual range of temperature. This region includes parts of North America, central Europe, and Asia, generally between latitudes 60°N and 40°N.

stopped when the sea eventually transgressed the entire area, giving subsequent rise to deposition of Clearwater Formation marine shales. Not all geologists are agreed upon the details of the formation and possible later biodegradation of tar sand deposits. Geographically close to the Athabasca deposit are the Cold Lake, Wabasca, and Peace River deposits. Underlying these Cretaceous tar sands are Devonian carbonate rocks (limestone and dolomite) impregnated with bitumen of essentially the same composition as the bitumen in the Cretaceous sands. See Fig. 1.

Along with other alternative sources of energy during the energy crisis of the 1970s, considerable attention was devoted to the exploitation of tar sands. Once, it was predicted that deposits in Canada could yield a light synthetic crude oil to the extent of a million barrels per day, or about one-third of Canada's petroleum requirements. Later, when serious environmental concern over fossil fuels was indicated, research turned essentially elsewhere. It was estimated in the late 1970s that tar sands reserves in the United States, mainly in Utah, would have the petroleum equivalent of 90 billion barrels.

It is interesting to note that tar sands worldwide contain the largest accumulations of liquid hydrocarbons in the earth's crust.

The seriousness of the Canadian tar sands effort is demonstrated by a view of a plant in Alberta as of about 1980. See Fig. 2.

TARTARIC ACID.   [CAS: 87-69-4]. $(CHOHCO_2H)_2$, formula weight 150.09, white crystalline solid with four physical isomers, three of which are optically active: (1) dextro- and (2) levotartaric acid, both with same mp 168–170°C and sp gr 1.760, (3) racemic acid (dextrolevo), mp 205–206°C, sp gr 1.697, and (4) mesotartaric acid (inactive), mp 159–160°C, sp gr 1.737. Racemic acid crystallizes with one molecule of H_2O. All forms decompose before reaching the boiling point at atmospheric pressure. All forms are soluble in H_2O, slightly soluble in alcohol, and essentially insoluble in ether. Tartaric acid is a primary example of optical isomerism and one of the earliest compounds studied in this regard. Tartaric acid is a dibasic acid with two series of salts and esters.

Tartrates (like citrates) in solution change silver of ammonio-silver nitrate into metallic silver. Potassium hydrogen tartrate and calcium tartrate, on account of their solubility characteristics, are of importance in the separation and recovery of tartaric acid. The former salt is readily converted into the latter, and the resulting calcium tartrate plus dilute sulfuric acid yields tartaric acid plus calcium sulfate, and the latter may be separated by filtration. Tartaric acid may be obtained by evaporation of the filtrate. Ester: Diethyl tartrate $COOC_2H_5(CHOH)_2COOC_2H_5$, melting point 17°C, boiling point 280°C. Tartaric acid may be obtained (1) from some natural products, e.g., in the juice of grapes and acid fruits, often in conjunction with citric or malic acid; potassium hydrogen tartrate, "argol," in the residue of wine vats, (2) by synthesis.

Tartaric acid is used: (1) in baking powders as potassium hydrogen tartrate ("cream of tartar") with sodium bicarbonate; (2) in medicine, e.g., potassium antimonyl tartrate ("Tartar emetic"); (3) in effervescent medicinal salts; (4) in blue printing as ferric tartrate; and (5) in silvering mirrors—ammonio-silver nitrate yielding a smooth deposit of silver. Sodium potassium tartrate ("Rochelle salt," $NaKC_4H_4O_6 \cdot 4H_2O$) is used in medicine, and in the preparation of Fehling's solution, which is an alkaline cupric solution made by mixing copper sulfate solution, sodium potassium tartrate solution and sodium hydroxide solution, and is used as an oxidizing reagent in the case of many organic compounds, such as glucose and reducing sugars, and aldehydes, with which cuprous oxide, red to yellow precipitate, is formed.

See also **Isomerism**.

TAUBE, HENRY (1915–).   A Canadian-born chemist who won the Nobel prize for Chemistry in 1983 for his pioneering work in inorganic chemistry and the study of electron-transfer reactions, particularly of metal complexes. Known as an outstanding teacher, he is admired and respected by students and colleagues for his work at Stanford University.

TAU CYCLE.   See **Carbohydrates**.

TAU PARTICLE.   Discovered in 1975, the tau particle is a lepton with a mass of 1.8 GeV, almost twice that of the proton. Like other leptons, the tau particle is considered as pointlike. See also **Particles (Subatomic)**.

TCA CYCLE.   (tricarboxylic acid cycle; Krebs cycle or citric acid cycle). A series of enzymatic reactions occurring in living cells of aerobic organisms, the net result of which is the conversion of pyruvic acid, formed by anaerobic metabolism of carbohydrates, into carbon dioxide and water. The metabolic intermediates are degraded by combination of decarboxylation and dehydrogenation. It is the major terminal pathway of oxidation in animal, bacterial, and plant cells. Recent research indicates that the TCA cycle may have predated life on earth and may have provided the pathway for formation of amino acids.

TECHNETIUM.   [CAS: 7440-26-8]. Chemical element symbol Tc, at. no. 43, at. wt. 98.906, periodic table group 7, mp 2172°C, bp 4877°C, does not occur in nature. The present location of technetium in the periodic table was vacant for many years, during which time several claims to having found the element were made, but never confirmed. One such claimant termed the element masurium. Technetium has been detected in certain stars and this discovery must be resolved with current theories of stellar evolution and element synthesis.

^{97}Tc, the first isotope to be isolated, was extracted by Perrier and Segré in 1937 from molybdenum which had been bombarded with deuterons in the Berkeley cyclotron. The reaction was ^{96}Mo$(d, n)^{97}$Tc. The isotope with the longest half-life, ^{99}Tc (half-life = 2.12×10^5 years), is found in relatively large amounts among the fission products of uranium. It is also produced by neutron irradiation of ^{98}Mo, by the reaction

$$^{98}\text{Mo}(n, \gamma)^{99}\text{Mo}(\beta - \text{decay})^{99m}\text{Tc(isomeric transition)}^{99}\text{Tc}$$

Significant quantities have been isolated and considerably larger quantities could be made available if applications for it were developed. A U.S. government-owned invention available for licensing concerns a method for recovering technetium from nuclear fuel reprocessing waste solutions. ^{99}Tc has found some application in diagnostic medicine. Ingested soluble technetium compounds tend to concentrate in the liver and are valuable in labeling and in radiological examination of that organ, and this was the basis of the early medical uses. However, the ideal nuclear properties of ^{99m}Tc have led to expanded usage in medical diagnostics. By technetium labeling of suitable compounds (or blood serum components), diseases involving the circulatory system and organs other than the liver can be diagnosed.

In all, sixteen isotopes of technetium have been reported of mass numbers 92–105, 107, and 108.

Superconductivity has been observed in technetium metal and in alloys based on technetium with additions of Pd, Os, Rh, Ru, Sn, V, Ti, Re, W, or C.

A study of the chemistry of technetium shows it to have, as expected, properties intermediate between those of its homologues manganese and rhenium, the resemblance to the latter being perhaps greater than to the former. Like rhenium, technetium apparently exists in (IV), (VI), and (VII) oxidation states. Pure technetium metal has been prepared by passing hydrogen gas at 1,000°C over the sulfide obtained by precipitation with H_2S from HCl solution. The metal has been shown to have the same crystal structure as rhenium and the adjacent elements osmium and ruthenium. Among its compounds are the ditechnetium heptasulfide, Te_2S_7, readily precipitated by H_2S from oxidized solutions, the corresponding oxide, Tc_2O_7 produced directly from the elements at higher temperatures, which reacts with NH_3 to form ammonium pertechnate, NH_4TcO_4, and the hexachloro complex ion, $TcCl_6^{2-}$, which like the corresponding rhenium ion, has a magnetic moment corresponding to three unpaired electron spins.

In 1991, cardiologists (University of California, Los Angeles) reported the use of a new combination of mixtures for yielding images of healthy and damaged areas of the heart. This enables physicians to assess the effectiveness of clot-busting drugs and other cardiac therapies. Use of the technique as a preventive measure for determining persons at risk of sudden blood-flow blockages and thus sudden heart attacks also has been suggested. The product (DuPont-Merck) is called *technetium-99 m sestamibi*. Technetium-99 m is a tracer, and sestamibi is an effective "heart-seeking" compound. This technique is superior to use of the thallium radioisotope because of the requirements of thallium to process images within 30 minutes of injection. Technetium is not that time-sensitive.

ROBERT Q. BARR
Director, Technical Information, Climax Molybdenum Company
Greenwich, Connecticut

Additional Reading

Fackelmann, K.A.: "Diagnostic Duo Highlights Heart Damage," *Science News*, 4 (January 5, 1991).
Greenwood, N.N. and A. Earnshaw: *Chemistry of the Elements*, 2nd Edition, Butterworth-Heinemann, Inc., Woburn, MA, 1997.
Krebs, R.E.: *The History and Use of Our Earth's Chemical Elements: A Reference Guide*, Greenwood Publishing Group, Inc., Westport, CT, 1998.
Lewis, R.J. and N.I. Sax: *Sax's Dangerous Properties of Industrial Materials*, 10th Edition, John Wiley & Sons, Inc., New York, NY, 1999.
Lide, D.R.: *CRC Handbook of Chemistry and Physics*, 84th Edition, CRC Press, LLC., Boca Raton, FL, 2003.
Sinflet, J.H. "Bimetallic Catalysts," *Sci. Amer.*, **90** (March 1985).

TEKTITE.   A small (usually walnut-size), rounded, pitted, jet-black to olive-greenish or yellowish body of silicate glass of nonvolcanic origin, found usually in groups in several widely separated areas of the earth's surface and apparently bearing no relation to the associated geologic formations. Most tektites have uniformly high silica (68–82%) and very low water contents (average, 0.005%). Their composition is unlike that of obsidian and more like that of shale. They have various shapes, strongly suggesting modeling by aerodynamic forces and they average a few grams in weight. The largest found weighs 3.2 kilograms. Some authorities believe that tektites are of extraterrestrial origin, or alternatively the product of large hypervelocity meteorite impacts on terrestrial rocks. The term was proposed by Suess in 1900 who believed they were meteorites which at one time had undergone melting.

TELLURIUM.   [CAS: 13494-80-9]. Chemical element, symbol Te, at. no. 52, at. wt. 127.60, periodic table group 6, mp 450°C, bp 690°C, density 6.24 g/cm^3 (crystalline form at 25°C), 6.00 (amorphous form at 25°C). Elemental tellurium has a hexagonal crystal structure with trigonal symmetry. Tellurium is a silver-white brittle semi-metal, stable in air, and in boiling H_2O, insoluble in HCl, but dissolved by HNO_3 or aqua regia to form telluric acid. The element is dissolved by NaOH solution and combines with chlorine upon heating to form tellurium tetrachloride.

In observing a peculiar phase in gold ores of the Transylvania region, Franz Müller von Reichenstein first identified the element in 1782. There are several natural occurring isotopes ^{120}Te, ^{122}Te through ^{126}Te, ^{128}Te, and ^{130}Te. Nine radioactive isotopes have been identified ^{118}Te, ^{119}Te, ^{121}Te, ^{123}Te, ^{127}Te, ^{129}Te, and ^{131}Te through ^{133}Te. With exception of ^{123}Te which has a half-life something greater than 10^{13} years, all of the other radioactive isotopes have half-lives measurable in terms of minutes, hours, or days. In terms of abundance, tellurium does not appear on the list of the first 36 elements that occur in the Earth's crust and hence is relatively scarce. Terrestrial abundance is estimated on the order of 0.002 ppm. Tellurium is found in seawater to the estimated extent of about 95 pounds per cubic mile of seawater. First ionization potential 9.01 eV; second, 18.6 eV; third, 30.5 eV; fourth 37.7 eV; fifth, 59.95 eV. Oxidation potentials $H_2Te(aq) \longrightarrow Te + 2H^+ + 2e^-$, 0.69 V; $Te + 2H_2O \longrightarrow ReO_2(s) + 4H^+ + 4e^-$, −0.529 V; $TeO_2(s) + 4H_2O \longrightarrow H_6TeO_6(s) + 2H^+ + 2e^-$, −1.02 V; $Te^{2-} \longrightarrow Te + 2e^-$, 0.92 V; $Te + 6OH^- \longrightarrow TeO_3^{2-} + 3H_2O + 4e^-$, 0.02 V; $Te \longrightarrow Te^{4+} + 4e^-$, −0.564 V. Electronic configuration $1s^2 2s^2 2p^6 3s^2 ep^6 3d^{10} - 4s^2 4p^6 4d^{10} 5s^2 5p^4$. Other important physical properties of tellurium are given under **Chemical Elements**.

Tellurium occurs chiefly as telluride in gold, silver, copper, lead, and nickel ores in Colorado, California, Ontario, Mexico, and Peru, and infrequently as free tellurium and tellurite (tellurium dioxide, TeO_2). The anode mud from copper and lead refineries, or the flue dust from roasting telluride gold ores is treated by fusion with sodium nitrate and carbonate and the melt extracted with water. The resulting solution is acidified carefully with H_2SO_4, whereupon tellurium dioxide is precipitated, and the dioxide reduced to free tellurium by heating with carbon.

Uses

On the scale of most other commercial metals, the production of elemental tellurium is relatively limited—approximately $\frac{1}{2}$ million pounds annually. Commercial tellurium is marketed at a purity of about 99.7%, although much purer forms are obtainable—up to 99.999%. The application of tellurium and tellurium compounds as catalysts is expanding. Small quantities are used in various electronic components, including solar cells, infrared detectors, emitters, and thermoelectric generators. Tellurium also is sometimes used as a dopant for semiconductor devices. The metal has been used in primer fuses for explosives. The main applications have been in metallurgy. Small additions of tellurium improve the machinability of low-carbon steels, stainless steels, and copper. The metal stabilizes the carbide in cast irons. Tellurium also helps to control pinhole porosity in steel castings. The very small addition (0.05%) of tellurium to lead improves a number of the physical properties of lead sheet, foil, and other shapes. To some extent, tellurium has been used as a curing agent and accelerator in rubber compounds.

Chemistry and Compounds

Tellurium occurs in the same periodic classification as sulfur, selenium, and polonium. Tellurium, unlike sulfur and selenium, has only two allotropic forms. Due to its $5s^2 5p^4$ electron configuration, tellurium, like sulfur and selenium, forms many divalent compounds with covalent bonds and two lone pairs, and d-hybridization is quite common, to form compounds with tellurium oxidation states of $+4$ and $+6$.

Tellurium dioxide, [CAS: 7446-07-3], TeO_2, made directly from the element or by heating tellurous acid, H_2TeO_3, is a solid, subliming at $450°C$, insoluble in H_2O, which dissolves in acids and alkalis, exemplifying the increasing metallic character with atomic weight of the main group 6 elements. Tellurium dioxide accepts a proton from strong acids to form the ion $TeOOH^+$. Dehydration of telluric acid at $400°C$ produces TeO_3, tellurium trioxide. It is not nearly as reactive as sulfur trioxide and selenium trioxide, but reacts with alkali hydroxides to form tellurates.

Tellurous acid, H_2TeO_3, can exist only in very dilute aqueous solutions (due to insolubility of TeO_2). It is a weak acid (ionization constants 2×10^{-3} and 2×10^{-8}). The salts of tellurous acid, the tellurites, may often be formed by reaction of TeO_2 with metal salts. Telluric acid, H_6TeO_6, is prepared by oxidation of tellurium with strong oxidizing agents, such as 30% hydrogen peroxide or boiling HNO_3 and catalyst. Various values of the ionization constants of telluric acid have been reported, on the order of 10^{-7} and 10^{-11}, but the best values would appear to be $pK_{A1} = 7.7$, $pK_{A2} = 11.0$, $pK_{A3} = 14.5$. Telluric acid is a quite strong oxidizing agent, forming halogens from hydrohalides in solution (except hydrogen fluoride). The alkali metal tellurates have the composition $M_2H_4TeO_6$, although metal tellurates with all H's replaced exist, such as Hg_3TeO_6 and Zn_3TeO_6.

Hydrogen telluride is a stronger acid than H_2S (ionization constants 2.27×10^{-3} and 10^{-11} (?) at $18°C$) and is less readily obtained from tellurides than hydrogen sulfide from sulfides. Aluminum telluride, Al_2Te_3, requires heating with H_2O or dilute acids. In general the metal tellurides are prepared by direct combination of the elements. Those of the transition metals and the zinc, gallium and germanium families exhibit many instances of both well defined compounds and non-daltonide compositions, as well as substitutional solid solutions. Also six intermediate phases are found in the palladium-tellurium system, Pd_4Te, Pd_3Te, $Pd_{2.5}Te$, Pd_2Te, $PdTe$, and $PdTe_2$.

Tellurium hexafluoride, [CAS: 7783-80-4], TeF_6, the only clearly defined tellurium hexahalide, is formed directly from the elements at $150°C$, while at $0°C$ the product is mainly the decafluoride, Te_2F_{10}. TeF_6 is a relatively weak Lewis acid, forming complexes with pyridine and other nitrogen bases. Tellurium tetrahalides of all four halogens exist, the TeF_4 formed from $TeCl_2$ and fluorine, the $TeCl_4$ from $TeCl_2$ and chlorine, the $TeBr_4$ from bromotrifluoromethane, CF_3Br, and molten tellurium, and the TeI_4 directly from the elements. Tellurium dichloride, $TeCl_2$, is prepared by passing dichlorodifluoromethane, CF_2Cl_2, over molten tellurium. $TeCl_2$ is quite reactive, disproportionating to tellurium and $TeCl_4$, and useful in preparing other tellurium compounds. $TeBr_2$ is obtained by distillation of the mixture of tellurium and $TeBr_4$ obtained in the reaction between bromotrifluoromethane and tellurium. The tetrahalides of tellurium form many addition compounds with other halides.

Organotellurium compounds corresponding generally to those of sulfur and selenium are known. Although carbon ditelluride has not been prepared, COTe and CSTe have been.

Toxicity

Tellurium and compounds are toxic. Acceptable concentration limit for an 8-hour daily exposure to dust and fumes in air is 0.1 milligrams of tellurium per cubic meter of air. Even exposure at this level may cause what is termed "garlic breath." Proper ventilation, appropriate hygienic practices, and good housekeeping should be observed in handling tellurium. Although

elemental tellurium causes no apparent problems in contact handling, skin contact with soluble tellurium compounds must be avoided.

S. C. CARAPELLA, JR.
ASARCO Incorporated
South Plainfield, New Jersey

Additional Reading

Chizhikov, D.M. and V.P. Schastsivity: *Tellurium and Tellurides,* (Translated from the Russian by E.M. Elms), Collet's Wellingbourough, Northants, England, 1970. (A Classic Reference.)

Greenwood, N.N. and A. Earnshaw: *Chemistry of the Elements,* 2nd Edition, Butterworth-Heinemann, Inc., Woburn, MA, 1997.

Krebs, R.E.: *The History and Use of Our Earth's Chemical Elements: A Reference Guide,* Greenwood Publishing Group, Inc., Westport, CT, 1998.

Lewis, R.J., N.I. Sax: *Sax's Dangerous Properties of Industrial Materials,* 10th Edition, John Wiley & Sons, Inc., New York, NY, 1999.

Lide, D.R.: *CRC Handbook of Chemistry and Physics,* 84th Edition, CRC Press, LLC., Boca Raton, FL, 2003.

Staff: *ASM Handbook—Properties and Selection: Nonferrous Alloys and Pure Metals,* ASM International, Materials Park, OH, 1990.

TELOMERIZATION REACTIONS. In telomerization reactions, a polymerizable unsaturated compound (the taxogen) is reacted under polymerization conditions in the presence of radical-forming catalysts or promoters with a so-called telogen. During the reaction, the telogen is split into radicals that attach to the ends of the polymerizing taxogen and in some instances add on to the double bond of the taxogen and thereby form chains whose therminal groups are formed of the radicals from the telogen.

Organic compounds containing an olefinic double bond, such as thylene, propylene, hexene, octene, or styrene, are normally employed as taxogens. Many different types of compounds can be employed as telogens, for example, halogenated hydrocarbons, such as chloroform or carbon tetrachloride, halogen derivatives of cyanogen, such as cyanogen chloride, aldehydes, alcohols, and the like.

Radical-forming catalysts, such as organic peroxides, hydrogen peroxide, aliphatic azo compounds of the type of azoisobutyric acid nitrile, and redox systems are employed for telomerization reactions.

Telomerization reactions are as a rule carried out at an elevated temperature up to 250 degrees. When volatile reactants are used, the reaction is carried out under elevated pressures, e.g., between 20 and 1000 atmospheres.

TEMPERATURE. The thermal state of a body, considered, with reference to its ability to communicate heat to other bodies (J. C. Maxwell). There is a distinction between temperature and heat, as is evidenced by Helmholtz's definition of heat, (energy that is transferred from one body to another by a thermal process), whereby a thermal process is meant radiation, conduction, and/or convection.

Temperature is measured by such instruments as thermometers, pyrometers, thermocouples, etc., and by scales such as centigrade (Celsius), Fahrenheit, Rankine, Reaumur, and absolute (Kelvin).

See also **Absolute Temperature**; **Thermodynamics**; and **Temperature Scales and Standards**.

TEMPERATURE SCALES AND STANDARDS. That property of systems which determines whether they are in thermodynamic equilibrium. Two systems are in equilibrium when their temperatures (measured on the same temperature scale) are equal. The existence of the property defined as temperature is a consequence of the zeroth law of thermodynamics. The zeroth law of thermodynamics leads to the conclusion that in the case of all systems there exist functions of their independent properties x_i such that at equilibrium

$$\phi_a(x_{ia}) = \phi_b(x_{ib}) = \theta \tag{1}$$

where subscripts a and b refer to two systems a and b each described by n_a properties x_{ia}, and n_b properties x_{ib}, respectively. The hypersurface

$$\theta = \phi(x_i) = \text{constant}$$

is called an *isotherm*, and the pairs of hypersurfaces, Equation (1), at equilibrium are called *corresponding isotherms*.

In order to establish a *temperature scale* it is necessary to assign numerical values θ to these corresponding isotherms in an arbitrary manner, subject only to the condition that the resulting function shall be single-valued. A temperature scale is established by taking the following steps:

1. An arbitrary system is chosen (*thermometer*).
2. It is agreed to maintain $n - 1$ properties of the system constant, and to use the nth property (*thermometric property* $x_n = X$) as a measure of temperature θ.
3. A single-valued *thermometric function* is assumed. Usually the function is simple, for example

$$\theta = aX \tag{2}$$

or

$$\theta = AX + B. \tag{3}$$

The function usually contains one or several constants (a, A, B, etc.).

4. The values of the constants in the thermometric function are determined with reference to fixed thermometric points whose temperatures are arbitrarily assumed. The *fixed thermometric points* most frequently employed are: the *ice point, steam point* and *triple point* of water.

It is not surprising that there exist many different scales of temperature, so-called *empirical temperature scales*, because of the large amount of arbitrariness inherent in the choice.

A *Centigrade* (or *Celsius*) *temperature scale* is obtained by choosing the thermometric function, Equation (3), and assigning the following arbitrary values of temperature, θ, to the ice point (θ_i) and steam point (θ_s) respectively

$$\theta_i = 0^\circ \text{C}.$$

$$\theta_s = 100^\circ \text{C}.$$

Hence on a Centigrade scale

$$\theta = 100 \frac{X - X_i}{X_s - X_i} \tag{4}$$

(X_s measured at θ_s, X_i measured at θ_i).

A Fahrenheit temperature scale is obtained by using Equation (3) but with

$$\theta_i = 32^\circ \text{F}$$

$$\theta_s = 212^\circ \text{F}.$$

Hence

$$\theta = 32^\circ + 180^\circ \frac{X - X_i}{X_s - X_i}. \tag{5}$$

An *absolute scale* is obtained by choosing Equation (2). Depending on the value assigned to a, we obtain the *Kelvin* or *Rankine scale*. The *Kelvin absolute temperature scale* assigns to the triple point of water the value

$$\theta_3 = 273.16^\circ \text{K} \tag{6}$$

which thus becomes a *universal constant of physics*. Hence for the Kelvin scale

$$\theta = 273.16 \frac{X}{X_3} \tag{7}$$

where X_3 is measured at θ_3. For the *Rankine absolute temperature* scale we assume

$$\theta_3 = 273.16 \times 1.8^\circ \text{R} = 491.69^\circ \text{R}. \tag{8}$$

It is clear that by definition

$$1^\circ \text{K} = 1.8^\circ \text{R}, \tag{9}$$

and that on the Rankine scale

$$\theta = 491.69^\circ \frac{X}{X_3}. \tag{10}$$

Different empirical temperature scales will naturally differ from each other except at the respective fixed thermometric points. Even different scales of the same type (say different Centigrade scales) will differ at all temperatures, except the steam point and ice point, depending on the fortuitous properties of the system chosen as a thermometer. It is, therefore, necessary to remove these differences and to obtain a more universal scale. This has been achieved in two ways. The practical way of achieving uniformity is to lay down detailed rules concerning the thermometer (actually different thermometers depending on the range of temperatures to be measured). Such rules have been agreed on internationally and

constitute the *international temperature scale*. Another way is to derive a universal scale from the principles of thermodynamics. The latter is called a *thermodynamic temperature scale*. Some authors refer to it as the *absolute temperature scale* which may be a source of confusion with the Kelvin and Rankine scales described earlier.

The *thermodynamic temperature scale* T is defined by the second law of thermodynamics. It can be shown that the *thermodynamic temperature scale* is identical with the *perfect-gas temperature scale* defined as follows:

1. The system is a gas thermometer (filled with a *real* gas).
2. The thermometric property is the product pV extrapolated to zero pressure, i.e.,

$$r = \lim_{p \to 0} (pV) \tag{11}$$

Hence the *thermodynamic Kelvin scale* is given by

$$T = 273.16 \frac{r}{r_3} (\text{K abs.}) \tag{12}$$

the *thermodynamic Centigrade scale* is given by

$$t - 100 \frac{r - r_i}{r_s - r_i} (\text{degrees C}), \tag{13}$$

the *thermodynamic Rankine scale* is given by

$$T = 491.69 \frac{r}{r_3} (\text{degrees R abs.}), \tag{14}$$

and the *thermodynamic Fahrenheit scale* is given by

$$t = 32 + 180 \frac{r - r_i}{r_s - r_i} (\text{degrees F}). \tag{15}$$

The zeros on the Kelvin and Rankine scales coincide and are termed the *absolute zero of temperature*. The absolute zero of temperature cannot be achieved by any finite process, as stated in the third law of thermodynamics.

The relation between the Centigrade and Kelvin thermodynamic scales is determined by

$$T_i = 273.16^\circ \text{K} = 0^\circ \text{C} \tag{16}$$

and that between the Fahrenheit and Rankine scales is determined by

$$T_i = 491.69^\circ \text{R} = 32^\circ \text{F}. \tag{17}$$

Hence the absolute zero of temperature is at

$$-273.16^\circ \text{C or} - 459.69^\circ \text{F}. \tag{18}$$

The Comité Consultative of the International Committee of Weights and Measures selected 273.16°K as the value for the triple point of water. This set the ice-point at 273.15°K.

The relation between the *international temperature scale* and the thermodynamic temperature scale must be determined empirically with the aid of careful measurements involving gas thermometers.

See also **Units and Standards**.

TEMPERATURE TRANSFER STANDARD. A device for the transfer of a temperature scale from one standardizing laboratory to another. One form consists of a sample of a purified material, the freezing point of which (when realized by a prescribed technique) is reproducible within narrow limits. Materials commonly employed are metals, such as zinc and tin, and organic compounds, such as benzoic acid, phenol, naphthalene, and phthalic anhydride. Another form is a tungsten ribbon-filament lamp, characterized by a stable lamp current-brightness temperature relation. This device is particularly useful for temperatures above $1,050^\circ \text{C}$.

TENACITY. Strength per unit weight of a fiber or filament, expressed as g/denier. It is the rupture load divided by the linear density of the fiber. See also **Tensile Strength**.

TENORITE. A mineral oxide and ore of copper, CuO. Crystallizes in the monoclinic system. Hardness, 3.5; specific gravity, 6.45; color, gray to black with metallic luster. Named after M. Tenore (1780–1861), Naples.

TENSILE STRENGTH. The rupture strength (stress-strain product at break) per unit area of a material subjected to a specified dynamic load; it is usually expressed in pounds per square inch (psi). This definition applies to elastomeric materials and to certain metals.

See also **Tenacity**.

TENSIOMETER.

An apparatus for measuring the surface tension of a liquid by registering the force necessary to detach a metal ring from the surface.

TENSION TEST.

Next to hardness tests, tension tests are the most frequently used to determine the mechanical properties of metals. Tension test specimens necessarily vary in form with the product to be tested. A machined cylindrical specimen with threaded or shouldered ends for gripping is used when the material is sufficiently thick. Standard flat specimens are used for flatrolled products. While both types of specimens have a reduced central section to ensure breaking within a measured gage length, wires and certain special shapes such as steel reinforcing bars for concrete are tested in full section without preparation. Special cast test specimens are often attached to castings, or cast separately, and these are generally tested without machining.

The significant loads determined in the test are reported as unit stresses based on the area of the original section (Stress equals load divided by area). The elongation is expressed as percent increase in length of the gage-marked section. The initial gage length is generally 2 inches (5 centimeters), although an 8-inch (20.3-centimeter) gage length is used for certain flat specimens and other gage lengths are used for special specimens. The gage length should be specified when elongation is reported since percent elongation values are higher for short than for long gage lengths.

The elongation measured over a fixed gage length, and the reduction of area of the section at the fracture are measures of ductility. In cylindrical specimens, the area is readily determined from the final diameter at the fracture. The percent reduction of area is then determined as: original area minus final area, divided by original area.

Autographic load-deformation curves are often drawn during the test. From such a curve, the modulus of elasticity, proportional limit, and yield strength can be determined.

A typical curve has an essentially linear portion (*OA*) in which the deformation is proportional to the applied load. See Fig. 1. It follows that the unit stress (load divided by original area) is proportional to the unit strain (deformation divided by original gage length) in accordance with Hooke's Law. The numerical value of this ratio (e.g., in psi) is known as Young's Modulus or Modulus of Elasticity.

The maximum stress that is developed without deviation from proportionality of stress to strain is the proportional limit (the stress corresponding to load *A*). The maximum stress that can be applied without causing permanent deformation upon release of the load is the elastic limit. Usually, there is little difference between the proportional limit and the elastic limit. Both are dependent on the sensitivity of the measuring devices used and certain details of testing technique. For this reason, the yield strength is generally used as a practical measure of the elastic properties of metals.

The yield strength is the stress at which the stress-strain curve deviates from the initial straight line by a specified increment of strain. The yield strength corresponding to the load at *B* is based on the specified strain deviation or offset *e*. The value of *e* may be as low as 0.0001 inch (0.0025

millimeter) of gage length but the most commonly used value is 0.002 inch (0.05 millimeter), or 0.2% strain.

If the load should be released after reaching *B*, the load deformation relationship will follow the line *BE*, or a curve line terminating between *O* and *E*. Thus the permanent strain will be *e* or a somewhat smaller value. When the final or permanent strain is specified, the stress is known as the proof stress.

An alternate type of yield strength is based on a specified total extension under load, such as 0.5%. If the specified extension is *e'* the load *B'* determines the "extension-underload" yield strength. Load *B'* may be greater or less than *B*.

The tensile strength, or ultimate tensile strength, is the maximum stress developed in the tension test (load *C* divided by original area).

The breaking stress, corresponding to load *D*, is seldom determined or reported.

In loading tension specimens of many soft irons and steels, a point is reached where stretching continues without increase in load. The unit stress obtained by dividing this load, *F*, by the original area of the section is called the yield point. The elongation of the specimen at the yield point may reach 8% in some instances, after which the load will again increase to a maximum in the normal manner. Upper and lower yield points are indicated at *F*; both are used, but the upper yield point is influenced by variations in testing technique such as alignment of the specimen in the testing machine and speed of test. Yield points occur only rarely in the nonferrous metals.

Conventional stress-strain curves are necessarily similar to the load-deformation curves from which they are derived. True stress-strain curves can also be derived in which the stress is based on the actual or instantaneous area of the cross-section. Such curves do not have a maximum corresponding to *C*, but increase continuously to the breaking load.

TERBIUM.

[CAS: 7440-27-9]. Chemical element symbol Tb, at. no. 65, at. wt. 158.92, eighth in the Lanthanide Series in the periodic table, mp 1365°C, bp 3230°C, density 8.230 g/cm³ (20°C). Elemental terbium has a close-packed hexagonal crystal structure at 25°C. The pure metallic terbium is silver-gray in color, and is stable in ambient air conditions. When pure, the metal is malleable. There is one natural isotope of terbium ^{159}Tb. The isotope is not radioactive and has a low acute-toxicity rating. Seventeen artificial isotopes have been identified. Little is known concerning the characteristics of terbium alloys and intermetallic compounds. Average content of the earth's crust is estimated at 0.9 ppm terbium, making this element the second least abundant of the rare-earth elements. Even at this level, however, terbium is potentially more available than antimony, bismuth, cadmium, or mercury. The element was first identified by C.G. Mosander in 1843. Electronic configuration of the ground state is mixed:

$$1s^2 2s^2 2p^6 3s^2 3p^6 3d^{10} 4s^2 4p^6 4d^{10} 4f^8 5s^2 5p^6 5d^1 6s^2.$$

Ionic radius Tb³⁺ 0.93 Å, Tb⁴⁺ 0.76 Å. Metallic radius 1.783 Å. First ionization potential 5.84 eV; second 11.52 eV. Other physical properties of terbium are given under **Rare-Earth Elements and Metals**. See also **Chemical Elements**.

Terbium occurs in apatite and xenotime and is derived from these minerals as a minor coproduct in the processing of yttrium. Processing involves organic ion-exchange or solvent extraction operations. Elemental terbium is produced by calcium reduction of anhydrous TbF_3 in a reactor under an inert atmosphere. Both the oxides and the metal are available at 99.9% purity.

To date, the uses for terbium have been quite limited. Terbium-activated lanthanum oxysulfide Tb:La_2O_2S is a phosphor finding use as an image intensifier for x-ray screens. Terbium-activated indium borate Tb:$InBO_3$ phosphor emits an intense narrow green light (5,450–5,500 Å) and has found use in information display systems where there are high ambient-light conditions. Future color television tubes may use terbium-activated yttrium silicate Tb:Y_2SiO_5 or yttrium phosphate Tb:YPO_4 green phosphors. They appear to be highly efficient post-deflection focused phosphors for elimination of the need for a shadow mask in a television tube. Although terbium oxide may be used as a stain for ceramics, the compound does not color glass. In soda-lime glass, a small quantity of terbium provides a strong green-blue fluorescence under ultraviolet radiation. Another important use of terbium is in amorphous Tb-Co(fe) alloys for magnetic recording and information storage.

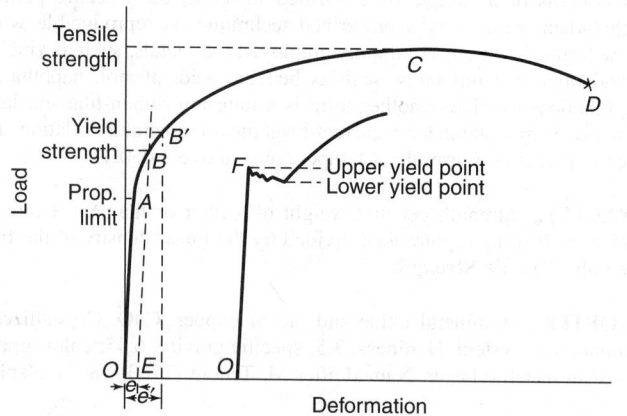

Fig. 1.   Stress-strain diagram

Note: This entry was revised and updated by K. A. Gschneidner, Jr., Director, and B. Evans, Assistant Chemist, Rare-earth Information Center, Institute for Physical Research and Technology, Iowa State University, Ames, IA.

TEREPHTHALIC ACID. [CAS: 100-21-0]. $C_6H_4(COOH)_2$, formula weight 166.13, crystalline solid sublimes upon heating, sp gr 1.510. The compound is almost insoluble in H_2O, only slightly soluble in warm alcohol, and insoluble in ether. Terephthalic acid (TPA) is a high-tonnage chemical, widely used in the production of synthetic materials, notably polyester fibers (poly-(ethylene terephthalate)).

There are several processes for making terephthalic acid on a large scale:

(1) Benzoic acid, phthalic acid and other benzene-carboxylic acids in the form of alkali-metal salts, comprise the chargestock. In a first step, the alkali-metal salts (usually potassium) are converted to terephthalates when heated to a temperature exceeding 350°C. The dried potassium salts (of benzoic acid or o- or isophthalic acid) are heated in anhydrous form to approximately 420°C in an inert atmosphere (CO_2) and in the presence of a catalyst (usually cadmium benzoate, phthalate, oxide, or carbonate). The corresponding zinc compounds also have been used as catalysts. In a following step, the reaction products are dissolved in H_2O and the terephthalic acid precipitated out with dilute H_2SO_4. The yield of terephthalic acid ranges from 95 to 98%.

(2) Toluene, formaldehyde, HCl, calcium hydroxide, and HNO_3 comprise the chargestock. In step 1 of this process, the toluene is reacted with concentrated HCl at about 70°C along with paraformaldehyde. This accomplishes chloromethylation of approximately 98% of the toluene. In step 2, saponification of the chloromethyltoluene is effected with lime and H_2O under pressure and at about 125°C. The product is methylbenzyl alcohol. In step 3, the methylbenzyl alcohol is oxidized with HNO_3 (dilute) under a pressure of about 20 atmospheres and at a temperature of about 170°C. The main products are o-phthalic acid in HNO_3 solution and insoluble terephthalic acid.

(3) Paraxylene and air comprise the chargestock. These materials, along with a proprietary catalyst and solvent, are fed to a liquid-phase oxidation reactor, operated at moderate pressure and temperature. The reaction is:

$$C_6H_4(CH_3)_2 + 3\ O_2 \longrightarrow C_6H_4(COOH)_2 + 2\ H_2O.$$

The design details of these processes are proprietary. There are several other processes which essentially are variations of the foregoing descriptions. See also **Intermediate (Chemical)**; **Phthalic Acid**; and **Phthalic Anhydride**.

TERPENE ALCOHOL. A generic name for an alcohol related to or derived from a terpene hydrocarbon, such as terpineol or borneol.

See also **Terpenes and Terpenoids**.

TERPENELESS OIL. An essential oil from which the terpene components have been removed by extraction and fractionation, either alone or in combination. The optical activity of the oil is thus reduced. The terpeneless grades are much more highly concentrated than the original oil (15–30 times). Removal of terpenes is necessary to inhibit spoilage, particularly of oils derived from citrus sources. On atmospheric oxidation the specific terpenes form compounds that impair the value of the oil; for example, d-limonene oxidazes to carvone and γ-terpinene to p-cymene. Terpeneless grades of citrus oils are commercially available.

TERPENES AND TERPENOIDS. The class of organic compounds known as terpenes is characterized by the presence of the repeating carbon skeleton of isoprene:

These compounds are widely distributed in nature. The name *terpene* used properly refers to the hydrocarbons which are exact multiples of the skeletal isoprene unit. However the name "terpene," sometimes used loosely, includes not only hydrocarbons but also other functional types of naturally occurring organic compounds which contain the reoccurring

isoprene skeleton. In the strictest sense, the names *terpenoid* or *isoprenoid* should be used instead of the more loosely applied usage of terpene.

Terpenoids are divided into subclasses as follows:

Subclass name	No. of carbon atoms	No. of isoprene skeletal units
Hemiterpenoids	C_5	1
Monoterpenoids	C_{10}	2
Sesquiterpenoids	C_{15}	3
Diterpenoids	C_{20}	4
Sesterterpenoids	C_{25}	5
Triterpenoids	C_{30}	6
Tetraterpenoids	C_{40}	8
Polyterpenoids	$(C_5)^n$	n

Together they possess a wide variety of functional groups and structures. Nearly every common functional group is represented. Acyclic, monocyclic and polycyclic structures are observed. The greatest structural variation within a single subclass is to be found among sesquiterpenoids.

The combination of skeletal isoprene units in a regular fashion was exemplified early in the study of terpenoids by nearly all of these compounds. On the basis of this regularity the *regular isoprene rule* was formulated and was taken to mean that terpenoids would possess structures built from a regular "head-to-tail" arrangement of isoprene units. However, as structures of more terpenoids were elucidated, departures from the regular rule were observed. In time "irregular" structures were also accommodated through postulated rearrangement of a regular isoprenoid chain to the "irregular" isoprene skeleton. Rearrangements could occur subsequent to or concomitant with natural cyclization. Thus, the adaptation of the *regular isoprene rule* now finds expression in the *biogenetic isoprene rule*, a rule which is supported experimentally. Particularly significant examples of naturally occurring compounds conforming to the biogenetic isoprene rule are lanosterol, a triterpenoid alcohol associated with cholesterol in wool fat, and gibberelic acid, an important plant-growth regulating substance which is the product of a diterpenoid precursor.

The various terpenoid subclasses are not equally distributed in nature. Representatives of the low-molecular weight end of the terpenoid spectrum are seldom encountered as stable isolable natural products. Isoprene itself has not been detected in plants or animals, but the existence of two highly reactive hemiterpenoid substances in living cells is well established. These are the isomeric γ, γ-dimethylallyl pyrophosphate and isopentenyl pyrophosphate, the two being ubiquitous in living organisms and represent the fundamental isoprene building block in terpenoid biogenesis. One source of evidence for the existence of these hemiterpenoids is the presence of the γ, γ-dimethylallyl unit as a substituent of other classes of natural products, often phenols. The origin of these truly vital hemiterpenoids has been found to be mevalonic acid (3,5-dihydroxy-3-methylpentonoic acid), a six-carbon acid produced by the coenzyme A-assisted condensation of three moles of acetic acid. Decarboxylation and dehydration of mevalonic acid pyrophosphate are known to give the five-carbon unit of isopentenyl pyrophosphate. Isopentenyl pyrophosphate and γ, γ-dimethylallyl pyrophosphate are not only the links between the various subclasses of terpenoids but also biogenetically connect seemingly unrelated plant constituents such as steroids and some types of phenolics and alkaloids.

Monoterpenoids and to a lesser extent sesquiterpenoids are the chief components of the volatile oils readily obtained by the distillation of leaves, wood, and blossoms of a broad array of plants. Sesquiterpenoids are among the most universally distributed natural products. Iridolactone, a monoterpenoid, occurs as a defensive secretion of an ant species belonging to the genus *Iridomyrmex*. The iridoids, which is the general name given to the structural type exemplified by iridolactone, make up an important group of monoterpenoids. Another representative of this group is loganin which along with the amino acid tryptophan provides the carbon atoms of a group of indole alkaloids. Medically important quinine and reserpine are members of this group of alkaloids. The "resin acids" are a group of diterpenoid carboxylic acids which form the major nonvolatile part of natural resins often obtained from conifers. Examples are abietic, pimaric and isopimaric acids. Sesterterpenoids, the most recently discovered terpenoid subclass, are produced by insects and fungi. The tricyclic cereoplasteric acid has been isolated from the waxy coating secreted by the insect *Cereoplastes albolineatus*. The fungus responsible for the leaf spot disease in corn produces the acyclic geranylnerolidol and the tricyclic ophiobola-7, 18-dien-3α-ol. Triterpenoids are found mainly in plants where they occur in resins and plant sap as free triterpenoids, esters or glycosides. A few are observed in animal sources, for example, the acyclic squalene and the tricyclic lanosterol. The connection between squalene and lanosterol is a vitally important one since these two triterpenoids, along with the hemiterpenoid isopentenyl pyrophosphate, are intermediates in the mevalonic acid based biogenesis of steroids. Tetraterpenoids are frequently referred to as *carotenoids*. These constitute a group of natural pigments containing long systems of conjugated carbon-carbon double bonds which are responsible for their color. β-Carotene is the principal pigment of the carrot but this pigment has been isolated from other plant sources also. See also **Carotenoids**. Finally the presence of polyisoprenoids (natural rubber and gutta-percha) in nature shows that the isoprene unit, like the simple sugar and amino acids, has been used to form linear macromolecules.

The use of terpenoids, usually as mixtures prepared from plants, dates from antiquity. The several "essential oils" produced by distillation of plant parts contained the plant "essences." These oils have been employed in the preparation of perfumes, flavorings, and medicinals. Examples are: oils of clove (local anesthetic in toothache), lemon (flavoring), lavender (perfume), and juniper (diuretic). Usually essential oil production depends on a simple technology which often involves steam distillation of plant material. The perfume industry of Southern France uses somewhat more sophisticated procedures in the isolation of natural flower oils since these oils are heat sensitive. The separation of oils from citrus fruit residues in California and Florida is done by machine.

The oleoresinous exudate or "pitch" of many conifers, but mainly pines, is the raw material for the major products of the naval stores industry. The oleoresin is produced in the epithelial cells which surround the resin canals. When the tree is wounded the resin canals are cut. The pressure of the epithelial cells forces the oleoresin to the surface of the wound where it is collected. The oleoresin is separated into two fractions by steam distillation. The volatile fraction is called "gum turpentine" and contains chiefly a mixture of monoterpenes but a smaller amount of sesquiterpenes is present also. The nonvolatile "gum rosin" consists mainly of the diterpenoid resin acids and smaller amounts of esters, alcohols and steroids. "Wood turpentine," "wood rosin" and a fraction of intermediate volatility, "pine oil" are obtained together by gasoline extraction of the chipped wood of old pine stumps. "Pine oil" is largely a mixture of the monoterpenoids terpineol, borneol and fenchyl alcohol. "Sulfate turpentine" and its nonvolatile counterpart, "tall oil," are isolated as by-products of the kraft pulping process. "Tall oil" consists of nearly equal amounts of saponified fatty acid esters and resin acids.

Turpentine is used in syntheses by the chemical and pharmaceutical industries. It also is used as a paint thinner and as a component of polishes and cleaning compounds. Pine oil finds application as a penetrant, wetting agent and preservative, especially by the textile and paper industries, and as an inexpensive deodorant and disinfectant in specialty products. The resin acids are used in the production of ester gum, "Glyptal" resins and are indispensable in paper sizing.

A few individual terpenoids, as well as less expensive mixtures of these compounds, find practical applications. Some examples are: the diterpenoid Vitamin A, the sesquiterpenoid santonin (as an anthelmintic), and the pyrethrins, pyretholone esters of the monoterpenoid chrysanthemic acid (used as an insecticide). A number of sesquiterpenoid lactones of the germacranolide, guaianolide and elemanolide types have shown promise as tumor inhibitors.

The different terpenoid content of plants has served as a finger printing method helpful in botanical identification, especially in cases where differentiation by morphological characteristics has failed. The striking difference in the chemotaxonomy of Jeffery and ponderosa pines serves as an example. The turpentine from the former species consists almost entirely of the paraffinic hydrocarbon *n*-heptane. Turpentine from ponderosa pine consists largely of the monoterpenes β-pinene and Δ^3-carene.

Besides being of considerable biochemical and botanical interest and of importance in the industrial arts, the food, perfumary, and pharmaceutical trades, the terpenoids have been also a continuing challenge to the organic chemist. The earliest work on the volatile oils was very difficult since the oils were usually complex mixtures and as a consequence individual compounds were isolated as liquids of uncertain purity. Physical constants such as molecular refraction and melting points of solid derivatives, especially those from which the compound could be regenerated were of importance in structural investigations. Organic chemistry relied heavily on oxidative degradation techniques. Dehydrogenation of cyclic terpenoids to aromatic systems and the synthesis of these aromatics played an important role in structure determination. In recent times much of the previous difficulty of obtaining pure samples of terpenoids has been overcome through the use of various chromatographic techniques. Gas-liquid phase chromatography has been used to good advantage in separating both microgram quantities and much larger amounts of the more volatile monoterpenoids and sesquiterpenoids. Much of the type of information formerly obtained only be degradative procedures and dehydrogenations can now be obtained through mass spectrometry. Through nuclear magnetic resonance, infrared and ultraviolet spectrometry, structural information can be obtained in a small fraction of the time that was formerly needed to gain the same amount of information.

Acid-promoted cyclization of acylic terpenoids is common. Geraniol, or more readily its trans isomer nerol, can be cyclized with acids to *p*-menthane derivatives. More importantly, citral, 2,6-dimethyl-2-octen-8-al, when condensed with acetone gives pseudoionone. The latter when cyclized with acid gives a mixture of α- and β-isomers which in turn are used in the preparation of perfumes, as an intermediate in a number of industrial syntheses of Vitamin A, and also in a commercial synthesis of the plant hormone abscisic acid. Polyclic terpenoids are prone to rearrangement of the carbon skeleton. The acid catalyzed rearrangement of the monoterpene camphene to derivatives of isobornyl alcohol are well known and have been the subject of extensive theoretical studies. Diterpenoids and triterpenoids undergo "backbone" rearrangement through the migration of hydride and methyl groups. Frequently these migrations are stereospecific and are acid promoted.

Studies of terpenoid chemistry have also involved syntheses. Several of the complex sesquiterpenoid structures have been confirmed or, in some cases, correctly established through synthesis. Many elegant new general synthetic methods have been developed as a result of attempts to synthesize terpenoids. The chemistry of terpenoids present a continually growing area of chemical research, perhaps the equal of any in complexity, subtlety, and variety.

ROBERT T. LaLONDE
State University of New York
Syracuse, New York

Additional Reading

Harborne, J.B., and F.A. Tomas-Barberan: *Ecological Chemistry and Biochemistry of Plant Terpenoids,* Oxford University Press, New York, NY, 1991.

Ho, Tse-Lok: *Enantioselective Synthesis: Natural Products from Chiral Terpenes,* John Wiley & Sons, Inc., New York, NY, 1992.

Sukh, D., A.S. Gupta, B.A. Nagasampagi, and S.A. Patwardhan: *Handbook of Terpenoids: Terpenoids,* CRC Press LLC., Boca Raton, FL, 1989.

Towers, G.H., and H.A. Stafford: *Biochemistry of the Mevalonic Acid Pathway to Terpenoids,* Kluwer Academic Publishers, Norwell, MA, 1990.

TESTING (Chemical). Identification of a substance by means of reagents, chromatography, spectroscopy, melting and boiling point determination, etc.

See also **Analysis (Chemical).**

TESTING (Physical). Application of any procedure whose object is to determine the physical properties of a material. There are four major categories of tests: (1) Those that are direct measurements of a property, e.g., tensile strength. (2) Those that subject the material to actual service conditions; these often require a long period of time, e.g., shelf life of foods and corrosion of metals. (3) Accelerated tests, which require specially designed equipment that simulates service conditions on an exaggerated scale; in these, only a few hours are necessary to duplicate years of service life, e.g., oxygen bomb aging of elastomers. (4) Nondestructive testing by N-ray of radiography. Elaborate standard testing procedures are established by the American Society for Testing and Materials: http://www.astm.org.

The more common types of tests are as follows:

- Abrasion (elastomers, textiles).
- Adhesion (glues, resins).
- Aging (elastomers, plastics, leather, food products).

- Color stability (pigments, organic dyes) (exposure).
- Corrosion (metals, alloys) (exposure).
- Dielectric (electrical tapes, plastics, glass).
- FLammability (textiles, fibers, paper, plastics).
- Flash point of combustible liquids (Tag closed cup TCC, Cleveland open cup COC, open cup OC.
- Hardness (metals, elastomers, plastics) (Brinell, Rockwell, Shore penetration).
- High temperature (elastomers, adhesives).
- Impact strength (composites, glass cement).
- Sun-cracking (paints, varnishes, elastomers) (exposure).
- Tear (paper, rubber, textiles).
- Tensile strength (fibers, elastomers, paper, textiles, metals).
- Viscosity (lubricants) (Saybolt, Engler).

See also **Nondestructive Testing (NDT)**.

TESTING (Physiological). Determination of the toxicity of a substance or product by administering it to laboratory animals in controlled dosages, by mouth, skin application, or injection. Materials commonly subjected to such evaluation are pharmaceuticals, pesticides, and foods. Extensive testing programs are required before such products are approved for human use.

TESTOSTERONE. See **Hormones**; **Steroids**.

TETRADYMITE. A mineral, bismuth tellurium sulfide, corresponding to the formula Bi_2Te_2S. It is rhombohedral. Tetradymite occurs usually in gold quartz veins. It is found in Norway, Sweden, England, Bolivia, British Columbia; and in the United States, in Virginia, North Carolina, Georgia, Montana, Colorado, and elsewhere. It derives its name from the Greek word meaning fourfold, in reference to the double twin crystals occasionally developed.

TETRAHEDRITE. A mineral of the composition, $(Cu, Fe)_{12}As_4S_{13}$, isomorphous with tennantite. The color ranges from steel-gray to iron-black. The mineral frequently contains cobalt, lead, mercury, nickel, silver, or zinc in replacement of the copper. Tetrahedrite usually occurs in tetrahedral crystals associated with copper ores. The mineral is considered an important copper ore and sometimes is a valuable ore for silver. The mineral sometimes is referred to as *fahlore, gray copper ore*, and *stylotypite*.

THALLIUM. [CAS: 7440-28-0]. Chemical element symbol Tl, at. no. 81, at. wt. 204.38, periodic table group 3, mp 303.5°C, bp 1447–1467°C, density 11.85 g/cm^3 (20°C). Elemental thallium has a hexagonal close-packed crystal structure normally, but also exhibits a face-centered cubic crystal structure.

Thallium metal is bluish-gray upon fresh exposure, changing to dark gray on standing, this oxidation increased with temperature above 25°C; soft, and may be easily cut with a knife. It is malleable but of low tenacity, so that it must be extruded to form wire; HNO_3 is the best solvent; forms alloys with many metals, e.g., mercury, cadmium, zinc, silver, copper, magnesium.

The element was first identified by Sir William Crookes spectrographically in 1861. While seeking tellurium, Crookes observed the characteristic bright green lines in the emission spectrum of thallium. At just about the same time, A. Lamy identified the element. Thallium occurs naturally as [203]Tl and [205]Tl. Eleven radioactive isotopes have been identified [198]Tl through [202]Tl, [204]Tl, and [206]Tl through [210]Tl. With exception of [204]Tl which has a half-life of 4.07 years, the other isotopes have relatively short half-lives expressed in minutes, hours, and days. See also **Radioactivity**. Thallium is not considered an abundant element, estimates of occurrence in the earth's crust ranging from 0.3 to 3.0 ppm. In a list of 65 chemicals found in seawater, thallium does not appear. First ionization potential 6.106 eV; second, 20.32 eV; third, 29.7 eV. Oxidation potentials $Tl \longrightarrow Tl^+ + e^-$, 0.336 V; $Tl^+ \longrightarrow Tl^{3+} + 2e^-$, −1.25 V; $Tl + OH^- \longrightarrow TlOH + e^-$, 0.3445 V; $TlOH + 2OH^- \longrightarrow Tl(OH)_3 + 2e^-$, 0.05 V. Electron configuration $1s^2 2s^2 2p^6 3s^2 3p^6 3d^{10} 4s^2 4p^6 4d^{10} 4f^{14} 5d^2 5p^6 5d^{10} 6s^2 6p^1$. Other important physical properties of thallium are given under **Chemical Elements**.

Thallium occurs in small amounts in pyrite, zinc blende, and hematite of certain localities, and in a few rare minerals in Sweden and Macedonia.

For the recovery of thallium from the flue dust of pyrite burners, the dust is boiled with H_2O, allowed to stand some time, filtered, and HCl added to the filtrate, whereupon crude thallous chloride is precipitated. This is purified by further treatment, and thallium metal obtained (1) by electrolysis of the sulfate solution or (2) by fusion of the chloride with sodium cyanide and carbonate.

Uses

Because thallium is recovered from smelting lead and zinc concentrates, it is available to fulfill any new uses up to several thousand pounds per year. To date, practical applications have been relatively limited. Thallium-activated sodium iodide crystals find use in photomultiplier tubes. It has been learned that thallium bromoiodide crystals transmit infrared radiation and that crystals of thallium oxysulfide detect infrared radiation. A combination of these crystals has been used in military communication systems. Because of their density, both thallous formate and thallous malonate have been used in the preparation of heavy-liquid sink-float solutions used in the gravity separation of minerals. Mixtures of thallium, arsenic, sulfur, and selenium form low-melting-point glasses for encapsulation of semiconductors has been under investigation. It has been found that the addition of small amounts of thallium to the counterelectrode alloy used in selenium rectifiers will improve the performance of the rectifiers. Claims have been made that the addition of a thallium salt to absorb traces of oxygen in tungsten-filament incandescent lamps will increase lamp life. Also, it has been shown that the addition of thallium to various glass formulations will improve optical properties and increase the refractive index.

For a number of years, thallium sulfate had been used in rodenticides. Some use of thallium has been made in connection with alloys for low-temperature applications, particularly for switches, seals, and thermometers. The ternary eutectic mercury-thallium-indium alloy has a freezing point of −63.3°C, while the binary eutectic mercury-thallium alloy has a freezing point of −60°C. These freezing points are considerably lower than that of mercury usually used for similar applications at higher temperatures. Mercury freezes at −38.87°C.

Toxicity

Thallium and thallium compounds are toxic and skin contact must be avoided. Impervious gloves and aprons should be worn and excellent ventilation and masks should be provided where dusts and fumes may be present.

Chemistry and Compounds

Oxidation states: thallous, Tl^+; thallic, Tl^{3+}.

Because of low oxidation potential of thallium to form Tl^+, thallium is quite reactive, dissolving slowly in most dilute mineral acids to form thallium(I) solutions. The thallium(I) halides are insoluble in water, but thallium trihalides are soluble; the latter are formed by treatment of the thallium(I) halide in solution with the corresponding halogen. Thallium(III) iodide, however, does not exist, TlI_3 being $[Tl^+][I_3^-]$.

Thallium(III) compounds are readily reduced to the thallium(I) state (see difference in oxidation potentials above) and are thus fairly strong oxidizing agents. Thallium(I) compounds resemble those of the alkali metals in many respects, including a soluble, strong basic hydroxide (TlOH) ($K_B = 0.14$), a soluble carbonate (Tl_2CO_3), the formation of well crystallized salts, including those with complex anions, the formation of polysulfides (Tl_2S_5), and polyiodides (thus TlI_3 contains the monovalent metal ion, like rubidium iodide, RbI_3 and cesium iodide, CsI_3). Thallium(I) ion resembles silver in forming insoluble halides, sulfide and chromate. The thallium(I) ion forms only weak complexes (probably because of its larger size and low charge) but the thallium(III) ion forms strong ones. There are four complex chloro ions $[TlCl_4]^-$, $[TlCl_5]^{2-}$, $[TlCl_6]^{3-}$, and $[Tl_2Cl_9]^{3-}$, the last having the six-coordinated structure

$$\begin{bmatrix} Cl & & Cl & & Cl \\ Cl-Tl & \cdot\cdot\cdot & Cl-Tl & \cdot\cdot\cdot & Cl \\ Cl & & Cl & & Cl \end{bmatrix}^{3-}$$

The complex compounds include the chelates, such as the oxine chelate, and also such compounds as $Tl(TlCl_4)$, $Tl_3(TlBr_6)$, $TlCl_3 \cdot 3NH_3$, $14Rb_3TlBr_6 \cdot 16H_2O$ (here the presence of the ion $[Tl(H_2O)_8]^+$ has been shown), oxalates, such as $H[Tl(C_2O_4)_2]$ (dioxalatothallic acid),

K2[Tl(C₂O₄)₂(NO₂)₂] ·H₂O and a number of complex hydrides, as well as the unstable binary hydride TlH₃.

The most readily prepared organometallic compounds are the dialkyl ones of the type R_2TlX, where X is an acid radical accompanying the ion $[R_2Tl]^+$. The trialkyl compounds of the type TlR_3 are immediately decomposed by H_2O, giving RH and R_2TlOH, in which the thallium atom is isoelectronic with the mercury atom in R_2Hg, and which is a strong base: $(C_2H_5)_2TlOH$, $K_B = 0.90$.

S. C. CARAPELLA, JR.
ASARCO Incorporated
South Plainfield, New Jersey

Additional Reading

Carter, G.F. and D.E. Paul: *Materials Science and Engineering,* ASM International, Materials Park, OH, 1991.

Greenwood, N.N. and A. Earnshaw: *Chemistry of the Elements,* 2nd Edition, Butterworth-Heinemann, Inc., Woburn, MA, 1997.

Hermann, A.M. and J.V. Yakhmi: *Thallium-Based High-Temperature Super Conductors,* Marcel Dekker, Inc., New York, NY, 1994.

Krebs, R.E.: *The History and Use of Our Earth's Chemical Elements: A Reference Guide,* Greenwood Publishing Group, Inc., Westport, CT, 1998.

Lewis, R.J. and N.I. Sax: *Sax's Dangerous Properties of Industrial Materials,* 10th Edition, John Wiley & Sons, Inc., New York, NY, 1999.

Lide, D.R.: *CRC Handbook of Chemistry and Physics,* 84th Edition, CRC Press, LLC., Boca Raton, FL, 2003.

Nriagu, J.O.: *Thallium in the Environment,* Vol. 30, John Wiley & Sons, Inc., New York, NY, 1998.

Staff: *ASM Handbook—Properties and Selection: Nonferrous Alloys and Pure Metals,* ASM International, Materials Park, OH, 1990.

THEORETICAL PLATE. Any contacting device in a fractionating column, such as packing, grids, or screens, that effects the same degree of separation of vapor from liquid as one simple distillation. A column that gives the same separation as 10 successive simple distillations is considered to have 10 theoretical plates. The effectiveness of a fractionating column is measured in terms of theoretical plates. As many as 100 theoretical plates are used in laboratory and industrial operation. The total column height divided by the number of theoretical plates is known as HETP (height equivalent to a theoretical plate). This concept is also used in chromatographic techniques.

THERMAL CRACKING.   See **Cracking Process**.

THERMAL EXPANSION COEFFICIENT.   The change in volume per unit volume per degree change in temperature (cubical coefficient). For isotropic solids the expansion is equal in all directions, and the cubical coefficient is about three times the linear coefficient of expansion. These coefficients vary with temperature, but for gases at constant pressure the coefficient of volume expansion is nearly constant and equals 0.00367 for each degree Celsius at any temperature.

THERMAL INSULATION.   See **Insulation (Thermal)**.

THERMAL RADIATION.   All bodies that are not at absolute zero emit radiation excited by the thermal agitation of their molecules or atoms, whether there are other causes of excitation or not. This thermal radiation ranges in wavelength from the longest infrared to the shortest ultraviolet rays, its spectral energy distribution, however, depending upon the nature of the body and upon its temperature. The total emissive power of a surface at any temperature is the rate at which it emits energy of all wavelengths and in all directions, per unit area of radiating surface. The flux density (per unit solid angle) in various directions obeys the cosine emission law approximately; but strictly only in the case of a black body.

THERMIONIC CONVERSION.   The process whereby electrons released by thermionic emission are collected and utilized as electric current. The simplest example of this is provided by a vacuum tube, in which the electrons released from a heated anode are collected at the cathode or plate. Used as a method of producing electrical power for spacecraft.

THERMIONIC EMISSION.   Direct ejection of electrons as the result of heating and material, which raises electron energy beyond the binding energy that holds the electron in the material.

THERMITE.   A mixture of ferric oxide and powdered aluminum, usually enclosed in a metal cylinder and used as an incendiary bomb, invented by the German chemist Hans Goldschmidt around 1900. On ignition by a ribbon of magnesium, the reaction produces a temperature of 2200°C, which is sufficient to soften steel. This is typical of some oxide/metal reactions that provide their own oxygen supply and thus are very difficult to stop.

THERMOCHEMISTRY.   That aspect of chemistry which deals with the heat changes which accompany chemical reactions and processes, the heat produced by them, and the influence of temperature and other thermal quantities upon them. It is closely related to chemical thermodynamics. The heat of formation of a compound is the heat absorbed when it is formed from its elements in their standard states. An exothermic reaction evolves heat; and endothermic reaction requires heat for initiation.

THERMOCOUPLE.   In 1821, Seebeck discovered that an electric current flows in a continuous circuit of two metals if the two junctions are at different temperatures, as shown in Fig. 1. A and B are two metals, T_1 and T_2 are the temperatures of the junctions. I is the thermoelectric current. A is thermoelectrically positive to B if T_1 is the colder junction. In 1834, Peltier found that current flowing across a junction of dissimilar metals causes heat to be absorbed or liberated. The direction of heat flow reverses if current flow is reversed. Rate of heat flow is proportional to current but depends upon both temperature and the materials at the junction. Heat transfer rate is given by PI, where P is the Peltier coefficient in watts per ampere, or the Peltier emf in volts. Many studies of the characteristics of thermocouples have led to the formulation of three fundamental laws:

1. *Law of the Homogeneous Circuit.* An electric current cannot be sustained in a circuit of a single homogeneous metal; however it may vary in section, by the application of heat alone.
2. *Law of Intermediate Metals.* If in any circuit of solid conductors the temperature is uniform from any point P through all the conducting matter to a point Q, the algebraic sum of the thermoelectromotive forces in the entire circuit is totally independent of this intermediate matter and is the same as if P and Q were put in contact.
3. *Law of Successive or Intermediate Temperatures.* The thermal emf developed by any thermocouple of homogeneous metals with its junctions at any two temperatures T_1 and T_3 is the algebraic sum of the emf of the thermocouple with one junction at T_1 and the other at any other temperature T_2 and the emf of the same thermocouple with its junctions at T_2 and T_3. See Fig. 2.

Common thermocouple wire combinations used in industry are listed in Table 1. A choice of different metals is needed to fulfill a broad range of temperatures as well as for oxidizing or reducing conditions in use. The temperature-thermal emf curves for common types of thermocouples are

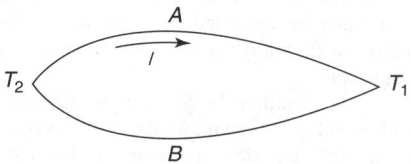

Fig. 1.   Simple thermocouple circuit

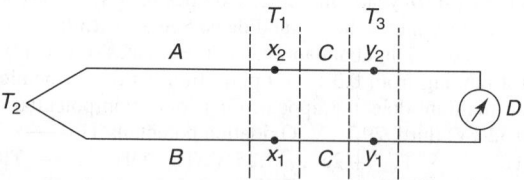

Fig. 2.   Law of Intermediate Metals makes it possible to use "foreign" wires to connect thermocouple to measuring instrument. Thermocouple materials A and B can be connected to the instrument by use of connecting materials C and D. If the temperatures at X_1 and X_2 are both at T_1 and if temperatures at Y_1 and Y_2 are both at T_3, the emf of the circuit will be independent of materials C and D

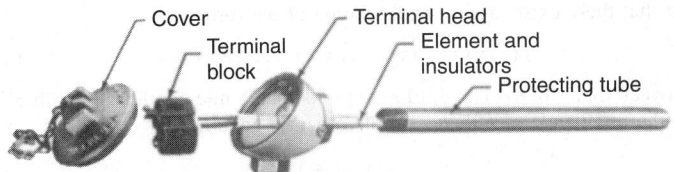

TABLE 1. COMMONLY USED THERMOCOUPLES AND TEMPERATURE RANGES

ANSI type	Positive element	Negative element	Normal temperature range	
			°F	°C
B	Platinum 20% Rhodium	Platinum 6% Rhodium	1,600–3,100	870–1,700
E	Chromel	Constantan	32–1,600	0–870
J	Iron	Constantan	32–1,400	0–760
K	Chromel	Alumel	32–2,300	0–1,260
R	Platinum 13% Rhodium	Platinum	32–2,700	0–1,480
S	Platinum 10% Rhodium	Platinum	32–2,700	0–1,480
T	Copper	Constantan	−300– +700	−180– +370

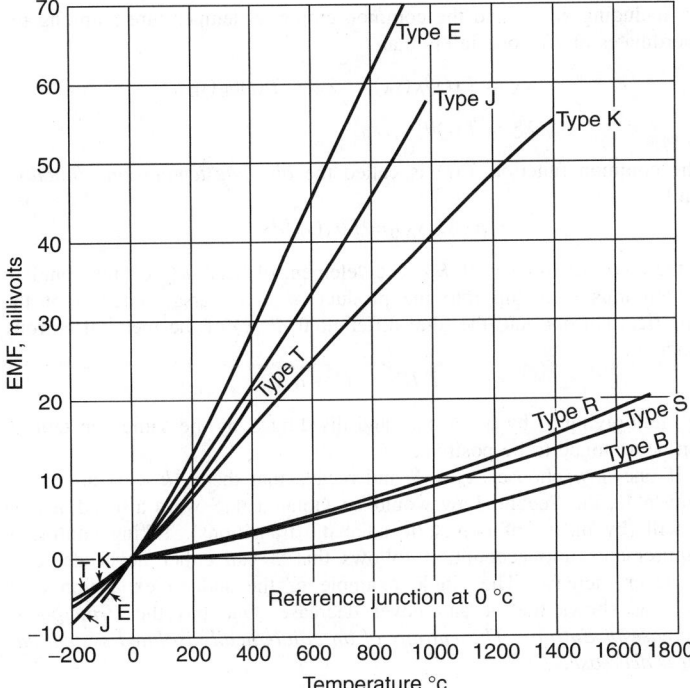

Fig. 3.   Temperature-thermal emf curves for common types of thermocouples. (*Honeywell*)

given in Fig. 3. The hot junction of a thermocouple may be joined by any means that will ensure good electrical continuity when in use. Commonly, the two wires are twisted together and either welded or silver-soldered. Simple clamping of wires together provides adequate connection for short-term use in clean atmospheres at lower temperatures. For industrial applications, the thermocouple is usually placed within a protecting tube. A typical assembly is shown in Fig. 4. For lower temperatures, carbon steel may be used. As the temperature goes up, wrought iron, stainless steel, nickel, nickel-chromium-iron, fused silica, silica-alumina, silicon carbide, alumina, and beryllia may be used. Beryllia protecting tubes will withstand operating temperatures of up to 4,000°F (~2,200°C). For some applications, disposable-tip thermocouples have been developed. These are particularly effective for high-temperature molten-metal temperature measurements.

The emf developed by a thermocouple depends upon the temperature of both the measuring and reference junctions. Thus, to determine temperature, the following data must be known: (1) the calibration data for the particular thermocouple; (2) the measured emf; and (3) the temperature of the reference junction. In laboratory cases, the reference junction can be maintained at the freezing temperature of water. However, in most modern instruments, the ambient temperature of the reference junction is sensed, and the correction is incorporated in the measurement circuitry.

Multiple thermocouples may be used in parallel and connected to a single instrument. A typical application is a fire-warning system. Multiple

thermocouples also may be connected in series. This is a means for obtaining the average temperature of an object. Also two or more thermocouples may be connected in series so that the emf outputs of the couples are additive. In application installations, thermopiles sometimes are used to detect the presence or absence of the pilot flame and cause a relay to turn off the main gas supply valve.

Additional Reading

Jackson, D.A. and A.E. Mushin: *Thermocouples*, in Process Instruments and Controls Handbook, 3rd Edition (D.M. and G.D. Considine, Eds.), The McGraw-Hill Companies, New York, NY, 1985.
Kerlin, T.W.: *Practical Thermocouple Thermometry*, ISA, Research Triangle Park, NC, 1998.
Pollock, D.D.: *Thermocouples: Theory and Properties*, CRC Press, LLC., Boca Raton, FL, 1991.
Staff: *Temperature Measurement Handbook*, Omega Press, Stamford, CT, 1982.
Staff: *Manual on the Use of Thermocouples in Temperature Measurements*, STP 470B, American Society for Testing and Materials, Philadelphia, PA, 1993.

THERMODYNAMICS.   Classical thermodynamics is a theory which on the basis of four main laws and some ancillary assumptions deals with general limitations exhibited by the behavior of macroscopic systems. Phenomenologically it takes no cognizance of the atomic constitution of matter. All *mechanical* concepts such as kinetic energy or work are presupposed. Thermodynamics is motivated by the existence of dissipative mechanical systems. A *thermodynamic system* K may be thought of as a collection of bodies in bulk; when its condition is found to be unchanging in time (on a reasonable time scale) it is *in equilibrium*. It is then characterized by the values of a finite set of, say, n physical quantities, it being supposed that none of these is redundant. Such a set of quantities constitutes the *coordinates* of K, denoted by $x(= x_1, \ldots, x_n)$. Any set of values of these is a *state* $\mathscr{G}$ of K. In virtue of these definitions, K is in a state only when it is in equilibrium. The passage of K from a state $\mathscr{G}1$ to a state $\mathscr{G}'$ is a *transition* of K. A transition is *quasi-static* if in its course it goes through a continuous sequence of states, and if the forces that do work on the system are just those which hold it in equilibrium. A transition is *reversible* if there exists a second transition that restores the initial state, the final condition of the surroundings of K being the same as the initial condition. Reversible transitions are assumed to be quasi-static.

An enclosure, such that the equilibrium of a system contained within it can only be disturbed by mechanical means, is *adiabatic*, otherwise it is *diathermic*. For instance, stirring, or the passage of an electric current, constitute "mechanical means." A system K_0 in an adiabatic enclosure is *adiabatically isolated*, but this does not preclude mechanical interactions with the surroundings. Its transitions are then called adiabatic.

For the time being, the masses of all substances present will be supposed fixed, and to achieve simplicity it will be given that (1) there are no substances present whose properties depend on their previous histories, and (2) capillary forces as well as long-range interactions are absent. Further, it will be supposed that of the n coordinates of K, just $n - 1$ have geometrical character (*deformation coordinates*, e.g., volumes of enclosures), so that the work done by K in a quasi-state transition is

$$\int dW = \int \sum_{k=1}^{n-1} P_k(x)dx_k \tag{1}$$

Such a system will be called a *standard system* ($n - 1$ enclosures in diathermic contact, each containing a simple fluid, may serve as example, x_n being any one of the pressures).

The Zeroth Law

Suppose two systems $K_A(x)$ and $K_B(y)$ to be in mutual diathermic contact. Experience shows that the states $\mathscr{G}_A$ and $\mathscr{G}_B$ cannot be assigned arbitrarily, but that there exists a necessary relation of the form

$$f(x; y) \equiv f(x_1, \ldots, x_n; y_1, \ldots, y_m) = 0 \qquad (2)$$

between them. If K_C is a third system, its diathermic equilibrium with K_B on the one hand, or with K_A on the other, is governed by conditions

$$g(y; z) = 0 \qquad (3)$$

and

$$h(z; x) = 0 \qquad (4)$$

respectively. That these three functions are not independent is expressed by the *Zeroth Law: If each of two systems is in equilibrium with a third system then they are in equilibrium with each other.* It follows that any two of Equations (2) through (4) imply the third, i.e., they must be equivalent to equations of the form

$$\xi(x) = \eta(y) = \zeta(z) \qquad (5)$$

Thus, with each system there is now associated a function, its *empirical temperature function*, such that two systems can be in equilibrium if and only if their *empirical temperatures* (i.e., the values of their empirical temperature functions) are equal. Write $t = \xi(x)$; so that one has the *equation of state* of K_A. Also, t may be introduced in place of any one of the x_k. Note that the empirical temperature is not uniquely determined since $t_A = t_B$ may be replaced by $\phi(t_A) = \phi(t_B)$ where the function ϕ is monotonic but otherwise arbitrary: one has a choice of *temperature scales*. For a system not in equilibrium, temperature is not defined.

The First Law

It is obvious that one can do mechanical work upon a system (say by stirring) while its initial and final states are the same. (Nothing is being said about the surroundings!) In this sense mechanical energy is not conserved. One might, however, hope that it is conserved at least in a restricted class of transitions. That this is so is asserted by the *First Law: The work W_0 done by a system K_0 in an adiabatic transition depends on the terminal states alone.* Thus, if $\mathscr{G}'(x')$, $\mathscr{G}''(x'')$ are the terminal states

$$W_0 = F(x'; x'')$$

If $\mathscr{G}'''(x''')$ is a third state, and the previous transition proceeds via $\mathscr{G}'''$, W_0 must not depend on x''', i.e.,

$$F(x'; x''') + F(x'''; x'') \equiv F(x'; x'')$$

It follows that there must exist a function $U(x)$, defined to within an arbitrary additive constant, such that

$$F(x'; x'') = U(x') - U(x'')(= -\Delta U, \text{ say})$$

$U(x)$ is the *internal energy function* of K. (To make sure that U is in fact defined for all states, one assumes that *some* adiabatic transition always exists between any pair of given states.) The energy of a compound standard system is the sum of the energies of its constituent standard systems. Further, U must be a monotonic function of t, and it is convenient to choose the scale of t such that $\partial U/\partial t > 0$.

When the transition from $\mathscr{G}'$ to $\mathscr{G}''$ is adiabatic, $W_0 + \Delta U$ vanishes by definition of U. If the transition is not adiabatic and W is the work done by K, the quantity

$$\Delta U + W(= Q, \text{ say}) \qquad (6)$$

will in general fail to vanish. Q is then called the *heat absorbed* by K. Every element of a quasistatic adiabatic transition is subject to $dQ = 0$, i.e., by Equations (1) and (6), to the differential equation

$$\sum_{k=1}^{n-1} \left(P_k(x) + \frac{\partial U(x)}{\partial x_k} \right) dx_k + \frac{\partial U}{\partial t} dt = 0 \qquad (7)$$

The Second Law

Experiment shows that if $\mathscr{G}'$ and $\mathscr{G}''$ are arbitrarily prescribed states, then it may be that no adiabatic transition from $\mathscr{G}'$ to $\mathscr{G}''$ exists. When this is the case one says that $\mathscr{G}''$ is *inaccessible* from $\mathscr{G}'$, but $\mathscr{G}'$ is then accessible from $\mathscr{G}''$, as has been already assumed. The states may of course happen to be mutually accessible. The existence of states adiabatically inaccessible from a given state is asserted precisely by the *Second Law: In every neighborhood of any state $\mathscr{G}$ of an adiabatically isolated system there are states inaccessible from $\mathscr{G}$.* (This formulation of the Second Law

is known as the *Principle of Carathéodory.*) A fortiori this law applies to quasistatic transitions, i.e., those which satisfy Equation (7). It asserts there are states $\mathscr{G}''$ near $\mathscr{G}'$ such that no functions $x_k(t)$ exist which satisfy Equation (7) and whose values when $t = t''$ are just x_k'', $(k = 1, \ldots, n-1)$. It is merely a mathematical problem (the Theorem of Carathéodory) to prove that this is the case if and only if there exist functions $\lambda(x)$ and $s(x)$, $(x_n \equiv t)$ such that the left-hand member is identically equal to λds, where ds is the total differential of s. Thus, the Second Law entails that

$$dQ = dU + dW = \lambda ds \qquad (8)$$

(dQ is of course not a total differential), s is called the *empirical entropy function* of K. It is not uniquely determined, since it may be replaced by any monotonic function of s. If two standard systems K_A and K_B in diathermic contact make up a compound system K_C; $dQ_C = dQ_A + dQ_B$, i.e., because of Equation (8),

$$\lambda_A ds_A + \lambda_B ds_B = \lambda_C ds_C$$

By including s_A, s_B and the common empirical temperature t among the coordinates of K_C, one infers that

$$\lambda_A = T(t)\theta_A(s_A), \quad \lambda_B = T(t)\theta_B(s_B),$$

$$\lambda_C = T(t)\theta(s_A, s_B)$$

The common function $T(t)$ is called the *absolute temperature function*, while

$$S_A(s_A) = \int \theta_A(s_A) ds_A$$

is the *metrical entropy* of K_A The "element of heat" dQ of any standard system thus splits up into the product of a universal function of the empirical entropy and the total differential $dS(x)$ of the metrical entropy function:

$$TdS = dU + dW \qquad (9)$$

By multiplying T by a constant and dividing S by the same constant, T can be arranged to be positive.

If one now chooses $x_n = S$ and recalls that the $x_k(k < n)$ are freely adjustable, the Second Law would be violated if S were also adjustable at will (by means of non-static adiabatic transitions). Taking continuity requirements into account, it follows that S can either never decrease or never increase. The single example of the sudden expansion of a real gas shows that it can never decrease. One has the *Principle of Increase of Entropy: The entropy of an adiabatically isolated system can never decrease.*

The Third Law

It is known from experiment that for given values of the deformation coordinates, the energy function has a lower bound U_0. The question rises whether the entropy S has an analogous property. It is found in practice that the specific heats $\partial U/\partial T$ of all substances appear to go to zero at least linearly with T as $T \to 0$. This ensures that the function S goes to a finite limit S_0 as $T \to 0$. Experiment shows, however, further that as $T \to 0$, the derivatives of S with respect to the deformation coordinates also go to zero. In contrast with U_0, S_0 has therefore the remarkable property that it is independent of the deformation coordinates. One thus arrives at the *Third Law: The entropy of any given system attains the same finite least value for every state of least energy.* One immediate consequence of this is that the so-called *classical ideal gas* (the product of whose volume V and pressure P is proportional to T, and whose energy is a function of T only) cannot exist in nature. Further, no system can have its absolute temperature reduced to zero. The Third Law is therefore a statement about the properties of functions, not of systems, at $T = 0$.

The practical applications of the theory just outlined divide themselves into two broad classes: (1) Those which are based on the existence and properties of the functions U and S and some others related to them—all "thermodynamic identities" being merely the integrability condition for the total differentials of these functions; and (2) those which are based on the Principle of Increase of Entropy: the entropy of the actual state of an adiabatically enclosed system being greater than that of any neighboring "virtual" state.

The most important of the auxiliary functions just mentioned are the *Helmholtz Function*:

$$F = U - TS \qquad (10)$$

the *Gibbs Functions*:

$$G = U - TS + \sum_{k=1}^{n-1} P_k x_k \tag{11}$$

the *Enthalpy*:

$$H = U + \sum_{k=1}^{n-1} P_k x_k \tag{12}$$

sometimes called *thermodynamic potentials*. Then, e.g.,

$$dF = -SdT - dW$$

F therefore contains all available quantitative information about K, since

$$S = -\frac{\partial F}{\partial G}, \text{ and } P_k = -\frac{\partial F}{\partial x_k} \tag{13}$$

The same is true of G for instance, since

$$S = -\frac{\partial G}{\partial T}, \text{ and } x_k = -\frac{\partial G}{\partial P_k}$$

F and G are naturally taken as functions of $x_1, \ldots, x_{n-1}, T$ and of $P_1, \ldots, P_{n-1}, T$, respectively. At times one speaks of F as the "Helmholtz free energy" and of G as the "Gibbs free energy." In an *isothermal* reversible transition, the amount W of work done by a system is equal not to the decrease of its energy U but to the decrease $-\Delta F$ of its (Helmholtz) free energy. In the presence of internal sources of irreversibility

$$W < -\Delta F$$

In considering physicochemical equilibria, that is to say, if one is interested in the internal constitution of a system in equilibrium when changes of phase and chemical reactions are admitted, one introduces the *constitutive coordinates* n_i^α; this being the number of moles of the ith constituent C_i in the αth phase. The definitions of Equations (10) through (12) remain unaltered, for the n_i^α do not enter into the description of the interaction of the system with its surroundings. Let an amount dn_i^α of C_i be introduced quasistatically into the αth phase of the system. The work done on K shall be $\mu_i^\alpha dn_i$. The quantity μ_i^α so defined is the *chemical potential* of C_i in the αth phase. It is in general a function of all the coordinates of K. Then, identically.

$$dG = \sum_{k=1}^{n-1} x_k dP_k - SdT + \sum_i \sum_a \mu_i^\alpha dn_i^\alpha$$

Integrability conditions such as

$$\partial \mu_i^\alpha / \partial T = -\partial S / \partial n_i^\alpha$$

are applications of the first kind. On the other hand, the minimal property of G, derived from the maximal property of S, requires that

$$\sum_i \sum_a \mu_i^\alpha dn_i^\alpha = 0$$

when all virtual states differ only in the values of the constitutive coordinates. If the system is chemically inert, the dn_i^α are subject only to the requirements of the conservation of matter. One then concludes that if there are c constituents and p phases, i.e., $n + pc$ coordinates in all, then the number f of these to which arbitrary values may be assigned is

$$f = c - p + n$$

This typical application of the second kind is the Gibbs Phase Rule (for inert systems). This rule is often stated merely for systems with only two external coordinates ($n = 2$, e.g., $x_i = P, x_2 = T$). There must then be no internal partitions within the system, nor may it, for instance, contain magnetic substances in the presence of external magnetic fields.

The beauty and power of phenomenological thermodynamics lies just in the generality and paucity of its basic laws which hold independently of any assumptions concerning the microscopic structure of the systems which they govern. Its quantitative content is limited to conditions of equilibrium. Its conceptual framework is too narrow to permit the description of the temporal behavior of systems, except to the extent that it makes it possible to decide which, of any pair of states of an adiabatically enclosed system, must have been the earlier state.

Statistical thermodynamics seeks to remedy these deficiencies by making specific assumptions about the microscopic structure of the system K, and relating its macroscopic behavior to that of its atomic constituents. K is then to be regarded as an *assembly* of a very large number of particles, which, on a non-quantal level, is a mechanical system with, say, N degrees of freedom. A *microstate* of K is a set of values of its N coordinates and its N conjugate momenta. It is out of the question to measure all these at a given time. One therefore constructs a *representative ensemble* ε_K of K, which is an abstract collection of a very large number of identical copies of K. At any time t, the members of ε_K will be in different microstates. Let the fractional number of members of the ensemble whose microstates lie in the range dp, dq about p, be $\phi dp dq$. Then ϕ is the *probability-in-phase*, and with $d\Gamma = dpdq$

$$\int \phi d\Gamma = 1 \tag{14}$$

The reason for this terminology is implicit in the *Postulate: The probability that a given assembly K will, at time t, be in a microstate lying in the range $d\Gamma$ about p, q, is equal to the probability $\phi d\Gamma$ that the microstate of a member of ε_K selected at random at time t, lies in the same range.*

The mean value $\langle f \rangle$ of a dynamical quantity f is defined to be

$$\langle f \rangle = \int f \phi d\Gamma$$

If N is sufficiently large, fluctuations about the mean will usually be negligible.

When K is in equilibrium, ϕ must be constant in time, and this will be the case if it is a function of the (time-independent) Hamiltonian H of K. Ensemble averages are now assumed to coincide with temporal averages. When, in particular, K is in diathermic equilibrium with its surroundings one can show that ϕ must have the form

$$\phi = \exp[(\phi - H)/\phi] \tag{15}$$

where ϕ and θ are independent of p, q. Then

$$\theta \langle \ln \theta \rangle = \phi - \langle H \rangle \tag{16}$$

and, because of Equation (14)

$$d \int \exp[(\phi - H)/\theta] d\Gamma = 0 \langle d[(\phi - H)/\theta]$$

where d refers to a variation of the macroscopic coordinates of K. Using Equation (16) and its variation, the relation

$$-\theta d \langle \ln \phi \rangle = d \langle H \rangle - \langle dH \rangle \tag{17}$$

follows. Now $\langle H \rangle (= \overline{U}$, say) is the total energy of the assembly, while $\langle dH \rangle$ is the average of the change of the potential energy, i.e., the work $-dW$ done by the external forces on K. If one writes

$$\overline{S} = -k \langle \ln \phi \rangle$$

where k is a constant, Equation (17) becomes

$$k^{-1} \theta d\overline{S} = d\overline{U} + dW$$

This is identical with the phenomenological relation of Equation (9) if one formally identifies S with $\overline{S} U$ with $\overline{U}$ and θ with kT. In this way, contact with the phenomenological theory has been established, and the quantities characteristic of the one theory have been *correlated* with that of the other. With this correlation, or interpretation, ϕ becomes F. However, because of Equations (14) and (15)

$$F = -kT \ln \int \exp(-H/kT) d\Gamma$$

so that if only H is known, the integral on the right (the *partition function*), and thus F, can be calculated. The equation of state of a real gas can thus in principle be obtained from a knowledge of the forces operating within the assembly. This illustrates how the additional information put into the theory yields a correspondingly greater output. Phenomenologically, such an equation of state might be written as

$$PV = \sum_{n=1}^{\infty} B_n(T) V^{1-n}$$

but here each of the *virial coefficients* $B_1, B_2, \ldots$ must be measured separately.

See also **Heat**; and **Heat Transfer**.

Additional Reading

Carter, A.H.: *Classical and Statistical Thermodynamics,* Prentice-Hall, Inc., Upper Saddle River, NJ, 2000.

Cengel, Y.A. and M.A. Boles: *Thermodynamics: An Engineering Approach,* 4th Edition, The McGraw-Hill Companies, Inc., New York, NY, 2001.

Granet, I., and M. Bluestein: *Thermodynamics and Heat Power,* Prentice Hall, Inc., Upper Saddle River, NJ, 2003.

Hudson, J.B.: *Thermodynamics of Materials: A Classical and Statistical Synthesis,* John Wiley & Sons, Inc., Hoboken, NJ, 2004.

Koretsky, M.: *Engineering and Chemical Thermodynamics,* John Wiley & Sons, Inc., New York, NY, 2003.

Mansoori, G.A.: *Thermodynamics: The Application of Classical and Statistical Thermodynamics to the Prediction of Equilibrium Properties,* Taylor & Francis, Inc., Philadelphia, PA, 1991.

Russell, L.D.: *Classical Thermodynamics,* Oxford University Press, Inc., New York, NY, 1995.

Sandler, S.L.: *Chemical and Engineering Thermodynamics,* 3rd Edition, John Wiley & Sons, Inc., New York, NY, 1998.

Sonntag, R.E. and G.J. Van Wylen: *Introduction to Thermodynamics: Classical and Statistical,* 3rd Edition, John Wiley & Sons, Inc., New York, NY, 1991.

THERMOELECTRIC COOLING. Like conventional refrigeration systems, thermoelectric systems obey the same basic laws of thermodynamics. Both in principle and result, thermoelectric cooling has much in common with conventional refrigeration methods. In a conventional refrigeration system, the main working parts are the freezer, condenser, and compressor. The freezer surface is where the liquid refrigerant boils, changes to vapor, and absorbs heat energy. The compressor circulates the refrigerant above ambient level. The condenser helps to discharge the absorbed heat into surrounding ambient. In thermoelectric refrigeration, the refrigerant in both liquid and vapor forms is replaced by two dissimilar conductors. The freezer surface becomes cold through absorption of energy by electrons as they pass from one semiconductor to another, instead of energy absorption by the refrigerant as it changes from liquid to vapor. The compressor is replaced by a direct current power source which pumps the electrons from one semiconductor to another. A heat sink replaces the conventional condenser fins, discharging the accumulated heat energy from the system.

The components of a thermoelectric cooler are indicated by the cross section of a typical unit shown in Fig. 1. Thermoelectric coolers such as this are actually small *heat pumps* that operate on the physical principles well established over a century ago. Semiconductor materials with dissimilar characteristics are connected electrically in series and thermally in parallel, so that two junctions are created. The semiconductor materials are *n*- and *p*-type and are so named because either they have more electrons than necessary to complete a perfect molecular lattice structure (*n*-type), or not enough electrons to complete a lattice structure (*p*-type). The extra electrons in the *n*-type material and the holes left in the *p*-type material are called *carriers* and they are the agents that move the heat energy from the cold to the hot junction.

Heat absorbed at the cold junction is pumped to the hot junction at a rate proportional to carrier current passing through the circuit and the number of couples. Good thermoelectric semiconductor materials, such as bismuth telluride, greatly impede conventional heat conduction from hot to cold areas, yet provide an easy flow for the carriers. In addition, these materials have carriers with a capacity for carrying more heat. Only since the refinement of semiconductor materials in the early 1950s has thermoelectric refrigeration been considered practical for many applications.

In practical use, couples are combined in a module where they are connected in series electrically and in parallel thermally. See Fig. 2. Normally, a module is the smallest component available. The user can tailor quantity, size, or capacity of the module to fit exact requirements without procuring more total capacity than is actually required. Modules are available in a variety of sizes, shapes, operating currents, operating voltages, number of couples, and ranges of heat-pumping levels. The present trend is toward a larger number of couples operating at a low current.

Thermoelectric coolers find three basic categories of applications. (1) Use in electronic components; (2) in temperature control units; and (3) in medical and laboratory instruments.

Modules normally contain from 2 to 71 couples with ceramic-metal laminate plates. If modules are to be used in cooling chambers of large components, a total surface area of virtually any size can be made by placing the appropriate number of modules side by side.

The interfaces at the cold junction and the hot junction must be constructed to transfer heat in and out of the module with little difference in temperature. This is accomplished with metal-ceramic laminate plates that give strength and permit good thermal bonding between the two interfaces. The outer plate surface is usually tinned to facilitate soldering to heat sinks. Where soldering is not practical, as in the case of thermal expansion differences, heat transfer grease is recommended. Epoxy bonding agents are available where a more permanent solderless bond is required.

The single-stage module is capable of pumping heat where the difference in temperature of the cold junction and hot junction is 70°C or less; however, in those applications requiring higher delta T_s, the modules can be cascaded. Cascading is a mechanical stacking of the modules so that the cold junction of one module becomes the heat sink for a smaller module placed on top. In addition to the heat pumped by any given stage, the next lower stage must also pump the heat resulting from the input power to that upper stage. Consequently, each succeeding stage must be larger and larger from the top of the cascade downward.

With any given set of heat sink and cold spot temperatures, there exists an optimum heat-pumping capacity or "size" ratio between each adjacent pair of stages. The optimum size ratio increases as the overall delta T increases, but decreases as the number of stages increases. It is not necessarily a constant from stage to stage, even with delta T and number of stages fixed. True optimization of a cascade design requires accurate temperature-dependent data on the thermoelectric materials in combination with a computerized numerical design theory.

Applications requiring low-temperature thermoelectric coolers usually have strict limitations on available power. Therefore, it is not practical to fabricate and stock numerous different cascades that can be optimized for only one set of conditions. On the other hand, fully optimized prototypes involve engineering and manufacturing costs that may prove uneconomical for some applications. Therefore, a low-cost alternative approach has been developed for responding to such requirements. Standard cascades are fabricated by assembling *partials* of standard modules. The number of

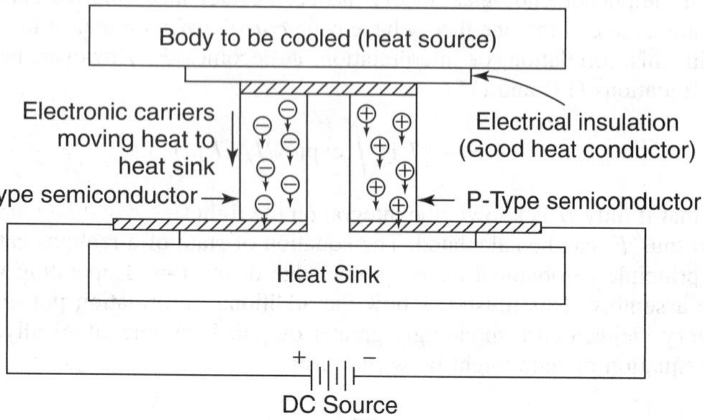

Fig. 1. Cross section of thermoelectric cooler

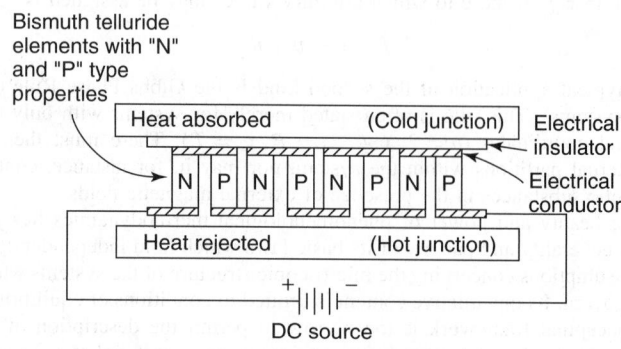

Fig. 2. Thermoelectric module assembly. Elements are electrically in series, thermally in parallel

different standard cascades is virtually unlimited due to "free" variables, such as number of stages, couple distribution, and the basic building block module. In order to determine the best standard cascade for a given application, the desired hot-side temperature, cold-side temperature, and thermal load are entered into a computer system. The result is a list of numerous standard coolers, which meet these specifications with various combinations of input power and cost. Generally, the lowest input power devices are of higher cost and vice versa.

The heat sink design is very important. The heat sink must carry heat away with minimum rise of temperature. It should be stressed that all the thermoelectric cooler does is to move energy from the load to the heat sink where it must be dissipated to another medium, the latter required to be cooler than the hot junction.

Power supply capabilities for thermoelectric coolers range from the simple open-loop direct current supply with a switch to sophisticated feedback systems with close temperature regulation and fast response. The only limitation on the supply is that ripple be maintained at a point lower than 10 to 15%. Open-loop systems will generally contain a transformer, rectifiers, choke, and chassis with heat sink for the rectifiers. In feedback systems, a thermistor is used to sense temperature at the cold junction. This signal is compared with the desired temperature setting to obtain an error signal.

THERMOELECTRICITY. Electricity produced directly by applying a temperature difference to various parts of electrically conducting or semiconducting materials. Usually two dissimilar materials are used, and the points of contact are kept at different temperatures (Peltier effect). Many temperature-measuring devices (thermocouples, thermopiles) work on this principle, since the voltage is proportional to the temperature difference. Metallic conductors are usually used for these "thermometers," which produce a rather small current. A newer use for the effect is as a source of electrical energy, i.e., a means of direct conversion of heat into electricity (or vice versa) without the use of a generator (or motor). The materials used for these thermoelectric couples are semiconductors (e.g., tellurium, zinc antimonide; lead, bismuth, and germanium tellurides; samarium sulfide) or thermoelectric alloys, all of which produce relatively large currents. Several of these "cells" are then hooked in series much like the cells of a battery.

THERMOFOR PROCESS. A moving-bed catalytic cracking process in which petroleum vapor is passed up through a reactor countercurrent to a flow of small beads or catalyst. The deactivated catalyst then passes through a regenerator and is recirculated.

THERMOGRAVIMETRIC ANALYSIS. This analytical technique (TGA), also sometimes referred to as thermogravimetry, is a method whereby the weight and temperature of a sample under test are continuously recorded as the sample is heated at a constant and linear rate. The heating, usually within the range from ambient up to 1,100 or 1,500°C, is achieved by a furnace, which surrounds, but does not touch, the sample. Thus, the sample remains freely suspended from the balancing mechanism, which is actuated as the sample mass alters in response to chemical reactions produced as its temperature is progressively increased. Weight-loss curves are preferably produced by means of an X-Y recorder. Of course, the term *weight-loss* must be used with reservation because in some atmospheres a sample may gain weight. Indirectly, TGA information can be applied to studies of rates of reaction and energies of activation for the reaction, sublimation, or vaporization of chemical compounds and minerals. Weight gains are caused by the adsorption and absorption of gases by solid samples, direct reaction as in oxidation, corrosion, recarbonation, and hydration, or the reaction of gases to produce solids. Conversely, weight losses are produced by the desorption of gases, dehydration, vaporization, sublimation, gaseous desolvation, and gas-liberating reactions of both organic and inorganic substances. Specific fields that have widely used TGA include: studies involving the thermal stability of minerals and mineral mixtures; the pyrolysis of coals, petroleum, and cellulose; the analysis of soils; roasting and calcining; thermochemical reactions of ceramics and cements; dehydration hygroscopicity studies; solid-state reactions; effect of radiation on various materials; corrosion studies; and the detection of short-lived unstable intermediate compounds.

THERMOMETER. An instrument that measures temperature. A thermometer may take advantage of one of several physical properties of materials that change when they are subjected to a change in temperature. Liquids, gases, and solids expand with increasing temperature; decrease in volume with decreasing temperature. Thus, there are liquid-in-glass thermometers, which depend upon the volumetric relation of mercury or other liquids with temperature. The difference in the energy radiating from materials at various temperatures is the basis of optical and radiation pyrometers. Bimetallic thermometers depend upon the differing expansiveness of different metals. Changes in electrical resistance with temperature are utilized in resistance thermometers and thermistors. Thermocouples depend upon the Seebeck, Peltier, and Thomson effects, wherein the emf in an electrical circuit comprising dissimilar metals bears a relationship to the temperature difference between a cold junction and the temperature being measured.

Several kinds of thermometers are described in this volume. See alphabetical index.

THERMOMETER (Filled-System). A representative filled-system thermometer is shown in Fig. 1. The temperature-sensing element (bulb) contains a fluid that changes its volume or pressure with temperature. The pressure-sensitive element (bourdon) responds to these changes by delivering a motion or force to a device that transduces the signal to a usable form. This is commonly a mechanical linkage which drives a pointer or pen, but may be a pneumatic or electric device which transmits the temperature signal over long distances. These signals frequently are used for process control purposes.

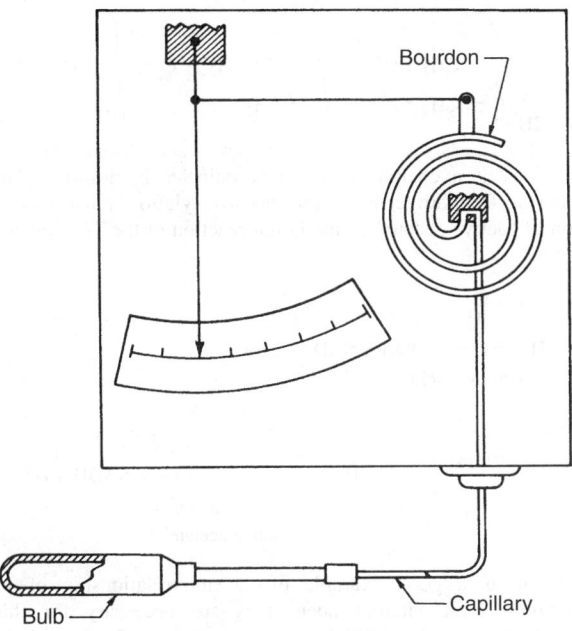

Fig. 1. Filled-system thermometer

Filled-system thermometers may be placed into one of two fundamental categories: (1) Those in which the bourdon responds to a volume change; and (2) those in which the bourdon responds to a *pressure change*. Those that respond to volume changes are completely filled with a liquid. The variation in liquid expansiveness with temperature is greater than that of the bulb metal, the net volume change being communicated to the bourdon. In this design, an internal-system pressure change is not of primary importance. In those systems that respond to pressure changes, the bulb is either filled with a gas, or partially filled with a volatile liquid. Changes in gas or vapor pressure with changes in bulb temperature are communicated to the bourdon. The bourdon will increase in volume with increase in pressure, but in this design, this effect is not of primary importance.

In liquid-filled thermal systems, mercury or various organic fills are used. In vapor-filled systems, a number of hydrocarbons, including ethane, ethyl chloride, ethyl ether, chlorobenzene, and propane, among others, are used. Nitrogen or other fully-dried and purified gases may be used in a gas thermometer.

THERMONUCLEAR FUSION REACTORS. See **Lithium; Nuclear Power Technology.**

THERMOPLASTIC. A high polymer that softens when exposed to heat and returns to its original condition when cooled to room temperature. Natural substances that exhibit this behavior are crude rubber and a number of waxes; however, the term is usually applied to synthetics such as polyvinyl chloride, nylons, fluorocarbons, linear polyethylene, polyurethane prepolymer, polystyrene, polypropylene, and cellulosic and acrylic resins.

See also **Plastics**.

THERMOSET. A high polymer that solidifies or "sets" irreversibly when heated. This property is usually associated with a cross-linking reaction of the molecular constituents induced by heat or radiation, as with proteins, and in the baking of dough. In many cases, it is necessary to add "curing" agents such as organic peroxides or (in the case of rubber) sulfur. For example, linear polyethylene can be cross-linked to a thermosetting material by either radiation or chemical reaction. Phenolics, alkyds, amino resins, polyesters, epoxies, and silicones are usually considered to be thermosetting, but the term also applies to materials in which additive-induced cross-linking is possible, e.g., natural-rubber.

THERMOSETTING RESINS. See **Plastics**.

THIAMINE (Vitamin B₁). Some earlier designations for this substance included aneurin, antineuritic factor, antiberiberi factor, and oryzamin. Thiamine is metabolically active as thiamine pyrophosphate (TPP), the formula of which is:

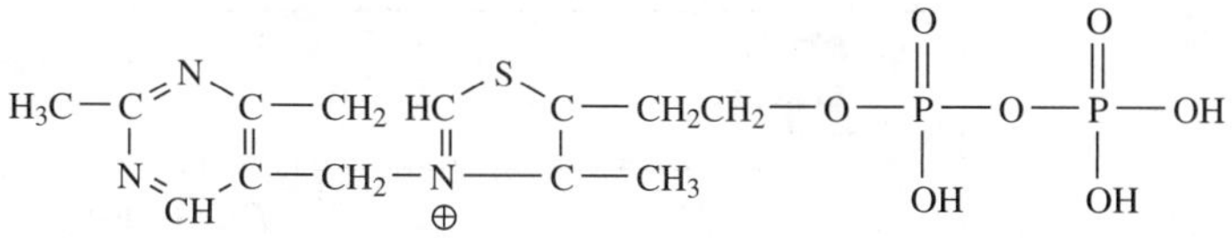

TPP functions as a coenzyme which participates in decarboxylation of α-keto acids. Dehydrogenation and decarboxylation must precede the formation of "active acetate" in the initial reaction of the TCA cycle (citric acid cycle):

This reaction is a good example of the interrelationship of vitamin B coenzymes. Four vitamin coenzymes are necessary for this one reaction: (1) thiamine (in TPP) for decarboxylation; (2) nicotinic acid in nicotinamide adenine dinucleotide (NAD); (3) riboflavin in flavin adenine dinucleotide (FAD); and (4) pantothenic acid in coenzyme A (CoA) for activation of the acetate fragment.

TPP also mediates the oxidative decarboxylation of α-ketoglutaric acid, another intermediate of carboxydrate metabolism in the citric acid cycle. The nutritional requirement for thiamine increases as dietary carbohydrate increases because of a greater demand for TPP.

The structure of thiamine hydrochloride is:

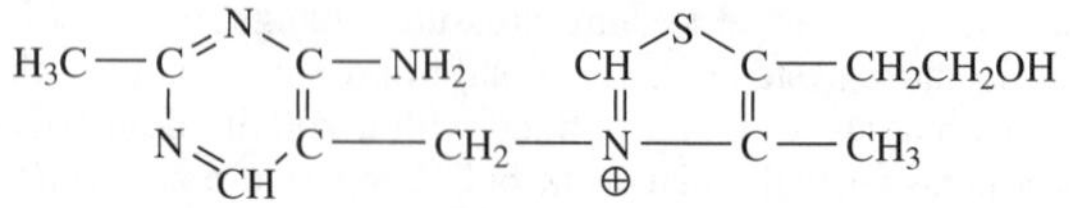

In this form and as other salts, such as thiamine mononitrate, the vitamin is available as a dietary supplement.

Diseases and disorders resulting from a deficiency of thiamine include beriberi, opisthotonos (in birds), polyneuritis, hyperesthesia, bradycardia, and edema. Rather than a specific disease, beriberi may be described as a clinical state resulting from a thiamine deficiency. In body cells, thiamine pyrophosphate is required for removing carbon dioxide from various substances, including pyruvic acid. Actually, this is accomplished by a decarboxylase of which thiamine pyrophosphate is a part. Where

thiamine is deficient, the process of oxidation necessary for converting food into energy is impeded, causing a variety of manifestations throughout the body.

In so-called *dry* beriberi, pathologic alterations in neurons and nerve fibers occurs, leading in some instances to degeneration of peripheral nerves. This condition is termed peripheral neuritis, generally affecting the nerves in the arms and legs. There often is altered skin sensitivity to touch in the extremities and pain on pressure over large nerves. There is a gradual loss of muscle strength, which may lead to paralysis of a limb. In *wet* beriberi, there is a lessening of strength of the heart muscles. There is enlargement of the heart, dyspnoea, increased pulse rate, palpitation, and edema. Pathologically, degenerative changes are found in the nervous tissue, heart muscle, and gastrointestinal tract. In later stages, marked enlargement of the heart and liver may be noted.

In one form of thiamine deficiency, Wernicke's syndrome may be noted. Therein is paralysis, or weakness of the muscles that causes motion of the eyeball. Closely associated with thiamine deficiency are dietary problems of alcoholism. The psychotic disturbances of alcoholism, including delirium tremens, frequently respond to thiamine and other B complex vitamins. Injections of thiamine often produce dramatic improvements in persons suffering from beriberi. Beriberi sometimes occurs in infants who are breast-fed by mothers who suffer a thiamine deficiency. Beriberi remains of concern in the Orient where polished rice is a dietary staple.

In cattle, a thiamine deficiency causes podioencephalomalcia (PEM), characterized by blindness, decreased feed intake, incoordination, failure of rumen to contract, spasms, and paralysis. In swine, a deficiency retards growth and sometimes causes cyanosis (insufficient oxygen in blood), enlarged heart, accompanied by fatty degeneration of heart muscles. Chicks suffer from paralysis of peripheral nerves, causing polyneuritis (head drawn back).

Distribution and Sources

Relatively few natural foods are considered high in thiamine content.

High thiamine content (1,000–10,000 micrograms/100 grams). Ham, rice bran, soybean flour, wheat germ, yeast.

Medium thiamine content (100–1000 micrograms/100 grams). Almond, asparagus, barley, brazil nut, bean (kidney, lima, snap, soy, wax), beef, beet greens, broccoli, Brussels sprouts, carp, cashew, cauliflower, chicory, chestnut, chicken, clam, cod, corn (maize), dandelion greens, eggs, endive, gooseberry, groundnut (peanut), hazelnut, kale, kohlrabi, leek, lentil (dry), lobster, mackerel, milk, mushroom, oats, oyster, parsley, pea, pecan, plum, pork, potato, prune (dry), raisin (dry), rice (brown), salmon, turkey, veal, walnut, watercress.

Low thiamine content (10–100 micrograms/100 grams). Apple, apricot, artichoke, avocado, banana, beet, berry (black-, blue-, cran-, rasp-, straw-), cabbage, carrot, celery, cheeses, cherry, coconut, cucumber, currant, date (dry), eggplant, fig, flounder, grape, grapefruit, haddock, halibut, herring, lemon, lettuce, melons, orange, parsnip, peach, pear, pepper (sweet), pike, pineapple, prune, sardine, scallop, shrimp, tangerine, trout, tuna, turnip.

Commercial thiamine dietary supplements are prepared by synthesis: Pyrimidine + thiazole nuclei synthesized separately and then condensed; also build on pyrimidine with acetamidine. Precursors in the biosynthesis of thiamine include thiazole and pyrimidine pyrophosphate, with thiamine phosphate as an intermediate. In plants, production sites are found in grain and cereal germ.

Bioavailability of Thiamine

Factors which contribute to a lessening of thiamine bioavailability include: (1) cooking, inasmuch as the vitamin is heat labile and water soluble; (2) presence of certain enzymes in food, such as thiaminase for vitamin breakdown; (3) destruction by calcium carbonate, dibasic potassium phosphate, and manganous sulfate; (4) destruction by nitrites and sulfites; (5) diuresis and gastrointestinal diseases; and (6) presence of live yeasts and alkalis. An increase in availability can result from: (1) presence of cellulose in diet, which increases intestinal synthesis; (2) storage capacity in heart, liver, and kidney; and (3) stimulation of bacterial synthesis in intestine (normally none).

Antagonists of thiamine include pyrithiamine, oxythiamine, and 2-*n*-butyl homologue. Synergists include vitamins B₂, B₆, B₁₂, and niacin, pantothenic acid, and somatotrophin (growth hormone).

Unusual features of thiamine as observed by some researchers include: (1) it exerts a hormonal function in plants, controlling root growth; (2) it aids phosphorylation in liver, dephosphorylation in kidney; (3) it easily poisoned by heavy metals, acetyl iodide; (4) plant and animal cocarboxylases are identical; (5) it exerts a diuretic effect and it is constipating; (6) it can be allergenic on injection; (7) it is not available from intestinal bacteria; and (8) blood contains most cocarboxylase in leukocytes.

Thiamine is soluble in water and easily destroyed by heat. These two properties account for appreciable losses of thiamine from processed and stored foods. An acid medium favors the retention of thiamine, whereas an alkaline medium is detrimental to retention.

Determination of Thiamine

Bioassay methods include yeast fermentation; polyneuritic rate of cure in rat; bacterial metabolism. Physicochemical methods include thiochrome fluorescence; polarography; chromatography; absorption in neutral and acid solutions.

THIAZIDES. See **Diuretic Agents**.

THIN. A nontechnical word used by scientists with a variety of meanings. (1) In electronic metallurgy a thin film is a vapor-deposited coating having a thickness of only a single atom; such *monatomic* films, e.g., thorium on tungsten, are used in electronic devices such as cathodes. (2) A coating or film of fatty acid on water that is one molecule thick (about 200 Å) is called *monomolecular* film. (3) In thin-layer chromatography the term applies to a specially prepared mixture of adsorbents spread on a glass slide to a thickness of 1/100 inch. (4) The word is also used in the sense of a liquid of low viscosity, as in paint thinner and thin-boiling starch.

THIN FILMS. The term *thin films* is used for a wide variety of physical structures. Self-supporting solid sheets usually are called foils when thinned from thicker material by such methods as rolling, beating, or etching, and films when obtained by stripping a deposited layer from its substrate. Supported thin films are deposited on planar or (in special cases) curved substrates by such methods as vacuum evaporation, cathode sputtering, electroplating, electroless plating, spraying, and various chemical surface reactions in a controlled atmosphere or electrolyte. Thicknesses of such supported films range from less than an atomic monolayer to a few micrometers (1 μm = 10^{-4} centimeter). Thin films not forming a continuous sheet are called "island films." Particularly noble metals may condense as islands of considerable thickness (up to ~10^2 micrometer).

In scientific studies and technical applications, the use of well-controllable deposition methods such as vacuum evaporation and cathode sputtering are generally preferred. The film structure is markedly influenced by such deposition parameters as substrate composition and surface structure, source and substrate temperatures, deposition rate, and composition and pressure of the ambient atmosphere (where applicable). In general, the structure of films is more disordered than the corresponding bulk material. Smaller grains, higher dislocation concentrations, and deviations from stoichiometry are typical, and films approach bulk structure only as a limiting case. Under certain growth conditions, films exhibit preferential crystal orientations or even epitaxy. (Epitaxy means that the film structure is determined by the crystal structure and orientation of the underlying substrate.)

Solid thin films are common study objects in most phases of solid state physics. They supply the samples for the study of general structural and physical properties of solid matter where special beam methods require small quantities of material or extremely thin layers. For example, thin films are used in transmission electron microscopy and diffraction, neutron diffraction, ultraviolet spectroscopy, and X-ray diffraction and spectroscopy. Thin films represent the best means for studying physical effects, where these effects are caused by the extreme thinness of the material itself. Examples are the rotational switching of ferromagnetic films, electron tunneling phenomena, electromagnetic skin effects of various kinds, and certain optical interference phenomena. Films also are convenient vehicles for the investigation of nucleation and crystal growth, and for states of extremely disturbed thermodynamic equilibrium.

Presently, films find three major industrial uses: the decorative finishing of plastics, optical coatings of various kinds (mainly antireflection coatings, reflection increasing films, multilayer interference filters, and fluorescent coatings), and in electronic components.

Nucleation, Growth and Mechanical Properties of Films

In vacuum evaporation, molecules or atoms of thermal energy are deposited at a uniform angle of incidence and under well-defined environmental conditions. Most nucleation and growth studies, therefore, have been made on evaporated films. A particle approaching the substrate enters close to its surface a field of attracting short-range London forces with an exchange energy proportional to—$1/r^6$. At a still shorter distance r, repulsive forces proportional to $e^{-r/\text{constant}}$ resist the penetration of the electron clouds of the surface atoms. Due to the atomic or crystalline structure of the substrate, this potential field exhibits periodicity or quasiperiodicity in the substrate plane. The freshly condensed particles migrate over the surface with a jump frequency $i_D \propto \exp(-Q_D/kT)$, or desorb with a frequency $i_{ad} \propto \exp(-Q_{ad}/kT)$, where the activation energy Q_D is often approximately one-fourth of Q_{ad}. Permanent condensation occurs in most cases at distinct nucleation centers which may consist of deep potential wells of the substrate, clusters of condensed particles, or previously deposited "seed" particles of a different material. The number of nuclei formed in the second case is strongly temperature and rate dependent.

Most metals always condense in crystalline form, but the grain size is extremely small at low temperatures (in the order of a few micrometers) and increases markedly with increasing substrate temperatures. Grain size decreases with increasing deposition rates. The condensation of amorphous or quasi-liquid phases at low temperatures has been observed for such metals as antimony and bismuth and a few dielectrics. Some of these materials, on annealing, pass through otherwise unobserved and probably metastable phases.

Stresses of considerable magnitude are often observed in deposited films. The main causes of these stresses are a mismatch of expansion coefficients between substrate and film, enclosed impurity atoms, a high concentration of lattice defects and in very thin films, a variety of surface effects. Often, the stresses resulting from lattice defects can be minimized by the choice of a higher substrate temperature during deposition, or they can be reduced by a post-deposition anneal. Metal films frequently exhibit tensile strengths that are considerably larger than those of the corresponding bulk materials.

Thin-film Optics

Deposited metal mirrors probably represent the oldest optical application of films. High-quality mirrors usually are produced by the vacuum evaporation of aluminum on an appropriately shaped glass substrate. Often, a glow-discharge cleaning of the substrate or a chromium undercoat is first applied to increase the adhesion of the aluminum. After deposition, the aluminum is protected by anodic oxidation or an evaporated overcoat of SiO, SiO_2, or Al_2O_3.

For SiO, maximum reflectance in the visible spectral region is achieved at a thickness of about 1400 micrometers. Rapid SiO evaporation reduces the reflectance at shorter wavelengths.

Single or multilayer coatings find increasing use as optical interference filters. These film stacks may consist solely of transparent films of different refractive indices n_f, or a combination of absorbing and nonabsorbing layers. Common low-index materials for glass coatings in the visible region of the spectrum are MgF_2 (n_f = 1.32 to 1.37), and cryolite Na_3AlF_6 (n_f = 1.28 to 1.34); high-index materials are SiO (n_f = 1.97), ZnS ($n_f \approx$ 2.34), TiO_2 (n_f = 2.66 to 2.69) and CeO_2 (n_f = 2.2 to 2.4). The indices are given for the sodium D-line. Various semiconductors are used for infrared coatings.

At each air-film, film-film, or film-substrate interface, the incident light amplitude is split into a reflected and a transmitted fraction according to the Fresnel coefficients

$$f_{j-1} = (\hat{N}_{j-1} - \hat{N}_j)/(\hat{N}_{j-1} + \hat{N}_j) \text{ and } g_{j-1} = 2\hat{N}_{j-1}/(\hat{N}_{j-1} + \hat{N}_j)$$

where j and $j-1$ denote the number of the optical layer counted from the side of the incident beam, $\hat{N}_j = N/\cos\Theta_j$ for p polarization or $\hat{N}_j = N_j/\cos\Theta_j$ for s polarization is the effective refractive index, and

$N_j = n_j - ik_j$ the refractive index of the j layer.

$$\cos \Theta_j = \sqrt{\left(\sqrt{p_j^2 + q_j^2} + p_j\right)/2} - i\sqrt{\left(\sqrt{p_j^2 + q_j^2} + p_j\right)/2}$$

$$p_j = 1 + (k_j^2 - n_j^2)[n_0 \sin \theta_0/(n_j^2 + k_j^2)]^2$$

$$q_j = -2n_j k_j [n_0 \sin \theta_0/(n_j^2 + k_j^2)]^2$$

The symbol θ_0 is the angle of incidence in the incident medium.

For nonabsorbing film stacks ($k_i = 0; i = 1, 2 \ldots, m + 1$), the overall reflectance and transmittance may be obtained by summing the multiple coherent reflections between the film boundaries. A more general treatment based on electromagnetic theory yields for amplitude reflectance and transmittance the recursions formulas

$$r_{(j-1)-} = (f_{j-1} + r_{j-} \exp(-2i\hat{\Phi}))/(1 + f_{j-1} r_{j-} \exp(-2i\hat{\Phi}_j))$$

and

$$t_{(j-1)-} = (g_{j-1} t_{j-} \exp(-i\hat{\Phi}_j))/(1 + f_{j-1} r_{j-} \exp(-2i\hat{\Phi}_j))$$

$\hat{\Phi}_j = \Phi_j \cos \Theta_j$ is the effective phase thickness. $\Phi_j = (2\pi/\lambda) \cdot N_j l_j$ where λ is the wavelength in vacuo, and l_j is the geometrical film thickness. The recursion is started on the side of emergence, using the initial conditions $r_{m-} = f_m$ and $t_{m-} = g_m$. Intensities are given by $R = |r_{0-}|^2$ and $T = (RN_{m+1}/n_0)|t_{0-}|^2$ where R denotes "real part of." If A_j is the absorption in the layer j, $R + T + \Sigma_j A_j = 1$.

A single antireflection coating of $\lambda/4$ optical thickness $n_f l_f$ yields zero reflectance at $n_f = \sqrt{n_{glass}}$. A double layer coating of $\lambda/4$ films requires $n_2/n_1 = \sqrt{n_g}$. The transmission of a Fabry-Perot interference filter consisting of a dielectric spacer layer between two partially reflecting metal films is given by $I/I_0 = [(1 + A/T)^2 + (4R/T^2) \sin^2(\delta - 1)]^{-1}$ where $\Phi = 2\pi nl \cos \theta/\lambda$. R, T, and A are the reflection, transmission, and absorption coefficients of the reflecting layers. The refractive index and thickness of the spacer film are n and l. θ is the angle of refraction in the spacer, and δ the phase change for reflections at the spacer-metal film interfaces. $(I/I_0)_{max} = (T/(1 - R))^2$ and $(I/I_0)_{min} = (T/(1 + R))^2$. The band pass half-width is $\Delta\lambda_{1/2}; \lambda(1 - R)/m\pi R^{1/2}$ for the interference order $m(m\pi = \Phi)$. More complex coatings and filters, and their various applications, cannot be discussed here. It should be mentioned, however, that films play a very important role today in the accurate determination of the optical constants of many materials, but particularly of metals.

Thin-Film Electronics

Deposited dielectric film materials are SiO, MgF$_2$, ZnS, and various organic compounds. Thin capacitive layers in the 100 to 500 micrometers thickness region are often produced by the anodization of tantalum and aluminum to Ta$_2$O$_5$ or Al$_2$O$_3$, respectively. The breakdown strength and dielectric constant of films approach bulk values, but might be reduced by surface roughness, structural faults, and lower density. According to the Lorentz-Lorenz formula, the dielectric constant D changes with reduced density ρ as $dD/d\rho = 3C/(1 - C\rho)^2$, where C is a constant depending on the material. On metal-dielectric-metal films, quantum mechanical tunneling through the dielectric film becomes observable below a dielectric thickness of about 100 micrometers. For applied voltages less than the metal-insulator work function ϕ, the tunneling current density J is proportional to the applied voltage V, demonstrating that the low-voltage tunneling resistance is ohmic. $J = (qV/h^2 s) \times (2m^*\phi)^{1/2} \exp[-(4\pi s/h)(2m^*\phi)^{1/2}]$. At high applied voltages ($qV > \phi$), the current increases very rapidly:

$$J = (q^2 V^2/8\pi h\phi_s^2) \exp[-(8\pi s/3hqV)(2m^*)^{1/2}\phi^{3/2}].$$

s is the insulator thickness, m^* the electronic effective mass, and q the electron charge. Thicker dielectric films may exhibit in high fields appreciable Schottky or avalanche currents when they are greatly disordered.

Polycrystalline metal films generally show, due to their low structural order, a larger resistivity than the bulk material. According to Matthiessen's rule, the total resistivity can be expressed as $\rho = \rho(t) + \rho(i)$ where $\rho(t)$ is the temperature-dependent resistivity associated with scattering by lattice vibrations, and $\rho(i)$ is a temperature-independent resistivity caused by impurity or imperfection scattering. Very thin specimens with a thickness comparable to the electron mean free path show a $\rho(i)$ rapidly increasing with decreasing thickness. This increase is caused by an increasing contribution of non-specular electron scattering at the film surfaces. By annealing a metal film, $\rho(i)$ might be reduced permanently. A large $\rho(i)$ results in a small temperature coefficient α.

Many known superconductors can be deposited as superconductive films.

Through thin-film experiments, the energy gap in semiconductors can be measured, and material parameters, such as the penetration depth of magnetic fields, can be studied at dimensions less than the coherence range.

Studies of semiconductor films have shown many facets. The properties of epitaxial films have mainly been investigated on Ge and Si, and to a lesser degree on III–V compounds. Much work has been done on polycrystalline II–VI films, particularly with regard to the stoichiometry of the deposits, doping and post-deposition treatments, conductivity and carrier mobility, photo-conductance, fluorescence, electroluminescence, and metal-semiconductor junction properties. Among other semiconductors, selenium, tellurium, and a few transition metal oxides have found some interest.

Film resistors, capacitors, and interconnected R-C networks on planar glass or ceramic substrates are finding widespread industrial use. Common resistor materials are carbon, nichrome, and tin oxide in individual components; and nichrome, tantalum, tantalum nitride, SiO-chromium cermet, and cermet glazes in planar networks. Gold, copper, aluminum, or tantalum is used for termination lands, connection leads, and capacitor plates. SiO, MgF$_2$, and Ta$_2$O$_5$ serve as film capacitor dielectrics and crossover insulation. The geometrical configuration of the desired component or circuit pattern is obtained either by deposition through mechanical masks or by removing from a continuous sheet the undesired portions after the deposition process is completed. This removal is frequently accomplished by a combination of photolitho-graphic and etch processes.

The minimum length l and width w of a resistor are calculated from the given resistance R, the sheet resistance R in ohms per square, dissipated power P, and permissible power dissipation per square inch P by use of the formulas $w = \sqrt{(P \cdot R)/P \cdot R}$ and $l = wR/R$. The capacitance of film capacitors is given by $C = 0.225D(N - 1)A/t$, where C is the capacitance in picofarads, D the dielectric constant, N the number of plates, A the area in square inches, and t the dielectric thickness in inches.

In retrospect, it is gratifying to note how much progress in thin-film electronics was made prior to a more penetrating and fundamental understanding of surface phenomena. It is only relatively recently that such experimental tools as angle-resolved photoelectron spectroscopy, synchrotron far-ultraviolet and X-ray spectroscopies, and back-scattering and channeling of energetic ions, have been used. New computational methods are now available for calculating detailed maps of the electron distribution at a surface. It has been found that the surface geometry is either a relaxed version of the bulk structure, or a reconstructed arrangement with symmetry wholly different from the bulk. A comparison of the electron spectroscopy results with theoretical predictions of surface state energy spectra based on a particular surface model allows confirmation of the assumed atomic arrangement. Together, the theoretical and experimental approaches have provided the first complete description of surfaces, including identification of the atomic species present, their atomic arrangement, and the distribution of valence electrons in space and energy. It is quite possible that these approaches will lead to a detailed picture at the atomic level of the interface structure, electron states, charges, reconstruction, and the related junction electronic properties of semiconductor-semiconductor and metal-semiconductor interfaces.

Various barrier layer diodes have exhibited impressive rectification ratios, but limited breakdown strength and low speed due to their large specific capacitance. Of the many film transistor concepts studied, the insulated gate field effect device has been promising. Its structure consists of a minute metal-dielectric-semiconductor capacitor. The semiconductor strip carries current between two terminals called source and drain. A field applied between metal "gate" and source modulates the semiconductor conductance and consequently the source-drain current. Usable semiconductor materials with a sufficiently low concentration of interface states are CdS, CdSe, and tellurium. These devices exhibit pentodelike characteristics with voltage gains ranging from 2.5 at 60 megahertz to 8.5 at 2.5 megahertz. The gain-bandwidth product GB, which is equal to the transconductance divided by 2π times the gate capacitance, reaches values of about 20 megahertz. It is determined by $GB = \mu_d V_D/2\pi L^2$, where μ_d is the effective drift mobility of the electrons, V_D the source-drain potential, and L the source-drain spacing, which is usually selected between 5 and 50 micrometers.

An outgrowth of prior thin-film technology and of basic surface science research has been molecular beam epitaxy—the MBE formation of compound semiconductor films. The MBE technology involves the use of separate atomic and molecular beams from multiple thermal sources in high vacuum which irradiate a substrate at intensities selected to grow films having the desired composition and doping. The ability to achieve slow growth rates, together with independent control of the separate beam sources, permits the fabrication of semiconductor junction profiles, both in doping and in composition, with a precision approaching that of a single atomic layer. To date MBE has been used to prepare films and layer structures involving a number of GaAs and $Ga_xAl_{1-x}As$ devices. Included in such devices are varactor diodes having highly controlled hyperabrupt capacitance-voltage characteristics, IMPATT diodes, microwave mixer diodes, Schottky barrier field-effect transistors (FETs), injection lasers, optical waveguides, and integrated optical structures. Some authorities believe that the potential for MBE in solid-state electronics may be greatest for microwave and optical solid-state devices as well as for circuits where submicrometer layer structures are required. A recently demonstrated MBE GaAs Schottky barrier diode cryogenic mixer with a noise temperature of 315 K at 102 gigahertz is exemplary of the potential of MBE technology for millimeter wave electronics.

MBE superlattice structures also are very promising. These superlattice structures, with periodicities of 50–100 micrometers, show negative resistance characteristics attributed to resonant tunneling into the quantized energy states associated with the narrow potential wells formed by the layers. Detailed studies have shown that the potential well distributions may be controlled and positioned to a precision of a few atomic layers.

Thin-film technology has also played an important role in developing Josephson superconducting devices, which offer outstanding advantages in constructing ultrahigh-speed computers. These are tunnel-junction type devices.

Thin-film and surface phenomena are fundamental to the successful development, production, and use of solid-state devices. The research in this area is extensive. See also **Molecular and Supermolecular Electronics**.

Magnetic Films

Magnetic thin films of nickel-iron (usually deposited at an 80:20 composition by weight) exhibit a number of unusual properties, which have led to many experimental and theoretical studies, as well as to important applications in binary storage and switching, magnetic amplifiers, and magneto-optical Kerr-effect displays.

Such "Permalloy" films have two stable states of magnetization, corresponding to positive and negative remanence. When deposited in a magnetic field or at an oblique angle, they exhibit uniaxial anisotropy. In practice, this anisotropy shows some dispersion, since it results from the alignment of local lattice disturbances. The stable states result from the minimization of the free energy $E = MH_L \cos\theta - MH_T \sin\theta + K\sin^2\theta$, where the last term represents the anisotropy energy, and θ is the angle between the magnetization M and the easy axis. From an inspection of the derivatives of this equation follows the hard-direction straight-line and the easy-direction square hysteresis loops of aniso-tropic films. In the latter case, the magnetization is always either $+M$ or $-M$, and the change occurs at $H_L = \pm H_K$. The transitions from unstable to stable states occur at $\partial^2 E/\partial\theta^2 = 0$, resulting in a critical curve $H_L^{2/3} + H_T^{2/3} = H_K^{2/3}$ which has the form of an asteroid enclosing the origin (see also **Magnetism**).

An important feature of magnetic films is the high speed with which the state of magnetization can be reversed. Dependent on film properties and magnetic fields, three modes of magnetization reversal occur: Domain wall motion, incoherent rotations, and the extremely fast coherent rotation of the magnetization. Wall motion switching is expected when the driving fields are smaller than the critical values. During the past decade, magnetic garnet films have gained prominence in a number of research and industrial applications.

See also **Magnetic Materials**.

Additional Reading

Brundle, C.R. and S. Wilson: *Encyclopedia of Materials Characterization: Surfaces, Interfaces, Thin Films,* Butterworth-Heinemann, Inc., Woburn, MA, 1992.

Cohen, E.D.: "Coatings: Going Below the Surface," *Chem. Eng. Progress,* 19 (September 1990).

Elshabini-Riad, A.R. and F.D. Barlow: *Thin Film Technology Handbook,* The McGraw-Hill Companies, Inc., New York, NY, 1996.

Ferendeci, A.M.: *Physical Foundations of Solid State and Electron Devices,* The McGraw-Hill Companies, Inc., New York, NY, 1991.

Fink, D.G. and H.W. Beaty: *Standard Handbook for Electrical Engineers,* 14th Edition, The McGraw-Hill Companies, Inc., New York, NY, 1999.

Feldman, L.C. and J.W. Mayer: *Fundamentals of Surface Thin Film,* Prentice-Hall, Inc., Upper Saddle River, NJ, 1998.

Karim, A. and S. Kumar: *Polymer Surfaces, Interfaces and Thin Films,* World Scientific Publishing Company, Inc., River Edge, NJ, 1999.

Lisenskey, G.C., et al.: "Electro-Optical Evidence for the Chelate Effect at Semiconductor Surfaces," *Science,* 840 (May 18, 1990).

Matacotta, F.C. and G. Ottaviani: *Science and Technology of Thin Films,* World Scientific Publishing Company, Inc., River Edge, NJ, 1995.

Mittal, K.L. and P. Kumar: *Emulsions, Foams, and Thin Films,* Marcel Dekker, Inc., New York, NY, 2000.

Ohring, M.: *The Materials Science of Thin Films,* Harcourt Brace & Company, San Diego, CA, 1991.

Sayer, M. and K. Sreenivas: "Ceramic Thin Films: Fabrication and Applications," *Science,* 1056 (March 2, 1990).

Scriven, L.E. and W.J. Suszynski: "Take a Closer Look at Coating Problems," *Chem. Eng. Progress,* 24 (September 1990).

Staff: *Future Directions in Thin Film,* World Scientific Publishing Company Inc., River Edge, NJ, 1997.

Staff: *Range of Critical Temperatures Observed for Superconductive Elements in Thin Films Condensed Usually at Low Temperatures,* in Handbook of Chemistry and Physics, CRC Press, Boca Raton, Florida, 73rd Edition (1992–1993).

Stuart, R.V.: *Vacuum Technology, Thin Films and Sputtering: An Introduction,* Academic Press, Inc., San Diego, CA, 1983.

Venables, J.A.: *Introduction to Surface and Thin Film Processes,* Cambridge University Press, New York, NY, 2000.

Vossen, J.L. and W. Kern: *Thin Film Processes II,* Academic Press, Inc., San Diego, CA, 1991.

Williams, E.D. and N.C. Bartelt: Thermodynamics of Surface Morphology, *Science,* 393 (January 25, 1991).

THIN-LAYER CHROMATOGRAPHY (TLC). A micro type of chromatography. The thin layer (0.01 inch) is the adsorbent, usually a special silica gel spread on glass or incorporated in a plastic film. Single drops of the solutions to be investigated are placed along one edge of the glass plate, and this edge then dipped into a solvent. The solvent carries the constituents of the original test drops up the thin layer in a selective separation, so that a comparison with known standards and various identifying tests may be made on the spots formed.

See also **Thin**.

THINNER. A hydrocarbon (naphtha) or oleoresinous solvent (turpentine) used to reduce the viscosity of paints to appropriate working consistency usually just before application. In this sense, a thinner is a liquid diluent, except that it has active solvent power on the dissolved resin.

THIOALCOHOLS. See **Mercaptans**.

THIO- AND DITHIOCARBAMIC ACIDS. See **Herbicides; Insecticide**.

THIOCYANATES AND ISOTHIOCYANATES. See **Herbicides; Insecticide**.

THIOCYANIC ACID. [CAS: 463-56-9]. Aqueous solution of hydrogen thiocyanate, HSCN, formula weight 59.08, yellow solid below mp 5°C, unstable gas at room temperature. The acid is moderately stable only when dilute and cold. The salts of this acid are known as thiocyanates.

Thiocyanic acid is formed by reaction of barium thiocyanate solution and dilute sulfuric acid, and filtering off barium sulfate, or by the action of hydrogen sulfide on silver thiocyanate, filtering off silver sulfide.

Sodium, potassium, barium, or calcium thiocyanate may be made by reaction of sulfur and the corresponding cyanide by heating to fusion. Ammonium thiocyanate (plus ammonium sulfide) may be made by reaction of ammonia and carbon disulfide, a reaction which probably accounts for the presence of ammonium thiocyanate in the products of the destructive distillation of coal. This reaction corresponds to the formation of ammonium cyanate from ammonia and carbon dioxide.

Silver, lead, copper(I), and thallium(I) thiocyanates are insoluble and mercury(II), bismuth, and tin(II) thiocyanates slightly soluble. All of these are soluble in excess of soluble (e.g., ammonium) thiocyanate, forming complexes. Iron(III) thiocyanate gives a blood-red solution, used in detecting either Fe(III) or thiocyanate in solution, and is extracted from water by amyl alcohol. It is not formed in the presence of fluoride, phosphate and other strongly complexing ions.

When thiocyanic acid is treated with certain oxidizing agents, e.g., nitric acid, sulfuric acid and hydrocyanic acid are formed, but the action of lead tetraacetate on the acid, or of bromine in ether on lead(II) thiocyanate, gives thiocyanogen ("Rhodan") NCSSCN, a yellow, volatile oil, mp about −3°C, which polymerizes irreversibly at room temperature to insoluble, brick-red parathiocyanogen (NCS)$_x$. Thiocyanogen reacts with organic compounds like a free halogen. It liberates iodine from iodides. In water it is rapidly hydrolyzed to sulfuric and hydrocyanic acids. When thiocyanic acid is treated with reducing agents, e.g., aluminum and dilute hydrochloric acid, hydrogen sulfide plus carbon plus ammonium chloride are formed.

Esters

Ethyl thiocyanate $C_2H_5 \cdot SCN$, colorless liquid, bp 142°C. Formed by reaction (1) of potassium thiocyanate and potassium ethyl sulfate, (2) of cyanogen chloride and ethane-thiol. Oxidizable with fuming nitric acid to ethyl sulfonic acid $C_2H_5 \cdot SO_2OH$, and reducible with zinc and dilute sulfuric acid to ethane thiol C_2H_5SH. Ethyl isothiocyanate $C_2H_5 \cdot NCS$, colorless, odorous liquid, bp 132°C. Formed by reaction of ethyl amine and carbon disulfide (cf. the formation of ammonium thiocyanate from ammonia and carbon disulfide). Reducible to ethyl amine $C_2H_5NH_2$ plus methylene sulfide CH_2S. Allyl isothiocyanate ("mustard oil") $C_3H_5 \cdot NCS$ liquid, bp 151°C, odor of mustard, and causes blisters in contact with the skin.

THIOETHERS.

Hydrogen sulfide yields two classes of organic compounds: (1) hydrosulfides, and (2) sulfides. The sulfides are termed thioethers. A more general term, *thiols*, also is used. This term not only embraces thioethers, but also covers thioalcohols, sulfhydrates, and thiophenols.

Ethyl sulfide $(C_2H_5)_2S$, one of the better known thioethers, is an odorous, inflammable liquid, mp −102.1°C, bp 91.6°C, sp gr 0.837. The compound is insoluble in H_2O and soluble in alcohol and ether. It is prepared by distilling ethyl potassium sulfate with potassium sulfide. Chemically, ethyl sulfide behaves much like the ethers. For example, none of the hydrogen atoms can be displaced by metals and generally the compound is very inert. Additional thioethers can be prepared in a similar manner with the corresponding proper ingredients. Upon oxidation with HNO_3, thioethers are converted to sulfones. The latter are stable crystalline substances. An example is ethyl sulfone $(C_2H_5)_2SO_2$.

THIOKOL RUBBERS.   See **Elastomers**.

THIOPHENE.

[CAS: 110-02-1]. $\langle(CH:CH)_2\rangle S$, formula weight 84.13, colorless liquid resembling benzene in odor, mp −30°C, bp 84°C, sp gr 1.070. Thiophene and its derivatives closely resemble benzene and its derivatives in physical and chemical properties. Thiophene is present in coal tar and is recovered in the benzene distillation fraction (up to about 0.5% of the benzene present). Its removal from benzene is accomplished by mixing with concentrated sulfuric acid, soluble thiophene sulfonic acid being formed. Thiophene gives a characteristic blue coloration with isatin in concentrated sulfuric acid.

Thiophene may be formed (1) by passing ethyl sulfide (diethyl sulfide) through a red-hot tube, (2) by reduction of sodium succinate and phosphorus trisulfide. Chlorine and bromine yield chloro- and bromo-substitution products, respectively, cold fuming nitric acid yields thiophene sulfonic acid. Thiophene aldehyde $C_4H_3S \cdot CHO$, liquid, bp 198°C, resembles benzaldehyde chemically rather than furfural. The corresponding primary alcohol and carboxylic acid are known. By comparison, where the sulfur atom of thiophene is occupied by oxygen, furane is the resulting compound. Where the sulfur atom of thiophene is occupied by a nitrogen group (NH), pyrrole is the resulting compound.

Benzothiophene $C_6H_4 \cdot (CH)_2S$ is a solid, mp 31°C, bp 221°C, with physical and chemical properties that resemble naphthalene. By comparison, where the sulfur atom of benzothiophene is occupied by oxygen, the resulting compound is benzofurane (coumarone). Where the sulfur atom of benzothiophene is occupied by a nitrogen group (NH), indole is the resulting compound.

THIOUREA.

[CAS: 62-56-6]. $(NH_2)_2CS$, formula weight 76.12, white crystalline solid, mp 180–182°C, decomposes before boiling at atmospheric pressure, sp gr 1.405. Thiourea is moderately soluble in H_2O, soluble in alcohol, and slightly soluble in ether. Sometimes referred to as thiocarbamide, sulfurea, and sulfocarbamide, thiourea may be considered chemically analogous to urea and is oxidized to urea by cold potassium permanganate solution. The compound is easily hydrolyzed to NH_3, CO_2, and H_2S. Upon long heating below the melting point, thiourea is transformed to ammonium thiocyanate. Thiourea is attractive for plastics manufacture because of the greater ease with which substitution can be made on the sulfur atom of thiourea than on the oxygen atom of urea.

Thiourea is formed by heating ammonium thiocyanate at 170°C. After about an hour, 25% conversion is achieved. With HCl, thiourea forms thiourea hydrochloride; with mercuric oxide, thiourea forms a salt; and with silver chloride, it forms a complex salt.

Symmetrical diphenyl thiourea (thiocarbanilide) $(C_6H_5NH)_2CS$ is a solid, mp 154°C. When heated with concentrated HCl, the compound yields aniline plus phenylisocyanate. Formed by the reaction of aniline and CS_2, symmetrical diethylthiourea $(C_2H_5NH)_2CS$ is a solid, mp 77°C.

In addition to its use in plastics manufacture, thiourea is used in some photographic processes and photocopying papers; in organic synthesis as an intermediate (drugs, dyes, cosmetics); in rubber accelerators; and as a mold inhibitor.

THIXOTROPY.

The ability of certain colloidal gels to liquefy when agitated (as by shaking or ultrasonic vibration) and to return to the gel form when at rest. This is observed in some clays, paints, and printing inks that flow freely on application of slight pressure, as by brushing or rolling. Suspensions of bentonite clay in water display this property, which is desirable in oil-well drilling fluids.

See also **Rheology**.

THOMSON, J. J. (1856–1940).

Joseph John Thomson was an English physicist. At age fourteen, his father sent him to Owens College for preparatory scientific training. His attendance here was important to his career because this college had an outstanding science faculty and it also offered many experimental physics courses.

Thomson earned his engineering degree from Owens. Then he attended Trinity College of Cambridge University and studied mathematics and theoretical physics. When he graduated he began working in the Cavendish Laboratory at Cambridge and by age twenty-seven became its director. Thomson's main research was on the conduction of electricity through gases. After, Roentgen's discovery of X-rays in 1895, Thomson started working with Rutherford and found that passing X-rays through gases greatly increased their ability to conduct electricity. Much of Thomson's further research dealt with the composition of cathode rays. He believed that cathode rays were streams of tiny charged particles. His work concluded that the atom was not the fundamental unit of matter. He devised a model of the atom incorporating his theory of *corpuscles*. Later, the name "electron" was adopted.

In 1906, Thomson received the Nobel Prize for Physics "in recognition of the great merits of his theoretical and experimental investigations on the conduction of electricity by gases." Between the years of 1906 and 1914, Thomson studied canal rays and worked on separating the different kinds of atoms and atomic groupings present in them. In 1903, Thomson proposed a discontinuous theory of light, with light rays being composed of separate particles rather than continuous streams, and later Einstein developed the photon theory of light.

Thomson's work revolutionized scientific understanding of the atom and ushered in a new era in physical science. He is also remembered for his excellent teaching at Cavendish Laboratory.

See also **Electron Theory**.

J.M.I.

THOMSON PARABOLA METHOD.

The method of investigating the charge-to-mass ratio of positive ions in which the ions are acted upon by electric and magnetic fields applied in the same direction normal to the path of the ions. It can be shown that ions of a given charge-to-mass ratio but different velocities will be deflected so as to form a parabola.

THOMSON PRINCIPLE.

The hypothesis that, if thermodynamically reversible and irreversible processes take place simultaneously in a system, the laws of thermodynamics may be applied to the reversible process while ignoring for this purpose the creation of entropy due to the irreversible process. Applied originally by Thomson to the case of thermoelectric effects. Also used in the treatment of electrochemical cells, thermal diffusion.

THORIANITE. This mineral of thorium oxide, ThO_2, is isomorphous with uraninite and occurs in black, nearly opaque cubic crystals in Ceylon and in Madagascar. Often containing rare-earth metals and uranium, the ore is strongly radioactive. Because of its radioactivity, it is valuable in helping to date the relative ages of rocks in which it occurs.

THORITE. The mineral thorite is a silicate of the rare element thorium and corresponds to the formula $ThSiO_4$. It is tetragonal and exhibits a prismatic cleavage. The original thorite was black in color with a specific gravity of 4.4–4.8. A variety orangite, so called from its orange-yellow color, has a specific gravity of 5.19–5.40. It has been found partly altered to thorite. Uranothorite contains uranium oxide. Thorite occurs in Norway in augite syenites. Thorite and orangite occur in Sweden, and orangite and uranothorite are found in Madagascar. Uranothorite is found in Ontario.

THORIUM. [CAS: 7440-29-1]. Chemical element symbol Th, at. no. 90, at. wt. 232.038, radioactive metal of the Actinide Series, mp 1750°C, bp 4790°C, density 11.5–11.9 g/cm³ (17°C). Thorium metal is dark gray, dissolves in HCl, is made passive in HNO_3, and is not affected by fusion with alkalis. The element combines with chlorine or sulfur at 450°C; with hydrogen or nitrogen at 650°C. All thorium-containing substances are radioactive. The element was discovered by J.J. Berzelius in 1829. The electronic configuration

$$1s^2 2s^2 2p^6 3s^2 3p^6 3d^{10} 4s^2 4p^6 4d^{10} 4f^{14} 5s^2 5p^6 5d^{10} 6s^2 6p^6 6d^2 7s^2.$$

Ionic radii Th^{3+} 1.08 Å, Th^{4+} 0.95 Å (Zachariasen). Metallic radius 1.797_5 Å. First ionization potential 5.7 eV; second, 16.2 eV; third, 29.4 eV. Oxidation potentials Th $\longrightarrow$ $Th^{4+} + 4e^-$, 1.90 V; Th + $4OH^- \longrightarrow ThO_2 + 2H_2O + 4e^-$, 2.48 V. See also **Chemical Elements**.

The isotopes of thorium include mass numbers 223–234. ^{232}Th has a half-life of 1.39×10^{10} years. See also **Radioactivity**. It emits an alpha-particle and forms meso-thorium 1 (radium-228), which is also radioactive, having a half-life of 6.7 years, emitting a beta-particle. Since ^{232}Th captures slow neutrons to form, by a series of nuclear reactions, ^{233}U which is fissionable, thorium can be used as a fuel for nuclear reactors of the breeder type. Thorium occurs in earth minerals, an average content estimated at about 12 ppm. Findings of the *Apollo 11* space flight indicated that thorium concentrations in some lunar rocks are about the same as the concentrations in terrestrial basalts.

Thorium occurs in monazite sand in Brazil, India, North and South Carolina; this ore contains 3–9% thorium oxide, and is the chief source; thorium is also found in thorite containing about 60% oxide and in thorianite, about 80% oxide. When heated with concentrated H_2SO_4 the minerals form thorium sulfate, from which, by a series of reactions, thorium nitrate, the chief commercial compound, is obtained.

Thorium has the oxidation state of (IV) in all of its important compounds. Its oxide, ThO_2, and its hydroxide are entirely basic. The nature of the ions present in a number of solutions of the soluble compounds is not known with certainty. Complex ions involving sulfate are suggested by the increased solubility of the sulfate in solutions of the acid sulfates. Similarly, other complex ions are suggested by the solubility of the carbonate in excess alkali carbonate and of the oxalate in ammonium oxalate. Such ready complex ion formation is consistent with the high positive charge of the thorium-(IV) ion.

Although the exact extent is not known accurately, hydrolysis of various salts is known to occur. Since the hydroxide is not precipitated it is assumed that the hydrolysis product is some ion on the form $Th(OH)_2^{++}$ or $ThOH^{3+}$. The solution chemistry of thorium is made more complicated because of the hydrolytic phenomena observed and the polynuclear complex ions that are formed at low acidities and higher thorium concentrations.

Studies of the complex ions formed by Th^{4+} with various complexing anions have given much information. For example, the equilibria and ionic species involved in the chloride complexing of aqueous thorium have been studied through the method of measuring the distribution between H_2O and benzene containing thenoyltrifluoroacetone. The conclusion: that there is successive complexing involving the species $ThCl^{3+}$, $ThCl_2^{2+}$, $ThCl_3^+$ and $ThCl_4$. Similarly, all the intermediate chelate complex ions between thorium and acetylacetone exist in aqueous solution of proper acidity.

Thorium dioxide (face-centered cubic structure) is very insoluble in H_2O, but dissolves in acids to yield salts.

Thorium forms one series of halides, another one of oxyhalides, and also a series of double or complex halides. In general, stability of these compounds toward heat decreases as the atomic weight of the halogen increases. These compounds are often isostructural with the corresponding compounds of other actinide elements in the (IV) oxidation state.

Thorium metal reacts with hydrogen at moderately elevated temperatures to yield two hydrides: (1) ThH_2, which has a pseudotetragonal body-centered unit containing two metal atoms, isomorphous or pseudoisomorphous with thorium carbide, zirconium hydride, and zirconium carbide, ThC_2, ZrH_2, and ZrC_2; and (2) a hydride of approximate composition $ThH_{3.75}$ or ThH_4, possessing a unique cubic structure unrelated to that of the parent metal.

Thorium sulfide, ThS_2, is obtained by the action of H_2S or sulfur on thorium metal. The oxysulfide, ThOS, has been obtained in several ways, one of which is by the action of CS_2 on thorium dioxide at elevated temperatures. At 800°C and under pressure, sulfur combines with thorium to yield compounds with approximately the formulas ThS, Th_2S_3, and Th_3S_7. The first two have semimetallic properties and may be employed as ceramics for use with highly electropositive metals, whereas the last appears to be a polysulfide.

Anhydrous thorium sulfate, $Th(SO_4)_2$, is obtained by the action of concentrated H_2SO_4 on thoria (ThO_2). A solution of this salt deposits crystals of $Th(SO_4)_2 \cdot 9H_2O$ at about 15°C, $Th(SO_4)_2 \cdot 8H_2O$ near 24°C, and $Th(SO_4)_2 \cdot 4H_2O$ around 45°C. At 100°C other hydrates change to $Th(SO_4)_2 \cdot 2H_2O$. In aqueous solution, the salt is considerably hydrolyzed to an oxysulfate—for instance, $ThOSO_4 \cdot H_2O$.

Thorium nitrate, CAS: 13823-29-5, $Th(NO_3)_4 \cdot 12H_2O$, is obtained by dissolving thorium hydroxide in HNO_3.

Thorium orthophosphate, $Th_3(PO_4)_4 \cdot 4H_2O$, is precipitated by adding a solution of sodium phosphate to an acidic solution of a thorium salt. Thorium pyrophosphate, $ThP_2O_7 \cdot 2H_2O$, precipitates when an acidic solution of thorium nitrate is treated with one of tetrasodium pyrophosphate.

Thorium has been used as a fuel for nuclear reactors since it is a fertile material for the generation of fissionable uranium-233. Some experts have estimated that the energy available from the world's reserves of thorium is greater than all of the remaining fossil fuels (coal and petroleum) and of all of the remaining uranium, combined. Thorium oxide is used for gas mantles. The oxide also helps to control grain size in tungsten filaments and strengthens nickel alloys (TD nickel). Thorium is also used as an alloying addition in magnesium technology and as a deoxidant for molybdenum, iron, and other metals. Several applications for thorium are found in electronic technology.

Thorium oxide has a high refractive index and low dispersion and thus finds use in high-quality camera and scientific instrument lenses. Thorium oxide also is used as a catalyst in the conversion of ammonia to nitric acid, in petroleum cracking, and in sulfuric acid production.

Handling

^{232}Th is sufficiently reactive to expose a photographic plate within a few hours. Thorium disintegrates with the production of thoron (^{220}radon). The latter is an alpha emitter and a radiation hazard. Areas where thorium is stored should be well ventilated and all precautions in the handling of thorium materials must be taken.

Additional Reading

Elvers, B., S. Hawkins, and W.E. Russey: *Ullmann's Encyclopedia of Industrial Chemistry: Thorium and Thorium Compounds to Vitamins,* 5th Edition, John Wiley & Sons, Inc., New York, NY, 1997.

Finlayson-Dutton, G.: "Tinkering with Glass and Ceramic Structures," *Science,* 627 (August 10, 1990).

Greenwood, N.N. and A. Earnshaw: *Chemistry of the Elements,* 2nd Edition, Butterworth-Heinemann, Inc., Woburn, MA, 1997.

Krebs, R.E.: *The History and Use of Our Earth's Chemical Elements: A Reference Guide,* Greenwood Publishing Group, Inc., Westport, CT, 1998.

LaTourrette, T.Z., A.K. Kennedy, and G.J. Wasserburg: "Thorium-Uranium Fractionation by Garnet: Evidence for a Deep Source and Rapid Rise of Oceanic Basalts," *Science,* 739 (August 6, 1993).

Lewis, R.J. and N.I. Sax: *Sax's Dangerous Properties of Industrial Materials,* 10th Edition, John Wiley & Sons, Inc., New York, NY, 1999.

Lide, D.R.: *CRC Handbook of Chemistry and Physics,* 84th Edition, CRC Press, LLC., Boca Raton, FL, 2003.

Smith, J.F. et al.: *Thorium Preparation and Properties,* Iowa State University Press, Ames, Iowa, 1975.

Staff: International Atomic Energy Agency, *Utilization of Thorium in Power Reactors,* Bernan Associates, Lanham, MD, 1996.

THORIUM OXIDE.   See **Thorianite**.

THORIUM SERIES.   See **Radioactivity**.

THREE-PHASE EQUILIBRIUM.   For every pure, chemically stable substance there is a certain temperature and pressure at which it can exist in all three states or phases, solid, liquid, and vapor, each phase being in equilibrium with each of the others. At higher temperatures and pressures than those at this so-called "triple point," the liquid and vapor states may attain equilibrium; solid-vapor equilibrium is possible at lower temperatures and pressures; and solid liquid equilibrium can be obtained at higher pressures and at lower or higher temperatures according as the substance contracts or expands upon melting. These three equilibria may be represented by three temperature-pressure graphs, which converge at the triple point. Figure 1 illustrates the case of water.

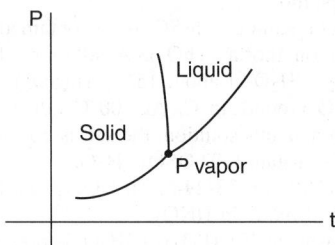

Fig. 1.   Triple-point (*P*) on temperature-pressure diagram

In 1954, the thermodynamic temperature scale (i.e., the absolute Kelvin scale was redefined by setting the triple point temperature for water equal to exactly 273.16 K).

THREONINE.   See **Amino Acids**.

THROMBIN.   A proteolytic enzyme that catalyzes the conversion of fibrinogen to fibrin and thus is essential in the clotting mechanism of blood. It is present in the blood in the form of prothrombin under normal conditions; when bleeding begins, the prothrombin is converted to thrombin, which in turn activates the formation of fibrin.

THULIUM.   [CAS: 7440-30-4]. Chemical element, symbol Tm, at. no. 69, at. wt. 168.934, twelfth in the Lanthanide Series in the periodic table, mp 1545°C, bp 1950°C, density 9.321 g/cm³ (20°C). Elemental thulium has a close-packed hexagonal crystal structure at 25°C. The pure metallic thulium is gray in color, with no evidence of tarnishing up to a temperature of 200°C. Above 200°C, the element combines with oxygen, sulfur, nitrogen, carbon, and hydrogen and will form intermetallic compounds with most metals. At higher temperatures, halogen gases react vigorously with the element to form trihalides. There is one natural isotope of thulium ^{169}Tm. Seventeen artificial isotopes have been produced. Average content of the earth's crust is estimated at 0.48 ppm thulium, making this element the least abundant of the rare-earth elements. Even at this level, however, thulium is potentially more available than antimony, bismuth, cadmium, or mercury. The element was first identified by P.T. Cleve in 1879. Electronic configuration

$$1s^2 2s^2 2p^6 3s^2 3p^6 3d^{10} 4s^2 4p^6 4d^{10} 4f^{12} 5s^2 5p^6 5d^1 6s^2.$$

Ionic radius Tm³⁺ 0.880 Å. Metallic radius 1.746 Å. First ionization potential 6.18 V; second 12.05 V. Other physical properties of thulium are given under **Rare-Earth Elements and Metals**.

Thulium occurs in apatite and xenotime and is derived from these minerals as a minor coproduct in the processing of yttrium. Processing involves organic ion-exchange, liquid-liquid, or solid-liquid, techniques. Prior to the development of cation exchange resins capable of separating the chemically similar rare earths, thulium was practically unavailable in

pure form. Thulium metal is made by the direct reduction of thulium oxide by lanthanum metal at high temperature in a vacuum.

Important scientific and industrial applications for thulium and its compounds remain to be developed. In particular, the photoelectric, semiconductor, and thermoelectric properties of the element and compounds, particularly behavior in the near-infrared region of the spectrum, are being studied. Thulium has been used in phosphors, ferrite bubble devices, and catalysts. Irradiated thulium (^{169}Tm) is used in a portable x-ray unit.

Note: This entry was revised and updated by K. A. Gschneidner, Jr., Director, and B. Evans, Assistant Chemist, Rare-earth Information Center, Institute for Physical Research and Technology, Iowa State University, Ames, IA.

Additional Reading

Lewis, R.J. and N.I. Sax: *Sax's Dangerous Properties of Industrial Materials,* 10th Edition, John Wiley & Sons, Inc., New York, NY, 1999.

Lide, D.R.: *CRC Handbook of Chemistry and Physics*, 84th Edition, CRC Press, LLC., Boca Raton, FL, 2003.

THYROID HORMONES.   See **Hormones**; **Iodine (In Biological Systems)**.

TIN.   [CAS: 7440-31-5]. Chemical element, symbol Sn, at. no. 50, at. wt. 118.69, periodic table group 4, mp 231.97°C, bp 2,270°C, density 7.29 g/cm³ (white tin at 15°C), 5.77 (gray tin at 13°C), 6.97 (liquid at mp). There are two allotropic forms of tin: (1) the more common soft white beta tin has a body-centered tetragonal crystal form, (2) the brittle gray alpha tin has a diamond-type cubic crystal form. The cubic form, α-tin, stable below 18°C, is an intrinsic semiconductor, and is gray. At 161°C, white tin undergoes a transition to rhombic or γ-tin.

Tin is a silver-white metal with a bluish tinge, softer than zinc and harder than lead. It is malleable, ductile at 100°C; can be powdered at 200°C, and upon exposure to temperatures below 18°C, it crumbles to a grayish powder due to the "tin pest," which is caused by the transformation of white to gray tin (the reverse transformation may be brought about by heating gray tin to about 100°C). When a bar of tin is bent a marked creaking sound is emitted due to the friction of the crystals. Tin: is not oxidized on exposure to air at ordinary temperatures; burns to stannic oxide when heated to high temperatures in air or oxygen; is soluble in HCl to form stannous chloride; is converted by concentrated HNO₃ into soluble beta-stannic acid; is soluble in aqua regia to form stannic chloride; is soluble in NaOH solution slowly to form sodium stannite and hydrogen gas; and reacts with chlorine to form volatile stannic chloride. Discovery, prehistoric.

Tin has the largest number of naturally occurring isotopes ^{112}Sn, ^{114}Sn through ^{120}Sn, ^{122}Sn, and ^{124}Sn. Five radioactive isotopes have been identified ^{111}Sn, ^{113}Sn, ^{121}Sn, ^{123}Sn, and ^{125}Sn. With exception of ^{121}Sn, which has a half-life of about 5 years, the half-lives of the other isotopes are comparatively short, expressed in minutes and days. Tin occurs in the Earth's crust to the extent of about 40 grams/ton. It is estimated that a cubic mile of seawater contains about 15 tons of tin. First ionization potential 7.332 eV; second, 14.52 eV; third, 30.49 eV; fourth, 40.57 eV. Oxidation potential Sn ⟶ Sn²⁺ + 2e⁻, 0.406 V; Sn²⁺ ⟶ Sn⁴⁺ + 2e⁻, −0.14 V;   HSnO₂⁻ + 3OH⁻ + H₂O ⟶ Sn(OH)₆²⁻ + 2e⁺, 0.96 V; Sn + 3OH⁻ ⟶ HSnO₂⁻ + H₂O + 2e⁻, 0.79 V. Electronic configuration $1s^2 2s^2 2p^6 3s^2 3p^6 3d^{10} 4s^2 4p^6 4d^{10} 5s^2 5p^2$.

Other important physical properties of tin are given under **Chemical Elements**.

Tin occurs as oxide (cassiterite, tin stone, stannic oxide, SnO₂), obtained commercially in Malaysia, Indonesia, Thailand, and Bolivia. The ore is concentrated and then roasted to oxide (83–88% stannic oxide). The product is treated in a blast furnace and crude tin recovered. Refining is conducted by electrolysis, or by fractional fusion.

Tin also occurs as complex sulfidic ores. The economic working of these ores is essentially confined to Bolivia. The ores include SnS₂ · Cu₂S · FeS (stannite), SnS (herzenbergite, SnS · PbS (teallite), 2SnS₂ · Sb₂S₃ · 5PbS (franckeite), Sn₆Pb₆Sb₂S₁₁ (cylindrite), 2SnS₂ · 2PbS · 2(FeZn)S · Sb₂S₃ (plumbostannite), and 4Ag₂S · SnS₂ (canfieldite).

Secondary tin is an important source of the metal. Tinplate scrap may be detinned electrolytically or chemically. The alkaline chemical process is the most widely used and involves a caustic solution, which contains an oxidizing agent to remove both tin and the underlying iron-tin alloy from the steel. The solution formed then is either (1) crystallized to form sodium stannate, (2) electrolyzed to recover tin metal, or (3) acidified with CO_2, H_2SO_4, or acidic gases to precipitate hydrated tin oxide. There are several secondary tin smelters in the United States, but only one primary smelter. The main primary tin smelters are located in the United Kingdom, Malaysia, and Thailand.

Uses. Not including former Soviet Bloc nations, world consumption of tin is in excess of 187,000 metric tons annually. Principal uses are: (1) tinplate, 35%; (2) solder, 24%; (3) bronze, 9%; (4) other alloys, 8%; (5) tinning, 4%; (6) chemicals, 5%; (7) other uses, 15%. In the United States, tin consumption in 1979 was: (1) tinplate, 29%; (2) solder, 29%; (3) bronze and brass, 14%; (4) chemicals, 8%; (5) other alloys, 9%; (6) tinning, 4%; (7) other uses, 7%. In the United States, the major portion of tinplate is used in the making of cans. The advantages of tin for cans and food-processing equipment include its nontoxic nature, resistance to corrosive attack by acids and other aqueous solutions, and when combined with other metals, strength.

Tin Plate. Tin coating may be applied to steel by (1) electroplating, usually as part of a high-speed, continuous process, or (2) by dipping cut sheets in a bath of molten tin. Electrolytic tin plate essentially is a sandwich in which the central core is strip steel. This core is thoroughly cleaned in a pickling solution prior to electroplating. The actual plating occurs as the strip moves through horizontal or vertical tanks containing electrolyte. The moving strip then is heated as it passes between high-frequency electric induction coils, whereupon the tin coating melts and flows to form a lustrous coat. The average thickness of tin on the end-product sheet is 0.00003 inch (0.0008 millimeter) on each side. A complex system of instrumentation is used to control process conditions and to inspect the moving sheet for any perforations in the plate. In hot-dip tinning, individual steel sheets are pickled and washed. A layer of hot palm oil is maintained on top of the molten tin bath to prevent oxidation of the molten tin by air and to prevent the molten tin from freezing too rapidly on the plate, thus providing a more even coating with a high luster.

Terne Plate. This is a sheet-steel product that is coated with an alloy of tin and lead. The coatings range from 50–50 mixtures of lead and tin to as low as 12% tin and 88% lead. Plate used for roofing normally is about 25% tin and 75% lead. In addition to roofing, terne plate is used in the manufacture of gasoline tanks for automotive vehicles, oil cans, and containers for solvents, resins, etc.

Tin-bearing solders are considered soft solders as contrasted with the hard solders which contain substantial quantities of silver. However, small quantities of silver are added to some tin solders to increase strength of the resulting joint and to adjust working temperature range. Antimony also is added in some cases for the latter purpose. Wiping solders usually have a tin content ranging from 35 to 40%. Solders used in automotive-body work require a wide plastic (working temperature) range. Solders with a low tin content (below 25%) are generally used. For very low-temperature melting, bismuth and cadmium also may be added to tin-lead solders.

Alloys. Tin is widely used as both a major and minor ingredient of alloy metals. These applications are summarized in Tables 1, 2, and 3. Phosphor bronzes (Table 3) actually contain very little phosphorus, ranging from 0.03 to 0.50%, and hence the alloys are poorly designated. Tin bronzes is the better term. High-silicon bronzes contain about 2.8% tin; low-silicon bronzes about 2.0% tin. Gun metals are tin bronze casting alloys with a 5–10% zinc content. Some wrought copper-base alloys contain tin: (1) Inhibited Admiralty metal, 1% tin; (2) manganese bronze, 1% tin; (3) naval brass, 0.75% tin, (4) leaded naval brass, 0.75% tin. See also **Copper**.

Copper-nickel-tin alloys (UNS C72500, 2700, and 2900) are *spinodal*[1] materials that can be age hardened after forming. They combine tensile strengths as high as 1380 MPa (200×10^3 psi) with resistance to oxidation, stress relaxation, fatigue, and stress-corrosion cracking. For use in round-wire form, these alloys are challenging the traditional position held by phosphor bronzes for electronic leads, contact pins, and sockets.

[1] Spinodal structure is a fine homogeneous mixture of two phases that form by the growth of composition waves in a solid solution during suitable heat treatment. The phases of a spinodal differ in composition from each other and from the parent phase, but have the same crystal structure as the parent phase.

TABLE 1. REPRESENTATIVE SOFT SOLDERS

Composition, %				Temperature		Working temperature
Tin	Lead	Antimony	Silver	Solidus	Liquidus	freezing range
80	20	0	0	183°C	203°C	20°C
70	30	0	0	183	192	9
60	40	0	0	183	189	6
50	50	0	0	183	216	33
49	50	1	0	186	210	24
40	60	0	0	183	234	51
39	60	1	0	186	230	44
30	70	0	0	183	252	69
29	70	1	0	186	252	66
20	80	0	0	183	273	90
20	78.75	0	1.25	180	270	90
19	80	1	0	185	273	88
10	88.50	0	1.50	178	290	112
10	90	0	0	183	297	114
5	95	0	0	270	311	41
2.5	97.5	0	0	301	319	18

TABLE 2. REPRESENTATIVE BABBITT METALS

Ingredients	[a]SAE 10 %	SAE 11 %	SAE 12 %	SAE 13 %	SAE 14 %	SAE 15 %
Tin	90	86	88.25	4.5–5.5	9.25–10.75	0.9–1.25
Antimony	4–5	6–7.5	7–8.5	9.25–10.75	14–16	14.5–15.5
Lead	0.35	0.35	0.35	86	76	Remainder
Copper	4–5	5–6.5	2.25–3.75	0.50	0.50	0.6
Iron	0.08	0.08	0.08	—	—	—
Arsenic	0.10	0.10	0.10	0.60	0.60	0.8–1.10
Bismuth	0.08	0.08	0.08	—	—	—

[a] Society of Automotive Engineers.

In flatwire form, they compete with beryllium copper for eyeglass frames, circuit boards, and electronic-contact clips. The alloys also have been used for rivets, self-threading screws, and a variety of cold-headed parts.

Chemistry and Compounds

Tin forms two series of compounds: tin(II) or stannous compounds and tin(IV) or stannic compounds. Tin(II) oxide, SnO, insoluble in water, is formed by precipitation of an SnO hydrate from an $SnCl_2$ solution with alkali and later treatment in water (near the boiling point and at constant pH). It is amphiprotic, but only slightly acid, forming stannites slowly with strong alkalis. Sodium stannite is conveniently prepared from tin(II) chloride: $SnCl_2 + 3NaOH \longrightarrow Na[Sn(OH)_3] + 2NaCl$. Tin(IV) oxide, SnO_2, is much more acidic; it readily reacts with NaOH to form stannate ions, $Sn(OH)_6^{2-}$. In fact, no hydroxide of the formula $Sn(OH)_4$ has ever been obtained. The metal metastannates, e.g., $M^{II}SnO_3$, are generally made by fusion methods and have three-dimensional polymeric anions in which each tin atom is surrounded octahedrally by six oxygen atoms. There are, however, two forms of stannic acid, H_2SnO_3. The α-stannic acid is a white, gelatinous precipitate obtained by treating $SnCl_4$ with NH_4OH. The α-stannic acid (also called metastannic acid) is a white powder obtained by action of concentrated HNO_3 on tin; unlike the α-form it is insoluble in concentrated acids and alkali metal hydroxides.

Tin forms dihalides and tetrahalides with all of the common halogens. These compounds may be prepared by direct combination of the elements, the tetrahalides being favored. Like the halides of the lower main group 4 elements, all are essentially covalent. Their hydrolysis requires, therefore, an initial step consisting of the coordinative addition of two molecules of water, followed by the loss of one molecule of HX, the process being repeated until the end product $H_2Sn(OH)_6$ is obtained. The most significant commercial tin halides are stannous chloride, stannic chloride, and stannous fluoride.

The increasingly electropositive character of main group 4 as tin is reached is evident from the fact that its hydrides are much less stable than

TABLE 3. REPRESENTATIVE TIN-BEARING BRONZES

	Phosphor bronzes				
	1.25% Tin	4% Tin[c]	5% Tin	8% Tin	10% Tin
Copper[a]	98.75%	88.00%	95.00%	92.00%	90.00%
Melting point, °C	1,077	1,000	1,050	1,027	1,000
Tensile strength, 1,000 psi					
Hard sheet	65	58	81	93	100
Soft sheet	40	44	47	55	66
Rockwell hardness					
Hard sheet	75B	68B	87B	93B	97B
Soft sheet	60F	65F	73F	75F	55B
Electrical conductivity % IACS[b]	48	19	18	13	11
Thermal conductivity Btu$(ft^2)(ft)(°F)$ at 68°F	120	50	47	36	29
Major Uses	Electrical contact wire Messenger cable Flexible metal hose Pole-line hardware	Bearings Bushings Gears Pinions Shafts Screw-machine products Washers Valve parts	Bearings Bellows Bourdons Gears Rivets Springs Wire cloth Truss wire	Bearings Bellows Bourdons Fasteners Washers Springs Switch parts Chemical hardware	Heavy bars, plates Bridge and expansion plates Heavy springs

[a] Small amounts of zinc, lead, iron, antimony, and phosphorus also present.

[b] International Annealed Copper Standard.

[c] Free-cutting phosphor bronze.

those of silicon and germanium. Known are Sn_2H_6 and SnH_4, which is obtained by hydrolysis of magnesium stannide, Mg_2Sn, or by electrolysis of a phosphoric acid solution with a tin cathode.

Among the other inorganic compounds of tin are the:

Nitrates. Stannous nitrate $Sn(NO_3)_2$, white solid, by reaction of tin metal and dilute HNO_3 and crystallization, soluble in water with slight excess of HNO_3.

Sulfates. [CAS: 7488-55-3], Stannous sulfate, a white powder soluble in water and H_2SO_4, is obtained commercially by action of H_2SO_4 on $SnCl_4$ or Sn. Stannic sulfate may be formed by the solution of stannic hydroxide in dilute H_2SO_4, or by action of oxidizing agents on stannous salts.

Sulfides. Stannous sulfide SnS, dark brown precipitate, by reaction of stannous salt solution and H_2S, insoluble in sodium sulfide solution but soluble in sodium polysulfide solution, forming sodium thiostannate; stannic sulfide SnS_2, yellow precipitate, by reaction of stannic salt solution and H_2S, soluble in sodium sulfide solution, forming sodium thiostannate.

Organometallic Compounds

In common with the other elements of main group 4, tin forms many organometallic compounds; the range of possible combinations is virtually limitless. They include:

(1) *Tetraorganotins*, R_4Sn, prepared either by alkylation of tin halides with Grignard Reagents or alkyl lithium; by reaction of an organic halide with a tin-sodium alloy; by direct reaction of tin with an organic halide; or by reaction of stannic chloride with alkyl aluminum compounds.

(2) *Organotin halides*, $RSnX_3$, R_2SnX_2, and R_3SnX, prepared by disproportionation of the tetraorganotin with stannic halide or by direct alkylation of stannic halide.

(3) *Organotin oxides*, R_2SnO or $(R_3Sn)_2O$, prepared by treatment of the organotin halides with alkali.

(4) *Stannoic acids*, RSnOOH.

The organotin halides and oxides are usually the intermediates used in the synthesis of other organotin derivatives, such as the organotin carboxylates, organotin sulfur-derivatives, organotin hydroxides, etc. The most significant commercial organotins include dibutyltin and dioctyltin carboxylates and sulfur derivatives, used as polyvinyl chloride (PVC) stabilizers and as catalysts in polymer systems; *bis*(tributyltin)oxide, triphenyltin fluoride, and tributyltin fluoride, used as antifoulants for marine paints, fungicides, bactericides, sanitizing agents, and wood preservatives; tricyclohexyltin hydroxide as an insecticide; triphenyltin hydroxide and triphenyltin acetate as agricultural fungicides; and dibutyltin dilaurate as a poultry anthelmintic.

Biochemical Aspects of Tin

Scientists at the University of Maryland have found that sediment microflora (from Chesapeake Bay sediments) can produce dimethyltin and trimethyltin species from inorganic Sn(IV) compounds. The results were consistent with a geocycle of tin proposed by Ridley et al. in 1977. The findings support the hypothesis that tin can be biotransformed in an estuarine environment.

Researchers J. Versieck and L. Vanballenberghe (University Hospital, Ghent, Belgium) have observed, "Tin has chemical properties offering potentials for a biological function. The element has a tendency to form truly covalent linkages as well as coordination complexes; hence, it was hypothesized that it could well contribute to the tertiary structure of proteins or other biologically important macromolecules, such as nucleic acids. The oxidation-reduction potential of $Sn^{2+} \rightleftharpoons Sn^{4+}$ being at 0.13 V, well within the range of physiological oxidation-reduction reactions, it was also speculated that the element could function as the active site of metalloenzymes."

During the late 1960s, several experiments led to the conclusion that tin was indispensable for the growth of rats fed purified amino acid diets in trace-element-controlled isolators. However, no additional evidence for biological essentiality has been added since that time. Information on the tin content of foods is meager. Estimates have shown that the intake is less than 1 milligram per day. However, because of contamination from packaging, it has been estimated that this figure occasionally could rise to 50 mg/day. The aforementioned researchers have developed what appears to be a reliable method for determining tin in human blood serum by radiochemical neutron activation analysis.

Additional Reading

Davis, J.R.: *Metals Handbook,* 2nd Edition, ASM International, Materials Park, OH, 1998.

Franklin, A.D., J.S. Olin, and T.A. Wertime (editors): *The Search for Ancient Tin,* Smithsonian Institution, Washington, DC, 1978.

Greenwood, N.N. and A. Earnshaw: *Chemistry of the Elements,* 2nd Edition, Butterworth-Heinemann, Inc., Woburn, MA, 1997.

Hallas, L.E., Means, J.C., and J.J. Cooney: "Methylation of Tin by Estuarine Microorganisms," *Science*, **215**, 1505–1507 (1982).

Krebs, R.E.: *The History and Use of Our Earth's Chemical Elements: A Reference Guide,* Greenwood Publishing Group, Inc., Westport, CT, 1998.

Lewis, R.J. and N.I. Sax: *Sax's Dangerous Properties of Industrial Materials,* 10th Edition, John Wiley & Sons, Inc., New York, NY, 1999.

Lide, D.R.: *CRC Handbook of Chemistry and Physics,* 84th Edition, CRC Press, LLC., Boca Raton, FL, 2003.

Meyer, C.: "Ore Metals Through Geologic History," *Science*, **227**, 1421–1428 (1985).

Patai, S.E.: *The Chemistry of Organic Germanium, Tin and Lead Compounds,* John Wiley & Sons, Inc., New York, NY, 1995.

Staff: "Forecast '91 Metals," *Advanced Materials & Processes*, 24 (January 1991).

Staff: Various publications on tin and its compounds, including *Tin and Its Uses* (quarterly): *Tin International* (monthly); and statistical publications on tin (periodically), International Tin Research Institute, Middlesex, England.

Staff: *Annual Review of the World Tin Industry,* Rayner-Harwill Ltd., London, published yearly.

Staff: *Tin Chemicals for Industry,* Tin Research Institute, Greenford, Middlesex, UB6 7AQ, England (published periodically).

Versieck, J. and L. Vanballenberghe: "Determination of Tin in Human Blood Serum by Radiochemical Neutron Activation Analysis," *Analytical Chemistry*, 1143 (June 1, 1991).

Web References

International Tin Research Institute (ITRI) Ltd: http://www.itri.co.uk/index.htm

Tin: http://me.mit.edu/2.01/Taxonomy/Characteristics/Tin.html

USGS Minerals Information, Tin: http://minerals.usgs.gov/minerals/pubs/commodity/tin/

TISELIUS, ARNE W. K. (1902–1971). A Swedish biochemist who won the Nobel prize for chemistry in 1948, for his research on electrophoresis and adsorption analysis, especially for his discoveries concerning the complex nature of the serum proteins. His work also involved virus isolation and synthesis of blood plasma. He earned degrees from the University of Uppsala and Princeton University, as well as a multitude of honorary degrees.

TISHCHENKO REACTION. Formation of esters from aldehydes by an oxidation-reduction process in the presence of aluminum or sodium alkoxides.

TITANITE. A yellow or brown calcium silicotitanite, $CaTiSiO_5$, having a waxy luster, and often containing niobium (columbium), chromium, fluorine, and other elements. Titanite occurs in wedge-shaped monoclinic crystals, usually as an accessory mineral in granitic rocks and in calcium-rich metamorphic rocks. See also **Sphene**.

TITANIUM. [CAS: 7440-32-6]. Chemical element, symbol Ti, at. no. 22, at. wt. 47.9, periodic table group 4, mp 1650–1670°C, bp 3,287°C, density 4.507 g/cm^3 (20°C). Below 885°C, elemental titanium has a hexagonal closepacked crystal structure; above this temperature, it has a body-centered cubic crystal structure. Compact titanium is a white metal, when cold it is brittle and may be powdered, but at red heat may be forged and drawn into wire. Titanium exhibits some passivity in air due to formation of coatings of oxide or nitride. At 610°C, titanium reacts with oxygen to form titanium dioxide; at 800°C, it reacts with nitrogen to form titanium nitride. Upon heating with chlorine, the metal forms titanium tetrachloride. Cold, dilute H_2SO_4 readily dissolves the metal to form titanous sulfate. Hot, concentrated H_2SO_4 yields titanic sulfate. The element was first identified by Gregor in 1789 and later named titanium by Klaproth (1795). A metal of 95% purity was not produced until 1887 when it was made by the reduction of titanium tetrachloride with sodium. The first commercial uses date back to 1860 when ferrotitanium was used as an alloying element in steel and a bit later as a deoxidizer in the production of steel. There are five natural isotopes of the metal ^{46}Ti through ^{50}Ti, and three radioactive isotopes have been identified ^{44}Ti, ^{45}Ti, and ^{51}Ti, the latter two with relatively short half-lives measured in minutes and hours. ^{44}Ti has a half-life of approximately 10^3 years. Titanium is relatively abundant, ranking 8th in the list of chemical elements occurring in the Earth's crust. Titanium ranks 35th among the elements in terms of content in seawater, with an estimated 5 tons of titanium per cubic mile (1.1 kilograms per cubic kilometer) of seawater. First ionization potential 6.83 eV; second, 13.60 eV; third, 27.6 eV; fourth, 44.66 eV. Oxidation potential Ti + 2H$_2$O $\longrightarrow$ TiO$_2$ +

$4H^+ + 4e^-$, 0.95 V; Ti $\longrightarrow$ Ti^2 + 2e$^-$, 1.75 V; Ti^{2+} $\longrightarrow$ Ti^{3+} + e$^-$, 0.37 V. Electronic configuration $1s^2 2s^2 2p^6 3s^2 3p^6 3d^2 4s^2$.

Other physical properties of titanium are given under **Chemical Elements**.

Titanium occurs in practically all rocks and is an important constituent of many minerals. Only rutile TiO_2, however, is of commercial importance. The most important sources of this mineral are the sand dunes of Australia and Florida. Presently, Australia furnishes over 80% of the rutile requirements. Projects are underway to beneficiate (reduce) the other major potential titanium source, e.g., ilmenite. The known reserves of ilmenite $FeTiO_3$ are estimated 50 × greater than those of rutile. For mining the sand deposits for rutile, large floating dredge concentrators are used. Gravity concentration, followed by magnetic and electrostatic separation, yield a raw rutile of about 95% TiO_2 content. See also **Ilmenite**; and **Rutile**.

Production of titanium metal first involves the preparation of $TiCl_4$, a colorless liquid. Rutile and coke are charged into a continuous chlorinator. Upon the addition of chlorine gas, $TiCl_4$ is yielded in an exothermic reaction. To separate the metal, the $TiCl_4$, in a separate process, is reacted with molten magnesium metal pigs at about 50°C. The products are magnesium chloride $MgCl_2$ and titanium metal sponge. The by-product $MgCl_2$ is electrolyzed and the resulting magnesium and chlorine are recycled in the process. In another process, sodium metal is used instead of magnesium. And in still another process, the $TiCl_4$ may be electrolyzed.

Uses

The major uses for titanium are in various alloys, although unalloyed titanium finds some application. Titanium alloys are classified as alpha, alpha-beta, or beta, determined by the phases present in the alloy at room temperature. The alpha alloys usually result when the main elements present are the alpha stabilizers, e.g., oxygen, nitrogen, hydrogen, and carbon. Alpha-beta alloys and beta alloys contain increasing amounts of beta stabilizers, mainly vanadium, molybdenum, iron, chromium, manganese, tantalum, and niobium (columbium). The alpha-beta class of alloys normally has great room-temperature strength and may be heat treated. The annealed beta alloys show poor thermal stability over about 230°C, but do have good formability and weldability. The beta alloys may be age heat treated wherein some alpha phase is precipitated and this results in a very high room-temperature strength. The complexity of titanium alloys is brought about by the fact that the element is allotropic and undergoes a phase transformation at about 885°C, changing from one crystalline form to another as mentioned at the start of this entry. The variations in strength and percent elongation for the three major types of alloys and for pure titanium are given in Table 1.

Many diversified applications have been found for titanium and its alloys. Moreover, the number of these applications tends to increase steadily as greater production and improved processes reduce costs. At the present time, titanium is still an expensive material and is only used where its light weight, high strength, and corrosion resistance justify its cost. Aeronautical and missile design engineers find titanium and its alloys to be materials whose light weight and high strength, particularly at elevated temperatures (600°C), give them many applications in aircraft and missile construction. About 99% of all titanium materials are used in these fields.

Titanium and its alloys are widely used in compressor blades, turbine disks and many other forged parts of the jet engine. Here they offer resistance to high temperature, as well as weight-saving. The latter quality is increasing their use in the structural airplane parts, ranging from engines and air frames to skin and fastenings. Titanium sheet finds application in shroud assemblies, cable shrouds and ammunition tracks. Titanium alloy sheet is formed into ribs for use as stiffeners, as well as fuselage frames and bulk heads. Other uses of titanium in aircraft include channel sections, flat rubbing strips, landing gear doors, hydraulic lines, baffles, tail cones, longerons, etc. Other uses of titanium alloys include bulk heads, ducts, fire walls, etc.

The light weight of titanium and its alloys, coupled with their corrosion resistance, has brought them into use in ships, especially naval ships. Here many investigations show the important advantages of the metal and its alloys as wet exhaust muffles for submarine diesel engines, and as meter disks, and heat exchanger tubes which offer improved service for widespread use in salt water. Military applications of titanium extend from cannon and guided missiles to light-weight armor-plate for tanks. These materials offer other weight savings in other parts of military vehicles, such

TABLE 1. TITANIUM ALLOYS

Alloy	Tensile strength Psi	Tensile strength Mpa	Yield strength psi	Yield strength MPa	Percent elongation
PURE TITANIUM					
High purity (99.9%)					
Annealed	34,000	237	20,000	138	54
Commercial purity (99.0%)	79,000	545	63,000	435	27
ALPHA ALLOY					
Ti-5Al-2.5 Sn					
Annealed	125,000	863	120,000	828	18
ALPHA-BETA ALLOY					
Ti-6Al-4V					
Annealed	135,000	932	120,000	828	11
Heat treated	170,000	1173	150,000	1035	7
BETA ALLOY					
Ti-3Al-13V-11Cr					
Heat treated	180,000	1242	170,000	1173	6

Classification of alloys by application:

Airframe Alloys
Ti-75A, Ti-SAl-2.5Sn, Ti-6Al-6V-2Sn, Ti-6Al-4 V, Ti-7Al-4Mo,
Ti-4Al-3Mo-IV, Ti-8Mn, Ti-13V-11Cr-3Al

Engine Alloys
Ti-8Al-1Mo-IV, Ti-SAl-2.5Sn, Ti-6Al-4V, Ti-6Al-2Sn-4Zr-2Mo,
Ti-6Al-2Sn-4Zr-6Mo

Corrosion-Resistant Alloys
Ti-35A, Ti-50A, Ti-65A, Ti-0.2Pd, Ti-2Ni

as piston rods and transmissions, which may extend to the transportation industries generally.

Throughout the chemical industry, titanium is used extensively both in plant and in laboratory. Among important present-day applications are heat exchangers, autoclave heads, autoclave coils for cooling and heating, chemical processing racks, and valves and tanks where corrosion resistance is necessary.

Advancements in Titanium Technology

Increasingly, titanium alloys are competing with nickel-base alloys on the basis of cost, strength, and corrosion resistance. The alloy Ti-3Al-2.5 V, for example, is finding expanded use in the process industries because of its resistance to mildly reducing chloride environments.

Authorities in the field do not believe that the demand for titanium in the aerospace industry will be adversely affected by the increasing use of polymer-matrix composites in airframes and engines. Titanium has been replaced outright by composite materials in only a few aerospace applications. In those instances, the primary considerations have been weight reduction and "stealth" characteristics. By contrast, titanium has been selected instead of composites for several applications where the primary considerations have been titanium's superior stiffness and toughness as well as its multidirectional strength characteristics. Titanium also has gained favor because of its compatibility with composites. An example of this compatibility is found in titanium-aluminide foils. These are used to fabricate honeycomb structures. Fiber-reinforced alpha 2 titanium aluminide-matrix composites are fabricated by consolidating a foil/fiber/foil layup, using hot isostatic processing. The reinforcements are carbon-coated silicon-carbide continuous fibers.

Near-Net-Shape Processing

Even greater use of titanium in aircraft has been limited because of its relatively high cost (as compared with aluminum, by a factor of $10\times$ to $20\times$) and also because of its lower machinability. Gains are being made by titanium on both of these counts, however, by initially making titanium parts very close to the shape and dimensions of the desired end product (near-net-shape or NNS). These gains have been made through advanced titanium processing techniques and include precision casting, hot isostatic pressing, blended elemental (BE) powder techniques, and precision forging.

Increasingly, titanium alloys are competing with nickel-based alloys on the basis of cost, strength, and corrosion resistance.

Titanium Diboride

TiB_2 cermet is an extremely hard material. For several years, there was no successful process for depositing the material on parts for achieving wear resistance. A process was introduced by Montreal Carbide Co. in 1990, however, that deposits microspheres of TiB_2 in a metallic matrix. Using a conventional plasma-spray technique, a series of cermet coatings, in which the ceramic phase is finely and uniformly dispersed, yields a wear-resistant surface. TiB_2 is synthesized during spraying by a reaction between powders of titanium-bearing alloys and boron-bearing alloys. Due to the rapid solidification, the TiB_2 crystals that form are finely dispersed in a metallic matrix, which originated from the alloys used in the process.

Chemistry and Compounds

Due to its $3d^2 4s^2$ electron configuration, titanium forms tetravalent compounds readily, although the Ti^{4+} ion does not exist as such in aqueous solution, except at very low or high pH values, the common cation being hydrated TiO^{2+} (or more probably $Ti(OH)_2^{2+}$). Many of the tetravalent compounds are largely covalent. There are also Ti(III) and a few Ti(II) compounds, the latter being very easily oxidized. Titanium dioxide, TiO_2, is well known both as a mineral, of which three structural forms exist, and as an industrial product obtained from ferrous titanate, $FeTiO_3$ ores or by oxidation of tin(IV) chloride, $TiCl_4$. See also **Titanium Dioxide.** Moreover, the precipitate obtained by action of alkali metal hydroxides upon solutions of tetravalent titanium is a hydrated oxide. The latter is readily soluble in acids to form oxysalts, which are usually formulated in terms of the TiO^{2+} ion, without including its water of hydration, e.g., as $NaTiOPO_4$. The hydrated TiO^{2+} ion is not amphiprotic, in that it does not dissolve in alkali hydroxides. However, it does react on fusion with alkali carbonates to form such compounds as M_2TiO_3 and $M_2Ti_2O_5$, these compounds having been shown to be mixed oxides rather than titanates. The alkaline earth titanates have the face-centered perovskite structure, and barium titanate, widely used for its electrical properties, has been produced in other crystalline forms.

Lower oxides of titanium, Ti_2O_3 and TiO, have been produced by reduction of TiO_2.

All four of the common halogens form tetrahalides of titanium, $TiCl_4$ being a liquid at ordinary temperatures, while TiF_4, $TiBr_4$, and TiI_4 are solids. They are readily hydrolyzed, yielding as end products TiO_2 and the hydrogen halide, in the case of $TiCl_4$ an intermediate addition product of the type $H_2O \cdot TiCl_4$ is considered to be formed. This is in accordance with the behavior of $TiCl_4$ and $TiBr_4$ as Lewis acids to form such unstable adducts, not only with water, but with oxygen-function organic compounds. Likewise, titanium chelates are formed with oxygen donor compounds such as acetylacetone.

The dihalides of titanium, formed by reduction of the tetrahalides, are vigorous reducing agents and unstable; $TiCl_2$ is inflammable in air. The trihalides, though more stable than the dihalides, are effective reducing agents. Ti(III) occurs in aqueous solutions as $Ti(H_2O)_6^{3+}$.

Normal oxyacid salts of titanium are unknown, but many basic salts, formulated as stated above, in terms of TiO^{2+}, though more or less hydrated, have been prepared.

Like the oxide, halides and sulfide, the nitride, boride, and carbide of titanium(IV) can be made by heating the elements together at high temperatures. The last three compounds are alloy-like in character, they can vary in composition without becoming unstable and they are extremely hard.

The halogen complexes are the most stable complex ions of titanium. The hexafluorotitanate ion, TiF_6^{2-} is very stable, as are the peroxo-complexes, containing $-Ti-O-O-$. The $TiCl_6^{2-}$ and $TiBr_6^{2-}$ complexes are less stable, except in concentrated solutions of the hydrogen halides. A number of compounds of the $TiCl_5^{2-}$ ion are known, especially of the higher alkali metals, e.g., $M_2TiCl_5 \cdot H_2O$.

Additional Reading

Bauccio, M.L.: *ASM Engineered Materials Reference Book,* 2nd Edition, ASM International, Materials Park, OH, 1994.

Brady, G.S., H.R. Clauser, and J.A. Vaccari: *Materials Handbook,* 14th Edition, McGraw-Hill Professional Book Group, New York, NY, 1996.

Carter, G.F. and D.E. Paul: *Materials Science and Engineering,* ASM International, Materials Park, OH, 1991.

Chiles, J.R.: "Titanium" *Smithsonian,* 86 (May 1987).

Collings, E.W. and G. Welsch: *Materials Properties Handbook: Titanium Alloys,* ASM International, Materials Park, OH, 1995.

Copley, S.M.: *Applied General and Nonferrous Physical Metallurgy,* Encyclopedia of Materials Science and Engineering, MIT Press, Cambridge, MA, 1986.

Davis, J.R.: *ASM Materials Engineering Dictionary,* ASM International, Materials Park, OH, 1992.

Donachie, M.J.: *Titanium: A Technical Guide,* 2nd Edition, ASM International, Materials Park, OH, 2000.

Gauthier, M.M.: *Engineered Materials Handbook,* ASM International, Materials Park, OH, 1995.

Greenwood, N.N. and A. Earnshaw: *Chemistry of the Elements,* 2nd Edition, Butterworth-Heinemann, Inc., Woburn, MA, 1997.

Jha, S.C., et al.: "Titanium-Aluminide Foils," *Advanced Materials & Processes,* 87 (April 1991).

Krebs, R.E.: *The History and Use of Our Earth's Chemical Elements: A Reference Guide,* Greenwood Publishing Group, Inc., Westport, CT, 1998.

Kubel, E.J., Jr.: "Titanium Near-Net-Shape Technology Shaping Up," *Advanced Materials & Processes,* 46 (February 1987).

Lide, D.R.: *CRC Handbook of Chemistry and Physics,* 84th Edition, CRC Press, LLC., Boca Raton, FL, 2003.

Nelson, O.E.: "Titanium Staves Off Composites," *Advanced Materials & Processes,* 18 (June 1991).

Square, M.: "Titanium Boride Cermet: New Wear-Resistant Coating," *Advanced Materials & Processes,* 117 (April 1990).

Staff: "Navy Studies Titanium Applications," *Advanced Materials & Processes,* 14 (September 1989)

Staff: *ASM Handbook—Properties and Selection: Nonferrous Alloys and Pure Metals,* ASM International, Materials Park, OH, 1990.

Staff: "Titanium and Titanium Aluminides," *Advanced Materials & Processes,* 71 (April 1990).

Staff: "Titanium Forecast," *Advanced Materials & Processes,* 18 (January 1991).

Staff: "Profile of Titanium Manufacturers," *Advanced Materials & Processes,* 52 (June 1991)

TITANIUM DIOXIDE. [CAS: 13463-67-7]. TiO_2, formula weight 79.90, variously colored, depending upon source, but white when purified and sold in commerce. Decomposes at about 1,640°C before melting, density 4.26 g/cm^3, insoluble in H_2O, soluble in H_2SO_4 or alkalis. Titanium dioxide is a very high-tonnage material and is the principal white pigment of commerce. The compound has an exceptionally high refractive index, great inertness, and a negligible color, all qualities that make it close to an ideal white pigment. Annual production approximates two million metric tons, of which nearly one-half of this amount is produced in the United States. Major uses of TiO_2 pigments are: (1) paint, 60%, (2) paper, 14%, (3) plastics and floor coverings, 12%, (4) printing inks, 3%, and (5) various applications including rubber, ceramics, roofing granules, and textiles, 11%.

Two major processes are used for producing raw titanium dioxide pigment: (1) the sulfate process, a batch process accounting for over half of current production, introduced by European makers in the early 1930s; and (2) the chloride process, a continuous process, introduced in the late 1950s and accounting for most of the new plant construction since the mid-1960s. The sulfate process can handle both rutile and anatase, but the chloride process is limited to rutile.

In the sulfate process, ilmenite (45–60% TiO_2) or a slag rich in titanium (70% TiO_2) obtained from electric smelting of ilmenite, is the feedstock. The raw materials first are digested: $FeTiO_3 + 2H_2SO_4 \longrightarrow FeSO_4 + TiO \cdot SO_4 + 2H_2O$. In a second step, the concentrated liquor is nucleated, diluted with H_2O, and boiled until nearly all of the titanium has precipitated out in the form of flocculated titanium dioxide (anatase) hydrate: $TiO \cdot SO_4 + 2H_2O \longrightarrow TiO_2 \cdot H_2O + H_2SO_4$. After filtering, the cake is leached under reducing conditions to remove residual iron. Conditioning agents are added, after which the hydrate is dried and calcined in a rotary kiln at approximately 900°C: $TiO_2 \cdot H_2O \longrightarrow TiO_2 + H_2O$. The conditioning agents usually consist of a phosphate and a potassium salt, as well as zinc, antimony, and aluminum compounds. The purpose of these additions is to improve the final properties of the pigment, including color, photochemical stability, and dispersibility, as well as to catalyze the formation of rutile from the anatase hydrate.

In the chloride process, the feedstock must be high in titanium and low in iron. Mineral rutile (95% TiO_2) is best suited, but leucoxene (65% TiO_2) can be used. See also **Brookite.** An economical conversion of ilmenite for use as a chloride process feedstock has not been developed to date. The ore is mixed with coke and chlorinated at about 900°C in a fluidized bed. The principal product is titanium tetrachloride, but other impurities including iron also are chlorinated and thus must be removed by selective condensation and distillation. Up to this point, the process is similar to that of producing titanium metal as described under **Titanium.** By selective reduction prior to distillation, vanadium present is removed as $VOCl_3$. In the next step, the purified $TiCl_4$ reacts with oxygen at a temperature of about 1,000°C. The presence of $AlCl_3$ in this reaction promotes the formation of rutile instead of anatase. The two major steps are: (1) chlorination: $3TiO_2 + 4C + 6Cl_2 \longrightarrow 3TiCl_4 + 2CO + 2CO_2$; and (2) oxidation: $TiCl_4 + O_2 \longrightarrow TiO_2 + 2Cl_2$. The chlorine is recycled. The raw titanium dioxide product generally is neutralized by washing in an aqueous solution of proper pH.

Many grades of titanium dioxide pigments are offered commercially. They range in crystal structure (anatase or rutile), particle shape and size, the type of hydrous oxide coating applied, and the type and quantity of additives applied. Generally, the commercial pigments contain 80–99% TiO_2, the remainder of the formulation comprised of alumina and silica hydrates. Nonpigmentary grades of titanium dioxide for the glass, welding-rod, electroceramic, and vitreous-enamel industries contain 99% TiO_2.

TITRATION (Potentiometric). This analytical method is based in principle upon the Nernst equation, which may be written in the form

$$E = E^0 - \frac{0.059}{n} \log \frac{A_{ox}}{A_{red}}$$

where E is the measured electromotive force, E^0 is the standard value of the electromotive force (electrode potential) when the substances of the electrochemical reaction are in their standard states, n is the valence change (change in number of electrons per mole of the reactants) and the A-terms are the activities of the oxidized and reduced forms of the reactants. See also **Activity Coefficient.** Activities are proportional to concentrations, so that the concentration of one of the reactants may be determined if that of the other is known. Thus the concentration of Cu^{2+} ions in a solution could be found by use of an electrode of metallic copper (unstrained metals are assumed to be at unit activity). Or the concentration of Cl^- ions in a solution could be found by use of an electrode of (insoluble) silver chloride $AgCl$ deposited on an electrode of metallic silver. Or the concentrations in a solution containing two ions in different states of oxidation, such as Fe^{2+} and Fe^{3+}, could be found by the use of an inert electrode, such as one of platinum. The three reactions and corresponding forms of Nernst's equation (using the approximation of substituting concentrations for activities) are:

$$Cu^0 \Longleftrightarrow Cu^{2+} + 2e^-$$

$$E = E^0 - \frac{0.0591}{2} \log c_{Cu^{2+}}$$

$$Ag^+ + Cl^- \Longleftrightarrow AgCl \downarrow$$

$$E = E^0 - (0.0591) \log K_{sp} + (0.0591) \log c_{Cl^-}$$

(where K_{sp} is the solubility product of $AgCl$)

$$Fe^{2+} \Longleftrightarrow Fe^{3+} + e^-$$

$$E = E^0 - (0.0591) \log \frac{c_{Fe^{3+}}}{c_{Fe^{2+}}}$$

In constructing an electric cell for potentiometric titrations it is necessary, of course, to use a second electrode to complete the circuit, in addition to the measuring electrodes (commonly called *indicator electrodes*) described above. Ideally the second electrode would be a hydrogen electrode which (as explained in the entry on electrode potential) is the standard reference electrode for which the potential, in equilibrium with its ions, is defined as zero. Since it is awkward to use, other electrodes of known potential, such as the calomel electrode or the glass electrode, are commonly used as reference electrodes. The arrangement of the apparatus is as shown in Fig. 1.

The procedure in a potentiometric titration is to determine the potential of the indicator electrode after each addition of the titrating solution. This is done by closing the switch in the circuit shown in Fig. 1 just long enough to read the potential on the sensitive measuring device used, such as a potentiometer or vacuum tube voltmeter. Of course, very small additions of titrant are made as the expected endpoint is approached. Then the readings of potential are plotted against volume of titrant added, as shown in Fig. 2. Reactions suitable for determination by this method show sharply defined endpoints as pictured in the figure. To correspond to the true stoichiometric endpoint, certain conditions should be met, including reversibility of the reaction and allowing time for the electrodes to reach

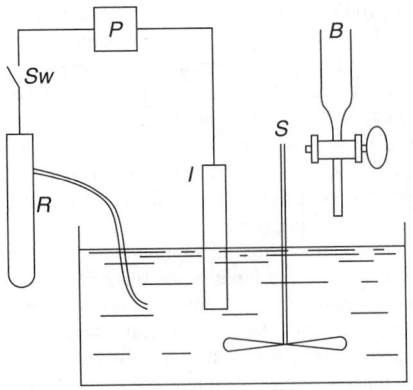

Fig. 1. Apparatus for potentiometric titration: (*R*) reference electrode; (*I*) indicating electrode; (*P*) potentiometer; (*B*) burette; (*S*) stirrer; (*Sw*) switch

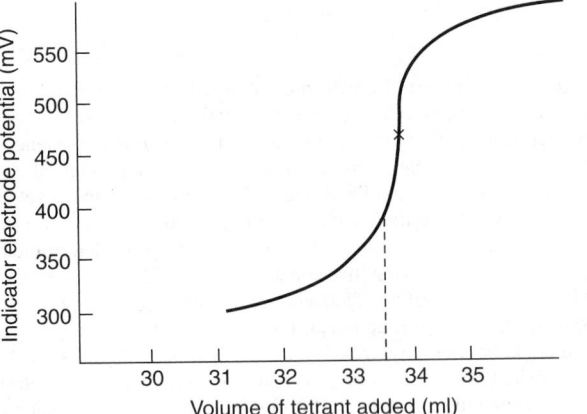

Fig. 2. Potentiometric titration curve

equilibrium before closing the switch to read the potential difference. While the description given was on the basis of titration with zero current drawn from the electrodes between readings, there is also a method in which a small constant current is drawn, not large enough to affect their potentials, but serving to eliminate some of the sources of error in the null method.

Types of titrations for which potentiometric methods of determining endpoints are particularly useful including titrations of halide mixtures, of various metals, of alkaloids, in non-aqueous solvents, and various titrations with oxidizing agents, such as permanganate, dichromate, iodate, and ceric sulfate.

TITRATION (Thermometric).   This technique consists of the detection and measurement of the change in temperature of a solution as the titrant is added to it, under as near adiabatic conditions as possible. Experimentally, the titrant is added from a constant-delivery burette into the titrate (solution to be titrated) which is contained in an insulated container such as a Dewar flask. The resultant temperature-volume (or time) curve thus obtained is similar to other titration curves, e.g., acid-base, in that the end point of the reaction can be readily ascertained. Since all reactions involve a detectable

endothermic or exothermic enthalpy change, the technique has potentially wide application in analytical chemistry, especially in those cases where other more common methods are not applicable.

Several idealized thermometric titration curves for an exothermic reaction are given in Fig. 1. A titration curve in which the titrant and the titrate are at the same initial temperature is illustrated in (a). The actual titration is preceded by a blank run, *CD*, in which no titrant is added to the titrate. At *D*, titrant is added, causing the temperature of the titrate to rise rapidly, reaching a maximum value at *E*. Beyond *E*, additional titrant causes no further change in the temperature of the titrate, hence, the horizontal excess reagent line, *EF*. The temperature rise of the titrate (and titration vessel), ΔT, is obtained by determining the vertical distance between the excess reagent line, *ED'*, and *CD*. In curve (b), the conditions of the titration were identical to those of (a) except that the titrant was at a higher temperature than the titrate, hence, the sloping excess reagent line, *EF*. For curve (c), conditions were also identical to the above except that the titrant was at a lower temperature than the titrate. The excess reagent line, *EF*, thus slopes in an opposite direction to that of curve (b). For an endothermic reaction, the curves would be identical to the above except that the temperature changes would be in an opposite direction.

TOCOPHEROLS.   See **Vitamin E.**

TODD, SIR ALEXANDER R. (1907–1997).   A British chemist who won the Nobel prize for chemistry in 1957. His diverse research and accomplishments involved phosphorylation and mechanisms of biological reactions concerning phosphates. Many of his studies concerned the structure of nucleic acids, nucleotides, nucleotidic coenzymes, as well as vitamins B_1, B_{12}, and E. Work in biological organic chemistry indicated that hemp plant could be used for production of narcotics. Todd had degrees awarded from Oxford, Frankfurt, and Glasgow, among others.

TODOROKITES.   Calcium-bearing manganese oxides found in terrestrial manganese ore depositions, in weathering products of manganese-bearing rocks, and in some manganese nodules. Todorokites are, in many instances, principal constituents of manganese nodules. Host copper and nickel within the modules is potentially of economic importance. Knowledge of todorokites is important for understanding how nodules form and how they concentrate transition elements from ocean waters. See **Ocean Resources (Mineral).**

Formerly believed to be a single phase, todorokite was discredited as a mineral by the International Mineralogical Association Commission on New Minerals and Mineral Names in 1970 when evidence was submitted showing it to be a complex mixture of several compounds. As reported by Turner and Buseck (1981), many mineralogists felt the decision was incorrect since recognizable x-ray diffraction patterns could be produced from todorokite collected from widespread deposits. X-ray patterns have not been adequate for structure determination. Further confusion is caused by the variable morphology of todorokite, which appears fibrous in some samples and platy in others. Todorokite material appears to have a range of related structures. High-resolution transmission electron microscopy reveals that terrestrial todorokites consist of tunnel structures of previously unreported dimensions and that these tunnel structures are intergrown coherently on a unit cell scale. As shown in Fig. 1, many types and degrees of disorder are evident. The widespread presence of disorder explains prior confusion with x-ray diffraction patterns and hence the difficulty to relate x-rays studies with specific structures. Turner and Buseck suggested a revised

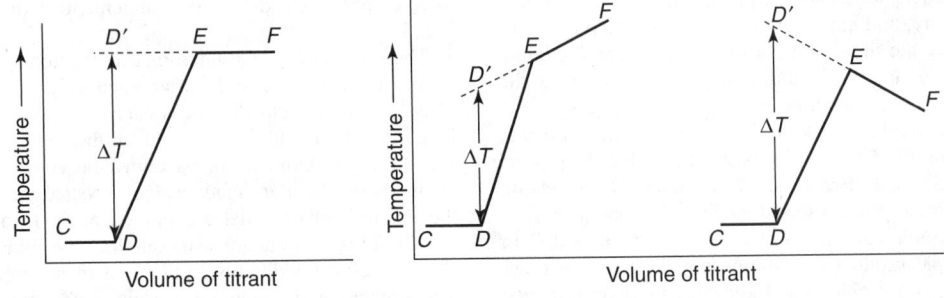

Fig. 1.   Thermometric curves for an exothermic reaction (ideal)

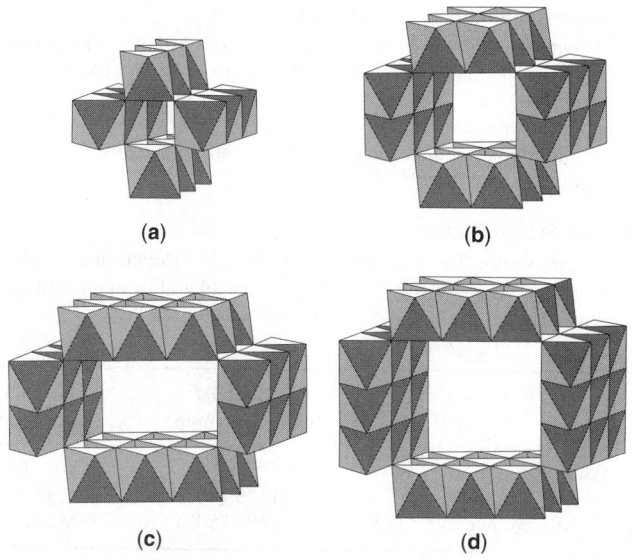

(a)   **(b)**

(c)   **(d)**

Fig. 1. Diagram of various manganese oxide tunnel structures. The common smallest unit of the structures is a $[Mn^{+3}, Mn^{+4}]_6$ octahedron. The octahedra are edge linked to form long chains that are corner linked to form the framework of tunnel minerals: (**a**) 1×1, pyrolusite, (**b**) 2×2, hollandite, (**c**) 2×3, romanechite and (**d**) 3×3, todorokite. The larger tunnel structures (hollandite, romanechite, todorokite) can accommodate large cations and water. There are many more possible tunnel structures and those with a common dimension (for instance the double chain sides of hollandite and romanechite) can intergrow on a unit cell level. As far as is now known, todorokite occurs only as an intergrowth of differently sized tunnel structures that have a triple chain in common (3×2, 3×3, 3×5, 3×6, etc.) The 3×3 structure is the most prevalent. (*Source: S. Turner*)

nomenclature of tunnel manganese oxides. This is described in *Science*, **212**, 1024–1027 (1981).

TOLUENE. Toluene, [CAS: 108-88-3], C_7H_8, is a colorless, mobile liquid with a distinctive aromatic odor somewhat milder than that of benzene. Prior to World War I, the main source of toluene was coke ovens. Petroleum became the source for toluene with the advent of catalytic reforming and the need for large quantities of toluene for use in aviation fuel during World War II. Since then, manufacture of toluene from petroleum sources has continued to increase, and manufacture from coke ovens and coaltar products has continued to decrease.

Toluene is generally produced along with benzene, xylenes, and C_9-aromatics by the catalytic reforming of C_6–C_9 naphthas. There have been, ca 1997, recent technological developments to produce benzene, toluene, and xylenes from pyrolysis of light hydrocarbons C_2–C_5, LPG, and naphthas. See also **Xylenes and Ethylbenzene**. About 85–90% of the toluene produced annually in the U.S. is not isolated, but is blended directly into the gasoline pool as a component of reformate and of pyrolysis gasoline.

Derivatives are formed by substitution of the hydrogen atoms of the methyl group, by substitution of the hydrogen atoms of the ring, and by addition to the double bonds. Substitutions on the methyl group are generally high-temperature, free-radical reactions. Thus, chlorination at ca 100°C, or in the presence of uv or other free-radical initiators, successively gives benzyl chloride, benzal chloride, and benzotrichloride. See also **Benzaldehyde**; and **Benzoic Acid**. With oxygen in the liquid phase, particularly in the presence of a catalyst, good yields of benzoic acid are obtained. In the presence of alkali metals, toluene is alkylated. With a lithium catalyst and a chelating compound, telomers are obtained with ethylene.

CH₃
|
[benzene ring] → CH₂—(C₂H₄)ₙ—CH₂CH₃
|
[benzene ring]

C_2H_4,Li,TMEDA
110°C

Additions to the double bonds results from both free-radical and catalytic reactions, e.g., chlorination at 0°C and hydrogenation. Usually, all three double bonds react. Substitution of the ring hydrogen atoms by electrophilic

attack takes place with the same reagents that react with benzene. Some of the common groups with which toluene can be substituted directly include

$$-Cl, \quad -Br, \quad -\overset{\overset{O}{\parallel}}{C}CH_3, \quad -SO_3H, \quad -NO_2, \quad -(C_nH_{2n+1}), \quad \text{and} \quad -CH_2Cl.$$

Under the same conditions, toluene reacts more rapidly than benzene. These reactivities and the related selectivity to the ortho and para positions can be explained in terms of the inductive effect of the methyl group. Toluene requires substitution by strongly negative groups, such as NO_2, to react with anions.

Substitution Reactions on the Methyl Group. The reactions that give substitution on the methyl group are generally high temperature and free-radical reactions. Thus, chlorination at ca 100°C, or in the presence of ultraviolet light and other free-radical initiators, successively gives benzyl chloride, benzal chloride, and benzotrichloride.

CH₃   CH₂Cl   CHCl₂   CCl₃
[benzene] →(Cl₂/hν) [benzene] + [benzene] + [benzene]

This oxidation reaction which yields benzoic acid is another example of this type of reaction.

In the presence of alkali metals such as potassium and sodium, toluene is alkylated with ethylene on the methyl group to yield, successively, normal propylbenzene, 3-phenylpentane, and 3-ethyl-3-phenylpentane.

In the formation of π-complexes with electrophiles such as silver ion, hydrogen chloride, and tetracyanoethylene, toluene differs from either benzene or the xylenes by a factor of less than two in relative basicity.

Properties

Physical and thermodynamic properties are given in Table 1. Toluene forms azeotropes with many hydrocarbons and most alcohols that boil in a similar range; all are minimum-boiling azeotropes. Toluene, water, and alcohols frequently form ternary azeotropes.

TABLE 1. PHYSICAL PROPERTIES OF TOLUENE

Property	Value
molecular weight	92.14
melting point, K	178.15
normal bp, K	383.75
critical temperature, K	591.80
critical pressure, MPa[a]	4.108
critical volume, L/(g · mol)	0.316
critical compressibility factor	0.264
acentric factor	0.262
flash point, K	278
autoignition temperature, K	809

GAS PROPERTIES, 298.15°K

H_f, kJ/mol[b]	50.17
G_f, kJ/mol[b]	122.2
C_p, J/(mol · K)[b]	104.7
H_{vap}, kJ/mol[b]	38.26
H_{comb}, kJ/mol[b]	−3734
viscosity, mPa · s(= cP)	0.00698
flammability limits, in air[c], vol%	
lower limit at 1 atm	1.2
upper limit at 1 atm	7.1

LIQUID PROPERTIES, 298.15°K

density, L/mol	9.38
C_p, J/(mol · K)[b]	156.5
viscosity, mPa · (= cP)	0.548
thermal conductivity, W/(m · K)	0.133
surface tension, mN · m(= dyn/cm)	27.9

SOLID PROPERTIES

density at 93.15°K, L/mol	11.18
C_p at 178.1 K, J/(mol · K)[b]	90.0
heat of fusion at 178.15°K, kJ/mol[b]	6.62

[a] To convert MPa to psi, multiply by 145.
[b] To convert J to cal, divide by 4.184.
[c] At 101.3 kPa (1 atm).

Because of the high electron density in the aromatic ring, toluene behaves as a base both in the formation of charge-transfer π-complexes and in the formation of sigma complexes. When only π-electrons are involved, toluene behaves much like benzene and xylene. When σ-bonds and complexes are involved, toluene reacts much faster than benzene and much slower than xylenes.

Manufacture and Processing

The principal source of toluene is catalytic reforming of refinery streams. This source accounts for ca 79% of the total toluene produced. An additional 16% is separated from pyrolysis gasoline produced in steam crackers during the manufacture of ethylene and propylene. The reactions taking place in catalytic reforming to yield aromatics are dehydrogenation or aromatization of cyclohexanes, dehydroisomerization of substituted cyclopentanes, and the cyclodehydrogenation of paraffins. The formation of toluene by these reactions is shown.

Of the main reactions, aromatization takes place most readily and proceeds ca 7 times as fast as the dehydroisomerization reaction and ca 20 times as fast as the dehydrocyclization. Hence, feeds richest in cycloparaffins are most easily reformed.

Because catalytic reforming is an endothermic reaction, most reforming units comprise about three reactors with reheat furnaces in between to minimize kinetic and thermodynamic limitations caused by decreasing temperature. There are three basic types of operations, i.e., semiregenerative, cyclic, and continuous. In the semiregenerative operation, feedstocks and operating conditions are controlled so that the unit can be maintained on-stream from 6 mo-2 yr before shutdown and catalyst regeneration. In cyclic operation, a swing reactor is employed so that one reactor can be regenerated while the other three are in operation. Regeneration, which may be as frequent as every 24 h, permits continuous operation at high severity. Since ca 1970, continuous units have been used commercially. In this type of operation, the catalyst is continuously withdrawn, regenerated, and fed back to the system.

Toluene, Benzene, and BTX Recovery. The composition of aromatics centers on the C_7- and C_8-fraction, depending somewhat on the boiling range of the feedstock used. Most catalytic reformate is used directly in gasoline. That part which is converted to benzene, toluene, and xylenes for commercial sale is separated from the unreacted paraffins and cycloparaffins or naphthenes by liquid–liquid extraction or by extractive distillation.

Proper choice of feedstocks and use of relatively severe operating conditions in the reformers produce streams high enough in toluene to be directly usable for hydrodemethylation to benzene without the need for extraction. Toluene is recovered from pyrolysis gasoline, usually by mixing the pyrolysis gasoline with reformate and processing the mixture in a typical aromatics extraction unit. Yields of pyrolysis gasoline and the toluene content depend on the feedstock to the steam-cracking unit. Pyrolysis gasoline is hydrotreated to eliminate dienes and styrene before processing to recover aromatics.

Emerging Technologies for Production of BTX from Light Hydrocarbons. Recent technological developments have centered on high temperature pyrolysis of light hydrocarbons C_2 to C_5, LPG, and naphtha to form aromatics in higher yields. Conversions were traditionally low because they were accompanied by a high degree of degradation to carbon and hydrogen. Recent improvements include modification of the thermal cracking process to produce higher yields of liquid products rich in aromatics and the extension of the catalytic hydroforming process to promote oligomerization and dehydrocyclization of the lower olefins. The common core of these developments is the use of shape-selective zeolite catalysts to promote the various reactions.

Specifications, Standards, and Quality Control

Toluene is marketed mostly as nitration and industrial grades. The generally accepted quality standards for the grades are given by ASTM D841 and D362, respectively.

Purity of toluene samples as well as the number, concentration, and identity of other components can be readily determined using standard gas chromatography techniques.

Health and Safety Factors

Permissible exposure limits established by the U.S. Department of Health and Human Services and the U.S. Department of Labor are summarized below, with the more restrictive levels proposed by NIOSH.

	OSHA, mg/m^3 (ppm)	NIOSH, mg/m^3 (ppm)
average during 8-h shift (TWA)	752 (200)	376 (100)
not to exceed	1129 (300)	
except for 10-min average (TLV)	1181 (500)	752 (200)

Toluene generally resembles benzene closely in its toxicological properties; however, it is devoid of benzene's chronic negative effects on blood formation.

Uses

About 90% of the toluene generated by catalytic reforming is blended into gasoline as a component of $>C_5$ reformate. The octane number $(R + M/2)$ of such reformates is typically in the range of 88.9–94.5, depending on severity of the reforming operation. Toluene itself has a blending octane number of 103–106, is exceeded only by oxygenated compounds such as methyl *tert*-butyl ether, ethanol, and methanol.

Toluene is a valuable blending component, particularly in unleaded premium gasolines. Although reformates are not extracted solely for the purpose of generating a high octane blending stock, the toluene that is co-produced when xylenes and benzene are extracted for use in chemicals, and that exceeds demands for use in chemicals, has a ready market as a blending component for gasoline.

Toluene is converted to benzene by hydrodemethylation either under thermal or catalytic conditions. Benzene produced from this source generally supplies 25–30% of the total benzene demand. The feedstock is usually extracted toluene, but some reformers are operated under sufficiently severe conditions or with selected feedstocks to provide toluene pure enough to be fed directly to the dealkylation unit without extraction.

Toluene is more important as a solvent than either benzene or xylene. Solvent use accounts for ca 14% of the total U.S. toluene demand for chemicals. About two-thirds of the solvent use is in paints and coatings; the remainder is in adhesives, inks, pharmaceuticals, and other formulated products utilizing a solvent carrier. Use of toluene as solvent in surface coatings has been declining, primarily because of various environmental and health regulations. It is being replaced by other solvents.

Potential Uses. Because much toluene is demethylated for use as benzene, considerable effort has been expended on developing processes in which toluene can be used in place of benzene to make directly from toluene the same products that are derived from benzene. Such processes both save the cost of demethylation and utilize the methyl group already on the ring. Most of this effort has been directed toward manufacture of styrene. An alternative approach is the manufacture of *para*-methylstyrene by selective ethylation of toluene, followed by dehydrogenation. Resins from this monomer are expected to displace polystyrene because of price and performance advantages.

Derivatives

Toluene Diisocyanate. Toluene diisocyanate is the basic raw material for production of flexible polyurethane foams. It is produced by the reaction in which toluene is dinitrated, the dinitrotoluene is hydrogenated to yield 2,4-diaminotoluene, and this diamine in turn is treated with phosgene to yield toluene 2,4-diisocyanate.

Benzoic Acid. Benzoic acid is manufactured from toluene by oxidation in the liquid phase using air and a cobalt catalyst. Typical conditions are 308–790 kPa (30–100 psi) and 130–160°C. The crude product is purified by distillation, crystallization, or both. Yields are generally >90 mol%, and product purity is generally >99%.

Benzyl Chloride. Benzyl chloride is manufactured by high temperature free-radical chlorination of toluene. The yield of benzyl chloride is maximized by use of excess toluene in the feed. More than half of the benzyl chloride produced is converted by butyl benzyl phthalate by reaction with monosodium butyl phthalate. The remainder is hydrolyzed to benzyl alcohol, which is converted to aliphatic esters for use in soaps, perfume, and flavors. Benzyl salicylate is used as a sunscreen in lotions and creams.

Disproportionation to Benzene and Xylenes. With acidic catalysts, toluene can transfer a methyl group to a second molecule of toluene to yield one molecule of benzene and one molecule of mixed isomers of xylene. This disproportionation is an equilibrium reaction. Disproportionation generates benzene from toluene and at the same time takes full advantage of the methyl group to generate a valuable product, i.e., xylene. Economic utility of the process is strongly dependent on the relative values of toluene, benzene, and the xylenes.

Vinyltoluene. Vinyltoluene is used as a resin modifier in unsaturated polyester resins. Its manufacture is similar to that of styrene; toluene is alkylated with ethylene, and the resulting ethyltoluene is dehydrogenated to yield vinyltoluene.

Toluenesulfonic Acid. Toluene reacts readily with fuming sulfuric acid to yield toluene–sulfonic acid. By proper control of conditions, *p*-toluenesulfonic acid is obtained. The primary use is for conversion, by fusion with NaOH, to *p*-cresol. The resulting high purity *p*-cresol is then alkylated with isobutylene to produce 2,6-di-*tert*-butyl-*p*-cresol (BHT), which is used as an antioxidant in foods, gasoline, and rubber. Mixed cresols can be obtained by alkylation of phenol and by isolation from certain petroleum and coal–tar process streams.

Benzaldehyde. Annual production of benzaldehyde requires ca 6,500–10,000 t (2–3 × 10⁶ gal) of toluene. It is produced mainly as by product during oxidation of toluene to benzoic acid, but some is produced by hydrolysis of benzal chloride. The main use of benzaldehyde is as a chemical intermediate for production of fine chemicals used for food flavoring, pharmaceuticals, herbicides, and dyestuffs.

Toluenesulfonyl Chloride. Toluene reacts with chlorosulfonic acid to yield both *o*- and *p*-toluenesulfonyl chlorides. The ortho isomer is converted to saccharin. The para isomer is used for preparation of specialty chemicals. Annual toluene requirements are ca 6500 t (2 × 10⁶ gal).

Miscellaneous Derivatives. Other derivatives of toluene, none of which is estimated to consume more than ca 3000 t (10⁶ gal) of toluene annually, are mono- and dinitrotoluene hydrogenated to amines; benzotrichloride and chlorotoluene, both used as dye intermediates; *tert*-butylbenzoic acid from *tert*-butyltoluene, used as a resin modifier, dodecyltoluene converted to a benzyl quaternary ammonium salt for use as a germicide; and biphenyl, obtained as by-product during demethylation, used in specialty chemicals. Toluene is also used as a denaturant in specially denatured alcohol (SDA) formulas 2-B and 12-A.

Acknowledgment

To R. A. Wilsak and M. E. Carrera (Amoco Chemical Co.) and O. C. Okoroafor (Cooper Union for the Advancement of Science and Arts).

E. DICKSON OZOKWELU
Amoco Chemical Company

Additional Reading

Stull, D.R.: and co-workers, *Chemical Thermodynamics of Hydrocarbon Compounds,* John Wiley & Sons, Inc., New York, NY, 1969, p. 368.
1996–97 Toluene–Xylenes Annuals, Dewitt & Co., Inc., Houston, TX, Jan. 1997.
Weissermel, K. and H. Arpe: *Industrial Organic Chemistry,* Verlag Chemie, New York, NY, 1978, pp. 288–289.
Wisniak, J. and A. Tamir: *Liquid–Liquid Equilibrium and Extraction,* Elsevier Scientific Publishing Co., Amsterdam, Pt. A, 1980; Pt. B, 1981; Suppl. 1, 1985; Suppl. 2, 1987.

TOPAZ. The mineral topaz is a silicate of aluminum and fluorine corresponding to the formula $Al_2SiO_4(F, OH)_2$. It is orthorhombic and its crystals are mostly prismatic terminated by pyramidal and other faces, the basal pinacoid being often present. Massive varieties are known. It has an easy and perfect basal cleavage, hence for this reason gems or fine specimens should be handled with care to avoid developing cleavage flaws. The fracture is conchoidal to uneven; hardness, 8; specific gravity, 3.4–3.6; luster, vitreous; color, of typical topaz, wine or straw-yellow but may be colorless, white, gray, green, blue or reddish-yellow; transparent to translucent. When heated, yellow topaz often becomes a reddish pink.

Topaz is found associated with the more acid rocks of the granite and rhyolite type and may occur with fluorite and cassiterite. Topaz comes from many localities, a few of which are: the former U.S.S.R. in the Urals and the Ilmen Mountains; the Czech Republic and Slovakia, Saxony, Norway, Sweden, Japan, Brazil and Mexico. In the United States, topaz has been found in Oxford County, Maine; Carroll County, New Hampshire; Fairfield County, Connecticut; El Paso and Chaffee Counties, Colorado; and in Texas, Utah and California.

The name *topaz* is derived from the Greek meaning to seek. It was the name of an island in the Red Sea that was difficult to find, and from which a yellow stone, now believed to be a yellowish-olivine, was obtained in ancient times. In the Middle Ages, any yellow stone was called topaz, but now the name is properly applied only to the species here described.

TOPOCHEMICAL REACTION. Any chemical reaction that is not expressible in stoichiometric relationships. Such reactions are characteristic of cellulose; they can take place only at certain sites on the molecule where reactive groups are available, i.e., in the amorphous areas or on the surfaces of the crystalline areas.

TORBERNITE. An ore of uranium with the composition, $Cu(UO_2)_2(PO_4)_2 \cdot 8 – 12H_2O$, green, radioactive, tetragonal, and isomorphous with autunite. Occurring in tabular crystals or in foliated form, the mineral is commonly a secondary mineral.

TOTH PROCESS. A process for production of aluminum metal that utilizes kaoliln and other high-alumina clays. The clay is chlorinated after calcination, and the aluminum chloride resulting is reacted with metallic manganese to yield aluminum and manganese chloride. The reaction occurs at the comparatively low temperature of 260°C. The manganese chloride is recovered as manganese metal and chlorine by oxidation and subsequent reduction, the manganese being recycled. This is a much cheaper and more efficient method than the Hall process, because it requires less energy input and does not utilize imported bauxite.

TOURMALINE. The mineral tourmaline is a complex silicate of aluminum and boron, but because of isomorphous replacements this mineral varies widely in chemical composition, iron, magnesium, and lithium entering into combination to a greater or less extent with the aluminum and boron. Its general formula is $(Na, Ca)(Mg, Fe^{2+}, Fe^{3+}, Al, Li)_3Al_6(BO_3)_3(Si_6O_{18})(OH, F)_4$. Tourmaline belongs to the hexagonal system, its crystals are usually prismatic, tending to be long and slender, often acicular. The crystals are ordinarily terminated with three faces of a rhombohedron and usually hemimorphic. The smaller crystals are frequently found in radial arrangement, and columnar masses are common. The prisms are usually three-, six-, or nine-sided with heavy vertical striations producing a rounded effect.

Tourmaline is essentially without cleavage; fracture, conchoidal to uneven; brittle; hardness, 7–7.5; specific gravity, 3.03–3.25; luster, vitreous inclining to resinous; color, in common tourmaline black, bluish-black, brown, blue, green, red or pink, and in the transparent varieties colorless (rare), various shades of rose and pink, greens, blues and browns. The color arrangement in tourmaline is of considerable interest; bicolored crystals are common and may be green at one end and pink at the other, or green on the outside, and pink within, which, in the case of transparent or translucent crystals, is very attractive.

The opaque black tourmaline is called schorl, a term which was applied to all tourmaline until 1703 when the word *tourmaline* was introduced, it being a corruption of the Ceylonese word, *turamali*. The origin of the word *schorl* is not known, but is perhaps Scandinavian, and is used to identify the iron-bearing black tourmalines; elbaites and liddicoatites tend to light shades of blue, red, green, and their bicolored combinations; the brown colored tourmalines of varying shades of dark brown to yellow to nearly colorless are called dravites and uvites (with the exception of the black tourmalines found at Pierrepont, New York, which have been identified as uvites); the completely colorless variety, achroite, falls within the elbaite group. Small tourmalines are found in granites and some gneisses.

Due to the mineralizing action of magmatic vapors, tourmaline is found particularly well developed in pegmatites, and as a contact metamorphic mineral. A few of the important localities are: the Ural Mountains; Bohemia; Saxony; the Island of Elba; Norway; Devonshire and Cornwall, England; Greenland; Madagascar. Magnificent elbaite crystals are obtained from Madagascar, Brazil, and Afghanistan; liddicoatite crystals from Madagascar. In the United States in Oxford and Androscoggin Counties, Maine; Grafton and Sullivan Counties, New Hampshire; Hampshire County, Massachusetts; Haddam and Fairfield Counties, Connecticut; St. Lawrence County, New York; Sussex County, New Jersey; Delaware County, Pennsylvania; and San Diego County, California.

See also terms listed under **Mineralogy**.

ELMER B. ROWLEY
Union College
Schenectady, New York

TOWNSEND AVALANCHE. A term used in gas-filled counter technology to describe a process which is essentially a cascade multiplication of ions. In this process an ion produces another ion by collision, and the new and original ions produce still others by further collisions, resulting finally in an "avalanche" of ions (or electrons). The terms "cumulative ionization" and "cascade" are also used to describe this process. It occurs in a nonself-maintained gas discharge, where ions have sufficient energy.

TOXICITY. The ability of a substance to cause damage to living tissue, impairment of the central nervous system, severe illness, or, in extreme cases, death when ingested, inhaled, or absorbed by the skin. The amounts required to produce these results vary widely with the nature of the substance and time of exposure to it. "Acute" toxicity refers to exposure of short duration, i.e., a single brief exposure; "chronic" toxicity refers to exposure of long duration, i.e., repeated or prolonged exposures.

The toxicity hazard of a material may depend on its physical state and on its solubility in water and acids. Some metals that are harmless in solid or bulk form are quite toxic as fume, powder, or dust. Many substances that are intensely poisonous are actually beneficial when administered in micro amounts, as in prescription drugs, e.g., strychnine.

Toxicity is objectively evaluated on the basis of test dosages made on experimental animals under controlled conditions. Most important of these are LD_{50} (lethal dose, 50%) and the LC_{50} (lethal concentration, 50%) tests, which include exposure of the animal to oral ingestion and inhalation of the material under test. A substance having an LD_{50} of less than 400 mg/kg of body weight is considered very toxic.

TOXICOLOGY. The technology of poisonous substances—their detection, and counteractions. Basic to this branch of science is the realization that chemical compounds vary in their danger to humans and their environment. Poisons can be simple or complex, inorganic or organic chemical compounds, bacterial or viral byproducts (toxins), animal-produced substances, such as venom—all of which produce ill effects on humans ranging from a low level of debilitation to almost instant death. Drugs used in countering diseases or physiological deficiencies can, in some doses, act as poisons. Many metals or elements are essential for life, but their body concentration for optimum health varies from element to element and depends somewhat on bodily weight. Once these optimum concentrations are exceeded, the metals or elements become contaminating, polluting, and often harmful.

Levels of toxicity ratings range from unknown, through low or light toxicity to moderate and severe. Exposure may also be acute, sub-acute, or chronic. Dusts, fumes, mists, vapors, gases or liquids may be absorbed through the skin, orally, or through the lungs. See also **Toxicity**.

Sources of information pertaining to toxic substances include local and national health organizations in many countries. Several treatises on the subject have been prepared, including the broad spectrum "Sax's Dangerous Properties of Industrial Materials," John Wiley & Sons, Inc., New York, NY, 2000. This book contains 20,000 entries, each of which gives physical, chemical, and toxicological data about potentially hazardous materials.

R.C.V.

TRACE AND RESIDUE ANALYSIS. Trace analysis is the detection of minute quantities of organic and inorganic materials. As of the mid-1990s, trace analysis is generally recognized as those determinations that represent around 0.0001%, i.e., at the parts per million (ppm) level, where 1 ppm

is equivalent to 1 μg/g. Ultratrace analysis, i.e., determination below trace analysis, corresponds to levels below ppm or $<\mu g/g$. Residue analysis is the analysis of material left from an operation, i.e., residual.

There are numerous applications for trace or ultratrace analyses in the chemical process industry. Environmental toxicology, in particular, is an area where determination of residues and traces of pesticides and other toxic substances is frequently employed.

Frontiers of Low Level Detection

Extremely low level detection work is being performed in analytical chemistry laboratories. Detection of rhodamine 6G at 50 yoctomole (50 × 10^{-24} mol) has been reported using a sheath flow cuvette for fluorescence detection following capillary electrophoresis. This represents 30 molecules of rhodamine, a highly fluorescent molecule. See also **Electrophoresis**.

Claims of single molecule detection in liquid samples have been made by combining the high sensitivity of laser-induced fluorescence (lif) and the spatial localization and imaging capabilities of optical microscopy. This technique combines confocal microscopy, diffraction-limited laser excitation, and a high efficiency detector. The probe volume is defined latitudinally by optical diffraction and longitudinally by spherical aberration. Using an unlimited excitation throughout and a low background level, this technique allows fluorescence detection of single rhodamine molecules at a signal-to-noise (S/N) ratio of approximately 10 in 1 ms. The use of confocal fluorescence microscopy can be extended to individual, fluorescently tagged biomolecules, including deoxynucleotides, whether single-stranded primers or double-stranded deoxyribonucleic acid (DNA).

Analysis of single mammalian cells by capillary electrophoresis has been reported using on-column derivatization and laser-induced fluorescence detection. Radioactive tracers are powerful tools for trace detection. A method of labeling proteins using ^{99m}Tc has been described. Radiotracer imaging agents have been used for mapping sympathetic nerves of the heart. The radioiodination of analogues of a calichemicin constituent have been employed as a possible brain-imaging agent.

Samples

Sampling. A sample used for trace or ultratrace analysis should always be representative of the bulk material. The principal considerations are determination of population or the whole from which the sample is to be drawn, procurement of a valid gross sample, and reduction of the gross sample to a suitable sample for analysis (see **Sampling**).

Sample Preparation. Sample contamination must be prevented throughout the sampling procedures. No significant changes should occur in the sample when it is being held for analysis. Sample stabilization generally includes storage at low temperatures; however, any stabilization generally includes storage at low temperatures; however, any stabilization step should be validated. A review of sample composition and properties is advised. This would include number of compounds present, chemical structures (functionality) of compounds, molecular weights of compounds, pK_a values of compounds, uv spectra of compounds, nature of sample matrix (solvent, fillers, etc.), concentration range of compounds in samples of interest, and sample solubility. Significant sample losses can occur during this step because of very small volume losses to glass walls of the recovery containers, pipets, and other glassware.

Solid-phase microextraction (SPME), used as a sample introduction technique for high speed gc, utilizes small-diameter fused-silica fibers coated with polymeric stationary phase for sample extraction and concentration. SPME has been utilized for determination of pollutants in aqueous solution by the adsorption of analyte onto stationary-phase coated fused-silica fibers, followed by thermal desorption in the injection system of a capillary gas chromatograph. Full automation can be achieved using an autosampler.

Polycyclic aromatic hydrocarbons (PAHs) have been extracted from contaminated land samples by *supercritical fluid extraction* (SFE) with both pure and modified carbon dioxide. Removing an analyte from a matrix using SFE requires knowledge about the solubility of the solute, the rate of transfer of the solute from the solid to the solvent phase, and interaction of the solvent phase with the matrix. These factors collectively control the effectiveness of the SFE process, if not of the extraction process in general. The range of samples for which SFE has been applied continues to broaden. Applications have been in the environment, food, and polymers.

Sample preparation techniques that prevent or minimize pollution in analytical laboratories, improve target analyte recoveries, and reduce sample preparation costs were evaluated with regard to the *microwave-assisted extraction* (MAE) procedure for 187 compounds and four Aroclors listed in EPA Methods 8250, 8081, and 8141A. The results indicate that most of these compounds can be recovered in good yields from the matrices investigated. Comparative studies were performed to evaluate microwave digestion with conventional sample destruction procedures. These included the analysis of shellfish, meats, rocks, and soils. Generally, comparable accuracy at much shorter digestion time was found for the MAE vs the classical digestion method.

Sample Cleanup. The recoveries from a quick cleanup method for waste solvents based on sample filtration through a Florisil and sodium sulfate column are available. This method offers an alternative for analysts who need to confirm the presence or absence of pesticides or PCBs.

Method Validation

Statistically designed studies should be performed to determine accuracy, precision, and selectivity of the methodology used for trace or ultratrace analyses. The reliability requirements for these studies are that the data generated withstand interlaboratory comparisons.

The following principles should be used to establish a valid analytical method: A specific detailed description and protocol should be written (standard operating procedure (SOP)). Each step in the method should be investigated to determine the extent to which environmental, matrix, material, or procedural variables, from time of collection of material until the time of analysis and including the time of analysis, may affect the estimation of analyte in the matrix. A method should be validated for its intended use with an acceptable protocol. Wherever possible, the same matrix should be used for validation purposes. The concentration range over which the analyte will be determined must be defined in the method, on the basis of actual standard samples over the range (standard curve). It is necessary to use a sufficient number of standards to adequately define the relationship between concentration and response. Determination of accuracy and precision should be made by analysis of replicate sets of analyte samples of known concentration from equivalent matrix.

Methodologies

The commonly used methods for ultratrace analyses together with the accepted detection limits are given in Table 1.

Atomic absorption or emission spectrometric methods are commonly used for inorganic elements in a variety of matrices. A radiochemical neutron activation analysis technique for determination of 26 elements, including the emitting elements Th and U and Cu, Fe, K, Na, Ni, and Zn, has been developed. The radiochemical separation was performed by anion exchange on Dowex 1×8 column from HF and HF–NH_4F medium, leading to selective removal of the matrix-produced radionuclides ^{46}Sc, ^{47}Sc, ^{48}Sc, and nearly selective isolation of ^{239}Np and ^{233}Pa, the indicator radionuclides of U and Th, respectively. For K, Na, Th, and U, a limit of detection of 30, 0.05, 0.03, and 0.07 ng/g, respectively, was achieved.

The most commonly used approach in thin-layer chromatography (tlc) entails separations on a silica gel plate where the silica gel is coated as a thin layer on a glass plate. The plate is developed using the mobile phase

of choice after a sample has been applied to the starting line of the plate. Quantification is achieved directly by scanning the plate or indirectly by scraping and eluting the sample.

A rapid tlc immunoaffinity chromatographic method has been reported for quantitation in serum of an acute phase reactant, C-reactive protein (CRP), which can differentiate between viral and bacterial infections.

A number of compounds have been quantified by tlc or high performance thin-layer chromatography (hptlc) using absorption or fluorescence scanning densitometry.

Gas chromatography is a technique utilized for separating volatile substances (or those that can be made volatile) between two phases, one of which is a gas. Purge-and-trap methods are frequently used for trace analysis. Various detectors have been employed in trace analysis, the most commonly used being flame ionization and electron capture detectors.

High pressure liquid chromatography (hplc), frequently referred to as simply lc or as high performance liquid chromatography, is used in virtually all fields of chemistry. Nonvolatile or thermally labile compounds are best separated by hplc. Although techniques such as adsorption and ion-exchange chromatography have been used, the technique of choice is reversed-phase liquid chromatography (rplc). In rplc the stationary phase is nonpolar and the mobile phase is polar and its polarity can be suitably changed.

In hplc, detection and quantitation have been limited by availability of detectors. Using a uv detector set at 254 nm, the lower limit of detection is 3.5×10^{-11} g/mL for a compound such as phenanthrene. A fluorescence detector can increase the detectability to 8×10^{-12} g/mL. The same order of detectability can be achieved using amperometric, electron-capture, or photoionization detectors. Hplc is capable of routine determination at the nanogram range.

Increased use of liquid chromatography/mass spectrometry (lc/ms) for structural identification and trace analysis has become apparent. Thermospray lc/ms has been used to identify by-products in phenyl isocyanate precolumn derivatization reactions. Liquid chromatography/thermospray mass spectrometric characterization of chemical adducts of DNA formed during *in vitro* reaction has been proposed as an analytical technique to detect and identify those contaminants in aqueous environmental samples which have a propensity to be genotoxic, i.e., to covalently bond to DNA.

Using capillary hplc, femtomole amounts of recombinant DNA-derived human growth hormone (rhgh) have been successfully detected from solutions at nanomolar concentrations. A sample of rhgh that was recovered from rat serum was analyzed by capillary reversed-phase hplc, using both acidic- and neutral-pH mobile phases, as well as by capillary ion-exchange chromatography. Submicrogram amounts of rhgh were also analyzed by tryptic mapping, using capillary hplc, and the resulting peptides were identified by capillary lc/ms.

Capillary electrophoresis (ce) is an analytical technique that can achieve rapid high resolution separation of water-soluble components present in small sample volumes. The separations are generally based on the principle of electrically driven ions in solution. Selectivity can be varied by the alteration of pH, ionic strength, electrolyte composition, or by incorporation of additives. Typical examples of additives include organic solvents, surfactants, and complexation agents.

Supercritical fluid chromatography (sfc) combines the advantages of gc and hplc in that it allows the use of gc-type detectors when supercritical fluids are used instead of the solvents normally used in hplc. Carbon dioxide, *n*-petane, and ammonia are common supercritical fluids. For example, carbon dioxide employed at 7.38 MPa (72.9 atm) and 31.3°C has a density of 448 g/mL. Derivatization of primary and secondary amines using 9-fluorenylmethyl chloroformate to form a nonpolar, uv-absorbing derivative has been reported.

Immunoassays may be simply defined as analytical techniques that use antibodies or antibody-related reagents for selective determination of sample components. These make up some of the most powerful and widespread techniques used in clinical chemistry. The main advantages of immunoassays are high selectivity, low limits of detection, and adaptability for use in detecting most compounds of clinical interest. Because of their high selectivity, immunoassays can often be used even for complex samples such as urine or blood, with little or no sample preparation.

A two-site immunometric assay of undecapeptide substance P (SP) has been developed. A number of solid-phase automated immunoassay analyzers have been used for performing immunoassays. A number of immunoassay methods have been found useful for environmental analysis (see **Automated Instrumentation**).

TABLE 1. ULTRATRACE ANALYSES METHODS

Method	Minimum amount detected, g
mass spectrometry	
electron impact	10^{-12}
spark source	10^{-13}
ion scattering	10^{-15}
flame emission spectrometry	10^{-12}
liquid chromatography	
ultraviolet detection	10^{-11}
fluorescence detection	10^{-12}
gas chromatography	
flame ionization	10^{-12} to 10^{-14}
electron capture	10^{-13}
combination techniques	
liquid chromatography/mass spectrometry	10^{-12}
gas chromatography/mass spectrometry	10^{-12}
electron capture (negative)/ionization mass spectrometry	10^{-15}

Applications

Trace or ultratrace and residue analyses are widely used throughout chemical technology. Areas of environmental investigations, explosives, food, pharmaceuticals, and biotechnology rely particularly on these methodologies.

Environment. Detection of environmental degradation products of nerve agents directly from the surface of plant leaves using static secondary ion mass spectrometry (sims) has been demonstrated.

Some of the methods used for determination of organic pollutants in the environment follow. The most notable are polyaromatic hydrocarbons (PAHs) and volatile organic compounds (VOCs).

Analytes	Methods
n-alkanes	gc/fid
VOCs	prefraction and gc
PAHs	sample clean up and gc
bromoform and other bromine compounds	gc for flame-retardant compounds in air
nitrodibenzopyranone	isomeric separations based on gc
PCDDs and PCDFs	gc/ms
PAH	on-line lc/gc for PAH in air
chlorinated PAH	lc/gc/ms
nitrated PAH	hplc with electrochemical detection
azaarenes	column chromatography, cleanup, and tlc
toluene diisocyanate	micro lc
dimethyl sulfoxide or sulfone	atmospheric pressure chemical ionization (APCI) ms
VOCs	ion trap monitoring
volatile chlorine compounds	fiber-optic emission sensor-based on AA of chlorine
formaldehyde	monitoring tape: hydroxylamine sulfate, methyl yellow
methyl nitrite	nitrogen oxide-indicating tubes with ir detection
VOCs	ir-based methods for on-site analysis
isocyanate species	chemiluminescent techniques
benzene	photoionization detector
hydrazine	coulometric methods
nitrobenzene	piezoelectric sensors
fluorocarbon	metal oxide sensors
perchloroethylene	quartz balance and calorimetric transducer

A variety of organic and inorganic analytes have been analyzed in air and in water.

Explosives. Explosives can be detected using either radiation- or vapor-based detection. The aim of both methods is to respond specifically to the properties of the energetic material that distinguish it from harmless material of similar composition. These techniques are useful for detecting organic as well as inorganic explosives.

Food. Laws and regulations controlling contamination of food were once the province of religious organizations. As far back as the Dark Ages, government set standards. Significant changes have occurred only relatively recently, since the time analytical chemists could characterize most of the substances that comprise food and thus more effectively control contaminants in it. The reliability of data regarding medicated feeds, dairy products. See also **Millerite**, seafood, meat products, fruits and vegetables, and beverages has been reviewed.

The development of analytical strategies for the regulatory control of drug residues in food-producing animals has also been reviewed. Because of the complexity of biological matrices such as eggs, milk, meat, and drug feeds, well-designed off-line or on-line sample treatment procedures are essential. For example, methylmercury in fish was extracted and cleaned using column chromatography and then determined using flameless atomic absorption. A rapid and sensitive electrothermal AA spectrophotometric method using a combination of microwave digestion and palladium as a stabilizer was used for detecting Cd and Pb at sub ppm levels in vegetables and protein in foodstuffs.

Inorganic elements in food can be determined by atomic absorption (AA) methods. These methods have been extensively reviewed.

Pharmaceuticals. Examples of trace and ultratrace analyses of various drugs and pharmaceuticals have been provided throughout. The purity of the active ingredient, its content and availability in dosage form, therapeutic blood levels, delivery to target areas, elimination (urine, feces, and metabolites), and toxicity are always of importance.

A safety profile of a generic drug can differ from that of the brandname product because different impurities may be present in each of the drugs. Impurities can arise out of the manufacturing processes and may be responsible for adverse interactions that can occur.

The subject of impurity analysis of pharmaceutical compounds has been insufficiently addressed in the scientific literature. Many monographs in the *U.S. Pharmacopeia* have nonspecific assay methods. An attempt has been made to address this problem by focusing on specific methodologies and delineating origination and concentration of impurities found in pharmaceutical compounds.

Chiral separations have become of significant importance because the optical isomer of an active component can be considered an impurity. Optical isomers can have potentially different therapeutic or toxicological activities. The pharmaceutical literature is trying to address the issues pertaining to these compounds. Frequently separations can be accomplished by glc, hplc, or ce.

Biotechnology. Particular attention must be paid to the detection of DNA in all finished biotechnology products because of the possibility that such DNA could be incorporated into the human genome and thus become a potential oncogene. The absence of DNA at the picogram-per-dose level should be demonstrated in order to assure the safety of biotechnology products.

The isolation and purification of DNA and ribonucleic acid (RNA) restriction fragments are of great importance in the area of molecular biology. These fragments are the product of site-specific digestion of large pieces of DNA and RNA with enzymes called restriction endonucleases. The fragments may range in size from a few base pairs to tens of thousands of base pairs. An ion-exchange column can provide DNA and RNA separations within one hour, giving resolution equivalent to that obtained with gel electrophoresis. Nucleic acid fragments are then visualized, using on-line uv detection and sample loading from 500 ng to 50 mg.

Molecular biologists are utilizing hplc for characterization and purification of proteins, peptides, and antibodies.

Miscellaneous. Trace analyses have been performed for a variety of other materials.

Thermal neutron activation analysis has been used for archeological samples, such as amber, coins, ceramics, and glass; biological samples; and forensic samples, as well as human tissues, including bile, blood, bone, teeth, and urine; laboratory animals; geological samples, such as meteorites and ores; and a variety of industrial products.

SATINDER AHUJA
Ahuja Consulting

Additional Reading

Ahuja, S.: *Impurities Evaluation of Pharmaceuticals,* Marcel Dekker, New York, NY, 1998.
Ahuja, S.: *Trace and Ultratrace Analysis by HPLC,* John Wiley & Sons, Inc., New York, NY, 1992.
Ahuja, S.: *Chiral Separations Applications & Technology,* American Chemical Society, Washington, DC, 1997.
Kolla, P.: *Anal. Chem.* **67**, 184A (1995).

TRACE ELEMENT (Micronutrient).   An element essential to plant and animal nutrition in trace concentration, i.e., minute fractions of 1% (1000 ppm or less). Plants require iron, copper, boron, zinc, manganese, potassium, molybdenum, sodium, and chlorine. Animals require iron, copper, manganese, cobalt, selenium, and potassium. Such elements are also called *micronutrients*. Do not confuse with *tracer*.

TRACER.   A chemical entity (almost invariably radioactive and usually an isotope) added to the reacting elements or compounds in a chemical process, which can be traced through the process by appropriate detection methods, e.g., Geiger counter. Compounds containing tracers are often said to be "tagged" or "labeled." Carbon-14 is a commonly used tracer, and radioactive forms of iodine and sodium are also used. Many complex biochemical reactions have been examined in this way (e.g., photosynthesis). Nonradioactive deuterium (hydrogen isotope) is sometimes used, the detection being by molecular weight determination. Radioactive enzymes are also available for tracer studies, e.g., ribonuclease, pepsin, trypsiln, and others.

See also **Radioactivity**.

TRACHYTE. The name of an extrusive, igneous, fine-grained or porphyritic rock, the surface equivalent of syenite. Trachyte is predominant in alkali feldspar and usually contains biotite and augite. Trachyte is an old name, proposed by Brongiart in 1813, and has never been altered or supplanted. It is derived from the Greek, meaning rough.

TRANSACTINIDE ELEMENTS. See **Chemical Elements**.

TRANSACTINIUM EARTHS. A group of chemical elements more frequently termed the Actinides. In order of increasing atomic number, they include actinium, thorium, protactinium, uranium, neptunium, plutonium, americium, curium, berkelium, californium, einsteinium, fermium, mendelevium, nobelium, and lawrencium. See also **Actinide Contraction**.

TRANS-COMPOUND. See **Isomerism**.

TRANSFERENCE NUMBER (Transport Number). Of a given ion in an electrolyte, the transference number is the fraction of total current carried by that ion.

TRANSITION ELEMENT (Transition Metal). Any of a number of elements in which the filling of the outermost shell to eight electrons within a period is interrupted to bring the penultimate shell from 8 to 18 or 32 electrons. Only these elements can use penultimate shell orbitals as well as outermost shell orbitals in bonding. All other elements, called "major group" elements, can use only outermost shell orbitals in bonding. Transition elements include elements 21 through 29 (scandium through copper), 39 through 47 (yttrium through silver), 57 through 79 (lanthanum through gold) and all known elements from 89 (actinium) on. All are metals. Many are noted for exhibiting a variety of oxidation states and forming numerous complex ions, as well as processing extremely valuable properties in the metallic state.

TRANSITION TEMPERATURE. 1. An arbitrarily defined temperature within the temperature range in which metal fracture characteristics determined usually by notched tests are changing rapidly such as from primarily fibrous (shear) to primarily crystalline (cleavage) fracture.

2. The arbitrarily defined temperature in a range in which, the ductility of a material changes rapidly with temperature.

TRANSMUTATION. The natural or artificial transformation of atoms of one element into atoms of a different element as the result of a nuclear reaction. The reaction may be one in which two nuclei interact, as in the formation of oxygen from nitrogen and helium nuclei (β-particles), or one in which a nucleus reacts with an elementary particle such as a neutron or proton. Thus, a sodium atom and a proton form a magnesium atom. Radioactive decay, e.g., of uranium, can be regarded as a type of transmutation. The first transmutation was performed by the English physicist Rutherford in 1919.

TRANSURANIUM ELEMENTS. The chemical elements with an atomic number higher than 92 (uranium), commencing with 93 (neptunium) and through 110 (darmstadtium) frequently are termed Transuranium elements. Any additional elements that may be identified will be a part of this series. See also **Actinide Contraction** and **Chemical Elements**.

TRAPPED IONS. Charged particles, including electrons and atomic ions, can be stored for long periods of time (days are not uncommon) without the usual perturbations associated with confinement (for example, the frequency shifts associated with the collisions of ions with buffer gases in a more traditional optical pumping experiment). Ions that are stored in electromagnetic "traps" can provide the basis for extremely high-resolution spectroscopy. By using lasers, the kinetic energy of the ions can be cooled to millikelvin temperatures, thus suppressing Doppler frequency

shifts. Potential accuracies of frequency standards and clocks based on such experiments are anticipated to be better than one part in 10^{15}. The current time standard (cesium clock) has an accuracy of one part in 10^{13} or less. See also **Mass Spectrometry**.

Storage has mainly been achieved in four types of traps: (1) the radio frequency or Paul trap; (2) the Penning trap; (3) the Kingdon electrostatic trap; and (4) the magnetostatic (magnetic bottle) trap. The principles, advantages, and disadvantages of these traps are detailed by D.J. Wineland (*Science*, **226**, 395–400, Oct. 26, 1984).

TRAVERTINE. Carbonated waters dissolve large amounts of calcium carbonate, especially under high temperature. Such waters reaching the earth's surface as hot springs often deposit the calcium carbonate, in great quantities. This material is called travertine from the ancient name for Tivoli, Italy, where a very thick deposit occurs. Travertine may be compact, crystalline, fibrous or, if rapidly deposited, spongy and porous. The less compact varieties are known as tufa. Travertine is being formed at the Mammoth Hot Springs, Yellowstone National Park and at many other localities. A banded travertine used as an ornamental stone is called onyx marble or Mexican onyx.

TREMOLITE. [CAS: 1332-21-4]. The mineral tremolite is a calcium-magnesium silicate corresponding to the formula $Ca_2Mg_5Si_8O_{22}(OH)_2$, belonging to the amphibole group. The replacement of magnesium by ferrous iron causes tremolite to approach actinolite in composition. Tremolite is monoclinic, developing bladed prismatic crystals, but it is frequently found in compact columnar, granular, or fibrous masses. The perfect prismatic cleavage at angles of 56° and 124° typical of this group is to be noted; hardness, 5–6; specific gravity, 2.9–3.1; luster, vitreous to silky; color, varies from white or whitish-gray through shades of green or greenish-yellow; transparent to opaque. Tremolite is formed as a result of contact metamorphism and occurs in marbles, dolomites, and schists. It may alter to talc. Tremolite is found in Switzerland, in the St. Gotthard region, being named for the Tremola Valley, and is common elsewhere in Europe. In the United States it occurs in Maine, Pennsylvania, and New York. In Canada tremolite has been found in Quebec and Ontario.

Hexagonite is a pinkish-purple variety of tremolite which contains a small amount of manganese. So called because it was at first believed to be hexagonal. It has been since shown to be monoclinic, and is found in St. Lawrence County, New York. Some nephrite and asbestos is tremolite.

See also terms listed under **Mineralogy**.

TRIACETATE. See **Cellulose Ester Plastics (Organic)**; **Fibers**.

TRICARBOXYLIC ACID CYCLE. See **Carbohydrates**.

TRIDYMITE. The mineral tridymite is, like quartz, silicon dioxide, SiO_2, but is a high-temperature variety, probably stable above 870°. It has a conchoidal fracture; is brittle; hardness, 7; specific gravity, 2.28–2.33; vitreous luster; usually colorless and transparent. It is found chiefly in volcanic rocks of the more acidic types like rhyolite, tachyte, and andesite. It is not a particularly uncommon mineral, occurring in Germany, France, Italy, Japan, the Island of Martinique, Mexico, and in the United States in Wyoming and Washington. Tridymite is hexagonal but when heated to about 1,470°C passes into an isometric form, cristobalite, which was first noted in the andesitic lavas of the Cerro San Cristobal, Pachuca, Mexico, together with tridymite. Cristobalite has been found also in California and in Germany.

TRIPHENYLMETHANE AND RELATED DYES. Triphenylmethane dyes are of brilliant hue, exhibit high tinctorial strength, are relatively inexpensive, and may be applied to a wide range of substrates. They are seriously deficient in fastness properties, especially fastness to light and washing. Consequently, the use of triphenylmethane dyes on textiles such as wool, silk, and cotton has decreased as dyes from other classes with superior lightfastness and washfastness properties have become available. Interest in this class of dyes was revived with the introduction of polyacrylonitrile fibers. See also **Acrylonitrile Polymers**; and **Fibers**. Triphenylmethane dyes are readily adsorbed on this fiber and show surprisingly high lightfastness and washfastness properties, compared with the same dyes on natural fibers. However, the durability of acrylic fibers created an even greater demand for fastness properties.

The triarylmethane dyes are broadly classified into the triphenylmethanes (CI 42000–43875), diphenylnaphthylmethanes (CI 44000–44100), and miscellaneous triphenylmethane derivatives (CI 44500–44535). The triphenylmethanes are classified further on the basis of substitution in the aromatic nuclei, as follows: (1) diamino derivatives of triphenylmethane, i.e., dyes of the malachite green series (CI 42000–42175); (2) triamino derivatives of triphenylmethane, i.e., dyes of the fuchsine, rosaniline, or magenta series (CI 42500–42800); (3) aminohydroxy derivatives of triphenylmethane (CI 43500–43570); and (4) hydroxy derivatives of triphenylmethane, i.e., dyes of the rosolic acid series (CI 43800–43875). Monoaminotriphenylmethanes are known but they are not included in the classification because they have little value as dyes.

Chemically, the triarylmethane dyes are monomethine dyes with three terminal aryl systems of which one or more are substituted with primary, secondary, or tertiary amino groups or hydroxyl groups in the para position to the methine carbon atom. Additional substituents such as carboxyl, sulfonic acid, halogen, alkyl, and alkoxy groups may be present on the aromatic rings. The number, nature, and position of these substituents determine both the hue or color of the dye and the application class to which the dye belongs.

Structure

The triarylmethane dyes are generally considered to be resonance hybrids. However, for convenience, usually only one hybrid is indicated, as shown for crystal violet, CI Basic Violet 3 (1), for which $\lambda_{max} = 589$ nm.

(1)   (2)

The ortho hydrogen atoms surrounding the central carbon atom show considerable steric overlap. Therefore, it can be assumed that the three aryl groups in the dye are not coplanar, but are twisted in such a fashion that the shape of the dye resembles that of a three-bladed propeller. Substitution in the para position of the three aryl groups determines the hue of the dye. When only one amino group is present, as in fuchsonimine hydrochloride, $\lambda_{max} = 440$ nm (2), the shade is a weak orange-yellow.

When at least two or more amino groups are present in different rings, the resonance possibility is greatly increased, resulting in a much greater intensity of absorption and in a strong bathochromic shift to longer wavelengths, e.g., Doebner's violet (3), $\lambda_{max} = 562$ nm, which is a reddish violet, and pararosaniline (4) $\lambda_{max} = 538$ nm, which is a bluish violet. The amino derivatives of commercial value contain two or three amino groups.

(3)   (4)

A further strong bathochromic shift is observed as the basicity of the primary amines is increased by N-alkylation, e.g., malachite green, CI Basic Green 4, $\lambda_{max} = 621$ nm.

Chemical Properties

Although many triarylmethane dyes are prepared by the oxidation of leuco bases, they are usually destroyed by strong oxidizing agents. The triarylmethane dyes are extremely sensitive to photochemical oxidation, a fact which accounts for their poor lightfastness on natural fibers. There are many factors which affect the rate of fading (degradation) of the triarylmethane dyes on natural and synthetic fibers. Several studies have revealed that N-dealkylation occurs simultaneously with the cleavage and contributes to the photodegradation. N-Dealkylation is a general phenomenon in dye photochemistry.

Triarylmethane dyes are reduced readily to leuco bases with a variety of reagents. Reduction with titanium trichloride (Knecht method) is used for rapidly assaying triarylmethane dyes.

The direct sulfonation of alkylaminotriphenylmethane dyes gives mixtures of substituted products. Although dyes containing anilino or benzylamino groups give more selective substitution, a sulfonated intermediate such as 3[(N-ethyl-N-phenyl-amino)methyl]benzenesulfonic acid (ethylbenzylanilinesulfonic acid) is the preferred starting material.

Dyes containing highly alkylated amino groups are prepared from highly alkylated intermediates and not by direct alkylation of dyes carrying primary amino groups. 4, 4′, 4″-Triaminotriphenylmethane (pararosaniline) may, however, be N-phenylated with excess aniline and benzoic acid to give the greenish blue, N, N′, N″-triphenylaminotriphenylmethane hydrochloride, CI Solvent Blue 23, $\lambda_{max} = 586$ nm.

Triarylmethane dyes can be converted into two types of insoluble compounds, which are used industrially as pigments. Both are salts of triarylmethane dyes. Known as pigment lakes, these complexes provide clean, brilliant red and violet shades. These pigments are used in printing inks, especially packaging and special printing inks. The second type of pigments derived from triarylmethane dyes are known as alkali blues. The main use of the alkali blues is as shading pigments in inks based on carbon black, where an inexpensive blue component is needed to correct the natural brown tone of the base pigment. The main area of application is in printing inks, particularly offset, letterpress, and to a lesser extent in aqueous flexographic inks. They are used to color ribbons for typewriters and also to blue copy paper.

Manufacture

The preparation of tri arylmethane dyes proceeds through several stages; formation of the colorless leuco base in acid media, conversion to the colorless carbinol base by using an oxidizing agent, e.g., lead dioxide, manganese dioxide, or alkali dichromates, and formation of the dye by treatment with acid. The oxidation of the leuco base can also be accomplished with atmospheric oxygen in the presence of catalysts.

Aldehyde Method. This method is generally used for the preparation of diaminotriphenylmethane dyes or hydroxytriphenylmethane dyes. The central carbon atom is derived from an aromatic aldehyde or a substance capable of generating an aldehyde during the course of the condensation. For example, Malachite green is prepared by heating benzaldehyde under reflux with a slight excess of dimethylaniline in aqueous acid.

Ketone Method. In the ketone method, the central carbon atom is derived from phosgene. A diarylketone is prepared from phosgene and a tertiary arylamine and then condenses with another mole of a tertiary arylamine (same or different) in the presence of phosphorus oxychloride or zinc chloride. The dye is produced directly without an oxidation step. Thus, ethyl violet, CI Basic Violet 4, is prepared from 4,4′-bis(diethylamino)benzophenone with diethylaniline in the presence of phosphorus oxychloride. This reaction is very useful for the preparation of unsymmetrical dyes.

Diphenylmethane Base Method. In this method, the central carbon atom is derived from formaldehyde, which condenses with two moles of an arylamine to give a substituted diphenylmethane derivative. The methane base is oxidized with lead dioxide or manganese dioxide to the benzhydrol derivative. The reactive hydrols condense fairly easily with arylamines, sulfonated arylamines, and sulfonated naphthalenes. The resulting leuco base is oxidized in the presence of acid.

Benzotrichloride Method. The central carbon atom of the dye is supplied by the trichloromethyl group from p-chlorobenzotrichloride. Both symmetrical and unsymmetrical triphenylmethane dyes suitable for acrylic fibers are prepared by this method.

Health, Safety, and Environmental Information

In the 1960s, problems were encountered with the interpretation of toxicological studies on animals given triarylmethane dyes used as food colorants. Conflicting test data have been obtained on various triarylmethane dyes. Positive and negative results for the same dye in different assays have only led to more genotoxic studies. In the U. S. CI Food Violet 2 was delisted for use in food, drugs, and cosmetics in 1973. There are triarylmethane dyes which have also been delisted worldwide for use in food, e.g., Guinea Green B, CI Food Green 1 (CI 42085), and Violet BNP, CI Food Violet 3.

The triarylmethane dyes of the rosaniline family, e.g., fuchsine and crystal violet, show similar toxic responses in assays. They are moderately toxic after acute exposure, but the effects usually pass within a couple of days. These effects are no cause for alarm as long as the dyes are not permitted for food use or contact. There is evidence that the toxicological effects might be the result of impurities, e.g., aromatic amines, or of certain functional groups, notably amino substituents found in the dyes.

The toxicity of dyes to aquatic organisms has also been investigated, with most of the work done on fish. In these studies, over 3000 dyes in common use were tested, of which 27 dyes had a LC_{50} around 0.05 mg/L. Ten of these cases had triarylmethane structures.

Environmental Concerns

The main route by which dyes enter the environment is via wastewater, both from their manufacture and their use. Accurate data on dyes released into the environment are not available, although lists of materials released to the environment from the processes operated for the production of some triarylmethane dyes have been reported.

Uses

Present usage of triarylmethane dyes is confined mainly to nontextile applications. Substantial quantities are used in the preparation of organic pigments for printing inks, pastes, and for the paper printing trade, where cost and brilliance of shade are more important than lightfastness. Triarylmethane dyes and their colorless precursors, e.g., carbinols and lactones, are used extensively in heat-, light-, and pressure-sensitive recording materials for high speed photoduplicating and photoimaging systems and for the production of printing plates and integrated circuits. They are also used for specialty applications such as tinting automobile antifreeze solutions and toilet sanitary preparations, in the manufacture of carbon paper, in ink for typewriter ribbons, and ink jet printing for high speed computer printers.

In addition to the dyeing and printing of natural and acrylic fibers, triarylmethane dyes are suitable for the coloration of other substrates such as paper, ceramics, leather, fur, anodized aluminium, waxes, polishes, soaps, plastics, drugs, and cosmetics. Several triarylmethane dyes are used as food colorants and are manufactured under stringent processing controls.

Triarylmethane dyes can be used for the coloration of glass. Other high technology applications using triarylmethane dyes include electrophotography and optical data storage.

Related Dyes

Diphenylmethane Dyes. The diphenylmethane dyes are usually classed with the triarylmethane dyes. The dyes of this subclass are ketoimine derivatives, and only three such dyes are registered in the *Colour Index.* They are Auramine O CI Basic Yellow 2 (CI 41000), Auramine G, Basic Yellow 3 (CI 41005), and CI Basic Yellow 37 (CI 41001). These dyes are still used extensively for the coloration of paper and in the preparation of pigment lakes.

Phthaleins. Dyes of this class are usually considered to be triarylmethane derivatives. Phenolphthalein and phenol red are used extensively as indicators in colorimetric and titrimetric determinations. See also **Hydrogen-Ion Activity**.

Heteroarylmethane Dyes. Dyes of this class usually have either one or two heteroaryl groups attached to the methane carbon atom. Trihetarylmethane dyes are known and have been investigated for their pharmacological activity as well as their color characteristics. These types of triarylmethane dyes and their derivatives are used as color formers in thermoreactive and pressure-sensitive recording materials.

Triarylmethane Dyes with Near-Infrared Absorption. The long wavelength absorption bands of triarylmethane dyes can be shifted into the near-infrared region, but the dyes still remain colored because other absorption bands are shifted to or stay in the visible region. These types of triarylmethane dyes and their derivatives have been claimed as infrared absorbers for optical information recording media and security devices, and as organic photoconductors for use in lithographic plate production.

DEAN THETFORD
Zeneca Specialties

Additional Reading

Abrahart, E.N.: *Dyes and their Intermediates,* 2nd Edition, Chemical Publishing, New York, NY, 1977.

Gregory, P.: *High-Technology Applications of Organic Colorants,* Plenum Publishing Corp., New York, NY, 1990.

Rys, P. and H. Zollinger: *Fundamentals of the Chemistry and Applications of Dyes,* Wiley-Interscience, New York, NY, 1972.

Waring, D.R. and G. Hallas, eds.: *The Chemistry and Applications of Dyes,* Plenum Publishing Corp., New York, NY, 1990.

TRIPHYLITE. This mineral is a phosphate of lithium and ferrous iron, $LiFePO_4$. It crystallizes in the orthorhombic system but usually is characterized by large cleavable masses. The hardness is 4.5–5.0; specific gravity, 3.42–3.56, vitreous to resinous luster, translucent, and bluegray color. The mineral occurs as a rare primary mineral in granitic pegmatites and, when available in large quantities, is a source of lithium. Worldwide occurrences include Bavaria, Finland, Sweden, and in the United States, New Hampshire, Maine, and South Dakota.

TRIPLE BOND. A highly unsaturated linkage between the two carbon atoms of acetylenic compounds (alkynes), typified by acetylene (HC ≡ CH).

See also **Bond (Chemical)**; and **Chemical Elements**.

TRIPLE BOND (Carbon). See **Organic Chemistry**.

TRIPLE POINT. The temperature and pressure at which the solid, liquid, and vapor of a substance are in equilibrium with one another. Also applied to similar equilibrium between any three phases, i.e., two solids and a liquid, etc. The triple point of water is +0.072 C at 4.6 mm Hg; it is of special importance because it is the fixed point for the absolute scale of temperature.

See also **Three-Phase Equilibrium**.

TRIPOLITE. A term frequently used synonymously with diatomaceous earth, but more particularly with reference to diatomaceous earth from Tripoli in North Africa. See also **Diatom**; and **Diatomite**.

TRITIUM. The radioactive isotope of hydrogen, with a mass number 3, is termed tritium. It is one form of heavy hydrogen, the other form being deuterium. See also **Nuclear Power Technology**.

TRIVIAL NAME. The name applied by early chemists to a number of simple organic compounds, usually based on their sources or properties, e.g., acetone and acetic acid, from Latin *acetum* (vinegar), urea from urine, glucose and glycerol from Greek *glyc*-(sweet). Such names remained in common use regardless of the systematic nomenclature later developed.

TUNGSTEN. [CAS: 7440-33-7]. Chemical element, symbol W, at. no. 74, at. wt. 183.85, periodic table group 6, mp 3390–3420°C, bp 5,660°C, density 19.3 g/cm³. Two forms of metallic tungsten are known: α-tungsten, which has a body-centered cubic crystal structure, and β-tungsten, which has a face-centered cubic crystal structure. The metal exhibits the phenomenon of passivity so that it is quite resistant to chemical action even though it has a strong reducing action when a fresh surface is exposed, or in potentiometric titrations. Tungsten is a silver-white to steel-gray, brittle, hard metal; not oxidized by air at ordinary temperature, but burns at high temperature, best dissolved by a mixture of hydrofluoric and HNO_3 acids. Tungsten has four naturally occurring isotopes ^{180}W, and ^{182}W through ^{184}W. ^{180}W is radioactive with a half-life of approximately 3×10^{14} years. Six other radioactive isotopes have been identified ^{176}W through ^{178}W, ^{181}W, ^{185}W, ^{187}W, and ^{188}W. All have half-lives considerably less than 4 months in length. Tungsten does not occur in the free state and is a relatively scarce element, making up an estimated $7 \times 10^{-3}\%$ of the Earth's crust. The tungsten content of seawater is estimated at about 950 pounds per cubic mile. Although the tungsten mineral *wolframite* (iron manganese tungstate) was described as early as 1574, it was then mistaken as a mineral of tin. The term *tungsten* first appeared about 1758. K.W. Scheele identified tungstic oxide in 1781, after which the calcium tungstate mineral *scheelite* was named. The first metallic tungsten was produced by J.J. d'Elhuyar and F. d'Elhuyar in 1783 by the carbon reduction of the oxide. W.D. Coolidge obtained a patent in 1908 for making ductile tungsten wire for use in incandescent lamps. Divers authorities accredit Scheele or the d'Elhuyar brothers with the discovery of the element. With exception of the United States, the element generally is referred to as *wolfram*. First ionization potential of tungsten is 7.98 eV. Electronic configuration $1s^2 2s^2 2p^6 3s^2 3p^6 3d^{10} 4s^2 4p^6 4d^{10} 4f^{14} 5s^2 5p^6 5d^4 6s^2$. Ionic radius

$W^{+4}0.68$ Å, $W^{+6}0.65$ Å. Metallic radius 1.3704_5 Å. Other important physical properties of tungsten are given under **Chemical Elements**.

Usually tungsten minerals are found in pegmatites, sills, and batholiths. Minerals often accompanying tungsten minerals are cassiterite, quartz, feldspar, sulfides, arsenites, apatite, calcite, molybdenite, and bismuthinite. Several of these minerals are described under separate alphabetical entries. In order of decreasing magnitude, tungsten deposits occur in the People's Republic of China, the United States, Korea, Bolivia, Portugal, Burma, and Australia. Deposits also are found in at least ten other areas. In the United States, the most significant deposits are found in California, Nevada, South Carolina, Idaho, and Colorado. Tungsten concentration in the ores found in the United States run from 0.5% to 3% WO_3 (20 pounds 9 kilograms of WO_3 contains about 15.9 pounds 7.2 kilograms of W). High-purity tungsten metal is prepared by extracting the tungsten from the ore by use of a strong alkali hydroxide solution at the boiling point. The alkali-metal carbonate or hydroxide thus obtained then is fused to form the water-soluble, alkali-metal tungstate. Where NH_4OH is used, the product is ammonium tungstate (NaOH yields sodium tungstate). The compound is reduced to metal powder. Conversion of the powder to massive metal is done by pressing, sintering, and mechanical working at high temperatures. In another process, the ore is fused with sodium carbonate and nitrate to yield sodium tungstate. Reduction of the oxide is accomplished by heating with carbon or hydrogen, whereupon tungsten metal is yielded.

Uses

Approximately one-half of the tungsten produced is in the form of sintered tungsten carbides. These compounds are used for cutting tools and wear-resistant parts. About 15% of production is consumed for making wire used in lamps and also for various shapes used in aerospace and defense products. Another 15% of production is used for high-temperature alloys and powder metallurgy. Approximately 10% goes into high-speed tools. The remaining production goes into a wide variety of applications.

Tungsten carbide, WC, is extremely hard (9.5 on the Mohs scale; diamond = 10) and has a melting point of $2,870°C$. This combination of hardness and high-temperature stability makes it an excellent material for cutting tools. Additionally, the wear-resistant properties are excellent, accounting for the use of tungsten carbide for dies for hot and cold working of wire, rod and tubing, mining tools, snow-tire studs, and ball-point pens. For hard carbide tools and dies, tungsten carbide in the form of fine powder (1–10 micrometer particle size) is bonded with cobalt.

Special carbide tools also will often contain various percentages of titanium, tantalum, niobium (columbium), and hafnium carbides, along with the tungsten carbide. Chromium and vanadium carbides are also added to produce special, fine-grain-size grades of cemented tungsten carbide-cobalt materials. See Fig. 1.

For hard-facing applications, fused tungsten carbide is used. Tungsten also forms the ditungsten carbide W_2C, which has a melting point of

$2,860°C$. However, the term *tungsten carbide* usually refers to the mono compound. WC generally is made by combining tungsten metal powder with finely divided lampblack. The mixture then is heated to about $1,500°C$. A variety of tungsten powders are made which then are subjected to various powder metallurgy techniques to form numerous shapes with a wide range of characteristics. Tungsten carbide can also be manufactured by a so-called menstrum process, which employs calcium carbide and aluminum metal to reduce scheelite via a thermite reaction, with the tungsten carbide recovered by acid washing.

Tungsten wire, including pure (unalloyed), doped nonsag; (potassium silicate and aluminum chloride or nitrate doped), and thoriated and zirconiated types are used extensively in applications such as filaments for incandescent lamps, thermocouples, arc-lamp electrodes, electrochemical electrodes, and instrument springs. See Fig. 2. Tungsten disks are used for electrical contacts; tungsten is used in glass-to-metal seals, where the coefficient of thermal expansion of tungsten is close to that of hard borosilicate glass; and tungsten pads are used in connection with silicon semiconductors because of the high thermal conductivity of tungsten and good match of the coefficient of thermal expansion of tungsten with that of silicon.

Compositions of silver and tungsten and of copper and tungsten find application as electrical contacts where they are subject to severe arcing. As a shield or as containers for radioactive materials, heavy-metal alloys of tungsten alloyed with about 7% nickel and 4% copper are effective. The same alloys also find other uses where high density is required, as in gyroscope rotors, counterweights in aircraft, and self-winding watch parts. Alloys of cobalt, chromium, and tungsten also find use in cutting tools, dies, and wear-resistant parts. The function of tungsten in steel is that of forming stable carbides, strengthening ferrite, and refining the grain size for retaining high hardness at elevated temperatures—a requirement of highspeed steels.

Tungsten chemicals find limited use in inks, paints, enamels, dyes, and glass manufacture. Some tungsten compounds and their derivative phosphors find use in x-ray screens, television picture tubes, and luminescent light sources.

Chemistry and Compounds

In keeping with its $5d^46s^2$ electron configuration, tungsten forms many compounds in which its oxidation state is 6+, just as molybdenum does. It forms divalent and tetravalent compounds to about the same extent as molybdenum but its trivalent and pentavalent compounds are somewhat fewer. Its anion chemistry is closely akin to that of molybdenum.

Among the divalent compounds of tungsten, the diiodide, WI_2, dibromide, WBr_2, and the dichloride, WCl_2, are among the most clearly characterized; they all hydrolyze, although the iodide reacts only with warm water. Like molybdenum, tungsten(II) has a complex chloroion, $[W_6Cl_8]^{4+}$ which, however, is much more easily oxidized than its molybdenum analog.

Fig. 1. Cemented tungsten carbide compacts and mud nozzles for oil well drilling. (*Fansteel*)

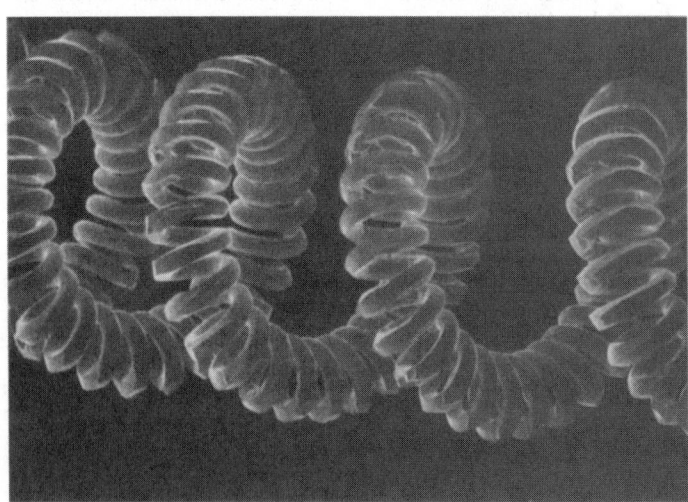

Fig. 2. Scanning electron photomicrograph of a nonsagging tungsten filament taken after several hundred hours of operation at 2500°C in a 60-watt light bulb. (*After Wittenaur, Nieh, and Wadsworth*)

Trivalent tungsten occurs rarely in simple compounds, other than certain high-temperature products such as one of the borides, WB, one of the phosphides, WP, and the complexes. Among the latter is the ion $[W_2Cl_9]^{3-}$ in which the two tungsten atoms participate in a ClWClWCl bridging structure, as they do in a W_2Cl_6 structure.

In addition to the tungsten(II) boride mentioned above, the element forms at least two other borides, W_2B and WB_2; it forms a similar series of phosphides, W_2P, WP, and WP_2 as well as WO_2 (brown oxide), W_4O_{11} (blue oxide), and WO_3 (yellow oxide), and two sulfides, WS_2 and WS_3. The tungsten(IV) oxide and sulfide are representative of the simple tetravalent compounds, which also include a tetrabromide, WBr_4, and tetraiodide, WI_4. Like the dihalides, these tetrahalides undergo hydrolysis quite readily.

Among the best known simple pentavalent tungsten compounds are the pentachloride, WCl_5, and the pentabromide, WBr_5. As is true of tungsten(IV), tungsten(V) forms complexes.

By far the greatest number of tungsten compounds are those in which the element is hexavalent. These include all common halides except the iodide, i.e., WF_6, WCl_6, as well as a number of oxyhalides, WOF_4, $WOCl_4$, WO_2Cl_2, $WOBr_4$, and WO_2Br_2, the trioxide, trisulfide, diboride, and diphosphide already mentioned, various complexes and organometallic compounds, and the anions.

Tungsten trioxide dissolves in hot alkali metal hydroxide solutions to yield in more or less hydrated form, the tungstate ion, WO_4^{2-}. However, the ionic species that exist in solution are more complex than mere hydration of the WO_4^{2-} would indicate, and this is especially true of the compounds obtained from such solutions. There are, however, two simple forms of the orthotungstic acid: H_2WO_4, which is precipitated upon addition of HCl to a hot tungstate solution, and $H_2WO_4 \cdot H_2O$, which is similarly obtained from a cold solution. Neutralization of a tungstate solution under most conditions yields, upon crystallization, much more complex salts. The acidic groups condense, with elimination of water, to form complexes, that can be crystallized as salts, which can be regarded as derived from "poly" acids. When such salts have only one kind of metal atom (e.g., W) in their anions, they are called *isopoly acids*; when they have more than one kind, they are called *heteropoly acids*. The latter group comprises an entire field of tungsten chemistry (as well as that of molybdenum and other elements.); the tungstophosphates are important in analysis and other applications. Other examples are the heteropoly acid salts formed by tungsten with oxyanions of boron, silicon, germanium, tin, arsenic, titanium, zirconium, and hafnium. In particular, the 6-series and the 12-series, containing, respectively, 6 and 12 tungsten atoms per molecule, have been extensively investigated.

Other interesting compounds are the "tungsten blues," complex oxides of colloidal nature, obtained by reduction of tungstates in alkaline solution. At higher temperatures, reduction of tungstates (of main group 1 and 2 elements) by alkali metal, hydrogen, zinc, tungsten, or electrolysis yields the semimetallic "tungsten bronzes," formulated as M_nWO_3, where M is the alkali metal and n is less than 1. They have cubical structures with W−O−W groups forming the sides, and the alkali or alkaline earth atom randomly located in the center of some of the cubes. The resulting extra electrons are considered to distribute over the entire structure, giving it metallic properties.

Tungsten forms many other complexes. Of particular interest are the octacyano complexes, containing eight cyanide, CN^- ions coordinated to a single tetravalent or pentavalent tungsten ion, $W(CN)_8^{4-}$ or $W(CN)_8^{3-}$, the latter being exceptionally stable, and forming octacyanotungstic(V) acid, $H_3[W(CN)_8] \cdot 6H_2O$, which is known in salts. Similar complexes are known for molybdenum, rhenium and osmium. The fluorocomplex of tungsten(VI) has the structure $[WF_8]^{2-}$ and forms salts with the higher alkali metals, potassium, rubidium and cesium.

Like molybdenum and chromium, tungsten forms a number of cyclo pentadienyl compounds which are also carbonyls, e.g., $C_5H_5W(CO)_3H$, $C_5H_5W(CO)_3CH_3$, $C_5H_5W(CO)_3C_5H_5$, $C_5H_5(CO)WW(CO)C_5H_5$, and $C_5H_5(CO)_3WW(CO)_3C_5H_5$. Tungsten(VI) also forms several other orga nometallic compounds, e.g., $W(OC_6H_5)_6$ and $W(OC_6H_4CH_3)_6$, as well as a simple carbonyl, $W(CO)_6$.

An excellent and comprehensive review of tungsten and its alloys is given in the Wittenauer reference listed.

Portions of this article were contributed by M. SCHUSSLER FANSTEEL Senior Scientist, North Chicago, Illinois

Additional Reading

Carter, G.F. and D.E. Paul: *Materials Science and Engineering,* ASM International, Materials Park, OH, 1991.

Farrar, L.C. and J.A. Shields, Jr.: "Tungsten and Tungsten-Copper Applications for Coal-Fired Magnetohydrodynamic Power Generation," *J. of Metals* (August 1992).

Greenwood, N.N. and A. Earnshaw: *Chemistry of the Elements,* 2nd Edition, Butterworth-Heinemann, Inc., Woburn, MA, 1997.

Krebs, R.E.: *The History and Use of Our Earth's Chemical Elements: A Reference Guide,* Greenwood Publishing Group, Inc., Westport, CT, 1998

Lassner, E. and W. Schubert: *Tungsten: Properties, Chemistry, Technology of the Element, Alloys, and Chemical Compounds,* Kluwer Academic Publishers, Norwell, MA, 1998.

Lewis, R.J. and N.I. Sax: *Sax's Dangerous Properties of Industrial Materials,* 10th Edition, John Wiley & Sons, Inc., New York, NY, 1999.

Lide, D.R.: *CRC Handbook of Chemistry and Physics,* 84th Edition, CRC Press, LLC., Boca Raton, FL, 2003.

Shin, K.E., et al.: "High-Temperature Properties of Particle-Strengthened W-Re," *J. of Metals* (August 1990).

Staff: *ASM Handbook—Properties and Selection: Nonferrous Alloys and Pure Metals,* ASM International, Materials Park, OH, 1990.

Wittenauer, J.P., T.G. Niehm, and J. Wadsworth: "Tungsten and Its Alloys," *Advanced Materials & Processes,* 28 (September 1992).

TURBIDIMETRY. The cloudiness in a liquid caused by the presence of finely divided, suspended material is termed *turbidity.* A turbidimeter measures this quality by determining the reduction in transmission of light that is caused by interposing a turbid solution between a light source and a detector, such as the eye or photocell. By using a known volume of solution in comparison with a standard, the instrument makes it possible to determine the mass effect, attributable to the number and size of the particles in the solution, and thus the quantitative amount of material present. Turbidimetric methods are similar to colorimetric procedures in that both involve measurement of the intensity of light transmitted through a medium. They differ in that the light intensity is attenuated by scattering in turbidimetry and by absorption in colorimetry. Both determinations may use similar instrumentation. Several designs of combined nephelometers, turbidimeters, and colorimeters are available.

Turbidimetry is applicable to the determination of suspended material in liquids encountered in nature and in manufacturing processes. Numerous uses are found in water treatment plants, sewage works, power and stream generating plants, beverage bottling plants, and petroleum refineries. The turbidity may be due to a single chemical substance, or it may be due to a combination of several components. For example, silica may be determined in the approximate concentrations of 0.1 to 150 parts per million, expressed as SiO_2. Sometimes, composite turbidities are expressed as equivalent to silica. Higher concentrations may be determined by dilution. Numerous applications are possible in which a turbidity is developed from the test sample under controlled conditions. A widely used determination is that of sulfate, after the addition of barium chloride to an unknown sample to form a suspension of barium sulfate. The procedure is particularly applicable within the concentration range of 0.2 to 100 parts per million sulfate. Routine procedures have been developed for the determination of sulfur in coal, oil, and other organic materials in which the sample first is fused in a sodium peroxide bomb prior to precipitation of the sulfur as barium sulfate.

The *Parr turbidimeter* is an extinction type instrument, which consists of a cylinder to contain the turbid suspension, a lamp filament of fixed intensity at the base, and an adjustable plunger through which visual observation is made. Measurement is made of the depth of turbid medium necessary to extinguish the image of the lamp filament. Standard suspensions are used to prepare a calibration curve, which is a plot of depth *vs.* concentration.

The *Hellige turbidimeter* is also a variable depth type of instrument using visual detection. A combination of vertical and horizontal illumination of the sample and a split ocular permit the eye to function merely to compare the intensities of two images simultaneously appearing in the ocular. This is a special form of double-beam operation, and the intensity emitted by the light source need not be extremely constant from one sample to another. Adjustment is made of a slit in the path of the direct, or vertical, illumination of the sample, and the calibration curve consists of a plot of this slit opening *vs.* concentration.

See also **Analysis (Chemical)**; **Nephelometry**; and **Photometers**.

TURBULENT FLOW. A condition of fluid flow in a closed conduit that is above a critical velocity. In this condition, there is a random, irregular motion of the fluid particles in directions transverse to the axis of the main flow. This is converse to the condition of laminar or streamline flow. See also **Laminar Flow**.

At velocities greater than the critical, the fluid velocity profile in the conduit is uniform across the conduit diameter except for a thin layer of fluid at the conduit wall. This *boundary layer* continues to move in laminar flow. In connection with flow measurement, most flowmeters have constant coefficients under turbulent flow conditions. Some flowmeters have the advantage of constant coefficients over Reynolds Number ranges encompassing both turbulent and laminar flows. See also **Fluid and Fluid Flow**; and **Reynolds Number**.

TURPENTINE.   See **Resins (Natural)**.

TURQUOISE.   The mineral turquoise is a hydrated phosphate of aluminum and copper. Its exact composition is doubtful, the formula may be expressed $CuAl_6(PO_4)_4(OH)_85 \cdot H_2O$; iron is often present. This mineral is found in minute triclinic crystals, but chiefly massive as seams and crusts. The fracture is conchoidal; hardness, 5–6; specific gravity, 2.6–2.8; luster, soft waxy; color, may be various shades of blue, bluish-green and green; essentially opaque. It takes a good polish and the sky-blue varieties have long been used as a gem material. Unfortunately, many beautiful blue stones in time change their color to some greenish hue, usually not attractive, rendering them practically valueless.

For hundreds of years turquoise has been mined in Iran, where it is found with limonite filling crevices in a brecciated trachyteporphyry, and because it found its way into Europe through Turkey, it became known as turquoise, from the French word *turque*, Turkish. Other mines were worked by the Egyptians in ancient times on the Sinai Peninsula. Turquoise is also found in Siberia, Turkestan, Saxony and France; and in the United States in Arizona, California and New Mexico. A blue stone that has passed for turquoise is in reality odontolite, from the Greek meaning tooth, usually fossil teeth or bones colored with iron phosphate. Odontolite is softer than true turquoise, has a somewhat higher specific gravity, 3.0–3.5, and may be distinguished by chemical tests, or by a microscopical examination which will reveal its organic structure.

TWINNING (Crystal).   A process in which a region in a crystal assumes an orientation which is symmetrically related to the basic orientation of the crystal. Usually, layers of atoms within this region are translated with respect to a basic plane (the twinning plane). Each atomic plane is displaced by a distance which is proportional to its distance from the twinning plane. Bands of metal (twin bands) thus assume a lattice structure which is the mirror image of the unchanged portion of the lattice. See also **Mineralogy**.

TWITCHELL PROCESS.   One of the very early methods for the hydrolysis of fats and oils on an industrial scale. The key to the process is *Twitchell's reagent*, which is prepared by interacting oleic acid, benzene, and H_2SO_4. The product sometimes is termed *sulfobenzenestearic acid*. The agent accelerates hydrolysis. Naphthalene may be used in place of benzene in the reagent with the assigned $C_{18}H_{35}O_2 \cdot C_{10}H_6SO_3H$.

See also **Vegetable Oils (Edible)**.

TYNDALL EFFECT.   A phenomenon first noticed by Faraday (1857). When a powerful beam of light is sent through a colloidal solution of high dispersity, the sol appears fluorescent and the light is polarized, the amount of polarization depending upon the size of the particles of the colloid. The polarization is complete, if the particles are much smaller than the wavelength of the radiation. See also **Nephelometry**.

TYNDALL, JOHN (1820–1893).   John Tyndall was a man of science: draftsman, surveyor, physics professor, mathematician, geologist, atmospheric scientist, public lecturer, and mountaineer. Throughout the course of his Irish and later, English life, he was able to express his thoughts in a manner none had seen or heard before. His ability to paint mental pictures for his audience enabled him to disseminate a popular knowledge of physical science that had not previously existed. Tyndall's original research on the radiative properties of gases as well as his work with other top scientists of his era opened up new fields of science and laid the groundwork for future scientific enterprises.

In January 1859, Tyndall began studying the radiative properties of various gases. Part of his experimentation included the construction of the first ratio spectrophotometer, which he used to measure the absorptive powers of gases such as water vapor, "carbonic acid" (now known as carbon dioxide), ozone, and hydrocarbons. Among his most important discoveries were the vast differences in the abilities of "perfectly colorless and invisible gases and vapors" to absorb and transmit radiant heat. He noted that oxygen, nitrogen, and hydrogen are almost transparent to radiant heat while other gases are quite opaque.

Tyndall's experiments also showed that molecules of water vapor, carbon dioxide, and ozone are the best absorbers of heat radiation, and that even in small quantities, these gases absorb much more strongly than the atmosphere itself. He concluded that among the constituents of the atmosphere, water vapor is the strongest absorber of radiant heat and is therefore the most important gas controlling Earth's surface temperature. He said, without water vapor, the Earth's surface would be "held fast in the iron grip of frost." He later speculated on how fluctuations in water vapor and carbon dioxide could be related to climate change.

Tyndall related his radiation studies to minimum nighttime temperatures and the formation of dew, correctly noting that dew and frost are caused by a loss of heat through radiative processes. He even considered London as a "heat island," meaning he thought that the city was warmer than its surrounding areas.

Over the course of his life, John Tyndall published numerous papers and essays on his scientific discoveries, as well as literature, religion, mountaineering, and travel. His accomplishments led him to receive five honorary doctorates and become a respected member of thirty-five scientific societies.

See also **Nephelometry**; and **Tyndall Effect**.

J.M.I.

TYROSINE.   See **Amino Acids**.

TYUYANUNITE.   An ore of uranium with the composition, $Ca(UO_2)_2(VO_4)_2 \cdot 5 - 8H_2O$, which occurs in yellow incrustations as a secondary mineral. The mineral is orthorhombic. It occurs as a secondary mineral as incrustations on limestones, and as disseminated impregnations in sandstones. Found abundantly in the Western United States, at Grants, New Mexico, and in Wyoming, Utah, Colorado, Nevada, Arizona and Texas. Also at Tyuya Muyan in Turkestan, the former U.S.S.R.

U

UGRANDITE. A group name for the calcium garnet minerals uvarovite, grossular, and andradite.

UINTAHITE. A black, shiny asphaltite, with a brown streak and conchoidal fracture, which is soluble in turpentine. It occurs primarily in the veins in the Uintas Basin, Utah.

ULEXITE. This mineral, a hydrated borate of sodium and calcium, $NaCaB_5O_9 \cdot 8H_2O$, is a product of crystallization in arid regions from shallow playas and lakes. Ulexite crystallizes in the triclinic system, but usually occurs in rounded masses of fine-fibered acicular crystals. The hardness is 2.5, specific gravity, 1.96, silky luster and white color. The mineral is found abundantly in Chile and Argentina and in Nevada and California in the United States. Ulexite is a source of boron.

ULMIN. One of a class of amorphous substances resulting from the decomposition of the cellulose and lignite tissues of plants. Ulmins represent on of the initial changes by which vegetable matter is converted into coal.

ULTRABASIC. A term proposed by Judd, in 1881, for exceedingly mafic igneous rocks composed largely, if not entirely, of the ferromagnesium minerals such as olivine and pyroxene. The limiting figure of total silica is approximately 45%, or barely sufficient to supply the needs of the basic silicates.

ULTRACENTRIFUGE. A high speed rotational separating device, usually of laboratory size, capable of developing a force of 250,000 times gravity. Its major uses are in research on molecular weight distribution, macromolecular structure and properties (proteins, nucleic acids, viruses) and separation of solutes from solutions. See also **Centrifugation**.

ULTRAFILTRATION. Ultrafiltration is a pressure-driven filtration separation occurring on a molecular scale. See also **Dialysis; Filtration; Hollow-Fiber Membranes; Membrane Separations Technology**; and **Reverse Osmosis**. Typically, a liquid including small dissolved molecules is forced through a porous membrane. Large dissolved molecules, colloids, and suspended solids that cannot pass through the pores are retained.

Ultrafiltration separations range from ca 1 to 100 nm. Above ca 50 nm, the process is often known as microfiltration. Below ca 2 nm, interactions between the membrane material and the solute and solvent become significant. That process, called reverse osmosis or hyperfiltration, is best described by solution–diffusion mechanisms.

Membrane-retained components are collectively called concentrate or retentate. Materials permeating the membrane are called filtrate, ultrafiltrate, or permeate.

Media

Most ultrafiltration membranes are porous, asymmetric, polymeric structures produced by phase inversion, i.e., the gelation or precipitation of a species from a soluble phase. See also **Membrane Separations Technology**. Membrane structure is a function of the materials used (polymer composition, molecular weight distribution, solvent system, etc) and the mode of preparation (solution viscosity, evaporation time, humidity, etc.). Commonly used polymers include cellulose acetates, polyamides, polysulfones, dynels (vinyl chloride–acrylonitrile copolymers) and poly(vinylidene fluoride).

Modification of the membranes affects the properties. Cross-linking improves mechanical properties and chemical resistivity. Fixed-charge membranes are formed by incorporating polyelectrolytes into polymer solution and cross-linking after the membrane is precipitated, or by substituting ionic species onto the polymer chain (e.g., sulfonation). Polymer grafting alters surface properties. Enzymes are added to react with permeable species and reduce fouling.

Polyelectrolyte complex membranes are phase-inversion membranes where polymeric anions and cations react during the gelation. Inorganic ultrafiltration membranes are formed by depositing particles on a porous substrate. Dynamic membranes are concentration–polarization layers formed *in situ* from the ultrafiltration of colloidal material analogous to a precoat in conventional filter operations. Track-etched membranes are made by exposing thin films (mica, polycarbonate, etc.) to fission fragments from a radiation source.

Process

Pore-flow models most accurately describe ultrafiltration processes. Other membrane transport mechanisms, which may occur simultaneously although generally at a much lower rate, include dialysis (diffusion), osmosis (solvent by osmotic gradient), anomalous osmosis (osmosis with a charged membrane), reverse osmosis (solvent by pressure gradient larger and opposite to osmotic gradient), electrodialysis (solute ions by electric field), piezodialysis (solute by pressure gradient), electroosmosis (solvent in electric field), Donnan effects, Knudsen flow, thermal effects, chemical reactions (including facilitated diffusion), and active transport.

When pure water is forced through a porous ultrafiltration membrane, Darcy's law states that the flow rate is directly proportional to the pressure gradient:

$$J = \frac{V}{A \cdot t} = \frac{K_m \Delta P}{\mu} \qquad (1)$$

where J is permeate flux in units of volume V per membrane area A, at time t, K_m is the membrane hydraulic permeability, μ is the fluid viscosity, and ΔP is the membrane pressure drop between the retentate and permeate. These parameters change with the membrane pressure history.

The addition of small membrane-permeable solutes to the water affects permeate transport in the following ways. (1) Solute–solvent interactions change the permeating fluid viscosity. (2) Solute adsorption reduces the apparent membrane-pore diameter. Because of high interfacial tension between water and certain materials, the water phase in the pores can be replaced. Surfactants suppress hydrophobic adsorption. Adsorption of permeate species is characterized by a lag in permeate concentration as a function of time. (3) The interfacial tension between the membrane-pore wall and the liquid affects permeate transport when the Debye screening length approaches (ca 10%) the membrane-pore size. (4) High surface tension on hydrophobic membranes forces water molecules to form large clusters in the pores. (5) Solvents, swelling agents, and plasticizers that diffuse into the polymer structure can change the apparent pore size (K_m in eq. 1) or increase the rate of long-term compaction. The rejection R of a solute is defined as:

$$R = 1 - \frac{C_{pi}}{C_{bi}} \qquad (2)$$

where C_p is the permeate concentration of species i and C_b is the concentration of that species in the retentate. There are two components of rejection. Observed rejection, R_o, is based on the concentration of the solute in the bulk solution, C_b. The intrinsic rejection, R_i, is based on the concentration of the solute on the surface of the membrane, C_w.

$$R_o = 1 - \frac{C_p}{C_b}$$

$$R_i = 1 - \frac{C_p}{C_w}$$

If the solute size is approximately the (apparent) membrane-pore size, it interferes with the pore dimensions. The solute concentration in the permeate first increases, then decreases with time. The point of maximum interference is further characterized as a minimum flux. If the solute size is greater than the pore dimensions, the solute is retained by mechanical sieving, forming a gel-polarization layer.

The gel usually has a much lower hydraulic permeability and smaller apparent pore size than the underlying membrane. (The gel layer and the concentration gradient between the gel layer and the bulk concentration are called the gel-polarization layer.)

The gel-layer thickness is limited by mass transport back into the solution bulk at the rate. At steady state,

$$JC_b = K \frac{dC}{dX} \tag{3}$$

where C_b is the bulk concentration of all retained species. Integration gives

$$J = K \cdot \ln \frac{C_g}{C_b} \tag{4}$$

In a static system, the gel-layer thickness rapidly increases and flux drops to uneconomically low values.

The gel-polarization layer has an hydraulic permeability of K_g. Equation 4 states that flux is independent of pressure, and K_g must therefore decrease with increasing pressure. Equation 1 becomes

$$J = \frac{\Delta P}{\mu \left(\frac{1}{K_m} + \frac{1}{K_g} \right)} = \frac{\Delta P}{\mu (R_m + R_g)} \tag{5}$$

where R_m and R_g are the hydraulic resistances of the membrane gel.

Flux is independent of pressure when the process flux is much less than the wafer flux ($K_g \ll K_m$). If $K_g > K_m$, the process is limited by the membrane water flux and flux would flatten out at low concentrations of solids.

Fouling. If the gel-polarization layer is not in hydrodynamic equilibrium with the fluid bulk, the membrane may be fouled. Fouling is caused either by adsorption of species on the membrane or on the surface of the pores, or by deposition of particles on the membrane or within the pores. Fouled systems are characterized as follows: flux is a function of total permeate production when hydrodynamic conditions are constant; if hydrodynamic conditions are changed, hydraulic permeability response of the gel layer is not reversible; and theoretical permeate flux (TPF) changes with time. A sensitive test for predicting fouling or process instability is to measure change in TPF after subjecting the system to process extremes.

Fouling is controlled by selection of proper membrane materials, pretreatment of feed and membrane, and operating conditions. Control and removal of fouling films is essential for industrial ultrafiltration processes.

When fouling is present or possible, ultrafiltration is usually operated at high liquid shear rates and low pressure to minimize the thickness of the gel polarization layer.

Cleaning. Fouling films are removed from the membrane surface by chemical and mechanical methods. Dissolved fouling material may pass into the membrane pores. Reprecipitation upon rinsing must be avoided. Membrane-swelling agents, such as hypochlorites, flushout material which may be lodged in the pores.

Certain applications require that the equipment meet FDA and USDA sanitary requirements. These requirements ensure that the products are not contaminated by extractables or microorganisms from the equipment. Special considerations are given to the design of such equipment.

Practical Aspects

The theoretical models cannot predict flux rates. Plant-design parameters must be obtained from laboratory testing, pilot-plant data, or in the case of established applications, performance of operating plants.

Flux is maximized when the upstream concentration is minimized. For any specific task, therefore, the most efficient (minimum membrane area) configuration is an open-loop system where retentate is returned to the feed tank. When the objective is concentration (e.g., enzyme), a batch system is employed. If the object is to produce a constant stream of uniform-quality permeate, the system may be operated continuously (e.g., electrocoating).

Open-loop systems have inherently long residence times which may be detrimental if the retentate is susceptible to degradation by shear or microbiological contamination. A feed-bleed or closed-loop configuration is a one-stage continuous membrane system. At steady state, the upstream concentration is constant at C_f. For concentration, a single-stage continuous system is the least efficient (maximum membrane area).

The single-pass system and the staged cascade have high flux at low residence time. Both trade the concentration dependence of the batch system on time for concentration dependence on position in the system. Thus, a uniform flux is maintained (assuming no fouling) allowing continuous process integration. In practice, the single-pass system is difficult to implement, and therefore most commercial systems are multistaged cascade. The more stages used, the closer the average flux approaches the batch flux.

Electroultrafiltration

Electroultrafiltration (EUF) combines forced-flow electrophoresis. See also **Electrophoresis** with ultrafiltration to control or eliminate the gel-polarization layer. Placing an electric field across an ultrafiltration membrane facilitates transport of retained species away from the membrane surface. Thus, the retention of partially rejected solutes can be dramatically improved. See also **Electrodialysis**.

Electroultrafiltration has been demonstrated on clay suspensions, electrophoretic paints, protein solutions, oil–water emulsions, and a variety of other materials.

Diafiltration

Diafiltration is an ultrafiltration process where water or an aqueous buffer is added to the concentrate and permeate is removed. The two steps may be sequential or simultaneous. Diafiltration improves the degree of separation between retained and permeable species.

Constant-volume batch diafiltration is the most efficient process mode. Sequential batch diafiltration is a series of dilution–concentration steps. Continuous diafiltration practiced in one or more stages of a cascade system has the same volume turnover relationship for overall recoveries as sequential batch diafiltration. The residence time however is dramatically reduced. If recovery of permeable solids is of primary importance, the permeate from the last stage may be used as diafiltration fluid for the previous stage. This countercurrent diafiltration arrangement results in higher permeate solids at the expense of increased membrane area.

Membrane Equipment

Commercial industrial ultrafiltration equipment first became available in the late 1960s. Since that time, the industry has focused on five different configurations.

Parallel-Leaf Cartridge. A parallel-leaf cartridge consists of several flat plates, each having membrane sealed to both sides. Cartridges are inserted in series into plastic or stainless-steel tubular pressure housings of square cross section. Feed flows parallel to the leaf surface. A permeate fitting secures each cartridge to the housing wall, which allows permeate egress and facilitates sealing between concentrate, atmosphere, and permeate channels.

Plate and Frame. Plate-and-frame systems consist of plates each with a membrane on both sides.

At least one hole near the perimeter of each plate connects the flow channels from one side of the plate to the other. The membrane is sealed around the hole to isolate the permeate from the concentrate. Permeate collects in a drain grid behind the membrane and exits from a withdrawal port on the frame perimeter.

Spiral Wound. A spiral-wound cartridge has two flat membrane sheets (skin side out) separated by a flexible, porous permeate drainage material.

Supported Tube. There are three types of supported tubular membranes: cast in place (integral with the support tube), cast externally and inserted into the tube (disposable linings), and dynamically formed membranes. The most common supported tubes are those with membranes cast in place.

Self-Supporting Tubes. Depending on the membrane material and operating pressure, self-supporting tubes are less than 2-mm ID; inside diameters as small as 0.04 mm are commercially available.

A large number of fibers are cut to length, and potted in epoxy resin at each end. The fiber bundle is shrouded in a cylinder which aids in permeate collection, reduces airborne contamination, and allows back pressing of the membrane. Hollow-fiber membranes have also found use in ultrafiltration.

Each of the membrane devices may be assembled by connecting the modules into combinations of series, parallel-flow paths, or both.

TABLE 1. ULTRAFILTRATION APPLICATIONS

Application	Process
electrophoretic paint	control of properties, recovery of solids from rinse systems
dairy wheys	protein recovery, concentration, purification, diafiltration
milk	cheese and yogurt mfg, 15–20% yield improvement, standardization
oil–water emulsions	concentration
effluents of wool, yarn scouring	lanolin recovery, pollution abatement
enzymes	concentration, purification
biological reactors	antibiotic mfg, alcohol fermentation, sewage treatment
vegetable proteins	
latex concentration	
production of pure[a] water	
pulp and paper	lignosulfonate sprn from spent liquor
blood and blood products	fractionation, purification
vaccines	conc, purification
biotechnology products	conc, purification

[a] Virus-free.

These assemblies are connected to pumps, valves, tanks, heat exchangers, instrumentation, and controls to provide complete systems.

Because of the broad differences between ultrafiltration equipment, the performance of one device cannot be used to predict the performance of another. Comparisons can only be made on an economic basis and only when the performance of each is known.

Uses

Applications of ultrafiltration are summarized in Table 1.

RALF KURIYEL
Millipore Corporation

Additional Reading

Cheryan, M.: *Ultrafiltration and Microfiltration Handbook,* CRC Press LLC., Boca Raton, FL, 1998.

Cheryan, M.: *Ultrafiltration Handbook,* CRC Press LLC., Boca Raton, FL, 1997.

Cooper, A. R. ed.: "Ultrafiltration Membranes and Applications, Proceedings of 178th National ACS Meeting," Washington, DC, 1979, Plenum Press, New York, NY, 1980.

Hwang, S. and K. Kammermeyer: *Membranes in Separations,* John Wiley & Sons, Inc., New York, NY, 1975; good study of membrane transport phenomenon.

Kesting, R. E.: *Synthetic Polymeric Membranes,* McGraw-Hill, New York, NY, 1971; good bibliographies.

Zeman, L. J., and A. L. Zydney: *Microfiltration and Ultrafiltration: Principles and Applications,* Marcel Dekker, Inc., New York, NY, 1997.

ULTRAMICROSCOPE.   The ultramicroscope is not an instrument of extraordinary magnifying power, as its name might suggest. The term has reference rather to a special system of illumination for very minute objects. Such objects as colloidal particles, fog drops, or smoke particles are held in liquid or gaseous suspension in an enclosure with an intensely black background (usually of the black-body type). They are illuminated by a convergent pencil of very bright light entering from one side and coming to focus in the field of view—the so-called "Tyndall cone" familiar in experiments on scattering. With this arrangement, objects too small to form visible images in the microscope produce small diffraction ring systems, which appear as minute bright specks on a dark field.

ULTRASONICS.   Sound waves above the frequency normally detectable by the human ear, that is, above 16 to 20 kHz are referred to as *ultrasonic waves*. The particles of matter transmitting a *longitudinal wave* move back and forth about mean positions in a direction parallel to the path of the wave. Alternate compressions and rarefactions in the transmitting material exist along the wave propagation direction. In *shear waves*, the particles move perpendicularly to the direction of wave propagation. In *surface waves*, in seismological studies and in waves through thin stock, the *Rayleigh* and *Lamb waves*, respectively, the particles undergo much more complex vibratory motions than in longitudinal and transverse waves. In most practical applications of ultrasonics, pulses or packets containing a number of oscillation cycles are sent through the solid or liquid

under investigation. See also **Cavitation**. A longitudinal wave pulse, when incident on the boundary between two materials having different sound velocities, is transformed into reflected and refracted shear and longitudinal waves. Snell's law governs the angles of reflection and refraction for both types of waves:

$$\frac{\sin\theta}{V} = \text{Constant}$$

where θ is the angle the beam makes with a plane normal to the intervening surface and V is the sound velocity. Therefore, in Fig. 1,

$$\text{Constant} = \frac{\sin\theta_1}{V_{L_1}} = \frac{\sin\theta_2}{V_{L_2}} = \frac{\sin\theta}{V_{S_2}} = \frac{\sin\phi_2}{V_{L_2}}$$

The practical application of ultrasonics requires effective transducers to change electrical energy into mechanical vibrations and vice versa. Transducers are usually piezoelectric, ferroelectric, or magnetostrictive. The application of a voltage across a piezoelectric crystal causes it to deform with an amplitude of deformation proportional to the voltage. Reversal of the voltage causes reversal of the mechanical strain. Quartz and synthetic ceramic materials are used.

Ferroelectric crystals are also electrostrictive. Barium titanate, for example, has an electrical mechanical conversion efficiency about 100 times that of quartz. Unlike the piezoelectric mode of oscillation, in ferroelectric crystals, application of a voltage in either direction across the crystal causes expansion of the crystal. This mode, however, can be converted to the piezoelectric by biasing the expansion in one direction, either by application of a strong dc field, or more commonly by cooling the ferroelectric crystal through its Curie temperature while it is under the influence of a strong electric field (on the order of 10^6 volts/meter).

Transducer crystals are normally cut to a resonant frequency, the thickness being one-half the acoustic wavelength. A bond between the crystal transducer and the specimen matches the acoustic impedance, and carries the acoustic power into the latter. Backing layers may be fixed to the rear surface of the transducer. These layers are selected to reflect power forward into the crystal and specimen in some applications. On the other hand, they may be selected to absorb power so as not to complicate signals received in material testing applications.

Ultrasonic Applications

An appreciation of useful applications of ultrasound dates back to the development of radar and sonar in the late 1930s and early 1940s World War II era. One of the very earliest uses was for detecting flaws in materials. Today ultrasound is one of several effective methods used in the field of nondestructive testing and inspection. Later, the use of ultrasound was extended to include a number of detectors, probes, and transducers for measuring such variables as fluid flow, liquid level, viscosity, density, proximity, and material thickness, among others. Most of these applications are described in other articles in this encyclopedia, notably **Nondestructive Testing (NDT)**, and Ultrasonic devices also are used in industrial processing applications, such as cleaning, drilling, emulsifying, soldering, and welding. Possible the most noteworthy of ultrasonic devices are found in the medical field for use in diagnostics where ultrasound vies with x-ray and other diagnostic procedures. Ultrasound is also applied in some security systems to detect intruders.

Ultrasonic Density Sensors.   A useful application is found in the density measurement of lime slurries for the purpose of adjusting the pH in acid

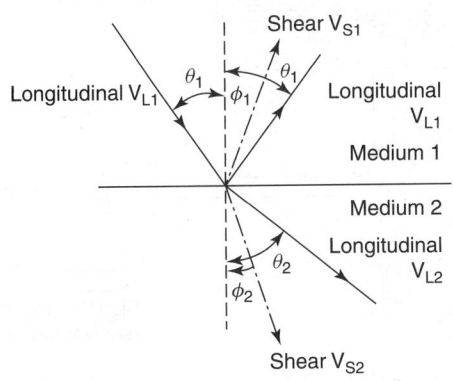

Fig. 1.   Reflection and refraction of ultrasonic waves

water neutralization processes. Where traditional measurement methods are not suited, control engineers will often turn to more sophisticated techniques, such as ultrasound.

Slurries are difficult to handle, have a strong tendency to settle out and to coat equipment with which they come in contact. An ultrasonic density control sensor can be fully immersed in an agitated slurry, thus avoiding coating and clogging. Use of the ultrasound device for controlling the specific gravity of slurries within ±0.01% of the desired value has been demonstrated. In the usual application, the specific gravity ranges between 1.05 and 1.10. Actually, the ultrasonic sensor measures the percentage of suspended solids in the slurry, thus providing a very close approximation of the true specific gravity. The suspended solids present in the slurry attenuate the ultrasonic beam, with the resulting electronic signal being proportional to the solids in suspension.

Ultrasonic Level Detectors. In the ultrasound system shown in Fig. 2, one transmitting sensor creates a sonic beam, the waves of which are picked up by a receiving sensor. This can be accomplished by a direct path or by the reflective waves from the transmitter, which strike a fluid surface and are reflected back to the receiving sensor. Sensors can be spaced as close as $\frac{1}{4}$ in. (6 mm) or as far apart as 10 ft (3 m) in a direct beam path. They can pipe their beams through tubing where sensor beams cannot be direct, but generally the tubing length is limited to several feet (few meters). The sensor's sound beam is unaffected by mist, smoke, dust, or fumes, since it is interrupted only by a solid or fluid entering the beam path. This type of system is often used in connection with measuring the level of dry bulk solids.

Ultrasonic Thermometers. These are usually designed to respond to the temperature dependence of sound speed. In special cases where only one particular temperature is of interest, such as the temperature of a phase change, or the recrystallization temperature of a substance, the temperature dependence of attenuation may be utilized. Ultrasonic thermometers have found applications in the range −80 to +250°C, where the so-called quartz thermometer offers resolution of 0.1 millidegree and linear superiority to platinum resistance thermometers.

Ultrasonic Pressure Transducers. Advantage is taken of the fact that pressure influences sound propagation in solids, liquids, and gases, but in different ways. In solids, applied pressure leads to so-called stress-induced anisotropy. In liquids, the effects of pressure are usually small (relative to effects in gases), but the frequency of relaxation peaks can be shifted significantly.

Sonochemistry. Where liquids and solids must react, researchers are finding that energy in the ultrasound range often promotes such reactions. K.S. Suslick (University of Illinois), who has pioneered in the field,

observes, "Ultrasound causes high-energy chemistry. It does so through the process of acoustic cavitation; the formation, growth and implosive collapse of bubbles in a liquid. During cavitational collapse, intense heating of the bubbles occurs. These localized hot spots have temperatures of roughly 5000°C, pressures of about 500 atmospheres, and lifetimes of a few microseconds. Shock waves from cavitation in liquid-solid slurries produce high-velocity interparticle collisions, the impact of which is sufficient to melt most metals." In recent research, Suslick also reports that ultrasound creates clean, highly reactive surfaces on metals. In both homogeneous and heterogeneous situations, ultrasound assists in initiating and enhancing a number of catalytic reactions.

In investigating the phenomenon of acoustic cavitation in homogeneous liquids, researchers find that the velocity of sound in liquids typically is about 1500 m/s. Ultrasound spans the frequencies of approximately 15 kHz to 10 MHz, with acoustic wavelengths of 10 to 0.01 cm. Researchers in sonochemistry stress that these are not molecular dimensions and thus no direct coupling of the acoustic field with chemical species takes place on a molecular level. Rather, the effects of ultrasound arise from a number of physical mechanisms, including the production of cavitation. Cavitation, originally investigated by Lord Rayleigh (1895), can produce inordinate local temperatures of 10,000 K and pressures up to 10,000 atmospheres when cavities collapse. These physical parameters indeed are conducive to promoting chemical reactivity. Considerable research over the years has been directed toward investigating the chemical effects of ultrasound on inorganic liquids, notably water. Little effort to date has been directed toward organic liquids.

In his summary of an excellent review article on sonochemistry (reference listed), Suslick points out, "Chemical applications of ultrasound are just beginning to emerge. The very high temperatures and very short times of cavitational collapse makes sonochemistry a unique interaction of energy and matter. In addition, ultrasound is well suited to industrial applications. Since the reaction liquid itself carries the sound, there is no barrier to its use with large volumes. In fact, ultrasound is already heavily used industrially for the physical processing of liquids, such as emulsification, solvent degreasing, solid dispersion, and sol formation. It is also extremely important in solids processing, including cutting, welding, cleaning and precipitation." See also **Cavitation**.

Ultrasonic Nondestructive Characterization (NDC) of Materials. The use of ultrasonics for the nondestructive characterization (testing) of metals and other materials was one of the first practical applications for ultrasound and dates back to the 1950s. The detailed properties of materials that can be measured with ultrasound include microstructure, surface characteristics, elastic properties, density, porosity, mechanical properties, process characteristics, and overt flaws. Ultrasound waves transmitted may take the form of longitudinal, shear, and surface waves. Wave characteristics include ultrasonic velocities and frequency dependence of ultrasonic attenuation/absorption. Specific reactions occurring within a tested material may include ultrasound reflection, transmission, refraction, diffraction, interference, scattering, and absorption. In applying ultrasound to testing, the characteristics of the radiation source that can be manipulated include wave type, frequency, bandwidth, pulse shape, and pulse size.

By comparison with other NDC methods, such as liquid penetrant examination, magnetic particle, eddy current testing, and radiography, the ultrasonic method is the only technique that is applicable to a wide range of materials.

The success of an ultrasonic NDC application depends upon the selection of the best-qualified transducer (i.e., one with optimum frequency response, pulse width and shape). Transducer characteristics can be customized through the use of the best-suited piezoelectric material, such as lead zirconate-lead titanate, lead metaniobates, polymer piezoelectrics, and other advanced ferro-electric materials.

Other Transducers. Ultrasound also has been used for the measurement of force, vibration, acceleration, interface location, position changes, differentiation between the composition of differing materials, grain size in metals, and evaluation of stress and strain and elasticity in materials. Sonic devices can used to detect gas leaks, and to count discrete parts by means of an interrupted sound beam. Frequently, an ultrasonic device can be applied where photoelectric devices are used. Particularly in situations where light-sensitive materials are being processed (hence presence of light must be avoided), ultrasonic devices may be the detectors of choice.

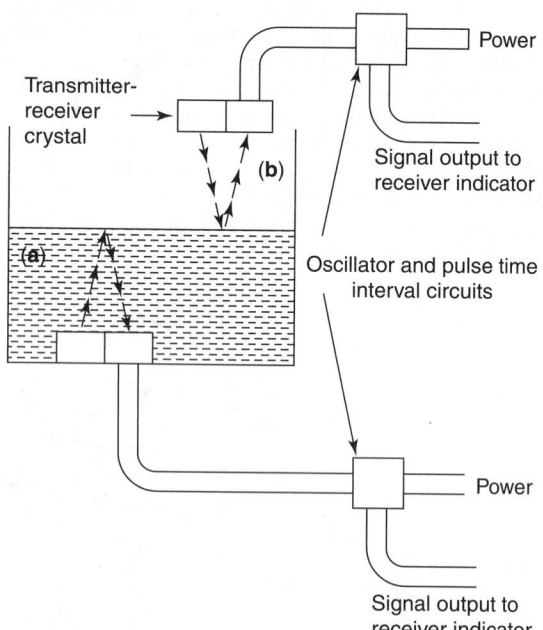

Fig. 2.   Continuous sonic-type level measuring device: (**a**) Liquid phase; (**b**) vapor phase

Ultrasonic Applications in Medicine

Ultrasonic imagery is one of the earlier tools used in the medical field for noninvasively probing body organs and tissues. Experience with this technique dates back at least to the 1970s.

The nonionizing character of ultrasonic radiation is particularly attractive, permitting the use of the technique repeatedly with a given patient at low risk. The sound frequency used for most diagnostic instruments ranges from 1 to 10 MHz. These frequencies are generated by piezoelectric transducers that reversibly convert electrical to vibratory mechanical (sound) energy. Short sound pulses propagate through the body, but a small portion of the energy is reflected back to the instrument where there are interfaces between tissues having different acoustic impedances.

Ultrasonic examination has become a well-established diagnostic tool in many abdominal diseases. In recent years, the degree of resolution attainable with ultrasonic equipment has been improved considerably, making more accurate diagnoses possible. As a consequence, transabdominal sonography has become increasingly important for diagnosing diseases of the gastrointestinal tract. It is noteworthy to observe that even acute appendicitis usually can be diagnosed by ultrasonic examination.

As noted by B. Limberg (University of Frankfurt), "With the use of sonography alone, it is impossible to detect diseases of the colon reliably, since the large bowel cannot be visualized in its entirety and detailed evaluation of wall structures and intra-aluminal lesions is difficult. The method can be improved considerably, however, by the retrograde instillation of water into the colon, a method known as *hydrocolonic sonography*. Studies have shown that this method permits not only the diagnosis of colonic tumors, but also that of inflammatory colonic diseases, such as Crohn's disease and ulcerative colitis." The detailed test of 300 patients in which conventional sonography and hydrocolonic sonography were used are given in the Limberg reference listed.

In order to facilitate surgery, preoperative imaging procedures, such as ultrasonography, computed tomography (CT), and angiography, have been used to localize the primary tumor in the case of pancreatic endocrine tumors. Because of their size, insulinomas and gastrinomas cannot be identified in up to 30% of patients. As pointed out by T. Rösch (Technical University of Munich), a large team made a comparative study involving some 37 patients. Some of the study results included: "The introduction of endoscopic ultrasonography has allowed high-resolution imaging of the pancreas that can distinguish structures as small as 2 to 3 mm in diameter. The accuracy of this procedure in diagnosing small pancreatic carcinomas has been reported to be close to 100 percent. The method also seems to be useful for the preoperative localization of small endocrine tumors of the pancreas, given the experience in this research. We report here the collective experience at six centers where endoscopic ultrasonography has been used for the preoperative localization of small tumors of the pancreas. Only patients with normal results on transabdominal ultrasonography and CT were included."

Ultrasound Assists Drug Implants. Plastic materials infused with drugs are used for a number of purposes where drugs are gradually diffused into the bloodstream. Such implants are used for contraceptions and various chemotherapies. The effects of exposure to ultrasound of biodegradable implant materials have been studied at several teaching hospitals with interesting results.

Human Perception of Ultrasonic Speech. The upper range of human air-conduction hearing has been estimated to be no higher than approximately 24,000 Hz, although there have been a conservative number of reports that indicate hearing well into the ultrasonic range, but only when signals are delivered by bone conduction. M.L. Lenhardt (Medical College of Virginia) and a team of researchers has reported that "Bone-conducted ultrasonic hearing has been found capable of supporting frequency discrimination and speech detection in normal, older hearing-impaired and profoundly deaf human subjects. When speech signals were modulated into the ultrasonic range, listening to words resulted in the clear perception of the speech stimuli and not a sense of high-frequency vibration. These data suggest that ultrasonic bone-conduction hearing has potential as an alternative communication channel in the rehabilitation of hearing disorders." Further details are given in the Lenhardt reference listed.

Additional Reading

Abramov, O.V.: *High-Intensity Ultrasonics: Theory and Industrial Applications,* Gordon & Breach Publishing Group, Newark, NJ, 1998.

Doktycz, S.J. and K.S. Suslick: "Interparticle Collisions Driven by Ultrasound," *Science,* 1067 (March 2, 1990).

Edward, I., E.I. Bluth, P.W. Ralls, P. Arger, and C. Benson: *Ultrasound: A Practical Approach to Clinical Problems,* Thieme Medical Publishers, Inc., New York, NY, 1999.

Evans, D.H. and N. McDicken: *Doppler Ultrasound: Physics, Instrumentation, and Signal Processing,* 2nd Edition, John Wiley & Sons, Inc., New York, NY, 2000.

Harness, J.K.: *Ultrasound in Surgical Practice: Basic Principles and Clinical Applications,* John Wiley & Sons, Inc., New York, NY, 1999.

Lenhardt, M.L., et al.: "Human Ultrasonic Speech Perception," *Science,* 82 (July 5, 1991).

Limberg, B.: "Diagnosis and Staging of Colonic Tumors by Conventional Abdominal Sonography as Compared with Hydrocolonic Sonography," *N. Eng. J. Med.,* 65 (July 9, 1992).

Nadel, A.S., et al.: "Absence of Needs for Amniocentesis in Patients with Elevated Levels of Maternal Serum Alpha-Fetoprotein and Normal Ultrasonographic Examinations," *N. Eng. J. Med.,* 557 (August 30, 1990).

Papadakis, E.P.: *Ultrasonic Instruments and Devices,* Academic Press, Inc., San Diego, CA, 2000.

Rennie, J.: "Ultrasound Speeds the Release of Drugs from Medical Implants," *Sci. Amer.,* 30 (April 1990).

Rifkin, M.D., et al.: "Comparison of Magnetic Resonance Imaging Ultrasonography in Staging Early Prostate Cancer," *N. Eng. J. Med.,* 621 (September 6, 1990).

Rösch, T., et al.: "Localization of Pancreatic Endocrine Tumors by Endoscopic Ultrasonography," *N. Eng. J. Med.,* 1721 (June 25, 1992).

Rose, J.L.: *Ultrasonic Waves in Solid Media,* Cambridge University Press, New York, NY, 1999.

Schmerr, L.W.: *Fundamentals of Ultrasonic Nondestructive Evaluation: A Modeling Approach,* Perseus Publishing, Boulder, CO, 1998.

Suslick, K.S., Editor: *Ultrasound. Its Chemical, Physical, and Biological Effects,* VCH Publishers, New York, NY, 1988.

Suslick, K.S.: "The Chemical Effects of Ultrasound," *Sci. Amer.,* 80 (February 1989).

Suslick, K.S.: "Sonochemistry," *Science,* 1439 (March 23, 1990).

Thurston, R.N., E. Papadakis: *Ultrasonic Instruments and Devices I: Reference for Modern Instrumentation, Techniques, and Technology,* Vol. 23, Academic Press, Inc., San Diego, CA, 1999.

Thurston, R.N., E.P. Papadakis, and A.D. Pierce: *Ultrasonic Instruments and Devices II: Reference for Modern Instrumentation, Techniques, and Technology,* Vol. 24, Academic Press, Inc., San Diego, CA, 1999.

Web Reference

Ultrasonic Industry Association: http://www.ultrasonics.org/

ULTRAVIOLET LASER CHEMISTRY. See **Photochemistry and Photolysis**.

ULTRAVIOLET RADIATION. This region of the electromagnetic spectrum is subdivided into: (1) the near-ultraviolet, 4,000 to 3,000 Å, present in sunlight, producing important biological effects, but not detectable by the human eye; (2) the middle-ultraviolet, 3,000 to 2,000 Å, not present in sunlight as it reaches the Earth's surface, but well transmitted through air; and (3) the long- or extreme-ultraviolet (XUV), 2,000 to 100 Å. The latter borders on x-radiation. The latter is also called the far-ultraviolet and it is not transmitted through air. The region between 2,000 and 1,350 Å is sometimes referred to as the Schumann region after its discoverer. The boundary between far-ultraviolet and x-rays is arbitrary.

Ultraviolet radiation is emitted by nearly all light sources to some degree. Generally, the higher the temperature of the source, or the more energetic the excitation, the shorter are the wavelengths produced. Tungsten lamps in quartz envelopes radiate in the ultraviolet in accordance with Planck's law, slightly modified by the emissivity function of tungsten. Because of its high temperature (3,800°K), the crater of an open carbon arc is an excellent source of ultraviolet radiation, extending to the air cutoff. Electrical discharges through gases produce intense ultraviolet emission, mainly in lines and bands. A widely used source is the quartz mercury arc. Magnetically compressed plasma, as produced by devices such as zeta and theta pinch and which reach an extremely high temperature, are also sources of highly ionized atoms and emission lines in the far-ultraviolet. Such radiation is also produced by a synchrotron.

Solids, liquids, and gases normally transmit effectively in the near-ultraviolet range, but become opaque in the middle or extreme ultraviolet range. For constructing lenses and prisms used in ultraviolet instruments, the unusual transmittance of crystal and fused quartz, fluorite, and lithium fluoride are an immense advantage. Gases vary considerably in their absorption characteristics. Oxygen molecules cause air to become opaque below about 1,850 Å. Molecular nitrogen is relatively transparent down to

1,000 Å. Hydrogen absorbs in the Lyman series lines, and in an ionization continuum beyond the series limit, 911.7 Å. Helium is the most ultraviolet-transparent of all gases. Absorption first takes place in the resonance lines, the longest lying at 584 Å, and in a continuum beyond the series limit, 504 Å. The more complex gaseous molecules, such as CO_2, NO, and N_2O are rather opaque throughout most of the far-ultraviolet. Water vapor commences to absorb at wavelengths below 1,850 Å.

As with all visible radiation, reflection occurs for ultraviolet. Reflectance becomes less as the wavelength decreases. Aluminum is the best reflector over much of the long-wavelength region, reaching 90% down to 2,000 Å (when the metal is properly prepared), and 80% down to 1,200 Å when the aluminum is coated with a thin layer of magnesium fluoride to prevent growth of aluminum oxide. Platinum is the best reflector below 1,000 Å, achieving 20% at 600 Å, but only about 4% at 300 Å.

The simplest way to detect and measure ultraviolet radiation is by using the fluorescence process, converting the ultraviolet into radiation that can be seen; or into the near-ultraviolet range, which can be easily photographed or measured with conventional photomultipliers. Materials used for the extreme ultraviolet include oil and sodium salicylate, the latter particularly valuable because its quantum efficiency of fluorescence is high and nearly independent of wavelength. Thus, an ordinary photomultiplier with a sodium salicylate coated glass window becomes a sensitive radiometer for use throughout the entire ultraviolet region.

Ionization chambers and Geiger counters can be used for detecting extreme ultraviolet radiation. Knowledge of the ionization efficiency of the gas makes it possible to use them for measurement of absolute energy. Ultraviolet radiation also can be detected with a thermocouple, thermopile, or bolometer.

The Sun emits strongly throughout the ultraviolet, but only the near-ultraviolet reaches the Earth's surface, wavelengths shorter than 2,900 Å being absorbed by a layer of ozone in the atmosphere. See also **Oxygen**. Possible disturbance of the ozone layer by various air-polluting chemicals is considered of major importance because of the possibility of eliminating this effect and thus exposing the Earth's surface to the shorter, more dangerous ultraviolet radiation. In addition to producing sunburn, exposure of the eye to ultraviolet can cause a painful burn of the cornea and conjunctivitis. Snow blindness is caused by reflection of intense sources of middle-ultraviolet radiation from snowfields and glaciers. Ultraviolet radiation also enters into the photochemical processes which contribute to the production of smog.

The use of ultraviolet lamps has been practiced for a number of years in some hospitals, schools, and factories to check the spread of respiratory infections, but their effectiveness is inconclusive. Possibly the most effective use of the characteristics of ultraviolet radiation is in optical and instrumentation applications. See also **Ultraviolet Spectrometers**.

For the use of ultraviolet lasers in chemistry, see also **Photochemistry and Photolysis**.

Additional Reading

Attwood, D.T.: *Soft X-Rays and Extreme Ultraviolet Radiation: Principles and Applications,* Cambridge University Press, New York, NY, 1999.

Cockell, C. and A.R. Blaustein: *Ecosystems, Evolution, and Ultraviolet Radiation,* Springer-Verlag, Inc., New York, NY, 2001.

Huffman, R.E.: *Atmosphere Ultraviolet Remote Sensing,* Academic Press, Inc., San Diego, CA, 1992.

Nilsson, A.: *Ultraviolet Reflections: Life under a Thinning Ozone Layer,* John Wiley & Sons, Inc., New York, NY, 1996.

ULTRAVIOLET SPECTROMETERS. Ultraviolet instruments are based upon the selective absorbance of ultraviolet radiation by various substances. The absorbance of a substance is directly proportional to the concentration of the substance which causes the absorption in accordance with the Lamber-Beer law (or simply Beer's law):

$$A = abc = \log \frac{I_0}{I} \log \frac{1}{T}$$

where A = absorbance; a = molar absorptivity, 1/(mole)(centimeter); b = path length, centimeters; c = concentration, moles/1; I_0 = intensity of radiation striking detector with nonabsorbing sample in light path; I = intensity of radiation striking detector with concentration c of absorbing sample in light path b; and T = transmittance = I/I_0.

For the vapor phase,

$$A = \frac{abc'}{2,450}$$

at 25°C and 760 torr pressure, where c' = volume percent or mole percent, or

$$A = \frac{abc'}{2,450} \times \frac{P + 14.7}{14.7} \times \frac{298}{t + 273}$$

at any temperature or pressure, where P = pressure, psig; and t = temperature °C.

For the liquid and solid phases.

$$c = \frac{c'' \times d}{\text{M.W.}} \times 10 = \text{moles/1}$$

where c'' = weight percent in liquid; d = density of liquid; and M.W. = molecular weight of material to be measured.

$$A = \frac{10abc''d}{\text{M.W.}}$$

The fundamental elements of an ultraviolet-absorption analyzer include: (a) a radiation source; (b) suitable optical filters; (c) a sample cell; and (d) an output meter. A transmittance measurement is made by calculating the ratio of the reading of the output with the sample in the cell to the reading with the cell empty (of ultraviolet-absorbing materials). The concentration can be calculated from the known absorptivity of the substances as previously demonstrated by the equations; or it may be determined by comparison with known samples.

Sources of ultraviolet radiation include: (a) tungsten-filament incandescent lamps; (b) tungsten-iodine cycle lamps with quartz envelopes; (c) mercury-vapor lamps; and (d) the zinc discharge lamp. Other types are available, but enjoy only limited application. The hydrogen or deuterium lamps are used in the laboratory, but are delicate and costly for process uses.

The analytical radiation in an ultraviolet analyzer must be as nearly monochromatic as possible in the interest of high linearity, long-time stability, and sustained accuracy. Monochromatic radiation is obtained by proper selection of sources, filters, and phototubes, each of which is selective in regard to the wavelengths that it respectively emits, transmits, or responds to. For lenses and windows, fused quartz is the most commonly used for windows and for some lenses. Corning 7910 glass and synthetic sapphire transmit throughout the near-ultraviolet range. Mirrors of rhodium or special alloys may be less reflecting in certain regions, but are preferred over silver and aluminum because of their high resistance to scratching and corrosion. Semitransparent mirrors for beamsplitting are made of special alloys or chromium coatings evaporated upon quartz or glass.

Vacuum phototubes are preferred as detectors over barrier-layer photocells because of their higher signal-to-noise ratio, greater stability, longer life, and freedom from fatigue. Simple tubes are preferred over multiplier types because they are less costly, are more stable, and can be used in simpler circuits.

The simplest ultraviolet-absorption analyzer is the *single beam* type. The output of this type of instrument will be affected by fluctuations and drift of the light source, dirt or bubbles in the sample cell, and any drift in the detector or detector circuit. Thus, single-beam instruments operate on relatively low sensitivity (high absorbance) levels to provide reasonably stable analyses. Improved single-beam instruments have found extensive applications where only a "go-no-go" or broad range measurement will suffice. The *split-beam* analyzer overcomes most of the foregoing shortcomings and is based upon a differential absorption measurement at two wavelengths. The optical diagram of a double-beam grating ultraviolet spectrophotometer is shown in Fig. 1. Each band of wavelengths, isolated by the grating monochromator, is split into two beams, which pass alternately through the sample and reference paths. The two beams are then recombined along a single path, but separated in time. Thus, the detector receives an alternating optical signal consisting of radiant power P through the sample and P_0 through the reference. This output is converted into an electrical signal that is related to the transmittance of the sample P/P_0.

One of the more important areas of use of ultraviolet instruments is the identification and determination of biologically active substances. Many components in body fluids can be determined either directly or through colorimetric methods. Drugs and narcotics can be measured both in the body as well as in formulations. Vitamin assay is another related activity. Nearly all metals and nonmetals can be determined through their ultraviolet absorption or by colorimetric methods. In recent years, ultraviolet instruments have been used extensively for the determination of air and water pollutants, such as aldehydes, phenolics, and ozone;

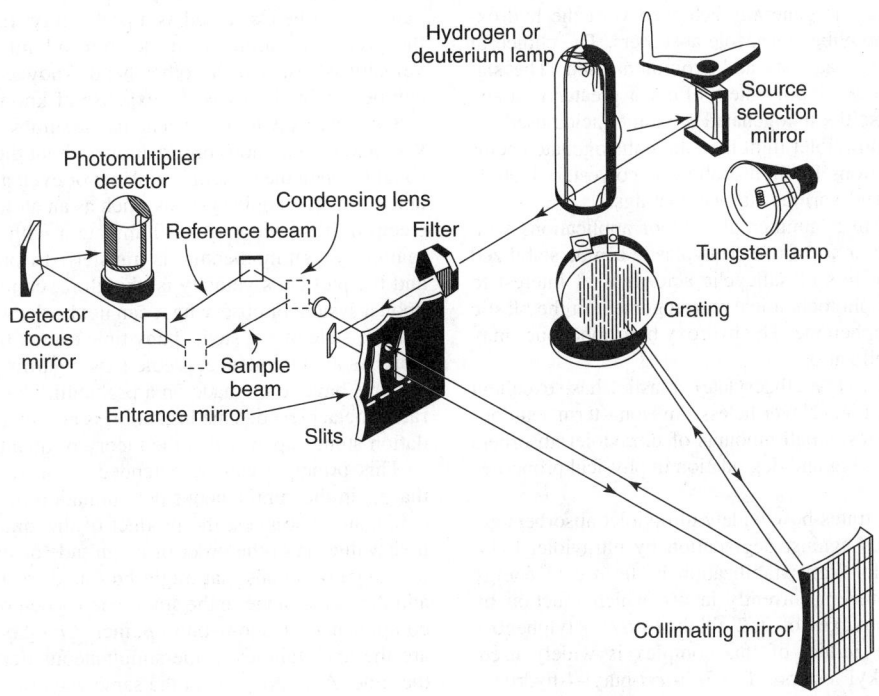

Fig. 1.   Optical path of double-beam grating ultraviolet spectrophotometer. (*Beckman*)

for the analysis of dyestuffs; for studies on polyaromatics and other carcinogenic substances; for the determination of food additives; for the analysis of petroleum fractions; and for the detection and determination of pesticide residues.

ULTRAVIOLET STABILIZERS.   When a polymer absorbs light, it may reradiate the absorbed energy at much longer wavelengths (heat), at slightly longer wavelengths (luminescence), or the energy may be transferred to another molecule. When none of these processes is operative, the absorbed light energy may cause bond breaking leading to degradation. By incorporating an ultraviolet absorbing compound into the plastic, it is possible to essentially eliminate all of the above processes, since the absorber even at concentrations as low as 0.5% can effectively compete for the incident ultraviolet radiation, thus protecting the plastic from degradation. A second approach to stabilization is to incorporate an additive which, though not an absorber in itself, can accept energy from the polymer substrate, and thus, leave the polymer intact. Since protection of plastics against light degradation can be achieved by these two mechanisms, the broader term *ultraviolet stabilizer* or *light stabilizer* is used to refer to such additives.

Ultraviolet absorbers continue to be the most widely used stabilizers. Such products must have long-term stability to ultraviolet light, be relatively nontoxic, heat stable, have little color, must not sensitize the substrate, and must be priced at levels which the plastics processor can tolerate. The principal classes of chemicals meeting these requirements at present are the 2-hydroxybenzophenones, and 2-(2'-hydroxyphenyl)benzotriazoles, substituted acrylates, and aryl esters. Typical compounds representative of these classes are 2-hydroxy-4-octoxybenzophenone, 2-(2'-hydroxy-5'-methylphenyl) benzotriazole, ethyl-2-cyano-3,3-diphenyl acrylate, dimethyl *p*-methoxybenzylidene malonate, and *p-tert*-octylphenyl salicylate.

The particular absorber to be used in a given application depends on several factors. One important criterion is whether the absorber will strongly absorb that portion of the ultraviolet spectrum responsible for degradation of the plastic under consideration. Compatibility, volatility, thermal stability, and interactions with other additives and fillers are other items that must be considered. When used in food wrappings, Food and Drug Administration approval must be obtained. While one or more of these considerations may rule out a given stabilizer or influence the choice of one class over another, the final selection must await the results of extensive accelerated and long-term tests.

At this point, it should be indicated that much effort has gone into the development of accelerated testing procedures. Many of the devices and techniques employed are based on knowledge gained in the evaluation of dyes, textiles, and rubber. For example, the carbon-arc Fade-O-Meter and the Xenon-arc Weather-O-Meter have been adapted from the dye field for use in plastics evaluation. Extensive use is also made of the fluorescent sunlamp, fluorescent blacklight, S-1 sunlamp, Hanovia lamp, and others. Such instruments are very useful for comparison of one stabilizer with others and for evaluating total stabilizing formulations in particular polymers. Nevertheless, no accelerated weathering device has yet been found which can accurately predict the outdoor weatherability of a broad range of polymers. Accelerated outdoor weathering is carried out in Phoenix, Arizona, where high levels of ultraviolet radiation occur and the temperature is high. To determine the lifetime under more humid conditions, tests are often conducted in the vicinity of Miami, Florida.

For extended outdoor applications, most polymers require some degree of light stabilization. There are wide variations in the inherent stability of different polymers ranging from less stable ones, such as polypropylene, to the highly light-stable poly(methyl methacrylate). Because of the dramatic growth of polyolefins, and particularly polypropylene, over the past several years, there has been an upsurge in requirements for ultraviolet absorbers. The hydroxybenzophenones, such as 2-hydroxy-4-octoxy benzophenone, have been widely used for stabilization of polypropylene. The benzotriazoles have also achieved commercial importance in this application. End uses of polypropylene requiring ultraviolet absorbers include upholstery fabrics, indoor-outdoor carpeting, lawn furniture, ropes, and various crates and boxes. Polyethylene is also stabilized with the hydroxy benzophenone absorbers. Applications include baskets, beverage cases, bags for fertilizer, and films for greenhouses.

Polystyrene light stabilization has been achieved with a variety of ultraviolet absorbers including the benzophenones, benzotriazoles, and salicylates. While yellowing of polystyrene occurs in many applications, it is particularly noticeable in diffusers used with fluorescent lights. This problem has been effectively solved by using ultraviolet light absorbers. In this instance, superior stabilization is achieved when the ultraviolet absorber is used in conjunction with specific antioxidants.

The hydroxybenzophenones, hydroxyphenylbenzotriazoles, and substituted acrylates are all used for stabilization of polyvinyl chloride. This polymer is growing at a substantial rate, and increasing uses are developing for light-stabilized grades. Among current uses, may be mentioned auto seat covers, floor tiles, light diffusers, vinyl-coated fabrics, siding, and exterior trim. Since the processing of polyvinyl chloride requires the use of a heat stabilizer, care must be exercised to avoid undesirable interactions between the heat and light stabilizers.

The stabilization of polyesters is generally achieved with the hydroxybenzophenone and hydroxyphenylbenzotriazole absorbers. The choice of absorber depends on the curing catalysts and promoters used. The stabilization of fire retardant grades of polyesters offers a greater problem than the standard grades because the halogenated monomer acids used are appreciably more sensitive to ultraviolet light than the unhalogenated acids (phthalic, isophthalic). Applications for light-stabilized polyesters include sheets for roofs and skylights and various surface coatings.

Cellulosic plastics are used in a number of outdoor applications with signs being one of the principal areas of use. This plastic can be stabilized reasonably well with the aryl esters of salicyclic acid. It is of interest to note that these esters undergo a photochemical rearrangement in the plastic to derivatives of hydroxy benzophenone. The hydroxy benzophenones may be added initially to effect stabilization.

As noted earlier, poly(methyl methacrylate) plastic has excellent resistance to ultraviolet radiation. Nevertheless, in long-term outdoor applications or in lighting fixtures, small amounts of ultraviolet absorbers are employed to retard the yellowing and degradation in physical properties which would otherwise occur.

The previous discussion illustrates how widely ultraviolet absorbers are used for stabilization of plastics against degradation by ultraviolet light. The second principal method for light stabilization is the use of energy transfer agents. Important stabilizers currently in use which function by this mechanism are nickel complexes of 2, 2'-thiobis(4-tert-octylphenol). For example, the butyl amine adduct of this complex is widely used. The nickel salts of mono alkyl esters of 3,5-di-tert-butyl-4-hydroxy-benzyl phosphonic acid are also useful stabilizers. When color is not a consideration, nickel dialkyl dithiocarbamates, and nickel acetophenone oximes may be used. Thus far, the nickel stabilizers have been used primarily in polypropylene and to some extent in polyethylene. They are especially useful in polypropylene fibers since stabilization by energy transfer is less dependent on sample thickness than is stabilization by ultraviolet absorption. Some of the nickel stabilizers have the further advantage that they act as dye acceptors and thus aid printing and dyeing of fibers and other items made from polyolefins.

Thus far, the discussion on ultraviolet stabilizers has been concerned only with their use for stabilization of plastics. While this is the principal use, the stabilizers which function by absorption are also widely used to prevent ultraviolet light from damaging furniture, clothing, and other articles. For example, a thin plastic film (6–10 mil) containing a high concentration of absorber is useful for covering a store window exposed to the sun. Such a film absorbs the ultraviolet radiation and thus prevents damage to the articles behind the film. The clarity and lack of color of the film permits customers to see the articles readily. Surface coatings containing ultraviolet absorbers are used in the same way to protect items such as flooring and furniture. Ultraviolet absorbing coatings may also be used to protect plastics, but generally, it is more practical to incorporate the absorber in the plastic itself. The incorporation of ultraviolet absorbers into plastic sunglasses is important for protecting the eye from ultraviolet radiation damage. Suntan lotions contain compounds such as monoglyceryl p-aminobenzoate that permit the longer wave lengths (330–400 nm) in the ultraviolet which cause tanning to pass through to the skin. At the same time, the more highly energetic short wave lengths (290–330 nm) which cause burning are strongly absorbed. When no tanning is desired, creams containing hydroxy benzophenones may be used, since these products remove a high percentage of the 290–400 nm radiation.

Highly satisfactory formulations have been developed for light stabilization of a wide range of polymers. Studies are continuing not only toward empirical development of superior stabilizing formulations, but also toward understanding the mechanisms of the degradation and inhibition process involved. This dual approach can be expected to yield products which will meet the increasingly severe demands that will result as plastics find their way into new outdoor uses.

W. B. HARDY
American Cyanamid Co.
Bound Brook, New Jersey

UNCERTAINTY PRINCIPLE. Also sometimes referred to as the *indeterminacy principle*, this was first stated by Heisenberg, Werner P, in connection with the position and momentum of an electron. In essence, the postulate states that it is impossible to determine simultaneously both the exact position and the exact momentum of an electron and thus these values must be expressed as a probability. If, for example, one determines the precise location of an electron, all information about the electron's velocity is lost. On the other hand, knowledge of the electron's velocity can be obtained only at the expense of knowledge of its location.

Classical Newtonian mechanics assumes that a physical system can be kept under continuous observation without thereby disturbing it. This is reasonable when the system is a planet or even a spinning top, but is unacceptable for microscopic systems, such as an atom. To observe the motion of an electron, it is necessary to illuminate it with light of ultrashort wavelength (gamma rays); momentum is transferred from the radiation to the electron and the particle's velocity is, therefore, continuously disturbed. The effect upon a system of observing it can not be determined exactly, and this means that the state of a system at any time cannot be known with complete precision. As a consequence, predictions regarding the behavior of microscopic systems have to be made on a probability basis and complete certainty can rarely be achieved. This limitation is accepted and is made one of the foundation stones upon which the theory of quantum mechanics is constructed.

This principle can be extended to other phenomena of a like nature, that is, in the simultaneous determination of the values of two canonically conjugated variables, the product of the smallest possible uncertainties in their values is of the order of magnitude of the Planck constant h. If Δq is the range of values that might be found for the coordinate q of a particle, and Δp is the range in the simultaneous determination of the corresponding component of its momentum p, then $\Delta p \cdot \Delta q \geq h$. Similarly, if ΔE and Δt are the uncertainties in the simultaneous determination of the energy and the time, $\Delta E \cdot \Delta t \geq h$. In the same way, the principle applies to any other pair of canonically conjugated variables. See also **Quantum Mechanics**.

UNITS AND STANDARDS. The General Conference on Weights and Measures, to which the United States adheres by treaty, has established the International System of Units, called SI units. The base quantities of this system and the corresponding units and symbols are:

Length	meter	m
Mass	kilogram	kg
Time	second	s
Electric current	ampere	A
Thermodynamic temperature	kelvin	K
Luminous intensity	candela	cd
Amount of substance	mole	mol

The units radian (rad) for plane angle and steradian (sr) for solid angle are described as supplementary units and are normally treated as though they were base units, although the corresponding quantities may be treated as dimensionless. The coherent SI unit system consists of the foregoing, plus all of the units derived from them by multiplication and division without introducing numerical factors.

TABLE 1. UNITS DERIVED FROM BASE SI QUANTITIES

Quantity	Name of SI derived unit	Symbol	Expressed in terms of SI base or derived units
frequency	hertz	Hz	$1 \text{ Hz} = 1 \text{ s}^{-1}$
force	newton	N	$1 \text{ N} = 1 \text{ kg} \cdot \text{m/s}^2$
pressure and stress	pascal	Pa	$1 \text{ Pa} = 1 \text{ N/m}^2$
work, energy, quantity of heat	joule	J	$1 \text{ J} = 1 \text{ N} \cdot \text{m}$
power	watt	W	$1 \text{ W} = 1 \text{ J/s}$
quantity of electricity	coulomb	C	$1 \text{ C} = 1 \text{ A} \cdot \text{s}$
electromotive force, potential difference	volt	V	$1 \text{ V} = 1 \text{ W/A}$
electric capacitance	farad	F	$1 \text{ F} = 1 \text{ A} \cdot \text{s/V}$
electric resistance	ohm	Ω	$1\Omega = 1 \text{ V/A}$
electric conductance	siemens	S	$1 \text{ S} = 1\Omega^{-1}$
flux of magnetic induction, magnetic flux	weber	Wb	$1 \text{ Wb} = 1 \text{ V} \cdot \text{s}$
magnetic flux density, magnetic induction	tesla	T	$1 \text{ T} = 1 \text{ Wb/m}^2$
inductance	henry	H	$1 \text{ H} = 1 \text{ V} \cdot \text{s/A}$
luminous flux	lumen	lm	$1 \text{ lm} = 1 \text{ cd} \cdot \text{sr}$
illuminance	lux	lx	$1 \text{ lx} = 1 \text{ lm/m}^2$

UNITS, NOT PART OF COHERENT SYSTEM, BUT GENERALLY ACCEPTED FOR USE WITH SI UNITS

Quantity	Name of unit	Unit symbol	Magnitude in SI units
time	minute	min	60 s
	hour	h	3600 s
	day	d	86400 s
plane angle	degree	°	$\pi/180$ rad
	minute	'	$\pi/10\,800$ rad
	second	"	$\pi/648\,000$ rad
Volume	litre	L	1 l = 1 dm^3
mass	tonne	T	1 t = 10^3 kg
energy	electronvolt	eV	approx. 1.60219×10^{-19} J
mass of an atom	atomic mass unit	u	approx. 1.66053×10^{-27} kg
length	astronomical unit	AU	149600×10^6 m
	parsec	pc	approx. 30857×10^{12} m

TABLE 2. STANDARD PREFIXES USED WITH SI UNITS

Factor by which the unit is multiplied	Prefix name	Symbol
10^{12}	tera	T
10^9	giga	G
10^6	mega	M
10^3	kilo	k
10^2	hecto	h
10	deca	Da
10^{-1}	deci	d
10^{-2}	centi	c
10^{-3}	milli	m
10^{-6}	Micro	m
10^{-9}	nano	n
10^{-12}	pico	p
10^{-15}	Femto	f
10^{-18}	atto	a

TABLE 3. PRINCIPAL UNITS—SYMBOLS, DEFINITIONS, DIMENSIONS

AMPERE (A). *The constant current that, if maintained in two straight parallel conductors that are of infinite length and negligible cross section and are separated from each other by a distance of 1 meter in a vacuum, will produce between these conductors a force equal to 2×10^{-7} newton per meter of length.* (The SI unit of electric current.)

AMPERE PER METER (A/m). *The magnetic field strength in the interior of an elongated uniformly wound solenoid which is excited with a linear current density in its winding of 1 ampere per meter of axial distance.* (The SI unit of magnetic field strength.)

AMPERE-HOUR (Ah). *The quantity of electricity represented by a current of 1 ampere flowing for 1 hour.*

ANGSTROM Å. *A unit of length equal to 10^{-10} meter.**

APOSTILB (asb). *A unit of luminance. One lumen per square meter leaves a surface whose luminance is 1 apostilb in all directions within a hemisphere.* (The candela per square meter is the preferred unit of luminance.)

ATMOSPHERE, STANDARD (atm). *A unit of pressure. One standard atmosphere equals 101,325 newtons per square meter.*

ATOMIC MASS UNIT, UNIFIED (u). *The atomic mass unit (unified) is 1/12th of the mass of an atom of the ^{12}C nuclide.* (Use of the prior atomic mass unit (amu), defined by reference to oxygen, is no longer preferred.)

BAR (bar). *A unit of pressure. One bar equals 100,000 newtons per square meter.*

BARN (b). *A unit of nuclear cross section. One barn equals 10^{-28} square meter.*

BARREL. (bbl). *A unit of volume. One barrel equals 9,702 cubic inches; or 0.15899 cubic meters.* (This is the standard barrel used for petroleum, etc. A different standard barrel is used for fruits, vegetables, and dry commodities.)

BAUD (Bd). *A unit of signaling speed. One baud equals one element per second.* (The signaling speed in bauds is equal to the reciprocal of the signal element length in seconds.)

BEL (B). *A dimensionless unit for expressing the ratio of two values of power, being the logarithm to the base 10 of the power ratio.* (The more commonly used unit, decibel (dB), is 10 times the logarithm to the base 10 of the power ratio. A bel is 10 decibel.)

BIT (b). *A unit of information, generally represented by a pulse. A bit is a binary digit, i.e., a 1 or 0 in computer technology.* (In information theory, the bit is the smallest possible unit of information.)

BIT PER SECOND (b/s). *A unit of signaling speed. A transference rate of 1 bit per second.*

BRITISH THERMAL UNIT (Btu). *A unit of heat. The heat required to warm 1 pound of pure water through an interval of 1 degree Fahrenheit.*

CALORIE (International Table) (cal$_{It}$). *A unit of heat. One International Table calorie equals 4.1868 joules.* (The 9th Conférence Générale des Poids et Mesures adopted the joule as the unit of heat.)

CALORIE (Thermochemical Calorie) (cal). *A unit of heat. One calorie equals 4.1840 joules.* (See foregoing note.)

CANDELA (cd). *The luminous intensity of 1/6000,000 of a square meter of a radiating cavity at the temperature of freezing platinum (2042 K).* (The SI unit of luminous intensity. The unit formerly was called the *candle*.)

CIRCULAR MIL (cmil). *The area of a circle whose diameter is 0.0001 inch. One circular mil equals $\pi/4 \cdot 10^{-6}$ square inches.*

COULOMB (C). *The quantity of electric charge which passes any cross section of a conductor in 1 second when the current is maintained constant at 1 ampere.* (The SI unit of electric charge.)

CURIE (Ci). *The unit of activity in the field of radiation dosimetry. One curie equals 3.7×10^{10} disintegrations per second.* (The activity of 1 gram of ^{226}Ra is slightly less than 1 curie.)

CYCLE (c). *An interval of space or time in which is completed 1 round of events or phenomena.*

CYCLE PER SECOND (Hz, c/s). *The number of cycles per second.* (The name hertz (Hz) is the accepted international term. The abbreviation Hz is preferred to c/s.)

DARCY (D). *A unit of permeability of a porous medium. One darcy equals 1 cP (cm/s)(cm/atm) equals 0.986923 square micrometers.* (A permeability of 1 darcy will allow the flow of 1 cubic centimeter per second of fluid of 1 centipoise viscosity through an area of 1 square centimeter under a pressure gradient of 1 atmosphere per centimeter.)

DAY (d). *A unit of time, the exact definition of which is dependent upon which system of time measurement is referred to, i.e., apparent solar time, mean solar time, universal time, apparent sidereal time, ephemeris time, or atomic time.* See **Time.** With exception of atomic time, the time base is referenced to rotation of the Earth. For general purposes, a day is considered the period taken for 1 revolution of the Earth about its axis.

DEGREE CELSIUS (°C). *One unit of temperature on the Celsius temperature scale*, which is derived from the thermodynamic of Kelvin scale of temperature and related by: Temperature (degrees Celsius) equals Temperature (Kelvin units) minus 273.15. See Temperature.

DEGREE FAHRENHEIT (°F). *One unit of temperature on the Fahrenheit temperature scale*, which is related to the Celsius temperature scale by: Temperature (degrees Fahrenheit) equals 1.8× (degrees Celsius) plus 32. See Temperature.

DEGREE RANKINE (°R). *One unit of temperature on the Rankine temperature scale*, which is related to the Fahrenheit temperature scale by: Temperature (degrees Rankine) equals Temperature (degrees Fahrenheit) plus 459.69. See **Temperature.**

DYNE (dyn). *A unit of force. One dyne equals the force necessary to give 1 gram mass an acceleration of 1 centimeter/(second)(second).* (The dyne is the unit of force in the CGS system.)

ELECTRONVOLT (eV). *A unit of energy. One electronvolt equals the energy acquired by an electron when it passes through a potential difference of 1 volt in a vacuum.* (One electronvolt equals 1.602×10^{-12} erg.)

ERG (erg). *A unit of energy. One erg equals 10^{-7} joule.* (Also, 1 erg equals the work done when a force of 1 dyne is applied through a distance of 1 centimeter. One foot-pound equals 13,560,000 ergs.)

FARAD (F). *The capacitance of a capacitor in which a charge of 1 coulomb produces a potential difference of 1 volt between the terminals.* (The SI unit of capacitance.)

FOOTCANDLE (fc). *A unit of luminance. One footcandle equals 1 lumen per square foot.* (The name *lumen per square foot* is recommended for this unit. The SI unit, lux (lumen per square meter), is preferred.)

FOOTLAMBERT (fL). *A unit of luminance. One lumen per square foot leaves a surface whose luminance is 1 footlambert in all directions within a hemisphere.* (If luminance is measured in English units, the candela per square inch is preferred. However, use of the SI unit, the candela per square meter, is generally accepted.)

GAL (Gal). *A unit of acceleration. One Gal equals 1 centimeter per second per second.*

GALLON (gal). Because the gallon, quart, and pint differ in the United States and the United Kingdom, the use of this unit and term is generally discouraged for scientific purposes. An imperial gallon equals 1.20095 U.S. gallons. One U.S. gallon equals 3.785×10^{-3} cubic meter.

GAUSS (G). *A unit of magnetic flux density, or magnetic induction. The ratio of the flux in any cross section to the area of that cross section, the cross section being taken normal to the direction of flow. One gauss equals 1 maxwell per square centimeter.* (The gauss is a unit of the CGS system. Use of the SI unit, the *tesla*, is preferred.)

(continued overleaf)

TABLE 3. (*continued*)

GILBERT (Gb). *A unit of magnetomotive force. One gilbert equals 0.4 π(ni), where (ni) is an ampere-turn.* (The gilbert is a unit of the CGS system. Use of the SI unit, the ampere (or ampere-turn), is preferred.)

GRAIN (gr). *A unit of mass. One grain equals 0.06480 gram.* (One ounce, avoidupois, equals 437.5 grains; 1 ounce, troy, equals 480 grains; 1 ounce, apothecaries', equals 480 grains. One pound, avoidupois, equals 7,000 grains.)

GRAM (g). *A unit of mass. One gram equals 1/1,000th kilogram.* (See also KILOGRAM in this list.)

HENRY (H). *A unit of inductance. The inductance of a circuit in which a current of 1 ampere induces a flux linkage of 1 weber.* (The SI unit of inductance.)

HERTZ (Hz). *A unit of frequency. One hertz equals a frequency of one cycle per second.* (The SI unit of frequency.)

HORSEPOWER (hp). The horsepower is considered an anachronism in science and technology. Use of the SI unit of power, the watt, is preferred. When used, 1 horsepower equals (1) 42.44 Btu/minute; (2) 33,000 foot-pounds/minute; or (3) 550 foot-pounds/second.

HOUR (h). *A unit of time. One hour equals 60 minutes, or 3,600 seconds.*

INCH (in). *A unit of length. One inch equals 2.540×10^{-2} meter.*

INCH OF MERCURY (inHg). *A unit of pressure. One inch of mercury equals 3,386.4 newtons per square miter.* (An inch of mercury also equals (1) 0.03342 atmosphere; (2) 1.133 feet of water; (3) 345.3 kilograms/square meter; (4) 70.73 pounds/square foot; or (5) 0.4912 pounds/square inch.

INCH OF WATER (in H$_2$O). *A unit of pressure. One inch of water equals 249.09 newtons per square meter.* (An inch of water also equals (1) 2.458×10^{-3} atmosphere; (2) 0.07355 inch of mercury; (3) 2.540×10^{-3} kilogram/square centimeter; (4) 0.5781 ounce/square inch; (5) 5.204 pounds/square foot; or (6) 0.03613 pound/square inch. The latter Figures hold for a temperature of 4°C.)

JOULE (J). *A unit of energy. The work done by 1 newton acting through a distance of 1 meter.* (The SI unit of energy. One joule equals 1 watt-second; equals 10^7 ergs; equals 10^7 dyne-centimeters.)

JOULE PER KELVIN (J/K). *A unit of heat capacity and entropy.*

KELVIN (K). *The basic unit of thermodynamic temperature. One kelvin is the fraction 1/273.16 of the thermodynamic temperature of the triple point of water.* (The term *degree Kelvin* was officially dropped in 1967. Thus, the symbol is K and not °K. Relationship of the Kelvin scale to the Celsius scale is given earlier in this list under DEGREE CELSIUS.)

KILOGRAM (kg). *A unit of mass and is based upon a cylinder of platinum-iridium alloy kept by the International Bureau of Weights and Measures at Paris.* A duplicate in the custody of the National Bureau of Standards at Washington is the mass standard for the United States. The kilogram is the only base unit still defined by an artifact. (A kilogram equals (1) 1,000 grams; (2) 2.205 pounds; (3) 9.842×10^{-4} long tons; or (4) 1.102×10^{-3} short tons.)

KNOT (kn). *A unit of speed. One knot equals 1 nautical mile per hour.* (A knot also equals 6,080.2 feet/hour; or 1.151 statute miles/hour.)

LAMBERT (L). *A unit of luminance. One lumen per square centimeter leaves a surface whose luminance is 1 lambert in all directions within a hemisphere.* (The candela per square meter is the preferred unit of luminance.)

LITER (l). *A unit of volume. One liter equals 10^{-3} cubic meter.* (A liter also equals (1) 1,000 cubic centimeters; (2) 0.03531 cubic foot; (3) 61.02 cubic inches; (4) 1.308×10^{-3} cubic yard; (5) 0.2642 U.S. liquid gallon; (6) 1.057 U.S. liquid quarts; or (7) 0.22 Imperial gallon.

LUMEN (lm). *A unit of luminous flux. The flux through a unit solid angle (steradian) from a uniform point source of 1 candela.* (The SI unit of luminous flux.)

LUMEN PER SQUARE FOOT (lm/ft^2). *A unit of illuminance and also a unit of luminous excitation.* (Use of the SI unit, lumen per square meter, is preferred.)

LUMEN PER SQUARE METER (lm/m^2). *A unit of luminous excitation.* (The SI unit of luminous excitation.)

LUMEN PER WATT (lm/W). *A unit of luminous efficacy.* (The SI unit of luminous efficacy.)

LUMEN SECOND (lm · s). *A unit of quantity of light.* (The SI unit of quantity of light.)

LUX (lx). *A unit of illuminance. One lux equals 1 lumen per square meter.* (The SI unit of illuminance.)

MAXWELL (Mx). *A unit of magnetic flux. The flux through a square centimeter normal to a field at 1 centimeter from a unit magnetic pole.*

METER (m). *A unit of length. Defined as 1,650,763.73 wavelengths in vacuum of the orange-red line of the spectrum of 86Kr(krypton).* (The SI unit of length. A meter also equals (1) 100 centimeters; (2) 3.281 feet; (3) 39.37 inches; (4) 0.001 kilometer; (5) 5.396×10^{-4} nautical mile; (6) 6.214×10^{-4} statute mile; or (7) 1.094 yards.)

MHO (mho). *A unit of conductance (and of admittance). The conductance of a conductor whose resistance is 1 ohm.* (The name *siemens* (S) also is used for this quantity.)

MICROMETER (μm). *A unit of length. One micrometer equals one-millionth of a meter.* (The term *micron* formerly used for this unit no longer is preferred.)

MICRON. See MICROMETER above.

MIL (mil). *A unit of length. One mil equals one-thousandth of an inch.*

MILE, STATUTE (mi). *A unit of length. One mile equals 5,280 feet.* (One statute mile also equals (1) 1.609 kilometers; (2) 1,760 yards; (3) 6.336×10^4 inches; or (4) 0.8684 nautical mile.)

MILE, NAUTICAL (nmi). *A unit of length. One nautical mile equals 1.1516 statute miles.* (One nautical mile also equals (1) 6,080.27 feet; (2) 1.853 kilometers; or (3) 2,027 yards.)

MINUTE, TIME (min). *A unit of time. One minute equals 60 seconds.* (Time also may be designated by means of superscripts, as in $9^h46^m30^s$, where there otherwise will be no confusion with abbreviations.)

MOLE (mol). *A unit of amount of substance. One mole is an amount of a substance, in specified mass units, equal to the molecular weight of that substance.* (The SI unit for amount of substance. Examples are the gram mole or the pound mole.)

NEPER (Np). *A dimensionless unit for expressing the ratio of two voltages, two currents, or two power values in a logarithmic manner. The number of nepers is the natural (Napierian) logarithm of the square root of the ratio of the two values being compared.* Thus, the neper uses the base of 2.71828 in contrast with the bel (or decibel) which uses the common-logarithm base of 10. One neper equals 8.686 decibels.

NEWTON (N). *A unit of force. One newton is the force that will impart an acceleration of 1 meter per second per second to a mass of 1 kilogram.* (The SI unit of force. One newton equals 10^5 dynes.)

NIT (nt). *A unit of luminance and is synonymous with candela per square meter.*

OERSTED (Oe). *A unit of magnetic field strength. The magnetic field produced at the center of a plane circular coil of 1 turn and of radius 1 centimeter, which carries a current of $(\frac{1}{2}\pi)$ abamperes.* (An abampere equals 10 amperes. The oersted is the CGS unit of magnetic field strength. Use of the SI unit, the ampere per meter, is preferred.)

OHM (Ω). *A unit of resistance (and of impedance). The resistance of a conductor such that a constant current of 1 ampere in it produces a voltage differences of 1 volt between its ends.* (The SI unit of resistance.)

PASCAL (Pa). *A unit of pressure or stress. One pascal equals 1 newton per square meter.*

PHON (phon). *A unit of loudness level. The pressure level in decibels of a pure 1,000 Hz tone.*

PHOT (ph). *A unit of illuminance. One phot equals 1 lumen per square centimeter.* (The phot is the CGS unit of illuminance. Use of the SI unit, the lux, is preferred.)

PINT (pt). Because the gallon, quart, and pint differ in the United States and the United Kingdom, the use of this unit and term is generally discouraged for scientific purposes. One U.S. pint equals: (1) 473.2 cubic centimeters; (2) 0.01671 cubic foot; (3) 28.87 cubic inches; (4) 4.732×10^{-4} cubic meter; (5) 6.189×10^{-4} cubic yard; (6) 0.125 U.S. gallon; (7) 0.4732 liter; or (8) 0.5 liquid U.S. quart.

POISE (P). *A unit of dynamic viscosity. The unit is expressed in dyne second per square centimeter.* The centipoise (cP) is more commonly used. The formal definition of viscosity arises from the concept put forward by Newton that under conditions of parallel flow, the shearing stress is proportional to the velocity gradient. If the force acting on each of two planes of area A parallel to each other, moving parallel to each other with a relative velocity V, and separated by a perpendicular distance X, be denoted by F, the shearing stress is F/A and the velocity gradient, which will be linear for a true liquid, is V/X. Thus, $F/A = \eta\, V/X$, where the constant η is the viscosity coefficient or dynamic viscosity of the liquid. The poise is the CGS unit of dynamic viscosity.

POUNDAL (pdl). *A unit of force. One poundal equals the force required to give a standard 1-pound body an acceleration of 1 foot per second per second.*

QUART (qt). Because the gallon, quart, and pint differ in the United States and the United Kingdom, the use of this unit and term is generally discouraged for scientific purposes. One U.S. quart equals: (1) 946.4 cubic centimeters; (2) 0.03342 cubic foot; (3) 57.75 cubic inches; (4) 9.464×10^{-4} cubic meter; (5) 1.238×10^{-3} cubic yard; (6) 0.25 U.S. gallon; or (7) 0.9463 liter.

RAD (rd). *A unit of absorbed dose in the field of radiation dosimetry. One rad equals the absorption of energy in any medium of 100 ergs per gram.*

RADIAN (rad). *A unit of plane angle. One radian equals the angle subtended at the center by a circular arc which is equal in length to the radius of the circle.* (The SI unit of plane angle.)

REM (rem). *A unit of dose equivalent in the field of radiation dosimetry. One rem equals the amount of ionizing radiation of any type which produces the same damage to humans as 1 roentgen of approximately 200 kilovolts x-radiation.* (The unit is abbreviation of Roentgen Equivalent Man.)

REVOLUTION PER MINUTE (r/min). Although use of rpm as an abbreviation is common, it should not be used as a symbol.

TABLE 3. (continued)

ROENTGEN (R). A unit of exposure in the field of radiation dosimetry. That quantity of x–or gamma-radiation such that the associated corpuscular emission per 0.001293 gram of dry air (equals 1 cubic centimeter at 0°C and 769 millimeters of mercury pressure) produces in air ions carrying 1 esu of quantity of electricity of either sign. (The emu (electrostatic unit) is a unit in the CGS system in which the statcoulomb is the charge that repels an exactly similar charge in a vacuum with a force of 1 dyne. One statcoulomb equals 3.3356×10^{-10} coulomb.)

SECOND (s). *A unit of time. The duration of 9,192,631,770 periods of the radiation corresponding to the transition between the two hyperfine levels of the ground state of the* ^{133}Cs *(cesium) atom. (The SI unit of time.)*

SLUG (slug). *A unit of mass. One slug equals 14.5959 kilograms.*

STERADIAN (sr). *A unit of solid angle. One steradian equals the solid angle subtended at the center by* $\frac{1}{4}\pi$ *of the surface area of a sphere of unit radius.*

STILB (sb). *A unit of luminance. One stilb equals 1 candela per square centimeter.*

STOKES (St). *A unit of kinematic viscosity. The centistokes (cSt) is more commonly used. Kinematic viscosity is the dynamic viscosity divided by the density. See POISE given previously in this list.*

TESLA (T). *A unit of magnetic flux density (magnetic induction). The magnetic flux density of a uniform field that produces a torque of 1 newton-meter on a plane current loop carrying 1 ampere and having a projected area of 1 square meter on the plane perpendicular to the field.* $T = N/A \cdot m$. *(The SI unit of magnetic flux density.)*

THERM (thm). *A unit of heat. One therm equals 100,000 British thermal units.*

TON (ton). A unit of weight. If not otherwise specified, a *short ton* equal to 2,000 pounds is assumed. A *long ton* equals 2,240 pounds. A metric ton equals 1,000 kilograms (2,205 pounds), also called *tonne* (t).

VAR (var). *A unit of reactive power. The reactive power at the port of entry of a single-phase two-wire circuit when the product of (a) the rms (root mean square) value in amperes of the sinusoidal current, (b) the rms value in volts of the voltage, and (c) the sine of the angular phase difference by which the voltage leads the current is equal to 1. (The SI unit of reactive power.)*

VOLT (V). *A unit of voltage. The voltage between 2 points of a conducting wire carrying a constant current of 1 ampere, when the power dissipated between these points is 1 watt. (The SI unit of voltage.)*

VOLTAMPERE (VA). *A unit of apparent power. The apparent power at the port of entry of a single-phase two-wire circuit when the product of (a) the rms (root mean square) value in amperes of the current and (b) the rms value in volts of the voltage is equal to 1. (The SI unit of apparent power.)*

WATT (W). *A unit of power. The watt equals 1 joule per second. (The SI unit of power.* One watt equals: (1) 3.4192 Btu/hour; (2) 0.05688 Btu/minute; (3) 10^7 ergs/second; (4) 44.27 foot-pounds/minute; (5) 0.7378 foot-pounds/second; (6) 1.341×10^{-3} horsepower; (7) 1.360×10^{-3} metric horsepower; (8) 0.01433 kilogram-calories/minute; or (9) 0.001 kilowatt.

WATT PER METER KELVIN (W/m · K). The SI unit of thermal conductivity.

WATT PER STERADIAN (W/sr). The SI unit of radiant intensity.

WATT PER STERADIAN SQUARE METER (W/Sr · m^2). The SI unit of radiance.

WATTHOUR (Wh). *A unit of energy. One watthour equals 3,600 joules.* (One watthour equals: (1) 3.413 Btu; (2) 3.60×10^{10} ergs; (3) 2,656 footpounds; (4) 859.85 gram-calories; (5) 1.341×10^{-3} horsepower-hour; (6) 0.8598 kilogram-calorie; (7) 367.2 kilogram-meters; or (8) 0.001 kilowatt-hour.

WEBER (Wb). *A unit of magnetic flux. The magnetic flux passing through an area of 1 square meter placed normal to a uniform magnetic field of magnetic flux density equal to 1 tesla.* $Wb = T \cdot m^2$. *(The SI unit of magnetic flux.)* If the flux linked by a circuit changes at a uniform rate of 1 weber per second, a voltage of 1 volt is induced in the circuit. $Wb = V \cdot s$.

---

*Although, officially, A (without small circle over it) may be used as an abbreviation or symbol for angstrom, to avoid possible confusion with the use of A for ampere, the Å symbol is used throughout this text.

A number of derived units have been given special names and symbols. See Table 1.

Decimal multiples of the coherent base and derived SI units are formed by attaching to these units the prefixes shown in Table 2.

The following units were accepted by the International Committee of Weights and Measures for use with the SI units for a transitional period: angstrom (1 Å = 10^{-10} m), barn (1b = 10^{-28} m²), bar (1bar = 10^5 Pa), standard atmosphere (1 atm = 101,325 Pa), curie (1Ci = 3.7×10^{10} s^{-1}), roentgen (1R = 2.58×10^{-4} C/kg), rad (1 rad = 10^2 J/kg).

The *cgs system of units*, based on the centimeter, gram, and second as units in mechanics, is a metric system which continues to be used in some branches of physics. In daily life, the customary units in the United States are those based on the foot, pound-force, and second, but these units are

TABLE 4. COMMON EQUIVALENTS AND CONVERSIONS

Approximate common equivalents		Conversions accurate to parts per million	
1 inch	25 millimeters	inches × 25.4*	millimeters
1 foot	0.3 meter	feet × 0.3048*	meters
1 yard	0.9 meter	yards × 0.9144*	meters
1 mile	1.6 kilometers	miles × 1.60934	Kilometers
1 square inch	6.5 square centimeters	square inches × 6.4516*	Square centimeters
1 square foot	0.09 square meter	square feet × 0.0929030	square meters
1 square yard	0.8 square meter	square yards × 0.836127	square meters
1 acre	0.4 hectare	acres × 0.404686	hectares
1 cubic inch	16 cubic centimeters	cubic inches × 16.3871	cubic centimeters
1 cubic foot	0.03 cubic meter	cubic feet × 0.0283168	cubic meters
1 cubic yard	0.8 cubic meter	cubic yards × 0.764555	cubic meters
1 quart (liquid)	0.9463 liter	quarts (liquid) × 0.946353	liters
1 gallon	0.004 cubic meter	gallons × 0.00378541	cubic meters
1 ounce (avoirdupois)	28 grams	ounces (avoirdupois) × 28.3495	grams
1 pound (avoirdupois)	0.45 kilogram	pounds (avoirdupois) × 0.453592	kilograms
1 horsepower	0.75 kilowatt	horsepower × 0.745700	kilowatts
1 millimeter	0.04 inch	millimeters × 0.0393701	inches
1 meter	3.3 feet	meters × 3.28084	feet
1 meter	1.1 yards	meters × 1.09361	yards
1 kilometer	0.6 mile	kilometers × 0.621371	miles
1 square centimeter	0.16 square inch	square centimeters × 0.155000	square inches
1 square meter	11 square feet	square meters × 10.7639	square feet
1 square meter	1.2 square yards	square meters × 1.19599	square yards
1 hectare	2.5 acres	hectares × 2.47105	acres
1 cubic centimeter	0.06 cubic inch	cubic centimeters × 0.0610237	cubic inches
1 cubic meter	35 cubic feet	cubic meters × 35.3147	cubic feet
1 gram	0.035 ounce (avoirdupois)	grams × 0.0352740	ounces (avoirdupois)
1 kilogram	2.2 pounds (avoirdupois)	kilograms × 2.20462	pounds (avoirdupois)
1 kilowatt	1.3 horsepower	kilowatts × 1.34102	horsepower

rarely used in physics except for the description of equipment (e.g., "a 2-inch pipe"). Considerable effort has been going forth in the United States and a number of other countries that are not accustomed to using the metric system to ultimately adopt it.

About one hundred of the most frequently used units are defined in Table 3. Common equivalents and conversions are given in Table 4.

Web References

General Conference on Weights and Measures: http://www.sizes.com/indexes.htm
General Tables of Units of Measurement: http://ts.nist.gov/ts/htdocs/230/235/appxc/appxc.htm

UPSILON PARTICLE. As of 1977, when the upsilon particle was discovered at the Fermi National Accelerator Laboratory, the particle was the heaviest to be identified. Discovery of upsilon prompted physicists to introduce a massive new quark, raising the number of quarks from four to five (but probably six). The upsilon has a mass three times greater than any subatomic entity previously detected. It was discovered in energetic collisions between protons and copper nuclei. With a mass at its lower energy state equivalent to 9.0 GeV and masses in excited states equivalent to 10 and 10.4 GeV, the upsilon particle has been interpreted by scientists as consisting of a massive new quark (fifth) bound to its antiquark. The experiment was later reinforced by research at the Deutsches Elektronen-Synchrotron (DESY), located near Hamburg. At Fermilab, the excited upsilon particle appeared as a resonance in the yield of muons generated in collisions between protons and nuclei. A discussion of the upsilon experiment is described by a principal scientist of the project, L.M. Lederman (*Sci. Amer.*, **239**(4), 72–80, 1978). See also **Particles (Subatomic)**.

URALITE. A metamorphic mineral. It is well established that pyroxene rocks may be metamorphosed into hornblende rocks. If the hornblende thus produced is fibrous and retains the original form of the pyroxene, it is called uralite, and the process by which the change is brought about chemical process which in many cases is accompanied by the generation chemical process which in many cases is accompanied by the generation of new minerals such as calcite, epidote, and magnetite. Uralite was first observed in rocks from the Ural Mountains, hence its name. See also **Hornblende**.

URANINITE. A mineral approximating the composition UO_2, but containing besides the higher oxide of uranium, UO_3, and oxides of lead, thorium, and rare earths. The uraninite usually occurs as cubic or cubo-octahedral crystals of specific gravity 7.5–10; when in masses of pitchy luster it is called pitchblende, specific gravity 6.5–9. All uraninites and pitchblende contain a minute amount of radium. It was in pitchblende obtained from the Joachimsthal in Czechoslovakia that Mme Curie discovered radium. Other localities for uraninite are in Saxony, Rumania, Norway, Cornwall, East Africa, and in the United States in the pegmatites of Connecticut Grafton Center, New Hampshire, North Carolina, and South Dakota, and in Gilpin County, Colorado. An important occurrence of pitchblende is at Great Bear Lake, Northwest Territories, Canada, where it has been found in large quantities associated with silver.

URANIUM. [CAS: 7440-61-1]. Chemical element symbol, U, at. no. 92, at. wt. 238.03, periodic table group (Actinides), mp 1,131 to 1.133°C, bp 3,818°C, density 18.9 g/cm³ (20°C). Uranium metal is found in three allotropic forms: (1) *alpha phase*, stable below 668°C, orthorhombic; (2) *beta phase*, existing between 668 and 774°C, tetragonal; and (3) *gamma phase*, above 774°C, body-centered cubic crystal structure. The gamma phase behaves most nearly that of a true metal. The alpha phase has several nonmetallic features in its crystallography. The beta phase is brittle. See also **Chemical Elements**.

Prior to the production of artificially-created elements, uranium was the highest in terms of atomic number and atomic weight. It was difficult to locate uranium in the periodic table of the elements, although chemically uranium resembles the elements of group 6b, namely, chromium, molybdenum, and tungsten. Subsequent to the production of the transuranium elements (atomic numbers 93 through 103), these elements, along with actinium (89), thorium (90), protactinium (91), and uranium (92) have been placed into the Actinide group of transition elements. They are similar in their mutual relations to the rare-earth group lanthanum (57) to lutetium (71). See also **Periodic Table of the Elements**.

Earthly abundance of uranium will be described shortly. In terms of presence in seawater, no significant concentrations have been reported. In terms of cosmic abundance, uranium also is very scarce. The study by Harold C. Urey (1952), in which silicon was given a base figure of 10,000, the concentration of uranium was represented by a figure of 0.0002.

Uranium is a white metal, ductile, malleable, and capable of taking a high polish, but tarnishes readily on exposure to the atmosphere. Finely divided uranium burns upon exposure to air, and the compact metal burns when heated in air at 170°C. Uranium metal slowly decomposes water at ordinary temperatures and rapidly at 100°C; is soluble in HCl and in HNO_3; and is unattacked by alkalis. Chemically related to chromium, molybdenum, and tungsten; and, like thorium, is radioactive. In the radioactive decomposition radium is formed. Discovered by Klaproth in 1789.

The element uranium found in nature consists of the three isotopes of mass numbers 238, 235, 234 with relative abundances 99.28, 0.71, and 0.006%, respectively.

The isotope ^{238}U is the parent of the natural uranium $4n + 2$ radioactive series, and the isotope ^{235}U is the parent of the natural actinium $4n + 3$ radioactive series.

The isotope ^{235}U has great importance because it undergoes the nuclear fission reaction with slow neutrons, and it has been separated in substantial amounts in nearly 100% isotopic composition.

Electronic configuration is $1s^2 2s^2 2p^6 3s^2 3p^6 3d^{10} 4s^2 4p^6 4d^{10} 4f^{14} 5s^2 5p^6 5d^{10} 5f^3 6s^2 6p^6 6d^1 7s^2$. Ionic radii U^{4+} 0.89 Å; U^{3+} 1.04 Å (Zachariasen). Metallic radius 1.4318 Å (805°C). Oxidation potential $U + 2H_2O \longrightarrow UO_2^{2+} + 4H^+ + 6e^-$, 0.82 V.

Uranium (^{233}U) is a fissionable isotope of uranium produced artificially by bombarding thorium-232 with neutrons. Used as an atomic fuel in molten salt reactor and is a possible fuel in breeder reactors. Half-life 1.62×10^5 years.

Uranium Reserves

Uranium has been known to be a distinct element since 1789. Apart from the small amount of its salts used in yellow pottery glazes, however, it remained more or less a laboratory curiosity until the 1920s. Then, the treatment of uranium ore, for the recovery of its radium (for the treatment of cancer), began in Eastern Europe and Zaire, followed by Canada in 1933. The separated uranium was mostly stockpiled or discarded.

After the development and successful explosion of the atomic bomb toward the end of World War II, an urgent search for workable uranium deposits was set in motion all over the world. The only high-grade deposits known to the western world were those in the countries just named as radium sources, but in view of the limited demand previously, serious exploration for uranium had never been undertaken. However, the offer of contracts by the U.S. Atomic Energy Commission, for fixed quantities at stated prices stimulated exploration for this hitherto largely ignored material.

Uranium is rather widely distributed throughout the world. See Table 1. Deposits vary markedly in richness.

Uranium Reserves

During the past few decades, the construction of new nuclear power reactors in the United States has been limited, although this does not hold for France and some other countries. As with any nonrenewable

TABLE 1. TYPES OF NATURAL URANIUM RESOURCES

Type of resource	Ore grade (ppm uranium)	Principal known locations
Vein deposits	10,000–30,000	Canada (near Great Bear Lake in the Northwest Territory) Western United States France Germany Russia Africa Australia China
Vein deposits (pegmatites, unconformity deposits)	2,000–10,000	Canada (Saskatchewan) Russia. Australia
Fossil placers, sandstones	200–2,000	Canada (Ontario) Western United States Brazil Chile Russia Japan Australia Africa
Shales, phosphates	10–100	United States (Florida) Morocco Sweden Russia
Pegmatites, other igneous and metamorphic	1–10	Canada (Ontario) deposits Greenland Brazil Spain Russia India Africa Australia

fuel, there is concern over some ultimate date in the future when the supply may approach exhaustion. As of the early 1990s, with a new period of calm prevailing pertaining to the need for construction of nuclear weaponry, forecasts of useable uranium reserves made in earlier years no longer hold, and revised forecasts are lacking. In terms of current usage of uranium, there are several avenues available for conserving the fuel.

(1) *Improvement of uranium efficiency in thermal reactors* which consume more fissionable material than they breed. If it is assumed that on average a typical light-water reactor (LWR) is operated at 75% total capacity (load factor = 0.75), it will consume about 6000 tons of U_3O_8 per gigawatt of electrical output GWe over an expected 30-year reactor life. During this period, the reactor will produce about 5 tons of fissile plutonium, which is discharged in the spent fuel. In the past, minimal consumption of uranium has not been a major design goal. Improvements in reactors (not requiring major alterations) could effect savings of from 10 to 15% in uranium consumption even without considering fuel recycling. For example, simply enriching the level of U-235 from 3 to 4.2% would allow the fuel to remain in the reactor for 5 years instead of the usual 3 years. As pointed out by Hafenmeister (California Polytechnic University), the longer residence time and the higher enrichment would allow a greater fraction of the U-235 to be used, increasing the *in situ* generation and burning of plutonium, while at the same time reducing the discharge of plutonium by 30% (from 5 to 3.5 tons over lifetime of the reactor). If the burn-up of fuel is increased from the traditional 30,000 to 50,000 megawatt-days per ton, the level of U-235 in the discharged fuel is reduced from 0.85% to 0.71%, and refueling can be considered either on a 12- or an 18-month cycle.

(2) *Design changes in new reactors can conserve uranium.* Traditional LWRs use control poisons such as boric acid in the reactor coolant as a means to reduce reactivity. This practice results in a waste of from 5 to 10% of available neutrons. Newer designs which would allow faster and more frequent refueling could reduce the need for such poisons and consequent loss of neutrons. Such changes could result in a saving of some 25% of the fuel required.

(3) *Reprocessing of spent fuel and blanket materials*, with the recovery of purified uranium and plutonium, could effect another fuel savings of nearly 20%. Such reprocessing not only would conserve the U_3O_8 supply, but would also alleviate a severe nuclear waste problem. As described in the entry on **Nuclear Power Technology**, the spent-fuel storage facilities at existing nuclear power facilities already have reached or are rapidly approaching their full capacity, and new storage systems must be developed. Processes for spent-fuel recovery are described in the next section of this article. The principal deterrent to reprocessing is concern with "nuclear proliferation." In the most straightforward reprocessing scheme, plutonium is recovered along with uranium. Plutonium is the primary ingredient of nuclear weapons. Contemporary nuclear power plants operating in a number of countries do not yield plutonium in a form useful for weapons production, whereas a fuel reprocessing scheme would.

(4) *Extending uranium supplies by using thorium in LWRs.* Instead of using boron as a poison in the coolant, 40% heavy water (D_2O) could be used along with thorium. Hafemeister observes that the heavier D_2O in this "spectral shift control reactor" would result in breeding and then burning more plutonium from the fertile U-238. It is estimated that the savings would be about 12% of the U_3O_8 when compared to the LWR on a conventional uranium cycle; and about 20% of the U-233 required for a LWR on a thorium cycle. The Canada Deuterium Uranium (CANDU) power reactors are described under "Heavy Water Reactor" in the entry on **Nuclear Power Technology**. This means of uranium conservation, of course, would require large capital expenditures.

(5) *Fast breeder reactors.* These reactors (FBRs) produce more fissionable material during operation than is originally furnished. A number of nations are interested in the FBR as a means of extending their available uranium. Plutonium is more valuable in an FBR because it raises breeding ratios by 10 to 20% when compared with U-235. But, as pointed out in the entry on **Nuclear Power Technology**, the FBR essentially remains in the design and development stage. In terms of nuclear proliferation, uranium has an advantage over plutonium because isotopic enrichment is necessary to obtain weapons-usable material from a fuel containing both U-233 and U-238, whereas only chemical separations are required to purify plutonium from a fuel containing plutonium and uranium.

Uranium Spent-Fuel Reprocessing

In the PUREX process, the spent fuel and blanket materials are dissolved in nitric acid to form nitrates of plutonium and uranium. These are separated chemically from the other fission products, including the highly radioactive actinides, and then the two nitrates are separated into two streams of partially purified plutonium and uranium. Additional processing will yield whatever purity of the two elements is desired. The process yields purified plutonium, purified uranium, and high-level wastes. See also "Radioactive Wastes" in the entry on **Nuclear Power Technology**. Because of the yield of purified plutonium, the PUREX process is most undesirable from a nuclear weapons proliferation standpoint.

In a modified PUREX process (sometimes called coprocessing), the plutonium and uranium are not handled separately as their nitrates, but rather they are processed together, yielding plutonium diluted with uranium. The diluted product is useful for nuclear reactors, but not directly usable for nuclear weapons.

In the CIVEX fuel reprocessing system, most of the uranium is separated and purified from the spent nuclear reactor fuel, while some of the fission fragments, some of the uranium, and all of the plutonium are handled together. The system thus yields a fuel containing some of all three materials. The fission products render the fuel unsafe for any but sophisticated handling for a period of at least one year. The plutonium does not exist in high concentrations. Hafemeister (1979) observes that the costs of remote fabrication for CIVEX make it potentially attractive only for nations with large breeder reactor programs. In addition, the fission products in CIVEX fuels would, to some extent, act as poisons in thermal reactors, reducing the flow of neutrons.

Processing of Uranium Ores

Preconcentration of uranium ores by methods based on gravity, magnetic, or electrical properties, or by flotation have not been generally successful because of the softness of the raw materials and the absence of differing physical properties among the constituents. Thus, treatment has involved the whole ore wherein an extractant has been used. The uranium is recovered from solution after appropriate solid-liquid separations. Either an acid or alkaline extractant can be used. The extraction is enhanced by use of fine grinding of the ore, by increasing the concentration of extractant, and by using higher temperatures and oxidizing agents. The uranium can be separated from the inert material by solid-liquid separation, filtration, or countercurrent washing. The uranium is recovered from solution by treatment with ion-exchange resin or solvent extraction. Alkaline leaching predominates over the acid process. The process is illustrated and briefly described in the Fig. 1. The uranium concentrate produced requires additional refining before fabrication into a fuel for nuclear reactors. Generally, this is accomplished by dissolving the product in HNO_3, filtering, and treating the uranyl nitrate solution by solvent extraction methods.

Chemistry of Uranium

Uranium has the four oxidation states, (III), (IV), (V), and (VI); the ions in aqueous solution are usually represented as U^{3+}, U^{4+}, UO_2^+, and UO_2^{2+}. The oxidation-reduction scheme, on the hydrogen scale (in which the potential for $\frac{1}{2}H_2 \longrightarrow H^+$ is taken as zero) is indicated in Fig. 2.

The UO_2^+ ion is unstable in solution and undergoes disproportionation to U^{4+} and UO_2^{2+}. A few solid compounds of this oxidation state are known, as for example, UF_5 and UCl_5.

The ion U^{3+} forms intense red solutions in H_2O and is oxidized by water at an appreciable rate. The rate of oxidation appears to increase with increasing ionic strength, although concentrated solutions are said to be stabilized by strong acids such as hydrochloric. Solutions of uranium(IV) are green, and uranium(VI) solutions, yellow.

The (IV) and (VI) are the important oxidation states and therefore the more important phases of the chemistry of uranium may be related to the two oxides UO_2 and UO_3, uranium dioxide and uranium trioxide. A series of salts such as the chloride and sulfate, UCl_4 and $U(SO_4)_2 \cdot 9H_2O$ is

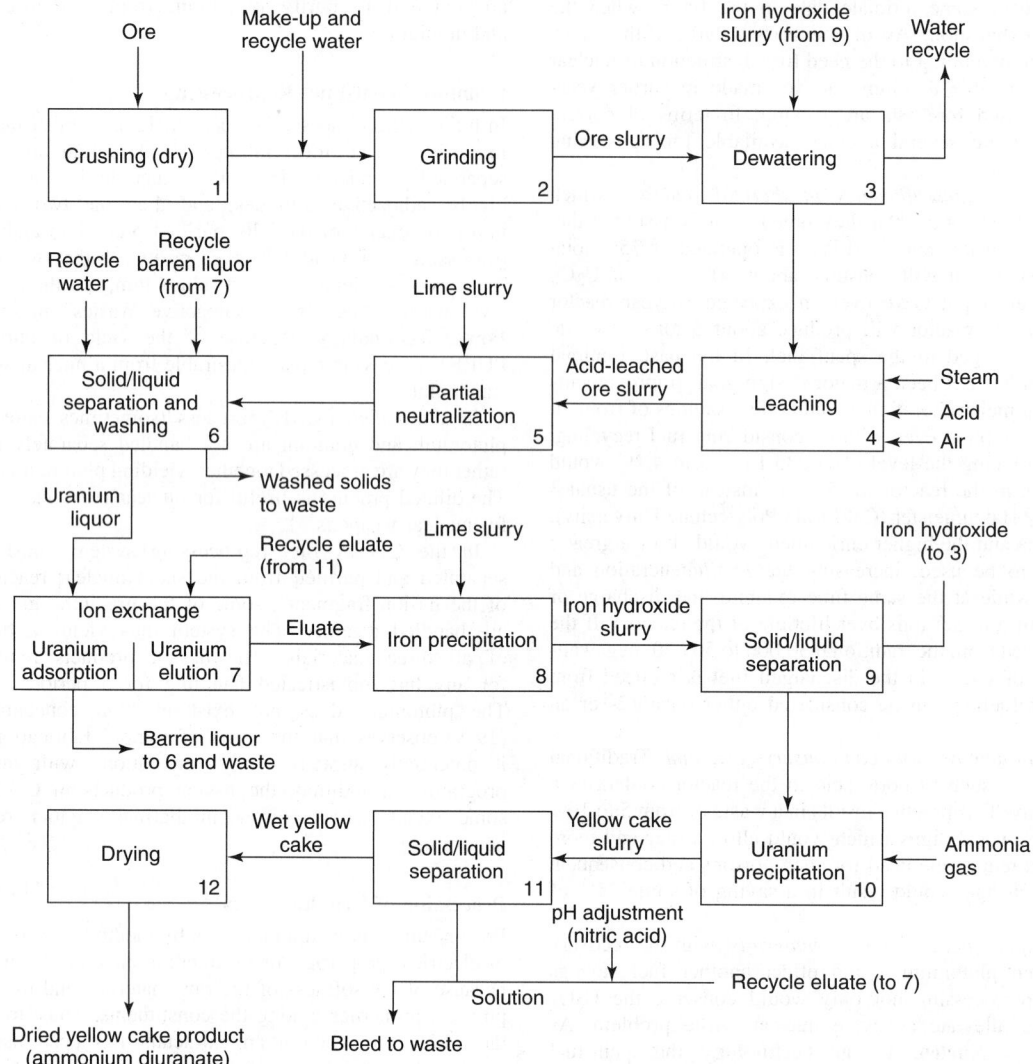

Fig. 1. Schematic flowsheet of uranium processing (acid leach and ion exchange) operation. Numbers refer to the numbers that appear in the boxes on the flowsheet. Operations (3), (6), (9), and (11) may be done by thickening or filtration. Most often, thickeners are used, followed by filters. The pH of the leach slurry (4) is elevated to reduce its corrosive effect and to improve the ion-exchange operation on the uranium liquor subsequently separated. In the ion exchange operation (7), resin contained in closed columns is alternately loaded with uranium and then eluted. The resin adsorbs the complex anions, such as $UO_2(SO_4)_3^{4-}$, in which the uranium is present in the leach solution. Ammonium nitrate is used for elution, obtained by recycling the uranium filtrate liquor after pH adjustment. Iron adsorbed with the uranium is eluted with it. Iron separation operation (8) is needed inasmuch as the iron hydroxide slurry is heavily contaminated with calcium sulfate and coprecipitated uranium salts. Therefore, the slurry is recycled to the watering stage (3). Washed solids from (6), the waste barren liquor from (7), and the uranium filtrate from (11) are combined. The pH is elevated to 7.5 by adding lime slurry before the mixture is pumped to the tailings disposal area. (*Rio Algom Mines Limited, Toronto*)

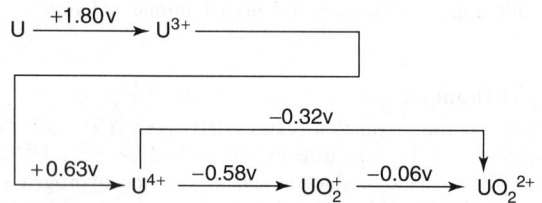

Fig. 2. Oxidation-reduction potentials of uranium ions (in 1-molar hydrochloric acid)

obtained from UO_2. The more common uranyl salts as $UO_2(NO_3)_2 \cdot 6H_2O$, UO_2Cl_2, and $UO_2SO \cdot nH_2O$ in which the UO_2^{2+} (uranyl ion) acts as a radical, are derived from UO_3. UO_3 is amphiprotic and forms a series of alkali and double alkali uranates and polyuranates of limited solubility, such as $Na_2U_2O_7$, $NaZn(UO_2)_3(C_2H_3O_2)_9 \cdot 6H_2O$, $NaMg(UO)_2)_3(C_2H_3O_2)_9 \cdot 6H_2O$, etc.

The element uranium also exhibits a formal oxidation number of (II) in a few solid compounds, semimetallic in nature, such as UO and US. No simple uranium ions of oxidation state (II) are known in solution.

In addition to three oxides, UO_2 (brown, cubic), U_3O_8 (greenish black, orthorhombic) and UO_3 (orange, hexagonal), which have been known for a long time, there are known to exist the monoxide, UO, and the pentoxide U_2O_5. There is also some evidence for the existence of U_4O_7 and U_6O_{17}. The phase relationships in the uranium-oxygen system are very complex because solid solutions are readily formed, so that it is possible to obtain uranium "oxides" with practically any composition intermediate between UO and UO_3, and with many crystal structures.

Uranyl peroxide, the formula of which is usually given as $UO_4 \cdot 2H_2O$, is formed by precipitation from solutions of uranyl nitrate by hydrogen peroxide. Alkali hydroxides, hydrogen peroxide, and sodium peroxide form soluble peroxyuranates, $Na_2UO_6 \cdot 4H_2O$ and $Na_4UO_8 \cdot 8H_2O$, when added to solutions of uranyl salts.

Two uranium carbides are known, the monocarbide, UC, and the dicarbide, UC_2. These can be prepared by direct reaction of carbon with molten uranium, or by reaction of carbon monoxide with metallic uranium at elevated temperatures. The sesquicarbide, U_2C_3, has been found to exist as a stable compound below about 1800°C and can be produced by heating a mixture of UC and UC_2 between 1,250 and 1,800°C.

Uranium and nitrogen form an extensive series of compounds that can be prepared by direct action of nitrogen on the metal. Uranium mononitride,

UN, is the lowest nitride of uranium. If the mononitride is treated with more nitrogen at atmospheric pressure, U_2N_3 is formed. With nitrogen under high pressure UN_2 can be prepared, but it is difficult to obtain samples of UN_2 that are completely free of UN.

Uranium metal reacts with hydrogen at 250–300°C to form a well-defined hydride, which resembles the rare-earth hydrides in many respects. The formula of this substance has been shown to be $UH_{3.00}$. The hydride undergoes decomposition with increasing temperature; the dissociation pressure of UH_3 is one atmosphere at 436°C.

Uranium tetrafluoride serves as a starting material for the preparation of the other fluorides. It is best prepared by hydrofluorination of uranium dioxide:

$$UO_2 + 4HF \xrightarrow{500°C} UF_4 + 2H_2O$$

Uranium trifluoride can be prepared by reduction of UF_4 with hydrogen at 1,000°C. Uranium hexafluoride, UF_6, white and orthorhombic, is best obtained by direct fluorination of UF_4, green and monoclinic, although any uranium compound will yield UF_6 by reaction with fluorine at elevated temperatures:

$$UF_4 + F_2 \xrightarrow{350°C} UF_6$$

The hexafluoride can also be prepared by the interesting reaction:

$$2UF_4 + O_2 \xrightarrow{900°C} UF_6 + UO_2F_2$$

The intermediate fluorides $U_2F_9(UF_{4.5})$, $U_4F_{17}(UF_{4.25})$ and UF_5 are prepared by reaction of solid UF_4 and gaseous UF_6 under appropriate conditions of temperature and pressure.

Uranium hexafluoride is probably the most interesting of the uranium fluorides. Under ordinary conditions, it is a dense, white solid with a vapor pressure of about 120 mm at room temperature. It can readily be sublimed or distilled, and it is by far the most volatile uranium compound known. Despite its high molecular weight, gaseous UF_6 is almost a perfect gas, and many of the properties of the vapor can be predicted from kinetic theory.

Uranium tetrachloride can be prepared by direct combination of chlorine with uranium metal or hydride; it can also be obtained by chlorination of uranium oxides with carbon tetrachloride, phosgene, sulfur chloride, or other powerful chlorinating agents. The trichloride is obtained by reaction of UCl_4 with hydrogen and the higher chlorides by reaction of UCl_4 and Cl_2. Uranium hexachloride, UCl_6 is a rather volatile, somewhat unstable substance. All of the uranium chlorides dissolve in or react readily with water to give solutions in which the oxidation state of the ion corresponds to that in the solid. All of the solid chlorides are sensitive to moisture and air.

The trichloride, tribromide and triiodide of uranium are obtained either by reaction of the elements or by treatment of UH_3 with the appropriate halogen acid. The thermal stability of the halides decreases as the atomic number of the halogen increases. No higher uranium bromides or iodides are known.

A series of oxyhalides of the type UO_2F_2, $UOCl_2$, UO_2Br_2, etc., are known. They are all water-soluble substances which become increasingly less stable in going from the oxyfluoride to the oxyiodide.

Uranyl ion forms complexes with many oxy anions. Both U(VI) and U(IV) compounds dissolve in alkali carbonate solutions with formation of carbonato complexes. Those of the larger alkali cations are only slightly soluble: $K_{sp} = 6 \times 10^{-5}$ for both $K_4[UO_2(CO_3)_3]$ and $(NH_4)_4[UO_2(CO_3)_3]$ $\cdot 2H_2O$.

Aqueous solutions of uranium(III), uranium(IV), and uranium(VI) are readily obtained. Solutions of uranium(III) are blood-red in appearance; hydrogen is slowly evolved with the formation of uranium(IV) is a strong reducing agent and is easily oxidized to uranyl ion by oxygen, peroxide, and numerous other oxidizing agents. Uranyl solutions in turn may be reduced to uranium(IV) with sodium dithionite, zinc or cadmium amalgams, or by electrochemical or photochemical means.

Separation of Isotopes

Several methods are available for the separation of isotopes, including gaseous diffusion, centrifugation, electromagnetic methods, thermal diffusion, electrolytic methods, distillation, and chemical-exchange methods. In the late 1960s, another, radically different process was added to the technology of isotope separation. This is known as laser enrichment and is described shortly. The separation of ^{235}U from ^{238}U represented the first large-scale isotope-separation operation and, after considerable study, the principal plant utilized gaseous diffusion.

The gaseous diffusion method of isotope separation is based upon the difference in the rate of diffusion of gases that differ in density. Since the rate of diffusion of a gas is inversely proportionate to the square root of its density, the lighter of two gases will diffuse more rapidly than the heavier. Therefore, the result of a partial diffusion process will be an enrichment of the partial product in the lighter component.

To separate isotopes by this process, they must be in the gaseous form. Therefore, the separation of isotopes of uranium required the conversion of the metallic uranium into a gaseous compound, for which purpose the hexafluoride, UF_6, was chosen. Since the atomic weight of fluorine is 19, the molecular weight of the hexafluoride of ^{235}U is $235 + (6 \times 19) = 349$, and the molecular weight of the hexafluoride of ^{238}U is $238 + (6 \times 19) = 352$. Since the rate of diffusion of a gas is inversely proportional to the square root of its density (mass per unit volume), the maximum separation factor for one diffusion process of the uranium isotopes is $\sqrt{352/349} = 1.0043$. Since only part of the gas can be allowed to diffuse, the actual separation factor is even less than this theoretical maximum.

From this small figure, it is apparent that many diffusion stages are necessary in the separation of ^{235}U from ^{238}U. The number originally calculated for the Oak Ridge plant was about 4,000. Other reasons are the small apertures demanded by diffusion processes (in this case less than .00001 centimeter in diameter), which reduce the rate of gas flow and demand a great barrier area for appreciable production.

The centrifugal method of isotope separation consists essentially of the passage of the mixture through a rapidly rotating force field, such as that of a rotating cylinder. If a current of mixed gases is passed into such a cylinder, moving parallel to the axis of rotation, the lighter gas will tend to concentrate near the axis, and the heavier gas, near the periphery. This is the principle of the cream separator; its successful application to separation of isotopes in the gaseous phase requires apparatus operating at very high speeds of rotation.

The electromagnetic method of isotope separation is based upon the principle of the mass spectrograph. As in that apparatus, a stream of charged particles is passed through a system of electric and magnetic fields. If the particles are ions of two or more isotopes of the same element, all bearing the same charge, the deflections produced by the fields will vary with the masses of the particles, and will thus provide a means for their separation. This method is especially effective for the separation of particles of a number of masses, and has been widely used for that purpose in research studies and in production-separation operations. The method is also used extensively in a number of research laboratories, particularly those of northern Europe, for the isotopic separation of individual radioactive nuclides that are to be used as sources in instruments, such as beta- and gamma-ray spectrometers, in which measurements are made of the characteristics of ionizing radiations.

The thermal diffusion method of isotope separation has broad application to liquid-phase as well as gaseous-phase separations. The apparatus widely used for this purpose consists of a vertical tube provided with an electrically heated central wire. The gaseous or liquid mixture containing the isotopes to be separated is placed in the tube, and heated by means of the wire. In such an apparatus two effects act to separate the isotopes. Thermal diffusion tends to concentrate the heavier isotopes in the cooler outer portions of the system, while the portions near the hot wire are enriched in the lighter isotopes. At the same time, thermal convection causes the hotter fluid near the hot wire to rise, while the cooler fluid in the outer portions of the system tends to fall. The overall result of these two effects causes the heavier isotopes to collect at the bottom of the tube and the lighter at the top, whereby both fractions may be withdrawn.

The electrolytic method of isotope separation is of importance not only because of its present day uses, but also because of its historical interest. It was by this method that G.N. Lewis and his co-workers at the University of California obtained practically pure deuterium. Since deuterium oxide had been shown to be present in ordinary water, the conclusion was drawn that water (or rather the dilute aqueous solution) from electrolytic cells used for the production of hydrogen and oxygen by continuous electrolysis of water, should be richer in the heavier isotope (deuterium having a mass number of 2, as against 1 for protium). Starting with such residual water from an electrolytic cell, it was found that by repeated electrolysis a small residue consisting almost entirely of deuterium oxide (D_2O) was obtained. This process is still used for the separation of pure hydrogen isotopes, as well as for other purposes.

The distillation method of isotope separation has also been the basis of important research contributions. In the work on the hydrogen isotopes, it preceded the electrolytic separation methods discussed above. Following the suggestion by Birge and Menzel, of the possible presence of deuterium in ordinary hydrogen to the extent of 1 part in about 4,500, Urey, Brickwedde and Murphy, in 1931, began their search for this isotope. By evaporating about 4,000 milliliters of liquid hydrogen to a volume of about 1 milliliter, Brickwedde obtained a residue that gave conclusive spectroscopic evidence of the presence of deuterium. The distillation method of separation has been responsible for many other research contributions. Among them may be mentioned separation of the isotopes of oxygen, of mercury, zinc, potassium and chlorine.

The chemical exchange methods of isotope separation are of value, not only for that purpose, but also because they provide a direct means for the study of chemical reactions. A well-known example of an isotopic chemical exchange is the heavy water equilibrium:

$$H_2 + D_2O \rightleftharpoons D_2 + H_2O$$

In this equation, the formula H_2 is used for the hydrogen isotope of mass number 1, which constitutes all but a small fraction of ordinary hydrogen; H_2O is the corresponding "light water"; D_2 is hydrogen of mass number 2 (deuterium) and D_2O is the corresponding heavy water. The double arrows indicate an equilibrium reaction, whereby under suitable conditions ordinary hydrogen reacts with heavy water to produce hydrogen of mass number 2 and light water. If one were to start either with the two reactants on the left of the arrows, or with the two on the right, the system at equilibrium would have all four present. However, in this system at equilibrium the reverse reaction predominates, so that the ratio of ^2H to ^1H (that is, the ratio of D_2O to H_2O) in the liquid phase is about three times as great as in the gas. Because of this differential reactivity, this method is useful in the separation of the two hydrogen isotopes.

Another equilibrium system is useful in the separation of ^{14}N and ^{15}N. It is represented by the equation:

$$^{15}NH_3 + {}^{14}NH_4NO_3 \rightleftharpoons {}^{14}NH_3 + {}^{15}NH_4NO_3$$

This exchange reaction is conducted by the countercurrent flow of ammonia gas and ammonium nitrate solution (in water). The forward reaction is favored, resulting in the concentration of the ^{15}N in the ammonium nitrate in solution. The multistate conduct of this reaction that is necessary for effective operation is accomplished by arranging later stages in which the enriched ammonium nitrate solution is divided into two parts. One part is treated with caustic soda to displace the enriched NH_3, which is then used in a second stage of the process with the other part of the NH_4NO_3 solution. Three or more stages may thus be used, until the desired concentration of ^{15}N has been effected.

Another method of isotope separation is by ion mobility, a process based on the difference in mobility of the ions in an electrolytic solution, under the influence of an electric field.

The laser process for enrichment of uranium dates back to the early 1970s. In addition to early and continuing development of this process at the Lawrence Livermore National Laboratory, pioneering efforts were made by Exxon Nuclear Corporation and Avco Corporation in early 1971. These efforts concentrated on an atomic vapor laser isotope separation approach. The process now in a reasonably late stage of development at the Lawrence laboratory is known by the acronym AVLIS (for atomic vapor laser isotope separation). The process differs radically from all other uranium enrichment approaches. The AVLIS processes uses a bank of very finely tuned lasers to create an electrical charge on uranium-235 atoms while leaving nonfissle uranium-238 atoms unchanged. Fundamentally, the process involves the firing of light from high-powered copper-vapor lasers into a stream of uranium atoms. Through tuning of the lasers, electrons will be stripped from some of the atoms, thus leaving positively charged uranium-235 ions. These are drawn to negatively charged plates. The uncharged uranium-238 atoms pass through the process unaffected. Advantages claimed for the process include a smaller capital investment to build and less costly to operate and maintain. Also, the construction of laser plants can be smaller and modular as compared with the former huge diffusion, centrifuge, and magnetic plants.

Health and Safety Factors

Exposure and Health Effects. Uranium is a general cellular poison, which can potentially affect any organ or tissue. Uranium and its compounds can be damaging due to chemical toxicity and by the injury caused by ionizing radiation. The chemical toxicity of uranium compounds depends on their solubility in biological media. Highly soluble and therefore highly transportable and toxic compounds include fluorides, chlorides, nitrates, and carbonates of uranium(VI); moderately transportable compounds include corresponding uranium(IV) compounds; slightly transportable compounds include oxides, hydrides, and carbides.

Uranium can enter the human body orally, by inhalation, and through the skin and mucous membranes. Uranium compounds, both soluble and insoluble, are absorbed most readily from the lungs. In the blood of exposed animals, uranium occurs in two forms in equilibrium with each other: as a nondiffusible complex with plasma proteins and as a diffusible bicarbonate complex.

Occupational Protection and Radiation Consideration. The main adverse factor during the mining and processing of uranium and uranium-containing minerals is airborne dust. Personal protection should be used. Finely divided uranium metal, some alloys, and uranium hydride are pyrophoric, therefore such materials should be handled in an inert atmosphere glovebox.

The toxicity of uranium caused by its radiation depends on the isotopes present. Natural uranium does not constitute an external radiation hazard since it emits mainly low energy α-radiation. It does, however present an internal radiation hazard if it enters the body by inhalation or ingestion. The concentration of 1 mg U/g biological tissue corresponds to an absorbed dose of 0.006 Sv per year. Large quantities of fissile isotopes, ^{233}U and ^{235}U, should be handled and stored appropriately to avoid a criticality hazard. Clear and relatively simple precautions, such as dividing quantities so that the minimum critical mass is avoided, following administrative controls, using neutron poisons, and avoiding critical configurations (or shapes), must be followed to prevent an extremely treacherous explosion.

Additional Reading

Bothwell, R.: *Nucleus: The History of Atomic Energy Limited,* University of Toronto, Toronto, Canada, 1988.

Golay, M.W. and N.E. Todreas: "Advanced Light-Water Reactors," *Sci. Amer.,* **82** (April 1990).

Golay, M.W.: "Longer Life for Nuclear Plants," *Technology Review (MIT),* 25 (May/June 1990).

Goldschmidt, B.: *Atomic Rivals,* Rutgers University Press, New Brunswick, NJ, 1990.

Grenwood, N.N. and A. Earnshaw: *Chemistry of the Elements,* 2nd Edition, Butterworth-Heinemann, Inc., Woburn, MA, 1997.

Hafemeister, D.W.: "Nonproliferation and Alternative Nuclear Technologies," *Technology Review (MIT),* **81**(3), 58–62 (1979).

Hotta, H.: "Recovery of Uranium from Seawater," *Oceanus,* 30 (Spring 1987).

Lewis, R.J. and N.I. Sax: *Sax'x Dangerous Properties of Industrial Materials,* 10th Edition, John Wiley & Sons, Inc., New York, NY, 2000.

Lide, D.R.: *CRC Handbook of Chemistry and Physics 2000–2001,* 81st Edition, CRC Press, LLC., Boca Raton, FL, 2000.

Marshall, E.: "Counting on New Nukes," *Science,* 1024 (March **2,** 1990).

Slovac, P., J.B. Flynn, and M. Layman: "Perceived Risk, Trust, and the Politics of Nuclear Waste," *Science,* 1603 (December 13, 1991).

Spinard, B.I.: "U.S. Nuclear Power in the Next Twenty Years," *Science,* 707 (December 12, 1988).

Suzuki, T.: "Japan's Nuclear Dilemma," *Technology Review (MIT),* 41 (October 1991).

URBAN WASTES (As Energy Source). See **Wastes as Energy Sources.**

UREA. [CAS: 57-13-6]. $H_2N \cdot CO \cdot NH_2$, formula weight 60.06, colorless crystalline solid, mp 132.7°C, sublimes unchanged under vacuum at its melting point, sp gr 1.335. Heating above the mp at atmospheric pressure causes decomposition, with the production of NH_3, isocyanic acid HNCO, cyanuric acid $(HNCO)_3$, biuret $NH_2CONHCONH_2$, and other products. Also known as carbamide, urea is very soluble in H_2O, soluble in alcohol, and slightly soluble in ether. The compound was discovered by Rouelle in 1773 as a constituent of urine. Historically, urea was the first organic compound to be synthesized from inorganic ingredients, accomplished by Wöhler in 1828. However, a century passed before the compound was manufactured on a large scale.

Because of the reactivity and versatility of its derivatives, urea is a very high-tonnage chemical. The compound and its derivatives are widely used in fertilizers, pharmaceuticals (e.g., barbiturates), and synthetic resins and plastics (urethanes). Although there are several chemical engineering approaches to the synthesis of urea, the principal reaction is that of combining NH_3 with CO_2 in a first step to form ammonium carbamate. In a second step, dehydrating the ammonium carbamate to yield urea: (1) $2NH_3 + CO_2 \longrightarrow NH_2COONH_4$, (2) $NH_2COONH_4 \longrightarrow NH_2CONH_2 + H_2O$. The processing is complicated because of the severe corrosiveness of the reactants, usually requiring reaction vessels that are lined with lead, titanium, zirconium, silver, or stainless steel. The second step of the process requires a temperature of about 200°C to effect the dehydration of the ammonium carbamate. The processing pressure ranges from 160 to 250 atmospheres. Only about one-half of the ammonium carbamate is dehydrated in the first pass. Thus, the excess carbamate, after separation from the urea, must be recycled to the urea reactor or used for other products, such as the production of ammonium sulfate.

Some of the reactions of urea and derivatives include: (1) as a weak mono-acid base, urea forms stable salts, such as urea nitrate $CO(NH_2)_2 \cdot HNO_3$ and urea oxalate $2CO(NH_2)_2 \cdot H_2C_2O_4$; (2) urea reacts with malonic acid to form barbituric acid $CO(NHCO)_2CH_2$, the derivatives of which are barbiturates (sedative drugs); (3) with alcohols, urea reacts to form urethanes; (4) with formaldehyde, urea forms ureaforms which can be used as slow-release fertilizers and also as ingredients for adhesives and plastics; (5) with hydrogen peroxide, urea forms a useful crystalline oxidizing agent; (6) with straight-chain alkanes, urea forms crystalline complexes (clathrates) which are used in the petroleum industry for separating straight- and branched-chain hydrocarbons; (7) when heated rapidly to about 350°C in a fluidized bed at atmospheric pressure, urea decomposes to isocyanic acid and NH_3. The latter products, when passed over a catalyst at 400°C, yield melamine $(NCH_2)_3$ which is the triamide of cyanuric acid and widely used in plastics; (8) with acids or bases, urea hydrolyzes, yielding NH_3

and CO_2. Hydrolysis in aqueous solutions is accelerated by the presence of urease (an enzyme). This reaction frequently is used for the quantitative determination of urea; (9) upon heating aqueous solutions of urea, biuret is formed. When crystallizing urea from aqueous solutions, the presence of about 5% biuret alters the crystals from long needles to short rhombic prisms, the latter greatly enhancing the handling properties of the final product. A content of up to 1.5% biuret is satisfactory for most fertilizer applications, although for citrus fruits, coffee plants, and cherry trees, the biuret content must remain below 0.3%. As a feed supplement for ruminants, pure biuret has proved advantageous because of the slower release rate of NH_3 from biuret as compared with urea. See also **Fertilizer**.

There are several different processes for making urea fertilizer. One process designed for energy savings is shown in Figs. 1 and 2.

Arginine-Urea Cycle (Ornithine Cycle)

In adult animals, including humans, the characteristic tissue-specific levels of different enzymes are maintained by a dynamic balance between the independently controlled rates of biosynthesis and degradation of each enzyme. A dynamic rather than a static system most likely emerged because it enables organisms to adapt to widely different nutritional conditions and other environmental changes. Depending upon the physiological state of the animal at a given moment, amino acids derived from the hydrolysis of exogenous or endogenous protein may be predominantly utilized for *synthesis* of tissue-specific proteins. Or, their carbon chains may be *metabolized* further to provide energy (ATP) or intermediates for synthesis of other cellular constituents. When the carbon chains of amino acids are utilized to provide energy, some provision must be made for disposal of the reduced nitrogen components. Animal tissues in general cannot tolerate accumulation of ammonia. Aquatic animals, which are surrounded

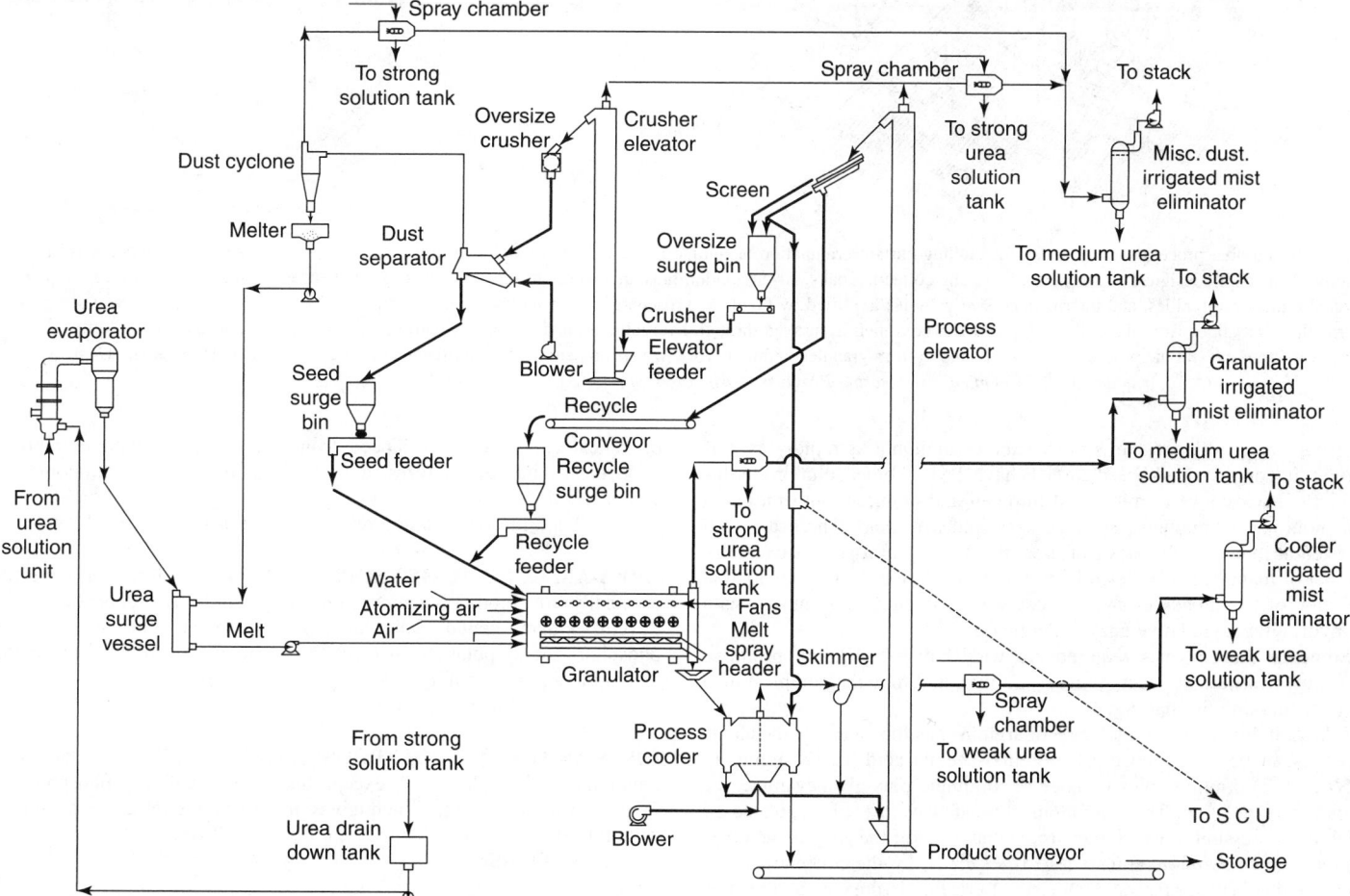

Fig. 1. A new process (*Urea Technologies*) developed for the Tennessee Valley Authority operates at considerable energy savings. Urea is produced in an overall exothermic reaction of ammonia and carbon dioxide at elevated pressure and temperature. In a highly exothermic reaction, ammonium carbamate is first formed as an intermediate compound, followed by its dehydration to urea and water, which is a slightly endothermic reaction. The conversion of CO_2 and NH_3 to urea depends on the ammonia-to-carbon dioxide ratio, temperature, and water-to-carbon dioxide ratio, among other factors. The new process makes maximum use of the heat created in the initial reaction, including heat recycling. (*Urea Technologies and Tennessee Valley Authority*)

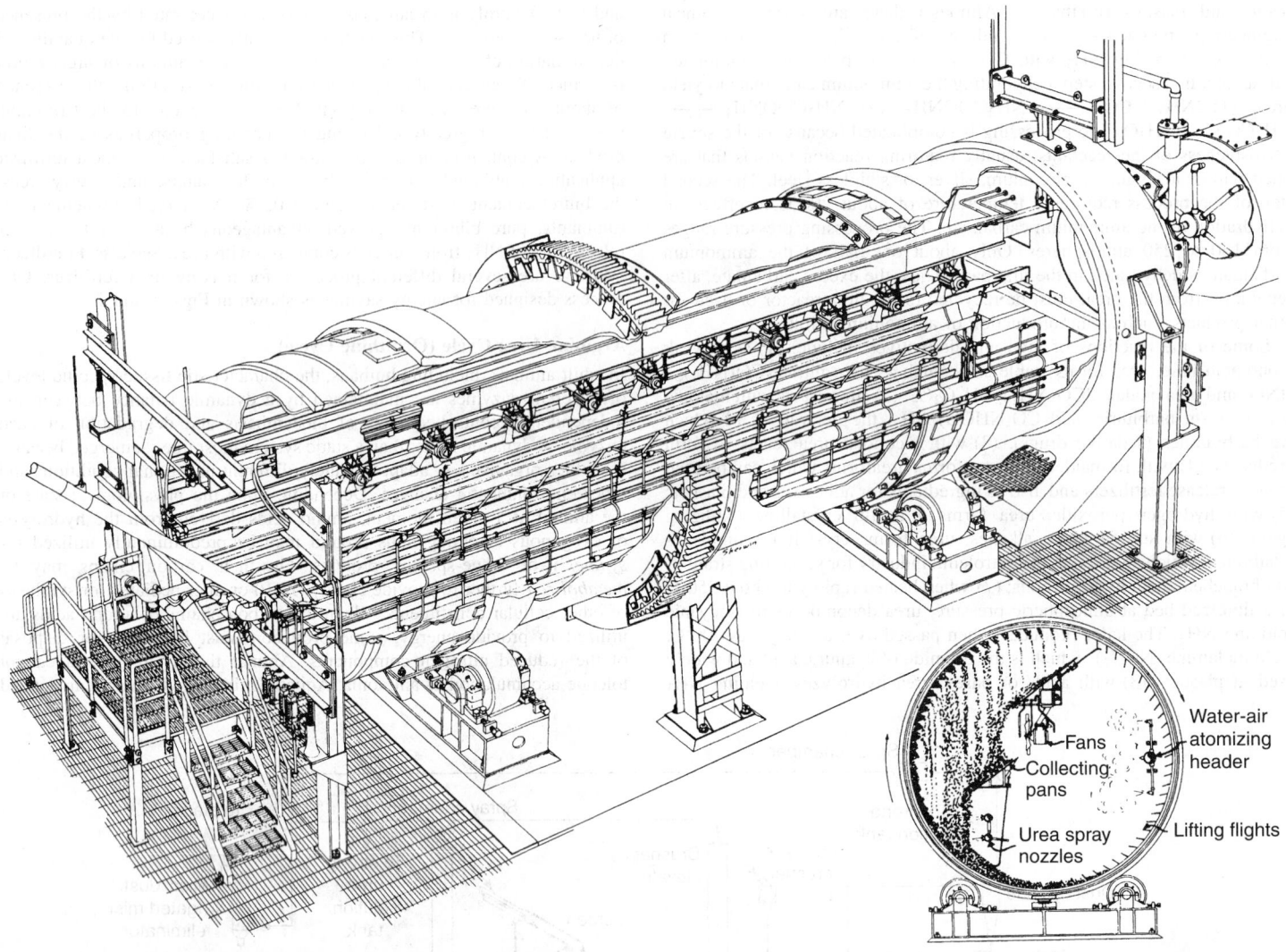

Fans
Collecting
pans
Water-air
atomizing
header
Urea spray
nozzles
Lifting flights

Fig. 2.   In the urea process shown in Fig. 1, a falling-curtain granulation technique is used. The granulator drum is shown here. The key to the process is the rotary drum, with specially designed components, including collecting pans, air-circulation fans, urea spray header, and water spray header. As the drum rotates, seed particles, recycled undersize granules, and intermediate-size granules are lifted by flights and discharged onto inclined collecting pans. The material forms a dense falling curtain of granules as it slides from the collecting pans. The urea melt is sprayed through hydraulic atomizing nozzles onto this falling curtain and quickly solidifies. Controls over several variables in the process make it possible to form granules within narrow size specifications. Further details of process are given by Kirkland (*R.W. Kirkland, "Energy-efficient route to granular urea," Chemical Engineering Progress, April, 1984, pp. 49–53, April 1984*)

by a convenient diluent, can simply excrete ammonia as rapidly as it is formed. In contrast, land-based animals have devised other solutions to this problem. They convert amino acid nitrogen and ammonia into nitrogen-rich, nontoxic compounds, such as *urea* and *uric acid*. These are then excreted at intervals. Synthesis of urea, the primary nitrogenous excretory product of mammals, is efficiently accomplished in the liver by combining a portion of the already established pathway of arginine biosynthesis with the hydrolytic degradative enzyme, arginase.

Some of the enzymes required are widely distributed, but *ornithine carbamoyltransferase* occurs only in the liver and thus the complete urea cycle occurs only in that organ.

In human liver, a given molecule of arginine has four possible metabolic fates. It can be converted to (1) argininosuccinic acid, or (2) argininyl-sRNA, or (3) ornithine plus urea, or (4) ornithine plus glycocyamine, the precursor of creatine. The flow along the pathway to (4) is regulated by feedback repression in which the steady-state level of the enzyme involved (arginine:glycine amidinotransferase) is regulated by the concentration of liver creatine. Thus, runaway synthesis of creatine is prevented. The flow along the pathway is regulated by supply and demand.

Experimentally, it has been observed that above a certain basal level, the quantity of urea excreted is proportional to the amount of ingested protein.

The enzyme urease was not discovered until 1926 (by Sumner). It was the first enzyme to be isolated as a crystalline protein. Sumner's accomplishment confirmed the then growing belief that enzymes, the

biological catalysts, were indeed from the chemical standpoint protein molecules. Urease catalyzes the cleavage of urea to ammonia and carbon dioxide.

For related information and references, see article on **Fertilizer**.

UREA-AMMONIUM ORTHOPHOSPHATE.   A fertilizer developed especially for food-deficient regions, particularly rice-dependent areas. Several grades contain all three primary plant nutrients (nitrogen, phosphorus, and potassium). Contains up to 60% nitrogen, phosphoric anhydride and potassium oxide.

See also **Fertilizer**.

UREA-AMMONIUM POLYPHOSPHATE.   A fertilizer similar to urea-ammonium othtrophosphate except that about half the phosphorus is in polyphosphate form, which gives improved sequestering action and solubility. It is excellent for use as a liquid fertilizer.

See also **Fertilizer**.

UREA AND THIOUREA DERIVATIVES.   See **Herbicides**; **Insecticide**.

UREAFORM.   A urea-formaldehyde reaction product that contains more than one molecule of urea per molecule of formaldehyde. It can be used as a fertilizer because of its high nitrogen content, its insolubility in water, and its gradual decomposition in the soil during the growing season to yield soluble nitrogen.

UREA-FORMALDEHYDE RESIN. An important class of amino resin. Urea and formaldehyde are united in a two-stage process in the presence of pyridine, ammonia, or certain alcohols with heat and control of pH to form intermediates (methylolurea, dimenthylolurea) that are mixed with fillers to produce molding powders. These are converted to thermosetting resins by further controlled heating and pressure in the presence of catalysts. These were first plastics that could be made in white, pastel, and colored products.

See also **Amino Acids; Melamine.**

UREASE. Enzyme present in low-percentages in jackbean and soybean; water soluble, its action is inhibited by heavy-metal ions. Its principal use is in the determination of urea in urine, blood, and other body fluids; it splits urea into ammonia and carbon dioxide or ammonium carbonate.

See also **Urea.**

URECH CYANOHYDRIN METHOD. Cyanohydrin formation by addition of alkali cyanide to the carbonyl group in the presence of acetic acid (Urech); or by reaction of the carbonyl compound with anhydrous hydrogen cyanide in the presence of basic catalyst (Ultee).

URECH HYDANTOIN SYNTHESIS. Formation of hydantoins from α-amino acids by treatment with potassium cyanate in a aqueous solution and heating of the salt of the intermediate hydantoic acid with 25% hydrochloric acid.

URETHANE. [CAS: 51-79-6]. $CO(NH_2)OC_2H_5$, also referred to as ethyl carbamate or ethyl urethane. Its structure is typical of the repeating unit in polyurethane resins. Colorless crystals or white powder, odorless, saltpeter-like taste, D 0.9862, mp 49C, bp 180C; solutions neutral to litmus, soluble in water, alcohol, ether, glycerol, and chloroform; slightly soluble in olive oil. Formed by the heating of ethanol and urea nitrate to 120–130°C or by the action of ammonia on ehtyl carbonate or ethyl chloroformate.

See also **Polyurethanes.**

URETHANE POLYMERS. The rapid formation of high molecular weight urethane polymers from liquid monomers, which occurs even at ambient temperature, is a unique feature of the polyaddition process, yielding products that range from cross-linked networks to linear fibers and elastomers. The enormous versatility of the polyaddition process allowed the manufacture of a myriad of products for a wide variety of applications.

Polyurethanes contain carbamate groups, $-NHCOO-$, also referred to as urethane groups, in their backbone structure. They are formed in the reaction of a diisocyanate with a macroglycol, a so-called polyol, or with a combination of a macroglycol and a short-chain diol extender. In the latter case, segmented block copolymers are generally produced. The macroglycols are based on polyethers, polyesters, or a combination of both. A linear polyurethane polymer has the structure of (**1**), whereas a linear segmented copolymer obtained from a diisocyanate, a macroglycol, and a diol extender, $HO(CH_2)_xOH$, has the structure of (**2**).

(**1**)

(**2**)

In addition to the linear thermoplastic polyurethanes obtained from difunctional monomers, branched or cross-linked thermoset polymers are made with higher functional monomers. Linear polymers have good impact strength, good physical properties, and excellent processibility, but, owing to their thermoplasticity, limited thermal stability. Thermoset polymers, on the other hand, have higher thermal stability but sometimes lower impact strength (rigid foams). The higher functionality is obtained with higher functional isocyanates (polymeric isocyanates), or with higher functional polyols. Cross-linking is also achieved by secondary reactions. Urea-modified segmented polyurethanes are manufactured from diisocyanates,

macroglycols, and diamine extenders. Urethane network polymers are also formed by trimerization of part of the isocyanate groups. This approach is used in the formation of rigid polyurethane-modified isocyanurate (PUIR) foams (**3**).

(**3**)

Formation and Properties

Polyurethane Formation. The polarization of the isocyanate group enhances the addition across the carbon-nitrogen double bond, which allows rapid formation of addition polymers from diisocyanates and macroglycols.

The liquid monomers are suitable for bulk polymerization processes. The reaction can be conducted in a mold (casting, reaction injection molding), continuously on a conveyor (block and panel foam production), or in an extruder (thermoplastic polyurethane elastomers and engineering thermoplastics). Also, spraying of the monomers onto the surface of suitable substrates provides insulation barriers or cross-linked coatings.

The polyaddition reaction is influenced by the structure and functionality of the monomers, including the location of substituents in proximity to the reactive isocyanate group (steric hindrance) and the nature of the hydroxyl group (primary or secondary). Impurities also influence the reactivity of the system.

The steric effects in isocyanates are best demonstrated by the formation of flexible foams from TDI. In the 2,4-isomer (**4**), the initial reaction occurs at the nonhindered isocyanate group in the 4-position. The unsymmetrically substituted ureas formed in the subsequent reaction with water are more soluble in the developing polymer matrix. Low density flexible foams are not readily produced from MDI or PMDI; enrichment of PMDI with the 2, 4'-isomer of MDI (**5**) affords a steric environment similar to the one in TDI, which allows the production of low density flexible foams that have good physical properties. The use of high performance polyols based on a copolymer polyol allows production of high resiliency (HR) slabstock foam from either TDI or MDI.

(**4**)

(**5**)

Tailoring of performance characteristics to improve processing and properties of polyurethane products requires the selection of efficient catalysts. In flexible foam manufacturing a combination of tin and tertiary amine catalysts are used in order to balance the gelation reaction (urethane formation) and the blowing reaction (urea formation). The tin catalysts used include dibutyltin dilaurate, dibutylbis(laurylthio)stannate, dibutyltinbis(isooctylmercapto acetate), and dibutyltinbis(isooctylmaleate). Strong bases, such as potassium acetate, potassium 2-ethylhexoate, or amine–epoxide combinations are the most useful trimerization catalysts.

The formation of cellular products also requires surfactants to facilitate the formation of small bubbles necessary for a fine-cell structure. The most effective surfactants are polyoxyalkylene–polysiloxane copolymers. The physical properties of polyurethanes are derived from their molecular structure and determined by the choice of building blocks as well as the supramolecular structures caused by atomic interaction between chains. The ability to crystallize, the flexibility of the chains, and spacing of polar groups are of considerable importance, especially in linear thermoplastic materials. In rigid cross-linked systems, e.g., polyurethane foams, other factors such as density determine the final properties.

Thermoplastic Polyurethanes. The unique properties of polyurethanes are attributed to their long-chain structure. In segmented polyether- and polyesterurethane elastomers, hydrogen bonds form between −NH−groups (proton donor) and the urethane carbonyl, polyether oxygen, or polyester carbonyl groups. The symmetrical MDI is more suitable for the preparation of segmented polyurethane elastomers having excellent physical properties. Segmented polyurethanes are also obtained from 2,6-TDI, but an economically attractive separation process for the TDI isomers has yet to be developed.

The melt viscosity of a thermoplastic polyurethane (TPU) depends on the weight-average molecular weight and is influenced by chain length and branching. TPUs are viscoelastic materials, which behave like a glassy, brittle solid, an elastic rubber, or a viscous liquid, depending on temperature and time scale of measurement. With increasing temperature, the material becomes rubbery because of the onset of molecular motion. At higher temperatures a free-flowing liquid forms. The melt temperature of a polyurethane is important for processibility. Melting should occur well below the decomposition temperature.

Thermoset Polyurethanes. The physical properties of rigid urethane foams are usually a function of foam density. A change in strength properties requires a change in density. Rigid polyurethane foams that have densities of <0.064 g/cm^5, used primarily for thermal insulation, are expanded with HCFCs, HFCs, or hydrocarbons. See also **Insulation (Thermal)**. Often water or a carbodiimide catalyst is added to the formulation to generate carbon dioxide as a coblowing agent. In addition to density, the strength of a rigid foam is influenced by the catalyst, surfactant, polyol, isocyanate, and the type of mixing. By changing the ingredients, foams can be made that have high modulus, low elongation, and some brittleness (friability), or relative flexibility and low modulus. See also **Foamed Plastics**.

The properties of thermoset flexible polyurethane foams are also related to density; load-bearing properties are likewise important. Under normal service temperatures, flexible foams exhibit rubber-like elasticity to deformations of short duration, but creep under long-term stress. Maximum tensile strength is obtained at densities of ca $0.024-0.030$ g/cm$^3 \cdot$ g/cm^3. The densities are controlled by the amount of water in the formulation and may range from 0.045 to 0.020 g/cm^3 by raising the amount of water from 2 to 5%. Auxiliary blowing agents are also used to reduce density and control hardness. The size and uniformity of the cells are controlled by the efficiency of mixing and the nucleation of the foam mix.

Hyperbranched polyurethanes are constructed using phenol-blocked trifunctional monomers in combination with 4-methylbenzyl alcohol for end capping. Polyurethane interpenetrating polymer networks (IPNs) are mixtures of two cross-linked polymer networks, prepared by latex blending, sequential polymerization, or simultaneous polymerization. IPNs have improved mechanical properties, as well as thermal stabilities, compared to the single cross-linked polymers. In pseudo-IPNs, only one of the involved polymers is cross-linked. Numerous polymers are involved in the formation of polyurethane-derived IPNs.

Raw Materials

Isocyanates. The commodity isocyanates TDI and PMDI are most widely used in the manufacture of urethane polymers. See also **Isocyantes, Organic**. The former is an 80:20 mixture of 2,4- and 2,6-isomers, respectively; the latter a polymeric isocyanate obtained by phosgenation of aniline−formaldehyde-derived polyamines. A coproduct in the manufacture of PMDI is 4, 4′-methylenebis(phenyl isocyanate) (MDI). The manufacture of TDI involves the dinitration of toluene, catalytic hydrogenation to the diamines, and phosgenation. Separation of the undesired 2,3-isomer is necessary because its presence interferes with polymerization. Polymeric isocyanates (PMDI) are crude products that vary in exact composition. The basic raw materials for the manufacture of PMDI and its coproduct MDI is benzene. Nitration and hydrogenation affords aniline (see **Amines; Aromatic Compound**). Reaction of aniline with formaldehyde in the presence of hydrochloric acid gives rise to the formation of a mixture of oligomeric amines, which are phosgenated to yield PMDI. The coproduct, MDI, is obtained by continuous thin-film vacuum distillation.

Urethanes obtained from aromatic diisocyanates undergo slow oxidation in the presence of air and light, causing discoloration, which is unacceptable in some applications. Polyurethanes obtained from aliphatic diisocyanates are color-stable, although it is necessary to add antioxidants and uv-stabilizers to the formulation to maintain the physical properties with

time. The least costly aliphatic diisocyanate is hexamethylene diisocyanate (HDI). Isophorone diisocyanate (IPDI) and its derivatives are also used in the formulation of rigid coatings; hydrogenated MDI (HMDI) and cyclohexane diisocyanate (CHDI) are used in the formulation of flexible coatings and polyurethane elastomers.

Masked or blocked diisocyanates are used in coatings applications. The blocked diisocyanates are storage-stable, nonvolatile, and easy to use in powder coatings. Blocked isocyanates are produced by reaction of the diisocyanate with blocking agents such as caprolactam, 3,5-dimethylpyrazole, phenols, oximes, acetoacetates, or malonates.

Polyether Polyols. Polyether polyols are addition products derived from cyclic ethers (Table 1). The alkylene oxide polymerization is usually initiated by alkali hydroxides, especially potassium hydroxide. In the base-catalyzed polymerization of propylene oxide, some rearrangement occurs to give allyl alcohol.

Polyether polyols are high molecular weight polymers that range from viscous liquids to waxy solids, depending on structure and molecular weight. Most commercial polyether polyols are based on the less expensive ethylene or propylene oxide or on a combination of the two. Block copolymers are manufactured first by the reaction of propylene glycol with propylene oxide to form a homopolymer.

With amine initiators the so-called self-catalyzed polyols are obtained, which are used in the formulation of rigid spray foam systems. The rigidity or stiffness of a foam is increased by aromatic initiators, such as Mannich bases derived from phenol, phenolic resins, toluenediamine, or methylenedianiline (MDA). In the manufacture of highly resilient flexible foams and thermoset RIM elastomers, graft or polymer polyols are used.

Polyester Polyols. Polyester polyols are based on saturated aliphatic or aromatic carboxylic acids and diols or mixtures of diols. The carboxylic acid of choice is adipic acid because of its favorable cost/performance ratio. For elastomers, linear polyester polyols of ca 2000 mol wt are preferred. Branched polyester polyols, formulated from higher functional glycols, are used for foam and coatings applications. Phthalates and terephthalates are also used.

In addition, polyester polyols are made by the reaction of caprolactone with diols. Poly(caprolactone diols) are used in the manufacture of thermoplastic polyurethane elastomers with improved hydrolytic stability. The hydrolytic stability of the poly(caprolactone diol)-derived TPUs is comparable to TPUs based on the more expensive long-chain diol adipates. Polyether/polyester polyol hybrids are synthesized from low molecular weight polyester diols, which are extended with propylene oxide.

Uses

Flexible Foam. Flexible slab or bun foam is poured by multicomponent machines at rates of >45 kg/min. One-shot pouring from traversing mixing heads is generally used. A typical formulation for furniture-grade foam having a density of 0.024 g/cm^3 includes a polyether triol, mol wt 3000; TDI; water; catalysts, i.e., stannous octoate in combination with a tertiary amine; and surfactant. Coblowing agents are often used to lower the density of the foam and to achieve a softer hand. Coblowing agents are methylene chloride, methyl chloroform, acetone, and CFC 11, but the last has been eliminated because of its ozone-depletion potential. Additive systems and new polyols are being developed to achieve softer low density

TABLE 1. COMMERCIAL POLYETHER POLYOLS

Product	Nominal functionality	Initiator	Cyclic ether[a]
poly(ethylene glycol) (PEG)	2	water or EG	EO
poly(propylene glycol) (PPG)	2	water or PG	PO
PPG/PEG[b]	2	water or PG	PO/EO
poly(tetramethylene glycol)	2	water	THF
glycerol adduct	3	glycerol	PO
trimethylolpropane adduct	3	TMP	PO
pentaerythritol adduct	4	pentaerythritol	PO
ethylenediamine adduct	4	ethylenediamine	PO
phenolic resin adduct	4	phenolic resin	PO
diethylenetriamine adduct	5	diethylenetriamine	PO
sorbitol adduct	6	sorbitol	PO/EO
sucrose adduct	8	sucrose	PO

[a] EO = ethylene oxide; PO = propylene oxide; THF = tetrahydrofuran.
[b] Random or block copolymer.

foams. Higher density (0.045 g/cm^3) slab or bun foam, also called high resiliency (HR) foam, is similarly produced, using polyether triols having molecular weight of 6000. The use of polymer polyols improves the load-bearing properties.

Flame retardants are incorporated into the formulations in amounts necessary to satisfy existing requirements. There are four main types of flexible slabstock foam: conventional, high resiliency, filled, and high load-bearing foam.

Most flexible foams produced are based on polyether polyols; ca 8–10% (15–20% in Europe) of the total production is based on polyester polyols. Flexible polyether foams have excellent cushioning properties, are flexible over a wide range of temperatures, and can resist fatigue, aging, chemicals, and mold growth. Polyester-based foams are superior in resistance to dry cleaning and can be flame-bonded to textiles.

Molded flexible foam products are becoming more popular. The bulk of the molded flexible urethane foam is employed in the transportation industry, where it is highly suitable for the manufacture of seat cushions, back cushions, and bucket-seat padding. TDI prepolymers were used in flexible foam molding in conjunction with polyether polyols. The need for heat curing has been eliminated by the development of cold-molded or high resiliency foams.

Semiflexible molded polyurethane foams are used in other automotive applications, such as instrument panels, dashboards, arm rests, head rests, door liners, and vibrational control devices. An important property of semiflexible foam is low resiliency and low elasticity, which results in a slow rate of recovery after deflection. The isocyanate used in the manufacture of semiflexible foams is PMDI, sometimes used in combination with TDI or TDI prepolymers. Both polyester as well as polyether polyols are used in the production of these water-blown foams.

Rigid Foams. Rigid polyurethane foam is mainly used for insulation. See also **Insulation (Thermal)**. The configuration of the product determines the method of production. Rigid polyurethane foam is produced in slab or bun form on continuous lines, or it is continuously laminated between either asphalt or tar paper, or aluminum, steel, and fiberboard, or gypsum facings. Rigid polyurethane products, for the most part, are self-supporting, which makes them useful as construction insulation panels and as structural elements in construction applications. Polyurethane can also be poured or frothed into suitable cavities, i.e., pour-in-place applications, or be sprayed on suitable surfaces.

Some formulations, particularly those for refrigerator and freezer insulation, are based on modified TDI (golden TDI) or TDI prepolymers, but these are being replaced by PMDI formulations. The polyols used include propylene oxide adducts of polyfunctional hydroxy compounds or amines (Table 1). The amine-derived polyols are used in spray foam formulations where high reaction rates are required. Crude aromatic polyester diols are often used in combination with the multifunctional polyether polyols. Blending of polyols of different functionality, Polyether–polyester polyol hybrids are also synthesized from low mol wt polyesters, which are subsequently propoxylated. Reactive or nonreactive fire retardants, containing halogen and phosphorus, are often added to meet the existing building code requirements. The most commonly used reactive fire retardants are Fyrol 6, chlorendic anhydride-derived diols, and tetrabromophthalate ester diols (PHT 4-Diol). Because the reactive fire retardants are combined with the polyol component, storage stability is important. Nonreactive fire retardants include halogenated phosphate esters, such as tris(chloroisopropyl) phosphate (TMCP) and tris(chloroethyl) phosphate (TCEP), and phosphonates, such as dimethyl methylphosphonate (DMMP). Also used are borax and melamine.

Because of the mandatory phaseout of CFCs by Jan. 1, 1996, it had become necessary to develop blowing agents that have a minimal effect on the ozone layer. As a short-term solution, two classes of blowing agents are considered: hydrochlorofluorocarbons (HCFCs) and hydrofluorocarbons (HFCs). For example, HCFC 141b, CH_3CCl_2F (bp 32°C), is a drop-in replacement for CFC-11, and HFC 134a, CF_3CH_2F (bp −26.5°C), was developed to replace CFC-12. HCFC 142b, CH_3CClF_2 (bp −9.2°C), is the blowing agent used in the 1990s. Addition of water or carbodiimide catalysts to the formulation generates carbon dioxide as a coblowing agent. Longer-range environmental considerations have prompted the use of hydrocarbons such as pentanes and cyclopentane as blowing agents.

From the onset of creaming to the end of the rise during the expansion process, the gas must be retained completely in the form of bubbles, which ultimately result in the closed-cell structure. Addition

of surfactants facilitates the production of very small uniform bubbles necessary for a fine-cell structure. The catalysts used in the manufacture of rigid polyurethane foams include tin and tertiary amine catalysts. Many of the rigid insulation foams produced in the 1990s are urethane-modified isocyanurate (PUIR) foams. In the formulation of poly(urethane isocyanurate) foams an excess of PMDI is used. The isocyanate index can range from 105 to 300 and higher. PUIR foams have a better thermal stability than polyurethane foams.

The formation of isocyanurates in the presence of polyols occurs via intermediate allophanate formation, i.e., the urethane group acts as a cocatalyst in the trimerization reaction. By combining cyclotrimerization with polyurethane formation, processibility is improved, and the friability of the derived foams is reduced. Modification of cellular polymers by incorporating amide, imide, oxazolidinone, or carbodiimide groups has been attempted but only the urethane-modified isocyanurate foams are produced in the 1990s. PUIR foams often do not require added fire retardants to meet most regulatory requirements. A typical PUIR foam formulation is shown in Table 2.

CASE Polyurethanes. CASE is the acronym for coatings, adhesives, sealants, and elastomers. Polyurethane coatings are mainly based on aliphatic isocyanates and acrylic or polyester polyols because of their outstanding weather-ability. For flexible elastomeric coatings, HMDI and IPDI are used with polyester polyols, whereas higher functional derivatives of HDI and IPDI with acrylic polyols are mainly used in the formulation of rigid coatings. Plastics coatings, textile coatings, and artificial leather are based on either aliphatic or aromatic isocyanates. For light-stable textile coatings, combinations of IPDI and IPDA (as chain extender) are used. The poly(urethane urea) coatings are applied either directly to the fabric or using transfer coating techniques. Microporous polyurethane sheets (poromerics) are used for shoe and textile applications. Polyurethane binder resins are also used to upgrade natural leather.

Blocked aliphatic isocyanates or their derivatives are used for one-component coating systems. Masked polyols are also used for this application. Water-borne polyurethane coatings are formulated by incorporating ionic groups into the polymer backbone. These ionomers are dispersed in water through neutralization. Ionic polymers are also formulated from TDI and MDI. Poly(urethane urea) and polyurea ionomers are obtained from divalent metal salts of *p*-aminobenzoic acid, MDA, dialkylene glycol, and 2,4-TDI. Polyurethane adhesives are known for excellent adhesion, flexibility, toughness, high cohesive strength, and fast cure rates. Two-component adhesives consist of an isocyanate prepolymer, which is cured with low equivalent weight diols, polyols, diamines, or polyamines. Such systems can be used neat or as solution. The two components are kept separately before application. Two-component polyurethane systems are also used as hot-melt adhesives.

Water-borne adhesives are preferred because of restrictions on the use of solvents. Low viscosity prepolymers are emulsified in water, followed by chain extension with water-soluble glycols or diamines. As cross-linker PMDI can be used. Water-borne polyurethane coatings are used for vacuum forming of PVC sheeting to ABS shells in automotive interior door panels, for the lamination of ABS/PVC film to treated polypropylene foam for use in automotive instrument panels, as metal primers for steering wheels, in flexible packaging lamination, as shoe sole adhesive, and as tie coats for polyurethane-coated fabrics. PMDI is also used as a binder for reconstituted wood products and as a foundry core binder.

Polyurethane sealant formulations use TDI or MDI prepolymers made from polyether polyols. The sealants contain 30–50% of the prepolymer; the remainder consists of pigments, fillers, plasticizers, adhesion promoters, and other additives.

TABLE 2. TYPICAL PUIR FOAM FORMULATION

Ingredients	Parts
PMDI (250 index)	208.7
Terate 203[a]	100.0
Dabco K-15	5.2
Dabco TMR 30	1.2
surfactant	2.0
HCFC 141b	35.0

[a] Crude aromatic polyester diol.

Polyurethane elastomers are either thermoplastic or thermoset polymers, depending on the functionality of the monomers used. Thermoplastic polyurethane elastomers are segmented block copolymers, comprising of hard- and soft-segment blocks. The soft-segment blocks are formed from long-chain polyester or polyether polyols and MDI; the hard segments are formed from short-chain diols, mainly 1,4-butanediol, and MDI. Thermoset polyurethanes are cross-linked polymers, which are produced by casting or reaction injection molding (RIM). For cast elastomers, TDI in combination with 3, 3'-dichloro-4, 4'-diphenylmethanediamine (MOCA) are often used.

Polyurethane engineering thermoplastics are also manufactured from MDI and short-chain glycols.

Segmented elastomeric polyurethane fibers (Spandex fibers) based on MDI have also been developed.

Recycling. The methods proposed for the recycling of polyurethanes include pyrolysis, hydrolysis, and glycolysis. Regrind from polyurethane RIM elastomers is used as filler in some RIM as well as compression molding applications. The RIM chips are also used in combination with rubber chips in the construction of athletic fields, tennis courts, and pavement of working roads of golf courses.

The use of rebound flexible foam for carpet underlay and for high load-bearing padding for furniture or for gymnasium mats is already a reality. Rebound flexible foam can also be used for sound dampening in cars. Rebounding of rigid foam particles with PMDI produces polyurethane particle boards. These boards are unaffected by water and are therefore used in furniture aboard ships. Rigid foam scrap is also used as filler in the manufacture of building products. The most convenient chemical recycling process, consists of glycolysis of solid polyurethane products. Heating of polyurethane scrap in a mixture of glycols and diethanolamine converts the cross-linked polymers into linear soluble oligomers via a transesterification process.

Health and Safety Factors

Fully cured polyurethanes present no health hazard; they are chemically inert and insoluble in water and most organic solvents. Dust can be generated in fabrication, and inhalation of the dust should be avoided. Polyether-based polyurethanes are not degraded in the human body, and are therefore used in biomedical applications. Some of the chemicals used in the production of polyurethanes, such as the highly reactive isocyanates and tertiary amine catalysts, must be handled with caution. The other polyurethane ingredients, polyols and surfactants, are relatively inert materials having low toxicity.

Isocyanates in general are toxic chemicals and require great care in handling. Respiratory effects are the primary toxicological manifestations of repeated overexposure to diisocyanates.

There are a multitude of governmental requirements for the manufacture and handling of isocyanates. The U.S. EPA mandates testing and risk management for TDI and MDI under Toxic Substance Control Administration (TSCA). Annual reports on emissions of both isocyanates are required by the EPA under SARA 313.

The liquid tertiary aliphatic amines used as catalysts in the manufacture of polyurethanes can cause contact dermatitis and severe damage to the eye. Inhalation can produce moderate to severe irritation.

Polyurethanes can be considered safe for human use. However, exposure to dust, generated in finishing operations, should be avoided. Polyurethanes are combustible. An approved fire-resistive thermal barrier must be applied over foam insulation on interior walls and ceilings. Under no circumstances should direct flame or excessive heat be allowed to contact polyurethane or polyisocyanurate foam.

Commercial Applications

The largest markets for flexible polyurethane foam are in the furniture, transportation, and bedding industries.

The bulk of the rigid polyurethane and polyisocyanurate foam is used in insulation. See also **Insulation (Thermal)**. More than half (60%) of the rigid foam consumed in 1994 was in the form of board or laminate; the remainder was used in pour-in-place and spray foam applications.

Polyurethane surface coatings are used wherever applications require abrasion resistance, skin flexibility, fast curing, good adhesion, and chemical resistance. See also **Coatings**. Synthetic leather products are also produced using a urethane binder. These poromeric materials are produced from textile-length fiber mats impregnated with DMF solutions of polyurethanes. Permeability to moisture vapor is the key property needed

in synthetic leather. In addition to shoe applications, poromerics are used for handbags, luggage, and apparel. Polyurethane films having oxygen and water permeability are applied in bandages and wound dressings and as artificial skin for burn victims.

Polyurethane elastomers are used in applications where toughness, flexibility, strength, abrasion resistance, and shock-absorbing qualities are required. Thermoplastic polyurethane elastomers and polyurethane engineering thermoplastics are molded or extruded to produce elastomeric products used as automobile parts, shoe soles, ski boots, roller skate and skateboard wheels, pond liners, cable jackets, and mechanical goods. Cast and RIM elastomers are used in auto fascia, bumper and fender extensions, printing and industrial rolls, industrial tires, and industrial and agricultural parts, such as oil well plugs and grain buckets. Elastomeric spandex fibers are used in hosiery and sock tops, girdles, brassieres, support hose, and swim wear. The use of spandex fibers in sport clothing is increasing.

HENRI ULRICH
Consultant

Additional Reading

Brandrup, J., D.R. Bloch, E.A. Grulke, E.H. Immergut, and A. Abe: *Polymer Handbook,* 2 Volume Set, 4th Edition, John Wiley & Sons, Inc., New York, NY, 2003.

Fried, J.: *Polymer Science and Technology,* 2nd Edition, Prentice Hall, Inc., Upper Saddle River, NJ, 2003.

Herrington, R. and K. Hook, eds.: *Flexible Polyurethane Foams,* Dow Chemical Company, Midland, MI, 1991.

Klempner, D., and K. Kurt, C. Frisch: *Advances in Urethane Science and Technology,* Rapra Technology Ltd., Cleveland, OH, 2002.

Kroschwitz, J.I.: *Encyclopedia of Polymer Science and Technology,* 12 Volume Set, 3rd Edition, John Wiley & Sons, Inc., Hoboken, NJ, 2004.

Oertel, G.: *Polyurethane Handbook,* 2nd Edition, Carl Hanser Publishers, Munich, Germany, 1993.

Stevens, M.P.: *Polymer Chemistry: An Introduction,* 3rd Edition, Oxford University Press, New York, NY, 1998.

Szycher, M.: *Handbook of Polyurethanes,* CRC Press LLC., Boca Raton, FL, 1999.

Ulrich, H.: *The Chemistry and Technology of Isocyanates,* John Wiley & Sons, Inc., New York, NY, 1996.

Woods, G.: *The ICI Polyurethanes Book,* John Wiley & Sons, Inc., New York, NY, 1987.

UREY, HAROLD C. (1894–1981). An American chemist who received the Nobel prize in chemistry in 1934 for his discovery of the heavy isotopes of hydrogen and oxygen. His discovery became and important factor in the development of nuclear fission and fusion and made possible the production of the transuranic element Pu. He was one was one of the leaders of the Manhattan Project, which constructed the first nuclear reactor at the University of Chicago and eventually produced the first atomic bomb. Obtaining his doctorate at the University of California in 1923, he taught at several leading universities, including Columbia where he discovered deuterium D oxide (heavy water), used as a moderator in early types of nuclear reactors. Later he devoted much study to the origin of the universe and the origin of life on earth. He was the author of many scientific treatises and made notable contributions to the cosmological theories.

See also **Nuclear Fission**; and **Nuclear Power Technology**.

URINE. The fluid secreted from the blood by the kidneys, stored in the bladder, and discharged by the urethra. In health, it is amber colored. About 1,250 milliliters of urine are excreted in 24 hours by normal humans, with specific gravities usually between 1.018 and 1.024 extremes: 1.003–1.040). Flow ranges from 0.5–20 milliliters/minute with extremes of dehydration and hydration. Maximum osmolar concentration is 1,400, compared to plasma osmolarity of 300. In diabetes insipidus, characterized by inadequate antidiuretic hormone (ADH) production, volumes of 15–25 liters/day of dilute urine may be formed. In addition to the substances listed in Table 1, there are trace amounts of purine bases and methylated purines, glucuronates, the pigments urochrome and urobilin, hippuric acid, and amino acids. In pathological states, other substances may appear: proteins (nephrosis); bile pigments and salts (biliary obstruction); glucose, acetone, acetoacetic acid and betahydroxybutyric acid (diabetes mellitus). The U/P ratios of the substances in the table vary widely because of differential handling by the kidney. Quantitative knowledge of glomerular filtration, tubular reabsorption, and secretion of these requires an understanding of the concept of renal plasma clearance.

TABLE 1. COMPOSITION OF 24-HOUR URINE IN THE NORMAL ADULT[a]

Substance	Amount (Grams)	U/P[b]
Urea	6.0–180.0 (nitrogen)	60.0
Creatinine	0.3–0.8 (nitrogen)	70.0
Ammonia	0.4–1.0 (nitrogen)	—
Uric acid	0.08–0.2 (nitrogen)	20.0
Sodium	2.0–4.0	0.8–1.5
Potassium	1.5–2.0	10.0–15.0
Calcium	0.1–0.3	—
Magnesium	0.1–0.2	—
Chloride	4.0–8.0	0.8–2.0
Bicarbonate	—	0.0–2.0
Phosphate	0.7–1.6 (phosphorus)	25.0
Inorganic sulfate	0.6–1.8 (sulfur)	50.0
Organic sulfate	0.06–0.2 (sulfur)	–

[a] Based upon data by White, Handler, Smith, and Stetten.
[b] U/P ratio = ratio of urinary to plasma concentration.

The rate at which a substance (X) is excreted in the urine is the product of its urinary concentration, U_x (milligram/milliliter), and the volume of urine per minute, V. The rate of excretion ($U_x V$) depends, among other factors, upon the concentration of X in the plasma, P_x (milligram/milliliter). It is therefore reasonable to relate $U_x V$ to P_x and this is called the clearance ratio: $(U_x \cdot V)/P_x$, or more generally, UV/P. This has the dimensions of volume and is in reality the smallest volume from which the kidneys can obtain the amount of X excreted per minute. The kidneys do not usually clear the plasma completely of X, but clear a larger volume incompletely. The clearance is therefore not a real, but a virtual volume. When substances are being cleared simultaneously, each has its own clearance rate, depending upon the amount absorbed from the glomerular filtrate or added by tubular secretion. The former will have the lower clearance, the latter the higher. Those cleared only by glomerular filtration will be intermediate, and their clearance will in effect measure the rate of glomerular filtration in milliliters/minute.

The best-known substance that can be infused into blood to provide a clearance equal to glomerular filtration rate is *inulin*, a polymer of fructose containing 32 hexose molecules (molecular weight 5,200). Strong evidence indicates that it is neither reabsorbed nor secreted, is freely filterable, is not metabolized, and has no physiological influences. Its clearance in humans is 120–130 milliliters/minute. This is taken to be the glomerular filtration rate (*GFR*) or C_F (amount of plasma water filtered through glomeruli/minute). Besides inulin in the dog and other vertebrates, creatinine, thiosulfate, ferrocyanide, and mannitol also fulfill these requirements.

Knowing the glomerular filtration rate permits quantification of the amount of any substance freely filtered (C_F(milliliters/minute) × P_x(milligrams/milliliter)). Subtracting from this one minute's excretion, $U_x V$, would give the amount reabsorbed in milligrams/minute. A classical example is the glucose mechanism. At normal plasma concentrations, none or a trace appears in the urine. When plasma glucose is elevated to about 180–200 milligram percent (the "threshold"), the amount appearing in the urine begins to increase. As concentration is raised more, the nephrons become progressively saturated until the rate of reabsorption becomes constant and maximal. This indication of saturation of the transport system is referred to as the T_m ("tubular maximum—T_{mG}"). In humans, T_{mG} has the value of 340 milligrams/minute. Absorption occurs in the proximal convoluted tubules.

Additional Reading

McBride, L.J.: *Textbook of Urinalysis and Body Fluids: A Clinical Approach,* Lippincott-Raven Publishers, Philadelphia, PA, 1997.

Ringsrud, K.M. and J.J. Linne: *Urinalysis and Body Fluids: A Colortext and Atlas,* Mosby-Year Book, Inc., St. Louis, MO, 1995.

Stamey, T.A. and R.W. Kindrachuk: *Urinary Sediment and Urinalysis: A Practical Guide for the Health Science Professional,* W.B. Saunders Company, Philadelphia, PA, 1996.

URONIC ACID. Any of a class of compounds similar to sugars but differing from them in that the terminal carbon has bee oxidized from an alcohol to carboxyl group. The most common are galacturonic acid and glucuronic acid.

V

VACCINE TECHNOLOGY. A vaccine is a preparation used to prevent a specific infectious disease by inducing immunity in the host against the pathogenic microorganism. The practice is also called immunization.

With the discovery and widespread use of antibiotics, beginning in the 1950s, the interest in vaccine research disappeared. It was anticipated that infectious diseases would no longer be a threat to human health. This expectation turned out to be too optimistic. Today, there are still numerous infection diseases, for which antibiotic has not been effective. The development of biotechnology and modern immunology created new opportunities for producing new antigens and vaccine research has become a primary focus in recent years. As a result, several vaccines such as Hepatitis B, Hepatitis A, *H. influenza*, and Varicella have been approved. A new vaccine against pertussis has been recently approved in the U.S.

Commercial Vaccines

Vaccines can be roughly categorized into killed vaccines and live vaccines. A killed vaccine can be (*1*) an inactivated, whole microorganism such as pertussis, (*2*) an inactivated toxin, called toxoid, such as diphtheria toxoid, or (*3*) one or more components of the microorganism commonly referred to as subunit vaccines.

Vaccines for human use are regulated by the FDA in the U.S. and Boards of Health in other countries. The manufacturing of vaccines requires adherence to strict current good manufacturing practices (cGMPs) and in the U.S. licenses for both the process and the facility where the vaccine is produced are required. The Center for Biologics Evaluation and Research (CBER); (http://www.fda.gov/cber/) is the branch of the FDA that regulates vaccines. Basic requirements are described in The Code of Federal Regulations (CFR):http://www.gpoaccess.gov/cfr/index.html

Vaccines for the General Population

Vaccines in this category protect children and adults from polio, diphtheria, tetanus, pertussis (whooping cough), measles (rubeola), mumps, rubella (German measles), hepatitis B, and haemophilus disease (meningitis, epiglotitis).

Poliomyelitis. Two vaccines are licensed for the control of poliomyelitis in the United States. The live, attenuated oral polio virus (OPV) vaccine can be used for the immunization of normal children. The killed or inactivated vaccine is recommended for immunization of adults at increased risk of exposure to poliomyelitis and of immunodeficient patients and their household contacts. Both vaccines protect against the three serotypes of poliomyelitis that cause disease.

Diphtheria, Tetanus, and Pertussis. These vaccines in combination (DTP) have been routinely used for active immunization of infants and young children since the 1940s. The recommended schedule calls for immunizations at 2, 4, and 6 months of age with boosters at 18 months and 4–5 years of age. Since 1993 these vaccines have been available in combination with a vaccine that protects against *Haemophilus* disease, thus providing protection against four bacterial diseases in one preparation. A booster immunization with diphtheria and tetanus only is recommended once every 10 years after the fifth dose.

Measles, Mumps, Rubella. Live, attenuated vaccines are used for simultaneous or separate immunization against measles, mumps, and rubella in children from around 15 months of age to puberty. Two doses, one at 12–15 months of age and the second at 4–6 or 11–12 years are recommended in the U.S.

Haemophilus influenza serotype b. Three vaccines are available for immunizing infants. Two of these vaccines are administered at 2, 4, and 6 months of age with a booster given at 12–15 months of age, and the third vaccine is administered at 2 and 4 months of age with a booster at 12–15 months of age.

Hepatitis B. Although Hepatitis B (Hep B) is not an infant disease, it is recommended for infant immunization to better control spread, because compliance with vaccine immunization programs is easier to achieve in an infant population. Infants receive immunizations at birth, 1–2 months, and a third dose at 6 months. Other schedules are available for immunization of adolescents and adults who have not previously received the vaccine.

Varicella. The varicella (chicken pox) vaccine was approved in April 1995 for immunization of children. A single dose at one year of age is recommended. In the future it may be combined with measles, mumps, and rubella.

Vaccines for Special Populations

Two vaccines that are in fairly widespread use in the adult population are vaccines that prevent viral influenza and pneumococcal pneumonia.

Influenza. The ACIP recommends annual influenza vaccination for all persons who are at risk from infections of the lower respiratory tract and for all older persons. Influenza viruses types A and B are responsible for periodic outbreaks of febrile respiratory disease.

Pneumococcal Polysaccharide. Pneumococcal polysaccharide vaccine may be used for immunization of persons two years of age or older who are at increased risk of pneumococcal disease.

Vaccines Being Developed

Despite the tremendous advances since the 1960s in the biomedical fields, there remains a large number of diseases that are endemic in many parts of the world. The Third World or developing countries bear the brunt of several of these, e.g., malaria, trypanosomiasis, and schistosomiasis. In developed countries, diseases such as herpes and gonorrhea are becoming increasingly prevalent. Some of these vaccines have been developed and licensed, whereas good progress is advancing in other areas. In the meantime, emerging exotic viruses such as HIV and drug-resistant pathogens continue to appear. There is an urgent need to expand vaccine R&D.

Meningitis. *Haemophilus influenze*, type b (Hib), *Streptococcus pneumoniae*, and *Neisseria meningitidis* are the major cause of meningitis in infants. Vaccines against Hib disease prepared using conjugate technology have been in use worldwide, and have been efficacious in eliminating the disease from the population. This same technology is being applied to the development of vaccines for *S. pneumoniae* and *N. meningitidis*.

S. pneumoniae has more than 80 sero-types. The current polysaccharide vaccine consists of 23 serotypes and covers about 87% of all pneumococcal diseases in the United States. Current vaccine development is based on conjugate technology and concentrates on the most prevalent 7–9 serotypes. Three multivalent vaccine candidates are in clinical trials. All are based on conjugating the polysaccharide to a T-dependent protein carrier. The results of phase I and II trials in infants have demonstrated the safety and immunogenicity of these vaccines. Phase III trials to demonstrate efficacy are in progress and final approval of this vaccine for infant immunization will be by the year 2000.

N. meningititidis also has several groups and serotypes. Most of the diseases are caused by groups A, B, and C. A multivalent polysaccharide vaccine consisting of types A, C, Y, and W$_{135}$ is available. However, like other polysaccharide vaccines, it is not immunogenic in infants. Conjugate vaccines against groups A and C are being developed, using different protein carries and conjugate chemistries. Clinical trials of these vaccines are in progress.

The capsular polysaccharide of group B meningococcus is not immunogenic in humans. Thus, a conjugate vaccine of the group B polysaccharide will not improve its efficacy, and this remains a major challenge in developing the vaccine against group B organisms.

Rotavirus. Rotavirus causes infant diarrhea, a disease which has major socio-economic impact. In developing countries it is the major cause of death in infants worldwide. In the U.S., diarrhea is still a primary cause of physician visits and hospitalization, although the mortality rate is relatively low. Two membrane proteins (VP4 and VP7) of the virus have been identified as protective epitopes and most vaccine development programs are based on these two proteins as antigens. Both live attenuated vaccines and subunit vaccines are being developed. By using the technique of viral gene re-assortant, a multivalent live attenuated vaccine for rotavirus has been developed. The vaccine candidates are generated by transferring the VP7 gene from a human rotavirus to Rhesus monkey (or bovine) rotavirus. On August 31, 1998 the FDA licensed a tetravalent rhesus rotavirus vaccine for use in infant. It will be well into the twenty-first century before a vaccine will be available.

Respiratory Syncytial Virus. RSV causes severe lower respiratory tract disease in infants. It is the major cause of hospitalization in the U. S. and it has a high mortality rate in neonates and other high risk populations, such as the geriatric population.

Both subunit and live, attenuated vaccine approaches are being developed for RSV. A candidate subunit vaccine based on the surface (F-) protein is being tested. Live attenuated vaccines for RSV are also being developed.

Parainfluenza. Parainfluenza viruses (PIV) also causes viral pneumonia in infants. It is similar to RSV, therefore similar approaches are being used for developing a vaccine. A live attenuated PIV-3 vaccine has been in clinical trial.

Otitis Media. Otitis media is thought to be caused by several bacteria, significantly *S. pneumoniae*, nontypable *H. influenza*, and *Moraxella catarrhalis*. Viruses such as influenza, RSV, and PIV may also play a role in the disease. The use of a pneumococcal vaccine is the first step in the development of an otitis media vaccine. Vaccines against nontypable *H. influenza* and *Moraxella* are at the development stages. Both vaccine candidates are derived from the exposed proteins of the bacteria.

Herpes Simplex. There are two types of herpes simplex virus (HSV) that infect humans. Type I causes orofacial lesions and 30% of the U.S. population suffers from recurrent episodes. Type II is responsible for genital disease and anywhere from $3 \times 10^4 - 3 \times 10^7$ cases per year (including recurrent infections) occur. The primary source of neonatal herpes infections, which are severe and often fatal, is the mother infected with type II. In addition, there is evidence to suggest that cervical carcinoma may be associated with HSV-II infection. Vaccine development is hampered by the fact that recurrent disease is common. Thus, natural infection does not provide immunity and the best method to induce immunity artificially is not clear. A much better understanding of the pathogenesis of the virus and virus-host interactions are required for the efficient development of the vaccine.

Influenza. Although current influenza vaccine (subunit split vaccine) has been in use yearly for the elderly, it is not recommended for the general population or infants. Improvements to increase or prolong the immunogenicity, reduce the side-effects (due to egg production procedure), and provide mass protection are still being pursued. One approach is to use a live, attenuated virus though cold adaptation. Subunit vaccines based on the surface proteins of virus are also being explored. It has been demonstrated that the two major protective antigens are haemagglutinin (HA) and neuraminidase (NA). The genes for these antigens have been cloned and expressed in baculovirus in insect cell culture.

Malaria. Malaria infection occurs in over 30% of the world's population and almost exclusively in developing countries. The majority of the disease in humans is caused by four different species of the malarial parasite. Vaccine development is problematic for several reasons. The parasites have a complex life cycle; malaria is difficult to grow in large quantities outside the natural host. Despite these difficulties, vaccine development has been pursued for many years. An overview of the state of the art is available.

Gonorrhea. Gonorrhea, caused by *Neisseria gonorrheae*, is the most commonly reported communicable disease in the United States. An increasing number of strains are becoming resistant to penicillin, the antibiotic that is usually used to treat this disease. Development of a vaccine is problematic because natural infection does not necessarily provide immunity. Studies are being carried out on various structural components of the gonococcal bacterium, including pili, outer membrane proteins, lipopolysaccharide, and the outer capsule, in an effort to develop a vaccine. One of the more promising approaches involves a vaccine made with pili. Human studies indicate that a pili vaccine stimulates antibody formation that is 50–100 times the prevaccination level and is effective in preventing disease after challenge.

Human Immunodeficiency Virus. HIV causes Acquired Immunodeficiency Syndrome (AIDS). HIV infects the cells of the human immune system, such as T-lymphocytes, monocytes, and macrophages. After a long period of latency and persistent infection, it results in the progressive decline of the immune system, and leads to full-blown AIDS, resulting in death.

The development of vaccine against HIV-1 has been a top priority of the national public health agencies and medical research institutes. The effort in developing a vaccine has not been as successful as expected. The main problem is the tremendous antigenic variability of the virus. An antigen derived from the cultured strain might not be the same as the clinical strain. Another problem is the fact that the virus infects the cells of the human immune system, making the design of the vaccine more complex. It will require certain combinations of immune responses to provide long-term protection or eliminate the virus from the host. So far, the proper immune mechanism for achieving this goal has not been identified, although it is generally agreed that a cell-mediated immune response (CMI) is essential. Up to the 1990s, most of the vaccine candidates have been derived from the surface proteins of the virus. Although these candidates all show immunogenicity and are protective in animal models, clinical studies of these proteins have not been able to demonstrate protection against disease. Efforts in development of the vaccine are being continued in the public and private research institutes.

Other Vaccines

There are many other diseases which do not have effective vaccines. These diseases are mostly regional in nature, epidemic in the developing world. Vaccines against parasites are also becoming critical to public health. Vaccines are being developed for Lyme disease, dengue, *Helicobacter pylori*, Japanese encephalitis, Equine encephalitis, Tickborne encephalitis, cholera, shigellas, schistosomiasis, group B streptococcus, and other sexually transmitted diseases.

Future Technology

Vaccines for many diseases are unavailable because of an inability to determine the appropriate method for vaccination or difficulty in obtaining large quantities of antigens. Advances in medical science and immunology have substantially improved the understanding of the design and delivery of antigens. Genetic engineering offers further advances in providing the techniques for construction and production of large quantities of antigens. Development of these fields has been responsible for the rapid advances of vaccinology. Development of new vaccines also requires different process technology for the production of antigens and preparation of delivery system for vaccines.

Genetic Engineering. Genetic engineering involves preparation of DNA fragments (passengers) coding for the substance of interest, inserting the DNA fragments into vectors (cloning vehicles), and introducing the recombinant vectors into living host cells where the passenger DNA fragments replicate and are expressed, i.e., transcribed and translated, to yield the desired substance.

Since the 1970s, genetic engineering has evolved to become the most powerful and routine tool in the study of immunology and the development of new vaccines. It offers new, and in some instances safer and more effective methods for production of vaccines of higher quality. It has allowed an efficient way for the study of construction of new attenuated live viral or bacterial vaccines. It can also be used to study the pathogenicity and immunology of viruses or bacteria. Recombinant hepatitis B vaccine is the first approved human vaccine based on a genetic engineering technique. The genetic engineering techniques can also be used to reduce the virulence of a pathogen which can then be used to produce vaccines. Live vectors are another application of genetic engineering. In this case, the genes from a pathogen are inserted into a vaccine vector, such as salmonella or vaccinia.

The use of naked DNA as a vaccine is the most recent development in this field. Since the demonstration of the possibility of genetic immune response by direct injection of DNA into muscle cells, the field is developing rapidly. Clinical trials for influenza, hepatitis, HIV-1, and herpes simplex are being initiated.

Adjuvants. Adjuvants are substances which can modify the immune response of an antigen. With better understanding of the functions of different arms of the immune system, it is possible to explore the effects of an adjuvant, such that the protective efficacy of a vaccine can be improved. At present, aluminum salt is the only adjuvant approved for use in human vaccines.

Peptide Vaccines. Development of a peptide vaccine is derived from the identification of the immunodominant epitope of an antigen. A polypeptide based on the amino acid sequence of the epitope can then by synthesized. Preparation of a peptide vaccine has the advantage of allowing for large-scale production of a vaccine at relatively low cost. It also allows for selecting the appropriate T- or B-cell epitopes to be included in the vaccine, which may be advantageous in some cases.

Process Technology

In the preparation of classical killed or toxoid vaccines, simple process technology was used. With the advance of new vaccines, far more sophisticated process technologies are needed. The desire to reduce side effects of vaccination requires processes which will yield antigens of extreme purity. The new regulation in cGMP requires consistent production procedures, and global competition also demands that the most efficient process technology be applied.

The basic process technology in vaccine production consists of fermentation for the production of antigen, purification of antigen, and formulation of the final vaccine. In bacterial fermentation, technology is well established. For viral vaccines, cell culture is the standard procedure. Different variations of cell line and process system are in use. For most of the live viral vaccine and other subunit vaccines, production is by direct infection of a cell substrate with the virus. Alternatively, some subunit viral vaccines can be generated by rDNA techniques and expressed in a continuous cell line or insect cells.

Development of conjugate and peptide vaccines requires the typical organic synthesis process and purification. This is a new area for vaccine technologists. Again, the main concern is to maintain the immunogenicity of the vaccine candidate during the chemical reaction and purification steps. Most of these procedures are proprietary.

Economic Aspects

Costs of vaccine manufacture vary according to the type of vaccine produced and how it is supplied. Live virus vaccines are generally less expensive because the quantitative mass to be given to the recipient is less than an inactivated or subunit vaccine. The purification process and yield and the number of strains or components in any given vaccine also affect the cost of manufacture. New vaccines often have a royalty cost, in addition to manufacturing and testing costs. Filling and packaging is often the most expensive part of the manufacturing process and the cost varies by how many doses are filled and packed into one unit.

Another important aspect of vaccine technology is the cost–benefit relationship between prevention vaccination and disease treatment. Generally the cost savings are high.

Liability for adverse reaction events associated in time with immunization have also played a principal role in vaccine economics. Prior to 1988, compensation for any adverse reaction associated in time with vaccination required that the vaccine recipient bring suit against the manufacturer or the health care provider that administered the vaccine. The uncertainty of numbers and costs associated with lawsuits contributed to the decline in the number of providers of routine childhood vaccines. The enactment in 1988 of the National Vaccine Injury Compensation Program was provided as a nonfault alternative to the tort system for resolving claims resulting from adverse reactions to mandated childhood vaccines, and has achieved its goal of providing compensation to those injured by rare adverse events associated with vaccination and providing some stability for the vaccine market.

CHIA-LUNG HSIEH
MARY B. RITCHEY
Wyeth-Lederle Vaccine and Pediatrics

Additional Reading

Ada, G.L. and A.J. Ramsay: *Vaccines, Vaccination and the Immune Response,* Lippincott-Raven Publishers, Philadelphia, PA, 1996.

Bazin, H. and E. Jenner: *The Eradication of Smallpox,* Academic Press, Inc., San Diego, CA, 2000.

Eby, R.: in M. Powell and M. Newman, eds., *Vaccine Design: The Subunit and Adjuvant Approach,* Plenum Press, New York, NY, 1995, Chapt. 31.

Ellis, R.: "The Application of rDNA Technology to Vaccines," in S.A. Plotkin and B. Fantini, eds., *Vaccinia, Vaccination and Vaccinology,* Elsevier, Paris, 1996.

Perez Tirse, J. and P.A. Gross: *Pharmaco. Economics,* **2**(3), 198 (1992).

Plotkin, S.A. and E.A. Mortimer: *Vaccines,* 2nd Edition, W. B. Saunders Co., Philadelphia, PA, 1994.

Kiyono, H., P.L. Ogra, and J.R. McGhee: *Mucosal Vaccines,* Academic Press, Inc., San Diego, CA, 1996.

Levine, M.M., G.C. Woodrow, J.B. Kaper, and G.S. Cobon: *New Generation Vaccines,* 2nd Edition, Marcel Dekker, Inc., New York, NY, 1997.

O'Hagan, D.T.: *Vaccine Adjuvants: Preparation Methods and Research Protocols,* Vol. 42, Humana Press, Totowa, NJ, 2000.

Orenstein, W.A. and R. Zorab: *Vaccines,* Harcourt Brace & Company, San Diego, CA, 1999.

Stanberry, L.R. and D. Bernstein: *Sexually Transmitted Diseases: Vaccines, Prevention and Control,* Academic Press, Inc., San Diego, CA, 2000.

Web References

American Public Health Association (APHA): http://www.apha.org/

Food and Drug Administration (FDA): Center for Biologics Evaluation and Research: http://www.fda.gov/cber/

Institute for Vaccine Safety: Johns Hopkins School of Public Health: http://www.vaccinesafety.edu/

National Institute of Allergy and Infectious Diseases: http://www.niaid.nih.gov

National Institute of Health (NIH): http://www.nih.gov

The Centers for Disease Control and Prevention: National Immunization Program: http://www.cdc.gov/nip/

The Immunization Gateway: Your Vaccine Fact-Finder: http://www.immu nofacts.com/

VACUUM. According to definition, a space entirely devoid of matter. The term is used in a relative sense in vacuum technology to denote gas pressures below the normal atmosphere pressure of 760 torr (1 torr = 1 millimeter of mercury). The degree or quality of the vacuum attained is indicated by the total pressure of the residual gases in the vessel that is pumped. Table 1 shows generally accepted terminology for denoting various degrees of vacuum, together with pertinent pressure ranges; the calculated molecular density (from the equation $p = nkT$, where p is the pressure n is the molecular density, i.e., number of molecules per cubic centimeter; k is the Boltzmann's constant; and T is the absolute temperature taken to be 293 K (20°C); and the mean free path λ from the approximate equation for air: $\lambda = 5/p$ centimeters, where p is the pressure in millitorr).

In the quantum field theories that describe the physics of elementary particles, the vacuum becomes somewhat more complex than previously defined. Even in empty space, matter can appear spontaneously as a result of fluctuations of the vacuum. It may be pointed out, for example, that an electron and a positron, or antielectron, can be created out of the void. Particles created in this way have only a fleeting existence; they are annihilated almost as soon as they appear, and their pressure can never be detected directly. They are called *virtual particles* in order to distinguish them from real particles. Thus, the traditional definition of vacuum (space with no real particles in it) holds. In their excellent paper, the aforementioned authors discuss how, near a superheavy atomic nucleus, empty space may become unstable, with the result that matter and antimatter can be created without any input of energy. The process may soon be observed experimentally.

Even when all matter and heat radiation have been removed from a region of space, the vacuum of classical physics remains filled with a distinctive pattern of electromagnetic fields. The discovery of a connection between thermal radiation and the structure of the classical vacuum reveals

TABLE 1. VARIOUS QUALITIES OF VACUUM AND PRESSURE RANGES

Quality of vacuum	Pressure range (torr)	Molecular density, n (molecules/cubic centimeter)	Mean free path, λ (centimeters)
Coarse or rough vacuum	760–1	$2.69 \times 10^{19} – 3.5 \times 10^{16}$	$6.6 \times 10^{-6} – 5 \times 10^{-3}$
Medium vacuum	$1–10^{-3}$	$3.5 \times 10^{16} – 3.5 \times 10^{13}$	$5 \times 10^{-3} – 5$
High vacuum	$10^{-3} – 10^{-7}$	$3.5 \times 10^{13} – 3.5 \times 10^{9}$	$5 – 5 \times 10^{4}$
Very high vacuum	$10^{-7} – 10^{-9}$	$3.5 \times 10^{9} – 3.5 \times 10^{7}$	$5 \times 10^{4} – 5 \times 10^{6}$
Ultrahigh vacuum	$<10^{-9}$	$<3.5 \times 10^{7}$	$>5 \times 10^{6}$

an unexpected unity in the laws of physics, but it also complicates the view of what was once considered simply *empty space*. But, even with its pattern of electric and magnetic fields, the vacuum remains the simplest state of nature. Perhaps this statement reflects more on the subtlety of nature than it does on the simplicity of the vacuum.

Vacuum Pumps

Two widely used vacuum pumps are the mechanical rotary oil-sealed pump and the vapor pump. The former provides a medium vacuum and works relative to the atmosphere. The vapor pump, on the other hand, provides a high or very high vacuum and operates relative to a medium vacuum provided by a rotary pump, referred to as a backing pump in this connection. Thus, the most widely used high-vacuum system able to establish an ultimate pressure of about 10^{-6} torr or below consists of a vapor pump backed by a rotary pump.

Several patterns of rotary oil-sealed pumps exist, but they have in common the fact that the volume between a rotor (or rotating plunger) and a stator is divided into two crescent-shaped sections, which are isolated from one another as regards the passage of gas. Further, they are furnished with an intake port and a discharge outlet valve to the atmosphere. One revolution of the rotor (speeds of 450 to 700 revolutions per minute are used), gas is swept from the intake port, compressed, and discharged to the atmosphere via the one-way outlet valve. The mechanism is immersed in a low-vapor-pressure oil for sealing and lubrication in a small pump; larger units have a separate oil reservoir and feed device. A spring-loaded vane type of rotary oil-sealed pump is shown in Fig. 1. A single-stage pump of this kind provides an ultimate pressure of about 10^{-2} torr; a two-stage one with two units in cascade will give an ultimate of about 10^{-4} torr. Rotary pumps with speeds from 20 to 20,000 liters/minute are commercially available, the smallest being driven by an 1/8-hp motor, the largest requiring a 40-hp motor.

These pumps handle permanent gases efficiently. Condensable vapors, e.g., water vapor, are not satisfactorily pumped because they may liquefy during the compression part of the rotation. To prevent this, gas ballast is a common provision whereby air from the atmosphere is admitted to the pump through a simple, adjustable screw valve to the region between the rotor and stator just before the discharge outlet valve. The amount of extra air admitted is readily adjusted to provide a compressed gas-vapor mixture, which opens the discharge valve before vapor condensation occurs. Gas ballasting will clearly increase significantly the ultimate pressure provided by the pump, but this is not important since the gas-ballast valve can be closed after initial pumping has removed most of the water vapor.

Vapor pumps are of two main types: vapor diffusion pumps and vapor ejector pumps. Both employ vapor (of either mercury or a low-vapor-pressure oil) issuing from a jet as a means of driving gas in the direction from the intake port to the discharge outlet, which is maintained at a medium vacuum by a backing rotary pump. In the diffusion pump (a two-stage design utilizing oil as the pump fluid), the vapor issuing from the top, first-stage jet is directed downward towards the backing region. Gas molecules from the intake port diffuse into the streaming vapor. The directed oil molecules collide with the gas molecules to give them velocity components toward the backing region. A large pressure gradient is thereby established in the pump so that the intake pressure may be over 100,000 times less than the backing pressure. The intake pressure may therefore be 10^{-6} torr or lower with a backing pressure of 10^{-1} torr.

In the diffusion pump, the vapor stream is not essentially influenced by the gas pumped. In the vapor ejector pump, however, the vapor stream is enabled by a higher boiler pressure to be denser and of greater speed with a higher intake pressure, so that the gas is entrained by the high-speed vapor. Viscous drag and turbulent mixing now carry the gas at initially supersonic speeds down a pump housing of diminishing cross section. The ejector pump is designed to operate with a maximum pumping speed at an intake pressure of 10^{-1} to 10^{-3} torr and with a backing pressure of 0.5 to 1 torr or more. The diffusion pump, on the other hand, is designed to have a fairly constant speed from 10^{-3} torr down to an ultimate 10^{-6} torr or much lower in a modern, bakeable stainless steel system.

An important mechanical pump, which operates in the same pressure region as the oil ejector, is the Roots pump, capable of very great speeds and requiring backing by a rotary oil–sealed pump.

A vacuum system may consist of a diffusion pump and a backing pump, together with baffles, cold traps and isolation valves. The cold trap is essential if a mercury vapor diffusion pump is used and is best filled with liquid nitrogen ($-196°C$); otherwise, the system will be exposed to the mercury vapor pressure, which is 10^{-3} torr at $18°C$.

Ultrahigh-vacuum systems with stainless steel traps and metal sealing gaskets, and bakeable (except for the pumps) for several hours to $450°C$, may be constructed to provide an ultimate pressure of 10^{-9} to 10^{-10} torr.

Other vacuum pumps include the sorption type based on the high gas take-up of charcoal or molecular sieve material at liquid nitrogen temperatures. Sorption pumps may be used in place of rotary pumps, with a desirable freedom from rotary pump oil vapor, especially in systems where the amount of gas to be handled is limited.

The chief rival to the vapor diffusion pump at present is the getterion pump of the Penning discharge type, sometimes called the sputter-ion pump, with electrodes of titanium metal. The principle of operation is illustrated by Fig. 2 where an egg-box type anode is situated between plane cathodes. The anode-cathode operating potential difference is of the order of 2 to 10 kV, and the magnetic flux density is about 3,000 gauss. The chief pumping action with active gases such as hydrogen, nitrogen and oxygen is to the anode which receives deposited titanium (which has very high gas affinity) sputtered from the cathodes under the action of the positive ion bombardment. Some gas, especially the inert gases like argon, is pumped to the cathodes.

A typical multicell pump of moderate size of this type has a pumping speed of about 250 liters/second. Much larger pumps with speeds of up to 5,000 liters/second are commercially available, as are small single-cell units with speeds of some 2 liters/second.

The sputter-ion pumps provide a vapor-free system giving a so-called dry vacuum, and they are often incorporated in plant with molecular sieve sorption as the backing pump. For medium-size laboratory plant able to provide ultrahigh vacuum they are most attractive. Probably their chief

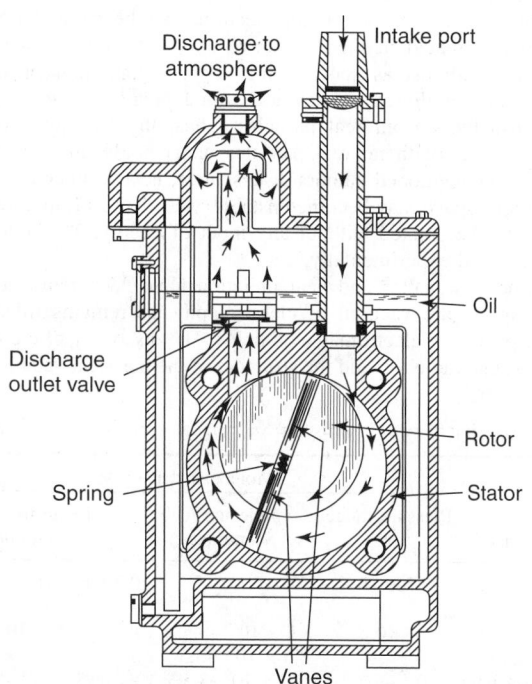

Fig. 1.   Spring-loaded vane-type rotary oil-sealed vacuum pump

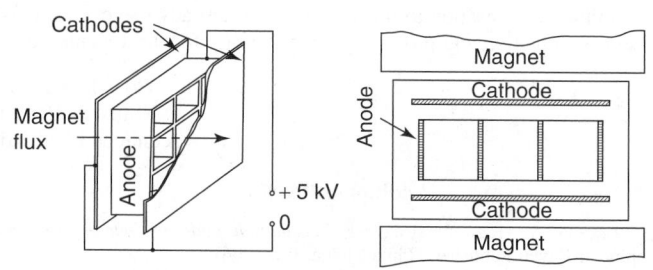

Fig. 2.   Sputter-ion vacuum pump

disadvantage is that the life of the pump is only about 40 hours at 10^{-3} torr, but this increases inversely with the pressure, so that it is 40,000 hours at 10^{-6} torr. At present, they are therefore strong rivals to the diffusion pump for plant where moderate amounts of gas are handled in the lower pressure ranges.

In cryogenic pumping, the provision is, within an initially evacuated system, a surface which is at such a low temperature that gas impinging on the surface will be condensed. For example, if the surface is maintained at the temperature of liquid helium ($-269°C$), all other gases have insignificantly low vapor pressures at this temperature and molecules of these gases impinging on the surface would remain there. A pumping speed for nitrogen of nearly 12 liters sec^{-1} cm^{-2} of cooled surface is hence theoretically possible. Liquid nitrogen, together with molecular sieve and other sorbent surfaces, and also liquid-hydrogen ($-253°C$, at which the vapor pressure of solid nitrogen is 10^{-10} torr) and liquid-helium cooled metallic surfaces, are being actively investigated for the possibility of providing very high pumping speeds (10^6 liters/sec is not out of question) in space simulators and other plants.

Vacuum Measurement

Subatmospheric pressure usually is expressed in reference to perfect vacuum or absolute zero pressure. Like absolute zero temperature (the concept is analogous), absolute zero pressure cannot be achieved, but it does provide a convenient reference datum. Standard atmospheric pressure is 14.695 psi absolute, 30 inches of mercury absolute, or 760 mmHg of density 13.595 g/cm³ where acceleration due to gravity is $g = 980.665$ cm/s². 1 mmHg, which equals 1 torr, is the most commonly used unit of absolute pressure. Derived units, the million or micrometer, representing 1/1000 of 1 mmHg or 1 torr, are also used for subtorr pressures.

In the MKS system of units, standard atmospheric pressure is 750 torr and is expressed as 100,000 Pa (N/m²) or 100 kPa. This means that 1 Pa is equivalent to 7.5 millitorr (1 torr = 133.3 pascal). Vacuum, usually expressed in inches of mercury, is the depression of pressure below the atmospheric level, with absolute zero pressure corresponding to a vacuum of 30 inches of mercury.

When specifying and using vacuum gages, one must constantly keep in mind that atmospheric pressure is *not* constant and that it also varies with elevation above sea level.

Vacuum gages can be either direct or indirect reading. Those that measure pressure by calculating the force exerted by incident particles of gas are direct reading, while instruments that record pressure by measuring a gas property that changes in a predictable manner with gas density are indirect reading.

The range of operation for these two classes of vacuum instruments is given in Table 2. Since the pressure range of interest in present vacuum technology extends from 760 to 10^{-13} torr (over 16 orders of magnitude), there is no single gage capable of covering such a wide range. The ranges of vacuum where specific types of gages are most applicable are shown in Fig. 3.

TABLE 2. RANGE OF OPERATION OF MAJOR VACUUM GAGES

Principle	Gage type	Range, torr
Direct reading	Force measuring:	
	Bourdon, bellows, manometer (oil and mercury),	$760-10^{-6}$
	McLeod capacitance (diaphragm)	760×10^{-6}
Indirect reading	Thermal conductivity:	
	Thermocouple (thermopile)	$10-10^{-3}$
	Pirani (thermistor)	$10-10^{-4}$
	Molecular friction	$10^{-2}-10^{-7}$
	Ionization:	
	Hot filament	$10-10^{-10}$
	Cold cathode	$10^{-2}-10^{-15}$

The operating principles of some vacuum gages, such as liquid manometers, bourdon, bellows, and diaphragm gages involving elastic members, were described earlier in this article. The remaining vacuum measurement devices include the thermal conductivity (or Pirani and thermocouple-type gages), the hot-filament ionization gage, the cold-cathode ionization gage (Philips), the spinning-rotor friction gage, and the partial-pressure analyzer.

Pirani or Thermocouple Vacuum Gage. Commercial thermal conductivity gages ordinarily should not be considered as precision devices. Within their rather limited but industrially important pressure range, they are outstandingly useful. The virtues of these gages include low cost, electrical indication readily adapted to remote readings, sturdiness, simplicity, and interchangeability of sensing elements. They are well adapted for uses where a single power supply and measuring circuit is used with several sensing elements located in different parts of the same vacuum system or in several different systems.

The working element of the gages consists of a metal wire or ribbon exposed to the unknown pressure and heated by an electric current. See Fig. 4. The temperature attained by the heater is such that the total rate of heat loss by radiation, gas convection, as thermal conduction, and thermal conduction through the supporting leads equals the electric power input to the element. Convection is unimportant and can be disregarded, but the heat loss by thermal conduction through the gas is a function of pressure. At pressures of approximately 10 torr and higher, the thermal conductivity of a gas is high and roughly independent of further pressure increases. Below about 1 torr, on the other hand, the thermal conductivity decreases with decreasing pressure, eventually in linear fashion, reaching zero at zero pressure. At pressures above a few torr, the cooling by thermal conduction limits the temperature attained by the heater to a relatively low value. As the pressure is reduced below a few hundred millitorr, the heater temperature rises, and at the lowest pressures, the heater temperature reaches an upper value established by heat radiation and by thermal conduction through the supporting leads.

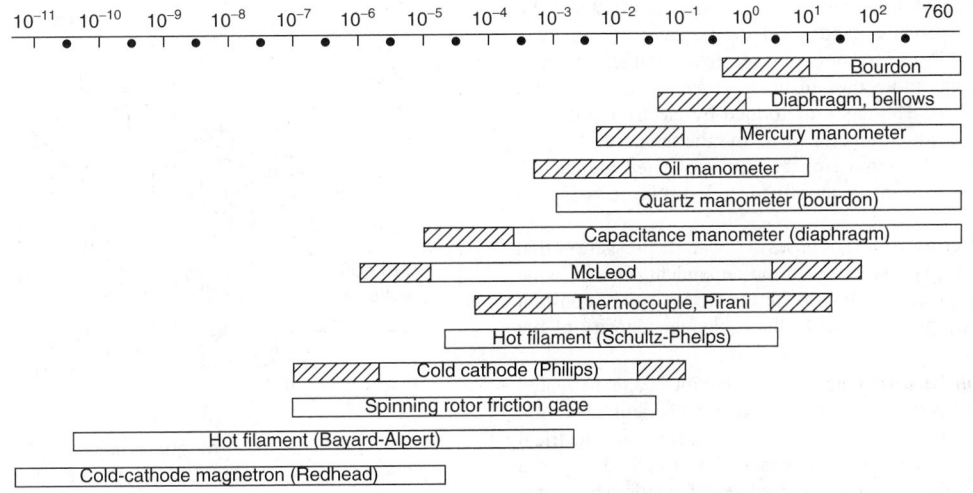

Fig. 3.   Vacuum gages versus measurement range

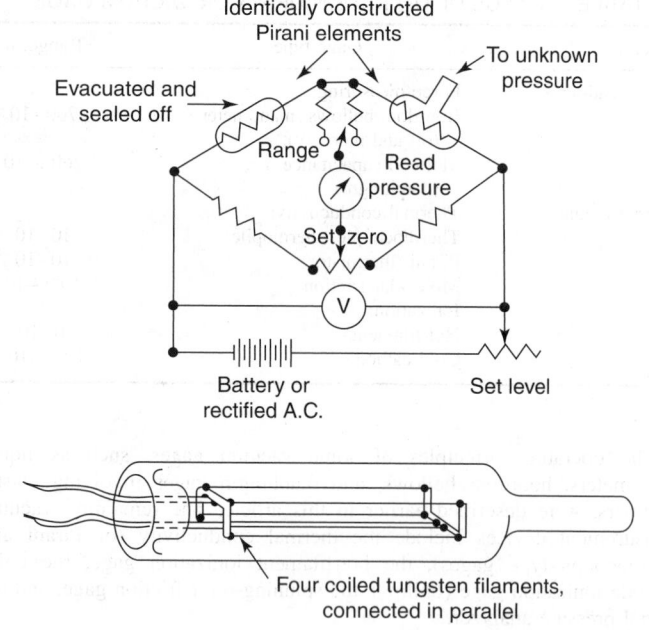

Fig. 4.   Pirani gage: (*Top*) Gage in fixed-voltage Wheatstone bridge. (*Bottom*) Sensing element

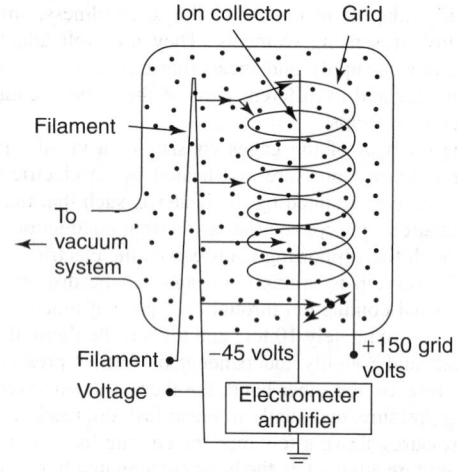

Fig. 5.   Hot-filament ionization gage. (*Bayard-Alpert type*)

Hot-Filament Ionization Vacuum Gage. This gage is the most widely used pressure-measuring device for the region from 10^{-2} to 10^{-11} torr. The operating principle of this gage is illustrated in Fig. 5.

A regulated electron current (typically about 10 mA) is emitted from a heated filament. The electrons are attracted to the helical grid by a dc potential of about +150 volts. In their passage from filament to grid, the electrons collide with gas molecules in the gage envelope, causing a fraction of them to be ionized. The gas ions formed by electron collisions are attracted to the central ion collector wire by the negative voltage on the collector (typically −30 volts). Ion currents collected are on the order of 100 mA/torr. This current is amplified and displayed using an electronic amplifier.

This ion current will differ for different gases at the same pressure—that is, a hot-filament ionization gage is composition-dependent. Over a wide range of molecular density, however, the ion current from a gas of constant composition will be directly proportional to the molecular density of the gas in the gage.

Cold-Cathode Ionization Vacuum Gage. This ingenious gage, in vented by Penning, possesses many of the advantages of the hot-filament ionization gage without being susceptible to burnout. Ordinarily, an electrical discharge between two electrodes in a gas cannot be sustained below a few millitorr pressure. To simplify a complicated set of relationships, this is because the "birthrate" of new electrons capable of sustaining ionization

is smaller than the "death rate" of electrons and ions. In the Philips gage this difficulty is overcome by the use of a collimating magnetic field, which forces the electrons to traverse a tremendously increased path length before they can reach the collecting electrode. In traversing this very long path, they have a correspondingly increased opportunity to encounter and ionize molecules of gas in the interelectrode region, even though this gas may be extremely rarefied. It has been found possible by this use of a magnetic field and appropriately designed electrodes, as indicated in Fig. 6, to maintain an electric discharge at pressures below 10^{-9} torr.

Comparison with the hot-filament ionization gage reveals that, in the hot-filament gage, the source of the inherently linear relationship between gas pressure (more exactly molecular density) and gage reading is the fact that the ionizing current is established and regulated independently of the resulting ion current. In the Philips gage this situation does not hold. Maintenance of the gas discharge current involves a complicated set of interactions in which electrons, positive ions, and photoelectrically effective x-rays all play a significant part. It is thus not surprising that the output current of the Philips gage is not perfectly linear with respect to pressure. Slight discontinuities in the calibration are also sometimes found, since the magnetic fields customarily used are too low to stabilize the gas discharge completely. Despite these objections, a Philips gage is a highly useful device, particularly where accuracy better than 10 or 20% is not required.

The Philips gage is composition-sensitive, but, unlike the situation with the hot-filament ionization gage, the sensitivity relative to some reference gas, such as air or argon, is not independent of pressure. Leak hunting with a Philips gage and a probe gas or liquid is a useful technique. Unlike the hot-filament ionization gage, the Philips gage does not involve the use of a high-temperature filament and consequently does not subject the gas to thermal stress. The voltages applied in the Philips gage are on the order of a few thousand volts, which is sufficient to cause some sputtering at the high-pressure end of the range. This results in a certain amount of gettering or enforced takeup of the gas by the electrodes and other part of the gage. Various design refinements have been used to facilitate periodic cleaning of the vacuum chamber and electrodes, since polymerized organic molecules are an ever-present contaminant.

The conventional cold-cathode (Philips) gage is used in the range from 10^{-2} to 10^{-7} torr. Redhead has developed a modified cold-cathode gage useful in the 10^{-6} to 10^{-12} torr range. See Fig. 7. The operating voltage is about 5000 volts in a 1-kG magnetic field.

Spinning-Rotor Friction Vacuum Gage. Although liquid manometers (U-tube, McLeod) serve as pressure standards for subatmospheric measurements (760 to 10^{-5} torr), and capacitance (also quartz) manometers duplicate this range as useful transfer standards, calibration at lower pressures depends on both volume expansion techniques and presumed linearity of the measuring system. The friction gage allows extension of the calibration range directly down to 10^{-7} torr. It measures pressure in a vacuum system by sensing the deceleration of a rotating steel ball levitating in a magnetic field. See Fig. 8.

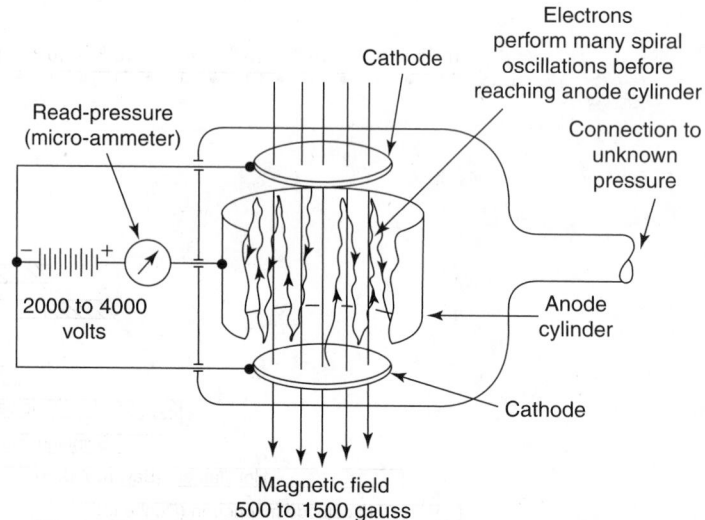

Fig. 6.   Philips cold-cathode ionization vacuum gage

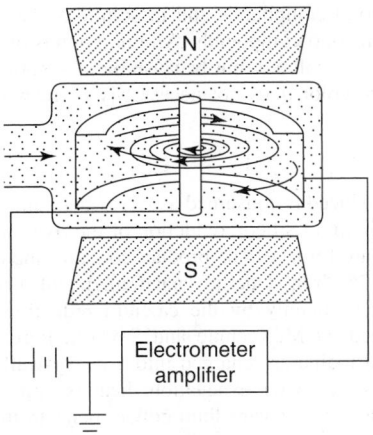

Fig. 7. Inverted magnetron, a cold-cathode gage, produces electrons by applying a high voltage to unheated electrodes. Electrons spiraling in toward the central electrode ionize gas molecules, which are collected on the curved cathode

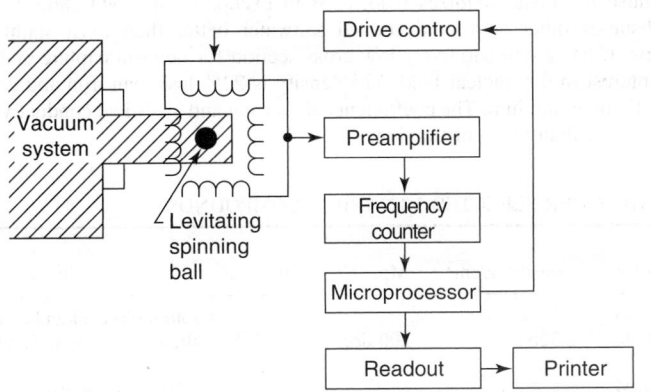

Fig. 8. Schematic diagram of spinning-rotor friction gage

TABLE 3. CHARACTERISTICS OF PARTIAL-PRESSURE ANALYZERS

Type	Minimum partial pressure, torr	Resolution,* au	Magnetic field
Magnetic sector	10^{-11}	20–150	Yes
Cycloidal	10^{-11}	100–150	Yes
Quadrupole	10^{-12}	100–300	No
Time of flight	10^{-12}	200	No

*Maximum mass number at which a mass number difference of 1 can be observed.

Partial-Pressure (Residual Gas) Analyzers. Many applications of high-vacuum technology are more concerned with the partial pressure of particular gas species than with total pressure. Also, "total" pressure gages generally give accurate readings only in pure gases. For these reasons partial-pressure analyzers are finding increasing application. These are basically low-resolution, high-sensitivity mass spectrometers that ionize a gas sample in a manner similar to that of a hot-filament ionization gage. The resulting ions are then separated in an analyzer section, depending on the mass-to-charge ratio of each ion. The ion current corresponding to one ion type is then collected, amplified, and displayed. Partial-pressure gages are very valuable diagnostic tools in both research and production work. The major types are listed in Table 3.

Additional Reading

Chanbers, A., R.K. Fitch, and B.S. Halliday: *Basic Vacuum Technology,* Iop Publishing, Philadelphia, PA, 1998.

Hoffman, D.M., J.H. Thomas, and B. Singh: *Handbook of Vacuum Science and Technology,* Morgan Kaufmann Publishers, Orlando, FL, 1997.

Lafferty, J.M.: *Foundations of Vacuum Science and Technology,* John Wiley & Sons, Inc., New York, NY, 1997.

Ohanlon, J.F.: *A User's Guide to Vacuum Technology,* 3rd Edition, John Wiley & Sons, Inc., New York, NY, 2003.

Santeler, D.J., D.W. Jones, D.H. Holkeboer, and F. Pagano: *Vacuum Technology and Space Simulation,* Springer-Verlag, Inc., New York, NY, 1997.

VACUUM DEPOSITION. The process of coating a base material by evaporating a metal under high vacuum and condensing it on the surface of the material to be coated, which is usually another metal or a plastic. Aluminum is most commonly used for this purpose. The coatings obtained range in thickness from 0.01 to as many as 3 mils. A vacuum of about one-millionth atmosphere is necessary. The process is used for jewelry, electronic components, decorative plastics, etc. Thermally evaporated metals and dielectric coatings can be effectively applied to glass by this method. It is also called vacuum coating and vacuum metallizing.

See also **Thin Films**.

VACUUM DISTILLATION. Distillation at pressure below atmospheric but not so low that it would be classed as molecular distillation. Since lowering the pressure also lowers the boiling point, vacuum distillation is useful for distilling high-boiling and heat-sensitive materials such as heavy distillates in petroleum, fatty acids, vitamins, etc.

See also **Petroleum**.

VALENCE. A whole number that represents of denotes the combining power of one element with another. By balancing these integral valence numbers in a given compound, the relative proportions of the elements present can be accounted for. If hydrogen and chlorine both have a valence of 1, oxygen a valence of 2, and nitrogen 3, the valence-balancing principle gives the formulas HCl, H_2, NH_3, Cl_2O, NCl_3, and N_2O_3, which indicate the relative numbers of atoms of these elements in compounds that they form with each other. In inorganic compounds it is necessary to assign either a positive or negative value to each valence number, so that valence balancing will give a zero sum by algebraic addition. Negative numbers are called polar valence numbers (-1 and -2). The valence of chlorine may be -1, $+1$, $+3$, $+5$, or $+7$, depending on the type of compound in which it occurs. In organic chemistry only nonpolar valence numbers are used.

See also **Bond (Chemical)**; **Chemical Elements**; **Coordination Number**; **Molecule**; **Oxidation and Oxidizing Agents**; and **Oxidation Number**.

VANADINITE. The mineral vanadinite corresponds to the formula $Pb_5(VO_4)_3Cl$, being composed of lead chloride and lead vanadate in the proportion of 90.2% of the former and 9.8% of the latter. It crystallizes in the hexagonal system, is usually prismatic, but the crystals are often skeletal or cavernous; it may be found in crusts. Its fracture is uneven; brittle; hardness, 2.75–3; specific gravity, 6.86; fresh fractures show a resinous luster; color, yellow, yellowish-brown, reddish-brown, and red; streak, white to yellowish; translucent to opaque. Vanadinite, not a common mineral, occurs as an alteration product in lead deposits. It is found in the Ural Mountains, Austria, Spain, Scotland, Morocco, the Transvaal, Argentina, and Mexico. In the United States it occurs in Arizona, New Mexico, and South Dakota. It is used as an ore of vanadium and to some extent of lead as well. It is interesting to note that this mineral was first described as a chromate upon its discovery in Mexico in 1801. It was not until the discovery of the element vanadium in 1830 that the true nature of this compound was known.

VANADIUM. [CAS: 7440-62-2]. Chemical element symbol V, at. no. 23, at. wt. 50.942, periodic table group 5, mp 1880–2000°C, bp 3380°C, density 6.10 g/cm^3. Elemental-vanadium has a body-centered cubic crystal structure. Vanadium is a silver-white, very hard (7 on the Mohs scale), oxidizes upon exposure to air, burns upon ignition to form the pentoxide V_2O_5, insoluble in HCl, slowly dissolves in hydrofluoric, HNO_3, or H_2SO_4 (hot, concentrated), or aqua regia. Insoluble in HaOH solution. The element was first reported by Andrés Manuel del Rio in 1801; later and separately reported by Nils Gabriel Sefstr "om in 1830. There are two naturally occurring isotopes ^{50}V (radioactive and with a half-life something greater than 10^{14} years and only present to the extent of 0.24% in natural substances) and ^{51}V (99.76 abundance percentage). Five other radioactive isotopes have been identified ^{46}V through ^{49}V and ^{52}V. ^{49}V has a half-life of 330 days. ^{48}V has a half-life of 16.1 days; the others have half-lives measured in seconds or minutes. Vanadium ranks 22nd among chemical elements occurring in the Earth's crust. An average composition of igneous rocks contains 0.017% V. It is estimated that vanadium occurs in seawater to the extent of about 9.5 tons per cubic mile (2.1 metric tons per cubic kilometer). First ionization potential 6.74 eV; second, 14.7 eV; third 29.6 eV; fourth 48.3 eV; fifth, 68.64 eV. Oxidation potentials: $V \longrightarrow V^{2+} + 2e^-$, 1.5 V, $V^{2+} \longrightarrow V^{3+} + e^-$, 0.255 V, $VO^{2+} + H_2O \longrightarrow VO_2^+ + 2H^+ + e^-$, -1.00 V. Electronic configuration $1s^2 2s^2 2p^6 3s^2 3p^6 3d^3 4s^2$.

Other important physical characteristics of vanadium are given under **Chemical Elements**.

Vanadium occurs as patronite, containing vanadium pentasulfide, in Peru, as carnotite, potassium uranyl vanadate, in Colorado and Utah, as vanadinite, lead vanadate, in Arizona, New Mexico, the Republic of South Africa and Zambia. See also **Carnotite**; and **Vanadinite**. Ships burning Venezuelan or Mexican petroleum fuel oil recover vanadium oxide from the boiler and stack dust. In Italy the refining of bauxite ore (for aluminum) yields vanadium, and in Germany some iron ores contain vanadium. Vanadium and radium ore is found in southwestern Colorado and southeastern Utah; in Arizona a complex ore of gold, silver, and lead contains vanadium and molybdenum; and extensive deposits of phosphate rock in Idaho yield tonnage quantities of vanadium. The sulfide ore is roasted to remove sulfur, and the residue fused with sodium carbonate, forming sodium vanadate. This last is extracted with H_2O and excess of H_2SO_4 is added, causing precipitation of vanadium pentoxide, which is later reduced by carbon or aluminum at high temperatures.

Physical Properties

Some properties of selected vanadium compounds are listed in Table 1. Vanadium, a typical transition element, displays well-characterized valence states of 2–5 in solid compounds and in solutions. Valence states of -1 and 0 may occur in solid compounds. All compounds of vanadium having unpaired electrons are colored, but because the absorption spectra may be complex, a specific color does not necessarily correspond to a particular oxidation state.

Uses

Most vanadium produced is consumed as ferrovanadium. Ferrovanadium is made by the aluminum or silicon reduction of the oxide in the presence of iron in an electric-arc furnace. The product contains about 85% vanadium, 12% carbon, and 2% iron. Ductile vanadium metal also is produced in significant quantities, mainly by the calcium reduction of the oxide in a process developed by McKechnie and Seybolt. Pure vanadium oxide, calcium metal, and iodine are charged into a heavy-walled steel cylinder, excluding all moisture. After evacuation, heat is applied to initiate the reaction. Molten droplets of vanadium collect beneath the calcium oxide-calcium iodide slag and there form a single button or regulus. Ductile vanadium metal produced in this manner has an analysis of 99.7% vanadium, 0.10% oxygen, 0.04% nitrogen, 0.008% hydrogen, 0.04% iron, and 0.03% carbon. The metal is soft and ductile, can be hot- and cold-worked easily. Any heating must be done in vacuum or an inert atmosphere because the metal oxidizes readily. With exception of HNO_3, the metal withstands other acids and aerated saltwater better than most stainless steels. It has a comparatively low cross section for neutron capture and is of interest in the nuclear field. The density is 22% less than iron and 28% greater than titanium. The coefficients of thermal and electrical conductivity are higher than those of titanium.

TABLE 1. PHYSICAL PROPERTIES OF SOME INDUSTRIAL AND OTHER SELECTED VANADIUM COMPOUNDS[a]

Compound	CAS Registry number	Formula	Appearance	Mol wt	Density, g/cm³	Mp, °C	Bp, °C	Solubility
vanadic acid, meta	[13470-24-1]	HVO₃	yellow scales	99.95				soluble in acid and alkali
ammonium metavanadate	[7803-55-6]	NH₄VO₃	white–yellowish or colorless crystals	116.98	2.326	200 dec		slightly soluble in H₂O
potassium metavanadate	[13769-43-2]	KVO₃	colorless crystals	134.04				soluble in hot H₂O
sodium metavanadate	[13718-26-8]	NaVO₃	colorless, mono-clinic prisms	121.93		630		soluble in H₂O
sodium orthovanadate	[13721-39-6]	Na₃VO₄	colorless, hex-agonal prisms	183.94		850–856		soluble in H₂O
sodium pyrovanadate	[13517-26-5]	Na₄V₂O₇	colorless, hexa-gonal prisms	305.84		632–654		soluble in H₂O
vanadium carbide	[12070-10-9]	VC	black cubic	62.95	5.77	2810	3900	insoluble in H₂O; soluble in HNO₃ with decomposition
vanadium nitride	[24646-85-3]	VN	black cubic	64.95	6.13	2320		soluble in *aqua regia*
vanadium(III) trichloride	[7718-98-1]	VCl₃	pink crystals, deliquescent	157.301	3.00	dec		soluble (deliquescent) in H₂O; soluble in methanol and ether
vanadium(IV) tetrachloride	[7632-51-1]	VCl₄	red-brown liquid	192.75	1.816	28 ± 2	148.5	soluble (deliquescent) in H₂O; soluble in methanol, ether, and chloroform
vanadium(V) oxytrichloride	[7727-18-6]	VOCl₃	yellow liquid	173.30	1.829	−77 ± 2	126.7	soluble (deliquescent) in H₂O; soluble in methanol, ether, acetone, and acid
vanadium(II) oxide	[12035-98-2]	VO	light green crystals	66.95	5.758	ignites		soluble in acid
vanadium(III) oxide	[1314-34-7]	V₂O₃	blue crystals	149.88	4.87	1970		soluble in HNO₃, HF, and alkali in presence of oxide
vanadium(IV) oxide	[12036-21-4]	VO₂	blue crystals	82.94	4.339	1967		soluble in acid and alkali
vanadium(V) oxide	[1314-62-1]	V₂O₅	yellow-red rhombohedra	181.88	3.357	690	1750 dec	slightly soluble in H₂O; soluble in acid and alkali
vanadium(IV) disilicide	[12039-87-1]	VSi₂	metallic prisms	107.11	4.42			soluble in HBr
vanadium(III) acetylacetonate	[13476-99-8]	V(C₅H₇O₂)₃	brown crystals	348.27	0.9–1.2	178–190		soluble in methanol, acetone, benzene, and chloroform
biscyclopentadienyl-vanadium chloride	[12083-18-6]	(C₅H₅)₂VCl₂	pale green crystals	252.04		<250 dec		soluble in methanol and chloroform

[a] Ref. 2. *Kirk–Othmer Encyclopedia of Chemical Technology.*

Most vanadium is used by the steel industry. The addition of vanadium to steel causes the formation of vanadium carbide. The carbides are very hard and wear-resistant; they maintain a fine dispersion. The addition of very small quantities (0.02–0.08%) of vanadium enhances strength and toughness of the resulting steel. Many structural, plate, bar, and pipe steels contain vanadium in these amounts. For little additional cost, in comparison with plain carbon steels, there is a marked increase in performance, including higher strength in the as-rolled condition without heat treatment. Often, manganese and copper are added in small quantities along with the vanadium. Some sheet steels that are used for deep-drawing as in auto and home-appliance parts contain vanadium to suppress aging. For such steels, ferrovanadium is added to rimming steels, resulting in a good, nonaging, deep-drawing steel at a smaller cost than for aluminum-killed deep-drawing steel. Some large steel forgings contain vanadium to the extent of 0.5 to 0.15% with the object of improving the mechanical properties of the forgings. Vanadium is particularly effective in raising the strength and ductility of large steel castings and forgings when added in a small percentage. A large number of tool steels contain vanadium to the extent of 0.10 to 5.00%. In these steels, vanadium insures the retention of hardness and cutting ability at the high temperatures resulting from the rapid cutting of metals. The use of vanadium in cast iron controls the size and distribution of graphite flakes and thus improves strength and wear resistance. One of the most popular of the titanium-base alloys contains 4% vanadium and 6% aluminum. The addition ingredient for this titanium alloy, produced in large quantities, is a base 40:60 vanadium-alloy.

The use of certain vanadium compounds as catalysts has been increasing. Vanadium oxytrichloride is a catalyst in making ethylene-propylene rubber. Ammonium metavanadate and vanadium pentoxide are used as oxidation catalysts, particularly in the production of polyamides, such as nylon, in the manufacture of H_2SO_4 by the contact process, in the production of phthalic and maleic anhydrides, and in numerous other oxidation reactions, such as alcohol to acetaldehyde, anthracene to anthraquinone, sugar to oxalic acid, and diphenylamine to carbazole. Vanadium compounds have been used for many years in the ceramics field for enamels and glazes. Colors are produced by various combinations of vanadium oxide and silica, zirconia, zinc, lead, tin, selenium, and cadmium. Vanadium intermediate compounds also are used in the making of aniline black used by the dye industry.

Chemistry and Compounds

The common oxidation states are: vanadous, V^{2+}, vanadic, V^{3+}, vanadyl, VO^{2+} or VO^{3+}; pervanadyl, VO_2^+; metavanadate, VO_3^+. There are also orthovanadates, VO_4^{3-}, pyrovanadates, $V_2O_7^{4-}$; and complex polyvanadates. The latter group includes di, tri-, tetra-, and octavanadates. The hexavanadates are regarded as oxyvanadium(V) pentavanadates, containing the group $V_5O_{16}^{7-}$. Vanadium ions also form heteropolyacids with acids of molybdenum, tungsten, arsenic, phosphorus, silicon and tin. The peroxy vanadium ions include the diperoxoortho-vanadate ions $[VO_2(O_2)_2]^{3-}$ and the peroxovanadium(V) ions, $[V(O_2)]^{3+}$.

Among the simpler compounds of vanadium are the following:

Fluorides: Vanadium trifluoride VF_3, green crystalline solid; vanadium tetrafluoride VF_4, brownish-yellow crystalline solid; vanadium pentafluoride VF_5, brownish-yellow crystalline solid, sublimes on heating.

Chlorides: Vanadium dichloride VCl_2, green crystalline solid, a strong reducing agent; vanadium trichloride VCl_3, pink crystalline solid; vanadium tetrachloride VCl_4, reddish-brown liquid, bp 148°C.

Bromide: Vanadium tribromide VBr_3, green crystalline solid.

Iodides: Vanadium diiodide VI_2, usually hydrated, green crystalline solid; vanadium triiodide VI_3, brown crystalline solid.

Oxyhalides of vanadium are common, including VOF_2, VOF_3, $VOCl$, $VOCl_2$, $VOCl_3$, $VOBr$ and $VOBr_3$.

Hydroxides: Vanadium dihydroxide $V(OH)_2$, brown precipitate by reaction of NaOH solution with hypovanadous acid (one of the most powerful of reducing agents) lavender solution; vanadous hydroxide $V(OH)_3$, green precipitate by reaction of NaOH solution with vanadous salt, green solution.

Oxides: Vanadium monoxide VO, gray solid; vanadium trioxide V_2O_3, black solid; vanadium dioxide VO_2, dark blue solid; vanadium pentoxide V_2O_5, orange to red solid. The last is the most important oxide; formed by the ignition in air of vanadium sulfide, or other oxide, or vanadium; used as a catalyzer, e.g., the reaction SO_2 gas plus oxygen of air to form sulfur trioxide, and the oxidation of naphthalene by air to form phthalic anhydride.

Bismuth vanadate, $BiVO_4$, exhibits a ferroelastic-paraelastic phase transition and has been the object of considerable investigation. Lattice dimensions of the compound have been determined under numerous different high-pressure or high-temperature conditions and all confirm this transition. R.M. Hazen and J.W.E. Mariathasan, key investigators in this area, have suggested that *in situ* determination of lattice parameters and crystal structures at combined temperature and pressure should be especially valuable in the documentation of nonquenchable, reversible phase transitions, such as that displayed by bismuth vanadate.

Sulfides: Vanadium monosulfide VS; vanadium trisulfide V_2S_3, most stable; vanadium pentasulfide V_2S_5.

Biological Systems: Very little is known of what roles V may play in biological systems. In 1984, scientists at the University of British Columbia found that vanadates stimulate glucose oxidation and transport in adipocytes, enhance glycogen synthesis in liver and diaphragm, and inhibit hepatic glucogenesis and intestinal glucose transport. Working with diabetic rats, the investigators found that vanadates appear to control the high blood glucose and prevent the decline in cardiac performance due to diabetes.

Health, Safety, and Environmental Considerations

Although most foods contain low concentrations of vanadium (<1 ng/g), food is the major source of exposure to vanadium for the general population. High air concentrations of vanadium occur in the occupation setting during boiler-cleaning operations as a result of the presence of vanadium oxides in the dust. The lungs absorb soluble vanadium compounds V_2O_5 well, but the absorption of vanadium salts from the gastrointestinal tract is poor. The excretion of vanadium by the kidneys is rapid with a biological half-life of 20–40 hours in the urine. Vanadium is probably an essential trace element, but a vanadium-deficiency disease has not been identified in humans. The estimated daily intake of the US population ranges from 10–60 micrograms V. Vanadyl sulfate is a common supplement used to enhance weight training in athletes at doses up to 60 mg/d. In vitro and animal studies indicate that vanadate and other vanadium compounds increase glucose transport activity and improve glucose metabolism. In general, the toxicity of vanadium compounds is low. Pentavalent compounds are the most toxic and the toxicity of vanadium compounds usually increases as the valence increases

Most of the toxic effects of vanadium compounds result from local irritation of the eyes and upper respiratory tract rather than systemic toxicity. The only clearly documented effect of exposure to vanadium dust is upper respiratory tract irritation characterized by rhinitis, wheezing, nasal hemorrhage, conjunctivitis, cough, sore throat, and chest pain. Case studies have described the onset of asthma after heavy exposure to vanadium compounds, but clinical studies to date have not detected an increased prevalence of asthma in workers exposed to vanadium.

Additional Reading

Carter, G.F. and D.E. Paul: *Materials Science and Engineering,* ASM International, Materials Park, OH, 1991.

Greenwood, N.N. and A. Earnshaw: *Chemistry of the Elements,* 2nd Edition, Butterworth-Heinemann, Inc., Woburn, MA, 1997.

Hazen, R.M. and J.W.E. Mariathasan: "Bismth Vanadate: A High-Pressure, High-Temperature Crystallographic Study of the Ferroelastic-Paraelastic Transition," *Science,* **216,** 991–992 (1982).

Lide, D.R.: *CRC Handbook of Chemistry and Physics 84th Edition,* CRC Press, LLC., Boca Raton, FL, 2004.

Nriagu, J.O.: *Vanadium in the Environment: Health Effects,* Vol. 2, John Wiley & Sons, Inc., New York, NY, 1998.

Staff: *ASM Handbook—Properties and Selection: Nonferrous Alloys and Pure Metals,* ASM International, Materials Park, OH, 1990.

Staff: *Properties and Selection: Iron, Steels, and High-Performance Alloys,* ASM International, Materials Park, OH, 1990.

Tracy, A.S. and D.C. Crans: *Vanadium Compounds: Chemistry, Biochemistry, and Therapeutic Applications,* Oxford University Press, Inc., New York, NY, 1998.

Web Reference

USGS Minerals Information: Vanadium: http://minerals.usgs.gov/minerals/pubs/commodity/vanadium/

VAN DER WAALS EQUATION.   A form of the equation of state, relating the pressure, volume, and temperature of a gas, and the gas constant. Van der Waals applied corrections for the reduction of total pressure by the attraction of molecules (effective at boundary surfaces) and for the effect of reduction of total volume by the volume of the molecules (but not the actual molecular volume). The equation takes the form

$$\left(P + \frac{a}{V^2}\right)(V - b) = RT$$

in which P is the pressure of the gas, V is the volume, T is the absolute temperature, R is the gas constant, and a and b are correction terms which have been evaluated and reported for many gases. See also **Characteristic Equation**.

VAN DER WAALS FORCES. Interatomic or intermolecular forces of attraction due to the interaction between fluctuating dipole moments associated with molecules not possessing permanent dipole moments. These dipoles result from momentary dissymmetry in the positive and negative charges of the atom or molecule, and on neighboring atoms or molecules. These dipoles tend to align in antiparallel direction and thus result in a net attractive force. This force varies inversely as the seventh power of the distance between ions.

VAN DER WAALS, JOHANNES DIDERIK (1837–1923). Van der Waals was a Dutch physicist, he was a professor of physics at the University of Amsterdam from 1877 to 1907. He was especially interested in thermodynamics and he won the 1910 Nobel Laureate in Physics for his work on gases and liquids, his most well known expressed as Van der Waals Equation. He also studied the attractive forces, which hold the atoms of molecules together and is named now Van der Waals forces. Van der Waals bonding, a crystalline structure with the weakest bonding, is also named in his honor.

See also **Characteristic Equation**; **Chemical Elements**; **Van Der Waals Equation**; and **Van Der Waals Forces**.

J.M.I.

VANILLIN. Vanillin, [CAS: 121–35–5]. $C_8H_8O_3$, a natural product, can be found as a glucoside (glucovanillin) in vanilla beans, at concentrations of about 2%. It can be extracted with water, alcohol, or other organic solvents. Approximately 250 by-products have been identified in natural vanilla, out of which 26 are present at levels in excess of 1 ppm. The balance of all these products contributes to the subtle taste of vanilla beans. The vanilla bean contains about 2% vanillin, but the 10% extract prepared from beans has several times the strength of a solution of 2% vanillin. The best known natural source of vanillin is the vanilla plant, *Vanilla planifolia* A., which belongs to the orchid family. It is cultivated mainly in Mexico, Madagascar, Reunion, Java, and Tahiti.

The long and expensive process of extracting vanillin from vanilla beans yields a product that has an inconsistent quality. The demand for this universally popular flavoring cannot be satisfied by vanilla beans alone. The consumption of naturally occurring vanilla has gradually given way to synthetic vanillin. Synthetic vanillin is identical to that contained in the pod, but differs in smell and flavor from natural vanillin as a result of the various compounds in the natural extract that do not exist in artificial vanillin. These other compounds represent only 2% of the extract; the remaining 98% is vanillin. Vanillin is the common name for 3-methoxy-4-hydroxybenzaldehyde (**1**).

(**1**)

Production

The manufacture of vanillin shows the progress made in the chemistry and chemical engineering of the substance. Most commercial vanillin is synthesized from guaiacol; the remainder is obtained by processing waste sulfite liquors. Preparation by oxidation of isoeugenol is of historical interest only.

Preparation from Guaiacol and Glyoxylic Acid. Several methods can be used to introduce an aldehyde group into an aromatic ring. Condensation of guaiacol (**2**) with glyoxylic acid (**3**), followed by oxidation of the resulting mandelic acid (**4**) to the corresponding phenylglyoxylic acid (**5**)

and decarboxylation continues to be a competitive industrial process for vanillin synthesis.

Preparation from Waste Sulfite Liquors. The starting material for vanillin production can also be the lignin present in sulfite wastes from the cellulose industry. The concentrated mother liquors are treated with alkali at elevated temperature and pressure in the presence of oxidants. The vanillin formed is separated from the by-products, particularly acetovanillone, 4-hydroxy-3-methoxyacetophenone, by extraction, distillation, and crystallization. In contrast to vanillin from lignin, the principal impurity found in vanillin from guaiacol is 5-methyl vanillin, typically present at levels of about 100 ppm, although levels as high as 3000 ppm have been found. This impurity is completely odorless.

No residual guaiacol can be found in vanillin produced by the guaiacol process. In contrast to vanillin from lignin, vanillin from guaiacol is extremely consistent in quality owing to the consistency of the supply source, and shows no variation in taste, odor, or color.

Specifications

The physical properties of Rhovanil Extra Pure vanillin of Rhône-Poulenc, the leading company in this area, are shown in Table 1.

Solubility. Solubility in water is less than 2%; the solubility in ethanol, is given by the ratio one part vanillin to two parts alcohol. Certain manufacturing processes require that the product be in liquid form.

Particle-Size Distribution. Particle size, crystal shape, and distribution of vanillin are important and greatly affect parameters such as taste, flavor, solubility, ease of dispersion in solvent, flowability of the powder, caking effect, and production of dust.

Taste and Flavor. The taste effect is generally sweet, but depends strongly on the base of preparation. For tasting purposes, vanillin is often evaluated in ice-cold milk with about 12% sugar. A concentration of 50 ppm in this medium is clearly perceptible. Vanilla is undoubtedly one of the most popular flavors; its consumption in the form of either vanilla extracts or vanillin is almost universal. The food flavor industry is the largest user of vanillin, an indispensable ingredient in chocolate, candy, bakery products, and ice cream. It is easy to smell a difference in the quality of vanillins from different origins, but it is normally difficult to taste the same difference, provided the various samples are of good quality.

Available Grades. Rhovanil Extra Pure is the trade name of the food-grade vanillin of Rhodia, worldwide leader in the diphenols area. The following grades are commercially available: Rhovanil Extra Pure crystallized, Rhovanil Fine Mesh, Rhovanil Free Flow, and Rhovanil Liquid.

Rhovanil Extra Pure is the standard mesh, multipurpose quality of food-grade extra pure vanillin. Rhovanil Fine Mesh, a specially calibrated extra pure vanillin that avoids demixing with other very fine dry ingredients such

TABLE 1. PHYSICAL PROPERTIES OF RHOVANIL EXTRA PURE VANILLIN

Property	Value
white to off-white nonhygroscopic crystalline powder	
mp, capillary, °C	81–83
assay, %	99.96 min
bulk density	~0.6
fp, °C	153
bp, °C	
at 101.3 kPa[a]	284–285
10 mm Hg	154
sublimation[b], °C	70

[a] To convert kPa to mm Hg, multiply by 7.5.
[b] At normal pressure.

as sucrose, flour, and dextrose, provides a faster dissolution rate at lower stirring, at lower temperature, in low acidity medium, or in viscous liquids.

Rhovanil Free Flow is obtained by adding an anticaking agent (0.5% max) to the extra pure vanillin. The flowability is increased, making it particularly suitable for self-dispensing equipment (instant beverage), while both mixability and dispersion/dissolution ratios remain as good as the standard Rhovanil Extra Pure vanillin.

Chemical Properties

Vanillin is a compound that possesses both a phenolic and an aldehydic group. It is capable of undergoing a number of different types of chemical reactions. Addition reactions are possible owing to the reactivity of the aromatic nucleus. On distillation at atmospheric pressure, vanillin undergoes partial decomposition with the formation of pyrocatechol. Exposure to air causes vanillin to oxidize slowly to vanillic acid. Reduction of vanillin by means of platinum black in the presence of ferric chloride gives vanillin alcohol in excellent yields. Because vanillin is a phenol aldehyde, it is stable to autooxidation and does not undergo the Cannizzarro reaction. All three functional groups in vanillin are highly reactive.

Applications

In flavor formulations, vanillin is used widely either as a sweetener or as a flavor enhancer, not only in imitation vanilla flavor, but also in butter, chocolate, and all types of fruit flavors, root beer, cream soda, etc. It is widely acceptable at different concentrations; 50–1000 ppm is quite normal in these types of finished products. Concentrations up to 20,000 ppm, i.e., one part in fifty parts of finished goods, are also used for direct consumption such as toppings and icings. Ice cream and chocolate are among the largest outlets for vanillin in the food and confectionery industries, and their consumption is many times greater than that of the perfume and fragrance industry.

When vanillin is not used as a single flavoring ingredient, it is a key part of flavor compounding. At least 30% of food-grade vanillin consumed in the world is through flavoring compounds.

In the industrial production of dry cookies, cakes, and pastries, the vanillin content ranges between 20 and 50 g per 100 kg of dough. Often, vanillin is added at the dry stage of dough preparation as the flour and sugar are being mixed. In fat-free recipes, it is possible to add and mix vanillin powder with eggs.

Vanillin is added, in powder form during the manufacturing process of chocolate, in average amounts of 20 g per 100 kg of the finished product. However, this amount varies according to the quality of the chocolate being made.

Main applications in confections are sugared almonds, caramel, nougat, and sweets. For sugared almonds and caramel, vanillin is mixed into the sugar in the dry phase of the recipe. For nougat, Vanillin is added during the liquid phase of manufacturing. In sweets, vanillin is added in the form of a 10% ethanol solution. Vanillin is used in flavored milk, desserts, yogurts, sorbets, and ice cream.

Vanillin sugar is prepared by dry mixing or impregnating the sugar with a vanillin alcohol solution and evaporating the alcohol. Vanillin confers a pleasant note to liqueur flavoring and improves the flavor of fortified wines by giving them a greatly enhanced bouquet. Vanillin is used as a palatability enhancer to make animal feed more appetizing by flavor-masking minerals with off-taste. Vanillin, a crystal, is the main constituent of the vanilla bean. Its importance can be illustrated by the fact that human preferences in fragrances and in flavors, as determined by various studies, comprise three main smells or tastes: rose, vanilla, and strawberry. Vanillin is used in perfumes and cosmetics as a fragrance and as a washing deodorant.

A flaked technical-grade vanillin, Vaniltek, to be used in pharmaceutical applications. The single largest use for vanillin is as a starting material for the manufacture of an antihypertensive drug having the chemical name of Methyldopa or L-3(3,4-dihydroxyphenyl)-2-methylalanine. Other drugs made from vanillin include L-Dopa, Trimethoprim Mebeverine, and Verazide for example.

Vanillin itself has some bacteriostatic properties and therefore has been used in formulations to treat dermatitis. Hydrazones of vanillin have been shown to have a herbicidal action similar to that of 2,4-D, and the zinc salts of dithiovanillic acid. A new potential use for vanillin is as a ripening agent to increase the yield of sucrose in sugarcane by the treatment of the cane crop a few weeks before harvest.

The antiultraviolet protection properties of vanillin have been patented and look promising for the plastics and cosmetics (suncreams) industries.

Health and Safety Factors

Vanillin is listed in the *Code of Federal Regulations* by the FDA as a Generally Recognized As Safe (GRAS) substance. The Council of Europe and the FAO/WHO Joint Expert Committee on Food Additives have both given vanillin an unconditional Acceptable Daily Intake (ADI) of 10 mg/kg.

Vanillin has a low potential for acute and chronic toxicity. Vanillin is known to cause allergic reactions in people previously sensitized to balsam of Peru, benzoic acid, orange peel, cinnamon, and clove, but vanillin itself is not an allergic sensitizer. Vanillin has been reported to be a bioantimutagen, demonstrating the ability to protect against mutagenic effects by enhancement of an error-free post-replication repair pathway.

LAWRENCE J. ESPOSITO
K. FORMANEK
G. KIENTZ
F. MAUGER
V. MAUREAUX
G. ROBERT
F. TRUCHET
Rhône-Poulenc

Additional Reading

Arctander, S. ed.: *Perfume and Flavor Chemicals (Aroma Chemicals),* Montclair, NJ, 1969.
Bedoukian, P. Z. ed.: *Perfumery and Flavoring Synthetics,* Allured Publishing Corp.
Bauer, K.: in D. Garbe, ed., *Common Fragrance and Flavor Materials: Preparation, Properties and Uses,* VCH, Weinheim, Germany, 1985.
Ullmann's Encyclopedia of Industrial Chemistry, 5th Edition, Vol. A 11, VCH, Weinheim, Germany, 1988, pp. 199–200.

VAN SLYKE REACTION. See **Amino Acids.**

VAN'T HOFF EQUATION. A relationship representing the variation with temperature (at constant pressure) of the equilibrium constant of a gaseous reaction in terms of the change in heat content, i.e., of the heat of reaction (at constant pressure). It has the form:

$$\frac{d \ln K_p}{dT} = \frac{\Delta H}{RT^2}$$

in which K_p is the equilibrium constant at constant pressure, T is absolute temperature, R is the gas constant, and ΔH is the standard change in heat content, or, for ideal gases, the change in heat content.

VAN'T HOFF, JACOBUS HENDRICUS (1852–1911). Van't Hoff was born in Rotterdam, the Netherlands. In 1869 he entered the Polytechnic School at Delft and obtained his technology diploma in 1871. His decision to follow a purely scientific career, however, came soon afterwards during vacation–work at a sugar factory when he anticipated for himself a dreary profession as a technologist. After having spent a year at Leyden, mainly for mathematics, he went to Bonn to work with A.F. Kekulé from autumn 1872 to spring 1873. This period was followed by another in Paris with A. Wurtz, when he attended a large part of the curriculum for 1873–1874. He returned to Holland in 1874 and received his Ph.D. from the University of Utrecht under E. Mulder. He did research on chemical dynamics and equilibrium. He was awarded the first Nobel Prize in Chemistry in 1901.

In his later years, Van't Hoff spent his research time trying to understand the action of enzymes as biological catalysts.

See also **Van't Hoff Equation**; and **Van't Hoff Law.**

J.M.I.

VAN'T HOFF LAW. A dissolved substance has the same osmotic pressure as the gas pressure it would exert in the form of an ideal gas occupying the same volume as that of the solution.

VAPOR. A substance in the gaseous state, but below its critical temperature, is called a vapor. If a pure liquid partly filling a closed container is allowed to stand, the space above it becomes filled with the vapor of the liquid, which then develops a pressure. This vapor pressure increases up to a certain limit, depending upon the temperature, where it becomes constant, and the space is then said to be saturated.

Such a body of vapor is not subject to all of the laws of gases. If the space occupied by it is diminished without change of temperature, there is no increase in pressure, but instead part of the vapor condenses. And if the temperature is raised, the pressure goes up not at a uniform but at an increasing rate, because of both the expansion of the liquid and the further evaporation from it. The relation of vapor to liquid takes on a curious aspect as the critical state is approached, in which the vapor and the liquid have equal density.

See also **Boiling Point**; and **Supersaturated Vapor**.

VAPOR DENSITY.   The density of a gas referred to the density of hydrogen or air as unity. If the density of hydrogen is taken as 2, the vapor density is approximately the molecular weight; if it is taken as one, the vapor density equals about half the molecular weight.

VAPORIZATION.   The change of a substance from the liquid or solid state to the gaseous state.

VAPORIZATION (Heat of).   The evaporation of a given mass of any liquid requires a definite quantity of heat, dependent upon the liquid and upon the temperature at which it evaporates. The quantity required per unit mass at a fixed temperature is called the heat of vaporization of the substance at that temperature. It may be measured by allowing the vapor to condense in a suitable calorimeter, the heat thus evolved, corrected for fall of temperature before and after condensation, being observed. (The heat evolved in condensing is equal to that absorbed when the liquid evaporates.) The result is often surprising. For example, the evaporation of water at the boiling point requires about 540 calories per gram, or more than five times the heat required to raise its temperature from freezing to boiling. The explanation is the large amount of energy necessary to separate the molecules against their cohesion, and the much smaller amount (about 7.4% of the whole) which is used in expanding the vapor against atmospheric pressure. At lower temperatures the value is still greater, because the cohesion is then more effective; with water, for each degree below the normal boiling point, about 0.6 calorie per gram must be added to the heat of vaporization. Trouton found that the heat of vaporization per mole for different liquids bears a nearly constant ratio to the absolute temperature of the boiling point.

A number of methods have been developed for measuring the heat of vaporization. In the Awberg and Griffith's method, the flow is continuous; the heat of evaporation of the liquid under investigation is transmitted to a stream of water flowing at constant rate and its increase in temperature is measured.

VAPOR PRESSURE.   The vapor pressure of a substance (solid or liquid) is the pressure exerted by its vapor when in equilibrium with the substance. For pure substances it depends only on the temperature. The simplest way to measure the vapor pressure of a substance is to introduce a small amount of it into the closed end of a barometer tube and note the decrease in the height of the barometer.

The vapor pressure of a solvent is lowered on dissolving the solute in it. This lowering for dilute solutions is proportional to the mole fraction of the solute (Raoult's Law). The lowering of the vapor pressure of the solution can be related to the lowering of the freezing point and the elevation of the boiling point. These phenomena serve as a basis for molecular weight determinations. If both components of the solution are volatile, each lowers the vapor pressure of the other and the ratios of the two substances in the liquid and vapor phase are not necessarily the same. Use is made of this fact to separate the two substances by distillation.

Equilibrium vapor pressure is the vapor pressure of a system in which two or more phases or a substance coexist in equilibrium. In meteorology, the reference is to water substance, unless otherwise specified. If the system consists of moist air in equilibrium with a plane surface of pure water or ice, the more specialized term *saturation vapor pressure* is usually employed, in which case, the vapor pressure is a function of temperature only. In the atmosphere, the system is complicated by the presence of impurities in liquid or solid water substance (see also **Raoult's Law**), drops or ice crystals or both, existing as aerosols; and, in general, the problem becomes one of nucleation. For example, the difference in vapor pressure over supercooled water drops and ice crystals is the basis for the Bergeron-Findeisen theory of precipitation formation.

VARIABLE (Process).   The quantity or characteristic that is the object of measurement in an instrumentation or automatic control system. Other terms used include measurement variable, instrumentation variable, and process variable. The latter term is commonly used in the manufacturing industries. Numerous ways to classify variables have been proposed—by methods of measurement, by end-measurement objectives, and so on. One of the most convenient and meaningful classifications is the physical and/or chemical nature of the variable, as follows:

Thermal Variables. These variables relate to the condition or character of a material dependent upon its thermal energy. Variables included are: *temperature, specific heat, thermal-energy variables* (enthalpy, entropy, etc.), and calorific value.

Radiation Variables. These variables relate to the emission, propagation, and absorption of energy through space or through a material in the form of waves; and by extension, corpuscular emission, propagation, and absorption. Variables included are: *nuclear radiation; electromagnetic radiation* (radiant heat, infrared, visible, and ultraviolet light; x- and cosmic rays; gamma radiation).

Force Variables, including: *total force, moment or torque*, and *force per unit area*, such as pressure, vacuum, and unit stress.

Rate Variables. These variables are concerned with the rate at which a body is moving toward or away from a fixed point. Time always is a component of a rate variable. Variables included are: *flow, speed, velocity,* and *acceleration.*

Quantity Variables. These variables relate to the total quantity of material that exists within specific boundaries. Variables include: *mass* and *weight.*

Physical Property Variables. These variables are concerned with the physical properties of materials with the exception of those properties which are related to chemical composition and direct mass and weight. Variables included are: *density* and *specific gravity, humidity, moisture content, viscosity, consistency*, and *structural characteristics*, such as hardness, ductility, and lattice structure.

Chemical-Composition Variables. These variables relate to the chemical properties and analysis of substances. A very abridged list of analysis variables would include: Identification and concentration of carbon dioxide, carbon monoxide, hydrogen, nitrogen, oxygen, water, hydrogen sulfide, nitrogen oxides, sulfur oxides, methane, ethylene, alcohol, and so on. Also included in this category is measurement of pH (hydrogen ion concentration) and redox measurements. See also **Analysis (Chemical)**.

Electrical Variables. Included here are those variables which are measured as the "product" of a process, as in the case of measuring the current and voltage of a generator, and also as part of an instrumentation system. Numerous transducers, of course, yield electrical signals that represent by inference some other variable quantity, such as a temperature or pressure. Variables in this class include: *electromotive force, electric current, resistance, conductance, inductance, capacitance*, and *impedance.*

Geometric Variables. These variables are related to position or dimension and relate to the fundamental standard of length. Variables include: *position, dimension, contour*, and *level* (as of a material in a tank or bin).

VARIOLITE.   A fine-grained basic rock that contains spherulites made up of fibers of feldspar and augite in radial development.

The spherulites themselves are known as varioles, and the texture of such rocks is said to be variolitic.

VARNISH.   (1) An organic protective coating similar to paint except that it does not contain colorant. It may be composed of a vegetable oil (linseed), tung, etc.), and solvent or of a synthetic or natural resin and solvent. In the first case the formation of the film is due to polymerization of the oil and in the second to evaporation of the solvent. "Long-oil" varnishes such as spar varnish have a high proportion of drying oil; "short-oil" types have a lower proportion, i.e., furniture varnishes. Spirit varnishes contain such solvents as methanol, toluene, ketones, etc. and often also thinners such as naphtha or other light hydrocarbon.

(2) A hard, tightly adherent deposit on the metal surfaces of automobile engines resulting from resinous oxidation products of gasoline and lubricating oils.

VEGETABLE OILS (Edible).

Vegetable oils are prepared from at least ten major oilseeds, plus a few other sources that may develop into high volume at some future date. There are both edible and nonedible vegetable oils. Technical and industrial vegetable oils, such as castorseed and tung oils, are used in nonfood applications. Some oils, such as linseed (flaxseed) and olive oil, depending upon the manner in which they have been treated and refined, find both food and nonfood applications. Other oils, such as soybean, sunflower, rapeseed, sesame, and safflower oils, are used predominantly in food processing and in the production of feedstuffs. Usually, in considering edible vegetable oils, the availability of animal fats, such as butterfat, lard, tallow, etc., and of fish and other marine oils, is noted because there is surprising versatility among all oils and fats, allowing significant substitutions in the same end-product. Perhaps this is best illustrated by the fact that coconut oil, corn (maize) oil, cottonseed oil, palm oil, groundnut (peanut) oil, safflower oil, and soybean oil all are or have been used in commercial margarines for the retail market. The processor of a given brand of margarine will select the oil or a combination of oils, considering price and availability, as well as desirable end-product characteristics (stick or brick; soft tub; diet or imitation; etc.), among other factors. The pressed cakes and meals remaining after extraction of oil from beans and seeds finds wide use in animal feedstuffs.

The consumption of fats and oils by persons in the United States has increased steadily since the early 1960s. In 1963, 46.3 pounds (21 kilograms) were consumed per capita per year, rising to 54.4 pounds (24.7 kilograms) in the late 1970s. During this period, butter dropped from 13.9% of total fats and oils to 7.7%; lard dropped from 12.9% to 4%; margarine increased from 19.4% to 20.2%; shortenings of various kinds increased from 27.2% to 30.6%; and salad and cooking oils increased markedly from 26.6% to 37.5%.

On a worldwide basis, as of the early 1980s, the various sources of edible oils, in terms of volume of production, rank as follows: Soybeans, 50.3%; cottonseed, 16.2%; groundnuts (peanuts), 11.4%; sunflower seeds, 7.9%; rapeseed, 6.9%; copra (coconut), 2.9%; flaxseed (linseed), 1.8%; sesame seed, 1.1%; palm kernel, 0.9%; and safflower seed, 0.6%.

Principal edible uses of these oils are found in cooking and salad oils; frying oils; margarine, mayonnaise, and salad dressings; bakery, cake mix, and pie shortenings; and whipped topping and other nondairy products, such as coffee creamers.

Definitions

Some of the common terms used in describing edible oils and fats are:

Saturated—the state in which all available valence bonds of an atom, especially carbon, are attached to other atoms. The straight-chain alkyls (paraffins) are typical saturated compounds. Several fatty acids found in oils and fats are listed in Table 1. Where no double- or triple-bonds are available, such compounds cannot be hydrogenated because there are no places for attaching additional hydrogen atoms.

Monosaturated—the state where one double bond is present in the compound. Under proper conditions of hydrogenation, an additional hydrogen atom can be added, thus resulting in a saturated compound.

Polyunsaturated—the state where two or more double bonds are present in the compound. Again, under proper conditions of hydrogenation, one or more additional hydrogen atoms can be added, thus resulting in a compound that is less unsaturated or that is saturated, depending upon the number of double bonds originally available and the degree of hydrogenation effected.

Hydrogenation—any reaction of hydrogen with an organic compound. In the case of fats and oils, this is the direct addition of hydrogen to double bonds of unsaturated molecules, resulting in a partially or a fully saturated product.

Hydrogenation can be accomplished with gaseous hydrogen under pressure in the presence of a catalyst (nickel, platinum, or palladium). The degree of saturation of an oil or its fatty acid substituents affects its fluidity (melting or softening point), density, refractive index, and general reactivity. Frequently, the degree of hydrogenation of an oil will be

TABLE 1. CHARACTERISTICS OF VEGETABLE OIL FATTY ACIDS

Principal fatty acid	Formula weight	Number of double bonds	Position of bonds	Number of carbon atoms	Formula
SATURATED COMPOUNDS					
Caproic	116.09	0	—	6	$C_5H_{11}COOH$
Caprylic	144.21	0	—	8	$C_7H_{15}COOH$
Capric	172.26	0	—	10	$C_9H_{19}COOH$
Lauric	200.31	0	—	12	$C_{11}H_{23}COOH$
Myristic	228.36	0	—	14	$C_{13}H_{27}COOH$
Palmitic	256.42	0	—	16	$C_{15}H_{31}COOH$
Stearic	258.47	0	—	20	$C_{19}H_{39}COOH$
UNSATURATED COMPOUNDS					
Palmitolenic	254.42	1		16	$C_{15}H_{29}COOH$
Linolenic	278.42	3	cis-9, cis-12, cis-15	18	$C_{17}H_{29}COOH$
Linoleic	180.44	2	cis-9, cis-12	18	$C_{17}H_{31}COOH$
Oleic	282.45	1	cis-9	18	$C_{17}H_{33}COOH$

determined by measuring its refractive index. Vegetable and fish oils can be hardened or solidified by catalytic hydrogenation. Partial hydrogenation clarifies some oils and makes them odorless. Fatty oils, such as oleic acid, are converted into stearic acid by hydrogenation. Coconut oil, groundnut (peanut) oil, and cottonseed oils can be made to appear, taste, and smell like lard; or they can be made to resemble tallow. Sometimes, hydrogenated oils are referred to as synthetic shortenings. Generally, hydrogenated oils have higher melting points and lower iodine values than the natural, untreated oils.

For many years, hydrogenation was a batch operation, but in recent years, continuous hydrogenation processes have been displacing the batch methods, especially in large-scale facilities. Hydrogenation pressures vary as well as temperatures, but the latter are usually in the 93–135°C range. It is not uncommon to mix two or more oils prior to hydrogenation. Considerably more detail on the hydrogenation process is given in the entry on **Hydrogenation**.

Deodorizing—crude oils tend to have undesirable odors and tastes. These are due to the presence of various volatile components. These are removed by passing steam through the heated oil under diminished pressure.

Winterizing—because of widespread use of refrigeration, salad oils and other end-products prepared from vegetable oils must not become cloudy or solidify at relatively low storage temperatures. Thus, most salad oils differ from cooking oils, in that the latter, if stored under refrigeration, may slowly solidify and be difficult to pour from a container. To prevent cloudiness and solidification at low temperatures, the oils are "winterized." This involves subjecting the oils to low temperatures in stages, during which time crystals of high-melting-point fat are formed. A final low temperature of about 5.5°C is reached in the process, at which temperature the oil is allowed to stand for a considerable time, during which considerable crystallization occurs. Slow cooling and gentle agitation ensure maximum removal of the high-melting-point fats. The crystalline material remaining after filtration is essentially stearine, with traces of wax, gums, and soaps. Improvements in the winterizing process during the past several years have reduced the time required from 3 to 6 days down to about 5 hours. The more recent processes involve solvent extraction and centrifugation.

Iodine Value (or Number)—the percentage of iodine that will be absorbed by a chemically unsaturated substance, such as vegetable oil, in a given time under specified conditions. The iodine value is a measure of degree of unsaturation. Two methods are described in the "Food Chemicals Codex:" the Hanus method and the Wijs method.

Saponification Number—the number of milligrams of potassium hydroxide required to hydrolyze one gram of sample of an ester (glyceride, fat) or mixture. This test is also described in the "Food Chemicals Codex."

Margarines

Regular margarines (by government regulations) contain at least 80% fat. The remaining content is ~16% water and small amounts (about

TABLE 2. CHARACTERISTICS AND PROPERTIES OF MAJOR VEGETABLE OILS

Property	Coconut oil	Cottonseed oil	Groundnut (peanut) oil	Pal oil	Rapeseed oil
Color	White	Pale-yellow or yellowish-brown to dark ruby-red, or black-red.	Yellow to greenish-yellow	Yellow-brown	Brown (raw); yellow (refined)
State (at room temperature)	Semisolid (nondrying oil)	Liquid (semidrying oil)	Liquid (nondrying)	Solid (buttery)	Liquid (viscous)
Odor	Slight	Slight (when refined)	Nutlike	Agreeable	Characteristic
Melting point	77°–90°F; 25°–32°C	Slightly below 32°F; 0°C# (before winterizing process)	23°–37°F; −5°–+3°C@	86°F; 30°C	328°F; 0°C$
Specific gravity	0.92	0.915–0.921	0.912–0.920	0.952	0.913–0.916
Saponification value	250–264	190–198	186–194	247.6	174
Iodine value	7–10	109–116	88–98	13.5	100.3
Principal constituents	Glycerides of fatty acids of approximate composition: Lauric acid 45–48% Myristic acid 17–20% Capric acid 6–8% Palmitic acid 5–9% Caprylic acid 5–7% Oleic acid 4–8% Stearic acid 2–5% Arachidic acid 1%± Palmitoleic acid 0.4%± Unsaturated acids 8% Saturated acids 92% (Approximate)	Glycerides of fatty acids of approximate composition: Linoleic acid 47–50% Palmitic acid 26–27% Oleic acid 18–19% Stearic acid 2%+ Palmitoleic acid 1%± Myristic acid 1%− Capric acid 0.5% Linolenic acid 0.4%± Lauric acid 0.4% Arachidic acid 0.2%± Unsaturated acids 73%	Glycerides of fatty acids of approximate composition: Oleic acid 40–52% Linoleic acid 25–37% Palmitic acid 8.5–10.5% Behenic acid 2.5%± Lignoceric acid 1.5%± Stearic acid 1.5–3.0% Arachidic acid 1.5–2.5% Myristic 0.2%± Unsaturated acids 82% Saturated acids 18% (Approximate) Saturated acids 27% (Approximate)	Triglycerides of fatty acids of approximate composition: Palmitic acid 37–47% Oleic acid 31–44% Lauric acid 4–8% Stearic acid 2.5–5.5% Myristic acid 0.8–1.5% Linoleic acid 1–3% Palmitoleic acid 0.2–0.4% Arachidic acid 0.2–0.4% Unsaturated acids 50% Saturated acids 50% (Approximate) There are some significant differences in composition of palm oil (from fleshy fruit pulp) and palm kernel oil.⊘	High in unsaturated acids, especially oleic, linoleic, and erucic acids, the latter being 40% in older varieties, but more recent varieties have a lower erucic acid content.
Soluble in	Alcohol, carbon disulfide, chloroform, ether. mmiscible in water.	Benzene, carbon disulfide, chloroform, ether. Slightlyform, soluble in alcohol.	Carbon disulfide, chloroform, ether, petroleum ether. Insoluble in alkalies, but saponified by alkali hydroxides with formation of soaps. Slightly soluble in alcohol. Insoluble in water.	Alcohol, carbon disulfide. chloroform, ether. Insoluble in water.	
Grades	Crude, refined, Ceyon, Manila	Crude, refined, prime summer yellow, bleachable, USP.	Crude, refined, edible, USP.	Crude, refined	
Derivation	Hydraulic press or expeller extraction from coconut meat, followed by alkali-refining, and bleaching.	From cotton seeds by hot-pressing or solvent extraction.	By pressing ground groundnut (peanut) meats or by extraction with hot or cold solvents. Purified by bleaching with fuller's earth or carbon. Hot-pressed oil may be allowed to stand to deposit stearin, prior to filtering.	Separation of fat from palm fruits by expression or centrifugation.	
Food uses	Margarine, hydrogenated shortenings, synthetic cocoa dietary supplements.	Margarine, shortening, salad oils and dressings, stabilizers.	Salad oils, mayonnaise, margarine Sometimes a substitutefor olive oil. Cooking oil.	Food shortenings, margarines, competes with soybean oil.	Salad dressings, margarine, substitute for soybean oil.
Special notations	See also entry on **Coconut**.	#Solidification range is 88°–95°F (31°–35°C) Flashpoint is 486°F (252°C) See also entry on **Cottonseed**.	@Flash point is 540°F (282°C) See also **Groundnut (Peanut)**.	⊘Approximate composition of palm kernel oil is: 50% lauric acid, 15% myristic acid, 16% oleic acid, 7% palmitic acid with lesser amounts of capric and caprylic acids. See also **Palm Oil**.	$Flash point is 325°F (617°F) Autoignition temperature is 836°F (447°C).

TABLE 2. (*continued*)

Property	Coconut oil	Cottonseed oil	Groundnut (peanut) oil	Pal oil	Rapeseed oil
Color	Straw (Nonyellowing)	Yellow	Pale Yellow	Pale Yellow	
State (at room temperature)	Liquid	Liquid	Liquid (fixed drying oil)	Liquid (semidrying)	
Odor	Almost odorless	Almost odorless	Characteristic	Pleasant	
Melting point	0°–+13 °F; –18°––25 °C	68°–77 °F; 20°–25 °C	77°–88 °F; 22° 31°	3° +°F; –16°––18 °C	
Specific gravity	0.923–0.927	0.9187	0.924–0.929	0.924–0.926	
Saponification value	186–193	188–193	190–193	186–194	
Iodine value	140–152	103–114	137–143	130–135	

Property	Safflower seed oil	Sesame seed oil	Soybean oil	Sunflower oil
Principal constituents	Basic fatty acid composition: Linoleic acid 78%± Oleic acid 13%± Stearic acid 3%± Palmitic acid 6%± Unsaturated acids 90% Saturated acids 10% (Approximate)	Basic fatty acid composition: Oleic acid 40%± Linoleic acid 44%± Other saturates 1%± Palmitic acid 9%± Stearic acid 4%± Other saturates 2%± Unsaturated acids 85% Saturated acids 15% (Approximate)	Basic fatty acid composition: Oleic acid 20–25% Linoleic acid 48–53% Linolenic acid 5–9% Stearic acid 3–5% Palmitic acid 9–12% Myristic acid 0.2%± Arachidic acid 0.2%± Palmitoleic acid 0.3% Lauric acid 0.1% Unsaturated acids 84% Saturated acids 16% (Approximate)	Basic fatty acid composition: Linoleic acid 60–75% Oleic acid 17–32% Palmitic acid 7–10% Stearic acid 4–8% Myristic acid 0.1% Arachidic acid 0.3% Linolenic acid 0.3%± Unsaturated acid 90% Saturated acids 10% (Approximate)
Soluble in		Benzene, carbon disulfide, chloroform, ether. Slightly soluble in alcohol.	Alcohol, carbon disulfide, chloroform, ether	Alcohol, carbon disulfide, chloroform, ether
Grades	Crude, refined	Edible should contain less than free fatty acids. Semi-refined, coast, USP.	Coast, refined (salad), crude, foots (for soapstock), clarified.	Crude, refined
Derivation	Hydraulic press or solvent extraction of seeds.	Pressing of seeds.	Oil for edible purposes is bleached with fuller's earth.	Expression from seeds.
Food uses	Dietetic foods, margarine, hydrogenated shortenings	Shortenings, salad oil, margarine	High-protein foods, margarine, salad dressings	Margarine, shortening
Special notations	Considered by some authorities as the most natural, nutritionally sound vegetable oil. See also **Safflower Seed Oil**.	≠Solidifying point is 23 °F (–5 °C) See also **Sesame Seed Oil**.	See also **Soybean Processing**.	See also **Safflower Seed Oil**.

4%) of skim or nonfat dry milk and salt. There are several unsalted margarines available. Small amounts of emulsifying agents consisting of lecithin and/or monoglycerides and diglycerides are contained in a number of commercially available margarines. The preservatives most commonly used include sodium benzoate, potassium sorbate, calcium disodium EDTA, isopropyl citrate, and citric acid. Most margarines in the United States are fortified with about 15,000 USP units of vitamin A and a few margarines also contain up to 2000 USP units of vitamin D.

Margarines produced in the United States resemble butter in their proximate composition. The great majority of margarines provide more polyunsaturated fatty acids than an equivalent weight of butter. Diet and imitation margarines were introduced about a decade ago and contain less than 80% fat. The fat content of light blends is about 60% and that of diet imitation margarines is about 40%. Lower fat content in these products is compensated for by higher water content.

A survey of some 40 margarines made in the United States in the late 1970s showed that stick margarines made from partially hydrogenated soybean and cottonseed oil were by far the largest category. Stick margarines made from corn (maize) oil and partially hydrogenated corn oil, and from partially hydrogenated soybean oil and liquid cottonseed oil ranked second and third, respectively. The survey showed that margarines with formulations that included liquid cottonseed oil or liquid and/or partially-hardened palm oil usually contained higher amounts of palmitic acid than did other margarines of the same type. The stearic acid content was reasonably constant for most hard and soft margarines. Margarines

made with coconut oil were extremely saturated and contained large amounts of lauric and myristic acids.

Properties of Major Edible Vegetable Oils

The major properties of nine of the principal edible vegetable oils are summarized in Table 2. For descriptions of the constituent acids, see also **Arachidic Acid**; **Capric Acid**; **Caproic Acid**; **Lauric Acid**; **Linoleic Acid**; **Linolenic Acid**; **Myristic Acid**; **Oleic Acid**; **Palmitic Acid**; and **Stearic Acid and Stearates**.

Unconventional Sources of Oils and Proteins

Research efforts continue to find new sources of both food and nonfood oils. Some investigators have reported on the prospects of various gourds indigenous to western North America. Such cucurbits are vigorous and highly drought resistant and produce large quantities of foliage and fruit containing seeds rich in protein and oil, with an extensive system of large, fleshy storage roots which contain starch. Varieties studied have included *Cucurbita foetidissima* and *C. digitata*. Also studied have been varieties of hibiscus (okra). As pointed out by the researchers, the nonconventional protein sources as well as oils would be welcome in a time of protein deficits.

Investigators also have described the potential of the *jojoba plant*, a native of the Sonoran desert of Mexico, Arizona, and California. The desert shrub produces seeds about the size of groundnuts (peanuts) which contain a liquid wax, frequently called *jojoba oil*. The oil is similar to sperm whale oil in its suitability for a number of industrial applications. Sperm whale was placed on the endangered species list by a number of countries, including the United States, in 1971. Prospects of an edible oil from jojoba are undetermined, but the meal remaining after pressing jojoba seeds contains 30–35% protein, and may have potential as a livestock feed. A new jojoba industry would mean a new, renewable resource that has not been exploited before and would offer economic relief to the people of the Sonoran desert region. Other researchers have recently shown a renewed interest in the lupine as a source of edible oils and proteins.

Additional Reading

Bockisch, M.: *Fats and Oils Handbook,* AOCS Press, Champaign, IL, 1998.

Gunstone, F.D. and D. Firestone: *Scientia Gras: A Select History of Fat Science and Technology,* AOCS Press, Champaign, IL, 2000.

O'Brien, R.D., W.E. Farr, and P.J. Wan: *Introduction to Fats and Oils Technology,* 2nd Edition, AOCS Press, Champaign, IL, 2000.

Przybylski, R. and B.E. McDonald: *Development and Processing of Vegetable Oils for Human Nutrition,* AOCS Press, Champaign, IL, 1995.

Widak, N.: *Physical Properties of Fats, Oils, and Emulsifiers,* AOCS Press, Champaign, IL, 2000.

VESUVIANITE. The mineral vesuvianite is a very complex silicate of calcium and aluminum with fluorine which may also contain varying amounts of boron, iron, lithium, magnesium, manganese, potassium, sodium and titanium. A suggested formula is $Ca_{10}Mg_2Al_4(SiO_4)_5$ $(Si_2O_7)_2(OH)_4$. Its tetragonal crystals are usually short, somewhat stoutish prisms, sometimes pyramids, but columnar to massive varieties are common. It is essentially without cleavage; fracture, uneven; brittle hardness, 6–7; specific gravity, 3.3–3.5; luster, vitreous to greasy resinous; color, commonly some shade of brown, green, or white, may be reddish, bluish or yellowish; may be transparent, but is usually translucent.

This mineral was formerly called idocrase, having been named Haüy from the Greek words meaning *form* and *mixture* because it resembled crystals of other species, scarcely a valid distinction. Werner gave it the name vesuvianite from Mt. Vesuvius where it was first found blocks of limestone appearing as inclusions in the lava. Vesuvianite not a constituent of the igneous rocks, but rather a contact metamorphic mineral resulting from the alteration of impure limestones and dolomites It is usually associated with diopside, wollastonite, epidote, grossularite and garnet. There are many localities worthy of mention among which are: The Urals, Central Europe, Rumania, Trentino and Monzoni in Italy, as well as at Mt. Vesuvius and Mt. Somma, Switzerland, Mexico and Japan. In the United States vesuvianite is found in Androscoggin and York Counties in Maine; Orange and Warren Counties, New York; Sussex County, New Jersey; Garland County, Arkansas; and Riverside and Tulare Counties in California.

VINEGAR. Vinegar is the liquid condiment or food flavoring used to give a sharp or sour taste to foods. It is also used as a preservative in pickling and as the sour component in many different sauces, dressings, and gravies.

Vinegar results from the action of the enzymes of bacteria of the genus *Acetobacter* and some others on dilute solutions of ethyl alcohol such as cider, wine, beer, or diluted distilled alcohol. See also **Ethyl Alcohol**. Most vinegars for table use, e.g., in the dressing of salads, derive from the acetic acid–bacterial fermentation of wine or cider. See also **Acetic Acid**. These latter, in turn, are produced by alcoholic fermentation (see also **Fermentation**) of dilute sugar solutions such as grape juice, apple juice, or malt. *Saccharomyces cerevisiae* is the yeast involved most frequently in alcoholic fermentation, i.e., in the enzymatic conversion of fermentable sugars to dilute alcoholic solutions. See also **Yeasts and Molds**. Although fruits and honey are used most frequently as sources of fermentable sugar for vinegar production, barley malt and, in the Orient, rice, after hydrolysis of starch, serve as primary sources. See also **Food Processing**.

Some raw materials used for vinegar production are listed in Table 1.

A number of factors govern the composition of vinegar: the nature of the raw material, the substances added to promote alcoholic fermentation and the growth and activity of *Acetobacter*, the procedure used for the acetification, and finally the aging, stabilization, and bottling operations.

Manufacture

Only since the mid-1800s has it been known that yeast and bacteria are the cause of fermentation and vinegar formation.

TABLE 1. RAW MATERIALS USED TO MAKE VINEGAR

Raw material
MAINLY SUGARY
jujube
sweet potato
dates
citrus
persimmon
pear
sugar cane
plum
tomato
kiwi fruit
pineapple
molasses
honey
palm sap
muscavado (brown sugar)
MAINLY STARCHY
potato, corn flour
soyabean
seaweed
rice
grain starch
VARIOUS
onions
bamboo grass
wood
whey
coconut water
vinasse (distillation residue)

Starch Hydrolysis and Alcoholic Fermentation. In general, because yeasts cannot utilize starch directly as a carbon source, the starch must first be hydrolyzed to sugar. Malt vinegars are made from malted barley or a mixture of malted barley with other starchy grains. Malt enzymes convert starch to sugars readily fermentable by *Saccharomyces* yeasts. In Japan, where vinegars are made from rice, a mixture of hydrolyzing enzymes produced by the fungus *Aspergillus oryzae* converts rice starches to sugars. Alcoholic fermentation is frequently conducted in two phases, although in a modern vinegar plant it can be conducted in one. The first phase is a vigorous fermentation during which the rapid evolution of carbon dioxide protects the alcoholic solution from air. The second or slower phase is fermentation of the residual sugar at a lower rate, during which, again, protection from air is required. In the first phase, 50–100 mg SO_2/L of sugarcontaining mash is added, followed after approximately 1 h by 1–3% of an actively fermenting, pure-culture starter of *Saccharomyces cerevisiae*. The fermentation process is monitored for disappearance of sugar and increase in temperature. Rates of alcoholic fermentation are highest at ca 25–30°C; higher temperatures tend to damage enzyme systems.

Both the fermentation of hexose sugars to ethanol and carbon dioxide and the oxidation of ethanol to acetic acid are exothermic (heat yielding) processes (see **Sugar**). Depending on the size of the fermenter and the rates of fermentation and aeration, loss of waste heat is apportioned among radiation, conduction, and vaporization of water and ethanol plus carbon dioxide. Although small fermenters may require no cooling, large ones require more cooling than occurs through natural radiation and conductance.

Acetic Acid. Ethyl alcohol is converted to acetic acid by air oxidation catalyzed by the enzymes within bacteria of the genus *Acetobacter*:

$$46 \text{ g } C_2H_5OH + 32 \text{ g } O_2 \longrightarrow 60 \text{ g } CH_3COOH + 18 \text{ g } H_2O$$
$$+ 487.2 \text{ kJ}(116.4 \text{ kcal})$$

One gram of ethanol should yield 1.304 g acetic acid. Practical yields are 77–85%. In contrast with the well-known Embden-Meyerhof-Parnass glycolysis pathway for the conversion of hexose sugars to alcohol, the steps in conversion of ethanol to acetic acid remain in some doubt.

Orleans Process. In the Orleans process, wine oxidizes slowly in a barrel where it is covered with a film of *Acetobacter*. Holes that are covered with screens to exclude insects are bored in each barrel head to permit access to air. Wine is added through the bung hole with a long-stemmed funnel below the surface of the bacterial film and without disturbing the film. Wines with 10–12% ethanol give vinegars of 8–10% acetic acid concentration. Orleans vinegars are characterized by a relatively high concentration of ethyl acetate, detected by its pleasantly strong fruity odor.

Modena-Style or Balsamic Vinegar Process. Balsamic vinegars are made in and around the city of Modena in central Italy and consist of two general types. The traditional product is made from juice of the Trebbiano grape concentrated to about 40% sugar by direct flame heating of the container. *Zygosaccharomyces, Saccharomyces cerevisiae,* and *Gluconobacter* in barrels convert the sugars to ethyl alcohol and gluconic acid as principal products. The ethyl alcohol is converted to acetic acid by the *Acetobacter* and *Gluconobacter.* Subsequent reactions form many taste and odor substances, among which the ester ethyl acetate is of key importance. Years of age in barrels result in a condiment having a harmonious balance of sweetness and tartness. Aceto Balsamico di Modena differs from the Tradizionale in that it is a blend of ordinary wine vinegar with concentrate. Lacking the complex character of Tradizionale and being younger, it is much less expensive.

Generator Process. Generators are packed with shavings of beech wood, which tend to curl and thus provide packing that does not consolidate but allows open spaces for the free flow of liquid and air. Beech wood does not contribute undesirable flavors or impurities to the vinegar. In the modern generator, a recirculating pump transfers the partially acetified alcoholic mixture from the bottom section of the generator to a distributing system at the top of the packed section. Cooling coils may be located in the packed section, but more frequently are placed at the bottom of the receiver section or are incorporated in the line for recirculating the liquid.

Various species and many strains of *Acetobacter* are used in vinegar production. Aeration rates, optimum temperatures and nutrient requirements vary with individual strains. In general, fermentation alcohol substrates require minimal nutrient supplementation while their addition is necessary for distilled alcohol substrates.

Adaptation of the surface-film growth procedure for producing antibiotics to an aerated submerged-culture process has been successful in making vinegar.

The most widely used submerged-culture oxidizer is the Frings acetator. It uses a bottom-driven hollow rotor turning in a field of stationary vanes arranged in such a way that the air which is drawn in is intimately mixed with the liquid throughout the whole bottom area of the tank.

Submerged-culture oxidizers are usually operated on a semicontinuous basis. In most cases, ca half the liquid in the tank is removed every 1–2 d, when the alcohol concentration has dropped to 0.1–0.2 vol %. The removed vinegar is replaced with wine or mash of richer ethanol and lower acetic acid concentration, giving a mixture in the tank of 5–6 vol % ethanol and 6–8 vol % of acetic acid. These are the optimum conditions for *Acetobacter* growth.

Foam production is most troublesome under conditions adverse to bacterial growth and thus can be minimized by keeping nutrient, ethanol, and acetic acid concentrations in the optimum ranges (see **Defoaming Agents**). Temperature and aeration rate are also critical.

Submerged culture oxidizers can also be operated on a continuous basis. Optimum production, however, is achieved by semicontinuous operation because the composition of vinegar desired in the withdrawal stream is so low in ethanol that vigorous bacterial growth is impeded. Bacterial concentrations up to 100×10^6 cells/cm^3 have been reported in generators making about 20% vinegars.

Fluidized-Bed Vinegar Reactors. Intimate contact of air with *Acetobacter* cells is achieved in fluidized-bed or tower-type systems. Air introduced through perforations in the bottom of each unit suspends the mixture of liquid and microorganisms within the unit. Air bubbles penetrating the bottom plate keep *Acetobacter* in suspension and active for the ethanol oxidation in the liquid phase.

Vinegars with High Concentrations of Acetic Acid. The U.S. regulations require at least 4 g acetic acid/100 cm^3 vinegar. Commercial vinegar and many quality table vinegars are significantly more concentrated. Submerged-culture oxidizers easily give acetic acid concentrations of 10–13 g/cm^3. Submerged-culture oxidizer techniques that produce vinegars with acetic acid concentrations ranging from 15–20% are now in commercial use. Continuous aeration, careful stepwise addition of ethanol as it is oxidized and careful control of temperature seem to be the keys to successful operation.

Vinegar Eels and Mother of Vinegar. Although esthetically undesirable, nematodes, known as vinegar eels, may actually be of some assistance in consuming dead bacteria from the surface of the packing material in the vinegar tank, and thus may aid in prolonging the operation of the system. Vinegar eels are removed from the raw vinegar by filtration and pasteurization before the vinegar is sold or used further in pickling or other processes.

Processing

Raw vinegars vary widely in stability and may contain materials that form cloudiness or deposits. Vinegars that carry a high and cloud suspension of bacterial cells are clarified and stabilized with bentonite and similar agents or activated carbon. Membrane filtration can be combined with aseptic bottling to provide a product free of all microorganisms. Wine, cider, and malt vinegars benefit from aging, which improves the character of the product. Vinegars bottled for table use and pickling usually are pasteurized at 77–78°C.

See also **Antimicrobial Agents (Foods).**

A. DINSMOOR WEBB
University of California, Davis

Additional Reading

Amerine, M.A. and co-workers: *Technology of Wine Making,* 4th Edition, AVI, Westport, Conn., 1980.
Ameyama, M. and S. Otsuka, eds.: *Science of Vinegar,* Asakura Publishing Co., Ltd., Tokyo, 1990.
Ebner, H. and H. Follmann: in G. Reed, ed., *Biotechnology,* Vol. 5, Verlag Chemie, Weinheim, Germany, 1983, pp. 425–446.
Ekundayo, J.A.: *Brit. Mycol. Soc. Symp. Ser. 3 (Fungal Biotechnol.),* 243–271, 1980.

Web Reference

The Vinegar Institute: http://www.versatilevinegar.org/

VINYL ACETAL POLYMERS.   Vinyl acetal polymers are made by the acid-catalyzed acetalization of poly(vinyl alcohol) with aldehydes.

Although many members of this class of resins have been made, only poly(vinyl formal) (PVF) and poly(vinyl butyral) (PVB) are made in significant commercial quantities.

Synthesis and Structure

Poly(vinyl acetals) are made from poly(vinyl alcohol) and aldehydes by acid-catalyzed addition–dehydration. The degree of acetalization and the conditions used during the reaction significantly affect product properties. Batch and continuous processes in both aqueous and organic media are used during manufacturing. In single-stage batch processes, hydrolysis of poly(vinyl acetate) and acetalization of the poly(vinyl alcohol) hydrolysis product are carried out in the same kettle at the same time. In two-stage batch processes, hydrolysis and acetalization take place in separate kettles.

Physical Properties

Unformulated poly(vinyl acetal) resins form hard, unpliable materials which are difficult to process without using solvents or plasticizers. Plasticizers aid resin processing, lower the glass-transition temperature, T_g, and can profoundly change other physical properties of the resins.

Poly(vinyl acetal)s can be formulated with other thermoplastic resins and with a variety of multifunctional cross-linkers. When cross-linking takes place the resin becomes thermoset. Thermosetting generally increases thermal stability, rigidity, and abrasion resistance, and improves resistance to solvents and to acids and bases. It also severely limits processibility by making the resin insoluble and impossible to extrude.

Health and Safety Factors

Representative unformulated PVB and PVF resins are practically nontoxic orally (rats) and no more than slightly toxic after skin application (rabbits).

Some forms of these products may contain sufficient fines to be considered nuisance dust and present dust explosion potential if sufficient quantities are dispersed in air. Unformulated PVB and PVF resins have flash points above 370°C. The lower explosive limit (lel) for PVB dust in air is about 20 g/m^2.

TABLE 1. PHYSICAL PROPERTIES OF BUTVAR RESINS

Property	Method	B-72	B-74	B-76	B-90	B-98
mol mass $\times 10^3$ (avg)	a	170–250	120–250	90–120	70–100	40–70
viscosity 15 wt %, Pa·s[b]	c	7–14	3–7	0.5–1	0.6–1.2	0.2–0.4
viscosity 10 wt %, Pa·s[b]	d	1.6–2.5	0.8–1.3	0.2–0.45	0.2–0.4	0.07–0.2
Ostwald soln viscosity, mPa·s[b] (= cP)	e	170–260	40–50	18–28	13–17	6–9
specific gravity, 23°C/23°C	ASTM D792-50	1.100	1.100	1.083	1.100	1.100
refractive index	ASTM D542-50	1.490	1.490	1.485	1.490	1.490
vinyl alcohol content, wt %		17–20	17–20	11–13	18–20	18–20
vinyl acetate content, wt %		0–2.5	0–2.5	0–1.5	0–1.5	0–2.5

[a] Determined by size exclusion chromatography in tetrahydrofuran with low angle light scattering.

[b] To convert Pa·s to P, multiply by 10.

[c] Measured in 60:40 toluene:ethanol at 25°C using a Brookfield viscometer.

[d] Measured in 95% ethanol at 25°C using an Ostwald-Cannon-Fenske viscometer.

[e] B-72 in 7.5 wt % anhydrous methanol at 20°C; B-76 and B-79 in 5.0 wt % SD 29 ethanol at 25°C; B-74, B-90, and B-98 in 6.0 wt % anhydrous methanol at 20°C, all using an Ostwald-Cannon-Fenske viscometer.

Poly(vinyl butyral)

Several grades are available that differ primarily in residual vinyl alcohol content and molecular weight. Both variables strongly affect solution viscosity, melt flow characteristics, and other physical properties. Some physical, properties of various grades of Monsanto's Butvar resins are listed in Table 1. In general, resin melt and solution viscosity increase with increasing molecular weight and vinyl alcohol content, whereas the tensile strength of materials made from PVB increases with vinyl alcohol content for a given molecular weight.

Commercially available PVB resins are generally soluble in lower molecular weight alcohols, glycol ethers, and certain mixtures of polar and nonpolar solvents. A common solvent for all of the Butvar resins is a combination of 60 parts of toluene and 40 parts of ethanol (95%) by weight.

PVB resins are also compatible with a limited number of plasticizers and resins. Plasticizers improve processibility, lower T_g, and increase flexibility and resiliency over a broad temperature range.

PVB combinations with the thermoplastic resins nitrocellulose or shellac have been used as sealers for wood finishing. In these applications the PVB component adds flexibility and adhesion. Tough, optically clear blends have been made with aliphatic polyurethanes. Thermosets are prepared with cross-linkers that form covalent bonds with hydroxyl groups.

Manufacture. PVBs are manufactured by a variety of two-stage heterogeneous processes. In one of these an alcohol solution of poly(vinyl acetate) and an acid catalyst are heated to 60–80°C with strong agitation. As the poly(vinyl alcohol) forms, it precipitates from solution. As the reaction approaches completion the reactants go into solution. When the reaction is complete, the catalyst is neutralized and the PVB is precipitated from solution with water, washed, and centrifuged and dried. Resin from this process has very low residual vinyl acetate and very low levels of gel from intermolecular acetalization.

In the second stage of a representative aqueous process, an aqueous solution of poly(vinyl alcohol) is heated with butyraldehyde and an acid catalyst. PVB precipitates from solution as it forms. Because PVB resin precipitates early in the reaction there is a tendency toward high levels of intermolecular acetalization. Cross-linking can be minimized by adding emulsifiers to control particle size or substances like ammonium thiocyanate or urea to improve the solubility of PVB in the aqueous phase.

Applications. During 1994, about 68,000 t of unplasticized PVB was manufactured worldwide. Of this, the overwhelming majority, about 66,000 t, was plasticized and extruded into sheet for use in laminated safety glass. Only about 2,300 t of unplasticized PVB was used for noninterlayer applications.

Plasticized PVB is uniquely suited for safety and security glazing applications. It is easily extruded into sheet. The laminated sheet exhibits high adhesion to glass, optical clarity, stability to sunlight, and high tear strength and impact-absorbing characteristics, all of which are demanded for safety glazing use.

Most laminated safety glazings are glass-PVB-glass trilayer composites. Adhesion of plasticized PVB to clean glass is very high.

Some categories of nonglazing uses for PVB are as follows: phenolic/adhesives; metal/glass binders; hard copy printing; and coatings/additives.

Poly(vinyl formal)

Estimated worldwide production of poly(vinyl formal) resin was about 2700 t in 1994. PVF resins are currently manufactured by Wacker Chemie (Pioloform F) in Germany and by Chisso (Vinylec) in Japan. Poly(vinyl formal) resins are free-flowing white powders with a poly(vinyl formal) content of about 81 wt %. Chemical resistance of poly(vinyl formal) to acids, bases, and aliphatic hydrocarbons is excellent, and chemical resistance to alcohols, aromatic hydrocarbons, esters, and ketones is good; however, chemical resistance to chlorinated solvents is rated poor. In general, PVF resins are soluble in a limited number of solvents and in certain mixtures of alcohols and aromatic hydrocarbons. The solubility of PVF resins in polar solvents increases with increasing proportions of residual vinyl acetate content. Increasing vinyl acetate content also reduces resin stiffness and tensile and impact strength. Solution viscosity increases with increasing vinyl alcohol content and with the average molecular weight of the resin.

PVF resins are generally compatible with phthalate, phosphate, adipate, and dibenzoate plasticizers, and with phenolic, melamine–formaldehyde, urea–formaldehyde, unsaturated polyester, epoxy, polyurethane, and cellulose acetate butylate resins. They are incompatible with polyamide, ethyl cellulose, and poly(vinyl chloride) resins.

Commercial PVF is manufactured by a single-stage batch process in acetic acid. The ratio of vinyl acetate and vinyl alcohol components in the acetal product is controlled by the ratio of acetic acid, water, and formaldehyde used.

Applications. PVF resins are used almost exclusively to make electric and magnetic wire insulation. In these applications the PVF resin component helps provide toughness, as well as abrasion and thermal resistance.

PVF resins have also been used in a variety of other applications, including conductive films, electrophotographic binders, as a component for inks and in membranes, photoimaging, solder masks, and reprographic toners.

JEROME W. KNAPCZYK
Solutia

Additional Reading

Brandrup, J., D. R. Bloch, E. A. Grulke, E. H. Immergut, and A. Abe: *Polymer Handbook,* 2 Volume Set, 4th Edition, John Wiley & Sons, Inc., New York, NY, 2003.

Butvar, Polyvinyl Butyral Resin, Technical Bulletin No. 8084A, Monsanto Chemical Co., St. Louis, Mo., 1991.

Farmer, P. H. and B. A. Jemmott: in I. Skeist, ed., *Handbook of Adhesives,* 3rd Edition, Van Nostrand Reinhold, New York, 1990, pp. 423–436.

Fried, J.: *Polymer Science and Technology,* 2nd Edition, Prentice Hall, Inc., Upper Saddle River, NJ, 2003.

Kroschwitz, J. I.: *Encyclopedia of Polymer Science and Technology,* 12 Volume Set, 3rd Edition, John Wiley & Sons, Inc., Hoboken, NJ, 2004.

Vinylec, Polyvinyl Formal Resins, Technical Bulletin, Chisso America Corp., New York, NY, 1994.

VINYL ACETATE POLYMERS. Vinyl acetate is a colorless, flammable liquid having an initially pleasant odor which quickly becomes sharp and irritating. Table 1 lists the physical properties of the monomer.

The most important chemical reaction of vinyl acetate is free-radical polymerization. The reaction is summarized as follows:

$$n\ CH_2{=}CHOCCH_3 \longrightarrow \left[CH_2{-}CH\right]_n$$

Vinyl acetate has been polymerized by bulk, suspension, solution, and emulsion methods. It copolymerizes readily with some monomers but not with others.

TABLE 1. SOME PROPERTIES AND CHARACTERISTICS OF VINYL ACETATE

Property	Value
formula weight	86.09
physical state	liquid
flammable limits in air (101.3 kPa[a]), vol %	LEL 2.6, UEL 13.4
flash point,°C	
Tag closed cup (ASTM D56)	−8
Tag open cup (ASTM D1310)	−4
autoignition temperature,°C	426.9
bp at 101.3 kpa[a],°C	72.7
relative evaporation rate (n − butyl acetate = 1)	8.9
vapor pressure at 20°C, kPa[a]	11.8
critical temp,°C	246
critical pressure, kPa[b]	3950
color	clear and colorless
sp gr, 20/20°C	0.934
vapor density (air = 1.00)	2.97
viscosity at 20°C	0.43 cps
fp,°C	−92.8
heat of combustion at 25°C	−495.0 kcal/mol
heat of vaporization (1 atm)	87.6 cal/g
heat of formation (liquid at 25°C), kJ/mol[c]	−349.4
heat of polarization, kJ/mol[c]	89.1
specific heat at 20°C (liquid)	0.46 cal/g°C
odor	sweetish smell in small quantities
reactivity	reactive with self and variety of other chemicals; stable when properly stored and inhibited
light sensitivity	light promotes polymerization
electrical conductivity at 23°C	2.6 × 104 pS/m(1 S = 1 mho)
refractive index, n_D^{20}	1.3953
surface tension at 20°C, mN/m(= dyn/cm)	23.6
coefficient of cubical expansion	0.00137/°C at 20°C

[a] To convert kPa to mm Hg, multiply by 7.5.
[b] To convert kPa to atm, divide by 101.3.
[c] To convert kJ/mol to kcal/mol, divide by 4.184.

TABLE 2. PHYSICAL CONSTANTS FOR POLY(VINYL ACETATE)

Property	Value
absorption of water at 20°C for 24–144 h, %	3–6
coefficient of thermal expansion, K^{-1}	
cubic	6.7×10^{-4}
linear, below T_g	7×10^{-5}
above T_g	22×10^{-5}
compressibility, cm³/(g·kPa)[a]	17.8×10^{-6}
decomposition temperature,°C	150
density at 20°C, g/cm³	1.191
dielectric constant at 50°C, at 2 MHz	3.5
dielectric dissipation factor at 50°C, at 2 MHz, tan δ	150
dielectric strength at 30°C, V/L	0.394
dipole moment at 20°C, C·m[b] per monomer unit	2.30
elongation at break, at 20°C and 0% rh, %	10–20
glass-transition temperature, T_g, °C	28–31
pressure dependence,°C/100 MPa[c]	0.22
hardness, at 20°C, Shore units	80–85
heat capacity, at 30°C, J/g[d]	1.465
heat distortion point,°C	50
heat of polymerization, kJ/mol[d]	87.5
refraction index at 20.7°C, n_D	1.4669
interfacial tension, mN/m (= dyn/cm)	
at 20°C with polyethlene	14.5
20°C with polystyrene	4.2
modulus of elasticity, GPa[c]	1.275–2.256
notched impact strength, J/m[e]	102.4
softening temperature,°C	35–50
surface resistance (ohm/cm)	5×10^{11}
surface tension, mN/m(= dyn/cm)	
at 20°C	36.5
180°C	25.9
tensile strength, MPa[c]	29.4–49.0
thermal conductivity, mW/(m · K)	159
Young's modulus, MPa[c]	600

[a] To convert kPa to atm, divide by 101.3.
[b] To convert C · m to debye, divide by 3.336×10^{-30}.
[c] To convert MPa to psi, multiply by 145; GPa to psi, multiply by 145,000.
[d] To convert J to cal, divide by 4.184.
[e] To convert J/m to lbf/in., divide by 53.38.

Hydrolysis of vinyl acetate is catalyzed by acidic and basic catalysts to form acetic acid and vinyl alcohol which rapidly tautomerizes to acetaldehyde. This rate of hydrolysis of vinyl acetate is 1000 times that of its saturated analogue, ethyl acetate, in alkaline media. Other chemical reactions which vinyl acetate may undergo are addition across the double bond, transesterification to other vinyl esters, and oxidation.

Vinyl acetate monomer is supplied in three grades, which differ in the amount of inhibitor they contain but otherwise have identical specifications. In storage vinyl acetate should be kept away from ignition sources. It should be stored in a cool environment away from heat, direct sunlight, oxidizing materials, and free-radical generating chemicals to avoid rapid uncontrolled polymerization. Bulk storage should be blanketed with dry nitrogen.

Health and Safety Aspects

NIOSH recommends a ceiling limit of 4 ppm for 15 minutes of exposure to vinyl acetate vapor. The ACGIH recommends an 8-hour TLV Time Weighted Average of 10 ppm and a 15 minute short-term exposure limit of 15 ppm to its vapor. Vinyl acetate is a severe eye and skin irritant, forming blisters on the skin, and redness, swelling, or corneal burns on the eyes. The vapor is irritating to the nose and throat, and high levels of exposure may result in pulmonary edema. Vinyl acetate has moderate acute toxicity if ingested.

Vinyl Acetate Polymers

Properties. Poly(vinyl acetate) (PVAc) polymer resins are manufactured in a variety of molecular weights. Some physical properties of the polymer are listed in Table 2. With increasing molecular weight, properties vary from viscous liquids to low melting solids to tough, horny materials. They are neutral, water-white to stray-colored, tasteless, odorless, and nontoxic. The resins have no sharply defined melting points but become softer with increasing temperature. Due to their solubility parameter, they are soluble in organic solvents, but are insoluble in the lower alcohols (excluding methanol), glycols, water, and nonpolar liquids.

The chemical properties of PVAc are those of an aliphatic ester. Thus, acidic or basic hydrolysis produces poly(vinyl alcohol) and acetic acid or the acetate of the basic cation.

Poly(vinyl acetate) emulsion films adhere well to most surfaces that have relatively high surface energy, (e.g., wood, paper, glass, and metal), and have good binding capacity for pigments and fillers. PVAc film can be laid down on a damp surface with trapped moisture gradually passing through the film without lifting or blistering it. Poly(vinyl acetate) polymers are environmentally friendly because they easily biodegrade.

Polymerization Processes. Vinyl acetate has been polymerized industrially by bulk, solution, suspension, and emulsion processes. Perhaps 90% of the material identified as poly(vinyl acetate) or copolymers that are predominantly vinyl acetate are made by emulsion techniques.

Emulsion Polymerization. Poly(vinyl acetate)-based emulsion polymers are produced by the polymerization of an emulsified monomer through free-radicals generated by an initiator system. An emulsion recipe, in general, contains monomer, water, protective colloid or surfactant, initiator, buffer, and perhaps a molecular weight regulator.

Many different combinations of surfactant and protective colloid are used in emulsion polymerizations of vinyl acetate as stabilizers.

The properties of the emulsion and the polymeric film depend to a large extent on the identity and quantity of the stabilizers. Poly(vinyl acetate) emulsions can be made with a surfactant alone or with a protective colloid alone, but the usual practice is to use a combination of the two. In general, the greater the quantity of stabilizers in a recipe, the smaller the particle size of the emulsion.

The initiators used in vinyl acetate polymerizations are the familiar free-radical types. Buffers are frequently added to emulsion recipes. Vinyl acetate emulsion polymerization recipes are usually buffered to pH 4–5. The pH of most commercially available emulsions is 4–6.

A chain-transfer agent is added to vinyl acetate polymerizations to control the polymer molecular weight.

A polymerization process may consist of simply charging all ingredients to the reactor, heating to reflux, and stirring until the reaction is over while controlling the heat removal at the reaction temperature using cooling systems. However, this simple procedure is seldom followed.

Industrially, polymerizations are carried out to over 99% conversion and thus there is no need to reduce the unreacted monomer unless very low levels are required to meet regulatory, product, or workplace requirements. Most poly(vinyl acetate) emulsions contain less than 0.5 wt % unreacted vinyl acetate. All of the processes are operated in conventional glass-lined or stainless steel kettles or reactors. Control of the process is important to ensure reproducibility of the product.

Bulk Polymerizations. In the bulk polymerization of vinyl acetate the viscosity increases significantly as the polymer forms making it difficult to remove heat from the process. Low molecular weight polymers have been made in this fashion. Continuous processes are known to be used for bulk polymerizations.

Suspension Polymerization. The suspension or pearl polymerization process has been used to prepare polymers for adhesive and coating applications and for conversion to poly(vinyl alcohol). Suspension polymerization are carried out with monomer-soluble initiators predominantly, with low levels of stabilizers. Continuous tubular polymerization of vinyl acetate in suspension yields stable dispersions of beads with narrow particle size distributions at high yields.

Solution Polymerization. Solution polymerization of vinyl acetate is carried out mainly as an intermediate step to the manufacture of poly(vinyl alcohol). A small amount of solution-polymerized vinyl acetate is prepared for the merchant market. When solution polymerization is carried out, the solvent acts as a chain-transfer agent, and depending on its transfer constant, has an effect on the molecular weight of the product. Continuous solution polymers of poly(vinyl acetate) in tubular reactors have been prepared at high yield and throughput.

Propagation. The rate of emulsion polymerization has been found to depend on initiator, monomer, and emulsifier concentrations. Vinyl acetate polymerizes chiefly in the usual head-to-tail fashion, but some of the monomers orient head-to-head and tail-to-tail as the chain grows. In vinyl acetate polymerizations, the molecular weights of the products increase with the extent of conversion: the ratio of weight-to-number-average-degree-of-polymerization also changes, becoming larger at higher conversions.

Chain Transfer. At the molecular scale, vinyl acetate polymerizations generally are understood as free-radical polymerizations, but are characterized in particular by a relatively large amount of chain transfer.

Grafting and Stabilizers. The degree of grafting of poly(vinyl acetate) (PVAc) on poly(vinyl alcohol) (PVA) and other stabilizers during emulsion polymerization strongly affects latex properties such as viscosity, rheology, and polymer solubility.

Copolymers. Vinyl acetate copolymerizes easily with a few monomers, e.g., ethylene, vinyl chloride, and vinyl neodecanoate, which have reactivity ratios close to its own. Block copolymers of vinyl acetate with methyl methacrylate, acrylic acid, acrylonitrile, and vinyl pyrrolidinone have been prepared by copolymerization in viscous conditions, with solvents that are poor solvents for the vinyl acetate macroradical.

Blends. Latex film properties are commonly modified through the blending of latexes, e.g., a "soft" polymer is made slightly harder by blending with a "hard" latex.

Specifications and Standards. Borax stability is an important property in adhesives, paper, and textile applications. Other emulsion properties tabulated by manufacturers include tolerance to specific solvents, surface tension, minimum film-forming temperature, dilution stability, freeze–thaw stability, percent soluble polymer, and molecular weight.

Poly(vinyl acetate) and its copolymers with ethylene are available as spray-dried emulsion solids with average particle sizes of 2–20 μm; the product can be reconstituted to an emulsion by addition of water or it can be added directly to formulations, e.g., concrete.

Uses. The uses of poly(vinyl acetate) adhesive are packaging and wood gluing. PVAc copolymer adhesives are finding application in more diverse areas such as construction and adhesion to more difficult to bond surfaces because of the range of adhesion and the flexibility that may be built into the polymer.

Poly(vinyl acetate) emulsions can be used in high speed gluing equipment. Poly(vinyl acetate) homopolymers adhere well to porous or cellulosic surfaces, e.g., wood, paper, cloth, leather, and ceramics. Homopolymer films tend to creep less than copolymer or terpolymer films. They are especially suitable in adhesives for high speed packaging operations.

Poly(vinyl acetate) dry resins and ethylene–vinyl acetate (EVA) copolymers are used in solvent adhesives, which can be applied by total industrial techniques, e.g., brushing, knife-coating, roller-coating, spraying, or dipping.

Paints prepared from poly(vinyl acetate) and its copolymers form flexible, durable films with good adhesion to clean surfaces, including wood, plaster, concrete, stone, brick, cinder blocks, asbestos board, asphalt, tar paper, wallboards, aluminum, and galvanized iron.

Poly(vinyl acetate) emulsions and resins have been used as the binder in coatings for paper and paperboard since 1955. The coatings may be clear, colored, or pigmented, and are glossy, odorless, tasteless, grease-proof, nonyellowing, and heat-sealable. Conventional paper-coating equipment is used.

The use of vinyl acetate copolymers as binding agents for nonwoven fabrics has grown rapidly. Poly(vinyl acetate) was first used in concrete in the 1940s as a thermoplastic polymer to strengthen the concrete matrix. Vinyl acetate resins are useful as antishrinking agents for glass fiber-reinforced polyester molding resins and as binders for numerous materials. Emulsions containing added poly(vinyl alcohol) and dichromate are used to make light-sensitive stencil screens for textile printing and ceramic decoration. Vinyl acetate polymers have long been used as chewing gum bases.

CAJETAN F. CORDEIRO
Air Products and Chemicals, Inc.

Additional Reading

Brandrup, J., D. R. Bloch, E. A. Grulke, E. H. Immergut, and A. Abe: *Polymer Handbook,* 2 Volume Set, 4th Edition, John Wiley & Sons, Inc., New York, NY, 2003.
Chemical Economics Handbook Marketing Research Report, Vinyl Acetate, (July 1993).
Fried, J.: *Polymer Science and Technology,* 2nd Edition, Prentice Hall, Inc., Upper Saddle River, NJ, 2003.
Kroschwitz, J. I.: *Encyclopedia of Polymer Science and Technology,* 12 Volume Set, 3rd Edition, John Wiley & Sons, Inc., Hoboken, NJ, 2004.
Lindemann, M. K.: in G. E. Ham, ed., *Vinyl Polymerization,* Vol. 1, Marcel Dekker, Inc., New York, NY, 1967, Part 1, Chapt. 4.

VINYL ALCOHOL POLYMERS. Poly(vinyl alcohol) (PVA), a poly-hydroxy polymer, is the largest-volume synthetic, water-soluble resin produced in the world. It is commercially manufactured by the hydrolysis of poly(vinyl acetate), because monomeric vinyl alcohol cannot be obtained in quantities and purity that makes polymerization to poly(vinyl alcohol) feasible.

The main uses of PVA are in textile sizing, adhesives, protective colloids for emulsion polymerization, fibers, production of poly(vinyl butyral), and paper sizing. Significant volumes are also used in the production of concrete additives and joint cements for building construction and water-soluble films for containment bags for hospital laundry, pesticides, herbicides, and fertilizers. Smaller volumes are consumed as emulsifiers for cosmetics, temporary protective film coatings, soil binding to control erosion, and photoprinting plates.

Physical Properties

The physical properties of poly(vinyl alcohol) are highly correlated with the method of preparation. The final properties are affected by the polymerization conditions of the parent poly(vinyl acetate), the hydrolysis conditions, drying, and grinding. Further, the term poly(vinyl alcohol) refers to an array of products that can be considered copolymers of vinyl acetate and vinyl alcohol. Representative properties are shown in Table 1.

The ability of PVA to crystallize is the single most important physical property of PVA as it controls water solubility, water sensitivity, tensile strength, oxygen barrier properties, and thermoplastic properties.

The glass-transition temperature, T_g, of fully hydrolyzed PVA has been determined to be 85°C for high molecular weight material. Poly(vinyl alcohol) is only soluble in highly polar solvents, such as water, dimethyl sulfoxide, acetamide, glycols, and dimethylformamide. The solubility in water is a function of degree of polymerization (DP) and hydrolysis.

TABLE 1. PHYSICAL PROPERTIES OF POLY(VINYL ALCOHOL)

Property	Value
appearance	white to ivory-white granular powder
specific gravity	1.27–1.31
tensile strength, MPa[a]	67–110[b]
elongation, %	0–300
specific heat, $J/(g \cdot K)$[c]	1.67
thermal conductivity, $W/(m \cdot K)$	0.2
T_g, K	358[b]
mp, K	503[b]
electrical resistivity, $\Omega \cdot cm$	$(3.1–3.8) \times 10^7$
refractive index, n_D^{20}	1.55
degree of crystallinity	0–0.54
storage stability (solid)	indefinite when protected from moisture
flammability	similar to paper
stability in sunlight	excellent

[a] To convert MPa to psi, multiply by 145.
[b] 98–99% hydrolyzed.
[c] To convert J to cal, divide by 4.184.

The viscosities of PVA solutions are mainly dependent on molecular weight and solution concentration. The viscosity increases with increasing degree of hydrolysis and decreases with increasing temperature.

The tensile strength of unplasticized PVA depends on degree of hydrolysis, molecular weight, and relative humidity. Tensile elongation of PVA is extremely sensitive to humidity and ranges from <10% when completely dry to 300–400% at 80% rh. Addition of plasticizer can double these values. Poly(vinyl alcohol) is virtually unaffected by hydrocarbons, chlorinated hydrocarbons, carboxylic acid esters, greases, and animal or vegetable oils. Resistance to organic solvents increases with increasing hydrolysis.

The oxygen-barrier properties of PVA at low humidity are the best of any synthetic resin. However, barrier performance deteriorates above 60% rh.

The surface tension of aqueous solutions of PVA varies with concentration, temperature, degree of hydrolysis, and acetate distribution on the PVA backbone. Surface tension decreases slightly as the molecular weight is reduced. The relationship between the intrinsic viscosity and molecular weight changes with degree of hydrolysis of the polymer.

Chemical Properties

Poly(vinyl alcohol) participates in chemical reactions in a manner similar to other secondary polyhydric alcohols. Of greatest commercial importance are reactions with aldehydes to form acetals, such as poly(vinyl butyral) and poly(vinyl formal).

Boric acid and borax form cyclic esters with poly(vinyl alcohol).

An unlimited number of organic esters can be prepared by reactions of poly(vinyl alcohol) employing standard synthesis. Ethers of poly(vinyl alcohol) are easily formed. Insoluble internal ethers are formed by the elimination of water, a reaction catalyzed by mineral acids and alkali.

Poly(vinyl alcohol) and aldehydes form products which find use in the manufacture of safety glass and as adhesives for hydrophilic surfaces.

Poly(vinyl alcohol) can be readily cross-linked using a multifunctional compound that reacts with hydroxyl groups. These types of reactions are of significant industrial importance as they provide ways to obtain improved water resistance of the poly(vinyl alcohol) or to increase the viscosity rapidly. The thermal decomposition of poly(vinyl alcohol) in the absence of oxygen occurs in two stages. The first stage begins at about 200°C and is mainly dehydration, accompanied by the formation of volatile products. Further heating to 400–500°C yields carbon and hydrocarbons.

Poly(vinyl alcohol) is one of the few truly biodegradable synthetic polymers; the degradation products are water and carbon dioxide. At least 55 species or varieties of microorganisms have been shown to degrade or participate in the degradation of PVA.

Manufacture

Poly(vinyl alcohol) can be derived from the hydrolysis of a variety of poly(vinyl esters), such as poly(vinyl acetate), poly(vinyl formate), and poly(vinyl benzoate), and of poly(vinyl ethers). However, all commercially produced poly(vinyl alcohol) is manufactured by the hydrolysis of poly(vinyl acetate). The manufacturing process can be viewed as one segment that deals with the polymerization of vinyl acetate and another that handles the hydrolysis of poly(vinyl acetate) to poly(vinyl alcohol).

Vinyl acetate is polymerized commercially using free-radical polymerization in either methanol or, in some circumstances, ethanol.

Poly(vinyl acetate) can be converted to poly(vinyl alcohol) by transesterification, hydrolysis, or aminolysis. Industrially, the most important reaction is that of transesterification, where a small amount of acid or base is added in catalytic amounts to promote the ester exchange.

Copolymers

Numerous vinyl alcohol copolymers have been prepared. Copolymers with ethylene and methacrylate are the only copolymers that have found sizable commercial utility. Ethylene–vinyl alcohol (EVOH) copolymers containing 20–30 mol % ethylene are used as an oxygen barrier in food packaging. Vinyl alcohol–methyl methacrylate copolymers are used as sizing agents in the textile industry. The presence of the methacrylate unit disrupts the crystallinity, making the product easier to remove during the desizing operation. The product is especially useful as an alkaline-resistant textile size.

Processing

Poly(vinyl alcohol) is not considered a thermoplastic polymer because the degradation temperature is below that of the melting point. Thus, industrial applications of poly(vinyl alcohol) are based on and limited by the use of water solutions.

Specifications and Standards

Poly(vinyl alcohol) is produced mainly in five molecular weight ranges. Industry practice expresses the molecular weight of a particular grade in terms of the viscosity of a 4% aqueous solution. An unlimited number of viscosities can be generated by blending the available molecular weights. Products having different degree of hydrolysis can also be blended to obtain a particular performance characteristic. Poly(vinyl alcohol) is an innocuous material having unlimited storage stability.

Health and Safety Factors

Poly(vinyl alcohol) is a nonhazardous material according to the American Standard for Precautionary Labeling of Hazardous Industrial Chemicals (ANSI 2129.1-1976). Extensive tests indicate a very low order of oral toxicity. Short-term inhalation of PVA dust has no known health significance, but can cause discomfort and should be avoided in accordance with industry standards for exposure to nuisance dust. During transport and handling, granular PVA may form an explosive mixture with air, which shows a low severity rating (Bureau of Mines Rating) of 0.1 on a scale in which coal dust has a rating of 1.0.

Uses

The main applications for PVA are in textile sizing, adhesives, polymerization stabilizers, paper coating, poly(vinyl butyral), and PVA fibers. In terms of percentage, and omitting the production of PVA not isolated prior to conversion into poly(vinyl butyral), the principal applications are textile sizes, at 30%; adhesives, including use as a protective colloid, at 25%; fibers, at 15%; paper sizes, at 15%, poly(vinyl butyral), at 10%; and others, at 5%, which include water-soluble films, nonwoven fabric binders, thickeners, slow-release binders for fertilizer, photoprinting plates, sponges for cosmetic, and health care applications.

F. LENNART MARTEN
Air Products and Chemicals, Inc.

Additional Reading

Brandrup, J., D. R. Bloch, E. A. Grulke, E. H. Immergut, and A. Abe: *Polymer Handbook*, 2 Volume Set, 4th Edition, John Wiley & Sons, Inc., New York, NY, 2003.

Finch, C. A. ed.: *Polyvinyl Alcohol*, John Wiley & Sons, Inc., New York, NY, 1973.

Finch, C. A. ed.: *Polyvinyl Alcohol Developments*, John Wiley & Sons, Inc., New York, NY, 1992.

Fried, J.: *Polymer Science and Technology*, 2nd Edition, Prentice Hall, Inc., Upper Saddle River, NJ, 2003.

Kroschwitz, J. I.: *Encyclopedia of Polymer Science and Technology*, 12 Volume Set, 3rd Edition, John Wiley & Sons, Inc., Hoboken, NJ, 2004.

Sakurada, I. *Poly(Vinyl Alcohol) Fibers*, Marcel Dekker, Inc., New York, NY, 1985.

N-VINYLAMIDE POLYMERS. *N*-Vinylamide-based polymers, especially the *N*-vinyllactams, such as poly(*N*-vinyl-2-pyrrolidinone) or simply polyvinylpyrrolidinone (PVP), continue to be of major importance to formulators of personal-care, pharmaceutical, agricultural, and industrial products because of desirable performance attributes and very low toxicity profiles. Because of hydrogen bonding of water to the amide group, many of the *N*-vinylamide homopolymers are water-soluble or dispersible. Like proteins, they contain repeating (but pendant) amide (lactam) linkages and share several protein-like characteristics. Many studies have actually employed PVP as a substitute for proteins, e.g., in simplifying the chemistry of the effects of radiation on polymers. Proteins are extremely complicated molecules with not only sequence distribution but tertiary bonding and structural complexity and it is an oversimplification to compare them to PVP, but the effects of radiation on PVP can be more readily studied. PVP can even be considered as a uniform synthetic protein-like analogue. By itself it does not enter into intermolecular hydrogen bonding, thus affording low viscosity concentrates, and also, unlike the proteins, PVP is soluble in polar solvents like alcohol. But even given these differences, the chemistry of PVP, the most commercially successful polymer of the class, is in many respects similar to that of proteins because of amide linkages sharing with them complexation to large anions such as polyphenols, anionic dyes, and surfactants. In addition to the ability to complex, PVP and its analogues along with a large assortment of copolymers are excellent film-formers. They exhibit the ability to interact with a variety of surfaces by hydrogen or electrostatic bonding, resulting in protective coatings and adhesive applications of commercial significance such as hair-spray fixatives, tablet binders, disintegrants, iodophors, antidye redeposition agents in detergents, protective colloids, dispersants, and solubilizers, among many others.

Monomers

N-Vinylamides and *N*-vinylimides can be prepared by reaction of amides and imides with acetylene, by dehydration of hydroxyethyl derivatives, by pyrolysis of ethylidenebisamides, or by vinyl exchange, among other methods; the monomers are stable when properly stored. Only *N*-vinyl-2-pyrrolidinone (VP) is of significant commercial importance. Vinylcaprolactam is available and is growing in importance, and vinyl formamide is available as a developmental monomer.

N-Vinyl-2-Pyrrolidinone. Commonly called vinylpyrrolidinone or VP, *N*-vinyl-2-pyrrolidinone is a clear, colorless liquid that is miscible in all proportions with water and most organic solvents. It can polymerize slowly by itself but can be easily inhibited by small amounts of ammonia, sodium hydroxide (caustic pellets), or antioxidants such as *N*, *N'*-di-*sec*-butyl-*p*-phenylenediamine. It is stable in neutral or basic aqueous solution but readily hydrolyzed in the presence of acid to form 2-pyrrolidinone and acetaldehyde. Properties are given in Table 1.

Commercially available VP is usually over 99% pure but does contain several methyl-substituted homologues and 2-pyrrolidinone. The vinylation of 2-pyrrolidinone is carried out under alkaline catalysis analogous to the vinylation of alcohols. 2-Pyrrolidinone is treated with ca 5% potassium hydroxide, then water and some pyrrolidinone are distilled at reduced pressure. A ca 1:1 mixture (by vol) of acetylene and nitrogen is heated at 150–160°C and ca 2 MPa (22 atm). Fresh 2-pyrrolidinone and catalyst are added continuously while product is withdrawn. Conversion is limited to ca 60% to avoid excessive formation of by-products. The *N*-vinyl-2-pyrrolidinone is distilled at 70–85°C at 670 Pa (5 mm Hg) and the yield is 70–80%.

One of the manufacturers, ISP, recommends that an appropriate workplace exposure limit be set at 0.1 ppm (vapor). In case of accidental eye contact, immediately flush with water for at least 15 minutes and seek medical attention.

Homopolymerization of *N*-Vinyl-2-Pyrrolidinone

Ammonia H₂O₂ Initiation. The lower molecular weight grades (K-15 and K-30) of PVP are prepared industrially with an ammonia/H₂O₂ initiation system. Such products are the standards for the pharmaceutical industry and conform to the various national pharmacopeias.

TABLE 1. PROPERTIES OF *N*-VINYL-2-PYRROLIDINONE (COMMERCIAL PRODUCTION)

Property	Value
mol wt	111
assay, %	98.5 (min)
moisture content, %	0.2 (max)
color (APHA)	100 (max)
vapor pressure, Pa[a]	
at 17°C	6.7
45°C	67
64°C	266
77°C	667
bp at 400 mm Hg, °C	193
fp, °C	13.5
flash point (open cup), °C	98.4
fire point, °C	100.5
viscosity at 25°C, mPa · s(= cP)	2.07
sp gr (25/4°C)	1.04
refractive index, n_D^{25}	1.511
solubility	miscible in H_2O and most organic solvents
uv spectrum	no significant absorption at wavelengths >220 nm

[a] To convert Pa to mm Hg, multiply by 0.0075.

Organic Peroxides and Azo Initiation. The H_2O_2/ammonia initiation system is not employed commercially in the manufacture of higher molecular weight homologues; they are prepared with organic initiators. Such polymerizations follow simple chain theory and are usually performed in water commercially. The rate of polymerization is at a maximum in aqueous media at pH 8–10 and at 75 wt % monomer. Polymerization rates follow the polarity and hydrogen bonding capability of the solvent.

Cationic Polymerization. VP polymerizes to low molecular weight (oligomers) with typical cationic initiators, such as boron trifluoride etherate. This reaction requires high concentrations, if not neat, of monomer and scrupulously anhydrous conditions for high yields; VP will readily hydrolyze to 2-pyrrolidinone and acetaldehyde even in the presence of trace moisture when catalyzed by strongly acidic reagents.

Proliferous Polymerization. Early attempts to polymerize VP anionically resulted in proliferous or "popcorn" polymerization. This was found to be a special form of free-radical addition polymerization, and not an example of anionic polymerization, as originally thought. VP contains a relatively acidic proton alpha to the pyrrolidinone carbonyl. In the presence of strong base such as sodium hydroxide, VP forms cross-linkers *in situ*. Both ethylidene vinyl pyrrolidinone (EVP) and ethylidene-bis-vinylpyrrolidinone (EBVP) are generated in about a 10:1 ratio, respectively. At the temperature required to generate these cross-linkers and when their concentration reaches some minimum level, usually a few percent, proliferous polymerization begins.

Crospovidones are produced commercially by two processes, i.e., *in situ* generation of cross-linker or addition of divinylimidazoline, and they are indistinguishable by ir. Both types exhibit a T_g of 190–195°C, which is not that much above the 175°C of high molecular weight, soluble PVP. Proliferous polymers prepared with easily hydrolyzed cross-linker containing an imine linkage do not further swell even when the cross-links are hydrolyzed. The crospovidones are unusually high molecular weight, highly chain-entangled polymers having covalent cross-links that most likely retard the termination reaction during polymerization and are not entirely responsible for the resulting mechanical properties, such as swell ratio.

The crospovidones are easily compressed when anhydrous but readily regain their form upon exposure to moisture. This is an ideal situation for use in pharmaceutical tablet disintegration and they have found commercial application in this technology. PVP strongly interacts with polyphenols, the crospovidones can readily remove them from beer, preventing subsequent interaction with beer proteins and the resulting formation of haze. The resin can be recovered and regenerated with dilute caustic.

PVP Hydrogels

Cross-linked versions of water-soluble polymers swollen in aqueous media are broadly referred to as hydrogels and have a growing commercial utility in such applications as oxygen-permeable soft contact lenses and controlled-release pharmaceutical drug delivery devices. Cross-linked PVP and selected copolymers fit this definition and are of interest because of the following structure/performance characteristics:

Structure	Performance	Benefit
nonionic	compatibility with other ingredients	stable formulation
pyrrolidinone	low toxicity	nonirritating/ nonthrombogenic
	complexation–actives/O_2	controlled release-transport
	high T_g	mechanical stability
	hydrolytic stability	storage-stable
ethylene backbone	nonbiodegradable, hydrolytic stability	resists biocontamination storage-stable
cross-links	swell volume/viscosity	mechanical stability/diffusion control

Cross-linked PVP can be prepared by several routes other than proliferous polymerization PPVP (crospovidones). Although a hydrogel, the swell volume of this type of polymer cannot be controlled over a large increment because the granular particles cannot be formed into larger uniform assemblies. These limitations can be overcome by the polymerization of VP in the presence of a few percent of suitable cross-linker utilizing standard free-radical initiation. The solution to this problem is to balance the reactivity ratios of the cross-linker and other comonomers with those of VP to obtain uniform copolymerization and cross-linking.

Cross-linked PVP can also be obtained by cross-linking the preformed polymer chemically (with persulfates, hydrazine, or peroxides) or with actinic radiation. If the starting PVP homopolymer is too low in molecular weight or too dilute, cyclization or cleavage is preferred.

Poly(*N*-Vinyl-2-Pyrrolidinone)

Poly(*N*-vinyl-2-pyrrolidinone) (PVP) is undoubtedly the best-characterized and most widely studied *N*-vinyl polymer. It derives its commercial success from its biological compatibility, low toxicity, film-forming and adhesive characteristics, unusual complexing ability, relatively inert behavior toward salts and acids, and thermal and hydrolytic stability.

Poly(*N*-vinyl-2-pyrrolidinone) is described in the *U.S. Pharmacopeia* as consisting of linear *N*-vinyl-2-pyrrolidinone groups of varying degrees of polymerization. The molecular weights of PVP samples are determined by size exclusion chromatography (sec), osmometry, ultracentrifugation, light-scattering, and solution viscosity techniques. The most frequently employed method of determining the molecular weight and reporting the molecular weight of PVP samples utilizes the sec/low angle light scattering (lalls) technique. A frequently used and commonly recognized method of distinguishing between different molecular weight grades of PVP is the *K* value. The relative viscosity is obtained with an Ostwald-Fenske or Cannon-Fenske capillary viscometer, and the *K* value is derived from Fikentscher's equation.

$$\log \frac{\eta_{rel}}{c} = \frac{75 K_0^2}{1 + 1.5 K_0 c} + K_0 \tag{1}$$

where $K = 1000\, K_0$, η_{rel} = relative viscosity, and c = concentration of the solution in g/100 mL. Solving directly for K, the Fikentscher equation is converted to:

$$K = [300c \log Z + (c + 1.5c \log Z)^2 + 1.5c \log Z - c]/(0.15c + 0.0003c^2)$$

where $Z = \eta_{rel}$.

Specifications for Pharmaceutical grade is given in Table 2.

The T_g of PVP is sensitive to residual moisture and unreacted monomer. It is even sensitive to how the polymer was prepared, suggesting that MWD, branching, and cross-linking may play a part. Polymers presumably with the same molecular weight prepared by bulk polymerization exhibit lower T_gs compared to samples prepared by aqueous solution polymerization, lending credence to an example of branching caused by chain-transfer to monomer.

TABLE 2. SPECIFICATIONS OF PHARMACEUTICAL PVP GRADES (POVIDONE)

Assay	Value (max)
K value	
10–15	85–115%[a]
16–90	90–107%[a]
moisture, %	5
pH[b]	3.0–7.0
residue on ignition, %	0.02
aldehydes, %[c]	0.02
N-vinyl-2-pyrrolidinone, %	0.20
lead, ppm	10
arsenic, ppm	1
nitrogen, %	11.5–12.8

[a] Of stated supplier's value.
[b] Of a 5% solution in distilled water.
[c] Calculated as acetaldehyde.

Molecular weight also plays a significant role in T_g, which increases to a limiting value of 180°C for high purity samples above K-90 in molecular weight. The following equation applies:

$$T_g(°C) = 175 - \frac{9685}{K^2}$$

One of PVP's more outstanding attributes is its solubility in both water and a variety of organic solvents. PVP is soluble in alcohols, acids, ethyl lactate, chlorinated hydrocarbons, amines, glycols, lactams, and nitroparaffins. PVP is insoluble in hydrocarbons, ethers, ethyl acetate, *sec*-butyl-4-acetate, 2-butanone, acetone, cyclohexanone, and chlorobenzene.

Complexation

The combination of electrostatic interaction (induced dipole—dipole interaction) with an increase in entropy resulting from the discharge of bound water is fundamental to PVP's ability to complex with a variety of large anions.

Other factors that can stabilize such a forming complex are hydrophobic bonding by a variety of mechanisms (Van der Waals, Debye, ion—dipole, charge-transfer, etc). Such forces complement the stronger hydrogen-bonding and electrostatic interactions.

Approximately a minimum $\overline{M}_n$ of 1 to 5,000 is required before complexation is no longer dependent on molecular weight for small anions such as KI_3 and 1-anilinonaphthaline-8-sulfonate (ANS).

Equilibrium dialysis studies indicate around 10 repeat VP units (base moles) are required to form favorable complexes. This figure can rise to several hundred for methyl orange and other anions depending on structure.

Iodine Complexes. The small molecule/PVP complex between iodine and PVP is probably the best-known example and can be represented as follows:

It is widely employed as a disinfectant in medicine (Povidone-iodine) because of its mildness, low toxicity, and water solubility. According to the *U.S. Pharmacopeia*, Povidone-iodine is a free-flowing, brown powder that contains from 9–12% available iodine. It is soluble in water and lower alcohols. When dissolved in water, the uncomplexed free iodine level is very low; however, the complexed iodine acts as a reservoir and by equilibrium replenishes the free iodine to the equilibrium level. This prevents free iodine from being deactivated because the free form is continually available at effective biocidal levels from this large reservoir. PVP will interact with other small anions and resembles serum albumin and other proteins in this regard. It can be "salted in" with anions such as NaSCN or "out" with Na_2SO_4 much like water-soluble proteins.

Phenolics. PVP readily complexes phenolics of all types to some degree, the actual extent depending on structural features such as number and orientation of hydroxyls and electron density of the associated aromatic system. A model has been proposed. Complexation with phenolics can result in reduced PVP viscosity and even polymer-complex precipitation.

One practical result of this strong interaction is the employment of PVP to remove unwanted phenolics such as bitter tanins from beer and wine.

Dyes. PVP is currently (ca 1997) employed in a variety of antidye redeposition detergents as a result of its strong interaction with fugitive anionic dyes. This interaction depends on the structure of the dye. Cationic dyes complex only if they also contain hydrogen-bonding functionality. Anionic dyes complex more easily, depending on the number of anionic groups, size of the aromatic nucleolus, and number and orientation of phenolic hydroxyl groups, etc.

Anionic Surfactants. PVP also interacts with anionic detergents, another class of large anions. The addition of PVP results in the formation of micelles at lower concentration than the critical micelle concentration (CMC) of the free surfactant the mechanism is described as a "necklace" of hemimicelles along the polymer chain, the hemimicelles being surrounded to some extent with PVP. The effective lowering of the CMC increases the surfactant's apparent activity at interfaces. PVP will increase foaming of anionic surfactants for this reason. Because of this interaction, PVP has found application in surfactant formulations.

Polymer/Polymer Complexes. PVP complexes with other polymers capable of interacting by hydrogen-bonding, ion-dipole, or dispersion forces. The interest in compatibility on a molecular level, an interesting phenomenon rarely found to exist between dissimilar polymers, is favored by the ability of PVP to form polymer/polymer complexes. Practical applications have been reported for PVP/cellulosics and PVP/polysulfones in membrane separation technology. Electrically conductive polymers of polyaniline are rendered more soluble and hence easier to process by complexation with PVP. Addition of small amounts of PVP to nylon 66 and 610 causes significant morphological changes, resulting in fewer but more regular spherulites.

Copolymerization

Copolymerizations can be conveniently carried out in aqueous solution or in a variety of solvents, depending on monomer/polymer solubilities. Various strategies have been employed to compensate for the divergence in reactivity ratios in order to form uniform (statistical) copolymers such as semibatch or mixed monomer feeds, the goal being to add the more reactive monomer at the rate at which it is being consumed. Clearly, if the difference in reactivity is too great, then the amount of more reactive monomer that can be uniformly incorporated is significantly reduced.

Poly(Vinylpyrrolidinone-co-Vinyl Acetate). The first commercially successful class of VP copolymers, poly(vinylpyrrolidinone-co-vinyl acetate) is currently manufactured in sizeable quantities. A wide variety of compositions and molecular weights are available as powders or as solutions in ethanol, isopropanol, or water (if soluble).

An important reason for the ongoing interest in these copolymers is that vinyl acetate reduces hydrophilicity so that applications that require less moisture-sensitive films such as those employed to set hair are less prone to plasticize and become tacky under high humidity conditions.

Desirable fixative properties superior to PVP homopolymer can be specified by judicious selection of the amount of vinyl acetate. Hair sprays are limited in the molecular weight of the resin because if they are too high the resulting viscosity of the formulation will result in a poor (coarse) spray pattern. Increasing the VP/VA ratio causes properties to increase in the direction shown by the arrows. Other applications for VP/VA copolymers are uses as water-soluble or remoistenable hot melt adhesives, pharmaceutical tablet coatings, binders, and controlled-release substrates.

Tertiary Amine-Containing Copolymers. Copolymers based on DMAEMA (dimethylaminoethyl methacrylate) in either free amine form or quaternized with diethyl sulfate or methyl chloride have achieved commercial significance as fixatives in hair-styling formulations, especially in the well-publicized "mousses" or as hair-conditioning shampoo additives.

The most successful of these products contain high ratios of VP to DMAEMA and are partially quaternized with diethyl sulfate (Polyquaternium 11). They afford very hard, clear, lustrous, nonflaking films on the hair that are easily removed by shampooing. More recently, copolymers with methylvinylimidazolium chloride (Polyquaternium 16) or MAPTAC (methacrylamidopropyltrimethyl ammonium chloride) (Polyquaternium 28) have been introduced. Unquaternized DMAEMA copolymers afford resins that are mildly cationic and less hydroscopic.

Copolymers Containing Carboxylic Groups. A new line of VP/acrylic acid copolymers in powdered form prepared by precipitation polymerization from heptane have been introduced commercially. A wide variety of

TABLE 3. PROPERTIES AND APPLICATIONS OF COMMERCIAL PVPs

Polymer	Properties/applications
	Homopolymers
PVP	film former, adhesive, binder, complexant, stabilizer, crystallization inhibitor, dye scavenger, detoxicant, viscosity modifier
	Cross-linked
proliferous polymerization	pharmaceutical tablet disintegrant, adsorbent for polyphenols (tanins), beverage clarification
	Copolymers
PVP/VA	film-forming adhesives for hair preparations, bio-adhesives, water-remoistenable or removable adhesives
PVP/DMAEMA	mildly cationic, hair styling aids and conditioners with strong hold; substantive, lustrous film-formers
PVP/DMAEMA DES quaternary	strongly cationic, substantive, hair fixative ingredients
PVP/imidazolinum quaternary	
PVP/styrene[a]	opacifier for personal care products; stable styrene emulsion
PVP/alpha-olefins[a]	surface-active film formers; waterproofing of sunscreens
	Terpolymers
VP/VCl/DMAEMA	cationic water-soluble hair styling aid
VP/tBMA/MA	hair fixatives

[a] Graft copolymers.

compositions and molecular weights are available, from 75/25 to 25/75 wt % VP/AA and from 20×10^3 to 250×10^3 molecular weights.

The copolymers are insoluble in water unless they are neutralized to some extent with base. They are soluble, however, in various ratios of alcohol and water, suggesting applications where delivery from hydroalcoholic solutions but subsequent insolubility in water is desired, such as in low volatile organic compound (VOC) hair-fixative formulations or tablet coatings. Unneutralized, their T_gs are higher than expected, indicating interchain hydrogen bonding.

Miscellaneous Copolymers. VP has been employed as a termonomer with various acrylic monomer—monomer combinations, especially to afford resins useful as hair fixatives.

Applications

An overview of the various product categories is given in Table 3.

ROBERT B. LOGIN
Sybron Chemicals Inc.

Additional Reading

Brandrup, J., D.R. Bloch, E.A. Grulke, E.H. Immergut, and A. Abe: *Polymer Handbook,* 2 Volume Set, 4th Edition, John Wiley & Sons, Inc., New York, NY, 2003.

Fried, J.: *Polymer Science and Technology,* 2nd Edition, Prentice Hall, Inc., Upper Saddle River, NJ, 2003.

Haaf, F., A. Sanner, F. Straub: *Polym. J.* **17**(1), 143 (1985).

Kirsh, Y.E.: *Prog. Polym. Sci.* **18**, 519–542 (1993).

Kroschwitz, J.I.: *Encyclopedia of Polymer Science and Technology,* 12 Volume Set, 3rd Edition, John Wiley & Sons, Inc., Hoboken, NJ, 2004.

Molyneuz P. and S. Vekavakaynondha: *J. Chem. Soc., Faraday Trans. 1* **82**, 291 (1986).

Robinson, B.V. and co-workers: PVP, *A Critical Review of the Kinetics and Toxicology of Polyvinylpyrrolidone (Povidone)*, Lewis, Chelsea, Mich., 1990.

VINYL CHLORIDE. Vinyl chloride, [CAS: 75-01-4] $CH_2=CHCl$, by virtue of the wide range of application for its polymers in both flexible and rigid forms, is a major commodity chemical in the U.S. and an important item of international commerce. Growth in vinyl chloride production is directly related to demand for its polymers and, on an energy-equivalent basis, rigid poly(vinyl chloride) (PVC) is one of the most energy-efficient construction materials available.

Vinyl chloride (also known as chloroethylene or chloroethene) is a colorless gas at normal temperature and pressure, but is typically handled as the liquid (bp $-13.4°C$). However, no human contact with the liquid is permissible. Vinyl chloride is an OSHA-regulated material.

Physical Properties

The physical properties of vinyl chloride are listed in Table 1. Vinyl chloride and water are nearly immiscible. Vinyl chloride is soluble in hydrocarbons, oil, alcohol, chlorinated solvents, and most common organic liquids.

Reactions

Polymerization. The most important reaction of vinyl chloride is its polymerization and copolymerization in the presence of a radical-generating initiator.

Substitution at the Carbon–Chlorine Bond. Vinyl chloride is generally considered inert to nucleophilic replacement compared to other alkyl halides. However, the chlorine atom can be exchanged under nucleophilic conditions in the presence of palladium and certain other metal chlorides and salts. Vinyl alcoholates, esters, and ethers can be readily produced from these reactions. Vinylmagnesium chloride (Grignard reagent) can be prepared from vinyl chloride and then used to make a variety of useful end products or intermediates by adding a vinyl anion to organic functional groups. Vinyl chloride similarly undergoes Grignard reactions with other organomagnesium halide compounds. Vinyllithium, another reactive intermediate, can be formed directly from vinyl chloride by means of a lithium dispersion. Vinyl chloride reacts with sulfides, thiols, alcohols, and oximes in basic media. Reaction of vinyl chloride with hydrogen fluoride over a chromia on alumina catalyst yields vinyl fluoride. The carbon–chlorine bond can also be activated at high temperatures.

Oxidation. The chlorine atom-initiated, gas phase oxidation of vinyl chloride yields 74% formyl chloride and 25% CO at high oxygen to Cl_2 ratios. It is unique among chloro olefin oxidations because CO is a major initial product and because the reaction proceeds by a nonchain path. The oxidation of vinyl chloride with oxygen in the gas phase proceeds by a nonradical path which, again, is unique among the chloro olefins. No C_2 carbonyl compounds are made; the major products are formyl chloride, CO, HCl, and formic acid. Oxidation of vinyl chloride with ozone in either the liquid or the gas phase gives formic acid and formyl chloride. Vinyl chloride can be completely oxidized to CO_2 and HCl using potassium permanganate in an aqueous solution at pH 10. The combustion of vinyl chloride in air produces mainly CO_2 and HCl, along with CO and a trace of phosgene.

Addition. Vinyl chloride undergoes a wide variety of addition reactions. Chlorine adds to vinyl chloride to form 1,1,2-trichloroethane by either an ionic or a radical path. Hydrogen halides add to vinyl chloride, usually to yield the 1,1-adduct. Many other vinyl chloride adducts can be formed under acid-catalyzed Friedel-Crafts conditions. Vinyl chloride can be hydrogenated to ethyl chloride and ethane over a platinum on alumina catalyst.

Photochemistry. Vinyl chloride is subject to photodissociation. Photoexcitation at 193 nm results in the elimination of HCl fragments and Cl atoms in an approximately 1.1:1 ratio. Both vinylidene and acetylene have been observed as photolysis products, as have H_2 molecules and H atoms.

Pyrolysis. Vinyl chloride is more stable than saturated chloroalkanes to thermal pyrolysis. That is why nearly all vinyl chloride made commercially comes from thermal dehydrochlorination of ethylene dichloride (EDC). When vinyl chloride is heated to 450°C, only small amounts of acetylene form. Decomposition of vinyl chloride via a free-radical chain process begins at approximately 550°C, and increases with increasing temperature. Acetylene, HCl, chloroprene, and vinylacetylene are formed in about 35% total yield at 680°C. At higher temperatures, tar and soot formation becomes increasingly important. When dry and in contact with metals, vinyl chloride does not decompose below 450°C. However, if water is present, vinyl chloride can corrode iron, steel, and aluminum because of the presence of trace amounts of HCl. This HCl may result from the hydrolysis of the peroxide formed between oxygen and vinyl chloride.

Manufacture

Vinyl chloride monomer was first produced commercially in the 1930s from the reaction of HCl with acetylene derived from calcium carbide. After ethylene became plentiful in the early 1950s, commercial processes were developed to produce vinyl chloride from ethylene and chlorine. These processes included direct chlorination of ethylene to form EDC, followed by pyrolysis of EDC to make vinyl chloride. However, because the EDC cracking process also produced HCl as a coproduct, the industry did not expand immediately, except in conjunction with acetylene-based technology. The development of ethylene oxychlorination technology in the late 1950s encouraged new growth in the vinyl chloride industry. In this process, ethylene reacts with HCl and oxygen to form EDC. Combining the component processes of direct chlorination, EDC pyrolysis, and oxychlorination provided the so-called balanced process for production of vinyl chloride from ethylene and chlorine with no net consumption or production of HCl.

Although a small fraction of the world's vinyl chloride capacity is still based on acetylene or mixed acetylene–ethylene feedstocks, nearly all production is conducted by the balanced process based on ethylene and chlorine. The reactions for each of the component processes are shown in equations 1–3 and the overall reaction is given by equation 4:

$$\text{Direct chlorination } CH_2{=}CH_2 + Cl_2 \longrightarrow ClCH_2CH_2Cl \qquad (1)$$

$$\text{EDC pyrolysis } 2\ ClCH_2CH_2Cl \longrightarrow 2\ CH_2{=}CHCl + 2\ HCl \qquad (2)$$

$$\text{Oxychlorination } CH_2{=}CH_2 + 2\ HCl + 1/2\ O_2 \longrightarrow ClCH_2CH_2Cl$$
$$+ H_2O \qquad (3)$$

$$\text{Overall reaction } 2\ CH_2{=}CH_2 + Cl_2 + 1/2\ O_2 \longrightarrow 2\ CH_2{=}CHCl$$
$$+ H_2O \qquad (4)$$

Direct chlorination of ethylene is usually conducted in liquid EDC in a bubble column reactor. Under typical process conditions, the reaction rate is controlled by mass transfer, with absorption of ethylene as the limiting factor. Ferric chloride is a highly selective and efficient catalyst for this reaction, and is widely used commercially. The direct chlorination process may be run with a slight excess of either ethylene or chlorine, depending on how effluent gases from the reactor are subsequently processed. Conversion of the limiting component is essentially 100%, and selectivity to EDC is greater than 99%. The direct chlorination reaction is exothermic ($\Delta H = -180$ kJ/mol for eq. 1) and requires heat removal for temperature control.

TABLE 1. PHYSICAL PROPERTIES OF VINYL CHLORIDE

Property	Value
mol wt	62.4985
melting point (1 atm), K	119.36
boiling point (1 atm), K	259.25
heat capacity at constant pressure, J/(mol · K)[a]	
vapor at 20°C	53.1
liquid at 20°C	84.3
critical temperature, K	432
critical pressure, MPa[b]	5.67
critical volume, cm^3/mol	179
critical compressibility	0.283
acentric factor	0.100107
dipole moment, C · m	4.84×10^{-30}
enthalpy of fusion (mp), kJ/mol[a]	4.744
enthalpy of vaporization (298.15 K), kJ/mol[a]	20.11
enthalpy of formation (298.15 K), kJ/mol[a]	28.45
Gibbs energy of formation (298.15 K), kJ/mol[a]	41.95
vapor pressure, kPa[b]	
−30°C	49.3
−20°C	78.4
−10°C	119
0°C	175
viscosity, mPa · s	
−40°C	0.345
−30°C	0.305
−20°C	0.272
−10°C	0.244
explosive limits in air, vol %	
lower limit	3.6
upper limit	33
autoignition temp, K	745

[a] To convert J to cal, divide by 4.184.

[b] To convert MPa, to psi, multiply by 145.

One widely used method involves operating the reactor at the boiling point of EDC, allowing the pure product to vaporize, and then either recovering heat from the condensing vapor or replacing one or more EDC fractionation column reboilers with the reactor itself.

In oxychlorination, ethylene reacts with dry HCl and either air or pure oxygen to produce EDC and water. While commercial oxychlorination processes may differ from one another to some extent because they were developed independently by many different vinyl chloride producers, in each case the reaction is carried out in the vapor phase in either a fixed- or fluidized-bed reactor containing a modified Deacon catalyst. Cupric chloride is usually the primary active ingredient of the catalyst, supported on a porous substrate such as alumina, silica–alumina, or diatomaceous earth. The oxychlorination reaction is highly exothermic ($\Delta H = -239$ kJ/mol for eq. 3) and requires heat removal for temperature control.

Fluidized bed reactors typically are vertical cylindrical vessels equipped with a support grid and feed sparger system for adequate fluidization and feed distribution, internal cooling coils for heat removal, and either external or internal cyclones to minimize catalyst carryover. Fluidization of the catalyst assures intimate contact between feed and product vapors, catalyst, and heat-transfer surfaces, and results in a uniform temperature within the reactor. Reaction heat can be removed by generating steam within the cooling coils or by some other heat-transfer medium.

Fixed-bed reactors resemble multitube heat exchangers, with the catalyst packed in vertical tubes held in a tubesheet at top and bottom. Reaction heat can be removed by generating steam on the shell side of the reactor or by some other heat-transfer fluid. However, temperature control is more difficult in a fixed-bed than in a fluidized-bed reactor because localized hot spots tend to develop in the tubes.

In the air-based oxychlorination process with either reactor type, ethylene and air are fed in slight excess of stoichiometric requirements to ensure high conversion of HCl and to minimize losses of excess ethylene that remains in the vent gas after product condensation. Under these conditions, typical feedstock conversions are 94–99% for ethylene and 98–99.5% for HCl with EDC selectivities of 94–97%. The use of oxygen instead of air in the oxychlorination process with either reactor type allows operation with excess ethylene and at lower temperatures. This enables recycling of unconverted ethylene, resulting in improved operating efficiency and product yield. An important advantage of oxygen-based oxychlorination technology over air-based operation is the drastic reduction in volume of the vent gas discharged. Since nitrogen is no longer present in the reactor feed streams, only a small amount of purge gas (about 2–5% of the vent gas volume for air-based operation) is vented.

Direct chlorination usually produces EDC with a purity greater than 99.5 wt %, so that, except for removal of the $FeCl_3$, little further purification is needed before it can be cracked. EDC from the oxychlorination process, however, is less pure and requires purification by distillation.

Thermal pyrolysis or cracking of EDC to vinyl chloride and HCl occurs as a vapor-phase, homogeneous, first-order, free-radical chain reaction. The endothermic cracking of EDC ($\Delta H = 71$ kJ/mol EDC reacted for eq. 2) is relatively clean at atmospheric pressure and at temperatures of 425–550°C. Commercial pyrolysis units, however, generally operate at gauge pressures of 1.4–3.0 MPa (200–435 psig) and at temperatures of 475–525°C to provide for better heat transfer and reduced equipment size, and to allow separation of HCl from vinyl chloride by fractional distillation at noncryogenic temperatures. EDC conversion per pass through the pyrolysis reactor is normally maintained at 53–63%. Cracking reaction selectivity to vinyl chloride of > 99% can be achieved at these conditions. Increasing cracking severity beyond this level gives progressively smaller increases in EDC conversion, with progressively lower selectivity to vinyl chloride. Higher conversion also increases pyrolysis tube coking rates and causes problems with downstream product purification. To minimize coke formation, it is necessary to quench or cool the pyrolysis reactor effluent quickly. Therefore, the hot effluent gases are normally quenched and partially condensed by direct contact with cold EDC in a quench tower. Alternatively, the pyrolysis effluent gases can first be cooled by heat exchange with cold liquid EDC furnace feed in a transfer line exchanger (TLE) prior to quenching in the quench tower.

Quenched pyrolysis product is typically distilled to remove first HCl and then vinyl chloride. The vinyl chloride is usually further treated to produce specification product, recovered HCl is sent to the oxychlorination process,

and unconverted EDC is purified for removal of by-products before it is recycled to the cracking furnace.

By-product disposal from vinyl chloride manufacturing plants is complicated by the need to process a variety of gaseous, organic liquid, aqueous, and solid streams, while ensuring that no chlorinated organic compounds are inadvertently released. Each class of by-product streams requires its own treatment and disposal system.

Vent streams from the different unit operations may contain traces (or more) of HCl, CO, methane, ethylene, chlorine, and vinyl chloride. The common treatment method is either incineration or catalytic combustion, followed by removal of HCl from the effluent gas. Organic liquid streams contain a variety of chlorinated compounds. When there is economic justification, these streams can be fractionated to recover specific, useful components, and the remainder subsequently incinerated and scrubbed to remove HCl. Alternative methods include catalytic oxidation or combustion, and may involve recycle of HCl to oxychlorination and recovery of the heat of combustion to make high pressure steam. Process water streams from vinyl chloride manufacture are typically steam-stripped to remove volatile organics, neutralized, and then treated in a conventional wastewater treatment process. Solid by-products include sludge from wastewater treatment, spent catalyst, and coke from the EDC pyrolysis process. These need to be disposed of in an environmentally sound manner, e.g., by sludge digestion, incineration, landfill, etc.

Environmental Considerations

Since the early 1980s, there has been much debate among environmental activist organizations, industry, and government about the impact of chlorine chemistry on the environment. One aspect of this debate involves the incidental manufacture and release of trace amounts of hazardous compounds such as polychlorinated dibenzodioxins, dibenzofurans, and biphenyls (PCDDs, PCDFs, and PCBs, respectively, but often referred to collectively as dioxins) during the production of chlorinated compounds like vinyl chloride. In 1994, the EPA released a review draft of its reassessment of the impact of dioxins in the environment on human health, which prompted speculation as to the amount of dioxins that might be attributed to chlorine-based industrial processes. The U.S. vinyl industry responded by committing to a voluntary characterization of dioxin levels in its products and in emissions from its facilities to the environment. The results of this study to date support the vinyl industry's position that it is a minor source of dioxins in the environment. In addition, a global benchmark study released recently by the American Society of Mechanical Engineers found no relationship between the chlorine content of waste and dioxin emissions from combustion processes.

Because of the toxicity of vinyl chloride, the EPA in 1975 proposed emission standards for vinyl chloride manufacture. This proposal was subsequently enacted as EPA Regulation 40 CFR 61, Subpart F. Compliance testing began in 1978. Environmental concerns and government regulations have prompted a major increase in the amount of add-on technology used in U.S. vinyl chloride production plants.

Technology Trends

The ethylene-based, balanced vinyl chloride process, which accounts for nearly all capacity worldwide, has been practiced by a variety of vinyl chloride producers since the mid-1950s. The technology is mature, so that the probability of significant changes is low. New developments in production technology will likely be based on incremental improvements in raw material and energy efficiency, environmental impact, safety, and process reliability.

More recent trends include widespread implementation of oxygen-based oxychlorination, further development of new catalyst formulations, a broader range of energy recovery applications, a continuing search for ways to improve conversion and minimize by-product formation during EDC pyrolysis, and chlorine source flexibility. The application of computer model-based process control and optimization is growing as a way to achieve even higher levels of feedstock and energy efficiency and plant process reliability.

Health and Safety Factors

Vinyl chloride is an OSHA-regulated substance. Current OSHA regulations impose a permissible exposure limit (PEL) to vinyl chloride vapors of no more than 1.0 ppm averaged over any 8-h period. Short-term exposure is limited to 5.0 ppm averaged over any 15-min period. Wherever exposure is above the OSHA limit, respirators are required. Contact with liquid vinyl chloride is prohibited.

Chronic exposure to vinyl chloride at concentrations of 100 ppm or more is reported to have produced Raynaud's syndrome, lysis of the distal bones of the fingers, and a fibrosing dermatitis. However, these effects are probably related to continuous intimate contact with the skin. Chronic exposure to large amounts of vinyl chloride gas over a period of many years is also reported to have produced a rare cancer of the liver (angiosarcoma) in a small number of workers.

Vinyl chloride also poses a significant fire and explosion hazard. It has a wide flammability range, from 3.6% to 33.0% by volume in air. Large fires of the compound are very difficult to extinguish, while vapors represent a severe explosion hazard.

Vinyl chloride is generally transported via pipeline, and in railroad tank cars and tanker ships. Because hazardous peroxides can form on standing in air, especially in the presence of iron impurities, vinyl chloride should always be handled and transported under an inert atmosphere.

Uses

Vinyl chloride has gained worldwide importance because of its industrial use as the precursor to PVC. It is also used in a wide variety of copolymers. The inherent flame-retardant properties, wide range of plasticized compounds, and low cost of polymers from vinyl chloride have made it a significant industrial chemical. About 95% of current vinyl chloride production worldwide ends up in polymer or copolymer applications. Vinyl chloride also serves as a starting material for the synthesis of a variety of industrial compounds. The primary nonpolymeric uses of vinyl chloride are in the manufacture of vinylidene chloride and tri- and tetrachloroethylene.

See also **Chlorinated Organics**.

JOSEPH A. COWFER
MAXIMILIAN B. GORENSEK
The Geon Company

Additional Reading

McPherson, R. W., C. M. Starks, and G. J. Fryar: *Hydrocarbon Process.* **58**(3), 75 (1979).

Shelton, L. G., D. E. Hamilton, and R. H. Fisackerly: in E. C. Leonard, ed., *Vinyl and Diene Monomers, P. 3,* Wiley-Interscience, New York, NY, 1971, pp. 1205–1289.

Weissermel, K. and H.-J. Arpe: *Industrial Organic Chemistry,* 2nd Edition, VCH Publishers, Inc., New York, 1993, pp. 215–218.

Wickson, E. J.: *Handbook of Polyvinyl Chloride Formulating,* John Wiley & Sons, Inc., New York, NY, 1993.

1997/98 World Vinyls Analysis, Chemical Market Associates, Inc., Houston, TX, May 1998.

VINYL CHLORIDE POLYMERS. Poly(vinyl chloride) (PVC), commanding large and broad uses in commerce, is second in volume only to polyethylene, having a volume sales in North America in 1995 of 6.2×10^9 kg(13.7×10^9 lb). Vinyl compounds usually contain close to 50% chlorine, which not only provides no fuel, but acts to inhibit combustion in the gas phase, thus supplying the vinyl with a high level of combustion resistance, useful in many building as well as electrical housings and electrical insulation applications.

PVC has a unique ability to be compounded with a wide variety of additives, making it possible to produce materials that range from flexible elastomers to rigid compounds, that are virtually unbreakable. Compounds are also made that have stiff melts for profile extrusion or low viscosity melts for thin-walled injection molding.

Produced by free radical polymerization, PVC has the structure of $-(CH_2CHCl)-_n$, where the degree of polymerization, n, ranges from 500 to 3500.

PVC Morphology

The principal type of polymerization of PVC is the suspension polymerization route. The morphology formed during polymerization strongly influences the processibility and physical properties.

In the suspension polymerization of PVC, droplets of monomer 30–150 μm in diameter are dispersed in water by agitation. A thin

membrane of a graft copolymer of poly(vinyl chloride) and poly(vinyl alcohol) is formed at the H_2O−monomer interface. Primary particles, 1 μm in diameter, deposit onto the membrane from the monomer side.

Mass-polymerized PVC also has a skin of compacted PVC primary particles very similar in thickness and appearance to the suspension-polymerized PVC skin. However, mass PVC does not contain the thinblock copolymer membrane.

In suspension PVC polymerization, droplets of polymerizing PVC, 30–150-μm dia agglomerate to form grains at 100–200-μm dia. With one droplet per grain, the shape is quite spherical. With several droplets making up the grain, the shape can be quite irregular and knobby. The grain shape plays an important role in determining grain packing and bulk density of a powder.

For both suspension and mass polymerizations at less than 2% conversion, PVC precipitates from its monomer as stable primary particles, slightly below 1-μm dia.

On an even smaller scale is the microdomain structure at 0.01-μm spacing. This is interpreted as a structure where the crystallites of about 0.01-μm spacing are tied together by molecules in the amorphous regions. Plasticizer swells the amorphous regions without dissolving the crystallites.

PVC has structure that is built upon structure which is, in turn, built upon even more structure. These many layers of structure are all important to performance and are interrelated. A summary of these structures is listed in Table 1.

The first step in processing is usually powder mixing in a high speed, intensive mixer. PVC resin, stabilizers, plasticizers, lubricants, processing aids, fillers, and pigments are added to the powder blend for distributive mixing.

In plasticized PVC, liquid plasticizers first fill the voids or pores in the PVC grains fairly rapidly during powder mixing. If a large amount of plasticizer is added, the excess plasticizer beyond the capacity of the pores initially remains on the surface of the grains, making the powder somewhat wet and sticky. Continued heating increases the diffusion rate of plasticizer into the PVC mass where the excess liquid is eventually absorbed and the powder dries.

PVC powder compounds are heated, sheared, and deformed during melt processing. During this process, the grains of PVC are broken down. A processing window of stable primary particles exists even with continued melt processing. The primary particle is about a billion molecules of PVC held together by a structure of crystallites and tie molecules.

The PVC crystallites are small, average 0.7 nm (3 monomer units), in the PVC chain direction, and are packed laterally to a somewhat greater extent.

PVC Fusion (Gelation). The PVC primary particle flow units (billion molecule bundles) can partially melt, freeing some molecules of PVC that can entangle at the flow unit boundary. These entangled molecules can recrystallize upon cooling, forming secondary crystallites, and tie the flow units together into a large three-dimensional structure. This process is known as fusion or gelation.

The strength created by the fusion process is strongly dependent on the previous processing temperature and the molecular weight of the PVC. PVC normally improves in properties with increasing fusion (or increasing melt temperature).

Plasticized PVC has the same structures as rigid PVC, except that plasticizer enters the amorphous phase of PVC and makes the tie molecules elastomeric. The grains break down to 1-μm primary particles which become the melt flow units. The crystallites are not destroyed by plasticizer. Table 2 provides a list of the PVC physical parameters.

Chemical Properties

The addition of vinyl monomer to a growing PVC chain can be considered to add in a head-to-tail fashion, resulting in a chlorine atom on every other carbon atom, i.e.,

$$-(CH_2CHClCH_2CHClCH_2CHCl)-_n$$

or in a head-to-head, tail-to-tail fashion, resulting in chlorine atoms on adjacent carbon atoms, i.e.,

$$-(CH_2CHClCHClCH_2CH_2CHClCHClCH_2)-_n$$

Dechlorination studies show that the head-to-tail structure is predominate.

TABLE 1. SUMMARY OF POLY(VINYL CHLORIDE) MORPHOLOGY

Feature	Size	Description
droplets	30–150 µm dia	dispersed monomer during suspension polymerization
membranes	0.01–0.02 µm thick	membrane at monomer–water interface in suspension PVC (usually graft copolymer of PVC and dispersant, such as poly(vinyl alcohol))
grains	100–200 µm dia	after polymerization, free-flowing powder usually made up of agglomerated droplets; in mass polymerization, is free-flowing powder
skins	0.5–5 µm thick	shell on grains made up of PVC deposited onto membrane during suspension polymerization; in mass polymerization, is PVC compacted on grain surface
primary particles	1 µm dia	formed as single polymerization site in both suspension and mass polymerization by precipitation of polymer from monomer; made up of over a billion molecules, it is often melt flow unit established during melt processing (in emulsion polymerization, it is emulsion particle)
agglomerates of primary particles	3–10 µm dia	formed during polymerization by merging of primary particles
domains	0.1 µm dia	formed under special conditions such as high temperature melting (205°C) followed by lower temperature mechanical work (140–150°C); water-phase polymerization also produces domain-sized structure
microdomains	0.01 µm spacing	crystallite spacing
secondary crystallinity	0.01 µm spacing	crystallinity reformed from amorphous melt and responsible for fusion (gelation)

Both saturated and unsaturated end groups can be formed during polymerization by chain transfer to monomer or polymer and by disproportionation.

PVC polymerization has a high chain-transfer activity to monomer; about 60% of the chains have unsaturated chain ends and the percentage of chain ends containing initiator fragments is low. Chain transfer to polymer leads to branching.

The addition of monomer fixes the tacticity of the previous monomer unit. The tacticity of PVC is nearly random, with syndiotacticity slightly favored at lower polymerization temperature.

Polymerization Kinetics of Mass and Suspension PVC. The polymerization kinetics of mass and suspension PVC are considered together because a droplet of monomer in suspension polymerization can be considered to be a mass polymerization in a very tiny reactor. During polymerization, the polymer precipitates from the monomer when the chain size reaches 10–20 monomer units. The precipitated polymer remains swollen with monomer, but has a reduced radical termination rate. This leads to a higher concentration of radicals in the polymer gel and an increased polymerization rate at higher polymerization conversion.

TABLE 2. PVC PHYSICAL PARAMETERS

PVC property	Value		
crystallographic data	orthorhombic, two monomer units/cell		
	a	*b*	*c*
commercial PVC, nm	1.06	0.54	0.51
single crystal, nm	1.024	0.524	0.508
crystallinity, %			
as polymerized	19		
from melt	4.9		
density (uncompounded), g/cc			
whole	1.39		
crystallites	1.53		
oxygen permeability, cc/(cm·s)cm² cm Hg	$238e^{-13.3/RT}$		
Poisson ratio (rigid PVC)	0.41		
refractive index	1.54		
glass-transition temperature, °C	83		
coefficient of linear thermal expansion (unplasticized), °C	7×10^{-5}		
specific heat	temp, °C		value, J/g °C[a]
rigid PVC	23		0.92
	80		1.45
plasticized PVC (50 phr DOP)	23		1.54
	80		1.75
thermal conductivity (unplasticized), J/(cm·s)°C	17.5×10^4		
dielectric strength			
kV/mil	0.5		
kV/mm	20		
solubility parameter, (J/cm³)^0.5	40.7 (av)		

[a] To convert J to cal, divide by 4.184.

Polymerization in two phases, the liquid monomer phase and the swollen polymer gel phase, forms the basis for kinetic descriptions of PVC polymerization.

Chain transfer to monomer is the main reaction controlling molecular weight and molecular weight distribution. PVC molecular weights are usually determined in the United States using inherent viscosity or relative viscosity measured according to ASTM D1243.

PVC Resin Manufacturing Processes

Mass Polymerization. Mass or bulk polymerization of PVC is normally difficult. At high conversions the mixture becomes extremely viscous, impeding agitation and heat removal, causing a high polymerization temperature and broad molecular weight distribution. A two-stage process overcomes these problems. The first stage of the process, which forms a skeleton seed grain for polymerization in a second stage, is carried out in a prepolymerizer with flat blade agitator and baffles to about 7–10% conversion. The number of grains remain constant throughout this polymerization.

Suspension Polymerization. Suspension polymerization is carried out in small droplets of monomer suspended in water. The monomer is first finely dispersed in water by vigorous agitation. Suspension stabilizers act to minimize coalescence of droplets by forming a coating at the monomer-water interface. The hydrophobic-hydrophilic properties of the suspension stabilizers are key to resin properties and grain agglomeration.

Kinetics of suspension PVC are identical to the kinetics of mass PVC.

Emulsion Polymerization. Emulsion polymerization takes place in a soap micelle where a small amount of monomer dissolves in the micelle. The initiator is water-soluble. Polymerization takes place when the radical enters the monomer-swollen micelle. Termination takes place in the growing micelle by the usual radical–radical interactions. The high solubility of vinyl chloride in water, 0.6 wt %, accounts for a strong deviation from true emulsion behavior. Also, PVC's insolubility in its own monomer accounts for such behavior as a rate dependence on conversion. Emulsions of up to 0.2-µm dia are sold in liquid form for water-based paints, printing inks, and finishes for paper and fabric. Other versions, 0.3–10-µm dia and dried by spray-drying or coagulation, are used as plastisol resins.

Microsuspension Polymerization. Microsuspension polymerization uses a monomer-soluble initiator. The monomer is homogenized in water along with emulsifiers or suspending agents to control the particle sizes. Microsuspension paste resins at 0.3–1-μm dia are used to make plastisols for flooring, seals, barriers, etc. These plastisols are also dispersions of PVC in liquid plasticizer and are cured by heating. Microsuspension blending resins at 10–100-μm dia are used as extenders to paste resins in plastisols.

Solution Polymerization. In solution polymerization, a solvent for the monomer is often used to obtain very uniform copolymers. Polymerization rates are normally slower than those for suspension or emulsion PVC.

Copolymerization. Vinyl chloride can be copolymerized with a variety of monomers. Vinyl acetate, the most important commercial comonomer, is used to reduce crystallinity, which aids fusion and allows lower processing temperatures. Copolymers are used in flooring and coatings. This copolymer sometimes contains maleic acid or vinyl alcohol (hydrolyzed from the poly(vinyl acetate)) to improve the coating's adhesion to other materials, including metals. Copolymers with vinylidene chloride are used as barrier films and coatings. Copolymers of vinyl chloride with acrylic esters in latex from are used as film formers in paint, nonwoven fabric binders, adhesives, and coatings. Copolymers with olefins improve thermal stability and melt flow, but at some loss of heat-deflection temperature.

Compounding

The additives found in PVC help make it one of the most versatile, cost-efficient materials in the world. Without additives, literally hundreds of commonly used PVC products would not exist. Many materials are useless until they undergo a similar modification process.

Stabilizers. Lead stabilizers, particularly tribasic lead sulfate, is commonly used in plasticized wire and cable compounds because of its good nonconducting electrical properties. Organotin stabilizers are commonly used for rigid PVC, including pipe, fittings, windows, siding profiles, packaging, and injection-molded parts.

Antimony tris(isooctylthioglycolate) has found use in pipe formulations at low levels. Barium–zinc stabilizers have found use in plasticized compounds, replacing barium–cadmium stabilizers. These are used in moldings, profiles, and wire coatings. Cadmium use has decreased because of environmental concerns surrounding certain heavy metals.

Calcium–zinc stabilizers are used in both plasticized PVC and rigid PVC for food contact where it is desired to minimize taste and odor characteristics. Applications include meat wrap, water bottles, and medical uses. Many stabilizers require costabilizers.

Impact Modifiers. Rubbery polymers are added to PVC to improve toughness. Rubbery particles are added to act as stress concentrators or multiple weak points, leading to crazing or shear-banding under impact load. This can result in cavitation and/or cold drawing, thus allowing the PVC to absorb large amounts of energy.

Processing Aids. PVC often flows in the form of billions of molecule primary particles. Processing aids glue these particles together before the PVC melts, thus acting as a fusion promoter. Processing aids also modify melt rheology by increasing melt elasticity and die swell; some by reducing melt viscosity and melt fracture.

Lubricants. A model for the lubrication mechanism has been developed that explains synergy between certain lubricants. This model treats lubricants as surface-active agents. Some lubricants have polar ends that are attracted to other polar ends and to polar PVC flow units and to polar metal surfaces. These also have nonpolar ends that are repelled by the polar groups. Synergy happens when nonpolar lubricants are added, which are attracted to the nonpolar ends and act as a slip layer.

Plasticizers. Solutions of PVC, prepared at elevated temperatures with high boiling solvents, possess unusual elastic properties when cooled to room temperature. Such solutions are flexible, elastic, and exhibit a high degree of chemical inertness and solvent resistance.

This unusual behavior results from unsolvated crystalline regions in the PVC that act as physical cross-links. These allow the PVC to accept large amounts of solvent (plasticizers) in the amorphous regions, lowering its T_g to well below room temperature, thus making it rubbery. PVC was, as a result, the first thermoplastic elastomer (TPE). This rubber-like material has stable properties over a wide temperature range.

Fillers. Fillers are used to improve strength and stiffness, to lower cost, and to control gloss. The most common filler is calcium carbonate.

Pigments. A variety of pigments are added to PVC to give color, including titanium dioxide and carbon black.

Ultraviolet Light Stabilizers. Both titanium dioxide and carbon black are strong ultraviolet light absorbers and effective in protecting the PVC. For ultraviolet light absorption in transparent PVC or for improvement of pigmented systems, various derivatives of benzotriazole are used.

Biocides. Although PVC itself and rigid PVC compounds are resistant to attack by microorganisms, plasticized PVC, in specific applications such as flashing and sealing boots on roofs, shower curtains, and swimming pools, may need protection. Many biocides, often containing arsenic compounds, are available for a balance of stability, compatibility, weatherability, and biocidal effectiveness.

Flame Retardants. Because PVC contains nearly half its weight of chlorine, it is inherently flame-retardant.

Foaming or Blowing Agents. Cellular PVC can be made by a variety of techniques, such as whipping air into a plastisol, incorporating a gas under pressure, incorporating a physical blowing agent into the melt, or using a chemical blowing agent which releases a gas when it decomposes with heat. The most common chemical blowing agent is 1, 1'-azobisdicarbonamide, which decomposes with heat to release nitrogen gas.

Uses

PVC is so versatile that it can be compounded for a wide range of properties and used in a wide variety of markets. Most of the products are durable goods and have long life spans. Its use in short-term, one-time-use products is limited.

Pipe and fittings, a principal market for PVC, are a prime example of PVC as an engineering thermoplastic.

PVC is accepted commercially as an excellent weathering material. PVC's chemical response to weathering is well understood so that compounds and products can be designed for satisfactory outdoor performance. Products include siding, windows, and doors.

Complex profiles require specialty manufacturing skills to build, maintain, and operate extrusion dies as well as cooling and sizing equipment to deliver the exact dimensions required. Cubed compound, where the PVC grains are already broken down, can be run faster on simple single-screw extruders on account of the typical low melt temperatures.

PVC has been used in wire and cable applications since World War II. The compounds are optimized for the requirements, including low temperature flexibility, high use temperature, especially low combustibility, weatherability, and high resistance to cutthrough.

Numerous housings, electrical enclosures, and cabinets are injection-molded from rigid PVC. These take advantage of PVC's outstanding UL flammability ratings and easy molding into thin-walled parts.

Health and Safety Factors

There are no significant health hazards arising from exposure to poly(vinyl chloride) at ambient temperature. At processing temperatures, most polymers emit fumes and vapors that may be irritating to the respiratory tract. Decomposition of plastics, e.g., through greatly elevated temperatures above normal operating temperatures, can result in personnel exposure to decomposition or combustion products. In the case of PVC compounds, such decomposition involves hydrogen chloride, which causes irritation of the respiratory tract, eyes, and skin.

Poly(vinyl chloride) resin has a flash point of approximately 391°C (735°F) and a self-ignition temperature of approximately 454°C (850°F) (ASTM D1929). In general, PVC burns with difficulty because a substantial amount of energy is required to break down the polymer. Poly(vinyl chloride) powder has a very low tendency to explode. In firefighting where PVC is involved, water, ABC dry chemical, or protein-type air foams should be sued as extinguishing media.

In the U.S., poly(vinyl chloride) is an EPA hazardous air pollutant under the Clean Air Act Section 112 (40 CFR 61) and is covered under the New Jersey Community Right-to-Know Survey: N.J. Environmental Hazardous Substances (EHS) List as "chloroethylene, polymer" with a reporting threshold of 225 kg (500 lb).

Environmental Considerations and Recycling

Over 30% of the chlorine produced on a global basis goes to make PVC. Chlorine makes PVC inherently flame-retardant. PVC is over 50% chlorine and, as a result, one of the most energy-efficient polymers. Chlorine makes PVC far more environmentally acceptable than other materials that are totally dependent on petrochemical feedstocks. In addition, recycling

PVC is easier because the chlorine in PVC acts as a marker, enabling automated equipment to sort PVC containers from other plastics in the waste stream.

Although vinyl is the world's second most widely used plastic, less than one-half percent by weight is found in the municipal solid waste stream. Most of that consists of vinyl packaging, bottles, blister packaging, and flexible film. This is because most vinyl applications are long-term uses, such as pipe and house siding, and are not disposed of quickly. Vinyl wastes are handled by all conventional disposal methods, i.e., recycling, landfilling, and incineration (including waste-to-energy).

A study sponsored by the American Society of Mechanical Engineers (ASME), involving the analysis of over 1700 test results from 155 large-scale, commercial incinerator facilities throughout the world, found no relationship between the chlorine content of waste and dioxin emissions from combustion processes. Instead, the study stated, the scientific literature is clear that the operating conditions of combustors are the critical factor in dioxin generation.

Incinerator scrubbing systems can remove about 99% of the hydrogen chloride generated by incinerating vinyl plastics and other chlorine-containing compounds and materials. Municipal incinerators are often targeted as a primary cause of acid rain. In fact, power plants burning fossil fuels, which produce sulfur dioxide and nitrogen oxide, are actually the leading cause of acid rain, along with automotive exhaust. In Europe and Japan, studies show that only about 0.02% of all acid rain can be traced to incineration of PVC.

Industrial scrap vinyl has been recycled for years, but in more recent years, post-consumer vinyl recycling is growing. In 1991, there were an estimated 1100 municipal recycling programs in place or planned in the United States that include vinyl.

In municipal recycling contamination occurs whether or not vinyl is present. Other resins are just as much a contamination problem as vinyl. Except for commingled plastics applications, different plastic materials cannot be mixed successfully in most recycled products applications. This is why it is crucial to separate efficiently one plastic from another.

It is not true that poly(ethylene terephthalate) (PET) and high density polyethylene (HDPE) packaging are listed as 1 and 2 in the Society of the Plastics Industry (SPI) recycling coding system because they are the most recyclable. The numbers assigned to each plastic in the SPI coding system are purely arbitrary and do not reflect the material's recyclability.

Vinyl plastics do not decompose in landfills and give off vinyl chloride monomer, because like all plastics, vinyl is an extremely stable landfill material. It resists chemical attack and degradation, and is so resistant to the conditions present in landfills that it is often used to make landfill liners. A recent study compared vinyl to a number of other packaging materials and found that vinyl consumed the least amount of energy, used the lowest level of fossil fuels, consumed the least amount of raw materials, and produced the lowest levels of carbon dioxide of any of the plastics studied.

It is not true that in a fire, vinyl is unusually hazardous and damaging. The real hazards in a fire are carbon monoxide and heat; these are especially a problem with other materials that readily burn. Because vinyl products contain chlorine, they are inherently flame-retardant and resist ignition. When it does burn, however, vinyl produces carbon monoxide, carbon dioxide, and hydrogen chloride. Of these, the most hazardous is carbon monoxide. Hydrogen chloride is an irritant gas that can be lethal at extremely high levels. However, research indicates that those levels are never reached or even approached in real fires.

JAMES W. SUMMERS
The Geon Company

Additional Reading

Brandrup, J., D. R. Bloch, E. A. Grulke, E. H. Immergut, and A. Abe: *Polymer Handbook,* 2 Volume Set, 4th Edition, John Wiley & Sons, Inc., New York, NY, 2003.

Davidson, J. A. and D. E. Witenhafer: *J. Polym. Sci.: Polym, Phys. Ed.* **18**, 51 (1980).

Fried, J.: *Polymer Science and Technology,* 2nd Edition, Prentice Hall, Inc., Upper Saddle River, NJ, 2003.

Kroschwitz, J. I.: *Encyclopedia of Polymer Science and Technology,* 12 Volume Set, 3rd Edition, John Wiley & Sons, Inc., Hoboken, NJ, 2004.

Lutz, J. T. Jr. and D. L. Dunkelberger: *Impact Modifiers for PVC; The History and Practice,* John Wiley & Sons, Inc., New York, NY, 1992.

Rigo, H. G. A. J. Chandler, and W. S. Lanier: *The Relationship Between Chlorine In Waste Streams and Dioxin Emissions From Waste Combustor Stacks,* CRTD, Vol. 36, The American Society of Mechanical Engineers, United Engineering Center, New York, 1995.

Summers, J. W.: *J. Vinyl Additive Technol.* **3**(2), 130 (1997).

VINYL ESTER RESINS.

The vinyl ester resins are a relatively recent addition[1] to thermosetting-polymer-chemistry. Superficially, they are similar to unsaturated polyester resins insofar as they contain ethylmic unsaturation and are cured through a free-radical mechanism, usually in the presence of a vinyl monomer, such as styrene. However, close examination of the chemistry and structure of the vinyl ester resins demonstrates several basic differences which lead to their unique characteristics.

Vinyl ester resins are manufactured through an addition reaction of an epoxy resin with an acrylic monomer, such as acrylic acid, methacrylic acid, or the half-ester product of an hydroxyalkyl acrylate and anhydride. In contrast, the polyester resins are condensation products of dibasic acids and polyhydric alcohols. The relatively low-molecular-weight precise polymer structure of the vinyl ester resins is in contrast to the high-molecular-weight random structure of the polyesters.

Of particular importance in describing the difference between these two families of resins are the locations of the reactive unsaturation. In the polyester resin, these groups are located along the backbone of the polymer with terminal hydroxyl or carboxylic acid groups. The vinyl ester resins contain no significant acidity but terminate in reactive vinyl ester groups. Because of the location of these reactive sites, the vinyl ester resins will homopolymerize as well as coreact with various vinyl monomers.

Resin Properties

Vinyl esters, because of their relatively low molecular weight and precise structure, can be characterized as low-viscosity, fast-wetting, consistent-reactivity products. Typical property profiles for some uncured vinyl ester resins are given in Table 1. Properties of some cured resins are given in Table 2. Typically, 6-month stability can be expected at 25°C (77°F) with decreased storage life under elevated temperatures. In general, anyone familiar with the proper storage and handling of unsaturated polyester resins and styrene monomer experiences no difficulty with these materials.

Cure Mechanism

The free-radical cure mechanism of the vinyl ester resins is well understood. In most respects, it is similar to that of the unsaturated polyester resins. To initiate the curing process, it is necessary to generate free radicals within the resin mass. Organic peroxides are the most common source of free radicals. These peroxides will decompose under the influence of elevated temperatures or chemical promoters, e.g., organometallics or tertiary amines, to form free radicals. Generation of free radicals also can be effected by ultraviolet or high-energy radiation applied directly to the resin system. The free radicals thus formed react to open the double bond

TABLE 1. PROPERTIES OF TYPICAL UNCURED VINYL ESTER RESINS

Property	Standard resin	Low viscosity resin
Monomer type	styrene	styrene
Level, %	45	45
Viscosity at 77°F (25°C), centipoises	550	200
Acid number	5	5
Specific gravity	1.04	1.04
SPI gel time (1% benzoyl peroxide), minutes		
at 82°C (180°F)	10	12
at 121°C (250°F)	1.4	1.5
Flash point (Tag open cup)		
°C	34	34
°F	93	93

Source: The Dow Chemical Company.

[1] The first literature reference was a patent issued in 1962 for a tooth-filling compound. Commercialization did not start until the late 1960s.

$$C=C-C-O \left[C-C-C-O-\text{(Ar)}-O \right]_8 C-C-C-O-C-C=C$$

(1) Typical vinyl ester resin

$$HO \left[R'-O-C-C=C-C-O \right] R'-O-C-C=C-C-OH$$

(2) Typical unsaturated polyester resin

TABLE 2. PROPERTIES OF TYPICAL CURED VINYL ESTER RESINS

CLEAR-CASTING PROPERTIES

Tensile strength, psi	12,000
Megapascals	83
Tensile modulus, psi	500,000
Megapascals	3,447
Ultimate elongation, %	5
Flexural strength, psi	18,000
Megapascals	124
Flexural modulus, psi	450,000
Megapascals	3,103
Yield compressive strength, psi	17,000
Megapascals	117
Compressive modulus, psi	350,000
Megapascals	2,413
Deflection at yield, %	7
Heat-distortion temperature,°C	101.7
°F	215.0
Barcol hardness	35

GLASS-REINFORCED LAMINATE PROPERTIES

Laminate thickness, inch	0.25
Millimeters	6.3
Fiber-glass content, %	30
Tensile strength, psi	19,000
Megapascals	131
Tensile modulus, psi	1,400,000
Megapascals	9,653
Flexural strength, psi	22,000
Megapascals	152
Flexural modulus, psi	1,000,000
Megapascals	6,895

Source: The Dow Chemical Company.

of the vinyl group. Once opened, the resin vinyl group is highly reactive and rapidly combines with several more vinyl groups available from both the unreacted resin and the monomer. This exothermic reaction is rapidly carried to completion, forming a 3-dimensional thermosetting network.

Applications

As might be expected from the wide variation in resin properties that can be built into the molecule, the vinyl ester resins find many applications. Chief among these are fiberglass-reinforced plastics, where the inherent characteristics of the vinyl ester resin provide a cost and performance advantage over other materials. The largest application is in the manufacture of corrosion-resistant reinforced plastic structures. Because of the reduced number of ester groups within the resin structure, the corrosion-resistant vinyl ester resins are less prone to attack by hydrolysis than the bisphenol A-fumaric acid polyesters. In addition, the resilience of the vinyl ester resins (4–6% ultimate tensile elongation) results in a fabricated part which is less prone to damage during shipping, field erection, and service.

All fabrication methods commonly used in the manufacture of reinforced plastics can be used with the vinyl ester resins. In those applications, such as filament winding and bag molding, where fast wetting is important, significant increases in output can be realized. One technique to which vinyl ester resins are particularly suited is the use of sheet molding compound (SMC). Here, the resin system (usually with a high filler loading) is combined in sheet form with the glass reinforcement and chemically thickened through the use of metal oxides, such as MgO. The SMC then is molded, usually in a matched-die molding operation, to give the desired final product, e.g., automotive parts, appliance housings, electrical structures, and panel configurations. Vinyl ester resins diluted with vinyl toluene monomer are used in the production of high-temperature electrical laminating systems. See Structures (1) and (2).

P. H. COOK
Dow Chemical U.S.A., an operating unit of The Dow
Chemical Company
Freeport, Texas

VINYL ETHER MONOMERS AND POLYMERS. Because of the strong electron-donating oxygen, the polymerization of vinyl ethers (VE) can be readily accomplished using cationic initiators, resulting in polymers and copolymers that have the potential for significant variety. However, only poly(methyl vinyl ether) (PMVE) achieved commercial success among the homopolymers, and its commercial importance has faded. Divinyl ethers are emerging as important ingredients in radiation-cured coatings, whereas copolymers of methyl vinyl ether (MVE) and maleic anhydride, easily prepared by free-radical initiation, continue to be valued as ingredients in personal care and pharmaceutical products.

Monomers

The most general commercial process for the manufacture of mono- and divinyl ethers, developed by Reppe in the 1930s at BASF, is by treating alcohols with acetylene under pressure of ≥ 6.8 atm (100 psi) at temperatures of 120–180°C in the presence of catalytic amounts of the corresponding metal alcoholate. The danger of handling acetylene under pressure in concentrated form requires sophisticated equipment and should only be attempted experimentally in an appropriately barricaded high pressure autoclave.

$$HC \equiv CH \xrightarrow{ROM} (ROCH = CHM) \xrightarrow{ROH} ROCH = CH_2 + ROM$$

Alternatively, thermal cracking of acetals or metal-catalyzed transvinylation can be employed. Some physical properties of the lower homologues of vinyl ether are presented in Table 1.

Reactions of Vinyl Ethers. Vinyl ethers undergo the typical reactions of activated carbon–carbon double bonds. A key reaction of VEs is acid-catalyzed hydrolysis to the corresponding alcohol and acetaldehyde, i.e., addition of water followed by decomposition of the hemiacetal. MVE is a reactive flammable gas and must be handled safely.

Homopolymerization

VEs such as MVE polymerize slowly in the presence of free-radical initiators to form low mol wt products of no commercial importance. Examples of anionic polymerization are unknown, whereas cationic initiation promotes rapid polymerization to high mol wt polymers in excellent yield and has been extensively studied.

A typical cationic polymerization is conducted with highly purified monomer free of moisture and residual alcohol, both of which act as inhibitors, in a suitably dry unreactive solvent such as toluene with a Friedel-Crafts catalyst, e.g., boron trifluoride, aluminum trichloride, and stannic chloride. Usually low temperatures (−40 to −70°C) are favored in order to prevent chain-transfer or side reactions.

The nature of the side chain R group exerts considerable influence on the reactivity of vinyl ethers toward cationic polymerization. The rate is fastest

TABLE 1. PHYSICAL PROPERTIES OF THE LOWER VINYL ETHERS

Property	Methyl	Ethyl	Isopropyl	n-Butyl	Isobutyl
odor	sweet, pleasant	pleasant	pleasant	pleasant	pleasant
bp, °C	5.5	35.6	55–56	94.3	83
fp, °C	−122	−115.3	−140	−112.7	−132.3
sp gr at 20/4°C	0.7511	0.753	0.753	0.778	0.767
refractive index, n_D^{25}	1.3947	1.3734	1.3829	1.3997	1.3946
sol in water at 20°C, wt %	0.97	0.039	0.6	0.1	0.1
flash point, °C	−56[a]	−18[b]		0.55	−9.4
heat of vaporization at 101.3 kPa		367		316	323

[a] Cleveland open cup.
[b] Tag open cup (ASTM D1310).

when the alkyl substituent is branched and electron-donating. Aromatic vinyl ethers are inherently less reactive and susceptible to side reactions.

Stereoregular Polymerization. In order to generate stereoregular (usually isotactic) polymers, the polymerization is conducted at low temperatures in nonpolar solvents. A variety of soluble initiators can produce isotactic polymers, but there are some initiators, e.g., $SnCl_4$, that produce atactic polymers under isotactic conditions. The nature of the pendant group can influence tacticity.

The low temperature limitation of homogeneous catalysis has been overcome with heterogeneous catalysts such as modified Ziegler-Natta solid-supported protonic acids and metal oxides.

It has been suggested that the mechanism of stereoregular vinyl ether polymerization heavily depends on the degree of association of the counterion with the growing terminal carbocation. An incoming monomer can approach the carbocation terminus from either the front-or back-side attack, which attack is prevalent depends on the tightness of the growing ion pair and the steric requirements of the particular vinyl ether monomer. Front-side attack is favored by a loose ion pair associated with polar solvents, whereas a back-side attack is favored by nonpolar solvents where a tight ion pair prevails.

Living Polymerization

Living polymerization is characterized by an increasing number-average molecular weight as the monomer is consumed. The rate of M_n increase is inversely proportional to the initial concentration of hydrogen iodide, not iodine, and the molecular weight distribution (MWD) of the polymer is very narrow throughout the course of the polymerization ($M_w/M_n < 1.1$). Thus, this type of polymerization can be stopped and started by consumption or addition of fresh monomer. It is similar to ethylene oxide/propylene oxide (EO/PO) anionic polymerization in this regard, but the initiation system is longer lived.

Living VE polymerization is usually terminated by addition of alcohols, phenols, amines, etc, that can replace iodide. Without some base present to neutralize generated HI, an aldehyde end group forms if moisture is present because of acid-catalyzed hydrolysis.

The living polymerization process offers enormous flexibility in the design of polymers. It is possible to control terminal functional groups, pendant groups, monomer sequencing along the main chain (including the order of addition and blockiness), steric structure, and spatial shape.

Homopolymer Properties

Physical properties, which depend on molecular weight, the nature of the alkyl group, the nature of the initiator, stereospecificity, and crystallinity, range from viscous liquids, through sticky liquids and rubbery solids, to brittle solids. Polyethers with long alkyl side chains are waxy, however, as the alkyl group in such cases dominates physical properties.

The glass-transition temperatures of the amorphous straight-chain alkyl vinyl ether homopolymers decrease with increasing length of the side chain. Also, the melting points of the semicrystalline poly(alkyl vinyl ether)s increase with increasing side-chain branching.

Commercial Aspects

No crystalline polymers are known to have been commercialized. This lack of commercial success results from the economically competitive situation concerning vinyl ether polymers versus other, more readily available polymers such as those based on acrylic and vinyl ester monomers.

Copolymerization

VEs do not readily enter into copolymerization by simple cationic polymerization techniques; instead, they can be mixed randomly or in blocks with the aid of living polymerization methods. Reactivity ratios must be taken into account if random copolymers, instead of mixtures of homopolymers, are to be obtained by standard cationic polymerization. VEs can also copolymerize by free-radical initiation with a variety of comonomers.

MVE/MAN Copolymers. Various mol wt grades of poly(methyl vinyl ether-*co*-maleic anhydride) (PMVEMA) are available. PMVEMA, supplied as a white, fluffy powder, is soluble in ketones, esters, pyridine, lactams, and aldehydes, and insoluble in aliphatic, aromatic, or halogenated hydrocarbons, as well as in ethyl ether and nitroparaffins. When the copolymer dissolves in water or alcohols, the anhydride group is cleaved, forming the polymers in free acid form or the half-esters of the corresponding alcohol, respectively.

International Specialty Products (ISP) supplies ethyl, isopropyl, and n-butyl half-esters of PMVEMA as 50% solutions in ethanol or 2-propanol. These half-esters do not dissolve in water but are soluble in dilute aqueous alkali and in aqueous alcoholic amine solutions. The main application for the half-esters is in hairsprays where they combine excellent hair-holding properties at high humidity without making the hair stiff or harsh. These half-esters are easily removed during shampooing, have a very low order of toxicity, and form tack-free films that exhibit good gloss, luster, and sheen (see **Enzyme Preparations**).

Health and Safety Factors

Poly(methyl vinyl ether-*co*-maleic anhydride) and their monoalkyl ester derivatives have been shown on rabbits to be neither primary irritants nor primary sensitizers to skin and eyes. The acute oral toxicities on white rats of the two copolymers are, respectively, 29 g/kg and 25 g/kg body weight.

Applications

Radiation-Curable Coatings. A wide variety of monovinyl and divinyl ethers are commercially available for this application, which allows the formulator greater latitude. For example, triethylene glycol divinyl ether (DVE-3) and 1,4-cyclohexanedimethanol divinyl ether (CHVE) can be combined as reactive diluents, with each contributing quite different properties to the subsequently cured coating. CHVE offers hard brittle films, whereas DVE-3 produces films that have greater flexibility.

Vinyl ethers can also be formulated with acrylic and unsaturated polyesters containing maleate or fumarate functionality. Because of their ability to form alternating copolymers by a free-radical polymerization mechanism, such formulations can be cured using free-radical photoinitiators. With acrylic monomers and oligomers, a hybrid approach has been taken using both simultaneous cationic and free-radical initiation.

Polymer-Polymer Compatibility. Frequently when polymers are mixed together they are immiscible because the combinatorial entropy of mixing is too small to overcome the enthalpy changes, which are usually positive. This small entropy of mixing is a result of the high mol wt nature of the component polymers. If the component polymers exhibit a specific interaction such as hydrogen bonding, Van der Waals, or electrostatic, etc, then miscibility can occur. In the case of PMVE–polystyrene, the blend presents a lower critical solution temperature (LCST) and the miscibility region depends on the molecular weight of the polymers. The interaction in this case is between the electrons of the ether groups and the aromatic polystyrene ring. In fact, PMVE can function as a diluent for isotactic polystyrene enhancing spherulite formation. Depending on the molecular weight and tacticity of the PMVE employed, separated regions of crystallized PMVE can function as reinforcement for polystyrene blends and offer improved plastic properties.

ROBERT B. LOGIN
Sybron Chemicals Inc.

Additional Reading

Brandrup, J., D.R. Bloch, E.A. Grulke, E.H. Immergut, and A. Abe: *Polymer Handbook,* 2 Volume Set, 4th Edition, John Wiley & Sons, Inc., New York, NY, 2003.

Cowie, J.M.G.: *Alternating Copolymers,* Plenum Publishing Corp., New York, NY, 1985.

Dougherty, J.A. and F.J. Vara: Proceedings of Radtech 88-North Americal Conference, New Orleans, LA, 1988.

Field, N.D. and D.H. Lorenz: in E.C. Leonard, ed., *Vinyl and Diene Monomer,* Part I, John Wiley & Sons, Inc., NY, 1970, p. 365.

Fried, J.: *Polymer Science and Technology,* 2nd Edition, Prentice Hall, Inc., Upper Saddle River, NJ, 2003.

Higashimura, T. and M. Sawamoto: in G. Allen and J. Bevington, eds., *Comprehensive Polymer Science,* Pergamon, Oxford, U.K., 1989, p. 673.

Kroschwitz, J.I.: *Encyclopedia of Polymer Science and Technology,* 12 Volume Set, 3rd Edition, John Wiley & Sons, Inc., Hoboken, NJ, 2004.

VINYLIDENE CHLORIDE MONOMER AND POLYMERS.

Vinylidene chloride copolymers' most valuable property is low permeability to a wide range of gases and vapors. From the beginning in 1939, the word Saran has been used for polymers with high vinylidene chloride content, and it is still a trademark of The Dow Chemical Company in some countries. Sometimes Saran and poly(vinylidene chloride) are used interchangeably in the literature.

Three types of comonomers are commercially important: vinyl chloride; acrylates, including alkyl acrylates and alkylmethacrylates; and acrylonitrile. When extrusion is the method of fabrication, the formulation includes plasticizers, stabilizers, and extrusion aids.

Monomer

Properties. Pure vinylidene chloride (1,1-dichloroethylene) is a colorless, mobile liquid with a characteristic sweet odor. Its properties are summarized in Table 1. Vinylidene chloride is soluble in most polar and nonpolar organic solvents. Its solubility in water (0.25 wt %) is nearly independent of temperature at 16–90°C.

Manufacture. Vinylidene chloride monomer can be conveniently prepared in the laboratory by the reaction of 1,1,2-trichloroethane with aqueous alkali:

$$2 \, CH_2ClCHCl_2 + Ca(OH)_2 \longrightarrow CH_2{=}CCl_2 + CaCl_2 + 2 \, H_2O$$

Vinylidene chloride (VDC) is prepared commercially by the dehydrochlorination of 1,1,2-trichloroethane with lime or caustic in slight excess (2–10%). A continuous liquid-phase reaction at 98–99°C yields ~90% VDC. Commercial grades contain 200 ppm of the monomethyl ether of hydroquinone (MEHQ).

For many polymerizations, MEHQ need not be removed; instead, polymerization initiators are added. Vinylidene chloride from which the inhibitor has been removed should be refrigerated in the dark at −10°C, under a nitrogen atmosphere, and in a nickel-lined or baked phenolic-lined storage tank. If not used within one day, it should be reinhibited.

Health and Safety Factors. Vinylidene chloride is highly volatile and, when free of decomposition products, has a mild, sweet odor. A single, brief exposure to a high concentration of vinylidene chloride vapor, e.g., 2000 ppm, rapidly causes intoxication, which may progress to unconsciousness on prolonged exposure. Vinylidene chloride is hepatotoxic, but does not appear to be a carcinogen. The liquid is irritating to the skin after only a few minutes of contact. In the presence of air or oxygen, uninhibited vinylidene chloride forms a violently explosive complex peroxide at temperatures as low as 40°C. Vinylidene chloride containing peroxides may be purified by being washed several times, either with 10 wt % sodium hydroxide at 25°C or with a fresh 5 wt % sodium bisulfite solution.

Polymerization

Homopolymerization. The free-radical polymerization of VDC has been carried out by solution, slurry, suspension, and emulsion methods. Slurry polymerizations are usually used only in the laboratory. The heterogeneity of the reaction makes stirring and heat transfer difficult; consequently, these reactions cannot be easily controlled on a large scale. Aqueous emulsion or suspension reactions are preferred for large-scale operations. The spontaneous polymerization of VDC, so often observed when the monomer is stored at room temperature, is caused by peroxides formed from the reaction of VDC with oxygen. Very pure monomer does not polymerize under these conditions. Heterogeneous polymerization is characteristic of a number of monomers, including vinyl chloride and acrylonitrile.

Emulsion and suspension reactions are doubly heterogeneous; the polymer is insoluble in the monomer and both are insoluble in water.

TABLE 1. PROPERTIES OF VINYLIDENE CHLORIDE MONOMER

Property	Value
odor	pleasant, sweet
color (APHA)	0–10
sol of monomer in H_2O at 25°C, wt %	0.25
sol of H_2O in monomer at 25°C, wt %	0.035
normal bp, °C	31.56
fp, °C	−122.56
flash point, °C	
Tag closed cup	−28
Tag open cup	−16
flammable limits in air (ambient conditions), vol %	6.5–15.5
autoignition temp, °C	513[a]
latent ΔH°_{v}, kJ/mol[b]	
at 25°C	26.48 ± 0.08
at normal bp	26.14 ± 0.08
latent ΔH_m fp, J/mol[b]	6514 ± 8
at 25°C, ΔH_p, kJ/mol[b]	−75.3 ± 3.8
ΔH_c, liquid monomer at 25°C, kJ/mol[b]	1095.9
ΔH_f, at 25°C, kJ/mol[b]	
liquid monomer	−25.1 ± 1.3
gaseous monomer	1.26 ± 1.26
C_p, at 25°C, J/(mol · K)[b]	
liquid monomer	111.27
gaseous monomer	67.03
T_c, °C	220.8
P_c, MPa[c]	5.21
V_c, cm³/mol	218
liquid density, at 20°C, g/cm³	1.2137
index of refraction at 20°C, n_D	1.42468
absolute viscosity at 20°C, mPa · s(= cP)	0.3302[d]

[a] Inhibited with methyl ether of hydroquinone.
[b] To convert J to cal, divide by 4.184.
[c] To convert MPa to atm, divide by 0.101.
[d] *P* measured from 6.7–104.7 kPa. To convert kPa to mm Hg, multiply by 7.5 (add 0.875 to the constant to convert $\log_{kPa}$ to $\log_{mm \, Hg}$).

The instability of PVDC is one of the reasons why ionic initiation of VDC polymerization has not been used extensively. Many of the common catalysts either react with the polymer or catalyze its degradation.

Copolymerization. The importance of VDC as a monomer results from its ability to copolymerize with other vinyl monomers. Bulk copolymerizations yielding high VDC-content copolymers are normally heterogeneous. During copolymerization, one monomer may add to the copolymer more rapidly than the other. Batch reactions carried to completion usually yield polymers of broad composition distribution. More often than not, this is an undesirable result.

Polymer Structure and Properties

The chemical composition of poly(vinylidene chloride) has been confirmed by various techniques, including elemental analysis, x-ray diffraction analysis, degradation studies, and ir, Raman, and nmr spectroscopy. The polymer chain is made up of vinylidene chloride units added head-to-tail:

$$-CH_2CCl_2-CH_2CCl_2-CH_2CCl_2-$$

Molecular weights of PVDC can be determined directly by dilute solution measurements in good solvents. Viscosity studies indicate that polymers having degrees of polymerization from 100 to more than 10,000 are easily obtained. Dimers and polymers having DP < 100 can be prepared by special procedures.

The crystal structure of PVDC is fairly well established. Several unit cells have been proposed. The unit cell contains four monomer units with two monomer units per repeat distance. The calculated density, 1.96 g/cm³, is higher than the experimental values, which are 1.80–1.94 g/cm³ at 25°C, depending on the sample. The melting temperature, T_m, of PVDC is independent of molecular weight above DP = 100. The properties of PVDC (Table 2) are usually modified by copolymerization.

TABLE 2. PROPERTIES OF POLY(VINYLIDENE CHLORIDE)

Property	Best value	Reported values
T_m, °C	202	198–205
T_g, °C	−17	−19 to −11
density at 25°C, g/cm^3		
amorphous	1.775	1.67–1.775
unit cell	1.96	1.949–1.96
crystalline		1.80–1.97
refractive index (crystalline), n_D	1.63	
ΔH_m, J/mol^a	6275	4600–7950

^{a} To convert J to cal, divide by 4.184.

The highly crystalline particles of PVDC precipitated during polymerization are aggregates of thin lamellar crystals.

Melting temperatures of as-polymerized powders are high, i.e., 198–205°C. As-polymerized PVDC does not have a well-defined glass-transition temperature because of its high crystallinity. The amorphous polymer has a glass-transition temperature of −17°C. Once melted, PVDC does not regain its as-polymerized morphology when subsequently crystallized.

Poly(vinylidene chloride) does not dissolve in most common solvents at ambient temperatures. Copolymers, particularly those of low crystallinity, are much more soluble. However, one of the outstanding characteristics of vinylidene chloride polymers is resistance to a wide range of solvents and chemical reagents. The insolubility of PVDC results less from its polarity than from its high melting temperature. It dissolves readily in a wide variety of solvents above 130°C.

Poly(vinylidene chloride) also dissolves readily in certain solvent mixtures. One component must be a sulfoxide or N,N-dialkylamide. Effective cosolvents are less polar and have cyclic structures.

Mechanical Properties. Because PVDC is difficult to fabricate into suitable test specimens, very few direct measurements of its mechanical properties have been made. Some characteristic properties of high VDC content, unplasticized copolymers are listed in Table 3. The performance of a given specimen is sensitive to morphology, including the amount and kind of crystallinity, as well as orientation. Tensile strength increases with crystallinity, whereas toughness and elongation decrease. Orientation, however, improves all three properties.

In cases where the copolymers have substantially lower glasstransition temperatures, the modulus decreases with increasing comonomer content.

The long side chains of the acrylate ester group can apparently act as internal plasticizers. Substitution of a carboxyl group on the polymer chain increases brittleness. Copolymers of VDC with N-alkylacrylamides are more brittle than the corresponding acrylates even when the side chains are long.

Vinylidene chloride polymers are more impermeable to a wider variety of gases and liquids than other polymers. For example, commercial copolymers are available with oxygen permeabilities of 0.05 nmol/m·s·GPa. This is a consequence of the combination of high density and high crystallinity in the polymer. An increase in either tends to reduce permeability. Permeability is affected by the kind and amounts of comonomer as well as crystallinity. A more polar comonomer, e.g., an AN comonomer, increases the water-vapor transmission more than VC when other factors are constant. All VDC copolymers, are very impermeable to

TABLE 3. MECHANICAL PROPERTIES OF HIGH VINYLIDENE CHLORIDE COPOLYMERS

Property	Range
tensile strength, MPa^a	
unoriented	34.5–69.0
oriented	207–414
elongation, %	
unoriented	10–20
oriented	15–40
softening range (heat distortion), °C	100–150
flow temp, °C	>185
brittle temp, °C	−10 to 10
impact strength, J/m^b	26.7–53.4

^{a} To convert MPa to psi, multiply by 145.
^{b} To convert J/m to ft·lbf/in., divide by 53.38 (see ASTM D256).

aliphatic hydrocarbons. Plasticizers increase permeability. However, water does not alter the permeability.

Degradation Chemistry

Vinylidene chloride polymers are highly resistant to oxidation, permeation of small molecules, and biodegradation, which makes them extremely durable under most use conditions. However, these materials are thermally unstable and, when heated above about 120°C, undergo degradative dehydrochlorination.

The principal steps in the thermal degradation of VDC polymer are formation of a conjugated polyene sequence followed by carbonization.

On being heated, the polymer gradually changes color from yellow to brown and finally to black. In general, the stability of the polymer reflects the method of preparation, with bulk > solution > suspension ≫ emulsion.

To some extent, the stability of VDC polymers is dependent on the nature of the comonomer present. Copolymers with acrylates degrade slowly. Copolymers with acrylonitrile or methacrylate undergo degradation more readily.

The degradation of VDC polymers in nonpolar solvents is comparable to degradation in the solid state. However, these polymers are unstable in many polar solvent. The rate of dehydrochlorination increases markedly with solvent polarity. This reaction is clearly unlike thermal degradation and may well involve the generation of ionic species as intermediates.

Stabilization. The ideal stabilizer system should (1) absorb or combine with evolved hydrogen chloride irreversibly under conditions of use, but not strip hydrogen chloride from the polymer chain; (2) act as a selective uv absorber; (3) contain a reactive dienophilic moiety capable of preventing discoloration by reacting with and disrupting the color-producing conjugated polymer sequences; (4) possess nucleophilicity sufficient for reaction with allylic dichloromethylene units; (5) possess antioxidant activity so as to prevent the formation of carbonyl groups and other chlorine-labilizing structures; (6) be able to scavenge chlorine atoms and other free radicals efficiently; and (7) chelate metals, e.g., iron, to prevent chlorine coordination and the formation of metal chlorides.

Commercial Methods of Polymerization and Processing

Emulsion polymerization and suspension polymerization are the preferred industrial processes. Either process is carried out in a closed, stirred reactor, which should be glass-lined and jacketed for heating and cooling. The reactor must be purged of oxygen, and the water and monomer must be free of metallic impurities to prevent an adverse effect on the thermal stability of the polymer.

Emulsion polymerization is used commercially to make vinylidene chloride copolymers. The principal advantages are high molecular weight polymers can be produced in reasonable reaction times, especially copolymers with vinyl chloride and monomer can be added during the polymerization to maintain copolymer composition control. The disadvantages of emulsion polymerization result from the relatively high concentration of additives in the recipe. The water-soluble initiators, activators, and surface-active agents generally cause the polymer to have greater water sensitivity, poorer electrical properties, and poorer heat and light stability.

Suspension polymerization of vinylidene chloride is used commercially to make molding and extrusion resins. The principal advantage is the use of fewer ingredients that might detract from the polymer properties. Stability is improved and water sensitivity is decreased. Extended reaction times and the difficult preparation of higher molecular weight polymers are disadvantages of the suspension process compared to the emulsion process, particularly for copolymers containing vinyl chloride.

The batch-suspension process does not compensate for composition drift, whereas constant-composition processes have been designed for emulsion or suspension reactions. It is more difficult to design controlled-composition processes by suspension methods.

Applications

Vinylidene chloride–vinyl chloride copolymers were originally developed for thermoplastic molding applications, and small amounts are still used for this purpose. Extrusion of VDC–VC copolymers is the main fabrication technique for filaments, films, rods, and tubing or pipe, and involves the same concerns for thermal degradation, streamlined flow, and noncatalytic materials of construction as described for injection-molding resins. A significant application for vinylidene chloride copolymer resins is in the construction of multilayer film and sheet. This permits the design of a packaging material with a combination of properties not obtainable in any single material. Rigid containers for food packaging can be made from coextruded sheet that contains a layer of a barrier polymer.

Vinylidene chloride polymers have several properties that are valuable in the coatings industry: excellent resistance to gas and moisture vapor transmission, good resistance to attack by solvents and by fats and oils, high strength, and the ability to be heat-sealed.

Vinylidene chloride polymers are often made in emulsion, but usually are isolated, dried, and used as conventional resins. Stable latices have been prepared and can be used directly for coatings. The principal applications for these materials are as barrier coatings on paper products and, more recently, on plastic films.

Vinylidene chloride emulsion copolymers are used in a variety of ignition-resistant binding applications.

R. A. WESSLING
D. S. GIBBS
P. T. DeLASSUS
B. E. OBI
The Dow Chemical Company
B. A. HOWELL
Central Michigan University

Additional Reading

Brandrup, J., D.R. Bloch, E.A. Grulke, E.H. Immergut, and A. Abe: *Polymer Handbook,* 2 Volume Set, 4th Edition, John Wiley & Sons, Inc., New York, NY, 2003.

Danforth, J.D.: in P.O. Klemchuk, ed., *Polymer Stabilization and Degradation,* American Chemical Society, Washington, DC, 1985, Chapt. 20, and references cited therein.

DeLassus, P.T.: *J. Vinyl Technol.* **1**, 14 (1979).

Fried, J.: *Polymer Science and Technology,* 2nd Edition, Prentice Hall, Inc., Upper Saddle River, NJ, 2003.

Kroschwitz, J.I.: *Encyclopedia of Polymer Science and Technology,* 12 Volume Set, 3rd Edition, John Wiley & Sons, Inc., Hoboken, NJ, 2004.

Wessling, R.A.: *Polyvinylidene Chloride,* Gordon & Breach, New York, NY, 1977.

VINYL PAINTS. See **Paint and Finish Removers**.

VINYLPYRIDINES. See **Pyridine and Derivatives**.

VIRTANEN, ARRTURI I. (1895–1973). A Finnish biochemist that won the Nobel prize in 1945. His work was primarily concerned with research in nutrition and agriculture. He made important discoveries regarding prevention of fodder spoilage and bacterial fermentation as well as nitrogen metabolism in plants. His Ph.D. was awarded at the University of Helsinki and followed by an illustrious career that included awards throughout Scandanavia.

VIRUS. Viruses are considered to be the smallest infectious agents capable of replicating themselves inside eukaryotic or prokaryotic cells. The majority of these extremely small infectious particles fall within a size range of about 0.02–0.25 micrometer and can only be visualized directly with the aid of an electron microscope.

In 1898, Loeffler and Frosch demonstrated that foot-and-mouth disease of cattle could be transferred by material passed through a filter capable of excluding bacteria. This new group of "organisms" subsequently became known as *filterable viruses.* Years of debate have centered around the question of whether viruses are living or nonliving and, although resolution of this is now considered to be simply a problem of semantics, several fundamental differences distinguish viruses from other organisms. Viruses, unlike true microorganisms, are not cells, do not replicate by binary fission, and contain a genome consisting of only one type of nucleic acid (DNA or RNA, double or single stranded). They contain no organelles, such as mitochondria or ribosomes (except for the Arena viruses, which contain cellular ribosomes). Some virions contain special enzymes, such as transcriptase required for initiation of the vital growth cycle, not present in host cells.

Smaller than viruses, however, are the particles known as *viroids,* which are nothing more than very short strands of RNA uncoated by protein as is a normal virus. Viroids are known to be responsible for several plant diseases and may also be involved in animal and rare human nerve diseases.

Very little is known about viroids except that when they are introduced into a host cell they replicate without the assistance of a helper virus. They are not translated into proteins and, presumably, their replication must rely entirely upon the enzyme systems of the host.

Viruses are obligatory intracellular parasites and, as such, cannot replicate in cell-free media. Therefore, the study of viruses is normally carried out with cultured cells. These are classified as: (1) *Primary cell cultures*; (2) *diploid cell strains*; and (3) *continuous cell lines.* Primary cell cultures are developed on tissue freshly removed from a plant or animal, contain several cell types capable of supporting replication of a wide range of viruses, but are limited to only a few cycles of cell division *in vitro.* Diploid cell strains contain cells of a single type which retain their original diploid chromosome number and are capable of up to about 100 divisions *in vitro.* Continuous cell lines consist of transformed or dedifferentiated cells of a single type which bear little resemblance to normal cells of that type. Continuous cell lines are capable of indefinite propagation *in vitro.*

Viral Structure. The mature virus particle, referred to as the *virion,* consists of a nucleic acid molecule(s) surrounded by a protein coat, the *capsid.* The capsid is composed of a number of *capsomeres* comprising one or more polypeptide chains and in some viruses surrounds a protein *core.* The capsid and enclosed nucleic acid together constitute the *nucleocapsid.* The viral capsid symmetry is characteristic of groups of virus and may be icosahedral (cubic) or helical. The icosahedron has 12 vertices and 20 faces, each an equilateral triangle. Nucleocapsids exhibiting helical symmetry consist of capsomeres and nucleic acid wound together into a spiral or helix. However, regardless of capsid symmetry, the actual virion may appear to be round, brick, or bullet-shaped. Some icosahedral and all helical viruses are enclosed in an outer *envelope* composed of lipoprotein, which is derived directly from the virus-modified cellular membrane during release of the nucleocapsid from the infected cell by a process called *budding.* Enveloped viruses can usually be inactivated by ether, chloroform, or bile salts.

Nucleic acid extracted from purified virus using phenol or dodecyl sulfate is easily destroyed by the homologous nucleases present in normal sera or tissues. DNA is destroyed by the enzyme deoxyribonuclease; RNA by ribonucleases. This provides one means of identifying the type of nucleic acid. The intact virus is not affected by these enzymes.

Bacteriophage, a virus infecting bacterial cells, has a structure somewhat different from those previously described. A *head* contains the nucleic acid and the viral DNA passes through a *tail* during the infection process. In the T-even phages (Fig. 1), the tail consists of a tube surrounded by a *sheath* and is connected to a thin *collar* at the head end and a plate at the tip end. The sheath is capable of contraction and the plate possesses *pins* and *tail fibers,* which are the organs of attachment of the bacteriophage to the wall of the host cell.

Some strains of the bacterium *Escherichia coli* harbor a dormant virus called lambda, which consists of a long molecule of DNA enclosed in protein. Exposure of such infected bacteria to ultraviolet light suddenly "switches on" these inactive lambda. The viruses proliferate and some 45 minutes after irradiation the bacteria burst, yielding a crop of new virus particles. If the bacteria are not irradiated, they grow normally, and rarely give rise spontaneously to viruses.

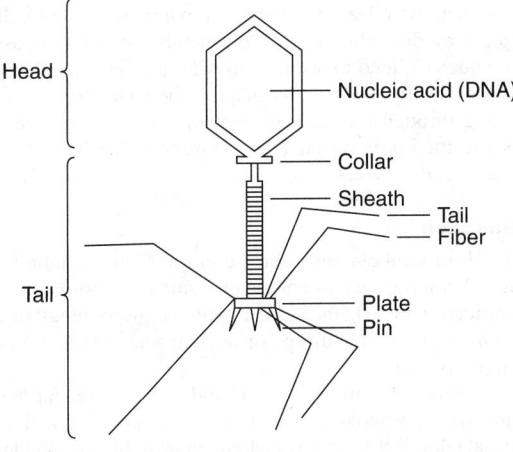

Fig. 1. T-even bacteriophage

Viral Replication. In contrast to eukaryotic and prokaryotic cells, which multiply by binary fission, viruses multiply by synthesis of their separate components, followed by assembly. Several stages are involved in viral replication:

(1) *Attachment or adsorption.* The virus becomes attached to the cell via specific receptors. Thus, cells lacking the receptors are resistant to attack.

(2) *Penetration.* With enveloped viruses, this step occurs when the virion's envelope fuses with the cellular membrane. Naked virions penetrate intact through the cellular plasma membrane and into the cell cytoplasm. Viruses may also enter the cell by cellular phagocytosis. Ordinarily, without a protein coat, viral nucleic acid is incapable of entering a cell, showing the importance of the coat in infectivity. The efficiency of infection with naked nucleic acid can be increased by the presence of basic polymers, such as DEAE-dextran, or by pretreating cells with hypertonic salt solution. Even under the most favorable conditions, however, the efficiency of infection is not more than 1% that of the corresponding intact virions.

(3) *Uncoating and eclipse.* Uncoating is detected by the lability of viral nucleic acids to nuclease after the artificial disruption of the cell. Eclipse is recognized by loss of infectivity of intracellular virions recovered from disrupted cells.

Once inside the cell, virulent viruses turn off cellular macromolecular synthesis and disaggregate cellular polyribosomes, thus favoring a shift to viral synthesis. These viruses cause the ultimate destruction of the infected cell. In contrast, moderate viruses may stimulate host DNA, mRNA, and protein synthesis—a phenomenon which may be of considerable importance in viral carcinogenesis.

In general, the DNA viruses multiply in the nucleus of the host cell. The viral DNA is transcribed in the nucleus and the resultant mRNA translated into proteins on cytoplasmic ribosomes. Depending upon the virus type, "early" or "late" proteins may be synthesized. These proteins may function as enzymes in replication of the viral DNA, as structural components of progeny virions, or as regulatory proteins. Replication of the viral DNA is semiconservative and, in general, depends upon viral proteins.

RNA viruses usually replicate in the cytoplasm and can be divided into five classes according to the nature of the RNA in the virion. *Class I* viruses contain a molecule of single-stranded RNA which acts as mRNA to be translated into viral proteins. The RNA is said to have plus *strand polarity.* The picornaviruses are an example. *Class II* viruses (e.g., paramyxoviruses) have a molecule of single-stranded RNA which cannot act as mRNA (minus strand polarity). A virion transcriptase synthesizes several complementary messenger molecules from which viral proteins are translated. *Class III* viruses (e.g., myxoviruses) contain single-stranded RNA of minus strand polarity, present in seven or more segments. A virion transcriptase transcribes each segment into a complementary messenger. *Class IV* viruses contain ten segments of double-stranded RNA which is transcribed into mRNA by a viral transcriptase. Representative of this group are the reoviruses. *Class V* viruses (e.g., leukoviruses) contain segmented single-stranded RNA of messenger polarity. Each RNA segment is transcribed into DNA by a *reverse transcriptase* present in the virion; mRNA is then transcribed from the DNA.

(4) *Assembly and release.* The assembly of the capsid and its association with nucleic acid is then followed by release of the virus from the cell. This may occur in different ways, depending upon the nature of the virus. Naked viruses may be released slowly and extruded without cell lysis, or released rapidly by disruption of the cell membrane. DNA viruses, which mature in the nucleus, tend to accumulate within infected cells over a long period. Enveloped viruses generally acquire their envelope and leave the cell by budding through the nuclear or cytoplasmic membrane at a point where virus-specified proteins have been inserted. The budding process is compatible with cell survival.

Viral Classification

Several methods of viral classification are in use. Classification based upon epidemiological criteria, such as enteric or respiratory viruses, is useful, but of more significance are schemes based upon the morphology of the virion (symmetry, envelope, etc.) and type of nucleic acid (DNA, RNA, number of strands, polarity, etc.).

The two groups of viruses, RNA and DNA, are further divided according to size, morphology, and biological and chemical properties. Thus, the icosahedral RNA viruses that are ether stable are divided into the picornaviruses and the reoviruses. The name picornavirus comes from *pico*

(meaning very small) and *rna* (indicating the type of nucleic acid). Included in the group are enteroviruses, such as polio, Coxsackie, foot-and-mouth, and echoviruses, among others, and also the rhinoviruses. The picornavirus capsid consists of 60 subunits each made up of four proteins, which change by mutation to yield antibody-resistant strains of cold and polio viruses. The reoviruses (*r*espiratory *e*nteric *o*rphan virus) cause inapparent infection in humans and other animals, and their relationship to spontaneous disease is uncertain. They are morphologically similar to the wound-tumor virus of clover, and a small cross-activity with this virus by means of complement fixation has been reported.

The arboviruses are those which multiply in both vertebrates and arthropods. The former serve as reservoirs and the latter primarily as vectors. Arbovirus is a somewhat arbitrary epidemiological classification, which contains several heterogeneous groups. The togaviruses contain such entities as Eastern and Western equine encephalitis viruses (EEE and WEE) and dengue and yellow fever viruses, which have mosquitoes as vectors. The arenaviruses comprise such agents as Lassa, Tacaribe, and lymphocytic choriomeningitis viruses. Arboviruses are dangerous and difficult to study. They appear to contain single-stranded RNA (positive polarity), are ether sensitive, and are relatively unstable. The capsids are suggestive of icosahedral symmetry.

Myxoviruses, orthomyxovirus, and paramyxoviruses are spherical or filamentous, enveloped single-stranded RNA viruses. The myxovirus group contains the influenza viruses which, in turn, have been separated into three distinct antigenic types, designated A, B, and C. The genome of myxoviruses, unlike that of the paramyxoviruses, is segmented. It is this characteristic that is responsible for the devastating influenza pandemics which have occurred periodically. Influenza A viruses have undergone three major antigenic shifts since 1933, and each new variant is able to successfully infect populations of individuals immune only to preexisting types. In the influenza pandemic of the winter of 1917–1918, over 20 million persons died worldwide, with better than one-half million fatal cases in the United States. Over 50 million cases of influenza were reported in the United States in the 1968–1969 winter. These cases were attributed to a hitherto unknown variant, first isolated in Hong Kong (hence named "Hong Kong flu"). Some 20,000 and possibly as many as 80,000 deaths resulted from this influenza invasion and the side effects it produced. In the 1972–1973 winter season, a much milder and minor variant, called the "London flu," caused well over 2000 deaths in the United States, particularly from complications such as pneumonia. When combined with influenza, pneumonia is the fifth most serious public health problem in the United States. In terms of absenteeism, it is the number one problem.

Only Type A influenza virus has been found to be capable of producing *pandemics.* The influenza A virus is identified as a medium-size RNA virus, some 110 nanometers in diameter and delimited by a membrane of lipids and polysaccharides derived from the host cell and virus-specific protein. Five distinct proteins have been identified, three of which are inside the virion. A schematic representation of the influenza virus emerging from a cell is given in Fig. 2.

It has been reported that the antigenic shifts are manifested in hemagglutinin and neuraminidase, two glycoproteins found on the surface of the influenza virion. It is suggested that the hemagglutinin binds the virus to the target cell and when the hemagglutinin function is inhibited

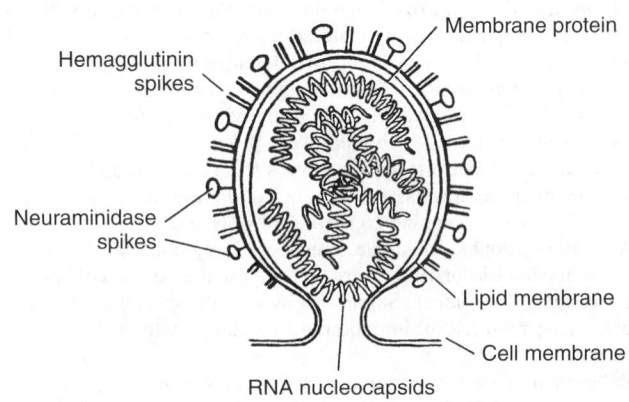

Fig. 2. Schematic representation of an influenza virus emerging from a cell. (*After Kilbourne*)

(as by an antibody), the virus is no longer infective. It is believed that the neuraminidase cleaves a glycoside bond in the host membrane. This action frees the newly formed virus from the cell and, if inhibited, will not reduce the infectivity of the virus, but will deter the spread of virus particles to other cells.

The emergence of new influenza subtypes appears to be too abrupt to be explained fully by conventional concepts of mutation and the full story may rest in the very nature of the segmented viral genome. If a host cell is infected by two different subtypes of influenza virus at the same time, the genes from the subtype may undergo random reassortment in the cell, resulting not only in production of the two original subtypes, but in production of one or several other subtypes as well. Each hybrid, of course, will have a different but full set of genes and recombination within the infected host can explain the large mutations that occur about once every decade. However, it is well established that only one influenza subtype can exist in humans at any given time. Also, that the emergence of a new subtype, such as the Hong Kong strain, is usually accompanied by the abrupt disappearance of the antecedent subtypes—thus allowing little, if any, opportunity for recombinations to occur within human cells. Of considerable interest, however, is the fact that several virus strains can exist simultaneously within animal hosts. In animals, the appearance of a new influenza virus strain is not necessarily accompanied by the disappearance of previously recognized strains. It has been established that there are at least two discrete subtypes of equine influenza, eight or more avian strains and two subtypes in swine. Thus, the postulation that recombination occurs in animals which share the general environs with humans. Some evidence of this may derive from the fact that most new subtypes appear to originate in Asia, where animals and humans commonly inhabit the same building.

The paramyxovirus group includes the causative viruses of mumps, measles, parainfluenza, Newcastle disease, canine distemper, and several other diseases. These viruses are generally larger than the myxoviruses, are enveloped and pleomorphic, and contain one molecule of single-stranded RNA.

The rhabdoviruses, causative agents of rabies in humans and other animals, are also enveloped and contain a single strand of RNA. A peculiar bullet-shaped morphology disguises the helical nucleocapsid.

The human immunodeficiency virus (HIV) which induces AIDS is a retrovirus. Its genetic material is RNA and it carries with it a reverse transcriptase which catalyzes transcription of viral RNA into double helical DNA that then integrates into the genome of the infected cell where it is known as a *provirus*. Transcription of this provirus produces new viral RNA and proteins. One characteristic of HIV is that its genome is significantly more complex than that of other known retroviruses; it possesses at least seven types of genes instead of the normal three. The virus replicates by budding off from a T lymphocyte to become a free infectious virus. See also **Immune System and Immunochemistry**.

The chemical nature of many viruses, which either do not grow well or do not lend themselves to purification, is unknown. The Riley lactic dehydrogenase virus is a nonpathogenic virus, which is recognized only by an increase in lactic dehydrogenase in the blood of infected mice. A lipovirus described by Chang causes marked degradation of infected cells and releases a lipogenic toxin dissociable from infectivity, which is capable of inducing fatty degeneration in other uninfected cells. A marked increase in the gamma globulin fraction of blood serum of mink infected with Aleutian mink disease is an indication of infection with a virus which causes a color change in the fur and often sickness and death.

Several groups of viruses, of importance in human disease, contain DNA. The adenoviruses, named for their original isolation from adenoid tissue, contain double-stranded DNA, have icosahedral symmetry and lack and envelope. This group of viruses that multiply in the nucleus of infected cells is usually associated with respiratory tract and eye infections, although it is now apparent that adenoviruses are not the etiological agents for the majority of acute viral respiratory infections. Although adenoviruses exhibit marked oncogenic (tumor causing) potential in animals, they are probably not oncogenic for humans.

The adenoviruses contain at least three protein moieties, and certain types are capable of inducing one or more new host antigens, such as tumor (T) antigens, the chemistry of which is presently unknown. The viral proteins can be separated by gel diffusion and correlated with results obtained by complement fixation. One moiety is the toxic protein that causes the host cell to degenerate. Another corresponds to the group antigen common to all 31 types of adenoviruses and the third is the type-specific protein.

The papovaviruses (*pap*-illoma, *po*-lyoma, *va*-cuolating agent, SV40) are small, nonenveloped, icosahedral viruses which also replicate in the cell nucleus. The virion contains double-stranded DNA. Apart from causing several forms of warts, this group of viruses is of interest as models for understanding mechanisms of viral carcinogenesis.

The major herpesviruses (Gr: *herpein* = to creep) that infect humans are herpes simplex (Type I: fever blisters; Type II: genital lesions), varicella (chickenpox), zoster (shingles), cytomegalovirus, and Epstein-Bar viruses. A number of viruses that infect lower animals also belong to this group of enveloped, icosahedral, double-stranded DNA viruses. Most members of this group tend to produce latent infections with periodic recurrent disease. Two examples are fever blisters caused by herpes simplex I, and shingles, the recurrent form of chickenpox. Cytomegalovirus causes a severe, often fatal, illness of newborns, usually affecting the salivary glands, brain, lungs, kidneys, and liver. Surpassing rubella virus, this is the most common viral cause of mental retardation. It has been estimated that cytomegalovirus (CMV) causes serious mental retardation of more than 3000 infants annually in the United States alone. In addition to mental retardation, the disease in infants may cause blindness and deafness. In about 90% of infants infected with CMV, the disease can be detected only through urine examination. In about 10% of the cases, the disease is typified by enlargement of the spleen and liver, blood abnormalities, and hepatitis. Microcephaly (abnormally small head) is also sometimes an indication. CMV causes enlargement of the affected cells (cytomegaly). The disease is found throughout the world and it is believed that congenital infections result from a primary infection of the mother during pregnancy. CMV, like herpesviruses, probably persist in a latent stage for long spans of time. Immunosuppressed patients, such as those suffering from cancer or recipients of organ transplants, are also prone to infections with CMV.

The Epstein-Barr viruses play an etiological role in infectious mononucleosis, an acute infectious disease that affects lymphoid tissue throughout the body. A strong association of this virus with Burkitt's lymphoma and perhaps nasopharyngeal carcinoma also has been observed.

The poxviruses are the largest and most complex viruses of vertebrates and contain a large, double-stranded DNA molecule. The virions are complex, brick-shaped particles, covered by several membrane layers of viral origin. Unlike other DNA viruses of mammals, poxviruses multiply in the cell cytoplasm. They can be divided into several groups on the basis of specific antigens, morphology, and natural hosts. *Group I* consists of mammalian viruses, such as variola (smallpox), vaccinia, cowpox, ectromelia, and monkeypox. Of this group, variola or smallpox has caused the greatest human morbidity and mortality. However, because the virus has no animal reservoir, and is spread chiefly by human contact, the World Health Organization was able to announce in 1980 that, because of massive immunization campaigns, smallpox has been completely eradicated. Since that announcement, remaining stores of the virus have been destroyed to prevent laboratory accidents, such as the one in 1979, which took the life of a scientist. *Group II* comprises the tumor-producing viruses, the fibroma and myxoma viruses.

The hepatitic viruses appear to fall into two different groups of small, icosahedral DNA viruses. *Type A* causes infectious hepatitis and is transmitted through the oral-intestinal route. *Type B* is transmitted by injection, usually of infected blood or its products.

Slow Viruses

During the last decade or two, there has been increasing speculation and some tentative evidence that so-called *slow viruses* may be operative and may be the underlying causes of a number of degenerative diseases, long poorly understood, such as multiple sclerosis and rheumatoid arthritis, among others. More recently, there have been increasing postulations of an association between viruses and diabetes. In fact, rather positive identification of slow viruses with some rare diseases has been established. Most investigators caution that the term "slow" should not necessarily be fully interpreted in terms of a virus per se, but equally if not completely with the manifestations of the virus. So-called slow virus infections are characterized by a long incubation period, followed by a protracted course of disease. The slowness may arise in some cases from the virus itself, but the slow pace also may be the result of weak but prolonged interactions between the virus and the host's immune system. It is also possible that these characterizations of slowness may not be attributable to viruses at all, but to some other unknown causative factors. Obviously as of this juncture, investigators are following a source of suspicion rather than a chain of hard evidence. Nevertheless, the case for the slow viruses is becoming increasingly convincing. The causative agents for at least four rare diseases, two in humans and two in animals, are sometimes referred to as "unconventional viruses."

One of these diseases in humans is *kuru*, encountered only in the Fore people and their neighbors in New Guinea. The disease for many years was considered a genetic disease. However, it has been established that the disease can be transmitted to chimpanzees by injection of extracts from the brains of human kuru victims into the brains of chimpanzees. Kuru is a neurological disease with brain lesions located mainly in the gray matter. The cerebral cortex takes on a spongy appearance. The other human disease is Creutzfeld-Jakob disease, rare but of worldwide distribution. It involves the premature development of the mental deterioration sometimes seen in old age. It also has been established that it is caused by a transmissible agent that can infect chimpanzees and lower primates. One of the animal diseases referred to is *scapie*, known for over two centuries as a fatal disease among sheep. The other animal disease is *transmissible mink encephalopathy*, first discovered in Wisconsin in the late 1940s. A puzzling aspect of the unconventional slow viruses is the fact that they cannot be observed with an electron microscope. Another puzzling aspect is their apparent lack of antigenicity. Although it has not been possible to demonstrate that any of these four "agents" will evoke production of antibodies, recent work has found fibrils in the brains of infected animals that are believed to be specific markers for the "unconventional" slow viruses and may indeed be the etiological agent. These unconventional slow viruses are not destroyed by ultraviolet radiation, and they are highly resistant to treatment with formalin or heat, but infectivity is destroyed by phenol or ether. Some investigators believe that these agents may incorporate a very small nucleus of the size range of the viroids (self-replicating infectious RNA molecules known to produce certain plant diseases).

Two slow infections of the human central nervous system—*progressive multifocal leukoencephalopathy* (PML) and *subacute sclerosing panencephalitis* (SSPE) are thought to be associated with conventional viruses. Although PML does not cause inflammation of the brain, it does produce demyelination, i.e., destruction of the layers of membranes surrounding nerve axons. Some investigators believe that the virus is a papovavirus (group of small viruses including human wart virus, simian virus 40, and the polyoma virus of mice). It is reasoned that in PML the virus destroys the cells needed for formation and maintenance of the myelin sheath. A conventional virus has been isolated from the brains of persons suffering from SSPE. An association between measles (in patients under two years of age) or immunization with a live measles virus vaccine and later development of SSPE has been shown. SSPE patients have unusually high titres of measles antibodies and affected brain cells have inclusions similar to those seen in measles infections.

Slow viruses are becoming increasingly suspect in the instances of much more common diseases, particularly the autoimmune diseases. An autoimmune disease may be defined as a disease wherein the immune system of the body does not direct its attack on an invading foreign substance, but instead at the body's own tissue. Many authorities consider rheumatoid arthritis and multiple sclerosis as autoimmune diseases. The precise causes of these diseases have remained obscure. Multiple sclerosis is a demyelinating disease and has variously been described as an autoimmune disease, a viral disease, or an autoimmune disease provoked by a virus. Epidemiological studies indicate that from 3 to 23 years may elapse between the time of exposure to the virus and the onset of symptoms. Further evidence points to involvement of a myxovirus. Measles virus is of this kind.

Possible Viral Connection to Diabetes. A Norwegian physician (J. Stang) in 1864 noted that diabetes developed in one of his patients within a short period after a mumps infection and was probably the first person to indicate a possible connection between viruses and diabetes. Over the years, numerous other connections have been attempted to relate diabetes with mumps, hepatitis, rubella, coxsackie, and influenza viruses, adenoviruses, enteroviruses, and cytomegalovirus. One of the presumptions made is that viruses are understood to replicate in the pancreas. Commencing in the late 1950s, more substantive evidence has been given. Reports from Sweden in 1958 link juvenile diabetes with mumps infection. Reports from New York State in 1974 relate closely the cycles of incidence of mumps and those of juvenile diabetes. The study was based upon investigation of records for the period 1946–1971. Tentative conclusions indicate an average lag period of about 3.8 years between onset of diabetes and exposure to mumps and it

is reasoned that this represents the time required for the virus to produce permanent damage to the pancreas. Other investigators have statistically linked diabetes to rubella (German measles). Some authorities suggest that the pancreas, along with other embryonic organs, may be damaged by the virus that causes congenital rubella. The records of nearly 3,000 juvenile diabetics treated at King's College Hospital in London (1955–1968) have been studied, and they reveal a seasonal pattern on the onset of juvenile diabetes, striking a low incidence in June and a high incidence in October. Without presenting the details, conclusions are suggested that an association of viral infections with the juvenile form of diabetes is evident. However, the relationship, if any, has not been determined in the case of the maturity-onset form of diabetes.

Viral Diagnosis and Vaccination

Viral Diagnosis. Three major approaches to identification of viruses are commonly used: (1) *Microscopy*. Viruses may be observed directly by electron microscopy; viral antigens may be recognized in infected tissue by immunofluorescence, using virus-specific antisera; virus-induced pathology may be identified by light microscopy. (2) *Virus isolation*. Provisional viral identification may be based upon cytopathic effects produced in cell cultures infected with virus present in tissues or secretions of the patient. (3) *Serology*. Antibodies specific for a particular virus may be identified in a patient's serum. A very sensitive, accurate, and recently developed diagnostic approach is radioimmunoassay, which involves the use of an isotope-labelled antibody or antigen.

Viral Vaccination. Vaccines, agents that elicit a specific antiviral immune response, have been very successful against smallpox, measles, rubella, poliomyelitis, and yellow fever, all of which are generalized diseases. Vaccines against diseases caused by respiratory tract viruses, where great antigenic diversity is found, have been less effective.

Vaccines may be prepared by rendering viruses harmless without affecting their immunogenicity. This can be done by either inactivating the virus, or by selecting avirulent mutants. The most successful vaccines are "living" avirulent viruses, which possess the advantage of multiplying in the host and which usually require only a single dose to be effective. This leads to prolonged immunological stimulation similar to that which occurs in natural infection. Live vaccines, however, are subject to a number of problems, such as genetic instability and contamination by extraneous viruses. Inactivated viruses are usually produced by treatment with formaldehyde, which destroys their infectivity. The major difficulty with inactivated viruses is the administration of sufficient viral antigen to induce a lasting immunity. In many cases, several injections must be given over a substantial period of time. The only inactivated viral vaccine in widespread use in humans is the influenza vaccine. The inactivated Salk polio vaccine has been largely replaced by the attenuated live-virus Sabin vaccine.

Interferon. Interferons (IFN) are proteins that evert virus nonspecific antiviral activities in cells through metabolic processes involving synthesis of both DNA and protein. The number of interferon-inducing substances has increased to include not only all of the major virus groups, but also bacterial and fungal products, nucleic acids, polymers, mitogens, and various low-molecular-weight substances. However, as interferons are induced by viruses and inhibit viral replication, viruses are usually considered to be natural inducers. The ability of viruses to induce interferon production depends upon the virus type. Some viruses, such as that responsible for Newcastle disease, are good inducers, while others, such as the adenoviruses, are regarded as poor inducers. Further, the type of cell used presents another factor in interferon production. In the whole animal, cells of the reticuloendothelial system are generally considered to be the major interferon producers. Recently, interferons have been classified into types on the basis of their antigenic specificities. Alpha and beta interferons (formerly called leukocyte and fibroblast, respectively) are acid stable and correspond to what have been called *Type I* IFNs (interferons). Gamma interferons (formerly called Immune) are acid labile and correspond to *Type II* IFNs.

Although interferon has been studied extensively for over a decade, the mechanism of its antiviral activity remains unclear. Considerable evidence exists to support the concept that interferon inhibits virus-specific protein synthesis, thus blocking viral replication in cells adjacent to the infected cell producing the interferon. There is no established reason to conclude, however, that interferon exerts antiviral action through a single mechanism.

Interferon is probably one of the most important early determinants of recovery from a number of viral diseases.

Recent work has centered upon use of interferon as a therapeutic agent in humans and animals. In humans, local application of monkey interferon is effective in reducing the severity of vaccinia virus skin infections. Recent results with herpes keratitis and chronic hepatitis are promising. Interferon appears to be active against oncogenic viruses in the treatment of such cancers as osteogenic sarcoma, and at present it is only the limited availability of interferon that prevents more extensive testing.

A number of virus diseases and virus related topics are described in this encyclopedia. Check alphabetical index for antiviral drugs, cancer research, chickenpox, common cold; coxsackie virus, dengue (breakbone fever), hepatitis, infectious mononucleosis, influenza, measles, mumps, Norwalk virus, poliomyelitis, rabies, Rift Valley fever, vaccinia, virus diseases (plants), and yellow fever.

ANN C. DeBALDO, PH.D.
Assoc. Prof., College of Public Health
University of South Florida
Tampa, Florida

Additional Reading

Ahmed, R. and I.S.Y. Chen: *Persistent Viral Infections,* John Wiley & Sons, Inc., New York, NY, 1999.

Campbell, I. and M. Buchmeier: *Neurovirology: Virus and the Brain (Advances in Virus Research,* Vol. 56, Academic Press, Inc., San Diego, CA, 2001.

Cann, A.J.: *Virus Culture: A Practical Approach,* Oxford University Press, Inc., New York, NY, 2000.

Goode, J.: *Gastroenteritis Viruses,* Vol. 238, Novartis Foundation Symposium, John Wiley & Sons, Inc., New York, NY, 2001.

Gosztonyi, I.G., M. Cooper, and R.W. Compans: *Mechanisms of Neuronal Damage in Virus Infections of the Nervous System,* Springer-Verlag, Inc., New York, NY, 2001.

Maramorosch, K., F.A. Murphy, and A.J. Shatkin: *Advances in Virus Research,* Vol. 54, Academic Press, Inc., San Diego, CA, 1999.

Montagnier, L.: *Virus: The Co-Discover of HIV Tracks Its Rampage and Charts the Future,* W.W. Norton Company, Inc., New York, NY, 1999.

Nowak, M.A. and R. May: *Virus Dynamics: Mathematical Principles of Immunology and Virology,* Oxford University Press, Inc., New York, NY, 2000.

Wagner, E.K. and M. Hewlett: *Basic Virology,* Blackwell Science, Inc., Malden, MA, 1999.

Zuckerman, A.J., J.R. Pattison, and J.E. Banatvala: *Principles and Practice of Clinical Virology,* 4th Edition, John Wiley & Sons, Inc., New York, NY, 2000.

VISCOELASTICITY. Mechanical behavior of material which exhibits viscous and delayed elastic response to stress in addition to instantaneous elasticity. Such properties can be considered to be associated with rate effects—time derivatives of arbitrary order of both stress and strain appearing in the constitutive equation—or hereditary or memory influences which include the history of the stress and strain variation from the undisturbed state. See also **Rheology**.

VISCOMETER (or Viscosimeter). A device for measuring the viscosity of a liquid. The types most widely used are the Engler, Saybolt, and Redwood, which indicate viscosity by the rate of flow of the test liquid through an orifice of standard diameter of the flow rate of a metal ball through a column of the liquid. Other types utilize the speed of a rotating spindle or vane immersed in the test liquid. The liquids commonly measured are lubricating oils and the like; heavier (non-Newtonian) liquids such as paints and paper coatings require more complex devices, e.g., Brookfield and Krebs-Stormer.

See also **Viscosity**.

VISCOSE PROCESS. The best-known process for making regenerated cellulose (rayon) by converting cellulose to the soluble xanthate, which can be spun into fibers and then reconverted to cellulose by treatment with acid. Wood pulp is steeped with 17–20% caustic soda; the resulting alkali cellulose is pressed to remove excess liquor and the soluble β- and γ-cellulose, and then shredded and aged. It is then treated with carbon disulfide and sodium hydroxide to form an orange, viscous solution of cellulose xanthate. After filtration and deaeration, this solution (viscose) is forced through minute spinnerette openings (or long slit dies in the case of cellophane) into a bath containing sulfuric acid and various salts such as sodium and zinc sulfate. The salts cause the viscose to gel immediately, forming a fiber or film of sufficient strength to permit it to be drawn through the bath under tension. At the same time the sulfuric acid decomposes the xanthate, converting the fibers to cellulose, in which form they are washed and dried.

See also **Xanthates**.

VISCOSITY. The internal resistance to flow exhibited by a fluid; the ratio of shearing stress to rate of shear. A liquid has a viscosity of one poise if a force of 1 dyne/square centimeter causes two parallel liquid surfaces one square centimeter in area and one centimeter apart to move past one another at a velocity of 1 cm/second. One poise equals 100 centipoises divided by the liquid density at the same temperature gives kinematic viscosity in centistokes (cs). One hundred centistokes equal on e stoke. To determine kinematic viscosity, the time is measured for an exact quantity of liquid to flow by gravity through a standard capillary. See also **Rheology**.

Water is the primary viscosity standard with an accepted viscosity at 20°C of 0.010019 poise. Hydrocarbon liquids such as hexane are less viscous. Molasses may have a viscosity of several hundred centistokes, while for a very heavy lubrication oil the viscosity may be 100 centistokes. There are many empirical methods for measuring viscosity.

See also **Saybolt Universal Viscosity**; and **Viscometer (or Viscosimeter)**.

VITAMIN. An organic compound that performs specific and necessary functions in humans, livestock, and other living organisms—even when it may be present in very small concentrations, at the milligram or microgram per 100 gram levels. The term *vitamine* was proposed by a Polish biochemist (Casimir Funk) in 1912 to designate substances required in trace amounts in the diet to prevent various nutritional-deficiency diseases. The principal vitamins are given in Table 1.

Nearly all vitamins are associated in some way with the normal growth function as well as with the maintenance and efficiency of living things. Various species are capable of synthesizing some of the vitamins from precursors that are present in the body. Synthesis is frequently by way of intestinal bacteria. In the case of vitamin D, substances in the skin combine with ultraviolet radiation from sunlight to yield the essential substance. Some vitamins, such as vitamin C, are specific, singular substances—in this case ascorbic acid. With other vitamins, there is a range of related compounds, as exemplified by the D, E, and K vitamins.

Because of inconsistencies in nomenclature, the B vitamins are not closely related as one might suspect. The B vitamins are different specific substances, and the use of the letter B to designate them indicates a degree of commonality that actually is not the case. Vitamin B_1 is thiamine, vitamin B_2 is riboflavin, vitamin B_6 is pyridoxine, vitamin B_{12} is cobalamin. Vitamins B_6 and B_{12}, for example differ markedly in function and structure. The alphabetical method of designation became complex and somewhat confusing as the various vitamins were recognized and studied over many years. During this period some substances were found to be identical with previously announced and described vitamins; or some substances were found not to be vitamins at all. Thus, over the years, the International Union of Pure and Applied Chemistry (I.U.P.A.C.) assigned new names to several of the vitamins.

The major vitamins are described in separate alphabetical entries in this book. Titles used for these entries have been selected on the basis of the most frequently used designations as of the early 1980s. In alphabetical order, the vitamins described in this book are: **Ascorbic Acid (Vitamin C); Biotin; Choline and Cholinesterase; Folic Acid; Inositol; Niacin; Pantothenic Acid; Vitamin B_2 (Riboflavin); Thiamine (Vitamin B_1); Vitamin A; Vitamin B_6 (Pyridoxine); Vitamin B_{12} (Cobalamin); Vitamin D; Vitamin E;** and **Vitamin K**.

The daily requirements of vitamins by humans are summarized in the entry on **Diet**. The relationship between hormones and vitamins is described in the entry on **Hormones**. Vitamins also are mentioned frequently in descriptions of various fruits, vegetables, and other food-stuffs throughout the book. Vitamins also figure prominently in discussions of some of the diseases, scores of which are described in this book.

Loss of Vitamins in Processing

During the last several years, much research has gone into determining the loss in effectiveness of vitamins as various foods are processed. It is a common tendency on the part of consumers to regard any fresh food as representing perfection in terms of nutritive, including vitamin, value

TABLE 1. COMPARATIVE LOSSES OF VITAMINS FROM VEGETABLES (CANNING AND FREEZE PROCESSING)

Method of preservation	Value	Loss of vitamins as compared with values of fresh cooke				
		Vitamin A	Thiamine (B$_1$)	Riboflavin (B$_2$)	Niacin	Ascorbic acid (C)
Frozen, cooked (boiled), drained	mean	12%	20%	24%	24%	26%
	range	0–50%	0–61%	0–45%	0–56%	0–78%
Canned, drained solids	mean	10%	67%	42%	49%	51%
	range	0–32%	56–83%	14–50%	30–65%	28–67%

TABLE 2. COMPARATIVE LOSSES OF VITAMINS FROM FRUITS

Method of preservation	Value	Loss of vitamins as compared with values of fresh products				
		vitamin A	Thiamine (B$_1$)	Riboflavin (B$_2$)	Niacin	Ascorbic acid (C)
Frozen (not thawed)	mean	37%	29%	17%	16%	18%
	range	0–78%	0–66%	0–67%	0–33%	0–50%
Canned, solids and liquids	mean	39%	47%	57%	42%	56%
	range	0–68%	22–67%	33–83%	25–60%	11–86%

and, conversely, to regard processed foods as nutritionally inferior. Under normal circumstances, these observations are true. Because fresh foods frequently are stored for several days at temperatures well above their freezing points, there are vitamin losses in unprocessed produce. Ascorbic acid content in vegeTables, for example, can severely degrade during improper storage. The degradation of vitamin values depends a great deal upon the type of food substance, the particular vitamin, and the manner in which the raw food is processed. The consumer today also is protected by vitamin-fortified foods, where vitamins have been added to compensate for losses during processing, or, in some cases, to generally enrich the foods nutritionally. Losses of vitamins from fruits and vegeTables during processing are tabulated in Tables 1 and 2.

Additional Reading

Ball, G.F.M.: *Fat-Soluble Vitamin Assays in Food Analysis: A Comprehensive Review,* Elsevier, New York, NY, 1989.

Ball, G.F.M.: *Bioavailability and Analysis of Vitamins in Foods,* Chapman & Hall, New York, NY, 1997.

Barinaga, M.: "Vitamin C Gets a Little Respect," *Science,* 374 (October 18, 1991).

Bender, D.A.: *Nutritional Biochemistry of the Vitamins,* 2nd Edition, Cambridge University Press, New York, NY, 2003.

Beardsley, T.: "Vitamin A and Its Cousins are Potent Regulators of Cells," *Sci. Amer.,* 16 (February 1991).

Blomhoff, R., et al.: "Transport and Storage of Vitamin A," *Science,* 399 (October 19, 1990).

Gaby, S.K., et al.: *Vitamin Intake and Health: A Scientific Review,* Marcel Dekker, New York, NY, 1991.

Litwack, G.: *Vitamins and Hormones,* Elsevier Science & Technology Books, New York, NY, 2004.

Navarra, T.: *Encyclopedia of Vitamins, Minerals, and Supplements,* 2nd Edition, Facts on File, Inc., New York, NY, 2004.

Staff: Academic Press: *Vitamins and Hormones,* Vol. 63, Academic Press, Inc., San Diego, CA, 2001.

Suttie, J.W., D.B. McCormic, and R. Rucker: *Handbook of Vitamins,* 3rd Edition, Marcel Dekker, Inc., New York, NY, 2001.

VITAMIN A. This substance also has been referred to as retinol, axerophthol, biosterol, vitamin A$_1$, anti-xerophthalmic vitamin, and anti-infective vitamin. The physiological forms of the vitamin include: Retinol (vitamin A$_1$) and esters; 3-dehydroretinol (vitamin A$_2$) and esters; 3-dehydroretinal (retinine-2); retinoic acid; neovitamin A; neo-b-vitamin A$_1$. The vitamin is required by numerous animal species. All vertebrates and some invertebrates convert plant dietary carotenoids in gut to vitamin A$_1$, which is absorbed. Most animal species store appreciable amounts of the vitamin in their livers, have low concentrations in the blood, and undetectable quantities in most other tissues. A deficiency of the vitamin produces a variety of symptoms, the most uniform being eye lesions, nerve degeneration, bone abnormalities, membrane keratinization, reproductive failure, and congenital abnormalities. Toxic symptoms from large doses of vitamin A are readily produced in animals and humans. Overdosage may cause irritability, nerve lesions, fatigue, insomnia, pain in bones and joints, exophthalmia, and mucous cell formation in keratinized membranes.

The principal physiological functions of this vitamin include growth, production of visual purple, maintenance of skin and epithelial cells, resistance to infection, gluconeogenesis, mucopolysaccharide synthesis, bone development, maintenance of myelin and membranes, maintenance of color and peripheral vision, maintenance of adrenal cortex and steroid hormone synthesis. Specific vitamin A deficiency diseases include xerophthalmia, nyctalopia, hemeralopia, keratomalacia, and hyperkeratosis.

In the rods of the retina, retinal is found combined with the protein *opsin,* the complex being called *rhodopsin* (visual purple). Although the entire series of reactions involved in dark vision has not been entirely worked out, the major steps in the cycle are quite clear. All-*trans*-retinol from the blood is oxidized by alcohol dehydrogenase (with NADP, nicotinamide adenine dinucleotide phosphate) to retinol which, in turn, is isomerized in the retina to 11-*cis*-retinal. This combines with opsin to form rhodopsin. On exposure to light, rhodopsin undergoes a sequence of changes with the eventual splitting off of retinal, which now has the all-*trans* configuration. This presumably can be reutilized in the retina by isomerization, or it can be reduced to retinol by alcohol dehydrogenase and returned to the circulation either as the free alcohol or as an ester.

The relatively recent observation, that retinoic acid can replace retinol or retinal for normal growth of animals, gave rise to further concepts in the biochemistry of vitamin A. Although retinoic acid cannot be demonstrated to be present normally in animal tissues, its formation by liver aldehyde dehydrogenase (NAD) and aldehyde oxidase has been accomplished, so that the molecule must be considered in the general scheme of vitamin A metabolism. When retinoic acid is given to animals as the only form of vitamin A, growth is normal, but the animals eventually become sterile and blind. This had led to the consideration that vitamin A may have at least three independent functions: (1) growth; (2) vision; and (3) reproduction.

The reversal of the oxidative pathway of vitamin A (retinol → retinal → retinoic acid) does not occur in the body. When retinoic acid is fed to animals, even in relatively large doses, there is no storage and, in fact, the molecule is rapidly metabolized and cannot be found several hours after administration. The metabolic products have not been fully identified. Several fractions from liver or intestine, isolated after administering retinoic acid marked with carbon-14, have been shown to have biological activity.

In 1912, Hopkins reported a factor in milk needed for the growth of rats. In 1913, Osborne and Mendel demonstrated that milk factor is fat soluble, and present in other fats also. McCollum and Davis, in 1913–1915, identified milk factor (fat-soluble A) in butter and egg yolk. In 1917, McCollum and Simmonds found xerophthalmia in rats due to lack of fat-soluble A. In 1920, Drummond renamed fat-soluble A, vitamin A. In 1930, Moore determined that carotene is a precursor of vitamin A. See also ???. During 1930–1937, Karrer et al. isolated and synthesized vitamin A. In 1935, Wald reported visual purple in retina to be a complex of protein and vitamin A.

Distribution and Sources. Provitamin carotenoids are contained in numerous foods, but of varying concentrations.

High vitamin A and procarotenoids content (10,000–76,000 I.U./100 grams).[1]. Carrot, dandelion green, kohlrabi, liver (beef, calf, chicken, pig, sheep), liver oil (cod, halibut, salmon, shark, sperm whale), mint, palm oil, parsley, spinach, turnip greens.

Medium vitamin A and procarotenoids content (1,000–10,000 I.U./100 grams). Apricot, beet greens, broccoli, butter, chard, cheese (except cottage), cherry (sour), chicory, chives, collards, cream, eel, egg yolk, endive, fennel, kale, kidney (beef, pig, sheep), leek greens, lettuce (butterhead and romaine), liver (pork), mango, margarine, melons (yellow), milk (dried), mustard, nectarine, peach, pumpkin, squash (acorn, butternut, hubbard), sweet potato, tomato, watercress, whitefish.

Low vitamin A and procarotenoids content (100–1,000 I.U./100 grams). Artichoke, asparagus, avocado, banana, bean (except kidney), berry

[1] One I.U. = 0.344 microgram vitamin A acetate = 0.3 microgram retinol.

(black-, blue-, boysen-, goose-, logan-, rasp-), Brussels sprouts, cabbage, carp, cashew, celery, cherry (sweet), clam, corn (maize), cowpea, cucumber, currant (red), grape, groundnut (peanut), hazelnut, herring, kumquat, leek, lentil (dry), lettuce, milk, okra, olive, orange, oyster, pea, pecan, pepper (sweet), pineapple, pistachio, plum, prune, rhubarb, rutabaga, salmon, sardine, squash (summer and zucchini), tangerine, walnut (black).

In higher plants, carotenoids are produced in green leaves. In animals, conversion of carotenoids to vitamin A occurs in the intestinal wall. Storage is in the liver; also kidney in rat and cat. Target tissues are retina, skin, bone, liver, adrenals, germinal epithelium. Commercial Vitamin A supplements are obtained chemically by extraction of fish liver; or synthetically from citral or β-ionone.

Bioavailability of Vitamin A. Factors which may cause a decrease in the availability of vitamin A include: (1) liver damage; (2) impaired intestinal conversion of carotenes; (3) impaired absorption (low bile); (4) loss in food preparation (cooking and frying—heat oxidation); (5) presence of antagonists; (6) illness, causing increased destruction and excretion of the vitamin. Increases in availability may result from: (1) storage in body (liver); (2) factors which stimulate intestinal conversion of carotenes—tetraiodothyronine (thyroxine), insulin; (3) absorption aids—bile, fat; and (4) dietary protein which mobilizes vitamin A from storage in liver.

Antagonists of vitamin A include sodium benzoate, bromobenzene, citral, oxidized derivatives of vitamin A, excessive concentrations of thyroxine, estrogens, vitamin E (as regards membrane permeability). Synergists include vitamins B₂, B₁₂, and E, ascorbic acid, thyroxine, testosterone, melanocyte-stimulating hormone (MSH), and somatotrophin growth hormone.

Unusual features of vitamin A as observed by some investigators include: (1) decreases serum cholesterol in large-quantity administration (chicks); (2) dietary protein required to mobilize liver reserves of vitamin A; (3) decreased quantities in tumors; (4) coenzyme Q₁₀ accumulates in A-deficient rat liver; (5) Ubichromenol-50 accumulation in A-deficient rat liver; (6) retinoic acid functions as vitamin A except for visual and reproductive functions; (7) anti-infection properties and anti-allergic properties; (8) decreases basal metabolism; (9) detoxification of poisons in the liver aided by vitamin A; and (10) vitamin A is involved in triose → glucose conversions.

Additional Reading

Ball, G.F.M.: *Bioavailability and Analysis of Vitamins In Foods,* Chapman & Hall, New York, NY, 1997.

Bender, D.A.: *Nutritional Biochemistry of the Vitamins,* 2nd Edition, Cambridge University Press, New York, NY, 2003.

Combs, G.F. Jr.: *The Vitamins: Fundamental Aspects in Nutrition and Health,* 2nd Edition, Academic Press, Inc., San Diego, CA, 1998.

Eitenmiller, R.R. and W.O. Landen: *Vitamin Analysis for the Health and Food Sciences,* CRC Press, LLC., Boca Raton, FL, 1998.

Litwack, G.: *Vitamins and Hormones,* Elsevier Science & Technology Books, New York, NY, 2004.

McDowell, L.R.: *Vitamins in Animal and Human Nutrition,* 2nd Edition, Iowa State University Press, Ames, IA, 2000.

Nau, H. and W.S. Blaner: *Retinoids: The Biochemical and Molecular Basis of Vitamin A and Retinoid Action,* Vol. 139, Springer-Verlag, Inc., New York, NY, 1999.

Navarra, T.: *Encyclopedia of Vitamins, Minerals, and Supplements,* 2nd Edition, Facts on File, Inc., New York, NY, 2004.

Newstrom, H.: *Nutrients Catalog: Vitamins, Minerals, Amino Acids, Macronutrients—Beneficial Use, Helpers, Inhibitors, Food Sources, Intake Recommendations, and Symptoms of over or under Use,* McFarland & Company, Inc., Publishers, Jefferson, NC, 1993.

Web Reference

Facts About Vitamin A and Carotenoids: http://www.cc.nih.gov/ccc/supplements/vita.html

VITAMIN B₂ (Riboflavin). Some earlier designations for this substance included vitamin G, lactoflavin, hepatoflavin, ovoflavin, verdoflavin. The chemical name is 6,7-dimethyl-9-d-l'ribityl isolloxazine. Riboflavin is a complex pigment with a green fluorescence. Riboflavin deficiency frequently accompanies pellagra and the typical lesions of both nicotinic acid and riboflavin deficiency are found in that disease. See also **Niacin.**

Riboflavin, like nicotinic acid, forms an oxidation enzyme and, as such, acts as an oxygen carrier to the cell. The structure of riboflavin is:

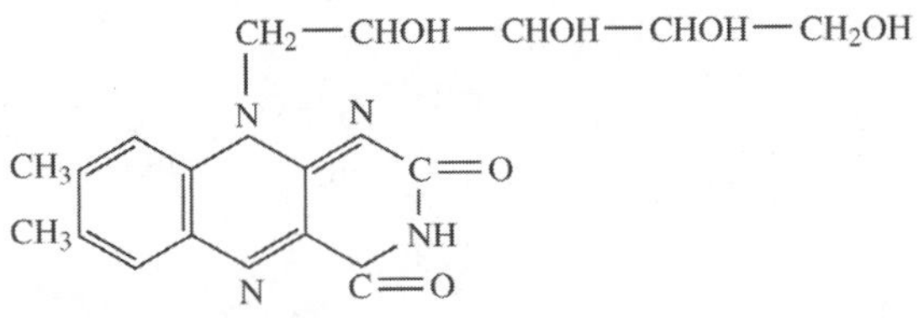

Disorders caused by a deficiency of riboflavin include anemia, cheilosis (a lip disorder); corneal vascularization, seborrheic dermatitis, and glossitis. Research leading to the current knowledge of riboflavin essentially commenced in 1917 when Emmet and McKim showed dietary growth factor for rats in rice polishings. In 1920, Emmet suggested the presence of several dietary growth factors in yeast concentrate, including the heat-stable component and B₁. The British Medical Research Council, in 1927, proposed that the designation B₂ be given to the heat-stable component. Warburg and Christian, in 1932, isolated yellow enzyme (containing riboflavin, FMN) from bottom yeast. In 1933, Kuhn isolated pure B₂ (riboflavin) from milk and recognized its growth-promoting activity. Several researchers (Kuhn et al.; Karrer et al.), in 1935, worked out the structure and synthesis of vitamin B₂, during which period it was named *riboflavin.* By 1954, Christie et al. had determined the structure and synthesized riboflavin dinucleotide (FAD).

Riboflavin has been shown to be a constituent of 2 coenzymes: (1) Flavin mononucleotide (FMN); and (2) flavin adenine dinucleotide (FAD). The structures are:

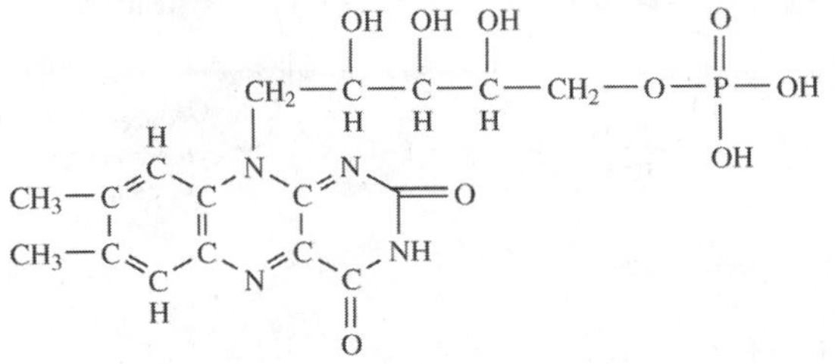

Flavin mononucleotide (FMN)

FMN was first identified as the coenzyme of an enzyme system that catalyzes the oxidation of the reduced nicotinamide coenzyme, NADPH (reduced NADP), to NADP (nicotinamide adenine dinucleotide phosphate). NADP is an essential coenzyme for glucose-6-phosphate dehydrogenase which catalyzes the oxidation of glucose-6-phosphate to 6-phosphogluconic acid. This reaction initiates the metabolism of glucose by a pathway other than the TCA cycle (citric acid cycle). The alternative route is known as the phosphogluconate oxidative pathway, or the hexose monophosphate shunt. The first step is:

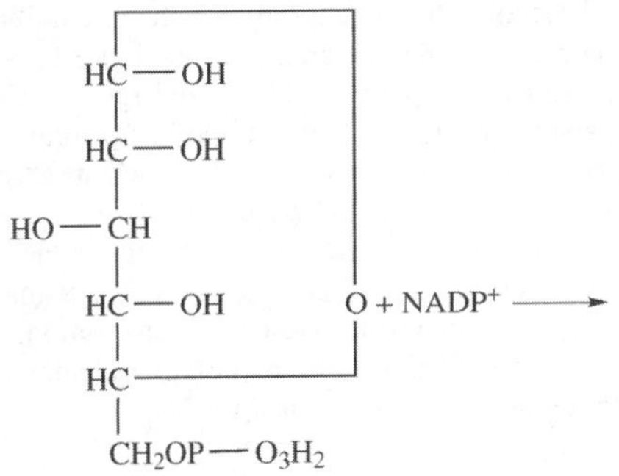

Glucose-6-phosphate

Most of the numerous other riboflavin-containing enzymes contain FAD. FAD is an integral part of the biological oxidation-reduction system where it mediates the transfer of hydrogen ions from NADH to the oxidized cytochrome system. FAD can also accept hydrogen ions directly from a metabolite and transfer them to either NAD, a metal ion, a heme derivative, or molecular oxygen. The various mechanisms of action of FAD are probably due to differences in protein apoenzymes to which it is bound.

The oxidized and reduced states of the flavin portion of FAD are:

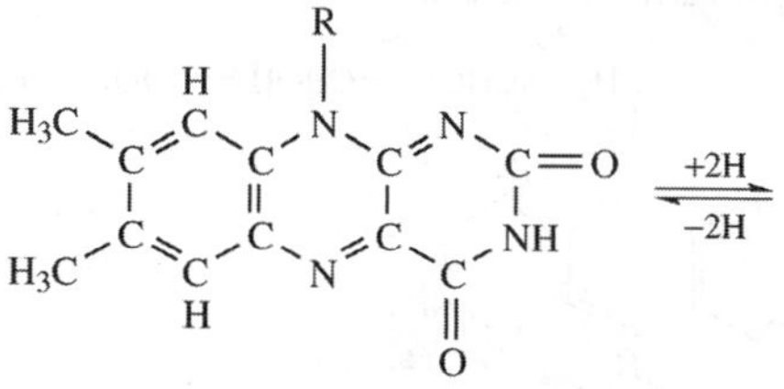

FAD
(oxidized)

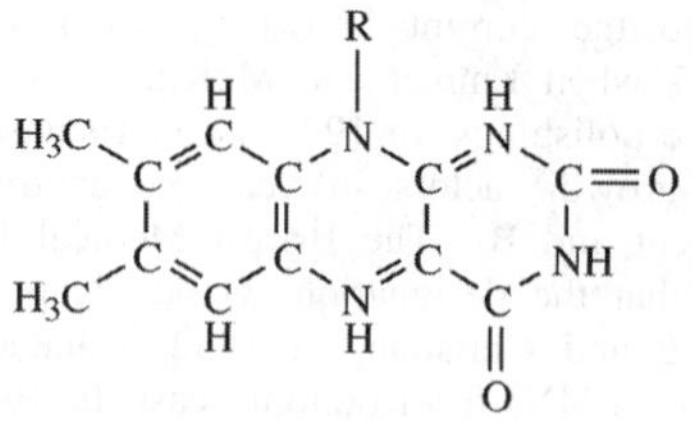

FADH$_2$
(reduced)

In the biological oxidation-reduction system, reduced NAD (i.e., ADH) is reoxidized to NAD by the riboflavin-containing coenzyme FAD as shown by:

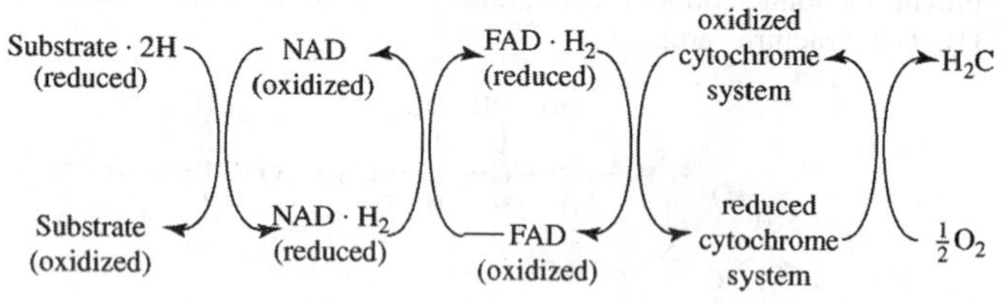

See also **Coenzymes**.

Distribution and Sources

Research indicates that all organisms require riboflavin. Endogenous sources exist in high plants, algae, some bacteria, and some fungi. All animals, some fungi and bacteria receive at least a partial supply of riboflavin from generation by intestinal bacteria. In the case of humans, there is a large dependence upon exogenous sources.

High riboflavin content (1000–10,000 micrograms/100 grams). Beef (kidneys, liver), calf (kidneys, liver), chicken (liver), pork (heart, kidneys, liver), sheep (kidneys, liver), yeast (killed)

Medium riboflavin content (100–1000 micrograms/100 grams). Almond (dry), asparagus, avocado, bacon, bean (kidney, lima, snap, wax), beef, beet greens, broccoli, Brussels sprouts, cashew, cauliflower, cheeses, chicken, chicory, corn (maize), cream, dandelion greens, eggs, endive, fish, goose, groundnut (peanut), kale, kohl-rabi, lamb, lentil (dry), milk, oats, parsley, parsnip, pea, pecan, pork, rice bran, soybean (dry), spinach, turkey, turnip greens, veal, walnut, wheat germ

Low riboflavin content (10–100 micrograms/100 grams). Apple, apricot, artichoke, banana, barley, beet, berry (black-, blue-, cran-, rasp-, straw-), cabbage, carrot, celery, cherry, coconut, cucumber, date (dry), eggplant, fig, grape, grapefruit, lettuce, melons, onion, orange, peach, pear, pepper (sweet), pineapple, plum, potato, radish, raisin (dry), rice, sweet potato, tangerine, tomato, turnip

Commercial riboflavin dietary supplements are prepared (1) by the fermentation process (bacteria or yeast); and (2) by chemical synthesis from alloxan, ribose, and *o*-xylene.

Precursors in the biosynthesis of riboflavin include purines, pyrimidines, and ribose. Intermediate in the synthesis is 6,7-dimethyl-8-ribityllumazine. In plants, riboflavin production sites are found in leaves, germinating seeds, and root nodules. Storage sites in animals are heart and liver, with small amounts in the kidneys. Riboflavin in overdose is essentially nontoxic to humans.

Bioavailability of Riboflavin

Factors which tend to decrease the availability of riboflavin include: (1) cooking, inasmuch as riboflavin is slightly soluble in water; (2) in some plant foods, availability is lower than might be expected because of bound forms; (3) decreased phosphorylation in intestines prevents absorption; (4) exposure of foods to sunlight; (5) enzymes required for breakdown are not present; (6) presence of gastrointestinal disease; and (7) diuresis. Riboflavin availability is increased by storage in heart, liver, and kidneys and by the presence of very actively producing intestinal bacteria.

Antagonists of riboflavin include isoriboflavin, lumiflavin, araboflavin, hydroxyethyl analogue, formyl methyl analogue, galactoflavin, and flavin-monosulfate. Synergists include vitamins A, B$_1$, B$_6$, and B$_{12}$, niacin, pantothenic acid, folic acid, biotin, tetraiodothyronine (thyroxine), insulin, and somatotrophin (growth hormone).

Determination of Riboflavin

Bioassay includes observance of the growth rate of rats; microbiological—*L. caseli*, and *L. mesenteroides*. Physicochemical methods include fluorimetry, paper electrophoresis, and polarography.

Unusual features of riboflavin as recorded by some researchers include: (1) High levels in liver inhibit tumor formation by azo compounds in animals; (2) free radicals are formed by light or dehydrogenation: flavine ⇌ semiquinone: dihydroflavin+; (3) free vitamin is found only in retina, urine, milk, and semen; (4) substitution of adenine by other purines and pyrimidines destroys activity of flavin adenine dinucleotide (FAD); (5) phosphorylation of vitamin in intestines allows absorption as flavin mononucleotide (FMN); (6) blood levels decrease during life in humans: (7) brain content remains constant; (8) available in plants as FMN and FAD; (8) very concentrated in bull semen.

Additional Reading

Ball, G.F.M.: *Bioavailability and Analysis of Vitamins In Foods,* Chapman & Hall, New York, NY, 1997.

Bender, D.A.: *Nutritional Biochemistry of the Vitamins,* 2nd Edition, Cambridge University Press, New York, NY, 2003.

Combs, G.F., Jr.: *The Vitamins: Fundamental Aspects in Nutrition and Health,* 2nd Edition, Academic Press, Inc., San Diego, CA, 1998.

Eitenmiller, R.R. and W.O. Landen: *Vitamin Analysis for the Health and Food Sciences,* CRC Press, LLC., Boca Raton, FL, 1998.

Litwack, G.: *Vitamins and Hormones,* Elsevier Science & Technology Books, New York, NY, 2004.

McDowell, L.R.: *Vitamins in Animal and Human Nutrition,* 2nd Edition, Iowa State University Press, Ames, IA, 2000.

Navarra, T.: *Encyclopedia of Vitamins, Minerals, and Supplements,* 2nd Edition, Facts on File, Inc., New York, NY, 2004.

Newstrom, H.: *Nutrients Catalog: Vitamins, Minerals, Amino Acids, Macronutrients—Beneficial Use, Helpers, Inhibitors, Food Sources, Intake Recommendations, and Symptoms of over or under Use,* McFarland & Company, Inc., Publishers, Jefferson, NC, 1993.

VITAMIN B$_6$ (Pyridoxine). Infrequently called adermine or pyridoxol, this vitamin participates in protein, carbohydrate, and lipid metabolism. The metabolically active form of B$_6$ is pyridoxal phosphate, the structures of which are:

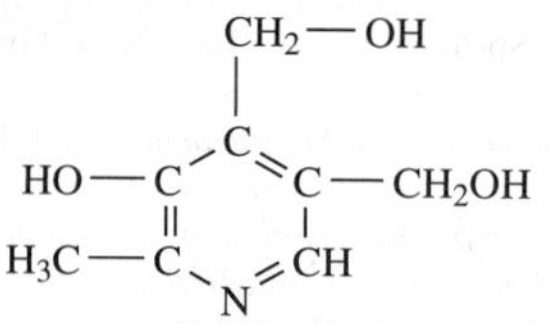

Pyridoxine

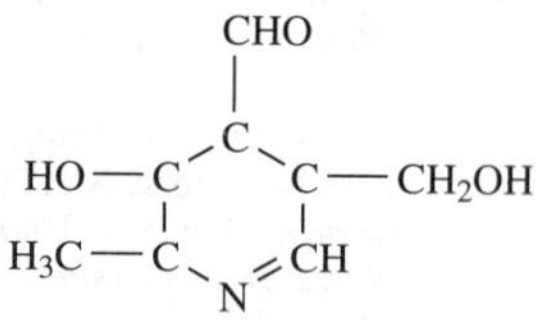

Pyridoxal

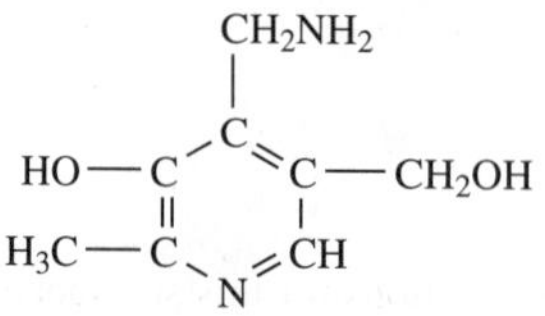

Pyridoxamine

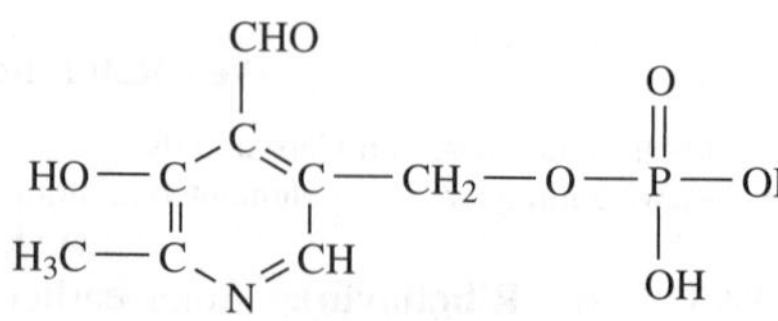

Pyridoxal phosphate

Pyridoxal phosphate enzymes mediate the nonoxidative decarboxylation of amino acids. This mechanism is of primary importance in bacteria, but it may be essential to proper function of the nervous system in humans by providing a pathway for the synthesis of a nerve impulse inhibitor, γ-amino-butyric acid from glutamic acid:

$$HOOC-CH_2CH_2CH-COOH$$
$$\mid$$
$$NH_2$$

Glutamic acid

$$\xrightarrow{\text{Pyridoxal phosphate}} HOOC-CH_2CH_2CH_2NH_2 + CO_2$$
$$\gamma\text{-Aminobutyric acid}$$

Pyridoxal phosphate is also a cofactor for transamination reactions. In these reactions, an amino group is transferred from an amino acid to an α-keto acid, thus forming a new amino acid and a new α-keto acid. Transamination reactions are important for the synthesis of amino acids from non-protein metabolites and for the degradation of amino acids for energy production. Since pyridoxal phosphate is intimately involved in amino acid metabolism, the dietary requirement for vitamin B₆ increases as the protein content of the diet increases.

The coenzyme especially participates in gluconeogenesis, production of neural hormones, bile acids, unsaturated fatty acids, and porphyrins.

A deficiency of the vitamin can result in lymphopenia, convulsions, dermatitis, irritability, and nervous disorders in humans. A deficiency in monkeys may cause arteriosclerosis, while in rats, acrodynia. Research indicates that all animals require vitamin B₆. Bacteria in intestines generate some of this vitamin, but relatively little is available to humans in this form. Endogenous sources are available to plants, fungi, and some bacteria.

In 1934, György cured a dermatitis in rats (not due to vitamins B₁ or B₂) with a yeast extract factor. In 1938, Lepkovsky isolated a similar factor from rice bran extract. In that same year, Keresztesy and Stevens isolated and crystallized pure B₆ from rice polishings. Also, in the same year, Kohn, Wendt, and Westphal synthesized pyridoxine and gave pyridoxine its present name. In the following year (1939), Stiller, Keresztesy, and Stevens established the structure of the vitamin. In 1945, Snell observed pyridoxal and pyridoxamine. The recognition of and establishment of B₆ requirements in humans was not achieved until 1953, by Snyderman et al.

In plants, the vitamin is present as pyridoxol-5-phosphate, pyridoxal-5-phosphate, or pyridoxamine-phosphate. In plants, production sites are found in fungi, cereal germ, and seeds.

Commercially, the vitamin is available as a dietary supplement in the compound pyridoxine hydrochloride. The compound can be synthesized by condensing ethoxy acetylacetone with cyanoacetamide (method of Harris and Folkers); or from oxazoles.

Distribution and Sources

Most fruits and vegetables are low in pyridoxine content, although most nuts are quite high. Cereals and a number of other substances have low-to-medium content.

High pyridoxine content (1,000–10,000 micrograms/100 grams). Groundnut (peanut), herring, liver (beef, calf, pork), molasses (black strap), rice (brown), salmon, walnut, wheat germ, yeast.

Medium pyridoxine content (100–1,000 micrograms/100 grams). Avocado, banana, barley, beef, Brussels sprouts, butter, cabbage, carrot, cauliflower, cod, corn (maize), eggs, flounder, grape, halibut, kale, lamb, mackerel, oats, pea, pear, pork, potato, rye, sardine, soybean, spinach, tomato, tuna, turnip, veal (brain, heart, kidney), whale, wheat, yam.

Low pyridoxine content (10–100 micrograms/100 grams). Apple, asparagus, bean, beet greens, cantaloupe, cheese, cherry, currant (red), grapefruit, lemon, lettuce, milk, onion, orange, peach, raisin, strawberry, watermelon.

Bioavailability of Pyridoxine

Factors which tend to decrease bioavailability of pyridoxine include: (1) Administration of isoniazid; (2) loss in cooking (estimated at 30–45%)—vitamin is water-soluble; (3) diuresis and gastrointestinal diseases; (4) irradiation. Availability can be increased by stimulating intestinal bacterial production (very small amount), and storage in liver. The target tissues of B₆ are nervous tissue, liver, lymph nodes, and muscle tissue. Storage is by muscle phosphorylase (skeletal muscle—small amount). It is estimated that 57% of the vitamin ingested per day is excreted. The vitamin exerts only limited toxicity for humans.

Precursors for biosynthesis of the vitamin include glycine, serine, or glycolaldehyde, although further research is required for further confirmation of these substances. Intermediates have not been identified. Antagonists of B₆ include 4-deoxypyridoxine, 4-methoxypyridoxine, toxopyrimidine, penicillamine, semicarbazide, and isoniazid. Synergists include ascorbic acid, biotin, epinephrine, folic acid, glucagon, niacin, norepinephrine, somatotrophin (growth hormone), and vitamins B₁, B₂, and E.

Determination of Vitamin B₆

As pointed out by investigators Gregory and Kirk (Department of Food Science and Human Nutrition, Michigan State University, East Lansing, Michigan), development of an adequate chemical procedure for the determination of biologically active forms of vitamin B₆ in foods has been a complex problem. Basic studies by Bonavita (1960), Toepfer et al. (1961), and Polansky et al. (1964) have demonstrated the feasibility of fluorometric measurement of pyridoxal (PAL), pyridoxamine (PAM), and pyrodixine (PIN) by conversion to PAL and reaction with potassium cyanide, forming the fluorophore 4-pyridoxic acid lactone. Various fluorometric methods have been applied to B₆ compounds in biological materials (Fujiita et al., 1955; Contractor and Shane, 1968; Loo and Badger, 1969; Takanashi et al., 1970; Fieldlerova and Davidek, 1974; Chin, 1975). The results of Chin suggested that interfering compounds may be present in the PAL fraction after column chromatographic separation of the B₆ analogs by the procedure of Toepfer and Lehmann (1961). In the Gregory-Kirk (1977) study, methods for improving chromatographic separation and fluorometric determination of vitamin B₆ compounds in foods were investigated. Their findings are presented in the reference indicated.

Traditionally, B₆ compounds also have been determined by bioassay, including rat and chicken growth assays.

Additional Reading

Ball, G.F.M.: *Bioavailability and Analysis of Vitamins In Foods,* Chapman & Hall, New York, NY, 1997.

Bender, D.A.: *Nutritional Biochemistry of the Vitamins,* 2nd Edition, Cambridge University Press, New York, NY, 2003.

Combs, G.F. Jr.: *The Vitamins: Fundamental Aspects in Nutrition and Health,* 2nd Edition, Academic Press, Inc., San Diego, CA, 1998.

Eitenmiller, R.R. and W.O. Landen: *Vitamin Analysis for the Health and Food Sciences,* CRC Press, LLC., Boca Raton, FL, 1998.

Litwack, G.: *Vitamins and Hormones,* Elsevier Science & Technology Books, New York, NY, 2004.

McDowell, L.R.: *Vitamins in Animal and Human Nutrition,* 2nd Edition, Iowa State University Press, Ames, IA, 2000.

Navarra, T.: *Encyclopedia of Vitamins, Minerals, and Supplements,* 2nd Edition, Facts on File, Inc., New York, NY, 2004.

Newstrom, H.: *Nutrients Catalog: Vitamins, Minerals, Amino Acids, Macronutrients—Beneficial Use, Helpers, Inhibitors, Food Sources, Intake Recommendations, and Symptoms of over or under Use,* McFarland & Company, Inc., Publishers, Jefferson, NC, 1993.

Web Reference

Facts About Vitamin B6:
http://www.cc.nih.gov/ccc/supplements/vitb6.html

VITAMIN B₁₂ (Cobalamin). Sometimes also called cyanocobalamin, this vitamin is one of the more recent of the major B complex vitamins to be fully identified, with its structure not definitized (by Hodkin et al.) until 1955. The vitamin is required by most vertebrates, some protozoa, bacteria, and algae. Principal physiological functions include: (1) Coenzyme in nucleic acid, protein, and lipid synthesis; (2) maintains growth; (3) participates in methylations; (4) maintains epithelial cells and nervous system (myelin sheath); (5) erythropoiesis (with folic acid); (6) leukopoiesis. Deficiency diseases or disorders include retarded growth; pernicious anemia; megaloblastic anemia; macrocytic, hyperchromic anemia; glossitis; spinal cord degeneration; and sprue. The major physiological forms of B₁₂ available include hydroxocobalamin (vitamin B₁₂ₐ) and aquocobalamin (vitamin B₁₂꜀).

In 1926, Minot and Murphy controlled pernicious anemia using liver. In 1944, Castle demonstrated intrinsic factor needed to control pernicious anemia with liver. Rickes et al., in 1948, isolated and crystallized factor in liver controlling pernicious anemia. In that same year, Smith and Parker crystallized and designated liver factor as vitamin B₁₂. West demonstrated, in 1948, clinical activity of vitamin B₁₂, and, in 1955, Hodgkin et al. determined the structure of the vitamin. This is shown in Structure 1. Vitamin B₁₂ is the only vitamin with a metal ion—in this case, cobalt.

Structure 1.   Vitamin B$_{12}$

Surrounding the cobalt is a macrocyclic corrin ring that is comprised of four nitrogen-containing, five-membered rings joined through three methylene bridges. There is a similarity between this corrin ring and the dihydroporphyrin (chlorin) ring of chlorophyll.

Absorption of Vitamin B$_{12}$

This vitamin is not synthesized in animals, but rather it results from the bacterial or fungal fermentation in the rumen, after which it is absorbed and concentrated during metabolism. Among the known vitamins, this exclusive microbial synthesis is of great interest. One of the major results of vitamin B$_{12}$ deficiency is pernicious anemia. This disease, however, usually does not result from a dietary deficiency of the vitamin, but rather by an absence of a glycoprotein ("gastric intrinsic factor") in the gastric juices that facilitates absorption of the vitamin in the intestine. Control of the diseases hence is either by injection of B$_{12}$ or by oral administration of the intrinsic factor, with or without the vitamin injection.

There are two separate and distinct mechanisms for absorption of vitamin B$_{12}$. One mechanism is active, the other passive; both operate simultaneously. The active process is physiologically more important, since it is operative primarily in the presence of the small (1–2 micrograms) quantities of vitamin B$_{12}$ made available for absorption from the average meal. This special mechanism, perhaps uniquely necessary for vitamin B$_{12}$ because of its large size and polar properties, operates as follows. The normal gastric mucosa secretes a substance, called the *intrinsic factor of Castle*, which combines with free vitamin B$_{12}$. The complex travels down the intestine to the ileum, where, in the presence of calcium and pH above 6, it attaches to "receptors" lining the wall of the ileal mucosa. Vitamin B$_{12}$ is then freed from intrinsic factor via a "releasing factor" mechanism of unknown nature, operating either at the surface of or within the ileal mucosal cell, and passes into the bloodstream. Thus, important requirements for normal absorption of vitamin B$_{12}$ from food are: (1) the vitamin must be freed from its peptide bonds in food; (2) the gastric mucosa must secrete an adequate quantity of intrinsic factor; (3) the ileal mucosa must be sufficiently normal both structurally and functionally so that vitamin B$_{12}$ may be absorbed across it.

Intrinsic factor is believed to be a glycoprotein or mucopolysaccharide with a molecular weight in the range of 50,000 and an end-group conformation like that of partly degraded blood group substance. The sole known role of intrinsic factor is to facilitate the transport of the large (molecular weight = 1,355) vitamin B$_{12}$ molecule across the wall of the ileal mucosa and into the bloodstream. Antibodies to intrinsic factor exist in the serum of approximately half of all patients with pernicious anemia.

The second mechanism for vitamin B$_{12}$ absorption is operative primarily in the presence of quantities of vitamin B$_{12}$ greater than those made available for absorption from the average diet (i.e., quantities greater than about 30 micrograms). This mechanism is a passive one, probably diffusion,

and most likely occurs along the entire length of the small intestine. It operates when patients with pernicious anemia (vitamin B$_{12}$ deficiency due to inadequate or absent intrinsic factor secretion of unknown cause) are treated with large quantities (500 micrograms or more daily) of oral vitamin B$_{12}$. Such treatment is probably better than treatment with oral hog intrinsic factor, to which refractoriness often develops, but it is not as certain as treatment with monthly injections of vitamin B$_{12}$.

Deficiency Effects

Further elucidating on the physiologic functions and deficiency disorders of vitamin B$_{12}$, this vitamin is required for DNA (deoxyribonucleic acid) synthesis and, therefore, is necessary in every reproducing cell in humans for maintenance of the ability to divide. The vitamin functions coenzymatically in the methylation of homocysteine to methionine. It is important in several isomerization reactions, and as a reducing agent, and is probably of special importance in enzymatic reduction of ribosides to deoxyribosides. It is involved in protein synthesis, partly via its role in the conversion of homocysteine to methionine; in fat and carbohydrate metabolism, partly via its role in the isomerization of succinate to methylmalonate (which then may be decarboxylated to propionate), and in folate metabolism. Where these two vitamins interrelate, vitamin B$_{12}$ appears to serve as a coenzyme and folate as a substrate; such is true in the vitamin B$_{12}$-mediated transfer of a methyl group from N^5-methyltetrahydrofolic acid to homocysteine, which is thereby converted to methionine.

Vitamin B$_{12}$ is one of the most potent nutrients known; the minimal daily requirement for absorption by the normal adult is probably in the range of 0.1 microgram. This equals, for example, 1/500th of the minimal daily adult folate requirement, which is in the range of 50 micrograms.

As with all nutritional deficiencies, lack of vitamin B$_{12}$ may arise from inadequate ingestion, absorption, or utilization, and from increased requirement or increased excretion. Deficiency of vitamin B$_{12}$ produces megaloblastic (large germ cell) anemia, damage to the alimentary tract (glossitis being the most striking feature), and neurologic damage. The most classic neurologic sign of vitamin B$_{12}$ deficiency is decreased ability to perceive the vibration of a tuning fork pressed against the ankles. This finding is associated with damage to the posterior and lateral columns of the spinal cord, and also with damage to the peripheral nerves. This damage occurs because vitamin B$_{12}$ deficiency results in gradual deterioration of the myelin sheath, which is followed by deterioration of the axon. These processes occur slowly over months to years, and during this stage are reversible by treatment with vitamin B$_{12}$. However, when the nerve nucleus finally deteriorates, the neurologic damage becomes irreversible.

Distribution and Sources of Vitamin B$_{12}$. Vegetables, fruits, seeds, and nuts have a very low content of this vitamin.
High vitamin B$_{12}$ content (50–500 micrograms/100 grams). Brain (beef), kidney (beef, lamb), liver (beef, calf, lamb, pork).
Medium Vitamin B$_{12}$ content (5–50 micrograms/100 grams). Clam, crab, egg yolk, heart (beef, chicken, rabbit), kidney (rabbit), liver (chicken, rabbit), oysters, sardine, salmon.
Low Vitamin B$_{12}$ content (0.5–5 micrograms/100 grams). Beef, cod, cheeses, chicken, eggs, flounder, haddock, halibut, lamb, lobster, milk, pork, scallops, shrimp, swordfish, tuna, whale.

Vitamin B$_{12}$ dietary supplements are often prepared commercially by the fermentation of *S. griseus, S. aureofaciens, Propionibacterium*; or as a by-product of antibiotic production.

Certain species of bacteria and actinomycetes biosynthesize vitamin B$_{12}$. Precursors for this synthesis include glycine-corrin nucleus; δ-aminolevulinic acid-corrin nucleus; and methionine-corrin nucleus. Intermediates during the synthesis include porphobilinogen, α-D-ribosides of benzimidazole; 5,6-dimethylbenzimidazole; and α-ribazole. Antagonists of vitamin B$_{12}$ include methylamide, ethylamide, anilide, lactone derivatives, pteridine, nicotinamide. Synergists include ascorbic acid, biotin, folic acid, pantothenic acid, thiamine, and vitamins A and E.

Bioavailability of Vitamin B$_{12}$

Factors which tend to decrease the availability of this vitamin include: (1) cooking losses, since the vitamin is heat labile; (2) cobalt deficiency in ruminants; (3) intestinal malabsorption or parasites; (4) lack of intrinsic factor; (5) intestinal disease; (6) aging; (7) vegetarian diet; (8) excretion in feces; (9) gastrectomy. Factors which help to increase availability include: (1) administration of sorbitol; (2) synthesis by intestinal bacteria (not normally); (3) reduced temperature; and (4) presence of food in the stomach.

Although vitamin B_{12} is essentially considered nontoxic, polycythemia has been reported from excessive dosages. From 30 to 60% of the vitamin is stored in the liver; the remainder is found in the kidneys, lungs, and spleen. Target tissues are the central nervous system, kidneys, myocardium, muscle, skin, and bone.

Unusual features of vitamin B_{12} observed by some investigators include: (1) the cyanide group is an artifact of preparation; (2) the only vitamin synthesized in appreciable amounts only by microorganisms (possible in tumors); (3) only vitamin with a metal ion; (4) works with glutathione; (5) glutathione content decreased on B_{12} deficiency; (6) mitosis retarded in B_{12} deficiency; (7) requires intrinsic factor (enzyme) for oral activity; (8) increases tumor size (Rous sarcoma); (9) diamagnetic properties; (10) no acidic or basic groups revealed on titration (no pKa).

Additional Sources of B_{12}

Fermented soybean and fish products have been found to contain B_{12} (Lee et al., 1958). Nutritionally significant amounts of B_{12} also were found in the Indonesian fermented products, *ontjom* and *tempeh* (Liem et al., 1977). The microbial production of vitamin B_{12} in *kimchi*, Korean fermented vegetables, including cabbage, has been reported (Lee et al., 1958; Kim et al., 1960). The strain producing the vitamin during the fermentation was identified as *Bacillus megaterium*. As reported by Ro, Woodburn, and Sandine (1979), Foods and Nutrition Department and Department of Microbiology, Oregon State University, Corvallis, Oregon, inoculation of fermented foods with strains known to produce vitamin B_{12} has been evaluated as a vitamin enrichment method. Soybean paste inoculated with *Bacillus megaterium* and fermented was found to contain increased vitamin levels (Choe et al., 1963; Ke et al., 1963). Propionibacterium species widely used in the industrial production of vitamin B_{12} (Wuest and Perlman, 1968) have been recommended for vitamin fortification of some dairy products. Karlin (1961) fortified *kefir* with vitamin B_{12} by the addition of *Propionibacterium* to the kefir grains. Kruglova (1963) prepared vitamin-enriched curds from pasteurized cow's milk by fermentation with equal parts of cultures of lactic acid and propionic acid bacteria (2.5% each). The curds had approximately 10 times more vitamin B_{12} than when produced in the usual way with only lactobacilli. In 1979, Ro, Woodburn, and Sandine undertook to increase the vitamin B_{12} content in the production of kimchi. Changes in the ascorbic acid content during the kimchi fermentation were also observed.

Determination of Vitamin B_{12}

Microbial (using *L. leichmanii, O. malhamensis, E. gracilis,* etc.) bioassay methods are used, as are checking the effects of curative doses on experimental animals (chick, rat, etc.). Physicochemical methods used include spectrophotometry, polarography, and isotope dilution.

Additional Reading

Ball, G.F.M.: *Bioavailability and Analysis of Vitamins In Foods,* Chapman & Hall, New York, NY, 1997.

Bender, D.A.: *Nutritional Biochemistry of the Vitamins,* 2nd Edition, Cambridge University Press, New York, NY, 2003.

Combs, G.F., Jr.: *The Vitamins: Fundamental Aspects in Nutrition and Health,* 2nd Edition, Academic Press, Inc., San Diego, CA, 1998.

Eitenmiller, R.R. and W.O. Landen: *Vitamin Analysis for the Health and Food Sciences,* CRC Press, LLC., Boca Raton, FL, 1998.

Litwack, G.: *Vitamins and Hormones,* Elsevier Science & Technology Books, New York, NY, 2004.

McDowell, L.R.: *Vitamins in Animal and Human Nutrition,* 2nd Edition, Iowa State University Press, Ames, IA, 2000.

Navarra, T.: *Encyclopedia of Vitamins, Minerals, and Supplements,* 2nd Edition, Facts on File, Inc., New York, NY, 2004.

Newstrom, H.: *Nutrients Catalog: Vitamins, Minerals, Amino Acids, Macronutrients—Beneficial Use, Helpers, Inhibitors, Food Sources, Intake Recommendations, and Symptoms of over or under Use,* McFarland & Company, Inc., Publishers, Jefferson, NC, 1993.

Web Reference

Facts About Vitamin B12:
http://www.cc.nih.gov/ccc/supplements/vitb12.html

VITAMIN D. Although the term "Vitamin D" is convenient to use in discussions of nutrition, this singular term is unsatisfactory when used in a strict biochemical context—because there are different substances, each of which is capable of performing vitamin D nutritional functions, namely, that of promoting growth, including bone growth, and preventing rickets in young animals. With reference to generalized terms used over the years in the development and refining of knowledge of related substances, such terms as *antirachitic vitamin, rachitamin, rachiasterol, cholecalciferol, activated 7-dehydrocholesterol,* etc., have been used. As pointed out later, some of these terms remain quite appropriate.

As a brief introductory summary, vitamin D substances perform the following fundamental physiological functions: (1) promote normal growth (via bone growth); (2) enhance calcium and phosphorus absorption from the intestine; (3) serve to prevent rickets; (4) increase tubular phosphorus reabsorption; (5) increase citrate blood levels; (6) maintain and activate alkaline phosphatase in bone; (7) maintain serum calcium and phosphorus levels. A deficiency of D substances may be manifested in the form of rickets, osteomalacia, and hypoparathyroidism. Vitamin D substances are required by vertebrates, who synthesize these substances in the skin when under ultraviolet radiation. Animals requiring exogenous sources include infant vertebrates and deficient adult vertebrates. Included there are vitamin D_2 (calciferol; ergocalciferol) and vitamin D_3 (activated 7-dehydrocholesterol; cholecalciferol).

The most important or at least the best-known members of the family of D vitamins are vitamin D_2 (calciferol), which is indicated in abbreviated form in Structure 1 and can be produced by ultraviolet irradiation of ergosterol, and vitamin D_3 [Structure 2], which may be produced by the irradiation of 7-dehydrocholesterol.

Nomenclature

Subscript numerals have a different connotation in connection with vitamin D substances than is true, for example, with B vitamins. Vitamins B_1, B_2, B_6, B_{12}, etc., represent individual substances which have little or no chemical resemblance to each other and perform different metabolic functions. The various vitamin D's, however, have very similar structures, differing only in the side chains, and perform the same functions.

Biochemical Requirements

There are several unique features exhibited by the D vitamins. First, they are not required nutritionally at all if the organism has access to ultraviolet light (which is present in sunlight). Some animals, kept away from ultraviolet light, require so little D vitamins that the need cannot be demonstrated using ordinary diets. Rats, for example, exhibit a need for D vitamins when the calcium/phosphorus ratio in the diet is about 5:1 but not when it is the more usual 1:1. Chickens, on the other hand, exhibit a need even when the calcium/phosphorus ratio is "normal" (1.5:1).

Different species of animals respond distinctively to the different members of the vitamin D family. The most striking example of this is the fact that vitamin D_2 (calciferol) has practically no vitamin D activity for chickens. Rats respond about equally to D_2 and D_3. Human beings respond both to D_2 and D_3. Information as to how various animals react to

Structure 1. Vitamin D_2

Structure 2. Vitamin D_3

the other less known forms of vitamin D is largely lacking and for practical reasons is not sought after.

Members of the vitamin D family are extremely difficult to isolate and identify in pure form from any source. Fish liver oils are rich sources, and vitamins D_2 and D_3 have been isolated from them. Most ordinary foods are such poor sources in terms of amounts present, that the presence of D vitamins in them has not been demonstrated. Sterols that can be converted into some form of vitamin D by ultraviolet light are, however, widespread, and it may be inferred that D vitamins are often present even when their presence has never been demonstrated.

The requirements of animals for D vitamins in terms of actual weight are extremely small. It is estimated that human beings need about 400 international units of vitamin D per day. Since an international unit of vitamin D corresponds to 0.025 microgram of crystalline vitamin D, this means that the daily human requirement is about 0.01 milligram. Foods can contain as little as 0.02 parts per million of vitamin D and yet furnish an ample supply on the basis of the foregoing estimate.

Excessive dosages of D vitamins have caused excessive calcification and damage (hypervitaminosis). The full story of vitamin D dosage remains obscure. It has been observed, for example, that some "susceptible" children do not respond to the usual doses, but require 5,000–10,000 units per day to keep them free from rickets. There are other children that are afflicted with "vitamin D-resistant rickets" who do not respond even to these high doses, but may do so when doses of the order of 500,000–1,000,000 units are administered. Although unclear, it would seem that in some individuals the vitamin D has difficulty in getting through to where it is needed.

For many years it has been recognized that all cells need calcium to function because their growth and development is related to changes in their intracellular calcium content. Reasoning further, it has been postulated that calcium may serve as a cellular regulatory agent. Growing interest has been shown by investigators, in a steroid that is derived from vitamin D and that regulates the amount of calcium in the animal's blood. This substance has been referred to as a hormone. It is 1,25-dihydroxyvitamin D_3 and is metabolized from vitamin D. In response to a skeletal need for calcium, the hormone is secreted by the kidney and transported to the intestine and bones. Many authorities believe that parathyroid hormone is involved in signaling the kidney to release 1,25-$(OH)_2D_3$. Hypoparathyroid patients lack parathyroid hormone and fail to make 1,25-$(OH)_2D_3$. The result is an abnormally low concentration of calcium in the blood, producing severe bone disease. DeLuca and associates have used 1,25$(OH)_2D_3$ along with calcium to correct deficits in serum calcium concentrations of a limited number of patients. Corticosteroid therapy of long duration is known to produce bone disease. Corticosteroids are frequently administered to persons with rheumatoid arthritis, systemic lupus erythematosis, and asthma, in addition to persons who have received transplants. Some investigators have found that large doses of vitamin D tend to overcome the adverse effects of the corticosteroids. Findings to date essentially are the results of clinical applications rather than based upon a more detailed knowledge of the molecular mechanisms that operate in the metabolism of D vitamins.

Chronology of Vitamin D Substances

In 1918, Mellanby produced experimental rickets in dogs. In 1919, Huldschinsky ameliorated rachitic symptoms in children with ultraviolet radiation. Hess, in 1922, showed that liver oils contain the same antirachitic factor as sunlight. In that same year, McCollum increased calcium deposition in rachitic rats with cod liver oil factor. In 1924, Steenbook and Hess demonstrated irradiated foods have antirachitic properties. It was in 1925 that McCollum named antirachitic factor as vitamin D. In 1931, Angus isolated crystalline vitamin D (calciferol). In 1936, Windaus isolated vitamin D3 (activated 7-dehydrocholesterol).

Rickets

Vitamin D deficiency (also calcium deficiency) produces a condition known in children as *rickets* and in adults as *osteomalacia*. The bones and teeth of children with rickets are poorly formed and soft. A child with rickets frequently has malformed limbs, especially bowlegs. Blood clotting may be impaired, and, in extreme cases, there may be disturbances of the nervous system. An improvement in the level of calcium in the diet, along with vitamin D or parathyroid extract when required, brings about a hardening of the bones, but leaves them misshapen if deformity has already occurred.

Adults, particularly pregnant or nursing women, also require vitamin D because calcium and phosphorus are continually dissolving from bones; and vitamin D is necessary for their utilization. Rickets is not to be confused with the entirely unrelated Rickeetsial group of diseases (Rocky Mountain fever, etc.) that are of virus origin.

Distribution and Sources

Fruits, nuts, and grains are not sources of vitamin D. Animal sources predominate.

> *High Vitamin D content* $(1,000 - 25 \times 10^6$ *I.U./100 grams*).[1] Liver oils from: Bonito, cod, halibut, herring, lingcod, sablefish, sea bass, soupfin shark, swordfish, tuna.
> *Medium Vitamin D content* (*100–1,000 I.U./100 grams*). Egg yolk, herring, kippers, lard, mackerel, margarine, pilchards, salmon, sardine, shrimp, tuna.
> *Low Vitamin D content* (*10–100 I.U./100 grams*). Beef, butter, cheeses, cod roe, cream, eggs, grain oils, halibut, horse meat, liver (beef, calf, lamb, pork), milk (vitamin D fortified),[2] veal, vegetable oils.

Bioavailability

Factors which tend to cause a decrease in available vitamin D substances include: (1) liver damage; (2) presence of antagonists; (3) presence of phytin in gut; (4) low bile salts in gut; (5) high pH in gut; (6) destruction of intestinal flora; and (7) excretion in feces. Factors that enhance availability include: (1) storage in liver and skin; (2) absorption aids, such as bile salts; (3) decrease in pH of lower intestine; and (4) irradiation by ultraviolet. Antagonists of vitamin D include toxisterol, phytin, phlorizin, cortisone, cortisol, thyrocalcitonin, and parathormone. Synergists include niacin, parathormone (concentration dependent), and somatotrophin (growth hormone).

Dosages exceeding 4000 I.U./day may cause varying degrees of toxicity in humans. Symptoms include anorexia, nausea, thirst, and diarrhea. There also may be polyuria, muscular weakness, and joint pains. Serum calcium increases and calcification of soft tissues (arteries, muscle) may commence. Arterial lesions and kidney injury have been noted in rats.

In the biosynthesis of vitamin D substances, precursors include cholesterol (skin + ultraviolet radiation) in animals; ergosterol (algae, yeast + ultraviolet radiation). Intermediates in the biosynthesis include preergocalciferol, tachysterol, and 7-dehydrocholesterol. Provitamins in very small quantities are generated in the leaves, seeds, and shoots of plants. In animals, the production site is the skin. Target tissues in animals are bone, intestine, kidney, and liver. Storage sites in animals are liver and skin.

Commercial vitamin D dietary supplements are prepared by the irradiation of ergosterol, 7-dehydrocholesterol; or by extraction of fish liver oils.

Unusual features of vitamin D substances noted by some investigators include: (1) vitamin has hormonal qualities due to internal synthesis; (2) vitamin D2 has little activity for chickens—various species differ in response to the vitamin; (3) vitamin D substances may play a role in aging calcification phenomena, especially in skin; (4) the vitamin can mimic rickets with a high-calcium–low-phosphorus diet; (5) the vitamin can mimic osteomalacia under same conditions; (6) the vitamin is absorbed through skin; (7) the vitamin activates transport of heavy metals by intestinal cells; (8) the vitamin has an exceptionally long half-life (days to weeks); (9) furred and feathered animals obtain some vitamin D as the result of grooming and licking; (10) fishes are believed to obtain vitamin D from marine invertebrates; (11) the vitamin has been found useful in the treatment of lead poisoning.

Determination of Vitamin D

Bioassay techniques involve testing rats on antirachitic qualities. An important physicochemical method involves reaction with antimony trichloride.

See also entries on **Calcium**; **Hormones**; **Lipids**; and **Phosphorus**.

---

[1] One I.U. = 0.025 microgram vitamin D_3.

[2] Milk is normally a poor source of vitamin D. Since milk forms a major part of the diet in many countries, particularly for children, the product is commonly fortified with vitamin D substances.

Additional Reading

Ball, G.F.M.: *Bioavailability and Analysis of Vitamins In Foods,* Chapman & Hall, New York, NY, 1997.

Bender, D.A.: *Nutritional Biochemistry of the Vitamins,* 2nd Edition, Cambridge University Press, New York, NY, 2003.

Combs, G.F., Jr.: *The Vitamins: Fundamental Aspects in Nutrition and Health,* 2nd Edition, Academic Press, Inc., San Diego, CA, 1998.

Eitenmiller, R.R. and W.O. Landen: *Vitamin Analysis for the Health and Food Sciences,* CRC Press, LLC., Boca Raton, FL, 1998.

Litwack, G.: *Vitamins and Hormones,* Elsevier Science & Technology Books, New York, NY, 2004.

McDowell, L.R.: *Vitamins in Animal and Human Nutrition,* 2nd Edition, Iowa State University Press, Ames, IA, 2000.

Navarra, T.: *Encyclopedia of Vitamins, Minerals, and Supplements,* 2nd Edition, Facts on File, Inc., New York, NY, 2004.

Newstrom, H.: *Nutrients Catalog: Vitamins, Minerals, Amino Acids, Macronutrients—Beneficial Use, Helpers, Inhibitors, Food Sources, Intake Recommendations, and Symptoms of over or under Use,* McFarland & Company, Inc., Publishers, Jefferson, NC, 1993.

Web Reference

Facts About Vitamin D: http://www.cc.nih.gov/ccc/supplements/vitd.html

VITAMIN E.   Sometimes referred to as the *antisterility vitamin,* factor X (an earlier designation), chemically vitamin E is alpha-tocopherol, the structure of which is:

Alpha-tocopherol

Active analogues and related compounds include: dl-α-Tocopherol; 1-α-tocopherol; esters (succinate, acetate, phosphate), and β, ζ_1, ζ_2-tocopherols. The principal physiological forms are D-a-tocopherol, tocopheronolactone, and their phosphate esters.

The physiological functions of vitamin E substances include: (1) bio logical antioxidant; (2) normal growth maintenance; (3) protects unsaturated fatty acids and membrane structures; (4) aids intestinal absorption of unsaturated fatty acids; (5) maintains normal muscle metabolism; (6) maintains integrity of vascular system and central nervous system; (7) detoxifying agent; and (8) maintains kidney tubules, lungs, genital structures, liver, and red blood cell membranes.

In livestock and laboratory animals, a deficiency of vitamin E substances may cause degeneration of reproductive tissues, muscular dystrophy, encephalomalacia, and liver necrosis. Considerable research is required to fully determine supplementation of livestock diets unless typical symptoms of a deficiency appear. Symptoms have appeared where there are selenium deficiencies in the soil and where there are excessive levels of nitrates in the soil. "White muscle" is the term used to describe a condition of muscular dystrophy in cattle.

In 1922, Evans and Bishop reported dietary factor "X" needed for normal rat reproduction. In that same year, Matill found dietary factor "X" in yeast and lettuce. Evans et al., in 1923, found factor "X" in alfalfa, butterfat, meat, oats, and wheat. The designation *factor "X"* was changed to *vitamin E* by Sure in 1924. In 1936, Evans et al. demonstrated that vitamin E belongs to the tocopherol family of compounds. During that year, these researchers isolated several active tocopherols and found a-tocopherol to be the most active of the number. Fernholz, in 1938, determined the structure of vitamin E. It was first synthesized by Karrer during that same year. During the interim between 1938 and 1956, several tocopherols were identified and studied. It was in 1956 that Green observed the eighth in the family of tocopherols.

The tocopherols were identified as naturally occurring oily substances and the first three were characterized as alpha, beta, and gamma forms, the biological activity of which decreased in that order.

Vitamin E substances are necessary for the normal growth of animals. Without vitamin E, the animals develop infertility, abnormalities of the central nervous system, and myopathies involving both skeletal and cardiac muscle. The antioxidant activity of the tocopherols is in reverse order to that of their vitamin activity. Muscular tissue taken from a deficient animal has an increased rate of oxygen utilization. The tocopherols are so widely distributed in natural foods that a spontaneous deficiency is infrequent unless diseases of the gastrointestinal or biliary system hinder absorption. Symptoms indicating a vitamin E deficiency include: (1) Red blood cell hemolysis, creatinuria, xanthomatosis and cirrhosis of gall bladder, steatorrhea (in young), cystic fibrosis of pancreases (in young), poorly developed muscles. Rats, dogs, monkeys, and chickens display muscular dystrophy; myocardial degeneration is observed in dogs and rabbits; resorption of fetus, degeneration of germ epithelium, disturbance of estrus cycle are observed in rats; hepatic necrosis is shown in rats; encephalomalacia and vascular degeneration is manifested in chickens.

Role of Vitamin E in Humans

The fundamental needs for vitamin E in humans have long been established. There are factors associated with this vitamin, however, that have created controversy and disagreement among highly qualified professional people. Although nearly every vitamin, at one time or other, has been used unwisely (in retrospect) in the treatment of human diseases, perhaps no other vitamin substance has aroused more discussion among clinicians than vitamin E. Because deficient animals develop a form of myopathy, it was natural to test the therapeutic efficacy of vitamin E in various forms of progressive muscular dystrophy and in diseases of the reproductive system. Enthusiastic claims have been made, and refuted, by investigators. From the standpoint of solid evidence, as of the early 1980s, the principal advantage of administering vitamin E lies exclusively in those instances where a vitamin E malabsorption syndrome exists. Associated with this fundamental situation are hemolytic anemia of premature infants; diseases caused by poor fat and oil absorption, and intermittent claudication (limping). A 1979 Institute of Food Technologists "Food Safety and Nutrition Panel" reported no incidence of vitamin E deficiency. Three underlying reasons were cited for this: (1) ample storage in adipose tissue; (2) slow elimination from the body; and (3) prevalence in foods. Significant amounts are present in vegetable oils and margarine (70% of the average daily intake), cereal products, fish, meat, eggs, dairy products, and leafy green vegetables.

Cure-all claims for the vitamin appear to stem from the vitamin's antioxidant properties and subsequent ability to neutralize harmful free radical products of oxidation. This had led to vitamin E administration for diseases of the circulatory, reproductive, and nervous systems, increased athletic and sexual endurance, and protection against aging and air pollution effects. Although some claims have been verified by animal studies, evidence is not conclusive for humans. Elderly individuals have resorted to vitamin E in hopes of slowing the aging process. The idea is not unfounded, for in the laboratory, the nutrient neutralizes radicals normally contributing to aging pigment formation. Neutralization within humans, however, remains unproven.

Distribution and Sources

Oily substances are, by far, the best natural sources of vitamin E.

High vitamin E content (50–300 milligrams/100 grams). Corn (maize) oil, cottonseed oil, margarine, safflower oil, soybean oil, wheat germ oil.

Medium vitamin E content (5–50 milligrams/100 grams). Alfalfa, apple seeds, asparagus, barley, cabbage, chocolate, coconut oil, groundnut (peanut), groundnut (peanut) oil, olive oil, rose hips, soybean (dry), spinach, wheat germ, yeast.

Low vitamin E content (0.5–5 milligrams/100 grams). Apple, bacon, bean (dry navy), beef, beef liver, blackberry, Brussels sprouts, butter, carrot, cauliflower, cheeses, coconut, corn (maize), corn (maize) meal, eggs, flour (whole wheat), kale, kohlrabi, lamb, lettuce, mustard, oats, oatmeal, olive, parsnip, pea, pear, pepper (sweet), pork, rice (brown), rye, sweet potato, turnip greens, veal, wheat.

Production sites for vitamin E biosynthesis occur in nuts, seeds, cereal germ, green leaves, legumes. Biosynthesis also occurs in some microorganisms. Precursors for biosynthesis include mevalonic acid and phenylalanine (probably these compounds with side chains). Considerably more research is required to pinpoint the exact precursors. Tocotrienol occurs as an intermediate in the biosynthesis.

Commercial production of vitamin E tocopherols is by way of molecular distillation from vegetable oils.

Antagonists of the tocopherols include α-tocopherol quinone, oxidants, cod liver oil, and thyroxine. Synergists include ascorbic acid, estradiol, somatotrophin (growth hormone), testosterone, and vitamins A, B_6, B_{12}, and K.

Bioavailability of Vitamin E

Factors which tend to reduce availability of the vitamin include: (1) presence of antagonists; (2) mineral oil ingestion; (3) presence of vitamin E oxidation products; (4) occurrence with other less active analogues; (5) excessive excretion in feces; (6) impaired fat absorption; (7) chemical binding in foods; (8) cooking losses (vitamin is heat and oxygen labile); (9) losses in frozen storage, steatorrhea, and variability of natural sources. Factors which may increase absorption include: (1) Storage of vitamin in adipose and muscle tissues; (2) esterification, which increases stability; (3) use of unprocessed fresh food sources; and (4) absorption aids, such as bile salts.

Storage sites for the tocopherols in the body include muscle and adipose tissues and the liver. Target tissues include the adrenals, pituitary, kidney, genital organs, muscles, liver, lungs, and bone marrow.

Unusual features of vitamin E substances as observed by various investigators include: (1) the vitamin may be involved in aging mechanisms by protecting unsaturated fatty acids and membranes against free radicals; (2) only D-isomers occur naturally; (3) vitamin E is replaceable by selenium salts in therapy of rat and pig liver necrosis, and chick exudative diathesis; (4) vitamin E is replaceable by coenzyme Q (see also **Coenzymes**) and antioxidants for certain symptoms of vitamin E deficiency, but not for all, e.g., red blood cell hemolysis, resorption gestation not affected; (5) species differences in response to vitamin E treatment of similar symptoms, e.g., muscular dystrophy—positive in rabbits, negative in humans; (6) other tocopherols are only slightly active as compared with vitamin E; (7) vitamin content is decreased in tumors.

Alpha-Tocopherol and Nitrosamine Formation

Because of the growing concern, commencing in the late 1970s, as regards the formation of *N*-nitrosamines, such as dimethylnitrosamine and *N*-nitrosopyrrolidine, upon cooking of certain meat products cured with sodium nitrite, a number of investigators began studies to find materials that may inhibit nitrosamine formation. Reporting in late 1978, W.J. Mergens and a team of investigators (Hoffmann–LaRoche Inc., Nutley, New Jersey) observed that *N*-nitrosopyrrolidine has been found in fried bacon, but not in raw bacon (Fazio et al., 1973; Fiddler et al., 1974), apparently because of the influence of heat in accelerating the reaction of nitrite with the amine group of proline or its decarboxylated product, pyrrolidine, formed in frying (Archer et al., 1976; Hwang and Rosen, 1976). The effect of ascorbic acid in inhibiting nitrosamine formation has been demonstrated by various workers both *in vitro* and *in vivo* (Mirvish et al., 1972, 1973; Kamm et al., 1973, 1975; Greenblatt, 1973; Ivankovic et al., 1973).

The promising contribution of adding tocopherol to bacon, along with sodium ascorbate, to inhibit nitrosamine formation undertaken by Mergens and associates is reported in detail in the Mergens et al. reference (1978).

Determination of Vitamin E

Bioassay methods include measurements of quantity required to prevent fetal resorption; and for red blood cell hemolysis (in rat). Measurements also are made of liver storage in the chick. Physicochemical methods used include colorimetric two-dimensional paper chromatography.

Additional Reading

Archer, M.C., et al.: *Nitrosamine Rofmation in the Presence of Carbonyl Compounds,* IARS Scientific Publication 14, International Agency for Research on Cancer, Lyon, France, 1976.

Ball, G.F.M.: *Bioavailability and Analysis of Vitamins in Foods,* Chapman & Hall, New York, NY, 1997.

Bender, D.A.: *Nutritional Biochemistry of the Vitamins,* 2nd Edition, Cambridge University Press, New York, NY, 2003.

Combs, G.F., Jr.: *The Vitamins: Fundamental Aspects in Nutrition and Health,* 2nd Edition, Academic Press, Inc., San Diego, CA, 1998.

Cort, W.M., W. Mergens, and A. Greene: "Stability of Alpha-and Gamma- Tocopherol: Fe^{3+} and Cu^{2+} Interactions," *J. Food Sci,* **43**, 3, 797–802 (1978).

Eitenmiller, R.R. and W.O. Landen: *Vitamin Analysis for the Health and Food Sciences,* CRC Press, LLC., Boca Raton, FL, 1998.

Fazio, T., et al.: "Nitrosopyrrolidine in Cooked Bacon," *J. Assoc. Offic. Anal. Chem.,* **56**, 919 (1973).

Fiddler, W., et al.: *Some Current Observations on the Occurrence and Formation of N-nitrosamines,* Proc., 18th Meeting Meat Res. Workers, Guelph, Ontario, Canada, 1972.

Greenblatt, M.: "Ascorbic Acid Blocking of Aminopyrine Nitrosation in NZO/BI Mice," *J. Nat. Cancer Inst.,* **50**, 1055 (1973).

Hwang, L.S. and J.D. Rosen: "Nitrosopyrrolidine Formation in Fried Bacon," *J. Agric. Food Chem.,* **24**, 1152 (1976).

Ivankovic, S., et al.: "Verhutung van Nitrosamidbedingtem Hydrocephalus durch Ascorbinsaure noch praenataler Gabe von Aethylharnstoff und Nitrite an Ratten," *Z. Krebsforsch.,* **79**, 145 (1973).

Kamm, J.J., et al.: "Protective Effect of Ascorbic Acid on Hepatotoxicity Caused by Sodium Nitrite plus Aminopyrine," *Proc., Nat. Acad. Sci.,* **70**, 747 (1973).

Kamm, J.J., et al.: "Inhibition of Amine-Nitrate Hepatotoxicity by Alpha-Tocopherol," *Toxical. Appl. Pharmacol.,* **41**, 575 (1977).

Litwack, G.: *Vitamins and Hormones,* Elsevier Science & Technology Books, New York, NY, 2004.

McDowell, L.R.: *Vitamins in Animal and Human Nutrition,* 2nd Edition, Iowa State University Press, Ames, IA, 2000.

Mergens, W.J., et al.: "Stability of Tocopherol in Bacon," *Food Technol.,* **32**, 11, 40–44, 52 (1978).

Navarra, T.: *Encyclopedia of Vitamins, Minerals, and Supplements,* 2nd Edition, Facts on File, Inc., New York, NY, 2004.

Newstrom, H.: *Nutrients Catalog: Vitamins, Minerals, Amino Acids, Macronutrients—Beneficial Use, Helpers, Inhibitors, Food Sources, Ilntake Recommendations, and Symptoms of over or under Use,* McFarland & Company, Inc., Publishers, Jefferson, NC, 1993.

Web Reference

Facts About Vitamin E: http://www.cc.nih.gov/ccc/supplements/vite.html

VITAMIN K. Sometimes referred to as the *antihemmorhagic vitamin,* and, earlier in its development, the prothrombin factor or Koagulations-vitamin, vitamin K is a substituted derivative of naphthoquinone and occurs in several forms. The designation *phylloquinone,* or K_1, refers to 2-methyl-3-phytyl-1,4 naphthoquinone; the designations *farnoquinone* and *prenyl-menaquinone,* or K_2, refer to 2-difarnesyl-3-methyl-1,4-naphthoquinone. *Menadione,* sometimes called oil-soluble vitamin K3, is 2-methyl-1,4-naphthoquinone. The structure of phylloquinone is:

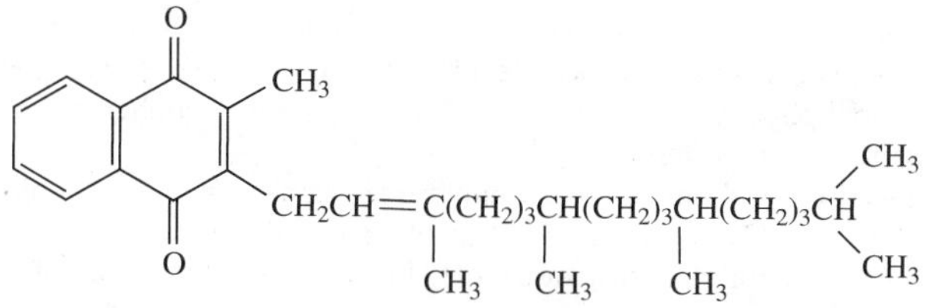

Generally, when vitamin K substances are absent or deficient in the diet of animals, including humans, a hemorrhagic disorder will appear. Young fowls that are allowed to continue on a deficient diet for extended periods will ultimately die of internal hemorrhage, or from extensive bleeding from small external wounds. Fowls experience difficulty in absorbing vitamin K from the intestine, whereas humans, rats, and dogs absorb it readily and normally obtain their requirement form intestinal bacteria without need of dietary supplementation. If, however, bacterial synthesis is inhibited by the use of sulfa drugs or certain antibiotics, the disease will develop, unless the diet is supplemented with some form of vitamin K. When there is a decrease in the amount of bile salts in the intestine, as in obstructive jaundice, vitamin K is absorbed in such small amounts that the disease will also ensue. The use of vitamin K also is suggested to control and prevent the disease in premature babies. Vitamin K_1 is also able to reverse the hemorrhagic condition resulting from the administration of dicumarol to animals.

It has been reported that vitamin K_1 and several of the vitamin K_2 homologues are capable of restoring electron transport in solvent-extracted or irradiated bacterial and mitochondrial preparations. Other reports suggest that vitamin K is concerned with the phosphorylation reactions accompanying oxidative phosphorylation. The capacity of these compounds to exist in several forms, e.g., quinone, quinol, chromanol, etc., appears to strengthen the proposal that links them to oxidative phosphorylation. Information has suggested that vitamin K acts to induce prothrombin synthesis. Since prothrombin has been shown to be synthesized only by liver parenchymal cells in the dog, it would appear that the proposed role for vitamin K is not specific for only prothrombin synthesis, but applicable to other proteins.

In 1929, Dam reported chicks on a synthetic diet develop hemorrhagic conditions. In 1935, Dam named vitamin K as the missing factor in synthetic diets. In that same year, Almquist and Stokstad demonstrated the presence of vitamin K in fish meal and alfalfa. In 1939, Dam and Karrer isolated vitamin K from alfalfa; and, in that same year, Doisy isolated K_1 from alfalfa, K_2 from fish meal, and demonstrated differences of the two substances. Also, in 1939, MacCorquodale, Cheney, and Fieser determined the structure of vitamin K_1. In that same year, Almquist and Klose synthesized vitamin K_1 for the first time. In 1941, Link et al. discovered dicoumarol, an anticoagulant and antagonist of vitamin K.

In addition to compounds previously mentioned, active analogues and related compounds include menadiol diphosphate, menadione bisulfite, phthicol, synkayvite, menadiol (vitamin K_4), and compounds designated as vitamins K_5, K_6, and K_7.

Many species require vitamin K. The vitamin is frequently administered to poultry via feedstuffs. Intestinal bacteria, normally functioning, supply the vitamin to the human body.

In the therapy of deep venous thrombosis, heparin is commonly administered. This drug takes effect immediately to prevent further thrombus formation. However, heparin is regarded as a hazardous drug and possibly may be the leading cause of drug-related deaths in hospitalized patients who are relatively well. Usually administered intravenously, preferably by pump-driven infusion at a constant rate rather than by intermittent injections, it sometimes may cause major bleeding, which is particularly hazardous if it is intracranial. The action of heparin can be terminated almost immediately by intravenous injection of protamine sulfate, but where there may be less urgency, vitamin K_1 may be used. The vitamin preparation may be administered intravenously, intramuscularly, or subcutaneously.

Vitamin K is also an antagonist of warfarin, which is sometimes used in rodenticides. Pets that have been exposed to warfarin-containing poisons may be saved from death by internal hemorrhaging through the immediate administration of vitamin K.

Vitamin K is sometimes used in the treatment of viral hepatitis.

It has been found that vitamin K analogues possess an ability to insert themselves into the oxygen-binding cleft of hemoglobin. This may result in hemolysis (dissolution of red blood corpuscles with liberation of their hemoglobin).

See also **Anticoagulants**.

Distribution and Sources

Some fruits, vegetables, and nuts, as well as meat products, contain good sources of K vitamins. Intestinal bacteria, M. phlei, synthesize it.

High vitamin K content (100–300 micrograms/100 grams). Beef kidney, beef liver, cabbage, cauliflower, pork, soybean, spinach.
Medium vitamin K content (10–100 micrograms/100 grams). Alfalfa, egg yolk, pine needles, potato, strawberry, tomato, wheat (bran, germ, whole).
Low vitamin K content (0–10 micrograms/100 grams). Carrot, corn (maize), milk, mushroom, parsley, pea.

Commercial production of vitamin K is by column chromatography of fish meal extracts. In biosynthesis, precursors include polyacetic acid (ring); acetate (side chain). Intermediates include dehydroquinic acid (ring); farnesol (side chain).

Bioavailability of Vitamin K

Factors which decrease availability of the vitamin include: (1) biliary obstruction; (2) liver damage—cirrhosis, toxins; (3) poor food preparation (vitamin is strong-acid, alkali, light, and reduction labile); (4) impaired lipid absorption in gut; (5) presence of antagonists; (6) ingestion of mineral oil; (7) sterilization of gut with antibiotics and sulfa drugs; and (8) excessive excretion in feces. Availability may be increased by way of storage in the liver and absorption aids, such as bile salts.

Antagonists of vitamin K substances include dicoumarol, sulfonamides, antibiotics, α-tocopherol quinone, dihydroxystearic acid glycide, salicylates, iodinin, warfarin. Synergists include ascorbic acid, somatotrophin (growth hormone), and vitamins A and E.

General symptoms of a vitamin K deficiency include hypoprothrombinemia, increased bleeding and hemorrhage, increased clotting time, and neonatal hemorrhage. Internal hemorrhage is a symptom in chicks. Usually the vitamin is nontoxic, but, in humans, very excessive dosages can cause thrombosis, vomiting, and porphyrinuria. Target tissues are liver and vascular system. Small quantities are stored in liver.

Determination of Vitamin K

A vitamin K deficient chick assay may be made; or physicochemical techniques, including polarographic methods, spectrophotometry of pure solutions, and prothrombin time determinations, may be used.

Additional Reading

Ball, G.F.M.: *Bioavailability and Analysis of Vitamins In Foods,* Chapman & Hall, New York, NY, 1997.
Bender, D.A.: *Nutritional Biochemistry of the Vitamins,* 2nd Edition, Cambridge University Press, New York, NY, 2003.
Combs, G.F., Jr.: *The Vitamins: Fundamental Aspects in Nutrition and Health,* 2nd Edition, Academic Press, Inc., San Diego, CA, 1998.
Eitenmiller, R.R. and W.O. Landen: *Vitamin Analysis for the Health and Food Sciences,* CRC Press, LLC., Boca Raton, FL, 1998.
Litwack, G.: *Vitamins and Hormones,* Elsevier Science & Technology Books, New York, NY, 2004.
McDowell, L.R.: *Vitamins in Animal and Human Nutrition,* 2nd Edition, Iowa State University Press, Ames, IA, 2000.
Navarra, T.: *Encyclopedia of Vitamins, Minerals, and Supplements,* 2nd Edition, Facts on File, Inc., New York, NY, 2004.
Newstrom, H.: *Nutrients Catalog: Vitamins, Minerals, Amino Acids, Macronutrients—Beneficial Use, Helpers, Inhibitors, Food Sources, Intake Recommendations, and Symptoms of over or under Use,* McFarland & Company, Inc., Publishers, Jefferson, NC, 1993.

Web Reference

MedlinePlus Medical Encyclopedia: Vitamin K: http://www.nlm.nih.gov/medlineplus/ency/article/002407.htm
Vitamin K: http://www.ctds.info/vitamink.html
Vitamin K, Linus Pauling Institute's Micronutrient Information: http://lpi.oregonstate.edu/infocenter/vitamins/vitaminK/

VITAMIN K (Blood Coagulation).   See **Anticoagulants**.

VITREOUS STATE.   When certain liquids are cooled fairly rapidly, crystals do not form at a definite temperature, but the viscosity of the liquid increases steadily until a glassy substance is obtained. A glass may be thought of as a disordered amorphous solid, or as a supercooled liquid, which only devitrifies into the crystalline state after extremely long standing. Glasses are optically isotropic, which explains their value in optical instruments. The property of forming a glass is possessed particularly by the oxides of silicon, boron, germanium, arsenic, phosphorus, etc., and by many organic compounds, especially those containing several hydroxyl groups per molecule. See Fig. 1.

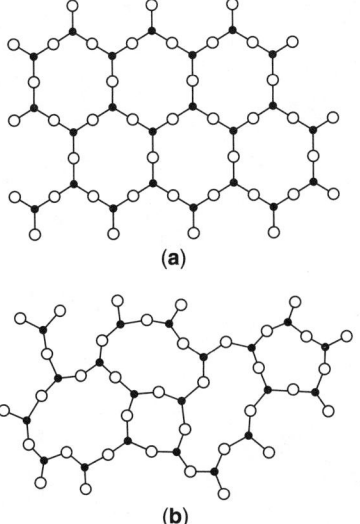

Fig. 1.   Two-dimensional diagram showing (**a**) an oxide of composition X_2O_3 in the crystalline form; and (**b**) the same oxide in the vitreous state

VITROPHYRE.   A volcanic glass carrying sporadic distinct crystals of feldspar and other minerals; in short, a porphyritic glass.

VIVIANITE. The mineral vivianite is a hydrous iron phosphate, $Fe_3(PO_4)_2 \cdot 8H_2O$, its monoclinic crystals are usually prismatic or blade-like but may be in massive forms. Vivianite has one perfect cleavage; hardness 1.5–2; specific gravity 2.58–2.68; luster, pearly on cleavage; faces, otherwise vitreous; colorless, when freshly exposed, but becoming blue or brownish with the alteration of the ferrous to ferric iron; transparent to translucent. Vivianite is an associate of pyrrhotite, pyrite and copper and tin ores. It is found also in clay beds forming the so-called blue iron earth which is common and of wide distribution in peat bogs. Vivianite is found in Rumania; Bavaria; Cornwall in England; and elsewhere in Europe; Australia; Bolivia; and Greenland. In the United States it occurs in New Jersey, Delaware, and Colorado. This mineral was named by Werner after the English mineralogist J.G. Vivian, its discoverer.

VOIDS. Empty spaces of molecular dimensions occurring between closely packed solid particles, as in powder metallurgy. Their presence permits barriers made by powder metallurgy techniques to act as diffusion membranes for separation of uranium isotopes in the gaseous diffusion process.

See also **Diffusion**.

VOLATILE. Having a low boiling or subliming temperature at ordinary pressure; in other words, having a high vapor pressure, as ether, camphor, naphthalene, iodine, chloroform, benzene; or methyl chloride.

VOLATILE OILS. The volatile oils are distinguished from the fixed oils by the fact that a drop of one of the former does not leave a spot on paper. Members of certain plant families, such as the *Labiatae*, contain a larger percentage of such oils than do other families. But volatile oils are in no sense restricted to any small group, nor are they found only in certain tissues. Sometimes, certain parts may be principally used for the oils, as the seeds of the *Umbelliferae*.

Various methods are used in extracting the oils from the plant tissue. Many are distilled with water or steam, the oil being carried over with the distillate. In others, as for example oil of bitter almonds, the oil develops in the tissues only after fermentation. It is then obtained by distillation. Another method, and one especially used for more delicate and valuable oils, is called "enfleurage." In this method the flowers containing the oil are spread as a thin layer over a layer of lard or olive oil. The latter absorbs the delicate oil in the flowers, after which distillation may separate the volatile oil from the other.

VOLATILE OILS (Flavoring).   See **Flavonoids**.

VOLATILE ORGANIC COMPOUNDS. Any hydrocarbon, except methane and ethane, with vapor pressure to or greater than 0.1 mm Hg.

VOLATILITY PRODUCT. The product of the concentrations of two or more ions or molecules that react to produce a volatile substance. The volatility product is analogous to the solubility product, except that, when it is exceeded, the substance escapes from the system by volatilization rather than precipitation. As with the solubility product, if any of the reacting ions or molecules have a numerical coefficient greater than one, then the concentration term of that ion or molecule is raised to the corresponding power.

VOLTAIC CELL. Two conductive metals of different potentials, in contact with an electrolyte, which generate an electric current. The original voltaic cell was composed of silver and zinc, with brine-moistened paper as electrolyte. Semisolid pastes are now used; electrodes may be lead, nickel, zinc, of cadmium.

See also **Solar Cell (Photovoltaic Cell)**.

VOLUME (Standard). The volume occupied by one gram molecular weight of a gas at 0°C and a pressure of 1 standard atmosphere.

VON BAEYER, ADOLF (1835–1917). A German chemist who received the Nobel prize for chemistry in 1905. He was recognized for his services in the advancement of organic chemistry and the chemical industry, through his work on organic dyes and hydroaromatic compounds. He was educated in Berlin under the direction of Bunsen and Kekule. He was a professor in Strasbourg and Munich. Many discoveries included barbituric acid and the molecular structure of indigo.

VULCANIZATION. A physiochemical change resulting from cross-linking the unsaturated hydrocarbon chain of polyisoprene (rubber) with sulfur, usually with application of heat. The precise mechanism that produces the network structure of the cross-linked molecules is not completely known. Sulfur is also used with unsaturated types of synthetic rubbers; some types require use of peroxides, metallic oxides, chlorinated quinones, or nitrobenzenes. Natural rubber can be vulcanized with selenium, organic peroxides, and quinone derivatives, but these have limited industrial use; high-energy radiation curing is and important innovation.

See also **Rubber (Natural)**.

W

WACKER REACTION. The oxidation of ethylene to acetaldehyde in the presence of palladium chloride and cupric chloride.

WAD. The mineral wad, sometimes called *bog manganese*, occurs in amorphous masses, and consists of mixtures of manganese oxides, MnO_2 and MnO, and oxides of other metals such as copper, lead, cobalt, and iron. It is bluish- to brownish-black, usually soft enough to soil the fingers and often porous and light. It is not a distinct mineral species.

WAFER (Silicon). See **Semiconductors**.

WAGNER-JAUREGG REACTION. Addition of maleic anhydride to diarylethylenes with formation of bis adducts that can be converted to aromatic ring systems.

WAGNER-MEERWIN REARRANGEMENT. Carbon-to-carbon migration of alkyl, aryl, or hydride ions. The original example is the acid-catalyzed rearrangement of camphene hydrochloride to isobornyl chloride.

WAKSMAN, SELMAN A (1888–1973). American microbiologist (Nobel prize winner in 1952) and professor at Rutgers University. He was the first to use the term *antibiotic* to designate the mold produced antibacterial substances discovered by Fleming in 1928. He became the outstanding authority in this field.

WALDEN INVERSION. Inversion of configuration of a chiral center in bimolecular nucleophilic substitution reactions. See also **Rearrangement (Organic Chemistry)**.

WALLACH, OTTO (1847–1931). German chemist who received the Nobel prize for chemistry in 1910 for recognition of his services to organic chemistry and the chemical industry by his pioneer work in the field of alicyclic compounds. His mentors were Hofmann and Wahler, and he worked at the University of Bonn under Kekule. He studied pharmacy and did work on terpenes, camphors, and essential oils. This was followed by research in aromatic oils, perfumes, and spices. His research of terpenes revealed their significance in sex hormones and vitamins. Ethereal oils and industrial uses were made possible by his work.

WARFARN. See **Anticoagulants**.

WASTE (Nuclear) Management. See **Nuclear Reactor**.

WASTES AND POLLUTION. The approach to reducing waste and abating pollution is complex and sometimes even controversial among the experts. Differences arise because not all of the facts are on hand. The basic chemistry and physics of the Earth's three waste "sinks"—the atmosphere, the hydrosphere, and the lithosphere—remain poorly understood. Although there is consensus among both technologists and the lay public that serious environmental pollution problems exist and indeed are worsening, there are differences pertaining to details and priorities. Environmental scientists and engineers are influenced by factors that are not exclusively scientific, but that are of a societal and economic nature as well. The populace in general is slow to accept changes in life-style preferences and habits, and commerce and business interests frequently resist the concept that environmental costs must be added to the other costs of doing business. Such conflicting factors minimize the key ingredient of attaining success—*dedication*.

Even in view of the aforementioned difficulties, impressive pollution abatement successes have been made, but unfortunately the pace of these programs has not kept up with the rate of the worsening environment. To illustrate the partial successes to date, one need only

to compare the environmental status in the advanced industrial nations with the environmental damage found in the former Soviet Bloc, where environmental problems essentially were ignored.

Pathways to Environmental Correction

From an idealistic viewpoint, one may place the efforts for restoring the purity of the natural environment as falling along three pathways. This is an approximation, and the pathways are not mutually exclusive. Table 1 essentially is included as a checklist of the numerous actions that are being taken.

Pathway 1 includes those actions that are directed more toward eliminating or drastically reducing waste production—that is, efforts made to correct the cause (waste) rather than the effect (pollution). Had actions along these lines been taken years ago, the massive pollution experienced today most likely would not have occurred.

Pathway 2 actions recognize that, with current technology and societal attitudes coupled with economic factors, a considerable amount of pollution must be accepted. Actions along this pathway, however, can reduce the amount of wastes produced and hence resulting pollution.

Pathway 3 actions recognize the current inevitability of mass pollution, but which are directed toward reducing the long-term effects of waste disposal. In the past, pollution has occurred in somewhat of a step-like fashion (i.e., creation of the wastes in the first place, sometimes followed by unscrupulous means taken to "hide" or simply "forget" abandoned wastes, followed by waste site clean up). These problems occur simply because wastes have been or are being disposed of improperly.

Establishment of Regulations

When it became apparent that environmental protection could not be accomplished strictly on a voluntary basis, several of the advanced industrial countries were forced to take regulatory actions (circa mid-1960s). The problem was too complex and not sufficiently understood at the outset to institute complete legislation at one time. Consequently in the United States, for example, numerous special acts were passed but stretched out for several years. See Table 2 on p. 3701. This has resulted in difficult compliance and enforcement procedures.

Numerous scholars of the environment have noted several major differences between the way some of the advanced European industrial nations and Japan approach the problems of waste reduction and pollution abatement, and the general approach adopted by the United States over the years. The European countries referred to here notably are the Netherlands, Sweden, and Germany and, of course, do not include the former Soviet Bloc countries.

In the United States, *regulation* is by far the predominant controlling tool. The U.S. system is highly legalized. In the European countries and Japan, regulation is but one tool used.

Government regulators in the United States infrequently provide specific professional assistance to a firm with pollution problems. In the European countries mentioned and in Japan, regulatory personnel frequently work closely and cooperatively with industrial personnel in seeking solutions. In fact, in some cases, government grants are made available for remedial implementation.

The principal incentive for regulators in the United States is that of *developing* and *enforcing* regulations. Less emphasis is directed toward finding better methods for achieving improved results. In connection with the Superfund program, regulators are credited when pollution sites are cleaned up, but emphasis is given to *initiating* the action, with less accountability required where the cleanup has been found inadequate.

TABLE 1. WASTE-POLLUTION OPTIONS AND STRATEGIES (ABRIDGED)

PATHWAY 1
ELIMINATE OR GROSSLY REDUCE WASTES

Increase Life Span of Products

Reduce "junking" frequency. Design for corrosion and wear resistance; easy maintainability. Discourage frequent styling changes simply in interest of increasing marketing appeal.

Design Energy Efficiency into Products and Processes

Traditional processes for converting energy resources (fuels) into electricity, for example, are major polluters. Thus, end products and end processes should consume minimum energy to overcome "hidden" pollution costs. Considerable progress has been made to design more efficient heating and cooling systems (at manufacturing and consuming levels). New electric motor designs consume less energy. Electric lamp efficiency is increasing.

Use Less Pollutive Energy Sources

Check fuel BTU content vs. pollution generated. Also, pretreat fuels and design equipment to increase combustion efficiency.

Evaluate Nontraditional Energy Sources and Conversion Processes

Select least pollutive of common fossil fuels. Consider the feasibility of geothermal, hydro, solar, biomass, and other substitute energy resources. Nuclear power scores high as a non-polluter, with the exception of the radioactivity wastes. A new generation of nuclear reactors is underway that will increase safe operation manyfold. Much research on nontraditional energy resources continues, but generally technical problems have slowed the pace of progress. These efforts are addressed elsewhere in this encyclopedia. Check index.

Search for New Products and Processes (Substitutes)

Many existing products either generate excessive pollutants during their manufacture, require large amounts of energy (hence hidden polluters), or are adversely pollutive in their own right. Some agricultural chemicals and refrigerants exemplify the latter property. The techniques of organic synthesis provide an avenue to substitute product and process development. The recent development of new refrigerants to replace chlorofluorocarbons (ozone problem) and biological insecticides are examples. Some progress has been made in developing less-pollutive fuels for internal combustion engines and the substitution of new technology for traditional motive power, such as electric- and solar-powered vehicles.

De-emphasize Throw-Away Products

Although somewhat justifiable for certain hospital and medical products, these are very hazardous in wastes. This practice should be reevaluated. Other throw-away items, such as pens and cameras are designed because of marketing motivations and contribute to the waste-disposal load.

Eliminate Product Frills and Frivolous Products

Packaging engineers in recent years have contributed tremendously to waste creation. A substantial portion of household and restaurant wastes, for example, consist of packaging and shipping materials. The marketplace is full of junk merchandise.

Safely Transport Products

Moving polluting products from the manufacturing source to a consuming destination poses an environmentally damaging threat. Product containers must be designed to withstand forceful damage. In connection with petrochemical products, containers range in size from oil drums and chemical-containing carboys to ocean-going oil tankers.

J.S. Hirschhorn (Congressional Office of Technology Assessment) as early as 1988 presented a scholarly summary of waste reduction as the ultimate key to pollution abatement, "Waste reduction is the only way to save industry some of the escalating costs of the current waste-management system." The direct costs of waste disposal have increased some 50 times just over the past few years. Hirschhorn listed six steps to waste reduction:

1. Transfer the economic motivation for waste reduction to those engaged in the manufacturing process.
2. Motivate employees by crediting their performance records by meeting waste-reduction timetables established by management and for proposing waste-reduction concepts.
3. Seek technical assistance from outside sources to gain new viewpoints and incentives.
4. Conduct and maintain a waste-reduction audit.
5. Make waste reduction a lasting part of corporate culture. Approach waste-reduction goals today as energy conservation was stressed a decade or so ago.
6. Initiate a corporate-wide waste-reduction educational program.

PATHWAY 2
RECYCLE WASTE MATERIALS

Design Containers for Recycling

Although not always desirable from a marketer's or consumer's viewpoint, throw-away containers of all kinds contribute massively to the waste-handling problem and to pollution.

Design Products and Components for Recycling

Whereas the aluminum beverage container cannot be reused as such, the aluminum in the can may be reprocessed. The recycling of aluminum is one of the current successes along these lines. The production of raw aluminum from ores consumes enormous amounts of electricity and thus contributes to pollution. Similarly, for years scrap and junk yards have specialized in recycling other metals and all manner of machine parts. From the viewpoint of pollution, this is an excellent practice. Nonmetallics have proved to be more difficult to recycle (most plastics, for example), but much technical progress in this area is underway. Recycled wood fibers in paper products has enjoyed much success. Product designers are in an excellent position to consider the recycling potential of materials after the useful life of the product itself has expired.

Design Processes for Recycling

Cooling water is a prime utility in manufacturing and most notably in the chemical and petrochemical industries. Excellent progress has been made in recycling water instead of continuously dipping into natural water reservoirs. Use is made of cooling ponds and relatively simple water treating at the plant site, thus bypassing pollutive procedures. This also avoids thermal pollution of water source.

Much more complex substances than water should be considered for recycling. These would include the reuse of solvents, cleaning compounds, and, in some instances, using traditional waste components as sources of raw materials. Recovery of valuable materials from wastes may prove less expensive than procuring the same substance from a supplier.

Consider Wastes as Energy Resources

Some industries that produce large amounts of combustible solid wastes have used such materials to augment solid fuels, such as coal, to generate utility steam and hot water. See article on **Wastes as Energy Sources**.

Louis J. Thibodeaux (Louisiana State University and Director, EPA-Sponsored Hazardous Waste Research Center), in 1990, outlined the four *natural laws* of hazardous waste:

1. In converting thermal energy to useful work, a certain amount of waste (thermal) energy must be discharged into the environment.
2. It is impossible to recycle waste completely. Recycling is one aspect of waste minimization, not a solution.
3. Some fraction of the energy and material needed to drive processes and make products will always be degraded to waste that will have to be disposed of in an environmentally acceptable manner, such as incineration or some version of the solidification/fixation process.
4. Small waste leaks are unavoidable and acceptable. Ecosystems can handle small infusions of hazardous substance. Such discharges must, however, be made small so that there will be no harmful effects either locally or globally.

PATHWAY 3

Note: The actual disposition of waste into one of the Earth's "waste sinks" is the least attractive of pollution handling procedures. To date, however, dumping of waste remains the most widely used practice. Progress along Pathways 1 and 2 contribute to a progressive reduction in the tonnage of waste to be disposed and in the long term will alleviate the pollution problem in a major way.

In the advanced countries of the world, waste disposal is no longer a simple matter of venting gases and vapors into the atmosphere, or of finding the nearest creek or river, or of creating a landfill.

The carefree disposal of waste that took place several years ago resulted in a public outcry and the creation of numerous, often quite complex regulations. Thus, the polluting source today must take a number of costly actions prior to the ultimate disposal of the waste.

WASTE DISPOSAL—RELOCATING THE WASTE

Classify and Characterize the Waste

Regulations vary considerably, depending upon the nature of the waste. Hazardous (toxic) wastes are treated as a separate category by federal, state, provincial, and municipal regulatory agencies.

TABLE 1. (continued)

Pretreat Wastes Prior to Disposal

In addition to rigid requirements for hazardous wastes, other regulations may require various forms of pretreatment, such as sorting wastes into various categories in the interest of handling efficiency at waste sites, incinerators, and so on. Although they require handling as wastes, biodegradable wastes pose a lesser threat to long-term pollution and often require less stringent regulation.

Select an Appropriate Depository

Gases, vapors, and airborne particulates, unless present in minor amounts, generally will require postproduction treatment before venting to the atmosphere. Post-treatment is also frequently required for the disposal of liquid and solids. The topic of disposal site selection for liquids and solids is complex because there are so many classes of materials, including such diverse wastes as sewage, public building and household wastes, medical and hospital wastes, packaging wastes, transportation vehicle wastes, office wastes, and agricultural wastes. The lists numbers into the hundreds of categories.

In the case of fluids and solids, the Earth's hydrosphere or lithosphere are the only sinks available. For solids, some geological formations are much more appropriate than others. Some areas may appear suitable, but are found to exist over an aquifer (essentially an underground stream), and hence pollution can occur in the lithosphere and then pass along to the hydrosphere.

In the past, a number of polluters have used temporary waste storage means, such as aboveground tanks. Storage of radioactive wastes at nuclear power facilities is another example. In-plant storage or nearby polluter-owned sites must meet all current pollution regulations. These practices have been costly in retrospect. They have comprised many of the targets of the so-called Superfund.

Consider a Professional Waste-Handling Firm

Expert assistance (applying mainly to liquid and solid wastes) is available, but extreme caution must be taken in selecting such assistance. There have been several instances of fraudulent practices that have led to disastrous pollution and resulted in strict legal judgments against the initiating polluter.

Run Continuous Checks on Waste Disposal Costs

Where costs continue to spiral, this may provide the incentive to initiate actions along Pathways 1 and 2, which can lower pollution costs in the long term.

Consider the Inevitable Conflicts

Severe regulations are reasonably clear in terms of what a polluter can and cannot do. But tradeoffs do remain. A basic triangle of conflicting forces—that is, Energy vs. Economy vs. Environment—is described in article on Electric Power Production and Distribution.

In Japan and the European countries mentioned, regulators are rewarded for eliminating waste streams and cleaning up pollution. Regulators contribute technical expertise. In some cases, research is conducted in government laboratories in an effort to solve a particular pollution problem. The emphasis is on cooperation, rather than adjudication.

In the United States, regulations are highly detailed, sometimes overburdened with detail. They are drawn up so that they can withstand litigation and with little practical latitude in enforcement. Industry has no greater access to the regulators than does any other interest. This, unfortunately, tends to create a climate of confrontation rather than one of cooperation.

TABLE 2. WASTES AND POLLUTANTS REGULATORY STATUTES

(United States—partial list)

Clean Air Act
Clean Water Act
Comprehensive Environmental Response, Compensation, and Liability Act (popularly known as the Superfund)
Federal Insecticide, Fungicide, and Rodenticide Act
Food, Drug, and Cosmetic Act
Hazardous Materials Transportation Act
National Environmental Policy Act
Occupational Safety and Health Act
Resource Conservation and Recovery Act
Safe Drinking Water Act
Superfund Amendments and Reauthorization Act
Toxic Substances Control Act

Note: Most of these statutes have been enacted since the early 1970s.

In the European countries mentioned and in Japan, polluting firms are not required to pay for past practices that did not break former law, with the exception of sites that the polluter continues to own. In the United States, industrial polluters are liable for cleanup costs for sites to which they have contributed waste, even though no laws were broken when the pollution was created.

A more thorough examination of these policy differences is given in the Beecher/Rappaport reference listed.

Particular Attention Given to Hazardous Wastes. In addition to toxicity, hazardous wastes include materials that may become chemically reactive, including ignitability and explosibility, or that may be corrosive. Some toxic materials require extensive pretreatment prior to dumping. See Table 3 on p. 3701.

With the growing use of throwaway products in the hospital and medical field, there is an increasing danger stemming from *infectious wastes*. The ultimate disposal of millions of needles and condoms becomes a part of municipal wastes. The virulence of microorganisms under such conditions is poorly understood. Throwaway diapers not only contribute immensely

TABLE 3. REGULATOR'S CHARACTERIZATION OF HAZARDOUS WASTES

Toxicity

Definition of Extract: The liquid component of a solid waste and deionized water at a pH 5.0 that has been in continuous contact with the solid phase of the waste for a minimum of 24 hours.

Permissible Upper Limit of Contaminant:

	Milligrams/Liter
Arsenic	5.0
Barium	100.0
Cadmium	1.0
Chromium	5.0
Lead	5.0
Mercury	0.2
Selenium	1.0
Silver	5.0
Endrin insecticide	0.02
Lindane insecticide	0.4
Methoxychlor insecticide	10.0
Toxaphene insecticide	0.5
2,4-D (2,4-Dichlorophenoxyacetic acid)	10.0
Silvex 2-(2,4–5 Trichlorophenoxy propionic acid)	1.0

Reactivity

Any substance that:

- Is normally unstable and readily undergoes violent changes with detonations.
- eacts violently with water.
- Reacts with water to generate toxic gases, vapors, or fumes in a quantity that is dangerous to human health or the environment.
- Is capable of detonating or undergoing an explosive reaction when subjected to a strong initiating source or if heated when confined.
- Is capable of detonation or explosive decomposition or reaction at standard temperatures and pressures.
- Is normally considered explosive and that meets transportation regulations.
- Forms potentially explosive mixtures with water.
- Is a cyanide or a sulfide-bearing material that, when exposed to a pH between 2 and 12.5, can generate toxic gases, vapors, or fumes that are dangerous to human health or the environment.

Ignitability

Any substance that

- Is a liquid with a flash point of less than 60°C (140°F).
- Is a solid and is capable of causing fire through friction, absorption of moisture, or spontaneous chemical changes that, when ignited, burns so vigorously as to create a hazard.
- Is a compressed gas or oxidizer that does not meet transportation regulations.

Corrosiveness

Any aqueous liquid with a pH less than or equal to 2.0 or greater than or equal to 12.5.

Any liquid that corrodes steel at a rate greater than 0.006 meter (1/4-inch) per year.

to the volume of wastes to be handled, but also contain hosts of living microorganisms, the survival rate of which, under disposal conditions, have not been documented. The common childhood intestinal pathogens, such as retoviruses, hepatitis A virus, and the protozoans *Giardia* and *Cryptospordium*, have not been ruled out.

Incineration of Hazardous Wastes

One authority has commented that the greatest public health danger with medical waste in the United States is *substandard* incineration practices at local hospitals. It has been estimated that in the early 1990s there were about 6000 substandard medical-waste incinerators throughout the United States. The congressional Office of Technology Assessment (OTA) has estimated that air emissions of dioxin and heavy metals from these incinerators average from 10 to 100 times more per gram of waste burned than emissions from well-controlled municipal waste incinerators. Also, hospital incinerators produce toxic remains in the ashes that can contaminate surface and groundwater when dumped in landfills.

The problem is exacerbated in large and crowded communities. For example, there are well over 50 hospital incinerators in New York City. Most local hospital incinerators are not equipped with acid-gas scrubbers, which convert harmful airborne substances into harmless calcium salts. Nor are most incinerators equipped with electrostatic precipitators to capture particles that have adsorbed toxic flue gases.

Authorities report that quite the contrary conditions exist in several parts of Europe, notably Switzerland and Germany. Legislation in the early 1980s in Germany mandated the closing of hospital incinerators, requiring that medical wastes be sent to regional facilities, at which the latest in technology is deployed. Incinerator operators are given special training and require certification of their skills. Particular precautions are used in feeding such incinerators to protect plant workers. The incinerated remains are disposed of in specially lined landfills.

Recently, the technology of autoclaving instead of incineration has been proposed. Disinfection is achieved through the use of high-pressure steam, which assists in breaking the refuse down and ready for compacting.

German authorities also mandate strict regulations over the transport of medical waste, disallowing the transport of foodstuffs in trucks that also are used to handle medical wastes. Hospitals also are required to "tag" all refuse, designating such categories as "office, cafeteria, and general," "infectious" (including pathological body parts, syringes, needles), "radioactive," and "dangerous to handle" (such as scalpels, which must be placed in unopenable containers).

Considerable design effort has been invested in the improvement of incineration systems. D.A. Tillman (Ebasco Environmental) and associates report that "Rotary kilns have become the incinerators of choice for eliminating hazardous wastes in accordance with the U.S. Resource Conservation and Recovery Act, Superfund, and related legislation."

The Ebasco team points out, "A rotary kiln is basically a rotating cylinder that, typically, is refractory lined. The cylinder is tilted slightly (i.e., 3°) and the feed material goes into the upper end. A heat source is applied to the material, usually by combusting liquid or gaseous fuel within the kiln. Gravity moves the material through the cylinder, and it is discharged from the lower end. When rotary kilns are used for hazardous waste incineration, the gaseous products from the kiln are ducted to a secondary combustion chamber for subsequent destruction." Several overall system design configurations are possible. Generalized configurations are illustrated in Figs. 1 and 2. See Tillman reference listed.

J.F. Mullen (Dorr-Oliver Inc.) reports that fluidized bed incinerators have been used for municipal sludge and industrial waste incineration since the early 1960s for a variety of wastes (petroleum tank bottoms, sludge from pharmaceutical, pulp and paper, and nylon manufacturing operations), waste plastics, waste oils, and solvents. Fluid beds were first considered for incinerating hazardous wastes in the 1980s.

Advantages claimed for fluid bed technology include efficient combustion, ease of control for handling a variety of feeds, and reasonably low capital and operating costs. Specific advantages claimed for incinerating hazardous wastes include fuel savings and lower emissions of nitrogen oxides (NO_x) and metals. The principles of a fluid bed incinerator are illustrated in Fig. 3. Among the primary organic hazardous constituents (POHCs) of waste, the fluid-bed has successfully demonstrated its effectiveness in achieving 99.99% efficiency in removal of aniline, carbon tetrachloride, chloroform, chlorobenzene, cresol, para-dichlorobenzene, methyl methacrylate, naphthalene, perchloroethylene, phenol, tetrachloroethane, 1,1,1-trichloroethane, trichloroethylene, and toluene.

Soil-Washing Technology

Tons of earth per hour can be sifted and scrubbed clean of hazardous materials by using what may be called a heavy-duty industrial *washing machine*. Soils contaminated with hazardous or radioactive materials may

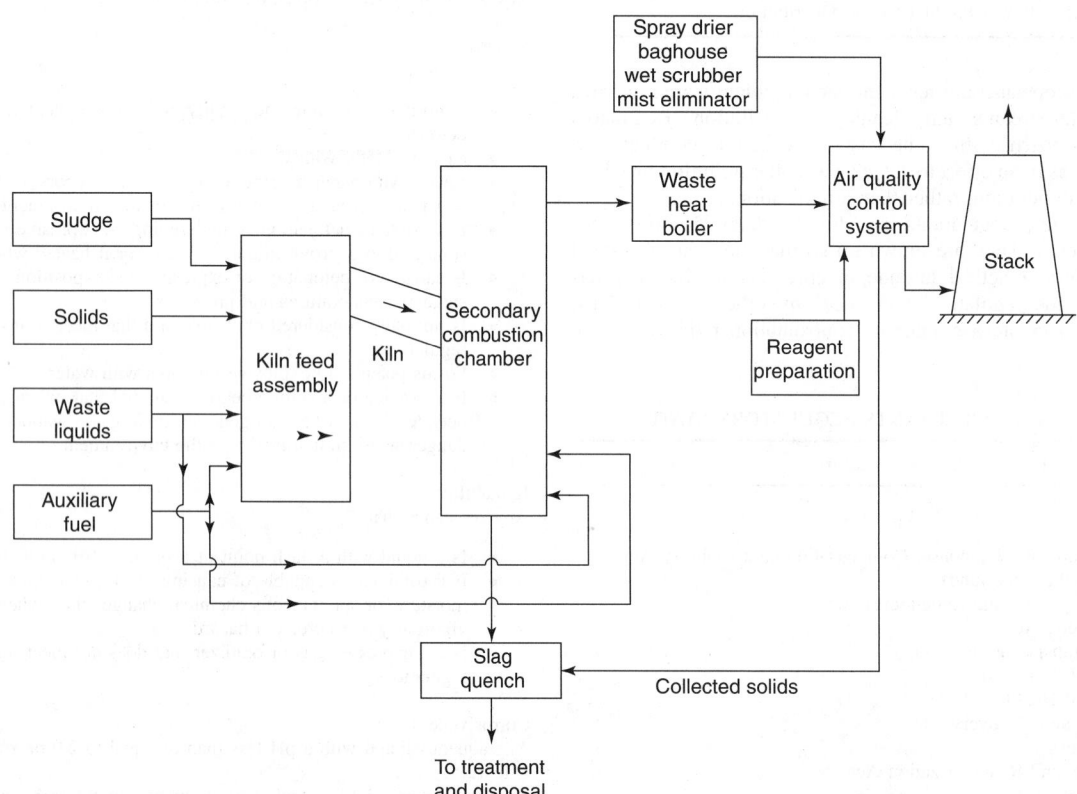

Fig. 1.   Simplified flowsheet of incineration train in a hazardous waste treatment complex. (*After Tillman*)

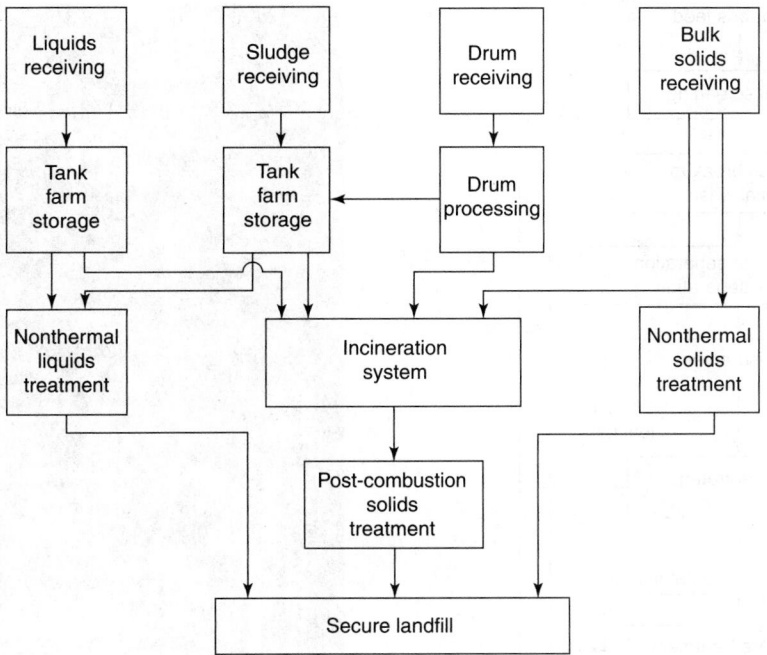

Fig. 2. Generalized flowsheet of a hazardous waste treatment system complex that utilizes a rotary kiln incineration system. (*After Tillman*)

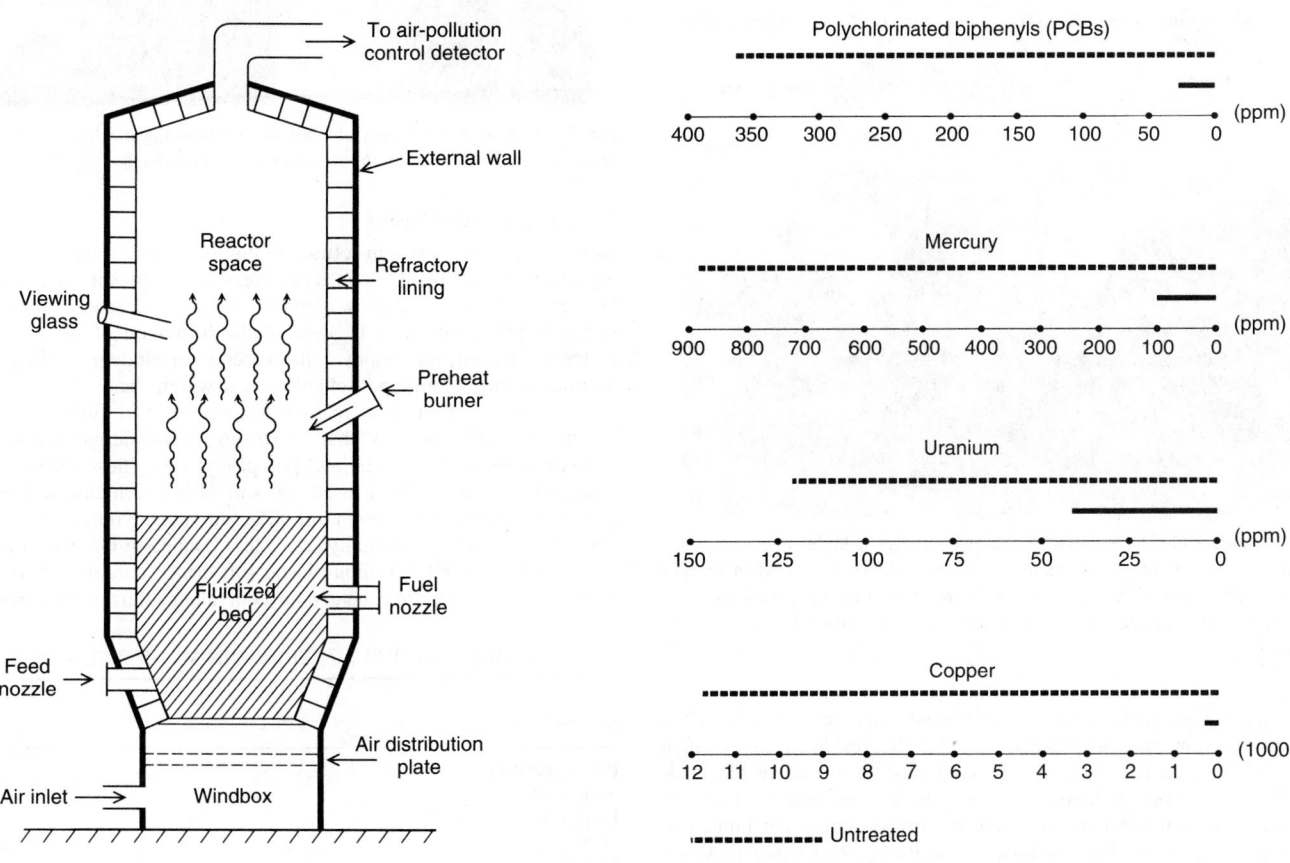

Fig. 3. Schematic diagram of a fluid-bed incinerator. Waste is injected into a bed of inert material that is fluidized by large quantities of air flowing upward through the unit. (*After Mullen*)

Fig. 4. Examples of performance of soil-washing process. (*Westinghouse Electric Corporation*)

be removed by washing. Contaminating material that "comes out in the wash" so to speak, such as copper, then can be recycled and reused by industry. Although not widely publicized, soil washing is not a new technology. It has been estimated that nearly 100,000 tons of contaminated soils are remedied in this manner each year in Europe. The bulk of these soils are sands contaminated with hydrocarbons and/or heavy metals. Westinghouse has developed a commercial unit.

The soil-washing process itself is environmentally benign. It is a closed-loop system, and thus no contaminants are discharged into the air or land. It is a permanent solution to many hazardous problems because it physically removes contaminants from the soil. Examples of tests made on the process are graphed in Fig. 4. Soil washers are mounted on truck trailers for mobility. A flow sheet of the process is given in Fig. 5.

Soil/debris feed

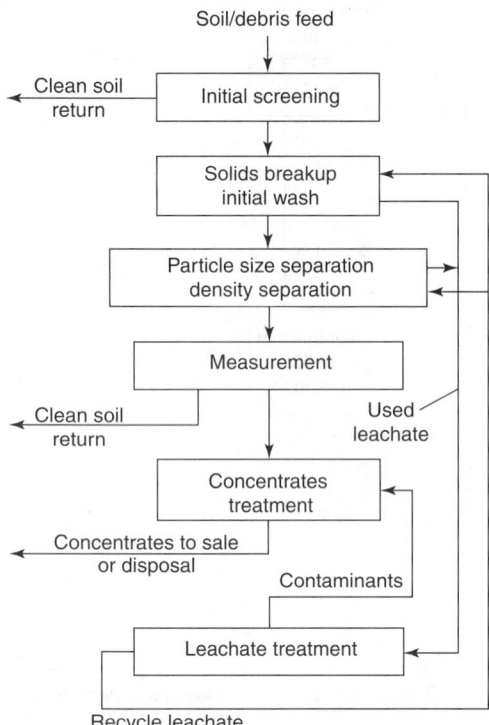

Fig. 5.   Soil-washing process flowsheet. (*Westinghouse Electric Corporation*)

Fig. 6.   Portable soil-washing machine shown here at a highway maintenance site used for removal of lead particles from the residue created when bridges and other metal structures are sand-blasted to remove old paint containing lead. There are numerous applications for such portable plants. (*Westinghouse Electric Corporation*)

In a particular application, Westinghouse unveiled in late 1992 equipment for removing lead particles from the residue created when bridges and other metal structures are sand-blasted to remove old lead-containing paint. This problem is commonly encountered by highway construction and maintenance personnel, but the system is not limited to such applications. It has been estimated that the method has achieved a two-thirds reduction in disposal costs as compared with burial or smelting without pretreatment. One of the first large-scale tests of the equipment was conducted by the Minnesota Department of Transportation. A portable soil-washing machine for highway work is shown in Figs. 6 and 7.

Soil washing is based on the use of water-based leachates, which are recycled continuously in the machine. The end products are a relatively clean, coarser soil fraction and a wash water containing finer soil particles and most contaminants. The small levels of organic contamination remaining in the coarser materials are removed readily with heat. The wash water, depending upon the specific contaminants, responds to various traditional treatment methods, such as bioremediation, air stripping, chemical precipitation, or membrane separation.

Fig. 7.   Operator at controls of portable soil-washing machine used at highway reconstruction site. (*Westinghouse Electric Corporation*)

Recycling Plastic Wastes

Although plastic wastes in refuse are highly visible to the public, and thus have caused considerable consumer pressures on plastics manufacturers to take anti-pollution measures, disposal of plastics creates special problems in municipal incinerators because of the formation of some toxic gases. In terms of dumpsite disposal, the absence of biodegradability results in long-term solids buildup. Actually, on a weight basis, however, plastics only comprise 7% of municipal wastes, as shown in Table 4.

The principal plastics that show up in municipal wastes are the polyethylenes, polystyrenes, and polypropylenes. These include polyethylene terephthalate (PET) used in soft drink containers, high-density polyethylene (HDPE), used in milk jugs, and polystyrene, used in fast-food containers, which, incidentally, were first banned in Oregon (1989).

As early as 1989, 7 billion plastic soft drink containers were produced in the United States and nearly an equivalent tonnage in Europe. As of

TABLE 4.   MAJOR MATERIALS IN MUNICIPAL WASTES

Material	Percent by weight
Paper products	40
Yard waste	18
Food waste (garbage)	12
Glass	8
Plastics	7
Steel/metals	7
Other	8
	100
Principal Classes of Plastics	
Low-density polyethylene (LDPE)	24
High-density polyethylene (HDPE)	19
Polystyrene	14.5
Polypropylene	14.5
Polypropylene terephthalate (PET)	4.5
Other plastics	23.5
	100.0

early 1991, it was reported that about 28% of the PET bottles produced in the United States were recycled, yielding over 20 million pounds of PET for subsequent use in making carpet yarn, fiberfill for clothing, nonfood containers, automobile parts, fencing, and industrial strapping. A doubling of these amounts was expected by 1995.

A principal plastic reclaiming process used was developed in the Netherlands by a resin manufacturing firm. In this process, after removal of labels and base cups by machine, the bottles (clear or green) are color sorted and granulated. The PET flake is then washed to remove glue. Closure material is separated by flotation from the PET. The remaining PET is dried and sifted for fines and then is ready for reuse.

In 1989, seven of the leading plastics producers in the United States formed the National Polystyrene Recycling Company, with a 1995 target for achieving a minimum of a 25% recycling rate for polystyrene. A pilot recycling center was set up in Leominster, Massachusetts.

A joint venture of a leading plastics manufacturer and a major waste management firm was established in 1990 to recycle PET and HDPE materials. The plan was that two existing plants would be joined by three additional recycling operations by 1994. Jointly, the plants would recycle about 200 million pounds of these plastics per year. Taken in perspective, however, this is a small quantity of the total of 1.5 billion pounds (PET) and 6.5 billion pounds (HDPE) disposed each year in the United States.

Still another leading plastics manufacturer commenced operation recently for recycling 400 million pounds/year of plastic film and rigid containers. Following a pilot-plant test run, a wide variety of polyethylene materials can be processed. These include polyethylene wrap, lawn and grocery bags, and containers (detergent, bleach, and motor oil), as well as plastic milk and juice bottles and PET soft drink and liquor containers. When in full operation, the plant will serve a 500-mile ($\sim$800-km) radius area. It is projected that new applications for the recycled materials will displace some of the requirements for virgin plastic and nonplastic materials. Food applications of recycled material are not in the current plans because of the special problems involved in altering the color and other physical properties of the recycled resins.

Plastic materials recycling poses a serious problem when plastic refuse is received in a commingled state. Because of the great variety of plastic products, only a comparatively few can be presorted before delivery to a recycling plant. These few exceptions would include plastic milk and other beverage containers. Impressive research has been underway at Rensselaer Polytechnic Institute, and a patent has been obtained for a process that dissolves shredded plastics (as described by a researcher), "one polymer at a time in a chip-filled vat" where a solvent (xylene) is used to dissolve five groups of plastics at five separate temperatures, ranging from room temperature to 138°C (280°F). For example, polystyrene dissolves first, while the other plastics remain unaffected. The top solubility temperature required is well below the boiling point of the xylene solvent. As pointed out by the researchers, the xylene polymer solution (for a specific polymer) drains to a separate part of the system, where it is heated under pressure to near the boiling point of xylene. Pressure is required to keep the xylene-polymer solution in liquid form. The solution then is sent through a valve into a vacuum chamber to undergo flash devolitization. The sudden change from high to low pressure, researchers say, causes the xylene to vaporize instantly, leaving behind the pure polymer.

After the recovered polymer is removed, the same xylenes are recompressed and cooled to return to its liquid state. It again is heated to a different temperature and reintroduced into the vat to dissolve another of the polymers, and the interim steps are repeated until all polymers present have been separated.

Genetic Engineering and Pollution

Research on the genetic engineering of microbes to degrade toxic wastes has been underway for at least two decades. Progress has been relatively slow, partially attributed to societal concerns over the possible release of new, untested microorganisms that in themselves could create environmental and health threats. As one researcher in the field has pointed out, "No one wants to release organisms before the possible consequences are known, but the possible consequences will remain unknown until the organisms are released."

A very cautious approach has been taken thus far concerning the possible use of engineered microbes in connection with the Superfund dumpsite cleanup program. Budget allocations for genetic engineering in the pollution field have not exceeded a few million dollars annually for the past several years.

Pseudomonas putida, a common soil microbe, has been the target of several researchers. A few years ago, scientists at the University Medical Center (Geneva, Switzerland) created a microbe that eats 4-ethylbenzoate (4-EB), a toxic synthetic chemical. The researchers attempted to find all the "right" genes and combine them into one organism. As early as 1980, researchers in the United States designed a microorganism to break down crude oil. This also involved research with *Pseudomonas* bacterial species. Other engineered microbes have been created that break down much of the sulfur in coal. Research at Johns Hopkins University also has yielded engineered microbes that can metabolize sulfur.

Microorganisms have been used successfully to clean up contaminated wastewater of such substances as pentachlorophenols (PCPs), polychlorinated biphenyls (PCBs), iron cyanide, and leachates containing chlorinated compounds, phenols, and formaldehydes. A dry cleaning solvent, tetrachlorodiphenylethane (TCE), which is a suspected carcinogen, has been degraded through the use of enzymes released by a strain of the bacterium, *Pseudomonas cetacia*. A wastewater treatment plant has used a mutant strain of *Pseudomonas*, which consumes mine-generated cyanide wastes. A process has been developed that uses naturally occurring microbes to concentrate phosphorus for easy removal. It has been estimated that 99.3% of phosphorus in wastewaters can be removed by the process. One firm has developed a bioreactor that uses an aerobic microorganism (naturally occurring in white-roto fungus) to break down toxic substances, such as 2-chlorophenol.

Other researchers have designed a microorganism for breaking down crude oil, and some experience has been gained from testing the product for cleaning up oil tanker spills. Several proprietary engineered microbes have been announced, but with little detail given. Research firms in the field are delaying commercialization, pending clarification of federal, state, and provincial regulations. See also **Water Pollution**.

Fermentation. In 1990, a bioprocessing research center was opened at Penn State University. A pilot plant demonstrates various processes for converting (recycling) agricultural and food processing wastes into dietary supplements for animals. The pilot plant is designed with versatility and flexibility for testing and processing a wide variety of substances. One example of agricultural and livestock wastes that can occur during abnormal situations is an estimated 1 million tons/day of chickens that die of natural causes and are buried at a site on the Delmarva Peninsula, which borders on the Chesapeake Bay. In extremely hot weather, this figure can increase to 4 million tons/day. Also, it has been estimated that 600 tons/day of fish wastes are dumped off Kodiak Island. The Netherlands is estimated to produce 100 million tons/year of poultry manure, twice the amount that the land can absorb naturally. Researchers at Penn State have applied for patents on one of the fermentation processes developed. A type of marine yeast converts protein-bearing waste into a slurry of water-soluble proteins. The process destroys all pathogens. Thus, when the slurry is dried, it is suitable as a dietary supplement. Researchers note that the same principles could be applied to cesspool wastes. Although centralized sewer systems collect *un*sanitary wastes in most urban areas today, there remain multi-thousands or millions of cesspools used in less-populated areas of the country.

Recycling/Regenerating Paper Wastes

Considerable progress has been made during the past decade for recycling paper wastes, which according to Table 4 constitute well over one-third of the typical municipal waste produced.

Investigators at Texas A&M University (Austin, Texas) are targeting on improved processes for regenerating paper wastes. One of the key problems is the cellulose content in newspapers and paper products. Cellulose is very difficult to "digest." The researchers have developed an ammonia fiber explosion (AFEX) technique that more efficiently utilizes enzymatic digestion of cellulose. Researchers explain that ground-up municipal waste is placed in a tank and soaked with ammonia for about one-half hour, after which high pressure is applied. When this pressure is released abruptly, cellulose fibers are literally blown apart, making it much easier for enzymes to digest them. The enzymes break down the cellulose into individual glucose molecules, after which yeast can convert them to ethanol. The researchers have noted an improvement of up to 150% in digestion as the result of the AFEX process. Researchers also forecast that 180 billion tons of municipal wastes could be converted into over 8 billion gallons of ethanol fuel for automotive consumption.

In many communities, prescribed routines for consumers to separate paper from other items of trash have been highly successful and, in fact, in recent years there has been more paper ready to recycle than there are facilities to process it. Consequently, a lot of newsprint and other paper products still go to the dumpsite. Paper in the household trash is a highly visible waste to the average consumer—hence, much cooperation from the populace has been evidenced. In recognition of an imminent "glut" of newsprint to recycle, some municipalities and states have lowered their targets. One example is Wisconsin, which had set a goal of 50% recycled fiber by 1995 to 17% by 2001.

As mentioned earlier in this article, there is a continuing conflict (triangle) of forces that come into play in numerous decisions that interlock the factors of energy, environment, and economy. Newsprint as of the early 1990s is an example. Publishers, particularly in the northeastern United States, can obtain paper of high quality and made of virgin fibers from nearby Canada at an attractive cost, not much in excess of the cost of recycled newsprint. Canadian paper is imported duty-free. Approximately 58% of the paper consumed in the United States (overall) is imported from Canadian mills from trees grown in Canada. Expected lower timber prices in the southeastern United States may make that region more competitive with Canada over the next decade or two.

Average use by newspapers of recycled fibers seldom exceeds 50% of the total. However, the publishers of the *Los Angeles Times* use approximately 80% recycled fiber.

Recycled newsprint does pose problems to publishers, notably those of books and magazines. Some virgin pulp is required to hold the paper together. Magazine publishers usually purchase what recycled paper they do use from pulp-substitute suppliers—that is, firms that process only selected used paper, such as envelope trim and cuttings, ledgers, business forms, and computer printout paper. A long-range recycling program also poses the problem of dealing with shorter fibers. Each time paper is reprocessed, the fibers are shortened. The present use of recycled fiber in newspapers is, on the average, about 20%.

The processing of recycled paper is essentially the same as the production of paper from virgin fibers once the feed slurry is made. See article on **Papermaking and Finishing**; and **Pulp (Wood) Production and Processing**.

In preparing the slurry for recycled newsprint, first all trash must be removed from the waste paper (some manual labor assisted by machine metal detectors, etc.). The paper then passes to a pulper, where water and some reagents are used to accomplish de-inking. In the pulper, the waste paper is shredded by rotating cutting blades to produce a slurry. This slurry is passed through a continuous pulper, similar to that used in making virgin paper slurry after the natural fibers have been processed in a digester. The slurry then passes through screens and onto a three-stage washer, where ink particles are fully removed and the paper adjusted for the proper consistency prior to being introduced onto the paper machine.

Considering the numerous errors that have appeared in the environmental literature, Stephen Strauss, a science writer for the *Toronto Globe and Mail*, points out how easy it is to draw conclusions regarding environmental measures when raw data have not been gathered, calculated, or presented with exacting care. Strauss makes the serious but amusing observation, "When historians of technology reflect on the final quarter of the twentieth century, they may well surmise that the archetypal public debate (over pollution) centered around the throw-away (paper or foam) cup."

See Strauss reference listed.

Several other articles in this encyclopedia address the topic of waste and pollution. See also **Pollution (Air)**; **Water Pollution**; and alphabetical index.

Additional Reading

Abelson, P.H.: "Remediation of Hazardous Waste Sites," *Science*, 901 (February 21, 1992).

Beecher, N. and A. Rappaport: "Hazardous Waste Management Policies Overseas," *Chem. Eng. Progress*, 30 (May 1990).

Bishop, P.L.: *Pollution Prevention: Fundamentals and Practice,* McGraw-Hill Higher Education, New York, NY, 1999.

Boerner, D.A.: "Recycling the Paper Forest," *Amer. Forests*, 37 (July/August 1990).

Bumble, S.: *Computer Simulated Plant Design for Waste Minimization/Pollution Prevention,* Lewis Publishers, Boca Raton, FL, 2000.

Cezeaus, A.: "East Meets West (Germany) to Look for Toxic Waste Sites," *Science*, 620 (February 8, 1991).

Crouch, M.S.: "Check Soil Contamination Easily," *Chem. Eng. Progress*, 41 (September 1990).

Davenport, G.B.: *The ABCs of Hazardous Waste Legislation, Chem. Eng. Progress*, 45 (May 1992).

Davis, M.L. and D.A. Cornwell: *Introduction to Environmental Engineering,* 3rd Edition, The McGraw-Hill Companies, Inc., New York, NY, 1997.

Davis, W.T. and A.J. Buonicore: *Air Pollution Engineering Manual,* John Wiley & Sons, Inc., New York, NY, 1997.

Davis, W.T.: *Air Pollution Engineering Manual,* 2nd Edition, John Wiley & Sons, Inc., New York, NY, 2000.

Dupont, R.R., K. Ganesan, and L. Theodore: *he Pollution Prevention the Waste Management Approach to the 21st Century,* Lewis Publishers, Boca Raton, FL, 1999.

Erb, J., E. Ortiz, and G. Woodside: "On-Line Characterization of Stack Emissions," *Chem. Eng. Progress*, 40 (May 1990).

Evanoff, S.P.: "Hazardous Waste Reduction in the Aerospace Industry," *Chem. Eng. Progress*, 51 (April 1990).

Garg, S.: *Introduction of Recombinant DNA-Engineered Organisms into the Environment: Key Issues,* National Academy Press, Washington, DC, 1987.

Garg, S. and D.P. Garg: "Genetic Engineering and Pollution Control," *Chem. Eng. Progress*, 46 (May 1990).

Gibbons, A.: "Making Plastics that Biodegrade," *Technology Review (MIT)*, 69 (February 1989).

Greenberg, R.A.: "Workshop Participants Focus on (Food) Packaging Waste Management," *Food Technology*, 42 (January 1991).

Hershkowitz, A.: "Without a Trace: Handling Medical Waste Safely," *Technology Review (MIT)*, 35 (August/September 1990).

Higgins, T.E.: *Pollution Prevention Handbook,* Lewils Publishers, Boca Raton, FL, 1995.

Hodge, C.A., N.N. Popovici: *Pollution Control in Fertilizer Production,* Marcel Dekker, Inc., New York, NY, 1994.

Hooker, L.: "Danger Below (Underground Aquifers)," *Chem. Eng. Progress*, 52 (May 1990).

Kamrin, M.A.: *Toxicology: A Primer,* Lewis Publishers, Boca Raton, FL, 1988.

Leaf, D.A.: "Acid Rain and the Clean Air Act," *Chem. Eng. Progress*, 25 (May 1990).

Lecomte, P., C. Mariotti: *Handbook of Diagnostic Procedures for Petroleum-Contaminated Sites,* John Wiley & Sons, Inc., New York, NY, 1999.

Lohr, L.: "Managing Solid Byproducts of Industrial Food Processing," *Food Review,* 21 (April–June 1991).

Loupe, D.E.: "To Rot or Not; Landfill Designers Argue the Benefits of Burying Garbage Wet vs Dry," *Science News*, 218 (October 6, 1990).

Majumdar, S.B.: "Regulatory Requirements and Hazardous Materials," *Chem. Eng. Progress*, 17 (May 1990).

Martin, A.M. et al.: "Control Odors from Chemical Process Industries," *Chem. Eng. Progress*, 51 (December 1992).

Morrow, D.R.: "Recycling of Plastic Packaging Materials," *Food Technology*, 89 (December 1989).

Mullen, J.F.: "Consider Fluid-Bed Incineration for Hazardous Waste Destruction," *Chem. Eng. Progress*, 50 (June 1992).

Nathanson, A.A.: *Basic Environmental Technology: Water Supply, Waste Management, and Pollution,* 2nd Edition, Prentice-Hall, Inc., Upper Saddle River, NJ, 1996.

Nathanson, J.A.: *Basic Environmental Technology,* 3rd Edition, Prentice-Hall, Inc., Upper Saddle River, NJ, 1999.

Nemerow, N.L.: *Zero Pollution for Industry: Waste Minimization through Industrial Complexes,* John Wiley & Sons, Inc., New York, NY, 1995.

Ostler, N.K.: *Introduction to Environmental Technology,* Prentice-Hall, Inc., Upper Saddle River, NJ, 1995.

Ostler, N.K., M. Malachowski, and T.A. Byrne: *Health Effects of Hazardous Materials,* Prentice-Hall, Inc., Upper Saddle River, NJ, 1996.

Ostler, N.K., J.T. Nielsen: *Waste Management Concepts,* Prentice-Hall, Inc., Upper Saddle River, NJ, 1997.

Powell, C.S.: "Plastic Goes Green (Recycled Plastics)," *Sci. Amer.*, 101 (August 1990).

Pszozola, D.E.: "Bottle Manufacturer Operates Plastic Recycling Plant," *Food Technology*, 54 (January 1991).

Rathje, W.L., L. Psihoyos: "Once and Future Landfills," *Nat'l. Geographic*, 116 (May 1991).

Renko, R.J.: "Minimize Operating Costs in Meeting Fume Emission Control Standards," *Chem. Eng. Progress*, 47 (October 1990).

Staff: "Environmental Protection, Safety, and Hazardous Waste Management," *Chem. Eng. Progress*, 15 (December 1988).

Staff: "Elements of Toxicology," *Chem. Eng. Progress*, 37 (August 1989).

Staff: "Plastic Recycling Plant in Philadelphia," *Chem. Eng. Progress*, 10 (February 1990).

Staff: "Nylon Meshes Well with the Environment," *Advanced Materials & Processes*, 6 (July 1990).

Staff: "Penn State Opens Pilot Plant for Biotechnology Companies," *Chem. Eng. Progress*, 9 (September 1990).

Staff: "Effective Management of Food Packing: From Production to Disposal," *Food Technology*, 225 (May 1991).

Staff: "Process Pushes the Upside of Garbage," *Chem. Eng. Progress*, 12 (October 1991).

Staff: "Solvent Sorts Out Plastics," *Chem. Eng. Progress*, 22 (November 1991).

Strauss, W.: "The Haze Around Environmental Audits," *Technology Review (MIT)*, 19 (April 1992).

Testin, R.F., P.J. Vergano: "Food Packaging," *Food Review*, 31 (April–June 1991).

Theodore, L., Y.C. McGuinn: *Pollution Prevention*, John Wiley & Sons, Inc., New York, NY, 1997.

Thibodeaux, L.G.: "The Four Natural Laws of Hazardous Waste," *Chem. Eng. Progress*, 7 (May 1990).

Tillman, D.A., A.J. Rossi, and K.M. Vick: "Rotary Incineration Systems for Solid Hazardous Wastes," *Chem. Eng. Progress*, 19 (July 1990).

Woodard, F.: *Industrial Waste Treatment Handbook*, Butterworth-Heinemann, Inc., Woburn, MA, 2001.

WASTES AS ENERGY SOURCES.   Initially, *biomass* was defined as the amount of living organisms in a particular area, stated in terms of the weight or volume of organisms per unit area or of the volume of the environment. This definition still applies very well to ecological and geophysical assessments of land areas or regions and depths of the seas and lakes.

In modern technology, the term also may be used to describe the exploitation of living terrestrial materials, such as plants or marine plants, all or parts of which may be combusted directly for the thermal energy that they yield or, more indirectly, as raw materials for processes that can convert the biomass into fuels.[1] Very generally, biomass may be considered the total amount of living matter within a given unit of area, volume, or mass. When biomaterials serve as foods or provide fibers and items of construction, for example, waste is created. These materials are typified by straw, sawdust, sewage sludge, and so on, which possess value as energy sources. That aspect of biomass is the topic of this article.

Generally, the pace of research and construction of facilities for transforming solid wastes into energy forms, such as methane, or to recover heat from combusting the wastes, slowed during the late 1980s and early 1990s. There are several causative factors, but of course a breakthrough could reverse these trends. In the 1970s, during the time of the oil embargo and energy crisis, finding new sources of energy was a major incentive. The fervor of the former energy programs has largely deteriorated in the presence of what is now considered by many a severe environmental crisis. This latter incentive has not been sufficient to foster extensive research programs in transforming garbage and other trash into energy because the *primary* advantages remain as an energy source and not as a means of pollution abatement. Nevertheless, progress has been made. Simply combusting municipal rubbish as a means of waste disposal is described in the preceding article.

Considerable research is being conducted on a variety of biomass feedstocks that contain cellulose, from which ethanol can be produced. Again, ethanol as a transportation fuel component has not been widely accepted, even though aggressively promoted in some areas. Brazil usually is cited as a prime example of progress in this area. L.R. Lynd (Dartmouth College) and a team of researchers have been studying the impacts of alternative fuel use on carbon dioxide (CO_2) accumulation, energy security, and economic effects on the United States as a whole. The team observes, "Production of ethanol from cellulosic biomass is believed to be an emerging energy technology with particularly great potential for the U.S. transportation sector. Research to improve conversion processes and to develop cellulosic energy crops is necessary to reduce costs and to increase production potential. Success can reasonably be expected in both these areas in light of the immature state of current technology and the powerful approaches available." In terms of their potential for yielding ethanol, in order of diminishing production potential, are agricultural wastes, forest sources, and municipal solid waste.

The majority of agricultural, commercial, industrial, and urban or municipal wastes are of a biological rather than mineral nature and thus fall under the umbrella of biomass. The simple burning of wood for heat illustrates one of the simplest ways to convert biomass to energy. All biomass represents an indirect form of solar energy. Biomass, as a source of energy, differs from coal, natural gas, and petroleum in one major way—biomass is renewable. Some potential biomass energy crops can be renewed as frequently as two or three times per year, depending upon location, while other materials such as trees have a renewable cycle of

several years. Anthropogenic wastes are renewed on a daily basis. Interest in biomass over the last several years has stemmed from the overall concern with ultimate exhaustion of nonrenewable energy sources, as well as gaining a degree of political independence by many nations that either do not have any fossil fuel resources, or that have insufficient supplies to maintain a strong economic and industrial position.

Urban Wastes as Energy Sources

Urban waste includes household, sewage, commercial, institutional, manufacturing, and demolition waste. The availability of this waste is directly related to the population living in urban areas of adequate size to support a given size system.

Manufacturing and processing wastes include all residuals generated from material inputs that leave the plant as product output. Office and packaging wastes associated with this sector are included in the urban waste sector. The majority of these wastes are from pulp and paper manufacturing, primary and secondary wood manufacturing, and the construction industry.

The energy recovery system selected dictates the extent that solid waste must be prepared. Some systems require nothing more than the removal of massive noncombustibles, such as kitchen appliances from the refuse, while other processes require extensive shredding, air classification, reshredding, and drying. In conjunction with fuel preparation, it is usually worthwhile to reclaim metals and glass for recycling.

One-stage shredding is often used to reduce waste to a nominal size as small as 1 inch (2.5 centimeters). When finer-sized fuel is required, a second shredding step is usually used after air classification has removed many of the noncombustibles. Both vertical and horizontal air classifiers depend on the heavy noncombustibles settling out by gravity in a moving air stream, while the lighter combustibles are pneumatically transferred through the air classifier. Denser combustibles, such as rubber and leather, may be removed with the heavy fraction, while some of the fine glass and metal foils are carried with the combustibles. Thus, desired separation may not always be achieved on one pass through an air classifier.

Some energy recovery systems require drying to remove excess moisture in the waste. This is required when sewage sludge is used as a fuel. Usually, waste heat from the total process can be used for the drying system.

Pyrolysis. In one system, municipal refuse is charged at the top of a shaft furnace and is pyrolyzed as it passes downward through the furnace. Oxygen enters the furnace through tuyeres near the furnace bottom and passes upward through a 1425 to 1650°C combustion zone. The products of combustion then pass through a pyrolysis zone and exit at about 93°C. The off-gas then passes through an electrostatic precipitator to remove flyash and oil formed during pyrolysis. The latter are recycled to the furnace combustion zone. The gas then passes through an acid absorber and a condenser. The clean fuel gas has a heating value of about 300 Btu/cubic foot (2670 Calories/cubic meter) and a flame temperature equivalent to that of natural gas. The solid waste that remains is a slag at the furnace bottom.

Biological Methane Production. This process involves the anaerobic digestion of a solid waste and water or sewage sludge slurry at 60°C for five days to produce a methane-rich gas. Solid waste is prepared by shredding and air classification, followed by blending with water to produce a mixture of 10 to 20% solids concentration. The slurry is heated and placed in a mixed digester at 60°C for 5 days detention. The digester gas is drawn off and separated into carbon dioxide and methane. The spent slurry from the digester is pumped through a heat exchanger to partially heat the incoming slurry prior to filtration. The filtrate is returned to the blender and the sludge is used for landfill. Heat addition to the refuse slurry is required to maintain the required digester temperature. The process is well suited for use on sewage sludge, animal manures, and other high-moisture-content solid wastes. It is estimated that the process can reduce the volume of volatile solids by 75%, while producing about 3000 cubic feet (85 cubic meters) of methane per ton of incoming solid waste. The major residue is used for landfill or incinerated. About 10% of the methane is required to heat the digester feed.

Direct Steam Process. One process uses a rotary kiln pyrolizer followed by an afterburner and boiler to produce steam from shredded waste. The pyrolysis process in a kiln is operated countercurrently. Solid waste enters at one end and pyrolyzed residue is discharged at the other. External fuel and air are introduced at the residue discharge area and combustion products and pyrolysis gases leave the kiln at the feed opening. This arrangement causes the solid waste to be exposed to progressively higher temperatures as it passes through the kiln. The kiln off-gases pass through a

---

[1] The generation of biomass from carbon dioxide is called "primary production" because it is the first fundamental step in turning inorganic material into organic compounds and cell constituents. This reduction of carbon dioxide uses sunlight as the source of energy. See also **Photosynthesis**.

refractory-lined afterburner into which air is introduced to allow complete combustion prior to passing through the waste heat boiler. A wet scrubber is used for air pollution control, while an induced draft fan is used to draw the gases through the system. One ton of solid waste, augmented by 1.25 million Btu (0.3 million Calories) from auxiliary fuel and 55 kilowatt-hours of electricity will produce about 4800 pounds (2177 kilograms) of steam at 330 psig (22.4 atmospheres) along with 200 pounds (91 kilograms) of char.

Waterwall Incinerators. These devices generate steam by burning unprepared solid waste on a grate and passing the hot products of combustion through a boiler. Numerous waterwall incinerators have been built in Europe and the United States. Unprepared refuse is taken from storage pits and charged directly into the incinerator feed hopper. From there, the refuse drops onto a feed chute and then is fed automatically onto the stoker by means of a hydraulic feed ram. Temperatures in the 870°C range effectively burn the solid waste. Before the flue gas enters the boiler, secondary air is added to produce a temperature near 1090°C. The boiler is constructed of membrane waterwalled tubes with extruded fins. After passing through the boiler, the gases travel through an economizer section and then into an electrostatic precipitator for particle removal. A typical 1000 tons/day (900 metric tons/day) waterwall incinerator produces about 300,000 pounds (136,080 kilograms) of steam per hour.

Additional engineering is required where toxic wastes may be present.

Principal Biomass Materials for Energy

Biomass-to-energy systems fall into two principal categories: (1) materials for direct combustion that will generate heat for processing, for warming living and working spaces, for steam and hence also for generating electricity; and (2) materials from which both fuels and chemicals can be obtained through biochemical or thermochemical conversion processes. Resulting fuels must have ample caloric content per unit of weight and thus rich in carbon and hydrogen and poor in the content of atoms, such as oxygen and nitrogen, which do not contribute to the caloric value of the fuel. In searching for new biomass raw materials, scientists have found it helpful to study the various biosynthetic pathways followed by plants from seed to maturation.

Direct Wood Burning. Wood was the major fuel of the United States until about 1886 when the consumption of coal equaled that of wood. Oil did not appear on the chart until about 1900 and gas in 1910–1920. The use of wood tapered off while other fuels climbed at amazing rates, but wood never ceased as a factor, even if small. It is interesting to note that about a million modern woodburning stoves are in use and that about 40% of the wood products industry is furnished by combusting bark and mill wastes. This amounts to about 1 quad (10^{15} Btu). Wood burning has been sufficiently extensive during the past few years to cause environmental concerns in some regions. Although wood has a low sulfur content and produces minimal amounts of nitrogen oxides even by the hottest fires (1370°C; 2500°F), it contains air pollutants in the form of particulates, gases, and tars. Environmentalists in the New England region have estimated that the 300,000–400,000 tons of wood burned per year (New Hampshire only), if the fuel is very dry red oak, will add 1000 tons of particulates to the air; if dry white pine is burned, the total may be over 5000 tons. Since a mixture of woods usually is used, the figure lies somewhere between the two aforementioned quantities.

Agricultural Wastes. In the absence of an energy crisis, with a few exceptions, attention to the use of agricultural wastes has waned. These materials continue to represent an essentially untouched source of energy in most countries. In the course of extensive scientific research in the early 1970s, at a time when renewable energy sources were being aggressively sought, evaluations were made of various materials for their caloric content and availability. Included were corn (maize), sorghum, wheat, sugar beats, sugarcane, pineapple, and cassava, among others. Several of these crops were considered as sources for the production of alcohol for admixture with gasoline (gasahol) as an automotive fuel. Brazil has had considerable success in this regard with sugarcane as a raw material.

Although burning and incineration offers means for disposing some waste products, these processes do contribute to air pollution and thus must meet the requirements now followed for the common fossil fuels.

Additional Reading

Abelson, P.H.: "Improved Yields of Biomass," *Science*, 1469 (June 14, 1991).

Corcoran, E.: "Dirty Business: How Companies are Seeking Their Fortunes in Garbage," *Sci. Amer.*, 98 (September 1989).

Klass, D.L.: *Biomass for Renewable Energy, Fuels, and Chemicals,* Harcourt Brace & Company, San Diego, CA, 1998.

Kumer, R., J.K. Van Sloun: "Purification (of Methane) by Adsorptive Separation," *Chem. Eng. Progress,* 34 (January 1989).

Lynd, L.R. et al.: "Fuel Ethanol from Cellulosic Biomass," *Science,* 1318 (March 15, 1991).

Monoastersky, R.: "Biomass Burning Ignites Concern," *Science News,* 196 (March 31, 1990).

Rowell, R.M., T.P. Schultzand, and R. Narayan: *Emerging Technologies for Materials and Chemicals from Biomass,* American Chemical Society, Washington, DC, 1992.

Saha, W.C., J. Woodward, and B.C. Saha: *Fuels and Chemicals from Biomass,* Vol. 666, American Chemical Society, Washington, DC, 1997.

Staff: "Waste Management: Technology Cuts through Emotional Myths," *Westinghouse Technology,* 15 (October 1988).

Turbak, G.: "Woodburning's New Age," *Amer. Forests,* 52 (November–December 1989).

Wereko-Brobby, C.Y., E.B. Hagen: *Biomass Conversion and Technology,* John Wiley & Sons, Inc., New York, NY, 1996.

Wright, J.D.: "Ethanol from Biomass by Enzymatic Hydrolysis," *Chem. Eng. Progress,* 62 (August 1988).

Wyman, C.E.: *Handbook on Bioethanol: Production and Utilization,* Taylor & Francis, Inc., Philadelphia, PA, 1996.

WATER. A colorless (blue in thick layers) liquid, H_2O or HOH, odorless, tasteless, melting point 0°C (one of the standard temperature points), boiling point 100°C at 760 millimeters of mercury pressure (another standard temperature point).

The boiling point of water increases with increasing pressure: 100.366°C at 770 mm; 120.6°C at 1,520 mm; 180.5°C at 7,600 mm. The boiling point decreases with decreasing pressure: 99.360°C at 750.0 mm; 99.255°C at 740.0 mm; 98.877°C at 730.0 mm; 81.7°C at 380 mm; 46.1°C at 76.0 mm.

At 0°C, the density of water is 0.99987 gram per milliliter. At 8°C, 0.99988; at 15°C, 0.99913; at 16°C, 0.99897; at 17.5°C, 0.99871; at 20°C, 0.99823; at 25°C, 0.99707; at 40°C, 0.99224; at 50°C, 0.99807; at 75°C, 0.97489; at 100°C, 0.95838; at 120°C, 0.9434.

The critical temperature of water is 374.15°C; critical pressure, 218.4 atmospheres; critical density, 0.323 gram per cubic centimeter.

The viscosity at 0°C is 0.01792 poise (dyne second per square centimeter), specific viscosity, 1.000. At 20°C the viscosity is 0.01005 poise, specific viscosity, 0.561. At 50°C, the viscosity is 0.00549 poise, specific viscosity, 0.307. At 75°C, the viscosity is 0.00380 poise, specific viscosity, 0.212. At 100°C, the viscosity is 0.00284 poise, specific viscosity, 0.158.

The surface tension of water against air at 0°C is 75.6 dynes per centimeter. At 10°C, the surface tension is 74.22; at 20°C, 72.75; at 30°C, 71.18; at 60°C, 66.18; and at 100°C, 58.9.

The specific heat of water is 1.000000 at 15°C (standard of specific heat). At 0°C, the specific heat is 1.00874; at 25°C, 0.99765; at 35°C, 0.99743 (minimum); at 50°C, 0.99829; at 65°C, 1.00001; at 80°C, 1.00239; at 100°C, 1.00645; at 120°C, 1.016; at 180°C, 1.04.

The electrical conductivity of water at 18°C is 0.04×10^{-6} reciprocal ohms (measurements of Kohlraush and Heydweiller, 1902); of pure water in equilibrium with air, 0.8×10^{-6}; of ordinary distilled water, about 5×10^{-6}.

The dielectric constant of water (specific inductive capacity) is 81.07 at 18°C.

Pure water, when free of dissolved gases, may be heated above 100°C (even up to 180°C) without boiling, but upon further heating, boiling with explosive violence may occur. Steam at 100°C occupies a volume 1,700 times greater than water at 100°C. Pure water, when not agitated, may be cooled somewhat below 0°C without freezing, but upon further cooling it congeals with an increase of volume (density of ice, 0.917) exerting great force, when confined, but if in intimate contact with water at atmospheric pressure, the freezing temperature is 0°C. The vapor pressure of ice and water is 4.579 millimeters at 1 atmosphere pressure and 0°C. The triple point (ice, water, and water vapor) occurs at a saturation vapor pressure of 6.11 millimeters and at a temperature of 273.16 K. When water is compressed to about 20,000 atmospheres and then cooled, other varieties of ice, all denser than water, are formed. Ice II is 12% denser; Ice III is 3% denser. At least six varieties of ice are known.

In ice I, ice II, and ice III, each oxygen atom is surrounded tetrahedrally by four other oxygen atoms, the difference between the three forms being largely in some distortion of the linkages, since the O—O distance varies little from the 2.76 value of ice I. Each oxygen atom has 2 hydrogen atoms quite close (about 1 Å) to it. At lower temperatures, and presumably higher pressures (forms V, VI, and VII, the water molecules with four hydrogen bonds are more in evidence.

Liquid water exhibits the same tendency toward increased bonding at lower temperatures. While individual water molecules have a nonlinear structure, there is association between H_2O molecules by hydrogen bonding, the degree of association being greater at lower temperatures. Based upon the statistical mechanical treatment of Frank and Wen, liquid water may be regarded as a mixture of hydrogen bonded clusters and unbonded molecules. Other research characterizes this model in terms of 5 species: unbonded molecules, tetrahydrogen bonded molecules in the interior of cluster; and surface molecules connected to the cluster by 1, 2, or 3 hydrogen bonds.

The chemical properties of water change with temperature at high temperatures. The reaction, $2H_2O \longleftrightarrow 2H_2 + O_2$, shows an appreciable shift to the right, reaching 0.8% at 2,000°C, and increasing rapidly above that temperature. At ordinary temperatures, the equilibrium, $2H_2O \longleftrightarrow H_3O^+OH^-$, is important because it enables water to act either as a proton donor or acceptor. With stronger acids, water can act as a proton acceptor:

$$HCl + H_2O \longleftrightarrow H_3O^+ + Cl^-$$

$$HBr + H_2O \longleftrightarrow H_3O^+ + Br^-$$

$$HSO_4^- + H_2O \longleftrightarrow H_3O^+ + SO_4^{2-}$$

With stronger bases, water can act as a proton onor:

$$H_2O + CO_3^{2-} \longrightarrow HCO_3^- + OH^-$$

$$H_2O + NH_3 \longrightarrow NH_4^+ + OH^-$$

Although the ions shown are written as CO_3^{2-}, HCO_3^-, Cl^-, etc., they of course are more or less solvated by the water, i.e., they have water molecules attached to them by ion-dipole bonds, since water is a polar compound. One of the most strongly marked properties of water is its behavior as an electrolytic solvent, which is due to its high dielectric constant. The energy of separation of two ions is an inverse function of the dielectric constant of the solvent. Some of the parameters of water are shown in diagram of Fig. 1.

Molecular Architecture

Greatly oversimplified, the water molecule may appear as shown in Fig. 2, which indicates the equilibrium position of the oxygen atom and the hydrogen atoms, i.e., the equilibrium position of the positive and negative charges of the molecule. Because of this orientation, the water molecule has a strong tendency to be oriented in an electrical field. The dipole moment depends upon the magnitude of the charge separation within the molecule, and in the water molecule, the separation is large. Thus, water may be described as having an exceptionally large dipole moment and consequently a large dielectric constant. On the basis of ascribing a dielectric constant of 1 for a vacuum, the dielectric constant of water is 80; i.e., in water, 2 electrical charges will attract or repel each other with only 1/80th as much strength as would be the case in a vacuum. This accounts, at least in part, for the remarkable ability of water to dissolve substances, particularly materials whose molecules are held together primarily by ionic bonding.

The bonding arrangements within the water molecule also account for the exceptional cohesive power exhibited in water's high surface tension and the outstanding ability of water to adhere strongly to a variety of materials (the property of wetting). Bonding also accounts for the manner in which water crystals, e.g., snowflakes, are formed and for the maximum density of water (4°C), below which water assumes less dense forms, causing ice to float. Bonding is responsible for the exceptional heat capacity and exceptionally high latent heats of fusion and evaporation of water.

The molecular behavior of various molecular types in electrolytes is shown in Fig. 3. This behavior is of particular significance to the role of water in biological systems. See also **Molecule**.

Although of continuing interest, research on the structure of water is exceptionally difficult. X-ray crystallography has provided a good picture of the stable hydrogen-bonded structure of ice, but thus far the structural imaging of liquid water has escaped investigators, and hence blackboard theorizing and computer modeling are the principal research techniques followed. In the *Journal of the American Chemical Society* (*JACS*) of May 20, 1992, water scientists present a complex new theory that may account for so many of the physical properties of liquid water.

Heavy water, also known as deuterium oxide, D_2O, is water in which the hydrogen of the water molecule consists entirely of the heavy-hydrogen isotope having a mass number of 2. The density of heavy water is 1.1076 at 20°C. Heavy water has been used as a moderator in nuclear reactors as well as a coolant.

Uses

Water is such a common substance that its importance and versatility are usually taken for granted. Included among the major ways in which water is important would be: (1) as a raw material for incorporation into final products without chemical change; (2) as a raw material for undergoing chemical change; (3) as a transport and conveyance medium with water acting as a solvent or carrier of solutions and suspensions in and out of reactions and physical-change operations—at an industrial as well as biochemical level; (4) as a heating and cooling medium over the wide temperature range from below normal freezing temperature (brine solutions, for example) to those of superheated steam; (5) as an energy-storage medium; (6) as a gathering medium for waste products; (7) as a cleaning medium; (8) as a shield against heat and nuclear radiation (heavy water); (9) as a convenient standard in terms of temperature, density, viscosity, and other units; and (10) with exception of a few situations where the presence of water is hazardous, as a fire-fighting medium.

Water Metabolism in Vertebrates

Those vertebrates that now inhabit land, seas, brackish and fresh waters have survived because they have developed homeostatic mechanisms that enable them to cope with considerable variation in the content and availability of water, sodium, potassium, and chloride in their external environment. These mechanisms prevent life-threatening changes in their internal environment by (1) assuring that the cells are bathed by fluid with the same osmotic concentrations as themselves; and (2) by preventing major qualitative changes in the intra- and extracellular content of these ions or water. Regardless of species, one is impressed not by the differences, but by the similarities in the ionic composition of their intra- and extracellular fluids. The water content of the fat-free tissues of all vertebrates ranges between 70 and 80%. Water diffuses freely along its concentration gradient (osmosis) throughout all body tissues. Therefore, any deviation of the osmotic pressure of intra- or extracellular fluids, by either withdrawal or addition of water, causes an immediate movement of

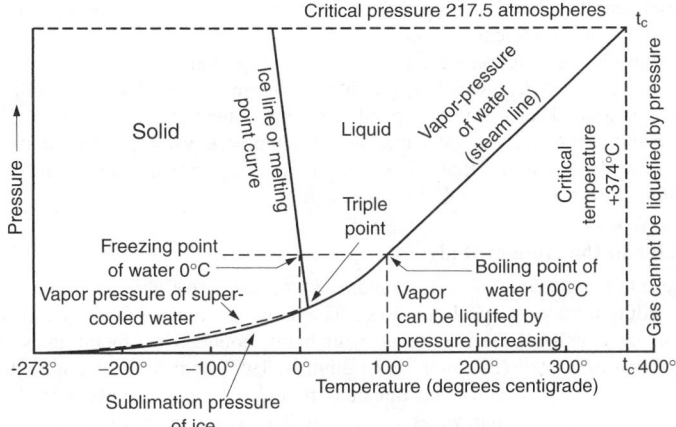

Fig. 1.   Pressure-temperature diagram for water

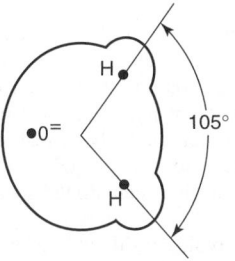

Fig. 2. Schematic representation of arrangement of electrical charges in water molecule

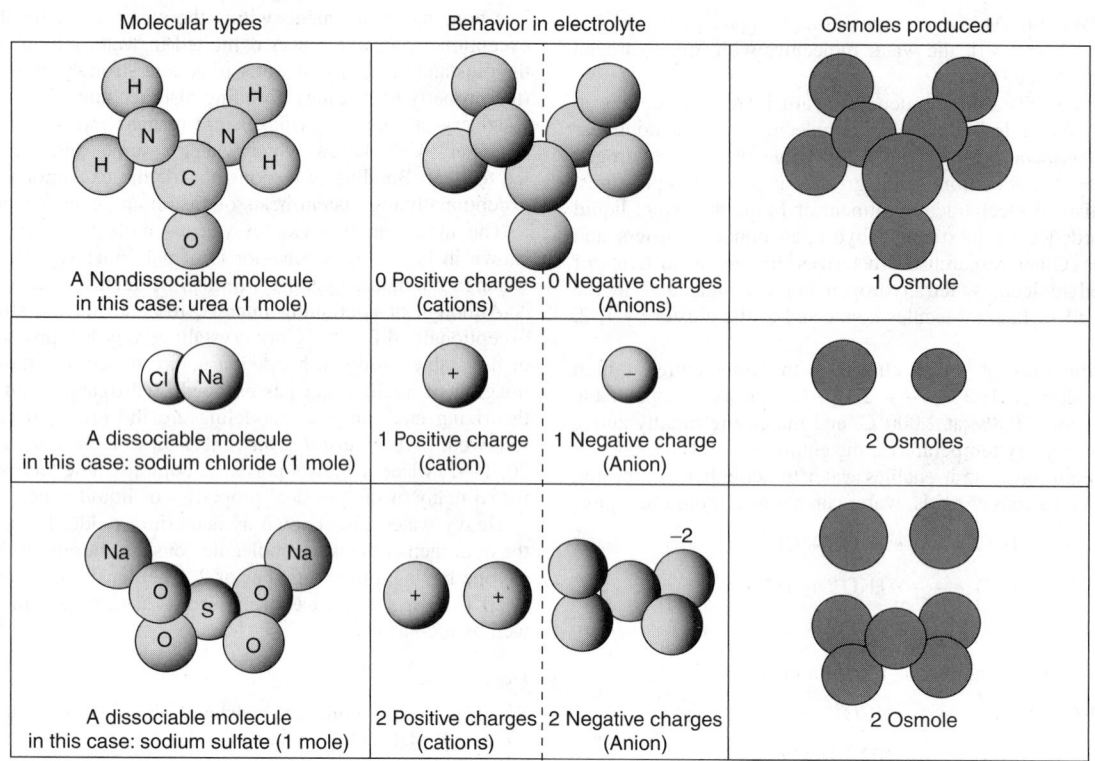

Fig. 3.   Behavior of various molecular types in electrolytes. (*After Maffy.*)

water from the more dilute to the more concentrated solution until osmotic equilibrium is reestablished.

Water is lost from the body of mammals by evaporation across the skin and in the expired air, urine, and feces. The more arid the environment, the more a mammal must be able to reduce water loss and tolerate longer periods of water dehydration and hypertonicity of its body fluids.

According to Chew, vertebrates fall into several groups in terms of how they maintain their water balance. *Fishes and amphibians* in fresh water are very hypertonic to their medium and must counteract a continual dilution of their body fluids. Water influx is reduced by the relative impermeability of the skin, and is balanced by diuresis. Electrolytes lost in this urine are replaced in food eaten and by absorption through gill surfaces (fishes) and skin (amphibians). *Marine elasmobranches* are unique in maintaining themselves slightly hypertonic to sea water by retention of urea (2±%) in their body fluids, making their osmotic pressure more than twice that of marine teleosts. *Marine teleosts* and terrestrial tetrapods face the continual problem of counteracting desiccation due to osmotic loss to a hypertonic medium or to evaporation. Mammals are the most effective of vertebrates in conserving urine water by concentrating the urine, which is achieved by reabsorption of water in the kidney tubules.

Terrestrial tetrapods adjust by avoidance of evaporative stress, reduction of evaporative and urinary water losses, and temporary toleration of hyperthermia or hypernatremia. Antidiuretic hormone (ADH) from the neurapophysis is very important in enhancing uptake of water through the skin (amphibians), reduction in glomerular filtration (amphibians, reptiles, birds), and increase in tubular reabsorption of water (mammals).

Water balance processes are best developed in species inhabiting deserts, where little drinking water is available and climatic conditions accentuate evaporation.

Certain toads and frogs survive in deserts, needing open water only for breeding, largely by remaining dormant during dry periods. Evaporation is greatly retarded in a cool damp burrow, and urine volume is reduced by 98–99% (filtration antidiuresis), but urine remains hypotonic. Urinary water may be recycled through the body by reabsorption from the bladder. Dormant animals tolerate a loss of 50–60% of their body water. They emerge during rains, and in their dehydrated state quickly reabsorb water through the skin.

Terrestrial reptiles also avoid considerable evaporation by being quiescent in burrows much of the time. Also, their skin is more impermeable than that of amphibians, although water is still lost in expired air. Hydrated lizards have low urine filtration rate (urine always hypotonic), and may become almost anuric when dehydrated. During dehydration, electrolyte wastes are retained in the body and tolerated in concentrations fatal to birds and mammals, until water is available for their excretion. A carnivorous diet (70±% water) provides adequate water intake while food is available. Water can be reabsorbed osmotically from the cloaca, reabsorption being particularly effective because of the nature of the principal nitrogenous waste, uric acid, which has a very low solubility. As uric acid precipitates in the cloaca, its osmotic effect is removed, and further water can then be absorbed by osmosis. This is probably the major value of uric acid excretion. Precipitated wastes are excreted en masse, with very little fluid loss.

Birds, being homeothermic, cannot reduce their evaporative loss by becoming dormant. Being diurnally active and exposed to radiant energy, they must often expend water for cooling, by panting. Consequently, in arid regions the distribution of birds is limited to areas within flying distance of water. Some water expenditure is avoided by allowing hyperthermia (up to 3°C) in the daytime.

Desert rodents lead the most water-independent life of all vertebrates. Kangaroo rats can so reduce their evaporation that they are able to maintain water balance on only metabolic water. Other species survive on only metabolic water plus free water in air-dry seeds. Respiratory water loss is reduced by cool nasal mucosal surfaces, which condense water from warm air coming from the lungs, before it can be expired. Skin impermeability involves a physical vapor barrier in the epidermis, plus unknown physiological factors.

Many larger mammals are exposed to daytime radiant energy and need to dissipate heat by sweating, panting or wetting themselves with saliva (marsupials). These water expenditures must be balanced periodically by drinking. A dehydrated camel is particularly physiologically adapted to store heat (rather than dissipate it by evaporation), undergoing a temperature rise of up to 6° in the daytime.

Water in the Human Body

The adult male human body contains about 60% (weight) of water and the adult female body about 50%. The large amount of body water is compartmentalized, each compartment being bounded by membranes. It has been estimated (Edelman and Leibman, 1959) that 55% of this water is contained within cells and that it is bounded by cell membranes. The remaining, extracellular water or fluids (ECF) is made up of a relatively small volume of plasma (7.5% of total body water in the vascular tree), with the remaining 37.5% in nonplasma and located outside the vascular

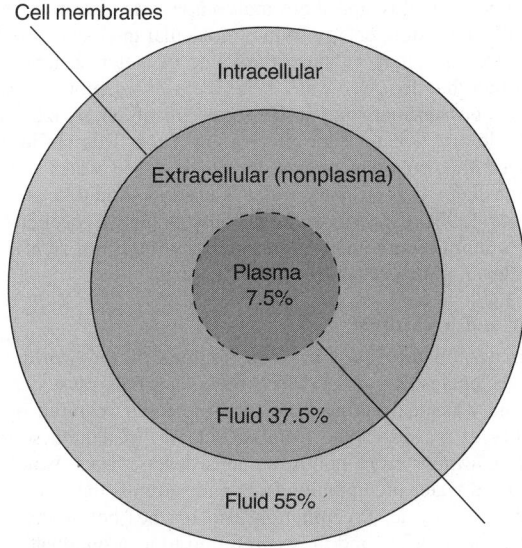

Fig. 4.   The three principal categories of body water

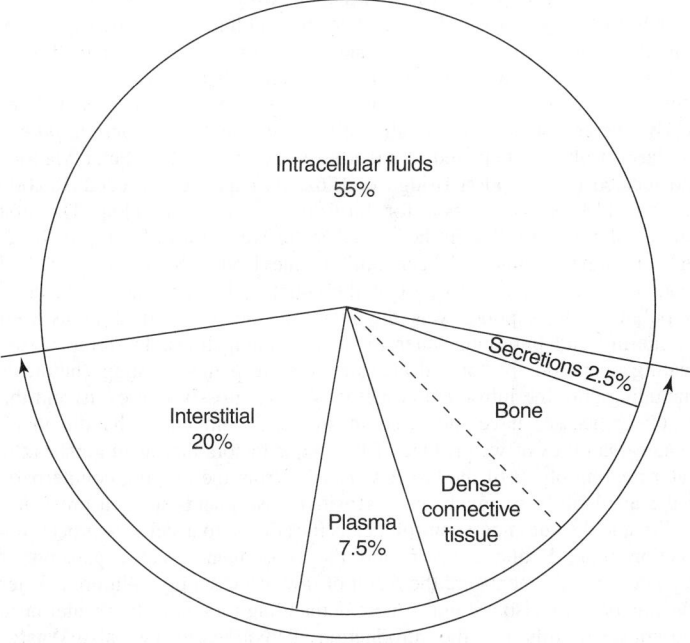

Fig. 5.   Pie chart showing approximate volumetric proportions of body fluids

tree. The latter includes interstitial water (20%), another 15% in bone and dense connective tissue, and 2.5% in secretions. These numbers are shown graphically in Figs. 4 and 5.

Two driving forces control the movement of water in the body, namely, hydrostatic pressure and osmotic pressure. Because transmembrane pressures are so low, it is not believed that hydrostatic pressure plays a role in the movement of water across cell membranes. On the other hand, hydrostatic pressure resulting from heart action creates a gradient of about 20 millimeters of mercury pressure across the capillary walls. The principles of osmosis are described in the entry on **Osmotic Pressure**. Normally, in describing osmotic pressure, reference is made to salt solutions of differing concentrations separated by a membrane. Concentrations are expressed in terms of solute in solvent (water). When thinking in terms of body water, one usually considers the addition of solute to the water as a dilution of the pure water rather than as an increase in solute content of the water. For example, pure water contains 55.5 moles per kilogram (about 55,500 mmoles/liter). Body fluids, that is, intracellular fluid (ICF) and extracellular fluid (ECF) contain approximately 99.5% water molecules and 0.5% solute molecules. This equals about 55,200 mmoles of water per liter; and 300 mmoles of solute per liter. Osmolality

is a measure of the concentration of osmoles present in the water. If the osmolality is low, the concentration of water is high; if the osmolality is high, the concentration of water is low. Movement of water across the cell membranes occurs because of differences in concentration, the movement always being from a phase of high water concentration to one of lower concentration. Movement is "downhill" so to speak and the motivating force is called osmotic pressure. This movement can be counteracted by the application of an opposing force, notably hydrostatic pressure.

It is interesting to note that the main cation present in ICF is the potassium ion, whereas the principal cation in ECF is the sodium ion. The role of potassium and sodium ions in the biological system is described in the entry on **Potassium and Sodium (In Biological Systems)**.

Water Intoxication. Either an increased intake of water or a decreased output of water can cause an excess of body water. Because healthy kidneys have outstanding ability to increase water excretion, a condition of water intoxication usually occurs because of a disability to excrete water (*hyponatremia*). Frequently impairment may not be the result of disease or damage to the kidneys per se, but rather due to faulty processing of stimuli by the kidneys. Excessive renal reabsorption of water can result from the action of antidiuretic hormone (ADH) or by the excessive reabsorption of sodium in the proximal tubules. Under such conditions, the excretion of water and sodium will be low. Retention of salt and water causes an expansion of the ECF, usually resulting in edema and effusions.

Water Deficiency. This condition occurs when water output exceeds intake. Water is continually lost by way of the lungs, skin, and kidneys and thus a deficiency of body water will occur if a critical minimal supply is not maintained. Decreased intake when water is available is uncommon. Very rarely, a brain malfunction may interfere with one's sense of thirst. Increased output of water can result from many causes. For example, a person with diabetes insipidus who lacks ADH (antidiuretic hormone) or a person whose kidneys do not respond normally to ADH, as in instances of nephrogenic diabetes insipidus, will increase water output. Other diseases which may cause excess excretion of water include osmotic diuresis, hypercalcemia, hypokalemia, chronic pyelonephritis, and sickle cell anemia, among others. Excessive water losses are also experienced in some cases with advanced age and in some burn cases. Two clinical features are good measures of dehydration—weight loss of the patient and an elevation of the serum sodium concentration. In situations of dehydration, the body initiates mechanisms which manipulate the transfer of water from one compartment to the next, retaining water in those cells and organs where it is most needed.

In cases of severe dehydration, rehydration must be brought about carefully and in steps. The usual practice is to make an initial estimate of water or electrolytes that require replacement and then to administer one-half the amount of the deficit, after which another series of measurements is made, followed by replacement of one-half of the estimate—until a satisfactory ultimate balance has been attained. If rehydration is not gradual, some organs, such as the brain, may take up water beyond normal requirements and this can result in cerebral edema. Also, in the case of acute dehydration, permanent brain damage can occur as the result of a shrinking brain tearing away vessels, causing cerebral hemorrhages.

Water and Macromolecules

Over eons of time, processes have appeared for making the great variety of macromolecules required by living organisms. Most of these developments have occurred in a water medium, or in one having a high water vapor content. So it is not surprising that the majority of the macromolecules involved in the life processes are hydrophilic, in different degrees. One may find minor exceptions in the cases of certain fats and lipids. It should be noted that in living organisms, the hydrophilic macromolecules of one kind often join with those of other kinds to produce the useful structures required in the life process, i.e., membranes.

Decades ago, biochemists recognized that it was difficult to remove all water from a large number of macromolecular materials. The term "bound water" was coined to explain the great affinity many of these materials showed for water, particularly the proteins. Biochemists were convinced that biological behavior, at least in part, resulted from the amount of "bound water" contained in the macromolecular structure. As an example, water held in plant structures so that it did not freeze in below-freezing temperatures was considered to be bound. At the time these ideas found favor, the method of lyophilization or quick freeze-drying had not been perfected. Lyophilization removes the bulk of the water held in biological substances without destroying their structures or their activity.

In more recent years, lyophilized proteins have been further dried to a constant weight in a high vacuum and then studied as they adsorbed water vapor. The heats of adsorption of the first water vapor molecules were considerably higher than the values obtained as the adsorption approached that of the saturated vapor. These results indicated that the first molecules to be adsorbed were on the most water-loving or active sites. Such adsorbed molecules on the higher-energy sites would be desorbed last in a high vacuum. Without going into considerable detail, the results of this line of research led to the conclusion that the earlier concept of "bound water" was unfounded.

Many biological systems depend in part on their degree of hydration. Most biological membranes are hydrophilic, but should the membrane be a multi-layered one, the hydration of the different layers may well differ markedly. A great many of the membranes used by living organisms are known to be selective in what passes through them. For many years the theory of the role of some membranes in pumping water into the cells they surround even against high osmotic gradients due to salt concentration has maintained among most biologists. The fascinating field of membrane biophysics has shed much light concerning the hydration of biological macromolecules.

As pointed out by Kolata, a primary difference between living and dead cells is that living cells selectively retain certain ions, such as potassium, and exclude others, such as sodium. The water in dead cells reflects the conditions of the solution around them. The conventional explanation is that this difference is due to ion "pumps" in membranes, pumps purported to use cell energy to transport some ions into and other ions out of the cell. There is another school of thought, however, which denies that such pumps exist. This school claims that ions are excluded from cells on the basis of their low solubilities in cellular water, except when specific charged sites with which the ions can associate are available. It is maintained that cell water has a different structure than either liquid water or ice and that it is this special structure that affects the solubility of various ions in it. This school also suggests that there is evidence that ion pumps are thermodynamically impossible, requiring more energy than is available to the cell. The school also claims that nuclear magnetic resonance (NMR) studies show that cell water is more structured than liquid water and less structured than ice. This, it is believed, would affect the solubilities of ions in the cell and could account for selective ion exclusions. A number of investigators, including advocates of pumps, have agreed that cell water may have some ordered structure that makes it different from liquid water. Most investigators do not question that cell water is likely to be structured, but do ask to what extent it is structured and what the physiological importance of this structure may be.

Degassed Water. It is interesting to note that during the past several years, Russian scientists have raised some intriguing points concerning water, not all of which have held up under intensive scrutiny. Some readers may recall the discussions of the early 1970s concerning the "discovery" of polywater or so-called anomalous water by some Russian scientists. Observations of this water did not meet with accepted criteria for physical properties. For a while, it was considered by some scientists to be of a polymeric nature, but as the result of subsequent numerous exchanges of views and a careful scrutiny of the water it was found to be ordinary water which had a large concentration of dissolved minerals.

In 1978, other Russian scientists proposed that meltwater (from freshly melted snows) carries certain biological properties not known in ordinary water. The Russian scientists theorized that meltwater retains some of the order that is characteristic of frozen water and that this increased order alters vital reaction rates within cells. This tends to tie in with the concept of structured water discussed briefly later.

In the course of investigation at the Institute of Fruit-Growing and Vine-Growing (Kazakhstan, Russia), investigators tested the relative growing rates, qualities, and yields of various plants subjected to meltwater, tap water, and boiled water. Although the plants responded in a superior way to the meltwater, as compared with tap water, it was found (more or less accidentally) that the plants responded even better to quickly cooled boiled water. Thus, the experimenters concluded that the superior action on plants derived from the fact that some of the waters tested (meltwater and quickly-cooled boiled water) contained less dissolved gases. In other words, the claim was made that any degassed water is superior when administered to plants. Igor Zelepukhin suggests that the conductivity of degassed water

is decreased considerably and there are comparative increases in density, viscosity, surface tension, energy of intermolecular interaction, and internal pressure, factors that may enhance the value of water. Zelephukhin thus observed further that degassed water bears a closer resemblance to the fluid in cells than does ordinary tap water. Experiments in utilizing degassed water for cattle and livestock are proceeding. Other Russian investigators have observed that concrete prepared with degassed water is from 8 to 10% stronger than when ordinary water is used. It should be stressed that, as of the early 1990s, these observations have not as yet been accepted by the general scientific community. Considerably more research is required toward the development of hard, convincing data.

Raw Water and Treatment

Aside from desalinators used in some regions of the world that have severe insufficiencies of rain and freshwater and thus must depend upon purified saline waters, drinking water and the water required by industry for a plurality of reasons come from two classes of natural sources. The first is *surface waters* from ponds, streams, rivers, lakes, waterfalls, and glaciers. Natural water precipitation (rain and snow) is the result of Earth's natural hydrologic cycle. Precipitation also reaches below the surface to collect or flow in aquifers and thus is referred to as groundwater. See also **Desalination**.

Prior to the great increases in world population and the development of extensive industrialization and modern agriculture, nature provided a number of built-in processes by which flowing streams essentially can be self-purifying, provided, of course, that pollution of any significant magnitude or intensity did not reach the natural water source. Today, with the exception of very few locales, such pristine conditions do not occur, and, in fact, some form of raw water treatment has existed for nearly five centuries. Traditionally called *waterworks*, or *pumping plants*, the earliest plant is believed to have been one designed by Peter Maurice and located near London Bridge in 1562. Its capacity exceeded 300,000 gallons (1136 cubic meters) for furnishing water to London. The first municipal plant installed in the United States was built in Boston in 1652. Only 15 more plants had been built in the United States by 1800, all located in the northeastern part of the country. The presence of bacteria in public water supplies was the major incentive for treating as well as filtering and pumping water to large municipalities. In terms of the treating operations performed at water treating plants, treating chemicals (including chlorine introduction methods) have grossly improved, and the various processes have increased in size and numbers to handle much greater quantities of water. One of the major factors that distinguishes the water treating plants of the last few decades from their earlier counterparts is the availability of much more sensitive instruments to measure water quality and the automatic control means introduced to accelerate processing reaction time. Public concern and the consequent great expansion of regulatory requirements are the result of raw water supply pollution. Water consumers today also are much more demanding for controlling water taste and odor. Controls over the introduction of toxic wastes (see also **Wastes and Pollution**) have been mandated.

In an average situation, water will be pumped from a natural source, passed through bar screens to remove any large pieces of debris, and moved on to a raw water storage tank for temporary holding. Treating boiler feedwater requires customized methods. See also **Water Treatment (Boiler)**. Raw water treatment plants are found in nearly all municipalities throughout the U.S., Canada, and other advanced countries. Traditionally, these plants have operated under strict regulations and guidelines. Additionally, some industrial firms will re-treat municipal water for a variety of special reasons. Ultrapure water, for example, is required by the semiconductor industry, and means used to obtain ultrapurity include reverse osmosis processes, decarbonators, distillation, deionization, ultraviolet sterilization, and ultrafiltration.

Principal Impurities

Substances that must be removed in order to meet required standards for municipal water fall into three general categories:

1. Suspended matter, color, and organic matter—sediment, turbidity, microorganisms, taste, odor, and other organic matter.
2. Dissolved mineral matter—bicarbonates, sulfate, chlorides of calcium, magnesium, and sodium. Small amounts of silica and alumina commonly are present. Other constituents frequently present

are iron, manganese, fluorides, nitrates, potassium, and sulfuric acid. (The presence of mine drainage components [common in some regions] and components of trade wastes are less frequently encountered today because of severe pollution restrictions. Severely toxic wastes also are controlled, with current regulations in most advanced industrialized nations mandating pretreatment to eliminate or neutralize all such substances prior to disposing of them in public reservoirs. However, municipal water treatment plant managers require the analysis of incoming raw water for the presence of toxic substances.)

3. Dissolved gases—Usually, these are present as oxygen, nitrogen, carbon dioxide, and, less frequently, hydrogen sulfide and methane.

Major Water Treating Operations

The principal operations required to remove and alleviate the undesirable components of raw water include the following.

Sedimentation

With waters containing large amounts of coarse, easily settled, suspended matter, sedimentation (plain sedimentation) is often of value in reducing the load on the filters and effecting economies in amounts of chemicals used for coagulation. Sedimentation may be carried out in sedimentation tanks or basins or in reservoirs. Detention periods vary over a wide range—from a few hours up to one or more months.

Coagulation

Coagulation is employed to form, by cataphoresis and entanglement, larger aggregates with the turbidity, color, microorganisms, and other organic matter present in the water. These larger particles, known as the "floc," may then be removed by filtration through a sand or Anthrafilt filter or by settling and filtration. The coagulant employed is either an aluminum or iron salt, usually the surface. Aluminum sulfate is the most widely used coagulant. Others are ferric sulfate, ferrous sulfate (must be oxidized by air or chlorine) and sodium aluminate. Most favorable pH values for aluminum coagulants usually range from 5.5 to 6.8 and for iron from 3.5 to 5.5 and above 9.0 but there are exceptions. Coagulation aids are ground clay (not too finely pulverized) and activated silica.

Filtration

Filtration is effected by flowing the coagulated or coagulated and settled water downward through a bed of fine filter sand or Anthrafilt is either a pressure type or gravity type filter. Flow rates in industrial practice range up to 3 gpm per sq. ft. (122 liters/min/sq. meter) of filter bed area while in municipal practice maximum flow rate is usually 2 gpm per sq. ft. (81.5 liters/min/sq. meter).

Chlorination

Chlorination is the most widely used disinfecting or sterilizing process. Where daily water requirements are not large, it is common practice to use a hypochlorite, but for large plants liquefied chlorine gas is used. Chlorination may be practiced before filtration (prechlorination), after filtration (postchlorination), or both before and after.

Taste and Odor Removal

Except for sulfur waters, most tastes and odors are organic in nature. Activated carbon is widely used for their removal. In powdered form, it may be added to the water being treated in coagulation and settling equipment. In such installations, aeration is frequently used as preliminary treatment. In granular form, it is used in filters (activated carbon filters or purifiers). As substances producing tastes and odors are usually extremely small in amount, activated carbon filters are frequently operated for 6 months to one or more years before replacement of bed is necessary.

Optional Treating Operations

These may include improving water quality for domestic and industrial purposes.

Hardness Removal (Water Softening)

Sodium Cation Exchanger (Zeolite) Process. This is the most widely used water-softening process in industrial, commercial, institutional and household applications. Hard water is softened by flowing it, usually downward, through a bed (2 feet to over 8 feet in thickness) of a granular or bead type sodium cation exchanger in either a pressure-type (most widely used) or gravity-type water softener. As water comes in contact with sodium cation exchanger, hardness (calcium and magnesium ions) is taken up and held by the exchanger which gives up to the water an equivalent amount of sodium ions. At end of softening run (4 to over 24 hours in industrial practice and 1 to over 2 weeks in household use), softener is cut out of service, regenerated and returned to service (1/2 to 11/2 hours). Regeneration is effected in 3 steps: (1) backwashing to cleanse and hydraulically regrade the bed (2) salting with specified amount of common salt (sodium chloride) solution, usually 10 to 15% in strength, which removes calcium and magnesium from the exchanger and restores sodium to it and (3) rinsing to remove calcium and magnesium chlorides and excess salt.

Hydrogen Cation Exchanger Process. Calcium, magnesium, sodium and other cations are removed by flowing water (usually downward) through a bed (2 feet to over 8 feet in thickness) in an acid-proof pressure-type (most widely used) or gravity-type shell. As water comes in contact with hydrogen cation exchanger, calcium, magnesium, sodium and other cations are taken up by the exchanger which gives up to the water an equivalent amount of hydrogen ions. At the end of operating run (4 to over 24 hours), unit is cut out of service, regenerated and returned to service (11/4 to 2 hours). Regeneration is effected in three steps: (1) backwashing to cleanse and hydraulically regrade the bed; (2) acid treatment with sulfuric or hydrochloric acid which removes metallic cations from the bed and restores hydrogen to it; and (3) rinsing to remove salts (sulfates or chlorides) and excess acid. The carbon dioxide formed from the bicarbonates may be reduced to below 5 to 10 ppm by aeration. The sulfuric and hydrochloric acids formed from chlorides and sulfates may be (1) neutralized with an alkali (usually caustic soda), (2) neutralized by sodium bicarbonate content of a sodium cation exchanger softened water (in which case aeration follows neutralization), or (3) removed by an anion exchanger.

Cold Lime (or Lime Soda) Process. Chemicals used may be (1) lime plus a coagulant or (2) lime plus soda ash plus a coagulant. Dosages vary according to composition of raw water and result desired such as (a) calcium alkalinity reduction, (b) calcium and magnesium alkalinity reduction, (c) reduction of total hardness without excess chemicals and (d) excess chemical treatment. Precipitates produced are calcium carbonate and magnesium hydroxide. Rated residuals without excess chemicals are 35 ppm for calcium and 33 ppm for magnesium, both expressed as calcium carbonate. Operating results will range between these and theoretical solubilities. With excess chemicals, total hardness may be lowered to 16 ppm. The process is best carried out in the sludge blanket-type of equipment in which the treated water is filtered upward through a suspended blanket of previously formed sludge. Detention periods range from one to two hours. Usually, treated water is filtered before going to service but where small amounts of turbidity are unobjectionable, filters may be omitted.

Hot Lime Soda Process. In this process, treatment with lime and soda ash is carried out at temperatures around the boiling point in closed, steel pressure tanks. Heating is usually accomplished with exhaust steam and pressures most widely used range from 5 to 10 psig, but higher pressures up to but seldom above 20 psig are also used. At these temperatures, the reactions proceed swiftly and precipitates formed are larger than those in cold lime soda process so no coagulant is needed. Detention period is one hour and de-aeration effected in primary heater is sufficient to lower dissolved oxygen content to 0.3 milliliter per liter which is sufficient for low-pressure boilers. For high-pressure boilers, either an integral or separate de-aerator is used and this will bring the dissolved oxygen down to less than 0.005 ml/l. With 20 to 30 ppm excess soda ash, the hardness will be reduced to 25 ppm. Softening to practically zero hardness may be effected by either (1) two-stage hot lime soda phosphate treatment in which effluent from hot lime soda softener is treated with sodium phosphate, or (2) the filtered effluent is passed through a sodium cation exchanger. Anthrafilt filters are usually employed with hot process softeners.

Fluoridation

See entry on **Fluorine.**

Demineralization (Deionization)

Metallic cations are removed by a hydrogen cation exchanger. Anions are removed by an anion exchanger. Depending on the hookup used, the carbon dioxide formed from the bicarbonates may be removed mechanically by an aerator, degasifier or vacuum de-aerator, or chemically by a strongly basic anion exchanger. Strongly basic anion exchangers will remove both strongly ionized acids, such as sulfuric and hydrochloric, and weakly ionized acids, such as silicic and carbonic. Weakly basic anion exchangers will remove only strongly ionized acids.

Iron and Manganese Removal

In clear, deep ground waters, iron and/or manganese may occur as soluble, colorless, divalent bicarbonates. These may be removed (1) by oxidation plus settling (if necessary) plus filtration (2) by cation exchange with sodium or hydrogen cation exchangers or (3) filtration through an oxidizing (manganese zeolite) filter. In (1) addition of an alkali or lime may be needed to build up the pH value so as to speed up the oxidation. Iron and/or manganese in organic (chelated) form may usually be removed by coagulation, settling and filtration. In acid waters, these metals may be removed by neutralization (plus increase of pH), aeration, settling and filtration.

Fluoride Removal

Fluorides may be reduced to below 1 ppm by filtration through a bed of a specially prepared, granular bone char (bone black). Regeneration is effected with caustic soda solution followed by treatment with dilute phosphoric acid.

Dissolved Gases

Oxygen and Nitrogen may be removed (1) hot in a deaerating heater (de-aerator) or (2) cold in a vacuum de-aerator. *Carbon dioxide* may be removed in (1) an aerator, (2) a de-aerating heater or (3) a vacuum de-aerator or it may be neutralized with lime or an alkali or by filtration through a bed of granular calcite. *Hydrogen sulfide* may be removed by (1) aeration followed by chlorination, (2) treatment with flue gas plus aeration followed by chlorination or (3) filtration through an oxidizing manganese zeolite filter (household use). If sulfur content and pH values are high, (1) may effect but little removal but, in some cases, with fairly long detention periods, sulfur bacteria may effect notable reductions.

Treatment of Wastewater (Sewage)

Wastewater may be defined as the spent or used water from a community or industry that contains dissolved or suspended matter. Toxic wastes must receive pretreatment by the polluter prior to introduction into a water reservoir or municipal used-water return lines. Most wastewater is 99.94% water by weight. The remaining 0.06% is material dissolved or suspended in the water. Water chemists differentiate *suspended solids* and *dissolved contaminants*. The concentration of dissolved or suspended matter usually is expressed as milligrams of pollutants per liter of water (mg/l), or as parts per million (ppm) (weight). On another scale, 1 ppm can be visualized as being equivalent to 1 minute of time in 1.9 years. Although pollutants may be so minute, innumerable studies have shown their adverse effects on human and other animal life.

A generally accepted estimate is that each individual, on a national (U.S.) average, contributes approximately 265 to 568 litres of water per day to a community's wastewater flow. While most people think of wastewater as only "sewage," wastewater also comes from other sources—commercial, industrial, and storm and ground water. Generally, each house or business has a pipe or sewer that carries the wastewater to the wastewater treatment plant. **Sanitary sewers** carry only domestic and industrial wastewater, while **combined sewers** carry wastewater and storm water runoff. Every reasonable effort is made to exclude storm (inflow) and ground (infiltration) water from the sanitary sewer system. These efforts are usually less successful on older sewer systems that leak.

The wastewater from the sewer system either flows by gravity or is pumped into the treatment plant. Usually, treatment consists of two major steps, primary and secondary, along with a process to dispose of solids removed during the two steps.

In **primary treatment**, the objective is to physically remove suspended solids from the wastewater, either by screening, settling, or floating. The major goal of **secondary treatment** is to biologically remove contaminants that are dissolved in wastewater. In secondary treatment, air is supplied to encourage the natural processes of growth of bacteria and other biological organisms to consume most of the waste. These organisms and other solids are then separated from the wastewater. Before discharge to the **receiving stream**, the water usually passes through a tank where a small amount of chemical (usually chlorine) is added to disinfect the treated water.

In primary and secondary treatment, solids are settled and removed for further processing. Solids, usually referred to as sludge, are normally processed in three steps—digestion, dewatering, and disposal. The digestion step reduces the volatile solids and prepares the sludge for further processing. Dewatering involves the application of a variety of processes that reduces the water content of the sludge and, in turn, its volume. The final step is the ultimate management of this treated material, or biosolids, which can be used beneficially through methods such as land application. See Figs. 6 and 7.

Screening removes large floating objects from the incoming wastewater stream. Treatment plant screens are sturdily built to withstand the flow of untreated wastewater for years at a time. Rags, wood, plastics, and other floating objects could clog pipes and disable treatment plant pumps if not removed at this point. Typically, screens are made of steel or iron bars set in parallel about one-half inch apart. Some treatment plants use a device known as a comminuter, which combines the functions of a screen with that of a grinder.

Sand, grit, and gravel flow through the screens to be picked up in the next stage of primary treatment—the grit chamber. Grit chambers are large tanks designed to slow the wastewater down just long enough for the grit to drop to the bottom. Grit is usually washed after its removal from the chamber and buried in a landfill.

After the flow passes out of the grit chamber, it enters a more sophisticated settling basin called a sedimentation tank. Sedimentation removes the solids that are too light to fall out in the grit chamber. Sedimentation tanks are designed to hold wastewater for several hours. During that time the suspended solids drift to the bottom of the tank, where they can be pushed into a large mass by mechanical scrapers and pumped out of the bottom of the tank. The solids removed at this point are called primary sludge. The primary sludge is usually pumped to a sludge digester for further treatment. During the sedimentation process, floatable substances, such as grease and oil, rise to the surface and are

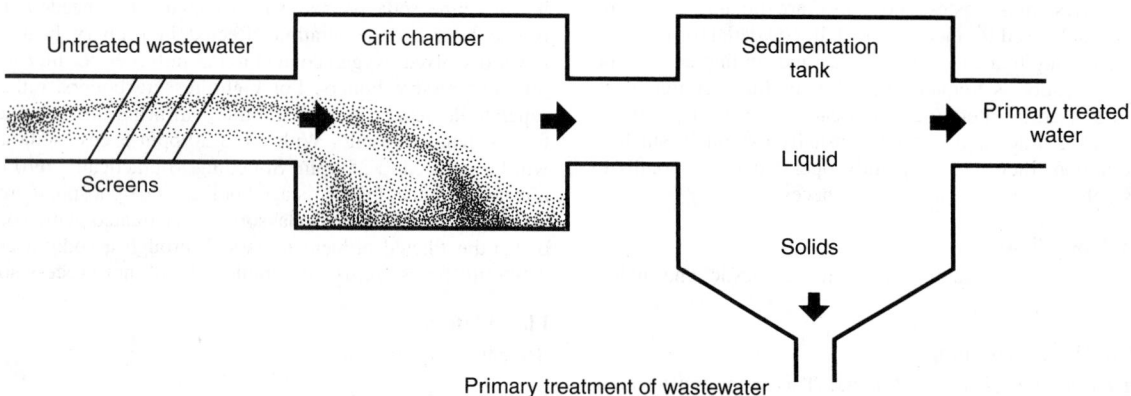

Fig. 6.   Wastewater entering a treatment plant receives primary treatment first. In this state, a series of operations removes most of the solids that can be screened out, will float, or will settle

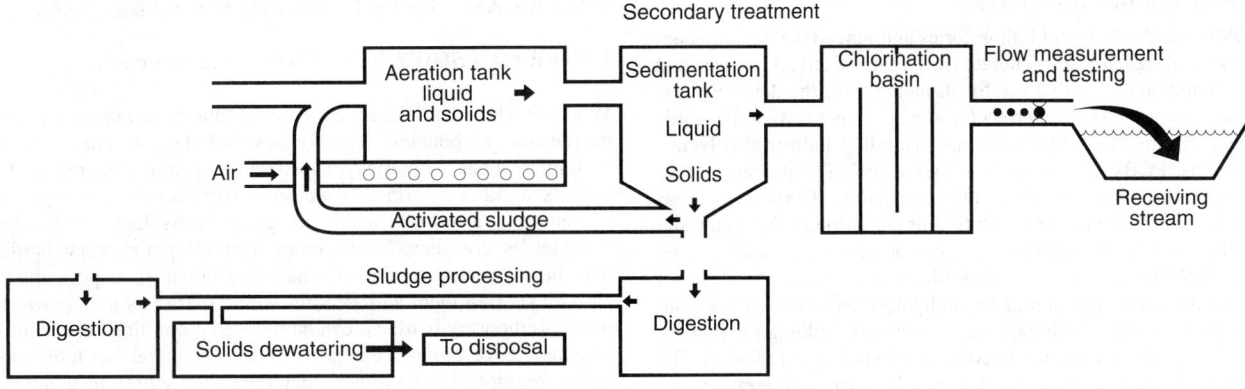

Fig. 7. Wastewater flowing out of primary treatment still contains some suspended solids and other solids that are dissolved in the water. In a natural stream, such substances are a source of food for protozoa, fungi, algae, and hundreds of varieties of bacteria. The secondary treatment stage is a highly controlled artificial environment in which these same microscopic organisms are allowed to work as fast and efficiently as they can. The microorganisms biologically convert the dissolved solids in the wastewater to suspended solids that will physically settle out at the end of secondary treatment

removed by a surface skimming system. The skimmed materials are either sent to the sludge digester for treatment along with the primary sludge or are incinerated. Sedimentation marks the end of primary treatment. At this point, most of the solids in the stream that can be removed by the purely physical processes of screening, skimming, and settling have been collected. An additional set of techniques using biological processes must be employed next. (*After Water Environment Federation.*)

There are several different ways to optimize biological conversion. Secondary treatment promotes the growth of millions of microorganisms, bringing them into close contact with the wastewater on which they feed. Care is taken to make sure that the temperature, oxygen level, and contact time support rapid and complete consumption of the dissolved wastes. The final products are carbon dioxide, water and more microorganisms. Three widely employed types of secondary treatment are common: activated sludge, trickling filters, and lagoons.

The most common is the activated sludge process. Activated sludge processes are much more tightly controlled than either trickling filters or lagoons. In this form of treatment, wastewater and microorganisms are mixed for a few hours in a large tank by constant aeration and agitation. Once the aeration is complete, the mixture of water and microorganisms flows to a sedimentation tank similar to the one used in primary treatment. The microorganisms and other solids settle to the bottom of the sedimentation tank. Since activated sludge is a continuous process, a portion of the settled solids (return activated sludge) are circulated back to the beginning of the process to serve as "seed" organisms. The part not needed for "seed" is commonly called waste activated sludge and is sent to a sludge digester for further treatment. (*After Water Environment Federation.*)

Sludge Handling

Treatment professionals refer to solids in general as **sludge**. Beneficial sludges are called "biosolids." But these solutions are not the thick, molasses-like substances that most people think of when they hear the word "sludge." Wastewater sludges are slurries of water and solids that are roughly 100 times more concentrated than untreated wastewater. That is, they contain about 3% solids compared to the .03% (or less) concentration of the initial flow into a treatment plant. The various techniques for handling these flows are designed to increase the solids concentration even further, to as much as 50%.

As a rule of thumb, higher degrees of wastewater treatment produce larger volumes of sludge. For example, primary treatment usually produces 2500 to 3500 gallons of sludge for every 1 million gallons of wastewater. Secondary treatment usually produces 15,000 to 25,000 gallons for every 1 million gallons treated. To try to dispose of or recycle such volumes of waste is practical. Thus, sludge handling methods are designed to remove as much water from the mixture as possible.

The spectrum of sludge handling techniques is divided into processes that condition, thicken, stabilize, and dewater the sludge flow. Conditioning operations usually employ chemicals or heat to make the sludge release water more easily. Thickening techniques use gravity, flotation, and chemicals to separate water from the solids. Conditioning and thickening are usually the first steps in handling primary and secondary sludges.

Stabilization converts the organic matter in the sludge so that the biosolids can be disposed of or used as a soil conditioner without posing a health hazard in the general environment. Sludge stabilization can occur with (**aerobic**) or without (*anaerobic*) oxygen in special tanks called digesters. Sometimes chemicals such as lime are used for stabilization.

Dewatering is done by mechanical means. Filters, centrifuges, and presses remove even more water from the biosolids. Biosolids dewatered by such equipment have the consistency of wet mud and can have a solids concentration of up to 20%. Other techniques, such as drying beds or special presses, can be used to dewater the sludge, producing up to 50% solids—about the consistency of dry soil. At the end of a sludge handling process, the concentrated solids can be placed in landfills, incinerated, applied to land, or composted for use as a soil conditioner.

Alternatives to Sludge Process

For some wastewater treating situations, the complexities of the sludge process may not be required. In such instances, trickling filters or lagoons may be used.

Trickling filters

are large beds of coarse, loosely packed material—rocks, wooden slats, or shaped plastic pieces—over which the wastewater is sprayed or spread. The surfaces of the filter material (also known as the "medium") become breeding grounds for the microorganisms that consume the wastes. A common trickling filter is a bed of stones 3 to 10 feet deep. Under the bed, a system of drains collects the treated wastewater and diverts it to a sedimentation tank or back over the filter medium for additional treatment. In the sedimentation tank, suspended solids settle and are pumped to a sludge digester. Trickling filters are relatively simple to construct and operate. Many communities in the United States rely on them for secondary treatment.

Lagoons are used by some communities to achieve secondary treatment. Lagoons generally treat the total wastewater from a community until the biological oxidation processes have consumed and converted most of the wastes present. This form of treatment depends heavily on the interaction of sunlight, algae, and oxygen. Sometimes the wastewater is aerated to speed the process, since these interactions are relatively slow. There is usually no sedimentation tank associated with a lagoon. Suspended solids settle to the bottom of the lagoon, where they remain or are removed every few years. Generally speaking, lagoons are simpler to operate than other forms of secondary treatment, but are less efficient.

At the end of a secondary treatment process, the wastewater is disinfected to remove disease-causing organisms. Usually, an agent such as chlorine is added to the stream of wastewater before it is discharged to receiving waters. Sometimes other techniques are used if the receiving waters are sensitive to the addition of chlorine.

An alternate to secondary and higher levels of treatment is land application of wastewater. The wastewater is usually sprayed over natural or specially sloped and seeded land. The wastewater seeps into the soil where natural solid microorganisms consume the wastes. The treated water is either used by plants, stays in the ground, or is collected and routed to a receiving stream.

Role of Sunlight as a Detoxifying Agent

In the late 1980s, Sandia National Laboratories announced the development of a solar-powered reactor to generate low-cost electrical power and recently have found an alternative use for it, namely, for the detoxification of polluted water. As stated by the project leader, C. Tyner, "We believe this process will destroy most organic materials, including industrial solvents, pesticides, dioxins, PCBs, and munitions chemicals." The process breaks down toxic chemicals into smaller, safer molecules. Current methods remove organic wastes from water by bubbling air through the water and thus volatilizing them for release into the atmosphere (not attractive) or by running the polluted water through carbon filters. The researchers have set up a troughlike arrangement (similar to sunlight collectors used for solar furnaces) along which runs a radiation-transparent tube holding the flowing water, which thus receives a maximum concentration of solar radiation. The future of the project will be determined largely by cost considerations.

See also **Wastes and Pollution**; **Water Pollution**; and **Water Resources**.

Additional Reading

Amato, I.: "A New Blueprint for Water's Architecture," *Science*, **1764** (June 26, 1992).

Beardsley, T.: "Mr. Clean: Sunlight Can Destroy Dangerous Chemicals," *Sci. Amer.*, **83** (June 1989).

Chew, R.M.: "Water Metabolism in Mammals," in *Physiological Mammalogy* (W.W. Mayer and R.G. Van Gelder, Eds.) Academic Press, Orlando, Florida, 1963.

Corbitt, R.A.: *Standard Handbook of Environmental Engineering*, 2nd Edition, The McGraw-Hill Companies, Inc., New York, NY, 1998.

Crompton, T.R.: *Determination of Metals in Natural and Treated Water*, Taylor & Francis, Inc., Philadelphia, PA, 2001.

Dale, J.E.: *Plants and Water*, Cambridge University Press, New York, NY, 2001.

Hauser, B.: *Fundamentals of Drinking Water*, Lewis Publishers, Boca Raton, FL, 2001.

Herschy, R.W. and R.W. Fairbridge: *Encyclopedia of Hydrology and Water Resources*, Chapman & Hall, New York, NY, 1998.

Horan, N.: *Biological Waste Water Treatment Systems*, 2nd Edition, John Wiley & Sons, Inc., New York, NY, 2001.

Kolata, G.B.: "Water Structure and Ion Binding: A Role in Cell Physiology?" *Science*, **192**, 1220–1222 (1976).

Martin, A.M.: "Use Biomonitoring Data to Reduce Effluent Toxicity," *Chem. Eng. Progress*, **43** (September 1992).

McLaughlin, L.A., H.S. McLaughlin, and K.A. Groff: "Develop an Effective Wastewater Treatment Strategy," *Chem. Eng. Progress*, **34** (September 1992).

Morra, M.: *Water Biotechnological Surface Science*, John Wiley & Sons, Inc., New York, NY, 2001.

Nollet, L.M.L.: *Handbook of Water Analysis*, Marcel Dekker, Inc., New York, NY, 2000.

Okoniewski, B.A.: "Remove VOCs from Wastewater by Air Stripping," *Chem. Eng. Progress*, **89** (December 1992).

Pankow, J.F.: *Aquatic Chemistry Concepts*, Lewis Publishers, Inc., Boca Raton, FL, 1991.

Patra, K.C.: *Hydrology and Water Resources Engineering*, CRC Press, LLC., Boca Raton, FL, 2000.

Schultz, G.A., E.T. Engman: *Remote Sensing in Hydrology and Water Management*, Springer-Verlag, Inc., New York, NY, 2000.

Staff: "Sun-Powered Chemical Reactor," *Chem. Eng. Progress*, **12** (September 1990).

Staff: "Structural Ice," *Advanced Materials & Processes*, **4** (December 1991).

Staff: Water Amer, American Waterworks Association, *The Drinking Water Dictionary*, McGraw-Hill Professional Book Group, New York, NY, 2001.

van der Leeden, F., F.L. Troise, and D.K. Todd: *The Water Encyclopedia*, 2nd Edition, Lewis Publishers, Inc., Boca Raton, FL, 1990.

WEF: *About Wastewater Treatment*, Water Environment Federation, Alexandria, VA, 1993.

WEF: *Clean Water for Today: What is Wastewater Treatment?* Water Environment Federation, Alexandria, VA, 1994.

Weinberg, C.J., R.H. Williams: "Energy from the Sun (Wind and Biomass)," *Sci. Amer.*, **147** (September 1990).

Web References

EPA Water: http://www.epa.gov/water/
Water Resources of the United States: http://water.usgs.gov/
Water Science for Schools (USGS): All about water: http://ga.water.usgs.gov/edu/

WATER (Desalination). See **Desalination**.

WATER (Electrolysis). See **Hydrogen (Fuel)**.

WATER GAS. See **Coal**; **Substitute Natural Gas (SNG)**.

WATER GAS SHIFT REACTION. See **Ammonia**.

WATER (Hard). Water containing low percentages of calcium and magnesium carbonates, bicarbonates, sulfates, or chlorides as a result of long contact with rocky substrates and soils. Degree of hardness is expressed either as grins per gallon or parts per million (ppm) of calcium carbonate (1 grain of $CaCO_3$ per gal is equivalent to 17.1 ppm). Up to 5 grains is considered soft, more than 30 grains very hard. Hardness may be temporary (carbonates and bicarbonates) or permanent (sulfates, chlorides). Treatment with zeolites is necessary to soften permanently hard water. Temporary hardness can be reduced by boiling. These impurities are responsible for boiler scale and corrosion of metals on long contact. Hard waters require use of synthetic detergents for satisfactory sudsing.

See also **Water Treatment (Boiler)**; and **Zeolite Group**.

WATER POLLUTION. Means for treating polluted waters are described in a prior entry on **Water**. This article is devoted to the current state of water pollution in the United States and some European countries, sources of pollution, and ways and means for preventing water pollution. This article also relates directly to articles on **Wastes and Pollution**; and a later article on **Water Resources**.

Water Pollution in the United States[1]

For an abridged analysis of the water pollution problem in the United States, a concerted look is taken at (1) the rivers and (2) groundwater.

Water Quality in the Rivers. River water serves as an excellent overall index of the pollution problem, because ultimately, most water, even consumed groundwater and a vast majority of the lakes, ponds, etc., in the long run ends up in a river (or series of rivers), and thence flows to the oceans. The majority of rivers are accessible with relative ease and thus add to the convenience of obtaining water samples for analysis.

Probably the most complete study of river water quality was completed by the U.S. Geological Survey, released in early 1987 and periodically updated. The initial survey was coordinated by Smith and Alexander (USGS) and Wolman (The Johns Hopkins University), including water quality records from two nationwide sampling networks. The network included over 300 locations on the major rivers of the United States. Twenty-four water quality parameters are measured. Originally, the two networks were comprised of: (1) the National Stream Quality Accounting Network (NASQUAN) and (2) the National Water Quality Surveillance System (NWQSS). Locations of stations are shown on the map in Fig. 1. The measured water-quality indicators include:

pH (hydrogen ion concentration)	Trace Elements
Alkalinity ($CaCO_3$)	Arsenic
Sulfate (SO_4)	Cadmium
Nitrate (total as N) (TN)	Chromium
Phosphorus (total as P) (TP)	Lead
Calcium	Iron
Magnesium	Manganese
Sodium	Mercury
Potassium	Selenium
Chloride	Zinc
Suspended sediment (SS)	
Fecal coliform bacteria	
Fecal Streptococcal bacteria	
Dissolved oxygen	
Dissolved-oxygen deficit (DOD)	

(1) The National Stream Quality Accounting Network (NASQAN), indicated by solid black dots in map; and (2) the National Water Quality Surveillance System (NWQSS), indicated by open black circles on map. Shown in outline are regional drainage basins. Abbreviations used for these basins are:

---

[1] It is interesting to note how interrelated the topics of water, air, and solids (soil) pollution are. Acid rain, for example, commences as an air pollutant and ends up as a soil and water pollutant. See also **Pollution (Air)**. Thus, water pollution may be direct or indirect. Because they have a mass, air pollutants ultimately fall to Earth's surface and thus pollute the oceans, bodies of freshwater, and the land.

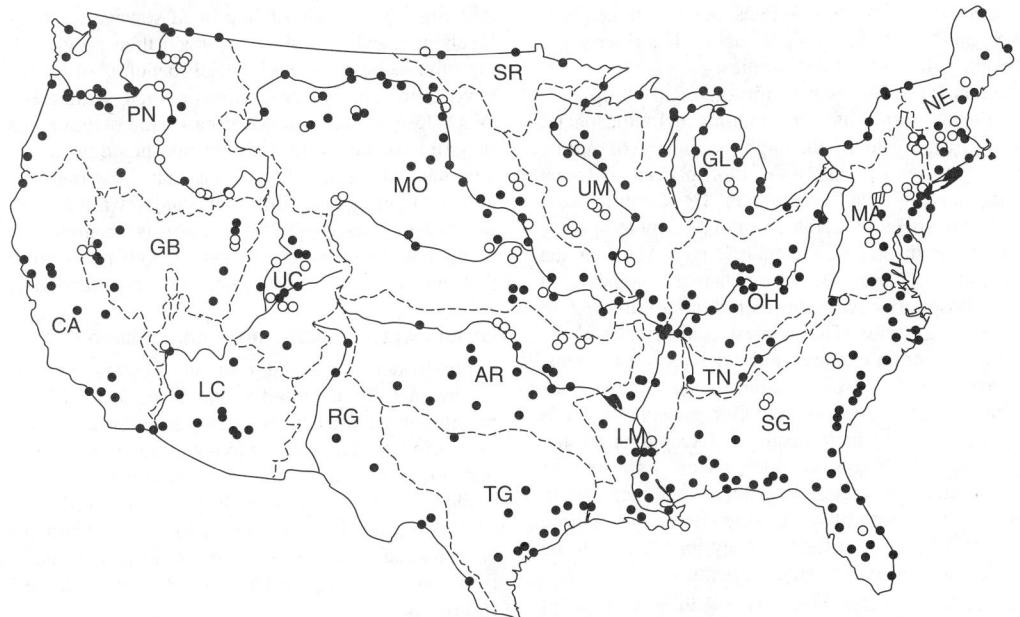

Fig. 1.   Analysis of major river waters based upon a sampling network comprised of over 300 stations and involving two systems

NE New England	UM Upper Mississippi	RG Rio Grande
MA Mid-Atlantic	LM Lower Mississippi	LC Lower Colorado
SG Southeast-Gulf	TG Texas-Gulf	UC Upper Colorado
TN Tennessee	AR Arkansas-Red	GB Great Basin
OH Ohio	MO Missouri	CA California
GL Great Lakes	SR Souris-Red-Rainy	PN Pacific Northwest

The largest rivers are shown as solid black lines. The NASQAN stations are located and associated with these rivers and their tributaries; the NWQSS stations are located along smaller rivers and usually near agricultural areas and some urban communities. (*U.S. Geological Survey.*)

Particularly noteworthy are widespread decreases in fecal bacteria and lead concentrations and widespread increases in nitrate, chloride, arsenic, and cadmium concentrations. Recorded increases in municipal waste treatment, use of salt on highways, and nitrogen fertilizer application, along with decreases in leaded gasoline consumption and regionally variable trends in coal production and combustion during the period, appear to be reflected in water-quality changes. In addition to data from the network of sampling stations, the researchers depended upon considerable ancillary data in their interpretation of the sampling station trend results.

Because of the passage of restrictive legislation, mainly over the past decade, improvements in river water quality can and should be expected. As discussed later, indirect pollution of landfills as they reach underground aquifers is an even more important concern and is less visible and hence less easy to police and control.

Since the passage of the Clean Water Act by the U.S. Congress, numerous improvements have been evidenced and thus tend to reinforce confidence that in many aspects the environment can be improved.

Trends in River Water Pollution

Biological Oxygen Demand (BOD). The Clean Water Act was passed by the U.S. Congress in 1972. In the decade that followed, municipal loads of BOD decreased an estimated 46% and industrial BOD loads decreased at least 71% nationwide. These reductions are impressive, especially in light of an increase in population (up 11% during the period) and an increase in the gross national product (GNP, adjusted for inflation) of 25%. Federal funding for upgrading municipal facilities peaked in 1980 and amounted to $35 billion for the decade.

Dissolved Oxygen Deficit (DOD). The sampling station report indicated a net decrease in DOD (thus an improvement in dissolved oxygen conditions). DOD declines were reported in the New England, Mid-Atlantic, Ohio, and Mississippi regional basins; increases were most frequently found in the Southeast. These data appear to confirm the success of point-source control efforts. Decreases in DOD were found most often (beyond expectations) where point sources dominated; increases

where nonpoint sources prevailed. The chronology of the station data also indicates that gains from industrial water treatment preceded those from modernization of municipal facilities.

Fecal Bacteria (coliform, FC; streptococcal, FS). During the study period, decreases in FC and FS were widespread. Major decreases were particularly evident in parts of the Gulf Coast, central Mississippi, and Columbia basins; significant decreases were also noted in the Arkansas-Red basin and along the Atlantic Ocean. During the study period, major efforts were made to control both municipal and agricultural sources of fecal bacteria.

Suspended Sediment (SS). These impurities, originating from nonpoint sources (notably agricultural) are probably the most damaging nonpoint sources of water pollution. E.H. Clark estimates the cost of the hydrologic impacts of soil erosion and related nutrients on aquatic ecosystems may increase. Fertilizer application rates increased by 68% during most of the testing period described here. The long-term history of fertilizer in the United States has been one of almost continuous increases in nitrogen and phosphorus application rates. As reported by Smith, et al., the extent to which changes in agricultural practice are reflected in trends of SS, phosphorus, and nitrogen concentrations in the nation's rivers has largely been a matter of conjecture because of the lack of systematic long-term studies.

Certain forms of land use, such as logging, traditionally have been associated with high rates of soil erosion. Thus, the study indicated large increases in SS in the Columbia (logging) and in the Arkansas-Red and Mississippi basins (agriculture) during the test period. Significant amounts of SS were also found along the Texas Gulf Coast, northern Florida, and scattered locations in northern Ohio, New England, and northern Minnesota. Declining rates of SS were indicated in the Missouri River basin and have been clearly traced to the effects of reservoir construction throughout that basin during the 1950s and 1960s.

Total Phosphorus (TP). Mainly as the result of controls over point sources of pollution, decreasing trends in TP were found in the Great Lakes and Upper Mississippi regions. Significant correlations in TP increases were found in connection with nonpoint sources (mainly agriculture) in several regions.

Total Nitrate (TN). Increases in TN were found to be widespread. Although also found to be significant in the Pacific Northwest and along the California coast, the great majority of increases in TN were found east of the 100th meridian (southward from North Dakota to southern Texas). Increases of TN were strongly associated with agricultural activity including fertilized acreage (as a percentage of basin area), livestock population density, and feedlot activity. Atmospheric deposition, in addition to agricultural runoff, was found to be a major source of nitrate in surface waters, East and northern Midwest. Separate studies of NOx emissions showed a general pattern of increasing rates, notably in the Ohio, Mid-Atlantic, Great Lakes, and Upper Mississippi basins. The river water quality study correlated well with the emission studies.

It is interesting to note that the differences in nitrogen and phosphorus trend patterns appear to be the result of three factors: (1) atmospheric deposition seems to play a large role in the high frequency of nitrate trends; (2) the low frequency of, and strong association patterns between phosphorus and SS trends suggests that increases in TP resulting from rises in agricultural activity have been moderated or delayed by temporary storage of sediment-ground TP in stream channels; and (3) during the study, point-source control efforts were focused much more intensely on TP than on TN because phosphorus was considered more limiting to eutrophication in freshwater ecosystems. The greatest consequence of the differences in TN and TP were seen in changes in the delivery of nutrients to coastal areas. Nitrate loads to East Coast estuaries, the Great Lakes, and the Gulf of Mexico have increased markedly, while phosphorus loads to coastal areas have changed little or have declined. Exceptions are the Gulf Coast and the Pacific Northwest basins, in which instances phosphorus loads to estuaries have increased in association with substantial increases in sediment loads. Researchers have stressed that increased delivery of nitrate to estuaries is of major concern because of the tendency for nitrogen to be limiting to eutrophication in many estuarine environments. See Table 1.

Salinity. Increasing trends in chloride, sulfate, and sodium were found in a large majority of the rivers tested. An average increase of 30% in salinity was found. Researchers attribute this increase to the following factors. (1) Increases in population in the basins surveyed—human wastes are a major source of chloride in populated basins. (2) Use of salt on highways increased by a factor of 12 between 1950 and 1980 and is considered a major contributor to total stream salinity. (The rates of highway salt use were particularly significant in the Ohio, Tennessee, lower Missouri, and Arkansas-Red basins.) (3) In the Missouri, Arkansas, and Tennessee basins, salinity increases were accurately correlated with changes in surface coal production during the testing period. By contrast, salinity decreased in the Upper Colorado Basin, an area historically plagued with salt problems. Decreases are attributed to control efforts and the effects of reservoir filling during the early 1970s.

Trace Elements. Moore and Ramamoorthy reported that trends in toxic element concentrations in surface waters have remained largely unknown despite rapidly increasing knowledge of the potential sources of toxic substances in aquatic systems. The study reported on analyses of several trace elements, including arsenic, cadmium, chromium, lead, iron, manganese, mercury, selenium, and zinc. Researchers found increases in dissolved arsenic and cadmium, particularly in those basins in the industrial Midwest. Atmospheric deposition is considered more significant than terrestrial sources of trace element contamination. Heit et al. confirmed this conclusion as based upon the results of lake sediment analyses in regions with high deposition of fossil-fuel combustion products. At many network sampling stations, dissolved lead concentrations showed a decrease, but despite significant declines in gasoline lead consumption, some sampling stations showed no decline. Lead concentrations were found particularly heavy in the Texas Gulf Coast region.

In deference to the excellence of the sampling study, researchers point out that additional sampling in selected smaller basins is needed to improve the ability to determine the effects of changes in point-source pollution.

TABLE 1. CHANGES AND TREND PATTERNS OF DELIVERY OF NUTRIENTS TO COASTAL AREAS OF THE UNITED STATES

Region	Change in load	
	Nitrate (%) total	Phosphorus (%) total
Northeast Atlantic Coast	32	−20
Long Island Sound/New York Bight	26	−1
Chesapeake Bay	29	−0.5
Southeast Atlantic Coast	20	12
Albemarle/Pamlico Sound	28	0
Gulf Coast	46	55
Great Lakes	36	−7
Pacific Northwest	6	34
California	−5	−5

Source: Smith/Alexander/Wolman 1987 reference listed.

Although the effects of improved sewage treatment on dissolved oxygen levels appear to be more localized than previously thought, it is possible that the ecological and social benefits of water-quality improvements have been large in proportion to their spatial extent. As pointed out by W.M. Leo, et al., individual case studies have demonstrated local effects of point-source pollution controls, but they do not provide an adequate national sample on which to base an assessment of the benefit of pollution abatement programs. Smith/Alexander/Wolman, in their report, suggest that in designing water-quality monitoring programs for the future, we should recognize the growing number of both point- and nonpoint-source issues that our economic and political systems must address.

Groundwater Quality and Contamination

Groundwater as used by humans consists of subsurface water, which occurs in fully saturated soils and geological formations. Nearly half the population of the United States uses groundwater from wells or springs as a primary source of drinking water. An estimated 75% of major cities depend on groundwater for most of their supply. Total fresh groundwater withdrawals (1980) were estimated at 88.5 billion gallons per day, of which 65% was used for agricultural irrigation. Numerous regulatory acts have been passed by the U.S. Congress in recent years to protect groundwater from pollution. These various acts are listed in article on **Wastes and Pollution**.

As early as the 1980s, the U.S. Environmental Protection Agency (EPA) proposed a national groundwater strategy—with the emphasis of targeting on prevention as contrasted with taking remedial action in treating known contaminated sites. Transport of groundwater is very slow. Flow rates are governed by hydraulic gradients and aquifer permeability, thus flows may range from a few centimeters to a meter or so per day. Contaminants usually mix with the water at a low rate. Sometimes natural ingredients tend to retard the actions of contaminants; in other cases the concentrations of contaminants in groundwater may exceed those found in surface water. Groundwater contamination occurs out of view and the effects of pollution may not be noted for weeks or months (even years in sparsely settled communities) after the initial cause(s). A wide variety of substances are involved in groundwater contamination. These include inorganic ions (chloride, nitrate, heavy metals), complex synthetic organic chemicals, and pathogens (viruses and bacteria). Thus, rarely does one find the same combination of variables when attacking a groundwater contamination problem.

The extent of groundwater is statistically very impressive: (1) of the water in the hydrologic cycle, 4% is groundwater, exceeded only by the seas and oceans; (2) groundwater volume at any given time exceeds that of the combined fresh water in lakes, streams, and rivers, of which an estimated 30% of the volume is furnished by groundwater; (3) during dry spells, groundwater furnishes a large majority of the low water flow of streams.

The manner in which waste disposal practices may interact with and contaminate the groundwater system is illustrated schematically in Fig. 2.

A study of groundwater contamination in ten selected states, sponsored by the Environmental Assessment Council of the Academy of Natural Sciences (Philadelphia) provides an excellent view of the sources and kinds of contamination that have been found in known, reported incidents of contamination. See Table 2.

One of the most serious cases of groundwater pollution occurred above the Great Miami Aquifer, which is a 2-mile (3.2-km) wide and 80-mile (129-km) long water basin that essentially follows the Great Miami River in southern Ohio and a section of which is located directly under the city of Hamilton. This aquifer is considered to be one of the Midwest's most productive sources of groundwater, making up about one-third of Ohio's groundwater. At points, the aquifer rises to within 20 feet (6 m) of ground level. A small, private firm established a waste-disposal operation in the 1970s at a time when regulations were minimal. Cleaning up the severely polluting operation commenced in May 1983 and was one of the first projects selected by the EPA to clean up after passage of the Superfund Act in late 1980. Cleanup contractors found 8600 drums, 30 storage tanks, and two open-top tanks on an approximately 10-acre (40,470–square-meter) area. Containers were found holding and leaking some 300,000 gallons (1,135,000 liters) of toxic wastes, including pesticides, rodenticides, waste oils, plastics and resins, acids, arsenic, and cyanide sludges. Among the chemicals found were DDT and PCBs. It was estimated in 1990 that environmental correction actions will require an additional 10 years at a multimillion dollar cost. Professionals in the cleanup field have estimated

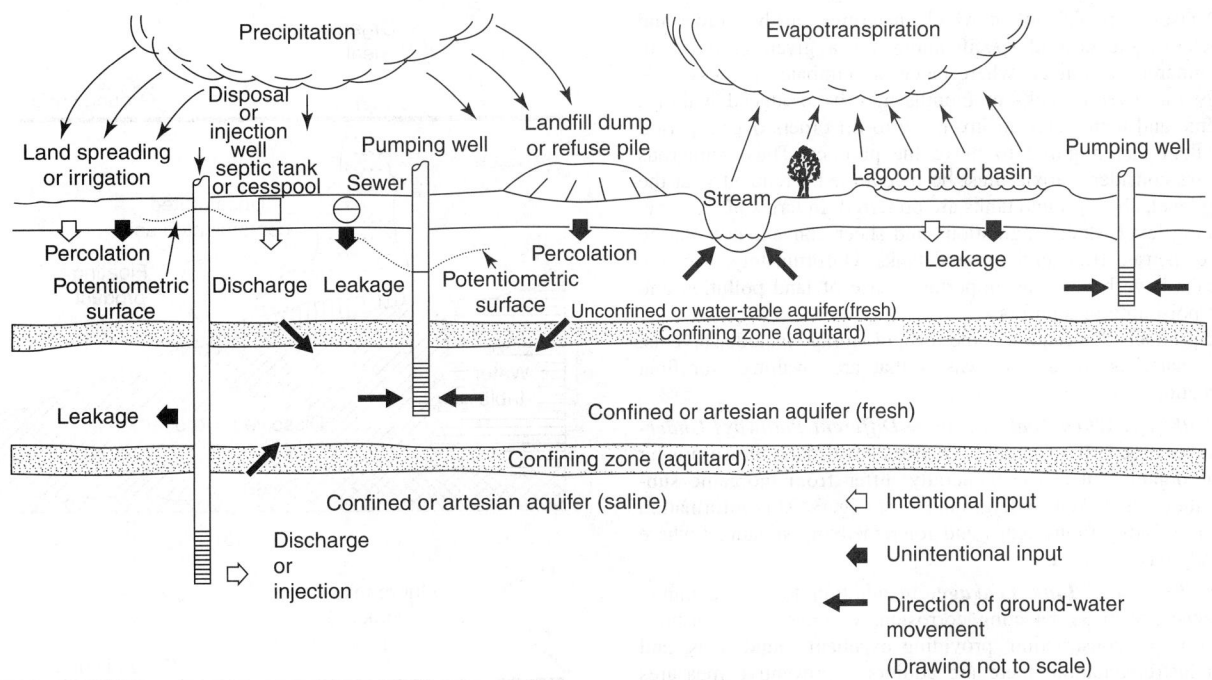

Fig. 2. Processes by which wastes can reach and contaminate groundwater systems. (*U.S. Environmental Protection Agency*)

TABLE 2. REPORTED INCIDENTS OF GROUNDWATER CONTAMINATION IN TEN SELECTED STATES (ACADEMY OF NATURAL SCIENCES, PHILADELPHIA)

State	Groundwater consumed (million gallons/day)	Source of contamination
Arizona	4800	Industrial wastes; landfill leachate; human and animal wastes
California	13,400–19,000	Saltwater intrusion; nitrates from agriculture; brines and other industrial and military wastes
Connecticut	116	Industrial wastes, petroleum products, human and animal wastes
Florida	3000	Chlorides from saltwater intrusion and agricultural return flow; industrial wastes; human and animal wastes
Idaho	5600	Human and animal wastes; industrial wastes; radioactive wastes
Illinois	1000	Human and animal wastes; landfill leachate; industrial wastes
Nebraska	5900	Irrigation and agriculture; human and animal wastes; industrial wastes
New Jersey	790	Industrial wastes; petroleum products; human and animal wastes
New Mexico	1500	Oil field brines; human and animal wastes; mine wastes
South Carolina	200	Petroleum products; industrial wastes; human and animal wastes

Note: Probably the most newsworthy, publicized incidents have stemmed from public and industrial dumping sites, covered in this summary under "landfill leachate." A class of pesticides most commonly found in groundwater is nematocides. They are particularly difficult because manufacturers design them to be both persistent and toxic. DBCP (1,2-dibromo-3-chloropropane) is a representative nematocide.

the existence of about 400 additional sites of serious groundwater pollution of the extent and nature of the site just described.

Remediation of Poisoned Aquifers. Means for correcting polluted underground water systems are limited and costly. To the knowledgeable would-be polluter, the costs and time involved serve as incentives for preventing such damage in the first place.

Containment. Once subterranean pollution has been detected for the first time, not only must the source(s) be located, but the extent of damage must be determined. This often involves the meticulous analysis of core samples and thorough geologic and hydrologic mapping of the poisoned volume

of earth and groundwater affected. Effective permanent remedial actions require an extensive and reliable data base. In some cases it may be found that, by way of constructing underground barriers, the polluted underground sections can be isolated and an aquifer rerouted. In such rerouting, it may be found that gravity flow underground may be insufficient, thus requiring pumps to assure flow through what, in essence, is a new "artery" for groundwater flow.

Where containment (isolation) procedures are not practical or fully adequate, extraction of pollutants may be required.

Extraction. In this process, sometimes referred to as the "pump and treat" method, polluted water is pumped to the surface, after which it is treated to remove toxic materials. Then the treated water is returned to the aquifer. Depending upon the extent of pollution, a large water-treating facility may be required adjacent the cleanup site, even though the facility may be required only for a comparatively short time of a few to several months.

Bioremediation. This process is gaining acceptance among site cleanup professionals. Wastes that have been pumped out of the aquifer are transferred to specially constructed tanks, to which fungi, bacteria, and other microbes are added, along with hydrogen peroxide, which furnishes oxygen. Also added are small amounts of nitrogen fertilizer, which acts as a nutrient for the microbes. Under these ideal conditions, the microorganisms secrete enzymes that transform immense quantities of substances such as benzene and trichloroethylene into harmless salts. A delicate balance must be maintained in providing just enough oxygen and nutrients, thus avoiding repollution of the wastes.

For cleaning up sites less extensive than the one just described, and thus requiring aboveground treating facilities for less time, portable plants are receiving consideration. New techniques also are being evaluated. In one system, a *wet oxidation* process converts waste sludge into clean water and sterile ash. Organic pollutants also are being subjected to a laser oxidation process that breaks molecular bonds in a photochemical reaction chamber. Low-frequency soundwaves and electric charges also are being tested to free pollutants from soils. In this process, after a positively charged anode is introduced to the ground, positively charged water and contaminants migrate toward a negatively charged withdrawal well, which acts as a cathode.

Above- and Underground Storage of Products and Wastes

For practical economic reasons, some potentially polluting products must be stored for indeterminate periods above ground. Chemical, petroleum, and petrochemicals are examples of the need for temporary holding. In the blending of gasolines, for example, refineries pump different grades of fuel to a "gasoline pool," out of which quantities can be drawn and blended to achieve the desired specifications for a given gasoline. In producing commercial chemicals where batch or semibatch processing is used, a supply for several weeks or months may be produced within a short time frame and then stored in inventory to fill orders over a period much longer than that required to make the product. Thus, numerous products that are considered toxic must be inventoried. Frequently, at the manufacturing level, aboveground tanks are preferred. In terms of gasoline, at the consuming level, safety regulations and sheer convenience require that fuels be dispensed from underground tanks. Unfortunately, over the years, tank leaks have become an important cause of land pollution and ultimately the poisoning of aquifers.

Other toxic products that require some form of storage include solvents, chemical raw materials (fluid) and wastes that are "waiting" for final disposal/destruction.

Pollutant Pathways When Leaked Follow Different Pathways Underground. Relatively recently, cleanup specialists have made the surprising discovery that organic substances generally differ from inorganic substances when they are leaked underground. See Fig. 3. This information is helpful for preventing future leaks and for remedying situations where pollution already has occurred.

Means for Preventing Tank Leakage. In addition to the "common sense" approaches, such as selecting corrosion, weather, and moisture-resistant materials of construction, providing excellent foundations, and leak-detection instrumentation, there are additional preventive measures that can be taken. Cathodic protection, for example, is described in an article on **Corrosion**.

Secondary containment of underground storage tanks is another means. EPA regulations now require secondary containment systems for hazardous substance underground storage. Many industrial firms now are turning to flexible membrane systems, which offer a number of advantages over their rigid counterparts. EPA regulations for new tank construction permit four options:

1. Fiberglass-reinforced plastic.
2. Steel with cathodic protection, using either dielectric coating, field installation by a corrosion expert, or impressed current system.
3. Metal without cathodic protection if site is determined to be noncorrosive.
4. Steel-fiberglass–reinforced plastic composite.

Options for existing tanks include:

1. Interior lining.
2. Cathodic protection if internal inspection is conducted and tank is less than 10 years old and is monitored monthly.

Landvaults. Approval of aboveground permanent landvaults sometimes can be obtained. They have a number of advantages:

1. A double or triple composite liner system provides maximum protection.
2. They permit future retrieval of wastes for improved treatment when such a process may be developed, and, inasmuch as the vault is fully sealed, the underground geology need only be strong, stable, and free of any tectonic history.

Pollution of the Seas and Oceans

Over the centuries, populations living along the tens of thousands of miles of coastline of the oceans and seas worldwide have dumped their refuse into these sinks of saline water. This was done with little if any reluctance or concern prior to the early 1900s. Ocean waters, without desalination, could not serve as drinking or irrigation water, and, inasmuch as these bodies of water are so tremendous as compared with the amounts of waste that may be added to them, there was no real sense of guilt when polluting them. Only within the last few decades has a public awareness of pollution of all types, including the oceans, developed. Modern means for exploring Earth's hydrosphere, from the surface to great depths, are relatively recent achievements. The American explorer, Charles Beebe, did not accomplish his remarkable descent to ocean depths until 1934, when he reached a depth of 3028 feet (923 m).

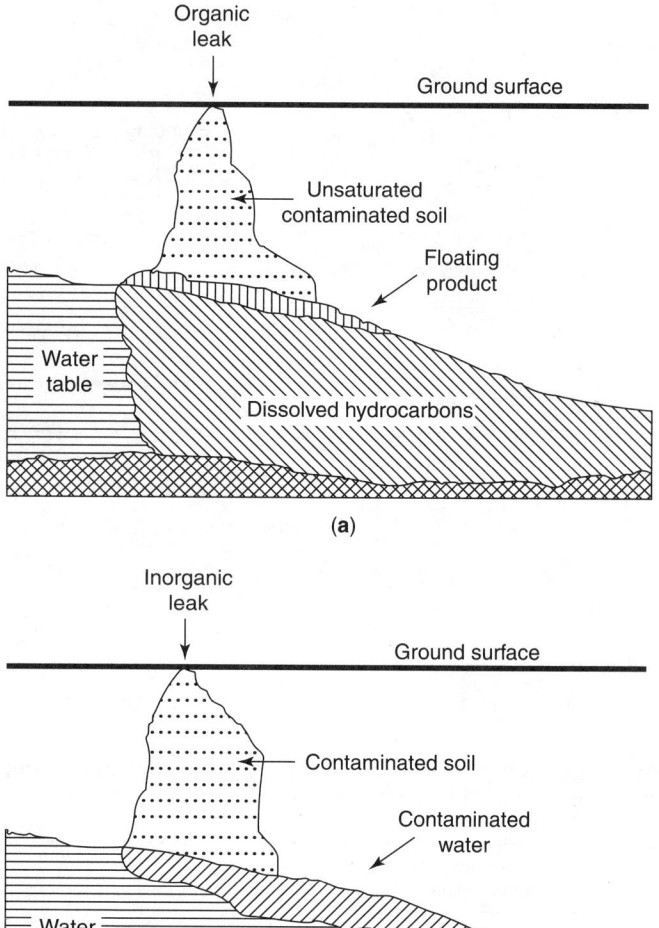

Fig. 3. Schematic diagrams illustrate how organic and inorganic fluids differ in how they permeate ground and water table: (**a**) Organic substances contaminate the soil to the surface of the groundwater table. If contaminant is lighter than water and only slightly soluble, such as a petroleum product, a substantial layer of product may be found floating on the groundwater surface. The groundwater will contain varying concentrations of dissolved contaminants, depending upon solubility. (**b**) Inorganic compounds, such as acidic wastes containing metals, may be attenuated in their flow. Soils tend to attenuate movement of positively charged ions (cationic) metals at various rates, depending on the cation exchange capacity of the soil. Clays have a high cation exchange capacity. Thus, extensive contamination of soils by metals is infrequent. In contrast, acids can increase movement through the soil

The effects of polluted waters on the proliferation and edibility of fishes and other forms of ocean life have been researched extensively in recent years, and this knowledge has created a vital incentive for reducing pollution. Just a few years ago lightweight refuse that rises to the ocean's surface and littered beaches created an environmental crisis and another incentive for ceasing ocean dumping of sewage and solid wastes into the oceans.

The U.S. Congress passed the Ocean Dumping Act in 1972. To continue with such dumping, permits were required and the polluter had to demonstrate to the Environmental Protection Agency that the materials to be dumped would not "unreasonably degrade or endanger human health or the marine environment." Radiological, chemical, and biological warfare agents and high-level radioactive wastes were fully banned. The dumping of dredged materials from navigable waters was put under the regulation of the U.S. Army Corps of Engineers. Dredged material is comprised of a mixture of sand, silt, and clay, but can include rock, gravel, organic matter, and contaminants derived from a wide range of agricultural, urban, and industrial sources. As reported by R.M. Engler (U.S. Army Corps of Engineers), "Contaminated or otherwise unacceptable dredged material accounts for only a small fraction of the total—less than 10 percent in the U.S. and globally." Clean dredged waste has a number of positive uses, including the enhancement of wetlands and aquatic and wildlife habitats, beach nourishment, offshore mound and island construction, agriculture mariculture uses, and as a construction aggregate. All requests for the ocean dumping of dredged materials had to be accompanied by a list of suggested alternative actions.

As will be noted in the article on **Wastes and Pollution**, much emphasis is given to reducing pollution by creating less waste. "No Waste," of course, is an impractical target, but "Much Less Waste" can be achieved globally within the next several years, even considering major population increases. But, even then, from an economical and practical viewpoint, some scientists do not believe that the oceans can escape, in the long term, playing a major role as a "waste sink." As of the mid-1990s, both scientific and lay opinions were strongly polarized. The oceans can accommodate huge annual volumes of nonhazardous, nonfloatable wastes. Much scientific research and engineering remains to be done in removing toxic materials prior to dumping and thus protect ocean life forms and the people who eat seafood. Detoxification is mandated now prior to land dumping. Also, much more knowledge of ocean characteristics is required in the way of selecting dumping sites.

Oceanologists and waste-handling professionals presently are attempting to answer the question, "How can society use the oceans for waste disposal without harming the marine environment or fisheries resources?" Much research remains, and numerous other questions remain unanswered.

The Boston Harbor Project

For nearly a century, scores of villages (later to become cities) developed around the city of Boston, Massachusetts. The Boston Harbor and associated rivers (Charles, Mystic, and Neponset) traditionally have been used for the disposal of sewage wastes. In 1904, Boston established a centralized area sewage disposal system, which because of population expansion, has multiplied in capacity manyfold. The cities that cluster around Boston found it easier and less costly to hook into the Boston central sewage system than to build their own treating and disposal facilities. The result over the past 20 years or so has been no surprise, and Boston Harbor became known as one of the most polluted in the world. Reconstruction of the Boston sewage system commenced during the same time frame as increased state and federal regulations governing sewage disposal. Although construction of a 9.5-mile (25-km) tunnel to carry sewage wastes out of the immediate Boston harbor area to Massachusetts Bay was nearing completion as of 1994, this project serves as an interesting example of how a major environmental project can become entangled in legislation and regulatory red tape and the involvement of citizen groups in environmental issues. An excellent background description is contained in the Spring 1993 issue of *Oceanus* magazine.

The waste-transporting tunnel was bored through bedrock that supports the seafloor of Boston Harbor and Massachusetts Bay. With exception of the tunnel bored under the English Channel, it was the major tunneling project to take place during the 1980s and 1990s. Toward the end of the tunnel, a series of 55 vertical effluent discharge pipes release the sewage into the bay.

Further Quantification of Ocean Data Needed

Factors[2] that remain to be studied concerning the oceans in terms of waste disposal include:

Physical—Diffusion, advection, sedimentation.
Chemical—Volatilization, neutralization, precipitation, flocculation, adsorption, desorption, dissolution, oxidation.
Benthic—Geochemical, biological.

Over the past several decades, oceanographers have found that the oceans and seas are not as uniform as once contemplated, ranging with regard to their latitudinal and longitudinal locations and also with depth.

[2] As suggested by I.W. Duedall (Florida Institute of Technology).

This suggests, then, that various areas of the ocean may be more suited to accept anthropogenic debris than others. For example, temperature profiles, depths, thermal gradients, and numerous other physical, biochemical, and life-sustaining qualities are known to exist.

An early appreciation of natural oceanic detritus is given by biologist Rachel Carson, who in 1950 observed in her book, *The Sea Around Us*, "When I think of the floor of the deep sea, the single overwhelming fact that possesses my imagination is the accumulation of sediments. I see always the steady, unremitting, downward drift of materials from above, flake upon flake.... For the sediments are the materials of the most stupendous 'snowfall' the earth has ever seen." Fine particles are sinking into the global oceans every microsecond of the day and night. Such particles may include "shreds and motes and globs of stuff"—dead and decayed remains of plants and animals, meteorites and other cosmic dust, old lobster molts, volcanic fallout, radioactive fallout, the pollen from flowers, grains of sand from the deserts—in summary, just about everything that is airborne contributes to oceanic detritus. The rate of passage of suspended material from the ocean's surface to the bottom of the sea varies widely, but some scientists estimate that for some particles it may take thousands of years to reach the ultimate deep-sea graveyard. It is only in recent years that some oceanologists have studied this phenomenon seriously. Some detritus never may reach the sea floor because it is consumed by various forms of ocean life as food. A. Aldredge and M. Silver (University of California) observe, "Marine snow particles are typically smaller than a pinhead. However, in Monterey Bay off California (and probably elsewhere), the particles may be 4 inches (10 cm) across. These particles are usually aggregates of many smaller particles that stick together, often with mucus.... The largest type of marine snow is the *giant house*, which can be 6 feet (1.8 meter) across. Each house is a blob of mucus that has been secreted by a zooplankter. Some investigators refer to them as "floating islands.... While a single zooplankter inhabits its balloon of mucus, there may be hundreds of copepods on its exterior, apparently grazing on the nourishing tidbits of *marine snow*.... We find typically that *marine snow* hosts organisms in very high concentrations.... There are certain groups of organisms that seem to be found only on these.... Among the unusual characters are rare species of copepod, certain protozoans and a unique assemblage of bacteria.... *Marine snow* is a major means by which material reaches the ocean floor, a major vehicle for the transport of organic matter down to the ocean's interior.... As these particles sink through the water, they change continuously, as do the communities living on them."

Scientists have observed that *marine snow* can be a nuisance. Because in some cases gas may be produced, particles rise and create a scum, and this can be dried by the sun to produce a surface of sufficient strength to permit seagulls to walk upon it. Such scum extends for many thousands of acres (hectares) in the Adriatic Sea, where it has become a menace to fishermen and the tourist trade. Such scums were reported as early as the 1700s.

Natural Pollution of the Oceans. Frequently overlooked is what may be termed "natural" pollution, which, when coupled with artificial (anthropogenic) pollution, contributes to the sum total of all pollutants found in fresh and ocean waters worldwide. Deep fissures in the ocean floor, fumaroles, and seamounts (underwater volcanoes) release megatons of sulfur-laden and other noxious gases into ocean water; other discontinuities in the ocean basins release vast quantities of crude oil and other hydrocarbons. Surface volcanoes are major contributors to atmospheric pollution, much of which ultimately affects Earth's hydrosphere. The present dissolved solids content of the oceans represents natural water pollution that has taken place ever since the land masses rose above sea level—through a constant erosion of soil.

Oil and Hydrocarbon Pollution of the Oceans

Petroleum is not a substance foreign to the marine environment. Natural seeps have been discharging petroleum hydrocarbons into the marine environment for millions of years, in amounts substantially greater than those resulting, for example, from present offshore production activities. About 200 submarine oil seeps have been identified worldwide. There is little doubt that many more exist. Petroleum has also continuously entered the seas as a result of erosion of uplifted sedimentary rocks containing trace amounts of petroleum hydrocarbons. There is also evidence of organisms living in the ocean that biologically produce hydrocarbons, ranging from gases (methane, ethane) through liquids, to solid paraffin waxes of high molecular weight.

These substances enter the marine environment from a variety of sources, both through natural phenomena and anthropogenic activities. In 1985, the National Academy of Sciences (NAS) published an assessment of petroleum pollution of the world's oceans and estimated that between 1.7 and 8.8 million metric tons per annum (mta) of oil enter the oceans. Within this range, 3.2 mta is regarded as the best single estimate—equivalent to about 0.1% of the total oil produced annually worldwide (about 3 billion metric tons).

Table 3 compares these estimates with those published by NAS in 1972. All categories except spills show about a 50% decrease. According to the NAS, part of the reduction in most categories can be attributed to refinements in estimating techniques. Other reductions, however, are attributed to efforts to reduce oil pollution, such as the international program to reduce tanker operational discharges. Because annual Figures for marine accidental spills vary significantly, an average over a number of years was used for both the 1975 and 1985 reports. The increase shown for such spills in the 1985 study reflects the influence of major incidents—such as the loss of the 220,000 dead-weight-ton tankers *Amoco Vadiz*, which occurred in the period (1975 through 1980)—used to calculate its annual average. The Alaskan oil spill (1989) is discussed later.

Most surface and near-surface open ocean waters contain petroleum hydrocarbons in the range of about 1 to 10 parts per billion (ppb), according to the NAS 1985 study. In deeper open ocean waters, the concentrations are

1 ppb or less. (One ppb is equal to about one drop of oil in 100,000 quarts of water.) Coastal waters, particularly those near populated and industrialized areas where the presence of oil is most likely, show higher levels (up to 100 ppb). Laboratory experiments conducted to determine the toxicity of crude oils or petroleum products indicate that concentrations from 10 to 100 times greater than those of coastal waters are required before measurable effects on marine organisms can be detected.

The process whereby an oil slick from a sudden spill is ultimately disposed has been the subject of intense research. Figure 4 depicts the processes occurring in the water, the overlying atmosphere, and the underlying bottom sediments. Figure 5 relates the time following discharge of oil into the sea to the various processes of movement and degradation. These processes include:

Drift. The movement of the center of a mass of oil on the surface of the ocean, drift, is caused by the combined action of wind, surface currents, waves, and tides. With larger amounts of oil, such as occur with accidental spills, the oil drift is largely independent of spill volume, spreading, or weathering. Since the "thick" portion of the slick drifts faster than the "thin" portion, a heavy oil accumulation forms the leading edge of an advancing slick. The drift process is always active—from the moment oil is released into the sea until the oil disappears from the surface. It is difficult to predict precisely all the complex interactions of oceanographic and meteorological factors that influence drift of an individual oil slick. However, field and laboratory observations of oil slick drift are surprisingly consistent.

Evaporation. The primary weathering process involved in the natural removal of oil from the sea is evaporation. It is particularly dominant soon after oil is released. Evaporation involves the transfer of hydrocarbon components from the liquid oil phase to the vapor phase. Estimates from major spills as well as experimental data indicate that evaporation may be responsible for the loss of up to 50% of a surface oil slick's volume during its life. Evaporation rates of oil at sea are determined by wind velocity, water and air temperatures, sea roughness, and oil composition. Some of the light, low-boiling hydrocarbons, such as benzene, toluene, and xylenes, which are rapidly lost through evaporation, are the most toxic. Thus, their removal decreases toxicity to marine life of the oil remaining on the surface.

Photooxidation. Much of the oil that evaporates is photo-oxidized in the atmosphere, but in the absence of good analytical data the contribution of this process cannot be estimated. It is reasonable to assume that some of the oil returns to the seas as atmospheric fallout.

Dissolution. This is another early process acting on spilled oil. Dissolution involves the transfer of oil compounds from a floating slick,

TABLE 3. PETROLEUM INPUTS TO THE OCEANS FROM DIFFERENT SOURCES

Source	1975 Study[1]	1985 Study[2]
	(Million metric tons annually)	
Municipal and Industrial Wastewater		
Discharges and Runoffs	2.5	1.0
Refinery Wastewater Discharges	0.2	0.1
Offshore Oil Production	0.08	0.05
Marine Transportation		
Tanker Operations	1.3	0.7
Accidental Spills	0.2	0.4
Other Maritime Activities	0.6	0.4
Natural Seeps and Erosion	0.6	0.25
Atmospheric Fallout	0.6	0.3
	6.1	3.2

[1] Petroleum in the Marine Environment, National Academy of Sciences, 1975.
[2] Oil in the Sea: Inputs, Fates, and Effects, National Academy of Sciences, 1985.

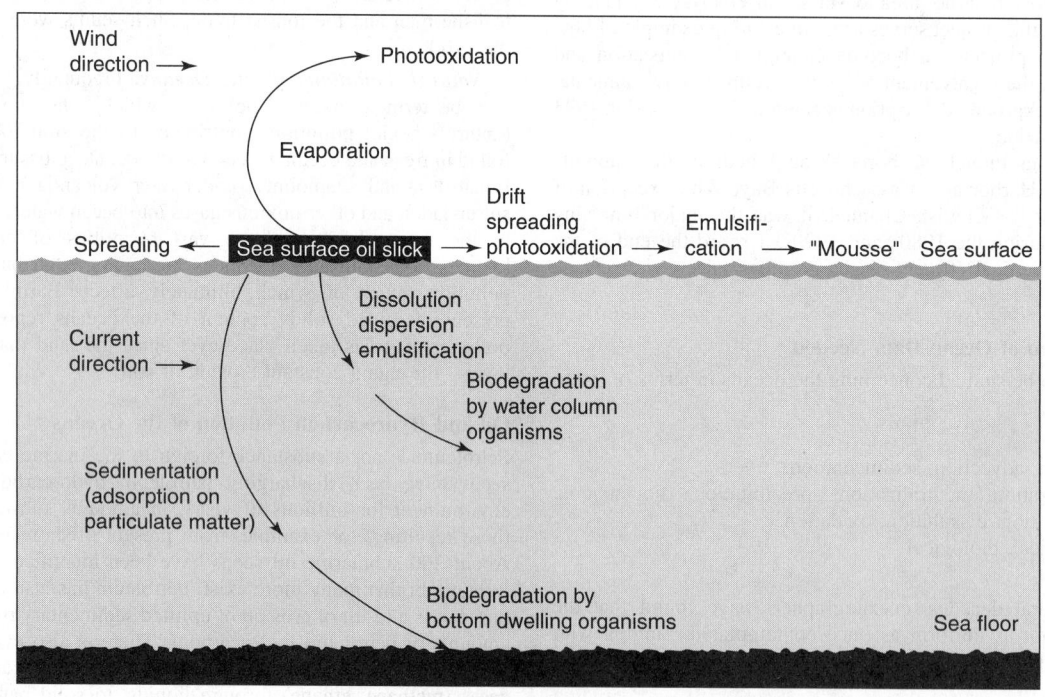

Fig. 4.   Processes that act upon an oil slick

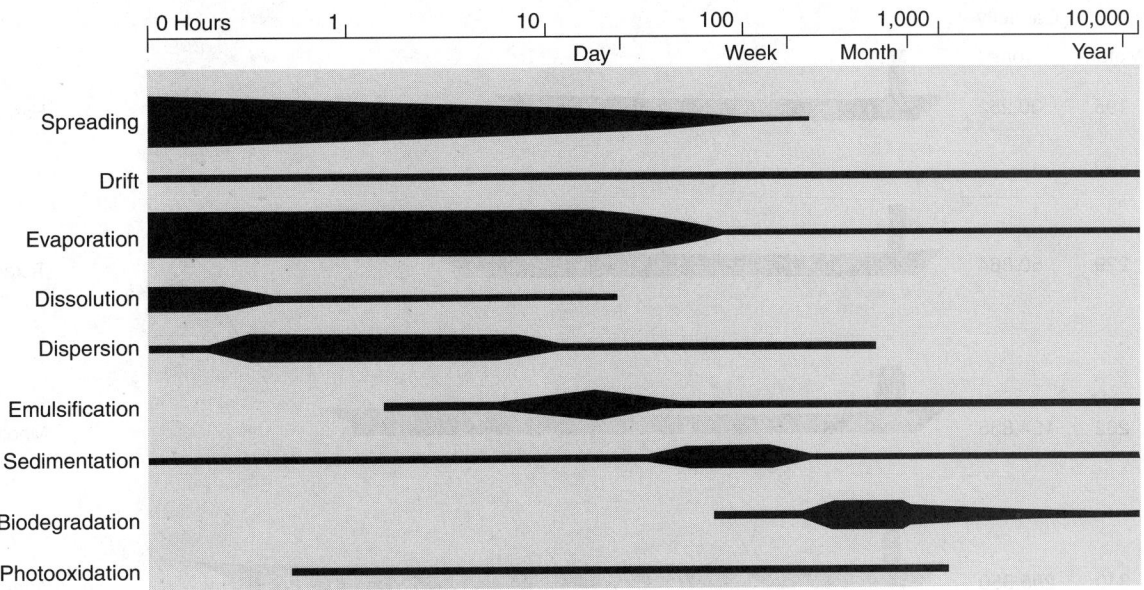

Line length—probable time span of any process.
Line width—relative magnitude of the process both through time and in relation to other contemporary processes.

Fig. 5.   Time span and relative magnitude of processes acting on spilled oil

or from dispersed oil droplets, into solution in the water phase. Lower-molecular-weight compounds tend to be the most soluble. It is unlikely, however, that dissolution significantly affects slick weathering as only a very small fraction of the oil dissolves. Evaporation proceeds much more rapidly than dissolution.

Dispersion. Oil also enters the water in forms larger than dissolved molecules. In natural dispersion, small droplets of oil (ranging in diameter from very small fractions of a millimeter to a few millimeters) are incorporated into the water in the form of a dilute oil-in-water suspension. Natural dispersion reaches a maximum rate in only a few hours (4 to 10) following a spill, but continues for some time. Oil dispersion is influenced by oil composition (e.g., wax and asphaltene content), density, viscosity, oil-water interfacial tension, and water turbulence. Crude oils and many petroleum products contain trace amounts of nitrogen, sulfur, and oxygen-bearing organic compounds, which can act as natural surfactants. These surfactants reduce the oil-water interfacial tension, allowing the oil to break up and to disperse into droplets more readily. Chemical dispersants may be applied to an oil slick to supplement the natural surfactants, thereby enhancing the dispersion process. The primary purpose for removing oil from the surface through dispersion is to enhance the degradation process. The increase in the surface area of dispersed oil droplets resulting from surfactant action accelerates the degradation of oil. Accelerating the dispersion process through the use of chemical dispersants also reduces the threat of floating oil stranding on a shoreline, where it can damage biota and property.

Emulsification. This is a water-in-oil process in which water is incorporated into the floating oil. Such emulsions, which may contain from 20 to 80% water, are often very viscous and referred to as "mousse." Mousse formation is highly dependent on oil composition. High levels of asphalt-type compounds, as well as waxes, appear to promote the formation of these emulsions. Ocean turbulence also accelerates mousse formation, although a fully developed, stable emulsion may be formed from some oils under relatively quiescent open-water conditions. Early treatment of spilled oil with chemical dispersants is an excellent way to prevent emulsification.

Sedimentation. Some organisms may ingest dispersed oil droplets in the water column and subsequently deposit them as fecal pellets. In some instances, this has been estimated to be a significant form of sedimentation.

Biodegradation. This is an important process for removing petroleum hydrocarbons from the marine environment. All surface waters, fresh or marine, contain natural populations of bacteria, yeast, and fungi capable of metabolizing and chemically degrading hydrocarbons through their normal life processes. These organisms are also primarily responsible for degrading most of the biologically produced hydrocarbons in the ocean. The rate

and extent of biodegradation depend on the abundance and variety of existing microorganisms, their predators, available oxygen and nutrients, temperature, and oil composition. Hydrocarbons, dissolved or dispersed in water, are the most easily degraded. Degradation of hydrocarbons contained in bottom sediments also occurs if oxygen is present. Emulsified oil (mousse) is slow to degrade because water is trapped within the emulsion and the nutrients and oxygen essential to biodegradation are kept out.

Invention of Oil-Eating Bacterium. It is interesting to note that the concept of genetically altering bacteria to transform crude oil into cattle feed was proposed by A.M. Chakrabarty (General Electric Co.) in the late 1960s. A patent was not granted until 1980 after considerable litigation. The concept is considered by many as the cornerstone of the biotech industry. Chakrabarty as of the 1990s continues in this field. Although applications in other areas of the petroleum industry may find applications for such bacteria, the primary interest in recent years has been in connection with oil tanker spill cleanups.

After the Alaskan oil spill of March 1989, approximately 70 miles (113 km) of beaches around Prince William Sound were sprayed with a fertilizer (*Inipol*) that had been invented by a French petroleum company (*Elf Aquataine*). The effort was made to stimulate the growth of naturally occurring bacteria (i.e., microorganisms that eat petroleum). This was the first large-scale test of bacteria for cleaning up an oil spill. Improvement was noted within a couple of weeks after the spraying. Research continues with the material to determine its effectiveness on shoreline rocks and pebbles. Early findings indicated that oil caught beneath the surface of rocks was consumed in 6 to 7 weeks. Researcher C. Oppenheimer (University of Texas) also developed microbial strains for producing fatty acids from oil, the product of which is more soluble in water. These compounds serve as food for plankton and other organisms.

Oil Tanker Spills

Usually the most publicized and one of the most dramatic examples of ocean water (and adjacent shoreline) pollution involves oil tanker or barge accidents. Most often, the saline waters of the oceans and seas are polluted, although there are instances where such accidents have occurred in fresh and brackish waters.

Oil spills present many different variations in the manner in which they develop and react to cleanup efforts. The Alaskan spill that occurred on March 24, 1989, in Prince William Sound was the most extensively researched to date. The cost of cleaning up the spill also exceeded all other tanker spill expenses to date. A study reported by the U.S. Forest Service estimates the final fate of the 10 million gallons as follows: evaporated, 35%; recovered, 17%; burned, 8%; biodegraded, 5%; and dispersed, 5%.

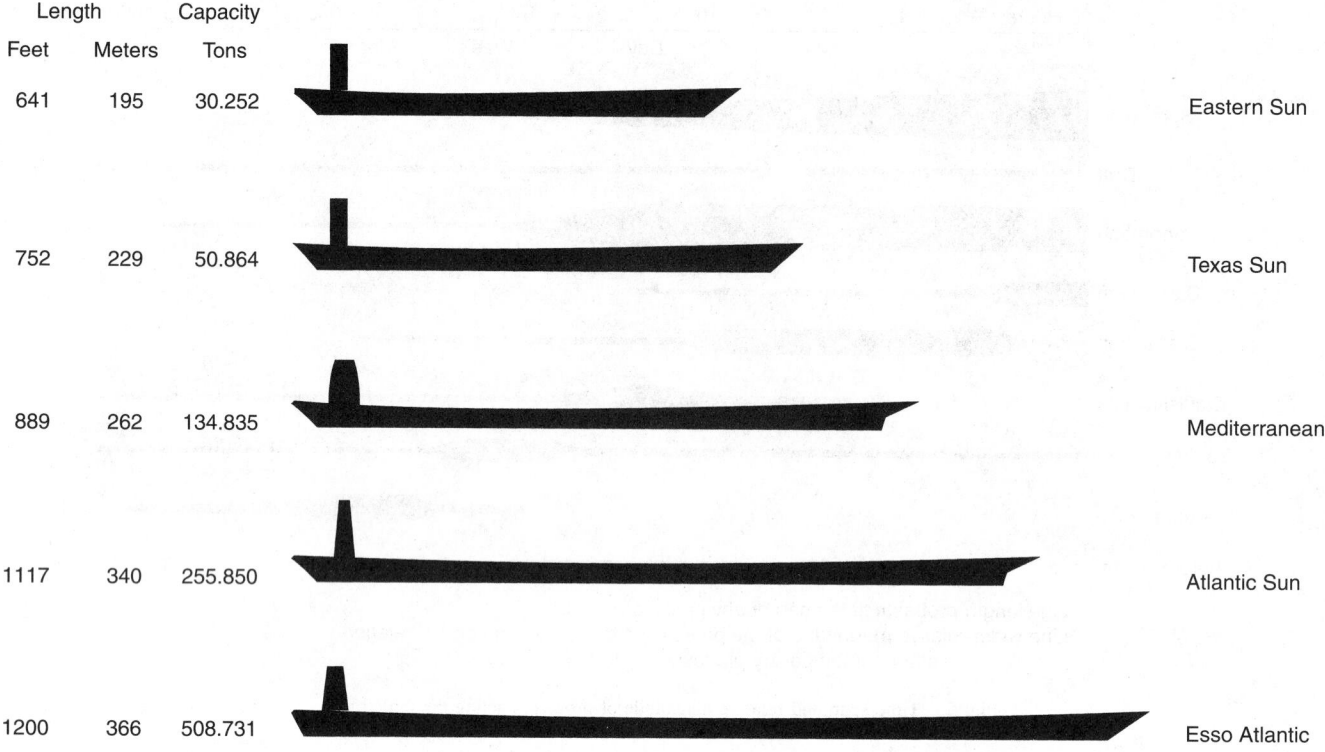

Length		Capacity		
Feet	Meters	Tons		
641	195	30.252		Eastern Sun
752	229	50.864		Texas Sun
889	262	134.835		Mediterranean
1117	340	255.850		Atlantic Sun
1200	366	508.731		Esso Atlantic

Fig. 6.   Trends in oil tanker design characteristics. (*After A. Dane*)

The total in the form of oil slicks on the Sound amounted to about 10% of the original spill and that on the shoreline about 18%. For many weeks after the spill, a fleet of specially equipped vessels mopped up about 120,000 gallons of crude oil per day. Other tanker spills have carried crude oil of somewhat different composition. The state of the sea, temperature, winds, and presence or absence of sunshine are among other variables that make each spill unique.

Tanker Design. One ship designer has observed that modern tankers are uniquely fragile and unwieldy vessels. The goal of tanker design is "to get as much cargo as you can into as little steel as possible and still have economical propulsion." Thus, the larger the ship, the easier to meet these specifications. Ultra-large crude carriers (ULCCs) required a distance of 3 miles (4.8 km) and about 20 minutes to stop from a top speed of 15 to 16 knots. Oil tankers have steadily increased in capacity and length over the last few decades. See Fig. 6.

Because of several major tanker incidents over the last few years and highlighted by the Alaskan spill, much design thought has gone into the "double-hull" concept. Indeed, this is not a new concept because two "skins" are used on the nearly 60,000 merchant vessels afloat—with the important *exception* of oil tankers. Double hulls have been standard on liquefied natural gas (LNG) ships for many years, and the design has been credited with preventing disasters at sea. The double-hull design enabled one LNG ship to sail many miles at its highest speed to the nearest port even though the outer hull had been torn open under several of the cargo tanks. None of the highly volatile cargo escaped.

The oil industry has objected to two hulls for the following reasons:

1. If oil leaked from the inner to the outer shell, the space in between could generate a vapor and thus be an explosion hazard.
2. When the outer hull was breached by an accident, water would fill the void, causing the ship to lose buoyancy and possibly go aground.

Even with these objections, there are some 530 oil tankers with double hulls, and these have been accident-free thus far.

Other improvements in tanker design have included more precise navigation systems that are customized to the vessel and the coarse of travel that it usually follows. In a single video readout are shown the ship's location and course with respect to shoreline, bottom contours, buoys, markers, and other ships. With another system, which includes radar reflectors located along the shoreline, the ship location can be determined within less than about 6 feet (1.8 meters). Another concept embraces use of a funnel that can be lowered from the ship immediately when a spill occurs and sucks the spilled oil back into the ship. A design of this kind has been under test in the Gulf of Mexico.

Radioactive Waste Dumping. Numerous proposed solutions for dumping radioactive wastes, including ocean burial, are described in the article on **Nuclear Power Technology**.

Additional Reading

Abel, P.D.: *Water Pollution: Biology,* 2nd Edition, Taylor & Francis, Inc., Philadelphia, PA, 1996.

Abelson, P.H.: "Oil Spills," *Science*, 629 (May 12, 1989).

Aubrey, D.G. and M.S. Connor: "Boston Harbor: Fallout Over the Outfall," *Oceanus*, 61 (Spring 1993).

Barinaga, M.: "Alaska Oil Spill: Health Risks Uncovered," *Science*, 463 (August 4, 1989).

Battle, J.B. and M. Lipeles: *Water Pollution,* 3rd Edition, Anderson Publishing Company, Cincinnati, OH, 1998.

Broadus, J.M.: "Tailoring Waste Disposal to Economic Realities," *Oceanus*, 707 (Summer 1990).

Cadwallader, M.: "Above-Ground Landvaults for Waste Containment," *Chem. Eng. Progress*, 9 (August 1989).

Capuzzo, J.E.M.: "Effects of Wastes on the Ocean: The Coastal Example," *Oceanus*, 39 (Summer 1990).

Clarke, E.H., II, A. Haverkamp, and W. Chapman: *Eroding Spo's* The Off-Farm Impacts, Conservation Foundation, Washington, DC, 1985.

Crawford, M.: "Bacteria Effective in Alaska Cleanup," *Science*, 1537 (March 30, 1990).

Curtis, C.E.: "Protecting the Oceans," *Oceanus*, 19 (Summer 1990).

Dane, A.: "America's Oil Tanker Mess," *Popular Mechanics*, 51 (November 1989).

Dane, A.: "Oil Slick Buster," *Popular Mechanics*, 58 (May 1990).

Dane, A.: "Learning from Disaster," *Popular Mechanics*, 94 (September 1991).

Duedall, I.W.: "A Brief History of Ocean Disposal," *Oceanus*, 29 (Summer 1990).

Eckenfelder, W.W., Jr.: *Industrial Water Pollution Control,* 3rd Edition, The McGraw-Hill Companies, Inc., New York, NY, 1999.

Erickson, D.: "Oil-Eating Bacterium that Spawned an Industry," *Sci. Amer.*, 88 (June 1990).

Grassle, F.: "Sludge Reaching Bottom at the 106 Site, Not Dispersing as Plan Predicted," *Oceanus*, 61 (Summer 1990).

Haberl, R., P. Cooper, R. Perfler, and J. Laber: *Wetland Systems for Water Pollution Control 1996*, Elsevier Science, New York, NY, 1997.

Hawley, T.M.: "Herculean Labors to Clean Wastewater," *Oceanus*, 772 (Summer 1990).

Helmer, R., I. Hespanhol: *Water Pollution Control: A Guide to the Use of Water Quality Management Principles,* Routledge, New York, NY, 1997.

Higgins, T.E. and W.D. Byers: "Leaking Underground Tanks: Conventional and Innovative Clean-Up Techniques," *Chem. Eng. Progress,* 12 (May 1989).

Hodgson, B. and N. Forbes: "Alaska's Big Spill: Can the Wilderness Heal?" *Nat'l. Geographic,* 5 (January 1990).

Hollister, C.D.: "Options for Waste: Space, Land, or Sea?" *Oceanus,* 13 (Summer 1990).

Holloway, M.: "Soiled Shores," *Sci. Amer.,* 102 (October 1991).

Holloway, M.: "Abyssal Proposal (Ocean Depths and Sewage Sludge)," *Sci. Amer.,* 30 (February 1992).

Hooker, L.: "Danger Below (Underground Aquifers)," *Chem. Eng. Progress,* 52 (May 1990).

Kistos, T.R., J.K.M. Bondareff: "Congress and Waste Disposal at Sea," *Oceanus,* 23 (Summer 1990).

Leo, W.M. et al.: Before and After Case Studies, EPA-430/9-007, Environmental Protection Agency, Washington, DC, 1984.

Levy, P.F.: "Sewer Infrastructure," *Oceanus,* 53 (Spring 1993).

Liptbak, B.G. and D.H. Liu: *Groundwater and Surface Water Pollution,* Lewis Publishers, Boca Raton, FL, 1999.

Marshall, E.: "Valdez: The Predicted Oil Spill," *Science,* 20 (April 7, 1989).

Mayer, J., S. McClurg: *Water Pollution,* Water Education Foundation, Sacramento, CA, 1996.

Moore, J.W., S. Ramamoorthy: *Heavy Metals in Natural Waters,* Springer-Verlag, Inc., New York, NY, 1984.

Noll, K.E., V. Gounaris, and Wain-sun Hou: *Adsorption Technology for Air and Water Pollution Control,* Lewis Publishers, Boca Raton, FL, 1991.

Peterson, S.: "Alternatives to the Big Pipe (Boston)," *Oceanus,* 71 (Spring 1993).

Schmitz, R.J.: *Introduction to Water Pollution Biology,* Buterworth-Heinemann, Inc., Woburn, MA, 1995.

Semonelli, C.T.: "Secondary Containment of Underground Storage Tanks," *Chem. Eng. Progress,* 78 (June 1990).

Smith, R.A., R.B. Alexander, and M.G. Wolman: "Water-Quality Trends in the Nation's Rivers," *Science,* 235, 1607–1615 (1987).

Spencer, D.W.: "The Ocean and Waste Management," *Oceanus,* 5 (Summer 1990).

Staff: *Prevention of Water Pollution by Agriculture and Related Activities,* Bernan Associates, Lanham, MD, 1993.

Staff: National Research Council, Ocean Studies Board, and Water Science Technology Staff, *Clean Coastal Waters: Understanding and Reducing the Effects of Nutrient Pollution,* National Academy Press, Washington, DC, 2000.

Stegeman, J.J.: "Detecting the Biological Effects of Deep-Sea Waste Disposal," *Oceanus,* 54 (Summer 1990).

Stone, R.: "Icy Inferno: Researchers Plan Oil Blaze in Arctic," *Science,* 1203 (September 13, 1991).

Viessman, W. and M.J. Hammer: *Water Supply and Pollution Control,* 6th Edition, Addison Wesley Longman, Inc., Reading, MA, 1998.

Wolman, M.G.: *Science,* **174**, 905 (1971).

WATER RESOURCES.

As pointed out by various authorities, major problems pertaining to water, in addition to pollution, include: (1) the heavy consumption, which began some years ago to limit the growth of various cities and of agriculture, particularly in the southwestern United States; (2) evaporation losses from reservoirs and storage ponds, particularly important in the arid western and west central sections of the United States, the control of which may be found to be more economic than that of developing new water sources; (3) lowering water Tables, again of major concern to the southwest and Pacific coastal regions of the United States, but also becoming a major factor near larger cities, and to some middle-Atlantic areas of the United States, a situation which is exerting considerable influence on the choice of new irrigation areas; (4) long-distance water transmission systems, which in the future may not be confined to the western United States; (5) waste water return to the oceans, which can become the largest and least expensive potential secondary water source and a very attractive source for many industrial uses; (6) salt water encroachment, which already is destroying some water sources and land; (7) watershed trash vegetation, the eradication of which can increase water yield, particularly in the southwestern United States; and (8) storage of seasonal and flood flows, which has been practiced for many years in most areas that are away from good lake or groundwater supplies, but which will require extension as the water problem becomes more severe in the less-arid areas.

The water supply problems of the United States image those of many other areas of the world. As is evident from Fig. 1, the eastern one-third of the United States, excepting a few areas and several of the major cities, does not generally run a seasonal water deficiency. Somewhat over one-third of the western United States, however, is characterized by no available surplus, and other areas by summer deficiency and winter surpluses.

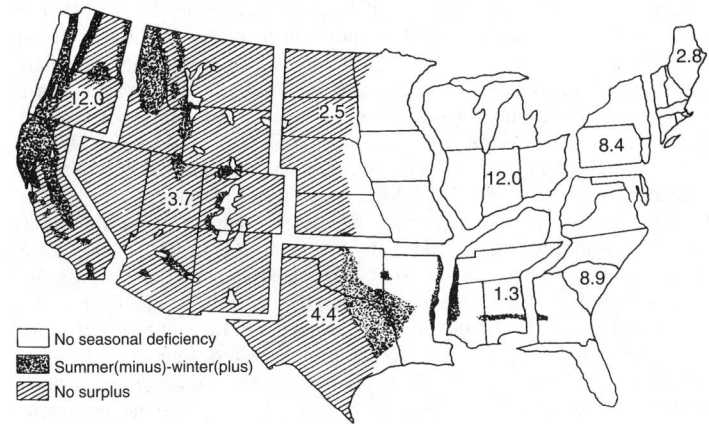

No seasonal deficiency
Summer(minus)-winter(plus)
No surplus

Fig. 1. Water supply versus population increases. Figures indicate expected increases (millions of people) by the year 2000. (*U.S. Department of the Interior*)

Water resources must be evaluated in the long term because unusually wet or dry winters (such as 1980–1981), while of near-crisis proportions, do not represent the average of a decade or more. What is important is to watch trends that appear to be consistent.

Widespread flooding in the midwestern United States during the summer of 1993 is regarded by some experts as a 1 in 100 year experience. Every century experiences a few weather phenomena that statistically contribute to the minimums and maximums of data, but do not necessarily predict any permanent changes in the weather picture for a given region. Final analyses of data from the 1993 flooding will not be completed for another year or two.

The environmental problem of finding additional sources of clean water have some parallels with the problem of finding new ways for disposing of wastes. These are indeed limited. Again, the subject of *waste reduction* enters into the picture. Whereas, fundamentally, pollution equates with cutting back on the production of wastes, water shortages equate markedly with reducing the waste of water. This, too, presents a variety of sociological preferences and concerns. Persons who live in communities where water shortages approach near-crisis proportions are fully aware of how inconvenient it is to cut back on water consumption. Over the years, a few imaginative suggestions have been made. Weather alteration in an effort to produce rain essentially has been abandoned by technologists. Dam building, in addition to creating hydroelectric power, also contributes to smoothing out the water supply for many regions. In recent years, however, a substantial citizen's movement against creating dams has arisen.

Although nearing the fantasy level, icebergs have not been ruled out technically as "portable and potable" sources of excellent water.

While desalination now serves a number of the arid regions of the world (see also **Desalination**) and the costs of desalination have been reduced because of improved processes, the economics of processing discourage the use of desalination in the less arid regions of the world. However, if the world's population continues to expand geometrically, the price per gallon or liter of drinking water, for example, may rise beyond belief.

Water from Icebergs

The concept of obtaining fresh water from icebergs dates back many years. One of the earlier proponents was Isaacs (Scripps Institute of Oceanography) who described the concept in the 1940s. The concept has been characterized by a cycling of interest over the years, but with little follow-through among world scientists and planners. In 1977, the first conference on iceberg utilization of major proportions was held at Iowa State University, with over 200 scientists, consultants, and representatives of private firms from 18 nations present to consider the technical, economic, environmental, and legal problems that may be involved in transporting and exploiting icebergs. In discussions over the last several years, most attention has been given to icebergs from the Antarctic rather than the Arctic region, principally those icebergs that break away from the Ross Ice Shelf.

Melted water from an iceberg is extremely pure, with only traces of rock fragments and with almost complete absence of trapped organic matter. Contamination has been estimated at one part per billion and thus far superior to other fresh water unless it is distilled. The Iowa State conference was financed in part by the National Science Foundation and Saudi Arabia.

In contemplating the possible use of icebergs as a future source of fresh water, it is interesting to note that although there are about 1.4 billion cubic kilometers of water on Earth, only about 9 million cubic kilometers (six-tenths of one per cent) of the total is both liquid and fresh. The useful supply tends to be about 10–20% of the total precipitation per year, or about 10,000–20,000 cubic kilometers.

The ice sheet in the Antarctic accumulates an input of precipitation equivalent to about 2000 cubic kilometers of water per year. About half of this total forms into tabular (flat) icebergs, each equivalent to about 200 cubic kilometers of water. The slabs are from 200 to 250 meters thick and up to about 0.5 kilometer wide. Greenland icebergs, in contrast, are more irregular in shape and size. Antarctic icebergs would have to be transported about 4,000 miles (6,436 kilometers) to the southwestern coast of Australia; about 9,500 miles (15,286 kilometers) to southern California; and about 12,000 miles (19,308 kilometers) to the Middle East. Several studies have been made as regards towing icebergs and the consensus is that the technology for this already exists. Towing of some icebergs already has been done in connection with oil drilling activities, where tugs are used to deflect Greenland-originated icebergs from the path of drillships and platforms. Also, many attempts have been made to destroy (break up) icebergs through the use of explosives to prevent the icebergs from drifting into major sea lanes. These experiments have largely been unsuccessful because one kilogram of explosive will break up only about 4 cubic meters of ice. Even though ice possesses a low strength and density, it reacts to blasting much like ordinary rock.

During the 1990s, the use of icebergs as a source of fresh water remained a topic of considerable conjecture. Topics discussed at the Iowa State Conference included: (1) use of earth resources satellites to find the most suitable icebergs for this purpose; (2) study of various transportation modes, such as tugs for pulling, semi-submersibles and submarines for pushing, and propellers mounted on icebergs to make them self-transportable; (3) ways to minimize melting during transport, e.g., stretching plastic sheets over the iceberg or spraying its surface with urethane foam; (4) investigation of legal problems relating to ownership of Antarctic ice; (5) the numerous environmental effects which such actions might provoke; (6) possible use of icebergs as energy sources, through harnessing of thermal and salinity gradients. If the concept takes hold, it would seem that the first tests would be conducted in the Southern Hemisphere.

For further details, see "The Iceberg Cometh," by W.F. Weeks and Malcolm Mellor, *Technol. Rev. (MIT)*, pp. 66–75 (August/September, 1979), and *Science*, **198**, 274–276 (1977).

Many of the scientific aspects of water resources are described in several entries in this encyclopedia. See also **Connate Water**; **Desalination**; **Groundwater** and **Water Pollution**.

WATER SOLUBLE POLYMERS. Water-soluble polymers find application in a wide variety of areas that include polymers as food sources, plasma substitutes, and as diluents in medical prescriptions. Other areas of importance for water-soluble polymers include detergents, cosmetics, sewage treatment, stabilizing agents in the production of commodity plastics, rheology modifiers in the various processes for petroleum, textile, paper, and latex coatings production. The water-soluble polymers discussed in this article have significant commercial impact.

Hydrophilic Groups. Water solubility can be achieved through hydrophilic units in the backbone of a polymer, such as O and N atoms that supply lone-pair electrons for hydrogen bonding to water. Solubility in water is also achieved with hydrophilic side groups (e.g., OH, NH_2, CO_2^-, SO_3^-). Truly unique in its ability to interact and promote water solubility is the $-O-CH_2-CH_2-$ group. The interactions of these groups with water and their placement in the polymer structure influence the water solubility of the polymer and its hydrodynamic volume.

Viscosity Efficiency. A majority of the applications of water-soluble polymers revolve around their role in increasing the viscosity of water solutions. The hydrodynamic volume of the polymer influences its viscosity efficiency, and thereby its ability to modify the rheology of an application formulation. The primary parameter influencing the hydrodynamic volume is the polymer's molecular weight. The second is in the conformational rigidity or extension of the polymer chain in solution. For example, carbohydrate polymers contain repeating ring structures that facilitate a more rigid structure than observed in non-ionic synthetic polymers. At a given molecular weight this effects a greater viscosity efficiency. The

rigidity also can be increased by inclusion of charged groups in both synthetic and carbohydrate polymers. Such groups lead to electrostatic repulsions and extension of the chain in deionized aqueous solutions and an increase in the hydrodynamic volume of the polymer. Greater rigidity also can be achieved in carbohydrate polymers if helical conformations are realized.

Carbohydrate Polymers

An anhydroglucose ring or glucopyranosyl unit (Fig. 1a) is the structural unit on which human metabolism is dependent. The anhydroglucose unit provides, through the four hydroxyl units, a diversity of polymer structures that can be formed through variations in positional bonding between the rings (i.e., $1 \rightarrow 2$, $1 \rightarrow 3$, $1 \rightarrow 4$, and $1 \rightarrow 6$). Examples of different positional bonding in carbohydrate polymers are illustrated in Figure 2.

Symmetrical $1 \rightarrow 4$ bonding provides the world's most abundant polymer, cellulose (Fig. 2a). The bonds linking hydroxyl units in the plane of the puckered ring are referred to as equatorial bonds. The glucopyranosyl unit is also present in amylose (Fig. 1b). The variance between cellulose and amylose (the latter is a component of starch) is in the interunit bonding between anhydroglucose rings. In cellulose, the rings are connected through equatorial–equatorial bonding (beta-linkage); in amylose the $1 \rightarrow 4$ inter-ring bonding is equatorial–axial (alpha-linkage; axial denotes a bond perpendicular to the general plane of the ring). The beta-linkage in cellulose, complemented by the intrahydrogen bonding among rings facilitates a linear projection of the polymer. This,

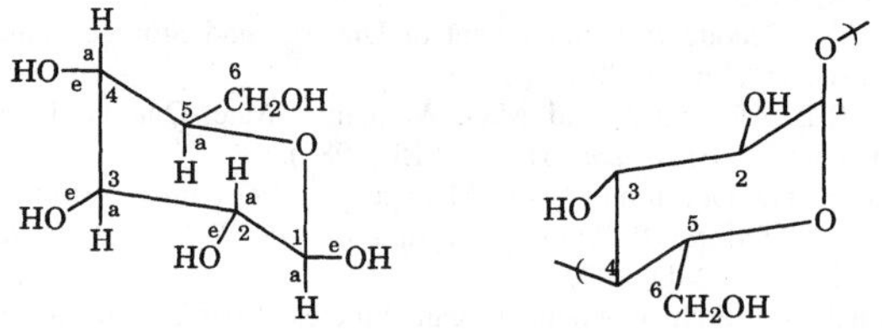

Fig. 1. (a) The glucopyranosyl basic unit, where e = equatorial and a = axial bond; (b) amylose with equatorial–axial interunit bonding (the C—H axial bonds have been omitted).

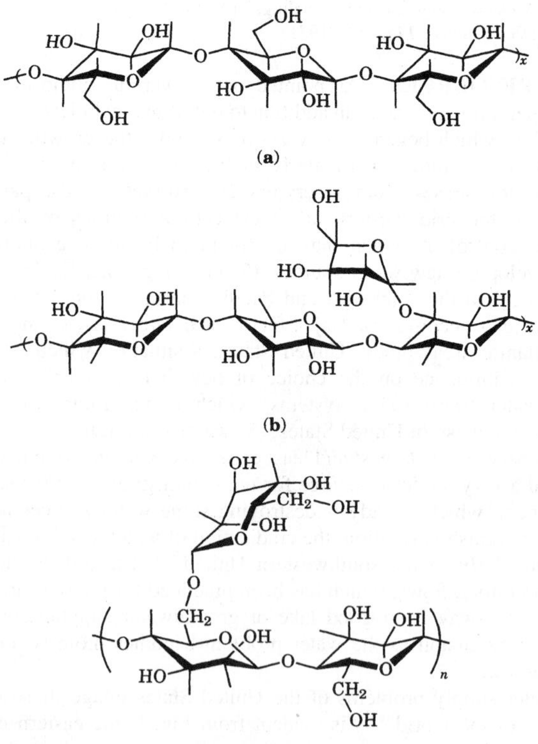

Fig. 2. Examples of carbohydrate polymers with interunit and branch position differences: (a) cellulose; (b) flaxseed gum; (c) guaran

complemented by interhydrogen bonding among polymer chains, provides crystallinity and a rigid structure. To transform the world's most abundant polymer, cellulose, into a water-soluble species requires replacement of some of the hydroxyls to disrupt the extensive hydrogen bonding.

Amylose (Fig. 1**b**), with alpha-interconnecting linkages, is soluble in hot water, (unlike cellulose), but it retrogrades in low-temperature aqueous solutions into a helical conformation and precipitates. Glucopyranose branching from the C-6 hydroxyl provides solubility at low temperatures. Thus this anomeric bonding difference (i.e., alpha-($1 \rightarrow 4$) instead of the beta ($1 \rightarrow 4$) linkage in cellulose) provides a readily accessible source of energy and is designated as a storage source in nature.

The anomeric difference (i.e., the alpha- and beta-linkages) between cellulose and amylose is important. Of greater importance in determining the aqueous solution properties of water-soluble polymers is the interunit bonding patterns between rings. As with amylose, branches from the C-6 position promote solubility in water. Two naturally occurring carbohydrate polymers with structural main chain similarity to cellulose (i.e., in the beta-($1 \rightarrow 4$) linkage), but with water solubility without derivatization due to side branch units, are the nonionic fraction of flaxseed gum and guaran (Fig. 2**b** and 2**c**, respectively).

Cellulose

Commercial Derivatization of Cellulose. Cellulose, the world's most abundant polymer, is derivatized for use in a variety of markets. Cellulose ethers are an important segment of water-soluble polymers.

Commercial derivatization of cellulose begins with the addition of sodium hydroxide to form alkali cellulose (AC) (eq. 1, R = carbohydrate).

$$ROH + NaOH \rightleftharpoons RO^-Na^+ + H_2O \qquad (1)$$

The AC may react with methyl chloride or alpha-chloroacetic acid via a direct displacement reaction (eq. 2). The derivatives would be methyl cellulose or carboxymethyl cellulose.

$$RO^-Na^+ + R'X \longrightarrow ROR' + NaX \qquad (2)$$

Alternatively, the AC may react with oxiranes e.g., ethylene oxide ($R'' =$ H) or propylene oxide ($R'' = CH_3$) (eq. 3).

$$RO^-Na^+ + CH_2 \overset{O}{-} CH \longrightarrow RO-CH_2 \overset{O^-Na^+}{-} CH \qquad (3)$$
$$\underset{R''}{\mid} \qquad\qquad \underset{R''}{\mid}$$

The derivatives are hydroxyethyl and hydroxypropyl cellulose. All four derivatives find numerous applications, and there are other reactants that can be added to cellulose, including the mixed addition of reactants leading to adducts of commercial significance. See also **Cellulose Ester Plastics (Organic).**

Biosynthesis. Although cellulose can be derivatized, such materials do not provide the optimum properties desired in many applications, such as retention of viscosity at higher solution temperatures, greater mechanical stability, and greater thickening efficiency. These properties can be approached with carbohydrate polymers in helical conformations, which can be achieved in some carbohydrate polymers prepared by fermentation processes. Yeasts have an advantage over fermentation processes using bacteria, because the fermentation can be conducted in low pH media. Yeast are also larger and easier to remove by filtration. However, the most successful commercial fermentation polymer is XCPS, synthesized by a bacteria, *Xanthomonas campestris*. In this polymer the main chain is simply cellulose, but with three pyranosyl rings branched from the C-3 position of every other repeating ring. This arrangement promotes a helical conformation. See also **Yeasts and Molds.**

Recently, two fermentation polymers have produced optimum properties through variations in positional and interunit bonding patterns: gellan and wellan.

Other Considerations. With multiple hydroxyl units on every repeating ring, most commercial carbohydrate particles are surface treated with glyoxal. After adequate dispersion of the particles in water, a small quantity of base solution is added to remove the acetal cross-linked structure. The individual particles then readily hydrate without the agglomeration of partially hydrated particles before complete hydration is achieved. The segmentally rigid and conformationally rigid (in helical polymers) ring structures provide a more viscous aqueous shear viscosity, but lower extensional solution viscosity for a given molecular weight. The

rigidity provides greater mechanical stability relative to synthetic polymers discussed below.

Synthetic Water-Soluble Polymers

Polyoxyethylene. Synthetic polymers with a variety of compositionally similar chemical structures are as follows. Based on polarity, poly(oxymethylene) (**1**) would be expected to be water soluble. It is a highly crystalline polymer used in engineering plastics but it is not water-soluble. Polyoxypropylene (PPO) (**2**) and poly(methyl vinyl ether) (PMVE) (**3**) have

$$\begin{array}{ccc} -\!\!\left(CH_2O\right)\!\!{}_{\overline{n}} & -\!\!\left(CH_2-CHO\right)\!\!{}_{\overline{n}} & -\!\!\left(CH_2-CH\right)\!\!{}_{\overline{n}} \\ & \underset{CH_3}{\mid} & \underset{OCH_3}{\mid} \\ (\mathbf{1}) & (\mathbf{2}) & (\mathbf{3}) \end{array}$$

identical chemical compositions; PMVE is water soluble up to a modest temperature ($\sim 50°C$), but PPO is not soluble in water. PPO is soluble only in the oligomeric form with less than 10 PO units. PO defines oxypropyl or propylene oxide monomer overcome the terminal hydroxyl groups facilitating water solubility when there are more than 10 repeating units. In view of the lack of water solubility of two compositionally similar polymers, the water solubility of polyoxyethylene POE (**4**) is notable. It occurs because of the unique interaction of water with the oxyethylene chain.

$$-\!\!\left(CH_2-CH_2O\right)\!\!{}_{\overline{n}}$$
$$(\mathbf{4})$$

Polyoxyethylene (POE) is synthesized employing different catalyst systems depending on the desired mol wt. For molecular weight below 20,000 (generally referred to as poly(ethylene glycol)s, PEGs), base or the Na^+ or K^+ alkoxides of methanol or butanol are used. Without the numerous hydroxyls present in carbohydrate polymers, POE particle structures cannot be treated to minimize their hydration prior to dissolution. The particle surfaces are covered with silica to minimize hydration and blocking of POE particles on storage. POE and PEGs of various molecular weights have found numerous applications.

Commodity Chain-Growth Polymers. Two of the largest commodity water-soluble polymers are poly(vinyl alcohol) (PVA) and polyacrylamide (PAM). They are prepared by the free-radical initiation of vinyl monomers, a chain-growth polymerization technique.

Poly(vinyl alcohol). The vinyl alcohol monomer is unstable and isomerizes to acetaldehyde. The polymer is obtained by the hydrolysis of poly(vinyl acetate). The vinyl acetate monomer produces a very high energy radical during chain-growth propagation. This accounts for the high chain transfer to monomer, a higher than normal head to head addition of propagating species, and grafting of some of the propagating species to polymer that is formed during an earlier stage of the polymerization. This leads to a more complex variation in structure than observed in PAM polymers, and this, with the differences that are realized with different methods of hydrolysis, can result in different Poly(vinyl alcohol) (PVA). These factors, on hydrolysis, lead to PVA below 50,000 molecular weights. See also **Vinyl Acetal Polymers.**

PVA is isomeric with POE; however, it can enter into hydrogen bonds with both water and with the other hydroxyl units of the repeating polymer chain, forming both inter- and intrahydrogen bonds. The extensive hydrogen bonding can lead to crystallinity, an occurrence that complicates its water solubility. Commercial PVA is essentially atactic. With most chain-growth polymers, crystallinity is associated with stereoregularity, but the small size of the hydroxyl substituent group promotes crystallinity even in the atactic polymer. For this reason poly(vinyl acetate) is seldom hydrolyzed completely; it is manufactured retaining three acetate levels: 25, 12, and 2 mole percents (these numbers represent averages). With higher acetate percentages the polymer is more surface active, which is important in its role as a suspending agent in poly(vinyl chloride) commodity resin production. Also, it is readily soluble in water at ambient temperatures. With low acetate percentages, PVA is difficult to dissolve unless the water temperature is high, and on cooling the low acetate PVA may precipitate.

Hydrolyzed Polyacrylamide. HPAM can be prepared by a free-radical process in which acrylamide is copolymerized with incremental amounts of acrylic acid or through homopolymerization of acrylamide followed by hydrolysis of some of the amide groups to carboxylate units. The carboxylated units, ionized, decrease adsorption on subterranean substrates, in proportion to the number of units, an important parameter in petroleum recovery processes. In waste treatment processes cationic acrylamide comonomer units are often used to increase adsorption and thereby flocculation of solids in wastewater. See also **Acrylamide Polymers**. Because the synthesis of HPAM can be conducted in water, the problem of dissolution in many applications is addressed by polymerizing the monomer in water-in-oil emulsions.

Poly(vinyl pyrrolidone). Another commercial polymer with significant usage is PVP. It was developed in World War II as a plasma substitute for blood. This monomer polymerizes faster in 50% water than it does in bulk, an abnormality inconsistent with general polymerization kinetics. This may be due to a complex with water that activates the monomer; it may also be related to the impurities in the monomer that are difficult to remove. See also **Vinyl Acetal Polymers**.

Poly(acrylic acid) and Poly(methacrylic acid). Poly(acrylic acid) (PAA) may be prepared by polymerization of the monomer with conventional free-radical initiators using the monomer either undiluted (with cross-linker for superadsorber applications) or in aqueous solution. Photochemical polymerization (sensitized by benzoin) of methyl acrylate in ethanol solution at $-78°C$ provides a syndiotactic form that can be hydrolyzed to syndiotactic PAA.

Cationic Water-Soluble Polymers

Cationic monomers are used to enhance adsorption on waste solids and facilitate flocculation. One of the first used in water treatment processes (**5**) is obtained by the cyclization of dimethyldiallylammonium chloride in $60-70$ wt % aqueous solution. Another cationic water-soluble polymer, poly(dimethylamine-*co*-epichlorohydrin), prepared by the step-growth

(5)

polymerization of dimethyl amine and epichlorohydrin, is also used for adsorption on clays.

Inorganic Water-Soluble Polymers

Two inorganic water-soluble polymers, both polyelectrolytes in their sodium salt forms, have been known for some time: poly(phosphoric acid) and poly(silicic acid). A more exciting inorganic water-soluble polymer with nonionic characteristics has been reported. This family of phosphazene polymers is prepared by the ring-opening polymerization of a heterocyclic monomer (**6**) followed by replacement of the chlorine atoms in the resultant polymer.

(6)

The hydrophobic inorganic backbone provides an easy route to variable water-solubilizing side nonionic or ionizing groups. These are stable to hydrolysis at room temperature.

One of the main advantages of water-soluble polyphosphazenes is the ease with which water-solubilizing side groups can sufficiently cross-link in a stable matrix with high energy radiation such as x-rays, gamma-rays, electron beams, or ultraviolet light.

The versatility of water-soluble polyphosphazenes is in the variations in the structures that can be prepared. Structures with a low glass-transition temperature backbone can be modified with a variety of versatile side units.

New Commercial Water-Soluble Polymers

Two recently developed water-soluble polymers have achieved limited market acceptance.

One product is poly(2-ethyl-2-oxazoline) (PEOX). It is prepared by the ring-opening polymerization of 2-ethyl-2-oxazoline with a cationic initiator. Most of the polymer's characteristics stem from its molecular structure, which like POE, promotes solubility in a variety of solvents in addition to water. It exhibits Newtonian rheology and is mechanically stable relative to other thermoplastics. It also forms miscible blends with a variety of other polymers. Another product is prepared from *N*-ethenylformamide formed from the reaction of acetaldehyde and formamide. The vinyl amide is polymerized with a free-radical initiator, then hydrolyzed.

The protonated form of poly(vinyl amine) (PVAm–HCl) has two advantages over many cationic polymers: high cationic charge densities are possible and the pendent primary amines have high reactivity. It has been applied in water treatment, paper making, and textiles.

Hydrophobe-Modification of Water-Soluble Polymers

Although many of the new water-soluble polymers discussed above have not achieved large-scale commercial acceptance, there is a class that has achieved outstanding success since the early 1980s: hydrophobically modified water-soluble polymers (HM-WSPs). They have filled certain voids in a number of applications that include cosmetic, paper, architectural, and original equipment manufacturing (OEM) coating areas and have found unsuspected application in the airplane de-icers market. The driving force for the development of HMWSPs is threefold in most application areas: 1. The achievement of high viscosities at low shear rates without high molecular weights. 2. Minimization of the elastic behavior of the fluid at high deformation rates that are present when high molecular weight water-soluble polymers are used. 3. Providing colloidal stability to disperse phases in aqueous media, not achievable with traditional water-soluble polymers.

Preparation of hydrophobically modified, water-soluble polymer in aqueous media by a chain-growth mechanism presents a unique challenge in that the hydrophobically modified monomers are surface active and form micelles.

The hydrophobe modification of acrylic acid represents an important class of hydrophobe-modified thickeners prepared by a chain-growth free-radical process. They differ slightly from other examples in that these products are generally cross-linked.

Hydrophobe-modification of hydroxyethylcellulose produces what should be considered model associative thickeners, for the distribution of hydroxyethyl units has been characterized. The commercial material, a Hercules product, contains three hydrophobes per chain.

HEUR associative thickeners are in effect poly(oxyethylene) polymers that contain terminal hydrophobe units. They can be synthesized via esterification with monoacids, tosylation reactions, or direct reaction with monoisocyanates.

J. Edward Glass
North Dakota State University

Additional Reading

Glass, J. E. ed.: *Water-Soluble Polymers: Beauty with Performance,* Advances in Chemistry Series 213, American Chemical Society, Washington, DC, 1986.

Glass, J. E. ed.: *Polymers in Aqueous Media, Performance through Association,* Advances in Chemistry 223, American Chemical Society, Washington, DC, 1989.

Glass, J. E. ed.: *Hydrophilic Polymers: Performance with Environmental Acceptance,* Advances in Chemistry Series 248, American Society, Washington, DC, 1995.

Molyneux, P.: *Water-Soluble Synthetic Polymers: Properties and Behavior,* Vols. I and II, CRC Press, Boca Raton, FL, 1982.

WATER TREATMENT (Boiler). One of the most critical and exacting requirements for pretreating water prior to industrial use is found in connection with the operation of modern boilers. Many of the basic principles of water treatment are encountered for this application.[1] The advantages of modern boilers can be realized to the fullest only if proper attention is given to water treatment. No boiler can operate efficiently or dependably if its heat-transfer surfaces are allowed to foul with scale or if corrosion is permitted to occur.

Water treatment must include conditioning of the:

1. Raw-water supply.
2. Condensate returns from process steam or turbines.
3. Boiler water.

Proper conditioning will result in:

[1] Abstracted, with permission, from "Steam/Its Generation and Use," 39th edition, Babcock & Wilcox, New York (1978).

1. Freedom from deposits on internal surfaces.
2. Absence of corrosion of internal surfaces.
3. Prevention of carry-over of boiler water solids into the steam, caused by foaming and/or high total dissolved solids.

Some definitions of the water terminology in various parts of the boiler cycle are desirable. Steam that is condensed and returned to the boiler system is termed condensate. Steam lost due to process requirements, blowdown or leakage out of the system, has to be replaced; the replacement water added to the system is termed makeup water. The condensate together with the makeup water comprise the feedwater to the boiler. In some plants only a small percentage of condensate is returned; in others, almost all the steam generated is recovered as condensate. Feedwater enters the boiler and is evaporated into steam, leaving behind solids to concentrate in the boiler water. If the concentration of solids in the boiler water exceeds certain limits, the quality of steam can be impaired by carry-over. Also, boiler-water solids may settle out on the boiler surfaces as sludge. The concentration of solids in the boiler water can be controlled by removing a portion of the water either intermittently or continuously. This bleeding of a portion of the boiler water from the drum is termed blowdown.

In Universal-Pressure boilers, there are no drums to concentrate the boiler-water salts and impurities, and blowdown is not utilized. Purification takes place by continuously passing all or part of the condensate through demineralizers in a process called condensate polishing.

The treatment of raw water, condensate, feedwater and boiler water, and the subjects of carry-over and steam purity are considered in detail in the sections which follow.

Raw-Water Treatment

Water never exists in the pure form. All natural waters contain varying amounts of dissolved and suspended matter. The type and amount of matter in water varies with the source, such as lake, river, well or rain, and also with the section of the country.

As rain, water brings into solution the atmospheric gases of oxygen, nitrogen and carbon dioxide. As it percolates through the soil, it dissolves and picks up many minerals harmful to boiler operation. Surface waters frequently contain organic matter that must be removed before the water is satisfactory for use in a boiler.

Suspended solids are those that do not dissolve in water and can be removed or separated by filtration. Examples of suspended solids are mud, silt, clay and some metallic oxides.

Dissolved solids are those which are in solution and cannot be removed by filtration. The major dissolved materials in water are silica, iron, calcium, magnesium and sodium. Metallic constituents occur in various combinations with bicarbonate, carbonate, sulfate and chloride radicals. In solution these materials divide into their component parts called ions, which carry an electrical charge. The metal ions carry a positive charge and are referred to as cations. The bicarbonate, carbonate, sulfate and chloride ions are negatively charged and are referred to as anions.

Scaling occurs when calcium or magnesium compounds in the water (*water hardness*) precipitate and adhere to boiler internal surfaces. These hardness compounds become less soluble as temperature increases, causing them to separate from solution. Scaling causes damage to heat-transfer surfaces by decreasing the heat-exchange capability. The result is overheating of tubes, followed by failure and equipment damage.

Porous deposits will allow concentration of boiler-water solids. This concentration of boiler-water solids, particularly if strong alkalies are present, will result in severe corrosion of the tube surfaces.

Since water impurities cause boiler problems, careful consideration must be given to the quality of water in the boiler. External treatment of water is required when the amount of one or more of the feedwater impurities is too high to be tolerated by the boiler system.

The selection of equipment for raw-water preparation should only be made after a careful analysis of the raw-water composition, quantity of makeup required, boiler type and operating pressure. Generally, the first step in the water processing involves coagulation and filtration of the suspended material. Natural settling in quiescent water will remove relatively coarse suspended solids. The required settling time depends on specific gravity, shape and size of particles and currents within the settling basin. This process can be speeded up by coagulation. Coagulation is the process by which finely divided materials are combined by the use of chemicals to produce large particles capable of rapid settling. Typical coagulant chemicals are alum and iron sulfate. The preliminary treatment involves chlorination of the water for the destruction of the organic matter. Several manufacturers offer equipment to operate on a completely automated basis, as illustrated in Fig. 1.

Following coagulation, settling and chlorination, the water should be passed through filters. Filtration removes the finely divided suspended particles not removed in the coagulation and settling tanks. Special equipment such as activated-charcoal filters may be necessary to remove the final traces of organic and excess chlorine. After the removal of the suspended material in the raw water, the hardness of scale-forming materials are still present in solution. Further treatment is required to remove these materials. This treatment consists of precipitating the hardness constituents and/or exchanging the hardness for non-hardness constituents in a process called ion exchange. Brief descriptions of each of these processes follow. It is recommended that assistance by obtained from a water consultant in order to select the best process and equipment for a specific installation.

Sodium-Cycle Softening. This process, called sodium zeolite softening, utilizes resin materials that have the property of exchanging the hardness constituents, calcium and magnesium, for sodium. The process continues until the sodium ions become depleted or, conversely, the resin capacity to absorb the calcium and magnesium no longer exists. When this occurs, the resin is said to be exhausted, and is regenerated by passing a solution of salt through it.

Water, after passing through the zeolite process, contains as much bicarbonate, sulfate and chloride as the raw water, only the calcium and magnesium having been exchanged for the sodium ions. There is no

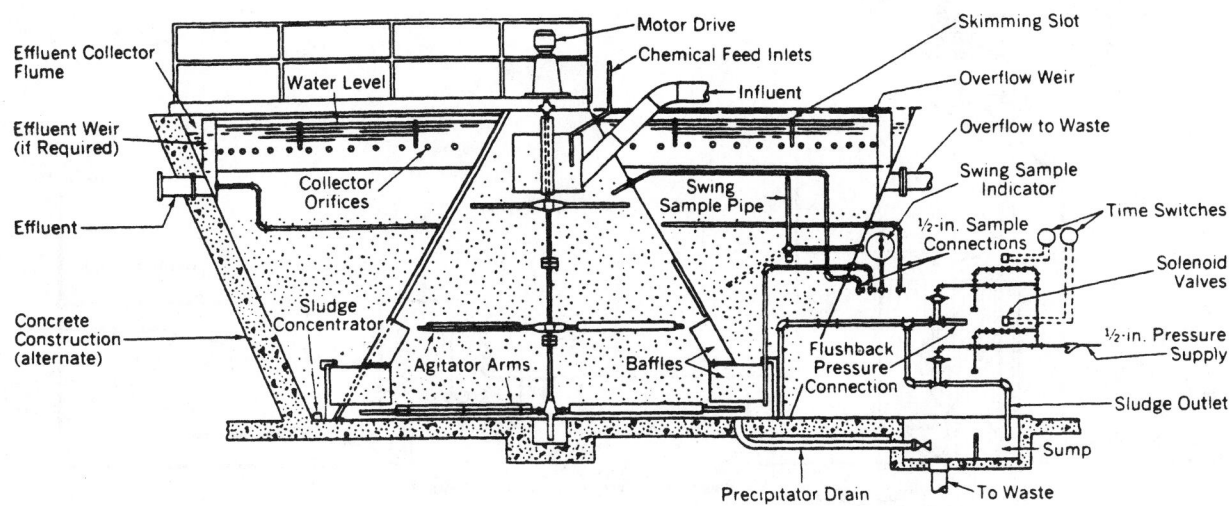

Fig. 1.   Sludge contact softener. (*Betz "Handbook of Industrial Water Conditioning."*)

reduction in the overall amount of dissolved solids and neither is there a reduction of alkalinity content. When it is necessary to reduce the amount of total dissolved solids, zeolite must be coupled with other methods such as hot lime zeolite softening. The reduction of alkalinity is discussed under *Hot Lime Zeolite–Split Stream Softening*.

Hot-Lime Zeolite Softening. In this process hydrated lime is employed to react with the bicarbonate alkalinity of the raw water. The precipitate is calcium carbonate and is filtered from the solution. To reduce silica, the natural magnesium of the raw supply can be precipitated as magnesium hydroxide, which acts as a natural absorbent for silica. These reactions are carried out in a vat or tank that is located just head of the zeolite softener tank. The effluent from this tank is filtered and then introduced into the zeolite softener. There is always some residual hardness leakage from the hot-process softener to be removed in the final zeolite process. The hot lime process operates at about 220°F (104°C). At this temperature the potential for the exchange of sodium for hardness ions is greater than at ambient temperature, and the result is a lower hardness effluent than is achieved at ambient temperatures. This system is shown schematically in Fig. 2.

Hot Lime Zeolite–Split Stream Softening. Many raw waters softened by the first two processes would contain more sodium bicarbonate than is acceptable for boiler feedwater purposes. Sodium bicarbonate will decompose in the boiler water to give caustic soda. Caustic soda in high concentrations is corrosive and promotes foaming. The American Boiler Manufacturers Association has adopted the standard that the alkalinity content should not exceed 20% of the total solids of the boiler water. Split stream softening provides a means for reducing the alkalinity content.

This requires a second zeolite tank that has a zeolite resin in the hydrogen form in addition to the usual tank with the resin in the sodium form. The two tanks are operated in parallel. In one tank, calcium and magnesium ions are replaced by hydrogen ions. The effluent from this tank with the resin in hydrogen form is on the acid side and has a lower total-solids content. The total flow can be proportioned between the two tanks to produce an effluent with any desired alkalinity as well as excellent hardness removal. When the hydrogen resin is exhausted, it is regenerated with acid.

Demineralization and Evaporation. At drum pressures over 1000 psi (68 atm), demineralization or evaporation of the makeup water is generally desirable. A water that closely approaches theoretical chemical purity can be obtained by either of these processes.

Evaporation as a source of purified water does not involve ion exchange. It is actually a distillation process, consisting of evaporation, leaving most of the solids behind, and recondensation of the purified water. While evaporated water is quite satisfactory, economics generally favors demineralization.

Demineralization, like the zeolite process, involves ion exchange. The metal ions are replaced with hydrogen ions by means of the process and equipment described for the hydrogen-zeolite system (see **Hot Lime Zeolite—Split Stream Softening**, previously described). In addition, the salt anions (bicarbonate, carbonate, sulfate and chloride) are replaced by hydroxide ions by means of a specially prepared resin saturated with hydroxide ions.

The two types of resins can be located in separate tanks. In this system, the two tanks are operated in series in a cation-anion sequence. The anion resin is regenerated after exhaustion with a solution of sodium hydroxide. The cation resin is regenerated with an acid, either hydrochloric or sulfuric. Some leakage of cations always occurs in a cation exchanger, resulting in leakage of alkalinity from the anion exchanger.

In another arrangement, known as the mixed-bed demineralizer, the two types of resins are mixed together in a single tank. In the mixed-bed demineralizer, cation and anion exchanges take place virtually simultaneously, resulting in a single irreversible reaction that goes to completion. Regeneration is possible in a mixed bed because the two resins can be hydraulically separated into distinct beds. The cation resin is approximately twice as dense as the anion resin. Resins can be regenerated in place or sluiced to external tanks for this purpose.

The raw water for drum-type boilers operating above 2000 psi (136 atm) drum pressure and for once-through units should be prepared by passing water through a mixed-bed demineralizer as a final step before adding to the cycle.

The effluent from demineralization is approximately neutral. With nearly all salts removed, the problem of chemical control of the boiler water is minimized.

Treatment of Condensate

In most cases, condensate does not require treatment prior to reuse. Makeup water is added directly to the condensate to form boiler feedwater. In some cases, however, especially where steam is used in industrial processes, the steam condensate is contaminated by corrosion products or by the in-leakage of cooling water or substances used in the process. Hence steps must be taken to reduce corrosion or to remove the undesirable substances before the condensate is recycled to the boiler as feedwater.

The presence of acidic gases in steam makes the condensate acidic with consequent corrosion of metal surfaces. In such cases, the corrosion rate can be reduced by feeding to the boiler water chemicals that produce alkaline gases in the steam. The addition of neutralizing and filming amines to boiler water or to condensate to minimize corrosion by condensate and feedwater is discussed later under *Control of* pH.

Many types of contaminants can be introduced to condensate by various industrial processes. They include liquids, such as oil and hydrocarbons, as well as all sorts of dissolved and suspended materials. Each installation must be studied for potential sources of contamination. The recommendations of a water consultant should be obtained to assist in determining corrective treatment.

Fig. 3 shows a condensate purification system used in a paper-mill boiler cycle. The resin beds not only remove dissolved impurities by ion exchange but also serve as filters to remove suspended solids. It is necessary to backwash and regenerate these resin beds periodically. Several types of condensate purification systems are available from various vendors. Some of these are capable of operation at temperatures as high as 300°F (149°C).

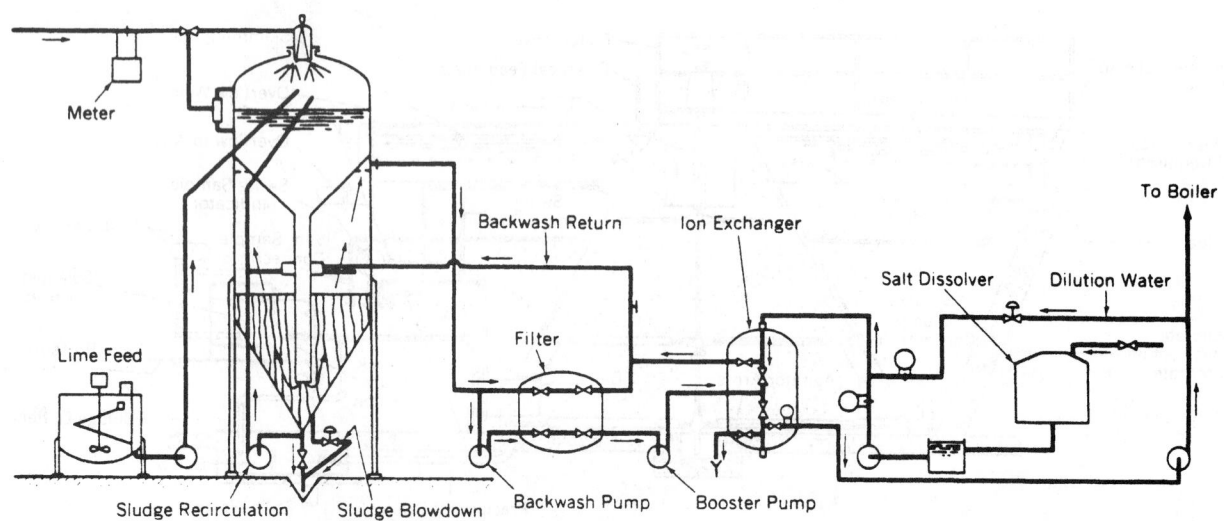

Fig. 2.   Flow sheet of typical hot lime zeolite softening process. (*Betz "Handbook of Industrial Water Conditioning."*)

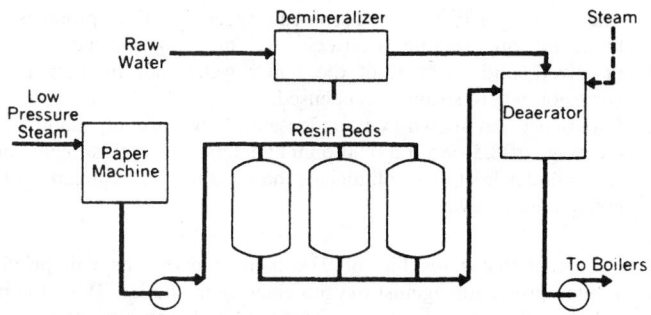

Fig. 3.   Condensate purification system. (*Cochrane Div., Crane Company.*)

1. Improved turbine capability and efficiency.
2. Shorter unit start-up time.
3. Protection from the effects of condenser leakage.
4. Longer intervals between acid cleanings.

Two types of condensate-polishing systems are available, both capable of removing suspended material, such as corrosion products, as well as ionized solids.

Deep-bed demineralizers operate at flow rates of 40–60 gpm per sq ft (1624–2441 liters/minute/square meter) of bed cross section. This type requires external regeneration facilities. The deep-bed system has a higher initial cost with possible lower operating costs, especially during initial unit start-up. Its greater capacity for removing ionized solids permits continued operation with small amounts of condenser leakage. The deep-bed system is usually operated with the cation resin in the hydrogen form, but the ammonium form can also be used. The cost of regenerating the resin in the ammonium form is greater than for the hydrogen form, but the time period between regenerations is much longer.

The cartridge-tubular type, such as the Powdex system, uses smaller amounts of disposable resins, eliminating the need for regeneration. The Powdex system uses cation resin in the ammonium form. Because of the many considerations involved, an evaluation of alternate types should be made before a system is selected for any given installation.

Use of the EMF with ion exchange equipment in a condensate purification system provides the ultimate in condensate polishing. Location of the EMF upstream of the resin beds offers the advantages of longer operating periods for the beds, thereby reducing the frequency of bed regeneration, and, subsequently, lesser costs for regeneration chemicals.

One such example of a condensate purification system is the use of ion exchange equipment as shown in Fig. 3 for a typical paper-mill boiler cycle. The resin beds not only remove dissolved impurities by ion exchange, but also serve as filters to remove suspended solids, such as products of corrosion. It is necessary to backwash and regenerate such resin beds periodically. These resin beds can be purchased for in-place regeneration or for regeneration in external tanks. External regeneration facilitates more efficient removal of suspended metal oxides from resin beds.

Where significant quantities of magnetic oxide or other magnetic species are present in the condensate-feedwater system, the application of an electro-magnetic filter (EMF) to effectively remove these suspended solids has been proved to be quite successful in both operation and performance. The EMF, Fig. 4, consists of a pressure vessel, coil, spheres, and a power control unit. The pressure vessel is constructed of a non-magnetic material and contains a featherbed of magnetizable spheres of approximately $\frac{1}{4}$-in. (6.5-mm) diam. The pressure vessel is surrounded by a magnetic coil which is supplied with direct current from the power control unit. Flow is upward through the filter both during operation and when flushing the filter, thereby minimizing and simplifying the piping and valving system. Some other advantages that the electro-magnetic filter offers are low pressure drop through the filter, minimum quantity of flush water, entire backwash process only takes several minutes and no chemicals are required in its use.

Laboratory analysis of flush water indicates that some non-magnetic iron oxide may also be retained by the EMF. The non-magnetic iron removal is believed to be due to the presence of magnetic/non-magnetic composite particles, which are magnetically attracted by the filter.

Condensate-Polishing Systems. Demineralizer systems, installed for the purpose of purifying condensate, are known as condensate-polishing systems. A condensate-polishing system is a requisite to maintain the purity required for satisfactory operation of once-through boilers (see **Water Treatment (Boiler)**). High-pressure drum-type boilers (over 2000 psi; 136 atm) can and do operate satisfactorily without condensate polishing. However, many utilities recognize the benefits of condensate polishing in high-pressure plants, including:

Treatment of Feedwater

The following discussion outlines what is required to produce and maintain the quality of feedwater recommended in Table 1.

The pre-boiler equipment, consisting of feedwater heaters, feed pumps and feed lines, is constructed of a variety of materials, including copper, copper alloys, carbon steel, and phosphor bronzes. To reduce corrosion, the makeup and condensate must be at the proper pH level and free of gases such as carbon dioxide and oxygen. The optimum pH level is that which introduces the least amount of iron and copper corrosion products into the boiler cycle. This optimum pH level should be established for each installation. It generally ranges between 8.0 and 9.5

Control of pH. The control of corrosion in the condensate system is generally accomplished by adding one of the following chemicals:

1. Neutralizing amines—ammonia, morpholine, cyclohexylamine and hydrazine.
2. Filming amines—octadecylamine acetate.

Neutralizing amines are volatile alkalizers that distill with the steam and neutralize acids that form in the condensate. Hydrazine, which is also an excellent oxygen scavenger, is included with the volatile alkalizers. It decomposes in the boiler, forming ammonia, hydrogen and nitrogen. The ammonia provides pH control in the condensate. The use of hydrazine as an oxygen scavenger is discussed later under *Chemical Scavenging of Oxygen by Hydrazine.*

Selection of the proper alkalizer should be considered for each plant to minimize the pickup of iron and copper in the condensate and feedwater. Optimum conditions should be established by tests during the early

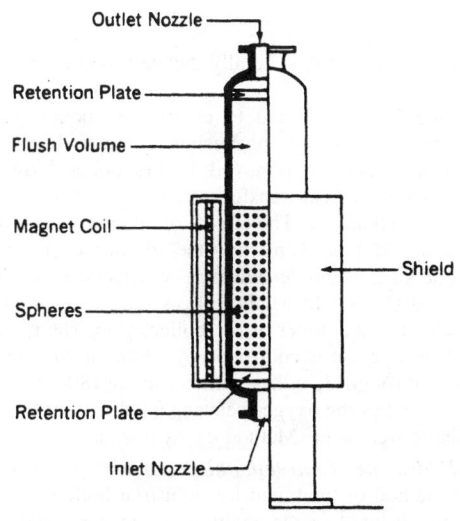

Fig. 4.   Sectional view of electromagnetic filter (EMF).

TABLE 1.   RECOMMENDED LIMITS OF SOLIDS IN BOILER FEEDWATER

Drum pressure	Below 600 psi (41 atm)	600 to 1,000 psi (41–68 atm)	1,000 to 2,000 psi (68–136 atm)	Over 2,000 psi (136 atm)
Total solids, ppm			0.15	0.05
Total hardness (as ppm CaCO₃)	0	0	0	0
Iron, ppm	0.1	0.05	0.01	0.01
Copper, ppm	0.05	0.03	0.005	0.002
Oxygen, ppm	0.007	0.007	0.007	0.007
pH	8.0–9.5	8.0–9.5	8.5–9.5	8.5–9.5
Organic	0	0	0	0

operation of the unit. Factors to be considered in selecting the alkalizer are steam temperature, makeup requirements, and carbon-dioxide concentration in the condensate.

Condensate corrosion rates are increased by the partial pressure of carbon dioxide in the steam. Carbon dioxide originates in the breakdown of carbonates in the boiler water. If steam has a high carbon-dioxide content, filming amines should be considered as a corrective. Filming amines reduce corrosion rates by forming a protective coating on the surfaces contacted by the steam and condensate. Since this is a surface phenomenon, the amount of metal surface to be protected is more important economically than the concentration of gas in the steam. Control of filming amine feed rate is critical. The protective film will not form if the feed is insufficient while excessive feed of this waxlike substance can plug the flow passages of the equipment.

In units with a high percentage of makeup feed, it may not be necessary to add chemicals specifically for pH control. When the makeup is treated by the lime-soda process or a lime-soda-zeolite system, the effluent is normally within the recommended pH range or slightly higher. If pH exceeds the recommended limits, the high alkalinity may lead to foaming and carry-over. For corrective action, see **Hot Lime Zeolite—Split Stream Softening**, previously discussed.

Control of Oxygen. The presence of gases, particularly oxygen, leads to corrosion of the boiler and cycle equipment. This type of attack will occur in an operating boiler as well as an improperly stored idle boiler. The consequent effect of dissolved oxygen in feedwater is pitting of the internal surfaces. This is most prevalent in the economizer, the steam drum and the supply tubes. The pitting may be general or selective. In either case, if allowed to proceed unchecked, it will adversely affect the reliability of the unit and shorten its service life.

The most logical approach to the prevention of corrosion by gases is to expel them from the system at the first opportunity. The usual method is by means of a deaerating heater. This equipment must be kept in prime operating order over the complete load range. If the deaerator operates under vacuum at low loads, the entrance of air must be prevented. Oxygen concentrations at the deaerator outlet should be consistently less than 0.007 ppm. As a further assurance against the destructive effect of dissolved oxygen, a residual quantity of an oxygen-scavenging compound should be maintained in the system.

Chemical scavenging of oxygen by sodium sulfite. Most operators use sodium sulfite for the chemical scavenging of oxygen. Fig. 5 shows the recommended sulfite concentration as a function of boiler pressure. The amount of sulfite that can be safely carried decreases as pressure increases. At the high temperatures associated with the higher pressures, sulfite decomposes into acidic gases that can cause increased corrosion. Consequently, sulfite should not be used at pressures greater than 1800 psi. On boilers having spray attemperation, the sodium sulfite should be added after the attemperator take-off point. About 8 ppm is required to remove 1 ppm of oxygen.

Chemical scavenging of oxygen by hydrazine. Hydrazine is an alternate scavenger, offering two principal advantages:

1. The decomposition and dissolved-oxygen reaction products of hydrazine are volatile. Consequently, they do not increase the dissolved-solids content of the boiler water, nor do they cause corrosion where steam is condensed.
2. Experience has shown that condensate pH will usually stabilize in the range of 8.5–9.5 if a 0.06-ppm hydrazine residual is maintained at the boiler inlet. This eliminates the need for pH treatment of the condensate-feedwater.

It is apparent that a residual of 0.06 ppm of hydrazine will provide only a limited protection against oxygen entering the boiler. Thus it is not practicable to utilize this scavenger as a substitute for an airtight system.

Changes in Feedwater Treatment. Any changes in feedwater treatment or boiler water conditions can have troublesome results. Changes should therefore be made gradually and with close observation. For instance, if sulfite treatment is to be replaced by hydrazine, initial dosage should be small and changes in the iron and copper concentration in the feedwater should be carefully monitored. If iron and copper concentrations in the feedwater and boiler water increase significantly, load should be reduced and blowdown increased. It may require days or weeks for conditions to stabilize, so results must be observed and evaluated over a significant period.

Treatment of Boiler Water for Natural-Circulation Units

Direct treatment of boiler water, usually referred to as internal treatment, is used (1) to prevent scale formations caused by hardness constituents, and (2) to provide pH control to prevent corrosion. Treatment that is incorrect or inadequate in either respect can lead to tube failures and result in costly unscheduled outages. The permissible limits on contaminants entering the boiler and also on treatment chemicals that can be added to the boiler decrease with rising boiler pressures.

Fig. 6 shows the relationship between dissolved solids in boiler water and solids in steam at various drum operating pressures. This correlation agrees reasonably well with both laboratory and field data. If a boiler-water total-solids concentration of 15 ppm is assumed, Fig. 6 indicates that 15 ppb solids would be expected in the steam at 2400 psi (163 atm) drum pressure, while at 2800 psi (190 atm) drum pressure about 75 ppb would be expected in the steam. Fig. 7 indicates the great reduction in silica concentration in boiler water that must occur as pressures increase if silica in the steam is to be limited to 20 ppb. Experience has demonstrated that a concentration of 20 ppb will pass through the superheater and turbine without deposition. This curve is valid for a boiler water pH of 9.5. At the higher pressures, boiler water additives must be reduced to low levels in order to avoid deposits on turbine parts.

There are four methods of internal treatment in common use on natural-circulation drum-type boilers:

1. Phosphate-hydroxide (conventional-treatment).
2. Coordinated phosphate.
3. Chelant.
4. Volatile.

The method of treatment is generally dictated by the pressure range of the unit.

Methods 1 and 2 are intended to control the boiler water pH and to precipitate the calcium and magnesium compounds as a flocculent sludge, so that they can be removed in the boiler blowdown rather than being deposited on heat-transfer surfaces. Method 1 maintains an excess of hydroxide alkalinity. The effects of alkalinity are discussed later under *Steam Purity.* Method 3 involves the addition of a complex metal-chelant compound such as ethylenediamine-tetraacetic acid (Na_4EDTA) or nitrilotriacetic acid (NTA). In Method 4, as the name implies, no solid chemicals are added to the boiler or pre-boiler cycle. The pH of the boiler water and condensate cycle is controlled by adding a volatile amine.

In Methods 1 through 3, either sulfite (up to 1800 psi; 122 atm) or hydrazine can be used as the oxygen scavenger. Above 1800 psi (122 atm), or with the volatile treatment (Method 4), hydrazine is used.

Phosphate-Hydroxide (Conventional-Treatment) Method. This is the most prevalent method of treatment for industrial boilers operating below 1000 psi (68 atm). It involves the addition of phosphate and caustic to the boiler water. Caustic is added in sufficient quantity to maintain a pH of

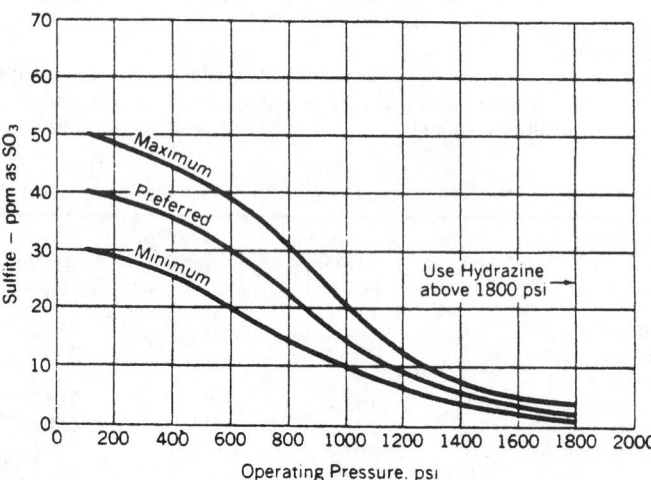

Fig. 5.   Usually recommended sulfite residual in boiler water

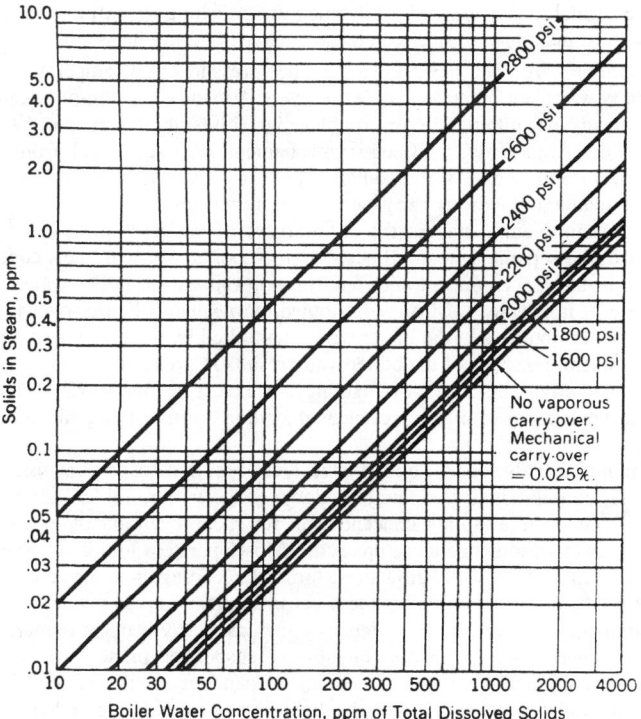

Fig. 6. Solids in steam versus dissolved solids in boiler water

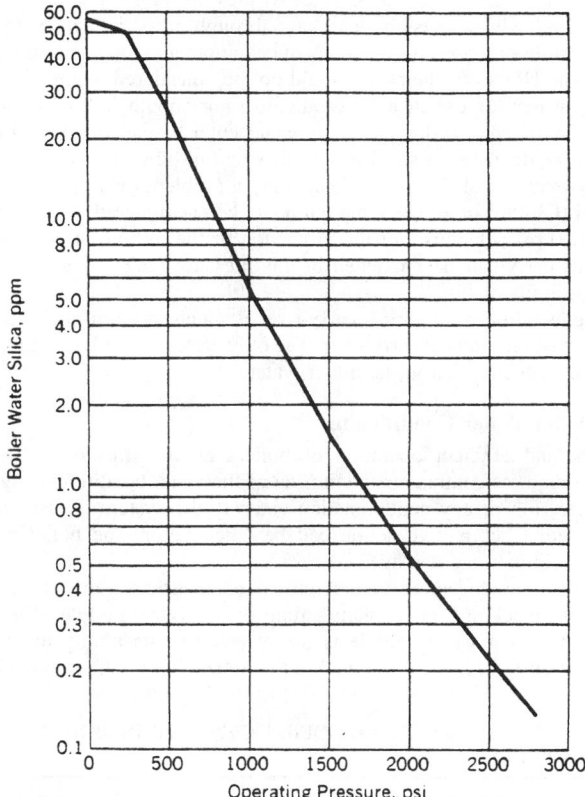

Fig. 7. Recommended maximum silica concentration in boiler water at pH 9.5 (drum-type boilers)

10.5 to 11.2. A boiler treated with caustic and phosphate is less sensitive to upsets than with other methods of feedwater control.

The primary purpose of phosphate addition is to precipitate the hardness constituents. The calcium reacts with phosphate under the proper pH conditions to precipitate calcium phosphate as calcium hydroxyapatite, $Ca_{10}(PO_4)_6(OH)_2$. This is a flocculent precipitate that tends to be less adherent to boiler surfaces than simple tricalcium phosphate, which is

precipitated below 10.2 pH. Caustic reacts with magnesium to form magnesium hydroxide or brucite, $Mg(OH)_2$. This precipitate is formed in preference to magnesium phosphate at a pH above 10.5 as it is less adherent.

The recommended phosphate concentration for a given boiler operating pressure is shown in Fig. 8. At the higher pressures, comparatively low phosphate residuals must be maintained in order to avoid appreciable phosphate "hideout." Hideout is the term used to identify the phenomenon of the temporary disappearance of phosphate in the boiler water upon increase in load and its reappearance upon load reduction. The recommended alkalinity as a function of pressure is given in Fig. 9.

Phosphate hideout does not appear to be as important below 1500 psi (102 atm) and even at this pressure, phosphate concentrations of 12 to 25 ppm as PO_4 can be carried without appreciable hideout. Either sulfite or hydrazine may be used to scavenge oxygen.

Coordinated Phosphate Method. In this method of treatment, no free caustic is maintained in the boiler water. Fig. 10 shows the phosphate concentration versus the resulting pH when trisodium phosphate is dissolved in water. Recent laboratory tests show that the crystals which precipitate from a concentrated solution of trisodium phosphate at elevated temperatures contain disodium phosphate and that the supernatant liquid is rich in sodium hydroxide. The sodium hydroxide can destroy the magnetite protective film on boiler surfaces. To assure that no free caustic is present, a boiler-water phosphate concentration that corresponds to a sodium-to-phosphate mole-ratio of 2.6 is recommended above 1000 psi (68 atm), as

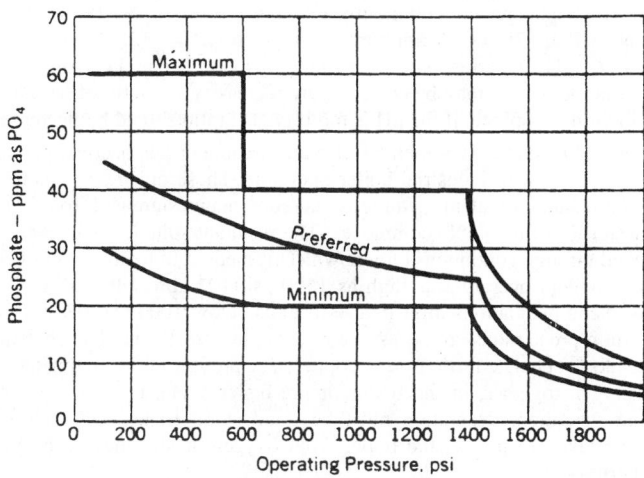

Fig. 8. Recommended phosphate concentration in boiler water at various boiler operating pressures (phosphate-hydroxide treatment)

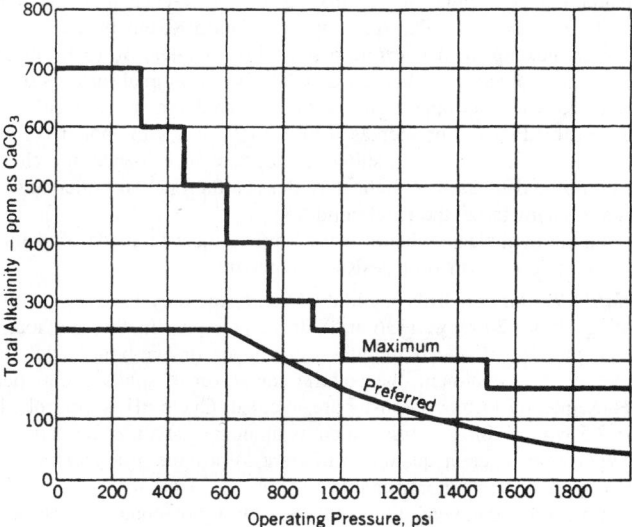

Fig. 9. Recommended alkalinity of boiler water at various boiler operating pressures (phosphate-hydroxide treatment)

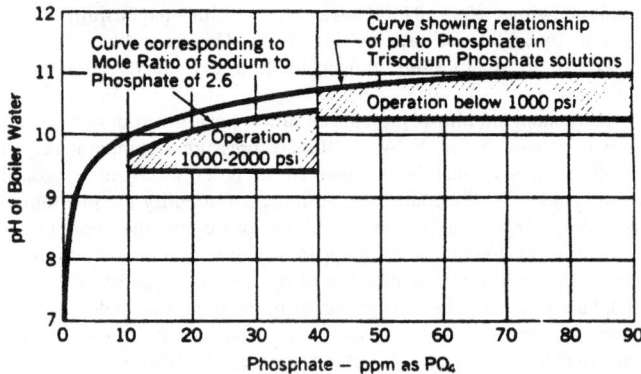

Fig. 10.   Recommended phosphate content of boiler water for drum boilers, using coordinated phosphate treatment

shown in Fig. 10. The precaution against free hydroxide alkalinity is less critical in boilers operating below 1000 psi (68 atm). The shaded areas on Fig. 10 indicate the recommended operating range of PO_4 and the resulting pH for boiler pressures to 2000 psi (136 atm). For drum pressures from 2000 to 2835 psi (136–192 atm) the boiler water should contain from 3 to 10 ppm Na_3PO_4 with corresponding pH of 9.0 to 9.7.

When using the regular commercial grades of chemicals, caution should be used in calculating the weights needed to provide the proper mole ratios. Commercial phosphates commonly are in the form of $Na_3PO_4 \cdot 12H_2O$ and $Na_2HPO_4 \cdot 7H_2O$. A mixture of 65% $Na_3PO_4 \cdot 12H_2O$ and 35% $Na_2HPO_4 \cdot 7H_2O$ corresponds to a mole ratio of Na to PO_4 of 2.6. If the pH is too low, it may be corrected by increasing the ratio of trisodium to disodium phosphate. If the pH is too high, the ratio should be decreased.

Use of Chelants. This method of water treatment has become popular in recent years with industrial boiler operators. These organic agents react with the residual divalent metal ions, calcium, magnesium and iron, in the feedwater to form soluble complexes. The resultant soluble complexes are removed through continuous blowdown. This method of treatment has been used in boilers operating as high as 1500 psi (102 atm) although present B&W recommendations limit its use to units below 1000 psi (68 atm).

Certain precautions are necessary in using this treatment. The chelating agents do not chelate ferric iron or copper. The presence of chelating agents and oxygen together, in the boiler or pre-boiler cycle, must be avoided. During operation, deaeration must be good at all times, and measures must be taken to protect the boiler from oxygen at all times during off-line periods.

Experience indicates that it is difficult to control chelant feed based on chelant residual in the boiler water. Excess chelant will attack clean boiler surfaces. B&W therefore recommends that chelant feed be based on known quantities of hardness and iron present in the feedwater, with the objective of maintaining a residual approximating 1 ppm of chelant in the boiler water. To protect the boiler from upsets resulting from heat-exchanger leakage or makeup plant overrun, a phosphate residual of 15 to 30 ppm should be maintained in the boiler water. The boiler internal surfaces should be inspected whenever opportunity permits. If sludge deposits accumulate, the chelant feed should be increased by 1 to 2 ppm. If the boiler is found to be exceptionally clean and shiny surfaces are in evidence, the chelant feed should be decreased. A light gray dust on the internal boiler surfaces appears to characterize the ideal condition.

For handling chelants, chemical feed piping must be made of stainless steel or some other corrosion-resistant material.

Volatile Treatment. This method of treatment may be used for units operating above 2000 psi (136 atm) drum pressure. In this method, no solid chemicals are added to either the boiler or pre-boiler cycle. By eliminating solid treatment, the volatile carry-over of solids is eliminated and consequently turbine deposits are avoided. Cycle pH is controlled at 9.0 to 9.5 with a volatile amine such as ammonia. Hydrazine is added as an oxygen scavenger in quantity sufficient to provide a concentration of 20 to 30 ppb at the economizer inlet.

With volatile treatment, the feedwater must not contain hardness of condenser-leak constituents. Since no phosphate is present to remove hardness, any contamination assumes major importance. Prompt detection and remedial action is required. Failure to take such action endangers the future availability of the unit. A condensate-polishing system in the cycle is the best insurance against condenser leakage and hardness constituents.

Steam Purity. The trend toward higher pressures and temperatures in steam power plant practice imposes a severe demand on steam-purification equipment for elimination of troublesome solids in the steam. Carry-over may result from ineffective mechanical separation and from the vaporization of boiler-water salts. Total carry-over is the sum of the mechanical and vaporous carry-over of all impurities.

Mechanical carry-over is the entrainment of small droplets of boiler water in the separated steam. Since entrained boiler-water droplets contain solids in the same concentration and proportions as the boiler water, the amount of impurities in steam contributed by mechanical carry-over is the sum of all impurities in the boiler water multiplied by the moisture content of the steam. Foaming of the boiler water results in gross mechanical carry-over. The common causes of foaming are excessive boiler-water solids, excessive alkalinity or the presence of certain forms of organic matter, such as oil.

Maintaining dissolved solids at the level required to prevent foaming requires continuous or periodic blowdown of the boiler. Table 2 gives the recommended total solids concentration for the prevention of excessive carryover at various operating pressures. Most operators find it convenient and advisable to run well below these limits. Exceeding them may endanger the superheater, the turbine, or the process application.

High boiler-water alkalinity tends to increase carryover, particularly in the presence of an appreciable quantity of suspended matter. This effect may be corrected by various methods, dependent on the cause of the high alkalinity. For example, if trisodium phosphate is being added to the boiler water, a less alkaline phosphate, such as disodium or monosodium phosphate will help in reducing alkalinity.

The presence of oil in boiler water is intolerable, as it causes foaming and carry-over. Steps should be taken to prevent its entry into the boiler through the feedwater system or leakage through pressure seals and joints. Organic antifoaming agents are a recent development with some successful application. However, their use should not be considered a cure-all.

Spray water for use in a spray attemperator should be of the highest quality. Solids entrained in the spray water enter the steam and can cause troublesome deposits on superheater tubes and turbine blades.

Carry-over of volatile silica is generally a problem only at pressures of 1000 psi (68 atm) or above, although it can be encountered at pressures as low as 600 psi (41 atm). For the protection of the turbine, it is important that silica carryover be prevented in this pressure range by adherence to the silica limits of Fig. 7.

The prevention of vaporous carry-over is much more difficult than the correction of mechanical carry-over. The only effective method is to reduce the solids concentration in the boiler water.

Controls for Water Conditioning

The safe and efficient operation of boilers at pressures over 1000 psi (68 atm) requires continuous monitoring of the water conditioning system. Early detection of any contamination entering the system is essential, so that immediate corrective action can be taken before the boiler and its related equipment are damaged.

Electrical conductance, the reciprocal of resistance, affords a rapid means of checking for contamination in a water sample. Electrical conductance of a water sample is the measure of its ability to conduct an electric current. It can be related to the ionizable dissolved solids in the

TABLE 2. LIMITS FOR TOTAL SOLIDS CONTENT IN BOILER WATER (DRUM BOILERS)

Drum pressure		Total solids (ppm)
(psi)	(atm)	
0–300	0–20.4	3500
301–450	20.5–30.6	3000
451–600	30.7–40.8	2500
601–750	40.9–51.0	2000
751–900	51.1–61.2	1500
901–1000	61.3–68.0	1250
1001–1500	68.1–102.0	1000
1501–2000	102.1–136.1	750
over 2000	over 136.1	15

water. A single instrument will measure and record important conductivities of the water from as many as twenty different locations in the system. The electrical conductivity signal can be used to actuate alarm systems or to operate equipment in the water system. The micromho (1×10^{-6} mho) is normally the unit of measurement. For most salts in low concentrations, 2 micromhos is equal to 1 ppm concentration when corrected to 77°F (25°C).

Ammonia or amines used for pH control affect the conductivity. To obtain an accurate indication of solids, a cation ion exchanger removes the volatile alkalizers and converts the salts to their corresponding acids. Seven micromhos are equivalent to 1 ppm concentration for most salts.

For boilers with operating pressures over 1000 psi (68 atm) cation conductivity of the condensate should normally run between 0.2 and 0.5 micromhos. A reading above this limit indicates the presence of condenser leakage or contamination from some other source. The source of the contamination should be investigated and remedied at the first opportunity. However, when a cation conductivity limit of 1.0 is reached, the internal water treatment and blowdown must be changed appropriately.

Dissolved oxygen should be monitored at the condensate pump discharge and the deaerator outlet. Sulfite or hydrazine can be used for oxygen scavenging. Over 1800 psi (122 atm) drum pressure, sulfite should not be used; only hydrazine is recommended. Sulfite or hydrazine can be added to the condensate on a manual or automated basis.

Feedwater pH is monitored at the condensate pump discharge and the economizer inlet. Chemical-injection pumps are usually adjusted manually to maintain the proper pH for the conventional and coordinated phosphate water-treatment systems. Where volatile water treatment is used, pH can be controlled automatically by using conductivity to transmit signals to the ammonia injection pumps. It is generally preferable to use conductivity rather than pH to transmit signals to the ammonia pumps. Conductivity equipment has been found to be more reliable for this purpose and the linear, rather than logarithmic relationship to concentration, enables better control. Ammonia should be added at the hotwell effluent, or, if condensate polishing is used, at the effluent of the demineralizing system.

Hydrogen should be monitored at the economizer inlet and the superheater outlet. A hydrogen analyzer-recorder can actuate an alarm when the hydrogen concentration of the feedwater or steam deviates from the safe value, which is specified for the plant. Deviation from the normal hydrogen concentration can indicate that corrosion is taking place within the water-steam system.

Automated equipment is commercially available for the continuous on-stream analysis of the critical constituents of the boiler water, such as hardness, phosphate, iron, copper and silica. Most laboratory analytical procedures that depend on the development of a color, and then measuring the intensity of that color to indicate the concentration of the constituent in the water sample, can be put on an automatic basis.

Water Treatment for Universal-Pressure Boilers

Satisfactory operation of the once-through boiler and associated turbine requires that the total solids in the feedwater be less than 0.05 ppm. Table 3 lists recommended maximum limits for feedwater contaminants and typical values obtained during operation.

Recommended limits should be low because all solids in the feedwater will either deposit in the boiler or be carried over with the steam to the turbine. Consequently, water-treatment chemicals must be volatile. All cycles should have condensate-polishing systems to meet the limits shown in Table 3. A schematic diagram is shown in Fig. 11. Laboratory tests as well as field studies show that high-flow-rate condensate-polishing systems [25 to 50 gal per min per sq ft (1015–2030 liters/minute/square meter) of cross-sectional bed area] perform as filters of suspended material and ionized particles. Ammonia is added to control the pH in the system. Fig. 12 indicates the amount of ammonia required, in terms of ppm or solution conductivity, to give a certain pH in the system. Hydrazine is added to the cycle for oxygen scavenging.

Most of the iron entering the boiler originates in the condensate-feedwater cycle downstream of the polishing demineralizers or in the shell side of feedwater heaters where drips bypass the polishing demineralizers. Studies on a number of installations with carbon-steel feedwater heaters have shown that iron pickup can be minimized by operating with feedwater pH in the range of 9.3 to 9.5. The best pH for minimizing iron pickup

TABLE 3. RECOMMENDED LIMITS OF SOLIDS IN FEEDWATER FOR UNIVERSAL-PRESSURE BOILERS

	Maximum limit	Typical concentrations
Total solids	0.050 ppm	0.020 ppm
Silica as SiO_2	0.020 ppm	0.002 ppm
Iron as Fe	0.010 ppm	0.003 ppm
Copper as Cu	0.002 ppm	0.001 ppm
Oxygen as O_2	0.007 ppm	0.002 ppm
Hardness	0.0 ppm	0.0 ppm
Carbon dioxide	0.0 ppm	not measured
Organic	0.0 ppm	0.002 ppm
pH	9.2–9.5	9.45

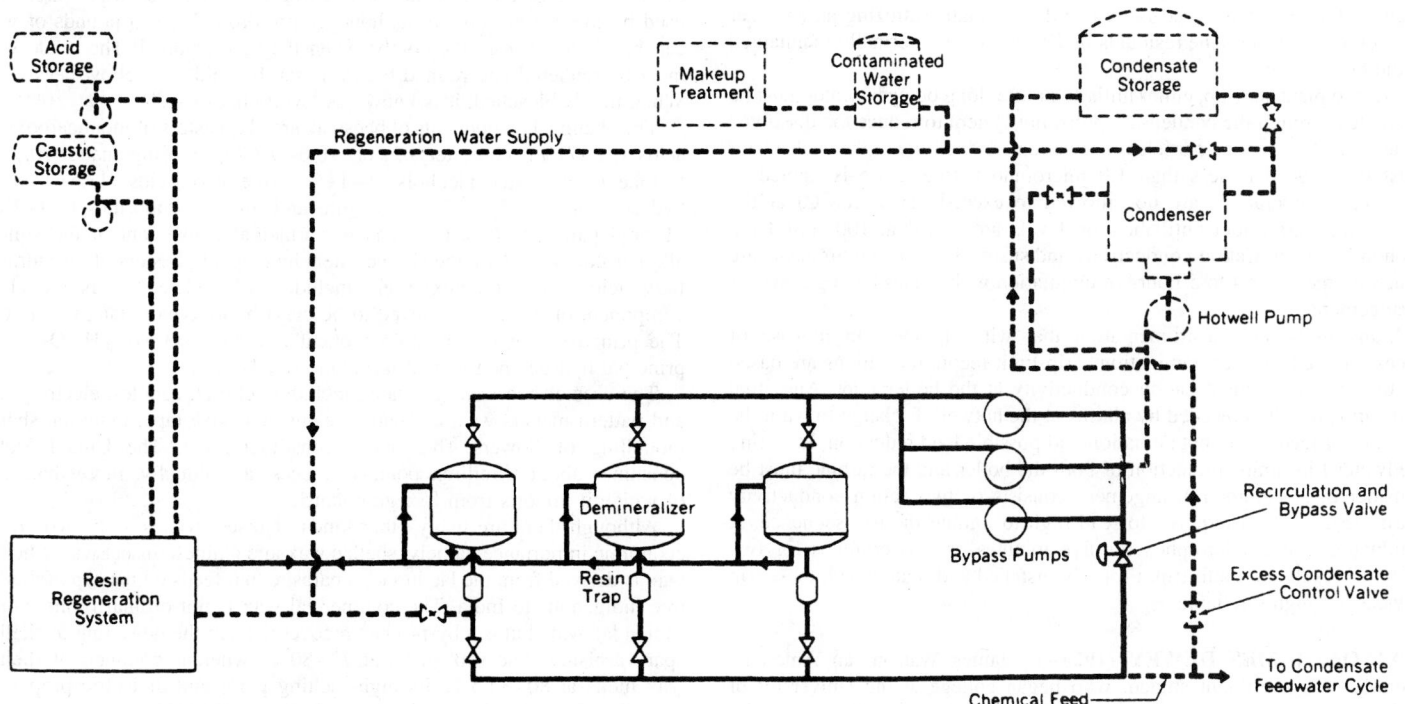

Fig. 11.   Schematic diagram of condensate-polishing system with high-quality makeup treatment (four-bed ion exchange or equivalent)

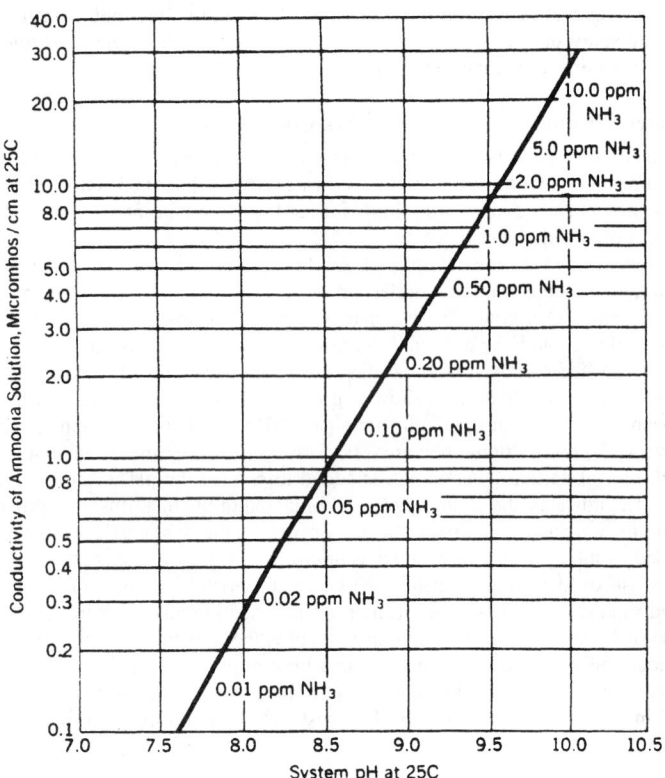

Fig. 12. Theoretical relationship between conductivity and pH for ammonia solutions

should be determined for each cycle during the first several months of operation.

Ammonia is injected downstream of the condensate polishers and controlled from a sample taken far enough downstream of the injection point to assure good mixing. Hydrazine is generally fed at the exit of the condensate-polishing system and/or at the boiler feed-pump suction. Automatic controls are available to regulate the positive displacement pumps that meter ammonia and hydrazine introduction. The signal to the pump-controller for ammonia usually comes from a specific conductivity-recording instrument that is compensated for temperature changes of the cycle water. Hydrazine feed is frequently automatic, utilizing an analyzer and controller. Hydrazine residuals of 10–20 ppb are normally maintained at the boiler inlet.

Prior to plant start-up, either initially or after long outages, water must be circulated through the condensate-polishing system to reduce the dissolved material and suspended particles. The cation conductivity of the cycle water must be reduced to less than 1.0 micromho before a fire is lighted in the unit. Temperatures are not allowed to exceed 550°F(288°C) at the convection pass outlet until the iron levels are less than 100 ppb at the economizer inlet. Cation-conductivity and suspended-iron requirements are generally met after 4 to 5 hours of circulation with cycles having a bypass arrangement.

Many units have instrumentation that will trip the unit in case of excessive feedwater contamination. Trip-limit recommendations are based on the measurement of cation conductivity at the boiler inlet. An actual unit trip is usually preceded by alarms at the hotwell discharge to warn the operator of feedwater contamination and possible load reduction. In setting feedwater trip limits, protection of both the boiler and the turbine must be considered. A common arrangement consists of two cation conductivity alarm devices, both required to read high to initiate the trip sequence. A conductivity of 2 micromhos for five minutes or 5 micromhos for two minutes, results in a unit trip. Properly installed and maintained, these trip devices are highly reliable.

WATSON, JAMES DEWEY (1928–). James Watson, an American chemist, was a brilliant student who began college at the University of Chicago at the young age of 15. Since he was young, Watson had a passion for bird watching and this interest was probably a factor in his fascination with genetics. He is known for his research contributions in the field of genetics.

He began working with Francis Crick in 1950 at the Cavendish laboratories. In 1953, Watson and Crick, using the photographs of Rosalind Franklin, which exposed crystallized molecules from the nucleus, identified the material that biologists were viewing in the nucleus as DNA. Watson and Crick created a three-dimensional structure DNA model, which provided scientists with a valuable tool in the study of heredity. In 1962 Watson-Crick were awarded the Nobel Prize for their work.

Watson taught at Harvard and CalTech and in 1968 he became the directorship for Cold Spring Harbor Laboratory. Under his directorship the lab became a leading research center in the world for molecular biology. In 1988, Watson became head of the Human Genome Project.

See also **Crick, Francis Harry Compton (1916–2004)**; and **Human Genome Project**.

J.M.I.

WAVELLITE. The mineral wavellite is a hydrous phosphate of aluminum, formula $Al_3(PO_4)_2(OH)_3 \cdot 5H_2O$. It is orthorhombic but crystals are of rare occurrence as it is ordinarily found in crusts or radial aggregates, sometimes fibrous. Its hardness is 3.25–4; specific gravity, 2.36; may be of various colors, gray, blue, green, yellow, black, or colorless. It has a vitreous luster, and is translucent. This mineral is of secondary origin, probably formed by waters bearing phosphoric acid which have acted on aluminum minerals. Wavellite is found in Saxony, Bavaria, Devonshire, from where it was originally described; and in the United States in Chester and Cumberland Counties, Pennsylvania; and Montgomery and Garland Counties, Arkansas. It was named after its discoverer, Dr. Wavel.

WAXES. The English term *wax* is derived from the Anglo-Saxon *weax*, which was the name applied to the natural material gleaned from the honeycomb of the bee. In modern times the term *wax* has taken on a broader significance, as it is generally applied to all waxlike solids, natural or synthetic, and to liquids when they are composed of monohydric alchol esters. Unlike the ordinary oils of animal and vegetable origin and the animal tallows, the waxes, with a few exceptions, are free from glycerides, which are common constituents of oils and fats. Bayberry wax is a vegetable tallow which happens to have all the physical characteristics of a wax, and has always been classed as such.

Animal and Vegetable Waxes

The most important insect wax from an economic viewpoint is *beeswax*, secreted by the hive-bee. Wax scales are secreted by eight wax glands on the underside of the abdomen of the worker bee. These wax wafers are used by the bee in building its honeycomb. From $1\frac{1}{2}$ to 3 pounds of wax can be obtained from the combs when they are scraped. The crude wax must be rendered and refined before it can be sold as "yellow beeswax." When this is bleached, it is known as "white beeswax."

The chemical components of beeswax are alkyl esters of monocarboxylic acids (71–72%), cholesteryl esters (0.6–0.8%), coloring matter (0.3%), lactone (0.6%), free alcohols ($1–1\frac{1}{2}$%), free wax acids (13.5–14.5%), hydrocarbons (10.5–11.5%), moisture and mineral impurities (0.9–2%). Myricyl palmitate ($C_{46}H_{92}O_2$) is the principal constituent of the simple alkyl esters (49–53%); the simple esters include alkyl esters of unsaturated fatty acids. The complex esters include hydroxylated esters the chief component of which is believed to be ceryl hydroxypalmitate, $C_{42}H_{84}O_3$. The principal free wax acid component is cerotic acid ($C_{26}H_{52}O_2$). The principal hydrocarbon is hentriacontane ($C_{31}H_{64}$).

The uses of beeswax are many, including church candles, electrotypers and pattern makers wax, cosmetic creams, adhesive tape, munition shells, modelling of flowers, shoe paste constituent, etc. The United States consumes about 8 million pounds of beeswax annually, more than half of which it imports from foreign countries.

Although there are many other kinds of insect waxes, only two are of economic importance namely, shellac wax and Chinese insect wax. Shellac wax is derived from the lac insect, a parasite that feeds on the sap of the lac tree indigenous to India. The commercial wax is not ordinarily the native Indian lac wax, but is a by-product recovered from the dewaxing of shellac spar varnishes. Lac wax melts at 72–80°C, whereas commercial shellac wax melts at 80–84.5°C. Its high melting point and dielectric properties favor its use in the electrical industry for insulation. Chinese insect wax is the product of the scale insect.

The land animal waxes are either solid or liquid. *Woolwax*, derived from the wool of the sheep, is of great economic value. It is better known as anhydrous lanolin, and is of a stiff, soft, solid consistency. The only representative of liquid animal wax is "mutton bird oil" obtainable from the stomach of the mutton bird.

The unsaponifiables of woolwax, known as "woolwax alcohols," are in considerable demand by cosmetic and pharmaceutical industries. Woolwax has a great affinity for water, of which it will absorb 25 to 30%. Refined woolwax is kneaded with water to produce a water-white, colorless ointment, known as hydrous lanolin or "lanolin USP." Anhydrous lanolin is widely used in cosmetic creams, since it is readily absorbed by the skin. It is also used in leather dressings and shoe pastes, as a superfatting agent for toilet soap, as a protective coating for metals, etc. United States consumption of wool wax is about 1.5 million lb/year.

The marine animal waxes are both solid and liquid. The solid marine animal waxes are represented by a wax of considerable economic importance, namely *spermaceti*, derived from a concrete obtained from the head of the sperm whale. The liquid waxes of marine animals are represented by sperm oil obtained from the blubber and cavities in the head of the sperm whale. Spermaceti is the wax used in the candle which defines our unit of candle power; it is used chiefly as a base for ointments, cerates, etc. Sperm oil contains a considerable amount of esters made up of unsaturated alcohols and acids, both of which are susceptible to hydrogenation. Hydrogenated sperm oil is the equivalent of spermaceti wax and harder than the commercial pressed spermaceti. Both yield cetyl alcohol as the unsaponifiable. There is a fairly large demand for cetyl alcohol in the manufacture of lipstick, shampoo, and other cosmetics. Sperm oil itself is an excellent lubricant for lubricating spindles of cotton and woolen mills, or wherever there is need for a very light, limpid, nongumming lubricant.

The waxes obtained from plants occur in the leaves, stems, barks, fruit, flowers, and roots. The leaves of palm trees furnish wax of great economic importance. Particularly is this true of the product furnished by harvesting the leaves of the carnauba palm. The wax is removed from the leaves by sundrying, trenching, threshing and beating; the powdered wax is melted in a clay or iron pot over a fire, strained, cast into blocks, and broken into chunks for shipment from Brazil. *Carnauba wax* dissolves well in hot turpentine and/or naphtha, from which solvents it gels or cooling; it has a good solvent retention power. Its hardness, luster, and favorable behavior with solvents make it a highly valued ingredient in shoe pastes, floor polishes, carbon paper, etc. A small amount of carnauba, such as 2.5%, when added to paraffin will raise the melting point of the latter enormously (e.g., from 130 to 170°F; 54 to 77°C), making it a very useful ingredient in the production of inexpensive high-melting blended waxes. United States consumption is provided by imports from Brazil which amount to over 11 million lb/year.

The chemical composition of carnauba wax comprises 84–85% of alkyl esters of higher fatty acids. Of these esters only 8–9% (wax basis) are simple esters of normal acids. The other esters are acid esters 8–9%, diesters 19–21%, and esters of hydroxylated acids 50–53% (was basis) of which about one-third are unsaturated. It is the hydroxylated saturated esters that give carnauba its extreme hardness, whereas the esters of the hydroxylated unsaturated fatty acids produce the outstanding luster to polishes.

Ouricury, carandá, and raffia are commercial palm leaf waxes of lesser importance. Ouricury wax has a very high content of esters of hydroxylated carboxylic acids and is used as a substitute for carnauba in carbon papers, etc. Carandá and raffia waxes have a very low contents of these acids and make unsatisfactory substitutes for carnauba.

The most important wax obtainable from the stems of plants is *candelilla*, obtained in Mexico and the southwestern United States. To recover the wax the plant stalks are pulled up by the roots and boiled in acidulated water. On cooling, the congealed wax is removed from the surface of the water in the tank. The crude wax is given an additional refinement before it si placed on the market. Candelilla wax is brownish in color, and melts at 66–78°C. Most vegetable waxes are essentially alkyl esters of aliphatic acids; candelilla, on the other hand, contains 51 to 59% of hydrocarbons and less than 30% of esters. The chief hydrocarbon is hentriacontane ($C_{31}H_{64}$), common to other vegetable waxes. The hydrocarbons melt at 68°C, and the esters at 88–90°C. Candelilla is often used in conjunction with carnauba in leather dressings, floor waxes, etc. It is also used in sound records, electrical insulators, candle compositions, etc. Imports average about 1600 tons per year.

Because of the enormous tonnage of sugar cane processed in Cuba and elsewhere, it is possible to recover an appreciable tonnage of *sugarcane wax* as a by-product. The crude wax contains about one-third each of wax, resin, and oil, and hence needs considerable refinement by selective solvents before it can become of value for industrial use. The refined sugarcane wax is dull yellow in color, melts at 79–81°C and is hard and brittle. It has a durometer hardness of 85–96. It is chemically composed of 78–82% of wax esters, 14% free wax acids, 6–7% free alcohols, and 3–5% hydrocarbons. A proportion of the esters are sterols—sitosterol and stigmasterol—combined with palmitic acid, which are responsible for the good emulsification properties of the wax itself in the preparation of polishes and the like. The proportions of sterols in the refined wax is far less than in the crude wax.

Of waxes obtained from fruits, *japanwax* is the only one of great economic importance, particularly to the Asiatic countries. The wax occurs as a greenish coating on the kernels of the fruit of a small sumac-like tree. Japanwax is actually a vegetable tallow, since it is comprised of 90–91% of glycerides. Peculiarly the glycerides include 3–6.5% (wax basis) of alkyl esters of dicarboxylic acids as well as monocarboxylic acids. The chief dicarboxylic acid is known as japanic acid $[(CH_2)_{19}(COOH)_2]$ which is present with lower as well as higher homologs. The dicarboxylic acids have 19 to 23 carbon atoms, whereas the monocarboxylic acids of the simple glycerides present have 16 to 20 carbons.

The textile industries in the past have been large users of japanwax since it is a source of emulsifying softening agents. Other industries using japanwax include those engaged in the manufacture of rubber, soap, polishes, pomades, leather dressings, cordage, etc. Japanwax is a relatively soft but firm wax, which melts at 48.5–54.5°C. About 3000 tons are normally produced per year in China, and twice that amount in Japan.

Other fruit waxes include *bayberry wax*, used in making Christmas candles since the days of the Pilgrams. The wax of rice bran is coming into commercial use, but waxes of the cranberry, apple, grapefruit, etc. are only of academic interest.

Waxes from grasses include bamboo leaf wax, esparto wax, and hemp fiber wax. Esparto wax is a hard, tough wax with a melting point of 73–78°C, and is the most important grass wax. Most of the esparto wax produced is consumed in the British Isles. It is chiefly useful as a substitute for carnauba. Waxes obtained from roots of various species of plants are minute in quantity and of no economic importance.

Mineral Waxes

The fossil waxes are associated with fossil remains which have not been bituminized, that is, converted to hydrocarbons by geological change. A fossil wax, chemically speaking, is composed largely of saponifiables, such as wax acids and esters. Fossil waxes of nearly pure ester composition are occasionally found in fragments of prehistoric plant life, still in a state of preservation as to the original wax constituents. Not far removed from fossil wax of the pure ester composition is *montan wax*, a natural mineral wax which is essentially an ester wax that has undergone partial bituminization. Montan wax is commercially extracted from the nonasphaltic insoluble pyrobitumen with which it is associated, by means of selective solvents such as alsohol and benzene, or by means of benzene alone. Crude montan wax is black and contains about 30% of resins, which is reduced to 10% or less upon refinement. The chemical components of montan wax (deresinified) are alkyl esters of fatty acids (40%), alkyl esters of hydroxy fatty acids (18%), free wax acids (18%), free monohydric alcohols (3%), resins (<12%), and ketones (<10%). There are a number of industrial uses of montan wax: electrical insulation, leather finishes, polishes, carbon papers, shoe pastes, brewer's pitch, etc. The crude wax is also used as a basic material in the manufacture of many synthetic waxes where its montanic acid content is utilized in making derivatives.

The principal source of the montan wax consumed in the United States is imports in annual amounts of some 3.4 million lb from Germany and Czechoslovakia, where it is derived from lignites and brown coals. Also, one California company extracts it from lignite.

Peat wax has somewhat the same composition as montan wax. It has only been produced on a limited scale in Ireland, where it is processed from the native peat. It has asphaltic constituents that tend to make it incompletely miscible with paraffin waxes.

The earth waxes are naturally occurring mineral waxes consisting of hydrocarbons with some oxygenated resinous bodies which can be eliminated by ordinary refining procedures. The earth wax of great economic importance is *ozocerite*, which originally was called ceresin wax. Important sources are the Carpathian mountains in Europe and to a far lesser extent Utah. Chemically speaking, ozocerite has hydrocarbons of a type different from those found in paraffin wax, giving it unique physical properties. The melting point of pure Galician ozocerite is 73°C. It is less soluble in organic solvents than paraffin. When added to paraffin wax in amounts of 15% or thereabouts, it will reduce the paraffin crystals to micro size and improve the tensile strength. Crude imported ozocerite has a dielectric constant of 2.37–2.43, the refined 2.03, and domestic (Utahwax) 2.63. Ozocerite is used in the electrical industry, paste polishes, cosmetics, wax flowers, crayons, etc. Of all the waxes it has the greatest affinity for oil.

Petroleum Waxes

Petroleum is the largest single source of hydrocarbon waxes. The largest single use of petroleum waxes is in paper coatings which require about 53% of the total. The second largest use is in candles. The third greatest use in electrical equipment. In contrast, in the early 1950's the largest single use for petroleum waxes was in the manufacture of paper containers for dairy products, and the second was in waxed wrappers for bread. Both outlets are now dominated by plastics.

Crude petroleums differ greatly in both the nature of their hydrocarbons (paraffinic, aromatic, naphthenic, etc.), as well as in their available content of wax. Wax distillates are obtainable with the batch-type, continuous-type, and pipe-still processes, but not from the cracking process. There are, broadly speaking, three principal types of wax encountered in crude oil, namely *paraffin wax, slop wax*, and *petrolatum*. The ordinary procedure in producing slack wax is to pump the paraffin distillate at a temperature of 80–100°F (26.7–38°C) to the paraffin sheds (wax plant), where it is allowed to repose in tanks to promote settling at a temperature between 0 and 32°F (−17.8–0°C). It is then pumped through a bank of cooling units (wax chillers) to hydraulic presses, which squeeze out the wax from the chilled distillate. The product is a soft solid known as *slack wax*. Slack wax finds uses in the industries, but most of it is sweated, pressed, and further refined to produce the various grades of fully refined paraffin wax of commerce. Some of the slack may be "pudged" to the extent that it still contains several percent of oil; it is then called *scale wax*. Most of the scale waxes produced have a melting point (drop) of 126–130°F (52–54°C) (ASTM) and are used in waterproofing thread in the fabrication of cotton duck and canvas, waxing kraft papers, builders' papers, cement bag stock, roofers' felt, car liners, and match splints. Scale wax of very low oil content and higher melting point is used in the manufacture of crayons.

Paraffin wax contains 14 hydrocarbons ranging from $C_{18}H_{38}$ to $C_{32}H_{66}$, solidifying between 27.0 and 68.9°C (80.5 and 156.0°F; 26.9–68.8°C). *Petrolatum wax*, which is a microcrystalline wax, has hydrocarbons ranging from $C_{34}H_{70}$ to $C_{43}H_{88}$, inclusive. Its solidifying range is 71.0–83.8°C (159.7–182.7°F). *Slop wax* (by-product from the heavy distillate in the coking process) has 13 hydrocarbons ranging from C_{26} to C_{43}, solidifying between 55.7 and 83.3°C (132.2 and 182.0°F). *Rod wax* (collected from the sucker rods in the field) has 8 hydrocarbons ranging from C_{35} to C_{41}, solidifying at 73.9 to 82.5°C (165.0–180.5°F).

Fully refined paraffin wax as regularly offered in the market is graded according to its melting point. There are also special refined grades offered by some refining companies, such as the so-called hard block fully refined paraffins with melting points of 138–140°F (58.8–60°C), and 143–145°F (61.6–62.7°C). The tensile strength of a paraffin wax is greatly influenced by the oil content. Ordinarily a well-refined paraffin of 130°F (54°C) melting point will have a tensile strength of about 250 psi (1.7 MPa). The addition of 1 or 2% of an oil-absorbent wax of microcrystalline structure will increase the tensile strength to 350 psi (2.4 MPa).

Microcrystalline Waxes

Microcrystalline petroleum waxes are characterized not only by microcrystalline structure but by very high average molecular weight, manifested by a much higher viscosity than that of paraffin wax. The chlorophyll present in plants is considered to be a microcrystalline wax.

Microcrystalline waxes are obtained as by-products from (a) the dewaxing of "lube oil raffinates," (b) the deoiling of petrolatum produced from deasphaltic residual oil, or (c) the deasphalting and deoiling of settlings of tanks holding crude oil in the oil field. These types of microcrystalline waxes are sometimes referred to as "motor-oil wax," "residual oil microcrystalline wax," and "tank-bottom microcrystalline wax," respectively. They have also been referred to as "micro wax," "petrolatum wax," and "petroleum ceresin," respectively.

The *micro waxes* are graded with 145–150°F (63–66°C) and 160–165°F (71–74°C) ASTM melting points, and are refined by selective solvent extraction from the crude wax, a "mobile slurry." A yield of 25–27% of refined wax of the lower melting point is claimed from S.A.E. 20 motor-oil distillate. These waxes are paraffin-like. *Petrolatum waxes* are of a 145–175°F (63–80°C) melting point range. The *petroleum ceresins* which are refined from deposits taken from tanks near the wells, called lease tanks, or in the refinery storage, have melting points which range between 165 and 195°F (74–91°C). In the solvent dewaxing processes the solvents for effectively separating the microcrystalline waxes vary with the refinery methods and the character of the feed stock.

A microcrystalline wax derived from petroleum may be defined as a solid hydrocarbon mixture, of average molecular weight range of 490 to 800, considerably higher than that of paraffin wax, which is 350 to 420. The viscosity (SUS at 210°F; 99°C) of a microcrystalline wax is within the range of 45 to 120 seconds. The lower limit corresponds to 5.75 and the upper limit to 25.1 centistokes at 210°F (99°C). The penetration value (ASTM) is of wide variation, namely 3 to 33, although sticky oily laminating waxes are encountered with as high a penetration as 60. Microcrystalline waxes have an occluded oil content which is not easily set free as it is in paraffin waxes. Therefore, a microcrystalline wax which has an oil content of 1 to 4% is virtually a dry wax. A microcrystalline wax which shows a penetration 20 to 30, which is desirable for many needs, will have an oil content of 5.5 to 10.5%.

When a microcrystalline wax is added to melted paraffin it acts like a solute with paraffin as the solvent; the melting point of the blend is greatly elevated, and the crystallization of the paraffin is depressed. The behavior is that of a two-phase system until about 15% of the microcrystalline wax has been added. With the addition of 15 percent of the petrolatum wax (M.P. 188°F; 86.6°C) the melting point of the paraffin is elevated from 130 to 160°F (54 to 71°C). For many industrial uses microcrystalline wax is admixed with paraffin wax.

The uses of microcrystalline waxes include adhesives, barrel lining, beater size for paper stocks, beer can lining, carbon papers, cheese coatings, cosmetic creams, drinking cups, electrical insulation, floor wax, fruit coating, glass fabric impregnation, heat sealing compounds, laminants for paper, ordnance packing, paper milk bottles, shoe and leather treatments, vegetable coatings, wax emulsions, wax figures and toys, and other miscellaneous purposes.

Synthetic Waxes

These include the following types:

(1) Long-chain polymers of ethylene with OH or other stop-length groupings at end of chain. An example is polyethylene wax of about 2000 molecular weight.

(2) Long-chain polymers of ethylene oxide combined with a dihydric alcohol, namely polyoxyethylene glycol, ("Carbo-wax").

(3) Chlorinated naphthalenes, ("Halowaxes").

(4) Waxy polyol ether-esters, as for example, polyoxyethylene sorbitol.

(5) Synthetic hydrocarbon waxes prepared by the water-gas synthesis in which carbon monoxide (CO) is reduced by hydrogen (H_2) under pressure, at a predescribed temperature, by means of a catalytic agent. (Fischer-Tropsch waxes "F-T 200" and "F-T 300").

(6) Wax-like ketones, straight-chain and cyclic: (a) Symmetrical ketones produced by the catalytic treatment of the higher fatty acids. (b) Unsymmetrical ketones produced by the Friedel-Crafts' condensation of fatty acids and the like with cyclic hydrocarbons. Examples of the straight-chain ketones are laurone, palmitone, and stearone, and of the cyclic ketones are phenoxyphenyl heptadecyl ketone.

(7) Amide derivatives of fatty acids. The length of the chain may be increased by heating the fatty acid, e.g., stearic acid, with an amino alcohol.

(8) Imide (*N*) condensation products that are wax-like are those of the condensation reaction of one mole of phthalic anhydride with one mole of a primary aliphatic amine to produce a phthalimide. Phthalimide waxes are used in polishes and carbon paper.

(9) Polyoxyethylene fatty acid esters are produced by the reaction of polyethylene glycols with fatty acids. The commercial products are waxy solids which include "Carbowax 4000 (Mono) Stearate." Some of the products act as plasticizers and lubricants for plastics. The polyethylene glycols are soluble in water.

(10) Miscellaneous synthetic waxes (unclassified).

In addition to the above waxes there is a group of synthetic wax-like emulsifiable materials extensively employed in the industries. They are the polyhydric alcohol fatty acid esters, such as ethylene glycol monostearate, glyceryl monostearate, glycerol distearate, and a number of others.

ALBIN H. WARTH
Cape May, New Jersey

WEERMAN DEGRADATION. Formation of an aldose with one less carbon atom from and aldonic acid by a Hoffmann-type rearrangement of the corresponding amid. This is a general reaction of α-hydroxycarboxylic acids.

WEIGHTING AGENT. (1) In soft drink technology, an oil or oil-soluble compound of high specific gravity, such as a brominated olive oil, which is added to citrus flavoring oils to raise the specific gravity of the mixture to about 1.00, so that stable emulsions with water can be made for flavoring. (2) In the textile industry a compound used both to deluster and lower the cost of a fabric, at the same time improving its "hand" or feeling. Zinc acetylacetonate, clays, chalk, etc. are used.

WERNER, A. (1866–1919). A native of Switzerland, Werner was awarded the Nobel prize for his development of the concept of the coordination theory of valence, which he advanced in 1893. His ideas revolutionized the approach to the structure of inorganic compounds and in recent years have permeated this entire area of chemistry. The term *Werner complex* has largely been replaced by "coordination compound".

See also **Coordination Compounds**.

WESSELY-MOSER REARRANGEMENT. Rearrangement of flavones and flavanones possessing the 5-hydroxyl groug, through fission of the heterocyclic ring and reclosure of the intermediate diaroylmethanes in the alternate direction.

WETTING AGENT. A surface-active agent that, when added to water, causes it to penetrate more easily into, or to spread over the surface of, another material by reducing the surface tension of the water. Soaps, alcohols, and fatty acids are examples.

See also **Detergents**.

WHEY PROTEIN. See **Proteins**.

WICHTERLE REACTION. Modification of the Robinson annellation reaction in which 1,3dichloro-*cis*-2-butene is used instead of methyl vinyl ketone

WIDMAN-STOERMER SYNTHESIS. The synthesis of cinnolines by cyclization of diazotized *o*-aminoarylethylenes at room temperature.

WIELAND, HEINRICH O. (1877–1957). A German chemist who won the Nobel prize for chemistry in 1927. His research included work on bile acids, organic radicals, nitrogen compounds, toxic substances, and chemical oxidation, as well as the discovery of the structure of cholesterol. He received his Ph.D. from the University of Munich.

WIGNER FORCE. Short-range nuclear force of nonexchange type postulated phenomenologically as part of the interaction between nucleons. Postulated exchange forces are Bartlett, Heisenberg, and Majorana forces.

WIGNER NUCLIDES. A special case of mirror nuclides. Pairs of odd-mass number isobars for which the atomic number and the neutron number differ by one, and in which the numbers of protons and neutrons are so related that each member of the pair would be transformed into the other by exchanging all neutrons for protons and vice versa.

WILKINSON, GEOFFREY (1921–1996). A British organic chemist who won the Nobel prize for chemistry in 1973 with Ernst Otto Fischer, for their pioneering work, performed independently, on the chemistry of the organometallic, so called sandwich compounds. He was a professor at the University of California and Harvard before returning as professor of inorganic chemistry at the University of London.

WILLEMITE. The mineral willemite is a zinc silicate, Zn_2SiO_4, occurring in hexagonal prisms, as masses or scattered grain. It is a brittle mineral with conchoidal fracture; hardness, 5.5; specific gravity, 3.9–4.2; subvitreous luster; usually some shade of yellow, yellowish-green, green, or reddish-brown, but may be colorless, white, or blue to nearly black; transparent to opaque. Much willemite is strongly fluorescent in yellow or yellowish-green hues. Willemite occurs associated with other zinc materials in Belgium, Algeria, Zaire, South West Africa, and Greenland. In the United States, except for three occurrences, one in Colorado, one in New Mexico, and one in Utah, Sussex County, New Jersey, is the only locality in the United States for willemite and is the only one in which that mineral is found in quantity. Here it is found associated with zincite and franklinite, forming an important ore of zinc. It was named by the French mineralogist, Michel Lévy, in honor of King William the First of the Netherlands.

WILLGERODT REACTION. This reaction, discovered in 1887, is conducted by heating a ketone, for example $ArCOCH_3$, with an aqueous solution of yellow ammonium sulfide (sulfur dissolved in ammonium sulfide), and results in formation of an amide derivative of an arylacetic acid and in some reduction of the ketone. The dark reaction mixture usually is refluxed with

$$ArCOCH_3 + (NH_4)Sx \longrightarrow ArCH_2CONH_2 + ArCH_2CH_3$$

alkali to effect hydrolysis of the amide, and the arylacetic acid is recovered from the alkaline solution. Although the yields are not high, the process sometimes offers the most satisfactory route to an arylacetic acid, as in the preparation of 1-acenaphthylacetic acid from 1-acetoacenaphthene, a starting material made in 45% yield by acylation of the hydrocarbon with acetic acid and liquid hydrogen fluoride. The product is obtained in better yield and is more easily purified than that from an alternate process consisting in hypochlorite oxidation, conversion to the acid chloride, and Arndt-Eistert reaction.

$$C_{12}H_9COCH_3 \xrightarrow[78\%]{1.KOCl\ 2.SOCl_2}$$
1-Acetoacenaphthene

$$C_{12}H_9COCl \xrightarrow{CH_2N_2}$$

$$[C_{12}H_9COCHN_2] \xrightarrow[64\%,\ from\ acid\ chloride]{Ag_2O,\ Na_2S_2O_3}$$
Diazo ketone

$$C_{12}H_9CH_2COOH$$
1-Acenaphthylacetic acid

A modification of the Willgerodt reaction that simplifies the procedure by obviating the necessity of a sealed tube or autoclave consists in refluxing the ketone with a high-boiling amine and sulfur (Schwenk, 1942). Morpholine, so named because of a relationship to an early erroneous partial formula suggested for morphine, is suitable and is made technically by dehydration of diethanolamine. The reaction is conducted in the absence of water, and the reaction product is not the amide but the thioamide; this, however, undergoes hydrolysis in the same manner to the arylacetic acid.

L. F. and MARY FIESER
Harvard University
Cambridge, Massachusetts

WILLIAMSON SYNTHESIS. An organic method for preparing ethers by the interaction of an alkylhalide with a sodium alcoholate (or phenolate).

WILLSTATER, RICHARD (1872–1942). A German chemist who won the Nobel prize for chemistry in 1915 or his work on plant pigments, especially chlorophyll. His education was at the University of Munich where he studied and taught before going to Zurich, Switzerland. He researched chlorophylls and pigments of plants and the relationship of cornflower blue to rose red. Work included the study of alkaloids including cocaine, tropine, and atropine. His work perfected the process of chromatographic partition.

WILZBACH PROCEDURE. Exposure of organic compounds to tritium gas yields tritiated products of high activity without extensive radiation damage. Concentrations of tritium ranging from 1 to 90 millicuries per gram have been obtained with quite varied compounds.

WINDAUS, ADOLF (1876–1959). A German chemist who won the Nobel prize for chemistry in 1928. His work involved the study of steroids and the effect of ultraviolet light activity, ergosterol, and vitamin D_2. He also researched digitalis and histamine. Although he studied medicine, he received his doctorate in chemistry at the University of Freiburg.

WITHERITE. The mineral witherite is barium carbonate, $BaCO_3$, crystallizing in the orthorhombic system. It is interesting to note that at 811°C it changes to the hexagonal system, and at 982°C it appears to become isometric. It has a rather imperfect prismatic cleavage; uneven fracture; hardness 3–3.7; specific gravity, 4.29; luster, vitreous to resinous; color, white to yellowish or grayish; streak, white; transparent to translucent. Witherite is found in veins, and often is associated with galena, as at Alston Moor, Cumberland, England. Associated with barite at Freiberg, Saxony, and at Lexington, Kentucky. Named in honor of Dr. William Withering, an English botanist.

WITTIG, GEORGE (1897–1987). A University of Heidelberg professor who won the Nobel prize for chemistry in 1979 along with Herbert C. Brown of Purdue. Wittig's research showed that phosphorous ylids react with ketones and aldehydes to form alkenes. This reaction is used a great deal in the synthesis of pharmaceuticals and other complex organic substances.

WITTIG REACTION. This reaction provides an excellent method for the conversion of a carbonyl compound to an olefin:

$$(C_6H_5)_3 \xrightarrow{CH_3Br}$$

Triphenylphosphine

$$(C_6H_5)_3\overset{+}{P}CH_3(Br^-) \xrightarrow[-C_6H_6-LiBr]{C_6H_5Li}$$

Methyltriphenyl-
phosphonium bromide

$$(C_6H_5)_3P = CH_2 \longleftrightarrow (C_6H_5)_3\overset{+}{P} - \overset{-}{C}H_2$$

Wittig reagent

$$(C_6H_5)_3P \overset{\cdot\cdot}{-} CH_2 \longrightarrow \begin{array}{c}(C_6H_5)_3P + CH_2 \\ \| \quad \| \\ O \quad C(C_6H_5)_3 \end{array}$$

with $(C_6H_5)_2C = O$ and $\overset{..}{O} \overset{..}{-} C(C_6H_5)_2$

The reagent is unstable and so is generated in the presence of the carbonyl compound by dehydrohalogenation of the alkyltriphenylphosphonium bromide with phenyllithium in dry ether in a nitrogen atmosphere. There are various modifications, such as the phosphonate, in which diethylbenzylphosphonate, cinnamaldehyde, and sodium methoxide yield 1,4-diphenylbutadiene.

Dr. Georg Wittig, University of Heidelberg, was awarded the 1979 Nobel Prize for chemistry for his work in organic synthesis. Dr. Herbert C. Brown of Purdue University also participated in the joint award, but for separate work in organic synthesis.

More details on the early development of the Wittig reaction, dating back to the 1940s, is given in *Science*, **207**, 42–44 (1980).

WOHL DEGRADATION. Method for the conversion of an aldose into an aldose with one less carbon atom by the reversal of the cyanohydrin synthesis. In the Wohl method, the nitrile group is eliminated by treatment with ammoniacal silver oxide.

WÖHLER, FRIEDRICH (1800–1882). A native of Germany, Wöhler working along with Jöns Jakob Berzelius, placed the qualitative analysis of minerals on a firm foundation. In the early 19th century, chemists still thought it impossible to synthesize organic compounds.

Wöhler is regarded as a pioneer in organic chemistry as a result of his (accidental) synthesizing urea in 1828, which electrified the scientific community. Until 1828, it was believed that organic substances could only be formed under the influence of the vital force in the bodies of animals and plants. Wöhler proved by the artificial preparation of urea from inorganic materials that this view was false. Urea synthesis was integral for

biochemistry because it showed that a compound known to be produced only by biological organisms could be produced in a laboratory, under controlled conditions, from inanimate matter.

Wöhler was also a co-discoverer of beryllium and silicon, as well as the synthesis of calcium carbide, among others. In 1834, Wöhler and Liebig published an investigation of the oil of bitter almonds. They proved by their experiments that a group of carbon, hydrogen, and oxygen atoms can behave like an element, take the place of an element, and can be exchanged for elements in chemical compounds. Thus the foundation was laid of the doctrine of compound radicals, a doctrine which had a profound influence on the development of chemistry.

Wöhler's discoveries had great influence on the theory of chemistry. Journals from 1820 to 1881 contained contributions from him. The sum of his work is absolutely overwhelming.

WÖHLER SYNTHESIS. Classical synthesis of urea by heating an aqueous solution of ammonium cyanate extended to preparation of urea derivatives.

WOHLWILL PROCESS. The official process of the U.S. mints for refining gold. It consists of subjecting gold anodes to electrolysis in a hot solution of hydrochloric acid containing gold chloride, the solution being continuously agitated with compressed air.

WOLFFENSTEIN-BOTERS REACTION. Simultaneous oxidation and nitration of aromatic compounds to nitrophenols with nitric acid or the higher oxides of nitrogen in the presence of a mercury salt as catalyst. Hydroxynitration of benzene yields picric acid.

WOLFF-KISHNER REACTION. This method of reduction was discovered independently in Germany (Wolff, 1912) and in Russia (Kishner, 1911). A ketone (or aldehyde) is converted into the hydrazone, and this derivative is heated in a sealed tube or an autoclave with sodium ethoxide in absolute ethanol.

$$\begin{array}{c} \diagup \\ C=O \diagdown \end{array} \xrightarrow{H_2NNH_2} \begin{array}{c}\diagup \\ C=NNH_2 \diagdown \end{array} \xrightarrow[200°]{NaOC_2H_5,} $$

$$\begin{array}{c}\diagup \\ CH_2 + N_2 \diagdown \end{array}$$

After preliminary technical improvements, Huang Minlon (1946) introduced a modified procedure by which the reduction is conducted on a large scale at atmospheric pressure with efficiency and economy. The ketone is refluxed in a high-boiling water-miscible solvent (usually di- or triethylene glycol) with the aqueous hydrazine and sodium hydroxide to form the hydrazone; water is then allowed to distil from the mixture till the temperature rises to a point favorable for decomposition of the hydrazone (200°); and the mixture is refluxed for three or four hours to complete the reduction.

L. F. and MARY FIESER
Harvard University
Cambridge, Massachusetts

WOLFF REARRANGEMENT. Rearrangement of diazoketone to ketenes by action of heat, light or some metallic catalyst. The rearrangement is the key step in the Arndt-Eistert synthesis.

WOLFRAMITE. The mineral wolframite, tungstate of iron and manganese, is an isomorphous mixture of tungstate of iron, $FeWo_4$, and tungstate of manganese, $MnWO_4$, the amounts being variable. The pure iron tungstate is called ferberite and the manganese tungstate, hübnerite. It has been proposed that the name ferberite be applied to mixtures of not less than 80% $FeWO_4$ and not more than 20% $MnWO_4$, and that the term hübnerite be given to mixtures of not less than 80% $MnWO_4$ but not more than 20% $FeWO_4$. Wolframite would thus include the minerals of intermediate composition, and its formula would be written $(Fe^{2+}, Mn)WO_4$. Wolframite is monoclinic, usually appearing in tabular, columnar or bladed crystals, sometimes quite large; also may be massive. Its hardness is 4–4.5; specific gravity, 7.371; color, gray, reddish-brown, brown or black; streak, reddish-brown to black; luster, submetallic; opaque, occasionally magnetic.

Wolframite is found associated with apatite, cassiterite, quartz, and fluorite; in granites and pegmatites. Often with scheelite, $CaWO_4$, and sometimes as a pseudomorph after that mineral. Wolframite is found in the Czech Republic and Slovakia, Rumania, Saxony, Cornwall in England, New South Wales, Bolivia; and in the United States at Trumbull, Connecticut; Luna and Lincoln Counties, New Mexico. It also occurs in small quantities in Nevada and Utah.

Ferberite, which has monoclinic tabular crystals, sometimes massive, resembles wolframite, and occurs in Spain and in Boulder County, Colorado. Superb crystals of Ferberite in association with apatite and arsenopyrite crystals occur at Panasqueira, Portugal.

Hübnerite crystals are monoclinic, often long fibrous or bladed, may be massive; resembles wolframite and is found in Peru, the Black Hills, South Dakota; San Juan County, Colorado; White Pine County, Nevada, and Lemhi County, Idaho.

WOLLASTONITE. The mineral wollastonite is calcium metasilicate, $CaSiO_3$, which is found as tabular or short prismatic triclinic crystals. This mineral has a hardness of 4.5–5; specific gravity, 2.87–3.09; color, white to gray, rarely green to colorless; luster, vitreous to pearly; transparent to translucent. Wollastonite in the main is formed by the action of contact metamorphic processes on limestones at relatively high temperature ($600°C+$). Its common associates, diopside, vesuvianite, garnet and epidote, suggest this origin. Some of the more important localities are in the copper mines of Rumania, in the lavas of Monte Somma and Vesuvius, in Finland and Mexico. In the United States it is found in Essex and Lewis Counties, New York; Keweenaw County, Michigan; and Riverside County, California. Wollastonite was named in honor of the English chemist, William Hyde Wollaston.

WOOD. A vascular tissue which occurs in all higher plants. The most important commercial sources of wood are the gymnosperms, or softwood trees and the dicotyledonous angiosperms, or hardwood trees. Botanically, wood serves the plant as supporting and conducting tissue, and it also contains certain cells which serve in the storage of food. The trunks and branches of trees and shrubs are composed of wood, except for the very narrow cylinder of pith in the center and the bark which covers the outside. Botanists refer to wood by its Greek name, *xylem*.

A new cylinder of wood is laid down each year around the previously formed wood in the tree. This new growth originates in the cambium, a very narrow growing layer, which elaborates both the wood and the bark, and which separates these two tissues from each other. Each year's growth of wood forms a new concentric ring in the woody-stem, as viewed in cross-section. These are termed the annual rings. Each of these has an inner part (toward the pith) which is laid down in the early growing season and is termed the spring wood and an outer layer (towards the bark) which is laid down later and is known as summer wood. The cell walls of the latter are often thicker, forming a denser structure than the spring wood.

Most of the cells of wood are long, narrow hollow fibers and tubular-shaped cells arranged with their long axes parallel to the axis of the tree trunk. Certain food storage cells lie in radial bands, termed wood rays, which are perpendicular to the tree axis. The walls of this complex system of plant cells form the basic framework and material of all wood substance. All wood substance is composed of two basic chemical materials, *lignin*, and a polysaccharidic system, which is termed *holocellulose*. The latter embraces *cellulose* and the *hemicelluloses*, a mixture of pentosans, hexosans and polyuronides, and in some instances small amounts of pectic materials. Wood cell wall tissue also always retains small amounts of mineral matter (ash).

The outer portion of the cell wall, known as the primary wall, is heavily lignified. The intercellular substance, termed the middle lamella, is mainly lignin. The lignin of the middle lamella and primary walls thus serves as a matrix in which the cells are imbedded. Dissolution and removal of the lignin results in separation of the wood fibers. This is the underlying principle in the manufacture of chemical pulps from wood for paper or other cellulose products. **Papermaking and Finishing**; and **Pulp (Wood) Production and Processing**.

About one-fourth or more of the lignin is in the middle lamella-primary wall complex. The remainder is within the holocellulose system of the cell walls.

Besides the cell wall tissue, which is the basic material of all wood substance, wood contains a variety of materials, many of which may be extracted by selected solvents. These extraneous components lie mainly within the cavities (lumen of the cells and on the surfaces of the cell walls. These "extraparietal substances" include a wide range of chemically different materials, such as essential oils, aliphatic hydrocarbons, fixed oils, resin acids, resinols, tannins, phytosterols, alkaloids, dyes, proteins, water-soluble carbohydrates, cyclitols, and salts of organic acids. The amount and composition of these extraneous substances vary greatly. The occurrence of certain of these substances is often very specific as to genera or species. Generally, however, the total amount of the extraneous components is only a few percent of the total weight of the wood.

The chemical composition of the extractive-free wood, i.e., the cell wall substance varies less than do the extractives. However, it is by no means constant, there being major differences between hardwoods and softwoods, and often between different genera and even between species. There is even some variability in chemical composition within the same log. For example, there is more lignin in the thinner-walled spring wood than in the summer wood, and more lignin in wood ray cells than in the tracheids or fibers. The heart-wood tissue often contains greater deposits of extraneous (extractive) components than the sapwood. There are major differences between most softwoods (conifers) and most temperate zone hardwoods (broad-leaf trees). Usually the softwoods have greater amounts of extraneous components extractable by organic solvents. Generally the lignin content of softwoods is higher than that of hardwoods, *viz.*, in the order of 25–30% compared to 17–24%. Also there is a major difference in the pentosan content between these two groups of woods, the hardwoods containing usually about 17–22% and the softwoods about 8–14%. There are exceptions to these generalizations, however.

The components of the cell wall substance of wood are exceedingly difficult to separate. Separations are rarely complete and generally bring about drastic chemical changes, especially in the lignin and molecular size degradation of the polysaccharides. To a considerable extent the components are apparently interpenetrating polymer systems. The long-chain linear polysaccharides tend to be parallel to the fiber axis and to form areas of varying degrees of crystallinity, as shown by x-rays and other physical and chemical properties. The lignin appears to be an amorphous tridimensional polymer.

Wood forms one of the world's most important chemical raw materials. It is the primary source of cellulose for the pulp and paper and cellulose industries. These industries are well up in the group of 10 major industries of the United States. For paper, rayon, films, lacquers, explosives and plastics, which comprise the greatest chemical uses of wood, it is the cellulose component (plus certain amounts of hemicellulose) of wood that is of value. The lignin forms a major industrial waste as a by-product of the paper and cellulose industries. Its major use is in its heat value in the recovery of alkaline pulping chemicals. A variety of minor uses for lignin have been developed, such as for the manufacture of vanillin, adhesives, plastics, oil-well drilling compounds and fillers for rubber.

Wood wastes from the lumber and woodworking industries form a great potential source of sugars and alcohol by acid hydrolysis of its polysaccharides followed by fermentation. Wood hydrolysis processes, however, are not yet economically competitive with other sources of sugars and alcohol in this country and many other areas of the world.

Wood is also an industrial source of charcoal, tannin, rosin, turpentine, and various other essential oils and pharmaceutical products.

EDWIN C. JAHN
State College of Forestry
Syracuse, New York

WOOD (As Energy Source). See **Wastes as Energy Sources**.

WOOD PRESERVATIVE. A material applied to wood to prevent its destruction by fungi, wood-boring insects, marine borers and fire. A common characteristic of these materials is toxicity to those organisms that attack wood, or in the case of fire retardants the ability to control combustion in terms defined by the Underwriters Laboratory. In addition, a satisfactory wood preservative must also (a) be capable of penetrating wood, (b) remain in the wood for extended periods without losing its effectiveness due to chemical breakdown, (c) be harmless to humans and animals, (d) be noncorrosive and, (e) be available in quantity at a reasonable cost. For certain uses, the preservative may be required to be colorless, odorless, nonswelling and paintable.

The principal wood preservatives in use today are classified as (a) preservative oils, (b) toxic chemicals in organic solvents, and (c) water-soluble salts.

The most important of the preservative oils is coal-tar creosote and its solutions in the form of creosote-coal tar and creosote petroleum. Coal-tar creosote is defined by the American Wood Preservers Association as—"A distillate of coal tar produced by high-temperature carbonization of bituminous coal; it consists principally of liquid and solid aromatic hydrocarbons and contains appreciable quantities of tar acids and tar bases; it is heavier than water and has a continuous boiling range of at least 125°C, beginning at about 200°C." This material is one of the oldest wood preservatives and is regarded by many as the best substance known for protection against all forms of wood-destroying organisms. For normal service conditions, minimum retentions of 8–10 lb/ft^3(128–161 kg/m^3) of wood penetrated are sufficient. For extreme conditions, retentions as high as 35 lb/ft^3(561 kg/m^3) are desirable. Currently, coal-tar creosote and creosote solutions are used as preservatives for about 60% of all wood products treated.

Of the large number of toxic chemicals that are oil-soluble, only three are recognized by the American Wood Preservers Association and only one of these is a commercially important wood preservative, pentachlorophenol (C_6Cl_5OH). Although "penta" is effective against fungi and insects, it will not protect against marine borers, and hence cannot be used as a wood preservative for salt water installations.

The two most common solvents for carrying penta into the wood are a relatively high-boiling No. 2 fuel oil and a very low-boiling liquified petroleum gas, butane. When fuel oil is used it remains in the wood and although the end product is brighter and cleaner than creosote-treated wood, it has a somewhat oily character and cannot readily be painted. When LP gas is used as the solvent, the butane is recovered and the penta is deposited in the wood as a dry crystalline material, hence the wood retains its color and is readily paintable.

Water-borne preservatives are divided into two categories. One group which includes acid copper chromate, chromated zinc chloride, copperized chromated zinc arsenate and fluorchrome-arsenate-phenol is used where the wood is not subjected to excessive leaching. The second group, ammoniacal copper arsenite and three types of chromated copper arsenate which react to become practically water insoluble, are used at about 0.6 lb/ft^3(9.6 kg/m^3) when wood is placed in ground contact under severe service conditions.

Water-borne preservatives penetrate wood easily and the solution presents no problem in flammability or health hazards. Disadvantages of water soluble preservatives include the swelling and shrinking of the treated material, reduction in bending strength and stiffness as a result of failure to redry following treatment, and less protection against weathering and mechanical wear than provided by either preservative oils or oil soluble chemicals in which the solvent remains in the wood.

In addition to the general preservative categories discussed there is a fourth group known as "proprietary preservatives" which are composed of various combinations of toxic materials and solvents. These are sold under trade names and in some cases are protected by patents.

While wood preservatives can be applied by simple means such as brushing, spraying and cold soaking, well over 90% of all commercial wood treatment is by one of the "pressure processes." Although the details of the individual processes vary, the type of equipment and general procedure used are similar. Treatment takes place in closed cylinders 6–9 feet (1.8-2.7 m) in diameter and up to 180 feet (54 m) in length. The wood to be treated is placed in the retort, submerged in the preservative and subjected to pressures in the order of 200 psi (13.6 atm). The preservative is often at an elevated temperature and in some cases the wood is given a preliminary pressure or vacuum period prior to admitting the preservative into the retort.

The most satisfactory nonpressure process is known as "thermal" treatment. The wood to be treated is immersed in a preservative at an elevated temperature. This causes the air in the wood cells to expand so that when the wood is transferred into a preservative bath of lower temperature, the air contracts forming a partial vacuum and atmospheric pressure forces the liquid into the wood.

Although the fire retardant treatment of wood is essentially the injection of water soluble salts into the wood by pressure treating methods, it deserves separate mention if for no other reason than the higher salt retention (4lbs/cu. ft. vs. 0.6 lbs/cu. ft.) (64 kg/m^3 vs 9.6 kg/m^3) required. Most fire retardant treatments are proprietary and are specified by the Underwriters Laboratory on the basis of flame spread ratings. The latter's list of building materials will give details of the ratings including those of a recently introduced leach-resistant fire retardant suitable for outdoor service. Common fire-retardant chemicals include diammonium phosphate, ammonium sulfate, sodium tetraborate and boric acid which are used in various combinations.

WILLIAM T. NEARN
Weyehaeuser Company
Seattle, Washington

WOOD PULP. See **Pulp (Wood) Production and Processing**.

WOODWARD-HOFFMANN RULES. Rules that predict the stereo-chemical course of concerted reactions in terms of the symmetry of the interacting molecular orbitals.

WOODWARD, ROBERT B. (1917–1979). An American chemist born in Quincy, MA, and widely regarded as one of the world's leading synthetic organic chemists. After receiving his doctorate from M.I.T. (the youngest student in the history of the institute to do so) he joined the Harvard faculty in 1937 as instructor and attained full professorship in 1950. He was recipient of the Nobel prize in 1965 for his brilliant work in synthesizing complex organic compounds, among them quinine, cholesterol, chlorophyll, reserpine, and cobalamin (vitamin B_{12}). When he died in 1979, his synthesis of the antibiotic erythromycin was virtually complete, and finished by his associate's two years later. He was director of the Woodward Research Institute in Basel, Switzerland, and a member of the governing board of the Weizmann Institute in Israel.

WOOL. The natural, highly crimped fiber from sheep, wool is one of the oldest fibers from the standpoint of use in textiles. Minute scales on the surface of the fibers allow them to interlock and are responsible for the ability of the fiber to *felt*, a phenomenon responsible for felt cloth and mill-finished worsteds. Crimpiness in wool is due to the open formation of the scales. Fine merino wool has 24 crimps per inch (~10 per centimeter). Luster of the fiber depends upon the size and smoothness of the scales. The basic wool protein, *keratin*, comprises molecular chains that are linked with sulfur. When sulfur is fed to sheep in areas deficient of the element, the quality of the wool improves. Wool fibers that fall below 3 inches (7.5 centimeters) in length are known as *clothing wool*; fibers 3–7 inches (7.5–17.8 centimeters) long are referred to as *combing wools*. The wool-fiber diameter ranges from 0.0025 to 0.005 inch (0.06–0.13 millimeter). See also **Fibers**.

WROUGHT IRON. A ferrous material aggregated from a solidifying mass of pasty particles of highly refined metallic iron, with which, and without subsequent fusion, is included a minutely and uniformly distributed quantity of slag. This definition of wrought iron indicates that it is a material made of two components; one, iron of a high degree of purity, the other, slag (chiefly silicate of iron). In the finished product the slag is distributed through the iron in threads and fibers, of which there is an enormous number. The slag imparts to the wrought iron a fibrous structure, quite different from the crystalline structure of cast metals. Wrought iron has made for itself a name as a metal which has resistance to corrosion, and which is exceptionally suitable for structural purposes where the structure is subject to shock. Wrought iron also can be readily worked, forged, machined, welded, galvanized, etc. Among the many applications of wrought iron might be mentioned tubes, pipes, and tanks.

WULFENITE. The mineral wulfenite is lead molybdate corresponding to the formula $PbMoO_4$, analyses showing that a part of the lead may be replaced by calcium. Wulfenite crystallizes in the tetragonal system usually in thin tabular forms, but is also found massive. It is a brittle mineral; hardness, 2.75–3; specific gravity, 6.5–7; luster, adamantine to resinous; color, yellowish to green or red, may be whitish or grayish; transparent to translucent. Wulfenite is a secondary mineral found in association with other lead minerals such as galena, and pyromorphite. It is believed to have been formed, at least in part, by the action of waters containing molybdenum salts on cerussite, anglesite, and pyromorphite.

Especially important localities are in the former Yugoslav Republics, the Czech Republic and Slovakia, Morocco, Zaire, New South Wales and Mexico. In the United States it has been found in Phoenixville, Pennsylvania, and in the Organ Mountains, New Mexico; Yuma County,

Arizona; Box Elder and Salt Lake Counties, Utah; and in Clark and Eureka Counties, Nevada. Wulfenite was named in honor of F.X. von Wülfen, an Austrian mineralogist of the eighteenth century.

WÜRTZ-FITTIG-FRANKLAND REACTION.

Sodium metal was used by Würtz as reagent for the preparation of paraffin hydrocarbons by treating alkyl iodide in ethereal solution, thus:

$$
\begin{array}{llll}
C_2H_5I & Na & & C_2H_5 \\
C_2H_5I & Na & \text{(Ether)} & | \\
& & & C_2H_5
\end{array}
$$

C$_2$H$_5$I	Na		C$_2$H$_5$		NaI
C$_2$H$_5$I	Na	(Ether) →	$\vert$ C$_2$H$_5$		NaI
Ethyl iodide	Sodium		Normal-butane		Sodium iodide

The method has been applied to the preparation of paraffin hydrocarbons as high in the series as hexacontane $C_{60}H_{122}$. The alkyl radicals may be the same or different in the iodide or iodides taken.

Sodium metal was also used similarly by Fittig as reagent for the preparation of hydrocarbons by treating aryl bromide or iodide in the presence of dry ether, thus:

C$_6$H$_5$Br	Na		C$_6$H$_5$		NaBr
C$_6$H$_5$Br	Na	(Ether) →	$\vert$ C$_6$H$_5$		NaBr
Phenyl iodide	Sodium		Biphenyl	Sodium bromide

When alkyl iodide and aryl bromide are taken, the hydrocarbon is of the mixed alkyl-aryl type, thus:

CH$_3$I	Na		CH$_3$	NaI
C$_6$H$_5$Br	Na	(Ether) →	$\vert$ C$_6$H$_5$	NaBr
Methyl iodide	Sodium		Toluene	Sodium iodide
Phenyl bromide				Sodium bromide

CH$_3$I	Na		C$_6$H$_4$ ⟨CH$_3$ (1) / CH$_3$ (4)⟩	NaI
C$_6$H$_4$ ⟨CH$_3$ (1) / Br (4)⟩	Na	(Ether) →		NaBr
Methyl iodide	Sodium		Para-xylene	Sodium iodide
Para-bromo-toluene				Sodium bromide

The method has been applied to the preparation of substituted benzene hydrocarbons containing as many as four alkyl-groups (durene, $C_6H_2(CH_3)_4(1,2,4,5)$ and isodurene, $C_6H_2(CH_3)_4(1,2,3,5)$).

Frankland introduced the use of zinc instead of sodium to accomplish similar reactions. See also **Fitting Reaction**; and **Organic Chemistry**.

WURTZITE.

A mineral zinc sulfide, (Zn, Fe)S, similar to sphalerite. Crystallizes in the hexagonal system. Hardness, 3.5–4; specific gravity, 3.98; color, brownish-black with resinous luster. Named after Adolphe Würtz, France.

X

XALSTOCITE. A pink to rose-pink variety of grossular garnet. Sometimes also called landerite or rosolite.

XANTHAN GUM. A very high-molecular-weight polysaccharide produced by pure culture fermentation of glucose by *Xanthamonas campestris*. The substance is readily soluble in hot or cold water, imparting a high viscosity at low concentrations. The solutions are pseudoplastic, with viscosity decreasing rapidly as shear rate increases. Heat, acid, and salt have little effect on the stability of its solutions. The substance is compatible with most other hydrocolloids, including starch. Xanthan gum undergoes a unique gel reaction with locust bean gum to produce a synergistic increase in viscosity. The gum is used as a thickening, suspending, emulsifying, and stabilizing agent in foods and has a number of important nonfood uses as well.

In 1974, the Northern Regional Research Center (Peoria, Illinois) of the U.S. Department of Agriculture and the Kelco Company were joint recipients of the Institute of Food Technologists award for the development and commercialization of xanthan gum. As early as 1956, researchers discovered unusual water-thickening abilities of a substance produced by the bacterium *Xanthamonas campestris*. They chemically identified the substance and named it *Polysaccharide B-1459* after the culture number. A food additive regulation for its use (when it was renamed xanthan gum) was issued in 1959. The process was brought into commercial production in 1964.

Glucose from starch is fermented by the aforementioned bacteria to produce xanthan gum, which is recovered by precipitation with isopropyl alcohol, then washed, dried, and milled. Among several applications, xanthan gum is used in pourable salad dressings for its emulsifying properties; in frozen foods for its freeze-thaw stabilizing effect; in juice drinks for its suspending properties; and in creamed cottage cheese for its stabilizing properties, pseudoplasticity, and mouthfeel. The gum makes a stable cream dressing that clings to the curd, but shears easily and thus does not have a gummy texture. The gum is also used as a stabilizer for frozen desserts and as a suspending agent in liquid feed supplements for cattle feeding and in milk replacers for calves. It also has been found that xanthan gum makes the proteins in nonwheat breads more extensible and thus produces better breads.

For references see the list in the entry on **Colloid Systems**. Other gums and mucilages are described in the entry on **Gums and Mucilages**.

XANTHATES. The salts of the *O*-esters of carbonodithioic acids and the corresponding *O,S*-diesters are xanthates. The free acids decompose on standing.

Properties

The free xanthic acids are unstable, colorless, or yellow oils, and may decompose with explosive violence. They are soluble in the common organic solvents and are slightly soluble in water: methyl xanthic acid at 0°C, 0.05 mol/L; ethyl xanthic acid at 0°C, 0.02 mol/L; and *n*-butyl xanthic acid at 0°C, 0.0008 mol/L. Values for the dissociation constant for ethyl xanthic acid are $(2.0-3.0) \times 10^{-2}$. Potentiometric determinations for C_1-C_8 xanthic acids show a decreasing acid strength with increasing molecular weight. The alkali metal salts, in contrast to the free acids, are relatively stable solids, are pale yellow when pure, and sometimes have a disagreeable odor.

When exposed to air, the sodium salts tend to take up moisture and form dihydrates. The alkali metal xanthates are soluble in water, alcohols, the lower ketones, pyridine, and acetonitrile (Table 1). They are not particularly soluble in nonpolar solvents.

The heavy metal salts, in contrast to the alkali metal salts, have lower melting points and are more soluble in organic solvents. They are slightly

TABLE 1. SOLUBILITIES OF SOME ALKALI-METAL XANTHATES

Xanthate	Solvent	Solubility, g/100 g soln			
		0°C	10°C	20°C	35°C
sodium ethyl	water	40.8	46.0	52.0	
potassium *n*-propyl	water	43.0			58.0
	n-propyl alcohol	1.9			8.9
sodium *n*-propyl	water	17.6			43.3
	n-propyl alcohol	10.2			22.5
potassium isopropyl	water	16.6			37.2
	isopropyl alcohol				2.0
	IPA-H$_2$O azeotrope			6.9	
sodium isopropyl	water	12.1			37.9
		24.5	27.3	30.5	
		24.0	27.5	31.0	37.5
	isopropyl alcohol				19.0
potassium *n*-butyl	water	32.4			47.9
	n-butyl alcohol				36.5
sodium *n*-butyl	water	20.0			76.2
	n-butyl alcohol				39.2
potassium isobutyl	water	10.7			47.7
	isobutyl alcohol	1.6			6.2
sodium isobutyl	water	11.2			33.4
		46.2	48.2	50.5	
		44.0	49.0	51.0	57.3
	isobutyl alcohol	1.2			20.5
sodium *sec*-butyl	water	29.4	34.0	38.8	
potassium isoamyl	water	28.4			53.5
		16.9	26.0	35.0	
		28.5	39.0	45.6	52.5
	isoamyl alcohol	10.9			15.5

soluble in water, alcohol, aliphatic hydrocarbons, and ethyl ether. Alkalies stabilize xanthate solutions somewhat and the solutions readily decompose at acidic pHs.

Reactions. The chemistry of the xanthates is essentially that of the dithio acids.

Peroxyxanthates. A new factor in the theory and practice of flotation was found in the Mount Isa, Australia, copper flotation solution. Secondary butyl perxanthate was formed by the reaction of the xanthate with hydrogen peroxide in dilute alkaline aqueous solution and was found to be identical to a substance from the flotation solution. The perxanthate was isolated as the ammonium salt.

Preparation and Manufacture

The alkali metal xanthates are generally prepared from the reaction of sodium or potassium hydroxide with an alcohol and carbon disulfide.

Many of the heavy metal xanthates have been prepared from aqueous solutions of the alkali metal xanthates and the water-soluble compound of the heavy metal desired.

Alkali Metal Xanthates. The commercially available xanthates are prepared from various primary or secondary alcohols. The alkyl group varies from C_2 to C_5 and the alkali metal may be sodium or potassium.

The alkali metal xanthates are fairly safe to handle. The standard precautions of rubber gloves, dust mask, and goggles are sufficient when handling the solid or the solution.

Under regulations for the enforcement of the Federal Insecticide, Fungicide, and Rodenticide Act, products containing over 50 wt % sodium isopropyl xanthate must bear the label "Caution. Irritating dust. Avoid breathing dust, avoid contact with skin and eyes". Rubber goods in repeated contact with food may contain diethyl xanthogen disulfide not to exceed 5 wt % of the rubber products.

Xanthate drums should be kept as cool and dry as possible. Protection from moisture is the most important factor. A combination of moisture and hot weather causes sodium ethyl xanthate to ignite spontaneously.

Environmental Concerns. Concern for the well-being of the environment has resulted in studies on the effects of mining chemicals, including the xanthates, on various aquatic organisms worldwide. In a thorough and detailed study of three typical mill operations, it was concluded that residual organic flotation reagents do not seem to present widespread problems in effluent disposal.

GUY H. HARRIS
University of California at Berkeley

Additional Reading

Gattow, G. and W. R. Bahrendt: "Topics in Sulfur Chemistry," Vol. 2: *Carbon Sulfides and Their Inorganic and Complex Chemistry,* Georg Thieme, Stuttgart, Germany, 1977.

Harris, G. H.: *Reagents in Mineral Technology,* Surfactant Science Series, Vol. 27, Marcel Dekker, Inc., New York, NY, 1988, pp. 371–383.

Herbst, W., and K. Hunger: *Industrial Organic Pigments: Production, Properties, Applications,* John Wiley & Sons, Inc., Hoboken, NJ, 2004.

Leja, J.: *Surface Chemistry of Froth Flotation,* Plenum Press, New York, NY, 1982.

Rao, S. R.: *Xanthates and Related Compounds,* Marcel Dekker, Inc., New York, NY, 1971.

XANTHENE DYES. Xanthene dyes are those containing the xanthylium (**1a**) or dibenzo-γ-pryan nucleus (xanthene) (**1b**) as the chromophore with amino or hydroxy groups meta to the oxygen as the usual auxochromes. They are important because of their brilliant hues; shades between greenish yellows to dark violets and blues are obtainable, the most important being reds and pinks. Xanthenes are often fluorescent, which adds to their strength and brightness; but, as is often the case with fluorescent dyes, they have lower light-fastness compared to other chromophores. Their use is concentrated on those areas in which light-fastness is relatively unimportant compared to economy (e.g., paper dyes) or where lightfastness can be achieved by modification. They are used for the direct dyeing of wool and silk and mordant dyeing of cotton. Paper, leather, woods, food, drugs, and cosmetics are dyed with xanthene dyes. See also **Dyes.**

Brilliant insoluble lakes are used in paints and varnishes. Recent applications for xanthene dyes include use in ink-jet printers, as markers in biological and medical research, and even as insecticides.

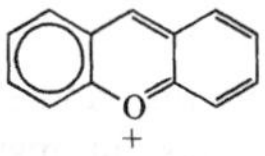

(1a) 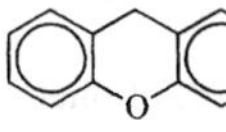   (1b)

Xanthenes date from 1871 when von Bayer synthesized fluorescein (**2**).

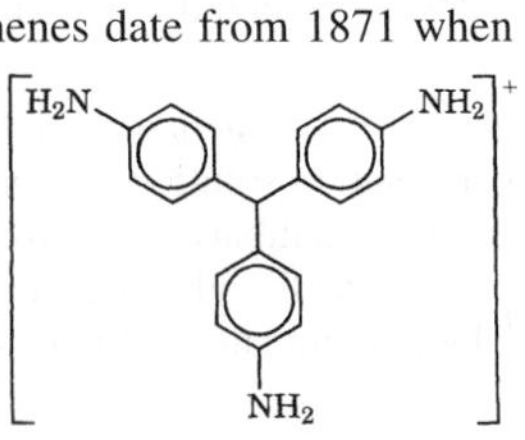

(3) 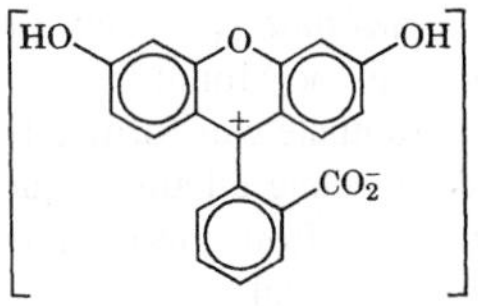   (2)

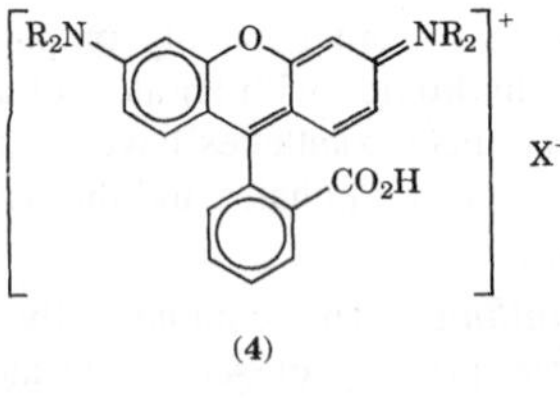

(4)

The xanthene dyes may be classified into two main groups: diphenylmethane derivatives, called pyronines, and triphenylmethane derivatives (e.g., (**3**)), which are mainly phthaleins made from phthalic anhydride condensations. A third much smaller group of rosamines (9-phenylxanthenes) is prepared from substituted benzaldehydes. The phthaleins may be further subdivided into the following: fluoresceins (hydroxy substituted); Rhodamines (amino substituted), e.g., (**4**); and mixed hydroxy/amino substituted.

Diphenylmethane Derivatives

Pyronines. Pyronines are diphenylmethane derivatives synthesized by the condensation of *m*-dialkylaminophenols with formaldehyde, followed by oxidation of the xanthene derivative to the corresponding xanthydrol, which in the presence of acid forms the dye (**5**). If R is methyl, the dye is pyronine G (CI 45005); if R is ethyl, pyronine B (CI 45010) is obtained.

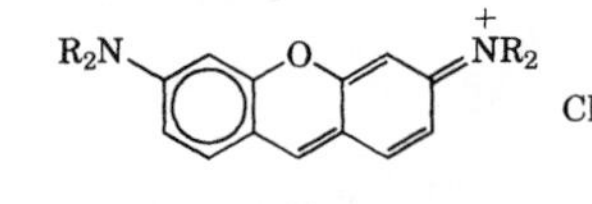

(5)

Succineins. Succineins are carboxyethyl-substituted pyronines made by substituting succinic anhydride for formaldehyde in the basic synthesis.

Triphenylmethane Derivatives

Amino-Derivatives

Rhodamines. Rhodamines are commercially the most important aminoxanthenes. If phthalic anhydride is used in place of formaldehyde in the above condensation reaction with *m*-dialkylaminophenol, a triphenylmethane analogue, 9-phenylxanthene, is produced. Historically, these have been called rhodamines. Rhodamine B (Basic Violet 10, CI 45170) (**6**) is usually manufactured by the condensation of two moles of *m*-diethylaminophenol with phthalic anhydride.

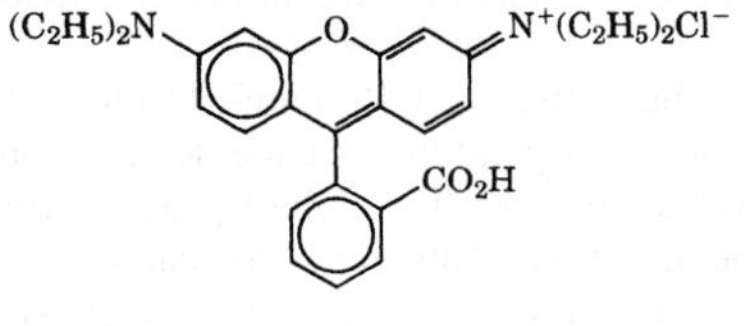

(6)

The rhodamines described thus far are basic rhodamines. They are used primarily for the dyeing of paper and the preparation of lakes for use as pigments. They are also used in the dyeing of silk and wool where brilliant shades with fluorescent effects are required, but where light-fastness is unimportant. Many new uses for rhodamine dyes have been reported. For example, when vacuum-sublimed onto a video disk, Rhodamine B loses its color to form a clear stable film which becomes permanently colored on exposure to uv light. This can be used in optical recording for computer storage or video recording. Acid rhodamines are made by the introduction of the sulfonic acid group to the aminoxanthene base.

Some acid rhodamines are used for silk and wool. Highly substituted acid rhodamines have been reported for fiber-reactive dye applications.

Rosamines. Rosamines are 9-phenylxanthene derivatives prepared from substituted benzaldehydes instead of phthalic anhydride. Sulforhodamine B (Acid Red 52; CI 45100) is the most important rosamine.

Hydroxyl Derivatives.. The building block of most hydroxyl-substituted xanthenes, or fluorones, is fluorescein (**2**). The sodium or potassium salt of fluorescein, commonly called uranine (CI 45350), is used for dyeing wool and silk brilliant yellow shades. However, the principal use of fluorescein is as an intermediate for more highly substituted hydroxyxanthenes.

Aminohydroxy Derivatives

Aminohydroxy-substituted xanthenes are of little commercial importance. They are synthesized by condensing one mole of *m*-dialkylaminophenol with phthalic anhydride, and then condensing that product with an appropriately substituted phenol.

Miscellaneous Derivatives

Two additional xanthene analogues are termed fluorescent brighteners. Fluorescent Brightener 74 (CI 45550) and Fluorescent Brightener 155 (CI 45555) are used in the formulation of solid dielectric compositions for application in high voltage cables to prevent conductive treeing.

Health and Safety Factors, Toxicology

Xanthene dyes have not exhibited health or safety properties warranting special precautions; however, standard chemical labeling instructions are required.

PAUL WIGHT
Zeneca Specialties

Additional Reading

Gregory, P.: *High-Technology Applications of Organic Colorants,* Plenum Publishing Corp., New York, NY, 1990.

Herbst, W., and K. Hunger: *Industrial Organic Pigments: Production, Properties, Applications,* John Wiley & Sons, Inc., Hoboken, NJ, 2004.

Lubs, H.A. ed.: *The Chemistry of Synthetic Dyes and Pigments,* American Chemical Society Monograph Series, Reinhold Publishing Corp., New York, NY, 1955.

Ventkataraman, K. ed.: *The Chemistry of Synthetic Dyes,* Vol. 2, Academic Press Inc., New York, NY, 1952.

XANTHOPHYLLS. See **Carotenoids**.

XENOBLAST. A term proposed by Becke in 1903 for metamorphic crystals with undeveloped crystal faces.

XENOCRYST. A term proposed by Sollas in 1894 for crystals, usually corroded, which are foreign to the magma from which the igneous rock in which they occur has crystallized.

XENOLITH. A fragment, large or small, of a foreign rock included in an igneous mass. The term is derived from the Greek, meaning stranger and stone. Xenoliths, both large and small, are best displayed at the contacts or margins of batholiths.

XENOMORPHIC. The texture or fabric of an igneous rock having or characterized by crystals not bounded by their own crystal faces and which have their form impressed upon them by preexisting adjacent mineral crystals.

XENON. [CAS: 7440-63-3]. Chemical element, symbol Xe, at. no. 54, at. wt. 131.30, periodic table group 18 (inert or noble gases), mp. $-112°C$, bp $-107.1 \pm 2.5°C$, density 3.5 g/cm^3 (liquid at $-109°C$). Specific gravity compared with air is 4.561. Xenon was found by Travers and Ramsay in 1898 when they were experimenting with liquefied air. Solid xenon has a face-centered cubic crystal structure. At standard conditions, xenon is a colorless, odorless gas and does not form stable compounds with any other element. Due to its low valence forces, xenon does not form diatomic molecules, except in discharge tubes. It does form compounds under highly favorable conditions, as excitation in discharge tubes, or pressure in the presence of a powerful dipole. Xenon forms a hydrate much more readily than argon, at a pressure slightly above 1 atmosphere pressure at $0°C$. The element also forms additional compounds with a number of organic substances, such as $Xe \cdot 2C_6H_5OH$ with phenol, which has a dissociation pressure of 1 atmosphere at $4°C$. See also **Chemical Elements**.

In 1962, the compound xenon platinum hexafluoride was synthesized by Bartlett. Later in the same year, Classen confirmed the synthesis and prepared the first binary compound of an inert gas, xenon tetrafluoride, a stable crystalline compound, mp about $90°C$. The compound was prepared by heating a 5:1 mixture of fluorine and xenon to $400°C$, then cooling it rapidly to room temperature. Since the original research, additional xylene-fluroine compounds have been reported, including XeF_2, XeF_4, XeF_6, XeF_6, $XeSiF_6$, XeO_2F_2, and Na_4XeO_6, as well as the hydrate. By heating xenon and fluorine above 10 atmospheres (up to 170 atmospheres), at $250°C$, Weinstock, Weaver, and Krop obtained XeF_5 and XeF_6 in an equilibrium mixture. D.F. Smith, S.M. Williamson, and C.W. Koch have reported the preparation of XeO_3 from the hexafluoride and tetrafluoride. XeO_3 is a white, crystalline, explosive compound.

Xenon also forms compounds, possibly clathrates, with certain substances in nonstoichiometric proportions. Crystalline compounds with benzene or hydroquinone, formed under 40 atmospheres pressure, contain about 26% xenon by weight. Alkaline hydrolysis of XeF_6 produces salts of octavalent xenon. No persistent divalent or tetravalent compounds are found in aqueous solution, but the former is intermediate in hydrolysis of the fluorides, and the latter in reactions of XeO_3 with XeF_2, H_2O_2, and various organics.

Xenon occurs in the atmosphere to the extent of approximately 0.00087%, making it the least abundant of the rare of noble gases in the atmosphere. In terms of abundance, xenon does not appear on lists of elements in the earth's crust because it does not exist in stable compounds under normal conditions. However, xenon because of its limited solubility in H_2O, is found in seawater to the extent of approximately 950 pounds per cubic mile (103 kilograms per cubic kilometer). Commercial xenon is derived from air by liquefaction and fractional distillation. There are nine natural isotopes ^{124}Xe, ^{126}Xe, ^{128}Xe, through ^{132}Xe, ^{134}Xe, and ^{136}Xe, and seven radioactive isotopes ^{123}Xe, ^{125}Xe, ^{127}Xe, ^{133}Xe, ^{135}Xe, ^{137}Xe and ^{138}Xe, all with relatively short half-lives, the longest ^{127}Xe with a half-life of about 36 days. See also Radioactivity. First ionization potential, 12.127 eV; second, 21.1 eV; third, 32.0 eV. Van der Waals radius 2.20 Å Electronic configuration $1s^22s^22p^63s^23p^63d^{10}4s^24p^64d^{10}5s^25p^6$.

Xenon is one of the elements of interstellar matter that is found in some meteorites. As reported by Lewis and Anders (1983), at least three types of xenon are present in carbonaceous chondrites. Two are abundant but controversial; the third is rare, but easy to explain. In 1964, Reynolds and Turner (University of California at Berkeley) examined the C2 chondrite Renazzo and were seeking xenon 129 (from radioactive decay of iodine 129). A controlled, step-heating process was used. In addition to finding xenon-129, the investigators found that in the fractions released between 600 and $1100°C$, the heavy isotopes of Xe ranging in mass from 131 through 136 were present. They found that these isotopes were enriched by as much as 6% with respect to primordial xenon. This enrichment increased from isotope 131 to isotope 136, as it does in the xenon formed by the fission of uranium and other heavy elements. Reynolds and Turner thus suggested that the new xenon component came from the fission of some extinct heavy element that had once been present in the meteorite. It was also noted that xenon 124 and 126 were also enriched in the meteorite, but inasmuch as these isotopes do not form by fission, their enrichment was not fully explained. The two sets of xenon components have been named H (heavy) and L (light) and, although apparently of different origins, they have proved inseparable in meteorites. Xenon also has been detected in some of the planetary atmospheres, such as that of Mars. Check various entries on planets in this book.

The gas finds principal application in special electronic devices and lamps. Xenon, in a vacuum tube, produces a beautiful blue glow when excited by an electrical discharge. Xenon lamps have been developed which provide a constant light (described as sunlight-plus-north-sky light) even when there are significant voltage changes. Thus, the lamps do not require voltage regulators. For a given wattage, xenon lamps have been found to deliver a greater light output. For example, an 800-watt xenon lamp will produce 2,000 lumens as compared with a 1,000-watt incandescent lamp that produces only about 200 lumens. Xenon also has found application in certain lasers. It has been found that xenon produces mild anesthesia, but cannot be used for surgery because the quantity required would cause asphyxiation. Xenon is used in bubble chambers, probes, and other applications where its high molecular weight is of advantage in the atomic energy field. Potentially, xenon is of interest as a gas for ion engines. The perxenates have been used as oxidizing agents in analytical chemistry. Xenon 133 and 135 are produced by neutron irradiation in air-cooled nuclear reactors. Xenon 133 has been found useful as a radioisotope in various studies.

Additional Reading

Anders, E.: "Noble Gases in Meteorites Evidence for Presolar Matter and Superheavy Elements," *Proceedings of the Royal Society of London,* Series A, Vol. **374**, No. 1757, 207–238 (February 4, 1981).

Anderson, D.L.: "Composition of the Earth," *Science,* 367 (January 20, 1989).

Birgeneau, R.J. and P.M. Horn: "Two-dimensional Rare Gas Solids," *Science,* **232**, 329–336 (1986).

Greenwood, N.N. and A. Earnshaw: *Chemistry of the Elements,* 2nd Edition, Butterworth-Heinemann, Inc., Woburn, MA, 1997.

Krebs, R.E.: *The History and Use of Our Earth's Chemical Elements: A Reference Guide,* Greenwood Publishing Group, Inc., Westport, CT, 1998.

Lagowski, J.J.: *MacMillan Encyclopedia of Chemistry,* Vol. 1, Macmillan Library Reference, New York, NY, 1997.

Lewis, R.S. and E. Anders: "Interstellar Matter in Meteorites," *Sci. Amer.,* **249**(2), 66–77 (August 1983).

Lide, D.R.: *CRC Handbook of Chemistry and Physics 2000–2001,* 84th Edition, CRC Press, LLC., Boca Raton, FL, 2003.

Parker, P.: *McGraw-Hill Encyclopedia of Chemistry,* 2nd Edition, The McGraw-Hill Companies, Inc., New York, NY, 1993.

Stwertka, A. and E. Stwertka: *A Guide to the Elements,* Oxford University Press, Inc., New York, NY, 1998.

XENOTOPIC. The fabric of a crystalline sedimentary rock in which the majority of the constituent crystals are anhedral. Fabric found in evaporites, chemically deposited cement, and recrystallized limestone or dolomite.

XI PARTICLE. A hyperon with a rest-mass energy of about 1318.4 MeV, an isospin quantum number $\frac{1}{2}$, an angular momentum spin quantum number $\frac{1}{2}$, and a strangeness quantum number 2. Symbol, Ξ.

X-RAY. In 1895, W. Roentgen of Würzburg, Germany discovered x-rays accidentally while experimenting with a Crookes tube. Roentgen observed the fluorescence of a barium platinocyanide screen that happened to lie near the tube and traced the effect to something that emanated from the spot where the cathode rays struck the tube wall. Putting the tube in a pasteboard box made no difference; so it was not light or ultraviolet radiation that caused the fluorescence. Investigation rapidly followed and relatively soon, x-rays were being used by surgeons to examine the bones of living people. X-ray photos are also used in industry. See Fig. 1.

It was soon found that x-rays arise wherever cathode rays encounter solids; that "targets" of high atomic weight yield more copious X-rays; and that the greater the speed of the cathode particles, the more penetrating, or the "harder" the X-rays are. Special tubes were designed for producing x-rays. The earlier tubes were of the Crookes type, depending on the conduction of ionized gas. Those most used now are thermionic, of the Coolidge type, with a hot-wire cathode operating in a high vacuum; a construction which permits the passage of very high-speed electrons under voltage control, the quantity of them, and the intensity of the resulting X-rays, being regulated by the filament temperature.

Early experiments indicated that x-rays are something essentially different from light, an erroneous conclusion based upon the failure to observe regular reflection, refraction, or diffraction. We now know that this failure was due to the extremely short wavelength of the rays, the range of which extends from the extreme ultraviolet into the gamma-ray region, that is, from 10^{-7} to 10^{-9} centimeter. It is also known that X-rays are produced: (1) when electrons, accelerated in a vacuum, strike a target and lose kinetic energy in passing through the strong electric fields surrounding the target nuclei, thus giving rise to bremsstrahlung and resulting in a continuous x-ray spectrum; and (2) by the transitions of atoms from higher energy states to K, L, ... energy states, thus giving rise to characteristic x-rays. The term x-rays is not used to refer to the characteristic radiation from an element of atomic number Z less than 10, since the wavelengths of such radiation exceed those in the x-ray range. However, every element has its characteristic X-ray spectrum, when used as a target, although according to the Duane and Hunt law the radiation also depends on the accelerating voltage.

There are three principal means of detecting x-rays: the fluorescent effect, the photographic effect, and the ionizing effect. The only method at first available for distinguishing radiation of different wavelengths was to measure their penetration or their absorption coefficient in various substances. The discovery of the X-ray diffraction or grating effect of crystals, by von Laue, Friedrich, and Knipping, in 1912, made it possible to analyze the rays and measure their wavelengths very much as light is studied with the spectroscope. When X-rays of given wavelength are incident upon a crystal turned in various directions, the layers of atoms, at certain angles of incidence, reflect wave trains in phase with each other which, if caught on a photographic plate, produce a "Laue pattern." While the matter is not as simple as in the case of light incident on a diffraction grating, it is nevertheless possible to interpret such patterns in somewhat the same way as a line spectrum, and to deduce the wavelength from it. A unit convenient for expressing X-ray wavelengths is the "X-unit," which is 10^{-11} centimeter or 0.001 angstrom. See also **X-Ray Analysis**.

See also list of references at the end of **X-Ray Analysis**.

Additional Reading

Michette, A.G. and A. Pfauntsch: *X-Rays: The First Hundred Years,* John Wiley & Sons, Inc., New York, NY, 1996.

X-RAY ANALYSIS.   X-rays occupy that portion of the electromagnetic spectrum between 0.01 and 100 angstroms (Å). Their range of approximate quantum energy is from 2×10^{-6} to 2×10^{-10} erg, or from 106 to 100 eV. Important X-ray analytical methods are based upon: (1) fluorescence; (2) emission; (3) absorption; and (4) diffraction. These methods are used qualitatively and quantitatively to determine the element content of complex mixtures and to determine exactly the atomic arrangement and spacings of crystalline materials. See also **Ion Microprobe Mass Analyzer.**

Source of X-rays

X-rays are emitted by atoms that are bombarded with energetic electrons. This results from two separate effects: (1) deceleration of high-speed electrons as they pass through matter, and (2) ionization of individual atoms which abruptly stop the electrons. The first effect results in a continuous-type spectrum; the second effect results in characteristic line spectra.

Continuous Spectrum

The bulk of X-radiation arising from electron bombardment is the continuous spectrum. If an individual electron is abruptly decelerated, but not necessarily stopped, in passing through or near the electric field of a target atom, the electron will lose some energy DE, which appears as an X-ray photon of frequency $v = \Delta E/h$, where h is Planck's constant. An electron may experience several such decelerations before it is finally stopped, emitting x-ray photons of widely different energy and wavelength. A few electrons will be stopped in a single process, losing their entire energy and emitting an X-ray photon having the exact energy of the incident electron.

X-Ray Spectral Lines

These result when the incident electrons knock orbital electrons out of an atom. If an ejected electron is from one of the inner orbits of the

Fig. 1.   Industrial-type x-ray photo of a Polaroid camera

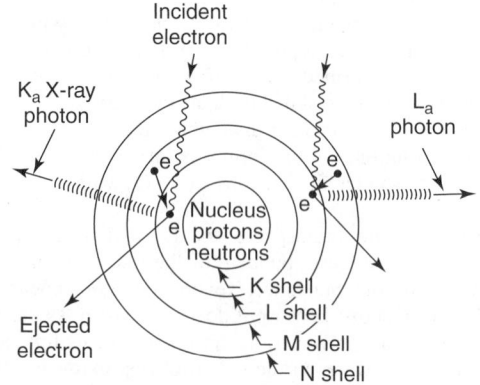

Fig. 1.   Origin of X-ray spectra due to electron bombardment

atom (Fig. 1), an electron from an outer shell will fall to the inner orbit to fill the vacancy. The decrease in potential energy of this electron in approaching the nucleus results in the emission of an X-ray photon having an energy exactly equal to that lost by the electron. The wavelength λ for such photons is related to ΔE by $\lambda = ch/\Delta E$, where c is the velocity of electromagnetic energy and h is Planck's constant. Because the energy of orbital electrons is quantized, the X-ray photons can have only certain definite wavelengths which are characteristic of the atom. This situation is somewhat analogous to the more familiar ultraviolet and visible-emission spectra of materials, the difference being that the optical spectra are the result of electron transitions between energy levels of just the outermost electrons of the atoms.

X-rays resulting from an electron transition filling an electron vacancy in the innermost shell of an atom are known as K x-rays or K lines; those from the L shell are known as L lines, and so on.

Generation of X-rays

An important component in an X-ray analytical device is an x-radiation generator. A high-vacuum Coolidge type tube, wherein electrons are emitted from a heated tungsten filament and accelerated by a high voltage to an anode (target) is a common source of x-rays. See Fig. 2.

A wide variety of tubes is available. All high-power (high-current) commercial tubes employ a water-cooled anode. Tubes of this type have been built with ratings up to 10 kilowatts.

Detection of X-rays

Detectors include (1) Geiger-Mueller tube, (2) ionization chambers, (3) scintillation counters, (4) proportional counter, (5) electron-multiplier tubes, and (6) nondispersive detectors using cooled lithium-drifted Si detectors. See Fig. 3.

X-ray Crystallography

X-rays penetrating below the surface of crystalline materials are scattered by the individual parallel layers of atoms; each atomic layer acts as a new, although weak, source of X-rays. To be reinforced in a given direction at an angle θ (Fig. 4), the spacing d between crystal planes must be rigorously related to the wavelength of the radiation. At a given angle, X-rays of one definite wavelength will be constructively reinforced. These variables are related by Bragg's law.

$$n\lambda = 2d \sin \theta$$

where n is an integer. Note that θ is measured relative to the crystal face rather than to the perpendicular.

In addition to fulfilling this wavelength requirement, the energy will be diffracted only at an angle equal to the angle of the incident x-rays, independent of wavelength; otherwise, destructive interference is possible, and the "reflected" energy is negligible.

X-Ray Analyzer, Electron-Microprobe

Advantages of this analytical instrument include: (1) analysis can be confined to very small (microsamples) amounts of materials; (2) the particular material to be analyzed need not be physically separated from its surrounding materials, as is often required with many analytical methods; and (3) through the development of associated instrumentation, diagnostic techniques, and information displays, the method can be quite fast. Limits of detection in solid solution are from approximately 0.005 to 0.5%, depending upon the elements and sample matrixes involved. See Fig. 5. Concentrations as low as 10^{-16} gram may be measured.

Mainly used for metallurgical studies, nonmetallics also may be analyzed when samples are properly prepared. Biological applications include tooth and bone samples, cytochemical problems and staining techniques, physiochemical problems, and studies in pathology. Relative weight-fraction-detection limits for most elements in biological specimens are in the general range of 0.01 to 0.10%. Electronics industry applications include studies of diffusion phenomena, electrical-contact surfaces, interfaces on transistors, and microcircuitry analysis.

As shown by Fig. 6, electrons from an electron gun are directed to the sample through an electron optical system. Once the electron beam

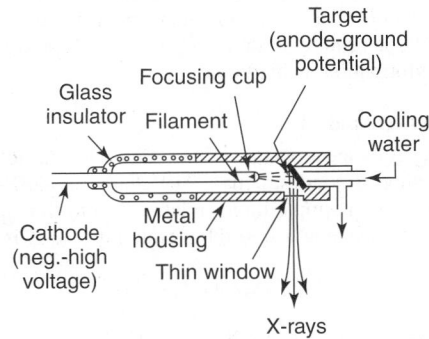

Fig. 2. High-voltage, high-vacuum X-ray tube

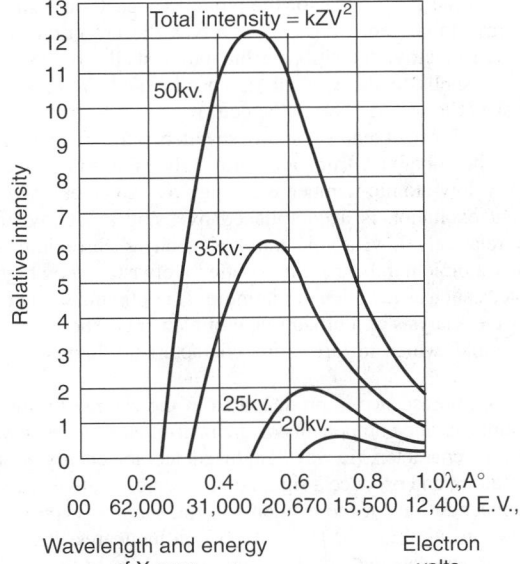

Fig. 3. Continuous X-ray spectrum of tungsten ($Z = 74$) at various tube voltages

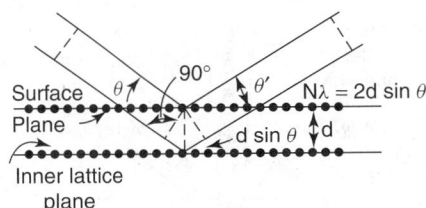

Fig. 4. Reflection of X-rays from internal crystal plane

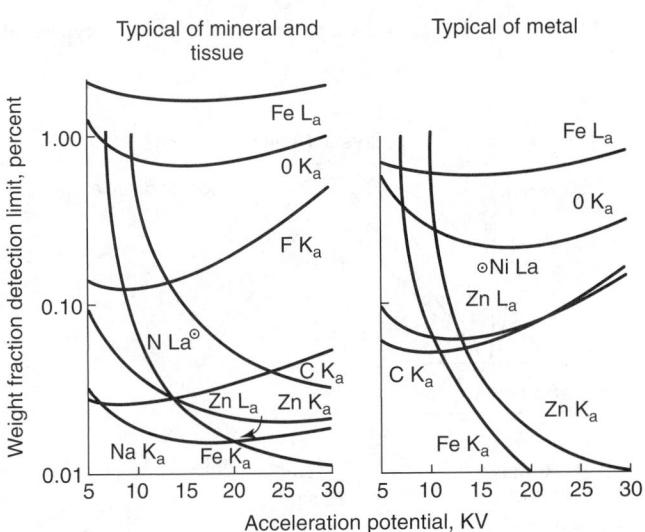

Fig. 5. Typical weight-fraction detecting limits of electron microprobe X-ray analyzer

strikes the sample, a number of signal sources are activated, including (1) high-energy backscattered electrons, (2) low-energy secondary electrons, (3) cathodoluminescence, and (4) x-rays. Some heat also is generated within the sample. Volume d_3 of the specimen is that *volume from which X-rays are emitted.*

The X-rays produced may be detected nondispersively by a proportional counter whose output may be separated as a function of energy by a pulse-height analysis system into the various wavelength components. Better detection sensitivities, however, can be obtained through the use of a fully focusing diffracting-crystal spectrometer in conjunction with a proportional detector and the necessary pulse-height analyzer. As shown by Fig. 7, the necessary condition for fully focusing optics is to have the x-ray source, the crystal, and the detector slit all placed on a common circle. This geometry requires that the diffracting-crystal planes be bent to the diameter of the Rowland circle.

Spherical aberration at the detector slit is minimized by further grinding the crystal surface to fit the radius of the Rowland circle. With the resultant

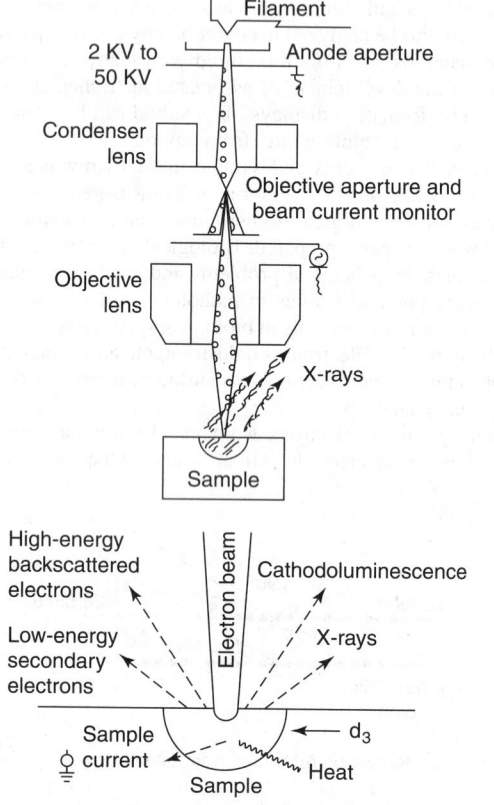

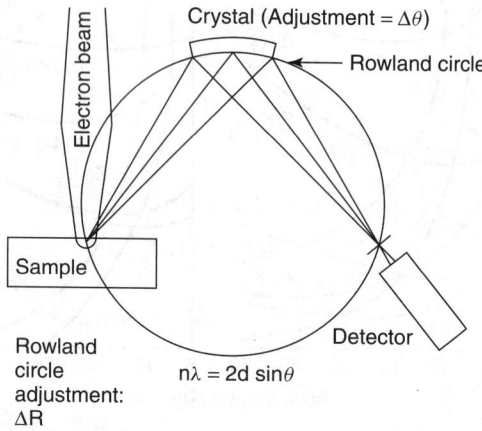

Fig. 6. Electron gun and probe-forming lens system of an integrated electron probe

Fig. 7. Geometry of a fully focusing diffracting-crystal spectrometer

Johansson optics, the crystal radius is fixed, and the 2θ range of the spectrometer is scanned by moving the crystal radially away from the source and, at the same time, rotating it into the detector to achieve a true focus throughout the spectrometer range. X-rays particularly of a wavelength greater than 2 Å, and electrons are highly absorbed in an air atmosphere. Thus, the spectrometer must be enclosed in a vacuum of the order of 10^{-5} torr. The present wavelength range of interest extends from approximately 1 to 100 Å. Diffracting crystals to cover this range must provide broad wavelength coverage, high diffraction efficiencies for high peak-count intensities, good resolution, and good resulting peak-to-background ratios. Crystals that meet these objectives include lithium fluoride, ammonium dihydrogen phosphate (ADP), ethylenediamine *d*-tartrate. (EDT), quartz, and sodium chloride.

Detectors

Of the three commonly used X-ray detectors—(1) Geiger counter, (2) scintillation counter, and (3) proportional counter—the latter is used most frequently for electron-probe microanalysis. In the wavelengths from 1 to 10 Å, sealed proportional counters may be used. For longer-wavelength analysis—in the range from 10 to 93 Å—the thinnest possible detector window is required to limit spectral attenuation. Nitrocellulose windows have proved successful. Nondispersive detection systems using cooled Li-drifted Si are also applicable.

An *optical microscope* is required in the system to provide the analyst with a means of reference to identify various sample areas for analysis. Sample stages may hold single or multiple samples and are provided with means for moving the sample in x, y, and z planes without breaking the system vacuum. After the point of interest is located on the specimen, the data may be read out in a number of ways: (1) quantitative and semiquantitative information may be obtained by processing the x-ray detector signal through a rate meter to a strip-chart or X–Y recorder; (2) scaler systems also provide direct readout of quantitative data integration; and (3) for operational convenience, a data translator and typewriter or teletype printout system may be connected directly to record digital-counter information as hard copy.

X-ray Fluorescence Analysis

One of several types of spectrochemical techniques now used for laboratory analysis. The method is nondestructive. The characteristic X-ray spectrum of each element bears a simple direct relationship to the atomic number. The relation of the wavelength λ to the atomic number Z is

$$\frac{1}{\lambda} a Z^2 \text{ (Mosley's law)}$$

Since the X-ray spectral lines come from the inner electrons of the atoms, the lines are not related to the chemical properties of the elements or to the compounds in which they may reside. Because the characteristics of the X-ray spectra are associated with energies released through transitions of electrons within the inner shells of the atom, the spectra are simple. Most practical X-ray fluorescence analysis involves the detection of radiation release through electron transitions from outer shells to the K shell (K spectra), outer shells to the L shell (L spectra) and, in very few cases, from outer shells to the M shell (M spectra).

The simplest form of energy source available for commercial instrumentation is that obtained from an X-ray tube. For samples containing predominantly low-atomic-number elements, as in cement raw mix, the most efficient excitation is accomplished by using an X-ray tube target material of relatively low atomic number, such as chromium. Elements having higher atomic numbers are most effectively excited by high-atomic-number targets, such as tungsten or platinum. An optimum target material is rhodium for the analysis of a broad range of elements. The X-ray tube irradiates the sample, which in turn emits characteristic fluorescent radiation of its atoms.

Once X-ray fluorescence is produced from the sample by means of an X-ray tube, appropriate components of the instrument (Fig. 8) separate this radiation into its characteristic wavelengths, detect the energy emitted from each excited atom, and produce a signal that is representative of the number of atoms (concentration) of the elements in the sample. Typical excitation conditions (x-ray tube) are 50 kilovolts, 35 milliamperes. Bragg's law of X-ray diffraction is satisfied by the condition $N\lambda = 2d \sin q$, where λ is the wavelength and θ is the angle of incidence and diffraction of X-rays from a crystal whose lattice spacing is defined by d and N, the order of harmonic of the diffraction. For almost all fluorescence analysis, $N = 1$. The usable x-ray spectrum normally extends from 0.1 to about 20 Å. A helium or vacuum path is required for x-ray analysis of elements with atomic numbers lower than $Z \cong 24$ (wavelengths longer than 2.3 Å).

Instrumentation can provide for simultaneous elemental analysis using fixed, preselected X-ray detection channels and scanners or a goniometer to provide one or more channels that can be tuned to a wide wavelength coverage. Up to 30 monochromator positions are possible. Typical crystal materials covering the practical wavelength range of 1 to 20 Å are lithium fluoride, silicon oxide, sodium chloride, EDT, and ADP. Optimum analytical data for the elements of interest are obtained by using fully focusing Johansson curved and ground crystals.

An optical diagram of a Johansson curved-crystal spectrometer is given in Fig. 9. Each spectrometer of an x-ray quantometer may be equipped with optimum crystal-detector combinations for specific determinations in a wide variety of matrixes, including steel, aluminum, copper-base materials, ores, cement, and slags—in both liquid and solid states.

The diffracted X-radiation is detected by Geiger, proportional, or semiproportional detectors. See Fig. 10. The detector of each monochromator generates pulses, which are a measure of the intensity of radiation of each wavelength. The pulses are filtered through a discriminator in order to avoid undesired interferences. Pulses shrinking due to an increase of frequency of pulses is automatically compensated. Collected pulses are transferred to a computer for processing and output. See Fig. 11.

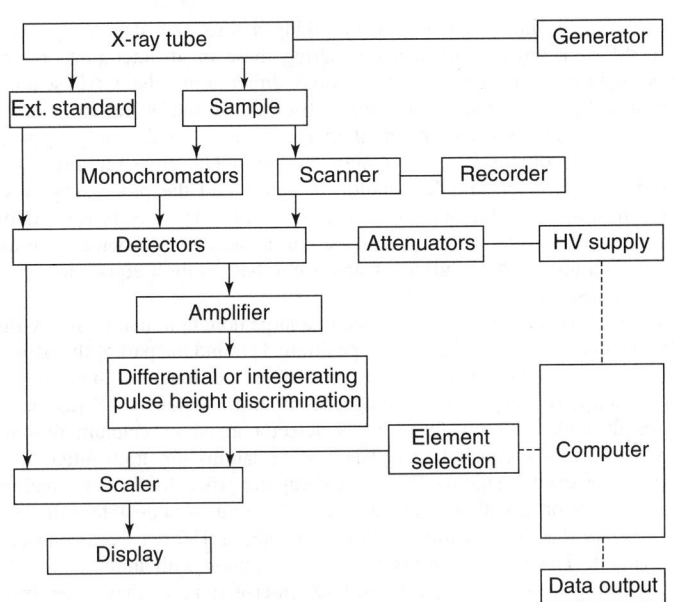

Fig. 11.   System configuration of X-ray fluorescence analyzer

The transition from laboratory to automated instrument to achieve high-speed continuous analysis of dry or wet materials primarily involves the sample-handling and presentation hardware.

Limits of detectability for the desired elemental analyses vary depending upon the matrix, elements, methods of sample preparation, and quality of instrumentation applied. Generally, these are on the order of 1 to 100 parts per million. The limit of detectability, however, is only one criterion in evaluating methods of analysis. The time of analysis is important, particularly in production and process control laboratories. In multi-element spectrometers, it is possible to perform as many as 30 simultaneous elemental determinations in from 20 to 120 seconds, depending upon the material being analyzed.

W. G. SHEQUEN, P. E.

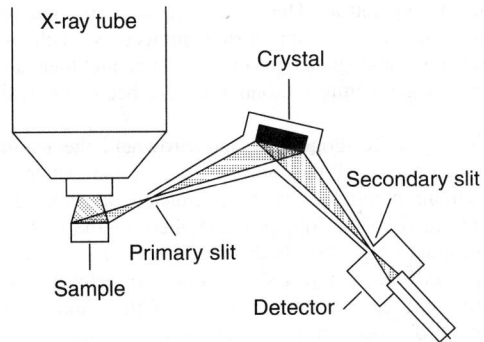

Fig. 8.   Optical path for one monochromator used in an industrial X-ray fluorescence analyzer

Additional Reading

Dunitz, J.D.D.: *X-Ray Analysis and the Structure of Organic Molecules,* John Wiley & Sons, Inc., New York, NY, 1995.

Gilfrich, J.V., et al.: *Advances in X-Ray Analysis,* Perseus Publishing, Boulder, CO, 1998.

Hammond, C.: *The Basics of Crystallography and Diffraction,* Oxford University Press, Inc., New York, NY, 1998.

Janssens, K.H., F. Adams, and A. Rindby: *Microscopic X-Ray Fluorescence Analysis,* John Wiley & Sons, New York, NY, 2000.

Jenkins, R., D.K. Smith, and V.E. Buhrke: *A Practical Guide for the Preparation of Specimens for X-Ray Fluorescence and X-Ray Diffraction Analysis,* John Wiley & Sons, Inc., New York, NY, 1997.

Lifshin, E.: *X-Ray Characterization of Materials,* John Wiley & Sons, Inc., New York, NY, 1999.

Suryanarayana, C. and M.G. Norton: *X-Ray Diffraction,* Perseus Publishing, Boulder, CO, 1998.

Warwick, T. and D. Atwood: *X-Ray Microscopy,* American Institute of Physics, College Park, MD, 2000.

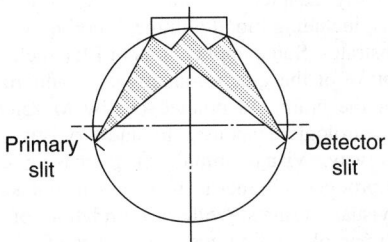

Fig. 9.   Optical diagram of Johansson curved-crystal spectrometer

X-RAY SCAN AND OTHER MEDICAL IMAGERY. Introduction of computerized axial tomography in the form of the X-ray cat scanner (CAT) was a major event in the history of diagnostic medicine. The technology is also referred simply as computerized tomography (CT).

Tomography as a technique of X-ray photography by which a single selected plane is photographed, with the outline of structures in other planes eliminated, has been known for many years. X-rays lose energy in direct proportion to the density of the structure through which they pass. Thus dense structures of the body, such as bone, absorb much of the X-radiation and therefore in a traditional X-ray negative appear as light images—because less radiation has reached the film. The CAT scanner utilizes this same general principle (more or less energy absorbed, depending upon density of intervening tissue). But, instead

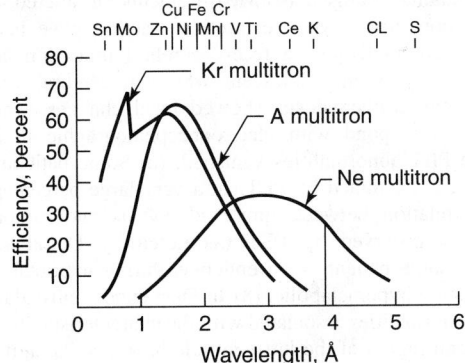

Fig. 10.   Relative efficiency of X-radiation detectors

of exposing a single X-ray film, or taking a series of single exposures from different angles and then comparing these in an attempt to create some semblance of the object in three dimensions, the CAT scanner indicates the exact amount of energy absorbed by the object from many different angles—so that the final image as seen by the radiologist is based upon a composite of contiguous one-millimeter cross sections. Film is replaced by electronic X-radiation detectors and the processing of so much information obtained in such a short time span is processed by a computer. A rapid rate of calculation is needed to measure tissue density, requiring the solution of about one-half million equations every few minutes.

The CAT scanner has been likened to a large doughnut that houses x-ray tube and detectors. The doughnut is positioned around the part of the patient to be examined. As the patient lies on a table, the doughnut, housed in a large gantry, rotates, creating images as it turns. When the X-ray beam passes through the body, it strikes a detector in which calcium fluoride crystals scintillate (flash light). These scintillations are transmitted to a computer as electric signals. While the computer processes the information, registering it on a cathode ray tube, the doughnut rotates a few degrees and repeats the process until it has completed a 180-degree arc around the patient. The scanner forms a grid of readings with innumerable X-rays creating a matrix of nearly 100,000 intersections as they enter from various angles. The average amount of X-radiation absorbed by a patient during a CAT scan is estimated to be about one-fifth that absorbed in X-ray exposures in a typical executive physical given by many clinics and hospitals over 40 years of age.

The result is an X-ray image that gives the illusion of three dimensionality. Because the CAT beam is rotated around the body, it can image organs that overlap and are therefore obscured under conventional x-rays or radiograms. By using several hundred X-ray detectors to produce one exposure, the CAT is an order of magnitude more sensitive to slight gradations in density than radiographs, which frequently do not allow the practitioner to distinguish between tissues of approximately the same density.

Prior to the CAT scanner, normal X-ray photographs of the brain tended to be blurred beyond recognition by the skull. Since the first application of the CAT scanner to brain imaging in 1972, the instrument has come to be considered an indispensable neurophysiological tool. Prior to the availability of the CAT technique, an image of the brain's complex vascular system was obtainable only by means of an angiogram or arteriogram. Patients were given an injection of a dye directed to a specific site. In the diagnosis of some brain diseases, it was necessary to make a pneumoencephalogram ("air scan"), in which gas was injected into the lower spine and flowed upward to fill and outline brain cavities prior to x-ray procedures. These were both dangerous and painful procedures and required up to 2 or 3 days of hospitalization just for diagnostic purposes. The CAT can provide a detailed cross section of the brain in as brief a span as five minutes. Air scans can be eliminated and the need for angiograms has been sharply reduced. Rapid, usually accurate diagnosis of brain disorders and injuries has saved lives by reducing exploratory surgery and by greatly shortening delays in commencing treatment.

The image of an 80–90% underexposed medical radiograph can be increased to readable density and contrast by autoradiographic image intensification. The technique consists of combining the image silver of the radiograph with a radioactive compound, thiourea labeled with sulfur-35, and then making an autoradiograph from the activated negative.

Minimizing the X-ray dose received by patients during medical examinations and maximizing the quality of the radiographs are subjects of concern to the medical profession and the public. Some conflict is inherent in the two objectives because higher quality radiographs, i.e., those which convey more information to the physician, usually require higher exposure levels. Recent gains in quality or exposure reductions, or both, are due to developments, such as computer processing, electrostatic imaging systems, improvements in intensifying screens, and scatter rejection techniques.

Radiographs which are normally classified as "badly underexposed" actually contain most of the information which was intended to be recorded by the original exposure. Autoradiographic intensification effectively retrieves this information by increasing the image density and contrast to readable levels. The intensification occurs on an autoradiograph made from the underexposed film after the original image silver has been chemically combined with a radioactive isotope.

Other Photon Imaging Techniques

Since introduction of X-ray CAT scan technology, other photon imaging approaches have been developed. These concepts augment and, in some instances, are alternatives to the X-ray CAT scan. The more recent developments include positron emission tomography, K-edge dichromography, the use of synchrotron radiation, angiographic imaging, and nuclear magnetic resonance imaging. The latter topic is described in the article on **Nuclear Magnetic Resonance (NMR) and Magnetic Resonance Imaging (MRI)**. The recent use of synchrotron radiation is described under **Particles (Subatomic)**.

Positron Emission Tomography (PET). In PET, image construction is based on the location and intensity of gamma rays emitted in the region of a *neutron-poor* isotope. Planar images formed by computer PET result from the attenuation coefficients of the tissues that intercept a transmitted x-ray beam. Neutron-poor isotopes undergo radioactive decay by the process $P^+ \rightarrow N + e^+ + v$, where P^+ is a nuclear proton, N is a neutron, e^+ is a positron, and v is a neutrino. A neutron-poor isotope, such as ^{11}C will undergo beta decay, in which a proton becomes a neutron and a positron and a neutrino are ejected from the nucleus. Within a short distance, the positron encounters an electron, upon which the two annihilate each other and give rise to a pair of gamma-ray photons that depart at an angle of about $180°$, each carrying an energy of 0.511 MeV.

The major isotopes used in PET are ^{15}O (half-life = 20 min), ^{13}N (half-life = 10 min), ^{11}C (half-life = 20 min), and ^{18}F (half-life = 110 min). Half-life spans are approximate. The foregoing isotopes are produced by a nearby cyclotron. They are either administered promptly, or are rapidly incorporated into appropriate molecules, such as metabolic substrates, substrate analogues, or drugs, which are then administered. Minicyclotrons for generating radionuclides are becoming available as of the late 1980s.

As pointed out by Ter-Pogossian and Brownell, the gamma photons derived from the decay of the isotopes within the patient's body are sensed by a circumferential array of collimated detectors, the circuitry of which is designed so that opposite members of the ring are coupled. A signal is recorded only when both members of the detector pair sense coincidental photons. By using a slight time difference in the activation of the detectors, one can locate the source of the photons on the basis of time-of-flight differences from an eccentrically positioned emitter. Data are fed into a computer, which generates the image based on location and source intensity. Tissue attenuation is taken into account. Spatial resolution is about 0.5 cm. Only minute amounts of tracers are needed. The radiation dose is small.

PET is particularly adapted to kinetic analysis of physiologic and biochemical events, including blood volume, blood flow, and consumption of oxygen and substrates. Some applications of PET include the following. (1) Identifying uptake of the glucose analogue ^{18}F-fluorodeoxyglucose in various regions of the brain. As pointed out by Mazziotta and Reivich, the laterality of neurologic responses to auditory stimulation has been correlated in this way. Visual stimuli of increasing complexity have been observed to produce symmetric increases in uptake in the primary and associative visual cortices, with a correlation of visual pathway abnormalities with neurologic findings. (2) Studies of Alzheimer's disease patients indicate dysfunction in temporal-parietal regions and in the structures near the third ventricle, as reported by de Leon and Friedland. (3) Studies of schizophrenic patients have suggested subtle local blood flow and metabolic changes. (4) Measurements of altered blood flow, oxygen utilization, and oxygen extraction fraction have been made in patients who have undergone a recent cerebral infarction, as noted by Baron. (5) A study of an adolescent who had cerebral vasculitis due to systemic lupus erythematosus showed local changes, when revealed by PET, that correspond with electroencephalographic findings. After remission, the PET abnormalities vanished. (6) Some patients with brain tumors have been examined by PET. In a very large percentage of cases, a positive correlation between tumor glycolysis and tumor histologic grade has been observed by PET (as noted by DeLaPaz). (7) Kuhl reports that in some patients with epilepsy, during interictal periods, the involved zones are hypometabolic. (8) In Duchenne's muscular dystrophy, biochemical abnormalities associated with the characteristic involvement of the posterolateral region of the left ventricle have been identified by PET. There are many other examples of the contributions of PET to patient diagnosis and biochemical research.

Digital Subtraction Angiography. Routine roentgenograms do not reveal vascular structures because blood vessels and surrounding soft tissues attenuate x-ray beams in the same way, thus not revealing any distinction between the two. Conventional arteriography requires high concentrations of iodine-containing contrast agents to be injected directly into an artery. The attenuation of the x-rays through Compton and photoelectric interactions with iodine provides the required contrast between vessels and surrounding tissues. An alternative technique known as digital subtraction angiography (DSA) requires considerably lower concentrations of contrast agents. DSA is now widely practiced by giving a peripheral or central intravenous injection of contrast materials. Most often, the agents will be administered by a pump-driven device. In this method, an x-ray image is acquired on an area detector, such as a fluoroscope screen, or on a scanning line detector. Data then are digitized, amplified, and transferred to a computer.

An optimal approach to DSA angiography is that of using monochromatic x-ray beams at energy levels just above and just below the K-shell absorption edge of iodine. Image data above the K-edge include information arising from Compton and photoelectric interactions of x-ray beams with iodine and with atoms in the molecules of the soft tissues and bones of the body. The image recorded just below the K-edge includes virtually the same attenuation data except for additional information that arises from photoelectric absorption by K-shell electrons in iodine atoms. The logarithmic subtraction of the two images almost totally suppresses signals arising from soft tissue and bone, as observed by Rubenstein and Hughes.

Additional Reading

Baron, J.C., et al.: "Comparison Study of CT and Positron Emission Tomographic Data," *Amer. J. of Neurological Research*, 536 (April 1983).

Brownell, G.L., et al.: "Positron Tomography and Nuclear Magnetic Resonance Imaging," *Science*, **215**, 619–626 (1982).

DeLa Paz, R.L., et al.: "Positron Emission Tomographic Study of Suppression of Gray-Matter Glucose Utilization by Brain Tumors," *Amer. J. of Neurological Research*, 826, (April 1983).

DeLeon, M., A.E. George, and et al.: "Regional Correlation of PET and CT in Senile Dementia of the Alzheimer Type," *Amer. J. of Neurological Research*, 533, (April 1983).

Friedland, R.P., et al.: "Regional Cerebral Metabolic Alterations in Dementia of the Alzheimer Type: Positron Emission Tomography," *J. Comput. Assist. Tomogr.*, 590 (July 1983).

Kuhl, D.E., et al.: "Epileptic Patterns of Local Cerebral Metabolism and Perfusion in Human Determined by Emission Computed Tomography," *Ann Neurol.*, 348 (August 1980).

Mazziotta, J.C., et al.: "Local Cerebral Glucose Metabolic Response—Studies in Human Subjects with Positron CT," *Human Neurobiology*, 11 (February 1983).

Reivich, M., et al.: "Positron Emission Tomographic Studies," *Human Neurobiology*, 25 (February 1983).

Rubenstein, E., E.B. Hughes, L.E. Cambell, and et al.: "Synchrotron Radiation and Its Application to Digital Subtraction Angiography," *Conf. On Digital Radiography*, **314**, 42, Bellingham, WA, 1981.

Ter-Pogossian, M.M., M.E. Raichle, and B.E. Sobel: "Positron-Emission Tomography," *Sci. Amer.*, 171–178 (October 1980).

XYLENE. [CAS: 1330-20-7]. $C_6H_4(CH_3)_2$, formula weight 106.16. There are three xylenes, ortho-, meta-, and para-xylene. Sometimes referred to as dimethylbenzenes, the xylenes have the following key physical properties:

o-Xylene	mp	−25°C	bp	144°C	sp gr 0.881
m-Xylene		−47.4°C		139°C	0.867
p-Xylene		13.2°C		138.5°C	0.861

All of these compounds are insoluble in H_2O, soluble in alcohol, and *o*-xylene and *m*-xylene are miscible in all proportions with ether; *p*-xylene is very soluble in ether.

The xylenes are very high-tonnage industrial chemicals and are raw materials or intermediate materials for numerous synthetic fibers, resins, and plastics. See also **Xylene Polymers.** A large amount of *p*-xylene goes into polyester fiber production, while substantial quantities of *o*-xylene are consumed by the manufacture of phthalic anhydride. The prime source of xylenes are petroleum refinery reformate streams in conjunction with benzene and toluene extraction. The xylenes occur mixed in these streams.

When naphtha or naphthenic gasoline fractions are catalytically reformed, they usually yield a C_8 aromatics stream that is comprised of mixed xylenes and ethylbenzene. It is possible to separate the ethylbenzene and *o*-xylene by fractionation. It is uneconomic to separate the *m*- and *p*-xylenes in this manner because of the closeness of their boiling points. To accomplish the separation, a Werner-type complex for selective absorption of *p*-xylene from the feed mixture may be used. Or, because of the widely different freezing points of the two xylene isomers, a process of fractional crystallization may be used. To boost the *p*-xylene yield, the filtrate from the crystallization step can be catalytically isomerized.

XYLENE POLYMERS. In a process capable of producing pinhole-free coatings of outstanding conformality and thickness uniformity through the unique chemistry of *p*-xylylene (PX) (**1**), a substrate is exposed to a controlled atmosphere of pure gaseous monomer. The coating process is best described as a vapor deposition polymerization (VDP). The monomer molecule is thermally stable, but kinetically very reactive toward polymerization with other molecules of its kind. Although it is stable as a rarified gas, upon condensation it polymerizes spontaneously to produce a coating of high molecular weight, linear poly(*p*-xylylene) PPX (**2**).

In the commercial Gorham process, PX is generated by the thermal cleavage of its stable dimer, *cyclo*-di-*p*-xylylene (DPX), a [2.2]paracyclophane (**3**). In many instances, substituents attached to the paracyclophane framework are carried through the process unchanged, ultimately becoming substituents of the polymer in the coating.

The PPXs formed as coatings in the Gorham process are referred to generically as the parylenes.

The parylene process has certain similarities with vacuum metallizing. The principal distinction is that truly conformal parylene coatings are deposited even on complex, three-dimensional substrates, such as on sharp points and in hidden or recessed areas. Vacuum metallizing, on the other hand, is a line-of-sight coating technology.

The *p*-xylylene species plays a central role in the coating process itself as well as in the making of the dimers which are used as feed-stocks for the coating process. Polymers and dimers have both been made from precursor *p*-xylene compounds (**4**) featuring a variety of X and Y leaving groups.

Gorham Process Monomers

The eight-carbon monomer PX is generated in the first stage of the parylene process by heating gaseous dimer as it passes through a high temperature zone.

Chemical Evidence for PX Monomer. Establishing early on that PX is indeed the pyrolysis product, rather than the molecule formed by breaking only one of the original dibenzyl bonds, the dimer diradical (**5**), proved to be an important development.

When the pyrolysis gases are quenched with a molar excess of iodine vapor, a yield of greater than 50% *p*-xylylene diiodide is recovered. The observation of this effect offered the first direct chemical support for the idea that DPX pyrolysis results in PX (**1**).

Monomer Properties. Despite difficulties involved in studying it owing to its great reactivity, a great deal is known about the structure of the parylene. The eight-carbon framework of the monomer PX is planar. The molecule is diamagnetic, i.e., all electron spins are paired in the ground state (spectroscopically, a singlet). The PX molecule is a conjugated tetra olefin whose particular arrangement gives it extreme reactivity at its end carbons (**6**).

(**6**)

A particularly useful property of the PX monomer is its enthalpy of formation. Using a semiempirical molecular orbital technique, the heat of formation of *p*-xylylene has been computed to be 234.8 kJ/mol (56.1 kcal/mol).

Successful p-Xylylene VDP Monomers. Within the limits mentioned above, it is frequently possible, and often desirable, to modify the *p*-xylylene monomer by attaching to it certain substituents. Limitations on such modifications lie in the three areas: reactivity, performance in the coater (deposition equipment) and cost.

Other, Related Processes

VDP processes using means other than the pyrolytic cleavage of DPX (Gorham process) to generate the reactive monomer are also known, although none are practiced commercially at present.

Dimer

In contrast to the extreme reactivity of the monomeric PX (**1**) generated from it, the dimer DPX (**3**) feedstock for the parylene process is an exceptionally stable compound. At present only three dimers are commercially available: DPXN, DPXC, and DPXD, which give rise to Parylene N, Parylene C, and Parylene D, respectively.

The unsubstituted C-16 hydrocarbon, [2.2]paracyclophane (**3**), is DPXN. Both DPXC and DPXD are prepared from DPXN by aromatic chlorination and differ only in the extent of chlorination; DPXC has an average of one chlorine atom per aromatic ring and DPXD has an average of two.

Manufacture. For the commercial production of DPXN (di-*p*-xylylene) (**3**), two principal synthetic routes have been used: the direct pyrolysis of *p*-xylene (**4**, X = Y = H) and the 1,6-Hofmann elimination of ammonium (HNR_3^+) from a quaternary ammonium hydroxide (**4**, X = H, Y = NR_3^+). Most of the routes to DPX share a common strategy: PX is generated at a controlled rate in a dilute medium, so that its conversion to dimer is favored over the conversion to polymer.

Purification. Unsubstituted di-*p*-xylylene (DPXN) is readily purified by recrystallization from xylene. It is a colorless, highly crystalline solid. The principal impurity is polymer, which is insoluble in the recrystallization solvent and easily removed by hot filtration. In purifying DPXC and DPXD, care is taken not to disturb the homologue composition, so that product uniformity is maintained.

Properties. The DPXs are all crystalline solids; melting points and densities are given in Table 1. Their solubility in aromatic hydrocarbons is limited.

The structure of DPXN was determined from x-ray diffraction studies. There is considerable strain energy in the buckled aromatic rings and distorted bond angles. The release of this strain energy is the principal reason for success in the preparation of monomer in the parylene process.

Polymer

The linear polymer of PX, poly(*p*-xylylene) (PPX) (**2**), is formed as a VDP coating in the parylene process. The energetics of the polymerization set it apart from all other known polymerizations and enable it to proceed as a vapor deposition polymerization.

Thermodynamic Considerations. On the basis of the value for the enthalpy of formation of *p*-xylylene, the enthalpy of polymerization, can be estimated.

Polymerization Mechanism. The physical processes of condensation and diffusion must be considered along with the *p*-xylylene polymerization chemistry for a proper understanding of what happens microscopically during vapor deposition polymerization. These processes point to an important distinction between VDP and vacuum metallization, i.e., that in the latter, adsorption is followed by a surface reorganization of the existing deposited material, and diffusion of incoming species through the bulk is nonexistent. In most parylene depositions, a coating forms from gaseous monomer under steady-state conditions.

The monomer is consumed by two chemical reactions: initiation, in which new polymer molecules are generated, and propagation, in which existing polymer molecules are extended to higher molecular weight. In steady-state VDP, both reactions proceed continuously inside polymeric coating, in the reaction zone just behind the growth interface.

The concentration of monomer within the coating decreases approximately exponentially with distance from the growth interface. With this decrease in monomer concentration, the rates of initiation and propagation reactions also decrease. Moving back into the polymer from the growth interface, through the reaction zone where polymer is being manufactured, a region in which the polymer formation is essentially complete is gradually entered. Under conditions prevailing during a typical deposition, the characteristic depth of the reaction zone is a few hundred nanometers, and the maximum concentration of monomer, i.e., the concentration at the growth interface, is of the order of a few tenths percent by weight. Thus the parylene polymerization takes place just behind the growth interface in a medium that is best described as a slightly swollen, solid polymer.

During the vapor deposition process, the polymer chain ends remain truly alive, ceasing to grow only when they are so far from the growth interface that fresh monomer can no longer reach them. No specific termination chemistry is needed.

Polymer Properties. The single most important feature of the parylenes, that feature which dominates the decision for their use in any specific situation, is the vapor deposition polymerization (VDP) process by which they are applied. VDP provides the room temperature coating process and produces the films of uniform thickness, having excellent thickness control, conformality, and purity. The engineering properties of commercial parylenes once they have been formed are given in Table 2.

The most important mode of degradation for parylenes is oxidative chain scission. Oxidative degradation limits the use of parylenes at elevated temperatures in many common applications. Conventional antioxidants, incorporated during or after VDP, can extend the life of the parylenes at elevated temperatures.

The oxidation of parylene appears to be enhanced by ultraviolet radiation.

The bulk barrier properties of parylenes are among the best of organic polymeric coatings.

TABLE 2. TYPICAL ENGINEERING PROPERTIES OF COMMERCIAL PARYLENES

Property	Parylene N	Parylene C	Parylene D
density, g/cm³	1.110	1.289	1.418
refractive index, n_D^{23}	1.661	1.639	1.669
tensile modulus, GPa[a]	2.4	3.2	2.8
tensile strength, MPa[b]	45	70	75
yield strength, MPa[b]	42	55	60
elongation to break, %	30	200	10
yield elongation, %	2.5	2.9	3
Rockwell hardness	R85	R80	
Thermal			
melting point, °C	420	290	380
heat capacity at 25°C, J/(g·K)[c]	1.3	1.0	
thermal conductivity at 25°C, W/(m·K)	0.12	0.082	
dielectric constant			
1 kHz	2.65	3.10	2.82
1 MHz	2.65	2.95	2.80
dissipation factor			
1 kHz	0.0002	0.019	0.003
1 MHz	0.0006	0.013	0.002

[a] To convert GPa to psi, multiply by 145,000.
[b] To convert MPa to psi, multiply by 145.
[c] To convert J to cal, divide by 4.184.

TABLE 1. PROPERTIES OF PARYLENE DIMERS

Dimer	Melting point, °C	Density, g/cm³
DPXN	284[a]	1.22
DPXC	140–160[b]	1.30
DPXD	170–195[b]	1.45

[a] Decomposes.
[b] Mixture of homologues and their isomers.

The parylenes do not absorb visible light, and absorb only at the shorter wavelength, high energy end of the ultraviolet range. Films and coatings are colorless in the visible, becoming opaque to sufficiently short wavelength uv light.

The crystallinity of the parylenes determines two of their most important practical characteristics: mechanical strength at elevated temperatures and solvent resistance. The crystallinity of parylenes is confined to small-submicrometer domains that are randomly dispersed throughout an amorphous continuum. Because the crystalline domains are much more resistant to permeation than the amorphous phase, they retain their reinforcing structural role even in the presence of permeants in the amorphous phase, thus giving the parylenes their resistance to solvent attack.

At temperatures below the melting of the crystallites, the parylenes resist all attempts to dissolve them.

Applications

Because the parylenes are generally insoluble in most solvents, even at elevated temperatures, they cannot be used as solvent-based coatings; neither can they be cast as films nor spun as fibers from solution. Because of their high crystalline melting points, melt-working (molding, extrusion, calendering, etc.) is also difficult. Yet it is often just these features of solvent resistance and high temperature mechanical strength that constitute the advantages of PPX materials.

The most important application of parylenes is as a conformal coating for printed wiring assemblies. These coatings provide excellent chemical resistance, and resistance to fungal attack. In addition, they exhibit stable dielectric properties over a wide range of temperatures.

The use of parylenes as a hybrid circuit coating is based on much the same rationale as its use in circuit boards. A significant distinction lies in obtaining adhesion to the ceramic substrate material, the success of which determines the eventual performance of the coated part. Adhesion to the which must be achieved using adhesion promoters, such as the organosilanes.

Parylenes are superior candidates for dielectrics in high quality capacitors. Their dielectric constant and loss remain constant over a wide temperature range. The thermistor sensing probe of a disposable bathythermograph is coated with parylene. This instrument is used to chart the ocean water temperature as a function of depth.

Parylene is used in the manufacture of high quality miniature stepping motors, such as those used in wristwatches, and as a coating for the ferrite cores of pulse transformers, magnetic tape-recording heads, and miniature inductors.

Parylene's use in the medical field is linked to electronics. For example, as a protective conformal coating on pacemaker circuitry.

As books age, the paper of their pages becomes brittle. A relatively thin coating of parylene can make these embrittled pages stronger.

By separating the coating from the substrate after deposition, the unique coating features of parylenes, especially continuity and thickness control and uniformity, can be imparted to a freestanding film. Applications include optical beam splitters, a window for a micrometeoroid detector, a detector cathode for an x-ray streak camera, and windows for x-ray proportional counters.

Parylenes can be used for contamination control, that is, securing small particles to prevent them from damaging a surface in a sealed unit; barrier coating; coating for corrosion control; and as dry lubricants.

Health and Safety

Provided the vacuum pump exhaust is appropriately vented and suitable caution is observed in cleaning out the cold trap, the VDP parylene process has an inherently low potential for operator contact with hazardous chemicals. Before using the process chemicals, operators must read and understand the current Material Safety Data Sheets, which are available from the manufacturers.

W. F. BEACH
Alpha Metals

Additional Reading

Cheremisinoff, N.P.: *Condensed Encyclopedia of Polymer Engineering Terms,* Elsevier Science & Technology Books, New York, NY, 2001.
Errede, L.A. and M. Szwarc: *Q. Rev. Chem. Soc.* **12**, 301 (1958).
Kroschwitz, J.I.: *Concise Encyclopedia of Polymer Science and Engineering,* John Wiley & Sons, Inc., New York, NY, 1998.
Mark, H.F.: *Encyclopedia of Polymer Science and Technology,* Part 2 (Four Volume Set), John Wiley & Sons, Inc., New York, NY, 2003.
Salamone, J.C.: *Concise Polymeric Materials Encyclopedia,* CRC Press LLC., Boca Raton, FL, 1998.
Szwarc, M.: *Polym. Eng. Sci.* **16**(7), 473 (1976).

XYLENES AND ETHYLBENZENE.

Xylenes and ethylbenzene (EB) are C_8 aromatic isomers having the molecular formula C_8H_{10}. The xylenes consist of three isomers: *o*-xylene (OX), *m*-xylene (MX), and *p*-xylene (PX). These differ in the positions of the two methyl groups on the benzene ring. The molecular structures are shown below.

o-xylene (OX)   *m*-xylene (MX)

p-xylene (PX)   ethylbenzene (EB)

Sources and Uses

The term mixed xylenes describes a mixture containing the three xylene isomers and usually EB. Commercial sources of mixed xylenes include catalytic reformate, pyrolysis gasoline, toluene disproportionation product, and coke-oven light oil. Ethylbenzene is present in all of these sources except toluene disproportionation product. Catalytic reformate is the product obtained from catalytic reforming processes.

Pyrolysis gasoline is a by-product of steam cracking of hydrocarbon feeds in ethylene. Coke over light oil is a by product of the manufacture of coke for the steel industry.

Properties

Because of their similar molecular structures, the three xylenes and EB exhibit many similar properties (see Table 1).

Chemical reactions that the xylenes participate in include (*1*) migration of the methyl groups, (*2*) reaction of the methyl groups, (*3*) reaction of the aromatic ring, and (*4*) complex formation.

Migration of the Methyl Groups. Reactions that involve migration of the methyl groups include isomerization, disproportionation, and dealkylation. The interconversion of the three xylene isomers via isomerization is catalyzed by acids.

Reactions of the Aromatic Ring. The reactions of the aromatic ring of the C_8 aromatic isomers are generally electrophilic substitution reactions. All of the classical electrophilic substitution reactions are possible, but in most instances they are of little practical significance. See also **Friedel-Crafts Reaction**.

Complex Formation. All four C_8 aromatic isomers have a strong tendency to form several different types of complexes. Complexes with electrophilic agents are utilized in xylene separation.

Manufacture of Xylenes

The initial manufacture of mixed xylenes and the subsequent production of high purity PX and OX consists of a series of stages in which (*1*) the mixed xylenes are initially produced; (*2*) PX and/or OX are separated from the mixed xylenes stream; and (*3*) the PX- (and perhaps OX-) depleted xylene stream is isomerized back to an equilibrium mixture of xylenes and then recycled back to the separation step. These steps are discussed below.

Mixed Xylenes Production Via Reforming. Two principal methods for producing xylenes are catalytic reforming and toluene disproportionation. In reforming, a light fraction from a straight run petroleum fraction or from an isocracker is fed to a catalytic reformer. This is followed by heart-cutting and extraction. The mixed xylenes stream must then be processed further to produce high purity PX and/or OX. However, because of the close

TABLE 1. PHYSICAL PROPERTIES FOR C_8 AROMATIC COMPOUNDS

Property	p-Xylene	m-Xylene	o-Xylene	Ethylbenzene
molecular weight	106.167	106.167	106.167	106.167
density at 25°C, g/cm^3	0.8610	0.8642	0.8802	0.8671
boiling point, °C	138.37	139.12	144.41	136.19
freezing point, °C	13.263	−47.872	−25.182	−94.975
refractive index at 25°C	1.4958	1.4971	1.5054	1.4959
surface tension, mN/m (= dyn/cm)	28.27	31.23	32.5	31.50
dielectric constant at 25°C	2.27	2.367	2.568	2.412
dipole moment of liquid [a]Cm·	0	0.30	0.51	0.36
critical properties:				
critical density, mmol/cm^3	2.64	2.66	2.71	2.67
critical volume, cm^3/mol	379.0	376.0	369.0	374.0
critical pressure, MPa[b]	3.511	3.535	3.730	3.701
critical temperature, °C	343.05	343.90	357.15	343.05
thermodynamic properties:				
C_s at 25°C, J/(mol · K)[c]	181.66	183.44	188.07	185.96
S_s at 25°C, J/(mol · K)[c]	247.36	253.25	246.61	255.19
$H_o − H_o$ at 25°C, J/mol[c]	44.641	40.616	42.382	40.219
$−(G_s − H_o/T)$ at 25°C, J/(mol · K)[c]	97.633	117.03	104.46	120.29
heats of transition, J/(mol · K)[c]				
vaporization at 25°C	42.036	42.036	43.413	42.226
formation at 25°C	−24.43	−25.418	−24.439	−12.456

[a] To convert Cm· to D, divide by 3.336×10^{-30}.
[b] To convert MPa to psi, multiply by 145.
[c] To convert J to cal, divide by 4.184.

boiling points of PX and MX, using distillation to produce high purity PX is impractical. Instead, other separation methods such as crystallization and adsorption are used.

Xylenes Production Via Toluene Transalkylation and Disproportionation. The toluene that is produced from processes such as catalytic reforming can be converted into xylenes via transalkylation and disproportionation. Toluene disproportionation is defined as the reaction of 2 mol of toluene to produce 1 mol of xylene and 1 mol of benzene. Toluene transalkylation is defined as the reaction of toluene with C_9 or higher aromatics to produce xylenes.

Separation Processes for PX. There are essentially two methods that are currently used commercially to separate and produce high purity PX: (*1*) crystallization and (*2*) adsorption. A third method, a hybrid crystallization/adsorption process, has been successfully field-demonstrated.

Low temperature fractional crystallization was the first and for many years the only commercial technique for separating PX from mixed xylenes. PX has a much higher freezing point than the other xylene isomers. Thus, upon cooling, a pure solid phase of PX crystallizes first. Eventually, upon further cooling, a temperature is reached where solid crystals of another isomer also form. This is called the eutectic point. The solid PX crystals are typically separated from the mother liquor by filtration or centrifugation.

Adsorption represents the second and newer method for separating and producing high purity PX. In this process, adsorbents such as molecular sieves are used to produce high purity PX by preferentially removing PX from mixed xylene streams. Separation is accomplished by exploiting the differences in affinity of the adsorbent for PX, relative to the other C_8 isomers. The adsorbed PX is subsequently removed from the adsorbent by displacement with a desorbent.

In 1994, IFP and Chevron announced the development of a hybrid process that reportedly combines the best features of adsorption and crystallization. In this process, the adsorbent bed is used to initially produce PX of 90–95% purity. The PX product from the adsorption section is then further purified in a small single-stage crystallizer and the filtrate is recycled back to the adsorption section. It is reported that ultra-high (99.9 + %) purity PX can be produced easily and economically with this scheme.

MX Separation Process. The Mitsubishi Gas–Chemical Company (MGCC) has commercialized a process for separating and producing high purity MX. This process is based on the formation of a complex between MX and HF–BF$_3$. MX is the most basic xylene, and its complex with HF–BF$_3$ is the most stable.

MX of >99% purity can be obtained with the MGCC process with <1% MX left in the raffinate by phase separation of hydrocarbon layer from the complex-HF layer. The latter undergoes thermal decomposition, which liberates the components of the complex.

Xylene Isomerization. After separation of the preferred xylenes, i.e., PX or OX, using the adsorption or crystallization processes discussed herein, the remaining raffinate stream, which tends to be rich in MX, is typically fed to a xylenes isomerization unit in order to further produce the preferred xylenes. Isomerization units are fixed-bed catalytic processes that are used to produce a close-to-equilibrium mixture of the xylenes. The catalysts are also designed to convert EB to either xylenes, benzene and lights, or benzene and diethylbenzene.

Health and Safety Factors

The xylene isomers are flammable liquids and should be stored in approved closed containers with appropriate labels and away from heat and open flames. The vapor can travel along the ground to an ignition source. In the event of fire, foam, carbon dioxide, and dry chemical are preferred extinguishers. The xylenes are mildly toxic. They are mild skin irritants, and skin protection and the cannister-type masks are recommended.

Uses

The majority of xylenes, which are mostly produced by catalytic reforming or petroleum fractions, are used in motor gasoline. The majority of the xylenes that are recovered for petrochemicals use are used to produce PX and OX. PX is the most important commercial isomer.

Almost all of the OX that is recovered is used to produce phthalic anhydride, the phthalic anhydride is a basic building block for plasticizers used in flexible PVC resins, for polyester resins used in glass-reinforced plastics, and for alkyd resins used for surface coatings. Some of the mixed xylenes that are produced are used as solvents in the paints and coatings industry. However, this use has declined.

WILLIAM J. CANNELLA
Chevron Research & Technology Co

Additional Reading

Chemical Economic Handbook, SRI International, Menlo Park, CA, 1996.
Igarashi, Y. and T. Ueno: *A New Xylene Separation Process,* ACS Meeting, Atlantic City, NJ, 1968.
Jeanneret, J. J.: in R. A. Meyers, ed., *Handbook of Petroleum Refining Processes,* 2nd Edition, McGraw-Hill Book Co., Inc., New York, NY, 1997.

XYLITOL. See **Sweeteners.**

Y

YAG AND YIG. Synthetic yttrium aluminum and yttrium iron garnets, respectively. These materials were developed in the mid-1960s. They are pressed and sintered polycrystalline ceramics and are made by a solid-state reaction of Y_2O_3 with iron oxide or aluminum oxide. Garnets operate in microwave bandpass (filters) circulators and isolators in telephone, radar, and space-communication networks. The original electronic use led to the development of single-crystal yttrium aluminum garnets which approach the brilliance and hardness of diamond. Yttrium oxide is the base for neodymium-doped laser crystals. See also **Neodymium**; **Rare-Earth Elements and Metals**; and **Yttrium**.

YEASTS AND MOLDS. These are very important plant organisms that make both positive and negative contributions to mammalian life processes. Their plus and minus values are particularly noted in connection with the production and storage of food products.

Taxonomy

Plants that lack true roots, stems, or leaves, and that are without highly-organized conducting systems are called *simple plants*, and they make up the phylum *Thallophyta*. The two subdivisions of *Thallophyta* are *algae* and *fungi*. Algae have chlorophyll; fungi do not. Fungi utilize carbohydrates that are synthesized by green plants. Fungi are classified as *parasites* or *saprophytes*, the latter obtaining food from nonliving organic material. The science of fungi is *mycology*. See also **Fungus**.

The fungi include: *Slime molds* (*Myxomycetes*); *algal fungi* (*Phycomycetes*); *sac fungi* (*Ascomycetes*); and *club fungi* (*Basidiomycetes*); among others. The yeasts and most of the molds with which food products are associated in some way are of the *Ascomycetes* variety.

The varied interface between the *Ascomycetes* and foods ranges from their very negative, parasitic habit on certain crops (exemplified by Chestnut blight fungus; or ergot on grains) and their causative involvement in certain foodborne diseases to their positive use in connection with the production (notably by fermentation) of major food products, such as bread and wine.

The term *yeast* is used to describe a relatively small number of *Ascomycetes* fungi (a few hundred— Lodder et al. (1970) classified 349 fermentative species, of which relatively few are used industrially), as compared with the vast number of other fungi (several thousand) that have been identified.

The term *mold* does not have a clear cut mycological definition. While the yeasts have a preponderantly positive value in food production and utilization, the molds make some positive contributions (some cheeses, etc.), but they participate in many more negative ways (leaf molds on plants; blue-green molds on fruits; machinery mold; causes of some foodborne diseases; etc.). Some authorities identify all or most mold as *saprophytes*, as previously defined. This identification, however, is not fully satisfactory in terms of the present loose ways in which the word mold is used in the professional literature. Molds do play an important role in the *biodegradation* of unwanted substances.

Yeasts

The importance of the economically useful yeasts can be attributed to two main factors: (1) *fermentation*—the transformation of simple sugars and other organic chemicals to other, more desirable chemicals; and (2) *respiratory (oxidative) metabolism*—the great capacity of some yeasts for a protein synthesis during growth in richly aerated media containing a wide variety of carbonaceous and nitrogenous nutrients. Thus, yeasts serve in many ways: (1) As living cells, they are biocatalysts in the production of bread, wine, beer, distilled beverages, among other important food products. (2) As dried, nonfermentative whole cells or hydrolyzed cell matter, yeasts contribute nutrition and flavor to human diets and animal rations. (3) As producers of vitamins and other biochemicals, yeasts are a rich source of enzymes, coenzymes, nucleic acids, nucleotides, sterols, and metabolic intermediates. See also **Coenzymes**; **Enzyme**; and **Enzyme Preparations**. (4) As a versatile biochemical tool, yeasts aid research studies in nutrition, enzymology, and molecular biology.

Background

It is estimated that the arts of making wine, leavened bread, and beer were practiced more than 4000 years ago. The phenomena for producing these foods were attributed to "yeast" at an early age. In many languages, the word for yeast describes the visible effects of fermentation, as observed in the expansion of bread dough and the accumulation of froth or barm on the surface of fermenting juices and mashes. Historically, it has been reported that yeast cells were first seen in a droplet of beer mounted on a slide in a crude microscope used by van Leeuwenhoek in 1680. He found globular bodies, but was not aware that they were living forms. For nearly two centuries the theory of spontaneous generation dominated thought and research on the causes of fermentation and disease. In 1818, Erxleben described beer yeast as a living vegetable matter responsible for fermentation. In the following twenty years, yeasts were shown to reproduce by budding and, in 1837, Meyen named yeast *Saccharomyces*, or "sugar fungus." By 1839, Schwann observed "endospores" in yeast cells, later named ascospores by Reess. As early as 1857, Pasteur proved the biological nature of fermentation and later, in 1876, Pasteur demonstrated that yeast can shift its metabolism from a fermentative to an oxidative pathway when subjected to aeration. This shift, then named the Pasteur effect, is especially characteristic of bakers' yeast (*Saccharomyces cerevisiae*) and is applied in the large-scale production of yeasts.

Botanically, yeasts form a heterogeneous group of saprophytic forms of life occurring naturally on the surface of fruits, in honey, exudates of trees, and in soil. They are disseminated by airborne dusts, insects, and animals. Typically, industrial yeasts are generally oval, microscopic, unicellular organisms. In addition to lacking chlorophyll, they also lack locomotion. They reproduce vegetatively by budding and sexually by spore formation (ascospores) within the mother cell or ascus. These properties place them in the family *Endomycetacea* of the class *Ascomycetes*, as previously mentioned. Among the most important industrial yeasts are:

Saccharomyces cerevisiae (alcoholic beverages and bread)
S. cerevisiae var. *ellipsoideus; S. bayanus*; and *S. beticus* (wines)
S. uvarium (formerly called *S. carlsbergensis*) (beer, ale, etc.)
Kluyveromyces fragilis (formerly called *S. fragilis*) (whey disposal)

Food and feed yeast production employ several molds in the family *Cryptococcaceae: Candida utilis, C. tropicalis*, and *C. japonica*, which are cultivated on plant wastes (wood sugars, molasses, stillage), and C. lipolytica, which converts hydrocarbons to yeast protein.

Properties of Yeasts

The cell structure of *S. cerevisiae*, as observed in the optical microscope, reveals a rigid cell wall, a colorless, granular cytoplasm, and one or more vacuoles. Dimensions of a typical bakers' yeast are about 4 to 6 by 7 to 10 micrometers. Electron microscopy of ultrathin sections of a yeast cell show the microstructures, including: birth and bud scars on the cell wall, plasmalemma (cytoplasmic membrane), nucleus, mitochondria, vacuoles, fat globules, cytoplasmic matrix and volution or polyphosphate bodies. The cell walls of bakers' yeast contain 30–35% glucan (yeast cellulose), 30% mannan (yeast gum), which is bound to protein (about 7%), 1–2% chitin, 8–13% lipid material, plus inorganic components, largely phosphates.

Gross chemical composition of compressed bakers' yeast is approximately 70% moisture. The dry matter is made up of 55% protein (N × 6.25), 6% ash, 1.5% fat, and the remainder mostly polysaccharides, including about 15% glycogen and 8% trehalose.

Food yeast, molasses-grown, is dried to about 5% moisture and has the same chemical composition as bakers' yeast. In terms of micrograms per gram of yeast, the vitamin content is: 165 thiamine; 100 riboflavin; 590 niacin; 20 pyridoxine; 13 folacin; 100 pantothenic acid; 0.6 biotin; 160 para-aminobenzoic acid; 2710 choline; and 3000 inositol. Yeast crude protein contains 80% amino acids; 12% nucleic acids; and 8% ammonia. The latter components lower the true protein content to 40% of the dry cell weight.

Yeast protein is easily digested (87%) and provides amino acids essential to human nutrition. Most commercial yeasts show the following pattern of amino acids, among others, as percent of protein: 8.2% lysine; 5.5% valine; 7.9% leucine; 2.5% methionine; 4.5% phenylalanine; 1.2% tryptophan; 1.6% cystine; 4% histidine; 5% tyrosine; and 5% arginine. The usual therapeutic dose of dried yeast is 40 grams/day, which supplies significant daily needs of thiamine, riboflavin, niacin, pyridoxine, and general protein.

The ash content of food yeasts ranges from 6 to 8% (dry basis), consisting mainly of calcium, phosphorus, and potassium. Contained in quantities of less than 1% are magnesium, sulfur, and sodium. At the microgram level are included iron, copper, lead, manganese, and iodine.

Triglycerides, lecithin, and ergosterol are the main constituents of yeast lipid (fat). Oleic and palmitic acids predominate in yeast fat. These resemble the composition of common vegetable fats. Ergosterol, the precursor of calciferol (vitamin D_2) varies from 1 to 3% of yeast dry matter.

Metabolic Activity

This is generally associated with the familiar alcoholic fermentation in which theoretically 100 parts of glucose are converted to 51.1 parts of ethyl alcohol (ethanol), 48.9 parts of carbon dioxide (CO_2), and heat. In addition, however, the anaerobic reaction also yields minor byproducts in small amounts—mainly glycerol, succinic acid, higher alcohols (fusel oil), 2,3-butanediol, and traces of acetaldehyde, acetic acid, and lactic acid. Fusel oil is a mixture of alcohols, including n-propyl, n-butyl, isobutyl, amyl, and isoamyl alcohols.

Respiratory activity of oxidative dissimilation is characteristic of many species of yeasts. During aerobic growth, sugar is oxidized to carbon dioxide and water, with release of large amounts of energy (about 680 kcal when complete oxidation occurs). Aerobiosis produces a variety of byproducts, some in unusually high concentration, such as acetic acid, succinic acid, zymonic acid, polyhydric alcohols (glycerol, erythritol, etc), extracellular lipids, carotenoid pigments in shades of red and yellow, black pigment (melanin), and capsular polysaccharides (phosphomannan).

Production

Well over 85,000 tons (76,500 metric tons) of yeast dry matter are produced in the United States alone each year. About 75% of this is in the form of bakers' yeast, the remaining 25% represents about equal amounts of food yeast and feed yeast. This production issues from four types of manufacture: (1) Bakers' yeast is grown batchwise in aerated molasses solutions. (2) *Candida utilis* is obtained from wood pulp mill spent liquid. (3) *K. fragilis* is grown batchwise in cottage cheese whey. (4) Dried yeast is recovered as spent beer yeast. Worldwide production of all types of food and feed yeast is estimated at more than 450,000 dry tons (405,000 metric tons) per year.

The process for growing bakers' yeast is a model system for the propagation of microorganisms. The process commences with a laboratory culture of a pure strain of *S. cerevisiae*. Seed yeast is developed in successively larger volumes of nutrient solutions, beginning with a Pasteur flask and ending in a fermenting tank containing as much as 40,000 gallons (1514 hectoliters) of sterilized and diluted molasses maintained at 30°C. During the highly aerated growth period, minerals are added, pH is adjusted to 4.5, and diluted molasses is continuously fed in proportion to the increase in cell mass. Under ideal conditions, yeast cells may double in number every 2.5 hours, converting more than half (56.7%) of the sugar supplied to cell components. Biosynthesis of cell matter requires an equal amount of oxygen. To produce 100 weight units of yeast dry matter with 50% protein content requires about 400 weight units of molasses, 25 weight units of aqua ammonia, 15 weight units of ammonium sulfate, and 7 weight units of monobasic ammonium phosphate. For each 100 pounds of dry yeast,

75,000 cubic feet of air are required; for 100 kilograms of dry yeast, 4,683 cubic meters of air are needed.

Fermentation Processes

The biochemistry of alcoholic fermentation involves a series of internal enzyme-mediated oxidation-reduction reactions in which glucose is degraded via the Embden-Meyerhof-Parnas pathway. See also **Carbohydrates**; and **Glycolysis**.

Some typical reactions performed by yeasts are listed in the Table 1.

Post-Processing Spoilage of Food by Yeasts

Microbiological spoilage of food is a competitive process occurring among yeasts, bacteria, and molds. Yeasts normally play a small role in spoilage, because they constitute only a small portion of the initial population, because they grow slowly in comparison with most bacteria, and because their growth may be limited by metabolic substances produced by bacteria. Evidence does not show food poisoning as caused by the presence of spoilage yeasts in foods. The byproducts of metabolism are not considered toxic and, while there are a few yeasts that may be considered pathogenic, they are not known to be responsible for foodborne infections or intoxications (Walker, 1977; Peppler, 1977). The metabolism of yeasts can result in the development of unnatural flavors and odors and changes in pH because of the utilization of organic acids important in fermented foods. Many yeasts are capable of utilizing lactic, acetic, and citric acids, which are essential for production of flavor and for preservation of some foods. Decreased concentration of these acids causes an increased pH and produces conditions that favor growth of spoilage bacteria. Normally, several factors interact to establish conditions that favor growth and subsequent spoilage of foods by yeasts. Examples of food spoilage by yeasts are given in the lower half of Table 1.

Although yeasts are abundant in nature, notably on leaves and in the soil, they do not compete with bacteria and molds as sources of major problems. But, in assessing the potential for spoilage by yeasts, one must consider those factors which are favorable or unfavorable to the growth and multiplication of yeast populations. Availability of oxygen is important. No yeast is known that can grow under strictly anaerobic conditions; thus, all require oxygen to be present in some proportion. Temperature is also an important factor. Yeasts are not heat resistant in the sense that bacterial endospores are heat resistant. Most yeasts cannot withstand temperatures above 65–70°C. The majority of yeast species have an optimum temperature for growth between 20 and 30°C. Even though yeasts are very resistant to low temperatures and can survive frozen storage, they normally are not a major problem in spoilage of frozen or refrigerated foods. In refrigerated foods, psychrotrophic bacteria rapidly outnumber the yeasts, the latter constituting a minor part of the initial population. However, where antibiotics or ionizing radiation have been used to reduce or inhibit bacterial growth, then yeast population may predominate. If the definition of a *psychrophilic* organism is one that has an optimum temperature below 20°C, then there are a number of strains of psychrophilic yeasts.

The survival and growth of all organisms is dependent on the presence of some water in the environment. Yeasts generally require more moisture for growth than molds, but less than bacteria. Systems containing high concentrations of sugar or salt have low a_w values or high osmotic pressure (Walker, 1977). Organisms growing in such systems are usually referred to as being *osmophilic*. A number of investigators have not been able to find any yeast that is clearly osmophilic. Windisch (1969) suggested the term osmotolerant to describe some yeasts. It is important to note that yeasts that normally will not grow in a high-sugar environment may appear, if the substance has been exposed to the air sufficiently to absorb water and dilute spots or edges of the material, thus favoring yeast growth. Species such as *Saccharomyces rouxii*, *S. rouxii* var. *polymorphus*, and *S. mellis* (Lodder, 1952) are capable of spoiling high-sugar foods, such as honey, maple sugar, sugar cane syrups, molasses, fruit syrups, candy, crystallized fruits, jams, jellies, and dried fruits. It is not uncommon for dried figs and prunes to be covered with a white coating of yeast during storage. The foods have a typical fermented odor and may contain gas pockets. This coating normally is a mixture of sugar and yeasts, with the principal yeast being *Schizosaccharomyces octosporus*.

Molds

The most critical factor in mold growth is the availability of sufficient moisture. Molds are widely distributed throughout nature. Chemicals for

TABLE 1. REACTIONS IN WHICH YEASTS PARTICIPATE[1]

Type of reaction	Number of known reactions	Examples of reactions
Reduction	156	Diacetyl to acetoin (in beer); cinnamic aldehyde to cinnamic alcohol.
Decarboxylation	21	Malic acid to lactic acid (in wines); amino acids to amines (histamine and tyramine accumulate in soft cheeses because of surface growth of *Torulopsis candida* and *Debaryomyces kloeckera*).
Deamination	17	Examples of deamination and decarboxylation include conversion of amino acids to fusel oil (leucine to isoamyl alcohol, isoleucine to amyl alcohol, and phenylalanine to phenyl ethanol). Fusel oil formation is a normal function of all yeast fermentations (in alcoholic beverages, levels range from trace to 2200 parts per million). Deamination: Glutamic acid to gamma-OH-butyric acid (*S. cerevisiae*).
Oxidation	14	Acids, alcohols, sugars, hydrocarbons. Also stepwise: alcohol to aldehyde or acid (sake yeast).
Esterification	10	Ethyl acetate (*Hansenula anomala*).
Condensation	9	Acetaldehyde to acetoin; acetaldehyde to pyruvic acid to alpha-acetolactic acid.
Hydrolysis	5	Starch hydrolysis: By *Endomycopsis fibuligera*; artichoke starch (insulin) by *Kluyveromyces fragilis*.
Amination	1	

Examples of Lesser-known Properties and Characteristics of Yeasts

Type of reaction	Yeasts involved	Examples
Lipolysis	*Candida lipolytica; C. rugosa; Torulopsis sphaerica*	Mainly in butter, margarine, and cheese.
Proteolysis	*C. lipolytica; T. sphaerica*	Especially on soft cheese surfaces.
Pectinolysis	*Saccharomyces kluyveri; Kluyveromyces fragilis; Hansenula anomala*	Softening of olives and cherries in brines, followed by formation of gas pockets in fruit. Strains of wine yeasts contain polygalacturonases which, during fermentation of grape juice, participate in the solubilization of pectin.
Acid formation	Species of Brettanomyces, Hansenula, Pichia, Saccharomyces	As a contaminant in wines, Brettanomyces spp. Forms a higher concentration of volatile acids (also isobutyric and isovaleric acids) than *S. cerevisiae*. Pichia species and other yeasts are responsible for acetic acid production in brines of domestic green olives; not lactobacilli, as assumed for years (Vaughn et al., 1976).
Pigmentation	*Rhodotorula glutinis.* Sporobolomyces spp.	Carotenoids (pink, red); *Rhodotorula glutinis* causes pink sauerkraut and discolors the surface of high-moisture cheeses.
Esterification	Species of Hansenula, Kluyveromyces, Brettanomyces	Ethyl acetate and ethyl lactate in cottage cheese and shredded Mozzarella cheese. See also **Milk and Milk Products**.
Turbidity formation	In wine: *Saccharomyces bailii, S. chevalieri*, Brettanomyces spp. In beer: *S. diastaticus, S. bayanus*	In soft drinks; in wine: *S. bailli, S. chevalieri*; in beer: *S. diastaticus, S. bayannus*.

[1] Based, in part, upon Peppler et al. (1977).

the destruction of molds, including those in the soil, are mentioned in the entry on **Antimicrobial Agents (Foods)**. In addition to food spoilage, various molds can be destructive of various materials, notably of substances prepared from animal skins, such as leather, book covers, shoes, and materials prepared from vegetable substances, such as paper and wood products. In warm, humid areas, mold (mildew) grows on wooden structures, including painted surfaces unless these have been chemically treated. In the case of foods, molds cause innumerable problems during the full cycle of food production.

Storage temperature of food is less important than the presence of moisture, since fungi can grow and produce toxins over a wider span of temperatures than can any other microorganisms. Most species are able to grow at an a_w of 0.8 to 0.88; while xerophilic types can grow at an a_w of 0.65 to 0.75. A relative humidity of from 70 to 90% establishes suitable moisture equilibrium for initiation of mold growth and toxin formation in numerous food products. See also **Activity Coefficient**.

Unfortunately for the would-be consumer of food, fungal contamination is not always immediately apparent and thus may not seem to be cause either for the rejection of food by humans or for the rejection of feedstuffs by livestock. Concentrations of mold organisms are not always easy to

see or to smell. Further, the relative stability of most mycotoxins to heat precludes the use of cooking as a detoxifying procedure. Among the most aggressive of the molds are species of *Candida, Aspergillus, Rhizopus*, and *Mucor*. Several of the fungi are not pathogenic to healthy humans, but may be virulent pathogens in debilitated persons, or those treated with broad-spectrum antibacterial drugs, or immunosuppressive measures. An example of this type of organism is *Cryptococcus neoformans* (Davis, 1973). Many of the fungi produce disease through infection rather than through mycotoxin production, but this usually requires predisposing factors.

Milner and Geddes pointed out in 1946 that the aflatoxigenic organism *Aspergillus flavus* is a common soil fungus throughout the world, and aflatoxin-contaminated food has been noted in many countries for a number of years (Wogan, 1966). Crops, such as groundnuts (peanuts), in intimate contact with soils, are likely to contact the necessary mold inoculum for aflatoxin production, resulting in toxin formation upon storage in air.

Several investigators have shown that toxic strains of *Fusarium, Cladosporium, Penicillium*, and *Mucor* can sporulate and grow at temperatures well below 0°C. It also has been shown that toxin formation can be associated with overwintering of grain in the field. In some varieties of barley, such as Siri and Mala, the incidence of *Aspergillus* and

Penicillium species is related solely to temperature and moisture storage conditions and is not correlated to the percentage of unripe grains. Low humidities tend to predispose hosts to invasion by molds by causing them to lose turgor. Mold is the cause of serious problems in connection with rice. The predominant molds present in wild rice during fermentation curing are *Mucor*, aflatoxigenic and nontoxigenic *Aspergillus*, *Penicillium*, and *Rhizopus* species. Most of the B_1 aflatoxin is in the hulls and the level of the toxin can be reduced by parching, which serves to reduce, but not fully destroy molds.

Contamination of soybeans with *Aspergillus flavus* is found in approximately 50% of commercial samples. Fortunately, the incidence of aflatoxin from this route is quite low. Moisture at the time of maturity, development of the seed in a closed pod, and binding of zinc by phytic acid are suggested as reasons for resistance of soybeans to aflatoxin production.

The most active sugar contaminants are molds, particularly *Aspergillus*. Penicillium growth on cheddar cheese results in a pH gradient, increasing to a pH of 8 at the surface. Mycotoxins diffuse through foods and are not removed when the surface molds are removed. Therefore, molds of unknown variety or origin should not be ingested. On the other hand, many of the common molds found as contaminants of Western foods are used in Asian fermented foods and beverages, principally as sources of flavor.

Degradation of Aflatoxin

Numerous researchers have investigated a variety of chemical and physical means for degrading aflatoxin. These include the use of irradiation, heat, acids and bases, oxidizing agents, bisulfite, and biological agents. The practical application of ultraviolet radiation to date has not proved successful. Aflatoxin is quite heat stable. Numerous investigators have concluded that although heat can degrade aflatoxins, it is not an effective and economically feasible means for inactivating these toxins when present in foods or feeds.

Among oxidizers used have been sodium hypochlorite, potassium permanganate, and hydrogen peroxide. Bisulfite is commonly used in the wet milling of corn (maize) and in processing wines, fruit juices, jams, and dried fruits.

Machinery Mold

This mold generally refers to the buildup of the organism *Geotrichum candidum* on food-contact factory equipment in processing plants. The term *dairy mold* is used in the processing of milk to identify this mold.

Since passage of the original Federal Food and Drugs Act of 1906 (United States), the presence of machinery mold in any food processing plant is a violation of regulations. Antimicrobial agents have made it possible to keep machinery clear of this mold. The mold can be particularly bothersome in tomato and pineapple processing plants. See also **Antimicrobial Agents (Foods)**.

Additional Reading

Adams, A., D.E. Gottschling, C. Kaiser, and T. Stearns: *Methods in Yeast Gene tics, 1997,* Cold Spring Harbor Laboratory Press, Cold Spring Harbor, NY, 1999.

Barnett, J.A., D. Yarrow, and R.W. Payne: *Yeasts Characteristics and Identification,* Cambridge University Press, New York, NY, 2000.

Davis, B.D., et al.: *Microbiology,* Harper and Row, Hagerstown, Maryland, 1973.

Deak, T. and L.R. Beuchat: *Handbook of Food Spoilage Yeasts,* CRC Press, LLC., Boca Raton, FL, 1996.

Fantes, P. and J. Beggs: *The Yeast Nucleus,* Oxford University Press, Inc., New York, NY, 2000.

Jones, E.W., et al.: *The Molecular and Cellular Biology of the Yeast Saccharomyces: Gene Expression,* Vol. 2, Cold Spring Harbor Laboratory Press, Cold Spring Harbor, NY, 1999.

Kurtzman, C.P. and J.W. Fell: *Yeasts: A Taxonomic Study,* Elsevier Science, New York, NY, 1998.

Lodder, J., C.P. Kurtzman, and J.W. Fell: *The Yeasts, A Taxonomic Study,* 4th Edition, Elsevier Science, New York, NY, 1998.

Milner, M. and W.F. Geddes: "Grain Storage Studies. III. The Relation between Moisture Content, Mold Growth, and Respiration of Soybeans," *Cereal Chem.,* **23,** 225 (1946).

Peppler, H.J.: "Yeast Properties Adversely Affecting Food Fermentations," *Food Technol.,* **34,** 2, 62–65 (1977).

Pringle, J.R., J.R. Broach, and E.W. Jones: *Molecular and Cellular Biology of the Yeast Saccharomyces,* Vol. 3, Cold Spring Harbor Laboratory Press, Cold Spring Harbor, NY, 2000.

Rose, A.H. and J.S. Harrison: *The Yeasts,* Vol. 5, Academic Press, Inc., San Diego, CA, 1997.

Walker, H.W.: "Spoilage of Food by Yeasts," *Food Technol.,* **31,** 2, 57–61 (1977).

Walker, G.M.: *Yeast Physiology and Biotechnology,* John Wiley & Sons, Inc., New York, NY, 1998.

Windisch, S.: *Studies on Osmotolerant Yeasts,* in "Yeasts" (A. Kocva-Kratochi flova, editor), Vyadavatel'stvo Slovenskej Akadmine Vied Bratislava, 1969.

Wogan, G.N.: "Chemical Nature and Biological Effects of the Aflatoxins," *Bacteriol. Rev.,* **30,** 460 (1966).

YEAST INHIBITORS.   See **Antimicrobial Agents (Foods)**.

YTTERBIUM.   [CAS: 7440-64-4]. Chemical element symbol Yb, at. no. 70, at. wt. 173.04, thirteenth in the Lanthanide Series in the periodic table, mp 819°C, bp 1196°C, density 6.966 g/cm³ (20°C). Elemental ytterbium has a face-centered cubic crystal structure at 25°C. The pure metallic ytterbium is silver-gray in color and is stable in moist or dry air up to 200°C, after which oxidation occurs. The metal is readily dissolved by dilute and concentrated mineral acids. The metal dissolves in liquid NH_3 to yield a dark-blue color. There are seven natural isotopes ^{168}Yb, ^{170}Yb through ^{174}Yb, and ^{176}Yb. Ten artificially-produced isotopes have been identified. Ytterbium is one of the least abundant elements of the rare-earth group and 53rd among all elements occurring in the earth's crust. The element was first identified by J.D.G. Marignac in 1878. Electronic configuration

$$1s^2 2s^2 2p^6 3s^2 3p^6 3d^{10} 4s^2 4p^6 4d^{10} 4f^{13} 5s^2 5p^6 5d^1 6s^2.$$

Ionic radius Yb^{2+} 1.00 Å; Yb^{+3} 0.88 Å. Metallic radius 1.940Å. First ionization potential 6.25 eV; second 12.18 eV. Other important physical properties of ytterbium are given under **Rare-Earth Elements and Metals**.

The principal sources of ytterbium are euxenite, gadolinite, monazite, and xenotime, the latter being the most important. Ytterbium is separated from a mixture of yttrium and the heavy Lanthanides by using the sodium amalgam reduction technique. Ytterbium metal is obtained by heating a mixture of lanthanum metal and ytterbium oxide under high vacuum. The ytterbium sublimes and is collected on condenser plates whereas the lanthanum is oxidized to the sesquioxide.

To date, the major uses of ytterbium have been in applied and fundamental research. The element and its compounds have been used in magnetic "bubble" domain devices (ytterbium orthoferrite), in phosphors to convert infrared to visible light, in lasers, and radioisotope ^{169}Yb has found application in portable industrial and medical radiographic units.

See references listed at ends of entries on **Chemical Elements**; and **Rare-Earth Elements and Metals**.

Note: This entry was revised and updated by K. A. Gschneidner, Jr., Director, and B. Evans, Assistant Chemist, Rare-earth Information Center, Institute for Physical Research and Technology, Iowa State University, Ames, Iowa.

Additional Reading

Gschneidner, K.A., Jr., B.J. Beaudry, and J. Capellen: *Rare Earth Metals,* pp. 720–732 in *Metals Handbook, 10th edition, Vol. 2, Properties and Selection: Nonferrous Alloys and Special Purpose Materials,* ASM International, Metals Park, OH, 1990.

Lide, D.R.: *CRC Handbook of Chemistry and Physics 2000–2001,* 84th Edition, CRC Press, LLC., Boca Raton, FL, 2003.

YTTRIUM.   [CAS: 7440-65-5]. Chemical element symbol Y, at. no. 39, at. wt. 88.905, periodic table group 3, mp 1522°C, bp 3,338°C, density 4.469 g/cm³ (20°C). Most of the properties of yttrium are similar to those of the heavy rare-earth elements, falling between gadolinium and erbium. Elemental yttrium has a close-packed hexagonal crystal structure at 25°C. The pure metallic yttrium is silver-gray in color, retaining a luster in air up to about 400°C, above which it oxidizes to Y_2O_3. The metal is dissolved by most mineral acids, but is relatively inert in a 1:1 mixture of concentrated HNO_3 and 48% hydrofluoric acid. Yttrium is capable of working common metallurgical fabrication procedures. The metal is immiscible with liquid or solid uranium metal. It has a low thermal-neutron-absorption cross section and a low acute-toxicity rating. The natural isotope of yttrium is ^{89}Y. Fourteen artificial isotopes have been identified. See also **Radioactivity**. In terms of abundance, yttrium is present on the average of 33 ppm in the earth's crust and potentially is as plentiful as cobalt. The element was first identified by Fredrich Wohler in 1828. Electronic configuration $1s^2 2s^2 2p^6 3s^2 3p^6 3d^{10} 4s^2 4p^6 4d^1 5s^2$. Ionic radius Y^3+ 0.900 Å. Metallic radius 1.801 Å. First ionization potential 6.38 eV;

second 12.24 eV. Other important physical properties of yttrium are given under **Chemical Elements**; and **Rare-Earth Elements and Metals**.

Residues from uranium mining operations in Canada have been a major source of yttrium. Xenotime (YPO_4) found in Malaysia is another source, as well as the ion-adsorption clay minerals in China. Some apatite deposits are unusually rich in yttrium and it also is found in gadolinite, euxenite, and samarskite.

In recovering yttrium, mixed rare-earth minerals or wastes are dissolved in HNO_3 or H_2SO_4. Liquid-liquid organic ion-exchange solvent extraction cells then separate a pure yttrium fraction, usually precipitated as an oxalate and then calcined to the oxide. The metal is obtained by metallothermic reduction, using calcium mixed with YCl_3 or YF_3 in a sealed retort at a temperature in excess of 1,550°C. Alloys of yttrium and cobalt, or yttrium and magnesium have been deposited out of a molten electrolyte BaF_2-LiF-YF_3.

The major uses of yttrium are as phosphors in color television, computer monitors, trichromatic fluorescent lights, x-ray intensifying screens and temperature sensors. Yttrium oxide stabilized zirconium oxide is used as oxygen sensors for automobile engines and in iron making, wear-resistant and corrosion-resistant cutting tools, abrasives, high temperature refractories for continuous casing nozzles, jet engine coatings, simulated diamond gemstones (i.e. cubic zirconia), and as a solid electrolyte in solid oxide fuel cells (SOFC). Yttrium aluminum garnets are used as laser crystals for industrial cutting and welding, medical and dental surgery, temperature and distance sensing, photochemistry, photoluminescence, digital communication and non-linear optics. $YBa_2Cu_3O_{7-x}$ is an important high temperature ceramic superconductor with a 90 K superconducting transition temperature. It is primarily used in the form of thin films for superconducting electronics (such as SQUIDS, MRI and NMR coils, wireless communication subsystems and digital instruments) and as superconducting wires and tapes for power transmission lines, motors and generators, transformers, current limiters, magnetic separation, research magnet systems and current leads. Yttrium-iron garnets find use as microwave bandpass filters, circulators, and isolators in electronic and communications circuitry. Y_2O_3 is used as a metallurgical dispersion hardening agent in nickel-based superalloys made by powder metallurgy techniques. Yttrium metal also is specified in several cobalt-base superalloys in which it improves hot corrosion (sulfidation) resistance at high temperatures. When used in iron-chromium-aluminum alloys, yttrium improves workability and adds resistance to sag when the alloys are used as electrical-heating elements. The metal also is applied as cladding to rotating turbine engine parts to obtain superior oxidation resistance. Yttrium also has been used in permanent magnets, YCo_5, and shows great promise. These magnets are second only to $PrCo_5$ as the most powerful permanent magnet materials developed to date, far exceeding alnico and other more conventional materials.

Chemistry and Compounds

Yttrium hydroxide, $Y(OH)_3$ is precipitated by NH_4OH from solutions of yttrium salts. It differs in properties from the lanthanide hydroxides, both structurally and chemically, in its ability to absorb atmospheric CO_2. Yttrium oxide, Y_2O_3, is obtained by heating the hydroxide or oxy-acid salts; it forms mixed oxides when heated with other oxides, such as Fe_2O_3 and TiO_2. Yttrium also forms a peroxide, Y_4O_9, obtained in hydrated form by treatment of yttrium solutions with hydrogen peroxide.

All four halides are known. The fluoride, YF_3, is readily formed by action of a fluoride on an yttrium nitrate solution. There is an oxyfluoride, YOF, formed by high-temperature, low-pressure heating of the mixed oxide and fluoride. Complex fluorides, containing $[YF_6]^{3-}$ exist; the cryolite minerals are of this composition. The group, $-YF_4$ is found in double salts formed by YF_3 and the alkali fluorides. The chloride, YCl_3, is formed as a hydrate by action of HCl solution upon the hydroxide. It may be dehydrated by slow heating; rapid heating gives the oxychloride, YOCl. The bromide, YBr_3, is formed as a hydrate by action of HBr upon the hydroxide. Its dehydration requires heating under vacuum. The iodide, YI_3 is best prepared from the anhydrous chloride, by reacting with HI and I_2.

Both normal and mixed carbonates are known. The former is precipitated, as $Y_2(CO_3)_3 \cdot 3H_2O$ from Y^{3+} solutions by alkali metal carbonates, which in excess dissolve the precipitate to form a soluble hydrated double carbonate. The oxycarbonate is also a double molecule, $3Y_2(CO_3)_3 \cdot 2Y(OH)_3$, formed by action of CO_2, upon the hydroxide.

The nitrate, $(Y(NO_3)_3)$ exists as a number of hydrates. The hexahydrate formed by action of HNO_3 upon the hydroxide, is dehydrated at 100°C to give the trihydrate and the anhydrous salt. However, other hydrates are known, as well as double nitrates (especially with the lanthanide elements) and oxynitrates, of the general formula, $xY_2O_3 \cdot yN_2O_5 \cdot zH_2O$.

Yttrium hydroxide forms a hydrated sulfate with H_2SO_4, which is dehydrated on heating. It forms double sulfates with alkali and ammonium sulfates.

The carbide, formed from the oxide and carbon in the electric furnace, appears to have the composition, YC_2. It yields acetylene and other hydrocarbons upon hydrolysis.

See references listed at ends of entries on **Chemical Elements**; and **Rare-Earth Elements and Metals**.

Note: This entry was revised and updated by K. A. GSCHNEIDNER, Jr., Director, and B. EVANS, Assistant Chemist, Rare-earth Information Center, Institute for Physical Research and Technology, Iowa State University, Ames, Iowa.

Additional Reading

Gschneidner, K.A., Jr., B.J. Beaudry, and J. Capellen: *Rare Earth Metals,* pp. 720–732 in "Metals Handbook," 10th Edition, Vol. 2, Properties and Selection: Nonferrous Alloys and Special Purpose Materials, ASM International, Metals Park, Ohio (1990).

Gschneidner, K.A., Jr.: *Physical Properties of the Rare Earth Elements,* pp. 4–112 to 4–121 in "CRC Handbook of Chemistry and Physics," 77th Edition, CRC Press, Boca Raton, Florida (1996–1997).

Hammond, C.R.: *The Elements,* pp. 4–1 to 4–34 in "CRC Handbook of Chemistry and Physics," 77th Edition, CRC Press, Boca Raton, Florida (1996–1997).

Z

ZEEMAN EFFECT. An effect of a moderately intense magnetic field upon the structure of the spectrum lines of a gas when subjected to its influence. The phenomenon, sought unsuccessfully by Faraday and finally observed by Zeeman in 1896, consists in the splitting up of each line into two or more components. In the simpler cases, when the source is viewed at right angles to the field, there are three components, of which the middle one has the same frequency as the unmodified line. This component is plane-polarized to vibrate parallel with the field, while the two side components vibrate at right angles to the field. When the source is viewed in the direction of the field, there are only two components, displaced in opposite directions, and circularly polarized in opposite senses. These phenomena constitute the so-called "normal" Zeeman effect. See also **Polarized Light**.

With most lines, however, an anomalous Zeeman effect is observed and the number of components is greater, in some cases reaching twelve or fifteen. They are symmetrically arranged and symmetrically polarized. The displacements, as in the simpler case, are proportional to the magnetic field intensity H, and are always expressible, in wave numbers, as rational multiples of the displacement in the normal effect, which is $4.67 \times 10^{-5} H$ (reciprocal centimeter), a quantity known as the "Lorentz unit." The Zeeman effects observed in sun spots give valuable information as to the magnetic conditions in those areas.

Closely related to the Zeeman effect are two others, the Paschen-Back effect, produced by very strong magnetic fields, and the Back-Goudsmit effect, observed with the spectra of elements having a nuclear magnetic moment, such as bismuth. See also **Chemical Elements**; **Electron Theory**; **Paschen-Back Effect**; and **Stark Effect**.

ZEEMAN, ERIK CHRISTOPHER (1925–). Zeeman was an English mathematician. His doctoral work was in pure mathematics and he received his Ph.D. in 1954 for a thesis on knots and all the algebra you need to actually prove the existence of knots. He did research in topology, which is a type of geometry that examines the properties of shapes in many dimensions. His best known work was in catastrophe theory. His work has consequences for a broad range of fields from weather to psychiatry. Zeeman also made contributions in the development of the chaos theory.

Zeeman is not only known for this mathematical contribution but also for being an excellent educator. He spent much of his life teaching mathematics both to adults and children. He built the mathematics department at University of Warwick in England and made it an internationally known research center. He gave many mathematical talks on radio and television to reach a broader audience feeling that many gifted were amongst the educationally deprived population. He was honored by Queen Elizabeth II knighting him in 1991 for his role in advancing mathematics education. See also **Chemical Elements**; **Electron Theory**; and **Zeeman Effect**.

J.M.I

ZEISBERG CONCENTRATOR. A nitric acid concentrator, consisting of a packed tower, into which weak nitric acid vapors are introduced. Some steam is admitted into the bottom, and strong nitric acid vapors are discharged from the top to a condenser.

ZEISEL REACTION. The demethylation of an organic compound by treatment with hydriodic acid, which leaves a hydroxy group in place of the methoxy group, and forms methyl iodide, which may readily be determined quantitatively, as in the study of the amount of methoxy groups present.

ZEOLITE GROUP. To the zeolite group of minerals belong a number of hydrous silicates of aluminum which also ordinarily contain sodium or calcium, but rarely they may carry barium, strontium, magnesium, and potassium. These minerals are not related crystal- lographically as they occur in the isometric, orthorhombic, hex- agonal, and monoclinic systems, but they are all characterized by the presence of water, up to 10 or 20%, which is easily released with the application of heat. They are all rather soft minerals, hardness, 3.5–5.5; of low specific gravity, 2.0–2.5, and they will decompose readily upon treatment with acid, most of them yielding a gelatinous mass.

The easy fusion, together with the rapid expulsion of water, is responsible for the name of this interesting group; it is derived from the Greek words to boil and a stone, hence zeolite, "a boiling stone." The zeolites are secondary minerals, usually found filling fissures and cavities in the more basic igneous rocks as basalt, and gabbro, but occasionally in the more acidic types as granite or in gneisses. The following members of the zeolite group are described under **Analcime**; **Chabazite**; **Harmotome**; **Heulandite**; **Natrolite**; **Phillipsite**; **Scolecite**; and **Stilbite**.

See also **Molecular Sieves**.

ZEOLITE SOFTENING (Boiler Water). See **Water Treatment (Boiler)**.

ZEOLYTIC CATALYST. See **Petroleum**.

ZEREVITINOV DETERMINATION. A method of analysis of organic compounds containing active hydrogen atoms, as in hydroxy, carboxy, or imino groups, by reaction with methyl magnesium halide to yield methane quantitatively: $ROH + CH_3MgX \longrightarrow CH_4 + ROMgX$.

ZERO-ORDER REACTION. A reaction that has a constant rate, as in the case of certain gases reacting on the surface of a solid when it is almost entirely covered. Under such conditions, the reaction rate may be independent of pressure, and thus, the process is kinetically of zero order.

ZERO POINT ENERGY. The kinetic energy remaining in a substance at the absolute zero point of temperature. According to quantum mechanics, a simple harmonic oscillator does not have a stationary state of zero kinetic energy. The ground state has still one half quantum, $h\nu$, of energy, and the motion corresponding thereto. This agrees with the uncertainty principle, which does not permit the oscillator particle to be absolutely at rest exactly at the origin. In solids the zero-point energy is distributed in the normal modes of lattice vibration, and may be an appreciable term in the binding energy of the crystal, especially in hydrogen, helium, rare gases, etc. The motion may be observed in x-ray diffraction, but does not contribute to electronic resistivity.

ZETA POTENTIAL. The potential across the interface of all solids and liquids. Specifically, the potential across the diffuse layer of ions surrounding a charged colloidal particle, which is largely responsible for colloidal stability. Discharge of the zeta potential, accompanied by precipitation of the colloid, occurs by addition of polyvalent ions of sign opposite to that of the colloidal particles. Zeta potentials can be calculated from electrophoretic mobilities, i.e., the rates at which colloidal particles travel between charged electrodes placed in the solution.

ZIEGLER CATALYST. A type of stereospecific catalyst, usually a chemical complex derived from a transition metal halide and a metal hydride or a metal alkyl. The transition metal may be any of those in groups IV to VIII of the periodic table; the hydride or alkyl metals are those of groups I, II, and III. Typical, titanium chloride is added to aluminum alkyl in a hydrocarbon solvent to form a dispersion or precipitate of the catalyst complex. These catalysts usually operate at atmospheric pressure and are used to convert ethylene to linear polyethylene and also in stereospecific polymerization of propylene to crystalline polypropylene (Ziegler process).

ZIEGLER, KARL (1898–1973).

A German chemist who won the Nobel prize for chemistry in 1963 with Giulio Natta for their discoveries in the field of the chemistry and technology of high polymers. A great deal of his work was concerned with the chemistry of carbon compounds and development of plastics. A recipient of the Swinburne medal from the Plastics Institute of London in 1964. After studying at Marburg he was a professor at Heidelberg.

ZIEGLER METHOD.

Cyclization of dinitriles at high dilution in dialkyl ether in the presence of ether-soluble metal alkylanilide and hydrolysis of the resultant imino-nitril with formation of macrocyclic ketones in good yields.

ZIEGLER-NATTA POLYMERIZATION.

Polymerization of vinyl monomers under mild conditions using aluminum alkyls and $TiCl_4$ (or other transition element halide) catalyst to give a stereoregulated, or tactic, polymer. These polymers, in which the stereochemistry of the chain is not random have very useful physical properties.

ZIMMERMANN REACTION.

The reaction that occurs between methylene ketones and aromatic polynitro compounds in the presence of alkali. When applied to 17-oxosteroids, the colored compounds formed can be used for the quantitative determination of 17-oxosteroids.

ZINC.

[CAS: 7440-66-6]. Chemical element, symbol Zn, at. no. 30, at. wt. 65.38, periodic table group 12, mp 419.58°C, bp 907°C, density 7.1 g/cm³. Elemental zinc has a close-packed hexagonal crystal structure. There are five stable isotopes ^{64}Zn, ^{66}Zn through ^{68}Zn, and ^{70}Zn. Six radioactive isotopes have been identified ^{62}Zn, ^{63}Zn, ^{65}Zn, ^{69}Zn, ^{71}Zn, and ^{72}Zn. With exception of ^{65}Zn which has a half-life of 245 days, the half-lives of the other isotopes are measured in minutes and hours.

Zinc is a bluish-white metal, malleable and ductile at 150°C, but at 180°C it changes rapidly so that at 205°C it may be easily powdered; remains lustrous in dry air but is slightly tarnished in moist air or in water; burns upon heating to vaporization with a bluish flame, forming zinc oxide; soluble in acids—slowly when pure but rapidly on contact with copper or platinum; soluble in alkalies. Discovery prehistoric.

Zinc ranks 27th in order of abundance of the chemical elements in the earth's crust, an estimated 0.004% content of igneous rocks on an average basis. It is estimated that a cubic mile of seawater contains about 48 tons of zinc. First ionization potential 9.391 eV; second, 17.89 eV. Oxidation potential $Zn \longrightarrow Zn^{2+} + 2e^-$, 0.762 V; $Zn + 4OH^- \longrightarrow ZnO_2^{2-} + 2H_2O + 2e^-$, 1.216 V. Electron configuration $1s^2 2s^2 2p^6 3s^2 3p^{10} 4s^2$. Ionic radius Zn^{+2} 0.75 Å. Metallic radius 1.3324_5 Å. Other physical properties of zinc are described under **Chemical Elements.**

Zinc and lead usually occur together in nature as sulfides. Earlier separation processes involved the fine grinding of the combined sulfides and then treating the particles with chemical reagents to cause one sulfide to be preferentially wetted and thus the two sulfides separated by the froth flotation process. In a first stage, the lead sulfide is floated while the zinc sulfide sinks to the bottom of the tank. In the second stage, the process is reversed and the zinc sulfide is floated. Gangue and other nonmetals collect at the bottom of the tank. The separated sulfides are dewatered to a 6–7% moisture content and are referred to as the zinc concentrate and the lead concentrate.

A major zinc ore is ZnS (sphalerite) which frequently occurs with the major lead ore PbS (galena). The lead-zinc ores usually contain recoverable quantities of copper, silver, antimony, and bismuth as well, Major deposits of this type are worked in Australia, the United States, Canada, Mexico, Peru, the former Yugoslav Republics, and the former Soviet Union. Two other important zinc ores are $ZnCO_3$ (smithsonite) and iron-zinc-manganese oxide (franklinite). Several of these minerals are described under separate alphabetical entries.

Extractive Metallurgy of Zinc[1]

As of the early 1980s, zinc production throughout the world is based almost exclusively on two processes:

1. Electrolytic Process—in which oxidized zinc concentrates are leached in sulfuric acid and then electrolyzed to plate SHG (special high-grade zinc metal) on the cathode and to regenerate the acid on the anode.
2. ISP (Imperial Smelting Process)—a combined lead-zinc process in which oxidized concentrates are reduced with coke in a shaft furnace and the zinc vapor collected in a lead splash condenser.

The older retort and electrothermic processes have largely been replaced because of environmental and economic reasons.

The ISP evolved to fill a very special niche in nonferrous metallurgy because of its capability of treating lead-zinc concentrates which may also contain appreciable amounts of copper. The concentrate is normally oxidized in a sintering machine to produce a feed for the blast furnace where the zinc oxide is reduced with coke. Some effort has been underway to develop a hot briquetting operation to produce a suitable feed without sintering. Other efforts to improve the economic competitiveness of the process include air preheat and the use of an oxygen-enriched blast to reduce coke consumption.

Although the electrolytic zinc process can trace its industrial history back over 60 years, only during the last 15 to 20 years has it become the industry standard, commanding a large portion of the free world's zinc capacity. While the electrolytic zinc process has been varied to meet the demands of the particular feed, the flow sheet always contains the basic steps of roasting, leaching, solution purification, and electrolysis. The zinc concentrate is oxidized with air to produce acid-soluble zinc oxide (calcine), and sulfur dioxide-containing off gas suitable for conversion to acid, as well as byproduct steam. The Vieille-Montagne/Lurgi fluidized bed roaster is the industry standard.

The roaster product is leached with spent electrolyte (sulfuric acid) under near-neutral conditions to dissolve most of the zinc, copper, and cadmium, but little of the iron. The leach residue solids are releached in hot, strong acid to dissolve more zinc, since it attacks the otherwise insoluble zinc ferrites. The iron which is also dissolved in this second leach is then precipitated as jarosite, goethite, or hematite. The development of these iron precipitation techniques permitted the use of the hot, strong acid leach and an increase in zinc extraction from about 87% to greater than 95%. Simultaneously, the hot acid leach frequently generates a leach residue rich enough in lead and silver to provide significant byproduct value, as well as increased recovery of cadmium and copper.

The neutral solution is purified to remove impurities more noble than zinc, e.g., cadmium, copper, cobalt, nickel, arsenic, antimony, and germanium. The purification is accomplished by cementation in two or more steps with the addition of zinc dust. Generally, at least one cementation step is conducted at high temperature with arsenic, antimony, or copper-arsenic added. Cadmium is usually recovered in the metallic state and copper, nickel, and cobalt are recovered as sludges if present in sufficient quantities.

Zinc is extracted from the purified solution in cells using lead/silver alloy anodes and aluminum cathodes at a current density of 38–60 amperes/square foot (400–650 amperes/square meter). The product is normally SHG zinc, particularly if strontium carbonate is added and/or lead/silver anodes of greater than 0.596 silver content are used. After deposition of 24 to 72 hours, the cathodes are removed from the cells and the zinc is stripped by automatic machines in modern plants, melted, and cast for market. The move to automated handling of large cathodes was a major factor in lowering the overall labor requirement in producing zinc.

The following technical developments have led to the acceptance of the electrolytic process:

1. Adoption of high-capacity, low-labor fluidized-bed roasters
2. Adoption of continuous schemes for leaching and solution purification, allowing more automation and lower operating costs
3. Improved raw material utilization via hot acid leaching and iron precipitation as previously described
4. Construction of higher-capacity plants with larger equipment
5. Adoption of mechanized cathode handling/stripping with dramatically lower labor demands
6. Improved byproduct recovery via the hot acid leach

With exception of leach residue (jarosite, goethite, etc.) disposal, the process is environmentally sound.

[1] Information for this topic furnished by C.O. Bounds, St. Joe Minerals Corporation, Monaca, Pennsylvania.

The electrolytic process is inherently a somewhat energy inefficient (being based on electrical rather than directly on fossil energy) and capital-intensive operation. A return to pyrometallurgical smelting is conceivable if the environmental concerns are addressed and a noncoke (or at least low-coke) process is developed. New hydrometallurgical developments include the near-commercial Sherritt-Gordon pressure leach to eliminate roasting and the generation of sulfur-dioxide-rich gases, and laboratory experiments with chloride-based leaching/electrolysis.

Production. Principal producers of slab zinc include the former U.S.S.R., Japan, Canada, the United States, Germany (West), Australia, Belgium, France, Poland, Italy, Spain, Mexico, and China. The first four countries mentioned account for well over half of the production. Countries with limited production include Finland, the Netherlands, the former Yugoslav Republics, Bulgaria, and North Korea.

Uses of Zinc

Some concept of the major uses for zinc can be gleaned from Table 1. Slab zinc is available in three grades, as specified by the American Society for Testing and Materials (Table 2).

Zinc Coating. Constituting the largest single use of the metal, zinc coating is accomplished mainly by dipping the product in molten zinc or by electroplating (electrogalvanizing). Hot-dip galvanizing employs chiefly the less pure grades of zinc. The life of galvanized material is proportional to the thickness of the coating, and recent developments in both electrogalvanizing and hot-dip galvanizing have been toward application of heavy coatings that will withstand deformation without peeling.

Because of its relatively high electropotential (position in the emf series of metals), zinc can provide electrolytic protection against corrosion of several common metals, notably products made of iron and steel. Advantage of this characteristic also is taken in the use of zinc as the anode material for a number of types of batteries, power packs, and fuel cells. In providing a protective coating over ferrous metals, the attack of corrosive materials on the zinc produces a relatively inert reaction-product film, which deters destruction of the underlying zinc and base metal. When the coating is broken, as may result from mechanical means such as scratching, abrading, etc., the zinc, having a higher electropotential than the ferrous metals, slowly is expended in furnishing the required protective current. Thus, serious corrosion is delayed for a long period, providing long useful life to the zinc-coated products except in the most adverse cases, or where a poor selection of construction materials was initially made.

The most widely used form of zinc coating is effected by hot-dip galvanizing in which steel sheets, coils, structurals, hardware, wire, and other forms are dipped in a bath of molten zinc. The zinc readily adheres to a previously cleaned (pickled and thoroughly washed) iron or steel surface. The thickness of the coating is controlled by manipulating process temperatures, time the underlying metals are in contact with the molten bath, and mechanical means used. Products that are commonly hot-dip galvanized include roofing, siding, transmission towers, highway guard-rails, light poles, culverts, and fencing. Other forms of zinc coating include flame-spraying or metallizing, flake galvanizing, sherardizing or cementation, plasma arc spraying, vacuum metallizing, and also painting with materials that contain zinc pigments.

Electrogalvanizing makes it possible to apply a zinc coating to such products as steel strip, wire, conduits, hardware, etc. in a high-speed fashion through electroplating. One-side zinc coating of steel sheet for automotive applications by electrodeposition is a notable innovation in recent years. One advantage of electroplating is that the products to be coated are not subject to thermal conditions that may alter dimensions and shape. Electrodeposits range in thickness from 0.00015 inch (0.004 millimeter) to 0.001 inch (0.25 millimeter). Sherardizing is accomplished by heating the ferrous materials to be coated at a temperature of about 350°C in contact with zinc dust in a closed vessel. The resulting coating consists of iron-zinc alloys. It has been found that sherardized coatings match the corrosion resistance of electrogalvanized or hot-dip coatings of equivalent thickness.

Die Castings. The major use of zinc as a structural material is in alloys for pressure die casting. Development of the modern zinc die-casting alloys was directly related to use of Special High Grade zinc, with the addition of particular alloying constituents held within close limits and control of impurities. For die castings it is essential to use this extra pure grade of zinc to ensure extremely low iron, lead, cadmium, and tin contents. It is also necessary to limit these same impurities in the metals added to make the desired zinc alloy composition. Only by such control can zinc die castings be produced that are stable in dimensions and properties.

The impurities lead, cadmium, and tin, if present in castings in amounts greater than the established maximums (0.005% lead; 0.004% cadmium; 0.003% tin), cause subsurface network corrosion. These limits are close to critical values. Iron is held to 0.10% maximum to prevent excessive skimming losses and machining problems.

Zinc alloys are low in cost of metal per casting, are easy to die cast, are cast at low temperatures, have greater strength than all other die-casting metals except the copper alloys, lend themselves to casting within close dimensional limits, permit the thinnest sections yet produced, and are machined at minimum cost. Their resistance to surface corrosion is adequate in a wide range of applications. Prolonged contact with moisture results in formation of white corrosion products, but surface treatments can be applied that largely prevent formation of such products.

Limiting service conditions for standard zinc-base die-casting alloys are as follows: At temperatures slightly above 95°C (200°F) their tensile strength is reduced 30% and their hardness 40%. At subzero temperatures, some embrittlement occurs, but impact strength is still in the same range as that of aluminum and magnesium die-casting alloys at normal service temperatures. At room temperature, impact strength of zinc die castings is much higher than that of aluminum or magnesium die castings or iron sand castings.

All die castings have at least a light flash at the die parting, and those requiring movable cores will have some flash around the cores. Flash is

TABLE 1. CONSUMPTION OF SLAB ZINC IN THE UNITED STATES

Application of zinc		Percent of total tonnage consumed
Galvanizing		40.8
Sheet and strip	26.5	
wire and wire rope	2.3	
tube and pipe	4.1	
fittings (tube and pipe)	0.6	
Tanks and containers	0.3	
structural shapes	0.3	
Fasteners	0.5	
pole-line hardware	0.4	
fencing, wire cloth and netting	1.5	
other	2.4	
In Brass and Bronze		13.8
Sheet, strip, and plate	6.6	
rod and wire	5.0	
tube	0.6	
casting and billets	0.2	
copper-base ingots	0.6	
Other copper-base products	0.8	
In Zinc-base Alloys		28.0
die casting alloy	27.4	
dies and rod alloy	0.1	
slush and sand casting alloy	0.5	
Other Uses		17.4
Rolled zinc	2.2	
zinc oxide	3.5	
other applications	11.7	
Total		100.0

Source: U.S. Bureau of Mines.

TABLE 2. GRADES OF SLAB ZINC

Grade of zinc	Composition (Percent)			
	Lead (maximum)	Iron (maximum)	Cadmium (maximum)	Zinc (minimum by difference)
Special High Grade	0.003	0.003	0.003	99.990
High Grade	0.03	0.02	0.02	99.90
Prime Western	1.4	0.05	0.20	98.0

Notes: When specified for use in manufacture of rolled zinc or brass, aluminum is held to 0.005% maximum.
Tin in Special High Grade zinc is held to 0.001% maximum.
Aluminum in Prime Western zinc is held to 0.05% maximum.
Source: American Society for Testing and Materials.

also formed around ejector pins at the points at which they make contact with the casting.

Although it often is cheaper to cut threads than to cast them, for many pieces cast threads usually are more economical. Male threads usually are made with a parting parallel to the axis; this leaves a flash at the parting. The flash can be removed with a shaving tool in some instances, but in others a chasing operation is necessary.

Zinc die castings are invariably cast within quite close dimensional limits, but some machining is commonly required in addition to removal of flash, even though it may consist only of such simple operations as punching, drilling reaming, or tapping of holes. Zinc die castings can be soldered or welded, but ordinarily neither technique is used except for special applications or repair.

Many of the finishes applied to other types of metal products can also be applied to zinc die castings, although some differences in formulation as well as occasional differences in method of application may be desirable. The types of finishes applicable to zinc die castings include: mechanical finishes (buffed, polished, brushed, and tumbled); electrodeposited finishes (copper, nickel, chromium, brass, silver, and black nickel); chemical finishes (chromate, phosphate, molybdate and black nickel); and organic finishes (enamel, lacquer, paint and varnish, and plastic finishes). Electrodeposited coatings of virtually any metal capable of electrodeposition can be applied to zinc die castings.

The automotive industry uses by far the largest number of zinc alloy die castings. Zinc makes the average automobile last longer—17 pounds of zinc protect it from rust. Another 20 pounds are used to make zinc die cast parts like door handles, locks, carburetors, bodies for fuel pumps, windshield-wiper parts, speedometer frames, grilles, horns, heaters, parts for hydraulic brakes, and each tire contains about 1/2 pound of zinc, which is needed to cure rubber. The electrical industry probably uses a larger diversity of die castings than the automotive industry. Such parts are used in washing machines, oil burners, stokers, motor housings, vacuum cleaners, electric clocks, and kitchen equipment and utensils. Zinc die castings are used in business machines—typewriters, recording machines, picture projectors, vending machines, accounting machines, cash registers, cameras, slicing machines, garbage disposers, gasoline pumps, hoists, and drink mixers. Building hardware, padlocks, toys, and novelties also consume a substantial percentage of the total production of zinc-base die castings.

It is interesting to note that Gutenberg (1440s) used metal-based inks in printing the early Bibles. Zinc was used, along with copper and lead. These inks replaced those made of soot, which were prone to evanescence.

Much more detail on zinc metal products and applications will be found in the Horvick (1979) reference listed. Zinc alloys with copper are described under **Copper**.

Chemistry and Compounds

In virtually all of its compounds zinc exhibits the $+2$ oxidation state, although compounds of zinc(I) have been reported in the gaseous phase. Zinc is readily oxidized in the presence of hydroxide ions, e.g., by H_2O, this behavior being attributed to the stability of the $Zn(OH)_4^{2-}$ ion. Like other transition group 2 elements zinc has a marked tendency to form covalent structures, e.g., ZnO and ZnS.

Zinc oxide, formed by oxidation of the metal, dissolves in acids to yield Zn^{2+} ions, and in alkalies to form $Zr(OH)_4^{2-}$ ions. It reacts slowly with moist CO_2 to form the oxycarbonate, $5ZnO \cdot 2CO_2 \cdot 4H_2O$. Addition of an alkali metal hydroxide to a solution of a zinc salt does not precipitate the hydroxide, producing instead hydroxyzincate salt or a precipitate of flocculant zinc oxide. However, the hydroxide can be obtained from a sodium zincate solution on dilution and standing. Zinc hydroxide is amphiprotic, yielding Zn^{2+} ions with acids and $Zn(OH)_4^{2-}$ ions with (excess of) alkali hydroxides. Zinc peroxide is produced by treating a zinc chloride solution with sodium peroxide at a pH of 9.5. It is unstable, decomposing slowly on standing. Zinc oxide is used extensively in rubber, paints, and chemicals. Expanding uses include exterior latex paints, particularly alkyd-modified latex, the photocopy paper field, and as a substitute for mercury in mildew and fungus prevention formulations.

Unlike the other zinc halides, zinc fluoride, ZnF_2, is only slightly soluble in cold water. The anhydrous halides are prepared by direct union of the elements. In solution, zinc chloride, bromide, and iodide exhibit anomalous conductance properties attributed to undissociated molecules and complex ions. On heating these solutions, halogen acids, HX, are evolved, leaving oxyhalides in the fused residue. The zinc halides readily form double salts with halides of elements of main groups 1, 2, and sometimes 3 and 4. See also **Bromine**.

Zinc oxycarbonate is formed by the reaction of suspensions of the oxide or hydroxide with CO_2; to produce the normal carbonate, a very rapid stream of carbon dioxide and, usually, a somewhat higher pH is required.

Zinc also forms both nitrates and oxynitrates (in various hydrated forms) but the oxynitrate is less stable. It is formed by heating the hexanitrate, or treating the nitrate solution with NH_3.

Zinc forms a wide variety of other salts, many by reaction with the acids, though some can only be obtained by fusing the oxides together. The salts include arsenates (ortho, pyro, and meta), the borate, bromate, chlorate, chlorite, various chromates, cyanide, iodate, various periodates, permanganate, phosphates (ortho, pyro, meta, various double phosphates), the selenate, selenites, various silicates, fluosilicate, sulfate, sulfite, and thiocyanate.

Zinc sulfide, selenide, and telluride are more pronouncedly covalent than the oxide. They can be prepared from the elements, or in the case of first two, by the action of H_2S or hydrogen selenide upon zinc solutions. Zinc nitride, Zn_3N_2, prepared from zinc dust and NH_3, hydrolyzes readily to NH_3 and the oxide. Two zinc phosphides are known; Zn_3P_2, formed by heating the elements, yields phosphine with acids; ZnP_2 and $ZnHP$ have also been prepared.

One of the features of the chemistry of zinc is the fact that it is among the elements having a large number of complex compounds, mostly with coordination numbers of four and tetrahedral, but some with coordination numbers of 6, such as those of ethylenediamine which contain the ion $[Zn(en)_3]_{2+}$. The large number of halogen double salts have already been cited. The marked donor ability of oxygen toward zinc is evident from the number of basic salts, the existence of the zincates (containing ZnO_2^{2-} and also $Zn(OH)_4^{2-}$, the latter in strong alkali), and the formation of such chelate complexes as acetylacetonates and dioxalato complexes. Sulfur is a better donor than oxygen, so that addition compounds of the type $(R_2S)_2ZnX_2$ are formed from dialkyl sulfides and zinc halides, while thiourea forms chelate complexes containing $[Zn(th)^2]^{2+}$. The ready reactions with ammonia, as with amines, give large numbers of complexes; those with ammonia include diammines, triammines and tetrammines, containing $[Zn(NH_3)_2]^{2+}$, $[Zn(NH_3)_3]^{2+}$ and $[Zn(NH_3)_4]^{2+}$ respectively.

Prominent among the carbon donor complexes are the cyanides, principally compounds of $[Zn(CN)_4]^{2-}$, although $[Zn(CN)_3]^-$ is also known. Other carbon donor complexes are the triethyl and tetraethyl complexes, which are readily electrolyzed.

Biological Aspects of Zinc. See also **Zinc (In Biological Systems)**.

Additional Reading

Bauccio, M.L.: *ASM Engineered Materials Reference Book,* 2nd Edition, ASM International, Materials Park, OH, 1994.

Brooks, C.R.: *Heat Treatment, Structure and Properties of Nonferrous Alloys,* ASM International, Materials Park, OH, 1989.

Craig, B.D. and D.S. Anderson: *Handbook of Corrosion Data,* 2nd Edition, ASM International, Materials Park, OH, 1995.

Davis, J.R.: *Metals Handbook,* 2nd Edition, ASM International, Materials Park, OH, 1998.

Dutrizac, J.: *Lead-Zinc 2000,* The Minerals, Metals & Materials Society, Warrendale, PA, 2000.

Greenwood, N.N. and A. Earnshaw: *Chemistry of the Elements,* 2nd Edition, Butterworth-Heinemann, Inc., Woburn, MA, 1997.

Guruswamy, S.: Staff, International Lead Zinc Research Organization, Marcel Dekker, Inc., New York, NY, 1999.

Hewitt, K. and T. Wall: *Zinc Industry,* Woodhead Publishing, Ltd., Cambridge, UK, 2000.

Horvick, E.W.: *Properties of Pure Zinc,* Metals Handbook, 9th Edition, Vol. 2, ASM International, Metals Park, OH, 1979.

Kubel, E.J., Jr.: "Expanding Horizons for Zinc-Aluminum Alloys," *Advanced Materials & Processes,* **51** (July 1987).

Lalo, J.: "Pennsylvania's Dead Mountain (Zinc Smelter)," *Amer. Forests,* **54** (March–April 1988).

Lide, D.R.: *CRC Handbook of Chemistry and Physics*, 84th Edition, CRC Press, LLC., Boca Raton, FL, 2003.

Loffler, H.: *Structure and Structure Development of Al-Zn Alloys,* John Wiley & Sons, Inc., New York, NY, 1995.

Parker, P.: *McGraw-Hill Encyclopedia of Chemistry,* 2nd Edition, The McGraw-Hill Companies, Inc., New York, NY, 1993.

Porter, F.C.: *Zinc Handbook: Properties, Processing, and Use in Design,* Marcel Dekker, Inc., New York, NY, 1991.

Sousa, L.J.: "The Changing World of Metals," *Advanced Materials & Processes*, 27 (September 1988).

Staff: *Properties and Selection: Nonferrous Alloys and Special-Purpose Materials*, Vol. 2, 10th Edition, ASM International, Materials Park, OH, 1991.

Staff: *Alloy Phase Diagrams Handbook*, Vol. 3, ASM International, Materials Park, OH, 1992.

Staff: *Nonferrous Metals–Nickel, Cobalt, Lead, Tin, Zinc, Cadmium, Precious, Reactive, Refractory Metals and Alloys,* Vol. 4, American Society for Testing & Materials, West Conshohocken, PA, 1997.

Vander Voort, G.: *Atlas of Time-Temperature Diagrams for Nonferrous Alloys*, ASM International, Materials Park, OH, 1991.

Zhang, X.G.: *Corrosion and Electrochemistry of Zinc*, Kluwer Academic Publishers, Norwell, MA, 1996.

Web References

International Lead Zinc Research Organization: http://www.ilzro.org/

The American Zinc Association: http://www.zinc.org/

The Minerals, Metals & Materials Society: http://www.tms.org/Tables

USGS Minerals Information Zinc: http://minerals.usgs.gov/minerals/pubs/commodity/zinc/

ZINC (In Biological Systems). Zinc was one of the first of the trace elements known to be essential for both plants and animals, and yet problems of zinc nutrition are still of pressing importance. Evidence of zinc deficiency in crops is being recognized in new areas and the use of zinc in fertilizers has increased steadily in recent years. A dry, cracked condition of the skin of pigs (*parakeratosis*) caused by a zinc-deficiency has been a problem to pork producers. Diseases and syndromes that have been attributed to zinc deficiency in the human diet include loss of appetite, loss of sense of taste, and delayed healing of burns and other wounds. It is interesting to note that application of zinc-containing ointments to promote healing is an old practice in human medicine. Laboratory animals deficient of zinc may be subject to serious reproductive problems, including infertility of males, failure of conception or implantation of the embryo, difficult births, and deformed offspring. The extent to which zinc deficiency is a primary cause of reproductive problems in farm animals and humans is not thoroughly understood, and research continues.

Patients of both sexes with sickle cell anemia have been reported to be zinc deficient. Administration of zinc to males with this disease has been shown to improve the hypogonadism and the short stature associated with this deficiency. Administration of small dosages of zinc sulfate is part of the therapy in treating certain types of sickle cell anemia. Zinc deficiency in crops is frequently observed where fields have been graded to smooth them so that irrigation water can be applied more uniformly. Where the topsoil is cut away from small areas of these fields, such crops as corn (maize) and beans may be stunted and many leaves will be white instead of the usual green. If zinc fertilizers, supplying as little as 10 pounds of zinc per acre (11.2 kilograms per hectare) are applied, bumper crops may be grown on these soils. Citrus trees are commonly fertilized with zinc.

When zinc fertilizers are used on soils deficient in zinc, crop production may be increased even though the zinc concentration in the plant tissues and especially in the seed show no increase. With higher levels of zinc fertilization, the zinc concentration in plants may increase. Some evidence shows that the value of food and feed crops as sources of dietary zinc can be improved by using zinc fertilizers at rates exceeding those required for optimal plant growth. However, very high rates of zinc fertilization can depress crop yields.

The zinc contained in plants is not fully utilized by animals. Diets high in calcium and phosphorus have been associated with poor digestibility of dietary zinc. Diets with large amounts of soy protein are particularly likely to require extra zinc fortification for livestock. Meat is an important source of zinc for human diets. Where supplementation of zinc is indicated, zinc sulfate, zinc oxide, and zinc carbonate are commonly used.

Zinc in Metabolism

Zinc was first recognized as a trace element—then referred to as a growth factor for Aspergillus niger—by Raulin (1869). Evidence for a specific biochemical role of zinc was first obtained by Keilin and Mann in 1940, when the metal was shown to be a stoichiometric component of bovine carbonic anhydrases. The findings of many other zinc-containing enzymes during the interim have indicated a diverse biological role for zinc.

The high affinity of zinc for nitrogeneous and sulfur-containing ligands seems chiefly responsible for the occurrence of zinc in a wide variety of biological compounds, such as proteins, amino acids, nucleic acids, and porphyrins. Operationally, the enzymes affected by zinc can be considered in two groups: (1) zinc metalloenzymes; and (2) zinc metal-enzyme complexes. Zinc metalloenzymes incorporate zinc so firmly in the protein matrix that they can generally be considered as an entity. Under reasonably mild conditions, the metal and protein moiety are isolated together and exhibit an integral stoichiometric relationship. On the same basis, a strict correlation is preserved between metal and enzyme activity, allowing the inferential identification of a specific biological function of zinc *in vivo*. Zinc metal-enzyme complexes, in contrast, comprise enzymes which are activated in vitro by the addition of zinc ions. The loose association and the relative lack of metal ion specificity render it difficult in many cases to assign specific biological significance to zinc *in vivo*.

In zinc metalloenzymes, zinc is a selective stoichiometric constituent and is essential for catalytic activity. It is frequently present in numerical correspondence with the number of active enzymatic sites, coenzyme binding sites, or enzyme subunits. Removal of zinc results in loss of activity. Inhibition by metal complexing agents is a characteristic feature of zinc metalloenzymes. However, no direct relationship holds between the inhibitory effectiveness of these agents and their affinity for ionic zinc. Although zinc is the only constituent of zinc metalloenzymes *in vivo*, it can be replaced by other metals in vitro, such as cobalt, nickel, iron, manganese, cadmium, mercury, and lead, as in the case of carboxy-peptidases.

Zinc is a ubiquitous component of animal and plant tissue. In vertebrates, most organs, including pancreas, contain 20–30 micrograms of zinc per gram of wet tissue. Liver, voluntary muscle, and bone hold about double this amount. Zinc contents ranging from 100 to 1000 micrograms/gram weight have been measured in islet tissues of certain teleost fishes. Correlation between zinc content and insulin storage suggest a parallelism, but evidence is wanting for zinc-insulin complexes *in vivo*. The highest zinc content determined among mammals is found in the *tapetium lucidum cellulosum* of adult for seals—up to 150,000 micrograms/gram. Human blood contains 7–8 micrograms zinc/milliliter. About 12% of this is present in serum, 3% in leukocytes, and 85% in erythrocytes. In these compartments, zinc occurs as part of zinc proteins and zinc metalloenzymes. In erythrocytes, zinc is correlated to carbonic anhydrase activity.

In 1991, B.C. Cunningham, M.J. Mulkerrin, and J.A. Wells (Genetech) reported that size-exclusion chromatography and sedimentation equilibrium studies demonstrated that zinc ion (Zn^{2+}) induced the dimerization of human growth hormone (HGH). Scatchard analysis of $^{65}An^{2+}$ binding to HGH shows that two ZN^{2+} ions are associated per dimer of HGH in a cooperative fashion. Cobalt (II) can substitute for Zn^{2+} in the hormone dimer and gives a visible spectrum characteristic of cobalt coordinated in a tetrahedral fashion by oxygen- and nitrogen-containing ligands. Human growth hormone is synthesized and secreted into storage granules before its release from the anterior pituitary.

In 1991, R.E. Klevit (University of Washington) reported on the recognition of DNA by Cys_2, His_2 zinc fingers. The "zinc finger" protein motif was so named because of the tandemly repeating pattern observed in the amino acid sequence of the transcription factor TFIIA. Klevit points out, "According to the original hypothesis, each 30-residue sequence is an independently folded unit that binds a zinc ion and is responsible for sequence-specific DNA binding."

Underwood (1977) reported that high levels of dietary zinc interfere with the normal absorption and metabolism of several minerals. Earlier, Stewart and Magee (1964) reported that experience with laboratory animals indicated that 7500 parts per million (ppm) zinc in a purified diet lowered the concentrations of calcium and phosphorus in bone. Other investigators showed that animals fed the same level of zinc were anemic and had deformed, fragile erythrocytes. The livers contained reduced amounts of ferritin and lower concentrations of iron in the ferritin. Whanger and Weswig (1971) showed that in rats fed 2000 and 4000 ppm zinc, liver copper concentration was decreased. The most sensitive responses to zinc have been obtained with animals receiving minimally required or suboptimal levels of copper Campbell and Mills, 1974; and Murthy et al., 1974. Based upon these earlier findings, Hamilton et al. (1979) investigated possible zinc interference with copper, iron, and manganese in young Japanese quail (*Coturnix coturnix japonica*). The objective of the investigators was to identify the minimal level of excess dietary zinc that would produce physiological and metabolic deviations from normal and to define some of the most sensitive zinc-mineral interactions. Because other workers had found a sensitive zinc-copper antagonism, the Hamilton team studied the effects of supplemental zinc at copper levels of marginal deficiency. Data from the study showed that adequacy of copper intake is important when supplemental zinc is consumed either as a dietary supplement or in foods fortified with zinc. It was reported that results of the findings may be important for the general human population, whose dietary intake of many minerals, including copper, does not usually exceed the requirement or may be marginally deficient (Milne et al., 1978; Harland et al., 1978; Klevay, 1978). See also **Copper (In Biological Systems)**.

Zinc Poisoning: Soluble zinc salts are usually the etiologic agent in cases of zinc poisoning. The poisonous nature of zinc may be described as astringent, corrosive, and emetic. Sources of food poisoning include zinc-coated galvanized containers, pots, cans, tubs, and acids that convert zinc into soluble salts. Products that may be so involved include lemonade, cooked apples, mashed potatoes, spinach, chicken and tomatoes, and fruit punch. Zinc in combination with various organic and inorganic radicals (dimethyl dithiocarbamate, arsenate, etc.) can be quite poisonous.

Additional Reading

Campbell, J.K. and C.F. Mills: "Effects of Dietary Cadmium and Zinc on Rats Maintained on Diets Low in Copper," *Proc. Nutr. Soc.,* **33**(1), 15A (Abstract) (1974).

Considine, D.M. and G.D. Considine: *Foods and Food Production Encyclopedia,* Van Nostrand Reinhold, New York, NY, 1983

Cunningham, B.C., M.G. Mulkerrin, and J.A. Wells: "Dimerization of Human Growth Hormone by Zinc," *Science,* 545 (August 2, 1991).

Gray, P.: *The Encyclopedia of the Biological Sciences,* 2nd Edition, Krieger Publishing Company, Melbourne, FL, 1981.

Hamilton R.P., et al.: "Zinc Interference with Copper, Iron and Manganese in Young Japanese Quail," *J. Food Sci.,* **44**(3), 738–741 (1979).

Harland, B., L. Prosky, and J. Vanderveen: "Nutritional Adequacy of Current Levels of Zn, et al., in the American Food Supply for Adults, Infants, and Toddlers," in *Trace Element Metabolism in Man and Animals* (M. Kirchgessner, editor), p. 311, Institut fur Ernahrungsphysiologie, Technische Universitat Munchen, Freising-Weihenstephan, Germany, 1978.

Herrmann, B.W.A.: *Copper, Silver, Gold, Zinc, Cadmium, and Mercury,* Vol. 5, Thieme Medical Publishers, Inc., New York, NY, 1999.

Klevay, L.M.: "Dietary Copper and Copper Requirements in Man," in *Trace Element Metabolism Man and Animals* (M. Kirchgessner, editor), p. 307, Institut fur Ernahrungsphysiologie Technische Universitat Munchen, Freising-Weihenstephan, Germany, 1978.

Klevit, R.E.: "Recognition of DNA by Cys$_2$, His$_2$ Zinc Fingers," *Science,* 1367 (September 20, 1991).

Lewis, R.J. and N.I. Sax: *Sax'x Dangerous Properties of Industrial Materials,* 10th Edition, John Wiley & Sons, Inc., New York, NY, 2000.

Mills, C.F.: *Zinc in Human Biology,* Springer-Verlag, Inc., New York, NY, 1989.

Milne, D.B., et al.: "Dietary Intakes of Copper, Zinc, and Manganese by Military Personnel," *Fed. Proc.,* **37**, 894 (Abstract) (1978).

Murthy, L., et al.: "Interrelationships of Zinc and Copper Nurtriture in the Rat," *J. Nutrition,* **104**, 1458 (1974).

Prasad, A.S.: *Biochemistry of Zinc,* Kluwer Academic Publishers, Norwell, MA, 1993.

Rainsford, K.D., et al.: *Copper and Zinc in Inflammatory and Degenerative Diseases,* Kluwer Academic Publishers, Norwell, MA, 1998.

Robson, A.D.: *Zinc in Soils and Plants,* Kluwer Academic Publishers, Norwell, MA, 1993.

Sarkar, B.: *Genetic Response to Metals,* Marcel Dekker, Inc., New York, NY, 1991.

Schollmerich, J. and J.D. Kruse-Jarres: *Zinc and Diseases of the Digestive Tract,* Kluwer Academic Publishers, Norwell, MA, 1997.

Stewartm, A.K. and A.C. Magee: "Effect of Zinc Toxicity on Calcium, Phosphorus and Magnesium Metabolism of Young Rats," *J. Nutrition,* **82**, 287 (1964).

Underwood, E.J. and W. Mertz: *Trace Elements in Human and Animal Nutrition,* 5th Edition, Academic Press, Inc., San Diego, CA, 1990.

Whanger, P.D. and P.H. Weswig: "Effect of Supplementary Zinc on the Intracellular Distribution of Hepatic Copper in Rats," *J. Nutrition,* **101**, 1093 (1971).

ZINC BLENDE. See **Sphalerite Blende**.

ZINC CARBONATE. See **Smithsonite**.

ZINCITE. This mineral, (Zn, Mn)O, is an ore of zinc and occurs in considerable quantities at Franklin Furnace, New Jersey, where it is associated with willemite and franklinite. Its hexagonal crystals are rare, as it usually occurs massive, foliated, or in coarse to fine grains. When the crystals are observable, it reveals a perfect cleavage parallel to the base of the prism. The fracture is conchoidal. The mineral has a hardness of 4; specific gravity, 5.684; luster, subadamantine to vitreous; orange-yellow

streak; color, red to orange-yellow; translucent to opaque. Zincite also has been found in Poland, Tuscany, Spain, Saxony, and Tasmania.

ZINC OXIDE PIGMENT. See **Paint and Finish Removers**.

ZINC PHOSPHATE COATINGS. See **Conversion Coatings**.

ZIRCON. This mineral is zirconium silicate, $ZrSiO_4$, and is the chief ore of zirconium. Zircon occurs in square tetragonal prisms, although sometimes it assumes pyramidal or irregular forms. The mineral may be found in some beach and river placer deposits, but generally it is associated as an accessory mineral in siliceous igneous rocks, crystalline limestones, schists, and gneisses—and in sedimentary rocks derived from the foregoing. Zircon is without good cleavage; is brittle, with a conchoidal fracture; hardness, 7.5; specific gravity, 3.7–4.7; luster, adamantine, brilliant; color, green, yellow-green, golden-yellow, red, red-brown, brown, and blue. The name zircon derives from an Arabic word *zarqun,* meaning vermilion, or perhaps from the old Persian *zargun,* meaning golden-colored.

Zircon occurs in the Ural Mountains; Trentino, Monte Somma, and Vesuvius, Italy; Arendal, Norway; Ceylon, India; Thailand; at the Kimberley mines, Republic of South Africa; Madagascar; and in Canada in Renfrew County, Ontario, and Grenville, Quebec. In the United States, zircon is found at Litchfield, Maine; Chesterfield, Massachusetts; in Essex, Orange, and St. Lawrence Counties, New York; Henderson County, North Carolina; the Pikes Peak district of Colorado; and Llano County, Texas.

Gem quality crystals from Ceylon (Sri Lanka) have been known for many years. They range from colorless to brownish orange, yellow, dark red, from light reddish-violet. Heat treated zircons provide a beautiful stone of light blue color. Colorless stones are used as a diamond substitute.

ZIRCONIUM. [CAS: 7440-67-7]. Chemical element symbol Zr, at. no. 40, at. wt. 91.22, periodic table group 4, mp 1,853°C, bp 4,376°C, density 6.44 g/cm^3, 6.47 g/cm^3 (single crystal). Metallic zirconium is allotropic. Up to about 863°C, the alpha phase (hexagonal close-packed) is stable; above this temperature, the metal assumes the beta phase (body-centered cubic). The most common impurity, oxygen, tends to stabilize the alpha phase.

Zirconium metal exhibits passivity in air due to the formation of adherent coatings of oxide or nitride. Even without the coating, it is resistant to the action of weak acids and acid salts, but dissolves in HCl (warm) or H_2SO_4 slowly, and more rapidly if F is present, forming compounds of ZrO^{2+} ions, or fluorozirconates in the last case.

Crystalline zirconium of high purity is a white, soft, ductile, and malleable metal, but that of 99% purity, when obtained at high temperature, is hard and brittle. Pure zirconium has a combination of properties which make it a valuable structural material for nuclear reactors. In addition to low neutron capture, zirconium has good strength at high temperatures, corrosion resistance to high velocity coolants, avoidance of formation of high activity isotopes, and resistance to mechanical damage from neutron radiation. Amorphous zirconium is a bluish-black powder. At about 500°C zirconium burns in air; heated in hydrogen forms hydride; heated in nitrogen a nitride; and heated in chlorine the tetrachloride. On the laboratory scale, zirconium metal may be produced by the reduction of the chloride, oxide, or potassium zirconium fluoride with sodium metal.

Zirconium was first identified by Klaproth in 1789. The first crude powder was made by Berzelius in 1824 by reducing potassium fluorozirconate with potassium. A sample with a purity of 98% was not produced until the 1950s. There are five natural isotopes ^{90}Zr through ^{92}Zr, ^{94}Zr, and ^{96}Zr. Six radioactive isotopes have been identified ^{87}Zr through ^{89}Zr, ^{93}Zr, ^{95}Zr, and ^{97}Zr. With exception of 93Zr with a half-life of 1.1×10^6 years, the half-lives of the other isotopes are expressed in minutes, hours, or days. Zirconium is ranked 19th in abundance of the chemical elements occurring in the earth's crust with an estimated average content of zirconium in igneous rocks of 0.026%.

First ionization potential 6.95 eV; second, 13.97 eV; third, 24.00 eV; fourth, 33.83 eV. Oxidation potentials $Zr + 2H_2O \longrightarrow ZrO_2 + 4H^+ + 4e^-$, 1.43 V; $Zr + 4OH^- \longrightarrow ZrO(OH)_2 + H_2O + 4e^-$, 2.32 V. Electronic configuration $1s^2 2s^2 2p^6 3s^2 3p^6 3d^{10} 4s^2 4p^6 4d^2 5s^2$. Ionic radius Zr^{+4} 0.80 Å. Metallic radius 1.5895 Å. Other important physical properties of zirconium are given under **Chemical Elements**.

The most important ore for production of zirconium metal is zircon $ZrSiO_4$, which occurs in several regions in the form of a beach sand, often mixed with silica, ilmenite, and rutile. A floating-dredge technique is used in the mining operation. Early phases of beneficiation often take place on the dredge. Nearly all of the silica is separated by means of spiral concentrators, with the ilmenite and rutile removed by magnetic and electrostatic separators. The purest concentrates are used for metal production; others for refractories. Direct chlorination of the ore is the most modern method of extraction. In a simple reaction, water-soluble zirconium tetrachloride is yielded. Liquid-liquid extraction in several stages is required for the removal of hafnium. With this process, zirconium containing less than 50 ppm hafnium can be produced. Ammonium thiocyanate is used to complex the zirconium while hafnium is extracted by a methyl ethyl ketone solvent. In a similar system, HNO_3 is used in the aqueous phase and tributyl phosphate as the solvent. After separation, the two metals are precipitated as their sulfates or hydroxides. Calcination yields a pure ZrO_2. For production of pure metal, the pure ZrO_2 is chlorinated to $ZrCl_4$. Sometimes this is sublimed for additional purification. The zirconium tetrachloride in the gaseous phase is reacted with molten magnesium, forming zirconium metal and magnesium chloride. There are several minor variations of these processes. For example, sodium may replace magnesium.

Uses

The most important application of zirconium is in the formulation of the base metal in an alloy of 98% zirconium, 1.5% tin, 0.35% iron-chromium-nickel, and 0.15% oxygen. This alloy is widely used in water-cooled nuclear reactors because of its excellent corrosion resistance up to about $350°$ in H_2O, and its low neutron cross section. Currently, about 90% of the zirconium produced is used for this application. The excellent corrosion resistance of zirconium to both strong acids and alkalis, particularly its resistance to strong caustic solutions at all concentrations and temperatures, is attracting increasing attention for application in chemical processing equipment. See also **Nuclear Power Technology**.

Chemistry and Compounds

Due to its $4d^2 5s^2$ electron configuration, zirconium forms tetravalent compounds readily, although the Zr^{4+} ion does not exist as such in aqueous solution, except at very low pH values, the common cation being hydrated ZrO^{2+} (or $Zr(OH)_2^{2+}$). Many of the tetravalent compounds are partly covalent. There are also less stable Zr(III) compounds. The remarkably close similarity in chemical properties to those of hafnium is due to the identical outer electron configuration ($5d^2 6s^2$ for Hf) and the almost identical ionic radii ($Hf^{4+} = 0.86$ Å) this relatively low value for Hf^{4+} being due to the lanthanide contraction.

Zirconium oxide, ZrO_2 is widely known, both as a mineral, baddeleyite, and as an industrial product obtained from zircon, $ZiSO_4$. Moreover, the precipitate obtained by action of alkali hydroxides upon solutions of tetravalent zirconium is a hydrated oxide. The latter is readily soluble in acids to form oxysalts, which are usually formulated in terms of the ZrO^{2+} ion, without including its water of hydration, e.g., as $ZrO(H_2PO_4)_2$. The hydrated ZrO^{2+} ion is not amphiprotic; it does not dissolve in alkali hydroxides. While it does react on alkali carbonate fusions, the compounds formed have been shown to be mixed oxides rather than zirconates.

All four of the halides of zirconium and the common halogens are known and are solids at ordinary temperatures. They are readily hydrolyzed, to form hydrated oxyhalides such as $ZrOCl_2 \cdot 8H_2O$ or $ZrOBr_2 \cdot 8H_2O$. The tetraiodide yields both $ZrOI \cdot 8H_2O$ and $ZrI(OH)_3 \cdot 3H_2O$, and this last composition probably represents best the structure of all these compounds. Like titanium tetrahalides, $ZrCl_4$ and $ZrBr_4$ act as Lewis Acids to form adducts, though of lesser stability, with H_2O and oxygen-function compounds, such as alcohols, ethers, and carboxy compounds generally. These include chelates formed with such compounds as 1,2-dihdroxy-benzene and acetylacetone, in which oxygen atoms act as electron donors. Zirconium also forms very stable complexes with $POCl_3$.

The trihalides of zirconium, like the dihalides of titanium, are extremely strong reducing agents, reacting even with H_2O.

Zirconium nitride, Zr_3N_4, is made by ammoniating the tetrachloride to yield $Zr(NH_3)4Cl_4$, which yields the nitride on heating. The nitride, like the boride and carbide, are alloy-like in character, with high fusing points, extreme hardness, and subject to considerable variation in composition. Thus Zr_3N_4 may vary in composition to ZrN without material change in its properties.

Unlike titanium, zirconium forms a few normal tetravalent salts, such as a tetranitrate and a tetrasulfate, as well as its more common basic salts. However, the normal salts readily undergo hydrolysis to form the basic salts.

Like titanium, zirconium forms halogen complexes, the most stable of which is the hexafluoride, ZrF_6^{2-}, as well as $ZrCl_6^{2-}$, and $ZrBr_6^{2-}$, which are less stable, except in concentrated solutions of the hydrogen halides. Zirconium also forms stable peroxy complexes, containing

$$-\!\!\!\!\overset{\diagdown}{\underset{\diagup}{Zr}}\!\!-\!O\!-\!O\!-$$

Unlike titanium, zirconium forms a heptafluoride ion, ZrF_7^{3-}, which is quite stable.

Additional Reading

Amato, I.: "Exploring the New Material World (Zirconium)," *Science*, 644 (May 3, 1991).
Anderson, D.L.: "Composition of the Earth," *Science*, 367 (January 20, 1989).
Bauccio, M.L.: *ASM Engineered Materials Reference Book,* 2nd Edition, ASM International, Materials Park, OH, 1994.
Brooks, C.R.: *Heat Treatment, Structure and Properties of Nonferrous Alloys,* ASM International, Materials Park, OH, 1989.
Craig, B.D.: *Handbook of Corrosion Data,* 2nd Edition, ASM International, Materials Park, OH, 1995.
Elvers, B., et al.: *Water to Zirconium and Zirconium Compounds,* Vol. 28, John Wiley & Sons, Inc., New York, NY, 1996.
Greenwood, N.N. and A. Earnshaw: *Chemistry of the Elements,* Butterworth-Heinemann, Inc., Woburn, MA, 1997.
Krebs, R.E.: *The History and Use of Our Earth's Chemical Elements: A Reference Guide,* Greenwood Publishing Group, Inc., Westport, CT, 1998.
Lewis, R.J. and N.I. Sax: *Sax's Dangerous Properties of Industrial Materials,* 10th Edition, John Wiley & Sons, Inc., New York, NY, 2000.
Lide, D.R.: *CRC Handbook of Chemistry & Physics,* 84th Edition, CRC Press, LLC., Boca Raton, FL, 2003.
Ondik, H.M. and H.F. McMurdie: *Phase Diagrams for Zirconium and Zirconia Systems,* The American Ceramic Society, Westerville, OH, 1998.
Parker, P.: *McGraw-Hill Encyclopedia of Chemistry,* 2nd Edition, The McGraw-Hill Companies, New York, NY, 1993.
Staff: *Properties and Selection: Nonferrous Alloys and Special-Purpose Materials,* ASM International, Materials Park, OH, 1991.
Staff: *Alloy Phase Diagrams Handbook,* ASM International, Materials Park, OH, 1992.
Vander Voort, G., Editor: *Atlas of Time-Temperature Diagrams for Nonferrous Alloys,* ASM International, Materials Park, OH, 1991.

ZOISITE. This mineral is a hydrous aluminum silicate corresponding to the formula $Ca_2Al_3(Si_3O_{12})(OH)$, crystallizing in the orthorhombic system. Clinozoisite (monoclinic) is its isomorphous counterpart. Zoisite occurs as prismatic crystals, usually deeply striated vertically, and as compact or columnar masses. Perfect prismatic cleavage; brittle; uneven to conchoidal fracture; hardness, 6.5–7; specific gravity, 3.355; luster, vitreous to pearly; transparent to translucent; color, grayish white, green, pink (the manganese-rich variety, thulite), and blue to purple (tanzanite).

Zoisite occurs in crystalline schists which are products of regional metamorphism of basic igneous rocks rich in plagioclase, the calcium-rich feldspar; also in argillaceous calcareous sandstones, thulite from quartz veins, pegmatites, and metamorphosed impure limestones and dolomites.

Tanzanite is the blue to purple variety of zoisite and represents a recent discovery of this heretofore unknown variety from Tanzania. It occurs here as excellent transparent crystals from which fine gems have been cut.

ZONE REFINING. One of a class of techniques known as fractional solidification, in which a separation is brought about by crystallization of a melt without solvent being added. See also **Crystallization**. A massive solid is formed slowly and a sizable temperature gradient is imposed at the solid—liquid interface.

Zone refining can be applied to the purification of almost every type of substance that can be melted and solidified, e.g., elements, organic compounds, and inorganic compounds. Because the solid—liquid phase equilibria are not favorable for all impurities, zone refining often is combined with other techniques to achieve ultrahigh purity.

Because of the high costs of zone refining its application thus far have been limited to laboratory reagents and valuable chemicals such as electronic materials. The cost arises primarily from the low processing rates, handling, and high energy consumption owing to the large temperature gradients needed.

ZSIGMONDY REAGENT. A reagent for colloids which is a red colloidal solution of metallic gold obtained by reducing auric chloride by formaldehyde in the presence of an alkali. When mixed with sodium chloride, this reagent becomes blue because of an agglomeration of the particles of gold, but this color change is prevented by the presence of an adequate amount of certain other colloids. They can be classified according to the amount required to prevent the color change.

ZSIGMONDY, RICHARD (1865–1929). A native of Austria, Zsigmondy received the Nobel prize in chemistry in 1925 for his demonstration of the heterogeneous nature of colloid solutions and for the methods he used, which have since become fundamental in modern colloid chemistry. His most important contribution to chemistry was his invention of the ultramicroscope (with Siedentoph) in 1903.

See also **Tyndall Effect**; and **Ultramicroscope**.

ZWITTERION. An ion carrying charges of opposite sign, which thus constitutes an electrically neutral molecule with a dipole moment; looking like a positive ion at one end and a negative ion at the other. Most aliphatic amino acids form such dipolar ions, hence react with both strong acids and strong bases.

ZYMOLYTIC REACTION. A chemical reaction catalyzed by an enzyme, especially a reaction involving bond rupture or splitting, usually a hydrolysis.

INDEX

Page references in **bold type** refer to primary articles. Page references followed by t indicate material in tables.

A

A477, producing organism, 118t
A42867, producing organism, 118t
A47934, producing organism, 118t
A51568A, producing organism, 118t
A80407 A, producing organism, 118t
A82846 A, producing organism, 118t
AAAS (American Association for the Advancement of Science), 1
AAD-216 complex, producing organism, 118t
AAD-609 complex, producing organism, 118t
AAJ-271, producing organism, 118t
AB-65, producing organism, 118t
Abaca, **1**, 632t
Abherent, **1**
Ablating material, **1**
Ablation, **1**
Above-ground storage of products and wastes, 1729–1730
Abrasion, **1**
Abrasion pH, **1–2, 1261**
Abrasives, **2**, 745–746
Absolute, **3**
Absolute humidity, 793
Absolute temperature, **3**
Absolute temperature scale, 1599
Absolute vacuum, 3
Absolute zero, **3**, 1599
Absorptimetry, **3**
Absorption (Process), **3–4**
Absorption band, **3**
Absorption coefficient, **3**
Absorption spectroscopy, **4–5**
Absorption spectrum, **5**
Absorption towers, 4
ABS resins, 21–22
Abundance, **5**
7-ACA, properties, 113t
Acacia gum, 748t
Acaricide, **5**
Acceleration due to gravity, *versus* altitude, 157t
Accelerator, **5**
Accutane, chemotherapeutic, 359t
Acetaldehyde [CAS: 75–07–0], **5–6**, 48
    formation from acetylene, 8
Acetaldehyde cyanohydrin, 465t
Acetal group, **6**
Acetal resins, **1436**
Acetaminophen, 93t
Acetate, fiber-dye property requirements, 521t
Acetate dye, **6**
Acetate fiber
    generic designation and definition, 621t
    physical properties, 1139t
Acetate fibers, **624–626**
Acetate kinase, metal chelate enzyme, 323t
Acetates, 6–7
Acetic acid, carboxylic acid, 294
Acetic acid [CAS: 64–19–7], **6–7**, 1674–1675
    acidulant, 14
Acetoacetic ester condensation, 7
Acetobacter, vinegar fermentation, 1674–1675
Acetone, as parent structure compound, 1090t

Acetone [CAS: 67–64–1], **7**, 899
    gas separation, 39t
    physical properties, 900t
Acetone chloride, 7
Acetone cyanohydrin, 7, 465t
Acetonephenyl-hydrazone, 7
Acetonesemicarbazone, 7
Acetone-sodiumbisulfite, 7
Acetonitrile, 75
Acetophenone, physical properties, 900t
Acetorphan, 92t
Acetoxime, 7
Acetylacetone, physical properties, 900t
Acetyl benzoyl peroxide, properties, 1237t
Acetyl *tert*-butanesulfonyl peroxide, properties, 1237t
Acetyl chloride, physical properties, 366t
AcetylCoA, 282
Acetyl cyclohexanesulfonyl peroxide, properties, 1237t
Acetyl cyclohexyl carbonyl peroxide, properties, 1237t
Acetylene
    combustion constants, 422t
    combustion reactions, 423t
    heating value, 686t
    ignition temperature in air, 425t
Acetylene [CAS: 74–86–2], **7–8**, 55
Acetylene dibromide, 8
Acetylene dichloride, 8
    physical properties, 366t
Acetylene series, **8**
Acetylene tetrabromide, 8
Acetylene tetrachloride, 8
Acetylenic hydrocarbons, 55
Acetyl *sec*-heptanesulfonyl peroxide, properties, 1237t
Acetyl(1-methycyclohexane)-sulfonylperoxide, properties, 1237t
Acetylornithinase, metal chelate enzyme, 323t
Acetyl propionyl peroxide, properties, 1237t
2-Acetyl-pyrroline, as key odor/tase compound, 649
Acetylsalicylic acid [CAS: 50–78–2], **8**, 1455
Achlorhydria, **8**
Achondrite meteorites, 600
Achromycin, year of discovery, 130t
Acid-amino acid ligases, 571t
Acid-ammonia ligases, 571t
Acid-base metabolism, 8
Acid-base regulation (Blood), **8**
Acid-base theories, 12
Acid deposition, 9, 10–11
Acid dyes, 512t, 517
    applications, 518
    dyeing process, 522, 523
Acidic solvent, **8**
Acidification, 12
    of lakes and streams, 9
Acidimetry, **8**
Acidity, **8–9**
Acid lead, 923t
Acid number, **9**
Acidosis, 9
Acid radicals, 31
Acid rain, **9–12**, 1329

Acid salt, 1456
Acids and bases, **12–13**
Acid sludge, excess air for combustion, 426t
Acid spoil, 1439t
Acid sulfite pulping process, 1379
Acid-thiol ligases, 571t
Acidulants and alkalizers (Foods), **13–14**
Acifluorfen, environmental health advisories, 771t
Acinetobacter, amphenicol susceptibility, 115t
Acmite-aegerine, **14**
Acne vulgaris, 99
A35512 complex, producing organism, 118t
A40926 complex, producing organism, 118t
A41030 complex, producing organism, 118t
Aconitase, 282t
    metal chelate enzyme, 323t
Aconitic acid, 282t
Acoustic cavitation, 1525
Acquired Immunodeficiency Syndrome (AIDS), 1660
Acree's reaction, **14**
Acrilan, 621t
Acrolein, 14–15
    physical properties, 17t
Acrolein and derivatives, **14–15**
Acrylamide, **15**
    physical properties, 15t, 17t
Acrylamide polymers, **15–16**
Acrylates, monomer reactivity and suitable initiator for, 838t
Acrylates and methacrylates, **16–17**
Acrylic acid and derivatives, **17–18**
Acrylic acid [CAS: 79–10–7], **19–20**
Acrylic acid nitrile, 20
Acrylic adhesives, load-bearing capabilities, 32t
Acrylic anhydride, physical properties, 17t
Acrylic emulsion polymers, 20
Acrylic ester monomers, 18
Acrylic ester polymers, **18–19**
Acrylic fiber
    generic designation and definition, 621t
    physical properties, 1139t
    physical properties of staple, 626t
    worldwide capacity, 628t
Acrylic fibers, **626–629**
    dyeing, 523
    fiber-dye property requirements, 521t
Acrylic latexes, 20
Acrylic plastics, **19–20**
Acrylic sealants, 1463
Acrylonitrile barrier polymer, diffusion and solubility coefficients for oxygen and carbon dioxide, 173t
Acrylonitrile-butadiene-styrene (ABS) polymers, 21–22
    physical properties of structural foams, 665t
Acrylonitrile [CAS: 107–13–1], **20–21**, 75
    monomer reactivity and suitable initiator for, 838t
    polymerization with acrylic monomers, 17
Acrylonitrile polymers, **21–22**
Acrylontrile-butadiene-styrene resins, **1436**
Acryloyl chloride, physical properties, 17t
Actamycins, biological activity, 108t

1781

Actaplanin (A4696 complex), producing organism, 118t
ACTH (adrenocorticotropic hormone) [CAS: 9002–60–2], 23
Actin, 23
Actinide contraction, 23
Actinides, 23t
    important binary compounds, 25–26t
    oxidation states, 24t
    physical properties of metals, 24t
    principal characteristics of elements, 327–328t
Actininides and transactinides, 23–24
Actinium
    abundance, 330t
    electronic structure, 337t
    interatomic distance, 342t
    ionic crystal radius, 341t
    principal characteristics, 327t
Actinium [CAS: 7440–34–8], 23t, 26–27, 27
    active deposit, 29
    binary compound properties, 25–26t
    oxidation states, 24t
    physical properties, 24t
Actinium series, 332
Actinoidin A, producing organism, 118t
Actinoidin A2, producing organism, 118t
Actinoidin B, producing organism, 118t
Actinolite, 27, 149
Actinomyces bovis, gram-positive, 168t
Actinomycin D, activity and producing organism, 128t
Actinomyosin, 23
Actinon, 27
Activated alumina, 36
    for liquid-phase adsorption, 40
Activated carbon
    commercial gas adsorption processes, 39t
    for liquid-phase adsorption, 40
Activated 7-dehydrocholesterol, 1703
Activated sludge, 28–29
    aeration in, 42
    for secondary water treatment, 1725
Activation (Molecular), 28–29
Activator, 29
Active center, 29
Active deposit, 29
Activity coefficient, 29–31
Activity (Radioactivity), 31
Activity series, 31
Acyclic hydrocarbons, 1170–1171
Acyl, 31
Acylase, metal chelate enzyme, 323t
Acylation, 31
Acyl group, analysis, 97
Acyltransferases, 571t
Adamantine compound, 31
Adams, Roger (1889–1971), 31
Addition compound, 13, 427
Addition polymerization, 1341
Additive color process, 31–32
Additive compound, 427
Adenine [CAS: 73–24–5], 32
Adenosine, constituent of human blood, 242t
Adenosine [CAS: 58–61–7], 32
Adenosine diphosphate (ADP), 32
Adenosine phosphates, 32
Adenosine phosphokinase, metal chelate enzyme, 323t
Adenosine triphosphate (ATP), 32
    constituent of human blood, 242t
    constituent of human plasma or serum, 244t
Adenoviruses, 1695

Adenylic acid (adenosine monophosphate, AMP), 32
Adermine, 1700
Adherend, 32
Adhesion, 32
Adhesive bond, 32
Adhesive joint, 32
Adhesives, 32–33
    load-bearing capabilities, 32t
Adiabatic process, 34
Adipic acid
    acidulant, 14
    monomer for nylon resins, 1333t
Adipic acid [CAS: 124–04–9], 34–35
Adipic ketone, physical properties, 900t
Adipoyl bis(acetyl peroxide), properties, 1237t
Adjuvants, 1661
Adozelesin, DNA interactive agent, 357t
Adrenal cortex hormones, 785, 790
Adrenal cortical hormones, 35
Adrenaline, 35, 49t, 51
    where produced, structure, and principal functions, 786t
Adrenal medula hormones, 35–36
Adrenamine, where produced, structure, and principal functions, 786t
Adrenin, where produced, structure, and principal functions, 786t
Adrenocorticotropic hormone (ACTH), 23
    where produced, structure, and principal functions, 786t
Adrenocorticotropin, where produced, structure, and principal functions, 786t
Adriamycin, DNA interactive agent, 357t
Adsorbents
    for gas separation, 39t, 40
    for liquid separation, 40–41, 41t
Adsorption, 36–38
    factors governing use of regeneration method, 38t
Adsorption: gas separation, 39–40
Adsorption: liquid separation, 40–42
Adsorption chromatography, 38
Adsorption columns, 37–38
Adsorption equilibrium, 36–37
Advanced Technology Development Program, 741
Aegerine, 14
Aeration, 42
Aerial infared photography, 1293
Aerobacter aerogenes, gram-negative, 168t
Aerogels, 42–44
Aerosols, 44
Aesculin, 734t
Affinity, 44
Aflatoxin, 1769
    degradation, 1770
Agar, 44, 748t
Agar-agar, 44, 748t
Agate, 44
    hardness, 1008t
    heat effect, 708t
Agent orange, 44
Agglomerating coals, 390t
Agglomeration, 44–45
Agglutination, 45
Agglutinin, 45
Aggregate, 45
Aggregation, 45
Agitated-pan dryers, 509t
Agribusiness database, 769t
Agricola database, 769t
Agricultural wastes, use as fuel, 1718
Agrochemicals handbook, 769t

AIChE (American Institute of Chemical Engineers), 45
Air, 45–46
    absolute index, 1426t
    composition, 424t
    density, 474t
    specific heat at constant pressure, 757t
    thermal conductivity, 758t
Air-aluminum cells, 186t
Air cell, cell reactions, energy content, and manufacturers, 179t
Air-cooled heat exchanger, 760
Air-iron cells, 186t, 189
AIRMoN (Atmospheric Integrated Research Monitoring Network), 11
Air pollution, 1324–1330
Air sparging, 205t
Air stripping, 205t
Air-zinc cells, 186t
Alabandite, 46
Alabaster, 46
    hardness, 1008t
Alachlor, environmental health advisories, 771t
Alamethicin, activity and producing organism, 128t
Alanine
    first isolation and isoelectric point, 77t
    manufacture, 76t
    structural classification, 79t
    synthesis from products of carbohydrate metabolism, 282t
Alberite, 46
Albinism, genetic influence, 716
Albite, 607
    abrasion pH, 1t
Albomycin, activity and producing organism, 128t
Albumin, 46, 1000
    target of medical diagnostic test, 975t
Alchemy, 46
Alcholism, and folic acid deficiency, 668
Alcohol, 46
Alcoholate, 46
Alcohol dehydrogenase
    metal chelate enzyme, 323t
    source and function, 986t
Alcoholic fermentation, yeast in, 1767, 1768
Alcoholism, and achlorhydria, 8
Alcohols, 46–48, 1173–1174
Alcoholysis, 48
Aldehyde, 1169t
Aldehyde dehydrogenase, metal chelate enzyme, 323t
Aldehyde-lyases, 571t
Aldehydes, 48
Alder, Kurt (1902–1958), 48
Aldocortin, where produced, structure, and principal functions, 786t
Aldohexoses, 279t
Aldol condensation, 47, 48
Aldoses, 899
Aldosterone, 35
    where produced, structure, and principal functions, 786t
Aldoximes, 48
ALDOX process, 47
Alexandrite, 48, 708
Alfalfa, susceptibility to iron deficiency, 874t
Alfisols, 1497t
Algae, 1767
    good source of protein, 1373
Algal fungi, 1767
Algicide, 48
Algin, 48

Alginic acid, 748t
Alicyclic compound, 427
Alicyclic hydrocarbons, 1171–1172
Aliphatic alcohols, 46
Aliphatic compound, 48, 427
Aliphatic diamines and higher amines, **482–484**
Alizarin Yellow R, pH range and color change of indicator, 825t
Alkali, **48**
Alkalic rocks, 49
Alkali metals, **48**
Alkali metal xanthates, 1755
Alkaline cell
 cell reactions, energy content, and manufacturers, 179t
 characteristics, 184t
Alkaline earths, **48–49**
Alkaline phosphate, target of medical diagnostic test, 975t
Alkali rocks, 49
Alkalizers (Foods), **13–14**
Alkaloids, **49–52**
Alkalosis, **52**
Alkane, **52**
Alkaptonuria, genetic influence, 716
Alkene, **52**, 1169t
Alkeran, DNA alkylating/cross-linking agent, 356t
Alkyd resins, **52–55**
Alkyl, **55**
OO-tert-Alkyl *O*-alkyl monoperoxycarbonates, organic peroxide initiators, 841t
Alkylaluminum compounds, alcohol production from, 47
Alkylate, 1255
Alkylation, **55**, 1179
 catalyst application, 304t, 305
 in petroleum refining, 1255
Alkylbenzenes, biodegradable, 207t
tert-Alkyl hydroperoxides, organic peroxide initiators, 841t
tert-Alkyl peroxyesters, organic peroxide initiators, 841t
Alkyne, 1169t
Alkynes, 8, **55**
Allanite, **55**
Allelopathic substance, **55**
Allergenicity, 709
Allobar, **56**
Allochromatic, **56**
Allomerism, **56**
Allotropes, 334
Alloy phase diagrams, 56–57
Alloys, **56–58**
AllWave fiber, 1155
Allyl acetate, 60
Allyl alcohol and monoallyl derivatives, **59–60**
Allylamine, 60
Allyl chloride, 60
 physical properties, 366t
Allylene, 55
Allyl ester resins, **60**
Allyl glycidyl ether, 60
Allyl methacrylate, 60
Allyl rearrangement, 1424
Almandine, 705
Alnico 9, magnetic properties, 956t
Alpha (*α*), **60–61**
Alpha decay, 61, 1406
Alpha emitter, 61
Alpha naphtholbenzein, pH range and color change of indicator, 825t
Alpha particle, **61**

Alpha ray emission, 1407
Altitude, atmosphere-altitude relationship, 155–157
Altman, Sidney (1939–), 61
Alum, **61**
 abrasion pH, 1t
Alumina, 61–62, 65
 commercial gas adsorption processes, 39t
 dielectric permittivity, 852t
 plasma synthesis, 318t
Alumina-silica ceramic fiber, insulation material, 856t
Aluminides, 70–71
Aluminizing, 763t
Aluminohydrides, 796
Aluminosilicates, 65
Aluminum
 abundance, 5t, 330t, 1132t
 constituent of human plasma or serum, 244t
 corrosion resistant alloys, 446t
 covalent radius, 344t
 density, 474t
 electronic structure, 336t
 hardness, 755t
 interatomic distance, 342t
 ionic crystal radius, 341t
 major sources of industrial minerals, 1011t
 modulus and rigidity, 538t
 nuclides (isotopes and isobars), 331t
 principal characteristics, 327t
 residence time in sea water, 1133t
 specific heat at constant pressure, 757t
 standard electrode potential, 31t
 ultraviolet reflectance, 1640
Aluminum acetate, 65
Aluminum-air cell, cell reactions and energy content (in development), 180t
Aluminum alloys, 58
 hardness, 755t
Aluminum alloys and engineered materials, **65–71**, 67t, 69t, 70t, 71t
Aluminum ammonium sulfate, 86
 acidulant, 14
Aluminum bronze, 440t
Aluminum carbide, 65
Aluminum [CAS: 7429–90–5], **61–65**
Aluminum casting alloys, 68–69, 70t
Aluminum chloride, 65
Aluminum composites, 64, 68
Aluminum electroplating, 71
Aluminum fluoride, 65
Aluminum hydride, physical properties, 795t
Aluminum hydroxide, 64–65
Aluminum-lithium alloys, 64, 69–70
Aluminum nitrate, 65
Aluminum nitride, plasma synthesis, 318t
Aluminum oleate, 65
Aluminum oxide, 65
 dielectric permittivity, 852t
Aluminum palmitate, 65
Aluminum perchlorate, 64
Aluminum powder metallurgy, 64, 71
Aluminum selenate, 64
Aluminum silicate, dielectric permittivity, 852t
Aluminum stearate, 65
Aluminum sulfate, 64, 65
Aluminum trifluoride, 64
Aluminum wrought alloys, 69t
Alumstone, 72
Alunite, **72**
AM374, producing organism, 118t
Amalgam, **72**
Amatol, 87

Amber, **72**, 558
 heat effect, 708t
Ambergris, **72**
Amblygonite, **72**
American Association for the Advancement of Science (AAAS), 1
American Association of Scientific Workers (AASW), **73**
American Association of Textile Chemists and Colorists (AATCC), **73**
American Carbon Society (ACS), **73**
American Ceramic Society (ACerS), **73**
American Chemical Society (ACS), **73**
American Institute of Chemical Engineers (AIChE), 45
American Institute of Chemists (AIC), **73**
American National Standards Institute (ANSI), **73**
American Oil Chemist's Society (AOCS), **73**
American Petroleum Institute (API), **73**
American Society for Metals (ASM), **73**
American Society for Testing and Materials (ASTM), **73**
Americium, 23t
 binary compound properties, 25–26t
 electronic structure, 337t
 interatomic distance, 342t
 ionic crystal radius, 341t
 oxidation states, 24t
 physical properties, 24t
 principal characteristics, 327t
Americium [CAS: 7440–35–9], **72–73**
Amethyst, **74**, 708
Ametryn, environmental health advisories, 771t
Amide, 1169t
Amides, **74**
Amide synthetases, 571t
Amidine-lyases, 571t
Amidinotransferases, 571t
Amination, **74**, 1178
 yeast participation, 1769t
Amine, 1169t
Amines, **74–75**
Amino acid abnormalities, genetic influence, 716
Amino acid racemization dating, 1415–1416
Aminoacid-RNA ligases, 571t
Amino acids, 12, **75–80**
 constituent of human blood, 242t
 essential/nonessential, 75, 75t, 76t, 1375t
 first isolations, 77–78t
 production, 76t, 79–80
 and protein structure, 1373–1375
 representative essential amino acid patterns for selected foods, 75t
 structural classification, 79t
 synthesis from products of carbohydrate metabolism, 282t
α-Aminoacyl-peptide hydrolases, 571t
Aminoacyltransferases, 571t
Aminobenzene, 101
p-Aminodiphenyl, definitely established workplace carcinogen, 297t
Aminoethylpiperazine, physical properties, 483t
N-(2-Aminoethyl)-1,3-propylenediamine, physical properties, 483t
Aminoglycoside antibiotics, **107**
Aminoguanidine, 747t
7-Aminoheptanoic acid, monomer for nylon resins, 1333t
4-Aminoisoquinoline, 1401
9-Aminononanoic acid, monomer for nylon resins, 1333t
2-Aminophenol, 81–82

3-Aminophenol, 81–82

4-Aminophenol, 81–82

Aminophenols, **81–82**

Aminoplasts, microcapsule material, 997t

Amino resins, **80–81**
  alkyd resins with, 54

Aminotransferases, 571t

11-Aminoundecanoic acid, monomer for nylon
      resins, 1333t

Amitrole, environmental health advisories, 771t

Ammines, **82**

Ammonia
    catalyst application in production, 304t, 306
    combustion constants, 422t
    constituent of human blood, 242t
    gas separation, 39t
    in 24 hour urine sample, 1657t
    measurement by NDIR analysis, 837t
    as parent structure compound, 1090t
    use in paper recycling, 1715
    van der Waals characteristic equation
          coefficients, 321t

Ammonia [CAS: 7664–41–7], **82–86**

Ammonia-lyases, 571t

Ammonium acetate [CAS: 631–61–8], 86

Ammonium benzoate [CAS: 1863–63–4], 86

Ammonium bicarbonate, alkalizer, 14

Ammonium borate [CAS: 11128–98–6], 86

Ammonium bromide [CAS: 12124–97–9], 86

Ammonium carbonate [CAS: 506–87–6], 86

Ammonium chloride [CAS: 12125–02–9], **86**

Ammonium chloroplatinate [CAS: 16919–58–7],
      86

Ammonium compounds, **86**

Ammonium cyanate, 86, 464

Ammonium dichromate [CAS: 7789–09–5], 86

Ammonium dihydrogen phosphate (ADP), for
      diffracting crystal spectrometer, 1760

Ammonium fluoride [CAS: 12125–01–8], 86

Ammonium hydrogen oxalate, 1185

Ammonium hydroxide, alkalizer, 14

Ammonium hydroxide [CAS: 1336–21–6], **87**

Ammonium iodide [CAS: 12027–06–4], 86

Ammonium ions, target of medical diagnostic test,
      975t

Ammonium iron(III) oxalate, 1185

Ammonium linoleate, 86

Ammonium nitrate [CAS: 6484–52–2], **87**

Ammonium nitrite [CAS: 13446–48–5], 86

Ammonium oxalate, 1185

Ammonium oxalate [CAS: 1113–38–8], 86

Ammonium perchlorate, 1222

Ammonium perchlorate [CAS: 7790–98–9], 86

Ammonium periodate, 86

Ammonium persulfate [CAS: 7727–54–0], 86

Ammonium phosphates, **87–88**
    acidulant, 14

Ammonium phosphomolybdate [CAS:
      12026–66–3], 86

Ammonium rhodanate, 86

Ammonium salicylate, 86

Ammonium sulfate [CAS: 7783–20–2], **88**

Ammonium sulfide [CAS: 12124–99–1], 86

Ammonium sulfocyanide, 86

Ammonium tartrate [CAS: 3164–29–2], 86

Ammonium tetraborate, 86

Ammonium thiocyanate [CAS: 1762–95–4], 86

Ammonolysis, 74, 1180

Ammoxidation, catalyst application, 304t, 305

Amorphous, **88**

Amorphous nylon, gas permeability, 172t

Amorphous silicon, 1470–1471

Amosite, **88**, 149–150
    physical and chemical properties, 150t

Amoxycillin, year of discovery/market introduction,
      106t

Ampere (A), 1643t

Ampere-hour (Ah), 1643t

Ampere per meter (A/m), 1643t

Amperometer, **88**

Amphetamine, 49t, **88**

Amphibole, **88**

Amphibolite, **88–89**

Amphipathy, **89**

Amphiphilic, **89**

Amphiprotic, **89**

Ampholyte, **89**

Amphomycin, activity and producing organism, 128t

Amphoteric surfactants, 1586

Ampicillin, year of discovery/market introduction,
      106t

Ampoule, **89**

Ampule, 89

AMS, environmental health advisories, 771t

Amsacrine, topoisomerase interactive, 357t

Amsidyl, topoisomerase interactive, 357t

Amygdalin, 734t

Amygdaloid, **89**

Amygdules, 89

Amyl, **89**

n-Amyl alcohol, physical properties, 89t

sec-Amyl alcohol, physical properties, 89t

tert-Amyl alcohol, physical properties, 89t

Amyl alcohols, **89–90**

Amylase, target of medical diagnostic test, 975t

α-Amylase, metal chelate enzyme, 323t

Amylases, 306

1-Amylene, melting range, 1546t

tert-Amyl hydroperoxide, physical properties, 1231t

Amyloidosis, genetic influence, 716

tert-Amyl peroxyacetate, properties, 1239t

Anabasine, 49t

Anaerobe, **90**

Analcime, **91**

Analgesics, chiral, 1268

Analgesics, antipyretics, and antiinflammatory
      agents, **91–93**

Analysis (Chemical), **93–96**

Analysis (Organic Chemical), **96**

Analytical chemistry, 93–94

Analytical instrumentation, 95–96

Analyzer (Reaction-Product), **97–98**

Analyzer (Reagent-Tape), **98**

Anamorphism, **98**

Anaphylaxis, **98**

Ana-position, **98**

Anatase, **98**

Andalusite, **98**
    hardness, 1008t

Andersen's disease, genetic influence, 716

Andesine, 607

Andesite, **98**

Andradite, 705, 1635

Androgens, **98–99**

Androsterone, 98

Anesthesia, 99–100

Anesthetics, **99–101**, 100t

Anfinsen, Christian B. (1916–1995), **101**

ANFO explosive, 87

Angelsite, **101**

Angiographic imaging, 1762

Ångstrom, **101**, 1643t

Angstrom, 1643t

Anhalidine, 49t

Anhydride, **101**

Anhydrite, **101**
    abrasion pH, 1t

Anhydrous, **101**

Aniline, 75, **101**
    as parent structure compound, 1090t

Animal feeds, genetic engineering, 710

Animal waxes, 1746–1747

Anion, **101**

Anionic acids and bases, 13

Anionic initiators, **838–839**

Anionic surfactants, 1584–1585

Anise, **102**

Anise oil, 647t

Anisodesmic structure, **102**

Anisotropic medium, **102**

Annabergite, **102**

Annatto food colors, **102**

Annealing, **102–103**

Anode, **104**

Anodic oxidation, **104**

Anodize, **104**

Anorthic crystal system, crystallographic elements,
      455t

Anorthite, 607

Anorthosite, **104**

Ansaid, 93t

Ansamacrolide antibiotics, **107–109**

Ansamitocins, biological activity, 108t

Ansathiazins, biological activity, 108t

Ansatrienins, biological activity, 108t

Antacids, **104**

Anthocyanins, **104,** 733

Anthophyllite, **105**

Anthracene [CAS: 120–12–7], **105**

Anthracite, characteristics, 390t

Anthracite coal, ignition temperature in air, 425t

Anthraquinol, 105

Anthraquinone dyes, **516–518**

Anthraxolite, 105

Anthrocyanins, 734t

Anthrone, 105

Anthroquinone [CAS: 84–65–1], 105

Antibiotic resistance markers, 710

Antibiotics, **105–106**
    chiral, 1268

Antibiotics: aminoglycosides, **107**

Antibiotics: ansamacrolides, **107–109**

Antibiotics: cephalosporins, **112–114**

Antibiotics: chloramphenicol and analogues,
      114–116

Antibiotics: elfamycins, **116–117**

Antibiotics: glycopeptides (dalbaheptides), **117–119**

Antibiotics: ß-lactamase inhibitors, **109–111**

Antibiotics: ß-lactams, **111–112**

Antibiotics: lincosaminides, **119–120**

Antibiotics: macrolides, **120–122**

Antibiotics: monobactams, **122–123**

Antibiotics: nucleosides and nucleotides, **123–124**

Antibiotics: oligosaccharides, **124–125**

Antibiotics: pencillins and others, **125–127**

Antibiotics: peptides, **127**

Antibiotics: polyethers, **127–129**

Antibiotics: tetracyclines, **129–131**

Antibody, **131–132**

Anticaking agents, **132**

Anticancer chemotherapeutics, **354–359**

Anti-CD3 Mab (Monoclonal Antibody), annual
      sales, 222t

Anticoagulant drugs, chiral, 1268

Anticoagulants, **132–134**

Antidiuretic hormone, where produced, structure,
      and principal functions, 789t

Antidiuretic hormone (ADH), 1656
Antidote, **134**
Antifouling agents, **134**
Antifreeze agents, **134**
Antigen, **134**
Antihemmorhagic vitamin, 1706
Antihistamine, **134–135**
Antihypertensive drugs, chiral, 1267–1268
Antiinfective vitamin, 1698
Antiinflammatory agents, 91–93
Antiknock agents, **134**
Antimatter rockets, 1449
Antimers, 79
Antimetabolites, **135**
Antimicrobial agents (Foods), **135–138**
Antimicrobial drugs, chiral, 1268
Antimony
    abundance, 330t, 1132t
    covalent radius, 344t
    electronic structure, 336t
    interatomic distance, 342t
    ionic crystal radius, 341t
    nuclides (isotopes and isobars), 331t
    principal characteristics, 327t
Antimony [CAS: 7440–36–0], **138–139**
Antineoplastic drugs, chiral, 1269
Antioparticles, **140**
Antioxidants, **139–140**
Antiproton, **140–141**
Antipyretic, **141**
Antipyretics, 91–93
Antirachitic vitamin, 1703
Antisterility vitamin, 1705
Antitoxin, **141**
Antiviral agents, **141–145**
Anti-xerophthalmic vitamin, 1698
Antlerite, **145**
Antonoff rule, **145**
Apatite, **145**
    hardness, 2t, 755t
API gravity, 1243
Aplite, **145**
AP600 nuclear power plant (Westinghouse),
    1119–1122
Apomorphine corydine, 49t
Apophyllite, **145–146**
Aporphone alkaloids, 49t
Apostilb (asb), 1643t
Apparent molar quantity, **146**
Applied research, **146**
Aqua regia, **146**
Aquifer bioremediation, 205t
Aquifers, 746
    remediation of poisoned, 1729
Aquifer sparging, 205t
Arabic gum, 748t
L-α-Arabinofuranosidase, therapeutic enzyme, 574t
Arachidic acid, in major vegetable oils, 1672–1673t
Arachidic acid [CAS: 506–30–9], **146**
Aragonite, **146**
    hardness, 1008t
ARALL laminates, 71
ARamid ALuminum Laminates (ARALL), 71
Aramid fiber
    generic designation and definition, 621–622t
    physical properties, 1139t
Arboviruses, 1694
Arbutin, 734t
Archaeometry, **146**
Ardacin, producing organism, 118t
Arenaviruses, 1694
Areosol air pollution, 1325–1326

Argentite, **146**
Argillite, **146**
Arginase, metal chelate enzyme, 323t
L-Arginase, therapeutic enzyme, 574t
Arginine
    first isolation and isoelectric point, 78t
    manufacture, 76t
    structural classification, 79t
Argininemia, genetic influence, 716
Arginine-urea Cycle, 1651–1652
Arginine vasopressin, where produced, structure, and
    principal functions, 789t
Argininosuccinicaciduria, genetic influence, 716
Argon
    abundance, 330t
    electronic structure, 336t
    nuclides (isotopes and isobars), 331t
    percentage in air, 46t
    principal characteristics, 327t
    in Texas natural gas, 1054t
Argon [CAS: 7440–37–1], **146–147**
Aridicin, producing organism, 118t
Aridisols, 1497t
Armco iron, hardness, 755t
Aromatic alcohols, 46
Aromatic compound, 427
Aromatic compounds, **147**
Aromatic hydrocarbons, 1172–1173
Aromatics, 1169t
    gas separation, 39t
Aromatization, catalyst application, 304t
Arrhenius-Ostwald acid-base theory, 12
Arsenate minerals, 1012
Arsenic
    abundance, 330t
    covalent radius, 344t
    crystal structure, 455
    definitely established workplace carcinogen, 297t
    electronic structure, 336t
    in hazardous waste, 1711t
    interatomic distance, 342t
    ionic crystal radius, 341t
    principal characteristics, 327t
    trace mineral nutrient, 1005
    water quality indicator, 1726
Arsenical copper, 438t
Arsenic [CAS: 7440–38–2], **147–148**
Arsenopyrite, **149**
Arsine, physical properties, 795t
Arterenol, where produced, structure, and principal
    functions, 788t
Artesian wells, 746
Arvin, therapeutic enzyme, 574t
Aryl-alphyl ketones, 899
Arylamides, 74
Arylsulfatase, therapeutic enzyme, 574t
Asbestos, **149–151**
    definitely established workplace carcinogen, 297t
    hardness, 1008t
Asbolite, cobalt source, 410t
Ascomycetes, 1767
Ascorbic acid, 756
    constituent of human blood, 242t
Ascorbic acid [CAS: 50–81–7], **151–152**
Ash, **153**
Asparaginase, chemotherapeutic, 359t
L-Asparaginase, therapeutic enzyme, 574t
Asparagine
    first isolation and isoelectric point, 77t
    manufacture, 76t
    structural classification, 79t
Aspartase, metal chelate enzyme, 323t

Aspartic acid
    first isolation and isoelectric point, 78t
    manufacture, 76t
    structural classification, 79t
Aspartylglycosaminuria, genetic influence, 716
Aspergillus, 1769
Aspergillus flavus, 1769, 1770
Asphalt, in representative U.S. crude oils, 1242t
Asphalt-base greases, 743
Asphalt [CAS: 8052–42–4], **153**
Asphaltium, 153
Aspirin, 8, 92
Aspirin [CAS: 50–78–2], **153**
Associated compound, 427
Association (Chemical), **153–154**
Association of Official Analytical Chemists
    (AOAC), **154**
Astatine, 754
    abundance, 1132t
    electronic structure, 337t
    ionic crystal radius, 341t
    principal characteristics, 327t
Astatine [CAS: 7440–68–8], **154**
Aston, Francis William (1877–1945), 154
Aston whole number rule, **154**
Astrochemistry, **154–155**
Asulam, environmental health advisories, 771t
Asymmetric top, **155**
Asymmetry (Chemical), **155**
Atacamite, **155**
Athermal transformation, **155**
Atmolysis, **155**
Atmosphere, standard (atm), 1643t
Atmosphere-altitude relationship, 155–157
Atmosphere (Earth), **155–158**
Atmospheric ozone, 1193
Atom, **158–159,** 1209
Atomic-absorption photometer, 1294
Atomic disintegration, **159**
Atomic energy, **159**
Atomic energy levels, **159**
Atomic frequency, **159**
Atomic heat, **159,** 758
Atomic heat of formation, **159**
Atomic mass, **159–160**
Atomic mass unit, unified UNIFIED (u), 1643t
Atomic number, **160**
Atomic orbitals, 1164–1165
Atomic percent, **160**
Atomic plane, **160**
Atomic radius, 340
Atomic species, **160**
Atomic spectra, **160**
Atomic spectroscopy, **160**
Atomic structure, 334–339
Atomic weight, **159–160**
Atom switch, 1470
ATPase, metal chelate enzyme, 323t
Atrazine, environmental health advisories, 771t
Atrial natiuretic factor, 784
Atropine, 49t, 49–50
Attar of rose, 647t
Atto, 1643t
Attrition mills, **160–161**
Augite, **161**
Augmentin, year of discovery/market introduction,
    106t
Aureomycin, year of discovery, 130t
Auric, **161**
Aurin, pH range and color change of indicator, 825t
Aurodox, formula and producing organism, 116t
Aurous, **161**

Austenic stainless steels, 58
Austenite, **161**
Autocatalysis, **161**
Autodeposition, **161**
Autoionization, **161**
Autolysis, **161**
Automated instrumentation: clinical chemistry, **161–163**
Automated instrumentation: hematology, **163–164**
Automotive coatings, and acid rain, 10
Autoxidation, **165**
Autunite, **165**
Auxins, 1313–1314
AVLIS (atomic vapor laser isotope separation), 1650
Avogadro constant, **165**
Avogadro law, **165**
Avoparcin (LL-AV290 complex), producing organism, 118t
Awamycins, biological activity, 108t
Axerophthol, 1698
Axinite, **165**
Ayfivin, activity and producing organism, 128t
5-Azacitidine, antimetabolite, 355t
Azdimycin, formula and producing organism, 116t
Azelaic acid, monomer for nylon resins, 1333t
Azeotropes, effect on adsorption, 37
Azeotropic system, **165**
Azides, **165**
Azines, **165**
Azithromycin, commercial products, 121t
Azo and diazo compounds, **165–166**
Azobis[2-acetoxy-2-propane], commercial azo initiator, 841t
2,2′-Azobis[2-amidinopropane]dihydrochloride, commercial azo initiator, 841t
4,4′-Azobis[4-cyanovaleric acid], commercial azo initiator, 841t
1,1′-Azobis[cyclohexanecarbonitrile], commercial azo initiator, 841t
2,2′-Azobis[2,4-dimethyl]-pentanenitrile, commercial azo initiator, 841t
2,2′-Azobis[isobutyronitrile], commercial azo initiator, 841t
2,2′-Azobis[4-methoxy-2,4-dimethyl]-pentanenitrile, commercial azo initiator, 841t
2,2′-Azobis[2-methylbutyronitrile], commercial azo initiator, 841t
Azo compound, 1169t
Azo dyes, 512t, 513
    applications, 519
Aztreonam, year of discovery/market introduction, 106t
Azurite, **166**

B

Babbits, 923t
Babbitts, antimony content, 139t
Babingtonite, **167**
Bacillomycin F, activity and producing organism, 128t
Bacillus anthracis, gram-positive, 168t
Bacillus catarrhalis, ß-lactamase-producing bacterium, 110t
Bacillus fragilis, ß-lactamase-producing bacterium, 110t
Bacilysin, activity and producing organism, 128t
Bacitracin
    activity and producing organism, 128t
    year of discovery/market introduction, 106t
Backdonation, of electrons by metals, 28
Back-Goudsmit effect, **167,** 1773

Backscattering, **167**
Bacteria, **167–169**
    commercial microbial transformations used to produce steroid hormones, 1550
Bacteriophage lambda, 1693–1694
Bacteriophages, 1693–1694
Baddeleyite, 1779
    physical properties, 1428t
Baekeland, L. H. (1863–1944), **170**
Baeyer-Villiger oxidation, 1179
Baffle, **170**
Bagasse, **170**
    excess air for combustion, 426t
Bainite, **170**
Bakers' yeast, 1767, 1768
Balance, **170**
Ball mills, **170–171**
Balsamic vinegar, 1675
Band theory, of solids, 1519–1520
Banting, Sir Frederick (1891–1941), **171**
Bar (bar), 1643t
Barbiturates, **171**
Bardhan–Sengupta synthesis, **171**
Barff process, **171**
Barite, **171**
    abrasion pH, 1t
    hardness, 1008t
Barium, 49
    abundance, 5t, 330t, 1132t
    covalent radius, 344t
    electronic structure, 336t
    in hazardous waste, 1711t
    interatomic distance, 342t
    ionic crystal radius, 341t
    nuclides (isotopes and isobars), 331t
    principal characteristics, 327t
    residence time in sea water, 1133t
    standard electrode potential, 31t
Barium [CAS: 7440–39–3], **171–172**
Barium glass, 1163t
Barium hydride, physical properties, 795t
Barium superoxide, 1580
Barium titanate
    dielectric permittivity, 852t
    ultrasonic applications, 1637
Barkers, 1379–1380
Barley, susceptibility to iron deficiency, 874t
Barlow rule, **172**
Barn (b), 1643t
Barometric pressure, **172**
Barrel (bbl), 1643t
Barrier (Moisture), **172**
Barrier layer, **172**
Barrier polymers, **172–175**
Bartlett force, 1749
Barton, Derek H. R. (1918–1998), **175**
Baryons, **175,** 1396
Barytocalcite, **175**
Basal metabolism, **175–176**
Bases, 12–13
Basic dyes, 512t
    applications, 519
Basic oxide, **176**
Basic salt, **176,** 1456
Basophils, 242t
Bastnasite, **176**
Batteries, **176–180**
Batteries: lead-acid, **181–182**
Batteries: other, **182–183**
Batteries: primary cells, **183–185**
Batteries: secondary cells, **185–189**
Battery grid metal, 923t

Batts, insulation material, 857
Bauxite, 65, **189–190**
    physical properties, 1428t
Bayberry wax, 1746, 1747
Bayer process, **190**
Bearing alloy, antimony content, 139t
Bebeering, 49t
Becke test, **190**
Beckmann, Ernest (1853–1923), **190**
Beckmann method, **190**
Beckmann rearrangement, 1424
Becquerel, Antoine Henri (1852–1908), **190**
Becquerel effect, **190**
Beer-making, 1767
Beer's law, **191**
Beeswax, 1746
Beet sugar, 1564–1565
Beggiatoa, **191**
Beilstein, F. P. (1838–1906), **191**
Beilstein's text, **191**
Bel (B), 1643t
Bénard convection cells, **191**
Bending alloy, 923t
Benedict solution, **191**
Benefin, environmental health advisories, 771t
Bensulide, environmental health advisories, 771t
Bentazon, environmental health advisories, 771t
Bentonite, **191**
Benzaldehyde, from toluene, 1625
Benzaldehyde [CAS: 100–52–7], **191**
Benzaldehyde cyanohydrin, 465t
Benzedrine, 88
Benzene
    biodegradable, 207t
    combustion constants, 422t
    definitely established workplace carcinogen, 297t
    heating value, 686t
    as parent structure compound, 1090t
    from toluene, 1625
Benzene [CAS: 71–43–2], **191–192**
Benzenesulfonic acid, physical properties, 1568t
Benzidine, definitely established workplace carcinogen, 297t
Benzidine rearrangement, 1424
Benzil rearrangement, 1424
Benzine, **192**
Benzoic acid
    carboxylic acid, 294
    from toluene, 1624–1625
Benzoic acid [CAS: 65–85–0], **192**
Benzoin, **192–193**
Benzophenone, physical properties, 900t
1,2-Benzoquinone, physical properties, 1402t
1,4-Benzoquinone, 1401
    physical properties, 1402t
Benzothiazolylrhodanines, sensitizer for semiconductors, 535t
Benzothiazolylstyryl dye, sensitizer for semiconductors, 535t
Benzoyl chloride, physical properties, 366t
Benzyl acetate, physical properties, 585t
Benzyl alcohol [CAS: 100–51–6], **193**
Benzyl benzoate, dye carrier, 512t
Benzyl benzoate [CAS: 120–51–4], **193–194**
Benzyl chloride
    physical properties, 366t
    from toluene, 1625
Benzyl cinnamate, physical properties, 585t
Benzyldimethyloctadecylammonium chloride, 1398
Benzyne, **194**
Benzyne cycloaddition, 1178
Berberine, 49t

Berberine alkaloids, 49t
Berg, Paul (1926–), **194**
Bergamot oil, **194,** 647t
Bergius, Frederick (1884–1949), **194**
Bergius process, **194**
Bergius-Willstatter saccharification process, **194**
Bergmann azlactone peptide synthesis, **194**
Bergmann degradation, **194**
Berkelium, 23t
    binary compound properties, 25–26t
    electronic structure, 337t
    interatomic distance, 342t
    ionic crystal radius, 341t
    oxidation states, 24t
    physical properties, 24t
    principal characteristics, 327t
Berkelium [CAS: 7440–40–6], **194–195**
Berries, susceptibility to iron deficiency, 874t
Berthellot, Claude Louis (1748–1822), **195**
Berthelot, Pierre Eugene Marcellin (1827–1907), **195**
Berthelot equation, **195**
Beryl, **195,** 708
    dielectric permittivity, 852t
    hardness, 1008t
    heat effect, 708t
Beryllium, 48
    abundance, 330t, 1132t
    corrosion resistant alloys, 446t
    covalent radius, 344t
    crystal structure, 455
    definitely established workplace carcinogen, 297t
    electronic structure, 336t
    interatomic distance, 342t
    ionic crystal radius, 341t
    nuclides (isotopes and isobars), 331t
    principal characteristics, 327t
Beryllium [CAS: 7440–41–7], **195–197**
Beryllium-copper, 439t
Beryllium hydride, physical properties, 795t
Berzelius, J. J. (1779–1848), **197**
Bessel function, **197**
Best, Charles H. (1899–1978), **197**
Beta decay, **198,** 1406
Beta radiation, 1407
Beta-ray chemical analyzers, **198**
Bettendorf's reagent, **198**
Betterton-Kroll process, **198**
Bialaphos, activity and producing organism, 128t
Bicarbonate
    constituent of human blood, 242t
    constituent of human plasma or serum, 244t
    in 24 hour urine sample, 1657t
Bicozamycin, activity and producing organism, 128t
Bicyclic alkane, 1169t
Bicyclomycin, activity and producing organism, 128t
Bile, **198–199**
Bile acids, constituent of human blood, 242t
Bilirubin, target of medical diagnostic test, 975t
Bimetal thermometer, **200**
Binary, **200**
Binary alloy, 56
Binary compound, 428
Binders, use in food processing, 671t
Binding energy, **200–202**
Bioactive barrier, 205t
Bioactive zone, 205t
Bioaugmentation, 205t
Bioavailability. *See also* specific Mineral
    micronutrients
    pharmaceuticals, 1263
    vitamin A, 1699

vitamin B$_2$, 1700
vitamin B$_6$, 1701
vitamin B$_{12}$, 1702–1703
vitamin D, 1704
vitamin E, 1706
vitamin K, 1707
Biochemical individuality, **202**
Biochemical nomenclature, 1090–1091
Biochemical oxygen demand (BOD), **202**
Biochemistry, **202,** 711
Biodegradability, **202–203**
Bioelectrochemistry, **203**
Biofilm reactor, 205t
Biofiltration, 205t
Biofluffing, 205t
Biogenesis, **928**
Biogeochemistry, **203**
Bioinorganic chemistry, **203**
Bioleaching, 205t
Biological Abstracts, 769t
Biological energy transfer, **203**
Biological fluidized bed, 205t
Biology, **203**
Bioluminescence, **203–204,** 946t
Biomass, as fuel, 1717–1718
Biomimetic chemistry, **204**
Biopile, 205t
Biopolymers, **204**
Bioremediation, **205–210**
Biosensors, **210–211**
BIOSIS previews, 769t
Bioslurping, 205t
Biosolids, 1725
Biosparging, 205t
Biosterol, 1698
Biostimulation, 205t
Biotechnology, microgravity applications, 741
Biotechnology (bioprocess engineering), **211–233**
    timeline of major events, 212–216t
Biotin, **235–236**
    constituent of human blood, 242t
Biotite, **236**
    abrasion pH, 1t
    mineralogical properties, 993t
Biotransformation, 205t
Biotrickling filter, 205t
Bioventing, 205t
Biowall, 205t
Biphenyl, **236–237**
    dye carrier, 512t
N,N′-Bis-(3-aminopropyl)-ethylenediamine, physical
    properties, 483t
Bis-benzylisoquinoline alkaloids, 49t
Bis[chloromethyl]ether, definitely established
    workplace carcinogen, 297t
S-Bis(2,3-dibromopropyl ether), flame retardant, 640t
2,2-Bis[4,4-di(*t*-butylperoxy)cyclohexyl]propane,
    boiling point, 1234t
Bis(2-hydroxyethyl ether), flame retardant, 640t
1,3-Bis(hydroxymethyl)-5,5-dimethylhydantoin, 794
Bismaleimide polymers, **237**
Bismuth
    abundance, 330t, 1132t
    electronic structure, 337t
    interatomic distance, 342t
    ionic crystal radius, 341t
    nuclides (isotopes and isobars), 331t
    principal characteristics, 327t
Bismuth [CAS: 7440–69–9], **237–238**
Bismuthinite, **239**
Bismuth solder, 238t
Bit (b), 1643t

Bit per second (b/s), 1643t
Bitter almond oil, 647t
Bitter patterns, **239**
Bitumen, **239**
Bituminous coal, ignition temperature in air, 425t
Black body, **239**
Blackbody, 760
Black liquor, excess air for combustion, 426t
Blagden law, **239**
Blanket insulation, insulation material, 857
Blast-furnace gas, excess air for combustion, 426t
Bleaching agents, **239–240**
Blending, **1013–1016**
Blenoxane, DNA interactive agent, 357t
Bleomycins, activity and producing organism, 128t
Bleomycin sulfate USP, DNA interactive agent, 357t
Block glass, insulation material, 856t
Block insulation, insulation material, 857
Blood, **240–247**
Blood buffering, 8
Blood plasma, 242t, 244
Bloodstone, **248**
Bloom, **248**
Blue asbestos, **451**
Blue glow, **248**
Blue quark, 1397
BMY-25067, DNA alkylating/cross-linking agent, 356t
Board insulation, insulation material, 857
Bodying and bulking agents (Foods), **248–249**
Bog manganese, 1709
Bohrium, 24, 333
    electronic structure, 337t
    principal characteristics, 328t
Boiler water treatment, 1738–1746, **1738–1746**
Boiling curve, **249**
Boiling point, **249**
Boiling point constant, **249–250**
Boiling point elevation, **250**
Boiling water nuclear reactor, 1102–1106
Boise de rose oil, 647t
Boltzmann, Ludwig (1844–1906), **250**
Boltzmann's distribution law, **250–251**
Boltzmann's H-theorem, 903
Boltzmann transport equation, **251**
Bond (Chemical), **251**
Bond angle, 346
Bonded abrasives, 2
Bond energies, 343t, 346
Boracite, **251**
Borane, **251**
Borarsane, 254
Borate minerals, 1012
Borates, 253–254
Borax, **251**
    in water-soluble flame retardant formulation, 642t
Borazine, 254
Borehole bioreactor, 205t
Boric acid, in water-soluble flame retardant
    formulation, 642t
Boric acid [CAS: 1303–86–2], 253
Boriding, 763t
Born, Max (1882–1970), **252**
Bornite, **251–252**
Born-Oppenheimer approximation, **252**
Borohydrides, 796
Boron
    abundance, 330t, 1132t
    covalent radius, 344t
    electronic structure, 336t
    interatomic distance, 342t

Boron (*continued*)
  ionic crystal radius, 341t
  nuclides (isotopes and isobars), 331t
  plant micronutrient, 613t
  plant micronutrient deficiencies, soils of U.S., 616t
  principal characteristics, 327t
  trace mineral nutrient, 1002t, 1004
Boron carbide
  hardness, 755t
  plasma synthesis, 318t
Boron [CAS: 7440–42–8], **252–254**
Boron halides, 254
Boron nitride, 254
  dielectric permittivity, 852t
  plasma synthesis, 318t
Boron sulfide, 254
Boron trifluoride, as acid, 13
Borosilicate glass, 1163t
Borosilicate glasses, 725
Bosons, **254**
Boston Harbor Project, 1731
Bostonite, **254**
Bottom quark, 1397
Bottromycin, activity and producing organism, 128t
Botulism, 90
Bouguer and Lambert law, **254–255**
Boulangerite, **255**
Bournonite, **255**
Boyle, Robert (1627–1691), **255**
Boyle-Charles law, **255**
Boyle's law, **255**
Bragg, Sir William Lawrence (1890–1971), **256**
Bragg's curve, **255**
Bragg's law, **255,** 1759
Bragg spectrometer, **255–256**
Bragg's rule, **256**
Branched alkane, 1169t
Brass, representative alloys, 440t
Brayton cycle, 1511–1512
Brazing, **256–257**
Brewster angle, **257**
Brick clay, analysis, 316t
Bridged hydrocarbon ring systems, 1173
Brightness meter, 1294–1295
Brilliant cut, 707
Brinase, therapeutic enzyme, 574t
Brinell hardness scale, 755
Brinell hardness test, **257**
Britannia metal, antimony content, 139t
British thermal unit (Btu), 1643t
Brochantite, **257**
Bromacil, environmental health advisories, 771t
Bromelain, therapeutic enzyme, 574t
Brominated polystyrene, low molecular weight, flame retardant, 640t
Bromination, 1180
Bromine, 754
  absolute index, 1426t
  abundance, 330t, 1132t
  constituent of human plasma or serum, 244t
  covalent radius, 344t
  electronic structure, 336t
  ionic crystal radius, 341t
  nuclides (isotopes and isobars), 331t
  principal characteristics, 327t
  standard electrode potential, 31t
Bromine [CAS: 7726–95–6], **257–259**
Bromoform, 7
Bromopicrin, physical properties, 1081t
Bronze, 440t
  representative alloys, 440t

Brookite, **259**
Broom corn, 632t
Broom root, 632t
Brown, Herbert C. (1912–), **259–260**
Brown coal, **929–930**
Brownian motion, **260**
Brucella abortus, gram-negative, 168t
Brucella suis, gram-negative, 168t
Brucine, 49t
Brucite, **261**
Brush-heap structure, 706
Bryamycin, activity and producing organism, 128t
Buchner, Eduard (1860–1917), **261**
Buckminsterfullerene (Buckyballs), **261**
Buckyballs, **261**
Budding, viruses, 1693
Buffer (Chemical), **261**
Buffering capacity, of lakes, 9
Bufotenine, 754
Bulking agents (Foods), **248–249**
Bulk polymerization, 1342
Bunsen, Robert Wilhelm (1811–1899), **262**
Bupivacaine, 100t
Buprenex, 93t
Buprenorphine, 93t
Burette, **262**
Busulfan USP, DNA alkylating/cross-linking agent, 356t
cis-1,4-Butadiene, melting range, 1546t
trans-1,4-Butadiene, melting range, 1546t
1,2-Butadiene, physical properties, 263t
1,3-Butadiene, 262
  physical properties, 263t
Butadiene [CAS: 106–99–0], **262**
Butane
  heating value, 686t
  measurement by NDIR analysis, 837t
n-Butane
  combustion constants, 422t
  physical properties, 263t
  in Texas natural gas, 1054t
1,4-Butanedicarboxylic acid, 34
Butanedioic acid, physical properties, 490t
1,4-Butanediol, 46
  from allyl alcohol, 59
2,3-Butanedione, physical properties, 900t
Butanesulfonic acid, physical properties, 1568t
2-Butanone
  permeation in selected barrier polymers, 174t
  physical properties, 900t
cis-2-Butene, physical properties, 263t
n-Butene, combustion constants, 422t
trans-2-Butene, physical properties, 263t
1-Butene, physical properties, 263t, 1150t
3-buten-2-one, physical properties, 900t
Buthidazole, environmental health advisories, 771t
Butorphanol, 93t
2-(2-Butoxyethoxy)ethyl acetate, physical properties, 585t
2-Butoxyethyl acetate, physical properties, 585t
Butvar resins, 1676t
Butyl acetate, physical properties, 585t
sec-Butyl acetate, physical properties, 585t
t-Butyl acetate, physical properties, 585t
Butyl acrylate, physical properties, 585t
n-Butyl acrylate
  physical properties, 18t
  physical properties of polymers, 18t
sec-Butyl acrylate, physical properties of polymers, 18t
t-Butyl acrylate, physical properties, 18t
tert-Butyl acrylate, physical properties of polymers, 18t

n-Butyl alcohol, physical properties, 262t
sec-Butyl alcohol, physical properties, 262t
tert-Butyl alcohol, physical properties, 262t
Butyl alcohols, **262–263**
9-Butyl-10-anthrylpropionic acid, Langmuir-Blodgett films made from, 1018
Butylate, environmental health advisories, 771t
tert-Butyl-benzeneperoxy sulfonate, properties, 1239t
Butyl benzoate, dye carrier, 512t
Butyl butyrate, physical properties, 585t
tert-Butyl 2-carboxy-peroxy-benzoate, properties, 1239t
tert-Butyl *p*-chloro-benzeneperoxy sulfonate, properties, 1239t
tert-Butyl *tert*-cumyl peroxide, physical properties, 1232t
tert-Butyl *N,N*-dimethyl peroxycarbamate, properties, 1239t
tert-Butyldioxytriethylplumbane, physical properties, 1232t
1-Butylene, melting range, 1546t
Butylenes, **263–264**
n-Butyl ether, physical properties, 587t
Butyl ethyl ketone, physical properties, 900t
tert-Butyl 2-ethyl-peroxy-haxonoate, properties, 1239t
Butyl formate, physical properties, 585t
O,O-tert-Butyl *O*-hydrogen monoperoxymaleate, properties, 1239t
n-Butyl hydroperoxide, physical properties, 1231t
sec-Butyl hydroperoxide, physical properties, 1231t
tert-Butyl hydroperoxide, physical properties, 1231t
tert-Butyl 1-hydroxybutyl peroxide, boiling point, 1234t
tert-Butyl 1-hydroxyethyl peroxide, boiling point, 1234t
tert-Butyl 2-hydroxyethyl peroxide, physical properties, 1232t
tert-Butyl hydroxymethyl peroxide, boiling point, 1234t
O, O-tert-Butyl *O*-isopropyl monoperoxycarbonate, properties, 1239t
Butyl methacrylate, 17, 20
  physical properties, 988t
tert-Butyl *p*-methoxy-benzeneperoxy sulfonate, properties, 1239t
tert-Butyl *p*-methyl-benzeneperoxy sulfonate, properties, 1239t
tert-Butyl methyl peroxide, physical properties, 1232t
tert-Butyl peroxyacetate, properties, 1239t
tert-Butyl peroxybenzoate, properties, 1239t
tert-Butyl peroxydecanoate, properties, 1239t
tert-Butyl peroxypivalate, properties, 1239t
tert-Butyl peroxystearate, properties, 1239t
tert-Butylperoxytrimethylsilane, physical properties, 1232t
tert-Butylperoxytrimethylstannane, physical properties, 1232t
sec-Butyl perxanthate, 1755
Butyl sealants, 1463
Butyl stearate, physical properties, 585t
tert-Butyl triethylgermanium peroxide, physical properties, 1232t
n-Butyl vinyl ether, physical properties, 1690t
n-Butyl xanthic acid, 1755
Bytownite, 607

C
CAB Abstracts, 769t
Cable sheath, 923t

Cabochon, 706

Cadmium
  abundance, 330t, 1132t
  corrosion resistant alloys, 446t
  covalent radius, 344t
  definitely established workplace carcinogen, 297t
  electronic structure, 336t
  EPA pretreatment standard for aqueous
      discharge, 985t
  in hazardous waste, 1711t
  interatomic distance, 342t
  ionic crystal radius, 341t
  nuclides (isotopes and isobars), 331t
  principal characteristics, 327t
  standard electrode potential, 31t
  water quality indicator, 1726
Cadmium arsenide, 266
Cadmium carbonate, 266
Cadmium [CAS: 7440–43–9], **265–266**
Cadmium-copper, 439t
Cadmium electroplating, 557t
Cadmium nitride, 266
Cadmium oxide, 266
Cadmium selenide, colorant for glass, 725t
Cadmium sulfide, 266
  colorant for glass, 725t
Cadmium telluride, 266
Caffeine, 50
Cairngorm stone, **267**
Calaverite, **267**
Calcalkalic rocks, 49
Calcareous spoil, 1439t
Calcination, **267**
Calcined magnesia, physical properties, 1428t
Calcite, **267**
  abrasion pH, 1t
  hardness, 2t, 755t
Calcium, 49
  abundance, 5t, 330t, 1132t
  biologically active chelates, 323t
  constituent of human plasma or serum, 244t
  covalent radius, 344t
  electronic structure, 336t
  essential mineral nutrient, 1001–1002, 1002t
  hard water, 1726
  in 24 hour urine sample, 1657t
  interatomic distance, 342t
  ionic crystal radius, 341t
  nuclides (isotopes and isobars), 331t
  plant micronutrient, 613t
  principal characteristics, 327t
  residence time in sea water, 1133t
  standard electrode potential, 31t
  target of medical diagnostic test, 975t
  water quality indicator, 1726
Calcium (In Biological Systems), **270–273**
Calcium acetate, 6
Calcium acetate [CAS: 62–54–4], 268
Calcium alginate, 748t
Calcium aluminate [CAS: 065997–16-], 268
Calcium aluminate cement, physical properties,
    1428t
Calcium aluminosilicates, 268
Calcium arsenate [CAS: 7778–44–1], 268
Calcium arsenite [CAS: 52740–16–6], 268
Calcium borates, 268
Calcium bromide [CAS: 7789–41–5], 268
Calcium carbide [CAS: 75–20–7], 268
Calcium carbonate
  alkalizer, 14
  dielectric permittivity, 852t
Calcium carbonate [CAS: 1317–65–3], 268

Calcium carbonate [CAS: 1337–65–3], **269–270**
Calcium [CAS: 7440–70–2], **267–269**
Calcium chloride [CAS: 10043–52–4], 268, **270**
Calcium chromate [CAS: 13765–19–0], 268
Calcium citrate, acidulant, 14
Calcium cyanamide [CAS: 156–62–7], 268
Calcium fluoride [CAS: 7789–75–5], 268
Calcium formate [CAS: 544–17–2], 268
Calcium furoate, 268
Calcium gluconate, acidulant, 14
Calcium hydride, physical properties, 795t
Calcium hydride [CAS: 7789–78–8], 268
Calcium hypochlorite [CAS: 7778–54–3], 269
Calcium hypophosphite [CAS: 7789–79–9], 269
Calcium iodide [CAS: 10102–68–8], 269
Calcium lactate [CAS: 814–80–2], 269
Calcium malate [CAS: 17482–42–7], 269
Calcium nitrate [CAS: 10124–37–5], 269
Calcium oxalate, 269, 1185
Calcium oxide, alkalizer, 14
Calcium oxide [CAS: 1305–78–8], 269
Calcium ricinoleate, as release agent, 1435t
Calcium silicate, insulation material, 857
Calcium silicates, 269
Calcium soap greases, 743
Calcium sulfate, 749
Calcium sulfate [CAS: 10101–11–4], **273–274**
Calcium sulfate [CAS: 10101–41–4], 269
Calcium sulfide [CAS: 20548–54–3], 269
Calcium sulfite [CAS: 10257–55–3], 269
Calcium superoxide, 1580
Calcium tartrate, 269
Calco-uranite, 165
Caliche, **274**
Californium, 23t
  binary compound properties, 25–26t
  electronic structure, 337t
  interatomic distance, 342t
  ionic crystal radius, 341t
  oxidation states, 24t
  physical properties, 24t
  principal characteristics, 327t
Californium [CAS: 7440–71–3], **274**
Calorescence, **274**
Calorie (International Table) (cal), 1643t
Calorie (Thermochemical Calorie) (cal), 1643t
Calorimetry, **274–276**
Calorizing, **276**
Calvin, Melvin (1911–1997), **276**
Camphor, **276**
Camphor oil, 647t
Canaga oil, 647t
Cancrinite, **276**
Candela (cd), 1643t
Candelilla wax, 1747
Candida, 1769
Candida japonica, 1767
Candida tropicalis, 1767
Candida utilis, 176, 1768
Cane sugar, 1563–1564
Cannizzaro, Stanislao (1826–1910), **276**
Cannizzaro reaction, **276**
Cantala, 632t
Cape ruby, 705
Capillarity, **276–277**
Capillary, **277**
Capillary condensation, 36
Capped steels, 58
Capreomycins, activity and producing organism,
    128t
Capric acid [CAS: 334–48–5], **277**
Caproic acid

fatty acid in vegetable oil, 1671t
  in major vegetable oils, 1672–1673t
Caproic acid [CAS: 142–61–1], **277**
Caprolactam, monomer for nylon resins, 1333t
Caprolactam [CAS: 105–60–2], **277**
Caprylic acid
  fatty acid in vegetable oil, 1671t
  in major vegetable oils, 1672–1673t
Capryllactam, monomer for nylon resins, 1333t
Capsid, 1693
Capsomeres, 1693
Caranda wax, 1747
Carat, 707
Caraway oil, 647t
Carbamates, **277**
Carbamazepine, target of medical diagnostic test,
    975t
Carbamide, 1650
Carbamidine, 747
Carbamoyltransferases, 571t
Carbanion, **277**
Carbene, **277**
Carbene addition, 1177–1178
Carbenicillin, year of discovery/market introduction,
    106t
Carbides, **277**
Carbogels, 43
Carbohydrate abnormalities, genetic influence, 716
Carbohydrates, **278–284**
Carbohydrate water-soluble polymers, 1736–1737
Carbon
  abundance, 5t, 330t, 1132t
  amount in pure irons, 877t
  combustion constants, 422t
  combustion reactions, 423t
  covalent radius, 344t
  electronic structure, 336t
  heating value, 686t
  ignition temperature in air (coals), 425t
  ionic crystal radius, 341t
  nuclides (isotopes and isobars), 331t
  polymorphic forms, 1009t
  principal characteristics, 327t
Carbonado, **289**
Carbon aerogels, 42
Carbonate, **289**
Carbonate minerals, 1012
Carbon black, **289**
Carbon boride, 254
Carbon/carbon composites, 739
Carbon-carbon lyases, 571t
Carbon [CAS: 7440–44–0], **284–288**
Carbon dioxide
  absolute index, 1426t
  combustion constants, 422t
  constituent of human blood, 242t
  crystal structure, 455
  density, 474t
  diffusion and solubility coefficients in barrier
      polymers, 173t
  gas separation, 39t
  girbotol adsorption, 723
  as greenhouse gas, 743–744
  mass peaks and relative intensities, 973t
  measurement by NDIR analysis, 837t
  percentage in air, 46t
  permeability in barrier polymers, 172t
  removal from raw water, 1724
  target of medical diagnostic test, 975t
  in Texas natural gas, 1054t
Carbon dioxide [CAS: 124–38–9], **290–291**
Carbon dioxide lasers, wavelength range, 910t

Carbon disulfide [CAS: 75–15–0], **291–292**
Carbon fiber
  generic designation and definition, 622t
  physical properties, 1139t
Carbon group, **292–293**
Carbon-halide lyases, 571t
Carbonic anhydrase
  metal chelate enzyme, 323t
  source and function, 986t
Carbonitriding, **293**, 763t
Carbonium ion, **293**
Carbon molecular sieves, commercial gas adsorption
    processes, 39t
Carbon monoxide
  air pollution, 1328
  combustion constants, 422t
  combustion reactions, 423t
  gas separation, 39t
  heating value, 686t
  ignition temperature in air, 425t
  mass peaks and relative intensities, 973t
  measurement by NDIR analysis, 837t
Carbon monoxide [CAS: 630–08–0], **293**
Carbon-nitrogen lyases, 571t
Carbon steels, 57–58, 58, 775
Carbon suboxide, **293–294**, 898
Carbon-sulfur lyases, 571t
Carbon tetrabromide, physical properties, 286t
Carbon tetrachloride
  physical properties, 286t
  physical properties, 366t
Carbon tetrachloride [CAS: 56–23–5], **294**
Carbon tetrafluoride, physical properties, 286t
Carbon tetraiodide, physical properties, 286t
Carbonyl group, analysis, 97
Carbon-zinc cell, characteristics, 184t
Carbon-zinc cells, 183
  cell reactions, energy content, and
      manufacturers, 179t
Carboplatin USP, DNA alkylating/cross-linking
    agent, 356t
Carborane, **294**
Carbo-wax, 1748
Carboxylic acid, 1169t, 1173
Carboxylic acids, 12, **294–295**
Carboxylic ester hydrolases, 571t
Carboxyl-lyases, 571t
Carboxyltransferases, 571t
Carboxypeptidase, metal chelate enzyme, 323t
Carburetion, **295**
Carburetor, 295
Carburizing, **295–296**, 763t
Carcinogens, **296–300**
Cardiac glycosides, 735
Carmustine USP, DNA alkylating/cross-linking
    agent, 356t
Carnallite, **300**
Carnauba wax, 1747
  as release agent, 1435t
  use in food processing, 671t
Carnelian, **300**
  heat effect, 708t
Carnosinase, metal chelate enzyme, 323t
Carnot cycle, 3, **300**
Carnotite, **300**
Carnot theorems, **301**
Caroa, 632t
Carob-bean gum, 748t
Carotenoids, 1602
Carothers, Wallace H. (1896–1937), **301**
Carpaine, 49t
Carrageenan, 748t

Carrier, **301**
Carrier (Food Additive), **301**
Carrolite, cobalt source, 410t
Cartridge brass, 440t
Cascade, **301**
CAS (Chemical Abstracts Service Registry number),
    301
Case hardening, **301**
Casein, 1000
Casein [CAS: 9005–46–3], **301**
CASE polyurethanes, 1655–1656
Cassia oil, 647t
Cassiterite, **301**
Cast ferrous metals, 57–58
Casting, **301–303**
Cast iron, magnetic properties, 955t
Castor oil, **303**
  as release agent, 1435t
Castorseed oil, 1671
Catalase, metal chelate enzyme, 323t
Catalysis, **303–306**
Catalytically cracked gasoline, 1255
Catalytic converter, **307**
Catalytic cracker, 1255
Catalytic cracking, 304t, 304–305
Catalytic reformate, 1255
Catalytic reformer, 1255
Catalytic reforming, 304t, 305
Catenation compound, 428
Cathode, **307**
Cathodoluminescence, 946t
Cation, **307**
Cationic acids and bases, 13
Cationic initiators, **839–840**
CAT scanner, 1761–1762
Cat's-eye, **307**
Caustic (Chemical), **307**
Cavendish, Henry (1731–1810), **308**
Cave stalagmite/stalactite, **1536**
Cavitand, **308**
Cavitatation, **308–309**, 1525
Cavitation, 1638
Cech, R. Thomas (1947–), **309**
Cedar leaf oil, 647t
Cedarwood oil, 647t
Cefotaxime, year of discovery/market introduction,
    106t
Cefoxitin, year of discovery/market introduction,
    106t
Celestite, **309**
Celion, 622t
Cellophane
  dielectric permittivity, 852t
  as release agent, 1435t
Cellular elastomers, insulation material, 857
Cellular glass, insulation material, 856t, 857
Cellular polystyrene, insulation material, 857
Cellular polyurethane, insulation material, 857
Cellular rubber, insulation material, 856t
Cellulose, 1751
  dyeing, 512–522
  heating value, 686t
  use in food processing, 671t
  as water-soluble polymer, 1737
Cellulose acetate, 311
  physical properties of rigid foamed plastics, 664t
  production methods for cellular, 664t
Cellulose acetate butyrate, 311
  outdoor stability, 990t
Cellulose acetate propionate, 311
Cellulose acetates, ultrafiltration media application,
    1635

Cellulose [CAS: 9004–34–6], **309–310**
Cellulose cotton, dielectric permittivity, 852t
Cellulose diacetate, dyeing, 524
Cellulose ester fibers, **629–630**
Cellulose ester plastics (Organic), **310–312**
Cellulose esters, dyeing, 524
Cellulose triacetate, 311
  dielectric permittivity, 852t
  dyeing, 524
  fiber-dye property requirements, 521t
Cellulosic-acrylic fiber blends, dyeing, 524
Cellulosic adhesives, 33
Cellulosic fibers, insulation material, 856t, 857
Cellulosic-nylon fiber blends, dyeing, 524
Cellulosic-polyester fiber blends, dyeing, 524
Cellulosics (Applications), **312**
Cell wall, wood, 1751
Celsius, Ander, **312**
Celsius temperature scale, **312**, 1599
Cement, **312–314**
Cementite, **314**
Center for the History of Chemistry (CHOC),
    314–315
Center of mass, **971**
Centi, 1643t
Centigrade temperature scale, 3, 1599
Centrifugation, **315–316**
Centrifuge, 315–316
Cephalexin
  properties, 113t
  year of discovery/market introduction, 106t
Cephalosporin antibiotics, **112–114**
Cephalosporin C
  properties, 113t
  year of discovery/market introduction, 106t
Cephalosporin N, year of discovery/market
    introduction, 106t
Cephalosporin P, year of discovery/market
    introduction, 106t
Cephalothin
  properties, 113t
  year of discovery/market introduction, 106t
Cephaloxidine, year of discovery/market
    introduction, 106t
Cephamycin C, properties, 113t
Ceramics, **316–318**
Cerium
  abundance, 330t, 1132t
  electronic structure, 336t
  interatomic distance, 342t
  ionic crystal radius, 341t
  nuclides (isotopes and isobars), 331t
  principal characteristics, 327t
  properties, 1421t, 1422t
Cermets, 383
Cerubidine, DNA interactive agent, 357t
Ceruloplasmin, source and function, 986t
Cerussite, **319**
Cervilaxin, where produced, structure, and principal
    functions, 788t
Cesium, 48
  abundance, 330t, 1132t
  covalent radius, 344t
  electronic structure, 336t
  interatomic distance, 342t
  ionic crystal radius, 341t
  nuclides (isotopes and isobars), 331t
  principal characteristics, 327t
Cesium [CAS: 7440–46–2], **319–320**
Cesium chloride, crystal structure, 455
Cesium-137 gamma radiation, 703
Cesium hydride, physical properties, 795t

Cesium superoxide, 1580
Cgs system of units, 1645
Chabazite, **320**
   composition, 1034t
Chadwick, James (1891–1974), **320**
Chain-growth polymerization, 1346
Chalcanthite, **320**
Chalcedony, **320–321**, 893
   heat effect, 708t
Chalcocite, **321**
Chalcopyrite, **321**
Chalk, **321**
Chamosite, 871t
Chappius bands, 3
Characteristic equation, **321**
Charcoal, ignition temperature in air, 425t
Chardonnet, H. (1839–1924), **321**
Charge-mass ratio, **321**
Charles law, **321–322**
Charm, quarks, 1397
Charnockite, **322**
Chelates and chelation, **322–323**
Chelating agents, for boiler water treatment, 1744
Chemical Abstracts search, 769t
Chemical Abstracts Service Registry Number (CAS), **324**
Chemical activity, 29
Chemical affinity, **324**
Chemical analysis, 93–96
Chemical association, **153–154**
Chemical bond, **251**
Chemical composition, **324–326**
Chemical-composition variables, 1670
Chemical compounds, **427–429**
Chemical decomposition, **471**
Chemical degradation, **471–472**
Chemical elements, **326–347**
   abundance, 330t
   electronic structure, 336t
   principal characteristics, 327–328t
Chemical equation, **347–348**
Chemical equilibrium, **348–349**
Chemical formula, **349–350**
Chemical indicators, **825**
Chemical intermediate, **858**
Chemical Manufacturer's Association (CMA), **350**
Chemical microscopy, **999**
Chemical oxygen demand, 1192
Chemical potential, 29, **350–351**
Chemical reaction rate, **351–354**
Chemical synthesis, **1591–1592**
Chemiluminescence, 946t, 946–947
Chemisorption, 36, **354**
Chemosphere, 156
Chemotherapeutics, anticancer, **354–359**
ChemTrec, 350
Chert, **359**, 893
Chicken pox vaccine, 1659
Chiller, 760
China clay, **359**
China jute, 632t
Chinese insect wax, 1746
Chiral pharmaceuticals, **1265–1269**
Chiral separations, **359–364**
Chloral, physical properties, 366t
Chloral hydrate, physical properties, 366t
Chloramben, environmental health advisories, 771t
Chlorambucil USP, DNA alkylating/cross-linking agent, 356t
Chloramphenicol, year of discovery/market introduction, 106t
Chloramphenicol antibiotics, **114–116**

Chlorargyrite, **364**
Chlordane, definitely established workplace carcinogen, 297t
Chlorendic acid, flame retardant, 640t
Chloride
   constituent of human plasma or serum, 244t
   in 24 hour urine sample, 1657t
   target of medical diagnostic test, 975t
   water quality indicator, 1726
Chloride anodes, 982
Chloride (Biological aspects), **364–366**
Chlorimuron, ethyl, environmental health advisories, 771t
Chlorinated organics, **366–370**
Chlorinated rubber, alkyd resins with, 54
Chlorination, 1180, 1722
Chlorination (Water), **370–371**
Chlorination-amination, 1178
Chlorine, 754
   abundance, 5t, 330t, 1132t
   covalent radius, 344t
   density, 474t
   electronic structure, 336t
   essential mineral nutrient, 1002t, 1003
   ionic crystal radius, 341t
   nuclides (isotopes and isobars), 331t
   principal characteristics, 327t
   reagent for sulfonation, with sulfur dioxide, 1566t
   standard electrode potential, 31t
Chlorine [CAS: 7782-50-5], **371–372**
Chlorine pentafluoride, as propellant, 1447t
Chlorine trifluoride, as propellant, 1447t
Chlorinity, **373**
Chlorite, **373**
Chloritoid, **373**
Chloroacetic acid, physical properties, 366t
Chlorobenzene, physical properties, 366t
2-Chloro-1,4-benzoquinone, physical properties, 1402t
Chlorofluorocarbons (CFCs), **373**
   as greenhouse gas, 744
Chloroform, 7, 100t
   physical properties, 366t
2-Chloro-1-hydroperoxycyclohexanol, 1234t
3-Chloro-1,2-naphthoquinone, physical properties, 1402t
Chloroorienticins A, B, C, D, E (PA-45052), producing organism, 118t
m-Chloroperoxybenzoic acid, 1236t
o-Chlorophenol, physical properties, 366t
p-Chlorophenol, physical properties, 366t
Chlorophylls, **373–374**
Chloropicrin, physical properties, 1081t
Chloropolysporin A, producing organism, 118t
Chloropolysporin B, producing organism, 118t
Chloropolysporin C, producing organism, 118t
Chloroprene rubber, physical properties of flexible foamed plastics, 665t
Chloroprocaine, 100t
Chloropropham, environmental health advisories, 771t
Chlorosulfuric acid, reagent for sulfonation, 1566t
Chlorotrianisene USP, hormonal therapy, 358t
Chloroxuron, environmental health advisories, 771t
Chlorsulfuron, environmental health advisories, 771t
Chlortetracycline, year of discovery/market introduction, 106t
Chlortetracyline, year of discovery, 130t
Cholecalciferol, 1703
Cholera vaccine, 1660
Cholesterol, 733–734

constituent of human blood, 242t
   target of medical diagnostic test, 975t
Cholesterol [CAS: 57-88-5], **373–374**
Cholesterol esterase, properties as diagnostic enzyme, 976t
Cholesterol esters, constituent of human blood, 242t
Cholesterol oxidase, properties as diagnostic enzyme, 976t
Choline, **374–375**
   constituent of human blood, 242t
Choline acylase, metal chelate enzyme, 323t
Cholinesterase, **374–375**
m-Chlorophenol, physical properties, 366t
Chondrite meteorites, 599–600
Chondrodite, **375**
Chromate minerals, 1012
Chromates, 382–383
   definitely established workplace carcinogen, 297t
Chromatin, **375**
Chromatography, **375–380**
Chromel-alumel thermocouples, temperature range, 1605t
Chromel-constantan thermocouples, temperature range, 1605t
Chrome ore, physical properties, 1428t
Chromic acid, purification, 863t
Chromic oxide, colorant for glass, 725t
Chromite, **380**
Chromium
   abundance, 330t, 1132t
   biologically active chelates, 323t
   corrosion resistant alloys, 446t
   covalent radius, 344t
   electronic structure, 336t
   EPA pretreatment standard for aqueous discharge, 985t
   in hazardous waste, 1711t
   interatomic distance, 342t
   ionic crystal radius, 341t
   major sources of industrial minerals, 1011t
   nuclides (isotopes and isobars), 331t
   physical properties of high temperature, 774t
   principal characteristics, 327t
   standard electrode potential, 31t
   trace mineral nutrient, 1002t, 1005
   water quality indicator, 1726
Chromium ammonium sulfate, 86
Chromium [CAS: 7440-47-3], **380–383**
Chromium-copper, 439t
Chromium electroplating, 557t
Chromium plating, 381–382
Chromium(VI), definitely established workplace carcinogen, 297t
Chromizing, **383**, 763t
Chromogenic couplers, **383**
Chromophore, **383**
Chromophoric electrons, **383**
Chrysoberyl, **384**
Chrysocolla, **384**
Chrysotile, 149–150, **384**
   physical and chemical properties, 150t
Chymotrypsin, therapeutic enzyme, 574t
Cilofungin, activity and producing organism, 128t
Cinchona alkaloids, 49t
Cinchonine, 49t
Conical chemistry, **388**
Cinnabar, **384**
Cinnamon oil, 647t
Circular dichrograph, 1295
Circular mil (cmil), 1643t
Circulins, activity and producing organism, 128t
Cirtullinemia, genetic influence, 716

Cisplatin USP, DNA alkylating/cross-linking agent, 356t
Cis-trans isomerases, 571t
Cistron, 712
Citric acid, 282t
  acidulant, 14
  carboxylic acid, 294
  purification, 863t
Citric acid [CAS: 77–92–9], **384**
Citric acid cycle, 1596–1597
Citrine, **384**, 708
Citrobacter
  amphenicol susceptibility, 115t
  ß-lactamase-producing bacterium, 110t
Citronella oil, 647t
Citrus, susceptibility to iron deficiency, 874t
CIVEX fuel reprocessing system, 1647
Clad aluminum alloys, 67–68
Claims/U.S. patent abstracts, 769t
Clarifying agents, **384–385**
Clarithromycin, commercial products, 121t
Classifying (Process), **385–386**
Clathrate compound, 428
Clausius-Clapeyron relationship, 39
Clays, **385–388**
Clean Air Act, 1711t
  Acid Rain program, 9, 11–12
Clean Air Status and Trends Network (CASTNET), 11
Clean Water Act, 1711t
Cleerspan, 624t
Clindamycin, year of discovery/market introduction, 106t
Clinical chemistry, automated instrumentation applications, **161–163**
Clinoptilolite, composition, 1034t
Clinoril, 93t
Clinozoisite, **388**, 1779
Clopyralid, environmental health advisories, 771t
Closed-loop bioremediation, 205t
Clostridium botulinum, 90
  and food processing, 672
Clostridium butyricum, gram-positive, 168t
Clostridium septicum, gram-positive, 168t
Clostridium sordelli, gram-positive, 168t
Clostridium tetani, gram-positive, 168t
Clostridium welchii, gram-positive, 168t
Clothing wools, 1752
Cloud point, **388**
Clove oil, 647t
Club fungi, 1767
CNS depressants, chiral, 1268
Coacervation, **388–389**
Coagulation, 45, **389**
  in water treatment, 1723
Coagulation factor VIII, therapeutic enzyme, 574t
Coagulation (Hofmeister series), **389**
Coal, **389–400**
  excess air for combustion, 426t
  revegetation after mining, 1439–1440
Coal conversion (Clean Coal) processes, **400–406**
Coal-tar analgesics, 92–93
Coal tar-creosote, wood preservative, 1752
Coarse vacuum, 1661t
Coar tar and derivatives, **406–409**
Coated abrasives, 2
Coating agents (Foods), **409**
Coatings, **1196–1203**
CoA-transferases, 571t
Cobalamin, 1701–1703
Cobalt
  abundance, 330t, 1132t

biologically active chelates, 323t
constituent of human plasma or serum, 244t
corrosion resistant alloys, 446t
covalent radius, 344t
electronic structure, 336t
interatomic distance, 342t
ionic crystal radius, 341t
major sources of industrial minerals, 1011t
nuclides (isotopes and isobars), 331t
physical properties of high temperature, 774t
principal characteristics, 327t
trace mineral nutrient, 1002t, 1005
Cobalt [CAS: 7440–48–4], **409–412**
Cobaltiferous pyrite, cobalt source, 410t
Cobalt (In Biological Systems), **412**
Cobaltite, 412
Cobalt-nickel, thin-film applications, 958t
Cobalt-nickel-phosphorus, thin-film applications, 958t
Cobalt oxide, colorant for glass, 725t
Cobalt-phosphorus, thin-film applications, 958t
Cobalt superalloys, 776
Cocaine, 49t, 50
Coconut oil, 1671
  characteristics and properties, 1672–1673t
  higher alcohol manufacture from, 47
Codamine, 49t
Codeine, 49t, 50, 91, 1041
Coenzymes, **412–415**
Coir, 632t
Coke-oven gas, excess air for combustion, 426t
Colchicine, 50–51
Cold-cathode ionization vacuum gage, 1664
Cold fusion, 699
Cold-rolled steel, magnetic properties, 955t
Cold-worked steel, **415**
Colemanite, **415**
Colistin, activity and producing organism, 128t
Collagen, **415**
Collagen abnormalities, genetic influence, 716
Collagenase, therapeutic enzyme, 574t
Colloid chemistry, **415**
Colloid systems, **415–419**
Color, quarks, 1397
Color addition, 31–32
Colorants (Food), **419–420**
Color centers, **420**
Color-difference meter, 1295
Colored glass, 725–726
Color film, 1291–1292
Colorimeter, 1295
Colorimetry, **420**
Colour Index (CI), 512, 527
Columbite, **420**
Columbium
  abundance, 330t
  corrosion resistant alloys, 446t
Combined sewers, 1724
Combing wools, 1752
Combustion, in microgravity, 741
Combustion (Fuels), **420–427**
Common ion effect, **427**
Comparative biology, **427**
Compet, 623t
Complex compound, 428
Composite materials, **427**
Composting, 205t
Compound (Chemical), **427–429**
Comprehensive Environmental Response, Compensation, and Liability Act (Superfund), 1711t
Compression (gas), **429–430**

Compton effect, gamma rays from, 704
Computational chemistry, **430**
Computer-assisted molecular modeling (CAMM), 1028
Computer graphics, molecular modeling application, 1028
Computerized tomography (CT), 1761–1762
Compuional methods, for molecular modeling, 1028–1029
Conalbumin, source and function, 986t
Conant, James Bryant (1893–1978), **430**
Concentration (Chemical), **430**
Concentration (Process), **430–431**
Concrete, **431**
Condensate, **431**
Condensation, yeast participation, 1769t
Condensation compound, 428
Condensation curve, 249
Condensation polymerization, 1181
Condensing enzyme, 282t
Conduction heat transfer, 758–759
Congenital adrenal plasia, genetic influence, 716
Coniferin, 734t
Coniine, 49t
Conjugate acid, 12
Conjugate base, 12
Conjugated double bonds, 346
Connate water, **431–432**
Conservation laws and symmetry, **432–434**
Consolute liquids, **434**
Consolute temperature, **434**
Constantan, thermal conductivity, 758t
Constituent, **434**
Constructed wetland, 205t
Contact angle, 1584
Contact potential difference, **435**
Contractile proteins, **434–435**, 1374
Contractility, **434–435**
Convection, **435**
Convection heat transfer, 759
Conversion, **435**
Conversion coatings, **435–436**
Conversion ratio, **436**
Coolant, **436**
Coordination compounds, **437**
Coordination number, **437**
Coordination polyhedra, **437**
Copaline, **437**
Copalite, **437**
Copper
  abundance, 330t, 1132t
  amount in pure irons, 877t
  biologically active chelates, 323t
  constituent of human plasma or serum, 244t
  corrosion resistant alloys, 446t
  crystal structure, 455
  density, 474t
  electronic structure, 336t
  EPA pretreatment standard for aqueous discharge, 985t
  hardness of cold-rolled, 755t
  interatomic distance, 342t
  ionic crystal radius, 341t
  major sources of industrial minerals, 1011t
  modulus and rigidity, 538t
  nuclides (isotopes and isobars), 331t
  plant micronutrient, 613t
  plant micronutrient deficiencies, soils of U.S., 616t
  principal characteristics, 327t
  purification, 863t
  residence time in sea water, 1133t

standard electrode potential, 31t
thermal conductivity, 758t
trace mineral nutrient, 1002t, 1004
Copper acetate, 6
Copper alloys, 58
Copper-cadmium cells, 186t
Copper carbonate, 441
Copper [CAS: 7440-50-8], **437-441**
Copper chromate, wood preservative, 1752
Copper-constantan thermocouples, temperature
range, 1605t
Copper glance, 321
Copper (In Biological Systems), **442-443**
Copper-lead cells, 186t
Copper-zinc cells, rechargeable, 186t
Coquimbite, abrasion pH, 1t
Corbino effect, 752
Cordierite, **443**
Corey, Elias James (1928-), **443**
Cork, **443**
density, 474t
Corn, susceptibility to iron deficiency, 874t
Cornforth, John (1917-), **443**
Cornmint oil, 647t
Corn oil, 1671
Corrosion, **443-446**
Corrosion embrittlement, **446**
Corrosion resistant alloys, 446t
Corsite, **446**
Corticosteroids, 35
Corticosterone, 35
Cortisol, 35
commercial microbial transformations used to
produce, 1550t
where produced, structure, and principal
functions, 786t
Cortisone, 35, 93
Corttrell, Frederick G. (1877-1948), **447**
Corundum, **446,** 708
abrasive properties, 745t
hardness, 2t, 755t, 1008t
heat effect, 708t
irradiation, 709t
Corynebacterium diphtheriae, gram-positive, 168t
Cosmegen, DNA interactive agent, 357t
Cotton, **446,** 632t
fiber-dye property requirements, 521t
generic designation and definition, 624t
physical properties of staple, 626t
susceptibility to iron deficiency, 874t
Cottonseed oil, 1671
characteristics and properties, 1672-1673t
good source of protein, 1373
Coulomb (C), 1643t
Coulometer, **446-447**
Coupling, 1180
Coupling (Chamical), **447**
Covalent bonds, 345-346
Covalent compound, 428
Covellite, **447**
CPT-11, topoisomerase interactive, 357t
Cracking, 1179
Cram, Donald Jmaes, **449**
Creatine kinase, target of medical diagnostic test,
975t
Creatinine
in 24 hour urine sample, 1657t
target of medical diagnostic test, 975t
Creep (Metals), **449-450**
Creking process, **447-449**
Cresex process, 41t
Creslan, 621t

Creutzfeld-Jakob disease, 1696
Crick, Francis Harry Compton (1916-), **449-450**
Crin vegetal, 632t
Cristobalite, crystal structure, 455
Critical composition, **450**
Critical concentration, **450**
Critical density, **450**
Critical humidity, 793
Critical mass, **450,** 1102
Critical point, **450**
Critical solution temperature, **450**
Critical temperature, **450**
Critical volume, **450**
Crocidolite, 149-150
physical and chemical properties, 150t
Crocidolite (Blue asbestos), **451**
Crocoite, **451**
Crospovidones, 1680
Crotonylene, 55
Crucible, **451**
Crude oil pipelines, 1259-1260
Cryochemistry, **451**
Cryogenic materials, **451**
Cryogenic pumping, 1663
Cryogenics, **451-452**
Cryohydrate, **452**
Cryohydric point, **452**
Cryolite, **452-453**
Cryopump, **453**
Cryoscope, **453**
Cryoscopic constant, **453**
Cryotron, **453**
Cryptococcus neoformans, 1769
Cryptocrystalline, **453**
Cryptopine, 49t
Cryptopine alkaloids, 49t
Crystal, **453-459**
Crystal (Homometric Pairs), **461**
Crystal (Isomorphous), **461**
Crystal (Mixed), **461**
Crystal field theory, **459-461**
Crystal habit, **461,** 751
Crystal lattice energy, 921
Crystallization, **461-462**
Crystal oscillator, **461-462**
Crystal phases, **462**
Crystal pickup, **462**
CT metal, antimony content, 139t
Cubic boron nitride
abrasive properties, 745t
hardness, 2t
Cubic crystal system, crystallographic elements, 455t
Cubic zirconia, 708
Cummingtonite, **462**
Cumulative excitation, **462**
α-Cumyl hydroperoxide, physical properties, 1231t
tert-Cumyl peroxyacetate, properties, 1239t
tert-Cumyl peroxybenzoate, properties, 1239t
tert-Cumyl peroxypivalate, properties, 1239t
Cunico I, thin-film applications, 958t
Cunico II, thin-film applications, 958t
Cunife I, thin-film applications, 958t
Cunife II, thin-film applications, 958t
Cupric acetate, 441
Cupric acetoarsenite, 441
Cupric acid orthoarsenite, 441
Cupric chloride, 441
Cupric hydroxide, 441
Cupric oxide, 441
Cupric sulfate, 441
Cuprite, **462**
crystal structure, 455

Cupronickels, 440t
Cuprous acetylide, 8
Cuprous cyanide, 441
Cuprous iodide, 441
Cuprous oxide, 441
colorant for glass, 725t
Curie, Marie (1867-1934), **462**
Curie, Pierre (1859-1906), **462**
Curie (Ci), 1643t
Curie point, **462**
Curie temperature, **462**
Curie-Weiss law, **462**
Curium, 23t
binary compound properties, 25-26t
electronic structure, 337t
interatomic distance, 342t
ionic crystal radius, 341t
oxidation state, 24t
physical properties, 24t
principal characteristics, 327t
Curium [CAS: 7440-51-9], **462-463**
Curl, Robert F., Jr. (1933-), **463**
Cut rubber thread, 631
physical properties, 631t
Cyamelide, **464**
Cyanamide [CAS: 420-04-2], **464**
Cyanamides, **464**
Cyanazine, environmental health advisories, 771t
Cyanic acid and related compounds, **464**
Cyanin, 734t
Cyanoacetic acid, 964
Cyanoacrylates, monomer reactivity and suitable
initiator for, 838t
2-Cyanoacrylic ester polymers, 18-19
Cyanocobalamin, 1701
Cyanogen [CAS: 460-19-5], **465**
Cyanohydrins, 48, **465,** 899
Cyanuric acid, 464
Cybotaxis, **466**
Cycle (c), 1643t
Cycle oils, 1255
Cycle per second (Hz, c/s), 1643t
Cyclic compound, 428
Cyclic imines, 820-821
Cycloheptanone, physical properties, 900t
2,5-Cyclohexadiene-1,4-dione, 1401
Cyclohexane diisocyanate (CHDI), 1654
Cyclohexanol, 47, **466**
Cyclohexanone, **466**
physical properties, 900t
Cyclohexanone cyanohydrin, 465t
Cyclohexasilanyl, 1477
Cyclohexyl acrylate, physical properties of polymers,
18t
Cyclohexyl hydroperoxide, physical properties,
1231t
Cyclo-ligases, 571t
Cycloolefins, polymerization, 1149
Cyclopentane, in Texas natural gas, 1054t
Cyclopentanone, physical properties, 900t
Cyclopentanoperhydrophenanthrene, 35
Cyclophosphamide USP, DNA
alkylating/cross-linking agent, 356t
Cyclopropane, 100t
Cyclosilicate minerals, 1013
Cyclosporin A, activity and producing organism,
128t
Cyclosporine, activity and producing organism, 128t
Cyclotron, **466**
Cymex process, 41t
Cysteine
first isolation and isoelectric point, 77t

Cysteine (*continued*)
manufacture, 76t
structural classification, 79t
Cystine
first isolation and isoelectric point, 77t
manufacture, 76t
production, 80
in selected foods, 75t
structural classification, 79t
Cystinuria, genetic influence, 716
Cytarabine USP, antimetabolite, 355t
Cytochrome *c*, 466–467
Cytochrome oxidase, 466–467
Cytochromes, **466–467**
metal chelate enzyme, 323t
Cytogenetics, 711
Cytology, 711
Cytosar, antimetabolite, 355t
Cytotoxic chemicals, **467**
Cytoxan, DNA alkylating/cross-linking agent, 356t

D

Dacarbazine USP, DNA alkylating/cross-linking
agent, 356t
Dacite, **469**
Dacron, 623t
Dactinomycin, activity and producing organism, 128t
Dactinomycin USP, DNA interactive agent, 357t
Daguerreotype, 1289
Dairy mold, 1770
Dalapon, environmental health advisories, 771t
Dalbaheptide antibiotics (glycopeptides), **117–119**
Dalton, John (1766–1844), **469**
Dalton law, **469**
Danburite, **469**
Daptomycin, activity and producing organism, 128t
Darcy (D), 1643t
Darcy's law, **469**
Darmstadtium, 24, 334
electronic structure, 337t
principal characteristics, 328t
Darvon, 93t
Datolite, **469**
Daturine, 49
Daughter element, **470**
Daunorubicin hydrochloride USP, DNA interactive
agent, 357t
Davy, Sir Humphrey (1778–1829), **469**
Day (d), 1643t
DCPA, environmental health advisories, 771t
2,4-D (2,4-Dichlorophenoxyacetic acid), in
hazardous waste, 1711t
Deacon process, **469**
Dealkylation, catalyst application, 305
Deaminase, therapeutic enzyme, 574t
Deamination, yeast participation, 1769t
Deaeration, **469**
De Broglie, Louis-Victor (1892–1987), **469–470**
De Broglie wavelength, **470**
Debye, Peter J. W. (1884–1966), **470**
Debye-Falkenhagen effect, **470**
Debye-Hückel limiting law, **470**
Debye-Sears effect, **470**
Debye theory of specific heat, **470**
Deca, 1643t
Decabromo-diphenyl oxide, flame retardant, 640t
Decanedioic acid, physical properties, 490t
Decaplanin, producing organism, 118t
Decarboxylation, yeast participation, 1769t
Decarburization, **470**
Decay product, **470**

1-Decene, physical properties, 1150t
Deci, 1643t
Declomycin, year of discovery, 130t
Decolorizing agents, **470–471**
Decomposition (Chemical), **471**
Deep inelastic scattering, 333
Deepwater oil production, 1252
Defoaming agents, **471**
Degasification, **471**
Degassed water, 1722
Degeneracy, **471**
Degradation (Chemical), **471–472**
Degree Celsius (C), 1643t
Degree Fahrenheit (F), 1643t
Degree Rankine (R), 1643t
Degrees of freedom, **679**
Dehumidification, **472**
Dehydration (Chemical), **472**
Dehydrochlorination, 1180
Dehydrocyclization, catalyst application, 304t
Dehydrogenation, **472**, 1180
catalyst application, 304t, 305
Dehydroisoandrosterone, 98
Deionizing, **472**
Deisenhofer, Johann (1943–), **472**
Deliquescence, **472–473**
Delphinin, 734t
Demal solution, **473**
Demantoid, 705
Demeclocycline, year of discovery, 130t
Demerol, 93t
N-Demethylefrotomycin, formula and producing
organism, 116t
N-Demethyl-vancomycin, producing organism, 118t
Demineralization, **473**
Demulsification, **473**
Dendrite, **473**
Dengue vaccine, 1660
Dengue virus, 1694
Dense glass, 1163t
Densitometer, 1295
Density, **473–474**
Deodorizing, oils, 1671
Deoxidized copper, 438t
Deoxidizing agent, **474**
Deoxyribonuclease
metal chelate enzyme, 323t
therapeutic enzyme, 574t
Deoxyribonucleic acid (DNA), **474**, 712–713
structure, 1125
Dephlegmation, **475**
Depo-provera, hormonal therapy, 358t
Deptomycin, activity and producing organism, 128t
Derivative (Chemical), **475**
Derived units, 1642t
Derwent world patents index, 769t
Desalination, **475–478**, 1735
Desflurane, 100t
Desorption, **478**
Desoxycorticosterone, 35
Destruxin B, activity and producing organism, 128t
Detergents, **478–481**
Deuterium, **481–482**
Deuteron, **482**
Devitrification, **482**
Dew point, **482**
Dexamethasone, 93
Dextrins, use in food processing, 671t
Dextrorotatory, 78, 1543
Dextrose, purification, 863t
Dezincification, **482**
Dhurrin, 734t

Diabetes, possible viral roles, 1696
Diabetes mellitus, genetic influence, 716
Diacetone alcohol, 7
physical properties, 900t
Diacetoneamine, 7
Diacetyl
as key odor/tase compound, 649
physical properties, 900t
Diacetylmorphine, 51
Diacetyl peroxide, properties, 1237t
Diacyl peroxides, organic peroxide initiators, 841t
Diafiltration, 1636
Diagenesis, **482**
Di-*tert*-alkyl peroxides, organic peroxide initiators,
841t
Dialkyl peroxydicarbonates, organic peroxide
initiators, 841t
Di(*tert*-alkylperoxy)ketals, organic peroxide
initiators, 841t
Diallage, **482**
Diallyl phthalate, dye carrier, 512t
Dialysis, **482**
Diamagnetism, **482**
Diamines and higher amines, aliphatic, **482–484**
1,3-Diaminopropane, physical properties, 483t
Diammonium phosphate, 87–88
in water-soluble flame retardant formulation,
642t
Diammonium sodium cobaltinitrite, 86
Diamond, **484–486**, 708, 1009t
abrasive properties, 745t
absolute index, 1426t
crystal structure, 455
hardness, 2t, 755t
heat effect, 708t
irradiation, 709t
Diamond anvil high pressure cell, **486–488**
Diamond Pyramid hardness, 756
Di-*tert*-amyl peroxide, physical properties, 1232t
2,2-Di(*t*-amylperoxy)propane, boiling point, 1234t
Dianthrol, 105
Dianthrone, 105
Diaphragm cell, **489**
Diaryl ketones, 899
Diaspore, **489**
Diastovaricins, biological activity, 108t
Diatomaceous earth, hardness, 1008t
Diatomaceous silica, insulation material, 856t, 857
Diatomic gases, 334
Diatomite, **489–490**
abrasive properties, 745t
Diazo compound, 1169t
Diazo compounds, **165–166**
Dibasic acids, 13
Di(benzenesulfonyl) peroxide, properties, 1237t
Di(benzocyclobutene-4-carbonyl) peroxide,
properties, 1237t
Dibenzoyl peroxide, properties, 1237t
Dibenzyl peroxydicarbonate, properties, 1237t
Dibromo-ethyldibromocyclohexane, flame retardant,
640t
Di(4-*tert*-butylcyclohexyl) peroxydicarbonate,
properties, 1237t
Di-*tert*-butyl diperoxyadipate, properties, 1239t
Di-*tert*-butyl diperoxycarbonate, properties,
1239t
Di-*tert*-butyl diperoxyoxalate, properties,
1239t
Di-*tert*-butyl diperoxyphthalate, properties,
1239t
Di-*tert*-butyl peroxide, physical properties,
1232t

2,2-Di(t-butylperoxy)butane, boiling point, 1234t
1,1-Di(t-butylperoxy)cyclohexane, boiling point, 1234t
Dibutyl peroxydicarbonate, properties, 1237t
Di-sec-butyl peroxydicarbonate, properties, 1237t
1,4-Di(2-tert-butylperoxyisopropyl)benzene, physical properties, 1232t
2,2-Di(t-butylperoxy)propane, boiling point, 1234t
Dibutyl phthalate, physical properties, 585t
Dicalcium phosphate dihydrate, properties of leavening acid, 925t
Dicamba, environmental health advisories, 771t
Di(2-carboxybenzoyl) peroxide, properties, 1237t
Di(4-carboxybutyryl) peroxide, properties, 1237t
Dicarboxylic acids, **490–491**
Di(3-carboxypropionyl) peroxide, properties, 1237t
Dichlobenil, environmental health advisories, 771t
Di(chloroacetyl) peroxide, properties, 1237t
m-Dichlorobenzene, physical properties, 366t
o-Dichlorobenzene
    dye carrier, 512t
    physical properties, 366t
p-Dichlorobenzene, physical properties, 366t
2,5-Dichloro-1,4-benzoquinone, physical properties, 1402t
Di(4-chlorobenzoyl) peroxide, properties, 1237t
1,3-Dichloro-cis-2-butene, 1749
2,3-Dichloro-5,6-dicyano-1,4-benzoquinone, physical properties, 1402t
2,3-Dichloro-1,4-naphthoquinone, physical properties, 1402t
Dichromates, 382–383
Dicinnamoyl peroxide, properties, 1237t
Diclofenac, 93t
Dictamine, 49t
Dicumyl peroxide, physical properties, 1232t
Di(cyclohexylcarbonyl) peroxide, properties, 1237t
Dicyclohexyl peroxydicarbonate, properties, 1237t
Didecanoyl peroxide, properties, 1237t
Didodecanoyl peroxide, properties, 1237t
Die, **491**
Die casting, **491**
Dieldrin, definitely established workplace carcinogen, 297t
Dielectric heating, **491–492**
Dielectric theory, **492–493**
Diels, Otto P. H. (1876–1954), **493**
Diels-Alder reaction, 48, **493**, 1177
Dienes, monomer reactivity and suitable initiator for, 838t
Diesel fuel, in representative U.S. crude oils, 1242t
Dietary fiber, **617**
Diethanolamine, 587
Diethoxyaluminum tert-cumyl peroxide, physical properties, 1232t
Diethyl adipate, physical properties, 585t
Diethyl carbonate, physical properties, 585t
3,6-Diethyl-3,6-dimethyl-1,2,4,5-tetraoxane, 1234t
Diethylene glycol, physical properties, 590t
Diethyleneglycol-bis(allylcarbonate), 60
Diethylene glycol monostearate, as release agent, 1435t
Diethylenetriamine, physical properties, 483t
Diethylenetriamine adduct, commercial polyether polyol, 1654t
Di(2-ethylhexyl) adipate, physical properties, 585t
Di(2-ethylhexyl) peroxydicarbonate, properties, 1237t
Di(2-ethylhexyl) phthalate, physical properties, 585t
Diethyl ketone, physical properties, 900t
Diethyl malonate [CAS: 105-53-3], 963

Diethylnitrosamine, 1082
Diethyl peroxide, physical properties, 1232t
Diethyl peroxydicarbonate, properties, 1237t
Diethyl phthalate
    dye carrier, 512t
    physical properties, 585t
Diethylstilbestrol diphosphate USP, hormonal therapy, 358t
Diffraction, **493**
Diffraction grating, **494**
Diffusion, **494–495**
Diffusion current, **495**
Diffusion potential, **495**
Diffusion pumps, 1663
Diflunisal, 93t
Di(2-furanylcarbonyl) peroxide, properties, 1237t
Digester, 205t
Digester (Process), **495**
Digesters, for wood pulp, 1380–1381
Digestion, **495**
Digipurpurin, 735
Digitalin, 734t
Digitalis, **495**, 735
Digital subtraction angiography, 1763
Digitonin, 734t, 735
Digitoxin, 734t
Diheptanoyl peroxide, properties, 1237t
Dihexadecanoyl peroxide, properties, 1237t
Dihexadecyl peroxydicarbonate, properties, 1237t
Dihydric alcohols, 46
9,10-Dihydro-9,10-epidi-oxyanthracene, physical properties, 1232t
Dihydrofollicular hormone, where produced, structure, and principal functions, 787t
Dihydrofolliculin, where produced, structure, and principal functions, 787t
Di(hydrogenated tallow)alkyldimethylammonium chloride, 1398
Dihydromocimycin, formula and producing organism, 116t
Di(2-hydroperoxy-2-butyl) peroxide, 1234t
1,1-Dihydroperoxycyclododecane, 1234t
Di(1-hydroperoxycyclohexyl) peroxide, 1234t
Dihydrotheelin, where produced, structure, and principal functions, 787t
Dihydroxyacetone phosphate, 283
Di(1-hydroxycyclohexyl) peroxide, 1234t
3,5-Dihydroxy-3,5-dimethyl-1,2-dioxolane, 1234t
Di(1-hydroxyethyl) peroxide, 1233t
Di(hydroxymethyl) peroxide, 1233t
Di(1-hydroxynonyl) peroxide, 1233t
Di(1-hydroxyoctyl) peroxide, 1233t
Di(1-hydroxypentyl) peroxide, 1233t
Dihydroxy-phenylalanine, manufacture, 76t
Di(isobornyl) peroxydicarbonate, properties, 1237t
Diisobutyl ketone, physical properties, 900t
Diisobutyryl peroxide, properties, 1237t
p-Diisopropylbenzene dihydroperoxide, physical properties, 1231t
p-Diisopropylbenzene mono-hydropero, physical properties, 1231t xide
Diisopropyl ketone, physical properties, 900t
Diisopropyl malonate, 963
Diisopropyl peroxide, physical properties, 1232t
Diisopropyl peroxydicarbonate, properties, 1237t
Diketene, 899
Dilatancy, **495**
Dilation number, **495**
Dilatometer, **495**
Diluent, **495**
Dimer acids, **495–496**
    monomer for nylon resins, 1333t

Di(methanesulfonyl) peroxide, properties, 1237t
Dimethyl adipate, physical properties, 585t
Dimethylallylamine, 60
Dimethylaminoethyl methacrylate, copolymers with PVP, 1682
2-Dimethylaminoethyl methacrylate, physical properties, 988t
Dimethylaminopropylamine, physical properties, 483t
Dimethyl-2,2'-azobis-[2-methylpropionate], commercial azo initiator, 841t
Dimethylbenzenes, 1763, 1764–1765
2,5-Dimethyl-1,4-benzoquinone, physical properties, 1402t
Di(2-methylbenzoyl) peroxide, properties, 1237t
cis-1,4-(2,3-Dimethyl butadiene), melting range, 1546t
trans-1,4-(2,3-Dimethyl butadiene), melting range, 1546t
3,3-Dimethyl-2-butanone, physical properties, 900t
Dimethyl carbonate, physical properties, 585t
2,5-Dimethyl-2,5-di(benzoyl-peroxy)-hexane, properties, 1239t
2,5-Dimethyl-2,5-di(tert-butyl-peroxy) hexane, physical properties, 1232t
2,5-Dimethyl-2,5-di(tert-butylperoxy)-3-hexyne, physical properties, 1232t
2,5-Dimethyl-2,5-dihydroperoxyhexane, physical properties, 1231t
2,5-Dimethyl-2,5-dihydroperoxy-3-hexyne, physical properties, 1231t
1,1-Dimethylethylnitrate, 1082
1,1-Dimethylethyl nitrite, 1082
2,6-Dimethyl-4-heptanone, physical properties, 900t
Dimethyl isophthalate, physical properties, 585t
Dimethylketene, 898
Dimethyl ketone, 7
Dimethyl maleate, physical properties, 585t
Dimethyl malonate [CAS: 108-59-8], 963
Dimethyl p-methoxybenzylidene malonate, 1641
Dimethyl oxalate, physical properties, 585t
2,4-Dimethyl-3-pentanone, physical properties, 900t
4,4-Dimethyl-1-pentene, melting range, 1546t
Dimethyl peroxide, physical properties, 1232t
Dimethyl phthalate
    dye carrier, 512t
    physical properties, 585t
2,2-Dimethyl-1-propanol, physical properties, 89t
1,1-Dimethylpropynyl hydroperoxide, physical properties, 1231t
Dimethyl sulfoxide, 1569–1570
Dimethyl terephthalate
    dye carrier, 512t
    physical properties, 585t
1,5-Dimethyl-6,7,8-trioxabicyclo[3.2.1]octane, boiling point, 1235t
3,3-Dimethyl-1,2,4-trioxolane, boiling point, 1235t
3,5-Dimethyl-1,2,4-trioxolane, boiling point, 1235t
Di(2-naphthaleny carbonyl) peroxide, properties, 1237t
Dinicotinoyl peroxide, properties, 1237t
Dinitramine, environmental health advisories, 771t
1,3-Dinitrobenzene, 1082
    physical properties, 1081t
Di(4-nitrobenzoyl) peroxide, properties, 1237t
2,4-Dinitrotoluene, physical properties, 1081t
Dinonanoyl peroxide, properties, 1237t
Dinoseb, environmental health advisories, 771t
Dioctadecanoyl peroxide, properties, 1237t
Dioctanoyl peroxide, properties, 1237t
Diode lasers, wavelength range, 910t
Dionin, 1041

Diopside, **496–497**
    abrasion pH, 1t
Dioptase, **497**
Diorite, **497**
1,2-Dioxane, physical properties, 1232t
Dioxin, 93, **497**
Dioxybis[triethylgermane], physical properties, 1232t
Dioxybis[triethylstannane], physical properties, 1232t
Dioxybis[trimethylsilane], physical properties, 1232t
Dipentanoyl peroxide, properties, 1237t
Dipeptide hydrolases, 571t
Diperoxydecanedioic acid, 1236t
Diperoxyhexanedioic acid, 1236t
Diperoxynonanedioic acid, 1236t
Diphenamide, environmental health advisories, 771t
Di(2-phenoxyethyl) peroxydicarbonate, properties, 1237t
Di(phenylacetyl) peroxide, properties, 1237t
1,3-Diphenylguanidine, 747t
Diphenylketene, 898
Diphenyl ketone, physical properties, 900t
Diphenylmethane 4,4'-diisocyanate, 464
Diphenylmethane dyes, 1630
Diphenylnitrosamine, 1082
    physical properties, 1082t
Diphenyl oxide; dye carrier, 512t
3,5-Diphenyl-1,2,4-trioxolane, boiling point, 1235t
Diphtheria, tetanus, and pertussis (DPT) vaccine, 1659
Diplococcus pneumoniae, gram-positive, 168t
Dipole moment, **497**
Dipotassium sodium cobaltnitrite, 1361
Dipropionyl peroxide, properties, 1237t
Dipropyl disulfide, permeation in selected barrier polymers, 174t
Di-*n*-propyl peroxydicarbonate, properties, 1237t
Diquatop, environmental health advisories, 771t
Direct dyes, 512t
    applications, 519
    dyeing process, 521
Disaccharides, 279t
Disilanoxy, 1477
Disilanyl, 1477
Disilanylamino, 1477
Disilanylene, 1477
Disilanylthio, 1477
Disilazanoxy, 1477
Disilazanyl, 1477
Disilazanylamino, 1477
Disiloxanoxy, 1477
Disiloxanyl, 1477
Disiloxanylamino, 1477
Disiloxanylthio, 1477
Disilthianoxy, 1477
Disilthianyl, 1477
Disilthianylthio, 1477
Disinfectant, **497**
Dislocation, **497**
Disodium hydrogen phosphite, 1492
Dispersants, **497–498**
Disperse dyes, 512t, 517
    applications, 519
    dyeing process, 523, 524
Dispersion, **498–499, 1303**
Displacement desorption, 38t
Displacement series, 31
Dissociation, **499–501**
Dissolved oxygen, water quality indicator, 1726
Dissolved-oxygen deficit, water quality indicator, 1726
Dissolved oxygen (DO), **497**
Distamycin A, activity and producing organism, 128t

Distillation, **501–504**
    in petroleum refining, 1255–1257
Disulfuric acid, 13
Diterpenoids, 1601
Ditetradecyl peroxydicarbonate, properties, 1237t
Di(2-thienylcarbonyl) peroxide, properties, 1237t
Di(*p*-toluenesulfonyl) peroxide, properties, 1237t
Ditran, 754
Di(2,2,2-trichloro-1-hydroxyethyl) peroxide, 1233t
Di(*cis*-3,3,5-trimethylcyclohexyl) peroxydicarbonate, properties, 1237t
Di(3,5,5-trimethylhexanoyl) peroxide, properties, 1237t
Diuretic agents, **504–506**
Diuron, environmental health advisories, 771t
cyclo-Di-*p*-xylylene (DPX), 1763
Di-*p*-xylylene (DPXN), 1764
DNA replication, 716
Dnase, annual sales, 222t
DNA viruses, 1693, 1694
Docetaxol, tubulin interactive, 358t
Dodecanedioic acid
    monomer for nylon resins, 1333t
    physical properties, 490t
1-Dodecene, physical properties, 1150t
Dodecyl acrylate, physical properties of polymers, 18t
Dodecyl stearate, physical properties, 585t
Dog, body weight-metabolic rate relationship, 176t
Dolerite, **506**
Dolobid, 93t
Dolomite, **506**
    abrasion pH, 1t
    hardness, 1008t
    physical properties, 1428t
Domain structure, **506–507**
Double-beam grating UV spectrophotometer, 1641.
Double-beta decay, 1407
Double bonds, 346
Double salt, **507,** 1456
Doublet, **507**
Down quark, 1397
Doxorubicin USP, DNA interactive agent, 357t
Doxycycline, year of discovery, 130t
DPNH-cytochrome c reductase, source and function, 986t
Dromoran, 93t
Drug discovery, 1270
Drug elimination, 1270–1271
Drug metabolism, 1270
Drug receptors, 1270, 1271–1272
Drugs. *See* Pharmaceuticals
    target of medical diagnostic test, 975t
Drying (Process), **508–511**
Drying oils, **507–508**
Dubnium, 24, 333
    electronic structure, 337t
    principal characteristics, 328t
Ductile iron, 57
Dulong and Petit law of specific heats, **511**
Dumortierite, **511**
Duranickel, 58
Duranickel 301, 1071t
Du Vigneaud, Vincent (1901–1978), **511**
Dye and dye intermediates, **512–516**
Dye carriers, **511–512,** 512t
Dyeing process, 520–521
Dye lasers, wavelength range, 910t
Dyes: anthraquinone, **516–518**
Dyes: application and evaluation, **518–527**
Dyes: environmental chemistry, **527–528**
Dyes: natural, **529–531**

Dyes: reactive, **531–533**
Dyes: sensitizing, **533–535**
Dyne (dyn), 1643t
Dynels, ultrafiltration media application, 1635
Dysprosium [CAS: 7429–91–6], **535**
    abundance, 330t
    electronic structure, 336t
    interatomic distance, 342t
    ionic crystal radius, 341t
    nuclides (isotopes and isobars), 331t
    principal characteristics, 327t
    properties, 1421t, 1422t

E

e (transcendental number), **587**
Earth atmosphere, **155–158**
Eberbach hardness scale, 756
Eberthella typhi, gram-negative, 168t
Ebonite, production methods for cellular, 664t
Ebulliometer, **537**
Ebullioscopic constant, 250
Ebullism, **537**
Ecgonine, 49t
Echinocandin B, activity and producing organism, 128t
Echo-planar imaging, 1100
Eclogite, **537**
Ecology, **537**
Eddy current nondestructive testing, 1094
Edeine A$_1$, activity and producing organism, 128t
Edible oil, **537**
Edible vegetable oils, 1671–1674
Efflorescence, **537**
Effusion, **537**
Efrotomycin, formula and producing organism, 116t
Ehlers-Damlos syndrome, genetic influence, 716
Ehrlich, Paul (1854–1915), **537**
Eicosanedioic acid, physical properties, 490t
Eicosanoid, **537**
1-Eicosene, physical properties, 1150t
Eigen, manfred, **537**
Einsteinium, 23t
    binary compound properties, 25–26t
    electronic structure, 337t
    interatomic distance, 342t
    ionic crystal radius, 341t
    oxidation states, 24t
    physical properties, 24t
    principal characteristics, 327t
Einsteinium [CAS: 7429–92–7], **538**
Elaidinization, **538**
Elasticity, **538–540**
Elastomeric fibers, **630–631**
    dyeing, 525
Elastomers, **540–541**
Elastostrictive materials, 1485
Eldisine, tubulin interactive, 358t
Electret, **541**
Electrical variables, 1670
Electric insulation, **851**
Electride, **541**
Electrocapillarity, **541**
Electrochemical equivalent, 543
Electrochemical machining (ECM), **541–542**
Electrochemistry, **542–543**
Electroconductive polymers, **1346–1347**
Electrocortin, where produced, structure, and principal functions, 786t
Electrode, **543–544**
Electrodeposition, **544**
Electrodialysis, **544–546**

Electrodynamics, 1393
Electroforming, **546**
Electrogalvanizing, 703
Electroluminescence, **546,** 946t
Electrolysis, 542–543, **546**
Electrolysis-type chemical analyzer, **546**
Electrolyte, **546**
Electrolytes, target of medical diagnostic test, 975t
Electrolytic cell, **546**
Electrolytic conductivity and resistivity
    measurements, **546–549**
Electrolytic tough-pitch copper, 438t
Electromagnetic radiation, **549**
Electromagnetic separation, **549**
Electromagnetic spectrum, **549**
Electromagnet iron, magnetic properties, 955t
Electrometallurgy, 987
Electromotive series, 31
Electron, **549–550,** 1209
Electron affinity, **550**
Electron capture (negative)/ionization mass
    spectrometry, for ultratrace analysis, 1627t
Electron diffraction, **550**
Electronegativity, **550–551**
Electron emission, **551**
Electroneutrality, **551**
Electron gas, **551**
Electron-microprobe x-ray analyzer, 1759–1760
Electron microscope, **551–552**
Electron-multiplier tubes, 1759
Electrons
    backdonation by metals, 28
    gemstone irradiation, 708t
    x-ray production from, 1758
Electron (photoelectron) spectroscopy, **553**
Electron theory, **553–554**
Electron volt, **554**
Electronvolt (eV), 1643t
Electrophile, 1127
    activation, 28
Electrophilic reaction, **554**
Electrophoresis, **554–556**
Electroplating, **556–557**
Electropolishing, **557**
Electrorheological materials, 1484–1485
Electrosol, **557**
Electrostatic precipitator, **557–558**
Electrostrictive materials, 1484
Electrotype, 923t
Electroultrafiltration, 1636
Electrovalence, 345
Electrovalent compound, 428
Electrum, **558**
Element 111, 24
Element 112, 24
Element 113, 24
Element 114, 24
Element 115, 24
Element 116, 24
Element 117, 24
Element 118, 24
Elephant, body weight-metabolic rate relationship,
    176t
Elfamycin antibiotics, **116–117**
Ellipsometer, 1295
Elovich equation, 37
Elsamitrucin tartrate, topoisomerase interactive, 357t
Elspar, chemotherapeutic, 359t
Elution, **558**
Elutriation, **558**
Embden-Meyerhof pathway, 281
Embrittlement, **558**

Emcyt, hormonal therapy, 358t
Emde degradation, **558**
Emerald, **558**
Emery
    abrasive properties, 745t
    hardness, 1008t
Emetine, 49t, 51
Emission spectroscopy, **558**
EMMI database, 769t
Emulsion adhesives, load-bearing capabilities, 32t
Emulsion-based adhesives, 33
Emulsion polymerization, **1342**
Emulsions, **559–560**
Enamels, porcelain or vitreous, **560–561**
Enantiomers, 79
Enantiomorphs, 79
Enantiotropy, **562**
Enargite, **562**
Endorphins, **565–566**
Endoscopic ultrasonography, 1639
Endothall, environmental health advisories, 771t
Endothermic compound, 428
Endrin insecticide, in hazardous waste, 1711t
Enduracidins, activity and producing organism, 128t
Energy, **562–564,** 756
Energy level, **564**
    selection rules, **1463–1464**
Energy state terms, **564–565**
Enfleurage, 1708
Enflurane, 100t
Enhanced oil recovery, 1252–1253
Enkephalins, **565–566**
Enniatin A, activity and producing organism, 128t
Enolase, metal chelate enzyme, 323t
Enramycin, activity and producing organism, 128t
Enrichment, **566**
Enstatite, **566**
Enterobacter
    amphenicol susceptibility, 115t
    ß-lactamase-producing bacterium, 110t
Enteroviruses, 1694
Enthalpy, **566–568**
Entisols, 1497t
Entropy, **568**
Enviomycin, activity and producing organism, 128t
Enviroline, 769t
Environmental chemistry, **568**
Environmental fallout, 10
Environmental Protection Agency (EPA), acid rain,
    10
Enzyme, **568–572**
Enzyme inhibitors, **573**
Enzyme preparations, **573–574**
Enzymes
    as catalysts, 303, 306
    classification of types, 571t
    metal chelates, 323t
    target of medical diagnostic test, 975t
Enzymes in organic synthesis, **575–577**
Enzyme therapeutic, **574–575**
Eosin, sensitizer for semiconductors, 535t
Eosinophils, 242t
Ephedra, 51
Ephedrine, 49t, 51
Epichlorohydrin, **577**
    from allyl alcohol, 59
    physical properties, 366t
Epidiorite, **577**
Epidote, **577**
Epimerases, 571t
Epinephrine, 35, 51
    where produced, structure, and principal
      functions, 786t

Episodic acidification, 9
Epitaxy, **577**
Epogen, unit value and relative production quantity,
    232t
Epoxies, dielectric permittivity, 852t
Epoxy adhesives, load-bearing capabilities, 32t
Epoxy compound, 428
Epoxy resin, production methods for cellular, 664t
Epoxy resins, **577–579**
Epsomite, **579**
Epstein-Barr virus, 1695
EPTC, environmental health advisories, 771t
Equation of state, **579–580**
Equilibrium, **580–581**
Equilibrium diagram, **581**
Equilibrium vapor pressure, 1670
Equine encephalitis vaccine, 1660
Equine encephalitis viruses, 1694
Equivalent electrons, **581**
Erbium
    abundance, 330t
    electronic structure, 337t
    interatomic distance, 342t
    ionic crystal radius, 341t
    nuclides (isotopes and isobars), 331t
    principal characteristics, 327t
    properties, 1421t, 1422t
Erbium [CAS: 7440–52–0], **581–582**
Erbium-doped optical fibers, 1157–1159
Eremomycin, producing organism, 118t
Erg (erg), 1643t
Erionite
    composition, 1034t
    pore structure, 41t
Ernst, Richard R. (1933–), **582**
Erysipelothrix muriseptica, gram-positive, 168t
Erythrite, **582**
Erythrocytes, 241, 242t
Erythromycin
    commercial products, 121t
    year of discovery/market introduction, 106t
Erythromycin-11,12-carbonate, commercial products,
    121t
Erythromycin stinoprate, commercial products, 121t
Erythromycin topical solution, commercial products,
    121t
Erythropoietin
    annual sales, 222t
    unit value and relative production quantity, 232t
Erythrose-4-phosphate, 283
Erythrosin, sensitizer for semiconductors, 535t
Escherichia coli
    amphenicol susceptibility, 115t
    b-lactamase-producing bacterium, 110t
    gram-negative, 168t
Esparto wax, 1747
Esperamicin A1, DNA interactive agent, 357t
Essences, **644–652**
Essential amino acids, 75, 75t, 76t, 1375t
Essential oils, **1136–1137**
Essera, 623t
Ester, 1169t, 1174
Esterification, **582–584,** 1181
    yeast participation, 1769t
Esters, organic, **584–586**
Estrace, hormonal therapy, 358t
Estradiol, where produced, structure, and principal
    functions, 787t
Estradiol USP, hormonal therapy, 358t
Estramustine phosphate sodium USP, hormonal
    therapy, 358t
Etamycin A, activity and producing organism, 128t

Etamycin B, activity and producing organism, 128t
Ethanal, 5
Ethane
    combustion constants, 422t
    combustion reactions, 423t
    gas separation, 39t
    heating value, 686t
    ignition temperature in air, 425t
    measurement by NDIR analysis, 837t
    physical properties, 47t
    in Texas natural gas, 1054t
Ethane [CAS: 74–84–0], **586**
Ethanedioic acid, physical properties, 490t
1,2-Ethanediol, 733
Ethanesulfonic acid, physical properties, 1568t
Ethanoic acid, 6
Ethanol, 588
    physical properties, 47t
Ethanolamines, 74–75
Ethanolamines [CAS: 141–43–5], **586–587**
Ethepon growth modifier, 1314
Ether, 587, 1174
Ethers, **587–588**
Ethine, 7
2-(2-Ethoxyethoxy)ethyl acetate, physical properties, 585t
2-Ethoxyethyl acetate, physical properties, 585t
2-Ethoxyethyl acrylate, physical properties of polymers, 18t
Ethyl acetate, 6
    physical properties, 585t
Ethyl acrylate
    physical properties, 18t, 585t
    physical properties of polymers, 18t
Ethyl alcohol
    combustion constants, 422t
    density, 474t
    heating value, 686t
    specific heat at constant pressure, 757t
Ethyl alcohol [CAS: 64–17–5], **588**
Ethyl amyl ketone, physical properties, 900t
Ethylbenzene, 1765
    manufacture, 1555
    physical properties, 1766t
Ethylbenzene hydroperoxide, physical properties, 1231t
Ethyl benzoate, physical properties, 585t
2-Ethylbutyl acrylate, physical properties of polymers, 18t
Ethyl butyrate
    permeation in selected barrier polymers, 174t
    physical properties, 585t
Ethyl carbamate, 1653
Ethyl cellulose, **588–589**
Ethylcellulose, microcapsule material, 997t
Ethyl chloride, physical properties, 366t
Ethyl cyanate, 464
Ethyl cyanoacetate, 964
Ethyl-2-cyano-3,3-diphenyl acrylate, 1641
Ethyl cyanurate, 464
Ethylene
    combustion constants, 422t
    combustion reactions, 423t
    formation from acetylene, 8
    heating value, 686t
    ignition temperature in air, 425t
    measurement by NDIR analysis, 837t
    monomer reactivity and suitable initiator for, 838t
    plant growth effects, 1314
    water-vapor transmission rate, 173t
Ethylenebis, as release agent, 1435t

Ethylenebisdibromonorbornanedicarboximide, flame retardant, 640t
Ethylenebistetra-bromophthalimide, flame retardant, 640t
Ethylene [CAS: 74–85–1], **589–590**
Ethylene cyanohydrin, 465t
Ethylenediamine, physical properties, 483t
Ethylenediamine adduct, commercial polyether polyol, 1654t
Ethylenediamine ditartrate (EDT), for diffracting crystal spectrometer, 1760
Ethylene dichloride, physical properties, 366t
Ethylene glycol [CAS: 107–21–1], 46, **590**, 733
    physical properties, 590t
    polyol for alkyd synthesis, 53t
Ethylene glycol diacetate, physical properties, 585t
Ethylene monochloride, 8
Ethylene oxide, as propellant, 1447t
Ethylene oxide [CAS: 75–21–8], **590**
    definitely established workplace carcinogen, 297t
    measurement by NDIR analysis, 837t
Ethylene-vinyl acetate copolymers, **590**
Ethylene-vinyl alcohol copolymer
    diffusion and solubility coefficients for oxygen and carbon dioxide, 173t
    gas permeability, 172t
    permeation of flavor/aroma compounds, 174t
    water-vapor transmission rate, 173t
Ethylene-vinyl alcohol (EVOH) copolymers, 1679
Ethyl ether, 100t, 587
    physical properties, 587
Ethyl 3-ethoxypropionate, physical properties, 585t
Ethyl formate, physical properties, 585t
Ethyl hexanoate, permeation in selected barrier polymers, 174t
2-Ethylhexyl acetate, physical properties, 585t
2-Ethylhexylacrylate, 17
2-Ethyl hexylacrylate, 20
2-Ethylhexyl acrylate
    physical properties, 18t, 585t
    physical properties of polymers, 18t
Ethyl hydroperoxide, physical properties, 1231t
Ethylidene diethyl ether, 6
Ethylidene dimethyl ether, 6
Ethyl isobutyrate, physical properties, 585t
Ethyl isocyanate, 464
Ethyl isocyanurate, 464
Ethyl methacrylate, 16–17, 19
    physical properties, 988t
Ethyl-2-methyl butyrate, as key odor/tase compound, 649
Ethyl 2-methylbutyrate, permeation in selected barrier polymers, 174t
Ethylnitrate, 1082
Ethylnitrite, 1082
3-Ethyl-3-pentyl hydroperoxide, physical properties, 1231t
Ethyl propyl ketone, physical properties, 900t
Ethyl salicylate, physical properties, 585t
Ethyl stearate, physical properties, 585t
Ethyl urethane, 1653
Ethyl vinyl ether, physical properties, 1690t
Ethyl xanthic acid, 1755
Ethyne, 7
Etidocaine, 100t
Etoposide phosphate, topoisomerase interactive, 357t
Etoposide USP, topoisomerase interactive, 357t
Eucalyptus oil, 647t
Euclase, **590**
Eudiometer, **590**
Eulexin, hormonal therapy, 358t
Europium

abundance, 330t
    electronic structure, 336t
    interatomic distance, 342t
    ionic crystal radius, 341t
    nuclides (isotopes and isobars), 331t
    principal characteristics, 327t
    properties, 1421t, 1422t
Europium [CAS: 7440–53–1], **590–591**
Eutectic, **591**
Eutectoid, **591**
Evaporation, **591–592**
Evaporite, **593**
Even-even nuclei, **593**
Exchange degeneracy, **593**
Excited state, **593**
Excitonic matter, 1517
Exosmosis, **593**
Exothermic compound, 428
Exotic nuclei, 1409–1410
Experimental breeding, 710
Explosives, **593–594**
Ex-situ bioremediation, 205t
Extender, **594**
Extracellular exudates, 747
Extraction (liquid-liquid), **594–598**
Extraction (liquid-solid), **598–599**
Extractive metallurgy, **599**
Extra dense glass, 1163t
Extraterrestrial materials, **599–600**
Extreme-ultraviolet, 1639
Extruded latex thread, 631
    physical properties, 631t
Extrusion, **601**
Extrusive, **601**
Erythromycin acistrate, commercial products, 121t
Erythromycin estolate, commercial products, 121t
Erythromycin ethyl succinate, commercial products, 121t
Erythromycin glucceptate, commercial products, 121t
Erythromycin lactobionate, commercial products, 121t
Erythromycin ophthalmic ointment, commercial products, 121t
Erythromycin stearate, commercial products, 121t

F

Fabry's disease, genetic influence, 716
Face-centered cubic structure, **603**
Factor group, 747
Factor VIII, annual sales, 222t
Factumycin, formula and producing organism, 116t
Facultative anaerobes, 90
Fahrenheit temperature scale, 3, **603,** 1599
Fallout (Radioactive), **603**
Familial goiter, genetic influence, 716
Fanconi's syndrome, genetic influence, 716
Faraday, Michael (1791–1867), **603**
Farad (F), 1643t
Far-ultraviolet, 1639
Fast breeder nuclear reactor, 1114–1119
Fast breeder reactors, 1647
Fat, **603**
Fatigue (Metals), **603–604**
Fats
    constituent of human blood, 242t
    industrial chemicals derived from renewable, 230t
    in vegetable oils, 1671t, 1671–1674
Fatty acid ester plasticizers, 1316
Fatty acids, 48, **604**
    constituent of human blood, 242t

Langmuir-Blodgett films made from, 1018
in vegetable oils, 1671t
Fatty alcohol, **604**
Fatty amine, **604**
Fatty ester, **604**
Faujasite
composition, 1034t
pore structure, 41t
Fecal coliform bacteria, water quality indicator, 1726
Fecal Streptococcal bacteria, water quality indicator, 1726
Federal Insecticide, Fungicide, and Rodenticide Act, 1711t
Feeder (Gravimetric), **604–605**
Feeder (Vibratory), **605**
Feeder (Volumetric), **605–606**
Feedstocks, **606**
Fehling's solution, **606**
Feldene, 93t
Feldspar, **606–607**
analysis, 316t
hardness, 755t
Felsite, **607**
Felt, 1752
Female sex hormones, 791
where produced, structure, and principal functions, 787t
Femic, **607**
Femto, 1643t
Fentanyl, 92t
Fergusonite, **607**
Fermentation, **607–608**, 1180
amino acids, 80
pollution, 1715
vinegar, 1674–1675
yeast in, 1767, 1768
Fermi, **608**
Fermi, Enrico (1901–1954), **608**
Fermions, 254, **608**
Fermi resonance, **608**
Fermi selection rules, **609**
Fermi surface, **609–610**
Fermium, 23t
binary compound properties, 25–26t
electronic structure, 337t
interatomic distance, 342t
ionic crystal radius, 341t
oxidation states, 24t
principal characteristics, 327t
Fermium [CAS: 7440–72–4], **610–611**
Ferredoxin, source and function, 986t
Ferric acetate, 6
Ferric ammonium alum, 61
Ferric ammonium sulfate, 86
Ferric nitrocarburizing, 763t
Ferrimagnetism, **611**
Ferrimycin A, activity and producing organism, 128t
Ferrites, **611**
magnetic properties, 956t
Ferritic stainless steel, 58
Ferritin, source and function, 986t
Ferrochromium, **611**
Ferroelectric effect, **611–612**
Ferroelectric materials, **612**
Ferromagnetism, 611, **612**
Ferromanganese, **612**
Ferromolybdenum, **612**
Ferroniobium, **612**
Ferrophosphorus, **612**
Ferrosilicon, **612**
Ferrotitanium, **612**
Ferrotungsten, **612**

Ferrovanadium, **612**, 1666
Ferroxdure, magnetic properties, 956t
Ferrozirconium, **612**
Fertile material, **612**
Fertilizer, **612–616**
Fiber (Dietary), **617**
Fiber glass, **617–620**
Fiberglass, definitely established workplace carcinogen, 297t
Fiber-optic systems, **1153–1163**
Fiber-reactive dyes. *See* Reactive dyes
Fiber-reinforced composites, **620**
Fibers, **620–624**
generic designations and definitions of synthetic, 621–624t
Fibers: acetate, **624–626**
Fibers: acrylic, **626–629**
Fibers: cellulose esters, **629–630**
Fibers: elastomeric, **630–631**
Fibers: vegetable, **631–633**
Fibrilon, 623t
Fibrinogen, constituent of human blood, 242t
Fibro, 624t
Fictile, **633**
Field beans, susceptibility to iron deficiency, 874t
Fieser, Louis F. (1899–1977), **633**
Filament, **633**
Filled-system thermometer, **1609**
Filler, **633**
Fillers, use in food processing, 671t
Film, **633**
Filterable viruses, 1693
Filter medium, **633**
Filtration, **633–636**
Fine solder, 923t
Fine structure, **636**
Finishing cement, insulation material, 857
Fire-brick, **636**
Fireclay, **637**
Fire clay, analysis, 316t
Fireclay, physical properties, 1428t
Firming agents (Foods), **637**
First law of thermodynamics, 1606
Fischer, Edmond H. (1920–1992), **637**
Fischer, Emil (1852–1919), **637**
Fischer, Ernst Otto (1918–), **637**
Fischer, Hans (1881–1945), **637**
Fischer-Tropsch process, **637**
Fischer-Tropsch synthesis, **637**
Fischer-Tropsch wax, as release agent, 1435t
Fish kills, 9
Fish protein concentrate, 1373
Fittig reaction (Fitting), **637**
Fixed bed, **637**
Fixed-bed bioreactor, 205t
Fixed oils, **637–638**
Flame emission spectrometry, for ultratrace analysis, 1627t
Flame hardening, **638**
Flame photometer, 1295
Flame photometry and spectrometry, **638**
Flame-retarding agents, **638–642**
Flashing (Thermal), **643**
Flash point, **643**
Flavanthrone, sensitizer for semiconductors, 535t
Flavonoids, **643**
Flavor enhancers and potentiators, **643–644**
Flavors, quarks, 1397
Flavors and essences, **644–652**
Flax, 632t
generic designation and definition, 624t
susceptibility to iron deficiency, 874t

Flaxseed oil, 1671
Flexible polyurethane foams, 1654–1655
Flexible PVC, 1357
Flint, **652**
hardness, 1008t
Flint clays, physical properties, 1428t
Flocculating agents, **652–654**
Flocculation, 45
Flooding, in kinetic measurements, 901
Flory, Paul J. (1910–1986), **652**
Flotation, **654**, 987
Floxuridine USP, antimetabolite, 355t
Flu, 1694–1695
Fluchloralin, environmental health advisories, 771t
Flucortolone, commercial microbial transformations used to produce, 1550t
Fluid, **654–657**
Fluid flow, **654–657**
Fluid friction, **657**
Fluidization, **657**
Fluid parcel, **657–658**
Fluometuron, environmental health advisories, 771t
Fluorescein, 1756
sensitizer for semiconductors, 535t
Fluorescence, **658**
Fluorescent Brightener 74 (CI 45550), 1756
Fluorescent Brightener 155 (CI 45555), 1756
Fluorescent brighteners, 512t
applications, 519
Fluoridation, 659
Fluorides, removal from raw water, 1724
Fluorimeter, 1295
Fluorine, 754
abundance, 330t, 1132t
constituent of human plasma or serum, 244t
covalent radius, 344t
electronic structure, 336t
ionic crystal radius, 341t
nuclides (isotopes and isobars), 331t
principal characteristics, 327t
standard electrode potential, 31t
trace mineral nutrient, 1002t, 1003
Fluorine [CAS: 7782–41–4], **658–659**
Fluorine (In Biological Systems), **659–660**
Fluorite, **660**
hardness, 755t
Fluoroaluminates, 64
Fluoroantimonic acid, 13
Fluorocarbon, **660–661**
Fluorocarbon fiber
generic designation and definition, 622t
physical properties, 1139t
Fluoroelastomer, **661**
Fluorometers, **661**
Fluorones, 1756
Fluorophlogopite, dielectric permittivity, 852t
Fluoroplastics, **661–662**
Fluorosilicates, 1476
Fluorouracil, antimetabolite, 355t
Fluoroxene, 100t
Flurbiprofen, 93t
Fluridone, environmental health advisories, 771t
Fluroxypyr, environmental health advisories, 771t
Flutamide, hormonal therapy, 358t
Flu vaccine, 1659
Flux (Slag), **662**
Flux (Solder), **662**
Foam, **662–663**
Foamed plastics, **663–667**
Fog, **668**
Fog seeding, 668
Fog tracks, **668**

Folic acid, **668–669**

Follicle ripening hormone, where produced, structure, and principal functions, 787t

Follicle-stimulating hormone (FSH), where produced, structure, and principal functions, 787t

Follotropin, where produced, structure, and principal functions, 787t

Food, Drug, and Cosmetic Act, 1711t

Food, drug, and cosmetic dyes, 512t, 531

Food additives, **669–671**
    carriers, 301

Foodc colorants, **419–420**

Food dehydration, 673

Food preservation, 672–673

Food processing, **671–673**

Foods
    antimicrobial agents, **135–138**
    bodying and bulking agents, **248–249**
    coating agents, **409**
    firming agents, **637**
    fortifying with amino acids, 75
    in municipal waste, 1714t
    spoilage by yeasts, 1768

Food toxicant, naturally occurring, **673–675**

Foolish seedling disease, in rice, 722

Footlambert (fL), 1643t

Forage sorghum, susceptibility to iron deficiency, 874t

Forbes' disease, genetic influence, 716

Forced-circulation reboiler, 760

Force variables, 1670

Forensic chemistry, **675–676**

Forests, acid rain, 10

Forging, **676**

Formaldehyde, 48
    definitely established workplace carcinogen, 297t
    purification, 863t

Formaldehyde [CAS: 50–00–0], **676–677**

Formaldehyde cyanohydrin, 465t

Formamide, **678**

Formestanef, hormonal therapy, 358t

Formic acid [CAS: 64–18–6], **678**

Formic hydrogenylase, metal chelate enzyme, 323t

Formyltransferases, 571t

Forsterite, dielectric permittivity, 852t

Fortrel, 623t

Fossil waxes, 1747

Fouling, in ultrafiltration, 1636

Fractional distillation, **678**

Fractional quantized Hall effect, 753

Fractionation, **678**

Francium, 48
    electronic structure, 337t
    interatomic distance, 342t
    ionic crystal radius, 341t
    principal characteristics, 327t

Francium [CAS: 7440–73–5], **679**

Franklinite, **679**

Frasch process, 679

Fraunhofer diffraction, 493

Fraunhofer holograms, 782

Fraunhofer lines, 5

Free convection, 759

Free-cutting copper, 438t

Freedom (Degrees of), **679**

Free electron theory of metals, **679**

Free energy, **679–680**

Free energy change, **680**

Free machining, **680**

Free-machining steels, 58

Free radical, **680**

Free radical addition, 1177

Free-radical initiators, **840–842**

Free-radical polymerization, **1343–1344**

Free radicals, activation, 28–29

Free rotation, **680**

Free volume, **680**

Freeze-concentrating, **680–681**

Freeze-drying, 672–673, **681–684**
    food, 672–673

Freeze-preserving, **684–685**

Freezing point, **685**

Freezing-point depression, **685**

Fresnal diffraction, 493

Fresnel holograms, 782

Freundlich adsorption isotherm, 37

Friction (Mechanical), **685**

Friedel-Crafts alkylation, 1178

Friedel-Crafts reaction, **685–686**

Fructokinase, metal chelate enzyme, 323t

Fructose, constituent of human blood, 242t

Fructose-1,6-diphosphate, 283

Fructose intolerance, genetic influence, 716

Fructose-6-phosphate, 283

FUDR, antimetabolite, 355t

Fuel, **686–687**

Fuel cells, **687–692**

Fuel combustion, **420–427**

Fuel gas, 1255

Fuel oil, 1255
    excess air for combustion, 426t

Fugacity, **692**

Fukui, Kenichi (1918–1998), **692**

Fuller's earth, **692**

Fulminic acid, 464

Fumarase, 282t

Fumaric acid, 282t
    acidulant, 14

Fumaric acid [CAS: 110–17–8], **962–963**

Fume, **692**

Fuming sulfuric acid (oleum), 13

Functional dyes, 517

Functional group analysis, 97

Fungi, 1767

Fungicides, **692–693**

Funk, Casimir (1884–1848), **693**

Furan, as parent structure compound, 1090t

Furan and related compounds, **693–694**

Furan [CAS: 110–00–9], 693–694

Furfural, 693

Furfuraldehyde [CAS: 98–01–1], **694**

Furfuryl mercaptan, as key odor/tase compound, 649

Furoic acid, carboxylic acid, 294

Fused alumina
    abrasive properties, 745t
    hardness, 755t
    physical properties, 1428t

Fused mullite, physical properties, 1428t

Fused silica, dielectric permittivity, 852t

Fused zirconia, hardness, 755t

Fused zirconia/alumina, hardness, 2t

Fusible lead alloys, 923t
    bismuth in, 238t

Fusidic acid, year of discovery/market introduction, 106t

Fusion (Heat of), **700**

Fusion (Nuclear), **700, 1097–1098**

Fusion (Phase change), **700**

Fusion energy, **694–699**

G

Gabbro, **701**

Gadolinium
    abundance, 330t
    electronic structure, 336t
    interatomic distance, 342t
    ionic crystal radius, 341t
    nuclides (isotopes and isobars), 331t
    principal characteristics, 327t
    properties, 1421t, 1422t

Gadolinium [CAS: 7440–54–2], **701**

Gadolinium oxide, 701

Gadolinium oxysulfide, 701

Gahnite, **701**

Galactin, where produced, structure, and principal functions, 788t

Galactokinase deficiency, genetic influence, 716

Galactosemia, genetic influence, 716

α-Galactosidase, therapeutic enzyme, 574t

ß-Galactosidase, therapeutic enzyme, 574t

Galena, hardness, 1008t

Galena [CAS: 1314–87–0], **701**

Gal (Gal), 1643t

Galipoline, 49t

Gallite, 702

Gallium
    abundance, 330t, 1132t
    covalent radius, 344t
    electronic structure, 336t
    interatomic distance, 342t
    ionic crystal radius, 341t
    principal characteristics, 327t

Gallium arsenide [CAS: 1303–00–0], 702

Gallium arsenide semiconductors, 1469–1470

Gallium [CAS: 7440–55–3], **702**

Galvanizing, **703**, 1775

Gametogenic hormone, where produced, structure, and principal functions, 787t

Gametokinetic hormone, where produced, structure, and principal functions, 787t

Gamma globulin, **703**

Gamma radiation, **703**, 1407

Gamma ray detectors, 704

Gamma rays, 703–704
    gemstone irradiation, 708t

Gamma ray spectra, 704

Gamma-ray spectroscopy, **703–704**

Gangliosides, **704–705**

Gangue, **705**, 987

Ganister, physical properties, 1428t

Garlic oil, 647t

Garnet, **705**
    abrasive properties, 745t
    hardness, 755t, 1008t

Garnierite, **705–706**

Gas, **706**

Gas bulk separation, 39t

Gas carbonitriding, 763t

Gas carburizing, 763t

Gas chromatography, for ultratrace analysis, 1627t

Gas chromatography/mass spectrometry, for ultratrace analysis, 1627t

Gas compression, **429–430**

Gas constant, **706**

Gaseous carbon, 7

Gaseous elements, 334

Gas hydrate, **706**

Gas law, 706

Gas-liquid absorption, 3–4

Gas nitriding, 763t

Gas oil, 1255
    in representative U.S. crude oils, 1242t
    in representative world crude oils, 1243t

Gasoline, 55
    density, 474t

ignition temperature in air, 425t
in representative U.S. crude oils, 1242t
in representative world crude oils, 1243t
Gasoline pool, 1255, 1730
Gas-phase adsorption, 39–40
Gas phase polymerization, 1342
Gas purification, 39t
Gattermann aldehyde synthesis, **706**
Gattermann-Koch reaction, **706**
Gaucher's disease, genetic influence, 716
Gauss (G), 1643t
Gay-Lussac, Joseph Louis (1778–1850), **706**
Gay-Lussac's law, **706**
Geiger-Mueller tube, 1759, 1760
Geiger-Muller counter, ultraviolet radiation, 1640
Gel, **706**
Gelatin, **706–707**
microcapsule material, 997t
purification, 863t
Gelatin-gum arabic, microcapsule material, 997t
Gelatin photographic film emulsions, 1290
Gelation, 1515–1516
alkyd resins, 53
Geldanamycins, biological activity, 108t
Gelsemine, 49t
Gemcitabine, antimetabolite, 355t
Gemstones, **707–709**
heating-enhanced, 708t
radiation-enhanced, 709t
Gemstone treatment, 708t, 708–709, 709t
Gene mutations, 714–715
General purpose ABS, physical properties, 22t
Genes, 712
and disease, 715–716
Gene therapy, 717
Genetically modified foods, 709
Genetic code, 712
Genetic engineering, **709–710**
and pollution, 1715
vaccines, 1660–1661
Genetic recombination, 712
Genetics and gene science (Classical), **710–720**
chronology of progress, 711t
Gentamicin, year of discovery/market introduction, 106t
Geochemistry, **721**
Geocorona, 155
Geocronite, **721**
Geometrical isomers, 1542
Geometric variables, 1670
Geopotential altitude, 157t
Geotrichum candidum, 1770
Geranium oil, 647t
Germane, 722
physical properties, 795t
Germanium
abundance, 330t, 1132t
covalent radius, 344t
electronic structure, 336t
interatomic distance, 342t
ionic crystal radius, 341t
nuclides (isotopes and isobars), 331t
principal characteristics, 327t
Germanium [CAS: 7440–56–4], **721–722**
Germanium hydroxide, 721
Gersdorffite, **722**
cobalt source, 410t
Gettering, **722**
Getters, 319–320
Giauque, William F. (1895–1982), **722**
Gibberellic acid, **722–723**
Gibberellins, **722–723**, 1314

Gibbs, Josiah Willard (1839–1903), **723**
Gibbs-Duhem equation, **723**
Gibbs-Konovalov theorems, **723**
Gibbs paradox, **723**
Giga, 1643t
Gilbert, Walter (1932–), **723**
Gilbert (Gb), 1644t
Gilding brass, 440t
Gilsonite, **723**
insulation material, 856t, 857
Girbotol adsorption, **723**
Glass, **723–728**
absolute index, 1426t
composition of commercial, 725t
density, 474t
in municipal waste, 1714t
radioactive dating, 1415
Glass blocks, **728–730**
Glass blowing, 727–728
Glass-ceramics, 725
Glass fiber
generic designation and definition, 622t
insulation material, 856t
physical properties, 1139t
Glass making, 726–727
Glass-melting furnaces, 726
Glass tubing, 726–727
Glassy metals, **730–732**
Glauberite, **732**
Glauconite, **732**
abrasion pH, 1t
Glaucophane, **732**
Global Warming Potential, 744
Globomycin, activity and producing organism, 128t
Globulins, **732**
Glospan, 624t
Glossmeter, 1295
Glovebox Flight Program, 741
Glucagon, 791
where produced, structure, and principal functions, 787t
Glucoamylase, 306
Glucocerebrosidase
annual sales, 222t
therapeutic enzyme, 574t
Glucocorticoids, 35
Glucono-δ-lactone
acidulant, 14
properties of leavening acid, 925t
Glucose
constituent of human blood, 242t
target of medical diagnostic test, 975t
Glucose isomerase, 306
Glucose oxidase, properties as diagnostic enzyme, 976t
Glucose-6-phosphate, 283
Glucose-6-phosphate dehydrogenase, 283
Glucose-6-phosphate dehydrogenase deficiency, genetic influence, 716
Glucose-1-P kinase, metal chelate enzyme, 323t
α-Glucosidase, therapeutic enzyme, 574t
ß-Glucosidase, therapeutic enzyme, 574t
Glufosinate, environmental health advisories, 771t
Glumamycin, activity and producing organism, 128t
Glumatic acid, first isolation and isoelectric point, 78t
Gluons, 1397
Glutamic acid
manufacture, 76t
structural classification, 79t
synthesis from products of carbohydrate metabolism, 282t

Glutamic dehydrogenase, metal chelate enzyme, 323t
L-Glutaminase, therapeutic enzyme, 574t
Glutamine
first isolation and isoelectric point, 77t
manufacture, 76t
structural classification, 79t
Glyceraldehyde-3-phosphate, 283
Glyceric kinase, metal chelate enzyme, 323t
Glycerin, purification, 863t
Glycerine, 732
absolute index, 1426t
density, 474t
Glycerol adduct, commercial polyether polyol, 1654t
Glycerol [CAS: 56–81–5], **732**
polyol for alkyd synthesis, 53t
Glyceryl triacetate, physical properties, 585t
Glyceryl tripropionate, physical properties, 585t
Glycidyl methacrylate, physical properties, 988t
Glycine
first isolation and isoelectric point, 77t
manufacture, 76t
structural classification, 79t
Glycol, 46, 732
Glycols, physical properties, 590t
Glycolysis, 281, **733**
Glycopeptide antibiotics (dalbaheptides), **117–119**
Glycoside hydrolases, 571t
Glycosides, **733**
Glycosides (Steroid), **733–735**
Glycosyltransferases, 571t
Glycyl alcohol, 732
Glyphosate, environmental health advisories, 771t
Goethite, **735,** 871t
Gold
abundance, 330t, 1132t
corrosion resistant alloys, 446t
density, 474t
electronic structure, 337t
interatomic distance, 342t
ionic crystal radius, 341t
nuclides (isotopes and isobars), 331t
principal characteristics, 327t
Gold amalgam, 72
Gold [CAS: 7440–57–5], **735–737**
colorant for glass, 725t
major sources of industrial minerals, 1011t
standard electrode potential, 31t
Gold halides, 737
Gold number, **737**
Gold oxides, 736–737
Goldschmidt detinning process, **737,** 738
Goldschmidt reduction process, **738**
Gold-silicon alloys, phase diagrams, 57
Gonadotropic hormones, 790
Gonadotropin 1, where produced, structure, and principal functions, 787t
Gonorrhea vaccine, 1660
Goodyear, Charles (1800–1860), **738**
Gore-Tex, 622t
Goserelin USP, hormonal therapy, 358t
Gout, genetic influence, 716
GPIIb/IIIa Mab, annual sales, 222t
Graham, Thomas (1805–1869), **738**
Graham law, **738**
Grain, **738**
Grain boundary, **738**
Grain (gr), 1644t
Grain refiner, **738**
Grain size, **738**
Grain sorghum, susceptibility to iron deficiency, 874t
Gram-atom, **738**
Gram-equivalent, 738

Gram (g), 1644t
Gramicidin S, activity and producing organism, 128t
Gramicidins, activity and producing organism, 128t
Gram-mole, 738
Gram-molecular weight, **738**
Gram-negative bacteria, ß-lactamase-producing
    bacteria, 110t
Gram-negative organisms, 738
Gram-positive bacteria, ß-lactamase-producing
    bacteria, 110t
Gram-positive organisms, 738
Gram stain, **738**
Granite, **738–739**
Granitoid, **739**
Granular leukocytes, 242t
Granulocyte colony stimulating factor (GCSF)
    annual sales, 222t
    unit value and relative production quantity, 232t
Granulocyte macrophage colony stimulating factor
    (GMCSF)
    annual sales, 222t
    unit value and relative production quantity, 232t
Granulocytes, 242t, 243
Grapefruit oil, 647t
Grapes, susceptibility to iron deficiency, 874t
Graphic granite, 739
Graphite, 1009t
    crystal structure, 455
    hardness, 1008t
    as release agent, 1435t
    thermal conductivity, 758t
Graphite composites, 739
Grasses, susceptibility to iron deficiency, 874t
Gravel, **740**
    physical properties, 1428t
Gravimetric analysis, 94, 740
Gravimetric feeder, **604–605**
Gravity and microgravity, **740–743**
Gray iron, 57
Grease, **743**
Greenalite, 871t
Green gold, 736
Greenhouse effect, **743**
Greenhouse gases, **743–744**
Greenockite, **744**
Greenstone, **744**
Greisen, **744**
Grignard reactions, **744–745**, 1179
Grignard reagents, 744–745
Grinding and polishing agents, **745–746**
Grinding fluids, 2
Grinding wheels, 2
    average particle size of abrasive used, 746t
Griseofulvin, year of discovery/market introduction,
    106t
Grossular, 1635
Grossularite, 705
Groundnut oil, 1671
    characteristics and properties, 1672–1673t
Groundnuts, susceptibility to iron deficiency, 874t
Groundwater, **746**
    contamination and quality, 1728–1729
Group, **746–747**
Growth hormone, 790
    where produced, structure, and principal
        functions, 789t
Graphite [CAS: 7440–44–0], **739–740**
Grundmann aldehyde synthesis, **747**
Grunerite, 462, 871t
Guanidine [CAS: 113–00–8], **747**
Guanylurea, 747t
Guanylyl cyclase receptors, 1272

Guar gum, 748t
Guayule, 1451
Guiac gum, 748t
Gum arabic, microcapsule material, 997t
Gum resins, 1438
Gums, use in food processing, 671t
Gums and mucilages, 748t, **748–749**
Gutzeit test, **749**
Gypsum, forms and composition, 273t
Gypsum [CAS: 10101–41–4], **749–750**
    abrasion pH, 1t
    hardness, 755t

H
Haber, Fritz (1868–1934), **751**
Habit, **751**
Hadrons, **751**, 1396
Haemophilus influenzae, ß-lactamase-producing
    bacterium, 110t
Haemophilus influenza serotype b vaccine, 1659
Hafnium
    abundance, 330t
    corrosion resistant alloys, 446t
    covalent radius, 344t
    electronic structure, 337t
    interatomic distance, 342t
    ionic crystal radius, 341t
    nuclides (isotopes and isobars), 331t
    principal characteristics, 327t
Hafnium [CAS: 7440–58–6], **751**
Hafnium nitride, plasma synthesis, 318t
Hair, fiber-dye property requirements, 521t
Half-life, 703, 1406
Halide minerals, 1011
Halides, **751–752**
Halite (Rock Salt), **752**
Hall, Charles Martin (1863–1914), **752**
Hall coefficient, 752
Hall effect and quantized Hall effect, **752–753**
Hälleflinta, **753**
Hall generator, 753
Hall process, **753**
Hallucinogens, **753–754**
Halocarbon, **754**
Halogenated compounds, 1174–1175
Halogenated hydantoins, 793–794
Halogen group, **754**
Halomycins, biological activity, 108t
Halothane, 100t
Halowax, 1748
Hamilton, Alice (1869–1970), **754**
Hammett acidity function, 13
Hann, Otto (1879–1968), **754**
Harden, Sir Arthur (1865–1940), **755**
Hardenability of steel, **754–755**
Hardening of metals, **755**
Hard lead, 923t
    antimony content, 139t
Hardness, **755–756**
Hardness removal (water softening), 1723
Hardness scales, 2t, 755t
Hard water, **1726**
Hargreaves process, **756**
Harmotome, **756**
Hartley bands, 3
Hassel, Odd (1897–1981), **756**
Hassium, 24, 333–334
    electronic structure, 337t
    principal characteristics, 328t
Hastelloy, 58
Hastelloy B, 1071t

Hastelloy F, 1071t
Hauptman, Herbert A. (1917–), **756**
Haworth, Sir Walter N. (1883–1950), **756**
Hazardous Materials Transportation Act, 1711t
Hazardous waste incineration, 1712
Heat, **756–758**
Heat capacity, 757, 758
Heat capacity equation (Einstein), **758**
Heat death, of universe, 903
Heat distortion resistant ABS, 22t
Heat dump, 758
Heat exchanger, **758,** 760–761
Heat measurement, 756–757
Heat of fusion, **700**
Heat of sublimation, **1558**
Heat of vaporization, 1670
Heat sink, **758**
Heats of adsorption, 39
Heat storage, 762
Heat transfer, **758–762**
Heat-treatable aluminum alloys, 67
Heat treating, **763–764**
Heat treatment, gemstones
Heavy gas oil, 1255
Heavy-ion fusion, 699
Heavy water, **764**
Heavy water nuclear reactor, 1113–1114
Hecto, 1643t
Heisenberg, Werner Karl (1901–1976), **764**
Heisenberg force, 1749
Helicobacter pylori vaccine, 1660
Heliogravure, 1289
Heliotrope, 248
Helium
    absolute index, 1426t
    abundance, 1132t
    electronic structure, 336t
    nuclides (isotopes and isobars), 331t
    percentage in air, 46t
    principal characteristics, 327t
    in Texas natural gas, 1054t
    van der Waals characteristic equation
        coefficients, 321t
Helium-cadmium lasers, wavelength range, 910t
Helium [CAS: 7440–59–7], **764–765**
Helium liquefaction, 765
Helium-neon lasers, wavelength range, 910t
Helleborein, 734t
Hellige turbidimeter, 1633
Helvecardin A, producing organism, 118t
Helvecardin B, producing organism, 118t
Hematite, **766,** 871t
Hematology, automated instrumentation applications,
    163–164
Heme, 767
Hemicellulose, **766–767**
Hemicelluloses, 1751
Hemicyanine, organic superlattice, 1023
Hemimorphite, **767**
Hemiterpenoids, 1601
Hemochromatosis, genetic influence, 716
Hemocuprein, source and function, 986t
Hemocyanin, source and function, 986t
Hemoglobin, **767–768**
    constituent of human blood, 242t
    target of medical diagnostic test, 975t
Hemoglobinometer, 1295
Hemoglobinopathies, genetic influence, 716
Hemophilus pertussis, gram-negative, 168t
Hemovanadin, source and function, 986t
Hemp, 632t, **768**
Heneicomycin, formula and producing organism,
    116t

Heneicosanedioic acid, physical properties, 490t
Henequen, 632t
Henry (H), 1644t
Henry's Law, 36
Henry's law, **768**
Heparin [CAS: 9005-49-6], **768**
Hepatitic viruses, 1695
Hepatitis B vaccine, 1659
    annual sales, 222t
Hepatocuprein, source and function, 986t
Hepatoflavin, 1699
Hepatolenticular degeneration, genetic influence, 716
Heptadecanedioic acid, physical properties, 490t
Heptadecene, biodegradable, 207t
1-Heptadecene, physical properties, 1150t
n-Heptane, heating value, 686t
Heptanedioic acid, physical properties, 490t
2-Heptanone, physical properties, 900t
3-Heptanone, physical properties, 900t
1-Heptene, physical properties, 1150t
Heptyl acrylate, physical properties of polymers, 18t
2-Heptyl acrylate, physical properties of polymers, 18t
n-Heptyl hydroperoxide, physical properties, 1231t
Herbicides, **768-773**
    pyridine derivatives, 1385-1386
Herbimycins, biological activity, 108t
Herculon, 623t
Hereditary spherocytosis, genetic influence, 716
Heroin, 51, 91, 1041
Herpes simplex vaccine, 1660
Herpesviruses, 1695
Herschbach, Dudley R. (1932-), **773**
Hers's disease, genetic influence, 716
Herzberg, Gerhard (1904-1999), **773**
Herzberg bands, 3
Hesperidin, 734t
Hessite, **773**
Heterocyclic compound, 428
Heterocyclic compounds, 1176
Heterogeneous catalysts, 303
Heterogenite, cobalt source, 410t
Heterogenous, **773**
Heteromolybdates, **773**
Heteropolyacids, **773**
Hetz (Hz), 1644t
Heulandite, **773**
Hevesy, Georg de (1885-1966), **773**
Hexachlorocyclopentadiene, flame retardant, 640t
Hexadecane, biodegradable, 207t
Hexadecanedioic acid, physical properties, 490t
1-Hexadecene, physical properties, 1150t
Hexadecyl acrylate, physical properties of polymers, 18t
Hexadecyl stearate, physical properties, 585t
Hexagonal crystal system, crystallographic elements, 455t
Hexamethylene diamine, monomer for nylon resins, 1333t
Hexamethylenediamine, physical properties, 483t
Hexamethylene diisocyanate (HDI), 1654
Hexamethylene tetramine, 75
3,3,6,6,9,9-Hexamethyl-1,2,4,5,7,8-hexoxononane, 1234t
Hexamine [CAS: 100-97-0], **773-774**
2,5-Hexanedione, physical properties, 900t
Hexane
    measurement by NDIR analysis, 837t
    in Texas natural gas, 1054t
n-Hexane
    combustion constants, 422t
    heating value, 686t

Hexanedioic acid, 34
    physical properties, 490t
Hexanesulfonic acid, physical properties, 1568t
Hexanol, permeation in selected barrier polymers, 174t
2-Hexanone, physical properties, 900t
3-Hexanone, physical properties, 900t
7,8,15,16,23,24-Hexaoxatrispiro[5.2.5.2.5.2] tetracosane, 1234t
Hexazinone, environmental health advisories, 771t
trans-2-Hexenal, permeation in selected barrier polymers, 174t
1-Hexene, physical properties, 1150t
Hexokinase, metal chelate enzyme, 323t
Hexoses, 279t
Hexosyltransferases, 571t
sec-Hexyl acetate, physical properties, 585t
Hexyl acrylate, physical properties of polymers, 18t
Heyrovsky, Jaroslav (1890-1967), **773**
HG-factor, where produced, structure, and principal functions, 787t
High-alloy irons, 57
High-alloy steels, 58
High density polyethylene, 1140t, 1142-1143
    diffusion and solubility coefficients for oxygen and carbon dioxide, 173t
    gas permeability, 172t
    permeation of flavor/aroma compounds, 174t
    water-vapor transmission rate, 173t
High-density polyethylene (HDPE)
    in municipal waste, 1714t
    physical properties of structural foams, 665t
Higher amines, aliphatic, **482-484**
Higher olefin polymers, 1148-1151
High impact polystyrene, physical properties of structural foams, 665t
High polymers, 1341
High pressure jet cutting, 2
High-pressure technology, 1368
High-quartz, 1397
High solids alkyd resins, 54
High-speed tool steel, hardness, 755t
High-strength low-alloy steels, 58
High-temperature, high-strength, iron-base alloys, 58
High temperature alloys, **774-776**
High-temperature gas-cooled nuclear reactor, 1109-1113
High vacuum, 1661t
High volatile A bituminous, characteristics, 390t
High volatile B bituminous, characteristics, 390t
High volatile C bituminous, characteristics, 390t
Hildebrand, Joel (1891-1983), **777**
Hildebrand rule, **777**
Hinsberg reaction, **777**
Hinshelwood, Sir Cyril N. (1897-1968), **777**
Histamine, **777-778**
    constituent of human blood, 242t
Histamine antagonists, **777-778**
Histamine receptors, **777-778**
Histamine release, 777
Histidine
    first isolation and isoelectric point, 78t
    manufacture, 76t
    production, 80
    structural classification, 79t
Histidinemia, genetic influence, 716
Histones, **778**
Histosols, 1497t
Hodgkin, Dorothy C. (1910-1994), **778**
Hoffman, Roald (1937-), **778**
Hofmann, August Wilhelm (1818-1892), **778**
Hofmann degradation, 778

Hofmann isonitrile synthesis, **778**
Hofmann-Loffler-Freytag reaction, **778**
Hofmann rearrangement, 1424
Hofmann rule, **779**
Hofmann's reaction, **779**
Hofmeister series, **389**
Hollocellulose, 1751
Hollow-fiber membranes, **779-780**
Holmium
    abundance, 330t
    electronic structure, 337t
    interatomic distance, 342t
    ionic crystal radius, 341t
    nuclides (isotopes and isobars), 331t
    principal characteristics, 327t
    properties, 1421t, 1422t
Holmium [CAS: 7440-60-0], **780-781**
Holograms, 781-782
Holography, **781-783**
Holohedral crystal, **782**
Homocyclic compound, 428
Homocystinuria, genetic influence, 716
Homogeneous, **783-784**
Homogeneous catalysts, 303
Homogenizing, **784**
Homolaudanosine, 49t
Homologous series, **784**
Homometric crystal pairs, **461**
Homopolymer, **784**
Hong Kong flu, 1694-1695
Hooke's law, **784**
Hopfield bands, 3
Hormones, **784-792**
Hornblende, **792**
    abrasion pH, 1t
Hornblendite, **792**
Horse, body weight-metabolic rate relationship, 176t
Horsepower (hp), 1644t
Hot-dip galvanizing, 703
Hot filament ionization vacuum gage, 1664
Hot melt adhesives, load-bearing capabilities, 32t
Hot-melt adhesives, 33
Hot working, **792**
Hour (h), 1644t
Hrd resins, 1437-1438
Huber, Robert (1937-), **792**
Huggins bands, 3
Human, body weight-metabolic rate relationship, 176t
Human Genome Project, 716-717
    major events in, 718-719t
Human growth hormone
    annual sales, 222t
    unit value and relative production quantity, 232t
Human Immunodeficiency Virus (HIV) vaccine, 1660
Human von Willebrand factor (vWF), 243-244
Humectants, use in food processing, 671t
Humectants and moisture-retaining agents, **792-793**
Hume-Rothery rules, **793**
Humidity, **793**
    influence on corrosion, 445
Hunter's syndrome, genetic influence, 716
Hurler's syndrome, genetic influence, 716
Hyatt, John Wesley (1837-1920), **793**
Hydantoin and derivatives, **793-794**
Hydracrylic acid, carboxylic acid, 294
Hydrastine, 49t
Hydrate, **794**
Hydration, 1181
Hydraulic fluid, **794**

Hydrazine
    as parent structure compound, 1090t
    as propellant, 1447t
Hydrazine [CAS: 302–01–2], **794–795**
Hydrazino group, analysis, 97
Hydrazoic acid [CAS: 7782–79–8], **795**
Hydrazones, 48, **795**, 899
Hydrea, antimetabolite, 355t
Hydrides, **795–796**
Hydroboration, **797**
Hydroboration-oxidation, 1178
Hydrocarbons, 1170–1181
    air pollution, 1328
    oxidation, 47
    pollution in oceans, 1731–1732
Hydrochloric acid [CAS: 7647–01–0], 12, **797**
    acidulant, 14
Hydrochlorofluorocarbons (HCFCs), as greenhouse gas, 744
Hydrocolloid, **797**
Hydrocolonic sonography, 1639
Hydrocortisone, where produced, structure, and principal functions, 786t
Hydrocracked gasoline, 1255
Hydrocracker, 1255
Hydrodealkylation (HDA), **797**
Hydrofluorocarbons (HFCs), as greenhouse gas, 744
Hydroforming, **797**
Hydroformylation, 1181
Hydrogem cyanide [CAS: 74–90–8], **804**
Hydrogen, 48
    abundance, 5t, 330t, 1132t
    catalyst application in production, 304t, 306
    combustion constants, 422t
    combustion reactions, 423t
    covalent radius, 344t
    density, 474t
    electronic structure, 336t
    fuel properties, 801t
    heating value, 686t
    ignition temperature in air, 425t
    ionic crystal radius, 341t
    nuclides (isotopes and isobars), 331t
    percentage in air, 46t
    principal characteristics, 327t
    standard electrode potential, 31t
    in Texas natural gas, 1054t
    van der Waals characteristic equation coefficients, 321t
Hydrogen (Fuel), **799–804**
Hydrogen-annealed iron, magnetic properties, 955t
Hydrogenated vegetable oils, microcapsule material, 997t
Hydrogenation, **799**, 1177, 1180
    catalyst application, 304t, 305
    fats and oils, 1671
Hydrogen bond, 346
Hydrogen [CAS: 1333–74–0], **797–799**
    specific heat at constant pressure, 757t
Hydrogen chloride, 797
Hydrogen-ion activity, **804–806**
Hydrogenolysis, 1181
Hydrogen-oxygen cell, 189
Hydrogen peroxide
    as parent structure compound, 1090t
    as propellant, 1447t
Hydrogen peroxide [CAS: 7722–84–1], **806–807**
Hydrogen scale, **807**
Hydrogen sulfide, 1571
    combustion constants, 422t
    combustion reactions, 423t
    girbotol adsorption, 723

removal from raw water, 1724
    in Texas natural gas, 1054t
Hydrogen sulfide [CAS: 7783–06–4], **807**
Hydroisoandrosterone, 98
Hydrolases, 571t
Hydro-lyases, 571t
Hydrolysis, **807**, 1180, 1525
    yeast participation, 1769t
Hydrolyzed polyacrylamide, water-soluble polymer, 1737–1738
Hydrometallurgy, 987
Hydromorphone, 92t
Hydronium ion, **807**
Hydrophilic, **807**, 947
Hydrophobic, **807**, 947
Hydroquinones, **807–808**
Hydrotreater, 1255
Hydrotreating, **808**
    catalyst application, 304t, 305
Hydrous calcium silicate, insulation material, 856t
Hydroxide minerals, 1011
17-ß-Hydroxy-4-androsten-3-one, where produced, structure, and principal functions, 789t
m-Hydroxybenzoic acid, 1455
p-Hydroxybenzoic acid, 1455
Hydroxycarboxylic acids, **808–809**
17-Hydroxycorticosterone, where produced, structure, and principal functions, 786t
1-Hydroxycyclohexyl 1-hydroperoxycyclohexyl peroxide, 1234t
Hydroxy dicarboxylic acids, **810–811**
3-Hydroxy-1,1-dimethylbutyl hydroperoxide, physical properties, 1231t
3-Hydroxy-1,1-dimethyl-butyl peroxyneodecanoate, properties, 1239t
1-Hydroxyethyl ethyl peroxide, boiling point, 1234t
1-Hydroxyethyl hydroperoxide, 1233t
2-Hydroxyethyl methacrylate, physical properties, 988t
1-Hydroxyethyl methyl peroxide, boiling point, 1234t
Hydroxylamine [CAS: 7803–49–8], **809–810**
Hydroxylases, 571t
Hydroxyl group, analysis, 97
Hydroxyl group alcohols, 46
Hydroxyl radical, and acid-base chemistry, 13
1-Hydroxymethyl-5,5-dimethylhydantoin, 794
Hydroxymethyl hydroperoxide, melting point, 1233t
Hydroxymethyl methyl peroxide, boiling point, 1234t
4-Hydroxy-4-methyl-2 pentanone, physical properties, 900t
2-(2′-Hydroxy-5′-Methylphenyl) benzotriazole, 1641
Hydroxymethyltransferases, 571t
1-Hydroxynonyl hydroperoxide, 1233t
2-Hydroxy-4-octoxybenzophenone, 1641
1-Hydroxyoctyl hydroperoxide, 1233t
1-Hydroxypentyl hydroperoxide, 1233t
Hydroxyproline
    first isolation and isoelectric point, 78t
    manufacture, 76t
    production, 80
    structural classification, 79t
2-Hydroxypropyl methacrylate, physical properties, 988t
Hydroxyurea USP, antimetabolite, 355t
Hygrine, 49t
Hygrometry and psychometry, **811–814**
Hygroscopic, **814**
Hyperammonemia, genetic influence, 716
Hyperconjugation, **814**
Hypereutectic alloy, **814**
Hyperfine structure, **814**

Hyperglycemic-glycogenolytic factor, where produced, structure, and principal functions, 787t
Hyperlipoproteinemias, genetic influence, 716
Hyperons, **814**
Hyperprolinemia, genetic influence, 716
Hypersorption, **814**
Hypersthene, **814–815**
Hypobaric (controlled-atmosphere) systems, **815**
Hypochlorination, 1180
Hypochlorites, **815–816**
Hypoeutectic alloy, **816**
Hypofluorite, **816**
Hypoiodites, 816
Hypoiodous acid and hypoiodites, **816**
Hyponitrites, 816
Hyponitrous acid and hyponitrites, **816**
α-Hypophamine, where produced, structure, and principal functions, 788t
Hypophosphates, 816
Hypophosphites, 816
Hypophosphoric acid, 1279
Hypophosphoric acid and hypophosphates, **816**
Hypophosphorous acid, 1279
Hypophosphorous acid and hypophosphites, **816**
Hypophyseal growth hormone, where produced, structure, and principal functions, 789t
Hyposulfites, 816
Hyposulfurous acid and hyposulfites, **816**
Hypothesis, 816
Hysteresis, **816–817**

I
Ibuprofen, 93, 93t
-IC, 819
ICE, 819
Ice, 1718–1719
    absolute index, 1426t
    thermal conductivity, 758t
Icebergs, 1735–1736
Ice point, **819**
Idaein, 734t
Idamycin, DNA interactive agent, 357t
Idarubicin hydrochloride, DNA interactive agent, 357t
-IDE, 819
Ideal gas law, **819–820**
Ideal system, **820**
Idiopathic hyperbilirubinemia, genetic influence, 716
Ifex, DNA alkylating/cross-linking agent, 356t
Ifosphamide USP, DNA alkylating/cross-linking agent, 356t
Ilamycin A, activity and producing organism, 128t
Illium G, 58, 1071t
Ilmenite, **820**, 871t
Imazamethabenz, environmental health advisories, 771t
Imazapyr, environmental health advisories, 771t
Imazaquin, environmental health advisories, 771t
Imazethapyr, environmental health advisories, 771t
2,4-Imidazolidinedione, 793
Imides, **820**
Imine, 1169t
Imines, cyclic, **820–821**
Imino acids, first isolation and isoelectric point, 78t
Iminobispropylamine, physical properties, 483t
Imino compounds, **821**
Iminourea, 747
Imipenem, year of discovery/market introduction, 106t
Immune system and immunochemistry, **821–823**

Immunochemistry, 821–823
Inceptisols, 1497t
Inch (in), 1644t
Inch of mercury (inHg), 1644t
Inch of water (inH2O), 1644t
Incineration, **824**
Inclusion compound, 428
Inclusion compounds, **824–825**
Inco, 58
Incoloy 800, 1071t
Inconel, 58
Inconel 600, 1071t
Indeterminacy principle, 1642
Indican, 734t
Indicator (Chemical), **825**
Indigo, dyeing process, 522
Indium
    abundance, 330t, 1132t
    corrosion resistant alloys, 446t
    covalent radius, 344t
    electronic structure, 336t
    interatomic distance, 342t
    ionic crystal radius, 341t
    nuclides (isotopes and isobars), 331t
    principal characteristics, 327t
Indium [CAS: 7440–74–6], **825–826**
Indocin, 93t
Indoles, **826**
Indomethacin, 93t
Indoor radon, 1417–1487
Industrial biotechnology, **826–830**
Industrial plasters, 750
Inert, **830**
Inert gases, 706, **830**
Infared film, 1292–1293
Inflammation, 91
Influenza A viruses, 1694
Influenza B viruses, 1694
Influenza C viruses, 1694
Influenza vaccine, 1659, 1660
Information retrieval, **830–832**
Infrared absorption bands, 3
Infrared radiation, **832–837**
Ingot, **838**
Inhibitor, **838**
Initiators (Anionic), **838–839**
Initiators (Cationic), **839–840**
Initiators (Free-Radical), **840–842**
Injection-molded SAN, 21
Inner compound, 428
Inner salt, 1456
Inorganic acids, 12
Inorganic bases, 12
Inorganic chemistry, **842**
Inorganic compound, 428
Inorganic high-polymers, **843–844**
Inorganic nomenclature, 1088–1089
Inorganic pigments, **1305–1308**
Inorganic polymers, **1347–1348**
Inosilicate minerals, 1013
Inositol, **844**
Insecticide, **844–848**
Insecticide and pesticide technology, **848–851**
Insects, as plant-growth control agents, 772–773
Instrumental analysis, **1532**
Insulating cement, insulation material, 857
Insulating concrete, insulation material, 857
Insulation (Electric), **851**
Insulation (Thermal), **851–857**
Insulators, electronic structure, 1519
Insulin, 791, **857**
    annual sales, 222t

unit value and relative production quantity, 232t
    where produced, structure, and principal
        functions, 787t
Intercalation compounds, **857**
Interface, **857–858**
Interfacial forces, 662, 1583–1584
Interferon, 1696–1697
    annual sales, 222t
Interleukin-2, annual sales, 222t
Intermediate (Chemical), **858**
Intermediate-moisture foods, **858**
Intermetallic compound, 428
Intermetallic compounds, **858**
Intermetallics, high temperature, 774t
Internal conversion coefficient, **436**
International Space Station, microgravity research,
    742
International Thermonuclear Experimental Reactor
    (ITER), 699
Interplanetary dust, 600
Interstitial, **858**
Interstitial compound, 428
Intrinsic bioremediation, 205t
Inulin, 1657
In vitro, **858**
In vivo, **858**
In-well bioreactor, 205t
Iodine, 754
    abundance, 330t, 1132t
    constituent of human plasma or serum, 244t
    covalent radius, 344t
    electronic structure, 336t
    ionic crystal radius, 341t
    nuclides (isotopes and isobars), 331t
    principal characteristics, 327t
    standard electrode potential, 31t
    trace mineral nutrient, 1002t, 1004
Iodine [CAS: 7553–56–2], **858–860**
Iodine (In Biological Systems), **860–861**
Iodine value, 1671
Iodoform, 7
Ion, **861–862**
Ion channels, 1272
Ion-exchange resins, **862–864**
Ion exclusion, **865**
Ionic compound, 428
Ionic crystal, **865**
Ionic equilibrium, **865**
Ionic mobility, **865**
Ionic polymerization, 1346
Ionic radius, 340–341
Ion implantation, **865**
Ionization, **866**
Ionization chambers, 1759
    ultraviolet radiation, 1640
Ionized gases, **866**
Ionizing particles, **866**
Ionizing radiation, **1406,** 1407
    definitely established workplace carcinogen, 297t
Ion lasers, wavelength range, 910t
Ion microprobe mass analyzer, **866–868**
Ion nitriding, 763t
Ionomers, **868–869**
Ionosphere, 156
Ion retardation, **869**
Ipatieff, Vladimir N. (1890–1952), **869**
Ipecac, 51
Iridescence, **869**
Iridium
    abundance, 330t
    electronic structure, 337t
    interatomic distance, 342t

ionic crystal radius, 341t
    nuclides (isotopes and isobars), 331t
    principal characteristics, 327t
    properties, 1317t
Iridium [CAS: 7439–88–5], **869–870**
Iridium oxide anodes, 982
Iron
    abundance, 330t, 1132t
    biologically active chelates, 323t
    constituent of human plasma or serum, 244t
    covalent radius, 344t
    density, 474t
    electronic structure, 336t
    interatomic distance, 342t
    ionic crystal radius, 341t
    nuclides (isotopes and isobars), 331t
    plant micronutrient, 613t
    plant micronutrient deficiencies, soils of U.S.,
        616t
    principal characteristics, 327t
Iron-air cells, 186t, 189
Iron alloys, 57–58
Iron [CAS: 7439–89–6], **870–876**
    abundance, 5t
    magnetic properties of annealed and
        alloys, 955t
    major sources of industrial minerals, 1011t
    physical properties of high temperature, 774t
    removal from raw water, 1724
    specific heat at constant pressure, 757t
    standard electrode potential, 31t
    thin-film applications, 958t
    trace mineral nutrient, 1002t, 1003
    water quality indicator, 1726
Iron chromite, colorant for glass, 725t
Iron-constantan thermocouples, temperature range,
    1605t
Iron metals, alloys, and steels, **877–887**
Iron meteorites, 600
Iron-nickel, thin-film applications, 958t
Iron-nickel-chromium, thin-film applications, 958t
Iron-nickel superalloys, 776
Iron oxide
    colorant for glass, 725t
    corrosion resistant alloys, 446t
Iron pyrites, 871t
Iron sulfide, colorant for glass, 725t
Irradiation, gemstones, 708t, 708–709
Irreversibility, 902–903
Isentropic process, 34
Isoamyl acetate, physical properties, 585t
isoamyl alcohol, physical properties, 89t
Isoamylethylene, melting range, 1546t
Isobars, 330–332, 331t
Isobornyl acrylate, physical properties of polymers,
    18t
Isobutane
    combustion constants, 422t
    physical properties, 263t
    in Texas natural gas, 1054t
Isobutene, combustion constants, 422t
Isobutyl acetate, physical properties, 585t
Isobutyl acrylate
    physical properties, 18t
    physical properties of polymers, 18t
Isobutyl alcohol, physical properties, 262t
Isobutylene, physical properties, 263t
Isobutylethylene, melting range, 1546t
Isobutyl heptyl ketone, physical properties, 900t
Isobutyl isobutyrate, physical properties, 585t
3-Isobutyl pyrazine, as key odor/tase compound,
    649

Isobutyl vinyl ether
    melting range, 1546t
    physical properties, 1690t
Isocitric acid, 282t
Isocitricdehydrogenase, metal chelate enzyme, 323t
Isocyanates, polyurethanes from, 1654
Isocyanates, organic, **886–888**
Isocyanurate laminate, physical properties of rigid
        foamed plastics, 664t
Isodesmic structure, **888**
Isodiaphere, **888**
Isodimorphism, **888**
Isoelectric point, 76, 77t, 79
Isoflurane, 100t
Isoleucine
    first isolation and isoelectric point, 77t
    manufacture, 76t
    in selected foods, 75t
    structural classification, 79t
Isomerases, 571t
Isomerism, **888–889**
Isomerization, **889**, 1179
    catalyst application, 305
Isomer (Nuclear), **888**
Isomesityl oxide, physical properties, 900t
Isomorphous crystal, **461**
Isonitrile, 1169t
Isoparaffins, gas separation, 39t
Isopelletierine, 49t
Isopentane
    combustion constants, 422t
    in Texas natural gas, 1054t
Isophorone, physical properties, 900t
ß-Isophorone, physical properties, 900t
Isophorone diisocyanate (IPDI), 1654
Isoprene, **889–890**
cis-1,4-Isoprene, melting range, 1546t
trans-1,4-Isoprene, melting range, 1546t
Isopropyl acetate, physical properties, 585t
Isopropyl acrylate, physical properties of polymers,
        18t
Isopropyl alcohol [CAS: 67–63–0], 890
Isopropyl ether, physical properties, 587t
Isopropylethylene, melting range, 1546t
Isopropyl hydroperoxide, physical properties, 1231t
Isopropylnitrate, 1082
Isopropylnitrite, 1082
Isopropyl vinyl ether, physical properties, 1690t
Isoquinoline, 1401–1403
    physical properties, 1400t
Isoquinoline alkaloids, 49t
Isotactic 1,2-butadiene, melting range, 1546t
Isothebaine, 49t
Isothermal, 890
Isotone, **890**
Isotope, **890–891**
Isotopes, 330–332, 331t
Isotretinoin USP, chemotherapeutic, 359t
Isoxaben, environmental health advisories, 771t
Istle, 632t
Iturins A, activity and producing organism, 128t
Izupeptin A, producing organism, 118t
Izupeptin B, producing organism, 118t

J

Jacobsen rearrangement, **893**
Jacquemart's reagent, **893**
Jade, **893**
Jadeite, **893**
Jamesonite, **893**
Janovsky reaction, **893**

Japanese encephalitis, 1660
Japanwax, 1747
Japp-Klingemann reaction, **893**
Jarosite, **893**
Jasper, **893**
Jatrophone, **893**
Jelly, **893**
Jenner, Edward (1749–1823), **893**
Jet cutting, 2
Jet fuel, **893**
Jetset, **893**
Jewelry bronze, 440t
JH (methyl-cis-10,11-epoxy-7-ethyl-3,
        11-dimethyl-trans,trans-2,6-tridecadienoate),
        894
Joliet-Curie, Frederick (1900–1958), **894**
Joliet-Curie, Irene (1897–1956), **894**
Jones oxidation, **894**
Josamycin, commercial products, 121t
Josephson, Brian David (1940–), **894**
Josephson effect, 894
Josephson junctions, 894
Josephson tunnel-junction, **894**
Joule, James Prescott (1818–1889), **894**
Joule (J), 1644t
Joule law, **894**
Joule per Kelvin (J/K), 1644t
Joule-Thomson coefficient, 894
Joule-Thomson effect, **894–895**, 902
Joule-Thomson inversion temperature, 894
Jourdan-Uiimann-Goldberg synthesis, **895**
J particle, 1397
Jute, 632t
Juvenile hormones, **895**

K

K-288, producing organism, 118t
Kanamycin
    antibiotic resistance markers, 710
    year of discovery/market introduction, 106t
Kanglemycins, biological activity, 108t
Kaolin, 897
    analysis, 316t
    physical properties, 1428t
    as release agent, 1435t
Kaolinite, **895**
    abrasion pH, 1t
    hardness, 1008t
Kapok, 632t
Karats, 735
Karaya gum, 748t
Karbutilate, environmental health advisories, 771t
Karle, Jerome (1918–), **897**
Karrer, Paul (1889–1971), **897**
Katamorphism, 98
Kauri-butanol value, **897**
K-edge dichromography, 1762
Kekule, August (1829–1896), **897**
Kelvin (K), 1644t
Kelvin temperature scale, 3, 1599
Kenaf, 632t
Kendrew, John C. (1917–1997), **897**
Keratin, **897**, 1752
Keratinase, **897**
Kerma, **897**
Kernite, **897**
Kerogen, **898**
Kerosene
    in representative U.S. crude oils, 1242t
    in representative world crude oils, 1243t
Kerosine, ignition temperature in air, 425t

Ketene lamp, 898
Ketenes, **898–899**, 899
Ketimine, **899**
Ketoacid-lyases, 571t
Ketoaciduria, genetic influence, 716
α-Ketoglutaric acid, 282t
α-Ketoglutaric acid dehydrogenase, 282t
Ketohexoses, 279t
Ketone, 1169t, 1174
Ketone bodies, constituent of human blood, 242t
Ketone peroxides, organic peroxide initiators, 841t
Ketones, **899–901**
    aldol condensation, 48
    physical properties of selected, 900t
Ketone splitting, 7
Ketonic acids, 899
Ketonic hydrolysis, 899
Ketoprofen, 93t
Ketoses, 899
Ketoximes, 899
Kevlar, 622t
Keyes equation, **901**
Keyes process, **901**
Kibdelin, producing organism, 118t
Kidneys
    acid-base regulation in, 8
    role in sodium/potassium balance, 1363–1364
Killed steels, 58
Kilo, 1643t
Kilogram (kg), 1644t
Kimberlite, **901**
Kinetic measurements, **901–902**
Kinetic polarimetry, 1322
Kinetics, adsorption, 37
Kinetic theory, **902–903**
Kinetins, 1314
Kinins, **903**
Kirromycin, formula and producing organism, 116t
Kirrothricin, formula and producing organism, 116t
Kishner cyclopropane synthesis, **903**
Kistiakowsky, George B. (1900–1982), **903–904**
Kjeldhal test, **904**
Klebsiella, amphenicol susceptibility, 115t
Klebsiella pneumoniae
    gram-negative, 168t
    ß-lactamase-producing bacterium, 110t
Klein-Gordon equation, 1395
Klug, Aaron S. (1926–), **904**
Kluyveromyces fragilis, 1767, 1768
Knoop hardness, 2t
Knot (kn), 1644t
Koagulationsvitamin, 1706
Kodel, 623t
Komarowsky reaction, **904**
Konowaloff rule, **904**
Kopp's law, **904**
Krabbe's disease, genetic influence, 716
Kraft pulping process, 1379
Krebs citric acid cycle, 281, 282t
Krebs cycle, 1596–1597
Kroll process, **904**
Kroto, Sir Harold W. (1939–), **904**
Krypton
    abundance, 330t, 1132t
    electronic structure, 336t
    nuclides (isotopes and isobars), 331t
    percentage in air, 46t
    principal characteristics, 327t
Krypton [CAS: 7439–90–9], **904**
K shell, 335, 335t
    electronic structure of the elements, 336–337t
Kucherov reaction, **905**

Kuhn, Richard (1900–1967), **904–905**
Kuhn-Winterstein reaction, **905**
Kuru, 1696
KW2149, DNA alkylating/cross-linking agent, 356t
Kyanite, **905**, 1009t
    physical properties, 1428t

L

L681,217, formula and producing organism, 116t
Labradorite, 607
    abrasion pH, 1t
Lacquer, **907**
ß-Lactam antibiotics, **111–112**
ß-Lactamase inhibitor antibiotics, **109–111**
Lactase, metal chelate enzyme, 323t
Lactase deficiencies, genetic influence, 716
Lactate dehydrogenase, target of medical diagnostic
    test, 975t
Lactic acid
    acidulant, 14
    constituent of human blood, 242t
Lactic acid [CAS: 50–21–5], **907–908**
Lactofen, environmental health advisories, 771t
Lactoflavin, 1699
Lactogen, where produced, structure, and principal
    functions, 788t
Lactogenic hormone, where produced, structure, and
    principal functions, 788t
Lactone, **908**
ß-Lactones, monomer reactivity and suitable initiator
    for, 838t
Lactose, 1000
    purification, 863t
Lactose [CAS: 63–42–3], **908**
Ladder polymer, **908**
Lakes, acidification, 9
Lambda particle, **908**
Lambert (L), 1644t
Lamb waves, 1637
Laminar flow, **908**
Laminate, **908**
Lampblack, **908**
Lana, 623t
Landerite, 1755
Land-farming, 205t
Land reclamation, 1440
Land treatment, 205t
Langmuir, Irving (1881–1957), **908–909**
Langmuir adsorption isotherm, 36–37
Langmuir-Blodgett films, 1017–1026
Langmuir-Blodgett film sensors, 1024–1026
Lanolin, 1747
    as release agent, 1435t
Lanthanide contraction, **909**
Lanthanides, principal characteristics of elements,
    327–328t
Lanthanide series, **909**
Lanthanum
    abundance, 330t, 1132t
    covalent radius, 344t
    electronic structure, 336t
    interatomic distance, 342t
    ionic crystal radius, 341t
    nuclides (isotopes and isobars), 331t
    principal characteristics, 327t
    properties, 1421t, 1422t
Lanthanum [CAS: 7439–91–0], **909**
Large scale integration, 1017
Laser chemistry, 1285–1286
Laser femtochemistry, 1286
Laser glasses, 725

Lasers, **909–919**
    optical fiber applications, 1156–1157
Latent heat, **920**
Laterite, **920**
Latex, **920**
Latex foam, production methods for cellular, 664t
Latex foam rubber, physical properties of flexible
    foamed plastics, 665t
Latex technology, **920–921**
Lattice compounds, **921**
Lattice energy of crystal, **921**
Lattice water, 794
Laue pattern, 1758
Lauric acid
    fatty acid in vegetable oil, 1671t
    in major vegetable oils, 1672–1673t
Lauric acid [CAS: 143–07–7], **921**
Laurolactam, monomer for nylon resins, 1333t
Lauryl methacrylate, physical properties, 988t
Lava, **921**
Lavender oil, 647t
Laviosier, Antoine Laurent (1743–1794), **921–922**
Lawrence, Ernest O. (1901–1958), **921**
Lawrencium, 23t
    binary compound properties, 25–26t
    electronic structure, 337t
    oxidation states, 24t
    principal characteristics, 327t
Lawrencium [CAS: 22537–19–5], **921**
Lawsonite, **921**
Lazulite, **921–922**
Lazurite, **922**
Leaching, 987
Lead
    abundance, 330t, 1132t
    constituent of human plasma or serum, 244t
    corrosion resistant alloys, 446t
    density, 474t
    electronic structure, 337t
    EPA pretreatment standard for aqueous
        discharge, 985t
    in hazardous waste, 1711t
    interatomic distance, 342t
    ionic crystal radius, 341t
    major sources of industrial minerals, 1011t
    nuclides (isotopes and isobars), 331t
    principal characteristics, 327t
    residence time in sea water, 1133t
    specific heat at constant pressure, 757t
    standard electrode potential, 31t
    water quality indicator, 1726
Lead acetate, 6
Lead-acid batteries, **181–182**
Lead base babbits, 923t
Lead [CAS: 7439–92–1], **922–925**
Lead diecasting alloys, antimony content, 139t
Leaded brass, 440t
Lead glasses, 725
Leather beating enzymes, 306
Leavened bread, 1767
Leavening agents, **925–926**
Le Châtelier's Principle, **926**
Lecithin, **926–927**
    constituent of human blood, 242t
Leduc's rulke, **927**
Lee, Yuan T. (1936–), **927**
Lehn, Jean-Marie Pierre (1939–), **927**
Lelanché cell, 179t
    cell reactions, energy content, and
        manufacturers, 179t
Leloir, Luis F. (1906–1987), **927**
Lemongrass oil, 647t

Lemon oil, 647t
Length of unused bed (LUB), adsorption columns,
    37
Lentaron, hormonal therapy, 358t
Lepidocrocite, 871t
Lepidolite, **927**
    mineralogical properties, 993t
Leptons, **927**
Lesch-Nyhan syndrome, genetic influence, 716
Leucine
    first isolation and isoelectric point, 77t
    manufacture, 76t
    production, 80
    in selected foods, 75t
    structural classification, 79t
Leucite, **927–928**
    abrasion pH, 1t
Leuprolide acetate USP, hormonal therapy, 358t
Leupron, hormonal therapy, 358t
Levarterenol, where produced, structure, and
    principal functions, 788t
Levarternol, 35–36
Levene-Hudson phenylhydrazide rule, **928**
Levorenine, where produced, structure, and principal
    functions, 786t
Levorotatory, 78, 1543
Levorotatory compounds, **928**
Levorphanol, 92t, 93t, 1041
Lewis, Gilbert N. (1875–1946), **928**
Lewis, Warren P. (1882–1974), **928**
Lewis acid, **928**
Lewis base, **928**
Lewis electron theory, **928**
Lewis salt, 13
Leyden temperature scale, **928**
Libby, Willard P. (1908–1980), **928**
Lidocaine, 100t
Liebig, Justus Von (1803–1873), **928**
Life, origin (Biogenesis), **928**
Ligand, **928–929**, 1542–1543
Ligases, 571t
Light distillate, 1255
Light-emitting diode, optical fiber applications, 1157
Light glass, 1163t
Light-sensitive materials, 1485
Light stabilizer, 1641
Light straight-run, 1255
Light-water reactor (LWR), 1647
Lignaloe, 647t
Lignin, **929**, 1751
Lignite A, characteristics, 390t
Lignite and brown coal, **929–930**
Lignite B, characteristics, 390t
Lime, **930–931**
Lime oil, 647t
Limestone, **930–931**
Limestone [CAS: 1317–65–3], 930–931
D-Limonene, permeation in selected barrier
    polymers, 174t
Limonite, 871t, **931**
Lincomycin, year of discovery/market introduction,
    106t
Lincosaminide antibiotics, **119–120**
Lindane insecticide, in hazardous waste, 1711t
Linear accelerator, 1215
Linear alkane, 1169t
Linear low density polyethylene, 1140t, 1143–1146
Linen, fiber-dye property requirements, 521t
Linnaeite, cobalt source, 410t
Linoleic acid, **931**
Linolenic acid, **931**
    fatty acid in vegetable oil, 1671t

Linotype, 923t
Linseed oil, 1671
    fatty acid composition, 507t
Linuron, environmental health advisories, 771t
Lipase
    metal chelate enzyme, 323t
    target of medical diagnostic test, 975t
    therapeutic enzyme, 574t
Lipids, **931**
    constituent of human blood, 242t
Lipmann, Fritz (1899–1986), **932**
Lipowitz' alloy, bismuth in, 238t
Lippmann hologram, 781
Lipscomb, William N. (1919–), **932**
Liquation, **932**
Liquefied petroleum gas LPG, 1255
Liquid, **932**
Liquid carburizing, 763t
Liquid Chromatography, 41
Liquid chromatography, **932–933**
    for ultratrace analysis, 1627t
Liquid chromatography/mass spectrometry, for
    ultratrace analysis, 1627t
Liquid crystalline materials, **933–935**
Liquid crystal polymers, **935**
Liquid crystals, **935–937**
Liquid cyaniding, 763t
Liquid junction, **937**
Liquid-liquid absorption, 4–5
Liquid-liquid extraction, **594–598**
Liquid-phase adsorption, 40–42
Liquid-solid extraction, **598–599**
Liquid state, **937–941**
Liquidus curve, 57
Liter (l), 1644t
Lithium, 48
    abundance, 330t, 1132t
    covalent radius, 344t
    electronic structure, 336t
    interatomic distance, 342t
    ionic crystal radius, 341t
    nuclides (isotopes and isobars), 331t
    principal characteristics, 327t
    residence time in sea water, 1133t
    standard electrode potential, 31t
Lithium-acid cell, cell reactions, energy content, and
    manufacturers, 180t
Lithium-aluminum/iron sulfide cell, 182–183
    cell reactions and energy content (in
        development), 180t
Lithium-carbon cell, cell reactions, energy content,
    and manufacturers, 180t
Lithium-carbon monofluoride cell, cell reactions,
    energy content, and manufacturers, 179t
Lithium [CAS: 7439-93-2], **941–943**
Lithium-copper oxide cell, cell reactions, energy
    content, and manufacturers, 179t
Lithium-drifted Si detectors, 1759
Lithium fluoride, for diffracting crystal spectrometer,
    1760
Lithium hydride, physical properties, 795t
Lithium-iodine cell, cell reactions, energy content,
    and manufacturers, 179t
Lithium-iron sulfide cell, cell reactions, energy
    content, and manufacturers, 179t
Lithium-manganese dioxide cell
    cell reactions, energy content, and
        manufacturers, 179t
    rechargeable, 180t
Lithium-molybdenum disulfide cell, cell reactions,
    energy content, and manufacturers, 180t
Lithium soap greases, 743

Lithium-solid polymer electrolyte cells, 182
Lithium-sulfur dioxide cell, 185
    cell reactions, energy content, and
        manufacturers, 179t
Lithium-thionyl chloride cell, 185
    cell reactions, energy content, and
        manufacturers, 179t
Lithium-vanadium pentoxide (Li-V2O5) cell, 183
    cell reactions, energy content, and
        manufacturers, 180t
Lithium-vanadium triskaidekaoxide (Li-V6O13) cell,
    cell reactions and energy content (in
        development), 180t
Lithography, nanofabrication application, 1047t
Little, Arthur D. (1863–1935), **943**
LL-E1902ß, formula and producing organism, 116t
LL-E19020α, formula and producing organism, 116t
Local anesthetics, 100–101
Local probe-assisted synthesis, nanofabrication
    application, 1047t
Locust bean gum, 748t
Lometrexol sodium, antimetabolite, 355t
Lomustine USP, DNA alkylating/cross-linking agent,
    356t
London flu, 1694
Long- ultraviolet, 1639
Loose abrasives, 2
Loose-fill Insulation, insulation material, 857
Low alloy steels, 775
Low-alloy steels, 57–58, 58
Low brass, 440t
Low carbon steel, magnetic properties, 955t
Low density polyethylene, 1140t, 1140–1142
    diffusion and solubility coefficients for oxygen
        and carbon dioxide, 173t
    gas permeability, 172t
    permeation of flavor/aroma compounds, 174t
    water-vapor transmission rate, 173t
Low-density polyethylene (LDPE), in municipal
    waste, 1714t
Low-quartz, 1397
Lowry-Bronsted acid-base theory, 12
Low volatile bituminous, characteristics, 390t
LSD-25, 566, 754
L shell, 335, 335t
    electronic structure of the elements, 336–337t
Lubricant, **943**
Lubricating agents, **943–946**
Lucovorin, antimetabolite, 355t
Lumen (lm), 1644t
Lumen per square foot (lm/ft2), 1644t
Lumen per square meter (lm/m2), 1644t
Lumen per watt (lm/W), 1644t
Luminescence, **946–947**
Lunar rocks, 10111
Lupanine, 49t
Lupine alkaloids, 49t
Lustran-35, 21t
Luteinizing hormone (LH), where produced,
    structure, and principal functions, 787t
Luteoantine, where produced, structure, and
    principal functions, 787t
Luteosterone, where produced, structure, and
    principal functions, 788t
Luteotropin, where produced, structure, and principal
    functions, 787t
Lutetium
    abundance, 330t
    electronic structure, 337t
    interatomic distance, 342t
    ionic crystal radius, 341t
    nuclides (isotopes and isobars), 331t

    principal characteristics, 327t
    properties, 1421t, 1422t
Lutetium [CAS: 7439-94-3], **947**
Lux (lx), 1644t
LY146032, activity and producing organism, 128t
Lyases, 571t
Lycorine, 49t
Lycra, 624t
Lyme disease vaccine, 1660
Lymphocytes, 242t, 243
Lyocell fiber, generic designation and definition, 622t
Lyophilic, **947**
Lyophobic, **947**
Lyotropic series, **389**
Lysergic acid diethylamide (LSD), 566, 754
Lysine
    first isolation and isoelectric point, 78t
    manufacture, 76t
    in selected foods, 75t
    structural classification, 79t
    synthesis from products of carbohydrate
        metabolism, 282t
Lysodren, chemotherapeutic, 359t
Lysosomal storage abnormalities, genetic influence,
    716

M

Macbecins, biological activity, 108t
Macerate, **949**
Machinery mold, 1770
Macrolide antibiotics, **120–122**
Macromolecular nomenclature, 1091
Macromolecular science, **949**
Macromolecule, **949**
Macrophages, 242t, 243
Magic acid, 13
Magnesia, dielectric permittivity, 852t
Magnesite, **949**
    abrasion pH, 1t
    physical properties, 1428t
Magnesium, 49
    abundance, 5t, 330t, 1132t
    biologically active chelates, 323t
    constituent of human plasma or serum, 244t
    corrosion resistant alloys, 446t
    covalent radius, 344t
    electronic structure, 336t
    essential mineral nutrient, 1002t, 1003
    hard water, 1726
    in 24 hour urine sample, 1657t
    interatomic distance, 342t
    ionic crystal radius, 341t
    modulus and rigidity, 538t
    nuclides (isotopes and isobars), 331t
    plant micronutrient, 613t
    principal characteristics, 327t
    residence time in sea water, 1133t
    standard electrode potential, 31t
    water quality indicator, 1726
Magnesium(II), target of medical diagnostic test,
    975t
Magnesium acetate, 6
Magnesium alloys, 58
    hardness, 755t
Magnesium carbonate, alkalizer, 14
Magnesium [CAS: 7439-95-4], **949–953**
Magnesium ethyl bromide, 744
Magnesium hydride, physical properties, 795t
Magnesium hydroxide, alkalizer, 14
Magnesium (In Biological Systems), **953–955**
Magnesium monoperoxyphthalate hexahydrate, 1236t

Magnesium nitride, plasma synthesis, 318t
Magnesium oxide
    alkalizer, 14
    dielectric permittivity, 852t
Magnesium stearate, as release agent, 1435t
Magnetic ingot iron, magnetic properties, 955t
Magnetic iron pyrites, 871t
Magnetic materials, **955–958**
Magnetic particle nondestructive testing, 1093–1094
Magnetic pyrites, **1390**
Magnetic quantum number, 1396
Magnetic refrigeration, 1432–1434
Magnetic resonance imaging (MRI), **1099–1100**
Magnetic separation, **958–959**
Magnetic thin films, 1613
Magnetism, **959–961**
Magnetite, 871t, **961**
    hardness, 1008t
Magnetochemistry, **961**
Magnetohydrodynamics (MHD), **961**
Magnetorheological materials, 1485
Magnetostriction, **961**
Magnetostrictive materials, 1484
Majorana force, 1749
Majority carriers, 1467
Malachite, **961–962**
Malachite green, pH range and color change of
    indicator, 825t
Malaic acid dehydrogenase, 282t
Malaprade reaction, **962**
Malaria vaccine, 1660
Maleic acid, carboxylic acid, 294
Maleic acid [CAS: 110–16–7], **962–963**
Maleic anhydride, copolymerization with vinyl
    ethers, 1689
Maleic anhydride, maleic acid, and fumaric acid,
    962–963
Maleic anhydride [CAS: 108–31–61], **962–963**
Maleic hydrazide, 1314
Maleic hydrazide growth inhibitor, **963**
Male pseudobermaphroditism, genetic influence, 716
Male secondary sex characteristics, 98
Male sex hormones, 790
    where produced, structure, and principal
        functions, 786t
Malic acid, 282t
    acidulant, 14
Malleable iron, 57
Malonates, 963–964
Malonic acid and derivatives, **963–964**
Malonic acid [CAS: 141–82–2], **963**
Malononitrile, 964
Malt, **964–966**
Maltha, **966**
Maltodextrin, use in food processing, 671t
Maltodextrins, microcapsule material, 997t
Mammotropin, where produced, structure, and
    principal functions, 788t
Manganese
    abundance, 5t, 330t, 1132t
    amount in pure irons, 877t
    biologically active chelates, 323t
    corrosion resistant alloys, 446t
    covalent radius, 344t
    electronic structure, 336t
    interatomic distance, 342t
    ionic crystal radius, 341t
    major sources of industrial minerals, 1011t
    nuclides (isotopes and isobars), 331t
    plant micronutrient, 613t
    plant micronutrient deficiencies, soils of U.S.,
        616t

principal characteristics, 327t
    removal from raw water, 1724
    trace mineral nutrient, 1002t, 1004–1005
    water quality indicator, 1726
Manganese acetate, 6
Manganese aluminum germanide, thin-film
    applications, 958t
Manganese bismuth, thin-film applications, 958t
Manganese [CAS: 7439–96–5], **966–968**
Manganese gallium germanide, thin-film
    applications, 958t
Manganese (In Biological Systems), **968–969**
Manganese-zinc cells, 186t
Manganite, **969**
Mannacryl, 621t
Mannich reaction, **969**
Manometer, 1368
Many-body force, **969**
Marathon-Howard process, **969**
Marble, **969–970**
    hardness, 1008t
Marcasite, 871t, **970**, 1009t
Marcus, Rudolph A. (1923–), **970**
Marfan's syndrome, genetic influence, 716
Margarines, 1671, 1673
Marijuana, 768
Mark-Houwink equation, **970**
Markownikoff rule, **970**
Marquesa, 623t
Martensite, **970**
Martensitic stainless steel, 58
Martin, Archer J. P. (1910–2002), **970**
Marves, 623t
Maser, **970–971**
Mass, **971**
Mass (Center of), **971**
Mass defect, **971**
Mass number, **971**
Mass spectrometry, **971–974**
    for ultratrace analysis, 1627t
Masticatory substances, **974–975**
Material balance, 170
Matrix metal, 923t
Matter, 1209
Matter-antimatter symmetry, 1394
Matulane, DNA interactive agent, 357t
Mauritius, 632t
Maxell (Mx), 1644t
Maytansinoids, biological activity, 108t
MB isoenzyme (CK-MB), target of medical
    diagnostic test, 975t
McLeod gage, **975**
McMillan, Edwin M. (1907–1991), **975**
M43 complex, producing organism, 118t
Mean free path, 902
Mean square molecular speed, 902
Measles vaccine, 1659
Mechanical equivalent of heat, 757
Mechanical friction, **685**
Mechlorethamine hydrochloride, definitely
    established workplace carcinogen, 297t
Mechlorethamine hydrochloride USP, DNA
    alkylating/cross-linking agent, 356t
Meclofenamate, 93t
Meclomen, 93t
Medical diagnostic reagents, **975–976**
Medicine
    ultrasound applications, 1639
    x-ray scan and other medical imagery,
        1761–1762
Medium-alloy steels, 58
Medium density polyethylene, 1140t

Medium vacuum, 1661t
Medium volatile bituminous, characteristics, 390t
Medroxyprogesterone acetate USP, hormonal
    therapy, 358t
Meerwin-Ponndorf-Verley reduction, 1179
Mega, 1643t
Megace, hormonal therapy, 358t
Megaloblasts, 241–243
Megestrol acetate USP, hormonal therapy, 358t
Meisenheimer complex, 893
Meitnerium, 24, 334
    electronic structure, 337t
    principal characteristics, 328t
Melamine [CAS: 108–78–1], 75, **979**
Melanite, 705
Melanocyte-stimulating Hormone (MSH), where
    produced, structure, and principal functions,
    787t
Melanotropin, where produced, structure, and
    principal functions, 787t
Melanterite, abrasion pH, 1t
Mellophanic acid, carboxylic acid, 294
Melphalan USP, DNA alkylating/cross-linking agent,
    356t
Membrane separations technology, **976–978**
Mendelevium, 23t
    binary compound properties, 25–26t
    electronic structure, 337t
    interatomic distance, 342t
    ionic crystal radius, 341t
    oxidation states, 24t
    principal characteristics, 327t
Mendelevium [CAS: 7440–11–1], **978**
Mendeleyev, Dimitri (1834–1907), **978**
Meningitis vaccine, 1659
Meniscus, **978**
p-Menthane, physical properties, 1231t
Menthanediamine, physical properties, 483t
Mentha piperita, 647t
Menthone, permeation in selected barrier polymers,
    174t
Meperidine, 93t, 1041
Mepivacaine, 100t
Mercaptans, **978**
Mercaptol, 7
Mercaptopurine USP, antimetabolite, 355t
Mercury
    abundance, 330t, 1132t
    density, 474t
    electronic structure, 337t
    interatomic distance, 342t
    ionic crystal radius, 341t
    nuclides (isotopes and isobars), 331t
    principal characteristics, 327t
Mercury-cadmium cells, 186t
Mercury [CAS: 7439–97–6], **978–980**
    in hazardous waste, 1711t
    standard electrode potential, 31t
    thermal conductivity, 758t
    water quality indicator, 1726
Mercury cell, cell reactions, energy content, and
    manufacturers, 179t
Merrifield, R. Bruce (1921–), **981**
Mescaline, 566
Mesitylene, 7
Mesityl oxide, 7
    physical properties, 900t
Mesna USP, DNA alkylating/cross-linking agent,
    356t
Mesnex, DNA alkylating/cross-linking agent, 356t
Mesodesmic structure, **981**
Mesons, **981**, 1396

Mesosphere, 156, 157
Messenger RNA, 713, 1125
Meta-anthracite, characteristics, 390t
Metabolism, **981**
Metachromatic leukodystrophy, genetic influence, 716
Metal anodes, **981–982**
Metal creep, **449–450**
Metal fatigue, **603–604**
Metal hardening, 755
Metallic coatings, **982–985**
Metallic glasses, 58, 730–732
Metallic radius, 341–345, 342t
Metallobiomolecules, **985**
Metallocene catalysts, 1149
Metallocenes, **985–986**
Metallography, **986**
Metalloid, **986**
Metalloproteins, **986**
Metallothionein, source and function, 986t
Metallothioneins, **986**
Metallurgy, **986–987**
    extractive, **599**
    powder, **1364–1365**
Metal-matrix composites, aluminum-containing, 70
Metalorganic compound, 429
Metal oxide semiconductor field effect transistor (MOSFET), 1467
Metals, **987**
    electronic structure, 1518–1519
    in municipal waste, 1714t
Metaphosphoric acid, 1279
Metastable nuclei, **987**
Metastable state, **987**
Meteorites, 599–600
    krypton in, 904
    neon in, 1064
    xenon in, 1757
Meter (m), 1644t
Metglas, magnetic properties, 956t
Metglas 2605, 731t
Methacrylates, monomer reactivity and suitable initiator for, 838t
Methacrylic acid and derivatives, **987–989**
Methacrylic acid [CAS: 79–41–4], 16–17, **987–989**
Methacrylic polymers, **989–990**
Methacycline, year of discovery, 130t
Methadone, 1041
Methane
    combustion constants, 422t
    combustion reactions, 423t
    heating value, 686t
    ignition temperature in air, 425t
    in Texas natural gas, 1054t
Methane [CAS: 74–82–8], **990–991**
    as greenhouse gas, 743–744
    measurement by NDIR analysis, 837t
    percentage in air, 46t
    physical properties, 47t
Methanesulfonic acid, physical properties, 1568t
Methanogens, **991**
Methanol, 991–993
    catalyst application in production, 304t, 306
    physical properties, 47t
    purification, 863t
    substitute natural gas from, 1563
Methionine
    first isolation and isoelectric point, 77t
    manufacture, 76t
    in selected foods, 75t
    structural classification, 79t
Methotrexate, antimetabolite, 355t

Methoxychlor insecticide, in hazardous waste, 1711t
1-Methoxycyclohexyl hydroperoxide, physical properties, 1231t
2-Methoxyethyl acetate, physical properties, 585t
Methoxyflurane, 100t
3-Methoxy-4-hydroxybenzaldehyde, 1668
2-Methoxy-2-propyl hydroperoxide, physical properties, 1231t
Methyl acetate, physical properties, 585t
Methyl acrylate, 17, 20
    physical properties, 18t, 585t
    physical properties of polymers, 18t
Methyl alcohol
    combustion constants, 422t
    heating value, 686t
Methyl alcohol [CAS: 67–56–1], **991–993**
Methylallene, 262
Methyl amyl ketone, physical properties, 900t
Methyl anthranilate, physical properties, 585t
Methyl benzoate
    dye carrier, 512t
    physical properties, 585t
2-Methyl-1,4-benzoquinone, physical properties, 1402t
Methylbenzoylepgonine, 50
Methylbiphenyl, dye carrier, 512t
2-Methyl-1-butanol, physical properties, 89t
2-Methyl-2-butanol, physical properties, 89t
3-Methyl-1-butanol, physical properties, 89t
3-Methyl-2-butanol, physical properties, 89t
3-Methyl-2-butanone, physical properties, 900t
3-Methyl-2-buten-2-one, physical properties, 900t
Methyl tert-butyl ether (MTBE), 587–588
    physical properties, 587t
Methyl n-butyl ketone, physical properties, 900t
Methyl sec-butyl ketone, physical properties, 900t
Methyl butyrate, physical properties, 585t
Methyl chloride, physical properties, 366t
Methyl cresotinate, dye carrier, 512t
Methyl cyanoacetate, 964
1-Methylcyclohexyl hydroperoxide, physical properties, 1231t
4,4′-Methylenebis(phenyl isocyanate) (MDI), polyurethanes from, 1653, 1654
Methylene blue, sensitizer for semiconductors, 535t
Methylene chloride, physical properties, 366t
Methylene dimethyl ether, 6
Methyl-cis-10,11-epoxy-7-ethyl-3,11-dimethyl-trans,trans-2,6-tridecadienoate (JH), **894**
Methylethyl ketone, 899
Methyl ethyl ketone, 899
    physical properties, 900t
Methyl formate, physical properties, 585t
5-Methyl-2-hexanone, physical properties, 900t
Methyl hexyl ketone, physical properties, 900t
5-Methylhydantoin, 794
Methyl hydroperoxide, physical properties, 1231t
Methyl isoamyl ketone, physical properties, 900t
Methyl isobutyl ketone, 899
    physical properties, 900t
Methyl isobutyrate, physical properties, 585t
Methyl isocyanate, 464
Methyl isopropenyl ketone, physical properties, 900t
Methyl isopropyl ketone, physical properties, 900t
Methylketene, 898
Methyl methacrylate, 17, 20
Methylmethacrylate, dielectric permittivity, 852t
Methyl methacrylate
    melting range, 1546t
    physical properties, 585t, 988t
Methylmethane, 586
Methyl N-methyl-anthrailate, as key odor/tase compound, 649

Methylmorphine, 50
1-Methylnaphthalene, dye carrier, 512t
2-Methylnaphthalene, dye carrier, 512t
2-Methyl-1,4-naphthoquinone, physical properties, 1402t
N-Methylolhydantoins, 794
Methyl orange, pH range and color change of indicator, 825t
3-Methyl-2-pentanone, physical properties, 900t
4-Methyl-2-pentanone, physical properties, 900t
4-Methyl-3-penten-2-one, physical properties, 900t
4-Methyl-4-penten-2-one, physical properties, 900t
4-Methylperoxybenzoic acid, 1236t
Methylphenethylamine, 88
Methyl phenyl ketone, physical properties, 900t
Methyl propyl ketone, physical properties, 900t
Methylpyridinium iodide, 1398
Methyl-2-pyridyl ketone, as key odor/tase compound, 649
Methyl red, pH range and color change of indicator, 825t
Methyl salicylate
    permeation in selected barrier polymers, 174t
    physical properties, 585t
Methyl stearate, physical properties, 585t
Methyltheobromine, 50
Methyltransferases, 571t
Methyl vinyl ether
    physical properties, 1690t
    polymerization, 1689
Methyl vinyl ketone, physical properties, 900t
Methyl violet, pH range and color change of indicator, 825t
Methyl xanthic acid, 1755
Metolachlor, environmental health advisories, 771t
Metribuzin, environmental health advisories, 771t
Metsulfuron, methyl, environmental health advisories, 771t
Meyer reaction, **993**
Meyer-Schuster rearrangement, **993**
Mezlocillin, year of discovery/market introduction, 106t
Mho (mho), 1644t
Mica, **993–995**
    dielectric permittivity, 852t
    hardness, 1008t
    as release agent, 1435t
Mica (glass-bonded), dielectric permittivity, 852t
Micelle, **995**, 1584
Michel, Hartmut (1948–), **995**
Microchemistry, **995**
Microcline, abrasion pH, 1t
Microclusters, 1471
Micrococcin P$_1$, activity and producing organism, 128t
Microcontact printing, nanofabrication application, 1047t
Microcrsystalline waxes, 1748
Microcrystalline, **995**
Microcrystalline wax, as release agent, 1435t
Microelectronics, 1017
Microemulsions, **995–996**
Microencapsulation, **996–998**
Microfabrication, nanofabrication application, 1047t
Microgravity, 740–743
    effect on calcium in bones, 271
Micrometer (μm), 1644t
Micronutrient, **1628**
Microgravity and materials processing, **998–999**
Microscopy (Chemical), **999**
Microwave spectroscopy, **999**
Middle-ultraviolet, 1639

Midgley, Thomas, Jr., **999**
Miescher degradation, **1000**
Mignonac reaction, **1000**
Mild steel, hardness, 755t
Mile, nautical (nmi), 1644t
Mile, statute (mi), 1644t
Milk and milk products, **1000–1001**
Milk fat, 1000
Milk proteins, 1000
Milkweed floss, 632t
Miller indices, **1001**
Millerite, **1001**
Millet, susceptibility to iron deficiency, 874t
Milli, 1643t
Mil (mil), 1644t
Mimetite, **1001**
Mineral dressing, 987
Mineral fiber, insulation material, 856t, 857
Mineral fibers, definitely established workplace
    carcinogen, 297t
Mineral nutrients, **1001–1005**
Mineralocorticoids, 35
Mineralogy, **1005–1013**
    minerals described in this encyclopedia, 1012t
Mineral oil, use in food processing, 671t
Minerals, 987
    described in this encyclopedia, 1012t
Mineral waxes, 1747–1748
Minnesotaite, 871t
Minocin, year of discovery, 130t
Minocycline
    year of discovery, 130t
    year of discovery/market introduction, 106t
Minority carriers, 1467
Mint, susceptibility to iron deficiency, 874t
Minute (min), 1644t
Miokamycin, commercial products, 121t
Mirror nuclides, **1013**
Miscibility, **1013**
Mitchell, Peter D. (1920–1992), **1013**
Mithracin, DNA interactive agent, 357t
Mitomycin C USP, DNA alkylating/cross-linking
    agent, 356t
Mitotane USP, chemotherapeutic, 359t
Mitoxanthrone hydrochloride USP, DNA interactive
    agent, 357t
Mitsunobu reaction, **1013**
Mixed ether, 587
Mixed ketones, 899
Mixing and blending, **1013–1016**
MM 45289, producing organism, 118t
MM 47756, producing organism, 118t
MM 47761, producing organism, 118t
MM 47766, producing organism, 118t
MM 47767, producing organism, 118t
MM 49721, producing organism, 118t
MM 49728, producing organism, 118t
MM 55256, producing organism, 118t
MM 55260, producing organism, 118t
MM 55266, producing organism, 118t
MM 55267, producing organism, 118t
MM 55268, producing organism, 118t
Modacrylic fiber
    generic designation and definition, 622t
    physical properties of staple, 626t
Modena-style vinegar, 1675
Moderator, **1016**, 1102
Mohs' hardness scale, 2t, 755, 755t, **1016**
Moiety, **1016**
Moissan, Henri (1852–1907), **1016**
Moisture barrier, **172**
Moisture-retaining agents, 792–793

Molal concentration, **1016**
Molar concentration, **1016**
Molar heat, 758, **1016**
Molar (Stoichiometry), **1016**
Molds, **1767–1770**
Molecular activation, 28–29
Molecular and supermolecular electronics,
    1016–1026
Molecular beam, **1026**
Molecular biology, **1026–1027**
Molecular compound, 428
Molecular distillation, **1027–1028**
Molecular dynamics, 1029
Molecular modeling, **1028–1029**
Molecular orbitals, 1165–1166
Molecular recognition, **1030–1033**
Molecular self-assembly, nanofabrication
    application, 1047, 1047t
Molecular sieves, **1033–1036**
    pore structure, 41t
Molecular streaming, **1553**
Molecular templating, nanofabrication application,
    1047t
Molecular weight, **1036**
Molecular weight distribution, 1341
Molecule, **1036–1038**, 1209
Mole fraction (mol fraction), **1038**
Mole (mol), 1644t
Mole volume, **1038**
Molex process, 41t
Molina, Mario (1943–), **1038**
Mollisols, 1497t
Molybdate minerals, 1012
Molybdenite, **1038**
Molybdenum
    abundance, 330t, 1132t
    biologically active chelates, 323t
    corrosion resistant alloys, 446t
    covalent radius, 344t
    electronic structure, 336t
    interatomic distance, 342t
    ionic crystal radius, 341t
    nuclides (isotopes and isobars), 331t
    plant micronutrient, 613t
    plant micronutrient deficiencies, soils of U.S.,
        616t
    principal characteristics, 327t
    properties of refractory, 1075t
    residence time in sea water, 1133t
Molybdenum [CAS: 7439-98-7],
    1038–1040
    physical properties of high temperature, 774t
    trace mineral nutrient, 1002t, 1005
Molybdenum (In Biological Systems), **1040**
Monatomic gases, 334
Monazite, **1041**
Monel, 58
Monel 400, 1071t
Monoammonium phosphate, 87–88
Monoazole, 1390
Monobactam antibiotics, **122–123**
Monobasic, **1041**
Monobasic acids, 13
Monocalcium phosphate (anhydrous), properties of
    leavening acid, 925t
Monocalcium phosphate (monohydrate), properties
    of leavening acid, 925t
Monoclinic crystal system, crystallographic
    elements, 455t
Monocyclic alkane, 1169t
Monocytes, 242t, 243
Monoethanolamine, 587

Monomer, **1041**
Monomolecular layer, **1041**
Monoperoxyphthalic acid, 1236t
Monoperoxysuccinic acid, 1236t
Monosaccharides, 279t
Monosaturated fats, 1671
Monosialoganglioside, 705
Monoterpenoids, 1601
Monotype, 923t
Montan wax, 1747
Monte Carlo simulation, 1029
Moore, Standford (1913–1982), **1041**
Mordant, **1041**
Mordant dyes, 512t, 517
    applications, 519
    dyeing process, 522
Mordenite
    composition, 1034t
    pore structure, 41t
    preparation, 1035t
Morphine alkaloids, 49t
Morphine [CAS: 57-27-2], 49t, 91, **1041–1042**
Mosley, Henry (1887–1915), **1042**
Mosley's law, **1042**, 1760
Mossbauer effect, **1042–1043**
Mother of vinegar, 1675
Motrin, 93t
Mountain (soil type), 1497t
Moxalactam, year of discovery/market introduction,
    106t
mRNA, 712, 713, 1125
MSD A63A, formula and producing organism, 116t
M shell, 335, 335t
    electronic structure of the elements, 336–337t
Mucilages, 748t, **748–749**
Mucopolysaccharides, constituent of human blood,
    242t
Mucopolysaccharidoses, genetic influence, 716
Mucoproteins, constituent of human blood, 242t
Mucor, 1769
Mullikan, Robert S. (1896–1986), **1043**
Mullis, Kary Banks (1944–), **1043**
Mumps vaccine, 1659
Municipal wastes, materials in, 1714t
Muntz metal, 440t
Muon, **1043**
Muonium, **1043**
Mureidomycins, 124t
Muscovite
    abrasion pH, 1t
    dielectric permittivity, 852t
    mineralogical properties, 993t
Mustargen, DNA alkylating/cross-linking agent,
    356t
Mutamycin, DNA alkylating/cross-linking agent,
    356t
Mutations, 714–715
MVE/MAN copolymers, 1690
Mycobacillin, activity and producing organism,
    128t
Mycobacterium tuberculosis, gram-positive,
    168t
Mycotoxins, 676
Mycotrienins, biological activity, 108t
Myleran, DNA alkylating/cross-linking agent,
    356t
Mylosar, antimetabolite, 355t
Myofibrils, 23
Myosin, and actin, 23
Myristic acid
    fatty acid in vegetable oil, 1671t
    in major vegetable oils, 1672–1673t

Myristic acid [CAS: 544–63–8], **1043**
Myxoviruses, 1694

N

Nacreous pigment, **1045**
NADP/NTN (National Trends Network), 11
Nalbuphine, 93t
Naloxone, 91, 93t, 100
Naltrexone, 91, 93t, 100
Nano, 1643t
Nanobiology, **1045–1046**
Nanochemical synthesis, nanofabrication application, 1047t
Nanotechnology (Molecular), **1045–1047**
Naphtha, 1255
Naphthalene
    biodegradable, 207t
    as parent structure compound, 1090t
Naphthalene [CAS: 91–20–3], **1047–1048**
Naphthalene derivatives, **1049–1052**
1-Naphthalenesulfonic acid, physical properties, 1568t
2-Naphthalenesulfonic acid, physical properties, 1568t
4-Naphthaquinoneoxime, physical properties, 1082t
Naphthenic acids, **1052–1053**
Naphthomycins, biological activity, 108t
1,2-Naphthoquinone, physical properties, 1402t
1,4-Naphthoquinone, physical properties, 1402t
ß-Naphthylamine, definitely established workplace carcinogen, 297t
Napropamide, environmental health advisories, 771t
Naprosyn, 93t
Naproxen, 93t
Naptalam, environmental health advisories, 771t
Napthalene, combustion constants, 422t
Narcan, 93t
Narceine, 49t
Narcotics, 49
Narcotine, 49t
Narcotine alkaloids, 49t
Nascent, **1053**
National Acid Precipitation Assessment Program (NAPAP), 10
National Air Monitoring Stations (NAMS), acid rain, 11
National Atmospheric Deposition Network (NADP), 11
National Environmental Policy Act, 1711t
Natrolite, **1053**
Natural gas, excess air for combustion, 426t
Natta, Giulio (1903–1979), **1053–1054**
Natural attenuation, 205t
Natural convection, 759
Natural dyes, 512t, 529–531, **529–531**
Natural gas, **1054–1066**, 1243
    heating value, 686t
Natural resins, **1437–1438**
Natural rubber, **1449–1451**
    physical properties of flexible foamed plastics, 665t
    production methods for cellular, 664t
Navelbine, tubulin interactive, 358t
Near-ultraviolet, 1639
Nebengruppen, 746
Néel temperature, **1066**
Nef reaction, **1067**
Nef synthesis, **1067**
Negatol, **1067**
Negatron, **1067**

Neisseria gonorrhea, ß-lactamase-producing bacterium, 110t
Neisseria gonorrheae, gram-negative, 168t
Neisseria intracellularis, gram-negative, 168t
Nencki reaction, **1067**
Nenitzescu indole synthesis, **1067**
Neocarzinostatin, activity and producing organism, 128t
Neodymium
    abundance, 330t
    electronic structure, 336t
    interatomic distance, 342t
    ionic crystal radius, 341t
    nuclides (isotopes and isobars), 331t
    principal characteristics, 327t
    properties, 1421t, 1422t
Neodymium [CAS: 7440–00–8], **1067**
Neodymium oxide, colorant for glass, 725t
Neomycin
    antibiotic resistance markers, 710
    year of discovery/market introduction, 106t
Neon
    abundance, 330t, 1132t
    electronic structure, 336t
    nuclides (isotopes and isobars), 331t
    percentage in air, 46t
    principal characteristics, 327t
Neon [CAS: 7440–01–9], **1067–1068**
Neopentane, combustion constants, 422t
Neopentyl alcohol, physical properties, 89t
Neopentyl glycol, polyol for alkyd synthesis, 53t
Neo-Synephrine, 52
Neoviridogrisein II, activity and producing organism, 128t
Neper (Np), 1644t
Nepheline, **1064**
Nephelometer, 1295
Nephelometry, **1064**
Nephrite, 893
Neptunium, 23t
    binary compound properties, 25–26t
    electronic structure, 337t
    interatomic distance, 342t
    ionic crystal radius, 341t
    oxidation states, 24t
    physical properties, 24t
    principal characteristics, 327t
Neptunium [CAS: 7439–99–8], **1064–1065**
Nernst, Hermann Walther (1864–1941), **1065**
Nernst effect, **1065**
Nernst heat theorem, **1065**
Nernst Principle of super position, **1580**
Nernst-Thompson rule, **1065**
Neroli oil, 647t
Nesosilicate minerals, 1013
Netropsin, activity and producing organism, 128t
Neuraminidase, therapeutic enzyme, 574t
Neuroregulators, **1065**
Neurotransmitters, chiral, 1268–1269
Neutralization, 12, **1066**
Neutrino, **1066–1067**
Neutron, **1067–1069**, 1210
Neutron activation analysis, **1069**
Neutrons, gemstone irradiation, 708t
Neutrophils, 242t
Newton (N), 1644t
Niacin, loss from food operations, 1698t
Niacin [CAS: 59–67–6], **1069–1070**
Niccolite, crystal structure, 455
Nickel
    abundance, 330t, 1132t
    corrosion resistant alloys, 446t

covalent radius, 344t
definitely established workplace carcinogen, 297t
electronic structure, 336t
EPA pretreatment standard for aqueous discharge, 985t
interatomic distance, 342t
ionic crystal radius, 341t
major sources of industrial minerals, 1011t
nuclides (isotopes and isobars), 331t
physical properties of high temperature, 774t
principal characteristics, 327t
in representative U.S. crude oils, 1242t
in representative world crude oils, 1243t
residence time in sea water, 1133t
standard electrode potential, 31t
trace mineral nutrient, 1002t, 1005
Nickel Alloys, 58
Nickel-cadmium cell, 186
    cell reactions, energy content, and manufacturers, 180t
Nickel-cadmium cells, 186, 186t
Nickel [CAS: 7440–02–0], **1070–1073**
Nickel electroplating, 557t
Nickel-hydrogen cell, cell reactions, energy content, and manufacturers, 180t
Nickel-hydrogen cells, 186t, 188
Nickel (In Biological Systems), **1073–1074**
Nickeline, **1074**
Nickel-iron cell, cell reactions, energy content, and manufacturers, 180t
Nickel-iron cells, 186t, 187
Nickel-metal hydride cell, cell reactions, energy content, and manufacturers, 180t
Nickel oxalate, 1185
Nickel oxide, colorant for glass, 725t
Nickel silvers, 440t
Nickel superalloys, 775–776
Nickel-zinc cell, cell reactions and energy content (in development), 180t
Nickel-zinc cells, 186t, 188
Nicotine, 49t, 51
Nicotinic acid, 51
    constituent of human blood, 242t
Niemann-Pick disease, genetic influence, 716
Ni-moly steel, quenched, hardness, 755t
Niobic acid, **1074**
Niobium
    abundance, 330t, 1132t
    corrosion resistant alloys, 446t
    covalent radius, 344t
    electronic structure, 336t
    interatomic distance, 342t
    ionic crystal radius, 341t
    major sources of industrial minerals, 1011t
    nuclides (isotopes and isobars), 331t
    physical properties of high temperature, 774t
    principal characteristics, 327t
    properties of refractory, 1075t
Niobium [CAS: 7440–03–1], **1074–1076**
Niobium dihydride, physical properties, 796t
Niobium hydride, physical properties, 796t
Niobium nitride, plasma synthesis, 318t
Niter, **1076**
Nithromethane, as propellant, 1447t
Nit (nt), 1644t
Nitrate, water quality indicator, 1726
Nitrate caliche, 274
Nitrate minerals, 1012
Nitrate reductase, metal chelate enzyme, 323t
Nitrates, 1082
Nitration, **1076–1077**, 1181
Nitric acid, 12

in acid rain, 9
conductivity-concentration curve, 546
red fuming, as propellant, 1447t
Nitric acid [CAS: 7697-37-2], **1077-1078**
Nitriding, 763t, **1078**
Nitrile, 1169t
Nitrile barrier polymers, gas permeability, 172t
Nitrile barrier resins, water-vapor transmission rate, 173t
Nitriles, 74, **1078-1081**
Nitrites, 1082
Nitroalkenes, monomer reactivity and suitable initiator for, 838t
1,3-Nitroaniline, physical properties, 1081t
1-Nitroanthraquinone, physical properties, 1081t
4-Nitrobenzaldehyde, physical properties, 1081t
Nitrobenzene, 1082
physical properties, 1081t
4-Nitrobenzoic acid, physical properties, 1081t
4-Nitrobenzyl alcohol, physical properties, 1081t
Nitrobromoform, physical properties, 1081t
Nitrocellulose-based lacquers, alkyd resins with, 54
Nitrochloroform, physical properties, 1081t
Nitrochlorotoluene, physical properties, 366t
Nitro compound, 1169t
Nitro compounds, **1081-1082**
Nitrodimethylmethane, 1082
Nitroethane, 1082
Nitroethyl alcohol, physical properties, 1081t
Nitrofurane, physical properties, 1081t
Nitrogen
abundance, 330t, 1132t
amount in pure irons, 877t
combustion constants, 422t
constituent of human blood, 242t
covalent radius, 344t
electronic structure, 336t
gas separation, 39t
ionic crystal radius, 341t
mass peaks and relative intensities, 973t
nuclides (isotopes and isobars), 331t
percentage in air, 46t, 424t
permeability in barrier polymers, 172t
plant micronutrient, 613t
principal characteristics, 327t
removal from raw water, 1724
in Texas natural gas, 1054t
van der Waals characteristic equation coefficients, 321t
Nitrogen [CAS: 7727-37-9], **1082-1086**
Nitrogen fixation, 1085-1086
Nitrogen group, **1086-1087**
Nitrogen oxides (NO_x)
in acid rain, 9-12
air pollution, 1327-1328
gas separation, 39t
Nitrogen tetroxide, as propellant, 1447t
Nitroglycerine, 732
Nitro group, analysis, 97
Nitroguanidine, physical properties, 1081t
2-Nitronaphthalene, physical properties, 1081t
3-Nitrophenol, physical properties, 1081t
2-Nitropropane, 1082
physical properties, 1081t
Nitrosamine, 1082
Nitrosobenzene, physical properties, 1082t
Nitroso compound, 1169t
Nitroso compounds, **1081-1082**
2-Nitrosonaphthol-1, physical properties, 1082t
4-Nitrosonaphthol-1, physical properties, 1082t
1-Nitrosonaphthylamine-2, physical properties, 1082t
4-Nitrosophenol, physical properties, 1082t

4-Nitrosophenylaniline, physical properties, 1082t
Nitrosotrimethylmethane, 1082
Nitrostilbene, organic superlattice, 1023
2-Nitrotoluene, physical properties, 1081t
Nitrotrimethylmethane, 1082
Nitrourea, physical properties, 1081t
Nitrous oxide, 100t
as greenhouse gas, 743-744
measurement by NDIR analysis, 837t
percentage in air, 46t
N-Nitrosomethylaniline, physical properties, 1082t
Nobel, Alfred B. (1833-1896), **1087**
Nobelium, 23t
binary compound properties, 25-26t
electronic structure, 337t
interatomic distance, 342t
ionic crystal radius, 341t
oxidation states, 24t
principal characteristics, 327t
Nobelium [CAS: 10028-14-5], **1087**
Nobel Prizes, **1087-1088**
Noble gases, 706
Nolvadex, hormonal therapy, 358t
Nomenclature, **1088-1091**
Nomex, 622t
Nonadecanedioic acid, physical properties, 490t
1-Nonadecene, physical properties, 1150t
trans-2-cis-6-Nonadienal, as key odor/tase compound, 649
Nonagglomerating coals, 390t
Nonanedioic acid, physical properties, 490t
Nondestructive testing, **1092-1094**
1-Nonene, physical properties, 1150t
Nonessential amino acids, 75, 76t
Nonheterocyclic nitrogen compounds, 1175
Nonheterocyclic sulfur compounds, 1175-1176
Nonionic surfactants, 1585-1586
Nonpolar compound, 428-429
Nonpolar solvent, 1524
Nonsteroidal antiinflammatory drugs (NSAIDs), 92
chiral, 1268
Nonstoichiometric compound, 428
Nootkatone, as key odor/tase compound, 649
Noradrenaline, 35-36
where produced, structure, and principal functions, 788t
Norepinephrine, 35-36
where produced, structure, and principal functions, 787t
Norite, 701
Normal concentration, **1094**
Norrish, Ronald G. W. (1897-1978), **1094**
Northrup, John H. (1891-1987), **1095**
Noscapine, 91
Novantrone, DNA interactive agent, 357t
Novobiocin, year of discovery/market introduction, 106t
Novolacs, 1274-1275
Noxious gas, **1095**
N shell, 335, 335t
electronic structure of the elements, 336-337t
Nubain, 93t
Nucleaic acids, **1125-1126**
Nuclear chemistry, **1095**
Nuclear fission, **1095-1096,** 1101-1102
Nuclear force, 1210-1211
Nuclear forces, **1096-1097**
Nuclear fuels, 1101-1102
Nuclear fusion, **700,** **1097-1098**
Nuclear fusion technology, 694-699
Nuclear isomer, 888
Nuclear magnetic moment, **1098**

Nuclear magnetic resonance (NMR), **1098-1099**
Nuclear potential, **1101**
Nuclear power technology, **1101-1123**
Nuclear selection rules, **1464**
Nuclear spin, **1124**
Nuclear structure, **1124**
Nuclear transmutation, **1124-1125**
Nucleic acids, 12, **1127**
Nucleocapsid, 1693
Nucleonics, **1127**
Nucleons, **1127**
Nucleophile, **1127**
Nucleophiles, activation, 29
Nucleophilic reaction, **1127**
Nucleoproteins, **1127**
Nucleoside, **1127**
Nucleoside antibiotics, **123-124**
Nucleosynthesis, 329
Nucleotide antibiotics, **123-124**
Nucleotide phosphorus, constituent of human blood, 242t
Nucleotides, constituent of human blood, 242t
Nucleotidyltransferases, 571t
Nucleus, **1127**
Nuclides, 330-332, 331t
Nutmeg oil, 647t
Nylon
dielectric permittivity, 852t
dyeing, 523
fiber-dye property requirements, 521t
physical properties of structural foams, 665t
Nylon-6, 623t, 1127
gas permeability, 172t
water-vapor transmission rate, 173t
Nylon-6, 6, 34, 622-623t, 1127
water-vapor transmission rate, 173t
Nylon-6, 10, 1127
Nylon-11, 1127
water-vapor transmission rate, 173t
Nylon-12, 1127
water-vapor transmission rate, 173t
Nylon blends, dyeing, 525
Nylon [CAS: 63428-83-1], **1127-1128**
Nylon fiber
generic designation and definition, 622-623t
physical properties, 1139t
worldwide capacity, 628t
Nylon-6,6 fiber, physical properties of staple, 626t
Nylon-MXD6, gas permeability, 172t
Nylon resin, 1128
Nylon resins, **1332-1334**
Nystatin, year of discovery/market introduction, 106t

O

OA-7653A, producing organism, 118t
OA-7653B, producing organism, 118t
Oats, susceptibility to iron deficiency, 874t
Obligatory anaerobes, 90
Obsidian hydration rate, dating method, 1415
Occupational Safety and Health Act, 1711t
Occupational Safety and Health Administration (OSHA), **1129**
Ocean Dumping Act, 1730-1731
Ocean pollution, 1730-1734
Ocean resources (Mineral), **1129-1131**
Ocean thermal energy conversion, **1131-1132**
Ocean water, **1132-1134**
Ocher, **1134**
Octadecanedioic acid, physical properties, 490t
1-Octadecene, physical properties, 1150t
n-Octane, heating value, 686t

Octanedioic acid, physical properties, 490t
Octane number, **1134**
2-Octanone, physical properties, 900t
3-Octanone
 permeation in selected barrier polymers, 174t
 physical properties, 900t
1-Octene, physical properties, 1150t
1-Octene-3-ol, as key odor/tase compound, 649
Octreotide acetate USP, hormonal therapy, 358t
p-tert-Octylphenyl salicylate, 1641
Odor, **1135**
Odorant, **1135**
Odor masking, **1135**
Odor modification, **1135–1136**
Oersted (Oe), 1644t
Offshore drilling, 1247–1248
Ohm (ω), 1644t
Oil black, **1136**
Oil blue, **1136**
Oil cake, **1136**
Oil-eating bacteria, 1733
Oil gas, **1136**
Oiliness, **1136**
Oil of vitriol, 1572
Oils
 industrial chemicals derived from renewable,
  230t
 use in food processing, 671t
 in vegetable oils, 1671t, 1671–1674
Oils, essential, **1136–1137**
Oil shale, **1137–1138**, 1243
Oil slicks, 1732
Oil tanker spills, 1733–1734
Oil wells, 1245–1253
Olah, George A. (1927–), **1138**
Oleate, **1138**
Olefin fiber, generic designation and definition, 623t
Olefin fibers, **1138–1139**
 physical properties, 1139t
Olefin polymers, **1139–1151**
Oleic acid
 fatty acid in vegetable oil, 1671t
 in major vegetable oils, 1672–1673t
 as release agent, 1435t
Oleic acid [CAS: 112–80–1], **1151–1152**
Oleoresin, **1152**
Oleoresins, 1438
Oleum, **1152**
 reagent for sulfonation, 1566t
Olex process, 41t
Oleyl palmitamide, as release agent, 1435t
Olfaction, 1135
Oligoclase, 607
Oligonucleotides, synthesis, 1126
Oligosaccharide antibiotics, **124–125**
Olive oil, 1671
Olivine, **1152**
 abrasion pH, 1t
Omega minus particle, 1396
Oncovin, tubulin interactive, 358t
Onion oil, 647t
Onsager, Lars (1903–1976), **1152**
Onyx, 44
Opacimeter, 1295
Opacity, **1152**
Opal, 708, **1152**
 hardness, 1008t
Opioid agonists, 91
Opioid antagonists, 91–92
Opioid partial agonists, 91–92
Opioid reversal agents, 100
Opium, 91

Optical antipodes, 79
Optical brightener, **1152**
Optical crystal, **1152**
Optical emission spectrochemical analysis,
 1152–1153
Optical fibers, 1155
Optical fiber systems, **1153–1163**
Optical glass, **1163**
Optical isomer, **1164**
Optical microscope, in x-ray fluorescence analysis,
 1164
Optical rotation, **1164**
Orange flower oil, 647t
Orange oil, 647t
Orbital quantum number, 1396
Orbitals, **1164–1166**
Order-disorder theory, **1166–1168**
Ore, 987
Ores, **1168**
 origin of, 1167
Organelle, **1168**
Organic acids, 12
Organic chemical analysis, 96
Organic chemistry, **1168–1181**
Organic compound, 429
Organic nomenclature, 1089–1090, 1168–1170
Organic pigments, **1308–1313**
Organic polymers, **1348–1350**
Organic reactions, 1176–1181
Organic solid state, 1016–1017
Organic superlattice, 1023
Organoborane, **1182**
Organochlorine pesticides, definitely established
  workplace carcinogen, 297t
Organoclay, **1182**
Organoleptic, **1182**
Organometallic compound, 429
Organometallic compounds, **1182**
Organophosphorus compound, **1182**
Organopyrosilicate, **1182**
Organosilicon, **1182**
Organosol, **1182,** 1357
Organotin compounds, **1182**
Orienticin (PA 42867-A, B, C, D), producing
  organism, 118t
Origanum oil, 647t
Ornithine, 80
Ornithine carbamoyltransferase, 1652
Ornithine cycle, 1651–1652
Orpiment, **1182**
Orris oil, 647t
Orthoacetoxybenzoic acid, 8
Orthoclase, hardness, 755t
Ortho-hydrogen, 798
Orthomyxovirus, 1694
Orthophosphoric acid, 1279
Orthorhombic crystal system, crystallographic
  elements, 455t
Ortho-state, **1182**
Orudis, 93t
Oryzalin, environmental health advisories,
 771t
OSHA (Occupational Safety and Health
  Administration), **1129**
O shell, electronic structure of the elements,
 336–337t
Osmium
 abundance, 330t
 electronic structure, 337t
 interatomic distance, 342t
 ionic crystal radius, 341t
 nuclides (isotopes and isobars), 331t

 principal characteristics, 327t
 properties, 1317t
Osmium [CAS: 7440–04–2], **1182–1183**
Osmophilic organism, 1768
Osmosis, **1183–1184**
Osmotic coefficient, **1184**
Osmotic pressure, **1184**
Osteogenesis imperfecta, genetic influence, 716
Osteomalacia, 1704
Ostwald, Wilhelm (1853–1923), **1184**
Otitis media vaccine, 1660
Ouricury wax, 1747
Outgassing, **1184**
Ovoflavin, 1699
Oxalacetic acid, 282t
Oxalic acid [CAS: 144–62–7], **1184–1186**
Oxalosis, genetic influence, 716
Oxamide, 1185
Oxidation, 1178, **1186**
 catalyst application, 304t, 305
 yeast participation, 1769t
Oxidation bases, 512t
Oxidation number, **1186**
Oxidation potential, **1186**
Oxidation-reduction indicator, **1186**
Oxidative cleavage, 1178
Oxidative coupling, **1186**
Oxidative-coupling polymerization, **1342–1343**
Oxidative phosphorylation, **1283**
Oxide, **1186**
Oxide minerals, 1011
Oxides, 1189
Oxidizing agents, **1186**
Oxidizing material, **1186**
Oxidoreductases, 571t
Oximes, **1186**
Oxime test (Rheinboldt), **1186**
Oximinotransferases, 571t
Oxiranes, monomer reactivity and suitable initiator
  for, 838t
Oxisols, 1497t
18-Oxocorticosterone, where produced, structure,
  and principal functions, 786t
Oxonium compounds, **1187**
OXO process, 47, 48
Oxo process, **1187–1188**
Oxy, **1188**
Oxyacid, **1188**
Oxyazo compounds, **1188**
Oxychlorination, catalyst application, 304t, 305
Oxygen
 absorption bands, 3
 abundance, 5t, 330t, 1132t
 activation, 29
 amount in pure irons, 877t
 combustion constants, 422t
 constituent of human blood, 242t
 covalent radius, 344t
 density, 474t
 diffusion and solubility coefficients in barrier
   polymers, 173t
 electronic structure, 336t
 gas separation, 39t
 ionic crystal radius, 341t
 nuclides (isotopes and isobars), 331t
 percentage in air, 46t, 424t
 permeability in barrier polymers, 172t
 principal characteristics, 327t
 reagent for sulfonation, with sulfur dioxide,
   1566t
 removal from raw water, 1724
 standard electrode potential, 31t

in Texas natural gas, 1054t
van der Waals characteristic equation
coefficients, 321t
Oxygen balance, **1191**
Oxygen [CAS: 7782–44–7], **1188–1191**
Oxygen cell, **1191**
Oxygen consumed (OC; COD; DOC), **1191–1192**
Oxygen debt, **1192**
Oxygen-evolving anodes, 982
Oxygen-free copper, 438t
Oxygen group, **1192**
Oxygen sink, **1192**
Oxyl process, **1192**
Oxymorphone, 92t
Oxytetracycline
year of discovery, 130t
year of discovery/market introduction, 106t
Oxytocic hormone, where produced, structure, and
principal functions, 788t
Oxytocin, **1192**
where produced, structure, and principal
functions, 788t
Ozocerite, **1192**, 1748
Ozone
absorption bands, 3
percentage in air, 46t
Ozone [CAS: 10028–15–6], 1188–1189,
1192–1194
Ozonolysis, **1194**
Ozonosphere, 156

P

Pacidamycins, 124t
Pack carburizing, 763t
Packed absorption tower, 4
Paclitaxol, tubulin interactive, 358t
Pain, 91
Paint and finish removers, **1195–1196**
Paints and coatings, **1196–1203**
Palladium
abundance, 330t
covalent radius, 344t
electronic structure, 336t
interatomic distance, 342t
ionic crystal radius, 341t
nuclides (isotopes and isobars), 331t
principal characteristics, 327t
properties, 1317t
Palladium [CAS: 7440–05–3], **1203–1204**
Palmarose oil, 647t
Palmitic acid
fatty acid in vegetable oil, 1671t
in major vegetable oils, 1672–1673t
as release agent, 1435t
Palmitic acid [CAS: 57–10–3], **1204**
Palmitoleic acid, in major vegetable oils,
1672–1673t
Palmitoleic acid [CAS: 373–49–9], **1204**
Palmitolenic acid, fatty acid in vegetable oil,
1671t
Palm oil, 1671
Palmyra palm, 632t
Pal oil, characteristics and properties,
1672–1673t
Palton III, 623t
Pandemics, 1694–1695
Paneth technique, **1204**
Pantothenic acid, constituent of human blood, 242t
Papain, therapeutic enzyme, 574t
Papaverine, 91
Papaverine alkaloids, 49t

Paper
in municipal waste, 1714t
waste recycling, 1715–1716
Papermaking and finishing, **1205–1208**
Papeverine, 49t
Papovaviruses, 1695
Paraffin, 44, **1208**
Paraffins, gas separation, 39t
Paraffin wax, 1748
as release agent, 1435t
use in food processing, 671t
Para-hydrogen, 798
Parainfluenza vaccine, 1660
Paraldehyde, 5
Paralenes, 1764–1765
Parallel-leaf cartridge ultrafilter, 1636
Paramagnetism, **1208**
Paramethasone, commercial microbial
transformations used to produce, 1550t
Para methyl red, pH range and color change of
indicator, 825t
Paramyxoviruses, 1694, 1695
Paraplatin, DNA alkylating/cross-linking agent, 356t
Paraquat, environmental health advisories, 771t
Parasites, 1767
Para-state, **1208**
Parathormone, where produced, structure, and
principal functions, 788t
Parathyroid hormone (PTH), where produced,
structure, and principal functions, 788t
Parathyroid hormones, 785
Parex process, 41t
Paris Green, 441
Parkes process, **1208**
Parr turbidimeter, **1195**, 1633
Partial condensers, 760
Partial pressure, **1208**
Particle, **1208**
Particle accelerator, 5, **1208**, 1213–1214
Particle generator, 1214–1215
Particles (Subatomic), **1209–1217**
Particle size, **1208**
Particulate air pollution, 1325–1326
Particulate matter, **1219**
Partile-pressure analyzers, 1665
Parvodicin, producing organism, 118t
Parylene C, physical properties, 1764t
Parylene D, physical properties, 1764t
Parylene N, physical properties, 1764t
Pascal (Pa), 1644t
Paschen-Back effect, **1219**, 1773
Passivity, **1219**
Pasteurella pestis, gram-negative, 168t
Pasteurization, **1219**
Pasteurm Louis (1822–1895), **1219**
Patchouli oil, 647t
Patentability, **1219**
Paterno-Buchi reaction, **1219**
Pathfinder element, **1219**
Pathway, **1219**
Patina, **1219**
Patterson function, 456
Pattinson process, **1219**
Pauli, Wolfgang Ernst (1900–1958), **1219**
Pauli exclusion principle, **1219**
Pauling, Linus Carl (1901–1994), **1219**
Peanut oil, 1671
characteristics and properties, 1672–1673t
Peanuts, susceptibility to iron deficiency, 874t
Pearl, irradiation, 709t
Pearlstone, **1226**
Pearson's solution, **1219**

Peat, **1220**
Peat wax, 1747
Pebble mill, **1220**
Pebble mills, **170–171**
Pechmanm pyrazole synthesis, **1220**
Pectinases, 306
Pectins, **1220–1221**
Pedersen, Charles John (1904–1989), **1221**
Pedigree analysis, 710–711
Pegmatite, **1221**
Pelargonin, 734t
Pellet, **1221**
Pellet mills, 45
Pellizzari reaction, **1221**
Pellotine, 49t
Pelouze synthesis, **1221**
Pencillins, **125–127**
Pendimethalin, environmental health advisories, 771t
Penetrant, **1221**
Penetrant nondestructive testing, 1093
Penicillin, year of discovery/market introduction,
106t
Penning discharge, **1221**
Penning effect, **1221**
Pentaborane, 254
Pentaboranze, 254
Pentabromobenzyl bromide, flame retardant, 640t
Pentachloroethane, physical properties, 366t
Pentachlorophenol, wood preservative, 1752
Pentadecanedioic acid, physical properties, 490t
1-Pentadecene, physical properties, 1150t
Pentaerythritol, polyol for alkyd synthesis, 53t
Pentaerythritol adduct, commercial polyether polyol,
1654t
Pentaethylenehexamine, physical properties, 483t
n-Pentane
combustion constants, 422t
heating value, 686t
in Texas natural gas, 1054t
Pentanedioic acid, physical properties, 490t
2,3-Pentanedione, physical properties, 900t
2,4-Pentanedione, physical properties, 900t
Pentanesulfonic acid, physical properties, 1568t
1-Pentanol, physical properties, 89t
2-Pentanol, physical properties, 89t
3-Pentanol, physical properties, 89t
2-Pentanone, physical properties, 900t
3-Pentanone, physical properties, 900t
Pentazocine, 93t
n-Pentene, combustion constants, 422t
1-Pentene, physical properties, 1150t
Pentlandite, **1221**
cobalt source, 410t
Pentosan, **1221**
Pentose phosphate cycle, 281, 283
Pentose phosphate epimerase, 283
Pentose phosphate isomerase, 283
Pentoses, 279t
Pentosuria, genetic influence, 716
Pentosyltransferases, 571t
Pentothenic acid, **1204–1205**
Pentyl, 89
Pentyl acetate, physical properties, 585t
Peppermint oil, 647t
Peptide antibiotics, **127**
Peptide hydrolases, 571t
Peptide synthetases, 571t
Peptide vaccines, 1661
Peptidomimetics, chiral, 1269
Peptidyl-amino acid hydrolases, 571t
Peptidyl-peptide hydrolases, 571t
Peptization, **1221**

Peptone, 1221
Perchlorates, **1221–1223**
Perchloric acid, **1221–1223**
Perfect gas, **1223**
Perfluorocarbons (PFCs), as greenhouse gas, 744
Perfluoro-di-*tert*-butyl peroxide, physical properties, 1232t
Perfluoro dimethyl peroxide, physical properties, 1232t
Perfluorolauric acid, as release agent, 1435t
Perfume, **1223**
Periclase, hardness, 755t
Peridotite, **1223**
Perilla oil, fatty acid composition, 507t
Periodic law, **1224**
Periodic table, **1224–1225**
Peritectic temperature, **1226**
Perkin, Sir William Henry (1838–1907), **1226**
Perlite, **1226**
    insulation material, 856t, 857
Permafil, **1226**
Permanent-press resin, **1226**
Permanent set, **1472**
Permeation, **1226**
Permenorm, **1226**
Permutation, **1226**
Permutation group, **1226**
Perovskite, **1226**
    crystal structure, 455
Peroxidase
    metal chelate enzyme, 323t
    properties as diagnostic enzyme, 976t
Peroxide, 1174
Peroxides and peroxide compounds (Inorganic), **1226–1229**
Peroxides and peroxide compounds (Organic), **1229–1240**
Peroxyacetic acid, 1236t
Peroxybenzoic acid, 1236t
Peroxybutyric acid, 1236t
Peroxycinnamic acid, 1236t
Peroxydecanoic acid, 1236t
Peroxydodecanoic acid, 1236t
Peroxyformic acid, 1236t
Peroxy group, analysis, 97
Peroxyhexadecanoic acid, 1236t
Peroxyhexanoic acid, 1236t
Peroxynonanoic acid, 1236t
Peroxyoctadecanoic acid, 1236t
Peroxyoctanoic acid, 1236t
Peroxypropionic acid, 1236t
Peroxytetradecanoic acid, 1236t
Peroxyxanthates, 1755
Perrhenic acid, **1240**
Perrin rule, **1240**
Perthite, **1240**
Perturbation, **1240**
Perutz, Max F. (1914–2002), **1240**
Pesticides, **848–851**
    pyridine derivatives, 1386
Petalite, **1240**
Petrochemicals, **1240–1242**
Petrolatum, **1242**, 1748
Petroleum, **1242–1254**
    pollution in oceans, 1731–1732
Petroleum ceresins, 1748
Petroleum exploration, 1245–1247
Petroleum exploratory drilling, 1247–1248
Petroleum refining, **1255–1260**
Petroleum reserves, 1254
Petroleum terminology, 1257–1259
Petroleum waxes, 1748

Petroleum wells, 1245–1253
Pettigrain oil, 647t
Pewter, antimony content, 139t
pH (Abrasion), **1–2, 1261**
Pharmaceuticals, **1262–1265**
    chiral, **1265–1269**
    pyridine derivatives, 1386–1388
Pharmacodynamics, **1269–1273**
Pharmacokinetics, 1263
Phase diagram (Metallurgy), **1273**
Phase rule, **1273**
Phenacite, **1273**
Phenanthrene, biodegradable, 207t
Phenazocine, 92t, 93t, 1041
Phenelfamycin A, formula and producing organism, 116t
Phenelfamycin B, formula and producing organism, 116t
Phenelfamycin C, formula and producing organism, 116t
Phenelfamycin D, formula and producing organism, 116t
Phenelfamycin E, formula and producing organism, 116t
Phenelfamycin F, formula and producing organism, 116t
ß-Phenethyl alcohol, **193**
Phenobarbitone, target of medical diagnostic test, 975t
Phenoclast, **1273**
Phenocryst, **1273**
Phenol, **1273–1274**
    as parent structure compound, 1090t
Phenolase, metal chelate enzyme, 323t
Phenolate process, **1274**
Phenol coefficient, **1274**
Phenol-furfural resin, **1274**
Phenolic adhesives, load-bearing capabilities, 32t
Phenolic resin, production methods for cellular, 664t
Phenolic resin adduct, commercial polyether polyol, 1654t
Phenolic resins, **1274–1275**
Phenolics, **1275**
    dielectric permittivity, 852t
Phenolphthalein, pH range and color change of indicator, 825t
Phenol Red, pH range and color change of indicator, 825t
Phenolsulfonphthalein, pH range and color change of indicator, 825t
Phenylacetone, physical properties, 900t
Phenylalanine
    first isolation and isoelectric point, 77t
    manufacture, 76t
    in selected foods, 75t
    structural classification, 79t
Phenylalanine hydroxylase, metal chelate enzyme, 323t
Phenylamine, 75, 101
1-Phenyl-2-aminopropane, 88
Phenylephrine hydrochloride, 52
Phenylephrine tyramine, 49t
2-Phenylethanol, 193
Phenyl ethyl ketone, physical properties, 900t
Phenylhydrazine, 794–795
Phenylhydrazones, 795
Phenyl isocyanate, 464
Phenylketonuria, genetic influence, 716
1-Phenyl-2-methylaminopropanol, 51
o-Phenylphenol, dye carrier, 512t
p-Phenylphenol, dye carrier, 512t
1-Phenyl-2-propanone, physical properties, 900t

Phenytoin, target of medical diagnostic test, 975t
pH (Hydrogen Ion Concentration), **1261–1262**
Phillipsite, **1275**
    composition, 1034t
Phlogopite, **1275–1276**
    mineralogical properties, 993t
Phloridzin, 734t
Phonons, **1276**
Phon (phon), 1644t
Phormium, 632t
Phorone, 7
Phosgenite, **1276**
Phosphate, target of medical diagnostic test, 975t
Phosphate minerals, 1012
Phosphate plasticizers, 1316
Phosphate rock, **1276**
Phosphates, in 24 hour urine sample, 1657t
Phosphazene, **1276**
Phosphine, 1279
6-Phosphogluconic acid, 283
6-Phosphogluconic acid dehydrogenase, 283
3-Phosphoglycerate kinase, metal chelate enzyme, 323t
Phospholipids, **1276**
Phosphonitrile, **1276**
Phosphor bronze, 440t
Phosphorescence, **1277**
Phosphoric acid, acidulant, 14
Phosphoric acid [CAS: 7664–38–2], **1276–1277**
Phosphoric diester hydrolases, 571t
Phosphoric monoester hydrolases, 571t
Phosphorus, abundance, 1132t
Phosphorous acid, 1279
Phosphors, **1277**
Phosphorus
    abundance, 5t, 330t
    amount in pure irons, 877t
    constituent of human plasma or serum, 244t
    covalent radius, 344t
    electronic structure, 336t
    essential mineral nutrient, 1002, 1002t
    interatomic distance, 342t
    ionic crystal radius, 341t
    nuclides (isotopes and isobars), 331t
    plant micronutrient, 613t
    principal characteristics, 327t
Phosphorus [CAS: 7723–14–0], **1277–1281**
Phosphorus (In Biological Systems), **1281–1283**
Phosphorylation (Oxidative), **1283**
Phosphorylation (Photosynthetic), **1283–1284**
Phosphotransferases, 571t
Photochemistry, **1284–1286**
Photoelectric constant, **1287**
Photoelectric effect, **1287–1288**
    gamma rays from, 704
Photoelectron spectroscopy, **553**
Photoemission, **1288**
Photographic chemistry, **1288**
Photographic emulsions, 1290–1291
Photographic grade, **1288**
Photography and imagery, **1288–1294**
Photoionization, **1294**
Photoisomerization, 1178
Photolithography, nanofabrication application, 1047t
Photoluminescence, 946t
Photolysis, **1284–1286**
Photometers, **1294–1295**
Photometric analysis, **1295**
Photomultipliers, **1288**
Photon, **1295–1296**
Photoneutron, **1296**
Photonics, **1295–1296**

Photons, 1393
Photonuclear reaction, **1296**
Photopolymer, **1296**
Photosensitive glass, **1296**
Photosynthesis, **1296–1297**
Photosynthetic phosphorylation, **1283–1284**
Photovoltaic cells, 687t, **1298–1300,** 1513
Phot (ph), 1644t
pH-sensitive materials, 1485
Phthalate plasticizers, 1315–1316
Phthalic acid [CAS: 88–99–3], **1300**
Phthalic anhydride [CAS: 85–44–9], **1300–1301**
Phthalide isoquinoline alkaloids, 49t
Phthalocyanine, sensitizer for semiconductors, 535t
Phthalocyanine compounds, **1301–1302**
Phthalocyanines, sensitizer for semiconductors, 535t
Phycocolloid, **1302**
Phyllosilicate minerals, 1013
Physical adsorption, 36
Physical chemistry, **1302**
Physical property variables, 1670
Physostigmine, 49t
Phytochemistry, **1302**
Phytoextraction, 205t
Phytofiltration, 205t
Phytoremediation, 205t
Phytostabilization, 205t
pi (π) bond, **1302**
Pickling, **1302**
Picloram, environmental health advisories, 771t
Pico, 1643t
Picornaviruses, 1694
Picric acid, physical properties, 1081t
Pidgeon process, **1302–1303**
Piezochemistry, **1303**
Piezoelectric effect, **1303**
Piezoelectricity, **1303**
Piezoelectric materials, 1484
Pig iron, **1303**
Pigmentation (Plants), **1304–1305**
Pigment dispersions, **1303–1304**
Pigments, 512t
Pigments (Inorganic), **1305–1308**
Pigments (Organic), **1308–1313**
Pilocarpine, 49t
Pilot plant, **1313**
Pimelic ketone, physical properties, 900t
Pinacolone, physical properties, 900t
Pinacol rearrangement, 1424
Pinane, physical properties, 1231t
Pineapple, 632t
Pineneedle oil, 647t
Pine oil, 647t
Pinner reaction, **1313**
Pint (pt), 1644t
Pion, 1210
Pipe insulation, insulation material, 857
Piperacillin, year of discovery/market introduction, 106t
Piperazine, physical properties, 483t
Pipette, **1313**
Pirani gage, 1663–1664
Piria reaction, **1313**
Piroxicam, 93t
Pissava, 632t
Pitch, 1255, **1313**
Pitchblende, **1313**
Pitocin, where produced, structure, and principal functions, 788t
Pitressin, where produced, structure, and principal functions, 789t
Pituitary hormones, 790

Pitzer equation, **1313**
p*K*, **1313**
Plant growth hormones, 722–723
Plant growth modification and regulation, **1313–1314**
Plant hormones, 791–792
Plant pathogens, 772–773
Plasma, 695, **1314**
Plasma (Particle), **1314–1315**
Plasma frequency, **1314**
Plasmid, **1315**
Plasmids, 715
Plaster, 749
Plastic deformation, **1315**
Plastic film, **1315**
Plastic flow, **1315–1316**
Plastic foams, insulation material, 856t
Plasticity, **1315**
Plasticizers, **1315–1316**
Plastic pipe, **1316**
Plastics, **1316**
  in municipal waste, 1714t
Plastids, **1317**
Plastisol, **1317,** 1357
Plate and Frame ultrafiltration, 1636
Plate glass, 727
Platforming, **1317**
Platinol, DNA alkylating/cross-linking agent, 356t
Platinum
  abundance, 330t
  corrosion resistant alloys, 446t
  density, 474t
  electronic structure, 337t
  interatomic distance, 342t
  ionic crystal radius, 341t
  major sources of industrial minerals, 1011t
  nuclides (isotopes and isobars), 331t
  principal characteristics, 327t
  properties, 1317t
  ultraviolet reflectance, 1640
Platinum [CAS: 7440–06–4], **1317–1319**
Platinum group, 1317
Platinum-iridium anodes, 982
Platinum-platinum thermocouples, temperature range, 1605t
Plicamycin USP, DNA interactive agent, 357t
Plutonium, 23t
  binary compound properties, 25–26t
  electronic structure, 337t
  interatomic distance, 342t
  ionic crystal radius, 341t
  oxidation states, 24t
  physical properties, 24t
  principal characteristics, 327t
Plutonium [CAS: 7440–07–5], **1319–1320**
Pneumatic conveyor dryers, 509t
Pneumococcal polysaccharide vaccin, 1659
Pneumoencephalogram (air scan), 1762
Pneumokonioses, **1320–1321**
Point mutation, 714–715
Poise (P), 1644t
Poison, **1321**
Polanyi, John C. (1929–), **1321**
Polar, **1321**
Polar compound, 429
Polarimeter, 1295, **1321,** 1321–1322
Polarimetry, **1321–1322**
Polarized ionic bond, 345
Polarized light, **1322–1323**
Polar molecule, **1323**
Polarographic analyzers, **1323**
Polar solvent, 1524

Polar valence, 345
Pole figure, **1323**
Poliomyelitis vaccine, 1659
Polish, **1323–1324**
Polishing agents, **745–746**
Pollucite, **1324**
Pollution, 1709–1716
Pollution (Air), **1324–1330**
Pollution abstracts, 769t
Polonium
  abundance, 330t
  electronic structure, 337t
  interatomic distance, 342t
  ionic crystal radius, 341t
  principal characteristics, 327t
Polonium [CAS: 7440–02–06], **1331**
Polonovski reaction, **1331**
Polyacetylene, **1331–1332**
Polyacrylamide, physical properties, 15t
Polyacrylamides, 15–16
Poly(acrylic acid), water-soluble polymer, 1738
Polyallomer resins, **1332**
Polyamide, microcapsule material, 997t
Polyamide-imide resins, **1334**
Polyamide resins, **1332–1334**
Polyamides
  dyeing, 523
  ultrafiltration media application, 1635
Polyarylates, **1334**
Polybasic acids, 13
Polybasite, **1334**
Polybenzimidazole fiber
  generic designation and definition, 623t
  physical properties, 1139t
Polybenzimidazoles, **1334**
Polybutene, 1150
Poly(*n*-butyl acrylate), physical properties, 18t
Poly(*sec*-butyl acrylate), physical properties, 18t
Poly(*tert*-butyl acrylate), physical properties, 18t
Polybutylene resins, **1335**
Polybutylene terephthalate polyesters, **1335**
Polycarbonate
  dielectric permittivity, 852t
  outdoor stability, 990t
  physical properties of structural foams, 665t
  water-vapor transmission rate, 173t
Polycarbonates, **1335–1336**
Polychlorinated biphenyls (PCBs), definitely established workplace carcinogen, 297t
Polycondensation polymerization, 1341
Polycyclic alkane, 1169t
Poly(cyclohexyl acrylate), physical properties, 18t
Polycyclo-hexylene-dimethylene terephthalate, **1336–1337**
Polycystic ovary syndrome, 99
Polydextrose, use in food processing, 671t
Poly(dibromostyrene), flame retardant, 640t
Polydimethylsiloxane, **1337**
  as release agent, 1435t
Poly(dodecyl acrylate), physical properties, 18t
Polyelectrolytes, **1337**
Polyene, **1337**
Polyester, physical properties of structural foams, 665t
Polyester-acrylic fiber blends, dyeing, 525
Polyester fiber
  dyeing, 523–524
  fiber-dye property requirements, 521t
  generic designation and definition, 623t
  physical properties, 1139t
  physical properties of staple, 626t
  worldwide capacity, 628t

Polyester fibers, **1337–1338**

Polyester film, **1338**

Polyester-nylon fiber blends, dyeing, 525

Polyester polyols, 1654

Polyester resins, **1338**

Polyether, physical properties of rigid foamed plastics, 664t

Polyether antibiotics, **127–129**

Polyether-etherketone, **1338**

Polyetherimide, **1338**

Polyether polyols, 1654

Poly(2-ethoxyethyl acrylate), physical properties, 18t

Poly(ethyl acrylate), physical properties, 18t

Poly(2-ethylbutyl acrylate), physical properties, 18t

Polyethylene, 1139–1140, **1338–1339**

dielectric permittivity, 852t

physical properties of flexible foamed plastics, 665t

production methods for cellular, 664t

Poly(ethylene glycol), commercial polyether polyol, 1654t

Polyethylene terephthalate

dielectric permittivity, 852t

in municipal waste, 1714t

Poly(ethylene terephthalate)

diffusion and solubility coefficients for oxygen and carbon dioxide, 173t

gas permeability, 172t

water-vapor transmission rate, 173t

Polyethylene wax, as release agent, 1435t

Poly(2-ethylhexyl acrylate), physical properties, 18t

Poly(2-ethyl-2-oxazoline), water-soluble polymer, 1738

Poly(fluoroethers), as release agent, 1435t

Polyhalite, **1338–1339**

Poly(heptyl acrylate), physical properties, 18t

Poly(2-heptyl acrylate), physical properties, 18t

Poly(hexadecyl acrylate), physical properties, 18t

Poly(hexyl acrylate), physical properties, 18t

Polyimide adhesives, load-bearing capabilities, 32t

Polyimides, **1339–1341**

Poly(isobornyl acrylate), physical properties, 18t

Poly(isobutyl acrylate), physical properties, 18t

cis-Polyisoprene, 1449

Poly(isopropyl acrylate), physical properties, 18t

Polymeric isocyanates, 1654

Polymerization, **1341–1342**

catalyst application, 304t

Polymerization (Emulsion), **1342**

Polymerization (Oxidative-coupling), **1342–1343**

Polymerization (Radical), **1343–1344**

Polymers, 1169t, **1345–1346**

Polymers (Electroconductive), **1346–1347**

Polymers (Inorganic), **1347–1348**

Polymers (Organic), **1348–1350**

Polymers (Water-soluble), **1350, 1736–1738**

Poly(methacrylic acid), water-soluble polymer, 1738

Polymethine dyes, **1350–1352**

Poly(methyl acrylate), physical properties, 18t

Polymethylbenzenes, **1352–1353**

Poly(methyl methacrylate), outdoor stability, 990t

Polymethyl(nonafluorohexyl)-siloxane, as release agent, 1435t

Poly(4-methyl-1-pentene), 1150

Poly(methyl vinyl ether), 1689

Polymorphism, **1353**

Polymorphonuclear leukocytes, 242t

Polymycin A, activity and producing organism, 128t

Polymyxin, year of discovery/market introduction, 106t

Polymyxin E, activity and producing organism, 128t

Polymyxins, activity and producing organism, 128t

Polynuclear aromatic hydrocarbons (PAHs), 298

definitely established workplace carcinogen, 297t

Polynucleotide phosphorylase, metal chelate enzyme, 323t

Polyol, **1353**

Polyolefin, **1353**

Polyolefin fiber, physical properties of staple, 626t

Polyorganosilicate graft polymer, **1353**

Polyoxalkylene, as release agent, 1435t

Polyoxyethylene, 1737

Poly(pentabromo-benzylacrylate), flame retardant, 640t

Polypeptide, **1353**

Polypeptide hormonal dysfunctions, genetic influence, 716

Polyphenol oxidase, source and function, 986t

Poly(propyl acrylate), physical properties, 18t

Polypropylene, 1146–1147

dielectric permittivity, 852t

diffusion and solubility coefficients for oxygen and carbon dioxide, 173t

fiber-dye property requirements, 521t

gas permeability, 172t

in municipal waste, 1714t

permeation of flavor/aroma compounds, 174t

physical properties of flexible foamed plastics, 665t

physical properties of structural foams, 665t

as release agent, 1435t

water-vapor transmission rate, 173t

Polypropylene [CAS: 9003–07–0], **1354**

Polypropylene fiber, worldwide capacity, 628t

Poly(propylene glycol), commercial polyether polyol, 1654t

Polysaccharides, 279t

constituent of human blood, 242t

Polystyrene, 1556–1557

dielectric permittivity, 852t

gas permeability, 172t

insulation material, 856t

in municipal waste, 1714t

physical properties of rigid foamed plastics, 664t

production methods for cellular, 664t

water-vapor transmission rate, 173t

Polystyrene [CAS: 9003–53–6], **1354–1355**

Polysulfide sealants, 1463

Polysulfone, **1355,** 1567–1568

Polysulfones, ultrafiltration media application, 1635

Polyterpenoids, 1601

Polytetrafluoroethylene, 662

dielectric permittivity, 852t

production methods for cellular, 664t

as release agent, 1435t

Poly(tetramethylene glycol), commercial polyether polyol, 1654t

Polytropic processes, **1355–1356**

Polyunsaturated fats, 1671

Polyurea, microcapsule material, 997t

Polyurethane, 1653–1656

insulation material, 856t

physical properties of flexible foamed plastics, 665t

physical properties of rigid foamed plastics, 664t

physical properties of structural foams, 665t

production methods for cellular, 664t

Polyurethane adhesives, load-bearing capabilities, 32t

Polyurethane-modified isocyanurate (PUIR) foams, 1653, 1655

typical formulation, 1655t

Polyurethanes [CAS: 9009–54–5], **1356**

Poly(vinyl acetate)

dielectric permittivity, 852t

physical properties, 1677t

Poly(vinyl alcohol), 1678–1679

physical properties, 1679t

as release agent, 1435t

water-soluble polymer, 1737

Polyvinyl alkyl ethers, **1356**

Poly(vinyl butyral), 1675, 1676

Poly(vinyl chloride), 1682, 1685–1688

dielectric permittivity, 852t

diffusion and solubility coefficients for oxygen and carbon dioxide, 173t

gas permeability, 172t

insulation material, 856t

physical properties, 1686t

physical properties of rigid foamed plastics, 664t

production methods for cellular, 664t

water-vapor transmission rate, 173t

Poly(vinyl chloride) [CAS: 9002–86–2], **1356–1357**

Poly(vinyl fluoride), dielectric permittivity, 852t

Poly(vinyl formal), 1675, 1676

Poly(vinylidene chloride), 1691–1693

physical properties, 1692t

Poly(vinylidene chloride) [CAS: 9002–86–2], **1357–1358**

Poly(vinylidene fluoride), **1358**

ultrafiltration media application, 1635

Polyvinylpyrrolidinone, 1680

Poly(N-vinyl-2-pyrrolidinone), 1680, 1681–1682

Poly(vinylpyrrolidinone-co-vinyl acetate), 1682

Poly(vinyl pyrrolidone), water-soluble polymer, 1738

Polywater, 1722

Poly(p-xylylene) (PPX), 1763–1765

Pomeranz-Fritsch reaction, **1358**

Pompe's disease, genetic influence, 716

Population genetics, 711–712

Population inversion, 1158

Porcelain, **1358**

Porcelain, Zircon, **1358**

Porcelain enamel, **1358**

Porcelain enamels, **560–561**

Pore, **1358**

Poromeric, **1358**

Porphyrias, genetic influence, 716

Porphyrin, **1358–1359**

Porphyrins, synthesis from products of carbohydrate metabolism, 282t

Porphyry, **1359**

Porter, George (1920–2002), **1359**

Portland cement, 750

Positive-negative acid-base theory, 12

Positron, **1359**

Positron emission tomography (PET), 1762

Positronium, **1359**

Potassium, 48

abundance, 5t, 330t, 1132t

constituent of human plasma or serum, 244t

covalent radius, 344t

electronic structure, 336t

essential mineral nutrient, 1002t, 1002–1003

in 24 hour urine sample, 1657t

interatomic distance, 342t

ionic crystal radius, 341t

nuclides (isotopes and isobars), 331t

plant micronutrient, 613t

principal characteristics, 327t

residence time in sea water, 1133t

standard electrode potential, 31t

water quality indicator, 1726

Potassium (In Biological Systems), **1362–1364**

Potassium acid tartrate

acidulant, 14

properties of leavening acid, 925t

Potassium alginate, 748t
Potassium alum, 61
Potassium bicarbonate, alkalizer, 14
Potassium bromate, 1361
Potassium n-butyl xanthate, 1755t
Potassium carbonate, 1361
    alkalizer, 14
Potassium [CAS: 7440-09-7], **1360-1362**
Potassium chlorate, 1361
Potassium chloride, 1361
Potassium chloroplatinate, 1361
Potassium chromate, 1361
Potassium cyanate, 1361
Potassium cyanide, 1361
Potassium dichromate, 1361
Potassium dihydrogen phosphate, crystal structure,
    455
Potassium hydride, physical properties, 795t
Potassium hydrogen oxalate, 1185
Potassium hydroxide, 12, 1361
    alkalizer, 14
Potassium hypophosphite, 1361
Potassium iodate, 1361
Potassium isoamyl xanthate, 1755t
Potassium isobutyl xanthate, 1755t
Potassium isopropyl xanthate, 1755t
Potassium manganate, 1361
Potassium nitrate, 1361
Potassium nitrite, 1361
Potassium oxalate, 1185
Potassium perchlorate, 1361
Potassium periodate, 1361
Potassium permanganate, 1361
Potassium persulfate, 1362
Potassium n-propyl xanthate, 1755t
Potassium silicate, 1362
Potassium sulfate, 1362
Potassium sulfide, 1362
Potassium sulfite, 1362
Potassium superoxide, 1580
Potassium thiocarbonate, 1362
Potassium thiocyanate, 1362
Potassium titanate fiber, insulation material, 856t
Potatoes, susceptibility to iron deficiency, 874t
Potentiator, **1364**
Potentiometric titration, **1621-1622**
Pound, Robert (1919-), **1364**
Poundal (pdl), 1644t
Pour point, **1364**
Pour point depressant, **1364**
Powder, **1364**
Powder metallurgy, **1364-1365**
Poxviruses, 1695
ppb (parts per billion), **1366**
ppm (parts per million), **1366**
Prandtl number, **1366**
Praseodymium
    abundance, 330t
    electronic structure, 336t
    interatomic distance, 342t
    ionic crystal radius, 341t
    nuclides (isotopes and isobars), 331t
    principal characteristics, 328t
    properties, 1421t, 1422t
Praseodymium [CAS: 7440-10-0], **1366**
Precipitate, **1366**
Precipitation hardening, **1366-1367**
Precursor, **1367**
Prednisolone, 93
    commercial microbial transformations used to
        produce, 1550t
Prednylidene, commercial microbial transformations
    used to produce, 1550t

Preferential, **1367**
Pregel, Fritz (1869-1930), **1367**
Prehnite, **1367**
Preionization, **161**
Prelog, Vladimir (1906-1998), **1367**
Premix molding, **1367**
Prepolymer, **1367**
Preservative, **1367**
Pressure, **1367-1369**
Pressure measurement, 1368
Pressure sensitive adhesives, load-bearing
    capabilities, 32t
Pressure-sensitive adhesives, 33
Pressure swing adsorption, 38t
Pressure-swing adsorption, 40
Pressurized water nuclear reactor, 1106-1109, 1120
Priestley, Joseph (1733-1804), **1369**
Prigogine, Ilya (1917-2003), **1369**
Prills, **1369**
Prilocaine, 100t
Primary alcohols, 46
Primary amides, 74
Primary amines, 74, 75
Primary amino group, analysis, 97
Primary wastewater treatment, 1724
Prinadol, 93t
Principal quantum number, 1396
Pristinamycin IA, activity and producing organism,
    128t
Pristinamycin IIA, activity and producing organism,
    128t
Procaine, 100t
Procarbazine hydrochloride USP, DNA interactive
    agent, 357t
Process-activating proteins, 1374
Process analyzers, 94
Progesterone, where produced, structure, and
    principal functions, 788t
Progestin, where produced, structure, and principal
    functions, 788t
Progressive multifocal leukoencephalopathy, 1696
Prolactin LTH, where produced, structure, and
    principal functions, 788t
Prolan A, where produced, structure, and principal
    functions, 787t
Prolinase, metal chelate enzyme, 323t
Proline
    first isolation and isoelectric point, 78t
    manufacture, 76t
    structural classification, 79t
Promethium
    electronic structure, 336t
    interatomic distance, 342t
    ionic crystal radius, 341t
    nuclides (isotopes and isobars), 331t
    principal characteristics, 328t
    properties, 1421t, 1422t
Promethium [CAS: 7440-12-2], **1369**
Prometon, environmental health advisories, 771t
Prometryn, environmental health advisories, 771t
Promoter, **1369**
Pronamide, environmental health advisories, 771t
Proof, **1369**
Propachlor, environmental health advisories, 771t
Propaldehyde, 48
Propane
    combustion constants, 422t
    measurement by NDIR analysis, 837t
    in Texas natural gas, 1054t
Propane [CAS: 74-98-6], **1369**
Propanedioic acid, physical properties, 490t
Propanesulfonic acid, physical properties, 1568t

Propanetriol, 732
Propanil, environmental health advisories, 771t
Propanone, 7
2-Propanone, physical properties, 900t
Propazine, environmental health advisories, 771t
2-Propenal, 14-15
2-Propeneamide, 15
Propenoic acid, 17
Propenoic acid nitrile, 20
Propham, environmental health advisories, 771t
Propionic acid, carboxylic acid, 294
Propionylerythromycin mercaptosuccinate,
    commercial products, 121t
Propiophenone, physical properties, 900t
Proportional counter, 1759, 1760
Proportional detectors, gamma rays, 704
Propoxyphene, 93t
Propyl acetate, physical properties, 585t
Propyl acrylate, physical properties of polymers, 18t
n-Propyl alcohol, heating value, 686t
Propyl butyrate, permeation in selected barrier
    polymers, 174t
Propylene
    combustion constants, 422t
    melting range, 1546t
Propylene cyanohydrin, 465t
1,2-Propylenediamine, physical properties, 483t
Propylene nitrile, 20
Prostaglandins, **1369-1370**
Protactinium, 23t
    abundance, 330t, 1132t
    binary compound properties, 25-26t
    electronic structure, 337t
    interatomic distance, 342t
    ionic crystal radius, 341t
    oxidation states, 24t
    physical properties, 24t
    principal characteristics, 328t
Protactinium [CAS: 7440-13-13], **1370**
Protease, **1371**
Proteases, 306
Protective coating, **1371**
Protein hydrolysate, **1371**
Protein in cerebrospinal fluid, target of medical
    diagnostic test, 975t
Proteins, **1371-1377**
    constituent of human blood, 242t
Proteus, amphenicol susceptibility, 115t
Proteus vulgaris, gram-negative, 168t
Prothrombin factor, 1706
Protium, **1378**
Proton, 1210, **1378**
Proton-proton reaction, **1378**
Protopine, 49t
Proustite, **1378**
Providencia, amphenicol susceptibility, 115t
Provirus, 1695
Provitamin, **1378**
Pseudogenes, 715
Pseudohypoparathyroidism, genetic influence, 716
Pseudomonas, amphenicol susceptibility, 115t
Pseudomonas aeruginosa, gram-negative, 168t
Pseudomones aeruginosa, ß-lactamase-producing
    bacterium, 110t
Pseudomorphs, **1378**
Pseudopelletierine, 49t
Pseudopotential theory, 346-347
Pseudoxanthoma elasticum, genetic influence, 716
P shell, electronic structure of the elements,
    336-337t
Psilocin, 754
Psilocybin, 566, 754

Psilomelane, **1378**
Psi particle, **1378,** 1397
Psychrophilic organism, 1768
Ptomaine, **1379**
Pulp (Wood) production and processing, **1379–1382**
Pultrusion, **1382**
Pumice, **1382**
    abrasive properties, 745t
    hardness, 1008t
Pump and treat, 205t
Pumping plants, 1722
PUREX process, 1647
Purines, constituent of human blood, 242t
Purines [CAS: 120–73–0], **1382**
Purinethol, antimetabolite, 355t
PVC resins, 1686–1687
PVP hydrogels, 1680–1681
Pycnometer, **1382**
Pyrargyrite, **1382**
Pyrazoles, **1382–1384**
Pyrazolines, **1382–1384**
Pyrazolones, **1382–1384**
Pyresis, 91
Pyrex, thermal conductivity, 758t
Pyrex (Corning 7740), dielectric permittivity, 852t
Pyridine, as parent structure compound, 1090t
Pyridine and derivatives, **1384–1385**
Pyridine [CAS: 110–86–1], **1384–1385**
Pyridinium tetracyanoquinodimethane molecular
        system, 1023
Pyridoxine, 1700–1701
Pyridoxol, 1700
ß-Pyridyl-α-N-methylpyrrolidine, 51
Pyrite, 871t, 1009t, **1389**
    crystal structure, 455
    hardness, 1008t
Pyrogenetic minerals, **1389**
Pyrolusite, **1389**
Pyrolysis, **1389**
Pyrometer, **1389**
Pyrometric cones, **1389–1390**
Pyromorphite, **1390**
Pyromucic acid, carboxylic acid, 294
Pyronine B (CI 45010), 1756
Pyronine G (CI 45005), 1756
Pyronines, 1756
Pyrope, 705
Pyrophoric material, **1390**
Pyrophosphoric acid, 1279
Pyrophosphotransferases, 571t
Pyrophyllite, **1390**
Pyrotechnics, **1390**
Pyroxene, **1390**
Pyrrhotite, 871t, **1390**
Pyrrole and related compounds, **1390–1391**
Pyrrole [CAS: 109–97–7], **1390–1391**
Pyruvate kinase deficiency, genetic influence, 716
Pyruvic acid, 282
    constituent of human blood, 242t

Q

Qionones, **1401–1403**
Q shell, electronic structure of the elements,
        336–337t
Quad, **1393**
Quadruple point, **1393**
Qualitative chemical analysis, 94
Quantitative chemical analysis, 94
Quantity variables, 1670
Quantized Hall effect, 753
Quantum chemistry, **1393**

Quantum chromodynamics, 1397
Quantum-effect devices, 1470
Quantum efficiency, **1393**
Quantum electrodynamics, **1393**
Quantum mechanics, **1393–1395**
Quantum number, **1396**
Quantum simulation, 1029
Quantum theory of spectra, **1396**
Quarks, **1396–1397**
Quart (qt), 1644t
Quartz, 708
    abrasion pH, 1t
    abrasive properties, 745t
    analysis, 316t
    crystal structure, 455
    for diffracting crystal spectrometer, 1760
    hardness, 755t
    heat effect, 708t
    irradiation, 709t
    ultrasonic applications, 1637
Quartz [CAS: 14808–60–7], **1397–1398**
Quartz glass, thermal conductivity, 758t
Quartzite, **1398**
Quartz porphyry, **1398**
Quasicrystal, 459
Quasicrystals, aluminum alloys, 71
Quaternary ammonium compounds, 74, **1398–1399**
Quelet reaction, **1399**
Quenching, **1399**
Quercitrin, 734t
Quinacridone, sensitizer for semiconductors, 535t
Quinidine, 49t
Quinine, 49t, 52
Quinoline, **1399–1401**
    physical properties, 1400t
Quinoline alkaloids, 49t
Quinoline and isoquinolines, **1399–1401**
Quinoline dyes, 1401
Quinoline Yellow A, 1401
Quinomycin C, activity and producing organism,
        128t
4-Quinoneoxime, physical properties, 1082t
1,2-Quinones, 1401
1,4-Quinones, 1401
Quotient group, 747

R

Rabbit, body weight-metabolic rate relationship, 176t
Racemases, 571t
Racemate, 78
Racemic modification, 78
Racemization, **1405**
Rachiasterol, 1703
Rachitamin, 1703
Racking, **1405**
Radian (rad), 1644t
Radiation, **1405**
Radiation (Biochemistry), **1405–1406**
Radiation heat transfer, 759–760, **1604**
Radiation (Industrial), **1406**
Radiation (Ionizing), **1406**
Radiation variables, 1670
Radical polymerization, **1343–1344**
Radioactivation analysis, 1410–1411
Radioactive dating, 1413–1414
Radioactive elements, 332
Radioactive fallout, **603**
Radioactive waste, **1406**
Radioactive waste disposal, 1122–1123
Radioactive wastes, ocean dumping, 1734
Radioactivity, **1406–1416**

Radiocarbon dating, 1414–1415
Radiochemistry, **1416**
Radiographic nondestructive testing, 1092
Radium, 49
    abundance, 330t, 1132t
    definitely established workplace carcinogen, 297t
    electronic structure, 337t
    interatomic distance, 342t
    ionic crystal radius, 341t
    principal characteristics, 328t
Radium [CAS: 7440–14–4], **1416–1417**
Radon
    abundance, 1132t
    definitely established workplace carcinogen, 297t
    electronic structure, 337t
    principal characteristics, 328t
Radon [CAS: 10043–92–2], **1417–1487**
Rad (rd), 1644t
Raffia, 632t
Raffia wax, 1747
Raman spectrometry, **1418–1419**
Ramie, 632t
Ramsey, Sir William (1852–1916), **1419**
Rankine cycle, 1511
Rankine temperature scale, 3, **1419,** 1588
Raoult's Law, **1419,** 1670
Rapeseed oil, 1671
    characteristics and properties, 1672–1673t
    good source of protein, 1373
Rapid solidification processing, 1364–1365
Rare-earth elements, **1419–1424**
Rare-earth metals, **1419–1424**
Rare gas, **1424**
Rare gases, 706
Rare gas halide lasers, wavelength range, 910t
Rat, body weight-metabolic rate relationship, 176t
Rate variables, 1670
Rayleigh waves, 1637
Rayon, physical properties, 1139t
Rayon fiber, generic designation and definition,
        623–624t
Reaction-product analyzer, 97–98
Reactive dyes, 512t, 517, 531–533, **531–533**
    applications, 519
    dyeing process, 521, 522
Reagent-tape analyzer, 98
Realgar, **1424**
Rearrangement (Organic chemistry), **1424**
Reaumur, **1425**
Reboilers, 760
Recalescence, **1425**
Reclaiming, **1425**
Recombinant DNA technology, 715
Recombination, **1425**
Rectification, **1438**
Recycling
    aluminum, 62
    paper wastes, 1715–1716
    plastic wastes, 1714–1715
    polyurethanes, 1656
Red brass, 440t
Red fuming nitric acid, as propellant, 1447t
Red gold, 736
Red quark, 1397
Red tide, **1425**
Reduced crude, 1255
Reduction, 1178
    yeast participation, 1769t
Refinery gas, excess air for combustion, 426t
Refining, **1425**
Reflection hologram, 781
Reflective insulation, insulation material, 857

Reflectometer, 1295
Reflux, **1425**
Reformatsky reaction, **1425–1426**
Reforming, **1426**
Refraction, **1426**
Refractive index, **1426–1427**
Refractomers, **1427–1428**
Refractories, **1428–1430**
Refractory metals, 776
Refrigerant, **1430**
Refrigeration, **1430–1434**
Regelation, **1434**
Regeneration, **1434**
Regenerative adsorbents, 38
Regular crystal system, crystallographic elements,
    455t
Reinforced plastics, **1434**
Reinforcing agent, **1434**
Relative humidity, 793
Relaxin, where produced, structure, and principal
    functions, 788t
Release agents, **1434–1435**
Releasin, where produced, structure, and principal
    functions, 788t
Rem (rem), 1644t
Remsen, Ira (1846–1927), 1436
Renal glycosuria, genetic influence, 716
Renal tubular acidosis, genetic influence, 716
Rene 41, 58
Reoviruses, 1694
Repellent, 1436
Reppe chemistry, 8
Reppe process, 1436
Residual fuel oil, 1255
Residual gas analyzers, 1665
Residuum, in representative world crude oils, 1243t
Resins, use in food processing, 671t
Resins: acetal, **1436**
Resins: acrylontrile-butadiene-styrene, **1436**
Resins: natural, **1437–1438**
Resins: synthetic, **1438**
Resols, 1275
Resonance, **1438**
Resonance fluorescence, 4
Resonance hybrid, 1438
Resorcinol-formaldehyde gels, 42
Resource Conservation and Recovery Act, 1711t
Respiratory syncytial virus (RSV) vaccine, 1660
Retinol, 1698
Retorting, **1438–1439**
Retrovirus, 1695
Reverse osmosis, **1441**
Reverse transcriptase, 715
Revegetation, **1439–1440**
Revolutions per minute (r/min), 1644t
Reynolds number, **1441**
Rhabdoviruses, 1695
Rheinboldt oxime test, **1186**
Rhenium
    covalent radius, 344t
    electronic structure, 337t
    interatomic distance, 342t
    ionic crystal radius, 341t
    nuclides (isotopes and isobars), 331t
    principal characteristics, 328t
Rhenium [CAS: 7440-15-5], **1441–1443**
Rheocasting, 68
Rheology, **1443–1444**
Rhizofiltration, 205t
Rhizopus, 1769
Rhodamine B, sensitizer for semiconductors, 535t
Rhodamine B (Basic Violet 10, CI 45170), 1756

Rhodamines, 1756
Rhodium
    abundance, 330t
    covalent radius, 344t
    electronic structure, 336t
    interatomic distance, 342t
    ionic crystal radius, 341t
    nuclides (isotopes and isobars), 331t
    principal characteristics, 328t
    properties, 1317t
Rhodium [CAS: 7440-16-6], **1445**
Rhodium-platinum thermocouples, temperature
    range, 1605t
Rhodium-rhodium thermocouples, temperature
    range, 1605t
Rhodochrosite, **1445**
Rhodonite, **1445**
Rhombic crystal system, crystallographic elements,
    455t
Rhombohedral crystal system, crystallographic
    elements, 455t
Riboflavin, 1699–1700
    constituent of human blood, 242t
Ribonuclease, **1445–1446**
    therapeutic enzyme, 574t
Ribonucleic acid (RNA), 712, 713–714
    constituent of human blood, 242t
    structure, 1125–1126
Ribosomal RNA, 713, 1125
Ribosome, **1446**
Ribulose-5-phosphate, 283
Rice, susceptibility to iron deficiency, 874t
Rice bran wax, 1747
Richards, Theodore W. (1868–1928), **1446**
Ricinine, 49t
Rickets, 1704
Ridgeway's hardness scale, 2t
Riebeckite, **1446**
Rifamycins, biological activity, 108t
Righi-Leduc effect, 752
Rigid polyurethane foams, 1655
Rigid vinyls, 1357
Rimmed steels, 58
Ristocetin A, producing organism, 118t
Ristocetin B, producing organism, 118t
Ristomycin A, producing organism, 118t
Ristomycin B, producing organism, 118t
Rivers, water quality, 1726–1728
RNA viruses, 1693, 1694
Roasting (minerals), 987
Robinson, Sir Robert (1886–1975), **1446**
Rock crystal, 708
Rocket propellants, **1446–1449**
Rock salt, 752
    absolute index, 1426t
Rockwell hardness scale, 755
Rod mills, **170–171**
Rod wax, 1748
Roentgen, W., x-rays, 1758
Roentgen (R), 1645t
Rokitamycin, commercial products, 121t
Rolling drum, 45
Rondomycin, year of discovery, 130t
Rontgen, Wilhelm Conrad (1845–1923), **1449**
Roots pump, 1663
Rosamines, 1756
Rose bengal, sensitizer for semiconductors, 535t
Roselle, 632t
Rosemary oil, 647t
Rose metal, 238t
Rose oil, 647t
Rosolic acid, pH range and color change of
    indicator, 825t

Rosolite, 1755
Rotary dryers, 509t
Rotating biological contractor, 205t
Rotation axis, **1449**
Rotavirus vaccine, 1660
Rough vacuum, 1661t
Rowland, Frank Sherwood (1927–), **1449**
Roxel process, **1449**
Roxithromycin, commercial products, 121t
rRNA, 713, 1125
Rubber
    insulation material, 856t
    production methods for cellular, 664t
Rubber (Natural), **1449–1451**
Rubber (Synthetic), **1452**
Rubber based adhesives, load-bearing capabilities,
    32t
Rubella vaccine, 1659
Rubidium, 48
    abundance, 330t, 1132t
    constituent of human plasma or serum, 244t
    covalent radius, 344t
    electronic structure, 336t
    interatomic distance, 342t
    ionic crystal radius, 341t
    nuclides (isotopes and isobars), 331t
    principal characteristics, 328t
    residence time in sea water, 1133t
Rubidium [CAS: 7440-17-7], **1452**
Rubidium hydride, physical properties, 795t
Rubidium superoxide, 1580
Ruby, 708
    heat effect, 708t
Rufomycin A, activity and producing organism, 128t
Russell-Saunders coupling, 338
Ruthenium
    abundance, 330t
    covalent radius, 344t
    electronic structure, 336t
    interatomic distance, 342t
    ionic crystal radius, 341t
    nuclides (isotopes and isobars), 331t
    principal characteristics, 328t
    properties, 1317t
Ruthenium [CAS: 7440-18-8], **1453**
Ruthenium chloride, 1453
Ruthenium-titanium oxide anodes, 981–982
Rutherford, Ernest (1871–1937), **1453–1454**
Rutherfordium, 24, 333
    principal characteristics, 328t
Rutile, 708, **1454**
    crystal structure, 455
Ruzicka, Leopold (1887–1976), **1454**
Ryton, 624t

S
Sabatier, Paul (1854–1941), **1455**
Sabatier-Senderens reaction, **1455**
Saccharification, 194
Saccharimeter, **1455**
Saccharomyces bayanus, 1767
Saccharomyces beticus, 1767
Saccharomyces cerevisiae, 1767–1768
Saccharomyces uvarium, 1767
Sac fungi, 1767
Safe Drinking Water Act, 1711t
Safflower oil, 1671
    fatty acid composition, 507t
Sal ammoniac, 86
Salicyl alcohol, 1455
Salicylic acid and related compounds, **1455–1456**

Salicylic acid [CAS: 69–72–7], **1455–1456**
Salmonella
    amphenicol susceptibility, 115t
    and food processing, 672
    ß-lactamase-producing bacterium, 110t
Salmonella enteriditis, gram-negative, 168t
Salt, **1456**
Salt bath, **1456**
Salting out, **1456**
Salt nitriding, 763t
Saltpeter, **1456**
Samarium
    abundance, 330t
    electronic structure, 336t
    interatomic distance, 342t
    ionic crystal radius, 341t
    nuclides (isotopes and isobars), 331t
    principal characteristics, 328t
    properties, 1421t, 1422t
Samarium [CAS: 7440–19–9], **1456–1457**
Sandalwood oil, 647t
Sandostatin, hormonal therapy, 358t
Sandstone, **1457**
Sanger, Frederick (1918–), **1457**
Sanitary sewers, 1724
Sansevieria, 632t
Sapogenins, 734
Saponification, **1457**
Saponification number, 1671
Saponin, 734t
Sapphire, 708
    hardness, 755t
    heat effect, 708t
Saprophytes, 1767
Saran, 624t, 1691
Sardonyx, 44
Sarex process, 41t
Sarsoline, 49t
Sassafras oil, 647t
Satellite energy collectors, 1513
Saturated compound, 429
Saturated fats, 1671
Saturated fatty acids, 1671t
Saturation, **1457–1458**
Saturation vapor pressure, 1670
Saybolt universal viscosity, **1458**
SB22484, factor 3, formula and producing organism, 116t
SB22484, factor 4, formula and producing organism, 116t
Scale, **1458**
Scandium
    abundance, 330t, 1132t
    covalent radius, 344t
    electronic structure, 336t
    interatomic distance, 342t
    ionic crystal radius, 341t
    nuclides (isotopes and isobars), 331t
    principal characteristics, 328t
    properties, 1421t, 1422t
Scandium [CAS: 7440–20–2], **1458**
Scanning probe microscopy, 1045
Scanning tunneling microscope (STM), **1458–1461**
Scapie, 1696
Scapolite, **1461**
Scavenging, **1461**
Scheele, C. W. (1742–1786), **1461**
Scheele's Green, 441
Scheelite, **1461**
Schele's syndrome, genetic influence, 716
Schist, **1461**
Schistosomiasis vaccine, 1660

Schmidlin ketene synthesis, **1461**
Schmidt reaction, **1462**
Schorigin reaction, **1462**
Schotten-Baumann reaction, **1462**
Schrodinger equation, 1395
Schumann region, 1639
Schumann-Runge bands, 3
Schweitzer's reagent, **1462**
Scintillation counters, 1759, 1760
Scintillation detectors, gamma rays, 704
Sclanidine, 49t
Sclerometer, **1462**
Scleroscope, 756
Scolecite, **1462**
Scopolamine, 49t
Scorodite, **1462**
Screw-conveyor dryers, 509t
Seaboard process, **1462**
Seaborg, Glenn T., **1462**
Seaborgium, 24, 333
    electronic structure, 337t
    principal characteristics, 328t
Sealants, **1462–1463**
Sealing glasses, 728
Season cracking, **1463**
Sebacic acid, monomer for nylon resins, 1333t
Secondary alcohols, 46
Secondary amides, 74
Secondary amines, 74, 75
Secondary wastewater treatment, 1724
Second law of thermodynamics, 1606
Second (s), 1645t
Sedimentation, **1463**
SEF (modacrylic staple), 622t
Seger cone, **1463**
Selar PA 3426, gas permeability, 172t
Selection rules (Energy levels), **1463–1464**
Selection rules (Nuclear), **1464**
Selectivity, adsorbents, 36
Selenium
    abundance, 330t, 1132t
    colorant for glass, 725t
    covalent radius, 344t
    electronic structure, 336t
    in hazardous waste, 1711t
    interatomic distance, 342t
    ionic crystal radius, 341t
    nuclides (isotopes and isobars), 331t
    principal characteristics, 328t
    trace mineral nutrient, 1002t, 1004
    water quality indicator, 1726
Selenium [CAS: 7782–49–2], **1464–1465**
Self-propagating high-temperature synthesis (powder metallurgy), 1365
Self-recognition. molecules, 1033
Semenov, Nikolai N. (1896–1986), **1465**
Semianthracite, characteristics, 390t
Semibituminous coal, ignition temperature in air, 425t
Semicarbazones, 899, **1465**
Semiconductor doping, 1467
Semiconductors, **1466–1471**
    electronic structure, 1519
    spectral sensitizers, 535t
Semikilled steels, 58
Semipermeable diaphragm, **1471**
Semipermeable membrane, **1471**
Sendust alloy, thin-film applications, 958t
Sepiolite, **1471–1472**
Sequencing batch reactor, 205t
Sequential analysis, **1472**
Serandite, **1472**

Serine
    first isolation and isoelectric point, 77t
    manufacture, 76t
    structural classification, 79t
L-Serine dehydratase, therapeutic enzyme, 574t
Sernyl, 754
Serpentine, **1472**
    hardness, 1008t
Serratia, amphenicol susceptibility, 115t
Sesame oil, 1671
Sesquiterpenoids, 1601
Sesterterpenoids, 1601
Set (Permanent), **1472**
Sevoflurane, 100t
Sewage treatment, 1724–1725
Sexually transmitted diseases, vaccines, 1660
SF 1293, activity and producing organism, 128t
Shale, **1472**
Shape-memory alloys, aluminum-containing, 71
Shear waves, 1637
Sheeting dryers, 509t
Sheetrock, 750
Shell, **1472**
Shellac, **1472**
Shellac wax, 1746
Shell-and-tube heat exchangers, 761
Sherardizing, **1472**
Shigella
    amphenicol susceptibility, 115t
    ß-lactamase-producing bacterium, 110t
Shigella paradysenteriae, gram-negative, 168t
Shigellas vaccine, 1660
Shrew, body weight-metabolic rate relationship, 176t
Shuttle-Mir microgravity program, 741
Sialons, 70
Sickle cell hemoglobin, 768
Siderite, 871t, **1472**
Silane, physical properties, 795t
Silanes, 1476
Silica
    commercial gas adsorption processes, 39t
    definitely established workplace carcinogen, 297t
    glassy state, 724
    hardness, 755t
    for liquid-phase adsorption, 40
    plasma synthesis, 318t
    as release agent, 1435t
    use in food processing, 671t
Silica aerogels, 43, 44t
Silica gel, 36
Silica glass, 725
Silicate minerals, 1012
Silicates, 1476
Silicates (Soluble), **1472–1473**
Silicic, **1473**
Silicification, **1473**
Silicoaluminates, 65
Silicon
    abundance, 5t, 330t, 1132t
    amount in pure irons, 877t
    constituent of human plasma or serum, 244t
    covalent radius, 344t
    electronic structure, 336t
    interatomic distance, 342t
    ionic crystal radius, 341t
    nuclides (isotopes and isobars), 331t
    principal characteristics, 328t
    residence time in sea water, 1133t
    trace mineral nutrient, 1002t, 1004
Silicon bronze, 440t
Silicon carbide, 1474
    abrasive properties, 745t

hardness, 2t, 755t
plasma synthesis, 318t
Silicon [CAS: 7440–21–3], **1473–1479**
Silicon dioxide, 1475–1476
Silicone, dielectric permittivity, 852t
Silicone resins, **1480–1481**
alkyd resins with, 54
Silicones, 1476
production methods for cellular, 664t
Silicone sealants, 1462
Siliconizing, 763t
Silicon nitride, plasma synthesis, 318t
Silicon-silicon double bond, 1477
Silicon solar cells, 1298–1300
Silk
dyeing, 522
fiber-dye property requirements, 521t
generic designation and definition, 624t
Sillimanite, 1009t, **1481**
physical properties, 1428t
Siloxanes, monomer reactivity and suitable initiator
for, 838t
Siloxy, 1477
Silver
abundance, 330t, 1132t
corrosion resistant alloys, 446t
covalent radius, 344t
density, 474t
electronic structure, 336t
EPA pretreatment standard for aqueous
discharge, 985t
in hazardous waste, 1711t
interatomic distance, 342t
ionic crystal radius, 341t
major sources of industrial minerals, 1011t
nuclides (isotopes and isobars), 331t
principal characteristics, 328t
residence time in sea water, 1133t
standard electrode potential, 31t
sterling, **1546**
Silver acetylide, 8
Silver amalgam, 72
Silver-bearing copper, 438t
Silver-cadmium cells, 186t, 188
Silver [CAS: 7440–22–4], **1482–1483**
Silver cell, cell reactions, energy content, and
manufacturers, 179t
Silver-hydrogen cells, 186t, 188–189
Silver-iron cells, 186t, 188
Silver oxide cell, 184t
Silver-zinc cell, cell reactions, energy content, and
manufacturers, 180t
Silver-zinc cells, 186t, 187–188
Silvex 2-(2,4–5 Trichlorophenoxy propionic acid),
in hazardous waste, 1711t
Silyl, 1477
Silylamino, 1477
Silylene, 1477
Silylidyne, 1477
Silylthio, 1477
Simazine, environmental health advisories, 771t
Simple distillation, **1483**
Simple plants, 1767
Single-cell protein, 1373
Sintered alumina, physical properties, 1428t
Sintered magnesium aluminate, physical properties,
1428t
Sintered mullite, physical properties, 1428t
Sintering, 45
Sisal, 632t
Sitosterol, 733
SI units, 1642–1645

Sizing compound, **1483**
Skraup quinoline synthesis, 1179
Skutterudite, **1483**
cobalt source, 410t
Slack, **1484**
Slag, **1484**
Slag flux, **662**
Slate, **1484**
Slime molds, 1767
Slop wax, 1748
Slow viruses, 1695–1696
Sludge, **1484**, 1725
Slug (slug), 1645t
Smalley, Richard E. (1943–), **1484**
Smart catalysts, 1485
Smart gels, 1485
Smart materials, **1484–1485**
Smart memory alloys, 1485
Smart polymers, 1485
Smell, 1135
Smelting, 987, **1486**
Smithsonite, **1486**
Smog, **1486**
Smoke, **1486**
Smokeless powder, **1486**
Smoky Mountain haze, 1327
Smoky quartz, 708
Soaps, **1486–1489**
Soda ash, 1490–1491
Soda-lime-silica glasses, 724–725
Sodalite, **1489**
Sodamide, 1490
Soda nitre, **1489**
Soddy, Frederick (1877–1965), **1489**
Sodium, 48
abundance, 5t, 330t, 1132t
constituent of human plasma or serum, 244t
covalent radius, 344t
electronic structure, 336t
essential mineral nutrient, 1002t, 1002–1003
in 24 hour urine sample, 1657t
interatomic distance, 342t
ionic crystal radius, 341t
nuclides (isotopes and isobars), 331t
principal characteristics, 328t
residence time in sea water, 1133t
standard electrode potential, 31t
target of medical diagnostic test, 975t
water quality indicator, 1726
Sodium (In Biological Systems), **1362–1364**
Sodium acetate, 6
Sodium acid pyrophosphate
acidulant, 14
properties of leavening acid, 925t
Sodium alginate, 748t
Sodium aluminate, 65, 1490
Sodium aluminosilicate, 1476, 1490
Sodium aluminum fluoride, 64
Sodium aluminum phosphate (anhydrous), properties
of leavening acid, 925t
Sodium aluminum phosphate hydrate, properties of
leavening acid, 925t
Sodium aluminum sulfate, properties of leavening
acid, 925t
Sodium bicarbonate, alkalizer, 14
Sodium bisulfite, reagent for sulfonation,
1566t
Sodium bisulfite, hydroperoxide catalyst, reagent for
sulfonation, 1566t
Sodium bromide, 1490
Sodium n-butyl xanthate, 1755t
Sodium sec-butyl xanthate, 1755t

Sodium carbonate, 1490–1491
alkalizer, 14
Sodium [CAS: 7440–23–5], **1489–1493**
Sodium caseinate, use in food processing, 671t
Sodium chlorate, 1491
Sodium chloride, 1491, **1493–1494**
conductivity-concentration curve, 546
crystal structure, 455
dielectric permittivity, 852t
for diffracting crystal spectrometer, 1760
Sodium chromate, 1491
Sodium chrome alum, 61
Sodium citrate, 1491
Sodium cyanate, 464
Sodium cyanide, 1491
Sodium dichromate, 1491
Sodium dithionate, 1491
Sodium ethyl xanthate, 1755t
Sodium fluoride, 1491
Sodium fluorosilicate, 1491
Sodium formate, 1491
Sodium hydride, 1491
physical properties, 795t
Sodium hydroxide, 12, 1491
alkalizer, 14
conductivity-concentration curve, 546
Sodium hypochlorite, 1491
Sodium hypophosphite, 1491
Sodium hyposulfite, 1491
Sodium iodide, 1491–1492
Sodium isobutyl xanthate, 1755t
Sodium isopropyl xanthate, 1755t
Sodium manganate, 1492
Sodium nitrate, 1492
Sodium nitrite, 1492
Sodium nitrobenzeneazosalicylate, pH range and
color change of indicator, 825t
Sodium oleate, 1492
Sodium oxalate, 1185, 1492
Sodium palmitate, 1492
Sodium permanganate, 1492
Sodium phenate, 1492
Sodium phosphate dodecahydrate, in water-soluble
flame retardant formulation, 642t
Sodium n-propyl xanthate, 1755t
Sodium pyrophosphate, acidulant, 14
Sodium salicylate, 1492
Sodium sesquicarbonate, alkalizer, 14
Sodium silicate, 1492
Sodium soap greases, 743
Sodium stearate, 1492
Sodium sulfate, 1492
Sodium sulfide, 1492
Sodium sulfite, 1492
reagent for sulfonation, 1566t
Sodium-sulfur cell, 183
cell reactions and energy content (in
development), 180t
Sodium superoxide, 1580
Sodium tartarate, 1492
Sodium thiosulfate, 1492
Sodium thiosulfate [CAS: 7772–98–7], **1494–1495**
Sodium tungstate, 1492
Soft, **1495**
Softener, **1495**
Soft solder, 923t
Soil, **1495–1498**
Soil-absent (soil type), 1497t
Soil chemistry, **1498–1499**
Soil classification system (U.S.D.A.), 1497t
Soil heaping, 205t
Soil-vapor extraction, 205t

Soil-washing technologies, 1712–1714
Solar cells, 1298–1300, 1513
Solar concentrators, 1505–1506
Solar energy, **1499–1513**
    heat treating application, 763–764
Solar furnaces, 1506–1507, 1512
Solar heating, 1501–1505
Solar One, 1508–1510
Solder, 923t
    antimony content, 139t
Solder flux, **662**
Solder glasses, 725
Sol-gel chemistry, 42
Sol-gel glass, 729–730
Sol-gel technology, **1514–1516**
Solid, **1516–1517**
Solid-state chemistry, **1517**
Solid state electronics, 1017
Solid state lasers, wavelength range, 910t
Solid-state physics, **1517**
Solid superacids, 13
Solidus curve, **1520**
Solidus line, 57
Solion, **1520**
Solubility, **1520**
Solubility product, **1520**
Solubilization, **1520–1521**
Solution polymerization, 1342
Solutions, **1521–1524**
Solvent, **1524**
Solvent action, 1524
Solvent-based adhesives, 33
Solvent dyes, 512t
Solvent extraction, **1524–1525**
Solvolysis, **1525**
Somatotrophic hormone, where produced, structure,
    and principal functions, 789t
Somatotropin, where produced, structure, and
    principal functions, 789t
Sommelet reaction, **1525**
Sonn-Muller method, **1525**
Sonocatalysis, 1526
Sonochemistry, **1525–1526**, 1638
Sonoluminescence, 1525–1526
Sonolysis, **1527**
Sorbex processes, 41
Sorbitol, purification, 863t
Sorbitol adduct, commercial polyether polyol, 1654t
Sorosilicate minerals, 1013
Sorption, **1527**
Soybean oil, 1671
    fatty acid composition, 507t
Soybeans, susceptibility to iron deficiency, 874t
Space chemistry, **1527**
Space processing, **1527–1528**
Spandex fiber, 631
    generic designation and definition, 624t
    physical properties, 631t
Sparteine, 49t
Specific gravity, **1528–1530**
Specific heat, 757, 758, **1530**
Specific humidity, 793
Specific volume, **1531**
Spectacle glass, 1163t
Spectra 900/1000, 623t
Spectrochemical analysis (Visible), **1531**
Spectro instruments, **1531**
Spectrophotometer, 1295
Spectropolarimetry, 1322
Spectroscope, **1531–1532**
Spectroscopy (Instrumental Analysis), **1532**
Spectrum, **1532**

Spent fuel reprocessing, 1647
Spermaceti, 1747
Sperm oil, 1747
Sperrylite, **1532**
Sperry process, **1532**
Spessartine, 705
Sphalerite blende, **1532**
Sphene, **1532**
Sphingomyelin, constituent of human blood, 242t
Spike oil, 647t
Spinel, 708, **1532**
    crystal structure, 455
Spinel-cobalt oxide anodes, 982
Spinning-rotor friction vacuum gage, 1664
Spin quantum number, 1396
Spiradoline, 92t
Spiro hydrocarbon ring systems, 1173
Spodosols, 1497t
Spodumene, **1532–1533**
    irradiation, 709t
Spoils revegetation, 1439–1440
Spray drying, **1533–1534**
Sprays, **1534–1535**
Spray-type agglomerator, 45
Sputtering, **1535**
Sputter-ion vacuum pumps, 1663–1664
Squids (super conducting quantum interferometric
    devices), 894
Stabilizer, **1535–1536**
Stadol, 93t
Stainless steel, hardness of annealed, 755t
Stainless steels, 58
Stalactite, **1536**
Stalagmite, **1536**
Stallimycin, activity and producing organism, 128t
Standard conditions, **1536**
Standard electrode potentials, 31t
Standard state, **1536**
Standard volume, **1708**
Stanley, Wendell M. (1904–1971), **1536**
Stannane, physical properties, 795t
Stannite, **1536**
Staphylococcus aureus
    amphenicol susceptibility, 115t
    gram-positive, 168t
    ß-lactamase-producing bacterium, 110t
Staphylococcus epidermidis, ß-lactamase-producing
    bacterium, 110t
Starch
    industrial chemicals derived from, 229t
    use in food processing, 671t
Starch-based adhesives, 33
Starch [CAS: 9005–25–8], **1536–1538**
Stark effect, **1538**
State, **1538**
State and Local Air Monitoring Stations (SLAMS),
    acid rain, 11
Statistical mechanics, **1539**
Statistical thermodynamics, 1607–1608
Staurolite, **1539**
Steam, **1539–1540**
Steam-tube rotary dryers, 509t
Stearamide, as release agent, 1435t
Stearates, **1540**
Stearic acid
    fatty acid in vegetable oil, 1671t
    in major vegetable oils, 1672–1673t
Stearic acid [CAS: 57–11–4], **1540**
Stearone, **1540**
Steatite, dielectric permittivity, 852t
Steel, 57–58, **877–887**
    hardenability, 754–755

hardness, 755t
    modulus and rigidity, 538t
    in municipal waste, 1714t
Stefan-Boltzmann Law, 760
Stellite, hardness, 755t
Stendomycin, activity and producing organism, 128t
Stengel process, **1540**
Step-growth polymerization, 1346
Stephanite, **1540**
Stephen aldehyde synthesis, **1540**
Steradian (sr), 1645t
Stereochemistry, **1540–1542**
Stereoisomerism, 350, **1542–1545**
Stereoregular polymers, **1545–1546**
Stereospecific polymerization, 1346
Stereotype, 923t
Sterling silver, **1546**
Steroidal saponins, 734–735
Steroid glycosides, 733–735
Steroid hormonal dysfunctions, genetic influence,
    716
Steroids, **1546–1551**
Sterolins, 733–734
Stevens rearrangement, **1552**
Stibnite, **1552**
Stigmasterol, 733
Stilbite, **1552**
Stilb (sb), 1645t
Stilphostrol, hormonal therapy, 358t
Stilpnomelane, 871t
Stoichiometry, **1552–1553**
Stoke's law, **1553**
Stokes (St), 1645t
Storage battery, **1553**
Strain, 538–539
Strain theory, **1553**
Strand polarity, viruses, 1694
Strangeness, quarks, 1397
Stratosphere, 156
Streaming (Molecular), **1553**
Streams, acidification, 9
Streptococcus fecalis, gram-positive, 168t
Streptococcus hemolyticus, gram-positive, 168t
Streptococcus lactis, gram-positive, 168t
Streptococcus pneumonae, amphenicol susceptibility,
    115t
Streptococcus salivarius, gram-positive, 168t
Streptococcus viridans, gram-positive, 168t
Streptogramin A, activity and producing organism,
    128t
Streptogramin B, activity and producing organism,
    128t
Streptokinase, therapeutic enzyme, 574t
Streptomycin, year of discovery/market introduction,
    106t
Streptothricin A, activity and producing organism,
    128t
Streptovaricins, biological activity, 108t
Streptozocin USP, DNA alkylating/cross-linking
    agent, 356t
Stress, 538–539
Stromatolite, **1553**
Strontianite, **1553**
Strontium, 49
    abundance, 330t, 1132t
    covalent radius, 344t
    electronic structure, 336t
    interatomic distance, 342t
    ionic crystal radius, 341t
    nuclides (isotopes and isobars), 331t
    principal characteristics, 328t
Strontium [CAS: 7440–24–6], **1553–1554**

Strontium hydride, physical properties, 795t
Strontium superoxide, 1580
Strontrium, residence time in sea water, 1133t
Structural adhesives, 33
    load-bearing capabilities, 32t
Structural genese, 712
Structural isomers, 1542
Structural proteins, 1374
Strychnine, 49t, 52
Styrene, **1554–1556**
    melting range, 1546t
    polymerization with acrylic monomers, 17
Styrene-acrylonitrile (SAN), 21
Styrene-butadiene rubber, **1556–1557**
Styrene-butadiene-rubber, physical properties of
    flexible foamed plastics, 665t
Styrene-maleic anhyride, **1557**
Styrenes, monomer reactivity and suitable initiator
    for, 838t
Sub, **1557–1558**
Subacute sclerosing panencephalitis, 1696
Subbituminous A, characteristics, 390t
Subbituminous B, characteristics, 390t
Subbituminous C, characteristics, 390t
Suberic acid, monomer for nylon resins, 1333t
Subgroups, 746
Sublimation, **1557–1558**
Sublimation (Heat of), **1558**
Substitute natural gas (SNG), **1558–1563**
    heating value, 686t
Substitution, 1178
Subtilin, activity and producing organism, 128t
Subtractive color process, **1563**
Succineins, 1756
Succinic acid, 282t
    acidulant, 14
Succinic acid dehydrogenase, 282t
α-Succinic acid thiokinase, 282t
Sucrose, purification, 863t
Sucrose adduct, commercial polyether polyol, 1654t
Sudangrass, susceptibility to iron deficiency, 874t
Sugar, **1563–1565**
Sugar beets, susceptibility to iron deficiency, 874t
Sugarcane wax, 1747
Sugars, industrial chemicals derived from, 229t
Sulfamation, 1567
Sulfamic acid, reagent for sulfonation, 1566t
Sulfanilamide, 1565
Sulfate
    in 24 hour urine sample, 1657t
    water quality indicator, 1726
Sulfate minerals, 1012
Sulfation, 1567
Sulfhydryl group, analysis, 97
Sulfide minerals, 1011
Sulfometuron, environmental health advisories, 771t
Sulfonamide drugs, **1565**
Sulfonamides, 74
Sulfonation, 1180, **1565–1566**
Sulfone polymers, **1567–1568**
Sulfonic acids, **1568–1569**
Sulforhodamine B (Acid Red 52; CI 45100), 1756
Sulfotransferases, 571t
Sulfoxides, **1569–1570**
Sulfur
    abundance, 5t, 330t, 1132t
    amount in pure irons, 877t
    combustion constants, 422t
    combustion reactions, 423t
    constituent of human plasma or serum, 244t
    content of petroleum, 1243
    covalent radius, 344t

electronic structure, 336t
essential mineral nutrient, 1002, 1002t
ignition temperature in air, 425t
ionic crystal radius, 341t
nuclides (isotopes and isobars), 331t
plant micronutrient, 613t
principal characteristics, 328t
    in representative U.S. crude oils, 1242t
    in representative world crude oils, 1243t
Sulfur (In Biological Systems), **1573–1575**
Sulfuranes, 1571
Sulfur [CAS: 7704-34-9], **1570–1572**
Sulfur dioxide
    in acid rain, 9–12
    air pollution, 1327
    combustion constants, 422t
    gas separation, 39t
    measurement by NDIR analysis, 837t
    reagent for sulfonation, 1566t
Sulfur dyes, 512t, 514
    applications, 519
    dyeing process, 522
Sulfur fiber, generic designation and definition, 624t
Sulfur hexafluoride, as greenhouse gas, 744
Sulfuric acid, 12, 13
    acidulant, 14
    conductivity-concentration curve, 546
    reagent for sulfonation, 1566t
Sulfuric acid [CAS: 7664-93-9], **1572–1573**
Sulfuric ester hydrolases, 571t
Sulfurous acid [CAS: 7782-99-2], **1575**
Sulfurtransferases, 571t
Sulfur trioxide, reagent for sulfonation, 1566t
Sulfuryl chloride, reagent for sulfonation, 1566t
Sulindac, 93t
Sulofenur, chemotherapeutic, 359t
Sunflower oil, 1671
    fatty acid composition, 507t
Sunn, 632t
Superabrasive wheels, 2
Superacids, 13
Superalloys, 56
    high temperature, 774t
Superconducting Super Collider, 1217
Superconductivity, **1576–1579**
Supercooling, 1579
Supercritical drying, aerogels, 42–43
Superfluidity, **1579–1580**
Superfund Amendments and Reauthorization Act,
    1711t
Superheater, 760
Superheavy elements, 332–334
Supermolecular electronics, **1016–1026**
Superoxide dismutase, 1001
    therapeutic enzyme, 574t
Superoxides, **1580**
Super position (Nernst Principle of), **1580**
Supersaturated vapor, **1580**
Supersaturation (Chemical), **1580**
Suprarenin, where produced, structure, and principal
    functions, 786t
Surface, **1581**
Surface chemistry, **1581–1582**
Surface hardenable steels, 58
Surface tension, **1582–1583**
Surface waters
    acid rain, 9–10
    treatment, 1722
Surface waves, 1637
Surfactants, **1583–1586**
Suspended sediment, water quality indicator,
    1726

Suspension, 1586
Suspension polymerization, 1342
Svedberg, Theodor (1884–1971), **1586**
Swamping catalyst procedure, 28
Swarts reaction, **1586**
Sweeteners, **1586–1591**
Syenite, **1591**
Sylvanite, **1591**
Sylvitea, **1591**
Symmetry, **1591**
    and conservation laws, **432–434**
Synchrotron, 1215–1216
Synchrotron radiation, 1216–1217
    medical imagery application, 1762
Syndiotactic 1,2-butadiene, melting range, 1546t
Syneresis, **1591**
Synge, Richard L. M. (1914–1994), **1591**
Synmonicin A, producing organism, 118t
Synmonicin B, producing organism, 118t
Synmonicon C (CWI-785), producing organism,
    118t
Syntex, 93t
Synthesis (Chemical), **1591–1592**
Synthesis gas, **1592**
    ammonia production from, 83–85
Synthetases, 571t
Synthetic abrasives, 745
Synthetic gemstones, 707
    manufacture, 708
Synthetic latex house paints, 54
Synthetic methyl acetone, 7
Synthetic narcotics, 1041
Synthetic resins, **1438**
Synthetic rubber, **1452**
Synthetic tausonite, 708
Synthetic waxes, 1748–1749
Synthetic zeolites, 1034–1036
    composition, 1034t
    preparation, 1035t

T
Tableting, 44–45
TACE, hormonal therapy, 358t
Tachylite, 1593
Tachylyte, **1593**
Taconite, 1593
Tacticity, **1593**
Tafel rearrangement, 1593
Talc
    hardness, 2t, 755t
    as release agent, 1435t
Talc [CAS: 14807-96-6], **1593**
Tallow, higher alcohol manufacture from, 47
Talwin, 93t
Tamm levels, **1593**
Tamoxifen citrate USP, hormonal therapy, 358t
Tannin, **1593**
Tannins, 733, 734t
Tantalite, **1594**
Tantalum
    abundance, 330t
    corrosion resistant alloys, 446t
    covalent radius, 344t
    electronic structure, 337t
    interatomic distance, 342t
    ionic crystal radius, 341t
    major sources of industrial minerals, 1011t
    nuclides (isotopes and isobars), 331t
    physical properties of high temperature, 774t
    principal characteristics, 328t
    properties of refractory, 1075t

Tantalum carbide
  hardness, 755t
  plasma synthesis, 318t
Tantalum [CAS: 7440-2-5]7, **1594-1595**
Tantalum hydride, physical properties, 796t
Tantalum nitride, plasma synthesis, 318t
Tanzanite, 1779
Tar acid, **1595**
Tarnish, **1595**
Tar sand, 1243
Tar sands, **1595-1596**
Tartaric acid
  acidulant, 14
  carboxylic acid, 294
Tartaric acid [CAS: 87-69-4], **1596**
Tarui's disease, genetic influence, 716
Taube, Henry (1915-), **1596**
Tau particle, **1596**
Tauromustine, DNA alkylating/cross-linking agent,
  356t
Taxol, tubulin interactive, 358t
Taxotere, tubulin interactive, 358t
Tay-Sachs, genetic influence, 716
TCA, environmental health advisories, 771t
TCA cycle, **1596-1597**
Tebuthiuron, environmental health advisories,
  771t
Technetium
  covalent radius, 344t
  electronic structure, 336t
  interatomic distance, 342t
  ionic crystal radius, 341t
  nuclides (isotopes and isobars), 331t
  principal characteristics, 328t
Technetium [CAS: 7440-26-8], **1597**
Tectosilicate minerals, 1013
Teflon, 622t
Teicoplanin (teichomycin complex), producing
  organism, 118t
Tektite, **1597**
Tellurium
  abundance, 330t
  covalent radius, 344t
  electronic structure, 336t
  interatomic distance, 342t
  ionic crystal radius, 341t
  nuclides (isotopes and isobars), 331t
  principal characteristics, 328t
Tellurium [CAS: 13494-80-9], **1597-1598**
Tellurium-copper, 439t
Telomerization reactions, **1598**
Telomycin, activity and producing organism, 128t
Temperature, **1598**
Temperature scales and standards, **1598-1599**
Temperature-swing adsorption, 40
Temperature transfer standard, **1599**
Tenacity, **1599**
Tencel, 622t
Teniposide, topoisomerase interactive, 357t
Tenorite, **1599**
Tensile strength, **1599-1600**
Tensiometer, **1600**
Tension test, **1600**
Tentoxin, activity and producing organism, 128t
Tera, 1643t
Terbacil, environmental health advisories, 771t
Terbium
  abundance, 330t
  electronic structure, 336t
  interatomic distance, 342t
  ionic crystal radius, 341t
  nuclides (isotopes and isobars), 331t

principal characteristics, 328t
  properties, 1421t, 1422t
Terbium [CAS: 7440-27-9], **1600-1601**
Terbutryn, environmental health advisories, 771t
Terephthalic acid [CAS: 100-21-0], **1601**
Ternary alloy, 56
Ternary critical point, **450**
Terpene alcohol, **1601**
Terpeneless oil, **1601**
Terpenes, **1601-1602**
  biodegradable, 207t
Terpenoids, **1601-1602**
Terphenyls, **236-237**
Terramycin, year of discovery, 130t
Tertiary alcohols, 46
Tertiary amides, 74
Tertiary amines, 74, 75
Teslac, hormonal therapy, 358t
Tesla (T), 1645t
Testicular feminization, genetic influence, 716
Testing (Chemical), **1602**
Testing (Physical), **1602-1603**
Testing (Physiological), **1603**
Testis, 98
Testolactone USP, hormonal therapy, 358t
Testosterone, 98, 99
  where produced, structure, and principal
    functions, 789t
Tetrabromo-bisphenol, flame retardant, 640t
Tetrabromobisphenol A, flame retardant, 640t
Tetrabromo-bisphenol A, flame retardant, 640t
Tetrabromo-bisphenol A carbonate oligomer,
  phenoxy end capped, flame retardant, 640t
Tetrabromocy-clooctane, flame retardant, 640t
Tetrabromodipentaerythritol, flame retardant, 640t
Tetra(*tert*-butylperoxy)silane, physical properties,
  1232t
Tetra-4-*tert*-butyl-10-phthalocyaninato silicon
  dichloride, Langmuir-Blodgett films made
  from, 1018
Tetracaine, 100t
2,3,5,6-Tetrachloro-1,4-benzoquinone, physical
  properties, 1402t
3,4,5,6-Tetrachloro-1,2-benzoquinone, physical
  properties, 1402t
Tetracycline, year of discovery, 130t
Tetracycline antibiotics, **129-131**
Tetradecabromo-diphenoxybenzene, flame retardant,
  640t
Tetradecane, physical properties, 47t
Tetradecanedioic acid, physical properties, 490t
Tetradecanol, physical properties, 47t
1-Tetradecene, physical properties, 1150t
Tetradymite, **1603**
Tetraethylenepentamine, physical properties, 483t
Tetragonal crystal system, crystallographic elements,
  455t
Tetrahedrite, **1603**
Tetrahydroaluminates, 65
Tetrahydrofolic acid, 668
Tetrahydrofuran, physical properties, 587t
1,2,3,4-Tetrahydronaphthalene, physical properties,
  1231t
Tetramethylammonium superoxide, 1580
1,1,3,3-Tetramethylbutyl hydroperoxide, physical
  properties, 1231t
3,3,6,6-Tetramethyl-1,2-dioxane, physical properties,
  1232t
3,3,5,5-Tetramethyl-1,2-dioxolane, physical
  properties, 1232t
3,3,6,6-Tetramethyl-1,2,4,5-tetroxane, 1234t
7,8,15,16-Tetraoxadispiro[5.2.5.2]hexadecane, 1234t

1,1,3,3-Tetraphenylguanidine, 747t
Tetraterpenoids, 1601
Tetroses, 279t
Thalassemias, genetic influence, 716
Thalium, abundance, 1132t
Thallium
  abundance, 330t
  electronic structure, 337t
  interatomic distance, 342t
  ionic crystal radius, 341t
  nuclides (isotopes and isobars), 331t
  principal characteristics, 328t
Thallium [CAS: 7440-28-0], **1603-1604**
Thebaine, 49t, 91
Theine, 50
Theophylline, target of medical diagnostic test, 975t
Theoretical plate, **1604**
Thermal conductivity, 758
Thermal cracker, 1255
Thermal expansion coefficient, **1604**
Thermal gasoline, 1255
Thermal insulation, **851-857**
Thermal radiation, 759-760, **1604**
Thermal swing adsorption, 38t
Thermal variables, 1670
Thermionic conversion, **1604**
Thermionic emission, **1604**
Thermite, **1604**
Thermal diffusivity, 759
Thermochemistry, **1604**
Thermocouple, **1604-1605**
Thermocouple vacuum gage, 1663-1664
Thermodynamics, **1605-1608**
Thermodynamic temperature scale, 1599
Thermoelectric cooling, **1608-1609**
Thermoelectricity, **1609**
Thermofor process, **1609**
Thermogravimetric analysis, **1609**
Thermometer, **1609**
Thermometer (Filled-system), **1609**
Thermometric titration, **1622**
Thermoplastic, **1610**
Thermoplastic polyurethanes, 1654
Thermoresponsive materials, 1485
Thermoset, **1610**
Thermoset polyurethanes, 1654
Thermosiphon reboiler, 760
Therm (thm), 1645t
Thiacarbocyanine, sensitizer for semiconductors,
  535t
Thiamides, 74
Thiamine, **1610-1611**
  constituent of human blood, 242t
Thiapyrylium dyes, sensitizer for semiconductors,
  535t
Thiiranes, monomer reactivity and suitable initiator
  for, 838t
Thin, **1611**
Thin film photovoltaics, 1300
Thin films, **1611-1613**
Thin-layer chromatography (TLC), **1613**
Thinner, **1613**
Thiobencarb, environmental health advisories, 771t
Thiocyanic acid [CAS: 463-56-9], **1613-1614**
Thioether hydrolases, 571t
Thioethers, **1614**
Thioindigo, sensitizer for semiconductors, 535t
Thiolester hydrolases, 571t
Thiopeptin, activity and producing organism, 128t
Thiophene, as parent structure compound, 1090t
Thiophene [CAS: 110-02-1], **1614**
Thiorphan, 92t

Thiosalicylic acid, 1456

Thiostrepton, activity and producing organism, 128t

Thiotepa, DNA alkylating/cross-linking agent, 356t

Thiotepa USP, DNA alkylating/cross-linking agent, 356t

Thiourea [CAS: 62–56–6], **1614**

Third law of thermodynamics, 1606–1607

Thixotropy, **1614**

Thomson, J. J., **1614**

Thomson parabola method, **1614**

Thomson particle, **1614–1615**

Thorianite, **1615**

Thorite, **1615**

Thorium, 23t

    abundance, 330t, 1132t

    active deposit, 29

    binary compound properties, 25–26t

    corrosion resistant alloys, 446t

    electronic structure, 337t

    as extender for uranium in LWRs, 1647

    interatomic distance, 342t

    ionic crystal radius, 341t

    oxidation states, 24t

    physical properties, 24t

    principal characteristics, 328t

Thorium [CAS: 7440–29–1], **1615**

Thorium series, 332

Three-phase equilibrium, **1616**

Threonine

    first isolation and isoelectric point, 77t

    manufacture, 76t

    in selected foods, 75t

    structural classification, 79t

L-Threonine, therapeutic enzyme, 574t

Thrombin, **1616**

Through-circulation dryers, 509t

Through hardenable steels, 58

Thulite, 1779

Thulium

    abundance, 330t

    electronic structure, 337t

    interatomic distance, 342t

    ionic crystal radius, 341t

    nuclides (isotopes and isobars), 331t

    principal characteristics, 328t

    properties, 1421t, 1422t

Thulium [CAS: 7440–30–4], **1616**

Thylakentrin, where produced, structure, and principal functions, 787t

Thyme oil, 647t

Thymol Blue, pH range and color change of indicator, 825t

Thymolphthalein, pH range and color change of indicator, 825t

Thymolsulfonphthalein, pH range and color change of indicator, 825t

Thyroid hormones, 785

Thyroid-stimulating hormone (TSH), where produced, structure, and principal functions, 789t

Thyrotrophic hormone, where produced, structure, and principal functions, 789t

Thyrotropic hormone, 790

Thyrotropin, where produced, structure, and principal functions, 789t

Thyroxine, where produced, structure, and principal functions, 789t

Tickborne encephalitis vaccine, 1660

Tiger's eye, heat effect, 708t

Tiger's-eye, **307**

Time-resolved photoacoustic calorimetry, 1286

Tin

    abundance, 330t, 1132t

    constituent of human plasma or serum, 244t

    corrosion resistant alloys, 446t

    covalent radius, 344t

    electronic structure, 336t

    interatomic distance, 342t

    ionic crystal radius, 341t

    nuclides (isotopes and isobars), 331t

    principal characteristics, 328t

    standard electrode potential, 31t

    trace mineral nutrient, 1002t, 1005

Tin base babbits, 923t

Tin [CAS: 7440–31–5], **1616–1618**

Tin electroplating, 557t

Tin-lead solders, antimony content, 139t

Tioguanine USP, antimetabolite, 355t

Tiselius, Arne W. K. (1902–1971), **1619**

Tishchenko reaction, **1619**

Tissue factor, 784

Tissue plasminogen activator, therapeutic enzyme, 574t

Tissue plasminogen activator (TPA)

    annual sales, 222t

    unit value and relative production quantity, 232t

Titanates, dielectric permittivity, 852t

Titanite, **1619**

Titanium

    abundance, 5t, 330t, 1132t

    corrosion resistant alloys, 446t

    covalent radius, 344t

    electronic structure, 336t

    interatomic distance, 342t

    ionic crystal radius, 341t

    major sources of industrial minerals, 1011t

    nuclides (isotopes and isobars), 331t

    principal characteristics, 328t

Titanium alloys, 58

Titanium carbide, 763t

    plasma synthesis, 318t

Titanium [CAS: 7440–32–6], **1619–1620**

Titanium diboride, 1620

Titanium dioxide, plasma synthesis, 318t

Titanium dioxide [CAS: 13463–67–7], **1621**

Titanium nitride, plasma synthesis, 318t

Titration (Potentiometric), **1621–1622**

Titration (Thermometric), **1622**

Titrimetric analysis, 94

Todd, Sir Alexander R. (1907–1997), **1622**

Todorokites, **1622–1623**

Togaviruses, 1694

Tokamak Fusion Test Reactor, 696–697, 698

Toluene

    biodegradable, 207t

    combustion constants, 422t

    heating value, 686t

    xylenes production from, 1766

Toluene [CAS: 108–88–3], **1623–1625**

Toluene diisocyanate (TDI), 1624

    polyurethanes from, 1653, 1654, 1655

    production, 464

Toluenesulfonic acid, 1625

p-Toluenesulfonic acid, physical properties, 1568t

Toluenesulfonyl chloride, 1625

Tolypomycins, biological activity, 108t

Tomography, 1761–1762

Tonephin, where produced, structure, and principal functions, 789t

Topaz, **1625**

    hardness, 2t, 755t

    irradiation, 709t

Topazolite, 705

Topochemical reaction, **1625**

Topotecan hydrochloride, topoisomerase interactive, 357t

Top quark, 1397

Torbernite, **1625**

Toth process, **1625**

Tourmaline, **1625–1626**

    hardness, 1008t

    irradiation, 709t

Townsend avalanche, **1626**

Toxaphene insecticide, in hazardous waste, 1711t

Toxicity, **1626**

Toxicology, **1626**

Toxic spoil, 1439t

Toxic Substances Control Act, 1711t

Toxline, 769t

Toxnet, 769t

Trace and residue analysis, **1626–1628**

Trace element (Micronutrient), **1628**

Trace mineral nutrients, 1001–1005, 1002t

Tracer, **1628–1629**

Tracer compound, 429

Trachyte, **1629**

Tragacanth gum, 748t

Transactinides, 23, 24, 332–334

    principal characteristics of elements, 328t

Transactinium earths, **1629**

Transaldolase, 283

Transferases, 571t

Transference number, **1629**

Transfer magnetoresistance, 752

Transferrin, source and function, 986t

Transfer RNA, 713–714, 1125

Transition element, **1629**

Transition metal, **1629**

Transition temperature, **1629**

Transketolase, 283

Transmissible mink encephalopathy, 1696

Transmutation, **1629**

Transport number, **1629**

Transport proteins, 1374

Transuranium elements, **1629**

Trapped ions, **1629**

Travertine, **1629**

Tray and compartment dryers, 509t

Tray dryers, 509t

Tree fruits, susceptibility to iron deficiency, 874t

Tremolite, 27, 149–150

    physical and chemical properties, 150t

Tremolite [CAS: 1332–21–4], **1629**

Trevira, 623t

Trexan, 93t

Triacetyloleandomycin, commercial products, 121t

Triallate, environmental health advisories, 771t

Triamcinolone, commercial microbial transformations used to produce, 1550t

Tribasic acids, 13

Triboluminescence, 946t

Tribomoneopentyl alcohol, flame retardant, 640t

2,4,6-Tribromophenol, flame retardant, 640t

Tribromophenyl allyl ether, flame retardant, 640t

Tribromostyrene, flame retardant, 640t

Tri (tert-butylperoxy)borane, physical properties, 1232t

Tricalcium phosphate, 269

Tricarboxylic acid cycle, 1596–1597

Tricarcillin, year of discovery/market introduction, 106t

1,3,5-Trichlorobenzene, dye carrier, 512t

Trichloroethylene, 100t

2,2,2-Trichloro-1-hydroxyethyl hydroperoxide, 1233t

Trickling filters, 1725

Triclinic crystal system, crystallographic elements, 455t

Triclopyr, environmental health advisories, 771t

ω-Tricosenoic acid, Langmuir-Blodgett films made from, 1018

Tridecanedioic acid, physical properties, 490t

1-Tridecene, physical properties, 1150t

Tridymite, **1629**
  crystal structure, 455

Trienomycins, biological activity, 108t

Triethanolamine, 587

Triethylenediamine, physical properties, 483t

Triethylene glycol, physical properties, 590t

Triethylenetetramine, physical properties, 483t

Triethylgallium, 702

3,6,9-Triethyl-3,6,9-trimethyl-1,2,4,5,7,8-hexoxononane, 1234t

Trifluoromethanesulfonic acid, physical properties, 1568t

Trifluralin, environmental health advisories, 771t

Triglycerides, target of medical diagnostic test, 975t

Trigonal crystal system, crystallographic elements, 455t

Trihydric alcohol, 46

Trilobine, 49t

Trim cooler, 760

Trimethylcarbinol, 7

3,3,5-Trimethylcyclohexanone, physical properties, 900t

3,5,5-Trimethyl-2-cyclohexen-1-one, physical properties, 900t

3,5,5-Trimethyl-3-cyclohexen-1-one, physical properties, 900t

Trimethylene glycol, 46

Trimethylgallium, 702

2,6,8-Trimethyl-4-nonanone, physical properties, 900t

Trimethylolethane, polyol for alkyd synthesis, 53t

Trimethylolpropane, polyol for alkyd synthesis, 53t

Trimethylolpropane adduct, commercial polyether polyol, 1654t

1,2,7-Trimethyl xanthine, 50

Trimetrexate, antimetabolite, 355t

1,3,5-Trinitrobenzene, 1082
  pH range and color change of indicator, 825t

2,4,7-Trinitrofluorenone, sensitizer for semiconductors, 535t

2,4,6-Trinitrophenol, physical properties, 1081t

Trinitrotoluene (TNT), physical properties, 1081t

1,2,4-Trioxolane, boiling point, 1235t

1,1,3-Triphenylguanidine, 747t

1,2,3-Triphenylguanidine, 747t

Triphenylmethane and related dyes, **1629–1631**

Triphosphoric monoester hydrolases, 571t

Triphylite, **1631**

Triple bond, **1631**

Triple bonds, 346

Triple point, **1631**

Tripoli, abrasive properties, 745t

Tripolite, **1631**

Trisaccharides, 279t

Tris-dibromopropylisocyanurate, flame retardant, 640t

Trisilanyl, 1477

Trisilanylene, 1477

Trisodium phosphate, 1492

Triterpenoids, 1601

Tritium, **1631**

Trivial name, **1631**

tRNA, 713–714, 1125

Troctolite, 701

Tropine, 49t

Tropine alkaloids, 49t

Troposphere, 156

Trypsin, therapeutic enzyme, 574t

Tryptophan
  first isolation and isoelectric point, 78t
  manufacture, 76t
  in selected foods, 75t
  structural classification, 79t

L-Tryptophanase, therapeutic enzyme, 574t

Tuberactinomycin N, activity and producing organism, 128t

Tukon hardness scale, 756

Tung oil, 1671
  fatty acid composition, 507t

Tungstate minerals, 1012

Tungsten
  abundance, 330t, 1132t
  covalent radius, 344t
  electronic structure, 337t
  interatomic distance, 342t
  ionic crystal radius, 341t
  nuclides (isotopes and isobars), 331t
  physical properties of high temperature, 774t
  principal characteristics, 328t
  properties of refractory, 1075t

Tungsten blues, 1633

Tungsten carbide, 1632
  hardness, 755t
  plasma synthesis, 318t

Tungsten [CAS: 7440–33–7], **1631–1633**

Tunnel dryers, 509t

Turbidimetry, **1633**

Turbulent flow, **1633–1634**

Turquoise, **1633–1634**

Twinning (Crystal), **1634**

Twitchell process, **1634**

Twitchell's reagent, **1634**

Tylenol, 93t

Tyndall, John (1820–1839), **1634**

Tyndall cone, 1636

Tyndall effect, **1634**

Type metals, antimony content, 139t

Tyril-880, 21t

Tyrocidines, activity and producing organism, 128t

Tyrosinase
  metal chelate enzyme, 323t
  source and function, 986t

L-Tyrosinase, therapeutic enzyme, 574t

Tyrosine
  first isolation and isoelectric point, 78t
  manufacture, 76t
  in selected foods, 75t
  structural classification, 79t

Tyrosine kinase receptors, 1272

Tyrosinemia, genetic influence, 716

Tyrothricin
  activity and producing organism, 128t
  year of discovery/market introduction, 106t

Tyuyanunite, **1634**

U

²³⁴U, 1646

²³⁵U, 1646, 1649–1650

²³⁸U, 1646, 1649–1650

U73,975, DNA interactive agent, 357t

Ugrandite, **1635**

Uintahite, **1635**

UK-68,597, producing organism, 118t

UK-69,753, formula and producing organism, 116t

UK-72,051, producing organism, 118t

Ulexite, **1635**

Ulmin, **1635**

Ultimate analysis, 97

Ultisols, 1497t

Ultraaccelerators, 5

Ultrabasic, **1635**

Ultracentifuge, **1635**

Ultracentrifuge, 315–316

Ultrafiltration, **1635–1636**

Ultrahigh molecular weight polyethylene, 1140t

Ultrahigh-strength steels, 58

Ultrahigh vacuum, 1661t

Ultramicroscope, **1636**

Ultrasonic density sensors, 1637–1638

Ultrasonic level detectors, 1638

Ultrasonic nondestructive characterization, 1638

Ultrasonic nondestructive testing, 1092–1093

Ultrasonic pressure transducers, 1638

Ultrasonics, **1637–1639**

Ultrasonic speech, human perception, 1639

Ultrasonic thermometers, 1638

Ultratrace mineral nutrients, 1001–1005, 1002t

Ultraviolet catastrophe, 1393

Ultraviolet film, 1293–1294

Ultraviolet radiation, **1639–1640**

Ultraviolet spectrometers, **1640–1641**

Ultraviolet stabilizers, **1641–1642**

Uncertainty principle, **1642**

Undecanedioic acid, physical properties, 490t

1-Undecene, physical properties, 1150t

Underground storage of products and wastes, 1729–1730

United States-Canada Air Quality Agreement, 11

Units and standards, **1642–1645**

Universal-pressure boiler water treatment, 1745–1746

Unphenelfamycin, formula and producing organism, 116t

Unresisted expansion, 34

Unsaturated compound, 429

Unsaturated fats, 1671

Unsaturated fatty acids, 1671t

UOP Sorbex, 41

Up quark, 1397

Upsilon particle, **1645–1646**

Uralite, **1646**

Urandium-radium series, 332

Uranine (CI 45350), 1756

Uraninite, **1646**

Uranium, 23t
  abundance, 330t, 1132t
  active deposit, 29
  binary compound properties, 25–26t
  electronic structure, 337t
  interatomic distance, 342t
  ionic crystal radius, 341t
  isotope separation, 1649–1650
  oxidation states, 24t
  physical properties, 24t
  principal characteristics, 328t
  purification, 863t

Uranium carbides, 1648

Uranium [CAS: 7440–61–1], **1646–1650**

Uranium chlorides, 1649

Uranium enrichment, **566**, 1649–1650

Uranium hexafluoride, 1649

Uranium oxides, 1648

Uranium spent-fuel reprocessing, 1647

Uranyl peroxide, 1648

Urate, target of medical diagnostic test, 975t

Urea
  constituent of human blood, 242t
  in 24 hour urine sample, 1657t

as parent structure compound, 1090t
    target of medical diagnostic test, 975t
Urea-ammonium orpthophosphate, **1652**
Urea-ammonium polyphosphate, **1652**
Urea [CAS: 57–13–6], **1650–1652**
Urea cycle abnormalities, genetic influence, 716
Ureaform, **1652–1653**
Urea-formaldehyde resin, production methods for
    cellular, 664t
Urea-formaldehyde resins, **1653**
Urease, **1653**
Urech cyanohydrin method, **1653**
Urech hydantoin synthesis, **1653**
Urena, 632t
Urethane [CAS: 51–79–6], **1653**
Urethane polymers, **1653–1656**
Urethane sealants, 1462
Urey, Harold C. (1894–1981), **1656**
Uric acid
    constituent of human blood, 242t
    in 24 hour urine sample, 1657t
    and purine metabolism, 1382
Uricase, metal chelate enzyme, 323t
Urine, **1656–1657**
Urokinase, therapeutic enzyme, 574t
Uronic acid, **1657**
Uteracon, where produced, structure, and principal
    functions, 788t
Uvarovite, 705, 1635

V

Vaccine technology, **1659–1661**
Vacuum, **1661–1665**
Vacuum carburizing, 763t
Vacuum deposition, **1665**
Vacuum distillation, **1665**
Vacuum electronics, 1017
Vacuum flasher, 1255
Vacuum measurement, 1663–1665
Vacuum pumps, 1662–1663
Vacuum rotary dryers, 509t
Vacuum tray dryers, 509t
Valence, 345–346, **1665**
Valine
    first isolation and isoelectric point, 77t
    manufacture, 76t
    in selected foods, 75t
    structural classification, 79t
    synthesis from products of carbohydrate
        metabolism, 282t
Valinomycin, activity and producing organism, 128t
Vanadate minerals, 1012
Vanadinite, **1665**
Vanadium
    abundance, 330t, 1132t
    biologically active chelates, 323t
    covalent radius, 344t
    electronic structure, 336t
    interatomic distance, 342t
    ionic crystal radius, 341t
    nuclides (isotopes and isobars), 331t
    physical properties of high temperature, 774t
    principal characteristics, 328t
    in representative U.S. crude oils, 1242t
    in representative world crude oils, 1243t
    trace mineral nutrient, 1002t, 1005
Vanadium [CAS: 7440–62–2], **1665–1667**
Vanadium hydride, physical properties, 796t
Vanadium nitride, plasma synthesis, 318t
Vancomycin, producing organism, 118t

Van Der Waals, Johannes Diderik (1837–1923),
    1668
Van der waals equation, **1667–1668**
Van der waals forces, **1668**
Van der Waals radius, 340
Vanillin [CAS: 121–35–5], **1668–1669**
Vanillin sugar, 1669
Van't Hoff, Jacobus Hendricus (1852–1911), **1669**
Van't Hoff equation, **1669**
Van't Hoff law, **1669**
Vapor, **1669–1670**
Vapor density, **1670**
Vapor heat exchanger, 760
Vaporization, **1670**
Vaporization (Heat of), **1670**
Vapor pressure, **1670**
Variable (Process), **1670**
Varicella vaccine, 1659
Variolite, **1670**
Varnish, **1670–1671**
Var (var), 1645t
Vasicine, 49t
Vasoconstrictine, where produced, structure, and
    principal functions, 786t
Vasophysin, where produced, structure, and principal
    functions, 789t
Vasopressin, where produced, structure, and
    principal functions, 789t
Vasotonin, where produced, structure, and principal
    functions, 786t
Vat dyes, 512t, 517–518
    applications, 519
    dyeing process, 521–522
Vegetable fibers, **631–633**
Vegetable oils (Edible), **1671–1674**
Vegetables, susceptibility to iron deficiency, 874t
Vegetable waxes, 1746–1747
Velban, tubulin interactive, 358t
Vepesid, topoisomerase interactive, 357t
Verdigris, 6
Verdoflavin, 1699
Vermiculite, insulation material, 857
Vernolate, environmental health advisories, 771t
Vertisols, 1497t
Very high vacuum, 1661t
Very large scale integration, 1017
Very low density polyethylene, 1140t
Vesuvianite, **1674**
Vetiver oil, 647t
Vibramycin, year of discovery, 130t
Vibrating-tray dryers, 509t
Vibratory feeder, **605**
Vibrio comma, gram-negative, 168t
Vicalloy II, thin-film applications, 958t
Vickers hardness scale, 756
Vinblastin sulfate USP, tubulin interactive, 358t
Vincristin sulfate USP, tubulin interactive, 358t
Vindesine sulfate, tubulin interactive, 358t
Vinegar, **1674–1675**
Vinegar acid, 6
Vineger eels, 1675
Vinorelbine, tubulin interactive, 358t
Vinyl acetal polymers, **1675–1676**
Vinyl acetate
    physical properties, 585t, 1677t
    polymerization, 1676–1678
Vinyl acetate polymers, **1676–1678**
Vinyl alcohol-methyl methacrylate copolymers, 1679
Vinyl alcohol polymers, **1678–1679**
N-Vinylamide polymers, **1680–1682**
N-Vinylamides, preparation, 1680
Vinylcaprolactam, 1680

Vinyl chloride
    definitely established workplace carcinogen, 297t
    physical properties, 366t
Vinyl chloride-acrylonitrile copolymers,
    ultrafiltration media application, 1635
Vinyl chloride [CAS: 75–01–4], **1682–1685**
Vinyl chloride polymers, **1685–1688**
Vinyl cyanide, 20
Vinyl ester polymers, **1687–1689**
Vinyl ether monomers and polymers, **1689–1690**
Vinyl ethers, polymerization, 1689–1690
Vinyl formamide, 1680
Vinylidene chloride, physical properties, 1691t
Vinylidene chloride copolymers
    diffusion and solubility coefficients for oxygen
        and carbon dioxide, 173t
    gas permeability, 172t
    permeation of flavor/aroma compounds, 174t
    water-vapor transmission rate, 173t
Vinylidene chloride monomer and polymers,
    1691–1693
Vinylidene cyanide, monomer reactivity and suitable
    initiator for, 838t
N-Vinylimides, preparation, 1680
Vinyl ketones, monomer reactivity and suitable
    initiator for, 838t
N-Vinyllactams, 1680
Vinyl polymerization, 1181, 1341
Vinylpyrrolidinone, 1680
N-Vinyl-2-pyrrolidinone, preparation, 1680
Vinyl resins, alkyd resins with, 54
Vinyltoluene, 1625
Viral replication, 1694
Viral vaccination, 1696–1697
Virginiamycin M_1, activity and producing organism,
    128t
Viridogrisein, activity and producing organism, 128t
Virion, 1693
Virtanen, Arrturi I. (1895–1973), **1693**
Virus, **1693–1697**
Viscoelasticity, **1697**
Viscometer (or Viscosimeter), **1697**
Viscose process, **1697**
Viscose rayon, fiber-dye property requirements, 521t
Viscosimeter, 1697
Viscosity, **1697**
Viscosity measurement, 1443
Visible spectrochemical analysis, **1531**
Vitamin, **1697–1698**
Vitamin A, 1697, **1697–1699**
    constituent of human blood, 242t
    loss from food operations, 1698t
Vitamin B_1, 1610–1611, 1697
    loss from food operations, 1698t
Vitamin B2, 1697
    loss from food operations, 1698t
Vitamin B_2 (Riboflavin), **1699–1700**
Vitamin B6, 1697
Vitamin B_6 (Pyridoxine), **1700–1701**
Vitamin B_{12}, 1697
    constituent of human blood, 242t
Vitamin B_{12} (Cobalamin), **1701–1703**
Vitamin B complex, 1697
Vitamin C, **151–152,** 756
    loss from food operations, 1698t
Vitamin D, 1697, **1703–1704**
    in milk, 1001
Vitamin D_2, constituent of human blood, 242t
Vitamin D-resistant rickets, genetic influence, 716
Vitamin E, 1697, **1705–1706**
    constituent of human blood, 242t
Vitamine, 1697

Vitamin G, 1699
Vitamin K, 1697, **1706–1707**
Vitamins
  loss in food processing, 1697–1698, 1698t
  relationship of hormones to, 786–789t
Vitreous enamels, **560–561**
Vitreous silica, hardness, 2t
Vitreous state, **1707**
Vitrophyre, **1708**
Vivianite, **1708**
Voids, 1708
Volatile, **1708**
Volatile oils, **1708**
Volatile organic compounds, **1708**
Volatile organic compounds (VOCs), in acid rain, 9
Volatility product, **1708**
Voltaic cell, **1708**
Voltampere (VA), 1645t
Voltaren, 93t
Volt (V), 1645t
Volume fraction, 1038
Volume (Standard), **1708**
Volumetric analysis, 94
Volumetric feeder, **605–606**
von Gierke's disease, genetic influence, 716
Vulcanization, 1450–1451, **1708**

W

Wacker reaction, **1709**
Wad, **1709**
Wagner-Jauregg reaction, **1709**
Wagner-Meerwin rearrangement, **1709**
Waksman, Selman A (1888–1973), **1709**
Walden inversion, 1424, **1709**
Wallach, Otto (1847–1931), **1709**
Wallboard, 750
Walnut oil, fatty acid composition, 507t
Walnuts, susceptibility to iron deficiency, 874t
Waspaloy, 58
Waste-heat boiler, 760
Wastes and energy sources, **1717–1718**
Wastes and pollution, **1709–1716**
Wastewater treatment, 1724–1725
  dyeing effluents, 527t
Water, **1718–1726**
  absolute index, 1426t
  combustion constants, 422t
  constituent of human blood, 242t
  content in saturated air, 46t
  density, 474t
  in human body, 1720–1721
  mass peaks and relative intensities, 973t
  measurement of vapor by NDIR analysis, 837t
  percent content of air at various temperatures
    and altitudes, 157t
  specific heat at constant pressure, 757t
  standard electrode potential, 31t
  thermal conductivity, 758t
  vapor transmission rate in barrier polymers, 173t
  in vertebrates, 1719–1720
Water deficiency, 1721
Water (Hard), **1726**
Water intoxication, 1721
Water of constitution, 794
Water pollution, **1726–1734**
Water-reducible alkyds, 54
Water resources, **1735–1736**
Water softening, 1723
Water-soluble polymers, **1350, 1736–1738**
Water table, 1735
Water treatment lagoons, 1725

Water treatment, 1722–1724
Water treatment (Boiler), **1738–1746**
Waterwall incinerators, 1718
Waterworks, 1722
Watson, James Dewey (1928–), **1746**
Watthour (Wh), 1645t
Watt per meter Kelvin (W/m*K), 1645t
Watt per steradian square meter (W/Sr*m2), 1645t
Watt per steradian (W/sr), 1645t
Watt (W), 1645t
Wavellite, **1746**
Waxes, **1746–1749**
  use in food processing, 671t
Weber (Wb), 1645t
Weed management agents, 772
Weerman degradation, **1749**
Weighting agent, **1749**
Weightlessness, 740
Weldalite, 70
Werner, A. (1866–1919), **1749**
Wessely-Moser rearrangement, **1749**
Wetting agent, **1749**
Wheat, susceptibility to iron deficiency, 874t
Whey, 1000
White glue, 33
White iron pyrites, 871t
Wichterle reaction, **1749**
Widman-Stoermer synthesis, **1749**
Wieland, Heinrich O. (1877–1957), **1749**
Wigner force, **1749**
Wigner nuclides, **1749**
Wilkinson, Geoffrey (1921–1996), **1749**
Willemite, **1749**
Willgerodt reaction, **1749**
Williamson synthesis, **1749**
Willstater, Richard (1872–1942), **1749**
Wilson's disease, genetic influence, 716
Wilzbach procedure, **1749**
Windaus, Adolf (1876–1959), **1750**
Wine-making, 1767
Winterizing, oils, 1671
Wiping solder, 923t
Witherite, **1750**
Wittig, George (1897–1987), **1750**
Wittig reaction, **1750**
Wohl degradation, **1750**
Wöhler, Friedrich (1800–1882), **1750**
Wöhler synthesis, **1750**
Wohlwill process, **1750**
Wolffenstein-Boters reaction, **1750**
Wolff-Kishner reaction, **1750**
Wolff rearrangement, **1750**
Wolframite, **1750–1751**
Wollastonite, **1751**
Wood, **1751**
  excess air for combustion, 426t
Wood burning, 1718
Wood dust, definitely established workplace
  carcinogen, 297t
Woodell's hardness scale, 2t
Wood fiber, insulation material, 857
Wood preservative, **1751–1752**
Wood pulp production and processing, **1379–1382**
Wood's metal, 923t
  bismuth in, 238t
Woodward, Robert B. (1917–1979), **1752**
Woodward-Hoffmann rules, **1752**
Wool, **1752**
  dyeing, 522
  fiber-dye property requirements, 521t
  generic designation and definition, 624t
  physical properties of staple, 626t

Wool-acrylic fiber blends, dyeing, 525
Wool-cellulosic fiber blends, dyeing, 525
Wool-nylon fiber blends, dyeing, 525
Wool-polyester fiber blends, dyeing, 525
Woolwax, 1747
Wrought ferrous metals, 58
Wrought iron, **1752**
Wrought nickel, 1071t
Wulfenite, **1752–1753**
Würtz-Fittig-Frankland reaction, **1753**
Wurtzite, **1753**
Wurzite, crystal structure, 455

X

Xalstocite, **1755**
Xanthan gum, **1755**
Xanthates, **1755–1756**
Xanthene dyes, **1756**
Xanthic acids, 1755
Xanthine oxidase
  metal chelate enzyme, 323t
  source and function, 986t
Xanthylium, 1756
Xenoblast, **1757**
Xenocryst, **1757**
Xenolith, **1757**
Xenomorphic, **1757**
Xenon
  abundance, 330t, 1132t
  electronic structure, 336t
  nuclides (isotopes and isobars), 331t
  percentage in air, 46t
  principal characteristics, 328t
Xenon 133, 1757
Xenon 135, 1757
Xenon [CAS: 7440–63–3], **1757**
Xenon platinum hexafluoride, 1757
Xenon tetrafluoride, 1757
Xenotime, 1771
Xenotropic, **1757**
Xi particle, **1758**
X-ray, **1758**
X-ray analysis, **1758–1761**
X-Ray analyzer, electron-microprobe, 1759–1760
X-ray cat scanner (CAT), 1761–1762
X-ray computed tomography, 1092
X-ray crystallography, 1759
X-ray detectors, 1759, 1760
X-ray fluorescence analysis, 1760–1761
X-ray generation, 1759
X-rays, gemstone irradiation, 708t
X-ray scan and other medical imagery, **1761–1763**
X-ray spectral lines, 1758–1759
Xylem, 1751
Xylene
  biodegradable, 207t
  combustion constants, 422t
  from toluene, 1625
m-Xylene
  manufacture, 1765–1766
  physical properties, 1763t, 1766t
o-Xylene
  manufacture, 1765–1766
  physical properties, 1763t, 1766t
p-Xylene
  liquid-phase adsorption, 41
  manufacture, 1765–1766
  physical properties, 1763t, 1766t
Xylene [CAS: 1330–20–7], **1763**
Xylene isomerization, 1766
Xylene polymers, **1763–1765**

Xylenes and ethylbenzene, **1765–1766**
p-Xylylene (PX), 1763–1765

Y

YAG and YIG, **1767**
Yard waste, in municipal waste, 1714t
Yeasts and molds, **1767–1770**
Yellow brass, 440t
    hardness of annealed and cold-rolled, 755t
Yellow fever virus, 1694
Yellow gold, 736
Yellow quark, 1397
Yield strength, 784
Ylang ylang, 647t
Yohimbine, 49t
Ytterbium
    abundance, 330t
    electronic structure, 337t
    interatomic distance, 342t
    ionic crystal radius, 341t
    nuclides (isotopes and isobars), 331t
    principal characteristics, 328t
    properties, 1421t, 1422t
Ytterbium [CAS: 7440–64–4], **1770**
Ytterbium orthoferrite, 1770
Yttrium
    abundance, 330t, 1132t
    covalent radius, 344t
    electronic structure, 336t
    interatomic distance, 342t
    ionic crystal radius, 341t
    nuclides (isotopes and isobars), 331t
    principal characteristics, 328t
    properties, 1421t, 1422t
Yttrium aluminum garnet (YAG), 1767, 1771
Yttrium carbide, 1771
Yttrium [CAS: 7440–65–5], **1770–1771**
Yttrium halides, 1771
Yttrium hydroxide, 1771
Yttrium iron garnet (YIG), 1767, 1771
Yttrium nitrate, 1771
Yttrium oxalate, 1185
Yttrium oxide, 1771
Yttrium peroxide, 1771

Z

Zanosar, DNA alkylating/cross-linking agent,
    356t
Zeeman, Erik Christopher (1925–), **1773**
Zeeman effect, **1773**
Zeisberg concentrator, **1773**
Zeisel reaction, **1773**
Zeolite A
    composition, 1034t
    pore structure, 41t
    preparation, 1035t
Zeolite group, **1773**
Zeolite L

composition, 1034t
    pore structure, 41t
Zeolite minerals, 1034
Zeolite omega
    composition, 1034t
    preparation, 1035t
Zeolites, 65, 1033–1036
    for gas adsorption, 39t
    for liquid adsorption, 41t
    for liquid-phase adsorption, 41
Zeolite X
    composition, 1034t
    preparation, 1035t
Zeolite Y
    composition, 1034t
    preparation, 1035t
Zeolite ZSM-5
    composition, 1034t
    pore structure, 41t
    preparation, 1035t
Zerevitinov determination, **1773**
Zero-order reaction, **1773**
Zero point energy, **1773**
Zeroth law of thermodynamics, 1605–1606
Zeta potential, **1773**
    role in dyeing, 520
Ziegler, Karl (1898–1973), **1774**
Ziegler catalyst, **1773–1774**
Ziegler method, **1774**
Ziegler-Natta catalysts, 1148–1149
Ziegler-Natta polymerization, **1774**
Ziegler process, 1774
Zimmermann reaction, **1774**
Zinc
    abundance, 330t, 1132t
    biologically active chelates, 323t
    constituent of human plasma or serum, 244t
    corrosion resistant alloys, 446t
    covalent radius, 344t
    electronic structure, 336t
    interatomic distance, 342t
    ionic crystal radius, 341t
    nuclides (isotopes and isobars), 331t
    plant micronutrient, 613t
    plant micronutrient deficiencies, soils of U.S.,
        616t
    principal characteristics, 328t
Zinc acetate, 6
Zinc-air cell, 184–185, 189
Zinc alloys, 58
Zinc arsenate, wood preservative, 1752
Zincblende, crystal structure, 455
Zinc-bromine cell, 183
    cell reactions and energy content (in
        development), 180t
Zinc [CAS: 7440–66–6], **1774–1776**
    major sources of industrial minerals, 1011t
    standard electrode potential, 31t
    trace mineral nutrient, 1002t, 1004
    water quality indicator, 1726

Zinc chloride, wood preservative, 1752
Zinc chloride cell, cell reactions, energy content, and
    manufacturers, 179t
Zinc-chlorine cell, 183
Zinc coating, 1775
Zinc cyanide, EPA pretreatment standard for
    aqueous discharge, 985t
Zinc die-casting alloys, 1775–1776
Zinc electroplating, 557t
Zinc fluoride, 1776
Zinc hydroxide, 1776
Zinc (In Biological Systems), **1777–1778**
Zincite, **1778**
Zinc-manganese dioxide cell, 184t
Zinc-mercuric oxide cell, 184
Zinc metalloenzymes, 1777–1778
Zinc nitride, 1776
Zinc oxide, 1776
    plasma synthesis, 318t
Zinc oxycarbonate, 1776
Zinc-oxygen cell, 189
Zinc peroxide, 1776
Zinc phosphide, 1776
Zinc poisoning, 1778
Zinc selenide, 1776
Zinc-silver oxide cell, 184t
Zinc spinel, 701
Zinc spraying, 703
Zinc stearate, as release agent, 1435t
Zinc sulfide, 1776
Zinc telluride, 1776
Zinostatin, activity and producing organism, 128t
Zircon, **1778**
    heat effect, 708t
    physical properties, 1428t
Zirconia, dielectric permittivity, 852t
Zirconium
    abundance, 330t
    corrosion resistant alloys, 446t
    covalent radius, 344t
    electronic structure, 336t
    interatomic distance, 342t
    ionic crystal radius, 341t
    nuclides (isotopes and isobars), 331t
    principal characteristics, 328t
Zirconium [CAS: 7440–67–7], **1778–1779**
Zirconium dioxide, plasma synthesis, 318t
Zirconium hexafluoride, 1779
Zirconium nitride, 1779
    plasma synthesis, 318t
Zirconium oxide, 1779
Zircon porcelain, **1358**
Zoisite, **1779**
    heat effect, 708t
Zoladex, hormonal therapy, 358t
Zone refining, **1779–1780**
Zsigmondy, Richard (1865–1929), **1780**
Zsigmondy reagent, **1780**
Zwitterion, **1780**
Zymolytic reaction, **1780**